MARCH'S ADVANCED ORGANIC CHEMISTRY

THE WILEY BICENTENNIAL—KNOWLEDGE FOR GENERATIONS

Each generation has its unique needs and aspirations. When Charles Wiley first opened his small printing shop in lower Manhattan in 1807, it was a generation of boundless potential searching for an identity. And we were there, helping to define a new American literary tradition. Over half a century later, in the midst of the Second Industrial Revolution, it was a generation focused on building the future. Once again, we were there, supplying the critical scientific, technical, and engineering knowledge that helped frame the world. Throughout the 20th Century, and into the new millennium, nations began to reach out beyond their own borders and a new international community was born. Wiley was there, expanding its operations around the world to enable a global exchange of ideas, opinions, and know-how.

For 200 years, Wiley has been an integral part of each generation's journey, enabling the flow of information and understanding necessary to meet their needs and fulfill their aspirations. Today, bold new technologies are changing the way we live and learn. Wiley will be there, providing you the must-have knowledge you need to imagine new worlds, new possibilities, and new opportunities.

Generations come and go, but you can always count on Wiley to provide you the knowledge you need, when and where you need it!

WILLIAM J. PESCE
PRESIDENT AND CHIEF EXECUTIVE OFFICER

PETER BOOTH WILEY
CHAIRMAN OF THE BOARD

MARCH'S ADVANCED ORGANIC CHEMISTRY

REACTIONS, MECHANISMS, AND STRUCTURE

SIXTH EDITION

Michael B. Smith
Professor of Chemistry

Jerry March
Professor of Chemistry

BICENTENNIAL
1807
WILEY
2007
BICENTENNIAL

WILEY-INTERSCIENCE
A JOHN WILEY & SONS, INC., PUBLICATION

Published by John Wiley & Sons, Inc., Hoboken, New Jersey
Published simultaneously in Canada

For general information on our other products and services or for technical support, please contact our Customer Care Department within the United States at (800) 762-2974, outside the United States at (317) 572-3993 or fax (317) 572-4002.

Wiley also publishes its books in a variety of electronic formats. Some content that appears in print may not be available in electronic formats. For more information about Wiley products, visit our web site at www.wiley.com.

Library of Congress Cataloging-in-Publication Data is available.

Smith, Michael B., March, Jerry
March's Advanced Organic Chemistry: Reactions, Mechanisms, and Structure, Sixth Edition

ISBN 13: 978-0-471-72091-1
ISBN 10: 0-471-72091-7

Printed in the United States of America

10 9 8 7 6 5 4 3 2 1

Organic chemistry is a vibrant and growing scientific discipline that touches a vast number of scientific areas. This sixth edition of "March's Advanced Organic Chemistry" has been thoroughly updated to reflect new areas of Organic chemistry, as well as new advances in well-known areas of Organic chemistry. Every topic retained from the fifth edition has been brought up to date. Changes include the addition of a few new sections, significant revision to sections that have seen explosive growth in that area of research, moving sections around within the book to better reflect logical and reasonable chemical classifications, and a significant rewrite of much of the book. More than 7000 new references have been added. As with the fifth edition, when older references were deleted and in cases where a series of papers by the same principal author were cited, all but the most recent were deleted. The older citations should be found within the more recent one or ones. The fundamental structure of the sixth edition is essentially the same as that of all previous ones, although acyl substitution reactions have been moved from chapter 10 to chapter 16, and many oxidation or reduction reactions have been consolidated into chapter 19.

Like the first five editions, the sixth is intended to be a textbook for a course in advanced organic chemistry taken by students who have had the standard undergraduate organic and physical chemistry courses.

The goal, as in previous editions is to give equal weight to the three fundamental aspects of the study of organic chemistry: reactions, mechanisms, and structure. A student who has completed a course based on this book should be able to approach the literature directly, with a sound knowledge of modern basic organic chemistry. Major special areas of organic chemistry: terpenes, carbohydrates, proteins, many organometallic reagents, combinatorial chemistry, polymerization and electrochemical reactions, steroids, etc. have been treated lightly or ignored completely. I share the late Professor March's opinion that these topics are best approached after the first year of graduate study, when the fundamentals have been mastered, either in advanced courses, or directly, by consulting the many excellent books and review articles available on these subjects. In addition, many of these topics are so vast, they are beyond the scope of this book.

The organization is based on reaction types, so the student can be shown that despite the large number of organic reactions, a relatively few principles suffice to explain nearly all of them. Accordingly, the reactions-mechanisms section of this book (Part 2) is divided into 10 chapters (10–19), each concerned with a different type of reaction. In the first part of each chapter the appropriate basic mechanisms are discussed along with considerations of reactivity and orientation, while the second part consists of numbered sections devoted to individual reactions, where the scope and the mechanism of each reaction are discussed. Numbered sections are used for the reactions. Since the methods for the preparation of individual classes of compounds (e.g., ketones, nitriles, etc.) are not treated all in one place, an index has been provided (Appendix B) by use of which all methods for the preparation of a given type of compound will be found. For each reaction, a list of *Organic Syntheses* references is given where they have been reported. Thus for many reactions the student can consult actual examples in *Organic Syntheses*. It is important to note that the numbers for each reaction *differ* from one edition to the other, and many of the sections in the fifth edition do not correlate with the fourth. A correlation table is included at the end of this Preface that directly correlates the sections found in the 5th edition with the new ones in the 6th edition.

The structure of organic compounds is discussed in the first five chapters of Part 1. This section provides a necessary background for understanding mechanisms and is also important in its own right. The discussion begins with chemical bonding and ends with a chapter on stereochemistry. There follow two chapters on reaction mechanisms in general, one for ordinary reactions and the other for photochemical reactions. Part 1 concludes with two more chapters that give further background to the study of mechanisms.

In addition to reactions, mechanisms, and structure, the student should have some familiarity with the literature of organic chemistry. A chapter devoted to this topic has been placed in Appendix A, though many teachers may wish to cover this material at the beginning of the course.

The IUPAC names for organic transformations are included, first introduced in the third edition. Since then the rules have been broadened to cover additional cases; hence more such names are given in this edition. Furthermore, IUPAC has now published a new system for designating reaction mechanisms (see p. 420), and some of the simpler designations are included.

In treating a subject as broad as the basic structures, reactions, and mechanisms of organic chemistry, it is obviously not possible to cover each topic in great depth. Nor would this be desirable even if possible. Nevertheless, students will often wish to pursue individual topics further. An effort has therefore been made to guide the reader to pertinent review articles and books published since about 1965. In this respect, this book is intended to be a guide to the secondary literature (since about 1965) of the areas it covers. Furthermore, in a graduate course, students should be encouraged to consult primary sources. To this end, more than 20,000 references to original papers have been included.

Although basically designed for a one-year course on the graduate level, this book can also be used in advanced undergraduate courses, but a one-year course in organic chemistry prior to this is essential, and a one year course in physical chemistry is strongly recommended. It can also be adapted, by the omission of a large part of its contents, to a one-semester course. Indeed, even for a one-year course, more is included than can be conveniently covered. Many individual sections can be easily omitted without disturbing continuity.

The reader will observe that this text contains much material that is included in first-year organic and physical chemistry courses, though in most cases it goes more deeply into each subject and, of course, provides references, which first-year texts do not. It has been my experience that students who have completed the first-year courses often have a hazy recollection of the material and greatly profit from a representation of the material if it is organized in a different way. It is hoped that the organization of the material on reactions and mechanisms will greatly aid the memory and the understanding. In any given course the teacher may want to omit some chapters because students already have an adequate knowledge of the material, or because there are other graduate courses that cover the areas more thoroughly. Chapters 1, 4, and 7 especially may fall into one of these categories.

This book is probably most valuable as a reasonably up-to-date reference work. Students preparing for qualifying examinations and practicing organic chemists will find that Part 2 contains a survey of what is known about the mechanism and scope of a large number of reactions, arranged in an orderly manner based on reaction type and on which bonds are broken and formed. Also valuable for reference purposes are the previously mentioned lists of reactions classified by type of compound prepared (Appendix B) and of all of the *Organic Syntheses* references to each reaction.

Anyone who writes a book such as this is faced with the question of which units to use, in cases where international rules mandate one system, but published papers use another. Two instances are the units used for energies and for bond distances. For energies, IUPAC mandates joules, and many journals do use this unit exclusively. However, organic chemists who publish in United States journals overwhelmingly use calories and this situation shows no signs of changing in the near future. Since previous editions of this book have been used extensively both in this country and abroad, I have now adopted the practice of giving virtually all energy values in both calories and joules. The question of units for bond distances is easier to answer. Although IUPAC does not recommend Ångstrom units, nearly all bond distances published in the literature anywhere in the world, whether in organic or in crystallographic journals, are in these units, though a few papers do use picometers. Therefore, I continue to use only Ångstrom units.

I would like to acknowledge the contributions of those chemists cited and thanked by Professor March in the first four editions. I especially thank George Majetich, Warren Hehre, and Amos B. Smith III for generous contributions to specialized sections in the book as well as reviewing those sections. I also thank the many people who have contributed comments or have pointed out errors in the 5th edition that were invaluable to putting together the 6th edition. I thank Cambridge-Soft Inc. for providing *ChemOffice*, with *ChemDraw*, which was used to prepare all reactions and several structures in this book. I thank Dr. Warren Hehre and Wavefunction, Inc. for providing MacSpartan, allowing the incorporation of Spartan 3D models for selected molecules and intermediates.

Special thanks are due to the Interscience division of John Wiley & Sons and to Dr. Darla Henderson without whose support the book would not have been completed. Special thanks are also given to Shirley Thomas and Rebekah Amos at

Wiley for their fine work as editors in turning the manuscript into the finished book. I also thank Ms. Jeannette Stiefel, for an excellent job of copy editing the manuscript. I gratefully acknowledge the work of the late Professor Jerry March, upon whose work this new edition is built, and who is responsible for the concept of this book and for carrying it through four very successful editions.

I encourage those who read and use the sixth edition to contact me directly with comments, errors, and with publications that might be appropriate for future editions. I hope that this new edition will carry on the tradition that Professor March began with the first edition.

My Email address is

michael.smith@uconn.edu

and my homepage is

http://orgchem.chem.uconn.edu/home/mbs-home.html

Finally, I want to thank my wife Sarah for her patience and understanding during the preparation of this manuscript. I also thank my son Steven for his support. Without their support, this work would not have been possible.

MICHAEL B. SMITH

June, 2006

5th edition ⟶ 6th edition

10-1 ⟶ 10-1	10-18 ⟶ 10-14	10-35 ⟶ 16-68
10-2 ⟶ 10-2	10-19 ⟶ 10-15	10-36 ⟶ 10-24
10-3 ⟶ 10-3	10-20 ⟶ 10-16	10-37 ⟶ 10-25
10-4 ⟶ 10-4	10-21 ⟶ 16-61	10-38 ⟶ 10-26
10-5 ⟶ 10-5	10-22 ⟶ 16-62	10-39 ⟶ 16-69
10-6 ⟶ 10-6	10-23 ⟶ 16-63	10-40 ⟶ 10-27
10-7 ⟶ 10-7	10-24 ⟶ 16-64	10-41 ⟶ 10-28
10-8 ⟶ 16-57	10-25 ⟶ 16-65	10-42 ⟶ 10-29
10-9 ⟶ 16-58	10-26 ⟶ 10-17	10-43 ⟶ 10-30
10-10 ⟶ 16-59	10-27 ⟶ 10-18	10-44 ⟶ 10-31
10-11 ⟶ 16-60	10-28 ⟶ 10-19	10-46 ⟶ 10-32
10-12 ⟶ 10-8	10-29 ⟶ 16-66	10-47 ⟶ 10-33
10-13 ⟶ 10-9	10-30 ⟶ 16-67	10-48 ⟶ 16-70
10-14 ⟶ 10-10	10-31 ⟶ 10-20	10-49 ⟶ 10-34
10-15 ⟶ 10-11	10-32 ⟶ 10-21	10-50 ⟶ 10-35
10-16 ⟶ 10-12	10-33 ⟶ 10-22	10-51 ⟶ 10-37
10-17 ⟶ 10-13	10-34 ⟶ 10-23	10-52 ⟶ 10-38

$10\text{-}53 \longrightarrow 10\text{-}39$

$10\text{-}54 \longrightarrow 10\text{-}40$

$10\text{-}55 \longrightarrow 16\text{-}72$

$10\text{-}56 \longrightarrow 16\text{-}73$

$10\text{-}57 \longrightarrow 16\text{-}74$

$10\text{-}58 \longrightarrow 16\text{-}75$

$10\text{-}59 \longrightarrow 16\text{-}76$

$10\text{-}60 \longrightarrow 16\text{-}77$

$10\text{-}61 \longrightarrow 10\text{-}41$

$10\text{-}62 \longrightarrow 10\text{-}42$

$10\text{-}63 \longrightarrow 10\text{-}36$

$10\text{-}64 \longrightarrow 10\text{-}42$

$10\text{-}65 \longrightarrow 10\text{-}43$

$10\text{-}66 \longrightarrow 10\text{-}44$

$10\text{-}67 \longrightarrow 10\text{-}45$

$10\text{-}68 \longrightarrow 10\text{-}46$

$10\text{-}69 \longrightarrow 10\text{-}47$

$10\text{-}70 \longrightarrow 10\text{-}48$

$10\text{-}71 \longrightarrow 10\text{-}49$

$10\text{-}72 \longrightarrow 10\text{-}50$

$10\text{-}73 \longrightarrow 10\text{-}51$

$10\text{-}74 \longrightarrow 10\text{-}52$

$10\text{-}75 \longrightarrow 10\text{-}53$

$10\text{-}76 \longrightarrow 10\text{-}54$

$10\text{-}77 \longrightarrow 16\text{-}79$

$10\text{-}78 \longrightarrow 16\text{-}80$

$10\text{-}79 \longrightarrow 19\text{-}53$

$10\text{-}80 \longrightarrow 19\text{-}57$

$10\text{-}81 \longrightarrow 19\text{-}54$

$10\text{-}82 \longrightarrow 19\text{-}58$

$10\text{-}83 \longrightarrow 19\text{-}66$

$10\text{-}84 \longrightarrow 19\text{-}56$

$10\text{-}85 \longrightarrow 19\text{-}35$

$10\text{-}86 \longrightarrow 19\text{-}59$

$10\text{-}87 \longrightarrow 19\text{-}67$

$10\text{-}88 \longrightarrow 19\text{-}70$

$10\text{-}89 \longrightarrow 19\text{-}39$

$10\text{-}90 \longrightarrow 19\text{-}40$

$10\text{-}91 \longrightarrow 19\text{-}41$

$10\text{-}92 \longrightarrow 10\text{-}55$

$10\text{-}93 \longrightarrow 10\text{-}56$

$10\text{-}94 \longrightarrow 10\text{-}57$

$10\text{-}95 \longrightarrow 10\text{-}58$

$10\text{-}96 \longrightarrow 10\text{-}59$

$10\text{-}98 \longrightarrow 10\text{-}61$

$10\text{-}99 \longrightarrow 10\text{-}63$

$10\text{-}100 \longrightarrow 10\text{-}60$

$10\text{-}101 \longrightarrow 10\text{-}64$

$10\text{-}102 \longrightarrow 10\text{-}65$

$10\text{-}103 \longrightarrow 10\text{-}66$

$10\text{-}104 \longrightarrow 10\text{-}67$

$10\text{-}105 \longrightarrow 10\text{-}68$

$10\text{-}106 \longrightarrow 10\text{-}70$

$10\text{-}107 \longrightarrow 10\text{-}71$

$10\text{-}108 \longrightarrow 10\text{-}72$

$10\text{-}109 \longrightarrow 10\text{-}73$

$10\text{-}110 \longrightarrow 10\text{-}74$

$10\text{-}111 \longrightarrow 10\text{-}75$

$10\text{-}112 \longrightarrow 10\text{-}76$

$10\text{-}113 \longrightarrow 10\text{-}77$

$10\text{-}114 \longrightarrow 16\text{-}81$

$10\text{-}115 \longrightarrow 16\text{-}82$

$10\text{-}116 \longrightarrow 16\text{-}83$

$10\text{-}117 \longrightarrow 16\text{-}84$

$10\text{-}118 \longrightarrow 16\text{-}85$

$10\text{-}119 \longrightarrow 16\text{-}86$

$10\text{-}120 \longrightarrow 16\text{-}87$

$10\text{-}121 \longrightarrow 16\text{-}88$

$10\text{-}122 \longrightarrow 16\text{-}89$

$10\text{-}123 \longrightarrow 16\text{-}90$

$10\text{-}124 \longrightarrow 16\text{-}100$

$10\text{-}125 \longrightarrow 16\text{-}101$

$10\text{-}126 \longrightarrow 16\text{-}102$

$10\text{-}127 \longrightarrow 16\text{-}103$

$10\text{-}128 \longrightarrow 16\text{-}104$

$10\text{-}129 \longrightarrow 16\text{-}105$

$11\text{-}1 \longrightarrow 11\text{-}1$

$11\text{-}2 \longrightarrow 11\text{-}2$

$11\text{-}3 \longrightarrow 11\text{-}3$

$11\text{-}4 \longrightarrow 11\text{-}4$

$11\text{-}5 \longrightarrow 11\text{-}5$

$11\text{-}6 \longrightarrow 11\text{-}6$

$11\text{-}7 \longrightarrow 11\text{-}7$

$11\text{-}8 \longrightarrow 11\text{-}8$

$11\text{-}9$

$11\text{-}10 \longrightarrow 11\text{-}9$

$11\text{-}11 \longrightarrow 11\text{-}10$

$11\text{-}12 \longrightarrow 11\text{-}11$

$11\text{-}13 \longrightarrow 11\text{-}15$

$11\text{-}14 \longrightarrow 11\text{-}17$

$11\text{-}15 \longrightarrow 11\text{-}18$

$11\text{-}19 \longrightarrow 11\text{-}19$

$11\text{-}20 \longrightarrow 11\text{-}20$

$11\text{-}21 \longrightarrow 11\text{-}21$

$11\text{-}22 \longrightarrow 11\text{-}12$

$11\text{-}23 \longrightarrow 11\text{-}13$

$11\text{-}24 \longrightarrow 11\text{-}14$

$11\text{-}25 \longrightarrow 11\text{-}22$

$11\text{-}26 \longrightarrow 11\text{-}23$

$11\text{-}27 \longrightarrow 11\text{-}24$

$11\text{-}28 \longrightarrow 11\text{-}25$

$11\text{-}29 \longrightarrow 11\text{-}26$

$11\text{-}30 \longrightarrow 11\text{-}27$

$11\text{-}31 \longrightarrow 11\text{-}28$

$11\text{-}32 \longrightarrow 11\text{-}29$

$11\text{-}33 \longrightarrow 11\text{-}30$

$11\text{-}34 \longrightarrow 11\text{-}31$

$11\text{-}35 \longrightarrow 11\text{-}32$

$11\text{-}36 \longrightarrow 11\text{-}33$

$11\text{-}37 \longrightarrow 11\text{-}34$

$11\text{-}38 \longrightarrow 11\text{-}35$

$11\text{-}39 \longrightarrow 11\text{-}36$

$11\text{-}40 \longrightarrow 11\text{-}37$

$11\text{-}41 \longrightarrow 11\text{-}38$

$11\text{-}42 \longrightarrow 11\text{-}39$

$11\text{-}43 \longrightarrow 11\text{-}40$

$11\text{-}44 \longrightarrow 11\text{-}41$

$12\text{-}1 \longrightarrow 12\text{-}1$

$12\text{-}2 \longrightarrow 12\text{-}2$

$12\text{-}3 \longrightarrow 12\text{-}3$

$12\text{-}4 \longrightarrow 12\text{-}4$

$12\text{-}5 \longrightarrow 12\text{-}5$

$12\text{-}6 \longrightarrow 12\text{-}6$

$12\text{-}7 \longrightarrow 12\text{-}7$

$12\text{-}8 \longrightarrow 12\text{-}8$

$12\text{-}9 \longrightarrow 12\text{-}10$

$12\text{-}10 \longrightarrow 12\text{-}11$

$12\text{-}11 \longrightarrow 12\text{-}12$

$12\text{-}12 \longrightarrow 12\text{-}13$

$12\text{-}13 \longrightarrow 12\text{-}14$

$12\text{-}14 \longrightarrow 12\text{-}16$

$12\text{-}15 \longrightarrow 12\text{-}18$

$12\text{-}16 \longrightarrow 12\text{-}19$

$12\text{-}17 \longrightarrow 12\text{-}20$

$12\text{-}18 \longrightarrow 10\text{-}69$

$12\text{-}19 \longrightarrow 12\text{-}21$

15-38 ⟶ 15-41
15-39 ⟶ 15-40
15-40 ⟶ 15-42
15-41 ⟶ 15-43
15-42 ⟶ 15-44
15-43 ⟶ 15-45
15-44 ⟶ 15-46
15-45 ⟶ 15-47
15-46 ⟶ 15-48
15-47 ⟶ 15-49
15-48 ⟶ 15-50
15-49 ⟶ 15-62
15-50 ⟶ 15-51
15-51 ⟶ 15-52
15-52 ⟶ 15-53
15-53 ⟶ 15-54
15-54 ⟶ 15-55
15-55 ⟶ 15-56
15-56 ⟶ 15-57
15-57 ⟶ 15-58
15-58 ⟶ 15-60
15-59 ⟶ 15-61
15-60 ⟶ 15-59
15-61 ⟶ 15-63
15-62 ⟶ 15-64
15-63 ⟶ 15-65
15-64 ⟶ 15-66

16-1 ⟶ 16-1
16-2 ⟶ 16-2
16-3 ⟶ 16-3
16-4 ⟶ 16-4
16-5 ⟶ 16-5
16-6 ⟶ 16-7
16-7 ⟶ 16-8
16-8 ⟶ 16-9
16-9 ⟶ 16-10
16-10 ⟶ 16-11
16-11 ⟶ 16-12
16-12 ⟶ 16-13
16-13 ⟶ 16-18
16-14 ⟶ 16-17
16-15 ⟶ 16-19
16-16 ⟶ 16-20
16-17 ⟶ 16-21

16-18 ⟶ 16-22
16-19 ⟶ 16-14
16-20 ⟶ 16-15
16-21 ⟶ 16-16
16-22 ⟶ 16-23
16-23 ⟶ 19-36
16-24 ⟶ 19-42
16-25 ⟶ 19-43
16-26 ⟶ 19-44
16-27 ⟶ 16-24
16-28 ⟶ 16-25
16-29 ⟶ 16-26
16-30 ⟶ 16-27
16-31 ⟶ 16-28
16-32 ⟶ 16-29
16-33 deleted - combined
with 10-115
16-34 ⟶ 16-30
16-35 ⟶ 16-31
16-36 ⟶ 16-32
16-37 ⟶ 16-33
16-38 ⟶ 16-34
16-39 ⟶ 16-35
16-40 ⟶ 16-36
16-41 ⟶ 16-38
16-42 ⟶ 16-41
16-43 ⟶ 16-42
16-44 ⟶ 16-39
16-45 ⟶ 16-40
16-46 ⟶ 16-43
16-47 ⟶ 16-44
16-48 ⟶ 16-45
16-49 ⟶ 16-50
16-50 ⟶ 16-51
16-51 ⟶ 16-52
16-52 ⟶ 16-53
16-53 ⟶ 16-54
16-54 ⟶ 16-55
16-55 ⟶ 16-56
16-56 ⟶ 16-91
16-57 ⟶ 16-6
16-58 ⟶ 16-92
16-59 ⟶ 16-93
16-60 ⟶ 16-94
16-61 ⟶ 16-46

16-62 ⟶ 16-48
16-63 ⟶ 16-95
16-64 ⟶ 16-96
16-65 ⟶ 16-97
16-66 ⟶ 16-98
16-67 ⟶ 16-99

17-1 ⟶ 17-1
17-2 ⟶ 17-2
17-3 ⟶ 17-4
17-4 ⟶ 17-5
17-5 ⟶ 17-6
17-6 ⟶ 17-7
17-7 ⟶ 17-8
17-8 ⟶ 17-9
17-9 ⟶ 17-10
17-10 ⟶ 17-11
17-11 ⟶ 17-12
17-12 ⟶ 17-13
17-13 ⟶ 17-14
17-14 ⟶ 17-15
17-15 ⟶ 17-16
17-16 ⟶ 17-17
17-17 ⟶ 17-18
17-18 ⟶ 17-19
17-19 ⟶ 17-3
17-20 ⟶ 17-20
17-21 ⟶ 17-21
17-22 ⟶ 17-22
17-23 ⟶ 17-23
17-24 ⟶ 17-24
17-25 ⟶ 17-25
17-26 deleted
combined with 17-25
17-27 ⟶ 17-26
17-28 ⟶ 17-27
17-29 ⟶ 17-28
17-30 ⟶ 17-29
17-31 deleted
combined with 17-30
17-32 ⟶ 17-30
17-33 ⟶ 17-31
17-34 ⟶ 17-32
17-35 ⟶ 17-33
17-36 ⟶ 17-34

$17\text{-}37 \longrightarrow 17\text{-}35$
$17\text{-}38 \longrightarrow 17\text{-}36$
$17\text{-}39 \longrightarrow 17\text{-}37$
$17\text{-}40 \longrightarrow 17\text{-}38$

$18\text{-}1 \longrightarrow 18\text{-}1$
$18\text{-}2 \longrightarrow 18\text{-}2$
$18\text{-}3 \longrightarrow 18\text{-}3$
$18\text{-}4 \longrightarrow 18\text{-}4$
$18\text{-}5 \longrightarrow 18\text{-}5$
$18\text{-}6 \longrightarrow 18\text{-}6$
$18\text{-}7 \longrightarrow 18\text{-}7$
$18\text{-}8 \longrightarrow 18\text{-}8$
$18\text{-}9 \longrightarrow 18\text{-}9$
$18\text{-}10 \longrightarrow 18\text{-}10$.
$18\text{-}11 \longrightarrow 18\text{-}11$
$18\text{-}12 \longrightarrow 18\text{-}12$
$18\text{-}13 \longrightarrow 18\text{-}13$
$18\text{-}14 \longrightarrow 18\text{-}14$
$18\text{-}15 \longrightarrow 18\text{-}15$
$18\text{-}16 \longrightarrow 18\text{-}16$
$18\text{-}17 \longrightarrow 18\text{-}17$
$18\text{-}18 \longrightarrow 18\text{-}18$
$18\text{-}19 \longrightarrow 18\text{-}19$
$18\text{-}20 \longrightarrow 18\text{-}20$
$18\text{-}21 \longrightarrow 18\text{-}21$
$18\text{-}22 \longrightarrow 18\text{-}22$
$18\text{-}23 \longrightarrow 18\text{-}23$
$18\text{-}24 \longrightarrow 18\text{-}24$
$18\text{-}25 \longrightarrow 18\text{-}25$
$18\text{-}26 \longrightarrow 18\text{-}26$
$18\text{-}27 \longrightarrow 18\text{-}27$
$18\text{-}28 \longrightarrow 18\text{-}28$
$18\text{-}29 \longrightarrow 18\text{-}29$
$18\text{-}30 \longrightarrow 18\text{-}30$
$18\text{-}31 \longrightarrow 18\text{-}31$
$18\text{-}32 \longrightarrow 18\text{-}32$
$18\text{-}33 \longrightarrow 18\text{-}33$

$18\text{-}34 \longrightarrow 18\text{-}34$
$18\text{-}35 \longrightarrow 18\text{-}35$
$18\text{-}36 \longrightarrow 18\text{-}36$
$18\text{-}37 \longrightarrow 18\text{-}37$
$18\text{-}38 \longrightarrow 18\text{-}38$
$18\text{-}39 \longrightarrow 18\text{-}39$
$18\text{-}40 \longrightarrow 18\text{-}40$
$18\text{-}42 \longrightarrow 18\text{-}42$
$18\text{-}43 \longrightarrow 18\text{-}43$
$18\text{-}44 \longrightarrow 18\text{-}44$

$19\text{-}1 \longrightarrow 19\text{-}1$
$19\text{-}2 \longrightarrow 19\text{-}2$
$19\text{-}3 \longrightarrow 19\text{-}3$
$19\text{-}4 \longrightarrow 19\text{-}4$
$19\text{-}5 \longrightarrow 19\text{-}5$
$19\text{-}6 \longrightarrow 19\text{-}6$
$19\text{-}7 \longrightarrow 19\text{-}7$
$19\text{-}8 \longrightarrow 19\text{-}8$
$19\text{-}9 \longrightarrow 19\text{-}9$
$19\text{-}10 \longrightarrow 19\text{-}10$
$19\text{-}11 \longrightarrow 19\text{-}11$
$19\text{-}12 \longrightarrow 19\text{-}12$
$19\text{-}13 \longrightarrow 19\text{-}13$
$19\text{-}14 \longrightarrow 19\text{-}17$
$19\text{-}15 \longrightarrow 19\text{-}15$
$19\text{-}16 \longrightarrow 19\text{-}18$
19-17 deleted
incorporated in 19-14
$19\text{-}18 \longrightarrow 19\text{-}19$
$19\text{-}19 \longrightarrow 19\text{-}20$
$19\text{-}20 \longrightarrow 19\text{-}21$
$19\text{-}21 \longrightarrow 19\text{-}22$
$19\text{-}22 \longrightarrow 19\text{-}25$
$19\text{-}23 \longrightarrow 19\text{-}27$
$19\text{-}24 \longrightarrow 19\text{-}28$
$19\text{-}25 \longrightarrow 19\text{-}30$
$19\text{-}26 \longrightarrow 19\text{-}26$

$19\text{-}27 \longrightarrow 19\text{-}29$
$19\text{-}28 \longrightarrow 19\text{-}31$
$19\text{-}29 \longrightarrow 19\text{-}24$
$19\text{-}30 \longrightarrow 19\text{-}32$
$19\text{-}31 \longrightarrow 19\text{-}33$
$19\text{-}32 \longrightarrow 19\text{-}34$
$19\text{-}33 \longrightarrow 19\text{-}61$
$19\text{-}34 \longrightarrow 19\text{-}37$
$19\text{-}35 \longrightarrow 19\text{-}64$
$19\text{-}36 \longrightarrow 19\text{-}62$
$19\text{-}37 \longrightarrow 19\text{-}63$
$19\text{-}38 \longrightarrow 19\text{-}38$
$19\text{-}39 \longrightarrow 19\text{-}65$
19-40 deleted
incorporated into 10-85
$19\text{-}41 \longrightarrow 19\text{-}45$
$19\text{-}42 \longrightarrow 19\text{-}46$
$19\text{-}43 \longrightarrow 19\text{-}47$
$19\text{-}44 \longrightarrow 19\text{-}48$
$19\text{-}45 \longrightarrow 19\text{-}50$
$19\text{-}46 \longrightarrow 19\text{-}51$
$19\text{-}47 \longrightarrow 19\text{-}71$
$19\text{-}48 \longrightarrow 19\text{-}68$
$19\text{-}49 \longrightarrow 19\text{-}72$
$19\text{-}50 \longrightarrow 19\text{-}60$
$19\text{-}51 \longrightarrow 19\text{-}49$
$19\text{-}52 \longrightarrow 19\text{-}73$
$19\text{-}53 \longrightarrow 19\text{-}74$
$19\text{-}54 \longrightarrow 19\text{-}75$
$19\text{-}55 \longrightarrow 19\text{-}76$
$19\text{-}56 \longrightarrow 19\text{-}77$
$19\text{-}57 \longrightarrow 19\text{-}78$
$19\text{-}58 \longrightarrow 19\text{-}79$
$19\text{-}59 \longrightarrow 19\text{-}80$
$19\text{-}60 \longrightarrow 19\text{-}81$
$19\text{-}61 \longrightarrow 19\text{-}82$
$19\text{-}62 \longrightarrow 19\text{-}83$
$19\text{-}63 \longrightarrow 19\text{-}84$

CONTENTS

■ BIOGRAPHICAL NOTE

Professor Michael B. Smith was born in Detroit, Michigan in 1946 and lived there until 1957. In 1957, he and his family moved to Madison Heights, Virginia, where he attended high school and then Ferrum Jr. College, where he graduated with an A.A in 1966. Professor Smith then transferred to Virginia Polytechnic Institute (Virginia Tech), and graduated with a B.S in chemistry in 1969. After working as an analytical chemist at the Newport News Shipbuilding and Dry Dock Co. (Tenneco) in Newport News, Virginia for three years, he began graduate studies at Purdue University under the mentorship of Professor Joseph Wolinsky. Professor Smith graduated with a Ph.D. in Organic chemistry in 1977. He then spent one year as a faculty research associate at the Arizona State University, in the Cancer Research Institute directed by Professor George R. Pettit. Professor Smith spent a second year doing postdoctoral work at the Massachusetts Institute of Technology under the mentorship of Professor Sidney Hecht. In 1979 Professor Smith began his independent academic career, where he now holds the rank of full professor.

Professor smith is the author of approximately 70 independent research articles, and is the author of 14 published books. The books include the 5th edition of March's Advanced Organic Chemistry (Wiley), volumes 6–11 of the Compendium of Organic Synthetic Methods (Wiley), Organic Chemistry a Two Semester Course (HarperCollins) into its 2nd edition, and Organic Synthesis (McGraw-Hill) through its 2nd edition. The 3rd edition of the Organic Synthesis book is due out in 2007, published by Wavefunction, Inc.

Professor Smith's current research involves the synthesis and structural verification of several bioactive lipids obtained from the dental pathogen *Porphyromonas gingivalis*. Another area of research examines the chemical reactivity of conducting polymers such as poly(ethylenedioxy)thiophene (PEDOT). Such polymers are supposed to be chemically inert but, in fact, induce a variety of chemical reactions, including Friedel-Crafts alkylation of aromatic compounds with alcohols. Another area of research involves the development of a dye-conjugate designed to target and image tumors, as well as the total synthesis of anti-cancer phenanthridone alkaloids such as pancratistatin.

Ac	Acetyl	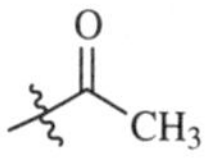
acac	Acetylacetonato	
AIBN	Azoisobutyronitrile	
aq.	Aqueous	

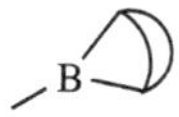

9-Borabicyclo[3.3.1]nonylboryl

9-BBN	9-Borabicyclo[3.3.1]nonane
BER	Borohydride exchange resin
BINAP	(2R,3S),2,2'-bis(diphenylphosphino)-1,1'-binapthyl
Bn	Benzyl
Bz	Benzoyl
BOC	*tert*-Butoxycarbonyl 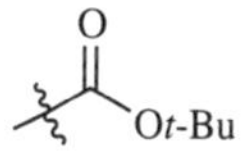
bpy (bipy)	2,2'-Bipyridyl
Bu	*n*-Butyl
CAM	Carboxamidomethyl
CAN	Ceric ammonium nitrate
c-	Cyclo-
cat.	Catalytic
Cbz	Carbobenzyloxy 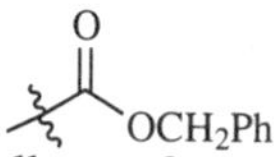
Chirald	(2S,3R)-(+)-4-dimethylamino-1,2-diphenyl-3-methylbutan-2-ol
Cod	1,5-Cyclooctadiene (ligand)
Cot	1,3,5,7-Cyclooctatetraene (ligand)
Cp	Cyclopentadienyl
CSA	Camphorsulfonic acid
CTAB	Cetyltrimethylammonium bromide

$-CH_2CH_2CH_2CH_3$

$(NH)_2Ce(NO_3)_6$

OCH_2Ph (O*t*-Bu)

$C_{16}H_{33}NMe_3^+Br^-$

Cy (*c*-C$_6$H$_{11}$)	Cyclohexyl 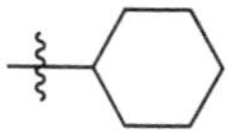
°C	Temperature in degrees Centigrade
DABCO	1,4-Diazobicyclo[2.2.2]octane
dba	Dibenzylidene acetone
DBE	1,2-Dibromoethane
DBU	1,8-Diazabicyclo[5.4.0]undec-7-ene
DBN	1,5-Diazabicyclo[4.3.0]non-5-ene
DCC	1,3-Dicyclohexylcarbodiimide
DCE	1,2-Dichloroethane

$BrCH_2CH_2Br$

$c\text{-}C_6H_{13}-N=C=N\text{-}c\text{-}C_6H_{13}$

$ClCH_2CH_2Cl$

DDQ	2,3-Dichloro-5,6-dicyano-1,4-benzoquinone	
% de	% Diasteromeric excess	
DEA	Diethylamine	$HN(CH_2CH_3)_2$
DEAD	Diethylazodicarboxylate	$EtO_2C-N=NCO_2Et$
Dibal-H	Diisobutylaluminum hydride	$(Me_2CHCH_2)_2AlH$
Diphos (dppe)	1,2-bis(Diphenylphosphino)ethane	$Ph_2PCH_2CH_2PPh_2$
Diphos-4 (dppb)	1,4-bis(Diphenylphosphino)butane	$Ph_2P(CH_2)_4PPh_2$
DMAP	4-Dimethylaminopyridine	
DMA	Dimethylacetamide	
DME	1,2-Dimethoxyethane	$MeOCH_2CH_2OMe$
DMF	N,N'-Dimethylformamide	

dmp   bis-[1,3-Di(*p*-methoxyphenyl)-1,3-propanedionato]

DMSO	Dimethyl sulfoxide	
dpm	Dipivaloylmethanato	
dppb	1,4-bis(Diphenylphosphino)butane $Ph_2P(CH_2)_4PPh_2$	
dppe	1,2-bis(Diphenylphosphino)ethane $Ph_2PCH_2CH_2CH_2PPh_2$	
dppf	bis(Diphenylphosphino)ferrocene	
dppp	1,3-bis(Diphenylphosphino)propane	$Ph_2P(CH_2)_3PPh_2$
dvb	Divinylbenzene	
e^-	Electrolysis	
% ee	% Enantiomeric excess	
EE	1-Ethoxyethyl	$EtO(Me)HCO-$
Et	Ethyl	$-CH_2CH_3$
EDA	Ethylenediamine	$H_2NCH_2CH_2NH_2$
EDTA	Ethylenediaminetetraacetic acid	
FMN	Flavin mononucleotide	
fod	*tris*-(6,6,7,7,8,8,8)-Heptafluoro-2,2-dimethyl-3,5-octanedionate	
Fp	Cyclopentadienyl-bis(carbonyl iron)	
FVP	Flash vacuum pyrolysis	
h	Hour (hours)	
hν	Irradiation with light	
1,5-HD	1,5-Hexadienyl	
HMPA	Hexamethylphosphoramide	$(Me_3N)_3P=O$
HMPT	Hexamethylphorous triamide	$(Me_3N)_3P$
*i*Pr	Isopropyl	$-CHMe_2$
IR	Infrared	
LICA (LIPCA)	Lithium cyclohexylisopropylamide	
LDA	Lithium diisopropylamide	$LiN(iPr)_2$
LHMDS	Lithium hexamethyl disilazide	$LiN(SiMe_3)_2$
LTMP	Lithium 2,2,6,6-tetramethylpiperidide	
MABR	Methylaluminum bis(4-bromo-2,6-di-*tert*-butylphenoxide)	

MAD	bis(2,6-Di-*tert*-butyl-4-methylphenoxy)methyl aluminum	
mCPBA	*meta*-Chloroperoxybenzoic acid	
Me	Methyl	$-CH_3$
MEM	β-Methoxyethoxymethyl	$MeOCH_2CH_2OCH_2-$
Mes	Mesityl	$2,4,6\text{-tri-Me-}C_6H_2$
MOM	Methoxymethyl	$MeOCH_2-$
Ms	Methanesulfonyl	CH_3SO_2-
MS	Molecular sieves (3 Å or 4 Å)	
MTM	Methylthiomethyl	CH_3SCH_2-
NAD	Nicotinamide adenine dinucleotide	
NADP	Sodium triphosphopyridine nucleotide	
Napth	Naphthyl ($C_{10}H_8$)	
NBD	Norbornadiene	
NBS	*N*-Bromosuccinimide	
NCS	*N*-Chlorosuccinimide	
NIS	*N*-Iodosuccinimide	
Ni(R)	Raney nickel	
NMP	*N*-Methyl-2-pyrrolidinone	
NY	New York	
NMR	Nuclear magnetic resonance	
Oxone	$2\ KHSO_5 \cdot KHSO_4 \cdot K_2SO_4$	
Ⓟ	Polymeric backbone	
PCC	Pyridinium chlorochromate	
PDC	Pyridinium dichromate	
PEG	Polyethylene glycol	
Ph	Phenyl	
PhH	Benzene	
PhMe	Toluene	
Phth	Phthaloyl	
pic	2-Pyridinecarboxylate	
Pip	Piperidyl	
PMP	4-Methoxyphenyl	
Pr	*n*-Propyl	$-CH_2CH_2CH_3$
Py	Pyridine	
quant.	Quantitative yield	
Red-Al	$[(MeOCH_2CH_2O)_2AlH_2]Na$	
sBu	*sec*-Butyl	$CH_3CH_2CH(CH_3)$
sBuLi	*sec*-Butyllithium	$CH_3CH_2CH(Li)CH_3$
Siamyl	Diisoamyl	$(CH_3)_2CHCH(CH_3)-$
TADDOL	α,α,α′α′-Tetraaryl-4,5-dimethoxy-1,3-dioxolane	
TASF	*tris*-(Diethylamino)sulfonium difluorotrimethyl silicate	
TBAF	Tetrabutylammonium fluoride	$n\text{-}Bu_4N^+F^-$
TBDMS	*tert*-Butyldimethylsilyl	$t\text{-}BuMe_sSi$
TBHP	*tert*-Butylhydroperoxide (*t*-BuOOH)	Me_3COOH

t-Bu	*tert*-Butyl	—C(CH$_3$)$_3$
TBS	*tert*-Butyl dimethylsilyl	*t*-BuMe$_2$Si
TEBA	Triethylbenzylammonium	Bn(CH$_3$)$_3$N$^+$
TEMPO	Tetramethylpiperdinyloxy free radical	
TFA	Trifluoroacetic acid	CF$_3$COOH
TFAA	Trifluoroacetic anhydride	(CF$_3$CO)$_2$O
Tf (OTf)	Triflate	—SO$_2$CF$_3$(—OSO$_2$CF$_3$)
THF	Tetrahydrofuran	
THP	Tetrahydropyran	
TMEDA	Tetramethylethylenediamine	Me$_2$NCH$_2$CH$_2$NMe$_2$
TMG	1,1,3,3-Tetramethylguanidine	
TMS	Trimethylsilyl	—Si(CH$_3$)$_3$
TMP	2,2,6,6-Tetramethylpiperidine	
TPAP	tetra-*n*-Propylammonium perruthenate	
Tol	Tolyl	4MeC$_6$H$_4$
Tr	Trityl	—CPh$_3$
TRIS	Triisopropylphenylsulfonyl	
Ts(Tos)	Tosyl = *p*-Toluenesulfonyl	4-MeC$_6$H$_4$
UV	Ultraviolet	
X$_c$	Chiral auxiliary	

MARCH'S ADVANCED
ORGANIC CHEMISTRY

This book contains 19 chapters. Chapters 10–19, which make up Part 2, are directly concerned with organic reactions and their mechanisms. Chapters 1–9 may be thought of as an introduction to Part 2. The first five chapters deal with the structure of organic compounds. These chapters discuss the kinds of bonding important in organic chemistry, the three-dimensional structure of organic molecules, and the structure of species in which the valence of carbon is less than 4. Chapters 6–9 are concerned with other topics that help to form a background to Part 2: acids and bases, photochemistry, the relationship between structure and reactivity, and a general discussion of mechanisms and the means by which they are determined.

This book contains 19 chapters. Chapters 10-19, which make up Part 2, are directly concerned with organic reactions and their mechanisms. Chapters 1-9 may be thought of as an introduction to Part 2. The first five chapters deal with the structure of organic compounds. These chapters discuss the kinds of bonding important in organic chemistry, the three-dimensional structure of organic molecules, and the structure of species in which the valence of carbon is less than 4. Chapter 6-9 are concerned with other topics that help to form a background to Part 2: acids and bases, photochemistry, the relationship between structure and reactivity, and a general discussion of mechanisms and the means by which they are determined.

CHAPTER 1

Localized Chemical Bonding

Localized chemical bonding may be defined as bonding in which the electrons are shared by two and only two nuclei. In Chapter 2, we will consider *delocalized bonding*, in which electrons are shared by more than two nuclei.

COVALENT BONDING[1]

Wave mechanics is based on the fundamental principle that electrons behave as waves (e.g., they can be diffracted) and that consequently a wave equation can be written for them, in the same sense that light waves, sound waves, and so on can be described by wave equations. The equation that serves as a mathematical model for electrons is known as the *Schrödinger equation*, which for a one-electron system is

$$\frac{\delta^2 \psi}{\delta x^2} + \frac{\delta^2 \psi}{\delta y^2} + \frac{\delta^2 \psi}{\delta z^2} + \frac{8\pi^2 m}{h^2}(E - V)\psi = 0$$

where m is the mass of the electron, E is its total energy, V is its potential energy, and h is Planck's constant. In physical terms, the function Ψ expresses the square root of the probability of finding the electron at any position defined by the coordinates x, y, and z, where the origin is at the nucleus. For systems containing more than one electron, the equation is similar, but more complicated.

---

[1]The treatment of orbitals given here is necessarily simplified. For much fuller treatments of orbital theory as applied to organic chemistry, see Matthews, P.S.C. *Quantum Chemistry of Atoms and Molecules*, Cambridge University Press, Cambridge, *1986*; Clark, T. *A Handbook of Computational Chemistry*, Wiley, NY, *1985*; Albright, T.A.; Burdett, J.K.; Whangbo, M. *Orbital Interactions in Chemistry*, Wiley, NY, *1985*; MacWeeny, R.M. *Coulson's Valence*, Oxford University Press, Oxford, *1980*; Murrell, J.N.; Kettle, S.F.A; Tedder, J.M. *The Chemical Bond*, Wiley, NY, *1978*; Dewar, M.J.S.; Dougherty. R.C. *The PMO Theory of Organic Chemistry*, Plenum, NY, *1975*; Zimmerman, H.E. *Quantum Mechanics for Organic Chemists*, Academic Press, NY, *1975*; Borden, W.T. *Modern Molecular Orbital Theory for Organic Chemists*, Prentice-Hall, Englewood Cliffs, NJ, *1975*; Dewar, M.J.S. *The Molecular Orbital Theory of Organic Chemistry*, McGraw-Hill, NY, *1969*; Liberles, A. *Introduction to Molecular Orbital Theory*, Holt, Rinehart, and Winston, NY, *1966*.

March's Advanced Organic Chemistry: Reactions, Mechanisms, and Structure, Sixth Edition, by Michael B. Smith and Jerry March
Copyright © 2007 John Wiley & Sons, Inc.

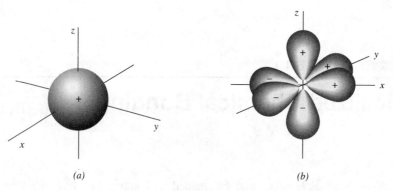

Fig. 1.1. (*a*) The 1*s* orbital. (*b*) The three 2*p* orbitals.

The Schrödinger equation is a differential equation, which means that solutions of it are themselves equations, but the solutions are not differential equations. They are simple equations for which graphs can be drawn. Such graphs, which are three-dimensional (3D) pictures that show the electron density, are called *orbitals* or electron clouds. Most students are familiar with the shapes of the *s* and *p* atomic orbitals (Fig. 1.1). Note that each *p* orbital has a *node*: A region in space where the probability of finding the electron is extremely small.[2] Also note that in Fig. 1.1 some lobes of the orbitals are labeled + and others −. These signs do not refer to positive or negative *charges*, since both lobes of an electron cloud must be negatively charged. They are the signs of the wave function Ψ. When two parts of an orbital are separated by a node, Ψ always has opposite signs on the two sides of the node. According to the Pauli exclusion principle, no more than two electrons can be present in any orbital, and they must have opposite spins.

Unfortunately, the Schrödinger equation can be solved exactly only for one-electron systems, such as the hydrogen atom. If it could be solved exactly for molecules containing two or more electrons,[3] we would have a precise picture of the shape of the orbitals available to each electron (especially for the important ground state) and the energy for each orbital. Since exact solutions are not available, drastic approximations must be made. There are two chief general methods of approximation: the molecular-orbital method and the valence-bond method.

In the molecular-orbital method, bonding is considered to arise from the overlap of atomic orbitals. When any number of atomic orbitals overlap, they combine to

[2]When wave-mechanical calculations are made according to the Schrödinger equation, the probability of finding the electron in a node is zero, but this treatment ignores relativistic considerations. When such considerations are applied, Dirac has shown that nodes do have a very small electron density: Powell, R.E. *J. Chem. Educ.* **1968**, *45*, 558. See also, Ellison, F.O. and Hollingsworth, C.A. *J. Chem. Educ.* **1976**, *53*, 767; McKelvey, D.R. *J. Chem. Educ.* **1983**, *60*, 112; Nelson, P.G. *J. Chem. Educ.* **1990**, *67*, 643. For a review of relativistic effects on chemical structures in general, see Pyykkö, P. *Chem. Rev.* **1988**, *88*, 563.

[3]For a number of simple systems containing two or more electrons, such as the H_2 molecule or the He atom, approximate solutions are available that are so accurate that for practical purposes they are as good as exact solutions. See, for example, Roothaan, C.C.J.; Weiss, A.W. *Rev. Mod. Phys.* **1960**, *32*, 194; Kolos, W.; Roothaan, C.C.J. *Rev. Mod. Phys.* **1960**, *32*, 219. For a review, see Clark, R.G.; Stewart, E.T. *Q. Rev. Chem. Soc.* **1970**, *24*, 95.

form an equal number of new orbitals, called *molecular orbitals*. Molecular orbitals differ from atomic orbitals in that they are clouds that surround the nuclei of two or more atoms, rather than just one atom. In localized bonding the number of atomic orbitals that overlap is two (each containing one electron), so that two molecular orbitals are generated. One of these, called a *bonding orbital*, has a lower energy than the original atomic orbitals (otherwise a bond would not form), and the other, called an *antibonding orbital*, has a higher energy. Orbitals of lower energy fill first. Since the two original atomic orbitals each held one electron, both of these electrons can now go into the new molecular *bonding* orbital, since any orbital can hold two electrons. The antibonding orbital remains empty in the ground state. The greater the overlap, the stronger the bond, although total overlap is prevented by repulsion between the nuclei. Figure 1.2 shows the bonding and antibonding orbitals that arise by the overlap of two 1s electrons. Note that since the antibonding orbital has a node between the nuclei, there is practically no electron density in that area, so that this orbital cannot be expected to bond very well. Molecular orbitals formed by the overlap of two atomic orbitals when the centers of electron density are on the axis common to the two nuclei are called σ (*sigma*) orbitals, and the bonds are called σ bonds. Corresponding antibonding orbitals are designated σ*. Sigma orbitals are formed not only by the overlap of two *s* orbitals, but also by the overlap of any of the kinds of atomic orbital (*s, p, d,* or *f*) whether the same or different, but the two lobes that overlap must have the same sign: a positive *s* orbital can form a bond only by overlapping with another positive *s* orbital or with a positive lobe of a *p, d,* or *f* orbital. Any *s* orbital, no matter what kind of atomic orbitals it has arisen from, may be represented as approximately ellipsoidal in shape.

Orbitals are frequently designated by their symmetry properties. The σ orbital of hydrogen is often written ψ_g. The *g* stands for *gerade*. A gerade orbital is one in which the sign on the orbital does not change when it is inverted through its center of symmetry. The σ* orbital is *ungerade* (designated ψ_u). An ungerade orbital changes sign when inverted through its center of symmetry.

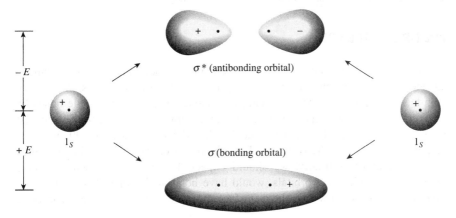

Fig. 1.2. Overlap of two 1s orbitals gives rise to a σ and a σ* orbital.

In molecular-orbital calculations, a wave function is formulated that is a linear combination of the atomic orbitals that have overlapped (this method is often called the *linear combination of atomic orbitals*, or LCAO). Addition of the atomic orbitals gives the bonding molecular orbital:

$$\psi = c_A \psi_A + c_B \psi_B \tag{1-1}$$

The functions ψ_A and ψ_B are the functions for the atomic orbitals of atoms A and B, respectively, and c_A and c_B represent weighting factors. Subtraction is also a linear combination:

$$\psi = c_A \psi_A - c_B \psi_B \tag{1-2}$$

This gives rise to the antibonding molecular orbital.

In the valence-bond method, a wave equation is written for each of various possible electronic structures that a molecule may have (each of these is called a *canonical form*), and the total ψ is obtained by summation of as many of these as seem plausible, each with its weighting factor:

$$\psi = c_1 \psi_1 + c_2 \psi_2 + \cdots \tag{1-3}$$

This resembles Eq. (1), but here each ψ represents a wave equation for an imaginary canonical form and each c is the amount contributed to the total picture by that form. For example, a wave function can be written for each of the following canonical forms of the hydrogen molecule:[4]

$$\text{H—H} \qquad \text{H}:^- \quad \text{H}^+ \qquad {}^+\text{H} \quad \text{H}:^-$$

Values for c in each method are obtained by solving the equation for various values of each c and choosing the solution of lowest energy. In practice, both methods give similar solutions for molecules that contain only localized electrons, and these are in agreement with the Lewis structures long familiar to the organic chemist. Delocalized systems are considered in Chapter 2.

MULTIPLE VALENCE

A univalent atom has only one orbital available for bonding. But atoms with a valence of 2 or more must form bonds by using at least two orbitals. An oxygen atom has two half-filled orbitals, giving it a valence of 2. It forms single bonds by the overlap of these with the orbitals of two other atoms. According to the principle of maximum overlap, the other two nuclei should form an angle of 90° with the oxygen nucleus, since the two available orbitals on oxygen are *p* orbitals, which are perpendicular. Similarly, we should expect that nitrogen, which has three mutually perpendicular *p* orbitals, would have bond angles of 90° when it forms three single bonds. However, these are not the observed bond angles. The bond

---

[4]In this book, a pair of electrons, whether in a bond or unshared, is represented by a straight line.

angles are,[5] in water, $104°27'$, and in ammonia, $106°46'$. For alcohols and ethers the angles are even larger (see p. 25). A discussion of this will be deferred to p. 25, but it is important to note that covalent compounds do have definite bond angles. Although the atoms are continuously vibrating, the mean position is the same for each molecule of a given compound.

HYBRIDIZATION

Consider the case of mercury. Its electronic structure is

$$[\mathbf{Xe}\text{ core}]4f^{14}5d^{10}6s^2$$

Although it has no half-filled orbitals, it has a valence of 2 and forms two covalent bonds. We can explain this by imagining that one of the $6s$ electrons is promoted to a vacant $6p$ orbital to give the excited configuration

$$[\mathbf{Xe}\text{ core}]4f^{14}5d^{10}6s^16p^1$$

In this state, the atom has two half-filled orbitals, but they are not equivalent. If bonding were to occur by the overlap of these orbitals with the orbitals of external atoms, the two bonds would not be equivalent. The bond formed from the $6p$ orbital would be more stable than the one formed from the $6s$ orbital, since a larger amount of overlap is possible with the former. A more stable situation is achieved when, in the course of bond formation, the $6s$ and $6p$ orbitals combine to form two new orbitals that *are* equivalent; these are shown in Fig. 1.3.

Since these new orbitals are a mixture of the two original orbitals, they are called *hybrid orbitals*. Each is called an *sp* orbital, since a merger of an *s* and a *p* orbital was required to form it. The *sp* orbitals, each of which consists of a large lobe and a very small one, are atomic orbitals, although they arise only in the bonding process and do not represent a possible structure for the free atom. A mercury atom forms

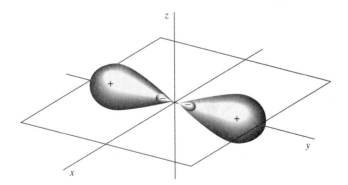

Fig. 1.3. The two *sp* orbitals formed by mercury.

[5]Bent, H.A. *Chem. Rev.* **1961**, *61*, 275, p. 277.

its two bonds by overlapping each of the large lobes shown in Fig. 1.3 with an orbital from an external atom. This external orbital may be any of the atomic orbitals previously considered (*s, p, d,* or *f*) or it may be another hybrid orbital, although only lobes of the same sign can overlap. In any of these cases, the molecular orbital that arises is called a σ orbital since it fits our previous definition of a σ orbital.

In general, because of mutual repulsion, equivalent orbitals lie as far away from each other as possible, so the two *sp* orbitals form an angle of 180°. This means that $HgCl_2$, for example, should be a linear molecule (in contrast to H_2O), and it is. This kind of hybridization is called *digonal hybridization.* An *sp* hybrid orbital forms a stronger covalent bond than either an *s* or a *p* orbital because it extends out in space in the direction of the other atom's orbital farther than the *s* or the *p* and permits greater overlap. Although it would require energy to promote a 6*s* electron to the 6*p* state, the extra bond energy more than makes up the difference.

Many other kinds of hybridization are possible. Consider boron, which has the electronic configuration

$$1s^2 2s^2 2p^1$$

yet has a valence of 3. Once again we may imagine promotion and hybridization:

$$1s^2 2s^2 2p^1 \xrightarrow{\text{promotion}} 1s^2 2s^1 2p_x^1 2p_y^1 \xrightarrow{\text{hybridization}} 1s^2 (sp^2)^3$$

In this case, there are three equivalent hybrid orbitals, each called sp^2 *(trigonal hybridization).* This method of designating hybrid orbitals is perhaps unfortunate since nonhybrid orbitals are designated by single letters, but it must be kept in mind that *each* of the three orbitals is called sp^2. These orbitals are shown in Fig. 1.4. The three axes are all in one plane and point to the corners of an equilateral triangle. This accords with the known structure of BF_3, a planar molecule with angles of 120°.

The case of carbon (in forming four single bonds) may be represented as

$$1s^2 2s^2 2p_x^1 2p_y^1 \xrightarrow{\text{promotion}} 1s^2 2s^1 2p_x^1 2p_y^1 2p_z^1 \xrightarrow{\text{hybridization}} 1s^2 (sp^3)^4$$

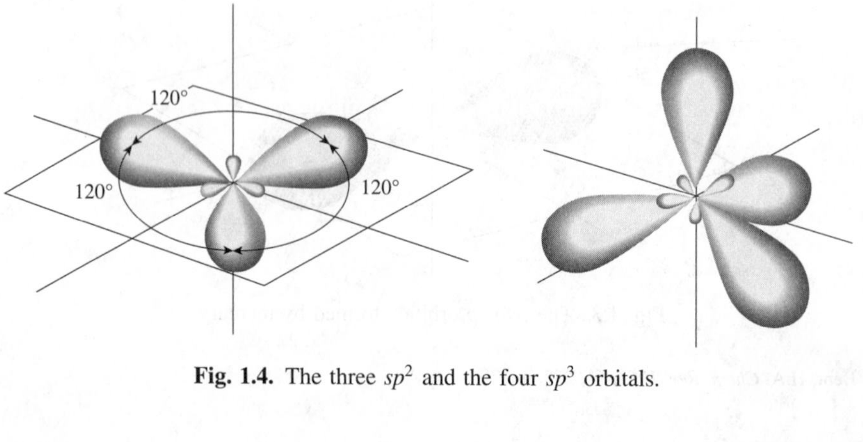

Fig. 1.4. The three sp^2 and the four sp^3 orbitals.

There are four equivalent orbitals, each called sp^3, which point to the corners of a regular tetrahedron (Fig. 1.4). The bond angles of methane would thus be expected to be $109°28'$, which is the angle for a regular tetrahedron.

Although the hybrid orbitals discussed in this section satisfactorily account for most of the physical and chemical properties of the molecules involved, it is necessary to point out that the sp^3 orbitals, for example, stem from only one possible approximate solution of the Schrödinger equation. The s and the three p atomic orbitals can also be combined in many other equally valid ways. As we shall see on p. 13, the four C—H bonds of methane do not always behave as if they are equivalent.

MULTIPLE BONDS

If we consider the ethylene molecule in terms of the molecular-orbital concepts discussed so far, we have each carbon using sp^2 orbitals to form bonds with the three atoms to which it is connected. These sp^2 orbitals arise from hybridization of the $2s^1$, $2p_x^1$, and $2p_y^1$ electrons of the promoted state shown on p. 8. We may consider that any carbon atom that is bonded to only three different atoms uses sp^2 orbitals for this bonding. Each carbon of ethylene is thus bonded by three σ bonds: one to each hydrogen and one to the other carbon. Each carbon therefore has another electron in the $2p_z$ orbital that is perpendicular to the plane of the sp^2 orbitals. The two parallel $2p_z$ orbitals can overlap sideways to generate two new orbitals, a bonding and an antibonding orbital (Fig. 1.5). Of course, in the ground state, both electrons go into the bonding orbital and the antibonding orbital remains vacant. Molecular orbitals formed by the overlap of atomic orbitals whose axes are parallel are called π orbitals if they are bonding and $π^*$ if they are antibonding.

In this picture of ethylene, the two orbitals that make up the double bond are not equivalent.[6] The σ orbital is ellipsoidal and symmetrical about the C—C axis. The π orbital is in the shape of two ellipsoids, one above the plane and one below. The plane itself represents a node for the π orbital. In order for the p orbitals to maintain maximum overlap, they must be parallel. This means that free rotation is not possible about the double bond, since the two p orbitals would have to reduce their overlap to allow one H—C—H plane to rotate with respect to the other. The six atoms of a double bond are therefore in a plane with angles that should be ~120°. Double bonds are shorter than the corresponding single bonds because maximum stability is obtained when the p orbitals overlap as much as possible. Double bonds between carbon and oxygen or nitrogen are similarly represented: they consist of one σ and one π orbital.

In triple-bond compounds, carbon is connected to only two other atoms and hence uses sp hybridization, which means that the four atoms are in a straight

---

[6]The double bond can also be pictured as consisting of two equivalent orbitals, where the centers of electron density point away from the C—C axis. This is the bent-bond or banana-bond picture. Support for this view is found in Pauling. L. *Theoretical Organic Chemistry, The Kekulé Symposium*, Butterworth, London, *1959*, pp. 2–5; Palke, W.E. *J. Am. Chem. Soc. 1986*, *108*, 6543. However, most of the literature of organic chemistry is written in terms of the σ–π picture, and we will use it in this book.

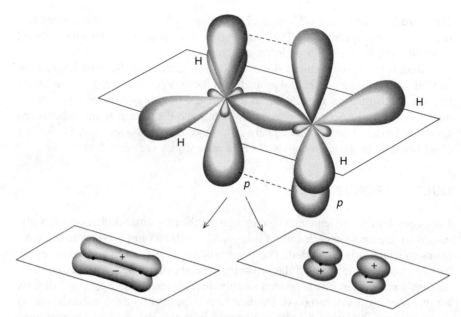

Fig. 1.5. Overlapping p orbitals form a π and a π^* orbital. The σ orbitals are shown in the upper figure. They are still there in the states represented by the diagrams below, but have been removed from the picture for clarity.

line (Fig. 1.6).[7] Each carbon has two p orbitals remaining, with one electron in each. These orbitals are perpendicular to each other and to the C—C axis. They overlap in the manner shown in Fig. 1.7 to form two π orbitals. A triple bond is thus composed of one σ and two π orbitals. Triple bonds between carbon and nitrogen can be represented in a similar manner.

Double and triple bonds are important only for the first-row elements carbon, nitrogen, and oxygen.[8] For second-row elements multiple bonds are rare and

Fig. 1.6. The σ electrons of acetylene.

[7]For reviews of triple bonds, see Simonetta, M.; Gavezzotti, A., in Patai, S. *The Chemistry of the Carbon-Carbon Triple Bond*, Wiley, NY, *1978*, pp. 1–56; Dale, J., in Viehe, H. G. *Acetylenes*, Marcel Dekker, NY, *1969*, pp. 3–96.

[8]This statement applies to the representative elements. Multiple bonding is also important for some transition elements. For a review of metal–metal multiple bonds, see Cotton, F.A. *J. Chem. Educ.* *1983*, *60*, 713.

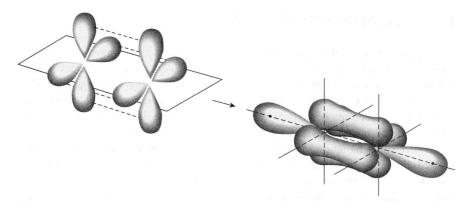

Fig. 1.7. Overlap of *p* orbitals in a triple bond for clarity, the σ orbitals have been removed from the drawing on the left, although they are shown on the right.

compounds containing them are generally less stable[9] because these elements tend to form weaker π bonds than do the first-row elements.[10] The only ones of any importance at all are C=S bonds, and C=S compounds are generally much less stable than the corresponding C=O compounds (however, see *pπ–dπ* bonding, p. $$$). Stable compounds with Si=C and Si=Si bonds are rare, but examples have been reported,[11] including a pair of cis and trans Si=Si isomers.[12]

[9]For a review of double bonds between carbon and elements other than C, N, S, or O, see Jutzi, P. *Angew. Chem. Int. Ed.* **1975**, *14*, 232. For reviews of multiple bonds involving silicon and germanium, see Barrau, J.; Escudié, J.; Satgé, J. *Chem. Rev.* **1990**, *90*, 283 (Ge only); Raabe, G.; Michl, J., in Patai, S. and Rappoport, Z. *The Chemistry of Organic Silicon Compounds, part 2*, Wiley: NY, **1989**, pp. 1015–1142; *Chem. Rev.* **1985**, *85*, 419 (Si only); Wiberg, N. *J. Organomet. Chem.* **1984**, *273*, 141 (Si only); Gusel'nikov, L.E.; Nametkin, N.S. *Chem. Rev.* **1979**, *79*, 529 (Si only). For reviews of C=P and C+P bonds, see Regitz, M. *Chem. Rev.* **1990**, *90*, 191; Appel, R.; Knoll, F. *Adv. Inorg. Chem.* **1989**, *33*, 259; Markovski, L.N.; Romanenko, V.D. *Tetrahedron* **1989**, *45*, 6019. For reviews of other second-row double bonds, see West, R. *Angew. Chem. Int. Ed.* **1987**, *26*, 1201 (Si=Si bonds); Brook, A.G.; Baines, K.M. *Adv. Organometal. Chem.* **1986**, *25*, 1 (Si=C bonds); Kutney, G.W.; Turnbull, K. *Chem. Rev.* **1982**, *82*, 333 (S=S bonds). For reviews of multiple bonds between heavier elements, see Cowley, A.H.; Norman, N.C. *Prog. Inorg. Chem.* **1986**, *34*, 1; Cowley, A.H. *Polyhedron* **1984**, *3*, 389; *Acc. Chem. Res.* **1984**, *17*, 386. For a theoretical study of multiple bonds to silicon, see Gordon, M.S. *Mol. Struct. Energ.* **1986**, *1*, 101.
[10]For discussions, see Schmidt, M.W.; Truong, P.N.; Gordon, M.S. *J. Am. Chem. Soc.* **1987**, *109*, 5217; Schleyer, P. von R.; Kost, D. *J. Am. Chem. Soc.* **1988**, *110*, 2105.
[11]For Si=C bonds, see Brook, A.G.; Nyburg, S.C.; Abdesaken, F.; Gutekunst, B.; Gutekunst, G.; Kallury, R.K.M.R.; Poon, Y.C.; Chang, Y.; Wong-Ng, W. *J. Am. Chem. Soc.* **1982**, *104*, 5667; Schaefer III, H.F. *Acc. Chem. Res.* **1982**, *15*, 283; Wiberg, N.; Wagner, G.; Riede, J.; Müller, G. *Organometallics* **1987**, *6*, 32. For Si=Si bonds, see West, R.; Fink, M.J.; Michl, J. *Science* **1981**, *214*, 1343; Boudjouk, P.; Han, B.; Anderson, K.R. *J. Am. Chem. Soc.* **1982**, *104*, 4992; Fink, M.J.; DeYoung, D.J.; West, R.; Michl, J. *J. Am. Chem. Soc.* **1983**, *105*, 1070; Fink, M.J.; Michalczyk, M.J.; Haller, K.J.; West, R.; Michl, J. *Organometallics* **1984**, *3*, 793; West, R. *Pure Appl. Chem.* **1984**, *56*, 163; Masamune, S.; Eriyama, Y.; Kawase, T. *Angew. Chem. Int. Ed.* **1987**, *26*, 584; Shepherd, B.D.; Campana, C.F.; West, R. *Heteroat. Chem.* **1990**, *1*, 1. For an Si=N bond, see Wiberg, N.; Schurz, K.; Reber, G.; Müller, G. *J. Chem. Soc. Chem. Commun.* **1986**, 591.
[12]Michalczyk, M.J.; West, R.; Michl, J. *J. Am. Chem. Soc.* **1984**, *106*, 821, *Organometallics* **1985**, *4*, 826.

PHOTOELECTRON SPECTROSCOPY

Although the four bonds of methane are equivalent according to most physical and chemical methods of detection (e.g., neither the nuclear magnetic resonances (NMR) nor the infrared (IR) spectrum of methane contains peaks that can be attributed to different kinds of C—H bonds), there is one physical technique that shows that the eight valence electrons of methane can be differentiated. In this technique, called *photoelectron spectroscopy,*[13] a molecule or free atom is bombarded with vacuum ultraviolet (UV) radiation, causing an electron to be ejected. The energy of the ejected electron can be measured, and the difference between the energy of the radiation used and that of the ejected electron is the *ionization potential* of that electron. A molecule that contains several electrons of differing energies can lose any one of them as long as its ionization potential is less than the energy of the radiation used (a single molecule loses only one electron; the loss of two electrons by any individual molecule almost never occurs). A photoelectron spectrum therefore consists of a series of bands, each corresponding to an orbital of a different energy. The spectrum gives a direct experimental picture of all the orbitals present, in order of their energies, provided that radiation of sufficiently high energy is used.[14] Broad

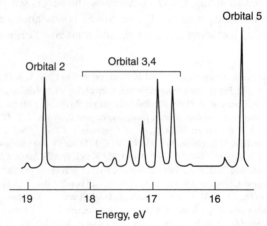

Fig. 1.8. Photoelectron spectrum of N_2.[15]

[13]Only the briefest description of this subject is given here. For monographs, see Ballard, R.E. *Photoelectron Spectroscopy and Molecular Orbital Theory,* Wiley, NY, *1978*; Rabalais, J.W., *Principles of Ultraviolet Photoelectron Spectroscopy,* Wiley, NY, *1977*; Baker, A.D.; Betteridge, D. *Photoelectron Spectroscopy,* Pergamon, Elmsford, NY, *1972*; Turner, D.W.; Baker, A.D..; Baker, C.; Brundle, C.R. *High Resolution Molecular Photoelectron Spectroscopy,* Wiley, NY, *1970*. For reviews, see Westwood, N.P.C. *Chem. Soc. Rev.* *1989*, *18*, 317; Carlson, T.A. *Annu. Rev. Phys. Chem. 1975*, *26*, 211; Baker, C.; Brundle, C.R.; Thompson, M. *Chem. Soc. Rev. 1972*, *1*, 355; Bock, H.; Mollère, P.D. *J. Chem. Educ. 1974*, *51*, 506; Bock, H.; Ramsey, B.G. *Angew. Chem. Int. Ed. 1973*, *12*, 734; Turner, D.W. *Adv. Phys. Org. Chem. 1966*, *4*, 31. For the IUPAC descriptive classification of the electron spectroscopies, see Porter, H.Q.; Turner, D.W. *Pure Appl. Chem. 1987*, *59*, 1343.
[14]The correlation is not perfect, but the limitations do not seriously detract from the usefulness of the method. The technique is not limited to vacuum UV radiation. Higher energy radiation can also be used.

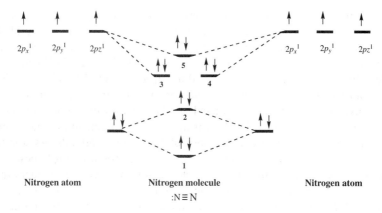

Fig. 1.9. Electronic structure of N_2 (inner-shell electrons omitted).

bands usually correspond to strongly bonding electrons and narrow bands to weakly bonding or nonbonding electrons. A typical spectrum is that of N_2, shown in Fig. 1.8.[15] The N_2 molecule has the electronic structure shown in Fig. 1.9. The two $2s$ orbitals of the nitrogen atoms combine to give the two orbitals marked 1 (bonding) and 2 (antibonding), while the six $2p$ orbitals combine to give six orbitals, three of which (marked 3, 4, and 5) are bonding. The three antibonding orbitals (not indicated in Fig. 1.9) are unoccupied. Electrons ejected from orbital 1 are not found in Fig. 1.8 because the ionization potential of these electrons is greater than the energy of the light used (they can be seen when higher energy light is used). The broad band in Fig. 1.8 (the individual peaks within this band are caused by different vibrational levels; see Chapter 7) corresponds to the four electrons in the degenerate orbitals 3 and 4. The triple bond of N_2 is therefore composed of these two orbitals and orbital 1. The bands corresponding to orbitals 2 and 5 are narrow; hence these orbitals contribute little to the bonding and may be regarded as the two unshared pairs of $\ddot{N}\equiv\ddot{N}$. Note that this result is contrary to that expected from a naive consideration of orbital roverlaps, where it would be expected that the two unshared pairs would be those of orbitals 1 and 2, resulting from the overlap of the filled $2s$ orbitals, and that the triple bond would be composed of orbitals 3, 4, and 5, resulting from overlap of the p orbitals. This example is one illustration of the value of photoelectron spectroscopy.

The photoelectron spectrum of methane[16] shows two bands,[17] at ~23 and 14 eV, and not the single band we would expect from the equivalency of the four C—H

---

[15]From Brundle, C.R.; Robin, M.B., in Nachod, F.C.; Zuckerman, J.J. *Determination of Organic Structures by Physical Methods, Vol. 3*, Academic Press, NY, *1971*, p. 18.

[16]Brundle, C.R.; Robin, M.B.; Basch, H. *J. Chem. Phys. 1970, 53*, 2196; Baker, A.D.; Betteridge, D.; Kemp, N.R.; Kirby, R.E. *J. Mol. Struct. 1971, 8*, 75; Potts, A.W.; Price, W.C. *Proc. R. Soc. London, Ser A 1972, 326*, 165.

[17]A third band, at 290 eV, caused by the $1s$ electrons of carbon, can also found if radiation of sufficiently high energy is used.

bonds. The reason is that ordinary sp^3 hybridization is not adequate to explain phenomena involving ionized molecules (e.g., the CH_4^+ radical ion, which is left behind when an electron is ejected from methane). For these phenomena it is necessary to use other combinations of atomic orbitals (see p. 9). The band at 23 eV comes from two electrons in a low-energy level (called the a_1 level), which can be regarded as arising from a combination of the $2s$ orbital of carbon with an appropriate combination of hydrogen $1s$ orbitals. The band at 14 eV comes from six electrons in a triply degenerate level (the t_2 level), arising from a combination of the three $2p$ orbitals of carbon with other combinations of $1s$ hydrogen orbitals. As was mentioned above, most physical and chemical processes cannot distinguish these levels, but photoelectron spectroscopy can. The photoelectron spectra of many other organic molecules are known as well,[18] including monocyclic alkenes, in which bands <10 eV are due to π-orbital ionization and those >10 eV originate from ionization of s-orbitals only.[19]

ELECTRONIC STRUCTURES OF MOLECULES

For each molecule, ion, or free radical that has only localized electrons, it is possible to draw an electronic formula, called a *Lewis structure*, that shows the location of these electrons. Only the valence electrons are shown. Valence electrons may be found in covalent bonds connecting two atoms or they may be unshared.[20] The student must be able to draw these structures correctly, since the position of electrons changes in the course of a reaction, and it is necessary to know where the electrons are initially before one can follow where they are going. To this end, the following rules operate:

1. The total number of valence electrons in the molecule (or ion or free radical) must be the sum of all outer-shell electrons "contributed" to the molecule by each atom plus the negative charge or minus the positive charge, for the case of ions. Thus, for H_2SO_4, there are 2 (one for each hydrogen) + 6 (for the sulfur) + 24 (6 for each oxygen) = 32; while for SO_4^{2-}, the number is also 32, since each atom "contributes" 6 plus 2 for the negative charge.

2. Once the number of valence electrons has been ascertained, it is necessary to determine which of them are found in covalent bonds and which are unshared. Unshared electrons (either a single electron or a pair) form part of the outer shell of just one atom, but electrons in a covalent bond are part of the outer shell of both atoms of the bond. *First-row atoms* (B, C, N, O, F) *can have a maximum of eight valence electrons*, and usually have this number, although some cases are known where a first-row atom has only six or seven.

[18]See Robinson, J.W., *Practical Handbook of Spectroscopy*, CRC Press, Boca Raton, FL, *1991*, p. 178.
[19]Novak, I.; Potts, A.W. *Tetrahedron* **1997**, *53*, 14713.
[20]It has been argued that although the Lewis picture of two electrons making up a covalent bond may work well for organic compounds, it cannot be successfully applied to the majority of inorganic compounds: Jørgensen, C.K. *Top. Curr. Chem.* **1984**, *124*, 1.

Where there is a choice between a structure that has six or seven electrons around a first-row atom and one in which all such atoms have an octet, it is the latter that generally has the lower energy and that consequently exists. For example, ethylene is

$$
\begin{array}{ccc}
\text{H}\quad\text{H} & & \text{H}\quad\text{H} & & \text{H}\quad\text{H} \\
\overset{\cdot}{\text{C}}=\text{C} & \text{and not} & \overset{\oplus}{\text{C}}-\overset{\cdot}{\text{C}}{:}^{\ominus} & \text{or} & \overset{\cdot}{\text{C}}-\text{C}\text{\textbullet} \\
\text{H}\quad\text{H} & & \text{H}\quad\text{H} & & \text{H}\quad\text{H}
\end{array}
$$

There are a few exceptions. In the case of the molecule O_2, the structure $:\overset{\cdot}{\text{O}}-\overset{\cdot}{\text{O}}:$ has a lower energy than $:\overset{\cdot\cdot}{\text{O}}=\overset{\cdot\cdot}{\text{O}}:$. Although first-row atoms are limited to 8 valence electrons, this is not so for second-row atoms, which can accommodate 10 or even 12 because they can use their empty d orbitals for this purpose.[21] For example, PCl_5 and SF_6 are stable compounds. In SF_6, one s and one p electron from the ground state $3s^2 3p^4$ of the sulfur are promoted to empty d orbitals, and the six orbitals hybridize to give six $sp^3 d^2$ orbitals, which point to the corners of a regular octahedron.

3. It is customary to show the formal charge on each atom. For this purpose, an atom is considered to "own" all unshared electrons, but only *one-half of the electrons in covalent bonds*. The sum of electrons that thus "belong" to an atom is compared with the number "contributed" by the atom. An excess belonging to the atom results in a negative charge, and a deficiency results in a positive charge. The total of the formal charges on all atoms equals the charge on the whole molecule or ion. Note that the counting procedure is not the same for determining formal charge as for determining the number of valence electrons. For both purposes, an atom "owns" all unshared electrons, but for outer-shell purposes it "owns" both the electrons of the covalent bond, while for formal-charge purposes it "owns" only one-half of these electrons.

Examples of electronic structures are (as mentioned in Ref. 4, an electron pair, whether unshared or in a bond, is represented by a straight line):

$$
\text{H}-\overset{\cdot\cdot}{\text{O}}-\overset{\overset{\displaystyle :\overset{\cdot\cdot}{\text{O}}-\text{H}}{|}}{\underset{+\ \overset{\cdot\cdot}{\text{O}}:^{-}}{\text{S}}}
\qquad
\text{H}-\overset{\overset{\displaystyle\text{H}}{|}}{\underset{\text{H}}{\text{C}}}\text{\textbullet}
\qquad
\text{H}_3\text{C}-\overset{\overset{\displaystyle\text{CH}_3}{|}}{\underset{\text{CH}_3}{\overset{+}{\text{N}}}}-\overset{\cdot\cdot}{\text{O}}:^{-}
\qquad
\overset{\text{H}_3\text{C}}{\underset{\text{H}_3\text{C}}{{>}}}\text{=}\overset{\cdot\cdot}{\text{N}}\overset{}{\underset{\text{CH}_3}{}}
\qquad
\text{F}-\overset{\overset{\displaystyle\text{F}}{}}{\underset{\text{F}}{\text{B}}}
$$

A coordinate-covalent bond, represented by an arrow, is one in which both electrons come from the same atom; that is, the bond can be regarded as being formed by the overlap of an orbital containing two electrons with an empty one. Thus trimethylamine oxide would be represented

$$
\text{H}_3\text{C}-\overset{\overset{\displaystyle\text{CH}_3}{|}}{\underset{\text{CH}_3}{\overset{+}{\text{N}}}}\rightarrow\overset{\cdot\cdot}{\text{O}}:^{-}
$$

[21]For a review concerning sulfur compounds with a valence shell larger than eight, see Salmond, W.G. *Q. Rev. Chem. Soc.* **1968**, *22*, 235.

For a coordinate-covalent bond the rule concerning formal charge is amended, so that both electrons count for the donor and neither for the recipient. Thus the nitrogen and oxygen atoms of trimethylamine oxide bear no formal charges. However, it is apparent that the electronic picture is exactly the same as the picture of trimethylamine oxide given just above, and we have our choice of drawing an arrowhead or a charge separation. Some compounds, for example, amine oxides, must be drawn one way or the other. It seems simpler to use charge separation, since this spares us from having to consider as a "different" method of bonding a way that is really the same as ordinary covalent bonding once the bond has formed.

ELECTRONEGATIVITY

The electron cloud that bonds two atoms is not symmetrical (with respect to the plane that is the perpendicular bisector of the bond) except when the two atoms are the same and have the same substituents. The cloud is necessarily distorted toward one side of the bond or the other, depending on which atom (nucleus plus electrons) maintains the greater attraction for the cloud. This attraction is called *electronegativity*;[22] and it is greatest for atoms in the upper-right corner of the periodic table and lowest for atoms in the lower-left corner. Thus a bond between fluorine and chlorine is distorted so that there is a higher probability of finding the electrons near the fluorine than near the chlorine. This gives the fluorine a partial negative charge and the chlorine a partial positive charge.

A number of attempts have been made to set up quantitative tables of electronegativity that indicate the direction and extent of electron-cloud distortion for a bond between any pair of atoms. The most popular of these scales, devised by Pauling, is based on bond energies (see p. 27) of diatomic molecules. It is rationalized that if the electron distribution were symmetrical in a molecule A—B, the bond energy would be the mean of the energies of A—A and B—B, since in these cases the cloud must be undistorted. If the actual bond energy of A—B is higher than this (and it usually is), it is the result of the partial charges, since the charges attract each other and make a stronger bond, which requires more energy to break. It is necessary to assign a value to one element arbitrarily (F = 4.0). Then the electronegativity of another is obtained from the difference between the actual energy of A—B and the mean of A—A and B—B (this difference is called Δ) by the formula

$$x_A - x_B = \sqrt{\frac{\Delta}{23.06}}$$

where x_A and x_B are the electronegativities of the known and unknown atoms and 23.06 is an arbitrary constant. Part of the scale derived from this treatment is shown in Table 1.1.

---

[22]For a collection of articles on this topic, see Sen, K.D.; Jørgensen, C.K. *Electronegativity* (Vol. 6 of *Structure and Bonding*); Springer: NY, *1987*. For a review, see Batsanov, S.S. *Russ. Chem. Rev.* *1968*, *37*, 332.

TABLE 1.1. Electronegativities of Some Atoms on the Pauling[23] and Sanderson[24] Scales

Element	Pauling	Sanderson	Element	Pauling	Sanderson
F	4.0	4.000	H	2.1	2.592
O	3.5	3.654	P	2.1	2.515
Cl	3.0	3.475	B	2.0	2.275
N	3.0	3.194	Si	1.8	2.138
Br	2.8	3.219	Mg	1.2	1.318
S	2.5	2.957	Na	0.9	0.835
I	2.5	2.778	Cs	0.7	0.220
C	2.5	2.746			

Other treatments[25] have led to scales that are based on different principles, for example, the average of the ionization potential and the electron affinity,[26] the average one-electron energy of valence-shell electrons in ground-state free atoms,[27] or the "compactness" of an atom's electron cloud.[24] In some of these treatments electronegativities can be calculated for different valence states, for different hybridizations (e.g., sp carbon atoms are more electronegative than sp^2, which are still more electronegative than sp^3),[28] and even differently for primary, secondary, and tertiary carbon atoms. Also, electronegativities can be calculated for groups rather than atoms (Table 1.2).[29]

Electronegativity information can be obtained from NMR spectra. In the absence of a magnetically anisotropic group[30] the chemical shift of a ^1H or a ^{13}C nucleus is approximately proportional to the electron density around it and hence to the electronegativity of the atom or group to which it is attached. The greater the electronegativity of the atom or group, the lower the electron density around the proton, and the further downfield the chemical shift. An example of the use of this correlation is found in the variation of chemical shift of the *ring* protons in the series

[23]Taken from Pauling, L. *The Nature of the Chemical Bond*, 3rd ed.; Cornell University Press: Ithaca, NY, p. 93, except for the value for Na, which is from Sanderson, R.T. *J. Am. Chem. Soc.* **1983**, *105*, 2259; *J. Chem. Educ.* **1988**, *65*, 112, 223.

[24]See Sanderson, R.T. *J. Am. Chem. Soc.* **1983**, *105*, 2259; *J. Chem. Educ.* **1988**, *65*, 112, 223.

[25]For several sets of electronegativity values, see Huheey, J.E. *Inorganic Chemistry*, 3rd ed., Harper and Row: NY, **1983**, pp. 146–148; Mullay, J., in Sen, K.D.; Jørgensen, C.K. *Electronegativity* (Vol. 6 of *Structure and Bonding*), Springer, NY, **1987**, p. 9.

[26]Mulliken, R.S. *J. Chem. Phys.* **1934**, *2*, 782; Iczkowski, R.P.; Margrave, J.L. *J. Am. Chem. Soc.* **1961**, *83*, 3547; Hinze, J.; Jaffé, H.H. *J. Am. Chem. Soc.* **1962**, *84*, 540; Rienstra-Kiracofe, J.C.; Tschumper, G.S.; Schaefer III, H.F.; Nandi, S.; Ellison, G.B. *Chem. Rev.* **2002**, *102*, 231.

[27]Allen, L.C. *J. Am. Chem. Soc.* **1989**, *111*, 9003.

[28]Walsh, A.D. *Discuss. Faraday Soc.* **1947**, *2*, 18; Bergmann, D.; Hinze, J., in Sen, K.D.; Jørgensen, C.K. *Electronegativity* (Vol. 6 of *Structure and Bonding*), Springer, NY, **1987**, pp. 146–190.

[29]Inamoto, N.; Masuda, S. *Chem. Lett.* **1982**, 1003. For a review of group electronegativities, see Wells, P.R. *Prog. Phys. Org. Chem.* **1968**, *6*, 111. See also Bratsch, S.G. *J. Chem. Educ.*, **1988**, *65*, 223; Mullay, J. *J. Am. Chem. Soc.* **1985**, *107*, 7271; Zefirov, N.S.; Kirpichenok, M.A.; Izmailov, F.F.; Trofimov, M.I. *Dokl. Chem.* **1987**, *296*, 440; Boyd, R.J.; Edgecombe, K.E. *J. Am. Chem. Soc.* **1988**, *110*, 4182.

[30]A magnetically anisotropic group is one that is not equally magnetized along all three axes. The most common such groups are benzene rings (see p. 55) and triple bonds.

TABLE 1.2. Some Group Electronegativites Relative to H = 2.176.[29]

CH$_3$	2.472	CCl$_3$	2.666
CH$_3$CH$_2$	2.482	C$_6$H$_5$	2.717
CH$_2$Cl	2.538	CF$_3$	2.985
CBr$_3$	2.561	C≡N	3.208
CHCl$_2$	2.602	NO$_2$	3.421

toluene, ethylbenzene, isopropylbenzene, *tert*-butylbenzene (there is a magnetically anisotropic group here, but its effect should be constant throughout the series). It is found that the electron density surrounding the ring protons decreases[31] in the order given.[32] However, this type of correlation is by no means perfect, since all the measurements are being made in a powerful field, which itself may affect the electron density distribution. Coupling constants between the two protons of a system $-\underset{|}{C}H\underset{|}{C}H-X$ have also been found to depend on the electronegativity of X.[33]

When the difference in electronegativities is great, the orbital may be so far over to one side that it barely covers the other nucleus. This is an *ionic bond*, which is seen to arise naturally out of the previous discussion, leaving us with basically only one type of bond in organic molecules. Most bonds can be considered intermediate between ionic and covalent. We speak of percent ionic character of a bond, which indicates the extent of electron-cloud distortion. There is a continuous gradation from ionic to covalent bonds.

DIPOLE MOMENT

The *dipole moment* is a property of the molecule that results from charge separations like those discussed above. However, it is not possible to measure the dipole moment of an individual bond within a molecule; we can measure only the total moment of the molecule, which is the vectorial sum of the individual bond moments.[34] These individual moments are roughly the same from molecule to molecule,[35] but this constancy is by no means universal. Thus, from the dipole moments of toluene and nitrobenzene (Fig. 1.10)[36] we should expect the moment of *p*-nitrotoluene to be ~4.36 D.

[31]This order is opposite to that expected from the field effect (p. 19). It is an example of the Baker–Nathan order (p. 96).

[32]Moodie, R.B.; Connor, T.M.; Stewart, R. *Can. J. Chem.* **1960**, *38*, 626.

[33]Williamson, K.L. *J. Am. Chem. Soc.* **1963**, *85*, 516; Laszlo, P.; Schleyer, P.v.R. *J. Am. Chem. Soc.* **1963**, *85*, 2709; Niwa, J. *Bull. Chem. Soc. Jpn.* **1967**, *40*, 2192.

[34]For methods of determining dipole moments and discussions of their applications, see Exner, O. *Dipole Moments in Organic Chemistry*; Georg Thieme Publishers: Stuttgart, **1975**. For tables of dipole moments, see McClellan, A.L. *Tables of Experimental Dipole Moments*, Vol. 1; W.H. Freeman: San Francisco, **1963**; Vol. 2, Rahara Enterprises: El Cerrito, CA, **1974**.

[35]For example, see Koudelka, J.; Exner, O. *Collect. Czech. Chem. Commun.* **1985**, *50*, 188, 200.

[36]The values for toluene, nitrobenzene, and *p*-nitrotoluene are from MacClellan, A.L., *Tables of Experimental Dipole Moments*, Vol. 1, W.H. Freeman, San Francisco, **1963**; Vol. 2, Rahara Enterprises, El Cerrito, CA, **1974**. The values for phenol and *p*-cresol were determined by Goode, E.V.; Ibbitson, D.A. *J. Chem. Soc.* **1960**, 4265.

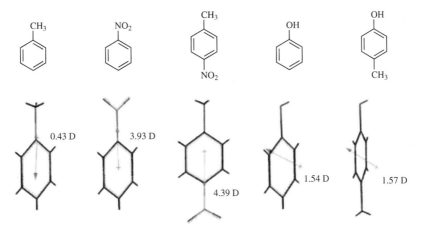

Fig. 1.10. Some dipole moments, in debye units, measured in benzene. In the 3D model, the arrow indicates the direction of the dipole moment for the molecule, pointing to the negative part of the molecule.[36]

The actual value 4.39 D is reasonable. However, the moment of *p*-cresol (1.57 D) is quite far from the predicted value of 1.11 D. In some cases, molecules may have substantial individual bond moments but no total moments at all because the individual moments are canceled out by the overall symmetry of the molecule. Some examples are CCl_4, *trans*-1,2-dibromoethene, and *p*-dinitrobenzene.

Because of the small difference between the electronegativities of carbon and hydrogen, alkanes have very small dipole moments, so small that they are difficult to measure. For example, the dipole moment of isobutane is 0.132 D[37] and that of propane is 0.085 D.[38] Of course, methane and ethane, because of their symmetry, have no dipole moments.[39] Few organic molecules have dipole moments >7 D.

INDUCTIVE AND FIELD EFFECTS

The C—C bond in ethane has no polarity because it connects two equivalent atoms. However, the C—C bond in chloroethane is polarized by the presence of the electronegative chlorine atom. This polarization is actually the sum of two effects. In the first of these, the C-1 atom, having been deprived of some of its electron density by the

$$\overset{\delta+}{^{1}CH_3} \longrightarrow \overset{\delta+}{^{2}CH_2} \longrightarrow \overset{\delta-}{Cl}$$

[37]Maryott, A.A.; Birnbaum, G. *J. Chem. Phys.* **1956**, *24*, 1022; Lide Jr., D.R.; Mann, D.E. *J. Chem. Phys.* **1958**, *29*, 914.

[38]Muenter, J.S.; Laurie, V.W. *J. Chem. Phys.* **1966**, *45*, 855.

[39]Actually, symmetrical tetrahedral molecules like methane do have extremely small dipole moments, caused by centrifugal distortion effects; these moments are so small that they can be ignored for all practical purposes. For CH_4 μ is $\sim 5.4 \times 10^{-6}$ *D*: Ozier, I. *Phys. Rev. Lett.* **1971**, *27*, 1329; Rosenberg, A.; Ozier, I.; Kudian, A.K. *J. Chem. Phys.* **1972**, *57*, 568.

greater electronegativity of Cl, is partially compensated by drawing the C—C electrons closer to itself, resulting in a polarization of this bond and a slightly positive charge on the C-2 atom. This polarization of one bond caused by the polarization of an adjacent bond is called the *inductive effect*. The effect is greatest for adjacent bonds but may also be felt farther away; thus the polarization of the C—C bond causes a (slight) polarization of the three methyl C—H bonds. The other effect operates not through bonds, but directly through space or solvent molecules, and is called the *field effect*.[40] It is often very difficult to separate the two kinds of effect, but it has been done in a number of cases, generally by taking advantage of the fact that the field effect depends on the geometry of the molecule but the inductive effect depends only on the nature of the bonds. For example, in isomers **1** and **2**[41] the inductive effect of the chlorine atoms on the position of the electrons in the COOH group (and hence on the

1	**2**
$pK_a = 6.07$	$pK_a = 5.67$

acidity, see Chapter 8) should be the same since the same bonds intervene; but the field effect is different because the chlorines are closer in space to the COOH in **1** than they are in **2**. Thus a comparison of the acidity of **1** and **2** should reveal whether a field effect is truly operating. The evidence obtained from such experiments is overwhelming that field effects are much more important than inductive effects.[42] In most cases, the two types of effect are considered together; in this book, we will not attempt to separate them, but will use the name *field effect* to refer to their combined action.[43]

Functional groups can be classified as electron-withdrawing $(-I)$ or electron-donating $(+I)$ groups relative to hydrogen. This means, for example, that NO_2, a $-I$ group, will draw electrons to itself more than a hydrogen atom would if it

---

[40]Roberts, J.D.; Moreland, Jr., W.T. *J. Am. Chem. Soc.* **1953**, *75*, 2167.

[41]This example is from Grubbs, E.J.; Fitzgerald, R.; Phillips, R.E.; Petty, R. *Tetrahedron* **1971**, *27*, 935.

[42]For example, see Dewar, M.J.S.; Grisdale, P.J. *J. Am. Chem. Soc.* **1962**, *84*, 3548; Stock, L.M. *J. Chem. Educ.*, **1972**, *49*, 400; Golden, R.; Stock, L.M. *J. Am. Chem. Soc.* **1972**, *94*, 3080; Liotta, C.; Fisher, W.F.; Greene Jr., G.H.; Joyner, B.L. *J. Am. Chem. Soc.* **1972**, *94*, 4891; Wilcox, C.F.; Leung, C. *J. Am. Chem. Soc.* **1968**, *90*, 336; Butler, A.R. *J. Chem. Soc. B* **1970**, 867; Rees, J.H.; Ridd, J.H.; Ricci, A. *J. Chem. Soc. Perkin Trans. 2* **1976**, 294; Topsom, R.D. *J. Am. Chem. Soc.* **1981**, *103*, 39; Grob, C.A.; Kaiser, A.; Schweizer, T. *Helv. Chim. Acta* **1977**, *60*, 391; Reynolds, W.F. *J. Chem. Soc. Perkin Trans. 2* **1980**, 985; *Prog. Phys. Org. Chem.* **1983**, *14*, 165-203; Adcock, W.; Butt, G.; Kok, G.B.; Marriott, S.; Topsom, R.D. *J. Org. Chem.* **1985**, *50*, 2551; Schneider, H.; Becker, N. *J. Phys. Org. Chem.* **1989**, *2*, 214; Bowden, K.; Ghadir, K.D.F. *J. Chem. Soc. Perkin Trans. 2* **1990**, 1333. Inductive effects may be important in certain systems. See, for example, Exner, O.; Fiedler, P. *Collect. Czech. Chem. Commun.* **1980**, *45*, 1251; Li, Y.; Schuster, G.B. *J. Org. Chem.* **1987**, *52*, 3975.

[43]There has been some question as to whether it is even meaningful to maintain the distinction between the two types of effect: see Grob, C.A. *Helv. Chim. Acta* **1985**, *68*, 882; Lenoir, D.; Frank, R.M. *Chem. Ber.* **1985**, *118*, 753; Sacher, E. *Tetrahedron Lett.* **1986**, *27*, 4683.

TABLE 1.3. Field Effects of Various Groups Relative to Hydrogen[a]

+I		−I	
O^-	NR_3^+	COOH	OR
COO^-	SR_2^+	F	COR
CR_3	NH_3^+	Cl	SH
CHR_2	NO_2	Br	SR
CH_2R	SO_2R	I	OH
CH_3	CN	OAr	$C\equiv CR$
D	SO_2Ar	COOR	Ar
			$C\equiv CR_2$

[a]The groups are listed approximately in order of decreasing strength for both −I and +I groups.

occupied the same position in the molecule.

$$O_2N \longleftarrow CH_2 \longleftarrow Ph$$
$$H \longrightarrow CH_2 \longrightarrow Ph$$

Thus, in α-nitrotoluene, the electrons in the N—C bond are farther away from the carbon atom than the electrons in the H—C bond of toluene. Similarly, the electrons of the C—Ph bond are farther away from the ring in α-nitrotoluene than they are in toluene. Field effects are always comparison effects. We compare the −I or +I effect of one group with another (usually hydrogen). It is commonly said that, compared with hydrogen, the NO$_2$ group is electron-withdrawing and the O$^-$ group electron-donating or electron releasing. However, there is no actual donation or withdrawal of electrons, though these terms are convenient to use; there is merely a difference in the position of electrons due to the difference in electronegativity between H and NO$_2$ or between H and O$^-$.

Table 1.3 lists a number of the most common −I and +I groups.[44] It can be seen that compared with hydrogen, most groups are electron withdrawing. The only electrondonating groups are groups with a formal negative charge (but not even all these), atoms of low electronegativity (Si,[45] Mg, etc., and perhaps alkyl groups). Alkyl groups[46] were formerly regarded as electron donating, but many examples of behavior have been found that can be interpreted only by the conclusion that alkyl groups are electron withdrawing compared with hydrogen.[47] In accord with this is the value of 2.472 for the group electronegativity of CH$_3$ (Table 1.2) compared with 2.176 for H. We will see that when an alkyl group is attached to an unsaturated or trivalent carbon (or other atom), its behavior is best explained by assuming it is +I (see, e.g., pp. 239, 251, 388, 669), but when it is connected to a saturated atom, the results are not as clear,

[44]See also Ceppi, E.; Eckhardt, W.; Grob, C.A. *Tetrahedron Lett.* **1973**, 3627.
[45]For a review of field and other effects of silicon-containing groups, see Bassindale, A.R.; Taylor. P.G., in Patai, S.; Rappoport, Z. *The Chemistry of Organic Silicon Compounds*, pt. 2, Wiley, NY, **1989**, pp. 893–963.
[46]For a review of the field effects of alkyl groups, see Levitt, L.S.; Widing, H.F. *Prog. Phys. Org. Chem.* **1976**, *12*, 119.
[47]See Sebastian, J.F. *J. Chem. Educ.* **1971**, *48*, 97.

and alkyl groups seem to be $+I$ in some cases and $-I$ in others[48] (see also p. 391). Similarly, it is clear that the field-effect order of alkyl groups attached to unsaturated systems is tertiary > secondary > primary > CH_3, but this order is not always maintained when the groups are attached to saturated systems. Deuterium is electron-donating with respect to hydrogen.[49] Other things being equal, atoms with sp bonding generally have a greater electron-withdrawing power than those with sp^2 bonding, which in turn have more electron-withdrawing power than those with sp^3 bonding.[50] This accounts for the fact that aryl, vinylic, and alkynyl groups are $-I$. Field effects always decrease with increasing distance, and in most cases (except when a very powerful $+I$ or $-I$ group is involved), cause very little difference in a bond four bonds away or more. There is evidence that field effects can be affected by the solvent.[51]

For discussions of field effects on acid and base strength and on reactivity, see Chapters 8 and 9, respectively.

BOND DISTANCES[52]

The distances between atoms in a molecule are characteristic properties of the molecule and can give us information if we compare the same bond in different molecules. The chief methods of determining bond distances and angles are X-ray diffraction (only for solids), electron diffraction (only for gases), and spectroscopic methods, especially microwave spectroscopy. The distance between the atoms of a bond is not constant, since the molecule is always vibrating; the measurements obtained are therefore average values, so that different methods give different results.[53] However, this must be taken into account only when fine distinctions are made.

Measurements vary in accuracy, but indications are that similar bonds have fairly constant lengths from one molecule to the next, though exceptions are known.[54] The variation is generally less than 1%. Table 1.4 shows

[48]See, for example, Schleyer, P. von.R.; Woodworth, C.W. *J. Am. Chem. Soc.* **1968**, *90*, 6528; Wahl Jr., G.H.; Peterson Jr., M.R. *J. Am. Chem. Soc.* **1970**, *92*, 7238. The situation may be even more complicated. See, for example, Minot, C.; Eisenstein, O.; Hiberty, P.C.; Anh, N.T. *Bull. Soc. Chim. Fr.* **1980**, II-119.

[49]Streitwieser Jr., A.; Klein, H.S. *J. Am. Chem. Soc.* **1963**, *85*, 2759.

[50]Bent, H.A. *Chem. Rev.* **1961**, *61*, 275, p. 281.

[51]See Laurence, C.; Berthelot, M.; Lucon, M.; Helbert, M.; Morris, D.G.; Gal, J. *J. Chem. Soc. Perkin Trans. 2* **1984**, 705.

[52]For tables of bond distances and angles, see Allen, F.H.; Kennard, O.; Watson, D.G.; Brammer, L.; Orpen, A.G.; Taylor, R. *J. Chem. Soc. Perkin Trans. 2* **1987**, S1–S19 (follows p. 1914); Tables of Interatomic Distances and Configurations in Molecules and Ions *Chem. Soc. Spec. Publ.* No. 11, **1958**; Interatomic Distances Supplement *Chem. Soc. Spec. Publ.* No. 18, **1965**; Harmony, M.D. Laurie, V.W.; Kuczkowski, R.L.; Schwendeman, R.H.; Ramsay, D.A.; Lovas, F.J.; Lafferty, W.J.; Maki, A.G. *J. Phys. Chem. Ref. Data* **1979**, *8*, 619–721. For a review of molecular shapes and energies for many small organic molecules, radicals, and cations calculated by molecular-orbital methods, see Lathan, W.A.; Curtiss, L.A.; Hehre, W.J.; Lisle, J.B.; Pople, J.A. *Prog. Phys. Org. Chem.* **1974**, *11*, 175. For a discussion of substituent effects on bond distances, see Topsom, R.D. *Prog. Phys. Org. Chem.* **1987**, *16*, 85.

[53]Burkert, U.; Allinger, N.L. *Molecular Mechanics*; ACS Monograph 177, American Chemical Society, Washington, **1982**, pp. 6–9; Whiffen, D.H. *Chem. Ber.* **1971**, *7*, 57–61; Stals, J. *Rev. Pure Appl. Chem.* **1970**, *20*, 1, pp. 2–5.

[54]Schleyer, P.v.R.; Bremer, M. *Angew. Chem. Int. Ed.* **1989**, *28*, 1226.

TABLE 1.4. Bond Lengths between sp^3 Carbons in Some Compounds

C—C bond in	Reference	Bond length, Å
Diamond	55	1.544
C_2H_6	56	1.5324 ± 0.0011
C_2H_5Cl	57	1.5495 ± 0.0005
C_3H_8	58	1.532 ± 0.003
Cyclohexane	59	1.540 ± 0.015
tert-Butyl chloride	60	1.532
n-Butane to *n*-heptane	61	$1.531 - 1.534$
Isobutane	62	1.535 ± 0.001

distances for single bonds between two sp^3 carbons. However, an analysis of C—OR bond distances in >2000 ethers and carboxylic esters (all with sp^3 carbon) shows that this distance increases with increasing electron withdrawal in the R group and as the C changes from primary to secondary to tertiary.[63] For these compounds, mean bond lengths of the various types ranged from 1.418 to 1.475 Å. Certain substituents can also influence bond length. The presence of a silyl substituent β- to a C—O (ester) linkage can lengthen the C—O, thereby weakening it.[64] This is believed to result from σ-σ* interactions in which the C—Si σ-bonding orbital acts as the donor and the C—O σ* orbitals acts as the receptor.

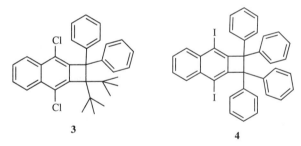

3   **4**

Although a typical carbon—carbon single bond has a bond length of ~1.54 Å, certain molecules are known that have significantly longer bond lengths.[65] Calculations

[55]Lonsdale, K. *Phil. Trans. R. Soc. London* **1947**, *A240*, 219.

[56]Bartell, L.S.; Higginbotham, H.K. *J. Chem. Phys.* **1965**, *42*, 851.

[57]Wagner, R.S.; Dailey, B.P. *J. Chem. Phys.* **1957**, *26*, 1588.

[58]Iijima, T. *Bull. Chem. Soc. Jpn.* **1972**, *45*, 1291.

[59]Tables of Interatomic Distances, Ref. 52.

[60]Momany, F.A.; Bonham, R.A.; Druelinger, M.L. *J. Am. Chem. Soc.* **1963**, *85*, 3075; also see, Lide, Jr., D.R.; Jen, M. *J. Chem. Phys.* **1963**, *38*, 1504.

[61]Bonham, R.A.; Bartell, L.S.; Kohl, D.A. *J. Am. Chem. Soc.* **1959**, *81*, 4765.

[62]Hilderbrandt, R.L.; Wieser, J.D. *J. Mol. Struct.* **1973**, *15*, 27.

[63]Allen, F.H.; Kirby, A.J. *J. Am. Chem. Soc.* **1984**, *106*, 6197; Jones, P.G.; Kirby, A.J. *J. Am. Chem. Soc.* **1984**, *106*, 6207.

[64]White, J.M.; Robertson, G.B. *J. Org. Chem.* **1992**, *57*, 4638.

[65]Kaupp, G.; Boy, J *Angew. Chem. Int. Ed.* **1997**, *36*, 48.

have been done for unstable molecules that showed them to have long bond lengths, and an analysis of the X-ray structure for the photoisomer of [2.2]-tetrabenzoparacyclophane (see Chapter 2) showed a C—C bond length of 1.77 Å.[66,65] Long bond lengths have been observed in stable molecules such as benzocyclobutane derivatives.[67] A bond length of 1.729 Å was reliably measured in 1,1-di-*tert*-butyl-2, 2-diphenyl-3,8-dichlorocyclobutan[*b*]naphthalene, **3**.[68] X-ray analysis of several of these derivations confirmed the presence of long C—C bonds, with **4** having a confirmed bond length of 1.734 Å.[69]

Bond distances for some important bond types are given in Table 1.5.[70] As can be seen in this table, carbon bonds are shortened by increasing *s* character.

TABLE 1.5. Bond distances[a]

Bond Type	Length, Å	Typical Compounds
C—C		
sp^3-sp^3	1.53	
sp^3-sp^2	1.51	Acetaldehyde, toluene, propene
sp^3-sp	1.47	Acetonitrile, propyne
sp^2-sp^2	1.48	Butadiene, glyoxal, biphenyl
sp^2-sp	1.43	Acrylonitrile, vinylacetylene
$sp-sp$	1.38	Cyanoacetylene, butadiyne
C=C		
sp^2-sp^2	1.32	Ethylene
sp^2-sp	1.31	Ketene, allenes
$sp-sp$[71]	1.28	Butatriene, carbon suboxide
C≡C[72]		
$sp-sp$	1.18	Acetylene
C–H[73]		
sp^3-H	1.09	Methane
sp^2-H	1.08	Benzene, ethylene
$sp-H$[74]	1.08	**HCN**, acetylene

[66]Ehrenberg, M. *Acta Crystallogr.* **1966**, *20*, 182.

[67]Toda, F.; Tanaka, K.; Stein, Z.; Goldberg, I *Acta Crystallogr., Sect. C* **1996**, *52*, 177.

[68]Toda, F.; Tanaka, K.; Watanabe, M.; Taura, K.; Miyahara, I.; Nakai, T.; Hirotsu, K. *J. Org. Chem.* **1999**, *64*, 3102.

[69]Tanaka, K.; Takamoto, N.; Tezuka, Y.; Kato, M.; Toda, F. *Tetrahedron* **2001**, *57*, 3761.

[70]Except where noted, values are from Allen, F.H.; Kennard, O.; Watson, D.G.; Brammer, L.; Orpen, A.G.; Taylor, R. *J. Chem. Soc. Perkin Trans. 2* **1987**, S1-S19 (follows p. 1914). In this source, values are given to three significant figures.

[71]Costain, C.C.; Stoicheff, B.P. *J. Chem. Phys.* **1959**, *30*, 777.

[72]For a full discussion of alkyne bond distances, see Simonetta, M.; Gavezzotti, A, in Patai, S. *The Chemistry of the Carbon–Carbon Triple Bond,* Wiley, NY, *1978*.

[73]For an accurate method of C–H bond distance determination, see Henry, B.R. *Acc. Chem. Res.* **1987**, *20*, 429.

[74]Bartell, L.S.; Roth, E.A.; Hollowell, C.D.; Kuchitsu, K.; Young, Jr., J.E. *J. Chem. Phys.* **1965**, *42*, 2683.

TABLE 1.5. (*continued*)

Bond Type	Length, Å	Typical Compounds
C–O		
sp^3–O	1.43	Dimethyl ether, ethanol
sp^2–O	1.34	Formic acid
C=O		
sp^2–O	1.21	Formaldehyde, formic acid
sp–O[59]	1.16	**CO_2**
C–N		
sp^3–N	1.47	Methylamine
sp^2–N	1.38	Formamide
C=N		
sp^2–N	1.28	Oximes, imines
C≡N		
sp–N	1.14	**HCN**
C–S		
sp^3–S	1.82	Methanethiol
sp^2–S	1.75	Diphenyl sulfide
sp–S	1.68	CH_3SCN
C=S		
sp–S	1.67	**CS_2**

C–halogen[75]	**F**	**Cl**	**Br**	**I**
sp^3–halogen	1.40	1.79	1.97	2.16
sp^2–halogen	1.34	1.73	1.88	2.10
sp–halogen	1.27[76]	1.63	1.79[77]	1.99[77]

[a]The values given are average lengths and do not necessarily apply exactly to the compounds mentioned.[70]

This is most often explained by the fact that, as the percentage of s character in a hybrid orbital increases, the orbital becomes more like an s orbital and hence is held more tightly by the nucleus than an orbital with less s character. However, other explanations have also been offered (see p. 39), and the matter is not completely settled.

Indications are that a C—D bond is slightly shorter than a corresponding C—H bond. Thus, electron-diffraction measurements of C_2H_6 and C_2D_6 showed a C—H bond distance of 1.1122 ± 0.0012 Å and a C—D distance of 1.1071 ± 0.0012 Å.[56]

BOND ANGLES

It might be expected that the bond angles of sp^3 carbon would always be the tetrahedral angle $109°28'$, but this is so only where the four groups are identical, as in

[75]For reviews of carbon-halogen bonds, see Trotter, J., in Patai, S. *The Chemistry of the Carbon–Halogen Bond*, pt. 1, Wiley, NY, *1973*, pp. 49–62; Mikhailov, B.M. *Russ. Chem. Rev. 1971*, *40*, 983.
[76]Lide, Jr., D.R. *Tetrahedron 1962*, *17*, 125.
[77]Rajput, A.S.; Chandra, S. *Bull. Chem. Soc. Jpn. 1966*, *39*, 1854.

methane, neopentane, or carbon tetrachloride. In most cases, the angles deviate a little from the pure tetrahedral value. For example, the C—C—Br angle in 2-bromo-propane is 114.2°.[78] Similarly, slight variations are generally found from the ideal values of 120 and 180° for sp^2 and sp carbon, respectively. These deviations occur because of slightly different hybridizations, that is, a carbon bonded to four other atoms hybridizes one s and three p orbitals, but the four hybrid orbitals thus formed are generally not exactly equivalent, nor does each contain exactly 25% s and 75% p character. Because the four atoms have (in the most general case) different electronegativities, each makes its own demand for electrons from the carbon atom.[79] The carbon atom supplies more p character when it is bonded to more electronegative atoms, so that in chloromethane, for example, the bond to chlorine has somewhat more than 75% p character, which of course requires that the other three bonds have somewhat less, since there are only three p orbitals (and one s) to be divided among the four hybrid orbitals.[80] Of course, in strained molecules, the bond angles may be greatly distorted from the ideal values (see p. 216).

For oxygen and nitrogen, angles of 90° are predicted from p^2 bonding. However, as we have seen (p. 6), the angles of water and ammonia are much larger than this, as are the angles of other oxygen and nitrogen compounds (Table 1.6); in fact, they are much closer to the tetrahedral angle of 109°28′ than to 90°. These facts have

TABLE 1.6. Oxygen, Sulfur, and Nitrogen Bond Angles in Some Compounds

Angle	Value	Compound	Reference
H—O—H	104°27′	Water	5
C—O—H	107–109°	Methanol	59
C—O—C	111°43′	Dimethyl ether	81
C—O—C	124° ± 5°	Diphenyl ether	82
H—S—H	92.1°	H_2S	82
C—S—H	99.4°	Methanethiol	82
C—S—C	99.1°	Dimethyl sulfide	83
H—N—H	106°46′	Ammonia	5
H—N—H	106°	Methylamine	84
C—N—H	112°	Methylamine	83
C—N—C	108.7°	Trimethylamine	85

[78]Schwendeman, R.H.; Tobiason, F.L. *J. Chem. Phys.* **1965**, *43*, 201.

[79]For a review of this concept, see Bingel, W.A.; Lüttke, W. *Angew. Chem. Int. Ed.* **1981**, *20*, 899.

[80]This assumption has been challenged: see Pomerantz, M.; Liebman, J.F. *Tetrahedron Lett.* **1975**, 2385.

[81]Blukis, V.; Kasai, P.H.; Myers, R.J. *J. Chem. Phys.* **1963**, *38*, 2753.

[82]Abrahams, S.C. *Q. Rev. Chem. Soc.* **1956**, *10*, 407.

[83]Iijima, T.; Tsuchiya, S.; Kimura, M. *Bull. Chem. Soc. Jpn.* **1977**, *50*, 2564.

[84]Lide, Jr., D.R. *J. Chem. Phys.* **1957**, *27*, 343.

[85]Lide, Jr., D.R.; Mann, D.E. *J. Chem. Phys.* **1958**, *28*, 572.

led to the suggestion that in these compounds oxygen and nitrogen use sp^3 bonding, that is, instead of forming bonds by the overlap of two (or three) p orbitals with $1s$ orbitals of the hydrogen atoms, they hybridize their $2s$ and $2p$ orbitals to form four sp^3 orbitals and then use only two (or three) of these for bonding with hydrogen, the others remaining occupied by unshared pairs (also called *lone pairs*). If this description is valid, and it is generally accepted by most chemists today,[86] it becomes necessary to explain why the angles of these two compounds are in fact not $109°28'$ but a few degrees smaller. One explanation that has been offered is that the unshared pair actually has a greater steric requirement than a pair in a bond, since there is no second nucleus to draw away some of the electron density and the bonds are thus crowded together. However, most evidence is that unshared pairs have smaller steric requirements than bonds[87] and the explanation most commonly accepted is that the hybridization is not pure sp^3. As we have seen above, an atom supplies more p character when it is bonded to more electronegative atoms. An unshared pair may be considered to be an "atom" of the lowest possible electronegativity, since there is no attracting power at all. Consequently, the unshared pairs have more s and the bonds more p character than pure sp^3 orbitals, making the bonds somewhat more like p^2 bonds and reducing the angle. As seen in Table 1.6, oxygen, nitrogen, and sulfur angles generally increase with decreasing electronegativity of the substituents. Note that the explanation given above cannot explain why some of these angles are *greater* than the tetrahedral angle.

BOND ENERGIES[88,89]

There are two kinds of bond energy. The energy necessary to cleave a bond to give the constituent radicals is called the *dissociation energy* D. For example, D for $H_2O \rightarrow HO + H$ is 118 kcal mol^{-1} (494/mol). However, this is not taken as the energy of the O—H bond in water, since D for $H-O \rightarrow H + O$ is 100 kcal mol^{-1} (418 kJ mol^{-1}). The average of these two values, 109 kcal mol^{-1} (456 kJ mol^{-1}), is taken as the *bond energy* E. In diatomic molecules, of course, $D = E$.

---

[86] An older theory holds that the bonding is indeed p^2, and that the increased angles come from repulsion of the hydrogen or carbon atoms. See Laing, M., *J. Chem. Educ.* **1987**, *64*, 124.

[87] See, for example, Pumphrey, N.W.J.; Robinson, M.J.T. *Chem. Ind. (London)* **1963**, 1903; Allinger, N.L.; Carpenter, J.G.D.; Karkowski, F.M. *Tetrahedron Lett.* **1964**, 3345; Jones, R.A.Y.; Katritzky, A.R.; Richards, A.C.; Wyatt, R.J.; Bishop, R.J.; Sutton, L.E. *J. Chem. Soc. B* **1970**, 127; Blackburne, I.D.; Katritzky, A.R.; Takeuchi, Y. *J. Am. Chem. Soc.* **1974**, *96*, 682; *Acc. Chem. Res.* **1975**, *8*, 300; Aaron, H.S.; Ferguson, C.P. *J. Am. Chem. Soc.* **1976**, *98*, 7013; Anet, F.A.L.; Yavari, I. *J. Am. Chem. Soc.* **1977**, *99*, 2794; Vierhapper, F.W.; Eliel, E.L. *J. Org. Chem.* **1979**, *44*, 1081; Gust, D.; Fagan, M.W. *J. Org. Chem.* **1980**, *45*, 2511. For other views, see Lambert, J.B.; Featherman, S.I. *Chem. Rev.* **1975**, *75*, 611; Crowley, P.J.; Morris, G.A.; Robinson, M.J.T. *Tetrahedron Lett.* **1976**, 3575; Breuker, K.; Kos, N.J.; van der Plas, H.C.; van Veldhuizen, B. *J. Org. Chem.* **1982**, *47*, 963.

[88] Blanksby, S.J.; Ellison, G.B. *Acc. Chem. Res.* **2003**, *36*, 255.

[89] For reviews including methods of determination, see Wayner, D.D.M.; Griller, D. *Adv. Free Radical Chem. (Greenwich, Conn.)* **1990**, *1*, 159; Kerr, J.A. *Chem. Rev.* **1966**, *66*, 465; Benson, S.W. *J. Chem. Educ.* **1965**, *42*, 520; Wiberg, K.B., in Nachod, F.C.; Zuckerman, J.J. *Determination of Organic Structures by Physical Methods*, Vol. 3, Academic Press, NY, **1971**, pp. 207–245.

				kcal	**kJ**
$C_2H_{6\,(gas)}$	$+ 3.5\ O_2$	$= 2\ CO_{2\,(gas)}$	$+ 3\ H_2O_{(liq)}$	+372.9	+1560
	$2\ CO_{2\,(gas)}$	$= 2\ C_{(graphite)}$	$+ 2\ O_{2\,(gas)}$	-188.2	-787
	$3\ H_2O_{(liq)}$	$= 3\ H_{2\,(gas)}$	$+ 1.5\ O_{2\,(gas)}$	-204.9	-857
	$3\ H_{2\,(gas)}$	$= 6\ H_{(gas)}$		-312/5	-1308
	$2\ C_{(graphite)}$	$= 2\ C_{(gas)}$		-343.4	-1437

$C_2H_{6\,(gas)}$	$= 6\ H_{(gas)}$	$+ 2\ C_{(gas)}$	-676.1 kcal	-2829 kJ

Fig. 1.11. Calculation of the heat of atomization of ethane at 25°C.

The D values may be easy or difficult to measure, and they can be estimated by various techniques.[90] When properly applied, "Pauling's original electronegativity equation accurately describes homolytic bond dissociation enthalpies of common covalent bonds, including highly polar ones, with an average deviation of (1.5 kcal mol^{-1} [$\approx$6.3 kJ mol^{-1}] from literature values)."[91] Whether measured or calculated, there is no question as to what D values mean. With E values the matter is not so simple. For methane, the total energy of conversion from CH_4 to C + 4H (at 0 K) is 393 kcal mol^{-1} (1644 kJ mol^{-1}).[92] Consequently, E for the C—H bond in methane is 98 kcal mol^{-1} (411 kJ mol^{-1}) at 0 K. The more usual practice, though, is not to measure the heat of atomization (i.e., the energy necessary to convert a compound to its atoms) directly but to calculate it from the heat of combustion. Such a calculation is shown in Figure 1.11.

Heats of combustion are very accurately known for hydrocarbons.[93] For methane the value at 25°C is 212.8 kcal mol^{-1} (890.4 kJ mol^{-1}), which leads to a heat of atomization of 398.0 kcal mol^{-1} (1665 kJ mol^{-1}) or a value of E for the C—H bond at 25°C of 99.5 kcal mol^{-1} (416 kJ mol^{-1}). This method is fine for molecules like methane in which all the bonds are equivalent, but for more complicated molecules assumptions must be made. Thus for ethane, the heat of atomization at 25°C is 676.1 kcal mol^{-1} or 2829 kJ mol^{-1} (Fig. 1.11), and we must decide how much of this energy is due to the C—C bond and how much to the six C—H bonds. Any assumption must be artificial, since there is no way of actually obtaining this information, and indeed the question has no real meaning. If we make the assumption that E for each of the C—H bonds is the same as E for the C—H bond in methane (99.5 kcal mol^{-1} or 416 kJ mol^{-1}), then 6× 99.5 (or 416) = 597.0 (or 2498), leaving 79.1 kcal mol^{-1} (331 kJ mol^{-1}) for the C—C bond. However, a similar calculation for propane gives a value of 80.3 (or 336) for the

[90]Cohen, N.; Benson, S.W. *Chem. Rev.* **1993**, *93*, 2419; Korth, H.-G.; Sicking, W. *J. Chem. Soc. Perkin Trans. 2* **1997**, 715.

[91]Matsunaga, N.; Rogers, D.W.; Zavitsas, A.A. *J. Org. Chem*, **2003**, *68*, 3158.

[92]For the four steps, D values are 101 to 102, 88, 124, and 80 kcal mol^{-1} (423–427, 368, 519, and 335 kJ mol^{-1}), respectively, though the middle values are much less reliable than the other two: Knox, B.E.; Palmer, H.B. *Chem. Rev.* **1961**, *61*, 247; Brewer, R.G.; Kester, F.L. *J. Chem. Phys.* **1964**, *40*, 812; Linevsky, M.J. *J. Chem. Phys.* **1967**, *47*, 3485.

[93]For values of heats of combustion of large numbers of organic compounds: hydrocarbons and others, see Cox, J.D.; Pilcher, G., *Thermochemistry of Organic and Organometallic Compounds*, Academic Press, NY, **1970**; Domalski, E.S. *J. Phys. Chem. Ref. Data* **1972**, *1*, 221–277. For large numbers of heats-of-formation values (from which heats of combustion are easily calculated) see Stull, D.R.; Westrum, Jr., E.F.; Sinke, G.C. *The Chemical Thermodynamics of Organic Compounds*, Wiley, NY, **1969**.

C—C bond, and for isobutane, the value is 81.6 (or 341). A consideration of heats of atomization of isomers also illustrates the difficulty. E values for the C—C bonds in pentane, isopentane, and neopentane, calculated from heats of atomization in the same way, are (at 25°C) 81.1, 81.8, and 82.4 kcal mol^{-1} (339, 342, 345 kJ mol^{-1}), respectively, even though all of them have twelve C—H bonds and four C—C bonds.

These differences have been attributed to various factors caused by the introduction of new structural features. Thus isopentane has a tertiary carbon whose C—H bond does not have exactly the same amount of *s* character as the C—H bond in pentane, which for that matter contains secondary carbons not possessed by methane. It is known that D values, which *can* be measured, are not the same for primary, secondary, and tertiary C—H bonds (see Table 5.3). There is also the steric factor. Hence, it is certainly not correct to use the value of 99.5 kcal mol^{-1} (416 kJ mol^{-1}) from methane as the E value for all C—H bonds. Several empirical equations have been devised that account for these factors; the total energy can be computed[94] if the proper set of parameters (one for each structural feature) is inserted. Of course, these parameters are originally calculated from the known total energies of some molecules which contain the structural feature.

Table 1.7 gives E values for various bonds. The values given are averaged over a large series of compounds. The literature contains charts that take account of

TABLE 1.7. Bond Energy E Values at 25°C for Some Important Bond Types[95a]

Bond	kcal mol^{-1}	kJ mol^{-1}	Bond	kcal mol^{-1}	kJ mol^{-1}
O—H	110–111	460–464	C—S[96]	61	255
C—H	96–99	400–415	C—I	52	220
N—H	93	390			
S—H	82	340	C≡C	199–200	835
			C=C	146–151	610–630
C—F	—	—	C—C	83–85	345–355
C—H	96–99	400–415			
C—O	85–91	355–380	C≡N	204	854
C—C	83–85	345–355	C=O	173–81	724–757
C—Cl	79	330			
C—N[97]	69–75	290–315	C=N[97]	143	598
C—Br	66	275	O—O[98]	42.9	179.6 ± 4.5

[a]The E values are arranged within each group in order of decreasing strength. The values are averaged over a large series of compounds.

[94]For a review, see Cox, J.D.; Pilcher, G. *Thermochemistry of Organic and Organometallic Compounds*, Academic Press, NY, *1970*, pp. 531–597. See also, Gasteiger, J.; Jacob, P.; Strauss, U. *Tetrahedron 1979, 35*, 139.
[95]These values, except where noted, are from Lovering, E.G.; Laidler, K.J. *Can. J. Chem. 1960, 38*, 2367; Levi, G.I.; Balandin, A.A. *Bull. Acad. Sci. USSR, Div. Chem. Sci. 1960*, 149.
[96]Grelbig, T.; Pötter, B.; Seppelt, K. *Chem. Ber. 1987, 120*, 815.
[97]Bedford, A.F.; Edmondson, P.B.; Mortimer, C.T. *J. Chem. Soc. 1962*, 2927.
[98]The average of the values obtained was $DH°$(O—O). dos Santos, R.M.B.; Muralha, V.S.F.; Correia, C.F.; Simões, J.A.M. *J. Am. Chem. Soc. 2001, 123*, 12670.

hybridization (thus an sp^3 C—H bond does not have the same energy as an sp^2 C—H bond).[99] Bond dissociation energies, both calculated and experientially determined, are constantly being refined. Improved values are available for the O—O bond of peroxides,[100] the C—H bond in alkyl amines,[101] the N—H bond in aniline derivatives,[102] the N—H bond in protonated amines,[103] the O—H bond in phenols,[104] the C—H bond in alkenes,[105] amides and ketones,[106] and in CH_2X_2 and CH_3X derivatives (X = COOR, C=O, SR, NO_2, etc.),[107] the O—H and S—H bonds of alcohols and thiols,[108] and the C—Si bond of aromatic silanes.[109] Solvent plays a role in the E values. When phenols bearing electron-releasing groups are in aqueous media, calculations show that the bond dissociation energies of decrease due to hydrogen-bonding interactions with water molecules, while electron-withdrawing substituents on the phenol increase the bond dissociation energies.[110]

Certain generalizations can be derived from the data in Table 1.7.

1. There is a correlation of bond strengths with bond distances. A comparison of Tables 1.5 and 1.7 shows that, in general, *shorter bonds are stronger bonds*. Since we have already seen that increasing *s* character shortens bonds (p. 24), it follows that bond strengths increase with increasing *s* character. Calculations show that ring strain has a significant effect on bond dissociation energy, particularly the C—H bond of hydrocarbons, because it forces the compound to adopt an undesirable hybridization.[111]

2. Bonds become weaker as we move down the Periodic Table. Compare C—O and C—S, or the carbon–halogen bonds C—F, C—Cl, C—Br, C—I. This is a consequence of the first generalization, since bond distances must increase as we go down the periodic table because the number of inner electrons increases. However, it is noted that "high-level *ab initio* molecular-orbital calculations confirm that the effect of alkyl substituents on R—X bond dissociation energies varies according to the nature of X (the stabilizing

[99]Cox, J.D.; Pilcher, G. *Thermochemistry of Organic and Organometallic Compounds*, Academic Press, NY, *1970*, pp. 531–597; Cox, J.D. *Tetrahedron* *1962*, *18*, 1337.

[100]Bach, R.D.; Ayala, P.Y.; Schlegel, H.B. *J. Am. Chem. Soc.* *1996*, *118*, 12758.

[101]Wayner, D.D.M.; Clark, K.B.; Rauk, A.; Yu, D.; Armstrong, D.A. *J. Am. Chem. Soc.* *1997*, *119*, 8925. For the α C—H bond of tertiary amines, see Dombrowski, G.W.; Dinnocenzo, J.P.; Farid, S.; Goodman, J.L. Gould, I.R. *J. Org. Chem.* *1999*, *64*, 427.

[102]Bordwell, F.G.; Zhang, X.-M.; Cheng, J.-P. *J. Org. Chem.* *1993*, *58*, 6410. See also, Li, Z.; Cheng, J.-P. *J. Org. Chem.* *2003*, *68*, 7350.

[103]Liu, W.-Z.; Bordwell, F.G. *J. Org. Chem.* *1996*, *61*, 4778.

[104]Lucarini, M.; Pedrielli, P.; Pedulli, G.F.; Cabiddu, S.; Fattuoni, C. *J. Org. Chem.* *1996*, *61*, 9259. For the O—H *E* of polymethylphenols, see de Heer, M.I.; Korth, H.-G.; Mulder, P. *J. Org. Chem.* *1999*, *64*, 6969.

[105]Zhang, X.-M. *J. Org. Chem.* *1998*, *63*, 1872.

[106]Bordwell, F.G.; Zhang, X.-M.; Filler, R. *J. Org. Chem.* *1993*, *58*, 6067.

[107]Brocks, J.J.; Beckhaus, H.-D.; Beckwith, A.L.J.; Rüchardt, C. *J. Org. Chem.* *1998*, *63*, 1935.

[108]Hadad, C.M.; Rablen, P.R.; Wiberg, K.B. *J. Org. Chem.* *1998*, *63*, 8668.

[109]Cheng, Y.-H.; Zhao, X.; Song, K.-S.; Liu, L.; Guo, Q.-X. *J. Org. Chem.* *2002*, *67*, 6638.

[110]Guerra, M.; Amorati, R.; Pedulli, G.F. *J. Org. Chem.* *2004*, *69*, 5460.

[111]Feng, Y.; Liu, L.; Wang, J.-T.; Zhao, S.-W.; Guo, Q.X. *J. Org. Chem.* *2004*, *69*, 3129; Song, K.-S.; Liu, L.; Guo, Q.X. *Tetrahedron* *2004*, *60*, 9909.

influence of the ionic configurations to increase in the order Me < Et < *i*-Pr < *t*-Bu, accounting for the *increase* (rather than expected decrease) in the R—X bond dissociation energies with increasing alkylation in the R—OCH$_3$, R—OH, and R—F molecules. This effect of X can be understood in terms of the increasing contribution of the ionic R$^+$X$^-$ configuration for electronegative X substituents."[112]

3. Double bonds are both shorter and stronger than the corresponding single bonds, but not twice as strong, because π overlap is less than σ overlap. This means that a σ bond is stronger than a π bond. The difference in energy between a single bond, say C—C, and the corresponding double bond is the amount of energy necessary to cause rotation around the double bond.[113]

---

[112]Coote, M.L.; Pross, A.; Radom, L. *Org. Lett.* **2003**, *5*, 4689.
[113]For a discussion of the different magnitdues of the bond energies of the two bonds of the double bond, see Miller, S.I. *J. Chem. Educ.* **1978**, *55*, 778.

CHAPTER 2

Delocalized Chemical Bonding

Although the bonding of many compounds can be adequately described by a single Lewis structure (p. 14), this is not sufficient for many other compounds. These compounds contain one or more bonding orbitals that are not restricted to two atoms, but that are spread out over three or more. Such bonding is said to be *delocalized*.[1] In this chapter, we will see which types of compounds must be represented in this way.

The two chief general methods of approximately solving the wave equation, discussed in Chapter 1, are also used for compounds containing delocalized bonds.[2] In the valence-bond method, several possible Lewis structures (called *canonical forms*) are drawn and the molecule is taken to be a weighted average of them. Each Ψ in Eq. (1.3), Chapter 1,

$$\Psi = c_1\psi_1 + c_1\psi_1 + \cdots$$

represents one of these structures. This representation of a real structure as a weighted average of two or more canonical forms is called *resonance*. For benzene the canonical forms are **1** and **2**. Double-headed arrows ($\leftrightarrow$) are used to indicate resonance. When the wave equation is solved, it is found that the energy value obtained by considering that **1** and **2** participate equally is lower than that for **1** or **2** alone. If **3**, **4**, and **5** (called *Dewar structures*) are also considered, the value

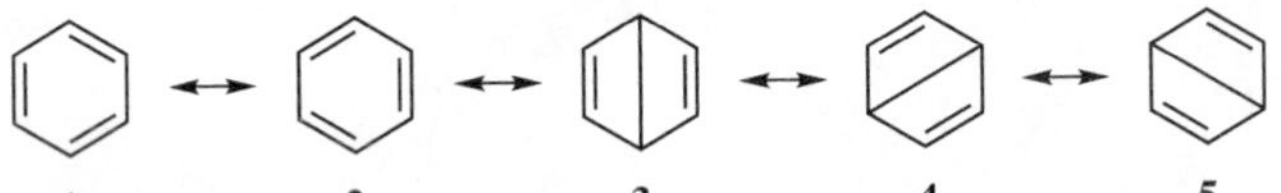

1 **2** **3** **4** **5**

[1]The classic work on delocalized bonding is Wheland, G.W. *Resonance in Organic Chemistry*; Wiley, NY, ***1955***.

[2]There are other methods. For a discussion of the free-electron method, see Streitwieser Jr., A. *Molecular Orbital Theory for Organic Chemists*; Wiley, NY, ***1961***, pp. 27–29. For the nonpairing method, in which benzene is represented as having three electrons between adjacent carbons, see Hirst, D.M.; Linnett, J.W. *J. Chem. Soc.* ***1962***, 1035; Firestone, R.A. *J. Org. Chem.* ***1969***, *34*, 2621.

is lower still. According to this method, **1** and **2** each contribute 39% to the actual molecule and the others 7.3% each.[3] The carbon–carbon bond order is 1.463 (not 1.5, which would be the case if only **1** and **2** contributed). In the valence-bond method, the *bond order* of a particular bond is the sum of the weights of those canonical forms in which the bonds is double plus 1 for the single bond that is present in all of them.[4] Thus, according to this picture, each C—C bond is not halfway between a single and a double bond but somewhat less. The energy of the actual molecule is obviously less than that of any one Lewis structure, since otherwise it would have one of those structures. The difference in energy between the actual molecule and the Lewis structure of lowest energy is call the *resonance energy*. Of course, the Lewis structures are not real, and their energies can only be estimated.

Qualitatively, the resonance picture is often used to describe the structure of molecules, but quantitative valence-bond calculations become much more difficult as the structures become more complicated (e.g., naphthalene, and pyridine). Therefore, the molecular-orbital method is used much more often for the solution of wave equations.[5] If we look at benzene by this method (qualitatively), we see that each carbon atom, being connected to three other atoms, uses sp^2 orbitals to form σ bonds, so that all 12 atoms are in one plane. Each carbon has a p orbital (containing one electron) remaining and each of these can overlap equally with the two adjacent p orbitals. This overlap of six orbitals (see Fig. 2.1) produces six new orbitals, three of which (shown) are bonding. These three (called π orbitals) all occupy approximately the same space.[6] One of the three is of lower energy than the other two, which are degenerate. They each have the plane of the ring as a node and so are in two parts, one above and one below the plane. The two orbitals of higher energy (Fig. 2.1*b* and *c*) also have another node. The six electrons that occupy this torus-shaped cloud are called the *aromatic sextet*. The carbon–carbon bond order for benzene, calculated by the molecular-orbital method, is 1.667.[7]

For planar unsaturated and aromatic molecules, many molecular-orbital calculations (*MO calculations*) have been made by treating the σ and π electrons separately. It is assumed that the σ orbitals can be treated as localized bonds and the

---

[3]Pullman, A. *Prog. Org. Chem.* **1958**, *4*, 31, p. 33.

[4]For a more precise method of calculating valence-bond orders, see Clarkson, D.; Coulson, C.A.; Goodwin, T.H. *Tetrahedron* **1963**, *19*, 2153. See also Herndon, W.C.; Párkányi, C. *J. Chem. Educ.* **1976**, *53*, 689.

[5]For a review of how MO theory explains localized and delocalized bonding, see Dewar, M.J.S. *Mol. Struct. Energ.*, **1988**, *5*, 1.

[6]According to the explanation given here, the symmetrical hexagonal structure of benzene is caused by both the σ bonds and the π orbitals. It has been contended, based on MO calculations, that this symmetry is caused by the σ framework alone, and that the π system would favor three localized double bonds: Shaik, S.S.; Hiberty, P.C.; Lefour, J.; Ohanessian, G. *J. Am. Chem. Soc.* **1987**, *109*, 363; Stanger, A.; Vollhardt, K.P.C. *J. Org. Chem.* **1988**, *53*, 4889. See also Cooper, D.L.; Wright, S.C.; Gerratt, J.; Raimondi, M. *J. Chem. Soc. Perkin Trans. 2* **1989**, 255, 263; Jug, K.; Köster, A.M. *J. Am. Chem. Soc.* **1990**, *112*, 6772; Aihara, J. *Bull. Chem. Soc. Jpn.* **1990**, *63*, 1956.

[7]The molecular-orbital method of calculating bond order is more complicated than the valence-bond method. See Pullman, A. *Prog. Org. Chem.* **1958**, *4*, 31, p. 36; Clarkson, D.; Coulson, C.A.; Goodwin, T.H. *Tetrahedron* **1963**, *19*, 2153.

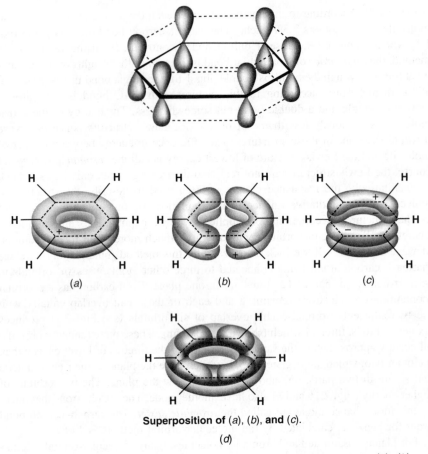

Superposition of (a), (b), and (c).

(d)

Fig. 2.1. The six p orbitals of benzene overlap to form three bonding orbitals, (a), (b), and (c). The three orbitals superimposed are shown in (d).

calculations involve only the π electrons. The first such calculations were made by Hückel; such calculations are often called *Hückel molecular-orbital* (HMO) *calculations*.[8] Because electron–electron repulsions are either neglected or averaged out in the HMO method, another approach, the *self-consistent field* (SCF), or *Hartree–Fock*, method, was devised.[9] Although these methods give many useful results for

[8]See Yates, K. *Hückel Molecular Orbital Theory*, Academic Press, NY, *1978*; Coulson, C.A.; O'Leary, B.; Mallion, R.B. *Hückel Theory for Organic Chemists*, Academic Press, NY, *1978*; Lowry, T.H.; Richardson, K.S. *Mechanism and Theory in Organic Chemistry*, 3rd ed., Harper and Row, NY, *1987*, pp. 100–121.
[9]Roothaan, C.C.J. *Rev. Mod. Phys.* *1951*, *23*, 69; Pariser, R.; Parr, R.G. *J. Chem. Phys.* *1952*, *21*, 466, 767; Pople, J.A. *Trans. Faraday Soc,.* *1953*, *49*, 1375, *J. Phys. Chem.* *1975*, *61*, 6; Dewar, M.J.S. *The Molecular Orbital Theory of Organic Chemistry*; McGraw-Hill, NY, *1969*; Dewar, M.J.S., in *Aromaticity, Chem. Soc. Spec. Pub.* no. 21, *1967*, pp. 177–215.

planar unsaturated and aromatic molecules, they are often unsuccessful for other molecules; it would obviously be better if all electrons, both σ and π, could be included in the calculations. The development of modern computers has now made this possible.[10] Many such calculations have been made[11] using a number of methods, among them an extension of the Hückel method (EHMO)[12] and the application of the SCF method to all valence electrons.[13]

One type of MO calculation that includes all electrons is called *ab initio*.[14] Despite the name (which means "from first principles") this type does involve assumptions, though not very many. It requires a large amount of computer time, especially for molecules that contain more than about five or six atoms other than hydrogen. Treatments that use certain simplifying assumptions (but still include all electrons) are called *semiempirical* methods.[15] One of the first of these was called CNDO (Complete Neglect of Differential Overlap),[16] but as computers have become more powerful, this has been superseded by more modern methods, including MINDO/3 (Modified Intermediate Neglect of Differential Overlap),[17] MNDO (Modified Neglect of Diatomic Overlap),[17] and AM1 (Austin Model 1), all of which were introduced by M.J. Dewar and co-workers.[18] Semiempirical calculations are generally regarded as less accurate than *ab initio* methods,[19] but are much faster and cheaper. Indeed, calculations for some very large molecules are possible only with the semiempirical methods.[20]

Molecular-orbital calculations, whether by *ab initio* or semiempirical methods, can be used to obtain structures (bond distances and angles), energies (e.g., heats of formation), dipole moments, ionization energies, and other properties of molecules,

---

[10]For discussions of the progress made in quantum chemistry calculations, see Ramsden, C.A. *Chem. Ber.* *1978*, *14*, 396; Hall, G.G. *Chem. Soc. Rev.* *1973*, *2*, 21.

[11]For a review of molecular-orbital calculatons on saturated organic compounds, see Herndon, W.C. *Prog. Phys. Org. Chem.* *1972*, *9*, 99.

[12]Hoffmann, R. *J. Chem. Phys.* *1963*, *39*, 1397. See Yates, K. *Hückel Molecular Orbital Theory*, Academic Press, NY, *1978*, pp. 190–201.

[13]Dewar, M.J.S. *The Molecular Orbital Theory of Chemistry*, McGraw-Hill, NY, *1969*; Jaffé, H.H. *Acc. Chem. Res.* *1969*, *2*, 136; Kutzelnigg, W.; Del Re, G.; Berthier, G. *Fortschr. Chem. Forsch.* *1971*, *22*, 1.

[14]Hehre, W.J.; Radom, L.; Schleyer, P.v.R.; Pople, J.A. *Ab Initio Molecular Orbital Theory*, Wiley, NY, *1986*; Clark, T. *A Handbook of Computational Chemistry*, Wiley, NY, *1985*, pp. 233–317; Richards, W.G.; Cooper, D.L. *Ab Initio Molecular Orbital Calculations for Chemists*, 2nd ed., Oxford University Press: Oxford, *1983*.

[15]For a review, see Thiel, W. *Tetrahedron* *1988*, *44*, 7393.

[16]Pople, J.A.; Santry, D.P.; Segal, G.A. *J. Chem. Phys.* *1965*, *43*, S129; Pople, J.A.; Segal, G.A. *J. Chem. Phys.* *1965*, *43*, S136; *1966*, *44*, 3289; Pople, J.A.; Beveridge, D.L. *Approximate Molecular Orbital Theory*; McGraw-Hill, NY, *1970*.

[17]For a discussion of MNDO and MINDO/3, and a list of systems for which these methods have been used, with references, see Clark, T. *A Handbook of Computational Chemistry*, Wiley, NY, *1985*, pp. 93–232. For a review of MINDO/3, see Lewis, D.F.V. *Chem. Rev.* *1986*, *86*, 1111.

[18]First publications are, MINDO/3: Bingham, R.C.; Dewar, M.J.S.; Lo, D.H. *J. Am. Chem. Soc.* *1975*, *97*, 1285; MNDO: Dewar, M.J.S.; Thiel, W. *J. Am. Chem. Soc.* *1977*, *99*, 4899; AM1: Dewar, M.J.S.; Zoebisch, E.G.; Healy, E.F.; Stewart, J.J.P. *J. Am. Chem. Soc.* *1985*, *107*, 3902.

[19]See, however, Dewar, M.J.S.; Storch, D.M. *J. Am. Chem. Soc.* *1985*, *107*, 3898.

[20]Clark, T. *A Handbook of Computational Chemistry*, Wiley, NY, *1985*, p. 141.

ions, and radicals: not only of stable ones, but also of those so unstable that these properties cannot be obtained from experimental measurements.[21] Many of these calculations have been performed on transition states (p. 302); this is the only way to get this information, since transition states are not, in general, directly observable. Of course, it is not possible to check data obtained for unstable molecules and transition states against any experimental values, so that the reliability of the various MO methods for these cases is always a question. However, our confidence in them does increase when (*1*) different MO methods give similar results, and (*2*) a particular MO method works well for cases that can be checked against experimental methods.[22]

Both the valence-bond and molecular-orbital methods show that there is delocalization in benzene. For example, each predicts that the six carbon–carbon bonds should have equal lengths, which is true. Since each method is useful for certain purposes, we will use one or the other as appropriate. Recent *ab initio, SCF* calculations confirms that the delocalization effect acts to strongly stabilize symmetric benzene, consistent with the concepts of classical resonance theory.[23]

Bond Energies and Distances in Compounds Containing Delocalized Bonds

If we add the energies of all the bonds in benzene, taking the values from a source like Table 1.7, the value for the heat of atomization turns out to be less than that actually found in benzene (Fig. 2.2). The actual value is $1323 \text{ kcal mol}^{-1}$ (5535 kJ mol^{-1}). If we use E values for a C=C double bond obtained from cyclohexene ($148.8 \text{ kcal mol}^{-1}$; $622.6 \text{ kJ mol}^{-1}$), a C—C single bond from cyclohexane ($81.8 \text{ kcal mol}^{-1}$, 342 kJ mol^{-1}), and C–H bonds from methane ($99.5 \text{ kcal mol}^{-1}$, 416 kJ mol^{-1}), we get a total of $1289 \text{ kcal mol}^{-1}$ (5390 kJ mol^{-1}) for structure **1** or **2**. By this calculation the resonance energy is 34 kcal mol^{-1} (145 kJ mol^{-1}). Of course, this is an arbitrary calculation since, in addition to the fact that we are calculating a heat of atomization for a nonexistent structure (**1**), we are forced to use E values that themselves do not have a firm basis in reality. The actual C—H bond energy for benzene has been measured to be $113.5 \pm 0.5 \text{ kcal mol}^{-1}$ at 300 K and estimated to be $112.0 \pm 0.6 \text{ kcal mol}^{-1}$ (469 kJ mol^{-1}) at 0 K.[24] The resonance energy can never be measured, only estimated, since we can measure the heat of atomization of the real molecule but can only make an intelligent guess at that of the Lewis structure of lowest energy.

---

[21]Another method of calculating such properies is molecular mechanics (p. $$$).

[22]Dias, J.R. *Molecular Orbital Calculations Using Chemical Graph Theory*, Spring-Verlag, Berlin, *1993*.

[23]Glendening, E.D.; Faust, R.; Streitwieser, A.; Vollhardt, K.P.C.; Weinhold, F. *J. Am. Chem.Soc.* *1993*, *115*, 10952.

[24]Davico, G.E.; Bierbaum, V.M.; DePuy, C.H.; Ellison, G.B.; Squires, R.R. *J. Am. Chem. Soc.* *1995*, *117*, 2590. See also Barckholtz, C.; Barckholtz, T.A.; Hadad, C.M. *J. Am. Chem. Soc.* *1999*, *121*, 491; Pratt, D.A.; DiLabio, G.A.; Mulder, P.; Ingold, K.U. *Acc. Chem. Res.* *2004*, *37*, 334.

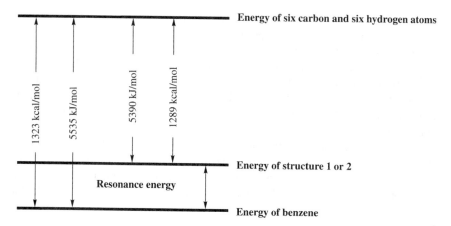

Fig. 2.2. Resonance energy in benzene.

Another method frequently used for estimation of resonance energy involves measurements of heats of hydrogenation.[25] Thus, the heat of hydrogenation of cyclohexene is 28.6 kcal mol^{-1} (120 kJ mol^{-1}), so we might expect a hypothetical **1** or **2** with three double bonds to have a heat of hydrogenation of about 85.8 kcal mol^{-1} (360 kJ mol^{-1}). The real benzene has a heat of hydrogenation of 49.8 kcal mol^{-1} (208 kJ mol^{-1}), which gives a resonance energy of 36 kcal mol^{-1} (152 kJ mol^{-1}). By any calculation the real molecule is more stable than a hypothetical **1** or **2**.

The energies of the six benzene orbitals can be calculated from HMO theory in terms of two quantities, α and β. The parameter α is the amount of energy possessed by an isolated $2p$ orbital before overlap, while β (called the *resonance integral*) is an energy unit expressing the degree of stabilization resulting from π-orbital overlap. A negative value of β corresponds to stabilization, and the energies of the six orbitals are (lowest to highest): $\alpha + 2\beta$, $\alpha + \beta$, $\alpha + \beta$, $\alpha - \beta$, $\alpha - \beta$, and $\alpha - 2\beta$.[26] The total energy of the three occupied orbitals is $6\alpha + 8\beta$, since there are two electrons in each orbital. The energy of an ordinary double bond is $\alpha + \beta$, so that structure **1** or **2** has an energy of $6\alpha + 6\beta$. The resonance energy of benzene is therefore 2β. Unfortunately, there is no convenient way to calculate the value of β from molecular-orbital theory. It is often given for benzene as about 18 kcal mol^{-1} (76 kJ mol^{-1}); this number being one-half of the resonance energy calculated from heats of combustion or hydrogenation. Using modern *ab initio* calculations, bond resonance energies for many aromatic hydrocarbons other than benzene have been reported.[27]

---

[25]For a review of heats of hydrogenation, with tables of values, see Jensen, J.L. *Prog. Phys. Org. Chem.* **1976**, *12*, 189.

[26]For the method for calculating these and similar results given in this chapter, see Higasi, K.; Baba, H.; Rembaum, A. *Quantum Organic Chemistry*, Interscience, NY, **1965**. For values of calculated orbital energies and bond orders for many conjugated molecules, see Coulson, C.A.; Streitwieser, Jr., A. *Dictionary of π Electron Calculations*, W.H. Freeman, San Francisco, **1965**.

[27]Aihara, J-i. *J. Chem. Soc. Perkin Trans 2* **1996**, 2185.

Isodesmic and homodesmotic reactions are frequently used for the study of aromaticity from the energetic point of view.[28] However, the energy of the reactions used experimentally or in calculations may reflects only the relative aromaticity of benzene and not its absolute aromaticity. A new homodesmotic reactions based on radical systems predict an absolute aromaticity of $29.13 \, \text{kcal mol}^{-1}$ ($121.9 \, \text{kJ mol}^{-1}$) for benzene and an absolute antiaromaticity of $40.28 \, \text{kcal mol}^{-1}$ ($168.5 \, \text{kJ mol}^{-1}$) for cyclobutadiene at the MP4(SDQ)/6-31G-(d,p) level.[29]

We might expect that in compounds exhibiting delocalization the bond distances would lie between the values gives in Table 1.5. This is certainly the case for benzene, since the carbon–carbon bond distance is $1.40 \, \text{Å}$,[30] which is between the $1.48 \, \text{Å}$ for an sp^2–sp^2 C—C single bond and the $1.32 \, \text{Å}$ of the sp^2–sp^2 C=C double bond.[31]

Kinds of Molecules That Have Delocalized Bonds

There are four main types of structure that exhibit delocalization:

1. *Double (or Triple) Bonds in Conjugation.*[32] The double bonds in benzene are conjugated, of course, but the conjugation exists in acyclic molecules such as butadiene. In the molecular orbital picture (Fig. 2.3), the overlap of four orbitals gives two bonding orbitals that contain the four electrons and two vacant antibonding orbitals. It can be seen that each orbital has one more node than the one of next lower energy. The energies of the four orbitals are (lowest to highest): $\alpha + 1.618\beta$, $\alpha + 0.618\beta$, $\alpha - 0.618\beta$, and $\alpha - 1.618\beta$; hence the total energy of the two occupied orbitals is $4\alpha + 4.472\beta$. Since the energy of two isolated double bonds is $4\alpha + 4\beta$, the resonance energy by this calculation is 0.472β.

In the resonance picture, these structures are considered to contribute:

$$CH_2=CH-CH=CH_2 \leftrightarrow \overset{\oplus}{C}H_2-CH=CH-\overset{\ominus}{C}H_2 \leftrightarrow \overset{\ominus}{C}H_2-CH=CH-\overset{\oplus}{C}H_2$$

$$\quad\;\; \mathbf{6} \qquad\qquad\qquad \mathbf{7} \qquad\qquad\qquad \mathbf{8}$$

[28]Hehre, W.J.; Ditchfield, R.; Radom, L.; Pople, J.A. *J. Am. Chem.Soc.* **1970**, *92*, 4796; Hehre, W.J.; Radom, L.; Pople, J.A. *J. Am. Chem. Soc.* **1971**, *93*, 289; George, P.; Trachtman, M.; Bock, C.W.; Brett, A.M. *Theor. Chim. Acta*, **1975**, *38*, 121; George, P.; Trachtman, M.; Bock, C.W.; Brett, A.M. *J. Chem. Soc. Perkin Trans. 2* **1976**, 1222; George, P.; Trachtman, M.; Brett, A.M. Bock, C.W.; *Tetrahedron* **1976**, *32*, 317; George, P.; Trachtman, M.; Brett, A.M.; Bock, C.W. *J. Chem. Soc. Perkin Trans. 2* **1977**, 1036.
[29]Suresh, C.H.; Koga, N. *J. Org. Chem.* **2002**, *67*, 1965.
[30]Bastiansen, O.; Fernholt, L.; Seip, H.M.; Kambara, H.; Kuchitsu, K. *J. Mol. Struct.* **1973**, *18*, 163; Tamagawa, K.; Iijima, T.; Kimura, M. *J. Mol. Struct.* **1976**, *30*, 243.
[31]The average C—C bond distance in aromatic rings is $1.38 \, \text{Å}$: Allen, F.H.; Kennard, O.; Watson, D.G.; Brammer, L.; Orpen, A.G.; Taylor, R. *J. Chem. Soc. Perkin Trans. 2* **1987**, p. S8.
[32]For reviews of conjugation in open-chain hydrocarbons, see Simmons, H.E. *Prog. Phys. Org. Chem.* **1970**, *7*, 1; Popov, E.M.; Kogan, G.A. *Russ. Chem. Rev.* **1968**, *37*, 119.

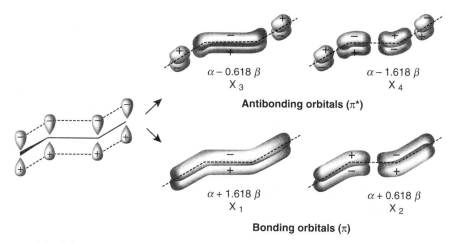

$\alpha - 0.618\,\beta$
X$_3$

$\alpha - 1.618\,\beta$
X$_4$

Antibonding orbitals (π^*)

$\alpha + 1.618\,\beta$
X$_1$

$\alpha + 0.618\,\beta$
X$_2$

Bonding orbitals (π)

Fig. 2.3. The four π-orbitals of butadiene, formed by overlap of four p orbitals.

In either picture, the bond order of the central bond should be >1 and that of the other carbon–carbon bonds <2, although neither predicts that the three bonds have equal electron density. Molecular-orbital bond orders of 1.894 and 1.447 have been calculated.[33]

The existence of delocalization in butadiene and similar molecules has been questoned. The bond lengths in butadiene are 1.34 Å for the double bonds and 1.48 Å for the single bond.[34] Since the typical single-bond distance of a bond that is not adjacent to an unsaturated group is 1.53 Å (p. 26), it has been argued that the shorter single bond in butadiene provides evidence for resonance. However, this shortening can also be explained by hybridization changes (see p. 26); and other explanations have also been offered.[35] Resonance energies for butadienes, calculated from heats of combustion or hydrogenation, are only about 4 kcal mol^{-1} (17 kJ mol^{-1}), and these values may not be entirely attributable to resonance. Thus, a calculation from heat of atomization data gives a resonance energy of 4.6 kcal mol^{-1} (19 kJ mol^{-1}) for cis-1,3-pentadiene, and −0.2 kcal mol^{-1} (−0.8 kJ mol^{-1}), for 1,4-pentadiene. These two compounds, each of which possesses two double bonds, two C–C single bonds, and eight C–H bonds, would seem to offer as similar a comparison as we could make of a conjugated with a nonconjugated compound, but they are nevertheless not strictly comparable. The former has three sp^3 C–H and five sp^2 C–H bonds, while the latter has two and six, respectively. Also, the two single C–C bonds

[33]Coulson, C.A. *Proc. R. Soc. London, Ser. A* **1939**, *169*, 413.
[34]Marais, D.J.; Sheppard, N.; Stoicheff, B.P. *Tetrahedron* **1962**, *17*, 163.
[35]Bartell, L.S. *Tetrahedron* **1978**, *34*, 2891, *J. Chem. Educ.* **1968**, *45*, 754; Wilson, E.B. *Tetrahedron* **1962**, *17*, 191; Hughes, D.O. *Tetrahedron* **1968**, *24*, 6423; Politzer, P.; Harris, D.O. *Tetrahedron* **1971**, *27*, 1567.

of the 1,4-diene are both sp^2–sp^3 bonds, while in the 1,3-diene, one is sp^2–sp^3 and the other sp^2–sp^2. Therefore, it may be that some of the already small value of 4 kcal mol^{-1} (17 kJ mol^{-1}) is not resonance energy but arises from differing energies of bonds of different hybridization.[36]

Although bond distances fail to show it and the resonance energy is low, the fact that butadiene is planar[37] shows that there is some delocalization, even if not as much as previously thought. Similar delocalization is found in other conjugated systems (e.g., C=C–C=O[38] and C=C–C=N), in longer systems with three or more multiple bonds in conjugation, and where double or triple bonds are conjugated with aromatic rings. Diynes such as 1,3-butadiyne (**9**) are another example of conjugated molecules. Based on calculations, Rogers et al. reported that the conjugation stabilization of 1,3-butadiyne is zero.[39] Later calculations concluded that consideration of hyperconjugative interactions provides a more refined measure of conjugative stabilization.[40] When this measure is used, the conjugation energies of the isomerization and hydrogenation reactions considered agree with a conjugative stabilization of 9.3 (0.5 kcal mol^{-1} for diynes and 8.2 (0.1 kcal mol^{-1} for dienes.

$$H–C\equiv C–C\equiv C–H$$

9

2. *Double (or Triple) Bonds in Conjugation with a p Orbital on an Adjacent Atom.* Where a *p* orbital is on an atom adjacent to a double bond, there are three parallel *p* orbitals that overlap. As previously noted, it is a general rule that the overlap of *n* atomic orbitals creates *n* molecular orbitals, so overlap of a *p* orbital with an adjacent double bond gives rise to three new orbitals, as

[36]For negative views on delocalization in butadiene and similar molecules, see Dewar, M.J.S.; Gleicher, G.J. *J. Am. Chem. Soc.* **1965**, *87*, 692; Brown, M.G. *Trans. Faraday Soc.* **1959**, *55*, 694; Somayajulu, G.R. *J. Chem. Phys.* **1959**, *31*, 919; Mikhailov, B.M. *J. Gen. Chem. USSR* **1966**, *36*, 379. For positive views, see Miyazaki, T.; Shigetani, T.; Shinoda, H. *Bull. Chem. Soc. Jpn.* **1971**, *44*, 1491; Berry, R.S. *J. Chem. Phys.* **1962**, *30*, 936; Kogan, G.A.; Popov, E.M. *Bull. Acad. Sci. USSR Div. Chem. Sci.* **1964**, 1306; Altmann, J.A.; Reynolds, W.F. *J. Mol. Struct.*, **1977**, *36*, 149. In general, the negative argument is that resonance involving excited structures, such as **7** and **8**, is unimportant. See rule 6 on p. $$$. An excellent discussion of the controversy is found in Popov, E.M.; Kogan, G.A. *Russ. Chem. Rev.* **1968**, *37*, 119, pp. 119–124.
[37]Marais, D.J.; Sheppard, N.; Stoicheff, B.P. *Tetrahedron* **1962**, *17*, 163; Fisher, J.J.; Michl, J. *J. Am. Chem. Soc.* **1987**, *109*, 1056; Wiberg, K.B.; Rosenberg, R.E.; Rablen, P.R. *J. Am. Chem. Soc.* **1991**, *113*, 2890.
[38]For a treatise on C=C–C=O systems, see Patai, S.; Rappoport, Z. *The Chemistry of Enones*, two parts; Wiley, NY, *1989*.
[39]Rogers, D.W.; Matsunaga, N.; Zavitsas, A.A.; McLafferty, F.J.; Liebman, J.F. *Org. Lett.* **2003**, *5*, 2373; Rogers, D.W.; Matsunaga, N.; McLafferty, F.J.; Zavitsas, A.A.; Liebman, J.F. *J. Org. Chem.* **2004**, *69*, 7143.
[40]Jarowski, P.D.; Wodrich, M.D.; Wannere, C.S.; Schleyer, P.v.R.; Houk, K.N. *J. Am. Chem. Soc.* **2004**, *126*, 15036.

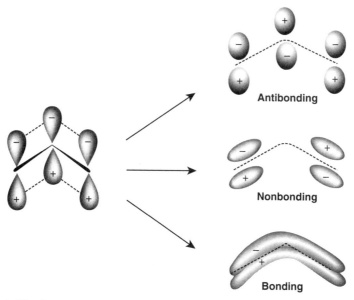

Fig. 2.4. The three orbitals of an allylic carbon, formed by overlap of three *p* orbitals.

shown in Fig. 2.4. The middle orbital is a *nonbonding orbital* of zero bonding energy. The central carbon atom does not participate in the nonbonding orbital.

There are three cases: the original *p* orbital may have contained two, one, or no electrons. Since the original double bond contributes two electrons, the total number of electrons accommodated by the new orbitals is four, three, or two. A typical example of the first situation is vinyl chloride CH_2=CH–Cl. Although the *p* orbital of the chlorine atom is filled, it still overlaps with the double bond (see **10**). The four electrons occupy the two molecular orbitals of lowest energies. This is our first example of resonance involving overlap between unfilled orbitals and a *filled* orbital. Canonical forms for vinyl chloride are shown in **11**.

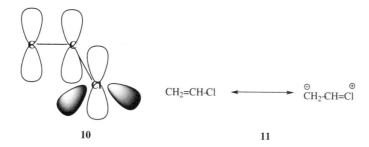

Any system containing an atom that has an unshared pair and that is directly attached to a multiple-bond atom can show this type of delocalization.

Another example is the carbonate ion:

$$^{\ominus}O \diagdown_{C} \diagup^{O^{\ominus}} \quad \longleftrightarrow \quad ^{\ominus}O-C^{\diagup O} \quad \longleftrightarrow \quad O^{\diagdown}_{C}-O^{\ominus}$$
$$\underset{O}{\overset{\parallel}{}} \qquad\qquad \underset{O_{\ominus}}{} \qquad\qquad \underset{O_{\ominus}}{}$$

The bonding in allylic carbanions, for example, $CH_2=CH-CH_2^-$, is similar.

The other two cases, where the original p orbital contains only one or no electron, are generally found only in free radicals and cations, respectively. Allylic free radicals have one electron in the nonbonding orbital. In allylic cations this orbital is vacant and only the bonding orbital is occupied. The orbital structures of the allylic carbanion, free radical, and cation differ from each other, therefore, only in that the nonbonding orbital is filled, half-filled, or empty. Since this is an orbital of zero bonding energy, it follows that the bonding π energies of the three species relative to electrons in the $2p$ orbitals of free atoms are the same. The electrons in the nonbonding orbital do not contribute to the bonding energy, positively or negatively.[41]

By the resonance picture, the three species may be described as having double bonds in conjugation with, respectively, an unshared pair, an unpaired electron, and an empty orbital as in the allyl cation **12** (see Chapter 5).

$$CH_2=CH-\overset{\ominus}{C}H_2 \quad \longleftrightarrow \quad \overset{\ominus}{C}H_2-CH=CH_2$$

$$CH_2=CH-\overset{\bullet}{C}H_2 \quad \longleftrightarrow \quad \overset{\bullet}{C}H_2-CH=CH_2$$

$$CH_2=CH-\overset{\oplus}{C}H_2 \quad \longleftrightarrow \quad \overset{\oplus}{C}H_2-CH=CH_2$$

12

3. *π-Allyl and Other η-Complexes.* In the presence of transition metals, delocalized cations are stabilized by donating electrons to the metal.[42] In a C—Metal bond, such as H_3C-Fe, the carbon donates (shares) one electron with them metal, and is considered to be a one-electron donor. With a π-bond, such as that found in ethylene, both electrons can be donated to the metal to

[41]It has been contended that here too, as with the benzene ring (Ref. 6), the geometry is forced upon allylic systems by the σ framework, and not the π system: Shaik, S.S.; Hiberty, P.C.; Ohanessian, G.; Lefour, J. *Nouv. J. Chim.*, *1985*, 9, 385. It has also been suggested, on the basis of ab initio calculations, that while the allyl cation has significant resonance stabilization, the allyl anion has little stabilization: Wiberg, K.B.; Breneman, C.M.; LePage, T.J. *J. Am. Chem. Soc.* *1990*, 112, 61.

[42]Crabtree, R.H. *The Organometallic Chemistry of the Transition Metals*, Wiley-Interscience, NY, *2005*; Hill, A.F. *Organotransition Metal Chemistry*, Wiley Interscience, Canberra, *2002*.

form a complex such as **14** by reaction of Wilkinson's catalyst (**13**) with an alkene and hydrogen gas,[43] and the π-bond is considered to be a two-electron donor. In these two cases, the electron donating ability of the group coordinated to the metal (the ligand) is indicated by terminology η^1, η^2, η^3, and so on, for a one-, two-, and three-electron donor, respectively.

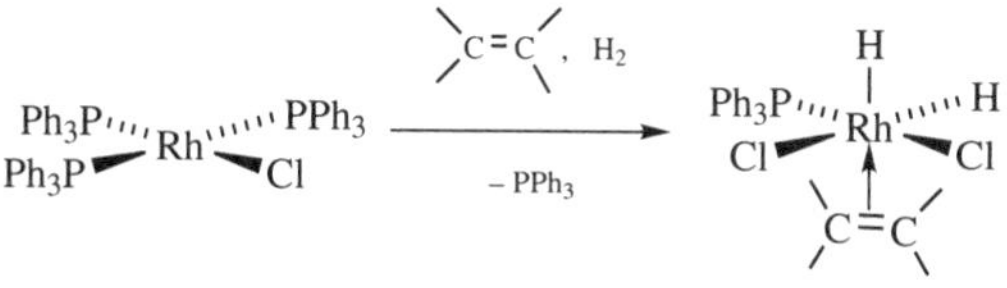

13

Wilkinson's catalyst

14

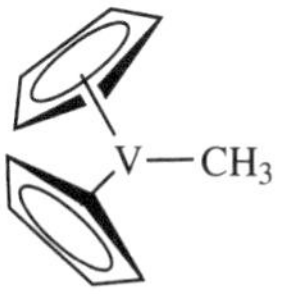

15 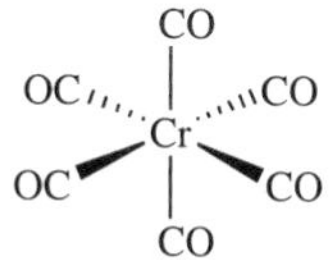   **16** 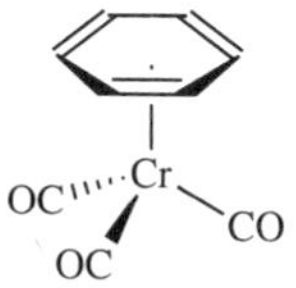   **17**    **18**

Ligands can therefore be categorized as η-ligands according to their electron donation to the metal. A hydrogen atom (as in **14**) or a halogen ligand (as in **13**) are η^1 ligands and an amine (NR_3), a phosphine (PR_3, as in **13**, **14**, and **18**), CO (as in **16** or **17**), an ether (OR_2) or a thioether (SR_2) are η^2 ligands. Hydrocarbon ligands include alkyl (as the methyl in **15**) or aryl with a C—metal bond (η^1), alkenes or carbenes (η^2, see p. 116), π-allyl (η^3), conjugated dienes such as 1,3-butadiene (η^4), cyclopentadienyl (η^5, as in **15** and see p 63), and arenes or benzene (η^6).[44] Note that in the formation of **14** from **13**, the two electron donor alkene displaces a two-electron donor phosphine. Other typical complexes include chromium hexacarbonyl $Cr(CO)_6$ (**16**), with six η^2-CO ligands; η^6-$C_6H_6Cr(CO)_3$ (**18**), and *tetrakis*-triphenylphosphinopalladium (0), **17**, with four η^2-phosphine ligands.

In the context of this section, the electron-delocalized ligand π-allyl (**12**) is an η^3 donor and it is well known that allylic halides react with $PdCl_2$ to form a *bis*-η^3-complex **19** (see the 3D model **20**).[45] Complexes, such as **19**, react with nucleophiles to give the corresponding coupling product (**10–60**).[46] The

[43]Jardine, F.H., Osborn, J.A.; Wilkinson, G.; Young, G.F. *Chem. Ind. (London) 1965*, 560; Imperial Chem. Ind. Ltd., *Neth. Appl. 6,602,062* [*Chem. Abstr., 66*: 10556y *1967*]; Bennett, M.A.; Longstaff, P.A. *Chem. Ind. 1965*, 846.

[44]Davies, S.G. *Organotransition Metal Chemistry*, Pergamon, Oxford, *1982*, p. 4.

[45]Trost, B.M.; Strege, P.E.; Weber, L.; Fullerton, T.J.; Dietsche, T.J. *J. Am. Chem. Soc. 1978*, *100*, 3407.

[46]Trost, B.M.; Weber, L.; Strege, P.E.; Fullerton, T.J.; Dietsche, T.J. *J. Am. Chem. Soc., 1978 100*, 3416.

reaction of allylic acetates or carbons and a catalytic amount of palladium (0) compounds also lead to an η^3-complex that can react with nucleophiles.[47]

$$[\text{PdCl}(\pi\text{-allyl})]_2 = [\text{PdCl}(\eta^3 C_3 H_5)]_2 =$$

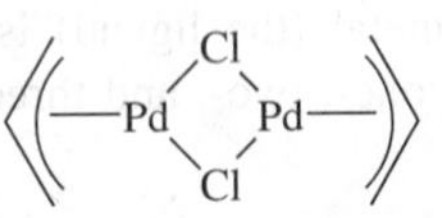

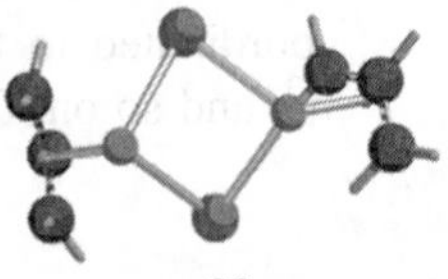

19

20

4. *Hyperconjugation.* The type of delocalization called *hyperconjugation*, is discussed on p. 95.

We will find examples of delocalization that cannot be strictly classified as belonging to any of these types.

Cross Conjugation[48]

In a cross-conjugated compound, three groups are present, two of which are not conjugated with each other, although each is conjugated with the third. Some examples[49] are benzophenone (**21**), triene **22** and divinyl ether **23**. Using the

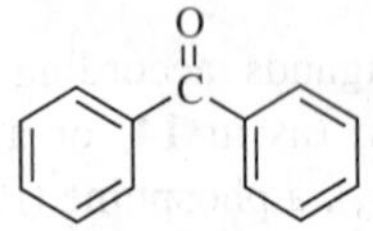

21

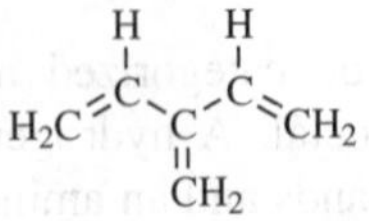

22

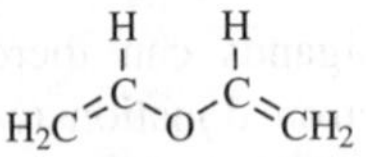

23

molecular-orbital method, we find that the overlap of six *p* orbitals in **22** gives six molecular orbitals, of which the three bonding orbitals are shown in Fig. 2.5, along with their energies. Note that two of the carbon atoms do not participate in the $\alpha + \beta$ orbital. The total energy of the three occupied orbitals is $6\alpha + 6.900\beta$, so the resonance energy is 0.900β. Molecular-orbital bond orders are 1.930 for the C-1,C-2 bond, 1.859 for the C-3,C-6 bond and 1.363 for the C-2,C-3 bond.[49] Comparing these values with those for butadiene (p. 39), we see that the C-1,C-2 bond contains more and the C-3,C-6 bond less double-bond character than the double bonds in butadiene. The resonance picture supports this conclusion, since each C-1,C-2 bond is double in three of the five canonical forms, while the C-3,C-6 bond is double in only one. In most cases, it is easier to treat cross-conjugated

[47]Melpolder, J.B.; Heck, R.F. *J. Org. Chem.* **1976**, *41*, 265; Trost, B.M.; Verhoeven, T.R. *J. Am. Chem. Soc.,* **1976**, *98*, 630; **1978**, *100*, 3435; Takahashi, K.; Miyake, A.; Hata, G. *Bull Chem. Soc. Jpn.* **1970**, *45*, 230,1183; Trost, B.M.; Verhoeven, T.R. *J. Org. Chem.* **1976**, *41*, 3215; Trost, B.M.; Verhoeven, T.R. *J. Am. Chem. Soc.* **1980**, *102*, 4730.

[48]For a discussion, see Phelan, N.F.; Orchin, M. *J. Chem. Educ.* **1968**, *45*, 633.

[49]Compound **22** is the simplest of a family of cross-conjugated alkenes, called dendralenes. For a review of these compounds, see Hopf, H. *Angew. Chem. Int. Ed.* **1984**, *23*, 948.

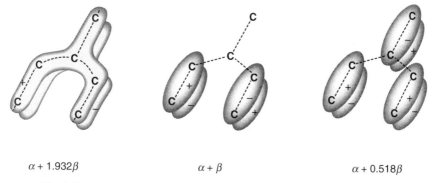

$$\alpha + 1.932\beta \qquad\qquad \alpha + \beta \qquad\qquad \alpha + 0.518\beta$$

Fig. 2.5. The three bonding orbitals of 3-methylelene-1,4-pentadiene (**22**).

molecules by the molecular-orbital method than by the valence-bond method.

One consequence of this phenomenon is that the cross-conjugated C=C unit has a slightly longer bond length that the noncross conjugated bond. In **24**, for example, the cross-conjugated bond is ∼0.01 Å longer.[50] The conjugative effect of a C=C or C≡C unit can be measured. An ethenyl substituent on a conjugated enone contributes 4.2 kcal mol^{-1} and an ethynyl substituent has a more variable effect but contributes ∼2.3 kcal mol^{-1}.[51]

The phenomenon of homoconjugation is related to cross-conjugation in that there are C=C units in close proximity, but not conjugated one to the other. Homoconjugation arises when the termini of two orthogonal π-systems are held in close proximity by being linked by a spiro-tetrahedral carbon atom.[52] Spiro[4.4]nonatetraene (**25**)[53] is an example and it known that the HOMO (p. 1208) of **25** is raised relative to cyclopentadiene, whereas the LUMO is unaffected[54] Another example

[50]Trætteberg, M.; Hopf, H. *Acta Chem. Scand. B* **1994**, *48*, 989.

[51]Trætteberg, M.; Liebman, J.F.; Hulce, M.; Bohn, A.A.; Rogers, D.W. *J. Chem. Soc. Perkin Trans. 2* **1997**, 1925.

[52]Simons, H.E.; Fukunaga, R. *J. Am. Chem. Soc.* **1967**, *89*, 5208; Hoffmann, R.; Imamura, A.; Zeiss, G.D. *J. Am. Chem. Soc.* **1967**, *89*, 5215; Durr, H.; Gleiter, R. *Angew. Chem. Int. Ed.* **1978**, *17*, 559.

[53]For the synthesis of this molecule, see Semmelhack, M.F.; Foos, J.S.; Katz, S. *J. Am. Chem. Soc.* **1973**, *95*, 7325.

[54]Raman, J.V.; Nielsen, K.E.; Randall, L.H.; Burke, L.A.; Dmitrienko, G.I. *Tetrahedron Lett.* **1994**, *35*, 5973.

is **26**, where there are bond length distortions caused by electronic interactions between the unsaturated bicyclic moiety and the cyclopropyl moiety.[55] It is assumed that cyclopropyl homoconjugation is responsible for this effect.

The Rules of Resonance

We have seen that one way of expressing the actual structure of a molecule containing delocalized bonds is to draw several possible structures and to assume that the actual molecule is a hybrid of them. These canonical forms have no existence except in our imaginations. The molecule does *not* rapidly shift between them. It is *not* the case that some molecules have one canonical form and some another. All the molecules of the substance have the same structure. That structure is always the same all the time and is a weighted average of all the canonical forms. In drawing canonical forms and deriving the true structures from them, we are guided by certain rules, among them the following:

1. All the canonical forms must be bona fide Lewis structures (see p. 14). For example, none of them may have a carbon with five bonds.
2. The positions of the nuclei must be the same in all the structures. This means that when we draw the various canonical forms, all we are doing is putting in the *electrons* in different ways. For this reason, shorthand ways of representing resonance are easy to devise:

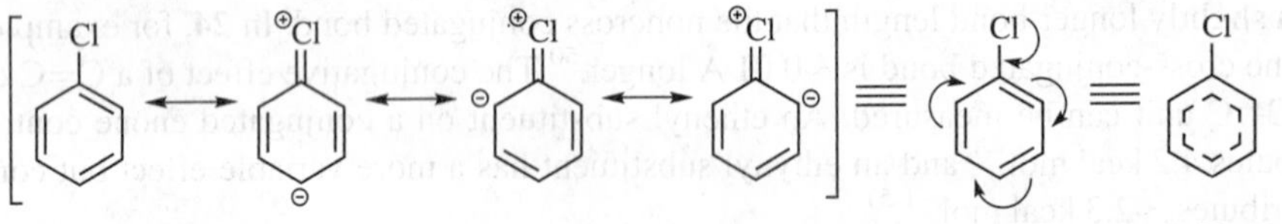

The resonance interaction of chlorine with the benzene ring can be represented as shown in **27** or **28** and both of these representations have been used in the literature to save space. However, we will not use the curved-arrow method of **27** since arrows will be used in this book to express the actual movement of electrons in reactions. We will use representations like **28** or else write out the canonical forms. The convention used in dashed-line formulas like **28** is that bonds that are present in all canonical forms are drawn as solid lines while bonds that are not present in all forms are drawn as dashed lines. In most resonance, σ bonds are not involved, and only the π or unshared electrons are put in, in different ways. This means that if we write one canonical form for a molecule, we can then write the others by merely moving π and unshared electrons.

[55]Haumann, T.; Benet-Buchholz, J.; Klärner, F.-G.; Boese, R. *Liebigs Ann. Chem.* **1997**, 1429.

3. All atoms taking part in the resonance, that is, covered by delocalized electrons, must lie in a plane or nearly so (see p. 48). This, of course, does not apply to atoms that have the same bonding in all the canonical forms. The reason for planarity is maximum overlap of the p orbitals.

4. All canonical forms must have the same number of unpaired electrons. Thus $^{\bullet}CH_2{-}CH{=}CH{-}CH_2{\bullet}$ is not a valid canonical form for butadiene.

5. The energy of the actual molecule is lower than that of any form, obviously. Therefore, delocalization is a stabilizing phenomenon.[56]

6. All canonical forms do not contribute equally to the true molecule. Each form contributes in proportion to its stability, the most stable form contributing most. Thus, for ethylene, the form $^{+}CH_2{-}CH_2^{-}$ has such a high energy compared to $CH_2{=}CH_2$ that it essentially does not contribute at all. We have seen the argument that such structures do not contribute even in such cases as butadiene.[36] Equivalent canonical forms, such as **1** and **2**, contribute equally. The greater the number of significant structures that can be written and the more nearly equal they are, the greater the resonance energy, other things being equal.

It is not always easy to decide relative stabilities of imaginary structures; the chemist is often guided by intuition.[57] However, the following rules may be helpful:

a. Structures with more covalent bonds are ordinarily more stable than those with fewer (cf. **6** and **7**).

b. Stability is decreased by an increase in charge separation. Structures with formal charges are less stable than uncharged structures. Structures with more than two formal charges usually contribute very little. An especially unfavorable type of structure is one with two like charges on adjacent atoms.

c. Structures that carry a negative charge on a more electronegative atom are more stable than those in which the charge is on a less electronegative atom. Thus, **30** is more stable than **29**. Similarly, positive charges are best carried on atoms of low electronegativity.

$$\begin{array}{ccc}
\underset{\textbf{29}}{H_2\overset{\ominus}{C}{-}\underset{\overset{\|}{O}}{C}{\diagdown}^{H}} & \longleftrightarrow & \underset{\textbf{30}}{H_2C{=}\underset{\overset{|}{O_\ominus}}{C}{\diagdown}^{H}} \qquad \underset{\textbf{31}}{\overset{H{-\!\!-\!\!-\!\!-\!\!-}H}{\underset{H{\diagup}{}^{C=C}{\diagdown}H}{{}_{H}{\diagdown}{}{\diagup}{}^{H}}}}
\end{array}$$

d. Structures with distorted bond angles or lengths are unstable, for example, the structure **31** for ethane.

---

[56]It has been argued that resonance is not a stabilizing phenomenon in all systems, especially in acyclic ions: Wiberg, K.B. *Chemtracts: Org. Chem.* **1989**, *2*, 85. See also, Siggel, M.R.; Streitwieser Jr., A.; Thomas, T.D. *J. Am. Chem. Soc.* **1988**, *110*, 8022; Thomas, T.D.; Carroll, T.X.; Siggel, M.R. *J. Org. Chem.* **1988**, *53*, 1812.

[57]A quantitative method for weighting canonical forms has been proposed by Gasteiger, J.; Saller, H. *Angew. Chem. Int. Ed.* **1985**, *24*, 687.

The Resonance Effect

Resonance always results in a different distribution of electron density than would be the case if there were no resonance. For example, if **32** were the actual structure of aniline, the two unshared electrons of the nitrogen would reside

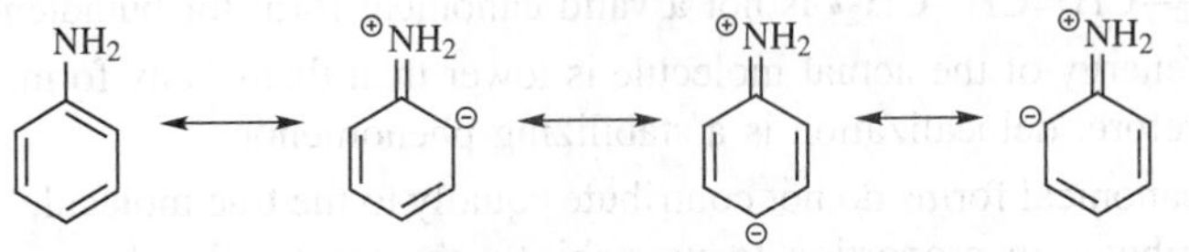

32

entirely on that atom. The structure of **32** can be represented as a hybrid that includes contributions from the canonical forms shown, indicating that the electron density of the unshared pair does not reside entirely on the nitrogen, but is spread over the ring. This decrease in electron density at one position (and corresponding increase elsewhere) is called the *resonance* or *mesomeric effect*. We loosely say that the NH_2 contributes or donates electrons to the ring by a resonance effect, although no actual contribution takes place. The "effect" is caused by the fact that the electrons are in a different place from that we would expect if there were no resonance. In ammonia, where resonance is absent, the unshared pair *is* located on the nitrogen atom. As with the field effect (p. 20), we think of a certain molecule (in this case ammonia) as a substrate and then see what happens to the electron density when we make a substitution. When one of the hydrogen atoms of the ammonia molecule is replaced by a benzene ring, the electrons are "withdrawn" by the resonance effect, just as when a methyl group replaces a hydrogen of benzene, electrons are "donated" by the the field effect of the methyl. The idea of donation or withdrawal merely arises from the comparison of a compound with a closely related one or a real compound with a canonical form.

Steric Inhibition of Resonance and the Influences of Strain

Rule 3 states that all the atoms covered by delocalized electrons must lie in a plane or nearly so. Many examples are known where resonance is reduced or prevented because the atoms are sterically forced out of planarity.

Bond lengths for the *o*- and *p*-nitro groups in picryl iodide are quite different.[58] Distance *a* in **33** is 1.45 Å, whereas *b* is 1.35 Å. This phenomenon can be explained if the oxygens of the *p*-nitro group are in the plane of the ring and thus in resonance with it, so that *b* has partial double-bond character, while the oxygens of the *o*-nitro

---

[58]Wepster, B.M. *Prog. Stereochem.* **1958**, 2, 99, p. 125. For another example of this type of steric inhibition of resonance, see Exner, O.; Folli, U.; Marcaccioli, S.; Vivarelli, P. *J. Chem. Soc. Perkin Trans. 2* **1983**, 757.

groups are forced out of the plane by the large iodine atom.

The Dewar-type structure for the central ring of the anthracene system in **34** is possible only because the 9,10 substituents prevent the system from being planar.[59] **34** is the actual structure of the molecule and is not in resonance with forms like **35**, although in anthracene itself, Dewar structures and structures like **35** both contribute. This is a consequence of rule 2 (p. 46). In order for a **35**-like structure to contribute to resonance in **34**, the nuclei would have to be in the same positions in both forms.

Even the benzene ring can be forced out of planarity.[60] In [5]paracyclophane (**36**),[61] the presence of a short bridge (this is the shortest para bridge known for a benzene ring) forces the benzene ring to become boat-shaped. The parent **36** has so far not proven stable enough for isolation, but a UV spectrum was obtained and showed that the benzene ring was still aromatic, despite the distorted ring.[62] The 8,11-dichloro analog of **36** is a stable solid, and X-ray diffraction showed

[59]Applequist, D.E.; Searle, R. *J. Am. Chem. Soc.* **1964**, *86*, 1389.

[60]For a review of planarity in aromatic systems, see Ferguson, G.; Robertson, J.M. *Adv. Phys. Org. Chem.* **1963**, *1*, 203.

[61]For a monograph, see Keehn, P.M.; Rosenfeld, S.M. *Cyclophanes*, 2 vols., Academic Press, NY, **1983**. For reviews, see Bickelhaupt, F. *Pure Appl. Chem.* **1990**, *62*, 373; Vögtle, F.; Hohner, G. *Top. Curr. Chem.* **1978**, *74*, 1; Cram, D.J.; Cram, J.M. *Acc. Chem. Res.* **1971**, *4*, 204; Vögtle, F.; Neumann, P. reviews in *Top. Curr. Chem.* **1983**, *113*, 1; **1985**, *115*, 1.

[62]Jenneskens, L.W.; de Kanter, F.J.J.; Kraakman, P.A.; Turkenburg, L.A.M.; Koolhaas, W.E.; de Wolf, W.H.; Bickelhaupt, F.; Tobe, Y.; Kakiuchi, K.; Odaira, Y. *J. Am. Chem. Soc.* **1985**, *107*, 3716. See also Tobe, Y.; Kaneda, T.; Kakiuchi, K.; Odaira, Y. *Chem. Lett.* **1985**, 1301; Kostermans, G.B.M.; de Wolf, W.E.; Bickelhaupt, F. *Tetrahedron Lett.* **1986**, *27*, 1095; van Zijl, P.C.M.; Jenneskens, L.W.; Bastiaan, E.W.; MacLean, C.; de Wolf, W.E.; Bickelhaupt, F. *J. Am. Chem. Soc.* **1986**, *108*, 1415; Rice, J.E.; Lee, T.J.; Remington, R.B.; Allen, W.D.; Clabo Jr., D.A.; Schaefer III, H.F. *J. Am. Chem. Soc.* **1987**, *109*, 2902.

that the benzene ring is boat-shaped, with one end of the boat bending ~27° out of the plane, and the other ~12°.[63] This compound too is aromatic, as shown by UV and NMR spectra. [6]Paracyclophanes are also bent,[64] but in [7]paracyclophanes the bridge is long enough so that the ring is only moderately distorted. Similarly, [n,m]paracyclophanes (37), where n and m are both 3 or less (the smallest yet prepared is [2.2]paracyclophane), have bent (boat-shaped) benzene rings. All these compounds have properties that depart significantly from those of ordinary benzene compounds. Strained paracyclophanes exhibit both π- and σ-strain, and the effect of the two types of strain on the geometry is approximately additive.[65] In "belt" cyclophane 38,[66] the molecule has a pyramidal structure with C_3 symmetry rather than the planar structure found in [18]-annulene. 1,8-Dioxa[8](2,70-pyrenophane (39)[67] is another severely distorted aromatic hydrocarbon, in which the bridge undergoes rapid pseudo-rotation (p. 212). A recent study showed that despite substantial changes in the hybridization of carbon atoms involving changes in the σ-electron structure of pyrenephane, such as 39, the aromaticity of the system decreases slightly and regularly upon increasing the bend angle θ from 0 to 109.2°.[68] Heterocyclic paracyclophane analogs have been prepared, such as the report of [2.n](2,5)pyridinophanes.[69]

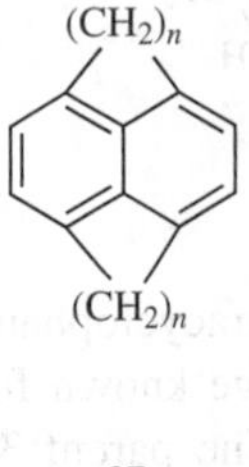

37

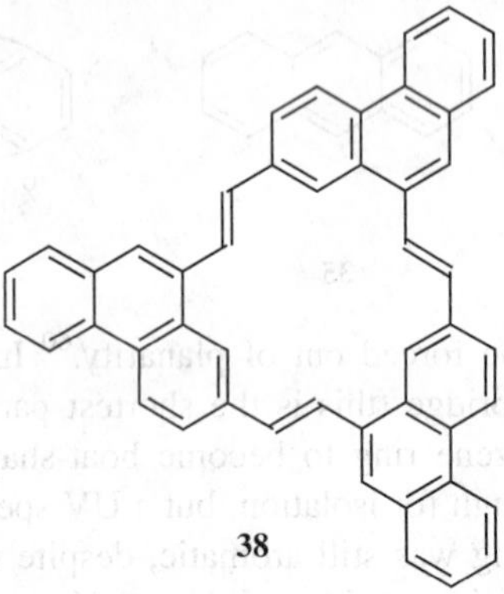

38

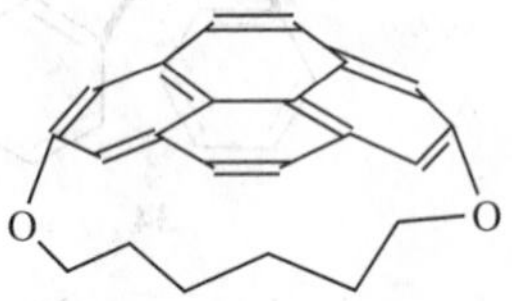

39

[63]Jenneskens, L.W.; Klamer, J.C.; de Boer, H.J.R.; de Wolf, W.H.; Bickelhaupt, F.; Stam, C.H. *Angew. Chem. Int. Ed.* **1984**, *23*, 238.

[64]See, for example, Liebe, J.; Wolff, C.; Krieger, C.; Weiss, J.; Tochtermann, W. *Chem. Ber.* **1985**, *118*, 4144; Tobe, Y.; Ueda, K.; Kakiuchi, K.; Odaira, Y.; Kai, Y.; Kasai, N. *Tetrahedron* **1986**, *42*, 1851.

[65]Stanger, A.; Ben-Mergui, N.; Perl, S. *Eur. J. Org. Chem.* **2003**, 2709.

[66]Meier, H.; Müller, K. *Angew. Chem. Int. Ed.*, **1995**, *34*, 1437.

[67]Bodwell, G.J.; Bridson, J.N.; Houghton, T.J.; Kennedy, J.W.J.; Mannion, M.R. *Angew. Chem. Int. Ed.*, **1996**, *35*, 1320.

[68]Bodwell, G.J.; Bridson, J.N.; Cyranski, M.K.; Kennedy, J.W.J.; Krygowski, T.M.; Mannion, M.R.; Miller, D.O. *J. Org. Chem.* **2003**, *68*, 2089; Bodwell, G.J.; Miller, D.O.; Vermeij, R.J. *Org. Lett.* **2001**, *3*, 2093

[69]Funaki, T.; Inokuma, S.; Ida, H.; Yonekura, T.; Nakamura, Y.; Nishimura, J. *Tetrahedron Lett.* **2004**, *45*, 2393.

There are many examples of molecules in which benzene rings are forced out of planarity, including 7-circulene (**40**),[70] 9,8-diphenyltetrabenz[*a,c,h,j*]anthracene (**41**),[71] and **42**[72] (see also p. 230). These have been called tormented aromatic systems.[73] The "record" for twisting an aromatic π-electron system appears to be 9,10,11,12,13,14,15,16-octaphenyldibenzo[*a,c*]naphthacene (**43**),[74] which has an end-to-end twist of 105°. This is >1.5 times as great as that observed in any previous polyaromatic hydrocarbon.

Perchlorotriphenylene has been reported in the literature and said to show severe molecular twisting, however, recent work suggests this molecule has not actually been isolated with perchlorofluorene-9-spirocyclohexa-2′,5′-diene being formed instead.[75] The X-ray structure of the linear [3]phenylene (benzo[3,4]cyclobuta-[1,2-b]biphenylene, **44**) has been obtained, and it shows a relatively large degree of bond alternation while the center distorts to a cyclic bis-allyl frame.[76]

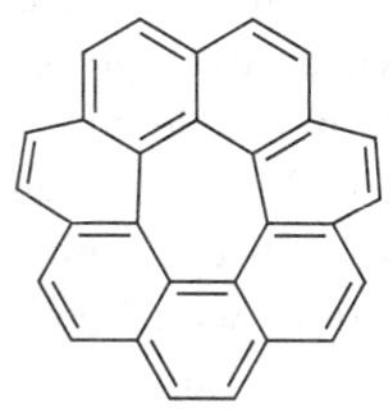

40

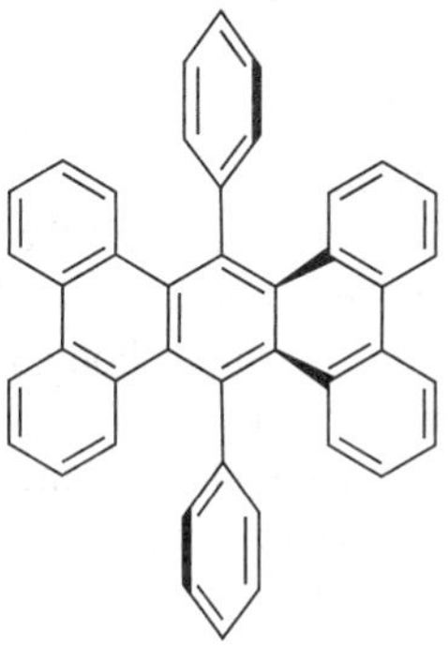

41

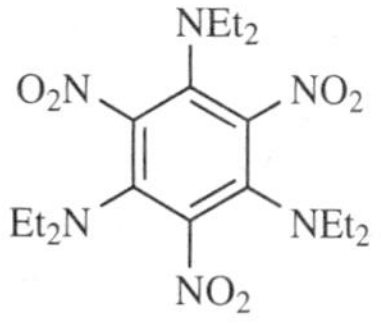

42

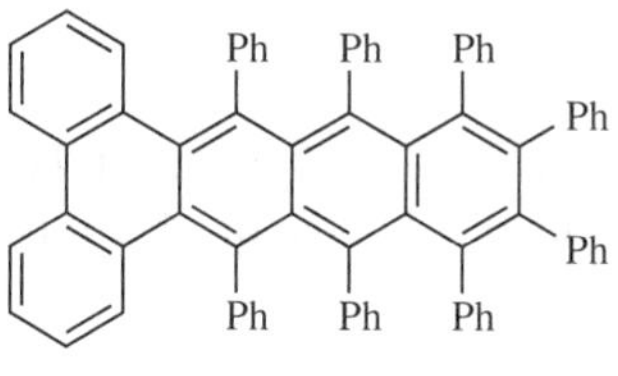

43

[70]Yamamoto, K.; Harada, T.; Okamoto, Y.; Chikamatsu, H.; Nakazaki, M.; Kai, Y.; Nakao, T.; Tanaka, M.; Harada, S.; Kasai, N. *J. Am. Chem. Soc.* **1988**, *110*, 3578.

[71]Pascal, Jr., R.A.; McMillan, W.D.; Van Engen, D.; Eason, R.G. *J. Am. Chem. Soc.* **1987**, *109*, 4660.

[72]Chance, J.M.; Kahr, B.; Buda, A.B.; Siegel, J.S. *J. Am. Chem. Soc.* **1989**, *111*, 5940.

[73]Pascal, Jr., R.A. *Pure Appl. Chem.* **1993**, *65*, 105.

[74]Qiao, X.; Ho, D.M.; Pascal Jr., R.A. *Angew. Chem. Int. Ed.,* **1997**, *36*, 1531.

[75]Campbell, M.S.; Humphries, R.E.; Munn, N.M. *J. Org. Chem.* **1992**, *57*, 641.

[76]Schleifenbaum, A.; Feeder, N.; Vollhardt, K.P.C. *Tetrahedron Lett.* **2001**, *42*, 7329.

It is also possible to fuse strained rings on benzene, which induces great strain on the benzene ring. In **45**, the benzene ring is compressed by the saturated environment of the tetrahydropyran units. In this case, the strain leads to distortion of the benzene ring in **45** into a *boat* conformation.[77] Benzocyclopropene (**46**) and benzocyclobutene (**47**) are also molecules where the small annellated ring induces great strain on the benzene ring. In these cases, bonds of annellation and those adjacent to it are strained.

44

45

46

47

Strain-induced bond localization was introduced in 1930 by Mills and Nixon[78] and is commonly referred to as the *Mills–Nixon effect* (see Chapter 11, p. 677). Ortho-fused aromatic compounds, such as **46**, are known as cycloproparenes[79] and are highly strained. Cyclopropabenzene (**46**) is a stable molecule with a strain energy of 68 kcal mol^{-1} (284.5 kJ mol^{-1}).[80] and the annellated bond is always the shortest, although in **47** the adjacent bond is the shortest.[81] In cycloproparenes, there is the expectation of partial aromatic bond localization, with bond length alternation in the aromatic ring.[82] When the bridging units are saturated, the benzene ring current is essentially unchanged, but annelation with one or more cyclobutadieno units disrupts the benzene ring current.[83] The chemistry of the cycloproparenes is dominated by the influence of the high strain energy. When fused to a benzene ring, the bicyclo[1.1.0]butane unit also leads to strain-induced localization of aromatic π-bonds.[84]

pπ–dπ Bonding: Ylids

We have mentioned (p. 10) that, in general, atoms of the second row of the Periodic table do not form stable double bonds of the type discussed in Chapter 1

[77]Hall, G.G *J. Chem. Soc. Perkin Trans. 2* **1993**, 1491.

[78]Mills, W. H.; Nixon, I.G. *J. Chem. Soc.* **1930**, 2510.

[79]Halton, B. *Chem. Rev.* **2003**, *103*, 1327; Halton, B. *Chem. Rev.* **1989**, *89*, 1161, and reviews cited therein.

[80]Billups, W.E.; Chow, W.Y.; Leavell, K.H.; Lewis, E.S.; Margrave, J.L.; Sass, R.L.; Shieh, J.J.; Werness, P.G.; Wood, J.L. *J. Am. Chem. Soc.* **1973**, *95*, 7878.; Apeloig, Y.; Arad, D. *J. Am. Chem. Soc.* **1986**, *108*, 3241.

[81]Boese, R.; Bläser, D.; Billups, W.E.; Haley, M.M.; Maulitz, A.H.; Mohler, D.L.; Vollhardt, K.P.C. *Angew. Chem. Int. Ed.*, **1994**, *33*, 313.

[82]Halton, B. *Pure Appl. Chem.* **1990**, *62*, 541; Stanger, A. *J. Am. Chem. Soc.* **1998**, *120*, 12034; Maksić, Z.B.; Eckert-Maksić, M.; Pfeifer, K.-H. *J. Mol. Struct.* **1993**, *300*, 445; Mó, M.; Yáñez, M.; Eckert-Maksić, M.; Maksić, Z.B. *J. Org. Chem.* **1995**, *60*, 1638; Eckert-Maksić, M.; Glasovac, Z.; Maksić, Z.B.; Zrinski, I. *J. Mol. Struct. (THEOCHEM)* **1996**, *366*, 173; Baldridge, K.K.; Siegel, J.S. *J. Am. Chem. Soc.* **1992**, *114*, 9583.

[83]Soncini, A.; Havenith, R.W.A.; Fowler, P.W.; Jenneskens, L.W.; Steiner, E. *J. Org. Chem.* **2002**, *67*, 4753

[84]Cohrs, C.; Reuchlein, H.; Musch, P.W.; Selinka, C.; Walfort, B.; Stalke, D.; Christl, M. *Eur. J. Org. Chem.* **2003**, 901.

(π bonds formed by overlap of parallel p orbitals). However, there is another type of double bond that is particularly common for the second-row atoms, sulfur and phosphorus. For example, such a double bond is found in the compound H_2SO_3,

as written on the left. Like an ordinary double bond, this double bond contains one s orbital, but the second orbital is not a π orbital formed by overlap of half-filled p orbitals; instead it is formed by overlap of a filled p orbital from the oxygen with an empty d orbital from the sulfur. It is called a $p\pi$–$d\pi$ orbital.[85] Note that we can represent this molecule by two canonical forms, but the bond is nevertheless localized, despite the resonance. Some other examples of $p\pi$–$d\pi$ bonding are Nitrogen

Phosphine oxides   Sulfones

Hypophorphoros acid   Sulfoxides

analogs are known for some of these phosphorus compounds, but they are less stable because the resonance is lacking. For example, amine oxides, analogs of phosphine oxides, can only be written R_3N^+–O^-. The $p\pi$–$d\pi$ canonical form is impossible since nitrogen is limited to eight outer-shell electrons.

In all the examples given above, the atom that donates the electron pair is oxygen and, indeed, oxygen is the most common such atom. But in another important class of compounds, called *ylids*, this atom is carbon.[86] There are three main types of ylids phosphorus,[87] nitrogen,[88] and sulfur ylids,[89] although

---

[85]For a monograph, see Kwart, H.; King, K. *d-Orbitals in the Chemistry of Silicon, Phosphorus, and Sulfur*; Springer, NY, **1977**.

[86]For a monograph, see Johnson, A.W. *Ylid Chemistry*; Academic Press, NY, **1966**. For reviews, see Morris, D.G., *Surv. Prog. Chem.* **1983**, *10*, 189; Hudson, R.F. *Chem. Br.*, **1971**, *7*, 287; Lowe, P.A. *Chem. Ind. (London)* **1970**, 1070. For a review on the formation of ylids from the reaction of carbenes and carbenoids with heteroatom lone pairs, see Padwa, A.; Hornbuckle, S.F. *Chem. Rev.* **1991**, *91*, 263.

[87]Although the phosphorus ylid shown has three R groups on the phosphorus atom, other phosphorus ylids are known where other atoms, for example, oxygen, replace one or more of these R groups. When the three groups are all alkyl or aryl, the phosphorus ylid is also called a phosphorane.

[88]For a review of nitrogen ylids, see Musker, W.K. *Fortschr. Chem. Forsch.* **1970**, *14*, 295.

[89]For a monograph on sulfur ylids, see Trost, B.M.; Melvin Jr., L.S. *Sulfur Ylids*; Academic Press, NY, **1975**. For reviews, see Fava, A, in Bernardi, F.; Csizmadia, I.G.; Mangini, A. *Organic Sulfur Chemistry*; Elsevier, NY, **1985**, pp. 299–354; Belkin, Yu.V.; Polezhaeva, N.A. *Russ. Chem. Rev.* **1981**, *50*, 481; Block, E. in Stirling, C.J.M. *The Chemistry of the Sulphonium Group*, part 2, Wiley, NY, **1981**, pp. 680–702; Block, E. *Reactions of Organosulfur Compounds*; Academic Press, NY, **1978**, pp. 91–127.

arsenic,[90] selenium, and so on, ylids are also known. Ylids may be defined as compounds in which a positively charged atom from group 15 or 16 of the Periodic table is connected to a carbon atom carrying an unshared pair of electrons. Because of $p\pi$–$d\pi$ bonding, two canonical forms can be written for phosphorus and sulfur, but there is only one for nitrogen ylids. Phosphorus ylids are much more stable than nitrogen ylids (see also p. 810). Sulfur ylids also have a low stability.

$$R-\overset{\underset{\displaystyle R}{|}}{\overset{\displaystyle R}{P}}=CR_2 \quad\longleftrightarrow\quad R-\overset{\underset{\displaystyle R}{|}}{\overset{\displaystyle R}{\overset{\oplus}{P}}}-\overset{\ominus}{C}R_2 \qquad \overset{R}{\underset{R}{\diagdown}}S=CR_2 \quad\longleftrightarrow\quad \overset{R}{\underset{R}{\diagdown}}\overset{\oplus}{S}-\overset{\ominus}{C}R_2 \qquad R-\overset{\underset{\displaystyle R}{|}}{\overset{\displaystyle R}{\overset{\oplus}{N}}}-\overset{\ominus}{C}R_2$$

Phosphorus ylids   Sulfur ylids   Nitrogen ylids

In almost all compounds that have $p\pi$–$d\pi$ bonds, the central atom is connected to four atoms or three atoms and an unshared pair and the bonding is approximately tetrahedral. The $p\pi$–$d\pi$ bond, therefore, does not greatly change the geometry of the molecule in contrast to the normal π bond, which changes an atom from tetrahedral to trigonal. Calculations show that nonstabilized phosphonium ylids have nonplanar ylidic carbon geometries whereas stabilized ylids have planar ylidic carbons.[91]

AROMATICITY[92]

In the nineteenth century, it was recognized that aromatic compounds[93] differ greatly from unsaturated aliphatic compounds,[94] but for many years chemists

[90]For reviews of arsenic ylids, see Lloyd, D.; Gosney, I.; Ormiston, R.A. *Chem. Soc. Rev.* **1987**, *16*, 45; Yaozeng, H.; Yanchang, S. *Adv. Organomet. Chem.* **1982**, *20*, 115.

[91]Bachrach, S.M. *J. Org. Chem.* **1992**, *57*, 4367.

[92]Krygowski, T.M.; Cyrañski, M.K.; Czarnocki, Z.; Häfelinger, G.; Katritzky, A.R. *Tetrahedron* **2000**, *56*, 1783; Simkin, B.Ya.; Minkin, V.I.; Glukhovtsev, M.N., in *Advances in Heterocyclic Chemistry*, Vol. 56, Katritzky, A.R., Ed., Academic Press, San Diego, **1993**, pp 303–428; Krygowski, T.M.; Cyranski, M.K. *Chem. Rev.* **2001**, *101*, 1385; Katritzky, A.R.; Jug, K.; Oniciu, D.C. *Chem. Rev.* **2001**, *101*, 1421; Katritzky, A.R.; Karelson, M.; Wells, A.P. *J. Org. Chem.* **1996**, *61*, 1619. See also Cyranski, M.K.; Krygowski, T.M.; Katritzky, A.R.; Schleyer, P.v.R. *J. Org. Chem.* **2002**, *67*, 1333.

[93]For books on Aromaticity, see Lloyd, D. *The Chemistry of Conjugated Cyclic Compounds*, Wiley, NY, **1989**; *Non-Benzenoid Conjugated Carbocyclic Compounds*, Elsevier, NY, **1984**; Garratt, P.J. *Aromaticity*, Wiley, NY, **1986**; Balaban, A.T.; Banciu, M.; Ciorba, V. *Annulenes, Benzo-, Hetero-, Homo-Derivatives and their Valence Isomers*, 3 vols., CRC Press, Boca Raton, FL **1987**; Badger, G.M. *Aromatic Character and Aromaticity*, Cambridge University Press, Cambridge, **1969**; Snyder, J.P. *Nonbenzenoid Aromatics*, 2 vols., Academic Press, NY, **1969–1971**; Bergmann, E.D.; Pullman, B. *Aromaticity, Pseudo-Aromaticity, and Anti-Aromaticity*, Israel Academy of Sciences and Humanities, Jerusalem, **1971**; *Aromaticity*; *Chem. Soc. Spec. Pub.* No. 21, **1967**. For reviews, see Gorelik, M.V. *Russ. Chem. Rev.* **1990**, *59*, 116; Stevenson, G.R. *Mol. Struct. Energ.*, **1986**, *3*, 57; Sondheimer, F. *Chimia*, **1974**, *28*, 163; Cresp, T.M.; Sargent, M.V. *Essays Chem.* **1972**, *4*, 91; Figeys, H.P. *Top. Carbocyclic Chem.* **1969**, *1*, 269; Garratt, P.J.; Sargent, M.V. papers in, *Top. Curr. Chem.* **1990**, 153 and *Pure Appl. Chem.* **1980**, *52*, 1397.

[94]For an account of the early history of Aromaticity, see Snyder, J.P., in Snyder, J.P. *Nonbenzenoid Aromatics*, Vol. 1, Academic Press, NY, **1971**, pp. 1–31. See also Balaban, A.T. *Pure Appl. Chem.* **1980**, *52*, 1409.

were hard pressed to arrive at a mutually satisfactory definition of aromatic character.[95] Qualitatively, there has never been real disagreement. Definitions have taken the form that aromatic compounds are characterized by a special stability and that they undergo substitution reactions more easily than addition reactions. The difficulty arises because these definitions are vague and not easy to apply in borderline cases. Definitions of aromaticity must encompass molecules ranging form polycyclic conjugated hydrocarbons,[96] to heterocyclic compounds[97] of various ring sizes, to reactive intermediates. In 1925 Armit and Robinson,[98] recognized that the aromatic properties of the benzene ring are related to the presence of a closed loop of electrons, the *aromatic sextet* (aromatic compounds are thus the arch examples of delocalized bonding), but it still was not easy to determine whether rings other than the benzene ring possessed such a loop. With the advent of magnetic techniques, most notably NMR, it is possible to determine experimentally whether or not a compound has a closed ring of electrons; aromaticity can now be defined as the *ability to sustain an induced ring current*. A compound with this ability is called **diatropic**. Although this definition also has its flaws,[99] it is the one most commonly accepted today. There are several methods of determining whether a compound can sustain a ring current, but the most important one is based on NMR chemical shifts.[100] In order to understand this, it is necessary to remember that, as a general rule, the value of the chemical shift of a proton in an NMR spectrum depends on the electron density of its bond; the greater the density of the electron cloud surrounding or partially surrounding a proton, the more upfield is its chemical shift (a lower value of δ). However, this rule has several exceptions; one is for protons in the vicinity of an aromatic ring. When an external magnetic field is imposed upon an aromatic ring (as in an NMR instrument), the closed loop of aromatic electrons circulates in a diamagnetic ring current, which sends out a field of its own. As can be seen in Fig. 2.6, this induced field curves around and in the area of the proton is parallel to the external field, so the field "seen" by the aromatic protons is greater than it would have been in the absence of the diamagnetic ring current. The protons are moved downfield (to higher δ) compared to where they would be if electron

[95]For a review of the criteria used to define aromatic character, see Jones, A.J. *Pure Appl. Chem.* **1968**, *18*, 253. For methods of assigning Aromaticity, see Jug, K.; Köster, A.M. *J. Phys. Org. Chem.* **1991**, *4*, 163; Zhou, Z.; Parr, R.G. *J. Am. Chem. Soc.* **1989**, *111*, 7371; Katritzky, A.R.; Barczynski, P.; Musumarra, G.; Pisano, D.; Szafran, M. *J. Am. Chem. Soc.* **1989**, *111*, 7; Schaad, L.J.; Hess, Jr., B.A. *J. Am. Chem. Soc.* **1972**, *94*, 3068, *J. Chem. Educ.* **1974**, *51*, 640. See also, Bird, C.W. *Tetrahedron* **1985**, *41*, 1409; **1986**, *42*, 89; **1987**, *43*, 4725.
[96]Randic, M. *Chem. Rev.* **2003**, *103*, 3449.
[97]Balaban, A.T.; Oniciu, D.C.; Katritzky, A.R. *Chem. Rev.* **2004**, *104*, 2777.
[98]Armit, J.W.; Robinson; R. *J. Chem. Soc.* **1925**, *127*, 1604.
[99]Jones, A.J. *Pure Appl. Chem.* **1968**, *18*, 253, pp. 266–274; Mallion, R.B. *Pure Appl. Chem.* **1980**, *52*, 1541. Also see, Schleyer, P.v.R.; Jiao, H. *Pure Appl. Chem.* **1996**, *68*, 209.
[100]For a review of NMR and other magnetic properties with respect to aromaticity, see Haddon, R.C.; Haddon, V.R.; Jackman, L.M. *Fortschr. Chem. Forsch.* **1971**, *16*, 103. For an example of a magentic method other than NMR, see Dauben Jr., H.J.; Wilson, J.D.; Laity, J.L., in Snyder, J.P. *Nonbenzenoid Aromatics*, Vol. 2, Academic Press, NY, **1971**, pp. 167–206.

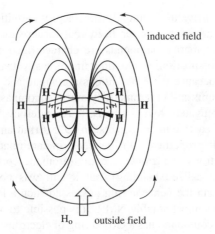

Fig. 2.6. Ring current in benzene.

density were the only factor. Thus ordinary alkene hydrogens are found at ~5–6 δ, while the hydrogens of benzene rings are located at ~7–8 δ. However, if there

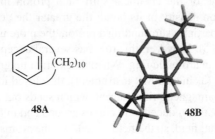

48A   **48B**

were protons located above or within the ring, they would be subjected to a *decreased* field and should appear at lower δ values than normal CH_2 groups (normal δ for CH_2 is ~1–2). The nmr spectrum of [10]paracyclophane (**48A**) showed that this was indeed the case[101] and that the CH_2 peaks were shifted to lower δ the closer they were to the middle of the chain. Examination of **48B** shows that a portion of the methylene chain is positioned directly over the benzene ring, making it subject to the anisotropy shift mentioned above.

It follows that aromaticity can be determined from an NMR spectrum. If the protons attached to the ring are shifted downfield from the normal alkene region, we can conclude that the molecule is diatropic, and hence aromatic. In addition, if the compound has protons above or within the ring (we shall see an example of the latter on p. 90), then if the compound is diatropic, these will be shifted upfield.

---

[101]Waugh, J.S.; Fessenden, R.W. *J. Am. Chem. Soc.* **1957**, *79*, 846. See also, Shapiro, B.L.; Gattuso, M.J.; Sullivan, G.R. *Tetrahedron Lett.* **1971**, 223; Pascal, Jr., R.A.; Winans, C.G.; Van Engen, D. *J. Am. Chem. Soc.* **1989**, *111*, 3007.

One drawback to this method is that it cannot be applied to compounds that have no protons in either category, for example, the dianion of squaric acid (p. 92). Unfortunately, ^{13}C NMR is of no help here, since these spectra do not show ring currents.[102]

Antiaromatic systems exhibit a *paramagnetic* ring current,[103] which causes protons on the outside of the ring to be shifted *upfield* while any inner protons are shifted *downfield*, in sharp contrast to a diamagnetic ring current, which causes shifts in the opposite directions. Compounds that sustain a paramagnetic ring current are called *paratropic*; and are prevalent in four- and eight-electron systems. As with aromaticity, we expect that antiaromaticity will be at a maximum when the molecule is planar and when bond distances are equal. The diamagnetic and paramagnetic effects of the ring currents associated with aromatic and antiaromatic compounds (i.e., shielding and deshielding of nuclei) can be measured by a simple and efficient criterion known as nucleus independent chemical shift (NICS).[104] The aromatic–antiaromatic ring currents reflect the extra π-effects that the molecules experience. The unique near zero value of NICS at the cyclobutadiene ring center is due to cancelation by large and opposite anistropic components.[105]

There are at least four theoretical models for aromaticity, which have recently been compared and evaluated for predictive ability.[106] The *Hess–Schaad model*[107] is good for predicting aromatic stability of benzenoid hydrocarbons, but does not predict reactivity. The *Herndon model*[108] is also good for predicting aromatic stability, but is unreliable for benzenoidicity and does not predict reactivity. The *conjugated-circuit model*[109] is very good for predicting aromatic stability, but not reactivity, and the *hardness model*[110] is best for predicting kinetic stability. Delocalization energy of π-electrons has also been used as an index for aromaticity in polycyclic aromatic hydrocarbons.[111] The claims for linear relationships between aromaticity and energetics, geometries, and magnetic criteria were said to be *invalid* for any representative set of heteroaromatics in which the number of heteroatoms varies.[112]

It should be emphasized that the old and new definitions of aromaticity are not necessarily parallel. If a compound is diatropic and therefore aromatic under the

[102]For a review of ^{13}C NMR spectra of aromatic compounds, see Günther, H.; Schmickler, H. *Pure Appl. Chem.* **1975**, *44*, 807.

[103]Pople, J.A.; Untch, K.G. *J. Am. Chem. Soc.* **1966**, *88*, 4811; Longuet-Higgins, H.C. in Garratt, P.J. *Aromaticity*, Wiley, NY, **1986**, pp. 109–111.

[104]Schleyer, P.v.R.; Maerker, C.; Dransfeld, A.; Jiao, H.; Hommes, N.J.R.v.E. *J. Am. Chem. Soc.* **1996**, *118*, 6317.

[105]Schleyer, P.v.R.; Manoharan, M.; Wang, Z.-X.; Kiran, B.; Jiao, H.; Puchta, R.; Hommes, N.J.R.v.E. *Org. Lett.* **2001**, *3*, 2465

[106]Plavić, D.; Babić, D.; Nikolić, S.; Trinajstić, N. *Gazz. Chim. Ital.*, **1993**, *123*, 243.

[107]Hess, Jr., B.A.; Schaad, L.J. *J. Am. Chem. Soc.* **1971**, *93*, 305.

[108]Herndon, W.C. *Isr. J. Chem.* **1980**, *20*, 270.

[109]Randić,M. *Chem.Phys.Lett.* **1976**, *38*, 68.

[110]Zhou, Z.; Parr, R.G. *J. Am. Chem. Soc.* **1989**, *111*, 7371; Zhou, Z.; Navangul, H.V. *J. Phys. Org. Chem.* **1990**, *3*, 784.

[111]Behrens, S.; Köster, A.M.; Jug, K. *J. Org. Chem.* **1994**, *59*, 2546.

[112]Katritzky, A.R.; Karelson, M.; Sild, S.; Krygowski, T.M.; Jug, K. *J. Org. Chem.* **1998**, *63*, 5228.

new definition, it is more stable than the canonical form of lowest energy, but this does not mean that it will be stable to air, light, or common reagents, since *this* stability is determined not by the resonance energy, but by the difference in free energy between the molecule and the transition states for the reactions involved; and these differences may be quite small, even if the resonance energy is large. A unified theory has been developed that relates ring currents, resonance energies, and aromatic character.[113] Note that aromaticity varies in magnitude relatively and sometimes absolutely with the molecular environment, which includes the polarity of the medium.[114]

The vast majority of aromatic compounds have a closed loop of six electrons in a ring (the aromatic sextet), and we consider these compounds first.[115] Note that a "formula Periodic table" for the benzenoid polyaromatic hydrocarbons has been developed.[116]

Six-Membered Rings

Not only is the benzene ring aromatic, but so are many heterocyclic analogs in which one or more heteroatoms replace carbon in the ring.[117] When nitrogen is the heteroatom, little difference is made in the sextet and the unshared pair of the nitrogen does not participate in the aromaticity. Therefore, derivatives such as *N*-oxides or pyridinium ions are still aromatic. However, for nitrogen heterocycles there are more significant canonical forms (e.g., **49**) than for benzene. Where oxygen or sulfur is the heteroatom, it must be present in its ionic form (**50**) in order to possess the valence of 3 that participation in such a system demands. Thus, pyran (**51**) is not aromatic, but the pyrylium ion (**49**) is.[118]

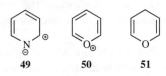

49   **50**   **51**

[113]Haddon, R.C. *J. Am. Chem. Soc.* *1979*, *101*, 1722; Haddon, R.C.; Fukunaga, T. *Tetrahedron Lett.* *1980*, *21*, 1191.
[114]Katritzky, A.R.; Karelson, M.; Wells, A.P. *J. Org. Chem.* *1996*, *61*, 1619.
[115]Values of molecular-orbital energies for many aromatic systems, calculated by the HMO method, are given in Coulson, C.A.; Streitwieser, Jr., A. *A Dictonary of π Electron Calculations*, W.H. Freeman, San Francisco, *1965*. Values calculated by a variation of the SCF method are given by Dewar, M.J.S.; Trinajstic, N. *Collect. Czech. Chem. Commun.* *1970*, *35*, 3136, 3484.
[116]Dias, J.R. *Chem. Br.* *1994*, 384.
[117]For reviews of Aromaticity of heterocycles, see Katritzky, A.R.; Karelson, M.; Malhotra, N. *Heterocycles* *1991*, *32*, 127.
[118]For a review of pyrylium salts, see Balaban, A.T.; Schroth, W.; Fischer, G. *Adv. Heterocycl. Chem.* *1969*, *10*, 241.

In systems of fused six-membered aromatic rings,[119] the principal canonical forms are usually not all equivalent. Compound **52** has a central double bond and is thus different from the other two canonical forms of naphthalene, which are equivalent to each other.[120] For naphthalene, these are the only forms that can be drawn

52

without consideration of Dewar forms or those with charge separation.[121] If we assume that the three forms contribute equally, the 1,2 bond has more double-bond character than the 2,3 bond. Molecular-orbital calculations show bond orders of 1.724 and 1.603, respectively, (cf. benzene, 1.667). In agreement with these predictions, the 1,2 and 2,3 bond distances are 1.36 and 1.415 Å, respectively,[122] and ozone preferentially attacks the 1,2 bond.[123] This nonequivalency of bonds, called *partial bond fixation*,[124] is found in nearly all fused aromatic systems. In phenanthrene, where the 9,10 bond is a single bond in only one of five forms (**53**), bond fixation becomes extreme and this bond is readily attacked by many reagents:[125] It has been observed that increased steric crowding leads to an increase in Dewar-benzene type structures.[126]

53

[119]For books on this subject, see Gutman, I.; Cyvin, S.J. *Introduction to the Theory of Benzenoid Hydrocarbons*, Springer, NY, *1989*; Dias, J.R. *Handbook of Polycyclic Hydrocarbons, Part A: Benzenoid Hydrocarbons*, Elsevier, NY, *1987*; Clar, E. *Polycyclic Hydrocarbons*, 2 vols., Academic Press, NY, *1964*. For a "Periodic table" that systematizes fused aromatic hydrocarbons, see Dias, J.R. *Acc. Chem. Res.* *1985*, *18*, 241; *Top. Curr. Chem.* *1990*, *253*, 123; *J. Phys. Org. Chem.* *1990*, *3*, 765.

[120]As the size of a given fused ring system increases, it becomes more difficult to draw all the canonical forms. For discussions of methods for doing this, see Herndon, W.C. *J. Chem. Educ.* *1974*, *51*, 10; Cyvin, S.J.; Cyvin, B.N.; Brunvoll, J.; Chen, R. *Monatsh. Chem.* *1989*, *120*, 833; Fuji, Z.; Xiaofeng, G.; Rongsi, C. *Top. Curr. Chem.* *1990*, *153*, 181; Wenchen, H.; Wenjie, H. *Top. Curr. Chem.* *1990*, *153*, 195; Sheng, R. *Top. Curr. Chem.* *1990*, *153*, 211; Rongsi, C.; Cyvin, S.J.; Cyvin, B.N.; Brunvoll, J.; Klein, D.J. *Top. Curr. Chem.* *1990*, *153*, 227, and references cited in these papers. For a monograph, see Cyvin, S.J.; Gutman, I. *Kekulé Structures in Benzenoid Hydrocarbons*; Springer, NY, *1988*.

[121]For a modern valence bond description of naphthalene, see Sironi, M.; Cooper, D.L.; Gerratt, J.; Raimondi, M. *J. Chem. Soc. Chem. Commun.* *1989*, 675.

[122]Cruickshank, D.W.J. *Tetrahedron* *1962*, *17*, 155.

[123]Kooyman, E.C. *Recl. Trav. Chim. Pays-Bas*, *1947*, *66*, 201.

[124]For a review, see Efros, L.S. *Russ. Chem. Rev.* *1960*, *29*, 66.

[125]See also Lai, Y. *J. Am. Chem. Soc.* *1985*, *107*, 6678.

[126]Zhang, J.; Ho, D.M.; Pascal Jr., R.A. *J. Am. Chem. Soc.* *2001*, *123*, 10919.

In general, there is a good correlation between bond distances in fused aromatic compounds and bond orders. Another experimental quantity that correlates well with the bond order of a given bond in an aromatic system is the NMR coupling constant for coupling between the hydrogens on the two carbons of the bond.[127]

The resonance energies of fused systems increase as the number of principal canonical forms increases, as predicted by rule 6 (p. 47).[128] Thus, for benzene, naphthalene, anthracene, and phenanthrene, for which we can draw, respectively, two, three, four, and five principal canonical forms, the resonance energies are, respectively, 36, 61, 84, and 92 kcal mol^{-1} (152, 255, 351, and 385 kJ mol^{-1}), calculated from heat-of-combustion data.[129] Note that when phenanthrene, which has a total resonance energy of 92 kcal mol^{-1} (385 kJ mol^{-1}), loses the 9,10 bond by attack of a reagent, such as ozone or bromine, two complete benzene rings remain, each with 36 kcal mol^{-1} (152 kJ mol^{-1}) that would be lost if benzene was similarly attacked. The fact that anthracene undergoes many reactions across the 9,10 positions can be explained in a similar manner. Resonance energies for fused systems can be estimated by counting canonical forms.[130]

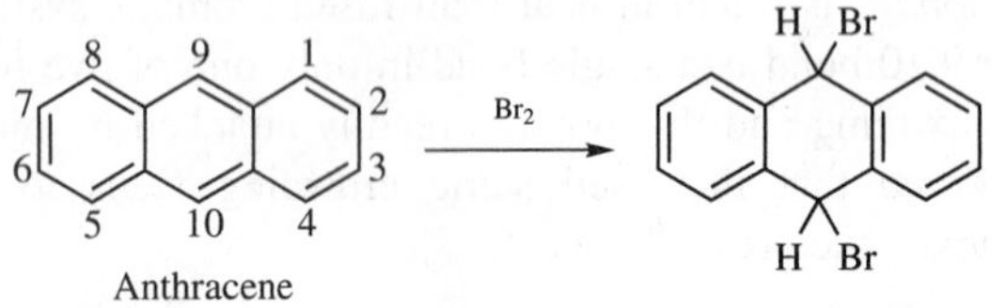

Anthracene

Not all fused systems can be fully aromatic. Thus for phenalene (**54**) there is no way double bonds can be distributed so that each carbon has one single and one double bond.[131] However, phenalene is acidic and reacts with potassium methoxide to give the corresponding anion (**55**), which is completely aromatic. So are the corresponding radical and cation, in which the resonance energies are the same (see p. 68).[132]

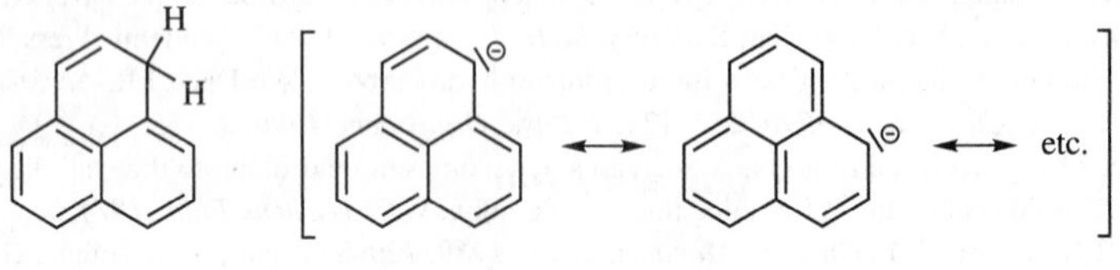

54   **55**

[127]Jonathan, N.; Gordon, S.; Dailey, B.P. *J. Chem. Phys.* **1962**, *36*, 2443; Cooper, M.A.; Manatt, S.L. *J. Am. Chem. Soc.* **1969**, *91*, 6325.

[128]See Herndon, W.C.; Ellzey Jr., M.L. *J. Am. Chem. Soc.* **1974**, *96*, 6631.

[129]Wheland, G.W. *Resonance in Organic Chemistry*, Wiley, NY, **1955**, p. 98.

[130]Swinborne-Sheldrake, R.; Herndon, W.C. *Tetrahedron Lett.* **1975**, 755.

[131]For reviews of phenalenes, see Murata, I. *Top. Nonbenzenoid Aromat. Chem.* **1973**, *1*, 159; Reid, D.H. *Q. Rev. Chem. Soc.* **1965**, *19*, 274.

[132]Pettit, R. *J. Am. Chem. Soc.* **1960**, *82*, 1972.

Molecules that contain fused rings, such as phenanthrene or anthracene, are generally referred to as linear or angular polyacenes. In a fused system, there are not six electrons for each ring.[133] In naphthalene, if one ring is to have six, the other must have only four. One way to explain the greater reactivity of the ring system of naphthalene compared with benzene is to regard one of the naphthalene rings as aromatic and the other as a butadiene system.[134] This effect can become extreme, as in the case of triphenylene.[135] For this compound, there are eight canonical forms like **56**, in which none of the three bonds marked *a* is a double bond and only one form (**57**) in which at least one of them is double. Thus the molecule behaves as if the 18 electrons were distributed so as to give each of the outer rings a sextet, while the middle ring is "empty." Since none of the outer rings need share

56   **57**

any electrons with an adjacent ring, they are as stable as benzene; triphenylene, unlike most fused aromatic hydrocarbons, does not dissolve in concentrated sulfuric acid and has a low reactivity.[136] This phenomenon, whereby some rings in fused systems give up part of their aromaticity to adjacent rings, is called *annellation* and can be demonstrated by UV spectra[119] as well as reactivities. In general, an increase of size of both linear and angular polyacenes is associated with a substantial edecrease in their aromaticity, with a greater decrease for the linear polyacenes.[137]

A six-membered ring with a circle is often used to indicate an aromatic system, and this will be used from time to time. Kekulé structures, those having the C=C units rather than a circle, are used most often in this book. Note that one circle can be used for benzene, but it would be misleading to use two circles for naphthalene, for example, because that would imply 12 aromatic electrons, although naphthalene has only 10.[138]

Five-, Seven-, and Eight-Membered Rings

Aromatic sextets can also be present in five- and seven-membered rings. If a five-membered ring has two double bonds, and the fifth atom possesses an unshared pair

[133]For discussions of how the electrons in fused aromatic systems interact to form $4n + 2$ systems, see Glidewell, C.; Lloyd, D. *Tetrahedron* **1984**, *40*, 4455, *J. Chem. Educ.* **1986**, *63*, 306; Hosoya, H. *Top. Curr. Chem.* **1990**, *153*, 255.

[134]Meredith, C.C.; Wright, G.F. *Can. J. Chem.* **1960**, *38*, 1177.

[135]For a review of triphenylenes, see Buess, C.M.; Lawson, D.D. *Chem. Rev.* **1960**, *60*, 313.

[136]Clar, E.; Zander, M. *J. Chem. Soc.* **1958**, 1861.

[137]Cyrański, M.K.; Stępień, B.T.; Krygowski, T.M. *Tetrahedron* **2000**, *56*, 9663.

[138]See Belloli, R. *J. Chem. Educ.* **1983**, *60*, 190.

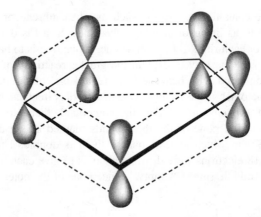

Fig. 2.7. Overlap of five *p* orbitals in molecules such as pyrrole, thiophene, and the cyclopentadienide ion

of electrons, the ring has five *p* orbitals that can overlap to create five new orbitals: three bonding and two antibonding (Fig. 2.7). There are six electrons for these orbitals: the four *p* orbitals of the double bonds each contribute one and the filled orbital contributes the other two. The six electrons occupy the bonding orbitals and

Pyrrole   Thiophene   Furan

constitute an aromatic sextet. The heterocyclic compounds pyrrole, thiophene, and furan are the most important examples of this kind of aromaticity, although furan has a lower degree of aromaticity than the other two.[139] Resonance energies for these three compounds are, respectively, 21, 29, and 16 kcal mol^{-1} (88, 121, and 67 kJ mol^{-1}).[140] The aromaticity can also be shown by canonical forms, for example, for pyrrole:

A

---

[139]The order of aromaticity of these compounds is benzene > thiophene > pyrrole > furan, as calculated by an Aromaticity index based on bond distance measurements. This index has been calculated for five- and six-membered monocyclic and bicyclic heterocycles: Bird, C.W. *Tetrahedron* **1985**, *41*, 1409; **1986**, *42*, 89; **1987**, *43*, 4725.
[140]Wheland, G.W. *Resonance in Organic Chemistry*, Wiley, NY, **1955**, p 99. See also, Calderbank, K.E.; Calvert, R.L.; Lukins, P.B.; Ritchie, G.L.D. *Aust. J. Chem.* **1981**, *34*, 1835.

In contrast to pyridine, the unshared pair in canonical structure **A** in pyrrole is needed for the aromatic sextet. This is why pyrrole is a much weaker base than pyridine.

The fifth atom may be carbon if it has an unshared pair. Cyclopentadiene has unexpected acidic properties ($pK_a \approx 16$) since on loss of a proton, the resulting carbanion is greatly stabilized by resonance although it is quite reactive. The cyclopentadienide ion is usually represented as in **58**. Resonance in this ion is greater than in pyrrole, thiophene, and furan, since all five forms are equivalent. The resonance energy for **58** has been estimated to be 24–27 kcal mol^{-1} (100–113 kJ mol^{-1}).[141]

That all five carbons are equivalent has been demonstrated by labeling the starting compound with ^{14}C and finding all positions equally labeled when cyclopentadiene was regenerated[142] As expected for an aromatic system, the cyclopentadienide ion is diatropic[143] and aromatic substitutions on it have been successfully carried out.[144] Average bond order has been proposed as a parameter to evaluate the aromaticity of these rings, but there is poor correlation with non-aromatic and antiaromatic systems.[145] A model that relies on calculating relative aromaticity from appropriate molecular fragments has also been developed.[146] Bird devised the aromatic index (I_A, or aromaticity index),[147] which is a statistical evaluation of the extent of ring bond order, and this has been used as a criterion of aromaticity. Another bond-order index was proposed by Pozharskii,[148] which goes back to the work of Fringuelli and co-workers.[149] Absolute hardness (see p. 377), calculated from molecular refractions for a range of aromatic and heteroaromatic compounds, shows good linear correlation with aromaticity.[150] Indene and fluorene are also acidic ($pK_a \approx 20$ and 23, respectively), but less so than cyclopentadiene, since annellation causes the electrons to be less available to the five-membered ring. On the other hand, the acidity of 1,2,3,4,5-pentakis(trifluoromethyl)cyclopentadiene (**59**) is greater than that of nitric acid,[151] because of the electron-

[141]Bordwell, F.G.; Drucker, G.E.; Fried, H.E. *J. Org. Chem.* **1981**, *46*, 632.

[142]Tkachuk, R.; Lee, C.C. *Can. J. Chem.* **1959**, *37*, 1644.

[143]Bradamante, S.; Marchesini, A.; Pagani, G. *Tetrahedron Lett.* **1971**, 4621.

[144]Webster, O.W. *J. Org. Chem.* **1967**, *32*, 39; Rybinskaya, M.I.; Korneva, L.M. *Russ. Chem. Rev.* **1971**, *40*, 247.

[145]Jursic, B.S. *J. Heterocycl. Chem.* **1997**, *34*, 1387.

[146]Hosmane, R.S.; Liebman, J.F. *Tetrahedron Lett.* **1992**, *33*, 2303.

[147]Bird, C.W. *Tetrahedron* **1985**, *41*, 1409; *Tetrahedron* **1992**, *48*, 335; *Tetrahedron* **1996**, *52*, 9945.

[148]Pozharskii, A.F. *Khimiya Geterotsikl Soedin* **1985**, 867.

[149]Fringuelli, F. Marino, G.; Taticchi, A.; Grandolini, G. *J. Chem. Soc. Perkin Trans. 2* **1974**, 332.

[150]Bird, C.W. *Tetrahedron* **1997**, *53*, 3319; *Tetrahedron* **1998**, *54*, 4641.

[151]Laganis, E.D.; Lemal, D.M. *J. Am. Chem. Soc.* **1980**, *102*, 6633.

withdrawing effects of the trifluoromethyl groups (see p. 381). Modifications of the Bird and Pozharskii systems have been introduced that are particularly useful for five-membered ring heterocycles.[152] Recent work introduced a new local aromaticity measure, defined as the mean of Bader's electron delocalization index (DI)[153] of para-related carbon atoms in six-membered rings.[154]

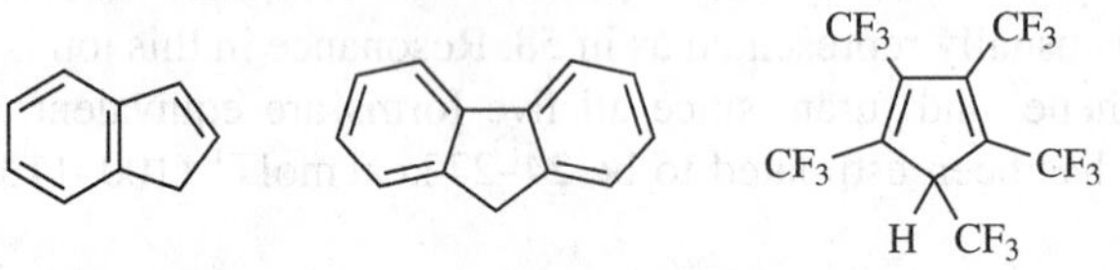

Indene   Fluorene   **59**

As seen above, acidity of compounds can be used to study the aromatic character of the resulting conjugate base. In sharp contrast to cyclopentadiene (see p. 63) is cycloheptatriene (**60**), which has no unusual acidity. This would be hard to explain without the aromatic sextet theory, since, on the basis of resonance forms or a simple

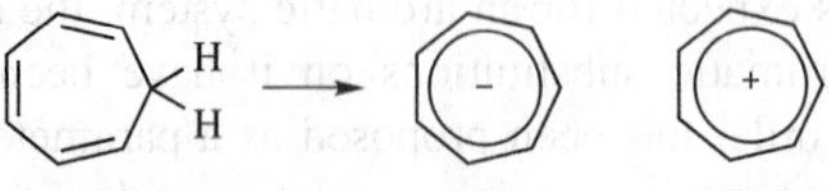

60   **61**   **62**

consideration of orbital overlaps, **61** should be as stable as the cyclopentadienyl anion (**58**). While **61** has been prepared in solution,[155] it is less stable than **58** and far less stable than **62**, in which **60** has lost not a proton, but a hydride ion. The six double-bond electrons of **62** overlap with the empty orbital on the seventh carbon and there is a sextet of electrons covering seven carbon atoms. The cycloheptatrienyl cations (known as the *tropylium ion*, **62**) is quite stable.[156] Tropylium bromide (**63**), which could be completely covalent if the electrons of the bromine were sufficiently attracted to the ring, is actually an ionic compound:[157] Many substituted tropylium ions have been prepared to probe the aromaticity, structure, and reactivity of such systems.[158] Just as with **58**, the equivalence of the carbons

[152]Kotelevskii, S.I.; Prezhdo, O.V. *Tetahedron* **2001**, *57*, 5715.

[153]See Bader, R.F.W. *Atoms in Molecules: A Quantum Theory*, Clarendon, Oxford, **1990**; Bader, R.F.W. *Acc. Chem. Res.* **1985**, *18*, 9; Bader, R.F.W. *Chem. Rev.* **1991**, *91*, 893.

[154]Poater, J.; Fradera, X.; Duran, M.; Solà, M. *Chem. Eur. J.* **2003**, *9*, 400; 1113.

[155]Dauben Jr., H.J.; Rifi, M.R. *J. Am. Chem. Soc.* **1963**, *85*, 3041; also see Breslow, R.; Chang, H.W. *J. Am Chem. Soc.* **1965**, *87*, 2200.

[156]For reviews, see Pietra, F. *Chem. Rev.* **1973**, *73*, 293; Bertelli, D.J. *Top. Nonbenzenoid Aromat. Chem.* **1973**, *1*, 29; Kolomnikova, G.D.; Parnes, Z.N. *Russ. Chem. Rev.* **1967**, *36*, 735; Harmon, K.H., in Olah, G.A.; Schleyer, P.v.R. *Carbonium Ions*, Vol. 4, Wiley, NY, 1973, pp. 1579–1641.

[157]Doering, W. von E.; Knox, L.H. *J. Am. Chem. Soc.* **1954**, *76*, 3203.

[158]Pischel, U.; Abraham, W.; Schnabel, W.; Müller, U. *Chem. Commun.* **1997**, 1383. See Komatsu, K.; Nishinaga, T.; Maekawa, N.; Kagayama, A.; Takeuchi, K. *J. Org. Chem.* **1994**, *59*, 7316 for a tropylium dication.

in **62** has been demonstrated by isotopic labeling.[159] The aromatic cycloheptatrienyl cations $C_7Me_7^+$ and $C_7Ph_7^+$ are known,[160] although their coordination complexes
with transition metals have been problematic, possibly because they assume a boat-
like rather than a planar conformation[161]

63

Another seven-membered ring that shows some aromatic character is tropone
(**64**). This molecule would have an aromatic sextet if the two C=O electrons stayed
away from the ring and resided near the electronegative oxygen atom. In fact,
tropones are stable compounds, and tropolones (**65**) are found in nature.[162] However, analyses of dipole moments, NMR spectra, and X-ray diffraction measurements show that tropones and tropolones display appreciable bond alternations.[163]

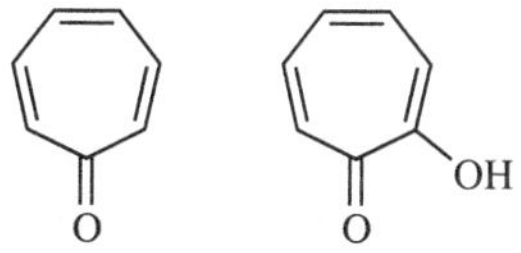

64   **65**

These molecules must be regarded as essentially non-aromatic, although with some
aromatic character. Tropolones readily undergo aromatic substitution, emphasizing
that the old and the new definitions of aromaticity are not always parallel. In sharp
contrast to **64**, cyclopentadienone (**66**) has been isolated only in an argon matrix
<38 K.[164] Above this temperature it dimerizes. Many earlier attempts to prepare
it were unsuccessful.[165] As in **64**, the electronegative oxygen atom draws electron
to itself, but in this case it leaves only four electrons and the molecule is

[159]Vol'pin, M.E.; Kursanov, D.N.; Shemyakin, M.M.; Maimind, V.I.; Neiman, L.A. *J. Gen. Chem. USSR*
1959, *29*, 3667.
[160]Takeuchi, K.; Yokomichi, Y.; Okamoto, K. *Chem. Lett.* *1977*,1177; Battiste, M.A. *J. Am. Chem. Soc.*
1961, *83*, 4101.
[161]Tamm, M.; Dreßel, B.; Fröhlich, R. *J. Org. Chem.* *2000*, *65*, 6795.
[162]For reviews of tropones and tropolones, see Pietra, F. *Acc. Chem. Res.* *1979*, *12*, 132; Nozoe, T. *Pure*
Appl. Chem. *1971*, *28*, 239.
[163]Bertelli, D.J.; Andrews, Jr., T.G. *J. Am. Chem. Soc.* *1969*, *91*, 5280; Bertelli, D.J.; Andrews Jr., T.G.;
Crews, P.O. *J. Am. Chem. Soc.* *1969*, *91*, 5286; Schaefer, J.P.; Reed, L.L. *J. Am. Chem. Soc.* *1971*, *93*,
3902; Watkin, D.J.; Hamor, T.A. *J. Chem. SOc. B 1971*, 2167; Barrow, M.J.; Mills, O.S.; Filippini, G. *J.*
Chem. Soc. Chem. Commun. *1973*, 66.
[164]Maier, G.; Franz, L.H.; Lanz, K.; Reisenauer, H.P. *Chem. Ber.* *1985*, *118*, 3196.
[165]For a review of cyclopentadienone derivatives and of attempts to prepare the parent compound, see
Ogliaruso, M.A.; Romanelli, M.G.; Becker, E.I. *Chem. Rev.* *1965*, *65*, 261.

unstable. Some derivatives of **66** have been prepared.[130]

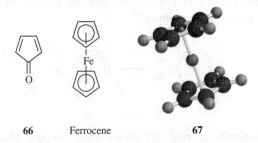

66   Ferrocene   **67**

Another type of five-membered aromatic compound is the *metallocenes* (also called *sandwich compounds*), in which two cyclopentadienide rings form a sandwich around a metallic ion. The best known of these is ferrocene, where the η^5-coordination of the two cyclopentadienyl rings to iron is apparent in the 3D model **67**. Other sandwich compounds have been prepared with Co, Ni, Cr, Ti, V, and many other metals.[166] As a reminder (see p. 43), the η terminology refers to π-donation of electrons to the metal (η^3 for π-allyl systems, η^6 for coordination to a benzene ring, etc.), and η^5 refers to donation of five π-electrons to the iron. Ferrocene is quite stable, subliming $>100°C$ and unchanged at $400°C$. The two rings rotate freely.[167] Many aromatic substitutions have been carried out on metallocenes.[168] Metallocenes containing two metal atoms and three cyclopentadienyl rings have also been prepared and are known as *triple-decker sandwiches*.[169] Even tetradecker, pentadecker, and hexadecker sandwiches have been reported.[170]

The bonding in ferrocene may be looked upon in simplified molecular-orbital terms as follows.[171] Each of the cyclopentadienide rings has five molecular orbitals: three filled bonding and two empty antibonding orbitals (p. 62). The outer

[166]For a monograph on metallocenes, see Rosenblum, M. *Chemistry of the Iron Group Metallocenes*, Wiley, NY, *1965*. For reviews, see Lukehart, C.M. *Fundamental Transition Metal Organometallic Chemistry*, Brooks/Cole, Monterey, CA, *1985*, pp. 85–118; Lemenovskii, D.A.; Fedin, V.P. *Russ. Chem. Rev. 1986*, *55*, 127; Sikora, D.J.; Macomber, D.W.; Rausch, M.D. *Adv. Organomet. Chem. 1986*, *25*, 317; Pauson, P.L. *Pure Appl. Chem. 1977*, *49*, 839; Nesmeyanov, A.N.; Kochetkova, N.S. *Russ. Chem. Rev. 1974*, *43*, 710; Shul'pin, G.B.; Rybinskaya, M.I. *Russ. Chem. Rev. 1974*, *43*, 716; Perevalova, E.G.; Nikitina, T.V. *Organomet. React., 1972*, *4*, 163; Bublitz, D.E.; Rinehart Jr., K.L. *Org. React., 1969*, *17*, 1; Leonova, E.V.; Kochetkova, N.S. *Russ. Chem. Rev. 1973*, *42*, 278; Rausch, M.D. *Pure Appl. Chem. 1972*, *30*, 523. For a bibliography of reviews on metallocenes, see Bruce, M.I. *Adv. Organomet. Chem. 1972*, *10*, 273, pp. 322–325.

[167]For a discussion of the molecular structure, see Haaland, A. *Acc. Chem. Res. 1979*, *12*, 415.

[168]For a review on aromatic substitution on ferrocenes, see Plesske, K. *Angew. Chem. Int. Ed. 1962*, *1*, 312, 394.

[169]For a review, see Werner, H. *Angew. Chem. Int. Ed. 1977*, *16*, 1.

[170]See, for example, Siebert, W. *Angew. Chem. Int. Ed. 1985*, *24*, 943.

[171]Rosenblum, M. *Chemistry of the Iron Group Metallocnes*, Wiley, NY, *1965*, pp. 13–28; Coates, G.E.; Green, M.L.H.; Wade, K. *Organometallic Compounds*, 3rd ed., Vol. 2, Methuene, London, *1968*, pp. 97–104; Grebenik, P.; Grinter, R.; Perutz, R.N. *Chem. Soc. Rev. 1988*, *17*, 453; 460.

shell of the Fe atom possesses nine atomic orbitals, that is, one $4s$, three $4p$, and five $3d$ orbitals. The six filled orbitals of the two cyclopentadienide rings overlap with the s, three p, and two of the d orbitals of the Fe to form twelve new orbitals, six of which are bonding. These six orbitals make up two ring-to-metal triple bonds. In addition, further bonding results from the overlap of the empty antibonding orbitals of the rings with additional filled d orbitals of the iron. All told, there are 18 electrons (10 of which may be considered to come from the rings and 8 from iron in the zero oxidation state) in nine orbitals; six of these are strongly bonding and three weakly bonding or nonbonding.

The tropylium ion has an aromatic sextet spread over seven carbon atoms. An analogous ion, with the sextet spread over eight carbon atoms, is 1,3,5,7-tetra-methylcyclooctatetraene dictation (**68**). This ion, which is stable in solution at $-50°C$, is diatropic and approximately planar. The dication **68** is not stable above about $-30°C$.[172]

68

Other Systems Containing Aromatic Sextets

Simple resonance theory predicts that pentalene (**69**), azulene (**70**), and heptalene (**71**) should be aromatic, although no nonionic canonical form can have a double bond at the ring junction. Molecular-orbital calculations show that azulene should be stable but not the other two, and this is borne out by experiment. Heptalene has been prepared,[173] but reacts readily with oxygen, acids, and bromine, is easily hydrogenated, and polymerizes on standing. Analysis of its NMR spectrum shows

69   **70**   **71**

[172]This and related ions were prepared by Olah, G.A.; Staral, J.S.; Liang, G.; Paquette, L.A.; Melega, W.P.; Carmody, M.J. *J. Am. Chem. Soc.* **1977**, *99*, 3349. See also Radom, L.; Schaefer III, H.F. *J. Am. Chem. Soc.* **1977**, *99*, 7522; Olah, G.A.; Liang, G. *J. Am. Chem. Soc.* **1976**, *98*, 3033; Willner, I.; Rabinovitz, M. *Nouv. J. Chim.*, **1982**, *6*, 129.

[173]Dauben, Jr., H.J.; Bertelli, D.J. *J. Am. Chem. Soc.* **1961**, *83*, 4659; Vogel, E.; Königshofen, H.; Wassen, J.; Müllen, K.; Oth, J.F.M. *Angew. Chem. Int. Ed.* **1974**, *13*, 732; Paquette, L.A.; Browne, A.R.; Chamot, E. *Angew. Chem. Int. Ed.* **1979**, *18*, 546. For a review of heptalenes, see Paquette, L.A. *Isr. J. Chem.* **1980**, *20*, 233.

that it is not planar.[174] The 3,8-dibromo and 3,8-dicarbomethoxy derivatives of **71** are stable in air at room temperature but are not diatropic.[175] A number of methylated heptalenes and dimethyl 1,2-heptalenedicarboxylates have also been prepared and are stable nonaromatic compounds.[176] Pentalene has not been prepared,[177] but the hexaphenyl[178] and 1,3,5-tri-*tert*-butyl derivatives[179] are known. The former is air sensitive in solution. The latter is stable, but X-ray diffraction and photoelectron spectral data show bond alternation.[180] Pentalene and its methyl and dimethyl derivatives have been formed in solution, but they dimerize before they can be isolated.[181] Many other attempts to prepare these two systems have failed.

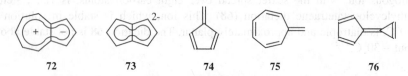

| **72** | **73** | **74** | **75** | **76** |

In sharp contrast to **69** and **71**, azulene, a blue solid, is quite stable and many of its derivatives are known.[182] Azulene readily undergoes aromatic substitution. Azulene may be regarded as a combination of **58** and **62** and, indeed, possesses a dipole moment of 0.8 D (see **72**).[183] Interestingly, if two electrons are added to pentalene, a stable dianion (**73**) results.[184] It can be concluded that an aromatic system of electrons will be spread over two rings only if 10 electrons (not 8 or 12) are available for aromaticity. $[n, m]$-Fluvalenes ($n \neq m$, where fulvalene is **74**) as well as azulene are known to shift their π-electrons due to the influence of dipolar aromatic resonance structures.[185] However, calculations showed that

[174]Bertelli, D.J., in Bergmann, E.D.; Pullman, B. *Aromaticity, Pseudo-Aromaticity, and Anti-Aromaticity*, Israel Academy of Sciences and Humanities, Jerusalem, *1971*, p. 326. See also Stegemann, J.; Lindner, H.J. *Tetrahedron Lett.* *1977*, 2515.

[175]Vogel, E.; Ippen, J. *Angew. Chem. Int. Ed.* *1974*, *13*, 734; Vogel, E.; Hogrefe, F. *Angew. Chem. Int. Ed.* *1974*, *13*, 735.

[176]Hafner, K.; Knaup, G.L.; Lindner, H.J. *Bull. Soc. Chem. Jpn.* *1988*, *61*, 155.

[177]Metal complexes of pentalene have been prepared: Knox, S.A.R.; Stone, F.G.A. *Acc. Chem. Res.* *1974*, *7*, 321.

[178]LeGoff, E. *J. Am. Chem. Soc.* *1962*, *84*, 3975. See also Hafner, K.; Bangert, K.F.; Orfanos, V. *Angew. Chem. Int. Ed.* *1967*, *6*, 451; Hartke, K.; Matusch, R. *Angew. Chem. Int. Ed.* *1972*, *11*, 50.

[179]Hafner, K.; Süss, H.U. *Angew. Chem. Int. Ed.* *1973*, *12*, 575. See also Hafner, K.; Suda, M. *Angew. Chem. Int. Ed.* *1976*, *15*, 314.

[180]Kitschke, B.; Lindner, H.J. *Tetrahedron Lett.* *1977*, 2511; Bischof, P.; Gleiter, R.; Hafner, K.; Knauer, K.H.; Spanget-Larsen, J.; Süss, H.U. *Chem. Ber.* *1978*, *111*, 932.

[181]Bloch, R.; Marty, R.A.; de Mayo, P. *J. Am. Chem. Soc.* *1971*, *93*, 3071; *Bull. Soc. Chim. Fr.*, *1972*, 2031; Hafner, K.; Dönges, R.; Goedecke, E.; Kaiser, R. *Angew. Chem. Int. Ed.* *1973*, *12*, 337.

[182]For a review on azulene, see Mochalin, V.B.; Porshnev, Yu.N. *Russ. Chem. Rev.* *1977*, *46*, 530.

[183]Tobler, H.J.; Bauder, A.; Günthard, H.H. *J. Mol. Spectrosc.*, *1965*, *18*, 239.

[184]Katz, T.J.; Rosenberger, M.; O'Hara, R.K. *J. Am. Chem. Soc.* *1964*, *86*, 249. See also, Willner, I.; Becker, J.Y.; Rabinovitz, M. *J. Am. Chem. Soc.* *1979*, *101*, 395.

[185]Möllerstedt, H.; Piqueras, M.C.; Crespo, R.; Ottosson, H. *J. Am. Chem. Soc.* *2004*, *126*, 13938.

dipolar resonance structures contribute only 5% to the electronic structure of heptafulvalene (**75**), although 22–31% to calicene (**76**).[186] Based on Baird's theory,[187] these molecules are influenced by aromaticity in both the ground and excited states, therefore acting as aromatic "chameleons." This premise was confirmed in work by Ottosson and co-workers.[185] Aromaticity indexes for various substituted fulvalene compounds has been reported.[188]

Alternant and Nonalternant Hydrocarbons[189]

Aromatic hydrocarbons can be divided into alternant and nonalternant hydrocarbons. In alternant hydrocarbons, the conjugated carbon atoms can be divided into two sets such that no two atoms of the same set are directly linked. For convenience, one set may be starred. Naphthalene is an alternant and azulene a nonalternant hydrocarbon:

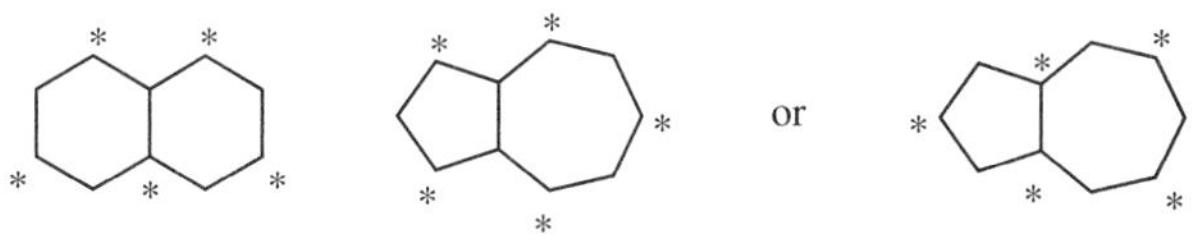

In alternant hydrocarbons, the bonding and antibonding orbitals occur in pairs; that is, for every bonding orbital with an energy $-E$ there is an antibonding one with energy $+E$ (Fig. 2.8[190]). Even-alternant hydrocarbons are those with an even number of conjugated atoms, that is, an equal number of starred and unstarred atoms. For these hydrocarbons, all the bonding orbitals are filled and the π electrons are uniformly spread over the unsaturated atoms.

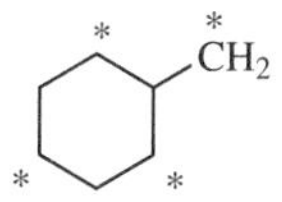

As with the allylic system, odd-alternant hydrocarbons (which must be carbocations, carbanions, or radicals) in addition to equal and opposite bonding and antibonding orbitals also have a nonbonding orbital of zero energy. When an odd number of orbitals overlap, an odd number is created. Since orbitals of alternant hydrocarbons occur in $-E$ and $+E$ pairs, one orbital can have no partner and must therefore have zero bonding energy. For example, in the benzylic system the cation has an unoccupied nonbonding orbital, the free radical has one electron there and the carbanion two (Fig. 2.9). As with the allylic system, all three species have the same bonding energy. The charge distribution (or unpaired-electron distribution)

[186]Scott, A.P.; Agranat, A.; Biedermann, P.U.; Riggs, N.V.; Radom, L. *J. Org. Chem.* **1997**, *62*, 2026.
[187]Baird, N.C. *J. Am. Chem. Soc.* **1972**, *94*, 4941.
[188]Stepien, B.T.; Krygowski, T.M.; Cyranski, M.K. *J. Org. Chem.* **2002**, *67*, 5987.
[189]For discussions, see Jones, R.A.Y. *Physical and Mechanistic Organic Chemistry*, 2nd ed.; Cambridge University Press, Cambridge, **1984**, pp. 122–129; Dewar, M.J.S. *Prog. Org. Chem.* **1953**, *2*, 1.
[190]Taken from Dewar, M.J.S *Prog. Org. Chem.* **1953**, *2*, 1, p. 8.

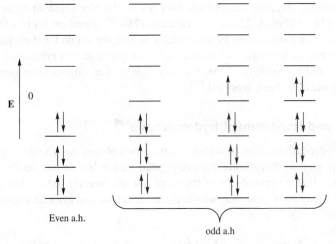

Fig. 2.8. Energy levels in odd- and even-alternant hydrocarbons.[190] The arrows represent electrons. The orbitals are shown as having different energies, but some may be degenerate.

over the entire molecule is also the same for the three species and can be calculated by a relatively simple process.[189]

For nonalternant hydrocarbons the energies of the bonding and antibonding orbitals are not equal and opposite and charge distributions are not the same in cations, anions, and radicals. Calculations are much more difficult but have been carried

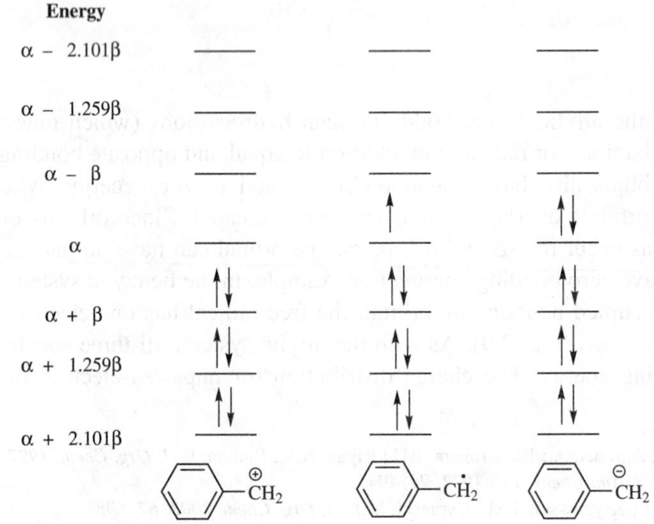

Energy

$\alpha - 2.101\beta$

$\alpha - 1.259\beta$

$\alpha - \beta$

α

$\alpha + \beta$

$\alpha + 1.259\beta$

$\alpha + 2.101\beta$

Fig. 2.9. Energy levels for the benzyl cation, free radical, and carbanion. Since α is the energy of a p-orbital (p. 36), the nonbonding orbital has no bonding energy.

out.[191] Theoretical approaches to calculate topological polarization and reactivity of these hydrocarbons have been reported.[192]

Aromatic Systems with Electron Numbers Other Than Six

Ever since the special stability of benzene was recognized, chemists have been thinking about homologous molecules and wondering whether this stability is also associated with rings that are similar but of different sizes, such as cyclobutadiene (**77**), cyclooctatetraene (**78**), cyclodecapentaene (**79**)[193], and so on. The general

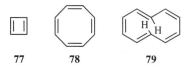

name *annulene* is given to these compounds, benzene being [6]annulene, and **77–79** being called, respectively, [4], [8], and [10]annulene. By a naïve consideration of resonance forms, these annulenes and higher ones should be as aromatic as benzene. Yet they proved remarkably elusive. The ubiquitous benzene ring is found in thousands of natural products, in coal and petroleum, and is formed by strong treatment of many noncyclic compounds. None of the other annulene ring systems has ever been found in nature and, except for cyclooctatetraene, their synthesis is not simple. Obviously, there is something special about the number six in a cyclic system of electrons.

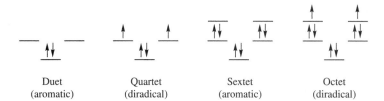

Hückel's rule, based on molecular-orbital calculations,[194] predicts that electron rings will constitute an aromatic system only if the number of electrons in the ring is of the form $4n + 2$, where n is zero or any position integer. Systems that contain $4n$ electrons are predicted to be nonaromatic. The rule predicts that

[191]Peters, D. *J. Chem. Soc.* **1958**, 1023, 1028, 1039; Brown, R.D.; Burden, F.R.; Williams, G.R. *Aust. J. Chem.* **1968**, *21*, 1939. For reviews, see Zahradnik, R., in Snyder, J.P. *Nonbenzenoid Aromatics* vol. 2, Academic Press, NY, **1971**, pp. 1–80; Zahradnik, R. *Angew. Chem. Int. Ed.* **1965**, *4*, 1039.

[192]Langler, R.F. *Aust. J. Chem.* **2000**, *53*, 471; Fredereiksen, M.U.; Langler, R.F.; Staples, M.A.; Verma, S.D. *Aust. J. Chem.* **2000**, *53*, 481.

[193]The cyclodecapentaene shown here is the cis–trans–cis–cis–trans form. For other stereoisomers, see p. 79.

[194]For reviews of molecular-orbital calculations of nonbenzenoid cyclic conjugated hydrocarbons, see Nakajima, T. *Pure Appl. Chem.* **1971**, *28*, 219; *Fortschr. Chem. Forsch.* **1972**, *32*, 1.

rings of 2, 6, 10, 14, and so on, electrons will be aromatic, while rings of 4, 8, 12, and so on, will not be. This is actually a consequence of Hund's rule. The first pair of electrons in an annulene goes into the π orbital of lowest energy. After that the bonding orbitals are degenerate and occur in pairs of equal energy. When there is a total of four electrons, Hund's rule predicts that two will be in the lowest orbital but the other two will be unpaired, so that the system will exist as a diradical rather than as two pairs. The degeneracy can be removed if the molecule is distorted from maximum molecular symmetry to a structure of lesser symmetry. For example, if **77** assumes a rectangular rather than a square shape, one of the previously degenerate orbitals has a lower energy than the other and will be occupied by two electrons. In this case, of course, the double bonds are essentially separate and the molecule is still not aromatic. Distortions of symmetry can also occur when one or more carbons are replaced by heteroatoms or in other ways.[195]

In the following sections systems with various numbers of electrons are discussed. When we look for aromaticity we look for (*1*) the presence of a diamagnetic ring current; (*2*) equal or approximately equal bond distances, except when the symmetry of the system is disturbed by a heteroatom or in some other way; (*3*) planarity; (*4*) chemical stability; (*5*) the ability to undergo aromatic substitution.

Systems of Two Electrons[196]

Obviously, there can be no ring of two carbon atoms though a double bond may be regarded as a degenerate case. However, in analogy to the tropylium ion, a three-membered ring with a double bond and a positive charge on the third atom (the *cyclopropenyl cation*) is a $4n + 2$ system and hence is expected to show aromaticity. The unsubstituted **80** has been prepared,[197] as well as several derivatives, e.g.,

$$\left[\triangledown^{\oplus} \longleftrightarrow {}^{\oplus}\triangledown \longleftrightarrow \triangledown^{\oplus} \right] \equiv \triangledown^{\oplus}$$

80

the trichloro, diphenyl, and dipropyl derivatives, and these are stable despite the angles of only 60°. In fact, the tripropylcyclopropenyl,[198] tricyclopropylcyclopropenyl,[199] chlorodipropylcyclopropenyl,[200] and chloro-bisdialkylaminocyclopropenyl[201] cations are among the most stable carbocations known, being stable

[195]For a discussion, see Hoffmann, R. *Chem. Commun.* **1969**, 240.

[196]For reviews, see Billups, W.E.; Moorehead, A.W., in Rappoport *The Chemistry of the Cyclopropyl Group*, pt. 2, Wiley, NY, **1987**, pp. 1533–1574; Potts, K.T.; Baum, J.S. *Chem. Rev.* **1974**, *74*, 189; Yoshida, Z. *Top. Curr. Chem.* **1973**, *40*, 47; D'yakonov, I.A.; Kostikov, R.R. *Russ. Chem. Rev.* **1967**, *36*, 557; Closs, G.L. *Adv. Alicyclic Chem.* **1966**, *1*, 53, pp. 102–126; Krebs, A.W. *Angew. Chem. Int. Ed.* **1965**, *4*, 10.

[197]Farnum, D.G.; Mehta, G.; Silberman, R.G. *J. Am. Chem. Soc.* **1967**, *89*, 5048; Breslow, R.; Groves, J.T. *J. Am. Chem. Soc.* **1970**, *92*, 984.

[198]Breslow, R.; Höver, H.; Chang, H.W. *J. Am. Chem. Soc.* **1962**, *84*, 3168.

[199]Komatsu, K.; Tomioka, K.; Okamoto, K. *Tetrahedron Lett.* **1980**, *21*, 947; Moss, R.A.; Shen, S.; Krogh-Jespersen, K.; Potenza, J.A.; Schugar, H.J.; Munjal, R.C. *J. Am. Chem. Soc.* **1986**, *108*, 134.

[200]Ito, S.; Morita, N.; Asao, T. *Tetrahedron Lett.* **1992**, *33*, 3773.

[201]Taylor, M.J.; Surman, P.W.J.; Clark, G.R. *J. Chem. Soc. Chem. Commun.* **1994**, 2517.

even in water solution. The tri-*tert*-butylcyclopropenyl cation is also very stable.[202] In addition, cyclopropenone and several of its derivatives are stable

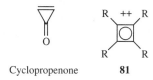

Cyclopropenone   **81**

compounds,[203] in accord with the corresponding stability of the tropones.[204] The ring system **80** is nonalternant and the corresponding radical and anion (which do not have an aromatic duet) have electrons in antibonding orbitals, so that their energies are much higher. As with **58** and **62**, the equivalence of the three carbon atoms in the triphenylcyclopropenyl cation has been demonstrated by ^{14}C labeling experiments.[205] The interesting dications **81** (R = Me or Ph) have been prepared,[206] and they too should represent aromatic systems of two electrons.[207]

Systems of Four Electrons: Antiaromaticity

The most obvious compound in which to look for a closed loop of four electrons is cyclobutadiene (**77**).[208] Hückel's rule predicts no aromatic character here, since 4 is not a number of the form $4n + 2$. There is a long history of attempts to prepare this compound and its simple derivatives, and those experiments fully bear out Hückel's prediction. Cyclobutadienes display none of the characteristics that would lead us to call them aromatic, and there is evidence that a closed loop of four electrons is actually *antiaromatic*.[209] If such compounds simply lacked aromaticity, we would expect

[202]Ciabattoni, J.; Nathan III, E.C. *J. Am. Chem. Soc.* **1968**, *90*, 4495.

[203]See, for example, Kursanov, D.N.; Vol'pin, M.E.; Koreshkov, Yu.D. *J. Gen. Chem. USSR* **1960**, *30*, 2855; Breslow, R.; Oda, M. *J. Am. Chem. Soc.* **1972**, *94*, 4787; Yoshida, Z.; Konishi, H.; Tawara, Y.; Ogoshi, H. *J. Am. Chem. Soc.* **1973**, *95*, 3043; Ciabattoni, J.; Nathan III, E.C. *J. Am. Chem. Soc.* **1968**, *90*, 4495.

[204]For a reveiw of cyclopropenones, see Eicher, T.; Weber, J.L. *Top. Curr. Chem. Soc.* **1975**, *57*, 1. For discussions of cyclopropenone structure, see Shäfer, W.; Schweig, A.; Maier, G.; Sayrac, T.; Crandall, J.K. *Tetrahedron Lett.* **1974**, 1213; Tobey, S.W., in Bergmann, E.D.; Pullman, B. *Aromaticity, Pseudo-Aromaticity, and Anti-Aromaticity*, Israel Academy of Sciences and Humanities, Jerusalem, *1971*, pp. 351–362; Greenberg, A.; Tomkins, R.P.T.; Dobrovolny, M.; Liebman, J.F. *J. Am. Chem. Soc.* **1983**, *105*, 6855.

[205]D'yakonov, I.A.; Kostikov, R.R.; Molchanov, A.P. *J. Org. Chem. USSR* **1969**, *5*, 171; *1970*, *6*, 304.

[206]Freedman, H.H.; Young, A.E. *J. Am. Chem. Soc.* **1964**, *86*, 734; Olah, G.A.; Staral, J.S. *J. Am. Chem. Soc.* **1976**, *98*, 6290. See also Lambert, J.B.; Holcomb, A.G. *J. Am. Chem. Soc.* **1971**, *93*, 2994; Seitz, G.; Schmiedel, R.; Mann, K. *Synthesis*, *1974*, 578.

[207]See Pittman Jr., C.U.; Kress, A.; Kispert, L.D. *J. Org. Chem.* **1974**, *39*, 378. See, however, Krogh-Jespersen, K.; Schleyer, P.v.R.; Pople, J.A.; Cremer, D. *J. Am. Chem. Soc.* **1978**, *100*, 4301.

[208]For a monograph, see Cava, M.P.; Mitchell, M.J. *Cyclobutadiene and Related Compounds*; Academic Press, NY, *1967*. For reviews, see Maier, G. *Angew. Chem. Int. Ed.* **1988**, *27*, 309; *1974*, *13*, 425–438; Bally, T.; Masamune, S. *Tetrahedron* **1980**, *36*, 343; Vollhardt, K.P.C. *Top. Curr. Chem.* **1975**, *59*, 113.

[209]For reviews of antiaromaticity, see Glukhovtsev, M.N.; Simkin, B.Ya.; Minkin, V.I. *Russ. Chem. Rev.* **1985**, *54*, 54; Breslow, R. *Pure Appl. Chem.* **1971**, *28*, 111; *Acc. Chem. Res.* **1973**, *6*, 393.

them to be about as stable as similar nonaromatic compounds, but both theory and experiment show that they are *much less stable*.[210] An antiaromatic compound may be defined as a compound that is destabilized by a closed loop of electrons.

After years of attempts to prepare cyclobutadiene, the goal was finally reached by Pettit and co-workers.[211] It is now clear that **77** and its simple derivatives are extremely unstable compounds with very short lifetimes (they dimerize by a Diels–Alder reaction; see **15-60**) unless they are stabilized in some fashion, either at ordinary temperatures embedded in the cavity of a hemicarcerand[212] (see the structure of a carcerand on p. 128), or in matrices at very low temperatures (generally under 35 K). In either of these cases, the cyclobutadiene molecules are forced to remain apart from each other, and other molecules cannot get in. The structures of **77** and some of its derivatives have been studied a number of times using the low-temperature matrix technique.[213] The ground-state structure of **77** is a rectangular diene (not a diradical) as shown by the ir spectra of **77** and deuterated **77** trapped in matrices,[214] as well as by a photoelectron spectrum.[215] Molecular-orbital calculations agree.[216] The same conclusion was also reached in an elegant experiment in which 1,2-dideuterocyclobutadiene was generated. If **77** is a rectangular diene, the dideutero compound should exist as two isomers:

The compound was generated (as an intermediate that was not isolated) and two isomers were indeed found.[217] The cyclobutadiene molecule is not static, even in the matrices. There are two forms (**77a** and **77b**), which rapidly interconvert.[218]

---

[210]For a discussion, see Bauld, N.L.; Welsher, T.L.; Cessac, J.; Holloway, R.L. *J. Am. Chem. Soc.* **1978**, *100*, 6920.

[211]Watts, L.; Fitzpatrick, J.D.; Pettit, R. *J. Am. Chem. Soc.* **1965**, *87*, 3253, **1966**, *88*, 623. See also, Cookson, R.C.; Jones, D.W. *J. Chem. Soc.* **1965**, 1881.

[212]Cram, D.J.; Tanner, M.E.; Thomas, R. *Angew. Chem. Int. Ed.* **1991**, *30*, 1024.

[213]See, for example, Lin, C.Y.; Krantz, A. *J. Chem. Soc. Chem. Commun.* **1972**, 1111; Chapman, O.L.; McIntosh, C.L.; Pacansky, J. *J. Am. Chem. Soc.* **1973**, *95*, 614; Maier, G.; Mende, U. *Tetrahedron Lett.* **1969**, 3155. For a review, see Sheridan, R.S. *Org. Photochem.* **1987**, *8*, 159; pp. 167–181.

[214]Masamune, S.; Souto-Bachiller, F.A.; Machiguchi, T.; Bertie, J.E. *J. Am. Chem. Soc.* **1978**, *100*, 4889.

[215]Kreile, J.; Münzel, N.; Schweig, A.; Specht, H. *Chem. Phys. Lett.* **1986**, *124*, 140.

[216]See, for example, Borden, W.T.; Davidson, E.R.; Hart, P. *J. Am. Chem. Soc.* **1978**, *100*, 388; Kollmar, H.; Staemmler, V. *J. Am. Chem. Soc.* **1978**, *100*, 4304; Jafri, J.A.; Newton, M.D. *J. Am. Chem. Soc.* **1978**, *100*, 5012; Ermer, O.; Heilbronner, E. *Angew. Chem. Int. Ed.* **1983**, *22*, 402; Voter, A.F.; Goddard III, W.A. *J. Am. Chem. Soc.* **1986**, *108*, 2830.

[217]Whitman, D.W.; Carpenter, B.K. *J. Am. Chem. Soc.* **1980**, *102*, 4272. See also Whitman, D.W.; Carpenter, B.K. *J. Am. Chem. Soc.* **1982**, *104*, 6473.

[218]Carpenter, B.K. *J. Am. Chem. Soc.* **1983**, *105*, 1700; Huang, M.; Wolfsberg, M. *J. Am. Chem. Soc.* **1984**, *106*, 4039; Dewar, M.J.S.; Merz, Jr., K.M.; Stewart, J.J.P. *J. Am. Chem. Soc.* **1984**, *106*, 4040; Orendt, A.M.; Arnold, B.R.; Radziszewski, J.G.; Facelli, J.C.; Malsch, K.D.; Strub, H.; Grant, D.M.; Michl, J. *J. Am. Chem. Soc.* **1988**, *110*, 2648. See, however, Arnold, B.R.; Radziszewski, J.G.; Campion, A.; Perry, S.S.; Michl, J. *J. Chem. Soc.* **1991**, *113*, 692.

Note that there is experimental evidence that the aromatic and antiaromatic characters of neutral and dianionic systems are measurably increased via deuteration.[219]

77a   **77b**   **82**

There are some simple cyclobutadienes that are stable at room temperature for varying periods of time. These either have bulky substituents or carry certain other stabilizing substituents such as seen in tri-*tert*-butylcyclobutadiene (**83**).[220] Such compounds are relatively stable because dimerization is sterically hindered. Examination of the NMR spectrum of **83** showed that the ring proton ($\delta = 5.38$) was shifted *upfield*, compared with the position expected for a nonaromatic proton, for example, cyclopentadiene. As we will see (pp. 89–90), this indicates that the compound is antiaromatic.

83   etc.

The other type of stable cyclobutadiene has two electron-donating and two electron-withdrawing groups,[221] and is stable in the absence of water.[222] An example is **58**. The stability of these compounds is generally attributed to the resonance shown, a type of resonance stabilization called the *push–pull or captodative effect*,[223] although it has been concluded from a photoelectron spectroscopy study that second-order bond fixation is more important.[224] An X-ray crystallographic study of **83** has shown[225] the ring to be a distorted square with bond lengths of 1.46 Å and angles of 87° and 93°.

[219]For experiments with [16]-annulene (see p 82), see Stevenson, C.D.; Kurth, T.L. *J. Am. Chem. Soc.* **1999**, *121*, 1623

[220]Masamune, S.; Nakamura, N.; Suda, M.; Ona, H. *J. Am. Chem. Soc.* **1973**, *95*, 8481; Maier, G.; Alzérreca, A. *Angew. Chem. Int. Ed.* **1973**, *12*, 1015. For a discussion, see Masamune, S. *Pure Appl. Chem.* **1975**, *44*, 861.

[221]The presence of electron-donating and -withdrawing groups on the same ring stabilizes 4n systems and destabilizes 4n + 2 systems. For a review of this concept, see Gompper, R.; Wagner, H. *Angew. Chem. Int. Ed.* **1988**, *27*, 1437.

[222]Neuenschwander, M.; Niederhauser, A. *Chimia*, **1968**, *22*, 491, *Helv. Chim. Acta*, **1970**, *53*, 519; Gompper, R.; Kroner, J.; Seybold, G.; Wagner, H. *Tetrahedron* **1976**, *32*, 629.

[223]Manatt, S.L.; Roberts, J.D. *J. Org. Chem.* **1959**, *24*, 1336; Breslow, R.; Kivelevich, D.; Mitchell, M.J.; Fabian, W.; Wendel, K. *J. Am. Chem. Soc.* **1965**, *87*, 5132; Hess Jr., B.A.; Schaad, L.J. *J. Org. Chem.* **1976**, *41*, 3058.

[224]Gompper, R.; Holsboer, F.; Schmidt, W.; Seybold, G. *J. Am. Chem. Soc.* **1973**, *95*, 8479.

[225]Lindner, H.J.; von Ross, B. *Chem. Ber.* **1974**, *107*, 598.

It is clear that simple cyclobutadienes, which could easily adopt a square planar shape if that would result in aromatic stabilization, do not in fact do so and are not aromatic. The high reactivity of these compounds is not caused merely by steric strain, since the strain should be no greater than that of simple cyclopropenes, which are known compounds. It is probably caused by antiaromaticity.[226]

84

The cyclobutadiene system can be stabilized as a η^4-complex with metals,[227] as with the iron complex **84** (see Chapter 3), but in these cases electron density is withdrawn from the ring by the metal and there is no aromatic quartet. In fact, these cyclobutadiene–metal complexes can be looked upon as systems containing an aromatic duet. The ring is square planar,[228] the compounds undergo aromatic substitution,[229] and nmr spectra of monosubstituted derivatives show that the C-2 and C-4 protons are equivalent.[229]

85   **86**   **87**

Other systems that have been studied as possible aromatic or antiaromatic four-electron systems include the cyclopropenyl anion (**86**), the cyclopentadienyl cation (**87**).[230] With respect to **86**, HMO theory predicts that an unconjugated **85** (i.e., a single canonical form) is more stable than a conjugated **86**,[231] so that **85** would actually lose stability by forming a closed loop of four electrons. The HMO theory

[226]For evidence, see Breslow, R.; Murayama, D.R.; Murahashi, S.; Grubbs, R. *J. Am. Chem. Soc.* **1973**, *95*, 6688; Herr, M.L. *Tetrahedron* **1976**, *32*, 2835.

[227]For reviews, see Efraty, A. *Chem. Rev.* **1977**, *77*, 691; Pettit, R. *Pure Appl. Chem.* **1968**, *17*, 253; Maitlis, P.M. *Adv. Organomet. Chem.* **1966**, *4*, 95; Maitlis, P.M.; Eberius, K.W., in Snyder, J.P. *Nonbenzenoid Aromatics*, vol. 2, Academic Press, NY, **1971**, pp. 359–409.

[228]Dodge, R.P.; Schomaker, V. *Acta Crystallogr.* **1965**, *18*, 614; *Nature (London)* **1960**, *186*, 798; Dunitz, J.D.; Mez, H.C.; Mills, O.S.; Shearer, H.M.M. *Helv. Chim. Acta*, **1962**, *45*, 647; Yannoni, C.S.; Ceasar, G.P.; Dailey, B.P. *J. Am. Chem. Soc.* **1967**, *89*, 2833.

[229]Fitzpatrick, J.D.; Watts, L.; Emerson, G.F.; Pettit, R. *J. Am. Chem. Soc.* **1965**, *87*, 3255. For a discussion, see Pettit, R. *J. Organomet. Chem.* **1975**, *100*, 205.

[230]For a review of cyclopentadienyl cations, see Breslow, R. *Top. Nonbenzenoid Aromat. Chem.* **1973**, *1*, 81.

[231]Clark, D.T. *Chem. Commun.* **1969**, 637; Glukhovtsev, M.N.; Simkin, B.Ya.; Minkin, V.I. *Russ. Chem. Rev.* **1985**, *54*, 54; Breslow, R. *Pure Appl. Chem.* **1971**, *28*, 111; *Acc. Chem. Res.* **1973**, *6*, 393.

is supported by experiment. Among other evidence,

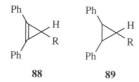

| | 88 | 89 |

it has been shown that **88** (R = COPh) loses its proton in hydrogen-exchange reactions ~6000 times more slowly than **89** (R = COPh).[232] Where R = CN, the ratio is ~10,000.[233] This indicates that **88** are much more reluctant to form carbanions (which would have to be cyclopropenyl carbanions) than **89**, which form ordinary carbanions. Thus the carbanions of **88** are less stable than corresponding ordinary carbanions. Although derivatives of cyclopropenyl anion have been prepared as fleeting intermediates (as in the exchange reactions mentioned above), all attempts to prepare the ion or any of its derivatives as relatively stable species have so far met with failure.[234]

In the case of **87**, the ion has been prepared and has been shown to be a diradical in the ground state,[235] as predicted by the discussion on p. 73.[236] Evidence that **87** is not only nonaromatic, but also antiaromatic comes from studies on **90** and **92**.[237] When **90** is treated with silver perchlorate in propionic acid, the molecule is rapidly solvolyzed (a reaction in which the intermediate **91** is formed; see Chapter 5). Under the same conditions, **92** undergoes no solvolysis at all; that is, **87** does not form. If **87** were merely nonaromatic, it should be about as stable as **91** (which of course has no resonance stabilization at all). The fact that it is so much more reluctant to form indicates that **87** is much less stable than **91**. It is noted that under certain conditions, **91** can be generated solvolytically.[238]

| | | | |
| **90** | **91** | **92** | **87** |

[232]Breslow, R.; Brown, J.; Gajewski, J.J. *J. Am. Chem. Soc.* **1967**, *89*, 4383.

[233]Breslow, R.; Douek, M. *J. Am. Chem. Soc.* **1968**, *90*, 2698.

[234]See, for example, Breslow, R.; Cortés, D.A.; Juan, B.; Mitchell, R.D. *Tetrahedron Lett.* **1982**, *23*, 795. A triphenylcyclopropyl anion has been prepared in the gas phase, with a lifetime of 1–2 s: Bartmess, J.E.; Kester, J.; Borden, W.T.; Köser, H.G. *Tetrahedron Lett.* **1986**, *27*, 5931.

[235]Saunders, M.; Berger, R.; Jaffe, A.; McBride, J.M.; O'Neill, J.; Breslow, R.; Hoffman Jr., J.M.; Perchonock, C.; Wasserman, E.; Hutton, R.S.; Kuck, V.J. *J. Am. Chem. Soc.* **1973**, *95*, 3017.

[236]Derivatives of **87** show similar behavior. Volz, H. *Tetrahedron Lett.* **1964**, 1899; Breslow, R.; Chang, H.W.; Hill, R.; Wasserman, E. *J. Am. Chem. Soc.* **1967**, *89*, 1112; Gompper, R.; Glöckner, H. *Angew. Chem. Int. Ed.* **1984**, *23*, 53.

[237]Breslow, R.; Mazur, S. *J. Am. Chem. Soc.* **1973**, *95*, 584. For further evidence, see Lossing, F.P.; Treager, J.C. *J. Am. Chem. Soc.* **1975**, *97*, 1579. See also, Breslow, R.; Canary, J.W. *J. Am. Chem. Soc.* **1991**, *113*, 3950.

[238]Allen, A.D.; Sumonja, M.; Tidwell, T.T. *J. Am. Chem. Soc.* **1997**, *119*, 2371.

It is strong evidence for Hückel's rule that **86** and **87** are not aromatic while the cyclopropenyl cation (**80**) and the cyclopentadienyl anion (**58**) are, since simple resonance theory predicts no difference between **86** and **80** or **87** and **58** (the same number of equivalent canonical forms can be drawn for **86** as for **80** and for **87** as for **58**).

78a

Systems of Eight Electrons

Cyclooctatetraene[239] ([8]annulene, **78a**) is not planar, but tub-shaped.[240] Therefore we would expect that it is neither aromatic nor antiaromatic, since both these conditions require overlap of parallel p orbitals. The reason for the lack of planarity is that a regular octagon has angles of 135°, while sp^2 angles are most stable at 120°. To avoid the strain, the molecule assumes a nonplanar shape, in which orbital overlap is greatly diminished.[241] Single- and double-bond distances in **78** are, respectively, 1.46 and 1.33 Å, which is expected for a compound made up of four individual double bonds.[240] The reactivity is also what would be expected for a linear polyene. Reactive intermediates can be formed in solution. Dehydrohalogenation of bromocyclooctatetraene at $-100°C$ has been reported, for example, and trapping by immediate electron transfer gave a stable solution of the [8]annulyne anion radical.[242]

The cyclooctadiendiynes **93** and **94** are planar conjugated eight-electron systems (the four extra triple-bond electrons do not participate), which nmr evidence show to be antiaromatic.[243] There is evidence that part of the reason for the lack of planarity in **78** itself is that a planar molecular would have to be antiaromatic.[244] The cycloheptatrienyl anion (**61**) also has eight electrons, but does not behave like an aromatic system.[151] The bond lengths for a series of molecules containing the cycloheptatrienide anion have recently been published.[245] The NMR spectrum

[239]For a monograph, see Fray, G.I.; Saxton, R.G. *The Chemistry of Cyclooctatetraene and its Derivatives*; Cambridge University Press: Cambridge, *1978*. For a review, see Paquette, L.A. *Tetrahedron* *1975*, *31*, 2855. For reviews of heterocyclic 8π systems, see Kaim, W. *Rev. Chem. Intermed.* *1987*, *8*, 247; Schmidt, R.R. *Angew. Chem. Int. Ed.* *1975*, *14*, 581.

[240]Bastiansen, O.; Hedberg, K.; Hedberg, L. *J. Chem. Phys.* *1957*, *27*, 1311.

[241]The compound perfluorotetracyclobutacyclooctatetraene has been found to have a planar cyclooctatetraene ring, although the corresponding tetracyclopenta analog is nonplanar: Einstein, F.W.B.; Willis, A.C.; Cullen, W.R.; Soulen, R.L. *J. Chem. Soc. Chem. Commun.* *1981*, 526. See also, Paquette, L.A.; Wang, T.; Cottrell, C.E. *J. Am. Chem. Soc.* *1987*, *109*, 3730.

[242]Peters, S.J.; Turk, M.R.; Kiesewetter, M.K.; Stevenson, C.D. *J. Am. Chem. Soc.* *2003*, *125*, 11264.

[243]For a review, see Huang, N.Z.; Sondheimer, F. *Acc. Chem. Res.* *1982*, *15*, 96. See also, Dürr, H.; Klauck, G.; Peters, K.; von Schnering, H.G. *Angew. Chem. Int. Ed.* *1983*, *22*, 332; Chan, T.; Mak, T.C.W.; Poon, C.; Wong, H.N.C.; Jia, J.H.; Wang, L.L. *Tetrahedron* *1986*, *42*, 655.

[244]Figeys, H.P.; Dralants, A. *Tetrahedron Lett.* *1971*, 3901; Buchanan, G.W. *Tetrahedron Lett.* *1972*, 665.

[245]Dietz, F.; Rabinowitz, M.; Tadjer, A.; Tyutyulkov, N. *J. Chem. Soc. Perkin Trans. 2* *1995*, 735.

of the benzocycloheptatrienyl anion (**95**) shows that, like **82**, **93**, and **94**, this compound is antiaromatic.[246] A new antiaromatic compound 1,4-biphenylene quinone (**96**) was prepared, but it rapidly dimerizes due to instability.[247]

| 93 | 94 | 95 | 96 |

Systems of Ten Electrons[248]

There are three geometrically possible isomers of [10]annulene: the all-*cis* (**97**), the mono-*trans* (**98**), and the *cis–trans–cis–cis–trans* (**79**). If Hückel's rule applies, they should be planar. But it is far from obvious that the molecules would adopt a planar

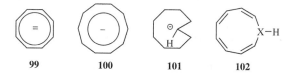

| 79 | 97 | 98 |

shape, since they must overcome considerable strain to do so. For a regular decagon (**97**) the angles would have to be 144°, considerably larger than the 120° required for sp^2 angles. Some of this strain would also be present in **98**, but this kind of strain is eliminated in **79** since all the angles are 120°. However, it was pointed out by Mislow[249] that the hydrogens in the 1 and 6 positions should interfere with each other and force the molecule out of planarity.

| 99 | 100 | 101 | 102 |

Compounds **97** and **98** have been prepared[250] as crystalline solids at −80°C. The NMR spectra show that all the hydrogens lie in the alkene region and it was concluded that neither compound is aromatic. Calculations on **98** suggest that

[246]Staley, S.W.; Orvedal, A.W. *J. Am. Chem. Soc.* **1973**, *95*, 3382.

[247]Kiliç, H.; Balci, M. *J. Org. Chem.* **1997**, *62*, 3434.

[248]For reviews, see Kemp-Jones, A.V.; Masamune, S. *Top. Nonbenzenoid Aromat. Chem.* **1973**, *1*, 121; Masamune, S.; Darby, N. *Acc. Chem. Res.* **1972**, *5*, 272; Burkoth, T.L.; van Tamelen, E.E., in Snyder,J.P. *Nonbenzenoid Aromaticity*, Vol. 1, Academic Press, NY, **1969**, pp. 63–116; Vogel, E., in Garratt, P.J. Aromaticity, Wiley, NY, **1986**, pp. 113–147.

[249]Mislow, K. *J. Chem. Phys.* **1952**, *20*, 1489.

[250]Masamune,S.; Hojo, K.; Bigam, G.; Rabenstein,D.L. *J. Am. Chem. Soc.* **1971**, *93*, 4966. [10]Annulenes had previously been prepared, but it was not known which ones: van Tamelen, E.E.; Greeley, R.H. *Chem. Commun.* **1971**, 601; van Tamelen, E.E.; Burkoth, T.L.; Greeley, R.H. *J. Am. Chem. Soc.* **1971**, *93*, 6120.

it may indeed be aromatic, although the other isomers are not.[251] It is known that the Hartree–Fock (HF) method incorrectly favors bond-length-alternating structures for [10]annulene, and aromatic structures are incorrectly favored by density functional theory. Improved calculations predict that the twist conformation is lowest in energy, and the naphthalene-like and heart-shaped conformations lie higher than the twist by 1.40 and 4.24 kcal mol^{-1}, respectively.[252] From ^{13}C and proton (H^1) nmr spectra it has been deduced that neither is planar. However, that the angle strain is not insurmountable has been demonstrated by the preparation of several compounds that have large angles, but that are definitely planar 10-electron aromatic systems. Among these are the dianion **99**, the anions **100** and **101**, and the azonine **102**.[253] Compound **99**[254] has angles of ~135°, while **100**[255] and **101**[256] have angles of ~140°, which are not very far from 144°. The inner proton in **101**[257] (which is the mono-trans isomer of the all-cis **100**) is found far upfield in the NMR (−3.5 δ). For **97** and **98**, the cost in strain energy to achieve planarity apparently outweighs the extra stability that would come from an aromatic ring. To emphasize the delicate balance between these factors, we may mention that the oxygen analog of **102** (X = O, oxonin) and the N-carbethoxy derivative of **102** (X = CH) are nonaromatic and nonplanar, while **102** (X = N) is aromatic and planar.[258] Other azaannulenes are known, including Vogel's 2,7-methanoazaannulene,[259] as well

[251] Sulzbach, H.M.; Schleyer, P.v.R.; Jiao, H.; Xie, Y.; Schaefer III, H.F. *J. Am. Chem. Soc.* **1995**, *117*, 1369. Also see, Sulzbach, H.M.; Schaefer III, H.F.; Klopper, W.; Lüthi, H.P. *J. Am. Chem. Soc.* **1996**, *118*, 3519 for a discussion of Aromaticity calculations for [10]annulene.

[252] King, R.A.; Crawford, T.D.; Stanton, J.F.; Schaefer, III, H.F. *J. Am. Chem. Soc.* **1999**, *121*, 10788.

[253] For reviews of **102** (X = N) and other nine-membered rings containing four double bonds and a hetero atom (heteronins), see Anastassiou, A.G. *Acc. Chem. Res.* **1972**, *5*, 281, *Top. Nonbenzenoid Aromat. Chem.* **1973**, *1*, 1, *Pure Appl. Chem.* **1975**, *44*, 691. For a review of heteroannulenes in general, see Anastassiou; Kasmai, H.S. *Adv. Heterocycl. Chem.* **1978**, *23*, 55.

[254] Katz, T.J. *J. Am. Chem. Soc.* **1960**, *82*, 3784, 3785; Goldstein, M.J.; Wenzel, T.T. *J. Chem. Soc. Chem. Commun.* **1984**, 1654; Garkusha, O.G.; Garbuzova, I.A.; Lokshin, B.V.; Todres, Z.V. *J. Organomet. Chem.* **1989**, *371*, 279. See also, Noordik, J.H.; van den Hark, T.E.M.; Mooij, J.J.; Klaassen, A.A.K. *Acta Crystallogr. Sect. B.* **1974**, *30*, 833; Goldberg, S.Z.; Raymond, K.N.; Harmon, C.A.; Templeton, D.H. *J. Am. Chem. Soc.* **1974**, *96*, 1348; Evans, W.J.; Wink, D.J.; Wayda, A.L.; Little, D.A. *J. Org. Chem.* **1981**, *46*, 3925; Heinz, W.; Langensee, P.; Müllen, K. *J. Chem. Soc. Chem. Commun.* **1986**, 947.

[255] Katz, T.J.; Garratt, P.J. *J. Am. Chem. Soc.* **1964**, *86*, 5194; LaLancette, E.A.; Benson, R.E. *J. Am. Chem. Soc.* **1965**, *87*, 1941; Simmons, H.E.; Chesnut, D.B.; LaLancette, E.A. *J. Am. Chem. Soc.* **1965**, *87*, 982; Paquette, L.A.; Ley, S.V.; Meisinger, R.H.; Russell, R.K.; Oku, M. *J. Am. Chem. Soc.* **1974**, *96*, 5806; Radlick, P.; Rosen, W. *J. Am. Chem. Soc.* **1966**, *88*, 3461.

[256] Anastassiou, A.G.; Gebrian, J.H. *Tetrahedron Lett.* **1970**, 825.

[257] Boche, G.; Weber, H.; Martens, D.; Bieberbach, A. *Chem. Ber.* **1978**, *111*, 2480. See also, Anastassiou, A.G.; Reichmanis, E. *Angew. Chem. Int. Ed.* **1974**, *13*, 728; Boche, G.; Bieberbach, A. *Tetrahedron Lett.* **1976**, 1021.

[258] Anastassiou, A.G.; Gebrian, J.H. *J. Am. Chem. Soc.* **1969**, *91*, 4011; Chiang, C.C.; Paul, I.C.; Anastassiou, A.G.; Eachus, S.W. *J. Am. Chem. Soc.* **1974**, *96*, 1636.

[259] Vogel, E.; Roth, H.D. *Angew. Chem. Int. Ed.* **1964**, *3*, 228; Vogel, E.; Biskup, M.; Pretzer, W.; Böll, W.A. *Angew. Chem. Int. Ed.* **1964**, *3*, 642.; Vogel, E.; Meckel, M.; Grimme, W. *Angew. Chem. Int. Ed.* **1964**, *3*, 643; Vogel, E.; Pretzer, W.; Böll, W.A. *Tetrahedron Lett.* **1965**, 3613; Sondheimer, F.; Shani, A. *J. Am. Chem. Soc.* **1964**, *86*, 3168; Shani, A.; Sondheimer, F. *J. Am. Chem. Soc.* **1967**, *89*, 6310; Bailey, N.A.; Mason, R. *J. Chem. Soc. Chem. Commun.* **1967**, 1039.

as 3,8-methanoaza[10]annulene,[260] and their alkoxy derivatives.[261] Calculations for aza[10]annulene concluded that the best olefinic twist isomer is 2.1 kcal mol^{-1} (8.8 kJ mol^{-1}) more stable than the aromatic form,[262] and is probably the more stable form.

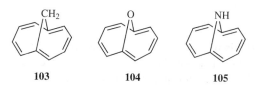

103   **104**   **105**

So far, **79** has not been prepared despite many attempts. However, there are various ways of avoiding the interference between the two inner protons. The approach that has been most successful involves bridging the 1 and 6 positions.[263] Thus, 1,6-methano[10]annulene (**103**)[264] and its oxygen and nitrogen analogs, **104**[265] and **105**,[266] have been prepared and are stable compounds that undergo aromatic substitution and are diatropic.[267] For example, the perimeter protons of **103** are found at 6.9–7.3 δ, while the bridge protons are at −0.5 δ. The crystal structure of **103** shows that the perimeter is nonplanar, but the bond distances are in the range 1.37–1.42 Å.[268] It has therefore been amply demonstrated that a closed loop of 10 electrons is an aromatic system, although some molecules that could conceivably have such a system are too distorted from planarity to be aromatic. A small distortion from planarity (as in **103**) does not prevent aromaticity, at least in part because the s orbitals so distort themselves as to maximize the favorable (parallel) overlap

[260]Schäfer-Ridder, M.; Wagner, A.; Schwamborn, M.; Schreiner, H.; Devrout, E.; Vogel, E. *Angew. Chem. Int. Ed.* **1978**, *17*, 853.; Destro, R.; Simonetta, M.; Vogel, E. *J. Am. Chem. Soc.* **1981**, *103*, 2863.

[261]Vogel, E. Presented at the 3rd International Symposium on Novel Aromatic Compounds (ISNA 3), San Francisco, Aug *1977*; Gölz, H.-J.; Muchowski, J.M.; Maddox, M.L. *Angew. Chem. Int. Ed.* **1978**, *17*, 855; Schleyer, P.v.R.; Jiao, H.; Sulzbach, H.M.; Schaefer III H.F. *J. Am. Chem. Soc.* **1996**, *118*, 2093.

[262]Bettinger, H.F.; Sulzbach, H.M.; Schleyer, P.v.R.; Schaefer III, H.F. *J. Org. Chem.* **1999**, *64*, 3278.

[263]For reviews of bridged [10]-, [14]-, and [18]annulenes, see Vogel, E. *Pure Appl. Chem.* **1982**, *54*, 1015; *Isr. J. Chem.* **1980**, *20*, 215; *Chimia*, *1968*, *22*, 21; Vogel, E.; Günther, H. *Angew. Chem. Int. Ed.* **1967**, *6*, 385.

[264]Vogel, E.; Roth, H.D. *Angew. Chem. Int. Ed.* **1964**, *3*, 228; Vogel, E.; Böll, W.A. *Angew. Chem. Int. Ed.* **1964**, *3*, 642; Vogel, E.; Böll, W.A.; Biskup, M. *Tetrahedron Lett.* **1966**, 1569.

[265]Vogel, E.; Biskup, M.; Pretzer, W.; Böll, W.A. *Angew. Chem. Int. Ed.* **1964**, *3*, 642; Shani, A.; Sondheimer, F. *J. Am. Chem. Soc.* **1967**, *89*, 6310; Bailey, N.A.; Mason, R. *Chem. Commun.* **1967**, 1039.

[266]Vogel, E.; Pretzer, W.; Böll, W.A. *Tetrahedron Lett.* **1965**, 3613. See also, Vogel, E.; Biskup, M.; Pretzer, W.; Böll, W.A. *Angew. Chem. Int. Ed.* **1964**, *3*, 642.

[267]For another type of bridged diatropic [10]annulene, see Lidert, Z.; Rees, C.W. *J. Chem. Soc. Chem. Commun.* **1982**, 499; Gilchrist, T.L.; Rees, C.W.; Tuddenham, D. *J. Chem. Soc. Perkin Trans. 1* **1983**, 83; McCague, R.; Moody, C.J.; Rees, C.W. *J. Chem. Soc. Perkin Trans. 1* **1984**, 165, 175; Gibbard, H.C.; Moody, C.J.; Rees, C.W. *J. Chem. Soc. Perkin Trans. 1* **1985**, 731, 735.

[268]Bianchi, R.; Pilati, T.; Simonetta, M. *Acta Crystallogr., Sect. B* **1980**, *36*, 3146. See also Dobler, M.; Dunitz, J.D. *Helv. Chim Acta*, **1965**, *48*, 1429.

of p orbitals to form the aromatic 10-electron loop.[269]

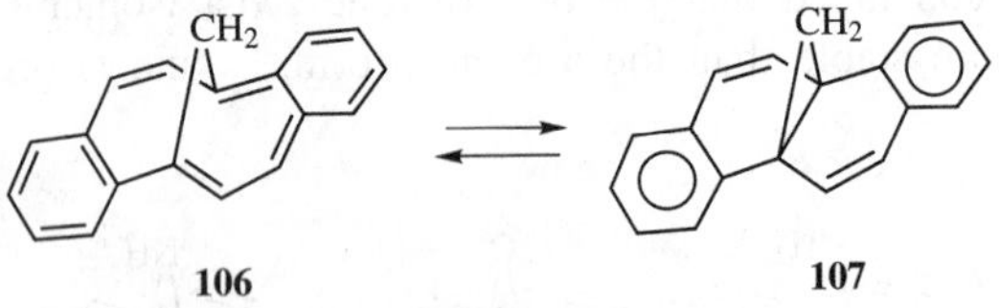

106   **107**

In **106**, where **103** is fused to two benzene rings in such a way that no canonical form can be written in which both benzene rings have six electrons, the aromaticity is reduced by annellation, as shown by the fact that the molecule rapidly converts to the more stable **107**, in which both benzene rings can be fully aromatic[270] (this is similar to the cycloheptatriene–norcaradiene conversions discussed on p. 1664).

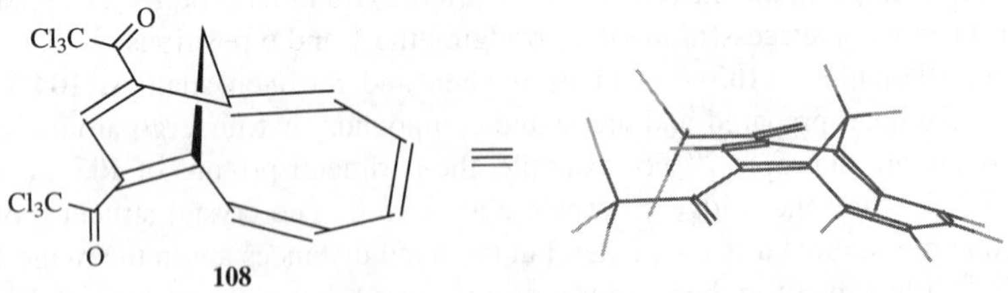

108

Molecules can sustain significant distortion from planarity and retain their aromatic character. 1,3-Bis(trichloroacetyl)homoazulene (**108**) qualifies as aromatic using the geometric criterion that there is only a small average deviation from the C—C bond length in the [10]annulene perimeter.[271] X-ray crystal structure shows that the 1,5-bridge distorts the [10]-annulene π-system away from planarity (see the 3D model) with torsion angles as large as 42.2° at the bridgehead position, but **108** does not lose its aromaticity.

Systems of More than Ten Electrons: $4n + 2$ Electrons[272]

Extrapolating from the discussion of [10]annulene, we expect larger $4n + 2$ systems to be aromatic if they are planar. Mislow[249] predicted that [14]annulene (**109**)

[269]For a discussion, see Haddon, R.C. *Acc. Chem. Res.* **1988**, *21*, 243.

[270]Hill, R.K.; Giberson, C.B.; Silverton, J.V. *J. Am. Chem. Soc.* **1988**, *110*, 497. See also, McCague, R.; Moody, C.J.; Rees, C.W.; Williams, D.J. *J. Chem. Soc. Perkin Trans. 1* **1984**, 909.

[271]Scott, L.T.; Sumpter, C.A.; Gantzel, P.K.; Maverick, E.; Trueblood, K.N. *Tetrahedron* **2001**, *57*, 3795.

[272]For reviews of annulenes, with particular attention to their nmr spectra, see Sondheimer, F. *Acc. Chem. Res.* **1972**, *5*, 81–91, *Pure Appl. Chem.* **1971**, *28*, 331, *Proc. R. Soc. London. Ser. A*, **1967**, *297*, 173; Sondheimer, F.; Calder, I.C.; Elix, J.A.; Gaoni, Y; Garratt, P.J.; Grohmann, K.; di Maio, G.; Mayer, J.; Sargent, M.V.; Wolovsky, R. in Garratt, P.G. *Aromaticity*, Wiley, NY, **1986**, pp. 75–107; Haddon, R.C.; Haddon, V.R.; Jackman, L.M. *Fortschr. Chem. Forsch.* **1971**, *16*, 103. For a review of annulenoannulenes (two annulene rings fused together), see Nakagawa, M. *Angew. Chem. Int. Ed.* **1979**, *18*, 202. For a review of reduction and oxidation of annulenes; that is, formation of radical ions, dianions, and dications, see Müllen, K. *Chem. Rev.* **1984**, *84*, 603. For a review of annulene anions, see Rabinovitz, M. *Top. Curr. Chem.* **1988**, *146*, 99. Also see Cyvin, S.J.; Brunvoll, J.; Chen, R.S.; Cyvin, B.N.; Zhang, F.J. *Theory of Coronoid Hydrocarbons II*, Springer-Verlag, Berlin, **1994**.

would possess the same type of interference as **79**, although in lesser degree. This is

109

borne out by experiment. Compound **109** is aromatic (it is diatropic; inner protons at 0.00 δ, outer protons at 7.6 δ),[273] but is completely destroyed by light and air in 1 day. X-ray analysis shows that although there are no alternating single and double bonds, the molecule is not planar.[274] A number of stable bridged [14]annulenes have been prepared,[275] for example, *trans*-15,16-dimethyldihydropyrene (**110**),[276] *syn*-1,6:8,13-diimino[14]annulene (**111**),[277] and *syn*- and *anti*-1,6:8,13-bis(methano[14]annulene) (**112** and **113**).[278] The dihydropyrene **110**

110   **111**   **112**   **113**

(and its diethyl and dipropyl homologs) is undoubtedly aromatic: the π perimeter is approximately planar;[279] the bond distances are all 1.39–1.40 Å; and the

[273]Gaoni, Y.; Melera, A.; Sondheimer, F.; Wolovsky, R. *Proc. Chem. Soc.* **1964**, 397.

[274]Bregman, J. *Nature (London)* **1962**, *194*, 679; Chiang, C.C.; Paul, I.C. *J. Am. Chem. Soc.* **1972**, *94*, 4741. Another 14-electron system is the dianion of [12]annulene, which is also apparently aromatic though not planar: Oth, J.F.M.; Schröder, G. *J. Chem. Soc. B*, **1971**, 904. See also Garratt, P.J.; Rowland, N.E.; Sondheimer, F. *Tetrahedron* **1971**, *27*, 3157; Oth, J.F.M.; Müllen, K.; Königshofen, H.; Mann, M.; Sakata, Y.; Vogel, E. *Angew. Chem. Int. Ed.* **1974**, *13*, 284. For some other 14-electron aromatic systems, see Anastassiou, A.G.; Elliott, R.L.; Reichmanis, E. *J. Am. Chem. Soc.* **1974**, *96*, 7823; Wife, R.L.; Sondheimer, F. *J. Am. Chem. Soc.* **1975**, *97*, 640; Ogawa, H.; Kubo, M.; Saikachi, H.*Tetrahedron Lett.* **1971**, 4859; Oth, J.F.M.; Müllen, K.; Königshofen, H.; Wassen, J.; Vogel, E. *Helv. Chim. Acta*, **1974**, *57*, 2387; Willner, I.; Gutman, A.L.; Rabinovitz, M. *J. Am. Chem. Soc.* **1977**, *99*, 4167; Röttele, H.; Schröder, G. *Chem. Ber.* **1982**, *115*, 248.

[275]For a review, see Vogel, E. *Pure Appl. Chem.* **1971**, *28*, 355.

[276]Boekelheide, V.; Phillips, J.B. *J. Am. Chem. Soc.* **1967**, *89*, 1695; Boekelheide, V.; Miyasaka, T. *J. Am. Chem. Soc.* **1967**, *89*, 1709. For reviews of dihydropyrenes, see Mitchell, R.H. *Adv. Theor. Interesting Mol.* **1989**, *1*, 135; Boekelheide, V. *Top. Nonbenzoid Arom. Chem.* **1973**, *1*, 47; *Pure Appl. Chem.* **1975**, *44*, 807.

[277]Vogel, E.; Kuebart, F.; Marco, J.A.; Andree, R.; Günther, H.; Aydin, R. *J. Am. Chem. Soc.* **1983**, *105*, 6982; Destro, R.; Pilati, T.; Simonetta, M.; Vogel, E. *J. Am. Chem. Soc.* **1985**, *107*, 3185, 3192. For the di-*O*- analog of **102**, see Vogel, A.; Biskup, M.; Vogel, E.; Günther, H. *Angew. Chem. Int. Ed.* **1966**, *5*, 734.

[278]Vogel, E.; Sombroek, J.; Wagemann, W. *Angew. Chem. Int. Ed.* **1975**, *14*, 564.

[279]Hanson, A.W. *Acta Crystallogr.* **1965**, *18*, 599, 1967, *23*, 476.

molecule undergoes aromatic substitution[276] and is diatropic.[280] The outer protons are found at 8.14–8.67 δ, while the CH_3 protons are at −4.25 δ. Other nonplanar aromatic dihydropyrenes are known.[281] Annulenes **111** and **112** are also diatropic,[282] although X-ray crystallography indicates that the π periphery in at least **111** is not quite planar.[283] However, **113**, in which the geometry of the molecule greatly reduces the overlap of the *p* orbitals at the bridgehead positions with adjacent *p* orbitals, is definitely not aromatic,[284] as shown by NMR spectra[278] and X-ray crystallography, from which bond distances of 1.33–1.36 Å for the double bonds and 1.44–1.49 Å for the single bonds have been obtained.[285] In contrast, all the bond distances in **111** are ∼1.38–1.40 Å.[283]

Another way of eliminating the hydrogen interferences of [14]annulene is to introduce one or more triple bonds into the system, as in dehydro[14]annulene (**114**).[286] All five known dehydro[14]annulenes are diatropic, and **87** can be nitrated or sulfonated.[287] The extra electrons of the triple bond do not form part of the aromatic system, but simply

| 114 | 115 | 116 |

exist as a localized bond. There has been a debate concerning the extent of delocalization in dehydrobenzoannulenes,[288] but there is evidence for a weak, but discernible ring current.[289] 3,4,7,8,9,10,13,14-Octahydro[14]annulene (**116**) has been

[280]A number of annellated derivatives of **110** are less diatropic, as would be expected from the discussion on p. $$$: Mitchell, R.H.; Williams, R.V.; Mahadevan, R.; Lai, Y.H.; Dingle, T.W. *J. Am. Chem. Soc.* *1982*, *104*, 2571 and other papers in this series.

[281]Bodwell, G.J.; Bridson, J.N.; Chen, S.-L.; Poirier, R.A. *J. Am. Chem. Soc.* *2001*, *123*, 4704; Bodwell, G.J.; Fleming, J.J.; Miller, D.O. *Tetrahedron* *2001*, *57*, 3577.

[282]As are several other similarly bridged [14]annulenes; see, for example, Flitsch, W.; Peeters, H. *Chem. Ber.* *1973*, *106*, 1731; Huber, W.; Lex, J.; Meul, T.; Müllen, K. *Angew. Chem. Int. Ed.* *1981*, *20*, 391; Vogel, E.; Nitsche, R.; Krieg, H. *Angew. Chem. Int. Ed.* *1981*, *20*, 811; Mitchell, R.H.; Anker, W. *Tetrahedron Lett.* *1981*, *22*, 5139; Vogel, E.; Wieland, H.; Schmalstieg, L.; Lex, J. *Angew. Chem. Int. Ed.* *1984*, *23*, 717; Neumann, G.; Müllen, K. *J. Am. Chem. Soc.* *1986*, *108*, 4105.

[283]Ganis, P.; Dunitz, J.D. *Helv. Chim. Acta,* *1967*, *50*, 2369.

[284]For another such pair of molecules, see Vogel, E.; Nitsche, R.; Krieg, H. *Angew. Chem. Int. Ed.* *1981*, *20*, 811. See also, Vogel, E.; Schieb, T.; Schulz, W.H.; Schmidt, K.; Schmickler, H.; Lex, J. *Angew. Chem. Int. Ed.* *1986*, *25*, 723.

[285]Gramaccioli, C.M.; Mimun, A.; Mugnoli, A.; Simonetta, M. *Chem. Commun.* *1971*, 796. See also, Destro, R.; Simonetta, M. *Tetrahedron* *1982*, *38*, 1443.

[286]For a review of dehydroannulenes, see, Nakagawa, M. *Top. Nonbenzenoid Aromat. Chem.* *1973*, *1*, 191.

[287]Gaoni, Y.; Sondheimer, F. *J. Am. Chem. Soc.* *1964*, *86*, 521.

[288]Balaban, A.T.; Banciu, M.; Ciorba, V. *Annulenes, Benzo-, Hetero-, Homo- Derivatives and their Valence Isomers*, Vols. 1–3, CRC Press, Boca Raton, FL, *1987*; Garratt, P.J. *Aromaticity*, Wiley, NY, *1986*; Minkin, V.I.; Glukhovtsev, M.N.; Simkin, B.Ya. *Aromaticity and Antiaromaticity*, Wiley, NY, *1994*.

[289]Kimball, D.B.; Wan, W.B.; Haley, M.M. *Tetrahdron Lett.* *1998*, *39*, 6795; Bell, M.L.; Chiechi, R.C.; Johnson, C.A.; Kimball, D.B.; Matzger, A.J.; Wan, W.B.; Weakley, T.J.R.; Haley, M.M. *Tetahedron* *2001*, *57*, 3507; Wan, W.B.; Chiechi, R.C.; Weakley, T.J.R.; Haley, M.M. *Eur. J. Org. Chem.* *2001*, 3485.

prepared, for example, and the evidence supported its aromaticity.[290] This study suggested that increasing benzoannelation of the parent, **116**, led to a step-down in aromaticity, a result of competing ring currents in the annulenic system.

[18]Annulene (**115**) is diatropic:[291] the 12 outer protons are found at $\sim\delta = 9$ and the 6 inner protons at $\sim\delta = -3$. X-ray crystallography[292] shows that it is nearly planar, so that interference of the inner hydrogens is not important in annulenes this large. Compound **115** is reasonably stable, being distillable at reduced pressures, and undergoes aromatic substitutions.[293] The C–C bond distances are not equal, but they do not alternate. There are 12 inner bonds of ~ 1.38 Å and 6 outer bonds of ~ 1.42 Å.[292] Compound **115** has been estimated to have a resonance energy of $\sim 37 \, \text{kcal mol}^{-1}$ ($155 \, \text{kJ mol}^{-1}$), similar to that of benzene.[294]

The known bridged [18]annulenes are also diatropic[295] as are most of the known dehydro[18]annulenes.[296] The dianions of open and bridged [16]annulenes[297] are also 18-electron aromatic systems,[298] and there are dibenzo[18]annulenes.[299]

[22]Annulene[300] and dehydro[22]annulene[301] are also diatropic. A dehydro-benzo[22]annulene has been prepared that has eight C≡C units, is planar and possesses a weak induced ring current.[302] In the latter compound there are 13 outer protons at 6.25–8.45 δ and 7 inner protons at 0.70–3.45 δ. Some aromatic bridged

[290]Bodyston, A.J.; Haley, M.M. *Org. Lett.* **2001**, *3*, 3599; Boydston, A.J.; Haley, M.M.; Williams, R.V.; Armantrout, J.R. *J. Org. Chem.* **2002**, *67*, 8812.

[291]Jackman, L.M.; Sondheimer, F.; Amiel, Y.; Ben-Efraim, D.A.; Gaoni, Y.; Wolovsky, R.; Bothner-By, A.A. *J. Am. Chem. Soc.* **1962**, *84*, 4307; Gilles, J.; Oth, J.F.M.; Sondheimer, F.; Woo, E.P. *J. Chem. Soc. B*, **1971**, 2177. For a thorough discussion, see Baumann, H.; Oth, J.F.M. *Helv. Chim. Acta*, **1982**, *65*, 1885.

[292]Bregman, J.; Hirshfeld, F.L.; Rabinovich, D.; Schmidt, G.M.J. *Acta Crystallogr.*, **1965**, *19*, 227; Hirshfeld, F.L.; Rabinovich, D. *Acta Crystallogr.*, **1965**, *19*, 235.

[293]Sondheimer, F. *Tetrahedron* **1970**, *26*, 3933.

[294]Oth, J.F.M.; Bünzli, J.; de Julien de Zélicourt, Y. *Helv. Chim. Acta*, **1974**, *57*, 2276.

[295]For some examples, see DuVernet, R.B.; Wennerström, O.; Lawson, J.; Otsubo, T.; Boekelheide, V. *J. Am. Chem. Soc.* **1978**, *100*, 2457; Ogawa, H.; Sadakari, N.; Imoto, T.; Miyamoto, I.; Kato, H.; Taniguchi, Y. *Angew. Chem. Int. Ed.* **1983**, *22*, 417; Vogel, E.; Sicken, M.; Röhrig, P.; Schmickler, H.; Lex, J.; Ermer, O. *Angew. Chem. Int. Ed.* **1988**, *27*, 411.

[296]Okamura, W.H.; Sondheimer, F. *J. Am. Chem. Soc.* **1967**, *89*, 5991; Ojima, J.; Ejiri, E.; Kato, T.; Nakamura, M.; Kuroda, S.; Hirooka, S.; Shibutani, M. *J. Chem. Soc. Perkin Trans. 1* **1987**, 831; Sondheimer, F. *Acc. Chem. Res.* **1972**, *5*, 81. For two that are not, see Endo, K.; Sakata, Y.; Misumi, S. *Bull. Chem. Soc. Jpn.* **1971**, *44*, 2465.

[297]For a review of this type of polycyclic ion, see Rabinovitz, M.; Willner, I.; Minsky, A. *Acc. Chem. Res.* **1983**, *16*, 298.

[298]Mitchell, R.H.; Boekelheide, V. *Chem. Commun.* **1970**, 1557; Oth, J.F.M.; Baumann, H.; Gilles, J.; Schröder, G. *J. Am. Chem. Soc.* **1972**, *94*, 3948. See also Brown, J.M.; Sondheimer, F. *Angew. Chem. Int. Ed.* **1974**, *13*, 337; Cresp, T.M.; Sargent, M.V. *J. Chem. Soc. Chem. Commun.* **1974**, 101; Schröder, G.; Plinke, G.; Smith, D.M.; Oth, J.F.M. *Angew. Chem. Int. Ed.* **1973**, *12*, 325; Rabinovitz, M.; Minsky, A. *Pure Appl. Chem.* **1982**, *54*, 1005.

[299]Michels, H.P.; Nieger, M.; Vögtle, F. *Chem. Ber.* **1994**, *127*, 1167.

[300]McQuilkin, R.M.; Metcalf, B.W.; Sondheimer, F. *Chem. Commun.* **1971**, 338.

[301]McQuilkin, R.M.; Sondheimer, F. *J. Am. Chem. Soc.* **1970**, *92*, 6341; Iyoda, M.; Nakagawa, M. *J. Chem. Soc. Chem. Commun.* **1972**, 1003. See also, Akiyama, S.; Nomoto, T.; Iyoda, M.; Nakagawa, M. *Bull. Chem. Soc. Jpn.* **1976**, *49*, 2579.

[302]Wan, W.B.; Kimball, D.B.; Haley, M.M. *Tetrahedron Lett.* **1998**, *39*, 6795.

[22]annulenes are also known.[303] [26]Annulene has not yet been prepared, but several dehydro[26]annulenes are aromatic.[304] Furthermore, the dianion of 1,3,7,9,13,-15,19,21-octadehydro[24]annulene is another 26-electron system that is aromatic.[305] Ojima and co-workers have prepared bridged dehydro derivatives of [26], [30], and [34] annulenes.[306] All of these are diatropic. The same workers prepared a bridged tetradehydro[38]annulene,[306] which showed no ring current. On the other hand, the dianion of the cyclophane, **117**, also has 38 perimeter electrons, and this species is diatropic.[307]

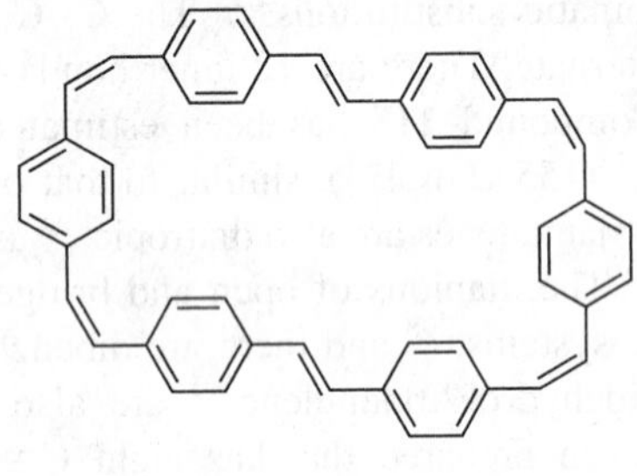

117

There is now no doubt that $4n + 2$ systems are aromatic if they can be planar, although **97** and **113** among others, demonstrate that not all such systems are in fact planar enough for aromaticity. The cases of **109** and **111** prove that absolute planarity is not required for aromaticity, but that aromaticity decreases with decreasing planarity.

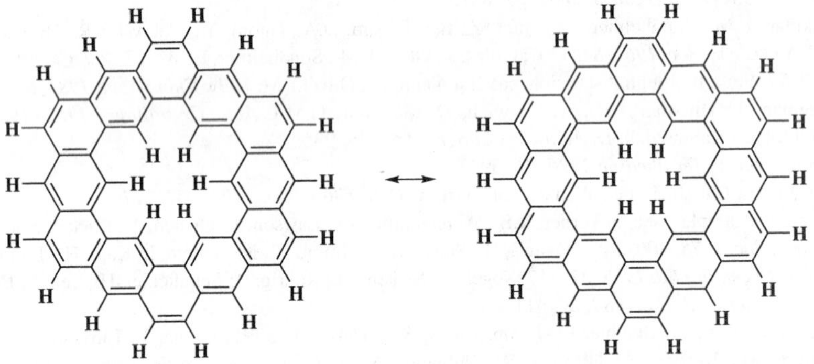

118a　　　　　　　　　　**118b**

[303]For example see Broadhurst, M.J.; Grigg, R.; Johnson, A.W. *J. Chem. Soc. Perkin Trans. 1* **1972**, 2111; Ojima, J.; Ejiri, E.; Kato, T.; Nakamura, M.; Kuroda, S.; Hirooka, S.; Shibutani, M. *J. Chem. Soc. Perkin Trans. 1* **1987**, 831; Yamamoto, K.; Kuroda, S.; Shibutani, M.; Yoneyama, Y.; Ojima, J.; Fujita, S.; Ejiri, E.; Yanagihara, K. *J. Chem. Soc. Perkin Trans. 1* **1988**, 395.

[304]Metcalf, B.W.; Sondheimer, F. *J. Am. Chem. Soc.* **1971**, *93*, 5271; Iyoda, M.; Nakagawa, M. *Tetrahedron Lett.* **1972**, 4253; Ojima, J.; Fujita, S.; Matsumoto, M.; Ejiri, E.; Kato, T.; Kuroda, S.; Nozawa, Y.; Hirooka, S.; Yoneyama, Y.; Tatemitsu, H. *J. Chem. Soc. Perkin Trans. 1* **1988**, 385.

[305]McQuilkin, R.M.; Garratt, P.J.; Sondheimer, F. *J. Am. Chem. Soc.* **1970**, *92*, 6682. See also, Huber, W.; Müllen, K.; Wennerström, O. *Angew. Chem. Int. Ed.* **1980**, *19*, 624.

[306]Ojima, J.; Fujita, S.; Matsumoto, M.; Ejiri, E.; Kato, T.; Kuroda, S.; Nozawa, Y.; Hirooka, S.; Yoneyama, Y.; Tatemitsu, H. *J. Chem. Soc., Perkin Trans. 1* **1988**, 385.

[307]Müllen, K.; Unterberg, H.; Huber, W.; Wennerström, O.; Norinder, U.; Tanner, D.; Thulin, B. *J. Am. Chem. Soc.* **1984**, *106*, 7514.

The proton NMR (^{1}H NMR) spectrum of **118** (called kekulene) showed that in a case where electrons can form either aromatic sextets or larger systems, the sextets are preferred.[308] There was initial speculation that kekulene might be *superaromatic*, that is, it would show enhanced aromatic stabilization. Recent calculations suggest that there is no enhanced stabilization.[309] The 48 π electrons of **118** might, in theory, prefer structure **118a**, where each ring is a fused benzene ring, or **118b**, which has a [30]annulene on the outside and an [18]annulene on the inside. The ^{1}H NMR spectrum of this compound shows three peaks at δ = 7.94, 8.37, and 10.45 in a ratio of 2:1:1. It is seen from the structure that **118** contains three groups of protons. The peak at 7.94 δ is attributed to the 12 ortho protons and the peak at 8.37 δ to the six external para protons. The remaining peak comes from the six inner protons. If the molecule preferred **118b**, we would expect to find this peak upfield, probably with a negative δ, as in the case of **115**. The fact that this peak is far downfield indicates that the electrons prefer to be in benzenoid rings. Note that in the case of the dianion of **117**, we have the opposite situation. In this ion, the 38-electron system is preferred even though 24 of these must come from the six benzene rings, which therefore cannot have aromatic sextets.

119

120

121

Phenacenes are a family of "graphite ribbons," where benzene rings are fused together in an alternating pattern. Phenanthrene is the simplest member of this family and other members include the 22-electron system picene (**119**); the 26-electron system fulminene (**120**); and the larger member of this family, the 30 electron [7]-phenancene, with seven rings (**121**).[310] In the series benzene to heptacene, reactivity increases although acene resonance energies per π electron are nearly constant. The inner rings of the "acenes" are more reactive, and calculations shown that those rings are more aromatic than the outer rings, and even more aromatic than benzene itself.[311]

[308]Staab, H.A.; Diederich, F. *Chem. Ber.* **1983**, *116*, 3487; Staab, H.A.; Diederich, F.; Krieger, C.; Schweitzer, D. *Chem. Ber.* **1983**, *116*, 3504. For a similar molecule with 10 instead of 12 rings, see Funhoff, D.J.H.; Staab, H.A. *Angew. Chem. Int. Ed.* **1986**, *25*, 742.

[309]Jiao, H.; Schleyer, P.v.R. *Angew. Chem. Int. Ed.*, **1996**, *35*, 2383.

[310]Mallory, F.B.; Butler, K.E.; Evans, A.C.; Mallory, C.W. *Tetrahedron Lett.* **1996**, *37*, 7173.

[311]Schleyer, P.v.R.; Manoharan, M.; Jiao, H.; Stahl, F. *Org. Lett.* **2001**, *3*, 3643.

A super ring molecule is formed by rolling a polyacene molecule into one ring with one edge benzene ring folding into the other. These are called cyclopolyacenes or cyclacenes.[312] Although the *zigzag* cyclohexacenes (**122**) are highly aromatic (this example is a 22-electron system), the linear cyclohexacenes (e.g., the 24 electron **123**) are much less aromatic.[313]

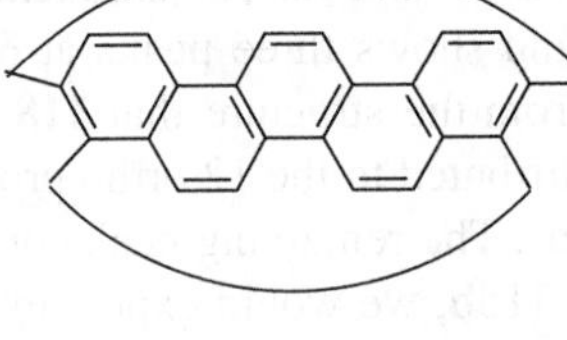

122

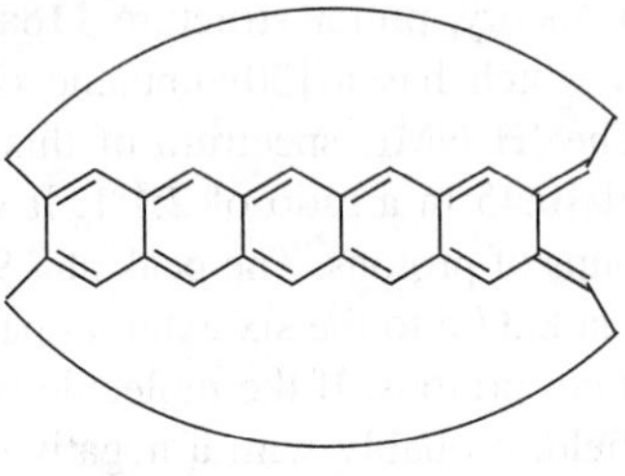

123

Systems of More Than Ten Electrons: $4n$ Electrons[224]

As we have seen (p. 74), these systems are expected to be not only nonaromatic, but actually antiaromatic.

124   **125**

The [12]annulene **124** has been prepared.[314] In solution, **124** undergoes rapid conformational mobility (as do many other annulenes),[315] and above $-150°C$ in this partiuclar case, all protons are magnetically equivalent. However, at $-170°C$ the mobility is greatly slowed and the three inner protons are found at $\sim 8\ \delta$ while the nine outer protons are at $\sim 6\ \delta$. Interaction of the "internal" hydrogens in annulene **124** leads to nonplanarity. Above $-50°C$, **124** is unstable and rearranges to **125**. Several bridged

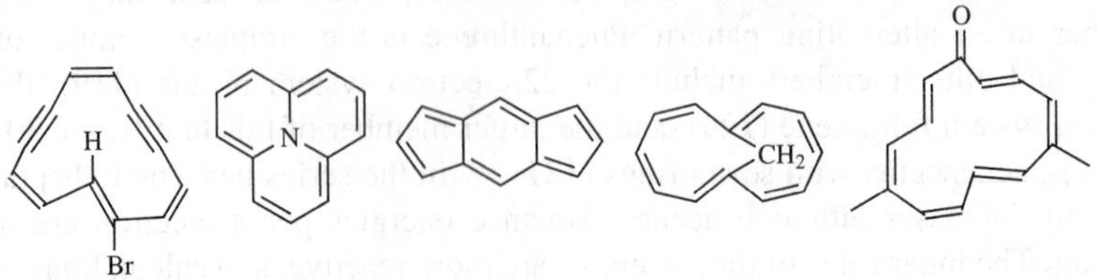

126   **127**   **128**   **129**   **130**

[312]Ashton, P.R.; Issacs, N.S.; Kohnke, F.H.; Slawin, A.M.Z.; Spencer, C.M.; Stoddart, J.F.; Williams, D.J. *Angew. Chem. Int. Ed.* **1988**, *27*, 966; Ashton, P.R.; Brown, G.R.; Issacs, N.S.; Giuffrida, D.; Kohnke, F.H.; Mathias, J.P.; Slawin, A.M.Z.; Smith, D.R.; Stoddart, J.F.; Williams, D.J. *J. Am. Chem. Soc.* **1992**, *114*, 6330; Ashton, P.R.; Girreser, U.; Giuffrida, D.; Kohnke, F.H.; Mathias, J.P.; Raymo, F.M.; Slawin, A.M.Z.; Stoddart, J.F.; Williams, D.J. *J. Am. Chem. Soc.* **1993**, *115*, 5422.
[313]Aihara, J-i. *J. Chem. Soc. Perkin Trans. 2* **1994**, 971.
[314]Oth, J.F.M.; Röttele, H.; Schröder, G. *Tetrahedron Lett.* **1970**, 61; Oth, J.F.M.; Gilles, J.; Schröder, G. *Tetrahedron Lett.* **1970**, 67.
[315]For a review of conformational mobility in annulenes, see Oth, J.F.M. *Pure Appl. Chem.* **1971**, *25*, 573.

and dehydro[12]annulenes are known, for example, 5-bromo-1,9-didehydro[12]annu-lene (**126**),[316] cycl[3.3.3]azine (**127**),[317] *s*-indacene (**128**),[318] and 1,7-methano[12]annu-lene (**129**).[319] *s*-Indacene is a planar, conjugated system perturbed by two cross-links, and studies showed that the low-energy structure has *localized* double bonds. In these compounds, both hydrogen interference and conformational mobility are prevented. In **127–129**, the bridge prevents conformational changes, while in **126** the bromine atom is too large to be found inside the ring. The NMR spectra show that all four compounds are paratropic, the inner proton of **126** being found at 16.4 δ. The dication of **112**[320] and the dianion of **103**[321] are also 12-electron paratropic species. An inter-esting 12-electron [13]-annulenone has recently been reported. 5,10-Dimethyl[13]an-nulenone (**130**) is the first monocyclic annulene larger than tropane,[322] and a linearly fused benzodehydro[12]annulene system has been reported.[323]

The results for [16]annulene are similar. The compound was synthesized in two different ways,[324] both of which gave **131**, which in solution is in equili-brium with **132**. Above −50°C there is conformational mobility, resulting in the magnetic equivalence of all protons, but at −130°C the compound is clearly paratropic: there are 4 protons at 10.56 δ and 12 at 5.35 δ. In the solid state, where the compound exists entirely as **131**, X-ray crystallography[325] shows that the molecules are nonplanar with almost complete bond alternation: the single bonds are 1.44–1.47 Å and the double bonds 1.31–1.35 Å. A number of dehydro and bridged [16]annulenes are also paratropic,[326] as are [20]annulene[327] and

[316]Untch, K.G.; Wysocki, D.C. *J. Am. Chem. Soc.* **1967**, *89*, 6386.

[317]Farquhar, D.; Leaver, D. *Chem. Commun.* **1969**, 24. For a review, see Matsuda, Y.; Gotou, H. *Heterocycles* **1987**, *26*, 2757.

[318]Hertwig, R.H.; Holthausen, M.C.; Koch, W.; Maksić, Z.B. *Angew. Chem. Int. Ed.* **1994**, *33*, 1192.

[319]Vogel, E.; Königshofen, H.; Müllen, K.; Oth, J.F.M. *Angew. Chem. Int. Ed.* **1974**, *13*, 281. See also, Mugnoli, A.; Simonetta, M. *J. Chem. Soc. Perkin Trans. 2* **1976**, 822; Scott, L.T.; Kirms, M.A.; Günther, H.; von Puttkamer, H. *J. Am. Chem. Soc.* **1983**, *105*, 1372; Destro, R.; Ortoleva, E.; Simonetta, M.; Todeschini, R. *J. Chem. Soc. Perkin Trans. 2* **1983**, 1227.

[320]Müllen, K.; Meul, T.; Schade, P.; Schmickler, H.; Vogel, E. *J. Am. Chem. Soc.* **1987**, *109*, 4992. This paper also reports a number of other bridged paratropic 12-, 16-, and 20-electron dianions and dications. See also Hafner, K.; Thiele, G.F. *Tetrahedron Lett.* **1984**, *25*, 1445.

[321]Schmalz, D.; Günther, H. *Angew. Chem. Int. Ed.* **1988**, *27*, 1692.

[322]Higuchi, H.; Hiraiwa, N.; Kondo, S.; Ojima, J.; Yamamoto, G. *Tetrahedron Lett.* **1996**, *37*, 2601.

[323]Gallagher, M.E.; Anthony, J.E. *Tetrahedron Lett.* **2001**, *42*, 7533.

[324]Schröder, G.; Oth, J.F.M. *Tetrahedron Lett.* **1966**, 4083; Oth, J.F.M.; Gilles, J. *Tetrahedron Lett.* **1968**, 6259; Calder, I.C.; Gaoni, Y.; Sondheimer, F. *J. Am. Chem. Soc.* **1968**, *90*, 4946. For monosubstituted [16]annulenes, see Schröder, G.; Kirsch, G.; Oth, J.F.M. *Chem. Ber.* **1974**, *107*, 460.

[325]Johnson, S.M.; Paul, I.C.; King, G.S.D. *J. Chem. Soc. B* **1970**, 643.

[326]For example, see Calder, I.C.; Garratt, P.J.; Sondheimer, F. *J. Am. Chem. Soc.* **1968**, *90*, 4954; Murata, I.; Okazaki, M.; Nakazawa, T. *Angew. Chem. Int. Ed.* **1971**, *10*, 576; Ogawa, H.; Kubo, M.; Tabushi, I. *Tetrahedron Lett.* **1973**, 361; Nakatsuji, S.; Morigaki, M.; Akiyama, S.; Nakagawa, M. *Tetrahedron Lett.* **1975**, 1233; Elix, J.A. *Aust. J. Chem.* **1969**, *22*, 1951; Vogel, E.; Kürshner, U.; Schmickler, H.; Lex, J.; Wennerström, O.; Tanner, D.; Norinder, U.; Krüger, C. *Tetrahedron Lett.* **1985**, *26*, 3087.

[327]Metcalf, B.W.; Sondheimer, F. *J. Am. Chem. Soc.* **1971**, *93*, 6675. See also Oth, J.F.M.; Woo, E.P.; Sondheimer, F. *J. Am. Chem. Soc.* **1973**, *95*, 7337; Nakatsuji, S.; Nakagawa, M. *Tetrahedron Lett.* **1975**, 3927; Wilcox, Jr., C.F.; Farley, E.N. *J. Am. Chem. Soc.* **1984**, *106*, 7195.

[24]annulene.[328] However, a bridged tetradehydro[32]annulene was atropic.[306]

| 131 | 132 | 133 | 134 |

Both pyracyclene (**133**)[329] (which because of strain is stable only in solution) and dipleiadiene (**134**)[330] are paratropic, as shown by NMR spectra. These molecules might have been expected to behave like naphthalenes with outer bridges, but the outer π frameworks (12 and 16 electrons, respectively) constitute antiaromatic systems with an extra central double bond. With respect to **133**, the $4n + 2$ rule predicts pyracylene to be "aromatic" if it is regarded as a 10-π-electron naphthalene unit connected to two 2-π-electron etheno systems, but "antiaromatic" if it is viewed as a 12-π-electron cyclododecahexaene periphery perturbed by an internal cross-linked etheno unit.[331] Recent studies have concluded on energetic grounds that **133** is a "borderline" case, in terms of aromaticity–antiaromaticity character.[329] Dipleiadiene appears to be antiaromatic.[330]

The fact that many $4n$ systems are paratropic, even though they may be nonplanar and have unequal bond distances, indicates that if planarity were enforced, the ring currents might be even greater. That this is true is dramatically illustrated by the NMR spectrum of the dianion of **110**[332] (and its diethyl and dipropyl homologs).[333] We may recall that in **110**, the outer protons were found at 8.14–8.67 δ with the methyl protons at -4.25 δ. For the dianion, however, which is forced to have approximately the same planar geometry, but now has 16 electrons, the outer protons are shifted to about -3 δ while the methyl protons are found at $\sim$21 δ, a shift of $\sim$25 δ! We have already seen where the converse shift was made, when [16]annulenes that were antiaromatic were converted to 18-electron dianions that were aromatic.[254] In these cases, the changes in nmr chemical shifts were almost

[328]Calder, I.C.; Sondheimer, F. *Chem. Commun.* **1966**, 904. See also, Stöckel, K.; Sondheimer, F. *J. Chem. Soc. Perkin Trans. 1* **1972**, 355; Nakatsuji, S.; Akiyama, S.; Nakagawa, M. *Tetrahedron Lett.* **1976**, 2623; Yamamoto, K.; Kuroda, S.; Shibutani, M.; Yoneyama, Y.; Ojima, J.; Fujita, S.; Ejiri, E.; Yanagihara, K. *J. Chem. Soc., Perkin Trans. 1* **1988**, 395.

[329]Trost, B.M.; Herdle, W.B. *J. Am. Chem. Soc.* **1976**, *98*, 4080.

[330]Vogel, E.; Neumann, B.; Klug, W.; Schmickler, H.; Lex, J. *Angew. Chem. Int. Ed.* **1985**, *24*, 1046.

[331]Diogo, H.P.; Kiyobayashi, T.; Minas da Piedade, M.E.; Burlak, N.; Rogers, D.W.; McMasters, D.; Persy, G.; Wirz, J.; Liebman, J.F. *J. Am. Chem. Soc.* **2002**, *124*, 2065.

[332]For a review of polycyclic dianions, see Rabinovitz, M.; Cohen, Y. *Tetrahedron* **1988**, *44*, 6957.

[333]Mitchell, R.H.; Klopfenstein, C.E.; Boekelheide, V. *J. Am. Chem. Soc.* **1969**, *91*, 4931. For another example, see Deger, H.M.; Müllen, K.; Vogel, E. *Angew. Chem. Int. Ed.* **1978**, *17*, 957.

as dramatic. Heat-of-combustion measures also show that [16]annulene is much less stable than its dianion.[334]

We can therefore conclude that $4n$ systems will be at a maximum where a molecule is constrained to be planar (as in **86** or the dianion of **110**) but, where possible, the molecule will distort itself from planarity and avoid equal bond distances in order to reduce. In some cases, such as cyclooctatraene, the distortion and bond alternation are great enough to be completely avoided. In other cases, for example, **124** or **131**, it is apparently not possible for the molecules to avoid at least some p-orbital overlap. Such molecules show evidence of paramagnetic ring currents, although the degree of is not as great as in molecules such as **86** or the dianion of **110**.

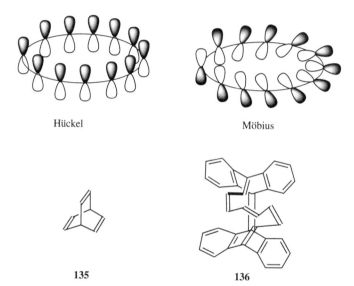

Hückel   Möbius

135   136

The concept of "Möbius aromaticity" was conceived by Helbronner in 1964[335] when he suggested that large cyclic [4n]annulenes might be stabilized if the π-orbitals were twisted gradually around a Möbius strip. This concept is illustrated by the diagrams labeled Hückel, which is a destabilized [4n] system, in contrast to the Möbius model, which is a stabilized [4n] system.[336] Zimmerman generalized this idea and applied the "Hückel–Möbius concept" to the analysis of ground-state systems, such as barrelene (**135**).[337] In 1998, a computational reinterpretation of existing experimental evidence for $(CH)_9^+$ as a Möbius

[334]Stevenson, G.R.; Forch, B.E. *J. Am. Chem. Soc.* **1980**, *102*, 5985.
[335]Heilbronner, E. *Tetrahedron Lett.* **1964**, 1923.
[336]Kawase, T; Oda, M. *Angew. Chem. Int. Ed.*, **2004**, *43*, 4396.
[337]Zimmerman, H.E. *J. Am. Chem. Soc.* **1966**, 88, 1564.; Zimmerman, H.E. *Acc. Chem. Res.* **1972**, *4*, 272.

aromatic cyclic annulene with $4n$ π-electrons was reported.[338] A recent computational study predicted several Möbius local minima for [12]-, [16]-, and [20]annulenes.[339] A twisted [16]annulene has been prepared and calculations suggested it should show Möbius aromaticity.[340] High-performance liquid chromatography (HPLC) separation of isomers gave **136**, which the authors concluded is Möbius aromatic.

Other Aromatic Compounds

We will briefly mention three other types of aromatic compounds.

1. *Mesoionic Compounds.*[341] These compounds cannot be satisfactorily represented by Lewis structures not involving charge separation. Most of them contain five-membered rings. The most common are the *sydnones*, stable aromatic compounds that undergo aromatic substitution when R′ is hydrogen.

Sydnone

2. *The Dianion of Squaric Acid.*[342] The stability of this system is illustrated by the fact that the pK_1 of squaric acid[343] is $\sim$1.5 and the pK_2 is $\sim$3.5,[344] which means that even the second proton is given up much more readily than the proton of acetic acid, for example.[345] The analogous three-,[346]

[338]Mauksch, M.; Gogonea, V.; Jiao, H.; Schleyer, P.v.R. *Angew. Chem. Int. Ed.*, *1998*, *37*, 2395.

[339]Castro, C.; Isborn, C.M.; Karney, W.L.; Mauksch, M.; Schleyer, P.v.R. *Org. Lett. 2002*, *4*, 3431.

[340]Ajami, D.; Oeckler, O.; Simon, A.; Herges, R. *Nature (London) 2003*, *426*, 819.

[341]For reviews, see Newton, C.G.; Ramsden, C.A. *Tetrahedron 1982*, *38*, 2965; Ollis, W.D.; Ramsden, C.A. *Adv. Heterocycl. Chem. 1976*, *19*, 1; Ramsden, C.A. *Tetrahedron 1977*, *33*, 3203; Yashunskii, V.G.; Kholodov, L.E. *Russ. Chem. Rev. 1980*, *49*, 28; Ohta, M.; Kato, H., in Snyder, J.P. *Nonbenzenoid Aromaticity*, Vol. 1, Academic Press, NY, *1969*, pp. 117–248.

[342]West, R.; Powell, D.L. *J. Am. Chem. Soc. 1963*, *85*, 2577; Ito, M.; West, R. *J. Am. Chem. Soc. 1963*, *85*, 2580.

[343]For a review of squaric acid and other nonbenzenoid quinones, see Wong, H.N.C.; Chan, T.; Luh, T., in Patai, S.; Rappoport, Z. *The Chemistry of the Quinonoid Compounds*, Vol. 2, pt. 2, Wiley, NY, *1988*, pp. 1501–1563.

[344]Ireland, D.T.; Walton, H.F. *J. Phys. Chem. 1967*, *71*, 751; MacDonald, D.J. *J. Org. Chem. 1968*, *33*, 4559.

[345]There has been a controversy as to whether this dianion is in fact aromatic. See Aihara, J. *J. Am. Chem. Soc. 1981*, *103*, 1633.

[346]Eggerding, D.; West, R. *J. Am. Chem. Soc. 1976*, *98*, 3641; Pericás, M.A.; Serratosa, F. *Tetrahedron Lett. 1977*, 4437; Semmingsen, D.; Groth, P. *J. Am. Chem. Soc. 1987*, *109*, 7238.

five-, and six-membered ring compounds are also known.[347]

3. *Homoaromatic Compounds*. When cyclooctatetraene is dissolved in concentrated H_2SO_4, a proton adds to one of the double bonds to form the homotropylium ion **137**.[348] In this species, an aromatic sextet is spread over seven carbons, as in the tropylium ion. The eighth carbon is an sp^3 carbon and so cannot take part in the aromaticity. The NMR spectra show the presence of a diatropic ring current: H_b is found at $\delta = -0.3$; H_a at 5.1 δ; H_1 and H_7 at 6.4 δ; H_2–H_6 at 8.5 δ. This ion is an example of a *homoaromatic* compound, which may be defined as a compound that contains one or more[349] sp^3-hybridized carbon atoms in an otherwise conjugated cycle.[350]

137

In order for the orbitals to overlap most effectively so as to close a loop, the sp^3 atoms are forced to lie almost vertically above the plane of the

[347]For a monograph, see West, R. *Oxocarbons*; Academic Press, NY, *1980*. For reviews, see Serratosa, F. *Acc. Chem. Res. 1983*, *16*, 170; Schmidt, A.H. *Synthesis 1980*, 961; West, R. *Isr. J. Chem. 1980*, *20*, 300; West, R.; Niu, J., in Snyder, J.P. *Nonbenzenoid Aromaticity*, Vol. 1, Academic Press, NY, *1969*, pp. 311–345, and in Zabicky, J. *The Chemistry of the Carbonyl Group*, Vol. 2, Wiley, NY, *1970*, pp. 241–275; Maahs, G.; Hegenberg, P. *Angew. Chem. Int. Ed. 1966*, *5*, 888.

[348]Rosenberg, J.L.; Mahler, J.E.; Pettit, R. *J. Am. Chem. Soc. 1962*, *84*, 2842; Keller, C.E.; Pettit, R. *J. Am. Chem. Soc. 1966*, *88*, 604, 606; Winstein, S.; Kreiter, C.G.; Brauman, J.I. *J. Am. Chem. Soc. 1966*, *88*, 2047; Haddon, R.C. *J. Am. Chem. Soc. 1988*, *110*, 1108. See also, Childs, R.F.; Mulholland, D.L.; Varadarajan, A.; Yeroushalmi, S. *J. Org. Chem. 1983*, *48*, 1431. See also, Alkorta, I.; Elguero, J.; Eckert-Maksić, M.; Maksić, Z.B. *Tetrahedron 2004*, *60*, 2259.

[349]If a compound contains two such atoms it is bishomoaromatic; if three, trishomoaromatic, and so on. For examples see Paquette, L.A. *Angew. Chem. Int. Ed. 1978*, *17*, 106.

[350]For reviews, see Childs, R.F. *Acc. Chem. Res. 1984*, *17*, 347; Paquette, L.A. *Angew. Chem. Int. Ed. 1978*, *17*, 106; Winstein, S. *Q. Rev. Chem. Soc. 1969*, *23*, 141; Garratt, P.J. *Aromaticity*, Wiley, NY, *1986*, pp. 5–45; and in Olah, G.A.; Schleyer, P.v.R. *Carbonium Ions*, Wiley, NY, Vol. 3, *1972*, the reviews by Story, P.R.; Clark, Jr., B.C. 1007–1098, pp. 1073–1093; Winstein, S. 965–1005. (The latter is a reprint of the *Q. Rev. Chem. Soc.* review mentioned above.)

aromatic atoms.[351] In **137**, H_b is directly above the aromatic sextet, and so is shifted far upfield in the nmr. All homoaromatic compounds so far discovered are ions, and it is questionable[352] as to whether homoaromatic character can exist in uncharged systems.[353] Homoaromatic ions of 2 and 10 electrons are also known.

New conceptual applications to 3D homoaromatic systems with cubane, dodecahedrane, and adamantane frameworks has been presented.[354] This concept includes families of spherical homoaromatics with both 2 and 8 mobile electrons. Each set has complete *spherical homoaromaticity*, that is, all the sp^2 carbon atoms in a highly symmetrical frameworks are separated by one or two sp^3-hybridized atoms.

4. *Fullerenes.* Fullerenes are a family of aromatic hydrocarbons based on the parent buckminsterfullerene (**138**; C_{60})[355] that have a variety of very interesting properties.[356] Molecular-orbital calculations showed that "fullerene aromaticity lies within $2 \, kcal \, mol^{-1}$ ($8.4 \, kJ \, mol^{-1}$) per carbon of a hypothetical ball of rolled up graphite.[357] Another class of polynuclear aromatic hydrocarbons are the *buckybowls*, which are essentially fragments of **138**. Corannulene (**139**)[358] (also called 5-circulene), for example, is the simplest curved-surface hydrocarbon possessing a carbon framework that is identified with the buckminsterfullerene

[351]Calculations show that only ~60% of the chemical shift difference between H_a and H_b is the result of the aromatic ring current, and that even H_a is shielded; it would appear at $\delta \sim 5.5$ without the ring current: Childs, R.F.; McGlinchey, M.J.; Varadarajan, A. *J. Am. Chem. Soc.* **1984**, *106*, 5974.

[352]Houk, K.N.; Gandour, R.W.; Strozier, R.W.; Rondan, N.G.; Paquette, L.A. *J. Am. Chem. Soc.* **1979**, *101*, 6797; Paquette, L.A.; Snow, R.A.; Muthard, J.L.; Cynkowski, T. *J. Am. Chem. Soc.* **1979**, *101*, 6991. See however, Liebman, J.F.; Paquette, L.A.; Peterson, J.R.; Rogers, D.W. *J. Am. Chem. Soc.* **1986**, *108*, 8267.

[353]Examples of uncharged homoantiaromatic compounds have been claimed: Wilcox, Jr., C.F.; Blain, D.A.; Clardy, J.; Van Duyne, G.; Gleiter, R.; Eckert-Maksic, M. *J. Am. Chem. Soc.* **1986**, *108*, 7693; Scott, L.T.; Cooney, M.J.; Rogers, D.W.; Dejroongruang, K. *J. Am. Chem. Soc.* **1988**, *110*, 7244.

[354]Chen, Z.; Haijun Jiao, H.; Andreas Hirsch, A.; Schleyer, P.v.R. *Angew. Chem. Int. Ed.*, **2002**, *41*, 4309

[355]Billups, W.E.; Ciufolini, M.A. *Buckminsterfullerenes*, VCH, NY, **1993**; Taylor, R. *The Chemistry of Fullerenes*, World Scientific, River Edge, NJ, Singapore, **1995**; Aldersey-Williams, H. *The Most Beautiful Molecule: The Discovery of the Buckyball*, Wiley, NY, **1995**; Baggott, J.E. *Perfect Symmetry: the Accidental Discovery of Buckminsterfullerene*, Oxford University Press, Oxford, NY, **1994**. Also see Kroto, H.W.; Heath, J.R.; O'Brien, S.C.; Curl, R.F.; Smalley, R.E. *Nature (London)* **1985**, *318*, 162.

[356]Smalley, R.E. *Acc. Chem. Res.* **1992**, *25*, 98; Diederich, F.; Whetten, R.L. *Acc. Chem. Res.* **1992**, *25*, 119; Hawkins, J.M. *Acc. Chem. Res.* **1992**, *25*, 150; Wudl, F. *Acc. Chem. Res.* **1992**, *25*, 157; McElvany, S.W.; Ross, M.M.; Callahan, J.H. *Acc. Chem. Res.* **1992**, *25*, 162; Johnson, R.D.; Bethune, D.S.; Yannoni, C.S. *Acc. Chem. Res.* **1992**, *25*, 169.

[357]Warner, P.M. *Tetrahedron Lett.* **1994**, *35*, 7173.

[358]Barth, W.E.; Lawton, R.G. *J. Am. Chem. Soc.* **1971**, *93*, 1730; Scott, L.T.; Hashemi, M.M.; Meyer, D.T.; Warren, H.B. *J. Am. Chem. Soc.* **1991**, *113*, 7082.

surface. It has been synthesized by Scott,[352] and several other groups.[359] Corannulene is a flexible molecule, with a bowl-to-bowl inversion barrier of ~ 10–11 kcal mol^{-1} (41.8–46.0 kJ mol^{-1}).[360] Benzocorannulenes are known,[361] and other bowl-shaped hydrocarbons include acenaphtho[3,2,1,8-*ijklm*]diindeno[4,3,2,1-*cdef*-1′,2′,3′,4′*pqra*]triphenylene.[362] The inversion barrier to buckybowl inversion has been lowered by such benzannelation of the rim.[363] Other semibuckminsterfullerenes include C_{2v}-$C_{30}H_{12}$ and C_3-$C_{30}H_{12}$.[358] Larger fullerenes include C_{60},C_{80}, C_{84}, and fullerenes are known that contain an endohedral metal, such as scandium or even Sc_3N.[364] Synthetic methods often generate mixtures of fullerenes that must be separated, as in the report of new methods for separating C_{84}-fullerenes.[365] A homofullerene has been prepared.[366]

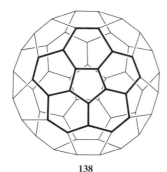

138

139

HYPERCONJUGATION

All of the delocalization discussed so far involves π electrons. Another type, called *hyperconjugation*, involves σ electrons.[367] When a carbon attached

[359]Borchardt, A.; Fuchicello, A.; Kilway, K.V.; Baldridge, K.K.; Siegel, J.S. *J. Am. Chem. Soc.* *1992*, *114*, 1921; Liu, C.Z.; Rabideau, P.W. *Tetrahedron Lett.* *1996*, *37*, 3437.

[360]Biedermann, P.U.; Pogodin, S.; Agranat, I. *J. Org. Chem.* *1999*, *64*, 3655; Rabideau, P.W.; Sygula, A. *Acc. Chem. Res.* *1996*, *29*, 235; Mehta, G.; Panda, G. *Chem. Comm.*, *1997*, 2081; Rabideau, P.W.; Abdourazak, A.H.; Folsom, H.E.; Marcinow, Z.; Sygula, A.; Sygula, R. *J. Am. Chem. Soc.* *1994*, *116*, 7891; Hagan, S.; Bratcher, M.S.; Erickson, M.S.; Zimmermann, G.; Scott, L.T. *Angew. Chem. Int. Ed.*, *1997*, *36*, 406. See also, Dinadayalane, T.C.; Sastry, G.N. *Tetrahedron 2003*, *59*, 8347.

[361]Dinadayalane, T.C.; Sastry, G.N. *J. Org. Chem.* *2002*, *67*, 4605.

[362]Marcinow, Z.; Grove, D.I.; Rabideau, P.W. *J. Org. Chem.* *2002*, *67*, 3537.

[363]Marcinow, Z.; Sygula, A.; Ellern, D.A.; Rabideau, P.W. *Org. Lett.* *2001*, *3*, 3527.

[364]Stevenson, S.; Rice, G.; Glass, T.; Harich, K.; Cromer, F.; Jordan, M.R.; Craft, J.; Hadju, E.; Bible, R.; Olmstead, M.M.; Maitra, K.; Fisher, A.J.; Balch, A.L.; Dorn, H.C. *Nature (London) 1999*, *401*, 55.

[365]Wang, G.-W.; Saunders, M.; Khong, A.; Cross, R.J. *J. Am. Chem. Soc.* *2000*, *122*, 3216.

[366]Kiely, A.F.; Haddon, R.C.; Meier, M.S.; Selegue, J.P.; Brock, C.P.; Patrick, B.O.; Wang, G.-W.; Chen, Y. *J. Am. Chem. Soc.* *1999*, *121*, 7971.

[367]For monographs, see Baker, J.W. *Hyperconjugation*, Oxford University Press, Oxford, *1952*; Dewar, M.J.S. *Hyperconjugation*, Ronald Press, NY, *1962*. For a review, see de la Mare, P.B.D. *Pure Appl. Chem.* *1984*, *56*, 1755.

forms there is no bond at all between the carbon and hydrogen. The effect of **140** on the actual molecule is that the electrons in the C—H bond are closer to the carbon than they would be if **140** did not contribute at all.

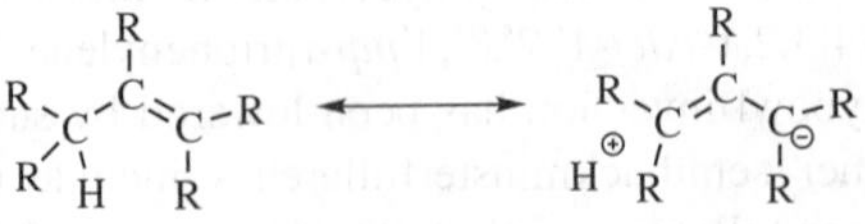

140

Hyperconjugation in the above case may be regarded as an overlap of the σ orbital of the C—H bond and the π orbital of the C—C bond, analogous to the π–π orbital overlap previously considered. As might be expected, those who reject the idea of resonance in butadiene (p. 39) believe it even less likely when it involves no-bond structures.

The concept of hyperconjugation arose from the discovery of apparently anomalous electron-release patterns for alkyl groups. By the field effect alone, the order of electron release for simple alkyl groups connected to an unsaturated system is *tert*-butyl > isopropyl > ethyl > methyl, and this order is observed in many phenomena. Thus, the dipole moments in the gas phase of $PhCH_3$, PhC_2H_5, $PhCH(CH_3)_2$, and $PhC(CH_3)_3$ are, respectively, 0.37, 0.58, 0.65, and 0.70 D.[368]

However, Baker and Nathan[369] observed that the rates of reaction with pyridine of *para*-substituted benzyl bromides (see reaction **10-31**) were opposite that expected from electron release by the field effect. That is, the methyl-substituted compound reacted fastest and the *tert*-butyl-substituted compounded reacted slowest.

This came to be called the *Baker–Nathan effect* and has since been found in many processes. Baker and Nathan explained it by considering that hyperconjugative forms contribute to the actual structure of toluene:

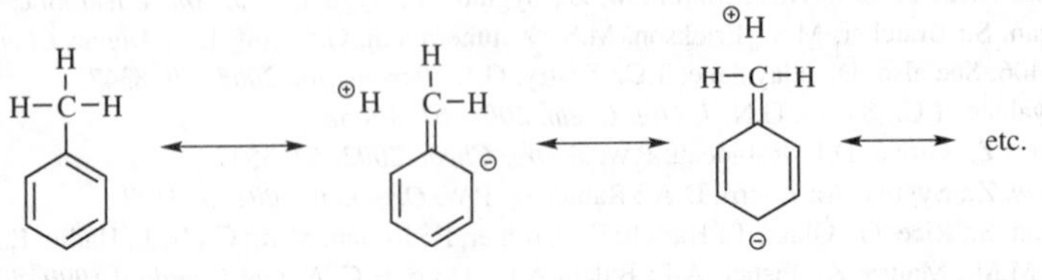

For the other alkyl groups, hyperconjugation is diminished because the number of C—H bonds is diminished and in *tert*-butyl there are none; hence, with

[368]Baker, J.W.; Groves, L.G. *J. Chem. Soc.* **1939**, 1144.
[369]Baker, J.W.; Nathan, W.S. *J. Chem. Soc.* **1935**, 1840, 1844.

respect to this effect, methyl is the strongest electron donor and *tert*-butyl is the weakest.

However, the Baker–Nathan effect has now been shown not to be caused by hyperconjugation, but by differential solvation.[370] This was demonstrated by the finding that in certain instances where the Baker–Nathan effect was found to apply in solution, the order was completely reversed in the gas phase.[371] Since the molecular structures are unchanged in going from the gas phase into solution, it is evident that the Baker–Nathan order in these cases is not caused by a structural feature (hyperconjugation), but by the solvent. That is, each alkyl group is solvated to a different extent.[372]

There is a large body of evidence against hyperconjugation in the ground states of neutral molecules.[373] A recent study of the one-bond coupling constants for the aromatic system **141**, however, appears to provide the first structural evidence for hyperconjugation in a neutral ground state.[374] In hyperconjugation

M = C, Si, Ge, Sn

141   X = NO₂, CN, H, Me, OMe

in the ground state of neutral molecules, which Muller and Mulliken call *sacrificial hyperconjugation*,[375] the canonical forms involve not only no-bond resonance, but also a charge separation not possessed by the main form (see **141**). For carbocations and free radicals[376] and for excited states of molecules,[377] there is evidence that hyperconjugation is important. In free radicals and carbocations, the canonical

[370]This idea was first suggested by Schubert, W.M.; Sweeney, W.A. *J. Org. Chem.* **1956**, *21*, 119.

[371]Hehre,W.J.; McIver, Jr., R.T.; Pople, J.A.; Schleyer, P.v.R. *J. Am. Chem. Soc.* **1974**, *96*, 7162; Arnett, E.M.; Abboud, J.M. *J. Am. Chem. Soc.* **1975**, *97*, 3865; Glyde, E.; Taylor, R. *J. Chem. Soc. Perkin Trans. 2* **1977**, 678. See also, Taylor, R. *J. Chem. Res. (S)*, **1985**, 318.

[372]For an opposing view, see Cooney, B.T.; Happer, D.A.R. *Aust. J. Chem.* **1987**, *40*, 1537.

[373]For some evidence in favor, see Laube, T.; Ha, T. *J. Am. Chem. Soc.* **1988**, *110*, 5511.

[374]Lambert, J.B.; Singer, R.A. *J. Am. Chem. Soc.* **1992**, *114*, 10246.

[375]Muller, N.; Mulliken, R.S. *J. Am. Chem. Soc.* **1958**, *80*, 3489.

[376]Symons, M.C.R. *Tetrahedron* **1962**, *18*, 333.

[377]Rao, C.N.R.; Goldman, G.K.; Balasubramanian, A. *Can. J. Chem.* **1960**, *38*, 2508.

forms display no more charge separation than the main form. Muller and Mulliken call this *isovalent hyperconjugation*: Even here the main form contributes more to the hybrid than the others.

TAUTOMERISM[378]

There remains one topic to be discussed in our survey of chemical bonding in organic compounds. For most compounds, all the molecules have the same structure, whether or not this structure can be satisfactorily represented by a Lewis formula. But for many other compounds there is a mixture of two or more structurally distinct compounds that are in rapid equilibrium. When this phenomenon, called *tautomerism*,[379] exists, there is a rapid shift back and forth among the molecules. In most cases, it is a proton that shifts from one atom of a molecule to another.

Keto–Enol Tautomerism[380]

A very common form of tautomerism is that between a carbonyl compound containing an a hydrogen and its enol form:[381] Such equilibria are pH dependent, as in the case of 2-acetylcyclohexanone.[382]

Keto form   Enol form

In simple cases ($R^2 = H$, alkyl, OR, etc.) the equilibrium lies well to the left (Table 2.1). The reason can be seen by examining the bond energies in Table 1.7.

[378]Baker, J.W. *Tautomerism*; D. Van Nostrand Company, Inc., New York, *1934*; Minkin, V.I.; Olekhnovich, L.P.; Zhdanov, Y.A. *Molecular Design of Tautomeric Compounds*, D. Reidel Publishing Co.: Dordrecht, Holland, *1988*.
[379]For reviews, see Toullec, J. *Adv. Phys. Org. Chem.* *1982*, *18*, 1; Kołsov, A.I.; Kheifets, G.M. *Russ. Chem. Rev.* *1971*, *40*, 773; *1972*, *41*, 452–467; Forsén, S.; Nilsson, M., in Zabicky, J. *The Chemistry of the Carbonyl Group*, Vol. 2, Wiley, NY, *1970*, pp. 157–240.
[380]The mechanism for conversion of one tautomer to another is discussed in Chapter 12 (reaction **12-3**).
[381]Capponi, M.; Gut, I.G.; Hellrung, B.; Persy, G.; Wirz, J. *Can. J. Chem.* *1999*, *77*, 605. For a treatise, see Rappoport, Z. *The Chemistry of Enols*, Wiley, NY, *1990*.
[382]Iglesias, E. *J. Org. Chem*, *2003*, *68*, 2680.

TABLE 2.1. The Enol Content of Some Carbonyl Compounds

Compound	Enol Content, %	References
Acetone	6×10^{-7}	383
$PhCOCH_3$	1.1×10^{-6}	384
Cyclopentanone	1×10^{-6}	385
CH_3CHO	6×10^{-5}	386
Cyclohexanone	4×10^{-5}	385
Butanal	5.5×10^{-4}	387
$(CH_3)_2CHCHO$	1.4×10^{-2}	388,387
Ph_2CHCHO	9.1	389
CH_3COOEt	No enol found[a]	385
CH_3COCH_2COOEt	8.4	390
$CH_3COCH_2COCH_3$	80	322
$PhCOCH_2COCH_3$	89.2	385
$EtOOCCH_2COOEt$	7.7×10^{-3}	385
$N{\equiv}C{-}CH_2COOEt$	2.5×10^{-1}	385
Indane-1-one	3.3×10^{-8}	391
Malonamide	No enol found	392

[a]Less than 1 part in 10 million.

The keto form differs from the enol form in possessing a C—H, a C—C, and a C=O bond, where the enol has a C=C, a C—O, and an O—H bond. The approximate sum of the first three is $359\,kcal\,mol^{-1}$ ($1500\,kJ\,mol^{-1}$) and of the second three is $347\,kcal\,mol^{-1}$ ($1452\,kJ\,mol^{-1}$). The keto form is therefore thermodynamically more stable by $\sim12\,kcal\,mol^{-1}$ ($48\,kJ\,mol^{-1}$) and enol forms cannot normally be isolated.[393] In certain cases, however, a larger amount of the enol form is present,

[383]Tapuhi, E.; Jencks, W.P. *J. Am. Chem. Soc.* **1982**, *104*, 5758; Chiang, Y.; Kresge, A.J.; Tang, Y.S.; Wirz, J. *J. Am. Chem. Soc.* **1984**, *106*, 460. See also, Hine, J.; Arata, K. *Bull. Chem. Soc. Jpn.* **1976**, *49*, 3089; Guthrie, J.P. *Can. J. Chem.* 1979, *57*, 797, 1177; Dubois, J.E.; El-Alaoui, M.; Toullec, J. *J. Am. Chem. Soc.* 1981, *103*, 5393; Toullec, J. *Tetrahedron Lett.* 1984, *25*, 4401; Chiang, Y.; Kresge, A.J.; Schepp, N.P. *J. Am. Chem. Soc.* **1989**, *111*, 3977.

[384]Keeffe, J.R.; Kresge, A.R.; Toullec, J. *Can. J. Chem.* **1986**, *64*, 1224.

[385]Gero, A. *J. Org. Chem.* **1954**, *19*, 469, 1960; Keeffe, J.R., Kresge, A.J.; Schepp, N.P. *J. Am. Chem. Soc.* **1990**, *112*, 4862; Iglesias, E. *J. Chem. Soc. Perkin Trans. 2* **1997**, 431. See these papers for values for other simple compounds.

[386]Chiang, Y.; Hojatti, M.; Keeffe, J.R.; Kresge, A.J.; Schepp, N.P.; Wirz, J. *J. Am. Chem. Soc.* **1987**, *109*, 4000.

[387]Bohne, C.; MacDonald, I.D.; Dunford, H.B. *J. Am. Chem. Soc.* **1986**, *108*, 7867.

[388]Chiang, Y.; Kresge, A.J.; Walsh, P.A. *J. Am. Chem. Soc.* **1986**, *108*, 6314.

[389]Chiang, Y.; Kresge, A.J.; Krogh, E.T. *J. Am. Chem. Soc.* **1988**, *110*, 2600.

[390]Moriyasu, M.; Kato, A.; Hashimoto, Y. *J. Chem. Soc. Perkin Trans. 2* 1986, 515. For enolization of β-ketoamides, see Hynes, M.J.; Clarke, E.M. *J. Chem. Soc. Perkin Trans. 2* **1994**, 901.

[391]Jefferson, E.A.; Keeffe, J.R.; Kresge, A.J. *J. Chem. Soc. Perkin Trans. 2* **1995**, 2041.

[392]Williams, D.L.H.; Xia, L. *J. Chem. Soc. Chem. Commun.* **1992**, 985.

[393]For reviews on the generation of unstable enols, see Kresge, A.J. *Pure Appl. Chem.* **1991**, *63*, 213; Capon, B., in Rappoport, Z. *The Chemistry of Enols*, Wiley, NY, *1990*, pp. 307–322.

and it can even be the predominant form.[394] There are three main types of the more stable enols:[395]

1. Molecules in which the enolic double bond is in conjugation with another double bond. Some of these are shown in Table 2.1. As the table shows, carboxylic esters have a much smaller enolic content than ketones. In molecules like acetoacetic ester (**142**), the enol is also stabilized by internal hydrogen bonding, which is unavailable to the keto form:

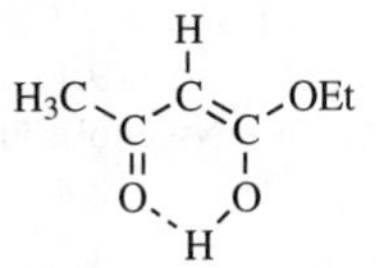

142

2. Molecules that contain two or three bulky aryl groups.[396] An example is 2,2-dimesitylethenol (**143**). In this case the keto content at equilibrium is only 5%.[397] In cases such as this, steric hindrance (p. 230) destabilizes the keto form. In **143**, the two aryl groups are ~120° apart, but in **144** they must move closer together (~109.5°). Such compounds are often called *Fuson-type enols*.[398] There is one example of an amide with a bulky aryl group [*N*-methyl bis(2,4,6-triisopropylphenyl)acetamide] that has a measurable enol content, in sharp contrast to most amides.[399]

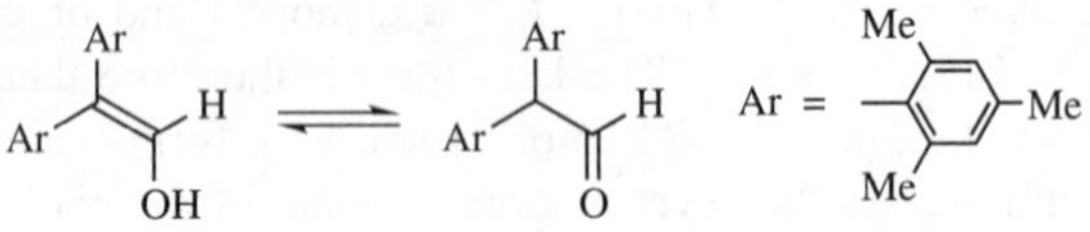

143   **144**

---

[394]For reviews of stable enols, see Kresge, A.J. *Acc. Chem. Res.* **1990**, *23*, 43; Hart, H.; Rappoport, Z.; Biali, S.E., in Rappoport, Z. *The Chemistry of Enols*, Wiley, NY, **1990**, pp. 481–589; Hart, H. *Chem. Rev*, **1979**, *79*, 515; Hart, H.; Sasaoka, M. *J. Chem. Educ.* **1980**, *57*, 685.

[395]For some examples of other types, see Pratt, D.V.; Hopkins, P.B. *J. Am. Chem. Soc.* **1987**, *109*, 5553; Nadler, E.B.; Rappoport, Z.; Arad, D.; Apeloig, Y. *J. Am. Chem. Soc.* **1987**, *109*, 7873.

[396]For a review, see Rappoport, Z.; Biali, S.E. *Acc. Chem. Res.* **1988**, *21*, 442. For a discussion of their structures, see Kaftory, M.; Nugiel, D.A.; Biali, D.A.; Rappoport, Z. *J. Am. Chem. Soc.* **1989**, *111*, 8181.

[397]Biali, S.E.; Rappoport, Z. *J. Am. Chem. Soc.* **1985**, *107*, 1007. See also, Kaftory, M.; Biali, S.E.; Rappoport, Z. *J. Am. Chem. Soc.* **1985**, *107*, 1701; Nugiel, D.A.; Nadler, E.B.; Rappoport, Z. *J. Am. Chem. Soc.* **1987**, *109*, 2112; O'Neill, P.; Hegarty, A.F. *J. Chem. Soc. Chem. Commun.* **1987**, 744; Becker, H.; Andersson, K. *Tetrahedron Lett.* **1987**, *28*, 1323.

[398]First synthesized by Fuson, R.C.; see, for example, Fuson, R.C.; Southwick, P.L.; Rowland, S.P. *J. Am. Chem. Soc.* **1944**, *66*, 1109.

[399]Frey, J.; Rappoport, Z. *J. Am. Chem. Soc.* **1996**, *118*, 3994.

3. Highly fluorinated enols, such as **145**.[400]

In this case, the enol form is not more stable than the keto form (**146**). The enol form is less stable, and converts to the keto form upon prolonged heating). It can, however, be kept at room temperature for long periods of time because the tautomerization reaction (**12-3**) is very slow, owing to the electron-withdrawing power of the fluorines.

Frequently, when the enol content is high, both forms can be isolated. The pure keto form of acetoacetic ester melts at $-39°C$, while the enol is a liquid even at $-78°C$. Each can be kept at room temperature for days if catalysts, such as acids or bases, are rigorously excluded.[401] Even the simplest enol, vinyl alcohol $CH_2=CHOH$, has been prepared in the gas phase at room temperature, where it has a half-life of ~30 min.[402] The enol $Me_2C=CCHOH$ is indefinitely stable in the solid state at $-78°C$ and has a half-life of ~24 h in the liquid state at $25°C$.[403] When both forms cannot be isolated, the extent of enolization is often measured by NMR.[404]

[400]For a review, see Bekker, R.A.; Knunyants, I.L. *Sov. Sci. Rev. Sect. B* **1984**, *5*, 145.

[401]For an example of particularly stable enol and keto forms, which could be kept in the solid state for more than a year without significant interconversion, see Schulenberg, J.W. *J. Am. Chem. Soc.* **1968**, *90*, 7008.

[402]Saito, S. *Chem. Phys. Lett.* **1976**, *42*, 399. See also, Capon, B.; Rycroft, D.S.; Watson, T.W.; Zucco, C. *J. Am. Chem. Soc.* **1981**, *103*, 1761; Holmes, J.L.; Lossing, F.P. *J. Am. Chem. Soc.* **1982**, *104*, 2648; McGarrity, J.F.; Cretton, A.; Pinkerton, A.A.; Schwarzenbach, D.; Flack, H.D. *Angew. Chem. Int. Ed.* **1983**, *22*, 405; Rodler, M.; Blom, C.E.; Bauder, A. *J. Am. Chem. Soc.* **1984**, *106*, 4029; Capon, B.; Guo, B.; Kwok, F.C.; Siddhanta, A.K.; Zucco, C. *Acc. Chem. Res.* **1988**, *21*, 135.

[403]Chin, C.S.; Lee, S.Y.; Park, J.; Kim, S. *J. Am. Chem. Soc.* **1988**, *110*, 8244.

[404]Cravero, R.M.; González-Sierra, M.; Olivieri, A.C. *J. Chem. Soc. Perkin Trans. 2* **1993**, 1067.

The extent of enolization[405] is greatly affected by solvent,[406] concentration, and temperature. Lactone enols, for example, have been shown to be stable in the gas phase, but unstable in solution.[407] Thus, acetoacetic ester has an enol content of 0.4% in water and 19.8% in toluene.[408] In this case, water reduces the enol concentration by hydrogen bonding with the carbonyl, making this group less available for internal hydrogen bonding. As an example of the effect of temperature, the enol content of pentan-2,4-dione, $CH_3COCH_2COCH_3$, was found to be 95, 68, and 44%, respectively, at 22, 180, and 275°C.[409] When a strong base is present, both the enol and the keto form can lose a proton. The resulting anion (the *enolate ion*) is the same in both cases. Since **147** and **148** differ only in placement of electrons, *they* are not tautomers, but canonical forms. The true structure of the enolate ion is a hybrid of **147** and **148** although **148** contributes more, since in this form the negative charge is on the more electronegative atom.

Other Proton-Shift Tautomerism

In all such cases, the anion resulting from removal of a proton from either tautomer is the same because of resonance. Some examples are:[410]

1. *Phenol–Keto Tautomerism.*[411]

Phenol   **Cyclohexadienone**

For most simple phenols, this equilibrium lies well to the side of the phenol, since only on that side is there aromaticity. For phenol itself, there is no evidence for the existence of the keto form.[412] However, the keto form

[405]For a review of keto–enol equilibrium constants, see Toullec, J. in Rappoport, Z. *The Chemistry of Enols*, Wiley, NY, *1990*, pp. 323–398.

[406]For an extensive study, see Mills, S.G.; Beak, P. *J. Org. Chem.* *1985*, *50*, 1216. For keto–enol tautomerism in aqueous alcohol solutions, see Blokzijl, W.; Engberts, J.B.F.N.; Blandamer, M.J. *J. Chem. Soc. Perkin Trans. 2* *1994*, 455; For theoretical calculations of keto–enol tautomerism in aqueous solutions, see Karelson, M.; Maran, U.; Katritzky, A.R. *Tetrahedron* *1996*, *52*, 11325.

[407]Tureček, F.; Vivekananda, S.; Sadílek, M.; Poláš ek, M. *J. Am. Chem. Soc,.* *2002*, *124*, 13282.

[408]Meyer, K.H. *Leibigs Ann. Chem.* *1911*, *380*, 212. See also, Moriyasu, M.; Kato, A.; Hashimoto, Y. *J. Chem. Soc. Perkin Trans. 2* *1986*, 515.

[409]Hush, N.S.; Livett, M.K.; Peel, J.B.; Willett, G.D. *Aust. J. Chem.* *1987*, *40*, 599.

[410]For a review of the use of X-ray crystallography to determine tautomeric forms, see Furmanova, N.G. *Russ. Chem. Rev.* *1981*, *50*, 775.

[411]For reviews, see Ershov, V.V.; Nikiforov, G.A. *Russ. Chem. Rev.* *1966*, *35*, 817; Forsén, S.; Nilsson, M., in Zabicky, J. *The Chemistry of the Carbonyl Group*, Vol. 2, Wiley, NY, *1970*, pp. 168–198.

[412]Keto forms of phenol and some simple derivatives have been generated as intermediates with very short lives, but long enough for spectra to be taken at 77 K. Lasne, M.; Ripoll, J.; Denis, J. *Tetrahedron Lett.* *1980*, *21*, 463. See also, Capponi, M.; Gut, I.; Wirz, J. *Angew. Chem. Int. Ed.* *1986*, *25*, 344.

becomes important and may predominate: (*1*) where certain groups, such as a second OH group or an N=O group, are present;[413] (*2*) in systems of fused aromatic rings;[414] (*3*) in heterocyclic systems. In many heterocyclic compounds in the liquid phase or in solution, the keto form is more stable,[415] although in the vapor phase the positions of many of these equilibria are reversed.[416] For example, in the equilibrium between 4-pyridone (**149**) and 4-hydroxypyridine (**150**), **149** is the only form detectable in ethanolic solution, while **150** predominates in the vapor phase.[416] In other heterocycles, the hydroxy-form predominates. 2-Hydroxypyridone (**151**) and pyridone-2-thiol (**153**)[417] are in equilibrium with their tautomers, 2-pyridone **152** and pyridine-2-thione **154**, respectively. In both cases, the most stable form is the hydroxy tautomer, **151** and **153**.[418]

2. *Nitroso–Oxime Tautomerism.*

The equiblirum shown for formaldehyde oxime and nitrosomethane illustrates this process.[419] In molecules where the products are stable, the equilibrium lies far to the right, and as a rule nitroso compounds are stable only when there is not a hydrogen.

[413]Ershov, V.V.; Nikiforov, G.A. *Russ. Chem. Rev.* **1966**, *35*, 817. See also, Highet, R.J.; Chou, F.E. *J. Am. Chem. Soc.* **1977**, *99*, 3538.

[414]See, for example, Majerski, Z.; Trinajstić, N. *Bull. Chem. Soc. Jpn.* **1970**, *43*, 2648.

[415]For a monograph on tautomerism in heterocyclic compounds, see Elguero, J.; Marzin, C.; Katritzky, A.R.; Linda, P. *The Tautomerism of Heterocycles*, Academic Press, NY, **1976**. For reviews, see Katritzky, A.R.; Karelson, M.; Harris, P.A. *Heterocycles* **1991**, *32*, 329; Beak, P. *Acc. Chem. Res.* **1977**, *10*, 186; Katritzky, A.R. *Chimia*, **1970**, *24*, 134.

[416]Beak, P.; Fry, Jr., F.S.; Lee, J.; Steele, F. *J. Am. Chem. Soc.* **1976**, *98*, 171.

[417]Moran, D.; Sukcharoenphon, K.; Puchta, R.; Schaefer III, H.F.; Schleyer, P.v.R.; Hoff, C.D. *J. Org. Chem.* **2002**, *67*, 9061.

[418]Parchment, O.G.; Burton, N.A.; Hillier, I.H.; Vincent, M.A. *J. Chem. Soc. Perkin Trans. 2* **1993**, 861.

[419]Long, J.A.; Harris, N.J.; Lammertsma, K. *J. Org. Chem.* **2001**, *66*, 6762.

3. *Aliphatic Nitro Compounds Are in Equilibrium with Aci Forms.*

Nitro form	Aci form

The nitro form is much more stable than the aci form in sharp contrast to the parallel case of nitroso–oxime tautomerism, undoubtedly because the nitro form has resonance not found in the nitroso case. Aci forms of nitro compounds are also called nitronic acids and azinic acids.

4. *Imine–Enamine Tautomerism.*[420]

$$R_2CH—CR=NR \quad \rightleftharpoons \quad R_2C=CR—NHR$$

Imine	Enamine

Enamines are normally stable only when there is no hydrogen on the nitrogen ($R_2C=CR–NR_2$). Otherwise, the imine form predominates.[421] The energy of various imine–enamine tautomers has been calculated.[422] In the case of 6-aminofulvene-1-aldimines, tautomerism was observed in the solid state, as well as in solution.[423]

5. *Ring-Chain Tautomerism.* Ring-chain tautomerism[424] occurs in sugars (aldehyde vs. the pyranose or furanose structures), and in γ-oxocarboxylic acids.[425] In benzamide carboxaldehyde, **156**, whose ring-chain tautomer is **155**, the equilibrium favors the cyclic form (**156**).[426] Similarly, benzoic acid 2-carboxyaldehyde (**157**) exists largely as the cyclic form (**158**).[427] In these latter cases, and in many others, this tautomerism influences chemical reactivity. Conversion of **157** to an ester, for example, is difficult since most standard methods lead to the OR derivative of **158** rather than the ester of **157**. Ring-chain tautomerism also occurs in spriooxathianes,[428] and in

[420]For reviews, see Shainyan, B.A.; Mirskova, A.N. *Russ. Chem. Rev.* **1979**, *48*, 107; Mamaev, V.P.; Lapachev, V.V. *Sov. Sci. Rev. Sect. B.* **1985**, *7*, 1. The second review also includes other closely related types of tautomerization.

[421]For examples of the isolation of primary and secondary enamines, see Shin, C.; Masaki, M.; Ohta, M. *Bull. Chem. Soc. Jpn.* **1971**, *44*, 1657; de Jeso, B.; Pommier, J. *J. Chem. Soc. Chem. Commun.* **1977**, 565.

[422]Lammertsma, K.; Prasad, B.V. *J. Am. Chem. Soc.* **1994**, *116*, 642.

[423]Sanz, D.; Perez-Torralba, M.; Alarcon, S.H.; Claramunt, R.M.; Foces-Foces, C.; Elguero, J. *J. Org. Chem.* **2002**, *67*, 1462.

[424]For a monograph, see Valters, R.E.; Flitsch, W. *Ring-Chain Tautomerism*, Plenum, NY, **1985**. For reviews, see Valters, R.E. *Russ. Chem. Rev.* **1973**, *42*, 464; **1974**, *43*, 665; Escale, R.; Verducci, J. *Bull. Soc. Chim. Fr.*, **1974**, 1203.

[425]Fabian, W.M.F.; Bowden, K. *Eur. J. Org. Chem.* **2001**, 303.

[426]Bowden, K.; Hiscocks, S.P.; Perjéssy, A. *J. Chem. Soc. Perkin Trans. 2* **1998**, 291.

[427]Ring chain tautomer of benzoic acid 2-carboxaldehdye.

[428]Terec, A.; Grosu, I.; Muntean, L.; Toupet, L.; Plé, G.; Socaci, C.; Mager, S. *Tetrahedron* **2001**, *57*, 8751; Muntean, L.; Grosu, I.; Mager, S.; Plé, G.; Balog, M. *Tetrahedron Lett.* **2000**, *41*, 1967.

decahydroquinazolines, such as **159** and **160**,[429] as well as other 1,3-hetero-cycles.[430]

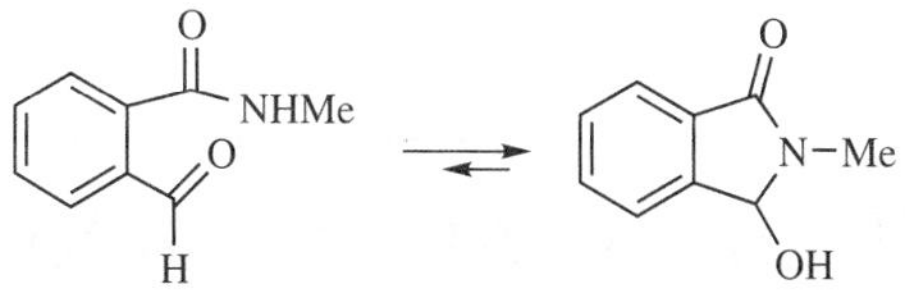

155   156

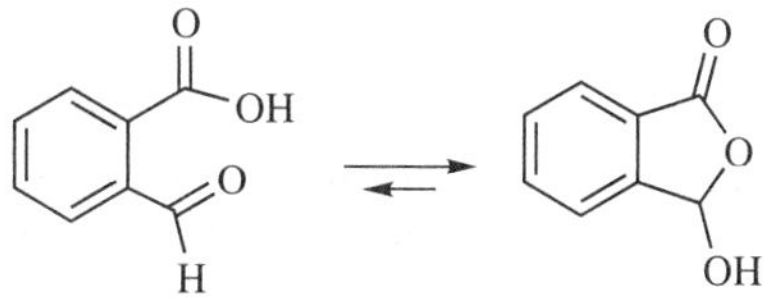

157   158

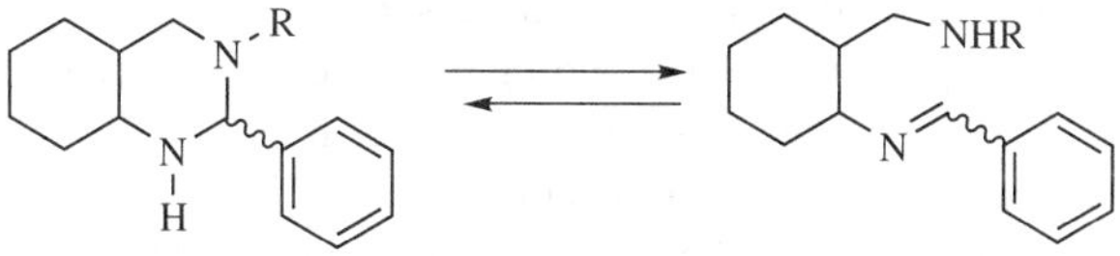

159   160

There are many other highly specialized cases of proton-shift tautomerism, including an internal Michael reaction (see **15-24**) in which 2-(2,2-dicyano-1-methylethenyl)benzoic acid (**161**) exists largely in the open chain form rather an its tautomer (**162**) in the solid state, but in solution there is an increasing amount of **162** as the solvent becomes more polar.[431]

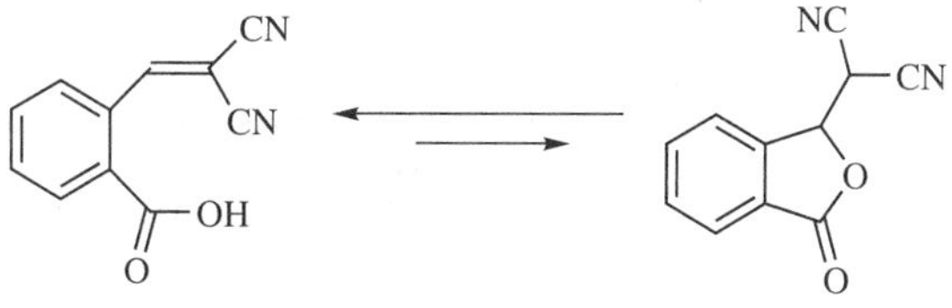

161   162

Valence Tautomerism

This type of tautomerism is discussed on p. 105.

[429]Lazar, L.; Goblyos, A.; Martinek, T.A.; Fulop, F. *J. Org. Chem.* **2002**, *67*, 4734.
[430]Lázár, L.; Fülöp, F. *Eur. J. Org. Chem.* **2003**, 3025.
[431]Kolsaker, P.; Arukwe, J.; Barcóczy, J.; Wiberg, A.; Fagerli, A.K. *Acta Chem. Scand. B* **1998**, *52*, 490.

██████ **CHAPTER 3**

Bonding Weaker than Covalent

In the first two chapters, we discussed the structure of molecules each of which is an aggregate of atoms in a distinct three-dimensional (3D) arrangement held together by bonds with energies on the order of $50–100\,\text{kcal mol}^{-1}$ ($200–400\,\text{kJ mol}^{-1}$). There are also very weak attractive forces *between* molecules, on the order of a few tenths of a kilocalorie per mole. These forces, called van der Waals forces, are caused by electrostatic attractions, such as those between dipole and dipole, induced dipole, and induced dipole, and are responsible for liquefaction of gases at sufficiently low temperatures. The bonding discussed in this chapter has energies of the order of $2–10\,\text{kcal mol}^{-1}$ ($9–40\,\text{kJ mol}^{-1}$), intermediate between the two extremes, and produces clusters of molecules. We will also discuss compounds in which portions of molecules are held together without any attractive forces at all.

HYDROGEN BONDING

A *hydrogen bond* is a bond between a functional group A—H and an atom or group of atoms B in the same or a different molecule.[1] With exceptions to be noted later, hydrogen bonds are *assumed to form only when A is oxygen, nitrogen, or fluorine and when B is oxygen, nitrogen, or fluorine.*[2] The oxygen may be singly or doubly

---

[1] For a treatise, see Schuster, P.; Zundel, G.; Sandorfy, C. *The Hydrogen Bond*, 3 vols., North-Holland Publishing Co.: Amsterdam, The Netherlands, *1976*. For a monograph, see Joesten, M.D.; Schaad, L.J. *Hydrogen Bonding*; Marcel Dekker, NY, *1974*. For reviews, see Meot-Ner, M. *Mol. Struct. Energ. 1987*, *4*, 71; Deakyne, C.A. *Mol. Struct. Energ. 1987*, *4*, 105; Joesten, M.D. *J. Chem. Educ. 1982*, *59*, 362; Gur'yanova, E.N.; Gol'dshtein, I.P.; Perepelkova, T.I. *Russ. Chem. Rev. 1976*, *45*, 792; Pimentel, G.C.; McClellan, A.L. *Annu. Rev. Phys. Chem. 1971*, *22*, 347; Kollman, P.A.; Allen, L.C. *Chem. Rev. 1972*, *72*, 283; Huggins, M.L. *Angew. Chem. Int. Ed. 1971*, *10*, 147; Rochester, C.H., in Patai, S. The Chemistry of the Hydroxyl Group, pt. 1; Wiley, NY, *1971*, pp. 327–392, 328–369. See also Hamilton, W.C.; Ibers, J.A. *Hydrogen Bonding in Solids*, W.A. Benjamin, NY, *1968*. Also see, Chen, J.; McAllister, M.A.; Lee, J.K.; Houk, K.N. *J. Org. Chem. 1998*, *63*, 4611 for a discussion of short, strong hydrogen bonds.

[2] The ability of functional groups to act as hydrogen bond acids and bases can be obtained from either equilibrium constants for 1:1 hydrogen bonding or overall hydrogen bond constants. See Abraham, M.H.; Platts, J.A. *J. Org. Chem. 2001*, *66*, 3484.

---

bonded and the nitrogen singly, doubly, or triply bonded. The bonds are usually represented by dotted or dashed lines, as shown in the following examples:

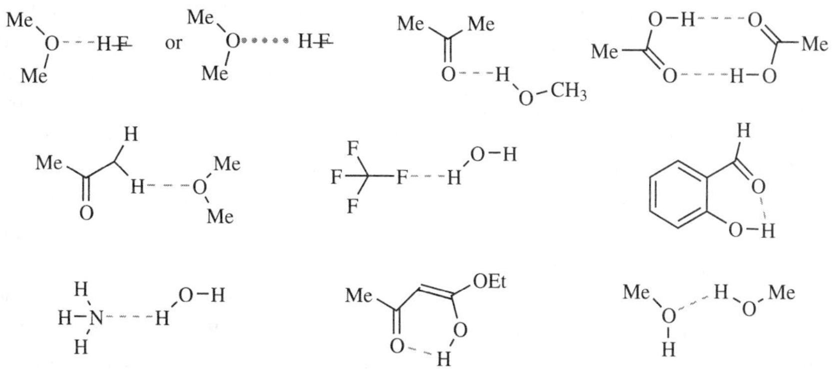

Hydrogen bonds can exist in the solid[3] and liquid phases and in solution.[4] Many organic reactions that will be discussed in later chapters can be done in aqueous media,[5] and their efficacy is due, in part, to the hydrogen bonding nature of aqueous media.[6] Even in the gas phase, compounds that form particularly strong hydrogen bonds may remain associated.[7] Acetic acid, for example, exists in the gas phase as a dimer, as shown above, except at very low pressures.[8] In solution and in the liquid phase, hydrogen bonds rapidly form and break. The mean lifetime of the $NH_3 \cdots H_2O$ bond is 2×10^{-12} s.[9] Except for a few very strong hydrogen bonds,[10] such as the $FH \cdots F^-$ bond (which has an energy of ~ 50 kcal mol^{-1} or 210 kJ mol^{-1}), the strongest hydrogen bonds are the $FH \cdots F$ bond and the bonds connecting one carboxylic acid with another. The energies of these bonds are in the range of 6–8 kcal mol^{-1} or 25–30 kJ mol^{-1} (for carboxylic acids, this refers to the energy of each bond). In general, short contact hydrogen bonds between fluorine and HO or NH are rare.[11] Other $OH \cdots O$ and $NH \cdots N$ bonds[12] have energies of 3–6 kcal mol^{-1} (12–25 kJ mol^{-1}).

[3]Steiner, T. *Angew. Chem. Int. Ed.* **2002**, *41*, 48. See also Damodharan, L.; Pattabhi, V. *Tetrahedron Lett.* **2004**, *45*, 9427.

[4]See Nakahara, M.; Wakai, C. *Chem. Lett.* **1992**, 809 for a discussion of monomeric and cluster states of water molecules in organic solvents due to hydrogen bonding.

[5]Li, C.-J.; Chen, T.-H. *Organic Reactions in Aqueous Media*, Wiley, NY, **1997**.

[6]Li, C.-J. *Chem. Rev.* **1993**, *93*, 2023.

[7]For a review of energies of hydrogen bonds in the gas phase, see Curtiss, L.A.; Blander, M. *Chem. Rev.* **1988**, *88*, 827.

[8]For a review of hydrogen bonding in carboxylic acids and acid derivatives, see Hadži, D.; Detoni, S., in Patai, S. *The Chemistry of Acid Derivatives*, pt. 1, Wiley, NY, **1979**, pp. 213–266.

[9]Emerson, M.T.; Grunwald, E.; Kaplan, M.L.; Kromhout, R.A. *J. Am. Chem. Soc.* **1960**, *82*, 6307.

[10]For a review of very strong hydrogen bonding, see Emsley, J. *Chem. Soc. Rev.* **1980**, *9*, 91.

[11]Howard, J.A.K.; Hoy, V.J.; O'Hagan, D.; Smith, G.T. *Tetrahedron* **1996**, *52*, 12613.

[12]For an *ab initio* study of diamine hydrogen bonds see Sorensen, J.B.; Lewin, A.H.; Bowen, J.P. *J. Org. Chem.* **2001**, *66*, 4105.

The intramolecular O–H•••N hydrogen bond in hydroxy amines is also rather strong.[13]

To a first approximation, the strength of hydrogen bonds increases with increasing acidity of A—H and basicity of B, but the parallel is far from exact.[14] A quantitative measure of the strengths of hydrogen bonds has been established, involving the use of an α scale to represent hydrogen-bond donor acidities and a β scale for hydrogen-bond acceptor basicities.[15] The use of the β scale, along with another parameter, ξ, allows hydrogen-bond basicities to be related to proton-transfer basicities (pK values).[16] A database has been developed to locate all possible occurrences of bimolecular cyclic hydrogen-bond motifs in the Cambridge Structural Database,[17] and donor–acceptor as well as polarity parameters have been calculated for hydrogen-bonding solvents.[18]

When two compounds whose molecules form hydrogen bonds with each other are both dissolved in water, the hydrogen bond between the two molecules is usually greatly weakened or completely removed,[19] because the molecules generally form hydrogen bonds with the water molecules rather than with each other, especially since the water molecules are present in such great numbers. In amides, the oxygen atom is the preferred site of protonation or complexation with water.[20] In the case of dicarboxylic acids, arguments have been presented that there is little or no evidence for strong hydrogen bonding in aqueous solution,[21] although recent studies concluded that strong, intramolecular hydrogen bonding can exist in aqueous acetone solutions (0.31 mole-fraction water) of hydrogen maleate and hydrogen cis-cyclohexane-1,2-dicarboxylate.[22]

Many studies have been made of the geometry of hydrogen bonds,[23] and the evidence shows that in most (though not all) cases, the hydrogen is on or near the

[13]Grech, E.; Nowicka-Scheibe, J.; Olejnik, Z.; Lis, T.; Pawêka, Z.; Malarski, Z.; Sobczyk, L. J. Chem. Soc., Perkin Trans. 2 **1996**, 343. See Steiner, T. J. Chem. Soc., Perkin Trans. 2 **1995**, 1315 for a discussion of hydrogen bonding in the crystal structure of α-amino acids.

[14]For reviews of the relationship between hydrogen-bond strength and acid-base properties, see Pogorelyi, V.K.; Vishnyakova, T.B. Russ. Chem. Rev. **1984**, 53, 1154; Epshtein, L.M. Russ. Chem. Rev. **1979**, 48, 854.

[15]For reviews, see Abraham, M.H.; Doherty, R.M.; Kamlet, M.J.; Taft, R.W. Chem. Br. **1986**, 551; Kamlet, M.J.; Abboud, J.M.; Taft, R.W. Prog. Phys. Org. Chem. **1981**, 13, 485. For a comprehensive table and α and β values, see Kamlet, M.J.; Abboud, J.M.; Abraham, M.H.; Taft, R.W. J. Org. Chem. **1983**, 48, 2877. For a criticism of the β scale, see Laurence, C.; Nicolet, P.; Helbert, M. J. Chem. Soc., Perkin Trans. 2 **1986**, 1081. See also Nicolet, P.; Laurence, C.; Luçon, M. J. Chem. Soc., Perkin Trans. 2 **1987**, 483; Abboud, J.M.; Roussel, C.; Gentric, E.; Sraidi, K.; Lauransan, J.; Guihéneuf, G.; Kamlet, M.J.; Taft, R.W. J. Org. Chem. **1988**, 53, 1545; Abraham, M.H.; Grellier, P.L.; Prior, D.V.; Morris, J.J.; Taylor, P.J. J. Chem. Soc., Perkin Trans. 2 **1990**, 521.

[16]Kamlet, M.J.; Gal, J.; Maria, P.; Taft, R.W. J. Chem. Soc., Perkin Trans. 2 **1985**, 1583.

[17]Allen, F.H.; Raithby, P.R.; Shields, G.P.; Taylor, R. Chem. Commun. **1998**, 1043.

[18]Joerg, S.; Drago, R.S.; Adams, J. J. Chem. Soc., Perkin Trans. 2 **1997**, 2431.

[19]Stahl, N.; Jencks, W.P. J. Am. Chem. Soc. **1986**, 108, 4196.

[20]Scheiner, S.; Wang, L. J. Am. Chem. Soc. **1993**, 115, 1958.

[21]Perrin, C.L. Annu. Rev. Phys. Org. Chem. **1997**, 48, 511.

[22]Lin, J.; Frey, P.A. J. Am. Chem. Soc. **2000**, 122, 11258.

[23]For reviews, see Etter, M.C. Acc. Chem. Res. **1990**, 23, 120; Taylor, R.; Kennard, O. Acc. Chem. Res. **1984**, 17, 320.

straight line formed by A and B.[24] This is true both in the solid state (where X-ray crystallography and neutron diffraction have been used to determine structures),[25] and in solution.[26] It is significant that the vast majority of intramolecular hydrogen bonding occurs where *six-membered rings* (counting the hydrogen as one of the six) can be formed, in which linearity of the hydrogen bond is geometrically favorable, while five-membered rings, where linearity is usually not favored (though it is known), are much rarer. A novel nine-membered intramolecular hydrogen bond has been reported.[27]

In certain cases, X-ray crystallography has shown that a single H–A can form simultaneous hydrogen bonds with two B atoms (*bifurcated* or *three-center hydrogen bonds*). An example is an adduct (**1**) formed from pentane-2,4-dione (in its enol form; see p. 98) and diethylamine, in which the O–H hydrogen simultaneously bonds[28] to an O and an N (the N–H hydrogen forms a hydrogen bond with the O of another pentane-2,4-dione molecule).[29] On the other hand, in the adduct (**2**) formed from 1,8-biphenylenediol and hexamethylphosphoramide (HMPA), the B atom (in this case oxygen) forms simultaneous hydrogen bonds with two A•••H hydrogens.[30] Another such case is found in methyl hydrazine carboxylate **3**.[31] Except for the special case of FH•••F⁻ bonds (see p. 107), the hydrogen is not equidistant between A and B. For example, in ice the O–H distance is 0.97 Å, while the H•••O distance is 1.79 Å.[32] A theoretical study of the vinyl alcohol–vinyl alcoholate system concluded the hydrogen bonding is strong, but asymmetric.[33] The hydrogen bond in the enol of malonaldehyde, in organic solvents, is asymmetric with the hydrogen atom closer to the basic oxygen atom.[34] There is recent evidence, however, that symmetrical hydrogen bonds to carboxylates should be regarded as two-center rather than three-center hydrogen bonds, since the criteria traditionally used to infer three-center hydrogen bonding are inadequate for carboxylates.[35] There is

[24]See Stewart, R. *The Proton: Applications to Organic Chemistry*; Academic Press, NY, *1985*, pp. 148–153.

[25]A statisical analysis of X-ray crystallographic data has shown that most hydrogen bonds in crystals are nonlinear by ~10–15°: Kroon, J.; Kanters, J.A.; van Duijneveldt-van de Rijdt, J.G.C.M.; van Duijneveldt, F.B.; Vliegenthart, J.A. *J. Mol. Struct.* *1975*, *24*, 109. See also, Ceccarelli, C.; Jeffrey, G.A.; Taylor, R. *J. Mol. Struct.* *1981*, *70*, 255; Taylor, R.; Kennard, O.; Versichel, W. *J. Am. Chem. Soc.* *1983*, *105*, 5761; *1984*, *106*, 244.

[26]For reviews of a different aspect of hydrogen-bond geometry: the angle between A•••H•••B and the rest of the molecule, see Legon, A.C.; Millen, D.J. *Chem. Soc. Rev.* *1987*, *16*, 467, *Acc. Chem. Res.* *1987*, *20*, 39.

[27]Yoshimi, Y.; Maeda, H.; Sugimoto, A.; Mizuno, K. *Tetrahedron Lett.* *2001*, *42*, 2341.

[28]Emsley, J.; Freeman, N.J.; Parker, R.J.; Dawes, H.M.; Hursthouse, M.B. *J. Chem. Soc., Perkin Trans. 1* *1986*, 471.

[29]For some other three-center hydrogen bonds, see Taylor, R.; Kennard, O.; Versichel, W. *J. Am. Chem. Soc.* *1984*, *106*, 244; Jeffrey, G.A.; Mitra, J. *J. Am. Chem. Soc.* *1984*, *106*, 5546; Staab, H.A.; Elbl, K.; Krieger, C. *Tetrahedron Lett.* *1986*, *27*, 5719.

[30]Hine, J.; Hahn, S.; Miles, D.E. *J. Org. Chem.* *1986*, *51*, 577.

[31]Caminati, W.; Fantoni, A.C.; Schäfer, L.; Siam, K.; Van Alsenoy, C. *J. Am. Chem. Soc.* *1986*, *108*, 4364.

[32]Pimentel, G.C.; McClellan, A.L. *The Hydrogen Bond*; W.H. Freeman: San Francisco, *1960*, p. 260.

[33]Chandra, A.K.; Zeegers-Huyskens, T., *J. Org. Chem.* *2003*, *68*, 3618.

[34]Perrin, C.L.; Kim, Y.-J. *J. Am. Chem. Soc.* *1998*, *120*, 12641.

[35]Görbitz, C.H.; Etter, M.C. *J. Chem. Soc., Perkin Trans. 2* *1992*, 131.

also an example of cooperative hydrogen bonding (O–H•••C≡C–H•••Ph) in crystalline 2-ethynyl-6,8-diphenyl-7H-benzocyclohepten-7-ol (**4**).[36]

Hydrogen bonding has been detected in many ways, including measurements of dipole moments, solubility behavior, freezing-point lowering, and heats of mixing, but one important way is by the effect of the hydrogen bond on IR.[37] The IR frequencies of groups, such as O–H or C=O, are shifted when the group is hydrogen bonded. Hydrogen bonding always moves the peak toward lower frequencies, for both the A—H and the B groups, though the shift is greater for the former. For example, a free OH group of an alcohol or phenol absorbs at $\sim$3590–3650 cm^{-1}, while a hydrogen-bonded OH group is found $\sim$50–100 cm^{-1} lower.[38] In many cases, in dilute solution, there is partial hydrogen bonding, that is, some OH groups are free and some are hydrogen bonded. In such cases, two peaks appear. Infrared spectroscopy can also distinguish between inter- and intramolecular hydrogen bonding, since intermolecular peaks are intensified by an increase in concentration while intramolecular peaks are unaffected. Other types of spectra that have been used for the detection of hydrogen bonding include Raman, electronic,[39] and NMR.[40] Since hydrogen bonding involves a rapid movement of protons from one atom to another, nmr records an average value. Hydrogen bonding can be detected because it usually produces a chemical shift to a lower field. For example, carboxylic acid–carboxylate systems arising from either mono- or diacids generally exhibit a downfield resonance (16–22 ppm), which indicates "strong" hydrogen bonding

---

[36]Steiner, T.; Tamm, M.; Lutz, B.; van der Maas, J. *Chem. Commun.* **1996**, 1127.

[37]For reviews of the use of ir spectra to detect hydrogen bonding, see Symons, M.C.R. *Chem. Soc. Rev.* **1983**, *12*, 1; Egorochkin, A.N.; Skobeleva, S.E. *Russ. Chem. Rev.* **1979**, *48*, 1198; Tichy, M. *Adv. Org. Chem.* **1965**, *5*, 115; Ratajczak, H.; Orville-Thomas, W.J. *J. Mol. Struct.* **1968**, *1*, 449. For a review of studies by ir of the shapes of intramolecular hydrogen-bonded compounds, see Aaron, H.S. *Top. Stereochem.* **1979**, *11*, 1. For a review of the use of rotational spectra to study hydrogen bonding, see Legon, A.C. *Chem. Soc. Rev.* **1990**, *19*, 197.

[38]Tichy, M. *Adv. Org. Chem.* **1965**, *5*, 115 contains a lengthy table of free and intramolecularly hydrogen-bonding peaks.

[39]For a discussion of the effect of hydrogen bonding on electronic spectra, see Lees, W.A.; Burawoy, A. *Tetrahedron* **1963**, *19*, 419.

[40]For a review of the use of nmr to detect hydrogen bonding, see Davis, Jr., J.C.; Deb, K.K. *Adv. Magn. Reson.* **1970**, *4*, 201. Also see, Kumar, G.A.; McAllister, M.A. *J. Org. Chem.* **1998**, *63*, 6968, which shows the relationship between ^{1}H NMR chemical shift and hydrogen bond strength.

in anhydrous, aprotic solvents.[41] Hydrogen bonding changes with temperature and concentration, and comparison of spectra taken under different conditions also serves to detect and measure it. As with IR spectra, intramolecular hydrogen bonding can be distinguished from intermolecular by its constancy when the concentration is varied. The spin–spin coupling constant across a hydrogen bond, obtained by NMR studies, has been shown to provide a "fingerprint" for hydrogen-bond type.[42]

Hydrogen bonds are important because of the effects they have on the properties of compounds, among them:

1. Intermolecular hydrogen bonding raises boiling points and frequently melting points.

2. If hydrogen bonding is possible between solute and solvent, this greatly increases solubility and often results in large or even infinite solubility where none would otherwise be expected.

3. Hydrogen bonding causes lack of ideality in gas and solution laws.

4. As previously mentioned, hydrogen bonding changes spectral absorption positions.

5. Hydrogen bonding, especially the intramolecular variety, changes many chemical properties. For example, it is responsible for the large amount of enol present in certain tautomeric equilibria (see p. 98). Also, by influencing the conformation of molecules (see Chapter 4), it often plays a significant role in determining reaction rates.[43] Hydrogen bonding is also important in maintaining the 3D structures of protein and nucleic acid molecules.

Besides oxygen, nitrogen, and fluorine, there is evidence that weaker hydrogen bonding exists in other systems.[44] Although many searches have been made for hydrogen bonding where A is carbon,[45] only three types of C–H bonds have been found that are acidic enough to form weak hydrogen bonds.[46] These are found in terminal alkynes, $RC{\equiv}CH$,[47] chloroform and some other halogenated alkanes, and HCN. Sterically unhindered C–H groups ($CHCl_3$, CH_2Cl_2, $RC{\equiv}CH$) form short contact hydrogen bonds with carbonyl acceptors, where there is a significant preference for coordination with the conventional carbonyl lone-pair direction.[48]

---

[41]Bruck, A.; McCoy, L.L.; Kilway, K.V. *Org. Lett.* **2000**, *2*, 2007.

[42]Del Bene, J.E.; Perera, S.A.; Bartlett, R.J. *J. Am. Chem. Soc.* **2000**, *122*, 3560.

[43]For reviews of the effect of hydrogen bonding on reactivity, see Hibbert, F.; Emsley, J. *Adv. Phys. Org. Chem.* **1990**, *26*, 255; Sadekov, I.D.; Minkin, V.I.; Lutskii, A.E. *Russ. Chem. Rev.* **1970**, *39*, 179.

[44]For a review, see Pogorelyi, V.K. *Russ. Chem. Rev.* **1977**, *46*, 316.

[45]For a monograph on this subject, see Green, R.D. *Hydrogen Bonding by C–H Groups*; Wiley, NY, **1974**. See also Taylor, R.; Kennard, O. *J. Am. Chem. Soc.* **1982**, *104*, 5063; Harlow, R.L.; Li, C.; Sammes, M.P. *J. Chem. Soc., Perkin Trans. 1* **1984**, 547; Nakai, Y.; Inoue, K.; Yamamoto, G.; Ō ki, M. *Bull. Chem. Soc. Jpn.* **1989**, *62*, 2923; Seiler, P.; Dunitz, J.D. *Helv. Chim. Acta* **1989**, *72*, 1125.

[46]For a theoretical study of weak hydrogen-bonds, see Calhorda, M.J. *Chem. Commun.* **2000**, 801.

[47]For a review, see Hopkinson, A.C., in Patai, S. *The Chemistry of the Carbon–Carbon Triple Bond*, pt. 1, Wiley, NY, **1978**, pp. 75–136. See also DeLaat, A.M.; Ault, B.S. *J. Am. Chem. Soc.* **1987**, *109*, 4232.

[48]Streiner, T.; Kanters, J.A.; Kroon, J. *Chem. Commun.* **1996**, 1277.

Weak hydrogen bonds are formed by compounds containing S–H bonds.[49] There has been much speculation regarding other possibilities for B. There is evidence that Cl can form weak hydrogen bonds,[50] but Br and I form very weak bonds if at all.[51] However, the *ions* Cl^-, Br^-, and I^- form hydrogen bonds that are much stronger than those of the covalently bonded atoms.[52] As we have already seen, the FH•••F$^-$ bond is especially strong. In this case, the hydrogen is equidistant from the fluorines.[53] Similarly, a sulfur atom[49] can be the B component in weak hydrogen bonds,[54] but the $^-$SH ion forms much stronger bonds.[55] There are theoretical studies of weak hydrogen bonding.[56] Hydrogen bonding has been directly observed (by NMR and IR) between a negatively charged carbon (see Carbanions, Chapter 5) and an OH group in the same molecule.[57] Another type of molecule in which carbon is the B component are isocyanides, $R–^+N{\equiv}C^-$ which form rather strong hydrogen bonds.[58] There is evidence that double and triple bonds, aromatic rings,[59] and even cyclopropane rings[60] may be the B component of hydrogen bonds, but these bonds are very weak. An interesting case is that of the *in*-bicyclo[4.4.4]-1-tetradecyl cation **5** (see in–out isomerism, p. 189). The NMR and IR spectra show that the actual structure of this ion is **6**, in which both the A and the B component of the hydrogen bond is a carbon.[61] These are sometimes

[49]For reviews of hydrogen bonding in sulfur-containing compounds, see Zuika, I.V.; Bankovskii, Yu.A. *Russ. Chem. Rev.* **1973**, *42*, 22; Crampton, M.R., in Patai, S. *The Chemistry of the Thiol Group*, pt. 1; Wiley, NY, **1974**, pp. 379–396; Pogorelyi, V.K. *Russ. Chem. Rev.* **1977**, *46*, 316.

[50]For a review of hydrogen bonding to halogens, see Smith, J.W., in Patai, S. *The Chemistry of the Carbon-Halogen Bond*, pt. 1; Wiley, NY, **1973**, pp. 265–300. See also, Bastiansen, O.; Fernholt, L.; Hedberg, K.; Seip, R. *J. Am. Chem. Soc.* **1985**, *107*, 7836.

[51]West, R.; Powell, D.L.; Whatley, L.S.; Lee, M.K.T.; Schleyer, P.v.R. *J. Am. Chem. Soc.* **1962**, *84*, 3221; Fujimoto, E.; Takeoka, Y.; Kozima, K. *Bull. Chem. Soc. Jpn.* **1970**, *43*, 991; Azrak, R.G.; Wilson, E.B. *J. Chem. Phys.* **1970**, *52*, 5299.

[52]Allerhand, A.; Schleyer, P.v.R. *J. Am. Chem. Soc.* **1963**, *85*, 1233; McDaniel, D.H.; Vallée, R.E. *Inorg. Chem.* **1963**, *2*, 996; Fujiwara, F.Y.; Martin, J.S. *J. Am. Chem. Soc.* **1974**, *96*, 7625; French, M.A.; Ikuta, S.; Kebarle, P. *Can. J. Chem.* **1982**, *60*, 1907.

[53]A few exceptions have been found, where the presence of an unsymmetrical cation causes the hydrogen to be closer to one fluorine than to the other: Williams, J.M.; Schneemeyer, L.F. *J. Am. Chem. Soc.* **1973**, *95*, 5780.

[54]Vogel, G.C.; Drago, R.S. *J. Am. Chem. Soc.* **1970**, *92*, 5347; Mukherjee, S.; Palit, S.R.; De, S.K. *J. Phys. Chem.* **1970**, *74*, 1389; Schaefer, T.; McKinnon, D.M.; Sebastian, R.; Peeling, J.; Penner, G.H.; Veregin, R.P. *Can. J. Chem.* **1987**, *65*, 908; Marstokk, K.; Møllendal, H.; Uggerrud, E. *Acta Chem. Scand.* **1989**, *43*, 26.

[55]McDaniel, D.H.; Evans, W.G. *Inorg. Chem.* **1966**, *5*, 2180; Sabin, J.R. *J. Chem. Phys.* **1971**, *54*, 4675.

[56]Calhorda, M.J. *Chem. Commun.* **2000**, 801.

[57]Ahlberg, P.; Davidsson, O.; Johnsson, B.; McEwen, I.; Rönnqvist, M. *Bull. Soc. Chim. Fr.* **1988**, 177.

[58]Ferstandig, L.L. *J. Am. Chem. Soc.* **1962**, *84*, 3553; Allerhand, A.; Schleyer, P.v.R. *J. Am. Chem. Soc.* **1963**, *85*, 866.

[59]For example, see Bakke, J.M.; Chadwick, D.J. *Acta Chem. Scand. Ser. B* **1988**, *42*, 223: Atwood, J.L.; Hamada, F.; Robinson, K.D.; Orr, G.W.; Vincent, R.L. *Nature (London)* **1991**, *349*, 683.

[60]Joris, L.; Schleyer, P.v.R.; Gleiter, R. *J. Am. Chem. Soc.* **1968**, *90*, 327; Yoshida, Z.; Ishibe, N.; Kusumoto, H. *J. Am. Chem. Soc.* **1969**, *91*, 2279.

[61]McMurry, J.E.; Lectka, T.; Hodge, C.N. *J. Am. Chem. Soc.* **1989**, *111*, 8867. See also, Sorensen, T.S.; Whitworth, S.M. *J. Am. Chem. Soc.* **1990**, *112*, 8135.

called 3-center–2-electron C–H–C bonds.[62] A technique called generalized population analysis has been developed to study this type of multicenter bonding.[63]

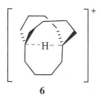

A weak ($\sim$1.5 kcal mol^{-1}) and rare C–H$\cdots$O=C hydrogen bond has been reported in a class of compounds known as a [6]semirubin (a dipyrrinone).[64] There is also evidence for a C–H$\cdots$N/CH$\cdots$OH bond in the crystal structures of α,β-unsaturated ketones carrying a terminal pyridine subunit,[65] and for R$_3$N$^+$–C–H$\cdots$O=C hydrogen bonding.[66]

Deuterium also forms hydrogen bonds; in some systems these seem to be stronger than the corresponding hydrogen bonds; in others, weaker.[67]

Weak hydrogen bonds can be formed between an appropriate hydrogen and a π bond, both with alkenes and with aromatic compounds. For example, IR data in dilute dichloromethane suggests that the predominant conformation for bis (amide) **7** contains an N–H$\cdots\pi$ hydrogen bond involving the C=C unit.[68] The strength of an intramolecular π-facial hydrogen bond between an NH group and an aromatic ring in chloroform has been estimated to have a lower limit of -4.5 ± 0.5 kcal mol^{-1}(-18.8 kJ mol^{-1}).[69] A neutron diffraction study of crystalline 2-ethynyladamantan-2-ol (**8**) shows the presence of an unusual O–H$\cdots\pi$

[62]McMurry, J.E.; Lectka, T. *Acc. Chem. Res.* **1992**, *25*, 47.

[63]Ponec, R.; Yuzhakov, G.; Tantillo, D.J. *J. Org. Chem.* **2004**, *69*, 2992.

[64]Huggins, M.T.; Lightner, D.A. *J. Org. Chem.* **2001**, *66*, 8402.

[65]Mazik, M.; Bläser, D.; Boese, R. *Tetrahedron 2001*, *57*, 5791.

[66]Cannizzaro, C.E.; Houk, K.N. *J. Am. Chem. Soc.* **2002**, *124*, 7163.

[67]Dahlgren Jr., G.; Long, F.A. *J. Am. Chem. Soc.* **1960**, *82*, 1303; Creswell, C.J.; Allred, A.L. *J. Am. Chem. Soc. 1962*, *84*, 3966; Singh, S.; Rao, C.N.R. *Can. J. Chem. 1966*, *44*, 2611; Cummings, D.L.; Wood, J.L. *J. Mol. Struct. 1974*, *23*, 103.

[68]Gallo, E.A.; Gelman, S.H. *Tetrahedron Lett.* **1992**, *33*, 7485.

[69]Adams, H.; Harris, K.D.M.; Hembury, G.A.; Hunter, C.A.; Livingstone, D.; McCabe, J.F. *Chem. Commun. 1996*, 2531. See Steiner, T.; Starikov, E.B.; Tamm, M. *J. Chem. Soc., Perkin Trans. 2 1996*, 67 for a related example with 5-ethynyl-5*H*-dibenzo[*a,d*]cyclohepten-5-ol.

hydrogen bond, which is short and linear, as well as the more common O–H•••O and C–H•••O hydrogen bonds.[70]

π–π INTERACTIONS

The π–π interactions are fundamental to many supramolecular organization and recognition processes.[71] There are many theoretical and experimental studies that clearly show the importance of π–π interactions.[72] Perhaps the simplest prototype of aromatic π–π interactions is the benzene dimer.[73] Within dimeric aryl systems such as this, possible π–π interactions are the sandwich and T-shaped interactions shown. It has been shown that all substituted sandwich dimers bind more strongly than benzene dimer, whereas the T-shaped configurations bind more or less favorably depending on the substituent.[74] Electrostatic, dispersion, induction, and exchange-repulsion contributions are all significant to the overall binding energies.[74]

Sandwich

T-shaped (1)

T-shaped (2)

The π-electrons of aromatic rings can interact with charged species, yielding strong cation–π interactions dominated by electrostatic and polarization effects.[75] Interactions with CH units is also possible. For CH–π interactions in both alkyl- and aryl-based model systems, dispersion effects dominate the interaction, but the electrostatics term is also relevant for aryl CH–π interactions.[76]

[70]Allen, F.H.; Howard, J.A.K.; Hoy, V.J.; Desiraju, G.R.; Reddy, D.S.; Wilson, C.C. *J. Am. Chem. Soc.* **1996**, *118*, 4081.

[71]Meyer, E.A.; Castellano, R.K.; Diederich, F. *Angew. Chem. Int. Ed.* **2003**, *42*, 1210.

[72]Tsuzuki, T.; Uchimaru, T.; Tanabe, K. *J. Mol. Struct. (THEOCHEM)* **1994**, *307*, 107; Hobza, P.; Selzle, H.L.; Schlag, E.W. *J. Phys. Chem.* **1996**, *100*, 18790; Tsuzuki, S.; Lüthi, H.P. *J. Chem. Phys.* **2001**, *114*, 3949; Steed, J.M.; Dixon, T.A.; Klemperer, W. *J. Chem. Phys.* **1979**, *70*, 4940.; Arunan, E.; Gutowsky, H.S. *J. Chem. Phys.* **1993**, *98*, 4294; Law, K.S.; Schauer, M.; Bernstein, E.R. *J. Chem. Phys.* **1984**, *81*, 4871; Felker, P.M.; Maxton, P.M.; Schaeffer, M.W. *Chem. Rev.* **1994**, *94*, 1787; Venturo, V.A.; Felker, P.M. *J. Chem. Phys.* **1993**, *99*, 748; Tsuzuki, S.; Honda, K.; Uchimaru, T.; Mikami, M.; Tanabe, K. *J. Am. Chem. Soc.* **2002**, *124*, 104; Hobza, P.; Jurečka, P. *J. Am. Chem. Soc.* **2003**, *125*, 15608.

[73]Sinnokrot, M.O.; Valeev, E.F.; Sherrill, C.D. *J. Am. Chem. Soc.* **2002**, *124*, 10887.

[74]Sinnokrot, M.O.; Sherrill, C.D. *J. Am. Chem. Soc.* **2004**, *126*, 7690

[75]Lindeman, S.V.; Kosynkin, D.; Kochi, J.K. *J. Am. Chem. Soc.* **1998**, *120*, 13268; Ma, J.C.; Dougherty, D.A. *Chem. Rev.* **1997**, *97*, 1303; Dougherty, D.A. *Science* **1996**, *271*, 163; Cubero, E.; Luque, F.J.; Orozco, M. *Proc. Natl. Acad. Sci. U.S.A.* **1998**, *95*, 5976.

[76]Ribas, J.; Cubero, E.; Luque, F. J.; Orozco, M. *J. Org. Chem.* **2002**, *67*, 7057.

Detection of π–π interactions has largely relied on NMR-based techniques, such as chemical shifts variations,[77] and Nuclear Overhauser Effect Spectroscopy (NOESY) or Rotating-Frame NOE Spectroscopy (ROESY).[78] Diffusion-ordered NMR spectroscopy (DOSY) has also been used to detect π–π stacked complexes.[79]

ADDITION COMPOUNDS

When the reaction of two compounds results in a product that contains all the mass of the two compounds, the product is called an *addition compound*. There are several kinds. In the rest of this chapter, we will discuss addition compounds in which the molecules of the starting materials remain more or less intact and weak bonds hold two or more molecules together. We can divide them into four broad classes: electron donor–acceptor complexes, complexes formed by crown ethers and similar compounds, inclusion compounds, and catenanes.

Electron Donor–Acceptor (EDA) Complexes[80]

In *EDA complexes*,[81] there is always a donor and an acceptor molecule. The donor may donate an unshared pair (an *n* donor) or a pair of electrons in a π orbital of a double bond or aromatic system (a π donor). One test for the presence of an EDA complex is the electronic spectrum. These complexes generally exhibit a spectrum (called a *charge-transfer spectrum*) that is not the same as the sum of the spectra of the two individual molecules.[82] Because the first excited state of the complex is relatively close in energy to the ground state, there is usually a peak in the visible or near-uv region and EDA complexes are often colored. Many EDA complexes are unstable and exist only in solutions in equilibrium with their components, but others are stable solids. In most EDA complexes the donor and acceptor molecules are present in an integral ratio, most often 1:1, but complexes with nonintegral ratios are also known. There are several types of acceptor molecules; we will discuss complexes formed by two of them.

---

[77]Petersen, S.B.; Led, J.J.; Johnston, E.R.; Grant, D.M. *J. Am. Chem. Soc.* **1982**, *104*, 5007.

[78]Wakita, M.; Kuroda, Y.; Fujiwara, Y.; Nakagawa, T. *Chem. Phys. Lipids* **1992**, *62*, 45.

[79]Viel, S.; Mannina, L.; Segre, A. *Tetrahedron Lett.* **2002**, *43*, 2515. See also, Ribas, J.; Cubero, E.; Luque, F.J.; Orozco, M. *J. Org. Chem.* **2002**, *67*, 7057.

[80]For monographs, see Foster, R. *Organic Charge-Transfer Complexes*, Academic Press, NY, **1969**; Mulliken, R.S.; Person, W.B. *Molecular Complexes*, Wiley, NY, **1969**; Rose, J. *Molecular Complexes*, Pergamon, Elmsford, NY, **1967**. For reviews, see Poleshchuk, O.Kh.; Maksyutin, Yu.K. *Russ. Chem. Rev.* **1976**, *45*, 1077; Banthorpe, D.V. *Chem. Rev.* **1970**, *70*, 295; Kosower, E.M. *Prog. Phys. Org. Chem.* **1965**, *3*, 81; Foster, R. *Chem. Br.* **1976**, *12*, 18.

[81]These have often been called *charge-transfer complexes*, but this term implies that the bonding involves charge transfer, which is not always the case, so that the more neutral name EDA complex is preferable. See Mulliken, R.S.; Person, W.B. *J. Am. Chem. Soc.* **1969**, *91*, 3409.

[82]For examples of EDA complexes that do not show charge-transfer spectra, see Bentley, M.D.; Dewar, M.J.S. *Tetrahedron Lett.* **1967**, 5043.

1. *Complexes in Which the Acceptor Is A Metal Ion and the Donor an Alkene or an Aromatic Ring* (*n* donors do not give EDA complexes with metal ions but form covalent bonds instead).[83] Many metal ions form complexes, that are often stable solids, with alkenes, dienes (usually conjugated, but not always), alkynes, and aromatic rings. The donor (or ligand) molecules in these complexes are classified by the prefix *hapto*[84] and/or the descriptor η^n (the Greek letter eta), where *n* indicates how many atoms the ligand uses to bond with the metal.[85] The generally accepted picture of the bonding in these complexes,[86] first proposed by Dewar,[87] can be

illustrated by the ethylene complex with silver, **9**, in which the alkene unit forms an η^2-complex with the silver ion (the alkene functions as a 2-electron donating ligand to the metal). There is evidence of π-complexation of Na^+ by $C=C$.[88]

---

[83]For monographs, see Collman, J.P.; Hegedus, L.S.; Norton, J.R.; Finke, R.G. *Principles and Applications of Organotransition Metal Chemistry*, 2nd ed, University Science Books, Mill Valley, CA, *1987*; Alper, H. *Transition Metal Organometallics in Organic Synthesis*, 2 vols., Academic Press, NY, *1976*, *1978*; King, R.B. *Transition-Metal Organic Chemistry*, Academic Press, NY, *1969*; Green, M.L.H. *Organometallic Compounds*, Vol. 2, Methuen, London, *1968*; For general reviews, see Churchill, M.R.; Mason, R. *Adv. Organomet. Chem. 1967*, *5*, 93; Cais, M., in Patai, S. *The Chemistry of Alkenes*, Vol. 1, Wiley, NY, *1964*, pp. 335–385. Among the many reviews limited to certain classes of complexes are transition metals–dienes, Nakamura, A. *J. Organomet. Chem. 1990*, *400*, 35; metals-cycloalkynes and arynes, Bennett, M.A.; Schwemlein, H.P. *Angew. Chem. Int. Ed. 1989*, *28*, 1296; metals-pentadienyl ions, Powell, P. *Adv. Organomet. Chem. 1986*, *26*, 125; complexes of main-group metals, Jutzi, P. *Adv. Organomet. Chem. 1986*, *26*, 217; intramolecular complexes, Omae, I. *Angew. Chem. Int. Ed. 1982*, *21*, 889; transition metals–olefins and acetylenes, Pettit, L.D.; Barnes, D.S. *Fortschr. Chem. Forsch. 1972*, *28*, 85; Quinn, H.W.; Tsai, J.H. *Adv. Inorg. Chem. Radiochem. 1969*, *12*, 217; Pt- and Pd-olefins and acetylenes, Hartley, F.R. *Chem. Rev. 1969*, *69*, 799; silver ions-olefins and aromatics, Beverwijk, C.D.M.; van der Kerk, G.J.M.; Leusink, J.; Noltes, J.G. *Organomet. Chem. Rev. Sect. A 1970*, *5*, 215; metals-substituted olefins, Jones, R. *Chem. Rev. 1968*, *68*, 785; transition metals-allylic compounds, Clarke, H.L. *J. Organomet. Chem. 1974*, *80*, 155; transition metals–arenes, Silverthorn, W.E. *Adv. Organomet. Chem. 1976*, *14*, 47; metals-organosilicon compounds, Haiduc, I.; Popa, V. *Adv. Organomet. Chem. 1977*, *15*, 113; metals–carbocations, Pettit, L.D.; Haynes, L.W., in Olah, G.A.; Schleyer, P.v.R. *Carbonium Ions*, Vol. 5, Wiley, NY, *1976*, pp. 2263–2302; metals-seven-and eight-membered rings, Bennett, M.A. *Adv. Organomet. Chem. 1966*, *4*, 353. For a list of review articles on this subject, see Bruce, M.I. *Adv. Organomet. Chem. 1972*, *10*, 273, pp. 317–321.

[84]For a discussion of how this system originated, see Cotton, F.A. *J. Organomet. Chem. 1975*, *100*, 29.

[85]Another prefix used for complexes is μ (mu), which indicates that the ligand bridges two metal atoms.

[86]For reviews, see Pearson, A.J. *Metallo-organic Chemistry*, Wiley, NY, *1985*; Ittel, S.D.; Ibers, J.A. *Adv. Organomet. Chem. 1976*, *14*, 33; Hartley, F.R. *Chem. Rev. 1973*, *73*, 163; *Angew. Chem. Int. Ed. 1972*, *11*, 596.

[87]Dewar, M.J.S. *Bull. Soc. Chim. Fr. 1951*, *18*, C79.

[88]Hu, J.; Gokel, G.W.; Barbour, L.J. *Chem. Commun. 2001*, 1858.

In the case of the silver complex, the bond is not from one atom of the $C=C$ unit to the silver ion, but from the π center such that two electrons are transferred from the alkene to the metal ion.[89] Ethene has two π-electrons and is a dihapto or η^2 ligand, as are other simple alkenes. Similarly, benzene has six π-electrons and is a hexahapto or η^6 ligand. Ferrocene (**10**) has two cyclopentadienyl ligands (each is a five-electron donor or an η^5 ligand), and ferrocene is properly called bis(η^5-cyclopentadienyl)iron(II). This system can be extended to compounds in which only a single σ bond connects the organic group to the metal, for example, C_6H_5–Li (a monohapto or η^1 ligand), and to complexes in which the organic group is an ion, for example, π-allyl complexes, such as **11**, in which the allyl ligand is trihapto or η^3. Note that in a compound such as allyllithium, where a σ bond connects the carbon to the metal, the allyl group is referred to as monohapto or η^1.

$$CH_2{=}CH-CH_2-Li$$

Allyllithium

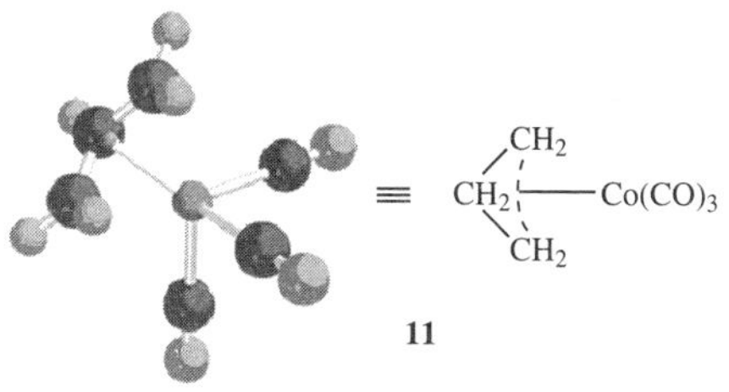

11

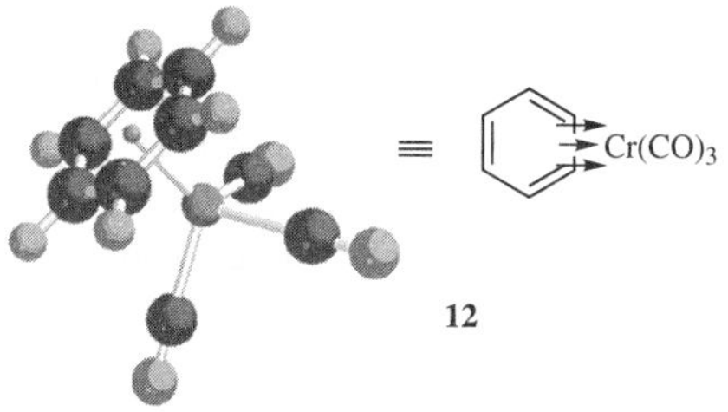

$$\begin{array}{c} CH_2 \\ | \\ CH_2{-}\!\!-Co(CO)_3 \\ | \\ CH_2 \end{array}$$

$\equiv$ ⟨benzene⟩$Cr(CO)_3$

12

As mentioned, benzene is an η^6 ligand that forms complexes with silver and other metals.[90] When the metal involved has a coordination number >1, more than one donor molecule (ligand) participates. The CO group is a common ligand (a two-electron donating or η^2 ligand), and in metal complexes the CO group is classified as a metal carbonyl. Benzenechromium tricarbonyl (**12**) is a stable compound[91] that illustrates both benzene and carbonyl ligands. Three arrows are shown to represent the six-electron donation (an η^6 ligand), but the accompanying model gives a clearer picture of the bonding. Cyclooctatetraene is an eight-electron donating or η^8 ligand that also forms complexes with metals. Metallocenes (see **10**) may be considered a special case of this type of complex, although the bonding in

---

[89]For a discussion of how the nature of the metal ion affects the stability of the complex, see p. $$$.
[90]For a monograph, see Zeiss, H.; Wheatley, P.J.; Winkler, H.J.S. *Benzenoid Metal Complexes*; Ronald Press, NY, *1966*.
[91]Nicholls, B.; Whiting, M.C. *J. Chem. Soc. 1959*, 551. For reviews of arene–transition-metal complexes, see Uemura, M. *Adv. Met.-Org. Chem. 1991, 2*, 195; Silverthorn, W.E. *Adv. Organomet. Chem. 1975, 13*, 47.

metallocenes is much stronger.

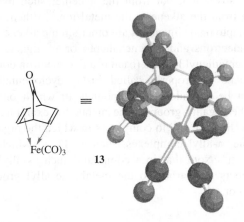

13

In a number of cases, alkenes that are too unstable for isolation have been isolated in the form of metal complexes. As example is norbornadienone, which was isolated in the form of its iron–tricarbonyl complex (**13**),[92] where the norbornadiene unit is an η^4 ligand, and each of the carbonyl units are η^2 ligands. The free dienone spontaneously decomposes to carbon monoxide and benzene (see reaction **17-28**).

2. *Complexes in Which the Acceptor Is an Organic Molecule.* Picric acid, 1,3,5-trinitrobenzene, and similar polynitro compounds are the most important of these.[93] Picric acid forms addition compounds with many

Picric acid

aromatic hydrocarbons, aromatic amines, aliphatic amines, alkenes, and other compounds. These addition compounds are usually solids with definite melting points and are often used as derivatives of the compounds in question. They are called picrates, though they are not salts of picric acid, but addition compounds. Unfortunately, salts of picric acid are also called picrates. Similar complexes are formed between phenols and quinones (quinhydrones).[94]

[92]Landesberg, J.M.; Sieczkowski, J. *J. Am. Chem. Soc.* **1971**, *93*, 972.

[93]For a review, see Parini, V.P. *Russ. Chem. Rev.* **1962**, *31*, 408; for a review of complexes in which the acceptor is an organic cation, see Kampar, V.E. *Russ. Chem. Rev.* **1982**, *51*, 107; also see Ref. 80.

[94]For a review of quinone complexes, see Foster, R.; Foreman, M.I., in Patai, S. *The Chemistry of the Quinonoid Compounds*, pt. 1, Wiley, NY, **1974**, pp. 257–333.

Alkenes that contain electron-withdrawing substituents also act as acceptor molecules, as do carbon tetrahalides[95] and certain anhydrides.[96] A particularly strong alkene acceptor is tetracyanoethylene.[97]

The bonding in these cases is more difficult to explain than in the previous case, and indeed no really satisfactory explanation is available.[98] The difficulty is that although the donor has a pair of electrons to contribute (both n and π donors are found here), the acceptor does not have a vacant orbital. Simple attraction of the dipole-induced dipole type accounts for some of the bonding,[99] but is too weak to explain the bonding in all cases;[100] for example, nitromethane, with about the same dipole moment as nitrobenzene, forms much weaker complexes. Some other type of bonding clearly must also be present in many EDA complexes. The exact nature of this bonding, called *charge-transfer bonding*, is not well understood, but it presumably involves some kind of donor–acceptor interaction.

Crown Ether Complexes and Cryptates[101]

Crown ethers are large-ring compounds containing several oxygen atoms, usually in a regular pattern. Examples are 12-crown-4 (**14**; where 12 is the size of the ring

---

[95]See Blackstock, S.C.; Lorand, J.P.; Kochi, J.K. *J. Org. Chem.* **1987**, *52*, 1451.

[96]For a review of anhydrides as acceptors, see Foster, R., in Patai, S. *The Chemistry of Acid Derivatives*, pt. 1, Wiley, NY, **1979**, pp. 175–212.

[97]For a review of complexes formed by tetracyanoethylene and other polycyano compounds, see Melby, L.R., in Rappoport, Z. *The Chemistry of the Cyano Group*, Wiley, NY, **1970**, pp. 639–669. See also, Fatiadi, A.J. *Synthesis* **1987**, 959.

[98]For reviews, see Bender, C.J. *Chem. Soc. Rev.* **1986**, *15*, 475; Kampar, E.; Neilands, O. *Russ. Chem. Rev.* **1986**, *55*, 334; Bent, H.A. *Chem. Rev.* **1968**, *68*, 587.

[99]See, for example, Le Fevre, R.J.W.; Radford, D.V.; Stiles, P.J. *J. Chem. Soc. B* **1968**, 1297.

[100]Mulliken, R.S.; Person, W.B. *J. Am. Chem. Soc.* **1969**, *91*, 3409.

[101]For a treatise, see Atwood, J.L.; Davies, J.E.; MacNicol, D.D. *Inclusion Compounds*, 3 vols.; Academic Press, NY, **1984**. For monographs, see Weber, E. et al., *Crown Ethers and Analogs*, Wiley, NY, **1989**; Vögtle, F. *Host Guest Complex Chemistry I, II, and III* (*Top. Curr. Chem. 98, 101, 121*); Springer, Berlin, **1981**, **1982**, **1984**; Vögtle, F.; Weber, E. *Host Guest Complex Chemistry/Macrocycles*, Springer, Berlin, **1985** [this book contains nine articles from the *Top. Curr. Chem.* vols. just mentioned]; Hiraoka, M. *Crown Compounds*, Elsevier, NY, **1982**; De Jong, F.; Reinhoudt, D.N. *Stability and Reactivity of Crown-Ether Complexes*, Academic Press, NY, **1981**; Izatt, R.M.; Christensen, J.J. *Synthetic Multidentate Macrocyclic Compounds*, Academic Press, NY, **1978**. For reviews, see McDaniel, C.W.; Bradshaw, J.S.; Izatt, R.M. *Heterocycles*, **1990**, *30*, 665; Sutherland, I.O. *Chem. Soc. Rev.* **1986**, *15*, 63; Sutherland, I.O., in Takeuchi, Y.; Marchand, A.P. *Applications of NMR Spectroscopy to Problems in Stereochemistry and Conformational Analysis*, VCH, NY, **1986**; Franke, J.; Vögtle, F. *Top. Curr. Chem.* **1986**, *132*, 135; Cram, D.J. *Angew. Chem. Int. Ed.* **1986**, *25*, 1039; Gutsche, C.D. *Acc. Chem. Res.* **1983**, *16*, 161; Tabushi, I.; Yamamura, K. *Top. Curr. Chem.* **1983**, *113*, 145; Stoddart, J.F. *Prog. Macrocyclic Chem.* **1981**, *2*, 173; De Jong, F.; Reinhoudt, D.N. *Adv. Phys. Org. Chem.* **1980**, *17*, 279; Vögtle, E.; Weber, E., in Patai, S. *The Chemistry of Functional Groups, Supplement E*, Wiley, NY, **1980**, pp. 59–156; Poonia, N.S. *Prog. Macrocyclic Chem.* **1979**, *1*, 115; Reinhoudt, D.N.; De Jong, F. *Prog. Macrocyclic Chem.* **1979**, *1*, 157; Cram, D.J.; Cram, J.M. *Acc. Chem. Res.* **1978**, *11*, 8, *Science* **1974**, *183*, 803; Knipe, A.C. *J. Chem. Educ.* **1976**, *53*, 618; Gokel, G.W.; Durst, H.D. *Synthesis* **1976**, 168; *Aldrichimica Acta* **1976**, *9*, 3; Lehn, J.M. *Struct. Bonding (Berlin)* **1973**, *16*, 1; Christensen, J.J.; Eatough, D.J.; Izatt, R.M. *Chem. Rev.* **1974**, *74*, 351; Pedersen, C.J.; Frensdorff, H.K. *Angew. Chem. Int. Ed.* **1972**, *11*, 16.

and 4 represents the number of coordinating atoms, here oxygen),[102] dicyclohexa-no-18-crown-6 (**15**), and 15-crown-5 (**16**). These compounds have the property[103] of forming complexes with positive ions, generally metallic ions (though not usually ions of transition metals) or ammonium and substituted ammonium ions.[104] The crown ether is called the *host* and the ion is the *guest*. In most cases, the ions are held tightly in the center of the cavity.[105] Each crown ether binds different ions, depending on the size of the cavity. For example, **14** binds Li^+ [106] but not K^+,[107] while **15** binds K^+ but not Li^+.[108] Similarly, **15** binds Hg^{2+}, but not Cd^{2+} or Zn^{2+}, and Sr^{2+} but not Ca^{2+}.[109] 18-Crown-5 binds alkali and ammonium cations >1000 times weaker than 18-crown-6, presumably because the larger 18-crown-6 cavity involves more hydrogen bonds.[110] The complexes can frequently be prepared as well-defined sharp-melting solids.

14   **15**   **16**

For a monograph on the synthesis of crown ethers, see Gokel, G.W.; Korzeniowski, S.H. *Macrocyclic Polyether Synthesis*, Springer, NY, *1982*. For reviews, see Krakowiak, K.E.; Bradshaw, J.S.; Zamecka-Krakowiak, D.J. *Chem. Rev. 1989*, *89*, 929; Jurczak, J.; Pietraszkiewicz, M. *Top. Curr. Chem. 1986*, *130*, 183; Gokel, G.W.; Dishong, D.M.; Schultz, R.A.; Gatto, V.J. *Synthesis 1982*, 997; Bradshaw, J.S.; Stott, P.E. *Tetrahedron 1980*, *36*, 461; Laidler, D.A.; Stoddart, J.F., in Patai, S. *The Chemistry of Functional Groups, Supplement E*, Wiley, NY, *1980*, pp. 3–42. For reviews of acyclic molecules with similar properties, see Vögtle, E. *Chimia 1979*, *33*, 239; Vögtle, E.; Weber, E. *Angew. Chem. Int. Ed. 1979*, *18*, 753. For a review of cryptands that hold two positive ions, see Lehn, J.M. *Pure Appl. Chem. 1980*, *52*, 2441. The 1987 Nobel Prize in Chemistry was awarded to Charles J. Pedersen, Donald J. Cram, and Jean-Marie Lehn for their work in this area. The three Nobel lectures were published in two journals (respectively, CJP, DJC, J-ML): *Angew. Chem. Int. Ed. 1988*, *27* pp. 1021, 1009, 89; and *Chem. Scr. 1988*, *28*, pp. 229, 263, 237. See also the series *Advances in Supramolecular Chemistry*.

[102]Cook, F.L.; Caruso, T.C.; Byrne, M.P.; Bowers, C.W.; Speck, D.H.; Liotta, C. *Tetrahedron Lett. 1974*, 4029.

[103]Discovered by Pedersen, C.J. *J. Am. Chem. Soc. 1967*, *89*, 2495, 7017. For an account of the discovery, see Schroeder, H.E.; Petersen, C.J. *Pure Appl. Chem. 1988*, *60*, 445.

[104]For a monograph, see Inoue, Y.; Gokel, G.W. *Cation Binding by Macrocycles*, Marcel Dekker, NY, *1990*.

[105]For reviews of thermodynamic and kinetic data for this type of interaction, see Izatt, R.M.; Bradshaw, J.S.; Nielsen, S.A.; Lamb, J.D.; Christensen, J.J.; Sen, D. *Chem. Rev. 1985*, *85*, 271; Parsonage, N.G.; Staveley, L.A.K., in Atwood, J.L.; Davies, J.E.; MacNicol, D.D. *Inclusion Compounds*, Vol. 3, Academic Press, NY, *1984*, pp. 1–36.

[106]Anet, F.A.L.; Krane, J.; Dale, J.; Daasvatn, K.; Kristiansen, P.O. *Acta Chem. Scand. 1973*, *27*, 3395.

[107]Certain derivatives of 14-crown-4 and 12-crown-3 show very high selectivity for Li^+ compared to the other alkali metal ions. See Bartsch, R.A.; Czech, B.P.; Kang, S.I.; Stewart, L.E.; Walkowiak, W.; Charewicz, W.A.; Heo, G.S.; Son, B. *J. Am. Chem. Soc. 1985*, *107*, 4997; Dale, J.; Eggestad, J.; Fredriksen, S.B.; Groth, P. *J. Chem. Soc., Chem. Commun. 1987*, 1391; Dale, J.; Fredriksen, S.B. *Pure Appl. Chem. 1989*, *61*, 1587.

[108]Izatt, R.M.; Nelson, D.P.; Rytting, J.H.; Haymore, B.L.; Christensen, J.J. *J. Am. Chem. Soc. 1971*, *93*, 1619.

[109]Kimura, Y.; Iwashima, K.; Ishimori, T.; Hamaguchi, H. *Chem. Lett. 1977*, 563.

[110]Raevsky, O.A.; Solov'ev, V.P.; Solotnov, A.F.; Schneider, H.-J.; Rüdiger, V. *J. Org. Chem. 1996*, *61*, 8113.

Apart from their obvious utility in separating mixtures of cations,[111] crown ethers have found much use in organic synthesis (see the discussion on p. 510). Chiral crown ethers have been used for the resolution of racemic mixtures (p. 138). Although crown ethers are most frequently used to complex cations, amines, phenols, and other neutral molecules have also been complexed[112] (see p. 189 for the complexing of anions).[113] Macrocycles containing nitrogen (aza-crown ethers) or sulfur atoms (thiacrown ethers),[114] such as **17** and **18**,[115] have complexing properties similar to other crown ethers, as do mixed heteroatom crown ethers such as **19**,[116] **20**,[117] or **21**.[118]

[111]Crown ethers have been used to separate isotopes of cations, for example, ⁴⁴Ca from ⁴⁰Ca. For a review, see Heumann, K.G. *Top. Curr. Chem.* **1985**, *127*, 77.

[112]For reviews, see Vögtle, F.; Müller, W.M.; Watson, W.H. *Top. Curr. Chem.* **1984**, *125*, 131; Weber, E. *Prog. Macrocycl. Chem.* **1987**, *3*, 337; Diederich, F. *Angew. Chem. Int. Ed.* **1988**, *27*, 362.

[113]A neutral molecule (e.g., urea) and a metal ion (e.g., Li⁺) were made to be joint guests in a macrocyclic host, with the metal ion acting as a bridge that induces a partial charge on the urea nitrogens: van Staveren, C.J.; van Eerden, J.; van Veggel, F.C.J.M.; Harkema, S.; Reinhoudt, D.N. *J. Am. Chem. Soc.* **1988**, *110*, 4994. See also, Rodrigue, A.; Bovenkamp, J.W.; Murchie, M.P.; Buchanan, G.W.; Fortier, S. *Can. J. Chem.* **1987**, *65*, 2551; Fraser, M.E.; Fortier, S.; Markiewicz, M.K.; Rodrigue, A.; Bovenkamp, J.W. *Can. J. Chem.* **1987**, *65*, 2558.

[114]For reviews of sulfur-containing macroheterocycles, see Voronkov, M.G.; Knutov, V.I. *Sulfur Rep.* **1986**, *6*, 137, *Russ. Chem. Rev.* **1982**, *51*, 856. For a review of those containing S and N, see Reid, G.; Schröder, M. *Chem. Soc. Rev.* **1990**, *19*, 239.

[115]For a review of **17** and its derivatives, see Chaudhuri, P.; Wieghardt, K. *Prog. Inorg. Chem.* **1987**, *35*, 329. N-Aryl-azacrown ethers are known, see Zhang, X.-X.; Buchwald, S.L. *J. Org. Chem.* **2000**, *65*, 8027.

[116]Gersch, B.; Lehn, J.-M.; Grell, E. *Tetrahedron Lett.* **1996**, *37*, 2213.

[117]Newcomb, M.; Gokel, G.W.; Cram, D.J. *J. Am. Chem. Soc.* **1974**, *96*, 6810.

[118]Graf, E.; Lehn, J.M. *J. Am. Chem. Soc.* **1975**, *97*, 5022; Ragunathan, K.G.; Shukla, R.; Mishra, S.; Bharadwaj, P.K. *Tetrahedron Lett.* **1993**, *34*, 5631.

Bicyclic molecules like **20** can surround the enclosed ion in three dimensions, binding it even more tightly than the monocyclic crown ethers. Bicyclics and cycles of higher order[119] are called *cryptands* and the complexes formed are called *cryptates* (monocyclic compunds are sometimes called cryptands). When the molecule contains a cavity that can accommodate a guest molecule, usually through hydrogen-bonding interactions, it is sometimes called a cavitand.[120] The tricyclic cryptand **21** has 10 binding sites and a spherical cavity.[93] Another molecule with a spherical cavity (though not a cryptand) is **22**, which complexes Li^+ and Na^+ (preferentially Na^+), but not K^+, Mg^{2+}, or Ca^{2+}.[121] Molecules such as these, whose cavities can be occupied only by spherical entities, have been called *spherands*.[77] Other types are *calixarenes*,[122] for example, **23**.[123] Spherand-type calixarenes are known.[124] There is significant hydrogen bonding involving the phenolic OH units in [4]calixarenes, but this diminishes as the size of the cavity increases in larger ring calixarenes.[125] There are also calix[6]arenes,[126] which have been shown to have conformational isomers (see p. 195) in equilibrium (cone vs. alternate) that can sometimes be isolated:[127] calix[8]arenes,[128] azacalixarenes,[129] homooxacalixarenes,[130]

[119]For reviews, see Potvin, P.G.; Lehn, J.M. *Prog. Macrocycl. Chem.* **1987**, *3*, 167; Kiggen, W.; Vögtle, F. *Prog. Macrocycl. Chem.* **1987**, *3*, 309; Dietrich, B., in Atwood, J.L.; Davies, J.E.; MacNicol, D.D. *Inclusion Compounds*, Vol. 2, Academic Press, NY, **1984**, pp. 337–405; Parker, D. *Adv. Inorg. Radichem.* **1983**, *27*, 1; Lehn, J.M. *Acc. Chem. Res.* **1978**, *11*, 49, *Pure Appl. Chem.* **1977**, *49*, 857.

[120]Shivanyuk, A.; Spaniol, T.P.; Rissanen, K.; Kolehmainen, E.; Böhmer, V. *Angew. Chem. Int. Ed.* **2000**, *39*, 3497.

[121]Cram, D.J.; Doxsee, K.M. *J. Org. Chem.* **1986**, *51*, 5068; Cram, D.J. *CHEMTECH* **1987**, 120, *Chemtracts: Org. Chem.* **1988**, *1*, 89; Bryany, J.A.; Ho, S.P.; Knobler, C.B.; Cram, D.J. *J. Am. Chem. Soc.* **1990**, *112*, 5837.

[122]Shinkai, S. *Tetrahedron* **1993**, *49*, 8933.

[123]For monographs, see Vicens, J.; Böhmer, V. *Calixarenes: A Versatile Class of Macrocyclic Compounds*, Kluver, Dordrecht, **1991**; Gutsche, C.D. *Calixarenes*; Royal Society of Chemistry: Cambridge, **1989**. For reviews, see Gutsche, C.D. *Prog. Macrocycl. Chem.* **1987**, *3*, 93; *Top. Curr. Chem.* **1984**, *123*, 1. Also see Geraci, C.; Piatelli, M.; Neri, P. *Tetrahedron Lett.* **1995**, *36*, 5429; Deng, G.; Sakaki, T.; Kawahara, Y.; Shinkai, S. *Tetrahedron Lett.* **1992**, *33*, 2163; Zhong, Z.-L.; Chen, Y.-Y.; Lu, X.-R. *Tetrahedron Lett.* **1995**, *36*, 6735; No, K.; Kim, J.E.; Kwon, K.M. *Tetrahedron Lett.* **1995**, *36*, 8453.

[124]Agbaria, K.; Aleksiuk, O.; Biali, S.E.; Böhmer, V.; Frings, M.; Thondorf, I. *J. Org. Chem.* **2001**, *66*, 2891. For the stereochemistry of such compounds, see Agbaria, K.; Biali, S.E.; Böhmer, V.; Brenn, J.; Cohen, S.; Frings, M., Grynszpan, F.; Harrowfield, J.Mc B.; Sobolev, A.N.; Thondorf, I. *J. Org. Chem.* **2001**, *66*, 2900.

[125]Cerioni, G.; Biali, S.E.; Rappoport, Z. *Tetrahedron Lett.* **1996**, *37*, 5797. For a synthesis of calix[4]arene see Molard, Y.; Bureau, C.; Parrot-Lopez, H.; Lamartine, R.; Regnourf-de-Vains, J.-B. *Tetrahedron Lett.* **1999**, *40*, 6383.

[126]Otsuka, H.; Araki, K.; Matsumoto, H.; Harada, T.; Shinkai, S. *J. Org. Chem.* **1995**, *60*, 4862.

[127]Neri, P.; Rocco, C.; Consoli, G.M.L.; Piatelli; M. *J. Org. Chem.* **1993**, *58*, 6535; Kanamathareddy, S.; Gutsche, C.D. *J. Org. Chem.* **1994**, *59*, 3871.

[128]Cunsolo, F.; Consoli, G.M.L.; Piatelli; M.; Neri, P. *Tetrahedron Lett.* **1996**, *37*, 715; Geraci, C.; Piatelli, M.; Neri, P. *Tetrahedron Lett.* **1995**, *36*, 5429.

[129]Miyazaki, Y.; Kanbara, T.; Yamamoto, T. *Tetrahedron Lett.* **2002**, *43*, 7945; Khan, I.U.; Takemura, H.; Suenaga, M.; Shinmyozu, T.; Inazu, T. *J. Org. Chem.* **1993**, *58*, 3158.

[130]Masci, B. *J. Org. Chem.* **2001**, *66*, 1497. For dioxocalix[4]arenes, see Seri, N.; Thondorf, I.; Biali, S.E. *J. Org. Chem.* **2004**, *69*, 4774. For tetraoxacalix[3]arenes, see Tsubaki, K.; Morimoto, T.; Otsubo, T.; Kinoshita, T.; Fuji, K. *J. Org. Chem.* **2001**, *66*, 4083.

and calix[9–20]arenes.[131] Note that substitution of the unoccupied "meta" positions immobilizes calix[4]arenes and substantially reduces the conformational mobility (see p. 211) in calix[8]arenes.[132] Amide-bridged calix[4]arenes[133] calix[4]azulene,[134] and quinone-bridged calix[6]arenes[135] are known, and diammoniumcalix[4]arene has been prepared.[136] Enantiopure calix[4]resorcinarene derivatives are known,[137] and water soluble calix[4]arenes have been prepared.[138] There are also a variety of calix[n]-crown ethers,[139] some of which are cryptands.[140]

calix[4]arene, n = 1
calix[6]arene, n = 3
calix[8]arene, n = 5

23

24

Other molecules include *cryptophanes*, for example, **24**,[141] *hemispherands* (an example is **25**[142]), and *podands*.[143] The last-named are host compounds in which two or more arms come out of a central structure. Examples are **26**[144] and **27**[145] and the latter molecule binds simple cations, such as Na^+, K^+, and Ca^{2+}. *Lariat ethers* are compounds containing a crown ether ring with one or more side chains

[131]Stewart, D.R.; Gutsche, C.D. *J. Am. Chem. Soc.* **1999**, *121*, 4136.

[132]Mascal, M.; Naven, R.T.; Warmuth, R. *Tetrahedron Lett.* **1995**, *36*, 9361.

[133]Wu, Y.; Shen, X.-P.; Duan, C.-y.; Liu, Y.-i.; Xu, Z. *Tetrahedron Lett.* **1999**, *40*, 5749.

[134]Colby, D.A.; Lash, T.D. *J. Org. Chem.* **2002**, *67*, 1031.

[135]Akine, S.; Goto, K.; Kawashima, T. *Tetrahedron Lett.* **2000**, *41*, 897.

[136]Aeungmaitrepirom, W.; Hagège, A.; Asfari, Z.; Bennouna, L.; Vicens, J.; Leroy, M. *Tetrahedron Lett.* **1999**, *40*, 6389.

[137]Page, P.C.B.; Heaney, H.; Sampler, E.P. *J. Am. Chem. Soc.* **1999**, *121*, 6751.

[138]Shimizu, S.; Shirakawa, S.; Sasaki, Y.; Hirai, C. *Angew. Chem. Int. Ed.* **2000**, *39*, 1256.

[139]Stephan, H.; Gloe, K.; Paulus, E.F.; Saadioui, M.; Böhmer, V. *Org. Lett.* **2000**, *2*, 839; Asfari, Z.; Thuéry, P.; Nierlich, M.; Vicens, J. *Tetrahedron Lett.* **1999**, *40*, 499; Geraci, C.; Piattelli, M.; Neri, P. *Tetrahedron Lett.* **1996**, *37*, 3899; Pappalardo, S.; Petringa, A.; Parisi, M.F.; Ferguson, G. *Tetrahedron Lett.* **1996**, *37*, 3907.

[140]Pulpoka, B.; Asfari, Z.; Vicens, J. *Tetrahedron Lett.* **1996**, *37*, 6315.

[141]For reviews, see Collet, A. *Tetrahedron* **1987**, *43*, 5725, in Atwood, J.L.; Davies, J.E.; MacNicol, D.D. *Inclusion Compounds*, Vol. 1, Academic Press, NY, **1984**, pp. 97–121.

[142]Lein, G.M.; Cram, D.J. *J. Am. Chem. Soc.* **1985**, *107*, 448.

[143]For reviews, see Kron, T.E.; Tsvetkov, E.N. *Russ. Chem. Rev.* **1990**, *59*, 283; Menger, F.M. *Top. Curr. Chem.* **1986**, *136*, 1.

[144]Tümmler, B.; Maass, G.; Weber, E.; Wehner, W.; Vögtle, F. *J. Am. Chem. Soc.* **1977**, *99*, 4683.

[145]Vögtle, F.; Weber, E. *Angew. Chem. Int. Ed.* **1974**, *13*, 814.

that can also serve as ligands, for example, **28**.[146] There is also a class of ortho cyclo-phanes that are crown ethers (see **29**) and have been given the name *starands*.[147]

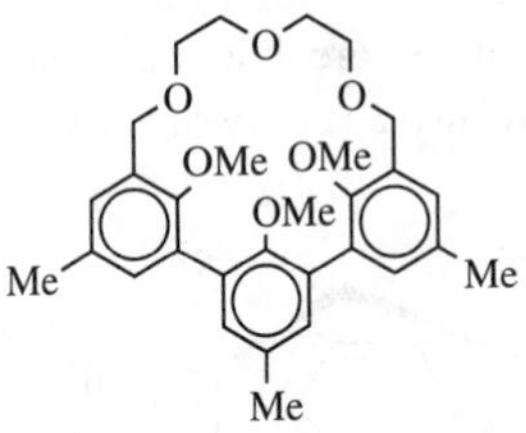

25

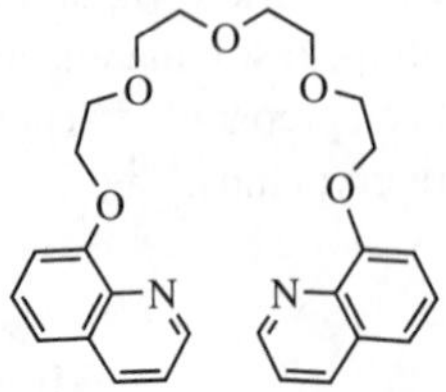

26

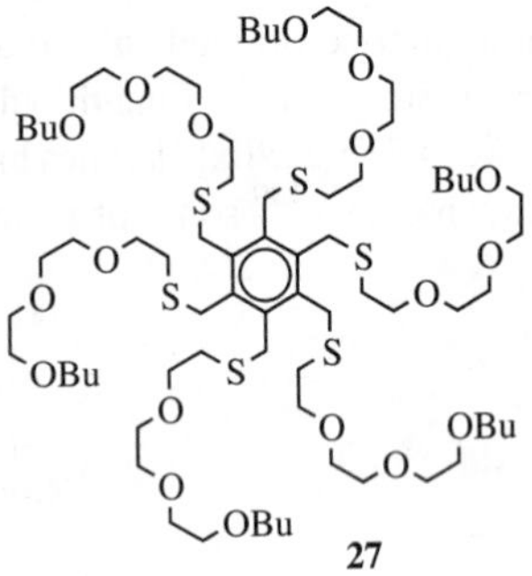

27

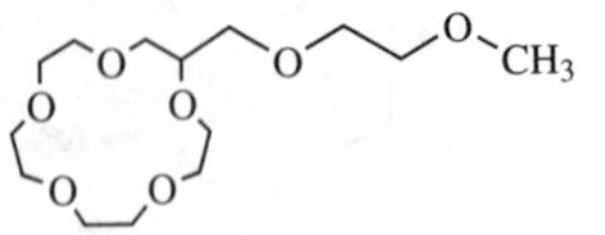

28

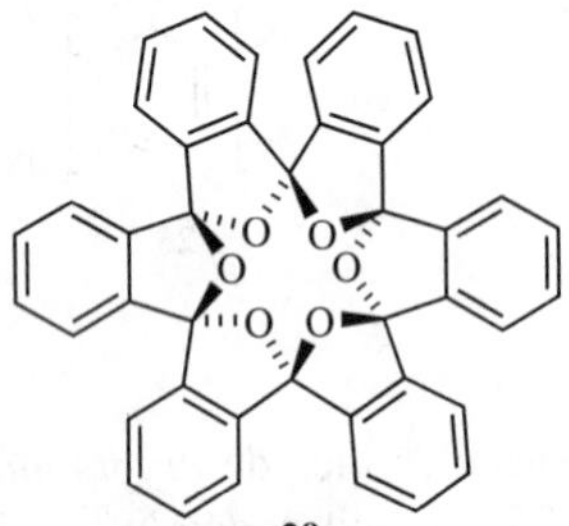

29

The bonding in these complexes is the result of ion-dipole attractions between the heteroatoms and the positive ions. The parameters of the host–guest interactions can sometimes be measured by NMR.[148]

As we have implied, the ability of these host molecules to bind guests is often very specific, often linked to the hydrogen-bonding ability of the host,[149] enabling the host to pull just one molecule or ion out of a mixture. This is called *molecular recognition*.[150] In general, cryptands, with their well-defined 3D cavities, are better for this than monocyclic crown ethers or ether derivatives. An example is the host **30**, which selectively binds the dication **31** ($n = 5$) rather than **31** ($n = 4$), and **31** ($n = 6$) rather than **31** ($n = 7$).[151] The host **32**, which is water soluble, forms 1:1 complexes with neutral aromatic hydrocarbons, such as pyrene and fluoranthene,

[146]See Gatto, V.J.; Dishong, D.M.; Diamond, C.J. *J. Chem. Soc., Chem. Commun.* **1980**, 1053; Gatto, V.J.; Gokel, G.W. *J. Am. Chem. Soc.* **1984**, *106*, 8240; Nakatsuji, Y.; Nakamura, T.; Yonetani, M.; Yuya, H.; Okahara, M. *J. Am. Chem. Soc.* **1988**, *110*, 531.

[147]Lee, W.Y.; Park, C.H. *J. Org. Chem.* **1993**, *58*, 7149.

[148]Wang, T.; Bradshaw, J.S.; Izatt, R.M. *J. Heterocyclic Chem.* **1994**, *31*, 1097.

[149]Fujimoto, T.; Yanagihara, R.; Koboyashi, K.; Aoyama, Y. *Bull. Chem. Soc. Jpn.* **1995**, *68*, 2113.

[150]For reviews, see Rebek Jr., J. *Angew. Chem. Int. Ed.* **1990**, *29*, 245; *Acc. Chem. Res.* **1990**, *23*, 399; *Top. Curr. Chem.* **1988**, *149*, 189; Diederich, F. *J. Chem. Educ.* **1990**, *67*, 813; Hamilton, A.D. *J. Chem. Educ.* **1990**, *67*, 821; Raevskii, O.A. *Russ. Chem. Rev.* **1990**, *59*, 219.

[151]Mageswaran, R.; Mageswaran, S.; Sutherland, I.O. *J. Chem. Soc., Chem. Commun.* **1979**, 722.

and even (though more weakly) with biphenyl and naphthalene, and is able to transport them through an aqueous phase.[152]

Of course, it has long been known that molecular recognition is very important in biochemistry. The action of enzymes and various other biological molecules is extremely specific because these molecules also have host cavities that are able to recognize only one or a few particular types of guest molecules. It is only in recent years that organic chemists have been able to synthesize nonnatural hosts that can also perform crude (compared to biological molecules) molecular recognition. The macrocycle **33** has been used as a catalyst, for the hydrolysis of acetyl phosphate and the synthesis of pyrophosphate.[153]

30   **31**   **32**

No matter what type of host, the strongest attractions occur when combination with the guest causes the smallest amount of distortion of the host.[154] That is, a fully preorganized host will bind better than a host whose molecular shape must change in order to accommodate the guest.

33

[152]Diederich, F.; Dick, K. *J. Am. Chem. Soc.* **1984**, *106*, 8024; Diederich, F.; Griebe, D. *J. Am. Chem. Soc.* **1984**, *106*, 8037. See also Vögtle, F.; Müller, W.M.; Werner, U.; Losensky, H. *Angew. Chem. Int. Ed.* **1987**, *26*, 901.
[153]Hosseini, M.W.; Lehn, J.M. *J. Am. Chem. Soc.* **1987**, *109*, 7047. For a discussion, see Mertes, M.P.; Mertes, K.B. *Acc. Chem. Res.* **1990**, *23*, 413.
[154]See Cram, D.J. *Angew. Chem. Int. Ed.* **1986**, *25*, 1039.

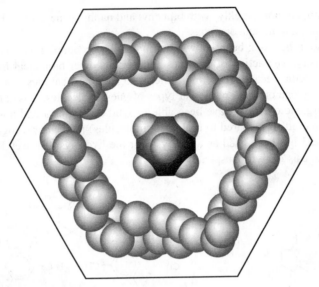

Fig. 3.1. Guest molecule in a urea lattice.[157]

Inclusion Compounds

This type of addition compound is different from either the EDA complexes or the crown ether type of complexes previously discussed. Here, the host forms a crystal lattice that has spaces large enough for the guest to fit into. There is no bonding between the host and the guest except van der Waals forces. There are two main types, depending on the shape of the space.[155] The spaces in *inclusion compounds* are in the shape of long tunnels or channels, while the other type, often called *clathrate*,[156] or *cage compounds* have spaces that are completely enclosed. In both types, the guest molecule must fit into the space and potential guests that are too large or too small will not go into the lattice, so that the addition compound will not form.[157]

One important host molecule among the inclusion compounds is urea.[158] Ordinary crystalline urea is tetragonal, but when a guest is present, urea crystallizes in a hexagonal lattice, containing the guest in long channels (Fig. 3.1).[157]

---

[155]For a treatise that includes both types, see Atwood, J.L.; Davies, J.E.; MacNicol, D.D. *Inclusion Compounds*, Vols. 1–3, Academic Press, NY, *1984*. For reviews, see Weber, E. *Top. Curr. Chem. 1987, 140*, 1; Gerdil, R. *Top. Curr. Chem. 1987, 140*, 71; Mak, T.C.W.; Wong, H.N.C. *Top. Curr. Chem. 1987, 140*, 141. For a review of channels with helical shapes, see Bishop, R.; Dance, I.G. *Top. Curr. Chem. 1988, 149*, 137.

[156]For reviews, see Goldberg, I. *Top. Curr. Chem. 1988, 149*, 1; Weber, E.; Czugler, M. *Top. Curr. Chem. 1988, 149*, 45; MacNicol, D.D.; McKendrick, J.J.; Wilson, D.R. *Chem. Soc. Rev. 1978, 7*, 65.

[157]This picture is taken from a paper by Montel, G. *Bull. Soc. Chim. Fr. 1955*, 1013.

[158]For a review of urea and thiourea inclusion compounds, see Takemoto, K.; Sonoda, N., in Atwood, J.L.; Davies, J.E.; MacNicol, D.D. *Inclusion Compounds*, Vol. 2, Academic Press, NY, *1984*, pp. 47–67.

The hexagonal type of lattice can form only when a guest molecule is present, showing that van der Waals forces between the host and the guest, while small, are essential to the stability of the structure. The diameter of the channel is ~5 Å, and which molecules can be guests is dependent only on their shapes and sizes and not on any electronic or chemical effects. For example, octane and 1-bromooctane are suitable guests for urea, but 2-bromooctane, 2-methylheptane, and 2-methyloctane are not. Also both dibutyl maleate and dibutyl fumarate are guests; neither diethyl maleate or diethyl fumarate is a guest, but dipropyl fumarate is a guest and dipropyl maleate is not.[159] In these complexes, there is usually no integral molar ratio (though by chance there may be). For example, the octane/urea ratio is 1:6.73.[160] A deuterium quadrupole echo spectroscopy study of a urea complex showed that the urea molecules do not remain rigid, but undergo 180° flips about the $C=O$ axis at the rate of >10^6 sec^{-1} at 30°C.[161]

The complexes are solids, but are not useful as derivatives, since they melt with decomposition of the complex at the melting point of urea. They are useful, however, in separating isomers that would be quite difficult to separate otherwise. Thiourea also forms inclusion compounds though with channels of larger diameter, so that *n*-alkanes cannot be guests but, for example, 2-bromooctane, cyclohexane, and chloroform readily fit.

The most important host for clathrates is hydroquinone.[162] Three molecules, held together by hydrogen bonding, make a cage in which fits one molecule of guest. Typical guests are methanol (but not ethanol), SO_2, CO_2, and argon (but not neon). One important use is the isolation of anhydrous hydrazine as complex.[163] Its highly explosive nature makes the preparation of anhydrous hydrazine by distillation of aqueous hydrazine solutions difficult and dangerous. The inclusion complex can be readily isolated and reactions done in the solid state, such as the reaction with esters to give hydrazides (reaction **16-75**).[163] In contrast to the inclusion compounds, the crystal lattices here can exist partially empty. Another host is water. Usually six molecules of water form the cage and many guest molecules, among them Cl_2, propane, and methyl iodide, can fit. The water clathrates, which are solids, can normally be kept only at low temperatures; at room temperature, they decompose.[164] Another inorganic host is sodium chloride (and some other alkali halides), which can encapsulate organic molecules, such as benzene, naphthalene, and diphenylmethane.[165]

[159]Radell, J.; Connolly, J.W.; Cosgrove Jr., W.R. *J. Org. Chem.* **1961**, *26*, 2960.
[160]Redlich, O.; Gable, C.M.; Dunlop, A.K.; Millar, R.W. *J. Am. Chem. Soc.* **1950**, *72*, 4153.
[161]Heatom, N.J.; Vold, R.L.; Vold, R.R. *J. Am. Chem. Soc.* **1989**, *111*, 3211.
[162]For a review, see MacNicol, D.D., in Atwood, J.L.; Davies, J.E.; MacNicol, D.D. *Inclusion Compounds*, Vol. 2, Academic Press, NY, **1984**, pp. 1–45.
[163]Toda, F.; Hyoda, S.; Okada, K.; Hirotsu, K. *J. Chem. Soc., Chem. Commun.* **1995**, 1531.
[164]For a monograph on water clathrates, see Berecz, E.; Balla-Achs, M. *Gas Hydrates*; Elsevier, NY, **1983**. For reviews, see Jeffrey, G.A., in Atwood, J.L.; Davies, J.E.; MacNicol, D.D. *Inclusion Compounds*, Vol. 1, Academic Press, NY, **1984**, pp. 135–190; Cady, G.H. *J. Chem. Educ.* **1983**, *60*, 915; Byk, S.Sh.; Fomina, V.I. *Russ. Chem. Rev.* **1968**, *37*, 469.
[165]Kirkor, E.; Gebicki, J.; Phillips, D.R.; Michl, J. *J. Am. Chem. Soc.* **1986**, *108*, 7106.

Among other hosts[166] for inclusion and/or clathrate compounds are deoxycholic acid,[167] cholic acid,[168] anthracene compounds, such as **34**,[169] dibenzo-24-crown-8,[170] and the compound **35**, which has been called a *carcerand*.[171] When carcerand-type molecules trap ions or other molecules (called guests), the resulting complex is called a carciplex.[172] It has been shown that in some cases, the motion of the guest within the carciplex is restricted.[173]

Hydroquinone   cholic acid (R = OH)   **34**   Perhydrotriphenylene
deoxycholic acid (R = H)

35

[166]See also Toda, F. *Pure App. Chem.* **1990**, *62*, 417, *Top. Curr. Chem.* **1988**, *149*, 211; **1987**, *140*, 43; Davies, J.E.; Finocchiaro, P.; Herbstein, F.H., in Atwood, J.L.; Davies, J.E.; MacNicol, D.D. *Inclusion Compounds*, Vol. 2, Academic Press, NY, **1984**, pp. 407–453.

[167]For a review, see Giglio, E., in Atwood, J.L.; Davies, J.E.; MacNicol, D.D. *Inclusion Compounds*, Vol. 2, Academic Press, NY, **1984**, pp. 207–229.

[168]See Miki, K.; Masui, A.; Kasei, N.; Miyata, M.; Shibakami, M.; Takemoto, K. *J. Am. Chem. Soc.* **1988**, *110*, 6594.

[169]Barbour, L.J.; Caira, M.R.; Nassimbeni, L.R. *J. Chem. Soc., Perkin Trans. 2* **1993**, 2321. Also see, Barbour, L.J.; Caira, M.R.; Nassimbeni, L.R. *J. Chem. Soc., Perkin Trans. 2* **1993**, 1413 for a dihydroanthracene derivative that enclathrates diethyl ether.

[170]Lämsä, M.; Suorsa, T.; Pursiainen, J.; Huuskonen, J.; Rissanen, K. *Chem. Commun.* **1996**, 1443.

[171]Sherman, J.C.; Knobler, C.B.; Cram, D.J. *J. Am. Chem. Soc.* **1991**, *113*, 2194.

[172]Kurdistani, S.K.; Robbins, T.A.; Cram, D.J. *J. Chem. Soc., Chem. Commun.* **1995**, 1259; Timmerman, P.; Verboom, W.; van Veggel, F.C.J.M.; van Duynhoven, J.P.M.; Reinhoudt, D.N. *Angew. Chem. Int. Ed.* **1994**, *33*, 2345; van Wageningen, A.M.A.; Timmerman, P.; van Duynhoven, J.P.M.; Verboom, W.; van Veggel, F.C.J.M.; Reinhoudt, D.N. *Chem. Eur. J.* **1997**, *3*, 639; Fraser, J.R.; Borecka, B.; Trotter, J.; Sherman, J.C. *J. Org. Chem.* **1995**, *60*, 1207; Place, D.; Brown, J.; Deshayes, K. *Tetrahedron Lett.* **1998**, *39*, 5915. See also: Jasat, A.; Sherman, J.C. *Chem. Rev.* **1999**, *99*, 931.

[173]Chapman, R.G.; Sherman, J.C. *J. Org. Chem.* **2000**, *65*, 513.

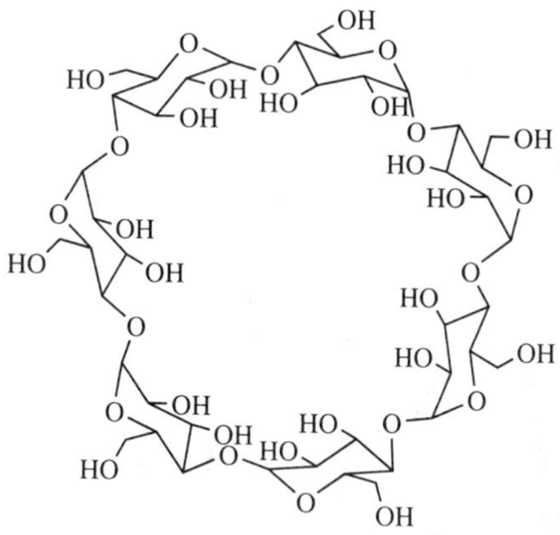

Fig. 3.2. β-Cyclodextrin.

Cyclodextrins

There is one type of host that can form both channel and cage complexes. This type is called *cyclodextrins* or *cycloamyloses*.[174] The host molecules are made up of six, seven, or eight glucose units connected in a large ring, called, respectively, α-, β-, or γ-cyclodextrin (Fig. 3.2 shows the β or seven-membered ring compound). The three molecules are in the shape of hollow truncated cones (Fig. 3.3) with primary OH groups projecting from the narrow side of the cones and secondary OH group from the wide side. As expected for carbohydrate molecules, all of them are soluble in water and the cavities normally fill with water molecules held in place by hydrogen bonds (6, 12, and 17 H_2O molecules for the α, β, and γ forms, respectively), but the insides of the cones are less polar than the outsides, so that nonpolar organic molecules readily displace the water. Thus the cyclodextrins form 1:1 cage complexes with many guests, ranging in size from the noble gases to large organic molecules. A guest molecule must not be too large or it will not fit, though many stable complexes are known in which one end of the guest molecule protrudes from the cavity (Fig. 3.4). On the other hand, if the guest is too small, it may go through the bottom hole (though some small polar molecules, e.g., methanol, do form complexes in which the cavity also contains some water molecules). Since the cavities of the three cyclodextrins are of different sizes (Fig. 3.3), a large variety of guests can be

---

[174]For a monograph, see Bender, M.L.; Komiyama, M. *Cyclodextrin Chemistry*, Springer, NY, *1978*. For reviews, see, in Atwood, J.L.; Davies, J.E.; MacNicol, D.D. *Inclusion Compounds*, Academic Press, NY, *1984*, the reviews, by Saenger, W. Vol. 2, 231–259, Bergeron, R.J. Vol. 3, 391–443, Tabushi, I. Vol. 3, 445–471, Breslow, R. Vol. 3, 473–508; Croft, A.P.; Bartsch, R.A. *Tetrahedron 1983*, *39*, 1417; Tabushi, I.; Kuroda, Y. *Adv. Catal.*, *1983*, *32*, 417; Tabushi, I. *Acc. Chem. Res. 1982*, *15*, 66; Saenger, W. *Angew. Chem. Int. Ed. 1980*, *19*, 344; Bergeron, R. *J. Chem. Ed. 1977*, *54*, 204; Griffiths, D.W.; Bender, M.L. *Adv. Catal. 1973*, *23*, 209.

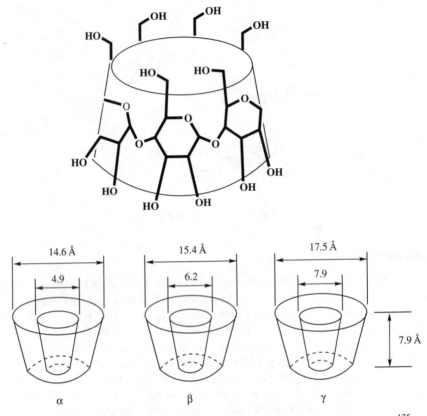

Fig. 3.3. Shape and dimensions of the α-, β-, and γ-cyclodextrin molecules.[175]

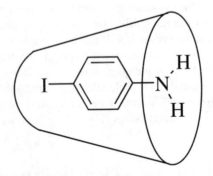

Fig. 3.4. Schematic drawing of the complex of α-cyclodextrin and *p*-iodoaniline.[176]

[175]Szejtli, J., in Atwood, J.L.; Davies, J.E.; MacNicol, D.D. *Inclusion Compounds*, Vol. 3, Academic Press, NY, *1984*, p. 332; Nickon, A.; Silversmith, E.F. *The Name Game*, Pergamon, Elmsford, NY, p. 235.
[176]Modified from Saenger, W.; Beyer, K.; Manor, P.C. *Acta Crystallogr. Sect. B 1976*, *32*, 120.

accommodated. Since cyclodextrins are nontoxic (they are actually small starch molecules), they are now used industrially to encapsulate foods and drugs.[177]

The cyclodextrins also form channel-type complexes, in which the host molecules are stacked on top of each other, like coins in a row.[178] For example, α-cyclodextrin (cyclohexaamylose) forms cage complexes with acetic, propionic, and butyric acids, but channel complexes with valeric and higher acids. Capped cyclodextrins are known.[179]

Catenanes and Rotaxanes[180]

These compounds contain two or more independent portions that are not bonded to each other by any valence forces but nevertheless must remain linked. [n]-*Catenanes* (where n corresponds to the number of linked rings) are made up of two or more rings held together as links in

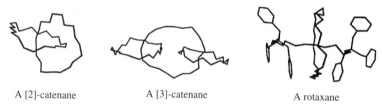

A [2]-catenane   A [3]-catenane   A rotaxane

a chain, while in *rotaxanes* a linear portion is threaded through a ring and cannot get away because of bulky end groups. Among several types of bulky molecular units, porphyrin units have been used to cap rotaxanes[181] as have C_{60} fullerenes.[182] [2]-Rotaxanes and [2]-catenanes are quite common, and [3]-catenanes are known having rather robust amide linkages.[183] More intricate variants, such as oligocatenanes,[184] molecular necklaces (a cyclic oligorotaxane in which a number of small rings are threaded onto a large ring),[185] and cyclic daisy chains (an interwoven chain in which each monomer unit acts as a donor and an acceptor for a threading

[177]For reviews, see Pagington, J.S. *Chem. Br.* **1987**, *23*, 455; Szejtli, J., in Atwood, J.L.; Davies, J.E.; MacNicol, D.D. *Inclusion Compounds*, Vol. 3, Academic Press, NY, **1984**, pp. 331–390.

[178]See Saenger, W. *Angew. Chem. Int. Ed.* **1980**, *19*, 344.

[179]Engeldinger, E.; Armspach, D.; Matt, D. *Chem. Rev.* **2003**, *103*, 4147.

[180]For a monograph, see Schill, G. *Catenanes, Rotaxanes, and Knots*, Academic Press, NY, **1971**. For a review, see Schill, G., in Chiurdoglu, G. *Conformational Analysis*, Academic Press, NY, **1971**, pp. 229–239.

[181]Solladié, N.; Chambron, J.-C.; Sauvage, J.-P. *J. Am. Chem. Soc.* **1999**, *121*, 3684.

[182]Sasabe, H.; Kihara, N.; Furusho, Y.; Mizuno, K.; Ogawa, A.; Takata, T. *Org. Lett.* **2004**, *6*, 3957.

[183]Safarowsky, O.; Vogel, E.; Vögtle, F. *Eur. J. Org. Chem.* **2000**, 499.

[184]Amabilino, D.B.; Ashton, P.R.; Boyd, S.E.; Lee, J.Y.; Menzer, S.; Stoddart, J.F.; Williams, D.J. *Angew. Chem. Int. Ed.* **1997**, *36*, 2070; Amabilino, D.B.; Ashton, P.R.; Balzani, V.; Boyd, S.E.; Credi, A.; Lee, J.Y.; Menzer, S.; Stoddart, J.F.; Venturi, M.; Williams, D.J. *J. Am. Chem. Soc.* **1998**, *120*, 4295.

[185]Chiu, S.-H.; Rowan, S.J.; Cantrill, S.J.; Ridvan, L.; Ashton, R.P.; Garrell, R.L.; Stoddart, J.-F. *Tetrahedron* **2002**, *58*, 807; Whang, D.; Park, K.-M.; Heo, J.; Ashton, P.R.; Kim, K. *J. Am. Chem. Soc.* **1998**, *120*, 4899; Roh, S.-G.; Park, K.-M.; Park, G.-J.; Sakamoto, S.; Yamaguchi, K.; Kim, K. *Angew. Chem. Int. Ed.* **1999**, *38*, 638.

interaction)[186] are known. Ring-in-ring complexes have also been reported.[187] Molecular thread, ribbon, and belt assemblies have been synthesized.[188] Rotaxanes have been used as the basis for molecular switches,[189] and a rotaxane eciplex has been generated that may have applications to molecular-scale photonic devices.[190]

Transitional isomers are possible in [2]-rotaxanes.[191] Catenanes and rotaxanes can be prepared by statistical methods or directed syntheses.[192] Catenanes can contain heteroatoms and heterocyclic units. In some cases, the catenane exists in equilibrium with the cyclic-non-catenane structures and in some cases this exchange is thought to proceed by ligand exchange and a Möbius strip mechanism.[193] An example of a statistical synthesis of a rotaxane is a reaction where a compound **A** is bonded at two positions to another compound **B** in the presence of a large ring **C**. It is hoped that some **A** molecules would by chance be threaded through **C** before combining with the two **B** molecules, so that some rotaxane (**D**) would be formed along with the normal product **E**.[194] In a directed synthesis,[195] the separate parts of the molecule are held together by other bonds that are later cleaved.

$$\vdash\!X \;+\; X\!-\!X \;+\; X\!\dashv \;+\; \bigcirc \;\longrightarrow\; \mid\!\bigcirc\!\mid \;+\; \mid\!\!-\!\!\mid \;+\; \bigcirc$$

$$\quad \textbf{B} \qquad \textbf{A} \qquad \textbf{B} \qquad \textbf{C} \qquad\qquad \textbf{D} \qquad \textbf{E} \qquad \textbf{C}$$

Rotation of one unit through the other catenanes is complex, often driven by making and breaking key hydrogen bonds or π–π interactions. In the case of the

[186]For example, see Ashton, P.R.; Baxter, I.; Cantrill, S.J.; Fyfe, M.C.T.; Glink, P.T.; Stoddart, J.F.; White, A.J.P.; Williams, D.J. *Angew. Chem. Int. Ed.* **1998**, *37*, 1294; Hoshino, T.; Miyauchi, M.; Kawaguchi, Y.; Yamaguchi, H.; Harada, A. *J. Am. Chem. Soc.* **2000**, *122*, 9876; Onagi, H.; Easton, C.J.; Lincoln, S.F. *Org. Lett.* **2001**, *3*, 1041; Cantrill, S.J.; Youn, G.J.; Stoddart, J.F.; Williams, D.J. *J. Org. Chem.* **2001**, *66*, 6857.
[187]Chiu, S.-H.; Pease, A.R.; Stoddart, J.F.; White, A.J.P.; Williams, D.J. *Angew. Chem. Int. Ed.* **2002**, *41*, 270.
[188]Schwierz, H.; Vögtle, F. *Synthesis* **1999**, 295.
[189]Jun, S.I.; Lee, J.W.; Sakamoto, S.; Yamaguchi, K.; Kim, K. *Tetrahedron Lett.* **2000**, *41*, 471; Elizarov, A.M.; Chiu, S.-H.; Stoddart, J.-F. *J. Org. Chem.* **2002**, *67*, 9175.
[190]MacLachlan, M.J.; Rose, A.; Swager, T.M. *J. Am. Chem. Soc.* **2001**, *123*, 9180.
[191]Amabilino, D.B.; Ashton, P.R.; Boyd, S.E.; Gómez-López, M.; Hayes, W.; Stoddart, J.F. *J. Org. Chem.* **1997**, *62*, 3062.
[192]For discussions, see Schill, G. *Catenanes, Rotaxanes, and Knots*, Academic Press, NY, **1971**. For a review, see Schill, G., in Chiurdoglu, G. *Conformational Analysis*, Academic Press, NY, **1971**, pp. 229–239; Walba, D.M. *Tetrahedron* **1985**, *41*, 3161.
[193]Fujita, M.; Ibukuro, F.; Seki, H.; Kamo, O.; Imanari, M.; Ogura, K. *J. Am. Chem. Soc.* **1996**, *118*, 899.
[194]Schemes of this type were carried out by Harrison, I.T.; Harrison, S. *J. Am. Chem. Soc.* **1967**, *89*, 5723; Ogino, H. *J. Am. Chem. Soc.* **1981**, *103*, 1303. For a different kind of statistical syntheszis of a rotaxane, see Harrison, I.T. *J. Chem. Soc., Perkin Trans. 1* **1974**, 301; Schill, G.; Beckmann, W.; Schweikert, N.; Fritz, H. *Chem. Ber.* **1986**, *119*, 2647. See also Agam, G.; Graiver, D.; Zilkha, A. *J. Am. Chem. Soc.* **1976**, *98*, 5206.
[195]For a directed synthesis of a rotaxane, see Schill, G.; Zürcher, C.; Vetter, W. *Chem. Ber.* **1973**, *106*, 228.

isophthaloyl [2]-catenane, **36**, the rate-determining steps do not necessarily correspond to the passage of the bulkiest groups.[196]

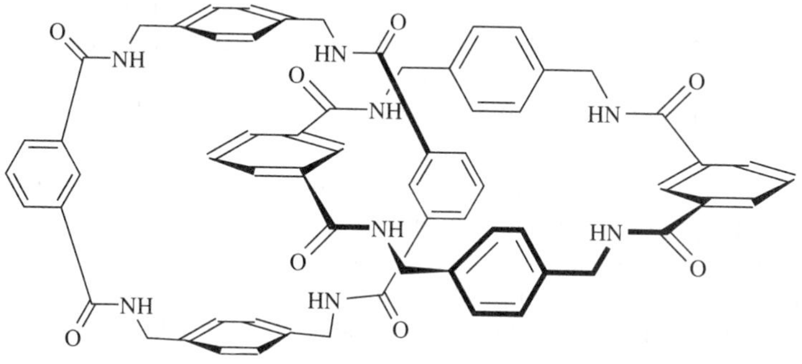

36

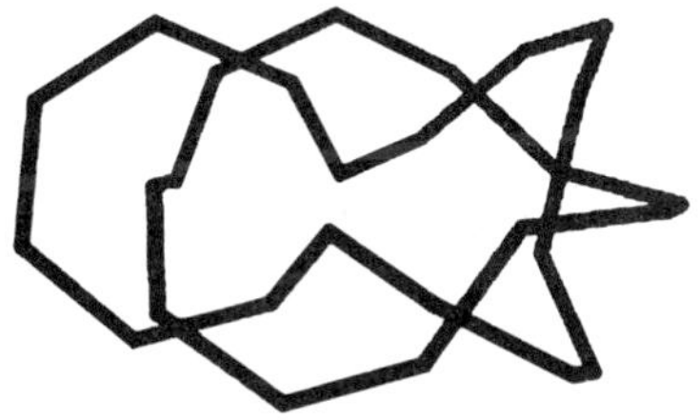

37 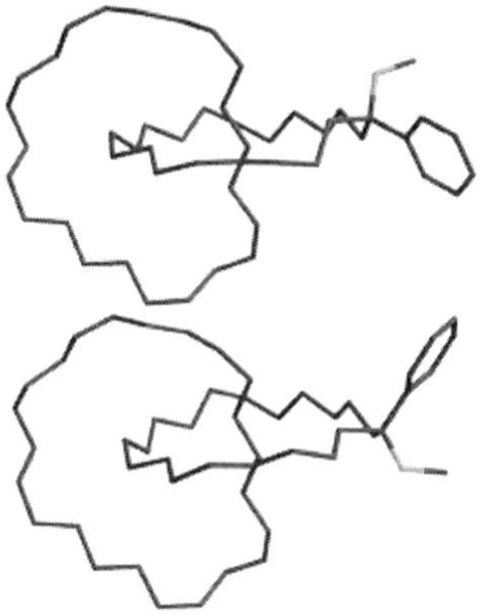   **38**

Singly and doubly interlocked [2]-catenanes[197] can exist as *topological stereo-isomers*[198] (see p. 163 for a discussion of diastereomers). Catenanes **37** and **38** are such stereoisomers, and would be expected to have identical mass spectra. Analysis showed that **37** is more constrained and cannot readily accommodate an excess of energy during the mass spectrometry ionization process and, hence, breaks more easily.

Catenanes, molecular knots, and other molecules in these structural categories can exist as enantiomers. In other words, stereoisomers can be generated in some cases. This phenomenon was first predicted by Frisch and Wassermann,[199] and the

---

[196]Deleuze, M.S.; Leigh, D.A; Zerbetto, F. *J. Am. Chem. Soc.* **1999**, *121*, 2364.
[197]For the synthesis of a doubly interlocking [2]-catenane, see Ibukuro, F.; Fujita, M.; Yamaguchi, K.; Sauvage, J.-P. *J. Am. Chem. Soc.* **1999**, *121*, 11014.
[198]See Lukin, O.; Godt, A.; Vögtle, F. *Chem. Eur. J.* **2004**, *10*, 1879.
[199]Frisch, H.L.; Wasserman, E. *J. Am. Chem. Soc.* **1961**, *83*, 3789.

first stereoisomeric catenanes and molecular knots were synthesized by Sauvage et al.[200] [2, 3]-Enantiomeric resolution has been achieved.[201] A chiral [3]-rotaxane containing two achiral wheels, mechanically bonded has been reported,[202] generating a cyclodiastereomeric compound,[8], and the enantiomers were separated using chiral HPLC. The terms cycloenantiomerism and cyclodiastereomerism were introduced by Prelog et al.[203] This stereoisomerism occurs in cyclic arrangements of several centrally chiral elements in combination with an orientation of the macrocycle.[202]

A rotaxane can also be an inclusion compound.[204] The molecule contains bulky end groups (or "stoppers," such as triisopropylsilyl groups, iPr_3Si-) and a chain that consists of a series of $-O-CH_2CH_2-O-$ groups, but also contains two benzene rings. The ring (or bead) around the chain is a macrocycle containing two benzene rings and four pyridine rings, and is preferentially attracted to one of the benzene rings in the chain. The benzene moiety serves as a "station" for the "bead." However, symmetry of the chain can make the two "stations" equivalent, so that the "bead" is equally attracted to them, and the "bead" actually moves back and forth rapidly between the two "stations," as shown by the temperature dependence of the NMR spectrum.[205] This molecule has been called a *molecular shuttle*. A copper(I) complexed rotaxane has been prepared with two fullerene (see p. 94) stoppers.[206]

Another variation of these molecules are called molecular knots, such as **39**, where the ● represents a metal [in this case, copper(I)].[207] This is particularly interesting since knotted forms of deoxyribonuclic acid (DNA) have been reported.[208]

[200]*Molecular Catenanes, Rotaxanes and Knots* (Eds.: Sauvage, J.-P.; Dietrich-Buchecker, C.O, Wiley-VCH, Weinheim, *1999*; Ashton, P.R.; Bravo, J.A.; Raymo, F.M.; Stoddart, J.F.; White, A.J.P.; Williams, D. *J. Eur. J. Org. Chem.* *1999*, 899; Mitchell, D.K.; Sauvage, J.-P. *Angew. Chem. Int. Ed.* *1988*, 27, 930; Nierengarten, J.-F.; Dietrich-Buchecker, C.O.; Sauvage, J.-P. *J. Am. Chem. Soc.* *1994*, *116*, 375; Walba, D.M. *Tetrahedron* *1985*, *41*, 3161; Chen, C.-T.; Gantzel, P.; Siegel, J.S.; Baldridge, K.K.; English, R.B.; Ho, D.M. *Angew. Chem. Int. Ed.* *1995*, *34*, 2657.

[201]Kaida, T.; Okamoto, Y.; Chambron, J.-C.; Mitchell, D.K.; Sauvage, J.-P. *Tetrahedron Lett.* *1993*, *34*, 1019.

[202]Schmieder, R.; Hübner, G.; Seel, C.; Vögtle, F. *Angew. Chem. Int. Ed.* *1999*, *38*, 3528.

[203]Prelog, V.; Gerlach, H. *Helv. Chim. Acta* *1964*, *47*, 2288; Gerlach, H.; Owtischinnkow, J.A.; Prelog, V. *Helv. Chim. Acta* *1964*, *47*, 2294; Eliel, E.L. *Stereochemie der Kohlenstoffverbindungen*, Verlag Chemie, Weinheim, *1966*; Eliel, E.L.; Wilen, S.H.; Mander, L.N. *Stereochemistry of Organic Compounds*, Wiley, NY, *1994*, pp. 1176–1181; Chorev, M.; Goodman, M. *Acc. Chem. Res.* *1993*, 26, 266 ; Mislow, K. *Chimia*, *1986*, *40*, 395.

[204]For an example, see Anelli, P.L.; Spencer, N.; Stoddart, J.F. *J. Am. Chem. Soc.* *1991*, *113*, 5131.

[205]Anelli, P.L.; Spencer, N.; Stoddart, J.F. *J. Am. Chem. Soc.* *1991*, *113*, 5131. For a review of the synthesis and properties of molecules of this type, see Philp, D.; Stoddart, J.F. *Synlett* *1991*, 445.

[206]Diederich, F.; Dietrich-Buchecker, C.O.; Nierengarten, S.-F.; Sauvage, J.-P. *J. Chem. Soc., Chem. Commun.* *1995*, 781.

[207]Dietrich-Buchecker, C.O.; Nierengarten, J.-F.; Sauvage, J.-P. *Tetrahedron Lett.* *1992*, *33*, 3625. See Dietrich-Buchecker, C.O.; Sauvage, J.-P. *Angew. Chem. Int. Ed.* *1989*, *28*, 189 and Dietrich-Buchecker, C.O.; Guilhem, J.; Pascard, C.; Sauvage, J.-P. *Angew. Chem. Int. Ed.* *1990*, *29*, 1154 for the synthesis of other molecular knots.

[208]Liu, L.F.; Depew, R.E.; Wang, J.C. *J. Mol. Biol.* *1976*, *106*, 439.

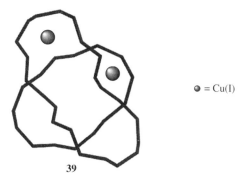

● = Cu(I)

39

Cucurbit[*n*]uril-Based Gyroscane

A new molecule known as gyroscane has been prepared, and proposed as a new supramolecular form.[209] The class of compounds known as cucurbit[*n*]urils, abbreviated Q_n (**40**),[210] are condensation products of glycoluril and formaldehyde. These macrocycles can act as molecular hosts. The new "supramolecular form is one in which a smaller macrocycle, Q5, is located inside a larger macrocycle, Q10, with facile rotation of one relative to the other in solution (see **41**).[210] The image of a ring rotating independently inside another ring, which resembles a gyroscope, suggests the name gyroscane for this new class of supramolecular system."[210]

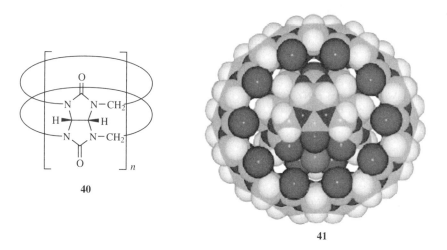

40

41

[209]Day, A.I.; Blanch, R.J.; Arnold, A.P.; Lorenzo, S.; Lewis, G.R.; Dance, I. *Angew. Chem. Int. Ed.* **2002**, *41*, 275.

[210]Freeman, W.A.; Mock, W.L.; Shih, N.Y. *J. Am. Chem. Soc.* **1981**, *103*, 7367; Cintas, P. *J. Inclusion Phenom.* **1994**, *17*, 205; Mock, W.L. *Top. Curr. Chem.* **1995**, *175*, 1; Mock, W.L., in *Comprehensive Supramolecular Chemistry*, Vol. 2, Atwood, J.L.; Davies, J.E.D.; MacNicol, D.D.; Vogtle, F. (Eds.), Pergamon, Oxford, **1996**, pp. 477–493; Day, A.; Arnold, A.P.; Blanch, R.J.; Snushall, B. *J. Org. Chem.* **2001**, *66*, 8094; Kim, J.; Jung, I.-S.; Kim, S.-Y.; Lee, E.; Kang, J.-L.; Sakamoto, S.; Yamaguchi, K.; Kim, K. *J. Am. Chem. Soc.* **2000**, *122*, 540.

CHAPTER 4

Stereochemistry

In Chapters 1–3, we discussed electron distribution in organic molecules. In this chapter, we discuss the 3D structure of organic compounds.[1] The structure may be such that *stereoisomerism*[2] is possible. Stereoisomers are compounds made up of the same atoms bonded by the same sequence of bonds, but having different 3D structures that are not interchangeable. These 3D structures are called *configurations*.

OPTICAL ACTIVITY AND CHIRALITY

Any material that rotates the plane of polarized light is said to be *optically active*. If a pure compound is optically active, the molecule is nonsuperimposable on its mirror image. If a molecule is superimposable on its mirror image, the compound does not rotate the plane of polarized light; it is *optically inactive*. The property

---

[1]For books on this subject, see Eliel, E.L.; Wilen, S.H.; Mander, L.N. *Stereochemistry of Organic Compounds*, Wiley-Interscience, NY, *1994*; Sokolov, V.I. *Introduction to Theoretical Stereochemistry*, Gordon and Breach, NY, *1991*; Bassindale, A. *The Third Dimension in Organic Chemistry*, Wiley, NY, *1984*; Nógrádi, M. *Sterochemistry*, Pergamon, Elmsford, NY, *1981*; Kagan, H. *Organic Stereochemistry*, Wiley, NY, *1979*; Testa, B. *Principles of Organic Stereochemistry*, Marcel Dekker, NY, *1979*; Izumi, Y.; Tai, A. *Stereo-Differentiating Reactions*, Academic Press, NY, Kodansha Ltd., Tokyo, *1977*; Natta, G.; Farina, M. *Stereochemistry*, Harper and Row, NY, *1972*; Eliel, E.L. *Elements of Stereochemistry*, Wiley, NY, *1969*; Mislow, K. *Introduction to Stereochemistry*, W. A. Benjamin, NY, *1965*. Two excellent treatments of stereochemistry that, though not recent, contain much that is valid and useful, are Wheland, G.W. *Advanced Organic Chemistry*, 3rd ed., Wiley, NY, *1960*, pp. 195–514; Shriner, R.L.; Adams, R.; Marvel, C.S. in Gilman, H. *Advanced Organic Chemistry*; Vol. 1, 2nd ed., Wiley, NY, *1943*, pp. 214–488. For a historical treatment, see Ramsay, O.B. *Stereochemistry*, Heyden & Son, Ltd., London, *1981*.

[2]The IUPAC 1974 Recommendations, Section E, Fundamental Stereochemistry, give definitions for most of the terms used in this chapter, as well as rules for naming the various kinds of stereoisomers. They can be found in *Pure Appl. Chem. 1976*, *45*, 13 and in *Nomenclature of Organic Chemistry*, Pergamon, Elmsford, NY, *1979* (the "Blue Book").

March's Advanced Organic Chemistry: Reactions, Mechanisms, and Structure, Sixth Edition, by Michael B. Smith and Jerry March
Copyright © 2007 John Wiley & Sons, Inc.

of nonsuperimposability of an object on its mirror image is called *chirality*. If a molecule is not superimposable on its mirror image, it is *chiral*. If it is superimposable on its mirror image, it is *achiral*. The relationship between optical activity and chirality is absolute. No exceptions are known, and many thousands of cases have been found in accord with it (however, see p. 141). The ultimate criterion, then, for optical activity is chirality (nonsuperimposability on the mirror image). This is both a necessary and a sufficient condition.[3] This fact has been used as evidence for the structure determination of many compounds, and historically the tetrahedral nature of carbon was deduced from the hypothesis that the relationship might be true. Note that parity violation represents an essential property of particle and atomic handedness, and has been related to chirality.[4]

If a molecule is nonsuperimposable on its mirror image, the mirror image must be a different molecule, since superimposability is the same as identity. In each case of optical activity of a pure compound there are two and only two isomers, called *enantiomers* (sometimes *enantiomorphs*), which differ in structure only in the left and right handedness of their orientations (Fig. 4.1). Enantiomers have identical[5] physical and chemical properties except in two important respects:

1. They rotate the plane of polarized light in opposite directions, although in equal amounts. The isomer that rotates the plane to the left (counterclockwise)

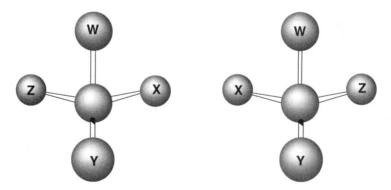

Fig. 4.1. Enantiomers.

---

[3]For a discussion of the conditions for optical activity in liquids and crystals, see O'Loane, J.K. *Chem. Rev.* **1980**, *80*, 41. For a discussion of chirality as applied to molecules, see Quack, M. *Angew. Chem. Int. Ed.* **1989**, *28*, 571.
[4]Avalos, M.; Babiano, R.; Cintas, P.; Jiménez, J.L.; Palacios, J.C. *Tetrahedron Asymmetry* **2000**, *11*, 2845.
[5]Interactions between electrons, nucleons, and certain components of nucleons (e.g., bosons), called *weak interactions*, violate parity; that is, mirror-image interactions do not have the same energy. It has been contended that interactions of this sort cause one of a pair of enantiomers to be (slightly) more stable than the other. See Tranter, G.E. *J. Chem. Soc. Chem. Commun.* **1986**, 60, and references cited therein. See also Barron, L.D. *Chem. Soc. Rev.* **1986**, *15*, 189.

is called the *levo isomer* and is designated (−), while the one that rotates the plane to the right (clockwise) is called the *dextro isomer* and is designated (+). Because they differ in this property they are often called *optical antipodes*.

2. They react at different rates with other chiral compounds. These rates may be so close together that the distinction is practically useless, or they may be so far apart that one enantiomer undergoes the reaction at a convenient rate while the other does not react at all. This is the reason that many compounds are biologically active while their enantiomers are not. Enantiomers react at the same rate with achiral compounds.[6]

In general, it may be said that enantiomers have identical properties in a symmetrical environment, but their properties may differ in an unsymmetrical environment.[7] Besides the important differences previously noted, enantiomers may react at different rates with achiral molecules if an optically active *catalyst* is present; they may have different solubilities in an optically active *solvent*; they may have different indexes of refraction or absorption spectra *when examined with circularly polarized light*, and so on. In most cases, these differences are too small to be useful and are often too small to be measured.

Although pure compounds are always optically active if they are composed of chiral molecules, mixtures of equal amounts of enantiomers are optically inactive since the equal and opposite rotations cancel. Such mixtures are called *racemic mixtures*[8] or *racemates*.[9] Their properties are not always the same as those of the individual enantiomers. The properties in the gaseous or liquid state or in solution usually are the same, since such a mixture is nearly ideal, but properties involving the solid state,[10] such as melting points, solubilities, and heats of fusion, are often different. Thus racemic tartaric acid has a melting point of 204–206°C and a solubility in water at 20°C of 206 g L^{-1}, while for the (+) or the (−) enantiomer, the corresponding figures are 170°C and 1390 g L^{-1}. The separation of a racemic mixture into its two optically active components is called *resolution*. The presence of optical activity always proves that a given compound is chiral, but its absence does not prove that the compound is achiral. A compound that is optically inactive may be achiral, or it may be a racemic mixture (see also, p. 142).

---

[6]For a reported exception, see Hata, N. *Chem. Lett.* **1991**, 155.

[7]For a review of discriminating interactions between chiral molecules, see Craig, D.P.; Mellor, D.P. *Top. Curr. Chem.* **1976**, *63*, 1.

[8]Strictly speaking, the term *racemic mixture* applies only when the mixture of molecules is present as separate solid phases, but in this book we shall use this expression to refer to any equimolar mixture of enantiomeric molecules, liquid, solid, gaseous, or in solution.

[9]For a monograph on the properties of racemates and their resolution, see Jacques, J.; Collet, A.; Wilen, S.H. *Enantiomers, Racemates, and Resolutions*, Wiley, NY, **1981**.

[10]For a discussion, see Wynberg, H.; Lorand, J.P. *J. Org. Chem.* **1981**, *46*, 2538, and references cited therein.

Dependence of Rotation on Conditions of Measurement

The *amount* of rotation α is not a constant for a given enantiomer; it depends on the length of the sample vessel, the temperature, the solvent[11] and concentration (for solutions), the pressure (for gases), and the wavelength of light.[12] Of course, rotations determined for the same compound under the same conditions are identical. The length of the vessel and the concentration or pressure determine the number of molecules in the path of the beam and a is linear with this. Therefore, a number is defined, called the *specific rotation* $[\alpha]$, which is

$$[\alpha] = \frac{\alpha}{lc} \quad \text{for solutions} \qquad [\alpha] = \frac{\alpha}{ld} \quad \text{for pure compounds}$$

where α is the observed rotation, l is the cell length in decimeters, c is the concentration in grams per milliliter, and d is the density in the same units. The specific rotation is usually given along with the temperature and wavelength, in this manner: $[\alpha]_{546}^{25}$. These conditions must be duplicated for comparison of rotations, since there is no way to put them into a simple formula. The expression $[\alpha]_D$ means that the rotation was measured with sodium D light; that is, $\lambda = 589$ nm. The molar rotation $[M]_\lambda^t$ is the specific rotation times the molecular weight divided by 100.

It must be emphasized that although the value of a changes with conditions, the molecular structure is unchanged. This is true even when the changes in conditions are sufficient to change not only the amount of rotation, but even the direction. Thus one of the enantiomers of aspartic acid, when dissolved in water, has $[\alpha]_D$ equal to $+4.36°$ at 20°C and $-1.86°$ at 90°C, although the molecular structure is unchanged. A consequence of such cases is that there is a temperature at which there is *no* rotation (in this case 75°C). Of course, the other enantiomer exhibits opposite behavior. Other cases are known in which the direction of rotation is reversed by changes in wavelength, solvent, and even concentration.[13] In theory, there should be no change in $[\alpha]$ with concentration, since this is taken into account in the formula, but associations, dissociations, and solute–solvent interactions often cause nonlinear behavior. For example, $[\alpha]_D^{24}$ for $(-)$-2-ethyl-2-methylsuccinic acid in $CHCl_3$ is $-5.0°$ at $c = 16.5$ g 100 mL^{-1} (0.165 g mL^{-1}), $-0.7°$ at $c = 10.6$, $+1.7°$ at $c = 8.5$, and $+18.9°$ at $c = 2.2$.[14] Note that the concentration is sometimes reported in g 100 mL^{-1} (as shown) or as g dL^{-1} (decaliters) rather than the standard grams per milliliter (g mL^{-1}). One should always check the concentration term to be certain. Noted that calculation of the optical rotation of (R)-$(-)$-3-chloro-1-butene found a remarkably large dependence on the C=C–C–C torsional angle.[15]

---

[11]A good example is found, in Kumata, Y.; Furukawa, J.; Fueno, T. *Bull. Chem. Soc. Jpn.* **1970**, *43*, 3920.
[12]For a review of polarimetry, see Lyle, G.G.; Lyle, R.E., in Morrison, J.D. *Asymmetric Synthesis*, Vol. 1, Academic Press, NY, **1983**, pp. 13–27.
[13]For examples, see Shriner, R.L.; Adams, R.; Marvel, C.S., in Gilman, H. *Advanced Organic Chemistry*, Vol. 1, 2nd ed. Wiley, NY, **1943**, pp. 291–301.
[14]Krow, G.; Hill, R.K. *Chem. Commun.* **1968**, 430.
[15]Wiberg, K. B.; Vaccaro, P. H.; Cheeseman, J. R. *J. Am. Chem. Soc.* **2003**, *125*, 1888.

However, the observed rotations are a factor of 2.6 smaller than the calculated values, independent of both conformation and wavelength from 589 to 365 nm.

What Kinds of Molecules Display Optical Activity?

Although the ultimate criterion is, of course, nonsuperimposability on the mirror image (chirality), other tests may be used that are simpler to apply but not always accurate. One such test is the presence of a *plane of symmetry.*[16] A plane of symmetry[17] (also called a *mirror plane*) is a plane passing through an object such that the part on one side of the plane is the exact reflection of the part on the other side (the plane acting as a mirror). *Compounds possessing such a plane are always optically inactive,* but there are a few cases known in which compounds lack a plane of symmetry and are nevertheless inactive. Such compounds possess a *center of symmetry,* such as in α-truxillic acid, or an *alternating axis of symmetry* as in **1**.[18] A center of symmetry[17] is a point within an object such that a straight line drawn from any part or element of the object to the center and extended an equal distance on the other side encounters an equal part or element. An alternating axis of symmetry[17] of order *n* is an axis such that when an object containing such an axis is rotated by 360°/*n* about the axis and then reflection is effected across a plane at right angles to the axis, a new object is obtained that is indistinguishable from the original one. Compounds that lack an alternating axis of symmetry are always chiral.

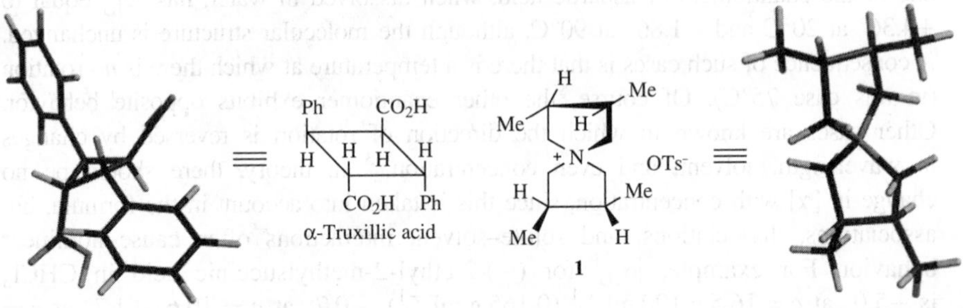

α-Truxillic acid

1

A molecule that contains just one *chiral (stereogenic) carbon atom* (defined as a carbon atom connected to four different groups; also called an *asymmetric carbon atom*) is always chiral, and hence optically active.[19] As seen in Fig. 4.1, such a

---

[16]For a theoretical discussion of the relationship between symmetry and chirality, including parity violation (Ref. 5), see Barron L.D. *Chem. Soc. Rev.* **1986**, *15*, 189.

[17]The definitions of plane, center, and alternating axis of symmetry are taken from Eliel, E.L. *Elements of Stereochemistry*, Wiley, NY, *1969*, pp. 6,7. See also Lemière, G.L.; Alderweireldt, F.C. *J. Org. Chem.* **1980**, *45*, 4175.

[18]McCasland, G.E.; Proskow, S. *J. Am. Chem. Soc.* **1955**, *77*, 4688.

[19]For discussions of the relationship between a chiral carbon and chirality, see Mislow, K.; Siegel, J. *J. Am. Chem. Soc.* **1984**, *106*, 3319; Brand, D.J.; Fisher, J. *J. Chem. Educ.* **1987**, *64*, 1035.

molecule cannot have a plane of symmetry, whatever the identity of W, X, Y, and Z, as long as they are all different. However, the presence of a chiral carbon is neither a necessary nor a sufficient condition for optical activity, since optical activity may be present in molecules with no chiral atom[20] and since some molecules with two or more chiral carbon atoms are superimposable on their mirror images, and hence inactive. Examples of such compounds will be discussed subsequently.

Optically active compounds may be classified into several categories.

1. *Compounds with a Stereogenic Carbon Atom.* If there is only one such atom, the molecule must be optically active. This is so no matter how slight the differences are among the four groups. For example, optical activity is present in

$$BrH_2CH_2CH_2CH_2CH_2CH_2C \diagdown \diagup CH_2CH_2CH_2CH_2CH_2Br$$
$$\underset{\underset{CH_3}{|}}{CH}$$

Optical activity has been detected even in cases,[21] such as 1-butanol-1-*d*, where one group is hydrogen and another deuterium.[22]

$$CH_3CH_2CH_2 \overset{\overset{H}{|}}{\underset{\underset{D}{|}}{—C—}} OH$$

However, the amount of rotation is greatly dependent on the nature of the four groups, in general increasing with increasing differences in polarizabilities among the groups. Alkyl groups have very similar polarizabilities[23] and the optical activity of 5-ethyl-5-propylundecane is too low to be measurable at any wavelength between 280 and 580 nm.[24]

2. *Compounds with Other Quadrivalent Stereogenic Atoms.*[25] Any molecule containing an atom that has four bonds pointing to the corners of a tetrahedron will be optically active if the four groups are different. Among atoms in this category are Si,[26] Ge, Sn,[27] and N (in quaternary salts or

[20]For a review of such molecules, see Nakazaki, M. *Top. Stereochem.* **1984**, *15*, 199.

[21]For reviews of compounds where chirality is due to the presence of deuterium or tritium, see Barth, G.; Djerassi, C. *Tetrahedron* **1981**, *24*, 4123; Arigoni, D.; Eliel, E.L. *Top. Stereochem.* **1969**, *4*, 127; Verbit, L. *Prog. Phys. Org. Chem.* **1970**, *7*, 51. For a review of compounds containing chiral methyl groups, see Floss, H.G.; Tsai, M.; Woodard, R.W. *Top. Stereochem.* **1984**, *15*, 253.

[22]Streitwieser, Jr., A.; Schaeffer, W.Γ. *J. Am. Chem. Soc.* **1956**, *78*, 5597.

[23]For a discussion of optical activity in paraffins, see Brewster, J.H. *Tetrahedron* **1974**, *30*, 1807.

[24]Ten Hoeve, W.; Wynberg, H. *J. Org. Chem.* **1980**, *45*, 2754.

[25]For reviews of compounds with asymmetric atoms other than carbon, see Aylett, B.J. *Prog. Stereochem.* **1969**, *4*, 213; Belloli, R. *J. Chem. Educ.* **1969**, *46*, 640; Sokolov, V.I.; Reutov, O.A. *Russ. Chem. Rev.* **1965**, *34*, 1.

[26]For reviews of stereochemistry of silicon, see Corriu, R.J.P.; Guérin, C.; Moreau, J.J.E., in Patai, S.; Rappoport, Z. *The Chemistry of Organic Silicon Compounds*, pt. 1, Wiley, NY, **1989**, pp. 305–370, *Top. Stereochem.* **1984**, *15*, 43; Maryanoff, C.A.; Maryanoff, B.E., in Morrison, J.D. *Asymmetric Synthesis*, Vol. 4, Academic Press, NY, **1984**, pp. 355–374.

[27]For reviews of the stereochemistry of Sn and Ge compounds, see Gielen, M. *Top. Curr. Chem.* **1982**, *104*, 57; *Top. Stereochem.* **1981**, *12*, 217.

N-oxides).[28] In sulfones, the sulfur bonds with a tetrahedral array, but since two of the groups are always oxygen, no chirality normally results. However, the preparation[29] of an optically active sulfone (**2**) in which one oxygen is ^{16}O and the other ^{18}O illustrates the point that slight differences in groups are all that is necessary. This has been taken even further with the preparation of the ester **3**, both enantiomers of which have been prepared.[30] Optically active chiral phosphates **4** have similarly been made.[31]

      2   **3**   **4**

3. *Compounds with Tervalent Stereogenic Atoms.* Atoms with pyramidal bonding[32] might be expected to give rise to optical activity if the atom is connected to three different groups, since the unshared pair of electrons is analogous to a fourth group, necessarily different from the others. For example, a secondary or tertiary amine where X, Y, and Z are different would be expected to be chiral and thus resolvable. Many attempts have been made to resolve such compounds, but until 1968 all of them failed because of *pyramidal inversion*, which is a rapid oscillation of the unshared pair from

one side of the XYZ plane to the other, thus converting the molecule into its enantiomer.[33] For ammonia, there are 2×10^{11} inversions every second. The inversion is less rapid in substituted ammonia derivatives[34] (amines,

[28]For a review, see Davis, F.A.; Jenkins, Jr., R.H., in Morrison, J.D. *Asymmetric Synthesis*, Vol. 4, Academic Press, NY, *1984*, pp. 313–353. The first resolution of a quaternary ammonium salt of this type was done by Pope, W, J.; Peachey, S.J. *J. Chem. Soc.* **1899**, *75*, 1127.

[29]Stirling, C.J.M. *J. Chem. Soc.* **1963**, 5741; Sabol, M.A.; Andersen, K.K. *J. Am. Chem. Soc.* **1969**, *91*, 3603; Annunziata, R.; Cinquini, M.; Colonna, S. *J. Chem. Soc. Perkin Trans. 1* **1972**, 2057.

[30]Lowe, G.; Parratt, M.J. *J. Chem. Soc. Chem. Commun.* **1985**, 1075.

[31]Abbott, S.J.; Jones, S.R.; Weinman, S.A.; Knowles, J.R. *J. Am. Chem. Soc.* **1978**, *100*, 2558; Cullis, P.M.; Lowe, G. *J. Chem. Soc. Chem. Commun.* **1978**, 512. For a review, see Lowe, G. *Acc. Chem. Res.* **1983**, *16*, 244.

[32]For a review of the stereochemistry at trivalent nitrogen, see Raban, M.; Greenblatt, J., in Patai, S. *The Chemistry of Functional Groups, Supplement F*, pt. 1, Wiley, NY, **1982**, pp. 53–83.

[33]For reviews of the mechanism of, and the effect of structure on, pyramidal inversion, see Lambert, J.B. *Top. Stereochem.* **1971**, *6*, 19; Rauk, A.; Allen, L.C.; Mislow, K. *Angew. Chem. Int. Ed.* **1970**, *9*, 400; Lehn, J.M. *Fortschr. Chem. Forsch.* **1970**, *15*, 311.

[34]For example, see Stackhouse, J.; Baechler, R.D.; Mislow, K. *Tetrahedron Lett.* **1971**, 3437, 3441.

amides, etc.). The interconversion barrier for endo vesus exo methyl in
N-methyl-2-azabicyclo[2.2.1]heptane, for example, is 0.3 kcal.[35] In this case,
torsional strain plays a significant role, along with angle strain, in determining
inversion barriers. Two types of nitrogen atom invert particularly slowly,
namely, a nitrogen atom in a three-membered ring and a nitrogen atom
connected to another atom bearing an unshared pair. Even in such compounds, however, for many years pyramidal inversion proved too rapid to
permit isolation of separate isomers. This goal was accomplished[28] only
when compounds were synthesized in which both features are combined: a
nitrogen atom in a three-membered ring connected to an atom containing an
unshared pair. For example, the two isomers of 1-chloro-2-methylaziridine (**5**
and **6**) were separated and do not interconvert at room temperature.[36] In
suitable cases this barrier to inversion can result in compounds that are
optically active solely because of a chiral tervalent nitrogen atom. For
example, **7** has been resolved into its separate enantiomers.[37] Note that in
this case too, the nitrogen is connected to an atom with an unshared pair.
Conformational stability has also been demonstrated for oxaziridines,[38]
diaziridines (e.g., **8**)[39] triaziridines (e.g., **9**),[40] and 1,2-oxazolidines (e.g.,
10)[41] even although in this case the ring is five membered. However, note that
the nitrogen atom in **10** is connected to two oxygen atoms.

  Another compound in which nitrogen is connected to two oxygens
is **11**. In this case, there is no ring at all, but it has been resolved
into (+) and (−) enantiomers ($[\alpha]_D^{20} \approx \pm 3°$).[42] This compound and

---

[35]Forsyth, D.A.; Zhang, W.; Hanley, J.A. *J. Org. Chem.* **1996**, *61*, 1284. Also see Adams, D.B. *J. Chem. Soc. Perkin Trans. 2* **1993**, 567.

[36]Brois, S.J. *J. Am. Chem. Soc.* **1968**, *90*, 506, 508. See also Shustov, G.V.; Kadorkina, G.K.; Kostyanovsky, R.G.; Rauk, A. *J. Am. Chem. Soc.* **1988**, *110*, 1719; Lehn, J.M.; Wagner, J. *Chem. Commun.* **1968**, 148; Felix, D.; Eschenmoser, A. *Angew. Chem. Int. Ed.* **1968**, *7*, 224; Kostyanovsky, R.G.; Samoilova, Z.E.; Chervin, I.I. *Bull. Acad. Sci. USSR Div. Chem. Sci.* **1968**, 2705, *Tetrahedron Lett.* **1969**, 719. For a review, see Brois, S.J. *Trans. N.Y. Acad. Sci.* **1969**, *31*, 931.

[37]Schurig, V.; Leyrer, U. *Tetrahedron: Asymmetry* **1990**, *1*, 865.

[38]Boyd, D.R. *Tetrahedron Lett.* **1968**, 4561; Boyd, D.R.; Spratt, R.; Jerina, D.M. *J. Chem. Soc. C* **1969**, 2650; Montanari, F.; Moretti, I.; Torre, G. *Chem. Commun.* **1968**, 1694; **1969**, 1086; Bucciarelli, M.; Forni, A.; Moretti, I.; Torre, G.; Brückner, S.; Malpezzi, L. *J. Chem. Soc. Perkin Trans. 2* **1988**, 1595. See also Mannschreck, A.; Linss, J.; Seitz, W. *Liebigs Ann. Chem.* **1969**, *727*, 224; Forni, A.; Moretti, I.; Torre, G.; Brückner, S.; Malpezzi, L.; Di Silvestro, G.D. *J. Chem. Soc. Perkin Trans. 2* **1984**, 791. For a review of oxaziridines, see Schmitz, E. *Adv. Heterocycl. Chem.* **1979**, *24*, 63.

[39]Shustov, G.V.; Denisenko, S.N.; Chervin, I.I.; Asfandiarov, N.L.; Kostyanovsky, R.G. *Tetrahedron* **1985**, *41*, 5719 and cited references. See also Mannschreck, A.; Radeglia, R.; Gründemann, E.; Ohme, R. *Chem. Ber.* **1967**, *100*, 1778.

[40]Hilpert, H.; Hoesch, L.; Dreiding, A.S. *Helv. Chim. Acta* **1985**, *68*, 1691, **1987**, *70*, 381.

[41]Müller, K.; Eschenmoser, A. *Helv. Chim. Acta* **1969**, *52*, 1823; Dobler, M.; Dunitz, J.D.; Hawley, D.M. *Helv. Chim. Acta* **1969**, *52*, 1831.

[42]Kostyanovsky, R.G.; Rudchenko, V.F.; Shtamburg, V.G.; Chervin, I.I.; Nasibov, S.S. *Tetrahedron* **1981**, *37*, 4245; Kostyanovsky, R.G.; Rudchenko, V.F. *Doklad. Chem.* **1982**, *263*, 121. See also Rudchenko, V.F.; Ignatov, S.M.; Chervin, I.I.; Kostyanovsky, R.G. *Tetrahedron* **1988**, *44*, 2233.

several similar ones reported in the same paper are the first examples of

compounds whose optical activity is solely due to an acyclic tervalent chiral nitrogen atom. However, **11** is not optically stable and racemizes at 20°C with a half-life of 1.22 h. A similar compound (**11**, with OCH$_2$Ph replaced by OEt) has a longer half-life, 37.5 h at 20°C.

In molecules in which the nitrogen atom is at a bridgehead, pyramidal inversion is of course prevented. Such molecules, if chiral, can be resolved even without the presence of the two structural features noted above. For example, optically active **12** (Tröger's base) has been prepared.[43] Phosphorus inverts more slowly and arsenic still more slowly.[44] Nonbridgehead phosphorus,[45] arsenic, and antimony compounds have also been resolved, for example, **13**.[46] Sulfur exhibits pyramidal bonding in sulfoxides, sulfinic

[43]Prelog, V.; Wieland, P. *Helv. Chim. Acta* **1944**, *27*, 1127.

[44]For reviews, see Yambushev, F.D.; Savin, V.I. *Russ. Chem. Rev.* **1979**, *48*, 582; Gallagher, M.J.; Jenkins, I.D. *Top. Stereochem.* **1968**, *3*, 1; Kamai, G.; Usacheva, G.M. *Russ. Chem. Rev.* **1966**, *35*, 601.

[45]For a review of chiral phosphorus compounds, see Valentine, Jr., D.J., in Morrison, J.D. *Asymmetric Synthesis*, Vol. 4, Academic Press, NY, **1984**, pp. 263–312.

[46]Horner, L.; Fuchs, H. *Tetrahedron Lett.* **1962**, 203.

esters, sulfonium salts, and sulfites. Examples of each of these have been resolved.[47] An interesting example is $(+)$-$Ph^{12}CH_2SO^{13}CH_2Ph$, a sulfoxide in which the two alkyl groups differ only in ^{12}C versus ^{13}C, but which has $[\alpha]280 = +0.71°$.[48] A computational study indicates that base-catalyzed inversion at sulfur in sulfoxides is possible via a tetrahedral intermediate.[49]

4. *Suitably Substituted Adamantanes.* Adamantanes bearing four different substituents at the bridgehead positions are chiral and optically active and **14**, for example, has been resolved.[50] This type of molecule is a kind of expanded tetrahedron and has the same symmetry properties as any other tetrahedron.

5. *Restricted Rotation Giving Rise to Perpendicular Disymmetric Planes.* Certain compounds that do not contain asymmetric atoms are nevertheless chiral because they contain a structure that can be schematically represented as in Fig. 4.2. For these compounds, we can draw two perpendicular planes neither of which can be bisected by a plane of symmetry. If either plane could be so bisected, the

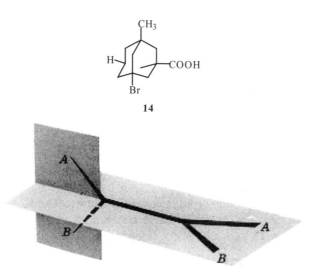

14

Fig. 4.2. Perpendicular disymmetric planes.

[47]For reviews of chiral organosulfur compounds, see Andersen, K.K., in Patai, S. Rappoport, Z. Stirling, C. *The Chemistry of Sulphones and Sulphoxides*, Wiley, NY, *1988*, pp. 55–94; and, in Stirling, C.J.M. *The Chemistry of the Sulphonium Group*, pt. 1, Wiley, NY, *1981*, pp. 229–312; Barbachyn, M.R.; Johnson, C.R., in Morrison, J.D. *Asymmetric Synthesis* Vol. 4, Academic Press, NY, *1984*, pp. 227–261; Cinquini, M.; Cozzi, F.; Montanari, F., in Bernardi, F.; Csizmadia, I.G.; Mangini, A. *Organic Sulfur Chemistry*; Elsevier, NY, *1985*, pp. 355–407; Mikoł ajczyk, M.; Drabowicz, J. *Top. Stereochem. 1982*, *13*, 333.
[48]Andersen, K.K.; Colonna, S.; Stirling, C.J.M. *J. Chem. Soc. Chem. Commun. 1973*, 645.
[49]Balcells, D.; Maseras, F.; Khiar, N. *Org. Lett. 2004*, *6*, 2197.
[50]Hamill, H.; McKervey, M.A. *Chem. Commun. 1969*, 864; Applequist, J.; Rivers, P.; Applequist, D.E. *J. Am. Chem. Soc. 1969*, *91*, 5705.

molecule would be superimposable on its mirror image, since such a plane would be a plane of symmetry. These points will be illustrated by examples.

Biphenyls containing four large groups in the ortho positions cannot freely rotate about the central bond because of steric hindrance.[51] For example, the activation energy (rotational barrier) for the enantiomerization process was determined, $\Delta G^{\ddagger} = 21.8 \pm 0.1 \, \text{kcal mol}^{-1}$, for the chiral 2-carboxy-2'-methoxy-6-nitrobiphenyl.[52] In such compounds, the two rings are in perpendicular planes. If either ring is symmetrically substituted, the molecule has a plane of symmetry. For example, consider the biaryls:

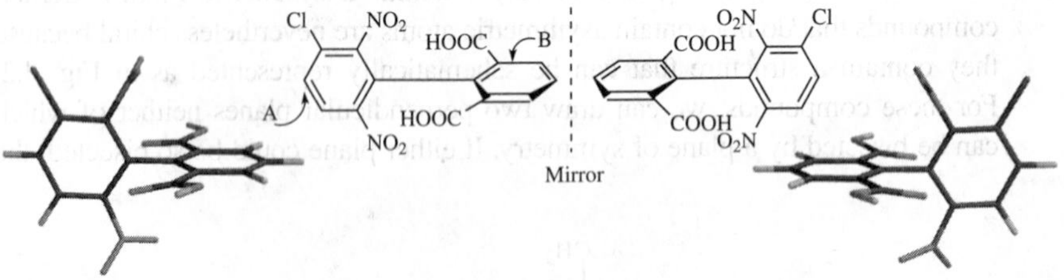

Mirror

Ring B is symmetrically substituted. A plane drawn perpendicular to ring B contains all the atoms and groups in ring A; hence, it is a plane of symmetry and the compound is achiral. On the other hand, consider:

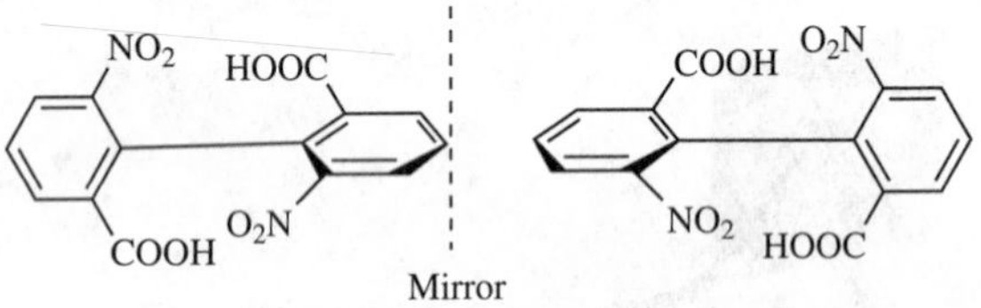

Mirror

There is no plane of symmetry and the molecule is chiral; many such compounds have been resolved. Note that groups in the para position cannot cause lack of symmetry. Isomers that can be separated only because rotation about single bonds is prevented or greatly slowed are called *atropisomers*.[53] 9,9'-Bianthryls also show hindered rotation and exhibit atropisomers.[54]

It is not always necessary for four large ortho groups to be present in order for rotation to be prevented. Compounds with three and even two groups, if large enough, can have hindered rotation and, if suitably substituted, can be resolved. An example is biphenyl-2,2'-bis-sulfonic acid.[55] In some cases, the groups may be large enough to slow rotation greatly but not to prevent it

[51]When the two rings of a biphenyl are connected by a bridge, rotation is of course impossible. For a review of such compounds, see Hall, D.M. *Prog. Stereochem.* **1969**, *4*, 1.

[52]Ceccacci, F.; Mancini, G.; Mencarelli, P.; Villani, C. *Tetrahedron Asymmetry* **2003**, *14*, 3117.

[53]For a review, see Ō ki, M. *Top. Stereochem.* **1983**, *14*, 1.

[54]Becker, H.-D.; Langer, V.; Sieler, J.; Becker, H.-C. *J. Org. Chem.* **1992**, *57*, 1883.

[55]Patterson, W.I.; Adams, R. *J. Am. Chem. Soc.* **1935**, *57*, 762.

completely. In such cases, optically active compounds can be prepared that

slowly racemize on standing. Thus, **15** loses its optical activity with a half-life of 9.4 min in ethanol at 25°C.[56] Compounds with greater rotational stability can often be racemized if higher temperatures are used to supply the energy necessary to force the groups past each other.[57]

Atropisomerism occurs in other systems as well, including monopyrroles.[58] Sulfoxide **16**, for example, forms atropisomers with an interconversion barrier with its atropisomer of 18–19 kcal mol^{-1}.[59] The atropisomers of hindered naphthyl alcohols, such as **17** exist as the *sp*-atropisomer (**17a**) and the *ap*-atropisomer (**17b**).[60] Atropisomers can also be formed in organometallic compounds, such as the bis(phosphinoplatinum) complex (see **18**), generated by reaction with R-BINAP (see p. 1801).[61]

[56]Stoughton, R.W.; Adams, R. *J. Am. Chem. Soc.* **1932**, *54*, 4426.

[57]For a monograph on the detection and measurement of restricted rotations, see Ō ki, M. *Applications of Dynamic NMR Spectroscopy to Organic Chemistry*, VCH, NY, **1985**.

[58]Boiadjiev, S.E.; Lightner, S.A. *Tetrahedron Asymmetry* **2002**, *13*, 1721.

[59]Casarini, D.; Foresti, E.; Gasparrini, F.; Lunazzi, L.; Macciantelli, D.; Misiti, D.; Villani, C. *J. Org. Chem.* **1993**, *58*, 5674.

[60]Casarini, D.; Lunazzi, L.; Mazzanti, A. *J. Org. Chem.* **1997**, *62*, 3315.

[61]Alcock, N.W.; Brown, J.M.; Pérez-Torrente, J.J. *Tetrahedron Lett.* **1992**, *33*, 389. See also, Mikami, K.; Aikawa, K.; Yusa, Y.; Jodry, J.J.; Yamanaka, M. *Synlett* **2002**, 1561.

It is possible to isolate isomers in some cases, often due to restricted rotation. In 9,10-bis(trifluorovinyl)phenanthrene (**19**) torsional diastereomers (see p. 163) are formed. The value of K for interconversion of **19a** and **19b** is 0.48, with $\Delta G° = 15.1$ kcal mol^{-1}.[62] The ability to isolate atropisomers can depend on interactions with solvent, as in the isolation of atropisomeric colchicinoid alkaloids, which have been isolated, characterized, and their dichroic behavior described.[63]

In allenes, the central carbon is *sp* bonded. The remaining two *p* orbitals are perpendicular to each other and each overlaps with the *p* orbital of one adjacent carbon atom, forcing the two remaining bonds of each carbon into perpendicular planes. Thus allenes fall into the category represented by Fig. 4.2: Like biphenyls, allenes are chiral only if both sides are unsymmetrically substituted.[64] For example,

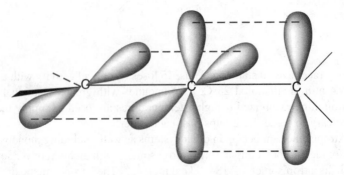

These cases are completely different from the cis–trans isomerism of compounds with one double bond (p. 182). In the latter cases, the four groups are all in one plane, the isomers are not enantiomers, and neither is chiral, while in allenes the groups are in two perpendicular planes and the isomers are a pair of optically active enantiomers.

$$A_{\prime\prime}{}_{B}C=C=C{\overset{A}{\underset{B}{\diagdown}}} \qquad {\overset{A}{\underset{B}{\diagup}}}C=C=C{\overset{\prime\prime A}{\underset{B}{\diagdown}}}$$

Mirror

When three, five, or any *odd* number of cumulative double bonds exist, orbital overlap causes the four groups to occupy one plane and cis–trans isomerism is observed. When four, six, or any *even* number of cumulative double bonds

[62]Dolbier Jr., W.R.; Palmer, K.W. *Tetrahedron Lett.* **1992**, *33*, 1547.
[63]Cavazza, M.; Zandomeneghi, M.; Pietra, F. *Tetrahedron Lett.* **2000**, *41*, 9129.
[64]For reviews of allene chirality, see Runge, W., in Landor, S.R. *The Chemistry of the Allenes*, Vol. 3, Academic Press, NY, **1982**, pp. 579–678, and, in Patai, S. *The Chemistry of Ketenes, Allenes, and Related Compounds*, pt. 1, Wiley, NY, **1980**, pp. 99–154; Rossi, R.; Diversi, P. *Synthesis* **1973**, 25.

exist, the situation is analogous to that in the allenes and optical activity is possible. Compound **20** has been resolved.[65]

Among other types of compounds that contain the system illustrated in Fig. 4.2 and that are similarly chiral if both sides are dissymmetric are spiranes (e.g., **21**) and compounds with exocyclic double bonds (e.g., **22**). Atropisomerism exists in (1,5)-bridgedcalix[8]arenes (see p. 123).[66]

6. *Chirality Due to a Helical Shape.*[67] Several compounds have been prepared that are chiral because they have a shape that is actually helical and can therefore be left or right handed in orientation. The entire molecule is usually less than one full turn of the helix, but this does not alter the possibility of left and right handedness. An example is hexahelicene,[68] in which one side of the molecule must lie above the other because of crowding.[69] The rotational barrier for helicene is $\sim$22.9 kcal mol^{-1}, and is significantly higher when substituents are present.[70] It has been shown that the dianion of helicene retains its chirality.[71] Chiral discrimination of helicenes is possible.[72] 1,16-Diazo[6]helicene has also been prepared and, interestingly, does not act as a proton sponge (see p. 386) because the helical structure leaves the basic nitrogen atoms too far apart. Heptalene is another compound that is not planar (p. 67). Its twisted structure makes it

[65]Nakagawa, M.; Shing ū, K.; Naemura, K. *Tetrahedron Lett.* **1961**, 802.
[66]Consoli, G.M.L.; Cunsolo, F.; Geraci, C.; Gavuzzo, E.; Neri, P. *Org. Lett.* **2002**, *4*, 2649.
[67]For a review, see Meurer, K.P.; Vögtle, F. *Top. Curr. Chem.* **1985**, *127*, 1. See also Laarhoven, W.H.; Prinsen, W.J.C. *Top. Curr. Chem.* **1984**, *125*, 63; Martin, R.H. *Angew. Chem. Int. Ed.* **1974**, *13*, 649.
[68]Newman, M.S.; Lednicer, D. *J. Am. Chem. Soc.* **1956**, *78*, 4765. Optically active heptahelicene has also been prepared, as have higher helicenes: Martin, R.H.; Baes, M. *Tetrahedron* **1975**, *31*, 2135; Bernstein, W.J.; Calvin, M.; Buchardt, O. *J. Am. Chem. Soc.* **1972**, *94*, 494, **1973**, *95*, 527; Defay, N.; Martin, R.H. *Bull. Soc. Chim. Belg.* **1984**, *93*, 313. Even pentahelicene is crowded enough to be chiral: Goedicke, C.; Stegemeyer, H. *Tetrahedron Lett.* **1970**, 937: Bestmann, H.J.; Roth, W. *Chem. Ber.* **1974**, *107*, 2923.
[69]For reviews of the helicenes, see Laarhoven, W.H.; Prinsen, W.J.C. *Top. Curr. Chem.* **1984**, *125*, 63; Martin, R.H. *Angew. Chem. Int. Ed.* **1974**, *13*, 649.
[70]Janke, R.H.; Haufe, G.; Würthwein, E.-U.; Borkent, J.H. *J. Am. Chem. Soc.* **1996**, *118*, 6031.
[71]Frim, R.; Goldblum, A.; Rabinovitz, M. *J. Chem. Soc. Perkin Trans. 2* **1992**, 267.
[72]Murguly, E.; McDonald, R.; Branda, N.R. *Org. Lett.* **2000**, *2*, 3169.

chiral, but the enantiomers rapidly interconvert.[73]

Hexahelicene   Heptalene   *trans*-Cyclooctene   **23**

24a   **24b**

trans-Cyclooctene (see also, p. 184) also exhibits helical chirality because the carbon chain must lie above the double bond on one side and below it on the other.[74] Similar helical chirality also appears in fulgide **23**[75] and dispiro-1,3-dioxane, **24**, shows two enantiomers, **24a** and **24b**.[76]

7. *Optical Activity Caused by Restricted Rotation of Other Types.* Substituted paracyclophanes may be optically active[77] and **25**, for example, has been resolved.[78] In this case, chirality results because the benzene ring cannot rotate in such a way that the carboxyl group goes through the alicyclic ring. Many chiral layered cyclophanes, (e.g., **26**) have been prepared.[79] Another cyclophane[80] with a different type of chirality is [12][12]para-cyclophane (**27**), where the chirality arises from the relative orientation of the two rings attached to the central benzene ring.[81] An aceytlenic cyclophane was shown to have helical chirality.[82] Metallocenes substituted with at least two different groups on one ring are also chiral.[83]

[73]Staab, H.A.; Diehm, M.; Krieger, C. *Tetrahedron Lett.* **1994**, *35*, 8357.

[74]Cope, A.C.; Ganellin, C.R.; Johnson Jr., H.W.; Van Auken, T.V.; Winkler, H.J.S. *J. Am. Chem. Soc.* **1963**, *85*, 3276. Also see Levin, C.C.; Hoffmann, R. *J. Am. Chem. Soc.* **1972**, *94*, 3446.

[75]Yokoyama, Y.; Iwai, T.; Yokoyama, Y.; Kurita, Y. *Chem. Lett.* **1994**, 225.

[76]Grosu, I.; Mager, S.; Plé, G.; Mesaros, E. *Tetrahedron* **1996**, *52*, 12783.

[77]For an example, see Rajakumar, P.; Srisailas, M. *Tetrahedron* **2001**, *57*, 9749.

[78]Blomquist, A.T.; Stahl, R.E.; Meinwald, Y.C.; Smith, B.H. *J. Org. Chem.* **1961**, *26*, 1687. For a review of chiral cyclophanes and related molecules, see Schlögl, K. *Top. Curr. Chem.* **1984**, *125*, 27.

[79]Nakazaki, M.; Yamamoto, K.; Tanaka, S.; Kametani, H. *J. Org. Chem.* **1977**, *42*, 287. Also see Pelter, A.; Crump, R.A.N.C.; Kidwell, H. *Tetrahedron Lett.* **1996**, *37*, 1273. for an example of a chiral [2.2]paracyclophane.

[80]For a treatise on the quantitative chirality of helicenes, see Katzenelson, O.; Edelstein, J.; Avnir, D. *Tetrahedron Asymmetry* **2000**, *11*, 2695.

[81]Chan, T.-L.; Hung, C.-W.; Man, T.-O.; Leung, M.-k. *J. Chem. Soc. Chem. Commun.* **1994**, 1971.

[82]Collins, S.K.; Yap, G.P.A.; Fallis, A.G. *Org. Lett.* **2000**, *2*, 3189.

[83]For reviews on the stereochemistry of metallocenes, see Schlögl, K. *J. Organomet. Chem.* **1986**, *300*, 219, *Top. Stereochem.* **1967**, *1*, 39; *Pure Appl. Chem.* **1970**, *23*, 413.

Several hundred such compounds have been resolved, one example

being **28**. Chirality is also found in other metallic complexes of suitable geometry.[84] For example, fumaric acid–iron tetracarbonyl (**29**)
has been resolved.[85] 1,2,3,4-Tetramethylcyclooctatetraene (**30**) is also
chiral.[86] This molecule, which exists in the tub form (p. 71), has

Perchlorotriphenylamine

neither a plane nor an alternating axis of symmetry. Another compound
that is chiral solely because of hindered rotation is the propeller-shaped
perchlorotriphenylamine, which has been resolved.[87] The 2,5-dideuterio

[84]For reviews of such complexes, see Paiaro, G. *Organomet. Chem. Rev. Sect. A* **1970**, *6*, 319.

[85]Paiaro, G.; Palumbo, R.; Musco, A.; Panunzi, A. *Tetrahedron Lett.* **1965**, 1067; also see Paiaro, G.;
Panunzi, A. *J. Am. Chem. Soc.* **1964**, *86*, 5148.

[86]Paquette, L.A.; Gardlik, J.M.; Johnson, L.K.; McCullough, K.J. *J. Am. Chem. Soc.* **1980**, *102*, 5026.

[87]Okamoto, Y.; Yashima, E.; Hatada, K.; Mislow, K. *J. Org. Chem.* **1984**, *49*, 557. For a conformational
study concerning stereomutation of the helical enantiomers of trigonal carbon diaryl-substituted
compounds by dynamic NMR, see Grilli, S.; Lunazzi, L.; Mazzanti, A.; Casarini, D.; Femoni, C. *J. Org.
Chem.* **2001**, *66*, 488.

derivative (**31**) of barrelene is chiral, although the parent hydrocarbon and the monodeuterio derivative are not. Compound **25** has been prepared in optically active form[88] and is another case where chirality is due to isotopic substitution.

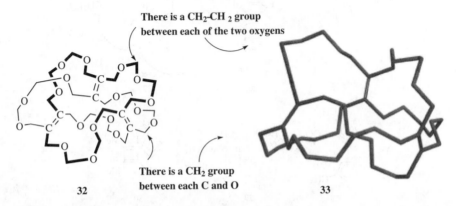

There is a CH$_2$-CH$_2$ group between each of the two oxygens

There is a CH$_2$ group between each C and O

32   33

The main molecular chain in compound **32** has the form of a Möbius strip (see Fig. 15.7 and 3D model **33**).[89] This molecule has no stereogenic carbons, nor does it have a rigid shape a plane nor an alternating axis of symmetry. However, **32** has been synthesized and has been shown to be chiral.[90] Rings containing 50 or more members should be able to exist as knots (**34**, and see **39** on p. 133 in Chapter 3). Such a knot would be nonsuperimposable on its mirror image. Calixarenes,[91] crown ethers,[92] catenanes, and rotaxanes (see p. 131) can also be chiral if suitably substituted.[93] For example, **40** and **41** are nonsuperimposable mirror images.

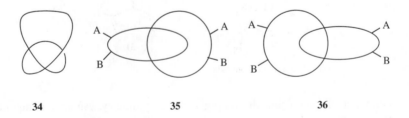

34   35   36

[88]Lightner, D.A.; Paquette, L.A.; Chayangkoon, P.; Lin, H.; Peterson, J.R.*J. Org. Chem.* **1988**, *53*, 1969.

[89]For a review of chirality in Möbius-strip molecules catenanes, and knots, see Walba, D.M. *Tetrahedron* **1985**, *41*, 3161.

[90]Walba, D.M.; Richards, R.M.; Haltiwanger, R.C. *J. Am. Chem. Soc.* **1982**, *104*, 3219.

[91]Iwanek, W.; Wolff, C.; Mattay, J. *Tetrahedron Lett.* **1995**, *36*, 8969.

[92]de Vries, E.F.J.; Steenwinkel, P.; Brussee, J.; Kruse, C.G.; van der Gen, A. *J. Org. Chem.* **1993**, *58*, 4315; Pappalardo, S.; Palrisi, M.F. *Tetrahedron Lett.* **1996**, *37*, 1493; Geraci, C.; Piattelli, M.; Neri, P. *Tetrahedron Lett.* **1996**, *37*, 7627.

[93]For a discussion of the stereochemistry of these compounds, see Schill, G. *Catenanes, Rotaxanes, and Knots*; Academic Press, NY, *1971*, pp. 11–18.

Creation of a Stereogenic Center

Any structural feature of a molecule that gives rise to optical activity may be called a *stereogenic center* (the older term is chiral center) In many reactions, a new chiral center is created, for example,

$$CH_3CH_2COOH \ + \ Br_2 \ \xrightarrow{P} \ CH_3CHBrCOOH$$

If the reagents and reaction conditions are all symmetrical, the product must be a racemic mixture. No optically active material can be created if all starting materials and conditions are optically inactive.[94] This statement also holds when one begins with a racemic mixture. Thus racemic 2-butanol, treated with HBr, must give racemic 2-bromobutane.

The Fischer Projection

For a thorough understanding of stereochemistry it is useful to examine molecular models (like those depicted in Fig. 4.1). However, this is not feasible when writing on paper or a blackboard. In 1891, Emil Fischer greatly served the interests of chemistry by inventing the Fischer projection, a method of representing tetrahedral carbons on paper. By this convention, the model is held so that the two bonds in front of the paper are horizontal and those behind the paper are vertical.

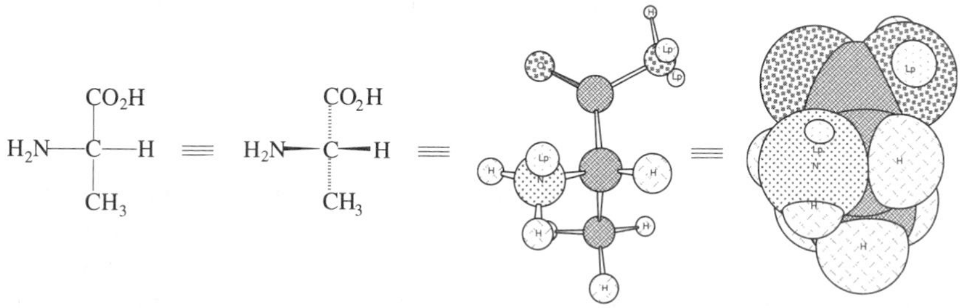

In order to obtain proper results with these formulas, it should be remembered that they are projections and must be treated differently from the models in testing for superimposability. Every plane is superimposable on its mirror image; hence with these formulas there must be added the restriction that they may not be taken out of the plane of the blackboard or paper. Also, they may not be rotated 90°, although 180° rotation is permissible:

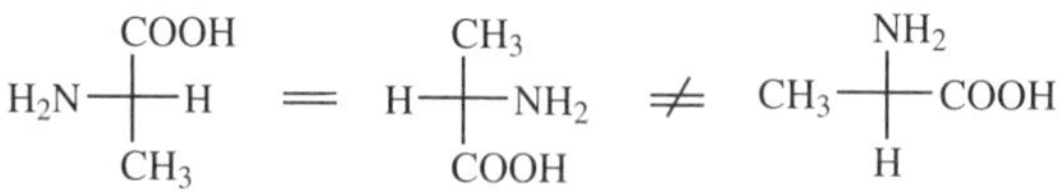

[94]There is one exception to this statement. In a very few cases, racemic mixtures may crystalize from solution in such a way that all the (+) molecules go into one crystal and the (−) molecules into another. If one of the crystals crystallizes before the other, a rapid filtration results in optically active material. For a discussion, see Pincock, R.E.; Wilson, K.R. *J. Chem. Educ.* **1973**, *50*, 455.

It is also permissible to keep any one group fixed and to rotate the other three clockwise or counterclockwise (because this can be done with models):

$$
\begin{array}{c}
\text{COOH} \\
\text{H}_2\text{N}\!\!-\!\!\!\!\underset{\text{CH}_3}{|}\!\!-\!\!\text{H}
\end{array}
\; = \;
\begin{array}{c}
\text{COOH} \\
\text{H}_3\text{C}\!\!-\!\!\!\!\underset{\text{H}}{|}\!\!-\!\!\text{NH}_2
\end{array}
\; = \;
\begin{array}{c}
\text{COOH} \\
\text{H}\!\!-\!\!\!\!\underset{\text{NH}_2}{|}\!\!-\!\!\text{CH}_3
\end{array}
\; = \;
\begin{array}{c}
\text{CH}_3 \\
\text{H}_2\text{N}\!\!-\!\!\!\!\underset{\text{H}}{|}\!\!-\!\!\text{COOH}
\end{array}
$$

However, the *interchange* of any two groups results in the conversion of an enantiomer into its mirror image (this applies to models as well as to the Fischer projections).

With these restrictions Fischer projections may be used instead of models to test whether a molecule containing asymmetric carbons is superimposable on its mirror image. However, there are no such conventions for molecules whose chirality arises from anything other than chiral atoms; when such molecules are examined on paper, 3D pictures must be used. With models or 3D pictures there are no restrictions about the plane of the paper.

Absolute Configuration

Suppose we have two test tubes, one containing (−)-lactic acid and the other the (+) enantiomer. One test tube contains **37** and the other **38**. How do we know which is which? Chemists in the early part of the twentieth century pondered this problem and

$$
\begin{array}{c}
\text{COOH} \\
\text{H}\!\!-\!\!\!\!\underset{\text{CH}_3}{|}\!\!-\!\!\text{OH} \\
\textbf{37}
\end{array}
\qquad
\begin{array}{c}
\text{COOH} \\
\text{HO}\!\!-\!\!\!\!\underset{\text{CH}_3}{|}\!\!-\!\!\text{H} \\
\textbf{38}
\end{array}
\qquad
\begin{array}{c}
\text{CHO} \\
(+)\;\text{H}\!\!-\!\!\!\!\underset{\text{CH}_2\text{OH}}{|}\!\!-\!\!\text{OH} \\
\textbf{39}
\end{array}
\qquad
\begin{array}{c}
\text{CHO} \\
(-)\;\text{HO}\!\!-\!\!\!\!\underset{\text{CH}_2\text{OH}}{|}\!\!-\!\!\text{H} \\
\textbf{40}
\end{array}
$$

decided that they could not know: for lactic acid or any other compound. Therefore Rosanoff proposed that one compound be chosen as a standard and a configuration be arbitrarily assigned to it. The compound chosen was glyceraldehyde because of its relationship to the sugars. The (+) isomer was assigned the configuration shown in **39** and given the label D. The (−) isomer, designated to be **39**, was given the label L. Once a standard was chosen, other compounds could then be related to it. For example, (+)-glyceraldehyde, oxidized with mercuric oxide, gives (−)-glyceric acid:

$$
\begin{array}{c}
\text{CHO} \\
(+)\;\text{H}\!\!-\!\!\!\!\underset{\text{CH}_2\text{OH}}{|}\!\!-\!\!\text{OH}
\end{array}
\quad \xrightarrow{\text{HgO}} \quad
\begin{array}{c}
\text{COOH} \\
(-)\;\text{H}\!\!-\!\!\!\!\underset{\text{CH}_2\text{OH}}{|}\!\!-\!\!\text{OH}
\end{array}
$$

Since it is highly improbable that the configuration at the central carbon changed, it can be concluded that (−)-glyceric acid has the same configuration as (+)-glyceraldehyde and therefore (−)-glyceric acid is also called D. This example emphasizes that molecules with the same configuration need not rotate the plane of polarized light in the same direction. This fact should not surprise us when we remember that the same compound can rotate the plane in opposite directions under different conditions.

Once the configuration of the glyceric acids was known (in relation to the glyceraldehydes), it was then possible to relate other compounds to either of these, and each time a new compound was related, others could be related to *it*. In this way, many thousands of compounds were related, indirectly, to D- or L-glyceraldehyde, and it was determined that **37**, which has the D configuration, is the isomer that rotates the plane of polarized light to the left. Even compounds without asymmetric atoms, such as biphenyls and allenes, have been placed in the D or L series.[95] When a compound has been placed in the D or L series, its *absolute configuration* is said to be known.[96]

In 1951, it became possible to determine whether Rosanoff's guess was right. Ordinary X-ray crystallography cannot distinguish between a D and a L isomer, but by use of a special technique, Bijvoet was able to examine sodium rubidium tartrate and found that Rosanoff had made the correct choice.[97] It was perhaps historically fitting that the first true absolute configuration should have been determined on a salt of tartaric acid, since Pasteur made his great discoveries on another salt of this acid.

In spite of the former widespread use of D and L to denote absolute configuration, the method is not without faults. The designation of a particular enantiomer as D or L can depend on the compounds to which it is related. Examples are known where an enantiomer can, by five or six steps, be related to a known D compound, and by five or six other steps, be related to the L enantiomer of the same compound. In a case of this sort, an arbitrary choice of D or L must be used. Because of this and other flaws, the DL system is no longer used, except for certain groups of compounds, such as carbohydrates and amino acids.

The Cahn–Ingold–Prelog System

The system that has replaced the DL system is the *Cahn–Ingold–Prelog* system, in which the four groups on an asymmetric carbon are ranked according to a set of sequence rules.[98] For our purposes, we confine ourselves to only a few

---

[95]The use of small *d* and *l* is now discouraged, since some authors used it for rotation, and some for configuration. However, a racemic mixture is still a *dl* mixture, since there is no ambiguity here.

[96]For lists of absolute configurations of thousands of compounds, with references, mostly expressed as (*R*) or (*S*) rather than D or L, see Klyne, W.; Buckingham, J. *Atlas of Stereochemistry*, 2nd ed., 2 vols., Oxford University Press: Oxford, *1978*; Jacques, J.; Gros, C.; Bourcier, S.; Brienne, M.J.; Toullec, J. *Absolute Configurations* (Vol. 4 of Kagan *Stereochemistry*), Georg Thieme Publishers, Stuttgart, *1977*.

[97]Bijvoet, J.M.; Peerdeman, A.F.; van Bommel, A.J. *Nature (London) 1951*, *168*, 271. For a list of organic structures whose absolute configurations have been determined by this method, see Neidle, S.; Rogers, D.; Allen, F.H. *J. Chem. Soc. C 1970*, 2340.

[98]For descriptions of the system and sets of sequence rules, see *Pure Appl. Chem. 19767*, *45*, 13; *Nomenclature of Organic Chemistry*, Pergamon, Elmsford, NY, *1979* (the Blue Book); Cahn, R.S.; Ingold, C.K.; Prelog, V. *Angew. Chem. Int. Ed. 1966*, *5*, 385; Cahn, R.S. *J. Chem. Educ. 1964*, *41*, 116; Fernelius, W.C.; Loening, K.; Adams, R.M. *J. Chem. Educ. 1974*, *51*, 735. See also, Prelog, V.; Helmchen, G. *Angew. Chem. Int. Ed. 1982*, *21*, 567. Eliel, E.L.; Wilen, S.H.; Mander, L.N. *Stereochemistry of Organic Compounds*, Wiley-Interscience, NY, *1994*, pp. 101–147. Also see, Smith, M.B. *Organic Synthesis*, 2nd ed., McGraw-Hill, NY, *2001*, pp. 13–20.

of these rules, which are sufficient to deal with the vast majority of chiral compounds.

1. Substituents are listed in order of decreasing atomic number of the atom directly joined to the carbon.

2. Where two or more of the atoms connected to the asymmetric carbon are the same, the atomic number of the second atom determines the order. For example, in the molecule $Me_2CH–CHBr–CH_2OH$, the CH_2OH group takes precedence over the Me_2CH group because oxygen has a higher atomic number than carbon. Note that this is so even although there are two carbons in Me_2CH and only one oxygen in CH_2OH. If two or more atoms connected to the second atom are the same, the third atom determines the precedence, and so on.

3. All atoms except hydrogen are formally given a valence of 4. Where the actual valence is less (as in nitrogen, oxygen, or a carbanion), phantom atoms (designated by a subscript $_0$) are used to bring the valence up to four. These phantom atoms are assigned an atomic number of zero and necessarily rank lowest. Thus the ligand $–HNHMe_2$ ranks higher than $–NMe_2$.

4. A tritium atom takes precedence over deuterium, which in turn takes precedence over ordinary hydrogen. Similarly, any higher isotope (e.g., ^{14}C) takes precedence over any lower one.

5. Double and triple bonds are counted as if they were split into two or three single bonds, respectively, as in the examples in Table 4.1 (note the treatment of the phenyl group). Note that in a $C=C$ double bond, the two carbon atoms are *each* regarded as being connected to two carbon atoms and that one of the latter is counted as having three phantom substituents.

As an exercise, we shall compare the four groups in Table 4.1. The first atoms are connected, respectively, to (H, O, O), (H, C, C), (C, C, C), and (C, C, C). That is enough to establish that –CHO ranks first and $–CH=CH_2$ last, since even one

TABLE 4.1. How Four Common Groups Are Treated in the Cahn–Ingold–Prelog System

Group	Treated as If It Were	Group	Treated as If It Were

oxygen outranks three carbons and three carbons outrank two carbons and a hydrogen. To classify the remaining two groups we must proceed further along the chains. We note that $-C_6H_5$ has two of its (C, C, C) carbons connected to (C, C, H), while the third is ($_{000}$) and is thus preferred to $-C{\equiv}CH$, which has only one (C, C, H) and two ($_{000}$)s.

By application of the above rules, some groups in descending order of precedence are COOH, COPh, COMe, CHO, $CH(OH)_2$, *o*-tolyl, *m*-tolyl, *p*-tolyl, phenyl, $C{\equiv}CH$, *tert*-butyl, cyclohexyl, vinyl, isopropyl, benzyl, neopentyl, allyl, *n*-pentyl, ethyl, methyl, deuterium, and hydrogen. Thus the four groups of glyceraldehyde are arranged in the sequence: OH, CHO, CH_2OH, H.

Once the order is determined, the molecule is held so that the lowest group in the sequence is pointed away from the viewer. Then if the other groups, in the order listed, are oriented clockwise, the molecule is designated (*R*), and if counterclockwise, (*S*). For glyceraldehyde, the (+) enantiomer is (*R*):

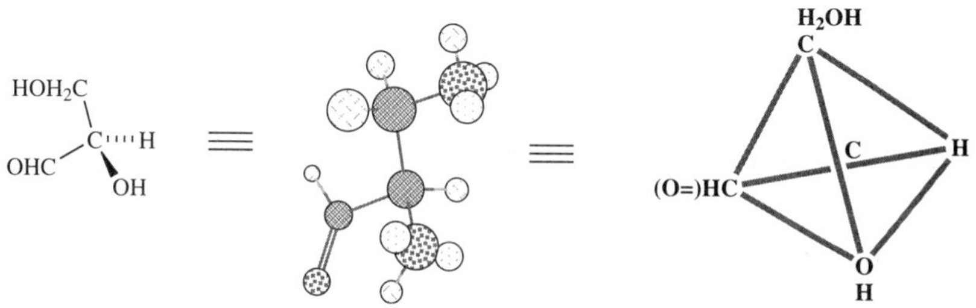

Note that when a compound is written in the Fischer projection, the configuration can easily be determined without constructing the model.[99] If the lowest ranking group is either at the top or the bottom (because these are the two positions pointing away from the viewer), the (*R*) configuration is present if the other three groups in descending order are clockwise, for example,

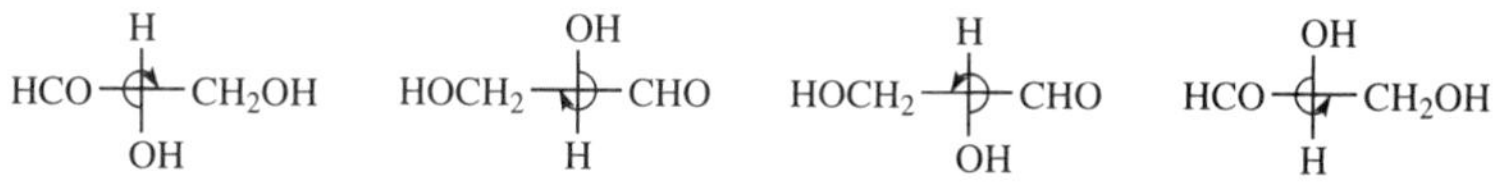

(*R*)-Glyceraldehyde (*S*)-Glyceraldehyde

If the lowestranking group is not at the top or bottom, one can simply interchange it with the top or bottom group, bearing in mind that in so doing, one is inverting the configuration, for example:

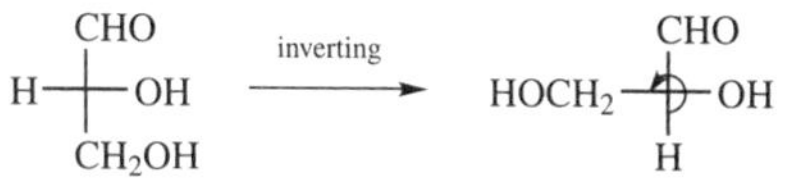

(*S*)-Glyceraldehyde

Therefore the original compound was (*R*)-glyceraldehyde.

[99]For a discussion of how to determine (*R*) or (*S*) from other types of formula, see Eliel, E.L. *J. Chem. Educ.* ***1985***, *62*, 223.

The Cahn–Ingold–Prelog system is unambiguous and easily applicable in most cases. Whether to call an enantiomer (*R*) or (*S*) does not depend on correlations, but the configuration must be known before the system can be applied and this does depend on correlations. The Cahn–Ingold–Prelog system has also been extended to chiral compounds that do not contain stereogenic centers, but have a chiral axis.[100] Compounds having a chiral axis include unsymmetrical allenes, biaryls that exhibit atropisomerism (see p. 146), and alkylidene cyclohexane derivatives, molecular propellers and gears, helicenes, cyclophanes, annulenes, *trans*-cycloalkenes, and metallocenes. A series of rules have been proposed to address the few cases where the rules can be ambiguous, as in cyclophanes and other systems.[101]

| Allenes | Biaryls | Alkylidenecyclohexanes |

Methods of Determining Configuration[102]

In all the methods,[103] it is necessary to relate the compound of unknown configuration to another whose configuration is known. The most important methods of doing this are

1. Conversion of the unknown to, or formation of the unknown from, a compound of known configuration without disturbing the chiral center. See the glyceraldehyde–glyceric acid example above (p. 154). Since the chiral

[100]Eliel, E.L.; Wilen, S.H.; Mander, L.N. *Stereochemistry of Organic Compounds*, Wiley, NY, *1994*, pp. 1119–1190. For a discussion of these rules, as well as for a review of methods for establishing configurations of chiral compounds not containing a stereogenic center, see Krow, G. *Top. Stereochem. 1970*, *5*, 31.

[101]Dodziuk, H.; Mirowicz, M. *Tetrahedron Asymmetry 1990*, *1*, 171; Mata, P.; Lobo, A.M.; Marshall, C.; Johnson, A.P. *Tetrahedron Asymmetry 1993*, *4*, 657; Perdih, M.; Razinger, M. *Tetrahedron Asymmetry 1994*, *5*, 835.

[102]For a monograph, see Kagan, H.B. *Determination of Configuration by Chemical Methods* (Vol. 3 of Kagan, H.B. *Stereochemistry*), Georg Thieme Publishers: Stuttgart, *1977*. For reviews, see Brewster, J.H., in Bentley, K.W.; Kirby, G.W. *Elucidation of Organic Structures by Physical and Chemical Methods*, 2nd ed. (Vol. 4 of Weissberger, A. *Techniques of Chemistry*), pt. 3, Wiley, NY, *1972*, pp. 1–249; Klyne, W.; Scopes, P.M. *Prog. Stereochem. 1969*, *4*, 97; Schlenk Jr., W. *Angew. Chem. Int. Ed. 1965*, *4*, 139. For a review of absolute configuration of molecules in the crystalline state, see Addadi, L.; Berkovitch-Yellin, Z.; Weissbuch, I.; Lahav, M.; Leiserowitz, L. *Top. Stereochem. 1986*, *16*, 1.

[103]Except the X-ray method of Bijvoet.

center was not disturbed, the unknown obviously has the same configuration as the known. This does not necessarily mean that if the known is (R), the unknown is also (R). This will be so if the sequence is not disturbed, but not otherwise. For example, when (R)-1-bromo-2-butanol is reduced to 2-butanol without disturbing the chiral center, the product is the (S) isomer, even although the configuration is unchanged, because CH_3CH_2 ranks lower than $BrCH_2$, but higher than CH_3.

2. Conversion at the chiral center if the mechanism is known. Thus, the S_N2 mechanism proceeds with inversion of configuration at an asymmetric carbon (see p. 426) It was by a series of such transformations that lactic acid was related to alanine:

See also, the discussion on p. 427.

3. Biochemical methods. In a series of similar compounds, such as amino acids or certain types of steroids, a given enzyme will usually attack only molecules with one kind of configuration. If the enzyme attacks only the L form of eight amino acids, say, then attack on the unknown ninth amino acid will also be on the L form.

4. Optical comparison. It is sometimes possible to use the sign and extent of rotation to determine which isomer has which configuration. In a homologous series, the rotation usually changes gradually and in one direction. If the configurations of enough members of the series are known, the configurations of the missing ones can be determined by extrapolation. Also certain groups contribute more or less fixed amounts to the rotation of the parent molecule, especially when the parent is a rigid system, such as a steroid.

5. The special X-ray method of Bijvoet gives direct answers and has been used in a number of cases.[86]

6. One of the most useful methods for determining enantiomeric composition is to derivatize the alcohol with a chiral nonracemic reagent and examine the ratio of resulting diastereomers by gas chromatography (gc).[104] There are many derivatizing agents available, but the most widely used are derivatives of α-methoxy-α-trifluoromethylphenyl acetic acid (MTPA, Mosher's acid,

[104]Parker, D. *Chem. Rev.* **1991**, *91*, 1441.

41).[105] Reaction with a chiral nonracemic alcohol (R*OH, where R* is a group containing a stereogenic center) generates a Mosher's ester (**42**) that can be analyzed for diastereomeric composition by ^{1}H or ^{19}F NMR, as well as by chromatographic techniques.[106] Alternatively, complexation with lanthanide shift reagents allow the signals of the MTPA ester to be resolved and used to determine enantiomeric composition.[107] This nmr method, as well as other related methods,[108] are effective for determining the absolute configuration of an alcohol of interest (R*OH).[109] Two, of many other reagents that have been developed to allow the enantiopurity of alcohols and amines to be determined include **43** and **44**. Chloromethyl lactam **43** reacts with R*OH or R*NHR (R*NH$_2$),[110] forming derivatives that allow analysis by ^{1}H NMR and **44** reacts with alkoxides (R*O$^-$)[111] to form a derivative that can be analyzed by ^{31}P NMR. For a more detailed discussion of methods to determine optical purity (see p. 179).

43   **44**

7. Other methods have also been used for determining absolute configuration in a variety of molecules, including optical rotatory dispersion,[112] circular dichroism,[113,114] and asymmetric synthesis (see p. 166). Optical rotatory dispersion (ORD) is a measurement of specific rotation, [α], as a function of wavelength.[115] The change of specific rotation [α] or molar rotation [Φ]

[105]Dale, J. A.; Dull, D.L.; Mosher, H. S. *J. Org. Chem.* **1969**, 34, 2543; Dale, J.A.; Mosher, H.S. *J. Am. Chem. Soc.* **1973**, 95, 512.

[106]See Mori, K.; Akao, H. *Tetrahedron Lett.* **1978**, 4127; Plummer, E.L.; Stewart, T.E.; Byrne, K.; Pearce, G.T.; Silverstein, R.M. *J. Chem. Ecol.* **1976**, 2, 307. See also Seco, J.M.; Quiñoá, E.; Riguera, R. *Tetrahedron Asymmetry* **2000**, 11, 2695.

[107]Yamaguchi, S.; Yasuhara, F.; Kabuto, K. *Tetrahedron* **1976**, 32, 1363; Yasuhara, F.; Yamaguchi, S. *Tetrahedron Lett.* **1980**, 21, 2827; Yamaguchi, S.; Yasuhara, F. *Tetrahedron Lett.* **1977**, 89.

[108]Latypov, S.K.; Ferreiro, M.J.; Quiñoá, E.; Riguera, R. *J. Am. Chem. Soc.* **1998**, 120, 4741; Latypov, S.K.; Seco, J.M.; Quiñoá, E.; Riguera, R. *J. Org. Chem.* **1995**, 60, 1538.

[109]Seco, J.M.; Quiñoá, E.; Riguera, R. *Chem. Rev.* **2004**, 104, 17.

[110]Smith, M.B.; Dembofsky, B.T.; Son, Y.C. *J. Org. Chem.* **1994**, 59, 1719; Latypov, S.K.; Riguera, R.; Smith, M.B.; Polivkova, J. *J. Org. Chem.* **1998**, 63, 8682.

[111]Alexakis, A.; Mutti, S.; Mangeney, P. *J. Org. Chem.* **1992**, 57, 1224.

[112]See Ref. 268 for books and reviews on optical rotatory dispersion and CD. For predictions about anomalous ORD, see Polavarapu, P.L.; Zhao, C. *J. Am. Chem. Soc.* **1999**, 121, 246.

[113]Gawroński, J.; Grajewski, J. *Org. Lett.* **2003**, 5, 3301. See Ref. 268.

[114]For a determination of the absolute configuration of chiral sulfoxides by vibrational circular dichroism spectroscopy, see Stephens, P.J.; Aamouche, A.; Devlin, F.J.; Superchi, S.; Donnoli, M.I.; Rosini, C. *J. Org. Chem.* **2001**, 66, 3671.

[115]Eliel, E.L.; Wilen, S.H.; Mander, L.N. *Stereochemistry of Organic Compounds*, Wiley, NY, **1994**, pp. 1203, 999–1003.

with wavelength is measured, and a plot of either versus wavelength is often related to the sense of chirality or the substance under consideration. In general, the absolute value of the rotation increases as the wavelength decreases. The plot of circular dichroism (CD) is the differential absorption of left and right circularly polarized radiation by a nonracemic sample, taking place only in spectral regions in which absorption bands are found in the isotropic or visible electronic spectrum.[116] The primary application of both ORD and CD is for the assignment of configuration or conformation.[117] Configurational and conformational analysis have been carried out using infrared and vibrational circular dichroism (VCD) spectroscopies.[118]

In one example of the use of these techniques, one of the more effective methods for derivatizing 1,2-diols is the method employing dimolybdenum tetraacetate [$Mo_2(AcO)_4$] developed by Snatzke and Frelek.[119] Exposure of the resulting complex to air leads, in most cases, to a significant induced CD spectrum (known as ICD). The method can be used for a variety of 1,2-diols.[120]

8. Kishi and co-worker's[121] developed an NMR database of various molecules in chiral solvents, for the assignment of relative and absolute stereochemistry without derivatization or degradation. Kishi referred to this database as a "universal NMR database."[122] The diagram provided for diols **45** illustrates the method. The graph presents the difference in carbon chemical shifts between the average and the values for **45** (100 MHz) in DMBA (N,α-dimethylbenzylamine). Spectra were recorded in both enantiomers of the solvent, where the solid bar was recorded in (R)-DMBA and the shaded bar in (S)-DMBA. The X- and Y-axes represent carbon number and $\Delta\delta$ ($\delta_{45a-h} - \delta_{ave}$ in ppm), respectively. The graphs are taken from "the ^{13}C NMR database in (R)- and (S)-DMBA as a deviation in chemical shift for each carbon of a given diastereomer from the average chemical shift of the carbon in question. Each diastereomer exhibits an almost identical NMR profile for (R)- and (S)-DMBA but shows an NMR profile distinct and differing from the other diastereomers, demonstrating that the database in (R)- and/or (S)-DMBA can

[116]Eliel, E.L.; Wilen, S.H.; Mander, L.N. *Stereochemistry of Organic Compounds*, Wiley, NY, *1994*, pp. 1195, 1003–1007.

[117]Eliel, E.L.; Wilen, S.H.; Mander, L.N. *Stereochemistry of Organic Compounds*, Wiley, NY, *1994*, pp. 1007–1071; Nakanishi, K.; Berova, N.; Woody, R.W. *Circular Dichroism: Principles and Applications*, VCH, NY, *1994*; Purdie, N.; Brittain, H.G. *Analytical Applications of Circular Dichroism*, Elsevier, Amsterdam, The Netherlands, *1994*.

[118]Devlin, F.J.; Stephens, P.J.; Osterle, C.; Wiberg, K.B.; Cheeseman, J.R.; Frisch, M.J. *J. Org. Chem.* *2002*, *67*, 8090.

[119]Frelek, J.; Geiger, M.; Voelter, W. *Curr. Org. Chem. 1999*, *3*, 117–146 and references cited therein.; Snatzke, G.; Wagner, U.; Wolff, H. P. *Tetrahedron 1981*, *37*, 349; Frelek, J.; Snatzke, G. *Fresenius J. Anal. Chem. 1983*, *316*, 261; Frelek, J.; Pakulski, Z.; Zamojski, A. *Tetrahedron: Asymmetry 1996*, *7*, 1363; Frelek, J.; Ikekawa, N.; Takatsuto, S.; Snatzke, G. *Chirality 1997*, *9*, 578.

[120]Di Bari, L.; Pescitelli, G.; Pratelli, C.; Pini, D.; Salvadori, P. *J. Org. Chem. 2001*, *66*, 4819.

[121]Kobayashi, Y.; Hayashi, N.; Tan, C.-H.; Kishi, Y. *Org. Lett. 2001*, *3*, 2245; Hayashi, N.; Kobayashi, Y.; Kishi, Y. *Org. Lett. 2001*, *3*, 2249; Kobayashi, Y.; Hayashi, N.; Kishi, Y. *Org. Lett. 2001*, *3*, 2253.

[122]Kobayashi, Y.; Tan, C.-H.; Kishi, Y. *J. Am. Chem. Soc. 2001*, *123*, 2076.

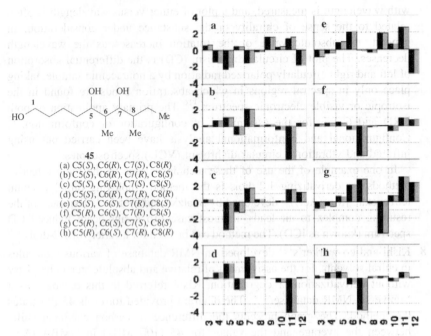

Fig. 4.3. Proton NMR analysis for assignment of stereochemistry.

be used for prediction of the relative stereochemistry of structural motifs in an intact form."[123]

A ^{1}H NMR analysis method has been developed that leads to the assignment of the stereochemistry of β-hydroxy ketones, by visual inspection of the ABX patterns for the (R)-methylene unit of the β-hydroxyketones.[124] Since β-hydroxy ketones are derived from the aldol reaction (see p. 1339), this new method is particularly useful in organic synthesis. A method has also been developed that uses ^{13}C NMR to determine the relative stereochemistry of 2,3-dialkylpentenoic acids.[125]

The Cause of Optical Activity

The question may be asked: Just why does a chiral molecule rotate the plane of polarized light? Theoretically, the answer to this question is known and in a greatly simplified form may be explained as follows.[126]

[123]Kobayashi, Y.; Hayashi, N.; Tan, C.-H.; Kishi, Y. *Org. Lett.* **2001**, *3*, 2245.

[124]Roush, W.R.; Bannister, T.D.; Wendt, M.D.; VanNieuwenhze, M.S.; Gustin, D.J.; Dilley, G.J.; Lane, G.C.; Scheidt, K.A.; Smith III, W.J. *J. Org. Chem.* **2002**, *67*, 4284.

[125]Hong, S.-p.; McIntosh, M.C. *Tetrahedron* **2002**, *57*, 5055.

[126]For longer, nontheoretical discussions, see Eliel, E.L.; Wilen, S.H.; Mander, L.N. *Stereochemistry of Organic Compounds*, Wiley-Interscience, NY, **1994**, pp. 93–94, 992–999; Wheland, G.W. *Advanced Organic CHemistry*, 3rd ed., Wiley, NY, **1960**, pp. 204–211. For theoretical discussions, see Caldwell, D.J.; Eyring, H. *The Theory of Optical Activity* Wiley, NY, **1971**; Buckingham, A.D.; Stiles, P.J. *Acc. Chem. Res.* **1974**, *7*, 258; Mason, S.F. *Q. Rev. Chem. Soc.* **1963**, *17*, 20.

Whenever any light hits any molecule in a transparent material, the light is slowed because of interaction with the molecule. This phenomenon on a gross scale is responsible for the refraction of light and the decrease in velocity is proportional to the refractive index of the material. The extent of interaction depends on the polarizability of the molecule. Plane-polarized light may be regarded as being made up of two kinds of circularly polarized light. Circularly polarized light has the appearance (or would have, if one could see the wave) of a helix propagating around the axis of light motion, and one kind is a left- and the other is a right-handed helix. As long as the plane-polarized light is passing through a symmetrical region, the two circularly polarized components travel at the same speed. However, a chiral molecule has a different polarizability depending on whether it is approached from the left or the right. One circularly polarized component approaches the molecule, so to speak, from the left and sees a different polarizability (hence on a gross scale, a different refractive index) than the other and is slowed to a different extent. This would seem to mean that the left- and right-handed circularly polarized components travel at different velocities, since each has been slowed to a different extent. However, it is not possible for two components of the same light to be traveling at different velocities. What actually takes place, therefore, is that the faster component "pulls" the other toward it, resulting in rotation of the plane. Empirical methods for the prediction of the sign and amount of rotation based on bond refractions and polarizabilities of groups in a molecule have been devised,[127] and have given fairly good results in many cases.

In liquids and gases, the molecules are randomly oriented. A molecule that is optically inactive because it has a plane of symmetry will very seldom be oriented so that the plane of the polarized light coincides with the plane of symmetry. When it is so oriented, that particular molecule does not rotate the plane, but all others not oriented in that manner do rotate the plane, even though the molecules are achiral. There is no net rotation because, even though the molecules are present in large numbers and randomly oriented, there will always be another molecule later on in the path of the light that is oriented exactly opposite and will rotate the plane back again. Even although nearly all molecules rotate the plane individually, the total rotation is zero. For chiral molecules, however (if there is no racemic mixture), no opposite orientation is present and there is a net rotation.

An interesting phenomenon was observed when the CD of chiral molecules was measured in achiral solvents. The chiral solvent contributed as much as 10–20% to the CD intensity in some cases. Apparently, the chiral compound can induce a solvation structure that is chiral, even when the solvent molecules themselves are achiral.[128]

---

[127]Brewster, J.H. *Top. Stereochem.* *1967*, *2*, 1, *J. Am. Chem. Soc.* *1959*, *81*, 5475, 5483, 5493; Davis, D.D.; Jensen, F.R. *J. Org. Chem.* *1970*, *35*, 3410; Jullien, F.R.; Requin, F.; Stahl-Larivière, H. *Nouv. J. Chim.*, *1979*, *3*, 91; Sathyanarayana, B.K.; Stevens, E.S. *J. Org. Chem.* *1987*, *52*, 3170; Wroblewski, A.E.; Applequist, J.; Takaya, A.; Honzatko, R.; Kim, S.; Jacobson, R.A.; Reitsma, B.H.; Yeung, E.S.; Verkade, J.G. *J. Am. Chem. Soc.* *1988*, *110*, 4144.
[128]Fidler, J.; Rodger, P.M.; Rodger, A. *J. Chem. Soc. Perkin Trans. 2* *1993*, 235.

MOLECULES WITH MORE THAN ONE STEREOGENIC CENTER

When a molecule has two stereogenic centers, each has its own configuration and can be classified (R) or (S) by the Cahn–Ingold–Prelog method. There are a total of four isomers, since the first center may be (R) or (S) and so may the second. Since a molecule can have only one mirror image, only one of the other three can be the enantiomer of **A**. This is **B** [the mirror image of an (R) center is *always* an (S) center]. Both **C** and **D** are a second pair of enantiomers and the relationship of **C** and **D**

to **A** and **B** is designated by the term *diastereomer*. Diastereomers may be defined as *stereoisomers that are not enantiomers*. Since **C** and **D** are enantiomers, they must have identical properties, except as noted on p. 138; the same is true for **A** and **B**. However, the properties of **A** and **B** are not identical with those of **C** and **D**. They have different melting points, boiling points, solubilities, reactivity, and all other physical, chemical, and spectral properties. The properties are usually *similar*, but not *identical*. In particular, diastereomers have different specific rotations; indeed one diastereomer may be chiral and rotate the plane of polarized light while another may be achiral and not rotate at all (an example is presented below).

It is now possible to see why, as mentioned on p. 138, enantiomers react at different rates with other chiral molecules, but at the same rate with achiral molecules. In the latter case, the activated complex formed from the (R) enantiomer and the other molecule is the mirror image of the activated complex formed from the (S)

The three stereoisomers of tartaric acid

enantiomer and the other molecule. Since the two activated complexes are enantiomeric, their energies are the same and the rates of the reactions in which they are formed must be the same (see Chapter 6). However, when an (R) enantiomer reacts with a chiral molecule that has, say, the (R) configuration, the activated complex has two chiral centers with configurations (R) and (R), while the activated complex formed from the (S) enantiomer has the configurations (S) and (R). The two activated complexes are diastereomeric, do not have the same energies, and consequently are formed at different rates.

Although four is the maximum possible number of isomers when the compound has two stereogenic centers (chiral compounds without a chiral carbon, or with one chiral carbon and another type of stereogenic center, also follow the rules described here), some compounds have fewer. When the three groups on one chiral atom are the same as those on the other, one of the isomers (called a *meso* form) has a plane of symmetry, and hence is optically inactive, even though it has two chiral carbons. Tartaric acid is a typical case. There are only three isomers of tartaric acid: a pair of enantiomers and an inactive meso form. For compounds that have two chiral atoms, meso forms are found only where the four groups on one of the chiral atoms are the same as those on the other chiral atom.

CH₃	CH₃	CH₃	CH₃
(S) H—OH	(S) H—OH	(S) H—OH	(R) HO—H
H—OH	HO—H	HO—H	H—OH
(R) H—OH	(R) H—OH	(S) HO—H	(R) H—OH
CH₃	CH₃	CH₃	CH₃
meso	meso	*dl* Pair	

In most cases with more than two stereogenic centers, the number of isomers can be calculated from the formula 2^n, where n is the number of chiral centers, although in some cases the actual number is less than this, owing to meso forms.[129] An interesting case is that of 2,3,4-pentanetriol (or any similar molecule). The middle carbon is not asymmetric when the 2- and 4-carbons are both (R) (or both S), but is asymmetric when one of them is (R) and the other is (S). Such a carbon is called a *pseudoasymmetric* carbon. In these cases, there are four isomers: two meso forms and one *dl* pair. The student should satisfy themselves, remembering the rules governing the use of the Fischer projections, that these isomers are different, that the *meso* forms are superimposable on their mirror images, and that there are no other stereoisomers. Two diastereomers that have a different configuration at only one chiral center are called *epimers*.

In compounds with two or more chiral centers, the absolute configuration must be separately determined for each center. The usual procedure is to determine the configuration at one center by the methods discussed on pp. 158–162 and then to relate the configuration at that center to the others in the molecule. One method is X-ray crystallography, which, as previously noted, cannot be used to determine the absolute configuration at any stereogenic center, but which does give relative configurations of all the stereogenic centers in a molecule and hence the absolute configurations of all once the first is independently determined. Other physical and chemical methods have also been used for this purpose.

The problem arises how to name the different stereoisomers of a compound when there are more than two.[2] Enantiomers are virtually always called by the same name, being distinguished by (R) and (S) or D and L or ($+$) or ($-$). In the early days of organic chemistry, it was customary to give each pair of enantiomers a different name or at least a different prefix (such as *epi-*, *peri-*, etc.). Thus the aldohexoses are called glucose, mannose, idose, and so on, although they are all 2,3,4,5,6-pentahydroxyhexanal (in their open-chain forms). This practice was partially due to lack of knowledge

---

[129]For a method of generating all stereoisomers consistent with a given empirical formula, suitable for computer use, see Nourse, J.G.; Carhart, R.E.; Smith, D.H.; Djerassi, C. *J. Am. Chem. Soc.* **1979**, *101*, 1216; **1980**, *102*, 6289.

about which isomers had which configurations.[130] Today it is customary to describe *each chiral position* separately as either (R) or (S) or, in special fields, to use other symbols. Thus, in the case of steroids, groups above the "plane" of the ring system are designated β, and those below it α. Solid lines are often used to depict β groups and dashed lines for a groups. An example is

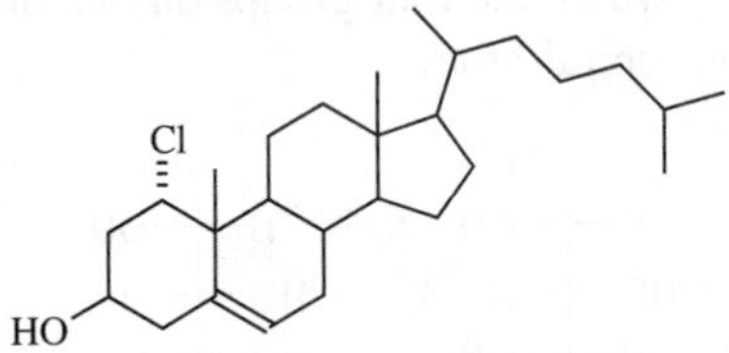

1α-Chloro-5-cholesten-3β-ol

For many open-chain compounds, prefixes are used that are derived from the names of the corresponding sugars and that describe the whole system rather than each chiral center separately. Two such common prefixes are erythro- and threo-, which are applied to systems containing two asymmetric carbons when two of the groups

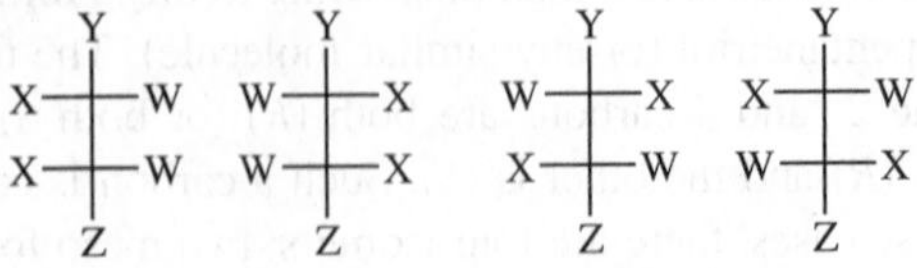

Erythro *dl* pair   Threo *dl* pair

are the same and the third is different.[131] The erythro pair has the identical groups on the same side when drawn in the Fischer convention, and if Y were changed to Z, it would be meso. The threo pair has them on opposite sides, and if Y were changed to Z, it would still be a *dl* pair. Another system[132] for designating stereoisomers[133] uses the terms syn and anti. The "main chain" of the molecule is drawn in the common zigzag manner. Then, if two non-hydrogen substituents are on the same side of the plane defined by the main chain, the designation is syn; otherwise it is anti.

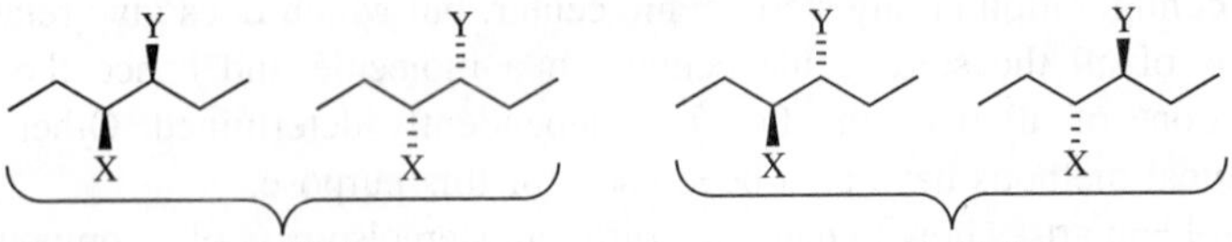

syn *dl* Pair   anti *dl* Pair

---

[130]A method has been developed for the determination of stereochemistry in six-membered chairlike rings using residual dipolar couplings. See Yan, J.; Kline, A. D.; Mo, H.; Shapiro, M. J.; Zartler, E. R. *J. Org. Chem.* **2003**, *68*, 1786.

[131]For more general methods of designating diastereomers, see Carey, F.A.; Kuehne, M.E. *J. Org. Chem.* **1982**, *47*, 3811; Boguslavskaya, L.S. *J. Org. Chem. USSR* **1986**, *22*, 1412; Seebach, D.; Prelog, V. *Angew. Chem. Int. Ed.* **1982**, *21*, 654; Brewster, J.H. *J. Org. Chem.* **1986**, *51*, 4751. See also Tavernier, D. *J. Chem. Educ.* **1986**, *63*, 511; Brook, M.A. *J. Chem. Educ.* **1987**, *64*, 218.

[132]For still another system, see Seebach, D.; Prelog, V. *Angew. Chem. Int. Ed.* **1982**, *21*, 654.

[133]Masamune, S.; Kaiho, T.; Garvey, D.S. *J. Am. Chem. Soc.* **1982**, *104*, 5521.

Asymmetric Synthesis

Organic chemists often wish to synthesize a chiral compound in the form of a single enantiomer or diastereomer, rather than as a mixture of stereoisomers. There are two basic ways in which this can be done.[134] The first way, which is more common, is to begin with a single stereoisomer, and to use a synthesis that does not affect the stereogenic center (or centers), as in the glyceraldehyde–glyceric acid example on p. 154. The optically active starting compound can be obtained by a previous synthesis, or by resolution of a racemic mixture (p. 172), but it is often more convenient to obtain it from Nature, since many compounds, such as amino acids, sugars, and steroids, are present in Nature in the form of a single enantiomer or diastereomer. These compounds are regarded as a *chiral pool;* that is, readily available compounds that can be used as starting materials.[135]

The other basic method is called *asymmetric synthesis,*[136] or *stereoselective synthesis.* As mentioned earlier, optically active materials cannot be created from

[134]For a monograph that covers both ways, including a list of commercially available optically active starting compounds, see Morrison, J.D.; Scott, J.W. *Asymmetric Synthesis* Vol. 4, Academic Press, NY, *1984*. For a monograph covering a more limited area, see Williams, R.M. *Synthesis of Optically Active α-Amino Acids*, Pergamon, Elmsford, NY, *1989*. For reviews on both ways, see Crosby, J. *Tetrahedron 1991*, *47*, 4789; Mori, K. *Tetrahedron 1989*, *45*, 3233.

[135]For books on the synthesis of optically active compounds starting from natural products, see Coppola, G.M.; Schuster, H.F. *Asymmetric Synthesis*, Wiley, NY, *1987* (amino acids as starting compounds); Hanessian, S. *Total Synthesis of Natural Products: The Chiron Approach*, Pergamon, Elmsford, NY, *1983* (mostly carbohydrates as starting compounds). For reviews, see Jurczak, J.; Pikul, S.; Bauer, T. *Tetrahedron 1986*, *42*, 447; Hanessian, S. *Aldrichimica Acta 1989*, *22*, 3; Jurczak, J.; Gotebiowski, A. *Chem. Rev. 1989*, *89*, 149.

[136]For a treatise on this subject, see Morrison, J.D. *Asymmetric Synthesis* 5 vols. [Vol. 4 coedited by Scott, J.W.], Academic Press, NY, *1983–1985*. For books, see Nógrádi, M. *Stereoselective Synthesis*, VCH, NY, *1986*; Eliel, E.L.; Otsuka, S. *Asymmetric Reactions and Processes in Chemistry*, American Chemical Society, Washington, *1982*; Morrison, J.D.; Mosher, H.S. *Asymmetric Organic Reactions*, Prentice-Hall, Englewood Cliffs, NJ, *1971*, paperback reprint, American Chemical Society, Washington, *1976*; Izumi,Y.; Tai, A. *Stereo-Differentiating Reactions*, Academic Press, NY, Kodansha Ltd. Tokyo, *1977*. For reviews, see Ward, R.S. *Chem. Soc. Rev. 1990*, *19*, 1; Whitesell, J.K. *Chem. Rev. 1989*, *89*, 1581; Fujita, E.; Nagao, Y. *Adv. Heterocycl. Chem. 1989*, *45*, 1; Kochetkov, K.A.; Belikov, V.M. *Russ. Chem. Rev. 1987*, *56*, 1045; Oppolzer, W. *Tetrahedron 1987*, *43*, 1969; Seebach, D.; Imwinkelried, R.; Weber, T. *Mod. Synth. Methods 1986*, *4*, 125; ApSimon, J.W.; Collier, T.L. *Tetrahedron 1986*, *42*, 5157; Mukaiyama, T.; Asami, M. *Top. Curr. Chem. 1985*, *127*, 133; Martens, J. *Top. Curr. Chem. 1984*, *125*, 165; Duhamel, L.; Duhamel, P.; Launay, J.; Plaquevent, J. *Bull. Soc. Chim. Fr. 1984*, II-421; Mosher, H.S.; Morrison, J.D. *Science, 1983*, *221*, 1013; Schöllkopf, U. *Top. Curr. Chem. 1983*, *109*, 65; Quinkert, G.; Stark, H. *Angew. Chem. Int. Ed. 1983*, *22*, 637; Tramontini, M. *Synthesis 1982*, 605; Drauz, K.; Kleeman, A.; Martens, J. *Angew. Chem. Int. Ed. 1982*, *21*, 584; Wynberg, H. *Recl. Trav. Chim. Pays-Bas 1981*, *100*, 393; Bartlett, P.A. *Tetrahedron 1980*, *36*, 2; Valentine, Jr., D.; Scott, J.W. *Synthesis 1978*, 329; Kagan, H.B.; Fiaud, J.C. *Top. Stereochem. 1978*, *10*, 175; ApSimon, J., in Bentley, K.W.; Kirby, G.W. *Elucidation of Organic Structures by Physical and Chemical Methods*, 2nd ed. (Vol. 4 of Weissberger, A. *Techniques of Chemistry*), pt. 3, Wiley, NY, *1972*, pp. 251–408; Boyd, D.R.; McKervey, M.A. *Q. Rev. Chem. Soc, 1968*, *22*, 95; Goldberg, S.I. *Sel. Org. Transform. 1970*, *1*, 363; Klabunovskii, E.I.; Levitina, E.S. *Russ. Chem. Rev. 1970*, *39*, 1035; Inch, T.D. *Synthesis 1970*, 466; Mathieu, J.; Weill-Raynal, J. *Bull. Soc. Chim. Fr. 1968*, 1211; Amariglio, A.; Amariglio, H.; Duval, X. *Ann. Chim. (Paris) [14] 1968*, *3*, 5; Pracejus, H. *Fortschr. Chem. Forsch. 1967*, *8*, 493; Velluz, L.; Valls, J.; Mathieu, J. *Angew. Chem. Int. Ed. 1967*, *6*, 778.

inactive starting materials and conditions, except in the manner previously noted.[94] However, when a new stereogenic center is created, the two possible configurations need not be formed in equal amounts if anything is present that is not symmetric. We discuss asymmetric synthesis under four headings:

1. *Active Substrate.* If a new chiral center is created in a molecule that is already optically active, the two diastereomers are not (except fortuitously) formed in equal amounts. The reason is that the direction of attack by the reagent is determined by the groups already there. For certain additions to the carbon–oxygen double bond of ketones containing an asymmetric α carbon, *Cram's rule* predicts which of two diastereomers will predominate (diastereoselectivity).[137,138] The reaction of **46**, which has a stereogenic center at the α-carbon, and HCN can generate two possible diastereomers,

47 and 48. If **46** is observed along its axis, it may be represented as in **49** (see p. 197), where S, M, and L stand for small, medium, and large, respectively. The oxygen of the carbonyl orients itself between the small- and the medium-sized groups. The rule is that the incoming group preferentially attacks on the side of the plane containing the small group. By this rule, it can be predicted that **48** will be formed in larger amounts than **47**.

| | **49** | | Major product | | Minor product |

Another model can be used to predict diastereoselectivity, which assumes reactant-like transition states and that the separation of the incoming group

[137]Leitereg, T.J.; Cram, D.J. *J. Am. Chem. Soc.* **1968**, *90*, 4019. For discussions, see Salem, L. *J. Am. Chem. Soc.* **1973**, *95*, 94; Anh, N.T. *Top. Curr. Chem*, **1980**, *88*, 145, 151–161; Eliel, E.L., in Morrison, J.D. *Asymmetric Synthesis*, Vol. 2, Academic Press, NY, **1983**, pp. 125–155. See Smith, R.J.; Trzoss, M.; Bühl, M.; Bienz, S. *Eur. J. Org. Chem.* **2002**, 2770.
[138]For reviews, see Eliel, E.L. *The Stereochemistry of Carbon Compounds*, McGraw-Hill, NY, **1962**, pp. 68–74. For reviews of the stereochemistry of addition to carbonyl compounds, see Bartlett, P.A. *Tetrahedron* **1980**, *36*, 2, pp. 22–28; Ashby, E.C.; Laemmle, J.T. *Chem. Rev.* **1975**, *75*, 521; Goller, E.J. *J. Chem. Educ.* **1974**, *51*, 182; Toromanoff, E. *Top. Stereochem.* **1967**, *2*, 157.

and any electronegative substituent at the α-carbon is greatest. Transition state models **50** and **51** are used to predict diastereoselectivity in what is known as the *Felkin–Ahn Model*.[139] The so-called Cornforth model has also been presented as a model for carbonyl addition.[140]

Many reactions of this type are known, and in some the extent of favoritism approaches 100% (for an example see reaction **12-12**).[141] The farther away the reaction site is from the chiral center, the less influence the latter has and the more equal the amounts of diastereomers formed.

In a special case of this type of asymmetric synthesis, a compound (**52**) with achiral molecules, but whose crystals are chiral, was converted by UV light to a single enantiomer of a chiral product (**53**).[142]

It is often possible to convert an achiral compound to a chiral compound by (*1*) addition of a chiral group; (*2*) running an asymmetric synthesis, and (*3*) cleavage of the original chiral group. An example is conversion of the achiral 2-pentanone to the chiral 4-methyl-3-heptanone, **55**.[143] In this case, >99% of the product was the (*S*) enantiomer. Compound **54** is called a *chiral auxiliary* because it is used to induce asymmetry and is then removed.

[139]Chérest, M.; Felkin, H.; Prudent, N. *Tetrahedron Lett.* *1968*, 2199.

[140]Evans, D.A.; Siska, S.J.; Cee, V.J. *Angew. Chem. Int. Ed.* *2003*, *42*, 1761.

[141]For other examples and references to earlier work, see Eliel, E.L., in Morrison, J.D. *Asymmetric Synthesis*, Vol. 2, Academic Press, NY, *1983*, pp. 125–155; Eliel, E.L.; Koskimies, J.K.; Lohri, B. *J. Am. Chem. Soc.* *1978*, *100*, 1614; Still, W.C.; McDonald, J.H. *Tetrahedron Lett.* *1980*, *21*, 1031; Still, W.C.; Schneider, J.A. *Tetrahedron Lett.* *1980*, *21*, 1035.

[142]Evans, S.V.; Garcia-Garibay, M.; Omkaram, N.; Scheffer, J.R.; Trotter, J.; Wireko, F. *J. Am. Chem. Soc.* *1986*, *108*, 5648; Garcia-Garibay, M.; Scheffer, J.R.; Trotter, J.; Wireko, F. *Tetrahedron Lett.* *1987*, *28*, 4789. For an earlier example, see Penzien, K.; Schmidt, G.M.J. *Angew. Chem. Int. Ed.* *1969*, *8*, 608.

[143]Enders, D.; Eichenauer, H.; Baus, U.; Schubert, H.; Kremer, K.A.M. *Tetrahedron* *1984*, *40*, 1345.

One diastereomer
predominates

2. *Active Reagent.* A pair of enantiomers can be separated by an active reagent
that reacts faster with one of them than it does with the other (this is also a
method of resolution). If the absolute configuration of the reagent is known,
the configuration of the enantiomers can often be determined by a knowledge
of the mechanism and by seeing which diastereomer is preferentially

formed.[144] Creation of a new chiral center in an inactive molecule can also be
accomplished with an active reagent, although it is rare for 100% selectivity
to be observed. An example[145,146] is the reduction of methyl benzoylformate

[144]See, for example, Horeau, A. *Tetrahedron Lett.* **1961**, 506; Marquet, A.; Horeau, A. *Bull. Soc. Chim. Fr.*
1967, 124; Brockmann Jr., H.; Risch, N. *Angew. Chem. Int. Ed.* **1974**, *13*, 664; Potapov, V.M.; Gracheva,
R.A.; Okulova, V.F. *J. Org. Chem. USSR* **1989**, *25*, 311.

[145]Meyers, A.I.; Oppenlaender, T. *J. Am. Chem. Soc.* **1986**, *108*, 1989. For reviews of asymmetric
reduction, see Morrison, J.D. *Surv. Prog. Chem.* **1966**, *3*, 147; Yamada, S.; Koga, K. *Sel. Org. Transform.*
1970, *1*, 1. See also, Morrison, J.D. *Asymmetric Synthesis*, Vol. 2, Academic Press, NY, **1983**.

[146]For reviews, see, in Morrison, J.D. *Asymmetric Synthesis* Vol. 5, Academic Press, NY, **1985**, the reviews by
Halpern, J. pp. 41–69, Koenig, K.E. pp. 71–101, Harada, K. pp. 345–383; Ojima, I.; Clos, N.; Bastos, C.
Tetrahedron **1989**, *45*, 6901, pp. 6902–6916; Jardine, F.H. in Hartley, F.R. *The Chemistry of the Metal-Carbon
Bond*, Vol. 4, Wiley, NY, **1987**, pp. 751–775; Nógrádi, M. *Stereoselective Synthesis*, VCH, NY, **1986**, pp. 53–
87; Knowles, W.S. *Acc. Chem. Res.* **1983**, *16*, 106; Brunner, H. *Angew. Chem. Int. Ed.* **1983**, *22*, 897;
Klabunovskii, E.I. *Russ. Chem. Rev.* **1982**, *51*, 630; Č aplar, V.; Comisso, G.; Š unjić, V. *Synthesis* **1981**, 85;
Morrison, J.D.; Masler, W.F.; Neuberg, M.K. *Adv. Catal.* **1976**, *25*, 81; Kagan, H.B. *Pure Appl. Chem.* **1975**,
43, 401; Bogdanović, B. *Angew. Chem. Int. Ed.* **1973**, *12*, 954. See also Brewster, J.H. *Top. Stereochem.* **1967**,
2, 1, *J. Am. Chem. Soc.* **1959**, *81*, 5475, 5483, 5493; Davis, D.D.; Jensen, F.R. *J. Org. Chem.* **1970**, *35*, 3410;
Jullien, F.R.; Requin, F.; Stahl-Larivière, H. *Nouv. J. Chim.* **1979**, *3*, 91; Sathyanarayana, B.K.; Stevens, E.S.
J. Org. Chem. **1987**, *52*, 3170; Wroblewski, A.E.; Applequist, J.; Takaya, A.; Honzatko, R.; Kim, S.;
Jacobson, R.A.; Reitsma, B.H.; Yeung, E.S.; Verkade, J.G. *J. Am. Chem. Soc.* **1988**, *110*, 4144.

with optically active *N*-benzyl-3-(hydroxymethyl)-4-methyl-1,4-dihydropyr-
idine (**56**) to produce mandelic acid that contained ~97.5% of the (*S*)-(+)
isomer and 2.5% of the (*R*)-(−) isomer (for another example, see p. 1079).
Note that the other product, **57**, is not chiral. Reactions like this, in which one
reagent (in this case **56**) gives up its chirality to another, are called *self-
immolative*. In this intramolecular example:

chirality is transferred from one atom to another in the same molecule.[147]

A reaction in which an inactive substrate is converted selectively to one of
two enantiomers is called an *enantioselective* reaction, and the process is
called *asymmetric induction*. These terms apply to reactions in this category
and in categories 3 and 4.

When an optically active substrate reacts with an optically active reagent
to form two new stereogenic centers, it is possible for both centers to be
created in the desired sense. This type of process is called *double asymmetric
synthesis*[148] (for an example, see p. 1349).

3. *Active Catalyst or Solvent.*[149] Many such examples are present in the literature, among them reduction of ketones and substituted alkenes to optically
active (though not optically pure) secondary alcohols and substituted alkanes
by treatment with hydrogen and a chiral homogeneous hydrogenation catalyst
(reactions **16-23** and **15-11**),[150] the treatment of aldehydes or ketones with
organometallic compounds in the presence of a chiral catalyst (see reaction
16-24), and the conversion of alkenes to optically active epoxides by
treatment with a hydroperoxide and a chiral catalyst (see reaction **15-50**).
In some instances, notably in the homogeneous catalytic hydrogenation of
alkenes (reaction **15-11**), the ratio of enantiomers prepared in this way is as
high as 98:2.[151] Other examples of the use of a chiral catalyst or solvent are

---

[147]Goering, H.L.; Kantner, S.S.; Tseng, C.C. *J. Org. Chem.* **1983**, *48*, 715.

[148]For a review, see Masamune, S.; Choy, W.; Petersen, J.S.; Sita, L.R. *Angew. Chem. Int. Ed.* **1985**, *24*, 1.

[149]For a monograph, see Morrison, J.D. *Asymmetric Synthesis*, Vol. 5, Academic Press, NY, **1985**. For
reviews, see Tomioka, K. *Synthesis* **1990**, 541; Consiglio, G.; Waymouth, R.M. *Chem. Rev.* **1989**, *89*, 257;
Brunner, H., in Hartley, F.R. *The Chemistry of the Metal-Carbon Bond*, Vol. 5, Wiley, NY, **1989**, pp. 109–
146; Noyori, R.; Kitamura, M. *Mod. Synth. Methods* **1989**, *5*, 115; Pfaltz, A. *Mod. Synth. Methods* **1989**, *5*,
199; Kagan, H.B. *Bull. Soc. Chim. Fr.* **1988**, 846; Brunner, H. *Synthesis* **1988**, 645; Wynberg, H. *Top.
Stereochem.* **1986**, *16*, 87.

[150]For reviews of these and related topics, see Zief, M.; Crane, L.J. *Chromatographic Separations*, Marcel
Dekker, NY, **1988**; Brunner, H. *J. Organomet. Chem.* **1986**, *300*, 39; Bosnich, B.; Fryzuk, M.D. *Top.
Stereochem.* **1981**, *12*, 119.

[151]See Vineyard, B.D.; Knowles, W.S.; Sabacky, M.J.; Bachman, G.L.; Weinkauff, D.J. *J. Am. Chem. Soc.*
1977, *99*, 5946; Fryzuk, M.D.; Bosnich, B. *J. Am. Chem. Soc.* **1978**, *100*, 5491.

the conversion of chlorofumaric acid (in the form of its diion) to the $(-)$-*threo* isomer of the di-ion of chloromalic acid by treatment with H_2O and the enzyme fumarase,[152] and the preparation of optically active aldols (aldol condensation, see reaction **16-35**) by the condensation of enolate anions with optically active substrates.[153]

$(-)$-*threo* Isomer

4. *Reactions in the Presence of Circularly Polarized Light.*[154] If the light used to initiate a photochemical reaction (Chapter 7) of achiral reagents is circularly polarized, then, in theory, a chiral product richer in one enantiomer might be obtained. However, such experiments have not proved fruitful. In certain instances, the use of left and right circularly polarized light *has* given products with opposite rotations[155] (showing that the principle is valid), but up to now the extent of favoritism has always been <1%.

Methods of Resolution[156]

A pair of enantiomers can be separated in several ways, of which conversion to diastereomers and separation of these by fractional crystallization is the most often used. In this method and in some of the others, both isomers can be recovered, but in some methods it is necessary to destroy one.

[152]Findeis, M.A.; Whitesides, G.M. *J. Org. Chem.* **1987**, *52*, 2838. For a monograph on enzymes as chiral catalysts, see Réty, J.; Robinson, J.A. *Stereospecificity in Organic Chemistry and Enzymology*, Verlag Chemie: Deerfield Beach, FL, **1982**. For reviews, see Klibanov, A.M. *Acc. Chem. Res.* **1990**, *23*, 114; Jones, J.B., *Tetrahedron* **1986**, *42*, 3351; Jones, J.B., in Morrison, J.D. *Asymmetric Synthesis*, Vol. 5, Academic Press, NY, **1985**, pp. 309–344; Svedas, V.; Galaev, I.U. *Russ. Chem. Rev.* **1983**, *52*, 1184. See also, Simon, H.; Bader, J.; Günther, H.; Neumann, S.; Thanos, J. *Angew. Chem. Int. Ed.* **1985**, *24*, 539.

[153]Heathcock, C.H.; White, C.T. *J. Am. Chem. Soc.* **1979**, *101*, 7076.

[154]For a review, See Buchardt, O. *Angew. Chem. Int. Ed.* **1974**, *13*, 179. For a discussion, see Barron L.D. *J. Am. Chem. Soc.* **1986**, *108*, 5539.

[155]See, for example, Bernstein, W.J.; Calvin, M.; Buchardt, O. *J. Am. Chem. Soc.* **1972**, *94*, 494; **1973**, *95*, 527, *Tetrahedron Lett.* **1972**, 2195; Nicoud, J.F.; Kagan, J.F. *Isr. J. Chem.* **1977**, *15*, 78. See also Zandomeneghi, M.; Cavazza, M.; Pietra, F. *J. Am. Chem. Soc.* **1984**, *106*, 7261.

[156]For a monograph, see Jacques, J.; Collet, A.; Wilen, S.H. *Enantiomers, Racemates, aand Resolutions*, Wiley, NY, **1981**. For reviews, see Wilen, S.H.; Collet, A.; Jacques, J. *Tetrahedron* **1977**, *33*, 2725; Wilen, S.H. *Top. Stereochem.* **1971**, *6*, 107; Boyle, P.H. *Q. Rev. Chem. Soc.* **1971**, *25*, 323; Buss, D.R.; Vermeulen, T. *Ind. Eng. Chem.* **1968**, *60* (8), 12. Eliel, E.L.; Wilen, S.H.; Mander, L.N. *Stereochemistry of Organic Compounds*, Wiley-Interscience, NY, **1994**, pp. 297–424.

1. *Conversion to Diastereomers*. If the racemic mixture to be resolved contains a carboxyl group (and no strongly basic group), it is possible to form a salt with an optically active base. Since the base used is, say, the (*S*) form, there will be a mixture of two salts produced having the configurations (*SS*) and (*RS*). Although the acids are enantiomers, the salts are diastereomers and have different properties. The property most often used for separation is differential solubility. The mixture of diastereomeric salts is allowed to crystallize from a suitable solvent. Since the solubilities are different, the initial crystals formed will be richer in one diastereomer. Filtration at this point will already have achieved a partial resolution. Unfortunately, the difference in solubilities is rarely if ever great enough to effect total separation with one crystallization. Usually, fractional crystallizations must be used and the process is long and tedious. Fortunately, naturally occurring optically active bases (mostly alkaloids) are readily available. Among the most commonly used are brucine, ephedrine, strychnine, and morphine. Once the two diastereomers have been separated, it is easy to convert the salts back to the free acids and the recovered base can be used again.

Most resolution is done on carboxylic acids and often, when a molecule does not contain a carboxyl group, it is converted to a carboxylic acid before resolution is attempted. However, the principle of conversion to diastereomers is not confined to carboxylic acids, and other functional groups[157] may be coupled to an optically active reagent.[158] Racemic bases can be converted to diastereomeric salts with active acids. Alcohols[159] can be converted to diastereomeric esters, aldehydes to diastereomeric hydrazones, and so on. Amino alcohols have been resolved using boric acid and chiral

---

[157]For summaries of methods used to resolve particular types of compounds, see Boyle, P.H. *Q. Rev. Chem. Soc.* **1971**, *25*, 323; Eliel, E.L.; Wilen, S.H.; Mander, L.N. *Stereochemistry of Organic Compounds*, Wiley-Interscience, NY, **1994**, pp. 322–424.

[158]For an extensive list of reagents that have been used for this purpose and of compounds resolved, see Wilen, S.H. *Tables of Resolving Agents and Optical Resolutions*, University of Notre Dame Press, Notre Dame, IN, **1972**.

[159]For a review of resolution of alcohols, see Klyashchitskii, B.A.; Shvets, V.I. *Russ. Chem. Rev.* **1972**, *41*, 592.

bipaphthols.[160] Phosphine oxides[161] and chiral calix[4]arenes[162] have been resolved. Chiral crown ethers have been used to separate mixtures of enantiomeric alkyl- and arylammonium ions, by the formation of diastereomeric complexes[163] (see also category 3, below). Even hydrocarbons can be converted to diastereomeric inclusion compounds,[164] with urea. Urea is not chiral, but the cage structure is.[165] Racemic unsaturated hydrocarbons have been resolved as inclusion complex crystals with a chiral host compound derived from tartaric acid.[166] *trans*-Cyclooctene (p. 150) was resolved by conversion to a platinum complex containing an optically active amine.[167]

Fractional crystallization has always been the most common method for the separation of diastereomers. When it can be used, binary phase diagrams for the diastereomeric salts have been used to calculate the efficiency of optical resolution.[168] However, it is tediousness and the fact that it is limited to solids prompted a search for other methods. Fractional distillation has given only limited separation, but GC[169] and preparative

[160]Periasamy, M.; Kumar, N. S.; Sivakumar, S.; Rao, V. D.; Ramanathan, C. R.; Venkatraman, L. *J. Org. Chem.* **2001**, *66*, 3828.

[161]Andersen, N.G.; Ramsden, P.D.; Che, D.; Parvez, M.; Keay, B.A. *Org. Lett.* **1999**, *1*, 2009; Andersen, N.G.; Ramsden, P.D.; Che, D.; Parvez, M.; Keay, B.A. *J. Org. Chem.* **2001**, *66*, 7478.

[162]Caccamese, S.; Bottino, A.; Cunsolo, F.; Parlato, S.; Neri, P. *Tetrahedron Asymmetry* **2000**, *11*, 3103.

[163]See, for example, Kyba, E.B.; Koga, K.; Sousa, L.R.; Siegel, M.G.; Cram, D.J. *J. Am. Chem. Soc.* **1973**, *95*, 2692; Slingenfelter, D.S.; Helgeson, R.C.; Cram, D.J. *J. Org. Chem.* **1981**, *46*, 393; Pearson, D.P.J.; Leigh, S.J.; Sutherland, I.O. *J. Chem. Soc. Perkin Trans. 1* **1979**, 3113; Bussman, W.; Lehn, J.M.; Oesch, U.; Plumeré, P.; Simon, W. *Helv. Chim. Acta* **1981**, *64*, 657; Davidson, R.B.; Bradshaw, J.S.; Jones, B.A.; Dalley, N.K.; Christensen, J.J.; Izatt, R.M.; Morin, F.G.; Grant, D.M. *J. Org. Chem.* **1984**, *49*, 353. See also Toda, F.; Tanaka, K.; Omata, T.; Nakamura, K.; Ōshima, T. *J. Am. Chem. Soc.* **1983**, *105*, 5151.

[164]For reviews of chiral inclusion compounds, including their use for resolution, see Prelog, V.; Kovačević, M.; Egli, M. *Angew. Chem. Int. Ed.* **1989**, *28*, 1147; Worsch, D.; Vögtle, F. *Top. Curr. Chem.* **1987**, *140*, 21; Toda, F. *Top. Curr. Chem.* **1987**, *140*, 43; Stoddart, J.F. *Top. Stereochem.* **1987**, *17*, 207; Sirlin, C. *Bull. Soc. Chim. Fr.* **1984**, II-5–40; Arad-Yellin, R.; Green, B.S.; Knossow, M.; Tsoucaris, G., in Atwood; Davies; MacNicol *Inclusion Compounds*, Vol. 3; Academic Press, NY, **1984**, pp. 263–295; Stoddart, J.F. *Prog. Macrocyclic Chem.* **1981**, *2*, 173; Cram, D.J.; Helgeson, R.C.; Sousa, L.R.; Timko, J.M.; Newcomb, M.; Moreau, P.; DeJong, F.; Gokel, G.W.; Hoffman, D.H.; Domeier, L.A.; Peacock, S.C.; Madan, K.; Kaplan, L. *Pure Appl. Chem.* **1975**, *43*, 327.

[165]See Schlenk Jr., W. *Liebigs Ann. Chem.* **1973**, 1145, 1156, 1179, 1195. Inclusion complexes of tri-*o*-thymotide can be used in a similar manner: see Arad-Yellin, R.; Green, B.S.; Knossow, M.; Tsoucaris, G. *J. Am. Chem. Soc.* **1983**, *105*, 4561.

[166]Miyamoto, H.; Sakamoto, M.; Yoskioka, K.; Takaoka, R.; Toda, F. *Tetrahedron Asymmetry* **2000**, *11*, 3045.

[167]Cope, A.C.; Ganellin, C.R.; Johnson, Jr., H.W.; Van Auken, T.V.; Winkler, H.J.S. *J. Am. Chem. Soc.* **1963**, *85*, 3276. For a review, see Tsuji, J. *Adv. Org. Chem.* **1969**, *6*, 109, see p. 220.

[168]Amos, R.D.; Handy, N.C.; Jones, P.G.; Kirby, A.J.; Parker, J.K.; Percy, J.M.; Su, M.D. *J. Chem. Soc. Perkin Trans. 2* **1992**, 549.

[169]See, for example, Casanova, J.; Corey, E.J. *Chem. Ind. (London)* **1961**, 1664; Gil-Av, E.; Nurok, D. *Proc. Chem. Soc.* **1962**, 146; Gault, Y.; Felkin, H. *Bull. Soc. Chim. Fr.* **1965**, 742; Vitt, S.V.; Saporovskaya, M.B.; Gudkova, I.P.; Belikov, V.M. *Tetrahedron Lett.* **1965**, 2575; Westley, J.W.; Halpern, B.; Karger, B.L. *Anal. Chem.* **1968**, *40*, 2046; Kawa, H.; Yamaguchi, F.; Ishikawa, N.*Chem. Lett.* **1982**, 745.

liquid chromatography[170] have proved more useful. In many cases, they have supplanted fractional crystallization, especially where the quantities to be resolved are small.[171]

2. *Differential Absorption.* When a racemic mixture is placed on a chromatographic column, if the column consists of chiral substances, then in principle the enantiomers should move along the column at different rates and should be separable without having to be converted to diastereomers.[171] This has been successfully accomplished with paper, column, thin-layer,[172] and gas and liquid chromatography.[173] For example, racemic mandelic acid has been almost completely resolved by column chromatography on starch.[174] Many workers have achieved separations with gas and liquid chromatography by the use of columns packed with chiral absorbents.[175] Columns packed with chiral materials are now commercially available and are capable of separating the enantiomers of certain types of compounds.[176]

3. *Chiral Recognition.* The use of chiral hosts to form diastereomeric inclusion compounds was mentioned above. But in some cases it is possible for a host to form an inclusion compound with one enantiomer of a racemic guest, but not the other. This is called *chiral recognition.* One enantiomer fits into the chiral host cavity, the other does not. More often, both diastereomers are formed, but one forms more rapidly than the other, so that if the guest is

[170]For example, See Pirkle, W.H.; Hauske, J.R. *J. Org. Chem.* **1977**, *42*, 1839; Helmchen, G.; Nill, G. *Angew. Chem. Int. Ed.* **1979**, *18*, 65; Meyers, A.I.; Slade, J.; Smith, R.K.; Mihelich, E.D.; Hershenson, F.M.; Liang, C.D. *J. Org. Chem.* **1979**, *44*, 2247; Goldman, M.; Kustanovich, Z.; Weinstein, S.; Tishbee, A.; Gil-Av, E. *J. Am. Chem. Soc.* **1982**, *104*, 1093.

[171]For monographs on the use of liquid chromatography to effect resolutions, see Lough, W.J. *Chiral Liquid Chromatography*; Blackie and Sons: London, **1989**; Krstulović, A.M. *Chiral Separations by HPLC*; Ellis Horwood: Chichester, **1989**; Zief, M.; Crane, L.J. *Chromatographic Separations*, Marcel Dekker, NY, **1988**. For a review, see Karger, B.L. *Anal. Chem.* **1967**, *39* (8), 24A.

[172]Weinstein, S. *Tetrahedron Lett.* **1984**, *25*, 985.

[173]For monographs, see Allenmark, S.G. *Chromatographic Enantioseparation*, Ellis Horwood, Chichester, **1988**; König, W.A. *The Practice of Enantiomer Separation by Capillary Gas Chromatography*, Hüthig, Heidelberg, **1987**. For reviews, see Schurig, V.; Nowotny, H. *Angew. Chem. Int. Ed.* **1990**, *29*, 939; Pirkle, W.H.; Pochapsky, T.C. *Chem. Rev.* **1989**, *89*, 347, *Adv. Chromatogr.*, **1987**, *27*, 73; Okamoto, Y. *CHEMTECH 1987*, 176; Blaschke, G. *Angew. Chem. Int. Ed.* **1980**, *19*, 13; Rogozhin, S.V.; Davankov, V.A. *Russ. Chem. Rev.* **1968**, *37*, 565. See also many articles in the journal *Chirality*.

[174]Ohara, M.; Ohta, K.; Kwan, T. *Bull. Chem. Soc. Jpn.* **1964**, *37*, 76. See also, Blaschke, G.; Donow, F. *Chem. Ber.* **1975**, *108*, 2792; Hess, H.; Burger, G.; Musso, H. *Angew. Chem. Int. Ed.* **1978**, *17*, 612.

[175]See, for example, Gil-Av, E.; Tishbee, A.; Hare, P.E. *J. Am. Chem. Soc.* **1980**, *102*, 5115; Hesse, G.; Hagel, R. *Liebigs Ann. Chem.* **1976**, 996; Schlögl, K.; Widhalm, M. *Chem. Ber.* **1982**, *115*, 3042; Koppenhoefer, B.; Allmendinger, H.; Nicholson, G. *Angew. Chem. Int. Ed.* **1985**, *24*, 48; Dobashi, Y.; Hara, S. *J. Am. Chem. Soc.* **1985**, *107*, 3406, *J. Org. Chem.* **1987**, *52*, 2490; Konrad, G.; Musso, H. *Liebigs Ann. Chem.* **1986**, 1956; Pirkle, W.H.; Pochapsky, T.C.; Mahler, G.S.; Corey, D.E.; Reno, D.S.; Alessi, D.M. *J. Org. Chem.* **1986**, *51*, 4991; Okamoto, Y.; Aburatani, R.; Kaida, Y.; Hatada, K. *Chem. Lett.* **1988**, 1125; Ehlers, J.; König, W.A.; Lutz, S.; Wenz, G.; tom Dieck, H. *Angew. Chem. Int. Ed.* **1988**, *27*, 1556; Hyun, M.H.; Park, Y.; Baik, I. *Tetrahedron Lett.* **1988**, *29*, 4735; Schurig, V.; Nowotny, H.; Schmalzing, D. *Angew. Chem. Int. Ed.* **1989**, *28*, 736; Ôi, S.; Shijo, M.; Miyano, S. *Chem. Lett.* **1990**, 59; Erlandsson, P.; Marle, I.; Hansson, L.; Isaksson, R.; Pettersson, C.; Pettersson, G. *J. Am. Chem. Soc.* **1990**, *112*, 4573.

[176]See, for example, Pirkle, W.H.; Welch, C.J. *J. Org. Chem.* **1984**, *49*, 138.

removed it is already partially resolved (this is a form of kinetic resolution, see category 6). An example is use of the chiral crown ether **58** partially to resolve the racemic amine salt **59**.[177] When an aqueous solution of **59** was

$$Ph-\overset{\overset{\displaystyle Me}{|}}{\underset{\underset{\displaystyle H}{|}}{C}}-NH_3^+ \qquad PF_6^-$$

58   **59**

mixed with a solution of optically active **58** in chloroform, and the layers separated, the chloroform layer contained about twice as much of the complex between **58** and (*R*)-**59** as of the diastereomeric complex. Many other chiral crown ethers and cryptands have been used, as have been cyclodextrins,[178] cholic acid,[179] and other kinds of hosts.[164] Of course, enzymes are generally very good at chiral recognition, and much of the work in this area has been an attempt to mimic the action of enzymes.

4. *Biochemical Processes*.[180] Biological molcules may react at different rates with the two enantiomers. For example, a certain bacterium may digest one enantiomer, but not the other. Pig liver esterase has been used for the selective cleavage of one enantiomeric ester.[181] This method is limited, since it is necessary to find the proper organism and since one of the enantiomers is destroyed in the process. However, when the proper organism is found, the method leads to a high extent of resolution since biological processes are usually very stereoselective.

5. *Mechanical Separation*.[182] This is the method by which Pasteur proved that racemic acid was actually a mixture of (+)- and (−)-tartaric acids.[183] In the case of racemic sodium ammonium tartrate, the enantiomers crystallize

[177]Cram, D.J.; Cram, J.M. *Science* **1974**, *183*, 803. See also, Yamamoto, K.; Fukushima, H.; Okamoto, Y.; Hatada, K.; Nakazaki M. *J. Chem. Soc. Chem. Commun.* **1984**, 1111; Kanoh, S.; Hongoh, Y.; Katoh, S.; Motoi, M.; Suda, H. *J. Chem. Soc. Chem. Commun.* **1988**, 405; Bradshaw, J.S.; Huszthy, P.; McDaniel, C.W.; Zhu, C.Y.; Dalley, N.K.; Izatt, R.M.; Lifson, S. *J. Org. Chem.* **1990**, *55*, 3129.

[178]See, for example, Hamilton, J.A.; Chen, L. *J. Am. Chem. Soc.* **1988**, *110*, 5833.

[179]See Miyata, M.; Shibakana, M.; Takemoto, K. *J. Chem. Soc. Chem. Commun.* **1988**, 655.

[180]For a review, see Sih, C.J.; Wu, S. *Top. Stereochem.* **1989**, *19*, 63.

[181]For an example, see Gais, H.-J.; Jungen, M.; Jadhav, V. *J. Org. Chem.* **2001**, *66*, 3384.

[182]For reviews, see Collet, A.; Brienne, M.; Jacques, J. *Chem. Rev.* **1980**, *80*, 215; *Bull. Soc. Chim. Fr.* **1972**, 127; **1977**, 494. For a discussion, see Curtin, D.Y.; Paul, I.C. *Chem. Rev.* **1981**, *81*, 525 pp. 535–536.

[183]Besides discovering this method of resolution, Pasteur also discovered the method of conversion to diastereomers and separation by fractional crystallization and the method of biochemical separation (and, by extension, kinetic resolution).

separately: all the (+) molecules going into one crystal and all the (−) into another. Since the crystals too are nonsuperimposable, their appearance is not identical and a trained crystallographer can separate them with twee-zers.[184] However, this is seldom a practical method, since few compounds crystallize in this manner. Even sodium ammonium tartrate does so only when it is crystallized <27°C. A more useful variation of the method, although still not very common, is the seeding of a racemic solution with something that will cause only one enantiomer to crystallize.[185] An inter-esting example of the mechanical separation technique was reported in the isolation of heptahelicene (p. 150). One enantiomer of this compound, which incidentally has the extremely high rotation of $[\alpha]_D^{20} = +6200°$, spontaneously crystallizes from benzene.[186] In the case of 1,1′-binaphthyl, optically active crystals can be formed simply by heating polycrystalline racemic samples of the compound at 76–150°C. A phase change from one crystal form to another takes place.[187] Note that 1,1′-binaphthyl is one of the few compounds that can be resolved by the Pasteur tweezer method. In some cases resolution can be achieved by enantioselective crystal-lization in the presence of a chiral additive.[188]

60

Spontaneous resolution has also been achieved by sublimation. In the case of the norborneol derivative **60**, when the racemic solid is subjected to sublimation, the (+) molecules condense into one crystal and the (−)

[184]This is a case of optically active materials arising from inactive materials. However, it may be argued that an optically active investigator is required to use the tweezers. Perhaps a hypothetical human being constructed entirely of inactive molecules would be unable to tell the difference between left- and right-handed crystals.

[185]For a review of the seeding method, see Secor, R.M. *Chem. Rev.* **1963**, *63*, 297.

[186]Martin, R.H; Baes, M. *Tetrahedron* **1975**, *31*, 2135. See also, Wynberg, H.; Groen, M.B. *J. Am. Chem. Soc.* **1968**, *90*, 5339. For a discussion of other cases, see McBride, J.M.; Carter, R.L. *Angew. Chem. Int. Ed.* **1991**, *30*, 293.

[187]Wilson, K.R.; Pincock, R.E. *J. Am. Chem. Soc.* **1975**, *97*, 1474; Kress, R.B.; Duesler, E.N.; Etter, M.C.; Paul, I.C.; Curtin, D.Y. *J. Am. Chem. Soc.* **1980**, *102*, 7709. See also, Lu, M.D.; Pincock, R.E. *J. Org. Chem.* **1978**, *43*, 601; Gottarelli, G.; Spada, G.P. *J. Org. Chem.* **1991**, *56*, 2096. For a discussion and other examples, see Agranat, I.; Perlmutter-Hayman, B.; Tapuhi, Y. *Nouv. J. Chem.* **1978**, *2*, 183.

[188]Addadi, L.; Weinstein, S.; Gati, E.; Weissbuch, I.; Lahav, M. *J. Am. Chem. Soc.* **1982**, *104*, 4610. See also, Weissbuch, I.; Addadi, L.; Berkovitch-Yellin, Z.; Gati, E.; Weinstein, S.; Lahav, M.; Leiserowitz, L. *J. Am. Chem. Soc.* **1983**, *105*, 6615.

molecules into another.[189] In this case, the crystals are superimposable, unlike the situation with sodium ammonium tartrate, but the investigators were able to remove a single crystal, which proved optically active.

6. *Kinetic Resolution.*[190] Since enantiomers react with chiral compounds at different rates, it is sometimes possible to effect a partial separation by stopping the reaction before completion. This method is very similar to the asymmetric syntheses discussed on p. 147. A method has been developed to evaluate the enantiomeric ratio of kinetic resolution using only the extent of substrate conversion.[191] An important application of this method is the resolution of racemic alkenes by treatment with optically active diisopinocampheylborane,[192] since alkenes do not easily lend themselves to conversion to diastereomers if no other functional groups are present. Another example

| Racemic | (R)-Enantiomer | (S)-Enantiomer |

61

is the resolution of allylic alcohols, such as **61** with one enantiomer of a chiral epoxidation agent (see **15-50**).[193] In the case of **61**, the discrimination was extreme. One enantiomer was converted to the epoxide and the other was not, the rate ratio (hence the selectivity factor) being >100. Of course, in this method only one of the enantiomers of the original racemic mixture is obtained, but there are at least two possible ways of getting the other: (*1*) use of the other enantiomer of the chiral reagent; (*2*) conversion of the product to the starting compound by a reaction that preserves the stereochemistry.

[189]Paquette, L.A.; Lau, C.J. *J. Org. Chem.* **1987**, *52*, 1634.

[190]For reviews, see Kagan, H.B.; Fiaud, J.C. *Top. Stereochem.* **1988**, *18*, 249; Ward, R.S. *Tetrahedron Asymmetry* **1995**, *6*, 1475; Pellissier, H. *Tetrahedron* **2003**, *59*, 8291.

[191]Lu, Y.; Zhao, X.; Chen, Z.-N. *Tetrahedron Asymmetry* **1995**, *6*, 1093.

[192]Brown, H.C.; Ayyangar, N.R.; Zweifel, G. *J. Am. Chem. Soc.* **1964**, *86*, 397.

[193]Martin, V.S.; Woodard, S.S.; Katsuki, T.; Yamada, Y.; Ikeda, M.; Sharpless, K.B. *J. Am. Chem. Soc.* **1981**, *103*, 6237. See also, Kobayashi, Y.; Kusakabe, M.; Kitano, Y.; Sato, F. *J. Org. Chem.* **1988**, *53*, 1586; Kitano, Y.; Matsumoto, T.; Sato, F. *Tetrahedron* **1988**, *44*, 4073; Carlier, P.R.; Mungall, W.S.; Schröder, G.; Sharpless, K.B. *J. Am. Chem. Soc.* **1988**, *110*, 2978; Discordia, R.P.; Dittmer, D.C. *J. Org. Chem.* **1990**, *55*, 1414. For other examples, see Miyano, S.; Lu, L.D.; Viti, S.M.; Sharpless, K.B. *J. Org. Chem.* **1985**, *50*, 4350; Paquette, L.A.; DeRussy, D.T.; Cottrell, C.E. *J. Am. Chem. Soc.* **1988**, *110*, 890; Weidert, P.J.; Geyer, E.; Horner, L. *Liebigs Ann. Chem.* **1989**, 533; Katamura, M.; Ohkuma, T.; Tokunaga, M.; Noyori, R. *Tetrahedron: Assymetry* **1990**, *1*, 1; Hayashi, M.; Miwata, H.; Oguni, N. *J. Chem. Soc. Perkin Trans. 2* **1991**, 1167.

Kinetic resolution of racemic allylic acetates[194] has been accomplished via asymmetric dihydroxylation (p. 1166), and 2-oxoimidazolidine-4-carboxylates have been developed as new chiral auxiliaries for the kinetic resolution of amines.[195] Reactions catalyzed by enzymes can be utilized for this kind of resolution.[196]

7. *Deracemization.* In this type of process, one enantiomer is converted to the other, so that a racemic mixture is converted to a pure enantiomer, or to a mixture enriched in one enantiomer. This is not quite the same as the methods of resolution previously mentioned, although an outside optically active substance is required. To effect the deracemization two conditions are necessary: (*1*) the enantiomers must complex differently with the optically active substance; (*2*) they must interconvert under the conditions of the experiment. When racemic thioesters were placed in solution with a specific optically active amide for 28 days, the solution contained 89% of one enantiomer and 11% of the other.[197] In this case, the presence of a base (Et$_3$N) was necessary for the interconversion to take place. Biocatalytic deracemization processes induce deracemization of chiral secondary alcohols.[198] In a specific example, *Sphingomonas paucimobilis* NCIMB 8195 catalyzes the efficient deracemization of many secondary alcohols in up to 90% yield of the (*R*)-alcohol.[199]

Optical Purity[200]

Suppose we have just attempted to resolve a racemic mixture by one of the methods described in the previous section. How do we know that the two enantiomers we have obtained are pure? For example, how do we know that the (+) isomer is not contaminated by, say, 20% of the (−) isomer and vice versa? If we knew the value of [α] for the pure material ($[\alpha]_{max}$), we could easily determine the purity of our sample by measuring its rotation. For example, if $[\alpha]_{max}$ is +80° and our (+) enantiomer contains 20% of the (−) isomer, [α] for the sample will be +48°.[201]

[194]Lohray, B.B.; Bhushan, V. *Tetrahedron Lett.* **1993**, *34*, 3911.

[195]Kubota, H.; Kubo, A.; Nunami, K. *Tetrahedron Lett.* **1994**, *35*, 3107.

[196]For example, see Nakamura, K.; Inoue, Y.; Ohno, A. *Tetrahedron Lett.* **1994**, *35*, 4375; Mohr, P. Rösslein, L.; Tamm, C. *Tetrahedron Lett.* **1989**, *30*, 2513; Kazlauskas, R.J. *J. Am. Chem. Soc.* **1989**, *111*, 4953; Schwartz, A.; Madan, P.; Whitesell, J.K.; Lawrence, R.M. *Org. Synth.*, **69**, 1; Francalanci, F.; Cesti, P.; Cabri, W.; Bianchi, D.; Martinengo, T.; Foá, M. *J. Org. Chem.* **1987**, *52*, 5079.

[197]Pirkle, W.H.; Reno, D.S. *J. Am. Chem. Soc.* **1987**, *109*, 7189. For another example, see Reider, P.J.; Davis, P.; Hughes, D.L.; Grabowski, E.J.J. *J. Org. Chem.* **1987**, *52*, 955.

[198]Stecher, H.; Faber, K. *Synthesis* **1997**, 1.

[199]Allan, G. R.; Carnell, A. J. *J. Org. Chem.* **2001**, *66*, 6495.

[200]For a review, see Raban, M.; Mislow, K. *Top. Stereochem.* **1967**, *2*, 199.

[201]If a sample contains 80% (+) and 20% (−) isomer, the (−) isomer cancels an equal amount of (+) isomer and the mixture behaves as if 60% of it were (+) and the other 40% inactive. Therefore the rotation is 60% of 80° or 48°. This type of calculation, however, is not valid for cases in which [α] is dependent on concentration (p. 139); see Horeau, A.*Tetrahedron Lett.* **1969**, 3121.

We define *optical purity* as

$$\text{Percent optical purity} = \frac{[\alpha]_{obs}}{[\alpha]_{max}} \times 100$$

Assuming a linear relationship between $[\alpha]$ and concentration, which is true for most cases, the optical purity is equal to the percent excess of one enantiomer over the other:

$$\text{Optical purity} = \text{percent enantiomeric excess} = \frac{[R] - [S]}{[R] + [S]} \times 100 = (\%R) - (\%S)$$

But how do we determine the value of $[\alpha]_{max}$? It is plain that we have two related problems here; namely, what are the optical purities of our two samples and what is the value of $[\alpha]_{max}$. If we solve one, the other is also solved. Several methods for solving these problems are known.

One of these methods involves the use of NMR[202] (see p. 161). Suppose we have a nonracemic mixture of two enantiomers and wish to know the proportions. We convert the mixture into a mixture of diastereomers with an optically pure reagent and look at the NMR spectrum of the resulting mixture, for example,

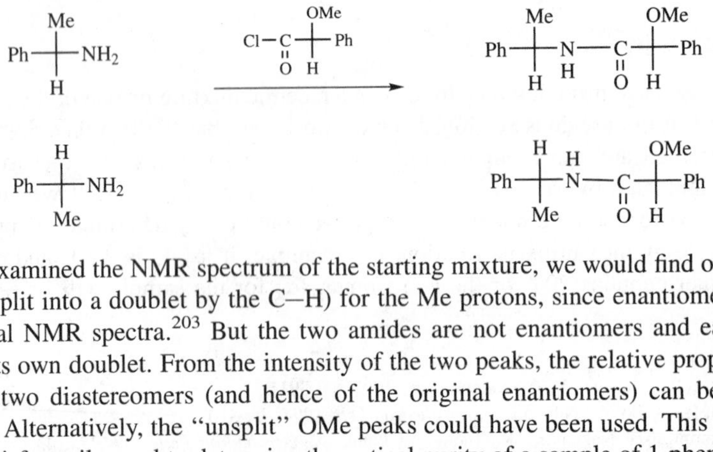

If we examined the NMR spectrum of the starting mixture, we would find only one peak (split into a doublet by the C—H) for the Me protons, since enantiomers give identical NMR spectra.[203] But the two amides are not enantiomers and each Me gives its own doublet. From the intensity of the two peaks, the relative proportions of the two diastereomers (and hence of the original enantiomers) can be determined. Alternatively, the "unsplit" OMe peaks could have been used. This method was satisfactorily used to determine the optical purity of a sample of 1-phenylethylamine (the case shown above),[204] as well as other cases, but it is obvious that

---

[202]Raban, M.; Mislow, K. *Tetrahedron Lett.* **1965**, 4249, **1966**, 3961; Jacobus, J.; Raban, M. *J. Chem. Educ.* **1969**, 46, 351; Tokles, M.; Snyder, J.K. *Tetrahedron Lett.* **1988**, 29, 6063. For a review, see Yamaguchi, S., in Morrison, J.D. *Asymmetric Synthesis*, Vol. 1, Academic Press, NY, **1983**, pp. 125–152. See also Raban, M.; Mislow, K. *Top. Stereochem.* **1967**, 2, 199.

[203]Though enantiomers give identical nmr spectra, the spectrum of a single enantiomer may be different from that of the racemic mixture, even in solution. See Williams, T.; Pitcher, R.G.; Bommer, P.; Gutzwiller, J.; Uskoković, M. *J. Am. Chem. Soc.* **1969**, 91, 1871.

[204]Raban, M.; Mislow, K. *Top. Stereochem.* **1967**, 2, 199, see pp. 216–218.

sometimes corresponding groups in diastereomeric molecules will give NMR signals that are too close together for resolution. In such cases, one may resort to the use of a different optically pure reagent. The ^{13}C NMR can be used in a similar manner.[205] It is also possible to use these spectra to determine the absolute configuration of the original enantiomers by comparing the spectra of the diastereomers with those of the original enantiomers.[206] From a series of experiments with related compounds of known configurations it can be determined in which direction one or more of the ^{1}H or ^{13}C NMR peaks are shifted by formation of the diastereomer. It is then assumed that the peaks of the enantiomers of unknown configuration will be shifted the same way.

A closely related method does not require conversion of enantiomers to diastereomers, but relies on the fact that (in principle, at least) enantiomers have different NMR spectra *in a chiral solvent*, or when mixed with a chiral molecule (in which case transient diastereomeric species may form). In such cases, the peaks may be separated enough to permit the proportions of enantiomers to be determined from their intensities.[207] Another variation, which gives better results in many cases, is to use an achiral solvent but with the addition of a *chiral lanthanide shift reagent* such as *tris*[3-trifluoroacetyl-*d*-camphorato]europium(III).[208] Lanthanide shift reagents have the property of spreading NMR peaks of compounds with which they can form coordination compounds, for example, alcohols, carbonyl compounds, and amines. Chiral lanthanide shift reagents shift the peaks of the two enantiomers of many such compounds to different extents.

Another method, involving GC,[209] is similar in principle to the NMR method. A mixture of enantiomers whose purity is to be determined is converted by means of an optically pure reagent into a mixture of two diastereomers. These diastereomers are then separated by GC (p. 172) and the ratios determined from the peak areas.

[205]For a method that relies on diastereomer formation without a chiral reagent, see Feringa, B.L.; Strijtveen, B.; Kellogg, R.M. *J. Org. Chem.* **1986**, *51*, 5484. See also, Pasquier, M.L.; Marty, W. *Angew. Chem. Int. Ed.* **1985**, *24*, 315; Luchinat, C.; Roelens, S. *J. Am. Chem. Soc.* **1986**, *108*, 4873.

[206]See Dale, J.A.; Mosher, H.S. *J. Am. Chem. Soc.* **1973**, *95*, 512; Rinaldi, P.L. *Prog. NMR Spectrosc.*, **1982**, *15*, 291; Faghih, R.; Fontaine, C.; Horibe, I.; Imamura, P.M.; Lukacs, G.; Olesker, A.; Seo, S. *J. Org. Chem.* **1985**, *50*, 4918; Trost, B.M.; Belletire, J.L.; Godleski, S.; McDougal, P.G.; Balkovec, J.M.; Baldwin, J.J.; Christy, M.E.; Ponticello, G.S.; Varga, S.L.; Springer, J.P. *J. Org. Chem.* **1986**, *51*, 2370.

[207]For reviews of nmr chiral solvating agents, see Weisman, G.R., in Morrison, J.D. *Asymmetric Synthesis*, Vol. 1, Academic Press, NY, **1983**, pp. 153–171; Pirkle, W.H.; Hoover, D.J. *Top. Stereochem.* **1982**, *13*, 263. For literature references, see Sweeting, L.M.; Anet, F.A.L. *Org. Magn. Reson.* **1984**, *22*, 539. See also, Pirkle, W.H.; Tsipouras, A. *Tetrahedron Lett.* **1985**, *26*, 2989; Parker, D.; Taylor, R.J. *Tetrahedron* **1987**, *43*, 5451.

[208]Sweeting, L.M.; Crans, D.C.; Whitesides, G.M. *J. Org. Chem.* **1987**, *52*, 2273. For a monograph on chiral lanthanide shift reagents, see Morrill, T.C. *Lanthanide Shift Reagents in Stereochemical Analysis*, VCH, NY, **1986**. For reviews, see Fraser, R.R., in Morrison, J.D. *Asymmetric Synthesis*, Vol. 1, Academic Press, NY, **1983**, pp. 173–196; Sullivan, G.R. *Top. Stereochem.* **1978**, *10*, 287.

[209]Charles, R.; Fischer, G.; Gil-Av, E. *Isr. J. Chem.* **1963**, *1*, 234; Halpern, B.; Westley, J.W. *Chem. Commun.* **1965**, 246; Vitt, S.V.; Saporovskaya, M.B.; Gudkova, I.P.; Belikov, V.M. *Tetrahedron Lett.* **1965**, 2575; Guetté, J.; Horeau, A. *Tetrahedron Lett.* **1965**, 3049; Westley, J.W.; Halpern, B. *J. Org. Chem.* **1968**, *33*, 3978.

Once again, the ratio of diastereomers is the same as that of the original enantiomers. High-pressure liquid chromatography has been used in a similar manner and has wider applicability.[210] The direct separation of enantiomers by gas or liquid chromatography on a chiral column has also been used to determine optical purity.[211]

Other methods[212] involve isotopic dilution,[213] kinetic resolution,[214] ^{13}C NMR relaxation rates of diastereomeric complexes,[215] and circular polarization of luminescence.[216]

CIS–TRANS ISOMERISM

Compounds in which rotation is restricted may exhibit cis–trans isomerism.[217] These compounds do not rotate the plane of polarized light (unless they also happen to be chiral), and the properties of the isomers are not identical. The two most important types are isomerism resulting from double bonds and that resulting from rings.

Cis–Trans Isomerism Resulting from Double Bonds

It has been mentioned (p. 10) that the two carbon atoms of a $C=C$ double bond and the four atoms directly attached to them are all in the same plane and that rotation around the double bond is prevented. This means that in the case of a molecule $WXC=CYZ$, stereoisomerism exists when $W \neq X$ and $Y \neq Z$. There are two and

$$\underset{X}{\overset{W}{\diagdown}}C=C\underset{Z}{\overset{Y}{\diagup}} \qquad \underset{X}{\overset{W}{\diagdown}}C=C\underset{Y}{\overset{Z}{\diagup}}$$

62   **63**

only two isomers (**62** and **63**), each superimposable on its mirror image unless one of the groups happens to carry a stereogenic center. Note that **62** and **63** are diastereomers, by the definition given on p. 155. There are two ways to name

---

[210]For a review, see Pirkle, W.H.; Finn, J., in Morrison, J.D. *Asymmetric Synthesis*, Vol. 1, Academic Press, NY, *1983*, pp. 87–124.

[211]For reviews, see in Morrison, J.D. *Asymmetric Synthesis*, Vol. 1, Academic Press, NY, *1983*, the articles by Schurig, V. pp. 59–86 and Pirkle, W.H.; Finn, J. pp. 87–124.

[212]See also Leitich, J. *Tetrahedron Lett. 1978*, 3589; Hill, H.W.; Zens, A.P.; Jacobus, J. *J. Am. Chem. Soc. 1979*, *101*, 7090; Matsumoto, M.; Yajima, H.; Endo, R. *Bull. Chem. Soc. Jpn. 1987*, *60*, 4139.

[213]Berson, J.A.; Ben-Efraim, D.A. *J. Am. Chem. Soc. 1959*, *81*, 4083. For a review, see Andersen, K.K.; Gash, D.M.; Robertson, J.D. in Morrison, J.D. *Asymmetric Synthesis*, Vol. 1, Academic Press, NY, *1983*, pp. 45–57.

[214]Horeau, A.; Guetté, J.; Weidmann, R. *Bull. Soc. Chim. Fr. 1966*, 3513. For a review, see Schoofs, A.R.; Guetté, J., in Morrison, J.D. *Asymmetric Synthesis*, Vol. 1, Academic Press, NY, *1983*, pp. 29–44.

[215]Hofer, E.; Keuper, R. *Tetrahedron Lett. 1984*, *25*, 5631.

[216]Eaton, S.S. *Chem. Phys. Lett. 1971*, *8*, 251; Schippers, P.H.; Dekkers, H.P.J.M. *Tetrahedron 1982*, *38*, 2089.

[217]Cis-trans isomerism was formerly called *geometrical isomerism*.

such isomers. In the older method, one isomer is called cis and the other trans. When W = Y, **62** is the cis and **63** the trans isomer. Unfortunately, there is no easy way to apply this method when the four groups are different. The newer method, which can be applied to all cases, is based on the Cahn–Ingold–Prelog system (p. 155). The two groups at each carbon are ranked by the sequence rules. Then that isomer with the two higher ranking groups on the same side of the double bond is called (*Z*) (for the German word *zusammen* meaning *together*); the other is (*E*) (for *entgegen* meaning *opposite*).[218] A few examples are shown. Note that the (*Z*) isomer is not necessarily the one that would be called cis under the older system (e.g., **64**, and **65**). Like *cis* and *trans*, (*E*) and (*Z*) are used as prefixes; for example, **65** is called (*E*)-1-bromo-1,2-dichloroethene.

(*E*)	(*Z*)	(*Z*)	(*E*)	(*E*)
		64	**65**	

This type of isomerism is also possible with other double bonds, such as C=N,[219] N=N, or even C=S,[220] although in these cases only two or three groups are connected to the double-bond atoms. In the case of imines, oximes, and other C=N compounds, if W = Y, **66** may be called syn and **67** anti, although (*E*) and (*Z*) are often used here too.[221] In azo compounds, there is no ambiguity. Compound **68** is always *syn* or (*Z*) regardless of the nature of W and Y.

		(*Z*)	(*E*)
66	**67**	**68**	

If there is more than one double bond[222] in a molecule and if W ≠ X and Y ≠ Z for each, the number of isomers in the most general case is 2^n, although this number may be decreased if some of the substituents are the same, as in

---

[218]For a complete description of the system, see *Pure Appl. Chem.* **19767**, *45*, 13; *Nomenclature of Organic Chemistry*, Pergamon, Elmsford, NY, *1979* (the Blue Book).

[219]For reviews of isomerizations about C=N bonds, see, in Patai, S. *The Chemistry of the Carbon–Nitrogen Double Bond*; Wiley, NY, *1970*, the articles by McCarty, C.G., 363–464 (pp. 364–408), and Wettermark, G. 565–596 (pp. 574–582).

[220]King, J.F.; Durst, T. *Can. J. Chem.* *1966*, *44*, 819.

[221]A mechanism has been reported for the acid-catalyzed (*Z/E*) isomerization of imines. See Johnson, J.E.; Morales, N.M.; Gorczyca, A.M.; Dolliver, D.D.; McAllister, M.A. *J. Org. Chem.* *2001*, *66*, 7979.

[222]This rule does not apply to allenes, which do not show cis–trans isomerism at all (see p. 148).

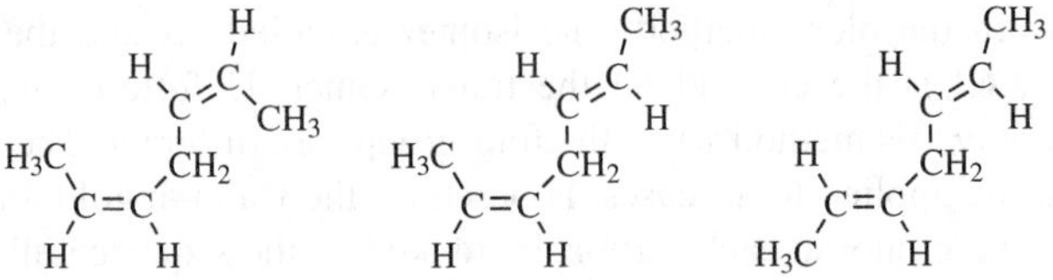

cis–cis or
(Z, Z)

cis–trans or
(Z, E)

trans–trans or
(E, E)

When a molecule contains a double bond and an asymmetric carbon, there are four isomers, a cis pair of enantiomers and a trans pair:

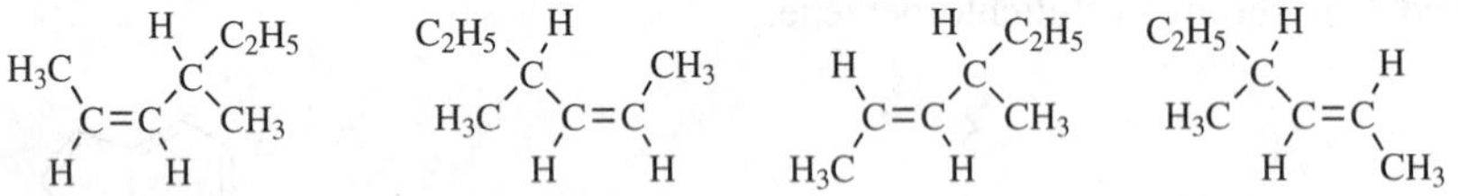

(Z) or cis *dl* pair

(E) or trans *dl* pair

Double bonds in small rings are so constrained that they must be cis. From cyclopropene (a known system) to cycloheptene, double bonds in a stable ring cannot be trans. However, the cyclooctene ring is large enough to permit trans double bonds to exist (see p. 151), and for rings larger than 10- or 11-membered, trans isomers are more stable[223] (see also, p. 225).

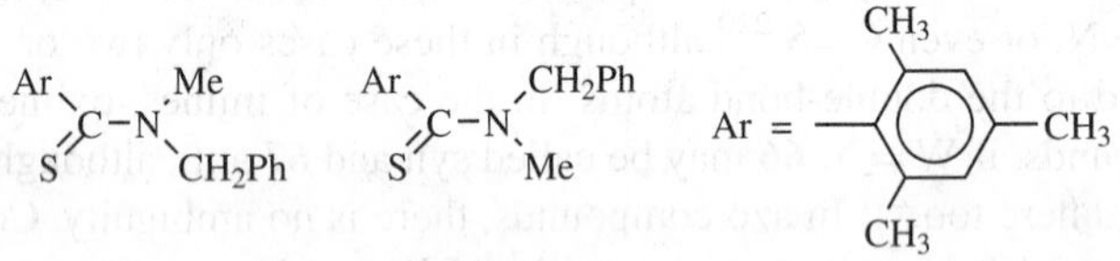

69   **70**

$Ar = $

In a few cases, single-bond rotation is so slowed that cis and trans isomers can be isolated even where no double bond exists[224] (see also p. 230). One example is *N*-methyl-*N*-benzylthiomesitylide (**69** and **70**),[225] the isomers of which are stable in the crystalline state but interconvert with a half-life of ∼25 h in CDCl$_3$ at 50°C.[226] This type of isomerism is rare; it is found chiefly in certain amides and thioamides, because resonance gives the single bond some double-bond character and slows rotation.[53] (For other examples of restricted rotation about single bonds, see pp. 230–233).

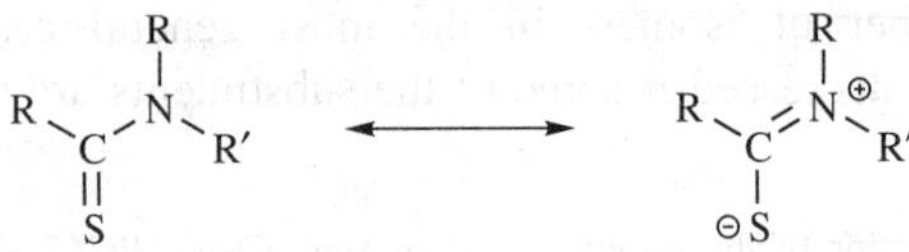

[223]Cope, A.C.; Moore, P.T.; Moore, W.R. *J. Am. Chem. Soc.* **1959**, *81*, 3153.

[224]For a review, see Ōki, M. *Applications of Dynamic NMR Spectroscopy to Organic Chemistry*, VCH, NY, **1985**, pp. 41–71.

[225]Mannschreck, A. *Angew. Chem. Int. Ed.* **1965**, *4*, 985. See also, Toldy, L.; Radics, L. *Tetrahedron Lett.* **1966**, 4753; Völter, H.; Helmchen, G. *Tetrahedron Lett.* **1978**, 1251; Walter, W.; Hühnerfuss, H. *Tetrahedron Lett.* **1981**, *22*, 2147.

[226]This is another example of atropisomerism (p. 145).

Conversely, there are compounds in which nearly free rotation is possible around what are formally C=C double bonds. These compounds, called *push–pull* or *captodative* ethylenes, have two electron-withdrawing groups on one carbon and two electron-donating groups on the other (**71**).[227] The contribution of di-ionic

71

A, B = electron withdrawing
C, D = electron donating

canonical forms, such as the one shown decreases the double-bond character and allows easier rotation. For example, compound **72** has a barrier to rotation of 13 kcal mol^{-1} (55 kJ mol^{-1}),[228] compared to a typical value of $\sim$62–65 kcal mol^{-1} (260–270 kJ mol^{-1}) for simple alkenes.

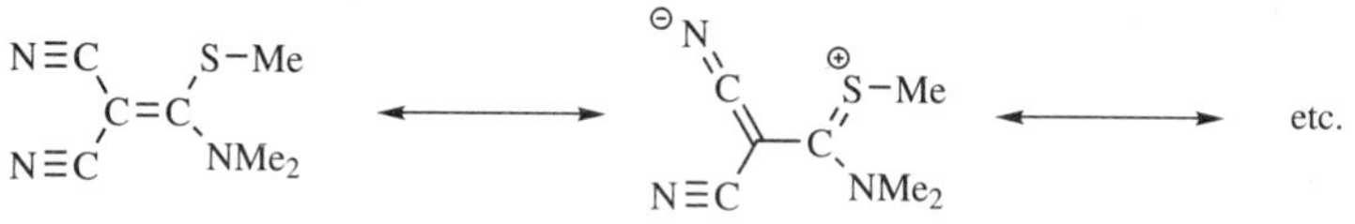

72

Since they are diastereomers, cis–trans isomers always differ in properties; the differences may range from very slight to considerable. The properties of maleic acid are so different from those of fumaric acid (Table 4.2) that it is not surprising that they have different names. Since they generally have more symmetry than cis isomers, trans isomers in most cases have higher melting points and lower

TABLE 4.2. Some Properties of Maleic and Fumaric Acids

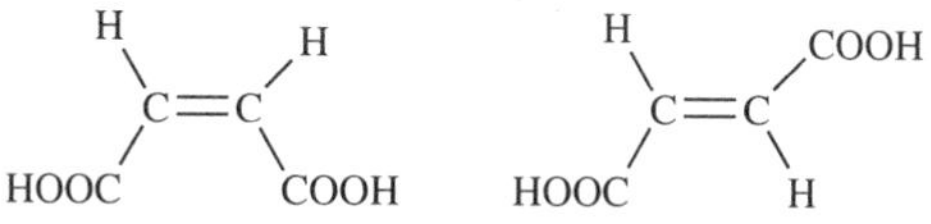

Maleic acid   Fumaric acid

Property	Maleic Acid	Fumaric Acid
Melting point, °C	130	286
Solubility in water at 25°C, g L^{-1}	788	7
K_1 (at 25°C)	1.5×10^{-2}	1×10^{-3}
K_2 (at 25°C)	2.6×10^{-7}	3×10^{-5}

[227]For reviews, see Sandström, J. *Top. Stereochem.* **1983**, *14*, 83; Ōki, M. *Applications of Dynamic NMR Spectroscopy to Organic Chemistry*, VCH, NY, **1985**, pp. 111–125.
[228]Sandström, J.; Wennerbeck, I. *Acta Chem. Scand. Ser. B*, **1978**, *32*, 421.

solubilities in inert solvents. The *cis* isomer usually has a higher heat of combustion, which indicates a lower thermochemical stability. Other noticeably different properties are densities, acid strengths, boiling points, and various types of spectra, but the differences are too involved to be discussed here.

It is also important to note that *trans*-alkenes are often more stable than *cis*-alkenes due to diminished steric hindrance (p. 232), but this is not always the case. It is known, for example, that *cis*-1,2-difluoroethene is thermodynamically more stable than *trans*-1,2-difluoroethene. This appears to be due to delocalization of halogen lone-pair electrons and an antiperiplanar effect between vicinal antiperiplanar bonds.[229]

Cis–Trans Isomerism of Monocyclic Compounds

Although rings of four carbons and larger are not generally planar (see p. 211), they will be treated as such in this section, since the correct number of isomers can be determined when this is done[230] and the principles are easier to visualize (see p. 204).

The presence of a ring, like that of a double bond, prevents rotation. Cis and trans isomers are possible whenever there are two carbons on a ring, each of which is substituted by two different groups. The two carbons need not be adjacent. Examples are

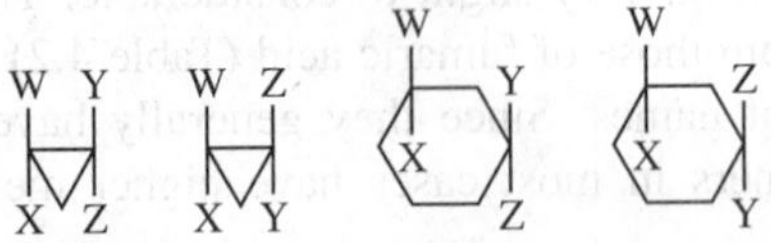

In some cases, the two stereoisomers can interconvert. In *cis*- and *trans*-disubstituted cyclopropanones, for example, there is reversible interconversion that favors the more stable trans isomer. This fluxional isomerization occurs via ring opening to an unseen oxyallyl *valence bond* isomer.[231]

As with double bonds, cis and trans isomers are possible, but the restrictions are that W may equal Y and X may equal Z, but W may not equal X and Y may not equal Z. There is an important difference from the double-bond case: The

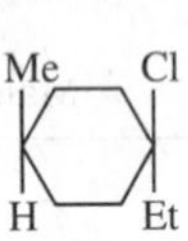

73

substituted carbons are sterogenic carbons. This means that there are not *only* two isomers. In the most general case, where W, X, Y, and Z are all different,

---

[229]Yamamoto, T.; Tomoda, S. *Chem. Lett.* **1997**, 1069.

[230]For a discussion of why this is so, see Leonard, J.E.; Hammond, G.S.; Simmons, H.E. *J. Am. Chem. Soc.* **1975**, *97*, 5052.

[231]Sorensen, T.S.; Sun, F. *J. Chem. Soc. Perkin Trans. 2* **1998**, 1053.

there are four isomers since neither the cis nor the trans isomer is superimposable on its mirror image. This is true regardless of ring size or which carbons are involved, except that in rings of even-numbered size when W, X, Y, and Z are at opposite corners, no chirality is present, for example, **73**. In this case, the substituted carbons are *not* chiral carbons. Note also that a plane of symmetry exists in such compounds. When W = Y and X = Z, the cis isomer is always superimposable on its mirror image, and hence is a meso compound, while the trans isomer consists of a *dl* pair, except in the case noted above. Again, the cis isomer has a plane of symmetry while the trans does not.

Me─┐ ┌─Me   Me─┐ ┌─H   H─┐ ┌─Me
H─┘ └─H   H─┘ └─Me   Me─┘ └─H

cis meso   trans *dl* pair

Rings with more than two differently substituted carbons can be dealt with on similar principles. In some cases, it is not easy to tell the number of isomers by inspection.[105] The best method for the student is to count the number n of differently substituted carbons (these will usually be asymmetric, but not always, e.g., in **73**), and then to draw 2^n structures, crossing out those that can be superimposed on others (usually the easiest method is to look for a plane of symmetry). By this means, it can be determined that for 1,2,3-cyclohexanetriol there are two meso compounds and a *dl* pair; and for 1,2,3,4,5,6-hexachlorocyclohexane there are seven meso compounds and a *dl* pair. The drawing of these structures is left as an exercise for the student.

Similar principles apply to heterocyclic rings as long as there are carbons (or other ring atoms) containing two different groups.

Cyclic stereoisomers containing only two differently substituted carbons are named either cis or trans, as previously indicated. The (Z, E) system is not used for cyclic compounds. However, cis–trans nomenclature will not suffice for compounds with more than two differently substituted atoms. For these compounds, a system is used in which the configuration of each group is given with respect to a reference group, which is chosen as the group attached to the lowest numbered ring member bearing a substituent giving rise to cis–trans isomerism. The reference group is indicated by the symbol *r*. Three stereoisomers named according to this system are *c*-3,*c*-5-dimethylcyclohexan-*r*-1-ol (**74**), *t*-3,*t*-5-dimethylcyclohexan-*r*-1-ol (**75**), and *c*-3,*t*-5-dimethylcyclohexan-*r*-1-ol (**76**). The last example demonstrates the rule that when there are two otherwise equivalent ways of going around the ring, one chooses the path that gives the cis designation to the first substituent after the reference. Another example is *r*-2,*c*-4-dimethyl-*t*-6-ethyl-1,3-dioxane (**77**).

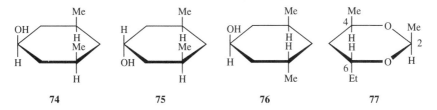

74   **75**   **76**   **77**

Cis–Trans Isomerism of Fused and Bridged Ring Systems

Fused bicyclic systems are those in which two rings share two and only two atoms. In such systems, there is no new principle. The fusion may be cis or trans, as illustrated by *cis-* and *trans-*decalin. However, when the rings are small enough, the trans configuration is impossible and the junction must be cis. The smallest trans junction that has been prepared when one ring is four membered is a four–five junction; *trans*-bicyclo[3.2.0]heptane (**78**) is known.[232] For the bicyclo[2.2.0] system

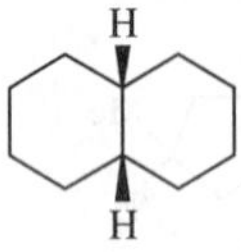

| *cis*-Decalin | 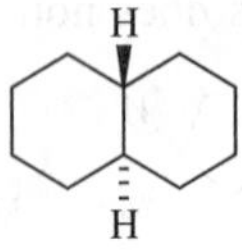 *trans*-Decalin | 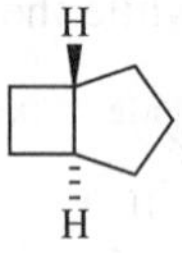 **78** | 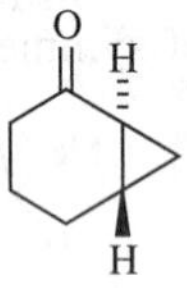 **79** |

(a four–four fusion), only cis compounds have been made. The smallest known trans junction when one ring is three-membered is a six–three junction (a bicyclo[4.1.0] system). An example is **79**.[233] When one ring is three membered and the other eight membered (an eight–three junction), the *trans*-fused isomer is more stable than the corresponding *cis*-fused isomer.[234]

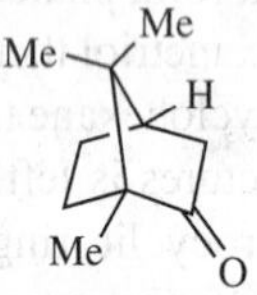

Camphor

In *bridged* bicyclic ring systems, two rings share more than two atoms. In these cases, there may be fewer than 2^n isomers because of the structure of the system. For example, there are only two isomers of camphor (a pair of enantiomers), although it has two chiral carbons. In both isomers, the methyl and hydrogen are *cis*. The *trans* pair of enantiomers is impossible in this case, since the bridge *must*

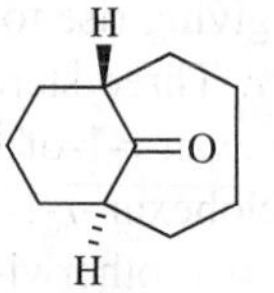

80

be *cis*. The smallest bridged system so far prepared in which the bridge is trans is the [4.3.1] system; the trans ketone **80** has been prepared.[235] In this case there

[232]Meinwald, J.; Tufariello, J.J.; Hurst, J.J. *J. Org. Chem.* **1964**, *29*, 2914.

[233]Paukstelis, J.V.; Kao, J. *J. Am. Chem. Soc.* **1972**, *94*, 4783. For references to other examples, see Dixon, D.A.; Gassman, P.G. *J. Am. Chem. Soc.* **1988**, *110*, 2309.

[234]Corbally, R.P.; Perkins, M.J.; Carson, A.S.; Laye, P.G.; Steele, W.V. *J. Chem. Soc. Chem. Commun.* **1978**, 778.

[235]Winkler, J.D.; Hey, J.P.; Williard, P.G. *Tetrahedron Lett.* **1988**, *29*, 4691.

are four isomers, since both the *trans* and the *cis* (which has also been prepared) are pairs of enantiomers.

When one of the bridges contains a substituent, the question arises as to how to name the isomers involved. When the two bridges that do *not* contain the substituent are of unequal length, the rule generally followed is that the prefix endo- is used when the substituent is closer to the longer of the two unsubstituted bridges; the prefix exo- is used when the substituent is closer to the shorter bridge; for example,

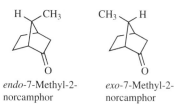

exo-2-Norborneol	*endo*-2-Norborneol

When the two bridges not containing the substituent are of equal length, this convention cannot be applied, but in some cases a decision can still be made; for example, if one of the two bridges contains a functional group, the endo isomer is the one in which the substituent is closer to the functional group:

endo-7-Methyl-2- norcamphor	*exo*-7-Methyl-2- norcamphor

Out–In Isomerism

Another type of stereoisomerism, called *out–in* isomerism (or *in–out*),[236] is found in salts of tricyclic diamines with nitrogen at the bridgeheads. In medium-sized bicyclic ring systems, *in–out* isomerisim is possible,[237] and the bridgehead nitrogen atoms adopt whichever arangement is more stable.[238] If we focus attention on the nitrogne lone pairs, 1,4-diazabicyclo[2.2.2]octane (**81**) favors the *out–out* isomer, 1,6-diazabicyclo[4.4.4]tetradecane (**82**) the *in,in*,[239] 1,5-diazabicyclo[3.3.3]undecane (**83**) has nearly planar nitrogen atoms,[240] and 1,9-diazabicyclo[7.3.1]tridecane (**84**) is *in,out*.[241] One can also focus on the NH unit in the case of ammonium salts.

[236]See Alder, R. *Acc. Chem. Res.* **1983**, *16*, 321.

[237]Alder, R.W.; East, S.P. *Chem. Rev.* **1996**, *96*, 2097.

[238]Alder, R.W. *Tetrahedron* **1990**, *46*, 683.

[239]Alder, R.W.; Orpen, A.G.; Sessions, R.B. *J. Chem. Soc., Chem. Commun.* **1983**, 999.

[240]Alder, R.W.; Goode, N.C.; King, T.J.; Mellor, J.M.; Miller, B.W. *J. Chem. Soc., Chem. Commun.* **1976**, 173; Alder, R.W.; Arrowsmith, R.J.; Casson, A.; Sessions, R.B.; Heilbronner, E.; Kovac, B.; Huber, H.; Taagepera, M. *J. Am. Chem. Soc.* **1981**, *103*, 6137.

[241]Alder, R.W.; Heilbronner, E.; Honegger, E.; McEwen, A.B.; Moss, R.E.; Olefirowicz, E.; Petillo, P.A.; Sessions, R.B.; Weisman, G.R.; White, J.M.; Yang, Z.-Z. *J. Am. Chem. Soc.* **1993**, *115*, 6580.

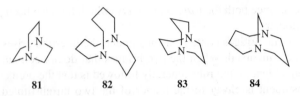

81   **82**   **83**   **84**

In the examples **85–87**, when k, l, and $m > 6$, the N—H bonds can be inside the molecular cavity or outside, giving rise to three isomers, as shown. Simmons and Park[242] isolated several such isomers with k, l, and m varying from 6 to 10. In the 9,9,9 compound, the cavity of the in-in isomer is large enough to encapsulate a

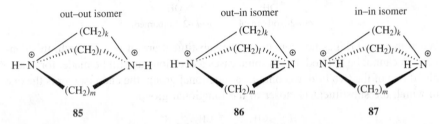

out–out isomer   out–in isomer   in–in isomer

85   **86**   **87**

chloride ion that is hydrogen bonded to the two N—H groups. The species thus formed is a cryptate, but differs from the cryptates discussed at p. 119 in that there is a negative rather than a positive ion enclosed.[243] Even smaller ones (e.g., the 4,4,4 compound) have been shown to form mono-inside-protonated ions.[244] In compound **88**, which has four quaternary nitrogens, a halide ion has been encapsulated without a hydrogen being present on a nitrogen.[245] This ion does not display *in–out* isomerism. *Out–in* and *in–in* isomers have also been prepared in analogous all-carbon tricyclic systems.[246]

It is known that chiral phosphanes are more pyramidal and that inversion is more difficult, usually requiring temperatures well over 100°C for racemization.[247] Alder

[242]Simmons, H.E.; Park, C.H. *J. Am. Chem. Soc.* *1968*, *90*, 2428; Park, C.H.; Simmons, H.E. *J. Am. Chem. Soc.* *1968*, *90*, 2429, 2431; Simmons, H.E.; Park, C.H.; Uyeda, R.T.; Habibi, M.F. *Trans. N.Y. Acad. Sci.* *1970*, *32*, 521. See also, Dietrich, B.; Lehn, J.M.; Sauvage, J.P. *Tetrahedron* *1973*, *29*, 1647; Dietrich, B.; Lehn, J.M.; Sauvage, J.P.; Blanzat, J.*Tetrahedron* *1973*, *29*, 1629.

[243]For reviews, see Schmidtchen, F.P.; Gleich, A.; Schummer, A. *Pure. Appl. Chem.* *1989*, *61*, 1535; Pierre, J.; Baret, P. *Bull. Soc. Chim. Fr.* *1983*, II-367. See also, Hosseini, M.W.; Lehn, J. *Helv. Chim. Acta* *1988*, *71*, 749.

[244]Alder, R.W.; Moss, R.E.; Sessions, R.B. *J. Chem. Soc. Chem. Commun.* *1983*, 997, 1000; Alder, R.W.; Orpen, A.G.; Sessions, R.B. *J. Chem. Soc. Chem. Commun.* *1983*, 999; Dietrich, B.; Lehn, J.M.; Guilhem, J.; Pascard, C. *Tetrahedron Lett.* *1989*, *30*, 4125; Wallon, A.; Peter-Katalinić, J.; Werner, U.; Müller, W.M.; Vögtle, F. *Chem. Ber.* *1990*, *123*, 375.

[245]Schmidtchen, F.P.; Müller, G. *J. Chem. Soc. Chem. Commun.* *1984*, 1115. See also, Schmidtchen, F.P. *J. Am. Chem. Soc.* *1986*, *108*, 8249; *Top. Curr. Chem.* *1986*, *132*, 101.

[246]Park, C.H.; Simmons, H.E. *J. Am. Chem. Soc.* *1972*, *94*, 7184; Gassman, P.G.; Hoye, R.C. *J. Am. Chem. Soc.* *1981*, *103*, 215; McMurry, J.E.; Hodge, C.N. *J. Am. Chem. Soc.* *1984*, *106*, 6450; Winkler, J.D.; Hey, J.P.; Williard, P.G. *J. Am. Chem. Soc.* *1986*, *108*, 6425.

[247]See Baechler, R.D.; Mislow, K. *J. Am. Chem. Soc.* *1970*, *92*, 3090; Rauk, A.; Allen, L.C.; Mislow, K. *Angew. Chem. Int. Ed.* *1970*, *9*, 400.

and Read found that deprotonation of bis(phosphorane) **89** (which is known to have an *in–out* structure with significant P–P bonding) leads to a rearrangement and the *out–out* diphosphane **90**.[248] Reprotonation gives **89**,[249] with inversion at the non-protonated phosphorus atom occurring at room temperature.

88	**89**	**90**

Enantiotopic and Diastereotopic Atoms, Groups, and Faces[250]

Many molecules contain atoms or groups that appear to be equivalent, but with a close inspection will show to be actually different. We can test whether two atoms are equivalent by replacing each of them in turn with some other atom or group. If the new molecules created by this process are identical, the original atoms are equivalent; otherwise they are not. We can distinguish three cases.

1. In the case of malonic acid $CH_2(COOH)_2$, propane CH_2Me_2, or any other molecule of the form CH_2Y_2,[251] if we replace either of the CH_2 hydrogens by a group Z, the identical compound results. The two hydrogens are thus equivalent. Equivalent atoms and groups need not, of course, be located on the same carbon atom. For example, all the chlorine atoms of hexachlorobenzene are equivalent as are the two bromine atoms of 1,3-dibromopropane.

2. In the case of ethanol CH_2MeOH, if we replace one of the CH_2 hydrogens by a group Z, we get one enantiomer of the compound ZCHMeOH (**91**), while replacement of the other hydrogen gives the *other* enantiomer (**92**). Since the

[248]Alder, R.W.; Read, D. *Angew. Chem. Int. Ed.* **2000**, *39*, 2879.

[249]Alder, R.W.; Ellis, D.D.; Gleiter, R.; Harris, C.J.; Lange, H.; Orpen, A.G.; Read, D.; Taylor, P.N. *J. Chem. Soc., Perkin Trans. 1* **1998**, 1657.

[250]These terms were coined by Mislow. For lengthy discussions of this subject, see Eliel, E.L. *Top. Curr. Chem.* **1982**, *105*, 1, *J. Chem. Educ.* **1980**, *57*, 52; Mislow, K.; Raban, M. *Top. Stereochem.* **1967**, *1*, 1. See also, Ault, A. *J. Chem. Educ.* **1974**, *51*, 729; Kaloustian, S.A.; Kaloustian, M.K. *J. Chem. Educ.* **1975**, *52*, 56; Jennings, W.B. *Chem. Rev.* **1975**, *75*, 307.

[251]In the case where Y is itself a chiral group, this statement is only true when the two Y groups have the same configuration.

two compounds that result upon replacement of H by Z (**91** and **92**) are not

$$
\underset{\textbf{91}}{\underset{Z}{\overset{OH}{\underset{H}{\bigwedge}}}Me} \quad \longleftarrow \quad \underset{H}{\overset{OH}{\underset{H}{\bigwedge}}}Me \quad \longrightarrow \quad \underset{\textbf{92}}{\underset{H}{\overset{OH}{\underset{Z}{\bigwedge}}}Me}
$$

identical but enantiomeric, the hydrogens are *not* equivalent. We define as *enantiotopic* two atoms or groups that upon replacement with a third group give enantiomers. In any symmetrical environment the two hydrogens behave as equivalent, but in a dissymmetrical environment they may behave differently. For example, in a reaction with a chiral reagent they may be attacked at different rates. This has its most important consequences in enzymatic reactions,[252] since enzymes are capable of much greater discrimination than ordinary chiral reagents. An example is found in the Krebs cycle, in biological organisms, where oxaloacetic acid (**93**) is converted to α-oxoglutaric

$$
\begin{array}{ccc}
\begin{array}{l}
4\,\overset{*}{C}OOH \\
\mid \\
3\,CH_2 \\
\mid \\
2\,C{=}O \\
\mid \\
1\,COOH
\end{array}
&
\xrightarrow{\text{enzymes}}
&
\begin{array}{l}
COOH \\
\mid \\
CH_2 \\
\mid \\
HO{-}C{-}COOH \\
\mid \\
CH_2 \\
\mid \\
COOH
\end{array}
&
\xrightarrow{\text{enzymes}}
&
\begin{array}{l}
1\,\overset{*}{C}OOH \\
\mid \\
2\,C{=}O \\
\mid \\
3\,CH_2 \\
\mid \\
4\,CH_2 \\
\mid \\
5\,COOH
\end{array}
\\
\textbf{93} & & \textbf{94} & & \textbf{95}
\end{array}
$$

acid (**95**) by a sequence that includes citric acid (**94**) as an intermediate. When **93** is labeled with ^{14}C at the 4 position, the label is found only at C-1 of **95**, despite the fact that **94** is not chiral. The two CH_2COOH groups of **94** are enantiotopic and the enzyme easily discriminates between them.[253] Note that the X atoms or groups of any molecule of the form CX_2WY are always enantiotopic if neither W nor Y is chiral, although enantiotopic atoms and groups may also be found in other molecules, for example, the hydrogen atoms in 3-fluoro-3-chlorocyclopropene (**96**). In this case, substitution of an H by a group Z makes the C-3 atom asymmetric and substitution at C-1 gives the opposite enantiomer from substitution at C-2.

$$
\underset{F \quad Cl}{\overset{H \quad \quad H}{\bigtriangledown}}
$$

96

[252]For a review, see Benner, S.A.; Glasfeld, A.; Piccirilli, J.A. *Top. Stereochem.* **1989**, *19*, 127. For a nonenzymatic example, see Job, R.C.; Bruice, T.C. *J. Am. Chem. Soc.* **1974**, *96*, 809.

[253]The experiments were carried out by Evans, Jr., E.A.; Slotin, L. *J. Biol. Chem.* **1941**, *141*, 439; Wood, H.G.; Werkman, C.H.; Hemingway, A.; Nier, A.O. *J. Biol. Chem.* **1942**, *142*, 31. The correct interpretation was given by Ogston, A.G. *Nature (London)* **1948**, *162*, 963. For discussion, see Hirschmann, H., in Florkin, M.; Stotz, E.H. *Comprehensive Biochemistry*, Vol. 12, pp. 236–260, Elsevier, NY, **1964**; Cornforth, J.W. *Tetrahedron* **1974**, *30*, 1515; Vennesland, B. *Top. Curr. Chem.* **1974**, *48*, 39; Eliel, E.L. *Top. Curr. Chem.*, **1982**, *105*, 1, pp. 5–7, 45–70.

The term *prochiral*[254] is used for a compound or group that has two enantiotopic atoms or groups, for example, CX_2WY. That atom or group X that would lead to an R compound if preferred to the other is called *pro-(R)*. The other is *pro-(S)*; for example,

$$H^2 = pro\text{-}(S)$$
$$H^1 = pro\text{-}(R)$$

3. Where two atoms or groups in a molecule are in such positions that replacing each of them in turn by a group Z gives rise to diastereomers, the atoms or groups are called *diastereotopic*. Some examples are the CH_2 groups of 2-chlorobutane (**97**), vinyl chloride (**98**), and chlorocyclopropane (**99**) and the

two alkenyl hydrogens of **100**. Diastereotopic atoms and groups are different in any environment, chiral or achiral. These hydrogens react at different rates with achiral reagents, but an even more important consequence is that in nmr spectra, diastereotopic hydrogens theoretically give different peaks and split each other. This is in sharp contrast to equivalent or enantiotopic hydrogens, which are indistinguishable in the NMR, except when chiral solvents are used, in which case enantiotopic (but not equivalent) protons give different peaks.[255] The term *isochronous* is used for hydrogens that are indistinguishable in the NMR.[256] In practice, the NMR signals from diastereotopic protons are often found to be indistinguishable, but this is merely because they are very close together. Theoretically they are distinct, and they have been resolved in many cases. When they appear together, it is sometimes possible to resolve them by the use of lanthanide shift reagents (p. 181) or by changing the solvent or concentration. Note that X atoms or groups CX_2WY are diastereotopic if either W or Y is chiral.

[254]Hirschmann, H.; Hanson, K.R. *Tetrahedron* **1974**, *30*, 3649.

[255]Pirkle, W.H. *J. Am. Chem. Soc.* **1966**, *88*, 1837; Burlingame, T.G.; Pirkle, W.H. *J. Am. Chem. Soc.* **1966**, *88*, 4294; Pirkle, W.H.; Burlingame, T.G. *Tetrahedron Lett.* **1967**, 4039.

[256]For a review of isochronous and nonisochronous nuclei in the nmr, see van Gorkom, M.; Hall, G.E. *Q. Rev. Chem. Soc.* **1968**, *22*, 14. For a discussion, see Silverstein, R.M.; LaLonde, R.T. *J. Chem. Educ.* **1980**, *57*, 343.

Just as there are enantiotopic and diastereotopic atoms and groups, so we may distinguish *enantiotopic and diastereotopic faces* in trigonal molecules. Again, we have three cases: (*1*) In formaldehyde or acetone (**101**), attack by an achiral reagent A from either face of the molecule gives rise to the same transition state and product; the two faces are thus equivalent. (*2*) In butanone or acetaldehyde (**102**), attack by an achiral A at one face gives a transition state and product that are the enantiomers of those arising from attack at the other face. Such faces are enantiotopic. As we have already seen (p. 153), a

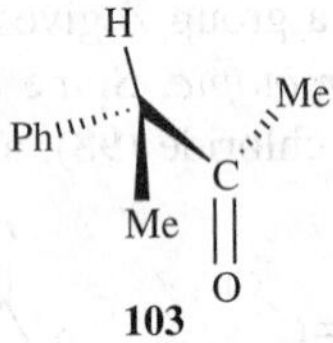

103

racemic mixture must result in this situation. However, attack at an enantiotopic face by a chiral reagent gives diastereomers, which are not formed in equal amounts. (*3*) In a case like **103**, the two faces are obviously not equivalent and are called diastereotopic. Enantiotopic and diastereotopic faces can be named by an extension of the Cahn–Ingold–Prelog system.[210] If the three groups as arranged by the sequence rules have the order $X > Y > Z$, that face in which the groups in this sequence are clockwise (as in **104**) is the *Re* face (from Latin *rectus*), whereas **105** shows the *Si* face (from Latin *sinister*).

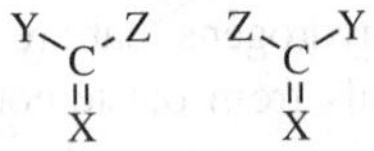

104   **105**

Note that new terminology has been proposed.[257] The concept of sphericity is used, and the terms homospheric, enantiospheric, and hemispheric have been coined to specify the nature of an orbit (an equivalent class) assigned to a coset representation.[258] Using these terms, prochirality can be defined: if a molecule has at least one enantiospheric orbit, the molecule is defined as being prochiral.[258]

Stereospecific and Stereoselective Syntheses

Any reaction in which only one of a set of stereoisomers is formed exclusively or predominantly is called a *stereoselective* synthesis.[259] The same term is used when a mixture of two or more stereoisomers is exclusively or predominantly formed at

[257]Fujita, S. *J. Org. Chem.* **2002**, *67*, 6055.

[258]Fujita, S. *J. Am. Chem. Soc.* **1990**, *112*, 3390.

[259]For a further discussion of these terms and of stereoselective reactions in general, see Eliel, E.L.; Wilen, S.H.; Mander, L.N. *Stereochemistry of Organic Compounds*, Wiley-Interscience, NY, *1994*, pp. 835–990.

the expense of other stereoisomers. In a *stereospecific* reaction, a given isomer leads to one product while another stereoisomer leads to the opposite product. All stereospecific reactions are necessarily stereoselective, but the converse is not true. These terms are best illustrated by examples. Thus, if maleic acid treated with bromine gives the *dl* pair of 2,3-dibromosuccinic acid while fumaric acid gives the meso isomer (this is the case), the reaction is stereospecific as well as stereoselective because two opposite isomers give two opposite isomers:

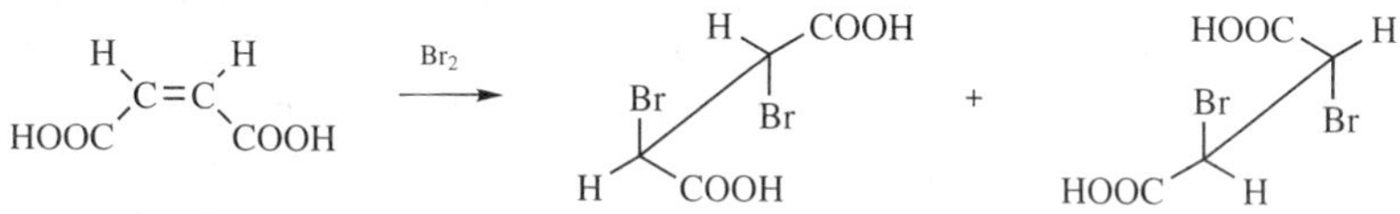

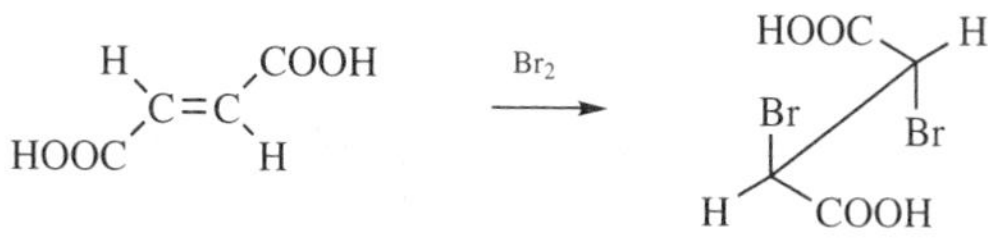

However, if both maleic and fumaric acid gave the *dl* pair or a mixture in which the *dl* pair predominated, the reaction would be stereoselective, but not stereospecific. If more or less equal amounts of *dl* and meso forms were produced in each case, the reaction would be nonstereoselective. A consequence of these definitions is that if a reaction is carried out on a compound that has no stereoisomers, it cannot be stereospecific, but at most stereoselective. For example, addition of bromine to methylacetylene could (and does) result in preferential formation of *trans*-1,2-dibromopropene, but this can be only a stereoselective, not a stereospecific reaction.

CONFORMATIONAL ANALYSIS

If two different 3D arrangements in space of the atoms in a molecule are interconvertible merely by free rotation about bonds, they are called *conformations*.[260] If they are not interconvertible, they are called *configurations*.[261] Configurations represent *isomers* that can be separated, as previously discussed in this chapter. Conformations represent *conformers*, which are rapidly interconvertible and thus

---

[260]For related discussions see Bonchev, D.; Rouvray, D.H. *Chemical Topology*, Gordon and Breach, Australia, *1999*.

[261]For books on conformational analysis see Dale, J. *Stereochemistry and Conformational Analysis*; Verlag Chemie: Deerfield Beach, FL, *1978*; Chiurdoglu, G. *Conformational Analysis*; Academic Press, NY, *1971*; Eliel, E.L.; Allinger, N.L.; Angyal, S.J.; Morrison, G.A. *Conformational Analysis*; Wiley, NY, *1965*; Hanack, M. *Conformation Theory*; Academic Press, NY, *1965*. For reviews, see Dale, J. *Top. Stereochem. 1976*, *9*, 199; Truax, D.R.; Wieser, H. *Chem. Soc. Rev. 1976*, *5*, 411; Eliel, E.L. *J. Chem. Educ. 1975*, *52*, 762; Bastiansen, O.; Seip, H.M.; Boggs, J.E. *Perspect. Struct. Chem. 1971*, *4*, 60; Bushweller, C.H.; Gianni, M.H., in Patai, S. *The Chemistry of Functional Groups, Supplement E*; Wiley, NY, *1980*, pp. 215–278.

nonseparable. The terms "conformational isomer" and "rotamer"[262] are sometimes used instead of "conformer." A number of methods have been used to determine conformations.[263] These include X-ray and electron diffraction, IR, Raman, UV, NMR,[264] and microwave spectra,[265] photoelectron spectroscopy,[266] supersonic molecular jet spectroscopy,[267] and optical rotatory dispersion and CD measurements.[268] Ring current NMR anisotropy has been applied to conformational analysis,[269] as has chemical shift simulation.[270] Some of these methods are useful only for solids. It must be kept in mind that the conformation of a molecule in the solid state is not necessarily the same as in solution.[271] Conformations can be *calculated* by a method called molecular mechanics (p. 213). A method was reported that characterized six-membered ring conformations as a linear combination of ideal basic conformations.[272] The term absolute conformation has been introduced for molecules for which one conformation is optically inactive but, by internal rotation about a $C(sp^3)-C(sp^3)$ bond, optically active conformers are produced.[273]

[262]Ōki, M. *The Chemistry of Rotational Isomers*, Springer-Verlag, Berlin, *1993*.

[263]For a review, see Eliel, E.L.; Allinger, N.L.; Angyal, S.J.; Morrison, G.A. *Conformational Analysis*, Wiley, NY, *1965*, pp. 129–188.

[264]For monographs on the use of NMR to study conformational questions, see Ōki, M. *Applications of Dynamic NMR Spectroscopy to Organic Chemistry*, VCH, NY, *1985*; Marshall, J.L. *Carbon–Carbon and Carbon–Proton NMR Couplings*, VCH, NY, *1983*. For reviews, see Anet, F.A.L.; Anet, R., in Nachod, F.C.; Zuckerman, J.J. *Determination of Organic Structures by Physical Methods*, Vol. 3, Academic Press, NY, *1971*, pp. 343–420; Kessler, H. *Angew. Chem. Int. Ed. 1970*, *9*, 219; Ivanova, T.M.; Kugatova-Shemyakina, G.P. *Russ. Chem. Rev. 1970*, *39*, 510; Anderson, J.E. *Q. Rev. Chem. Soc. 1965*, *19*, 426; Franklin, N.C.; Feltkamp, H. *Angew. Chem. Int. Ed. 1965*, *4*, 774; Johnson, Jr., C.S. *Adv. Magn. Reson. 1965*, *1*, 33. See also, Whitesell, J.K.; Minton, M. *Stereochemical Analysis of Alicyclic Compounds by C-13 NMR Spectroscopy*, Chapman and Hall, NY, *1987*.

[265]For a review see Wilson, E.B. *Chem. Soc. Rev. 1972*, *1*, 293.

[266]For a review, see Klessinger, M.; Rademacher, P. *Angew. Chem. Int. Ed. 1979*, *18*, 826.

[267]Breen, P.J.; Warren, J.A.; Bernstein, E.R.; Seeman, J.I. *J. Am. Chem. Soc. 1987*, *109*, 3453.

[268]For monographs, see Kagan, H.B. *Determination of Configurations by Dipole Moments, CD, or ORD* (Vol. 2 of Kagan, *Stereochemistry*), Georg Thieme Publishers, Stuttgart, *1977*; Crabbé, P. *ORD and CD in Chemistry and Biochemistry*, Academic Press, NY, *1972*, *Optical Rotatory Dispersion and Circular Dichroism in Organic Chemistry*, Holden-Day, San Francisco, *1965*; Snatzke, G. *Optical Rotatory Dispersion and Circular Dichroism in Organic Chemistry*, Sadtler Research Laboratories, Philadelphia, *1967*; Velluz, L.; Legrand, M.; Grosjean, M. *Optical Circular Dichroism*, Academic Press, NY, *1965*. For reviews, see Smith, H.E. *Chem. Rev. 1983*, *83*, 359; Håkansson, R., in Patai, S. *The Chemistry of Acid Derivatives*, pt. 1, Wiley, NY, *1979*, pp. 67–120; Hudec, J.; Kirk, D.N. *Tetrahedron 1976*, *32*, 2475; Schellman, J.A. *Chem. Rev. 1975*, *75*, 323; Velluz, L.; Legrand, M. *Bull. Soc. Chim. Fr. 1970*, 1785; Barrett, G.C., in Bentley, K.W.; Kirby, G.W. *Elucidation of Organic Structures by Physical and Chemical Methods*, 2nd ed. (Vol. 4 of Weissberger, A. *Techniques of Chemistry*), pt. 1, Wiley, NY, *1972*, pp. 515–610; Snatzke, G. *Angew. Chem. Int. Ed. 1968*, *7*, 14; Crabbé, P., in Nachod, F.C.; Zuckerman, J.J. *Determination of Organic Structures by Physical Methods*, Vol. 3, Academic Press, NY, *1971*, pp. 133–205; Crabbé, P.; Klyne, W. *Tetrahedron 1967*, *23*, 3449; Crabbé, P. *Top. Stereochem. 1967*, *1*, 93–198; Eyring, H.; Liu, H.; Caldwell, D. *Chem. Rev. 1968*, *68*, 525.

[269]Chen, J.; Cammers-Goodwin, A. *Eur. J. Org. Chem. 2003*, 3861.

[270]Iwamoto, H.; Yang, Y.; Usui, S.; Fukazawa, Y. *Tetrahedron Lett. 2001*, *42*, 49.

[271]See Kessler, H.; Zimmermann, G.; Förster, H.; Engel, J.; Oepen, G.; Sheldrick, W.S. *Angew. Chem. Int. Ed. 1981*, *20*, 1053.

[272]Bérces, A.; Whitfield, D.M.; Nukada, T. *Tetrahedron 2001*, *57*, 477.

[273]Ōki, M.; Toyota, S. *Eur. J. Org. Chem. 2004*, 255.

Conformation in Open-Chain Systems[274]

For any open-chain single bond that connects two sp^3 carbon atoms, an infinite number of conformations are possible, each of which has a certain energy associated with it. As a practical matter, the number of conformations is much less. If one ignores duplications due to symmetry, the number of conformations can be *estimated* as being greater than 3^n, where $n =$ the number of internal C—C bonds. *n*-Pentane, for example, has 11, *n*-hexane 35, *n*-heptane 109, *n*-octane 347, *n*-nonane 1101, and *n*-decane 3263.[275] For ethane there are two extremes, a conformation of highest and one of lowest potential energy, depicted in two ways as:

| Staggered | Eclipsed | Staggered | Eclipsed |

In *Newman projection formulas* (the two figures on the right), the observer looks at the C—C bond head on. The three lines emanating from the center of the circle represent the bonds coming from the front carbon, with respect to the observer.

The staggered conformation is the conformation of lowest potential energy for ethane. As the bond rotates, the energy gradually increases until the eclipsed conformation is reached, when the energy is at a maximum. Further rotation decreases the energy again. Fig. 4.4 illustrates this. The *angle of torsion*, which is a dihedral angle, is the angle between the X—C—C and the C—C—Y planes, as shown:

For ethane, the difference in energy is $\sim 2.9 \, \text{kcal mol}^{-1}$ ($12 \, \text{kJ mol}^{-1}$).[276] This difference is called the *energy barrier*, since in free rotation about a single bond there must be enough rotational energy present to cross the barrier every time two hydrogen atoms are opposite each other. There has been much speculation about the cause of the barriers and many explanations have been suggested.[277] It

---

[274]For a review, see Berg, U.; Sandström, J. *Adv. Phys. Org. Chem.* **1989**, *25*, 1. Eliel, E.L.; Wilen, S.H.; Mander, L.N. *Stereochemistry of Organic Compounds*, Wiley-Interscience, NY, *1994*, pp. 597–664. Also see, Smith, M.B. *Organic Synthesis*, 2nd ed., McGraw-Hill, NY, *2001*, pp. 32–37.

[275]Gotō, H.; Ō sawa, E.; Yamato, M. *Tetrahedron* **1993**, *49*, 387.

[276]Lide, Jr., D.R. *J. Chem. Phys.* **1958**, *29*, 1426; Weiss, S.; Leroi, G.E. *J. Chem. Phys.* **1968**, *48*, 962; Hirota, E.; Saito, S.; Endo, Y. *J. Chem. Phys.* **1979**, *71*, 1183.

[277]For a review of methods of measuring barriers, of attempts to explain barriers, and of values of barriers, see Lowe, J.P. *Prog. Phys. Org. Chem.* **1968**, *6*, 1. For other reviews of this subject, see Oosterhoff, L.J. *Pure Appl. Chem.* **1971**, *25*, 563; Wyn-Jones, E.; Pethrick, R.A. *Top. Stereochem.* **1970**, *5*, 205; Pethrick, R.A.; Wyn-Jones, E. *Q. Rev. Chem. Soc.* **1969**, *23*, 301; Brier, P.N. *J. Mol. Struct.* **1970**, *6*, 23; Lowe, J.P. *Science* , *1973*, *179*, 527.

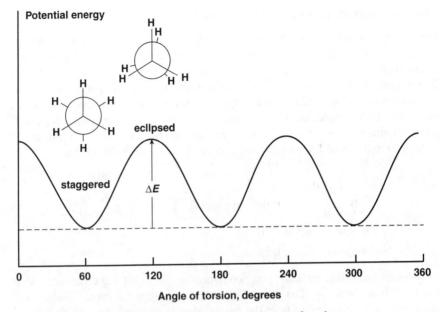

Fig. 4.4. Conformational energy diagram for ethane.

was concluded from molecular-orbital calculations that the barrier is caused by repulsion between overlapping filled molecular orbitals.[278] That is, the ethane molecule has its lowest energy in the staggered conformation because in this conformation the orbitals of the C—H bonds have the least amount of overlap with the C—H orbitals of the adjacent carbon.

At ordinary temperatures, enough rotational energy is present for the ethane molecule rapidly to rotate, although it still spends most of its time at or near the energy minimum. Groups larger than hydrogen cause larger barriers. When the barriers are large enough, as in the case of suitably substituted biphenyls (p. 146) or the diadamantyl compound mentioned on p. 201, rotation at room temperature is completely prevented and we speak of configurations, not conformations. Even for compounds with small barriers, cooling to low temperatures may remove enough rotational energy for what would otherwise be conformational isomers to become configurational isomers.

A slightly more complicated case than ethane is that of a 1,2-disubstituted ethane (YCH_2—CH_2Y or YCH_2—CH_2X),[279] such as *n*-butane, for which there are four extremes: a fully staggered conformation, called anti, trans, or antiperiplanar; another

[278]See Pitzer, R.M. *Acc. Chem. Res.* **1983**, *16*, 207. See, however, Bader, R.F.W.; Cheeseman, J.R.; Laidig, K.E.; Wiberg, K.B.; Breneman, C.*J. Am. Chem. Soc.* **1990**, *112*, 6350.
[279]For discussions of the conformational analysis of such systems, see Kingsbury, C.A. *J. Chem. Educ.* **1979**, *56*, 431; Wiberg, K.B.; Murcko, M.A. *J. Am. Chem. Soc.* **1988**, *110*, 8029; Allinger, N.L.; Grev, R.S.; Yates, B.F.; Schaefer III, H.F. *J. Am. Chem. Soc.* **1990**, *112*, 114.

staggered conformation, called *gauche* or *synclinal*; and two types of eclipsed

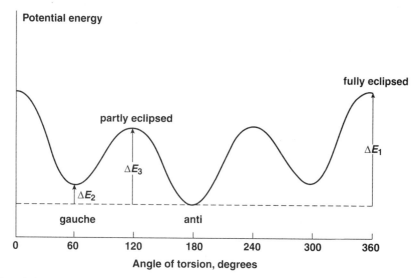

anti, trans, or   anticlinal   *gauche* or   synperiplanar
antiperiplanar   synclinal

106

conformations, called synperiplanar and anticlinal. An energy diagram for this system is given in Fig. 4.5. Although there is constant rotation about the central bond, it is possible to estimate what percentage of the molecules are in each conformation at a given time. For example, it was concluded from a consideration of dipole moment and polarizability measurements that for 1,2-dichloroethane in CCl_4 solution at 25°C ~70% of the molecules are in the anti and ~30% in the gauche conformation.[280] The corresponding figures for 1,2-dibromoethane are 89% *anti* and 11% *gauche*.[281] The eclipsed conformations are unpopulated and serve only as pathways from one staggered conformation to another. Solids normally consist of a single conformer.

Fig. 4.5. Conformational energy for $YCH_2{-}CH_2Y$ or $YCH_2{-}CH_2X$. For *n*-butane, $\Delta E_1 = 4\text{--}6$, $\Delta E_2 = 0.9$, and $\Delta E_3 = 3.4\ \text{kcal mol}^{-1}$ (17–25, 3.8, 14 kL mol^{-1}, respectively).

[280] Aroney, M.; Izsak, D.; Le Fèvre, R.J.W. *J. Chem. Soc.* **1962**, 1407; Le Fèvre, R.J.W.; Orr, B.J. *Aust. J. Chem.* **1964**, *17*, 1098.
[281] The *anti* form of butane itself is also more stable than the *gauche* form: Schrumpf, G. *Angew. Chem. Int. Ed.* **1982**, *21*, 146.

It may be observed that the *gauche* conformation of butane (**106**) or any other similar molecule is chiral. The lack of optical activity in such compounds arises from the fact that **106** and its mirror image are always present in equal amounts and interconvert too rapidly for separation.

For butane and for most other molecules of the forms YCH_2—CH_2Y and YCH_2—CH_2X, the *anti* conformer is the most stable, but exceptions are known. One group of exceptions consists of molecules containing small electronegative atoms, especially fluorine and oxygen. Thus 2-fluoroethanol,[282] 1,2-difluoroethane,[283] and 2-fluoroethyl trichloroacetate $(FCH_2CH_2OCOCCl_3)$[284] exist predominantly in the *gauche* form and compounds, such as 2-chloroethanol and 2-bromoethanol,[282] also prefer the *gauche* form. It has been proposed that the preference for the *gauche* conformation in these molecules is an example of a more general phenomenon, known as the *gauche effect*, that is, a tendency to adopt that structure that has the maximum number of *gauche* interactions between adjacent electron pairs or polar bonds.[285] It was believed that the favorable *gauche* conformation of 2-fluoroethanol was the result of intramolecular hydrogen bonding, but this explanation does not do for molecules like 2-fluoroethyl trichloroacetate and has in fact been ruled out for 2-fluoroethanol as well.[286] The effect of β-substituents in Y—C—C—OX systems, where Y = F or SiR_3 has been examined and there is a small bond shortening effect on C—OX that is greatest when OX is a good leaving group. Bond lengthening was also observed with the β-silyl substituent.[287] Other exceptions are known, where small electronegative atoms are absent. For example, 1,1,2,2-tetrachloroethane and 1,1,2,2-tetrabromoethane both prefer the *gauche* conformation,[288] even although 1,1,2,2-tetrafluoroethane prefers the *anti*.[289] Also, both 2,3-dimethylpentane and 3,4-dimethylhexane prefer the *gauche* conformation,[290] and 2,3-dimethylbutane shows no preference for either.[291] Furthermore, the solvent can exert a powerful

[282]Wyn-Jones, E.; Orville-Thomas, W.J. *J. Mol. Struct.* **1967**, *1*, 79; Buckley, P.; Giguère, P.A.; Yamamoto, D. *Can. J. Chem.* **1968**, *46*, 2917; Davenport, D.; Schwartz, M. *J. Mol. Struct.* **1978**, *50*, 259; Huang, J.; Hedberg, K. *J. Am. Chem. Soc.* **1989**, *111*, 6909.

[283]Klaboe, P.; Nielsen, J.R. *J. Chem. Phys.* **1960**, *33*, 1764; Abraham, R.J.; Kemp, R.H. *J. Chem. Soc. B* **1971**, 1240; Bulthuis, J.; van den Berg, J.; MacLean, C. *J. Mol. Struct.* **1973**, *16*, 11; van Schaick, E.J.M.; Geise, H.J.; Mijlhoff, F.C.; Renes, G. *J. Mol. Struct.* **1973**, *16*, 23; Friesen, D.; Hedberg, K. *J. Am. Chem. Soc.* **1980**, *102*, 3987; Fernholt, L.; Kveseth, K. *Acta Chem. Scand. Ser. A* **1980**, *34*, 163.

[284]Abraham, R.J.; Monasterios, J.R. *Org. Magn. Reson.* **1973**, *5*, 305.

[285]This effect is ascribed to nuclear electron attactive forces between the groups or unshared pairs: Wolfe, S.; Rauk, A.; Tel, L.M.; Csizmadia, I.G. *J. Chem. Soc. B* **1971**, 136; Wolfe, S. *Acc. Chem. Res.* **1972**, *5*, 102. See also, Phillips, L.; Wray, V. *J. Chem. Soc. Chem. Commun.* **1973**, 90; Radom, L.; Hehre, W.J.; Pople, J.A. *J. Am. Chem. Soc.* **1972**, *94*, 2371; Zefirov, N.S. *J. Org. Chem. USSR* **1974**, *10*, 1147; Juaristi, E. *J. Chem. Educ.* **1979**, *56*, 438.

[286]Griffith, R.C.; Roberts, J.D. *Tetrahedron Lett.* **1974**, 3499.

[287]Amos, R.D.; Handy, N.C.; Jones, P.G.; Kirby, A.J.; Parker, J.K.; Percy, J.M.; Su, M.D. *J. Chem. Soc. Perkin Trans. 2* **1992**, 549.

[288]Kagarise, R.E. *J. Chem. Phys.* **1956**, *24*, 300.

[289]Brown, D.E.; Beagley, B. *J. Mol. Struct.* **1977**, *38*, 167.

[290]Ritter, W.; Hull, W.; Cantow, H. *Tetrahedron Lett.* **1978**, 3093.

[291]Lunazzi, L.; Macciantelli, D.; Bernardi, F.; Ingold, K.U. *J. Am. Chem. Soc.* **1977**, *99*, 4573.

effect. For example, the compound 2,3-dinitro-2,3-dimethylbutane exists entirely in the gauche conformation in the solid state, but in benzene, the *gauche/anti* ratio is 79:21; while in CCl_4 the anti form is actually favored (*gauche/anti* ratio 42:58).[292] In many cases, there are differences in the conformation of these molecules between the gas and the liquid phase (as when X = Y = OMe) because of polar interactions with the solvent.[293]

In one case, two conformational isomers of a single aliphatic hydrocarbon, 3,4-di(1-adamantyl)-2,2,5,5-tetramethylhexane, have proven stable enough for isolation at room temperature.[294] The two isomers **107** and **108** were separately crystallized, and the structures proved by X-ray crystallography. (The actual dihedral angles are distorted from the 60° angles shown in the drawings, owing to steric hindrance between the large groups.)

All the conformations so far discussed have involved rotation about sp^3–sp^3 bonds. Many studies were also made of compounds with sp^3–sp^2 bonds.[295] For example, propanal (or any similar molecule) has four extreme conformations, two of which are called *eclipsing* and the other two *bisecting*. For propanal the eclipsing conformations have lower energy than the other two, with **109** favored over **110** by ~1 kcal mol^{-1} (4 kJ mol^{-1}).[296] As has already been pointed out (p. 184), for a few of these compounds, rotation is slow enough to permit cis–trans isomerism, although for simple compounds rotation is rapid. The cis conformer of acetic acid was produced in solid Ar,[297] and it was reported that acetaldehyde has a lower rotational barrier (~1 kcal mol^{-1} or 4 kJ mol^{-1}) than ethane.[298] Calculations have examined the rotational barriers around the CO and CC bonds

[292]Tan, B.; Chia, L.H.L.; Huang, H.; Kuok, M.; Tang, S. *J. Chem. Soc. Perkin Trans. 2* **1984**, 1407.
[293]Smith, G.D.; Jaffe, R.L.; Yoon, D.Y. *J. Am. Chem. Soc.* **1995**, *117*, 530. For an analysis of *N,N*-dimethylacetamide see Mack, H.-G.; Oberhammer, H. *J. Am. Chem. Soc.* **1997**, *119*, 3567.
[294]Flamm-ter Meer; Beckhaus, H.; Peters, K.; von Schnering, H.; Fritz, H.; Rüchardt, C. *Chem. Ber.* **1986**, *119*, 1492; Rüchardt, C.; Beckhaus, H. *Angew. Chem. Int. Ed.* **1985**, *24*, 529.
[295]For reviews, see Sinegovskaya, L.M.; Keiko, V.V.; Trofimov, B.A. *Sulfur Rep.* **1987**, *7*, 337 (for enol ethers and thioethers); Karabatsos, G.J.; Fenoglio, D.J. *Top. Stereochem.* **1970**, *5*, 167; Jones, G.I.L.; Owen, N.L. *J. Mol. Struct.* **1973**, *18*, 1 (for carboxylic esters). See also, Schweizer, W.B.; Dunitz, J.D. *Helv. Chim. Acta* **1982**, *65*, 1547; Chakrabarti, P.; Dunitz, J.D. *Helv. Chim. Acta* **1982**, *65*, 1555; Cossé-Barbi, A.; Massat, A.; Dubois, J.E. *Bull. Soc. Chim. Belg.* **1985**, *94*, 919; Dorigo, A.E.; Pratt, D.W.; Houk, K.N. *J. Am. Chem. Soc.* **1987**, *109*, 6591.
[296]Butcher, S.S.; Wilson Jr., E.B. *J. Chem. Phys.* **1964**, *40*, 1671; Allinger, N.L.; Hickey, M.J. *J. Mol. Struct.* **1973**, *17*, 233; Gupta, V.P. *Can. J. Chem.* **1985**, *63*, 984.
[297]Macoas, E. M. S.; Khriachtchev, L.; Pettersson, M.; Fausto, R.; Rasanen, M. *J. Am. Chem. Soc.* **2003**, *125*, 16188.
[298]Davidson, R.B.; Allen, L.C. *J. Chem. Phys.* **1971**, *54*, 2828.

in formic acid, ethanedial and glycolaldedyde molecules.[299]

| Eclipsing | Eclipsing | Bisecting | Bisecting |
| **109** | **110** | | |

Other carbonyl compounds exhibit rotation about sp^3–sp^3 bonds, including amides.[300] In N-acetyl-N-methylaniline, the cis conformation (**111**) is more stable than the *trans*- (**112**) by 3.5 kcal mol^{-1} (14.6 kJ mol^{-1}).[301] This is due to destabilization of (S) due to steric hindrance between two methyl groups, and to electronic repulsion between the carbonyl lone-pair electrons and the phenyl π-electrons in the twisted phenyl orientation.[301]

| | | |
| **111** | **112** | **113** |

A similar conformational analysis has been done with formamide derivatives,[302] with secondary amides,[303] and for hydroxamide acids.[304] It is known that thioformamide has a larger rotational barrier than formamide, which can be explained by a traditional picture of amide "resonance' that is more appropriate for the thioformamide than formamide itself.[305] Torsional barriers in α-keto amides have been reported,[306] and the C–N bond of acetamides,[307] thioamides,[308] enamides[309] carbamates (R$_2$N–CO$_2$R'),[310,311] and enolate anions derived

[299]Ratajczyk, T.; Pecul, M.; Sadlej, J. *Tetrahedron* **2004**, *60*, 179.

[300]Avalos, M.; Babiano, R.; Barneto, J.L.; Bravo, J.L.; Cintas, P.; Jiménez, J.L.; Palcios, J.C. *J. Org. Chem.* **2001**, *66*, 7275.

[301]Saito, S.; Toriumi, Y.; Tomioka, A.; Itai, A. *J. Org. Chem.* **1995**, *60*, 4715.

[302]Axe, F.U.; Renugopalakrishnan, V.; Hagler, A.T. *J. Chem. Res.* **1998**, 1. For an analysis of DMF see Wiberg, K.B.; Rablen, P.R.; Rush, D.J.; Keith, T.A. *J. Am. Chem. Soc.* **1995**, *117*, 4261.

[303]Avalos, M.; Babiano, R.; Barneto, J.L.; Cintas, P.; Clemente, F.R.; Jiménez, J.L.; Palcios, J.C. *J. Org. Chem.* **2003**, *68*, 1834.

[304]Kakkar, R.; Grover, R.; Chadha, P. *Org. Biomol. Chem.* **2003**, *1*, 2200.

[305]Wiberg, K.B.; Rablen, P.R. *J. Am. Chem. Soc.* **1995**, *117*, 2201.

[306]Bach, R.D.; Mintcheva, I.; Kronenberg, W.J.; Schlegel, H.B. *J. Org. Chem.* **1993**, *58*, 6135.

[307]Ilieva, S.; Hadjieva, B.; Galabov, B. *J. Org. Chem.* **2002**, *67*, 6210.

[308]Wiberg, K. B.; Rush, D. J. *J. Am. Chem. Soc.* **2001**, *123*, 2038; *J. Org. Chem.* **2002**, *67*, 826.

[309]Rablen, P.R.; Miller, D.A.; Bullock, V.R.; Hutchinson, P.H.; Gorman, J.A. *J. Am. Chem. Soc.* **1999**, *121*, 218.

[310]Menger, F.M.; Mounier, C.E. *J. Org. Chem.* **1993**, *58*, 1655.

[311]Deetz, M.J.; Forbes, C.C.; Jonas, M.; Malerich, J.P.; Smith, B.D.; Wiest, O. *J. Org. Chem.* **2002**, *67*, 3949.

from amides[312] have been examined. It is known that substituents influence rotational barriers.[313]

On p. 146, atropisomerism was possible when ortho substituents on biphenyl derivatives and certain other aromatic compounds prevented rotation about the Csp^3–Csp^3 bond. The presence of ortho substituents can also influence the conformation of certain groups. In **113**, R = alkyl the carbonyl unit is planar with the trans C=O···F conformer more stable when X = F. When X = CF$_3$, the cis and trans are planar and the trans predominates.[314] When R = alkyl there is one orthogonal conformation, but there are two interconverting nonplanar conformations when R = O-alkyl.[314] In 1,2-diacylbenzenes, the carbonyl units tend to adopt a twisted conformation to minimize steric interactions.[315]

Conformation in Six-Membered Rings[316]

For cyclohexane there are two extreme conformations in which all the angles are tetrahedral.[317] These are called the *boat* and the *chair* conformations and in each the ring is said to be *puckered*. The chair conformation is a rigid structure, but the boat form is flexible[318] and can easily pass over to a somewhat more stable form

Boat   Chair   Twist

known as the *twist* conformation. The twist form is $\sim$1.5 kcal mol^{-1} (6.3 kJ mol^{-1}) more stable than the boat because it has less eclipsing interaction (see p. 224).[319] The chair form is more stable than the twist form by $\sim$5 kcal mol^{-1} (21 kJ mol^{-1}).[320] In the vast majority of compounds containing a cyclohexane ring, the molecules exist almost entirely in the chair form.[321] Yet, it

[312]Kim, Y.-J.; Streitwieser, A.; Chow, A.; Fraenkel, G. *Org. Lett.* **1999**, *1*, 2069.

[313]Smith, B.D.; Goodenough-Lashua, D.M.; D'Souza, C.J.E.; Norton, K.J.; Schmidt, L.M.; Tung, J.C. *Tetrahedron Lett.* **2004**, *45*, 2747.

[314]Abraham, R.J.; Angioloni, S.; Edgar, M.; Sancassan, F. *J. Chem. Soc. Perkin Trans. 2* **1997**, 41.

[315]Casarini, D.; Lunazzi, L.; Mazzanti, A. *J. Org. Chem.* **1997**, *62*, 7592.

[316]For reviews, see Jensen, F.R.; Bushweller, C.H. *Adv. Alicyclic Chem.* **1971**, *3*, 139; Robinson, D.L.; Theobald, D.W. *Q. Rev. Chem. Soc.* **1967**, *21*, 314; Eliel, E.L. *Angew. Chem. Int. Ed.* **1965**, *4*, 761. Eliel, E.L.; Wilen, S.H.; Mander, L.N. *Stereochemistry of Organic Compounds*, Wiley-Interscience, NY, **1994**, pp. 686–753. Also see, Smith, M.B. *Organic Synthesis*, 2nd ed., McGraw-Hill, NY, **2001**, pp. 46–57.

[317]The C–C–C angles in cyclohexane are actually 111.5° [Davis, M.; Hassel, O. *Acta Chem. Scand.* **1963**, *17*, 1181; Geise, H.J.; Buys, H.R.; Mijlhoff, F.C. *J. Mol. Struct.* **1971**, *9*, 447; Bastiansen, O.; Fernholt, L.; Seip, H.M.; Kambara, H.; Kuchitsu, K. *J. Mol. Struct.* **1973**, *18*, 163], but this is within the normal tetrahedral range (see p. 26).

[318]See Dunitz, J.D. *J. Chem. Educ.* **1970**, *47*, 488.

[319]For a review of nonchair forms, see Kellie, G.M.; Riddell, F.G. *Top. Stereochem.* **1974**, *8*, 225.

[320]Margrave, J.L.; Frisch, M.A.; Bautista, R.G.; Clarke, R.L.; Johnson, W.S. *J. Am. Chem. Soc.* **1963**, *85*, 546; Squillacote, M.; Sheridan, R.S.; Chapman, O.L.; Anet, F.A.L. *J. Am. Chem. Soc.* **1975**, *97*, 3244.

[321]For a study of conformations in the cyclohexane series, see Wiberg, K. B.; Hammer, J. D.; Castejon, H.; Bailey, W. F.; DeLeon, E. L.; Jarret, R. M. *J. Org. Chem.* **1999**, *64*, 2085; Wiberg, K.B.; Castejon, H.; Bailey, W.F.; Ochterski, J. *J. Org. Chem.* **2000**, *65*, 1181.

is known that the boat or twist form exists transiently. An inspection of the chair form shows that six of its bonds are directed differently from the other six:

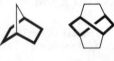

a = axial group
e = equatorial group

On each carbon, one bond is directed up or down and the other more or less in the "plane" of the ring. The up or down bonds are called *axial* and the others *equatorial*. The axial bonds point alternately up and down. If a molecule were frozen into a chair form, there would be isomerism in mono-substituted cyclohexanes. For example, there would be an equatorial methylcyclohexane and an axial isomer. However, it has never been possible to isolate isomers of this type at room temperature.[322] This proves the transient existence of the boat or twist form, since in order for the two types of methylcyclohexane to be nonseparable, there must be rapid interconversion of one chair form to another (in which all axial bonds become equatorial and vice versa) and this is possible only through a boat or twist conformation. Conversion of one chair form to another requires an activation energy of $\sim 10 \, \text{kcal mol}^{-1}$ ($42 \, \text{kJ mol}^{-1}$)[323] and is very rapid at room temperature.[324] However, by working at low temperatures, Jensen and Bushweller were able to obtain the pure equatorial conformers of chlorocyclohexane and trideuteriomethoxycyclohexane as solids and in solution.[325] Equatorial chlorocyclohexane has a half-life of 22 years in solution at $-160°C$.

In some molecules, the twist conformation is actually preferred.[326] Of course, in certain bicyclic compounds, the six-membered ring is forced to maintain a boat or twist conformation, as in norbornane or twistane.

Norbornane   Twistane

In mono-substituted cyclohexanes, the substituent normally prefers the equatorial position because in the axial position there is interaction between the substituent

[322]Wehle, D.; Fitjer, L. *Tetrahedron Lett.* **1986**, *27*, 5843, have succeeded in producing two conformers that are indefinitely stable in solution at room temperature. However, the other five positions of the cyclohexane ring in this case are all spirosubstituted with cyclobutane rings, greatly increasing the barrier to chair-chair interconversion.

[323]Jensen, F.R.; Noyce, D.S.; Sederholm, C.H.; Berlin, A.J. *J. Am. Chem. Soc.* **1962**, *84*, 386; Bovey, F.A.; Hood, F.P.; Anderson, E.W.; Kornegay, R.L. *J. Chem. Phys.* **1964**, *41*, 2041; Anet, F.A.L.; Bourn, A.J.R. *J. Am. Chem. Soc.* **1967**, *89*, 760. See also Strauss, H.L. *J. Chem. Educ.* **1971**, *48*, 221.

[324]For reviews of chair–chair interconversions, see Ōki, M. *Applications of Dynamic NMR Spectroscopy to Organic Chemistry*, VCH, NY, **1985**, pp. 287–307; Anderson, J.E. *Top. Curr. Chem.* **1974**, *45*, 139.

[325]Jensen, F.R.; Bushweller, C.H. *J. Am. Chem. Soc.* **1966**, *88*, 4279; Paquette, L.A.; Meehan, G.V.; Wise, L.D. **1969**, *91*, 3223.

[326]Weiser, J.; Golan, O.; Fitjer, L.; Biali, S.E. *J. Org. Chem.* **1996**, *61*, 8277.

and the axial hydrogens in the 3 and 5 positions, but the extent of this preference depends greatly on the nature of the group.[327] Alkyl groups have a greater preference for the equatorial postion than polar groups. For alkyl groups, the preference increases with size, although size seems to be unimportant for polar groups. Both the large $HgBr$[328] and $HgCl$[329] groups and the small F group have been reported to have little or no conformational preference (the HgCl group actually shows a slight preference for the axial position). Table 4.3 gives approximate values of the free energy required for various groups to go from the equatorial position to the axial (these are called *A* values),[330] although it must be kept in mind that they vary somewhat with physical state, temperature, and solvent.[331]

In disubstituted compounds, the rule for alkyl groups is that the conformation is such that as many groups as possible adopt the equatorial position. How far it is possible depends on the configuration. In a *cis*-1,2-disubstituted cyclohexane, one substituent must be axial and the other equatorial. In a *trans*-1,2 compound both may be equatorial or both axial. This is also true for 1,4-disubstituted cyclohexanes, but the reverse holds for 1,3 compounds: the trans isomer must have the *ae* conformation and the cis isomer may be *aa* or *ee*. For alkyl groups, the *ee* conformation predominates over the *aa*, but for other groups this is not necessarily so. For example, both *trans*-1,4-dibromocyclohexane and the corresponding dichloro compound have the *ee* and *aa* conformations about equally populated[332] and most *trans*-1,2-dihalocyclohexanes exist predominantly in the *aa* conformation.[333] Note that in the latter case the two halogen atoms are anti in the *aa* conformation, but *gauche* in the *ee* conformation.[334]

Since compounds with alkyl equatorial substituents are generally more stable, trans-1,2 compounds, which can adopt the *ee* conformation, are thermodynamically more stable than their cis-1,2 isomers, which must exist in the *ae* conformation. For the 1,2-dimethylcyclohexanes, the difference in stability is $\sim$2 kcal mol^{-1}

---

[327]For a study of thioether, sulfoxide and sulfone substituents, see Juaristi, E.; Labastida, V.; Antúnez, S. *J. Org. Chem.* **2000**, *65*, 969.

[328]Jensen, F.R.; Gale, L.H. *J. Am. Chem. Soc.* **1959**, *81*, 6337.

[329]Anet, F.A.L.; Krane, J.; Kitching, W.; Dodderel, D.; Praeger, D. *Tetrahedron Lett.* **1974**, 3255.

[330]Except where otherwise indicated, these values are from Jensen, F.R.; Bushweller, C.H. *Adv. Alicyclic Chem.* **1971**, *3*, 139. See also Schneider, H.; Hoppen, V. *Tetrahedron Lett.* **1974**, 579 and see Smith, M.B. *Organic Synthesis*, 2nd ed., McGraw-Hill, NY, **2001**, pp. 46–57.

[331]See, for example, Ford, R.A.; Allinger, N.L. *J. Org. Chem.* **1970**, *35*, 3178. For a critical review of the methods used to obtain these values, see Jensen, F.R.; Bushweller, C.H. *Adv. Alicyclic Chem.* **1971**, *3*, 139.

[332]Atkinson, V.A.; Hassel, O. *Acta Chem. Scand.* **1959**, *13*, 1737; Abraham, R.J.; Rossetti, Z.L. *Tetrahedron Lett.* **1972**, 4965, *J. Chem. Soc. Perkin Trans. 2* **1973**, 582. See also, Hammarström, L.; Berg, U.; Liljefors, T. *Tetrahedron Lett.* **1987**, *28*, 4883.

[333]Hageman, H.J.; Havinga, E. *Recl. Trav. Chim. Pays-Bas* **1969**, *88*, 97; Klaeboe, P. *Acta Chem. Scand.* **1971**, *25*, 695; Abraham, M.H.; Xodo, L.E.; Cook, M.J.; Cruz, R. *J. Chem. Soc. Perkin Trans. 2* **1982**, 1503; Samoshin, V.V.; Svyatkin, V.A.; Zefirov, N.S. *J. Org. Chem. USSR* **1988**, *24*, 1080, and references cited therein. *trans*-1,2-Difluorocyclohexane exists predominantly in the ee conformation: see Zefirov, N.S.; Samoshin, V.V.; Subbotin, O.A.; Sergeev, N.M. *J. Org. Chem. USSR* **1981**, *17*, 1301.

[334]For a case of a preferential diaxial conformation in 1,3 isomers, see Ochiai, M.; Iwaki, S.; Ukita, T.; Matsuura, Y.; Shiro, M.; Nagao, Y. *J. Am. Chem. Soc.* **1988**, *110*, 4606.

TABLE 4.3. Free-Energy Differences between Equatorial and Axial Substituents on a Cyclohexane Ring (A Values)[330]

	Approximate $-\Delta G°$,			Approximate $-\Delta G°$	
Group	kcal mol^{-1}	kJ mol^{-1}	Group	kcal mol^{-1}	kJ mol^{-1}
HgCl[330]	−0.25	−1.0	NO$_2$	1.1	4.6
HgBr	0	0	COOEt	1.1–1.2	4.6–5.0
D[335]	0.008	0.03	COOMe	1.27–1.31	5.3–5.5
CN	0.15–0.25	0.6–1.0	COOH	1.35–1.46	5.7–6.1
F	0.25	1.0	NH$_2$[336]	1.4	5.9
C≡CH	0.41	1.7	CH=CH$_2$[337]	1.7	7.1
I	0.46	1.9	CH$_3$[338]	1.74	7.28
Br	0.48–0.62	2.0–2.6	C$_2$H$_5$	~1.75	~7.3
OTs	0.515	2.15	i-Pr	~2.15	~9.0
Cl	0.52	2.2	C$_6$H$_{11}$[339]	2.15	9.0
OAc	0.71	3.0	SiMe$_3$[340]	2.4–2.6	10–11
OMe[341]	0.75	3.1	C$_6$H$_5$[342]	2.7	11
OH	0.92–0.97	3.8–4.1	t-Bu[343]	4.9	21

(8 kJ mol^{-1}). Similarly, *trans*-1,4 and *cis*-1,3 compounds are more stable than their stereoisomers.

An interesting anomaly is *all-trans*-1,2,3,4,5,6-hexaisopropylcyclohexane, in which the six isopropyl groups prefer the axial position, although the six ethyl groups of the corresponding hexaethyl compound prefer the equatorial position.[344] The alkyl groups of these compounds can of course only be all axial or all equatorial, and it is likely that the molecule prefers the all-axial conformation because of unavoidable strain in the other conformation.

Incidentally, we can now see, in one case, why the correct number of stereoisomers could be predicted by assuming planar rings, even although they are not planar (p. 186). In the case of both a *cis*-1,2-X,X-disubstituted and a *cis*-1,2-X,Y-disubstituted cyclohexane, the molecule is nonsuperimposable on its mirror image;

[335]Anet, F.A.L.; O'Leary, D.J. *Tetrahedron Lett.* **1989**, *30*, 1059.
[336]Buchanan, G.W.; Webb, V.L. *Tetrahedron Lett.* **1983**, *24*, 4519.
[337]Eliel, E.L.; Manoharan, M. *J. Org. Chem.* **1981**, *46*, 1959.
[338]Booth, H.; Everett, J.R. *J. Chem. Soc. Chem. Commun.* **1976**, 278.
[339]Hirsch, J.A. *Top. Stereochem.* **1967**, *1*, 199.
[340]Kitching, W.; Olszowy, H.A.; Drew, G.M.; Adcock, W. *J. Org. Chem.* **1982**, *47*, 5153.
[341]Schneider, H.; Hoppen, V. *Tetrahedron Lett.* **1974**, 579.
[342]Squillacote, M.E.; Neth, J.M. *J. Am. Chem. Soc.* **1987**, *109*, 198. Values of 2.59–2.92 kcal mol^{-1} were determined for 4-X-C$_6$H$_4$- substituents (X = NO$_2$, Cl, MeO) - see Kirby, A.J.; Williams, N.H. *J. Chem. Soc. Chem. Commun.* **1992**, 1285, 1286.
[343]Manoharan, M.; Eliel, E.L. *Tetrahedron Lett.* **1984**, *25*, 3267.
[344]Golan, O.; Goren, Z.; Biali, S.E. *J. Am. Chem. Soc.* **1990**, *112*, 9300.

neither has a plane of symmetry. However, in the former case (**114**) conversion of one chair form to the other (which of course happens rapidly) turns the molecule into its mirror image, while in the latter case (**115**) rapid interconversion does not give the mirror image but merely the conformer in which the original axial and equatorial substituents exchange places. Thus the optical inactivity of **114** is not due to a plane of symmetry, but to a rapid interconversion of the molecule and its mirror image. A similar situation holds for cis-1,3 compounds. However, for cis-1,4 isomers (both X,X and X,Y) optical inactivity arises from a plane of symmetry in both conformations. All-trans-1,2- and trans-1,3-disubstituted cyclohexanes are chiral (whether X,X or X,Y), while trans-1,4 compounds (both X,X and X,Y) are achiral, since all conformations have a plane of symmetry. It has been shown that the equilibrium is very dependent on both the solvent and the concentration of the disubstituted cyclohexane.[345] A theoretical study of the 1,2-dihalides showed a preference for the diaxial form with X = Cl, but predicted that the energy difference between diaxial and diequatorial was small when X = F.[346]

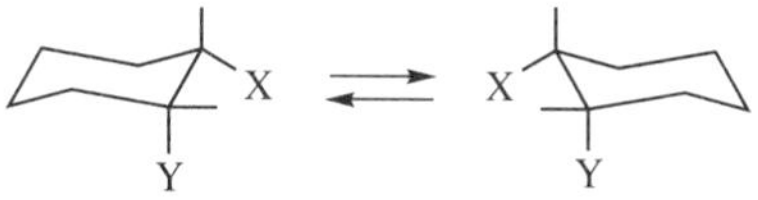

114

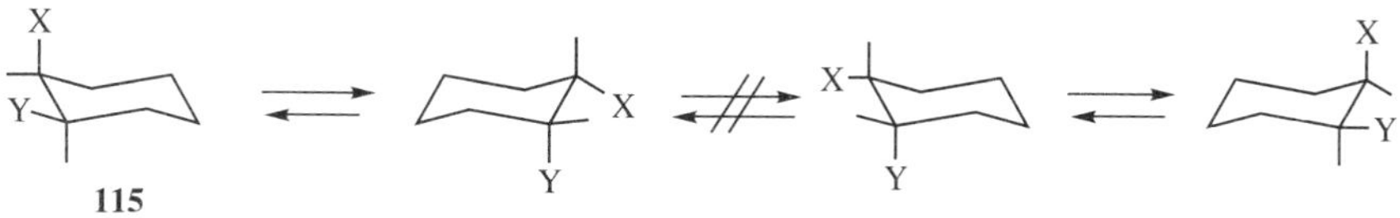

115

The conformation of a group can be frozen into a desired position by putting into the ring a large alkyl group (most often *tert*-butyl), which greatly favors the equatorial position.[347] It is known that silylated derivatives of *trans*-1,4- and *trans*-1,2-dihydroxycyclohexane, some monosilyloxycyclohexanes and some silylated sugars have unusually large populations of chair conformations with axial substituents.[348] Adjacent silyl groups in the 1,2-disubstituted series show a stabilizing interaction in all conformations, and this leads generally to unusually large axial populations.

[345]Abraham, R.J.; Chambers, E.J.; Thomas, W.A. *J. Chem. Soc. Perkin Trans. 2* **1993**, 1061.
[346]Wiberg, K. B. *J. Org. Chem.* **1999**, *64*, 6387.
[347]This idea was suggested by Winstein, S.; Holness, N.J. *J. Am. Chem. Soc.* **1955**, *77*, 5561. There are a few known compounds in which a *tert*-butyl group is axial. See, for example, Vierhapper, F.W. *Tetrahedron Lett.* **1981**, *22*, 5161.
[348]Marzabadi, C. H.; Anderson, J.E.; Gonzalez-Outeirino, J.; Gaffney, P.R.J.; White, C.G.H.; Tocher, D.A.; Todaro, L.J. *J. Am. Chem. Soc.* **2003**, *125*, 15163.

The principles involved in the conformational analysis of six-membered rings containing one or two trigonal atoms, for example, cyclohexanone and cyclohexene, are similar.[349-351] The barrier to interconversion in cyclohexane has been calculated to be 8.4–12.1 kcal mol^{-1}.[352] Cyclohexanone derivatives also assume a chair-conformation. Substituents at C-2 can assume an axial or equatorial position depending on steric and electronic influences. The proportion of the conformation with an axial X group is shown in Table 4.4 for a variety of substituents (X) in 2-substituted cyclohexanones.[353]

TABLE 4.4. Proportion of Axial Conformation in 2-Substituted Cyclohexanones, in CDCl$_3$.[353]

X	% Axial Conformation
F	17 ± 3
Cl	45 ± 4
Br	71 ± 4
I	88 ± 5
MeO	28 ± 4
MeS	85 ± 7
MeSe	(92)
Me$_2$N	44 ± 3
Me	(26)

[349]For a monograph, see Rabideau, P.W. *The Conformational Analysis of Cyclohexenes, Cyclohexadienes, and Related Hydroaromatic Compounds*, VCH, NY, *1989*. For reviews, see Vereshchagin, A.N. *Russ. Chem. Rev. 1983*, *52*, 1081; Johnson, F. *Chem. Rev. 1968*, *68*, 375. See also, Lambert, J.B.; Clikeman, R.R.; Taba, K.M.; Marko, D.E.; Bosch, R.J.; Xue, L. *Acc. Chem. Res. 1987*, *20*, 454.

[350]For books on conformational analysis see Dale, J. *Stereochemistry and Conformational Analysis*, Verlag Chemie, Deerfield Beach, FL, *1978*; Chiurdoglu, G. *Conformational Analysis*, Academic Press, NY, *1971*; Eliel, E.L.; Allinger, N.L.; Angyal, S.J.; Morrison, G.A. *Conformational Analysis*, Wiley, NY, *1965*; Hanack, M. *Conformation Theory*, Academic Press, NY, *1965*. For reviews, see Dale, J. *Top. Stereochem. 1976*, *9*, 199; Truax, D.R.; Wieser, H. *Chem. Soc. Rev. 1976*, *5*, 411; Eliel, E.L. *J. Chem. Educ. 1975*, *52*, 762; Bastiansen, O.; Seip, H.M.; Boggs, J.E. *Perspect. Struct. Chem. 1971*, *4*, 60; Bushweller, C.H.; Gianni, M.H., in Patai, S. *The Chemistry of Functional Groups, Supplement E*, Wiley, NY, *1980*, pp. 215–278.

[351]For reviews, see Jensen, F.R.; Bushweller, C.H. *Adv. Alicyclic Chem. 1971*, *3*, 139; Robinson, D.L.; Theobald, D.W. *Q. Rev. Chem. Soc. 1967*, *21*, 314; Eliel, E.L. *Angew. Chem. Int. Ed. 1965*, *4*, 761. Eliel, E.L.; Wilen, S.H.; Mander, L.N. *Stereochemistry of Organic Compounds*, Wiley-Interscience, NY, *1994*, pp. 686–753. Also see Smith, M.B. *Organic Synthesis*, 2nd ed., McGraw-Hill, NY, *2001*, pp. 53–55.

[352]Laane, J.; Choo, J. *J. Am. Chem. Soc. 1994*, *116*, 3889.

[353]Basso, E.A.; Kaiser, C.; Rittner, R.; Lambert, J.B. *J. Org. Chem. 1993*, *58*, 7865.

Conformation in Six-Membered Rings Containing Heteroatoms

In six-membered rings containing heteroatoms,[354] the basic principles are the same; that is, there are chair, twist, and boat forms, axial, and equatorial groups. The conformational equilibrium for tetrahydropyridines, for example, has been studied.[355] In certain compounds a number of new factors enter the picture. We deal with only two of these.[356]

1. In 5-alkyl-substituted 1,3-dioxanes, the 5-substituent has a much smaller preference for the equatorial position than in cyclohexane derivatives;[357] the *A* values are much lower. This indicates that the lone pairs on the oxygens have a smaller steric requirement than the C—H bonds in the corresponding cyclohexane derivatives. There is some evidence of an homoanomeric interaction in these systems.[358]

Similar behavior is found in the 1,3-dithianes,[359] and 2,3-disubstituted-1,4-dithianes have also been examined.[360] With certain non-alkyl substituents (e.g., F, NO_2, SOMe,[361] NMe_3^+) the axial position is actually preferred.[362]

2. An alkyl group located on a carbon α to a heteroatom prefers the equatorial position, which is of course the normally expected behavior, but a *polar* group in such a location prefers the *axial* position. An example of this

---

[354]For monographs, see Glass, R.S. *Conformational Analysis of Medium-Sized Heterocycles*, VCH, NY, **1988**; Riddell, F.G. *The Conformational Analysis of Heterocyclic Compounds*, Academic Press, NY, **1980**. For reviews, see Juaristi, E. *Acc. Chem. Res.* **1989**, *22*, 357; Crabb, T.A.; Katritzky, A.R. *Adv. Heterocycl. Chem.* **1984**, *36*, 1; Eliel, E.L. *Angew. Chem. Int. Ed.* **1972**, *11*, 739; *Pure Appl. Chem.* **1971**, *25*, 509; *Acc. Chem. Res.* **1970**, *3*, 1; Lambert, J.B. *Acc. Chem. Res.* **1971**, *4*, 87; Romers, C.; Altona, C.; Buys, H.R.; Havinga, E. *Top. Stereochem.* **1969**, *4*, 39; Bushweller, C.H.; Gianni, M.H., in Patai, S. *The Chemistry of Functional Groups, Supplement E*, Wley, NY, **1980**, pp. 232–274.

[355]Bachrach, S.M.; Liu, M. *Tetrahedron Lett.* **1992**, *33*, 6771.

[356]These factors are discussed by Eliel, E.L. *Angew. Chem. Int. Ed.* **1972**, *11*, 739.

[357]Riddell, F.G.; Robinson, M.J.T. *Tetrahedron* **1967**, *23*, 3417; Eliel, E.L.; Knoeber, M.C. *J. Am. Chem. Soc.* **1968**, *90*, 3444. See also Eliel, E.L.; Alcudia, F. *J. Am. Chem. Soc.* **1974**, *96*, 1939. See Cieplak, P.; Howard, A.E.; Powers, J.P.; Rychnovsky, S.D.; Kollman, P.A. *J. Org. Chem.* **1996**, *61*, 3662 for conformational energy differences in 2,2,6-trimethyl-4-alkyl-1,3-dioxane.

[358]Cai, J.; Davies, A.G.; Schiesser, C.H. *J. Chem. Soc. Perkin Trans. 2* **1994**, 1151.

[359]Hutchins, R.O.; Eliel, E.L. *J. Am. Chem. Soc.* **1969**, *91*, 2703. See also, Juaristi, E.; Cuevas, G. *Tetrahedron* **1999**, *55*, 359.

[360]Strelenko, Y.A.; Samoshin, V.V.; Troyansky, E.I.; Demchuk, D.V.; Dmitriev, D.E.; Nikishin, G.I.; Zefirov, N.S. *Tetrahedron* **1994**, *50*, 10107.

[361]Gordillo, B.; Juaristi, E.; Matínez, R.; Toscano, R.A.; White, P.S.; Eliel, E.L. *J. Am. Chem. Soc.* **1992**, *114*, 2157.

[362]Kaloustian, M.K.; Dennis, N.; Mager, S.; Evans, S.A.; Alcudia, F.; Eliel, E.L. *J. Am. Chem. Soc.* **1976**, *98*, 956. See also Eliel, E.L.; Kandasamy, D.; Sechrest, R.C. *J. Org. Chem.* **1977**, *42*, 1533.

phenomenon, known as the *anomeric effect*,[363] is the greater stability of α-glucosides over β-glucosides. A number of explanations have been offered

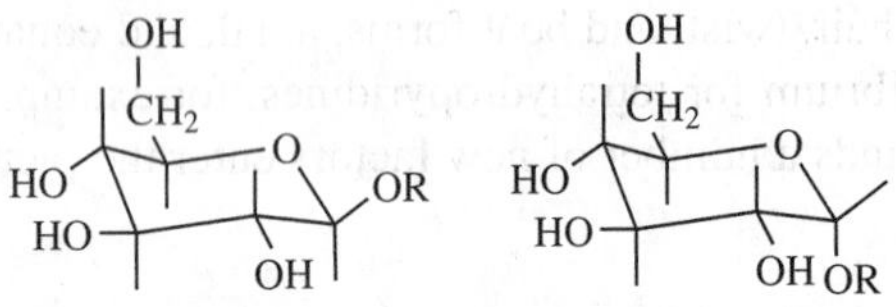

A β-glucoside
116

An α-glucoside
117

for the anomeric effect.[364] The one[365] that has received the most acceptance[366] is that one of the lone pairs of the polar atom connected to the carbon (an oxygen atom in the case of **117**) can be stabilized by overlapping with an antibonding orbital of the bond between the carbon and the other polar atom:

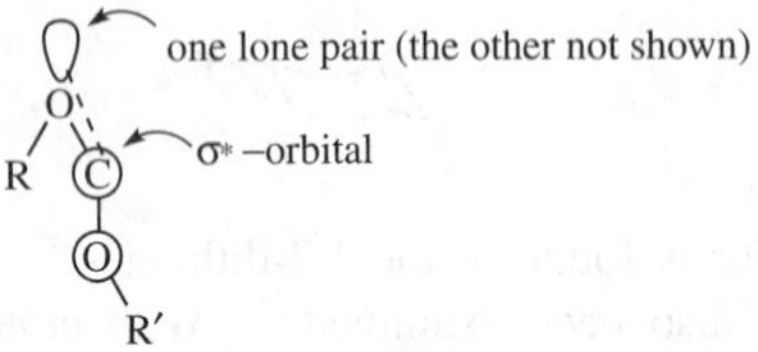

one lone pair (the other not shown)

σ* –orbital

This can happen only if the two orbitals are in the positions shown. The situation can also be represented by this type of hyperconjugation (called "negative hyperconjugation"):

$$R-O-\overset{|}{\underset{|}{C}}-O-R' \longleftrightarrow R-\overset{\oplus}{O}=\overset{|}{\underset{|}{C}} \quad \overset{\ominus}{O}-R'$$

It is possible that simple repulsion between parallel dipoles in **116** also plays a part in the greater stability of **117**. It has been shown that aqueous solvation effects reduce anomeric stabilization in many systems, particularly for tetrahydropyranosyls.[367] In contrast to cyclic acetals, simple acyclic acetlas

[363]For books on this subject, see Kirby, A.J. *The Anomeric Effect and Related Stereoelectronic Effects at Oxygen*, Springer, NY, *1983*; Szarek, W.A.; Horton, D. *Anomeric Effect*, American Chemical Society, Washington, *1979*. For reviews see Deslongchamps, P. *Stereoelectronic Effects in Organic Chemistry*, Pergamon, Elmsford, NY, *1983*, pp. 4–26; Zefirov, N.S. *Tetrahedron 1977*, *33*, 3193; Zefirov, N.S.; Shekhtman, N.M. *Russ. Chem. Rev. 1971*, *40*, 315; Lemieux, R.U. *Pure Appl. Chem. 1971*, *27*, 527; Angyal, S.J. *Angew. Chem. Int. Ed. 1969*, *8*, 157; Martin, J. *Ann. Chim. (Paris) [14]*, *1971*, *6*, 205.
[364]Juaristi, E.; Cuevas, G. *Tetrahedron 1992*, *48*, 5019.
[365]See Romers, C.; Altona, C.; Buys, H.R.; Havinga, E. *Top. Stereochem. 1969*, *4*, 39, see pp. 73–77; Wolfe, S.; Whangbo, M.; Mitchell, D.J. *Carbohydr. Res. 1979*, *69*, 1.
[366]For some evidence for this explanation, see Fuchs, B.; Ellencweig, A.; Tartakovsky, E.; Aped, P. *Angew. Chem. Int. Ed. 1986*, *25*, 287; Praly, J.; Lemieux, R.U. *Can. J. Chem. 1987*, *65*, 213; Booth, H.; Khedhair, K.A.; Readshaw, S.A. *Tetrahedron 1987*, *43*, 4699. For evidence against it, see Box, V.G.S. *Heterocycles 1990*, *31*, 1157.
[367]Cramer, C.J. *J. Org. Chem. 1992*, *57*, 7034; Booth, H.; Dixon, J.M.; Readshaw, S.A. *Tetrahedron 1992*, *48*, 6151.

rarely adopt the anomeric conformation, apparently because the eclipsed conformation better accommodates steric interactions of groups linked by relatively short carbon–oxygen bonds.[368] In all-cis-2,5-di-*tert*-butyl-1,4-cyclohexanediol, hydrogen bonding stabilizes the otherwise high-energy form[369] and 1,3-dioxane (**118**) exists largely as the twist conformation shown.[370] The conformational preference of 1-methyl-1-silacyclohexane (**121**) has been studied.[371] A strongly decreased activation barrier in silacyclohexane was observed, as compared to that in the parent ring, and is explained by the longer endocyclic Si–C bonds.

118   **119**   **120**   **121**

Second-row heteroatoms are known to show a substantial anomeric effect.[372] There appears to be evidence for a reverse anomeric effect in 2-aminotetrahydropyrans.[373] It has been called into question whether a reverse anomeric effect exists at all.[374] In **119**, the lone-pair electrons assume an axial conformation and there is an anomeric effect.[375] In **120**, however, the lone-pair electron orbitals are oriented gauche to both the axial and equatorial α-CH bond and there is no anomeric effect.[375]

Conformation in Other Rings[376]

Three-membered saturated rings are usually planar, but other three-membered rings can have some flexibility. Cyclobutane[377] is not planar but exists as in **122**, with an

[368]Anderson, J.E. *J. Org. Chem.* **2000**, *65*, 748.

[369]Stolow, R.D. *J. Am. Chem. Soc.* **1964**, *86*, 2170; Stolow, R.D.; McDonagh, P.M.; Bonaventura, M.M. *J. Am. Chem. Soc.* **1964**, *86*, 2165. For some other examples, see Camps, P.; Iglesias, C. *Tetrahedron Lett.* **1985**, *26*, 5463; Fitjer, L.; Scheuermann, H.; Klages, U.; Wehle, D.; Stephenson, D.S.; Binsch, G. *Chem. Ber.* **1986**, *119*, 1144.

[370]Rychnovsky, S.D.; Yang, G.; Powers, J.P. *J. Org. Chem.* **1993**, *58*, 5251.

[371]Arnason, I.; Kvaran, A.; Jonsdottir, S.; Gudnason, P. I.; Oberhammer, H. *J. Org. Chem.* **2002**, *67*, 3827.

[372]Juaristi, E.; Cuevas, G. *Tetrahedron* **1992**, *48*, 5109; Juaristi, E.; Tapia, J.; Mendez, R. *Tetrahedron* **1986**, *42*, 1253; Zefirov, N.S.; Blagoveschenskii, V.S.; Kazimirchik, I.V.; Yakovleva, O.P. *J. Org. Chem. USSR* **1971**, *7*, 599; Salzner, U.; Schleyer, P.v.R. *J. Am. Chem. Soc.* **1993**, *115*, 10231; Aggarwal, V.K.; Worrall, J.M.; Adams, H.; Alexander, R.; Taylor, B.F. *J. Chem. Soc. Perkin Trans. 1* **1997**, 21.

[373]Salzner, U.; Schleyer, P.v.R. *J. Org. Chem.* **1994**, *59*, 2138.

[374]Perrin, C.L. *Tetrahedron* **1995**, *51*, 11901.

[375]Anderson, J.E.; Cai, J.; Davies, A.G. *J. Chem. Soc. Perkin Trans. 2* **1997**, 2633. For some controversy concerning the anomeric effect a related system, see Perrin, C.L.; Armstrong, K.B.; Fabian, M.A. *J. Am. Chem.Soc.* **1994**, *116*, 715 and Salzner, U. *J. Org. Chem.* **1995**, *60*, 986.

[376]Eliel, E.L.; Wilen, S.H.; Mander, L.N. *Stereochemistry of Organic Compounds*, Wiley-Interscience, NY, **1994**, pp. 675–685 and 754–770.

[377]For reviews of the stereochemistry of four-membered rings, see Legon, A.C. *Chem. Rev.* **1980**, *80*, 231; Moriarty, R.M. *Top. Stereochem.* **1974**, *8*, 271; Cotton, F.A.; Frenz, B.A. *Tetrahedron* **1974**, *30*, 1587.

angle between the planes of ~35°.[378] The deviation from planarity is presumably caused by eclipsing in the planar form (see p. 219). Oxetane, in which eclipsing is

$$CH_2{-}O$$
$$|\quad\ \ |$$
$$CH_2{-}CH_2$$

122   Oxetane

less, is closer to planarity, with an angle between the planes of ~10°.[379] Cyclopentane might be expected to be planar, since the angles of a regular pentagon are 108°, but it is not so, also because of eclipsing effects.[380] There are two puckered conformations, the *envelope* and the *half-chair*. There is little energy difference between these two forms and many five-membered ring systems have conformations somewhere in between them.[381] Although in the envelope conformation one carbon is shown above the others, ring motions cause each of the carbons in

Envelope   Half-chair

rapid succession to assume this position. The puckering rotates around the ring in what may be called a *pseudorotation*.[382] In substituted cyclopentanes and five-membered rings in which at least one atom does not contain two substituents [e.g., tetrahydrofuran (THF), cyclopentanone, C_3 and C_7-mono- and disubstituted hexahydroazepin-2-ones (caprolactams),[383] and tetrahydrothiophene S-oxide[384]], one conformer may be more stable than the others. The barrier to planarity in cyclopentane has been reported to be 5.2 kcal mol^{-1} (22 kJ mol^{-1}).[385] Contrary to previous reports, there is only weak stabilization (<2 kcal mol^{-1}; <8 kJ mol^{-1}) of 3-, 4-, and 5-membered rings by *gem*–dialkoxycarbonyl substituents (e.g., COOR).[386]

123

[378]Dows, D.A.; Rich, N. *J. Chem. Phys.* **1967**, *47*, 333; Stone, J.M.R.; Mills, I.M. *Mol. Phys.* **1970**, *18*, 631; Miller, F.A.; Capwell, R.J.; Lord, R.C.; Rea, D.G. *Spectrochim. Acta Part A*, **1972**, *28*, 603. However, some cyclobutane derivatives are planar, at least in the solid state: for example, see Margulis, T.N. *J. Am. Chem. Soc.* **1971**, *93*, 2193.

[379]Luger, P.; Buschmann, J. *J. Am. Chem. Soc.* **1984**, *106*, 7118.

[380]For reviews of the conformational analysis of five-membered rings, see Fuchs, B. *Top. Stereochem.* **1978**, *10*, 1; Legon, A.C. *Chem. Rev.* **1980**, *80*, 231.

[381]Willy, W.E.; Binsch, G.; Eliel, E.L. *J. Am. Chem. Soc.* **1970**, *92*, 5394; Lipnick, R.L. *J. Mol. Struct.* **1974**, *21*, 423.

[382]Kilpatrick, J.E.; Pitzer, K.S.; Spitzer, R. *J. Am. Chem. Soc.* **1947**, *69*, 2438; Pitzer, K.S.; Donath, W.E. *J. Am. Chem. Soc.* **1959**, *81*, 3213; Durig, J.R.; Wertz, D.W. *J. Chem. Phys.* **1968**, *49*, 2118; Lipnick, R.L. *J. Mol. Struct.* **1974**, *21*, 411; Poupko, R.; Luz, Z.; Zimmermann, H. *J. Am. Chem. Soc.* **1982**, *104*, 5307; Riddell, F.G.; Cameron K.S.; Holmes, S.A.; Strange, J.H. *J. Am. Chem. Soc.* **1997**, *119*, 7555.

[383]Matallana, A.; Kruger, A.W.; Kingsbury, C.A. *J. Org. Chem.* **1994**, *59*, 3020.

[384]Abraham, R.J.; Pollock, L.; Sancassan, F. *J. Chem. Soc. Perkin Trans. 2* **1994**, 2329.

[385]Carreira, L.A.; Jiang, G.J.; Person, W.B.; Willis, Jr., J.N. *J. Chem. Phys.* **1972**, *56*, 1440.

[386]Verevkin, S.P.; Kümmerlin, M.; Beckhaus, H.-D.; Galli, C.; Rüchardt, C. *Eur. J. Org. Chem.* **1998**, 579.

Rings larger than six-membered are always puckered[387] unless they contain a large number of sp^2 atoms (see the section on strain in medium rings, p. 223). The energy and conformations of the alkane series cycloheptane to cyclodecane has been reported.[388] The conformation shown for oxacyclooctane (**123**), for example, appears to be the most abundant one.[389] The conformations of other large ring compounds have been studied, including 11-membered ring lactones,[390] 10- and 11-membered ring ketones,[391] and 11- and 14-membered ring lactams.[392] Dynamic NMR was used to determine the conformation large-ring cycloalkenes and lactones.[393] Note that axial and equatorial hydrogens are found only in the chair conformations of six-membered rings. In rings of other sizes the hydrogens protrude at angles that generally do not lend themselves to classification in this way,[394] although in some cases the terms "pseudo-axial" and "pseudo-equatorial" have been used to classify hydrogens in rings of other sizes.[395]

Molecular Mechanics[396]

Molecular mechanics Molecular Mechanics[397] describes a molecule in terms of a collection of bonded atoms that have been distorted from some idealized geometry due to non-bonded van der Waals (steric) and coulombic (charge–charge)

[387]For reviews of conformations in larger rings, see Arshinova, R.P. *Russ. Chem. Rev.* **1988**, *57*, 1142; Ounsworth, J.P.; Weiler, L. *J. Chem. Educ.* **1987**, *64*, 568; Ōki, M. *Applications of Dynamic NMR Spectroscopy to Organic Chemistry*, VCH, NY, **1985**, pp. 307–321; Casanova, J.; Waegell, B. *Bull. Soc. Chim. Fr.* **1975**, 911; Anet, F.A.L. *Top. Curr. Chem.* **1974**, *45*, 169; Dunitz, J.D. *Pure Appl. Chem.* **1971**, *25*, 495; *Perspect. Struct. Chem.* **1968**, *2*, 1; Tochtermann, W. *Fortchr. Chem. Forsch.* **1970**, *15*, 378; Dale, J. *Angew. Chem. Int. Ed.* **1966**, *5*, 1000. For a monograph, see Glass, R.S. *Conformational Analysis of Medium-Sized Heterocycles*, VCH, NY, **1988**. Also see the monographs by Eliel, E.L.; Allinger, N.L.; Angyal, S.J.; Morrison, G.A. *Conformational Analysis*; Wiley, NY, **1965**; Hanack, M. *Conformation Theory*, Academic Press, NY, **1965**.

[388]Wiberg, K.B. *J. Org. Chem* **2003**, *68*, 9322.

[389]Meyer, W.L.; Taylor, P.W.; Reed, S.A.; Leister, M.C.; Schneider, H.-J.; Schmidt, G.; Evans, F.E.; Levine, R.A. *J. Org. Chem.* **1992**, *57*, 291.

[390]Spracklin, D.K.; Weiler, L. *J. Chem. Soc. Chem. Commun.* **1992**, 1347; Ogura, H.; Furuhata, K.; Harada, Y.; Iitaka, Y. *J. Am. Chem. Soc.* **1978**, *100*, 6733; Ounsworth, J.P.; Weiler, L. *J. Chem. Ed.*, **1987**, *64*, 568; Keller, T.H.; Neeland, E.G.; Rettig, S.; Trotter, J.; Weiler, L. *J. Am. Chem. Soc.* **1988**, *110*, 7858.

[391]Pawar, D.M.; Smith, S.V.; Moody, E.M.; Noe, E.A. *J. Am. Chem. Soc.* **1998**, *120*, 8241.

[392]Borgen, G.; Dale, J.; Gundersen, L.-L.; Krivokapic, A.; Rise, F.; Øverås, A.T. *Acta Chem. Scand. B*, **1998**, *52*, 1110.

[393]Pawar, D.M.; Davids, K.L.; Brown, B.L.; Smith, S.V.; Noe, E.A. *J. Org. Chem.* **1999**, *64*, 4580; Pawar, D.M.; Moody, E.M.; Noe, E.A. *J. Org. Chem.* **1999**, *64*, 4586.

[394]For definitions of axial, equatorial, and related terms for rings of any size, see Anet, F.A.L.*Tetrahedron Lett.* **1990**, *31*, 2125.

[395]For a discussion of the angles of the ring positions, see Cremer, D. *Isr. J. Chem.* **1980**, *20*, 12.

[396]Thanks to Dr. Warren Hehre, Wavefunction, Inc., Irvine, CA. Personal communication. See Hehre, W.J. *A Guide to Molecular Mechanics and Quantum Chemical Calculations*, Wavefunction, Inc., Irvine, CA, **2003**, pp. 56–57.

[397]For a review, see Rappe, A.K.; Casewit, C.J. *Molecular Mechanics Across Chemistry*, University Science Books, Sausalito, CA, **1997**.

interactions. This approach is fundamentally different from molecular-orbital theory that is based on quantum mechanics and that make no reference whatsoever to chemical bonding. The success of molecular mechanics depends on the ability to represent molecules in terms of unique valence structures, on the notion that bond lengths and angles may be transferred from one molecule to another and on a predictable dependence of geometrical parameters on the local atomic environment.

The molecular mechanics energy of a molecule is given as a sum of contributions arising from distortions from ideal bond distances (stretch contributions), bond angles (bend contributions) and torsion angles (torsion contributions), together with contributions from nonbonded interactions. This energy is commonly referred to as a strain energy, meaning that it reflects the inherent strain in a real molecule relative to a hypothetical idealized (strain-free) form.

$$E^{\text{strain}} = E_A^{\text{stretch}} + E_A^{\text{bend}} + E_A^{\text{torsion}} + E_{AB}^{\text{nonbonded}} \tag{1}$$

Stretch and bend terms are most simply given in terms of quadratic (Hooke's law) forms:

$$E^{\text{stretch}}(r) = \frac{1}{2}k^{\text{stretch}}(r - r^{\text{eq}})^2 \tag{2}$$

$$E^{\text{bend}}(\alpha) = \frac{1}{2}k^{\text{bend}}(r - r^{\text{eq}})^2 \tag{3}$$

r and α are the bond distance and angle, respectively, and r^{eq} and α^{eq} are the ideal bond length and angle, respectively.

Torsion terms need to properly reflect the inherent periodicity of the particular bond involved in a rotation. For example, the threefold periodicity of the carbon-carbon bond in ethane may be represented by a simple cosine form.

$$E^{\text{torsion}}(\omega) = k^{\text{torsion3}}[1 - \cos 3(\omega - \omega^{\text{eq}})] \tag{4}$$

Ω is the torsion angle, ω^{eq} is the ideal torsion angle and k^{torsion} is a parameter. Torsion contributions to the strain energy will also usually need to include contributions that are onefold and twofold periodic. These can be represented in the same manner as the threefold term.

$$E^{\text{torsion}}(\omega) = k^{\text{torsion1}}[1 - \cos(\omega - \omega^{\text{eq}})] + k^{\text{torsion2}}[1 - \cos 2(\omega - \omega^{\text{eq}})]$$
$$+ k^{\text{torsion3}}[1 - \cos 3(\omega - \omega^{\text{eq}})] \tag{5}$$

Nonbonded interacations invovle a sum of van der Waals (VDW) interactions and coulombic interactions. The coulombic term accounts for charge–charge interactions.

$$E^{\text{nonbonded}}(r) = E^{\text{VDW}}(r) + E^{\text{coulombic}}(r) \tag{6}$$

The VDW is made up of two parts, the first to account for strong repulsion on nonbonded atoms as the closely approach, and the second to account for weak long-range attraction, r is the nonbonded distance.

Molecular mechanics methods differ both in the form of the terms that make up the strain energy and in their detailed parameterization. Older methods, such as SYBYL,[398] use very simple forms and relatively few parameters, while newer methods such as MM3,[399] MM4,[400] and MMFF[401] use more complex forms and many more parameters. In general, the more complex the form of the strain energy terms and the more extensive the parameterization, the better will be the results. Of course, more parameters mean that more (experimental) data will be needed in their construction. Because molecular mechanics is not based on "physical fundamentals," but rather is essentially an interpolation scheme, its success depends on the availability of either experimental or high-quality theoretical data for parameterization. A corollary is that molecular mechanics would not be expected to lead to good results for "new" molecules, that is, molecules outside the range of their parameterization.

The two most important applications of molecular mechanics are geometry calculations on very large molecules, for example, on proteins, and conformational analysis on molecules for which there may be hundreds, thousands, or even tens of thousands of distinct structures. It is here that methods based on quantum mechanics are simply not (yet) practical. It should be no surprise that equilibrium geometries obtained from molecular mechanics are generally in good accord with experimental values. There are ample data with which to parameterize and evaluate the methods. However, because there are very few experimental data relating to the equilibrium conformations of molecules and energy differences among different conformations, molecular mechanics calculations for these quantities need to be viewed with a very critical eye. In time, high-quality data from quantum mechanics will provide the needed data and allow more careful parameterization (and assessment) than now possible.

The most important limitation of molecular mechanics is its inability to provide thermochemical data. The reason for this is that the mechanics strain energy is specific to a given molecule (it provides a measure of how much this molecule deviates from an ideal arrangement), and different molecules have different ideal arrangements. For example, acetone and methyl vinyl ether have different bonds and would be referenced to different standards. The only exception occurs for conformational energy differences or, more generally, for energy comparisons among molecules with exactly the same bonding, for example, cis- and trans-2-butene.

Because a molecular mechanics calculation reveals nothing about the distribution of electrons or distribution of charge in molecules, and because mechanics

---

[398]Clark, M.; Cramer III, R.D.; van Opdenbosch, N. *J. Computational Chem.* **1989**, *10*, 982.

[399]Allinger, N.L.; Li, F.; Yan, L. *J. Computational Chem.* **1990**, *11*, 855, and later papers in this series.

[400]Allinger, N.L.; Chen, K.; Lii, J.-H. *J. Computational Chem.* **1996**, *17*, 642, and later papers in this series.

[401]Halgren, T.A. *J. Computational Chem* **1996**, *17*, 490, and later papers in this series.

methods have not (yet) been parameterized to reproduce transition state geometries, they are of limited value in describing either chemical reactivity or product selectivity. There are, however, situations where steric considerations associated with either the product or reactants are responsible for trends in reactivity and selectivity, and here molecular mechanics would be expected to be of some value.

Because of the different strengths and limitations of molecular mechanics and quantum chemical calculations, it is now common practice to combine the two, for example, to use molecular mechanics to establish conformation (or at least a set of reasonable conformations) and then to quantum calculations to evaluate energy differences.

In practical terms, molecular mechanics calculations may easily be performed on molecules comprising several thousand atoms. Additionally, molecular mechanics calculations are sufficiently rapid to permit extensive conformational searching on molecules containing upwards of a hundred atoms. Modern graphical based programs for desktop computers make the methods available to all chemists.

STRAIN

Steric strain[402] exists in a molecule when bonds are forced to make abnormal angles. This results in a higher energy than would be the case in the absence of angle distortions. It has been shown that there is a good correlation between the $^{13}C-H$ coupling constants in NMR and the bond angles and bond force angles in strained organic molecules.[403] There are, in general, two kinds of structural features that result in sterically caused abnormal bond angles. One of these is found in small-ring compounds, where the angles must be less than those resulting from normal orbital overlap.[404] Such strain is called *small-angle strain*. The other arises when nonbonded atoms are forced into close proximity by the geometry of the molecule. These are called *nonbonded interactions*.

Strained molecules possess *strain energy*. That is, their potential energies are higher than they would be if strain were absent.[405] The strain energy for a particular molecule can be estimated from heat of atomization or heat of combustion data. A strained molecule has a lower heat of atomization than it would have if it were strain-free (Fig. 4.6). As in the similar case of resonance energies (p. 36), strain energies can not be known exactly, because the energy of a real molecule can be measured, but not the energy of a hypothetical unstrained model. It is also possible

[402]For a monograph, see Greenberg, A.; Liebman, J.F. *Strained Organic Molecules*, Academic Press, NY, *1978*. For reviews, see Wiberg, K.B. *Angew. Chem. Int. Ed. 1986*, *25*, 312; Greenberg, A.; Stevenson, T.A. *Mol. Struct. Energ.*, *1986*, *3*, 193; Liebman, J.F.; Greenberg, A. *Chem. Rev. 1976*, *76*, 311. For a review of the concept of strain, see Cremer, D.; Kraka, E. *Mol. Struct. Energ. 1988*, *7*, 65.

[403]Zhao, C.-Y.; Duan, W.-S.; Zhang, Y.; You, X.-Z. *J. Chem. Res. (S) 1998*, 156.

[404]Wiberg, K.B. *Accts. Chem. Res. 1996*, *29*, 229.

[405]For discussions, see Wiberg, K.B.; Bader, R.F.W.; Lau, C.D.H. *J. Am. Chem. Soc. 1987*, *109*, 985, 1001.

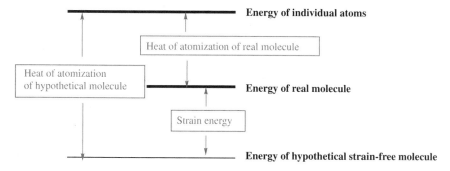

Fig. 4.6. Strain energy calculation.

to calculate strain energies by molecular mechanics, not only for real molecules, but also for those that cannot be made.[406]

Strain in Small Rings

Three-membered rings have a great deal of angle strain, since 60° angles represent a large departure from the tetrahedral angles. In sharp contrast to other ethers, ethylene oxide is quite reactive, the ring being opened by many reagents (see p. 496). Ring opening, of course, relieves the strain.[407] Cyclopropane,[408] which is even more strained[409] than ethylene oxide, is also cleaved more easily than would be expected for an alkane.[410] Thus, pyrolysis at 450–500°C converts it to propene, bromination gives 1,3-dibromopropane,[411] and it can be hydrogenated to propane (though at high pressure).[412] Other three-membered rings are similarly reactive.[413] Alkyl substituents influence the strain energy of small ring compounds.[414] *gem*-Dimethyl substitution, for example, "lowers the strain energy of cyclopropanes,

[406]For a review, see Rüchardt, C.; Beckhaus, K. *Angew. Chem. Int. Ed.* **1985**, *24*, 529. See also Burkert, U.; Allinger, N.L. *Molecular Mechanisms*, American Chemical Society, Washington, **1982**, pp. 169–194; Allinger, N.L. *Adv. Phys. Org. Chem.* **1976**, *13*, 1, pp. 45–47.

[407]For reviews of reactions of cyclopropanes and cyclobutanes, see Trost, B.M. *Top. Curr. Chem.* **1986**, *133*, 3; Wong, H.N.C.; Lau, C.D.H.; Tam, K. *Top. Curr. Chem.* **1986**, *133*, 83.

[408]For a treatise, see Rappoport, Z. *The Chemistry of the Cyclopropyl Group*, 2 pts., Wiley, NY, **1987**.

[409]For reviews of strain in cyclopropanes, see, in Rappoport, Z. *The Chemistry of the Cyclopropyl Group*, 2 pts, Wiley, NY, **1987**, the papers by Wiberg, K.B. pt. 1., pp. 1–26; Liebman, J.F.; Greenberg, A. pt. 2, pp. 1083–1119; Liebman, J.F.; Greenberg, A. *Chem. Rev.* **1989**, *89*, 1225.

[410]For reviews of ring-opening reactions of cyclopropanes, see Wong, H.N.C.; Hon, M.; Ts, C.e; Yip, Y.; Tanko, J.; Hudlicky, T. *Chem. Rev.* **1989**, *89*, 165; Reissig, H., in Rappoport, Z. *The Chemistry of the Cyclopropyl Group*, pt. 1, Wiley, NY, **1987**, pp. 375–443.

[411]Ogg Jr., R.A.; Priest, W.J. *J. Am. Chem. Soc.* **1938**, *60*, 217.

[412]Shortridge, R.W.; Craig, R.A.; Greenlee, K.W.; Derfer, J.M.; Boord, C.E. *J. Am. Chem. Soc.* **1948**, *70*, 946.

[413]For a review of the pyrolysis of three- and four-membered rings, see Frey, H.M. *Adv. Phys. Org. Chem.* **1966**, *4*, 147.

[414]Bach, R. D.; Dmitrenko, O. *J. Org. Chem.* **2002**, *67*, 2588.

cyclobutanes, epoxides, and dimethyldioxirane by 6–10 kcal mol^{-1} (25.42 kJ mol^{-1}) relative to an unbranched acyclic reference molecule."[414] The C—H bond dissociation energy also tends to increase ring strain in small-ring alkenes.[415]

There is much evidence, chiefly derived from NMR coupling constants, that the bonding in cyclopropanes is not the same as in compounds that lack small-angle strain.[416] For a normal carbon atom, one s and three p orbitals are hybridized to give four approximately equivalent sp^3 orbitals, each containing ~25% s character. But for a cyclopropane carbon atom, the four hybrid orbitals are far from equivalent. The two orbitals directed to the outside bonds have more s character than a normal sp^3 orbital, while the two orbitals involved in ring bonding have less, because the more p-like they are the more they resemble ordinary p orbitals, whose preferred bond angle is 90° rather than 109.5°. Since the small-angle strain in cyclopropanes is the difference between the preferred angle and the real angle of 60°, this additional p character relieves some of the strain. The external orbitals have ~33% s character, so that they are ~sp^2 orbitals, while the internal orbitals have ~17% s character, so that they may be called ~sp^5 orbitals.[417] Each of the three carbon–carbon bonds of cyclopropane is therefore formed by overlap of two sp^5 orbitals. Molecular-orbital calculations show that such bonds are not completely s in character. In normal C—C bonds, sp^3 orbitals overlap in such a way that the straight line connecting the nuclei becomes an axis about which the electron density is symmetrical. But in cyclopropane, the electron density is directed *away from* the ring.[418] Fig. 4.7 shows the direction of orbital overlap.[419] For cyclopropane, the angle (marked θ) is 21°. Cyclobutane exhibits the same phenomenon but to a lesser extent, θ being 7°.[419,418] Molecular-orbital calculations also show that the

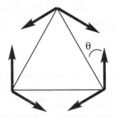

Fig. 4.7. Orbital overlap in cyclopropane. The arrows point toward the center of electron density.

[415]Bach, R.D.; Dmitrenko, O. *J. Am. Chem. Soc.* **2004**, *126*, 4444.

[416]For discussions of bonding in cyclopropanes, see Bernett, W.A. *J. Chem. Educ.* **1967**, *44*, 17; de Meijere, A. *Angew. Chem. Int. Ed.* **1979**, *18*, 809; Honegger, E.; Heilbronner, E.; Schmelzer, A. *Nouv. J. Chem.* **1982**, *6*, 519; Cremer, D.; Kraka, E. *J. Am. Chem. Soc.* **1985**, *107*, 3800, 3811; Slee, T.S. *Mol. Struct. Energ.* **1988**, *5*, 63; Casaarini, D.; Lunazzi, L.; Mazzanti, A. *J. Org. Chem.* **1997**, *62*, 7592.

[417]Randić, M.; Maksić, Z. *Theor. Chim. Acta* **1965**, *3*, 59; Foote, C.S. *Tetrahedron Lett.* **1963**, 579; Weigert, F.J.; Roberts, J.D. *J. Am. Chem. Soc.* **1967**, *89*, 5962.

[418]Wiberg, K.B. *Acc. Chem. Res.* **1996**, *29*, 229.

[419]Coulson, C.A.; Goodwin, T.H. *J. Chem. Soc.* **1962**, 2851; *1963*, 3161; Peters, D. *Tetrahedron* **1963**, *19*, 1539; Hoffmann, R.; Davidson, R.B. *J. Am. Chem. Soc.* **1971**, *93*, 5699.

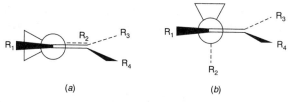

Fig. 4.8. Conformations of α-cyclopropylalkenes. Conformation (*a*) leads to maximum conjugation and conformation (*b*) to minimum conjugation.

maximum electron densities of the C—C σ orbitals are bent away from the ring, with $\theta = 9.4°$ for cyclopropane and 3.4° for cyclobutane.[420] The bonds in cyclopropane are called *bent bonds*, and are intermediate in character between σ and π, so that cyclopropanes behave in some respects like double-bond compounds.[421] For one thing, there is much evidence, chiefly from UV spectra,[422] that a cyclopropane ring is conjugated with an adjacent double bond and that this conjugation is greatest for the conformation shown in *a* in Fig. 4.8 and least or absent for the conformation shown in *b*, since overlap of the double-bond π-orbital with two of the *p*-like orbitals of the cyclopropane ring is greatest in conformation *a*. However, the conjugation between a cyclopropane ring and a double bond is less than that between two double bonds.[423] For other examples of the similarities in behavior of a cyclopropane ring and a double bond (see p. 212).

Four-membered rings also exhibit angle strain, but much less, and are less easily opened. Cyclobutane is more resistant than cyclopropane to bromination, and although it can be hydrogenated to butane, more strenuous conditions are required. Nevertheless, pyrolysis at 420°C gives two molecules of ethylene. As mentioned earlier (p. 212), cyclobutane is not planar.

Many highly strained compounds containing small rings in fused systems have been prepared,[424] showing that organic molecules can exhibit much more

[420]Wiberg, K.B.; Bader, R.F.W.; Lau, C.D.H. *J. Am. Chem. Soc.* **1987**, *109*, 985, 1001; Cremer, D.; Kraka, E. *J. Am. Chem. Soc.* **1985**, *107*, 3800, 1811.

[421]For reviews, see Tidwell, T.T., in Rappoport, Z. *The Chemistry of the Cyclopropyl Groups*, pt. 1, Wiley, NY, **1987**, pp. 565–632; Charton, M. in Zabicky, J. *The Chemistry of Alkenes*, Vol. 2, pp. 511–610, Wiley, NY, **1970**.

[422]See, for example, Cromwell, N.H.; Hudson, G.V. *J. Am. Chem. Soc.* **1953**, *75*, 872; Kosower, E.M.; Ito, M. *Proc. Chem. Soc.* **1962**, 25; Dauben, W.G.; Berezin, G.H. *J. Am. Chem. Soc.* **1967**, *89*, 3449; Jorgenson, M.J.; Leung, T. *J. Am. Chem. Soc.* **1968**, *90*, 3769; Heathcock, C.H.; Poulter, S.R. *J. Am. Chem. Soc.* **1968**, *90*, 3766; Tsuji, T.; Shibata, T.; Hienuki, Y.; Nishida, S. *J. Am. Chem. Soc.* **1978**, *100*, 1806; Drumright, R.E.; Mas, R.H.; Merola, J.S.; Tanko, J.M. *J. Org. Chem.* **1990**, *55*, 4098.

[423]Staley, S.W. *J. Am. Chem. Soc.* **1967**, *89*, 1532; Pews, R.G.; Ojha, N.D. *J. Am. Chem. Soc.* **1969**, *91*, 5769. See, however, Noe, E.A.; Young, R.M. *J. Am. Chem. Soc.* **1982**, *104*, 6218.

[424]For reviews discussing the properties of some of these as well as related compounds, see the reviews in *Chem. Rev.* **1989**, *89*, 975, and the following: Jefford, C.W. *J. Chem. Educ.* **1976**, *53*, 477; Seebach, D. *Angew. Chem. Int. Ed.* **1965**, *4*, 121; Greenberg, A.; Liebman, J.F. *Strained Organic Molecules*, Academic Press, NY, **1978**, pp. 210–220. For a review of bicyclo[*n.m.0*]alkanes, see Wiberg, K.B. *Adv. Alicyclic Chem.* **1968**, *2*, 185. Eliel, E.L.; Wilen, S.H.; Mander, L.N. *Stereochemistry of Organic Compounds*, Wiley-Interscience, NY, **1994**, pp. 771–811.

strain than simple cyclopropanes or cyclobutanes.[425] Table 4.5 shows a few of these compounds.[426] Perhaps the most interesting are cubane, prismane, and the substituted

tetrahedrane, since preparation of these ring systems had been the object of much endeavor. Prismane is tetracyclo[2.2.0.0^{2,6}.0^{3,5}]hexane and many derivatives are known,[427] including bis(homohexaprismane) derivatives.[428] The bicyclobutane molecule is bent, with the angle θ between the planes equal to $126 \pm 3°$.[429] The rehybridization effect, described above for cyclopropane, is even more extreme in this molecule. Calculations have shown that the central bond is essentially formed by overlap of two p orbitals with little or no s character.[430] *Propellanes* are compounds in which two carbons, directly connected, are also connected by three other bridges. [1.1.1]Propellane is in the table and it is the smallest possible propellane,[431] and is in fact more stable than the larger [2.1.1]propellane and [2.2.1]propellane, which have been isolated only in solid matrixes at low temperature.[432] The bicyclo[1.1.1]pentanes are obviously related to the propellanes except that the central connecting bond is missing, and several derivatives are known.[433] Even more complex systems are known.[434]

[425]For a useful classification of strained polycyclic systems, see Gund, P.; Gund, T.M. *J. Am. Chem. Soc.* *1981, 103*, 4458.

[426]For a computer program that generates IUPAC names for complex bridged systems, see Rücker, G.; Rücker, C. *Chimia,* *1990, 44*, 116.

[427]Gleiter, R.; Treptow, B.; Irngartinger, H.; Oeser, T. *J. Org. Chem.* *1994, 59*, 2787; Gleiter, R.; Treptow, B. *J. Org. Chem. 1993, 58*, 7740.

[428]Golobish, T.D.; Dailey, W.P. *Tetrahedron Lett. 1996, 37*, 3239.

[429]Haller, I.; Srinivasan, R. *J. Chem. Phys. 1964, 41*, 2745.

[430]Schulman, J.M.; Fisanick, G.J. *J. Am. Chem. Soc. 1970, 92*, 6653; Newton, M.D.; Schulman, J.M. *J. Am. Chem. Soc. 1972, 94*, 767.

[431]Wiberg, K.B.; Waddell, S.T. *J. Am. Chem. Soc. 1990, 112*, 2194; Seiler, S.T. *Helv. Chim. Acta 1990, 73*, 1574; Bothe, H.; Schlüter, A. *Chem. Ber. 1991, 124*, 587; Lynch, K.M.; Dailey, W.P. *J. Org. Chem. 1995, 60*, 4666. For reviews of small-ring propellanes, see Wiberg, K.B. *Chem. Rev. 1989, 89*, 975; Ginsburg, D., in Rappoport, Z *The Chemistry of the Cyclopropyl Group*, pt. 2, Wiley, NY, *1987*, pp. 1193–1221. For a discussion of the formation of propellanes, see Ginsburg, D. *Top. Curr. Chem. 1987, 137*, 1.

[432]Wiberg, K.B.; Walker, F.H.; Pratt, W.E.; Michl, J. *J. Am. Chem. Soc. 1983, 105*, 3638.

[433]Della, E.W.; Taylor, D.K. *J. Org. Chem. 1994, 59*, 2986.

[434]See Kuck, D.; Krause, R.A.; Gestmann, D.; Posteher, F.; Schuster, A. *Tetrahedron 1998, 54*, 5247 for an example of a [5.5.5.5.5.5]centrohexacycline.

TABLE 4.5. Some Strained Small-Ring Compounds

Structural Formula of Compound Prepared	Systematic Name of Ring System	Common Name If Any	Reference
	Bicyclo[1.1.0]butane	Bicyclobutane	435
	$\Delta^{1,4}$-Bicyclo[2.2.0]hexene		436
	Tricyclo[1.1.0.0^{2,4}]butane	Tetrahedrane	437
	Pentacyclo[5.1.0.0^{2,4}.0^{3,5}.0^{6,8}]-octane Tricyclo[1.1.1.0^{1,3}]-pentane	Octabisvalene a [1.1.1]propellane	438 364
	Tetradecaspiro[2.0.2.0.0.0.0.0.-2.0.2.0.0.0.2.0.2.0.0.1.0.0.2.0.2.-0.0.0]untriacontane	[15]-triangulane	439
	Tetracyclo[2.2.0.0^{2,6}.0^{3,5}]-hexane	Prismane	440

[435]Lemal, D.M.; Menger, F.M.; Clark, G.W. *J. Am. Chem. Soc. 1963*, *85*, 2529; Wiberg, K.B.; Lampman, G.M. *Tetrahedron Lett. 1963*, 2173. For reviews of preparations and reactions of this system, see Hoz, S., in Rappoport, Z *The Chemistry of the Cyclopropyl Group*, pt. 2, Wiley, NY, *1987*, pp. 1121–1192; Wiberg, K.B.; Lampman, G.M.; Ciula, R.P.; Connor, D.S.; Schertler, P.; Lavanish, J.M. *Tetrahedron 1965*, *21*, 2749; Wiberg, K.B. *Rec. Chem. Prog.*, *1965*, *26*, 143; Wiberg, K.B. *Adv. Alicyclic Chem. 1968*, *2*, 185. For a review of [n.1.1] systems, see Meinwald, J.; Meinwald, Y.C. *Adv. Alicyclic Chem. 1966*, *1*, 1.

[436]Casanova, J.; Bragin, J.; Cottrell, F.D. *J. Am. Chem. Soc. 1978*, *100*, 2264.

[437]Maier, G.; Pfriem, S.; Schäfer, U.; Malsch, K.; Matusch, R. *Chem. Ber. 1981*, *114*, 3965; Maier, G.; Pfriem, S.; Malsch, K.; Kalinowski, H.; Dehnicke, K. *Chem. Ber. 1981*, *114*, 3988; Irngartinger, H.; Goldmann, A.; Jahn, R.; Nixdorf, M.; Rodewald, H.; Maier, G.; Malsch, K.; Emrich, R. *Angew. Chem. Int. Ed. 1984*, *23*, 993; Maier, G.; Fleischer, F. *Tetrahedron Lett. 1991*, *32*, 57. For reviews of attempts to synthesize tetrahedrane, see Maier, G. *Angew. Chem. Int. Ed. 1988*, *27*, 309; Zefirov, N.S.; Koz'min, A.S.; Abramenkov, A.V. *Russ. Chem. Rev. 1978*, *47*, 163. For a review of tetrahedranes and other cage molecules stabilized by steric hindrance, see Maier, G.; Rang, H.; Born, D., in Olah, G.A. *Cage Hydrocarbons*, Wiley, NY, *1990*, pp. 219–259. See also, Maier, G.; Born, D. *Angew. Chem. Int. Ed. 1989*, *28*, 1050.

[438]Rücker, C.; Trupp, B. *J. Am. Chem. Soc. 1988*, *110*, 4828.

[439]Von Seebach, M.; Kozhushkov, S.I.; Boese, R.; Benet-Buchholz, J.; Yufit, D.S.; Howard, J.A.K.; de Meijere, A. *Angew. Chem. Int. Ed. 2000*, *39*, 2495.

[440]Katz, T.J.; Acton, N. *J. Am. Chem. Soc. 1973*, *95*, 2738. See also Viehe, H.G.; Merényi, R.; Oth, J.F.M.; Senders, J.R.; Valange, P. *Angew. Chem. Int. Ed. 1964*, *3*, 755; Wilzbach, K.E.; Kaplan, L. *J. Am. Chem. Soc. 1965*, *87*, 4004.

TABLE 4.5 (*Continued*)

Structural Formula of Compound Prepared	Systematic Name of Ring System	Common Name If Any	Reference
	Pentacyclo[4.2.0.0^{2,5}.0^{3,8}.0^{4,7}]octane	Cubane	441
	Pentacyclo[5.4.1.0^{3,1}.0^{5,9}.0^{8,11}]dodecane	4[Peristylane]	442
	Hexacyclo[5.3.0.0^{2,6}-.0^{3,10}.0^{4,9}.0^{5,8}]decane	Pentaprismane	443
	Tricyclo[3.1.1.1^{2,4}]octane	Diasterane	444
	Hexacyclo[4.4.0.0^{2,4}.0^{3,9}-.0^{5,8}.0^{7,10}]decane		445
	Nonacyclo[10.8.0^{2,11}.0^{4,9}-.0^{4,19}.0^{6,17}.0^{7,16}.0^{9,14}.0^{14,19}]eicosane	A double tetraesterane	446
	Undecacyclo[9.9.0.0^{1,5}-.0^{2,12}.0^{2,18}.0^{3,7}.0^{6,10}.0^{8,12}-.0^{11,15}.0^{13,17}.0^{16,20}]eicosane	Pagodane	447

[441]Barborak, J.C.; Watts, L.; Pettit, R. *J. Am. Chem. Soc.* **1966**, *88*, 1328; Hedberg, L.; Hedberg, K.; Eaton, P.E.; Nodari, N.; Robiette, A.G. *J. Am. Chem. Soc.* **1991**, *113*, 1514. For a review of cubanes, see Griffin, G.W.; Marchand, A.P. *Chem. Rev.* **1989**, *89*, 997.

[442]Paquette, L.A.; Fischer, J.W.; Browne, A.R.; Doecke, C.W. *J. Am. Chem. Soc.* **1985**, *105*, 686.

[443]Eaton, P.E.; Or, Y.S.; Branca, S.J.; Shankar, B.K.R. *Tetrahedron* **1986**, *42*, 1621. See also Dauben, W.G.; Cunningham Jr., A.F. *J. Org. Chem.* **1983**, *48*, 2842.

[444]Otterbach, A.; Musso, H. *Angew. Chem. Int. Ed.* **1987**, *26*, 554.

[445]Allred, E.L.; Beck, B.R. *J. Am. Chem. Soc.* **1973**, *95*, 2393.

[446]Hoffmann, V.T.; Musso, H. *Angew. Chem. Int. Ed.* **1987**, *26*, 1006.

[447]Rihs, G. *Tetrahedron Lett.* **1983**, *24*, 5857. See Mathew, T.; Keller, M.; Hunkler, D.; Prinzbach, H. *Tetrahedron Lett.* **1996**, *37*, 4491 for the synthesis of azapagodanes (also called azadodecahedranes).

In certain small-ring systems, including small propellanes, the geometry of one or more carbon atoms is so constrained that all four of their valences are directed to the same side of a plane (inverted tetrahedron), as in **124**.[448] An example is 1,3-dehydroadamantane, **125** (which is also a propellane).[449] X-ray crystallography of the 5-cyano derivative of **125** shows that the four carbon valences at C-1 and C-3 are all directed "into" the molecule and none point outside.[450] Compound **125** is quite reactive; it is unstable in air, readily adds hydrogen, water, bromine, or acetic acid to the C_1—C_3 bond, and is easily polymerized. When two such atoms are connected by a bond (as in **125**), the bond is very long (the C_1—C_3 bond length in the 5-cyano derivative of **125** is 1.64 Å), as the atoms try to compensate in this way for their enforced angles. The high reactivity of the C_1—C_3 bond of **125** is not only caused by strain, but also by the fact that reagents find it easy to approach these atoms since there are no bonds (e.g., C—H bonds on C-1 or C-3) to get in the way.

124   **125**

Strain in Other Rings[451]

In rings larger than four-membered, there is no small-angle strain, but there are three other kinds of strain. In the chair form of cyclohexane, which does not exhibit any of the three kinds of strain, all six carbon–carbon bonds have the two attached carbons in the gauche conformation. However, in five-membered rings and in rings containing from 7 to 13 carbons any conformation in which all the ring bonds are gauche contains transannular interactions, that is, interactions between the substituents on C-1 and C-3 or C-1 and C-4, and so on. These interactions occur because the internal space is not large enough for all the quasiaxial hydrogen atoms to fit without coming into conflict. The molecule can adopt other conformations in which this *transannular strain* is reduced, but then some of the carbon–carbon bonds must adopt eclipsed or partially eclipsed conformations. The strain resulting from eclipsed conformations is called *Pitzer strain*. For saturated rings from 3- to 13-membered (except for the chair form of cyclohexane) there is no escape from at least one of these two types of strain. In practice, each ring adopts conformations that minimize both sorts of strain as much as possible. For cyclopentane, as we have seen (p. 212), this means that the molecule is not planar. In rings larger than

---

[448]For a review, see Wiberg, K.B. *Acc. Chem. Res.* **1984**, *17*, 379.
[449]Scott, W.B.; Pincock, R.E. *J. Am. Chem. Soc.* **1973**, *95*, 2040.
[450]Gibbons, C.S.; Trotter, J. *Can. J. Chem.* **1973**, *51*, 87.
[451]For reviews, see Gol'dfarb, Ya.L.; Belen'kii, L.I. *Russ. Chem. Rev.* **1960**, *29*, 214; Raphael, R.A. *Proc. Chem. Soc.* **1962**, 97; Sicher, J. *Prog. Stereochem.* **1962**, *3*, 202.

9-membered, Pitzer strain seems to disappear, but transannular strain is still present.[452] For 9- and 10-membered rings, some of the transannular and Pitzer strain may be relieved by the adoption of a third type of strain, *large-angle strain*. Thus, C—C—C angles of 115–120° have been found in X-ray diffraction of cyclononylamine hydrobromide and 1,6-diaminocyclodecane dihydrochloride.[453]

126

Strain can exert other influences on molecules. 1-Aza-2-adamantanone (**126**) is an extreme case of a twisted amide.[454] The overlap of the lone pair electrons on nitrogen with the π-system of the carbonyl is prevented.[454] In chemical reactions, **126** reacts more or less like a ketone, giving a Wittig reaction (**16-44**) and it can form a ketal (**16-7**). A twisted biadamantylidene compound has been reported.[455]

127a   **127b**   **128**   **129**

The amount of strain in cycloalkanes is shown in Table 4.6,[456] which lists heats of combustion per CH_2 group. As can be seen, cycloalkanes larger than 13-membered are as strain-free as cyclohexane.

Transannular interactions can exist across rings from 8- to 11-membered and even larger.[457] Such interactions can be detected by dipole and spectral measurements. For example, that the carbonyl group in **127a** is affected by the nitrogen (**127b** is probably another canonical form) has been demonstrated by photoelectron spectroscopy, which shows that the ionization potentials of the nitrogen *n* and C=O π orbitals in **127** differ from those of the two comparison molecules **128** and **129**.[458] It is significant that when **127** accepts a proton, it goes to the

[452]Huber-Buser, E.; Dunitz, J.D. *Helv. Chim. Acta* **1960**, *43*, 760.

[453]Dunitz, J.D.; Venkatesan, K. *Helv. Chim. Acta* **1961**, *44*, 2033.

[454]Kirby, A.J.; Komarov, I.V.; Wothers, P.D.; Feeder, N. *Angew. Chem. Int. Ed.*, **1998**, *37*, 785. For other examples of twisted amides, see Duspara, P.A.; Matta, C.F.; Jenkins, S.I.; Harrison, P.H.M. *Org. Lett.* **2001**, *3*, 495; Madder, R.D.; Kim, C.-Y.; Chandra, P.P.; Doyon, J.B.; Barid Jr., T.A.; Fierke, C.A.; Christianson, D.W.; Voet, J.G.; Jain, A. *J. Org. Chem.* **2002**, *67*, 582.

[455]Okazaki, T.; Ogawa, K.; Kitagawa, T.; Takeuchi, K. *J. Org. Chem.* **2002**, *67*, 5981.

[456]Gol'dfarb, Ya.L.; Belen'kii, L.I. *Russ. Chem. Rev.* **1960**, *29*, 214, p. 218.

[457]For a review, see Cope, A.C.; Martin, M.M.; McKervey, M.A. *Q. Rev. Chem. Soc.* **1966**, *20*, 119.

[458]Spanka, G.; Rademacher, P. *J. Org. Chem.* **1986**, *51*, 592. See also, Spanka, G.; Rademacher, P.; Duddeck, H. *J. Chem. Soc. Perkin Trans. 2* 1988, 2119; Leonard, N.J.; Fox, R.C.; Ō ki, M. *J. Am. Chem. Soc.* **1954**, *76*, 5708.

TABLE 4.6. Heats of Combustion in the Gas Phase for Cycloalkanes, per CH$_2$ Group[456]

Size of Ring	$-\Delta H_c$, (g) kcal mol^{-1}	kJ mol^{-1}	Size of Ring	$-\Delta H_c$, (g) kcal mol^{-1}	kJ mol^{-1}
3	166.3	695.8	10	158.6	663.6
4	163.9	685.8	11	158.4	662.7
5	158.7	664.0	12	157.8	660.2
6	157.4	658.6	13	157.7	659.8
7	158.3	662.3	14	157.4	658.6
8	158.6	663.6	15	157.5	659.0
9	158.8	664.4	16	157.5	659.0

oxygen rather than to the nitrogen. Many examples of transannular reactions are known, including:

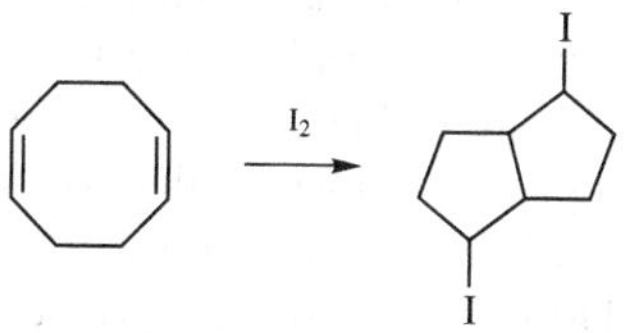

Ref. [459]

Ref. [460]

In summary, we can divide saturated rings into four groups, of which the first and third are more strained than the other two.[461]

1. *Small rings* (3- and 4-membered). Small-angle strain predominates.
2. *Common rings* (5-, 6-, and 7-membered). Largely unstrained. The strain that is present is mostly Pitzer strain.
3. *Medium rings* (8- to 11-membered). Considerable strain; Pitzer, transannular, and large-angle strain.
4. *Large rings* (12-membered and larger). Little or no strain.[462]

[459]Uemura, S.; Fukuzawa, S.; Toshimitsu, A.; Okano, M.; Tezuka, H.; Sawada, S. *J. Org. Chem.* **1983**, *48*, 270.

[460]Schläpfer-Dähler, M.; Prewo, R.; Bieri, J.H.; Germain, G.; Heimgartner, H. *Chimia* **1988**, *42*, 25.

[461]For a review on the influence of ring size on the properties of cyclic systems, see Granik, V.G. *Russ. Chem. Rev.* **1982**, *51*, 119.

[462]An example is the calculated strain of 1.4–3.2 kcal mol^{-1} in cyclotetradecane. See Chickos, J.S.; Hesse, D.G.; Panshin, S.Y.; Rogers, D.W.; Saunders, M.; Uffer, P.M.; Liebman, J.F. *J. Org. Chem.* **1992**, *57*, 1897.

Unsaturated Rings[463]

Double bonds can exist in rings of any size. As expected, the most highly strained are the three-membered rings. Small-angle strain, which is so important in cyclopropane, is even greater in cyclopropene[464] because the ideal angle is greater. In cyclopropane, the bond angle is forced to be 60°, ~50° smaller than the tetrahedral angle; but in cyclopropene, the angle, also ~60°, is now ~60° smaller than the ideal angle of 120°. Thus, the angle is cyclopropene is ~10° more strained than in cyclopropane. However, this additional strain is offset by a decrease in strain arising from another factor. Cyclopropene, lacking two hydrogens, has none of the eclipsing

Benzocyclopropene

strain present in cyclopropane. Cyclopropene has been prepared[465] and is stable at liquid-nitrogen temperatures, although on warming even to −80°C it rapidly polymerizes. Many other cyclopropenes are stable at room temperature and above.[464] The highly strained benzocyclopropene,[466] in which the cyclopropene ring is fused to a benzene ring, has been prepared[467] and is stable for weeks at room temperature, although it decomposes on distillation at atmospheric pressure.

As previously mentioned, double bonds in relatively small rings must be cis. A stable trans double bond[468] first appears in an eight-membered ring (*trans*-cyclooctene, p. 150), although the transient existence of *trans*-cyclohexene and cycloheptene has been demonstrated.[469] Above ~11 members, the trans isomer

[463]For a review of strained double bonds, see Zefirov, N.S.; Sokolov, V.I. *Russ. Chem. Rev.* **1967**, *36*, 87. For a review of double and triple bonds in rings, see Johnson, R.P. *Mol. Struct. Energ.* **1986**, *3*, 85.

[464]For reviews of cyclopropenes, see Baird, M.S. *Top. Curr. Chem.* **1988**, *144*, 137; Halton, B.; Banwell, M.G. in Rappoport, Z. *The Chemistry of the Cyclopropyl Group*, pt. 2, pp. Wiley, NY, **1987**, pp. 1223–1339; Closs, G.L. *Adv. Alicyclic Chem.* **1966**, *1*, 53; For a discussion of the bonding and hybridization, see Allen, F.H. *Tetrahedron* **1982**, *38*, 645.

[465]Dem'yanov, N.Ya.; Doyarenko, M.N. *Bull. Acad. Sci. Russ.* **1922**, *16*, 297, *Ber.* **1923**, *56*, 2200; Schlatter, M.J. *J. Am. Chem. Soc.* **1941**, *63*, 1733; Wiberg, K.B.; Bartley, W.J. *J. Am. Chem. Soc.* **1960**, *82*, 6375; Stigliani, W.M.; Laurie, V.W.; Li, J.C. *J. Chem. Phys.* **1975**, *62*, 1890.

[466]For reviews of cycloproparenes, see Halton, B. *Chem. Rev.* **1989**, *89*, 1161; **1973**, *73*, 113; Billups, W.E.; Rodin, W.A.; Haley, M.M. *Tetrahedron* **1988**, *44*, 1305; Halton, B.; Stang, P.J. *Acc. Chem. Res.* **1987**, *20*, 443; Billups, W.E. *Acc. Chem. Res.* **1978**, *11*, 245.

[467]Vogel, E.; Grimme, W.; Korte, S. *Tetrahedron Lett.* **1965**, 3625. Also see Anet, R.; Anet, F.A.L. *J. Am. Chem. Soc.* **1964**, *86*, 526; Müller, P.; Bernardinelli, G.; Thi, H.C.G. *Chimia* **1988**, *42*, 261; Neidlein, R.; Christen, D.; Poignée, V.; Boese, R.; Bläser, D.; Gieren, A.; Ruiz-Pérez, C.; Hübner, T. *Angew. Chem. Int. Ed.* **1988**, *27*, 294.

[468]For reviews of trans cycloalkenes, see Nakazaki, M.; Yamamoto, K.; Naemura, K. *Top. Curr. Chem.* **1984**, *125*, 1; Marshall, J.A. *Acc. Chem. Res.* **1980**, *13*, 213.

[469]Bonneau, R.; Joussot-Dubien, J.; Salem, L.; Yarwood, A.J. *J. Am. Chem. Soc.* **1979**, *98*, 4329; Wallraff, G.M.; Michl, J. *J. Org. Chem.* **1986**, *51*, 1794; Squillacote, M.; Bergman, A.; De Felippis, J. *Tetrahedron Lett.* **1989**, *30*, 6805.

is more stable than the cis.[223] It has proved possible to prepare compounds in which a trans double bond is shared by two cycloalkene rings (e.g., **130**). Such compounds have been called *[m.n]betweenanenes,* and several have been prepared with *m* and *n* values from 8 to 26.[470] The double bonds of the smaller betweenanenes, as might be expected from the fact that they are deeply buried within the bridges, are much less reactive than those of the corresponding *cis-cis* isomers.

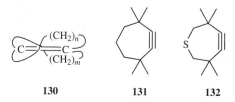

130   **131**   **132**

The smallest unstrained cyclic triple bond is found in cyclononyne.[471] Cyclooctyne has been isolated,[472] but its heat of hydrogenation shows that it is considerably strained. There have been a few compounds isolated with triple bonds in seven-membered rings. 3,3,7,7-Tetramethylcycloheptyne (**131**) dimerizes within 1 h at room temperature,[473] but the thia derivative **132**, in which the C—S bonds are longer than the corresponding C—C bonds in **131**, is indefinitely stable even at 140°C.[474] Cycloheptyne itself has not been isolated, although its transient existence has been shown.[475] Cyclohexyne[476] and its 3,3,6,6-tetramethyl derivative[477] have been trapped at 77 K, and in an argon matrix at 12 K, respectively, and IR spectra

[470]Nakazaki, M.; Yamamoto, K.; Yanagi, J. *J. Am. Chem. Soc.* **1979**, *101*, 147; Ceré, V.; Paolucci, C.; Pollicino, S.; Sandri, E.; Fava, A. *J. Chem. Soc. Chem. Commun.* **1980**, 755; Marshall, J.A.; Flynn, K.E. *J. Am. Chem. Soc.* **1983**, *105*, 3360. For reviews, see Nakazaki, M.; Yamamoto, K.; Naemura, K. *Top. Curr. Chem.* **1984**, *125*, 1; Marshall, J.A. *Acc. Chem. Res.* **1980**, *13*, 213. For a review of these and similar compounds, see Borden, W.T. *Chem. Rev.* **1989**, *89*, 1095.

[471]For reviews of triple bonds in rings, see Meier, H. *Adv. Strain Org. Chem.* **1991**, *1*, 215; Krebs, A.; Wilke, J. *Top. Curr. Chem.* **1983**, *109*, 189; Nakagawa, M., in Patai, S. *The Chemistry of the C≡C Triple Bond*, pt. 2; Wiley, NY, **1978**, pp. 635–712; Krebs, A. in Viehe, H.G. *Acetylenes*, Marcel Dekker, NY, **1969**, pp. 987–1062. For a list of strained cycloalkynes that also have double bonds, see Meier, H.; Hanold, N.; Molz, T.; Bissinger, H.J.; Kolshorn, H.; Zountsas, J. *Tetrahedron* **1986**, *42*, 1711.

[472]Blomquist, A.T.; Liu, L.H. *J. Am. Chem. Soc.* **1953**, *75*, 2153. See also, Bühl, H.; Gugel, H.; Kolshorn, H.; Meier, H. *Synthesis* **1978**, 536.

[473]Krebs, A.; Kimling, H. *Angew. Chem. Int. Ed.* **1971**, *10*, 509; Schmidt, H.; Schweig, A.; Krebs, A. *Tetrahedron Lett.* **1974**, 1471.

[474]Krebs, A.; Kimling, H. *Tetrahedron Lett.* **1970**, 761.

[475]Wittig, G.; Meske-Schüller, J. *Liebigs Ann. Chem.* **1968**, *711*, 65; Krebs, A.; Kimling, H. *Angew. Chem. Int. Ed.* **1971**, *10*, 509; Bottini, A.T.; Frost II, K.A.; Anderson, B.R.; Dev, V. *Tetrahedron* **1973**, *29*, 1975.

[476]Wentrup, C.; Blanch, R.; Briehl, H.; Gross, G. *J. Am. Chem. Soc.* **1988**, *110*, 1874.

[477]See Sander, W.; Chapman, O.L. *Angew. Chem. Int. Ed.* **1988**, *27*, 398; Krebs, A.; Colcha, W.; Müller, M.; Eicher, T.; Pielartzik, H.; Schnöckel, H. *Tetrahedron Lett.* **1984**, *25*, 5027.

have been obtained. Transient six-and even five-membered rings containing triple bonds have also been demonstrated.[478]

⬡ᵒᶜᵗᵃᵍᵒⁿ—CMe₃

133

A derivative of cyclopentyne has been trapped in a matrix.[479] Although cycloheptyne and cyclohexyne have not been isolated at room temperatures, Pt(0) complexes of these compounds have been prepared and are stable.[480] The smallest cyclic allene[481] so far isolated is 1-*tert*-butyl-1,2-cyclooctadiene **133**.[482] The parent 1,2-cyclooctadiene has not been isolated. It has been shown to exist transiently, but rapidly dimerizes.[483] The presence of the *tert*-butyl group apparently prevents this. The transient existence of 1,2-cycloheptadiene has also been shown,[484] and both 1,2-cyclooctadiene and 1,2-cycloheptadiene have been isolated in platinum complexes.[485] 1,2-Cyclohexadiene has been trapped at low temperatures, and its structure has been proved by spectral studies.[486] Cyclic allenes in general are less strained than their acetylenic isomers.[487] The cyclic cumulene 1,2,3-cyclononatriene has also been synthesized and is reasonably stable in solution at room temperature in the absence of air.[488]

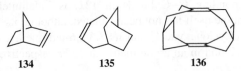

134   **135**   **136**

There are many examples of polycyclic molecules and bridged molecules that have one or more double bonds. There is flattening of the ring containing the C=C unit, and this can have a significant effect on the molecule. Norbornene (bicyclo[2.2.1]hept-2-ene; **134**) is a simple example and it has been calculated that it contains a distorted

[478]See, for example, Wittig, G. Mayer, U. *Chem. Ber.* **1963**, *96*, 329, 342; Wittig, G.; Weinlich, J. *Chem. Ber.* **1965**, *98*, 471; Bolster, J.M.; Kellogg, R.M. *J. Am. Chem. Soc.* **1981**, *103*, 2868; Gilbert, J.C.; Baze, M.E. *J. Am. Chem. Soc.* **1983**, *105*, 664.

[479]Chapman, O.L.; Gano, J.; West, P.R.; Regitz, M.; Maas, G. *J. Am. Chem. Soc.* **1981**, *103*, 7033.

[480]Bennett, M.A.; Robertson, G.B.; Whimp, P.O.; Yoshida, T. *J. Am. Chem. Soc.* **1971**, *93*, 3797.

[481]For reviews of cyclic allenes, see Johnson, R.P. *Adv. Theor. Interesting Mol.* **1989**, *1*, 401; *Chem. Rev.* **1989**, *89*, 1111; Thies, R.W. *Isr. J. Chem.* **1985**, *26*, 191; Schuster, H.F.; Coppola, G.M. *Allenes in Organic Synthesis*; Wiley, NY, **1984**, pp. 38–56.

[482]Price, J.D.; Johnson, R.P. *Tetrahedron Lett.* **1986**, *27*, 4679.

[483]See Marquis, E.T.; Gardner, P.D. *Tetrahedron Lett.* **1966**, 2793.

[484]Wittig, G.; Dorsch, H.; Meske-Schüller, J. *Liebigs Ann. Chem.* **1968**, *711*, 55.

[485]Visser, J.P.; Ramakers, J.E. *J. Chem. Soc. Chem. Commun.* **1972**, 178.

[486]Wentrup, C.; Gross, G.; Maquestiau, A.; Flammang, R. *Angew. Chem. Int. Ed.* **1983**, *22*, 542. 1,2,3-Cyclohexatriene has also been trapped: Shakespeare, W.C.; Johnson, R.P. *J. Am. Chem. Soc.* **1990**, *112*, 8578.

[487]Moore, W.R.; Ward, H.R. *J. Am. Chem. Soc.* **1963**, *85*, 86.

[488]Angus Jr., R.O.; Johnson, R.P. *J. Org. Chem.* **1984**, *49*, 2880.

π-face.[489] The double bond can appear away from the bridgehead carbon atoms, as in bicyclo[4.2.2]dec-3-ene (**135**) and that part of the molecule is flattened. In pentacyclo[8.2.1.1^{2,5}.1^{4,7}.1^{8,11}]hexadeca-1,7-diene (**136**), the C=C units are held in a position where there is significant π–π interactions across the molecule.[490]

Double bonds at the bridgehead of bridged bicyclic compounds are impossible in small systems. This is the basis of *Bredt's rule*,[491] which states that elimination to give a double bond in a bridged bicyclic system (e.g., **137**) always leads away from the bridgehead. This rule no longer applies when the rings are large enough. In

137

determining whether a bicyclic system is large enough to accommodate a bridgehead double bond, the most reliable criterion is the size of the ring in which the double bond is located.[492] Bicyclo[3.3.1]non-1-ene[493] (**138**) and bicyclo[4.2.1]non-1(8)ene[494] (**139**) are stable compounds. Both can be looked upon as derivatives of *trans*-cyclooctene, which is of course a known compound. Compound **138** has been shown to have a strain energy of the same order of magnitude

138   139   140

as that of *trans*-cyclooctene.[495] On the other hand, in bicyclo[3.2.2]non-1-ene (**140**), the largest ring that contains the double bond is *trans*-cycloheptene, which is as yet unknown. Compound **140** has been prepared, but dimerized before it could be isolated.[496] Even smaller systems ([3.2.1] and [2.2.2]), but with imine double

[489]Ohwada, T. *Tetrahedron* **1993**, *49*, 7649.

[490]Lange, H.; Schäfer, W.; Gleiter, R.; Camps, P.; Vázquez, S. *J. Org. Chem.* **1998**, *63*, 3478.

[491]For reviews, see Shea, K.J. *Tetrahedron* **1980**, *36*, 1683; Buchanan, G.L. *Chem. Soc. Rev.* **1974**, *3*, 41; Köbrich, G. *Angew. Chem. Int. Ed.* **1973**, *12*, 464. For reviews of bridgehead olefins, see Billups, W.E.; Haley, M.M.; Lee, G. *Chem. Rev.* **1989**, *89*, 1147; Warner, P.M. *Chem. Rev.* **1989**, *89*, 1067; Szeimies, G. *React. Intermed. (Plenum)* **1983**, *3*, 299; Keese, R. *Angew. Chem. Int. Ed.* **1975**, *14*, 528. Also see, Smith, M.B. *Organic Synthesis*, 2nd ed., McGraw-Hill, NY, **2001**, pp. 502–504.

[492]For a discussion and predictions of stability in such compounds, see Maier, W.F.; Schleyer, P.v.R. *J. Am. Chem. Soc.* **1981**, *103*, 1891.

[493]Marshall, J.A.; Faubl, H. *J. Am. Chem. Soc.* **1967**, *89*, 5965, **1970**, *92*, 948; Wiseman, J.R.; Pletcher, W.A. *J. Am. Chem. Soc.* **1970**, *92*, 956; Kim, M.; White, J.D. *J. Am. Chem. Soc.* **1975**, *97*, 451; Becker, K.B. *Helv. Chim. Acta* **1977**, *60*, 81. For the preparation of optically active **125**, see Nakazaki, M.; Naemura, K.; Nakahara, S. *J. Org. Chem.* **1979**, *44*, 2438.

[494]Wiseman, J.R.; Chan, H.; Ahola, C.J. *J. Am. Chem. Soc.* **1969**, *91*, 2812; Carruthers, W.; Qureshi, M.I. *Chem. Commun.* **1969**, 832; Becker, K.B. *Tetrahedron Lett.* **1975**, 2207.

[495]Lesko, P.M.; Turner, R.B. *J. Am. Chem. Soc.* **1968**, *90*, 6888; Burkert, U. *Chem. Ber.* **1977**, *110*, 773.

[496]Wiseman, J.R.; Chong, J.A. *J. Am. Chem. Soc.* **1969**, *91*, 7775.

bonds (**141–143**), have been obtained in matrixes at low temperatures.[497] These compounds are destroyed on warming. Compounds **141** and **142** are the first reported example of (*E–Z*) isomerism at a strained bridgehead double bond.[498]

(*E*) Isomer   (*Z*) Isomer

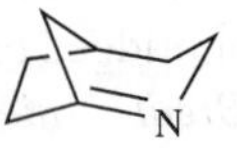

141   **142**   **143**

Strain Due to Unavoidable Crowding[499]

In some molecules, large groups are so close to each other that they cannot fit into the available space in such a way that normal bond angles are maintained. It has proved possible to prepare compounds with a high degree of this type of strain. For example, success has been achieved in synthesizing benzene rings containing *ortho-tert*-butyl groups. Two examples that have been prepared, of several, are 1,2,3-tri-*tert*-butyl compound **144**[500] and the 1,2,3,4-tetra-*tert*-butyl compound **145**.[501] That these molecules are strained is demonstrated by UV and IR spectra,

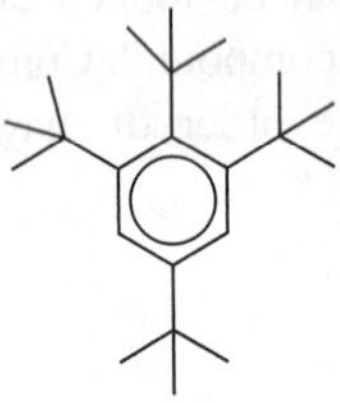

144 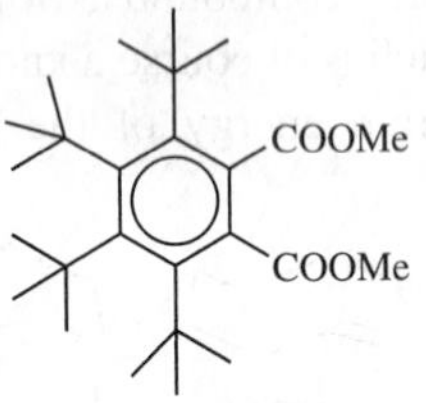   **145** 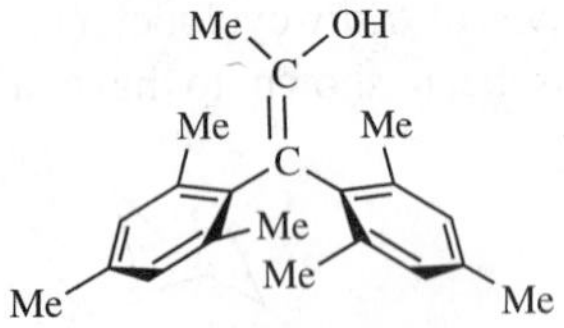   **146**

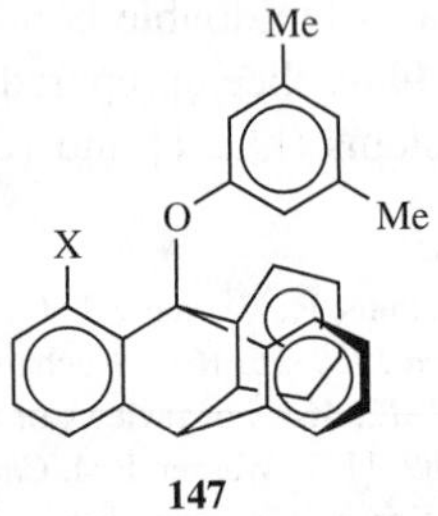

147

[497]Sheridan, R.S.; Ganzer, G.A. *J. Am. Chem. Soc.* **1983**, *105*, 6158; Radziszewski, J.G.; Downing, J.W.; Wentrup, C.; Kaszynski, P.; Jawdosiuk, M.; Kovacic, P.; Michl, J. *J. Am. Chem. Soc.* **1985**, *107*, 2799.

[498]Radziszewski, J.G.; Downing, J.W.; Wentrup, C.; Kaszynski, P.; Jawdosiuk, M.; Kovacic, P.; Michl, J. *J. Am. Chem. Soc.* **1985**, *107*, 2799.

[499]For reviews, see Tidwell, T.T. *Tetrahedron* **1978**, *34*, 1855; Voronenkov, V.V.; Osokin, Yu.G. *Russ. Chem. Rev.* **1972**, *41*, 616. For a review of early studies, see Mosher, H.S.; Tidwell, T.T. *J. Chem. Educ.* **1990**, *67*, 9. For a review of van der Waals radii, see Zefirov, Yu.V.; Zorkii, P.M. *Russ. Chem. Rev.* **1989**, *58*, 421.

[500]Arnett, E.M.; Bollinger, J.M. *Tetrahedron Lett.* **1964**, 3803.

[501]Maier, G.; Schneider, K. *Angew. Chem. Int. Ed.* **1980**, *19*, 1022. For another example, see Krebs, A.; Franken, E.; Müller, S. *Tetrahedron Lett.* **1981**, *22*, 1675.

which show that the ring is not planar in 1,2,4-tri-*tert*-butylbenzene, and by a comparison of the heats of reaction of this compound and its 1,3,5 isomer, which show that the 1,2,4 compound possesses ~22 kcal mol^{-1} (92 kJ mol^{-1}) more strain energy than its isomer[502] (see also, p. 1642). Since SiMe$_3$ groups are larger than CMe$_3$ groups, and it has proven possible to prepare C$_6$(SiMe$_3$)$_6$. This compound has a chair-shaped ring in the solid state, and a mixture of chair and boat forms in solution.[503] Even smaller groups can sterically interfere in ortho positions. In hexaisopropylbenzene, the six isopropyl groups are so crowded that they cannot rotate but are lined up around the benzene ring, all pointed in the same direction.[504] This compound is an example of a *geared molecule*.[505] The isopropyl groups fit into each other in the same manner as interlocked gears. Another example

is **146** (which is a stable enol).[506] In this case each ring can rotate about its C—aryl bond only by forcing the other to rotate as well. In the case of triptycene derivatives, such as **147**, a complete 360° rotation of the aryl group around the O—aryl bond requires the aryl group to pass over three rotational barriers; one of which is the C—X bond and other two the "top" C–H bonds of the other two rings. As expected, the C—X barrier is the highest, ranging from 10.3 kcal mol^{-1} (43.1 kJ mol^{-1}) for X = F to 17.6 kcal mol^{-1} (73.6 kJ mol^{-1}) for X = *tert*-butyl.[507] In another instance, it has proved possible to prepare cis and trans isomers of 5-amino-2,4,6-triiodo-*N,N,N′,N′*-tetramethylisophthalamide because there is no room for the CONMe$_2$ groups to rotate, caught as they are between two bulky iodine atoms.[508] The trans isomer is chiral and has been resolved, while the cis isomer is a meso form. Another

[502]Arnett, E.M.; Sanda, J.C.; Bollinger, J.M.; Barber, M. *J. Am. Chem. Soc.* **1967**, *89*, 5389; Krüerke, U.; Hoogzand, C.; Hübel, W. *Chem. Ber.* **1961**, *94*, 2817; Dale, J. *Chem. Ber.* **1961**, *94*, 2821. See also Barclay, L.R.C.; Brownstein, S.; Gabe, E.J.; Lee, F.L. *Can. J. Chem.* **1984**, *62*, 1358.

[503]Sakurai, H.; Ebata, K.; Kabuto, C.; Sekiguchi, A. *J. Am. Chem. Soc.* **1990**, *112*, 1799.

[504]Arnett, E.M.; Bollinger, J.M. *J. Am. Chem. Soc.* **1964**, *86*, 4730; Hopff, H.; Gati, A. *Helv. Chim. Acta* **1965**, *48*, 509; Siegel, J.; Gutiérrez, A.; Schweizer, W.B.; Ermer, O.; Mislow, K. *J. Am. Chem. Soc.* **1986**, *108*, 1569. For the similar structure of hexakis(dichloromethyl)benzene, see Kahr, B.; Biali, S.E.; Schaefer, W.; Buda, A.B.; Mislow, K. *J. Org. Chem.* **1987**, *52*, 3713.

[505]For reviews, see Iwamura, H.; Mislow, K. *Acc. Chem. Res.* **1988**, *21*, 175; Mislow, K. *Chemtracts: Org. Chem.* **1989**, *2*, 151; *Chimia*, **1986**, *40*, 395; Berg, U.; Liljefors, T.; Roussel, C.; Sandström, J. *Acc. Chem. Res.* **1985**, *18*, 80.

[506]Nugiel, D.A.; Biali, S.E.; Rappoport, Z. *J. Am. Chem. Soc.* **1984**, *106*, 3357.

[507]Yamamoto, G.; Ō ki, M. *Bull. Chem. Soc. Jpn.* **1986**, *59*, 3597. For reviews of similar cases, see Yamamoto, G. *Pure Appl. Chem.* **1990**, *62*, 569; Ō ki, M. *Applications of Dynamic NMR Spectroscopy to Organic Chemistry*, VCH, NY, **1985**, pp. 269–284.

[508]Ackerman, J.H.; Laidlaw, G.M.; Snyder, G.A. *Tetrahedron Lett.* **1969**, 3879; Ackerman, J.H.; Laidlaw, G.M. *Tetrahedron Lett.* **1969**, 4487. See also Cuyegkeng, M.A.; Mannschreck, A. *Chem. Ber.* **1987**, *120*, 803.

example of cis–trans isomerism resulting from restricted rotation about single bonds[509] is found in 1,8-di-*o*-tolylnapthalene[510] (see also, p. 182).

There are many other cases of intramolecular crowding that result in the distortion of bond angles. We have already mentioned hexahelicene (p. 150) and bent benzene rings (p. 48). The compounds tri-*tert*-butylamine and tetra-*tert*-butylmethane are as yet unknown. In the latter, there is no way for the strain to be relieved and it is questionable whether this compound can ever be made. In tri-*tert*-butylamine the crowding can be eased somewhat if the three bulky groups assume a planar instead of the normal pyramidal configuration. In tri-*tert*-butylcarbinol, coplanarity of the three *tert*-butyl groups is prevented by the presence of the OH group, and yet this compound has been prepared.[511] Tri-*tert*-butylamine should have less steric strain than tri-*tert*-butylcarbinol and it should be possible to prepare it.[512] The tetra-*tert*-butylphosphonium cation $(t\text{-Bu})_4P^+$ has been prepared.[513] Although steric effects are nonadditive in crowded molecules, a quantitative measure has been proposed by D. F. DeTar, based on molecular mechanics calculations. This is called *formal steric enthalpy* (FSE), and values have been calculated for alkanes, alkenes, alcohols, ethers, and methyl esters.[514] For example, some FSE values for alkanes are butane 0.00; 2,2,3,3-tetramethylbutane 7.27; 2,2,4,4,5-pentamethylhexane 11.30; and tri-*tert*-butylmethane 38.53.

The two carbon atoms of a C=C double bond and the four groups attached to them are normally in a plane, but if the groups are large enough, significant

[509]For a monograph on restricted rotation about single bonds, see Ōki, M. *Applications of Dynamic NMR Spectroscopy to Organic Chemistry*, VCH, NY, *1985*. For reviews, see Förster, H.; Vögtle, F. *Angew. Chem. Int. Ed. 1977*, *16*, 429; Ōki, M. *Angew. Chem. Int. Ed. 1976*, *15*, 87.

[510]Clough, R.L.; Roberts, J.D. *J. Am. Chem. Soc. 1976*, *98*, 1018. For a study of rotational barriers in this system, see Cosmo, R.; Sternhell, S. *Aust. J. Chem. 1987*, *40*, 1107.

[511]Bartlett, P.D.; Tidwell, T.T. *J. Am. Chem. Soc. 1968*, *90*, 4421.

[512]For attempts to prepare tri-*tert*-butylamine, see Back, T.G.; Barton, D.H.R. *J. Chem. Soc. Perkin Trans 1*, *1977*, 924. For the preparation of di-*tert*-butylmethylamine and other sterically hindered amines, see Kopka, I.E.; Fataftah, Z.A.; Rathke, M.W. *J. Org. Chem. 1980*, *45*, 4616; Audeh, C.A.; Fuller, S.E.; Hutchinson, R.J.; Lindsay Smith, J.R. *J. Chem. Res. (S), 1979*, 270.

[513]Schmidbaur, H.; Blaschke, G.; Zimmer-Gasser, B.; Schubert, U. *Chem. Ber. 1980*, *113*, 1612.

[514]DeTar, D.F.; Binzet, S.; Darba, P. *J. Org. Chem. 1985*, *50*, 2826, 5298, 5304.

deviation from planarity can result.[515] The compound tetra-*tert*-butylethene (**148**) has not been prepared,[516] but the tetraaldehyde **149**, which should have about the same amount of strain, has been made. X-ray crystallography shows that **149** is twisted out of a planar shape by an angle of 28.6°.[517] Also, the C=C double-bond distance is 1.357 Å, significantly longer than a normal C=C bond of 1.32 Å (Table 1.5). (*Z*)-1,2-Bis(*tert*-butyldimethylsilyl)-1,2-bis(trimethylsilyl)ethene (**150**) has an even greater twist, but could not be made to undergo conversion to the (*E*) isomer, probably because the groups are too large to slide past each other.[518] A different kind of double bond strain is found in tricyclo[4.2.2.2^{2,5}]dodeca-1,5-diene (**151**),[519] cubene (**152**),[520] and homocub-4(5)-ene (**153**).[521] In these molecules, the four groups on the double bond are all forced to be on one side

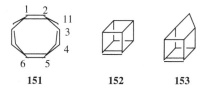

148	**149**	**150**

of the double-bond plane.[522] In **151**, the angle between the line C_1–C_2 (extended) and the plane defined by C_2, C_3, and C_{11} is 27°. An additional source of strain in this molecule is the fact that the two double bonds are pushed

151	**152**	**153**

into close proximity by the four bridges. In an effort to alleviate this sort of strain, the bridge bond distances (C_3–C_4) are 1.595 Å, which is considerably longer than the 1.53 Å expected for a normal sp^3–sp^3 C–C bond (Table 1.5). Compounds **152** and **153** have ***not*** been isolated, but have been generated as intermediates that were trapped by reaction with other compounds.[520,521]

[515]For reviews, see Luef, W.; Keese, R. *Top. Stereochem.* **1991**, *20*, 231; Sandström, J. *Top. Stereochem.* **1983**, *14*, 83, pp. 160–169.

[516]For a list of crowded alkenes that have been made, see Drake, C.A.; Rabjohn, N.; Tempesta, M.S.; Taylor, R.B. *J. Org. Chem.* **1988**, *53*, 4555. See also, Garratt, P.J.; Payne, D.; Tocher, D.A. *J. Org. Chem.* **1990**, *55*, 1909.

[517]Krebs, A.; Nickel, W.; Tikwe, L.; Kopf, J. *Tetrahedron Lett.* **1985**, *26*, 1639.

[518]Sakurai, H.; Ebata, K.; Kabuto, C.; Nakadaira, Y. *Chem. Lett.* **1987**, 301.

[519]Wiberg, K.B.; Matturo, M.G.; Okarma, P.J.; Jason, M.E. *J. Am. Chem. Soc.* **1984**, *106*, 2194; Wiberg, K.B.; Adams, R.D.; Okarma, P.J.; Matturo, M.G.; Segmuller, B. *J. Am. Chem. Soc.* **1984**, *106*, 2200.

[520]Eaton, P.E.; Maggini, M. *J. Am. Chem. Soc.* **1988**, *110*, 7230.

[521]Hrovat, D.A.; Borden, W.T. *J. Am. Chem. Soc.* **1988**, *110*, 7229.

[522]For a review of such molecules, see Borden, W.T. *Chem. Rev.* **1989**, *89*, 1095. See also, Hrovat, D.A.; Borden, W.T. *J. Am. Chem. Soc.* **1988**, *110*, 4710.

CHAPTER 5

Carbocations, Carbanions, Free Radicals, Carbenes, and Nitrenes

There are four types of organic species in which a carbon atom has a valence of only 2 or 3.[1] They are usually very short-lived, and most exist only as intermediates that are quickly converted to more stable molecules. However, some are more stable than others and fairly stable examples have been prepared of three of the four

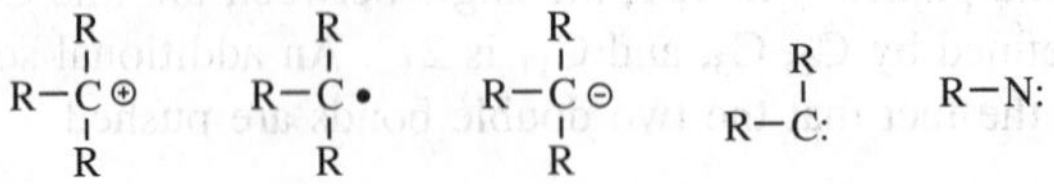

$$
\begin{array}{ccccc}
\underset{\substack{|\\R}}{\overset{\substack{R\\|}}{R-C^{\oplus}}} & \underset{\substack{|\\R}}{\overset{\substack{R\\|}}{R-C\bullet}} & \underset{\substack{|\\R}}{\overset{\substack{R\\|}}{R-C^{\ominus}}} & \overset{\substack{R\\|}}{\underset{\substack{|\\R-C:}}{}} & R-N: \\
\mathbf{A} & \mathbf{B} & \mathbf{C} & \mathbf{D} & \mathbf{E}
\end{array}
$$

types. The four types of species are *carbocations* (**A**), *free radicals* (**B**), *carbanions* (**C**), and *carbenes* (**D**). Of the four, only carbanions have a complete octet around the carbon. There are many other organic ions and radicals with charges and unpaired electrons on atoms other than carbon, but we will discuss only *nitrenes* (**E**), the nitrogen analogs of carbenes. Each of the five types is discussed in a separate section, which in each case includes brief summaries of the ways in which the species form and react. These summaries are short and schematic. The generation and fate of the five types are more fully treated in appropriate places in Part 2 of this book.

---

[1]For general references, see Isaacs, N.S. *Reactive Intermediates in Organic Chemistry*, Wiley, NY, *1974*; McManus, S.P. *Organic Reactive Intermediates*, Academic Press, NY, *1973*. Two serial publications devoted to review articles on this subject are *Reactive Intermediates (Wiley)* and *Reactive Intermediates (Plenum)*.

March's Advanced Organic Chemistry: Reactions, Mechanisms, and Structure, Sixth Edition, by Michael B. Smith and Jerry March
Copyright © 2007 John Wiley & Sons, Inc.

CARBOCATIONS[2]

Nomenclature

First, we must say a word about the naming of **A**. For many years these species were called "carbonium ions," although it was suggested[3] as long ago as 1902 that this was inappropriate because "-onium" usually refers to a covalency higher than that of the neutral atom. Nevertheless, the name "carbonium ion" was well established and created few problems[4] until some years ago, when George Olah and his co-workers found evidence for another type of intermediate in which there is a positive charge at a carbon atom, but in which the formal covalency of the carbon atom is five rather than three. The simplest example is the methanonium ion CH_5^+ (see p. 766). Olah proposed[5] that the name "carbonium ion" be reserved for pentacoordinated positive ions, and that **A** be called "carbenium ions." He also proposed the term "carbocation" to encompass both types. The International Union of Pure and Applied Chemistry (IUPAC) has accepted these definitions.[6] Although some authors still refer to **A** as carbonium ions and others call them carbenium ions, the general tendency is to refer to them simply as *carbocations*, and we will follow this practice. The pentavalent species are much rarer than **A**, and the use of the term "carbocation" for **A** causes little or no confusion.

Stability and Structure

Carbocations are intermediates in several kinds of reactions.[7] The more stable ones have been prepared in solution and in some cases even as solid salts, and X-ray crystallographic structures have been obtained in some cases.[8] The X-ray of the

[2]For a treatise, see Olah, G.A.; Schleyer, P.v.R. *Carbonium Ions*, 5 vols., Wiley, NY, *1968–1976*. For monographs, see Vogel, P. *Carbocation Chemistry*, Elsevier, NY, *1985*; Bethell, D.; Gold, V. *Carbonium Ions*, Academic Press, NY, *1967*. For reviews, see Saunders, M.; Jiménez-Vázquez, H.A. *Chem. Rev. 1991*, *91*, 375; Arnett, E.M.; Hofelich, T.C.; Schriver, G.W. *React. Intermed. (Wiley) 1987*, *3*, 189; Bethell, D.; Whittaker, D. *React. Intermed. (Wiley) 1981*, 2, 211; Bethell, D. *React. Intermed. (Wiley) 1978*, *1*, 117; Olah, G.A. *Chem. Scr. 1981*, 18, 97, *Top. Curr. Chem. 1979*, *80*, 19, *Angew. Chem. Int. Ed. 1973*, *12*, 173 (this review has been reprinted as Olah, G.A. *Carbocations and Electrophilic Reactions*, Wiley, NY, *1974*); Isaacs, N.S. *Reactive Intermediates in Organic Chemistry*, Wiley, NY, *1974*, pp. 92–199; McManus, S.P.; Pittman, Jr., C.U., in McManus, S.P. *Organic Reactive Intermediates*, Academic Press, NY, *1973*, pp. 193–335; Buss, V.; Schleyer, P.v.R.; Allen, L.C. *Top. Stereochem. 1973*, 7, 253; Olah, G.A.; Pittman Jr., C.U. *Adv. Phys. Org. Chem. 1966*, *4*, 305. For reviews of dicarbocations, see Lammertsma, K.; Schleyer, P.v.R.; Schwarz, H. *Angew. Chem. Int. Ed. 1989*, 28, 1321; Pagni, R.M. *Tetrahedron 1984*, *40*, 4161; Prakash, G.K.S.; Rawdah, T.N.; Olah, G.A. *Angew. Chem. Int. Ed. 1983*, *22*, 390. See also, the series *Advances in Carbocation Chemistry*.
[3]Gomberg, M. *Berchte 1902*, *35*, 2397.
[4]For a history of the term "carbonium ion," see Traynham, J.G. *J. Chem. Educ. 1986*, *63*, 930.
[5]Olah, G.A. *CHEMTECH 1971*, *1*, 566; *J. Am. Chem. Soc. 1972*, *94*, 808.
[6]Gold, V.; Loening, K.L.; McNaught, A.D.; Sehmi, P. *Compendium of Chemical Terminology: IUPAC Recommendations*, Blackwell Scientific Publications, Oxford, *1987*.
[7]Olah, G.A. *J. Org. Chem. 2001*, *66*, 5943.
[8]See Laube, T. *J. Am. Chem. 2004*, 126, 10904 and references cited therein. For the X-ray of a vinyl carbocation, see Müller, T.; Juhasz, M.; Reed, C.A. *Angew. Chem. Int. Ed. 2004*, *43*, 1543.

tert-butyl cation complexed with dichloromethane was reported,[9] for example, and is presented as **1** with the solvent molecules removed for clarity. An isolable dioxa-stabilized pentadienylium ion was isolated and its structure was determined by ^{1}H-, ^{13}C-NMR, mass spectrometry (MS), and IR.[10] A β-fluoro substituted 4-methoxyphenethyl cation has been observed directly by laser flash photolysis.[11] In solution, the carbocation may be free (this is more likely in polar solvents, in which it is solvated) or it may exist as an ion pair,[12] which means that it is closely associated with a negative ion, called a *counterion* or *gegenion*. Ion pairs are more likely in nonpolar solvents.

$$ \equiv \left[H_3C \overset{+}{-\!\!-\!\!-} \underset{CH_3}{\overset{\cdots\!\!\backslash CH_3}{}} \right] $$

1

Among simple alkyl carbocations[13] the order of stability is tertiary > secondary > primary. There are many known examples of rearrangements of primary or secondary carbocations to tertiary, both in solution and in the gas phase. Since simple alkyl cations are not stable in ordinary strong-acid solutions (e.g., H_2SO_4), the study of these species was greatly facilitated by the discovery that many of them could be kept indefinitely in stable solutions in mixtures of fluorosulfuric acid and antimony pentafluoride. Such mixtures, usually dissolved in SO_2 or SO_2ClF, are among the strongest acidic solutions known and are often called *super acids*.[14] The original experiments involved the addition of alkyl fluorides to SbF_5.[15]

$$ RF \ + \ SbF_5 \ \longrightarrow \ R^+ \ SbF_6^- $$

Subsequently, it was found that the same cations could also be generated from alcohols in super acid-SO_2 at $-60°C$[16] and from alkenes by the addition of a proton from super acid or HF–SbF_5 in SO_2 or SO_2ClF at low temperatures.[17] Even alkanes give carbocations in super acid by loss of H^-. For example,[18]

[9]Kato, T.; Reed, C.A. *Angew. Chem. Int. Ed.* **2004**, *43*, 2908.

[10]Lüning, U.; Baumstark, R. *Tetrahedron Lett.* **1993**, *34*, 5059.

[11]McClelland, R.A.; Cozens, F.L.; Steenken, S.; Amyes, T.L.; Richard, J.P. *J. Chem. Soc. Perkin Trans. 2* **1993**, 1717.

[12]For a treatise, see Szwarc, M. *Ions and Ion Pairs in Organic Reactions*, 2 vols., Wiley, NY, *1972–1974*.

[13]For a review, see Olah, G.A.; Olah, J.A., in Olah, G.A.; Schleyer, P.v.R. *Carbonium Ions*, Vol. 2, WIley, NY, *1969*, pp. 715–782. Also see Fǎrcaşiu, D.; Norton, S.H. *J. Org. Chem.* **1997**, *62*, 5374.

[14]For a review of carbocations in super acid solutions, see Olah, G.A.; Prakash, G.K.S.; Sommer, J., in *Superacids*, Wiley, NY, *1985*, pp. 65–175.

[15]Olah, G.A.; Baker, E.B.; Evans, J.C.; Tolgyesi, W.S.; McIntyre, J.S.; Bastien, I.J. *J. Am. Chem. Soc.* *1964*, *86*, 1360; Brouwer, D.M.; Mackor, E.L. *Proc. Chem. Soc.* *1964*, 147; Kramer, G.M. *J. Am. Chem. Soc.* *1969*, *91*, 4819.

[16]Olah, G.A.; Sommer, J.; Namanworth, E. *J. Am. Chem. Soc.* *1967*, *89*, 3576.

[17]Olah, G.A.; Halpern, Y. *J. Org. Chem.* *1971*, *36*, 2354. See also, Herlem, M. *Pure Appl. Chem.* *1977*, *49*, 107.

[18]Olah, G.A.; Lukas, J. *J. Am. Chem. Soc.* *1967*, *89*, 4739.

isobutane gives the *tert*-butyl cation

$$Me_3CH \xrightarrow{FSO_3H-SbF_6} Me_3\overset{\oplus}{C} \ SbF_5F\overset{\ominus}{SO_3} \ + \ H_2$$

No matter how they are generated, study of the simple alkyl cations has provided dramatic evidence for the stability order.[19] Both propyl fluorides gave the isopropyl cation; all four butyl fluorides[20] gave the *tert*-butyl cation, and all seven of the pentyl fluorides tried gave the *tert*-pentyl cation. *n*-Butane, in super acid, gave only the *tert*-butyl cation. To date, no primary cation has survived long enough for detection. Neither methyl nor ethyl fluoride gave the corresponding cations when treated with SbF_5. At low temperatures, methyl fluoride gave chiefly the methylated sulfur dioxide salt $(CH_3OSO)^+ \ SbF_6^-$,[21] while ethyl fluoride rapidly formed the *tert*-butyl and *tert*-hexyl cations by addition of the initially formed ethyl cation to ethylene molecules also formed.[22] At room temperature, methyl fluoride also gave the *tert*-butyl cation.[23] In accord with the stability order, hydride ion is abstracted from alkanes by super acid most readily from tertiary and least readily from primary positions.

The stability order can be explained by the polar effect and by hyperconjugation. In the polar effect, nonconjugated substituents exert an influence on stability through bonds (inductive effect) or through space (field effect). Since a tertiary carbocation has more carbon substituents on the positively charged carbon, relative to a primary, there is a greater polar effect that leads to great stability. In the hyperconjugation explanation,[24] we compare a primary carbocation with a tertiary. It should be made clear that "*the hyperconjugation concept arises solely from our model-building procedures.* When we ask whether hyperconjugation is important in a given situation, we are asking only whether the localized model is adequate for that situation at the particular level of precision we wish to use, or whether the model must be corrected by including some delocalization in order to get a good enough description."[25] Using the hyperconjugation model, is seen that the

[19]See Amyes, T.L.; Stevens, I.W.; Richard, J.P. *J. Org. Chem.* **1993**, *58*, 6057 for a recent study.

[20]The *sec*-butyl cation has been prepared by slow addition of *sec*-butyl chloride to SbF_5-SO_2ClF solution at $-110°C$ [Saunders, M.; Hagen, E.L.; Rosenfeld, J. *J. Am. Chem. Soc.* **1968**, *90*, 6882] and by allowing molecular beams of the reagents to impinge on a very cold surface [Saunders, M.; Cox, D.; Lloyd, J.R. *J. Am. Chem. Soc.* **1979**, *101*, 6656; Myhre, P.C.; Yannoni, C.S. *J. Am. Chem. Soc.* **1981**, *103*, 230].

[21]Peterson, P.E.; Brockington, R.; Vidrine, D.W. *J. Am. Chem. Soc.* **1976**, *98*, 2660; Calves, J.; Gillespie, R.J. *J. Chem. Soc. Chem. Commun.* **1976**, 506; Olah, G.A.; Donovan, D.J. *J. Am. Chem. Soc.* **1978**, *100*, 5163.

[22]Olah, G.A.; Olah, J.A., in Olah, G.A.; Schleyer, P.v.R. *Carbonium Ions*, Vol. 2, Wiley, NY, **1969**, p. 722.

[23]Olah, G.A.; DeMember, J.R.; Schlosberg, R.H. *J. Am. Chem. Soc.* **1969**, *91*, 2112; Bacon, J.; Gillespie, R.J. *J. Am. Chem. Soc.* **1971**, *91*, 6914.

[24]For a review of molecular-orbital theory as applied to carbocations, see Radom, L.; Poppinger, D.; Haddon, R.C., in Olah, G.A.; Schleyer, P.v.R. *Carbonium Ions*, Vol. 5, Wiley, NY, **1976**, pp. 2303–2426.

[25]Lowry, T.H.; Richardson, K.S. *Mechanism and Theory in Organic Chemistry, 3rd ed.*, HarperCollins, NY, **1987**, p. 68.

primary ion has only two hyperconjugative forms while the tertiary has six:

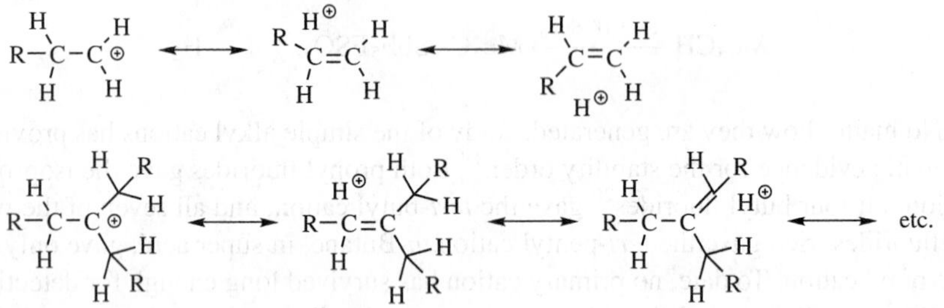

According to rule 6 for resonance contributors (p. 47), the greater the number of equivalent forms, the greater the resonance stability. Evidence used to support the hyperconjugation explanation is that the equilibrium constant for this reaction:

$$(CD_3)_3C^{\oplus} + (CH_3)_3CH \rightleftharpoons (CH_3)_3C^{\oplus} + (CD_3)_3CH \qquad K_{298} = 1.97 \pm 0.20$$

$$\mathbf{2} \qquad\qquad\qquad\qquad \mathbf{3}$$

is 1.97, showing that **3** is more stable than **2**.[26] Due to a β secondary isotope effect, there is less hyperconjugation in **2** than in **3** (see p. 324 for isotope effects).[27]

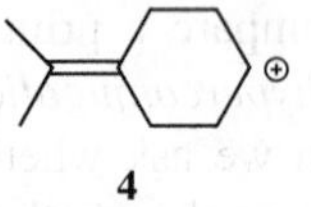

4

There are several structural types of delocalization, summarized in Table 5.1.[28] The stabilization of dimethylalkylidine cation **4** is an example of double hyperconjugation.[28,29]

The field effect explanation is that the electron-donating effect of alkyl groups increases the electron density at the charge-bearing carbon, reducing the net charge on the carbon, and in effect spreading the charge over the α carbons. It is a general rule that the more concentrated any charge is, the less stable the species bearing it will be.

The most stable of the simple alkyl cations is the *tert*-butyl cation. Even the relatively stable *tert*-pentyl and *tert*-hexyl cations fragment at higher temperatures to

---

[26]Meot-Ner, M. *J. Am. Chem. Soc.* **1987**, *109*, 7947.

[27]If only the field effect were operating, **2** would be more stable than **3**, since deuterium is electron-donating with respect to hydrogen (p. 23), assuming that the field effect of deuterium could be felt two bonds away.

[28]Lambert, J.B.; Ciro, S.M. *J. Org. Chem.* **1996**, *61*, 1940.

[29]Alabugin, I.V.; Manoharan, M. *J. Org. Chem.* **2004**, *69*, 9011.

TABLE 5.1. Structural Types of Delocalization[25]

Valence Structures	Abbreviation	Name
	$\pi\pi$	Simple conjugation
	$\sigma\pi$	Hyperconjugation
	$\pi\sigma$	Homoconjugation
	$\sigma\sigma$	Homohyperconjugation
	$\sigma\pi/\pi\pi$	Hyperconjugation/ conjugation
	$\sigma\pi/\sigma\pi$	Double hyperconjugation

produce the *tert*-butyl cation, as do all other alkyl cations with four or more carbons so far studied.[30] Methane,[31] ethane, and propane, treated with super acid, also yield *tert*-butyl cations as the main product (see reaction **12-20**). Even paraffin wax and polyethylene give *tert*-butyl cation. Solid salts of *tert*-butyl and *tert*-pentyl cations (e.g., Me_3C^+ SbF_6^-) have been prepared from super acid solutions and are stable below $-20°C$.[32]

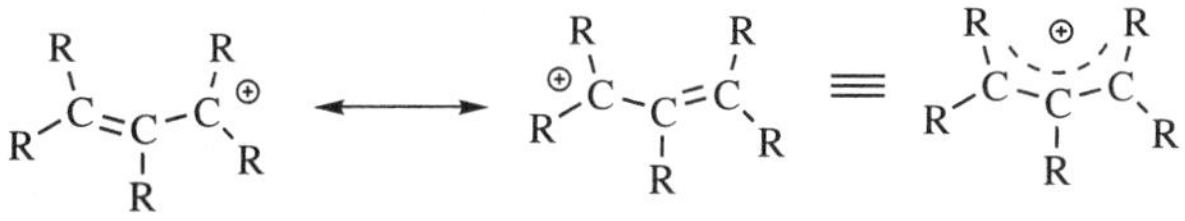

5

In carbocations where the positive carbon is in conjugation with a double bond, as in allylic cations (the allyl cation is **5**, R = H), the stability is greater because of increased delocalization due to resonance,[33] where the positive charge is spread over several atoms instead of being concentrated on one (see the molecular-orbital picture of this species on p. 41). Each of the terminal atoms has a charge of $\sim\frac{1}{2}$ (the charge is exactly $\frac{1}{2}$ if all of the R groups are the same). Stable cyclic and

---

[30]Olah, G.A.; Lukas, J. *J. Am. Chem. Soc.* **1967**, *89*, 4739; Olah, G.A.; Olah, J.A., in Olah, G.A.; Schleyer, P.v.R. *Carbonium Ions*, Vol. 2, Wiley, NY, **1969**, pp. 750–764.
[31]Olah, G.A.; Klopman, G.; Schlosberg, R.H. *J. Am. Chem. Soc.* **1969**, *91*, 3261. See also, Hogeveen, H.; Gaasbeek, C.J. *Recl. Trav. Chim. Pays-Bas* **1968**, *87*, 319.
[32]Olah, G.A.; Svoboda, J.J.; Ku, A.T. *Synthesis* **1973**, 492; Olah, G.A.; Lukas, J. *J. Am. Chem. Soc.* **1967**, *89*, 4739.
[33]See Barbour, J.B.; Karty, J.M. *J. Org. Chem.* **2004**, *69*, 648; Mo, Y. *J. Org. Chem.* **2004**, *69*, 5563 and references cited therein.

acyclic allylic-type cations[34] have been prepared by the solution of conjugated dienes in concentrated sulfuric acid, for example,[35]

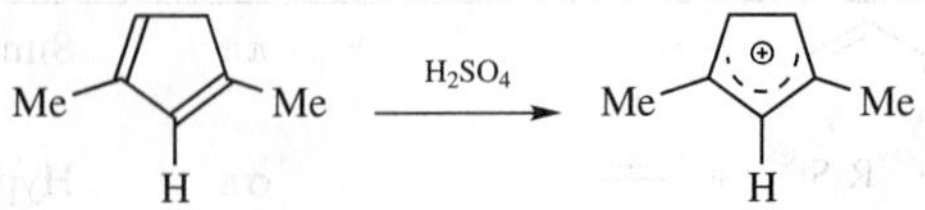

Stable allylic cations have also been obtained by the reaction between alkyl halides, alcohols, or alkenes (by hydride extraction) and SbF_5 in SO_2 or SO_2ClF.[36] Bis(allylic) cations[37] are more stable than the simple allylic type, and some of these have been prepared in concentrated sulfuric acid.[38] Arenium ions (p. 658) are familiar examples of this type. Propargyl cations ($RC\equiv CCR_2^+$) have also been prepared.[39]

Canonical forms can be drawn for benzylic cations,[40] similar to those shown above for allylic cations, for example,

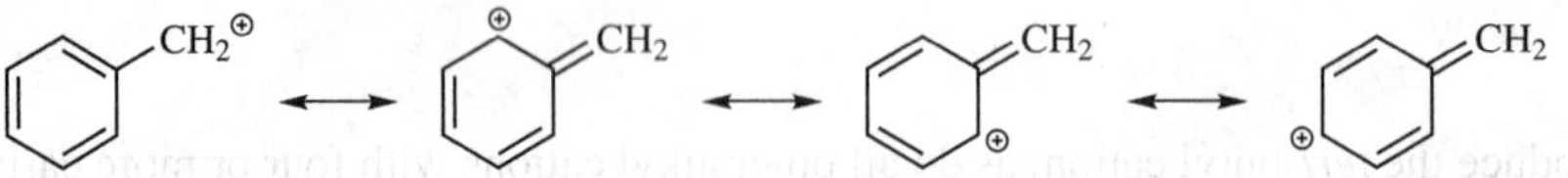

A number of benzylic cations have been obtained in solution as SbF_6^- salts.[41] Diarylmethyl and triarylmethyl cations are still more stable. Triphenylchloromethane ionizes in polar solvents that do not, like water, react with the ion. In SO_2, the equilibrium

$$Ph_3CCl \rightleftharpoons Ph_3C^{\oplus} + Cl^{\ominus}$$

has been known for many years. Both triphenylmethyl and diphenylmethyl cations have been isolated as solid salts[42] and, in fact, $Ph_3C^+ BF_4^-$ and related salts are available commercially. Arylmethyl cations are further stabilized if they have

[34]For reviews, see Deno, N.C., in Olah, G.A.; Schleyer, P.v.R. *Carbonium Ions*, Vol. 2, Wiley, NY, *1970*, pp. 783–806; Richey Jr., H.G., in Zabicky, J. *The Chemistry of Alkenes*, Vol. 2, Wiley, NY, *1970*, pp. 39–114.

[35]Deno, N.C.; Richey, Jr., H.G.; Friedman, N.; Hodge, J.D.; Houser, J.J.; Pittman, Jr., C.U. *J. Am. Chem. Soc. 1963*, *85*, 2991.

[36]Olah, G.A.; Spear, R.J. *J. Am. Chem. Soc. 1975*, *97*, 1539 and references cited therein.

[37]For a review of divinylmethyl and trivinylmethyl cations, see Sorensen, T.S., in Olah, G.A.; Schleyer, P.v.R. *Carbonium Ions*, Vol. 2, Wiley, NY, *1970*, pp. 807–835.

[38]Deno, N.C.; Pittman, Jr., C.U. *J. Am. Chem. Soc. 1964*, *86*, 1871.

[39]Pittman, Jr., C.U.; Olah, G.A. *J. Am. Chem. Soc. 1965*, *87*, 5632; Olah, G.A.; Spear, R.J.; Westerman, P.W.; Denis, J. *J. Am. Chem. Soc. 1974*, *96*, 5855.

[40]For a review of benzylic, diarylmethyl, and triarymethyl cations, see Freedman, H.H., in Olah, G.A.; Schleyer, P.v.R. *Carbonium Ions*, Vol. 4, Wiley, NY, *1971*, pp. 1501–1578.

[41]Olah, G.A.; Porter, R.D.; Jeuell, C.L.; White, A.M. *J. Am. Chem. Soc. 1972*, *94*, 2044.

[42]Volz, H.; Schnell, H.W. *Angew. Chem. Int. Ed. 1965*, *4*, 873.

electron-donating substituents in ortho or para positions.[43] Dications[44] and trica-
tions are also possible, including the particularly stable dication (**6**), where each
positively charged benzylic carbon is stabilized by two azulene rings.[45] A related
trication is known where each benzylic cationic center is also stabilized by two
azulene rings.[46]

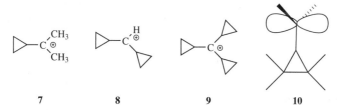

6

Cyclopropylmethyl cations[47] are even more stable than the benzyl type. Ion **9**
has been prepared by solution of the corresponding alcohol in 96% sulfuric acid,[48]
and **7**, **8**, and similar ions by solution of the alcohols in $FSO_3H-SO_2-SbF_5$.[49]
This special stability, which increases with each additional cyclopropyl group, is a

7	**8**	**9**	**10**

result of conjugation between the bent orbitals of the cyclopropyl rings (p. $$$)
and the vacant *p* orbital of the cationic carbon (see **10**). Nuclear magnetic resonance
and other studies have shown that the vacant *p* orbital lies parallel to the C-2,C-3
bond of the cyclopropane ring and not perpendicular to it.[50] In this respect, the

[43]Goldacre, R.J.; Phillips, J.N. *J. Chem. Soc.* **1949**, 1724; Deno, N.C.; Schriesheim, A. *J. Am. Chem. Soc.*
1955, *77*, 3051.

[44]Prakash, G.K.S. *Pure Appl. Chem.* **1998**, *70*, 2001.

[45]Ito, S.; Morita, N.; Asao, T. *Tetrahedron Lett.* **1992**, *33*, 3773.

[46]Ito, S.; Morita, N.; Asao, T. *Tetrahedron Lett.* **1994**, *35*, 751.

[47]For reviews, see, in Olah, G.A.; Schleyer, P.v.R. *Carbonium Ions*, Vol. 3, Wiley, NY, **1972**: Richey, Jr.,
H.G. pp. 1201–294; Wiberg, K.B.; Hess Jr., B.A.; Ashe III, A.H. pp. 1295–1345.

[48]Deno, N.C.; Richey, Jr., H.G.; Liu, J.S.; Hodge, J.D.; Houser, H.J.; Wisotsky, M.J. *J. Am. Chem. Soc.*
1962, *84*, 2016.

[49]Pittman Jr., C.U.; Olah, G.A. *J. Am. Chem. Soc.* **1965**, *87*, 2998; Deno, N.C.; Liu, J.S.; Turner, J.O.;
Lincoln, D.N.; Fruit, Jr., R.E. *J. Am. Chem. Soc.* **1965**, *87*, 3000.

[50]For example, see Ree, B.; Martin, J.C. *J. Am. Chem. Soc.* **1970**, *92*, 1660; Kabakoff, D.S.; Namanworth,
E. *J. Am. Chem. Soc.* **1970**, *92*, 3234; Buss, V.; Gleiter, R.; Schleyer, P.v.R. *J. Am. Chem. Soc.* **1971**, *93*,
3927; Poulter, C.D.; Spillner, C.J. *J. Am. Chem. Soc.* **1974**, *96*, 7591; Childs, R.F.; Kostyk, M.D.; Lock,
C.J.L.; Mahendran, M. *J. Am. Chem. Soc.* **1990**, *112*, 8912; Deno, N.C.; Richey Jr., H.G.; Friedman, N.;
Hodge, J.D.; Houser, J.J.; Pittman Jr., C.U. *J. Am. Chem. Soc.* **1963**, *85*, 2991.

geometry is similar to that of a cyclopropane ring conjugated with a double bond (p. 218). Cyclopropylmethyl cations are further discussed on pp. 459–463. The stabilizing effect just discussed is unique to cyclopropyl groups. Cyclobutyl and larger cyclic groups are about as effective at stabilizing a carbocation as ordinary alkyl groups.[51]

Another structural feature that increases carbocation stability is the presence, adjacent to the cationic center, of a heteroatom bearing an unshared pair,[52] for example, oxygen,[53] nitrogen,[54] or halogen.[55] Such ions are stabilized by resonance:

The methoxymethyl cation can be obtained as a stable solid, $MeOCH_2^+$ SbF_6^-.[56] Carbocations containing either α, β, or γ silicon atom are also stabilized,[57] relative to similar ions without the silicon atom. In super acid solution, ions such as CX_3^+ (X = Cl, Br, I) have been prepared.[58] Vinyl-stabilized halonium ions are also known.[59]

Simple acyl cations RCO^+ have been prepared[60] in solution and the solid state.[61] The acetyl cation CH_3CO^+ is about as stable as the *tert*-butyl cation (see, e.g., Table 5.1). The 2,4,6-trimethylbenzoyl and 2,3,4,5,6-pentamethylbenzoyl cations are especially stable (for steric reasons) and are easily formed in 96% H_2SO_4.[62] These

[51]Sorensen, T.S.; Miller, I.J.; Ranganayakulu, K. *Aust. J. Chem.* **1973**, *26*, 311.

[52]For a review, see Hevesi, L. *Bull. Soc. Chim. Fr.* **1990**, 697. For examples of stable solutions of such ions, see Kabus, S.S. *Angew. Chem. Int. Ed.* **1966**, *5*, 675; Dimroth, K.; Heinrich, P. *Angew. Chem. Int. Ed.* **1966**, *5*, 676; Tomalia, D.A.; Hart, H. *Tetrahedron Lett.* **1966**, 3389; Ramsey, B.; Taft, R.W. *J. Am. Chem. Soc.* **1966**, *88*, 3058; Olah, G.A.; Liang, G.; Mo, Y.M. *J. Org. Chem.* **1974**, *39*, 2394; Borch, R.F. *J. Am. Chem. Soc.* **1968**, *90*, 5303; Rabinovitz, M.; Bruck, D. *Tetrahedron Lett.* **1971**, 245.

[53]For a review of ions of the form R_2C^+—OR', see Rakhmankulov, D.L.; Akhmatdinov, R.T.; Kantor, E.A. *Russ. Chem. Rev.* **1984**, *53*, 888. For a review of ions of the form $R'C^+(OR)_2$ and $C^+(OR)_3$, see Pindur, U.; Müller, J.; Flo, C.; Witzel, H. *Chem. Soc. Rev.* **1987**, *16*, 75.

[54]For a review of such ions where nitrogen is the heteroatom, see Scott, F.L.; Butler, R.N., in Olah, G.A.; Schleyer, P.v.R. *Carbonium Ions*, Vol. 4, Wiley, NY, **1974**, pp. 1643–1696.

[55]See Allen, A.D.; Tidwell, T.T. *Adv. Carbocation Chem.* **1989**, *1*, 1. See also, Teberekidis, V.I.; Sigalas, M.P. *Tetrahedron* **2003**, *59*, 4749.

[56]Olah, G.A.; Svoboda, J.J. *Synthesis* **1973**, 52.

[57]For a review and discussion of the causes, see Lambert, J.B. *Tetrahedron* **1990**, *46*, 2677. See also, Lambert, J.B.; Chelius, E.C. *J. Am. Chem. Soc.* **1990**, *112*, 8120.

[58]Olah, G.A.; Heiliger, L.; Prakash, G.K.S. *J. Am. Chem. Soc.* **1989**, *111*, 8020.

[59]Haubenstock, H.; Sauers, R.R. *Tetrahedron* **2004**, *60*, 1191.

[60]For reviews of acyl cations, see Al-Talib, M.; Tashtoush, H. *Org. Prep. Proced. Int.* **1990**, *22*, 1; Olah, G.A.; Germain, A.; White, A.M., in Olah, G.A.; Schleyer, P.v.R. *Carbonium Ions*, Vol. 5, Wiley, NY, **1976**, pp. 2049–2133. For a review of the preparation of acyl cations from acyl halides and Lewis acids, see Lindner, E. *Angew. Chem. Int. Ed.* **1970**, *9*, 114.

[61]See, for example, Deno, N.C.; Pittman, Jr., C.U.; Wisotsky, M.J. *J. Am. Chem. Soc.* **1964**, *86*, 4370; Olah, G.A.; Dunne, K.; Mo, Y.K.; Szilagyi, P. *J. Am. Chem. Soc.* **1972**, *94*, 4200; Olah, G.A.; Svoboda, J.J. *Synthesis* **1972**, 306.

[62]Hammett, L.P.; Deyrup, A.J. *J. Am. Chem. Soc.* **1933**, *55*, 1900; Newman, M.S.; Deno, N.C. *J. Am. Chem. Soc.* **1951**, *73*, 3651.

ions are stabilized by a canonical form containing a triple bond (**12**), although the positive charge is principally located on the carbon,[63] so that **11** contributes more than **12**.

$$R\overset{\oplus}{-}C{=}O \quad\longleftrightarrow\quad R{-}C{\equiv}O^{\oplus}$$

$$\textbf{11} \qquad\qquad \textbf{12}$$

The stabilities of most other stable carbocations can also be attributed to resonance. Among these are the tropylium, cyclopropenium,[64] and other aromatic cations discussed in Chapter 2. Where resonance stability is completely lacking, as in the phenyl ($C_6H_5^+$) or vinyl cations,[65] the ion, if formed at all, is usually very short lived.[66] Neither vinyl[67] nor phenyl cation has as yet been prepared as a stable species in solution.[68] However, stable alkenyl carbocations have been generated on Zeolite Y.[69]

Various quantitative methods have been developed to express the relative stabilities of carbocations.[70] One of the most common of these, although useful only for relatively stable cations that are formed by ionization of alcohols in acidic solutions, is based on the equation[71]

$$H_R = pK_{R^+} - \log\frac{C_{R^+}}{C_{ROH}}$$

---

[63]Boer, F.P. *J. Am. Chem. Soc.* **1968**, *90*, 6706; Le Carpentier, J.; Weiss, R. *Acta Crystallogr. Sect. B*, **1972**, 1430. See also, Olah, G.A.; Westerman, P.W. *J. Am. Chem. Soc.* **1973**, *95*, 3706.

[64]See Komatsu, K.; Kitagawa, T. *Chem. Rev.* **2003**, *103*, 1371. Also see, Gilbertson, R.D.; Weakley, T.J.R.; Haley, M.M. *J. Org. Chem.* **2000**, *65*, 1422.

[65]For the preparation and reactivity of a primary vinyl carbocation see Gronheid, R.; Lodder, G.; Okuyama, T. *J. Org. Chem.* **2002**, *67*, 693.

[66]For a review of destabilized carbocations, see Tidwell, T.T. *Angew. Chem. Int. Ed.* **1984**, *23*, 20.

[67]Solutions of aryl-substituted vinyl cations have been reported to be stable for at least a short time at low temperatures. The NMR spectra was obtained: Abram, T.S.; Watts, W.E. *J. Chem. Soc. Chem. Commun.* **1974**, 857; Siehl, H.; Carnahan, Jr., J.C.; Eckes, L.; Hanack, M. *Angew. Chem. Int. Ed.* **1974**, *13*, 675. The l-cyclobutenyl cation has been reported to be stable in the gas phase: Franke, W.; Schwarz, H.; Stahl, D. *J. Org. Chem.* **1980**, *45*, 3493. See also, Siehl, H.; Koch, E. *J. Org. Chem.* **1984**, *49*, 575.

[68]For a monograph, see Stang, P.J.; Rappoport, Z.; Hanack, M.; Subramanian, L.R. *Vinyl Cations*, Academic Press, NY, 1979. For reviews of aryl and/or vinyl cations, see Hanack, M. *Pure Appl. Chem.* **1984**, *56*, 1819, *Angew. Chem. Int. Ed.* **1978**, *17*, 333; *Acc. Chem. Res.* **1976**, *9*, 364; Rappoport, Z. *Reactiv. Intermed. (Plenum)* **1983**, *3*, 427; Ambroz, H.B.; Kemp, T.J. *Chem. Soc. Rev.* **1979**, *8*, 353; Richey Jr., H.G.; Richey, J.M., in Olah, G.A.; Schleyer, P.v.R. *Carbonium Ions*, Vol. 2, Wiley, NY, **1970**, pp. 899–957; Richey Jr., H.G., in Zabicky, J. *The Chemistry of Alkenes*, Vol. 2, Wiley, NY, **1970**, pp. 42–49; Modena, G.; Tonellato, U. *Adv. Phys. Org. Chem.* **1971**, *9*, 185; Stang, P.J. *Prog. Phys. Org. Chem.* **1973**, *10*, 205. See also, Charton, M. *Mol. Struct. Energ.* **1987**, *4*, 271. For a computational study, see Glaser, R.; Horan, C. J.; Lewis, M.; Zollinger, H. *J. Org. Chem.* **1999**, *64*, 902.

[69]Yang, S.; Kondo, J.N.; Domen, K. *Chem. Commun.* **2001**, 2008.

[70]For reviews, see Bagno, A.; Scorrano, G.; More O'Ferrall, R.A. *Rev. Chem. Intermed.* **1987**, *7*, 313; Bethell, D.; Gold, V. *Carbonium Ions*, Academic Press, NY, **1967**, pp. 59–87.

[71]Deno, N.C.; Berkheimer, H.E.; Evans, W.L.; Peterson, H.J. *J. Am. Chem. Soc.* **1959**, *81*, 2344.

pK_{R^+} is the pK value for the reaction $R^+ + 2\ H_2O \rightleftharpoons ROH + H_3O^+$ and is a measure of the stability of the carbocation. The H_R parameter is an early obtainable measurement of the stability of a solvent (see p. 371) and approaches pH at low concentrations of acid. In order to obtain pK_{R^+}, for a cation R^+, one dissolves the alcohol ROH in an acidic solution of known H_R. Then the concentration of R^+ and ROH are obtained, generally from spectra, and pK_{R^+} is easily calculated.[72] A measure of carbocation stability that applies to less-stable ions is the dissociation energy $D(R^+-H^-)$ for the cleavage reaction $R - H \rightarrow R^+ + H^-$, which can be obtained from photoelectron spectroscopy and other measurements. Some values of $D(R^+-H^-)$ are shown in Table 5.2.[75] Within a given class of ion (primary, secondary, allylic, aryl, etc.), $D(R^+-H^-)$ has been shown to be a linear function of the logarithm of the number of atoms in R^+, with larger ions being more stable.[74]

        13   **14**

TABLE 5.2. R–H $\rightarrow$ R$^+$ + H$^-$ Dissociation Energies in the Gas Phase

Ion	$D(R^+-H^-)$		
	kcal mol^{-1}	kJ mol^{-1}	Reference
CH_3^+	314.6	1316	73
$C_2H_5^+$	276.7	1158	73
$(CH_3)_2CH^+$	249.2	1043	73
$(CH_3)_3C^+$	231.9	970.3	73
$C_6H_5^+$	294	1230	74
$H_2C=CH^+$	287	1200	74
$H_2C=CH-CH_2^+$	256	1070	74
Cyclopentyl	246	1030	74
$C_6H_5CH_2^+$	238	996	74
CH_3CHO	230	962	74

[72]For a list of stabilities of 39 typical carbocations, see Arnett, E.M.; Hofelich, T.C. *J. Am. Chem. Soc.* **1983**, *105*, 2889. See also, Schade, C.; Mayr, H.; Arnett, E.M. *J. Am. Chem. Soc.* **1988**, *110*, 567; Schade, C.; Mayr, H. *Tetrahedron* **1988**, *44*, 5761.

[73]Schultz, J.C.; Houle, F.A.; Beauchamp, J.L. *J. Am. Chem. Soc.* **1984**, *106*, 3917.

[74]Lossing, F.P.; Holmes, J.L. *J. Am. Chem. Soc.* **1984**, *106*, 6917.

[75]Hammett, L.P.; Deyrup, A.J. *J. Am. Chem. Soc.* **1933**, *55*, 1900; Newman, M.S.; Deno, N.C. *J. Am. Chem. Soc.* **1951**, *73*, 3651; Boer, F.P. *J. Am. Chem. Soc.* **1968**, *90*, 6706; Le Carpentier, J.; Weiss, R. *Acta Crystallogr. Sect. B*, **1972**, 1430. See also, Olah, G.A.; Westerman, P.W. *J. Am. Chem. Soc.* **1973**, *95*, 3706. See also, Staley, R.H.; Wieting, R.D.; Beauchamp, J.L. *J. Am. Chem. Soc.* **1977**, *99*, 5964; Arnett, E.M.; Petro, C. *J. Am. Chem. Soc.* **1978**, *100*, 5408; Arnett, E.M.; Pienta, N.J. *J. Am. Chem. Soc.* **1980**, *102*, 3329.

Since the central carbon of tricoordinated carbocations has only three bonds and no other valence electrons, the bonds are sp^2 and should be planar.[76] Raman, IR, and NMR spectroscopic data on simple alkyl cations show this to be so.[77] In methylcycohexyl cations, there are two chair conformations where the carbon bearing the positive charge is planar (**13** and **14**), and there is evidence that **14** is more stable due to a difference in hyperconjugation.[78] Other evidence is that carbocations are difficult to form at bridgehead atoms in [2.2.1] systems,[79] where they cannot be planar (see p. 435).[80] Bridgehead carbocations are known, however, as in [2.1.1]hexanes[81] and cubyl carbocations.[82] However, larger bridgehead ions can exist. For example, the adamantyl cation (**15**) has been synthesized, as the SF_6^- salt.[83] The relative stability of 1-adamantyl cations is influenced by the number and nature of substituents. For example, the stability of the 1-adamantyl cation increases with the number of isopropyl substituents at C-3, C-5 and C-7.[84] Among other bridgehead cations that have been prepared in super acid solution at $-78°C$ are the dodecahydryl cation (**16**)[85] and the 1-trishomobarrelyl cation (**17**).[86] In the latter

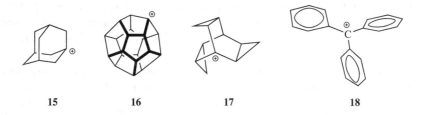

| **15** | **16** | **17** | **18** |

[76]For discussions of the stereochemistry of carbocations, see Henderson, J.W. *Chem. Soc. Rev.* **1973**, *2*, 397; Buss, V.; Schleyer, P.v.R.; Allen, L.C. *Top. Stereochem.* **1973**, *7*, 253; Schleyer, P.v.R., in Chiurdoglu, G. *Conformational Analysis*; Academic Press, NY, **1971**, p. 241; Hehre, W.J. *Acc. Chem. Res.* **1975**, *8*, 369; Freedman, H.H., in Olah, G.A.; Schleyer, P.v.R. *Carbonium Ions*, Vol. 4, Wiley, NY, **1974**, pp. 1561–574.

[77]Olah, G.A.; DeMember, J.R.; Commeyras, A.; Bribes, J.L. *J. Am. Chem. Soc.* **1971**, *93*, 459; Yannoni, C.S.; Kendrick, R.D.; Myhre, P.C.; Bebout, D.C.; Petersen, B.L. *J. Am. Chem. Soc.* **1989**, *111*, 6440.

[78]Rauk, A.; Sorensen, T.S.; Maerker, C.; de M. Carneiro, J.W.; Sieber, S.; Schleyer, P.v.R. *J. Am. Chem. Soc.* **1996**, *118*, 3761.

[79]For a review of bridgehead carbocations, see Fort, Jr., R.C., in Olah, G.A.; Schleyer, P.v.R *Carbonium Ions*, Vol. 4, Wiley, NY, **1974**, pp. 1783–1835.

[80]Della, E.W.; Schiesser, C.H. *J. Chem. Soc. Chem. Commun.* **1994**, 417.

[81]Åhman, J.; Somfai, P.; Tanner, D. *J. Chem. Soc. Chem. Commun.* **1994**, 2785.

[82]Della, E.W.; Head, N.J.; Janowski, W.K.; Schiesser, C.H. *J. Org. Chem.* **1993**, *58*, 7876.

[83]Schleyer, P.v.R.; Fort, Jr., R.C.; Watts, W.E.; Comisarow, M.B.; Olah, G.A. *J. Am. Chem. Soc.* **1964**, *86*, 4195; Olah, G.A.; Prakash, G.K.S.; Shih, J.G.; Krishnamurthy, V.V.; Mateescu, G.D.; Liang, G.; Sipos, G.; Buss, V.; Gund, T.M.; Schleyer, P.v.R. *J. Am. Chem. Soc.* **1985**, *107*, 2764. See also, Kruppa, G.H.; Beauchamp, J.L. *J. Am. Chem. Soc.* **1986**, *108*, 2162; Laube, T. *Angew. Chem. Int. Ed.* **1986**, *25*, 349.

[84]Takeuchi, K.; Okazaki, T.; Kitagawa, T.; Ushino, T.; Ueda, K.; Endo, T.; Notario, R. *J. Org. Chem.* **2001**, *66*, 2034.

[85]Olah, G.A.; Prakash, G.K.S.; Fessner, W.; Kobayashi, T.; Paquette, L.A. *J. Am. Chem. Soc.* **1988**, *110*, 8599.

[86]de Meijere, A.; Schallner, O. *Angew. Chem. Int. Ed.* **1973**, *12*, 399.

TABLE 5.3. The ^{13}C Chemical Shift Values, in Parts Per Million from $^{13}CS_2$ for the Charged Carbon Atom of Some Carbocations in $SO_2ClF–SbF_5$, $SO_2–FSO_3H–SbF_6$, or $SO_2–SbF_5$[90]

Ion	Chemical Shift	Temperature, °C	Ion	Chemical Shift	Temperature, °C
Et_2MeC^+	−139.4	−20	$C(OH)_3^+$	+28.0	−50
Me_2EtC^+	−139.2	−60	$PhMe_2C^+$	−61.1	−60
Me_3C^+	−135.4	−20	$PhMeCH^+$	−40[91]	
Me_2CH^+	−125.0	−20	Ph_2CH^+	−5.6	−60
Me_2COH^+	−55.7	−50	Ph_3C^+	−18.1	−60
$MeC(OH)_2^+$	−1.6	−30	$Me_2(cyclopropyl)C^+$	−86.8	−60
$HC(OH)_2^+$	+17.0	−30			

case, the instability of the bridgehead position is balanced by the extra stability gained from the conjugation with the three cyclopropyl groups.

Triarylmethyl cations (**18**)[87] are propeller shaped, although the central carbon and the three ring carbons connected to it are in a plane:[88] The three benzene rings cannot be all in the same plane because of steric hindrance, although increased resonance energy would be gained if they could.

An important tool for the investigation of carbocation structure is measurement of the ^{13}C NMR chemical shift of the carbon atom bearing the positive charge.[89] This shift approximately correlates with electron density on the carbon. The ^{13}C chemical shifts for a number of ions are given in Table 5.3.[90] As shown in this table, the substitution of an ethyl for a methyl or a methyl for a hydrogen causes a down-field shift, indicating that the central carbon becomes somewhat more positive. On the other hand, the presence of hydroxy or phenyl groups decreases the positive character of the central carbon. The ^{13}C chemical shifts are not always in exact order of carbocation stabilities as determined in other ways. Thus the chemical shift shows that the triphenylmethyl cation has a more positive central carbon than diphenylmethyl cation, although the former is more stable. Also, the 2-cyclopropyl-propyl and 2-phenylpropyl cations have shifts of −86.8 and −61.1, respectively, although we have seen that according to other criteria a cyclopropyl group is better

[87]For a review of crystal-structure determinations of triarylmethyl cations and other carbocations that can be isolated in stable solids, see Sundaralingam, M.; Chwang, A.K., in Olah, G.A.; Schleyer, P.v.R. *Carbonium Ions*, Vol. 5, Wiley, NY, *1976*, pp. 2427–2476.

[88]Sharp, D.W.A.; Sheppard, N. *J. Chem. Soc. 1957*, 674; Gomes de Mesquita, A.H.; MacGillavry, C.H.; Eriks, K. *Acta Crystallogr. 1965*, *18*, 437; Schuster, I.I.; Colter, A.K.; Kurland, R.J. *J. Am. Chem. Soc. 1968*, *90*, 4679.

[89]For reviews of the nmr spectra of carbocations, see Young, R.N. *Prog. Nucl. Magn. Reson. Spectrosc. 1979*, *12*, 261; Farnum, D.G. *Adv. Phys. Org. Chem. 1975*, *11*, 123.

[90]Olah, G.A.; White, A.M. *J. Am. Chem. Soc. 1968*, *90*, 1884; *1969*, *91*, 5801. For ^{13}C NMR data for additional ions, see Olah, G.A.; Donovan, D.J. *J. Am. Chem. Soc. 1977*, *99*, 5026; Olah, G.A.; Prakash, G.K.S.; Liang, G. *J. Org. Chem. 1977*, *42*, 2666.

than a phenyl group at stabilizing a carbocation.[91] The reasons for this discrepancy are not fully understood.[88,92]

Nonclassical Carbocations

These carbocations are discussed at pp. 450–455.

The Generation and Fate of Carbocations

A number of methods are available to generate carbocations, stable or unstable.

1. A direct ionization, in which a leaving group attached to a carbon atom leaves with its pair of electrons, as in solvolysis reactions of alkyl halides (see p. 480) or sulfonate esters (see p. 522):

$$R{-}X \longrightarrow R^{\oplus} + X^{\ominus} \quad \text{(may be reversible)}$$

2. Ionization after an initial reaction that converts one functional group into a leaving group, as in protonation of an alcohol to give an oxonium ion or conversion of a primary amine to a diazonium salt, both of which ionize to the corresponding carbocation:

$$R{-}OH \xrightarrow{\text{H}^+} R{-}OH_2 \longrightarrow R^{\oplus} + H_2O \quad \text{(may be reversible)}$$

$$R{-}NH_2 \xrightarrow{\text{HONO}} R{-}N_2 \longrightarrow R^{\oplus} + N_2$$

3. A proton or other positive species adds to one atom of an alkene or alkyne, leaving the adjacent carbon atom with a positive charge (see Chapters 11, 15).

4. A proton or other positive species adds to one atom of an C=X bond, where X = O, S, N in most cases, leaving the adjacent carbon atom with a positive charge (see Chapter 16). When X = O, S this ion is resonance stabilized, as shown. When X = NR, protonation leads to an iminium ion, with the charge localized on the

[91]Olah, G.A.; Porter, R.D.; Kelly, D.P. *J. Am. Chem. Soc.* **1971**, *93*, 464.

[92]For discussions, see Brown, H.C.; Peters, E.N. *J. Am. Chem. Soc.* **1973**, *95*, 2400; **1977**, *99*, 1712; Olah, G.A.; Westerman, P.W.; Nishimura, J. *J. Am. Chem. Soc.* **1974**, *96*, 3548; Wolf, J.F.; Harch, P.G.; Taft, R.W.; Hehre, W.J. *J. Am. Chem. Soc.* **1975**, *97*, 2902; Fliszár, S. *Can. J. Chem.* **1976**, *54*, 2839; Kitching, W.; Adcock, W.; Aldous, G. *J. Org. Chem.* **1979**, *44*, 2652. See also, Larsen, J.W.; Bouis, P.A. *J. Am. Chem. Soc.* **1975**, *97*, 4418; Volz, H.; Shin, J.; Streicher, H. *Tetrahedron Lett.* **1975**, 1297; Larsen, J.W. *J. Am. Chem. Soc.* **1978**, *100*, 330.

nitrogen. A silylated carboxonium ion, such as **19**, has been reported.[93]

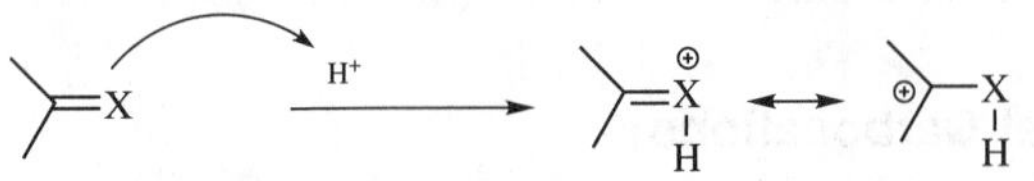

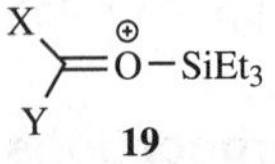

Formed by either process, carbocations are most often short-lived transient species and react further without being isolated. The intrinsic barriers to formation and reaction of carbocations has been studied.[94] Carbocations have been generated in zeolites.[95]

The two chief pathways by which carbocations react to give stable products are the reverse of the two pathways just described.

1. *The Carbocation May Combine with a Species Possessing an Electron Pair* (a Lewis acid–base reaction, see Chapter 8):

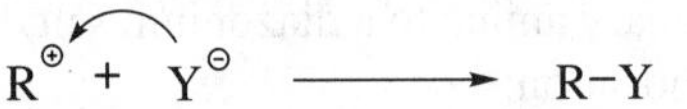

$$R^{\oplus} + Y^{\ominus} \longrightarrow R-Y$$

This species may be ^{-}OH, halide ion, or any other negative ion, or it may be a neutral species with a pair to donate, in which case, of course, the immediate product must bear a positive charge (see Chapters 10, 13, 15, 16). These reactions are very fast. A recent study measured k_s (the rate constant for reaction of a simple tertiary carbocation) to be 3.5×10^{12} s^{-1}.[96]

2. *The Carbocation May Lose a Proton* (or much less often, another positive ion) from the adjacent atom (see Chapters 11, 17):

Carbocations can also adopt two other pathways that lead not to stable products, but to other carbocations:

3. *Rearrangement.* An alkyl or aryl group or a hydrogen (sometimes another group) migrates with its electron pair to the positive center, leaving another positive charge behind (see Chapter 18):

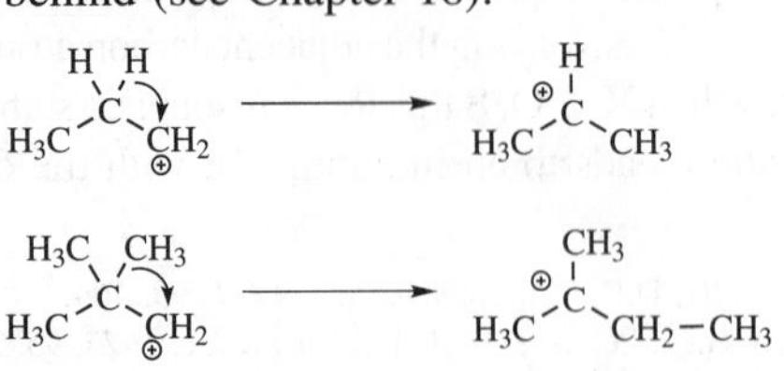

[93]Prakash, G.K.S.; Bae, C.; Rasul, G.; Olah, G.A. *J. Org. Chem.* **2002**, *67*, 1297.
[94]Richard, J.P.; Amyes, T.L.; Williams, K.B. *Pure. Appl. Chem.* **1998**, *70*, 2007.
[95]Song, W.; Nicholas, J. B.; Haw, J. F. *J. Am. Chem. Soc.* **2001**, *123*, 121.
[96]Toteva, M.M.; Richard, J.P. *J. Am. Chem. Soc.* **1996**, *118*, 11434.

A novel rearrangement has been observed. The 2-methyl-2-butyl-1-^{13}C cation (^{13}C-labeled *tert*-amyl cation) shows an interchange of the inside and outside carbons with a barrier of 19.5 ($\pm$2.0 kcal mol^{-1}).[97] Another unusual migratory process has been observed for the nonamethylcyclopentyl cation. It has been shown that "four methyl groups undergo rapid circumambulatory migration with a barrier $<$2 kcal mol^{-1} while five methyl groups are fixed to ring carbons, and the process that equalizes the two sets of methyls has a barrier of 7.0 kcal mol^{-1}."[98]

4. *Addition.* A carbocation may add to a double bond, generating a positive charge at a new position (see Chapters 11, 15):

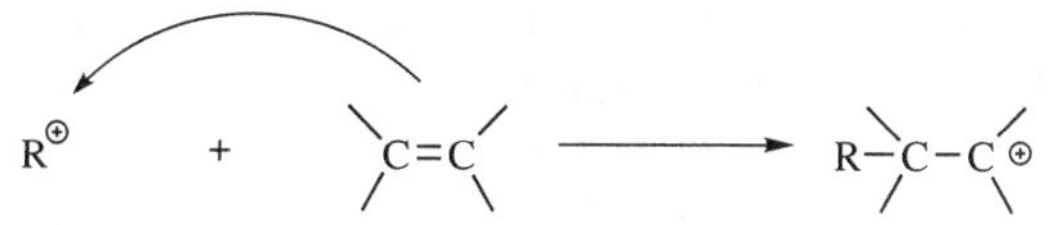

Whether formed by pathway 3 or 4, the new carbocation normally reacts further in an effort to stabilize itself, usually by pathway 1 or 2. However, **20** can add to another alkene molecule, and this product can add to still another, and so on. This is one of the mechanisms for vinyl polymerization.

CARBANIONS

Stability and Structure[99]

An *organometallic compound* is a compound that contains a bond between a carbon atom and a metal atom. Many such compounds are known, and organometallic chemistry is a very large area, occupying a borderline region between organic and inorganic chemistry. Many carbon–metal bonds (e.g., carbon–mercury bonds)

[97]Vrcek, V.; Saunders, M.; Kronja, O. *J. Am. Chem. Soc.* **2004**, *126*, 13703.

[98]Kronja, O.; Kohli, T.-P.; Mayr, H.; Saunders, M. *J. Am. Chem. Soc.* **2000**, *122*, 8067.

[99]For monographs, see Buncel, E.; Durst, T. *Comprehensive Carbanion Chemistry*, pts. A, B, and C; Elsevier, NY, **1980**, **1984**, **1987**; Bates, R.B.; Ogle, C.A. *Carbanion Chemistry*, Springer, NY, **1983**; Stowell, J.C. *Carbanions in Organic Synthesis*, Wiley, NY, **1979**; Cram, D.J. *Fundamentals of Carbanion Chemistry*, Academic Press, NY, **1965**. For reviews, see Staley, S.W. *React. Intermed. (Wiley)* **1985**, *3*, 19; Staley, S.W.; Dustman, C.K. *React. Intermed. (Wiley)* **1981**, *2*, 15; le Noble, W.J. *React. Intermed. (Wiley)* **1978**, *1*, 27; Solov'yanov, A.A.; Beletskaya, I.P. *Russ. Chem. Rev.* **1978**, *47*, 425; Isaacs, N.S. *Reactive Intermediates in Organic Chemistry*, Wiley, NY, **1974**, pp. 234–293; Kaiser, E.M.; Slocum, D.W., in McManus, S.P. *Organic Reactive Intermediates*, Academic Press, NY, **1973**, pp. 337–422; Ebel, H.F. *Fortchr. Chem. Forsch.* **1969**, *12*, 387; Cram, D.J. *Surv. Prog. Chem.* **1968**, *4*, 45; Reutov, O.A.; Beletskaya, I.P. *Reaction Mechanisms of Organometallic Compounds*, North Holland Publishing Co, Amsterdam, The Netherlands, **1968**, pp. 1–64; Streitwieser Jr., A.; Hammons, J.H. *Prog. Phys. Org. Chem.* **1965**, *3*, 41. For reviews of nmr spectra of carbanions, see Young, R.N. *Prog. Nucl. Magn. Reson. Spectrosc.* **1979**, *12*, 261. For a review of dicarbanions, see Thompson, C.M.; Green, D.L.C. *Tetrahedron* **1991**, *47*, 4223.

are undoubtedly covalent, but in bonds between carbon and the more active metals the electrons are closer to the carbon. Whether the position of the electrons in a given bond is close enough to the carbon to justify calling the bond ionic and the carbon moiety a carbanion depends on the metal, on the structure of the carbon moiety, and on the solvent and in some cases is a matter of speculation. In this section, we discuss carbanions with little reference to the metal. In the next section, we will deal with the structures of organometallic compounds.

By definition, every carbanion possesses an unshared pair of electrons and is therefore a base. When a carbanion accepts a proton, it is converted to its conjugate acid (see Chapter 8). The stability of the carbanion is directly related to the strength of the conjugate acid. The weaker the acid, the greater the base strength and the lower the stability of the carbanion.[100] By stability here we mean stability toward a proton donor; the lower the stability, the more willing the carbanion is to accept a proton from any available source, and hence to end its existence as a carbanion. Thus the determination of the order of stability of a series of carbanions is equivalent to a determination of the order of strengths of the conjugate acids, and one can obtain information about relative carbanion stability from a table of acid strengths like Table 8.1.

Unfortunately, it is not easy to measure acid strengths of very weak acids like the conjugate acids of simple unsubstituted carbanions. There is little doubt that these carbanions are very unstable in solution, and in contrast to the situation with carbocations, efforts to prepare solutions in which carbanions, such as ethyl or isopropyl, exist in a relatively free state have not yet been successful. Nor has it been possible to form these carbanions in the gas phase. Indeed, there is evidence that simple carbanions, such as ethyl and isopropyl, are unstable toward loss of an electron, which converts them to radicals.[101] Nevertheless, there have been several approaches to the problem. Applequist and O'Brien[102] studied the position of equilibrium for the reaction

$$RLi + R'I \rightleftharpoons RI + R'Li$$

in ether and ether–pentane. The reasoning in these experiments was that the R group that forms the more stable carbanion would be more likely to be bonded to lithium than to iodine. Carbanion stability was found to be in this order: vinyl > phenyl > cyclopropyl > ethyl > n-propyl > isobutyl > neopentyl > cyclobutyl > cyclopentyl. In a somewhat similar approach, Dessy and co-workers[103] treated a

---

[100]For a monograph on hydrocarbon acidity, see Reutov, O.A.; Beletskaya, I.P.; Butin, K.P. *CH-Acids*; Pergamon: Elmsford, NY, *1978*. For a review, see Fischer, H.; Rewicki, D. *Prog. Org. Chem. 1968*, *7*, 116.

[101]See Graul, S.T.; Squires, R.R. *J. Am. Chem. Soc. 1988*, *110*, 607; Schleyer, P.v.R.; Spitznagel, G.W.; Chandrasekhar, J. *Tetrahedron Lett. 1986*, *27*, 4411.

[102]Applequist, D.E.; O'Brien, D.F. *J. Am. Chem. Soc. 1963*, *85*, 743.

[103]Dessy, R.E.; Kitching, W.; Psarras, T.; Salinger, R.; Chen, A.; Chivers, T. *J. Am. Chem. Soc. 1966*, *88*, 460.

number of alkylmagnesium compounds with a number of alkylmercury compounds in tetrahydrofuran (THF), setting up the equilibrium

$$R_2Mg + R'_2Hg \rightleftharpoons R_2Hg + R'_2Mg$$

where the group of greater carbanion stability is linked to magnesium. The carbanion stability determined this way was in the order phenyl > vinyl > cyclopropyl > methyl > ethyl > isopropyl. The two stability orders are in fairly good agreement, and they show that stability of simple carbanions decreases in the order methyl > primary > secondary. It was not possible by the experiments of Dessy and co-workers to determine the position of *tert*-butyl, but there seems little doubt that it is still less stable. We can interpret this stability order solely as a consequence of the field effect since resonance is absent. The electron-donating alkyl groups of isopropyl result in a greater negative charge density at the central carbon atom (compared with methyl), thus decreasing its stability. The results of Applequist and O'Brien show that β branching also decreases carbanion stability. Cyclopropyl occupies an apparently anomalous position, but this is probably due to the large amount of *s* character in the carbanionic carbon (see p. 254).

A different approach to the problem of hydrocarbon acidity, and hence carbanion stability is that of Shatenshtein and co-workers, who treated hydrocarbons with deuterated potassium amide and measured the rates of hydrogen exchange.[104] The experiments did not measure *thermodynamic* acidity, since rates were measured, not positions of equilibria. They measured *kinetic* acidity, that is, which compounds gave up protons most rapidly (see p. 307 for the distinction between thermodynamic and kinetic control of product). Measurements of rates of hydrogen exchange enable one to compare acidities of a series of acids against a given base even where the positions of the equilibria cannot be measured because they lie too far to the side of the starting materials, that is, where the acids are too weak to be converted to their conjugate bases in measurable amounts. Although the correlation between thermodynamic and kinetic acidity is far from perfect,[105] the results of the rate measurements, too, indicated that the order of carbanion stability is methyl > primary > secondary > tertiary.[104]

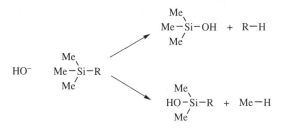

[104]For reviews, see Jones, J.R. *Surv. Prog. Chem.* **1973**, *6*, 83; Shatenshtein, A.I.; Shapiro, I.O. *Russ. Chem. Rev.* **1968**, *37*, 845.

[105]For example, see Bordwell, F.G.; Matthews, W.S.; Vanier, N.R. *J. Am. Chem. Soc.* **1975**, *97*, 442.

However, experiments in the gas phase gave different results. In reactions of $^-$OH with alkyltrimethylsilanes, it is possible for either R or Me to cleave. Since the R or Me comes off as a carbanion or incipient carbanion, the product ratio RH/MeH can be used to establish the relative stabilities of various R groups. From these experiments a stability order of neopentyl > cyclopropyl > *tert*-butyl > *n*-propyl > methyl > isopropyl > ethyl was found.[106] On the other hand, in a different kind of gas-phase experiment, Graul and Squires were able to observe CH_3^- ions, but not the ethyl, isopropyl, or *tert*-butyl ions.[107]

Many carbanions are far more stable than the simple kind mentioned above. The increased stability is due to certain structural features:

1. *Conjugation of the Unshared Pair with an Unsaturated Bond:*

In cases where a double or triple bond is located α to the carbanionic carbon, the ion is stabilized by resonance in which the unshared pair overlaps with the π electrons of the double bond. This factor is responsible for the stability of the allylic[108] and benzylic[109] types of carbanions:

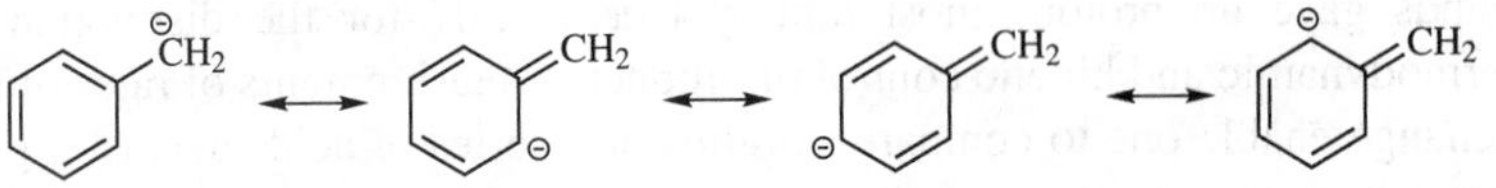

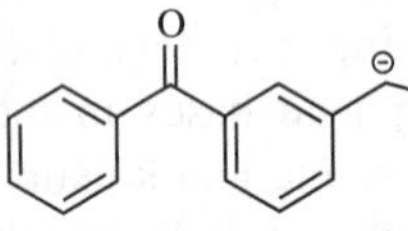

21

Diphenylmethyl and triphenylmethyl anions are still more stable and can be kept in solution indefinitely if water is rigidly excluded.[110]

---

[106]DePuy, C.H.; Gronert, S.; Barlow, S.E.; Bierbaum, V.M.; Damrauer, R. *J. Am. Chem. Soc.* **1989**, *111*, 1968. The same order (for *t*-Bu, Me, *i*Pr, and Et) was found in gas-phase cleavages of alkoxides (**12-41**): Tumas, W.; Foster, R.F.; Brauman, J.I. *J. Am. Chem. Soc.* **1984**, *106*, 4053.

[107]Graul, S.T.; Squires, R.R. *J. Am. Chem. Soc.* **1988**, *110*, 607.

[108]For a review of allylic anions, see Richey, Jr., H.G., in Zabicky, J. *The Chemistry of Alkenes*, Vol. 2, Wiley, NY, *1970*, pp. 67–77.

[109]Although benzylic carbanions are more stable than the simple alkyl type, they have not proved stable enough for isolation so far. The benzyl carbanion has been formed and studied in submicrosecond times; Bockrath, B.; Dorfman, L.M. *J. Am. Chem. Soc.* **1974**, *96*, 5708.

[110]For a review of spectrophotometric investigations of this type of carbanion, see Buncel, E.; Menon, B., in Buncel, E.; Durst, T. *Comprehensive Carbanion Chemistry*, pts. A, B, and C, Elsevier, NY, *1980*, *1984*, *1987*, pp. 97–124.

Condensed aromatic rings fused to a cyclopentadienyl anion are known to stabilize the carbanion.[111] X-ray crystallographic structures have been obtained for Ph_2CH^- and Ph_3C^- enclosed in crown ethers.[112] Carbanion **21** has a lifetime of several minutes (hours in a freezer at $-20\ °C$) in dry THF.[113]

Where the carbanionic carbon is conjugated with a carbon–oxygen or carbon–nitrogen multiple bond (Y = O or N), the stability of the ion is greater than that of the triarylmethyl anions, since these electronegative atoms are better capable of bearing a negative charge than carbon. However, it is questionable whether ions of this type should be called carbanions at all, since

in the case of enolate ions, for example, **23** contributes more to the hybrid than **22** although such ions react more often at the carbon than at the oxygen. In benzylic enolate anions such as **24**, the conformation of the enolate can be coplanar with the aromatic ring or bent out of plane if the strain is too great.[114] Enolate ions can also be kept in stable solutions. In the case of carbanions at a carbon α- to a nitrile, the "enolate" resonance form would be a ketene imine nitranion, but the existence of this species has been called into question.[115] A nitro group is particularly effective in stabilizing a negative charge on an adjacent carbon, and the anions of simple nitro alkanes can exist in water. Thus pK_a for nitromethane is 10.2. Dinitromethane is even more acidic ($pK_a = 3.6$).

In contrast to the stability of cyclopropylmethyl cations (p. 241), the cyclopropyl group exerts only a weak stabilizing effect on an adjacent carbanionic carbon.[116]

By combining a very stable carbanion with a very stable carbocation, Okamoto and co-workers[117] were able to isolate the salt **25**, as well as several

[111]Kinoshita, T.; Fujita, M.; Kaneko, H.; Takeuchi, K-i.; Yoshizawa, K.; Yamabe, T. *Bull. Chem. Soc. Jpn.* **1998**, *71*, 1145.

[112]Olmstead, M.M.; Power, P.P. *J. Am. Chem. Soc.* **1985**, *107*, 2174.

[113]Laferriere, M.; Sanrame, C.N.; Scaiano, J.C. *Org. Lett.* **2004**, *6*, 873.

[114]Eldin, S.; Whalen, D.L.; Pollack, R.M. *J. Org. Chem.* **1993**, *58*, 3490.

[115]Abbotto, A.; Bradamanti, S.; Pagani, G.A. *J. Org. Chem.* **1993**, *58*, 449.

[116]Perkins, M.J.; Peynircioglu, N.B. *Tetrahedron* **1985**, *41*, 225.

[117]Okamoto, K.; Kitagawa, T.; Takeuchi, K.; Komatsu, K.; Kinoshita, T.; Aonuma, S.; Nagai, M.; Miyabo, A. *J. Org. Chem.* **1990**, *55*, 996. See also, Okamoto, K.; Kitagawa, T.; Takeuchi, K.; Komatsu, K.; Miyabo, A. *J. Chem. Soc. Chem. Commun.* **1988**, 923.

similar salts, as stable solids. These are salts that consist entirely of carbon and hydrogen.

25

2. *Carbanions Increase in Stability with an Increase in the Amount of s Character at the Carbanionic Carbon.* Thus the order of stability is

$$RC\equiv C^- > R_2C=CH^- \sim Ar^- > R_3C-CH_2^-$$

Acetylene, where the carbon is sp hybridized with 50% s character, is much more acidic than ethylene[118] (sp^2, 33% s), which in turn is more acidic than ethane, with 25% s character. Increased s character means that the electrons are closer to the nucleus and hence of lower energy. As previously mentioned, cyclopropyl carbanions are more stable than methyl, owing to the larger amount of s character as a result of strain (see p. 218).

3. *Stabilization by Sulfur*[119] *or Phosphorus.* Attachment to the carbanionic carbon of a sulfur or phosphorus atom causes an increase in carbanion stability, although the reasons for this are in dispute. One theory is that there is overlap of the unshared pair with an empty d orbital[120] ($p\pi-d\pi$ bonding, see p. 52). For example, a carbanion containing the SO_2R group would be written

[118]For a review of vinylic anions, see Richey, Jr., H.G., in Zabicky, J. *The Chemistry of Alkenes*, Vol. 2, Wiley, NY, *1970*, pp. 49–56.

[119]For reviews of sulfur-containing carbanions, see Oae, S.; Uchida, Y., in Patai, S.; Rappoport, Z.; Stirling, C. *The Chemistry of Sulphones and Sulphoxides*, Wiley, NY, *1988*, pp. 583–664; Wolfe, S., in Bernardi, F.; Csizmadia, I.G.; Mangini, A. *Organic Sulfur Chemistry*, Elsevier, NY, *1985*, pp. 133–190; Block, E. *Reactions of Organosulfur Compounds*; Academic Press, NY, *1978*, pp. 42–56; Durst, T.; Viau, R. *Intra-Sci. Chem. Rep. 1973*, 7 (3), 63. For a review of selenium-stabilized carbanions, see Reich, H.J., in Liotta, D.C. *Organoselenium Chemistry*, Wiley, NY, *1987*, pp. 243–276.

[120]For support for this theory, see Wolfe, S.; LaJohn, L.A.; Bernardi, F.; Mangini, A.; Tonachini, G. *Tetrahedron Lett. 1983*, 24, 3789; Wolfe, S.; Stolow, A.; LaJohn, L.A. *Tetrahedron Lett. 1983*, 24, 4071.

However, there is evidence against d-orbital overlap; and the stabilizing effects have been attributed to other causes.[121] In the case of a PhS substituent, carbanion stabilization is thought to be due to a combination of the inductive and polarizability effects of the group, and d–$p\pi$ resonance and negative hyperconjugation play a minor role, if any.[122] An α silicon atom also stabilizes carbanions.[123]

4. *Field Effects.* Most of the groups that stabilize carbanions by resonance effects (either the kind discussed in 1 above or the kind discussed in paragraph 3) have electron-withdrawing field effects and thereby stabilize the carbanion further by spreading the negative charge, although it is difficult to separate the field effect from the resonance effect. However, in a nitrogen ylid R_3N^+—$^-CR_2$ (see p. 54), where a positive nitrogen is adjacent to the negatively charged carbon, only the field effect operates. Ylids are more stable than the corresponding simple carbanions. Carbanions are stabilized by a field effect if there is any heteroatom (O, N, or S) connected to the carbanionic carbon, provided that the heteroatom bears a positive charge in at least one important canonical form,[124] for example,

5. *Certain Carbanions are Stable because they are Aromatic* (see the cyclopentadienyl anion p. 63, and other aromatic anions in Chapter 2).

6. *Stabilization by a Nonadjacent π Bond.*[125] In contrast to the situation with carbocations (see pp. 450–455), there have been fewer reports of carbanions stabilized by interaction with a nonadjacent π bond. One that may be mentioned is **17**, formed when optically active camphenilone (**15**) was treated with a strong base (potassium *tert*-butoxide).[126] That **17** was truly formed was

[121]Bernardi, F.; Csizmadia, I.G.; Mangini, A.; Schlegel, H.B.; Whangbo, M.; Wolfe, S. *J. Am. Chem. Soc.* **1975**, *97*, 2209; Lehn, J.M.; Wipff, G. *J. Am. Chem. Soc.* **1976**, *98*, 7498; Borden, W.T.; Davidson, E.R.; Andersen, N.H.; Denniston, A.D.; Epiotis, N.D. *J. Am. Chem. Soc.* **1978**, *100*, 1604; Bernardi, F.; Bottoni, A.; Venturini, A.; Mangini, A. *J. Am. Chem. Soc.* **1986**, *108*, 8171.

[122]Bernasconi, C.F.; Kittredge, K.W. *J. Org. Chem.* **1998**, *63*, 1944.

[123]Wetzel, D.M.; Brauman, J.I. *J. Am. Chem. Soc.* **1988**, *110*, 8333.

[124]For a review of such carbanions, see Beak, P.; Reitz, D.B. *Chem. Rev.* **1978**, *78*, 275. See also, Rondan, N.G.; Houk, K.N.; Beak, P.; Zajdel, W.J.; Chandrasekhar, J.; Schleyer, P.v.R. *J. Org. Chem.* **1981**, *46*, 4108.

[125]For reviews, see Werstiuk, N.H. *Tetrahedron* **1983**, *39*, 205; Hunter, D.H.; Stothers, J.B.; Warnhoff, E.W., in de Mayo, P. *Rearrangements in Ground and Excited States*, Vol. 1, Academic Press, NY, **1980**, pp. 410–437.

[126]Nickon, A.; Lambert, J.L. *J. Am. Chem. Soc.* **1966**, *88*, 1905. Also see, Brown, J.M.; Occolowitz, J.L. *Chem. Commun.* **1965**, 376; Grutzner, J.B.; Winstein, S. *J. Am. Chem. Soc.* **1968**, *90*, 6562; Staley, S.W.; Reichard, D.W. *J. Am. Chem. Soc.* **1969**, *91*, 3998; Miller, B. *J. Am. Chem. Soc.* **1969**, *91*, 751; Werstiuk, N.H.; Yeroushalmi, S.; Timmins, G. *Can. J. Chem.* **1983**, *61*, 1945; Lee, R.E.; Squires, R.R. *J. Am. Chem. Soc.* **1986**, *108*, 5078; Peiris, S.; Ragauskas, A.J.; Stothers, J.B. *Can. J. Chem.* **1987**, *65*, 789; Shiner, C.S.; Berks, A.H.; Fisher, A.M. *J. Am. Chem. Soc.* **1988**, *110*, 957.

shown by the following facts: (*1*) A proton was abstracted: ordinary

        26   **27**   **28**

CH_2 groups are not acidic enough for this base; (*2*) recovered **26** was racemized: **28** is symmetrical and can be attacked equally well from either side; (*3*) when the experiment was performed in deuterated solvent, the rate of deuterium uptake was equal to the rate of racemization; and (*4*) recovered **26** contained up to three atoms of deuterium per molecule, although if **27** were the only ion, no more than two could be taken up. Ions of this type, in which a negatively charged carbon is stabilized by a carbonyl group two carbons away, are called *homoenolate ions*.

Overall, functional groups in the a position stabilize carbanions in the following order: $NO_2 > RCO > COOR > SO_2 > CN \sim CONH_2 > Hal > H > R$.

It is unlikely that free carbanions exist in solution. Like carbocations, they usually exist as either ion pairs or they are solvated.[127] Among experiments that demonstrated this was the treatment of $PhCOCHMe^- M^+$ with ethyl iodide, where M^+ was Li^+, Na^+, or K^+. The half-lives of the reaction were[128] for Li, 31×10^{-6}; Na, 0.39×10^{-6}; and K, 0.0045×10^{-6}, demonstrating that the species involved were not identical. Similar results[129] were obtained with Li, Na, and Cs triphenylmethides $Ph_3C^- M^+$.[130] Where ion pairs are unimportant, carbanions are solvated. Cram[99] has demonstrated solvation of carbanions in many solvents. There may be a difference in the structure of a carbanion depending on whether it is free (e.g., in the gas phase) or in solution. The negative charge may be more

[127]For reviews of carbanion pairs, see Hogen-Esch, T.E. *Adv. Phys. Org. Chem.* **1977**, *15*, 153; Jackman, L.M.; Lange, B.C. *Tetrahedron* **1977**, *33*, 2737. See also, Laube, T. *Acc. Chem. Res.* **1995**, *28*, 399.

[128]Zook, H.D.; Gumby, W.L. *J. Am. Chem. Soc.* **1960**, *82*, 1386.

[129]Solov'yanov, A.A.; Karpyuk, A.D.; Beletskaya, I.P.; Reutov, O.A. *J. Org. Chem. USSR* **1981**, *17*, 381. See also, Solov'yanov, A.A.; Beletskaya, I.P.; Reutov, O.A. *J. Org. Chem. USSR* **1983**, *19*, 1964.

[130]For other evidence for the existence of carbanionic pairs, see Hogen-Esch, T.E.; Smid, J. *J. Am. Chem. Soc.* **1966**, *88*, 307, 318; **1969**, *91*, 4580; Abatjoglou, A.G.; Eliel, E.L.; Kuyper, L.F. *J. Am. Chem. Soc.* **1977**, *99*, 8262; Solov'yanov, A.A.; Karpyuk, A.D.; Beletskaya, I.P.; Reutov, V.M. *Doklad. Chem.* **1977**, *237*, 668; DePalma, V.M.; Arnett, E.M. *J. Am. Chem. Soc.* **1978**, *100*, 3514; Buncel, E.; Menon, B. *J. Org. Chem.* **1979**, *44*, 317; O'Brien, D.H.; Russell, C.R.; Hart, A.J. *J. Am. Chem. Soc.* **1979**, *101*, 633; Streitwieser, Jr., A.; Shen, C.C.C. *Tetrahedron Lett.* **1979**, 327; Streitwieser, Jr., A. *Acc. Chem. Res.* **1984**, *17*, 353.

localized in solution in order to maximize the electrostatic attraction to the counterion.[131]

The structure of simple unsubstituted carbanions is not known with certainty since they have not been isolated, but it seems likely that the central carbon is sp^3 hybridized, with the unshared pair occupying one apex of the tetrahedron. Carbanions would thus have pyramidal structures similar to those of amines.

The methyl anion CH_3^- has been observed in the gas phase and reported to have a pyramidal structure.[132] If this is a general structure for carbanions, then any carbanion in which the three R groups are different should be chiral and reactions in which it is an intermediate should give retention of configuration. Attempts have been made to demonstrate this, but without success.[133] A possible explanation is that pyramidal inversion takes place here, as in amines, so that the unshared pair and the central carbon rapidly oscillate from one side of the plane to the other. There is, however, other evidence for the sp^3 nature of the central carbon and for its tetrahedral structure. Carbons at bridgeheads, although extremely reluctant to undergo reactions in which they must be converted to carbocations, undergo with ease reactions in which they must be carbanions and stable bridgehead carbanions are known.[134] Also, reactions at vinylic carbons proceed with retention,[135] indicating that the intermediate **29** has sp^2 hybridization and not the sp hybridization that would be expected in the analogous carbocation. A cyclopropyl anion can also hold its configuration.[136]

$$\underset{R}{\overset{R}{\diagdown}} C = \overset{R}{\underset{}{C_\ominus}}$$

29

[131]See Schade, C.; Schleyer, P.v.R.; Geissler, M.; Weiss, E. *Angew. Chem. Int. Ed.* **1986**, *21*, 902.

[132]Ellison, G.B.; Engelking, P.C.; Lineberger, W.C. *J. Am. Chem. Soc.* **1978**, *100*, 2556.

[133]Retention of configuration has never been observed with simple carbanions. Cram has obtained retention with carbanions stabilized by resonance. However, these carbanions are known to be planar or nearly planar, and retention was caused by asymmetric solvation of the planar carbanions (see p. $$$).

[134]For other evidence that carbanions are pyramidal, see Streitwieser, Jr., A.; Young, W.R. *J. Am. Chem. Soc.* **1969**, *91*, 529; Peoples, P.R.; Grutzner, J.B. *J. Am. Chem. Soc.* **1980**, *102*, 4709.

[135]Curtin, D.Y.; Harris, E.E. *J. Am. Chem. Soc.* **1951**, *73*, 2716, 4519; Braude, E.A.; Coles, J.A. *J. Chem. Soc.* **1951**, 2078; Nesmeyanov, A.N.; Borisov, A.E. *Tetrahedron* **1957**, *1*, 158. Also see, Miller, S.I.; Lee, W.G. *J. Am. Chem. Soc.* **1959**, *81*, 6313; Hunter, D.H.; Cram, D.J. *J. Am. Chem. Soc.* **1964**, *86*, 5478; Walborsky, H.M.; Turner, L.M. *J. Am. Chem. Soc.* **1972**, *94*, 2273; Arnett, J.F.; Walborsky, H.M. *J. Org. Chem.* **1972**, *37*, 3678; Feit, B.; Melamed, U.; Speer, H.; Schmidt, R.R. *J. Chem. Soc. Perkin Trans. 1* **1984**, 775; Chou, P.K.; Kass, S.R. *J. Am. Chem. Soc.* **1991**, *113*, 4357.

[136]Walborsky, H.M.; Motes, J.M. *J. Am. Chem. Soc.* **1970**, *92*, 2445; Motes, J.M.; Walborsky, H.M. *J. Am. Chem. Soc.* **1970**, *92*, 3697; Boche, G.; Harms, K.; Marsch, M. *J. Am. Chem. Soc.* **1988**, *110*, 6925. For a monograph on cyclopropyl anions, cations, and radicals, see Boche, G.; Walborsky, H.M. *Cyclopropane Derived Reactive Intermediates*, Wiley, NY, **1990**. For a review, see Boche, G.; Walborsky, H.M., in Rappoport, Z. *The Chemistry of the Cyclopropyl Group*, pt. 1, Wiley, NY, **1987**, pp. 701–808 (the monograph includes and updates the review).

Carbanions in which the negative charge is stabilized by resonance involving overlap of the unshared-pair orbital with the π electrons of a multiple bond are essentially planar, as would be expected by the necessity for planarity in resonance, although unsymmetrical solvation or ion-pairing effects may cause the structure to deviate somewhat from true planarity.[137] Cram and co-workers showed that where chiral carbanions possessing this type of resonance are generated, retention, inversion, or racemization can result, depending on the solvent (see p. 759). This result is explained by unsymmetrical solvation of planar or near-planar carbanions. However, some carbanions that are stabilized by adjacent sulfur or phosphorus, for example,

are inherently chiral, since retention of configuration is observed where they are generated, even in solvents that cause racemization or inversion with other carbanions.[138] It is known that in THF, PhCH(Li)Me behaves as a prochiral entity,[139] and **30** has been prepared as an optically pure α-alkoxylithium reagent.[140] Cyclohexyllithium **31** shows some configurationally stability, and it is known that isomerization is slowed by an increase in the strength of lithium coordination and by an increase in solvent polarity.[141] It is known that a vinyl anion is configurationally stable whereas a vinyl radical is not. This is due to the instability of the radical anion that must be an intermediate for conversion of one isomer of vinyllithium to the other.[142] The configuration about the carbanionic carbon, at least for some of the α-sulfonyl carbanions, seems to be planar,[143] and the inherent chirality is caused by lack of rotation about the C—S bond.[144]

30   **31**

[137]See the discussion, in Cram, D.J. *Fundamentals of Carbanion Chemistry*, Academic Press, NY, *1965*, pp. 85–105.

[138]Cram, D.J.; Wingrove, A.S. *J. Am. Chem. Soc. 1962, 84,* 1496; Goering, H.L.; Towns, D.L.; Dittmer, B. *J. Org. Chem. 1962, 27,* 736; Corey, E.J.; Lowry, T.H. *Tetrahedron Lett. 1965,* 803; Bordwell, F.G.; Phillips, D.D.; Williams, Jr., J.M. *J. Am. Chem. Soc. 1968, 90,* 426; Annunziata, R.; Cinquini, M.; Colonna, S.; Cozzi, F. *J. Chem. Soc. Chem. Commun. 1981,* 1005; Chassaing, G.; Marquet, A.; Corset, J.; Froment, F. *J. Organomet. Chem. 1982, 232,* 293. For a discussion, see Cram, D.J. *Fundamentals of Carbanion Chemistry*, Academic Press, NY, *1965*, pp. 105–113. Also see Hirsch, R.; Hoffmann, R.W. *Chem. Ber. 1992, 125,* 975.

[139]Hoffmann, R.W.; Rühl, T.; Chemla, F.; Zahneisen, T. *Liebigs Ann. Chem. 1992,* 719.

[140]Rychnovsky, S.D.; Plzak, K.; Pickering, D. *Tetrahedron Lett. 1994, 35,* 6799.

[141]Reich, H.J.; Medina, M.A.; Bowe, M.D. *J. Am. Chem. Soc. 1992, 114,* 11003.

[142]Jenkins, P.R.; Symons, M.C.R.; Booth, S.E.; Swain, C.J. *Tetrahedron Lett. 1992, 33,* 3543.

[143]Boche, G.; Marsch, M.; Harms, K.; Sheldrick, G.M. *Angew. Chem. Int. Ed. 1985, 24,* 573; Gais, H.; Müller, J.; Vollhardt, J.; Lindner, H.J. *J. Am. Chem. Soc. 1991, 113,* 4002. For a contrary view, see Trost, B.M.; Schmuff, N.R. *J. Am. Chem. Soc. 1985, 107,* 396.

[144]Grossert, J.S.; Hoyle, J.; Cameron, T.S.; Roe, S.P.; Vincent, B.R. *Can. J. Chem. 1987, 65,* 1407.

The Structure of Organometallic Compounds[145]

Whether a carbon–metal bond is ionic or polar-covalent is determined chiefly by the electronegativity of the metal and the structure of the organic part of the molecule. Ionic bonds become more likely as the negative charge on the metal-bearing carbon is decreased by resonance or field effects. Thus the sodium salt of acetoacetic ester has a more ionic carbon–sodium bond than methylsodium.

Most organometallic bonds are polar-covalent. Only the alkali metals have electronegativities low enough to form ionic bonds with carbon, and even here the behavior of lithium alkyls shows considerable covalent character. The simple alkyls and aryls of sodium, potassium, rubidium, and cesium[146] are nonvolatile solids[147] insoluble in benzene or other organic solvents, while alkyllithium reagents are soluble, although they too are generally nonvolatile solids. Alkyllithium reagents do not exist as monomeric species in hydrocarbon solvents or ether.[148] In benzene and cyclohexane, freezing-point-depression studies have shown that alkyllithium reagents are normally hexameric unless steric interactions favor tetrameric aggregates.[149] The NMR studies, especially measurements of $^{13}C-^{6}Li$ coupling, have also shown aggregation in hydrocarbon solvents.[150] Boiling-point-elevation studies have been performed in ether solutions, where alkyllithium reagents exist in two- to fivefold aggregates.[151] Even in the gas phase[152] and in

[145]For a monograph, see Elschenbroich, C.; Salzer, A. *Organometallics*, VCH, NY, *1989*. For reviews, see Oliver, J.P., in Hartley, F.R.; Patai, S. *The Chemistry of the Metal–Carbon Bond*, Vol. 2, Wiley, NY, *1985*, pp. 789–826; Coates, G.E.; Green, M.L.H.; Wade, K. *Organometallic Compounds*, 3rd ed., Vol. 1; Methuen: London, *1967*. For a review of the structures of organodialkali compounds, see Grovenstein, Jr., E., in Buncel, E.; Durst, T. *Comprehensive Carbanion Chemistry*, pt. C, Elsevier, NY, *1987*, pp. 175–221.
[146]For a review of X-ray crystallographic studies of organic compounds of the alkali metals, see Schade, C.; Schleyer, P.v.R. *Adv. Organomet. Chem.* *1987*, *27*, 169.
[147]X-ray crystallography of potassium, rubidium, and cesium methyls shows completely ionic crystal lattices: Weiss, E.; Sauermann, G. *Chem. Ber.* *1970*, *103*, 265; Weiss, E.; Köster, H. *Chem. Ber.* *1977*, *110*, 717.
[148]For reviews of the structure of alkyllithium compounds, see Setzer, W.N.; Schleyer, P.v.R. *Adv. Organomet. Chem.* *1985*, *24*, 353; Schleyer, P.v.R. *Pure Appl. Chem.* *1984*, *56*, 151; Brown, T.L. *Pure Appl. Chem.* *1970*, *23*, 447, *Adv. Organomet. Chem.* *1965*, *3*, 365; Kovrizhnykh, E.A.; Shatenshtein, A.I. *Russ. Chem. Rev.* *1969*, *38*, 840. For reviews of the structures of lithium enolates and related compounds, see Boche, G. *Angew. Chem. Int. Ed.* *1989*, *28*, 277; Seebach, D. *Angew. Chem. Int. Ed.* *1988*, *27*, 1624. For a review of the use of nmr to study these structures, see Günther, H.; Moskau, D.; Bast, P.; Schmalz, D. *Angew. Chem. Int. Ed.* *1987*, *26*, 1212. For monographs on organolithium compounds, see Wakefield, B.J. *Organolithium Methods*, Academic Press, NY, *1988*, *The Chemistry of Organolithium Compounds*, Pergamon, Elmsford, NY, *1974*.
[149]Lewis, H.L.; Brown, T.L. *J. Am. Chem. Soc.* *1970*, *92*, 4664; Brown, T.L.; Rogers, M.T. *J. Am. Chem. Soc.* *1957*, *79*, 1859; Weiner, M.A.; Vogel, G.; West, R. *Inorg. Chem.* *1962*, *1*, 654.
[150]Fraenkel, G.; Henrichs, M.; Hewitt, M.; Su, B.M. *J. Am. Chem. Soc.* *1984*, *106*, 255; Thomas, R.D.; Jensen, R.M.; Young, T.C. *Organometallics* *1987*, *6*, 565. See also, Kaufman, M.J.; Gronert, S.; Streitwieser, Jr., A. *J. Am. Chem. Soc.* *1988*, *110*, 2829.
[151]Wittig, G.; Meyer, F.J.; Lange, G. *Liebigs Ann. Chem.* *1951*, *571*, 167. See also, McGarrity, J.F.; Ogle, C.A. *J. Am. Chem. Soc.* *1985*, *107*, 1805; Bates, T.F.; Clarke, M.T.; Thomas, R.D. *J. Am. Chem. Soc.* *1988*, *110*, 5109.
[152]Brown, T.L.; Dickerhoof, D.W.; Bafus, D.A. *J. Am. Chem. Soc.* *1962*, *84*, 1371; Chinn, Jr., J.W.; Lagow, R.L. *Organometallics* *1984*, *3*, 75; Plavšić, D.; Srzić, D.; Klasinc, L. *J. Phys. Chem.* *1986*, *90*, 2075.

the solid state,[153] alkyllithium reagents exist as aggregates. X-ray crystallography has shown that methyllithium has the same tetrahedral structure in the solid state as in ether solution.[153] However, *tert*-butyllithium is monomeric in THF, although dimeric in ether and tetrameric in hydrocarbon solvents.[154] Neopentyllithium exists as a mixture of monomers and dimers in THF.[155]

The C—Mg bond in Grignard reagents is covalent and not ionic. The actual structure of Grignard reagents in solution has been a matter of much controversy over the years.[156] In 1929, it was discovered[157] that the addition of dioxane to an ethereal Grignard solution precipitates all the magnesium halide and leaves a solution of R_2Mg in ether; that is, there can be no RMgX in the solution since there is no halide. The following equilibrium, now called the *Schlenk equilibrium*, was proposed as the composition of the Grignard solution:

$$2 \ RMgX \ \rightleftharpoons \ R_2Mg \ + \ MgX_2 \ \rightleftharpoons \ R_2Mg{\cdot}MgX_2$$
$$\mathbf{32}$$

in which **32** is a complex of some type. Much work has demonstrated that the Schlenk equilibrium actually exists and that the position of the equilibrium is dependent on the identity of R, X, the solvent, the concentration, and the temperature.[158] It has been known for many years that the magnesium in a Grignard solution, no matter whether it is RMgX, R_2Mg, or MgX_2, can coordinate with two molecules of ether in addition to the two covalent bonds:

$$
\begin{array}{ccc}
OR'_2 & OR'_2 & OR'_2 \\
\uparrow & \uparrow & \uparrow \\
R{-}Mg{-}X & R{-}Mg{-}R & X{-}Mg{-}X \\
\uparrow & \uparrow & \uparrow \\
OR'_2 & OR'_2 & OR'_2
\end{array}
$$

Rundle and co-workers[159] performed X-ray diffraction studies on solid phenylmagnesium bromide dietherate and on ethylmagnesium bromide dietherate, which they obtained by cooling ordinary ethereal Grignard solutions until the

[153]Dietrich, H. *Acta Crystallogr.* **1963**, *16*, 681; Weiss, E.; Lucken, E.A.C. *J. Organomet. Chem.* **1964**, *2*, 197; Weiss, E.; Sauermann, G.; Thirase, G. *Chem. Ber.* **1983**, *116*, 74.

[154]Bauer, W.; Winchester, W.R.; Schleyer, P.v.R. *Organometallics* **1987**, *6*, 2371.

[155]Fraenkel, G.; Chow, A.; Winchester, W.R. *J. Am. Chem. Soc.* **1990**, *112*, 6190.

[156]For reviews, see Ashby, E.C. *Bull. Soc. Chim. Fr.* **1972**, 2133; *Q. Rev. Chem. Soc.* **1967**, *21*, 259; Wakefield, B.J. *Organomet. Chem. Rev.* **1966**, *1*, 131; Bell, N.A. *Educ. Chem.* **1973**, 143.

[157]Schlenk, W.; Schlenk Jr., W. *Ber.* **1929**, *62B*, 920.

[158]See Parris, G.; Ashby, E.C. *J. Am. Chem. Soc.* **1971**, *93*, 1206; Salinger, R.M.; Mosher, H.S. *J. Am. Chem. Soc.* **1964**, *86*, 1782; Kirrmann, A.; Hamelin, R.; Hayes, S. *Bull. Soc. Chim. Fr.* **1963**, 1395.

[159]Guggenberger, L.J.; Rundle, R.E. *J. Am. Chem. Soc.* **1968**, *90*, 5375; Stucky, G.; Rundle, R.E. *J. Am. Chem. Soc.* **1964**, *86*, 4825.

solids crystallized. They found that the structures were monomeric:

$$
\begin{array}{c}
\text{OEt}_2 \\
\uparrow \\
\text{R}\!-\!\text{Mg}\!-\!\text{Br} \qquad \text{R = ethyl, phenyl} \\
\uparrow \\
\text{OEt}_2
\end{array}
$$

These solids still contained ether. When ordinary ethereal Grignard solutions[160] prepared from bromomethane, chloromethane, bromoethane, and chloroethane were evaporated at $\sim100°C$ under vacuum so that the solid remaining contained no ether, X-ray diffraction showed *no* RMgX, but a mixture of R_2Mg and MgX_2.[161] These results indicate that in the presence of ether $RMgX\cdot2Et_2O$ is the preferred structure, while the loss of ether drives the Schlenk equilibrium to $R_2Mg + MgX_2$. However, conclusions drawn from a study of the solid materials do not necessarily apply to the structures in solution.

Boiling-point-elevation and freezing-point-depression measurements have demonstrated that in THF at all concentrations and in ether at low concentrations (up to ~0.1 *M*) Grignard reagents prepared from alkyl bromides and iodides are monomeric, that is, there are few or no molecules with two magnesium atoms.[162] Thus, part of the Schlenk equilibrium is operating but not the other

$$2\,RMgX \;\rightleftharpoons\; R_2Mg \;+\; MgX_2$$

part; that is, **32** is not present in measurable amounts. This was substantiated by ^{25}Mg NMR spectra of the ethyl Grignard reagent in THF, which showed the presence of three peaks, corresponding to EtMgBr, Et_2Mg, and $MgBr_2$.[163] That the equilibrium between RMgX and R_2Mg lies far to the left for "ethylmagnesium bromide" in ether was shown by Smith and Becker, who mixed 0.1 *M* ethereal solutions of Et_2Mg and $MgBr_2$ and found that a reaction occurred with a heat evolution of $3.6\,kcal\,mol^{-1}$ ($15\,kJ\,mol^{-1}$) of Et_2Mg, and that the product was *monomeric* (by boiling-point-elevation measurements).[164] When either solution was added little by little to the other, there was a linear output of heat until almost a 1:1 molar ratio was reached. Addition of an excess of either reagent gave no further heat output. These results show that at least under some conditions the Grignard reagent is largely RMgX (coordinated with solvent) but that the equilibrium can be driven to R_2Mg by evaporation of all the ether or by addition of dioxane.

[160]The constitution of alkylmagnesium chloride reagents in THF has been determined. See Sakamoto, S.; Imamoto, T.; Yamaguchi, K. *Org. Lett.* **2001**, *3*, 1793.

[161]Weiss, E. *Chem. Ber.* **1965**, *98*, 2805.

[162]Ashby, E.C.; Smith, M.B. *J. Am. Chem. Soc.* **1964**, *86*, 4363; Vreugdenhil, A.D.; Blomberg, C. *Recl. Trav. Chim. Pays-Bas* **1963**, *82*, 453, 461.

[163]Benn, R.; Lehmkuhl, H.; Mehler, K.; Rufińska, A. *Angew. Chem. Int. Ed.* **1984**, *23*, 534.

[164]Smith, M.B.; Becker, W.E. *Tetrahedron* **1966**, *22*, 3027.

For some aryl Grignard reagents it has proved possible to distinguish separate NMR chemical shifts for ArMgX and Ar$_2$Mg.[165] From the area under the peaks it is possible to calculate the concentrations of the two species, and from them, equilibrium constants for the Schlenk equilibrium. These data show[165] that the position of the equilibrium depends very markedly on the aryl group and the solvent but that conventional aryl Grignard reagents in ether are largely ArMgX, while in THF the predominance of ArMgX is less, and with some aryl groups there is actually more Ar$_2$Mg present. Separate nmr chemical shifts have also been found for alkyl RMgBr and R$_2$Mg in HMPA[166] and in ether at low temperatures.[167] When Grignard reagents from alkyl bromides or chlorides are prepared in triethylamine the predominant species is RMgX.[168] Thus the most important factor determining the position of the Schlenk equilibrium is the solvent. For primary alkyl groups the equilibrium constant for the reaction as written above is lowest in Et$_3$N, higher in ether, and still higher in THF.[169]

However, Grignard reagents prepared from alkyl bromides or iodides in ether at higher concentrations (0.5–1 M) contain dimers, trimers, and higher polymers, and those prepared from alkyl chlorides in ether at all concentrations are dimeric,[170] so that **32** is in solution, probably in equilibrium with RMgX and R$_2$Mg; that is, the complete Schlenk equilibrium seems to be present.

The Grignard reagent prepared from 1-chloro-3,3-dimethylpentane in ether undergoes rapid inversion of configuration at the magnesium-containing carbon (demonstrated by NMR; this compound is not chiral).[171] The mechanism of this inversion is not completely known. Therefore, in almost all cases, it is not possible to retain the configuration of a stereogenic carbon while forming a Grignard reagent.

Organolithium reagents (RLi) are tremendously important reagents in organic chemistry. In recent years, a great deal has been learned about their structure[172] in both the solid state and in solution. X-ray analysis of complexes of n-butyllithium with N,N,N',N'-tetramethylethylenediamine (TMEDA), THF, and 1,2-dimethoxyethane (DME) shows them to be dimers and tetramers [e.g., (BuLi·DME)$_4$].[173] X-ray analysis of isopropyllithium shows it to be a hexamer,

[165]Evans, D.F.; Fazakerley, V. *Chem. Commun.* **1968**, 974.

[166]Ducom, J. *Bull. Chem. Soc. Fr.* **1971**, 3518, 3523, 3529.

[167]Ashby, E.C.; Parris, G.; Walker, F. *Chem. Commun.* **1969**, 1464; Parris, G.; Ashby, E.C. *J. Am. Chem. Soc.* **1971**, *93*, 1206.

[168]Ashby, E.C.; Walker, F. *J. Org. Chem.* **1968**, *33*, 3821.

[169]Parris, G.; Ashby, E.C. *J. Am. Chem. Soc.* **1971**, *93*, 1206.

[170]Ashby, E.C.; Smith, M.B. *J. Am. Chem. Soc.* **1964**, *86*, 4363.

[171]Whitesides, G.M.; Witanowski, M.; Roberts, J.D. *J. Am. Chem. Soc.* **1965**, *87*, 2854; Whitesides, G.M.; Roberts, J.D. *J. Am. Chem. Soc.* **1965**, *87*, 4878. Also see, Witanowski, M.; Roberts, J.D. *J. Am. Chem. Soc.* **1966**, *88*, 737; Fraenkel, G.; Cottrell, C.E.; Dix, D.T. *J. Am. Chem. Soc.* **1971**, *93*, 1704; Pechhold, E.; Adams, D.G.; Fraenkel, G. *J. Org. Chem.* **1971**, *36*, 1368; Maercker, A.; Geuss, R. *Angew. Chem. Int. Ed.* **1971**, *10*, 270.

[172]For a computational study of acidities, electron affinities, and bond dissociation energies of selected organolithium reagents, see Pratt, L.M.; Kass, S.R. *J. Org. Chem.* **2004**, *69*, 2123.

[173]Nichols, M.A.; Williard, P.G. *J. Am. Chem. Soc.* **1993**, *115*, 1568.

(iPrLi)$_6$],[174] and unsolvated lithium aryls are tetramers.[175] α-Ethoxyvinyllithium [CH$_2$=C(OEt)Li] shows a polymeric structure with tetrameric subunits.[176] Aminomethyl aryllithium reagents have been shown to be chelated and dimeric in solvents such as THF.[177]

The dimeric, tetrameric, and hexameric structures of organolithium reagents[178] in the solid state is often retained in solution, but this is dependent on the solvent and complexing additives, if any. A tetrahedral organolithium compound is known,[179] and the X-ray of an α,α-dilithio hydrocarbon has been reported.[180] Phenyllithium is a mixture of tetramers and dimers in diethyl ether, but stoichiometric addition of THF, dimethoxyethane, or TMEDA leads to the dimer.[181] The solution structures of mixed aggregates of butyllithium and amino-alkaloids has been determined,[182] and also the solution structure of sulfur-stabilized allyllithium compounds.[183] Vinyllithium is an 8:1 mixture of tetramer:dimer in THF at $-90°$C, but addition of TMEDA changes the ratio of tetramer:dimer to 1:13 at $-80°$C.[184] Internally solvated allylic lithium compounds have been studied, showing the coordinated lithium to be closer to one of the terminal allyl carbons.[185] A relative scale of organolithium stability has been established,[186] and the issue of configurational stability of enantio-enriched organolithium reagents has been examined.[187]

Enolate anions are an important class of carbanions that appear in a variety of important reactions, including alkylation α- to a carbonyl group and the aldol (reaction **16-34**) and Claisen condensation (reaction **16-85**) reactions. Metal enolate anions of aldehydes, ketones, esters, and other acid derivatives exist as aggregates in ether solvents,[188] and there is evidence that the lithium enolate of

[174]Siemeling, U.; Redecker, T.; Neumann, B.; Stammler, H.-G. *J. Am. Chem. Soc.* **1994**, *116*, 5507.

[175]Ruhlandt-Senge, K.; Ellison, J.J.; Wehmschulte, R.J.; Pauer, F.; Power, P.P. *J. Am. Chem. Soc.* **1993**, *115*, 11353. For the X-ray structure of 1-methoxy-8-naphthyllithium see Betz, J.; Hampel, F.; Bauer, W. *Org. Lett.* **2000**, *2*, 3805.

[176]Sorger, K.; Bauer, W.; Schleyer, P.v.R.; Stalke, D. *Angew. Chem. Int. Ed.* **1995**, *34*, 1594.

[177]Reich, H.J.; Gudmundsson, B.O.; Goldenberg, W.S.; Sanders, A.W.; Kulicke, K.J.; Simon, K.; Guzei, I.A. *J. Am. Chem. Soc.* **2001**, *123*, 8067.

[178]For an *ab initio* correlation of structure with NMR, see Parisel, O.; Fressigne, C.; Maddaluno, J.; Giessner-Prettre, C. *J. Org. Chem.* **2003**, *68*, 1290.

[179]Sekiguchi, A.; Tanaka, M. *J. Am. Chem. Soc.* **2003**, *125*, 12684.

[180]Linti, G.; Rodig, A.; Pritzkow, H. *Angew. Chem. Int. Ed.* **2002**, *41*, 4503.

[181]Reich, H.J.; Green, D.P.; Medina, M.A.; Goldenberg, W.S.; Gudmundsson, B.Ö.; Dykstra, R.R.; Phillips. N.H. *J. Am. Chem. Soc.* **1998**, *120*, 7201.

[182]Sun, X.; Winemiller, M.D.; Xiang, B.; Collum, D.B. *J. Am. Chem. Soc.* **2001**, *123*, 8039. See also, Rutherford, J.L.; Hoffmann, D.; Collum, D.B. *J. Am. Chem. Soc.* **2002**, *124*, 264.

[183]Piffl, M.; Weston, J.; Günther, W.; Anders, E. *J. Org. Chem.* **2000**, *65*, 5942.

[184]Bauer, W.; Griesinger, C. *J. Am. Chem. Soc.* **1993**, *115*, 10871.

[185]Fraenkel, G.; Chow, A.; Fleischer, R.; Liu, H. *J. Am. Chem. Soc.* **2004**, *126*, 3983.

[186]Graña, P.; Paleo, M.R.; Sardina, F.J. *J. Am. Chem. Soc.* **2002**, *124*, 12511.

[187]Basu, A.; Thayumanavan, S. *Angew. Chem. Int. Ed.* **2002**, *41*, 717. See also, Fraenkel, G.; Duncan, J.H.; Martin, K.; Wang, J. *J. Am. Chem. Soc.* **1999**, *121*, 10538.

[188]Stork, G.; Hudrlik, P.F. *J. Am. Chem. Soc.* **1968**, *90*, 4464; Bernstein, M.P.; Collum, D.B. *J. Am. Chem. Soc.* **1993**, *115*, 789; Bernstein, M.P.; Romesberg, F.E.; Fuller, D.J.; Harrison, A.T.; Collum, D.B.; Liu, Q.Y.; Williard, P.G. *J. Am. Chem. Soc.* **1992**, *114*, 5100; Collum, D.B. *Acc. Chem. Res.* **1992**, *25*, 448.

isobutyrophenone is a tetramer in THF,[189] but a dimer in DME.[190] X-ray crystallography of ketone enolate anions have shown that they can exist as tetramers and hexamers.[191] There is also evidence that the aggregate structure is preserved in solution and is probably the actual reactive species. Lithium enolates derived from esters are as dimers in the solid state[192] that contain four tetrahydrofuran molecules. It has also been established that the reactivity of enolate anions in alkylation and condensation reactions is influenced by the aggregate state of the enolate. It is also true that the relative proportions of (*E*) and (*Z*) enolate anions are influenced by the extent of solvation and the aggregation state. Addition of LiBr to a lithium enolate anion in THF suppresses the concentration of monomeric enolate.[193] *Ab initio* studies confirm the aggregate state of acetaldehyde.[194] It is also known that α-Li benzonitrile [PhCH(Li)CN] exists as a dimer in ether and with TMEDA.[195] Mixed aggregates of *tert*-butyllithium and lithium *tert*-butoxide are known to be hexameric.[196]

It might be mentioned that matters are much simpler for organometallic compounds with less-polar bonds. Thus Et_2Hg and $EtHgCl$ are both definite compounds, the former a liquid and the latter a solid. Organocalcium reagents are also know, and they are formed from alkyl halides via a single electron-transfer (SET) mechanism with free-radical intermediates.[197]

The Generation and Fate of Carbanions

The two principal ways in which carbanions are generated are parallel with the ways of generating carbocations.

1. A group attached to a carbon leaves without its electron pair:

$$R—H \longrightarrow R^{\ominus} + H^{\oplus}$$

The leaving group is most often a proton. This is a simple acid–base reaction, and a base is required to remove the proton.[198] However, other

[189]Jackman, L.M.; Szeverenyi, N.M. *J. Am. Chem. Soc.* **1977**, *99*, 4954; Jackman, L.M.; Lange, B.C. *J. Am. Chem. Soc.* **1981**, *103*, 4494.

[190]Jackman, L.M.; Lange, B.C. *Tetrahedron* **1977**, *33*, 2737.

[191]Williard, P.G.; Carpenter, G.B. *J. Am. Chem. Soc.* **1986**, *108*, 462; Williard, P.G.; Carpenter, G.B. *J. Am. Chem. Soc.* **1985**, *107*, 3345; Amstutz, R.; Schweizer, W.B.; Seebach, D.; Dunitz, J.D. *Helv. Chim. Acta* **1981**, *64*, 2617; Seebach, D.; Amstutz, D.; Dunitz, J.D. *Helv. Chim. Acta* **1981**, *64*, 2622.

[192]Seebach, D.; Amstutz, R.; Laube, T.; Schweizer, W.B.; Dunitz, J.D. *J. Am. Chem. Soc.* **1985**, *107*, 5403.

[193]Abu-Hasanayn, F.; Streitwieser, A. *J. Am. Chem. Soc.* **1996**, *118*, 8136.

[194]Abbotto, A.; Streitwieser, A.; Schleyer, P.v.R. *J. Am. Chem. Soc.* **1997**, *119*, 11255.

[195]Carlier, P.R.; Lucht, B.L.; Collum, D.B. *J. Am. Chem. Soc.* **1994**, *116*, 11602.

[196]DeLong, G.T.; Pannell, D.K.; Clarke, M.T.; Thomas, R.D. *J. Am. Chem. Soc.* **1993**, *115*, 7013.

[197]Walborsky, H.M.; Hamdouchi, C. *J. Org. Chem.* **1993**, *58*, 1187.

[198]For a review of such reactions, see Durst, T., in Buncel, E.; Durst, T. *Comprehensive Carbanion Chemistry*, pt. B, Elsevier, NY, **1984**, pp. 239–291.

leaving groups are known (see Chapter 12):

$$R \overset{\frown}{\underset{\underset{O}{\overset{\parallel}{C}}}{}} \overset{\frown}{O^{\ominus}} \longrightarrow R^{\ominus} + CO_2$$

2. A negative ion adds to a carbon–carbon double or triple bond (see Chapter 15):

$$\overset{}{\underset{}{C}} = \overset{}{\underset{}{C}} \overset{\frown}{\longleftarrow} \quad \overset{Y^{\ominus}}{\longleftarrow} \longrightarrow \overset{\ominus}{\underset{}{C}} - \overset{}{\underset{}{C}} - Y$$

The addition of a negative ion to a carbon–oxygen double bond does not give a carbanion, since the negative charge resides on the oxygen.

The most common reaction of carbanions is combination with a positive species, usually a proton, or with another species that has an empty orbital in its outer shell (a Lewis acid–base reaction):

$$R^{\ominus} + Y \longrightarrow R-Y$$

Carbanions may also form a bond with a carbon that already has four bonds, by displacing one of the four groups (S_N2 reaction, see Chapter 10):

$$R^{\ominus} + \quad \overset{}{\underset{}{C}} \overset{\frown}{-X} \longrightarrow R - \overset{}{\underset{}{C}} + X^{\ominus}$$

Like carbocations, carbanions can also react in ways in which they are converted to species that are still not neutral molecules. They can add to double bonds (usually C=O double bonds; see Chapters 10 and 16),

$$R^{\ominus} + \quad \overset{}{\underset{\underset{O}{\overset{\parallel}{}}}{C}} \longrightarrow \overset{R}{\underset{O^{\ominus}}{C}}$$

or rearrange, although this is rare (see Chapter 18),

$$Ph_3C\overset{\ominus}{C}H_2 \longrightarrow Ph_2\overset{\ominus}{C}CH_2Ph$$

or be oxidized to free radicals.[199] A system in which a carbocation [Ph(p-Me$_2$NC$_6$H$_4$)$_2$C$^+$] oxidizes a carbanion [(p-NO$_2$C$_6$H$_4$)$_3$C$^-$] to give two free radicals, reversibly, so that all four species are present in equilibrium, has been demonstrated.[200,201]

---

[199]For a review, see Guthrie, R.D., in Buncel, E.; Durst, T. *Comprehensive Carbanion Chemistry*, pt. A, Elsevier, NY, *1980*, pp. 197–269.

[200]Arnett, E.M.; Molter, K.E.; Marchot, E.C.; Donovan, W.H.; Smith, P. *J. Am. Chem. Soc.* *1987*, *109*, 3788.

[201]Okamoto, K.; Kitagawa, T.; Takeuchi, K.; Komatsu, K.; Kinoshita, T.; Aonuma, S.; Nagai, M.; Miyabo, A. *J. Org. Chem.* *1990*, *55*, 996. See also, Okamoto, K.; Kitagawa, T.; Takeuchi, K.; Komatsu, K.; Miyabo, A. *J. Chem. Soc. Chem. Commun.* *1988*, 923.

Organometallic compounds that are not ionic, but polar-covalent behave very much as if they were ionic and give similar reactions.

FREE RADICALS

Stability and Structure[202]

A *free radical* (often simply called a *radical*) may be defined as a species that contains one or more unpaired electrons. Note that this definition includes certain stable inorganic molecules (e.g., NO and NO_2), as well as many individual atoms (e.g., Na and Cl). As with carbocations and carbanions, simple alkyl radicals are very reactive. Their lifetimes are extremely short in solution, but they can be kept for relatively long periods frozen within the crystal lattices of other molecules.[203] Many spectral[204] measurements have been made on radicals trapped in this manner. Even under these conditions the methyl radical decomposes with a half-life of 10–15 min in a methanol lattice at 77 K.[205] Since the lifetime of a radical depends not only on its inherent stability, but also on the conditions under which it is generated, the terms *persistent* and *stable* are usually used for the different senses. A stable radical is inherently stable; a persistent radical has a relatively long lifetime under the conditions at which it is generated, although it may not be very stable.

Radicals can be characterized by several techniques, such as mass spectrometry[206] or the characterization of alkoxycarbonyl radicals by Step-Scan Time-Resolved Infrared Spectroscopy.[207] Another technique makes use of the magnetic moment that is associated with the spin of an electron, which can be expressed by a quantum number of $\frac{1}{+2}$ or $\frac{1}{-2}$. According to the Pauli principle, any two electrons occupying the same orbital must have opposite spins, so the total magnetic

[202]For monographs, see Alfassi, Z.B. *N-Centered Radicals*, Wiley, Chichester, *1998*; Alfassi, Z.B. *Peroxyl Radicals*, Wiley, Chichester, *1997*; Alfassi, Z.B. *Chemical Kinetics of Small Organic Radicals*, 4 vols., CRC Press: Boca Raton, FL, *1988*; Nonhebel, D.C.; Tedder, J.M.; Walton, J.C. *Radicals*, Cambridge University Press, Cambridge, *1979*; Nonhebel, D.C.; Walton, J.C. *Free-Radical Chemistry*, Cambridge University Press, Cambridge, *1974*; Kochi, J.K. *Free Radicals*, 2 vols., Wiley, NY, *1973*; Hay, J.M. *Reactive Free Radicals*, Academic Press, NY, *1974*; Pryor, W.A. *Free Radicals*, McGraw-Hill, NY, *1966*. For reviews, see Kaplan, L. *React. Intermed. (Wiley)* **1985**, *3*, 227; *1981*, *2*, 251–314; *1978*, *1*, 163; Griller, D.; Ingold, K.U. *Acc. Chem. Res.* **1976**, *9*, 13; Huyser, E.S., in McManus, S.P. *Organic Reactive Intermediates*, Academic Press, NY, *1973*, pp. 1–59; Isaacs, N.S. *Reactive Intermediates in Organic Chemistry*, Wiley, NY, *1974*, pp. 294–374.
[203]For a review of the use of matrices to study radicals and other unstable species, see Dunkin, I.R. *Chem. Soc. Rev.* **1980**, *9*, 1; Jacox, M.E. *Rev. Chem. Intermed.* **1978**, *2*, 1. For a review of the study of radicals at low temperatures, see Mile, B. *Angew. Chem. Int. Ed.* **1968**, *7*, 507.
[204]For a review of infrared spectra of radicals trapped in matrices, see Andrews, L. *Annu. Rev. Phys. Chem.* **1971**, *22*, 109.
[205]Sullivan, P.J.; Koski, W.S. *J. Am. Chem. Soc.* **1963**, *85*, 384.
[206]Sablier, M.; Fujii, T. *Chem. Rev.* **2002**, *102*, 2855.
[207]Bucher, G.; Halupka, M.; Kolano, C.; Schade, O.; Sander, W. *Eur. J. Org. Chem.* **2001**, 545.

moment is zero for any species in which all the electrons are paired. In radicals, however, one or more electrons are unpaired, so there is a net magnetic moment and the species is paramagnetic. Radicals can therefore be detected by magnetic-susceptibility measurements, but for this technique a relatively high concentration of radicals is required.

A much more important technique is *electron spin resonance* (esr), also called *electron paramagnetic resonance* (epr).[208] The principle of esr is similar to that of nmr, except that electron spin is involved rather than nuclear spin. The two electron spin states ($m_s = \frac{1}{2}$ and $m_s = -\frac{1}{2}$) are ordinarily of equal energy, but in a magnetic field the energies are different. As in NMR, a strong external field is applied and electrons are caused to flip from the lower state to the higher by the application of an appropriate radio-frequency (rf) signal. Inasmuch as two electrons paired in one orbital must have opposite spins which cancel, an esr spectrum arises only from species that have one or more unpaired electrons (i.e., free radicals).

Since only free radicals give an esr spectrum, the method can be used to detect the presence of radicals and to determine their concentration.[209] Furthermore, information concerning the electron distribution (and hence the structure) of free radicals can be obtained from the splitting pattern of the esr spectrum (esr peaks are split by nearby protons).[210] Fortunately (for the existence of most free radicals is very short), it is not necessary for a radical to be persistent for an esr spectrum to be obtained. Electron spin resonance spectra have been observed for radicals with lifetimes considerably <1 s. Failure to observe an esr spectrum does not prove that radicals are not involved, since the concentration may be too low for direct observation. In such cases, the *spin trapping* technique can

---

[208]For monographs, see Wertz, J.E.; Bolton, J.R. *Electron Spin Resonance*; McGraw-Hill, NY, *1972* [reprinted by Chapman and Hall, NY, and Methuen, London, *1986*]; Assenheim, H.M. *Introduction to Electron Spin Resonance*, Plenum, NY, *1967*; Bersohn, R.; Baird, J.C. *An Introduction to Electron Paramagnetic Resonance*, W.A. Benjamin, NY, *1966*. For reviews, see Bunce, N.J. *J. Chem. Educ. 1987*, *64*, 907; Hirota, N.; Ohya-Nishiguchi, H., in Bernasconi, C.F. *Investigation of Rates and Mechanisms of Reactions*, 4th ed., pt. 2, Wiley, NY, *1986*, pp. 605–655; Griller, D.; Ingold, K.U. *Acc. Chem. Res. 1980*, *13*, 193; Norman, R.O.C. *Chem. Soc. Rev. 1980*, *8*, 1; Fischer, H., in Kochi, J.K. *Free Radicals*, Vol. 2, Wiley, NY, *1973*, pp. 435–491; Russell, G.A., in Nachod, F.C.; Zuckerman, J.J. *Determination of Organic Structures by Physical Methods*, Vol. 3; Academic Press, NY, *1971*, pp. 293–341; Rassat, A. *Pure Appl. Chem. 1971*, *25*, 623; Kevan, L. *Methods Free-Radical Chem. 1969*, *1*, 1; Geske, D.H. *Prog. Phys. Org. Chem. 1967*, *4*, 125; Norman, R.O.C.; Gilbert, B.C. *Adv. Phys. Org. Chem. 1967*, *5*, 53; Schneider, F.; Möbius, K.; Plato, M. *Angew. Chem. Int. Ed. 1965*, *4*, 856. For a review on the application of epr to photochemistry, see Turro, N.J.; Kleinman, M.H.; Karatekin, E. *Angew. Chem. Int. Ed. 2000*, *39*, 4437. For a review of the related ENDOR method, see Kurreck, H.; Kirste, B.; Lubitz, W. *Angew. Chem. Int. Ed. 1984*, *23*, 173. See also, Poole, Jr., C.P. *Electron Spin Resonance. A Comprehensive Treatise on Experimental Techniques*, 2nd ed., Wiley, NY, *1983*.
[209]Davies, A.G. *Chem. Soc. Rev. 1993*, *22*, 299.
[210]For reviews of the use of esr spectra to determine structures, see Walton, J.C. *Rev. Chem. Intermed. 1984*, *5*, 249; Kochi, J.K. *Adv. Free-Radical Chem. 1975*, *5*, 189. For esr spectra of a large number of free radicals, see Bielski, B.H.J.; Gebicki, J.M. *Atlas of Electron Spin Resonance Spectra*; Academic Press, NY, *1967*.

be used.[211] In this technique, a compound is added that is able to combine with very reactive radicals to produce more persistent radicals; the new radicals can be observed by esr. Azulenyl nitrones have been developed as chromotropic spin trapping agents.[212] The most important spin-trapping compounds are nitroso compounds, which react with radicals to give fairly stable nitroxide radicals:[213] $RN{=}O + R'^{\bullet} \rightarrow RR'N{-}O^{\bullet}$. An N-oxide spin trap has been developed [**33**; 2(diethylphosphino)-5,5-dimethyl-1-pyrroline-N-oxide], and upon trapping a reactive free radical, ^{31}P NMR can be used to identify it.[214] This is an effective technique, and short-lived species such as the oxiranylmethyl radical has been detected by spin trapping.[215] Other molecules have been used to probe the intermediacy of radicals via SET processes. They are called SET probes.[216]

33

Because there is an equal probability that a given unpaired electron will have a quantum number of $\frac{1}{+2}$ or $\frac{1}{-2}$, radicals are observed as a single line in an esr spectrum unless they interact with other electronic or nuclear spins or possess magnetic anisotropy, in which case two or more lines may appear in the spectrum.[217]

Another magnetic technique for the detection of free radicals uses an ordinary NMR instrument. It was discovered[218] that if an nmr spectrum is taken during the course of a reaction, certain signals may be enhanced, either in a positive or negative direction; others may be reduced. When this type of behavior, called *chemically*

[211]For reviews, see Janzen, E.G.; Haire, D.L. *Adv. Free Radical Chem. (Greenwich, Conn.)* *1990*, *1*, 253; Gasanov, R.G.; Freidlina, R.Kh. *Russ. Chem. Rev. 1987*, *56*, 264; Perkins, M.J. *Adv. Phys. Org. Chem. 1980*, *17*, 1; Zubarev, V.E.; Belevskii, V.N.; Bugaenko, L.T. *Russ. Chem. Rev. 1979*, *48*, 729; Evans, C.A. *Aldrichimica Acta 1979*, *12*, 23; Janzen, E.G. *Acc. Chem. Res. 1971*, *4*, 31. See also, the collection of papers on this subject in *Can. J. Chem. 1982*, *60*, 1379.
[212]Becker, D.A. *J. Am. Chem. Soc. 1996*, *118*, 905; Becker, D.A.; Natero, R.; Echegoyen, L.; Lawson, R.C. *J. Chem. Soc. Perkin Trans. 2 1998*, 1289. Also see, Klivenyi, P.; Matthews, R.T.; Wermer, M.; Yang, L.; MacGarvey, U.; Becker, D.A.; Natero, R.; Beal, M.F. *Experimental Neurobiology 1998*, *152*, 163.
[213]For a series of papers on nitroxide radicals, see *Pure Appl. Chem. 1990*, *62*, 177.
[214]Janzen, E.G.; Zhang, Y.-K. *J. Org. Chem. 1995*, *60*, 5441. For the preparation of a new but structurally related spin trap see Karoui, H.; Nsanzumuhire, C.; Le Moigne, F.; Tordo, P. *J. Org. Chem. 1999*, *64*, 1471.
[215]Grossi, L.; Strazzari, S. *Chem. Commun. 1997*, 917.
[216]Timberlake, J.W.; Chen, T. *Tetrahedron Lett. 1994*, *35*, 6043; Tanko, J.M.; Brammer Jr., L.E.; Hervas', M.; Campos, K. *J. Chem. Soc. Perkin Trans. 2 1994*, 1407.
[217]Harry Frank, University of Connecticut, Storrs, CT., Personal Communication.
[218]Ward, H.R.; Lawler, R.G.; Cooper, R.A. *J. Am. Chem. Soc. 1969*, *91*, 746; Bargon, J.; Fischer, H.; Johnsen, U. *Z. Naturforsch., Teil A 1967*, *22*, 1551; Bargon, J.; Fischer, H. *Z. Naturforsch., Teil A 1967*, *22*, 1556; Lepley, A.R. *J. Am. Chem. Soc. 1969*, *91*, 749; Lepley, A.R.; Landau, R.L. *J. Am. Chem. Soc. 1969*, *91*, 748.

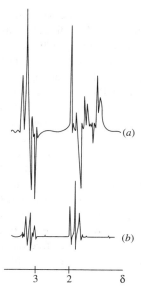

Fig. 5.1 (*a*) The NMR spectrum taken during reaction between EtI and EtLi in benzene (the region between 0.5 and 3.5 δ was scanned with an amplitude twice that of the remainder of the spectrum). The signals at 1.0–1.6 δ are due to butane, some of which is also formed in the reaction. (*b*) Reference spectrum of EtI.[221]

induced dynamic nuclear polarization[219] (CIDNP), is found in the nmr spectrum of the product of a reaction, it means that *at least a portion of that product was formed via the intermediacy of a free radical.*[220] For example, the question was raised whether radicals were intermediates in the exchange reaction between ethyl iodide and ethyllithium (reaction **12-39**):

$$EtI + EtLi \rightleftharpoons EtLi + EtI$$

Curve *a* in Fig. 5.1[221] shows an NMR spectrum taken during the course of the reaction. Curve *b* is a reference spectrum of ethyl iodide (CH$_3$ protons at δ = 1.85; CH$_2$ protons at δ = 3.2). Note that in curve *a* some of the ethyl iodide signals are

[219]For a monograph on CIDNP, see Lepley, R.L.; Closs, G.L. *Chemically Induced Magnetic Polarization,* Wiley, NY, *1973*. For reviews, see Adrian, F.J. *Rev. Chem. Intermed.* *1986*, 7, 173; Closs, G.L.; Miller, R.J.; Redwine, O.D. *Acc. Chem. Res.* *1985*, *18*, 196; Lawler, R.G.; Ward, H.R., in Nachod, F.C.; Zuckerman, J.J. *Determination of Rates and Mechanisms of Reactions*, Vol. 5, Academic Press, NY, *1973*, pp. 99–150; Ward, H.R., in Kochi, J.K. *Free Radicals*, Vol. 1, Wiley, NY, *1973*, pp. 239–273; *Acc. Chem. Res.* *1972*, 5, 18; Closs, G.L. *Adv. Magn. Reson.* *1974*, 7, 157; Lawler, R.G. *Acc. Chem. Res.* *1972*, 5, 25; Kaptein, R. *Adv. Free-Radical Chem.* *1975*, 5, 319; Bethell, D.; Brinkman, M.R. *Adv. Phys. Org. Chem.* *1973*, 10, 53.
[220]A related technique is called chemically induced dynamic electron polarization (CIDEP). For a review, see Hore, P.J.; Joslin, C.G.; McLauchlan, K.A. *Chem. Soc. Rev.* *1979*, 8, 29.
[221]Ward, H.R.; Lawler, R.G.; Cooper, R.A. *J. Am. Chem. Soc.* *1969*, *91*, 746.

enhanced; others go below the base line (*negative enhancement*; also called *emission*). Thus the ethyl iodide formed in the exchange shows CIDNP, and hence was formed via a free-radical intermediate. Chemically induced dynamic nuclear polarization results when protons in a reacting molecule become dynamically coupled to an unpaired electron while traversing the path from reactants to products. Although the presence of CIDNP almost always means that a free radical is involved,[222] its absence does not prove that a free-radical intermediate is necessarily absent, since reactions involving free-radical intermediates can also take place without observable CIDNP. Also, the presence of CIDNP does not prove that *all* of a product was formed via a free-radical intermediate, only that some of it was. It is noted that dynamic nuclear polarization (DNP) enhance signal intensities in NMR spectra of solids and liquids. In a contemporary DNP experiment, a diamagnetic sample is doped with a paramagnet and the large polarization of the electron spins is transferred to the nuclei via microwave irradiation of the epr spectrum.[223] Dynamic nuclear polarization has been used to examine biradicals.[224]

As with carbocations, the stability order of free radicals is tertiary > secondary > primary, explainable by field effects and hyperconjugation, analogous to that in carbocations (p. 235):

$$R-\overset{\overset{\displaystyle H}{|}}{\underset{\underset{\displaystyle H}{|}}{C}}-\overset{\overset{\displaystyle H}{|}}{\underset{\underset{\displaystyle H}{|}}{C}}\cdot \quad \longleftrightarrow \quad R-\overset{\overset{\displaystyle \cdot H}{|}}{\underset{\underset{\displaystyle H}{|}}{C}}=\overset{\overset{\displaystyle H}{|}}{\underset{\underset{\displaystyle H}{|}}{C} } \quad \longleftrightarrow \quad R-\overset{\overset{\displaystyle H}{|}}{\underset{\underset{\displaystyle \cdot H}{|}}{C}}=\overset{\overset{\displaystyle H}{|}}{\underset{\underset{\displaystyle H}{|}}{C}}$$

With resonance possibilities, the stability of free radicals increases;[225] some can be kept indefinitely.[226] Benzylic and allylic[227] radicals for which canonical forms can be drawn similar to those shown for the corresponding cations

$$2\ Ph_3C\cdot \quad \rightleftharpoons \quad Ph-\overset{\overset{\displaystyle Ph}{|}}{\underset{\underset{\displaystyle Ph}{|}}{C}}\overset{\displaystyle H}{-}\!\!\!\left\langle\!\!\!\bigcirc\!\!\!\right\rangle\!\!\!=C\!\!\overset{\displaystyle Ph}{\underset{\displaystyle Ph}{<}}$$

34

(pp. 239, 240) and anions (pp. 252) are more stable than simple alkyl radicals, but still have only a transient existence under ordinary conditions. However, the triphenylmethyl and similar radicals[228] are stable enough to exist in solution

---

[222]It has been shown that CIDNP can also arise in cases where para hydrogen (H_2 in which the nuclear spins are opposite) is present: Eisenschmid, T.C.; Kirss, R.U.; Deutsch, P.P.; Hommeltoft, S.I.; Eisenberg, R.; Bargon, J.; Lawler, R.G.; Balch, A.L. *J. Am. Chem. Soc.* **1987**, *109*, 8089.

[223]Wind, R.A.; Duijvestijn, M.J.; van der Lugt, C.; Manenschijn, A; Vriend, J. *Prog. Nucl. Magn. Reson. Spectrosc.* **1985**, *17*, 33.

[224]Hu, K.-N.; Yu, H.-h.; Swager, T.M.; Griffin, R.G. *J. Am. Chem. Soc.* **2004**, *126*, 10844.

[225]For a discussion, see Robaugh, D.A.; Stein, S.E. *J. Am. Chem. Soc.* **1986**, *108*, 3224.

[226]For a monograph on stable radicals, including those in which the unpaired electron is not on a carbon atom, see Forrester, A.R.; Hay, J.M.; Thomson, R.H. *Organic Chemistry of Stable Free Radicals*, Academic Press, NY, *1968*.

[227]For an electron diffraction study of the allyl radical, see Vajda, E.; Tremmel, J.; Rozsondai, B.; Hargittai, I.; Maltsev, A.K.; Kagramanov, N.D.; Nefedov, O.M. *J. Am. Chem. Soc.* **1986**, *108*, 4352.

[228]For a review, see Sholle, V.D.; Rozantsev, E.G. *Russ. Chem. Rev.* **1973**, *42*, 1011.

at room temperature, although in equilibrium with a dimeric form. The concentration of triphenylmethyl radical in benzene solution is $\sim 2\%$ at room temperature. For many years it was assumed that $Ph_3C\cdot$, the first stable free radical known,[229] dimerized to hexaphenylethane (Ph_3C-CPh_3),[230] but UV and NMR investigations have shown that the true structure is **34**.[231] Although triphenylmethyl-type radicals are stabilized by resonance:

$$Ph_3C\cdot \longleftrightarrow \quad =CPh_2 \quad \longleftrightarrow \quad \cdot\quad =CPh_2 \quad \longleftrightarrow \quad \text{etc.}$$

it is steric hindrance to dimerization and not resonance that is the major cause of their stability.[232] This was demonstrated by the preparation of the radicals **35** and **36**.[233] These radicals are electronically very similar, but **35**, being planar, has much less steric hindrance to dimerization than $Ph_3C\cdot$, while **36**, with six groups in ortho positions, has much more. On the other hand, the planarity of **35** means that

35   **36**

it has a maximum amount of resonance stabilization, while **36** must have much less, since its degree of planarity should be even less than $Ph_3C\cdot$, which itself is propeller shaped and not planar. Thus if resonance is the chief cause of the stability of $Ph_3C\cdot$, **36** should dimerize and **35** should not, but if steric hindrance is

---

[229]Gomberg, M. *J. Am. Chem. Soc.* **1900**, *22*, 757, *Ber.* **1900**, *33*, 3150.

[230]Hexaphenylethane has still not been prepared, but substituted compounds [hexakis(3,5-di-*tert*-butyl-4-biphenylyl)ethane and hexakis(3,5-di-*tert*-butylphenyl)ethane] have been shown by X-ray crystallography to be nonbridged hexaarylethanes in the solid state: Stein, M.; Winter, W.; Rieker, A. *Angew. Chem. Int. Ed.* **1978**, *17*, 692; Yannoni, N.; Kahr, B.; Mislow, K. *J. Am. Chem. Soc.* **1988**, *110*, 6670. In solution, both dissociate into free radicals.

[231]Lankamp, H.; Nauta, W.T.; MacLean, C. *Tetrahedron Lett.* **1968**, 249; Staab, H.A.; Brettschneider, H.; Brunner, H. *Chem. Ber.* **1970**, *103*, 1101; Volz, H.; Lotsch, W.; Schnell, H. *Tetrahedron* **1970**, *26*, 5343; McBride, J. *Tetrahedron* **1974**, *30*, 2009. See also, Guthrie, R.D.; Weisman, G.R. *Chem. Commun.* **1969**, 1316; Takeuchi, H.; Nagai, T.; Tokura, N. *Bull. Chem. Soc. Jpn.* **1971**, *44*, 753. For an example where a secondary benzilic radical undergoes this type of dimerization, see Peyman, A.; Peters, K.; von Schnering, H.G.; Rüchardt, C. *Chem. Ber.* **1990**, *123*, 1899.

[232]For a review of steric effects in free-radical chemistry, see Rüchardt, C. *Top. Curr. Chem.* **1980**, *88*, 1.

[233]Sabacky, M.J.; Johnson Jr., C.S.; Smith, R.G.; Gutowsky, H.S.; Martin, J.C. *J. Am. Chem. Soc.* **1967**, *89*, 2054.

the major cause, the reverse should happen. It was found[233] that **36** gave no evidence of dimerization, even in the solid state, while **35** existed primarily in the dimeric form, which is dissociated to only a small extent in solution,[234] indicating that steric hindrance to dimerization is the major cause for the stability of triarylmethyl radicals. A similar conclusion was reached in the case of $(NC)_3C\cdot$, which dimerizes readily although considerably stabilized by resonance.[235] Nevertheless, that resonance is still an important contributing factor to the stability of radicals is shown by the facts that (*1*) the radical $t\text{-}Bu(Ph)_2C\cdot$ dimerizes more than $Ph_3C\cdot$, while $p\text{-}PhCOC_6H_4(Ph_2)C\cdot$ dimerizes less.[236] The latter has more canonical forms than $Ph_3C\cdot$, but steric hindrance should be about the same (for attack at one of the two rings). (*2*) A number of radicals ($p\text{-}XC_6H_4)_3C\cdot$, with X = F, Cl, O_2N, CN, and so on do not dimerize, but are kinetically stable.[237] Completely chlorinated triarylmethyl radicals are more stable than the unsubstituted kind, probably for steric reasons, and many are quite inert in solution and in the solid state.[238]

Allylic radical are relatively stable, and the pentadienyl radical is particularly stable. In such molecules, $(E,E)\text{-}(E,Z)\text{-}$, and (Z,Z)-stereoisomers can form. It has been calculated that (Z,Z)-pentadienyl radical is $5.6\,\text{kcal mol}^{-1}(23.4\,\text{kJ mol}^{-1})$ less stable than (E,E)-pentadienyl radical.[239] 2-Phenylethyl radicals have been shown to exhibit bridging of the phenyl group.[240] It is noted that vinyl radical have (E)- and (Z)-forms and the inversion barrier from one to the other increases as the electronegativity of substituents increase.[241] Enolate radicals are also known.[242]

It has been postulated that the stability of free radicals is enhanced by the presence at the radical center of *both* an electron-donating and an electron-withdrawing group.[243] This is called the *push–pull* or *captodative effect* (see also, pp. 185). The effect arises from increased resonance, for example:

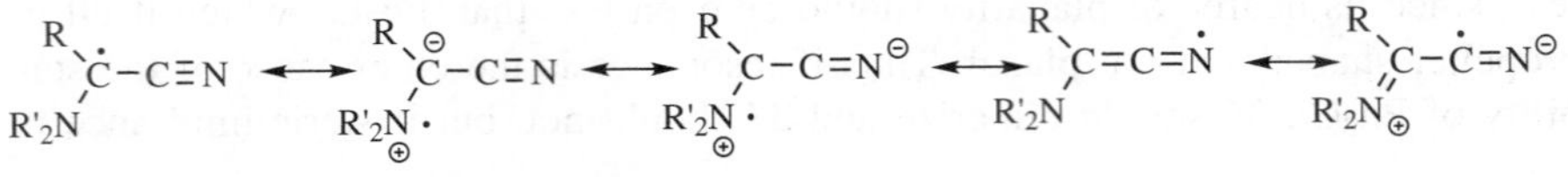

[234]Müller, E.; Moosmayer, A.; Rieker, A.; Scheffler, K. *Tetrahedron Lett.* **1967**, 3877. See also, Neugebauer, F.A.; Hellwinkel, D.; Aulmich, G. *Tetrahedron Lett.* **1978**, 4871.

[235]Kaba, R.A.; Ingold, K.U. *J. Am. Chem. Soc.* **1976**, 98, 523.

[236]Zarkadis, A.K.; Neumann, W.P.; Marx, R.; Uzick, W. *Chem. Ber.* **1985**, *118*, 450; Zarkadis, A.K.; Neumann, W.P.; Uzick, W. *Chem. Ber.* **1985**, *118*, 1183.

[237]Dünnebacke, D.; Neumann, W.P.; Penenory, A.; Stewen, U. *Chem. Ber.* **1989**, *122*, 533.

[238]For reviews, see Ballester, M. *Adv. Phys. Org. Chem.* **1989**, *25*, 267, pp. 354–405; *Acc. Chem. Res.* **1985**, *18*, 380. See also, Hegarty, A.F.; O'Neill, P. *Tetrahedron Lett.* **1987**, *28*, 901.

[239]Fort Jr., R.C.; Hrovat, D.A.; Borden, W.T. *J. Org. Chem.* **1993**, *58*, 211.

[240]Asensio, A.; Dannenberg, J.J. *J. Org. Chem.* **2001**, *66*, 5996.

[241]Galli, C.; Guarnieri, A.; Koch, H.; Mencarelli, P.; Rappoport, Z. *J. Org. Chem.* **1997**, *62*, 4072.

[242]Giese, B.; Damm, W.; Wetterich, F.; Zeltz, H.-G.; Rancourt, J.; Guindon, Y. *Tetrahedron Lett.* **1993**, *34*, 5885.

[243]For reviews, see Sustmann, R.; Korth, H. *Adv. Phys. Org. Chem.* **1990**, *26*, 131; Viehe, H.G.; Janousek, Z.; Merényi, R.; Stella, L. *Acc. Chem. Res.* **1985**, *18*, 148.

There is some evidence in favor[244] of the captodative effect, some of it from esr studies.[245] However, there is also experimental[246] and theoretical[247] evidence against it. There is evidence that while $FCH_2^{\bullet}$ and $F_2CH^{\bullet}$ are more stable than $CH_3^{\bullet}$, the radical $CF_3^{\bullet}$ is less stable; that is, the presence of the third F destabilizes the radical.[248]

Diphenylpicrylhydrazyl

39   **40**

Certain radicals with the unpaired electron not on a carbon are also very stable.[249] Radicals can be stabilized by intramolecular hydrogen bonding.[250]

[244]For a summary of the evidence, see Pasto, D.J. *J. Am. Chem. Soc.* **1988**, *110*, 8164. See also, Ashby, E.C. *Bull. Soc. Chim. Fr.* **1972**, 2133; *Q. Rev. Chem. Soc.* **1967**, *21*, 259; Wakefield, B.J. *Organomet. Chem. Rev.* **1966**, *1*, 131; Bell, N.A. *Educ. Chem.* **1973**, 143.

[245]See, for example, Korth, H.; Lommes, P.; Sustmann, R.; Sylvander, L.; Stella, L. *New J. Chem.* **1987**, *11*, 365; Sakurai, H.; Kyushin, S.; Nakadaira, Y.; Kira, M. *J. Phys. Org. Chem.* **1988**, *1*, 197; Rhodes, C.J.; Roduner, E. *Tetrahedron Lett.* **1988**, *29*, 1437; Viehe, H.G.; Merényi, R.; Janousek, Z. *Pure Appl. Chem.* **1988**, *60*, 1635; Creary, X.; Sky, A.F.; Mehrsheikh-Mohammadi, M.E. *Tetrahedron Lett.* **1988**, *29*, 6839; Bordwell, F.G.; Lynch, T. *J. Am. Chem. Soc.* **1989**, *111*, 7558.

[246]See, for example, Beckhaus, H.; Rüchardt, C. *Angew. Chem. Int. Ed.* **1987**, *26*, 770; Neumann, W.P.; Penenory, A.; Stewen, U.; Lehnig, M. *J. Am. Chem. Soc.* **1989**, *111*, 5845; Bordwell, F.G.; Bausch, M.J.; Cheng, J.P.; Cripe, T.H.; Lynch, T.-Y.; Mueller, M.E. *J. Org. Chem.* **1990**, *55*, 58; Bordwell, F.G.; Harrelson Jr., J.A. *Can. J. Chem.* **1990**, *68*, 1714.

[247]See Pasto, D.J. *J. Am. Chem. Soc.* **1988**, *110*, 8164.

[248]Jiang, X.; Li, X.; Wang, K. *J. Org. Chem.* **1989**, *54*, 5648.

[249]For reviews of radicals with the unpaired electron on atoms other than carbon, see, in Kochi, J.K. *Free Radicals*, Vol. 2, Wiley, NY, **1973**, the reviews by Nelson, S.F. pp. 527–593 (*N*-centered); Bentrude, W.G. pp. 595–663 (*P*-centered); Kochi, J.K. pp. 665–710 (*O*-centered); Kice, J.L. pp. 711–740 (*S*-centered); Sakurai, H. pp. 741–807 (Si, Ge, Sn, and Pb centered).

[250]Maki, T.; Araki, Y.; Ishida, Y.; Onomura, O.; Matsumura, Y. *J. Am. Chem. Soc.* **2001**, *123*, 3371.

Diphenylpicrylhydrazyl is a solid that can be kept for years, and stable neutral azine radicals have been prepared.[251] Nitroxide radicals were mentioned previously (p. 273),[252] and the commercially available TEMPO (2,2,6,6-tetramethylpiperidine-1-oxyl free radical, 37) is a stable nitroxyl radical used in chemical reactions such as oxidations.[253] or as a spin trap.[254] Nitroxyl radical 38 is a nitroxide radical so stable that reactions can be performed on it without affecting the unpaired electron[255] (the same is true for some of the chlorinated triarylmethyl radicals mentioned above[256]). Several nitrogen-containing groups are known to stabilize radicals, and the most effective radical stabilization is via spin delocalization.[257] A number of persistent N-tert-butoxy-1-aminopyrenyl radicals, such as 39, have been isolated as monomeric radical crystals (see 40, the X-ray crystal structure of 39),[258] and monomeric N-alkoxyarylaminyls have been isolated.[259] α-Trichloromethylbenzyl(tert-butyl)aminoxyl (41) is extremely stable.[260] In aqueous media it is stable for >30 days, and in solution in an aromatic hydrocarbon solvent it has survived for more than 90 days.[260] Although the stable nitroxide radicals have the α-carbon blocked to prevent radical formation there, stable nitroxide radicals are also known with hydrogen at the α-carbon,[261] and long-lived vinyl nitroxide radicals are known.[262] A stable organic radical lacking resonance stabilization has been prepared (42) and its X-ray crystal structure was

[251]Jeromin, G.E. *Tetrahedron Lett.* **2001**, *42*, 1863.

[252]For a study of the electronic structure of persistent nitroxide radicals see Novak, I.; Harrison, L.J.; Kovač, B.; Pratt, L.M. *J. Org. Chem.* **2004**, *69*, 7628.

[253]See Anelli, P.L.; Biffi, C.; Montanari, F.; Quici, S. *J. Org. Chem.* **1987**, *52*, 2559; Anelli, P.L.; Banfi, S.; Montanari, F.; Quici, S. *J. Org. Chem.* **1989**, *54*, 2970; Anelli, P.L.; Montanari, F.; Quici, S. *Org. Synth.* **1990**, *69*, 212; Fritz-Langhals, E. *Org. Process Res. Dev.* **2005**, *9*, 577. See also, Rychnovsky, S.D.; Vaidyanathan, R.; Beauchamp, T.; Lin, R.; Farmer, P.J. *J. Org. Chem.* **1999**, *64*, 6745.

[254]Volodarsky, L.B.; Reznikov, V.A.; Ovcharenko, V.I. *Synthetic Chemistry of Stable Nitroxides*, CRC Press: Boca Raton, FL, **1994**; Keana, J.F.W. *Chem. Rev.* **1978**, *78*, 37; Aurich, H.G. *Nitroxides. In Nitrones, Nitronates, Nitroxides*, Patai, S., Rappoport, Z., (Eds.), Wiley, NY, **1989**; Chapt. 4.

[255]Neiman, M.B.; Rozantsev, E.G.; Mamedova, Yu.G. *Nature* **1963**, *200*, 256. For reviews of such radicals, see Aurich, H.G., in Patai, S. *The Chemistry of Functional Groups, Supplement F*, pt. 1, Wiley, NY, **1982**, pp. 565–622 [This review has been reprinted, and new material added, in Breuer, E.; Aurich, H.G.; Nielsen, A. *Nitrones, Nitronates, and Nitroxides*, Wiley, NY, **1989**, pp. 313–399]; Rozantsev, E.G.; Sholle, V.D. *Synthesis* **1971**, 190, 401.

[256]See Ballester, M.; Veciana, J.; Riera, J.; Castañer, J.; Armet, O.; Rovira, C. *J. Chem. Soc. Chem. Commun.* **1983**, 982.

[257]Adam, W.; Ortega Schulte, C.M. *J. Org. Chem.* **2002**, *67*, 4569.

[258]Miura, Y.; Matsuba, N.; Tanaka, R.; Teki, Y.; Takui, T. *J. Org. Chem.* **2002**, *67*, 8764. For another stable nitroxide radical, see Huang, W.-l.; Chiarelli, R.; Rassat, A. *Tetrahedron Lett.* **2000**, *41*, 8787.

[259]Miura, Y.; Tomimura, T.; Matsuba, N.; Tanaka, R.; Nakatsuji, M.; Teki, Y. *J. Org. Chem.* **2001**, *66*, 7456.

[260]Janzen, E.G.; Chen, G.; Bray, T.M.; Reinke, L.A.; Poyer, J.L.; McCay, P.B. *J. Chem. Soc. Perkin Trans. 2* **1993**, 1983.

[261]Reznikov, V.A.; Volodarsky, L.B. *Tetrahedron Lett.* **1994**, *35*, 2239.

[262]Reznikov, V.A.; Pervukhina, N.V.; Ikorskii, V.N.; Ovcharenko, V.I; Grand, A. *Chem. Commun.* **1999**, 539.

obtained.[263]

41   **42**

Dissociation energies (D values) of R—H bonds provide a measure of the relative inherent stability of free radicals R.[264] Table 5.4 lists such values.[265] The higher the D value, the less stable the radical. Bond dissociation energies have also been reported for the C—H bond of alkenes and dienes[266] and for the C—H bond in radical precursors XYC—H, where X,Y can be H, alkyl, COOR, COR, SR, CN, NO_2, and so on.[267] Bond dissociation energies for the C—O bond in hydroperoxide radicals (ROO•) have also been reported.[268]

TABLE 5.4. The D_{298} Values for Some R—H Bonds.[265] Free-radical Stability is in the Reverse Order

	D	
R	kcal mol^{-1}	kJ mol^{-1}
Ph•[269]	111	464
CF_3•	107	446
CH_2=CH•	106	444
Cyclopropyl[270]	106	444
Me•	105	438
Et•	100	419

[263]Apeloig, Y.; Bravo-Zhivotovskii, D.; Bendikov, M.; Danovich, D.; Botoshansky, M.; Vakulrskaya, T.; Voronkov, M.; Samoilova, R.; Zdravkova, M.; Igonin, V.; Shklover, V.; Struchkov, Y. *J. Am. Chem. Soc.* **1999**, *121*, 8118.

[264]It has been claimed that relative D values do not provide such a measure: Nicholas, A.M. de P.; Arnold, D.R. *Can. J. Chem.* **1984**, *62*, 1850, 1860.

[265]Except where noted, these values are from Kerr, J.A., in Weast, R.C. *Handbook of Chemistry and Physics*, 69th ed.; CRC Press: Boca Raton, FL, **1988**, p. F-183. For another list of D values, see McMillen, D.F.; Golden, D.M. *Annu. Rev. Phys. Chem.* **1982**, *33*, 493. See also, Tsang, W. *J. Am. Chem. Soc.* **1985**, *107*, 2872; Holmes, J.L.; Lossing, F.P.; Maccoll, A. *J. Am. Chem. Soc.* **1988**, *110*, 7339; Holmes, J.L.; Lossing, F.P. *J. Am. Chem. Soc.* **1988**, *110*, 7343; Roginskii, V.A. *J. Org. Chem. USSR* **1989**, *25*, 403.

[266]Zhang, X.-M. *J. Org. Chem.* **1998**, *63*, 1872.

[267]Brocks, J.J.; Beckhaus, H.-D.; Beckwith, A.L.J.; Rüchardt, C. *J. Org. Chem.* **1998**, *63*, 1935.

[268]Pratt, D.A.; Porter, N.A. *Org. Lett.* **2003**, *5*, 387.

[269]For the infra-red of a matrix-isolated phenyl radical see Friderichsen, A.V.; Radziszewski, J.G.; Nimlos, M.R.; Winter, P.R.; Dayton, D.C.; David, D.E.; Ellison, G.B. *J. Am. Chem. Soc.* **2001**, *123*, 1977.

[270]For a review of cyclopropyl radicals, see Walborsky, H.M. *Tetrahedron* **1981**, *37*, 1625. See also, Boche, G.; Walborsky, H.M. *Cyclopropane Derived Reactive Intermediates*, Wiley, NY, **1990**.

$Me_3CCH_2\bullet$	100	418
$Pr\bullet$	100	417
$Cl_3C\bullet$	96	401
$Me_2CH\bullet$	96	401
$Me_3C\bullet$[271]	95.8	401
Cyclohexyl	95.5	400
$PhCH_2\bullet$	88	368
$HCO\bullet$	87	364
$CH_2{=}CH{-}CH_2\bullet$	86	361

There are two possible structures for simple alkyl radicals.[272] They might have sp^2 bonding, in which case the structure would be planar, with the odd electron in a p orbital, or the bonding might be sp^3, which would make the structure pyramidal and place the odd electron in an sp^3 orbital. The esr spectra of $\bullet CH_3$ and other simple alkyl radicals, as well as other evidence indicate that these radicals have planar structures.[273] This is in accord with the known loss of optical activity when a free radical is generated at a chiral carbon.[274] In addition, electronic spectra of the CH_3 and CD_3 radicals (generated by flash photolysis) in the gas phase have definitely established that under these conditions the radicals are planar or near planar.[275] IR spectra of $\bullet CH_3$ trapped in solid argon led to a similar conclusion.[276]

43a   43b

Despite the usual loss of optical activity noted above, asymmetric radicals can be prepared in some cases. For example, asymmetric nitroxide radicals are known.[277] An anomeric effect was observed in alkoxy radical **43**, where the ratio of **43a/43b** was 1:1.78.[278]

[271]This value is from Gutman, D. *Acc. Chem. Res.* **1990**, *23*, 375.

[272]For a review, see Kaplan, L., in Kochi, J.K. *Free Radicals*, Vol. 2, Wiley, NY, *1973*, pp. 361–434.

[273]See, for example, Cole, T.; Pritchard, D.E.; Davidson, N.; McConnell, H.M. *Mol. Phys.* **1958**, *1*, 406; Fessenden, R.W.; Schuler, R.H. *J. Chem. Phys.* **1963**, *39*, 2147; Symons, M.C.R. *Nature* **1969**, *222*, 1123, *Tetrahedron Lett.* **1973**, 207; Bonazzola, L.; Leray, E.; Roncin, J. *J. Am. Chem. Soc.* **1977**, *99*, 8348; Giese, B.; Beckhaus, H. *Angew. Chem. Int. Ed.* **1978**, *17*, 594; Ellison, G.B.; Engelking, P.C.; Lineberger, W.C. *J. Am. Chem. Soc.* **1978**, *100*, 2556. See, however, Paddon-Row, M.N.; Houk, K.N. *J. Am. Chem. Soc.* **1981**, *103*, 5047.

[274]There are a few exceptions. See p. $$$.

[275]Herzberg, G.; Shoosmith, J. *Can. J. Phys.* **1956**, *34*, 523; Herzberg, G. *Proc. R. Soc. London, Ser. A* **1961**, *262*, 291. See also, Tan, L.Y.; Winer, A.M.; Pimentel, G.C. *J. Chem. Phys.* **1972**, *57*, 4028; Yamada, C.; Hirota, E.; Kawaguchi, K. *J. Chem. Phys.* **1981**, *75*, 5256.

[276]Andrews, L.; Pimentel, G.C. *J. Chem. Phys.* **1967**, *47*, 3637; Milligan, D.E.; Jacox, M.E. *J. Chem. Phys.* **1967**, *47*, 5146.

[277]Tamura, R.; Susuki, S.; Azuma, N.; Matsumoto, A.; Todda, F.; Ishii, Y. *J. Org. Chem.* **1995**, *60*, 6820.

[278]Rychnovsky, S.D.; Powers, J.P.; LePage, T.J. *J. Am. Chem. Soc.* **1992**, *114*, 8375.

Evidence from studies on bridgehead compounds shows that although a planar configuration is more stable, pyramidal structures are not impossible. In contrast to the situation with carbocations, free radicals have often been generated at bridgeheads, although studies have shown that bridgehead free radicals are less rapidly formed than the corresponding open-chain radicals.[279] In sum, the available evidence indicates that although simple alkyl free radicals prefer a planar, or near-planar shape, the energy difference between a planar and a pyramidal free radical is not great. However, free radicals in which the carbon is connected to atoms of high electronegativity, for example, $\cdot CF_3$, prefer a pyramidal shape;[280] increasing the electronegativity increases the deviation from planarity.[281] Cyclopropyl radicals are also pyramidal.[282] Free radicals with resonance are definitely planar, although triphenylmethyl-type radicals are propeller-shaped,[283] like the analogous carbocations (p. 245). Radicals possessing simple alkyl substituents attached to the radical carbon (C$\cdot$) that have C^{sp^3}–C^{sp^3} bonds, and rotation about those bonds is possible. The internal rotation barrier for the t-butyl radical (Me$_3$C$\cdot$), for example, was estimated to be $\sim 1.4 \, kcal \, mol^{-1}$ ($6 \, kJ \, mol^{-1}$).[284]

A number of diradicals (also called biradicals) are known,[285] and the thermodynamic stability of diradicals has been examined.[286] Orbital phase theory has been applied to the development of a theoretical model of localized 1,3-diradicals, and used to predict the substitution effects on the spin preference and S–T gaps, and to design stable localized carbon-centered 1,3-diradicals.[287] When the unpaired electrons of a diradical are widely separated, for example, as in $\cdot CH_2CH_2CH_2CH_2\cdot$,

[279]Lorand, J.P.; Chodroff, S.D.; Wallace, R.W. *J. Am. Chem. Soc.* **1968**, *90*, 5266; Humphrey, L.B.; Hodgson, B.; Pincock, R.E. *Can. J. Chem.* **1968**, *46*, 3099; Oberlinner, A.; Rüchardt, C. *Tetrahedron Lett.* **1969**, 4685; Danen, W.C.; Tipton, T.J.; Saunders, D.G. *J. Am. Chem. Soc.* **1971**, *93*, 5186; Fort, Jr., R.C.; Hiti, J. *J. Org. Chem.* **1977**, *42*, 3968; Lomas, J.S. *J. Org. Chem.* **1987**, *52*, 2627.

[280]Fessenden, R.W.; Schuler, R.H. *J. Chem. Phys.* **1965**, *43*, 2704; Rogers, M.T.; Kispert, L.D. *J. Chem. Phys.* **1967**, *46*, 3193; Pauling, L. *J. Chem. Phys.* **1969**, *51*, 2767.

[281]For example, 1,1-dichloroalkyl radicals are closer to planarity than the corresponding 1,1-difluoro radicals, though still not planar: Chen, K.S.; Tang, D.Y.H.; Montgomery, L.K.; Kochi, J.K. *J. Am. Chem. Soc.* **1974**, *96*, 2201. For a discussion, see Krusic, P.J.; Bingham, R.C. *J. Am. Chem. Soc.* **1976**, *98*, 230.

[282]See Deycard, S.; Hughes, L.; Lusztyk, J.; Ingold, K.U. *J. Am. Chem. Soc.* **1987**, *109*, 4954.

[283]Adrian, F.J. *J. Chem. Phys.* **1958**, *28*, 608; Andersen, P. *Acta Chem. Scand.* **1965**, *19*, 629.

[284]Kubota, S.; Matsushita, M.; Shida, T.; Abu-Raqabah, A.; Symons, M.C.R.; Wyatt, J.L. *Bull. Chem. Soc. Jpn.* **1995**, *68*, 140.

[285]For a monograph, see Borden, W.T. *Diradicals*, Wiley, NY, **1982**. For reviews, see Johnston, L.J.; Scaiano, J.C. *Chem. Rev.* **1989**, *89*, 521; Doubleday, Jr., C.; Turro, N.J.; Wang, J. *Acc. Chem. Res.* **1989**, *22*, 199; Scheffer, J.R.; Trotter, J. *Rev. Chem. Intermed.* **1988**, *9*, 271; Wilson, R.M. *Org. Photochem.* **1985**, *7*, 339; Borden, W.T. *React. Intermed. (Wiley)* **1985**, *3*, 151; **1981**, *2*, 175; Borden, W.T.; Davidson, E.R. *Acc. Chem. Res.* **1981**, *14*, 69; Salem, L.; Rowland, C. *Angew. Chem. Int. Ed.* **1972**, *11*, 92; Salem, L. *Pure Appl. Chem.* **1973**, *33*, 317; Jones II, G. *J. Chem. Educ.* **1974**, *51*, 175; Morozova, I.D.; Dyatkina, M.E. *Russ. Chem. Rev.* **1968**, *37*, 376. See also, Döhnert, D.; Koutecky, J. *J. Am. Chem. Soc.* **1980**, *102*, 1789. For a series of papers on diradicals, see *Tetrahedron* **1982**, *38*, 735.

[286]Zhang, D.Y.; Borden, W.T. *J. Org. Chem.* **2002**, *67*, 3989.

[287]Ma, J.; Ding, Y.; Hattori, K.; Inagaki, S. *J. Org. Chem.* **2004**, *69*, 4245.

the species behaves spectrally like two doublets. When they are close enough for interaction or can interact through an unsaturated system as in trimethyl-enemethane,[288] they can have total spin numbers of $+1$, 0, or -1, since each

Trimethylenemethane

electron could be either $\frac{1}{+2}$ or $\frac{1}{-2}$. Spectroscopically they are called *triplets*,[289] since each of the three possibilities is represented among the molecules and gives rise to its own spectral peak. In triplet molecules the two unpaired electrons have the same spin. Not all diradicals have a triplet ground state. In 2,3-dimethylelecycohexane-1,4-diyl (**44**), the singlet and triplet states were found to be almost degenerate.[290] Some diradicals, such as **45**, are very stable with a triplet ground state.[291] Diradicals are generally short-lived species. The lifetime of **46** was measured to be <0.1 ns and other diradicals were found to have lifetimes in the 4–316-ns range.[292] Diradical **47** [3,5-di-*tert*-butyl-3'-(N-*tert*-butyl-N-aminoxy)-4-oxybiphenyl] was found to have a lifetime of weeks even in the presence of oxygen, and survived brief heating in toluene up to $\sim60°C$.[293] Radicals with both unpaired electrons on the same carbon are discussed under carbenes.

44　　　　**45**　　　　**46**　　　　**47**

[288]For reviews of trimethylenemethane, see Borden, W.T.; Davidson, E.R. *Ann. Rev. Phys. Chem.* **1979**, *30*, 125; Bergman, R.G., in Kochi, J.K. *Free Radicals*, Vol. 1, Wiley, NY, **1973**, pp. 141–149.

[289]For discussions of the triplet state, see Wagner, P.J.; Hammond, G.S. *Adv. Photochem.* **1968**, *5*, 21; Turro, N.J. *J. Chem. Educ.* **1969**, *46*, 2. For a discussion of esr spectra of triplet states, see Wasserman, E.; Hutton, R.S. *Acc. Chem. Res.* **1977**, *10*, 27. For the generation and observation of triplet 1,3-biradicals see Ichinose, N.; Mizuno, K.; Otsuji, Y.; Caldwell, R.A.; Helms, A.M. *J. Org. Chem.* **1998**, *63*, 3176.

[290]Matsuda, K.; Iwamura, H. *J. Chem. Soc. Perkin Trans. 2* **1998**, 1023. Also see, Roth, W.R.; Wollweber, D.; Offerhaus, R.; Rekowski, V.; Lenmartz, H.-W.; Sustmann, R.; Müller, W. *Chem. Ber.* **1993**, *126*, 2701.

[291]Inoue, K.; Iwamura, H. *Angew. Chem. Int. Ed.* **1995**, *34*, 927. Also see, Ulrich, G.; Ziessel, R.; Luneau, D.; Rey, P. *Tetrahedron Lett.* **1994**, *35*, 1211.

[292]Engel, P.S.; Lowe, K.L. *Tetrahedron Lett.* **1994**, *35*, 2267.

[293]Liao, Y.; Xie, C.; Lahti, P.M.; Weber, R.T.; Jiang, J.; Barr, D.P. *J. Org. Chem.* **1999**, *64*, 5176.

The Generation and Fate of Free Radicals[294]

Free radicals are formed from molecules by breaking a bond so that each fragment keeps one electron.[295,296] The energy necessary to break the bond is supplied in one of two ways.

1. *Thermal Cleavage*. Subjection of any organic molecule to a high enough temperature in the gas phase results in the formation of free radicals. When the molecule contains bonds with D values or $20\text{–}40 \, \text{kcal mol}^{-1}$ (80–170 kJ mol^{-1}), cleavage can be caused in the liquid phase. Two common examples are cleavage of diacyl peroxides to acyl radicals that decompose to alkyl radicals[297] and cleavage of azo compounds to alkyl radicals[298]

$$R-\overset{\overset{\displaystyle O}{\|}}{C}-O-O-\overset{\overset{\displaystyle }{}}{\underset{\underset{\displaystyle O}{\|}}{C}}-R \quad \xrightarrow{\Delta} \quad 2 \; R-\overset{\overset{\displaystyle O}{\|}}{C}-O\cdot \quad \xrightarrow{-CO_2} \quad 2 \; R^{\bullet}$$

$$R-N=N-R \quad \xrightarrow{\Delta} \quad 2 \; R^{\bullet} \; + \; N_2$$

2. *Photochemical Cleavage* (see p. 335). The energy of light of 600–300 nm is $48\text{–}96 \, \text{kcal mol}^{-1}$ (200–400 kJ mol^{-1}), which is of the order of magnitude of covalent-bond energies. Typical examples are photochemical cleavage of alkyl halides in the presence of triethylamine,[299] alcohols in the presence of mercuric oxide and iodine,[300] alkyl 4-nitrobenzenesulfenates,[301] chlorine, and of ketones:

$$Cl_2 \quad \xrightarrow{h\nu} \quad 2 \; Cl^{\bullet}$$

$$R-\overset{\overset{\displaystyle R}{}}{\underset{\underset{\displaystyle O}{\|}}{C}} \quad \xrightarrow[\text{vapor phase}]{h\nu} \quad R-\overset{\overset{\displaystyle \bullet}{}}{\underset{\underset{\displaystyle O}{\|}}{C}} \; + \; R^{\bullet}$$

The photochemistry of radicals and biradicals has been reviewed.[302]

[294]For a summary of methods of radical formation, see Giese, B. *Radicals in Organic Synthesis: Formation of Carbon-Carbon Bonds*; Pergamon: Elmsford, NY, *1986*, pp. 267–281. For a review on formation of free radicals by thermal cleavage, see Brown, R.F.C. *Pyrolytic Methods in Organic Chemistry*; Academic Press, NY, *1980*, pp. 44–61.

[295]It is also possible for free radicals to be formed by the collision of two nonradical species. For a review, see Harmony, J.A.K. *Methods Free-Radical Chem.* *1974*, *5*, 101.

[296]For a review of homolytic cleavage of carbon–metal bonds, see Barker, P.J.; Winter, J.N., in Hartley, F.R.; Patai, S. *The Chemistry of the Metal-Carbon Bond*, Vol. 2, Wiley, NY, *1985*, pp. 151–218.

[297]Chateauneuf, J.; Lusztyk, J.; Ingold, K.U. *J. Am. Chem. Soc.* *1988*, *110*, 2877, 2886; Matsuyama, K.; Sugiura, T.; Minoshima, Y. *J. Org. Chem.* *1995*, *60*, 5520; Ryzhkov, L.R. *J. Org. Chem.* *1996*, *61*, 2801. For a review of free radical mechanisms involving peroxides in solution, see Howard, J.A., in Patai, S. *The Chemistry of Peroxides*, Wiley, NY, *1983*, pp. 235–258. For a review of pyrolysis of peroxides in the gas phase, see Batt, L.; Liu, M.T.H. in the same volume, pp. 685–710.

[298]For a review of the cleavage of azoalkanes, see Engel, P.S. *Chem. Rev.* *1980*, *80*, 99. For summaries of later work, see Adams, J.S.; Burton, K.A.; Andrews, B.K.; Weisman, R.B.; Engel, P.S. *J. Am. Chem. Soc.* *1986*, *108*, 7935; Schmittel, M.; Rüchardt, C. *J. Am. Chem. Soc.* *1987*, *109*, 2750.

[299]Cossy, J.; Ranaivosata, J.-L.; Bellosta, V. *Tetrahedron Lett.* *1994*, *35*, 8161.

[300]Courtneidge, J.L. *Tetrahedron Lett.* *1992*, *33*, 3053.

[301]Pasto, D.J.; Cottard, F. *Tetrahedron Lett.* *1994*, *35*, 4303.

[302]Johnston, L.J. *Chem. Rev.* *1993*, *93*, 251.

Radicals are also formed from other radicals, either by the reaction between a radical and a molecule (which *must* give another radical, since the total number of electrons is odd) or by cleavage of a radical[303] to give another radical, for example,

$$Ph-\underset{\underset{O}{\overset{\|}{C}}}{}-O^{\bullet} \longrightarrow Ph\bullet + CO_2$$

Radicals can also be formed by oxidation or reduction, including electrolytic methods.

Reactions of free radicals either give stable products (termination reactions) or lead to other radicals, which themselves must usually react further (propagation reactions). The most common termination reactions are simple combinations of similar or different radicals:

$$R\bullet + R'\bullet \longrightarrow R{-}R'$$

Another termination process is disproportionation:[304]

There are four principal propagation reactions, of which the first two are most common:

$$2\,CH_3{-}CH_2^{\bullet} \longrightarrow CH_3{-}CH_3 + CH_2{=}CH_2$$

1. *Abstraction of Another Atom or Group, Usually a Hydrogen Atom* (see Chapter 14):

$$R\bullet + R'{-}H \longrightarrow R{-}H + R'\bullet$$

2. *Addition to a Multiple Bond* (see Chapter 15):

$$R\bullet + \overset{\diagdown}{\underset{\diagup}{C}}{=}\overset{\diagup}{\underset{\diagdown}{C}} \longrightarrow R{-}\overset{\diagdown}{\underset{\diagup}{C}}{-}\overset{\diagup}{\underset{\diagdown}{C}}\bullet$$

The radical formed here may add to another double bond and so on. This is one of the chief mechanisms for vinyl polymerization.

3. *Decomposition.* This can be illustrated by the decomposition of the benzoxy radical (above).

4. *Rearrangement*:

$$\underset{R}{\overset{R}{\underset{|}{R-\overset{\displaystyle C}{}}}}\!\!-\overset{\bullet}{C}H_2 \longrightarrow R-\overset{\overset{\textstyle R}{|}}{\overset{\bullet}{C}}-\underset{H_2}{C}{-}R$$

[303]For a deterimination of activation barriers in the homolytic cleavage of radicals and ion radicals see Costentin, C.; Robert, M.; Saveant, J.-M. *J. Am. Chem. Soc.* **2003**, *125*, 105.
[304]For reviews of termination reactions, see Pilling, M.J. *Int. J. Chem. Kinet.* **1989**, *21*, 267; Khudyakov, I.V.; Levin, P.P.; Kuz'min, V.A. *Russ. Chem. Rev.* **1980**, *49*, 982; Gibian, M.J.; Corley, R.C. *Chem. Rev.* **1973**, *73*, 441.

This is less common than rearrangement of carbocations, but it does occur (though not when R = alkyl or hydrogen; see Chapter 18). Perhaps the best-known rearrangement is that of cyclopropylcarbinyl radicals to a butenyl radical.[305] The rate constant for this rapid ring opening has been measured in certain functionalized cyclopropylcarbinyl radicals by picosecond radical kinetics.[306] Substituent effects on the kinetics of ring opening in substituted cyclopropylcarbinyl radicals has been studied.[307] "The cyclopropylcarbinyl radical has found an important application as a radical clock.[308] Various radical processes can be clocked by the competition of direct reaction with the cyclopropylcarbinyl radical (k_t) and opening of that radical to the 1-buten-4-yl radical (k_r) followed by trapping. Relative rates (k_t/k_r) can be determined from yields of 4-X-1-butene and cyclopropylcarbinyl products as a function of the radical trap[309] (X—Y) concentration. Absolute rate constants have been determined for a number of radicals with various radical traps by laser flash photolysis methods.[310] From these absolute rate constants, reasonably accurate values of k_t can be estimated, and with the relative rate (k_t/k_r), a value for k_r can be calculated. From the calibrated radical-clock reaction rate (k_r), rates (k_t) of other competing reactions can be determined from relative rate data (k_t/k_r)."[306] Other radical clocks are known.[311]

Free radicals can also be oxidized to carbocations or reduced to carbanions.[312]

---

[305]For a discussion of radical vs. radical anion character see Stevenson, J. P.; Jackson, W. F.; Tanko, J. M. *J. Am. Chem. Soc.* **2002**, *124*, 4271.

[306]LeTadic-Biadatti, M.-H.; Newcomb, M. *J. Chem. Soc. Perkin Trans. 2* **1996**, 1467. See also, Choi, S.-Y.; Horner, J.H.; Newcomb, M. *J. Org. Chem.* **2000**, *65*, 4447. For determination of k for rearrangement and for and competing reactions, see Cooksy, A. L.; King, H.F.; Richardson, W.H. *J. Org. Chem.* **2003**, *68*, 9441. For the ring opening of fluorinated cyclopropylcarbinyl systems see Tian, F.; Dolbier Jr., W.R. *Org. Lett.* **2000**, *2*, 835.

[307]Halgren, T.A.; Roberts, J.D.; Horner, J.H.; Martinez, F.N.; Tronche, C.; Newcomb, M. *J. Am. Chem. Soc.* **2000**, *122*, 2988.

[308]Griller, D.; Ingold, K.U. *Acc. Chem. Res.* **1980**, *13*, 317; Newcomb, M.; Choi, S.-Y.; Toy, P.H. *Can. J. Chem.* **1999**, *77*, 1123; Le Tadic-Biadatti, M.-H.; Newcomb, M. *J. Chem. Soc., Perkin Trans. 2* **1996**, 1467; Choi, S.Y.; Newcomb, M. *Tetrahedron* **1995**, *51*, 657; Newcomb, M. *Tetrahedron* **1993**, *49*, 1151; Newcomb, M.; Johnson, C.; Manek, M.B.; Varick, T.R. *J. Am. Chem. Soc.* **1992**, *114*, 10915; Nevill, S.M.; Pincock, J.A. *Can. J. Chem.* **1997**, *75*, 232.

[309]For an alkyl radical trap in aqueous medium see Barton, D.H.R.; Jacob, M.; Peralez, E. *Tetrahedron Lett.* **1999**, *40*, 9201.

[310]Choi, S.-Y.; Horner, J.H.; Newcomb, M. *J. Org. Chem.* **2000**, *65*, 4447; Engel, P.S.; He, S.-L.; Banks, J.T.; Ingold, K.U.; Lusztyk, J. *J. Org. Chem.* **1997**, *62*, 1210; Johnston, L.J.; Lusztyk, J.; Wayner, D.D.M.; Abeywickreyma, A.N.; Beckwith, A.L.J.; Scaiano, J.J.; Ingold, K.U. *J. Am. Chem. Soc.* **1985**, *107*, 4594; Chatgilialoglu, C.; Ingold, K.U.; Scaiano, J.J. *J. Am. Chem. Soc.* **1981**, *103*, 7739.

[311]For example, see Leardini, R.; Lucarini, M.; Pedulli, G.F.; Valgimigli, L. *J. Org. Chem.* **1999**, *64*, 3726.

[312]For a review of the oxidation and reduction of free radicals, see Khudyakov, I.V.; Kuz'min, V.A. *Russ. Chem. Rev.* **1978**, *47*, 22.

Radical Ions[313]

Several types of radical anions are known with the unpaired electron or the charge or both on atoms other than carbon. Examples include semiquinones[314] (**48**),

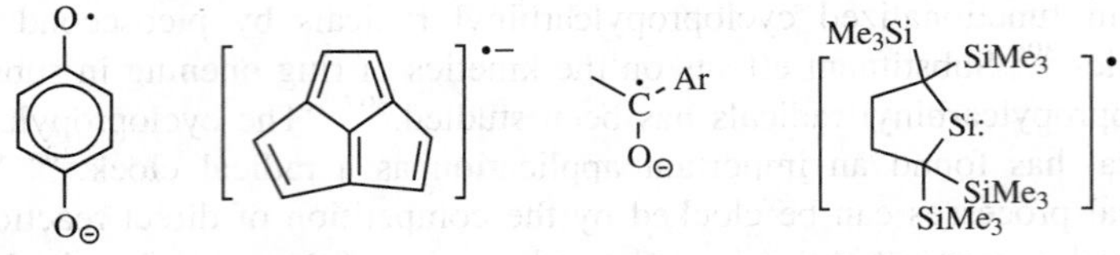

48   **49**   **50**   **51**

M = Cs, Rb, Na, Li

acepentalenes (**49**),[315] ketyls[316] (**50**) and the radical anion of the isolable dialkylsilylene **51**.[317] Reactions in which alkali metals are reducing agents often involve radical anion intermediates, for example, reaction **15-13**:

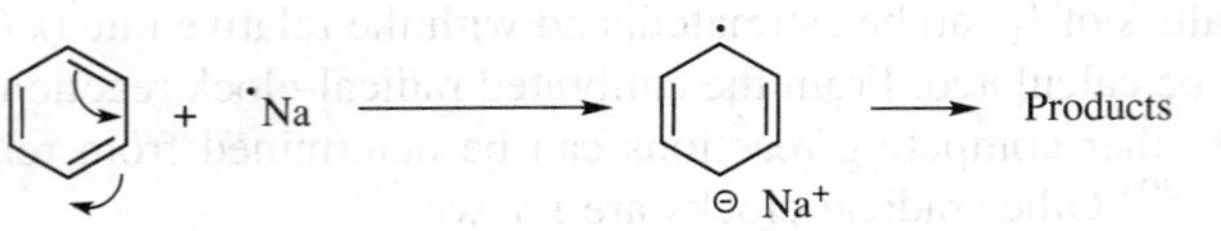

Several types of radical cation are also known.[318] Typical examples include alkyl azulene cation radicals (**52**),[319] trialkyl amine radical cations,[320]

[313]For a monograph, see Kaiser, E.T.; Kevan, L. *Radical Ions*, Wiley, NY, *1968*. For reviews, see Gerson, F.; Huber, W. *Acc. Chem. Res. 1987*, *20*, 85; Todres, Z.V. *Tetrahedron 1985*, *41*, 2771; Russell, G.A.; Norris, R.K., in McManus, S.P. *Organic Reactive Intermediates*; Academic Press, NY, *1973*, pp. 423–448; Holy, N.L.; Marcum, J.D. *Angew. Chem. Int. Ed. 1971*, *10*, 115; Bilevitch, K.A.; Okhlobystin, O.Yu. *Russ. Chem. Rev. 1968*, *37*, 954; Szwarc, M. *Prog. Phys. Org. Chem. 1968*, *6*, 322. For a related review, see Chanon, M.; Rajzmann, M.; Chanon, F. *Tetrahedron 1990*, *46*, 6193. For a series of papers on this subject, see *Tetrahedron 1986*, *42*, 6097.

[314]For a review of semiquinones, see Depew, M.C.; Wan, J.K.S., in Patai, S.; Rappoport, Z. *The Chemistry of the Quinonoid Compounds*, Vol. 2, pt. 2, Wiley, NY, *1988*, pp. 963–1018. For a discussion of the thermodynamic stability of aromatic radical anions see Huh, C.; Kang, C.H.; Lee, H.W.; Nakamura, H.; Mishima, M.; Tsuno, Y.; Yamataka, H. *Bull. Chem. Soc. Jpn. 1999*, *72*, 1083.

[315]de Meijere, A.; Gerson, F.; Schreiner, P.R.; Merstetter, P.; Schüngel, F.-M. *Chem. Commun. 1999*, 2189.

[316]For a review of ketyls, see Russell, G.A., in Patai, S.; Rappoport, Z. *The Chemistry of Enones*, pt. 1, Wiley, NY, *1989*, pp. 471–512. See Davies, A.G.; Neville, A.G. *J. Chem. Soc. Perkin Trans. 2 1992*, 163, 171 for ketyl and thioketyl cation radicals.

[317]Ishida, S.; Iwamoto, T.; Kira, M. *J. Am. Chem. Soc. 2003*, *125*, 3212. For bis(tri-*tert*-butylsilyl)silylene: triplet ground state silylene see Sekiguchi, A.; Tanaka, T.; Ichinohe, M.; Akiyama, K.; Tero-Kubota, S. *J. Am. Chem. Soc. 2003*, *125*, 4962.

[318]For reviews, see Roth, H.D. *Acc. Chem. Res. 1987*, *20*, 343; Courtneidge, J.L.; Davies, A.G. *Acc. Chem. Res. 1987*, *20*, 90; Hammerich, O.; Parker, V.D. *Adv. Phys. Org. Chem. 1984*, *20*, 55; Symons, M.C.R. *Chem. Soc. Rev. 1984*, *13*, 393; Bard, A.J.; Ledwith, A.; Shine, H.J. *Adv. Phys. Org. Chem. 1976*, *13*, 155.

[319]Gerson, F.; Scholz, M.; Hansen, H.-J.; Uebelhart, P. *J. Chem. Soc. Perkin Trans. 2 1995*, 215.

[320]de Meijere, A.; Chaplinski, V.; Gerson, F.; Merstetter, P.; Haselbach, E. *J. Org. Chem. 1999*, *64*, 6951.

1,2-bis(dialkylamino)benzenes radical cations, such as **53**,[321] dimethylsulfonium cation radicals (Me$_2$S$^{+\bullet}$),[322] N-alkyl substituted imine cation radicals (Ph$_2$C=NEt$^{\bullet+}$),[323] dibenzo[a,e]cyclooctene (**54**, a nonplanar cation radical),[324] and [n.n]paracyclophane cation radicals.[325] A twisted radical cation derived from bicyclo[2.2.2]oct-2-ene has been reported.[326]

| 52 | 53 | 54 |

CARBENES

Stability and Structure[327]

Carbenes are highly reactive species, practically all having lifetimes considerably under 1 s. With exceptions noted below (p. 289), carbenes have been isolated only by entrapment in matrices at low temperatures (77 K or less).[328] The parent species CH$_2$ is usually called *methylene*, although derivatives are more often named by the carbene nomenclature. Thus CCl$_2$ is generally known as dichlorocarbene, although it can also be called dichloromethylene.

---

[321]Neugebauer, F.A.; Funk, B.; Staab, H.A. *Tetrahedron Lett.* **1994**, *35*, 4755. See Stickley, K.R.; Blackstock, S.C. *Tetrahedron Lett.* **1995**, *36*, 1585 for a *tris*-diarylaminobenzene cation radical.

[322]Dauben, W.G.; Cogen, J.M.; Behar, V.; Schultz, A.G.; Geiss, W.; Taveras, A.G. *Tetrahedron Lett.* **1992**, *33*, 1713.

[323]Rhodes, C.J.; AgirBas H. *J. Chem. Soc. Perkin Trans. 2* **1992**, 397.

[324]Gerson, F.; Felder, P.; Schmidlin, R.; Wong, H.N.C. *J. Chem. Soc. Chem. Commun.* **1994**, 1659.

[325]Wartini, A.R.; Valenzuela, J.; Staab, H.A.; Neugebauer, F.A. *Eur. J. Org. Chem.* **1998**, 139.

[326]Nelson, S.F.; Reinhardt, L.A.; Tran, H.Q.; Clark, T.; Chen, G.-F.; Pappas, R.S.; Williams, F. *Chem. Eur. J.* **2002**, *8*, 1074.

[327]For monographs, see Jones, Jr., M.; Moss, R.A. *Carbenes*, 2 vols., Wiley, NY, **1973–1975**; Kirmse, W. *Carbene Chemistry*, 2nd ed.; Academic Press, NY, **1971**; Rees, C.W.; Gilchrist, T.L. *Carbenes, Nitrenes, and Arynes*, Nelson, London, **1969**. For reviews, see Minkin, V.I.; Simkin, B.Ya.; Glukhovtsev, M.N. *Russ. Chem. Rev.* **1989**, *58*, 622; Moss, R.A.; Jones, Jr., M. *React. Intermed. (Wiley)* **1985**, *3*, 45; **1981**, *2*, 59; **1978**, *1*, 69; Isaacs, N.S. *Reactive Intermediates in Organic Chemistry*, Wiley, NY, **1974**, pp. 375–407; Bethell, D. *Adv. Phys. Org. Chem.* **1969**, *7*, 153; Bethell, D., in McManus, S.P. *Organic Reactive Intermediates*, Academic Press, NY, **1973**, pp. 61–126; Closs, G.L. *Top. Stereochem.* **1968**, *3*, 193; Herold, B.J.; Gaspar, P.P. *Fortschr. Chem. Forsch.*, **1966**, *5*, 89; Rozantsev, G.G.; Fainzil'berg, A.A.; Novikov, S.S. *Russ. Chem. Rev.* **1965**, *34*, 69. For a theoretical study, see Liebman, J.F.; Simons, J. *Mol. Struct. Energ.* **1986**, *1*, 51.

[328]For example, see Murray, R.W.; Trozzolo, A.M.; Wasserman, E.; Yager, W.A. *J. Am. Chem. Soc.* **1962**, *84*, 3213; Brandon, R.W.; Closs, G.L.; Hutchison, C.A. *J. Chem. Phys.* **1962**, *37*, 1878; Milligan, D.E.; Mann, D.E.; Jacox, M.E.; Mitsch, R.A. *J. Chem. Phys.* **1964**, *41*, 1199; Nefedov, O.M.; Maltsev, A.K.; Mikaelyan, R.G. *Tetrahedron Lett.* **1971**, 4125; Wright, B.B. *Tetrahedron* **1985**, *41*, 1517. For reviews, see Zuev, P.S.; Nefedov, O.M. *Russ. Chem. Rev.* **1989**, *58*, 636; Sheridan, R.S. *Org. Photochem.* **1987**, *8*, 159, pp. 196–216; Trozzolo, A.M. *Acc. Chem. Res.* **1968**, *1*, 329.

The two nonbonded electrons of a carbene can be either paired or unpaired. If they are paired, the species is spectrally a *singlet*, while, as we have seen (p. 278), two unpaired electrons appear as a *triplet*. An ingenious method of distinguishing

between the two possibilities was developed by Skell,[329] based on the common reaction of addition of carbenes to double bonds to form cyclopropane derivatives (**15-51**). If the singlet species adds to *cis*-2-butene, the resulting cyclopropane should be the cis isomer since the movements of the two pairs of electrons should

occur either simultaneously or with one rapidly succeeding another. However, if the attack is by a triplet species, the two unpaired electrons cannot both go into a new covalent bond, since by Hund's rule they have parallel spins. So one of the unpaired electrons will form a bond with the electron from the double bond that has the opposite spin, leaving two unpaired electrons that have the same spin and therefore cannot form a bond at once but must wait until, by some collision process, one of the electrons can reverse its spin. During this time, there is free rotation about the C—C bond and a mixture of *cis*- and *trans*-1,2-dimethylcyclopropanes should result.[330]

The results of this type of experiment show that CH_2 itself is usually formed as a singlet species, which can decay to the triplet state, which consequently has a lower energy (molecular-orbital calculations[331] and experimental determinations show that the difference in energy between singlet and triplet CH_2 is $\sim$8–10 kcal mol^{-1} or 33–42 kJ mol^{-1} [332]). However, it is possible to prepare triplet CH_2 directly by a

[329]Skell, P.S.; Woodworth, R.C. *J. Am. Chem. Soc.* **1956**, *78*, 4496; Skell, P.S. *Tetrahedron* **1985**, *41*, 1427.
[330]These conclusions are generally accepted though the reasoning given here may be oversimplified. For discussions, see Closs, G.L. *Top. Stereochem.* **1968**, *3*, 193, pp. 203–210; Bethell, D. *Adv. Phys. Org. Chem.* **1969**, *7*, 153, pp. 194; Hoffmann, R. *J. Am. Chem. Soc.* **1968**, *90*, 1475.
[331]Richards, Jr., C.A.; Kim, S.-J.; Yamaguchi, Y.; Schaefer III, H.F. *J. Am. Chem. Soc.* **1995**, *117*, 10104.
[332]See, for example, Hay, P.J.; Hunt, W.J.; Goddard III, W.A. *Chem. Phys. Lett.* **1972**, *13*, 30; Dewar, M.J.S.; Haddon, R.C.; Weiner, P.K. *J. Am. Chem. Soc.* **1974**, *96*, 253; Frey, H.M.; Kennedy, G.J. *J. Chem. Soc. Chem. Commun.* **1975**, 233; Lucchese, R.R.; Schaefer III, H.F. *J. Am. Chem. Soc.* **1977**, *99*, 6765; Roos, B.O.; Siegbahn, P.M. *J. Am. Chem. Soc.* **1977**, *99*, 7716; Lengel, R.K.; Zare, R.N. *J. Am. Chem. Soc.* **1978**, *100*, 7495; Borden, W.T.; Davidson, E.R. *Ann. Rev. Phys. Chem.* **1979**, *30*, 125, see pp. 128–134; Leopold, D.G.; Murray, K.K.; Lineberger, W.C. *J. Chem. Phys.* **1984**, *81*, 1048.

photosensitized decomposition of diazomethane.[333] The CH_2 group is so reactive[334] that it generally reacts as the singlet before it has a chance to decay to the triplet state.[335] As to other carbenes, some react as triplets, some as singlets, and others as singlets or triplets, depending on how they are generated. There are, however, molecules that generate persistent triplet carbenes.[336] Indeed, remarkably stable diaryl triplet carbenes have been prepared.[337]

There is a limitation to the use of stereospecificity of addition as a diagnostic test for singlet or triplet carbenes.[338] When carbenes are generated by photolytic methods, they are often in a highly excited singlet state. When they add to the double bond, the addition is stereospecific; but the cyclopropane formed carries excess energy; that is, it is in an excited state. It has been shown that under certain conditions (low pressures in the gas phase) the excited cyclopropane may undergo cis-trans isomerization *after* it is formed, so that triplet carbene may seem to be involved although in reality the singlet was present.[339]

Studies of the IR spectrum of CCl_2 trapped at low temperatures in solid argon indicate that the ground state for this species is the singlet.[340] The geometrical structure of triplet methylene can be investigated by esr measurements,[341] since triplet species are diradicals. Such measurements made on triplet CH_2 trapped in matrices at very low temperatures (4 K) show that triplet CH_2 is a bent molecule, with an angle of $\sim 136°$.[342] Epr measurements cannot be made on singlet species, but from electronic spectra of CH_2 formed in flash photolysis of diazomethane it was concluded that singlet CH_2 is also bent, with an angle of $\sim 103°$.[343] Singlet CCl_2[286] and CBr_2[344] are also bent, with angles of 100 and 114°, respectively. It

---

[333]Kopecky, K.R.; Hammond, G.S.; Leermakers, P.A. *J. Am. Chem. Soc.* **1961**, *83*, 2397; **1962**, *84*, 1015; Duncan, F.J.; Cvetanović, R.J. *J. Am. Chem. Soc.* **1962**, *84*, 3593.

[334]For a review of the kinetics of CH_2 reactions, see Laufer, A.H. *Rev. Chem. Intermed.* **1981**, *4*, 225.

[335]Decay of singlet and triplet CH_2 has been detected in solution, as well as in the gas phase: Turro, N.J.; Cha, Y.; Gould, I.R. *J. Am. Chem. Soc.* **1987**, *109*, 2101.

[336]Tomioka, H. *Acc. Chem. Res.* **1997**, *30*, 315; Kirmse, W. *Angew. Chem. Int. Ed.* **2003**, *42*, 2117.

[337]Hirai, K.; Tomioka, H. *J. Am. Chem. Soc.* **1999**, *121*, 10213; Woodcock, H.L.; Moran, D.; Schleyer, P.v.R.; Schaefer III, H.F. *J. Am. Chem. Soc.* **2001**, *123*, 4331.

[338]For other methods of distinguishing singlet from triplet carbenes, see Hendrick, M.E.; Jones Jr., M. *Tetrahedron Lett.* **1978**, 4249; Creary, X. *J. Am. Chem. Soc.* **1980**, *102*, 1611.

[339]Rabinovitch, B.S.; Tschuikow-Roux, E.; Schlag, E.W. *J. Am. Chem. Soc.* **1959**, *81*, 1081; Frey, H.M. *Proc. R. Soc. London, Ser. A* **1959**, *251*, 575. It has been reported that a singlet carbene (CBr_2) can add nonstereospecifically: Lambert, J.B.; Larson, E.G.; Bosch, R.J. *Tetrahedron Lett.* **1983**, *24*, 3799.

[340]Andrews, L. *J. Chem. Phys.* **1968**, *48*, 979.

[341]The technique of spin trapping (p. 268) has been applied to the detection of transient triplet carbenes: Forrester, A.R.; Sadd, J.S. *J. Chem. Soc. Perkin Trans. 2* **1982**, 1273.

[342]Wasserman, E.; Kuck, V.J.; Hutton, R.S.; Anderson, E.D.; Yager, W.A. *J. Chem. Phys.* **1971**, *54*, 4120; Bernheim, R.A.; Bernard, H.W.; Wang, P.S.; Wood, L.S.; Skell, P.S. *J. Chem. Phys.* **1970**, *53*, 1280; **1971**, *54*, 3223.

[343]Herzberg, G.; Johns, J.W.C. *Proc. R. Soc. London, Ser. A* **1967**, *295*, 107, *J. Chem. Phys.* **1971**, *54*, 2276 and cited references.

[344]Ivey, R.C.; Schulze, P.D.; Leggett, T.L.; Kohl, D.A. *J. Chem. Phys.* **1974**, *60*, 3174.

has long been known that triplet aryl carbenes are bent.[345]

| Triplet methylene (136°) | Singlet methylene (103°) |

The most common carbenes are :CH_2 and: CCl_2,[346] but many others have been reported, [347] including heterocyclic carbenes, such as **55** (stabilized by the steric constraints of the ring geometry),[348] **56** (an aminocarbene without π conjugation),[349] bicyclo[2.2.2]octylidene, **57**,[350] alkylidene carbenes, such as **58**,[351] conformationally restricted cyclopropylcarbenes, such as **59**,[352] β-Silylcarbenes, such as **60**,[353] α-keto carbenes,[354] vinyl carbenes,[355] and chiral carbenoids.[356] In the case of **55** (R = Ph),[357] the precursor is a tetraaminoethylene, and when potassium hydride is present to preclude electrophilic catalysis, starting tetraaminoethylenes are recovered unchanged.

| **55** | **56** | **57** | **58** | **59** | **60** |

[345]Trozzolo, A.M.; Wasserman, E.; Yager, W.A. *J. Am. Chem. Soc.* **1965**, *87*, 129; Senthilnathan, V.P.; Platz, M.S. *J. Am. Chem. Soc.* **1981**, *103*, 5503; Gilbert, B.C.; Griller, D.; Nazran, A.S. *J. Org. Chem.* **1985**, *50*, 4738.

[346]For reviews of halocarbenes, see Burton, D.J.; Hahnfeld, J.L. *Fluorine Chem. Rev.* **1977**, *8*, 119; Margrave, J.L.; Sharp, K.G.; Wilson, P.W. *Fort. Chem. Forsch.* **1972**, *26*, 1, pp. 3–13.

[347]For reviews of unsaturated carbenes, see Stang, P.J. *Acc. Chem. Res.* **1982**, *15*, 348; *Chem. Rev.* **1978**, *78*, 383. For a review of carbalkoxycarbenes, see Marchand, A.P.; Brockway, N.M. *Chem. Rev.* **1974**, *74*, 431. For a review of arylcarbenes, see Schuster, G.B. *Adv. Phys. Org. Chem.* **1986**, *22*, 311. For a review of carbenes with neighboring hetero atoms, see Taylor, K.G. *Tetrahedron* **1982**, *38*, 2751.

[348]Denk, M.K.; Thadani, A.; Hatano, K.; Lough, A.J. *Angew. Chem. Int. Ed.* **1997**, *36*, 2607; Herrmann, W.A. *Angew. Chem. Int. Ed.* **2002**, *41*, 1290.

[349]Ye, Q.; Komarov, I.V.; Kirby, A.J.; Jones, Jr., M. *J. Org. Chem.* **2002**, *67*, 9288.

[350]Ye, Q.; Jones, Jr., M.; Chen, T.; Shevlin, P.B. *Tetrahedron Lett.* **2001**, *42*, 6979.

[351]Ohira, S.; Okai, K.; Moritani, T. *J. Chem. Soc. Chem. Commun.* **1992**, 721; Walsh, R.; Wolf, C.; Untiedt, S.; de Meijere, A. *J. Chem. Soc. Chem. Commun.* **1992**, 421, 422; Ohira, S.; Yamasaki, K.; Nozaki, H.; Yamato, M.; Nakayama, M. *Tetrahedron Lett.* **1995**, *36*, 8843. For dimethylvinylidene carbene, see Reed, S.C.; Capitosti, G.J.; Zhu, Z.; Modarelli, D.A. *J. Org. Chem.* **2001**, *66*, 287. For a review of akylidenecarbenes, see Knorr, R. *Chem. Rev.* **2004**, *104*, 3795.

[352]Fernamberg, K.; Snoonian, J.R.; Platz, M.S. *Tetrahedron Lett.* **2001**, *42*, 8761.

[353]Creary, X.; Butchko, M.A. *J. Org. Chem.* **2002**, *67*, 112.

[354]Bonnichon, F.; Richard, C.; Grabner, G. *Chem. Commun.* **2001**, 73.

[355]Zuev, P.S.; Sheridan, R.S. *J. Am. Chem. Soc.* **2004**, *126*, 12220.

[356]Topolski, M.; Duraisamy, M.; Rachoń, J.; Gawronski, J.; Gawronska, K.; Goedken, V.; Walborsky, H.M. *J. Org. Chem.* **1993**, *58*, 546.

[357]See Wanzlick, H.-W.; Schikora, E. *Angew. Chem.* **1960**, *72*, 494.

Flash photolysis of $CHBr_3$ produced the intermediate CBr.[358]

$$CHBr_3 \xrightarrow[\text{photolysis}]{\text{flash}} \cdot \bar{C}\text{-Br}$$

This is a *carbyne*. The intermediates CF and CCl were generated similarly from $CHFBr_2$ and $CHClBr_2$, respectively.

The Generation and Fate of Carbenes[359]

Carbenes are chiefly formed in two ways, although other pathways are also known.

1. In α elimination, a carbon loses a group without its electron pair, usually a proton, and then a group with its pair, usually a halide ion:[360]

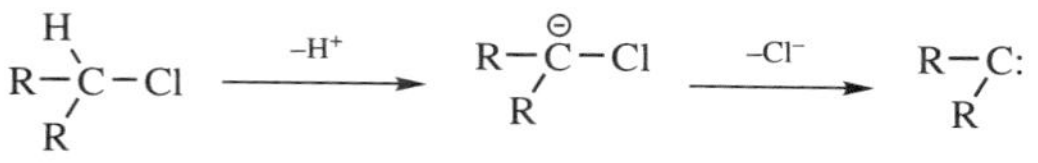

$$R\text{-}\underset{R}{\overset{H}{C}}\text{-Cl} \xrightarrow{-H^+} R\text{-}\underset{R}{\overset{\ominus}{C}}\text{-Cl} \xrightarrow{-Cl^-} R\text{-}\underset{R}{C}:$$

The most common example is formation of dichlorocarbene by treatment of chloroform with a base (see reaction **10-3**) and geminal alkyl dihalides with Me_3Sn^-,[361] but many other examples are known, such as

Ref. 362

$$CCl_3\text{-}COO^{\ominus} \xrightarrow{\Delta} CCl_2 + CO_2 + Cl^{\ominus}$$

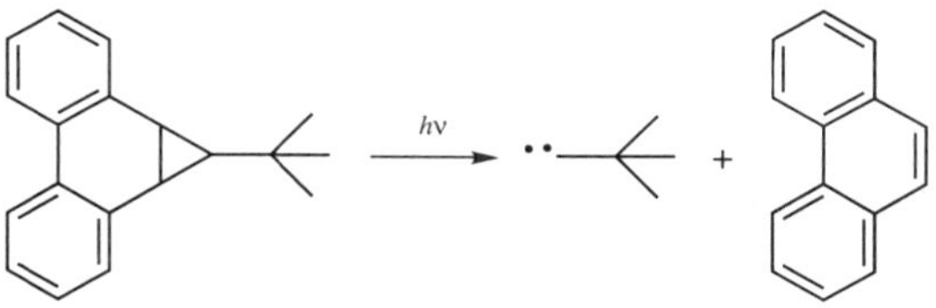

Ref. 363

2. Disintegration of compounds containing certain types of double bonds:

$$R_2C{=}Z \longrightarrow R_3C: + Z$$

[358]Ruzsicska, B.P.; Jodhan, A.; Choi, H.K.J.; Strausz, O.P. *J. Am. Chem. Soc.* **1983**, *105*, 2489.
[359]For reviews, see Jones Jr., M. *Acc. Chem. Res.* **1974**, *7*, 415; Kirmse, W., in Bamford, C.H.; Tipper, C.F.H. *Comprehensive Chemical Kinetics*, Vol. 9; Elsevier, NY, **1973**, pp. 373–415; Ref. 327. For a review of electrochemical methods of carbene generation, see Petrosyan, V.E.; Niyazymbetov, M.E. *Russ. Chem. Rev.* **1989**, *58*, 644.
[360]For a review of formation of carbenes in this manner, see Kirmse, W. *Angew. Chem. Int. Ed.* **1965**, *4*, 1.
[361]Ashby, E.C.; Deshpande, A.K.; Doctorovich, F. *J. Org. Chem.* **1993**, *58*, 4205.
[362]Wagner, W.M. *Proc. Chem. Soc.* **1959**, 229.
[363]Glick, H.C.; Likhotvovik, I.R.; Jones Jr., M. *Tetrahedron Lett.* **1995**, *36*, 5715; Stang, P.J. *Acc. Chem. Res.* **1982**, *15*, 348; *Chem. Rev.* **1978**, *78*, 383.

The two most important ways of forming :CH_2 are examples: the photolysis of ketene

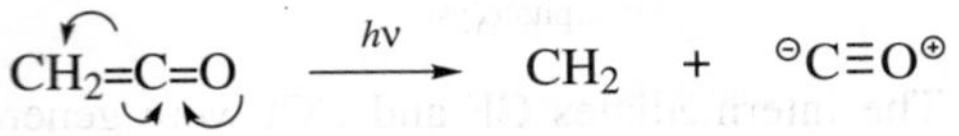

$$CH_2{=}C{=}O \xrightarrow{\ h\nu\ } CH_2 \ + \ {}^{\ominus}C{\equiv}O^{\oplus}$$

and the isoelectronic decomposition of diazomethane.[364]

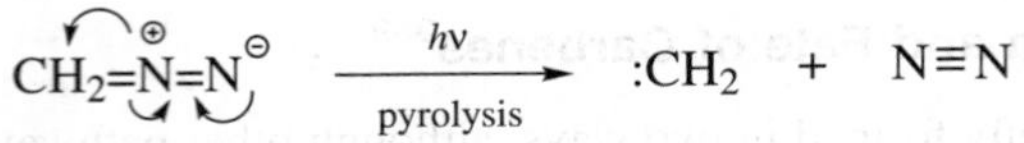

$$CH_2{=}N{=}N^{\ominus} \xrightarrow[\text{pyrolysis}]{\ h\nu\ } \ :CH_2 \ + \ N{\equiv}N$$

Diazirines[365] (isomeric with diazoalkanes) give carbenes,[366] but arylmethyl radicals have also been generated from diazirines.[367] In a different study, thermolysis of diaryloxydiazirines gave the anticipated carbene products, but photolysis gave both carbenes and aryloxy radicals by α-scission.[368]

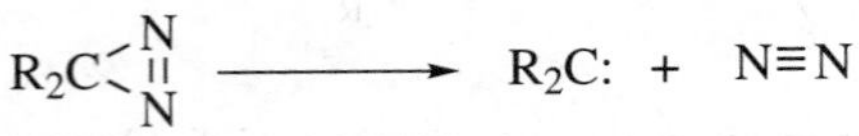

$$R_2C\!\!\underset{N}{\overset{N}{\left\langle\,\|\right.}} \longrightarrow R_2C: \ + \ N{\equiv}N$$

Because most carbenes are so reactive, it is often difficult to prove that they are actually present in a given reaction. The lifetime of formylcarbene was measured by transient absorption and transient grating spectroscopy to be 0.15–0.73 ns in dichloromethane.[369] In many instances where a carbene is *apparently* produced by an α elimination or by disintegration of a double-bond compound there is evidence that no free carbene is actually involved. The neutral term *carbenoid* is used where it is known that a free carbene is not present or in cases where there is doubt. α-Halo organometallic compounds, R_2CXM, are often called carbenoids because they readily give a elimination reactions[370] (e.g., see **12-39**).

The reactions of carbenes are more varied than those of the species previously discussed in this chapter. Solvent effects have been observed in carbene reactions. The selectivity of certain carbenes is influenced by the nature of the solvent.[371] the distribution of rearrangement products (see below) from *tert*-butylcarbene[372] are

[364]For a review, see Regitz, M.; Maas, G. *Diazo Compounds*, Academic Press, NY, *1986*, pp. 170–184.
[365]For syntheses, see Martinu, T.; Dailey, W.P. *J. Org. Chem.* *2004*, *69*, 7359; Likhotvorik, I.R.; Tae, E.L.; Ventre, C.; Platz, M.S. *Tetahedron Lett.* *2000*, *41*, 795.
[366]For a treatise, see Liu, M.T.H. *Chemistry of Diazirines*, 2 vols., CRC Press, Boca Raton, FL, *1987*. For reviews, see Liu, M.T.H. *Chem. Soc. Rev.* *1982*, *11*, 127; Frey, H.M. *Adv. Photochem.* *1966*, *4*, 225.
[367]Moss, R.A.; Fu, X. *Org. Lett.* *2004*, *6*, 3353.
[368]Fede, J.-M.; Jockusch, S.; Lin, N.; Moss, R.A.; Turro, N.J. *Org. Lett.* *2003*, *5*, 5027.
[369]Toscano, J.P.; Platz, M.S.; Nikolaev, V.; Cao, Y.; Zimmt, M.B. *J. Am. Chem. Soc.* *1996*, *118*, 3527.
[370]For a review, see Nefedov, O.M.; D'yachenko, A.I.; Prokof'ev, A.K. *Russ. Chem. Rev.* *1977*, *46*, 941.
[371]Tomioka, H.; Ozaki, Y.; Izawa, Y. *Tetrahedron* *1985*, *41*, 4987.
[372]Moss, R.A.; Yan, S.; Krogh-Jesperson, K. *J. Am. Chem. Soc.* *1998*, *120*, 1088.; Krogh-Jesperson, K.; Yan, S.; Moss, R.A. *J. Am. Chem. Soc.* *1999*, *121*, 6269.

influenced by changes in solvent.[373] It is known that singlet methylene forms a charge-transfer complex with benzene.[374] Solvent interactions for chlorophenylcarbene and fluorophenylcarbene, however, are weak.[375]

1. Additions to carbon–carbon double bonds have already been mentioned. Carbenes also add to aromatic systems, but the immediate products rearrange, usually with ring enlargement (see **15-65**). Additions of carbenes to other double bonds, such as C=N (**16-46** and **16-48**), and to triple bonds have also been reported.

2. An unusual reaction of carbenes is that of insertion into C–H bonds (**12-21**). Thus, :CH_2 reacts with methane to give ethane and with propane to give

n-butane and isobutane, as shown. Elimination to give an alkene is a competing side reaction in polar solvents, but this is suppressed in nonpolar solvents.[376] Simple alkyl carbenes, such as this, are not very useful for synthetic purposes, but do illustrate the extreme reactivity of carbene. However, carbenoids generated by rhodium catalyzed decomposition of diazoalkanes are very useful (p. 803) and have been used in a variety of syntheses. Treatment in the liquid phase of an alkane, such as pentane with carbene formed from the photolysis of diazomethane, gives the three possible products in statistical ratios[377] demonstrating that carbene is displaying no selectivity. For many years, it was a generally accepted principle that the lower the selectivity the greater the reactivity; however, this principle is no longer regarded as general because many exceptions have been found.[378] Singlet CH_2 generated by photolysis of diazomethane is probably the most reactive organic species known, but triplet CH_2 is somewhat less reactive, and other carbenes are still less reactive. The following series of carbenes of decreasing reactivity has

[373]Ruck, R.T.; Jones Jr., M. *Tetrahedron Lett.* **1998**, *39*, 2277.

[374]Khan, M.I.; Goodman, J.L. *J. Am. Chem. Soc.* **1995**, *117*, 6635.

[375]Sun, Y.; Tippmann, E.M.; Platz, M.S. *Org. Lett.* **2003**, *5*, 1305.

[376]Ruck, R.T.; Jones Jr., M. *Tetrahedron Lett.* **1998**, *39*, 2277.

[377]Doering, W. von E.; Buttery, R.G.; Laughlin, R.G.; Chaudhuri, N. *J. Am. Chem. Soc.* **1956**, *78*, 3224; Richardson, D.B.; Simmons, M.C.; Dvoretzky, I. *J. Am. Chem. Soc.* **1961**, *83*, 1934; Halberstadt, M.L.; McNesby, J.R. *J. Am. Chem. Soc.* **1967**, *89*, 3417.

[378]For reviews of this question, see Buncel, E.; Wilson, H. *J. Chem. Educ.* **1987**, *64*, 475; Johnson, C.D. *Tetrahedron* **1980**, *36*, 3461; *Chem. Rev.* **1975**, *75*, 755; Giese, B. *Angew. Chem. Int. Ed.* **1977**, *16*, 125; Pross, A. *Adv. Phys. Org. Chem.* **1977**, *14*, 69. See also, Ritchie, C.D.; Sawada, M. *J. Am. Chem. Soc.* **1977**, *99*, 3754; Argile, A.; Ruasse, M. *Tetrahedron Lett.* **1980**, *21*, 1327; Godfrey, M. *J. Chem. Soc. Perkin Trans. 2* **1981**, 645; Kurz, J.L.; El-Nasr, M.M.S. *J. Am. Chem. Soc.* **1982**, *104*, 5823; Srinivasan, C.; Shunmugasundaram, A.; Arumugam, N. *J. Chem. Soc. Perkin Trans. 2* **1985**, 17; Bordwell, F.G.; Branca, J.C.; Cripe, T.A. *Isr. J. Chem.* **1985**, *26*, 357; Formosinho, S.J. *J. Chem. Soc. Perkin Trans. 2* **1988**, 839; Johnson, C.D.; Stratton, B. *J. Chem. Soc. Perkin Trans. 2* **1988**, 1903. For a group of papers on this subject, see *Isr. J. Chem.* **1985**, *26*, 303.

been proposed on the basis of discrimination between insertion and addition reactions: $CH_2 > HCCOOR > PhCH > BrCH \sim ClCH.$[379] Dihalocarbenes generally do not give insertion reactions at all. Insertion of carbenes into other bonds has also been demonstrated, although not insertion into C—C bonds.[380]

Two carbenes that are stable at room temperature have been reported.[381] These are **61** and **62**. In the absence of oxygen and moisture, **61** exists as stable crystals with a melting point of 240–241°C.[382] Its structure was proved by X-ray crystallography.

61   **62**

3. It would seem that dimerization should be an important reaction of carbenes

$$R_2C\colon + R_2C\colon \longrightarrow R_2C{=}CR_2$$

but it is not, because the reactivity is so great that the carbene species do not have time to find each other and because the dimer generally has so much energy that it dissociates again. Apparent dimerizations have been observed, but it is likely that the products in many reported instances of "dimerization" do not arise from an actual dimerization of two carbenes but from attack by a carbene on a molecule of carbene precursor, for example,

$$R_2C\colon + R_2CN_2 \longrightarrow R_2C{=}CR_2 + N_2$$

[379]Closs, G.L.; Coyle, J.J. *J. Am. Chem. Soc.* **1965**, *87*, 4270.
[380]See, for example, Doering, W. von E.; Knox, L.H.; Jones, Jr., M. *J. Org. Chem.* **1959**, *24*, 136; Franzen, V. *Liebigs Ann. Chem.* **1959**, *627*, 22; Bradley, J.; Ledwith, A. *J. Chem. Soc.* **1961**, 1495; Frey, H.M.; Voisey, M.A. *Chem. Commun.* **1966**, 454; Seyferth, D.; Damrauer, R.; Mui, J.Y.; Jula, T.F. *J. Am. Chem. Soc.* **1968**, *90*, 2944; Tomioka, H.; Ozaki, Y.; Izawa, Y. *Tetrahedron* **1985**, *41*, 4987; Frey, H.M.; Walsh, R.; Watts, I.M. *J. Chem. Soc. Chem. Commun.* **1989**, 284.
[381]For a discussion, see Regitz, M. *Angew. Chem. Int. Ed.* **1991**, *30*, 674.
[382]Arduengo III, A.J.; Harlow, R.L.; Kline, M. *J. Am. Chem. Soc.* **1991**, *113*, 361.

4. Alkylcarbenes can undergo rearrangement, with migration of alkyl or hydrogen.[383] Indeed these rearrangements are generally so rapid[384] that additions to multiple bonds and insertion reactions, which are so common for CH_2, are seldom encountered with alkyl or dialkyl carbenes. Unlike rearrangement of the species previously encountered in this chapter, most rearrangements of carbenes directly give stable molecules. A carbene intermediate has been suggested for the isomerization of cyclopropane.[385] Some examples of carbene rearrangement are

The rearrangement of acylcarbenes to ketenes is called the Wolff rearrangement (reaction **18-8**). A few rearrangements in which carbenes rearrange to other carbenes are also known.[390] Of course, the new carbene must stabilize itself in one of the ways we have mentioned.

---

[383]For a probe of migratory aptitudes of hydrogen to carbenes see Locatelli, F.; Candy, J.-P.; Didillon, B.; Niccolai, G.P.; Uzio, D.; Basset, J.-M. *J. Am. Chem. Soc.* **2001**, *123*, 1658. For reviews of carbene and nitrene rearrangements, see Brown, R.F.C. *Pyrolytic Methods in Organic Chemistry*, Academic Press, NY, **1980**, pp. 115–163; Wentrup, C. *Adv. Heterocycl. Chem.* **1981**, *28*, 231; *React. Intermed. (Plenum)* **1980**, *1*, 263; *Top. Curr. Chem.* **1976**, *62*, 173; Jones, W.M., in de Mayo, P. *Rearrangements in Ground and Excited States*, Vol. 1, Academic Press, NY, **1980**, pp. 95–160; Schaefer III, H.F. *Acc. Chem. Res.* **1979**, *12*, 288; Kirmse, W. *Carbene Chemistry*, 2nd ed., Academic Press, NY, **1971**, pp. 457–496.
[384]The activation energy for the 1,2-hydrogen shift has been estimated at 1.1 kcal mol^{-1} (4.5 kJ mol^{-1}), an exceedingly low value: Stevens, I.D.R.; Liu, M.T.H.; Soundararajan, N.; Paike, N. *Tetrahedron Lett.* **1989**, *30*, 481. Also see, Pezacki, J.P.; Couture, P.; Dunn, J.A.; Warkentin, J.; Wood, P.D.; Lusztyk, J.; Ford, F.; Platz, M.S. *J. Org. Chem.* **1999**, *64*, 4456.
[385]Bettinger, H.F.; Rienstra-Kiracofe, J.C.; Hoffman, B.C.; Schaefer III, H.F.; Baldwin, J.E.; Schleyer, P.v.R. *Chem. Commun.* **1999**, 1515.
[386]Kirmse, W.; Doering, W. von E. *Tetrahedron* **1960**, *11*, 266. For kinetic studies of the rearrangement: Cl–C̄–CHR$_2$ → ClCH=CR$_2$, see Liu, M.T.H.; Bonneau, R. *J. Am. Chem. Soc.* **1989**, *111*, 6873; Jackson, J.E.; Soundararajan, N.; White, W.; Liu, M.T.H.; Bonneau, R.; Platz, M.S. *J. Am. Chem. Soc.* **1989**, *111*, 6874; Ho, G.; Krogh-Jespersen, K.; Moss, R.A.; Shen, S.; Sheridan, R.S.; Subramanian, R. *J. Am. Chem. Soc.* **1989**, *111*, 6875; LaVilla, J.A.; Goodman, J.L. *J. Am. Chem. Soc.* **1989**, *111*, 6877.
[387]Friedman, L.; Shechter, H. *J. Am. Chem. Soc.* **1960**, *82*, 1002.
[388]McMahon, R.J.; Chapman, O.L. *J. Am. Chem. Soc.* **1987**, *109*, 683.
[389]Friedman, L.; Berger, J.G. *J. Am. Chem. Soc.* **1961**, *83*, 492, 500.
[390]For a review, see Jones, W.M. *Acc. Chem. Res.* **1977**, *10*, 353.

5. The fragmentation reactions of alicyclic oxychlorocarbenes such as **63** and **64**[391] give substitution and elimination products. Menthyloxychlorocarbene, **63**, gave primarily the substitution product, whereas neomenthyloxychloro-carbene, **64**, gave primarily the elimination product, as shown. In this case, the substitution product is likely due to rearrangement of the chlorocar-bene.[392] It is known that fragmentation of nortricyclyloxychlorocarbene in pentane occurs by an S_Ni-like process to give nortricyclyl chloride.[393] In more polar solvents, fragmentation leads to nortricyclyl cation–chloride anion pair that gives nortricyclyl chloride and a small amount of *exo*-2-norbornenyl chloride. Fragmentation can also lead to radicals.[394]

6. Triplet carbenes can abstract hydrogen or other atoms to give free radicals, for example,

$$\cdot \overset{\cdot}{C}H_2 \; + \; CH_3CH_3 \; \longrightarrow \; \cdot CH_3 \; + \; \cdot CH_2CH_3$$

This is not surprising, since triplet carbenes are free radicals. But singlet carbenes can also give this reaction, although in this case only halogen atoms are abstracted, not hydrogen.[395]

[391]Moss, R.A.; Johnson, L.A.; Kacprzynski, M.; Sauers, R.R. *J. Org. Chem.* **2003**, *68*, 5114.

[392]A rearrangement product was noted for adamantylchlorocarbenes, possibly due to rearrangement of the chlorine atom from a chlorocarbene. See Yao, G.; Rempala, P.; Bashore, C.; Sheridan, R.S. *Tetrahedron Lett.* **1999**, *40*, 17.

[393]Moss, R.A.; Ma, Y.; Sauers, R.R.; Madni, M. *J. Org. Chem.* **2004**, *69*, 3628.

[394]Mekley, N.; El-Saidi, M.; Warkentin, J. *Can. J. Chem.* **2000**, *78*, 356.

[395]Roth, H.D. *J. Am. Chem. Soc.* **1971**, *93*, 1527, 4935, *Acc. Chem. Res.* **1977**, *10*, 85.

NITRENES

Nitrenes,[396] R—N, are the nitrogen analogs of carbenes, and most of what we have said about carbenes also applies to them. Nitrenes are too reactive for isolation under ordinary conditions,[397] although *ab initio* calculations show that nitrenes are more stable than carbenes with an enthalpy difference of 25–26 kcal mol^{-1} (104.7–108.8 kJ mol^{-1}).[398]

$$R-\ddot{\underset{\cdot\cdot}{N}}: \qquad R-\ddot{\underset{\cdot}{N}}\cdot$$

Singlet   Triplet

Alkyl nitrenes have been isolated by trapping in matrices at 4 K,[399] while aryl nitrenes, which are less reactive, can be trapped at 77 K.[400] The ground state of NH, and probably of most nitrenes,[401] is a triplet, although nitrenes can be generated in both triplet[402] and singlet states. In additions of EtOOC—N to C=C double bonds two species are involved, one of which adds in a stereospecific manner and the other not. By analogy with Skell's proposal involving carbenes (p. 284) these are taken to be the singlet and triplet species, respectively.[403]

The two principal means of generating nitrenes are analogous to those used to form carbenes.

1. *Elimination.* An example is

$$\underset{H}{\overset{R}{\diagdown}}N-OSO_2Ar \xrightarrow{\text{base}} R-N \ + \ B-H \ + \ ArSO_2^{\ominus}$$

---

[396]For monographs, see Scriven, E.F.V. *Azides and Nitrenes*, Academic Press, NY, **1984**; Lwowski, W. *Nitrenes*, Wiley, NY, **1970**. For reviews, see Scriven, E.F.V. *React. Intermed. (Plenum)* **1982**, *2*, 1; Lwowski, W. *React. Intermed. (Wiley)* **1985**, *3*, 305; **1981**, *2*, 315; **1978**, *1*, 197; *Angew. Chem. Int. Ed.* **1967**, *6*, 897; Abramovitch, R.A., in McManus, S.P. *Organic Reactive Intermediates*, Academic Press, NY, **1973**, pp. 127–192; Hünig, S. *Helv. Chim. Acta* **1971**, *54*, 1721; Belloli, R. *J. Chem. Educ.* **1971**, *48*, 422; Kuznetsov, M.A.; Ioffe, B.V. *Russ. Chem. Rev.* **1989**, *58*, 732 (N- and O-nitrenes); Meth-Cohn, O. *Acc. Chem. Res.* **1987**, *20*, 18 (oxycarbonylnitrenes); Abramovitch, R.A.; Sutherland, R.G. *Fortsch. Chem. Forsch.*, **1970**, *16*, 1 (sulfonyl nitrenes); Ioffe, B.V.; Kuznetsov, M.A. *Russ. Chem. Rev.* **1972**, *41*, 131 (N-nitrenes).

[397]McClelland, R.A. *Tetrahedron* **1996**, *52*, 6823.

[398]Kemnitz, C.R.; Karney, W.L.; Borden, W.T. *J. Am. Chem. Soc.* **1998**, *120*, 3499.

[399]Wasserman, E.; Smolinsky, G.; Yager, W.A. *J. Am. Chem. Soc.* **1964**, *86*, 3166. For the structure of CH$_3$–N:, as determined in the gas phase, see Carrick, P.G.; Brazier, C.R.; Bernath, P.F.; Engelking, P.C. *J. Am. Chem. Soc.* **1987**, *109*, 5100.

[400]Smolinsky, G.; Wasserman, E.; Yager, W.A. *J. Am. Chem. Soc.* **1962**, *84*, 3220. For a review, see Sheridan, R.S. *Org. Photochem.* **1987**, *8*, 159, pp. 159–248.

[401]A few nitrenes have been shown to have singlet ground states. See Sigman, M.E.; Autrey, T.; Schuster, G.B. *J. Am. Chem. Soc.* **1988**, *110*, 4297.

[402]For the direct detection of triplet alkyl nitrenes in solution via photolysis of α-azidoacetophenones see Singh, P.N.D.; Mandel, S.M.; Robinson, R.M.; Zhu, Z.; Franz, R.; Ault, B.S.; Gudmundsdottir, A.D. *J. Org. Chem.* **2003**, *68*, 7951.

[403]McConaghy, Jr., J.S.; Lwowski, W. *J. Am. Chem. Soc.* **1967**, *89*, 2357, 4450; Mishra, A.; Rice, S.N.; Lwowski, W. *J. Org. Chem.* **1968**, *33*, 481.

2. *Breakdown of Certain Double-Bond Compounds.* The most common method of forming nitrenes is photolytic or thermal decomposition of azides,[404]

$$R-N=N=N \xrightarrow{\Delta \text{ or } h\nu} R-N \ + \ N_2$$

The unsubstituted nitrene NH has been generated by photolysis of or electric discharge through NH_3, N_2H_4, or HN_3.

The reactions of nitrenes are also similar to those of carbenes.[405] As in that case, many reactions in which nitrene intermediates are suspected probably do not involve free nitrenes. It is often very difficult to obtain proof in any given case that a free nitrene is or is not an intermediate.

1. *Insertion* (see reaction **12-13**). Nitrenes, especially acyl nitrenes and sulfonyl nitrenes, can insert into C—H and certain other bonds, for example,

$$\underset{O}{\overset{R'}{\underset{\|}{C}}-N} + R_3CH \longrightarrow \underset{O}{\overset{R'}{\underset{\|}{C}}}-\overset{H}{\underset{|}{N}}-CR_3$$

2. *Addition to C=C Bonds* (see reaction **15-54**):

$$R-N \ + \ R_2C=CR_2 \longrightarrow \underset{R_2C-CR_2}{\overset{R}{\underset{|}{N}}}$$

3. *Rearrangements.*[383] Alkyl nitrenes do not generally give either of the two preceding reactions because rearrangement is more rapid, for example,

$$\underset{H}{\overset{R}{\underset{}{CH-N}}} \longrightarrow RHC=NH$$

Such rearrangements are so rapid that it is usually difficult to exclude the possibility that a free nitrene was never present at all, that is, that migration takes place at the same time that the nitrene is formed[406] (see p. 1606). However, the rearrangement of naphthylnitrenes to novel bond-shift isomers has been reported.[407]

[404]For reviews, see Dyall, L.K., in Patai, S.; Rappoport, Z. *The Chemistry of Functional Groups, Supplement D*, pt. 1, Wiley, NY, *1983*, pp. 287–320; Dürr, H.; Kober, H. *Top. Curr. Chem. 1976, 66*, 89; L'Abbé, G. *Chem. Rev. 1969, 69*, 345.

[405]For a discussion of nitrene reactivity, see Subbaraj, A.; Subba Rao, O.; Lwowski, W. *J. Org. Chem. 1989, 54*, 3945.

[406]For example, see Moriarty, R.M.; Reardon, R.C. *Tetrahedron 1970, 26*, 1379; Abramovitch, R.A.; Kyba, E.P. *J. Am. Chem. Soc. 1971, 93*, 1537.

[407]Maltsev, A.; Bally, T.; Tsao, M.-L.; Platz, M.S.; Kuhn, A.; Vosswinkel, M.; Wentrup, C. *J. Am. Chem. Soc. 2004, 126*, 237.

4. *Abstraction*, for example,

$$\overset{\bullet}{R-N} \;+\; R-H \;\longrightarrow\; R-\overset{\bullet}{\underset{\cdot\cdot}{N}}-H \;+\; R\cdot$$

5. *Dimerization*. One of the principal reactions of NH is dimerization to diimide N_2H_2. Azobenzenes are often obtained in reactions where aryl nitrenes are implicated:[408]

$$2\ Ar-N \;\longrightarrow\; Ar-N=N-Ar$$

It would thus seem that dimerization is more important for nitrenes than it is for carbenes, but again it has not been proved that free nitrenes are actually involved.

$$\underset{\textbf{65}}{\overset{\displaystyle R\diagdown \underset{\cdot\cdot}{N}\diagup R'}{}\ \oplus} \qquad \underset{\textbf{66}}{\overset{\displaystyle R\diagdown}{\underset{\displaystyle R'\diagup}{}}{=}\overset{\cdot\cdot}{N}\ \oplus}$$

At least two types of *nitrenium ions*,[409] the nitrogen analogs of carbocations, can exist as intermediates, although much less work has been done in this area than on carbocations. In one type (**65**), the nitrogen is bonded to two atoms (R or R′ can be H)[410] and in the other (**66**) to only one atom.[411] When R = H in **65** the species is a protonated nitrene. Like carbenes and nitrenes, nitrenium ions can exist in singlet or triplet states.[412]

[408]See, for example, Leyva, E.; Platz, M.S.; Persy, G.; Wirz, J. *J. Am. Chem. Soc.* **1986**, *108*, 3783.

[409]Falvey, D.E. *J. Phys. Org. Chem.* **1999**, *12*, 589; Falvey, D.E., in Ramamurthy, V., Schanze, K. *Organic, Physical, and Materials Photochemistry*, Marcel Dekker, NY, **2000**; pp. 249–284; Novak, M.; Rajagopal, S. *Adv. Phys. Org. Chem.* **2001**, *36*, 167; Falvey, D.E., in Moss, R.A., Platz, M.S., Jones, Jr., M. *Reactve Intermediate Chemistry*, Wiley-Interscience: Hoboken, NJ, **2004**; Vol. 1, pp. 593–650.

[410]Winter, A.H.; Falvey, D.E.; Cramer, C.J. *J. Am. Chem. Soc.*, **2004**, *126*, 9661.

[411]For reviews of **65**, see Abramovitch, R.A.; Jeyaraman, R., in Scriven, E.F.V. *Azides and Nitrenes*, Academic Press, NY, **1984**, pp. 297–357; Gassman, P.G. *Acc. Chem. Res.* **1970**, *3*, 26. For a review of **66**, see Lansbury, P.T., in Lwowski, W. *Nitrenes*, Wiley, NY, **1970**, pp. 405–419.

[412]Gassman, P.G.; Cryberg, R.L. *J. Am. Chem. Soc.* **1969**, *91*, 5176.

CHAPTER 6

Mechanisms and Methods of Determining Them

A mechanism is the actual process by which a reaction takes place: which bonds are broken, in what order, how many steps are involved, the relative rate of each step, and so on. In order to state a mechanism completely, we should have to specify the positions of all atoms, including those in solvent molecules, and the energy of the system, at every point in the process. A proposed mechanism must fit all the facts available. It is always subject to change as new facts are discovered. The usual course is that the gross features of a mechanism are the first to be known and then increasing attention is paid to finer details. The tendency is always to probe more deeply, to get more detailed descriptions.

Although for most reactions gross mechanisms can be written today with a good degree of assurance, no mechanism is known completely. There is much about the fine details that is still puzzling, and for some reactions even the gross mechanism is not yet clear. The problems involved are difficult because there are so many variables. Many examples are known where reactions proceed by different mechanisms under different conditions. In some cases, there are several proposed mechanisms, each of which completely explains all the data.

TYPES OF MECHANISM

In most reactions of organic compounds, one or more covalent bonds are broken. We can divide organic mechanisms into three basic types, depending on how the bonds break.

1. If a bond breaks in such a way that both electrons remain with one fragment, the mechanism is called *heterolytic*. Such reactions do not necessarily involve ionic intermediates, although they usually do. The important thing is that the electrons are never unpaired. For most reactions, it is convenient to call one reactant the *attacking reagent* and the other the *substrate*. In this book,

---

March's Advanced Organic Chemistry: Reactions, Mechanisms, and Structure, Sixth Edition, by Michael B. Smith and Jerry March
Copyright © 2007 John Wiley & Sons, Inc.

we will always designate as the substrate that molecule that supplies carbon to the new bond. When carbon–carbon bonds are formed, it is necessary to be arbitrary about which is the substrate and which is the attacking reagent. In heterolytic reactions, the reagent generally brings a pair of electrons to the substrate or takes a pair of electrons from it. A reagent that brings an electron pair is called a *nucleophile* and the reaction is *nucleophilic*. A reagent that takes an electron pair is called an *electrophile* and the reaction is *electrophilic*. In a reaction in which the substrate molecule becomes cleaved, part of it (the part not containing the carbon) is usually called the *leaving group*. A leaving group that carries away an electron pair is called a *nucleofuge*. If it comes away without the electron pair, it is called an *electrofuge*.

2. If a bond breaks in such a way that each fragment gets one electron, free radicals are formed and such reactions are said to take place by *homolytic* or *free-radical mechanisms*.

3. It would seem that all bonds must break in one of the two ways previously noted. But there is a third type of mechanism in which electrons (usually six, but sometimes some other number) move in a closed ring. There are no intermediates, ions or free radicals, and it is impossible to say whether the electrons are paired or unpaired. Reactions with this type of mechanism are called *pericyclic*.[1]

Examples of all three types of mechanisms are given in the next section.

TYPES OF REACTION

The number and range of organic reactions is so great as to seem bewildering, but actually almost all of them can be fitted into just six categories. In the description of the six types that follows, the immediate products are shown, although in many cases they then react with something else. All the species are shown without charges, since differently charged reactants can undergo analogous changes. The descriptions given here are purely formal and are for the purpose of classification and comparison. All are discussed in detail in Part 2 of this book.

1. *Substitutions.* If heterolytic, these can be classified as nucleophilic or electrophilic depending on which reactant is designated as the substrate and which as the attacking reagent (very often Y must first be formed by a previous bond cleavage).

   a. Nucleophilic substitution (Chapters 10, 13).

   $$A{-}X \ + \ Y \ \longrightarrow \ A{-}Y \ + \ X$$

---

[1]For a classification of pericyclic reactions, see Hendrickson, J.B. *Angew. Chem. Int. Ed.* **1974**, *13*, 47. Also see, Fleming, I. *Pericyclic Reactions*, Oxford University Press, Oxford, **1999**.

b. Electrophilic substitution (Chapters 11, 12).

$$A—X \ + \ Y \longrightarrow A—Y \ + \ X$$

c. Free-radical substitution (Chapter 14).

$$A—X \ + \ Y\bullet \longrightarrow A—Y \ + \ X\bullet$$

In free-radical substitution, Y• is usually produced by a previous free-radical cleavage, and X• goes on to react further.

2. *Additions to Double or Triple Bonds* (Chapters 15, 16). These reactions can take place by all three of the mechanistic possibilities.

a. Electrophilic addition (heterolytic).

$$A{=}B \ + \ Y{-}W \longrightarrow \underset{A-B}{\overset{Y}{|}} \ + \ W \longrightarrow \overset{W}{\underset{Y}{A{-}B}}$$

b. Nucleophilic addition (heterolytic).

$$A{=}B \ + \ Y{-}W \longrightarrow \underset{A-B}{\overset{Y}{|}} \ + \ W \longrightarrow \overset{W}{\underset{Y}{A{-}B}}$$

c. Free-radical addition (homolytic).

$$A{=}B \ + \ Y{-}W \xrightarrow{-W\bullet} \underset{A-B}{\overset{Y}{|}} \ + \ W{-}Y \longrightarrow \overset{W}{\underset{Y}{A{-}B}} \ + \ Y\bullet$$

d. Simultaneous addition (pericyclic).

$$\underset{A{=}B}{W{-}Y} \longrightarrow \overset{W}{\underset{Y}{A{-}B}}$$

The examples show Y and W coming from the same molecule, but very often (except in simultaneous addition) they come from different molecules. Also, the examples show the Y—W bond cleaving at the same time that Y is bonding to B, but often (again except for simultaneous addition) this cleavage takes place earlier.

3. *β Elimination* (Chapter 17).

$$\overset{W}{\underset{Y}{A{-}B}} \longrightarrow A{=}B \ + \ W \ + \ X^{\ominus}$$

These reactions can take place by either heterolytic or pericyclic mechanisms. Examples of the latter are shown on p. $$$. Free-radical β eliminations are extremely rare. In heterolytic eliminations W and X may or may not leave simultaneously and may or may not combine.

4. *Rearrangement* (Chapter 18). Many rearrangements involve migration of an atom or group from one atom to another. There are three types, depending on how many electrons the migrating atom or group carries with it.

a. Migration with electron pair (nucleophilic).

$$
\begin{array}{c}
\text{W} \\
\text{\textbackslash} \\
\text{A}-\text{B}
\end{array}
\longrightarrow
\begin{array}{c}
\text{W} \\
\text{/} \\
\text{A}-\text{B}
\end{array}
$$

b. Migration with one electron (free-radical).

$$
\begin{array}{c}
\text{W} \\
\text{\textbackslash} \\
\text{A}-\dot{\text{B}}
\end{array}
\longrightarrow
\begin{array}{c}
\text{W} \\
\text{/} \\
\dot{\text{A}}-\text{B}
\end{array}
$$

c. Migration without electrons (electrophilic; rare).

$$
\begin{array}{c}
\text{W} \\
\text{\textbackslash}\ \ \ .. \\
\text{A}-\ddot{\text{B}}
\end{array}
\longrightarrow
\begin{array}{c}
..\ \ \ \text{W} \\
\ \ \ /\\
\text{A}-\text{B}
\end{array}
$$

The illustrations show 1,2 rearrangements, in which the migrating group moves to the adjacent atom. These are the most common, although longer rearrangements are also possible. There are also some rearrangements that do not involve simple migration at all (see Chapter 18). Some of the latter involve pericyclic mechanisms.

5. *Oxidation and Reduction* (Chapter 19). Many oxidation and reduction reactions fall naturally into one of the four types mentioned above, but many others do not. For a description of oxidation–reduction mechanistic types, see p. 1704.

6. *Combinations of the above.* Note that arrows are used to show movement of *electrons.* An arrow always follows the motion of electrons and never of a nucleus or anything else (it is understood that the rest of the molecule follows the electrons). Ordinary arrows (double-headed) follow electron pairs, while single-headed arrows follow unpaired electrons. Double-headed arrows are also used in pericyclic reactions for convenience, although in these reactions we do not really know how or in which direction the electrons are moving.

THERMODYNAMIC REQUIREMENTS FOR REACTION

In order for a reaction to take place spontaneously, the free energy of the products must be lower than the free energy of the reactants; that is, ΔG must be negative. Reactions can go the other way, of course, but only if free energy is added. Like water on the surface of the earth, which only flows downhill and never uphill (though it can be carried or pumped uphill), molecules seek the lowest possible potential energy. Free energy is made up of two components, enthalpy H and entropy S. These quantities are related by the equation

$$\Delta G = \Delta H - T\Delta S$$

The enthalpy change in a reaction is essentially the difference in bond energies (including resonance, strain, and solvation energies) between the reactants and the products. The enthalpy change can be calculated by totaling the bond energies of all the bonds broken, subtracting from this the total of the bond energies of all the bonds formed, and adding any changes in resonance, strain, or solvation energies. Entropy changes are quite different, and refer to the disorder or randomness of the system. The less order in a system, the greater the entropy. The preferred conditions in Nature are *low* enthalpy and *high* entropy, and in reacting systems, enthalpy spontaneously decreases while entropy spontaneously increases.

For many reactions entropy effects are small and it is the enthalpy that mainly determines whether the reaction can take place spontaneously. However, in certain types of reaction entropy is important and can dominate enthalpy. We will discuss several examples.

1. In general, liquids have lower entropies than gases, since the molecules of gas have much more freedom and randomness. Solids, of course, have still lower entropies. Any reaction in which the reactants are all liquids and one or more of the products is a gas is therefore thermodynamically favored by the increased entropy; the equilibrium constant for that reaction will be higher than it would otherwise be. Similarly, the entropy of a gaseous substance is higher than that of the same substance dissolved in a solvent.

2. In a reaction in which the number of product molecules is equal to the number of reactant molecules, for example, $A + B \rightarrow C + D$, entropy effects are usually small, but if the number of molecules is increased, for example, $A \rightarrow B + C$, there is a large gain in entropy because more arrangements in space are possible when more molecules are present. Reactions in which a molecule is cleaved into two or more parts are therefore thermodynamically favored by the entropy factor. Conversely, reactions in which the number of product molecules is less than the number of reactant molecules show entropy decreases, and in such cases there must be a sizable decrease in enthalpy to overcome the unfavorable entropy change.

3. Although reactions in which molecules are cleaved into two or more pieces have favorable entropy effects, many potential cleavages do not take place because of large increases in enthalpy. An example is cleavage of ethane into two methyl radicals. In this case, a bond of $\sim 79\,\text{kcal mol}^{-1}$ ($330\,\text{kJ mol}^{-1}$) is broken, and no new bond is formed to compensate for this enthalpy increase. However, ethane can be cleaved at very high temperatures, which illustrates the principle that *entropy becomes more important as the temperature increases*, as is obvious from the equation $\Delta G = \Delta H - T\Delta S$. The

enthalpy term is independent of temperature, while the entropy term is directly proportional to the absolute temperature.

4. An acyclic molecule has more entropy than a similar cyclic molecule because there are more conformations (cf. hexane and cyclohexane). Ring opening therefore means a gain in entropy and ring closing a loss.

KINETIC REQUIREMENTS FOR REACTION

Just because a reaction has a negative ΔG does not necessarily mean that it will take place in a reasonable period of time. A negative ΔG is a *necessary*, but not a *sufficient*, condition for a reaction to occur spontaneously. For example, the reaction between H_2 and O_2 to give H_2O has a large negative ΔG, but mixtures of H_2 and O_2 can be kept at room temperature for many centuries without reacting to any significant extent. In order for a reaction to take place, *free energy of activation* $\Delta G^{\ddagger}$ must be added.[2] This situation is illustrated in Fig. 6.1,[3] which is an energy

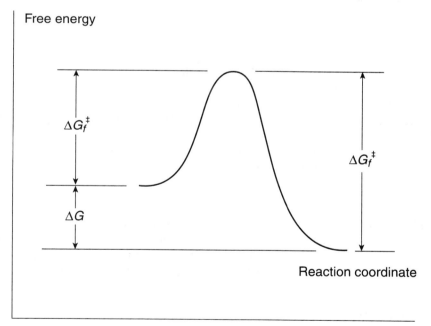

Fig. 6.1. Free-energy profile of a reaction without an intermediate where the products have a lower free energy than the reactants.

---

[2]For mixtures of H_2 and O_2 this can be done by striking a match.
[3]Strictly speaking, this is an energy profile for a reaction of the type $XY + Z \rightarrow X + YZ$. However, it may be applied, in an approximate way, to other reactions.

profile for a one-step reaction without an intermediate. In this type of diagram, the horizontal axis (called the *reaction coordinate*)[4] signifies the progression of the reaction. The parameter $\Delta G_f^{\ddagger}$ is the free energy of activation for the forward reaction. If the reaction shown in Fig. 6.1 is reversible, must be $>\Delta G_f^{\ddagger}$, since it is the sum of ΔG and $\Delta G_f^{\ddagger}$.

When a reaction between two or more molecules has progressed to the point corresponding to the top of the curve, the term *transition state* is applied to the positions of the nuclei and electrons. The transition state possesses a definite geometry and charge distribution but has no finite existence; the system passes through it. The system at this point is called an *activated complex*.[5]

In the *transition-state theory*[6] the starting materials and the activated complex are taken to be in equilibrium, the equilibrium constant being designated $K^{\ddagger}$. According to the theory, all activated complexes go on to product at the same rate (which, although at first sight surprising, is not unreasonable, when we consider that they are all "falling downhill") so that the rate constant (see p. 315) of the reaction depends only on the position of the equilibrium between the starting materials and the activated complex, that is, on the value of $K^{\ddagger}$. The parameter $\Delta G^{\ddagger}$ is related to $K^{\ddagger}$ by

$$\Delta G^{\ddagger} = -2.3 \, RT \, \log \, K^{\ddagger}$$

so that a higher value of $\Delta G^{\ddagger}$ is associated with a smaller rate constant. The rates of nearly all reactions increase with increasing temperature because the additional energy thus supplied helps the molecules to overcome the activation energy barrier.[7] Some reactions have no free energy of activation at all, meaning that $K^{\ddagger}$ is essentially infinite and that virtually all collisions lead to reaction. Such processes are said to be *diffusion-controlled*.[8]

Like ΔG, $\Delta G^{\ddagger}$ is made up of enthalpy and entropy components

$$\Delta G^{\ddagger} = \Delta H^{\ddagger} - \text{T}\Delta S^{\ddagger}$$

$\Delta H^{\ddagger}$, the *enthalpy of activation*, is the difference in bond energies, including strain, resonance, and solvation energies, between the starting compounds and the *transition state*. In many reactions, bonds have been broken or partially broken by the time the transition state is reached; the energy necessary for this is $\Delta H^{\ddagger}$. It is

[4]For a review of reaction coordinates and structure–energy relationships, see Grunwald, E. *Prog. Phys. Org. Chem.* **1990**, *17*, 55.

[5]For a discussion of transition states, see Laidler, K.J. *J. Chem. Educ.* **1988**, *65*, 540.

[6]For fuller discussions, see Kreevoy, M.M.; Truhlar, D.G. in Bernasconi, C.F. *Investigation of Rates and Mechanisms of Reactions*, 4th ed. (Vol. 6 of Weissberger, A. *Techniques of Chemistry*), pt. 1, Wiley, NY, **1986**, pp. 13–95; Moore, J.W.; Pearson, R.G. *Kinetics and Mechanism*, 3rd ed, Wiley, NY, **1981**, pp. 137–181; Klumpp, G.W. *Reactivity in Organic Chemistry*, Wiley, NY, **1982**; pp. 227–378.

[7]For a review concerning the origin and evolution of reaction barriers see Donahue, N.M. *Chem. Rev.* **2003**, *103*, 4593.

[8]For a monograph on diffusion-controlled reactions, see Rice, S.A. *Comprehensive Chemical Kinetics*, Vol. 25 (edited by Bamford, C.H.; Tipper, C.F.H.; Compton, R.G.); Elsevier: NY, **1985**.

true that additional energy will be supplied by the formation of new bonds, but if this occurs after the transition state, it can affect only ΔH and not $\Delta H^{\ddagger}$.

Entropy of activation, $\Delta S^{\ddagger}$, which is the difference in entropy between the starting compounds and the transition state, becomes important when two reacting molecules must approach each other in a specific orientation in order for the reaction to take place. For example, the reaction between a simple noncyclic alkyl chloride and hydroxide ion to give an alkene (reaction **17-13**) takes place only if, in the transition state, the reactants are oriented as shown.

Not only must the $^-$OH be near the hydrogen, but the hydrogen must be oriented anti to the chlorine atom.[9] When the two reacting molecules collide, if the $^-$OH should be near the chlorine atom or near R^1 or R^2, no reaction can take place. In order for a reaction to occur, the molecules must surrender the freedom they normally have to assume many possible arrangements in space and adopt only that one that leads to reaction. Thus, a considerable loss in entropy is involved, that is, $\Delta S^{\ddagger}$ is negative.

Entropy of activation is also responsible for the difficulty in closing rings[10] larger then six membered. Consider a ring-closing reaction in which the two groups that must interact are situated on the ends of a 10-carbon

chain. In order for reaction to take place, the groups must encounter each other. But a 10-carbon chain has many conformations, and in only a few of these are the ends of the chain near each other. Thus, forming the transition state requires a great loss of entropy.[11] This factor is also present, although less so, in closing rings of six members or less (except three-membered rings), but with rings of this size the

[9]As we will see in Chapter 17, with some molecules elimination is also possible if the hydrogen is oriented syn, instead of anti, to the chlorine atom. Of course, this orientation also requires a considerable loss of entropy.

[10]For discussions of the entropy and enthalpy of ring-closing reactions, see De Tar, D.F.; Luthra, N.P. *J. Am. Chem. Soc.* **1980**, *102*, 4505; Mandolini, L. *Bull. Soc. Chim. Fr.* **1988**, 173. For a related discussion, see Menger, F.M. *Acc. Chem. Res.* **1985**, *18*, 128.

[11]For reviews of the cyclization of acyclic molecules, see Nakagaki. R.; Sakuragi, H.; Mutai, K. *J. Phys. Org. Chem.* **1989**, 2, 187; Mandolini, L. *Adv. Phys. Org. Chem.* **1986**, *22*, 1. For a review of the cyclization and conformation of hydrocarbon chains, see Winnik, M.A. *Chem. Rev.* **1981**, *81*, 491. For a review of steric and electronic effects in heterolytic ring closures, see Valters, R. *Russ. Chem. Rev.* **1982**, *51*, 788.

TABLE 6.1. Relative Rate Constants at 50°C.[a] The rate for an eight-membered ring = 1 for the reaction.

Ring Size	Relative Rate
3	21.7
4	5.4×10^3
5	1.5×10^6
6	1.7×10^4
7	97.3
8	1.00
9	1.12
10	3.35
11	8.51
12	10.6
13	32.2
14	41.9
15	45.1
16	52.0
18	51.2
23	60.4

[a](Eight-membered ring = 1) for the reaction

$$Br(CH_2)_{n-2}CO_2^- \longrightarrow (CH_2)_{n-2}\overset{O}{\underset{O}{\diagdown}},$$

where $n =$ the ring size[12].

entropy loss is less than that of bringing two individual molecules together. For example, a reaction between an OH group and a COOH group in the same molecule to form a lactone with a five- or six-membered ring takes place much faster than the same reaction between a molecule containing an OH group and another containing a COOH group. although $\Delta H^{\ddagger}$ is about the same, $\Delta S^{\ddagger}$ is much less for the cyclic case. However, if the ring to be closed has three or four members, small-angle strain is introduced and the favorable $\Delta S^{\ddagger}$ may not be sufficient to overcome the unfavorable $\Delta H^{\ddagger}$ change. Table 6.1 shows the relative rate constants for the closing of rings of 3–23 members all by the same reaction.[12] Reactions in which the transition state has more disorder than the starring compounds, for example, the pyrolytic conversion of cyclopropane to propene, have positive $\Delta S^{\ddagger}$ values and are thus favored by the entropy effect.

Reactions with intermediates are two-step (or more) processes. In these reactions there is an energy "well." There are two transition states, each with an energy higher than the intermediate (Fig. 6.2). The deeper the well, the more stable the intermediate. In Fig. 6.2a, the second peak is higher than the first. The opposite situation

[12]The values for ring sizes 4, 5, and 6 are from Mandolini, L. *J. Am. Chem. Soc.* **1978**, *100*, 550; the others are from Galli, C.; Illuminati, G.; Mandolini, L.; Tamborra, P. *J. Am. Chem. Soc.* **1977**, *99*, 2591. See also, Illuminati, G.; Mandolini, L. *Acc. Chem. Res.* **1981**, *14*, 95. See, however, van der Kerk, S.M.; Verhoeven, J.W.; Stirling, C.J.M. *J. Chem. Soc. Perkin Trans. 2* **1985**, 1355; Benedetti, F.; Stirling, C.J.M. *J. Chem. Soc. Perkin Trans. 2* **1986**, 605.

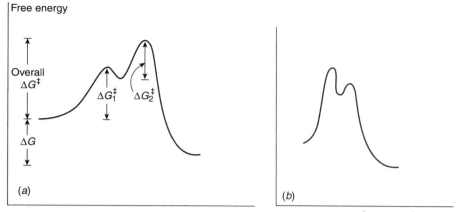

Fig. 6.2. (*a*) Free-energy profile for a reaction with an intermediate $\Delta G_1^{\ddagger}$ and $\Delta G_2^{\ddagger}$ are the free energy of activation for the first and second stages, respectively. (*b*) Free-energy profile for a reaction with an intermediate in which the first peak is higher than the second.

is shown in Fig. 6.2*b*. Note that in reactions in which the second peak is higher than the first, the overall $\Delta G^{\ddagger}$ is less than the sum of the $\Delta G^{\ddagger}$ values for the two steps. Minima in free-energy-profile diagrams (*intermediates*) correspond to real species, which have a finite although usually short existence. These may be the carbocations, carbanions, free radicals, etc., discussed in Chapter 5 or molecules in which all the atoms have their normal valences. In either case, under the reaction conditions they do not live long (because $\Delta G_2^{\ddagger}$ is small), but rapidly go on to products. Maxima in these curves, however, do not correspond to actual species but only to transition states in which bond breaking and/or bond making have partially taken place. Transition states have only a transient existence with an essentially zero lifetime.[13]

THE BALDWIN RULES FOR RING CLOSURE[14]

In previous sections, we discussed, in a general way, the kinetic and thermodynamic aspects of ring-closure reactions. J. E. Baldwin has supplied a more specific set of rules for certain closings of three- to seven-membered rings.[15] These rules

---

[13]Despite their transient existences, it is possible to study transition states of certain reactions in the gas phase with a technique called laser femtochemistry: Zewall, A.H.; Bernstein, R.B. *Chem. Eng. News* **1988**, *66*, No. 45 (Nov. 7), 24–43. For another method, see Collings, B.A.; Polanyi, J.C.; Smith, M.A.; Stolow, A.; Tarr, A.W. *Phys. Rev. Lett.* **1987**, *59*, 2551.

[14]See Smith, M.B. *Organic Synthesis,* 2nd ed., McGraw-Hill, NY, **2001**, pp. 517–523.

[15]Baldwin, J.E. *J. Chem. Soc. Chem. Commun.* **1976**, 734; Baldwin, J.E., in *Further Perspectives in Organic Chemistry* (*Ciba Foundation Symposium 53*), Elsevier North Holland, Amsterdam, The Netherlands, **1979**, pp. 85–99. See also, Baldwin, J.E.; Thomas, R.C.; Kruse, L.I.; Silberman, L. *J. Org. Chem.* **1977**, *42*, 3846; Baldwin, J.E.; Lusch, M.J. *Tetrahedron* **1982**, *38*, 2939; Anselme, J. *Tetrahedron Lett.* **1977**, 3615; Fountain, K.R.; Gerhardt, G. *Tetrahedron Lett.* **1978**, 3985.

distinguish two types of ring closure, called Exo and Endo, and three kinds of atoms at the starred positions: *Tet* for sp^3, *Trig* for sp^2, and *Dig* for sp. The following are Baldwin's rules for closing rings of three to seven members.

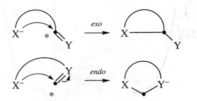

Rule 1. Tetrahedral systems
  (a) 3–7-*Exo–Tet* are all favored processes
  (b) 5–6-*Endo–Tet* are disfavored

Rule 2. Trigonal systems
  (a) 3–7-*Exo–Trig* are favored
  (b) 3–5-*Endo–Trig* are disfavored[16]
  (c) 6–7-*Endo–Trig* are favored

Rule 3. Digonal systems
  (a) 3–4-*Exo–Dig* are disfavored
  (b) 5–7-*Exo–Dig* are favored
  (c) 3–7-*Endo–Dig* are favored

"Disfavored" does not mean it cannot be done: only that it is more difficult than the favored cases. These rules are empirical and have a stereochemical basis. The favored pathways are those in which the length and nature of the linking chain enables the terminal atoms to achieve the proper geometries for reaction. The disfavored cases require severe distortion of bond angles and distances. Many cases in the literature are in substantial accord with these rules, and they important in the formation of five- and six-membered rings.[17]

Although Baldwin's rules can be applied to ketone enolates,[18] additional rules were added to make the terminology more specific.[19] The orientation of the orbital as it approaches the reactive center must be considered for determining

[16]For some exceptions to the rule in this case, see Trost, B.M.; Bonk, P.J. *J. Am. Chem. Soc.* **1985**, *107*, 1778; Auvray, P.; Knochel, P.; Normant, J.F. *Tetrahedron Lett.* **1985**, *26*, 4455; Torres, L.E.; Larson, G.L. *Tetrahedron Lett.* **1986**, *27*, 2223.

[17]Johnson, C.D. *Acc. Chem. Res.* **1997**, *26*, 476.

[18]Baldwin, J.E.; Kruse, L.I. *J. Chem. Soc. Chem. Commun.* **1977**, 233.

[19]Baldwin, J.E.; Lusch, M.J. *Tetrahedron* **1982**, *38*, 2939.

the correct angle of approach. Diagrams that illustrate the enolate rules are

ENOLENDO-EXOTET

ENOLEXO-EXOTET

ENOLENDO-EXOTRIG

ENOLEXO-EXOTRIG

The rules are

    (a) 6–7 enolendo–exo–tet reactions are favored.
    (b) 3–5 enolendo–exo–tet reactions are disfavored.
    (c) 3–7 enolexo–exo–tet reactions are favored.
    (d) 3–7 enolexo–exo–trig reactions are favored.
    (e) 6–7 enolendo–exo–trig reactions are favored.
    (f) 3–5 enolendo–exo–trig reactions are disfavored.

KINETIC AND THERMODYNAMIC CONTROL

There are many cases in which a compound under a given set of reaction conditions can undergo competing reactions to give different products:

Figure 6.3 shows a free-energy profile for a reaction in which **B** is thermodynamically more stable than **C** (ΔG_B is $> \Delta G_C$), but **C** is formed faster (lower $\Delta G^{\ddagger}$). If neither reaction is reversible, **C** will be formed in larger amount because it is formed faster. The product is said to be *kinetically controlled*. However, if the reactions are reversible, this will not necessarily be the case. If such a process is stopped well before the equilibrium has been established, the reaction will be kinetically controlled since more of the faster-formed product will be present.

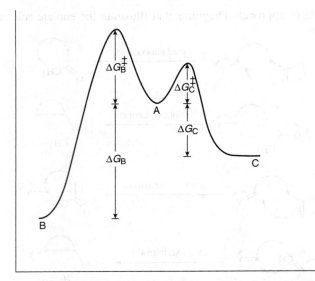

Fig. 6.3. Free-energy profile illustrating kinetic versus thermodynamic control of products. The starting compounds (**A**) can react to give either **B** or **C**.

However, if the reaction is permitted to approach equilibrium, the predominant or even exclusive product will be **B**. Under these conditions the **C** that is first formed reverts to **A**, while the more stable **B** does so much less. We say the product is *thermodynamically controlled*.[20] Of course, Fig. 6.3 does not describe all reactions in which a compound **A** can give two different products. In many cases the more stable product is also the one that is formed faster. In such cases, the product of kinetic control is also the product of thermodynamic control.

THE HAMMOND POSTULATE

Since transition states have zero lifetimes, it is impossible to observe them directly and information about their geometries must be obtained from inference. In some cases our inferences can be very strong. For example, in the S_N2 reaction (p. 426) between CH_3I and I^- (a reaction in which the product is identical to the starting compound), the transition state should be perfectly symmetrical. In most cases, however, we cannot reach such easy conclusions, and we are greatly aided by the *Hammond postulate*,[21] which states that for any single reaction step, *the geometry of the transition state for that step resembles the side to which it is closer*

[20]For a discussion of thermodynamic versus kinetic control, see Klumpp, G.W. *Reactivity in Organic Chemistry*, Wiley, NY, *1982*, pp. 36–89.
[21]Hammond, G.S. *J. Am. Chem. Soc. 1955*, *77*, 334. For a discussion, see Fǎrcaşiu, D. *J. Chem. Educ. 1975*, *52*, 76.

in free energy. Thus, for an exothermic reaction like that shown in Fig. 6.1, the transition state resembles the reactants more than the products, although not much more because there is a substantial $\Delta G^{\ddagger}$ on both sides. The postulate is most useful in dealing with reactions with intermediates. In the reaction illustrated in Fig. 6.2*a*, the first transition state lies much closer in energy to the intermediate than to the reactants, and we can predict that the geometry of the transition state resembles that of the intermediate more than it does that of the reactants. Likewise, the second transition state also has a free energy much closer to that of the intermediate than to the products, so that both transition states resemble the intermediate more than they do the products or reactants. This is generally the case in reactions that involve very reactive intermediates. Since we usually know more about the structure of intermediates than of transition states, we often use our knowledge of intermediates to draw conclusions about the transition states (e.g., see pp. 479, 1019).

MICROSCOPIC REVERSIBILITY

In the course of a reaction, the nuclei and electrons assume positions that at each point correspond to the lowest free energies possible. If the reaction is reversible, these positions must be the same in the reverse process, too. This means that the forward and reverse reactions (run under the same conditions) must proceed by the same mechanism. This is called the *principle of microscopic reversibility*. For example, if in a reaction **A** → **B** there is an intermediate **C**, then **C** must also be an intermediate in the reaction **B** → **A**. This is a useful principle since it enables us to know the mechanism of reactions in which the equilibrium lies far over to one side. Reversible photochemical reactions are an exception, since a molecule that has been excited photochemically does not have to lose its energy in the same way (Chapter 7).

MARCUS THEORY

It is often useful to compare the reactivity of one compound with that of similar compounds. What we would like to do is to find out how a reaction coordinate (and in particular the transition state) changes when one reactant molecule is replaced by a similar molecule. Marcus theory is a method for doing this.[22]

In this theory, the activation energy $\Delta G^{\ddagger}$ is thought of as consisting of two parts.

1. An *intrinsic* free energy of activation, which would exist if the reactants and products had the same ΔG°.[23] This is a kinetic part, called the *intrinsic barrier* $\Delta G_{int}^{\ddagger}$

2. A thermodynamic part, which arises from the ΔG° for the reaction.

---

[22]For reviews, see Albery, W.J. *Annu. Rev. Phys. Chem.* **1980**, *31*, 227; Kreevoy, M.M.; Truhlar, D.G., in Bernasconi, C.F. *Investigation of Rates and Mechanisms of Reactions*, 4th ed. (Vol. 6 of Weissberger, A. *Techniques of Chemistry*), pt. 1, Wiley, NY, **1986**, pp. 13–95.

[23]The parameter ΔG° is the standard free energy; that is, ΔG at atmospheric pressure.

The Marcus equation says that the overall $\Delta G^{\ddagger}$ for a one-step reaction is[24]

$$\Delta G^{\ddagger} = \Delta G^{\ddagger}_{int} + \frac{1}{2}\Delta G^{\Delta} + \frac{(\Delta G^{\Delta})^2}{16(\Delta G^{\ddagger}_{int} - w^R)}$$

where the term ΔG^{Δ} stands for

$$\Delta G^{\Delta} = \Delta G^{\circ} - w^R + w^P$$

w^R, a work term, is the free energy required to bring the reactants together and w^P is the work required to form the successor configuration from the products.

For a reaction of the type $AX + B \rightarrow BX$, the intrinsic barrier[25] $\Delta G^{\ddagger}_{int}$ is taken to be the average $\Delta G^{\ddagger}$ for the two symmetrical reactions

$$AX \;+\; A \;\longrightarrow\; AX \;+\; A \qquad \Delta G^{\ddagger}_{A,A}$$

$$BX \;+\; B \;\longrightarrow\; BX \;+\; B \qquad \Delta G^{\ddagger}_{B,B}$$

so that

$$\Delta G^{\ddagger}_{int} + \frac{1}{2}(\Delta G^{\ddagger}_{A,A} + \Delta G^{\ddagger}_{B,B})$$

One type of process that can successfully be treated by the Marcus equation is the S_N2 mechanism (p. 426)

$$R{-}X \;+\; Y \;\longrightarrow\; R{-}Y \;+\; X$$

When R is CH_3 the process is called *methyl transfer*.[26] For such reactions, the work terms w^R and w^P are assumed to be very small compared to ΔG°, and can be neglected, so that the Marcus equation simplifies to

$$\Delta G^{\ddagger} = \Delta G^{\ddagger}_{int} + \frac{1}{2}\Delta G^{\circ} + \frac{(\Delta G)^2}{16\Delta G^{\ddagger}_{int}}$$

The Marcus equation allows $\Delta G^{\ddagger}$ for $RX + Y \rightarrow RY + X$ to be calculated from the barriers of the two symmetrical reactions $RX + X \rightarrow RX + X$ and

[24]Albery, W.J.; Kreevoy, M.M. *Adv. Phys. Org. Chem.* **1978**, *16*, 87, pp. 98–99.

[25]For discussions of intrinsic barriers, see Lee, I. *J. Chem. Soc. Perkin Trans. 2* **1989**, 943, *Chem. Soc. Rev.* **1990**, *19*, 133.

[26]For a review of Marcus theory applied to methyl transfer, see Albery, W.J.; Kreevoy, M.M. *Adv. Phys. Org. Chem.* **1978**, *16*, 87. See also, Lee, I. *J. Chem. Soc., Perkin Trans. 2* **1989**, 943; Lewis, E.S.; Kukes, S.; Slater, C.D. *J. Am. Chem. Soc.* **1980**, *102*, 1619; Lewis, E.S.; Hu, D.D. *J. Am. Chem. Soc.* **1984**, *106*, 3292; Lewis, E.S.; McLaughlin, M.L.; Douglas, T.A. *J. Am. Chem. Soc.* **1985**, *107*, 6668; Lewis, E.S. *Bull. Soc. Chim. Fr.* **1988**, 259.

$RY + Y \rightarrow RY + Y$. The results of such calculations are generally in agreement with the Hammond postulate.

Marcus theory can be applied to any single-step process where something is transferred from one particle to another. It was originally derived for electron transfers,[27] and then extended to transfers of H^+ (see p. 372),

H^-,[28] and $H\bullet$,[29] as well as methyl transfers.

METHODS OF DETERMINING MECHANISMS

There are a number of commonly used methods for determining mechanisms.[30] In most cases, one method is not sufficient, and the problem is generally approached from several directions.

Identification of Products

Obviously, any mechanism proposed for a reaction must account for all the products obtained and for their relative proportions, including products formed by side reactions. Incorrect mechanisms for the von Richter reaction (reaction **13-30**) were accepted for many years because it was not realized that nitrogen was a major product. A proposed mechanism cannot be correct if it fails to predict the products in approximately the observed proportions. For example, any mechanism for the reaction

$$CH_4 + Cl_2 \xrightarrow{h\nu} CH_3Cl$$

that fails to account for the formation of a small amount of ethane cannot be correct (see **14-1**), and any mechanism proposed for the Hofmann rearrangement (**18-13**):

must account for the fact that the missing carbon appears as CO_2.

Determination of the Presence of an Intermediate

Intermediates are postulated in many mechanisms. There are several ways, none of them foolproof,[31] for attempting to learn whether or not an intermediate is present and, if so, its structure.

---

[27]Marcus, R.A. *J. Phys. Chem.* **1963**, *67*, 853, *Annu. Rev. Phys. Chem.* **1964**, *15*, 155; Eberson, L. *Electron Transfer Reactions in Organic Chemistry*; Springer: NY, **1987**.

[28]Kim, D.; Lee, I.H.; Kreevoy, M.M. *J. Am. Chem. Soc.* **1990**, *112*, 1889, and references cited therein.

[29]See, for example, Dneprovskii, A.S.; Eliseenkov, E.V. *J. Org. Chem. USSR* **1988**, *24*, 243.

[30]For a treatise on this subject, see Bernasconi, C.F. *Investigation of Rates and Mechanisms of Reactions*, 4th ed. (Vol. 6 of Weissberger, A. *Techniques of Chemistry*), 2 pts., Wiley, NY, **1986**. For a monograph, see Carpenter, B.K. *Determination of Organic Reaction Mechanisms*, Wiley, NY, **1984**.

[31]For a discussion, see Martin, R.B. *J. Chem. Educ.* **1985**, *62*, 789.

1. *Isolation of an Intermediate.* It is sometimes possible to isolate an intermediate from a reaction mixture by stopping the reaction after a short time or by the use of very mild conditions. For example, in the Neber rearrangement (reaction **18-12**)

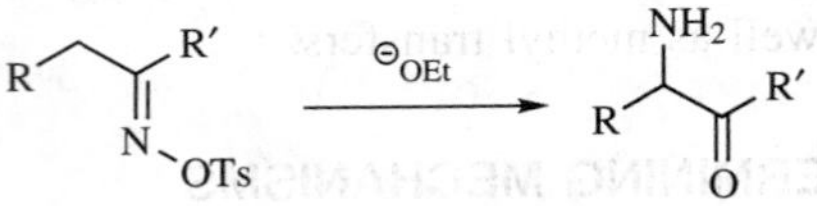

the intermediate **1** (an azirene)[32] has been isolated. If it can be shown that the isolated compound gives the same product when subjected to the reaction conditions and at a rate no slower than the starting compound, this constitutes strong evidence that the reaction involves that intermediate, although it is not conclusive, since the compound may arise by an alternate path and by coincidence give the same product.

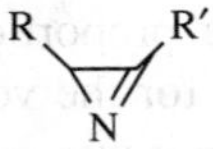

2. *Detection of an intermediate.* In many cases, an intermediate cannot be isolated, but can be detected by IR, NMR, or other spectra.[33] The detection by Raman spectra of NO_2^+ was regarded as strong evidence that this is an intermediate in the nitration of benzene (see **11-2**). Free radical and triplet intermediates can often be detected by esr and by CIDNP (see Chapter 5). Free radicals (as well as radical ions and EDA complexes) can also be detected by a method that does not rely on spectra. In this method, a double-bond compound is added to the reaction mixture, and its fate traced.[34] One possible result is cis–trans conversion. For example, *cis*-stilbene is isomerized to the trans isomer in the presence of RS• radicals, by this mechanism:

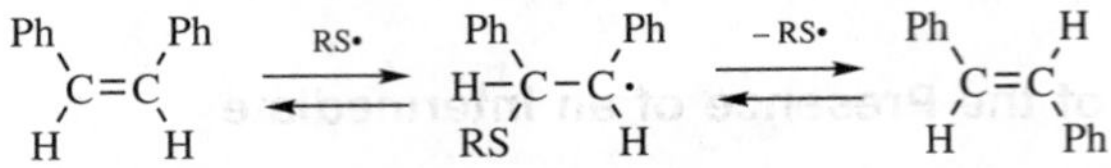

Since the trans isomer is more stable than the cis, the reaction does not go the other way, and the detection of the isomerized product is evidence for the presence of the RS• radicals.

---

[32]See Gentilucci, L.; Grijzen, Y.; Thijs, L.; Zwanenburg, B. *Tetrahedron Lett.* **1995**, *36*, 4665 for the synthesis of an azirene derivative.

[33]For a review on the use of electrochemical methods to detect intermediates, see Parker, V.D. *Adv. Phys. Org. Chem.* **1983**, *19*, 131. For a review of the study of intermediates trapped in matrixes, see Sheridan, R.S. *Org. Photochem.* **1987**, *8*, 159.

[34]For a review, see Todres, Z.V. *Tetrahedron* **1987**, *43*, 3839.

3. *Trapping of an Intermediate.* In some cases, the suspected intermediate is known to be one that reacts in a given way with a certain compound. The intermediate can then be trapped by running the reaction in the presence of that compound. For example, benzynes (p. 859) react with dienes in the Diels–Alder reaction (reaction **15-60**). In any reaction where a benzyne is a suspected intermediate, the addition of a diene and the detection of the Diels–Alder adduct indicate that the benzyne was probably present.

4. *Addition of a Suspected Intermediate.* If a certain intermediate is suspected, and if it can be obtained by other means, then under the same reaction conditions it should give the same products. This kind of experiment can provide conclusive negative evidence: if the correct products are not obtained, the suspected compound is not an intermediate. However, if the correct products are obtained, this is not conclusive since they may arise by coincidence. The von Richter reaction (reaction **13-30**) provides us with a good example here too. For many years, it had been assumed that an aryl cyanide was an intermediate, since cyanides are easily hydrolyzed to carboxylic acids (**16-4**). In fact, in 1954, *p*-chlorobenzonitrile was shown to give *p*-chlorobenzoic acid under normal von Richter conditions.[35] However, when the experiment was repeated with 1-cyanonaphthalene, no 1-naphthoic acid was obtained, although 2-nitronaphthalene gave 13% 1-naphthoic acid under the same conditions.[36] This proved that 2-nitronaphthalene must have been converted to 1-naphthoic acid by a route that does not involve 1-cyanonaphthalene. It also showed that even the conclusion that *p*-chlorobenzonitrile was an intermediate in the conversion of *m*-nitrochlorobenzene to *p*-chlorobenzoic acid must now be suspect, since it is not likely that the mechanism would substantially change in going from the naphthalene to the benzene system.

The Study of Catalysis[37]

Much information about the mechanism of a reaction can be obtained from a knowledge of which substances catalyze the reaction, which inhibit it, and which do neither. Of course, just as a mechanism must be compatible with the products, so must it be compatible with its catalysts. In general, catalysts perform their actions by providing an alternate pathway for the reaction in which $\Delta G^{\ddagger}$ is less than it would be without the catalyst. Catalysts do not change ΔG.

[35]Bunnett, J.F.; Rauhut, M.M.; Knutson, D.; Bussell, G.E. *J. Am. Chem. Soc.* **1954**, *76*, 5755.
[36]Bunnett, J.F.; Rauhut, M.M. *J. Org. Chem.* **1956**, *21*, 944.
[37]For treatises, see Jencks, W.P. *Catalysis in Chemistry and Enzymology*, McGraw-Hill, NY, **1969**; Bender, M.L. *Mechanisms of Homogeneous Catalysis from Protons to Proteins*, Wiley, NY, **1971**. For reviews, see Coenen, J.W.E. *Recl. Trav. Chim. Pays-Bas* **1983**, *102*, 57; and in Bernasconi, C.F. *Investigation of Rates and Mechanisms of Reactions*, 4th ed. (Vol. 6 of Weissberger, A. *Techniques of Chemistry*), pt. 1, Wiley, NY, **1986**, the articles by Keeffe, J.R.; Kresge, A.J. pp. 747–790; Haller, G.L.; Delgass, W.N. pp. 951–979.

Isotopic Labeling[38]

Much useful information has been obtained by using molecules that have been isotopically labeled and tracing the path of the reaction in that way. For example, in the reaction

$$RCOO^* + BrCN \longrightarrow RCN^*$$

does the CN group in the product come from the CN in the BrCN? The use of ^{14}C supplied the answer, since $R^{14}CO_2^-$ gave *radioactive* RCN.[39] This surprising result saved a lot of labor, since it ruled out a mechanism involving the replacement of CO_2 by CN (see reaction **16-94**). Other radioactive isotopes are also frequently used as tracers, but even stable isotopes can be used. An example is the hydrolysis of esters

$$\underset{R}{\overset{O}{\|}}\!\!-\!OR' + H_2O \longrightarrow \underset{R}{\overset{O}{\|}}\!\!-\!OH + ROH$$

Which bond of the ester is broken, the acyl–O or the alkyl–O bond? The answer is found by the use of $H_2^{18}O$. If the acyl–O bond breaks, the labeled oxygen will appear in the acid; otherwise it will be in the alcohol (see **16-59**). Although neither compound is radioactive, the one that contains ^{18}O can be determined by submitting both to mass spectrometry. In a similar way, deuterium can be used as a label for hydrogen. In this case, it is not necessary to use mass spectrometry, since ir and nmr spectra can be used to determine when deuterium has been substituted for hydrogen. Carbon-13 NMR is also nonradioactive: It can be detected by ^{13}C NMR.[40]

In the labeling technique, it is not generally necessary to use completely labeled compounds. Partially labeled material is usually sufficient.

Stereochemical Evidence[41]

If the products of a reaction are capable of existing in more than one stereoisomeric form, the form that is obtained may give information about the mechanism. For example, (+)-malic acid was discovered by Walden[42] to give (−)-chlorosuccinic acid when treated with PCl_5 and the (+) enantiomer when treated with $SOCl_2$,

[38]For reviews see Wentrup, C., in Bernasconi, C.F. *Investigation of Rates and Mechanisms of Reactions*, 4th ed. (Vol. 6 of Weissberger, A. *Techniques of Chemistry*), pt. 1, Wiley, NY, *1986*, pp. 613–661; Collins, C.J. *Adv. Phys. Org. Chem. 1964*, *2*, 3. See also, the series *Isotopes in Organic Chemistry*.

[39]Douglas, D.E.; Burditt, A.M. *Can. J. Chem. 1958*, *36*, 1256.

[40]For a review, see Hinton, J.; Oka, M.; Fry, A. *Isot. Org. Chem. 1977*, *3*, 41.

[41]For lengthy treatments of the relationship between stereochemistry and mechanism, see Billups, W.E.; Houk, K.N.; Stevens, R.V., in Bernasconi, C.F. *Investigation of Rates and Mechanisms of Reactions*, 4th ed. (Vol. 6 of Weissberger, A. *Techniques of Chemistry*), pt. 1, Wiley, NY, *1986*, pp. 663–746; Eliel, E.L. *Stereochemistry of Carbon Compounds*; McGraw-Hill: NY, *1962*; Newman, M.S. *Steric Effects in Organic Chemistry*, Wiley, NY, *1956*.

[42]Walden, P. *Ber. 1896*, *29*, 136; *1897*, *30*, 3149; *1899*, *32*, 1833.

showing that the mechanisms of these apparently similar conversions could not be the same (see pp. 427, 469). Much useful information has been obtained about nucleophilic substitution, elimination, rearrangement, and addition reactions from this type of experiment. The isomers involved need not be enantiomers. Thus, the fact that *cis*-2-butene treated with $KMnO_4$ gives *meso*-2,3-butanediol and not the racemic mixture is evidence that the two OH groups attack the double bond from the same side (see reaction **15-48**).

Kinetic Evidence[43]

The rate of a homogeneous reaction[44] is the rate of disappearance of a reactant or appearance of a product. The rate nearly always changes with time, since it is usually proportional to concentration and the concentration of reactants decreases with time. However, the rate is not always proportional to the concentration of all reactants. In some cases, a change in the concentration of a reactant produces no change at all in the rate, while in other cases the rate may be proportional to the concentration of a substance (a catalyst) that does not even appear in the stoichiometric equation. A study of which reactants affect the rate often tells a good deal about the mechanism.

If the rate is proportional to the change in concentration of only one reactant (**A**), the *rate law* (the rate of change of concentration of **A** with time t) is

$$\text{Rate} = \frac{-d[\mathbf{A}]}{dt} = k[\mathbf{A}]$$

where k is the *rate constant* for the reaction.[45] There is a minus sign because the concentration of **A** decreases with time. A reaction that follows such a rate law is called a *first-order reaction*. The units of k for a first-order reaction are s^{-1}. The rate of a *second-order reaction* is proportional to the concentration of two reactants, or to the square of the concentration of one:

$$\text{Rate} = \frac{-d[\mathbf{A}]}{dt} = k[\mathbf{A}][\mathbf{B}] \qquad \text{or} \qquad \text{Rate} = \frac{-d[\mathbf{A}]}{dt} = k[\mathbf{A}]^2$$

For a second-order reaction the units are $\text{L mol}^{-1}\,\text{s}^{-1}$ or some other units expressing the reciprocal of concentration or pressure per unit time interval.

---

[43]For the use of kinetics in determining mechanisms, see Connors, K.A. *Chemical Kinetics*, VCH, NY, *1990*; Zuman, P.; Patel, R.C. *Techniques in Organic Reaction Kinetics*, Wiley, NY, *1984*; Drenth, W.; Kwart, H. *Kinetics Applied to Organic Reactions*, Marcel Dekker, NY, *1980*; Hammett, L.P. *Physical Organic Chemistry*, 2nd ed.; McGraw-Hill: NY, *1970*, pp. 53–100; Gardiner, Jr., W.C. *Rates and Mechanisms of Chemical Reactions*, W.A. Benjamin, NY, *1969*; Leffler, J.E.; Grunwald, E. *Rates and Equilibria of Organic Reactions*, Wiley, NY, *1963*; Jencks, W.P. *Catalysis in Chemistry and Enzymology*, McGraw-Hill, NY, *1969*, pp. 555–614; Refs. 6 and 26
[44]A homogeneous reaction occurs in one phase. Heterogeneous kinetics have been studied much less.
[45]Colins, C.C.; Cronin, M.F.; Moynihan, H.A.; McCarthy, D.G. *J. Chem. Soc. Perkin Trans. 1* *1997*, 1267 for the use of Marcus theory to predict rate constants in organic reactions.

Similar expressions can be written for third-order reactions. A reaction whose rate is proportional to [**A**] and to [**B**] is said to be first order in **A** and in **B**, second order overall. A reaction rate can be measured in terms of any reactant or product, but the rates so determined are not necessarily the same. For example, if the stoichiometry of a reaction is $2\mathbf{A} + \mathbf{B} \rightarrow \mathbf{C} + \mathbf{D}$ then, on a molar basis, **A** must disappear twice as fast as **B**, so that $-d[\mathbf{A}]/dt$ and $-d[\mathbf{B}]/dt$ are not equal, but the former is twice as large as the latter.

The rate law of a reaction is an experimentally determined fact. From this fact, we attempt to learn the *molecularity*, which may be defined as the number of molecules that come together to form the activated complex. It is obvious that if we know how many (and which) molecules take part in the activated complex, we know a good deal about the mechanism. The experimentally determined rate order is not necessarily the same as the molecularity. Any reaction, no matter how many steps are involved, has only one rate law, but each step of the mechanism has its own molecularity. For reactions that take place in one step (reactions without an intermediate) the order is the same as the molecularity. A first-order, one-step reaction is always unimolecular; a one-step reaction that is second order in **A** always involves two molecules of **A**; if it is first order in **A** and in **B**, then a molecule of **A** reacts with one of **B**, and so on. For reactions that take place in more than one step, the order *for each step* is the same as the molecularity *for that step*. This fact enables us to predict the rate law for any proposed mechanism, although the calculations may get lengthy at times.[46] If any one step of a mechanism is considerably slower than all the others (this is usually the case), the rate of the overall reaction is essentially the same as that of the slow step, which is consequently called the *rate-determining* step.[47]

For reactions that take place in two or more steps, two broad cases can be distinguished:

1. The first step is slower than any subsequent step and is consequently rate determining. In such cases, the rate law simply includes the reactants that participate in the slow step. For example, if the reaction $\mathbf{A} + 2\mathbf{B} \rightarrow \mathbf{C}$ has the mechanism

$$\mathbf{A} + \mathbf{B} \xrightarrow{\text{slow}} \mathbf{I}$$

$$\mathbf{I} + \mathbf{B} \xrightarrow{\text{fast}} \mathbf{C}$$

where **I** is an intermediate, the reaction is second order, with the rate law

$$\text{Rate} = \frac{-d[\mathbf{A}]}{dt} = k[\mathbf{A}][\mathbf{B}]$$

---

[46]For a discussion of how order is related to molecularity in many complex situations, see Szabó, Z.G. in Bamford, C.H.; Tipper, C.F.H. *Comprehensive Chemical Kinetics*, Vol. 2; Elsevier: NY, *1969*, pp. 1–80.
[47]Many chemists prefer to use the term *rate-limiting step or rate-controlling step* for the slow step, rather than *rate-determining step*. See the definitions, in Gold, V.; Loening, K.L.; McNaught, A.D.; Sehmi, P. *IUPAC Compendium of Chemical Terminology;* Blackwell Scientific Publications: Oxford, *1987*, p. 337. For a discussion of rate-determining steps, see Laidler, K.J. *J. Chem. Educ. 1988*, 65, 250.

2. When the first step is not rate determining, determination of the rate law is usually much more complicated. For example, consider the mechanism

$$A + B \underset{k_{-1}}{\overset{k_1}{\rightleftharpoons}} I$$

$$I + B \overset{k_2}{\longrightarrow} C$$

where the first step is a rapid attainment of equilibrium, followed by a slow reaction to give **C**. The rate of disappearance of **A** is

$$\text{Rate} = \frac{-d[A]}{dt} = k_1[A][B] - k_{-1}[I]$$

Both terms must be included because **A** is being formed by the reverse reaction as well as being used up by the forward reaction. This equation is of very little help as it stands since we cannot measure the concentration of the intermediate. However, the combined rate law for the formation and disappearance of **I** is

$$\text{Rate} = \frac{-d[A]}{dt} = k_1[A][B] - k_{-1}[I] - k_2[I][B]$$

At first glance, we seem no better off with this equation, but we can make the assumption that *the concentration of I does not change with time*, since it is an intermediate that is used up (going either to **A** + **B** or to **C**) as fast as it is formed. This assumption, called the assumption of the *steady state*,[48] enables us to set $d[I]/dt$ equal to zero and hence to solve for [I] in terms of the measurable quantities [A] and [B]:

$$[I] = \frac{k_1[A][B]}{k_2[B] + k_{-1}}$$

We now insert this value for [I] into the original rate expression to obtain

$$\frac{-d[A]}{dt} = \frac{k_1 k_2 [A][B]^2}{k_2[B] + k_{-1}}$$

Note that this rate law is valid whatever the values of k_1, k_{-1}, and k_2. However, our original hypothesis was that the first step was faster than the second, or that

$$k_1[A][B] \gg k_2[I][B]$$

---

[48]For a discussion, see Raines, R.T.; Hansen, D.E. *J. Chem. Educ.* **1988**, *65*, 757.

Since the first step is an equilibrium

$$k_1[\mathbf{A}][\mathbf{B}] = k_{-1}[\mathbf{I}]$$

we have

$$k_{-1}[\mathbf{I}] \gg k_2[\mathbf{I}][\mathbf{B}]$$

Canceling [**I**], we get

$$k_{-1} \gg k_2[\mathbf{B}]$$

We may thus neglect $k_2[\mathbf{B}]$ in comparison with k_{-1} and obtain

$$\frac{-d[\mathbf{A}]}{dt} = \frac{k_1 k_2}{k_{-1}} [\mathbf{A}][\mathbf{B}]^2$$

The overall rate is thus third order: first order in **A** and second order in **B**. Incidentally, if the first step is rate determining (as was the case in the preceding paragraph), then

$$k_2[\mathbf{B}] \gg k_{-1} \qquad \text{and} \qquad \frac{-d[\mathbf{A}]}{dt} = k_1[\mathbf{A}][\mathbf{B}]$$

which is the same rate law we deduced from the rule that where the first step is rate determining, the rate law includes the reactants that participate in that step.

It is possible for a reaction to involve **A** and **B** in the rate-determining step, although only [**A**] appears in the rate law. This occurs when a large excess of **B** is present, say 100 times the molar quantity of **A**. In this case, the complete reaction of **A** uses up only 1 equivalent of **B**, leaving 99 equivalents. It is not easy to measure the change in concentration of **B** with time in such a case, and it is seldom attempted, especially when **B** is also the solvent. Since [**B**], for practical purposes, does not change with time, the reaction appears to be first order in **A** although actually both **A** and **B** are involved in the rate-determining step. This is often referred to as a *pseudo-first-order* reaction. Pseudo-order reactions can also come about when one reactant is a catalyst whose concentration does not change with time because it is replenished as fast as it is used up and when a reaction is conducted in a medium that keeps the concentration of a reactant constant, for example, in a buffer solution where H^+ or ^-OH is a reactant. Pseudo-first-order conditions are frequently used in kinetic investigations for convenience in experimentation and calculations.

What is actually being measured is the change in concentration of a product or a reactant with time. Many methods have been used to make such measurements.[49]

---

[49]For a monograph on methods of interpreting kinetic data, see Zuman, P.; Patel, R.C. *Techniques in Organic Reaction Kinetics*, Wiley, NY, *1984*. For a review of methods of obtaining kinetic data, see Batt, L. in Bamford, C.H.; Tipper, C.F.H. *Comprehensive Chemical Kinetics,* Vol. 1, Elsevier, NY, *1969*, pp. 1–111.

The choice of a method depends on its convenience and its applicability to the reaction being studied. Among the most common methods are

1. *Periodic or Continuous Spectral Readings.* In many cases, the reaction can be carried out in the cell while it is in the instrument. Then all that is necessary is that the instrument be read, periodically or continuously. Among the methods used are ir and uv spectroscopy, polarimetry, nmr, and esr.[50]

2. *Quenching and Analyzing.* A series of reactions can be set up and each stopped in some way (perhaps by suddenly lowering the temperature or adding an inhibitor) after a different amount of time has elapsed. The materials are then analyzed by spectral readings, titrations, chromatography, polarimetry, or any other method.

3. *Removal of Aliquots at Intervals.* Each aliquot is then analyzed as in method 2.

4. *Measurement of Changes in Total Pressure, for Gas-Phase Reactions.*[51]

5. *Calorimetric Methods.* The output or absorption of heat can be measured at time intervals.

Special methods exist for kinetic measurements of very fast reactions.[52]

In any case, what is usually obtained is a graph showing how a concentration varies with time. This must be interpreted[53] to obtain a rate law and a value of k. If a reaction obeys simple first- or second-order kinetics, the interpretation is generally not difficult. For example, if the concentration at the start is A_0, the first-order rate law

$$\frac{-d[\mathbf{A}]}{dt} = k[\mathbf{A}] \qquad \text{or} \qquad \frac{-d[\mathbf{A}]}{[\mathbf{A}]} = kdt$$

---

[50]For a review of esr to measure kinetics, see Norman, R.O.C. *Chem. Soc. Rev. 1979*, 8, 1.

[51]For a review of the kinetics of reactions in solution at high pressures, see le Noble, W.J. *Prog. Phys. Org. Chem. 1967*, 5, 207. For reviews of synthetic reactions under high pressure, see Matsumoto, K.; Sera, A.; Uchida, T. *Synthesis 1985*, 1; Matsumoto, K.; Sera, A. *Synthesis 1985*, 999.

[52]For reviews, see Connors, K.A. *Chemical Kinetics*, VCH, NY, *1990*, pp. 133–186; Zuman, P.; Patel, R.C. *Techniques in Organic Reaction Kinetics*, Wiley, NY, *1984*, pp. 247–327; Krüger, H. *Chem. Soc. Rev. 1982*, 11, 227; Hague, D.N. in Bamford, C.H.; Tipper, C.F.H. *Comprehensive Chemical Kinetics*, Vol. 1, Elsevier, NY, *1969*, pp. 112–179, Elsevier, NY, *1969*; Bernasconi, C.F. *Investigation of Rates and Mechanisms of Reactions*, 4th ed. (Vol. 6 of Weissberger, A. *Techniques of Chemistry*), pt. 2, Wiley, NY, *1986*,. See also, Bamford, C.H.; Tipper, C.F.H. *Comprehensive Chemical Kinetics*, Vol. 24, Elsevier, NY, *1983*.

[53]For discussions, much fuller than that given here, of methods for interpreting kinetic data, see Connors, K.A. *Chemical Kinetics*, VCH, NY, *1990*, pp. 17–131; Ritchie, C.D. *Physical Organic Chemistry*, 2nd ed., Marcel Dekker, NY, *1990*, pp. 1–35; Zuman, P.; Patel, R.C. *Techniques in Organic Reaction Kinetics*, Wiley, NY, *1984*; Margerison, D., in Bamford, C.H.; Tipper, C.F.H. *Comprehensive Chemical Kinetics*, Vol. 1, Elsevier, NY, *1969*, pp. 343–421; Moore, J.W.; Pearson, R.G. *Kinetics and Mechanism*, 3rd ed., Wiley, NY, *1981*, pp. 12–82; in Bernasconi, C.F. *Investigation of Rates and Mechanisms of Reactions*, 4th ed. (Vol. 6 of Weissberger, A. *Techniques of Chemistry*), pt. 1, Wiley, NY, *1986*, the articles by Bunnett, J.F. pp. 251–372, Noyes Pub., pp. 373–423, Bernasconi, C.F. pp. 425–485, Wiberg, K.B. pp. 981–1019.

can be integrated between the limits $t = 0$ and $t = t$ to give

$$-\ln\frac{[\mathbf{A}]}{\mathbf{A}_0} = kt \qquad \text{or} \qquad \ln[\mathbf{A}] = -kt + \ln\mathbf{A}_0$$

Therefore, if a plot of ln [**A**] against t is linear, the reaction is first order and k can be obtained from the slope. For first-order reactions, it is customary to express the rate not only by the rate constant k, but also by the *half-life*, which is the time required for one-half of any given quantity of a reactant to be used up. Since the half-life $t_{1/2}$ is the time required for [**A**] to reach $\mathbf{A}_0/2$, we may say that

$$\ln\frac{\mathbf{A}_0}{2} = kt_{1/2} + \ln\mathbf{A}_0$$

so that

$$t_{1/2} = \frac{\ln\left[\frac{\mathbf{A}_0}{\mathbf{A}_0/2}\right]}{k} = \frac{\ln 2}{k} = \frac{0.693}{k}$$

For the general case of a reaction first order in **A** and first order in **B**, second order overall, integration is complicated, but it can be simplified if equimolar amounts of **A** and **B** are used, so that $\mathbf{A}_0 = \mathbf{B}_0$. In this case,

$$\frac{-d[\mathbf{A}]}{dt} = k[\mathbf{A}][\mathbf{B}]$$

is equivalent to

$$\frac{-d[\mathbf{A}]}{dt} = k[\mathbf{A}]^2 \qquad \text{or} \qquad \frac{-d[\mathbf{A}]}{[\mathbf{A}]^2} = k\,dt$$

Integrating as before gives

$$\frac{1}{[\mathbf{A}]} - \frac{1}{\mathbf{A}_0} = kt$$

Thus, under equimolar conditions, if a plot of $1/[\mathbf{A}]$ against t is linear, the reaction is second order with a slope of k. It is obvious that the same will hold true for a reaction second order in **A**.[54]

Although many reaction-rate studies do give linear plots, which can therefore be easily interpreted, the results in many other studies are not so simple. In some cases, a reaction may be first order at low concentrations but second order at higher concentrations. In other cases, fractional orders are obtained, and even negative orders. The interpretation of complex kinetics often requires much skill and effort. Even where the kinetics are relatively simple, there is often a problem in interpreting the data because of the difficulty of obtaining precise enough measurements.[55]

[54]We have given the integrated equations for simple first- and second-order kinetics. For integrated equations for a large number of kinetic types, see Margerison, D., in Bamford, C.H.; Tipper C.F.H. *Comprehensive Chemical Kinetics,* Vol. 1, Elsevier, NY, *1969*, p. 361.

[55]See, Hammett, L.P. *Physical Organic Chemistry,* 2nd ed., McGraw-Hill, NY, *1970*, pp. 62–70.

Nuclear magnetic resonance spectra can be used to obtain kinetic information in a completely different manner from that mentioned on p. 319. This method, which involves the study of NMR line shapes,[56] depends on the fact that NMR spectra have an inherent time factor: If a proton changes its environment less rapidly than $\sim 10^3$ times/s, an NMR spectrum shows a separate peak for each position the proton assumes. For example, if the rate of rotation around

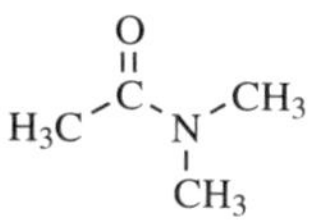

the C—N bond of *N,N*-dimethylacetamide is slower than 10^3 rotations per second, the two *N*-methyl groups each have separate chemical shifts since they are not equivalent, one being cis to the oxygen and the other trans. However, if the environmental change takes place more rapidly than $\sim 10^3$ times per second, only one line is found, at a chemical shift that is the weighted average of the two individual positions. In many cases, two or more lines are found at low temperatures, but as the temperature is increased, the lines coalesce because the interconversion rate increases with temperature and passes the 10^3 per second mark. From studies of the way line shapes change with temperature it is often possible to calculate rates of reactions and of conformational changes. This method is not limited to changes in proton line shapes but can also be used for other atoms that give nmr spectra and for esr spectra.

Several types of mechanistic information can be obtained from kinetic studies.

1. From the order of a reaction, information can be obtained about which molecules and how many take part in the rate-determining step. Such knowledge is very useful and often essential in elucidating a mechanism. For any mechanism that can be proposed for a given reaction, a corresponding rate law can be calculated by the methods discussed on pp. 316–320. If the experimentally obtained rate law fails to agree with this, the proposed mechanism is wrong. However, it is often difficult to relate the order of a reaction to the mechanism, especially when the order is fractional or negative. In addition, it is frequently the case that two or more proposed mechanisms for a reaction are kinetically indistinguishable, that is, they predict the same rate law.

2. Probably the most useful data obtained kinetically are the rate constants themselves. They are important since they can tell us the effect on the rate of

[56]For a monograph, see Ōki, M. *Applications of Dynamic NMR Spectroscopy to Organic Chemistry*, VCH, NY, *1985*. For reviews, see Fraenkel, G., in Bernasconi, C.F. *Investigation of Rates and Mechanisms of Reactions*, 4th ed. (Vol. 6 of Weissberger, A. *Techniques of Chemistry*), pt. 2, Wiley, NY, *1986*, pp. 547–604; Aganov, A.V.; Klochkov, V.V.; Samitov, Yu.Yu. *Russ. Chem. Rev.* *1985*, *54*, 931; Roberts, J.D. *Pure Appl. Chem.* *1979*, *51*, 1037; Binsch, G. *Top. Stereochem.* *1968*, *3*, 97; Johnson Jr., C.S. *Adv. Magn. Reson.* *1965*, *1*, 33.

a reaction of changes in the structure of the reactants (see Chapter 9), the solvent, the ionic strength, the addition of catalysts, and so on.

3. If the rate is measured at several temperatures, in most cases a plot of ln k against $1/T$ (T stands for absolute temperature) is nearly linear[57] with a negative slope, and fits the equation

$$\ln k = \frac{-E_a}{RT} + \ln A$$

where R is the gas constant and A is a constant called the *frequency factor*. This permits the calculation of E_a, which is the Arrhenius activation energy of the reaction. The parameter $\Delta H^{\ddagger}$ can then be obtained by

$$E_a = \Delta H^{\ddagger} + RT$$

It is also possible to use these data to calculate $\Delta S^{\ddagger}$ by the formula[58]

$$\frac{\Delta S^{\ddagger}}{4.576} = \log k - 10.753 - \log T + \frac{E_a}{4.576T}$$

for energies in calorie units. For joule units the formula is

$$\frac{\Delta S^{\ddagger}}{19.15} = \log k - 10.753 - \log T + \frac{E_a}{19.15T}$$

One then obtains $\Delta G^{\ddagger}$ from $\Delta G^{\ddagger} = \Delta H^{\ddagger} - T\Delta S^{\ddagger}$.

Isotope Effects

When a hydrogen in a reactant molecule is replaced by deuterium, there is often a change in the rate. Such changes are known as *deuterium isotope effects*[59] and are

---

[57]For a review of cases where such a plot is nonlinear, see Blandamer, M.J.; Burgess, J.; Robertson, R.E.; Scott, J.M.W. *Chem. Rev.* **1982**, *82*, 259.

[58]For a derivation of this equation, see Bunnett, J.F., in Bernasconi, C.F. *Investigation of Rates and Mechanisms of Reactions*, 4th ed. (Vol. 6 of Weissberger, A. *Techniques of Chemistry*), pt. 1, Wiley, NY, **1986**, p. 287.

[59]For a monograph, see Melander, L.; Saunders, Jr., W.H. *Reaction Rates of Isotopic Molecules*, Wiley, NY, **1980**. For reviews, see Isaacs, N.S. *Physical Organic Chemistry*, Longman Scientific and Technical, Essex, **1987**, pp. 255–281; Lewis, E.S. *Top. Curr. Chem.* **1978**, *74*, 31; Saunders, Jr., W.H. in Bernasconi, C.F. *Investigation of Rates and Mechanisms of Reactions*, 4th ed. (Vol. 6 of Weissberger, A. *Techniques of Chemistry*), pt. 1, Wiley, NY, **1986**, pp. 565–611; Bell, R.P. *The Proton in Chemistry*, 2nd ed.; Cornell University Press: Ithaca, NY, **1973**, pp. 226–296, *Chem. Soc. Rev.* **1974**, *3*, 513; Bigeleisen, J.; Lee, M.W.; Mandel, F. *Annu. Rev. Phys. Chem.* **1973**, *24*, 407; Wolfsberg, M. *Annu. Rev. Phys. Chem.* **1969**, *20*, 449; Saunders, Jr., W.H. *Surv. Prog. Chem.* **1966**, *3*, 109; Simon, H.; Palm, D. *Angew. Chem. Int. Ed.* **1966**, *5*, 920; Jencks, W.P. *Catalysis in Chemistry and Enzymology*, McGraw-Hill, NY, **1969**, pp. 243–281. For a review of temperature dependence of primary isotope effects as a mechanistic criterion, see Kwart, H. *Acc. Chem. Res.* **1982**, *15*, 401. For a review of the effect of pressure on isotope effects, see Isaacs, E.S. *Isot. Org. Chem.* **1984**, *6*, 67. For a review of isotope effects in the study of reactions in which there is branching from a common intermediate, see Thibblin, A.; Ahlberg, P. *Chem. Soc. Rev.* **1989**, *18*, 209. See also, the series *Isotopes in Organic Chemistry*.

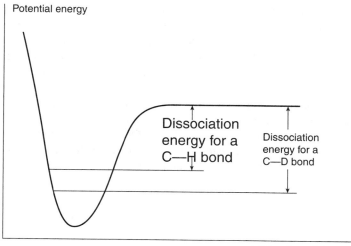

Fig. 6.4. A C—D bond has a lower zero point than does a corresponding C—H bond; thus the dissociation energy is higher.

expressed by the ratio k_H/k_D. The ground-state vibrational energy (called the zero-point vibrational energy) of a bond depends on the mass of the atoms and is lower when the reduced mass is higher.[60] Therefore, D—C, D—O, D—N bonds, and so on, have lower energies in the ground state than the corresponding H—C, H—O, H—N bonds, and so on. Complete dissociation of a deuterium bond consequently requires more energy than that for a corresponding hydrogen bond in the same environment (Fig. 6.4). If an H—C, H—O, or H—N bond is not broken at all in a reaction or is broken in a nonrate-determining step, substitution of deuterium for hydrogen causes no change in the rate (see below for an exception to this statement), but if the bond is broken in the rate-determining step, the rate must be lowered by the substitution.

This provides a valuable diagnostic tool for determination of mechanism. For example, in the bromination of acetone (reaction **12-4**)

$$CH_3COCH_3 + Br_2 \longrightarrow CH_3COCH_2Br$$

the fact that the rate is independent of the bromine concentration led to the postulate that the rate-determining step was prior tautomerization of the acetone:

In turn, the rate-determining step of the tautomerization involves cleavage of a C—H bond (see **12-3**). Thus there should be a substantial isotope effect if deuterated

[60]The reduced mass μ of two atoms connected by a covalent bond is $\mu = m_1 m_2/(m_1 + m_2)$.

acetone is brominated. In fact, k_H/k_D was found to be ~ 7.[61] Deuterium isotope effects usually range from 1 (no isotope effect at all) to ~ 7 or 8, although in a few cases, larger[62] or smaller values have been reported.[63] Values of $k_H/k_D < 1$ are called *inverse isotope effects*. Isotope effects are greatest when, in the transition state, the hydrogen is symmetrically bonded to the atoms between which it is being transferred.[64] Also, calculations show that isotope effects are at a maximum when the hydrogen in the transition state is on the straight line connecting the two atoms between which the hydrogen is being transferred and that for sufficiently nonlinear configurations they decrease to $k_H/k_D = 1$–2.[65] Of course, in open systems there is no reason for the transition state to be nonlinear, but this is not the case in many intramolecular mechanisms, for example, in a 1,2 migration of a hydrogen

Transition state

To measure isotope effects it is not always necessary to prepare deuterium-enriched starting compounds. It can also be done by measuring the change in deuterium concentration at specific sites between a compound containing deuterium in natural abundance and the reaction product, using a high-field NMR instrument.[66]

The substitution of tritium for hydrogen gives isotope effects that are numerically larger. Isotope effects have also been observed with other elements, but they are much smaller, ~ 1.02–1.10. For example, k_{12C}/k_{13C} for

$$Ph*CH_2Br \;+\; CH_3O^{\ominus} \xrightarrow{\;CH_3OH\;} Ph*CH_2OCH_3$$

[61]Reitz, O.; Kopp, J. *Z. Phys. Chem. Abt. A* **1939**, *184*, 429.

[62]For an example of a reaction with a deuterium isotope effect of 24.2, see Lewis, E.S.; Funderburk, L.H. *J. Am. Chem. Soc.* **1967**, *89*, 2322. The high isotope effect in this case has been ascribed to *tunneling* of the proton: because it is so small a hydrogen atom can sometimes get through a thin potential barrier without going over the top, that is, without obtaining the usually necessary activation energy. A deuterium, with a larger mass, is less able to do this. The phenomenon of tunneling is a consequence of the uncertainty principle. k_H/k_D for the same reaction is 79: Lewis, E.S.; Robinson, J.K. *J. Am. Chem. Soc.* **1968**, *90*, 4337. An even larger deuterium isotope effect (~ 50) has been reported for the oxidation of benzyl alcohol. This has also been ascribed to tunneling: Roecker, L.; Meyer, T.J. *J. Am. Chem. Soc.* **1987**, *109*, 746. For discussions of high isotope effects, see Kresge, A.J.; Powell, M.F. *J. Am. Chem. Soc.* **1981**, *103*, 201; Caldin, E.F.; Mateo, S.; Warrick, P. *J. Am. Chem. Soc.* **1981**, *103*, 202. For arguments that high isotope effects can be caused by factors other than tunneling, see McLennan, D.J. *Aust. J. Chem.* **1979**, *32*, 1883; Thibblin, A. *J. Phys. Org. Chem.* **1988**, *1*, 161; Kresge, A.J.; Powell, M.F. *J. Phys. Org. Chem.* **1990**, *3*, 55.

[63]For a review of a method for calculating the magnitude of isotope effects, see Sims, L.B.; Lewis, D.E. *Isot. Org. Chem.* **1984**, *6*, 161.

[64]Kwart, H.; Latimore, M.C. *J. Am. Chem. Soc.* **1971**, *93*, 3770; Pryor, W.A.; Kneipp, K.G. *J. Am. Chem. Soc.* **1971**, *93*, 5584; Bell, R.P.; Cox, B.G. *J. Chem. Soc. B* **1971**, 783; Bethell, D.; Hare, G.J.; Kearney, P.A. *J. Chem. Soc. Perkin Trans. 2* **1981**, 684, and references cited therein. See, however, Motell, E.L.; Boone, A.W.; Fink, W.H. *Tetrahedron* **1978**, *34*, 1619.

[65]More O'Ferrall, R.A. *J. Chem. Soc. B* **1970**, 785, and references cited therein.

[66]Pascal, R.A.; Baum, M.W.; Wagner, C.K.; Rodgers, L.R.; Huang, D. *J. Am. Chem. Soc.* **1986**, *108*, 6477.

is $1.053.^{67}$ Although they are small, heavy-atom isotope effects can be measured quite accurately and are often very useful.[68]

Deuterium isotope effects have been found even where it is certain that the C—H bond does not break at all in the reaction. Such effects are called *secondary isotope effects*,[69] the term *primary isotope effect* being reserved for the type discussed previously. Secondary isotope effects can be divided into α and β effects. In a β secondary isotope effect, substitution of deuterium for hydrogen β to the position of bond breaking slows the reaction. An example is solvolysis of isopropyl bromide:

$$(CH_3)_2CHBr \; + \; H_2O \; \xrightarrow{\;k_H\;} \; (CH_3)_2CHOH$$

$$(CD_3)_2CHBr \; + \; H_2O \; \xrightarrow{\;k_D\;} \; (CD_3)_2CHOH$$

where k_H/k_D was found to be $1.34.^{70}$ The cause of β isotope effects has been a matter of much controversy, but they are most likely due to hyperconjugation effects in the transition state. The effects are greatest when the transition state has considerable carbocation character.[71] Although the C—H bond in question is not broken in the transition state, the carbocation is stabilized by hyperconjugation involving this bond. Because of hyperconjugation, the difference in vibrational energy between the C—H bond and the C—D bond in the transition state is less than it is in the ground state, so the reaction is slowed by substitution of deuterium for hydrogen.

Support for hyperconjugation as the major cause of β isotope effects is the fact that the effect is greatest when D is anti to the leaving group[72] (because of the requirement that all atoms in a resonance system be coplanar, planarity of the D—C—C—X system would most greatly increase the hyperconjugation), and the fact that secondary isotope effects can be transmitted through unsaturated systems.[73] There is evidence that at least some β isotope effects are steric in

[67]Stothers, J.B.; Bourns, A.N. *Can. J. Chem.* **1962**, *40*, 2007. See also, Ando, T.; Yamataka, H.; Tamura, S.; Hanafusa, T. *J. Am. Chem. Soc.* **1982**, *104*, 5493.

[68]For a review of carbon isotope effects, see Willi, A.V. *Isot. Org. Chem.* **1977**, *3*, 237.

[69]For reviews, see Westaway, K.C. *Isot. Org. Chem.* **1987**, *7*, 275; Sunko, D.E.; Hehre, W.J. *Prog. Phys. Org. Chem.* **1983**, *14*, 205; Shiner, Jr., V.J., in Collins, C.J.; Bowman, N.S. *Isotope Effects in Chemical Reactions*, Van Nostrand-Reinhold, Princeton, NJ, **1970**, pp. 90–159; Laszlo, P.; Welvart, Z. *Bull. Soc. Chim. Fr.* **1966**, 2412; Halevi, E.A. *Prog. Phys. Org. Chem.* **1963**, *1*, 109. For a review of model calculations of secondary isotope effects, see McLennan, D.J. *Isot. Org. Chem.* **1987**, *7*, 393. See also, Sims, L.B.; Lewis, D.E. *Isot. Org. Chem.* **1984**, *6*, 161.

[70]Leffek, K.T.; Llewellyn, J.A.; Robertson, R.E. *Can. J. Chem.* **1960**, *38*, 2171.

[71]Bender, M.L.; Feng, M.S. *J. Am. Chem. Soc.* **1960**, *82*, 6318; Jones, J.M.; Bender, M.L. *J. Am. Chem. Soc.* **1960**, *82*, 6322.

[72]Shiner, Jr., V.J.; Jewett, J.G. *J. Am. Chem. Soc.* **1964**, *86*, 945; DeFrees, D.J.; Hehre, W.J.; Sunko, D.E. *J. Am. Chem. Soc.* **1979**, *101*, 2323. See also, Siehl, H.; Walter, H. *J. Chem. Soc. Chem. Commun.* **1985**, 76.

[73]Shiner, Jr., V.J.; Kriz, Jr., G.S. *J. Am. Chem. Soc.* **1964**, *86*, 2643.

origin[74] (e.g., a CD_3 group has a smaller steric requirement than a CH_3 group) and a field-effect explanation has also been suggested (CD_3 is apparently a better electron donor than CH_3[75]), but hyperconjugation is the most probable cause in most instances.[76] Part of the difficulty in attempting to explain these effects is their small size, ranging only as high as ~1.5.[77] Another complicating factor is that they can change with temperature. In one case,[78] k_H/k_D was 1.00 ± 0.01 at $0°C$, 0.90 ± 0.01 at $25°C$, and 1.15 ± 0.09 at $65°C$. Whatever the cause, there seems to be a good correlation between β secondary isotope effects and carbocation character in the transition state, and they are thus a useful tool for probing mechanisms.

The other type of secondary isotope effect results from a replacement of hydrogen by deuterium at the carbon containing the leaving group. These (called *secondary isotope effects*) are varied, with values so far reported[79] ranging from 0.87 to 1.26.[80] These effects are also correlated with carbocation character. Nucleophilic substitutions that do not proceed through carbocation intermediates (S_N2 reactions) have a isotope effects near unity.[81] Those that do involve carbocations (S_N1 reactions) have higher a isotope effects, which depend on the nature of the leaving group.[82] The accepted explanation for a isotope effects is that one of the bending C—H vibrations is affected by the substitution of D for H more or less strongly in the transition state than in the ground state.[83] Depending on the nature of the transition state, this may increase or decrease the rate of the reaction. The α isotope effects on S_N2 reactions can vary with concentration,[84] an

[74]Bartell, L.S. *J. Am. Chem. Soc.* **1961**, *83*, 3567; Brown, H.C.; Azzaro, M.E.; Koelling, J.G.; McDonald, G.J. *J. Am. Chem. Soc.* **1966**, *88*, 2520; Kaplan, E.D.; Thornton, E.R. *J. Am. Chem. Soc.* **1967**, *89*, 6644; Carter, R.E.; Dahlgren, L. *Acta Chem. Scand.* **1970**, *24*, 633; Leffek, K.T.; Matheson, A.F. *Can. J. Chem.* **1971**, *49*, 439; Sherrod, S.A.; Boekelheide, V. *J. Am. Chem. Soc.* **1972**, *94*, 5513.

[75]Halevi, E.A.; Nussim, M.; Ron, M. *J. Chem. Soc.* **1963**, 866; Halevi, E.A.; Nussim, M. *J. Chem. Soc.* **1963**, 876.

[76]Karabatsos, G.J.; Sonnichsen, G.; Papaioannou, C.G.; Scheppele, S.E.; Shone, R.L. *J. Am. Chem. Soc.* **1967**, *89*, 463; Kresge, A.J.; Preto, R.J. *J. Am. Chem. Soc.* **1967**, *89*, 5510; Jewett, J.G.; Dunlap, R.P. *J. Am. Chem. Soc.* **1968**, *90*, 809; Sunko, D.E.; Szele, I.; Hehre, W.J. *J. Am. Chem. Soc.* **1977**, *99*, 5000; Kluger, R.; Brandl, M. *J. Org. Chem.* **1986**, *51*, 3964.

[77]Halevi, E.A.; Margolin, Z. *Proc. Chem. Soc.* **1964**, 174. A value for k_{CH_3}/k_{CD_3} of 2.13 was reported for one case: Liu, K.; Wu, Y.W. *Tetrahedron Lett.* **1986**, *27*, 3623.

[78]Halevi, E.A.; Margolin, Z. *Proc. Chem. Soc.* **1964**, 174.

[79]A value of 2.0 has been reported in one case, for a cis–trans isomerization, rather than a nucleophilic substitution: Caldwell, R.A.; Misawa, H.; Healy, E.F.; Dewar, M.J.S. *J. Am. Chem. Soc.* **1987**, *109*, 6869.

[80]Shiner, Jr., V.J.; Buddenbaum, W.E.; Murr, B.L.; Lamaty, G. *J. Am. Chem. Soc.* **1968**, *90*, 418; Harris, J.M.; Hall, R.E.; Schleyer, P.v.R. *J. Am. Chem. Soc.* **1971**, *93*, 2551.

[81]For reported exceptions, see Tanaka, N.; Kaji, A.; Hayami, J. *Chem. Lett.* **1972**, 1223; Westaway, K.C. *Tetrahedron Lett.* **1975**, 4229.

[82]Willi, A.V.; Ho, C.; Ghanbarpour, A. *J. Org. Chem.* **1972**, *37*, 1185; Shiner Jr., V.J.; Neumann, A.; Fisher, R.D. *J. Am. Chem. Soc.* **1982**, *104*, 354; and references cited therein.

[83]Streitwieser, Jr., A.; Jagow, R.H.; Fahey, R.C.; Suzuki, S. *J. Am. Chem. Soc.* **1958**, *80*, 2326.

[84]Westaway, K.C.; Waszczylo, Z.; Smith, P.J.; Rangappa, K.S. *Tetrahedron Lett.* **1985**, *26*, 25.

effect attributed to a change from a free nucleophile to one that is part of an ion pair[85] (see p. 492). This illustrates the use of secondary isotope effects as a means of studying transition state structure. The γ secondary isotope effects have also been reported.[86]

Another kind of isotope effect is the *solvent isotope effect*.[87] Reaction rates often change when the solvent is changed from H_2O to D_2O or from ROH to ROD. These changes may be due to any of three factors or a combination of all of them.

1. The solvent may be a reactant. If an O—H bond of the solvent is broken in the rate-determining step, there will be a primary isotope effect. If the molecules involved are D_2O or D_3O^+ there may also be a secondary effect caused by the O—D bonds that are not breaking.

2. The substrate molecules may become labeled with deuterium by rapid hydrogen exchange, and then the newly labeled molecule may become cleaved in the rate-determining step.

3. The extent or nature of solvent–solute interactions may be different in the deuterated and nondeuterated solvents; this may change the energies of the transition state, and hence the activation energy of the reaction. These are secondary isotope effects. Two physical models for this third factor have been constructed.[88]

It is obvious that in many cases the first and third factors at least, and often the second, are working simultaneously. Attempts have been made to separate them.[89]

The methods described in this chapter are not the only means of determining mechanisms. In an attempt to elucidate a mechanism, the investigator is limited only by their ingenuity.

[85]Westaway, K.C.; Lai, Z. *Can. J. Chem.* **1988**, *66*, 1263.

[86]Leffek, K.T.; Llewellyn, J.A.; Robertson, R.E. *J. Am. Chem. Soc.* **1960**, *82*, 6315; *Chem. Ind. (London)* **1960**, 588; Werstiuk, N.H.; Timmins, G.; Cappelli, F.P. *Can. J. Chem.* **1980**, *58*, 1738.

[87]For reviews, see Alvarez, F.J.; Schowen, R.L. *Isot. Org. Chem.* **1987**, *7*, 1; Kresge, A.J.; More O'Ferrall, R.A.; Powell, M.F. *Isot. Org. Chem.* **1987**, *7*, 177; Schowen, R.L. *Prog. Phys. Org. Chem.* **1972**, *9*, 275; Gold, V. *Adv. Phys. Org. Chem.* **1969**, *7*, 259; Laughton, P.M.; Robertson, R.E., in Coetzee; Ritchie *Solute–Solvent Interactions*, Marcel Dekker, NY, **1969**, pp. 399–538. For a review of the effect of isotopic changes in the solvent on the properties of nonreacting solutes, see Arnett, E.M.; McKelvey, D.R., in Coetzee, J.F.; Ritchie, C.D. cited above, pp. 343–398.

[88]Bunton, C.A.; Shiner, Jr., V.J. *J. Am. Chem. Soc.* **1961**, *83*, 42, 3207, 3214; Swain, C.G.; Thornton, E.R. *J. Am. Chem. Soc.* **1961**, *83*, 3884, 3890. See also, Mitton, C.G.; Gresser, M.; Schowen, R.L. *J. Am. Chem. Soc.* **1969**, *91*, 2045.

[89]More O'Ferrall, R.A.; Koeppl, G.W.; Kresge, A.J. *J. Am. Chem. Soc.* **1971**, *93*, 9.

CHAPTER 7

Irradiation Processes in Organic Chemistry

Most reactions carried out in organic chemistry laboratories take place between molecules all of which are in their ground electronic states. In a *photochemical reaction*,[1] however, a reacting molecule has been previously promoted by absorption of light to an electronically excited state. A molecule in an excited state must lose its extra energy in some manner; it cannot remain in the excited condition for long. The subject of electronic spectra is closely related to photochemistry. A chemical reaction is not the only possible means of relinquishing the extra energy in a photochemical process. In this chapter, first we discuss electronically excited states and the processes of promotion to these states. Two other methods are available to facilitate chemical reactions: sonochemistry and microwave chemistry. Although the physical processes involved are not necessarily the same excitation processes observed in photochemistry, irradiation with ultrasound or with microwaves have a significant influence on chemical reactivity. For that reason, they are included in this chapter.

---

[1]There are many books on photochemistry. Some recent ones are Michl, J.; Bonačić-Koutecký, V. *Electronic Aspects of Organic Photochemistry*, Wiley, NY, *1990*; Scaiano, J.C. *Handbook of Organic Photochemistry*, 2 vols., CRC Press, Boca Raton, FL, *1989*; Coxon, J.M.; Halton, B. *Organic Photochemistry*, 2nd ed.; Cambridge University Press: Cambridge, *1987*; Coyle, J.D. *Photochemistry in Organic Synthesis*, Royal Society of Chemistry, London, *1986*, *Introduction to Organic Photochemistry*, Wiley, NY, *1986*; Horspool, W.M. *Synthetic Organic Photochemistry*, Plenum, NY, *1984*; Margaretha, P. *Preparative Organic Photochemistry*, *Top. Curr. Chem.* *1982*, *103*; Turro, N.J. *Modern Molecular Photochemistry*, W.A. Benjamin, NY, *1978*; Rohatgi-Mukherjee. K.K. *Fundamentals of Photochemistry*, Wiley, NY, *1978*; Barltrop, J.A.; Coyle, J.D. *Principles of Photochemistry*, Wiley, NY, *1978*. For a comprehensive older treatise, see Calvert, J.G.; Pitts, Jr., J.N. *Photochemistry*, Wiley, NY, *1966*. For a review of the photochemistry of radicals and carbenes, see Scaiano, J.; Johnston, L.J. *Org. Photochem.* *1989*, *10*, 309. For a history of photochemistry, see Roth, H.D. *Angew. Chem. Int. Ed.* *1989*, *28*, 1193. For a glossary of terms used in photochemistry, see Braslavsky, S.E.; Houk, K.N. *Pure Appl. Chem.* *1988*, *60*, 1055. See also, the series, *Advances in Photochemistry*, *Organic Photochemistry*, and *Excited States*.

March's Advanced Organic Chemistry: Reactions, Mechanisms, and Structure, Sixth Edition, by Michael B. Smith and Jerry March
Copyright © 2007 John Wiley & Sons, Inc.

PHOTOCHEMISTRY

Excited States and the Ground State

Electrons can move from the ground-state energy level of a molecule to a higher level (i.e., an unoccupied orbital of higher energy) if outside energy is supplied. In a photochemical process, this energy is in the form of light. Light of any wavelength has associated with it an energy value given by $E = h\nu$, where n is the frequency of the light ($n =$ velocity of light c divided by the wavelength λ) and h is Planck's constant. Since the energy levels of a molecule are quantized, the amount of energy required to raise an electron in a given molecule from one level to a higher one is a fixed quantity. Only light with exactly the frequency corresponding to this amount of energy will cause the electron to move to the higher level. If light of another frequency (too high or too low) is sent through a sample, it will pass out without a loss in intensity, since the molecules will not absorb it. However, if light of the correct frequency is passed in, the energy will be used by the molecules for electron promotion, and hence the light that leaves the sample will be diminished in intensity or altogether gone. A *spectrophotometer* is an instrument that allows light of a given frequency to pass through a sample and that detects (by means of a phototube) the amount of light that has been transmitted, that is, not absorbed. A spectrophotometer compares the intensity of the transmitted light with that of the incident light. Automatic instruments gradually and continuously change the frequency, and an automatic recorder plots a graph of absorption versus frequency or wavelength.

The energy of electronic transitions corresponds to light in the visible, UV, and far-UV regions of the spectrum (Fig. 7.1). Absorption positions are normally expressed in wavelength units, usually nanometers (nm).[2] If a compound absorbs in the visible, it is colored, possessing a color complementary to that which is absorbed.[3] Thus a compound absorbing in the violet is yellow. The far-uv region is studied by organic chemists less often than the visible or ordinary uv regions because special vacuum instruments are required owing to the fact that oxygen and nitrogen absorb in these regions.

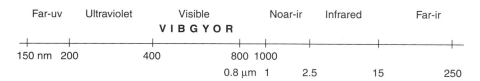

Fig. 7.1. The uv, visible, and ir portions of the electromagnetic spectrum.

---

[2]Formerly, millimicrons (mμ) were frequently used; numerically they are the same as nanometers.
[3]For monographs, see Zollinger, H. *Color Chemistry*, VCH, NY, *1987*; Gordon, P.F.; Gregory, P. *Organic Chemistry in Colour*, Springer, NY, *1983*; Griffiths, J. *Colour and Constitution of Organic Molecules*, Academic Press, NY, *1976*. See also, Fabian, J.; Zahradník, R. *Angew. Chem. Int. Ed. 1989*, 28, 677.

From these considerations it would seem that an electronic spectrum should consist of one or more sharp peaks, each corresponding to the transfer of an electron from one electronic level to another. Under ordinary conditions the peaks are seldom sharp. In order to understand why, it is necessary to realize that molecules are constantly vibrating and rotating and that these motions are also quantized. A molecule at any time is not only in a given electronic state but also in a given vibrational and rotational state. The difference between two adjacent vibrational levels is much smaller than the difference between adjacent electronic levels, and the difference between adjacent rotational levels is smaller still. A typical situation is shown in Fig. 7.2. When an electron moves from one electronic level to another, it moves from a given vibrational and rotational level within that electronic level to some vibrational and rotational level at the next electronic level. A given sample contains a large number of molecules, and even if all of them are in the ground electronic state, they are still distributed among the vibrational and rotational states (though the ground vibrational state V_0 is most heavily populated). This means that not just one wavelength of light will be absorbed, but a number of them close together, with the most probable transition causing the most intense peak. But in molecules containing more than a few atoms there are so many possible transitions and these are so close together that what is observed is a relatively broad band. The height of the peak depends on the number of molecules making the transition and is proportional to log ε, where ε is the *extinction coefficient*. The extinction coefficient can be expressed by $\varepsilon = E/cl$, where c is the concentration in moles per liter, l is the

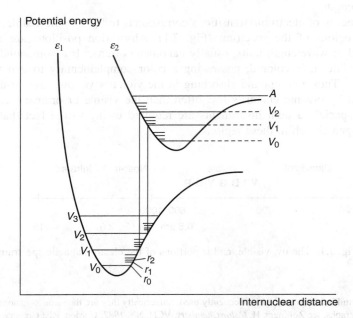

Fig. 7.2. Energy curves for a diatomic molecule. Two possible transitions are shown. When an electron has been excited to the point marked A, the molecule may cleave (p. 335).

cell length in centimeters, and $E = \log I_0/I$, where I_0 is the intensity of the incident light and I of the transmitted light. The wavelength is usually reported as λ_{max}, meaning that this is the top of the peak. Purely vibrational transitions, such as between V_0 and V_1 of E_1, which require much less energy, are found in the ir region and are the basis of ir spectra. Purely rotational transitions are found in the far-ir and microwave (beyond the far-ir) regions.

A UV or visible absorption peak is caused by the promotion of an electron in one orbital (usually a ground-state orbital) to a higher orbital. Normally, the amount of energy necessary to make this transition depends mostly on the nature of the two orbitals involved and much less on the rest of the molecule. Therefore, a simple functional group such as the C=C double bond always causes absorption in the same general area. A group that causes absorption is called a *chromophore*.

Singlet and Triplet States: "Forbidden" Transitions

In most organic molecules, all electrons in the ground state are paired, with each member of a pair possessing opposite spin as demanded by the Pauli principle. When one of a pair of electrons is promoted to an orbital of higher energy, the two electrons no longer share an orbital, and the promoted electron may, in principle, have the same spin as its former partner or the opposite spin. As we saw in Chapter 5, a molecule in which two unpaired electrons have the same spin is called a *triplet*,[4] while one in which all spins are paired is a *singlet*. Thus, at least in principle, for every excited singlet state there is a corresponding triplet state. In most cases, the triplet state has a lower energy than the corresponding singlet, which is in accord with Hund's rule. Therefore, a different amount of energy, and hence a different wavelength is required to promote an electron from the ground state (which is almost always a singlet) to an excited singlet than to the corresponding triplet state.

It would thus seem that promotion of a given electron in a molecule could result either in a singlet or a triplet excited state depending on the amount of energy added. However, this is often not the case because transitions between energy levels are governed by selection rules, which state that certain transitions are "forbidden." There are several types of "forbidden" transitions, two of which are more important than the others.

1. *Spin-Forbidden Transitions.* Transitions in which the spin of an electron changes are not allowed, because a change from one spin to the opposite involves a change in angular momentum and such a change would violate the law of conservation of angular momentum. Therefore, singlet–triplet and triplet–singlet transitions are forbidden, whereas singlet–singlet and triplet–triplet transitions are allowed.

2. *Symmetry-Forbidden Transitions.* Among the transitions in this class are those in which a molecule has a center of symmetry. In such cases, a $g \rightarrow g$ or

---

[4]See Kurreck, H. *Angew. Chem. Int. Ed.* **1993**, *32*, 1409 for a brief discussion of the triplet state in organic chemistry.

$u \rightarrow u$ transition (see p. 5) is "forbidden," while a $g \rightarrow u$ or $u \rightarrow g$ transition is allowed.

We have put the word "forbidden" into quotation marks because these transitions are not actually forbidden but only highly improbable. In most cases, promotions from a singlet ground state to a triplet excited state are so improbable that they cannot be observed, and it is safe to state that in most molecules only singlet–singlet promotions take place. However, this rule does break down in certain cases, most often when a heavy atom (e.g., iodine) is present in the molecule, in which cases it can be shown from spectra that singlet–triplet promotions are occurring.[5] Symmetry-forbidden transitions can frequently be observed, though usually with low intensity.

Types of Excitation

When an electron in a molecule is promoted (normally only one electron in any molecule), it usually goes into the lowest available vacant orbital, though promotion to higher orbitals is also possible. For most organic molecules, there are consequently four types of electronic excitation:

1. $\sigma \rightarrow \sigma^*$. Alkanes, which have no n or π electrons, can be excited only in this way.[6]
2. $n \rightarrow \sigma^*$. Alcohols, amines,[7] ethers, and so on, can also be excited in this manner.
3. $\pi \rightarrow \pi^*$. This pathway is open to alkenes as well as to aldehydes, carboxylic esters, and so on.
4. $n \rightarrow \pi^*$. Aldehydes, ketones, carboxylic esters, and so on, can undergo this promotion as well as the other three.

The four excitation types above are listed in what is normally the order of decreasing energy. Thus light of the highest energy (in the far uv) is necessary for $\sigma \rightarrow \sigma^*$ excitation, while $n \rightarrow \pi^*$ promotions are caused by ordinary uv light. However, the order may sometimes be altered in some solvents.

In 1,3-butadiene (and other compounds with two conjugated double bonds) there are two π and two π^* orbitals (p. 39). The energy difference between the higher $\pi(\chi_2)$ and the lower $\pi^*(\chi_3)$ orbital is less than the difference between the π and π^* orbitals of ethylene. Therefore 1,3-butadiene requires less energy than ethylene, and thus light of a higher wavelength, to promote an electron. This is a general phenomenon, and it may be stated that, in general, *the more conjugation in a molecule, the more the absorption is displaced toward higher wavelengths* (see Table 7.1).[8]

---

[5]For a review of photochemical heavy-atom effects, see Koziar, J.C.; Cowan, D.O. *Acc. Chem. Res.* **1978**, *11*, 334.

[6]An n electron is one in an unshared pair.

[7]For a review of the photochemistry of amines, see Malkin, Yu.N.; Kuz'min, V.A. *Russ. Chem. Rev.* **1985**, *54*, 1041.

[8]Bohlmann, F.; Mannhardt, H. *Chem. Ber.* **1956**, *89*, 1307.

TABLE 7.1. Ultraviolet Absorption[8] of CH$_3$–(CH=CH)$_n$–CH$_3$ for Some Values of n

n	nm
2	227
3	263
6	352
9	413

When a chromophore absorbs at a certain wavelength and the substitution of one group for another causes absorption at a longer wavelength, a *bathochromic shift* is said to have occurred. The opposite kind of shift is called *hypsochromic*.

Of the four excitation types listed above, the $\pi \rightarrow \pi^*$ and $n \rightarrow \pi^*$ are far more important in organic photochemistry than the other two. Compounds containing C=O groups can be excited in both ways, giving rise to at least two peaks in the UV.

As we have seen, a chromophore is a group that causes a molecule to absorb light. Examples of chromophores in the visible or UV are C=O, N=N,[9] Ph, and NO$_2$. Some chromophores in the far UV (beyond 200 nm) are C=C, C≡C, Cl, and OH. An *auxochrome* is a group that displaces (through resonance) and usually intensifies the absorption of a chromophore present in the same molecule. Groups, such as Cl, OH, and NH$_2$, are generally regarded as auxochromes since they shift (usually bathochromically) the uv and visible bands of chromophores, such as Ph or C=O (see Table 7.2).[10] Since auxochromes are themselves chromophores

TABLE 7.2. Some UV Peaks of Substituted Benzenes in Water, or Water With a Trace of Methanol (for Solubility)[a]

	Primary Band		Secondary Band	
	λ_{max}, nm	ε_{max}	λ_{max}, nm	ε_{max}
PhH	203.5	7,400	254	204
PhCl	209.5	7,400	263.5	190
PhOH	210.5	6,200	270	1,450
PhOMe	217	6,400	269	1,480
PhCN	224	13,000	271	1,000
PhCOOH	230	11,600	273	970
PhNH$_2$	230	8,600	280	1,430
PhO$^-$	235	9,400	287	2,600
PhAc	245.5	9,800		
PhCHO	249.5	11,400		
PhNO$_2$	268.5	7,800		

[a]Note how auxochromes shift and usually intensify the peaks.

[9]For a review of the azo group as a chromophore, see Rau, H. *Angew. Chem. Int. Ed.* 1973, *12*, 224.
[10]These values are from Jaffé, H.H.; Orchin, M. *Theory and Applications of Ultraviolet Spectroscopy*, Wiley, NY, *1962*, p. 257.

(to be sure, generally in the far-UV), it is sometimes difficult to decide which group in a molecule is an auxochrome and which a chromophore. For example, in acetophenone (PhCOMe) is the chromophore Ph or C=O? In such cases, the distinction becomes practically meaningless.

Nomenclature and Properties of Excited States

An excited state of a molecule can be regarded as a distinct chemical species, different from the ground state of the same molecule and from other excited states. It is obvious that we need some method of naming excited states. Unfortunately, there are several methods in use, depending on whether one is primarily interested in photochemistry, spectroscopy, or molecular-orbital theory.[11] One of the most common methods simply designates the original and newly occupied orbitals, with or without a superscript to indicate singlet or triplet. Thus the singlet state arising from promotion of a π to a π^* orbital in ethylene would be the $^1(\pi,\pi^*)$ state or the π,π^* singlet state. Another very common method can be used even in cases where one is not certain which orbitals are involved. The lowest energy excited state is called S_1, the next S_2, and so on, and triplet states are similarly labeled T_1, T_2, T_3, and so on. In this notation, the ground state is S_0. Other notational systems exist, but in this book we will confine ourselves to the two types just mentioned.

The properties of excited states are not easy to measure because of their generally short lifetimes and low concentrations, but enough work has been done for us to know that they often differ from the ground state in geometry, dipole moment and acid or base strength.[12] For example, acetylene, which is linear in the ground state, has a trans geometry in the excited state

$$\underset{H}{\overset{\displaystyle H}{C\equiv C}}$$

with $\sim sp^2$ carbons in the $^1(\pi,\pi^*)$ state.[13] Similarly, the $^1(\pi,\pi^*)$ and the $^3(\pi,\pi^*)$ states of ethylene have a perpendicular and not a planar geometry,[14] and the $^1(n,\pi^*)$ and $^3(n,\pi^*)$ states of formaldehyde are both pyramidal.[15] Triplet species tend to stabilize themselves by distortion, which relieves interaction between the

---

[11] For discussions of excited-state notation and other terms in photochemistry, see Pitts, Jr., J.N.; Wilkinson, F.; Hammond, G.S. *Adv. Photochem.* **1963**, *1*, 1; Porter, G.B.; Balzani, V.; Moggi, L. *Adv. Photochem.* **1974**, *9*, 147. See also, Braslavsky, S.E.; Houk, K.N. *Pure Appl. Chem.* **1988**, *60*, 1055.

[12] For reviews of the structures of excited states, see Zink, J.I.; Shin, K.K. *Adv. Photochem.* **1991**, *16*, 119; Innes, K.K. *Excited States* **1975**, *2*, 1; Hirakawa, A.Y.; Masamichi, T. *Vib. Spectra Struct.* **1983**, *12*, 145.

[13] Ingold, C.K.; King, G.W. *J. Chem. Soc.* **1953**, 2702, 2704, 2708, 2725, 2745. For a review of acetylene photochemistry, see Coyle, J.D. *Org. Photochem.* **1985**, *7*, 1.

[14] Merer, A.J.; Mulliken, R.S. *Chem. Rev.* **1969**, *69*, 639.

[15] Robinson, G.W.; Di Giorgio, V.E. *Can. J. Chem.* **1958**, *36*, 31; Buenker, R.J.; Peyerimhoff, S.D. *J. Chem. Phys.* **1970**, *53*, 1368; Garrison, B.J.; Schaefer III, H.F.; Lester, Jr., W.A. *J. Chem. Phys.* **1974**, *61*, 3039; Streitwieser, Jr., A.; Kohler, B. *J. Am. Chem. Soc.* **1988**, *110*, 3769. For reviews of excited states of formaldehyde, see Buck, H.M. *Recl. Trav. Chim. Pays-Bas* **1982**, *101*, 193, 225; Moule, D.C.; Walsh, A.D. *Chem. Rev.* **1975**, *75*, 67.

TABLE 7.3. **Typical Energies for Some Covalent Single Bonds (see Table 1.7) and the Corresponding Approximate Wavelengths**

Bond	E		nm
	kcal mol^{-1}	kJ mol^{-1}	
C–H	95	397	300
C–O	88	368	325
C–C	83	347	345
Cl–Cl	58	243	495
C–O	35	146	820

unpaired electrons. Obviously, if the geometry is different, the dipole moment will probably differ also and the change in geometry and electron distribution often results in a change in acid or base strength.[16] For example, the S_1 state of 2-naphthol is a much stronger acid (pK 3.1) than the ground state (S_0) of the same molecule (pK 9.5).[17]

Photolytic Cleavage

We have said that when a molecule absorbs a quantum of light, it is promoted to an excited state. Actually, that is not the only possible outcome. Because the energy of visible and UV light is of the same order of magnitude as that of covalent bonds (Table 7.3), another possibility is that the molecule may cleave into two parts, a process known as *photolysis*. There are three situations that can lead to cleavage:

1. The promotion may bring the molecule to a vibrational level so high that it lies above the right-hand portion of the E_2 curve (line *A* in Fig. 7.2). In such a case, the excited molecule cleaves at its first vibration.

2. Even where the promotion is to a lower vibrational level, one which lies wholly within the E_2 curve (e.g., V_1 or V_2), the molecule may still cleave. As Fig. 7.2 shows, equilibrium distances are greater in excited states than in the ground state. The *Franck–Condon principle* states that promotion of an electron takes place much faster than a single vibration (the promotion takes $\sim 10^{-15}$ s; a vibration $\sim 10^{-12}$ s). Therefore, when an electron is suddenly promoted, even to a low vibrational level, the distance between the atoms is essentially unchanged and the bond finds itself in a compressed condition like a pressed-in spring; this condition may be relieved by an outward surge that is sufficient to break the bond.

---

[16]For a review of acid–base properties of excited states, see Ireland, J.F.; Wyatt, P.A.H. *Adv. Phys. Org. Chem.* **1976**, *12*, 131.

[17]Weller, A. *Z. Phys. Chem. (Frankfurt am Main)* **1955**, *3*, 238, *Discuss. Faraday Soc.* **1959**, *27*, 28.

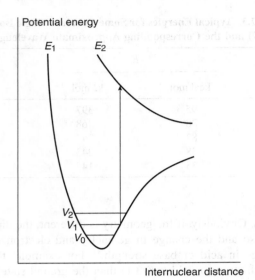

Fig. 7.3. Promotion to a dissociative state results in bond cleavage.

3. In some cases, the excited state is entirely dissociative (Fig. 7.3), that is, there is no distance where attraction outweighs repulsion, and the bond must cleave. An example is the hydrogen molecule, where a $\sigma \rightarrow \sigma*$ promotion always results in cleavage.

A photolytic cleavage can break the molecule into two smaller molecules or into two free radicals (see p. 343). Cleavage into two ions, though known, is much rarer. Once free radicals are produced by a photolysis, they behave like free radicals produced in any other way (Chapter 5) except that they may be in excited states, and this can cause differences in behavior.[18]

The Fate of the Excited Molecule: Physical Processes

When a molecule has been photochemically promoted to an excited state, it does not remain there for long. Most promotions are from the S_0 to the S_1 state. As we have seen, promotions from S_0 to triplet states are "forbidden." Promotions to S_2 and higher singlet states take place, but in liquids and solids these higher states usually drop very rapidly to the S_1 state ($\sim 10^{-13}$–10^{-11} s). The energy lost when an S_2 or S_3 molecule drops to S_1 is given up in small increments to the environment by collisions with neighboring molecules. Such a process is called an *energy cascade*. In a similar manner, the initial excitation and the decay from higher singlet states initially populate many of the vibrational levels of S_1, but these also cascade, down to the lowest vibrational level of S_1. Therefore, in most cases, the lowest

[18]Lubitz, W.; Lendzian, F.; Bittl, R. *Acc. Chem. Res.* **2002**, *35*, 313.

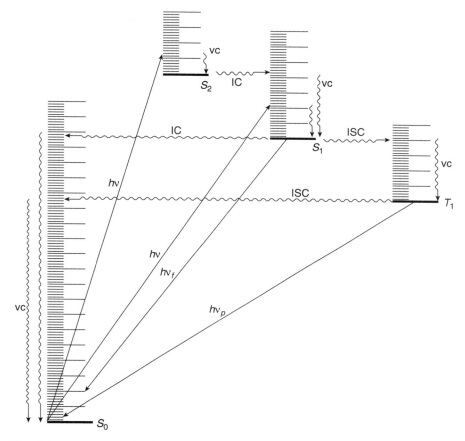

Fig. 7.4. Modified Jablonski diagram showing transitions between excited states and the ground state. Radiative processes are shown by straight lines, radiationless processes by wavy lines. vc = vibrational cascade; hv_f = fluorescence; hv_p = phosphorescence.

vibrational level of the S_1 state is the only important excited singlet state.[19] This state can undergo various physical and chemical processes. In the following list, we describe the physical pathways open to molecules in the S_1 and excited triplet states. These pathways are also shown in a modified Jablonski diagram (Fig. 7.4) and in Table 7.4.

1. A molecule in the S_1 state can cascade down through the vibrational levels of the S_0 state and thus return to the ground state by giving up its energy in small increments to the environment, but this is generally quite slow because the

[19]For a review of physical and chemical processes undergone by higher states, see Turro, N.J.; Ramamurthy, V.; Cherry, W.; Farneth, W. *Chem. Rev.* **1978**, *78*, 125.

TABLE 7.4. Physical Processes Undergone by Excited Molecules[a]

$S_0 + h\nu \rightarrow S_1^v$	Excitation
S_1^v	$\sim\!\sim\!\!\rightarrow \rightarrow S_1 + \Delta$ Vibrational relaxation
$S_1 \rightarrow S_1 + h\nu$	Fluorescence
S_1	$\sim\!\sim\!\!\rightarrow \rightarrow S_0 + \Delta$ Internal conversion
$S_1 \sim\!\sim\!\!\rightarrow T_1^v$	Intersystem crossing
$T_1^v \sim\!\sim\!\!\rightarrow T_1 + \Delta$	Vibrational relaxation
$T_1 \rightarrow S_0 + h\nu$	Phosphorescence
$T_1 \sim\!\sim\!\!\rightarrow S_0 + \Delta$	Intersystem crossing
$S_1 + A_{(S_0)} \rightarrow S_0 + A_{(S_1)}$	Singlet–singlet transfer (photosensitization)
$T_1 + A_{(S_0)} \rightarrow S_0 + A_{(T_1)}$	Triplet–triplet transfer (photosensitization)

[a]The superscript v indicates vibrationally excited state: excited states higher than S_1 or T_1 are omitted.

amount of energy is large. The process is called *internal conversion* (IC, see Fig. 7.4). Because it is slow, most molecules in the S_1 state adopt other pathways.[20]

2. A molecule in the S_1 state can drop to some low vibrational level of the S_0 state all at once by giving off the energy in the form of light. This process, which generally happens within 10^{-9} s, is called *fluorescence*. This pathway is not very common either (because it is relatively slow), except for small molecules, for example, diatomic, and rigid molecules, for example, aromatic. For most other compounds, fluorescence is very weak or undetectable. For compounds that do fluoresce, the fluorescence emission spectra are usually the approximate mirror images of the absorption spectra. This comes about because the fluorescing molecules all drop from the lowest vibrational level of the S_1 state to various vibrational levels of S_0, while excitation is from the lowest vibrational level of S_0 to various levels of S_1 (Fig. 7.5). The only peak in common is the one (called the 0–0 peak) that results from transitions between the lowest vibrational levels of the two states. In solution, even the 0–0 peak may be noncoincidental because the two states are solvated differently. Fluorescence nearly always arises from a $S_1 \rightarrow S_0$ transition, though azulene (p. $$$) and its simple derivatives are exceptions,[21] emitting fluorescence from $S_2 \rightarrow S_0$ transitions.

Because of the possibility of fluorescence, any chemical reactions of the S_1 state must take place very fast, or fluorescence will occur before they can happen.

[20]For a monograph on radiationless transitions, see Lin, S.H. *Radiationless Transitions*; Academic Press, NY, *1980*. For reviews, see Kommandeur, J. *Recl. Trav. Chim. Pays-Bas* **1983**, *102*, 421; Freed, K.F. *Acc. Chem. Res.* **1978**, *11*, 74.
[21]For other exceptions, see Gregory, T.A.; Hirayama, F.; Lipsky, S. *J. Chem. Phys.* **1973**, *58*, 4697; Sugihara, Y.; Wakabayashi, S.; Murata, I.; Jinguji, M.; Nakazawa, T.; Persy, G.; Wirz, J. *J. Am. Chem. Soc.* **1985**, *107*, 5894, and references cited therein. See also Turro, N.J.; Ramamurthy, V.; Cherry, W.; Farneth, W. *Chem. Rev.* **1978**, *78*, 125, see pp. 126–129.

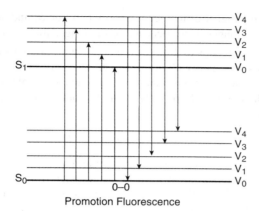

Promotion Fluorescence

Fig. 7.5. Promotion and fluorescence between S_1 and S_0 states.

3. Most molecules (though by no means all) in the S_1 state can undergo an intersystem crossing (ISC, see Fig. 7.4) to the lowest triplet state T_1.[22] An important example is benzophenone, of which ~100% of the molecules that are excited to the S_1 state cross over to the T_1.[23] Intersystem crossing from singlet to triplet is of course a "forbidden" pathway, since the angular-momentum problem (p. 331) must be taken care of, but this often takes place by compensations elsewhere in the system. Intersystem crossings take place without loss of energy. Since a singlet state usually has a higher energy than the corresponding triplet, this means that energy must be given up. One way for this to happen is for the S_1 molecule to cross to a T_1 state at a high vibrational level and then for the T_1 to cascade down to its lowest vibrational level (see Fig. 7.4). This cascade is very rapid (10^{-12} s). When T_2 or higher states are populated, they too rapidly cascade to the lowest vibrational level of the T_1 state.

4. A molecule in the T_1 state may return to the S_0 state by giving up heat (intersystem crossing) or light (this is called *phosphorescence*).[24] Of course, the angular-momentum difficulty exists here, so that both intersystem crossing and phosphorescence are very slow ($\sim 10^{-3}$–10^1 s). This means that T_1 states generally have much longer lifetimes than S_1 states. When they occur in the same molecule, phosphorescence is found at lower frequencies than fluorescence

---

[22]Intersystem crossing from S_1 to T_2 and higher triplet states has also been reported in some aromatic molecules: Li, R.; Lim, E.C. *Chem. Phys.* **1972**, *57*, 605; Sharf, B.; Silbey, R. *Chem. Phys. Lett.* **1970**, *5*, 314. See also, Schlag, E.W.; Schneider, S.; Fischer, S.F. *Annu. Rev. Phys. Chem.* **1971**, *22*, 465, pp. 490. There is evidence that ISC can also occur from the S_2 state of some molecules: Samanta, A. *J. Am. Chem. Soc.* **1991**, *113*, 7427. Also see, Tanaka, R.; Kuriyama, Y.; Itoh, H.; Sakuragi, H.; Tokumaru, K. *Chem. Lett.* **1993**, 1447; Ohsaku, M.; Koga, N.; Morokuma, K. *J. Chem. Soc. Perkin Trans. 2* **1993**, 71.
[23]Moore, W.M.; Hammond, G.S.; Foss, R.P. *J. Am. Chem. Soc.* **1961**, *83*, 2789.
[24]For a review of physical processes of triplet states, see Lower, S.K.; El-Sayed, M.A. *Chem. Rev.* **1966**, *66*, 199. For a review of physical and chemical processes of triplet states see Wagner, P.J.; Hammond, G.S. *Adv. Photochem.* **1968**, *5*, 21.

(because of the higher difference in energy between S_1 and S_0 than between T_1 and S_0) and is longer-lived (because of the longer lifetime of the T_1 state).

5. If nothing else happens to it first, a molecule in an excited state (S_1 or T_1) may transfer its excess energy all at once to another molecule in the environment, in a process called *photosensitization*.[25] The excited molecule (which we will call D for donor) thus drops to S_0 while the other molecule (A for acceptor) becomes excited:

$$D^* + A \longrightarrow A^* + D$$

Thus there are *two* ways for a molecule to reach an excited state: by absorption of a quantum of light or by transfer from a previously excited molecule.[26] The donor D is also called a *photosensitizer*. This energy transfer is subject to the *Wigner spin-conservation rule*, which is actually a special case of the law of conservation of momentum we encountered previously. According to the Wigner rule, the total electron spin does not change after the energy transfer. For example, when a triplet species interacts with a singlet these are some allowed possibilities:[27]

D*	**A**	**D**	**A***	

In all these cases, the products have three electrons spinning "up" and the fourth "down" (as do the starting molecules). However, formation of, say, two triplets ($\uparrow\downarrow + \downarrow\downarrow$) or two singlets ($\uparrow\downarrow + \uparrow\downarrow$), whether ground states or excited, would violate the rule.

In the two most important types of photosensitization, both of which are in accord with the Wigner rule, a triplet excited state generates another triplet and a singlet generates a singlet:

$$D_{T_1} + A_{S_0} \longrightarrow A_{T_1} + D_{S_0} \qquad \text{triplet–triplet transfer}$$

$$D_{S_1} + A_{S_0} \longrightarrow A_{S_1} + D_{S_0} \qquad \text{singlet–triplet transfer}$$

[25]For reviews, see Albini, A. *Synthesis*, **1981**, 249; Turro, N.J.; Dalton, J.C.; Weiss, D.S. *Org. Photochem.* **1969**, 2, 1.

[26]There is also a third way: in certain cases excited states can be produced directly in ordinary reactions. For a review, see White, E.H.; Miano, J.D.; Watkins, C.J.; Breaux, E.J. *Angew. Chem. Int. Ed.* **1974**, *13*, 229.

[27]For another table of this kind, see Calvert, J.G.; Pitts, Jr., J.N. *Photochemistry*, Wiley, NY, **1966**, p. 89.

Singlet–singlet transfer can take place over relatively long distances (e.g., 40 Å), but triplet transfer normally requires a collision between the molecules.[28] Both types of photosensitization can be useful for creating excited states when they are difficult to achieve by direct irradiation. Photosensitization is therefore an important method for carrying out photochemical reactions when a molecule cannot be brought to the desired excited state by direct absorption of light. Triplet–triplet transfer is especially important because triplet states are usually much more difficult to prepare by direct irradiation than singlet states (often impossible) and because triplet states, having longer lifetimes, are much more likely than singlets to transfer energy by photosensitization. Photosensitization can also be accomplished by electron transfer.[29]

In choosing a photosensitizer, one should avoid a compound that absorbs in the same region as the acceptor because the latter will then compete for the light.[30] For examples of the use of photosensitization to accomplish reactions, see **15-62** and **15-63**.

6. An excited species can be quenched. Qunching is the deactivation of an excited molecular entity intermolecularly by an external environmental influence (e.g., a quencher) or intramolecularly by a substituent through a nonradiative process.[31] When the external environmental influence (quencher) interferes with the behavior of the excited state after its formation, the process is referred to as dynamic quenching. Common mechanisms include energy transfer, charge transfer, and so on. When the environmental influence inhibits the excited state formation the process is referred to as static quenching. A quencher is defined as a molecular entity that deactivates (quenches) an excited state of another molecular entity, either by energy tranfer, electron transfer, or by a chemical mechanism.[31]

An example is the rapid triplet quenching of aromatic ketone triplets by amines, which is well known.[32] Alkyl and aryl thiols and thioethers also serve as quenchers in this system[33] In this latter case, the mechanism involves electron

---

[28]Long-range triplet-triplet transfer has been observed in a few cases: Bennett, R.G.; Schwenker, R.P.; Kellogg, R.E. *J. Chem. Phys.* **1964**, *41*, 3040; Ermolaev, V.L.; Sveshnikova, E.B. *Izv. Akad. Nauk SSSR, Ser. Fiz.* **1962**, *26*, 29 [*C. A.* **1962**, *57*, 1688], *Opt. Spectrosc. (USSR)* **1964**, *16*, 320.
[29]For a review, see Kavarno, G.J.; Turro, N.J. *Chem. Rev.* **1986**, *86*, 401. See also, Mariano, P.S. *Org. Photochem.* **1987**, *9*, 1.
[30]For a review of other complications that can take place in photosensitized reactions, see Engel, P.S.; Monroe, B.M. *Adv. Photochem.* **1971**, *8*, 245.
[31]Verhoeven, J.W. *Pure Appl. Chem.* **1996**, *68*, 2223 (see p 2268).
[32]See Aspari, P.; Ghoneim, N.; Haselbach, E.; von Raumer, M.; Suppan, P.; Vauthey, E. *J. Chem. Soc., Faraday Trans.* **1996**, *92*, 1689; Cohen, S.G.; Parola, A.; Parsons, Jr., G.H. *Chem. Rev.* **1973**, *73*, 141; Inbar, S.; Linschitz, H.; Cohen, S.G. *J. Am. Chem. Soc.* **1981**, *103*, 1048; Peters, K.S.; Lee, J. *J. Phys. Chem.* **1993**, *97*, 3761; von Raumer, M.; Suppan, P.; Haselbach, E. *Helv. Chim. Acta* **1997**, *80*, 719.
[33]Guttenplan, J.B.; Cohen, S.G. *J. Org. Chem.* **1973**, *38*, 2001; Inbar, S.; Linschitz, H.; Cohen, S.G. *J. Am. Chem. Soc.* **1982**, *104*, 1679; Bobrowski, K.; Marciniak, B.; Hug, G.L. *J. Photochem. Photobiol. A: Chem.* **1994**, *81*, 159; Wakasa, M.; Hayashi, H. *J. Phys. Chem.* **1996**, *100*, 15640.

transfer from the sulfur atom to the triplet ketone, and this is supported by theoretical calculations.[34] Aromatic ketone triplets are quenched by phenols and the photochemical reaction between aromatic ketones and phenols is efficient only in the presence of an acid catalyst.[35] Indirect evidence has been provided for involvement of the hydrogen-bonded triplet exciplex and for the role of electron transfer in this reaction.[36]

The Fate of the Excited Molecule: Chemical Processes

Although both excited singlet and triplet species can undergo chemical reactions, they are much more common for triplets, simply because these generally have much longer lifetimes. Excited singlet species, in most cases, have a lifetime of $<10^{-10}$ s and undergo one of the physical processes already discussed before they have a chance to react chemically. Therefore, photochemistry is largely the chemistry of triplet states.[37] Table 7.5[38] lists many of the possible chemical pathways that can be taken by an excited molecule.[39] The first four of these are unimolecular reactions; the others are bimolecular. In the case of bimolecular reactions, it is rare for two excited molecules to react with each other (because the concentration of excited molecules at any one time is generally low); reactions are between an excited molecule and an unexcited molecule of either the same or another species. The reactions listed in Table 7.5 are primary processes. Secondary reactions often follow, since the primary products are frequently radicals or carbenes; even if they are ordinary molecules, they are often in upper vibrational levels and so have excess energy. In almost all cases, the primary products of photochemical reactions are in their ground states, though exceptions are known.[40] Of the reactions listed in Table 7.5, the most common are cleavage into radicals (1), decomposition into molecules (2), and (in the presence of a suitable acceptor molecule) photosensitization (7), which we have already discussed. The following are some specific examples of reaction categories (1)–(6). Other examples are discussed in Part 2 of this book.[41]

[34]Marciniak, B.; Bobrowski, K.; Hug, G.L. *J. Phys. Chem.* **1993**, *97*, 11937.
[35]Becker, H.-D. *J. Org. Chem.* **1967**, *32*, 2115; *J. Org. Chem.* **1967**, *32*, 2124; *J. Org. Chem.* **1967**, *32*, 2140.
[36]Lathioor, E.C.; Leigh, W.J.; St. Pierre, M.J. *J. Am. Chem. Soc.* **1999**, *121*, 11984.
[37]For a review of the chemical reactions of triplet states, see Wagner, P.J.; Hammond, G.S. Wagner, P.J.; Hammond, G.S. *Adv. Photochem.* **1968**, *5*, 21. For other reviews of triplet states, see *Top. Curr. Chem.* **1975**, Vols. 54 and 55.
[38]Adapted from Calvert, J.G.; Pitts, Jr., J.N. *Photochemistry*, Wiley, NY, **1966**, p. 367.
[39]For a different kind of classification of photochemical reactions, see Dauben, W.G.; Salem, L.; Turro, N.J. *Acc. Chem. Res.* **1975**, *8*, 41. For reviews of photochemical reactions where the molecules are geometrically constrained, see Ramamurthy, V. *Tetrahedron* **1986**, *42*, 5753; Ramamurthy, V.; Eaton, D.F. *Acc. Chem. Res.* **1988**, *21*, 300; Turro, N.J.; Cox, G.S.; Paczkowski, M.A. *Top. Curr. Chem.* **1985**, *129*, 57.
[40]Turro, N.J.; Lechtken, P.; Lyons, A.; Hautala, R.T.; Carnahan, E.; Katz, T.J. *J. Am. Chem. Soc.* **1973**, *95*, 2035.
[41]For monographs on the use of photochemistry for synthesis, see Ninomiya, I.; Naito, T. *Photochemical Synthesis*, Academic Press, NY, **1989**; Coyle, J.D. *Photochemistry in Organic Synthesis*, Royal Society of Chemistry, London, **1986**; Schönberg, A. *Preparative Organic Photochemistry*, Springer, Berlin, **1968**.

.imary photochemical reactions^a of an excited molecule A—B—C^{38}

TABLE

	Reaction Type	Example Number
Re$\cdot \longrightarrow$ A—B$\bullet$ + C$\bullet$	Simple cleavage into radicals^{42}	(1)
$\longrightarrow$ E + F	Decomposition into molecules	(2)
$\cdot) \longrightarrow$ A—C—B	Intramolecular rearrangement	(3)
$\check{C}) \longrightarrow$ A—B—C$'$	Photoisomerization	(4)
—C) $\xrightarrow{RH}$ A—B—C—H + R$\bullet$	Hydrogen-atom abstraction	(5)
$\mathcal{B}$—C) $\longrightarrow$ (ABD)$_2$	Photodimerization	(6)
(A—B—C) $\xrightarrow{A}$ ABC + A*	Photosensitization	(7)

^{a}Examples are given in the text; the most common are (1), (2), and, in the presence of a suitable acceptor molecule, (7).

Category 1. *Simple Cleavage into Radicals.*^{43} Aldehydes and ketones absorb in the 230–330-nm region. This is assumed to result from an $n \rightarrow \pi^*$ singlet–singlet transition. The excited aldehyde or ketone can then cleave.^{44}

$$\underset{O}{\overset{\displaystyle R'\diagdown \underset{\|}{C} \diagup R}{}} \quad \xrightarrow{h\nu} \quad \underset{O}{\overset{\displaystyle R'\diagdown \underset{\|}{C}\bullet}{}} + R\bullet$$

When applied to ketones, this is called *Norrish Type I cleavage* or often just *Type I cleavage*. In a secondary process, the acyl radical R$'$—CO$\bullet$ can then lose CO to give R$'\bullet$ radicals. Another example of a category 1 process is cleavage of Cl$_2$ to give two Cl atoms. Other bonds that are easily cleaved by photolysis are the O—O bonds of peroxy compounds and the C—N bonds of aliphatic azo

---

^{42}For a polymer-supported reagent used for the photochemical generation of radicals in solution see DeLuca, L.; Giacomelli, G.; Porcu, G.; Taddei, M. *Org. Lett.* **2001**, *3*, 855.

^{43}For reviews, see Jackson, W.M.; Okabe, H. *Adv. Photochem.* **1986**, *13*, 1; Kresin, V.Z.; Lester Jr., W.A. *Adv. Photochem.* **1986**, *13*, 95.

^{44}For full discussions of aldehyde and ketone photochemistry, see Formosinho, S.J.; Arnaut, L.G. *Adv. Photochem.* **1991**, *16*, 67; Newton, R.F., in Coyle, J.D. *Photochemistry in Organic Synthesis*, Royal Society of Chemistry, London, **1986**, pp. 39–60; Lee, E.K.C.; Lewis, R.S. *Adv. Photochem.* **1980**, *12*, 1; Calvert, J.G.; Pitts, Jr., J.N. *Photochemistry*, Wiley, NY, **1966**, pp. 368–427; Coyle, J.D.; Carless, H.A. J. *Chem. Soc. Rev.* **1972**, *1*, 465; Pitts, Jr., J.N.; Wan, J.K.S., in Patai, S. *The Chemistry of the Carbonyl Group*, Wiley, NY, **1966**, pp. 823–916; Dalton, J.C.; Turro, N.J. *Annu. Rev. Phys. Chem.* **1970**, *21*, 499; Bérces, T. in Bamford, C.H.; Tipper, C.F.H. *Comprehensive Chemical Kinetics*, Vol. 5; Elsevier, NY, **1972**, pp. 277–380; Turro, N.J.; Dalton, J.C.; Dawes, K.; Farrington, G.; Hautala, R.; Morton, D.; Niemczyk, M.; Shore, N. *Acc. Chem. Res.* **1972**, *5*, 92; Wagner, P.J. *Top. Curr. Chem.* **1976**, *66*, 1; Wagner, P.J.; Hammond, G.S. *Adv. Photochem.* **1968**, *5*, 21, 87–129. For reviews of the photochemistry of cyclic ketones, see Weiss, D.S. *Org. Photochem.* **1981**, *5*, 347; Chapman, O.L.; Weiss, D.S. *Org. Photochem.* **1973**, *3*, 197; Morton, B.M.; Turro, N.J. *Adv. Photochem.* **1974**, *9*, 197. For reviews of the photochemistry of α-diketones, see Rubin, M.B. *Top. Curr. Chem.* **1985**, *129*, 1; **1969**, *13*, 251; Monroe, B.M. *Adv. Photochem.* **1971**, *8*, 77. For a review of the photochemistry of protonated unsaturated carbonyl compounds, see Childs, R.F. *Rev. Chem. Intermed.* **1980**, *3*, 285. For reviews of the photochemistry of C=S compounds, see Coyle, J.D. *Tetrahedron* **1985**, *41*, 5393; Ramamurthy, V. *Org. Photochem.* **1985**, *7*, 231. For a review of the chemistry of C=N compounds, see Mariano, P.S. *Org. Photochem.* **1987**, *9*, 1.

compounds R—N=N—R.[45] The latter is an important source of radica⎯
the other product is the very stable N_2.

Category 2. *Decomposition into Molecules.* Aldehydes (though not ge⎯
ketones) can also cleave in this manner:

$$R-\underset{\underset{O}{\overset{\|}{}}}{C}{\overset{H}{}} \xrightarrow{h\nu} R-H + CO$$

This is an extrusion reaction (see Chapter 17). In another example of a
process in category 2, aldehydes and ketones with a γ hydrogen can cleave in
still another way (a β elimination, see Chapter 17):

$$R_2HC-CR_2-CR_2-\underset{\overset{\|}{O}}{C}-R' \xrightarrow{h\nu} R_2C=CR_2 + R_2HC-\underset{\overset{\|}{O}}{C}-R'$$

This reaction, called *Norrish Type II cleavage*,[46] involves intramolecular
abstraction of the γ hydrogen followed by cleavage of the resulting diradical[47] (a secondary reaction) to give an enol that tautomerizes to the aldehyde
or ketone product.[48]

[45]For reviews of the photochemistry of azo compounds, see Adam, W.; Oppenländer, T. *Angew. Chem. Int. Ed.* **1986**, *25*, 661; Dürr, H.; Ruge, B. *Top. Curr. Chem.* **1976**, *66*, 53; Drewer, R.J., in Patai, S. *The Chemistry of the Hydrazo, Azo, and Azoxy Groups*, pt. 2, Wiley, NY, **1975**, pp. 935–1015.

[46]For thorough discussions of the mechanism, see Wagner, P.J., in de Mayo, P. *Rearrangements in Ground and Excited States*, Vol. 3, Academic Press, NY, **1980**, pp. 381–444; *Acc. Chem. Res.* **1971**, *4*, 168; Dalton, J.C.; Turro, N.J. *Annu. Rev. Phys. Chem.* **1970**, *21*, 499, 526–538. See Niu, Y.; Christophy, E.; Hossenlopp, J.M. *J. Am. Chem. Soc.* **1996**, *118*, 4188 for a new view of Norrish Type II elimination.

[47]For reviews of the diradicals produced in this reaction, see Wilson, R.M. *Org. Photochem.* **1985**, *7*, 339, 349–373; Scaiano, J.C.; Lissi, E.A.; Encina, M.V. *Rev. Chem. Intermed.* **1978**, *2*, 139. For a review of a similar process, where δ hydrogens are abstracted, see Wagner, P.J. *Acc. Chem. Res.* **1989**, *22*, 83.

[48]This mechanism was proposed by Yang, N.C.; Yang, D.H. *J. Am. Chem. Soc.* **1958**, *80*, 2913. Among the evidence for this mechanism is the fact that the diradical intermediate has been trapped: Wagner, P.J.; Zepp, R.G. *J. Am. Chem. Soc.* **1972**, *94*, 287; Wagner, P.J.; Kelso, P.A.; Zepp, R.G. *J. Am. Chem. Soc.* **1972**, *94*, 7480; Adam, W.; Grabowski, S.; Wilson, R.M. *Chem. Ber.* **1989**, *122*, 561. See also Caldwell, R.A.; Dhawan, S.N.; Moore, D.E. *J. Am. Chem. Soc.* **1985**, *107*, 5163.

Both singlet and triplet n,π^* states undergo the reaction.[49] The intermediate diradical can also cyclize to a cyclobutanol, which is often a side product. Carboxylic esters, anhydrides, and other carbonyl compounds can also give this reaction.[50] The photolysis of ketene to CH_2 (p. 288) is still another example of a reaction in category 2. Both singlet and triplet CH_2 are generated, the latter in two ways:

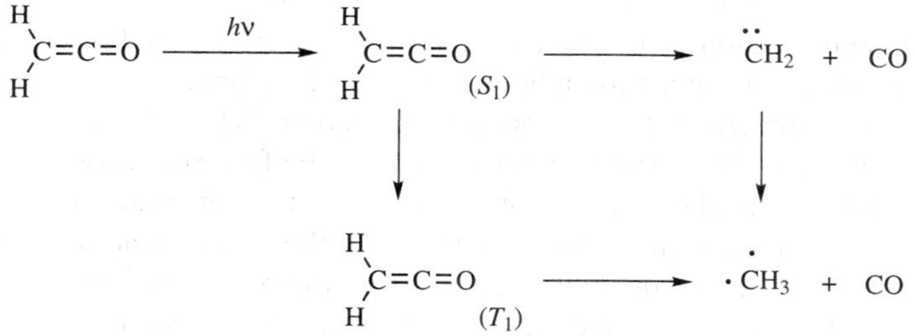

Reactions are known where *both* Norrish Type I and Norrish Type II reactions compete, and the substituents on and nature of the substrate will determine which leads to the major product.[51]

Category 3. *Intramolecular Rearrangement.* Two examples are the rearrangement of the trimesityl compound (**1**) to the enol ether (**2**),[52] and irradiation of *o*-nitrobenzaldehydes (**3**) to give *o*-nitrosobenzoic acids (**4**).[53]

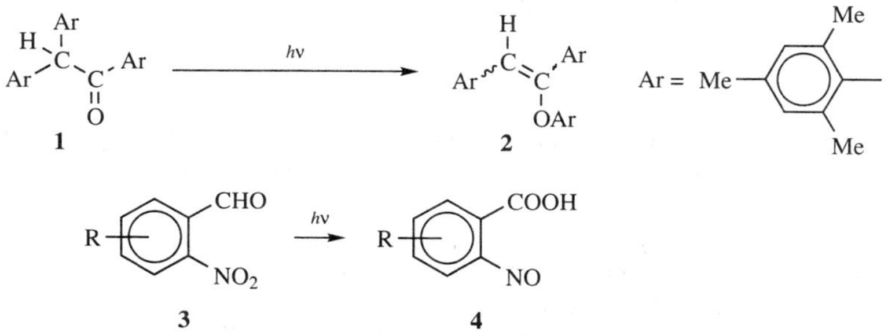

[49]Wagner, P.J.; Hammond, G.S. *J. Am. Chem. Soc.* **1965**, *87*, 4009; Dougherty, T.J. *J. Am. Chem. Soc.* **1965**, *87*, 4011; Ausloos, P.; Rebbert, R.E. *J. Am. Chem. Soc.* **1964**, *86*, 4512; Casey, C.P.; Boggs, R.A. *J. Am. Chem. Soc.* **1972**, *94*, 6457.

[50]For a review of the photochemistry of carboxylic acids and acid derivatives, see Givens, R.S.; Levi, N., in Patai, S. *The Chemistry of Acid Derivatives*, pt. 1, Wiley, NY, **1979**, pp. 641–753.

[51]See Hwu, J.R.; Chen, B.-L.; Huang, L.W.; Yang, T.-H. *J. Chem. Soc. Chem. Commun.* **1995**, 299 for an example.

[52]Hart, H.; Lin, L.W. *Tetrahedron Lett.* **1985**, *26*, 575; Wagner, P.J.; Zhou, B. *J. Am. Chem. Soc.* **1988**, *110*, 611.

[53]For a review of this and closely related reactions, see Morrison, H.A., in Feuer, H. *The Chemistry of the Nitro and Nitroso Groups*, pt. 1; Wiley, NY, **1969**, pp. 165–213, 185–191. For a review of photochemical rearrangements of benzene derivatives, see Kaupp, G. *Angew. Chem. Int. Ed.* **1980**, *19*, 243. See also, Yip, R.W.; Sharma, D.K. *Res. Chem. Intermed.* **1989**, *11*, 109.

Category 4. *Photoisomerization.* The most common reaction in this category is photochemical cis–trans isomerization.[54] For example, *cis*-stilbene can be converted to the trans isomer,[55] and the photoisomerization of *O*-methyl oximes is known.[56]

$$\begin{array}{ccc}
\underset{H}{\overset{Ph}{\diagdown}}C=C\underset{H}{\overset{Ph}{\diagup}} & \xrightarrow{\ h\nu\ } & \underset{H}{\overset{Ph}{\diagdown}}C=C\underset{Ph}{\overset{H}{\diagup}}
\end{array}$$

The isomerization takes place because the excited states, both S_1 and T_1, of many alkenes have a perpendicular instead of a planar geometry (p. 334), so cis–trans isomerism disappears upon excitation. When the excited molecule drops back to the S_0 state, either isomer can be formed. A useful example is the photochemical conversion of *cis*-cyclooctene to the much less stable trans isomer.[57] Another interesting example of this isomerization involves azo crown ethers. The crown ether **5**, in which the N=N bond is anti, preferentially binds NH_4^+, Li^+, and Na^+, but the syn isomer preferentially binds K^+ and Rb^+ (see p. 119). Thus, ions can be selectively put in or taken out of solution merely by turning a light source on or off.[58]

5

6

In another example, the trans azo compound **6** is converted to its cis isomer when exposed to light. In this case[59] the cis isomer is a stronger acid than the

[54] For reviews of cis–trans isomerizations, see Sonnet, P.E. *Tetrahedron* **1980**, *36*, 557; Schulte-Frohlinde, D.; Görner, H. *Pure Appl. Chem.* **1979**, *51*, 279; Saltiel, J.; Charlton, J.L., in de Mayo, P. *Rearrangements in Grund and Excited States*, Vol. 3, Academic Press, NY, **1980**, pp. 25–89; Saltiel, J.; Chang, D.W.L.; Megarity, E.D.; Rousseau, A.D.; Shannon, P.T.; Thomas, B.; Uriarte, A.K. *Pure Appl. Chem.* **1975**, *41*, 559; Saltiel, J.; D'Agostino, J.; Megarity, E.D.; Metts, L.; Neuberger, K.R.; Wrighton, M.; Zafiriou, O.C. *Org. Photochem.* **1979**, *3*, 1. For reviews of the photochemistry of alkenes, see Leigh, W.J.; Srinivasan, R. *Acc. Chem. Res.* **1987**, *20*, 107; Steinmetz, M.G. *Org. Photochem.* **1987**, *8*, 67; Adam, W.; Oppenländer, T. *Angew. Chem. Int. Ed.* **1986**, *25*, 661; Mattes, S.L.; Farid, S. *Org. Photochem.* **1984**, *6*, 233; Kropp, P.J. *Org. Photochem.* **1979**, *4*, 1; Morrison, H. *Org. Photochem.* **1979**, *4*, 143; Kaupp, G. *Angew. Chem. Int. Ed.* **1978**, *17*, 150. For a review of the photochemistry of allenes and cumulenes, see Johnson, R.P. *Org. Photochem.* **1985**, *7*, 75.
[55] For a review of the photoisomerization of stilbenes, see Waldeck, D.H. *Chem. Rev.* **1991**, *91*, 415.
[56] Kawamura, Y.; Takayama, R.; Nishiuchi, M.; Tsukayama, M. *Tetrahedron Lett.* **2000**, *41*, 8101.
[57] Deyrup, J.A.; Betkouski, M. *J. Org. Chem.* **1972**, *37*, 3561.
[58] Shinkai, S.; Nakaji, T.; Nishida, Y.; Ogawa, T.; Manabe, O. *J. Am. Chem. Soc.* **1980**, *102*, 5860. See also, Irie, M.; Kato, M. *J. Am. Chem. Soc.* **1985**, *107*, 1024; Akabori, S.; Kumagai, T.; Habata, Y.; Sato, S. *J. Chem. Soc. Perkin Trans. 1* **1989**, 1497; Shinkai, S.; Yoshioka, A.; Nakayama, H.; Manabe, O. *J. Chem. Soc. Perkin Trans. 2* **1990**, 1905. For a review, see Shinkai, S.; Manabe, O. *Top. Curr. Chem.* **1984**, *121*, 67.
[59] Haberfield, P. *J. Am. Chem. Soc.* **1987**, *109*, 6177.

trans. The trans isomer is dissolved in a system containing a base, wherein a liquid membrane separates two sides, one of which is illuminated, the other kept dark. On the illuminated side, the light converts the trans isomer to the cis. The cis isomer, being a stronger acid, donates its proton to the base, converting *cis*-ArOH to *cis*-ArO⁻. This ion migrates to the dark side, where it rapidly reverts to the *trans* ion, which reacquires a proton. Because each cycle forms one H_3O^+ ion in the illuminated compartment and one ^-OH ion in the dark compartment, the process reverses the normal reaction whereby these ions neutralize each other.[60] Thus the energy of light is used to do chemical work.[61] Another example of a category 4 reaction is the conversion of bicyclo[2.2.1]hept-2,5-diene to **7**.[54] The thermal isomerization of dibenzosemibullvalene **9** to the corresponding dibenzodihydropenta-lenofuran **8** in quantitative yield was known,[62] but in another example of a category 4 reaction the photochemical isomerization of **8** to **9** has now been reported.[63]

These examples illustrate that the use of photochemical reactions can make it very easy to obtain compounds that would be difficult to get in other ways. Reactions similar to these are discussed at **15-63**.

Category 5. Hydrogen-Atom Abstraction. When benzophenone is irradiated in isopropyl alcohol, the initially formed S_1 state crosses to the T_1 state, which abstracts hydrogen from the solvent to give the radical **10**. Radical **10** then

[60]Haberfield, P. *J. Am. Chem. Soc.* **1987**, *109*, 6178.

[61]For a review of instances where macrocycles change in response to changes in light, pH, temperature, and so on, see Beer, P.D. *Chem. Soc. Rev.* **1989**, *18*, 409. For an example not involving a macrocycle, see Feringa, B.L.; Jager, W.F.; de Lange, B.; Meijer, E.W. *J. Am. Chem. Soc.* **1991**, *113*, 5468.

[62]Sajimon, M.C.; Ramaiah, D.; Muneer, M.; Ajithkumar, E.S.; Rath, N.P.; George, M.V. *J. Org. Chem.* **1999**, *64*, 6347; Sajimon, M.C.; Ramaiah, D.; Muneer, M.; Rath, N.P.; George, M.V. *J. Photochem. Photobiol. A Chem.* **2000**, *136*, 209.

[63]Sajimon, M.C.; Ramaiah, D.; Thomas, K.G.; George, M.V. *J. Org. Chem.* **2001**, *66*, 3182.

abstracts another hydrogen to give benzhydrol (**11**) or dimerizes to benzpinacol (**12**):

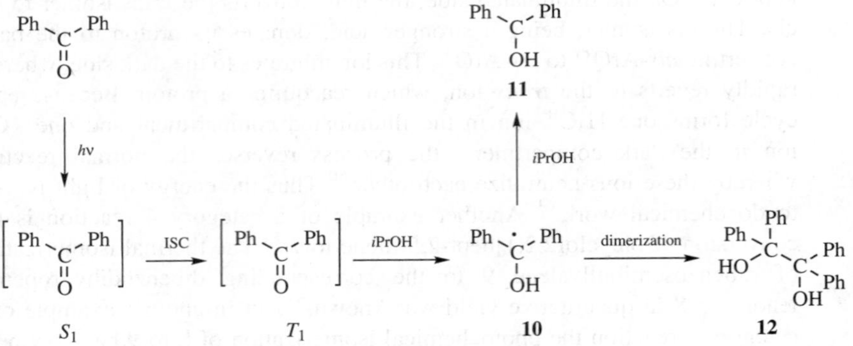

Ph–C(=O)–Ph →(hv) [Ph–C(=O)–Ph] S_1 →(ISC) [Ph–C(=O)–Ph] T_1 →(iPrOH) Ph–Ċ(OH)–Ph **10** →(dimerization) Ph$_2$C(OH)–C(OH)Ph$_2$ **12**

Ph$_2$C(OH) **11** ↑(iPrOH)

An example of intramolecular abstraction has already been given (p. $$$).

Category 6. *Photodimerization.* An example is dimerization of cyclopentenone:[64]

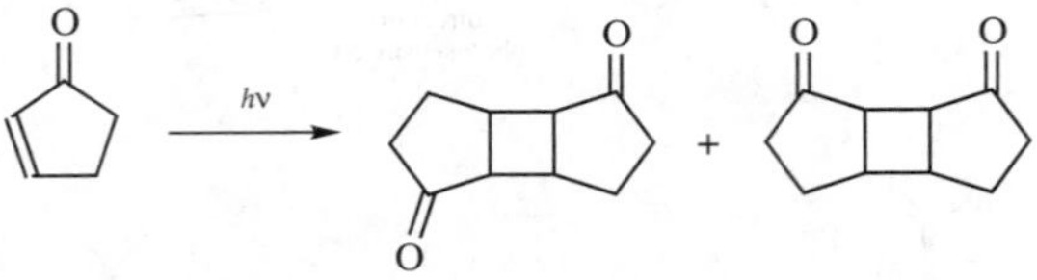

See reaction **15-63** for a discussion of this and similar reactions.

The Determination of Photochemical Mechanisms[65]

The methods used for the determination of photochemical mechanisms are largely the same as those used for organic mechanisms in general (Chapter 6): product identification, isotopic tracing, the detection and trapping of intermediates, and kinetics. There are, however, a few new factors: (*1*) there are generally many products in a photochemical reaction, as many as 10 or 15; (*2*) in measuring kinetics, there are more variables, since we can study the effect on the rate of the intensity or the wavelength of light; (*3*) in the detection of intermediates by spectra we can use the technique of *flash photolysis*, which can detect extremely short-lived intermediates.

In addition to these methods, there are two additional techniques.

1. The use of emission (fluorescence and phosphorescence) as well as absorption spectroscopy. From these spectra the presence of as well as the energy and lifetime of singlet and triplet excited states can often be calculated.

---

[64]Eaton, P.E. *J. Am. Chem. Soc.* **1962**, *84*, 2344, 2454, *Acc. Chem. Res.* **1968**, *1*, 50. For a review of the photochemistry of α,β-unsaturated ketones, see Schuster, D.I., in Patai, S.; Rappoport, Z. *The Chemistry of Enones*, pt. 2, Wiley, NY, **1989**, pp. 623–756.
[65]For a review, see Calvert, J.G.; Pitts, Jr., J.N. *Photochemistry*, Wiley, NY, **1966**, pp. 580–670.

2. The study of quantum yields. The *quantum yield* is the fraction of absorbed light that goes to produce a particular result. There are several types. A *primary quantum yield* for a particular process is the fraction of molecules absorbing light that undergo that particular process. Thus, if 10% of all the molecules that are excited to the S_1 state cross over to the T_1 state, the primary quantum yield for that process is 0.10. However, primary quantum yields are often difficult to measure. A *product quantum yield* (usually designated Φ) for a product P that is formed from a photoreaction of an initially excited molecule A can be expressed as

$$\Phi = \frac{\text{number of molecules of P formed}}{\text{number of quanta absorbed by A}}$$

Product quantum yields are much easier to measure. The number of quanta absorbed can be determined by an instrument called an *actinometer*, which is actually a standard photochemical system whose quantum yield is known. An example of the information that can be learned from quantum yields is the following. If the quantum yield of a product is finite and invariant with changes in experimental conditions, it is likely that the product is formed in a primary rate-determining process. Another example: in some reactions, the product quantum yields are found to be well over 1 (perhaps as high as 1000). Such a finding indicates a chain reaction (see p. $$$ for a discussion of chain reactions).

SONOCHEMISTRY

Sonochemistry (chemical events induced by exposure to ultrasound) occupies an important place in organic chemistry.[66] The chemical effects of high-intensity ultrasound were extensively studied in aqueous solutions for many years,[67] but is now applied to a variety of organic solvents. The origin of sonochemistry is acoustic cavitation: the creation, growth, and implosive collapse of gas vacuoles in solution by the sound field. Acoustic cavitation is the phenomenon by which intense ultrasonic waves induce the formation, oscillation, and implosion of gas

[66]Mason, T.J., Ed. *Advances in Sonochemistry*; JAI Press, NY, *1990-1994*; Vols. 1–3; Price, G.J., Ed. *Current Trends in Sonochemistry*, Royal Society of Chemistry, Cambridge, UK, *1992*; Suslick, K.S. *Science 1990*, 247, 1439; Suslick, K.S. *Ultrasound: Its Chemical, Physical, and Biological Effects*, VCH, NY, *1988*; Knapp, R.T.; Daily, J.W.; Hammitt, F.G. *Cavitation*, McGraw-Hill, NY, *1970*; Young, F.R. *Cavitation*, McGraw-Hill, NY, *1989*; Brennen, C.E. *Cavitation and Bubble Dynamics*, Oxford University Press, Oxford, UK, *1995*; Anbar, M. *Science 1968*, 161, 1343.
[67]Apfel, R.E., in Edmonds, P. *Methods in Experimental Physics*, Academic Press: New York, *1981*; Vol. 19; Margulis, M.A. *Russ. J. Phys. Chem. 1976*, 50, 1; Chendke, P.K.; Fogler, H.S. *Chem. Eng. J. 1974*, 8, 165; Makino, K.; Mossoba, M.M.; Riesz, P. *J. Am. Chem. Soc. 1982*, 104, 3537.

bubbles in liquids.[68] Liquids irradiated with high-power ultrasound undergo chemical decomposition and emit light.[69] These phenomena occur near the end of the collapse of bubbles expanded many times their equilibrium sizes. Chemistry (sonochemistry), light emission (sonoluminescence), and cavitation noise often accompany the process of acoustic cavitation.[70]

The collapse of gas vaculoes generates transient hot spots with local temperatures and pressures of several thousand degree K and hundreds of atmosphere. A sonochemical hot spot forms where the gas- and liquid-phase reaction zones have effective temperatures of 5200 and 1900 K, respectively.[71] The high temperatures and pressures that are achieved in the bubbles during the quasiadiabatic collapse[72] lead to the generation of chemistry and to the emission of light, most probably coming from molecular excited states and molecular recombination results. Note that work has been done that shows the commonly held view that bubbles are filled with saturated gas is inconsistent with a realistic estimate of condensation rates.[73] The alternative view of extensive solvent vapor supersaturation in bubbles uniformly heated to a few thousand K, depending on the conditions, is in accord with sonochemical rates and products.[74]

There is a correlation between sonochemical and sonoluminescence measurements, which is usually not observed. Sonoluminescence is the consequence that both the sonochemical production (under air) of oxidizing species and the emission of light reflect the variations of the primary sonochemical acts, which are themselves due to variations of the number of "active" bubbles.[75] Pulsed ultrasound in the high-frequency range (>1 MHz) is extensively used in medical diagnosis, and the effects of pulsed ultrasound in the 20-kHz range using an immersed titanium horn has been reported.[76]

The chemical effects of ultrasound have been studied for >50 years,[77] and applied to colloid chemistry in the 1940s.[78] Modern interest in the chemical uses

[68]Stottlemeyer, T.R.; Apfel, R.E. *J. Acoust. Soc. Am.* **1997**, *102*, 1413.

[69]Suslick, K.S.; Crum, L.A., in *Sonochemistry and Sonoluminescence, Handbook of Acoustics*; Crocker, M.J., Ed., Wiley, NY, **1998**, Chapt. 23; Leighton, T.G. *The Acoustic Bubble*, Academic Press, London, **1994**, Chapter 4, Brennen, C.E. *Cavitation and Bubble Dynamics*, Oxford University Press, **1995**; Chapts. 1–4. (4) Hua, I.; Hoffmann, M.R. *Environ. Sci. Technol.* **1997**, *31*, 2237.

[70]Suslick, K.S.; Didenko, Y.T.; Fang, M.M.; Hyeon, T.; Kolbeck, K.J.; McNamara, III, W.B.; Mdleleni, M.M.; Wong, M. *Philos. Trans. R. Soc. London A* **1999**, *357*, 335. For problems of sonochemistry and cavitation, see Margulis, M.A. *Ultrasonics Sonochemistry*, **1994**, *1*, S87.

[71]Suslick, K.S.; Hammerton, D.A.; Cline Jr., R.E. *J. Am. Chem. Soc.* **1986**, *108*, 5641.

[72]Didenko, Y.T.; McNamara III, W.B.; Suslick, K.S. *J. Am. Chem. Soc.* **1999**, *121*, 5817.

[73]Colussi, A. J.; Hoffmann, M.R. *J. Phys. Chem. A.* **1999**, *103*, 11336.

[74]Colussi, A.J.; Weavers, L.K.; Hoffmann, M.R. *J. Phys. Chem. A* **1998**, *102*, 6927; Hart, E.J.; Henglein, A. *J. Phys. Chem.* **1985**, *89*, 4342.

[75]Segebarth, N.; Eulaerts, O.; Reisse, J.; Crum, L. A.; Matula, T. J. *J. Phys. Chem. B.* **2002**, *106*, 9181.

[76]Dekerckheer, C.; Bartik, K.; Lecomte, J.-P.; Reisse, J. *J. Phys. Chem. A.* **1998**, *102*, 9177.

[77]Elpiner, I. E. *Ultrasound: Physical, Chemical, and Biological Effects*, Consultants Bureau, NY, **1964**.

[78]Sollner, K. *Chem. Rev.* **1944**, *34*, 371.

of ultrasound involve chemistry in both homogeneous[79] and heterogeneous[80] systems. Organic solvents, such as alkanes, support acoustic cavitation and the associated sonochemistry, and this leads to carbon–carbon bond cleavage and radical rearrangements, with the peak temperatures reached in such cavities controlled by the vapor pressure of the solvent.[81]

It is often difficult to compare the sonochemical results reported from different laboratories (the reproducibility problem in sonochemistry).[82] The sonochemical power irradiated into the reaction system can be different for different instruments. Several methods are available to estimate the amount of ultrasonic power entered into a sonochemical reaction,[82] the most common being calorimetry. This experiment involves measurement of the initial rate of a temperature rise produced when a system is irradiated by power ultrasound. It has been shown that calorimetric methods combined with the Weissler reaction can be used to standardize the ultrasonic power of individual ultrasonic devices.[83]

Sonochemistry has been used to facilitate or assist many organic reactions,[84] as well as other applications.[85] The scope of reactions studied is beyond this work, but some representative examples will be listed. Ultrasound has been used to promote lithiation of organic compounds,[86] for the generation of carbenes,[87] and

[79]Suslick, K.S.; Schubert, P.F.; Goodale, J.W. *J. Am. Chem. Soc.* **1981**, *103*, 7342; Lorimer, J.P.; Mason, T.J. *J. Chem. Soc., Chem. Commun.* **1980**,1135; Margulis, M.A. *Khim. Zh.* **1981**, 57; Nishikawa, S.; Obi, U.; Mashima, M. *Bull. Chem. Soc. Jpn.* **1977**, *50*, 1716; Yu, T.J.; Sutherland, R.G.; Verrall, R.E. *Can. J. Chem.* **1980**, *58*, 1909; Sehgal, C.; Sutherland, R.G.; Verrall, R.E. *J. Phys. Chem.* **1980**, *84*, 2920; Sehgal, C.; Yu, T.J.; Sutherland, R.G.; Verrall, R.E. *J. Phys. Chem.* **1982**, *86*,2982; Sehgal, C.M.; Wang, S.Y. *J. Am. Chem. Soc.* **1981**, *103*, 6606; Staas, W.H.; Spurlock, L.A. *J. Chem. Soc., Perkin Trans. 1* **1975**, 1675.
[80]Han, B.-H.; Boudjouk, P. *J. Org. Chem.* **1982**, *47*, 5030; Boudjouk, P.; Han, B.-H. *Tetrahedron Lett.* **1981**, *22*, 3813; Han, B.-H.; Boudjouk, P. *J. Org. Chem.* **1982**, *47*, 751; Boudjouk, P.; Han, B.-H.; Anderson, K.R. *J. Am. Chem. Soc.* **1982**, *104*, 4992; Boudjouk, P.; Han, B.-H. *J. Catal.* **1983**, *79*, 489; Fry, A.J.; Ginsburg, G.S. *J. Am. Chem. Soc.* **1979**, *101*, 3927; Kitazume, T.; Ishikawa, N. *Chem. Lett.* **1981**, 1679; Kristol, D.S.; Klotz, H.; Parker, R.C. *Tetrahedron Lett.* **1981**, *22*, 907; Lintner, W.; Hanesian, D. *Ultrasonics* **1977**, *15*, 21; Luche, J.-L.; Damiano, J. *J. Am. Chem. Soc.* **1980**, *102*, 7926; Moon, S.; Duchin, L.; Cooney, J.V. *Tetrahedron Lett.* **1979**, 3917; Racher, S.; Klein, P. *J. Org. Chem.* **1981**, *46*, 3558; Regen, S.L.; Singh, A. *J. Org. Chem.* **1982**, *47*, 1587; Kegelaers, Y.; Eulaerts, O.; Reisse, J.; Segebarth, N. *Eur. J. Org. Chem.* **2001**, 3683.
[81]Suslick, K.S. Gawienowski, J.J.; Schubert, P.F.; Wang, H.H. *J. Phys. Chem.* **1983**, *87*, 2299.
[82]Mason, T.J. *Practical Sonochemistry: User's Guide to Applications in Chemistry and Chemical Engineering*, Ellis Horwood, West Sussex, **1991**, pp. 43–46; Broeckaert, L.; Caulier, T.; Fabre, O.; Maerschalk, C.; Reisse, J.; Vandercammen, J.; Yang, D.H.; Lepoint, T.; Mullie, F. *Current Trends in Sonochemistry*, Price, G.J., Ed., Royal Society of Chemistry, Cambridge, **1992**, p. 8; Mason, T.J.; Lorimer, J.P.; Bates, D.M.; Zhao, Y. *Ultrasonics Sonochemistry* **1994**, *1*, S91; Mason, T.J.; Lorimer, J.P.; Bates, D.M. *Ultrasonics* **1992**, *30*, 40.
[83]Kimura, T.; Sakamoto, T.; Leveque, J.-M.; Sohmiya, H.; Fujita, M.; Ikeda, S.; Ando, T. *Ultrasonics Sonochemistry* **1996**, *3*, S157.
[84]*Synthetic Organic Sonochemistry*, Luche, J.-L. (Universite de Savoie, France), Plenum Press, NY. **1998**; Luche, J.-L. *Ultrasonics Sonochemistry* **1996**, *3*, S215; Bremner, D.H. *Ultrasonics Sonochemistry* **1994**, *1*, S119.
[85]Thompson, L.H.; Doraiswamy, L.K. *Ind. Eng. Chem. Res.* **1999**, *38*, 1215; Adewuyi, Y.G. *Ind. Eng. Chem. Res.* **2001**, *40*, 4681.
[86]Boudjouk, P.; Sooriyakumaran, R.; Han, B.H. *J. Org. Chem.* **1986**, *51*, 2818, and Ref. 1 therein.
[87]Regen, S.L.; Singh, A. *J. Org. Chem.* **1982**, *47*, 1587.

reactions of metal carbonyls where sonochemical ligand dissociation has been observed, which often produces multiple CO substitution.[88] The influence of ultrasound on phase-transfer catalyzed thioether synthesis has been studied.[89]

Sonochemistry has been applied to acceleration of the Reformatsky reaction,[90] Diels–Alder reactions,[91] the arylation of active methylene compounds[92] nucleophilic aromatic substitution of haloarenes,[93] and to hydrostannation and tin hydride reduction.[94] Other sonochemical applications involve the reaction of benzyl chloride and nitrobenzene,[95] a $S_{RN}1$ reaction in liquid ammonia at room temperature,[96] and Knoevenagel condensation of aromatic aldehydes.[97] Iodination of aliphatic hydrocarbons can be accelerated,[98] and oxyallyl cations have been prepared from α,α'-diiodoketones using sonochemistry.[99] Sonochemistry has been applied to the preparation of carbohydrate compounds.[100] When sonochemistry is an important feature of a chemical reaction, this fact will be noted in the reactions presented in Chapters 10–19.

MICROWAVE CHEMISTRY

In 1986, independent work by Gedye and co-workers[101] as well as Giguere and Majetich[102] reported the advantages of microwave irradiation for organic synthesis. Gedye described four different types of reactions were studied, including the hydrolysis of benzamide to benzoic acid under acidic conditions, and all reactions showed significant rate enhancements when compared to the same reactions done at reflux conditions.[103] Giguere and Majetich reported rate enhancements for microwave-promoted Diels–Alder, Claisen, and ene reactions. At this point,

[88]Suslick, K.S.; Goodale, J.W.; Schubert, P.F.; Wang, H.H. *J. Am. Chem. Soc.* **1983**, *105*, 5781.

[89]Wang, M.-L.; Rajendran, V. *J. Mol. Catalysis A: Chemical* **2005**, *244*, 237.

[90]Han, B.H.; Boudjouk, P. *J. Org. Chem.*, **1982**, *47*, 5030.

[91]Nebois, P.; Bouaziz, Z.; Fillion, H.; Moeini, L.; Piquer, Ma.J.A.; Luche, J.-L.; Riera, A.; Moyano, A.; Pericàs, M.A. *Ultrasonics Sonochemistry* **1996**, *3*, 7.

[92]Mečiarová, M.; Kiripolský, M.; Toma, Š *Ultrasonics Sonochemistry* **2005**, *12*, 401.

[93]Mečiarová, M.; Toma, S.; Magdolen, P. *Ultrasonics Sonochemistry* **2003**, *10*, 265.

[94]Nakamura, E.; Machii, D.; Inubushi, T. *J. Am. Chem. Soc.* **1989**, *111*, 6849.

[95]Vinatoru, M.; Stavrescua, R.; Milcoveanu, A.B.; Toma, M.; T.J. Mason, T.J. *Ultrasonics Sonochemistry* **2002**, *9*, 245.

[96]Manzo, P.G.; Palacios, S.M.; Alonso, R.A. *Tetrahedron Lett.* **1994**, *35*, 677.

[97]McNulty, J.; Steere, J.A.; Wolf, S. *Tetrahedron Lett.* **1998**, *39*, 8013.

[98]Kimura, T.; Fujita, M.; Sohmiya, H. Ando, T. *Ultrasonics Sonochemistry* **2002**, *9*, 205.

[99]Montaña, A.M.; Grima, P.M. *Tetrahedron Lett.* **2001**, *42*, 7809.

[100]Kardos, N.; Luche, J.-L. *Carbohydrate Res.* **2001**, *332*, 115.

[101]Gedye R.; Smith, F.; Westaway, K.; Ali, H.; Baldisera, L. *Tetrahedron Lett.* **1986**, *27*, 279; Gedye, R. N.; Smith, F. E.; Westaway, K. C. *Can. J. Chem.* **1987**, *66*, 17.

[102]Giguere, R.J.; Bray, T.; Duncan, S.M.; Majetich, G. *Tetrahedron Lett.* **1986**, *27*, 4945.

[103]Taken from Horeis, G.; Pichler, S.; Stadler, A.; Gössler, W.; Kappe, C.O. *Microwave-Assisted Organic Synthesis - Back to the Roots*, Fifth International Electronic Conference on Synthetic Organic Chemistry (ECSOC-5), **2001** (http://www.mdpi.org/ecsoc-5.htm).

>2000 publications[104] have appeared describing chemical synthesis promoted by microwave irradiation, including many review articles[105] and books.[106]

Microwaves are electromagnetic waves (see p. 329) and there are electric and magnetic field components. Charged particles start to migrate or rotate as the electric field is applied,[107] which leads to further polarization of polar particles. Because the concerted forces applied by the electric and magnetic components of microwaves are rapidly changing in direction ($2.4 \times 10^9/s$), warming occurs.[107] In general, the most common frequencies used for microwave dielectric heating[108] are 918 MHz and 2.45 GHz[109] (wavelengths of 33.3 and 12.2 cm, respectively), which are in the region between the IR and radiowave wavelengths in the electromagnetic spectrum. For chemical reactions done with microwave irradiation, rapid heating is usually observed, and if a solvent is used superheating of that solvent was always observed.[108] In the early days of microwave chemistry, reactions were often done in open vessels, but also in sealed Teflon or glass vessels using unmodified domestic household ovens.[110] Dielectric heating is direct so if the reaction matrix has a sufficiently large dielectric loss tangent, and contains molecules possessing a dipole moment, a solvent is not required. The use of dry-reaction microwave chemistry is increasingly popular.[111]

Microwave dielectric heating was initially categorized by thermal effects and nonthermal effects.[112] "Thermal effects are those which are caused by the different temperature regime which can be created due to microwave dielectric heating. Nonthermal effects are effects,[113] which are caused by effects specifically inherent to

---

[104]Kappe, C. O. *Angew. Chem. Int. Ed.* **2004**, *43*, 6250.

[105]Majetich, G.; Karen, W. in Kingston, H.M.; Haswell, S.J. *Microwave-Enhanced Chemistry: Fundamentals, Sample Preparation, and Applications.* American Chemical Society, Washington, DC, **1997**, p 772; Giguere, R.J. *Org. Synth.: Theory Appl.* **1989**, *1*, 103; Mingos, D.M.P.; Baghurst, D.R. *Chem. Soc. Rev.* **1991**, 20, 1; Abramovitch, R.A. *Org. Prep. Proced. Int.* **1991**, *23*, 683; Bose, A.K.; Manhas, M.S.; Banik, B.K.; Robb, E.W. *Res. Chem. Intermed.* **1994**, 20, 1; Majetich, G.; Hicks, R. *Res. Chem. Intermed.* **1994**, 20, 61; Strauss, C.R.; Trainor, R.W. *Aust. J. Chem.* **1995**, *48*, 1665; Caddick, S. *Tetrahedron* **1995**, *51*, 10403; Mingos, D M.P. *Res. Chem. Intermed.* **1994**, 20, 85; Berlan, J. *Rad. Phys. Chem.* **1995**, *45*, 581; Fini, A.; Breccia, A. *Pure Appl. Chem.* **1999**, *71*, 573.

[106]Kingston, H. M.; Haswell, S.J. *Microwave-Enhanced Chemistry. Fundamentals, Sample Preparation, and Applications*, American Chemical Society, **1997**; Loupy, A. *Microwaves in Organic Synthesis*, Wiley-VCH, Weinheim, **2002**; Hayes, B.L. *Microwave Synthesis: Chemistry at the Speed of Light*, CEM Publishing, Matthews, NC, **2002**; Lidström, P., Tierney, J.P. *Microwave-Assisted Organic Synthesis*, Blackwell Scientific, **2005** ; Kappe, C.O.; Stadler, A. *Microwaves in Organic and Medicinal Chemistry*, Wiley-VCH, Weinheim, **2005**.

[107]Galema, S.A. *Chem. Soc. Rev.* **1997**, *26*, 233.

[108]Gabriel, C.; Gabriel, S.; Grant, E. H.; Halstead, B. S. J.; Mingos, D. M. P. *Chem. Soc. Rev.* **1998**, *27*, 213.

[109]This frequencey is usually applied in domestic microwave ovens.

[110]Caddick, S. *Tetrahedron* **1995**, *51*, 10403.

[111]Loupy, A.; Petit, A.; Hamelin, J.; Texier-Boullet, F.; Jacquault, P.; Mathé, D. *Synthesis* **1998**, 1213; Varma, R. S. *Green Chem.* **1999**, 43; Kidawi, M. *Pure Appl. Chem.* **2001**, *73*, 147; Varma, R. S. *Pure Appl. Chem.* **2001**, *73*, 193.

[112]Langa, F.; de la Cruz, P.; de la Hoz, A.; Díaz-Ortiz, A.; Díez-Barra, E. *Contemp. Org. Synth.* **1997**, *4*, 373.

[113]See Kuhnert, N. *Angew. Chem. Int. Ed.* **2002**, *41*, 1863.

the microwaves and are not caused by different temperature regimes."[107] Some claimed special effects[114] in microwave chemistry, such as lowering of Gibbs energy of activation, but later study under careful temperature control indicated no special rate effects.[115] When conventional microwave ovens were used, temperature control was difficult particularly when reactions are carried out in closed reaction vessels. The main contributing factor to any rate acceleration caused by microwave dielectric heating seems to be due to a thermal effect. The thermal effect may be due to a faster initial heating rate or to the occurrence of local regions with higher temperatures.[107]

Conventional microwave ovens are used less often for microwave chemistry today. Microwave reactors for chemical synthesis are commercially available and widely used in academia and in industry. These instruments have built-in magnetic stirring, direct temperature control of the reaction mixture, shielded thermocouples or IR sensors, and the ability to control temperature and pressure by regulating microwave output power.

The applications of microwave chemistry to organic chemistry are too numerous to mention. A few representative examples will be given to illustrate the scope and utility. Microwave chemistry is widely used in synthesis.[116] Examples include the Heck reaction (reaction **13-10**),[117] the Suzuki reaction (reaction **13-12**),[118] the Sonogashira reaction (reaction **13-13**),[119] Ullman type couplings (reaction

[114]Laurent, R.; Laporterie, A.; Dubac, J.; Berlan, J.; Lefeuvre, S.; Audhuy, M. *J. Org. Chem.* **1992**, *57*, 7099, and references cited therein.

[115]Raner, K.D.; Strauss, C.R.; Vyskoc, F.; Mokbel, L. J. *Org. Chem.* **1993**, *58*, 950, and references cited therein.

[116]Abramovitch, R. A. *Org. Prep. Proced. Int.* **1991**, *23*, 685; Caddick, S. *Tetrahedron* **1995**, *51*, 10403; Strauss, C. R.; Trainor, R. W. *Aust. J. Chem.* **1995**, *48*, 1665; Bose, A. K.; Banik, B. K.; Lavlinskaia, N.; Jayaraman, M.; Manhas, M. S. *Chemtech* **1997**, *27*, 18; Lidström, P.; Tierney, J.; Wathey, B.; Westman, J. *Tetrahedron* **2001**, *57*, 9225; Larhed, M.; Moberg, C.; Hallberg, A. *Acc. Chem. Res.* **2002**, *35*, 717; Nüchter, M.; Ondruschka, B.; Bonrath, W.; Gum, A. *Green Chem.* **2004**, *6*, 128; Hayes, B.L. *Aldrichim. Acta* **2004**, *37*, 66.

[117]Larhed, M.; Moberg, C.; Hallberg, A. *Acc. Chem. Res.* **2002**, *35*, 717; Olofsson, K.; Larhed, M. in Lidström, P.; Tierney, J.P. *Microwave-Assisted Organic Synthesis*, Blackwell, Oxford, **2004**, Chapt. 2.; Andappan, M.M.S.; Nilsson, P.; Larhed, M. *Mol. Diversity* **2003**, *7*, 97.

[118]Nuteberg, D.; Schaal, W.; Hamelink, E.; Vrang, L.; Larhed, M. *J. Comb. Chem.* **2003**, *5*, 456; Miller, S.P.; Morgan, J.B.; Nepveux, F.J.; Morken, J.P. *Org. Lett.* **2004**, *6*, 131; Kaval, N.; Bisztray, K.; Dehaen, W.; Kappe, C.O.; Van der Eycken, E. *Mol. Diversity* **2003**, *7*, 125; Gong, Y.; He, W. *Heterocycles* **2004**, *62*, 851; Organ, M.G.; Mayer, S.; Lepifre, F.; N'Zemba, B.; Khatri, J. *Mol. Diversity* **2003**, *7*, 211; Luo, G.; Chen, L.; Pointdexter, G.S. *Tetrahedron Lett.* **2002**, *43*, 5739; Wu, T.Y.H.; Schultz, P.G.; Ding, S. *Org. Lett.* **2003**, *5*, 3587; Han, J.W.; Castro, J.C.; Burgess, K. *Tetrahedron Lett.* **2003**, *44*, 9359; Leadbeater, N.E.; Marco, M. *J. Org. Chem.* **2003**, *68*, 888; Bai, L.; Wang, J.-X.; Zhang, Y. *Green Chem.* **2003**, *5*, 615 ; Leadbeater, N.E.; Marco, M. *J. Org. Chem.* **2003**, *68*, 5660.

[119]Kaval, N.; Bisztray, K.; Dehaen, W.; Kappe, C.O.; Van der Eycken, E. *Mol. Diversity* **2003**, *7*, 125; Gong, Y.; He, W. *Heterocycles* **2004**, *62*, 851; Miljani, O.Š; Vollhardt, K.P.C.; Whitener, G.D. *Synlett* **2003**, 29; Petricci, E.; Radi, M.; Corelli, F.; Botta, M. *Tetrahedron Lett.* **2003**, *44*, 9181; Leadbeater, N.E.; Marco, M.; Tominack, B.J. *Org. Lett.* **2003**, *5*, 3919 ; Appukkuttan, P.; Dehaen, W.; Van der Eycken, E. *Eur. J. Org. Chem.* **2003**, 4713.

13-3),[120] cycloaddition reactions (reactions **15-58–15-66**),[121] dihydroxylation (reaction **15-48**),[122] and the Mitsunobu reaction (reaction **10-23**).[123] There are a multitude of other reactions types from earlier literature that can be found in the cited review articles. When microwave chemistry is an important feature of a chemical reaction, this fact will be noted in the reactions presented in Chapters 10–19.

[120]Wu, Y.-J.; He, H.; L'Heureux, A. *Tetrahedron Lett.* **2003**, *44*, 4217; Lange, J.H.M.; Hofmeyer, L.J.F.; Hout, F.A.S.; Osnabrug, S.J.M.; Verveer, P.C.; Kruse, C.G.; Feenstra, R.W. *Tetrahedron Lett.* **2002**, *43*, 1101.
[121]For example, see de la Hoz, A.; D'az-Ortis, A.; Moreno, A.; Langa, F. *Eur. J. Org. Chem.* **2000**, 3659; Van der Eycken, E.; Appukkuttan, P.; De Borggraeve, W.; Dehaen, W.; Dallinger, D.; Kappe, C.O. *J. Org. Chem.* **2002**, *67*, 7904 ; Pinto, D.C.G.A.; Silva, A.M.S.; Almeida, L.M.P.M.; Carrillo, J.R.; D'az-Ortiz, A.; de la Hoz, A.; Cavaleiro, J.A.S. *Synlett* **2003**, 1415.
[122]Dupau, P.; Epple, R.; Thomas, A.A.; Fokin, V.V.; Sharpless, K.B. *Adv. Synth. Catal.* **2002**, *344*, 421.
[123]Lampariello, L.R.; Piras, D.; Rodriquez, M.; Taddei, M. *J. Org. Chem.* **2003**, *68*, 7893; Raheem, I.T.; Goodman, S.N.; Jacobsen, E.N. *J. Am. Chem. Soc.* **2004**, *126*, 706.

CHAPTER 8

Acids and Bases

Two acid–base theories are used in organic chemistry today: the Brønsted theory and the Lewis theory.[1] These theories are quite compatible and are used for different purposes.[2]

BRØNSTED THEORY

According to this theory, an acid is defined as a *proton donor*[3] and a base as a *proton acceptor* (a base must have a pair of electrons available to share with the proton; this is usually present as an unshared pair, but sometimes is in a π orbital). An acid–base reaction is simply the transfer of a proton from an acid to a base. (Protons do not exist free in solution but must be attached to an electron pair.) When the acid gives up a proton, the species remaining still retains the electron pair to which the proton was formerly attached. Thus the new species, in theory at least, can reacquire a proton and is therefore a base. It is referred to as the *conjugate base* of the acid. All acids have a conjugate base, and all bases have a *conjugate acid*. All acid–base reactions fit the equation

$$A\text{-}H \quad + \quad B \quad \rightleftharpoons \quad A \quad + \quad B\text{-}H$$

$$\text{Acid}_1 \qquad\qquad \text{Base}_2 \qquad\qquad \text{Base}_1 \qquad \text{Acid}_2$$

---

[1]For monographs on acids and bases, see Stewart, R. *The Proton: Applications to Organic Chemistry*, Academic Press, NY, *1985*; Bell, R.P. *The Proton in Chemistry*, 2nd ed., Cornell University Press, Ithaca, NY, *1973*; Finston, H.L.; Rychtman, A.C. *A New View of Current Acid–Base Theories*, Wiley, NY, *1982*.
[2]For discussion of the historical development of acid–base theory, see Bell, R.P. *Q. Rev. Chem. Soc. 1947*, *1*, 113; Bell, R.P. *The Proton in Chemistry*, 1st ed., Cornell University Press, Ithaca, NY, *1959*, pp. 7–17.
[3]According to IUPAC terminology (Bunnett, J.F.; Jones, R.A.Y. *Pure Appl. Chem. 1988*, *60*, 1115), an acid is a *hydron* donor. IUPAC recommends that the term *proton* be restricted to the nucleus of the hydrogen isotope of mass 1, while the nucleus of the naturally occurring element (which contains ∼0.015% deuterium) be called the *hydron* (the nucleus of mass 2 has always been known as the *deuteron*). This accords with the naturally occurring negative ion, which has long been called the *hydride* ion. In this book, however, we will continue to use *proton* for the naturally occurring form, because most of the literature uses this term.

No charges are shown in this equation, but an acid always has a charge one positive unit higher than that of its conjugate base.

Acid strength may be defined as the tendency to give up a proton and *base strength* as the tendency to accept a proton. Acid–base reactions occur because acids are not equally strong. If an acid, say HCl, is placed in contact with the conjugate base of a weaker acid, say acetate ion, the proton will be transferred because the HCl has a greater tendency to lose its proton than acetic acid. That is, the equilibrium

$$HCl \; + \; CH_3COO^- \; \rightleftharpoons \; CH_3COOH \; + \; Cl^-$$

lies well to the right. On the other hand, treatment of acetic acid with chloride ion gives essentially no reaction, since the weaker acid already has the proton.

This is always the case for any two acids, and by measuring the positions of the equilibrium the relative strengths of acids and bases can be determined.[4] Of course, if the two acids involved are close to each other in strength, a measurable reaction will occur from both sides, though the position of equilibrium will still be over to the side of the weaker acid (unless the acidities are equal within experimental limits). In this manner, it is possible to construct a table in which acids are listed in order of acid strength (Table 8.1).[5] Next to each acid in Table 8.1 is shown its conjugate base. It is obvious that if the acids in such a table are listed in *decreasing* order of acid strength, the bases must be listed in *increasing* order of base strength, since the stronger the acid, the weaker must be its conjugate base. The pK_a values in Table 8.1 are most accurate in the middle of the table. They are much harder to measure[6] for very strong and very weak acids, and these values must be regarded as approximate. Qualitatively, it can be determined that $HClO_4$ is a stronger acid than H_2SO_4, since a mixture of $HClO_4$ and H_2SO_4 in 4-methyl-2-pentanone can be titrated to an $HClO_4$ end point without interference by H_2SO_4.[7] Similarly, $HClO_4$ can be shown to be stronger than HNO_3 or HCl. However, this is not quantitative, and the value of -10 in the table is not much more than an educated guess. The values for RNO_2H^+, $ArNO_2H^+$, HI, $RCNH^+$ and RSH_2^+ must also be regarded as highly speculative.[8] A wide variety of pK_a values has been reported for the conjugate acids of even such simple bases

[4]Although equilibrium is reached in most acid–base reactions extremely rapidly (see p. $$$), some are slow (especially those in which the proton is given up by a carbon) and in these cases time must be allowed for the system to come to equilibrium.
[5]Table 8.1 is a thermodynamic acidity scale and applies only to positions of equilibria. For the distinction between thermodynamic and kinetic acidity (see p. 367).
[6]For a review of methods of determining pK_a values, see Cookson, R.F. *Chem. Rev.* **1974**, *74*, 5.
[7]Kolthoff, I.M.; Bruckenstein, S., in Kolthoff, I.M.; Elving, P.J. *Treatise on Analytical Chemistry*, Vol. 1, pt. 1; Wiley, NY, *1959*, pp. 475–542, p. 479.
[8]For reviews of organic compounds protonated at O, N, or S, see Olah, G.A.; White, A.M.; O'Brien, D.H. *Chem. Rev.* **1970**, *70*, 561; Olah, G.A.; White, A.M.; O'Brien, D.H., in Olah, G.A.; Schleyer, P.V.R. *Carbonium Ions*, Vol. 4; Wiley, NY, *1973*, pp. 1697–1781.

as acetone[9] (-0.24 to -7.2), diethyl ether (-0.30 to -6.2), ethanol (-0.33 to -4.8), methanol (-0.34 to -4.9), and 2-propanol (-0.35 to -5.2), depending on the method used to measure them.[10] Very accurate values can be obtained only for acids weaker than hydronium ion and stronger than water.

A crystallographic scale of acidity has been developed, including the acidity of C—H compounds. Measuring the mean C—H•••O distances in crystal structures correlated well with conventional p$K_{a(DMSO)}$ values, where dimethyl sulfoxide = DMSO.[11] An *ab initio* study was able to correlate ring strain in strained hydrocarbons with hydrogen-bond acidity.[12]

The bottom portion of Table 8.1 consists of very weak acids (pK_a above ~ 17).[13] In most of these acids, the proton is lost from a carbon atom, and such acids are known as *carbon acids*. The pK_a values for such weak acids are often difficult to measure and are known only approximately. The methods used to determine the relative positions of these acids are discussed in Chapter 5.[14] The acidity of carbon acids is proportional to the stability of the carbanions that are their conjugate bases (see p. 249).

The extremely strong acids at the top of the table are known as *super acids* (see p. 236).[15] The actual species present in the FSO_3H—SbF_5 mixture are probably $H[SbF_5(SO_3F)]$ and $H[SbF_2(SO_3F)_4]$.[16] The addition of SO_3 causes formation of the still stronger $H[SbF_4(SO_3F)_2]$, $H[SbF_3(SO_3F)_3]$, and $H[(SbF_5)_2(SO_3F)]$.[16]

By the use of tables, such as Table 8.1, it is possible to determine whether a given acid will react with a given base. For tables in which acids are listed in order of decreasing strength, the rule is that *any acid will react with any base in the table that is below it but not with any above it*.[17] It must be emphasized that the order of

[9]For discussions of pK_a determinations for the conjugate acids of ketones, see Bagno, A.; Lucchini, V.; Scorrano, G. *Bull. Soc. Chim. Fr.* **1987**, 563; Toullec, J. *Tetrahedron Lett.* **1988**, *29*, 5541.

[10]Rochester, C.H. *Acidity Functions*; Academic Press, NY, **1970**. For discussion of the basicity of such compounds, see Liler, M. *Reaction Mechanisms in Sulfuric Acid*, Academic Press, NY, **1971**, pp. 118–139.

[11]Pedireddi, V.R.; Desiraju, G.R. *J. Chem. Soc. Chem. Commun.* **1992**, 988.

[12]Alkorta, I.; Campillo, N.; Rozas, I.; Elguero, J. *J. Org. Chem.* **1998**, *63*, 7759.

[13]For a monograph on very weak acids, see Reutov, O.A.; Beletskaya, I.P.; Butin, K.P. *CH-Acids*, Pergamon, NY, **1978**. For other discussions, see Cram, D.J. *Fundamentals of Carbanion Chemistry*, Academic Press, NY, **1965**, pp. 1–45; Streitwieser, Jr., A.; Hammons, J.H. *Prog. Phys. Org. Chem.* **1965**, *3*, 41. For a study of substituent effects of weak acids see Wiberg, K.B. *J. Org. Chem.* **2002**, *67*, 1613.

[14]For reviews of methods used to measure the acidity of carbon acids, see Jones, J.R. *Q. Rev. Chem. Soc.* **1971**, *25*, 365; Fischer, H.; Rewicki, D. *Prog. Org. Chem.* **1968**, *7*, 116; Reutov, O.A.; Beletskaya, I.P.; Butin, K.P. *CH-Acids*, Chapt. 1, Pergamon, NY, **1978** [an earlier version of this chapter appeared in *Russ. Chem. Rev.* **1974**, *43*, 17]; Ref. 6. For reviews on acidities of carbon acids, see Gau, G.; Assadourian, L.; Veracini, S. *Prog. Phys. Org. Chem.* **1987**, *16*, 237; in Buncel, E.; Durst, T. *Comprehensive Carbanion Chemistry*, pt. A, Elsevier, NY, **1980**, the reviews by Pellerite, M.J.; Brauman, J.I. pp. 55–96 (gas-phase acidities); and Streitwieser, Jr., A.; Juaristi, E.; Nebenzahl, L. pp. 323–381.

[15]For a monograph, see Olah, G.A.; Prakash, G.K.S.; Sommer, J. *Superacids*; Wiley, NY, **1985**. For a review, see Gillespie, R.J.; Peel, T.E. *Adv. Phys. Org. Chem.* **1971**, *9*, 1. For a review of solid super acids, see Arata, K. *Adv. Catal.* **1990**, *37*, 165. For a review of methods of measuring superacidity, see Jost, R.; Sommer, J. *Rev. Chem. Intermed.* **1988**, *9*, 171.

[16]Gillespie, R.J. *Acc. Chem. Res.* **1968**, *1*, 202.

[17]These reactions are equilibria. What the rule actually says is that the position of equilibrium will be such that the weaker acid predominates. However, this needs to be taken into account only when the acid and base are close to each other in the table (within ~ 2 pK units).

TABLE 8.1. The p K_a Values for Many Types of Acids. The values in boldface are exact values; the others are approximate, especially above 18 and below −2.[18]

Acid	Base	Approximate pK_a (relative to water)	References
Super Acids			
HF–SbF$_5$	SbF$_6^-$		19
FSO$_3$H–SbF$_5$–SO$_3$			16
FSO$_3$H–SbF$_5$			16,19
FSO$_3$H	FSO$_3^-$		16
RNO$_2$H$^+$	RNO$_2$	−12	20
ArNO$_2$H$^+$	ArNO$_2$	−11	20
HClO$_4$	ClO$_4^-$	−10	21
HI	I$^-$	−10	21
RCNH$^+$	RCN	−10	22
R–C–H $\overset{\shortparallel}{\underset{+OH}{}}$	R–C–H $\overset{\shortparallel}{\underset{O}{}}$	−10	23
H$_2$SO$_4$	HSO$_4^-$		
HBr	Br$^-$	−9	21

(*continued*)

[18]In this table we do not give pK_a values for individual compounds (with a few exceptions), only average values for functional groups. Extensive tables of pK values for many carboxylic and other acids and amines are given in Brown, H.C.; McDaniel, D.H.; Häflinger, O., in Braude, E.A.; Nachod, F.C. *Determination of Organic Structures by Physical Methods*, Vol. 1, Academic Press, NY, *1955*. Values for >5500 organic acids are given, in Serjeant, E.P.; Dempsey, B. *Ionisation Constants of Organic Acids in Aqueous Solution*, Pergamon, Elmsford NY, *1979*; Kortüm, G.; Vogel, W.; Andrussow, K. *Dissociation Constants of Organic Acids in Aqueous Solution*, Butterworth, London, *1961*. The index in the 1979 volume covers both volumes. Kortüm, G.; Vogel, W.; Andrussow, K. *Pure Appl. Chem. 1960*, *1*, 190 give values for 631 carboxylic acids and 110 phenols. Arnett, E.M. *Prog. Phys. Org. Chem. 1963*, *1*, 223 gives hundreds of values for very strong acids (very weak bases). Perrin, D.D. *Dissociation Constants of Organic Bases in Aqueous Solution*, Butterworth, London, *1965*, and Supplement, 1972 list pK values for >7000 amines and other bases. Collumeau, A. *Bull. Soc. Chim. Fr. 1968*, 5087 gives pK values for ~800 acids and bases. Bordwell, F.G. *Acc. Chem. Res. 1988*, *21*, 456 gives values for >300 acids in dimethyl sulfoxide. For inorganic acids and bases, see Perrin, D.D. *Ionisation Constants of Inorganic Acids and Bases in Aqueous Solution*, 2nd ed., Pergamon, Elmsford NY, *1982*; *Pure Appl. Chem. 1969*, 20, 133.
[19]Brouwer, D.M.; van Doorn, J.A. *Recl. Trav. Chim. Pays-Bas 1972*, *91*, 895; Gold,V.; Laali, K.; Morris, K.P.; Zdunek, L.Z. *J. Chem. Soc. Chem. Commun. 1981*, 769; Sommer, J.; Canivet, P.; Schwartz, S.; Rimmelin, P. *Nouv. J. Chim. 1981*, *5*, 45.
[20]Arnett, E.M. *Prog. Phys. Org. Chem. 1963*, *1*, 223, 324–325.
[21]Bell, R.P. *The Proton in Chemistry*, 2nd ed., Cornell University Press, Ithaca, NY, *1973*.
[22]Deno, N.C.; Gaugler, R.W.; Wisotsky, M.J. *J. Org. Chem. 1966*, *31*, 1967.
[23]Levy, G.C.; Cargioli, J.D.; Racela, W. *J. Am. Chem. Soc. 1970*, *92*, 6238. See, however, Brouwer, D.M.; van Doorn, J.A. *Recl. Trav. Chim. Pays-Bas 1971*, *90*, 1010.

TABLE 8.1. (*Continued*)

Acid	Base	Approximate pK_a (relative to water)	References
Ar—C(=OH⁺)—OR [24]	Ar—C(=O)—OR	−7.4	20
HCl	Cl⁻	−7	21
RSH₂+	RSH	−7	20
Ar—C(=OH⁺)—OH [24]	Ar—C(=O)—OH	−7	25
Ar—C(=OH⁺)—H	Ar—C(=O)—H	−7	26
R—C(=OH⁺)—R	R—C(=O)—R	−7	9,22,27
ArSO₃H	ArSO₃⁻	−6.5	28
R—C(=OH⁺)—OR [24]	R—C(=O)—OR	−6.5	20
ArOH₂⁺	ArOH	−6.4	29
R—C(=OH⁺)—OH [24]	R—C(=O)—OH	−6	20
Ar—C(=OH⁺)—R	Ar—C(=O)—R	−6	26,30
Ar—O(⁺)(H)—R	Ar—O—R	−6	29,31
CH(CN)₃	⁻C(CN)₃	−5	32

[24]Carboxylic acids, esters, and amides are shown in this table to be protonated on the carbonyl oxygen. There has been some controversy on this point, but the weight of evidence is in that direction. See, for example, Katritzky, A.R.; Jones, R.A.Y. *Chem. Ind. (London)* **1961**, 722; Ottenheym, J.H.; van Raayen, W.; Smidt, J.; Groenewege, M.P.; Veerkamp, T.A. *Recl. Trav. Chim. Pays-Bas* **1961**, *80*, 1211; Stewart, R.; Muenster, L.J. *Can. J. Chem.* **1961**, *39*, 401; Smith, C.R.; Yates, K. *Can. J. Chem.* **1972**, *50*, 771; Benedetti, E.; Di Blasio, B.; Baine, P. *J. Chem. Soc. Perkin Trans. 2* **1980**, 500; Ref. 8; Homer, R.B.; Johnson, C.D., in Zabicky, J. *The Chemistry of Amides*; Wiley, NY, **1970**, pp. 188–197. It has been shown that some amides protonate at nitrogen: see Perrin, C.L. *Acc. Chem. Res.* **1989**, *22*, 268. For a review of alternative proton sites, see Liler, M. *Adv. Phys. Org. Chem.* **1975**, *11*, 267.

[25]Stewart, R.; Granger, M.R. *Can. J. Chem.* **1961**, *39*, 2508.

[26]Yates, K.; Stewart, R. *Can. J. Chem.* **1959**, *37*, 664; Stewart, R.; Yates, K. *J. Am. Chem. Soc.* **1958**, *80*, 6355.

[27]Lee, D.G. *Can. J. Chem.* **1970**, *48*, 1919.

[28]Cerfontain, H.; Koeberg-Telder, A.; Kruk, C. *Tetrahedron Lett.* **1975**, 3639.

[29]Arnett, E.M.; Wu, C.Y. *J. Am. Chem. Soc.* **1960**, *82*, 5660; Koeberg-Telder, A.; Lambrechts, H.J.A.; Cerfontain, H. *Recl. Trav. Chim. Pays-Bas* **1983**, *102*, 293.

[30]Fischer, A.; Grigor, B.A.; Packer, J.; Vaughan, J. *J. Am. Chem. Soc.* **1961**, *83*, 4208.

[31]Arnett, E.M.; Wu, C.Y. *J. Am. Chem. Soc.* **1960**, *82*, 4999.

[32]Boyd, R.H. *J. Phys. Chem.* **1963**, *67*, 737.

TABLE 8.1. (*Continued*)

Acid	Base	Approximate pK_a (relative to water)	References
Ar_3NH^+	Ar_3N	−5	33
$\underset{+OH}{\overset{H-C-H}{\underset{\parallel}{}}}$	$\underset{O}{\overset{H-C-H}{\underset{\parallel}{}}}$	−4	34
$\underset{H}{\overset{R-\overset{+}{O}-R}{\underset{\mid}{}}}$	$R\diagdown O\diagup R$	−3.5	22,31,35
$R_3COH_2{}^+$	R_3COH	−2	35
$R_2CHOH_2{}^+$	R_2CHOH	−2	35,36
$RCH_2OH_2{}^+$	RCH_2OH	−2	22,35,36
H_3O^+	H_2O	**−1.74**	37
$\underset{+OH}{\overset{Ar-C-NH_2{}^{24}}{\underset{\parallel}{}}}$	$\underset{O}{\overset{Ar-C-NH_2}{\underset{\parallel}{}}}$	−1.5	38
HNO_3	$NO_3{}^-$	−1.4	21
$\underset{+OH}{\overset{R-C-NH_2\ ^{24}}{\underset{\parallel}{}}}$	$\underset{O}{\overset{R-C-NH_2}{\underset{\parallel}{}}}$	−0.5	38
$Ar_2NH_2{}^+$	Ar_2NH	1	33
$HSO_4{}^-$	$SO_4{}^{2-}$	**1.99**	39
HF	F^-	**3.17**	39
$HONO$	$NO_2{}^-$	**3.29**	39
$ArNH_3{}^+$	$ArNH_2$	3-5	40
$ArNR_2H^+$	$ArNR_2$	3-5	40
$RCOOH$	$RCOO^-$	4-5	40
$HCOCH_2CHO$	$HCOC-HCHO$	5	41
$H_2CO_3{}^{42}$	$HCO_3{}^-$	**6.35**	39
H_2S	HS^-	**7.00**	39
$ArSH$	ArS^-	6–8	43

(*continued*)

[33]Arnett, E.M.; Quirk, R.P.; Burke, J.J. *J. Am. Chem. Soc.* **1970**, *92*, 1260.

[34]McTigue.P.T.; Sime, J.M. *Aust. J. Chem.* **1963**, *16*, 592.

[35]Deno, N.C.; Turner, J.O. *J. Org. Chem.* **1966**, *31*, 1969.

[36]Lee, D.G.; Demchuk, K.J. *Can. J. Chem.* **1987**, *65*, 1769; Chandler, W.D.; Lee, D.G. *Can. J. Chem.* **1990**, *68*, 1757.

[37]For a discussion, see Campbell, M.L.; Waite, B.A. *J. Chem. Educ.* **1990**, *67*, 386.

[38]Cox, R.A.; Druet, L.M.; Klausner, A.E.; Modro, T.A.; Wan, P.; Yates, K. *Can. J. Chem.* **1981**, *59*, 1568; Grant, H.M.; McTigue, P.; Ward, D.G. *Aust. J. Chem.* **1983**, *36*, 2211.

[39]Bruckenstein, S.; Kolthoff, I.M., in Kolthoff, I.M.; Elving, P.J. *Treatise on Analytical Chemistry*, Vol. 1, pt. 1; Wiley, NY, **1959**, pp. 432–433.

[40]Brown, H.C.; McDaniel, D.H.; Häflinger, O., in Braude, E.A.; Nachod, F.C. *Determination of Organic Structures by Physical Methods*, Vol. 1; Academic Press, NY, **1955**, pp. 567–662.

[41]Pearson, R.G.; Dillon, R.L. *J. Am. Chem. Soc.* **1953**, *75*, 2439.

[42]This value includes the CO_2 usually present. The value for H_2CO_3 alone is 3.9, in Bell, R.P. *The Proton in Chemistry*, 2nd ed., Cornell University Press, Ithaca, NY, **1973**.

[43]Crampton, M.R., in Patai, S. *The Chemistry of the Thiol Group*, pt. 1, Wiley, NY, **1974**, pp. 396–410.

TABLE 8.1. (*Continued*)

Acid	Base	Approximate pK_a (relative to water)	References
$CH_3COCH_2COCH_3$[44]	$CH_3COC{-}HCOCH_3$	9	41
HCN	CN^-	9.2	45
NH_4^+	NH_3	**9.24**	39
ArOH	ArO^-	8–11	46
RCH_2NO_2	$RC{-}HNO_2$	10	47
R_3NH^+	R_3N	10–11	40
RNH_3^+	RNH_2	10–11	40
HCO_3^-	CO_3^{2-}	**10.33**	39
RSH	RS^-	10–11	43
$R_2NH_2^+$	R_2NH	11	40
$N{\equiv}CCH_2C{\equiv}N$	$N{\equiv}CC{-}HC{\equiv}N$	11	41,48
CH_3COCH_2COOR	$CH_3COC{-}HCOOR$	11	41
$CH_3SO_2CH_2SO_2CH_3$	$CH_3SO_2C{-}HSO_2CH_3$	12.5	49
$EtOOCCH_2COOEt$	$EtOOCC{-}HCOOEt$	13	41
CH_3OH	CH_3O^-	15.2	50,51
H_2O	OH^-	**15.74**	52
(cyclopentadiene)	(cyclopentadienyl anion)	16	53
RCH_2OH	RCH_2O^-	16	50
RCH_2CHO	$RC{-}HCHO$	16	54
R_2CHOH	R_2CHO^-	16.5	50
R_3COH	R_3CO^-	17	50
$RCONH_2$	$RCONH^-$	17	55
$RCOCH_2R$	$RCOC{-}HR$	19-20[56]	57

[44]See Bunting, J.W.; Kanter, J.P. *J. Am. Chem. Soc.* **1993**, *115*, 11705 for pK_a values of several β-ketone esters and amides.

[45]Perrin, D.D. *Ionisation Constants of Inorganic Acids and Bases in Aqueous Solution*, 2nd ed., Pergamon, Elmsford, NY, *1982*.

[46]Rochester, C.H., in Patai, S. *The Chemistry of the Hydroxyl Group*, pt. 1, Wiley, NY, *1971*, p. 374.

[47]Cram, D.J. *Chem. Eng. News* **1963**, *41* (No. 33, Aug. 19), 94.

[48]Bowden, K.; Stewart, R. *Tetrahedron* **1965**, *21*, 261.

[49]Hine, J.; Philips, J.C.; Maxwel, J.I. *J. Org. Chem.* **1970**, *35*, 3943. See also, Ang, K.P.; Lee, T.W.S. *Aust. J. Chem.* **1977**, *30*, 521.

[50]Reeve, W.; Erikson, C.M.; Aluotto, P.F. *Can. J. Chem.* **1979**, *57*, 2747.

[51]See also Mackay, G.I.; Bohme, D.K. *J. Am. Chem. Soc.* **1978**, *100*, 327; Olmstead, W.N.; Margolin, Z.; Bordwell, F.G. *J. Org. Chem.* **1980**, *45*, 3295.

[52]Harned, H.S.; Robinson, R.A. *Trans. Faraday Soc.* **1940**, *36*, 973.

[53]Streitwieser Jr., A.; Nebenzahl, L. *J. Am. Chem. Soc.* **1976**, *98*, 2188.

[54]Guthrie, J.P.; Cossar, J. *Can. J. Chem.* **1986**, *64*, 2470.

[55]Homer, R.B.; Johnson, C.D., in Zabicky, J. *The Chemistry of Amides*, Wiley, NY, *1970*, pp. 238–240.

[56]The pK_a of acetone in DMSO is reported to be 26.5. See Bordwell, F.G.; Zhang, X.-M. *Accts. Chem. Res.* **1997**, *26*, 510.

[57]Tapuhi, E.; Jencks, W.P. *J. Am. Chem. Soc.* **1982**, *104*, 5758; Guthrie, J.P.; Cossar, J.; Klym, A. *J. Am. Chem. Soc.* **1984**, *106*, 1351; Chiang, Y.; Kresge, A.J.; Tang, Y.S.; Wirz, J. *J. Am. Chem. Soc.* **1984**, *106*, 460.

TABLE 8.1. (*Continued*)

Acid	Base	Approximate pK_a (relative to water)	References
		20	58,59
		20.08[a]	60
		18.91[a]	60
		23	58,59
$ROOCCH_2R$	$ROOCC\text{-}HR$	24.5	41
$RCH_2C{\equiv}N$	$RC\text{-}HC{=}N$	25	41,61
$HC{\equiv}CH$	$HC{\equiv}CC^-$	25	62
Ph_2NH	PH_2N^-	24.95[b]	56
$EtOCOCH_3$	$EtOCOCH_2{}^-$	25.6	63
$PhNH_2$	$PhNH^-$	30.6[b]	56
Ar_3CH	Ar_3C^-	31.5	58,64
Ar_2CH_2	Ar_2CH^-	33.5	58,64
H_2	H^-	35	65
NH_3	$NH_2{}^-$	38	66
$PhCH_3$	$PhCH_2{}^-$	40	67
$CH_2{=}CHCH_3$		43	68
PhH	Ph^-	43	69

[58]Streitwieser, Jr., A.; Ciuffarin, E.; Hammons, J.H. *J. Am. Chem. Soc.* **1967**, *89*, 63.

[59]Streitwieser, Jr., A.; Hollyhead, W.B.; Pudjaatmaka, H.; Owens, P.H.; Kruger, T.L.; Rubenstein, P.A.; MacQuarrie, R.A.; Brokaw, M.L.; Chu, W.K.C.; Niemeyer, H.M. *J. Am. Chem. Soc.* **1971**, *93*, 5088.

[60]Streitwieser, A.; Wang, G.P.; Bors, D.A. *Tetrahedron* **1997**, *53*, 10103.

[61]For a review of the acidity of cyano compounds, see Hibbert, F., in Patai, S.; Rappoport, Z. *The Chemistry of Triple-bonded Functional Groups*, pt. 1; Wiley, NY, **1983**, pp. 699–736.

[62]Cram, D.J. *Fundamentals of Carbanion Chemistry*, Academic Press, NY, **1965**, p. 19. See also, Dessy, R.E.; Kitching, W.; Psarras, T.; Salinger, R.; Chen, A.; Chivers, T. *J. Am. Chem. Soc.* **1966**, *88*, 460.

[63]Amyes, T.L.; Richard J.P. *J. Am. Chem. Soc.* **1996**, *118*, 3129.

[64]Streitwieser, Jr., A.; Hollyhead W.B.; Sonnichsen, G.; Pudjaatmaka, H.; Chang, C.J.; Kruger. T.L. *J. Am. Chem. Soc.* **1971**, *93*, 5096.

[65]Buncel, E.; Menon, B. *J. Am. Chem. Soc.* **1977**, *99*, 4457.

[66]Buncel, E.; Menon, B. *J. Organomet. Chem.* **1977**, *141*, 1.

[67]Streitwieser, Jr., A.; Ni, J.X. *Tetrahedron Lett.* **1985**, *26*, 6317; Albrech, H.; Schneider, G. *Tetrahedron* **1986**, *42*, 4729.

[68]Boerth, D.W.; Streitwieser, Jr., A. *J. Am. Chem. Soc.* **1981**, *103*, 6443.

[69]Streitwieser, Jr., A.; Scannon, P.J.; Niemeyer, H.M. *J. Am. Chem. Soc.* **1972**, *94*, 7936.

TABLE 8.1. (*Continued*)

Acid	Base	Approximate pK_a (relative to water)	References
$CH_2=CH_2$	$CH_2=CH^-$	44	70
cyclo-C_3H_6	*c*-$C_3H_5^-$	46	71
CH_4[72]	CH_3^-	48	73
C_2H_6	$C_2H_5^-$	50	74
$(CH_3)_2CH_2$[72]	$(CH_3)_2CH^-$	51	74
$(CH_3)_3CH$[72]	$(CH_3)_3C^-$		75

[a]pK_a in THF.
[b]pK_a in DMSO.

acid strength in Table 8.1 applies when a given acid and base react without a solvent or, when possible, in water. In other solvents the order may be greatly different (see p. 392). In the gas phase, where solvation effects are completely or almost completely absent, acidity orders may also differ greatly.[76] For example, in the gas phase, toluene is a stronger acid than water and *tert*-butoxide ion is a weaker base than methoxide ion[77] (see also, pp. 390–394). It is also possible for the acidity order to change with temperature. For example, >50°C the order of base strength is BuOH > H$_2$O > Bu$_2$O; from 1 to 50°C the order is BuOH > Bu$_2$O > H$_2$O; while <1°C the order becomes Bu$_2$O > BuOH > H$_2$O.[78]

A hydrogen-bond basicity scale has been developed that can be used to determine the relative basicity of molecules. Table 8.2 gives the pK_{HB} values for several common heteroatom containing molecules. This is obtained from the protonated form (conjugated acid) of the molecule in question. The larger the number, the more basic is that compound. The basicity of aliphatic amines has been calculated,[79] the ion-pair

[70]Streitwieser, Jr., A.; Boerth, D.W. *J. Am. Chem. Soc.* **1978**, *100*, 755.

[71]This value is calculated from results given in Streitwieser, Jr., A.; Caldwell, R.A.; Young, W.R. *J. Am. Chem. Soc.* **1969**, *91*, 529. For a review of acidity and basicity of cyclopropanes, see Battiste, M.A.; Coxon, J.M., in Rappoport, Z. *The Chemistry of the Cyclopropyl Group*, pt. 1; Wiley, NY, **1987**, pp. 255–305.

[72]See Daasbjerg, K. *Acta Chem. Scand. B* **1995**, *49*, 878 for pK_a values of various hydrocarbons in DMF.

[73]This value is calculated from results given in Streitwieser, Jr., A.; Taylor, D.R. *J. Chem. Soc. D* **1970**, 1248.

[74]These values are based on those given in Cram, D.J. *Chem. Eng. News* **1963**, *41* (No. 33, Aug. 19), 94, but are corrected to the newer scale of Streitwieser, A.; Streitwieser, Jr., A.; Scannon, P.J.; Niemeyer, H.M. *J. Am. Chem. Soc.* **1972**, *94*, 7936; Streitwieser, Jr., A.; Boerth, D.W. *J. Am. Chem. Soc.* **1978**, *100*, 755.

[75]Breslow, R. and co-workers report a value of 71 [Breslow, R.; Grant, J.L. *J. Am. Chem. Soc.* **1977**, *99*, 7745], but this was obtained by a different method, and is not comparable to the other values in Table 8.1. A more comparable value is ~53. See also Juan, B.; Schwarz, J.; Breslow, R. *J. Am. Chem. Soc.* **1980**, *102*, 5741.

[76]For a review of acidity and basicity scales in the gas phase and in solution, see Gal, J.; Maria, P. *Prog. Phys. Org. Chem.* **1990**, *17*, 159.

[77]Brauman, J.I.; Blair, L.K. *J. Am. Chem. Soc.* **1970**, *92*, 5986; Bohme, D.K.; Lee-Ruff, E.; Young, L.B. *J. Am. Chem. Soc.* **1972**, *94*, 4608, 5153.

[78]Gerrard, W.; Macklen, E.D. *Chem. Rev.* **1959**, *59*, 1105. For other examples, see Calder, G.V.; Barton, T.J. *J. Chem. Educ.* **1971**, *48*, 338; Hambly, A.N. *Rev. Pure Appl. Chem.* **1965**, *15*, 87, 88.

[79]Caskey, D.C.; Damrauer, R.; McGoff, D. *J. Org. Chem.* **2002**, *67*, 5098.

basicity of amines in THF[80] and in water[81] has been determined, and the basicity of pyridine was examined.[82] Weaker bases have also been examined, and the basicity of carbonyl compound in carbon tetrachloride has been determined.[83]

A class or organic compounds termed *super bases* has been developed. Vinamidine type or Schwesinger proton sponges (see p. 386), **1**,[84] are dubbed super bases and are probably the most powerful organic neutral bases known. The pK_a ($pK_{BH}{}^+$) in MeCN was measure as 31.94. It has been shown that the pK_a values of strong neutral organic (super)bases in acetonitrile are well described by the density functional theory.[85] The fundamental type of proton sponge is 1,8-bis(dimethylamino)-naphthalene, **2** (see p. 386), with a $pK_{BH}{}^+$ of 18.18.[86] Other super base type compounds include amidinazines such as N^1,N^1-dimethyl-N^2-β-(2-pyridylethyl)-formamidine (**3**), $pK_{BH}{}^+$ in DMSO = 25.1,[87] 1,8-bis(tetramethylguanidino) naphthalene, **4**,[88] and quinolino[7,8-*h*]quinolines, such as **5** with a pK_{BH^+} = 12.8.[89]

[80]Streitwieser, A.; Kim, H.-J. *J. Am. Chem. Soc.* **2000**, *122*, 11783.

[81]Canle, L.M.; Demirtas, I.; Freire, A.; Maskill, H.; Mishima, M. *Eur. J. Org. Chem.* **2004**, 5031.

[82]Chmurzyński, L. *J. Heterocyclic Chem.* **2000**, *37*, 71.

[83]Carrasco, N.; González-Nilo, F.; Rezende, M.C. *Tetrahedron* **2002**, *58*, 5141.

[84]Schwesinger, R.; Mißfeldt, M.; Peters, K.; von Schnering, H.G. *Angew. Chem. Int. Ed.* **1987**, *26*, 1165; Schwesinger, R.; Schlemper, H.; Hasenfratz, Ch.; Willaredt, J.; Dimbacher, T.; Breuer, Th.; Ottaway, C.; Fletschinger, M.; Boele, J.; Fritz, H.; Putzas, D.; Rotter, H.W.; Bordwell, F.G.; Satish, A.V.; Ji, G.Z.; Peters, E.-M.; Peters, K.; von Schnering, H.G. *Liebigs Ann.* **1996**, 1055.

[85]Kovačević, B.; Maksić, Z. B. *Org. Lett.* **2001**, *3*, 1523.

[86]Alder, R.W.; Bowman, P.S.; Steele, W.R.S.; Winterman, D.R. *Chem. Commun.* **1968**, 723 ; Alder, R.W. *Chem. Rev.* **1989**, *89*, 1215.

[87]Raczyńska, E.D.; Darowska, M.; Dabkowska, I.; Decouzon, M.; Gal, J.-F.; Maria, P-C.; Poliart, C.D. *J. Org. Chem.* **2004**, *69*, 4023.

[88]Raab, V.; Kipke, J.; Gschwind, R.M.; Sundermeyer, J. *Chem. Eur. J.* **2002**, *8*, 1682.

[89]Staab, H.A.; Saupe, T. *Angew. Chem. Int. Ed.* **1988**, *27*, 865; Staab, H.A.; Höne, M.; Krieger, C. *Tetrahedron Lett.* **1988**, *29*, 1905; Staab, H.A.; Krieger, C.; Höne, M. *Tetrahedron Lett.* **1988**, *29*, 5629; Krieger, C.; Newsom, I.; Zirnstein, M.A.; Staab, H.A. *Angew. Chem. Int. Ed.* **1989**, *28*, 84 ; Zirnstein, M.A.; Staab, H.A. *Angew. Chem. Int. Ed.* **1987**, *26*, 460.

TABLE 8.2. pK_{HB} Values for Many Types of Bases

Base	Approximate pK_{HB}	Reference
N-Methyl-2-piperidone	2.60	90
$Et_2NCONEt_2$	2.43	90
N-Methyl-2-pyrrolidinone	2.38	90
$PhCONMe_2$	2.23	90
$HCONMe_2$	2.10	90
PhCONHMe	2.03	90
18-crown-6	1.98	91
HCONHMe	1.96	90
15-crown-5	1.82	91
12-crown-4	1.73	91
$PhOCONMe_2$	1.70	90
Et_2N-CN	1.63	92
Me_2N-CN	1.56	92
δ-Valerolactone	1.43	93
Oxetane	1.36	91
γ-Butyrolactone	1.32	93
Tetrahydrofuran (THF)	1.28	91
Cyclopentanone	1.27	94
t-BuOMe	1.19	91
Acetone	1.18	94
MeCOOEt	1.07	93
1,4-Dioxane	1.03	91
Et_2O	1.01	91
1,3-Dioxane	0.93	91
1-Methyloxirane	0.97	91
PhCOOMe	0.89	93
MeOCOOMe	0.82	93
PhCHO	0.78	94
Bu_2O	0.75	91
HCOOEt	0.66	93
Me_3CHO	0.65	94
Me_2NO_2	0.41	95
$MeNO_2$	0.27	95
$PhNO_2$	0.30	95
Furan	−0.40	91

[90]Le Questel, J.-Y.; Laurence, C.; Lachkar, A.; Helbert, M.; Berthelot, M. *J. Chem. Soc. Perkin Trans. 2* **1992**, 2091.

[91]Berthelot, M.; Besseau, F.; Laurence, C. *Eur. J. Org. Chem.* **1998**, 925.

[92]Berthelot, M.; Helbert, M.; Laurence, C.; LeQuestel, J.-Y.; Anvia, F.; Taft, R.W. *J. Chem. Soc. Perkin Trans. 2* **1993**, 625.

[93]Besseau, F.; Laurence, C.; Berthelot, M. *J. Chem. Soc. Perkin Trans. 2* **1994**, 485.

[94]Besseau, F.; Luçon, M.; Laurence, C.; Berthelot, M. *J. Chem. Soc. Perkin Trans. 2* **1998**, 101.

[95]Laurence, C.; Berthelot, M.; Luçon, M.; Morris, D.G. *J. Chem. Soc. Perkin Trans. 2* **1994**, 491.

THE MECHANISM OF PROTON-TRANSFER REACTIONS

Proton transfers between oxygen and nitrogen acids and bases are usually extremely fast.[96] In the thermodynamically favored direction they are generally diffusion controlled.[97] In fact, a *normal acid* is defined[98] as one whose proton-transfer reactions are completely diffusion controlled, except when the conjugate acid of the base to which the proton is transferred has a pK value very close (differs by $<\sim2$ pK units) to that of the acid. The normal acid–base reaction mechanism consists of three steps:

1. HA + I B ⇌ AH•••ᛁ•B

2. AH••••ᛁB ⇌ A•ᛁ••••HB

3. Aᛁ••••HB ⇌ AI + HB

The actual proton transfer takes place in the second step the first step is formation of a hydrogen-bonded complex. The product of the second step is another hydrogen-bonded complex, which dissociates in the third step.

However, not all such proton transfers are diffusion controlled. For example, if an internal hydrogen bond exists in a molecule, reaction with an external acid or base is often much slower.[99] In a case such as 3-hydroxypropanoic acid,

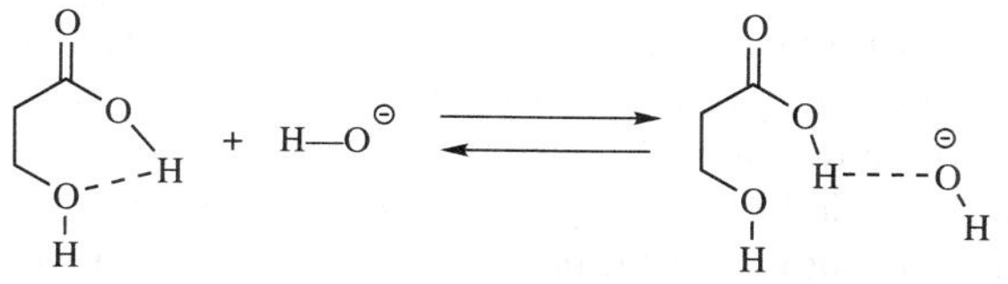

3-Hydroxypropanoic acid

the $^-$OH ion can form a hydrogen bond with the acidic hydrogen only if the internal hydrogen bond breaks. Therefore only some of the collisions between $^-$OH ions and 3-hydroxypropanoic acid molecules result in proton transfer. In many collisions, the $^-$OH ions will come away empty handed, resulting in a lower reaction rate. Note that this affects only the rate, not the equilibrium. Other systems are capable of hydrogen bonding, such as 1,2-diols. In the case of cyclohexane-1,2-diols, hydrogen bonding, ion–dipole interactions, polarizability, and stereochemistry all

---

[96]For reviews of such proton transfers, see Hibbert, F. *Adv. Phys. Org. Chem.* **1986**, *22*, 113; Crooks, J.E., in Bamford, C.H.; Tipper, C.F.H. *Chemical Kinetics*, Vol. 8; Elsevier, NY, **1977**, pp. 197–250.

[97]Kinetic studies of these very fast reactions were first carried out by Eigen. See Eigen, M. *Angew. Chem. Int. Ed.* **1964**, *3*, 1.

[98]See, for example, Hojatti, M.; Kresge, A.J.; Wang, W. *J. Am. Chem. Soc.* **1987**, *109*, 4023.

[99]For an example of a slow proton transfer from F_3CCOOH to $(PhCH_2)_3N$, see Ritchie, C.D.; Lu, S. *J. Am. Chem. Soc.* **1989**, *111*, 8542.

play a role in determining the acidity.[100] The presence of halogen atoms such as chlorine can lead to hydrogen-bonding effects.[101] Another factor that can create lower rates is a molecular structure in which the acidic proton is protected within a molecular cavity (e.g., the in–in and out–in isomers shown on p. 190). See also, the proton sponges mentioned on p. 386. Proton transfers between an acidic and a basic group within the same molecule can also be slow, if the two groups are too far apart for hydrogen bonding. In such cases, participation of solvent molecules may be necessary.

Proton transfers to or from a carbon atom[102] in most cases are much slower than those strictly between oxygen or nitrogen atoms. At least three factors can be responsible for this,[103] not all of them applying in every case:

1. Hydrogen bonding is very weak or altogether absent for carbon (Chapter 3).
2. Many carbon acids, upon losing the proton, form carbanions that are stabilized by resonance. Structural reorganization (movement of atoms to different positions within the molecule) may accompany this. Chloroform, HCN, and 1-alkynes do not form resonance-stabilized carbanions, and these[104] behave kinetically as normal acids.[105] It has been reported that carborane acids, such as $H(CHB_{11}H_5Cl_6)$, are the strongest isolable (Lewis-free) Brønsted acids known.[106]
3. There may be considerable reorganization of solvent molecules around the ion as compared to the neutral molecule.[107]

In connection with factors 2 and 3, it has been proposed[103] that any factor that stabilizes the product (e.g., by resonance or solvation) lowers the rate constant if it develops late on the reaction coordinate, but increases the rate constant if it develops early. This is called the *Principle of Imperfect Synchronization*.

Mechanisms of proton transfer have been studied for many compounds, including the reactions of acids with lactams,[108] amides with various bases,[109] and amines with alkoxide bases.[110]

[100]Chen, X.; Walthall, D.A.; Brauman, J.I. *J. Am. Chem. Soc.* **2004**, *126*, 12614.

[101]Abraham, M. H.; Enomoto, K.; Clarke, E. D.; Sexton, G. *J. Org. Chem.* **2002**, *67*, 4782.

[102]For reviews of proton transfers to and from carbon, see Hibbert, F., in Bamford, C.H.; Tipper, C.F.H. *Chemical Kinetics*, Vol. 8, Elsevier, NY, **1977**, pp. 97–196; Kreevoy, M.M. *Isot. Org. Chem.* **1976**, 2, 1; Leffek, K.T. *Isot. Org. Chem.* **1976**, 2, 89.

[103]See Bernasconi, C.F. *Tetrahedron* **1985**, *41*, 3219.

[104]Bednar, R.A.; Jencks, W.P. *J. Am. Chem. Soc.* **1985**, *107*, 7117, 7126, 7135; Kresge, A.J.; Powell, M.F. *J. Org. Chem.* **1986**, *51*, 822; Formosinho, S.J.; Gal, V.M.S. *J. Chem. Soc. Perkin Trans. 2* **1987**, 1655.

[105]Not all 1-alkynes behave as normal acids; see Aroella, T.; Arrowsmith, C.H.; Hojatti, M.; Kresge, A.J.; Powell, M.F.; Tang, Y.S.; Wang, W. *J. Am. Chem. Soc.* **1987**, *109*, 7198.

[106]Juhasz, M.; Hoffmann, S.; Stoyanov, E.; Kim, K.-C.; Reed, C.A. *Angew. Chem. Int. Ed.* **2004**, *43*, 5352.

[107]See Bernasconi, C.F.; Terrier, F. *J. Am. Chem. Soc.* **1987**, *109*, 7115; Kurz, J.L. *J. Am. Chem. Soc.* **1989**, *111*, 8631.

[108]Wang, W.; Cheng, P.; Huang, C.; Jong, Y. *Bull. Chem. Soc. Jpn.* **1992**, *65*, 562.

[109]Wang, W.-h.; Cheng, C.-c. *Bull. Chem. Soc. Jpn.* **1994**, *67*, 1054.

[110]Lambert, C.; Hampel, F.; Schleyer, P.v.R. *Angew. Chem. Int. Ed.* **1992**, *31*, 1209.

MEASUREMENTS OF SOLVENT ACIDITY[111]

When a solute is added to an acidic solvent it may become protonated by the solvent. This effect can lead to an enhancement of acidity, as in the effect of using formic acid rather than methanol.[112] An acidity scale has been reported for ionic liquids[113] (see p 415 for a discussion of ionic liquids), and the Lewis acidity of ionic liquids has been established using ir.[114] If the solvent is water and the concentration of solute is not very great, then the pH of the solution is a good measure of the proton-donating ability of the solvent. Unfortunately, this is no longer true in concentrated solutions because activity coefficients are no longer unity. A measurement of solvent acidity is needed which works in concentrated solutions and applies to mixed solvents as well. The Hammett acidity function[115] is a measurement that is used for acidic solvents of high dielectric constant.[116] For any solvent, including mixtures of solvents (but the proportions of the mixture must be specified), a value H_0 is defined as

$$H_{\mathrm{o}} = \mathrm{p}K_{\mathrm{BH_w}^+} - \log \frac{[\mathrm{BH}^+]}{[\mathrm{B}]}$$

H_0 is measured by using "indicators" that are weak bases (B) and so are partly converted, in these acidic solvents, to the conjugate acids BH^+. Typical indicators are o-nitroanilinium ion, with a pK in water of -0.29, and 2,4-dinitroanilinium ion, with a pK in water of -4.53. For a given solvent, $[\mathrm{BH}^+]/[\mathrm{B}]$ is measured for one indicator, usually by spectrophotometric means. Then, using the known pK in water (p$K_{\mathrm{BH_w}^+}$) for that indicator, H_0 can be calculated for that solvent system. In practice, several indicators are used, so that an average H_0 is taken. Once H_0 is known for a given solvent system, pK_a values in it can be calculated for any other acid-base pair.

The symbol H_0 is defined as

$$h_{\mathrm{o}} = \frac{a_{\mathrm{H}^+} f\mathrm{I}}{f_{\mathrm{HI}^+}}$$

[111]For fuller treatments, see Hammett, L.P. *Physical Organic Chemistry*, 2nd ed., McGraw-Hill, NY, *1970*, pp. 263–313; Jones, R.A.Y. *Physical and Mechanistic Organic Chemistry*, 2nd ed., Cambridge University Press, Cambridge, *1984*, pp. 83–93; Arnett, E.M.; Scorrano, G. *Adv. Phys. Org. Chem. 1976*, *13*, 83.

[112]Holt, J.; Karty, J.M. *J. Am. Chem. Soc. 2003*, *125*, 2797.

[113]Thomazeau, C.; Olivier-Bourbigou, H.; Magna, L.; Luts, S.; Gilbert, B. J. *Am. Chem. Soc. 2003*, *125*, 5264.

[114]Yang, Y.-l.; Kou, Y. *Chem. Commun. 2004*, 226.

[115]Hammett, L.P.; Deyrup, A.J. *J. Am. Chem. Soc. 1932*, *54*, 2721.

[116]For a monograph on acidity functions, see Rochester, C.H. *Acidity Functions*, Academic Press, NY, *1970*. For reviews, see Ref. 111; Cox, R.A.; Yates, K. *Can. J. Chem. 1983*, *61*, 2225; Boyd R.H., in Coetzee, J.F.; Ritchie, C.D. *Solute–Solvent Interactions*, Marcel Dekker, NY, *1969*, pp. 97–218; Vinnik, M.I. *Russ. Chem. Rev. 1966*, *35*, 802; Liler, M. *Reaction Mechanisms in Sulfuric Acid*, Academic Press, NY, *1971*, pp. 26–58.

where a_{H^+} is the activity of the proton and f_I and f_{HI^+} are the activity coefficients of the indicator and conjugate acid of the indicator,[117] respectively. The parameter H_0 is related to H_0 by

$$H_o = -\log h_o$$

so that H_0 is analogous to pH and H_0 to $[H^+]$, and indeed in dilute aqueous solution $H_o = pH$.

The parameter H_0 reflects the ability of the solvent system to donate protons, but it can be applied only to acidic solutions of high dielectric constant, mostly mixtures of water with acids, such as nitric, sulfuric, perchloric, and so on. It is apparent that the H_0 treatment is valid only when f_I/f_{HI^+} is independent of the nature of the base (the indicator). Since this is so only when the bases are structurally similar, the treatment is limited. Even when similar bases are compared, many deviations are found.[118] Other acidity scales[119] have been set up, including a scale for C—H acids,[120] among them H_- for bases with a charge of -1, H_R for aryl carbinols,[121] H_C for bases that protonate on carbon,[122] and H_A for unsubstituted amides.[123] It is now clear that there is no single acidity scale that can be applied to a series of solvent mixtures, irrespective of the bases employed.[124]

Although most acidity functions have been applied only to acidic solutions, some work has also been done with strongly basic solutions.[125] The H_- function, which is used for highly acidic solutions when the base has a charge of -1, can also be used for strongly basic solvents, in which case it measures the ability of these

---

[117]For a review of activity coefficient behavior of indicators in acid solutions, see Yates, K.; McClelland, R.A. *Prog. Phys. Org. Chem.* **1974**, *11*, 323.

[118]For example, see Kresge, A.J.; Barry, G.W.; Charles, K.R.; Chiang, Y. *J. Am. Chem. Soc.* **1962**, *84*, 4343; Katritzky, A.R.; Waring, A.J.; Yates, K. *Tetrahedron* **1963**, *19*, 465; Arnett, E.M.; Mach, G.W. *J. Am. Chem. Soc.* **1964**, *86*, 2671; Jorgenson, M.J.; Hartter, D.R. *J. Am. Chem. Soc.* **1963**, *85*, 878; Kreevoy, M.M.; Baughman, E.H. *J. Am. Chem. Soc.* **1973**, *95*, 8178; García, B.; Leal, J.M.; Herrero, L.A.; Palacios, J.C. *J. Chem. Soc. Perkin Trans. 2* **1988**, 1759; Arnett, E.M.; Quirk, R.P.; Burke, J.J. *J. Am. Chem. Soc.* **1970**, *92*, 1260.

[119]For lengthy tables of many acidity scales, with references, see Cox, R.A.; Yates, K. *Can. J. Chem.* **1983**, *61*, 2225. For an equation that is said to combine the vast majority of acidity functions, see Zalewski, R.I.; Sarkice, A.Y.; Geltz, Z. *J. Chem. Soc. Perkin Trans. 2* **1983**, 1059.

[120]See Vianello, R.; Maksić, Z.B. *Eur. J. Org. Chem.* **2004**, 5003.

[121]Deno, N.C.; Berkheimer, H.E.; Evans, W.L.; Peterson, H.J. *J. Am. Chem. Soc.* **1959**, *81*, 2344.

[122]Reagan, M.T. *J. Am. Chem. Soc.* **1969**, *91*, 5506.

[123]Yates, K.; Stevens, J.B.; Katritzky, A.R. *Can. J. Chem.* **1964**, *42*, 1957; Yates, K.; Riordan, J.D. *Can. J. Chem.* **1965**, *43*, 2328; Edward, J.T.; Wong, S.C. *Can. J. Chem.* **1977**, *55*, 2492; Liler, M.; Marković, D. *J. Chem. Soc. Perkin Trans. 2* **1982**, 551.

[124]Hammett, L.P. *Physical Organic Chemistry*, 2nd ed., McGraw-Hill, NY, **1970**, p. 278; Rochester, C.H. *Acidity Functions*, Academic Press, NY, **1970**, p. 21.

[125]For another approach to solvent basicity scales, see Catalán, J.; Gómez, J.; Couto, A.; Laynez, J. *J. Am. Chem. Soc.* **1990**, *112*, 1678.

solvents to abstract a proton from a neutral acid BH.[126] When a solvent becomes protonated, its conjugate acid is known as a *lyonium ion*.

Another approach to the acidity function problem was proposed by Bunnett and Olsen,[127] who derived the equation

$$\log \frac{[SH^+]}{[S]} + H_o = \phi(H_o + \log[H^+]) + pK_{SH^+}$$

where S is a base that is protonated by an acidic solvent. Thus the slope of a plot of $\log ([SH^+]/[S]) + H_0$ against $H_0 + \log[H^+]$ is the parameter ϕ, while the intercept is the pK_a of the lyonium ion SH^+ (referred to infinite dilution in water). The value of ϕ expresses the response of the equilibrium $S + H^+ \rightleftharpoons SH^+$ to changing acid concentration. A negative ϕ indicates that the log of the ionization ratio $[SH^+]/[S]$ increases, as the acid concentration increases, more rapidly than $-H_0$. A positive ϕ value indicates the reverse. The Bunnett–Olsen equation given above is a linear free-energy relationship (see p. 405) that pertains to acid-base equilibria. A corresponding equation that applies to kinetic data is

$$\log k_\psi + H_0 = \phi(H_o + \log[H^+]) + \log k_2^0$$

where k_ψ is the pseudo-first-order rate constant for a reaction of a weakly basic substrate taking place in an acidic solution and k_2^0 is the second-order rate constant at infinite dilution in water. In this case, ϕ characterizes the response of the reaction rate to changing acid concentration of the solvent. The Bunnett–Olsen treatment has also been applied to basic media, where, in a group of nine reactions in concentrated NaOMe solutions, no correlation was found between reaction rates and either H_- or stoichiometric base concentration but where the rates were successfully correlated by a linear free-energy equation similar to those given above.[128]

A treatment partially based on the Bunnett–Olsen one is that of Bagno Scorrano, and More O'Ferrall,[129] which formulates medium effects (changes in acidity of solvent) on acid–base equilibria. An appropriate equilibrium is chosen as reference, and the acidity dependence of other reactions compared with it, by use of the linear free-energy equation

$$\log \frac{K'}{K'_0} = m^* \log \frac{K}{K_o}$$

---

[126]For reviews, see Rochester, C.H. *Q. Rev. Chem. Soc.* **1966**, *20*, 511; Rochester, C.H. *Acidity Functions*, Academic Press, NY, **1970**, pp. 234–264; Bowden, K. *Chem. Rev.* **1966**, *66*, 119 (the last review is reprinted in (Coetzee, J.F.; Ritchie, C.D. *Solute–Solvent Interactions*, Marcel Dekker, NY, **1969**, pp. 186–215).

[127]Bunnett, J.F.; McDonald, R.L.; Olsen, F.P. *J. Am. Chem. Soc.* **1974**, *96*, 2855.

[128]More O'Ferrall, R.A. *J. Chem. Soc. Perkin Trans. 2* **1972**, 976.

[129]Bagno, A.; Scorrano, G.; More O'Ferrall, R.A. *Rev. Chem. Intermed.* **1987**, *7*, 313. See also, Sampoli, M.; De Santis, A.; Marziano, N.C. *J. Chem. Soc. Chem. Commun.* **1985**, 110; Cox, R.A. *Acc. Chem. Res.* **1987**, *20*, 27.

where the K values are the equilibrium constants for the following: K for the reaction under study in any particular medium; K' for the reference reaction in the same medium; K_0 for the reaction under study in a reference solvent; K'_0 for the reference reaction in the same reference solvent; and m^* is the slope of the relationship [corresponding to $(1-\phi)$ of the Bunnett–Olsen treatment]. This equation has been shown to apply to many acid–base reactions.

Another type of classification system was devised by Bunnett[130] for reactions occurring in moderately concentrated acid solutions. Log $k_\psi + H_0$ is plotted against log a_{H_2O}, where K_ψ is the pseudo-first-order rate constant for the protonated species and a_{H_2O} is the activity of water. Most such plots are linear or nearly so. According to Bunnett, the slope of this plot w tells something about the mechanism. Where w is between -2.5 and 0, water is not involved in the rate-determining step; where w is between 1.2 and 3.3, water is a nucleophile in the rate-determining step; where w is between 3.3 and 7, water is a proton-transfer agent. These rules hold for acids in which the proton is attached to oxygen or nitrogen.

A new acidity scale has been developed based on calorimetric measurement of N-methylimidazole and N-methylpyrrole in bulk solvents.[131] A revised version of this method was shown to give better results in some cases.[132] Another scale of solvent acidities was developed based on the hydrogen-bond donor acidities in aqueous DMSO.[133] It is noted that bond energies, acidities, and electron affinities are related in a thermodynamic cycle, and Kass and Fattahi have shown that by measuring two of these quantities the third can be found.[134]

ACID AND BASE CATALYSIS[135]

Many reactions are catalyzed by acids, bases, or both. In such cases, the catalyst is involved in a fundamental way in the mechanism. Nearly always the first step of such a reaction is a proton transfer between the catalyst and the substrate.

Reactions can be catalyzed by acid or base in two different ways, called *general* and *specific catalysis*. If the rate of an acid-catalyzed reaction run in a solvent S is

[130]Bunnett, J.F. *J. Am. Chem. Soc.* **1961**, *83*, 4956, 4968, 4973, 4978.

[131]Catalán, J.; Couto, A.; Gomez, J.; Saiz, J.L.; Laynez, J. *J. Chem. Soc. Perkin Trans. 2* **1992**, 1181.

[132]Abraham, M.H.; Taft, R.W. *J. Chem. Soc. Perkin Trans. 2* **1993**, 305.

[133]Liu, P.C.; Hoz, S.; Buncel, E. *Gazz. Chim. Ital.* **1996**, *126*, 31. See also Abraham, M.H.; Zhao, Y.J. *J. Org. Chem.* **2004**, *69*, 4677.

[134]Fattahi, A.; Kass, S.R. *J. Org. Chem.* **2004**, *69*, 9176.

[135]For reviews, see Stewart, R. *The Proton: Applications to Organic Chemistry*, Academic Press, NY, **1985**, pp. 251–305; Hammett, L.P. *Physical Organic Chemistry*, 2nd ed., McGraw-Hill, NY, **1970**, pp. 315–345; Willi, A.V., in Bamford, C.H.; Tipper, C.F.H. *Chemical Kinetics*, Vol. 8, Elsevier, NY, **1977**, pp. 1–95; Jones, R.A.Y. *Physical and Mechanistic Organic Chemistry*, 2nd ed., Cambridge University Press, Cambridge, **1984**, pp. 72–82; Bell, R.P. *The Proton in Chemistry*, 2nd ed., Cornell University Press, Ithaca, NY, **1973**, pp. 159–193; Jencks, W.P. *Catalysis in Chemistry and Enzymology*, McGraw-Hill, NY, **1969**, pp. 163–242; Bender, M.L. *Mechanisms of Homogeneous Catalysis from Protons to Proteins*; Wiley, NY, **1971**, pp. 19–144.

proportional to $[SH^+]$, the reaction is said to be subject to *specific acid catalysis*, the acid being the lyonium ion SH^+. The acid that is put into the solvent may be stronger or weaker than SH^+, but the rate is proportional only to the $[SH^+]$ that is actually present in the solution (derived from $S + HA \rightleftharpoons SH^+ + A^-$). The identity of HA makes no difference except insofar as it determines the position of equilibrium and hence the $[SH^+]$. Most measurements have been made in water, where SH^+ is H_3O^+.

In *general acid catalysis*, the rate is increased not only by an increase in $[SH^+]$, but also by an increase in the concentration of other acids (e.g., in water by phenols or carboxylic acids). These other acids increase the rate even when $[SH^+]$ is held constant. In this type of catalysis the strongest acids catalyze best, so that, in the example given, an increase in the phenol concentration catalyzes the reaction much less than a similar increase in $[H_3O^+]$. This relationship between acid strength of the catalyst and its catalytic ability can be expressed by the *Brønsted catalysis equation*[136]

$$\log k = \alpha \log K_\alpha + C$$

where k is the rate constant for a reaction catalyzed by an acid of ionization constant K_α. According to this equation, when $\log k$ is plotted against $\log K_\alpha$ for catalysis of a given reaction by a series of acids, a straight line should be obtained with slope and intercept C. Although straight lines are obtained in many cases, this is not always the case. The relationship usually fails when acids of different types are compared. For example, it is much more likely to hold for a group of substituted phenols than for a collection of acids that contains both phenols and carboxylic acids. The Brønsted equation is another linear free-energy relationship (see p. 405).

Analogously, there are *general* and *specific* (S^- from an acidic solvent SH) *base-catalyzed reactions*. The Brønsted law for bases is

$$\log k = \beta \log K_b + C$$

The Brønsted equations relate a rate constant k to an equilibrium constant K_a. In Chapter 6, we saw that the Marcus equation also relates a rate term (in that case $\Delta G^{\ddagger}$) to an equilibrium term ΔG°. When the Marcus treatment is applied to proton transfers[137] between a carbon and an oxygen (or a nitrogen), the simplified[138] equation (p. 309)

$$\Delta G^{\ddagger} = \Delta G^{\ddagger}_{int} + \frac{1}{2}\Delta G^\circ + \frac{(\Delta G^\circ)^2}{16\Delta G^{\ddagger}_{int}}$$

[136]For reviews, see Klumpp, G.W. *Reactivity in Organic Chemistry*; Wiley, NY, *1982*, pp. 167–179; Bell, R.P., in Chapman, N.B.; Shorter, J. *Correlation Analysis in Chemistry: Recent Advances*; Plenum Press, NY, *1978*, pp. 55–84; Kresge, A.J. *Chem. Soc. Rev. 1973*, 2, 475.

[137]For applications of Marcus theory to proton transfers, see Marcus, R.A. *J. Phys. Chem. 1968*, 72, 891; Kreevoy, M.M.; Konasewich, D.E. *Adv. Chem. Phys. 1971*, 21, 243; Kresge, A.J. *Chem. Soc. Rev. 1973*, 2, 475.

[138]Omitting the work terms.

where

$$\Delta G_{int}^{\ddagger} = \frac{1}{2}\left(\Delta G_{(0^{\ddagger},0)} + \Delta G_{(c^{\ddagger},c)}\right)$$

can be further simplified: Because proton transfers between oxygen and oxygen (or nitrogen and nitrogen) are much faster than those between carbon and carbon, $\Delta G_{(0,^{\ddagger}0)}$ is much smaller than $\Delta G_{(c,^{\ddagger}c)}$ and we can write[139]

$$\Delta G\ddagger = \frac{1}{2}\Delta G(c,^{\ddagger}c) + \frac{1}{2}\Delta G^{\circ} + \frac{(\Delta G^{\circ})^2}{8\Delta G_{(c,^{\ddagger}c)}}$$

Thus, if the carbon part of the reaction is kept constant and only the A of HA is changed (where A is an oxygen or nitrogen moiety), then $\Delta G^{\ddagger}$ is dependent only on ΔG°. Differentiation of this equation yields the Brønsted α:

$$\frac{d\Delta G\ddagger}{d\Delta G^{\circ}} = \alpha = \frac{1}{2}\left(1 + \frac{\Delta G^{\circ}}{2\Delta G_{(c\ddagger,c)}}\right)$$

The Brønsted law is therefore a special case of the Marcus equation.

A knowledge of whether a reaction is subject to general or specific acid catalysis supplies information about the mechanism. For any acid-catalyzed reaction we can write

Step 1 $A \overset{SH^+}{\rightleftharpoons} AH^+$
Step 2 $AH^+ \rightarrow$ products

If the reaction is catalyzed only by the specific acid SH^+, it means that step 1 is rapid and step 2 is rate-controlling, since an equilibrium has been rapidly established between A and the strongest acid present in the solution, namely, SH^+ (since this is the strongest acid that can be present in S). On the other hand, if step 2 is faster, there is no time to establish equilibrium and the rate-determining step must be step 1. This step is affected by all the acids present, and the rate reflects the sum of the effects of each acid (general acid catalysis). General acid catalysis is also observed if the slow step is the reaction of a hydrogen-bond complex A•••HB, since each complex reacts with a base at a different rate. A comparable discussion can be used for general and specific base catalysis.[140] Further information can be obtained from the values α and β in the Brønsted catalysis equations, since these are

[139]Albery, W.J. *Annu. Rev. Phys. Chem.* **1980**, *31*, 227, p. 244.
[140]For discussions of when to expect general or specific acid or base catalysis, see Jencks, W.P. *Acc. Chem. Res.* **1976**, *9*, 425; Stewart, R.; Srinivasan, R. *Acc. Chem. Res.* **1978**, *11*, 271; Guthrie, J.P. *J. Am. Chem. Soc.* **1980**, *102*, 5286.

approximate measures of the extent of proton transfer in the transition state. In most cases values of α and β are between 1 and 0. A value of α or β near 0 is generally taken to mean that the transition state resembles the reactants; that is, the proton has been transferred very little when the transition state has been reached. A value of α or β near 1 is taken to mean the opposite; that is, in the transition state the proton has been almost completely transferred. However, cases are known in which these generalizations are not followed,[141] and their theoretical basis has been challenged.[142] In general, the proton in the transition state lies closer to the weaker base.

LEWIS ACIDS AND BASES: HARD AND SOFT ACIDS AND BASES

At about the same time that Brønsted proposed his acid–base theory, Lewis put forth a broader theory. A base in the Lewis theory is the same as in the Brønsted one, namely, a compound with an available pair of electrons, either unshared or in a π orbital. A *Lewis acid*, however, is any species with a vacant orbital.[143] In a Lewis acid–base reaction the unshared pair of the base forms a covalent bond with the vacant orbital of the acid, as represented by the general equation

$$A + :B \longrightarrow A–B$$

in which charges are not shown, since they may differ. A specific example is

$$BF_3 + :NH_3 \longrightarrow F_3\overset{\ominus}{B} \to \overset{\oplus}{N}H_3$$

In the Brønsted picture, the acid is a proton donor, but in the Lewis picture the proton itself is the acid since it has a vacant orbital. A Brønsted acid becomes, in the Lewis picture, the compound that gives up the actual acid. The advantage of the Lewis theory is that it correlates the behavior of many more processes. For example, $AlCl_3$ and BF_3 are Lewis acids because they have only six electrons in the outer shell and have room for eight. Lewis acids $SnCl_4$ and SO_3 have eight, but their central elements, not being in the first row of the Periodic table, have room for 10 or 12. Other Lewis acids are simple cations, like Ag^+. The simple reaction

[141]See, for example, Bordwell, F.G.; Boyle, Jr., W.J. *J. Am. Chem. Soc.* **1972**, *94*, 3907; Davies, M.H. *J. Chem. Soc. Perkin Trans. 2* **1974**, 1018; Agmon, N. *J. Am. Chem. Soc.* **1980**, *102*, 2164; Murray, C.J.; Jencks, W.P. *J. Am. Chem. Soc.* **1988**, *110*, 7561.

[142]Pross, A.; Shaik, S.S. *New J. Chem.* **1989**, *13*, 427; Lewis, E.S. *J. Phys. Org. Chem.* **1990**, *3*, 1.

[143]For a monograph on Lewis acid–base theory, see Jensen, W.B. *The Lewis Acid–Base Concept*, Wiley, NY, **1980**. For a discussion of the definitions of Lewis acid and base, see Jensen, W.B. *Chem. Rev.* **1978**, *78*, 1.

$A + \bar{B} \rightarrow A{-}B$ is not very common in organic chemistry, but the scope of the Lewis picture is much larger because reactions of the types

$$A^1 \quad + \quad A^2{-}B \longrightarrow A^1{-}B \quad + \quad A^2$$
$$B^1 \quad + \quad A{-}B^2 \longrightarrow A{-}B^1 \quad + \quad B^2$$
$$A^1{-}B^1 \quad + \quad A^2{-}B^2 \longrightarrow A^1{-}B^2 \quad + \quad A^2{-}B^1$$

which are very common in organic chemistry, are also Lewis acid-base reactions. In fact, all reactions in which a covalent bond is formed through one species contributing a filled and the other a vacant orbital may be regarded as Lewis acid–base reactions. An *ab initio* analysis of the factors that determine Lewis versus Lowry–Brønsted acidity–basicity is available.[144]

When a Lewis acid combines with a base to give a negative ion in which the central atom has a higher than normal valence, the resulting salt is called an *ate complex*.[145] Examples are

$$Me_3B \quad + \quad LiMe \longrightarrow Me_4B^- \ Li^+$$
An ate complex

$$Ph_5Sb \quad + \quad LiPh \longrightarrow Ph_6Sb^- \ Li^+$$
An ate complex

The ate complexes are analogous to the onium salts formed when a Lewis base expands its valence, for example,

$$Me_3N \quad + \quad MeI \longrightarrow Me_4N^+ \ I^-$$
Onium salt

Far fewer quantitative measurements have been made of Lewis acid strength compared to that of Brønsted acids.[146] A simple table of Lewis acidities based on some quantitative measurement (e.g., that given for Brønsted acids in Table 8.1) is not feasible because Lewis acidity depends on the nature of the base and any solvent that can function as a base. For example, lithium perchlorate functions as a weak Lewis acid in ether.[147] Qualitatively, the following approximate sequence

[144]Rauk, A.; Hunt, I.R.; Keay, B.A. *J. Org. Chem.* **1994**, *59*, 6808.

[145]For a review of ate complexes, see Wittig, G. *Q. Rev. Chem. Soc.* **1966**, *20*, 191.

[146]For reviews of the quantitative aspects of Lewis acidity, see Satchell, D.P.N.; Satchell, R.S. *Q. Rev. Chem. Soc.* **1971**, *25*, 171; *Chem. Rev.* **1969**, *69*, 251. See also Maria, P.; Gal, J. *J. Phys. Chem.* **1985**, *89*, 1296; Larson, J.W.; Szulejko, J.E.; McMahon, T.B. *J. Am. Chem. Soc.* **1988**, *110*, 7604; Sandström, M.; Persson, I.; Persson, P. *Acta Chem. Scand.* **1990**, *44*, 653; Laszlo, P.; Teston-Henry, M. *Tetrahedron Lett.* **1991**, *32*, 3837.

[147]Springer, G.; Elam, C.; Edwards, A.; Bowe, C.; Boyles, D.; Bartmess, J.; Chandler, M.; West, K.; Williams, J.; Green, J.; Pagni, R.M.; Kabalka, G.W. *J. Org. Chem.* **1999**, *64*, 2202.

of acidity of Lewis acids of the type MX_n has been suggested, where X is a halogen atom or an inorganic radical: $BX_3 > AlX_3 > FeX_3 > GaX_3 > SbX_5 > SnX_4 > AsX_5 > ZnX_2 > HgX_2$.

The facility with which an acid–base reaction takes place depends of course on the strengths of the acid and the base. But it also depends on quite another quality, called the *hardness*[148] or *softness* of the acid or base.[149] Hard and soft acids and bases have these characteristics:

Soft Bases. The donor atoms are of low electronegativity and high polarizability and are easy to oxidize. They hold their valence electrons loosely.

Hard Bases. The donor atoms are of high electronegativity and low polarizability and are hard to oxidize. They hold their valence electrons tightly.

Soft Acids. The acceptor atoms are large, have low positive charge, and contain unshared pairs of electrons (*p* or *d*) in their valence shells. They have high polarizability and low electronegativity.

Hard Acids. The acceptor atoms are small, have high positive charge, and do not contain unshared pairs in their valence shells. They have low polarizability and high electronegativity.

A qualitative listing of the hardness of some acids and bases is given in Table 8.3.[150] The treatment has also been made quantitative,[151] with the following operational definition:

$$\eta = \frac{I - A}{2}$$

In this equation, η, the *absolute hardness*, is half the difference between I, the ionization potential, and A, the electron affinity.[152] The softness, σ, is the reciprocal of

[148]See Ayers, P.W.; Parr, R.G. *J. Am. Chem. Soc.* **2000**, *122*, 2010.

[149]Pearson, R.G. *J. Am. Chem. Soc.* **1963**, *85*, 3533; *Science*, **1966**, *151*, 172; Pearson, R.G.; Songstad, J. *J. Am. Chem. Soc.* **1967**, *89*, 1827. For a monograph on the concept, see Ho, T. *Hard and Soft Acids and Bases Principle in Organic Chemistry*; Academic Press, NY, **1977**. For reviews, see Pearson, R.G. *J. Chem. Educ.* **1987**, *64*, 561; Ho, T. *Tetrahedron* **1985**, *41*, 1, *J. Chem. Educ.* **1978**, *55*, 355; *Chem. Rev.* **1975**, *75*, 1; Pearson, R.G. in Chapman, N.B.; Shorter, J. *Advances in Linear Free-Energy Relationships*, Plenum Press, NY, **1972**, pp. 281–319; Pearson, R.G. *Surv. Prog. Chem.* **1969**, *5*, 1 [portions of this article slightly modified also appear in Pearson, R.G. *J. Chem. Educ.* **1968**, *45*, 581, 643]; Garnovskii, A.D.; Osipov, O.A.; Bulgarevich, S.B. *Russ. Chem. Rev.* **1972**, *41*, 341; Seyden-Penne, J. *Bull. Soc. Chim. Fr.* **1968**, 3871. For a collection of papers, see Pearson, R.G. *Hard and Soft Acids and Bases*, Dowden, Hutchinson, and Ross, Stroudsberg, PA, **1973**.

[150]Taken from larger listings, in Pearson, R.G. Ref. 149.

[151]Pearson, R.G. *Inorg. Chem.* **1988**, *27*, 734; *J. Org. Chem.* **1989**, *54*, 1423. See also, Orsky, A.R.; Whitehead M.A. *Can. J. Chem.* **1987**, *65*, 1970.

[152]For a computational study of proton and electron affinities see Sauers, R.R. *Tetrahedron* **1999**, *55*, 10013.

TABLE 8.3. Hard and Soft Acids and Bases[150]

Hard Bases	Soft Bases	Borderline Cases
H_2O	OH^- F^-	R_sS RSH RS^- $ArNH_2C_5H_5N$
AcO^-	SO_4^{2-}	Cl^- I^- R_3P $(RO)_3P$
	N_3^-	Br^-
CO_3^{2-}	NO_3^-	ROH CN^- RCN CO NO_2^-
RO^-	R_2O	NH_3 C_2H_4 C_6H_6
RNH_2	H^-	R^-

Hard Acids	Soft Acids	Borderline Cases
H^+	Li^+	Na^+ Cu^+ Ag^+ Pd^{2+} Fe^{2+}
	Co^{2+}	Cu^{2+}
K^+	Mg^{2+}	Ca^{2+} Pt^{2+} Hg^{2+} BH_3 Zn^{2+}
	Sn^{2+}	Sb^{3+}
Al^{3+}	Cr^{2+}	Fe^{3+} $GaCl_3$ I_2 Br_2 Bi^{3+}
	BMe_3	SO_2
BF_3	$B(OR)_3$	$AlMe_3$ CH_2 Carbenes
	R_3C^+	NO^+ GaH_3
$AlCl_3$	AlH_3	SO_3 $C_6H_5^+$
RCO^+	CO_2	
HX (hydrogen-bonding molecules)		

η. Values of η for some molecules and ions are given in Table 8.4.[153] Note that the proton, which is involved in all Brønsted acid–base reactions, is the hardest acid listed, with $\eta = \infty$ (it has no ionization potential). The above equation cannot be applied to anions, because electron affinities cannot be measured for them. Instead, the assumption is made that η for an anion X^- is the same as that for the radical $X^{\bullet}$.[154] Other methods are also needed to apply the treatment to polyatomic cations.[154]

Once acids and bases have been classified as hard or soft, a simple rule can be given: *hard acids prefer to bond to hard bases, and soft acids prefer to bond to soft bases (the HSAB principle)*.[155] The rule has nothing to do with acid or base *strength* but merely says that the product A—B will have extra stability if both A and B are hard or if both are soft. Another rule is that a soft Lewis acid and a soft Lewis base tend to form a covalent bond, while a hard acid and a hard base tend to form ionic bonds.

One application of the first rule given above is found in complexes between alkenes or aromatic compounds and metal ions (p. 376). Alkenes and aromatic rings

---

[153]Note that there is not always a strict correlation between the values in Table 8.3 and the categories of Table 8.2.

[154]Pearson, R.G. *J. Am. Chem. Soc.* **1988**, *110*, 7684.

[155]For proofs of this principle, see Chattaraj, P.K.; Lee, H.; Parr, R.G. *J. Am. Chem. Soc.* **1991**, *113*, 1855.

TABLE 8.4. Some Absolute Hardness Values in Electron Volts[151]

Cations		Molecules		Anions[b]	
Ion	η	Compound	η	Ion	η
H^+	∞	HF	11.0	F^-	7.0
Al^{3+}	45.8	CH_4	10.3	H^-	6.4
Li^+	35.1	BF_3	9.7	OH^-	5.7
Mg^{2+}	32.6	H_2O	9.5	NH_2^-	5.3
Na^+	21.1	NH_3	8.2	CN^-	5.1
Ca^{2+}	19.5	HCN	8.0	CH_3^-	4.9
K^+	13.6	$(CH_3)_2O$	8.0	Cl^-	4.7
Zn^{2+}	10.9	CO	7.9	$CH_3CH_2^-$	4.4
Cr^{3+}	9.1	C_2H_2	7.0	Br^-	4.2
Cu^{2+}	8.3	$(CH_3)_3N$	6.3	$C_6H_5^-$	4.1
Pt^{2+}	8.0	H_2S	6.2	SH^-	4.1
Sn^{2+}	7.9	C_2H_4	6.2	$(CH_3)_2CH^-$	4.0
Hg^{2+}	7.7	$(CH_3)_2S$	6.0	I^-	3.7
Fe^{2+}	7.2	$(CH_3)_3P$	5.9	$(CH_3)_3C^-$	3.6
Pd^{2+}	6.8	CH_3COCH_3	5.6		
Cu^+	6.3	C_6H_6	5.3		
		HI	5.3		
		C_5H_5N	5.0		
		C_6H_5OH	4.8		
		CH_2^a	4.7		
		C_6H_5SH	4.6		
		Cl_2	4.6		
		$C_6H_5NH_2$	4.4		
		Br_2	4.0		
		I_2	3.4		

[a]For singlet state.
[b]The same as for the corresponding radical.

are soft bases and should prefer to complex with soft acids. Thus, Ag^+, Pt^{2+}, and Hg^{2+} complexes are common, but complexes of Na^+, Mg^{2+}, or Al^{3+} are rare. Chromium complexes are also common, but in such complexes the chromium is in a low or zero oxidation state (which softens it) or attached to other soft ligands. In another application, we may look at this reaction:

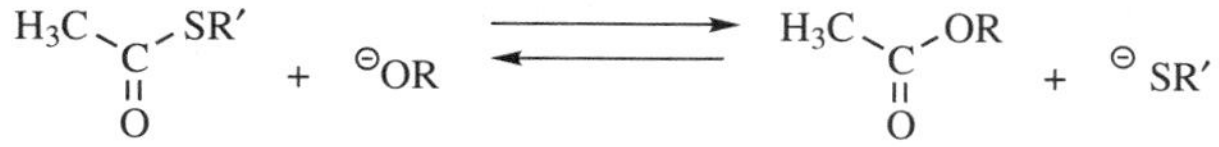

The HSAB principle predicts that the equilibrium should lie to the right, because the hard acid CH_3CO^+ should have a greater affinity for the hard base RO^- than for the soft base RS^-. Indeed, thiol esters are easily cleaved by RO^- or hydrolyzed by dilute base ($^-$OH is also a hard base).[156] Another application of the rule is discussed on p. 493.[157] The HSAB principles have been applied to analyze the reactivity of ketone and ester enolates,[158] and in analyzing catalyst selectivity in synthesis.[159]

THE EFFECTS OF STRUCTURE ON THE STRENGTHS OF ACIDS AND BASES[160]

The structure of a molecule can affect its acidity or basicity in a number of ways. Unfortunately, in most molecules two or more of these effects (as well as solvent effects) are operating, and it is usually very difficult or impossible to say how much each effect contributes to the acid or base strength.[161] Small differences in acidity or basicity between similar molecules are particularly difficult to interpret. It is well to be cautious when attributing them to any particular effect.

1. *Field Effects.* These were discussed on p. 19. In general, changes in substituents can have an effect on acidity. As an example of the influence of field effects on acidity, we may compare the acidity of acetic acid and nitroacetic acid:

$pK_a = 4.76$ $pK_a = 1.68$

The only difference in the structure of these molecules is the substitution of NO_2 for H. Since NO_2 is a strongly electron-withdrawing group, it withdraws electron density from the negatively charged COO^- group in the anion of

---

[156]Wolman, Y., in Patai, S. *The Chemistry of the Thiol Group*, pt. 2, Wiley, NY, *1974*, p. 677; Maskill, H. *The Physical Basis of Organic Chemistry*, Oxford University Press, Oxford *1985*, p. 159.

[157]See also Bochkov, A.F. *J. Org. Chem. USSR* *1986*, 22, 1830, 1837.

[158]Méndez, F.; Gázquez, J.L. *J. Am. Chem. Soc.* *1994*, *116*, 9298.

[159]Woodward, S. *Tetrahedron* *2002*, *58*, 1017.

[160]For a monograph, see Hine, J. *Structural Effects on Equilibria in Organic Chemistry*, Wiley, NY, *1975*. For reviews, see Taft, R.W. *Prog. Phys. Org. Chem.* *1983*, *14*, 247; Petrov, E.S. *Russ. Chem. Rev.* *1983*, *52*, 1144 (NH acids); Bell, R.P. *The Proton in Chemistry*, 2nd ed., Cornell University Press, Ithaca, NY, *1973*, pp. 86–110; Barlin, G.B.; Perrin, D.D., in Bentley, K.W.; Kirby, G.W. *Elucidation of Organic Structures by Physical and Chemical Methods*, 2nd ed. (Vol. 4 of Weissberger, A. *Techniques of Chemistry*), pt. 1; Wiley, NY, *1972*, pp. 611–676. For discussions, see Bolton, P.D.; Hepler, L.G. *Q. Rev. Chem. Soc.* *1971*, *25*, 521; Barlin, G.B.; Perrin, D.D. *Q. Rev. Chem. Soc.* *1966*, *20*, 75; Thirot, G. *Bull. Soc. Chim. Fr.* *1967*, 3559; Liler, M. *Reaction Mechanisms in Sulfuric acid*, Academic Press, NY, *1971*, pp. 59–144. For a monograph on methods of estimating pK values by analogy, extrapolation, and so on, see Perrin, D.D.; Dempsey, B.; Serjeant, E.P. *pKa Prediction for Organic Acids and Bases*, Chapman and Hall, NY, *1981*.

[161]The varying degrees by which the different factors that affect gas-phase acidities of 25 acids has been calculated: Taft, R.W.; Koppel, I.A.; Topsom, R.D.; Anvia, F. *J. Am. Chem. Soc.* *1990*, *112*, 2047.

nitroacetic acid (compared with the anion of acetic acid) and, as the pK_a values indicate, nitroacetic acid is ~1000 times stronger than acetic acid.[162] Any effect that results in electron withdrawal from a negatively charged center is a stabilizing effect because it spreads the charge. Thus, $-I$ groups increase the acidity of uncharged acids, such as acetic because they spread the negative charge of the anion. However, $-I$ groups also increase the acidity of any acid, no matter what the charge. For example, if the acid has a charge of $+1$ (and its conjugate base is therefore uncharged), a $-I$ group destabilizes the positive center (by increasing and concentrating the positive charge) of the acid, a destabilization that will be relieved when the proton is lost. In general, we may say that *groups that withdraw electrons by the field effect increase acidity and decrease basicity, while electron-donating groups act in the opposite direction*. Another example is the molecule $(C_6F_5)_3CH$, which has three strongly electron-withdrawing C_6F_5 groups and a pK_a of 16,[163] compared with Ph_3CH, with a pK_a of 31.5 (Table 8.1), an acidity enhancement of $\sim10^{15}$. Table 8.5 shows pK_a values for some acids. An approximate idea of field effects can be obtained from this table. In the case of the chlorobutyric acids note how the effect decreases with distance. It must be remembered, however, that field effects are not the sole cause of the acidity differences noted and that in fact solvation effects may be more important in many cases (see pp. 390–394).[164] The influence of various substituents on the acidity of acetic acid has been calculated,[165] Substituent effects for weak acids, such as phenols and benzyl alcohols, have been discussed.[166]

$$X = o, m, p$$

6

Field effects are important in benzoic acid derivatives, and the pK_a of the acid will vary with the nature and placement of the "X" group in **6**.[167] The pK_a of 3-OMe **6** is 5.55, but 4-OMe **6** is 6.02 in 50% aq. methanol,[168] compared with a pK_a of 5.67 when X = H. When X = 4-NO_2, the pK_a is 4.76 and 4-Br is 5.36.[157] The pK_a of 2,6-diphenylbenzoic acid is 6.39.[169]

[162]For a review of the enhancement of acidity by NO_2, see Lewis, E.S., in Patai, S. *The Chemistry of Functional Groups, Supplement F*, pt. 2, Wiley, NY, *1982*, pp. 715–729.

[163]Filler, R.; Wang, C. *Chem. Commun. 1968*, 287.

[164]For discussions, see Edward, J.T. *J. Chem. Educ. 1982*, *59*, 354; Schwartz, L.M. *J. Chem. Educ. 1981*, *58*, 778.

[165]Headley, A.D.; McMurry, M.E.; Starnes, S.D. *J. Org. Chem. 1994*, *59*, 1863.

[166]Wiberg, K.B. *J. Org. Chem. 2003*, *68*, 875.

[167]For calculated gas-phase acidities of substituted benzoic acids, see Wiberg, K. B. *J. Org. Chem. 2002*, *67*, 4787.

[168]DeMaria, P.; Fontana, A.; Spinelli, D.; Dell'Erba, C.; Novi, M.; Petrillo, G.; Sancassan, F. *J. Chem. Soc. Perkin Trans. 2 1993*, 649.

[169]Chen, C.-T.; Siegel, J.S. *J. Am. Chem. Soc. 1994*, *116*, 5959. Several 2,6-diaryl derivatives are also reported. See also Sotomatsu, T.; Shigemura, M.; Murata, Y.; Fujita, T. *Bull. Chem. Soc. Jpn. 1992*, *65*, 3157.

TABLE 8.5. The pK Values for Some Acids[40]

Acid	pK	Acid	pK
HCOOH	3.77	ClCH$_2$COOH	2.86
CH$_3$COOH	4.76	Cl$_2$CHCOOH	1.29
CH$_3$CH$_2$COOH	4.88	Cl$_3$COOH	0.65
CH$_3$(CH$_2$)$_n$COOH	4.82–4.95		
($n = 2$–7)		O$_2$NCH$_2$COOH	1.68
(CH$_3$)$_2$CHCOOH	4.86	(CH$_3$)$_3$N^{H}CH$_2$COOH	1.83
(CH$_3$)$_3$CCOOH	5.05	HOOCCH$_2$COOH	2.83
		PhCH$_2$COOH	4.31
FCH$_2$COOH	2.66		
ClCH$_2$COOH	2.86	$^-$OOCCH$_2$COOH	5.69
BrCH$_2$COOH	2.86		
ICH$_2$COOH	3.12	$^-$O$_3$SCH$_2$COOH	4.05
		HOCH$_2$COOH	3.83
ClCH$_2$CH$_2$CH$_2$COOH	4.52	H$_2$C=CHCH$_2$COOH	4.35
CH$_3$CHClCH$_2$COOH	4.06		
CH$_3$CH$_2$CHClCOOH	2.84		

2. *Resonance Effects.* Resonance that stabilizes a base but not its conjugate acid results in the acid having a higher acidity than otherwise expected and vice versa. An example is found in the higher acidity of carboxylic acids[170] compared with primary alcohols.

The RCOO$^-$ ion is stabilized by resonance not available to the RCH$_2$O$^-$ ion (or to RCOOH).[171] Note that the RCOO$^-$ is stabilized not only by the fact that there are two equivalent canonical forms, but also by the fact that the negative charge is spread over both oxygen atoms and is therefore less

[170]See Exner, O.; Čársky, P. *J. Am. Chem. Soc.* **2001**, *123*, 9564. See also, Liptak, M.D.; Shields, G.C. *J. Am. Chem. Soc.* **2001**, *123*, 7314.

[171]It has been contended that resonance delocalization plays only a minor role in the increased strength of carboxylic acids compared to alcohols, and the "...higher acidity of acids arises principally because the electrostatic potential of the acidic hydrogens is more positive in the neutral acid molecule...": Siggel, M.R.; Streitwieser, Jr., A.; Thomas, T.D. *J. Am. Chem. Soc.* **1988**, *110*, 8022; Thomas, T.D.; Carroll, T.X.; Siggel, M.R. *J. Org. Chem.* **1988**, *53*, 1812. For contrary views, see Exner, O. *J. Org. Chem.* **1988**, *53*, 1810; Dewar, M.J.S.; Krull, K.L. *J. Chem. Soc. Chem. Commun.* **1990**, 333; Perrin, D.D. *J. Am. Chem. Soc.* **1991**, *113*, 2865. See also, Godfrey, M. *Tetrahedron Lett.* **1990**, *31*, 5181.

concentrated than in RCH_2O^-. The same effect is found in other compounds containing a $C=O$ or CN group. Thus amides $RCONH_2$ are more acidic than amines RCH_2NH_2; esters $RCH_2\text{-}COOR'$ than ethers RCH_2CH_2OR'; and ketones RCH_2COR' than alkanes RCH_2CH_2R' (Table 8.1). The effect is enhanced when two carbonyl groups are attached to the same carbon (because of additional resonance and spreading of charge); for example, β-keto esters (see **7**) are more acidic than simple ketones or carboxylic esters (Table 8.1). The influence of substituents in the α-position of substituted ethyl acetate derivatives has been studied.[172] Extreme examples of this effect are found in the molecules tricyanomethane $(NC)_3CH$, with a pK_a of -5 (Table 8.1, p. 359), and 2-(dicyanomethylene)-1,1,3,3-tetracyanopropene$(NC)_2$ $C=C[CH(CN)_2]_2$, whose first pK_a is below -8.5 and whose second pK_a is -2.5.

Resonance effects are also important in aromatic amines. m-Nitroaniline is a weaker base than aniline, a fact that can be accounted for by the $-I$ effect of the nitro group. But p-nitroaniline is weaker still, though the $-I$ effect should be less because of the greater distance. It is noted that the pK_a values reported are those of the conjugate acid, the ammonium ion.[173] We can explain this result by taking into account the canonical form **A**. Because **A** contributes to the resonance hybrid,[174] the electron density of the unshared pair is lower in p-nitroaniline than in m-nitroaniline, where a canonical form such as **A** is impossible. The basicity is lower in the

| 4.60 | 2.47 | 1.11 | A |

para compound for two reasons, both caused by the same effect: (*1*) the unshared pair is less available for attack by a proton, and (2) when the conjugate acid is formed, the resonance stabilization afforded by **A** is no longer available because the previously unshared pair is now being shared by the proton. The acidity of phenols is affected by substituents in a similar manner.[175]

[172]Goumont, R.; Magnier, E.; Kizilian, E.; Terrier, F. *J. Org. Chem.* **2003**, *68*,6566.
[173]Smith, J.W. in Patai, S. *The Chemistry of the Amino Group;* Wiley, NY, **1968**, pp. 161–204.
[174]See, however, Lipkowitz, K.B. *J. Am. Chem. Soc.* **1982**, *104*, 2647; Krygowski, T.M.; Maurin, J. *J. Chem. Soc. Perkin Trans. 2* **1989**, 695.
[175]Liptak, M.D.; Gross, K.C.; Seybold, P.G.; Feldus, S.; Shields, G.C. *J. Am. Chem. Soc.* **2002**, *124*, 6421.

In general, resonance effects lead to the same result as field effects. That is, here too, electron-withdrawing groups increase acidity and decrease basicity, and electron-donating groups act in the opposite manner. As a result of both resonance and field effects, charge dispersal leads to greater stability.

3. *Periodic Table Correlations.* When comparing Brønsted acids and bases that differ in the position of an element in the periodic table:

a. Acidity increases and basicity decreases in going from left to right across a row of the Periodic table. Thus acidity increases in the order $CH_4 <$ $NH_3 < H_2O < HF$, and basicity decreases in the order $^-CH_3 > ^-NH_2 >$ $^-OH > ^-F$. This behavior can be explained by the increase in electronegativity upon going from left to right across the table. It is this effect that is responsible for the great differences in acidity between carboxylic acids, amides, and ketones: $RCOOH \gg RCONH_2 \gg RCOCH_3$.

b. Acidity increases and basicity decreases in going down a column of the periodic table, despite the decrease in electronegativity. Thus acidity increases in the order $HF < HCl < HBr < HI$ and $H_2O < H_2S$, and basicity decreases in the order $NH_3 > PH_3 > AsH_3$. This behavior is related to the size of the species involved. Thus, for example, F^-, which is much smaller than I^-, attracts a proton much more readily because its negative charge occupies a smaller volume and is therefore more concentrated (note that F^- is also much harder than I^- and is thus more attracted to the hard proton; see p. $$$). This rule does not always hold for positively charged acids. Thus, although the order of acidity for the group 16 hydrides is $H_2O < H_2S < H_2Se$, the acidity order for the positively charged ions is $H_3O^+ > H_3S^+ > H_3Se^+$.[176]

Lewis acidity is also affected by Periodic table considerations. In comparing acid strengths of Lewis acids of the form MX_n:[146]

c. Acids that require only one electron pair to complete an outer shell are stronger than those that require two. Thus $GaCl_3$ is stronger than $ZnCl_2$. This results from the relatively smaller energy gain in adding an electron pair that does not complete an outer shell and from the buildup of negative charge if two pairs come in.

d. Other things being equal, the acidity of MX_n decreases in going down the periodic table because as the size of the molecule increases, the attraction between the positive nucleus and the incoming electron pair is weaker. Thus BCl_3 is a stronger acid than $AlCl_3$.[177]

4. *Statistical Effects.* In a symmetrical diprotic acid, the first dissociation constant is twice as large as expected since there are two equivalent ionizable

---

[176]Taft, R.W. *Prog. Phys. Org. Chem.* **1983**, *14*, 247, see pp. 250–254.

[177]Note that Lewis acidity *decreases*, whereas Brønsted acidity *increases*, going down the table. There is no contradiction here when we remember that in the Lewis picture the actual acid in all Brønsted acids is the same, namely, the proton. In comparing, say, HI and HF, we are not comparing different Lewis acids but only how easily F^- and I^- give up the proton.

TABLE 8.6. Bases Listed in Increasing Order of Base Strength when Compared with Certain Reference Acids

Increasing Order of Base Strength[a]	Reference Acid			
	H$^+$ or BMe$_3$	BMe$_3$	B(CMe$_3$)$_3$	
	NH$_3$	Et$_3$N	Me$_3$N	Et$_3$N
	Me$_3$N	NH$_3$	Me$_2$NH	Et$_2$NH
	MeNH$_2$	Et$_2$NH	NH$_3$	EtNH$_2$
	Me$_2$NH	EtNH$_2$	MeNH$_2$	NH$_3$

[a]The order of basicity (when the reference acids were boranes) was determined by the measurement of dissociation pressures

hydrogens, while the second constant is only half as large as expected because the conjugate base can accept a proton at two equivalent sites. So K_1/K_2 should be 4, and approximately this value is found for dicarboxylic acids where the two groups are sufficiently far apart in the molecule that they do not influence each other. A similar argument holds for molecules with two equivalent basic groups.[178]

5. *Hydrogen Bonding.* Internal hydrogen bonding can greatly influence acid or base strength. For example, the pK for *o*-hydroxybenzoic acid is 2.98, while the value for the para isomer is 4.58. Internal hydrogen bonding between the OH and COO$^-$ groups of the conjugate base of the ortho isomer stabilizes it and results in an increased acidity.

6. *Steric Effects.* The proton itself is so small that direct steric hindrance is seldom encountered in proton transfers. Steric effects are much more common in Lewis acid–base reactions in which larger acids are used. Spectacular changes in the order of base strength have been demonstrated when the size of the acid was changed. Table 8.6 shows the order of base strength of simple amines when compared against acids of various size.[179] It can be seen that the usual order of basicity of amines (when the proton is the reference acid) can be completely inverted by using a large enough acid. The strain caused by formation of a covalent bond when the two atoms involved each have three large groups is called *face strain* or *F strain*.

Steric effects can indirectly affect acidity or basicity by affecting the resonance (see p. 48). For example, *o-tert*-butylbenzoic acid is ∼10 times as strong as the para isomer, because the carboxyl group is forced out of the plane by the *tert*-butyl group. Indeed, virtually all ortho benzoic acids are stronger than the corresponding para isomers, regardless of whether the group on the ring is electron-donating or electron-withdrawing.

[178]The effect discussed here is an example of a symmetry factor. For an extended discussion, see Eberson, L., in Patai, S. *The Chemistry of Carboxylic Acids and Esters*, Wiley, NY, *1969*, pp. 211–293.

[179]Brown, H.C. *J. Am. Chem. Soc. 1945, 67*, 378, 1452, *Boranes in Organic Chemistry*, Cornell University Press, Ithaca, NY, *1972*, pp. 53–64. See also, Brown, H.C.; Krishnamurthy, S.; Hubbard, J.L. *J. Am. Chem. Soc. 1978, 100*, 3343.

Steric effects can also be caused by other types of strain. 1,8-Bis(diethyla-mino)-2,7-dimethoxynaphthalene (**8**) is an extremely strong base for a tertiary amine (pK_a of the conjugate acid = 16.3; compare *N,N*-dimethylaniline, pK_a = 5.1), but proton transfers to and from the nitrogen are exceptionally slow; slow enough to be followed by a UV spectrophotometer.[180] Compound **8** is severely strained because the two nitrogen lone pairs are forced to be near each other.[181] Protonation relieves the strain: one lone pair is now connected to a hydrogen, which forms a hydrogen bond to the other lone pair (shown in **9**).

The same effects are found in 4,5-bis(dimethylamino)fluorene (**10**)[182] and 4,5-bis(dimethylamino)phenanthrene (**11**).[183] Compounds, such as **8**, **10**, and **11**, are known as *proton sponges*.[184] The basicity of proton sponge has been calculated as the sum of the proton affinity[152] of an appropriate reference monoamine, the strain released on protonation, and the energy of the intramolecular hydrogen bond formed on protonation.[185] Another type of proton sponge is quino[7,8-*h*]quinoline (**12**).[186] Protonation of this compound also gives a stable mono-protonated ion similar to **9**, but the steric hindrance found in **8**, **10**, and **11** is absent. Therefore, **12** is a much stronger base than quinoline (**13**) (pK_a values of the conjugate acids are 12.8 for **12** and 4.9 for **13**), but proton transfers are not abnormally slow. A cyclam-like macrocyclic

[180]Barnett, G.H.; Hibbert, F. *J. Am. Chem. Soc.* **1984**, *106*, 2080; Hibbert, F.; Simpson, G.R. *J. Chem. Soc. Perkin Trans. 2* **1987**, 243, 613.

[181]For a review of the effect of strain on amine basicities, see Alder, R.W. *Chem. Rev.* **1989**, *89*, 1215.

[182]Staab, H.A.; Saupe, T.; Krieger, C. *Angew. Chem. Int. Ed.* **1983**, *22*, 731.

[183]Saupe, T.; Krieger, C.; Staab, H.A. *Angew. Chem. Int. Ed.* **1986**, *25*, 451.

[184]For a review, see Staab, H.A.; Saupe, T. *Angew. Chem. Int. Ed.* **1988**, *27*, 865.

[185]Howard, S.T. *J. Am. Chem. Soc.* **2000**, *122*, 8238.

[186]Krieger, C.; Newsom, I.; Zirnstein, M.A.; Staab, H.A. *Angew. Chem. Int. Ed.* **1989**, *28*, 84. See also, Schwesinger, R.; Missfeldt, M.; Peters, K.; Schnering, H.G. von *Angew. Chem. Int. Ed.* **1987**, *26*, 1165; Alder, R.W.; Eastment, P.; Hext, N.M.; Moss, R.E.; Orpen, A.G.; White, J.M. *J. Chem. Soc. Chem. Commun.* **1988**, 1528; Staab, H.A.; Zirnstein, M.A.; Krieger, C. *Angew. Chem. Int. Ed.* **1989**, *28*, 86.

tetramine (**15**) was prepared by a coupling reaction of bispidine, and was shown to be a new class of proton sponge.[187]

| 12 | 13 | 14 | 15 |

Chiral Lewis acids are known. Indeed, an air stable and storable chiral Lewis acid catalyst has been prepared, a chiral zirconium catalyst combined with molecular sieves powder.[188] Association of a bulky silicon group with the bis(trifluoromethanesulfonyl)imide anion leads to enhancement of the electrophilic character of R_3SiNTf_2. The presence of a chiral substituent derived from (−)-myrtenal on the silicon atom led to a chiral silicon Lewis acid.[189]

Another type of steric effect is the result of an entropy effect. The compound 2,6-di-*tert*-butylpyridine is a weaker base than either pyridine or 2,6-dimethylpyridine.[190] The reason is that the conjugate acid (**14**) is less stable than the conjugate acids of non-sterically hindered pyridines. In all cases, the conjugate acids are hydrogen bonded to a water molecule, but in the case of **14** the bulky *tert*-butyl groups restrict rotations in the water molecule, lowering the entropy.[191]

The conformation of a molecule can also affect its acidity. The following pK_a values were determined for these compounds:[192]

	16	**17**	**18**	**19**
pK_a	13.3	11.2	15.9	7.3

Since ketones are stronger acids than carboxylic esters (Table 8.1), we are not surprised that **16** is a stronger acid than **18**. But cyclization of **16** to **17** increases the acidity by only 2.1 pK units while cyclization of **18** to **19**

[187]Miyahara, Y.; Goto, K.; Inazu, T. *Tetrahedron Lett.* **2001**, *42*, 3097.

[188]Ueno, M.; Ishitani, H.; Kobayashi, S. *Org. Lett.* **2002**, *4*, 3395.

[189]Mathieu, B.; de Fays, L.; Ghosez, L. *Tetrahedron Lett* **2000**, *41*, 9651

[190]Brown, H.C.; Kanner, B. *J. Am. Chem. Soc.* **1953**, *75*, 3865; **1966**, *88*, 986.

[191]Hopkins, Jr., H.P.; Janagirdar, D.V.; Moulik, P.S.; Aue, D.H.; Webb, H.M.; Davidson, W.R.; Pedley; M.D. *J. Am. Chem. Soc.* **1984**, *106*, 4341; Meot-Ner, M.; Smith, S.C. *J. Am. Chem. Soc.* **1991**, *113*, 862, and references cited therein. See also, Benoit, R.L.; Fréchette, M.; Lefebvre, D. *Can. J. Chem.* **1988**, *66*, 1159.

[192]Arnett, E.M.; Harrelson Jr., J.A. *J. Am. Chem. Soc.* **1987**, *109*, 809.

increases it by 8.6 units. Indeed, it has long been known that **19** (called *Meldrum's acid*) is an unusually strong

acid for a 1,3-diester. In order to account for this very large cyclization effect, molecular-orbital calculations were carried out two conformations of methyl acetate and of its enolate ion by two groups.[193] Both found that loss of a proton is easier by $\sim$5 kcal mol^{-1} (21 kJ mol^{-1}) for the syn than for the anti conformer of the ester. In an acyclic molecule like **18**, the preferred conformations are anti, but in Meldrum's acid (**19**) the conformation on both sides is constrained to be syn.

Facial differences in proton reactivity can lead to enantioselective deprotonation. A more common way to achieve enantioselective deprotonation is to use a chiral base and/or a chiral complexing agent. Enantioselective deprotonation in cyclic ketones has been studied.[194] Enantioselective deprotonation with heterodimer bases has been studied.[195]

When a Lewis acid coordinates to a base, the resulting complex can have conformational properties that influence reactivity. Coordination of SnCl$_4$ with aldehydes and esters, for example, leads to a complex where the conformation is determined by interactions of the C=O•••SnCl$_4$ unit with substituents attached to the carbonyl.[196]

7. *Hybridization.* An *s* orbital has a lower energy than a *p* orbital. Therefore the energy of a hybrid orbital is lower the more *s* character it contains. It follows that a carbanion at an *sp* carbon is more stable than a corresponding carbanion at an *sp^2* carbon. Thus HC≡C$^-$, which has more *s* character in its unshared pair than CH$_2$=CH$^-$ or CH$_3$CH$_2^-$ (*sp* vs. *sp^2* vs. *sp^3*, respectively), is a much weaker base. This explains the relatively high acidity of acetylenes and HCN. Another example is that alcohol and ether oxygens, where the unshared pair is *sp^3*, are more strongly basic than carbonyl oxygens, where the unshared pair is *sp^2* (Table 8.1).

A recent development in understanding the reactivity of bases has focused on their structures in solution and in the crystalline state. Due to the importance of dialkyl amide bases, there is a significant body of work, led by Williard and by Collum, that has attempted to understand the structures of

[193]Wang, X.; Houk, K.N. *J. Am. Chem. Soc.* **1988**, *110*, 1870; Wiberg, K.B.; Laidig, K.E. *J. Am. Chem. Soc.* **1988**, *110*, 1872.
[194]Majewski, M.; Wang, F. *Tetrahedron* **2002**, *58*, 4567.
[195]Amedjkouh, M. *Tetrahedron Asymm.* **2004**, *15*, 577.
[196]Gung, B.W.; Yanik, M.M. *J. Org. Chem.* **1996**, *61*, 947.

these reactive molecules. It is clear that they are aggregates. Note that the simplest member of the amide base family, lithium amide ($LiNH_2$) was shown to be monomeric and unsolvated, as determined using a combination of gas-phase synthesis and millimeter–submillimeter wave spectroscopy.[197] Note that monomeric $LiNH_2$ and $LiNMe_2$ are planar.[198] Lithium diisopropylamide ($LiNiPr_2$, LDA) was isolated from a THF solution and X-ray crystallography revealed a dimeric structure (**20**; R = iPr, S = THF) in the solid state.[199] Lithium diisopropylamide was also shown to be a dimer in solutions of THF[200] and/or HMPA (see **20**, R = iPr and S = THF, HMPA).[201] In the presence of HMPA, many derivatives **20** tend to be mixed aggregates.[202] Extremely hindered $LiNR_2$ (R = 2-adamantyl) are monomeric under all conditions.[203] In hydrocarbon solvents, lithium tetramethylpiperidide [LTMP, RR'NLi where RR' = $-CMe_2(CH_2)_3C(Me_2)-$] forms cyclic trimers and tetramers, with the tetrameric species predominating.[204] In THF, lithium hexamethyldisilazide [LHMDS, $(Me_3Si)_2NLi$] forms a five-coordinate tetrasolvate [$(Me_3Si)_2NLi(thf)_4$],[205] but in ether there is an equilibrium mixture of monomer and dimer.[206] A review is available that discusses the solution structures of amide bases $LiNR_2$.[207] Chiral lithium amide bases are known and they show similar behavior in solution.[208] Chelation effects are common in enantio-enriched amide bases, which also form aggregates.[209] The aggregation state of lithium phenylacetonitrile has been studied.[210] Dianion aggregates can be generated, and in the case of the lithiation reaction of N-silyl allylamine, X-ray structure determination showed the presence of three

[197]Grotjahn, D.B.; Sheridan, P.M.; Al Jihad, I.; Ziurys, L.M. *J. Am. Chem. Soc.* **2001**, *123*, 5489.

[198]Fressigné, C.; Maddaluno, J.; Giessner-Prettre, C.; Silvi, B. *J. Org. Chem.* **2001**, *66*, 6476.

[199]Williard P.G.; Salvino, J.M. *J. Org. Chem.* **1993**, *58*, 1. For a study of the oligomer structure of LDA at low ligand concentrations, see Rutherford, J.L.; Collum, D.B. *J. Am. Chem. Soc.* **2001**, *123*, 199.

[200]Ito, H.; Nakamura, T.; Taguchi, T.; Hanzawa, Y. *Tetrahedron Lett.* **1992**, *33*, 3769.

[201]Aubrecht, K.B.; Collum, D.B. *J. Org. Chem.* **1996**, *61*, 8674.

[202]Romesberg, F.E.; Collum, D.B. *J. Am.Chem. Soc.* **1994**, *116*, 9198, 9187. For a study of other mixed aggregates see Thomas, R.D.; Huang, J. *J. Am. Chem. Soc.* **1999**, *121*, 11239.

[203]Sakuma, K.; Gilchrist, J.H.; Romesberg, F.E.; Cajthami, C.E.; Collum, D.B. *Tetrahedron Lett.* **1993**, *34*, 5213.

[204]Lucht, B.L.; Collum, D.B. *J. Am. Chem. Soc.* **1994**, *116*, 7949.

[205]Lucht, B.L.; Collum, D.B. *J. Am. Chem. Soc.* **1995**, *117*, 9863. See also, Lucht, B.L.; Collum, D.B. *J. Am. Chem. Soc.* **1996**, *118*, 2217, 3529. See Romesberg, F.E.; Bernstein, M.P.; Gilchrist, J.H.; Harrison, A.T.; Fuller, D.J.; Collum, D.B. *J. Am. Chem. Soc.* **1993**, *115*, 3475 for the structure in HMPA.

[206]Lucht, B.L.; Collum, D.B. *J. Am. Chem. Soc.* **1994**, *116*, 6009.

[207]Collum, D.B. *Acc. Chem. Res.* **1993**, *26*, 227. For NMR studies of $LiNEt_2$ and ring laddering see Rutherford, J.L.; Collum, D.B. *J. Am. Chem. Soc.* **1999**, *121*, 10198.

[208]Hilmersson, G.; Davidsson, Ö. *J. Org. Chem.* **1995**, *60*, 7660. See also, O'Brien, P. *J. Chem. Soc. Perkin Trans. 1* **1998**, 1439. See also, Sott, R.; Grandander, J.; Dinér, P.; Hilmersson, G. *Tetrahedron Asymm.* **2004**, *15*, 267.

[209]Arvidsson, P.I.; Hilmersson, G.; Ahlberg, P. *J. Am. Chem. Soc.* **1999**, *121*, 183.

[210]Carlier, P.R.; Madura, J.D. *J. Org. Chem.* **2002**, *67*, 3832.

uniquely different aggregates.[211] A mixed aggregate is formed when the lithium enolate of a ketone is mixed with a lithium amide.[212]

$$R'''''\text{---}N\underset{R}{\overset{\overset{\displaystyle S}{\vdots}}{\underset{\underset{\displaystyle S}{\vdots}}{\underset{Li}{\overset{Li}{<}}}}N'''''R \qquad S = \text{solvent}$$

20

Similar information is available for other bases. Lithium phenoxide (LiOPh) is a tetramer in THF.[213] Lithium 3,5-dimethylphenoxide is a tetramer in ether, but addition of HMPA leads to dissociation to a monomer.[214]

Enolate anions are nucleophiles in reactions with alkyl halides (reaction **10-68**), with aldehydes and ketones (reactions **16-34**, **16-36**) and with acid derivatives (reaction **16-85**). Enolate anions are also bases, reacting with water, alcohols and other protic solvents, and even the carbonyl precursor to the enolate anion. Enolate anions exist as aggregates, and the effect of solvent on aggregation and reactivity of lithium enolate anions has been studied.[215] The influence of alkyl substitution on the energetics of enolate anions has been studied.[216]

THE EFFECTS OF THE MEDIUM ON ACID AND BASE STRENGTH

Structural features are not the only factors that affect acidity or basicity. The same compound can have its acidity or basicity changed when the conditions are changed. The effect of temperature (p. 364) has already been mentioned. More important is the effect of the solvent, which can exert considerable influence on acid and base strengths by differential solvation.[217] If a base is more solvated than its conjugate acid, its stability is increased relative to the conjugate acid. For example,

[211]Williard, P. G.; Jacobson, M. A. *Org. Lett.* **2000**, *2*, 2753. For the structure and bonding of dilithiodiamines see Pratt, L.M.; Mu, R. *J. Org. Chem.* **2004**, *69*, 7519.

[212]Sun, C.; Williard, P.G. *J. Am. Chem. Soc.* **2000**, *122* , 7829. See also, Pratt, L.M.; Streitwieser, A. *J. Org. Chem.* **2003**, *68*, 2830.

[213]Jackman, L.M.; Çizmeciyan, D.; Williard, P.G.; Nichols, M.A. *J. Am. Chem. Soc.* **1993**, *115*, 6262.

[214]Jackman, L.M.; Chen, X. *J. Am. Chem. Soc.* **1992**, *114*, 403.

[215]Streitwieser, A.; Juaristi, E.; Kim, Y.-J.; Pugh, J.K. *Org. Lett.* **2000**, *2*, 3839.

[216]Alconcel, L.S.; Deyerl, H.-J.; Continetti, R.E. *J. Am. Chem. Soc.* **2001**, *123*, 12675.

[217]For reviews of the effects of solvent, see Epshtein, L.M.; Iogansen, A.V. *Russ. Chem. Rev.* **1990**, *59*, 134; Dyumaev, K.M.; Korolev, B.A. *Russ. Chem. Rev.* **1980**, *49*, 1021. For a review of the effects of the solvent DMSO, see Taft, R.W.; Bordwell, F.G. *Acc. Chem. Res.* **1988**, *21*, 463. For determination of pK_a values of various compounds in acetonitrile see Heemstra, J.M.; Moore, J.S. *Tetrahedron* **2004**, *60*, 7287.

Table 8.6 shows that toward the proton, where steric effects are absent, methylamine is a stronger base than ammonia and dimethylamine is stronger still.[218] These results are easily explainable if one assumes that methyl groups are electron donating. However, trimethylamine, which should be even stronger, is a weaker base than dimethylamine or methylamine. This apparently anomalous behavior can be explained by differential hydration.[219] Thus, NH_4^+ is much better hydrated (by hydrogen bonding to the water solvent) than NH_3 because of its positive charge.[220] It has been estimated that this effect contributes ~11 pK units to the base strength of ammonia.[221] When methyl groups replace hydrogen, this difference in hydration decreases[222] until, for trimethylamine, it contributes only ~6 pK units to the base strength.[179] Thus two effects act in opposite directions, the field effect increasing the basicity as the number of methyl groups increases and the hydration effect decreasing it. When the effects are added, the strongest base is dimethylamine and the weakest is ammonia. If alkyl groups are electron donating, one would expect that in the gas phase,[223] where the solvation effect does not exist, the basicity order of amines toward the proton should be $R_3N > R_2NH > RNH_2 > NH_3$, and this has indeed been confirmed, for R = Me as well as R = Et and Pr.[224] Aniline

---

[218]For a review of the basicity of amines, see Smith, J.W., in Patai, S. *The Chemistry of the Amino Group*, Wiley, NY, *1968*, pp. 161–204.

[219]Trotman-Dickenson, A.F. *J. Chem. Soc.* *1949*, 1293; Pearson, R.G.; Williams, F.V. *J. Am. Chem. Soc.* *1954*, *76*, 258; Hall, Jr., H.K. *J. Am. Chem. Soc.* *1957*, *79*, 5441; Arnett, E.M.; Jones III, F.M; Taagepera, M.; Henderson, W.G.; Beauchamp, J.L.; Holtz, D.; Taft, R.W. *J. Am. Chem. Soc.* *1972*, *94*, 4724; Aue, D.H.; Webb, H.M.; Bowers, M.T. *J. Am. Chem. Soc.* *1972*, *94*, 4726; *1976*, *98*, 311, 318; Mucci, A.; Domain, R.; Benoit, R.L. *Can. J. Chem.* *1980*, *58*, 953. See also Drago, R.S.; Cundari, T.R.; Ferris, D.C. *J. Org. Chem.* *1989*, *54*, 1042.

[220]For discussions of the solvation of ammonia and amines, see Jones III, F.M.; Arnett, E.M. *Prog. Phys. Org. Chem.* *1974*, *11*, 263; Grunwald, E.; Ralph, E.K. *Acc. Chem. Res.* *1971*, *4*, 107.

[221]Condon, F.E. *J. Am. Chem. Soc.* *1965*, *87*, 4481, 4485.

[222]For two reasons: (*1*) the alkyl groups are poorly solvated by the water molecules, and (2) the strength of the hydrogen bonds of the BH$^+$ ions decreases as the basicity of B increases: Lau, Y.K.; Kebarle, P. *Can. J. Chem.* *1981*, *59*, 151.

[223]For reviews of acidities and basicities in the gas phase, see Liebman, J.F. *Mol. Struct. Energ.* *1987*, *4*, 49; Dixon, D.A.; Lias, S.G. *Mol. Struct. Energ.* 1987, *2*, 269; Bohme, D.K., in Patai, S. *The Chemistry of Functional Groups, Supplement F*, pt. 2, Wiley, NY, *1982*, pp. 731–762; Bartmess, J.E.; McIver, Jr., R.T., in Bowers, M.T. *Gas Phase Ion Chemistry*, Vol. 2, Academic Press, NY, *1979*, pp. 88–121; Kabachnik, M.I. *Russ. Chem. Rev.* *1979*, *48*, 814; Kebarle, P. *Annu. Rev. Phys. Chem.* *1977*, *28*, 445; Arnett, E.M. *Acc. Chem. Res.* *1973*, *6*, 404. For a comprehensive table of gas-phase basicities, see Lias, S.G.; Liebman, J.F.; Levin, R.D. *J. Phys. Chem. Ref. Data*, *1984*, *13*, 695. See also the tables of gas-phase acidities and basicities in the following articles, and their cited references: Meot-Ner, M.; Kafafi, S.A. *J. Am. Chem. Soc.* *1988*, *110*, 6297; Headley, A.D. *J. Am. Chem. Soc.* *1987*, *109*, 2347; McMahon, T.B.; Kebarle, P. *J. Am. Chem. Soc.* *1985*, *107*, 2612; *1977*, *99*, 2222, 3399; Wolf, J.F.; Staley, R.H.; Koppel, I.; Bartmess, J.E.; Scott, J.A.; McIver, Jr., R.T. *J. Am. Chem. Soc.* *1979*, *101*, 6046; Fujio, M.; McIver, Jr., R.T.; Taft, R.W. *J. Am. Chem. Soc.* *1981*, *103*, 4017; Lau, Y.K.; Nishizawa, K.; Tse, A.; Brown, R.S.; Kebarle, P. *J. Am. Chem. Soc.* *1981*, *103*, 6291.

[224]Munson, M.S.B. *J. Am. Chem. Soc.* *1965*, *87*, 2332; Brauman, J.I.; Riveros, J.M.; Blair, L.K. *J. Am. Chem. Soc.* *1971*, *93*, 3914; Briggs, J.P.; Yamdagni, R.; Kebarle, P. *J. Am. Chem. Soc.* *1972*, *94*, 5128; Aue, D.H.; Webb H.M.; Bowers, M.T. *J. Am. Chem. Soc.* *1972*, *94*, 4726; *1976*, *98*, 311, 318.

too, in the gas phase, is a stronger base than NH_3,[225] so its much lower basicity in aqueous solution (pK_a of $PhNH_3^+$ 4.60 compared with 9.24 for aqueous NH_4^+) is caused by similar solvation effects and not by resonance and field electron-withdrawing effects of a phenyl group. Similarly, pyridine[226] and pyrrole[227] are both much less basic than NH_3 in aqueous solution (pyrrole[228] is neutral in aqueous solution), but *more* basic in the gas phase. These examples in particular show how careful one must be in attributing relative acidities or basicities to any particular effect. Solvent has a significant influence on the Hammett reaction constant (p. 679), which influences the acidity of substituted benzoic acids.[229]

In the case of Lewis acids, protic solvents such as water or alcohol can strongly influence their reactivity, cause it to react via an alternative path to the one desired, or even cause decomposition. Recently, rare earth metal triflates were used to develop water tolerant Lewis acids that can be used in many organic reactions.[230]

For simple alcohols, the order of gas-phase *acidity* is completely reversed from that in aqueous solution. In solution, the acidity is in the order $H_2O > MeCH_2OH > Me_2CHOH > Me_3COH$, but in the gas phase the order is precisely the opposite.[231] Once again solvation effects can be invoked to explain the differences. Comparing the two extremes, H_2O and Me_3COH, we see that the OH^- ion is very well solvated by water while the bulky Me_3CO^- is much more poorly solvated because the water molecules cannot get as close to the oxygen. Thus in solution H_2O gives up its proton more readily. When solvent effects are absent, however, the intrinsic acidity is revealed and Me_3COH is a stronger acid than H_2O. This result demonstrates that simple alkyl groups cannot be simply regarded as electron donating. If methyl is an electron-donating group, then Me_3COH should be an intrinsically weaker acid than H_2O, yet it is stronger. A similar pattern is found with carboxylic acids, where simple aliphatic acids, such as propanoic, are stronger than acetic acid in the gas phase,[232] though weaker in aqueous solution (Table 8.5).

[225]Briggs, J.P.; Yamdagni, R.; Kebarle, P. *J. Am. Chem. Soc. 1972*, *94*, 5128; Dzidic, I. *J. Am. Chem. Soc. 1972*, *94*, 8333; Ikuta, S.; Kebarle, P. *Can. J. Chem. 1983*, *61*, 97.

[226]Taft, R.W.; Taagepera, M.; Summerhays, K.D.; Mitsky, J. *J. Am. Chem. Soc. 1973*, *95*, 3811; Briggs, J.P.; Yamdagni, R.; Kebarle, P. *J. Am. Chem. Soc. 1972*, *94*, 5128.

[227]Yamdagni, R.; Kebarle, P. *J. Am. Chem. Soc. 1973*, *95*, 3504.

[228]For a review of the basicity and acidity of pyrroles, see Catalan, J.; Abboud, J.L.M.; Elguero, J. *Adv. Heterocycl. Chem. 1987*, *41*, 187.

[229]Bartnicka, H.; Bojanowska, I.; Kalinowski, M.K. *Aust. J. Chem. 1993*, *46*, 31.

[230]Kobayashi, S. *Synlett, 1994*, 689.

[231]Baird, N.C. *Can. J. Chem. 1969*, *47*, 2306; Arnett, E.M.; Small, L.E.; McIver, Jr., R.T.; Miller, J.S. *J. Am. Chem. Soc. 1974*, *96*, 5638; Blair, L.K.; Isolani, P.C.; Riveros, J.M. *J. Am. Chem. Soc. 1973*, *95*, 1057; McIver, Jr., R.T.; Scott, J.A.; Riveros, J.M. *J. Am. Chem. Soc. 1973*, *95*, 2706. The alkylthiols behave similarly; gas-phase acidity increases with increasing group size while solution (aqueous) acidity decreases: Bartmess, J.E.; McIver Jr., R.T. *J. Am. Chem. Soc. 1977*, *99*, 4163.

[232]For a table of gas-phase acidities of 47 simple carboxylic acids, see Caldwell, G.; Renneboog, R.; Kebarle, P. *Can. J. Chem. 1989*, *67*, 611.

The evidence in these and other cases[233] is that alkyl groups can be electron donating when connected to unsaturated systems, but in other systems may have either no effect or may actually be electron withdrawing. The explanation given for the intrinsic gas-phase acidity order of alcohols as well as the basicity order of amines is that alkyl groups, because of their polarizability, can spread both positive and negative charges.[234] It has been calculated that even in the case of alcohols the field effects of the alkyl groups are still operating normally, but are swamped by the greater polarizability effects.[235] Polarizability effects on anionic centers are a major factor in gas-phase acid–base reactions.[236]

It has been shown (by running reactions on ions that are solvated in the gas phase) that solvation by even one molecule of solvent can substantially affect the order of basicities.[237]

An important aspect of solvent effects is the effect on the orientation of solvent molecules when an acid or base is converted to its conjugate. For example, consider an acid RCOOH converted to RCOO$^-$ in aqueous solution. The solvent molecules, by hydrogen bonding, arrange themselves around the COO$^-$ group in a much more orderly fashion than they had been arranged around the COOH group (because they are more strongly attracted to the negative charge). This represents a considerable loss of freedom and a decrease in entropy. Thermodynamic measurements show that for simple aliphatic and halogenated aliphatic acids in aqueous solution at room temperature, the entropy ($T\Delta S$) usually contributes much more to the total free-energy change ΔG than does the enthalpy ΔH.[238] Two examples are shown in Table 8.7.[239] Resonance and field effects of functional groups therefore affect the acidity of RCOOH in two distinct ways. They affect the enthalpy (electron withdrawing groups increase acidity by stabilizing RCOO$^-$ by charge dispersal), but they also affect the entropy (by lowering the charge on the COO$^-$ group and by changing the electron-density distribution in the COOH group, electron-withdrawing groups alter the solvent orientation patterns around both the acid and the ion, and consequently change ΔS).

[233]Brauman, J.I.; Blair, L.K. *J. Am. Chem. Soc.* **1971**, *93*, 4315; Kwart, H.; Takeshita, T. *J. Am. Chem. Soc.* **1964**, *86*, 1161; Fort, Jr., R.C.; Schleyer, P.v.R. *J. Am. Chem. Soc.* **1964**, *86*, 4194; Holtz, H.D.; Stock, L.M. *J. Am. Chem. Soc.* **1965**, *87*, 2404; Laurie, V.W.; Muenter, J.S. *J. Am. Chem. Soc.* **1966**, *88*, 2883.

[234]Brauman, J.I.; Blair, L.K. *J. Am. Chem. Soc.* **1970**, *92*, 5986; Munson, M.S.B. *J. Am.Chem. Soc.* **1965**, *87*, 2332; Brauman, J.I.; Riveros, J.M.; Blair, L.K. *J. Am. Chem. Soc.* **1971**, *93*, 3914; Huheey, J.E. *J. Org. Chem.* **1971**, *36*, 204; Radom, L. *Aust. J. Chem.* **1975**, *28*, 1; Aitken, E.J.; Bahl, M.K.; Bomben, K.D.; Gimzewski, J.K.; Nolan, G.S.; Thomas, T.D. *J. Am. Chem. Soc.* **1980**, *102*, 4873.

[235]Taft, R.W.; Taagepera, M.; Abboud J.M.; Wolf, J.F.; Defrees, D.J.; Hehre, W.J.; Bartmess, J.E.; McIver Jr., R.T. *J. Am. Chem. Soc.* **1978**, *100*, 7765. For a scale of polarizability parameters, see Hehre, W.J.; Pau, C.; Headley, A.D.; Taft, R.W.; Topsom, R.D. *J. Am. Chem. Soc.* **1986**, *108*, 1711.

[236]Bartmess, J.E.; Scott, J.A.; McIver Jr., R.T. *J. Am. Chem. Soc.* **1979**, *101*, 6056.

[237]Bohme, D.K.; Rakshit, A.B.; Mackay, G.I. *J. Am. Chem. Soc.* **1982**, *104*, 1100.

[238]Bolton, P.D.; Hepler, L.G. *Q. Rev. Chem. Soc.* **1971**, *25*, 521; Gerrard, W.; Macklen, E.D. *Chem. Rev.* **1959**, *59*, 1105. See also Wilson, B.; Georgiadis, R.; Bartmess, J.E. *J. Am. Chem. Soc.* **1991**, *113*, 1762.

[239]Bolton, P.D.; Hepler, L.G. *Q. Rev. Chem. Soc.* **1971**, *25*, 521; p. 529; Hambly, A.N.*Rev. Pure Appl. Chem.* **1965**, *15*, 87, p. 92.

TABLE 8.7. Thermodynamic Values for the Ionizations of Acetic and Chloroacetic Acids in H$_2$O at 25°C[238]

Acid	pK_a	ΔG		ΔH		$T\Delta S$	
		kcal mol^{-1}	kJ mol^{-1}	kcal mol^{-1}	kJ mol^{-1}	kcal mol^{-1}	kJ mol^{-1}
CH$_3$COOH	4.76	+6.5	+27	−0.1	−0.4	−6.6	−28
ClCH$_2$COOH	2.86	+3.9	+16	−1.1	−4.6	−5.0	−21
Cl$_3$CCOOH	0.65	+0.9	+3.8	+1.5	+6.3	+0.6	+2.5

A change from a protic to an aprotic solvent can also affect the acidity or basicity, since there is a difference in solvation of anions by a protic solvent (which can form hydrogen bonds) and an aprotic one.[240] The effect can be extreme: in DMF, picric acid is stronger than HBr,[241] though in water HBr is far stronger. This particular result can be attributed to size. That is, the large ion $(O_2N)_3C_6H_2O^-$ is better solvated by DMF than the smaller ion Br$^-$.[242] The ionic strength of the solvent also influences acidity or basicity, since it has an influence on activity coefficients.

In summary, solvation can have powerful effects on acidity and basicity. In the gas, phase the effects discussed in the previous section, especially resonance and field effects, operate unhindered by solvent molecules. As we have seen, electron-withdrawing groups generally increase acidity (and decrease basicity); electron-donating groups act in the opposite way. In solution, especially aqueous solution, these effects still largely persist (which is why pK values in Table 8.5 do largely correlate with resonance and field effects), but in general are much weakened, and occasionally reversed.[164]

[240]For a review, see Parker, A.J. *Q. Rev. Chem. Soc.* **1962**, *16*, 163.
[241]Sears, P.G.; Wolford, R.K.; Dawson, L.R. *J. Electrochem. Soc.* **1956**, *103*, 633.
[242]Miller, J.; Parker, A.J. *J. Am. Chem. Soc.* **1961**, *83*, 117.

CHAPTER 9

Effects of Structure and Medium on Reactivity

When the equation for a reaction of, say, carboxylic acids, is written, it is customary to use the formula RCOOH, which implies that all carboxylic acids undergo the reaction. Since most compounds with a given functional group do give more or less the same reactions, the custom is useful, and the practice is used in this book. It allows a large number of individual reactions to be classified together and serves as an aid both for memory and understanding. Organic chemistry would be a huge morass of unconnected facts without the symbol R. Nevertheless, it must be borne in mind that a given functional group does not always react the same way, regardless of what molecule it is a part of. The reaction at the functional group is influenced by the rest of the molecule. This influence may be great enough to stop the reaction completely or to make it take an entirely different course. Even when two compounds with the same functional group undergo the same reaction, the rates and/or the positions of equilibrium are usually different, sometimes slightly, sometimes greatly, depending on the structures of the compounds. The greatest variations may be expected when additional functional groups are present.

The effects of structure on reactivity can be divided into three major types: field, resonance (or mesomeric), and steric.[1] In most cases, two or all three of these are operating, and it is usually not easy to tell how much of the rate enhancement (or decrease) is caused by each of the three effects.

---

[1]For a monograph, see Klumpp, G.W. *Reactivity in Organic Chemistry*, Wiley, NY, *1982*. For a general theoretical approach to organic reactivity, see Pross, A. *Adv. Phys. Org. Chem. 1985*, *21*, 99.

March's Advanced Organic Chemistry: Reactions, Mechanisms, and Structure, Sixth Edition, by Michael B. Smith and Jerry March
Copyright © 2007 John Wiley & Sons, Inc.

395

RESONANCE AND FIELD EFFECTS

It is often particularly difficult to separate resonance and field effects; they are frequently grouped together under the heading of *electrical effects*.[2] Field effects were discussed on pp. 19–22. Table 1.3 contains a list of some $+I$ and $-I$ groups. As for resonance effects, on p. 48 it was shown how the electron density distribution in aniline is not the same as it would be if there were no resonance interaction between the ring and the NH_2 group. Most groups that contain an unshared pair on an atom connected to an unsaturated system display a similar effect; that is, the electron density on the group is less than expected, and the density on the unsaturated system is greater. Such groups are said to be electron donating by the resonance effect ($+M$ groups). Alkyl groups, which do not have an unshared pair, are also $+M$ groups, presumably because of hyperconjugation.

On the other hand, groups that have a multiple-bonded electronegative atom directly connected to an unsaturated system are $-M$ groups. In such cases, we can draw canonical forms in which electrons have been taken from the unsaturated system into the group, as in nitrobenzene, **1**. Table 9.1 contains a list of some $+M$ and $-M$ groups.

1

**TABLE 9.1. Some Groups with $+M$ and $-M$ Effects,
Not Listed in Order of Strength of Effect**[a]

$+M$		$-M$	
O^-	SR	NO_2	CHO
S^-	SH	CN	COR
NR_2	Br	COOH	SO_2R
NHR	I	COOR	SO_2OR
NH_2	Cl	$CONH_2$	NO
NHCOR	F	CONHR	Ar
OR	R	$CONR_2$	
OH	Ar		
OCOR			

[a]Ar appears in both lists because it is capable of both kinds of effect.

[2]For reviews of the study of electrical effects by ab initio mo methods, see Topsom, R.D. *Prog. Phys. Org. Chem.* **1987**, *16*, 125, *Mol. Struct. Energ.* **1987**, *4*, 235.

The resonance effect of a group, whether $+M$ or $-M$, operates only when the group is directly connected to an unsaturated system, so that, for example, in explaining the effect of the CH_3O group on the reactivity of the COOH in $CH_3OCH_2CH_2COOH$, only the field effect of the CH_3O need be considered. This is one way of separating the two effects. In p-methoxybenzoic acid both effects must be considered. The field effect operates through space, solvent molecules, or the σ bonds of a system, while the resonance effect operates through π electrons.

It must be emphasized once again that neither by the resonance nor by the field effect are any electrons actually being donated or withdrawn, though these terms are convenient (and we will use them). As a result of both effects, the electron-density distribution is not the same as it would be without the effect (see pp. 21, 48). One thing that complicates the study of these effects on the reactivity of compounds is that a given group may have an effect in the transition state that is considerably more or less than it has in the molecule that does not react.

An example will show the nature of electrical effects (resonance and field) on reactivity. In the alkaline hydrolysis of aromatic amides (reaction **16-60**), the rate-determining step is the attack of hydroxide ion at the carbonyl carbon:

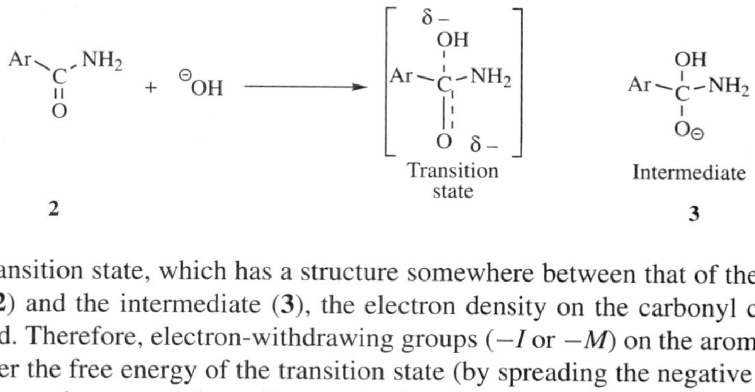

In the transition state, which has a structure somewhere between that of the starting amide (**2**) and the intermediate (**3**), the electron density on the carbonyl carbon is increased. Therefore, electron-withdrawing groups ($-I$ or $-M$) on the aromatic ring will lower the free energy of the transition state (by spreading the negative charge). These groups have much less effect on the free energy of **2**. Since G is lowered for the transition state, but not substantially for **2**, $\Delta G^{\ddagger}$ is lowered and the reaction rate is increased (Chapter 6). Conversely, electron-donating groups ($+I$ or $+M$) should decrease the rate of this reaction. Of course, many groups are $-I$ and $+M$, and for these it is not always possible to predict which effect will predominate.

STERIC EFFECTS

It occasionally happens that a reaction proceeds much faster or much slower than expected on the basis of electrical effects alone. In these cases it can often be shown that steric effects are influencing the rate. For example, Table 9.2 lists relative rates for the S_N2 ethanolysis of certain alkyl halides (see p. 426).[3] All these compounds

[3]Hughes, E.D. *Q. Rev. Chem. Soc.* **1948**, *2*, 107.

TABLE 9.2. Relative Rates of Reaction of RBr with Ethanol[3]

R	Relative Rate
CH_3	17.6
CH_3CH_2	1
$CH_3CH_2CH_2$	0.28
$(CH_3)_2CHCH_2$	0.030
$(CH_3)_3CCH_2$	4.2×10^{-6}

are primary bromides; the branching is on the second carbon, so that field-effect differences should be small. As Table 9.2 shows, the rate decreases with increasing β branching and reaches a very low value for neopentyl bromide. This reaction is known to involve an attack by the nucleophile from a position opposite to that of the bromine (see p. 426). The great decrease in rate can be attributed to *steric hindrance*, a sheer physical blockage to the attack of the nucleophile. Another example of steric hindrance is found in 2,6-disubstituted benzoic acids, which are difficult to esterify no matter what the resonance or field effects of the groups in the 2 or the 6 position. Similarly, once 2,6-disubstituted benzoic acids *are* esterified, the esters are difficult to hydrolyze.

Not all steric effects decrease reaction rates. In the hydrolysis of RCl by an S_N1 mechanism (see p. 433), the first step, which is rate determining, involves ionization of the alkyl chloride to a carbocation:

$$R - \underset{\underset{R}{|}}{\overset{\overset{R}{|}}{C}} - Cl \longrightarrow R - \overset{\overset{R}{\diagdown}}{\underset{\underset{R}{\diagup}}{C}}{}^{\oplus}$$

The central carbon in the alkyl chloride is sp^3 hybridized, with angles of ~109.5°, but when it is converted to the carbocation, the hybridization becomes sp^2 and the preferred angle is 120°. If the halide is tertiary and the three alkyl groups are large enough, they will be pushed together by the enforced tetrahedral angle, resulting in strain (see p. 232). This type of strain is called *B strain*[4] (for back strain), and it can be relieved by ionization to the carbocation.[5]

The rate of ionization (and hence the solvolysis rate) of a molecule in which there is B strain is therefore expected to be larger than in cases where B strain is not present. Table 9.3 shows that this is so.[6] Substitution of ethyl groups for the

[4]For a discussion, see Brown, H.C. *Boranes in Organic Chemistry*, Cornell University Press, Ithaca, NY, *1972*, pp. 114–121.
[5]For reviews of the effects of strain on reactivity, see Stirling, C.J.M. *Tetrahedron 1985*, *41*, 1613; *Pure Appl. Chem. 1984*, *56*, 1781.
[6]Brown, H.C.; Fletcher, R.S. *J. Am. Chem. Soc. 1949*, *71*, 1845.

**TABLE 9.3. Rates of Hydrolysis of Tertiary Alkyl
Chlorides at 25°C in 80% Aqueous Ethanol[6]**

Halide	Rate	Halide	Rate
Me$_3$Cl	0.033	Et$_3$CCl	0.099
Me$_2$EtCCl	0.055	Me$_2$(iPr)CCl	0.029
MeEt$_2$CCl	0.086	Me(iPr)$_2$CCl	0.45

methyl groups of *tert*-butyl chloride does not cause B strain; the increase in
rate is relatively small, and the rate smoothly rises with the increasing number
of ethyl groups. The rise is caused by normal field and resonance (hyperconjugation) effects. Substitution by one isopropyl group is not greatly different.
But with the second isopropyl group the crowding is now great enough to cause
B strain, and the rate is increased 10-fold. Substitution of a third isopropyl
group increases the rate still more. Another example where B strain increases
solvolysis rates is found with the highly crowded molecules tri-*tert*-butylcarbi-
nol, di-*tert*-butylneopentylcarbinol, *tert*-butyldineopentylcarbinol, and trineo-
pentylcarbinol, where rates of solvolysis of the *p*-nitrobenzoate esters
are faster than that of *tert*-butyl nitrobenzoate by factors of 13,000, 19,000,
68,000, and 560, respectively.[7]

Another type of strain, that can affect rates of cyclic compounds, is called
I strain (internal strain).[8] This type of strain results from changes in ring strain
in going from a tetrahedral to a trigonal carbon or vice versa. For example, as mentioned above, S$_N$1 solvolysis of an alkyl halide involves a change in the bond angle
of the central carbon from ~109.5 to ~120°. This change is highly favored in
1-chloro-1-methylcyclopentane because it relieves eclipsing strain (p. 223); thus
this compound undergoes solvolysis in 80% ethanol at 25°C, 43.7 times faster

	t-BuCl	Me Cl (cyclopentyl)	Me Cl (cyclohexyl)
Relative solvolysis rates	1.0	43.7	0.35

than the reference compound *tert*-butyl chloride.[9] In the corresponding cyclohexyl compound, this factor is absent because the substrate does not have
eclipsing strain (p. 223), and this compound undergoes the reaction at about

[7]Bartlett, P.D.; Tidwell, T.T. *J. Am. Chem. Soc.* **1968**, *90*, 4421.
[8]For a discussion, see Brown, H.C. *Boranes in Organic Chemistry*, Cornell University Press, Ithaca, NY, **1972**, pp. 105–107, 126–128.
[9]Brown, H.C.; Borkowski, M. *J. Am. Chem. Soc.* **1952**, *74*, 1894. See also, Brown, H.C.; Ravindranathan, M.; Peters, E.N.; Rao, C.G.; Rho, M.M. *J. Am. Chem. Soc.* **1977**, *99*, 5373.

one-third the rate of *tert*-butyl chloride. The reasons for this small decrease in rate are not clear. Corresponding behavior is found in the other direction, in changes from a trigonal to a tetrahedral carbon. Thus cyclohexanone undergoes addition reactions faster than cyclopentanone. Similar considerations apply to larger rings. Rings of 7–11 members exhibit eclipsing and transannular strain; and in these systems reactions in which a tetrahedral carbon becomes trigonal generally proceed faster than in open-chain systems.[10] I-Strain has been shown to be a factor in other reactions as well.[11]

Conformational effects on reactivity can be considered under the heading of steric effects,[12] though in these cases we are considering not the effect of a group X and that of another group X' upon reactivity at a site Y but the effect of the conformation of the molecule. Many reactions fail entirely unless the molecules are able to assume the proper conformation. An example is the rearrangement of *N*-benzoylnorephedrine. The two diastereomers of this compound behave very

differently when treated with alcoholic HCl. In one of the isomers, nitrogen-to-oxygen migration takes place, while the other does not react at all.[13] In order for the migration to take place, the nitrogen must be near the oxygen (*gauche* to it). When **4** assumes this conformation, the methyl and phenyl groups are anti to each other, which is a favorable position, but when **5** has the nitrogen gauche to the oxygen, the methyl must be *gauche* to the phenyl, which is so unfavorable that the reaction does not occur. Other examples are electrophilic additions to C=C double bonds (see p. 999) and E2 elimination reactions (see p. 1478). Also, many examples are known where axial and equatorial groups behave differently.[14]

In steroids and other rigid systems, a functional group in one part of the molecule can strongly affect the rate of a reaction taking place at a remote part of the

[10]See, for example, Schneider, H.; Thomas, F. *J. Am. Chem. Soc.* **1980**, *102*, 1424.

[11]Sands, R.D. *J. Org. Chem.* **1994**, *59*, 468.

[12]For reviews of conformational effects, see Green, B.S.; Arad-Yellin, R.; Cohen, M.D. *Top. Stereochem.* **1986**, *16*, 131; Ōki, M. *Acc. Chem. Res.* **1984**, *17*, 154; Seeman, J.I. *Chem. Rev.* **1983**, *83*, 83. See also Ōki, M.; Tsukahara, J.; Moriyama, K.; Nakamura, N. *Bull. Chem. Soc. Jpn.* **1987**, *60*, 223, and other papers in this series.

[13]Fodor, G.; Bruckner, V.; Kiss, J.; Óhegyi, G. *J. Org. Chem.* **1949**, *14*, 337.

[14]For a discussion, see Eliel, E.L. *Stereochemistry of Carbon Compounds*, McGraw-Hill, NY, **1962**, pp. 219–234.

same molecule by altering the conformation of the whole skeleton.

6   **7**

An example of this effect, called *conformational transmission*, is found in ergost-7-en-3-one (**6**) and cholest-6-en-3-one (**7**), where **7** condenses with benzaldehyde 15 times faster than **6**.[15] The reaction site in both cases is the carbonyl group, and the rate increases because moving the double bond from the 7 to the 6 position causes a change in conformation at the carbonyl group (the difference in the side chain at C-17 does not affect the rate).

QUANTITATIVE TREATMENTS OF THE EFFECT OF STRUCTURE ON REACTIVITY[16]

Suppose a reaction is performed on a substrate molecule that can be represented as XGY, where Y is the site of the reaction, X a variable substituent, and G a skeleton group to which X and Y are attached, and we find that changing X from H to CH_3 results in a rate increase by a factor, say, 10. We would like to know just what part of the increase is due to each of the effects previously mentioned. The obvious way to approach such a problem is to try to find compounds in which one or two of the factors are absent or at least negligible. This is not easy to do acceptably because factors that seem negligible to one investigator do not always appear so to another. The first attempt to give numerical values was that of Hammett.[17] For the cases of

[15]Barton, D.H.R.; McCapra, F.; May, P.J.; Thudium, F. *J. Chem. Soc.* **1960**, 1297.
[16]For monographs, see Exner, O. *Correlation Analysis of Chemical Data*, Plenum, NY, **1988**; Johnson, C.D. *The Hammett Equation*, Cambridge University Press, Cambridge, **1973**; Shorter, J. *Correlation Analysis of Organic Reactivity*, Wiley, NY, **1982**, *Correlation Analysis in Organic Chemistry*, Clarendon, N.B. Press, Oxford, **1973**; Chapman, N.B.; Shorter, J. *Correlation Analysis in Chemistry: Recent Advances*, Plenum, NY, **1978**, *Advances in Linear Free Energy Relationships*, Plenum, NY, **1972**; Wells, P.R. *Linear Free Energy Relationships*, Academic Press, NY, **1968**. For reviews, see Connors, K.A. *Chemical Kinetics*, VCH, NY, **1990**, pp. 311–383; Lewis, E.S., in Bernasconi, C.F. *Investigation of Rates and Mechanisms of Reactions* (Vol. 6 of Weissberger, A. *Techniques of Chemistry*), 4th ed., Wiley, NY, **1986**, pp. 871–901; Jones, R.A.Y. *Physical and Mechanistic Organic Chemistry*, 2nd ed., Cambridge University Press: Cambridge, **1984**, pp. 38–68; Charton, M. *CHEMTECH 1974*, 502, **1975**, 245; Hine, J. *Structural Effects in Organic Chemistry*, Wiley, NY, **1975**, pp. 55–102; Afanas'ev, I.B. *Russ. Chem. Rev.* **1971**, *40*, 216; Laurence, C.; Wojtkowiak, B. *Ann. Chim. (Paris)* **1970**, [*14*] *5*, 163. For a historical perspective, see Grunwald, E. *CHEMTECH 1984*, 698.
[17]For a review, see Jaffé, H.H. *Chem. Rev.* **1953**, *53*, 191.

m- and p-XC_6H_4Y, Hammett set up the equation

$$\log \frac{k}{k_0} = \sigma\rho$$

where k_0 is the rate constant or equilibrium constant for $X = H$, k is the constant for the group X, ρ is a constant for a given reaction under a given set of conditions, and σ is a constant characteristic of the group X. The equation is called the *Hammett equation.*

The value of ρ was set at 1.00 for ionization of XC_6H_4COOH in water at 25°C. The values of σ_m and σ_p were then calculated for each group (for a group X, σ is different for the meta and para positions). Once a set of σ values was obtained, ρ values could be obtained for other reactions from the rates of just two X-substituted compounds, if the σ values of the X groups were known (in practice, at least four well-spaced values are used to calculate ρ because of experimental error and because the treatment is not exact). With the ρ value thus calculated and the known σ values for other groups, rates can be predicted for reactions that have not yet been run.

The σ values are numbers that sum up the total electrical effects (resonance plus field) of a group X when attached to a benzene ring. The treatment usually fails for the ortho position. The Hammett treatment has been applied to many reactions and to many functional groups and correlates quite well an enormous amount of data. Jaffé's review article[17] lists ρ values for 204 reactions,[18] many of which have different ρ values for different conditions. Among them are reactions as disparate as the following:

Rate constants for

$ArCOOMe + OH^- \longrightarrow ArCOO^-$

$ArCH_2Cl + I^- \longrightarrow ArCH_2I$

$ArNH_2 + PhCOCl \longrightarrow ArNHCOPh$

$ArH + NO_2^+ \longrightarrow ArNO_2$

$ArCO_2OCMe_3 \longrightarrow$ Decomposition (a free-radical process)

Equilibrium constants for

$ArCOOH + H_2O \rightleftharpoons ArCOO^- + H_3O^+$

$ArCHO + HCN \rightleftharpoons ArCH(CN)OH$

---

[18]Additional ρ values are given in Wells, P.R. *Chem. Rev.* **1963**, *63*, 171 and van Bekkum, H.; Verkade, P.E.; Wepster, B.M. *Recl. Trav. Chim. Pays-Bas* **1959**, *78*, 821.

The Hammett equation has also been shown to apply to many physical measurements, including ir frequencies and nmr chemical shifts.[19] The treatment is reasonably successful whether the substrates are attacked by electrophilic, nucleophilic, or free-radical reagents, the important thing being that the mechanism be the same *within* a given reaction series.

However, there are many reactions that do not fit the treatment. These are mostly reactions where the attack is directly on the ring and where the X group can enter into direct resonance interaction with the reaction site in the transition state (i.e., the substrate is XY rather than XGY). For these cases, two new sets of σ values have been devised: σ^+ values (proposed by H.C. Brown) for cases in which an electron-donating group interacts with a developing positive charge in the transition state (this includes the important case of electrophilic aromatic substitutions; see Chapter 11), and σ values, where electron-withdrawing groups interact with a developing negative charge. Table 9.4 gives σ, σ^+, and σ^- values for some common X groups.[20] As shown in the table, σ is not very different from σ^+ for most electron-withdrawing groups. The values of σ_m^- are not shown in Table 9.4, since they are essentially the same as the σ_m values.

A positive value of σ indicates an electron-withdrawing group and a negative value an electron-donating group.[21] The constant ρ measures the susceptibility of the reaction to electrical effects.[22] Reactions with a positive ρ are helped by electron-withdrawing groups and vice versa. The following ρ values for the ionization of some carboxylic acids illustrate this:[23]

XC_6H_4-COOH	1.00	$XC_6H_4-CH=CH-COOH$	0.47
$XC_6H_4-CH_2-COOH$	0.49	$XC_6H_4-CH_2CH_2-COOH$	0.21

[19]For a review of Hammett treatment of nmr chemical shifts, see Ewing, D.F., in Chapman, N.B.; Shorter, J. *Correlation Analysis in Chemistry: Recent Advances*, Plenum, NY, *1978*, pp. 357–396.

[20]Unless otherwise noted, σ values are from Exner, O. in Chapman, N.B.; Shorter, J. *Correlation Analysis in Chemistry: Recent Advances*, Plenum, NY, *1978*, pp. 439–540, and σ^+ values from Okamoto,Y.; Inukai, T.; Brown, H.C. *J. Am. Chem. Soc. 1958*, *80*, 4969 and Brown, H.C.; Okamoto,Y. *J. Am. Chem. Soc. 1958*, *80*, 4979. σ^- values, except as noted, are from Jaffé, H.H. *Chem. Rev. 1953*, *53*, 191. Exner, O. pp. 439–540, has extensive tables giving values for >500 groups, as well as σ^+, σ^-, σ_I, σ_R^0, and E_s values for many of these groups. Other large tables of the various sigma values are found in Hansch, C.; Leo, A.; Taft, R.W. *Chem. Rev. 1991*, *91*, 165. For tables of σ_p, σ_m, σ^+, σ_I, and σ_R^0 values of many groups containing Si, Ge, Sn, and Pb atoms, see Egorochkin, A.N.; Razuvaev, G.A. *Russ. Chem. Rev. 1987*, *56*, 846. For values for heteroaromatic groups, see Mamaev, V.P.; Shkurko, O.P.; Baram, S.G. *Adv. Heterocycl. Chem. 1987*, *42*, 1.

[21]For discussions of the precise significance of σ, see Dubois, J.E.; Ruasse, M.; Argile, A. *J. Am. Chem. Soc. 1984*, *106*, 4840; Ruasse, M.; Argile, A.; Dubois, J.E. *J. Am. Chem. Soc. 1984*, *106*, 4846; Lee, I.; Shim, C.S.; Chung, S.Y.; Kim, H.Y.; Lee, H.W. *J. Chem. Soc. Perkin Trans. 2 1988*, 1919.

[22]Hine, J. *J. Am. Chem. Soc. 1960*, *82*, 4877.

[23]Binev, I.G.; Kuzmanova, R.B.; Kaneti, J.; Juchnovski, I.N. *J. Chem. Soc. Perkin Trans. 2 1982*, 1533.

TABLE 9.4. The σ, σ^+, and σ^- Values for Some Common Groups[20]

Group	σ_p	σ_m	σ_p^+	σ_m^+	σ_p^-
O⁻	-0.81^{24}	-0.47^{24}	-4.27^{25}	-1.15^{25}	
NMe₂	-0.63	10.10	-1.7		
NH₂	-0.57	-0.09	-1.3	-0.16	
OH	-0.38^{26}	0.13^{26}	-0.92^{27}		
OMe	-0.28^{26}	0.10	-0.78	0.05	
CMe₃	-0.15	-0.09	-0.26	-0.06	
Me	-0.14	-0.06	-0.31	-0.10^{28}	
H	0	0	0	0	0
Ph	0.05^{29}	0.05	-0.18	0^{29}	
COO⁻	0.11^{24}	0.02^{24}	-0.41^{25}	-0.10^{25}	
F	0.15	0.34	-0.07	0.35	
Cl	0.24	0.37	0.11	0.40	
Br	0.26	0.37	0.15	0.41	
I	0.28^{29}	0.34	0.14	0.36	
N=NPh³⁰	0.34	0.28	0.17		
COOH³¹	0.44	0.35	0.42	0.32	0.73
COOR	0.44	0.35	0.48	0.37	0.68
COMe	0.47	0.36			0.87
CF₃	0.53	0.46		0.57^{28}	
NH₃⁺	0.60^{25}	0.86^{26}			
CN³²	0.70	0.62	0.66	0.56	1.00
SO₂Me	0.73	0.64			
NO₂	0.81	0.71	0.79	0.73^{28}	1.27
NMe₃⁺	0.82^{33}	0.88^{33}	0.41	0.36	
N₂⁺	1.93^{34}	1.65^{34}	1.88^{34}		3^{35}

[24]Hine, J. *J. Am. Chem. Soc.* **1960**, *82*, 4877; Jones, R.A.Y. *Physical and Mechanistic Organic Chemistry*, 2nd ed., Cambridge University Press, Cambridge, **1984**, p. 42.

[25]See Hine, J. *J. Am. Chem. Soc.* **1960**, *82*, 4877.

[26]Matsui, T.; Ko, H.C.; Hepler, L.G. *Can. J. Chem.* **1974**, *52*, 2906.

[27]de la Mare, P.B.D.; Newman, P.A. *Tetrahedron Lett.* **1982**, *23*, 1305 give this value as -1.6.

[28]Amin, H.B.; Taylor, R. *Tetrahedron Lett.* **1978**, 267.

[29]Sjöström, M.; Wold, S. *Chem. Scr.* **1976**, *9*, 200.

[30]Byrne, C.J.; Happer, D.A.R.; Hartshorn, M.P.; Powell, H.K.J. *J. Chem. Soc. Perkin Trans. 2* **1987**, 1649.

[31]For a review of directing and activating effects of C=O, C=C, C=N, and C=S groups, see Charton, M., in Patai, S. *The Chemistry of Double-Bonded Functional Groups*, Vol. 2, pt. 1, Wiley, NY, **1989**, pp. 239–298.

[32]For a review of directing and activating effects of C≡N and C≡C groups, see Charton, M., in Patai, S.; Rappoport, Z. *The Chemistry of Functional Groups, Supplement C*, pt. 1, Wiley, NY, **1983**, pp. 269–323.

[33]McDaniel, D.H.; Brown, H.C. *J. Org. Chem.* **1958**, *23*, 420.

[34]Ustynyuk, Yu. A.; Subbotin, O.A.; Buchneva, L.M.; Gruzdneva, V.N.; Kazitsyna, L.A. *Doklad. Chem.* **1976**, *227*, 175.

[35]Lewis, E.S.; Johnson, M.D. *J. Am. Chem. Soc.* **1959**, *81*, 2070.

This example shows that the insertion of a CH_2 or a $CH=CH$ group diminishes electrical effects to about the same extent, while a CH_2CH_2 group diminishes them much more. A $\rho > 1$ would mean that the reaction is more sensitive to electrical effects than is the ionization of XC_6H_4COOH ($\rho = 1.00$).

Similar calculations have been made for compounds with two groups X and X′ on one ring, where the σ values are sometimes additive and sometimes not,[36] for other ring systems, such as naphthalene[37] and heterocyclic rings,[38] and for ethylenic and acetylenic systems.[39]

The Hammett equation is a *linear free-energy relationship* (*LFER*). This can be demonstrated as follows for the case of equilibrium constants (for rate constants a similar demonstration can be made with $\Delta G^{\ddagger}$ instead of ΔG). For each reaction, where X is any group,

$$\Delta G = -RT \ln K$$

For the unsubstituted case,

$$\Delta G_0 = -RT \ln K_0$$

The Hammett equation can be rewritten

$$\log K - \log K_0 = \sigma\rho$$

so that

$$\frac{-\Delta G}{2.3RT} + \frac{\Delta G_0}{2.3RT} = \sigma\rho$$

and

$$-\Delta G = \sigma\rho 2.3RT - \Delta G_0$$

---

[36]Stone, R.M.; Pearson, D.E. *J. Org. Chem.* **1961**, *26*, 257.

[37]Berliner, E.; Winikov, E.H. *J. Am. Chem. Soc.* **1959**, *81*, 1630; see also, Well, P.R.; Ehrenson, S.; Taft, R.W. *Prog. Phys.Org. Chem.* **1968**, *6*, 147.

[38]For reviews, see Charton, M. in Chapman, N.B.; Shorter, J. *Correlation Analysis in Chemistry: Recent Advances*, Plenum, NY, **1978**, pp. 175–268; Tomasik, P.; Johnson, C.D. *Adv. Heterocycl. Chem.* **1976**, *20*, 1.

[39]For reviews of the application of the Hammett treatment to unsaturated systems, see Ford, G.P.; Katritzky, A.R.; Topsom, R.D., in *Correlation Analysis in Chemistry: Recent Advances*, Plenum, NY, **1978**, pp. 269–311; Charton, M. *Prog. Phys. Org. Chem.* **1973**, *10*, 81.

For a given reaction under a given set of conditions, σ, R, T, and ΔG_0 are all constant, so that σ is linear with ΔG.

The Hammett equation is not the only LFER.[40] Some, like the Hammett equation, correlate structural changes in reactants, but the Grunwald–Winstein relationship (see p. 505) correlates changes in solvent and the Brønsted relation (see p. 373) relates acidity to catalysis. The Taft equation is a structure-reactivity equation that correlates only field effects.[41]

Taft, following Ingold,[42] assumed that for the hydrolysis of carboxylic esters, steric and resonance effects will be the same whether the hydrolysis is catalyzed by acid or base (see the discussion of ester-hydrolysis mechanisms, reaction **16-59**). Rate differences would therefore be caused only by the field effects of R and R' in RCOOR'. This is presumably a good system to use for this purpose because the transition state for acid-catalyzed hydrolysis (**8**) has a

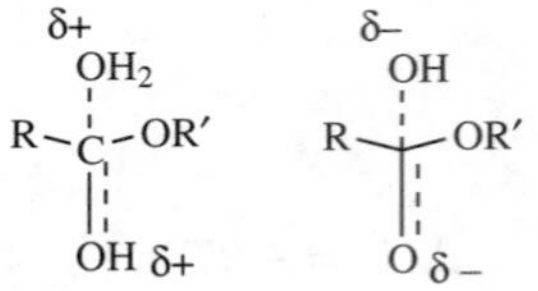

$$
\begin{array}{ccc}
\overset{\delta+}{\underset{}{\text{OH}_2}} & \quad & \overset{\delta-}{\underset{}{\text{OH}}} \\[2pt]
\text{R}\!-\!\overset{|}{\underset{|}{\text{C}}}\!-\!\text{OR}' & \quad & \text{R}\!-\!\overset{|}{\underset{|}{\text{}}}\!-\!\text{OR}' \\[2pt]
\underset{\delta+}{\text{OH}} & \quad & \underset{\delta-}{\text{O}} \\[4pt]
\mathbf{8} & & \mathbf{9}
\end{array}
$$

greater positive charge (and is hence destabilized by $-I$ and stabilized by $+I$ substituents) than the starting ester, while the transition state for base-catalyzed hydrolysis (**9**) has a greater negative charge than the starting ester. Field effects of substituents X could therefore be determined by measuring the rates of acid- and base-catalyzed hydrolysis of a series XCH_2COOR',[43] where R' is held constant.[38] From these rate constants, a value σ_I could be determined by the equation[44]

$$
\sigma_I + 0.181 \left[\log\!\left(\frac{k}{k_0}\right)_{\!B} - \log\!\left(\frac{k}{k_0}\right)_{\!A} \right]
$$

[40]For a discussion of physicochemical preconditions for LFERs, see Exner, O. *Prog. Phys. Org. Chem.* **1990**, *18*, 129.

[41]For reviews of the separation of resonance and field effects, see Charton, M. *Prog. Phys. Org. Chem.* **1981**, *13*, 119; Shorter, J. *Q. Rev. Chem. Soc.* **1970**, *24*, 433; *Chem. Ber.* **1969**, *5*, 269. For a review of field and inductive effects, see Reynolds, W.F. *Prog. Phys. Org. Chem.* **1983**, *14*, 165. For a review of field effects on reactivity, see Grob, C.A. *Angew. Chem. Int. Ed.* **1976**, *15*, 569.

[42]Ingold, C.K. *J. Chem. Soc.* **1930**, 1032.

[43]For another set of field-effect constants, based on a different premise, see Draffehn, J.; Ponsold, K. *J. Prakt. Chem.* **1978**, *320*, 249.

[44]The symbol σ_F is also used in the literature; sometimes in place of σ_I, and sometimes to indicate only the field (not the inductive) portion of the total effect (p. 19).

In this equation $(k/k_0)_B$ is the rate constant for basic hydrolysis of XCH_2COOR' divided by the rate constant for basic hydrolysis of CH_3COOR', $(k/k_0)_A$ is the similar rate-constant ratio for acid catalysis, and 0.181 is an arbitrary constant. σ_I is a substituent constant for a group X, substituted at a saturated carbon, that reflects only field effects.[45] Once a set of σ_I values was obtained, it was found that the equation

$$\sigma_I + 0.181 \left[\log\left(\frac{k}{k_0}\right)_B - \log\left(\frac{k}{k_0}\right)_A \right]$$

holds for a number of reactions, among them:[46]

$$RCH_2OH \longrightarrow RCH_2O^-$$

$$RCH_2Br \ + \ PhS^- \longrightarrow RCH_2SPh \ + \ Br^-$$

$$Acetone \ + \ I_2 \text{, catalyzed by } RCOOH \longrightarrow$$

$$o\text{-Substituted-}ArNH_2 \ + \ PhCOCl \longrightarrow ArNHCOPh$$

As with the Hammett equation, σ_I is constant for a given reaction under a given set of conditions. For very large groups the relationship may fail because of the presence of steric effects, which are not constant. The equation also fails when X enters into resonance with the reaction center to different extents in the initial and transition states. A list of some σ_I values is given in Table 9.5.[47] The σ_I values are about what we would expect for pure field-effect values (see p. 21) and are additive, as field effects (but not resonance or steric effects) would be expected to be. Thus, in moving a group one carbon down the chain, there is a decrease by a factor of 2.8 ± 0.5 (cf. the values of R and RCH_2 in Table 9.5 for $R = Ph$ and CH_3CO). An inspection of Table 9.5 shows that σ_I values for most groups are fairly close to the σ_m values (Table 9.4) for the same groups. This is not surprising, since σ_m values would be expected to arise almost entirely from field effects, with little contribution from resonance.

---

[45]There is another set of values (called σ^* values) that are also used to correlate field effects. These are related to σ_I values by $= \sigma_I(X) = 0.45\sigma$. We discuss only σ_I, and not σ^* values.

[46]Wells, P.R. *Chem. Rev.* **1963**, *63*, 171, p. 196.

[47]These values are from Bromilow, J.; Brownlee, R.T.C.; Lopez, V.O.; Taft, R.W. *J. Org. Chem.* **1979**, *44*, 4766, except that the values for NHAc, OH, and I are from Wells, P.R.; Ehrenson, S.; Taft, R.W. *Prog. Phys. Org. Chem.* **1968**, *6*, 147, the values for Ph and NMe_3^+ are from Taft, R.W.; Ehrenson, S.; Lewis, I.C.; Glick, R. *J. Am.Chem. Soc.* **1959**, *81*, 5352 and Taft, R.W.; Deno, N.C.; Skell, P.S. *Annu. Rev. Phys. Chem.* **1958**, *8*, 287, and the value for CMe_3 is from Seth-Paul, W.A.; de Meyer-van Duyse, A.; Tollenaere, J.P. *J. Mol. Struct.* **1973**, *19*, 811. The values for the CH_2Ph and CH_2COCH_3 groups were calculated from σ^* values by the formula given in reference 45. For much larger tables of σ_I and σ_R values, see Charton, M. *Prog. Phys. Org. Chem.* **1981**, *13*, 119. See also Ref. 20 and Taylor, P.J.; Wait, A.R. *J. Chem. Soc. Perkin Trans. 2* **1986**, 1765.

TABLE 9.5. The σ_I and σ_R^0 Values for Some Groups[47]

Group	σ_I	σ_R^0	Group	σ_I	σ_R^0
CMe_3	−0.07	−0.17	OMe	0.27	−0.42
Me	−0.05	−0.13	OH	0.27	−0.44
H	0	0	I	0.39	−0.12
$PhCH_2$	0.04		CF_3	0.42	0.08
NMe_2[48]	0.06	−0.55	Br	0.44	−0.16
Ph	0.10	−0.10	Cl	0.46	−0.18
CH_3COCH_2	0.10		F	0.50	−0.31
NH_2	0.12	−0.50	CN	0.56	0.08
CH_3CO	0.20	0.16	SO_2Me	0.60	0.12
COOEt	0.20	0.16	NO_2	0.65	0.15
NHAc	0.26	−0.22	NMe_3[49]	0.86	

Since σ_p values represent the sum of resonance and field effects, these values can be divided into resonance and field contributions if σ_I is taken to represent the field-effect portion.[50] The resonance contribution σ_R[51] is defined as

$$\sigma_R = \sigma_p - \sigma_I$$

As it stands, however, this equation is not very useful because the σ_R value for a given group, which should be constant if the equation is to have any meaning, is actually not constant but depends on the nature of the reaction.[52] In this respect, the σ_I values are much better. Although they vary with solvent in some cases, σ_I values are essentially invariant throughout a wide variety of reaction series. However, it is possible to overcome[53] the problem of varying σ_R values by using a special set of σ_R values, called

[48]For σ_R^0 values for some other NR_2 groups, see Korzhenevskaya, N.G.; Titov, E.V.; Chotii, K.Yu.; Chekhuta, V.G. *J. Org. Chem. USSR* **1987**, *28*, 1109.

[49]Although we give a σ_I value for NMe_3^+, (and *F* values for three charged groups in Table 9.6), it has been shown that charged groups (called polar substituents) cannot be included with uncharged groups (dipolar substituents) in one general scale of electrical substituent effects: Marriott, S.; Reynolds, J.D.; Topsom, R.D. *J. Org. Chem.* **1985**, *50*, 741.

[50]Roberts, J.D.; Moreland, Jr., W.T. *J. Am. Chem. Soc.* **1953**, *75*, 2167; Taft, R.W. *J. Am. Chem. Soc.* **1957**, *79*, 1045; *J. Phys. Chem.* **1960**, *64*, 1805; Taft, R.W.; Lewis, I.C. *J. Am. Chem. Soc.* **1958**, *80*, 2436; Taft, R.W.; Deno, N.C.; Skell, P.S. *Annu. Rev. Phys. Chem.* **1958**, *9*, 287, see pp. 290–293.

[51]For reviews of the σ_I and σ_R concept as applied to benzenes and naphthalenes, respectively, see Ehrenson, S.; Brownlee, R.T.C.; Taft, R.W. *Prog. Phys. Org. Chem.* **1973**, *10*, 1. See also, Taft, R.W.; Topsom, R.D. *Prog. Phys. Org. Chem.* **1987**, *16*, 1; Charton, M. *Prog. Phys. Org. Chem.* **1987**, *16*, 287.

[52]Taft, R.W.; Lewis, I.C. *J. Am. Chem. Soc.* **1959**, *81*, 5343; Reynolds, W.F.; Dais, P.; MacIntyre, D.W.; Topsom, R.D.; Marriott, S.; von Nagy-Felsobuki, E.; Taft, R.W. *J. Am. Chem. Soc.* **1983**, *105*, 378.

[53]For a different way of overcoming this problem, see Happer, D.A.R.; Wright, G.J. *J. Chem. Soc. Perkin Trans. 2* **1979**, 694.

σ_R^o,[54] that measure the ability to delocalize π electrons into or out of an unperturbed or "neutral" benzene ring. Several σ_R^o scales have been reported; the most satisfactory values are obtained from ^{13}C chemical shifts of substituted benzenes.[55] Table 9.5 lists some values of σ_R^o, most of which were obtained in this way.[56]

An equation such as

$$\log \frac{k}{k_0} = \rho_I \sigma_I + \rho_R \sigma_R^o$$

which treats resonance and field effects separately, is known as a *dual substituent parameter equation.*[57]

The only groups in Table 9.5 with negative values of σ_I are the alkyl groups methyl and *t*-butyl. There has been some controversy on this point.[58] One opinion is that σ_I values decrease in the series methyl, ethyl, isopropyl, *tert*-butyl (respectively, -0.046, -0.057, -0.065, -0.074).[59] Other evidence, however, has led to the belief that all alkyl groups have approximately the same field effect and that the σ_I values are invalid as a measure of the intrinsic field effects of alkyl groups.[60]

Another attempt to divide σ values into resonance and field contributions[61] is that of Swain and Lupton, who have shown that the large number of sets of σ values (σ_m, σ_p, σ_p^-, σ_p^+, σ_I, σ_R^o, etc., as well as others we have not mentioned) are not entirely independent and that linear combinations of two sets of new values F (which expresses the field-effect contribution) and R (the resonance contribution) satisfactorily express 43 sets of values.[62] Each set is expressed as

$$\sigma = fF + rR$$

[54]Taft, R.W.; Ehrenson, S.; Lewis, I.C.; Glick, R.E. *J. Am. Chem. Soc.* **1959**, *81*, 5352.

[55]Bromilow, J.; Brownlee, R.T.C.; Lopez, V.O.; Taft, R.W. *J. Org. Chem.* **1979**, *44*, 4766. See also, Marriott, S.; Topsom, R.D. *J. Chem. Soc. Perkin Trans. 2* **1985**, 1045.

[56]For a set of σ_R values for use in XY^+ systems, see Charton, M. *Mol. Struct. Energ.* **1987**, *4*, 271.

[57]There are also three-parameter equations. See, for example, de Ligny, C.L.; van Houwelingen, H.C. *J. Chem. Soc. Perkin Trans. 2* **1987**, 559.

[58]For a discussion, see Shorter, J. in Chapman, N.B.; Shorter, J. *Advances in Linear Free Energy Relationships*, Plenum, NY, **1972**, pp. 98–103.

[59]For support for this point of view, see Levitt, L.S.; Widing, H.F. *Prog. Phys. Org. Chem.* **1976**, *12*, 119; Taft, R.W.; Levitt, L.S. *J. Org. Chem.* **1977**, *42*, 916; MacPhee, J.A.; Dubois, J.E. *Tetrahedron Lett.* **1978**, 2225; Screttas, C.G. *J. Org. Chem.* **1979**, *44*, 3332; Hanson, P. *J. Chem. Soc. Perkin Trans. 2* **1984**, 101.

[60]For support for this point of view, see, for example, Ritchie, C.D. *J. Phys. Chem.* **1961**, *65*, 2091; Bordwell, F.G.; Drucker, G.E.; McCollum, G.J. *J. Org. Chem.* **1976**, *41*, 2786; Bordwell, F.G.; Fried, H.E. *Tetrahedron Lett.* **1977**, 1121; Charton, M. *J. Am. Chem. Soc.* **1977**, *99*, 5687; *J. Org. Chem.* **1979**, *44*, 903; Adcock, W.; Khor, T. *J. Org. Chem.* **1978**, *43*, 1272; DeTar, D.F. *J. Org. Chem.* **1980**, *45*, 5166; *J. Am. Chem. Soc.* **1980**, *102*, 7988.

[61]Yukawa and Tsuno have still another approach, also involving dual parameters: Yukawa, Y.; Tsuno, Y. *Bull. Chem. Soc. Jpn.* **1959**, *32*, 971. For a review and critique of this method, see Shorter, J., in Chapman, N.B.; Shorter, J. *Correlation Analysis in Chemistry: Recent Advances*, Plenum, NY, **1978**, pp. 119–173, 126–144. This article also discusses the Swain–Lupton and Taft σ_I, σ_R approaches. For yet other approaches, see Afanas'ev, I.B. *J. Org. Chem. USSR* **1981**, *17*, 373; *J. Chem. Soc. Perkin Trans. 2* **1984**, 1589; Ponec, R. *Coll. Czech. Chem. Commun.* **1983**, *48*, 1564.

[62]Swain, C.G.; Unger, S.H.; Rosenquist, N.R.; Swain, M.S. *J. Am. Chem. Soc.* **1983**, *105*, 492 and references cited therein.

TABLE 9.6. The F and R Values for Some Groups[63]

Group	F	R	Group	F	R
COO$^-$	−0.27	0.40	OMe	0.54	−1.68
Me$_3$C	−0.11	−0.29	CF$_3$	0.64	0.76
Et	−0.02	−0.44	I	0.65	−0.12
Me	−0.01	−0.41	Br	0.72	−0.18
H	0	0	Cl	0.72	−0.24
Ph	0.25	−0.37	F	0.74	−0.60
NH$_2$	0.38	−2.52	NHCOCH$_3$	0.77	−1.43
COOH	0.44	0.66	CN	0.90	0.71
OH	0.46	−1.89	NMe$_3^+$	1.54	
COOEt	0.47	0.67	N$_2^+$	2.36	2.81
COCH$_3$	0.50	0.90			

where f and r are weighting factors. Some F and R values for common groups are given in Table 9.6.[63] From the calculated values of f and r, Swain and Lupton calculated that the importance of resonance, $\% R$, is 20% for σ_m, 38% for σ_p, and 62% for σ_p^+.[64] This is another dual substituent parameter approach.

Taft was also able to isolate steric effects.[65] For the acid-catalyzed hydrolysis of esters in aqueous acetone, long (k/k_0) was shown to be insensitive to polar effects.[66] In cases where resonance interaction was absent, this value was proportional only to steric effects (and any others[67] that are not field or resonance). The equation is

$$\log \frac{k}{k_0} = E_S$$

[63]Taken from a much longer list in Swain, C.G.; Unger, S.H.; Rosenquist, N.R.; Swain, M.S. *J. Am. Chem. Soc.* **1983**, *105*, 492. Long tables of R and F values are also given in Hansch, C.; Leo, A.; Taft, R.W. *Chem. Rev.* **1991**, *91*, 165.

[64]The Swain–Lupton treatment has been criticized by Reynolds, W.F.; Topsom, R.D. *J. Org. Chem.* **1984**, *49*, 1989; Hoefnagel, A.J.; Oosterbeek, W.; Wepster, B.M. *J. Org. Chem.* **1984**, *49*, 1993; and Charton, M. *J. Org. Chem.* **1984**, *49*, 1997. For a reply to these criticisms, see Swain, C.G. *J. Org. Chem.* **1984**, *49*, 2005. A study of the rates of dediazoniation reactions (**13-32**) was more in accord with the Taft and Charton (see Charton, M. *Prog. Phys. Org. Chem.* **1981**, *13*, 119) σ_I and σ_R values than with the Swain–Lupton F and R values: Nakazumi, H.; Kitao, T.; Zollinger, H. *J. Org. Chem.* **1987**, *52*, 2825.

[65]For reviews of quantitative treatments of steric effects, see Gallo, R.; Roussel, C.; Berg, U. *Adv. Heterocycl. Chem.* **1988**, *43*, 173; Gallo, R. *Prog. Phys. Org. Chem.* **1983**, *14*, 115; Unger, S.H.; Hansch, C. *Prog. Phys. Org. Chem.* **1976**, *12*, 91.

[66]Another reaction used for the quantitative measurement of steric effects is the aminolysis of esters (**16-75**); De Tar, D.F.; Delahunty, C. *J. Am. Chem. Soc.* **1983**, *105*, 2734.

[67]It has been shown that E_s values include solvation effects: McClelland, R.A.; Steenken, S. *J. Am. Chem. Soc.* **1988**, *110*, 5860.

TABLE 9.7. The E_s, υ, and V^a Values for Some Groups[68]

Group	E_s	υ	$V^a \times 10^2$	Group	E_s	υ	$V^a \times 10^2$
H	0	0		Cyclohexyl	−2.03	0.87	6.25
F	−0.46	0.27	1.22	i-Bu	−2.17	0.98	5.26
CN	−0.51			sec-Bu	−2.37	1.02	6.21
OH	−0.55			CF_3	−2.4	0.91	3.54
OMe	−0.55		3.39	t-Bu	−2.78	1.24	7.16
NH_2	−0.61			NMe_3^+	−2.84		
Cl	−0.97	0.55	2.54	Neopentyl	−2.98	1.34	5.75
Me	−1.24	0.52	2.84	CCl_3	−3.3	1.38	6.43
Et	−1.31	0.56	4.31	CBr_3	−3.67	1.56	7.29
I	−1.4	0.78	4.08	$(Me_3CCH_2)_2CH$	−4.42	2.03	
Pr	−1.6	0.68	4.78	Et_3C	−5.04	2.38	
iPr	−1.71	0.76	5.74	Ph_3C	−5.92	2.92	

Some E_s values are given in Table 9.7,[68] where hydrogen is taken as standard, with a value of 0.[69] This treatment is more restricted than those previously discussed, since it requires more assumptions, but the E_s values are approximately in order of the size of the groups. Charton has shown that E_s values for substituents of types CH_2X, CHX_2, and CX_3 are linear functions of the van der Waals radii for these groups.[70]

Two other steric parameters are independent of any kinetic data. Charton's υ values are derived from van der Waals radii,[71] and Meyer's V^a values from the volume of the portion of the substituent that is within 0.3 nm of the reaction center.[72] The V^a values are obtained by molecular mechanics calculations based on the structure of the molecule. Table 9.7 gives υ and V^a values for some groups.[73] As can be seen in the table, there is a fair, but not perfect, correlation among the E_s, υ, and V^a values. Other sets of steric values, for example, E_s',[74] E_s^*,[75] Ω_s,[76] and f,[77] have also been proposed.[73]

[68]The E_s, υ, and V^a values are taken from longer tables in, respectively, Gallo, R.; Roussel, C.; Berg, U. *Adv. Heterocycl. Chem.* **1988**, *43*, 173; Gallo, R. *Prog. Phys. Org. Chem.* **1983**, *14*, 115; Unger, S.H.; Hansch, C. *Prog. Phys. Org. Chem.* **1976**, *12*, 91. Charton, M. *J. Am. Chem. Soc.* **1975**, *97*, 1552; *J. Org. Chem.* **1976**, *41*, 2217; and Meyer, A.Y. *J. Chem. Soc. Perkin Trans. 2* **1986**, 1567.
[69]In Taft's original work, Me was given the value 0. The E_s values in Table 9.7 can be converted to the orginal values by adding 1.24.
[70]Charton, M. *J. Am. Chem. Soc.* **1969**, *91*, 615.
[71]Charton, M. *J. Am. Chem. Soc.* **1975**, *97*, 1552; *J. Org. Chem.* **1976**, *41*, 2217. See also, Charton, M. *J. Org. Chem.* **1978**, *43*, 3995; Idoux, J.P.; Schreck, J.O. *J. Org. Chem.* **1978**, *43*, 4002.
[72]Meyer, A.Y. *J. Chem. Soc. Perkin Trans. 2* **1986**, 1567.
[73]For a discussion of the various steric parameters, see DeTar, D.F. *J. Org. Chem.* **1980**, *45*, 5166; *J. Am. Chem. Soc.* **1980**, *102*, 7988.
[74]MacPhee, J.A.; Panaye, A.; Dubois, J.E. *J. Org. Chem.* **1980**, *45*, 1164; Dubois, J.E.; MacPhee, J.A.; Panaye, A. *Tetrahedron* **1980**, *36*, 919. See also, Datta, D.; Sharma, G.T. *J. Chem. Res. (S)* **1987**, 422.
[75]Fellous, R.; Luft, R. *J. Am. Chem. Soc.* **1973**, *95*, 5593.
[76]Komatsuzaki, T.; Sakakibara, K.; Hirota, M. *Tetrahedron Lett.* **1989**, *30*, 3309; *Chem. Lett.* **1990**, 1913.
[77]Beckhaus, H. *Angew. Chem. Int. Ed.* **1978**, *17*, 593.

Since the Hammett equation has been so successful in the treatment of the effects of groups in the meta and para positions, it is not surprising that attempts have been made to apply it to ortho positions also.[78] The effect on a reaction rate or equilibrium constant of a group in the ortho position is called the *ortho effect*.[79] Despite the many attempts made to quantify ortho effects, no set of values has so far commanded general agreement. However, the Hammett treatment is successful for ortho compounds when the group Y in $o\text{-}XC_6H_4Y$ is separated from the ring; for example, ionization constants of $o\text{-}XC_6H_4OCH_2COOH$ can be successfully correlated.[80]

Linear free-energy relationships can have mechanistic implications. If $\log(k/k_0)$ is linear with the appropriate σ, it is likely that the same mechanism operates throughout the series. If not, a smooth curve usually indicates a gradual change in mechanism, while a pair of intersecting straight lines indicates an abrupt change,[81] though nonlinear plots can also be due to other causes, such as complications arising from side reactions. If a reaction series follows σ^+ or σ^- better than σ it generally means that there is extensive resonance interaction in the transition state.[82]

Information can also be obtained from the magnitude and sign of ρ. For example, a strongly negative ρ value indicates a large electron demand at the reaction center, from which it may be concluded that a highly electron-deficient center, perhaps an incipient carbocation, is involved. Conversely, a positive ρ value is associated with a developing negative charge in the transition state.[83] The $\sigma\rho$ relationship even applies to free-radical reactions, because free radicals can have some polar character (p. 939), though ρ values here are usually small (less than $\sim$1.5) whether positive or negative. Reactions involving cyclic transition states (p. 297) also exhibit very small ρ values.

EFFECT OF MEDIUM ON REACTIVITY AND RATE

There is no question that the solvent chosen for a given reaction has a profound influence on the course of that reaction. Protic versus aprotic solvents as well as polar versus nonpolar solvents can have effects ranging from solubility to solvent assisted ionization or stabilization of transition states. Reactions can

---

[78]For reviews, see Fujita, T.; Nishioka, T. *Prog. Phys. Org. Chem.* **1976**, *12*, 49; Charton, M. *Prog. Phys. Org. Chem.* **1971**, *8*, 235; Shorter, J., in Chapman, N.B.; Shorter, J. *Advances in Linear Free Energy Relaltionships*, Plenum, NY, **1972**, pp. 103–110. See also, Segura, P. *J. Org. Chem.* **1985**, *50*, 1045; Robinson, C.N.; Horton, J.L.; Fosheé, D.O.; Jones, J.W.; Hanissian, S.H.; Slater, C.D. *J. Org. Chem.* **1986**, *51*, 3535.

[79]This is not the same as the ortho effect discussed on p. $$$.

[80]Charton, M. *Can. J. Chem.* **1960**, *38*, 2493.

[81]For a discussion, see Schreck, J.O. *J. Chem. Educ.* **1971**, *48*, 103.

[82]See, however, Gawley, R.E. *J. Org. Chem.* **1981**, *46*, 4595.

[83]For another method of determining transition state charge, see Williams, A. *Acc. Chem. Res.* **1984**, *17*, 425.

also be done neat in one of the reactants, in the gas phase, on solid support or in the solid phase. Environmental friendly chemistry (green chemistry) is becoming increasingly important, and chemical reactions in nonpolluting (often non-organic) solvents is of particular interest.[84] This section will describe alternative reaction media as well as other medium-related things that influence chemical reactions.

HIGH PRESSURE

Acceleration of some chemical reactions is possible when high-pressure techniques are employed.[85,86] The effects on a given reaction can be predicted to a certain extent because the thermodynamic properties of solutions are well known. The rate of a reaction can be expressed in terms of the activation volume, $\Delta V^{\ddagger}$

$$\frac{\delta \ln k}{\delta p} = \frac{\Delta V^{\ddagger}}{RT}$$

so rate constants vary with pressure.[86] "The activation volume[87] is the difference in partial molal volume between the transition state and the initial state. From a synthetic point of view this could be approximated by the molar volume."[86] If the volume of activation is negative, the rate of the reaction will be accelerated by increasing pressure. As the pressure increases, the value of $\Delta V^{\ddagger}$ decreases and the system does not strictly obey equation (11.4) > 10 kbar (1 bar = 0.986924 atm = 1.1019716 kg cm^{-2}). If the transition state of a reaction involves bond formation, concentration of charge, or ionization, a negative volume of activation often results. Cleavage of a bond, dispersal of charge, or neutralization of the transition state and diffusion control lead to a positive volume of activation. Matsumoto summarized the reactions for which rate enhancement is expected at high pressure.[86]

1. Reactions in which the molecularity number (number of molecules) decreases when starting materials are converted to products: cycloadditions, condensations.
2. Reactions that proceed via cyclic transition states.
3. Reactions that take place through dipolar transition states.
4. Reactions with steric hindrance.

---

[84]For example, see Clark, J.H. *Green Chem.* **1999**, *1*, 1; Cave, G.W.V.; Raston, C.L.; Scott, J.L. *Chem. Commun.* **2001**, 2159.

[85]Jenner, G. *Tetrahedron* **2002**, *58*, 5185; Matsumoto, K.; Morris, A.R. *Organic Synthesis at High Pressure*, Wiley, New York, **1991**.

[86]Matsumoto, K.; Sera, A.; Uchida, T. *Synthesis* **1985**, 1; Matsumoto, K.; Sera, A. *Synthesis* **1985**, 999.

[87]See le Noble, W.J. *Progr. Phys. Org. Chem.* **1967**, *5*, 207; Isaacs, N.S. *Liquid Phase High Pressure Chemistry*, Wiley, Chichester, **1981**; Asano, T.; le Noble, W.J. *Chem. Rev.* **1978**, *78*, 407.

Many high-pressure reactions are done neat, but if a solvent is used, the influence of pressure on that solvent is important. The melting point generally increases at elevated pressures, and this influences the viscosity of the medium (the viscosity of liquids increases approximately two times per kilobar increase in pressure). Controlling the rate of diffusion of reactants in the medium is also important, leading to another influence of high pressure on reactivity.[86,88] In most reactions, pressure is applied (5–20 kbar) at room temperature and then the temperature is increased until reaction takes place. The temperature is lowered and the pressure is reduced to isolate the products.

WATER AND OTHER NONORGANIC SOLVENTS

Chemical reactions of organic substrates usually employs an organic solvent, such as a hydrocarbon, ether, dichloromethane, and so on. With the exception of small molecular weight molecules with polar functional groups and polyfunctional molecules or salts, organic chemicals have poor solubility in water. The first indication that water accelerated a reaction was in a patent by Hopff and Rautenstrauch in 1939,[89] who reported that yields in the Diels–Alder reaction (**15-60**) were enhanced in aqueous detergent solutions. In an early study, Berson showed a clear relationship between the endo/exo product ratio and solvent polarity, in the Diels–Alder reaction of cyclopentadiene and acrylates.[90] Breslow showed there was a hydrophobic acceleration for an intermolecular Diels–Alder reaction in which cyclopentadiene reacted with methyl vinyl ketone.[91] Clearly, there is an accelerating effect on some chemical reactions when done in water that is useful in organic chemistry.[92]

When nonpolar compounds are suspended in water their relative insolubility causes them to associate, diminishing the water–hydrocarbon interfacial area (a hydrophobic effect). This association is greater in water than in methanol and brings the reactive partners into close proximity, increasing the rate of reaction. Any additive that increases the hydrophobic effect will increase the rate.[91]

Carbon dioxide can be used as a reaction solvent when pressurized (supercritical carbon dioxide, scCO$_2$). Carbon dioxide is nontoxic, inexpensive, abundant, and easily recycled. These properties have made it attractive as an extraction solvent.[93] The low critical temperature of CO$_2$ (T_c) 31.1 °C ensures that scCO$_2$ is a safe solvent for many applications.[94] There are solubility issues that suggest scCO$_2$ is a rather polar solvent.[95] For example, many systems with hydrocarbon chains are

[88]Firestone, R.A.; Vitale, M.A. *J. Org. Chem.* **1981**, *46*, 2160.

[89]Hopff, H.; Rautenstrauch, C.W. *U.S. Patent* 2,262,002, **1939** [*Chem. Abstr. 36*: 10469, *1942*].

[90]Berson, J.A.; Hamlet, Z.; Mueller, W.A. *J. Am. Chem. Soc.* **1962**, *84*, 297.

[91]Rideout, D.; Breslow, R. *J. Am. Chem. Soc.* **1980**, *102*, 7816.

[92]Engberts, J.B.F.N.; Blandamer, M.J. *Chem. Commun.* **2001**, 1701; Lindström, U.M. *Chem. Rev.* **2002**, *102*, 2751; Ribe, S.; Wipf, P. *Chem. Commun.* **2001**, 299.

[93]See Raynie, D.E. *Anal. Chem.* **2004**, *76*, 4659.

[94]Subramaniam, B.; Rajewski, R.A.; Snavely, K. *J. Pharm. Sci.* **1997**, *86*, 885.

[95]Raveendran, P.; Ikushima, Y.; Wallen, S.L. *Acc. Chem. Res.* **2005**, *38*, 478.

not very soluble in CO_2.[96] Water/carbon dioxide emulsions have also been employed.[97]

The use of supercritical carbon dioxide ($ScCO_2$) has been explored in many reactions,[98] including catalysis.[99] Some applications of this technique include the electrochemical synthesis of conducting polymers[100] and highly cross-linked polymers[101] in $scCO_2$, the synthesis of octyl palmitate,[102] of carbonated fatty methyl esters,[103] and of methyl carbamates.[104] A carbonylation reaction was done is $ScCO_2$ in the course of a synthesis of trisubstituted cyclopentanes and cyclohexanes as key components of substance P antagonists.[105] A continuous flow acid catalyzed dehydration of alcohols was accomplished in $ScCO_2$.[106] Supercritical fluids are playing an increasingly important role in synthetic organic chemistry.[107]

Other supercritical fluids can be used for chemical reactions, such as supercritical ammonia in the synthesis of labeled guanidines.[108]

IONIC SOLVENTS

Environmentally friendly solvents, such as ionic liquids, is of great interest. It was discovered that some molecules form ionic liquids that are suitable as a medium for chemical reactions.[109] An ionic liquid is a salt in which the ions are poorly coordinated, usually leading to their being liquid $<100°C$ and sometimes at room temperature. In such ionic species, there is usually at least one ion with a delocalized charge whereas the other component is usually organic. This combination inhibits the formation of a stable crystal lattice. Both methylimidazolium and pyridinium ions form the basis of common ionic liquids that have been used in organic chemistry. One of the most common ionic solvents is 1-butyl-3-methylimidazolium as the hexafluorophosophate, **10**

[96]Consani, K.A.; Smith, R.D.J. *Supercrit. Fluids* **1990**, *3*, 51.

[97]Jacobson, G.B.; Lee Jr., C.T.; da Rocha, S.R.P.; Johnston, K.P. *J. Org. Chem.* **1999**, *64*, 1207; Jacobson, G.B.; Lee, Jr., C.T.; Johnston, K.P. *J. Org. Chem.* **1999**, *64*, 1201.

[98]Gopalan, A.D.; Wai, C.M.; Jacobs, H.K. *Supercritical Carbon Dioxide: Separations and Processes*, American Chemical Society (distributed by Oxford University Press), Washington, DC, **2003**; Beckman, E.J. *Ind. Eng. Chem. Res.* **2003**, *42*, 1598; Wang, S.; Kienzle, F. *Ind. Eng. Chem. Res.* **2000**, *39*, 4487.

[99]Leitner, W. *Acc. Chem. Res.* **2002**, *35*, 746.

[100]Anderson, P.E.; Badlani, R.N.; Mayer, J.; Mabrouk, P.A. *J. Am. Chem. Soc.* **2002**, *124*, 10284.

[101]Cooper, A.I.; Hems, W.P.; Holmes, A.B. *Macromolecules* **1999**, *32*, 2156.

[102]Madras, G.; Kumar, R.; Modak, J. *Ind. Eng. Chem. Res.* **2004**, *43*, 7697,1568.

[103]Doll, K.M.; Erhan, S.Z. *J. Agric. Food Chem.* **2005**, *53*, 9608.

[104]Selva, M.; Tundo, P.; Perosa, A.; Dall'Acqua, F. *J. Org. Chem.* **2005**, *70*, 2771.

[105]Kuethe, J.T.; Wong, A.; Wu, J.; Davies, I.W.; Dormer, P.G.; Welch, C.J.; Hillier, M.C.; Hughes, D.L.; Reider, P.J. *J. Org. Chem.* **2002**, *67*, 5993.

[106]Gray, W.K.; Smail, F.R.; Hitzler, M.G.; Ross, S.K.; Poliakoff, M. *J. Am. Chem. Soc.* **1999**, *121*, 10711.

[107]Oakes, R.S.; Clifford, A.A.; Rayner, C.M. *J. Chem. Soc., Perkin Trans. 1* **2001**, 917; Prajapati, D.; Gohain, M. *Tetrahedron* **2004**, *60*, 815.

[108]Jacobson, G.B.; Westerberg, G.; Markides, K.E.; Langstrom, B. *J. Am. Chem. Soc.* **1996**, *118*, 6868.

[109]Wasserscheid, P.; Keim, W. *Angew. Chem. Int. Ed.* **2000**, *39*, 3772; Earle, M.J.; Seddon, K.R. *Pure. Appl. Chem.* **2000**, *72*, 1391; Wasserscheid, P.; Welton, T. *Ionic Liquids in Synthesis*, Wiley-VCH, NY, **2002**; Adams, D.J.; Dyson, P.J.; Taverner, S.J. *Chemistry in Alternative Reaction Media*, Wiley, NY, **2003**.

(Bmim PF$_6$).[110] Hydrogenbutylimidazolium tetrafluoroborate (HBuIm, **11**) and 1,3-dibutylimidazolium, tetrafluoroborate (DiBuIm, **12**), for example,[111] have been reported to facilitate Diels–Alder reactions (**15-60**).[112] Pyridinium based ionic liquids, such as ethylpyridinium tetrafluoroborate (**13**), have also been used.[113]

Ionic solvents have been used to facilitate the Heck reaction (**13-9**),[114] the oxidation of alcohols with hypervalent iodine reagents (**19-3**),[115] and the catalytic asymmetric dihydroxylation of olefins (**15-48**) using a recoverable and reusable osmium/ligand.[116] Reactions in ionic liquids is a rapidly growing area of organic chemistry, including microwave reactions (see p. 354) in ionic solvents.[117] The development and use of ionic solvents is a growth area of organic chemistry.[118]

SOLVENTLESS REACTIONS

In some cases, it should be possible to accomplish a chemical transformation without the use of a solvent. Dry media reaction under microwaves is an important area of study (see p. 352).[119] There are several advantages of solventless reactions: (*1*) the possibility of direct formation of high purity compounds, (*2*) the possibility of sequential reactions, (*3*) fast kinetics, (*4*) lower energy usage, (*5*) minimal need for preformed salts and metal–metalloid complexes, (*6*) simplicity and low equipment cost, and (*7*) the possibility of avoiding functional group protection–deprotection.[120] Potential difficulties include the possibility of hot spots and runaway reactions, and difficulties in handling solid or highly viscous materials.[121] An example of this approach is the aldol condensation, where a single aldol product was obtained in high yield.[122] 3-Carboxylcoumarins have been produced via a solventless aldol.[120]

[110]Dupont, J.; Consorti, C.S.; Suarez, P.A.Z.; de Souza, R.F. *Org. Synth. Coll. Vol. X*, 184.

[111]For discussion of HBuIM and DiBuIm, see Harlow, K.J.; Hill, A.F.; Welton, T. *Synthesis* **1996**, 697; Holbrey, J.D.; Seddon, K.R. *J. Chem. Soc., Dalton Trans.* **1999**, 2133; Larsen, A.S.; Holbrey, J.D.; Tham, F.S.; Reed, C.A. *J. Am. Chem. Soc.* **2000**, *122*, 7264.

[112]Jaegar, D.A.; Tucker, C.E. *Tetrahedron Lett.* **1989**, *30*, 1785.

[113]See Xiao, Y.; Malhotra, S.V. *Tetrahedron Lett.* **2004**, *45*, 8339.

[114]Handy, S.T.; Okello, M.; Dickenson, G. *Org. Lett.* **2003**, *5*, 2513.

[115]Yadav, J.S.; Reddy, B.V.S.; Basak, A.K.; Narsaiah, A.V. *Tetrahedron* **2004**, *60*, 2131.

[116]Branco, L.C.; Afonso, C.A.M. *J. Org. Chem.* **2004**, *69*, 4381.

[117]See Leadbeater, N.E.; Torenius, H.M. *J. Org. Chem.* **2002**, *67*, 3145.

[118]For studies to expand the polarity range of ionic solvents see Dzyuba, S.V.; Bartsch, R.A. *Tetrahedron Lett.* **2002**, *43*, 4657.

[119]Kidwai, M. *Pure Appl. Chem.* **2001**, *73*, 147.

[120]Cave, G.W.V.; Raston, C.L.; Scott, J.L. *Chem. Commun.* **2001**, 2159; Toda, F.; Tanaka, K. *Chem. Rev.* **2000**, *100*, 1025.

[121]Raston, C.L. *Chemistry in Australia* **2004**, 10.

[122]Toda, F.; Tanaka, K.; Hamai, K. *J. Chem. Soc., Perkin Trans. 1* **1990**, 3207.

PART TWO

In Part 2 of this book, we will be directly concerned with organic reactions and their mechanisms. The reactions have been classified into 10 chapters, based primarily on reaction type: substitutions, additions to multiple bonds, eliminations, rearrangements, and oxidation–reduction reactions. Five chapters are devoted to substitutions; these are classified on the basis of mechanism as well as substrate. Chapters 10 and 13 include nucleophilic substitutions at aliphatic and aromatic substrates, respectively. Chapters 12 and 11 deal with electrophilic substitutions at aliphatic and aromatic substrates, respectively. All free-radical substitutions are discussed in Chapter 14. Additions to multiple bonds are classified not according to mechanism, but according to the type of multiple bond. Additions to carbon–carbon multiple bonds are dealt with in Chapter 15; additions to other multiple bonds in Chapter 16. One chapter is devoted to each of the three remaining reaction types: Chapter 17, eliminations; Chapter 18, rearrangements; Chapter 19, oxidation–reduction reactions. This last chapter covers only those oxidation–reduction reactions that could not be conveniently treated in any of the other categories (except for oxidative eliminations).

Each chapter in Part 2 consists of two main sections. The first section of each chapter (except Chapter 19) deals with mechanism and reactivity. For each reaction type the various mechanisms are discussed in turn, with particular attention given to the evidence for each mechanism and to the factors that cause one mechanism rather than another to prevail in a given reaction. Following this, each chapter contains a section on reactivity, including, where pertinent, a consideration of orientation and the factors affecting it.

The second main section of each chapter is a treatment of the reactions belonging to the category indicated by the title of the chapter. It is not possible to discuss in a book of this nature all or nearly all known reactions. However, an attempt has been made to include all the important reactions of standard organic chemistry that can be used to prepare relatively pure compounds in reasonable yields. In order to present a well-rounded picture and to include some reactions that are traditionally discussed in textbooks, a number of reactions that do not fit into the above category have been included. The scope of the coverage is apparent from the fact that more than 90% of the individual preparations given in *Organic Syntheses* are treated. However, certain special areas have been covered only lightly or not at all. Among these are electrochemical and polymerization reactions, and the preparation and reactions of heterocyclic compounds, carbohydrates, steroids, and compounds containing phosphorus, silicon, arsenic, boron, and mercury. The basic principles involved in these areas are of course no different from those in the areas more fully treated. Even with these omissions, however, some 580 reactions are treated in this book.

Each reaction is discussed in its own numbered section.[1] These are numbered consecutively within a chapter, each section number preceded by the chapter number, so that reaction **16-1** is the first reaction of Chapter 16 and reaction **13-21** is the twenty-first reaction of Chapter 13. The order in which the reactions are presented is not arbitrary, but is based on an orderly outline that depends on the type of reaction. Within each section, the scope and utility of the reaction are discussed and references are given to review articles, if any. If there are features of the mechanism that especially pertain to that reaction, these are also discussed within the section rather than in the first part of the chapter where the discussion of mechanism is more general.

IUPAC NOMENCLATURE FOR TRANSFORMATIONS

There has long been a need for a method of naming reactions. As most students know well, many reactions are given the names of their discoverers or those who popularized them (e.g., Claisen, Diels–Alder, Stille, Wittig, Cope, Dess–Martin). This is useful as far as it goes, but each name must be individually memorized, and there are many reactions that do not have such names. The IUPAC Commission on Physical Organic Chemistry has produced a *system* for naming not reactions, but transformations (a reaction includes all reactants; a transformation shows only the substrate and product, omitting the reagents). The advantages of a systematic method are obvious. Once the system is known, no memorization is required; the name can be generated directly from the equation. The system includes rules for naming eight types of transformation: substitutions, additions, eliminations, attachments and detachments, simple rearrangements, coupling and uncoupling, insertions and extrusions, and ring opening and closing. We give here only the most basic rules for the first three of these types, which, however, will suffice for naming many transformations.[2] The complete rules give somewhat different names for speech writing and indexing. In this book, we give only the speech-writing names.

---

[1] The classification of reactions into sections is, of course, to some degree arbitrary. Each individual reaction (e.g., $CH_3Cl + CN^- \rightarrow CH_3CN$ and $C_2H_5Cl + CN^- \rightarrow C_2H_5CN$) is different, and custom generally decides how we group them together. Individual preferences also play a part. Some chemists would say that $C_6H_5N_2^+ + CuCN \rightarrow C_6H_5CN$ and $C_6H_5N_2^+ + CuCl \rightarrow C_6H_5Cl$ are examples of the "same" reaction. Others would say that they are not, but that $C_6H_5N_2^+ + CuCl \rightarrow C_6H_5Cl$ and $C_6H_5N_2^+ + CuBr \rightarrow C_6H_5Br$ are examples of the "same" reaction. No claim is made that the classification system used in this book is more valid than any other. For another way of classifying reactions, see Fujita, S. *J. Chem. Soc., Perkin Trans. 2* **1988**, 597.

[2] For the complete rules, as so far published, see Jones, R.A.Y.; Bunnett, J.F. *Pure Appl. Chem.* **1989**, *61*, 725.

Substitutions

A name consists of the entering group, the syllable "de," and the leaving group. If the leaving group is hydrogen, it may be omitted (in all examples, the substrate is written on the left).

$$CH_3CH_2Br \; + \; CH_3O^- \longrightarrow CH_3CH_2-O-CH_3 \qquad \text{Metthoxy-de-hydrogenation}$$

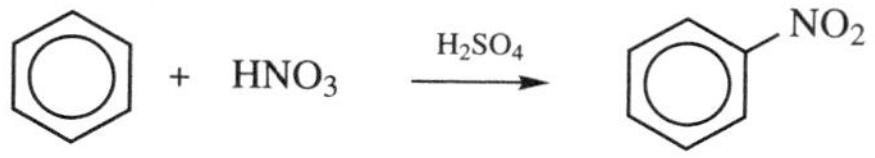

Nitro-de-hydrogenation
or Nitration

Multivalent substitutions are named by a modification of this system that includes suffixes, such as "bisubstitution" and "tersubstitution."

$$CH_2Cl_2 \; + \; 2\;EtO^- \longrightarrow CH_2(OEt)_2 \qquad \text{Diethoxy-de-chloro-bisubstitution}$$

$$CH_3CHO \; + \; Ph_3P{=}CH_2 \longrightarrow CH_3CH{=}CH_2 \qquad \text{Methylene-de-oxo-bisubstitution}$$

Hydroxy, oxo-de-nitrilo-tersubstitution

(Note: The nitrilo group is $\equiv N$.)

Additions

For simple 1,2-additions, the names of both addends are given followed by the suffix "addition." The addends are named in order of priority in the Cahn–Ingold–Prelog system (p. 155), the lower ranking addend coming first. Multivalent addition is indicated by "biaddition," and so on.

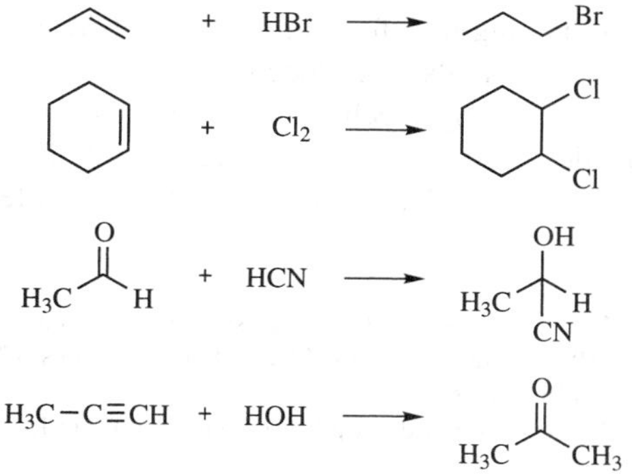

Hydro-bromo-addition

Dichloro-addition

O-Hydro-C-cyano-addition

Dihydro-oxo-biaddition

Eliminations are named the same way as additions, except that "elimination" is used instead of "addition."

In the reaction sections of this book, we will give IUPAC names for most transformations (these names will be printed in the same typeface used above), including examples of all eight types.[3] As will become apparent, some transformations require more rules than we have given here.[2] However, it is hoped that the simplicity of the system will also be apparent.

Two further notes: (*1*) Many transformations can be named using either of two reactants as the substrate. For example, the transformation methylene-de-oxo-bisubstitution above, can also be named ethylidene-de-triphenylphosphoranediyl-bisubstitution. In this book, unless otherwise noted, we will show only those names in which the substrate is considered to undergo the reactions indicated by the titles of the chapters. Thus the name we give to **11-11** (ArH + RCl → ArR) is alkyl-de-hydrogenation, not aryl-de-chlorination, though the latter name is also perfectly acceptable under the IUPAC system. (*2*) The IUPAC rules recognize that some transformations are too complex to be easily fitted into the system, so they also include a list of names for some complex transformations, which are IUPAC approved, but nonsystematic (for some examples, see reactions **12-44**, **18-34**).

IUPAC SYSTEM FOR SYMBOLIC REPRESENTATION OF MECHANISMS

In addition to providing a system for naming transformations, the IUPAC Commission on Physical Organic Chemistry has also produced one for representing mechanisms.[4] As we will see in Part 2, many mechanisms (though by no means all) are commonly referred to by designations, such as S_N2, $A_{AC}2$, $E1_{cB}$, and $S_{RN}1$, many of them devised by C.K. Ingold and his co-workers. While these

---

[3]For some examples, see: attachments (**18-27**, **19-29**), detachments (**19-72**), simple rearrangements (**18-7**, **18-29**), coupling (**10-56**, **19-34**), uncoupling (**19-9**, **19-75**), insertions (**12-21**, **18-9**), extrusions (**17-35**, **17-38**), ring opening (**10-14**, **10-35**), ring closing (**10-9**, **15-60**).

[4]Guthrie, R.D. *Pure Appl. Chem.* **1989**, *61*, 23. For a briefer description, see Guthrie, R.D.; Jencks, W.P. *Acc. Chem. Res.* **1989**, *22*, 343.

designations have been useful (and we will continue to use them in this book), the sheer number of them can be confusing, especially since the symbols do not give a direct clue to what is happening. For example, there is no way to tell directly from the symbols how S_N2' is related to S_N2 (see p. 426). The IUPAC system is based on a very simple description of bond changes.[5] The letter A represents formation of a bond (association); D the breaking of a bond (dissociation). These are *primitive changes*. The basic description of a mechanism consists of these letters, with subscripts to indicate where the electrons are going. In any mechanism, the *core atoms* are defined as (*1*) the two atoms in a multiple bond that undergoes addition, or (*2*) the two atoms that will be in a multiple bond after elimination, or (*3*) the single atom at which substitution takes place.

As an example of the system, this is how an $E1_{cB}$ mechanism (p. 1488) would be represented:

Overall designation: $A_nD_E + D_N$ (or $A_{xh}D_H + D_N$)
In this case, the overall reaction is

and the core atoms are the two carbons in boldface.

Step 1, First Symbol. A bond is being formed between O and H. Bond formation is represented by A. For this particular case, the system gives two choices for subscript. In any process, the subscript is N if a core atom is forming a bond to a nucleophile (A_N) or breaking a bond to a nucleofuge (D_N). If a noncore atom is doing the same thing, lowercase n is used instead. Since H and O are noncore atoms, the lowercase n is used, and the formation of the O—H bond is designated by A_n. However, because involvement of H^+ is so common in organic mechanisms, the rules allow an alternative. The subscript H or h may

---

[5]There are actually two IUPAC systems. The one we use in this book (Ref. 4) is intended for general use. A more detailed system, which describes every conceivable change happening in a system, and which is designed mostly for computer handling and storage, is given by Littler, J.S. *Pure Appl. Chem.* **1989**, *61*, 57. The two systems are compatible; the Littler system uses the same symbols as the Guthrie system, but has additional symbols.

replace N or n. The symbol xh denotes that the H^+ comes from or goes to an unspecified carrier atom X. Thus the term A_{xh} means that a bond is being formed between H (moving without electrons) and an outside atom, in this case O. The same subscript, xh, would be used if the outside atom were any other nucleophilic atom, say, N or S.

Step 1, Second Symbol. A bond is being broken between C and H. The symbol is D. In any process, the subscript is E if a core atom is forming a bond to an electrophile (A_E) or breaking a bond to an electrofuge (D_E). Since C is a core atom, the symbol here is D_E. Alternatively, the symbol could be D_H. The rules allow A_H or D_H to replace A_E or D_E if the electrophile or electrofuge is H^+. Because a core atom is involved in this primitive change the H in the subscript is capitalized.

Step 1, Combined Symbols. In step 1, two bond changes take place simultaneously. In such cases, they are written together with no space or punctuation:

$$A_nD_E \quad \text{or} \quad A_{xh}D_H$$

Step 2. Only one bond is broken in this step and no bonds are formed. (The movement of a pair of unshared electrons into the C—C bond, forming a double bond, is not designated by any symbol. In this system bond multiplicity changes are understood without being specified.) Thus the symbol is D. The broken bond is between a core atom (C) and a nucleofuge (Cl), so the designation is D_N.

Overall designation. This can be either $A_nD_N + D_N$ or $A_{xh}D_H + D_N$. The + symbol shows that there are two separate steps. If desired, rate-limiting steps can be shown by the symbol. In this case, if the first step is the slow step [old designation $(E1_{cB})_I$], the designation would be $A_nD_E + D_N$ or $A_{xh}D_H + D_N$.

For most mechanisms (other than rearrangements), there will be only two A or D terms with uppercase subscripts, and the nature of the reaction can be immediately recognized by looking at them. If both are A, the reaction is an addition; if both are D (as in $A_nD_E + D_N$) it is an elimination. If one is A and the other D, the reaction is a substitution.

Here, we have given only a brief description of the system. Other IUPAC designations will be shown in Part Two, where appropriate. For more details, further examples, and additional symbols, see Ref. 4.

ORGANIC SYNTHESES REFERENCES

At the end of each numbered section there is a list of *Organic Syntheses* references (abbreviated OS). With the exception of a few very common reactions (**12-3**, **12-23**, **12-24**, and **12-38**) the list includes *all* OS references for each reaction. The volumes of OS that have been covered are Collective Volumes **I–X** and individual volumes **80–81**. Where no OS references are listed at the end of a section, the reaction has not been reported in OS through volume **81**. These listings thus constitute a kind of index to OS.[6] Certain ground rules were followed in assembling these lists. A reaction in which two parts of a molecule independently undergo simultaneous reaction is listed under both reactions. Similarly, if two reactions happen (or might happen) rapidly in succession without the isolation of an intermediate, the reactions are listed in both places. For example, at OS **IV**, 266 is

$$\text{(cyclic ether)} \xrightarrow[\text{H}_2\text{SO}_4]{\text{POCl}_3} \text{Cl(CH}_2)_4\text{O(CH}_2)_4\text{Cl}$$

This reaction is treated as **10-49** followed by **10-12** and is listed in both places. However, certain reactions are not listed because they are trivial examples. An instance of this is the reaction found at OS **III**, 468:

$$\text{4-nitrophenol} + \text{CH}_2\text{(OMe)}_2 \xrightarrow[\text{H}_2\text{SO}_4]{\text{HCl}} \text{(2-chloromethyl-4-nitrophenol)}$$

This is a chloromethylation reaction and is consequently listed at **11-14**. However, in the course of the reaction formaldehyde is generated from the acetal. This reaction is not listed at **10-6** (hydrolysis of acetals), because it is not really a preparation of formaldehyde.

---

[6]Two indexes to *Organic Syntheses* have been published as part of the series. One of these, Liotta, D.C.; Volmer, M. *Organic Syntheses Reaction Guide*, Wiley, NY, *1991*, which covers the series through Vol. 68, is described on p. 1896. There are two others. One covers the series through Collective Vol. V, Shriner, R.L.; Shriner, R.H. *Organic Syntheses Collective Volumes I-V, Cumulative Indices*, Wiley, NY, *1976*. An updated version covers through Collective Vol. VIII: Freeman, J.P. *Organic Syntheses Collective Volumes I–VIII, Cumulative Indices*, Wiley: NY, *1995*. For an older index to *Organic Syntheses* (through Vol. 45), see Sugasawa, S.; Nakai, S. *Reaction Index of Organic Syntheses*, Wiley, NY, *1967*.

Aliphatic Substitution: Nucleophilic and Organometallic

In nucleophilic aliphatic substitution the attacking (electron donating) reagent (the nucleophile) brings an electron pair to the substrate, using this pair to form the new bond, and the leaving group (the nucleofuge) comes away with an electron pair:

$$R\overset{\frown}{-}X \ + \ Y: \ \longrightarrow \ R-Y \ + \ X:$$

This equation says nothing about charges. Nuclephile Y may be neutral or negatively charged; RX may be neutral or positively charged; so there are four charge types, examples of which are

Type I	$R-I$	$+$ OH^-	$\longrightarrow$	$R-OH$	$+$	I^-
Type II	$R-I$	$+$ NMe_3	$\longrightarrow$	$R-\overset{\oplus}{N}Me_3$	$+$	I^-
Type III	$R-\overset{\oplus}{N}Me_3$	$+$ OH^-	$\longrightarrow$	$R-OH$	$+$	NMe_3
Type IV	$R-\overset{\oplus}{N}Me_3$	$+$ H_2S	$\longrightarrow$	$R-\overset{\oplus}{S}H_2$	$+$	NMe_3

In all cases, Y must have an unshared pair of electrons, so that all nucleophiles are Lewis bases. When Y is the solvent, the reaction is called *solvolysis*. Nucleophilic substitution at an aromatic carbon is considered in Chapter 13.

Nucleophilic substitution at an alkyl carbon is said to *alkylate* the nucleophile. For example, the above reaction between RI and NMe_3 is an *alkylation* of trimethylamine. Similarly, nucleophilic substitution at an acyl carbon is an *acylation* of the nucleophile.

March's Advanced Organic Chemistry: Reactions, Mechanisms, and Structure, Sixth Edition, by Michael B. Smith and Jerry March
Copyright © 2007 John Wiley & Sons, Inc.

MECHANISMS

Several distinct mechanisms are possible for aliphatic nucleophilic substitution reactions, depending on the substrate, nucleophile, leaving group, and reaction conditions. In all of them, however, the attacking reagent carries the electron pair with it, so that the similarities are greater than the differences. Mechanisms that occur at a saturated carbon atom are considered first.[1] By far the most common are the S_N1 and S_N2 mechanisms.

The S_N2 Mechanism

The designation S_N2 stands for *substitution nucleophilic bimolecular*. The IUPAC designation (p. 420) is $A_N D_N$. In this mechanism, there is *backside attack*:[2] the nucleophile approaches the substrate from a position 180° away from the leaving group. The reaction is a one-step process with no intermediate (see, however, pp. 428–431 and 440). The C—Y bond is formed as the C—X bond is broken to generate transition state **1**.

$$\ominus Y \;+\; -\!\overset{|}{\underset{|}{C}}\!-X \;\longrightarrow\; Y\text{-}\text{-}\text{-}\overset{|}{\underset{|}{C}}\text{-}\text{-}\text{-}X \;\longrightarrow\; X\!-\!\overset{|}{\underset{|}{C}}\!- \;+\; X^{\ominus}$$

$$\mathbf{1}$$

The energy necessary to break the C—X bond is supplied by simultaneous formation of the C—Y bond. The position of the atoms at the top of the curve of free energy of activation is represented as transition state **1**. Of course, the reaction does not stop here since this is the transition state. The group X must leave as the group Y comes in, because at no time can the carbon have more than eight electrons in its outer shell. When the transition state is reached, the central carbon atom has gone from its initial sp^3 hybridization to an sp^2 state with an approximately perpendicular p orbital. One lobe of this p orbital overlaps with the nucleophile and the other with the leaving group. This is why a frontside S_N2 mechanism has never been observed. In a hypothetical frontside transition state, both the nucleophile and the leaving group would have to overlap with the same lobe of the p orbital. The backside mechanism involves the maximum amount of overlap throughout the course of the reaction. During the transition state the three nonreacting substituents and the central carbon are approximately coplanar. They will be exactly coplanar if both the entering and the leaving group are the same.

---

[1]For a monograph on this subject, see Hartshorn, S.R. *Aliphatic Nucleophilic Substitution*, Cambridge University Press, Cambridge, *1973*. For reviews, see Katritzky, A.R.; Brycki, B.E. *Chem. Soc. Rev. 1990*, *19*, 83; Richard, J.P. *Adv. Carbocation Chem. 1989*, *1*, 121; de la Mare, P.B.D.; Swedlund, B.E., in Patai, S. *The Chemistry of the Carbon–Halogen Bond*, pt. 1, Wiley, NY, *1973*, pp. 409–490. Streitwieser, A. *Solvolytic Displacement Reactions*, McGraw-Hill, NY, *1962*.

[2]See Sun, L.; Hase, W.L.; Song, K. *J. Am. Chem. Soc. 2001*, *123*, 5753.

There is a large amount of evidence for the S_N2 mechanism. First, there is the kinetic evidence. Since both the nucleophile and the substrate are involved in the rate-determining step (the only step, in this case), the reaction should be first order in each component, second order overall, and satisfy the rate expression, Eq. (10.1).

$$\text{Rate} = k[\text{RX}][\text{Y}] \tag{10.1}$$

This rate law has been found to apply. Note that the 2 in S_N2 stands for bimolecular. It must be remembered that this is not always the same as second order (see p. 315). If a large excess of nucleophile is present (for example, if it is the solvent) the mechanism may still be bimolecular, although the experimentally determined kinetics will be first order, Eq. (10.2).

$$\text{Rate} = k[\text{RX}] \tag{10.2}$$

As previously mentioned (p. 318), such kinetics are called *pseudo-first order*.

The kinetic evidence is a necessary but not a sufficient condition; we will meet other mechanisms that are also consistent with these data. Much more convincing evidence is obtained from the fact that the mechanism predicts inversion of configuration when substitution occurs at a chiral carbon and this has been observed many times. This inversion of configuration (see p. 158) that proceeds through transition state **1** is called the *Walden inversion* and was observed long before the S_N2 mechanism was formulated by Hughes and Ingold.[3]

At this point it is desirable for us to see just how it was originally proved that a given substitution reaction proceeds with inversion of configuration, even before the mechanism was known. Walden presented a number of examples[4] in which inversion *must* have taken place. For example, (+)-malic acid (**2**) could be converted to (+)-chlorosuccinic acid by thionyl chloride and to (−)-chlorosuccinic acid by phosphorus pentachloride.

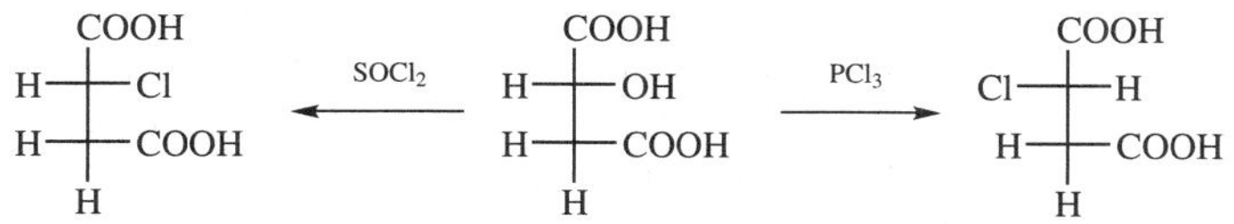

                                            2

---

[3]Cowdrey, W.A.; Hughes, E.D.; Ingold, C.K.; Masterman, S.; Scott, A.D. *J. Chem. Soc.* **1937**, 1252. The idea that the addition of one group and removal of the other are simultaneous was first suggested by Lewis, G.N., in *Valence and the Structure of Atoms and Molecules*, Chemical Catalog Company, NY, **1923**, p. 113. The idea that a one-step substitution leads to inversion was proposed by Olsen, A.R. *J. Chem. Phys.* **1933**, *1*, 418.

[4]Walden, P. *Berichte* **1893**, 26, 210; **1896**, 29, 133; **1899**, 32, 1855.

One of these must be an inversion and the other a retention of configuration, but the question is which is which? The signs of rotation are of no help in answering this question since, as we have seen (p. 154), rotation need not be related to configuration. Another example discovered by Walden is formation of **3** from **4**.[5]

A series of experiments designed to settle the matter of exactly where inversion takes place was performed by Phillips, Kenyon, and co-workers. In 1923, Phillips carried out the following cycle based on (+)-1-phenyl-2-propanol.[6]

In this cycle, (+)-1-phenyl-2-propanol is converted to its ethyl ether by two routes, path *AB* giving the (−) ether, and path *CD* giving the (+) ether. Therefore, at least one of the four steps must be an inversion. It is extremely unlikely that there is inversion in step *A*, *C*, or *D*, since in all these steps the C—O bond is unbroken, and in none of them could the oxygen of the bond have come from the reagent. There is therefore a high probability that *A*, *C*, and *D* proceeded with retention, leaving *B* as the inversion. A number of other such cycles were carried out, always with nonconflicting results.[7] These experiments not only definitely showed that certain specific reactions proceed with inversion, but also established the configurations of many compounds.

Walden inversion has been found at a primary carbon atom by the use of a chiral substrate containing a deuterium and a hydrogen atom at the carbon bearing the

[5]For a discussion of these cycles, see Kryger, L.; Rasmussen, S.E. *Acta Chem. Scand.* **1972**, *26*, 2349.
[6]Phillips, H. *J. Chem. Soc.* **1923**, *123*, 44. For analyses of such cycles and general descriptions of more complex ones, see Garwood, D.C.; Cram, D.J. *J. Am. Chem. Soc.* **1970**, *92*, 4575; Cram, D.J.; Cram, J.M. *Fortschr. Chem. Forsch.* **1972**, *31*, 1.
[7]See Kenyon, J.; Phillips, H.; Shutt, G.R. *J. Chem. Soc.* **1935**, 1663 and references cited therein.

leaving group.[8] Inversion of configuration has also been found for S_N2 reactions proceeding in the gas phase.[9] High-pressure mass spectrometry has been used to probe the energy surface for gas-phase S_N2 reactions, which have two transition states (a "loose" transitions state and a "tight" transition state).[10]

Another kind of evidence for the S_N2 mechanism comes from compounds with potential leaving groups at bridgehead carbons. If the S_N2 mechanism is correct, these compounds should not be able to react by this mechanism, since

the nucleophile cannot approach from the rear. Among the many known examples of unsuccessful reaction attempts at bridgeheads under S_N2 conditions[11] are treatment of the [2.2.2] system **5** with ethoxide ion[12] and treatment of the [3.3.1] system **6** with sodium iodide in acetone.[13] In these cases, open-chain analogs underwent the reactions readily. As a final example of evidence for the S_N2 mechanism, the reaction between optically active 2-octyl iodide and radioactive iodide ion may be mentioned:

$$C_6H_{13}CHMeI \ + \ ^*I^- \ \longrightarrow \ C_6H_{13}CHMe^*I \ + \ I^-$$

We expect racemization in this reaction, since if we start with the pure R isomer, at first each exchange will produce an S isomer, but with increasing concentration of S isomer, *it* will begin to compete for I^- with the (R) isomer, until at the end a racemic mixture is left. The point investigated was a comparison of the rate of inversion with the rate of uptake of radioactive $^*I^-$. It was found[14] that the rates were identical within experimental error:

Rate of inversion	$2.88 \pm 0.03 \times 10^{-5}$
Rate of exchange	$3.00 \pm 0.25 \times 10^{-5}$

[8]Streitwieser, Jr., A. *J. Am. Chem. Soc.* **1953**, *75*, 5014.

[9]Speranza, M.; Angelini, G. *J. Am. Chem. Soc.* **1980**, *102*, 3115 and references cited therein; Sauers, R.R. *J. Org. Chem.* **2002**, *67*, 1221; Kempf, B.; Hampel, N.; Ofial, A.R.; Mayr, H. *Chem. Eur. J.* **2003**, *9*, 2209. For a review of nucleophilic displacements in the gas phase, see Riveros, J.M.; José, S.M.; Takashima, K. *Adv. Phys. Org. Chem.* **1985**, *21*, 197.

[10]Li, C.; Ross, P.; Szulejko, J.E.; McMahon, T.B. *J. Am. Chem. Soc.* **1996**, *118*, 9360.

[11]For a review of bridgehead reactivity in nucleophilic substitution reactions, see Müller, P.; Mareda, J., in Olah, G.A. *Cage Hydrocarbons*, Wiley, NY, **1990**, pp. 189–217. For a review of reactions at bridgehead carbons, see Fort, Jr., R.C.; Schleyer, P.v.R. *Adv. Alicyclic Chem.* **1966**, *1*, 283.

[12]Doering, W. von E.; Levitz, M.; Sayigh, A.; Sprecher, M.; Whelan, Jr., W.P. *J. Am. Chem. Soc.* **1953**, *75*, 1008. Actually, a slow substitution was observed in this case, but not by an S_N2 mechanism.

[13]Cope, A.C.; Synerholm, M.E. *J. Am. Chem. Soc.* **1950**, *72*, 5228.

[14]Hughes, E.D.; Juliusburger, F.; Masterman, S.; Topley, B.; Weiss, J. *J. Chem. Soc.* **1935**, 1525.

What was actually measured was the rate of racemization, which is twice the rate of inversion, since each inversion creates, in effect, two racemic molecules. The significance of this result is that it shows that every act of exchange is an act of inversion.

Eschenmoser and co-workers have provided strong evidence that the transition state in an S_N2 reaction must be linear.[15] Base treatment of methyl α-tosyl-o-toluenesulfonate (**7**) gives the o-(1-tosylethyl)benzenesulfonate ion (**9**). The role of

the base is to remove the a proton to give the ion **8**. It might be supposed that the negatively charged carbon of **8** attacks the methyl group in an internal S_N2 process, but this is not the case. Cross-over experiments[15] (p. 736) have shown that the negatively charged carbon attacks the methyl group of another molecule rather than the nearby one in the same molecule, that is, the reaction is intermolecular and not intramolecular, despite the more favorable entropy of the latter pathway (p. 302). The obvious conclusion is that intramolecular attack does not take place because complete linearity cannot be attained. This behavior is in sharp contrast to that in cases in which the leaving group is not constrained (p. 446), where intramolecular S_N2 mechanisms operate freely.

There is evidence, both experimental and theoretical, that there are intermediates in at least some S_N2 reactions in the gas phase, in charge type I reactions, where a negative ion nucleophile attacks a neutral substrate.[16] Two energy minima, one before and one after the transition state appear in the reaction coordinate (Fig. 10.1).[17] The energy surface for the S_N2 Menshutkin reaction (p. 555) has been examined and it was shown that charge separation was promoted by the solvent.[18] An ab initio study of the S_N2 reaction at primary and secondary carbon centers has looked at the energy barrier (at the transition state) to the reaction.[19] These minima correspond to unsymmetrical ion–dipole complexes.[20] Theoretical calculations also show such minima in certain solvents (e.g., DMF), but not in water.[21] The

[15]Tenud, L.; Farooq, S.; Seibl, J.; Eschenmoser, A. *Helv. Chim. Acta* *1970*, *53*, 2059. See also, King, J.F.; McGarrity, M.J. *J. Chem. Soc., Chem. Commun.* *1979*, 1140.

[16]See Angel, L.A; Ervin, K.M. *J. Am. Chem. Soc.* *2003*, *125*, 1014.

[17]Taken from Chandrasekhar, J.; Smith, S.F.; Jorgensen, W.L. *J. Am. Chem. Soc.* *1985*, *107*, 154.

[18]Gao, J.; Xia, X. *J. Am. Chem. Soc.* *1993*, *115*, 9667.

[19]Lee, I.; Kim, C.K.; Chung, D.S.; Lee, B.-S. *J. Org. Chem.* *1994*, *59*, 4490.

[20]Pellerite, M.J.; Brauman, J.I. *J. Am. Chem. Soc.* *1980*, *102*, 5993; Wolfe, S.; Mitchell, D.J.; Schlegel, H.B. *J. Am. Chem. Soc.* *1981*, *103*, 7692; Evanseck, J.D.; Blake, J.F.; Jorgensen, W.L. *J. Am. Chem. Soc.* *1987*, *109*, 2349; Kozaki, T.; Morihashi, K.; Kikuchi, O. *J. Am. Chem. Soc.* *1989*, *111*, 1547; Jorgensen, W.L. *Acc. Chem. Res.* *1989*, *22*, 184.

[21]Chandrasekhar, J.; Jorgensen, W.L. *J. Am. Chem. Soc.* *1985*, *107*, 2974.

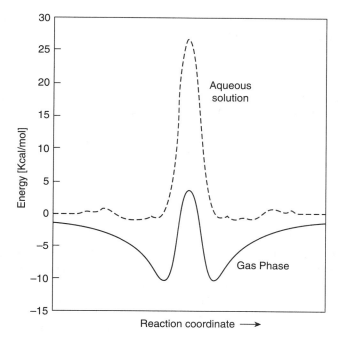

Fig. 10.1. Free-energy profile for the gas-phase (solid line) and aqueous solution (dashed line) S_N2 reaction between CH_3Cl and Cl^-, from molecular orbital calculations.[17]

S_N2 reactions can occur at atoms other than carbon, X (e.g., nitrogen or sulfur[22]), and analogous to the phenomenon observed for S_N2 reactions at carbon.[23] The valence of the element X, controls the intrinsic barrier for the reaction in accord with the properties seen in the Periodic table.[24]

For a list of some of the more important reactions that operate by the S_N2 mechanism, see Table 10.7.

Note that in some reactions, such as bromine transfer between carbanions via nucleophilic attack on bromine, anomalous kinetic behavior is observed. The largest rate constants are associated with bromine transfer between cyano-activated carbanions and the smallest relate to the removal of bromine from the nitromethane and nitroethane moieties.[25] The Brønsted plot ($\log k$ vs. ΔpK_a) for this reaction shows that unlike any normal Brønsted plot, which by definition displays a positive slope, the plot for $MeNO_2$ and $EtNO_2$ is negative. In deprotonation reactions of carbon compounds, the reactivity of nitroethane and nitromethane were shown to be anomalous.[26] In the series nitromethane, ethane, and isopropane, contrary

[22]See reactions **10-60–10-68** and Bachrach, S.M.; Gailbreath, B.D. *J. Org. Chem.* **2001**, *66*, 2005.

[23]Hoz, S.; Basch, H.; Wolk, J.L.; Hoz, T.; Rozental, E. *J. Am. Chem. Soc.* **1999**, *121*, 7724.

[24]Yi, R.; Basch, H.; Hoz, S. *J. Org. Chem.* **2002**, *67*, 5891.

[25]Grinblat, J.; Ben-Zion, M.; Hoz, S. *J. Am. Chem. Soc.* **2001**, *123*, 10738.

[26]Pearson, R.G.; Dillon, R.L. *J. Am. Chem. Soc.* **1953**, *75*, 2439.

to expectations, compounds with higher acidity undergo slower deprotonation (i.e., the Brønsted plot displays a negative slope).[27]

The S$_N$1 Mechanism

The most ideal version of the S$_N$1 mechanism (*substitutional nucleophilic unimolecular*) consists of two steps[28] (once again, possible charges on the substrate and nucleophile are not shown):

Step 1
$$R-X \underset{}{\overset{slow}{\rightleftarrows}} R^+ + X$$

Step 2
$$R^+ + Y \xrightarrow{fast} R-Y$$

The first step is a slow ionization of the substrate and is the rate-determining step. The second is a rapid reaction between the intermediate carbocation and the nucleophile. The reactive nature of the carbocation can be expressed by its electrophilic character, or electrophlicity. A theoretical discussion concerning the origin of the electrophilicity concept was proposed by Parr et al.[29] In general, a good electrophile was characterized by having a high value of electronegativity (or a high value of electronic chemical potential), and a low value of the chemical hardness. The effect of substitution has been studied[30] in the context of superelectrophilicity (where carbocations are generated in super acidic media). Solvent effects have also been studied.[31] Electrophlicity scales have been proposed using other carbocations.[32]

Returning to the S$_N$1 mechanism, ionization of a leaving group to form the carbocation is always assisted by the solvent,[33] since the energy necessary to break the bond is largely recovered by solvation of R$^+$ and of X. For example, the ionization of *t*-BuCl to *t*-Bu$^+$ and Cl$^-$ in the gas phase without a solvent requires 150 kcal mol^{-1} (630 kJ mol^{-1}). In the absence of a solvent, such a process simply would not take place, except at very high temperatures. In water, this ionization requires only 20 kcal mol^{-1} (84 kJ mol^{-1}). The difference is solvation energy. In

[27]Kresge, A.J. *Can. J. Chem.* **1974**, *52*, 1897; Yamataka, H.; Mustanir; Mishima, M. *J. Am. Chem. Soc.* **1999**, *121*, 10223.

[28]For a direct observation of the two steps see Mayr, H.; Minegishi, S. *Angew. Chem. Int. Ed.* **2002**, *41*, 4493.

[29]Parr, R. G.; Szentpály, L.V.; Liu, S. *J. Am. Chem. Soc.* **1999**, *121*, 1922.

[30]See Pérez, P. *J. Org. Chem.* **2004**, *69*, 5048.

[31]Pérez, P.; Toro-Labbé, A.; Contreras, R. *J. Am. Chem. Soc.* **2001**, *123*, 5527.

[32]Pérez, P.; Toro-Labbé, A.; Aizman, A.; Contreras, R. *J. Org. Chem.* **2002**, *67*, 4747; Parr, R.G.; Szentpály, L.-v.; Liu, S. *J. Am. Chem. Soc.* **1999**, *121*, 1922.

[33]For reviews of solvolysis, see Okamoto, K. *Adv. Carbocation Chem.* **1989**, *1*, 171; Blandamer, M.J.; Scott, J.M.W.; Robertson, R.E. *Prog. Phys. Org. Chem.* **1985**, *15*, 149; Robertson, R.E. *Prog. Phys. Org. Chem.* **1967**, *4*, 213. For a review of the solvolytic cleavage of *tert*-butyl substrates, see Dvorko, G.F.; Ponomareva, E.A.; Kulik, N.I. *Russ. Chem. Rev.* **1984**, *53*, 547.

cases where the role of the solvent is solely to assist in departure of the leaving group from the frontside, that is, where there is a complete absence of backside (S_N2) participation by solvent molecules, the mechanism is called *limiting* S_N1. There is kinetic and other evidence[34] that in pulling the leaving group X away from RX, two molecules of a protic solvent form weak hydrogen bonds with X.

$$R-X \begin{matrix} \diagup H-O-R \\ \diagdown H-O-R \end{matrix} \longrightarrow R^{\oplus}$$

In the IUPAC system, the S_N1 mechanism is $D_N + A_N$ or $D_N^{\ddagger} + A_N$ (where $\ddagger$ denotes the rate-determining step). The IUPAC designations for the S_N1 and S_N2 mechanisms thus clearly show the essential differences between them: A_ND_N indicates that bond breaking is concurrent with bond formation; $D_N + A_N$ shows that the former happens first.

In looking for evidence for the S_N1 mechanism, the first thought is that it should be a first-order reaction following the rate law:

$$\text{Rate} = k[\text{RX}] \tag{10.3}$$

Since the slow step involves only the substrate, the rate should be dependent only on the concentration of that. Although the solvent is necessary to assist in the process of ionization, it does not enter the rate expression, because it is present in large excess. However, the simple rate law given in Eq. (10.3) is not sufficient to account for all the data. Many cases are known where pure first-order kinetics are followed, but in many other cases more complicated kinetics are found. We can explain this by taking into account the reversibility of the first step. The X formed in this step competes with Y for the cation and the rate law must be modified as shown (see Chapter 6).

$$\text{RX} \underset{k_{-1}}{\overset{k_1}{\rightleftharpoons}} \text{R}^+ + \text{X}$$

$$\text{R}^+ + \text{Y} \xrightarrow{k_2} \text{RY} \tag{10.4}$$

$$\text{Rate} = \frac{k_1 k_2 [\text{RX}][\text{Y}]}{k_{-1}[\text{X}] + k_2[\text{Y}]}$$

At the beginning of the reaction, when the concentration of X is very small, $k_{-1}[\text{X}]$ is negligible compared with $k_2[\text{Y}]$ and the rate law is reduced to Eq. (10.3). Indeed, S_N1 reactions generally do display simple first-order kinetics in their initial stages. Most kinetic studies of S_N1 reactions fall into this category. In the later stages of S_N1 solvolyses, [X] becomes large and Eq. (10.4) predicts that the rate should

[34]Blandamer, M.J.; Burgess, J.; Duce, P.P.; Symons, M.C.R.; Robertson, R.E.; Scott, J.M.W. *J. Chem. Res. (S)* **1982**, 130.

decrease. This is found to be the case for diarylmethyl halides,[35] although not for
tert-butylhalides, which follow Eq. (10.3) for the entire reaction.[36] An explanation
for this difference is that tert-butylcations are less selective than the relatively stable
diarylmethyl type (p. 240). Although halide ion is a much more powerful nucleophile than water, there is much more water available since it is the solvent.[37] The
selective diphenylmethyl cation survives many collisions with solvent molecules
before combining with a reactive halide, but the less selective tert-butylion cannot
wait for a reactive but relatively rare halide ion and combines with the solvent.

If the X formed during the reaction can decrease the rate, at least in some cases,
it should be possible to *add* X from the outside and further decrease the rate in that
way. This retardation of rate by addition of X is called *common-ion effect* or the
mass-law effect. Once again, addition of halide ions decreases the rate for diphenylmethyl but not for tert-butylhalides.

One factor that complicates the kinetic picture is the *salt effect*. An increase in
ionic strength of the solution usually increases the rate of an S_N1 reaction (p. 501).
But when the reaction is of charge type II, where both Y and RX are neutral, so that
X is negatively charged (and most solvolyses are of this charge type), the ionic
strength increases as the reaction proceeds and this increases the rate. This effect
must be taken into account in studying the kinetics. Incidentally, the fact that the
addition of outside ions *increases* the rate of most S_N1 reactions makes especially
impressive the *decrease* in rate caused by the common ion.

Note that the pseudo-first-order rate law for an S_N2 reaction in the presence of
a large excess of Y [Eq. (10.1)] is the same as that for an ordinary S_N1 reaction
[Eq. (10.3)]. It is thus not possible to tell these cases apart by simple kinetic measurements. However, we can often distinguish between them by the common-ion
effect mentioned above. Addition of a common ion will not markedly affect the
rate of an S_N2 reaction beyond the effect caused by other ions. Unfortunately, as
we have seen, not all S_N1 reactions show the common-ion effect, and this test fails
for tert-butyl and similar cases.

Kinetic studies also provide other evidence for the S_N1 mechanism. One technique used ^{19}F NMR to follow the solvolysis of trifluoroacetyl esters.[38] If this
mechanism operates essentially as shown on p. 432, the rate should be the same
for a given substrate under a given set of conditions, *regardless of the identity of
the nucleophile or its concentration*. In one experiment that demonstrates this,
benzhydryl chloride (Ph$_2$CHCl) was treated in SO$_2$ with the nucleophiles fluoride
ion, pyridine, and triethylamine at several concentrations of each nucleophile.[39]
In each case the initial rate of the reaction was approximately the same when

[35]Benfey, O.T.; Hughes, E.D.; Ingold, C.K. *J. Chem. Soc.* *1952*, 2488.
[36]Bateman, L.C.; Hughes, E.D.; Ingold, C.K. *J. Chem. Soc.* *1940*, 960.
[37]In the experiments mentioned, the solvent was actually "70%" or "80%" aqueous acetone. The "80%"
aqueous acetone consists of 4 vol of dry acetone and 1 vol of water.
[38]Creary, X.; Wang, Y.-X. *J. Org. Chem.* *1992*, 57, 4761. Also see, Fărcaşiu, D.; Marino, G.; Harris, J.M.;
Hovanes, B.A.; Hsu, C.S. *J. Org. Chem.* *1994*, 59, 154.
[39]Bateman, L.C.; Hughes, E.D.; Ingold, C.K. *J. Chem. Soc.* *1940*, 1011.

corrections were made for the salt effect. The same type of behavior has been shown in a number of other cases, even when the reagents are as different in their nucleophilicities (see p. 490) as H_2O and HO^-.

It is normally not possible to detect the carbocation intermediate of an S_N1 reaction directly, because its lifetime is very short. However, in the case of 3,4-dimethoxydiphenylmethyl acetate (**10**) and certain other substrates in polar solvents it was possible to initiate the reaction photolytically, and under these conditions the UV spectra of the intermediate carbocations could be obtained,[40] providing additional evidence for the S_N1 mechanism.

Further evidence for the S_N1 mechanism is that reactions run under S_N1 conditions fail or proceed very slowly at the bridgehead positions[10] of [2.2.1] (norbornyl) systems[41] (e.g., 1-chloroapocamphane, **8**). If S_N1 reactions require carbocations

and if carbocations must be planar or nearly planar, then it is no surprise that bridgehead 1-norbornyl carbon atoms, which cannot assume planarity, do not become the seat of carbocations. As an example, **11**, boiled 21 h with 30% KOH in 80% ethanol or 48 h with aqueous ethanolic silver nitrate, gave no reaction in either case,[42] although analogous open-chain systems reacted readily. According to this theory, S_N1 reactions should be possible with larger rings, since near-planar carbocations might be expected there. This turns out to be the case. For example, [2.2.2] bicyclic systems undergo S_N1 reactions much faster than smaller bicyclic systems, although the reaction is still slower than with open-chain systems.[43] Proceeding to a still larger system, the bridgehead [3.2.2] cation **12** is actually stable enough to be kept in solution in SbF_5–SO_2ClF at temperatures below $-50°C$[44] (see also, p. 486). Other small bridgehead systems that undergo S_N1 reactions are the

[40]McClelland, R.A.; Kanagasabapathy, V.M.; Steenken, S. *J. Am. Chem. Soc.* ***1988***, *110*, 6913.

[41]For a review, see Fort, Jr., R.C., in Olah, G.A.; Schleyer, P.v.R. *Carbonium Ions*, Vol. 4; Wiley, NY, ***1973***, pp. 1783–1835.

[42]Bartlett, P.D.; Knox, L.H. *J. Am. Chem. Soc.* ***1939***, *61*, 3184.

[43]For synthetic examples, see Kraus, G.A.; Hon, Y. *J. Org. Chem.* ***1985***, *50*, 4605.

[44]Olah, G.A.; Liang, G.; Wiseman, J.R.; Chong, J.A. *J. Am. Chem. Soc.* ***1972***, *74*, 4927.

[3.1.1] (e.g., **13**)[45] and the cubyl (e.g., **14**)[46] systems. *Ab initio* calculations show that the cubyl cation, although it cannot be planar, requires less energy to form than the 1-norbornyl cation.[47] There are reactions where the cationic carbon is not coplanar with conjugating substituents (such as phenyl), and formation of the carbocation is more difficult but the reaction proceeds.[48]

Certain nucleophilic substitution reactions that normally involve carbocations can take place at norbornyl bridgeheads[49] (though it is not certain that carbocations are actually involved in all cases) if the leaving group used is of the type that cannot function as a nucleophile (and thus come back) once it has gone, and in the displacement of $ClCO_2$ in **15**. In this example,[50] chlorobenzene is the nucleophile (see **11-10**).

Additional evidence for the S_N1 mechanism, in particular, for the intermediacy of carbocations, is that solvolysis rates of alkyl chlorides in ethanol parallel carbocation stabilities as determined by heats of ionization measured in superacid solutions (p. 236).[51] It is important to note that some solvolysis reactions proceed by an S_N2 mechanism.[52]

Ion Pairs in the S_N1 Mechanism[53]

Like the kinetic evidence, the stereochemical evidence for the S_N1 mechanism is less clear-cut than it is for the S_N2 mechanism.[54] If there is a free carbocation, it is planar (p. 245), and the nucleophile should attack with equal facility from either

[45]Della, E.W.; Pigou, P.E.; Tsanaktsidis, J. *J. Chem. Soc., Chem. Commun.* **1987**, 833.

[46]Eaton, P.E.; Yang, C.; Xiong, Y. *J. Am. Chem. Soc.* **1990**, *112*, 3225; Moriarty, R.M.; Tuladhar, S.M.; Penmasta, R.; Awasthi, A.K. *J. Am. Chem. Soc.* **1990**, *112*, 3228.

[47]Hrovat, D.A.; Borden, W.T. *J. Am. Chem. Soc.* **1990**, *112*, 3227.

[48]Lee, I.; Kim, N.D.; Kim, C.K. *Tetrahedron Lett.* **1992**, *33*, 7881.

[49]Bartlett, P.D.; Knox, L.H. *J. Am. Chem. Soc.* **1939**, *61*, 3184; Clive, D.L.J.; Denyer, C.V. *Chem. Commun.* **1971**, 1112; White, E.H.; McGirk, R.H.; Aufdermarsh, Jr., C.A.; Tiwari, H.P.; Todd, M.J. *J. Am. Chem. Soc.* **1973**, *95*, 8107; Beak, P.; Harris, B.R. *J. Am. Chem. Soc.* **1974**, *96*, 6363.

[50]For a review of reactions with the OCOCl leaving group, see Beak, P. *Acc. Chem. Res.* **1976**, *9*, 230.

[51]Arnett, E.M.; Petro, C.; Schleyer, P.v.R. *J. Am. Chem. Soc.* **1979**, *101*, 522; Arnett, E.M.; Pienta, N.J. *J. Am. Chem. Soc.* **1980**, *102*, 3329; Arnett, E.M.; Molter, K.E. *Acc. Chem. Res.* **1985**, *18*, 339.

[52]Lee, I.; Lee, Y.S.; Lee, B.-S.; Lee, H.W. *J. Chem. Soc. Perkin Trans. 2* **1993**, 1441.

[53]For reviews of ion pairs in S_N reactions, see Beletskaya, I.P. *Russ. Chem. Rev.* **1975**, *44*, 1067; Harris, J.M. *Prog. Phys. Org. Chem.* **1974**, *11*, 89; Raber, D.J.; Harris, J.M.; Schleyer, P.v.R., in Szwarc, M. *Ions and Ion Pairs in Organic Reactions*, Vol. 2; Wiley, NY, **1974**, pp. 247–374.

[54]For an alternative view of the S_N1/S_N2 mechanism see Uggerud, E. *J. Org. Chem.* **2001**, *66*, 7084.

side of the plane, resulting in complete racemization. Although many first-order substitutions do give complete racemization, many others do not. Typically there is 5–20% inversion, although in a few cases, a small amount of retention of configuration has been found. These and other results have led to the conclusion that in many S_N1 reactions at least some of the products are not formed from free carbocations but rather from *ion pairs*. According to this concept,[55] S_N1 reactions proceed in this manner:

$$R\text{-}X \; \rightleftharpoons \; \underset{\textbf{16}}{R^+ X^-} \; \rightleftharpoons \; \underset{\textbf{17}}{R^+ \| X^-} \; \rightleftharpoons \; \underset{\textbf{18}}{R^+ \; + \; X^-}$$

where **16** is an *intimate, contact,* or *tight* ion pair, **17** a *loose,* or *solvent-separated* ion pair, and **18** the dissociated ions (each surrounded by molecules of solvent).[56] The reaction in which the intimate ion pair recombines to give the original substrate is referred to as *internal return*. The reaction products can result from attack by the nucleophile at any stage. In the intimate ion pair **16**, R^+ does not behave like the free cation of **18**. There is probably significant bonding between R^+ and X^- and asymmetry may well be maintained.[57] Here, X^- "solvates" the cation on the side from which it departed, while solvent molecules near **16** can only solvate it from the opposite side. Nucleophilic attack by a solvent molecule on **16** thus leads to inversion.

Ignoring the possibilities of elimination or rearrangement (see Chapters 17 and 18), a complete picture of the possibilities for solvolysis reactions[58] in a solvent SH is represented by following diagram,[59] although in any particular case it is unlikely that all these reactions occur:

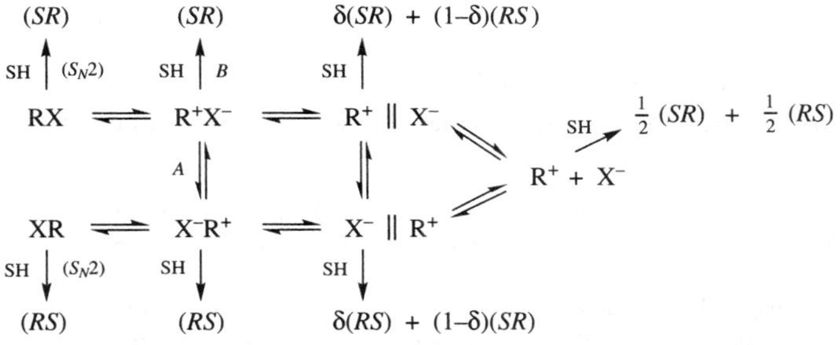

[55]Proposed by Winstein, S.; Clippinger, E.; Fainberg, A.H.; Heck, R.; Robinson, G.C. *J. Am. Chem. Soc.* **1956**, *78*, 328.

[56]For a review of the energy factors involved in the recombination of ion pairs, see Kessler, H.; Feigel, M. *Acc. Chem. Res.* **1982**, *15*, 2.

[57]Fry, J.L.; Lancelot, C.J.; Lam, L.K.M.; Harris, J.M.; Bingham, R.C.; Raber, D.J.; Hall, R.E.; Schleyer, P.v.R. *J. Am. Chem. Soc.* **1970**, *92*, 2538.

[58]For solvolysis of tertiary derivatives with a discussion of solvent participation versus solvation see Richard, J.P.; Toteva, M.M.; Amyes, T.L. *Org. Lett.* **2001**, *3*, 2225.

[59]Shiner, Jr., V.J.; Fisher, R.D. *J. Am. Chem. Soc.* **1971**, *93*, 2553.

In this scheme RS and SR represent enantiomers, and so on, and δ represents some fraction. The following are the possibilities: (*1*) Direct attack by SH on RX gives SR (complete inversion) in a straight S_N2 process. (*2*) If the intimate ion pair $R^+ X^-$ is formed, the solvent can attack at this stage. This can lead to total inversion if reaction A does not take place or to a combination of inversion and racemization if there is competition between A and B. (*3*) If the solvent-separated ion pair is formed, SH can attack here. The stereochemistry is not maintained as tightly and more racemization (perhaps total) is expected. (*4*) Finally, if free R^+ is formed, it is planar, and attack by SH gives complete racemization.

The ion-pair concept thus predicts that S_N1 reactions can display either complete racemization or partial inversion. The fact that this behavior is generally found is evidence that ion pairs are involved in many S_N1 reactions. There is much other evidence for the intervention of ion pairs,[60] includng ion-molecule pairs.[61]

1. The compound 2-octyl brosylate was labeled at the sulfone oxygen with ^{18}O and solvolyzed. The unreacted brosylate recovered at various stages of solvolysis had the ^{18}O considerably, although not completely, scrambled:[62]

In an intimate ion pair, the three oxygens become equivalent:

Similar results were obtained with several other sulfonate esters.[63] The possibility must be considered that the scrambling resulted from ionization

[60]For further evidence beyond that given here, see Winstein, S.; Baker, R.; Smith, S. *J. Am. Chem. Soc.* **1964**, *86*, 2072; Streitwieser, Jr., A.; Walsh, T.D. *J. Am. Chem. Soc.* **1965**, *87*, 3686; Sommer, L.H.; Carey, F.A. *J. Org. Chem.* **1967**, *32*, 800, 2473; Kwart, H.; Irvine, J.L. *J. Am. Chem. Soc.* **1969**, *91*, 5541; Harris, J.M.; Becker, A.; Fagan, J.F.; Walden, F.A. *J. Am. Chem. Soc.* **1974**, *96*, 4484; Bunton, C.A.; Huang, S.K.; Paik, C.H. *J. Am. Chem. Soc.* **1975**, *97*, 6262; Humski, K.; Sendijarević, V.; Shiner, Jr., V.J. *J. Am. Chem. Soc.* **1976**, *98*, 2865; Maskill, H.; Thompson, J.T.; Wilson, A.A. *J. Chem. Soc., Chem. Commun.* **1981**, 1239; McManus, S.P.; Safavy, K.K.; Roberts, F.E. *J. Org. Chem.* **1982**, *47*, 4388; McLennan, D.J.; Stein, A.R.; Dobson, B. *Can. J. Chem.* **1986**, *64*, 1201; Kinoshita, T.; Komatsu, K.; Ikai, K.; Kashimura, K.; Tanikawa, S.; Hatanaka, A.; Okamoto, K. *J. Chem. Soc. Perkin Trans. 2* **1988**, 1875; Ronco, G.; Petit, J.; Guyon, R.; Villa, P. *Helv. Chim. Acta* **1988**, *71*, 648; Kevill, D.N.; Kyong, J.B.; Weitl, F.L. *J. Org. Chem.* **1990**, *55*, 4304.
[61]Jia, Z.S.; Ottosson, H.; Zeng, X.; Thibblin, A. *J. Org. Chem.* **2002**, *67*, 182.
[62]Diaz, A.F.; Lazdins, I.; Winstein, S. *J. Am. Chem. Soc.* **1968**, *90*, 1904.
[63]Goering, H.L.; Jones, B.E. *J. Am. Chem. Soc.* **1980**, *102*, 1628; Yukawa, Y.; Morisaki, H.; Tsuji, K.; Kim, S.; Ando, T. *Tetrahedron Lett.* **1981**, *22*, 5187; Chang, S.; le Noble, W.J. *J. Am. Chem. Soc.* **1983**, *105*, 3708; Paradisi, C.; Bunnett, J.F. *J. Am. Chem. Soc.* **1985**, *107*, 8223; Fujio, M.; Sanematsu, F.; Tsuno, Y.; Sawada, M.; Takai, Y. *Tetrahedron Lett.* **1988**, *29*, 93.

of one molecule of $ROSO_2Ar$ to R^+ and $ArSO_2O^-$ followed by attack by the $ArSO_2O^-$ ion on *another* carbocation or perhaps on a molecule of $ROSO_2Ar$ in an S_N2 process. However, this was ruled out by solvolyzing unlabeled substrate in the presence of labeled $HOSO_2Ar$. These experiments showed that there was some intermolecular exchange (3–20%), but not nearly enough to account for the amount of scrambling found in the original experiments. Similar scrambling was found in solvolysis of labeled carboxylic esters $R-{}^{18}O-COR'$, where the leaving group is $R'COO^-$.[64] In this case also, the external addition of $RCOO^-$ did not result in significant exchange. However, it has been proposed that the scrambling could result from a concerted process, not involving ion-pair intermediates, and there is some evidence for this view.[65]

2. The *special salt effect*. The addition of $LiClO_4$ or LiBr in the acetolysis of certain tosylates produced an initial steep rate acceleration that then decreased to the normal linear acceleration (caused by the ordinary salt effect).[66] This is interpreted as follows: the ClO_4^- (or Br^-) traps the solvent-separated ion pair to give $R^+ \parallel ClO_4^-$ which, being unstable under these conditions, goes to product. Hence, the amount of solvent-separated ion pair that would have returned to the starting material is reduced, and the rate of the overall reaction is increased. The special salt effect has been directly observed by the use of picosecond absorption spectroscopy.[67]

3. We have previously discussed the possibilities of racemization or inversion of the *product* RS of a solvolysis reaction. However, the formation of an ion pair followed by internal return can also affect the stereochemistry of the *substrate* molecule RX. Cases have been found where internal return racemizes an original optically active RX, an example being solvolysis in aqueous acetone of α-*p*-anisylethyl *p*-nitrobenzoate,[68] while in other cases partial or complete retention is found, for example, solvolysis in aqueous acetone of *p*-chlorobenzhydryl *p*-nitrobenzoate.[69] Racemization of RX is presumably caused by the equilibrium pathway: $RX \rightleftharpoons R^+X^- \rightleftharpoons X^-R^+ \rightleftharpoons XR$. Evidence for ion pairs is that, in some cases where internal return involves racemization, it has been shown that such racemization is *faster* than solvolysis. For example, optically active *p*-chlorobenzhydryl chloride racemizes $\sim$30 times faster than it solvolyzes in acetic acid.[70]

[64]Goering, H.L.; Hopf, H. *J. Am. Chem. Soc.* **1971**, *93*, 1224, and references cited therein.
[65]Dietze, P.E.; Wojciechowski, M. *J. Am. Chem. Soc.* **1990**, *112*, 5240.
[66]Winstein, S.; Clippinger, E.; Fainberg, A.H.; Heck, R.; Robinson, G.C. *J. Am. Chem. Soc.* **1956**, *78*, 328; Winstein, S.; Klinedinst, Jr., P.E.; Clippinger, E. *J. Am. Chem. Soc.* **1961**, *83*, 4986; Cristol, S.J.; Noreen, A.L.; Nachtigall, G.W. *J. Am. Chem. Soc.* **1972**, *94*, 2187.
[67]Simon, J.D.; Peters, K.S. *J. Am. Chem. Soc.* **1982**, *104*, 6142.
[68]Goering, H.L.; Briody, R.G.; Sandrock, G. *J. Am. Chem. Soc.* **1970**, *92*, 7401.
[69]Goering, H.L.; Briody, R.G.; Levy, J.F. *J. Am. Chem. Soc.* **1963**, *85*, 3059.
[70]Winstein, S.; Gall, J.S.; Hojo, M.; Smith, S. *J. Am. Chem. Soc.* **1960**, *82*, 1010. See also, Shiner, Jr., V.J.; Hartshorn, S.R.; Vogel, P.C. *J. Org. Chem.* **1973**, *38*, 3604.

Molecular orbital calculations[71] made on *t*-BuCl show that the C—Cl distance in the intimate ion pair is 2.9 Å and the onset of the solvent-separated ion pair takes place at about 5.5 Å (cf. the C—Cl bond length of 1.8 Å).

In a few cases, S_N1 reactions have been found to proceed with partial retention (20–50%) of configuration. Ion pairs have been invoked to explain some of these.[72] For example, it has been proposed that the phenolysis of optically active α-phenylethyl chloride, in which the ether of net retained configuration is obtained, involves a four-center mechanism:

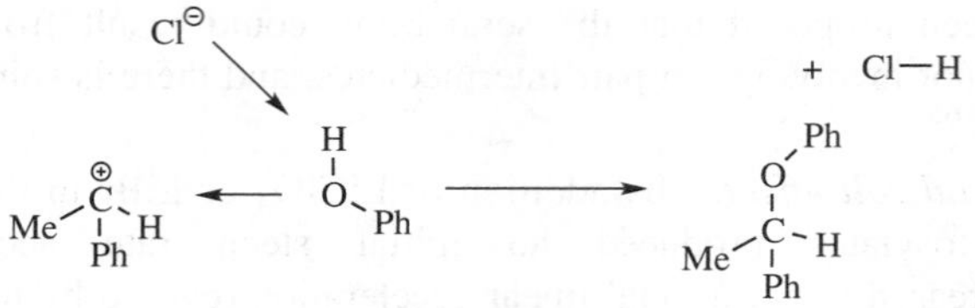

This conclusion is strengthened by the fact that partial retention was obtained in this system only with chloride or other neutral leaving groups; with leaving groups bearing a positive charge, which are much less likely to form hydrogen bonds with the solvent, no retention was found.[73] Partial retention can also arise when the ion pair is shielded at the backside by an additive such as acetonitrile, acetone, or aniline.[74]

The difference between the S_N1 and S_N2 mechanisms is in the timing of the steps. In the S_N1 mechanism, first X leaves, then Y attacks. In the S_N2 case, the two things happen simultaneously. One could imagine a third possibility: first the attack of Y and then the removal of X. This is not possible at a saturated carbon, since it would mean more than eight electrons in the outer shell of carbon. However, this type of mechanism is possible and indeed occurs at other types of substrate (p. 473; Chapter 13).

Mixed S_N1 and S_N2 Mechanisms

Some reactions of a given substrate under a given set of conditions display all the characteristics of S_N2 mechanisms; other reactions seem to proceed by S_N1 mechanisms, but cases are found that cannot be characterized so easily. There seems to be something in between, a mechanistic "borderline" region.[75] At least two broad

[71]Jorgensen, W.L.; Buckner, J.K.; Huston, S.E.; Rossky, P.J. *J. Am. Chem. Soc.* **1987**, *109*, 1891.

[72]Okamoto, K. *Pure Appl. Chem.* **1984**, *56*, 1797. For a similar mechanism with amine nucleophiles, see Lee, I.; Kim, H.Y.; Kang, H.K.; Lee, H.W. *J. Org. Chem.* **1988**, *53*, 2678; Lee, I.; Kim, H.Y.; Lee, H.W.; Kim, I.C. *J. Phys. Org. Chem.* **1989**, *2*, 35.

[73]Okamoto, K.; Kinoshita, T.; Shingu, H. *Bull. Chem. Soc. Jpn.* **1970**, *43*, 1545.

[74]Okamoto, K.; Nitta, I.; Dohi, M.; Shingu, H. *Bull. Chem. Soc. Jpn.* **1971**, *44*, 3220; Kinoshita, T.; Ueno, T.; Ikai, K.; Fujiwara, M.; Okamoto, K. *Bull. Chem. Soc. Jpn.* **1988**, *61*, 3273; Kinoshita, T.; Komatsu, K.; Ikai, K.; Kashimura, K.; Tanikawa, S.; Hatanaka, A.; Okamoto, K. *J. Chem. Soc. Perkin Trans. 2* **1988**, 1875.

[75]For an essay on borderline mechanisms in general, see Jencks, W.P. *Chem. Soc. Rev.* **1982**, *10*, 345.

theories have been devised to explain these phenomena. One theory holds that intermediate behavior is caused by a mechanism that is neither "pure" S_N1 nor "pure" S_N2, but some "in-between" type. According to the second theory, there is no intermediate mechanism at all, and borderline behavior is caused by simultaneous operation, in the same flask, of both the S_N1 and S_N2 mechanisms; that is, some molecules react by the S_N1, while others react by the S_N2 mechanism.

One formulation of the intermediate-mechanism theory is that of Sneen.[76] The formulation is in fact very broad and applies not only to borderline behavior but to all nucleophilic substitutions at a saturated carbon.[77] According to Sneen, all S_N1 and S_N2 reactions can be accommodated by one basic mechanism (the *ion-pair mechanism*). The substrate first ionizes to an intermediate ion pair that is then converted to products:

$$RX \xrightleftharpoons{k_1} R^+ \ X^- \xrightleftharpoons{k_2} \text{Products}$$

The difference between the S_N1 and S_N2 mechanisms is that in the former case the *formation* of the ion pair (k_1) is rate determining, while in the S_N2 mechanism its *destruction* (k_2) is rate determining. Borderline behavior is found where the rates of formation and destruction of the ion pair are of the same order of magnitude.[78] However, a number of investigators have asserted that these results could also be explained in other ways.[79]

There is evidence for the Sneen formulation where the leaving group has a positive charge. In this case, there is a cation–molecule pair $(RX^+ \rightarrow R^+X)$[80] instead of the ion pair that would be present if the leaving group were uncharged. Katritzky

---

[76]Sneen, R.A.; Felt, G.R.; Dickason, W.C. *J. Am. Chem. Soc.* **1973**, *95*, 638 and references cited therein; Sneen, R.A. *Acc. Chem. Res.* **1973**, *6*, 46.

[77]Including substitution at an allylic carbon; see Sneen, R.A.; Bradley, W.A. *J. Am. Chem. Soc.* **1972**, *94*, 6975; Sneen, R.A.; Carter, J.V. *J. Am. Chem. Soc.* **1972**, *94*, 6990; Bordwell, F.G.; Mecca, T.G. *J. Am. Chem. Soc.* **1975**, *97*, 123, 127; Bordwell, F.G.; Wiley, P.F.; Mecca, T.G. *J. Am. Chem. Soc.* **1975**, *97*, 132; Kevill, D.N.; Degenhardt, C.R. *J. Am. Chem. Soc.* **1979**, *101*, 1465.

[78]For evidence for this point of view, see Sneen, R.A.; Felt, G.R.; Dickason, W.C. *J. Am. Chem. Soc.* **1973**, *95*, 638 and references cited therein; Sneen, R.A. *Acc. Chem. Res.* **1973**, *6*, 46; Robbins, H.M. *J. Am. Chem. Soc.* **1972**, *94*, 7868; Graczyk, D.G.; Taylor, J.W. *J. Am. Chem. Soc.* **1974**, *96*, 3255; Peeters, H.L.; Anteunis, M. *J. Org. Chem.* **1975**, *40*, 312; Pross, A.; Aronovitch, H.; Koren, R. *J. Chem. Soc. Perkin Trans. 2* **1978**, 197; Blandamer, M.J.; Robertson, R.E.; Scott, J.M.W.; Vrielink, A. *J. Am. Chem. Soc.* **1980**, *102*, 2585; Stein, A.R. *Can. J. Chem.* **1987**, *65*, 363.

[79]See, for example, Gregory, B.J.; Kohnstam, G.; Queen, A.; Reid, D.J. *Chem. Commun.* **1971**, 797; Raber, D.J.; Harris, J.C.; Hall, R.E.; Schleyer, P.v.R. *J. Am. Chem. Soc.* **1971**, *93*, 4821; McLennan, D.J. *Acc. Chem. Res.* **1976**, *9*, 281; McLennan, D.J.; Martin, P.L. *Tetrahedron Lett.* **1973**, 4215; Raaen, V.F.; Juhlke, T.; Brown, F.J.; Collins, C.J. *J. Am. Chem. Soc.* **1974**, *96*, 5928; Gregoriou, G.A. *Tetrahedron Lett.* **1976**, 4605, 4767; Queen, A.; Matts, T.C. *Tetrahedron Lett.* **1975**, 1503; Stein, A.R. *J. Org. Chem.* **1976**, *41*, 519; Stephan, E. *Bull. Soc. Chim. Fr.* **1977**, 779; Katritzky, A.R.; Musumarra, G.; Sakizadeh, K. *J. Org. Chem.* **1981**, *46*, 3831. For a reply to some of these objections, see Sneen, R.A.; Robbins, H.M. *J. Am. Chem. Soc.* **1972**, *94*, 7868. For a discussion, see Klumpp, G.W. *Reactivity in Organic Chemistry*, Wiley, NY, **1982**, pp. 442–450.

[80]For ion-molecule pairs in other solvolysis reactions, see Thibblin, A. *J. Chem. Soc. Perkin Trans. 2* **1987**, 1629.

and co-workers found that when such a reaction was run at varying high pressures, there was a minimum in the plot of rate constant versus pressure.[81] A minimum of this sort usually indicates a change in mechanism, and the interpretation in this case was that the normal S_N2 mechanism operates at higher pressures and the cation–molecule mechanism at lower pressures.

An alternative view that also favors an intermediate mechanism is that of Schleyer and co-workers,[82] who believe that the key to the problem is varying degrees of nucleophilic solvent assistance to ion-pair formation. They have proposed an S_N2(intermediate) mechanism.[83]

Among the experiments that have been cited for the viewpoint that borderline behavior results from simultaneous S_N1 and S_N2 mechanisms is the behavior of 4-methoxybenzyl chloride in 70% aqueous acetone.[84] In this solvent, hydrolysis (that is, conversion to 4-methoxybenzyl alcohol) occurs by an S_N1 mechanism. When azide ions are added, the alcohol is still a product, but now 4-methoxybenzyl azide is another product. Addition of azide ions increases the rate of ionization (by the salt effect) but *decreases* the rate of hydrolysis. If more carbocations are produced but fewer go to the alcohol, then some azide must be formed by reaction with carbocations: an S_N1 process. However, the rate of ionization is always *less* than the total rate of reaction, so some azide must also form by an S_N2 mechanism.[84] Thus, the conclusion is that S_N1 and S_N2 mechanisms operate simultaneously.[85]

Some nucleophilic substitution reactions that seem to involve a "borderline" mechanism actually do not. Thus, one of the principal indications that a "borderline" mechanism is taking place has been the finding of partial racemization and partial inversion. However, Weiner and Sneen have demonstrated that this type of stereochemical behavior is quite consistent with a strictly S_N2 process. These workers studied the reaction of optically active 2-octyl brosylate in 75% aqueous dioxane, under which conditions inverted 2-octanol was obtained in 77% optical purity.[86] When

[81]Katritzky, A.R.; Sakizadeh, K.; Gabrielsen, B.; le Noble, W.J. *J. Am. Chem. Soc.* **1984**, *106*, 1879.

[82]Bentley, T.W.; Bowen, C.T.; Morten, D.H.; Schleyer, P.v.R. *J. Am. Chem. Soc.* **1981**, *103*, 5466.

[83]For additional evidence for this view, see Laureillard, J.; Casadevall, A.; Casadevall, E. *Tetrahedron* **1984**, *40*, 4921; *Helv. Chim. Acta* **1984**, *67*, 352; McLennan, D.J. *J. Chem. Soc. Perkin Trans. 2* **1981**, 1316. For evidence against the S_N2 (intermediate) mechanism, see Allen, A.D.; Kanagasabapathy, V.M.; Tidwell, T.T. *J. Am. Chem. Soc.* **1985**, *107*, 4513; Fărcaşiu, D.; Jähme, J.; Rüchardt, C. *J. Am. Chem. Soc.* **1985**, *107*, 5717; Dietze, P.E.; Jencks, W.P. *J. Am. Chem. Soc.* **1986**, *108*, 4549; Dietze, P.E.; Hariri, R.; Khattak, J. *J. Org. Chem.* **1989**, *54*, 3317; Coles, C.J.; Maskill, H. *J. Chem. Soc. Perkin Trans. 2* **1987**, 1083; Richard, J.P.; Amyes, T.L.; Vontor, T. *J. Am. Chem. Soc.* **1991**, *113*, 5871.

[84]Kohnstam, G.; Queen, A.; Shillaker, B. *Proc. Chem. Soc.* **1959**, 157; Amyes, T.L.; Richard, J.P. *J. Am. Chem. Soc.* **1990**, *112*, 9507. For other evidence supporting the concept of simultaneous mechanisms, see Pocker, Y. *J. Chem. Soc.* **1959**, 3939, 3944; Casapieri, P.; Swart, E.R. *J. Chem. Soc.* **1963**, 1254; Ceccon, A.; Papa, I.; Fava, A. *J. Am. Chem. Soc.* **1966**, *88*, 4643; Okamoto, K.; Uchida, N.; Saitô, S.; Shingu, H. *Bull. Chem. Soc. Jpn.* **1966**, *39*, 307; Guinot, A.; Lamaty, G. *Chem. Commun.* **1967**, 960; Queen, A. *Can. J. Chem.* **1979**, *57*, 2646; Richard, J.P.; Rothenberg, M.E.; Jencks, W.P. *J. Am. Chem. Soc.* **1984**, *106*, 1361; Richard, J.P.; Jencks, W.P. *J. Am. Chem. Soc.* **1984**, *106*, 1373, 1383; Katritzky, A.R.; Brycki, B.E. *J. Phys. Org. Chem.* **1988**, *1*, 1; Stein, A.R. *Can. J. Chem.* **1989**, *67*, 297.

[85]These data have also been explained as being in accord with the ion-pair mechanism: Sneen, R.A.; Larsen, J.W. *J. Am. Chem. Soc.* **1969**, *91*, 6031.

[86]Weiner, H.; Sneen, R.A. *J. Am. Chem. Soc.* **1965**, *87*, 287.

sodium azide was added, 2-octyl azide was obtained along with the 2-octanol, *but the latter was now 100% inverted*. It is apparent that, in the original case, 2-octanol was produced by two different processes: an S_N2 reaction leading to inverted product, and another process in which some intermediate leads to racemization or retention. When azide ions were added, they scavenged this intermediate, so that the entire second process now went to produce azide, while the S_N2 reaction, unaffected by addition of azide, still went on to give inverted 2-octanol. What is the nature of the intermediate in the second process? At first thought we might suppose that it is a carbocation, so that this would be another example of simultaneous S_N1 and S_N2 reactions. However, solvolysis of 2-octyl brosylate in pure methanol or of 2-octyl methanesulfonate in pure water, in the absence of azide ions, gave methyl 2-octyl ether or 2-octanol, respectively, *with 100% inversion of configuration*, indicating that the mechanism in these solvents was pure S_N2. Since methanol and water are more polar than 75% aqueous dioxane and since an increase in polarity of solvent increases the rate of S_N1 reactions at the expense of S_N2 (p. 500), it is extremely unlikely that any S_N1 process could occur in 75% aqueous dioxane. The intermediate in the second process is thus not a carbocation. It's nature is suggested by the fact that, in the absence of azide ions, the amount of inverted 2-octanol decreased with an increasing percentage of dioxane in the solvent. Thus the intermediate is an oxonium ion (**19**) formed by an S_N2 attack *by dioxane*. This ion is not a stable product but reacts with water in another S_N2 process to produce 2-octanol with retained configuration.

That part of the original reaction that resulted in retention of configuration[87] is thus seen to stem from two successive S_N2 reactions and not from any "borderline" behavior.[88]

SET MECHANISMS

In certain reactions where nucleophilic substitutions would seem obviously indicated, there is evidence that radicals and/or radical ions are actually involved.[89]

---

[87]According to this scheme, the configuration of the isolated RN_3 should be retained. It was, however, largely inverted, owing to a competing S_N2 reaction where N_3^- directly attacks ROBs.

[88]For other examples, see Streitwieser, Jr., A.; Walsh, T.D.; Wolfe, Jr., J.R. *J. Am. Chem. Soc.* **1965**, *87*, 3682; Streitwieser, Jr., A.; Walsh, T.D. *J. Am. Chem. Soc.* **1965**, *87*, 3686; Beronius, P.; Nilsson, A.; Holmgren, A. *Acta Chem. Scand.* **1972**, *26*, 3173. See also, Knier, B.L.; Jencks, W.P. *J. Am. Chem. Soc.* **1980**, *102*, 6789.

[89]Kornblum, N.; Michel, R.E.; Kerber, R.C. *J. Am. Chem. Soc.* **1966**, *88*, 5660, 5662; Russell, G.A.; Danen, W.C. *J. Am. Chem. Soc.* **1966**, *88*, 5663; Bank, S.; Noyd, D.A. *J. Am. Chem. Soc.* **1973**, *95*, 8203; Ashby, E.C.; Goel, A.B.; Park, W.S. *Tetrahedron Lett.* **1981**, *22*, 4209. For discussions of the relationship between S_N2 and SET mechanisms, see Lewis, E.S. *J. Am. Chem. Soc.* **1989**, *111*, 7576; Shaik, S.S. *Acta Chem. Scand.* **1990**, *44*, 205.

The first step in such a process is transfer of an electron from the nucleophile to the substrate to form a radical anion:

Step 1 $R-X$ + $\overline{Y}^-$ $\longrightarrow$ $R-X^{\bullet-}$ + $Y^{\bullet}$

Mechanisms that begin this way are called SET *(single electron transfer) mechanisms.*[90] Once formed, the radical ion cleaves:

Step 2 $R-X^{\bullet-}$ $\longrightarrow$ $R^{\bullet}$ + $\overline{X}^-$

The radicals formed in this way can go on to product by reacting with the $Y^{\bullet}$ produced in Step 1 or with the original nucleophilic ion Y^-, in which case an additional step is necessary:

Step 3 $R^{\bullet}$ + $Y^{\bullet}$ $\longrightarrow$ $R-Y$

or

Step 3 $R^{\bullet}$ + $\overline{Y}^-$ $\longrightarrow$ $R-Y^{\bullet-}$

Step 4 $R-Y^{\bullet-}$ + $R-X$ $\longrightarrow$ $R-Y$ + $R-X^{\bullet-}$

In the latter case, the radical ion $R-X^{\bullet-}$ is formed by Step 4 as well as by Step 1, so that a chain reaction (p. 936) can take place.

One type of evidence for an SET mechanism is the finding of some racemization. A totally free radical would of course result in a completely racemized product RY, but it has been suggested[91] that inversion can also take place in some SET processes. The suggestion is that in Step 1 the Y^- still approaches from the back side, even although an ordinary S_N2 mechanism will not follow, and that the radical $R\bullet$, once formed, remains in a solvent cage with $Y\bullet$ still opposite X^-, so that Steps 1, 2, and 3 can lead to inversion.

$\overline{Y}^-$ + $R-X$ $\longrightarrow$ $[Y^{\bullet}\,R-X^{\bullet-}]$ $\longrightarrow$ $[Y^{\bullet}R^{\bullet}\,X^-]$ $\longrightarrow$ $Y-R$ + $\overline{X}^-$
    Solvent   Solvent
    cage   cage

Reactions with SET mechanisms typically show predominant, although not 100%, inversion.

---

[90]For reviews, see Savéant, J. *Adv. Phys. Org. Chem.* **1990**, *26*, 1; Rossi, R.A.; Pierini, A.B.; Palacios, S.M. *J. Chem. Educ.* **1989**, *66*, 720; Ashby, E.C. *Acc. Chem. Res.* **1988**, *21*, 414; Chanon, M.; Tobe, M.L. *Angew. Chem. Int. Ed.* **1982**, *21*, 1. See also, Pross, A. *Acc. Chem. Res.* **1985**, *18*, 212; Chanon, M. *Acc. Chem. Res.* **1987**, *20*, 214. See Rossi, R.A.; Pierini, A.B.; Peñéñory, A.B. *Chem. Rev.* **2003**, *103*, 71.
[91]Ashby, E.C.; Pham, T.N. *Tetrahedron Lett.* **1987**, *28*, 3183; Daasbjerg, K.; Lund, T.; Lund, H. *Tetrahedron Lett.* **1989**, *30*, 493.

Other evidence cited[92] for SET mechanisms has been detection of radical or radical ion intermediates by esr[93] or CIDNP; the finding that such reactions can take place at 1-norbornyl bridgeheads;[94] and the formation of cyclic side products when the substrate has a double bond in the 5,6 position (such substrates are called *radical probes*).

Free radicals with double bonds in this position are known to cyclize readily (p. 1011).[95]

20   **21**

The SET mechanism is chiefly found, where X = I or NO$_2$ (see **10-67**). A closely related mechanism, the S$_{RN}$1, takes place with aromatic substrates (Chapter 13).[96] In that mechanism, the initial attack is by an electron donor, rather than a nucleophile. The S$_{RN}$1 mechanism has also been invoked for reactions of enolate anions with 2-iodobicyclo[4.1.0]heptane.[97] An example is the reaction of 1-iodobicyclo[2.2.1]heptane (**20**) with NaSnMe$_3$ or LiPPh$_2$, and some other nucleophiles, to give the substitution product.[98] Another is the reaction of bromo 4-bromoacetophenone (**21**) with Bu$_4$NBr in cumene.[99] The two mechanisms, S$_N$2 versus SET, have been compared and contrasted.[100] There are also reactions where it is reported that radical, carbanion, and carbene pathways occur simultaneously.[101]

[92]See also, Chanon, M.; Tobe, M.L. *Angew. Chem. Int. Ed.* **1982**, *21*, 1; Fuhlendorff, R.; Lund, T.; Lund, H.; Pedersen, J.A. *Tetrahedron Lett.* **1987**, *28*, 5335.

[93]See, for example Russell, J.A.; Pecoraro, J.M. *J. Am. Chem. Soc.* **1979**, *101*, 3331.

[94]Santiago, A.N.; Morris, D.G.; Rossi, R.A. *J. Chem. Soc., Chem. Commun.* **1988**, 220.

[95]For criticisms of this method for demonstrating SET mechanisms, see Newcomb, M.; Kaplan, J. *Tetrahedron Lett.* **1988**, *29*, 3449; Newcomb, M.; Curran, D.P. *Acc. Chem. Res.* **1988**, *21*, 206; Newcomb, M. *Acta Chem. Scand.* **1990**, *44*, 299. For replies to the criticism, see Ashby, E.C. *Acc. Chem. Res.* **1988**, *21*, 414; Ashby, E.C.; Pham, T.N.; Amrollah-Madjdabadi, A.A. *J. Org. Chem.* **1991**, *56*, 1596.

[96]In this book, we make the above distinction between the SET and S$_{RN}$1 mechanisms. However, many workers use the designation SET to refer to the S$_{RN}$1, the chain version of the SET, or both.

[97]Nazareno, M.A.; Rossi, R.A. *Tetrahedron* **1994**, *50*, 9267; Nazareno, M.A.; Rossi, R.A. *J. Org. Chem.* **1996**, *61*, 1645.

[98]Ashby, E.C.; Sun, X.; Duff, J.L. *J. Org. Chem.* **1994**, *59*, 1270.

[99]Haberfield, P. *J. Am. Chem. Soc.* **1995**, *117*, 3314.

[100]Shaik, S.S. *Acta Chem. Scand.* **1990**, *44*, 205.

[101]Ashby, E.C.; Park, B.; Patil, G.S.; Gadru, K.; Gurumurthy, R. *J. Org. Chem.* **1993**, *58*, 424.

The mechanisms so far considered can, in theory at least, operate on any type of saturated (or for that matter unsaturated) substrate. There are other mechanisms that are more limited in scope.

The Neighboring-Group Mechanism[102]

It is occasionally found with certain substrates that (*1*) the rate of reaction is greater than expected, and (*2*) the configuration at a chiral carbon is *retained* and not inverted or racemized. In these cases, there is usually a group with an unshared pair of electrons β to the leaving group (or sometimes farther away). The mechanism operating in such cases is called the *neighboring-group mechanism* and consists essentially of two S_N2 substitutions, each causing an inversion so the net result is retention of configuration.[103] In the first step of this reaction, the neighboring group acts as a nucleophile, pushing out the leaving group but still retaining attachment to the molecule. In the second step, the external nucleophile displaces the neighboring group by a backside attack:

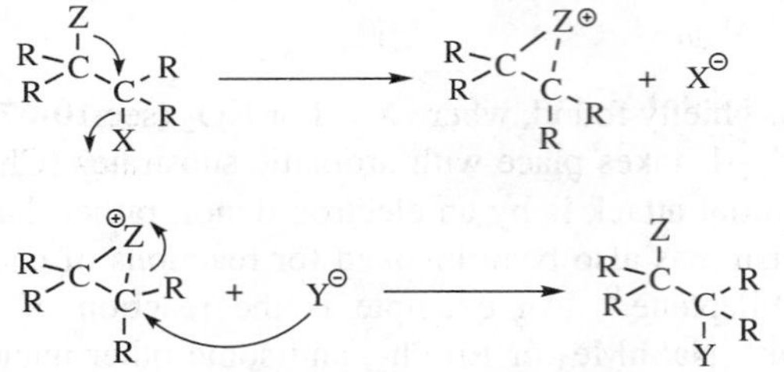

Step 1

Step 2

The reaction obviously must go faster than if Y were attacking directly, since if the latter process were faster, *it* would be happening. The neighboring group Z is said to be lending *anchimeric assistance*. The rate law followed in the neighboring-group mechanism is the first-order law shown in Eq. (10.2) or (10.3); that is, Y does not take part in the rate-determining step.

The reason attack by Z is faster than that by Y is that the group Z is more available. In order for Y to react, it must collide with the substrate, but Z is immediately available by virtue of its position. A reaction between the substrate and Y involves a large decrease in entropy of activation $(\Delta S^{\ddagger})$, since the reactants are far less free in the transition state than before. Reaction of Z involves a much smaller loss of $\Delta S^{\ddagger}$ (see p. 303).[104]

---

[102]For a monograph, see Capon, B.; McManus, S. *Neighboring Group Participation*, Vol. 1, Plenum, NY, **1976**.

[103]There is evidence that this kind of process can happen intermolecularly (e.g., $RX + Z^- \rightarrow RZ + Y^-$). In this case Z^- acts as a catalyst for the reaction $RX + Y^- \rightarrow RY$: McCortney, B.A.; Jacobson, B.M.; Vreeke, M.; Lewis, E.S. *J. Am. Chem. Soc.* **1990**, *112*, 3554.

[104]For a review of the energetics of neighboring-group participation, see Page, M.I. *Chem. Soc. Rev.* **1973**, *2*, 295.

It is not always easy to determine when a reaction rate has been increased by anchimeric assistance. In order to be certain, it is necessary to know what the rate would be without participation by the neighboring group. The obvious way to examine this question is to compare the rates of the reaction with and without the neighboring group, for example, $HOCH_2CH_2Br$ versus CH_3CH_2Br. However, this will certainly not give an accurate determination of the extent of participation, since the steric and field effects of H and OH are not the same. Furthermore, no matter what the solvent, the shell of solvent molecules that surrounds the polar protic OH group must differ greatly from that which surrounds the nonpolar H. Because of these considerations, it is desirable to have a large increase in the rate, preferably >50-fold, before a rate increase is attributed to neighboring-group participation.

The first important evidence for the existence of this mechanism was the demonstration that retention of configuration can occur if the substrate is suitable. It was shown that the threo *dl* pair of 3-bromo-2-butanol when treated with HBr gave *dl*-2,3-dibromobutane, while the erythro pair gave the meso isomer (**22**).[105]

threo *dl* Pair   *dl* Pair

erythro *dl* Pair   **22** (meso)

This indicated that retention had taken place. Note that both products are optically inactive and so cannot be told apart by differences in rotation. The meso and *dl* dibromides have different boiling points and indexes of refraction and were identified by these properties. Even more convincing evidence was that either of the two threo isomers alone gave not just one of the enantiomeric dibromides, but the *dl* pair. The reason for this is that the intermediate present after the attack by the neighboring group (**23**) is symmetrical, so the external nucleophile Br⁻ can attack

23

both carbon atoms equally well. Intermediate **23** is a bromonium ion, the existence of which has been demonstrated in several types of reactions.

[105]Winstein, S.; Lucas, H.J. *J. Am. Chem. Soc.* **1939**, *61*, 1576, 2845.

Although **23** is symmetrical, intermediates in most neighboring-group mechanisms are not, and it is therefore possible to get not a simple substitution product but a rearrangement. This will happen if Y attacks not the carbon atom from which X left, but the one to which Z was originally attached:

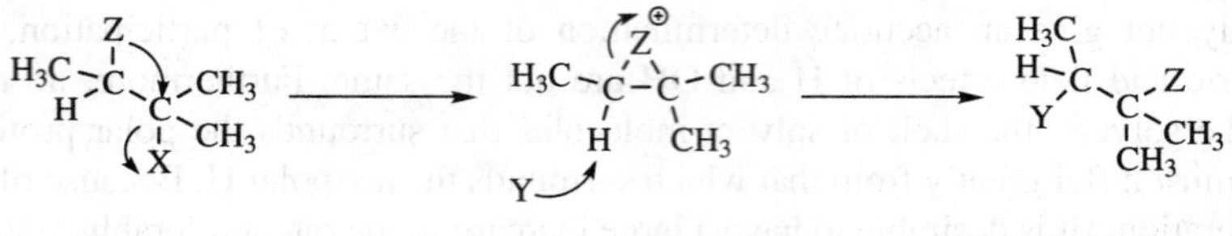

In such cases, substitution and rearrangement products are often produced together. For a discussion of rearrangements, see Chapter 18.

Another possibility is that the intermediate may be stable or may find some other way to stabilize itself. In such cases, Y never attacks at all and the product is cyclic. These are simple internal S_N2 reactions. Two examples are formation of epoxides and lactones:

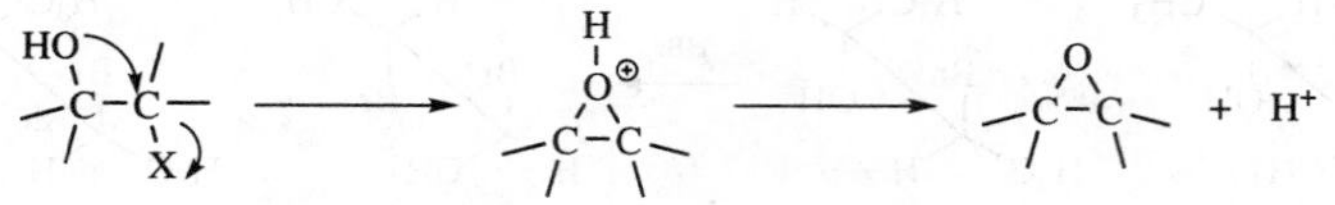

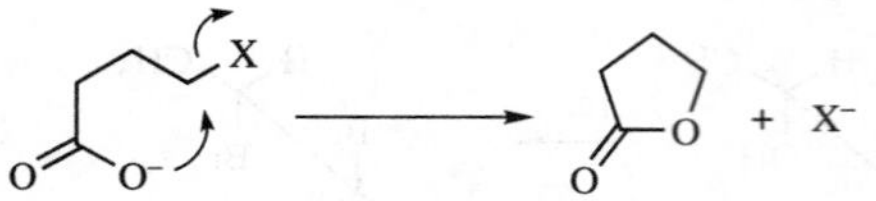

The fact that acetolysis of both 4-methoxy-1-pentyl brosylate (**24**) and 5-methoxy-2-pentyl brosylate (**25**) gave the same mixture of products is further evidence for participation by a neighboring group.[106] In this case, the intermediate **26** is common to both substrates.

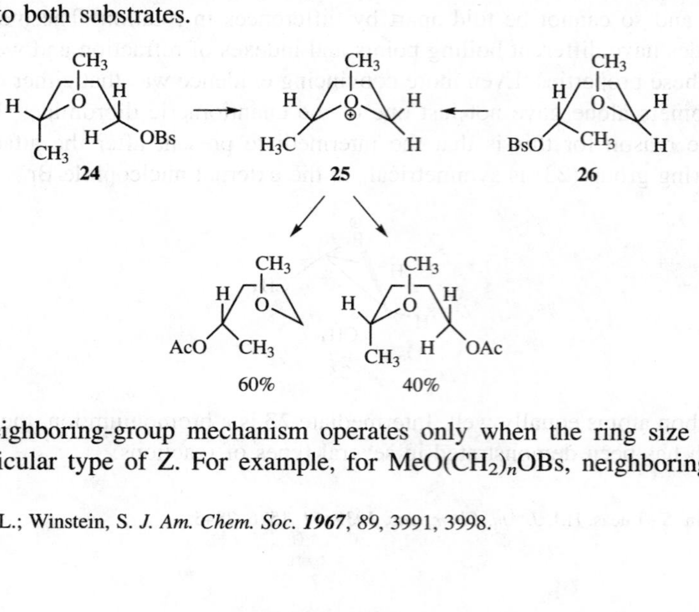

The neighboring-group mechanism operates only when the ring size is right for a particular type of Z. For example, for $MeO(CH_2)_nOBs$, neighboring-group

[106]Allred, E.L.; Winstein, S. *J. Am. Chem. Soc.* **1967**, *89*, 3991, 3998.

participation was important for $n = 4$ or 5 (corresponding to a five- or six-membered intermediate), but not for $n = 2$, 3, or 6.[107] However, optimum ring size is not the same for all reactions, even with a particular Z. In general, the most rapid reactions occur when the ring size is three, five, or six, depending on the reaction type. The likelihood of four-membered ring neighboring-group participation is increased when there are alkyl groups a or β to the neighboring group.[108]

The following are some of the more important neighboring groups: COO^- (but not COOH), COOR, COAr, OCOR,[109] OR, OH, O^-,[110] NH_2, NHR, NR_2, NHCOR, SH, SR, S^-,[111] SO_2Ph,[112] I, Br, and Cl. The effectiveness of halogens as neighboring groups decreases in the order $I > Br > Cl$.[113] The Cl is a very weak neighboring group and can be shown to act in this way only when the solvent does not interfere. For example, when 5-chloro-2-hexyl tosylate is solvolyzed in acetic acid, there is little participation by the Cl, but when the solvent is changed to trifluoroacetic acid, which is much less nucleophilic, neighboring-group participation by the Cl becomes the major reaction pathway.[114] Thus, Cl acts as a neighboring group *only when there is need for it* (for other examples of the *principle of increasing electron demand*, see below; p. 454).

A number of intermediates of halogen participation (halonium ions),[115] for example, **27** and **28**, have been prepared as stable salts in SbF_5-SO_2 or SbF_5-SO_2ClF solutions.[116] Some have even been crystallized. Attempts to prepare

---

[107]Allred, E.L.; Winstein, S. *J. Am. Chem. Soc.* **1967**, *89*, 4012.

[108]Eliel, E.L.; Clawson, L.; Knox, D.E. *J. Org. Chem.* **1985**, *50*, 2707; Eliel, E.L.; Knox, D.E. *J. Am. Chem. Soc.* **1985**, *107*, 2946.

[109]For an example of OCOR as a neighboring group where the ring size is seven membered, see Wilen, S.H.; Delguzzo, L.; Saferstein, R. *Tetrahedron* **1987**, *43*, 5089.

[110]For a review of oxygen functions as neighboring groups, see Perst, H. *Oxonium Ions in Organic Chemistry*; Verlag Chemie, Deerfield Beach, FL, **1971**, pp. 100–127. There is evidence that the oxygen in an epoxy group can also act as a neighboring group: Francl, M.M.; Hansell, G.; Patel, B.P.; Swindell, C.S. *J. Am. Chem. Soc.* **1990**, *112*, 3535.

[111]For a review of sulfur-containing neighboring groups, see Block, E. *Reactions of Organosulfur Compounds*, Academic Press, NY, **1978**, pp. 141–145.

[112]Lambert, J.B.; Beadle, B.M.; Kuang, K. *J. Org. Chem.* **1999**, *64*, 9241.

[113]Peterson, P.E. *Acc. Chem. Res.* **1971**, *4*, 407, and references cited therein.

[114]Peterson, P.E.; Bopp, R.J.; Chevli, D.M.; Curran, E.L.; Dillard, D.E.; Kamat, R.J. *J. Am. Chem. Soc.* **1967**, *89*, 5902. See also, Reich, I.L.; Reich, H.J. *J. Am. Chem. Soc.* **1974**, *96*, 2654.

[115]For a monograph, see Olah, G.A. *Halonium Ions*, Wiley, NY, **1975**. For a review, see Koster, G.F., in Patai, S.; Rappoport, Z. *The Chemistry of Functional Groups, Supplement D*, pt. 2, Wiley, NY, **1983**, pp. 1265–1351.

[116]See, for example Olah, G.A.; Peterson, P.E. *J. Am. Chem. Soc.* **1968**, *90*, 4675; Henrichs, P.M.; Peterson, P.E. *J. Am. Chem. Soc.* **1973**, *95*, 7449, *J. Org. Chem.* **1976**, *41*, 362; Olah, G.A.; Liang, G.; Staral, J. *J. Am. Chem. Soc.* **1974**, *96*, 8112; Vančik, H.; Percač, K.; Sunko, D.E. *J. Chem. Soc., Chem. Commun.* **1991**, 807.

four-membered homologs of **27** and **28** were not successful.[117] There is no evidence that F can act as a neighboring group.[113]

The principle that a neighboring group lends assistance in proportion to the need for such assistance also applies to differences in leaving-group ability. Thus, $p\text{-}NO_2C_6H_4SO_2O$ (the nosylate group) is a better leaving group than $p\text{-}MeC_6H_4\text{-}SO_2O$ (the tosylate group). Experiments have shown that the OH group in *trans*-2-hydroxycyclopentyl arenesulfonates **29** acts as a neighboring group when the leaving group is tosylate but not when it is nosylate, apparently because the nosylate group leaves so rapidly that it does not require assistance.[118]

Neighboring-Group Participation by π and σ Bonds: Nonclassical Carbocations[119]

For all the neighboring groups listed in the preceding section, the nucleophilic attack is made by an atom with an unshared pair of electrons. In this section, we consider neighboring-group participation by C=C π bonds and C—C and C—H σ bonds. There has been a great deal of controversy over whether such bonds can act as neighboring groups and about the existence and structure of the intermediates involved. These intermediates are called *nonclassical* (or *bridged*) carbocations. In classical carbocations (Chapter 5) the positive charge is localized on one carbon atom or delocalized by resonance involving an unshared pair of electrons or a double or triple bond in the allylic position. In a nonclassical carbocation, the positive charge is delocalized by a double or triple bond that is not in the allylic position or by a single bond. Examples are the 7-norbornenyl cation (**30**), the norbornyl cation

[117]Olah, G.A.; Bollinger, J.M.; Mo, Y.K.; Brinich, J.M. *J. Am. Chem. Soc.* **1972**, *94*, 1164.

[118]Haupt, F.C.; Smith, M.R. *Tetrahedron Lett.* **1974**, 4141.

[119]For monographs, see Olah, G.A.; Schleyer, P.v.R. *Carbonium Ions*, Vol. 3, Wiley, NY, **1972**; Bartlett, P.D. *Nonclassical Ions*, W.A. Benjamin, NY, **1965**. For reviews, see Barkhash, V.A. *Top. Curr. Chem.* **1984**, *116/117*, 1; Kirmse, W. *Top. Curr. Chem.* **1979**, *80*, 125, pp. 196–288; McManus, S.P.; Pittman, Jr., C.U., in McManus, S.P. *Organic Reactive Intermediates*, Academic Press, NY, **1973**, pp. 302–321; Bethell, D.; Gold, V. *Carbonium Ions*, Academic Press, NY, **1967**; pp. 222–282. For a related review, see Prakash, G.K.S.; Iyer, P.S. *Rev. Chem. Intermed.* **1988**, *9*, 65.

(**31**),[120] and the cyclopropylmethyl cation (**32**). A cyclopropyl group (as in **33**) is capable of stabilizing the norbornyl cation, inhibiting this rearrangement.[121] Carbocation **30** is called a *homoallylic* carbocation, because in **30a** there is one carbon atom between the positively charged carbon and the double bond. Many of these carbocations can be produced in more than one way if the proper substrates are chosen. For example, **31** can be generated by the departure of a leaving group from

34 or from **35**.[122] The first of these pathways is called the σ route to a nonclassical carbocation, because participation of a σ bond is involved. The second is called the π route.[123] The argument against the existence of nonclassical carbocations is essentially that the structures **30a–c** (or **31a**, **31b**, etc.) are not canonical forms but real structures and that there is rapid equilibration among them. This debate remains an active area of interest for some reactions.[124] In one study, the solvolysis and rearrangement of 2-bicyclo[3.2.2]nonanyl tosylate in methanol generated ethers derived from the 2-bicyclo[3.2.2]nonanyl and 2-bicyclo[3.3.1]nonanyl systems that were rationalized in terms of a classical carbocation.[125] Density functional and *ab initio* calculations indicated that the products of the 2-bicyclo[3.2.2]nonanyl tosylate solvolysis were found to have nonclassical structures.[126]

In discussing nonclassical carbocations, we must be careful to make the distinction between neighboring-group participation and the existence of nonclassical carbocations.[127] If a nonclassical carbocation exists in any reaction, then an ion with electron delocalization, as shown in the above examples, is a discrete reaction intermediate. If a carbon–carbon double or single bond participates in the departure of the leaving group to form a carbocation, it may be that a nonclassical carbocation is involved, but there is no necessary relation. In any particular case either or both of these possibilities can be taking place.

In the following pages, we consider some of the evidence bearing on the questions of the participation of π and s bonds and on the existence of nonclassical

[120]Sieber, S.; Schleyer, P.v.R.; Vančik, H.; Mesić, M.; Sunko, D.E. *Angew. Chem. Int. Ed.* **1993**, *32*, 1604; Schleyer, P.v.R.; Sieber, S. *Angew. Chem. Int. Ed.* **1993**, *32*, 1606.

[121]Herrmann, R.; Kirmse, W. *Liebigs Ann. Chem.* **1995**, 703.

[122]Lawton, R.G. *J. Am. Chem. Soc.* **1961**, *83*, 2399; Bartlett, P.D.; Bank, S.; Crawford, R.J.; Schmid, G.H. *J. Am. Chem. Soc.* **1965**, *88*, 1288.

[123]Winstein, S.; Carter, P. *J. Am. Chem. Soc.* **1961**, *83*, 4485.

[124]For example see Brunelle, P.; Sorensen, T.S.; Taeschler, C. *J. Org. Chem.* **2001**, *66*, 7294.

[125]Okazaki, T.; Terakawa, E.; Kitagawa, T.; Takeuchi, K. *J. Org. Chem.* **2000**, *65*, 1680.

[126]Smith, W. B. *J. Org. Chem.* **2001**, *66*, 376.

[127]This was pointed out by Cram, D.J. *J. Am. Chem. Soc.* **1964**, *86*, 3767.

carbocations,[128] although a thorough discussion is beyond the scope of this book.[89]

1. **C=C** *as a Neighboring Group.*[129] The most striking evidence that C=C can act as a neighboring group is that acetolysis of **36**-OTs is 10^{11} times faster than that of **37**-OTs and *proceeds with retention of configuration.*[130] The rate data alone do not necessarily prove that acetolysis of **36**-OTs involves a

nonclassical intermediate (**30d**), but it is certainly strong evidence that the C=C group assists in the departure of the OTs. Evidence that **30** is indeed a nonclassical ion comes from an NMR study of the relatively stable norbornadienyl cation (**38**). The spectrum shows that the 2 and 3 protons are not equivalent to the 5 and 6 protons.[131] Thus there is interaction between the charged carbon and one double bond, which is evidence for the existence of **30d**.[132] In the case of **36**, the double bond is geometrically fixed in an especially favorable position for backside attack on the carbon bearing the leaving group (hence the very large rate enhancement), but there is much evidence that other double bonds in the homoallylic position,[133] as well as in

[128]The arguments against nonclassical ions are summed up in Brown, H.C. *The Nonclassical Ion Problem*; Plenum, NY, *1977*. This book also includes rebuttals by Schleyer, P.v.R. See also, Brown, H.C. *Pure Appl. Chem.* *1982*, *54*, 1783.

[129]For reviews, see Story, P.R.; Clark, Jr., B.C., in Olah, G.A.; Schleyer, P.v.R. *Carbonium Ions*, Vol. 3, Wiley, NY, *1972*, pp. 1007–1060; Richey, Jr., H.G., in Zabicky, J. *The Chemistry of Alkenes*, Vol. 2; Wiley, NY, *1970*, pp. 77–101.

[130]Winstein, S.; Shatavsky, M. *J. Am. Chem. Soc.* *1956*, *78*, 592.

[131]Story, P.R.; Snyder, L.C.; Douglass, D.C.; Anderson, E.W.; Kornegay, R.L. *J. Am. Chem. Soc.* *1963*, *85*, 3630. For a discussion, see Story, P.R.; Clark, Jr., B.C., in Olah, G.A.; Schleyer, P.v.R. *Carbonium Ions*, Vol. 3, Wiley, NY, *1972*, pp. 1026–1041. See also, Lustgarten, R.K.; Brookhart, M.; Winstein, S. *J. Am. Chem. Soc.* *1972*, *94*, 2347.

[132]For further evidence for the nonclassical nature of **30**, see Brookhart, M.; Diaz, A.; Winstein, S. *J. Am. Chem. Soc.* *1966*, *88*, 3135; Richey, Jr., H.G.; Lustgarten, R.K. *J. Am. Chem. Soc.* *1966*, *88*, 3136; Gassman, P.G.; Doherty, M.M. *J. Am. Chem. Soc.* *1982*, *104*, 3742 and references cited therein; Laube, T. *J. Am. Chem. Soc.* *1989*, *111*, 9224.

[133]For examples, see Shoppee, C.W. *J. Chem. Soc.* *1946*, 1147; LeBel, N.A.; Huber, J.E. *J. Am. Chem. Soc.* *1963*, *85*, 3193; Closson, W.D.; Kwiatkowski, G.T. *Tetrahedron 1965*, *21*, 2779; Brown, H.C.; Peters, E.N.; Ravindranathan, M. *J. Am. Chem. Soc.* *1975*, *97*, 7449; Schleyer, P.v.R.; Bentley, T.W.; Koch, W.; Kos, A.J.; Schwarz, H. *J. Am. Chem. Soc.* *1987*, *109*, 6953; Fernández-Mateos, A.; Rentzsch, M.; Sánchez, L.R.; González, R.R. *Tetrahedron 2001*, *57*, 4873.

positions farther away,[134] can also lend anchimeric assistance, although generally with much lower rate ratios. One example of the latter is the compound β-(*syn*-7-norbornenyl)ethyl brosylate (**39**), which at 25°C undergoes acetolysis ~140,000 times faster than the saturated analog **40**.[135] Triple bonds[136] and allenes[137] can also act as neighboring groups.

We have already seen evidence that participation by a potential neighboring group can be reduced or eliminated if an outside nucleophile is present that is more effective than the neighboring group in attacking the central carbon (p. 450), or if a sufficiently good leaving group is present (p. 450). In another example of the principle of increasing electron demand, Gassman and co-workers have shown that neighboring-group participation can also be reduced if the stability of the potential carbocation is increased. They found that the presence of a *p*-anisyl group at the 7 position of **36** and **37** exerts a powerful leveling effect on the rate differences. Thus, solvolysis in acetone–water at 85°C of **38** was only ~2.5 times faster than that of the saturated

compound **42**.[138] Furthermore, both **41** and its stereoisomer **43** gave the same mixture of solvolysis products, showing that the stereoselectivity in the solvolysis of **36** is not present here. The difference between **41** and **36** is that in the case of **41** the positive charge generated at the 7 position in the transition state is greatly stabilized by the *p*-anisyl group. Apparently, the stabilization by the *p*-anisyl group is so great that further stabilization that would come from

[134]For examples, see LeNy, G. *C. R. Acad. Sci.* **1960**, *251*, 1526; Goering, H.L.; Closson, W.D. *J. Am. Chem. Soc.* **1961**, *83*, 3511; Bartlett, P.D.; Trahanovsky, W.S.; Bolon, D.A.; Schmid, G.H. *J. Am. Chem. Soc.* **1965**, *87*, 1314; Bly, R.S.; Swindell, R.T. *J. Org. Chem.* **1965**, *30*, 10; Marvell, E.N.; Sturmer, D.; Knutson, R.S. *J. Org. Chem.* **1968**, *33*, 2991; Cogdell, T.J. *J. Org. Chem.* **1972**, *37*, 2541; Ferber, P.H.; Gream, G.E. *Aust. J. Chem.* **1981**, *34*, 1051; Orlović, M.; Borčić, S.; Humski, K.; Kronja, O.; Imper, V.; Polla, E.; Shiner, Jr., V.J. *J. Org. Chem.* **1991**, *56*, 1874; Winstein, S.; Carter, P. *J. Am. Chem. Soc.* **1961**, *83*, 4485.
[135]Bly, R.S.; Bly, R.K.; Bedenbaugh, A.O.; Vail, O.R. *J. Am. Chem. Soc.* **1967**, *89*, 880.
[136]See, for example, Closson, W.D.; Roman, S.A. *Tetrahedron Lett.* **1966**, 6015; Hanack, M.; Herterich, I.; Vött, V. *Tetrahedron Lett.* **1967**, 3871; Lambert, J.B.; Papay, J.J.; Mark, H.W. *J. Org. Chem.* **1975**, *40*, 633; Peterson, P.E.; Vidrine, D.W. *J. Org. Chem.* **1979**, *44*, 891. For a review of participation by triple bonds and allylic groups, see Rappoport, Z. *React. Intermed. (Plenum)* **1983**, *3*, 440.
[137]Bly, R.S.; Koock, S.U. *J. Am. Chem. Soc.* **1969**, *91*, 3292, 3299; Von Lehman, T.; Macomber, R. *J. Am. Chem. Soc.* **1975**, *97*, 1531.
[138]Gassman, P.G.; Zeller, J.; Lamb, J.T. *Chem. Commun.* **1968**, 69.

participation by the C=C bond is not needed.[139] The use of a phenyl instead of a *p*-anisyl group is not sufficient to stop participation by the double bond completely, although it does reduce it.[140] These results permit us to emphasize our previous conclusion that *a neighboring group lends anchimeric assistance only when there is sufficient demand for it*.[141] The π-bond of a neighboring alkene group can assist solvolysis via π-participation.[142]

The ability of C=C to serve as a neighboring group can depend on its electron density. When the strongly electron-withdrawing CF_3 group was attached to a double bond carbon of **44**, the solvolysis rate was lowered by a factor of about 10^6.[143]

Relative Rates	
$R^1 = R^2 = H$	1.4×10^{12}
$R^1 = H, R^2 = CF_3$	1.5×10^6
$R^1 = R^2 = CF_3$	1

Mos = MeO—⟨ ⟩—$\overset{O_2}{\underset{}{S}}$—

44

A second CF_3 group had an equally strong effect. In this case, two CF_3 groups decrease the electron density of the C=C bond to the point that the solvolysis rate for **44** ($R^1 = R^2 = CF_3$) was about the same as (actually ∼17 times slower than) the rate for the saturated substrate **37** (X = OMos). Thus, the two CF_3 groups completely remove the ability of the C=C bond to act as a neighboring group.

2. *Cyclopropyl*[144] *as a Neighboring Group*.[145] On p. 217 we saw that the properties of a cyclopropane ring are in some ways similar to those of a double bond. Therefore it is not surprising that a suitably placed cyclopropyl ring can

H   OCOAr   H   OBs   BsO   H

Ar = *p*-NO₂C₆H₄

45   **46**   **47**

---

[139]Nevertheless, there is evidence from ^{13}C NMR spectra that some π participation is present, even in the cation derived from **38**: Olah, G.A.; Berrier, A.L.; Arvanaghi, M.; Prakash, G.K.S. *J. Am. Chem. Soc.* **1981**, *103*, 1122.

[140]Gassman, P.G.; Fentiman, Jr., A.F. *J. Am. Chem. Soc.* **1969**, *91*, 1545; **1970**, *92*, 2549.

[141]For a discussion of the use of the tool of increasing electron demand to probe neighboring-group activity by double bonds, sigma bonds, and aryl rings, see Lambert, J.B.; Mark, H.W.; Holcomb, A.G.; Magyar, E.S. *Acc. Chem. Res.* **1979**, *12*, 317.

[142]Malnar, I.; Jurić, S.; Vrček, V.; Gjuranović, Ž.; Mihalič, Z.; Kronja, O. *J. Org. Chem.* **2002**, *67*, 1490.

[143]Gassman, P.G.; Hall, J.B. *J. Am. Chem. Soc.* **1984**, *106*, 4267.

[144]In this section, we consider systems in which at least one carbon separates the cyclopropyl ring from the carbon bearing the leaving group. For a discussion of systems in which the cyclopropyl group is directly attached to the leaving-group carbon, see p. $$$.

[145]For a review, see Haywood-Farmer, J. *Chem. Rev.* **1974**, *74*, 315.

also be a neighboring group. Thus *endo-anti*-tricyclo[$3.2.1.0^{2,4}$]octan-8-yl
p-nitrobenzoate (**45**) solvolyzed $\sim 10^{14}$ times faster that the *p*-nitrobenzoate of
37-OH.[146] Obviously, a suitably placed cyclopropyl ring can be even more
effective[147] as a neighboring group than a double bond.[148] The need for suitable
placement is emphasized by the fact that **47** solvolyzed only about five times
faster than **37**-OBs,[149] while **46** solvolyzed three times *slower* than
37-OBs.[150] In the case of **45** and of all other cases known where cyclopropyl
lends considerable anchimeric assistance, the developing *p* orbital of the carbocation is orthogonal to the participating bond of the cyclopropane ring.[151] An
experiment designed to test whether a developing *p* orbital that would
be parallel to the participating bond would be assisted by that bond showed
no rate enhancement.[151] This is in contrast to the behavior of cyclopropane rings
directly attached to positively charged carbons, where the *p* orbital is parallel to
the plane of the ring (pp. 241, 464). Rate enhancements, although considerably
smaller, have also been reported for suitably placed cyclobutyl rings.[152]

3. *Aromatic Rings as Neighboring Groups*.[153] There is a great deal of evidence
that aromatic rings in the β position can function as neighboring

                    48   **49**   **50**

[146]Tanida, H.; Tsuji, T.; Irie, T. *J. Am. Chem. Soc.* *1967*, *89*, 1953; Battiste, M.A.; Deyrup, C.L.; Pincock,
R.E.; Haywood-Farmer, J. *J. Am. Chem. Soc.* *1967*, *89*, 1954.

[147]For a competitive study of cyclopropyl versus double-bond participation, see Lambert, J.B.;
Jovanovich, A.P.; Hamersma, J.W.; Koeng, F.R.; Oliver, S.S. *J. Am. Chem. Soc.* *1973*, *95*, 1570.

[148]For other evidence for anchimeric assistance by cyclopropyl, see Sargent, G.D.; Lowry, N.; Reich, S.D.
J. Am. Chem. Soc. *1967*, *89*, 5985; Battiste, M.A.; Haywood-Farmer, J.; Malkus, H.; Seidl, P.; Winstein, S.
J. Am. Chem. Soc. *1970*, *92*, 2144; Coates, R.M.; Yano, K. *Tetrahedron Lett.* *1972*, 2289; Masamune, S.;
Vukov, R.; Bennett, M.J.; Purdham, J.T. *J. Am. Chem. Soc.* *1972*, *94*, 8239; Gassman, P.G.; Creary, X. *J.
Am. Chem. Soc.* *1973*, *95*, 2729; Costanza, A.; Geneste, P.; Lamaty, G.; Roque, J. *Bull. Soc. Chim. Fr.*
1975, 2358; Takakis, I.M.; Rhodes, Y.E. *Tetrahedron Lett.* *1983*, *24*, 4959.

[149]Battiste, M.A.; Deyrup, C.L.; Pincock, R.E.; Haywood-Farmer, J. *J. Am. Chem. Soc.* *1967*, *89*, 1954;
Haywood-Farmer, J.; Pincock, R.E. *J. Am. Chem. Soc.* *1969*, *91*, 3020; Haywood-Farmer, J. *Chem. Rev.*
1974, *74*, 315; Haywood-Farmer, J. *Chem. Rev.* *1974*, *74*, 315.

[150]Haywood-Farmer, J.; Pincock, R.E.; Wells, J.I. *Tetrahedron* *1966*, *22*, 2007; Haywood-Farmer, J.;
Pincock, R.E. *J. Am. Chem. Soc.* *1969*, *91*, 3020. For some other cases where there was little or no rate
enhancement by cyclopropyl, see Wiberg, K.B.; Wenzinger, G.R. *J. Org. Chem.* *1965*, *30*, 2278; Sargent,
G.D.; Taylor, R.L.; Demisch, W.H. *Tetrahedron Lett.* *1968*, 2275; Rhodes, Y.E.; Takino, T. *J. Am. Chem.
Soc.* *1970*, *92*, 4469; Hanack, M.; Krause, P. *Liebigs Ann. Chem.* *1972*, *760*, 17.

[151]Gassman, P.G.; Seter, J.; Williams, F.J. *J. Am. Chem. Soc.* *1971*, *93*, 1673. For a discussion, see
Haywood-Farmer, J.; Pincock, R.E. *J. Am. Chem. Soc.* *1969*, *91*, 3020. See also, Chenier, P.J.; Jenson,
T.M.; Wulff, W.D. *J. Org. Chem.* *1982*, *47*, 770.

[152]For example, see Sakai, M.; Diaz, A.; Winstein, S. *J. Am. Chem. Soc.* *1970*, *92*, 4452; Battiste, M.A.;
Nebzydoski, J.W. *J. Am. Chem. Soc.* *1970*, *92*, 4450; Schipper, P.; Driessen, P.B.J.; de Haan, J.W.; Buck, H.M. *J. Am.
Chem. Soc.* *1974*, *96*, 4706; Ohkata, K.; Doecke, C.W.; Klein, G.; Paquette, L.A. *Tetrahedron Lett.* *1980*, *21*, 3253.

[153]For a review, see Lancelot, L.A.; Cram, D.J.; Schleyer, P.v.R., in Olah, G.A.; Schleyer, P.v.R.
Carbonium Ions, Vol. 3; Wiley, NY, *1972*, pp. 1347–1483.

groups.[154] Stereochemical evidence was obtained by solvolysis of L-*threo*-3-phenyl-2-butyl tosylate (**48**) in acetic acid.[155] Of the acetate product 96% was the threo isomer and only about 4% was erythro. Moreover, both the (+) and (−) threo isomers (**49 and 50**) were produced in approximately equal amounts (a racemic mixture). When solvolysis was conducted in formic acid, even less erythro isomer was obtained. This result is similar to that found on reaction of 3-bromo-2-butanol with HBr (p. 446) and leads to the conclusion that configuration is retained because phenyl acts as a neighboring group. However, evidence from rate studies is not so simple. If β-aryl groups assist the departure of the leaving group, solvolysis rates should be enhanced. In general, they are not. However, solvolysis rate studies in 2-arylethyl systems are complicated by the fact that, for primary and secondary systems, two pathways can exist.[156] In one of these (designated k_Δ), the aryl, behaving as a neighboring group, pushes out the leaving group to give a bridged ion, called a *phenonium ion* (**51**), and is in turn pushed out by the solvent SOH, so

the net result is substitution with retention of configuration (or rearrangement, if **51** is opened from the other side). The other pathway (k_s) is simple S_N2 attack by the solvent at the leaving-group carbon. The net result here is substitution with inversion and no possibility of rearrangement. Whether the leaving group is located at a primary or a secondary carbon, there is no cross-over between these pathways; they are completely independent.[157] (Both the k_Δ and k_s pathways are unimportant when the leaving group is at a tertiary carbon.) In these cases, the mechanism is S_N1 and open carbocations $ArCH_2CR_2^+$ are intermediates. This pathway is designated k_c. Which of the two pathways (k_s or k_Δ) predominates in any given case depends on the solvent and on the nature of the aryl group. As expected from the results we have seen for Cl as a neighboring group (p. 450), the k_Δ/k_s ratio is highest for solvents that are poor nucleophiles and so compete very poorly with the aryl

[154]Kevill, D.N.; D'Souza, M.J. *J. Chem. Soc. Perkin Trans. 2* **1997**, 257; Fujio, M.; Goto, N.; Dairokuno, T.; Goto, M.; Saeki, Y.; Okusako, Y.; Tsuno, Y. *Bull. Chem. Soc. Jpn.* **1992**, 65, 3072.

[155]Cram, D.J. *J. Am. Chem. Soc.* **1949**, 71, 3863; **1952**, 74, 2129.

[156]Brookhart, M.; Anet, F.A.L.; Cram, D.J.; Winstein, S. *J. Am. Chem. Soc.* **1966**, 88, 5659; Lee, C.C.; Unger, D.; Vassie, S. *Can. J. Chem.* **1972**, 50, 1371.

[157]Brown, H.C.; Kim, C.J.; Lancelot, C.J.; Schleyer, P.v.R. *J. Am. Chem. Soc.* **1970**, 92, 5244; Brown, H.C.; Kim, C.J. *J. Am. Chem. Soc.* **1971**, 93, 5765.

group. For several common solvents the k_Δ/k_s ratio increases in the order EtOH < CH_3COOH < HCOOH < CF_3COOH.[158] In accord with this, the following percentages of retention were obtained in solvolysis of 1-phenyl-2-propyl tosylate at 50°C: solvolysis in EtOH 7%, CH_3COOH 35%, HCOOH 85%.[158] This indicates that k_s predominates in EtOH (phenyl participates very little), while k_Δ predominates in HCOOH. Trifluoroacetic acid is a solvent of particularly low nucleophilic power, and in this solvent the reaction proceeds entirely by k_Δ;[159] deuterium labeling showed 100% retention.[160] This case provides a clear example of neighboring-group rate enhancement by phenyl: The rate of solvolysis of $PhCH_2CH_2OTs$ at 75°C in CF_3COOH is 3040 times the rate for CH_3CH_2OTs.[159]

With respect to the aromatic ring, the k_Δ pathway is electrophilic aromatic substitution (Chapter 11). We predict that groups on the ring that activate that reaction (p. 665) will increase, and deactivating groups will decrease, the rate of this pathway. This prediction has been borne out by several investigations. The *p*-nitro derivative of **48** solvolyzed in acetic acid 190 times slower than **48**, and there was much less retention of configuration; the acetate produced was only 7% threo and 93% erythro.[161] At 90°C, acetolysis of *p*-$ZC_6H_4CH_2CH_2OTs$ gave the rate ratios shown in Table 10.1.[162] Throughout this series k_s is fairly constant, as it should be since it is affected only by the rather remote field effect of Z. It is k_Δ that changes substantially as Z is changed from activating to deactivating. The evidence is thus fairly clear that participation by aryl groups depends greatly on the nature of the group. For some groups (e.g., *p*-nitrophenyl), in some solvents (e.g., acetic acid), there is essentially no neighboring-group participation at all,[163] while for others (e.g., *p*-methoxyphenyl), neighboring-group participation is substantial. The combined effect of solvent and structure is shown in Table 10.2, where the figures shown were derived by three different methods.[164] The decrease in neighboring-group effectiveness when aromatic rings are substituted by electron-withdrawing groups is reminiscent of the similar case of C=C bonds substituted by CF_3 groups (p. 454).

[158]Diaz, A.; Winstein, S. *J. Am. Chem. Soc.* **1969**, *91*, 4300. See also, Schadt, F.L.; Lancelot, C.J.; Schleyer, P.v.R. *J. Am. Chem. Soc.* **1978**, *100*, 228.

[159]Nordlander, J.E.; Kelly, W.J. *J. Am. Chem. Soc.* **1969**, *91*, 996.

[160]Jablonski, R.J.; Snyder, E.I. *J. Am. Chem. Soc.* **1969**, *91*, 4445.

[161]Thompson, J.A.; Cram, D.J. *J. Am. Chem. Soc.* **1969**, *91*, 1778. See also, Tanida, H.; Tsuji, T.; Ishitobi, H.; Irie, T. *J. Org. Chem.* **1969**, *34*, 1086; Kingsbury, C.A.; Best, D.C. *Bull. Chem. Soc. Jpn.* **1972**, *45*, 3440.

[162]Coke, J.L.; McFarlane, F.E.; Mourning, M.C.; Jones, M.G. *J. Am. Chem. Soc.* **1969**, *91*, 1154; Jones, M.G.; Coke, J.L. *J. Am. Chem. Soc.* **1969**, *91*, 4284. See also, Harris, J.M.; Schadt, F.L.; Schleyer, P.v.R.; Lancelot, C.J. *J. Am. Chem. Soc.* **1969**, *91*, 7508.

[163]The k_Δ pathway is important for *p*-nitrophenyl in CF_3COOH: Ando, T.; Shimizu, N.; Kim, S.; Tsuno, Y.; Yukawa, Y. *Tetrahedron Lett.* **1973**, 117.

[164]Lancelot, C.J.; Schleyer, P.v.R. *J. Am. Chem. Soc.* **1969**, *91*, 4291, 4296; Lancelot, C.J.; Harper, J.J.; Schleyer, P.v.R. *J. Am. Chem. Soc.* **1969**, *91*, 4294; Schleyer, P.v.R.; Lancelot, C.J. *J. Am. Chem. Soc.* **1969**, *91*, 4297.

TABLE 10.1. Approximate k_Δ/k_s Ratios for Acetolysis of p-ZC$_6$H$_4$CH$_2$CH$_2$OTs at 90°C[162]

Z	k_Δ/k_s
MeO	30
Me	11
H	1.3
Cl	0.3

TABLE 10.2. Percent of Product Formed by the k_Δ Pathway in Solvolysis of p-ZC$_6$H$_4$CH$_2$CH$_2$OTs[164]

Z	Solvent	Percent by k_Δ
H	CH$_3$COOH	35–38
H	HCOOH	72–79
MeO	CH$_3$COOH	91–93
MeO	HCOOH	99

Several phenonium ions have been prepared as stable ions in solution where they can be studied by NMR, among them **52**,[165] **53**,[166] and the unsubstituted **51**.[167] These were prepared[168] by the method shown for **51**: treatment of the corresponding β-arylethyl chloride with SbF$_5$–SO$_2$ at low temperatures. These conditions are even more extreme than the solvolysis in CF$_3$COOH mentioned earlier. The absence of any nucleophile at all eliminates

not only the k_s pathways, but also nucleophilic attack on **51**. Although **51** is not in equilibrium with the open-chain ion PhCH$_2$CH$_2^+$ (which is primary

[165]Olah, G.A.; Comisarow, M.B.; Namanworth, E.; Ramsey, B. *J. Am. Chem. Soc.* **1967**, *89*, 5259; Ramsey, B.; Cook, Jr., J.A.; Manner, J.A. *J. Org. Chem.* **1972**, *37*, 3310.

[166]Olah, G.A.; Comisarow, M.B.; Kim, C.J. *J. Am. Chem. Soc.* **1969**, *91*, 1458. See, however, Ramsey, B.; Cook, Jr., J.A.; Manner, J.A. *J. Org. Chem.* **1972**, *37*, 3310.

[167]Olah, G.A.; Spear, R.J.; Forsyth, D.A. *J. Am. Chem. Soc.* **1976**, *98*, 6284.

[168]For some others, see Olah, G.A.; Singh, B.P.; Liang, G. *J. Org. Chem.* **1984**, *49*, 2922; Olah, G.A.; Singh, B.P. *J. Am. Chem. Soc.* **1984**, *106*, 3265.

and hence unstable), **53** is in equilibrium with the open-chain tertiary ions $PhCMe_2C^+Me_2$ and PhC^+MeCMe_3, although only **53** is present in appreciable concentration. Proton and ^{13}C NMR show that **51**, **52**, and **53** are classical carbocations where the only resonance is in the six-membered ring. The three-membered ring is a normal cyclopropane ring that is influenced only to a relatively small extent by the positive charge on the adjacent ring. Nuclear magnetic resonance spectra show that the six-membered rings have no aromatic character, but are similar in structure to the arenium ions, for example,

54

54, that are intermediates in electrophilic aromatic substitution (Chapter 11). A number of phenonium ions, including **51**, have also been reported to be present in the gas phase, where their existence has been inferred from reaction products and from ^{13}C labeling.[169]

It is thus clear that β-aryl groups can function as neighboring groups.[170] Much less work has been done on aryl groups located in positions farther away from the leaving group, but there is evidence that these too can lend anchimeric assistance.[171]

4. *The Carbon–Carbon Single Bond as a Neighboring Group.*[172]

    a. *The 2-Norbornyl System.* In the investigations to determine whether a C–C σ bond can act as a neighboring group, by far the greatest attention

[169]Fornarini, S.; Muraglia, V. *J. Am. Chem. Soc.* **1989**, *111*, 873; Mishima, M.; Tsuno, Y.; Fujio, M. *Chem. Lett.* **1990**, 2277.

[170]For additional evidence, see Tanida, H. *Acc. Chem. Res.* **1968**, *1*, 239; Kingsbury, C.A.; Best, D.C. *Tetrahedron Lett.* **1967**, 1499; Braddon, D.V.; Wiley, G.A.; Dirlam, J.; Winstein, S. *J. Am. Chem. Soc.* **1968**, *90*, 1901; Tanida, H.; Ishitobi, H.; Irie, T. *J. Am. Chem. Soc.* **1968**, *90*, 2688; Brown, H.C.; Tritle, G.L. *J. Am. Chem. Soc.* **1968**, *90*, 2689; Bentley, M.D.; Dewar, M.J.S. *J. Am. Chem. Soc.* **1970**, *92*, 3996; Raber, D.J.; Harris, J.M.; Schleyer, P.v.R. *J. Am. Chem. Soc.* **1971**, *93*, 4829; Shiner, Jr., V.J.; Seib, R.C. *J. Am. Chem. Soc.* **1976**, *98*, 862; Faïn, D.; Dubois, J.E. *Tetrahedron Lett.* **1978**, 791; Yukawa, Y.; Ando, T.; Token, K.; Kawada, M.; Matsuda, K.; Kim, S.; Yamataka, H. *Bull. Chem. Soc. Jpn.* **1981**, *54*, 3536; Ferber, P.H.; Gream, G.E. *Aust. J. Chem.* **1981**, *34*, 2217; Fujio, M.; Goto, M.; Seki, Y.; Mishima, M.; Tsuno, Y.; Sawada, M.; Takai, Y. *Bull. Chem. Soc. Jpn.* **1987**, *60*, 1097. For a discussion of evidence obtained from isotope effects, see Scheppele, S.E. *Chem. Rev.* **1972**, *72*, 511, p. 522.

[171]Heck, R.; Winstein, S. *J. Am. Chem. Soc.* **1957**, *79*, 3105; Muneyuki, R.; Tanida, H. *J. Am. Chem. Soc.* **1968**, *90*, 656; Ouellette, R.J.; Papa, R.; Attea, M.; Levin, C. *J. Am. Chem. Soc.* **1970**, *92*, 4893; Jackman, L.M.; Haddon, V.R. *J. Am. Chem. Soc.* **1974**, *96*, 5130; Gates, M.; Frank, D.L.; von Felten, W.C. *J. Am. Chem. Soc.* **1974**, *96*, 5138; Ando, T.; Yamawaki, J.; Saito, Y. *Bull. Chem. Soc. Jpn.* **1978**, *51*, 219.

[172]For a review pertaining to studies of this topic at low temperatures, see Olah, G.A. *Angew. Chem. Int. Ed.* **1973**, *12*, 173, pp. 192–198.

has been paid to the 2-norbornyl system.[173] Winstein and Trifan found that solvolysis in acetic acid of optically active *exo*-2-norbornyl brosylate (**55**) gave a racemic mixture of the two exo acetates; no endo isomers were formed:[174]

Furthermore, **55** solvolyzed ~350 times faster than its endo isomer **58**. Similar high exo/endo rate ratios have been found in many other [2.2.1] systems. These two results—(*1*) that solvolysis of an optically active exo isomer gave only racemic exo isomers and (*2*) the high exo/endo rate ratio—were interpreted by Winstein and Trifan as indicating that the 1,6 bond assists in the departure of the leaving group and that a nonclassical intermediate (**59**) is involved. They reasoned that solvolysis of the endo isomer **58** is not assisted by the 1,6 bond because it is not in a favorable position for backside attack,

and that consequently solvolysis of **58** takes place at a "normal" rate. Therefore the much faster rate for the solvolysis of **55** must be caused by anchimeric assistance. The stereochemistry of the product is also explained by the intermediacy of **59**, since in **59** the 1 and 2 positions are equivalent and would be attacked by the nucleophile with equal facility, but only from the exo direction in either case. Incidentally, acetolysis of **58** also leads exclusively to the exo acetates (**56** and **57**), so that in this case Winstein and Trifan postulated that a classical ion (**60**)

[173]For reviews, see Olah, G.A.; Prakash, G.K.S.; Williams, R.E. *Hypercarbon Chemistry*, Wiley, NY, *1987*, pp. 157–170; Grob, C.A. *Angew. Chem. Int. Ed. 1982, 21*, 87; Sargent, G.D., in Olah, G.A.; Schleyer, P.v.R. *Carbonium Ions*, Vol. 3, Wiley, NY, *1972*, pp. 1099–1200; Sargent, G.D. *Q. Rev. Chem. Soc. 1966, 20*, 301; Gream, G.E. *Rev. Pure Appl. Chem. 1966, 16*, 25. For a closely related review, see Kirmse, W. *Acc. Chem. Res. 1986, 19*, 36. See also, Ref. 177.
[174]Winstein, S.; Clippinger, E.; Howe, R.; Vogelfanger, E. *J. Am. Chem. Soc. 1965, 87*, 376.

is first formed, and then converted to the more stable **59**. Evidence for this interpretation is that the product from solvolysis of **58** is not racemic but contains somewhat more **57** than **56** (corresponding to 3–13% inversion, depending on the solvent),[174] suggesting that when **60** is formed, some of it goes to give **57** before it can collapse to **59**.

The concepts of σ participation and the nonclassical ion **59** were challenged by H.C. Brown,[128] who suggested that the two results can also be explained by postulating that **55** solvolyzes without participation of the 1,6 bond to give the classical ion **60**, which is in rapid equilibrium with **61**. This

55   **60**   **61**

rapid interconversion has been likened to the action of a windshield wiper.[175] Obviously, in going from **60** to **61** and back again, **59** must be present, but in Brown's view it is a transition state and not an intermediate. Brown's explanation for the stereochemical result was that exclusive exo attack is a property to be expected from any 2-norbornyl system, not only for the cation but even for reactions not involving cations, because of steric hindrance to attack from the endo side. There is a large body of data that shows that exo attack on norbornyl systems is fairly general in many reactions. A racemic mixture will be obtained if **60** and **61** are present in equal amounts, since they are equivalent and exo attack on **60** and **61** gives, respectively, **57** and **56**. Brown explained the high exo/endo rate ratios by contending that it is not the endo rate that is normal and the exo rate abnormally high, but the exo rate that is normal and the endo rate abnormally *low*, because of steric hindrance to removal of the leaving group in that direction.[176]

A vast amount of work has been done[177] on solvolysis of the 2-norbornyl system in an effort to determine whether the 1,6 bond

[175]Another view is somewhere in between: There are two interconverting ions, but each is asymmetrically bridged: Bielmann, R.; Fuso, F.; Grob, C.A. *Helv. Chim. Acta* **1988**, *71*, 312; Flury, P.; Grob, C.A.; Wang, G.Y.; Lennartz, H.; Roth, W.R. *Helv. Chim. Acta* **1988**, *71*, 1017.

[176]For evidence against steric hindrance as the only cause of this effect, see Menger, F.M.; Perinis, M.; Jerkunica, J.M.; Glass, L.E. *J. Am. Chem. Soc.* **1978**, *100*, 1503.

[177]For thorough discussions, see Lenoir, D.; Apeloig, Y.; Arad, D.; Schleyer, P.v.R. *J. Org. Chem.* **1988**, *53*, 661; Grob, C.A. *Acc. Chem. Res.* **1983**, *16*, 426; Brown, H.C. *Acc. Chem. Res.* **1983**, *16*, 432; Walling, C. *Acc. Chem. Res.* **1983**, *16*, 448; Allred, E.L.; Winstein, S. *J. Am. Chem. Soc.* **1967**, *89*, 3991, 3998; Nordlander, J.E.; Kelly, W.J. *J. Am. Chem. Soc.* **1969**, *91*, 996. For commentary on the controversy, see Arnett, E.M.; Hofelich, T.C.; Schriver, G.W. *React. Intermed. (Wiley)* **1985**, *3*, 189, pp. 193–202.

participates and whether **59** is an intermediate. Most,[178] although not all,[179] chemists now accept the intermediacy of **59**.

Besides the work done on solvolysis of 2-norbornyl compounds, the 2-norbornyl cation has also been extensively studied at low temperatures; there is much evidence that under these conditions the ion is definitely nonclassical. Olah and co-workers have prepared the 2-norbornyl cation in stable solutions at temperatures below $-150°C$ in SbF_5-SO_2 and $FSO_3H-SbF_5-SO_2$, where the structure is static and hydride shifts are absent.[180] Studies by proton and ^{13}C NMR, as well as by laser Raman spectra and X-ray electron spectroscopy, led to the conclusion[181] that under these conditions the ion is nonclassical.[182] A similar result has been reported for the 2-norbornyl cation in the solid state where at 77 and even 5 K, ^{13}C NMR spectra gave no evidence of the freezing out of a single classical ion.[183]

Nortricyclane   **62**   **63**   **64**

[178]For some recent evidence in favor of a nonclassical **59**, see Arnett, E.M.; Petro, C.; Schleyer, P.v.R. *J. Am. Chem. Soc.* **1979**, *101*, 522: Albano, C.; Wold, S. *J. Chem. Soc. Perkin Trans. 2* **1980**, 1447; Wilcox, C.F.; Tuszynski, W.J. *Tetrahedron Lett.* **1982**, *23*, 3119; Kirmse, W.; Siegfried, R. *J. Am. Chem. Soc.* **1983**, *105*, 950; Creary, X.; Geiger, C.C. *J. Am. Chem. Soc.* **1983**, *105*, 7123; Chang, S.; le Noble, W.J. *J. Am. Chem. Soc.* **1984**, *106*, 810; Kirmse, W.; Brandt, S. *Chem. Ber.* **1984**, *117*, 2510; Wilcox, C.F.; Brungardt, B. *Tetrahedron Lett.* **1984**, *25*, 3403; Lajunen, M. *Acc. Chem. Res.* **1985**, *18*, 254; Sharma, R.B.; Sen Sharma, D.K.; Hiraoka, K.; Kebarle, P. *J. Am. Chem. Soc.* **1985**, *107*, 3747; Servis, K.L.; Domenick, R.L.; Forsyth, D.A.; Pan, Y. *J. Am. Chem. Soc.* **1987**, *109*, 7263; Lenoir, D.; Apeloig, Y.; Arad, D.; Schleyer, P.v.R. *J. Org. Chem.* **1988**, *53*, 661.

[179]For some evidence against a nonclassical **59** see Dewar, M.J.S.; Haddon, R.C.; Komornicki, A.; Rzepa, H. *J. Am. Chem. Soc.* **1977**, *99*, 377; Lambert, J.B.; Mark, H.W. *J. Am. Chem. Soc.* **1978**, *100*, 2501; Christol, H.; Coste, J.; Pietrasanta, F.; Plénat, F.; Renard, G. *J. Chem. Soc. (S) 1978*, 62; Brown, H.C.; Rao, C.G. *J. Org. Chem.* **1979**, *44*, 133, 3536; *1980*, *45*, 2113; Liu, K.; Yen, C.; Hwang, H. *J. Chem. Res.(S)* **1980**, 152; Werstiuk, N.H.; Dhanoa, D.; Timmins, G. *Can. J. Chem.* **1983**, *61*, 2403; Brown, H.C.; Ikegami, S.; Vander Jagt, D.L. *J. Org. Chem.* **1985**, *50*, 1165; Nickon, A.; Swartz, T.D.; Sainsbury, D.M.; Toth, B.R. *J. Org. Chem.* **1986**, *51*, 3736. See also, Brown, H.C. *Top. Curr. Chem.* **1979**, *80*, 1.

[180]The presence of hydride shifts (p. $$$) under solvolysis conditions has complicated the interpretation of the data.

[181]Olah, G.A. *Acc. Chem. Res.* **1976**, *9*, 41; Olah, G.A.; Liang, G.; Mateescu, G.D.; Riemenschneider, J.L. *J. Am. Chem. Soc.* **1973**, *95*, 8698; Saunders, M.; Kates, M.R. *J. Am. Chem. Soc.* **1980**, *102*, 6867; *1983*, *105*, 3571; Olah, G.A.; Prakash, G.K.S.; Saunders, M. *Acc. Chem. Res.* **1983**, *16*, 440. See also, Schleyer, P.v.R.; Lenoir, D.; Mison, P.; Liang, G.; Prakash, G.K.S.; Olah, G.A. *J. Am. Chem. Soc.* **1980**, *102*, 683; Johnson, S.A.; Clark, D.T. *J. Am. Chem. Soc.* **1988**, *110*, 4112.

[182]This conclusion has been challenged: Fong, F.K. *J. Am. Chem. Soc.* **1974**, *96*, 7638; Kramer, G.M. *Adv. Phys. Org. Chem.* **1975**, *11*, 177; Brown, H.C.; Periasamy, M.; Kelly, D.P.; Giansiracusa, J.J. *J. Org. Chem.* **1982**, *47*, 2089; Kramer, G.M.; Scouten, C.G. *Adv. Carbocation Chem.* **1989**, *1*, 93. See, however, Olah, G.A.; Prakash, G.K.S.; Farnum, D.G.; Clausen, T.P. *J. Org. Chem.* **1983**, *48*, 2146.

[183]Yannoni, C.S.; Macho, V.; Myhre, P.C. *J. Am. Chem. Soc.* **1982**, *104*, 907, 7380; *Bull. Soc. Chim. Belg.* **1982**, *91*, 422; Myhre, P.C.; Webb, G.G.; Yannoni, C.S. *J. Am. Chem. Soc.* **1990**, *112*, 8991.

Olah and co-workers represented the nonclassical structure as a corner-protonated nortricyclane (**62**); the symmetry is better seen when the ion is drawn as in **63**. Almost all the positive charge resides on C-1 and C-2 and very little on the bridging carbon C-6. Other evidence for the nonclassical nature of the 2-norbornyl cation in stable solutions comes from heat of reaction measurements that show that the 2-norbornyl cation is more stable (by $\sim$6–10 kcal mol^{-1} or 25–40 kJ mol^{-1}) than would be expected without the bridging.[184] Studies of ir spectra of the 2-norbornyl cation in the gas phase also show the nonclassical structure.[185] *Ab initio* calculations show that the nonclassical structure corresponds to an energy minimum.[186]

The spectra of other norbornyl cations have also been investigated at low temperatures. Spectra of the tertiary 2-methyl- and 2-ethylnorbornyl cations show less delocalization,[187] and the 2-phenylnorbornyl cation (**64**) is essentially classical,[188] as are the 2-methoxy-[189] and 2-chloronorbornyl cations.[190] We may recall (p. 242) that methoxy and halo groups also stabilize a positive charge. The ^{13}C NMR data show that electron-withdrawing groups on the benzene ring of **64** cause the ion to become less classical, while electron-donating groups enhance the classical nature of the ion.[191]

b. *The Cyclopropylmethyl System.* Apart from the 2-norbornyl system, the greatest amount of effort in the search for C—C participation has been devoted to the cyclopropylmethyl system.[192] It has long been known that cyclopropylmethyl substrates solvolyze with abnormally high rates and

[184]For some examples, see Hogeveen, H.; Gaasbeek, C.J. *Recl. Trav. Chim. Pays-Bas* **1969**, *88*, 719; Hogeveen, H. *Recl. Trav. Chim. Pays-Bas* **1970**, *89*, 74; Solomon, J.J.; Field, F.H. *J. Am. Chem. Soc.* **1976**, *98*, 1567; Staley, R.H.; Wieting, R.D.; Beauchamp, J.L. *J. Am. Chem. Soc.* **1977**, *99*, 5964; Arnett, E.M.; Pienta, N.; Petro, C. *J. Am. Chem. Soc.* **1980**, *102*, 398; Saluja, P.P.S.; Kebarle, P. *J. Am. Chem. Soc.* **1979**, *101*, 1084; Schleyer, P.v.R.; Chandrasekhar, J. *J. Org. Chem.* **1981**, *46*, 225; Lossing, F.P.; Holmes, J.L. *J. Am. Chem. Soc.* **1984**, *106*, 6917.

[185]Koch, W.; Liu, B.; DeFrees, D.J.; Sunko, D.E.; Vančik, H. *Angew. Chem. Int. Ed.* **1990**, *29*, 183.

[186]See, for example Koch, W.; Liu, B.; DeFrees, D.J. *J. Am. Chem. Soc.* **1989**, *111*, 1527.

[187]Olah, G.A.; DeMember, J.R.; Lui, C.Y.; White, A.M. *J. Am. Chem. Soc.* **1969**, *91*, 3958. See also, Laube, T. *Angew. Chem. Int. Ed.* **1987**, *26*, 560; Forsyth, D.A.; Panyachotipun, C. *J. Chem. Soc., Chem. Commun.* **1988**, 1564.

[188]Olah, G.A. *Acc. Chem. Res.* **1976**, *9*, 41; Farnum, D.G.; Mehta, G. *J. Am. Chem. Soc.* **1969**, *91*, 3256. See also, Schleyer, P.v.R.; Kleinfelter, D.C.; Richey, Jr., H.G. *J. Am. Chem. Soc.* **1963**, *85*, 479; Farnum, D.G.; Wolf, A.D. *J. Am. Chem. Soc.* **1974**, *96*, 5166.

[189]Nickon, A.; Lin, Y. *J. Am. Chem. Soc.* **1969**, *91*, 6861. See also, Montgomery, L.K.; Grendze, M.P.; Huffman, J.C. *J. Am. Chem. Soc.* **1987**, *109*, 4749.

[190]Fry, A.J.; Farnham, W.B. *J. Org. Chem.* **1969**, *34*, 2314.

[191]Olah, G.A.; Prakash, G.K.S.; Liang, G. *J. Am. Chem. Soc.* **1977**, *99*, 5683; Farnum, W.B.; Botto, R.E.; Chambers, W.T.; Lam, B. *J. Am. Chem. Soc.* **1978**, *100*, 3847. See also, Olah, G.A.; Berrier, A.L.; Prakash, G.K.S. *J. Org. Chem.* **1982**, *47*, 3903.

[192]For reviews, see, in Olah, G.A.; Schleyer, P.v.R. *Carbonium Ions*, Vol. 3, Wiley, NY, **1972**, the articles by Richey, Jr., H.G. pp. 1201–1294, and by Wiberg, K.B.; Hess, Jr., B.A.; Ashe III, A.J. pp. 1295–1345; Hanack, M.; Schneider, H. *Fortschr. Chem. Forsch.* **1967**, *8*, 554, *Angew. Chem. Int. Ed.* **1967**, *6*, 666; Sarel, S.; Yovell, J.; Sarel-Imber, M. *Angew. Chem. Int. Ed.* **1968**, *7*, 577.

that the products often include not only unrearranged cyclopropylmethyl, but also cyclobutyl and homoallylic compounds. An example is[193]

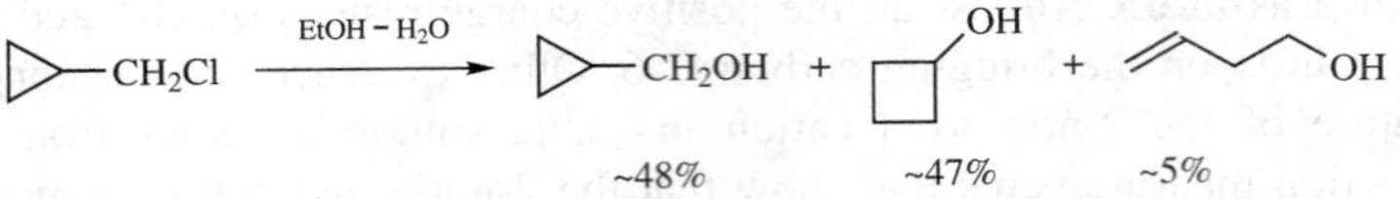

$$\sim48\% \qquad \sim47\% \qquad \sim5\%$$

Cyclobutyl substrates also solvolyze abnormally rapidly and give similar products. Furthermore, when the reactions are carried out with labeled substrates, considerable, although not complete, scrambling is observed. For these reasons, it has been suggested that a common intermediate (some kind of nonclassical intermediate, e.g., **32**, p. 450) is present in these cases. This common intermediate could then be obtained by three routes:

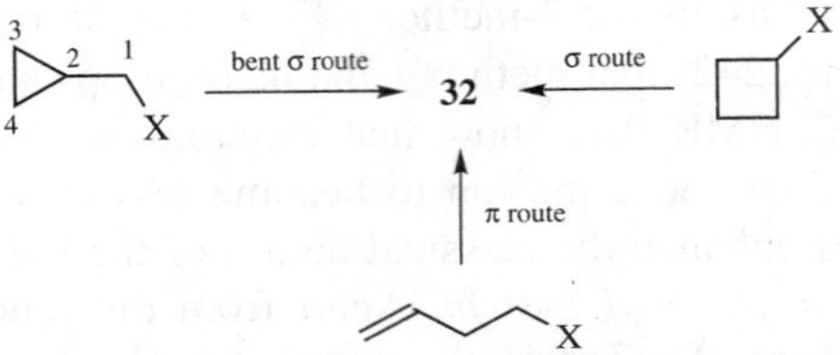

In recent years, much work has been devoted to the study of these systems, and it is apparent that matters are not so simple. Although there is much that is still not completely understood, some conclusions can be drawn.

i. In solvolysis of simple primary cyclopropylmethyl systems the rate is enhanced because of participation by the σ bonds of the ring.[194] The ion that forms initially is an unrearranged cyclopropylmethyl cation[195] that is *symmetrically* stabilized, that is, both the 2,3 and 2,4 σ bonds help stabilize the positive charge. We have already seen (p. 240) that a cyclopropyl group stabilizes an adjacent positive charge even better than a phenyl group. One way of representing the structure of this cation is as shown in **65**. Among the

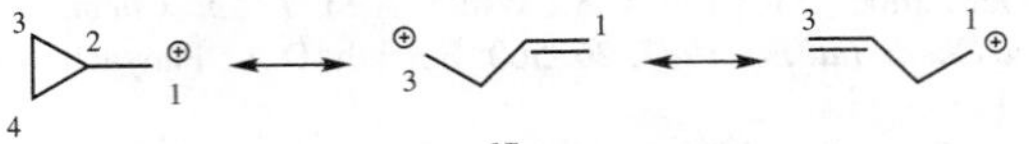

65

evidence that **65** is a symmetrical ion is that substitution of one or more methyl groups in the 3 and 4 positions increases the rate of solvolysis of cyclopropylcarbinyl 3,5-dinitrobenzoates by approximately a factor of 10 for *each* methyl group.[196] If only one of the σ bonds (say, the 2,3 bond)

[193]Roberts, D.D.; Mazur, R.H. *J. Am. Chem. Soc.* **1951**, *73*, 2509.

[194]See, for example, Roberts, D.D.; Snyder, Jr., R.C. *J. Org. Chem.* **1979**, *44*, 2860, and references cited therein.

[195]Wiberg, K.B.; Ashe III, A.J. *J. Am. Chem. Soc.* **1968**, *90*, 63.

[196]Schleyer, P.v.R.; Van Dine, G.W. *J. Am. Chem. Soc.* **1966**, *88*, 2321. See also, Kevill, D.N.; Abduljaber, M.H. *J. Org. Chem.* **2000**, *65*, 2548.

stabilizes the cation, then methyl substitution at the 3 position should increase the rate, and a second methyl group at the 3 position should increase it still more, but a second methyl group at the 4 position should have little effect.[197]

ii. The most stable geometry of simple cyclopropylmethyl cations is the bisected one shown on p. 240. There is much evidence that in systems where this geometry cannot be obtained, solvolysis is greatly slowed.[198]

iii. Once a cyclopropylmethyl cation is formed, it can rearrange to two other cyclopropylmethyl cations:

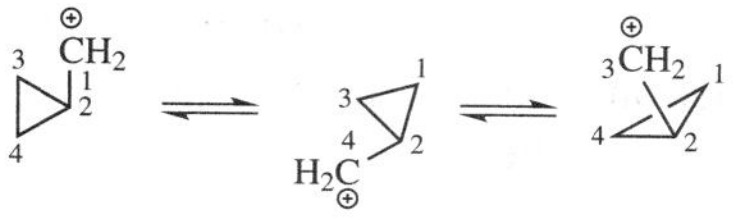

This rearrangement, which accounts for the scrambling, is completely stereospecific.[199] The rearrangements probably take place through a nonplanar cyclobutyl cation intermediate or transition state. The formation of cyclobutyl and homoallylic products from a cyclopropylmethyl cation is also completely stereospecific. These products may arise by direct attack of the nucleophile on **65** or on the cyclobutyl cation intermediate.[200] A planar cyclobutyl cation is ruled out in both cases because it would be symmetrical and the stereospecificity would be lost.

iv. The rate enhancement in the solvolysis of secondary cyclobutyl substrates is probably caused by participation by a bond leading directly to **65**, which accounts for the fact that solvolysis of cyclobutyl and of cyclopropylmethyl

substrates often gives similar product mixtures. There is no evidence that requires the cyclobutyl cations to be intermediates in most secondary cyclobutyl systems, although tertiary cyclobutyl cations can be solvolysis intermediates.

v. The unsubstituted cyclopropylmethyl cation has been generated in super acid solutions at low temperatures, where ^{13}C NMR spectra have led to

---

[197]For a summary of additional evidence for the symmetrical nature of cyclopropylmethyl cations, see Olah, G.A.; Schleyer, P.v.R. *Carbonium Ions*, Vol. 3, Wiley, NY, *1972*, the article by Wiberg, K.B.; Hess, Jr., B.A.; Ashe III, A.J. pp. 1300–1303.

[198]For example, see Ree, B.; Martin, J.C. *J. Am. Chem. Soc. 1970*, 92, 1660; Rhodes, Y.E.; DiFate, V.G. *J. Am. Chem. Soc. 1972*, 94, 7582. See, however, Brown, H.C.; Peters, E.N. *J. Am. Chem. Soc. 1975*, 97, 1927.

[199]Wiberg, K.B.; Szeimies, G. *J. Am. Chem. Soc. 1968*, 90, 4195; *1970*, 92, 571; Majerski, Z.; Schleyer, P.v.R. *J. Am. Chem. Soc. 1971*, 93, 665.

[200]Koch, W.; Liu, B.; DeFrees, D.J. *J. Am. Chem. Soc. 1988*, 110, 7325; Saunders, M.; Laidig, K.E.; Wiberg, K.B.; Schleyer, P.v.R. *J. Am. Chem. Soc. 1988*, 110, 7652.

the conclusion that it consists of a mixture of the bicyclobutonium ion **32** and the bisected cyclopropylmethyl cation **65**, in equilibrium with **32**.[201] Molecular-orbital calculations show that these two species are energy minima, and that both have nearly the same energy.[200]

c. *Methyl as a Neighboring Group.* Both the 2-norbornyl and cyclopropyl-methyl system contain a σ bond that is geometrically constrained to be in a particularly favorable position for participation as a neighboring group. However, there have been a number of investigations to determine whether a C—C bond can lend anchimeric assistance even in a simple open-chain compound, such as neopentyl tosylate. On solvolysis, neo-pentyl systems undergo almost exclusive rearrangement and **66** must lie on the reaction path, but the two questions that have been asked are (*1*) Is the departure of the

Neopentyl tosylate   **66**

leaving group concerted with the formation of the CH_3—C bond (i.e., does the methyl participate)? (*2*) Is **66** an intermediate or only a transition state? With respect to the first question, there is evidence, chiefly from isotope effect studies, that indicates that the methyl group in the neopentyl system does indeed participate,[202] although it may not greatly enhance the rate. As to the second question, evidence that **66** is an intermediate is that small amounts of cyclopropanes (10–15%) can be isolated in these reactions.[203] Cation **66** is a protonated cyclopropane and would give cyclopropane on loss of a proton.[204] In an effort to isolate a species that has structure **66**, the 2,3,3-trimethyl-2-butyl cation was prepared in super acid solutions at low temperatures.[205] However, ^{1}H and ^{13}C NMR, as well as Raman spectra,

[201]Staral, J.S.; Yavari, I.; Roberts, J.D.; Prakash, G.K.S.; Donovan, D.J.; Olah, G.A. *J. Am. Chem. Soc.* *1978*, *100*, 8016. See also, Olah, G.A.; Spear, R.J.; Hiberty, P.C.; Hehre, W.J. *J. Am. Chem. Soc. 1976*, *98*, 7470; Saunders, M.; Siehl, H. *J. Am. Chem. Soc. 1980*, *102*, 6868; Brittain, W.J.; Squillacote, M.E.; Roberts, J.D. *J. Am. Chem. Soc. 1984*, *106*, 7280; Siehl, H.; Koch, E. *J. Chem. Soc., Chem. Commun. 1985*, 496; Prakash, G.K.S.; Arvanaghi, M.; Olah, G.A. *J. Am. Chem. Soc. 1985*, *107*, 6017; Myhre, P.C.; Webb, G.G.; Yannoni, C.S. *J. Am. Chem. Soc. 1990*, *112*, 8992.

[202]For example, see Dauben, W.G.; Chitwood, J.L. *J. Am. Chem. Soc. 1968*, *90*, 6876; Ando, T.; Morisaki, H. *Tetrahedron Lett. 1979*, 121; Shiner, V.J.; Tai, J.J. *J. Am. Chem. Soc. 1981*, *103*, 436; Yamataka, H.; Ando, T.; Nagase, S.; Hanamura, M.; Morokuma, K. *J. Org. Chem. 1984*, *49*, 631. For an opposing view, see Zamashchikov, V.V.; Rudakov, E.S.; Bezbozhnaya, T.V.; Matveev, A.A. *J. Org. Chem. USSR 1984*, *20*, 11.

[203]Skell, P.S.; Starer, I. *J. Am. Chem. Soc. 1960*, *82*, 2971; Silver, M.S. *J. Am. Chem. Soc. 1960*, *82*, 2971; Friedman, L.; Bayless, J.H. *J. Am. Chem. Soc. 1969*, *91*, 1790; Friedman, L.; Jurewicz, A.T. *J. Am. Chem. Soc. 1969*, *91*, 1800, 1803; Dupuy, W.E.; Hudson, H.R.; Karam, P.A. *Tetrahedron Lett. 1971*, 3193; Silver, M.S.; Meek, A.G. *Tetrahedron Lett. 1971*, 3579; Dupuy, W.E.; Hudson, H.R. *J. Chem. Soc. Perkin Trans. 2 1972*, 1715.

[204]For further discussions of protonated cyclopropanes, see pp. $$$, $$$.

[205]Olah, G.A.; DeMember, J.R.; Commeyras, A.; Bribes, J.L. *J. Am. Chem. Soc. 1971*, *93*, 459.

showed this to be a pair of rapidly equilibrating open ions.

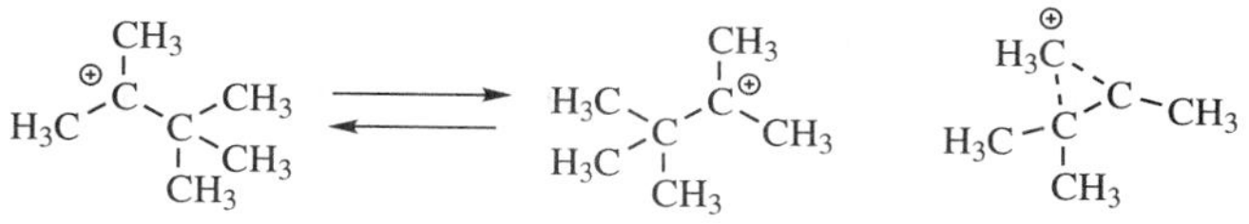

67

Of course, **67** must lie on the reaction path connecting the two open ions, but it is evidently a transition state and not an intermediate. However, evidence from X-ray photoelectron spectroscopy (ESCA) has shown that the 2-butyl cation is substantially methyl bridged.[206]

d. *Silylalkyl as a Neighboring Group.* Rates of solvolysis are enhanced in molecules that contain a silylalkyl or silylaryl group β- to the carbon bearing the leaving group. This is attributed to formation of a cyclic transition state involving the silicon.[207]

5. *Hydrogen as a Neighboring Group.* The questions relating to hydrogen are similar to those relating to methyl. There is no question that hydride can migrate, but the two questions are (*1*) Does the hydrogen participate in the

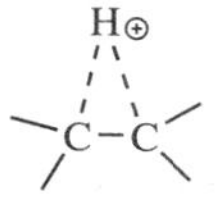

68

departure of the leaving group? (*2*) Is **68** an intermediate or only a transition state? There is some evidence that a β hydrogen can participate.[208] Evidence that **68** can be an intermediate in solvolysis reactions comes from

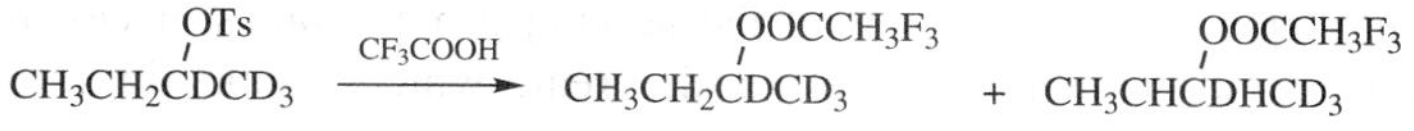

$$CH_3CH_2CDCD_3 \xrightarrow{CF_3COOH} CH_3CH_2CDCD_3 + CH_3CHCDHCD_3$$

69   **70**   **71**

a study of the solvolysis in trifluoroacetic acid of deuterated *sec*-butyl tosylate **69**. In this solvent of very low nucleophilic power, the products were

---

[206]Johnson, S.A.; Clark, D.T. *J. Am. Chem. Soc.* **1988**, *110*, 4112. See also, Carneiro, J.W.; Schleyer, P.v.R.; Koch, W.; Raghavachari, K. *J. Am. Chem. Soc.* **1990**, *112*, 4064.
[207]Fujiyama, R.; Munechika, T. *Tetrahedron Lett.* **1993**, *34*, 5907.
[208]See, for example, Shiner, Jr., V.J.; Jewett, J.G. *J. Am. Chem. Soc.* **1965**, *87*, 1382; Tichy, M.; Hapala, J.; Sicher, J. *Tetrahedron Lett.* **1969**, 3739; Myhre, P.C.; Evans, E. *J. Am. Chem. Soc.* **1969**, *91*, 5641; Inomoto, Y.; Robertson, R.E.; Sarkis, G. *Can. J. Chem.* **1969**, *47*, 4599; Shiner, V.J.; Stoffer, J.O. *J. Am. Chem. Soc.* **1970**, *92*, 3191; Krapcho, A.P.; Johanson, R.G. *J. Org. Chem.* **1971**, *36*, 146; Chuit, C.; Felkin, H.; Le Ny, G.; Lion, C.; Prunier, L. *Tetrahedron* **1972**, *28*, 4787; Stéhelin, L.; Kanellias, L.; Ourisson, G. *J. Org. Chem.* **1973**, *38*, 847, 851; Hirsl-Starševič, S.; Majerski, Z.; Sunko, D.E. *J. Org. Chem.* **1980**, *45*, 3388; Buzek, P.; Schleyer, P.v.R.; Sieber, S.; Koch, W.; Carneiro, J.W. de M.; Vančik, H.; Sunko, D.E. *J. Chem. Soc., Chem. Commun.* **1991**, 671; Imhoff, M.A.; Ragain, R.M.; Moore, K.; Shiner, V.J. *J. Org. Chem.* **1991**, *56*, 3542.

an equimolar mixture of **70** and **71**,[209] but *no* **72** or **73** was found. If this

$$\underset{CH_3CHDCHCD_3}{\overset{OOCCH_3F_3}{|}}$$

$$\underset{CH_3CDCH_2CD_3}{\overset{OOCCH_3F_3}{|}}$$

$$\underset{\overset{|}{H}\quad\overset{|}{D}}{H_3C-\overset{H\oplus}{\overset{.\,.\,.}{C}}-C-CD_3}$$

72   **73**   **74**

reaction did not involve neighboring hydrogen at all (pure S_N2 or S_N1), the product would be only **70**. On the other hand, if hydrogen does migrate, but only open cations are involved, then there should be an equilibrium among these four cations:

$$CH_3CH_2\overset{\oplus}{C}DCD_3 \;\rightleftharpoons\; CH_3\overset{\oplus}{C}HCDHCD_3 \;\rightleftharpoons\; CH_3CD\overset{\oplus}{H}CHCD_3 \;\rightleftharpoons\; CH_3\overset{\oplus}{C}DCH_2CD_3$$

leading not only to **70** and **71**, but also to **72** and **73**. The results are most easily compatible with the intermediacy of the bridged ion **74**, which can then be attacked by the solvent equally at the 2 and 3 positions. Attempts to prepare **68** as a stable ion in super acid solutions at low temperatures have not been successful.[208]

The S$_N$i Mechanism

In a few reactions, nucleophilic substitution proceeds with retention of configuration, even where there is no possibility of a neighboring-group effect. In the S_Ni mechanism (*substitution nucleophilic internal*), part of the leaving group must be able to attack the substrate, detaching itself from the rest of the leaving group in the process. The IUPAC designation is $D_N + A_ND_e$. The first step is the same as the very first step of the S_N1 mechanism dissociation into an intimate ion pair.[210] But in the second step part of the leaving group attacks, necessarily from the front since it is unable to get to the rear, which results in retention of configuration.

Step 1 $R\diagdown\!\!\diagup\!\!\underset{Cl}{\overset{O}{\underset{|}{S}}}\!\!=\!\!O \longrightarrow R^+ \quad \overset{\ominus}{\underset{Cl}{\overset{O}{\underset{|}{S}}}}\!\!=\!\!O$

Step 2 $R^+ \curvearrowleft \overset{\overset{\ominus}{O}}{\underset{Cl}{\overset{|}{S}}}\!\!=\!\!O \longrightarrow R\!-\!Cl \;+\; O\!\!=\!\!S\!\!=\!\!O$

[209]Dannenberg, J.J.; Goldberg, B.J.; Barton, J.K.; Dill, K.; Weinwurzel, D.H.; Longas, M.O. *J. Am. Chem. Soc.* **1981**, *103*, 7764. See also, Dannenberg, J.J.; Barton, J.K.; Bunch, B.; Goldberg, B.J.; Kowalski, T. *J. Org. Chem.* **1983**, *48*, 4524; Allen, A.D.; Ambidge, I.C.; Tidwell, T.T. *J. Org. Chem.* **1983**, *48*, 4527.
[210]Lee, C.C.; Finlayson, A.J. *Can. J. Chem.* **1961**, *39*, 260; Lee, C.C.; Clayton, J.W.; Lee, C.C.; Finlayson, A.J. *Tetrahedron* **1962**, *18*, 1395.

The example shown is the most important case of this mechanism yet discovered, since the reaction of alcohols with thionyl chloride to give alkyl halides usually proceeds in this way, with the first step in this case being $ROH + SOCl_2 \rightarrow ROSOCl$ (these alkyl chlorosulfites can be isolated).

Evidence for this mechanism is as follows: the addition of pyridine to the mixture of alcohol and thionyl chloride results in the formation of alkyl halide with *inverted* configuration. Inversion results because the pyridine reacts with ROSOCl to give $ROSONC_5H_5$ before anything further can take place. The Cl^- freed in this process now attacks from the rear. The reaction between alcohols and thionyl chloride is second order, which is predicted by this mechanism, but the decomposition by simple heating of ROSOCl is first order.[211]

The S_Ni mechanism is relatively rare. Another example is the decomposition of ROCOCl (alkyl chloroformates) into RCl and CO_2.[212]

Nucleophilic Substitution at an Allylic Carbon: Allylic Rearrangements

Allylic substrates rapidly undergo nucleophilic substitution reactions (see p. 482), but we discuss them in a separate section because they are commonly accompanied by a certain kind of rearrangement known as an *allylic rearrangement*.[213] When allylic substrates are treated with nucleophiles under S_N1 conditions, two products are usually obtained: the normal one and a rearranged one.

Two products are formed because an allylic type of carbocation is a resonance hybrid

$$R-CH=CH-\overset{\oplus}{C}H_2 \longleftrightarrow R-\overset{\oplus}{C}H-CH=CH_2$$

so that C-1 and C-3 each carry a partial positive charge and both are attacked by Y. Of course, an allylic rearrangement is undetectable in the case of symmetrical allylic cations, as in the case where $R = H$, unless isotopic labeling is used. This mechanism has been called the S_N1' mechanism. The IUPAC designation is $1/D_N + 3/A_N$, the numbers 1 and 3 signifying the *relative* positions where the nucleophile attacks and from which the nucleofuge leaves.

[211]Lewis, E.S.; Boozer, C.E. *J. Am. Chem. Soc.* **1952**, *74*, 308.

[212]Lewis, E.S.; Herndon, W.C.; Duffey, D.C. *J. Am. Chem. Soc.* **1961**, *83*, 1959; Lewis, E.S.; Witte, K. *J. Chem. Soc. B* **1968**, 1198. For other examples, see Hart, H.; Elia, R.J. *J. Am. Chem. Soc.* **1961**, *83*, 985; Stevens, C.L.; Dittmer, H.; Kovacs, J. *J. Am. Chem. Soc.* **1963**, *85*, 3394; Kice, J.L.; Hanson, G.C. *J. Org. Chem.* **1973**, *38*, 1410; Cohen, T.; Solash, J. *Tetrahedron Lett.* **1973**, 2513; Verrinder, D.J.; Hourigan, M.J.; Prokipcak, J.M. *Can. J. Chem.* **1978**, *56*, 2582.

[213]For a review, see DeWolfe, R.H., in Bamford, C.H.; Tipper, C.F.H. *Comprehensive Chemical Kinetics*, Vol. 9, Elsevier, NY, **1973**, pp. 417–437. For comprehensive older reviews, see DeWolfe, R.H.; Young, W.G. *Chem. Rev.* **1956**, *56*, 753; in Patai, S. *The Chemistry of Alkenes*, Wiley, NY, **1964**, the sections by Mackenzie, K. pp. 436–453 and DeWolfe, R.H.; Young, W.G. pp. 681–738.

As with other S_N1 reactions, there is clear evidence that S_N1' reactions can involve ion pairs. If the intermediate attacked by the nucleophile is a completely free carbocation, then, say,

$$CH_3CH=CHCH_2Cl \quad \text{and} \quad CH_3CHClCH=CH_2$$
$$\mathbf{75} \qquad\qquad\qquad \mathbf{76}$$

should give the same mixture of alcohols when reacting with hydroxide ion, since the carbocation from each should be the same. When treated with 0.8 M aq. NaOH at 25°C, **75** gave 60% $CH_3CH=CHCH_2OH$ and 40% $CH_3CHOHCH=CH_2$, while **76** gave the products in yields of 38 and 62%, respectively.[214] This phenomenon is called the *product spread*. In this case, and in most others, the product spread is in the direction of the starting compound. With increasing polarity of solvent,[215] the product spread decreases and in some cases is entirely absent. It is evident that in such cases the high polarity of the solvent stabilizes completely free carbocations. There is other evidence for the intervention of ion pairs in many of these reactions. When $H_2C=CHCMe_2Cl$ was treated with acetic acid, both acetates were obtained, but also some $ClCH_2CH=CMe_2$,[216] and the isomerization was faster than the acetate formation. This could not have arisen from a completely free Cl^- returning to the carbon, since the rate of formation of the rearranged chloride was unaffected by the addition of external Cl^-. All these facts indicate that the first step in these reactions is the formation of an unsymmetrical intimate ion pair that undergoes a considerable amount of internal return and in which the counterion remains close to the carbon from which it departed. Thus, **75** and **76**, for example, give rise to two *different* intimate ion pairs. The field of the anion polarizes the allylic cation, making the nearby carbon atom more electrophilic, so that it has a greater chance of attracting the nucleophile.[217]

Nucleophilic substitution at an allylic carbon can also take place by an S_N2 mechanism, in which case no allylic rearrangement usually takes place. However, allylic rearrangements can also take place under S_N2 conditions, by the following mechanism, in which the nucleophile attacks at the γ carbon rather than the usual position:[218]

The IUPAC designation is $3/1/A_ND_N$. This mechanism is a second-order allylic rearrangement; it usually comes about where S_N2 conditions hold but where

[214]DeWolfe, R.H.; Young, W.G. *Chem. Rev.* **1956**, *56*, 753 give several dozen such examples.
[215]Katritzky, A.R.; Fara, D.C.; Yang, H.; Tämm, K.; Tamm, T.; Karelson, M. *Chem. Rev.* **2004**, *104*, 175.
[216]Young, W.G.; Winstein, S.; Goering, H.L. *J. Am. Chem. Soc.* **1951**, *73*, 1958.
[217]For additional evidence for the involvement of ion pairs in S_N1' reactions, see Goering, H.L.; Linsay, E.C. *J. Am. Chem. Soc.* **1969**, *91*, 7435; d'Incan, E.; Viout, P. *Bull. Soc. Chim. Fr.* **1971**, 3312; Astin, K.B.; Whiting, M.C. *J. Chem. Soc. Perkin Trans. 2* **1976**, 1157; Kantner, S.S.; Humski, K.; Goering, H.L. *J. Am. Chem. Soc.* **1982**, *104*, 1693; Thibblin, A. *J. Chem. Soc. Perkin Trans. 2* **1986**, 313; Ref. 77.
[218]For a review of the S_N2' mechanism, see Magid, R.M. *Tetrahedron* **1980**, *36*, 1901, see pp. 1901–1910.

a substitution sterically retards the normal S_N2 mechanism. There are few well-established cases of the S_N2' mechanism on substrates of the type $C=C-CH_2X$, but compounds of the form $C=C-CR_2X$ give the S_N2' rearrangement almost exclusively[219] when they give bimolecular reactions at all. Increasing the size of the nucleophile can also increase the extent of the S_N2' reaction at the expense of the S_N2.[219] In certain cases, the leaving group can also have an affect on whether the rearrangement occurs. Thus $PhCH=CHCH_2X$, treated with $LiAlH_4$, gave 100% S_N2 reaction (no rearrangement) when $X = Br$ or Cl, but 100% S_N2' when $X = PPh_3^+$ Br^-.[220] The solvent also plays a role in some cases, with more polar solvents giving more S_N2' product.[221]

The S_N2' mechanism as shown above involves the simultaneous movement of three pairs of electrons. However, Bordwell has contended that there is no evidence that requires that this bond making and bond breaking be in fact concerted,[222] and that a true S_N2' mechanism is a myth. There is evidence both for[223] and against[224] this proposal. There is also a review of the S_N' reaction.[225]

The stereochemistry of S_N2' reactions has been investigated. It has been found that both syn[226] (the nucleophile enters on the side from which the leaving group departs) and anti[227] reactions can take place, depending on the nature of X and Y,[228] although the syn pathway predominates in most cases.

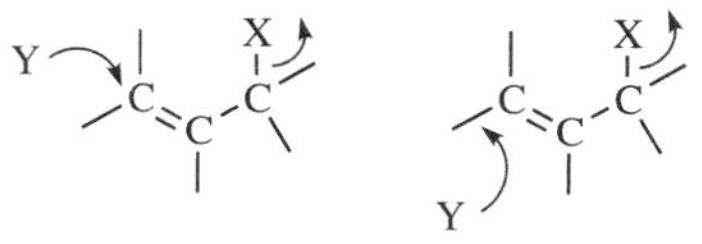

syn   anti

[219]Bordwell, F.G.; Clemens, A.H.; Cheng, J. *J. Am. Chem. Soc.* **1987**, *109*, 1773. Also see, Young, J.-j; Jung, L.-j.; Cheng, K.-m. *Tetrahedron Lett.* **2000**, *41*, 3411.

[220]Hirab, T.; Nojima, M.; Kusabayashi, S. *J. Org. Chem.* **1984**, *49*, 4084.

[221]Hirashita, T.; Hayashi, Y.; Mitsui, K.; Araki, S. *Tetrahedron Lett.* **2004**, *45*, 3225.

[222]Bordwell, F.G.; Mecca, T.G. *J. Am. Chem. Soc.* **1972**, *94*, 5829; Bordwell, F.G. *Acc. Chem. Res.* **1970**, *3*, 281, pp. 282–285. See also, de la Mare, P.B.D.; Vernon, C.A. *J. Chem. Soc. B* **1971**, 1699; Dewar, M.J.S. *J. Am. Chem. Soc.* **1984**, *106*, 209.

[223]See Uebel, J.J.; Milaszewski, R.F.; Arlt, R.E. *J. Org. Chem.* **1977**, *42*, 585.

[224]See Fry, A. *Pure Appl. Chem.* **1964**, *8*, 409; Georgoulis, C.; Ville, G. *J. Chem. Res. (S)* **1978**, 248; *Bull. Soc. Chim. Fr.* **1985**, 485; Meislich, H.; Jasne, S.J. *J. Org. Chem.* **1982**, *47*, 2517.

[225]Paquette, L.A.; Stirling, C.J.M. *Tetrahedron* **1992**, *48*, 7383.

[226]See, for example, Stork, G.; White, W.N. *J. Am. Chem. Soc.* **1956**, *78*, 4609; Jefford, C.W.; Sweeney, A.; Delay, F. *Helv. Chim. Acta* **1972**, *55*, 2214; Kirmse, W.; Scheidt, F.; Vater, H. *J. Am. Chem. Soc.* **1978**, *100*, 3945; Gallina, C.; Ciattini, P.G. *J. Am. Chem. Soc.* **1979**, *101*, 1035; Magid, R.M.; Fruchey, O.S. *J. Am. Chem. Soc.* **1979**, *101*, 2107; Bäckvall, J.E.; Vågberg, J.O.; Genêt, J.P. *J. Chem. Soc., Chem. Commun.* **1987**, 159.

[227]See, for example, Borden, W.T.; Corey, E.J. *Tetrahedron Lett.* **1969**, 313; Takahashi, T.T.; Satoh, J.Y. *Bull. Chem. Soc. Jpn.* **1975**, *48*, 69; Staroscik, J.; Rickborn, B. *J. Am. Chem. Soc.* **1971**, *93*, 3046; See also, Liotta, C. *Tetrahedron Lett.* **1975**, 523; Stork, G.; Schoofs, A.R. *J. Am. Chem. Soc.* **1979**, *101*, 5081.

[228]Stork, G.; Kreft III, A.F. *J. Am. Chem. Soc.* **1977**, *99*, 3850, 3851; Oritani, T.; Overton, K.H. *J. Chem. Soc., Chem. Commun.* **1978**, 454; Bach, R.D.; Wolber, G.J. *J. Am. Chem. Soc.* **1985**, *107*, 1352. See also, Chapleo, C.B.; Finch, M.A.W.; Roberts, S.M.; Woolley, G.T.; Newton, R.F.; Selby, D.W. *J. Chem. Soc. Perkin Trans. 1* **1980**, 1847; Stohrer, W. *Angew. Chem. Int. Ed.* **1983**, *22*, 613.

When a molecule has in an allylic position a nucleofuge capable of giving the S_Ni reaction, it is possible for the nucleophile to attack at the γ position instead of the α position. This is called the S_Ni' mechanism and has been demonstrated on 2-buten-1-ol and 3-buten-2-ol, both of which gave 100% allylic rearrangement

$$+ \ O=S=O$$

when treated with thionyl chloride in ether.[229] Ordinary allylic rearrangements (S_N1') or S_N2' mechanisms could not be expected to give 100% rearrangement in *both* cases. In the case shown, the nucleophile is only part of the leaving group, not the whole. But it is also possible to have reactions in which a simple leaving group, such as Cl, comes off to form an ion pair and then returns not to the position whence it came, but to the allylic position:

Most S_Ni' reactions are of this type.

Allylic rearrangements have also been demonstrated in propargyl systems, for example,[230]

(Reaction 19-67)

The product in this case is an allene,[231] but such shifts can also give triple-bond compounds or, if $Y = OH$, an enol will be obtained that tautomerizes to an α,β-unsaturated aldehyde or ketone.

[229]Young, W.G. *J. Chem. Educ.* **1962**, *39*, 456. For other examples, see Mark, V. *Tetrahedron Lett.* **1962**, 281; Czernecki, S.; Georgoulis, C.; Labertrande, J.; Prévost, C. *Bull. Soc. Chim. Fr.* **1969**, 3568; Lewis, E.S.; Witte, K. *J. Chem. Soc. B* **1968**, 1198; Corey, E.J.; Boaz, N.W. *Tetrahedron Lett.* **1984**, *25*, 3055.
[230]Vermeer, P.; Meijer, J.; Brandsma, L. *Recl. Trav. Chim. Pays-Bas* **1975**, *94*, 112.
[231]For reviews of such rearrangements, see Schuster, H.F.; Coppola, G.M. *Allenes in Organic Synthesis*, Wiley, NY, **1984**, pp. 12–19, 26–30; Taylor, D.R. *Chem. Rev.* **1967**, *67*, 317, pp. 324–328. For a palladium-catalyzed variation of this transformation see Larock, R.C.; Reddy, Ch.K. *Org. Lett.* **2000**, *2*, 3325.

When X = OH, this conversion of acetylenic alcohols to unsaturated aldehydes or ketones is called the *Meyer–Schuster rearrangement*.[232] The propargyl rearrangement can also go the other way; that is, 1-haloalkenes, treated with organocopper compounds, give alkynes.[233]

The S_N2' reaction has been shown to predominate in reactions of mixed cuprates (**10-57**) with allylic mesylates,[234] and in ring opening reactions of aziridines.[235] A related reaction is the opening of cyclopropylcarbinyl halides with organocuprates where the cyclopropane ring reacts similarly to the C=C unit of an alkene to give a homoallylic substituted product.[236] This latter reaction is interesting since the reaction of **77** with piperidine leads to the S_N2' product (**78**) in ~87% yield, but there is ~8% of the direct substitution product, **79**. Since the carbon bearing the bromine is very hindered, formation of **72** is somewhat unusual under these conditions. As Bordwell has suggested (see above), this may not be a true S_N2 process.

Nucleophilic Substitution at an Aliphatic Trigonal Carbon: The Tetrahedral Mechanism

All the mechanisms so far discussed take place at a saturated carbon atom. Nucleophilic substitution is also important at trigonal carbons, especially when the carbon is double bonded to an oxygen, a sulfur, or a nitrogen. These reactions are discussed in Chapter 16. Nucleophilic substitution at vinylic carbons is considered in the next section; at aromatic carbons in Chapter 13.

Nucleophilic Substitution at a Vinylic Carbon

Nucleophilic substitution at a vinylic carbon[237] is difficult (see p. 481), but many examples are known. The most common mechanisms are the tetrahedral mechanism

[232]For a review, see Swaminathan, S.; Narayanan, K.V. *Chem. Rev.* **1971**, *71*, 429. For discussions of the mechanism, see Edens, M.; Boerner, D.; Chase, C.R.; Nass, D.; Schiavelli, M.D. *J. Org. Chem.* **1977**, *42*, 3403; Andres, J.; Cardenas, R.; Silla, E.; Tapi, O. *J. Am. Chem. Soc.* **1988**, *110*, 666.

[233]Corey, E.J.; Boaz, N.W. *Tetrahedron Lett.* **1984**, *25*, 3059, 3063.

[234]Ibuka, T.; Taga, T.; Habashita, H.; Nakai, K.; Tamamura, H.; Fujii, N.; Chounan, Y.; Nemoto, H.; Yamamoto, Y. *J. Org. Chem.* **1993**, *58*, 1207.

[235]Wipf, P.; Fritch, P.C. *J. Org. Chem.* **1994**, *59*, 4875.

[236]Smith, M.B.; Hrubiec, R.T. *Tetrahedron* **1984**, *40*, 1457; Hrubiec, R.T.; Smith, M.B. *J. Org. Chem.* **1984**, *49*, 385; Hrubiec, R.T.; Smith, M.B. *Tetrahedron Lett.* **1983**, *24*, 5031.

[237]For reviews, see Rappoport, Z. *Recl. Trav. Chim. Pays-Bas* **1986**, *104*, 309; *React. Intermed. (Plenum)* **1983**, *3*, 427, *Adv. Phys. Org. Chem.* **1969**, *7*, 1; Shainyan, B.A. *Russ. Chem. Rev.* **1986**, *55*, 511; Modena, G. *Acc. Chem. Res.* **1971**, *4*, 73.

and the closely related *addition–elimination mechanism*. Both of these mechanisms are impossible at a saturated substrate. The addition–elimination mechanism

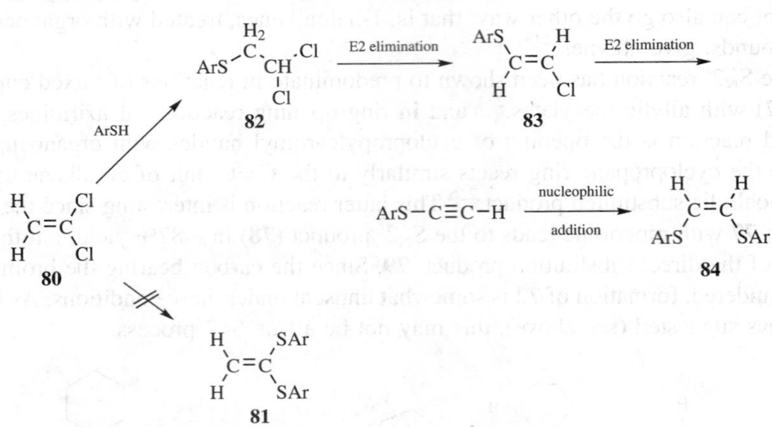

has been demonstrated for the reaction between 1,1-dichloroethene (**80**) and ArS⁻ catalyzed by ⁻OEt.[238] The product was not the 1,1-dithiophenoxy compound **81** but the "rearranged" compound **84**. Isolation of **82** and **83** showed that an addition–elimination mechanism had taken place. In the first step, ArSH adds to the double bond (nucleophilic addition, p. 1007) to give the saturated **82**. The second step is an E2 elimination reaction (p. 1478) to give the alkene **83**. A second elimination and addition give **84**.

The tetrahedral mechanism, often also called addition–elimination (AdN-E), takes place with much less facility than with carbonyl groups, since the negative charge of the intermediate must be borne by a carbon, which is less electronegative than oxygen, sulfur, or nitrogen:

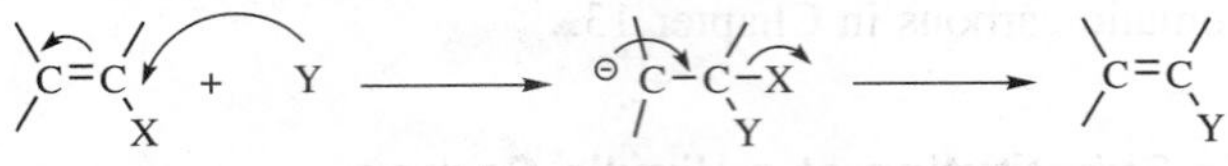

Such an intermediate can also stabilize itself by combining with a positive species. When it does, the reaction is nucleophilic addition to a C=C double bond (see Chapter 15). It is not surprising that with vinylic substrates addition and substitution often compete. For chloroquinones, where the charge is spread by resonance, tetra-hedral intermediates have been isolated:[239]

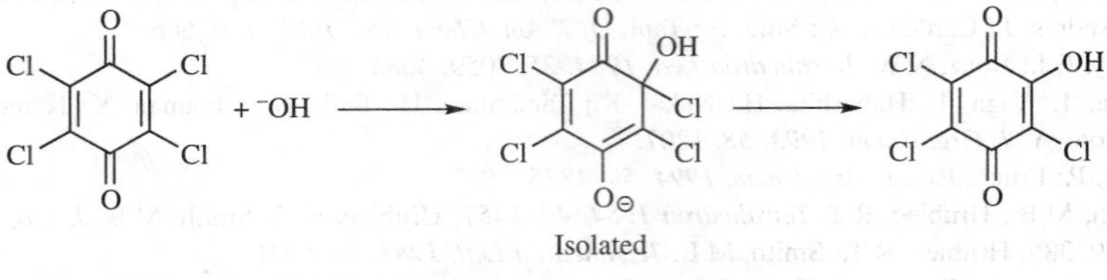

Isolated

[238]Truce, W.E.; Boudakian, M.M. *J. Am. Chem. Soc.* **1956**, *78*, 2748.
[239]Hancock, J.W.; Morrell, C.E.; Rhom, D. *Tetrahedron Lett.* **1962**, 987.

In the case of $Ph(MeO)C=C(NO_2)Ph + RS^-$, the intermediate lived long enough to be detected by UV spectroscopy.[240]

Since both the tetrahedral and addition–elimination mechanisms begin the same way, it is usually difficult to tell them apart, and often no attempt is made to do so. The strongest kind of evidence for the addition–elimination sequence is the occurrence of a "rearrangement," but of course the mechanism could still take place even if no rearrangement is found. Evidence[241] that a tetrahedral or an addition–elimination mechanism takes place in certain cases (as opposed, e.g., to an S_N1 or S_N2 mechanism) is that the reaction rate increases when the leaving group is changed from Br to Cl to F (this is called the *element effect*).[242] This clearly demonstrates that the carbon–halogen bond does not break in the rate-determining step (as it would in both the S_N1 and S_N2 mechanisms), because fluorine is by far the poorest leaving group among the halogens in both the S_N1 and S_N2 reactions (p. 496). The rate is faster with fluorides in the cases cited, because the superior electron-withdrawing character of the fluorine makes the carbon of the C—F bond more positive, and hence more susceptible to nucleophilic attack.

Ordinary vinylic substrates react very poorly if at all by these mechanisms, but substitution is greatly enhanced in substrates of the type $ZCH=CHX$, where Z is an electron-withdrawing group, such as HCO, RCO,[243] EtOOC, $ArSO_2$, NC, and F, since these β groups stabilize the carbanion:

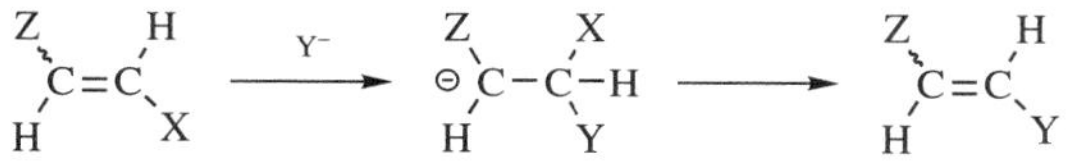

Many such examples are known. In most cases where the stereochemistry has been investigated, retention of configuration is observed,[244] but stereoconvergence [the same product mixture from an (E) or (Z) substrate] has also been observed,[245] especially where the carbanionic carbon bears two electron-withdrawing groups. Although rare, nucleophilic substitution with inversion has also been reported as in the intramolecular substitution of the C—Br bond of 2-bromobut-2-enylamines by the pendant nitrogen atom, giving 2-ethylene aziridines by way of stereochemical inversion.[246] It is not immediately apparent why the tetrahedral mechanism

[240]Bernasconi, C.F.; Fassberg, J.; Killion, Jr., R.B.; Rappoport, Z. *J. Am. Chem. Soc.* **1989**, *112*, 3169; *J. Org. Chem.* **1990**, *55*, 4568.

[241]Additional evidence comes from the pattern of catalysis by amines, similar to that discussed for aromatic substrates on p. 856. See Rappoport, Z.; Peled, P. *J. Am. Chem. Soc.* **1979**, *101*, 2682, and references cited therein.

[242]Beltrame, P.; Favini, G.; Cattania, M.G.; Guella, F. *Gazz. Chim. Ital.* **1968**, *98*, 380. See also, Solov'yanov, A.A.; Shtern, M.M.; Beletskaya, I.P.; Reutov, O.A. *J. Org. Chem. USSR* **1983**, *19*, 1945; Avramovitch, B.; Weyerstahl, P.; Rappoport, Z. *J. Am. Chem. Soc.* **1987**, *109*, 6687.

[243]For a review, see Rybinskaya, M.I.; Nesmeyanov, A.N.; Kochetkov, N.K. *Russ. Chem. Rev.* **1969**, *38*, 433.

[244]Rappoport, Z. *Adv. Phys. Org. Chem.* **1969**, *7*, see pp. 31–62; Shainyan, B.A. *Russ. Chem. Rev.* **1986**, *55*, 516. See also, Rappoport, Z.; Gazit, A. *J. Am. Chem. Soc.* **1987**, *109*, 6698.

[245]See Rappoport, Z.; Gazit, A. *J. Org. Chem.* **1985**, *50*, 3184, *J. Am. Chem. Soc.* **1986**, *51*, 4112; Park, K.P.; Ha, H. *Bull. Chem. Soc. Jpn.* **1990**, *63*, 3006.

[246]Shiers, J.J.; Shipman, M.; Hayes, J.-F.; Slawin, A.M.Z. *J. Am. Chem. Soc.* **2004**, *126*, 6868.

should lead to retention, but this behavior has been ascribed, on the basis of molecular orbital calculations, to hyperconjugation involving the carbanionic electron pair and the substituents on the adjacent carbon.[247]

Vinylic substrates are in general very reluctant to undergo S_N1 reactions, but they can be made to do so in two ways:[248] (*1*) By the use of an a group that stabilizes the vinylic cation. For example, α-aryl vinylic halides $ArCBr=CR'_2$ have often been shown to give S_N1 reactions.[249] The S_N1 reactions have also been demonstrated with other stabilizing groups: cyclopropyl,[250] vinylic,[251] alkynyl,[252] and an adjacent double bond ($R_2C=C=CR'X$).[253] (*2*) Even without a stabilization, by the use of a very good leaving group, OSO_2CF_3 (triflate).[254] The stereochemical outcome of S_N1 reactions at a vinylic substrate is often randomization,[255] that is, either a cis or a trans substrate gives a 1:1 mixture of cis and trans products, indicating that vinylic cations are linear. Another indication that vinylic cations prefer to be linear is the fact that reactivity in cycloalkenyl systems decreases with decreasing ring size.[256] However, a linear vinylic cation need not give random products.[257] The empty *p* orbital lies in the plane of the double bond,

so entry of the nucleophile can be and often is influenced by the relative size of R^1 and R^2.[258] It must be emphasized that even where vinylic substrates do give S_N1

[247]Apeloig, Y.; Rappoport, Z. *J. Am. Chem. Soc.* **1979**, *101*, 5095.

[248]For reviews of the S_N1 mechanism at a vinylic substrate, see Stang, P.J.; Rappoport, Z.; Hanack, H.; Subramanian, L.R. *Vinyl Cations*, Chapt. 5; Academic Press, NY, *1979*; Stang, P.J. *Acc. Chem. Res.* **1978**, *11*, 107; Rappoport, Z. *Acc. Chem. Res.* **1976**, *9*, 265; Subramanian, L.R.; Hanack, M. *J. Chem. Educ.* **1975**, *52*, 80; Hanack, M. *Acc. Chem. Res.* **1970**, *3*, 209; Modena, G.; Tonellato, U. *Adv. Phys. Org. Chem.* **1971**, *9*, 185, 231–253; Grob, C.A. *Chimia* **1971**, *25*, 87; Rappoport, Z.; Bässler, T.; Hanack, M. *J. Am. Chem. Soc.* **1970**, *92*, 4985.

[249]For a review, see Stang, P.J.; Rappoport, Z.; Hanack, H.; Subramanian, L.R. *Vinyl Cations*, Chapt. 6, Academic Press, NY, *1979*.

[250]Kelsey, D.R.; Bergman, R.G. *J. Am. Chem. Soc.* **1970**, *92*, 238; **1971**, *93*, 1941; Hanack, M.; Bässler, T.; Eymann, W.; Heyd, W.E.; Kopp, R. *J. Am. Chem. Soc.* **1974**, *96*, 6686.

[251]Grob, C.A.; Spaar, R. *Tetrahedron Lett.* **1969**, 1439; *Helv. Chim. Acta* **1970**, *53*, 2119.

[252]Hassdenteufel, J.R.; Hanack, M. *Tetrahedron Lett.* **1980**, 503. See also, Kobayashi, S.; Nishi, T.; Koyama, I.; Taniguchi, H. *J. Chem. Soc., Chem. Commun.* **1980**, 103.

[253]Schiavelli, M.D.; Gilbert, R.P.; Boynton, W.A.; Boswell, C.J. *J. Am. Chem. Soc.* **1972**, *94*, 5061.

[254]See, for example, Clarke, T.C.; Bergman, R.G. *J. Am. Chem. Soc.* **1972**, *94*, 3627; *1974*, *96*, 7934; Summerville, R.H.; Schleyer, P.v.R. *J. Am. Chem. Soc.* **1972**, *94*, 3629; *1974*, *96*, 1110; Hanack, M.; Märkl, R.; Martinez, A.G. *Chem. Ber.* **1982**, *115*, 772.

[255]Rappoport, Z.; Apeloig, Y. *J. Am. Chem. Soc.* **1969**, *91*, 6734; Kelsey, D.R.; Bergman, R.G. *J. Am. Chem. Soc.* **1970**, *92*, 238; *1971*, *93*, 1941.

[256]Pfeifer, W.D.; Bahn, C.A.; Schleyer, P.v.R.; Bocher, S.; Harding, C.E.; Hummel, K.; Hanack, M.; Stang, P.J. *J. Am. Chem. Soc.* **1971**, *93*, 1513.

[257]For examples of inversion, see Clarke, T.C.; Bergman, R.G. *J. Am. Chem. Soc.* **1972**, *94*, 3627; *1974*, *96*, 7934; Summerville, R.H.; Schleyer, P.v.R. *J. Am. Chem. Soc.* **1972**, *94*, 3629; *1974*, *96*, 1110.

[258]Maroni, R.; Melloni, G.; Modena, G. *J. Chem. Soc., Chem. Commun.* **1972**, 857.

reactions, the rates are generally lower than those of the corresponding saturated compounds.

Alkynyl cations are so unstable that they cannot be generated even with very good leaving groups. However, one way in which they have been generated was by formation of a tritiated substrate.

$$R-C\equiv C-T \xrightarrow{\beta \ \text{decay}} R-C\equiv C-{}^3He \xrightarrow[\text{fast}]{\text{very}} R-C\equiv C\oplus + {}^3He$$

When the tritium (half-life 12.26 years) decays it is converted to the helium-3 isotope, which, of course, does not form covalent bonds, and so immediately departs, leaving behind the alkynyl cation. When this was done in the presence of benzene, $RC=CC_6H_5$ was isolated.[259] The tritium-decay technique has also been used to generate vinylic and aryl cations.[260]

Besides the mechanisms already discussed, another mechanism, involving an *elimination–addition* sequence, has been observed in vinylic systems (a similar mechanism is known for aromatic substrates, p. 859). An example of a reaction involving this mechanism is the reaction of 1,2-dichloroethane with ArS^- and ^-OEt to produce **84**. The mechanism may be formulated as:

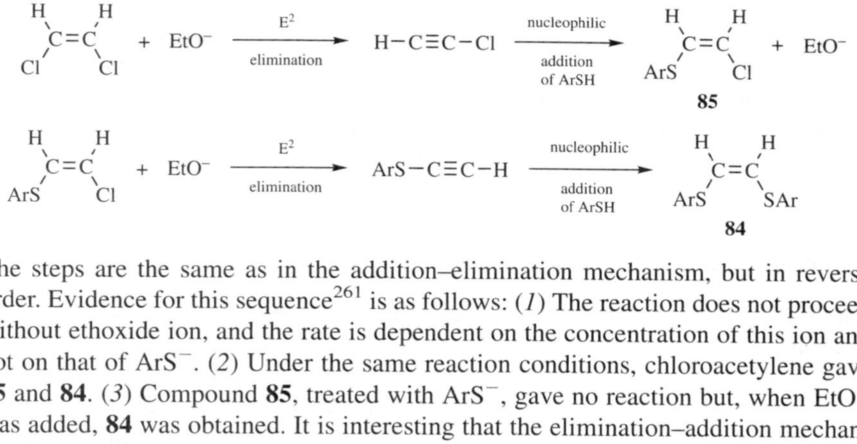

The steps are the same as in the addition–elimination mechanism, but in reverse order. Evidence for this sequence[261] is as follows: (*1*) The reaction does not proceed without ethoxide ion, and the rate is dependent on the concentration of this ion and not on that of ArS^-. (*2*) Under the same reaction conditions, chloroacetylene gave **85** and **84**. (*3*) Compound **85**, treated with ArS^-, gave no reaction but when EtO^- was added, **84** was obtained. It is interesting that the elimination–addition mechanism has even been shown to occur in five- and six-membered cyclic systems, where triple bonds are greatly strained.[262] Note that both the addition–elimination and elimination–addition sequences, as shown above, lead to overall retention of configuration, since in each case both addition and elimination are anti.

[259]Angelini, G.; Hanack, M.; Vermehren, J.; Speranza, M. *J. Am. Chem. Soc.* **1988**, *110*, 1298.

[260]For a review, see Cacace, F. *Adv. Phys. Org. Chem.* **1970**, *8*, 79. See also, Fornarini, S.; Speranza, M. *J. Am. Chem. Soc.* **1985**, *107*, 5358.

[261]Flynn, Jr., J.; Badiger, V.V.; Truce, W.E. *J. Org. Chem.* **1963**, *28*, 2298. See also, Shainyan, B.A.; Mirskova, A.N. *J. Org. Chem. USSR* **1984**, *20*, 885, 1989; **1985**, *21*, 283.

[262]Montgomery, L.K.; Clouse, A.O.; Crelier, A.M.; Applegate, L.E. *J. Am. Chem. Soc.* **1967**, *89*, 3453; Caubere, P.; Brunet, J. *Tetrahedron* **1971**, *27*, 3515; Bottini, A.T.; Corson, F.P.; Fitzgerald, R.; Frost II, K.A. *Tetrahedron* **1972**, *28*, 4883.

The elimination–addition sequence has also been demonstrated for certain reactions of saturated substrates, for example, $ArSO_2CH_2CH_2SO_2Ar$.[263] Treatment of this with ethoxide proceeds as follows:

$$ArSO_2CH_2CH_2SO_2Ar \xrightarrow[\text{E}^2 \text{ elimination}]{\text{EtO}^-} ArSO_2CH=CH_2 \xrightarrow[\text{addition}]{\text{EtO}^-} ArSO_2CH_2CH_2OEt$$

Mannich bases (see **16-19**) of the type $RCOCH_2CH_2NR_2$ similarly undergo nucleophilic substitution by the elimination–addition mechanism.[264] The nucleophile replaces the NR_2 group.

The simple S_N2 mechanism has never been convincingly demonstrated for vinylic substrates.[265]

Vinylic halides can react by a $S_{RN}1$ mechanism (p. 862) in some cases. An example is the $FeCl_2$-catalyzed reaction of 1-bromo-2-phenylethene and the enolate anion of pinacolone (t-$BuCOCH_2^-$), which gave a low yield of substitution products along with alkynes.[266]

REACTIVITY

A large amount of work has been done on this subject. although a great deal is known, much is still poorly understood, and many results are anomalous and hard to explain. In this section, only approximate generalizations are attempted. The work discussed here, and the conclusions reached, pertain to reactions taking place in solution. Some investigations have also been carried out in the gas phase.[267]

The Effect of Substrate Structure

The effect on the reactivity of a change in substrate structure depends on the mechanism.

1. *Branching at the α and β Carbons.* For the S_N2 mechanism, branching at either the α or the β carbon decreases the rate. Tertiary systems seldom[268]

[263]Kader, A.T.; Stirling, C.J.M. *J. Chem. Soc.* **1962**, 3686. For another example, see Popov, A.F.; Piskunova, Z.; Matvienko, V.N. *J. Org. Chem. USSR* **1986**, 22, 1299.

[264]For an example, see Andrisano, R.; Angeloni, A.S.; De Maria, P.; Tramontini, M. *J. Chem. Soc. C* **1967**, 2307.

[265]For discussions, see Miller, S.I. *Tetrahedron* **1977**, 33, 1211; Texier, F.; Henri-Rousseau, O.; Bourgois, J. *Bull. Soc. Chim. Fr.* **1979**, II-11; Rappoport, Z. *Acc. Chem. Res.* **1981**, 14, 7; Rappoport, Z.; Avramovitch, B. *J. Org. Chem.* **1982**, 47, 1397.

[266]Galli, C.; Gentili, P.; Rappoport, Z. *J. Org. Chem.* **1994**, 59, 6786; Galli, C.; Gentili, P. *J. Chem. Soc., Chem. Commun.* **1993**, 570.

[267]See, for example, DePuy, C.H.; Gronert, S.; Mullin, A.; Bierbaum, V.M. *J. Am. Chem. Soc.* **1990**, 112, 8650.

[268]For a reported example, see Edwards, O.E.; Grieco, C. *Can. J. Chem.* **1974**, 52, 3561.

TABLE 10.3. Average Relative S$_N$2 Rates for Some Alkyl Substrates[270]

R	Relative rate	R	Relative rate
Methyl	30	Isobutyl	0.03
Ethyl	1	Neopentyl	10^{-5}
Propyl	0.4	Allyl	40
Butyl	0.4	Benzyl	120
Isopropyl	0.025		

react by the S$_N$2 mechanism and neopentyl systems react so slowly as to make such reactions, in general, synthetically useless.[269] Table 10.3 shows average relative rates for some alkyl substrates.[270] The reason for these low rates is almost certainly steric.[271] The transition state **1** is more crowded when larger groups are close to the central carbon.

The tetrahedral mechanism for substitution at a carbonyl carbon is also slowed or blocked completely by α or β branching for similar reasons. Solvolysis in such systems is linked to relief of B-strain, but solvent participation can overshadow this as steric hindrance increases.[272] Severe steric strain can cause distortion from coplanarity in the carbocation intermediate,[273] although there seems to be no loss of resonance stability.[274] Adding electron-donating substituents to such molecules improves coplanarity in the cation.[275] For example, esters of the formula R$_3$CCOOR$'$ cannot generally be hydrolyzed by the tetrahedral mechanism (see **16-59**), nor can acids R$_3$CCOOH be easily esterified.[276] Synthetic advantage can be taken of this fact, for example, when in a molecule containing two ester groups only the less hindered one is hydrolyzed.

1

[269]The S$_N$2 reactions on neopentyl tosylates have been conveniently carried out in the solvents HMPA and DMSO: Lewis, R.G.; Gustafson, D.H.; Erman, W.F. *Tetrahedron Lett.* **1967**, 401; Paquette, L.A.; Philips, J.C. *Tetrahedron Lett.* **1967**, 4645; Anderson, P.H.; Stephenson, B.; Mosher, H.S. *J. Am. Chem. Soc.* **1974**, *96*, 3171.

[270]This table is from Streitwieser, A. *Solvolytic Displacement Reactions*, McGraw-Hill, NY, **1962**, p. 13. Also see, Table 9.2.

[271]For evidence, see Caldwell, G.; Magnera, T.F.; Kebarle, P. *J. Am. Chem. Soc.* **1984**, *106*, 959.

[272]Liu, K.-T.; Hou, S.-J.; Tsao, K.-L. *J. Org. Chem.* **1998**, *63*, 1360.

[273]Fujio, M.; Nomura, H.; Nakata, K.; Saeki, Y.; Mishima, M.; Kobayashi, S.; Matsushita, T.; Nishimoto, K.; Tsuno, Y. *Tetrahedron Lett.* **1994**, *35*, 5005.

[274]Fujio, M.; Nakata, K.; Kuwamura, T.; Nakamura, H.; Saeki, Y.; Mishima, M.; Kobayashi, S.; Tsuno, Y. *Tetrahedron Lett.* **1992**, *34*, 8309.

[275]Liu, K.T.; Tsao, M.-L.; Chao, I. *Tetrahedron Lett.* **1996**, *37*, 4173.

[276]For a molecular mechanics study of this phenomenon, see DeTar, D.F.; Binzet, S.; Darba, P. *J. Org. Chem.* **1987**, *52*, 2074.

TABLE 10.4. Relative Rates of Solvolysis of RBr in Two Solvents[277]

RBr Substrate	In 60% Ethanol at 55°C	In Water at 50°C
MeBr	2.08	1.05
EtBr	1.00	1.00
iPrBr	1.78	11.6
t-BuBr	2.41×10^4	1.2×10^6

For the S_N1 mechanism, a branching increases the rate, as shown in Table 10.4.[277] We can explain this by the stability order of alkyl cations (tertiary > secondary > primary). Of course, the rates are not actually dependent on the stability of the ions, but on the difference in free energy between the starting compounds and the transition states. We use the Hammond postulate (p. 308) to make the assumption that the transition states resemble the cations and that anything (e.g., a branching) that lowers the free energy of the ions also lowers it for the transition states. For simple alkyl groups, the S_N1 mechanism is important under all conditions only for tertiary substrates.[278] As previously indicated (p. 440), secondary substrates generally react by the S_N2 mechanism,[279] except that the S_N1 mechanism may become important at high solvent polarities. Table 10.4 shows that isopropyl bromide reacts less than twice as fast as ethyl bromide in the relatively nonpolar 60% ethanol (compare this with the 10^4 ratio for *tert*-butylbromide, where the mechanism is certainly S_N1), but in the more polar water the rate ratio is 11.6. The 2-adamantyl system is an exception; it is a secondary system that reacts by the S_N1 mechanism because backside attack is hindered for steric reasons.[280] Because there is no S_N2 component, this system provides an opportunity for comparing the pure S_N1 reactivity of secondary and tertiary substrates. It has been found that substitution of a methyl group for the a

---

[277]These values are from Streitwieser, A. *Solvolytic Displacement Reactions*, McGraw-Hill, NY, *1962*, p. 43, where values are also given for other conditions. Methyl bromide reacts faster than ethyl bromide (and in the case of 60% ethanol, ispropyl bromide) because most of it (probably all) reacts by the S_N2 mechanism.

[278]For a report of an S_N1 mechanism at a primary carbon, see Zamashchikov, V.V.; Bezbozhnaya, T.V.; Chanysheva, I.R. *J. Org. Chem. USSR* *1986*, 22, 1029.

[279]See Raber, D.J.; Harris, J.M. *J. Chem. Educ.* *1972*, 49, 60; Lambert, J.B.; Putz, G.J.; Mixan, C.E. *J. Am. Chem. Soc.* *1972*, 94, 5132; Nordlander, J.E.; McCrary, Jr., T.J. *J. Am. Chem. Soc.* *1972*, 94, 5133; Fry, J.L.; Lancelot, C.J.; Lam, L.K.M.; Harris, J.M.; Bingham, R.C.; Raber, D.J.; Hall, R.E.; Schleyer, P.v.R. *J. Am. Chem. Soc.* *1970*, 92, 2538; Dietze, P.E.; Jencks, W.P. *J. Am. Chem. Soc.* *1986*, 108, 4549; Dietze, P.E.; Hariri, R.; Khattak, J. *J. Org. Chem.* *1989*, 54, 3317.

[280]Fry, J.L.; Harris, J.M.; Bingham, R.C.; Schleyer, P.v.R. *J. Am. Chem. Soc.* *1970*, 92, 2540; Schleyer, P.v.R.; Fry, J.L.; Lam, L.K.M.; Lancelot, C.J. *J. Am. Chem. Soc.* *1970*, 92, 2542. See also, Pritt, J.R.; Whiting, M.C. *J. Chem. Soc. Perkin Trans. 2* *1975*, 1458. For an *ab initio* molecular-orbital study of the 2-adamantyl cation, see Dutler, R.; Rauk, A.; Sorensen, T.S.; Whitworth, S.M. *J. Am. Chem. Soc.* *1989*, 111, 9024.

hydrogen of 2-adamantyl substrates (thus changing a secondary to a tertiary system) increases solvolysis rates by a factor of $\sim10^8$.[281] Simple primary substrates react by the S_N2 mechanism (or with participation by neighboring alkyl or hydrogen), but not by the S_N1 mechanism, even when solvolyzed in solvents of very low nucleophilicity[282] (e.g., trifluoroacetic acid or trifluoroethanol[283]), and even when very good leaving groups (e.g., OSO_2F) are present[284] (see, however, p. 497).

For some tertiary substrates, the rate of S_N1 reactions is greatly increased by the relief of B strain in the formation of the carbocation (see p. 398). Except where B strain is involved, β branching has little effect on the S_N1 mechanism, except that carbocations with β branching undergo rearrangements readily. Of course, isobutyl and neopentyl are primary substrates, and for this reason react very slowly by the S_N1 mechanism, but not more slowly than the corresponding ethyl or propyl compounds.

To sum up, primary and secondary substrates generally react by the S_N2 mechanism and tertiary by the S_N1 mechanism. However, tertiary substrates seldom undergo nucleophilic substitution at all. Elimination is always a possible side reaction of nucleophilic substitutions (wherever a β hydrogen is present), and with tertiary substrates it usually predominates. With a few exceptions, nucleophilic substitutions at a tertiary carbon have little or no preparative value. However, tertiary substrates that can react by the SET mechanism (e.g., p-$NO_2C_6H_4CMe_2Cl$) give very good yields of substitution products when treated with a variety of nucleophiles.[285]

2. *Unsaturation at the α Carbon.* Vinylic, acetylenic,[286] and aryl substrates are very unreactive toward nucleophilic substitutions. For these systems, both the S_N1 and S_N2 mechanisms are greatly slowed or stopped altogether. One reason that has been suggested for this is that sp^2 (and even more, sp) carbon atoms have a higher electronegativity than sp^3 carbons and thus a greater attraction for the electrons of the bond. As we have seen (p. 388), an sp–H bond has a higher acidity than an sp^3–H bond, with that of an sp^2 H bond in

[281]Fry, J.L.; Engler, E.M.; Schleyer, P.v.R. *J. Am. Chem. Soc.* *1972*, *94*, 4628. See also, Gassman, P.G.; Pascone, J.M. *J. Am. Chem. Soc.* *1973*, *95*, 7801.

[282]For discussions and attempts to develop quantitative scales of solvent nucleophilicity see Minegishi, S.; Kobayashi, S.; Mayr, H. *J. Am. Chem. Soc.* *2004*, *126*, 5174; Catalan, J.; Diaz, C.; Garcia-Blanco, F. *J. Org. Chem.* *1999*, *64*, 6512; Bentley, T.W.; Llewellyn, G. *Prog. Phys. Org. Chem.* *1990*, *17*, 121; Kevill, D.N., in Charton, M. *Advances in Quantitative Structure-Property Relationships*, Vol. 1, JAI Press, Greenwich, CT, *1996*, pp. 81–115; Grunwald, E.; Winstein, S. *J. Am. Chem. Soc.* *1948*, *70*, 846; Winstein, S.; Fainberg, A.H.; Grunwald, E. *J. Am. Chem. Soc.* *1957*, *79*, 4146; Peterson, P.E.; Waller, F.J. *J. Am. Chem. Soc.* *1972*, *94*, 991; Schadt, F.L.; Bentley, T.W.; Schleyer, P.v.R. *J. Am. Chem. Soc.* *1976*, *98*, 7667.

[283]Dafforn, G.A.; Streitwieser, Jr., A. *Tetrahedron Lett.* *1970*, 3159.

[284]Cafferata, L.F.R.; Desvard, O.E.; Sicre, J.E. *J. Chem. Soc. Perkin Trans. 2* *1981*, 940.

[285]Kornblum, N.; Cheng, L.; Davies, T.M.; Earl, G.W.; Holy, N.L.; Kerber, R.C.; Kestner, M.M.; Manthey, J.W.; Musser, M.T.; Pinnick, H.W.; Snow, D.H.; Stuchal, F.W.; Swiger, R.T. *J. Org. Chem.* *1987*, *52*, 196.

[286]For a discussion of S_N reactions at acetylenic substrates, see Miller, S.I.; Dickstein, J.I. *Acc. Chem. Res.* *1976*, *9*, 358.

between. This is reasonable; the carbon retains the electrons when the proton is lost and an sp carbon, which has the greatest hold on the electrons, loses the proton most easily. But in nucleophilic substitution, the leaving group *carries off* the electron pair, so the situation is reversed and it is the sp^3 carbon that loses the leaving group and the electron pair most easily. It may be recalled (p. 24) that bond distances decrease with increasing s character. Thus the bond length for a vinylic or aryl C—Cl bond is 1.73 Å compared with 1.78 Å for a saturated C—Cl bond. Other things being equal, a shorter bond is a stronger bond.

Of course, we have seen (p. 476) that S_N1 reactions at vinylic substrates can be accelerated by α substituents that stabilize that cation, and that reactions by the tetrahedral mechanism can be accelerated by β substituents that stabilize the carbanion. Also, reactions at vinylic substrates can in certain cases proceed by addition–elimination or elimination–addition sequences (pp. 473, 476).

In contrast to such systems, substrates of the type RCOX are usually much *more* reactive than the corresponding RCH_2X. Of course, the mechanism here is almost always the tetrahedral one. Three reasons can be given for the enhanced reactivity of RCOX: (*1*) The carbonyl carbon has a sizable partial positive charge that makes it very attractive to nucleophiles. (*2*) In an S_N2 reaction, a σ bond must break in the rate-determining step, which requires more energy than the shift of a pair of π electrons, which is what happens in a tetrahedral mechanism. (*3*) A trigonal carbon offers less steric hindrance to a nucleophile than a tetrahedral carbon.

For reactivity in aryl systems, see Chapter 13.

3. *Unsaturation at the* β *Carbon.* The S_N1 rates are increased when there is a double bond in the β position, so that allylic and benzylic substrates react rapidly (Table 10.5).[287] The reason is that allylic (p. 239) and benzylic[288]

TABLE 10.5. Relative Rates for the S_N1 Reaction between ROTs and Ethanol at 25°C[285]

Group	Relative Rate
Et	0.26
iPr	0.69
$CH_2=CHCH_2$	8.6
$PhCH_2$	100
Ph_2CH	$\sim 10^5$
Ph_3C	$\sim 10^{10}$

[287]Streitwieser, A. *Solvolytic Displacement Reactions*, McGraw-Hill, NY, *1962*, p. 75. Actually, the figures for Ph_2CHOTs and Ph_3COTs are estimated from the general reactivity of these substrates.
[288]For a Grunwald-Winstein correlation analysis of the solvolysis of benzyl bromide, see Liu, K.-T.; Hou, I.-J. *Tetrahedron* **2001**, *57*, 3343.

(p. 240) cations are stabilized by resonance. As shown in Table 10.5, a second and a third phenyl group increase the rate still more, because these carbocations are more stable yet. Remember that allylic rearrangements are possible with allylic systems.

In general, S_N1 rates at an allylic substrate are increased by any substituent in the 1 or 3 position that can stabilize the carbocation by resonance or hyperconjugation.[289] Among these are alkyl, aryl, and halo groups.

86

The S_N2 rates for allylic and benzylic systems are also increased (see Table 10.3), probably owing to resonance possibilities in the transition state. Evidence for this in benzylic systems is that the rate of the reaction was 8000 times slower than the rate with $(PhCH_2)_2SEt^+$.[290] The cyclic **86** does not have the proper geometry for conjugation in the transition state.

Triple bonds in the β position (in propargyl systems) have about the same effect as double bonds.[291] Alkyl, aryl, halo, and cyano groups, among others, in the 3 position of allylic substrates increase S_N2 rates, owing to increased resonance in the transition state, but alkyl and halo groups in the 1 position decrease the rates because of steric hindrance.

4. α *Substitution.* Compounds of the formula ZCH_2X, where $Z = RO$, RS, or R_2N undergo S_N1 reactions very rapidly,[292] because of the increased resonance in the carbocation. These groups have an unshared pair on an atom directly attached to the positive carbon, which stabilizes the carbocation (p. 242). The field effects of these groups would be expected to decrease S_N1 rates (see Section 6, p. 485), so the resonance effect is far more important.

When Z in ZCH_2X is RCO,[293] HCO, ROCO, NH_2CO, NC, or F_3C,[294] S_N1 rates are decreased compared to CH_3X, owing to the electron-withdrawing field

[289]For a discussion of the relative reactivities of different allylic substrates, see DeWolfe, R.H.; Young, W.G., in Patai, S. *The Chemistry of Alkenes*, Wiley, NY, **1964**, pp. 683–688, 695–697.

[290]King, J.F.; Tsang, G.T.Y.; Abdel-Malik, M.M.; Payne, N.C. *J. Am. Chem. Soc.* **1985**, *107*, 3224.

[291]Hatch, L.F.; Chiola, V. *J. Am. Chem. Soc.* **1951**, *73*, 360; Jacobs, T.L.; Brill, W.F. *J. Am. Chem. Soc.* **1953**, *75*, 1314.

[292]For a review of the reactions of α-haloamines, sulfides, and ethers, see Gross, H.; Höft, E. *Angew. Chem. Int. Ed.* **1967**, *6*, 335.

[293]For a review of α-halo ketones, including reactivity, see Verhé, R.; De Kimpe, N., in Patai, S.; Rappoport, Z. *The Chemistry of Functional Groups, Supplement D*, pt. 1, Wiley, NY, **1983**, pp. 813–931. This review has been reprinted, and new material added, in De Kimpe, N.; Verhé, R. *The Chemistry of α-Haloketones, α-Haloaldehydes, and α-Haloimines*, Wiley, NY, **1988**, pp. 225–368.

[294]Liu, K.; Kuo, M.; Shu, C. *J. Am. Chem. Soc.* **1982**, *104*, 211; Gassman, P.G.; Harrington, C.K. *J. Org. Chem.* **1984**, *49*, 2258; Allen, A.D.; Girdhar, R.; Jansen, M.P.; Mayo, J.D.; Tidwell, T.T. *J. Org. Chem.* **1986**, *51*, 1324; Allen, A.D.; Kanagasabapathy, V.M.; Tidwell, T.T. *J. Am. Chem. Soc.* **1986**, *108*, 3470; Richard, J.P. *J. Am. Chem. Soc.* **1989**, *111*, 1455.

effects of these groups. Furthermore, carbocations[295] with an a CO or CN group are greatly destabilized because of the partial positive charge on the adjacent carbon (**87**). The S_N1 reactions have been carried out on such compounds,[296] but the rates are very low. For example, from a comparison of the solvolysis rates of **88** and **89**, a rate-retarding effect of $10^{7.3}$

was estimated for the C=O group.[297] However, when a different kind of comparison is made: $RCOCR'_2X$ versus HCR'_2X (where X = a leaving group), the RCO had only a small or negligible rate-retarding effect, indicating that resonance stabilization[298]

may be offsetting the inductive destabilization for this group.[299] For a CN group also, the rate-retarding effect is reduced by this kind of resonance.[300] A carbocation with an a COR group has been isolated.[301]

When S_N2 reactions are carried out on these substrates, rates are greatly increased for certain nucleophiles (e.g., halide or halide-like ions), but decreased or essentially unaffected by others.[302] For example, α-chloroacetophenone ($PhCOCH_2Cl$) reacts with KI in acetone at 75°C ~32,000 times faster than 1-chlorobutane,[303] but α-bromoacetophenone reacts with the nucleophile triethylamine 0.14 times as fast as iodomethane.[302] The reasons

[295]For reviews of such carbocations, see Bégué, J.; CharpentierMorize, M. *Acc. Chem. Res.* **1980**, *13*, 207; Charpentier-Morize, M. *Bull. Soc. Chim. Fr.* **1974**, 343.

[296]For reviews, see Creary, X. *Acc. Chem. Res.* **1985**, *18*, 3; Creary, X.; Hopkinson, A.C.; Lee-Ruff, E. *Adv. Carbocation Chem.* **1989**, *1*, 45; Charpentier-Morize, M.; Bonnet-Delpon, D. *Adv. Carbocation Chem.* **1989**, *1*, 219.

[297]Creary, X. *J. Org. Chem.* **1979**, *44*, 3938.

[298]**D**, which has the positive charge on the more electronegative atom, is less stable than **C**, according to rule c on p. 47, but it nevertheless seems to be contributing in this case.

[299]Creary, X. *J. Am. Chem. Soc.* **1984**, *106*, 5568. See, however, Takeuchi, K.; Yoshida, M.; Ohga,Y.; Tsugeno, A.; Kitagawa, T. *J. Org. Chem.* **1990**, *55*, 6063.

[300]Gassman, P.G.; Saito, K.; Talley, J.J. *J. Am. Chem. Soc.* **1980**, *102*, 7613.

[301]Takeuchi, K.; Kitagawa, T.; Okamoto, K. *J. Chem. Soc., Chem. Commun.* **1983**, 7. See also, Dao, L.H.; Maleki, M.; Hopkinson, A.C.; Lee-Ruff, E. *J. Am. Chem. Soc.* **1986**, *108*, 5237.

[302]Halvorsen, A.; Songstad, J. *J. Chem. Soc., Chem. Commun.* **1978**, 327.

[303]Bordwell, F.G.; Brannen, Jr., W.T. *J. Am. Chem. Soc.* **1964**, *86*, 4645. For some other examples, see Conant, J.B.; Kirner, W.R.; Hussey, R.E. *J. Am. Chem. Soc.* **1925**, *47*, 488; Sisti, A.J.; Lowell, S. *Can. J. Chem.* **1964**, *42*, 1896.

for this varying behavior are not clear, but those nucleophiles that form a "tight" transition state (one in which bond making and bond breaking have proceeded to about the same extent) are more likely to accelerate the reaction.[304]

When Z is SOR or SO_2R (e.g., α-halo sulfoxides and sulfones), nucleophilic substitution is retarded.[305] The S_N1 mechanism is slowed by the electron-withdrawing effect of the SOR or SO_2R group,[306] and the S_N2 mechanism presumably by the steric effect.

5. β *Substitution.* For compounds of the type ZCH_2CH_2X, where Z is any of the groups listed in the previous section as well as halogen[307] or phenyl, S_N1 rates are lower than for unsubstituted systems, because the resonance effects mentioned in Section 4 are absent, but the field effects are still there, although smaller. These groups in the β position do not have much effect on S_N2 rates unless they behave as neighboring groups and enhance the rate through anchimeric assistance,[308] or unless their size causes the rates to decrease for steric reasons.[309] It has been shown that silicon exerts a β-effect, and that tin exerts a γ-effect.[310] Silcon also exerts a γ-effect.[311]

6. *The Effect of Electron-Donating and Electron-Withdrawing Groups.* If substitution rates of series of compounds $p\text{-}ZC_6H_4CH_2X$ are measured, it is possible to study the electronic effects of groups Z on the reaction. Steric effects of Z are minimized or eliminated, because Z is so far from the reaction site. For S_N1 reactions electron-withdrawing Z decrease the rate and electron-donating Z increase it,[312] because the latter decrease the energy of the transition state (and of the carbocation) by spreading the positive charge, for example,

[304]For discussions of possible reasons, see McLennan, D.J.; Pross, A. *J. Chem. Soc. Perkin Trans. 2* **1984**, 981; Yousaf, T.I.; Lewis, E.S. *J. Am. Chem. Soc.* **1987**, *109*, 6137; Lee, I.; Shim, C.S.; Chung, S.Y.; Lee, I. *J. Chem. Soc. Perkin Trans. 2* **1988**, 975; Yoh, S.; Lee, H.W. *Tetrahedron Lett.* **1988**, *29*, 4431.

[305]Bordwell, F.G.; Jarvis, B.B. *J. Org. Chem.* **1968**, *33*, 1182; Loeppky, R.N.; Chang, D.C.K. *Tetrahedron Lett.* **1968**, 5414; Cinquini, M.; Colonna, S.; Landini, D.; Maia, A.M. *J. Chem. Soc. Perkin Trans. 2* **1976**, 996.

[306]See, for example, Creary, X.; Mehrsheikh-Mohammadi, M.E.; Eggers, M.D. *J. Am. Chem. Soc.* **1987**, *109*, 2435.

[307]See Gronert, S.; Pratt, L.M.; Mogali, S. *J. Am. Chem. Soc.* **2001**, *123*, 3081.

[308]For example, substrates of the type $RSCH_2CH_2X$ are so prone to the neighboring-group mechanism that ordinary S_N2 reactions have only recently been observed: Sedaghat-Herati, M.R.; McManus, S.P.; Harris, J.M. *J. Org. Chem.* **1988**, *53*, 2539.

[309]See, for example, Okamoto, K.; Kita, T.; Araki, K.; Shingu, H. *Bull. Chem. Soc. Jpn.* **1967**, *40*, 1913.

[310]Sugawara, M.; Yoshida, J.-i. *Bull. Chem. Soc. Jpn.* **2000**, *73*, 1253.

[311]Nakashima, T.; Fujiyama, R.; Fujio, M.; Tsuno, Y. *Bull. Chem. Soc. Jpn.* **1999**, *72*, 741, 1043; Nakashima, T.; Fujiyama, R.; Kim, H.-J.; Fujio, M.; Tsuno, Y. *Bull. Chem. Soc. Jpn.* **2000**, *73*, 429.

[312]Jorge, J.A.L.; Kiyan, N.Z.; Miyata, Y.; Miller, J. *J. Chem. Soc. Perkin Trans. 2* **1981**, 100; Vitullo, V.P.; Grabowski, J.; Sridharan, S. *J. Chem. Soc., Chem. Commun.* **1981**, 737.

while electron-withdrawing groups concentrate the charge. The Hammett $\sigma\rho$ relationship (p. 402) correlates fairly successfully the rates of many of these reactions (with σ^+ instead of σ). ρ values are generally about -4, which is expected for a reaction where a positive charge is created in the transition state.

For S_N2 reactions, no such simple correlations are found.[313] In this mechanism, bond breaking is about as important as bond making in the rate-determining step, and substituents have an effect on both processes, often in opposite directions. The unsubstituted benzyl chloride and bromide solvolyze by the S_N2 mechanism.[306]

For Z = alkyl, the Baker–Nathan order (p. 96) is usually observed both for S_N1 and S_N2 reactions.

In para-substituted benzyl systems, steric effects have been removed, but resonance and field effects are still present. However, Holtz and Stock studied a system that removes not only steric effects, but also resonance effects. This is the 4-substituted bicyclo[2.2.2]octylmethyl tosylate system (**90**).[314] In

$$Z-\underset{\mathbf{90}}{\longleftrightarrow}-CH_2OTs$$

this system, steric effects are completely absent owing to the rigidity of the molecules, and only field effects operate. By this means, Holtz and Stock showed that electron-withdrawing groups increase the rate of S_N2 reactions. This can be ascribed to stabilization of the transition state by withdrawal of some of the electron density.

For substrates that react by the tetrahedral mechanism, electron-withdrawing groups increase the rate and electron-donating groups decrease it.

7. *Cyclic Substrates.* Cyclopropyl substrates are extremely resistant to nucleo-philic attack.[315] For example, cyclopropyl tosylate solvolyzes $\sim10^6$ times more slowly than cyclobutyl tosylate in acetic acid at 60°C.[316] When such attack does take place, the result is generally not normal substitution (though exceptions are known,[317] especially when an a stabilizing group, such as aryl

---

[313]See Sugden, S.; Willis, J.B. *J. Chem. Soc.* **1951**, 1360; Baker, J.W.; Nathan, W.S. *J. Chem. Soc.* **1935**, 1840; Hayami, J.; Tanaka, N.; Kurabayashi, S.; Kotani, Y.; Kaji, A. *Bull. Chem. Soc. Jpn.* **1971**, *44*, 3091; Westaway, K.C.; Waszczylo, Z. *Can. J. Chem.* **1982**, *60*, 2500; Lee, I.; Sohn, S.C.; Oh, Y.J.; Lee, B.C. *Tetrahedron* **1986**, *42*, 4713.

[314]Holtz, H.D.; Stock, L.M. *J. Am. Chem. Soc.* **1965**, *87*, 2404.

[315]For reviews, see Friedrich, E.C., in Rappoport, Z. *The Chemistry of the Cyclopropyl Group*, pt. 1; Wiley, NY, **1987**, pp. 633–700; Aksenov, V.S.; Terent'eva, G.A.; Savinykh, Yu.V. *Russ. Chem. Rev. 1980*, *49*, 549.

[316]Roberts, J.D.; Chambers, V.C. *J. Am. Chem. Soc.* **1951**, *73*, 5034.

[317]For example, see Kirmse, W.; Schütte, H. *J. Am. Chem. Soc.* **1967**, *89*, 1284; Landgrebe, J.A.; Becker, L.W. *J. Am. Chem. Soc.* **1967**, *89*, 2505; Howell, B.A.; Jewett, J.G. *J. Am. Chem. Soc.* **1971**, *93*, 798; van der Vecht, J.R.; Steinberg, H.; de Boer, T.J. *Recl. Trav. Chim. Pays-Bas 1978*, *96*, 313; Engbert, T.; Kirmse, W. *Liebigs Ann. Chem. 1980*, 1689; Turkenburg, L.A.M.; de Wolf, W.H.; Bickelhaupt, F.; Stam, C.H.; Konijn, M. *J. Am. Chem. Soc.* **1982**, *104*, 3471; Banert, K. *Chem. Ber. 1985*, *118*, 1564; Vilsmaier, E.; Weber, S.; Weidner, J. *J. Org. Chem.* **1987**, *52*, 4921.

or alkoxy is present), but ring opening:[310]

$$\overset{3}{\underset{2}{\triangleright}}\!\!\!\triangleright\!\!\!-X \longrightarrow H_2C=C\overset{\diagup H}{\underset{\diagdown CH_2}{}} \overset{Y}{\longrightarrow} H_2C=C\overset{\diagup H}{\underset{\diagdown C\diagdown Y}{}}$$
$$\underset{\oplus}{} \qquad\qquad H_2$$

There is much evidence that the ring opening is usually concerted with the departure of the leaving group[318] (as in the similar case of cyclobutyl substrates, p. 465), from which we can conclude that if the 2,3 bond of the cyclopropane ring did not assist, the rates would be lower still. Strain plays a role in the ring-opening process.[319] It has been estimated[320] that without this assistance the rates of these already slow reactions would be further reduced by a factor of perhaps 10^{12}. For a discussion of the stereochemistry of the ring opening, see p. 1644. For larger rings, we have seen (p. 399) that, because of I strain, cyclohexyl substrates solvolyze slower than analogous compounds in which the leaving group is attached to a ring of 5 or of from 7 to 11 members.

8. *Bridgeheads.*[11] The S_N2 mechanism is impossible at most bridgehead compounds (p. 429). Nucleophilic attack in [1.1.1]-propellane has been reported, however.[321] In general, a relatively large ring is required for an S_N1 reaction to take place (p. 435).[322] The S_N1 reactions have been claimed to occur for 1-iodobicyclo[1.1.1]pentane via the bicyclo[1.1.1]pentyl cation,[323] but this has been disputed and the bicyclo[1.1.0]butyl carbinyl cation was calculated to be the real intermediate.[324] Solvolytic reactivity at bridgehead positions spans a wide range; for example, from $k = 4 \times 10^{-17}\,s^{-1}$

91   **92**

for **91** (very slow) to $3 \times 10^6\,s^{-1}$ for the [3.3.3] compound **92** (very fast);[325] a range of 22 orders of magnitude. Molecular mechanics calculations show that

[318]For example, see Schleyer, P.v.R.; Van Dine, G.W.; Schöllkopf, U.; Paust, J. *J. Am. Chem. Soc.* **1966**, *88*, 2868; DePuy, C.H.; Schnack, L.G.; Hausser, J.W. *J. Am. Chem. Soc.* **1966**, *88*, 3343; Jefford, C.W.; Wojnarowski, W. *Tetrahedron* **1969**, *25*, 2089; Hausser, J.W.; Uchic, J.T. *J. Org. Chem.* **1972**, *37*, 4087.

[319]See Wolk, J.L.; Hoz, T.; Basch, H.; Hoz, S. *J. Org. Chem.* **2001**, *66*, 915.

[320]Sliwinski, W.F.; Su, T.M.; Schleyer, P.v.R. *J. Am. Chem. Soc.* **1972**, *94*, 133; Brown, H.C.; Rao, C.G.; Ravindranathan, M. *J. Am. Chem. Soc.* **1978**, *100*, 7946.

[321]Sella, A.; Basch, H.; Hoz, S. *Tetrahedron Lett.* **1996**, *37*, 5573.

[322]For a review of organic synthesis using bridgehead carbocations, see Kraus, G.A.; Hon, Y.; Thomas, P.J.; Laramay, S.; Liras, S.; Hanson, J. *Chem. Rev.* **1989**, *89*, 1591.

[323]Adcock, J.L.; Gakh, A.A. *Tetrahedron Lett.* **1992**, *33*, 4875.

[324]Wiberg, K.B.; McMurdie, N. *J. Org. Chem.* **1993**, *58*, 5603.

[325]Bentley, T.W.; Roberts, K. *J. Org. Chem.* **1988**, *50*, 5852.

TABLE 10.6. List of Groups in Approximately Descending Order of Reactivity Toward S_N1 and S_N2 Reactions[a]

S_N1 Reactivity	S_N2 Reactivity
Ar_3CX	Ar_3CX
Ar_2CHX	Ar_2CHX
$ROCH_2X$, $RSCH_2X$, R_2NCH_2X	$ArCH_2X$
R_3CX	ZCH_2X
	$-\overset{\mid}{C}=\overset{\mid}{C}-CH_2X$
$ArCH_2X$	
$-\overset{\mid}{C}=\overset{\mid}{C}-CH_2X$	$RCH_2X \sim RCHDX \sim RCHDCH_2X$
R_2CHX	R_2CHX
$RCH_2X \sim R_3CCH_2X$	R_3CX
$RCHDX$	ZCH_2CH_2X
$RCHDCH_2X$	R_3CCH_2X
$-\overset{\mid}{C}=\overset{\mid}{C}-X$	$-\overset{\mid}{C}=\overset{\mid}{C}-X$
ZCH_2X	
ZCH_2CH_2X	ArX
ArX	Bridgehead-X
[2.2.1] Bridgehead-X	

[a]The Z group is RCO, HCO, ROCO, NH_2CO, NC, or a similar one.

S_N1 bridgehead reactivity is determined by strain changes between the substrate and the carbocation intermediate.[326]

9. *Deuterium Substitution.* Both α and β secondary isotope effects affect the rate in various ways (p. 324). The measurement of a secondary isotope effects provides a means of distinguishing between S_N1 and S_N2 mechanisms, since for S_N2 reactions the values range from 0.95 to 1.06 per α D, while for S_N1 reactions the values are higher.[327] This method is especially good because it provides the minimum of perturbation of the system under study; changing from α H to α D hardly affects the reaction, while other probes, such as changing a substituent or the polarity of the solvent, may have a much more complex effect.

Table 10.6 is an approximate listing of groups in order of S_N1 and S_N2 reactivity. Table 10.7 shows the main reactions that proceed by the S_N2 mechanism (if R = primary or, often, secondary alkyl).

[326]Bingham, R.C.; Schleyer, P.v.R. *J. Am. Chem. Soc.* **1971**, *93*, 3189; Müller, P.; Blanc, J.; Mareda, J. *Chimia* **1987**, *41*, 399; Müller, P.; Mareda, J. *Helv. Chim. Acta* **1987**, *70*, 1017; Bentley, T.W.; Roberts, K. *J. Org. Chem.* **1988**, *50*, 5852.
[327]Shiner, Jr., V.J.; Fisher, R.D. *J. Am. Chem. Soc.* **1971**, *93*, 2553. For a review of secondary isotope effects in S_N2 reactions, see Westaway, K.C. *Isot. Org. Chem.* **1987**, *7*, 275.

TABLE 10.7. The More Important Synthetic Reactions of Chapter 10 That Take Place by an S_N2 Mechanism.[a] Catalysts are not shown[b]

10-1	$RX + OH^- \longrightarrow ROH$
10-8	$RX + OR' \longrightarrow ROR'$

10-9

$$\underset{\overset{\displaystyle |}{OH}}{\overset{\overset{\displaystyle Cl}{|}}{>\!C\!-\!C\!<}} \longrightarrow -C\overset{\displaystyle O}{\diagup\!\diagdown}C-$$

10-10	$R-OSO_2OR'' + OR' \longrightarrow ROR'$
10-12	$2\,ROH \longrightarrow ROR$

10-14

$$-C\overset{\displaystyle O}{\underset{\overset{\displaystyle |}{}}{\diagup\!\diagdown}}C- + ROH \longrightarrow \underset{\overset{\displaystyle |}{OH}}{\overset{\overset{\displaystyle OR}{|}}{>\!C\!-\!C\!<}}$$

10-15	$R_3O^+ + R'OH \longrightarrow ROR'$
10-17	$RX + R'COO^- \longrightarrow R'COOR$
10-21	$RX + OOH^- \longrightarrow ROOH$
10-25	$RX + SH^- \longrightarrow RSH$
10-26	$RX + R'S^- \longrightarrow RSR'$
10-27	$RX + S_2^{2-} \longrightarrow RSSR$
10-30	$RX + SCN^- \longrightarrow RSCN$
10-31	$RX + R'_2NH \longrightarrow RR'_2N$
10-31	$RX + R'_3N \longrightarrow RR'_3N^+\ X^-$

10-35

$$-C\overset{\displaystyle O}{\underset{\overset{\displaystyle |}{}}{\diagup\!\diagdown}}C- + RNH_2 \longrightarrow \underset{\overset{\displaystyle |}{OH}}{\overset{\overset{\displaystyle NHR}{|}}{>\!C\!-\!C\!<}}$$

10-41	$RX + R'CONH^- \longrightarrow RNHCOR'$
10-42	$RX + NO_2^- \longrightarrow RNO_2 + RONO$
10-43	$RX + N_3^- \longrightarrow RN_3$
10-44	$RX + NCO^- \longrightarrow RNCO$
10-44	$RX + NCS^- \longrightarrow RNCS$
10-46	$RX + X' \longrightarrow RX'$
10-47	$R-OSO_2OR' + X^- \longrightarrow RX$
10-48	$ROH + PCl_5 \longrightarrow RCl$
10-49	$ROR' + 2HI \longrightarrow RI + R'I$

10-50

$$-C\overset{\displaystyle O}{\underset{\overset{\displaystyle |}{}}{\diagup\!\diagdown}}C- + HX \longrightarrow \underset{\overset{\displaystyle |}{OH}}{\overset{\overset{\displaystyle X}{|}}{>\!C\!-\!C\!<}}$$

10-51	$R-O-COR' + LiI \longrightarrow RI + R'COO^-$
10-57	$RX + R'_2CuLi \longrightarrow RR'$

10-65

$$-C\overset{\displaystyle O}{\underset{\overset{\displaystyle |}{}}{\diagup\!\diagdown}}C- + RMgX \longrightarrow \underset{\overset{\displaystyle |}{OH}}{\overset{\overset{\displaystyle R}{|}}{>\!C\!-\!C\!<}}$$

10-67	$RX + HC^-(CO_2R')_2 \longrightarrow RCH(CO_2R')_2$

(*continued*)

TABLE 10.7. (*Continued*)

10-68	RX + R''$\overset{\ominus}{C}$H—COR' ———→ RCR''—COR'	
10-70	RX + R'CHCOO⁻ ———→ RR'CHCOO⁻	

10-71 R-X + H—$\left\langle\begin{smallmatrix}S\\S\end{smallmatrix}\right\rangle^{\ominus}$ ———→ $\overset{R}{\underset{H}{\diagup}}\!\!\!\diagdown\!\left\langle\begin{smallmatrix}S\\S\end{smallmatrix}\right\rangle$

10-74	RX + RC≡C$^{\ominus}$ ———→ RC≡CR'	
10-75	RX + CN⁻ ———→ RCN	

^{a}(R = primary, often secondary, alkyl).
^{b}This is a schematic list only. Some of these reactions may also take place by other mechanisms and the scope may vary greatly. See the discussion of each reaction for details.

The Effect of the Attacking Nucleophile[328]

Any species that has an unshared pair (i.e., any Lewis base) can be a nucleophile, whether it is neutral or has a negative charge. The rates of S_N1 reactions are independent of the identity of the nucleophile, since it does not appear in the rate-determining step.[329] This may be illustrated by the effect of changing the nucleophile from H_2O to ⁻OH for a primary and a tertiary substrate. For methyl bromide, which reacts by an S_N2 mechanism, the rate is multiplied >5000 by the change to the more powerful nucleophile ⁻OH, but for *tert*-butylbromide, which reacts by an S_N1 mechanism, the rate is unaffected.[330] A change in nucleophile can, however, change the *product* of an S_N1 reaction. Thus solvolysis of benzyl tosylate in methanol gives benzyl methyl ether (the nucleophile is the solvent methanol). If the more powerful nucleophile Br⁻ is added, the rate is unchanged, but the product is now benzyl bromide.

For S_N2 reactions in solution, there are four main principles that govern the effect of the nucleophile on the rate, although the nucleophilicity order is not invariant, but depends on substrate, solvent, leaving group, and so on.

1. A nucleophile with a negative charge is always a more powerful nucleophile than its conjugate acid (assuming the latter is also a nucleophile). Thus ⁻OH is more powerful than H_2O, ⁻NH_2 more powerful than NH_3, and so on.

2. In comparing nucleophiles whose attacking atom is in the same row of the periodic table, nucleophilicity is approximately in order of basicity, although

[328]For a monograph, see Harris, J.M.; McManus, S.P. *Nucleophilicity*, American Chemical Society, Washington, DC, *1987*. For reviews, see Klumpp, G.W. *Reactivity in Organic Chemistry*; Wiley, NY, *1982*, pp. 145–167, 181–186; Hudson, R.F., in Klopman, G. *Chemical Reactivity and Reaction Paths*; Wiley, NY, *1974*, pp. 167–252.
[329]It is, however, possible to measure the rates of reaction of nucleophiles with fairly stable carbocations: see Ritchie, C.D. *Acc. Chem. Res. 1972, 5*, 348; Ritchie, C.D.; Minasz, R.J.; Kamego, A.A.; Sawada, M. *J. Am. Chem. Soc. 1977, 99*, 3747; McClelland, R.A.; Banait, N.; Steenken, S. *J. Am. Chem. Soc. 1986, 108*, 7023.
[330]Bateman, L.C.; Cooper, K.A.; Hughes, E.D.; Ingold, C.K. *J. Chem. Soc. 1940*, 925.

basicity is thermodynamically controlled and nucleophilicity is kinetically controlled. So an approximate order of nucleophilicity is $^-NH_2^- > RO^- > ^-OH > R_2NH > ArO^- > NH_3 >$ pyridine $> F^- > H_2O > ClO_4^-$, and another is $R_3C^- > R_2N^- > RO^- > F^-$ (see Table 8.1). This type of correlation works best when the structures of the nucleophiles being compared are similar, as with a set of substituted phenoxides. Within such a series, linear relationships can often be established between nucleophilic rates and pK values.[331]

3. Going down the Periodic table, nucleophilicity increases, although basicity decreases. Thus the usual order of halide nucleophilicity is $I^- > Br^- > Cl^- > F^-$ (as we will see below, this order is solvent dependent). Similarly, any sulfur nucleophile is more powerful than its oxygen analog, and the same is true for phosphorus versus nitrogen. The main reason for this distinction between basicity and nucleophilic power is that the smaller negatively charged nucleophiles are more solvated by the usual polar protic solvents; that is, because the negative charge of Cl^- is more concentrated than the charge of I^-, the former is more tightly surrounded by a shell of solvent molecules that constitute a barrier between it and the substrate. This is most important for protic polar solvents in which the solvent may be hydrogen bonded to small nucleophiles. Evidence for this is that many nucleophilic substitutions with small negatively charged nucleophiles are much more rapid in aprotic polar solvents than in protic ones[332] and that, in DMF, an aprotic solvent, the order of nucleophilicity was $Cl^- > Br^- > I^-$.[333] Another experiment was the use of $Bu_4N^+ X^-$ and LiX as nucleophiles in acetone, where X^- was a halide ion. The halide ion in the former salt is much less associated than in LiX. The relative rates with LiX were Cl^-, 1; Br^-, 5.7; I^-, 6.2, which is in the normal order, while with $Bu_4N^+ X^-$, where X^- is much freer, the relative rates were Cl^-, 68; Br^-, 18; I^-, 3.7.[334] In a further experiment, halide ions were allowed to react with the molten salt $(n\text{-}C_5H_{11})_4N^+ X^-$ at 180°C in the absence of a solvent.[335] Under these conditions, where the ions are unsolvated and unassociated, the relative rates were Cl^-, 620; Br^-, 7.7; I^-, 1. In the gas phase, where no solvent is present, an approximate order of nucleophilicity was found to be $^-OH > F^- \approx MeO^- > MeS^- \gg Cl^- > CN^- > Br^-$,[336]

[331]See, for example, Jokinen, S.; Luukkonen, E.; Ruostesuo, J.; Virtanen, J.; Koskikallio, J. *Acta Chem. Scand.* **1971**, *25*, 3367; Bordwell, F.G.; Hughes, D.L. *J. Org. Chem.* **1983**, *48*, 2206; *J. Am. Chem. Soc.* **1984**, *106*, 3234.

[332]Parker, A.J. *J. Chem. Soc.* **1961**, 1328 has a list of ~20 such reactions.

[333]Weaver, W.M.; Hutchison, J.D. *J. Am. Chem. Soc.* **1964**, *86*, 261; See also, Fuchs, R.; Mahendran, K. *J. Org. Chem.* **1971**, *36*, 730; Müller, P.; Siegfried, B. *Helv. Chim. Acta* **1971**, *54*, 2675; Liotta, C.; Grisdale, E.E.; Hopkins, Jr., H.P. *Tetrahedron Lett.* **1975**, 4205; Bordwell, F.G.; Hughes, D.L. *J. Org. Chem.* **1981**, *46*, 3570. For a contrary result in liquid SO_2, see Lichtin, N.N.; Puar, M.S.; Wasserman, B. *J. Am. Chem. Soc.* **1967**, *89*, 6677.

[334]Winstein, S.; Savedoff, L.G.; Smith, S.G.; Stevens, I.D.R.; Gall, J.S. *Tetrahedron Lett.* **1960**, no. 9, 24.

[335]Gordon, J.E.; Varughese, P. *Chem. Commun.* **1971**, 1160. See also, Ford, W.T.; Hauri, R.J.; Smith, S.G. *J. Am. Chem. Soc.* **1974**, *96*, 4316.

[336]Olmstead, W.N.; Brauman, J.I. *J. Am. Chem. Soc.* **1977**, *99*, 4219. See also, Tanaka, K.; Mackay, G.I.; Payzant, J.D.; Bohme, D.K. *Can. J. Chem.* **1976**, *54*, 1643.

providing further evidence that solvation[337] is responsible for the effect in solution.

However, solvation is not the entire answer since, even for *uncharged* nucleophiles, nucleophilicity increases going down a column in the periodic table. These nucleophiles are not so greatly solvated and changes in solvent do not greatly affect their nucleophilicity.[338] To explain these cases we may use the principle of hard and soft acids and bases (p. 375).[339] The proton is a hard acid, but an alkyl substrate (which may be considered to act as a Lewis acid toward the nucleophile considered as a base) is a good deal softer. According to the principle given on p. 380, we may then expect the alkyl group to prefer softer nucleophiles than the proton does. Thus the larger, more polarizable (softer) nucleophiles have a greater (relative) attraction toward an alkyl carbon than toward a proton.

4. The freer the nucleophile, the greater the rate.[340] We have already seen one instance of this.[334] Another is that the rate of attack by $(EtOOC)_2CBu^- Na^+$ in benzene was increased by the addition of substances (e.g., 1,2-dimethoxyethane, adipamide) that specifically solvated the Na^+ and thus left the anion freer.[341] In a nonpolar solvent, such as benzene, salts, such as $(EtOOC)_2CBu^- Na^+$, usually exist as ion-pair aggregations of large molecular weights.[342] Similarly, it was shown that the half-life of the reaction between $C_6H_5COCHEt^-$ and ethyl bromide depended on the positive ion: K^+, 4.5×10^{-3}; Na^+, 3.9×10^{-5}; Li^+, 3.1×10^{-7}.[343] Presumably, the potassium ion leaves the negative ion most free to attack most rapidly. Further evidence is that in the gas phase,[344] where nucleophilic ions are completely free, without solvent or counterion, reactions take place orders of magnitude faster than the same reactions in solution.[345] It has proven possible to measure the rates of reaction of ^-OH with methyl bromide in the gas phase, with ^-OH either unsolvated or solvated with one, two, or three molecules of water.[346] The rates were, with the number of water molecules

[337]See Kormos, B.L.; Cramer, C.J. *J. Org. Chem.* **2003**, *68*, 6375.

[338]Parker, A.J. *J. Chem. Soc.* **1961**, 4398.

[339]Pearson, R.G. *Surv. Prog. Chem.* **1969**, *5*, 1, pp. 21–38.

[340]For a review of the effect of nucleophile association on nucleophilicity, see Guibe, F.; Bram, G. *Bull. Soc. Chim. Fr.* **1975**, 933.

[341]Zaugg, H.E.; Leonard, J.E. *J. Org. Chem.* **1972**, *37*, 2253. See also, Solov'yanov, A.A.; Ahmed, E.A.A.; Beletskaya, I.P.; Reutov, O.A. *J. Org. Chem. USSR* **1987**, *23*, 1243; Jackman, L.M.; Lange, B.C. *J. Am. Chem. Soc.* **1981**, *103*, 4494.

[342]See, for example Williard, P.G.; Carpenter, G.B. *J. Am. Chem. Soc.* **1986**, *108*, 462.

[343]Zook, H.D.; Gumby, W.L. *J. Am. Chem. Soc.* **1960**, *82*, 1386. See also, Cacciapaglia, R.; Mandolini, L. *J. Org. Chem.* **1988**, *53*, 2579.

[344]For some other measurements of rates of S_N2 reactions in the gas phase, see Barlow, S.E.; Van Doren, J.M.; Bierbaum, V.M. *J. Am. Chem. Soc.* **1988**, *110*, 7240; Merkel, A.; Havlas, Z.; Zahradník, R. *J. Am. Chem. Soc.* **1988**, *110*, 8355.

[345]Olmstead, W.N.; Brauman, J.I. *J. Am. Chem. Soc.* **1977**, *99*, 4219.

[346]Bohme, D.K.; Raksit, A.B. *J. Am. Chem. Soc.* **1984**, *106*, 3447. See also, Hierl, P.M.; Ahrens, A.F.; Henchman, M.; Viggiano, A.A.; Paulson, J.F.; Clary, D.C. *J. Am. Chem. Soc.* **1986**, *108*, 3142.

in parentheses: (0) 1.0×10^{-9}; (1) 6.3×10^{-10}; (2) 2×10^{-12}; (3) 2×10^{-13} cm^3 molecule^{-1} s^{-1}. This provides graphic evidence that solvation of the nucleophile decreases the rate. The rate of this reaction in aqueous solution is 2.3×10^{-25} cm^3 molecule^{-1} s^{-1}. Similar results were found for other nucleophiles and other solvents.[347] In solution too, studies have been made of the effect of solvation of the nucleophile by a specific number of water molecules. When the salt $(n\text{-}C_6H_{13})_4N^+$ F$^-$ was allowed to react with n-octyl methanesulfonate, the relative rate fell from 822 for no water molecules to 96 for 1.5 water molecules to 1 for 6 water molecules.[348]

In Chapter 3, we saw that cryptands specifically solvate the alkali metal portion of salts like KF, KOAc, and so on. Synthetic advantage can be taken of this fact to allow anions to be freer, thus increasing the rates of nucleophilic substitutions and other reactions (see p. 509).

However, the four rules given above do not always hold. One reason is that steric influences often play a part. For example, the *tert*-butoxide ion Me_3CO^- is a stronger base than $^-$OH or $^-$OEt, but a much poorer nucleophile because its large bulk hinders it from closely approaching a substrate.

The following overall nucleophilicity order for S_N2 mechanisms (in protic solvents) was given by Edwards and Pearson:[349] $RS^- > ArS^- > I^- > CN^- > ^-OH > N_3^- > Br^- > ArO^- > Cl^- >$ pyridine $> AcO^- > H_2O$. A quantitative relationship[350] (the *Swain–Scott equation*) has been worked out similar to the linear free-energy equations considered in Chapter 9:[351]

$$\log \frac{k}{k_0} = sn$$

where n is the nucleophilicity of a given group, s is the sensitivity of a substrate to nucleophilic attack, and k_0 is the rate for H_2O, which is taken as the standard and for which n is assigned a value of zero. The parameter s is defined as 1.0 for methyl bromide. Table 10.8 contains values of n for some common nucleophiles.[352] The order is similar to that of Edwards and Pearson. The Swain–Scott equation can be derived from Marcus theory.[353]

[347]Bohme, D.K.; Raksit, A.B. *Can. J. Chem.* **1985**, *63*, 3007.
[348]Landini, D.; Maia, A.; Rampoldi, A. *J. Org. Chem.* **1989**, *54*, 328.
[349]Edwards, J.O.; Pearson, R.G. *J. Am. Chem. Soc.* **1962**, *84*, 16.
[350]Swain, C.G.; Scott, C.B. *J. Am. Chem. Soc.* **1953**, *75*, 141.
[351]This is not the only equation that has been devised in an attempt to correlate nucleophilic reactivity. For reviews of attempts to express nucleophilic power quantitatively, see Ritchie, C.D. *Pure Appl. Chem.* **1978**, *50*, 1281; Duboc, C., in Chapman, N.B.; Shorter, J. *Correlation Analysis in Chemistry: Recent Advances*, Plenum, NY, **1978**, pp. 313–355; Ibne-Rasa, K.M. *J. Chem. Educ.* **1967**, *44*, 89. See also, Hoz, S.; Speizman, D. *J. Org. Chem.* **1983**, *48*, 2904; Kawazoe, Y.; Ninomiya, S.; Kohda, K.; Kimoto, H. *Tetrahedron Lett.* **1986**, *27*, 2897; Kevill, D.N.; Fujimoto, E.K. *J. Chem. Res. (S)* **1988**, 408.
[352]From Wells, P.R. *Chem. Rev.* **1963**, *63*, 171, p. 212. See also, Koskikallio, J. *Acta Chem. Scand.* **1969**, *23*, 1477, 1490.
[353]Albery, W.J.; Kreevoy, M.M. *Adv. Phys. Org. Chem.* **1978**, *16*, 87, pp. 113–115.

TABLE 10.8. Nucleophilicities of Some Common Reagents[352]

Nucleophile	n	Nucleophile	n
^-SH	5.1	Br^-	3.5
^-CN	5.1	PhO^-	3.5
I^-	5.0	AcO^-	2.7
$PhNH_2$	4.5	Cl^-	2.7
^-OH	4.2	F^-	2.0
N_3^-	4.0	NO_3^-	1.0
Pyridine	3.6	H_2O	0.0

It is now evident that an absolute order of either nucleophilicity[354] or leaving-group ability, even in the gas phase where solvation is not a factor, does not exist, because they have an effect on each other. When the nucleophile and leaving group are both hard or both soft, the reaction rates are relatively high, but when one is hard and the other soft, rates are reduced.[344] Although this effect is smaller than the effects in paragraphs one and four above, it still prevents an absolute scale of either nucleophilicity or leaving-group ability.[355] There has been controversy as to whether the selectivity of a reaction should increase with decreasing reactivity of a series of nucleophiles, or whether the opposite holds. There is evidence for both views.[356]

For substitution at a carbonyl carbon, the nucleophilicity order is not the same as it is at a saturated carbon, but follows the basicity order more closely. The reason is presumably that the carbonyl carbon, with its partial positive charge, resembles a proton more than does the carbon at a saturated center. That is, a carbonyl carbon is a much harder acid than a saturated carbon. The following nucleophilicity order for these substrates has been determined:[357] $Me_2C=NO^- > EtO^- > MeO^- > {}^-OH > OAr^- > N_3^- > F^- > H_2O > Br^- \sim I^-$. Soft bases are ineffective at a carbonyl carbon.[358] In a reaction carried out in the gas phase with alkoxide nucleophiles OR^- solvated by only one molecule of an alcohol $R'OH$, it was found that both RO^- and $R'O^-$ attacked the formate substrate ($HCOOR''$) about equally, although in the unsolvated case, the more basic alkoxide is the better nucleophile.[359] In this study, the product ion R^2O^- was also solvated by one molecule of ROH or R'OH.

[354]However, for a general model of intrinsic nucleophilicity in the gas phase, see Pellerite, M.J.; Brauman, J.I. *J. Am. Chem. Soc.* **1983**, *105*, 2672.

[355]For reference scales for the characterization of cationic electrophiles and neutral nucleophiles see Mayr, H.; Bug, T.; Gotta, M.F.; Hering, N.; Irrgang, B.; Janker, B.; Kempf, B.; Loos, R.; Ofial, A.R.; Remennikov, G.; Schimmel, H. *J. Am. Chem. Soc.* **2001**, *123*, 9500.

[356]For discussions, see Dietze, P.; Jencks, W.P. *J. Am. Chem. Soc.* **1989**, *111*, 5880.

[357]Hudson, R.F.; Green, M. *J. Chem. Soc.* **1962**, 1055; Bender, M.L.; Glasson, W.A. *J. Am. Chem. Soc.* **1959**, *81*, 1590; Jencks, W.P.; Gilchrist, M. *J. Am. Chem. Soc.* **1968**, *90*, 2622.

[358]For theoretical treatments of nucleophilicity at a carbonyl carbon, see Buncel, E.; Shaik, S.S.; Um, I.; Wolfe, S. *J. Am. Chem. Soc.* **1988**, *110*, 1275, and references cited therein.

[359]Baer, S.; Stoutland, P.O.; Brauman, J.I. *J. Am. Chem. Soc.* **1989**, *111*, 4097.

If an atom containing one or more unshared pairs is adjacent to the attacking atom on the nucleophile, the nucleophilicity is enhanced.[360] Examples of such nucleophiles are HO_2^-, $Me_2C=NO^-$, NH_2NH_2, and so on. This is called the *alpha effect* (α-effect),[361] and a broader definition is a positive deviation exhibited by an α-nucleophile from a Brønsted type nucleophilicity plot, [362] where the reference (or normal) nucleophile is one that possesses the same basicity as the α-nucleophile, but does not deviate from the Brønsted-type plot. Several reviews of the α-effect have been published previously,[362,363]

Several possible explanations have been offered.[364] One is that the ground state of the nucleophile is destabilized by repulsion between the adjacent pairs of electrons;[365] another is that the transition state is stabilized by the extra pair of electrons;[366] a third is that the adjacent electron pair reduces solvation of the nucleophile.[367] Evidence supporting the third explanation is that there was no alpha effect in the reaction of HO_2^- with methyl formate in the gas phase,[368] although HO_2^- shows a strong alpha effect in solution. The α-effect has been demonstrated to be remarkably dependent on the nature of the solvent. [369] The α-effect is substantial for substitution at a carbonyl or other unsaturated carbon, at some inorganic atoms,[370] and for reactions of a nucleophile with a carbocation,[371] but is generally smaller or absent entirely for substitution at a saturated carbon.[372]

---

[360]Definition in the *Glossary of Terms used in Physical Organic Chemistry, Pure & Appl. Chem. **1979**, 51,* 1731.

[361]For reviews, see Grekov, A.P.; Veselov, V.Ya. *Russ. Chem. Rev. **1978**, 47,* 631; Fina, N.J.; Edwards, J.O. *Int. J. Chem. Kinet. **1973**, 5,* 1.

[362]Hoz, S.; Buncel, E. *Israel J. Chem. **1985**, 26,* 313.

[363]Grekov, A.P.; Veselov, V.Ya. *Russ. Chem. Rev. **1978**, 47,* 631; Fina, N.J.; Edwards, J.O. *Int. J. Chem. Kinet. **1973**, 5,* 1; Jencks, W.P. *Catalysis in Chemistry and Enzymology,* McGraw-Hill, New York, ***1969***; pp. 107–111.

[364]For discussions, see Wolfe, S.; Mitchell, D.J.; Schlegel, H.B.; Minot, C.; Eisenstein, O. *Tetrahedron Lett. **1982**, 23,* 615; Ho, S.; Buncel, E. *Isr. J. Chem. **1985**, 26,* 313.

[365]Buncel, E.; Hoz, S. *Tetrahedron Lett. **1983**, 24,* 4777. For evidence that this is not the sole cause, see Oae, S.; Kadoma, Y. *Can. J. Chem. **1986**, 64,* 1184.

[366]See Hoz, S. *J. Org. Chem. **1982**, 47,* 3545; Laloi-Diard, M.; Verchere, J.; Gosselin, P.; Terrier, F. *Tetrahedron Lett. **1984**, 25,* 1267.

[367]For other explanations, see Hudson, R.F.; Hansell, D.P.; Wolfe, S.; Mitchell, D.J. *J. Chem. Soc., Chem. Commun. **1985**,* 1406; Shustov, G.V. *Doklad. Chem. **1985**, 280,* 80. For a discussion, see Herschlag, D.; Jencks, W.P. *J. Am. Chem. Soc. **1990**, 112,* 1951.

[368]DePuy, C.H.; Della, E.W.; Filley, J.; Grabowski, J.J.; Bierbaum, V.M. *J. Am. Chem. Soc. **1983**, 105,* 2481; Buncel, E.; Um, I. *J. Chem. Soc., Chem. Commun. **1986**,* 595; Terrier, F.; Degorre, F.; Kiffer, D.; Laloi, M. *Bull. Soc. Chim. Fr. **1988**,* 415. For some evidence against this explanation, see Moss, R.A.; Swarup, S.; Ganguli, S. *J. Chem. Soc., Chem. Commun. **1987**,* 860.

[369]Buncel, E.; Um, I.-H. *Tetrahedron* **2004**, 60, 7801.

[370]For example, see Kice, J.L.; Legan, E. *J. Am. Chem. Soc. **1973**, 95,* 3912.

[371]Dixon, J.E.; Bruice, T.C. *J. Am. Chem. Soc. **1971**, 93,* 3248, 6592.

[372]Gregory, M.J.; Bruice, T.C. *J. Am. Chem. Soc. **1967**, 89,* 4400; Oae, S.; Kadoma, Y.; Yano, Y. *Bull. Chem. Soc. Jpn. **1969**, 42,* 1110; McIsaac, Jr., J.E.; Subbaraman, L.R.; Subbaraman, J.; Mulhausen, H.A.; Behrman, E.J. *J. Org. Chem. **1972**, 37,* 1037. See, however, Beale, J.H. *J. Org. Chem. **1972**, 37,* 3871; Buncel, E.; Wilson, H.; Chuaqui, C. *J. Am. Chem. Soc. **1982**, 104,* 4896; *Int. J. Chem. Kinet. **1982**, 14,* 823.

The Effect of the Leaving Group

1. *At a Saturated Carbon.* The leaving group comes off more easily the more stable it is as a free entity. This is usually inverse to its basicity, and the best leaving groups are the weakest bases. Thus iodide is the best leaving group among the halides and fluoride the poorest. Since XH is always a weaker base than X^-, nucleophilic substitution is always easier at a substrate RXH^+ than at RX. An example of this effect is that OH and OR are not leaving groups from ordinary alcohols and ethers, but can come off when the groups are protonated, that is, converted to ROH_2^+ or $RORH^+$.[373] Reactions in which the leaving group does not come off until it has been protonated have been called S_N1cA or S_N2cA, depending on whether after protonation the reaction is an S_N1 or S_N2 process (these designations are often shortened to A1 and A2). The cA stands for conjugate acid, since the substitution takes place on the conjugate acid of the substrate. The IUPAC designations for these mechanisms are, respectively, $A_h + D_N + A_N$ and $A_h + A_N D_N$; that is, the same designations as S_N1 and S_N2, with A_h to show the preliminary step. When another electrophile assumes the role of the proton, the symbol A_e is used instead. The ions ROH_2^+ and $RORH^+$ can be observed as stable entities at low temperatures in super acid solutions.[374] At higher temperatures they cleave to give carbocations.

It is obvious that the best nucleophiles (e.g., NH_2^-, ^-OH) cannot take part in S_N1cA or S_N2cA processes, because they would be converted to their conjugate acids under the acidic conditions necessary to protonate the leaving groups.[375] Because S_N1 reactions do not require powerful nucleophiles, but do require good leaving groups, most of them take place under acidic conditions. In contrast, S_N2 reactions, which do require powerful nucleophiles (which are generally strong bases), most often take place under basic or neutral conditions.

Another circumstance that increases leaving-group power is ring strain. Ordinary ethers do not cleave at all and protonated ethers only under

[373]For a review of ORH^+ as a leaving group, see Staude, E.; Patat, F., in Patai, S. *The Chemistry of the Ether Linkage*, Wiley, NY, *1967*, pp. 22–46.

[374]Olah, G.A.; O'Brien, D.H. *J. Am. Chem. Soc. 1967*, *89*, 1725; Olah, G.A.; Sommer, J.; Namanworth, E. *J. Am. Chem. Soc. 1967*, *89*, 3576; Olah, J.A.; Olah, G.A., in Olah, G.A.; Schleyer, P.v.R. *Carbonium Ions*, Vol. 2, Wiley, NY, *1970*, pp. 743–747.

[375]Even in the gas phase, NH_3 takes a proton from $CH_3OH_2^+$ rather than acting as a nucleophile: Okada, S.; Abe, Y.; Taniguchi, S.; Yamabe, S. *J. Chem. Soc., Chem. Commun. 1989*, 610.

strenuous conditions, but epoxides[376] (**93**) are cleaved quite easily and protonated epoxides (**94**) even more easily. Aziridines (**95**)[377] and episulfides (**96**) are also easily cleaved (see p. 518).[378]

Although halides are common leaving groups in nucleophilic substitution for synthetic purposes, it is often more convenient to use alcohols. Since OH does not leave from ordinary alcohols, it must be converted to a group that does leave. One way is protonation, mentioned above. Another is conversion to a reactive ester, most commonly a sulfonic ester. The sulfonic ester groups *tosylate*, *brosylate*, *nosylate*, and *mesylate* are better leaving groups

$R-OSO_2-\langle\bigcirc\rangle-CH_3$ $R-OSO_2-\langle\bigcirc\rangle-Br$

ROTs   ROBs

p-Toluenesulfonates   *p*-Bromobenzenesulfonates
Tosylates   Brosylates

$R-OSO_2-\langle\bigcirc\rangle-NO_2$ $R-OSO_2CH_3$

RONs   ROMs

p-Nitrobenzenesulfonates   Methanesulfonates
Nosylates   Mesylates

than halides and are frequently used.[379] Other leaving groups are still better, and compounds containing these groups make powerful alkylating agents. Among them are oxonium ions (ROR_2^+),[380] and the fluorinated compounds

$R-OSO_2CF_3$ $R-OSO_2C_4F_9$ $R-OSO_2CCH_2F_3$

ROTf   Nonafluorobutanesulfonates   2,2,2-Trifluoroethanesulfonates
Trifluoromethanesulfonates   Nonaflates   Tresylates
Triflates

---

[376]For a review of the reactions of epoxides, see Smith, J.G. *Synthesis* **1984**, 629. For a review of their synthesis and reactions, see Bartók, M.; Láng, K.L., in Patai, S. *The Chemistry of Functional Groups, Supplement E*, Wiley, NY, **1980**, pp. 609–681.

[377]See Kametani, T.; Honda, T. *Adv. Heterocycl. Chem.* **1986**, *39*, 181; Hu, X.E. *Tetrahedron* **2004**, *60*, 2701.

[378]There is evidence that relief of ring strain is not the only factor responsible for the high rates of ring opening of three-membered rings: Di Vona, M.L.; Illuminati, G.; Lillocci, C. *J. Chem. Soc. Perkin Trans. 2* **1985**, 1943; Bury, A.; Earl, H.A.; Stirling, C.J.M. *J. Chem. Soc., Chem. Commun.* **1985**, 393.

[379]Bentley, T.W.; Christl, M.; Kemmer, R.; Llewellyn, G.; Oakley, J.E. *J. Chem. Soc. Perkin Trans. 2* **1994**, 2531.

[380]For a monograph, see Perst, H. *Oxonium Ions in Organic Chemistry*; Verlag Chemie: Deerfield Beach, FL, **1971**, pp. 100–127. For reviews, see Perst, H., in Olah, G.A.; Schleyer, P.v.R. *Carbonium Ions*, Vol. 5, Wiley, NY, **1976**, pp. 1961–2047; Granik, V.G.; Pyatin, B.M.; Glushkov, R.G. *Russ. Chem. Rev.* **1971**, *40*, 747. For a discussion of their use, see Curphey, T.J. *Org. Synth. VI*, 1021.

triflates[381] and *nonaflates.*[381] *Tresylates* are ~400 times less reactive than triflates, but still ~100 times more reactive than tosylates.[382] Halonium ions ($RClR^+$, $RBrR^+$, RIR^+), which can be prepared in super acid solutions (p. 236) and isolated as solid SbF_6^- salts, are also extremely reactive in nucleophilic substitution.[383] Of the above types of compound, the most important in organic synthesis are tosylates, mesylates, oxonium ions, and triflates. The others have been used mostly for mechanistic purposes.

The leaving group ability of NH_2, NHR, and NR_2 are extremely poor,[384] but the leaving-group ability of NH_2 can be greatly improved by converting a primary amine RNH_2 to the ditosylate $RNTs_2$. The NTs_2 group has been successfully replaced by a number of nucleophiles.[385] Another way of converting NH_2 into a good leaving group has been extensively developed by Katritzky and co-workers.[386] In this method the amine is converted to a

pyridinium compound (**98**) by treatment with a pyrylium salt (frequently a 2,4,6-triphenylpyrylium salt, **97**).[387] When the salt is heated, the counterion acts as a nucleophile. In some cases, a non-nucleophilic ion, such as BF_4^-, is used as the counterion for the conversion **97 → 98**, and then Y^- is added to **98**. Among the nucleophiles that have been used successfully in this reaction are I^-, Br^-, Cl^-, F^-, ^-OAc, N_3^-, NHR_2, and H^-. Ordinary NR_2 groups are good leaving groups when the substrate is a Mannich base (these are compounds of the form $RCOCH_2CH_2NR_2$; see reaction **16-19**).[388] The elimination–addition mechanism applies in this case.

[381]For reviews of triflates, nonaflates, and other fluorinated ester leaving groups, see Stang, P.J.; Hanack, M.; Subramanian, L.R. *Synthesis* **1982**, 85; Howells, R.D.; McCown, J.D. *Chem. Rev.* **1977**, *77*, 69, pp. 85–87.

[382]Crossland, R.K.; Wells, W.E.; Shiner, Jr., V.J. *J. Am. Chem. Soc.* **1971**, *93*, 4217.

[383]Peterson, P.E.; Clifford, P.R.; Slama, F.J. *J. Am. Chem. Soc.* **1970**, *92*, 2840; Peterson, P.E.; Waller, F.J. *J. Am. Chem. Soc.* **1972**, *94*, 5024; Olah, G.A.; Mo, Y.K. *J. Am. Chem. Soc.* **1974**, *96*, 3560.

[384]For a review of the deamination of amines, see Baumgarten, R.J.; Curtis, V.A., in Patai, S. *The Chemistry of Functional Groups, Supplement F*, pt. 2, Wiley, NY, **1982**, pp. 929–997.

[385]For references, see Müller, P.; Thi, M.P.N. *Helv. Chim. Acta* **1980**, *63*, 2168; Curtis, V.A.; Knutson, F.J.; Baumgarten, R.J. *Tetrahedron Lett.* **1981**, *22*, 199.

[386]For reviews, see Katritzky, A.R.; Marson, C.M. *Angew. Chem. Int. Ed.* **1984**, *23*, 420; Katritzky, A.R. *Tetrahedron* **1980**, *36*, 679. For reviews of the use of such leaving groups to study mechanistic questions, see Katritzky, A.R.; Sakizadeh, K.; Musumarra, G. *Heterocycles* **1985**, *23*, 1765; Katritzky, A.R.; Musumarra, G. *Chem. Soc. Rev.* **1984**, *13*, 47.

[387]For discussions of the mechanism, see Katritzky, A.R.; Brycki, B. *J. Am. Chem. Soc.* **1986**, *108*, 7295, and other papers in this series.

[388]For a review of Mannich bases, see Tramontini, M. *Synthesis* **1973**, 703.

Probably the best leaving group is N_2 from the species RN_2^+, which can be generated in several ways,[389] of which the two most important are the treatment of primary amines with nitrous acid (see p. $$$ for this reaction)

$$RNH_2 + HONO \longrightarrow RN_2{}^+$$

and the protonation of diazo compounds[390]

$$R_2C{=}\overset{\oplus}{N}{=}\overset{\ominus}{N} + H^+ \longrightarrow R_2CHN_2{}^+$$

No matter how produced, RN_2^+ are usually too unstable to be isolable,[391] reacting presumably by the S_N1 or S_N2 mechanism.[392] Actually, the exact mechanisms are in doubt because the rate laws, stereochemistry, and products have proved difficult to interpret.[393] If there are free carbocations they should give the same ratio of substitution to elimination to rearrangements, and so on, as carbocations generated in other S_N1 reactions, but they often do not. "Hot" carbocations (unsolvated and/or chemically activated) that can hold their configuration have been postulated,[394] as have ion pairs, in which ^-OH (or ^-OAc, and so on, depending on how the diazonium ion is generated) is the counterion.[395] One class of aliphatic diazonium salts of which several

[389]For reviews, see Kirmse, W. *Angew. Chem. Int. Ed.* **1976**, *15*, 251; Collins, C.J. *Acc. Chem. Res.* **1971**, *4*, 315; Moss, R.A. *Chem. Eng. News* **1971**, *49*, 28 (No. 48, Nov. 22).

[390]For a treatise, see Regitz, M.; Maas, G. *Diazo Compounds*, Academic Press, NY, **1986**. For reviews of the reactions of aliphatic diazo compounds with acids, see Hegarty, A.F., in Patai, S. *The Chemistry of Diazonium and Diazo Groups*, pt. 2, Wiley, NY, 1978, pp. 511–591, 571–575; More O'Ferrall, R.A. *Adv. Phys. Org. Chem.* **1967**, *5*, 331. For review of the structures of these compounds, see Studzinskii, O.P.; Korobitsyna, I.K. *Russ. Chem. Rev.* **1970**, *39*, 834.

[391]Aromatic diazonium salts can, of course, be isolated (see Chapter 13), but only a few aliphatic diazonium salts have been prepared (see also, Weiss, R.; Wagner, K.; Priesner, C.; Macheleid, J. *J. Am. Chem. Soc.* **1985**, *107*, 4491). For reviews see Laali, K.; Olah, G.A. *Rev. Chem. Intermed.* **1985**, *6*, 237; Bott, K., in Patai, S.; Rappoport, Z. *The Chemistry of Functional Groups, Supplement C*, pt. 1, Wiley, NY, **1983**, pp. 671–697; Bott, K. *Angew. Chem. Int. Ed.* **1979**, *18*, 259. The simplest aliphatic diazonium ion $CH_3N_2^+$ has been prepared at $-120°C$ in superacid solution, where it lived long enough for an nmr spectrum to be taken: Berner, D.; McGarrity, J.F. *J. Am. Chem. Soc.* **1979**, *101*, 3135.

[392]For an example of a diazonium ion reacting by an S_N2 mechanism, see Mohrig, J.R.; Keegstra, K.; Maverick, A.; Roberts, R.; Wells, S. *J. Chem. Soc., Chem. Commun.* **1974**, 780.

[393]For reviews of the mechanism, see Manuilov, A.V.; Barkhash, V.A. *Russ. Chem. Rev.* **1990**, *59*, 179; Saunders, Jr., W.H.; Cockerill, A.F. *Mechanisms of Elimination Reactions*, Wiley, NY, **1973**, pp. 280–317; in Olah, G.A.; Schleyer, P.v.R. *Carbonium Ions*, Vol. 2, Wiley, NY, **1970**, the articles by Keating, J.T.; Skell, P.S. pp. 573–653; and by Friedman, L. pp. 655–713; White, E.H.; Woodcock, D.J., in Patai, S. *The Chemistry of the Amino Group*, Wiley, NY, **1968**, pp. 440–483; Ref. 389.

[394]Semenow, D.; Shih, C.; Young, W.G. *J. Am. Chem. Soc.* **1958**, *80*, 5472. For a review of "hot" or "free" carbocations, see Olah, G.A.; Schleyer, P.v.R. *Carbonium Ions*, Vol. 2, Wiley, NY, **1970**, the articles by Keating, J.T.; Skell, P.S. pp. 573–653.

[395]Collins, C.J. *Acc. Chem. Res.* **1971**, *4*, 315; Collins, C.J.; Benjamin, B.M. *J. Org. Chem.* **1972**, *37*, 4358; White, E.H.; Field, K.W. *J. Am. Chem. Soc.* **1975**, *97*, 2148; Cohen, T.; Daniewski, A.R.; Solash, J. *J. Org. Chem.* **1980**, *45*, 2847; Maskill, H.; Thompson, J.T.; Wilson, A.A. *J. Chem. Soc. Perkin Trans. 2* **1984**, 1693; Connor, J.K.; Maskill, H. *Bull. Soc. Chim. Fr.* **1988**, 342.

members have been isolated as stable salts are the cyclopropeniumyldiazonium salts:[396]

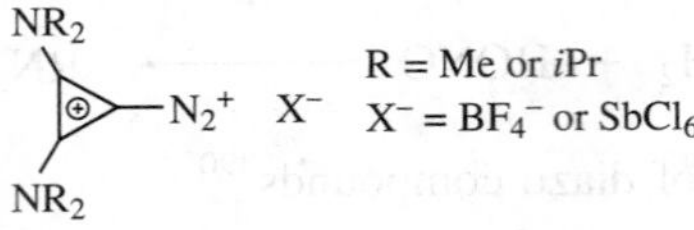

$$
\begin{array}{c}
NR_2 \\
\triangleright\!=\!\!\oplus\!\!-N_2^+ \quad X^- \\
NR_2
\end{array}
\qquad
\begin{array}{l}
R = Me \text{ or } iPr \\
X^- = BF_4^- \text{ or } SbCl_6^-
\end{array}
$$

Diazonium ions generated from ordinary aliphatic primary amines are usually useless for preparative purposes, since they lead to a mixture of products giving not only substitution by any nucleophile present, but also elimination and rearrangements if the substrate permits. For example, diazotization of n-butylamine gave 25% 1-butanol, 5.2% 1-chlorobutane, 13.2% 2-butanol, 36.5% butenes (consisting of 71% 1-butene, 20% *trans*-2-butene, and 9% *cis*-2-butene), and traces of butyl nitrites.[397]

In the S_N1cA and S_N2cA mechanisms (p. 496) there is a preliminary step, the addition of a proton, before the normal S_N1 or S_N2 process occurs. There are also reactions in which the substrate *loses* a proton in a preliminary step. In these reactions, there is a carbene intermediate.

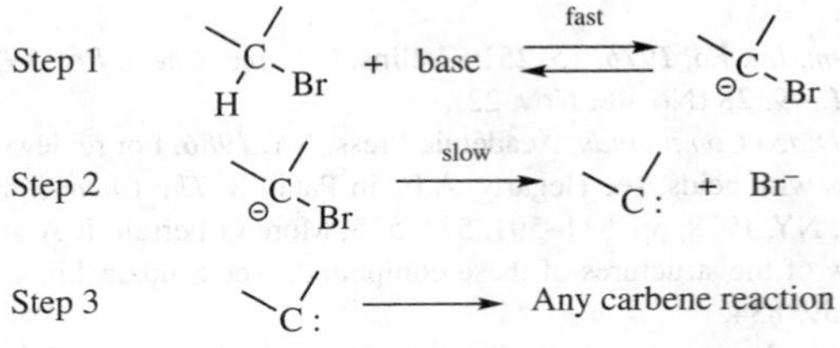

Step 1 $\quad \overset{\diagup}{\underset{H}{C}}\!-Br \;+\; base \;\underset{}{\overset{fast}{\rightleftharpoons}}\; \overset{\diagup}{\underset{\ominus}{C}}\!-Br$

Step 2 $\quad \overset{\diagup}{\underset{\ominus}{C}}\!-Br \;\overset{slow}{\longrightarrow}\; \overset{\diagup}{C}\!: \;+\; Br^-$

Step 3 $\quad \overset{\diagup}{C}\!: \;\longrightarrow\; \text{Any carbene reaction}$

Once formed by this process, the carbene may undergo any of the normal carbene reactions (see p. 287). When the net result is substitution, this mechanism has been called the S_N1cB (for conjugate base) mechanism.[398] Although the slow step is an S_N1 step, the reaction is second order; first order in substrate and first order in base.

Table 10.9 lists some leaving groups in approximate order of ability to leave. The order of leaving-group ability is about the same for S_N1 and S_N2 reactions.

2. At a Carbonyl Carbon. This reaction is discussed in Chapter 16.

[396]Weiss, R.; Wagner, K.; Priesner, C.; Macheleid, J. *J. Am. Chem. Soc.* **1985**, *107*, 4491.

[397]Whitmore, F.C.; Langlois, D.P. *J. Am. Chem. Soc.* **1932**, *54*, 3441; Streitwieser, Jr., A.; Schaeffer, W.D. *J. Am. Chem. Soc.* **1957**, *79*, 2888.

[398]Pearson, R.G.; Edgington, D.N. *J. Am. Chem. Soc.* **1962**, *84*, 4607.

TABLE 10.9. Leaving Groups Listed in Approximate Order of Decreasing Ability to Leave[a]

	Common Leaving Groups	
Substrate RX	At Saturated Carbon	At Carbonyl Carbon
RN_2^+	x	
$ROR_2'^+$		
$ROSO_2C_4F_9$		
$ROSO_2CF_3$	x	
$ROSO_2F$		
ROTs, etc.[b]	x	
RI	x	
RBr	x	
ROH_2^+	x (conjugate acid of alcohol)	
RCl	x	x (acyl halides)
$RORH^+$	x (conjugate acid of ether)	
$RONO_2$, etc.[b]		
$RSR_2'^{+400}$		
$RNR_3'^+$	x	
RF		
ROCOR$'^{401}$	x	x (anhydrides)
RNH_3^+		
ROAr^{402}		x (aryl esters)

(continued)

The Effect of the Reaction Medium[399]

The effect of solvent polarity[403] on the rate of S_N1 reactions depends on whether the substrate is neutral or positively charged.[404] For neutral substrates, which constitute the majority of cases, the more polar the solvent, the faster the reaction, since there is a greater charge in the transition state than in the starting compound (Table 10.10[405]) and the energy of an ionic transition state is reduced by polar solvents.

[399]For a monograph, see Reichardt, C. *Solvents and Solvent Effects in Organic Chemistry*, 2nd ed., VCH, NY, *1988*. For reviews, see Klumpp, G.W. *Reactivity in Organic Chemistry*, Wiley, NY, *1982*, pp. 186–203; Bentley, T.W.; Schleyer, P.v.R. *Adv. Phys. Org. Chem. 1977*, *14*, 1.

[400]For a review of the reactions of sulfonium salts, see Knipe, A.C., in Stirling, C.J.M. *The Chemistry of the Sulphonium Group*, pt. 1, Wiley, NY, *1981*, pp. 313–385. See also, Badet, B.; Julia, M.; Lefebvre, C. *Bull. Soc. Chim. Fr. 1984*, II-431.

[401]For a review of S_N2 reactions of carboxylic esters, where the leaving group is OCOR', see McMurry, J.E. *Org. React. 1976*, *24*, 187.

[402]Nitro substitution increases the leaving-group ability of ArO groups, and alkyl picrates [2,4,6-$ROC_6H_2(NO_2)_3$] react at rates comparable to tosylates: Sinnott, M.L.; Whiting, M.C. *J. Chem. Soc. B 1971*, 965. See also, Page, I.D.; Pritt, J.R.; Whiting, M.C. *J. Chem. Soc. Perkin Trans. 2 1972*, 906.

[403]Mu, L.; Drago, R.S.; Richardson, D.E. *J. Chem. Soc. Perkin Trans. 2, 1998*, 159; Fujio, M.; Saeki, Y.; Nakamoto, K.; Kim, S.H.; Rappoport, Z.; Tsuno, Y. *Bull. Chem. Soc. Jpn. 1996*, *69*, 751.

[404]Mitsuhashi, T.; Hirota, H.; Yamamoto, G. *Bull. Chem. Soc. Jpn. 1994*, *67*, 824; Bentley, T.W.; Llewellyn, G.; Ryu, Z.H. *J. Org. Chem. 1998*, *63*, 4654.

[405]This analysis is due to Ingold, C.K. *Structure and Mechanism in Organic Chemistry*, 2nd ed., Cornell University Press, Ithaca, NY, *1969*, pp. 457–463.

TABLE 10.9. (*Continued*)

	Common Leaving Groups	
Substrate RX	At Saturated Carbon	At Carbonyl Carbon
ROH		x (carboxylic acids)
ROR		x (alkyl esters)
RH		
RNH$_2$		x (amides)
RAr		
RR		

[a]Groups that are common leaving groups at saturated and carbonyl carbons are indicated.
[b]The substrates ROTs, and so on, includes esters of sulfuric and sulfonic acids in general, for example, ROSO$_2$OH, ROSO$_2$OR, ROSO$_2$R. The substrate RONO$_2$, and so on, includes inorganic ester leaving groups, such as ROPO(OH)$_2$ and ROB(OH)$_2$.

TABLE 10.10. Transition States for S$_N$1 Reactions of Charged and Uncharged Substrates, and for S$_N$2 Reactions of the Four Charge Types[405]

Reactants and Transition States		Charge in the Transition State Relative to Starting Materials	How an Increase in Solvent Polarity Affects the Rate
S$_N$2	Type I RX + Y$^-$ → Y$^{\delta -}$$\bullet\bullet$R$\bullet\bullet\bullet$X$^{\delta -}$	Dispersed	Small decrease
	Type II RX + Y$^-$ → Y$^{\delta +}$$\bullet\bullet\bullet$R$\bullet\bullet\bullet$X$^{\delta -}$	Increased	Large increase
	Type III RX + Y$^-$ → Y$^{\delta -}$$\bullet\bullet\bullet$R$\bullet\bullet\bullet$X$^{\delta +}$	Decreased	Large decrease
	Type IV RX + Y$^-$ → Y$^{\delta +}$$\bullet\bullet\bullet$R$\bullet\bullet\bullet$X$^{\delta +}$	Dispersed	Small decrease
S$_N$1	RX → R$^{\delta +}$$\bullet\bullet\bullet$X$^{\delta -}$	Increased	Large increase
	RX$^-$ → R$^{\delta -}$$\bullet\bullet\bullet$X$^{\delta -}$	Dispersed	Small decrease

However, when the substrate is positively charged, the charge is more spread out in the transition state than in the starting ion, and a greater solvent polarity slows the reaction. Even for solvents with about the same polarity, there is a difference between protic and aprotic solvents.[406] The S$_N$1 reactions of un-ionized substrates are more rapid in protic solvents, which can form hydrogen bonds with the leaving group. Examples of protic solvents are water,[407] alcohols, and carboxylic acids, while some polar aprotic solvents are DMF, dimethyl sulfoxide (DMSO),[408] acetonitrile, acetone, sulfur dioxide, and

---

[406]See, for example, Ponomareva, E.A.; Dvorko, G.F.; Kulik, N.I.; Evtushenko, N.Yu. *Doklad. Chem.* **1983**, *272*, 291.

[407]For a study of nucleophilic reactivities in water, see Bug, T.; Mayr, H. *J. Am. Chem. Soc.* **2003**, *125*, 12980. For a correlation of the Hammett equation and micellar effects see Brinchi, L.; DiProfio, P.; Germani, R.; Savelli, G.; Spreti, N.; Bunton, L.A. *Eur. J. Org. Chem.* **2000**, 3849.

[408]For reviews of reactions in dimethyl sulfoxide, see Buncel, E.; Wilson, H. *Adv. Phys. Org. Chem.* **1977**, *14*, 133; Martin, D.; Weise, A.; Niclas, H. *Angew. Chem. Int. Ed.* **1967**, *6*, 318.

hexamethylphosphoramide [(Me$_2$N)$_3$PO], HMPA.[409] An algorithm has been developed to accurately calculate dielectric screening effects in solvents.[410] S$_N$2 reactions have been done in ionic liquids (see p. 415),[411] and in supercritical carbon dioxide (see p. 414).[412]

For S$_N$2 reactions, the effect of the solvent[413] depends on which of the four charge types the reaction belongs to (p. 425). In types I and IV, an initial charge is dispersed in the transition state, so the reaction is hindered by polar solvents. In type III, initial charges are *decreased* in the transition state, so that the reaction is even more hindered by polar solvents. Only type II, where the reactants are uncharged but the transition state has built up a charge, is aided by polar solvents. These effects are summarized in Table 10.10.[405] Westaway has proposed a "solvation rule" for S$_N$2 reactions, which states that changing the solvent will not change the structure of the transition state for type I reactions, but will change it for type II reactions.[414] For S$_N$2 reactions also, the difference between protic and aprotic solvents must be considered.[415] For reactions of types I and III the transition state is more solvated in polar aprotic solvents than in protic ones,[416] while (as we saw on p. 490) the original charged nucleophile is less solvated in aprotic solvents[417] (the second factor is generally much greater than the first[418]). So the change from, say, methanol to DMSO should greatly increase the rate. As an example, the relative rates at 25°C for the reaction between MeI and Cl$^-$ were[332] in MeOH, 1; in HCONH$_2$ (still protic although a weaker acid), 12.5; in HCONHMe, 45.3; and HCONMe$_2$, 1.2×10^6. The change in rate in going from a protic to an aprotic solvent is also related to the *size* of the attacking anion. Small ions are solvated best in protic solvents, since hydrogen bonding is most important for them, while large anions are solvated best in aprotic solvents (protic solvents have highly developed structures held together by hydrogen bonds; aprotic solvents have much looser structures, and it is easier for a large anion to be fitted in). So the rate of attack by small anions is most greatly increased by the change from a protic to an aprotic solvent. This may have preparative significance. The review articles in Ref. 400 have lists of several dozen reactions of charge types I and III in which

[409]For reviews of HMPA, see Normant, H. *Russ. Chem. Rev.* **1970**, *39*, 457; *Bull. Soc. Chim. Fr.* **1968**, 791; *Angew. Chem. Int. Ed.* **1967**, *6*, 1046.

[410]Klamt, A.; Schüürmann, G. *J. Chem. Soc. Perkin Trans. 2* **1993**, 799.

[411]Wheeler, C.; West, K.N.; Liotta, C.L.; Eckert, C.A. *Chem. Commun.* **2001**, 887; Kim, D.W.; Song, C.E.; Chi, D.Y. *J. Org. Chem.* **2003**, *68*, 4281; Chiappe, C.; Pieraccini, D.; Saullo, P. *J. Org. Chem.* **2003**, *68*, 6710.

[412]DeSimone, J.; Selva, M.; Tundo, P. *J. Org. Chem.* **2001**, *66*, 4047.

[413]For microsolvation of S$_N$2 transition states see Craig, S.L.; Brauman, J.I. *J. Am. Chem. Soc.* **1999**, *121*, 6690.

[414]Westaway, K.C. *Can. J. Chem.* **1978**, *56*, 2691; Westaway, K.C.; Lai, Z. *Can. J. Chem.* **1989**, *67*, 345.

[415]For reviews of the effects of protic and aprotic solvents, see Parker, A.J. *Chem. Rev.* **1969**, *69*, 1; *Adv. Phys. Org. Chem.* **1967**, *5*, 173; *Adv. Org. Chem.* **1965**, *5*, 1; Madaule-Aubry, F. *Bull. Soc. Chim. Fr.* **1966**, 1456.

[416]However, even in aprotic solvents, the transition state is less solvated than the charged nucleophile: Magnera, T.F.; Caldwell, G.; Sunner, J.; Ikuta, S.; Kebarle, P. *J. Am. Chem. Soc.* **1984**, *106*, 6140.

[417]See, for example, Fuchs, R.; Cole, L.L. *J. Am. Chem. Soc.* **1973**, *95*, 3194.

[418]See, however, Haberfield, P.; Clayman, L.; Cooper, J.S. *J. Am. Chem. Soc.* **1969**, *91*, 787.

TABLE 10.11. Relative Rates of Ionization of *p*-Methoxyneophyl Toluenesulfonate in Various Solvents[419]

Solvent	Relative Rate	Solvent	Relative Rate
HCOOH	153	Ac$_2$O	0.020
H$_2$O	39	Pyridine	0.013
80% EtOH–H$_2$O	1.85	Acetone	0.0051
AcOH	1.00	EtOAc	6.7×10^{-4}
MeOH	0.947	THF	5.0×10^{-4}
EtOH	0.370	Et$_2$O	3×10^{-5}
Me$_2$SO	0.108	CHCl$_3$	
Octanoic acid	0.043	Benzene	Lower still
MeCN	0.036	Alkanes	
HCONMe$_2$	0.029		

yields are improved and reaction times reduced in polar aprotic solvents. Reaction types II and IV are much less susceptible to the difference between protic and aprotic solvents.

Since for most reactions S$_N$1 rates go up and S$_N$2 rates go down in solvents of increasing polarity, it is quite possible for the same reaction to go by the S$_N$1 mechanism in one solvent and the S$_N$2 in another. Table 10.11 is a list of solvents in order of ionizing power;[419] a solvent high on the list is a good solvent for S$_N$1 reactions. Trifluoroacetic acid, which was not studied by Smith, Fainberg, and Winstein, has greater ionizing power than any solvent listed in Table 10.11.[420] Because it also has very low nucleophilicity, it is an excellent solvent for S$_N$1 solvolyses. Other good solvents for this purpose are 1,1,1-trifluoroethanol CF$_3$CH$_2$OH, and 1,1,1,3,3,3-hexafluoro-2-propanol, (F$_3$C)$_2$CHOH.[421]

We have seen how the polarity of the solvent influences the rates of S$_N$1 and S$_N$2 reactions. The ionic strength of the medium has similar effects. In general, the addition of an external salt affects the rates of S$_N$1 and S$_N$2 reactions in the same way as an increase in solvent polarity, although this is not quantitative; different salts have different effects.[422] However, there are exceptions: although the rates of S$_N$1 reactions are usually increased by the addition of salts (this is called the *salt effect*), addition of the leaving-group ion often decreases the rate (the common-ion effect, p. 434). There is also the special salt effect of LiClO$_4$, mentioned on p. 439. In addition to these effects, S$_N$1 rates are also greatly accelerated when there are ions present that specifically help in pulling off the leaving group.[423] Especially

[419]Smith, S.G.; Fainberg, A.H.; Winstein, S. *J. Am. Chem. Soc.* **1961**, *83*, 618.

[420]Capon, B.; McManus, S. *Neighboring Group Participation*, Vol. 1; Plenum, NY, **1976**; Haywood-Farmer, J. *Chem. Rev.* **1974**, *74*, 315; Streitwieser, Jr., A.; Dafforn, G.A. *Tetrahedron Lett.* **1969**, 1263.

[421]Schadt, F.L.; Schleyer, P.v.R.; Bentley, T.W. *Tetrahedron Lett.* **1974**, 2335.

[422]See, for example, Duynstee, E.F.J.; Grunwald, E.; Kaplan, M.L. *J. Am. Chem. Soc.* **1960**, *82*, 5654; Bunton, C.A.; Robinson, L. *J. Am. Chem. Soc.* **1968**, *90*, 5965.

[423]For a review, see Kevill, D.N., in Patai, S.; Rappoport, Z. *The Chemistry of Functional Groups, Supplement D*, pt. 2, Wiley, NY, **1983**, pp. 933–984.

important are Ag^+, Hg^{2+}, and Hg_2^{2+}, but H^+ helps to pull off F (hydrogen bonding).[424] Even primary halides have been reported to undergo S_N1 reactions when assisted by metal ions.[425] This does not mean, however, that reactions in the presence of metallic ions invariably proceed by the S_N1 mechanism. It has been shown that alkyl halides can react with $AgNO_2$ and $AgNO_3$ by the S_N1 or S_N2 mechanism, depending on the reaction conditions.[426]

The effect of solvent has been treated quantitatively (for S_N1 mechanisms, in which the solvent pulls off the leaving group) by a linear free-energy relationship[427]

$$\log \frac{k}{k_0} = m\,Y$$

where m is characteristic of the substrate (defined as 1.00 for t-BuCl) and is usually near unity, Y is characteristic of the solvent and measures its "ionizing power," and k_0 is the rate in a standard solvent, 80% aqueous ethanol at 25°C. This is known as the Grunwald–Winstein equation, and its utility is at best limited. The Y values can of course be measured for solvent *mixtures* too, and this is one of the principal advantages of the treatment, since it is not easy otherwise to assign a polarity arbitrarily to a given mixture of solvents.[428] The treatment is most satisfactory for different proportions of a given solvent pair. For wider comparisons, the treatment is not so good quantitatively, although the Y values do give a reasonably good idea of solvolyzing power.[429] Table 10.12 contains a list of some Y values.[430]

Ideally, Y should measure only the ionizing power of the solvent, and should not reflect any backside attack by a solvent molecule in helping the nucleofuge

[424]For a review of assistance by metallic ions, see Rudakov, E.S.; Kozhevnikov, I.V.; Zamashchikov, V.V. *Russ. Chem. Rev.* **1974**, *43*, 305. For an example of assistance in removal of F by H^+, see Coverdale, A.K.; Kohnstam, G. *J. Chem. Soc.* **1960**, 3906.

[425]Zamashchikov, V.V.; Rudakov, E.S.; Bezbozhnaya, T.V.; Matveev, A.A. *J. Org. Chem. USSR* **1984**, *20*, 424. See, however, Kevill, D.N.; Fujimoto, E.K. *J. Chem. Soc., Chem. Commun.* **1983**, 1149.

[426]Kornblum, N.; Jones, W.J.; Hardies, D.E. *J. Am. Chem. Soc.* **1966**, *88*, 1704; Kornblum, N.; Hardies, D.E. *J. Am. Chem. Soc.* **1966**, *88*, 1707.

[427]Grunwald, E.; Winstein, S. *J. Am. Chem. Soc.* **1948**, *70*, 846.

[428]For reviews of polarity scales of solvent mixtures, see Reichardt, C. *Solvents and Solvent Effects in Organic Chemistry*, 2nd ed., VCH, NY, **1988**, pp. 339–405; Langhals, H. *Angew. Chem. Int. Ed.* **1982**, *21*, 724.

[429]For a criticism of the Y scale, see Abraham, M.H.; Doherty, R.M.; Kamlet, M.J.; Harris, J.M.; Taft, R.W. *J. Chem. Soc. Perkin Trans. 2* **1987**, 1097.

[430]Y values are from Fainberg, A.H.; Winstein, S. *J. Am. Chem. Soc.* **1956**, *78*, 2770, except for the value for CF_3CH_2OH, which is from Shiner, Jr., V.J.; Dowd, W.; Fisher, R.D.; Hartshorn, S.R.; Kessick, M.A.; Milakofsky, L.; Rapp, M.W. *J. Am. Chem. Soc.* **1969**, *91*, 4838. Y_{OTs} values are from Bentley, T.W.; Llewellyn, G. *Prog. Phys. Org. Chem.* **1990**, *17*, 143–144. Z values are from Kosower, E.M.; Wu, G.; Sorensen, T.S. *J. Am. Chem. Soc.* **1961**, *83*, 3147. See also, Larsen, J.W.; Edwards, A.G.; Dobi, P. *J. Am. Chem. Soc.* **1980**, *102*, 6780. $E_T(30)$ values are from Reichardt, C.; Dimroth, K. *Fortschr. Chem. Forsch.* **1969**, *11*, 1; Reichardt, C. *Angew. Chem. Int. Ed.* **1979**, *18*, 98; Laurence, C.; Nicolet, P.; Reichardt, C. *Bull. Soc. Chim. Fr.* **1987**, 125; Laurence, C.; Nicolet, P.; Lucon, M.; Reichardt, C. *Bull. Soc. Chim. Fr.* **1987**, 1001; Reichardt, C.; Eschner, M.; Schäfer, G. *Liebigs Ann. Chem.* **1990**, 57. Values for many additional solvents are given, in the last five papers. Many values from all of these scales are given, in Reichardt, C. *Solvents and Solvent Effects in Organic Chemistry*, 2nd ed.; VCH, NY, **1988**.

TABLE 10.12. The Y, Y_{OTs}, Z, and E_T (30) Values for Some Solvents[430]

Solvent	Y	Y_{OTs}	Z	E_T (30)
CF_3COOH		4.57		
H_2O	3.5	4.1	94.6	63.1
$(CF_3)_2CHOH$		3.82		65.3
HCOOH	2.1	3.04		
H_2O–EtOH (1:1)	1.7	1.29	90	55.6
CF_3CH_2OH	1.0	1.77		59.8
$HCONH_2$	0.6		83.3	56.6
80% EtOH	0.0	0.0	84.8	53.7
MeOH	−1.1	−0.92	83.6	55.4
AcOH	−1.6	−0.9	79.2	51.7
EtOH	−2.0	−1.96	79.6	51.9
90% dioxane	−2.0	−2.41	76.7	46.7
iPrOH	−2.7	−2.83	76.3	48.4
95% acetone	−2.8	−2.95	72.9	48.3
t-BuOH	−3.3	−3.74	71.3	43.9
MeCN		−3.21	71.3	45.6
Me_2SO			71.1	45.1
$HCONMe_2$		−4.14	68.5	43.8
Acetone			65.7	42.2
HMPA				40.9
CH_2Cl_2				40.7
Pyridine			64.0	40.5
$CHCl_3$			63.2	39.1
PhCl				37.5
THF				37.4
Dioxane				36.0
Et_2O				34.5
C_6H_6			54	34.3
PhMe				33.9
CCl_4				32.4
n-Octane				31.1
n-Hexane				31.0
Cyclohexane				30.9

to leave (nucleophilic assistance; k_s, p. 456). Actually, there is evidence that many solvents do lend some nucleophilic assistance,[431] even with tertiary substrates.[432] It was proposed that a better measure of solvent "ionizing power" would be a relationship based on 2-adamantyl substrates, rather than t-BuCl, since the structure of this system completely prevents backside nucleophilic assistance (p. 480). Such a

[431] A scale of solvent nucleophilicity (as opposed to ionizing power), called the N_T scale, has been developed: Kevill, D.N.; Anderson, S.W. *J. Org. Chem.* **1991**, *56*, 1845.

[432] For discussions, with references, see Kevill, D.N.; Anderson, S.W. *J. Am. Chem. Soc.* **1986**, *108*, 1579; McManus, S.P.; Neamati-Mazreah, N.; Karaman, R.; Harris, J.M. *J. Org. Chem.* **1986**, *51*, 4876; Abraham, M.H.; Doherty, R.M.; Kamlet, M.J.; Harris, J.M.; Taft, R.W. *J. Chem. Soc. Perkin Trans. 2* **1987**, 913.

scale, called Y_{OTs}, was developed, with m defined as 1.00 for 2-adamantyl tosylate.[433] Some values of Y_{OTs} are given in Table 10.12. These values, which are actually based on both 1- and 2-adamantyl tosylates (both are equally impervious to nucleophilic assistance and show almost identical responses to solvent ionizing power[434]) are called Y_{OTs} because they apply only to tosylates. It has been found that solvent "ionizing power" depends on the leaving group, so separate scales[435] have been set up for OTf,[436] Cl,[402] Br,[437] I,[438] and other nucleofuges,[439] all based on the corresponding adamantyl compounds. A new Y scale has been established based on benzylic bromides.[440] In part, this was done because benzylic tosylates did not give a linear correlation with the 2-adamantyl Y_{OTs} parameter.[441] This is substrate dependent, since solvolysis of 2,2,-dimethyl-1-phenyl-1-propanol tosylate showed no nucleophilic solvent participation.[442]

In order to include a wider range of solvents than those in which any of the Y values can be conveniently measured, other attempts have been made at correlating solvent polarities.[443] Kosower found that the position of the charge-transfer peak (see p. 115) in the UV spectrum of the complex (**99**) between iodide ion and

R = Me or Et

99   **100**

[433]Schadt, F.L.; Bentley, T.W.; Schleyer, P.v.R. *J. Am. Chem. Soc.* **1976**, *98*, 7667.

[434]Bentley, T.W.; Carter, G.E. *J. Org. Chem.* **1983**, *48*, 579.

[435]For a review of these scales, see Bentley, T.W.; Llewellyn, G. *Prog. Phys. Org. Chem.* **1990**, *17*, 121.

[436]Kevill, D.N.; Anderson, S.W. *J. Org. Chem.* **1985**, *50*, 3330. See also, Creary, X.; McDonald, S.R. *J. Org. Chem.* **1985**, *50*, 474.

[437]Bentley, T.W.; Carter, G.E. *J. Am. Chem. Soc.* **1982**, *104*, 5741. See also, Liu, K.; Sheu, H. *J. Org. Chem.* **1991**, *56*, 3021.

[438]Bentley, T.W.; Carter, G.E.; Roberts, K. *J. Org. Chem.* **1984**, *49*, 5183.

[439]See Bentley, T.W.; Roberts, K. *J. Org. Chem.* **1985**, *50*, 4821; Takeuchi, K.; Ikai, K.; Shibata, T.; Tsugeno, A. *J. Org. Chem.* **1988**, *53*, 2852; Kevill, D.N.; Hawkinson, D.C. *J. Org. Chem.* **1990**, *55*, 5394 and references cited therein.

[440]Fujio, M.; Saeki, Y.; Nakamoto, K.; Yatsugi, K.-i.; Goto, N.; Kim, S.H.; Tsuji, Y.; Rappoport, Z.; Tsuno, Y. *Bull. Chem. Soc. Jpn.* **1995**, *68*, 2603; Liu, K.-T.; Chin, C.-P.; Lin, Y.-S.; Tsao, M.-L. *J. Chem. Res. (S)* **1997**, 18.

[441]Fujio, M.; Susuki, T.; Goto, M.; Tsuji, Y.; Yatsugi, K.; Saeki, Y.; Kim, S.H.; Tsuno, Y. *Bull. Chem. Soc. Jpn.* **1994**, *67*, 2233.

[442]Tsuji, Y.; Fujio, M.; Tsuno, Y. *Tetrahedron Lett.* **1992**, *33*, 349.

[443]For reviews of solvent polarity scales, see Abraham, M.H.; Grellier, P.L.; Abboud, J.M.; Doherty, R.M.; Taft, R.W. *Can. J. Chem.* **1988**, *66*, 2673; Kamlet, M.J.; Abboud, J.M.; Taft, R.W. *Prog. Phys. Org. Chem.* **1981**, *13*, 485; Shorter, J. *Correlation Analysis of Organic Reactivity*, Wiley, NY, **1982**, pp. 127–172; Reichardt, C.; Dimroth, K. *Fortschr. Chem. Forsch.* **1969**, *11*, 1; Reichardt, C. *Angew. Chem. Int. Ed.* **1979**, *18*, 98; Abraham, M.H. *Prog. Phys. Org. Chem.* **1974**, *11*, 1; Koppel, I.A.; Palm, V.A., in Chapman, N.B.; Shorter, J. *Advances in Linear Free Energy Relationships*, Plenum, NY, **1972**, pp. 203–280; Ref. 443. See also, Chastrette, M.; Rajzmann, M.; Chanon, M.; Purcell, K.F. *J. Am. Chem. Soc.* **1985**, *107*, 1.

1-methyl- or 1-ethyl-4-carbomethoxypyridinium ion was dependent on the polarity of the solvent.[444] From these peaks, which are very easy to measure, Kosower calculated transition energies that he called Z values. These values are thus measures of solvent polarity analogous to Y values. Another scale is based on the position of electronic spectra peaks of the pyridinium-N-phenolbetaine (**100**) in various solvents.[445] Solvent polarity values on this scale are called $E_T(30)$[446] values. The $E_T(30)$ values are related to Z values by the expression[447]

$$Z = 1.41\,E_T(30) + 6.92$$

Table 10.12 shows that Z and $E_T(30)$ values are generally in the same order as Y values. Other scales, the π^* scale,[448] the π^*_{azo} scale,[449] and the Py scale,[450] are also based on spectral data.[451]

Carbon dioxide can be liquefied under high pressure (supercritical CO_2). Several reactions have been done using supercritical CO_2 as the medium, but special apparatus is required. This medium offers many advantages,[452] and some disadvantages, but is an interesting new area of research.

The effect of solvent on nucleophilicity has already been discussed (pp. 490–495).

Phase-Transfer Catalysis

A difficulty that occasionally arises when carrying out nucleophilic substitution reactions is that the reactants do not mix. For a reaction to take place the reacting molecules must collide. In nucleophilic substitutions the substrate is usually insoluble in water and other polar solvents, while the nucleophile is often an anion, which is soluble in water but not in the substrate or other organic solvents. Consequently, when the two reactants are brought together, their concentrations in the same phase are too low for convenient reaction rates. One way to overcome this

[444]Kosower, E.M.; Wu, G.; Sorensen, T.S. *J. Am. Chem. Soc.* **1961**, *83*, 3147. See also, Larsen, J.W.; Edwards, A.G.; Dobi, P. *J. Am. Chem. Soc.* **1980**, *102*, 6780.

[445]Dimroth, K.; Reichardt, C. *Liebigs Ann. Chem.* **1969**, *727*, 93. See also, Haak, J.R.; Engberts, J.B.F.N. *Recl. Trav. Chim. Pays-Bas* **1986**, *105*, 307.

[446]The symbol E_T comes from *energy, transition*. The (30) is used because the ion **100** bore this number in Dimroth, K.; Reichardt, C. *Liebigs Ann. Chem.* **1969**, *727*, 93. Values based on other ions have also been reported: See, for example, Reichardt, C.; Harbusch-Görnert, E.; Schäfer, G. *Liebigs Ann. Chem.* **1988**, 839.

[447]Reichardt, C.; Dimroth, K. *Fortschr. Chem. Forsch.* **1969**, *11*, p. 32.

[448]Kamlet, M.J.; Abboud, J.M.; Taft, R.W. *J. Am. Chem. Soc.* **1977**, *99*, 6027; Doherty, R.M.; Abraham, M.H.; Harris, J.M.; Taft, R.W.; Kamlet, M.J. *J. Org. Chem.* **1986**, *51*, 4872; Kamlet, M.J.; Doherty, R.M.; Abboud, J.M.; Abraham, M.H.; Taft, R.W. *CHEMTECH* **1986**, 566, and other papers in this series. See also, Doan, P.E.; Drago, R.S. *J. Am. Chem. Soc.* **1982**, *104*, 4524; Kamlet, M.J.; Abboud, J.M.; Taft, R.W. *Prog. Phys. Org. Chem.* **1981**, *13*, 485; Bekárek, V. *J. Chem. Soc. Perkin Trans. 2* **1986**, 1425; Abe, T. *Bull. Chem. Soc. Jpn.* **1990**, *63*, 2328.

[449]Buncel, E.; Rajagopal, S. *J. Org. Chem.* **1989**, *54*, 798.

[450]Dong, D.C.; Winnik, M.A. *Can. J. Chem.* **1984**, *62*, 2560.

[451]For a review of such scales, see Buncel, E.; Rajagopal, S. *Acc. Chem. Res.* **1990**, *23*, 226.

[452]Kaupp, G. *Angew. Chem. Int. Ed.* **1994**, *33*, 1452.

difficulty is to use a solvent that will dissolve both species. As we saw on p. 501, a dipolar aprotic solvent may serve this purpose. Another way, which is used very often, is *phase-transfer catalysis.*[453]

In this method, a catalyst is used to carry the nucleophile from the aqueous into the organic phase. As an example, simply heating and stirring a two-phase mixture of 1-chlorooctane for several days with aqueous NaCN gives essentially no yield of 1-cyanooctane. But if a small amount of an appropriate quaternary ammonium salt is added, the product is quantitatively formed in ~2 h.[454] There are two principal types of phase-transfer catalyst, although the action of the two types is somewhat different, the effects are the same. Both get the anion into the organic phase and allow it to be relatively free to react with the substrate.

1. *Quaternary Ammonium or Phosphonium Salts.* In the above-mentioned case of NaCN, the uncatalyzed reaction does not take place because the $^-$CN ions cannot cross the interface between the two phases, except in very low concentration. The reason is that the Na$^+$ ions are solvated by the water, and this solvation energy would not be present in the organic phase. The CN$^-$ ions cannot cross without the Na$^+$ ions because that would destroy the electrical neutrality of each phase. In contrast to Na$^+$ ions, quaternary ammonium (R_4N^+)[455] and phosphonium (R_4P^+) ions with sufficiently large R groups are poorly solvated in water and prefer organic solvents. If a small amount of such a salt is added, three equilibria are set up:

Organic phase   Q$^\oplus$ $^\ominus$CN + RCl $\xrightarrow{\;4\;}$ RCN + Q$^\oplus$ $^\ominus$Cl

Aqueous phase   Q$^\oplus$ $^\ominus$CN + Na$^\oplus$ Cl$^\ominus$ $\rightleftharpoons^{3}$ Na$^\oplus$ CN$^\ominus$ + Q$^\oplus$ Cl$^\ominus$

Q$^\oplus$ = R$_4$N$^\oplus$ or   R$_4$P$^\oplus$

The Na$^+$ ions remain in the aqueous phase; they cannot cross. The Q$^+$ ions do cross the interface and carry an anion with them. At the beginning of the reaction the chief anion present is $^-$CN. This gets carried into the organic phase (equilibrium 1) where it reacts with RCl to produce RCN and Cl$^-$. The Cl$^-$ then gets carried into the aqueous phase (equilibrium 2). Equilibrium 3, taking place entirely in the aqueous phase, allows Q^{+-}CN to be regenerated.

[453]For monographs, see Dehmlow, E.V.; Dehmlow, S.S. *Phase Transfer Catalysis,* 2nd ed., Verlag Chemie, Deerfield Beach, FL, *1983*; Starks, C.M.; Liotta, C. *Phase Transfer Catalysis,* Academic Press, NY, *1978*; Weber, W.P.; Gokel, G.W. *Phase Transfer Catalysis in Organic Synthesis,* Springer, NY, *1977.* For reviews, see Makosza, M. *Pure Appl. Chem.* **2000**, *72*, 1399; Montanari, F.; Landini, D.; Rolla, F. *Top. Curr. Chem.* **1982**, *101*, 147; Alper, H. *Adv. Organomet. Chem.* **1981**, *19*, 183; Dehmlow, E.V. *Chimia* **1980**, *34*, 12; Makosza, M. *Surv. Prog. Chem.* **1980**, *9*, 1; Sjöberg, K. *Aldrichimica Acta* **1980**, *13*, 55; Brändström, A. *Adv. Phys. Org. Chem.* **1977**, *15*, 267; Dockx, J. *Synthesis* **1973**, 441.
[454]Starks, C.M.; Liotta, C. *Phase Transfer Catalysis,* Academic Press, NY, *1978*, p. 2.
[455]Bis-quaternary ammonium salts have also been used: Lissel, M.; Feldman, D.; Nir, M.; Rabinovitz, M. *Tetrahedron Lett.* **1989**, *30*, 1683.

All the equilibria are normally reached much faster than the actual conversion of RCl to RCN, so the latter is the rate-determining step.

In some cases, the Q^+ ions have such a low solubility in water that virtually all remain in the organic phase.[456] In such cases the exchange of ions (equilibrium 3) takes place across the interface. Still another mechanism (*the interfacial mechanism*) can operate where $^-$OH extracts a proton from an organic substrate.[457] In this mechanism, the $^-$OH ions remain in the aqueous phase and the substrate in the organic phase; the deprotonation takes place at the interface.[458] Thermal stability of the quaternary ammonium salt is a problem, limiting the use of some catalysts. The trialkylacyl ammonium halide **101** is thermally stable, however, even at high reaction temperatures.[459] The use of molten quaternary ammonium salts as ionic reaction media for substitution reactions has also been reported.[460]

$$CH_3(CH_2)_n \overset{\overset{O}{\|}}{} \overset{\oplus}{N}Et_3 \ \overset{\ominus}{Cl} \qquad n = 8\text{--}14$$

101

2. *Crown Ethers and Other Cryptands.*[461] We saw in Chapter 3 that certain cryptands are able to surround certain cations. In effect, a salt-like KCN is converted by dicyclohexano-18-crown-6 into a new salt (**102**) whose anion is the same, but whose cation is now a much larger species with the positive

102   **103**   **104**

charge spread over a large volume and hence much less concentrated. This larger cation is much less solubilized by water than K^+ and much more attracted to organic solvents, although KCN is generally insoluble in organic solvents, the cryptate salt is soluble in many of them. In these cases we do not need an aqueous phase at all but simply add the salt to the organic phase.

[456]Landini, D.; Maia, A.; Montanari, F. *J. Chem. Soc., Chem. Commun.* **1977**, 112; *J. Am. Chem. Soc.* **1978**, *100*, 2796.

[457]For a review, see Rabinovitz, M.; Cohen, Y.; Halpern, M. *Angew. Chem. Int. Ed.* **1986**, *25*, 960.

[458]This mechanism was proposed by Makosza, M. *Pure Appl. Chem.* **1975**, *43*, 439. See also, Dehmlow, E.V.; Thieser, R.; Sasson, Y.; Pross, E. *Tetrahedron* **1985**, *41*, 2927; Mason, D.; Magdassi, S.; Sasson, Y. *J. Org. Chem.* **1990**, *55*, 2714.

[459]Bhalerao, U.T.; Mathur, S.N.; Rao, S.N. *Synth. Commun.* **1992**, *22*, 1645.

[460]Badri, M.; Brunet, J.-J.; Perron, R. *Tetrahedron Lett.* **1992**, *33*, 4435.

[461]For a review of this type of phase-transfer catalysis, see Liotta, C., in Patai, S. *The Chemistry of Functional Groups, Supplement E*, Wiley, NY, **1980**, pp. 157–174.

Suitable cryptands have been used to increase greatly the rates of reactions where F^-, Br^-, I^-, ^-OAc, and ^-CN are nucleophiles.[462] Certain compounds that are not cryptands can act in a similar manner. One example is the podand tris(3,6-dioxaheptyl)amine (**103**), also called TDA-1.[463] Another, not related to the crown ethers, is the pyridyl sulfoxide **104**.[464]

Both of the above-mentioned catalyst types get the anions into the organic phase, but there is another factor as well. There is evidence that sodium and potassium salts of many anions, even if they could be dissolved in organic solvents, would undergo reactions very slowly (dipolar aprotic solvents are exceptions) because in these solvents the anions exist as ion pairs with Na^+ or K^+ and are not free to attack the substrate (p. 492). Fortunately, ion pairing is usually much less with the quaternary ions and with the positive cryptate ions, so the anions in these cases are quite free to attack. Such anions are sometimes referred to as "naked" anions.

Not all quaternary salts and cryptands work equally well in all situations. Some experimentation is often required to find the optimum catalyst.

Although phase-transfer catalysis has been most often used for nucleophilic substitutions, it is not confined to these reactions. Any reaction that needs an insoluble anion dissolved in an organic solvent can be accelerated by an appropriate phase-transfer catalyst. We will see some examples in later chapters. In fact, in principle, the method is not even limited to anions, and a small amount of work has been done in transferring cations,[465] radicals, and molecules.[466] The reverse type of phase-transfer catalysis has also been reported: transport into the aqueous phase of a reactant that is soluble in organic solvents.[467] Microwave activated phase-transfer catalysis has been reported.[468]

The catalysts mentioned above are soluble. Certain cross-linked polystyrene resins, as well as alumina[469] and silica gel, have been used as insoluble phase-transfer catalysts. These, called *triphase catalysts*,[470] have the advantage of

[462]See, for example, Liotta, C.; Harris, H.P.; McDermott, M.; Gonzalez, T.; Smith, K. *Tetrahedron Lett.* *1974*, 2417; Sam, D.J.; Simmons, H.E. *J. Am. Chem. Soc.* *1974*, *96*, 2252; Durst, H.D. *Tetrahedron Lett.* *1974*, 2421.

[463]Soula, G. *J. Org. Chem.* *1985*, *50*, 3717.

[464]Furukawa, N.; Ogawa, S.; Kawai, T.; Oae, S. *J. Chem. Soc. Perkin Trans. 1* *1984*, 1833. See also, Fujihara, H.; Imaoka, K.; Furukawa, N.; Oae, S. *J. Chem. Soc. Perkin Trans. 1* *1986*, 333.

[465]See Armstrong, D.W.; Godat, M. *J. Am. Chem. Soc.* *1979*, *101*, 2489; Iwamoto, H.; Yoshimura, M.; Sonoda, T.; Kobayashi, H. *Bull. Chem. Soc. Jpn.* *1983*, *56*, 796.

[466]See, for example, Dehmlow, E.V.; Slopianka, M. *Chem. Ber.* *1979*, *112*, 2765.

[467]Mathias, L.J.; Vaidya, R.A. *J. Am. Chem. Soc.* *1986*, *108*, 1093; Fife, W.K.; Xin, Y. *J. Am. Chem. Soc.* *1987*, *109*, 1278.

[468]Deshayes, S.; Liagre, M.; Loupy, A.; Luche, J.-L.; Petit, A. *Tetrahedron* *1999*, *55*, 10851.

[469]Quici, S.; Regen, S.L. *J. Org. Chem.* *1979*, *44*, 3436.

[470]For reviews, see Regen, S.L. *Nouv. J. Chim.* *1982*, *6*, 629; *Angew. Chem. Int. Ed.* *1979*, *18*, 421. See also, Molinari, H.; Montanari, F.; Quici, S.; Tundo, P. *J. Am. Chem. Soc.* *1979*, *101*, 3920; Bogatskii, A.V.; Luk'yanenko, N.G.; Pastushok, V.N.; Parfenova, M.N. *Doklad. Chem.* *1985*, *283*, 210; Pugia, M.J.; Czech, B.P.; Czech, B.P.; Bartsch, R.A. *J. Org. Chem.* *1986*, *51*, 2945.

simplified product work-up and easy and quantitative catalyst recovery, since the catalyst can easily be separated from the product by filtration.

Influencing Reactivity by External Means

In many cases, reactions are slow. This is sometimes due to poor mixing or the aggregation state of one or more reactants. A powerful technique used to increase reaction rates is *ultrasound* (see p. 349). In this technique, the reaction mixture is subjected to high-energy sound waves, most often 20 KHz, but sometimes higher (a frequency of 20 KHz is about the upper limit of human hearing). When these waves are passed through a mixture, small bubbles form (*cavitation*). Collapse of these bubbles produces powerful shock waves that greatly increase the temperatures and pressures within these tiny regions, resulting in an increased reaction rate.[471] In the common instance where a metal, as a reactant or catalyst, is in contact with a liquid phase, a further effect is that the surface of the metal is cleaned and/or eroded by the ultrasound, allowing the liquid-phase molecules to come into closer contact with the metal atoms. Among the advantages of ultrasound is that it may increase yields, reduce side reactions, and permit the use of lower temperatures and/or pressures. The reaction of pyrrolidinone **105** with allyl bromide, under phase-transfer conditions, gave <10% of the *N*-allyl product, **106**. When the reaction was done under identical conditions, but with exposure to ultrasound (in an ultrasonic bath), the yield of **106** was 78%.[472] It has been postulated that ultrasound has its best results with reactions that proceed, at least partially, through free-radical intermediates.[473]

105   **106**

As noted in Chapter 7 (see p. 352), microwave irradiation is used extensively. Reaction times are greatly accelerated in many reactions, and reactions that took hours to be complete in refluxing solvents are done in minutes. Benzyl alcohol was converted to benzyl bromide, for example, using microwave irradiation (650 W) in only 9 min on a doped K10 Montmorillonite clay.[474] This is a growing and very useful technique.

The rate of many reactions can be increased by application of high pressure.[475] In solution, the rate of a reaction can be expressed in terms of the activation

[471]Reaction rates can also be increased by running reactions in a microwave oven. For reviews, see Mingos, D.M.P.; Baghurst, D.R. *Chem. Soc. Rev.* **1991**, *20*, 1; Giguere, R.J. *Org. Synth. Theory Appl.* **1989**, *1*, 103.

[472]Keusenkothen, P.F.; Smith, M.B. *Tetrahedron Lett.* **1989**, *30*, 3369.

[473]See Einhorn, C.; Einhorn, J.; Dickens, M.J.; Luche, J. *Tetrahedron Lett.* **1990**, *31*, 4129.

[474]Kad, G.-L.; Singh, V.; Kuar, K.P.; Singh, J. *Tetrahedron Lett.* **1997**, *38*, 1079.

[475]Matsumoto, K.; Morris, A.R. *Organic Synthesis at High Pressure*, Wiley, NY, **1991**; Matsumoto, K.; Sera, A.; Uchida, T. *Synthesis* **1985**, 1; Matsumoto, K.; Sera, A. *Ibid.*, **1985**, 999.

volume, $\Delta V^{\ddagger}$.[476]

$$\frac{\delta \ln k}{\delta p} = \frac{\Delta V^{\ddagger}}{RT}$$

The value of $\Delta V^{\ddagger}$ is the difference in partial molal volume between the transition state and the initial state, but it can be approximated by the molar volume.[476] Increasing pressure decreases the value of $\Delta V^{\ddagger}$ decreases and it $\Delta V^{\ddagger}$ is negative the reaction rate is accelerated. This equation is not strictly obeyed above 10 kbar. If the transition state of a reaction involves bond formation, concentration of charge, or ionization, a negative volume of activation often results. Cleavage of a bond, dispersal of charge, neutralization of the transition state and diffusion control lead to a positive volume of activation. Reactions for which rate enhancement is expected at high pressure include:[476]

1. Reactions in which the number of molecules decreases when starting materials are converted to products: cycloadditions, such as the Diels–Alder (**15-60**); condensations, such as the Knoevenagel condensation (**16-38**).
2. Reactions that proceed via cyclic transition states: Claisen (**18-33**) and Cope (**18-32**) rearrangements.
3. Reactions that take place through dipolar transition states: Menschutkin reaction (**10-31**), electrophilic aromatic substitution.
4. Reactions with steric hindrance.

Many high pressure reactions are done neat, but if a solvent is used, the influence of pressure on that solvent is important. The melting point generally increases at elevated pressures, which influences the viscosity of the medium (viscosity of liquids increases approximately two times per kilobar increase in pressure). Controlling the rate of diffusion of reactants in the medium is also important.[477] In most reactions, pressure is applied (5–20 kbar) at room temperature and then the temperature is increased until reaction takes place.

Ambident (Bidentant) Nucleophiles: Regioselectivity

Some nucleophiles have a pair of electrons on each of two or more atoms, or canonical forms can be drawn in which two or more atoms bear an unshared pair. In these cases, the nucleophile may attack in two or more different ways to give different products. Such reagents are called *ambident nucleophiles*.[478] In most cases, a nucleophile with two potentially attacking atoms can attack with either of them,

---

[476]le Noble, W.J. *Progr. Phys. Org. Chem.* **1967**, *5*, 207; Isaacs, N.S. *Liquid Phase High Pressure Chemistry*, Wiley, Chichester, **1981**; Asano, T.; le Noble, W.J. *Chem. Rev.* **1978**, *78*, 407.

[477]Firestone, R.A.; Vitale, M.A. *J. Org. Chem.* **1981**, *46*, 2160.

[478]For a monograph, see Reutov, O.A.; Beletskaya, I.P.; Kurts, A.L. *Ambident Anions*, Plenum, NY, **1983**. For a review, see Black, T.H. *Org. Prep. Proced. Int.* **1989**, *21*, 179.

depending on conditions, and mixtures are often obtained, although this is not always the case. For example, the nucleophile NCO^- usually gives only isocyanates RNCO and not the isomeric cyanates ROCN.[479] When a reaction can potentially give rise to two or more structural isomers (e.g., ROCN or RNCO), but actually produces only one, the reaction is said to be *regioselective*[480] (cf. the definitions of stereoselective, p. 194 and enantioselective, p. 171). Some important ambient nucleophiles are

1. *Ions of the Type* $\underline{\quad}CO-\overset{\ominus}{C}R-CO\underline{\quad}$. These ions, which are derived by removal of a proton from malonic esters, β-keto esters, β-diketones, and so on, are resonance hybrids:

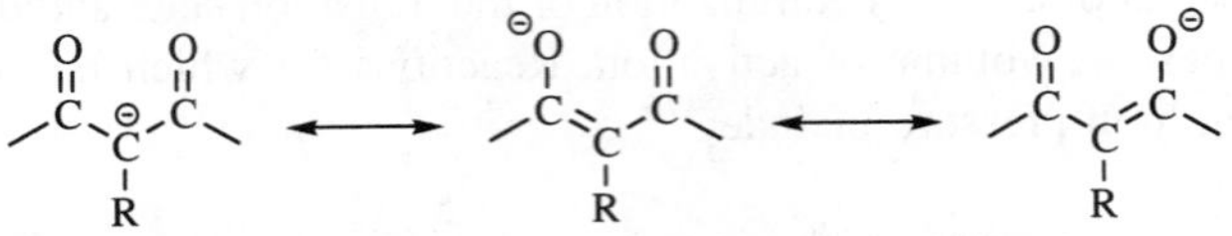

They can thus attack a saturated carbon with their carbon atoms (*C*-alkylation) or with their oxygen atoms (*O*-alkylation):

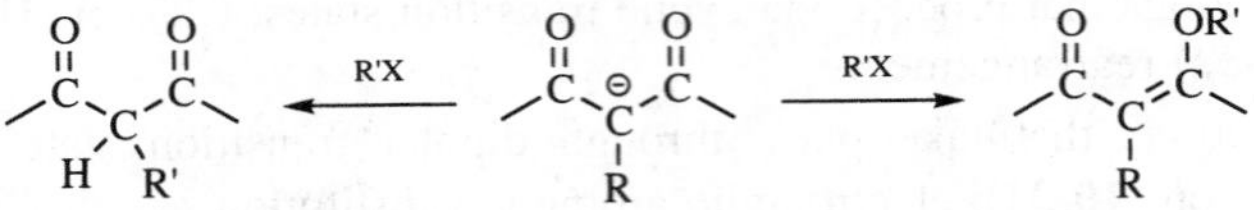

With unsymmetrical ions, three products are possible, since either oxygen can attack. With a carbonyl substrate the ion can analogously undergo *C*-acylation or *O*-acylation.

2. *Compounds of the Type* CH_3CO-CH_2-CO- *Can Give Up Two Protons,* if treated with 2 equivalents of a strong enough base, to give dicarbanions:

$$CH_3-CO-CH_2-CO- \xrightarrow{\text{2 equivalents of base}} \overset{\ominus}{C}H_2-CO-\overset{\ominus}{C}H-CO-$$
$$\textbf{107}$$

Such ions are ambident nucleophiles, since they have two possible attacking carbon atoms, aside from the possibility of attack by oxygen. In such cases, the attack is virtually always by the more basic carbon.[481] Since the hydrogen of a carbon bonded to two carbonyl groups is more acidic than that of a carbon bonded to just one (see Chapter 8), the CH group of **107** is less basic than the CH_2 group, so the latter attacks the substrate. This gives rise to a useful general principle: whenever we desire to remove a proton at a given

---

[479]Both cyanates and isocyanates have been isolated in treatment of secondary alkyl iodides with NCO^-: Holm, A.; Wentrup, C. *Acta Chem. Scand.* **1966**, *20*, 2123.

[480]This term was introduced by Hassner, A. *J. Org. Chem.* **1968**, *33*, 2684.

[481]For an exception, see Trimitsis, G.B.; Hinkley, J.M.; TenBrink, R.; Faburada, A.L.; Anderson, R.; Poli, M.; Christian, B.; Gustafson, G.; Erdman, J.; Rop, D. *J. Org. Chem.* **1983**, *48*, 2957.

position for use as a nucleophile, but there is a stronger acidic group in the molecule, it may be possible to take off both protons; if it is, then attack is always by the desired position since it is the ion of the weaker acid. On the other hand, if it is desired to attack with the more acidic position, all that is necessary is to remove just one proton.[482] For example, ethyl acetoacetate can be alkylated at either the methyl or the methylene group (**10-67**):

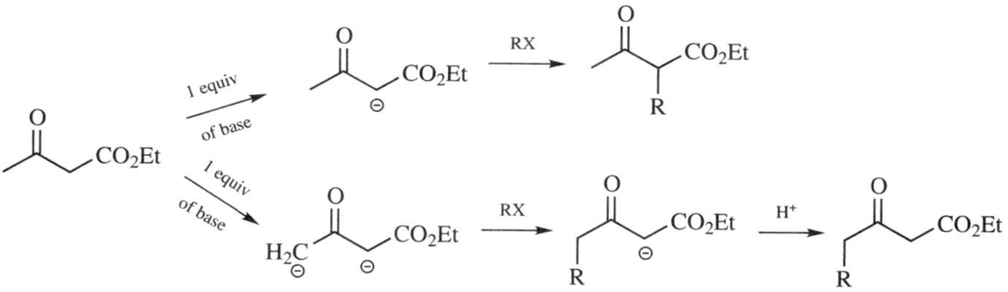

3. *The CN⁻ Ion.* This nucleophile can give nitriles RCN (**10-75**) or isocyanides RN≡C.

4. *The Nitrite Ion.* This ion can give nitrite esters R—O—N=O (**10-22**) or nitro compounds RNO_2 (**10-76**), which are not esters.

5. Phenoxide ions (which are analogous to enolate anions) can undergo *C*-alkylation or *O*-alkylation:

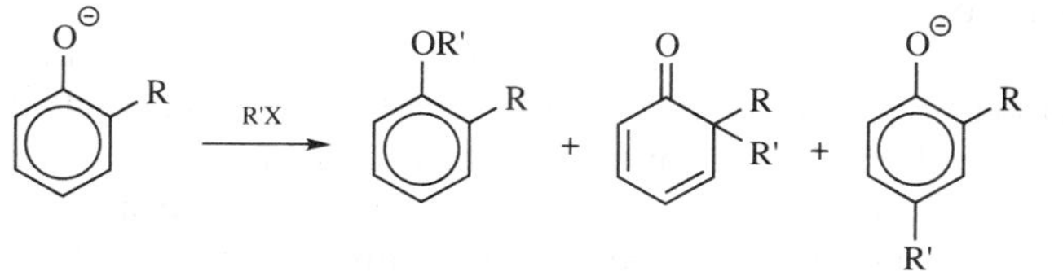

6. Removal of a proton from an aliphatic nitro compound gives a carbanion $(R_2\bar{C}{-}NO_2)$ that can be alkylated at oxygen or carbon.[483] *O*-Alkylation gives nitronic esters, which are generally unstable to heat but break down to give an oxime and an aldehyde or ketone.

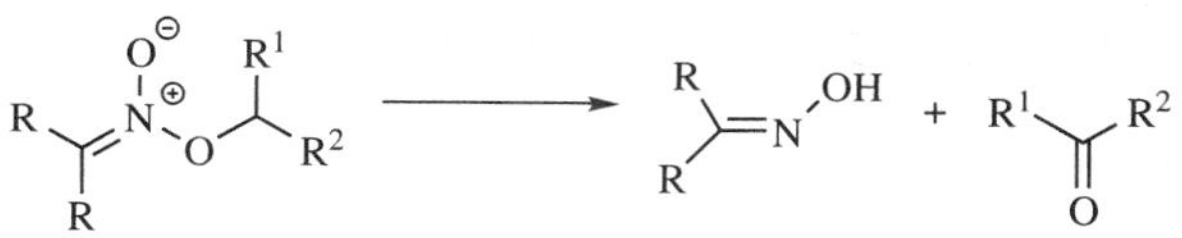

There are many other ambident nucleophiles.

---

[482]The use of this principle was first reported by Hauser, C.R.; Harris, C.M. *J. Am. Chem. Soc.* **1958**, *80*, 6360. It has since been applied many times. For reviews, see Thompson, C.M.; Green, D.L.C. *Tetrahedron* **1991**, *47*, 4223; Kaiser, E.M.; Petty, J.D.; Knutson, P.L.A. *Synthesis* **1977**, 509; Harris, T.M.; Harris, C.M. *Org. React.* **1969**, *17*, 155.

[483]For a review, see Erashko, V.I.; Shevelev, S.A.; Fainzil'berg, A.A. *Russ. Chem. Rev.* **1966**, *35*, 719.

It would be useful to have general rules as to which atom of an ambient nucleophile will attack a given substrate under a given set of conditions.[484] Unfortunately, the situation is complicated by the large number of variables. It might be expected that the more electronegative atom would always attack, but this is often not the case. Where the products are determined by thermodynamic control (p. 307), the principal product is usually the one in which the atom of higher basicity has attacked (i.e., $C > N > O > S$).[485] However, in most reactions, the products are kinetically controlled and matters are much less simple. Nevertheless, the following generalizations can be made, while recognizing that there are many exceptions and unexplained results. As in the discussion of nucleophilicity in general (p. 490), there are two major factors: the polarizability (hard–soft character) of the nucleophile and solvation effects.

1. The principle of hard and soft acids and bases states that hard acids prefer hard bases and soft acids prefer soft bases (p. 375). In an S_N1 mechanism, the nucleophile attacks a carbocation, which is a hard acid. In an S_N2 mechanism, the nucleophile attacks the carbon atom of a molecule, which is a softer acid. The more electronegative atom of an ambient nucleophile is a harder base than the less electronegative atom. We may thus make the statement: As the character of a given reaction changes from S_N1- to S_N2-like, an ambient nucleophile becomes more likely to attack with its less electronegative atom.[486] Therefore, changing from S_N1 to S_N2 conditions should favor C attack by ^-CN, N attack by NO_2^-, C attack by enolate or phenoxide ions, etc. As an example, primary alkyl halides are attacked (in protic solvents) by the carbon atom of the anion of CH_3COCH_2COOEt, while α-chloro ethers, which react by the S_N1 mechanism, are attacked by the oxygen atom. However, this does not mean that attack is by the less electronegative atom in all S_N2 reactions and by the more electronegative atom in all S_N1 reactions. The position of attack also depends on the nature of the nucleophile, the solvent, the leaving group, and other conditions. The rule merely states that increasing the S_N2 character of the transition state makes attack by the less electronegative atom more likely.

2. All negatively charged nucleophiles must of course have a positive counter-ion. If this ion is Ag^+ (or some other ion that specifically helps in removing the leaving group, p. 504), rather than the more usual Na^+ or K^+, then the transition state is more S_N1-like. Therefore the use of Ag^+ promotes attack at the more electronegative atom. For example, alkyl halides treated with NaCN

---

[484]For reviews, see Jackman, L.M.; Lange, B.C. *Tetrahedron* **1977**, *33*, 2737; Reutov, O.A.; Kurts, A.L. *Russ. Chem. Rev.* **1977**, *46*, 1040; Gompper, R.; Wagner, H. *Angew. Chem. Int. Ed.* **1976**, *15*, 321.

[485]For an example, see Bégué, J.; Charpentier-Morize, M.; Née, G. *J. Chem. Soc., Chem. Commun.* **1989**, 83.

[486]This principle, sometimes called *Kornblum's rule*, was first stated by Kornblum, N.; Smiley, R.A.; Blackwood, R.K.; Iffland, D.C. *J. Am. Chem. Soc.* **1955**, *77*, 6269.

generally give mostly RCN, but the use of AgCN increases the yield of isocyanides RNC.[487]

3. In many cases, the solvent influences the position of attack. The freer the nucleophile, the more likely it is to attack with its more electronegative atom, but the more this atom is encumbered by either solvent molecules or positive counterions, the more likely is attack by the less electronegative atom. In protic solvents, the more electronegative atom is better solvated by hydrogen bonds than the less electronegative atom. In polar aprotic solvents, neither atom of the nucleophile is greatly solvated, but these solvents are very effective in solvating cations. Thus in a polar aprotic solvent the more electronegative end of the nucleophile is freer from entanglement by both the solvent and the cation, so that a change from a protic to a polar aprotic solvent often increases the extent of attack by the more electronegative atom. An example is attack by sodium β-naphthoxide on benzyl bromide, which resulted in 95% *O*-alkylation in dimethyl sulfoxide and 85% *C*-alkylation in 2,2,2-trifluoroethanol.[488] Changing the cation from Li^+ to Na^+ to K^+ (in nonpolar solvents) also favors *O*- over *C*-alkylation[489] for similar reasons (K^+ leaves the nucleophile much freer than Li^+), as does the use of crown ethers, which are good at solvating cations (p. 119).[490] Alkylation of the enolate anion of cyclohexanone in the gas phase, where the nucleophile is completely free, showed only *O*-alkylation and no *C*-alkylation.[491]

4. In extreme cases, steric effects can govern the regioselectivity.[492]

Ambident Substrates

Some substrates (e.g., 1,3-dichlorobutane) can be attacked at two or more positions. We may call these *ambident substrates*. In the example given, there happen to be

---

[487]Actually, this reaction is more complicated than it seems on the surface; see Austad, T.; Songstad, J.; Stangeland, L.J. *Acta Chem. Scand.* **1971**, *25*, 2327; Carretero, J.C.; García Ruano, J.L. *Tetrahedron Lett.* **1985**, *26*, 3381.

[488]Kornblum, N.; Berrigan, P.J.; le Noble, W.J. *J. Chem. Soc.* **1963**, *85*, 1141; Kornblum, N.; Seltzer, R.; Haberfield, P. *J. Am. Chem. Soc.* **1963**, *85*, 1148. For other examples, see le Noble, W.J.; Puerta, J.E. *Tetrahedron Lett.* **1966**, 1087; Brieger, G.; Pelletier, W.M. *Tetrahedron Lett.* **1965**, 3555; Heiszwolf, G.J.; Kloosterziel, H. *Recl. Trav. Chim. Pays-Bas* **1970**, *89*, 1153, 1217; Kurts, A.L.; Masias, A.; Beletskaya, I.P.; Reutov, O.A. *J. Org. Chem. USSR* **1971**, *7*, 2323; Schick, H.; Schwarz, H.; Finger, A.; Schwarz, S. *Tetrahedron* **1982**, *38*, 1279.

[489]Kornblum, N.; Seltzer, R.; Haberfield, P. *J. Am. Chem. Soc.* **1963**, *85*, 1148; Kurts, A.L.; Beletskaya, I.P.; Masias, A.; Reutov, O.A. *Tetrahedron Lett.* **1968**, 3679. See, however, Sarthou, P.; Bram, G.; Guibe, F. *Can. J. Chem.* **1980**, *58*, 786.

[490]Smith, S.G.; Hanson, M.P. *J. Org. Chem.* **1971**, *36*, 1931; Kurts, A.L.; Dem'yanov, P.I.; Beletskaya, I.P.; Reutov, O.A. *J. Org. Chem. USSR* **1973**, *9*, 1341; Cambillau, C.; Sarthou, P.; Bram, G. *Tetrahedron Lett.* **1976**, 281; Akabori, S.; Tuji, H. *Bull. Chem. Soc. Jpn.* **1978**, *51*, 1197. See also, Zook, H.D.; Russo, T.J.; Ferrand, E.F.; Stotz, D.S. *J. Org. Chem.* **1968**, *33*, 2222; le Noble, W.J.; Palit, S.K. *Tetrahedron Lett.* **1972**, 493.

[491]Jones, M.E.; Kass, S.R.; Filley, J.; Barkley, R.M.; Ellison, G.B. *J. Am. Chem. Soc.* **1985**, *107*, 109.

[492]See, for example, O'Neill, P.; Hegarty, A.F. *J. Org. Chem.* **1987**, *52*, 2113.

two leaving groups in the molecule, but there are two kinds of substrates that are inherently ambident (unless symmetrical). One of these, the allylic type, has already been discussed (p. 469). The other is the epoxy (or the similar aziridine[493] or episulfide) substrate.[494]

Substitution of the free epoxide, which generally occurs under basic or neutral conditions, usually involves an S_N2 mechanism. Since primary substrates undergo S_N2 attack more readily than secondary, unsymmetrical epoxides are attacked in neutral or basic solution at the less highly substituted carbon, and stereospecifically, with inversion at that carbon. Under acidic conditions, it is the protonated epoxide that undergoes the reaction. Under these conditions the mechanism can be either S_N1 or S_N2. In S_N1 mechanisms, which favor tertiary carbons, we might expect that attack would be at the more highly substituted carbon, and this is indeed the case. However, even when protonated epoxides react by the S_N2 mechanism, attack is usually at the more highly substituted position.[495] Thus, it is often possible to change the direction of ring opening by changing the conditions from basic to acidic or vice versa. In the ring opening of 2,3-epoxy alcohols, the presence of Ti(O-iPr)$_4$ increases both the rate and the regioselectivity, favoring attack at C-3 rather than C-2.[496] When an epoxide ring is fused to a cyclohexane ring, S_N2 ring opening invariably gives diaxial rather than diequatorial ring opening.[497]

Cyclic sulfates (**108**), prepared from 1,2-diols, react in the same manner as epoxides, but usually more rapidly:[498]

108

[493]Chechik, V.O.; Bobylev, V.A. *Acta Chem. Scand. B* **1994**, *48*, 837.

[494]For reviews of S_N reactions at such substrates, see Rao, A.S.; Paknikar, S.K.; Kirtane, J.G. *Tetrahedron* **1983**, *39*, 2323; Behrens, C.H.; Sharpless, K.B. *Aldrichimica Acta* **1983**, *16*, 67; Enikolopiyan, N.S. *Pure Appl. Chem.* **1976**, *48*, 317; Fokin, A.V.; Kolomiets, A.F. *Russ. Chem. Rev.* **1976**, *45*, 25; Dermer, O.C.; Ham, G.E. *Ethylenimine and Other Aziridines*; Academic Press, NY, **1969**, pp. 206–273; Akhrem, A.A.; Moiseenkov, A.M.; Dobrynin, V.N. *Russ. Chem. Rev.* **1968**, *37*, 448; Gritter, R.J., in Patai, S. *The Chemistry of the Ether Linkage*, Wiley, NY, **1967**, pp. 390–400.

[495]Addy, J.K.; Parker, R.E. *J. Chem. Soc.* **1963**, 915; Biggs, J.; Chapman, N.B.; Finch, A.F.; Wray, V. *J. Chem. Soc. B* **1971**, 55.

[496]Caron M.; Sharpless, K.B. *J. Org. Chem.* **1985**, *50*, 1557. See also, Chong, J.M.; Sharpless, K.B. *J. Org. Chem.* **1985**, *50*, 1560; Behrens, C.H.; Sharpless, K.B. *J. Org. Chem.* **1985**, *50*, 5696.

[497]Murphy, D.K.; Alumbaugh, R.L.; Rickborn, B. *J. Am. Chem. Soc.* **1969**, *91*, 2649. For a method of overriding this preference, see McKittrick, B.A.; Ganem, B. *J. Org. Chem.* **1985**, *50*, 5897.

[498]Gao, Y.; Sharpless, K.B. *J. Am. Chem. Soc.* **1988**, *110*, 7538; Kim, B.M.; Sharpless, K.B. *Tetrahedron Lett.* **1989**, *30*, 655.

Reactions

The reactions in this chapter are classified according to the attacking atom of the nucleophile in the order O, S, N, halogen, H, C. For a given nucleophile, reactions are classified by the substrate and leaving group, with alkyl substrates usually considered before acyl ones. Nucleophilic substitutions at a sulfur atom are treated at the end.

Not all the reactions in this chapter are actually nucleophilic substitutions. In some cases, the mechanisms are not known with enough certainty even to decide whether a nucleophile, an electrophile, or a free radical is attacking. In other cases, conversion of one compound to another can occur by two or even all three of these possibilities, depending on the reagent and the reaction conditions. However, one or more of the nucleophilic mechanisms previously discussed do hold for the overwhelming majority of the reactions in this chapter. For the alkylations, the S_N2 is by far the most common mechanism, as long as R is primary or secondary alkyl. For the acylations, the tetrahedral mechanism is the most common.

OXYGEN NUCLEOPHILES

A. Attack by OH at an Alkyl Carbon

10-1   Hydrolysis of Alkyl Halides

Hydroxy-de-halogenation

$$RX \ + \ H_2O \ \longrightarrow \ ROH_2^+ \ \xrightarrow{\ -H^+\ } \ ROH \ + \ H^+$$

$$RX \ + \ OH^- \ \longrightarrow \ ROH$$

Alkyl halides can be hydrolyzed to alcohols. Hydroxide ion is usually required, although particularly active substrates such as allylic or benzylic alcohols can be hydrolyzed by water. Ordinary halides can also be hydrolyzed by water,[499] if the solvent is HMPA or *N*-methyl-2-pyrrolidinone,[500] or if the reaction is done in an ionic solvent.[501] In contrast to most nucleophilic substitutions at saturated carbons, this reaction can be performed on tertiary substrates without significant interference from elimination side reactions. Tertiary alkyl α-halocarbonyl compounds can be converted to the corresponding alcohol with silver oxide in aqueous acetonitrile.[502] The

[499]It has been proposed that the mechanism of the reaction of primary halides with water is not the ordinary S_N2 mechanism, but that the rate-determining process involves a fluctuation of solvent configuration: Kurz, J.L.; Kurz, L.C. *Isr. J. Chem.* **1985**, *26*, 339; Kurz, J.L.; Lee, J.; Love, M.E.; Rhodes, S. *J. Am. Chem. Soc.* **1986**, *108*, 2960.

[500]Hutchins, R.O.; Taffer, I.M. *J. Org. Chem.* **1983**, *48*, 1360.

[501]Kim, D.W.; Hong, D.J.; Seo, J.W.; Kim, H.S.; Kim, H.K.; Song, C.E.; Chi, D.Y. *J. Org. Chem.* **2004**, *69*, 3186.

[502]Cavicchioni, G. *Synth. Commun.* **1994**, *24*, 2223.

reaction is not frequently used for synthetic purposes, because alkyl halides are usually obtained from alcohols.

An indirect conversion of halides to alcohols involved triethylborane. The reaction of an α-iodo ester with BEt_3, followed by reaction with dimethyl sulfide in methanol, gave an α-hydroxy ester.[503]

Vinylic halides are unreactive (p. 473), but they can be hydrolyzed to ketones at room temperature with mercuric trifluoroacetate, or with mercuric acetate in either

$$ \underset{X}{\overset{R}{\diagdown}}C=C\diagup \quad \xrightarrow[\text{CF}_3\text{COOH}]{\text{Hg(OAc)}_2} \quad -\underset{\underset{H}{|}}{C}-\underset{\underset{O}{\|}}{C}\diagup^{R} $$

trifluoroacetic acid or acetic acid containing BF_3 etherate.[504] Primary bromides and iodides give alcohols when treated with bis(tributyltin)oxide $Bu_3Sn-O-SnBu_3$ in the presence of silver salts.[505]

OS **II**, 408; **III**, 434; **IV**, 128; **VI**, 142, 1037.

10-2   Hydrolysis of *gem*-Dihalides

Oxo-de-dihalo-bisubstitution

$$ R-\underset{\underset{X}{|}}{\overset{\overset{X}{|}}{C}}-R' \quad \xrightarrow[\text{H}^+ \text{ or OH}^-]{\text{H}_2\text{O}} \quad R-\underset{\underset{O}{\|}}{C}-R' $$

gem-Dihalides can be hydrolyzed with either acid or basic catalysis to give aldehydes or ketones.[506] Formally, the reaction may be regarded as giving R—C(OH)XR′, which is unstable and loses HX to give the carbonyl compound. For aldehydes derived from $RCHX_2$, strong bases cannot be used, because the product undergoes the aldol reaction (**16-34**) or the Cannizzaro reaction (**19-81**). A mixture of calcium carbonate and sodium acetate is effective,[507] and heating to 100°C in DMSO gives good yields.[508] Heating 1,1-dihaloalkenes ($C=CX_2$) with zinc and water leads to the corresponding methyl ketone.[509]

OS **I**, 95; **II**, 89, 133, 244, 549; **III**, 538, 788; **IV**, 110, 423, 807. Also see, OS **III**, 737.

[503]Kihara, N.; Ollivier, C.; Renaud, P. *Org. Lett.* **1999**, *1*, 1419.

[504]Martin, S.F.; Chou, T. *Tetrahedron Lett.* **1978**, 1943; Yoshioka, H.; Takasaki, K.; Kobayashi, M.; Matsumoto, T. *Tetrahedron Lett.* **1979**, 3489.

[505]Gingras, M.; Chan, T.H. *Tetrahedron Lett.* **1989**, *30*, 279.

[506]For a review, see Salomaa, P., in Patai, S. *The Chemistry of the Carbonyl Group*, Vol. 1, Wiley, NY, **1966**, pp. 177–210.

[507]Mataka, S.; Liu, G.-B.; Sawada, T.; Tori-i, A.; Tashiro, M. *J. Chem. Res. (S)* **1995**, 410.

[508]Li, W.; Li, J.; DeVincentis, D.; Masour, T.S. *Tetrahedron Lett.* **2004**, *45*, 1071.

[509]Wang, L.; Li, P.; Yan, J.; Wu, Z. *Tetrahedron Lett.* **2003**, *44*, 4685.

10-3 Hydrolysis of 1,1,1-Trihalides

Hydroxy,oxo-de-trihalo-tersubstitution

$$RCX_3 + H_2O \longrightarrow RCOOH$$

This reaction is similar to the previous one. The utility of the method is limited by the lack of availability of trihalides, although these compounds can be prepared by addition of CCl_4 and similar compounds to double bonds (**15-38**) and by the free-radical halogenation of methyl groups on aromatic rings (**14-1**). When the hydrolysis is carried out in the presence of an alcohol, a carboxylic ester can be obtained directly.[510] 1,1-Dichloroalkenes can also be hydrolyzed to carboxylic acids, by treatment with H_2SO_4. In general 1,1,1-trifluorides do not undergo this reaction,[511] although exceptions are known.[512]

Aryl 1,1,1-trihalomethanes can be converted to acyl halides by treatment with sulfur trioxide.[513]

Chloroform is more rapidly hydrolyzed with base than dichloromethane or carbon tetrachloride and gives not only formic acid, but also carbon monoxide.[514] Hine[515] has shown that the mechanism of chloroform hydrolysis is quite different from that of dichloromethane or carbon tetrachloride, although superficially the three reactions appear similar. The first step is the loss of a proton to give CCl_3^-, which then loses Cl^- to give dichlorocarbene CCl_2, which is hydrolyzed to formic acid or carbon monoxide.

This is an example of an S_N1cB mechanism (p. 500). The other two compounds react by the normal mechanisms. Carbon tetrachloride cannot give up a proton and dichloromethane is not acidic enough.

OS **III**, 270; **V**, 93. Also see, OS **I**, 327.

[510]See, for example, Le Fave, G.M.; Scheurer, P.G. *J. Am. Chem. Soc.* **1950**, *72*, 2464.

[511]Sheppard, W.A.; Sharts, C.M. *Organic Fluorine Chemistry*, W.A. Benjamin, NY, **1969**, pp. 410–411; Hudlický, M. *Chemistry of Organic Fluorine Compounds*, 2nd ed., Ellis Horwood, Chichester, **1976**, pp. 273–274.

[512]See, for example, Kobayashi, Y.; Kumadaki, I. *Acc. Chem. Res.* **1978**, *11*, 197.

[513]Rondestvedt Jr., C.S. *J. Org. Chem.* **1976**, *41*, 3569, 3574, 3576. For another method, see Nakano, T.; Ohkawa, K.; Matsumoto, H.; Nagai, Y. *J. Chem. Soc., Chem. Commun.* **1977**, 808.

[514]For a review, see Kirmse, W. *Carbene Chemistry*, 2nd ed., Academic Press, NY, **1971**, pp. 129–141.

[515]Hine, J. *J. Am. Chem. Soc.* **1950**, *72*, 2438. Also, see le Noble, W.J. *J. Am. Chem. Soc.* **1965**, *87*, 2434.

10-4   Hydrolysis of Alkyl Esters of Inorganic Acids

Hydroxy-de-sulfonyloxy-substitution, and so on.

$$ R\text{-}X \xrightarrow{\hspace{4cm}} R\text{-}OH $$

$$ X = OSO_2R' , OSO_2OH , OSO_2OR , OSO_2R' , OSOR' $$
$$ = ONO_2 , ONO , OPO(OH)_2 , OPO(OR')_2 , OB(OH)_2 \quad \text{and others} $$

Esters of inorganic acids, including those given above and others, can be hydrolyzed to alcohols. The reactions are most successful when the ester is that of a strong acid, but it can be done for esters of weaker acids by the use of hydroxide ion (a more powerful nucleophile) or acidic conditions (which make the leaving group come off more easily). When vinylic substrates are hydrolyzed, the products are aldehydes or ketones.

$$ R_2C{=}CH\text{-}X \xrightarrow{H_2O} R_2C{=}CH\text{-}OH \; \underset{\longleftarrow}{\hspace{1.5cm}} \; R_2CH\text{-}CHO $$

These reactions are all considered at one place because they are formally similar, but although some of them involve R—O cleavage and are thus nucleophilic substitutions at a saturated carbon, others involve cleavage of the bond between the inorganic atom and oxygen and are thus nucleophilic substitutions at a sulfur, nitrogen, etc. It is even possible for the same ester to be cleaved at either position, depending on the conditions. Thus benzhydryl p-toluenesulfinate ($Ph_2CHOSOC_6H_4CH_3$) was found to undergo C—O cleavage in $HClO_4$ solutions and S—O cleavage in alkaline media.[516] In general, the weaker the corresponding acid, the less likely is C—O cleavage. Thus, sulfonic acid esters $ROSO_2R'$ generally give C—O cleavage,[517] while nitrous acid esters RONO usually give N—O cleavage.[518] Esters of sulfonic acids that are frequently hydrolyzed are mentioned on p. 497. For hydrolysis of sulfonic acid esters, see also **16-100**.

OS **VI**, 852. See also, **VIII**, 50.

10-5   Hydrolysis of Diazoketones

Hydro,hydroxy-de-diazo-bisubstitution

$$ R \underset{O}{\overset{}{\diagup}}CHN_2 + H_2O \xrightarrow{H^+} R \underset{O}{\overset{}{\diagup}}\diagdown OH $$

[516]Bunton, C.A.; Hendy, B.N. *J. Chem. Soc.* *1963*, 627. For another example, see Batts, B.D. *J. Chem. Soc. B 1966*, 551.

[517]Barnard, P.W.C.; Robertson, R.E. *Can. J. Chem. 1961*, *39*, 881. See also, Drabicky, M.J.; Myhre, P.C.; Reich, C.J.; Schmittou, E.R. *J. Org. Chem. 1976*, *41*, 1472.

[518]For a discussion of the mechanism of hydrolysis of alkyl nitrites, see Williams, D.L.H. *Nitrosation*, Cambridge University Press, Cambridge, *1988*, pp. 162–163.

Diazoketones are relatively easy to prepare (see **16-89**). When treated with acid, they add a proton to give α-keto diazonium salts, which are hydrolyzed to the alcohols by the S_N1 or S_N2 mechanism.[519] Relatively good yields of α-hydroxy ketones can be prepared in this way, since the diazonium ion is somewhat stabilized by the presence of the carbonyl group, which discourages N_2 from leaving because that would result in an unstable α-carbonyl carbocation.

10-6   Hydrolysis of Acetals, Enol Ethers, and Similar Compounds[520]

The alkoxyl group OR is not a leaving group, so these compounds must be converted to the conjugate acids before they can be hydrolyzed. Although 100% sulfuric acid and other concentrated strong acids readily cleave simple ethers,[521] the only acids used preparatively for this purpose are HBr and HI (**10-49**). However, acetals, ketals, and ortho esters[522] are easily cleaved by dilute acids. These compounds are hydrolyzed with greater facility because carbocations of the type RO—CH— are greatly stabilized by resonance (p. 242). The reactions therefore

[519]Dahn, H.; Gold, H. *Helv. Chim. Acta* **1963**, *46*, 983; Thomas, C.W.; Leveson, L.L. *Int. J. Chem. Kinet.*, **1983**, *15*, 25. For a review of the acidpromoted decomposition of diazoketones, see Smith III, A.B; Dieter, R.K. *Tetrahedron* **1981**, *37*, 2407.

[520]For reviews, see Bergstrom, R.G., in Patai, S. *The Chemistry of Functional Groups, Supplement E*, Wiley, NY, **1980**, pp. 881–902; Cockerill, A.F.; Harrison, R.G., in Patai, S. *The Chemistry of Functional Groups, Supplement A*, pt. 1, Wiley, NY, **1977**, pp. 149–329; Cordes, E.H.; Bull, H.G. *Chem. Rev.* **1974**, *74*, 581; Cordes, E.H. *Prog. Phys. Org. Chem.* **1967**, *4*, 1; Salomaa, P., in Patai, S. *The Chemistry of the Carbonyl Group*, Vol. 1, Wiley, NY, **1966**, pp. 184–198; Pindur, U.; Müller, J.; Flo, C.; Witzel, H. *Chem. Soc. Rev.* **1987**, *16*, 75 (ortho esters); Cordes, E.H., in Patai, S. *The Chemistry of Carboxylic Acids and Esters*, Wiley, NY, **1969**, pp. 632–656 (ortho esters); DeWolfe, R.H. *Carboxylic Ortho Acid Derivatives*, Academic Press, NY, **1970**, pp. 134–146 (ortho esters); Rekasheva, A.F. *Russ. Chem. Rev.* **1968**, *37*, 1009 (enol ethers).

[521]Jaques, D.; Leisten, J.A. *J. Chem. Soc.* **1964**, 2683. See also, Olah, G.A.; O'Brien, D.H. *J. Am. Chem. Soc.* **1967**, *89*, 1725.

[522]For a review of the reactions of ortho esters, see Pavlova, L.A.; Davidovich, Yu.A.; Rogozhin, S.V. *Russ. Chem. Rev.* **1986**, *55*, 1026.

proceed by the S_N1 mechanism,[523] as shown for acetals:[524]

Hemiacetal

This mechanism (which is an S_N1cA or A1 mechanism) is the reverse of that for acetal formation by reaction of an aldehyde and an alcohol (**16-5**). Among the facts supporting the mechanism are[525] (*1*) The reaction proceeds with *specific* H_3O^+ catalysis (see p. 373). (*2*) It is faster in D_2O. (*3*) Optically active ROH are not racemized. (*4*) Even with *tert*-butylalcohol the R—O bond does not cleave, as shown by ^{18}O labeling.[526] (*5*) In the case of acetophenone ketals, the intermediate corresponding to **109** [$ArCMe(OR)_2$] could be trapped with sulfite ions (SO_3^{2-}).[527] (*6*) Trapping of this ion did not affect the hydrolysis rate,[527] so the rate-determining step must come earlier. (*7*) In the case of 1,1-dialkoxyalkanes, intermediates corresponding to **109** were isolated as stable ions in super acid solution at $-75°C$, where their spectra could be studied.[528] (*8*) Hydrolysis rates greatly increase in the order $CH_2(OR')_2 < RCH(OR')_2 < R_2C(OR')_2 < RC(OR')_3$, as would be expected for a carbocation intermediate.[529] Formation of **109** is usually the rate-determining step (as marked above), but there is evidence that at least in some cases this step is fast, and the rate-determining step is loss of R'OH from the protonated hemiacetal.[530] Rate-determining addition of water to **109** has also been reported.[531]

[523]For a review of the mechanisms of hydrolysis of acetals and thioacetals, see Satchell, D.P.N.; Satchell, R.S. *Chem. Soc. Rev.* **1990**, *19*, 55.

[524]Kreevoy, M.M.; Taft, R.W. *J. Am. Chem. Soc.* **1955**, *77*, 3146, 5590.

[525]For a discussion of these, and of other evidence, see Cordes, E.H. *Prog. Phys. Org. Chem.* **1967**, *4*, 1.

[526]Cawley, J.J.; Westheimer, F.H. *Chem. Ind. (London)* **1960**, 656.

[527]Young, P.R.; Jencks, W.P. *J. Am. Chem. Soc.* **1977**, *99*, 8238. See also, Jencks, W.P. *Acc. Chem. Res.* **1980**, *13*, 161; McClelland, R.A.; Ahmad, M. *J. Am. Chem. Soc.* **1978**, *100*, 7027, 7031; Young, P.R.; Bogseth, R.C.; Rietz, E.G. *J. Am. Chem. Soc.* **1980**, *102*, 6268. However, in the case of simple aliphatic acetals, **103** could not be trapped: Amyes, T.L.; Jencks, W.P. *J. Am. Chem. Soc.* **1988**, *110*, 3677.

[528]See White, A.M.; Olah, G.A. *J. Am. Chem. Soc.* **1969**, *91*, 2943; Akhmatdinov, R.T.; Kantor, E.A.; Imashev, U.B.; Yasman, Ya.B.; Rakhmankulov, D.L. *J. Org. Chem. USSR* **1981**, *17*, 626.

[529]For the influence of alkyl group size on the mechanism see Belarmino, A.T.N.; Froehner, S.; Zanette, D.; Farah, J.P.S.; Bunton, C.A.; Romsted, L.S. *J. Org. Chem.* **2003**, *68*, 706.

[530]Jensen, J.L.; Lenz, P.A. *J. Am. Chem. Soc.* **1978**, *100*, 1291; Finley, R.L.; Kubler, D.G.; McClelland, R.A. *J. Org. Chem.* **1980**, *45*, 644; Przystas, T.J.; Fife, T.H. *J. Am. Chem. Soc.* **1981**, *103*, 4884; Chiang, Y.; Kresge, A.J. *J. Org. Chem.* **1985**, *50*, 5038; Fife, T.H.; Natarajan, R. *J. Am. Chem. Soc.* **1986**, *108*, 2425, 8050; McClelland, R.A.; Sørensen, P.E. *Acta Chem. Scand.* **1990**, *44*, 1082.

[531]Toullec, J.; El-Alaoui, M. *J. Org. Chem.* **1985**, *50*, 4928; Fife, T.H.; Natarajan, R. *J. Am. Chem. Soc.* **1986**, *108*, 2425, 8050.

While the A1 mechanism shown above operates in most acetal hydrolyses, it has been shown that at least two other mechanisms can take place with suitable substrates.[532] In one of these mechanisms the second and third of the above steps are concerted, so that the mechanism is S_N2cA (or A2). This has been shown, for example, in the hydrolysis of 1,1-diethoxyethane, by isotope effect studies:[533]

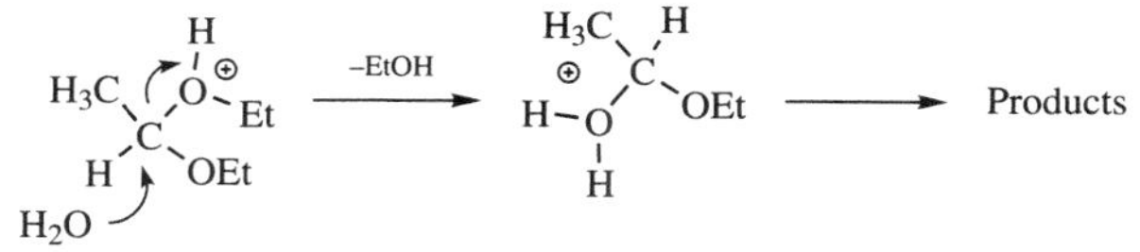

In the second mechanism, the first and second steps are concerted. In the case of hydrolysis of 2-(*p*-nitrophenoxy)tetrahydropyran, *general* acid catalysis was shown[534] demonstrating that the substrate is protonated in the rate-determining step (p. 373). Reactions in which a substrate is protonated in the rate-determining step are called AS_E2 reactions.[535] However, if protonation of the substrate were all that happens in the slow step, then the proton in the transition state would be expected to lie closer to the weaker base (p. 373). Because the substrate is a much weaker base than water, the proton should be largely transferred. Since the Brønsted coefficient was found to be 0.5, the proton was actually transferred only about halfway. This can be explained if the basicity of the substrate is increased by partial breaking of the C—O bond. The conclusion drawn is that steps 1 and 2 are concerted. The hydrolysis of ortho esters in most cases is also subject to general acid catalysis.[536]

The hydrolysis of acetals and ortho esters is governed by the stereoelectronic control factor discussed on p. 1258,[537] although the effect can generally be seen only in systems where conformational mobility is limited, especially in cyclic systems. There is evidence for synplanar stereoselection in the acid hydrolysis of

---

[532]For a review, see Fife, T.H. *Acc. Chem. Res.* **1972**, *5*, 264. For a discussion, see Wann, S.R.; Kreevoy, M.M. *J. Org. Chem.* **1981**, *46*, 419.

[533]Kresge, A.J.; Weeks, D.P. *J. Am. Chem. Soc.* **1984**, *106*, 7140. See also, Fife, T.H. *J. Am. Chem. Soc.* **1967**, *89*, 3228; Craze, G.; Kirby, A.J.; Osborne, R. *J. Chem. Soc. Perkin Trans. 2* **1978**, 357; Amyes, T.L.; Jencks, W.P. *J. Am. Chem. Soc.* **1989**, *111*, 7888, 7900.

[534]Fife, T.H.; Brod, L.H. *J. Am. Chem. Soc.* **1970**, *92*, 1681. For other examples, see Kankaanperä, A.; Lahti, M. *Acta Chem. Scand.* **1969**, *23*, 2465; Mori, A.L.; Schaleger, L.L. *J. Am. Chem. Soc.* **1972**, *94*, 5039; Capon, B.; Nimmo, K. *J. Chem. Soc. Perkin Trans. 2* **1975**, 1113; Eliason, R.; Kreevoy, M.M. *J. Am. Chem. Soc.* **1978**, *100*, 7037; Jensen, J.L.; Herold, L.R.; Lenz, P.A.; Trusty, S.; Sergi, V.; Bell, K.; Rogers, P. *J. Am. Chem. Soc.* **1979**, *101*, 4672.

[535]For a review of A-S_E2 reactions, see Williams Jr., J.M.; Kreevoy, M.M. *Adv. Phys. Org. Chem.* **1968**, *6*, 63.

[536]Chiang, Y.; Kresge, A.J.; Lahti, M.O.; Weeks, D.P. *J. Am. Chem. Soc.* **1983**, *105*, 6852, and references cited therein; Santry, L.J.; McClelland, R.A. *J. Am. Chem. Soc.* **1983**, *105*, 6138; Fife, T.H.; Przystas, T.J. *J. Chem. Soc. Perkin Trans. 2* **1987**, 143.

[537]See, for example, Kirby, A.J. *Acc. Chem. Res.* **1984**, *17*, 305; Bouab, O.; Lamaty, G.; Moreau, C. *Can. J. Chem.* **1985**, *63*, 816. See, however, Ratcliffe, A.J.; Mootoo, D.R.; Andrews, C.W.; Fraser-Reid, B. *J. Am. Chem. Soc.* **1989**, *111*, 7661.

acetals.[538] The mechanism of Lewis acid-mediated cleavage of chiral acetals is also known.[539]

Convenient reagents for acetals are wet silica gel[540] and Amberlyst-15 (a sulfonic acid-based polystyrene cation exchange resin).[541] Both cyclic and acyclic acetals and ketals can be converted to aldehydes or ketones under nonaqueous conditions by treatment with Montmorillonite K10 clay in various solvents,[542] with Lewis acids, such as $FeCl_3 \bullet 6\ H_2O$ in chloroform,[543] $Bi(OTf)_3 \bullet xH_2O$,[544] or 5% $Ce(OTf)_3$ in wet nitromethane.[545] Hydrolysis techniques include treatment with β-cyclodextrin in water,[546] Me_3SiI in CH_2Cl_2, or $CHCl_3$,[547] $LiBF_4$,[548] ceric ammonium nitrate in aqueous acetonitrile,[549] DDQ[550] in wet MeCN, or Magtrieve in chloroform.[551]

Although acetals, ketals, and ortho esters are easily hydrolyzed by acids, they are extremely resistant to hydrolysis by bases. An aldehyde or ketone can therefore be protected from attack by a base by conversion to the acetal or ketal (**16-5**), and then can be cleaved with acid. Pyridine–HF has also been used for this conversion.[552] Thioacetals, thioketals, *gem*-diamines, and other compounds that contain any two of the groups OR, OCOR, NR_2, NHCOR, SR, and halogen on the same carbon can also be hydrolyzed to aldehydes or ketones, in most cases, by acid treatment. Several $ArCH(OAc)_2$ derivatives were hydrolyzed to the aldehyde using Montmorillonite K10,[553] alumina with microwaves,[554] ceric ammonium nitrate on silica gel,[555] or by heating with CBr_4 in acetonitirle.[556] Thioacetals $RCH(SR')_2$ and thioketals

[538]Li, S.; Kirby, A.J.; Deslongchamps, P. *Tetrahedron Lett.* **1993**, *34*, 7757.

[539]Sammakia, T.; Smith, R.S. *J. Org. Chem.* **1992**, *57*, 2997.

[540]Huet, F.; Lechevallier, A.; Pellet, M.; Conia, J.M. *Synthesis* **1978**, 63. See Caballero, G.M.; Gros, E.G. *Synth. Commun.* **1995**, *25*, 395 for hydrolysis of hindered ketals with $CuSO_4$ on silica gel.

[541]Coppola, G.M. *Synthesis* **1984**, 1021.

[542]Li, T.-S.; Li, S.-H. *Synth. Commun.* **1997**, *27*, 2299; Gautier, E.C.L.; Graham, A.E.; McKillop, A.; Standen, S.T.; Taylor, R.J.K. *Tetrahedron Lett.* **1997**, *38*, 1881.

[543]Sen, S.E.; Roach, S.L.; Boggs, J.K.; Ewing, G.J.; Magrath, J. *J. Org. Chem.* **1997**, *62*, 6684.

[544]Carringan, M.D.; Sarapa, D.; Smith, R.C.; Wieland, L.C.; Mohan, R.S. *J. Org. Chem.* **2002**, *67*, 1027.

[545]Dalpozzo, R.; De Nino, A.; Maiuolo, L.; Procopio, A.; Tagarelli, A.; Sindona, G.; Bartoli, G. *J. Org. Chem.* **2002**, *67*, 9093.

[546]Krishnaveni, N. S.; Surendra, K.; Reddy, M. A.; Nageswar, Y. V. D.; Rao, K. R. *J. Org. Chem.* **2003**, *68*, 2018.

[547]Jung, M.E.; Andrus, W.A.; Ornstein, P.L. *Tetrahedron Lett.* **1977**, 4175. See also, Balme, G.; Goré, J. *J. Org. Chem.* **1983**, *48*, 3336.

[548]Lipshutz, B.H.; Harvey, D.F. *Synth. Commun.* **1982**, *12*, 267.

[549]Ates, A.; Gautier, A.; Leroy, B.; Plancher, J.M.; Quesnel, Y.; Markó, I.E. *Tetrahedron Lett.* **1999**, *40*, 1799.

[550]Tanemura, K.; Suzuki, T.; Horaguchi, T. *J. Chem. Soc., Chem. Commun.* **1992**, 979.

[551]Ko, J.-y.; Park, S.-T. *Tetrahedron Lett.* **1999**, *40*, 6025.

[552]Watanabe, Y.; Kiyosawa, Y.; Tatsukawa, A.; Hayashi, M. *Tetrahedron Lett.* **2001**, *42*, 4641.

[553]Li, T.-S.; Zhang, Z.-H.; Fu, C.-G. *Tetrahedron Lett.* **1997**, *38*, 3285.

[554]Varma, R.S.; Chatterjee, A.K.; Varma, M. *Tetrahedron Lett*, **1993**, *34*, 3207.

[555]Cotelle, P.; Catteau, J.-P. *Tetrahedron Lett.* **1992**, *33*, 3855.

[556]Ramalingam, T.; Srinivas, R.; Reddy, B.V.S.; Yadav, J.S. *Synth. Commun.* **2001**, *31*, 1091.

$R_2C(SR')_2$ are among those compounds generally resistant to acid hydrolysis.[557] Because conversion to these compounds (**16-11**) serves as an important method for protection of aldehydes and ketones, many methods have been devised to cleave them to the parent carbonyl compounds. Among reagents[558] used for this purpose are $HgCl_2$,[559] $FeCl_3 \cdot 6\ H_2O$,[560] cetyltrimethylammonium tribromide in dichloromethane,[561] *m*-chloroperoxybenzoic acid, and CF_3COOH in CH_2Cl_2,[562] Oxone® on wet alumina,[563] the Dess–Martin periodinane,[564] and DDQ in water under photolysis conditions,[565] and sodium nitrite in aqueous acetyl chloride.[566] Electrochemical methods have also been used.[567] Mixed acetals and ketals (RO—C—SR) can be hydrolyzed with most of the reagents mentioned above, including *N*-bromosuccinimide (NBS) in aqueous acetone,[568] and glyoxylic acid on Amberlyst 15 with microwave irradiation.[569]

Enol ethers are readily hydrolyzed by acids; the rate-determining step is protonation of the substrate.[570] However, protonation does not take place at the oxygen, but at the β carbon,[571] because that gives rise to the stable carbocation **110**.[572] After that the mechanism is similar to the A1 mechanism given above for the hydrolysis of acetals.

[557]Ali, M.; Satchell, D.P.N. *J. Chem. Soc. Perkin Trans. 2* **1992**, 219; **1993**, 1825; Ali, M.; Satchell, D.P.N.; Le, V.T. *J. Chem. Soc. Perkin Trans. 2* **1993**, 917.

[558]For references to other reagents, see Gröbel, B.; Seebach, D. *Synthesis* **1977**, 357, see pp. 359–367; Cussans, N.J.; Ley, S.V.; Barton, D.H.R. *J. Chem. Soc. Perkin Trans. 1* **1980**, 1654.

[559]Corey, E.J.; Erickson, B.W. *J. Org. Chem.* **1971**, *36*, 3553. For a mechanistic study, see Satchell, D.P.N.; Satchell, R.S. *J. Chem. Soc. Perkin Trans. 2* **1987**, 513.

[560]Kamal, A.; Laxman, E.; Reddy, P.S.M.M. *Synlett* **2000**, 1476.

[561]Mondal, E.; Bose, G.; Khan, A.T. *Synlett* **2001**, 785.

[562]Cossy, J. *Synthesis* **1987**, 1113.

[563]Ceccherelli, P.; Curini, M.; Marcotullio, M.C.; Epifano, F.; Rosati, O. *Synlett,* **1996**, 767.

[564]Langille, N.F.; Dakin, L.A.; Panek, J.S. *Org. Lett.* **2003**, *5*, 575. See also, Stork, G.; Zhao, K. *Tetrahedron Lett.* **1989**, *30*, 287.

[565]Mathew, L.; Sankararaman, S. *J. Org. Chem.* **1993**, *58*, 7576.

[566]Khan, A.T.; Mondal, E.; Sahu, P.R. *Synlett* **2003**, 377.

[567]See Schulz-von Itter, N.; Steckhan, E. *Tetrahedron* **1987**, *43*, 2475; Suda, K.; Watanabe, J.; Takanami, T. *Tetrahedron Lett.* **1992**, *33*, 1355.

[568]Karimi, B.; Seradj, H.; Tabaei, M.H. *Synlett* **2000**, 1798.

[569]Chavan, S.P.; Soni, P.; Kamat, S.K. *Synlett* **2001**, 1251.

[570]Jones, J.; Kresge, A. J. *Can. J. Chem.* **1993**, *71*, 38.

[571]Jones, D.M.; Wood, N.F. *J. Chem. Soc.* **1964**, 5400; Okuyama, T.; Fueno, T.; Furukawa, J. *Bull. Chem. Soc. Jpn.* **1970**, *43*, 3256; Kreevoy, M.M.; Eliason, R. *J. Phys. Chem.* **1969**, *72*, 1313; Lienhard, G.; Wang, T.C. *J. Am. Chem. Soc.* **1969**, *91*, 1146; Burt, R.A.; Chiang, Y.; Kresge, A.J.; Szilagyi, S. *Can. J. Chem.* **1984**, *62*, 74.

[572]See Chwang, W.K.; Kresge, A.J.; Wiseman, J.R. *J. Am. Chem. Soc.* **1979**, *101*, 6972.

Among the facts supporting this mechanism (which is an A-S_E2 mechanism because the substrate is protonated in the rate-determining step) are (*1*) the ^{18}O labeling shows that in $ROCH=CH_2$ it is the vinyl–oxygen bond and not the RO bond that cleaves;[573] (*2*) the reaction is subject to general acid catalysis;[574] (*3*) there is a solvent isotope effect when D_2O is used.[574] Enamines are also hydrolyzed by acids (see **16-2**); the mechanism is similar. Ketene dithioacetals $R_2C=C(SR')_2$ also hydrolyze by a similar mechanism, except that the initial protonation step is partially reversible.[575] Furans represent a special case of enol ethers that are cleaved by acid to give 1,4-diones.[576] Thus oxonium ions are cleaved by water to give an alcohol and an ether:

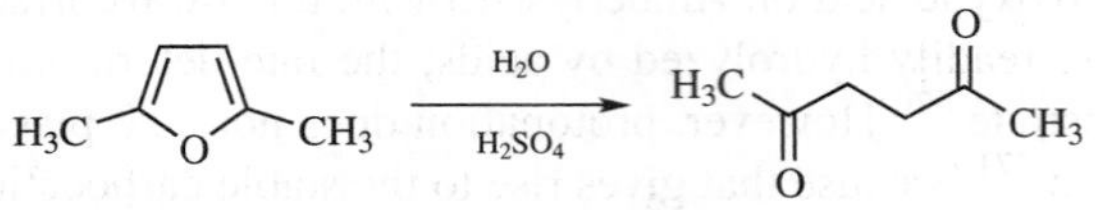

OS **I**, 67, 205; **II**, 302, 305, 323; **III**, 37, 127, 465, 470, 536, 541, 641, 701, 731, 800; **IV**, 302, 499, 660, 816, 903; **V**, 91, 292, 294, 703, 716, 937, 967, 1088; **VI**, 64, 109, 312, 316, 361, 448, 496, 683, 869, 893, 905, 996; **VII**, 12, 162, 241, 249, 251, 263, 271, 287, 381, 495; **VIII**, 19, 155, 241, 353, 373

10-7 Hydrolysis of Epoxides

(3)*OC-seco*-hydroxy-de-alkoxy-substitution

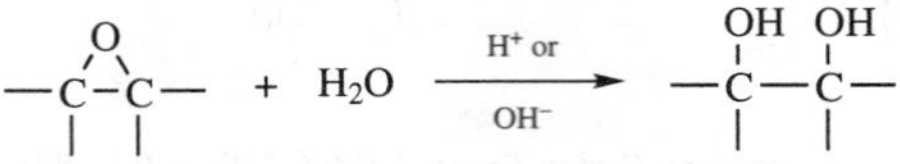

The hydrolysis of epoxides is a convenient method for the preparation of *vic*-diols. The reaction is catalyzed by acids or bases (see discussion of the mechanism on p. 518). Among acid catalysts, perchloric acid leads to minimal side reactions,[577] and 10% Bu_4NHSO_4 in water is effective.[578] Water reacts with epoxides in the presence of β-cyclodextrin to give the corresponding diol.[579] Dimethyl sulfoxide is a superior solvent for the alkaline hydrolysis of epoxides.[580] Water at 10 kbar and 60°C opens epoxides with high stereoselectivity,[581] and epoxide hydrolase

[573]Kiprianova, L.A.; Rekasheva, A.F. *Dokl. Akad. Nauk SSSR*, **1962**, *142*, 589.
[574]Fife, T.H. *J. Am. Chem. Soc.* **1965**, *87*, 1084; Salomaa, P.; Kankaanperä, A.; Lajunen, M. *Acta Chem. Scand.* **1966**, *20*, 1790; Kresge, A.J.; Yin, Y. *Can. J. Chem.* **1987**, *65*, 1753.
[575]For a review, see Okuyama, T. *Acc. Chem. Res.* **1986**, *19*, 370.
[576]Enzymatic hydrolysis of 2,5-dimethylfuran gave hex-3-en-2,5-dione. See Finlay, J.; McKervey, M.A.; Gunaratne, H.Q.N. *Tetrahedron Lett.* **1998**, *39*, 5651.
[577]Fieser, L.F.; Fieser, M. *Reagents for Organic Synthesis* Vol. 1, Wiley, NY, **1967**, p. 796.
[578]Fan, R.-H.; Hou, X.-L. *Org. Biomol. Chem.* **2003**, *1*, 1565.
[579]Reddy, M.A.; Reddy, L.R.; Bhanumthi, N.; Rao, K.R. *Org. Prep. Proceed. Int.* **2002**, *34*, 537.
[580]Berti, G.; Macchia, B.; Macchia, F. *Tetrahedron Lett.* **1965**, 3421.
[581]Kotsuki, H.; Kataoka, M.; Nishizawa, H. *Tetrahedron Lett.* **1993**, *34*, 4031.

opens epoxides with high enantioselectivity.[582] Cobalt salen [salen = bis(salicyli-dene)ethylenediamine] catalysts, in the presence of water, open epoxides with high stereoselectivity.[583] Photolysis of epoxy-ketones in the presence of 1,3-dimethylbenzimidazoline in AcOH/THF leads to β-hydroxy ketones.[584]

OS **V**, 414.

B. Attack by OR at an Alkyl Carbon

10-8   Alkylation With Alkyl Halides: The Williamson Reaction

Alkoxy-de-halogenation

$$RX \ + \ OR'^- \ \longrightarrow \ ROR'$$

The *Williamson reaction*, discovered in 1850, is still the best general method for the preparation of unsymmetrical or symmetrical ethers.[585] The reaction can also be carried out with aromatic R′, although *C*-alkylation is sometimes a side reaction (see p. 515).[586] The normal method involves treatment of the halide with alkoxide or aroxide ion prepared from an alcohol or phenol, although methylation using dimethyl carbonate has been reported.[587] It is also possible to mix the halide and alcohol or phenol directly with Cs_2CO_3 in acetonitrile,[588] or with solid KOH in Me_2SO.[589] The reaction can also be carried out in a dry medium,[590] on zeolite–HY[591] or neat[592] or in solvents[593] using microwave irradiation. Williamson ether synthesis in ionic liquids has also been reported.[594] The reaction is not successful for tertiary R (because of elimination), and low yields are often obtained with secondary R. Mono-ethers can be formed from diols and alkyl halides.[595] Many other

[582]Zhao, L.; Han, B.; Huang, Z.; Miller, M.; Huang, H.; Malashock, D.S.; Zhu, Z.; Milan, A.; Robertson, D.E.; Weiner, D.P.; Burk, M. J. *J. Am. Chem. Soc.* **2004**, *126*, 11156; See also, Pedragosa-Moreau, S.; Archelas, A.; Furstoss, R. *Tetrahedron Lett.* **1996**, *37*, 3319.

[583]Ready, J.M.; Jacobsen, E.N. *J. Am. Chem. Soc.* **2001**, *123*, 2687.

[584]Hasegawa, E.; Chiba, N.; Nakajima, A.; Suzuki, K.; Yoneoka, A.; Iwata, K. *Synthesis* **2001**, 1248. For a related reaction with NO, see Liu, Z.; Li, R.; Yang, D.; Wu, L. *Tetrahedron Lett.* **2004**, *45*, 1565.

[585]For a review, see Feuer, H.; Hooz, J., in Patai, S. *The Chemistry of the Ether Linkage*, Wiley, NY, **1967**, pp. 446–450, 460–468.

[586]For a list of reagents used to convert alcohols and phenols to ethers, see Larock, R.C. *Comprehensive Organic Transformations*, 2nd ed., Wiley-VCH, NY, **1999**, pp. 890–893.

[587]Ouk, S.; Thiebaud, S.; Borredon, E.; Legars, P.; Lecomte, L. *Tetrahedron Lett.* **2002**, *43*, 2661.

[588]Lee, J.C.; Yuk, J.Y.; Cho, S.H. *Synth. Commun.* **1995**, *25*, 1367.

[589]Benedict, D.A.; Bianchi, T.A.; Cate, L.A. *Synthesis* **1979**, 428; Johnstone, R.A.W.; Rose, M.E. *Tetrahedron* **1979**, *35*, 2169. See also, Loupy, A.; Sansoulet, J.; Vaziri-Zand, F. *Bull. Soc. Chim. Fr.* **1987**, 1027.

[590]Bogdal, D.; Pielichowski, J.; Jaskot, K. *Org. Prep. Proceed. Int.* **1998**, *30*, 427.

[591]Gadhwal, S.; Boruah, A.; Prajapati, D.; Sandhu, J.S. *Synth. Commun.* **1999**, *29*, 1921.

[592]Yuncheng, Y.; Yulin, J.; Jun, P.; Xiaohui, Z.; Conggui, Y. *Gazz. Chim. Ital.*, **1993**, *123*, 519.

[593]Paul, S.; Gupta, M. *Tetrahedron Lett.* **2004**, *45*, 8825.

[594]In bmim PF₆, 1-butyl-3-methylimidazolium hexafluorophosphate: Xu, Z.Y.; Xu, D.Q.; Liu, B.Y. *Org. Prep. Proceed. Int.* **2004**, *36*, 156.

[595]For an example, see Jha, S.C.; Joshi, N.N. *J. Org. Chem.* **2002**, *67*, 3897.

functional groups can be present in the molecule without interference. Ethers with one tertiary group *can* be prepared by treatment of an alkyl halide or sulfate ester (**10-10**) with a tertiary alkoxide $R'O^-$. Di-*tert*-butylether was prepared in high yield by direct attack by *t*-BuOH on the *tert*-butylcation (at $-80°C$ in SO_2ClF).[596] Di-*tert*-alkyl ethers in general have proved difficult to make, but they can be prepared in low-to-moderate yields by treatment of a tertiary halide with Ag_2CO_3 or Ag_2O.[597] Active halides, such as Ar_3CX, may react directly with the alcohol without the need for the more powerful nucleophile alkoxide ion.[598] Even tertiary halides have been converted to ethers in this way, with no elimination,[599] and hindered alcohols react as well.[600] Treatment of tertiary halides (R_3C-Cl) with zinc acetate and ultrasound leads to the corresponding acetate (R_3C-OAc) in a related reaction.[601] The mechanism is these cases is of course S_N1. *tert*-Butyl halides can be converted to aryl *tert*-butylethers by treatment with phenols and an amine, such as pyridine.[602] Aryl alkyl ethers can be prepared from alkyl halides by treatment with an aryl acetate (instead of a phenol) in the presence of K_2CO_3 and a crown ether.[603] It is possible to selectively alkylate the primary hydroxyl in a diol $HOCH_2CH(OH)R$ using a tin complex.[604] It is also possible to hydrogenate aldehydes and ketones (**19-36**) and trap the intermediate with an alcohol to form an ether.[605] The palladium-catalyzed displacement of allylic acetates with aliphatic alcohols has been shown to give the corresponding alkyl allyl ether.[606] The rhodium-catalyzed conversion of allylic carbonates to allylic benzyl ethers has also been reported.[607] Aryl ethers have been prepared using Mitsunobu conditions (see **10-17**).[608]

gem-Dihalides react with alkoxides to give acetals, and 1,1,1-trihalides give ortho esters.[609] Both aryl alkyl and dialkyl ethers can be efficiently prepared with

[596]Olah, G.A.; Halpern, Y.; Lin, H.C. *Synthesis* **1975**, 315. For another synthesis of di-*tert*-butyl ether, see Masada, H.; Yonemitsu, T.; Hirota, K. *Tetrahedron Lett.* **1979**, 1315.

[597]Masada, H.; Sakajiri, T. *Bull. Chem. Soc. Jpn.* **1978**, *51*, 866.

[598]For a review of reactions in which alcohols serve as nucleophiles, see Salomaa, P.; Kankaanperä, A.; Pihlaja, K., in Patai, S. *The Chemistry of the Hydroxyl Group*, pt. 1, Wiley, NY, **1971**, pp. 454–466.

[599]Biordi, J.; Moelwyn-Hughes, E.A. *J. Chem. Soc.* **1962**, 4291.

[600]Aspinall, H.C.; Greeves, N.; Lee, W.-M.; McIver, E.G.; Smith, P.M. *Tetrahedron Lett.* **1997**, *38*, 4679.

[601]Jayasree, J.; Rao, J.M. *Synth. Commun.* **1996**, *26*, 1103.

[602]Masada, H.; Oishi, Y. *Chem. Lett.* **1978**, 57. For another method, see Camps, F.; Coll, J.; Moretó, J.M. *Synthesis* **1982**, 186.

[603]Banerjee, S.K.; Gupta, B.D.; Singh, K. *J. Chem. Soc., Chem. Commun.* **1982**, 815.

[604]Boons, G.-J.; Castle, G.H.; Clase, J.A.; Grice, P.; Ley, S.V.; Pinel, C. *Synlett*, **1993**, 913.

[605]Bethmont, V.; Fache, F.; LeMaire, M. *Tetrahedron Lett.* **1995**, *36*, 4235.

[606]Nakagawa, H.; Hirabayashi, T.; Sakaguchi, S.; Ishii, Y. *J. Org. Chem.* **2004**, *69*, 3474; Haight, A.R.; Stoner, E.J.; Peterson, M.J.; Grover, V.K. *J. Org. Chem.* **2003**, *68*, 8092.

[607]Evans, P.A.; Leahy, D.K. *J. Am. Chem. Soc.* **2002**, *124*, 7882.

[608]Lepore, S.D.; He, Y. *J. Org. Chem.* **2003**, *68*, 8261.

[609]For a review of the formation of ortho esters by this method, see DeWolfe, R.H. *Carboxylic Ortho Acid Derivatives*, Academic Press, NY, **1970**, pp. 12–18.

the use of phase transfer catalysis (p. 511)[610] and with micellar catalysis.[611] Symmetrical benzylic ethers have been prepared by reaction of benzylic alcohols with Mg/I$_2$ followed by triflic anhydride.[612]

Hydroxy groups can be protected[613] by reaction of their salts with chloromethyl methyl ether.

$$RO^- \ + \ CH_3OCH_2Cl \longrightarrow ROCH_2OCH_3$$

This protecting group is known as MOM (methoxymethyl) and such compounds are called MOM ethers. The resulting acetals are stable to bases and are easily cleaved with mild acid treatment (**10-7**). Another protecting group, the 2-methoxyethoxymethyl group (the MEM group), is formed in a similar manner. Both MOM and MEM groups can be cleaved with dialkyl- and diarylboron halides, such as Me$_2$BBr.[614]

Aryl cyanates[615] can be prepared by reaction of phenols with cyanogen halides in the presence of a base: ArO$^-$ + ClCN → ArOCN + Cl$^-$.[616] This reaction has also been applied to certain alkyl cyanates.[617]

Most Williamson reactions proceed by the S$_N$2 mechanism, but there is evidence (see p. 446) that in some cases the SET mechanism can take place, especially with alkyl iodides.[618] Secondary alcohols have been converted to the corresponding methyl ether by reaction with methanol in the presence of ferric nitrate nonahydrate.[619] Vinyl ethers have been formed by coupling tetravinyl tin with phenols, in the presence of cupric acetate and oxygen.[620] The palladium-catalyzed coupling of vinyl triflates and phenols has also been reported.[621]

[610]For reviews, see Starks, C.M.; Liotta, C. *Phase Transfer Catalysis*, Springer, NY, *1978*, pp. 128–138; Weber, W.P.; Gokel, G.W. *Phase Transfer Catalysis in Organic Synthesis*, Springer, NY, *1977*, pp. 73–84. See also, Dueno, E.E.; Chu, F.; Kim, S.-I.; Jung, K.W. *Tetrahedron Lett.* *1999*, *40*, 1843; Eynde, J.J.V.; Mailleux, I. *Synth. Commun.* *2001*, *31*, 1. For the use of phase-transfer catalysis to convert one OH group of a diol or triol to a mono ether with selectivity, see de la Zerda, J.; Barak, G.; Sasson, Y. *Tetrahedron* *1989*, *45*, 1533.

[611]Juršić, B. *Tetrahedron* *1988*, *44*, 6677.

[612]Nishiyama, T.; Kameyama, H.; Maekawa, H.; Watanuki, K. *Can. J. Chem.* *1999*, *77*, 258.

[613]For other protecting groups for OH, see Wuts, P.G.M.; Greene, T.W. *Protective Groups in Organic Synthesis Vol. II*, Wiley, NY, *1991*, pp. 15–104; Wuts, P.G.M.; Greene, T.W. *Protective Groups in Organic Synthesis*, 3rd ed., Wiley, New York, *1999*. pp. 23–127; Corey, E.J.; Gras, J.; Ulrich, P. *Tetrahedron Lett.* *1976*, 809 and references cited therein.

[614]Guindon, Y.; Yoakim, C.; Morton, H.E. *J. Org. Chem.* *1984*, *49*, 3912. For other methods, see Williams, D.R.; Sakdarat, S. *Tetrahedron Lett.* *1983*, *24*, 3965; Hanessian, S.; Delorme, D.; Dufresne,Y. *Tetrahedron Lett.* *1984*, *25*, 2515; Rigby, J.H.; Wilson, J.Z. *Tetrahedron Lett.* *1984*, *25*, 1429.

[615]For reviews of alkyl and aryl cyanates, see Jensen, K.A.; Holm, A., in Patai, S. *The Chemistry of Cyanates and Their Thio Derivatives*, pt. 1, Wiley, NY, *1977*, pp. 569–618; Grigat, E.; Pütter, R. *Angew. Chem. Int. Ed.* *1967*, *6*, 206.

[616]Grigat, E.; Pütter, R. *Chem. Ber.* *1964*, *97*, 3012; Martin, D.; Bauer, M. *Org. Synth.* **VII**, 435.

[617]Kauer, J.C.; Henderson, W.W. *J. Am. Chem. Soc.* *1964*, *86*, 4732.

[618]Ashby, E.C.; Bae, D.; Park, W.; Depriest, R.N.; Su, W. *Tetrahedron Lett.* *1984*, *25*, 5107.

[619]Namboodiri, V.V.; Varma, R.S. *Tetrahedron Lett.* *2002*, *43*, 4593.

[620]Blouin, M.; Frenette, R. *J. Org. Chem.* *2001*, *66*, 9043.

[621]Willis, M.C.; Taylor, D.; Gillmore, A.T. *Chem. Commun.* *2003*, 2222.

OS **I**, 75, 205, 258, 296, 435; **II**, 260; **III**, 127, 140, 209, 418, 432, 544; **IV**, 427, 457, 558, 590, 836; **V**, 251, 258, 266, 403, 424, 684; **VI**, 301, 361, 395, 683; **VII**, 34, 386, 435; **VIII**, 26, 161, 155, 373; **80**, 227.

10-9   Epoxide Formation (Internal Williamson Ether Synthesis)

(3)*OC-cyclo*-Alkoxy-de-halogenation

This is a special case of **10-8**. The base removes the proton from the OH group and the resulting alkoxide subsequently attacks in an internal S_N2 reaction.[622] Many epoxides have been made in this way.[623] The course of the reaction can be influenced by neighboring group effects.[624] The method can also be used to prepare larger cyclic ethers: five- and six-membered rings.[625] Additional treatment with base yields the glycol (**10-7**). Thiiranes can be prepared by the reaction of α-chloro ketones with $(EtO)_2P(=O)$—SH and $NaBH_4$—Al_2O_3 with microwave irradiation.[626]

OS **I**, 185, 233; **II**, 256; **III**, 835; **VI**, 560; **VII**, 164, 356; **VIII**, 434.

10-10   Alkylation With Inorganic Esters

Alkoxy-de-sulfonyloxy-substitution

$$R-OSO_2OR'' \ + \ R'O^- \longrightarrow ROR$$

The reaction of alkyl sulfates with alkoxide ions is quite similar to **10-8** in mechanism and scope. Other inorganic esters can also be used. Methyl ethers of alcohols and phenols are commonly formd by treatment of alkoxides or aroxides with methyl sulfate. The alcohol or phenol can be methylated directly with dimethyl sulfate under various conditions.[627] Carboxylic esters sometimes give ethers when treated with alkoxides ($B_{AL}2$ mechanism, p. 1403) in a very similar process (see also, **16-64**). A related reaction heated **111** with alumina to give the corresponding benzofuran, **112**.[628]

[622]See, for example, Swain, C.G.; Ketley, A.D.; Bader, R.F.W. *J. Am. Chem. Soc.* **1959**, *81*, 2353; Knipe, A.C. *J. Chem. Soc. Perkin Trans. 2* **1973**, 589.

[623]For a review, see Berti, G. *Top. Stereochem.* **1973**, *7*, 93, pp. 187.

[624]Lang, F.; Kassab, D.J.; Ganem, B. *Tetrahedron Lett.* **1998**, *39*, 5903.

[625]See Kim, K.M.; Jeon, D.J.; Ryu, E.K. *Synthesis* **1998**, 835 for cyclization to an alkene in the presence of a catalytic amount of iodine. See Marek, I.; Lefrançois, J.-M.; Normant, J.-F. *Tetrahedron Lett.* **1992**, *33*, 1747 for a related reaction.

[626]Yadav, L.D.S.; Kapoor, R. *Synthesis* **2002**, 2344.

[627]Ogawa, H.; Ichimura, Y.; Chihara, T.; Teratani, S.; Taya, K. *Bull. Chem. Soc. Jpn.* **1986**, *59*, 2481; Cao, Y.-Q.; Pei, B.-G. *Synth. Commun.* **2000**, *30*, 1759.

[628]Mihara, M.; Ishino, Y.; Minakata, S.; Komatsu, M. *Synlett* **2002**, 1526.

The reaction of aliphatic alcohols and potassium organotrifluoroborate salts also gives ethers.[629]

tert-Butyl ethers (**113**) can be prepared by treating the compound *tert*-butyl2,2,2-trichloroacetimidate with an alcohol or phenol in the presence of boron trifluoride etherate.[630] Trichloroimidates can be used to prepare other ethers as well.[631] *tert*-Butyl ethers can be cleaved by acid-catalyzed hydrolysis.[632]

OS **I**, 58, 537; **II**, 387, 619; **III**, 127, 564, 800; **IV**, 588; **VI**, 737, 859, **VII**, 41. Also see, OS **V**, 431.

10-11   Alkylation With Diazo Compounds

Hydro,alkoxy-de-diazo-bisubstitution

$$CH_2N_2 \ + \ ROH \ \xrightarrow{HBF_4} \ CH_3OR$$

$$R_2CN_2 \ + \ ArOH \ \longrightarrow \ R_2CHOAr$$

Alcohols react with diazo compounds to form ethers, but diazomethane and diazo ketones are most readily available, giving methyl ethers or α-keto ethers,[633] respectively. With diazomethane[634] the method is expensive and requires great caution, but the conditions are mild and high yields are obtained. Diazomethane is used chiefly to methylate alcohols and phenols that are expensive or available in small amounts. Hydroxy compounds react better as their acidity increases; ordinary alcohols do not react at all unless a catalyst, such as HBF_4[635] or silica gel,[636] is present. The more acidic phenols react very well in the absence of a catalyst. The reaction of oximes, and ketones that have substantial enolic contributions,

[629]Quach, T.D.; Batey, R.A. *Org. Lett.* **2003**, *5*, 1381.

[630]Armstrong, A.; Brackenridge, I.; Jackson, R.F.W.; Kirk, J.M. *Tetrahedron Lett.* **1988**, *29*, 2483.

[631]Rai, A.N.; Basu, A. *Tetrahedron Lett.* **2003**, *44*, 2267.

[632]Lajunen, M.; Ianskanen-Lehti, K. *Acta Chem. Scand. B*, **1994**, *48*, 861.

[633]Pansare, S.V.; Jain, R.P.; Bhattacharyya, A. *Tetrahedron Lett.* **1999**, *40*, 5255.

[634]For a review of diazomethane, see Pizey, J.S. *Synthetic Reagents*, Vol. 2, Wiley, NY, **1974**, pp. 65–142.

[635]Neeman, M.; Caserio, M.C.; Roberts, J.D.; Johnson, W.S. *Tetrahedron* **1959**, *6*, 36.

[636]Ohno, K.; Nishiyama, H.; Nagase, H. *Tetrahedron Lett.* **1979**, 4405; Ogawa, H.; Hagiwara, H.; Chihara, T.; Teratani, S.; Taya, K. *Bull. Chem. Soc. Jpn.* **1987**, *60*, 627.

give *O*-alkylation to form, respectively, *O*-alkyl oximes and enol ethers. The mechanism[637] is as in **10-5**:

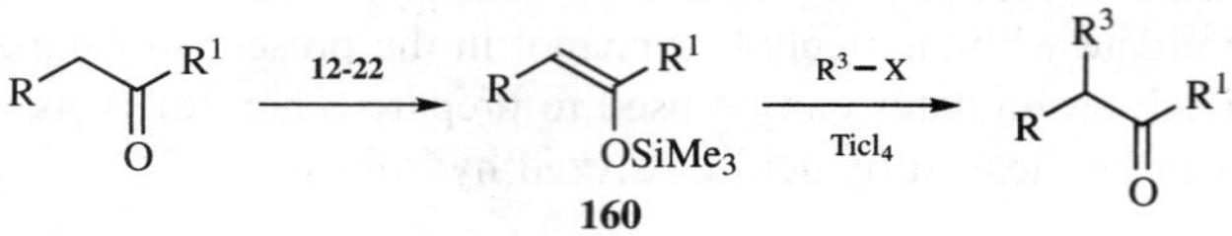

160

Diazoalkanes can also be converted to ethers by thermal or photochemical cleavage in the presence of an alcohol. These are carbene or carbenoid reactions.[638] Similar intermediates are involved when diazoalkanes react with alcohols in the presence of *t*-BuOCl to give acetals.[639]

$$ R_2CN_2 \ + \ 2\ R'OH \ \xrightarrow{\textit{t}\text{-BuOCl}} \ R_2C(OR')_2 $$

OS **V**, 245. Also see, OS **V**, 1099.

10-12   Dehydration of Alcohols

Alkoxy-de-hydroxylation

$$ 2\ ROH \ \xrightarrow{H_2SO_4} \ ROR \ + \ H_2O $$

The dehydration of alcohols to form symmetrical ethers[640] is analogous to **10-8** and **10-10**, but the species from which the leaving group departs is ROH_2^+ or $ROSO_2OH$. The former is obtained directly on treatment of alcohols with sulfuric acid and may go, by an S_N1 or S_N2 pathway, directly to the ether if attacked by another molecule of alcohol. On the other hand, it may, again by either an S_N1 or S_N2 route, be attacked by the nucleophile HSO_4^-, in which case it is converted to $ROSO_2OH$, which in turn may be attacked by an alcohol molecule to give ROR. Elimination is always a side reaction and, in the case of tertiary alkyl substrates, completely predominates. Good yields of ethers were obtained by heating diarylcarbinols [ArAr′CHOH → (ArAr′CH)$_2$O] with TsOH in the solid state.[641] Acids, such as Nafion-H with silyl ethers,[642] can be used in this transformation, and Lewis acids can be used with alcohols in some cases.[643]

---

[637]Kreevoy, M.M.; Thomas, S.J. *J. Org. Chem.* **1977**, *42*, 3979. See also, McGarrity, J.F.; Smyth, T. *J. Am. Chem. Soc.* **1980**, *102*, 7303.

[638]Bethell, D.; Newall, A.R.; Whittaker, D. *J. Chem. Soc. B* **1971**, 23; Noels, A.F.; Demonceau, A.; Petiniot, N.; Hubert, A.J.; Teyssié, P. *Tetrahedron* **1982**, *38*, 2733.

[639]Baganz, H.; May, H. *Angew. Chem. Int. Ed.* **1966**, *5*, 420.

[640]For a review, see Feuer, H.; Hooz, J., in Patai, S. *The Chemistry of the Ether Linkage*, Wiley, NY, **1967**, pp.457–460, 468–470.

[641]Toda, F.; Takumi, H.; Akehi, M. *J. Chem. Soc. Perkin Trans. 2* **1990**, 1270.

[642]Zolfigol, M.A.; Mohammadpoor-Baltork, I.; Habibi, D.; Mirjalili, B.B.F.; Bamoniri, A. *Tetrahedron Lett.* **2003**, *44*, 8165.

[643]For a reaction that used MeAl(NTf)$_2$, see Ooi, T.; Ichikawa, H.; Itagaki, Y.; Maruoka, K. *Heterocycles* **2000**, *52*, 575.

Mixed (unsymmetrical) ethers can be prepared if one group is tertiary alkyl and the other primary or secondary, since the latter group is not likely to compete with the tertiary group in the formation of the carbocation, while a tertiary alcohol is a very poor nucleophile.[644] If one group is not tertiary, the reaction of a mixture of two alcohols leads to all three possible ethers. Unsymmetrical ethers have been formed by treatment of two different alcohols with $MeReO_3$[645] or with $BiBr_3$.[646] Unsymmetrical ethers have been prepared under Mitsunobu conditions (**10-17**) with a polymer-supported phosphine and diethyl azodicarboxylate (DEAD).[647] Diols can be converted to cyclic ethers,[648] although the reaction is most successful for five-membered rings, but five-, six-, and seven-membered rings have been prepared.[649] Thus, 1,6-hexanediol gives mostly 2-ethyltetrahydrofuran. This reaction is also important in preparing furfural derivatives from aldoses, with concurrent elimination:

Phenols and primary alcohols form ethers when heated with dicyclohexylcarbodiimide[650] (see **16-63**). 1,2-Diols can be converted to epoxides by treatment with DMF dimethyl acetal, $(MeO)_2CHNMe_2$,[651] with diethyl azodicarboxylate, $EtOOCN=NCOOEt$, and Ph_3P,[652] with a dialkoxytriphenylphosphorane,[653] or with $TsCl^-Na^-OHPhCH_2NEt_3^+\ Cl^-$.[654]

OS **I**, 280; **II**, 126; **IV**, 25, 72, 266, 350, 393, 534; **V**, 539, 1024; **VI**, 887; **VIII**, 116. Also see, OS **V**, 721.

10-13 Transetherification

Hydroxy-de-alkoxylation and Alkoxy-de-hydroxylation

$$ROR' \ + \ R''OH \ \longrightarrow \ ROR'' \ + \ R'OH$$

The exchange of one alkoxy group for another is rare for *ethers* without a reactive R group, such as diphenylmethyl,[655] or by treatment of alkyl aryl ethers with

---

[644]See, for example, Jenner, G. *Tetrahedron Lett.* **1988**, *29*, 2445.

[645]Zhu, Z.; Espenson, J.H. *J. Org. Chem.* **1996**, *61*, 324.

[646]Boyer, B.; Keramane, E.-M.; Roque, J.-P.; Pavia, A.A. *Tetrahedron Lett.* **2000**, *41*, 2891.

[647]Lizarzaburu, M.E.; Shuttleworth, S. *Tetrahedron Lett.* **2002**, *43*, 2157.

[648]For a list of reagents, with references, see Larock, R.C. *Comprehensive Organic Transformations*, 2nd ed., Wiley-VCH, NY, **1999**, pp. 893–894.

[649]For an example, see Olah, G.A.; Fung, A.P.; Malhotra, R. *Synthesis* **1981**, 474.

[650]Vowinkel, E. *Chem. Ber.* **1962**, *95*, 2997; **1963**, *96*, 1702; **1966**, *99*, 42.

[651]Neumann, H. *Chimia*, **1969**, *23*, 267.

[652]Guthrie, R.D.; Jenkins, I.D.; Yamasaki, R.; Skelton, B.W.; White, A.H. *J. Chem. Soc. Perkin Trans. 1* **1981**, 2328 and references cited therein. For a review of diethyl azodicarboxylate-Ph_3P, see Mitsunobu, O. *Synthesis* **1981**, 1.

[653]Kelly, J.W.; Evans, Jr., S.A. *J. Org. Chem.* **1986**, *51*, 5490. See also, Hendrickson, J.B.; Hussoin, M.S. *Synlett*, **1990**, 423.

[654]Szeja, W. *Synthesis* **1985**, 983.

[655]Pratt, E.F.; Draper, J.D. *J. Am. Chem. Soc.* **1949**, *71*, 2846. Transetherification using $Fe(ClO_4)_3$ was reported. See Salehi, P.; Irandoost, M.; Seddighi, B.; Behbahani, F.K.; Tahmasebi, D.P. *Synth. Commun.* **2000**, *30*, 1743.

alkoxide ions: $ROAr + R'O^- \rightarrow ROR' + ArO^-$.[656] 3-(2-Benzyloxyethyl)-3-methyl-oxetane was transformed into 3-benzyloxymethyl-3-methyltetrahydrofuran by an internal transetherification catalyzed by $BF_3 \cdot OEt_2$.[657]

Acetals and ortho esters undergo transetherification readily,[658] as with the trasnfomraton of **114** to **115**.[659]

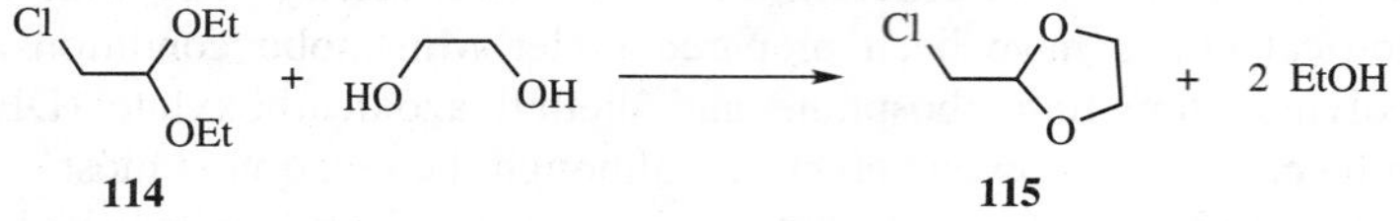

114   **115**

As seen in **10-6**, departure of the leaving group from an acetal gives a particularly stable carbocation. It is also possible to convert a dimethylketal directly to a dithiane by reaction with butane 1,4-dithiol on clay.[660] These are equilibrium reactions, and most often the equilibrium is shifted by removing the lower-boiling alcohol by distillation. Enol ethers can be prepared by treating an alcohol with an enol ester or a different enol ether, with mercuric acetate as a catalyst,[661] for example,

$$ROCH{=}CH_2 + R'OH \xrightarrow{Hg(OAc)_2} R'OCH{=}CH_2 + ROH$$

1,2-Diketones can be converted to α-keto enol ethers by treatment with an alkoxytrimethylsilane ($ROSiMe_3$).[662]

OS **VI**, 298, 491, 584, 606, 869; **VII**, 334; **VIII**, 155, 173. Also see, OS **V**, 1080, 1096.

10-14 Alcoholysis of Epoxides

(3)*OC-seco*-alkoxy-de-alkoxylation

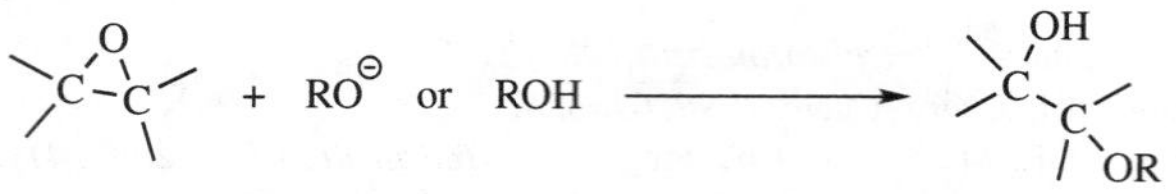

[656]Zoltewicz, J.A.; Sale, A.A. *J. Org. Chem.* **1970**, *35*, 3462.

[657]Itoh, A.; Hirose, Y.; Kashiwagi, H.; Masaki, Y. *Heterocycles* **1994**, *38*, 2165.

[658]For reviews, see Salomaa, P.; Kankaanperä, A.; Pihlaja, K., in Patai, S. *The Chemistry of the Hydroxyl Group*, pt. 1, Wiley, NY, **1971**, pp. 458–463; DeWolfe, R.H. *Carboxylic Ortho Acid Derivatives*, Academic Press, NY, **1970**, pp. 18–29, 146–148.

[659]McElvain, S.M.; Curry, M.J. *J. Am. Chem. Soc.* **1948**, *70*, 3781.

[660]Jnaneshwara, G.K.; Barahate, N.B.; Sudalai, A.; Deshpande, V.H.; Wakharkar, R.D.; Gajare, A.S.; Shingare, M.S.; Sukumar, R. *J. Chem. Soc. Perkin Trans. 1* **1998**, 965.

[661]Watanabe, W.H.; Conlon, L.E. *J. Am. Chem. Soc.* **1957**, *79*, 2828; Büchi, G.; White, J.D. *J. Am. Chem. Soc.* **1964**, *86*, 2884. For a review, see Shostakovskii, M.F.; Trofimov, B.A.; Atavin, A.S.; Lavrov, V.I. *Russ. Chem. Rev.* **1968**, *37*, 907. For a discussion of the mechanism, see Gareev, G.A. *J. Org. Chem. USSR* **1982**, *18*, 36.

[662]Ponaras, A.A.; Meah, M.Y. *Tetrahedron Lett.* **1986**, *27*, 4953.

This reaction is analogous to **10-7**. It may be acid (including Lewis acids[663]), base, or alumina[664] catalyzed, occur with electrolysis,[665] and may occur by either an S_N1 or S_N2 mechanism. Catalysts, such as $[Rh(CO)_2Cl]_2$,[666] $TiCl_3$ (OTf),[667] $Fe(ClO_4)_3$,[668] $Cu(BF_4)_2 \cdot n$ H_2O,[669] or $BiCl_3$,[670] have been used. β-Cyclodextrin has been used to promote the reaction with phenoxides in aqueous media.[671] Many of the β-hydroxy ethers produced in this way are valuable solvents, for example, diethylene glycol and Cellosolve. Reaction with thiols leads to hydroxy thioethers.[672] The reaction of alcohols with aziridines leads to β-amino ethers,[673] and reaction with thiols gives β-amino thioethers.[674] It has been shown that ring-opening of aziridines by phenols is promoted by tributylphosphine.[675]

Opening an epoxide by an alkoxide moiety can be done intramolecularly, and a new cyclic ether is generated. Ethers of various ring sizes can be produced depending on the length of the tether between the alkoxide unit and the epoxide. Specialized conditions are common, as in the conversion of **116** to **117**.[676] Another variant of this transformation used a cobalt–salen catalyst.[677] A specialized version has the alkoxide moiety on the carbon adjacent to the epoxide, leading to the *Payne rearrangement*, where a 2,3-epoxy alcohol is converted to an isomeric one, by treatment

[663]Iranpoor, N.; Tarrian, T.; Movahedi, Z. *Synthesis* **1996**, 1473; Iranpoor, N.; Salehi, P. *Synthesis* **1994**, 1152. See Moberg, C.; Rákos, L.; Tottie, L. *Tetrahedron Lett.* **1992**, *33*, 2191 for an example that generates a hydroxy ether with high enantioselectivity. Also see, Chini, M.; Crotti, P.; Gardelli, C.; Macchia, F. *Synlett*, **1992**, 673.

[664]See Posner, G.H.; Rogers, D.Z. *J. Am. Chem. Soc.* **1977**, *99*, 8208, 8214.

[665]Safavi, A.; Iranpoor, N.; Fotuhi, L. *Bull. Chem. Soc. Jpn.* **1995**, *68*, 2591.

[666]Fagnou, K.; Lautens, M. *Org. Lett.* **2000**, *2*, 2319.

[667]Iranpoor, N.; Zeynizadeh, B. *Synth. Commun.* **1999**, *29*, 1017.

[668]Salehi, P.; Seddighi, B.; Irandoost, M.; Behbahani, F.K. *Synth. Commun.* **2000**, *30*, 2967.

[669]Barluenga, J.; Vázquez-Villa, H.; Ballesteros, A.; González, J.M. *Org. Lett.* **2002**, *4*, 2817.

[670]Mohammadpoor-Baltork, I.; Tangestaninejad, S.; Aliyan, H.; Mirkhani, V. *Synth. Commun.*, **2000**, *30*, 2365.

[671]Surendra, K.; Krishnaveni, N.; Nageswar, Y.V.D.; Rao, K.R. *J. Org. Chem.* **2003**, *68*, 4994.

[672]Iida, T.; Yamamoto, N.; Sasai, H.; Shibasaki, M. *J. Am. Chem. Soc.* **1997**, *119*, 4783; Kesavan, V.; Bonnet-Delpon, D.; Bégué, J.-P. *Tetrahedron Lett.* **2000**, *41*, 2895; Fringuelli, F.; Pizzo, F.; Toroioli, S.; Vaccaro, L. *J. Org. Chem.* **2003**, *68*, 8248; Amantini, D.; Friguelli, F.; Pizzo, F.; Tortioli, S.; Vaccaro, L. *Synlett* **2003**, 2292.

[673]For a review, see Dermer, O.C.; Ham, G.E. *Ethlenimine and Other Aziridines*, Academic Press, NY, **1969**, pp. 224–227, 256–257.

[674]Wu, J.; Hou, X.-L.; Dai, L.-X. *J. Chem. Soc., Perkin Trans. 1* **2001**, 1314.

[675]Hou, X.-L.; Fan, R.-H.; Dai, L.-X. *J. Org. Chem.* **2002**, *67*, 5295.

[676]Matsumura, R.; Suzuki, T.; Sato, K.; Oku, K.-i.; Hagiwara, H.; Hoshi, T.; Ando, M.; Kamat, V.P. *Tetrahedron Lett.* **2000**, *41*, 7701. See also, Karikomi, M.; Watanabe, S.; Kimura, Y.; Uyehara, T. *Tetrahedron Lett.* **2002**, *43*, 1495.

[677]Wu, M.H.; Hansen, K.B.; Jacobsen, E.N. *Angew. Chem. Int. Ed.* **1999**, *38*, 2012.

with aqueous base:[678]

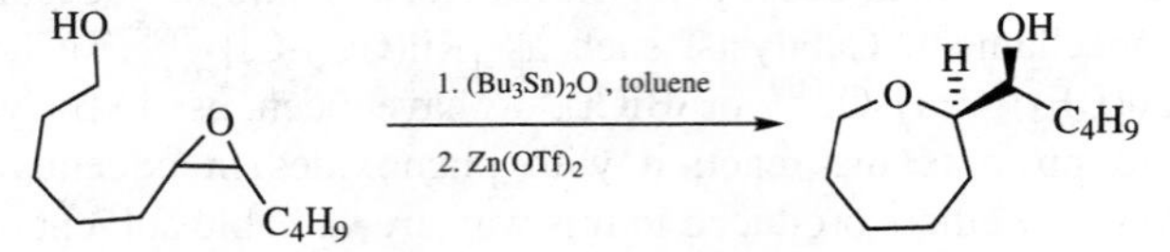

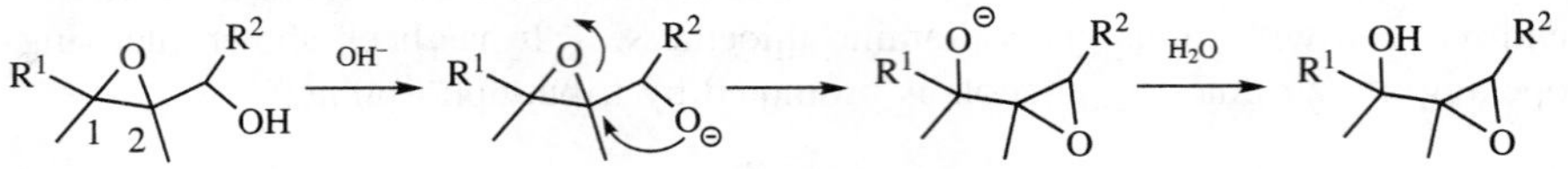

116   117

The reaction results in inverted configuration at C-2. Of course, the product can also revert to the starting material by the same pathway, so a mixture of epoxy alcohols is generally obtained.

Other nucleophilic oxygen or sulfur species have been shown to open epoxides. Examples include thiocyanate[679] and acetate via acetic anhydride and zeolite HY.[680] Epoxide react with sodium acetate and a cerium catalyst in detergent solutions to give hydroxy acetates.[681] In addition, N-tosylaziridines are opened by acetic acid in the presence of In(OTf)$_3$ to give N-tosylamino acetates.[682] The reaction of N-tosyl aziridines with 10% ceric ammonium nitrate in aqueous methanol leads to N-tosylamino alcohols,[683] and reaction with ethanol and 10% BF$_3$•OEt$_2$ gives N-tosyl ethers.[684] In the presence of Amberlyst 15, N-Boc (Boc = tert-butoxycarboxyl, —CO$_2$t-Bu) aziridines react with LiBr to give the corresponding bromo amide.[685]

10-15 Alkylation With Onium Salts

Alkoxy-de-hydroxylation

$$R_3O^+ \ + \ R'OH \ \longrightarrow \ ROR' \ + \ R_2O$$

Oxonium ions are excellent alkylating agents, and ethers can be conveniently prepared by treating them with alcohols or phenols.[686] Quaternary ammonium salts can sometimes also be used.[687]

OS **VIII**, 536.

[678]Payne, G.B. *J. Org. Chem.* **1962**, *27*, 3819; Behrens, C.H.; Ko, S.Y.; Sharpless, K.B.; Walker, F.J. *J. Org. Chem.* **1985**, *50*, 5687. See Yamazaki, T.; Ichige, T.; Kitazume, T. *Org. Lett.* **2004**, *6*, 4073.

[679]Sharghi, H.; Nasserri, M.A.; Niknam, K. *J. Org. Chem.* **2001**, *66*, 7287.

[680]Ramesh, P.; Reddy, V.L.N.; Venugopal, D.; Subrahmanya, M.; Venkateswarlu, Y. *Synth. Commun.* **2001**, *31*, 2599.

[681]Iranpoor, N.; Firouzabadi, H.; Safavi, A.; Shekarriz, M. *Synth. Commun.* **2002**, *32*, 2287.

[682]Yadav, J.S.; Reddy, B.V.S.; Sadashiv, K.; Harikishan, K. *Tetrahedron Lett.* **2002**, *43*, 2099.

[683]Chandrasekhar, S.; Narshihmulu, Ch.; Sultana, S.S. *Tetrahedron Lett.* **2002**, *43*, 7361.

[684]Prasad, B.A.B.; Sekar, G.; Singh, V.K. *Tetrahedron Lett.* **2000**, *41*, 4677.

[685]Righi, G.; Potini, C.; Bovicelli, P. *Tetrahedron Lett.* **2002**, *43*, 5867.

[686]Granik, V.G.; Pyatin, B.M.; Glushkov, R.G. *Russ. Chem. Rev.*, **1971**, *40*, 747, see p. 749.

[687]For an example, see Vogel, D.E.; Büchi, G.H. *Org. Synth.*, **66**, 29.

10-16   Hydroxylation of Silanes

Hydroxy-de-silylalkylation

$$R-SiR^1{}_2Ar \xrightarrow{F^-} R-SiR^{1'}{}_2F \xrightarrow{oxidation} R-OH$$

$$R-SiR^1{}_2SiR^2{}_3 \xrightarrow{F^-} R-SiR^1{}_2F \xrightarrow{oxidation} R-OH$$

Alkylsilanes can be oxidized, with the silyl unit converted to a hydroxy unit. This usually requires either an aryl group[688] or another silyl group[689] attached to silicon. It has been shown that a strained four-membered ring silane (a siletane) also gives the corresponding alcohol upon oxidation.[690] Treatment with a fluorinating agent, such as tetrabutylammonium fluoride or CsF replaces Ar or SiR$_3$ with F, which is oxidized with hydrogen peroxide or a peroxy acid to give the alcohol. This sequence is often called the *Tamao–Fleming oxidation*.[688] There are several variation in substrate that allow versatility in the initial incorporation of the silyl unit.[691] Hydroperoxide oxidation of a cyclic silane leads to a diol.[692]

C. Attack by OCOR at an Alkyl Carbon

10-17   Alkylation of Carboxylic Acid Salts

Acyloxy-de-halogenation

$$RX + R'COO^- \xrightarrow{HMPA} R'COOR$$

Sodium salts of carboxylic acids, including hindered acids, such as mesitoic, rapidly react with primary and secondary bromides and iodides at room temperature in dipolar aprotic solvents, especially HMPA, to give high yields of carboxylic esters.[693] The mechanism is S$_N$2. Several bases or basic media have been used to generate the carboxylate salt.[694] Sodium salts are often used, but potassium, silver, cesium,[695] and substituted ammonium salts have also been used. An important

[688]Kumada, M.; Tamao, K.; Yoshida, J.I. *J. Organomet. Chem.* **1982**, *239*, 115; Tamao, K.; Kakui, T.; Akita, M.; Iwahara, T.; Kanatani, R.; Yoshida, J.; Kumada, M. *Tetrahedron* **1983**, *39*, 983; Fleming, I.; Henning, R.; Plaut, H. *J. Chem. Soc., Chem. Commun.* **1984**, 29. For the protodesilylation step see Häbich, D.; Effenberger, F. *Synthesis* **1979**, 841. For the peroxyacid reaction see Buncel, E.; Davies, A.G. *J. Chem. Soc.* **1958**, 1550.
[689]Suginome, M.; Matsunaga, S.; Ito, Y. *Synlett*, **1995**, 941.
[690]Sunderhaus, J.D.; Lam, H.; Dudley, G.B. *Org. Lett.* **2003**, *5*, 4571.
[691]For examples see Matsumoto, Y.; Hayashi, T.; Ito, Y. *Tetrahedron* **1994**, *50*, 335; Uozumi, Y.; Kitayama, K.; Hayashi, T.; Yanagi, K.; Fukuyo, E. *Bull. Chem. Soc. Jpn.* **1995**, *68*, 713.
[692]Liu, D.; Kozmin, S.A. *Angew. Chem. Int. Ed.* **2001**, *40*, 4757.
[693]Parker, A.J. *Adv. Org. Chem.* **1965**, *5*, 1, 37; Alvarez, F.S.; Watt, A.N. *J. Org. Chem.* **1968**, *33*, 2143; Mehta, G. *Synthesis* **1972**, 262; Shaw, J.E.; Kunerth, D.C. *J. Org. Chem.* **1974**, *39*, 1968; Larock, R.C. *J. Org. Chem.* **1974**, *39*, 3721; Pfeffer, P.E.; Silbert, L.S. *J. Org. Chem.* **1976**, *41*, 1373.
[694]Bases include DBU (p. $$$): See Mal, D. *Synth. Commun.* **1986**, *16*, 331. Cs$_2$CO$_3$: Lee, J.C.; Oh, Y.S.; Cho, S.H.; Lee, J.I. *Org. Prep. Proceed. Int.* **1996**, *28*, 480. CsF-Celite: Lee, J.C.; Choi, Y. *Synth. Commun.* **1998**, *28*, 2021.
[695]See Dijkstra, G.; Kruizinga, W.H.; Kellogg, R.M. *J. Org. Chem.* **1987**, *52*, 4230.

variation uses phase-transfer catalysis,[696] and good yields of esters have been obtained from primary, secondary, benzylic, allylic, and phenacyl halides.[697] Without phase-transfer catalysts and in protic solvents, the reaction is useful only for fairly active R, such as benzylic and allylic, (S_N1 mechanism), but not for tertiary alkyl, since elimination occurs instead.[698] Solid-state procedures are available. Addition of the dry carboxylate salt and the halide to alumina as a solid support, and microwave irradiation gives the ester in a procedure that is applicable to long-chain primary halides.[699] A similar reaction of hexanoic acid and benzyl bromide on solid benzyltributylammonium chloride gave the ester with microwave irradiation.[700] Ionic liquid solvents have been shown to facilitate this alkylation reaction.[701]

The reaction of an alcohol and a carboxylate anion with diethyl azodicarboxylate EtOOCN=NCOOEt and Ph_3P[702] is called the *Mitsunobu esterification reaction*.[703] This reaction can also be considered as an S_N2. Other Mitsunobu catalysts are available,[704] and a polymer-bound phosphine has been used.[705] A renewable phosphine ligand has been developed.[706] Note that other functional groups, including azides[707] and thiocyanates[708] can be generated from alcohols using Mitsunobu conditions.

Lactones can be prepared from halo acids by treatment with base (see **16-63**). This has most often been accomplished with γ and δ lactones, but macrocyclic

---

[696]For reviews of phase-transfer catalysis of this reaction, see Starks, C.M.; Liotta, C. *Phase Transfer Catalysis*, Acaemic Press, NY, *1978*, pp. 140–155; Weber, W.P.; Gokel, G.W. *Phase Transfer Catalysis in Organic Synthesis Phase Transfer Catalysis in Organic Synthesis*, Springer, NY, *1977*, pp. 85–95.

[697]For an alternative method for phenacyl halides, see Clark, J.H.; Miller, J.M. *Tetrahedron Lett.* *1977*, 599.

[698]See, however, Moore, G.G.; Foglia, T.A.; McGahan, T.J. *J. Org. Chem.* *1979*, 44, 2425.

[699]Bram, G.; Loupy, A.; Majdoub, M.; Gutierrez, E.; Ruiz-Hitzky, E. *Tetrahedron* *1990*, 46, 5167. See Arrad, O.; Sasson, Y. *J. Am. Chem. Soc.* *1988*, 110, 185; Dakka, J.; Sasson, Y.; Khawaled, K.; Bram, G.; Loupy, A. *J. Chem. Soc., Chem. Commun.* *1991*, 853.

[700]Yuncheng, Y.; Yulin, J.; Dabin, G. *Synth. Commun.* *1992*, 22, 3109.

[701]Brinchi, L.; Germani, R.; Savelli, G. *Tetraheron Lett.* *2003*, 44, 2027, 6583. In bmim BF_4, 1-butyl-3-methylimidazolium tetrafluoroborate: Liu, Z.; Chen, Z.-C.; Zheng, Q.-G. *Synthesis* *2004*, 33.

[702]Mitsunobu, O.; Yamada, M. *Bull. Chem. Soc. Jpn.* *1967*, 40, 2380; Camp, D.; Jenkins, I.D. *Aust. J. Chem.* *1988*, 41, 1835.

[703]For discussions of the mechanism, see Ahn, C.; Correia, R.; DeShong, P. *J. Org. Chem.* *2002*, 67, 1751 and references cited therein. See also, Hughes, D.L. *Org. Prep. Proceed. Int.* *1996*, 28, 127; Dembinski, R. *Eur. J. Org. Chem.* *2004*, 2763; Dandapani, S.; Curran, D.P. *Chem. Eur. J.* *2004*, 10, 3131. For a discussion of microwave-promoted Mitsunobu reactions, see Steinreiber, A.; Stadler, A.; Mayer, S.F.; Faber, K.; Kappe, C.O. *Tetrahedron Lett.* *2001*, 42, 6283.

[704]See Tsunoda, T.; Yamamiya, Y.; Kawamura, Y.; Itô, S. *Tetrahedron Lett.* *1995*, 36, 2529; Tsunoda, T.; Nagaku, M.; Nagino, C.; Kawamura, Y.; Ozaki, F.; Hioki, H.; Itô, S. *Tetrahedron Lett.* *1995*, 36, 2531; Walker, M.A. *Tetrahedron Lett.* *1994*, 35, 665. For fluorous reactions and reagents, see Dandapani, S.; Curran, D.P. *Tetrahedron* *2002*, 58, 3855.

[705]Charette, A.B.; Janes, M.K.; Boezio, A.A. *J. Org. Chem.* *2001*, 66, 2178. See also, Elson, K.E.; Jenkins, I.D.; Loughlin, W.A. *Tetrahedron Lett.* *2004*, 45, 2491.

[706]Yoakim, C.; Guse, I.; O'Meara, J.A.; Thavonokham, B. *Synlett* *2003*, 473.

[707]For an example, see Papeo, G.; Poster, H.; Vianello, P.; Varasi, M. *Synthesis* *2004*, 2886.

[708]Iranpoor, N.; Firouzabadi, H.; Akhlaghinia, B.; Azadi, R. *Synthesis* *2004*, 92.

lactones (e.g., 11–17 members) have also been prepared in this way.[709] An interesting variation treated 2-ethylbenzoic acid with hypervalent iodine and then $I_2/h\nu$ to give the five-membered ring lactone.[710]

Copper(I) carboxylates give esters with primary (including neopentyl without rearrangement), secondary, and tertiary alkyl, allylic, and vinylic halides.[711] A simple S_N mechanism is obviously precluded in this case. Vinylic halides can be converted to vinylic acetates by treatment with sodium acetate if palladium(II) chloride is present.[712]

A carboxylic acid (not the salt) can be the nucleophile if F^- is present.[713] Mesylates are readily displaced, for example, by benzoic acid/CsF.[714] Dihalides have been converted to diesters by this method.[713] A COOH group can be conveniently protected by reaction of its ion with a phenacyl bromide ($ArCOCH_2Br$).[715] The resulting ester is easily cleaved when desired with zinc and acetic acid. Dialkyl carbonates can be prepared without phosgene (see **16-61**) by phase-transfer catalyzed treatment of primary alkyl halides with dry $KHCO_3$ and K_2CO_3.[716]

Other leaving groups can also be replaced by OCOR. Alkyl chlorosulfites (ROSOCl) and other derivatives of sulfuric, sulfonic, and other inorganic acids can be treated with carboxylate ions to give the corresponding esters. Treatment with oxalyl chloride allows displacement by carboxylate salts.[717] The use of dimethyl sulfate[718] or trimethyl phosphate[719] allows sterically hindered COOH groups to be methylated. The reaction of benzoic acid with aqueous lithium hydroxide and then dimethyl sulfate gave methyl benzoate.[720] Dimethyl carbonate in the presence of 1,8-diazabicyclo[5.4.0]undec-7-ene (DBU) has been used to prepare methyl esters.[721] With certain substrates, carboxylic acids are strong enough nucleophiles

[709]For example, see Galli, C.; Mandolini, L. *Org. Synth.* **VI**, 698; Kruizinga, W.H.; Kellogg, R.M. *J. Am. Chem. Soc.* **1981**, *103*, 5183; Kimura, Y.; Regen, S.L. *J. Org. Chem.* **1983**, *48*, 1533.

[710]Togo, H.; Muraki, T.; Yokoyama, M. *Tetrahedron Lett.* **1995**, *36*, 7089.

[711]Lewin, A.H.; Goldberg, N.L. *Tetrahedron Lett.* **1972**, 491; Klumpp, G.W.; Bos, H.; Schakel, M.; Schmitz, R.F.; Vrielink, J.J. *Tetrahedron Lett.* **1975**, 3429.

[712]Kohll, C.F.; van Helden, R. *Recl. Trav. Chim. Pays-Bas* **1968**, *87*, 481; Volger, H.C. *Recl. Trav. Chim. Pays-Bas* **1968**, *87*, 501; Yamaji, M.; Fujiwara, Y.; Asano, R.; Teranishi, S. *Bull. Chem. Soc. Jpn.* **1973**, *46*, 90.

[713]Clark, J.H.; Emsley, J.; Hoyte, O.P.A. *J. Chem. Soc. Perkin Trans. 1* **1977**, 1091; Ooi, T.; Sugimoto, H.; Doda, K.; Maruoka, K. *Tetrahedron Lett.* **2001**, *42*, 9245.

[714]Sato, T.; Otera, J. *Synlett*, **1995**, 336.

[715]Hendrickson, J.B.; Kandall, L.C. *Tetrahedron Lett.* **1970**, 343.

[716]Lissel, M.; Dehmlow, E.V. *Chem. Ber.* **1981**, *114*, 1210; Verdecchia, M.; Frochi, M.; Palombi, L.; Rossi, L. *J. Org. Chem.* **2002**, *67*, 8287. See also, Kadokawa, J.-i.; Habu, H.; Fukamachi, S.; Karasu, M.; Tagaya, H.; Chiba, K. *J. Chem. Soc., Perkin Trans. 1* **1999**, 2205.

[717]Barrett, A.G.M.; Braddock, D.C.; James, R.A.; Koike, N.; Procopiou, P.A. *J. Org. Chem.* **1998**, *63*, 6273.

[718]Grundy, J.; James, B.G.; Pattenden, G. *Tetrahedron Lett.* **1972**, 757.

[719]Harris, M.M.; Patel, P.K. *Chem. Ind. (London)* **1973**, 1002.

[720]Chakraborti, A.K.; Basak, A.; Grover, V. *J. Org. Chem.* **1999**, *64*, 8014. See also, Avila-Zárraga, J.G.; Martínez, R. *Synth. Commun.* **2001**, *31*, 2177.

[721]Shieh, W.-C.; Dell, S.; Repič, O. *Tetrahedron Lett.* **2002**, *43*, 5607.

for the reaction. Examples of such substrates are trialkyl phosphites $P(OR)_3$[722] and acetals of DMF.[723]

$$(RO)_2CHNMe_2 + R'COOH \longrightarrow R'COOR + ROH + HCONMe_2$$

This is an S_N2 process, since inversion is found at R. Another good leaving group is NTs_2 and ditosylamines react quite well with acetate ion in dipolar aprotic solvents:[724] $RNTs_2 + OAc^- \rightarrow ROAc$. Ordinary primary amines have been converted to acetates and benzoates by the Katritzky pyrylium–pyridinium method (p. 498).[725] Quaternary ammonium salts can be cleaved by heating with AcO^- in an aprotic solvent.[726] Oxonium ions can also be used as substrates:[727] $R_3O^+ + R'COO^- \rightarrow R'COOR + R_2O$. The reaction of potassium thioacetate with alkyl halides give dithiocarboxylic esters.[728]

In a variation of this reaction, alkyl halides can be converted to carbamates, by treatment with a secondary amine and K_2CO_3 under phase-transfer conditions.[729] The reaction of alcohols and alkyl halides can lead to carbonates.[730]

$$R-X + R'_2NH + K_2CO_3 \xrightarrow{Bu_4NH^+HSO_4^-} R \underset{O}{\overset{O}{\diagdown}} NR'_2$$

OS **II**, 5; **III**, 650; **IV**, 582; **V**, 580; **VI**, 273, 576, 698.

10-18   Cleavage of Ethers With Acetic Anhydride or Acid Halides

Acyloxy-de-alkoxylation

$$R-O-R' + Ac_2O \xrightarrow{FeCl_3} ROAc + R'OAc$$

Dialkyl ethers can be cleaved by treatment with anhydrous ferric chloride in acetic anhydride,[731] or with Me_3SiOTf in acetic anhydride.[732] In this reaction both R groups are converted to acetates and yields are moderate to high. Ethers

[722]Szmuszkovicz, J. *Org. Prep. Proceed. Int.* **1972**, *4*, 51.

[723]Vorbrüggen, H. *Angew. Chem. Int. Ed.* **1963**, *2*, 211; Brechbühler, H.; Büchi, H.; Hatz, E.; Schreiber, J.; Eschenmoser, A. *Angew. Chem. Int. Ed.* **1963**, *2*, 212.

[724]Andersen, N.H.; Uh, H. *Synth. Commun.* **1972**, *2*, 297; Curtis, V.A.; Schwartz, H.S.; Hartman, A.F.; Pick, R.M.; Kolar, L.W.; Baumgarten, R.J. *Tetrahedron Lett.* **1977**, 1969.

[725]See Katritzky, A.R.; Gruntz, U.; Kenny, D.H.; Rezende, M.C.; Sheikh, H. *J. Chem. Soc. Perkin Trans. 1* **1979**, 430.

[726]Wilson, N.D.V.; Joule, J.A. *Tetrahedron* **1968**, *24*, 5493.

[727]Raber, D.J.; Gariano Jr., P.; Brod, A.O.; Gariano, A.; Guida, W.C.; Guida, A.R.; Herbst, M.D. *J. Org. Chem.* **1979**, *44*, 1149.

[728]Zheng, T.-C.; Burkart, M.; Richardson, D.E. *Tetrahedron Lett.* **1999**, *40*, 603.

[729]Gómez-Parra, V.; Sánchez, F.; Torres, T. *Synthesis* **1985**, 282; *J. Chem. Soc. Perkin Trans. 2* **1987**, 695. For another method, with lower yields, see Yoshida, Y.; Ishii, S.; Yamashita, T. *Chem. Lett.* **1984**, 1571.

[730]Dueno, E.E.; Chu, F.; Kim, S.-I.; Jung, K.W. *Tetrahedron Lett.* **1999**, *40*, 1843. For the synthesis of cyclic carbonates see Yoshida, M.; Fujita, M.; Ishii, T.; Ihara, M. *J. Am. Chem. Soc.* **2003**, *125*, 4874.

[731]Ganem, B.; Small, Jr., V.M. *J. Org. Chem.* **1974**, *39*, 3728.

[732]Procopiou, P.A.; Baugh, S.P.D.; Flack, S.S.; Inglis, G.G.A. *Chem. Commun.* **1996**, 2625.

can also be cleaved by the mixed anhydride acetyl tosylate:[733]

$$R_2O \ + \ \underset{H_3C}{\overset{O}{\underset{}{\|}}}\!\!\!\!\!\! \overset{}{C}\text{-OTs} \quad\longrightarrow\quad \underset{H_3C}{\overset{O}{\underset{}{\|}}}\!\!\!\!\!\!\overset{}{C}\text{-OR} \ + \ ROTs$$

Epoxides give β-hydroxyalkyl carboxylates when treated with a carboxylic acid or a carboxylate ion and a suitable catalyst.[734] Tetrahydrofuran was opened to give *O*-acyl-4-iodo-1-butanol by treatment with acid chlorides and samarium halides[735] or BCl_3.[736] In a highly specialized transformation, the reaction of an epoxide with carbon dioxide and $ZnCl_2$ in an ionic liquid leads to a cyclic carbonate.[737] Epoxides react with CO and methanol in the presence of 10% of 3-hydroxypyridine and 5% of $Co_2(CO)_8$ to give a β-hydroxy methyl ester.[738]

OS **VIII**, 13.

10-19 Alkylation of Carboxylic Acids With Diazo Compounds

Hydro, **acyloxy-de-diazo-bisubstitution**

$$R_2CN_2 \ + \ R'COOH \quad\longrightarrow\quad R'COOCHR_2$$

Carboxylic acids can be converted to esters with diazo compounds in a reaction essentially the same as **10-11**. In contrast to alcohols, carboxylic acids undergo the reaction quite well at room temperature, since the reactivity of the reagent increases with acidity. The reaction is used where high yields are important or where the acid is sensitive to higher temperatures. Because of availability diazomethane (CH_2N_2)[634] is commonly used to prepare methyl esters, and diazo ketones are common. The mechanism is as shown in **10-11**.

OS **V**, 797.

D. Other Oxygen Nucleophiles

10-20 Formation of Oxonium Salts

$$RX \ + \ R_2O \quad\longrightarrow\quad R_3\overset{\oplus}{O} \ \overset{\ominus}{B}F_4 \ + \ AgX \qquad \textbf{Dialkyloxonio-de-halogenation}$$

$$RX \ + \ R_2'CO \quad\longrightarrow\quad R_2'C{=}\overset{\oplus}{O}{-}R \ \overset{\ominus}{B}F_4 \ + \ AgX$$

Alkyl halides can be alkylated by ethers or ketones to give oxonium salts, if a very weak, negatively charged nucleophile is present to serve as a counterion and a

[733]Karger, M.H.; Mazur, Y. *J. Am. Chem. Soc.* **1968**, *90*, 3878. See also, Coffi-Nketsia, S.; Kergomard, A.; Tautou, H. *Bull. Soc. Chim. Fr.* **1967**, 2788.

[734]See Otera, J.; Matsuzaki, S. *Synthesis* **1986**, 1019; Deardorff, D.R.; Myles, D.C. *Org. Synth.*, *67*, 114.

[735]Yu, Y.; Zhang, Y.; Ling, R. *Synth. Commun.* **1993**, *23*, 1973; Kwon, D.W.; Kim, Y.H.; Lee, K. *J. Org. Chem.* **2002**, *67*, 9488.

[736]Malladi, R.R.; Kabalka, G.W. *Synth. Commun.* **2002**, *32*, 1997.

[737]Li, F.; Xiao, L.; Xia, C.; Hu, B. *Tetrahedron Lett.* **2004**, *45*, 8307.

[738]Hinterding, K.; Jacobsen, E.N. *J. Org. Chem.* **1999**, *64*, 2164.

Lewis acid is present to combine with X^-.[739] A typical procedure consists of treating the halide with the ether or the ketone in the presence of $AgBF_4$ or $AgSbF_6$. The Ag^+ serves to remove X^- and the BF_4^- or SbF_6^- acts as the counterion. Another method involves treatment of the halide with a complex formed between the oxygen compound and a Lewis acid, for example, $R_2O\bullet BF_3 + RX \rightarrow R_3O^+ BF_4^-$, although this method is most satisfactory when the oxygen and halogen atoms are in the same molecule so that a cyclic oxonium ion is obtained. Ethers and oxonium ions also undergo exchange reactions:

$$2\,R_3O^+\,BF_4^- \;+\; 3\,R_2'O \;\rightleftharpoons\; 2\,R_3'O^+\,BF_4^- \;+\; 3\,R_2O$$

OS **V**, 1080, 1096, 1099; **VI**, 1019.

10-21   Preparation of Peroxides and Hydroperoxides

Hydroperoxy-de-halogenation

$$RX \;+\; {}^-OOH \longrightarrow ROOH$$

Hydroperoxides can be prepared by treatment of alkyl halides, esters of sulfuric or sulfonic acids, or alcohols with hydrogen peroxide in basic solution, where it is actually HO_2^+.[740] Sodium peroxide is similarly used to prepare dialkyl peroxides $(2\,RX + Na_2O_2 \rightarrow ROOR)$. Another method, which gives primary, secondary, or tertiary hydroperoxides and peroxides, involves treatment of the halide with H_2O_2 or a peroxide in the presence of silver trifluoroacetate.[741] Peroxides can also be prepared[742] by treatment of alkyl bromides or tosylates with potassium superoxide KO_2 in the presence of crown ethers (though alcohols may be side products[743]) and by the reaction between alkyl triflates and germanium or tin peroxide.[744] However, alkyl halides can be converted to symmetrical ethers by treatment with oxide ion generated *in situ* by a reaction between an organotin oxide and fluoride ion in the presence of a quaternary ammonium iodide or a crown ether.[745]

[739]Meerwein, H.; Hederich, V.; Wunderlich, K. *Arch. Pharm.* **1958**, *291/63*, 541. For a review, see Perst, H.*Oxonium Ions in Organic Chemistry*, Verlag Chemie, Deerfield Beach, VA, **1971**, pp. 22–39.

[740]For a review, see Hiatt, R., in Swern, D. *Organic Peroxides*, Vol. 2, Wiley, NY, **1971**, pp. 1–151. For a review of hydrogen peroxide, see Pandiarajan, K., in Pizey, J.S. *Synthetic Reagents*, Vol. 6, Wiley, NY, **1985**, pp. 60–155.

[741]Cookson, P.G.; Davies, A.G.; Roberts, B.P. *J. Chem. Soc., Chem. Commun.* **1976**, 1022. For another preparation of unsymmetrical peroxides, see Bourgeois, M.; Montaudon, E.; Maillard, B. *Synthesis* **1989**, 700.

[742]Johnson, R.A.; Nidy, E.G.; Merritt, M.V. *J. Am. Chem. Soc.* **1978**, *100*, 7960.

[743]Alcohols have also been reported to be the main products: San Filippo, Jr., J.; Chern, C.; Valentine, J.S. *J. Org. Chem.* **1975**, *40*, 1678; Corey, E.J.; Nicolaou, K.C.; Shibasaki, M.; Machida, Y.; Shiner, C.S. *Tetrahedron Lett.* **1975**, 3183.

[744]Salomon, M.F.; Salomon, R.G. *J. Am. Chem. Soc.* **1979**, *101*, 4290.

[745]Harpp, D.N.; Gingras, M. *J. Am. Chem. Soc.* **1988**, *110*, 7737.

Diacyl peroxides and acyl hydroperoxides can similarly be prepared[746] from acyl halides or anhydrides and from carboxylic acids.[747] Diacyl peroxides can

also be prepared by the treatment of carboxylic acids with hydrogen peroxide in the presence of dicyclohexylcarbodiimide,[748] H_2SO_4, methanesulfonic acid, or some other dehydrating agent. Mixed alkyl–acyl peroxides (peresters) can be made from acyl halides and hydroperoxides.

OS **III**, 619, 649; **V**, 805, 904; **VI**, 276.

10-22   Preparation of Inorganic Esters

Nitrosooxy-de-hydroxylation, and so on.

$$\text{ROH} + \text{HONO} \xrightarrow{\text{H+}} \text{RONO}$$
$$\text{ROH} + \text{HONO}_2 \xrightarrow{\text{H+}} \text{RONO}_2$$
$$\text{ROH} + \text{SOCl}_2 \longrightarrow \text{ROSOOR}$$
$$\text{ROH} + \text{POCl}_3 \longrightarrow \text{PO(OR)}_3$$
$$\text{ROH} + \text{SO}_3 \longrightarrow \text{ROSO}_2\text{OH}$$
$$\text{ROH} + (\text{CF}_3\text{SO}_2)_2\text{O} \longrightarrow \text{ROSO}_2\text{CF}_3$$

The above transformations show a few of the many inorganic esters that can be prepared by the reaction of an alcohol with an inorganic acid or, better, its acid halide or anhydride[749] These similar reactions are grouped together for convenience, but not all involve nucleophilic substitutions at R. The other possible pathway

[746]For a review of the synthesis and reactions of acyl peroxides and peresters, see Bouillon, G.; Lick, C.; Schank, K., in Patai, S. *The Chemistry of Peroxides*, Wiley, NY, *1983*, pp. 279–309. For a review of the synthesis of acyl peroxides, see Hiatt, R. Swern, D. *Organic Peroxides*, Vol. 2, Wiley, NY, *1971*, pp. 799–929.

[747]See Silbert, L.S.; Siegel, E.; Swern, D. *J. Org. Chem. 1962*, 27, 1336.

[748]Greene, F.D.; Kazan, J. *J. Org. Chem. 1963*, 28, 2168.

[749]For a review, see Salomaa, P.; Kankaanperä, A.; Pihlaja, K., in Patai, S. *The Chemistry of the Hydroxyl Group*, pt. 1, Wiley, NY, *1971*, pp. 481–497.

is nucleophilic substitution at the inorganic central atom, such as the attack of the alcohol oxygen at the electrophilic sulfur atom in **118**,[750] or a corresponding

$$\underset{\substack{\text{R}'}}{\overset{\substack{O\;\;\;O\\\diagdown\;\;\diagup\\ S}}{\diagdown}}\text{Cl} \longrightarrow \left[\underset{\substack{\text{R}'}}{\overset{\substack{O\;\;\;O\\\diagdown\;\;\diagup\\ S}}{\diagdown}}O^{\oplus} \right] \xrightarrow{\text{ROH}} \underset{\substack{\text{R}'}}{\overset{\substack{O\;\;\;O\\\diagdown\;\;\diagup\\ S}}{\diagdown}} \overset{H}{\underset{\substack{O\;\oplus\\ R}}{O}} \xrightarrow{-H^+} \underset{\substack{\text{R}'}}{\overset{\substack{O\;\;\;O\\\diagdown\;\;\diagup\\ S}}{\diagdown}}\text{OR}$$

118

S_N2-type process (see p. 1470). In such cases, there is no alkyl-O cleavage. Mono esters of sulfuric acid (alkylsulfuric acids), which are important industrially because their salts are used as detergents, can be prepared by treating alcohols with SO_3, H_2SO_4, $ClSO_2OH$, or SO_3 complexes.[751] It is possible to prepare a primary sulfonate ester such as tosylate, in the presence of a secondary alcohol unit when tosic acid reacts with a 1,2-diol in the presence of Fe^{3+}-Montmorillonite.[752] Polymer-bound reagents have been used to prepared sulfonate esters.[753] Phenolic triflate have been prepared using N,N-ditrifylaniline and K_2CO_3 under microwave irradiation.[754] Alkyl nitrites[755] can be conveniently prepared by an exchange reaction $ROH + R'ONO \rightarrow RONO + R'OH$, where $R = t$-Bu.[756] Primary amines can be converted to alkyl nitrates ($RNH_2 \rightarrow RONO_2$) by treatment with N_2O_4 at $-78°C$ in the presence of an excess of amidine base.[757] Mitsunobu conditions (**10-17**) can be used to prepare phosphate ester or phosphonate esters. The reaction can be done intramolecularly for prepare cyclic phosphonate esters.[758]

Alkyl halides are often used as substrates instead of alcohols. In such cases, the *salt* of the inorganic acid is usually used and the mechanism is nucleophilic substitution at the carbon atom. An important example is the treatment of alkyl halides with silver nitrate to form alkyl nitrates. This is used as a test for alkyl halides. In some cases, there is competition from the central atom. Thus nitrite ion is an ambident nucleophile that can give nitrites or nitro compounds (see **10-42**).[759] Dialkyl or aryl alkyl ethers can be cleaved with anhydrous sulfonic acids.[760]

$$\text{ROR}' + \text{R}''\text{SO}_2\text{OH} \longrightarrow \text{ROSO}_2\text{R}'' + \text{R}'\text{OH}$$

[750]For an example involving nitrite formation, see Aldred, S.E.; Williams, D.L.H.; Garley, M. *J. Chem. Soc. Perkin Trans. 2* **1982**, 777.

[751]For a review, see Sandler, S.R.; Karo, W. *Organic Functional Group Preparations*, 2nd ed., Vol 3; Academic Press, NY, **1989**, pp. 129–151.

[752]Choudary, B.M. Chowdari, N.S.; Kantam, M.L. *Tetraheron* **2000**, *56*, 7291.

[753]Vignola, N.; Dahmen, S.; Enders, D.; Bräse, S. *Tetrahedron Lett.* **2001**, *42*, 7833.

[754]Bengtson, A.; Hallberg, A.; Larhed, M. *Org. Lett.* **2002**, *4*, 1231.

[755]For a review of alkyl nitrites, see Williams, D.L.H. *Nitrosation*, Cambridge University Press, Cambridge, **1988**, pp. 150–172.

[756]Doyle, M.P.; Terpstra, J.W.; Pickering, R.A.; LePoire, D.M. *J. Org. Chem.* **1983**, *48*, 3379. For a review of the nitrosation of alcohols, see Williams, D.L.H. *Nitrosation*, Cambridge University Press, Cambridge, **1988**, pp. 150–156.

[757]Barton, D.H.R.; Narang, S.C. *J. Chem. Soc. Perkin Trans. 1* **1977**, 1114.

[758]Pungente, M.D.; Weiler, L. *Org. Lett.* **2001**, *3*, 643.

[759]For a review of formation of nitrates from alkyl halides, see Boguslavskaya, L.S.; Chuvatkin, N.N.; Kartashov, A.V. *Russ. Chem. Rev.* **1988**, *57*, 760.

[760]Klamann, D.; Weyerstahl, P. *Chem. Ber.* **1965**, *98*, 2070.

R'' may be alkyl or aryl. For dialkyl ethers, the reaction does not end as indicated above, since $R'OH$ is rapidly converted to $R'OR'$ by the sulfonic acid (reaction **10-12**), which in turn is further cleaved to $R'OSO_2R''$ so that the product is a mixture of the two sulfonates. For aryl alkyl ethers, cleavage always takes place to give the phenol, which is not converted to the aryl ether under these conditions. Ethers can also be cleaved in a similar manner by mixed anhydrides of sulfonic and carboxylic acids[761] (prepared as in **16-68**). β-Hydroxyalkyl perchlorates[762] and sulfonates can be obtained from epoxides.[763] Epoxides and oxetanes give α,ω-dinitrates when treated with N_2O_5.[764] Aziridines and azetidines react similarly, giving nitramine nitrates; for example, N-butylazetidine gave $NO_2OCH_2CH_2CH_2\text{-}N(Bu)NO_2$.[764]

OS **II**, 106, 108, 109, 112, 204, 412; **III**, 148, 471; **IV**, 955; **V**, 839; **VIII**, 46, 50, 616. Also see, OS **II**, 111.

10-23   Alcohols from Amines

Hydroxy-de-amination

$$RNH_2 \longrightarrow ROH$$

This is a rare transformation. A rather direct method was reported whereby a primary amine reacted with KOH in diethylene glycol at $210°C$.[765] The reaction of S-phenethylamine and the bis(sulfonyl chloride) of 1,2-benzenesulfonic acid, followed by KNO_2 and 18-crown-6 gave (R)-phenethyl alcohol in 70% yield and 40% enantiomeric excess (ee).[766]

10-24   Alkylation of Oximes[767]

A nitrone

Oximes can be alkylated by alkyl halides or sulfates. N-Alkylation is a side reaction, yielding a nitrone.[768] The relative yield of oxime ether and nitrone depends on the nature of the reagents, including the configuration of the oxime,

[761]Karger, M.H.; Mazur, Y. *J. Org. Chem.* **1971**, *36*, 532, 540.
[762]For a review of the synthesis and reactions of organic perchlorates, see Zefirov, N.S.; Zhdankin, V.V.; Koz'min, A.S. *Russ. Chem. Rev.* **1988**, *57*, 1041.
[763]Zefirov, N.S.; Kirin, V.N.; Yur'eva, N.M.; Zhdankin, V.V.; Kozmin, A.S. *J. Org. Chem. USSR* **1987**, *23*, 1264.
[764]Golding, P.; Millar, R.W.; Paul, N.C.; Richards, D.H. *Tetrahedron Lett.* **1988**, *29*, 2731, 2735.
[765]Rahman, S.M.A.; Ohno, H.; Tanaka, T. *Tetrahedron Lett.* **2001**, *42*, 8007.
[766]Sørbye, K.; Tautermann, C.; Carlsen, P.; Fiksdahl, A. *Tetraheron Asymmetry*, **1998**, *9*, 681.
[767]For a review of the chemistry of oximes see Ãbele, E.; Lukevics, E. *Org. Prep. Proceed. Int.* **2000**, *32*, 235.
[768]For a review of nitrones, see Torssell, K.B.G. *Nitrile Oxides, Nitrones, and Nitronates in Organic Synthesis*, VCH, NY, **1988**, pp. 75–93. For the synthesis of nitrones see Katritzky, A.R.; Cui, X.; Long, Q.; Yanga, B.; Wilcox, A.L.; Zhang, Y.-K. *Org. Prep. Proceed. Int.* **2000**, *32*, 175.

and on the reaction conditions.[769] For example, *anti*-benzaldoximes give nitrones, while the syn isomers give oxime ethers.[770]

OS **III**, 172; **V**, 1031. Also see, OS **V**, 269; **VI**, 199.

SULFUR NUCLEOPHILES

Sulfur compounds[771] are better nucleophiles than their oxygen analogs (p. 491), so in most cases these reactions take place faster and more smoothly than the corresponding reactions with oxygen nucleophiles. There is evidence that some of these reactions take place by SET mechanisms.[772]

10-25   Attack by SH at an Alkyl Carbon: Formation of Thiols[773]

Mercapto-de-halogenation

$$RX + H_2S \longrightarrow RSH_2^+ \longrightarrow RSH + H^+$$

$$RX + HS^- \longrightarrow RSH$$

Sodium sulfhydride (NaSH) is a much better reagent for the formation of thiols (mercaptans) from alkyl halides than H_2S and is used much more often. It is easily prepared by bubbling H_2S into an alkaline solution, but hydrosulfide on a supported polymer resin has also been used.[774] The reaction is most useful for primary halides. Secondary substrates give much lower yields, and the reaction fails completely for tertiary halides because elimination predominates. Sulfuric and sulfonic esters can be used instead of halides. Thioethers (RSR) are often side products.[775] The conversion can also be accomplished under neutral conditions by treatment of a primary halide with F^- and a tin sulfide, such as $Ph_3SnSSnPh_3$.[776] An indirect method for the preparation of a thiol is the reaction of an alkyl halide with thiourea to give an isothiuronium salt (**119**), and subsequent treatment with alkali or a

---

[769]For a review, see Reutov, O.A.; Beletskaya, I.P.; Kurts, A.L. *Ambident Anions*, Plenum, NY, *1983*, pp. 262–272.

[770]Buehler, E. *J. Org. Chem. 1967*, *32*, 261.

[771]For monographs on sulfur compounds, see Bernardi, F.; Csizmadia, I.G.; Mangini, A. *Organic Sulfur Chemistry*, Elsevier, NY, *1985*; Oae, S. *Organic Chemistry of Sulfur*, Plenum, NY, *1977*. For monographs on selenium compounds, see Krief, A.; Hevesi, L. *Organoselenium Chemistry I*, Springer, NY, *1988*; Liotta, D. *Organoselenium Chemistry*, Wiley, NY, *1987*.

[772]See Ashby, E.C.; Park, W.S.; Goel, A.B.; Su, W. *J. Org. Chem. 1985*, *50*, 5184.

[773]For a review, see Wardell, J.L., in Patai, S. *The Chemistry of the Thiol Group*, pt. 1; Wiley, NY, *1974*, pp. 179–211.

[774]Bandgar, B.P.; Sadavarte, V.S.; Uppalla, L.S. *Chem. Lett. 2000*, 1304.

[775]For a method of avoiding thioether formation, see Vasil'tsov, A.M.; Trofimov, B.A.; Amosova, S.V. *J. Org. Chem. USSR 1983*, *19*, 1197.

[776]Gingras, M.; Harpp, D.N. *Tetrahedron Lett. 1990*, *31*, 1397.

high-molecular-weight amine gives cleavage to the thiol.

$$
\underset{\underset{H_2N}{\overset{S}{\overset{\|}{C}}}\underset{NH_2}{}}{} + R-X \longrightarrow \underset{\underset{H_2N}{\overset{S-R}{\overset{\oplus}{C}}}\underset{NH_2}{}}{X^{\ominus}} \overset{^-OH}{\longrightarrow} R-S^{\ominus}
$$

$$\textbf{119}$$

Other indirect methods are treatment of the halide with silyl-thiols and KH, followed by treatment with fluoride ion and water,[777] and hydrolysis of Bunte salts (see **10-28**) is another method.

Thiols have also been prepared from alcohols. One method involves treatment with H_2S and a catalyst, such as Al_2O_3,[778] but this is limited to primary alcohols. Another method involves treatment with Lawesson's reagent (see **16-10**).[779] When epoxides are substrates, the products are β-hydroxy thiols.[780] Tertiary nitro compounds give thiols ($RNO_2 \rightarrow RSH$) when treated with sulfur and sodium sulfide, followed by amalgamated aluminum.[781]

OS **III**, 363, 440; **IV**, 401, 491; **V**, 1046; **VIII**, 592. Also see, OS **II**, 345, 411, 573; **IV**, 232; **V**, 223; **VI**, 620.

10-26   Attack by S at an Alkyl Carbon: Formation of Thioethers

Alkylthio-de-halogenation; Alkylthio-de-hydroxylation

$$R-X + R'-S^- \longrightarrow R-S-R'$$

$$R-OH + R'-SH \xrightarrow{\text{additives}} R-S-R'$$

Thioethers (sulfides) can be prepared by treatment of alkyl halides with salts of thiols (thiolate ions).[782] The R′ groups may be alkyl or aryl, and organolithium bases can be used to deprotonate the thiol.[783] As in **10-25**, RX cannot be a tertiary halide, and sulfuric and sulfonic esters can be used instead of halides. As in the Williamson reaction (**10-8**), yields are improved by phase-transfer catalysis.[784] Thiols can be reacted directly with alkyl halides in the presence of bases such as

---

[777]Miranda, E.I.; Díaz, M.J.; Rosado, I.; Soderquist, J.A. *Tetrahedron Lett.* **1994**, *35*, 3221; Rane, A.M.; Miranda, E.I.; Soderquist, J. *Tetrahedron Lett.* **1994**, *35*, 3225.

[778]Lucien, J.; Barrault, J.; Guisnet, M.; Maurel, R. *Nouv. J. Chim.* **1979**, *3*, 15.

[779]Nishio, T. *J. Chem. Soc., Chem. Commun.* **1989**, 205; Nishio, T. *J. Chem. Soc. Perkin Trans. 1* **1993**, 1113.

[780]For a review, see Wardell, J.L., in Patai, S. *The Chemistry of the Thiol Groups*, pt. 1, Wiley, NY, **1974**, pp. 246–251.

[781]Kornblum, N.; Widmer, J. *J. Am. Chem. Soc.* **1978**, *100*, 7086.

[782]For a review, see Peach, M.E., in Patai, S. *The Chemistry of the Thiol Groups*, pt. 2, Wiley, NY, **1974**, pp. 721–735.

[783]Yin, J.; Pidgeon, C. *Tetrahedron Lett.* **1997**, *38*, 5953.

[784]For a review of the use of phase transfer catalysis to prepare sulfur-containing compounds, see Weber, W.P.; Gokel, G.W. *Phase Transfer Catalysis in Organic Synthesis*, Springer, NY, **1977**, pp. 221–233.

DBU (p. 1531)[785] or CsF.[786] Neopentyl bromide was converted to Me_3CCH_2SPh in good yield by treatment with PhS^- in liquid NH_3 at $-33°C$ under the influence of light.[787] This probably takes place by an $S_{RN}1$ mechanism (see p. 862). Leaving groups other than chloride can be used, as in the ruthenium-catalyzed reaction of thiols with propargylic carbonates.[788] Vinylic sulfides can be prepared by treating vinylic bromides with PhS^- in the presence of a nickel complex,[789] with R_3SnSPh[790] or with $PhSLi$[791] in the presence of $Pd(PPh_3)_4$.

In some cases, alcohols can be converted to thioethers by reaction with thiols. Tertiary alcohols react with thiols in the presence of sulfuric acid to give thioethers, and the reaction works best with tertiary substrates.[792] This reaction is analogous to **10-12**. Thiophenol reacts with propargylic alcohols in the presence of a ruthenium catalysts to give propargylic thioethers.[793] Primary and secondary alcohols can be converted to alkyl aryl sulfides ($ROH \rightarrow RSAr$) in high yields by treatment with Bu_3P and an N-(arylthio)succinimide in benzene.[794] Primary alcohols reacted with benzylic thiols in the presence of PMe_3, 1,1'(azodicarbonyl)dipyridine (ADDP) and imidazole to give the thioether.[795] Thioethers RSR' can be prepared from an alcohol ROH and a halide $R'Cl$ by treatment with tetramethylthiourea $Me_2NC(=S)NMe_2$ followed by NaH.[796]

Thiolate ions are also useful for the demethylation of certain ethers,[797] esters, amines, and quaternary ammonium salts. Aryl methyl ethers[798] can be cleaved by heating with EtS^- in the dipolar aprotic solvent DMF: $ROAr + EtS^- \rightarrow ArO^- + EtSR$.[799] Carboxylic esters and lactones are cleaved (the lactones give ω-alkylthio carboxylic acids) with a thiol and $AlCl_3$ or $AlBr_3$.[800] Esters and lactones

[785]Ono, N.; Miyake, H.; Saito, T.; Kaji, A. *Synthesis* **1980**, 952. See also, Ferreira, J.T.B.; Comasseto, J.V.; Braga, A.L. *Synth. Commun.* **1982**, *12*, 595; Ando, W.; Furuhata, T.; Tsumaki, H.; Sekiguchi, A. *Synth. Commun.* **1982**, *12*, 627.; Feroci, M.; Inesi, A.; Rossi, L. *Synth. Commun.* **1999**, *29*, 2611.

[786]Shah, S.T.A.; Khan, K.M.; Heinich, A.M.; Voelter, W. *Tetrahedron Lett.* **2002**, *43*, 8281.

[787]Pierini, A.B.; Peñéñory, A.B.; Rossi, R.A. *J. Org. Chem.* **1985**, *50*, 2739.

[788]Kondo, T.; Kanda, Y.; Baba, A.; Fukuda, K.; Nakamura, A.; Wada, K.; Morisaki, Y.; Mitsudo, T.-a. *J. Am. Chem. Soc.* **2002**, *124*, 12960.

[789]Cristau, H.J.; Chabaud, B.; Labaudiniere, R.; Christol, H. *J. Org. Chem.* **1986**, *51*, 875.

[790]Carpita, A.; Rossi, R.; Scamuzzi, B. *Tetrahedron Lett.* **1989**, *30*, 2699. For another method, see Ogawa, T.; Hayami, K.; Suzuki, H. *Chem. Lett.* **1989**, 769.

[791]See Martínez, A.G.; Barcina, J.O.; Cerezo, A. de F.; Subramanian, L.R. *Synlett*, **1994**, 561.

[792]See Cain, M.E.; Evans, M.B.; Lee, D.F. *J. Chem. Soc.* **1962**, 1694.

[793]Inada, Y.; Nishibayashi, Y.; Hidai, M.; Uemura, S. *J. Am. Chem. Soc.* **2002**, *124*, 15172.

[794]Walker, K.A.M. *Tetrahedron Lett.* **1977**, 4475. See the references in this paper for other methods of converting alcohols to sulfides. See also, Cleary, D.G. *Synth. Commun.* **1989**, *19*, 737.

[795]Falck, J.R.; Lai, J.-Y.; Cho, S.-D.; Yu, J. *Tetrahedron Lett.* **1999**, *40*, 2903.

[796]Fujisaki, S.; Fujiwara, I.; Norisue, Y.; Kajigaeshi, S. *Bull. Chem. Soc. Jpn.* **1985**, *58*, 2429.

[797]For a review, see Evers, M. *Chem. Scr.* **1986**, *26*, 585.

[798]Certain other sulfur-containing reagents also cleave methyl and other ethers: see Hanessian, S.; Guindon, Y. *Tetrahedron Lett.* **1980**, *21*, 2305; Williard, P.G.; Fryhle, C.B. *Tetrahedron Lett.* **1980**, *21*, 3731; Node, M.; Nishide, K.; Fuji, K.; Fujita, E. *J. Org. Chem.* **1980**, *45*, 4275. For cleavage with selenium-containing reagents, see Evers, M.; Christiaens, L. *Tetrahedron Lett.* **1983**, *24*, 377. For a review of the cleavage of aryl alkyl ethers, see Tiecco, M. *Synthesis* **1988**, 749.

[799]Feutrill, G.I.; Mirrington, R.N. *Tetrahedron Lett.* **1970**, 1327, *Aust. J. Chem.* **1972**, *25*, 1719, 1731.

[800]Node, M.; Nishide, K.; Ochiai, M.; Fuji, K.; Fujita, E. *J. Org. Chem.* **1981**, *46*, 5163.

are similarly cleaved in high yield by phenyl selenide ion PhSe⁻.[801] Allylic sulfides
have been prepared by treating allylic carbonates ROCOOMe (R = an allylic group)
with a thiol and a Pd(0) catalyst.[802] A good method for the demethylation of quaternary ammonium salts consists of refluxing them with PhS⁻ in butanone:[803]

$$R_3\overset{\oplus}{N}Me \ + \ Ph\overset{\ominus}{S} \ \xrightarrow{\ \text{MeCOEt}\ } \ R_3N \ + \ PhSMe$$

A methyl group is cleaved more readily than other simple alkyl groups (such as
ethyl), although loss of these groups competes, but benzylic and allylic groups
cleave even more easily, and this is a useful procedure for the cleavage of benzylic
and allylic groups from quaternary ammonium salts, even if methyl groups are also
present.[804]

Symmetrical thioethers can also be prepared by treatment of an alkyl halide with
sodium sulfide.[805] Symmetrical thioethers have also been prepared by the reaction
of S(MgBr)₂ with allylic halides.[806]

$$2\ RX \ + \ Na_2S \ \longrightarrow \ RSR$$

This reaction can be carried out internally, by treatment of sulfide ions with 1,4-, 1,5-,
or 1,6-dihalides, to prepare five-, six-, and seven-membered[807] sulfur-containing
heterocyclic rings. Certain larger rings have also been closed in this way.[808] A
related variation converts epxoides to thiiranes with thiourea and LiBF₄ in
acetonitrile.[809]

gem-Dihalides can be converted to dithioacetals RCH(SR')₂,[810] and acetals have
been converted to monothioacetals $R_2C(OR')(SR^2)$,[811] and to dithioacetals.[812] The
combination of carbon disulfide and NaBH₄ converted 1,3-dibromopropane to 1,3-
dithiane.[813]

[801]Scarborough, Jr., R.M.; Smith III, A.B. *Tetrahedron Lett.* **1977**, 4361; Liotta, D.; Sunay, U.;
Santiesteban, H.; Markiewicz, W. *J. Org. Chem.* **1981**, *46*, 2605; Kong, F.; Chen, J.; Zhou, X. *Synth.
Commun.* **1988**, *18*, 801.
[802]Trost, B.M.; Scanlan, T.S. *Tetrahedron Lett.* **1986**, *27*, 4141; Goux, C.; Lhoste, P.; Sinou, D.
Tetrahedron Lett. **1992**, *33*, 8099; *Tetrahedron* **1994**, *50*, 10321.
[803]Shamma, M.; Deno, N.C.; Remar, J.F. *Tetrahedron Lett.* **1966**, 1375. For alternative procedures, see
Hutchins, R.O.; Dux, F.J. *J. Org. Chem.* **1973**, *38*, 1961; Posner, G.H.; Ting, J. *Synth. Commun.* **1974**, *4*,
355.
[804]Kametani, T.; Kigasawa, T.; Hiiragi, M.; Wagatsuma, N.; Wakisaka, K. *Tetrahedron Lett.* **1969**, 635.
[805]For another reagent, see Harpp, D.N.; Gingras, M.; Aida, T.; Chan, T.H. *Synthesis* **1987**, 1122.
[806]Nedugov, A.N.; Pavlova, N.N. *Zhur. Org. Khim.*, **1992**, *28*, 1401 (Engl. 1103).
[807]Tan, L.C.; Pagni, R.M.; Kabalka, G.W.; Hillmyer, M.; Woosley, J. *Tetrahedron Lett.* **1992**, *33*, 7709.
[808]See Hammerschmidt, E.; Bieber, W.; Vögtle, F. *Chem. Ber.* **1978**, *111*, 2445; Singh, A.; Mehrotra, A.;
Regen, S.L. *Synth. Commun.* **1981**, *11*, 409.
[809]Kazemi, F.; Kiasat, A.R.; Ebrahimi, S. *Synth. Commun.* **2003**, *33*, 595.
[810]See, for example, Wähälä, K.; Ojanperä, I.; Häyri, L.; Hase, T.A. *Synth. Commun.* **1987**, *17*, 137.
[811]Masaki, Y.; Serizawa, Y.; Kaji, K. *Chem. Lett.* **1985**, 1933; Sato, T.; Kobayashi, T.; Gojo, T.; Yoshida,
E.; Otera, J.; Nozaki, H. *Chem. Lett.* **1987**, 1661.
[812]Firouzabadi, H.; Iranpoor, N.; Hazarkhami, H. *J. Org. Chem.* **2001**, *66*, 7527, and references cited
therein; Ranu, B.C.; Das, A.; Samanta, S. *Synlett.* **2002**, 727.
[813]Wan,Y.; Kurchan, A.N.; Barnhurst, L.A.; Kutateladze, A.G. *Org. Lett.* **2000**, *2* , 1133.

When epoxides are substrates,[814] reaction with $PhSeSnBu_3/BF_3 \cdot OEt_2$[815] gives the corresponding β-hydroxy selenide in a manner analogous to that mentioned in **10-25**. Reaction of an epoxide with Ph_3SiSH followed by treatment with Bu_4NF gives hydroxy-thiols.[816] Epoxides can also be directly converted to episulfides[817] by treatment with a phosphine sulfide, such as Ph_3PS,[818] with thiourea and titanium tetraisopropoxide,[819] with NH_4SCN and $TiO(tfa)_2$,[820] with $(EtO)_2P(=O)H/S/Al_2O_3$,[821] with KSCN and $InBr_3$,[822] and with KSCN in ionic liquids.[823]

$$-\overset{O}{\overset{|}{C}}-\overset{|}{C}- \quad \xrightarrow[\text{NH}_2\text{CSNH}_2 \ + \ \text{Ti(O-i-Pr)}_4]{\text{Ph}_2\text{PS or}} \quad -\overset{S}{\overset{|}{C}}-\overset{|}{C}-$$

Alkyl halides, treated with thioethers, give sulfonium salts.[824] Other leaving groups have also been used for this purpose.[825]

Selenides (selenoethers) and tellurides can be prepared via RSe^- and RTe^- species,[826] and selenium and borohydride exchange resin followed by the halide give the selenoether.[827] The La/I_2-catalyzed reaction of diphenyl diselenide with primary alkyl iodides gave arylalkyl selenides,[828] and InI has been used with benzyl halides.[829] Diaryl selenides (Ar—Se—Ar') have been prepared by coupling aryl iodides with tin reagents ($ArSeSnR_3$) with a palladium(0) catalyst.[830]

[814]Chini, M.; Crotti, P.; Giovani, E.; Macchia, F.; Pineschi, M. *Synlett*, **1992**, 303.

[815]Nishiyama, Y.; Ohashi, H.; Itoh, K.; Sonoda, N. *Chem. Lett.* **1998**, 159.

[816]Brittain, J.; Gareau, Y. *Tetrahedron Lett.* **1993**, *34*, 3363.

[817]For a review of episulfides, see Fokin, A.V.; Kolomiets, A.F. *Russ. Chem. Rev.* **1975**, *44*, 138.

[818]Chan, T.H.; Finkenbine, J.R. *J. Am. Chem. Soc.* **1972**, *94*, 2880.

[819]Gao, Y.; Sharpless, K.B. *J. Org. Chem.* **1988**, *53*, 4114. For other methods, see Calò, V.; Lopez, L.; Marchese, L.; Pesce, G. *J. Chem. Soc., Chem. Commun.* **1975**, 621; Takido, T.; Kobayashi, Y.; Itabashi, K. *Synthesis* **1986**, 779; Bouda, H.; Borredon, M.E.; Delmas, M.; Gaset, A. *Synth. Commun.* **1987**, *17*; 943, **1989**, *19*, 491.

[820]Iranpoor, N.; Zeynizadeh, B. *Synth. Commun.* **1998**, *28*, 3913. See also, Tamami, B.; Kolahdoozan, M. *Tetrahedron Lett.* **2004**, *45*, 1535.

[821]Kaboudin, B.; Norouzi, H. *Tetrahedron Lett.* **2004**, *45*, 1283.

[822]Yadav, J.S.; Reddy, B.V.S.; Baishya, G. *Synlett.* **2003**, 396.

[823]Yadav, J.S.; Reddy, B.V.S.; Reddy, Ch.S.; Rajasekhar, K. *J. Org. Chem.* **2003**, *68*, 2525.

[824]For a review of the synthesis of sulfonium salts, see Lowe, P.A., in Stirling, C.J.M. *The Chemistry of the Sulphonium Group*, pt. 1, Wiley, NY, **1981**, pp. 267–312.

[825]See Badet, B.; Jacob, L.; Julia, M. *Tetrahedron* **1981**, *37*, 887; Badet, B.; Julia, M. *Tetrahedron Lett.* **1979**, 1101, and references cited in the latter paper.

[826]Brandsma, L.; Wijers, H.E. *Recl. Trav. Chim. Pays-Bas* **1963**, *82*, 68; Clarembeau, M.; Krief, A. *Tetrahedron Lett.* **1984**, *25*, 3625; Cohen, R.J.; Fox, D.L.; Salvatore, R.N. *J. Og. Chem.* **2004**, *69*, 4265. For a review of nucleophilic selenium, see Monahan, R.; Brown, D.; Waykole, L.; Liotta, D., in Liotta, D.C. *Organoselenium Chemistry*, Wiley, NY, **1987**, pp. 207–241.

[827]Yanada, K.; Fujita, T.; Yanada, R. *Synlett*, **1998**, 971.

[828]Nishino, T.; Okada, M.; Kuroki, T.; Watanabe, T.; Nishiyama, Y.; Sonoda, N. *J. Org. Chem.* **2002**, *67*, 8696. Zinc in aqueous media has also been used: see Bieber, L.W.; de Sá, A.C.P.F.; Menezes, P.H. Gonçalves, S.M.C. *Tetrahedron Lett.* **2001**, *42*, 4597.

[829]Ranu, B.C.; Mandal, T.; Samanta, S. *Org. Lett.* **2003**, *5*, 1439.; Ranu, B.C.; Mandal, T. *J. Org. Chem.* **2004**, *69*, 5793.

[830]Nishiyama, Y.; Tokunaga, K.; Sonoda, N. *Org. Lett.* **1999**, *1*, 1725.

OS **II**, 31, 345, 547, 576; **III**, 332, 751, 763; **IV**, 396, 667, 892, 967; **V**, 562, 780, 1046; **VI**, 5, 31, 268, 364, 403, 482, 556, 601, 683, 704, 737, 833, 859; **VII**, 453; **VIII**, 592. See also, OS **VI**, 776.

$$RI \ + \ R_2'S \ \longrightarrow \ R_2'\overset{\oplus}{S}R \ \ I^{\ominus}$$

10-27 Formation of Disulfides[831]

Dithio-de-dihalo-*aggre*-substitution

$$2\,RX \ + \ S_2{}^{2-} \ \longrightarrow \ RSSR \ + \ 2\,X^-$$

Disulfides can be prepared by treatment of alkyl halides with disulfide ions and also indirectly by the reaction of Bunte salts (see **10-28**) with acid solutions of iodide, thiocyanate ion, or thiourea,[832] or by pyrolysis or treatment with hydrogen peroxide. Alkyl halides also give disulfides when refluxed with sulfur and NaOH,[833] and with piperidinium tetrathiotungstate or piperidinium tetrathiomolybdate.[834] Other molybdenum compounds convert alkyl halides to disulfides, including $(BnNEt_3)_6Mo_7S_{24}$.[835]

There are no OS references, but a similar preparation of a polysulfide may be found in OS **IV**, 295.

10-28 Formation of Bunte Salts

Sulfonatothio-de-halogenation

$$RX + S_2O_3^{2-} \ \longrightarrow \ R{-}S{-}SO_3^- + X^-$$

Primary and secondary, but not tertiary, alkyl halides are easily converted to Bunte salts ($RSSO_3^-$) by treatment with thiosulfate ion.[836] Bunte salts can be hydrolyzed with acids to give the corresponding thiols[837] or converted to disulfides, tetrasulfides, or pentasulfides.[838]

OS **VI**, 235.

[831]For a discussion of disulfide exchange reactions, see Arisawa, M.; Yamaguchi, M. *J. Am. Chem. Soc.* **2004**, *125*, 6624.

[832]Milligan, B.; Swan, J.M. *J. Chem. Soc.* **1962**, 2712.

[833]Chorbadjiev, S.; Roumian, C.; Markov, P. *J. Prakt. Chem.* **1977**, *319*, 1036. For an example using microwave irradiation, see Wang, J.-X.; Gao, L.; Huang, D. *Synth. Commun.* **2002**, *32*, 963.

[834]Dhar, P.; Chandrasekaran, S. *J. Org. Chem.* **1989**, *54*, 2998.

[835]Polshettiwar, V.; Nivsarkar, M.; Acharya, J.; Kaushik, M.P. *Tetrahedron Lett.* **2003**, *44*, 887.

[836]For a review of Bunte salts, see Distler, H. *Angew. Chem. Int. Ed.* **1967**, *6*, 544–553.

[837]Kice, J.L. *J. Org. Chem.* **1963**, *28*, 957.

[838]Milligan, B.; Saville, B.; Swan, J.M. *J. Chem. Soc.* **1963**, 3608.

10-29 Alkylation of Sulfinic Acid Salts

Alkylsulfonyl-de-halogenation

$$RX + R'SO_2^- \longrightarrow R-SO_2-R' + X^-$$

Alkyl halides or alkyl sulfates, treated with the salts of sulfinic acids, give sulfones.[839] A palladium catalyzed reaction with a chiral complexing agent led to sulfones with modest asymmetric induction.[840] Alkyl sulfinates R'SO—OR may be side products.[841] Sulfonic acids themselves can be used, if DBU (p. 1530) is present.[842] Sulfonyl halides react with allylic halides in the presence of $AlCl_3^-Fe$[843] and wit benzyl hlaides in the presence of $Sm/HgCl_2$.[844] Sulfones have also been prepared by treatment of alkyl halides with tosylhydrazide.[845]

Vinyl sulfones were prepared from $PhSO_2Na$ and vinyl iodinium salts $C=C-I^+Ph$ BF_4^-.[846] Sulfinate esters (RS(=O)OR' were prepared from alcohols and sulfinyl chlorides, in the presence of Proton Sponge®.[847]

OS **IV**, 674; **IX**, 497. See also, OS **VI**, 1016.

10-30 Formation of Alkyl Thiocyanates

Thiocyanato-de-halogenation

$$RX + SCN^- \longrightarrow RSCN + X^-$$

Alkyl halides[848] or sulfuric or sulfonic esters can be heated with sodium or potassium thiocyanate to give alkyl thiocyanates,[849] although the attack by the analogous cyanate ion (**10-44**) gives exclusive *N*-alkylation. Primary amines can be converted to thiocyanates by the Katritzky pyrylium–pyridinium method (p. 498).[850] Tertiary

[839]For a review, see Schank, K., in Patai, S.; Rappoport, Z.; Stirling, C. *The Chemistry of Sulphones and Sulphoxides*, Wiley, NY, *1988*, pp. 165–231, 177–188.

[840]Eichelmann, H.; Gais, H.-J. *Tetrahedron Asymmetry*, *1995*, 6, 643.

[841]See, for example Meek, J.S.; Fowler, J.S. *J. Org. Chem. 1968*, 33, 3422; Kiełbasiński, P.; Żurawiński, R.; Drabowicz, J.; Mikołajczyk, M. *Tetrahedron 1988*, 44, 6687.

[842]Biswas, G.; Mal, D. *J. Chem. Res. (S) 1988*, 308.

[843]Saikia, P.; Laskar, D.D.; Prajapati, D.; Sandhu, J.S. *Chem. Lett. 2001*, 512.

[844]Zhang, J.; Zhang, Y. *J. Chem. Res. (S) 2001*, 516.

[845]Ballini, R.; Marcantoni, E.; Petrini, M. *Tetrahedron 1989*, 45, 6791.

[846]Ochiai, M.; Oshima, K.; Masaki, Y.; Kunishima, M.; Tani, S. *Tetrahedron Lett. 1993*, 34, 4829.

[847]Evans, J.W.; Fierman, M.B.; Miller, S.J.; Ellman, J.A. *J. Am. Chem. Soc. 2004*, 126, 8134.

[848]Renard, P.-Y.; Schwebel, H.; Vayron, P.; Leclerc, E.; Dias, S.; Mioskowski, C. *Tetrahedron Lett. 2001*, 42, 8479. For a variation involving *in situ* halogenation of active methylene compounds with formation of the thiocyanate, see Prakash, O.; Kaur, H.; Batra, H.; Rani, N.; Singh, S.P.; Moriarty, R.M. *J. Org. Chem. 2001*, 66, 2019. The reagent $Ph_3P(SCN)_2$ has also been used: see Iranpoor, N.; Firouzabadi, H.; Shaterian, H.R. *Tetrahedron Lett. 2002*, 43, 3439.

[849]For a review of thiocyanates, see Guy, R.G., in Patai, S. *The Chemistry of Cyanates and Their Thio Derivatives*, pt. 2; pp. 819–886, Wiley, NY, *1977*, pp. 819–886.

[850]Katritzky, A.R.; Gruntz, U.; Mongelli, N.; Rezende, M.C. *J. Chem. Soc. Perkin Trans. 1 1979*, 1953. For the conversion of primary alcohols to thiocyanates, see Tamura, Y.; Kawasaki, T.; Adachi, M.; Tanio, M.; Kita, Y. *Tetrahedron Lett. 1977*, 4417.

chlorides are converted to tertiary thiocyanates with $Zn(SCN)_2$ in pyridine and ultrasound.[851]

OS **II**, 366.

NITROGEN NUCLEOPHILES

A. Attack by NH_2, NHR, or NR_2 at an Alkyl Carbon

10-31   Alkylation of Amines

  Amino-de-halogenation (alkyl)

$$3\,RX + NH_3 \longrightarrow R_3N + RX \longrightarrow R_4N^+\ X^-$$
$$2\,RX + R'NH_2 \longrightarrow R_2R'N + RX \longrightarrow R_3R'N^+\ X^-$$
$$RX + R''R'NH_2 \longrightarrow RR'R''N + RX \longrightarrow R_2R'R''N^+\ X^-$$
$$RX + RR'R''N \longrightarrow RR'R''R''N^+\ X^-$$

The reaction between alkyl halides and ammonia or primary amines is not usually a feasible method for the preparation of primary or secondary amines, since they are stronger bases than ammonia and preferentially attack the substrate. However, the reaction is very useful for the preparation of tertiary amines[852] and quaternary ammonium salts. If ammonia is the nucleophile,[853] the three or four alkyl groups on the nitrogen of the product must be identical. If a primary, secondary, or tertiary amine is used, then different alkyl groups can be placed on the same nitrogen atom. The conversion of tertiary amines to quaternary salts is called the *Menshutkin reaction*.[854] It is sometimes possible to use this method for the preparation of a primary amine by the use of a large excess of ammonia or a secondary amine by the use of a large excess of primary amine. The use of ammonia in methanol with microwave irradiation has also been effective.[855] Microwave irradiation has also been used in reactions of aniline with allyl iodides.[856] A base other than the amine

[851]Bettadaiah, B.K.; Gurudutt, K.N.; Srinivas, P. *Synth. Commun.* **2003**, *33*, 2293.

[852]For reviews of this reaction, see Gibson, M.S., in Patai, S. *The Chemistry of the Amino Group*, Wiley, NY, **1968**, pp. 45–55; Spialter, L.; Pappalardo, J.A. *The Acyclic Aliphatic Tertiary Amines*, Macmillan, NY, **1965**, pp. 14–29.

[853]For a review of ammonia as a synthetic reagent, see Jeyaraman, R., in Pizey, J.S. *Synthetic Reagents*, Vol. 5, Wiley, NY, **1983**, pp. 9–83.

[854]For a discussion of solvent effects see Deleuze, M.S.; Leigh, D.A.; Zerbetto, F. *J. Am. Chem. Soc.* **1999**, *121*, 2364. For a review of stereoselectivity in this reaction see Bottini, A.T. *Sel. Org. Transform.* **1970**, *1*, 89. For a discussion of steric effects, see Persson, J.; Berg, U.; Matsson, O. *J. Org. Chem.* **1995**, *60*, 5037. For a review of quaternization of heteroaromatic rings, see Zoltewicz, J.A.; Deady, L.W. *Adv. Heterocycl. Chem.* **1978**, *22*, 71. See Shaik, S.; Ioffe, A.; Reddy, A.C.; Pross, A. *J. Am. Chem. Soc.* **1994**, *116*, 262 for a discussion of the transition state for this reaction.

[855]Saulnier, M.G.; Zimmermann, K.; Struzynski, C.P.; Sang, X.; Velaparthi, U.; Wittman, M.; Frennesson, D.B. *Tetrahedron Lett.* **2004**, *45*, 397.

[856]Romera, J.L.; Cid, J.M.; Trabanco, A.A. *Tetrahedron Lett.* **2004**, *45*, 8797.

can be added to facilitate the reaction. Sodium carbonate has been used,[857] as has lithium hydroxide.[858] Cesium hydroxide was successfully used as a base in the presence of molecular sieve 4 Å,[859] and cesium fluoride has been used with benzylic halides.[860] Potassium carbonate in DMSO has been used for the alkylation of aniline.[861] Bromides react faster than chlorides, and secondary amines reaction with 3-chloro-1-bromopropane via the bromide, in the presence of Zn and THF.[862]

The limitations of this approach can be seen in the reaction of a saturated solution of ammonia in 90% ethanol with ethyl bromide in a 16:1 molar ratio, under which conditions the yield of primary amine was 34.2% (at a 1:1 ratio the yield was 11.3%).[863] Alkyl amines can be one type of substrate that does give reasonable yields of primary amine (provided a large excess of NH_3 is used) are α-halo acids, which are converted to amino acids. N-Chloromethyl lactams also react with amines to give good yields to the N-aminomethyl lactam.[864] Primary amines can be prepared from alkyl halides by **10-43**, followed by reduction of the azide (**19-32**),[865] or by the Gabriel synthesis (**10-41**).

The immediate product in any particular step is the protonated amine, but it rapidly loses a proton to another molecule of ammonia or amine in an equilibrium process, for example,

$$RX + R_2NH \longrightarrow R_3\overset{\oplus}{N}H + R_2NH \rightleftharpoons R_3N + R_2\overset{\oplus}{N}H_2$$

When it is desired to convert a primary or secondary amine directly to the quaternary salt (*exhaustive alkylation*), the rate can be increased by the addition of a non-nucleophilic strong base that serves to remove the proton from $RR'NH_2^+$ or $RR'R^2NH^+$ and thus liberates the amine to attack another molecule of RX.[866]

The conjugate bases of ammonia and of primary and secondary amines (NH_2^-, RNH^- R_2N^-) are sometimes used as nucleophiles,[867] including amide bases generated from organolithium reagents and amines (R_2NLi).[868] This is in contrast to the

[857]Faul, M.M.; Kobierski, M.E.; Kopach, M.E. *J. Org. Chem.* **2003**, *68*, 5739.

[858]Cho, J.H.; Kim, B.M. *Tetrahedron Lett.* **2002**, *43*, 1273.

[859]Salvatore, R.N.; Nagle, A.S.; Schmidt, S.E.; Jung, K.W. *Org. Lett.* **1999**, *1*, 1893; Salvatore, R.N.; Schmidt, S.E.; Shin, S.I.; Nagle, A.S.; Worrell, J.H.; Jung, K.W. *Tetrahedron Lett.* **2000**, *41*, 9705.

[860]Hayat, S.; Rahman, A.-U.; Choudhary, M.I.; Khan, K.M.; Schumann, W.; Bayer, E. *Tetrahedron* **2001**, *57*, 9951.

[861]Srivastava, S.K.; Chauhan, P.M.S.; Bhaduri, A.P. *Synth. Commun.* **1999**, *29*, 2085; Jaisinghani, H.G.; Khadilkar, B.M. *Synth. Commun.* **1999**, *29*, 3693; Salvatore, R.N.; Nagle, A.S.; Jung, K.W. *J. Org. Chem.* **2002**, *67*, 674.

[862]Murty, M.S.R.; Jyothirmai, B.; Krishna, P.R.; Yadav, J.S. *Synth. Commun.* **2003**, *33*, 2483.

[863]Werner, E.A. *J. Chem. Soc.* **1918**, *113*, 899.

[864]Chen, P.; Suh, D.J.; Smith, M.B. *J. Chem. Soc. Perkin Trans. 1* **1995**, 1317; Deskus, J.; Fan, D.-p.; Smith. M.B. *Synth. Commun.* **1998**, *28*, 1649.

[865]See Kumar, H.M.S.; Anjaneyulu, S.; Reddy, B.V.S.; Yadav, J.S. *Synlett.* **1999**, 551.

[866]Sommer, H.Z.; Jackson, L.L. *J. Org. Chem.* **1970**, *35*, 1558; Sommer, H.Z.; Lipp, H.I.; Jackson, L.L. *J. Org. Chem.* **1971**, *36*, 824. See also, Chuang, T.-H.; Sharpless, K.B. *Org. Lett.* **2000**, *2*, 3555.

[867]For a discussion of the mechanism of the reaction between a primary halide and Ph_2NLi, see DePue, J.S.; Collum, D.B. *J. Am. Chem. Soc.* **1988**, *110*, 5524.

[868]Vitale, A.A.; Chiocconi, A.A. *J. Chem. Res. (S)* **1996**, 336.

analogous methods **10-1**, **10-8**, **10-25**, and **10-26**. Pyrrole is converted to
N-methylpyrrole with KOH, iodomethane in ionic liquids.[869] Primary alkyl, allylic,
and benzylic bromides, iodides, and tosylates react with sodium bis(trimethylsilyl)
amide to give derivatives that are easily hydrolyzed to produce amine salts in high
overall yields.[870] Primary arylamines are easily alkylated, but diaryl- and triarylamines are very poor nucleophiles. However, the reaction has been carried out with
diarylamines.[871] Sulfates or sulfonates can be used instead of halides. The reaction
can be carried out intramolecularly to give cyclic amines, with three-, five-, and six-
membered (but not four-membered) rings being easily prepared. Thus, 4-chloro-1-
aminobutane treated with base gives pyrrolidine, and 2-chloroethylamine gives aziridine[872] (analogous to **10-9**):

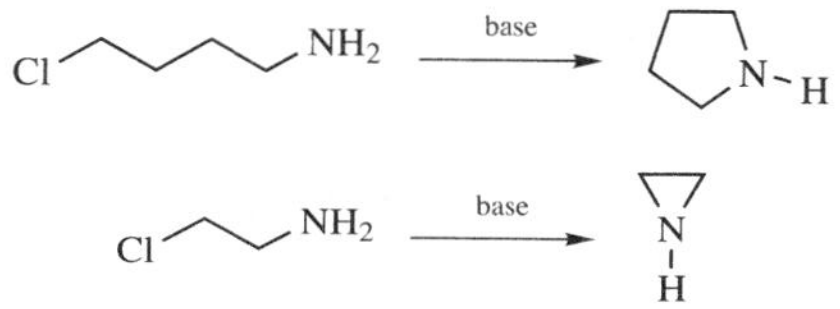

Reduction of N-(3-bromopropyl) imines gives a bromo-amine *in situ*, which
cyclizes to the aziridine.[873] Five-membered ring amines (pyrrolidines) can be prepared from alkenyl amines via treatment with N-chlorosuccinimide and then
Bu_3SnH.[874] Internal addition of amine to allylic acetates, catalyzed by $Pd(PPh_3)_4$,
leads to cyclic products via a S_N2' reaction.[875] Three-membered cyclic amines
(aziridines) can be prepared from chiral conjugated amides via bromination and
reaction with an amine.[876] Four-membered cyclic amines (azetidines) have been
prepared in a different way:[877]

This reaction was also used to close five-, six-, and seven-membered rings.

As usual, tertiary substrates do not give the reaction at all but undergo preferential elimination. However, tertiary (but not primary or secondary) halides R_3CCl
can be converted to primary amines R_3CNH_2 by treatment with NCl_3 and
$AlCl_3$[878] in a reaction related to **10-39**.

[869]In bmim PF_6, 1-butyl-3-methylimidazolium hexafluorophosphate: Le, Z.-G.; Chen, Z.-C.; Hu, Y.;
Zheng, Q.-G. *Synthesis* **2004**, 1951.
[870]Bestmann, H.J.; Wölfel, G. *Chem. Ber.* **1984**, *117*, 1250.
[871]Patai, S.; Weiss, S. *J. Chem. Soc.* **1959**, 1035.
[872]For a review of aziridine formation by this method, see Dermer, O.C.; Ham, G.E. *Ethylenimine and
Other Aziridines*, Academc Press, NY, **1969**, pp. 1–59.
[873]DeKimpe, N.; DeSmaele, D. *Tetrahedron Lett.*, **1994**, 35, 8023. Also see, De Kimpe, N.; Boelens, M.;
Piqueur, J.; Baele, J. *Tetrahedron Lett.* **1994**, 35, 1925.
[874]Tokuda, M.; Fujita, H.; Suginome, H. *J. Chem. Soc. Perkin Trans. 1* **1994**, 777.
[875]Grellier, M.; Pfeffer, M.; van Koten, G. *Tetrahedron Lett.* **1994**, 35, 2877.
[876]Garner, P.; Dogan, O.; Pillai, S. *Tetrahedron Lett.*,**1994**, 35, 1653.
[877]Juaristi, E.; Madrigal, D. *Tetrahedron* **1989**, 45, 629.
[878]Strand, J.W.; Kovacic, M.K. *J. Am. Chem. Soc.* **1973**, 95, 2977.

Amines can be *N*-alkylated by reaction with alcohols, in a sealed tube with microwave irradiation,[879] by ruthenium-catalyzed,[880] palladium-[881] or iridium-catalyzed[882] reactions. Heating indoles with benzylic alcohols in the presence of $Me_3P=CH(CN)$ give the *N*-benzylindole.[883] Heating an alcohol on γ-Al_2O_3 leads to an amine,[884] as does treatment with the amine, $SnCl_2$ and $Pd(PPh_3)_4$.[885] The palladium-catalyzed displacement of allylic acetates leads to allylic amines.[886] Chlorodiethylaluminum (Et_2AlCl), with a Cu(II) catalysts can be used to prepare *N*-ethylaniline derivatives.[887] *tert*-Butylamines can be prepared from isobutylene, HBr and the amine by heating a sealed tube.[888]

Phosphines behave similarly, and compounds of the type R_3P and $R_4P^+ X^-$ can be so prepared.[889] The reaction between triphenylphosphine and quaternary salts of nitrogen heterocycles in an aprotic solvent is probably the best way of dealkylating the heterocycles, for example,[890]

Primary amines can be prepared from alkyl halides by the use of hexamethylenetetramine[891] followed by cleavage of the resulting salt with ethanolic HCl. The method, called the *Delépine reaction*, is most successful for active halides such as allylic and benzylic halides and α-halo ketones, and for primary

A convenient way of obtaining secondary amines without contamination by primary or tertiary amines involves treatment of alkyl halides with the sodium or

[879]Jiang, Y.-L.; Hu, Y.-Q.; Feng, S.-Q.; Wu, J.-S.; Wu, Z.-W.; Yuan, Y.-C.; Liu, J.-M.; Hao, Q.-S.; Li, D.-P. *Synth. Commun.* **1996**, *26*, 161.

[880]Watanabe, Y.; Morisaki, Y.; Kondo, T.; Mitsudo, T. *J. Org. Chem.* **1996**, *61*, 4214.

[881]Yang, S.-C.; Yu, C.-L.; Tsai, Y.-C. *Tetrahedron Lett.* **2000**, *41*, 7097; Shue, Y.-J.; Yang, S.-C.; Lai, H.-C. *Tetrahedron Lett.* **2003**, *44*, 1481; Kimura, M.; Futamata, M.; Shibata, K.; Tamaru, Y. *Chem. Commun.* **2003**, 234.

[882]Takeuchi, R.; Ue, N.; Tanabe, K.; Yamashita, K.; Shiga, N. *J. Am. Chem. Soc.* **2001**, *123*, 9525; Fujita, K.-i.; Li, Z.; Ozeki, N.; Yamaguchi, R. *Tetrahedron Lett.* **2003**, *44*, 2687.

[883]Bombrun, A.; Casi, G. *Tetrahedron Lett.* **2002**, *43*, 2187.

[884]Valot, F.; Fache, F.; Jacquot, R.; Spagnol, M.; Lemaire, M. *Tetrahedron Lett.* **1999**, *40*, 3689. For a zeolite mediated reaction that uses methyl acetate, see Selva, M.; Tundo, P.; Perosa, A. *J. Org. Chem.* **2003**, *68*, 7374.

[885]Masuyama, Y.; Kagawa, M.; Kurusu, Y. *Chem. Lett.* **1995**, 1121.

[886]Kodama, H.; Taiji, T.; Ohta, T.; Furukawa, I. *Synlett* **2001**, 385; Feuerstein, M.; Laurenti, D.; Doucet, H.; Santelli, M. *Tetrahedron Lett.* **2001**, *42*, 2313; Watson, I.D.G.; Styler, S.A.; Yudin, A.K. *J. Am. Chem. Soc.* **2004**, *126*, 5086; Ohta, T.; Sasayama, H.; Nakajima, O.; Kurahashi, N.; Fujii, J.; Furukawa, I. *Tetrahedron Asymmetry* **2003**, *14*, 537. See also, Evans, P.A.; Robinson, J.E.; Moffett, K.K. *Org. Lett.* **2001**, *3*, 3269. For a titanium-catalyzed variation see Mahrwald, R.; Quint, S. *Tetrahedron Lett.* **2001**, *42*, 1655.

[887]Barton, D.H.R.; Doris, E. *Tetrahedron Lett.* **1996**, *37*, 3295.

[888]Gage, J.R.; Wagner, J.M. *J. Org. Chem.* **1995**, *60*, 2613.

[889]See Honaker, M.T.; Sandefur, B.J.; Hargett, J.L.; McDaniel, A.L.; Salvatore, R.N. *Tetrahedron Lett.* **2003**, *44*, 8373.

[890]For example, see Deady, L.W.; Finlayson, W.L.; Korytsky, O.L. *Aust. J. Chem.* **1979**, *32*, 1735.

[891]For a review of the reactions of this reagent, see Blažević, N.; Kolbah, D.; Belin, B.; Šunjić, V.; Kajfež, F. *Synthesis* **1979**, 161.

calcium salt of cyanamide NH_2—CN to give disubstituted cyanamides, which are then hydrolyzed and decarboxylated to secondary amines. Good yields are obtained when the reaction is carried out under phase-transfer conditions.[892] The R group may be primary, secondary, allylic, or benzylic. 1,ω-Dihalides give cyclic secondary amines. Aminoboranes react with sulfonate esters to give a derivative that can be hydrolyzed to a tertiary amine.[893] An aminyl-radical cyclization process was used to prepare cyclic amines.[894]

N-Silylalkyl amines are formed from amines by reaction with halotrialkylsilanes and a suitable base.[895] Amines react directly with triarylsilanes in the presence of Yb catalysts.[896]

OS **I**, 23, 48, 102, 300, 488; **II**, 85, 183, 290, 328, 374, 397, 419, 563; **III**, 50, 148, 254, 256, 495, 504, 523, 705, 753, 774, 813, 848; **IV**, 84, 98, 383, 433, 466, 582, 585, 980; **V**, 88, 124, 306, 361, 434, 499, 541, 555, 608, 736, 751, 758, 769, 825, 883, 985, 989, 1018, 1085, 1145; **VI**, 56, 75, 104, 106, 175, 552, 652, 704, 818, 967; **VIII**, 9, 152, 231, 358. Also see, OS **II**, 395; **IV**, 950; OS **V**, 121; OS **I**, 203.

For *N*-arylation of amines see **13-5**.

10-32   Replacement of a Hydroxy or Alkoxy by an Amino Group

Amino-de-hydroxylation and Amino-de-alkoxylation

$$R—OH \longrightarrow R—NH_2$$
$$Ar—OR' \longrightarrow R'—NH_2 + ArOH$$

Alcohols can be converted to alkyl halides, which then react with amines (**10-43**). Alcohols react with various amine reagents that give products convertible to the amine.[897] The conversion ROH → RNH_2 can be accomplished for primary and secondary alcohols by treatment with hydrazoic acid (HN_3), diisopropyl azodicarboxylate (*i*Pr—OOCN=NCOO—*i*Pr), and excess Ph_3P in THF, followed by water or aqueous acid.[898] This is a type of Mitsunobu reaction (see **10-17**). Other

[892]Jończyk, A.; Ochal, Z.; Makosza, M. *Synthesis* **1978**, 882.

[893]Thomas, S.; Huynh, T.; Enriquez-Rios, V.; Singaram, B. *Org. Lett.* **2001**, *3*, 3915.

[894]Crich, D.; Shirai, M.; Rumthao, S. *Org. Lett.* **2003**, *5*, 3767.

[895]Greene, T.W. *Protective Groups in Organic Synthesis* Wiley, NY, **1980**, p. 283; Wuts, P.G.M.; Greene, T.W. *Protective Groups in Organic Synthesis 2nd ed.*, Wiley, NY, **1991**, pp. 69–71; Wuts, P.G.M.; Greene, T.W. *Protective Groups in Organic Synthesis 3rd ed.*, Wiley, NY, **1999**; Pratt, J.R.; Massey, W.D.; Pinkerton, F.H.; Thames, S.F. *J. Org. Chem.* **1975**, 40, 1090.

[896]Takaki, K.; Kamata, T.; Miura, Y.; Shishido, T.; Takehira, K. *J. Org. Chem.* **1999**, *64*, 3891.

[897]See Laurent, M.; Marchand-Brynaert, J. *Synthesis* **2000**, 667; Jirgensons, A.; Kauss, V.; Kalvinsh, I.; Gold, M.R. *Synthesis* **2000**, 1709; Katritzky, A.R.; Huang, T.-B.; Voronkov, M.V. *J. Org. Chem.* **2001**, *66*, 1043; Cami-Kobeci, G.; Williams, J.M.J. *Chem. Commun.* **2004**, 1072. See also, Salehi, P.; Motlagh, A.R. *Synth. Commun.* **2000**, *30*, 671; Lakouraj. M.M.; Movassagh, B.; Fasihi, J. *Synth. Commun.* **2000**, *30*, 821.

[898]Fabiano, E.; Golding, B.T.; Sadeghi, M.M. *Synthesis* **1987**, 190. See also, Klepacz, A.; Zwierzak, A. *Synth. Commun.* **2001**, *31*, 1683.

alcohol-to-amine Mitsunobu reactions have also been reported.[899] Primary and secondary alcohols ROH (but not methanol) can be converted to tertiary amines,[900] $R_2'NR$, by treatment with the secondary amine $R_2'NH$ and $(t\text{-BuO})_3Al$ in the presence of Raney nickel.[901] The use of aniline gives secondary amines PhNHR. Allylic alcohols ROH react with primary ($R'NH_2$) or secondary ($R_2'NH$) amines in the presence of platinum or palladium complexes, to give secondary (RNHR') or tertiary (RNR_2') allylic amines.[902] Conversion of an allyic alcohol to the corresponsodning allylic crbonate, foolwed by reacatin with an N-tosylamine and lihtium hexamethyldisilazide, followedby by $Rh(PPh_3)_3Cl$ and $P(OMe)_3$, gives the N-tosylallylic amine.[903] α-Hydroxy phosphonates react with aniline on alumina with microwave irradiation.[904] The ruthenium-catalyzed reaction of amines and diols leads to cyclic amines.[905]

β-Amino alcohols give aziridines (**120**) when treated with triphenylphosphine dibromide in the presence of triethylamine.[906] The fact that inversion takes place at the OH carbon indicates that an S_N2 mechanism is involved, with $OPPh_3$ as the leaving group.

120

Alcohols can be converted to amines in an indirect manner.[907] The alcohols are converted to alkyloxyphosphonium perchlorates which in DMF successfully

[899]See, for example, Henry, J.R.; Marcin, L.R.; McIntosh, M.C.; Scola, P.M.; Harris Jr., G.D.; Weinreb, S.M. *Tetrahedron Lett.* **1989**, *30*, 5709; Edwards, M.L.; Stemerick, D.M.; McCarthy, J.R. *Tetrahedron Lett.* **1990**, *31*, 3417.

[900]For other methods of converting certain alcohols to secondary and tertiary amines, see Murahashi, S.; Kondo, K.; Hakata, T. *Tetrahedron Lett.* **1982**, *23*, 229; Baiker, A.; Richarz, W. *Tetrahedron Lett.* **1977**, 1937; *Helv. Chim. Acta* **1978**, *61*, 1169; *Synth. Commun.* **1978**, *8*, 27; Grigg, R.; Mitchell, T.R.B.; Sutthivaiyakit, S.; Tongpenyai, N. *J. Chem. Soc., Chem. Commun.* **1981**, 611; Arcelli, A.; Bui-The-Khai; Porzi, G. *J. Organomet. Chem.* **1982**, *235*, 93; Kelly, J.W.; Eskew, N.L.; Evans, Jr., S.A. *J. Org. Chem.* **1986**, *51*, 95; Huh, K.; Tsuji, Y.; Kobayashi, M.; Okuda, F.; Watanabe, Y. *Chem. Lett.* **1988**, 449.

[901]Botta, M.; De Angelis, F.; Nicoletti, R. *Synthesis* **1977**, 722.

[902]Atkins, K.E.; Walker, W.E.; Manyik, R.M. *Tetrahedron Lett.* **1970**, 3821; Tsuji, Y.; Takeuchi, R.; Ogawa, H.; Watanabe, Y. *Chem. Lett.* **1986**, 293.

[903]Evans, P.A.; Robinson, J.E.; Nelson, J.D. *J. Am. Chem. Soc.* **1999**, *121*, 6761.

[904]Kaboudin, B. *Tetrahedron Lett.* **2003**, *44*, 1051.

[905]Fujita, K.-i.; Fujii, T.; Yamaguchi, R. *Org. Lett.* **2004**, *6*, 3525.

[906]Okada, I.; Ichimura, K.; Sudo, R. *Bull. Chem. Soc. Jpn.* **1970**, *43*, 1185. See also, Pfister, J.R. *Synthesis* **1984**, 969; Suzuki, H.; Tani, H. *Chem. Lett.* **1984**, 2129; Marsella, J.A. *J. Org. Chem.* **1987**, *52*, 467.

[907]For some other indirect methods, see White, E.H.; Ellinger, C.A. *J. Am. Chem. Soc.* **1965**, *87*, 5261; Burgess, E.M.; Penton Jr., H.R.; Taylor, E.A. *J. Am. Chem. Soc.* **1970**, *92*, 5224; Hendrickson, J.B.; Joffee, I. *J. Am. Chem. Soc.* **1973**, *95*, 4083; Trost, B.M.; Keinan, E. *J. Org. Chem.* **1979**, *44*, 3451; Koziara, A.; Osowska-Pacewicka, K.; Zawadzki, S.; Zwierzak, A. *Synthesis* **1985**, 202; **1987**, 487.

monoalkylate not only secondary but also primary amines.[908]

$$ROH \xrightarrow[\text{2. NH}_4\text{ClO}_4]{\text{1. CCl}_4-\text{P(NMe}_2)_3} RO^{\oplus}P(NMe_2)_3 \ ^{\ominus}ClO_4 \xrightarrow[R'R''NH]{\text{DMF}} RR'R''N + OP(NMe_2)_3$$

Thus by this means secondary as well as tertiary amines can be prepared in good yields. Benzylic alcohols can be converted to an azide and then treated with triphenylphosphine to give the amine (**19-50**).[909]

Cyanohydrins can be converted to amines by treatment with ammonia. The use of primary or secondary amines instead of ammonia leads to secondary and tertiary cyanoamines, respectively. It is more common to perform the conversion of an aldehyde or ketone directly to the cyanoamine without isolation of the cyanohydrin (see **16-52**). α-Hydroxy ketones (acyloins and benzoins) behave similarly.[910]

A solution of the sodium salt of *N*-methylaniline in HMPA can be used to cleave the methyl group from aryl methyl ethers:[911] ArOMe + PhNMe⁻ → ArO⁻ + PhNMe₂. This reagent also cleaves benzylic groups. In a similar reaction, methyl groups of aryl methyl ethers can be cleaved with lithium diphenylphosphide, Ph₂PLi.[912] This reaction is specific for methyl ethers and can be carried out in the presence of ethyl ethers with high selectivity. Phenyl allyl ethers react with secondary amines in the presence of a palladium catalyst to give phenol and the tertiary allyl amine.[913]

OS **II**, 29, 231; **IV**, 91, 283; **VI**, 567, 788; **VII**, 501. Also see, OS **I**, 473; **III**, 272, 471.

10-33   Transamination

Alkylamino-de-amination

$$RNH_2 + R'NH^- \longrightarrow RR'NH + NH_2^-$$

Where the nucleophile is the conjugate base of a primary amine, NH₂ can be a leaving group. The method has been used to prepare secondary amines.[914] In another process, primary amines are converted to secondary amines in which

[908]Castro, B.; Selve, C. *Bull. Soc. Chim. Fr.* **1971**, 4368. For a similar method, see Tanigawa, Y.; Murahashi, S.; Moritani, I. *Tetrahedron Lett.* **1975**, 471.

[909]Reddy, G.V.S.; Rao, G.V.; Subrmanyam, R.V.K.; Iyengar, D.S. *Synth. Commun.* **2000**, *30*, 2233.

[910]For example, see Klemmensen, P.; Schroll, G.; Lawesson, S. *Ark. Kemi*, **1968**, *28*, 405.

[911]Loubinoux, B.; Coudert, G.; Guillaumet, G. *Synthesis* **1980**, 638.

[912]Ireland, R.E.; Walba, D.M. *Org. Synth. VI*, 567.

[913]Widehem, R.; Lacroix, T.; Bricout, H.; Monflier, E. *Synlett* **2000**, 722.

[914]Baltzly, R.; Blackman, S.W. *J. Org. Chem.* **1963**, *28*, 1158.

both R groups are the same $(2\ RNH_2 \rightarrow R_2NH + NH_3)$[915] by refluxing in xylene in the presence of Raney nickel.[916] Quaternary salts can be dealkylated with ethanolamine.[917]

$$R_4N^+ + NH_2CH_2CH_2OH \longrightarrow R_3N + \overset{\oplus}{R}NH_2CH_2CH_2OH$$

In this reaction, methyl groups are cleaved in preference to other saturated alkyl groups. A similar reaction takes place between a Mannich base (see **16-19**) and a secondary amine, where the mechanism is elimination–addition (see p. 477). See also, **19-5**.

OS V, 1018.

10-34   Alkylation of Amines With Diazo Compounds

Hydro,dialkylamino-de-diazo-bisubstitution

$$CR_2N_2 + R_2'NH \overset{BF_3}{\longrightarrow} CHR_2NR_2'$$

The reaction of diazo compounds with amines is similar to **10-11**.[918] The acidity of amines is not great enough for the reaction to proceed without a catalyst, but BF_3, which converts the amine to the F_3B-NHR_2' complex, enables the reaction to take place. Cuprous cyanide can also be used as a catalyst.[919] The most common substrate is diazomethane,[630] in which case this is a method for the methylation of amines. Ammonia has been used as the amine but, as in the case of **10-31**, mixtures of primary, secondary, and tertiary amines are obtained. Primary aliphatic amines give mixtures of secondary and tertiary amines. Secondary amines give successful alkylation. Primary aromatic amines also give the reaction, but diaryl or arylalkylamines react very poorly.

10-35   Reaction of Epoxides With Nitrogen Reagents

(3)*OC-seco*-Amino-de-alkoxylation

---

[915]In a similar manner, a mixture of primary amines can be converted to a mixed secondary amine. For a review of the mechanism, see Geller, B.A. *Russ. Chem. Rev.* **1978**, *47*, 297.

[916]De Angelis, F.; Grgurina, I.; Nicoletti, R. *Synthesis* **1979**, 70; See also, Ballantine, J.A.; Purnell, H.; Rayanakorn, M.; Thomas, J.M.; Williams, K.J. *J. Chem. Soc., Chem. Commun.* **1981**, 9; Arcelli, A.; Bui-The-Khai; Porzi, G. *J. Organomet. Chem.* **1982**, *231*, C31; Jung, C.W.; Fellmann, J.D.; Garrou, P.E. *Organometallics* **1983**, *2*, 1042; Tsuji, Y.; Shida, J.; Takeuchi, R.; Watanabe, Y. *Chem. Lett.* **1984**, 889; Bank, S.; Jewett, R. *Tetrahedron Lett.* **1991**, *32*, 303.

[917]Hünig, S.; Baron W. *Chem. Ber.* **1957**, *90*, 395, 403.

[918]Müller, E.; Huber-Emden, H.; Rundel, W. *Liebigs Ann. Chem.* **1959**, *623*, 34.

[919]Saegusa, T.; Ito, Y.; Kobayashi, S.; Hirota, K.; Shimizu, T. *Tetrahedron Lett.* **1966**, 6131.

The reaction between epoxides and ammonia[920] (or ammonium hydroxide)[921] is a general and useful method for the preparation of β-hydroxyamines. With epoxide derived from terminal alkenes, the reaction with ammonia gives largely the primary amine, but secondary and tertiary amine products are possible from the appropriate epoxide. The reaction of **121** with ammonium hydroxide with microwave irradiation, for example, gave **122**.[922] Ethanolamines, which are useful solvents

as well as synthetic precursors, are prepared by this reaction. Similar ring opening occurs with alkyl and aromatic amines.[923] For another way of accomplishing this conversion, see **10-40**. The reaction can be catalyzed with Yb(OTf)$_3$ and in the presence of (R)-BINOL (BINOL = 1,1′-bi-2-naphthol) gives amino alcohols with high asymmetric induction.[924] Many other metal-catalyzed ring-opening reactions have been reported.[925] Ring opening has been accomplished with aniline on silica gel.[926]

Primary and secondary amines give, respectively, secondary and tertiary amines (**121**). Aniline reacts with epoxides in the presence of aqueous β-cyclodextrin[927] in 5 M LiClO$_4$ in ether,[928] or in fluoro-alcohol solvents.[929] Aniline reacts with epoxides in the presence of a VCl$_3$ catalyst.[930] N-Boc-amine (H$_2$N—CO$_2$t-Bu) reacted

[920]For an example, see McManus, S.P.; Larson, C.A.; Hearn, R.A. *Synth. Commun.* **1973**, *3*, 177; Charrada, B.; Hedhli, A.; Baklouti, A. *Tetrahedron Lett.* **2000**, *41*, 7347.

[921]Pastó, M.; Rodríguez, B.; Riera, A.; Pericàs, M.A. *Tetrahedron Lett.* **2003**, *44*, 8369.

[922]Lindström, U.M.; Olofsson, B.; Somfai, P. *Tetrahedron Lett.* **1999**, *40*, 9273.

[923]See Harrack, Y.; Pujol, M.D. *Tetrahedron Lett.* **2002**, *43*, 819; Steiner, D.; Sethofer, S.G.; Goralski, C.T.; Singaram, B. *Tetrahedron Asymmetry* **2002**, *13*, 1477. For a reaction catalyzed by LiBr, see Chakraborti, A.K.; Rudrawar, S.; Kondaskar, A. *Eur. J. Org. Chem.* **2004**, 3597.

[924]Hou, X.-L.; Wu, J.; Dai, L.-X.; Xia, L.-J.; Tang, M.-H. *Tetrahedron Asymmetry* **1998**, *9*, 1747.

[925]Examples include, **Sn(OTf)$_2$**: Sekar, G.; Singh, V.K. *J. Org. Chem.* **1999**, *64*, 287; **CeCl$_3$-NaI**: Reddy, L.R.; Reddy, M.A.; Bhanumathi, N.; Rao, K.R. *Synthesis* **2001**, 831; **Zr catalysts**: Curini, M.; Epifano, F.; Marcotullio, M.C.; Rosati, O. *Eur. J. Org. Chem.* **2001**, 4149 and Charkraborti, A.K.; Kondaskar, A. *Tetrahedron Lett.* **2003**, *44*, 8315; **LiNTf$_2$**: Cossy, J.; Bellosta, V.; Hamoir, C.; Desmurs, J.-R. *Tetrahedron Lett.* **2002**, *43*, 7083; **Bi compounds**: Ollevier, T.; Lavie-Compin, G. *Tetrahedron Lett.* **2002**, *43*, 7891 and **2004**, *45*, 49; **ZnCl$_2$**: Pachón, L.D.; Gamez, P.; van Brussel, J.J.M.; Reedijk, J. *Tetrahedron Lett.* **2003**, *44*, 6025; **InBr$_3$**: Rodríguez, J.R.; Navarro, A. *Tetrahedron Lett.* **2004**, *45*, 7495; **SmI$_2$(thf)$_2$**: Carrée, F.; Gil, R.; Collin, J. *Tetrahedron Lett.* **2004**, *45*, 7749; **CoCl$_2$**: Sundararajan, G.; Viyayakrishna, K.; Varghese, B. *Tetrahedron Lett.* **2004**, *45*, 8253.

[926]Chakraborti, A.K.; Rudrawar, S.; Kondaskar, A. *Org. Biomol. Chem.* **2004**, *2*, 1277.

[927]Reddy, L.R.; Reddy, M.A.; Chanumathi, N.; Rao, K.R. *Synlett* **2000**, 339.

[928]Heydar, A.; Mehrdad, M.; Malecki, A.; Ahmadi, N. *Synthesis* **2004**, 1563.

[929]Das, U.; Crousse, B.; Kesavan, V.; Bonnet-Delpon, D.; Bégue, J.P. *J. Org. Chem.* **2000**, *65*, 6749.

[930]Sabitha, G.; Reddy, G.S.K.K.; Reddy, K.B.; Yadav, J.S. *Synthesis* **2003**, 2298.

with epoxides in the presence of a cobalt–salen catalyst to give the amido alcohol.[931] Solvent free reactions using a catalytic amount of $SnCl_4$ are known.[932] Tetrahydropyrimidones can be used to mediate the addition of indole to epoxides.[933] Amide bases react differently with epoxides. Lithium tetramethylpiperidide (LTMP), for example, reacted with epoxides, but the product was the corresponding enamine.[934] This latter reaction follows a very different mechanism. Initial formation of the lithio-epoxide is followed by rearrangement to give the aldehyde,[935] and subsequent reaction with the amine by-product of the lithiation leads to the enamine.

An indirect method for generating an amino alcohol (**124**) is to open an epoxide with azide to give the azido-alcohol **123**,[936] and subsequent reduction (**19-50**) gives the amine group.[937] Sodium azide and Oxone® react with epoxides to give an azido-alcohol.[938] Under Mitsunobu conditions (**10-17**), epoxides are converted to 1,2-diazides with HN_3.[939] The reaction of trimethylsilyl azide and an epoxide was reported using an ionic solvent.[940] The cerium ammonium nitrate catalyzed reaction of epoxides and sodium azide, for example, gave the azido alcohol with selectivity for the azide group on the more substituted position.[941] Cerium chloride has also been used, giving the azide on the less substituted carbon.[942] Manganese–salen complexes, immobilized on mesoporous material has also been used to mediate the ring opening of epoxides by azide.[943] In the presence of $AlCl_3$ in water at pH 4, sodium azide reacts with epxoy acids to give the β-azido-α-hydroxycarboxylic acid.[944] Silylazides can be used as well.[945]

123   **124**

[931]Bartoli, G.; Bosco, M.; Carlone, A.; Locatelli, M.; Mechiorre, P.; Sambri, L. *Org. Lett.* **2004**, *6*, 3973.

[932]Zhao, P.-Q.; Xu, L.-W.; Xia, C.-G. *Synlett* **2004**, 846.

[933]Fink, D.M. *Synlett* **2004**, 2394.

[934]Hodgson, D.M.; Bray, C.D.; Kindon, N.D. *J. Am. Chem. Soc.* **2004**, *126*, 6870.

[935]Yanagisawa, A.; Yasue, K.; Yamamoto, H. *J. Chem. Soc., Chem. Commun.* **1994**, 2103.

[936]Kazemi, F.; Kiasat, A.R.; Ebrahimi, S. *Synth. Commun.* **2003**, *33*, 999. For a reaction done under phase-transfer conditions, see Tamami, B.; Mahdavi, H. *Tetrahedron Lett.* **2001**, *42*, 8721.

[937]Larock, R.C. *Comprehensive Organic Transformations*, 2nd ed., Wiley–VCH, NY, **1999**, p. 815.

[938]Sabitha, G.; Babu, R.S.; Reddy, M.S.K.; Yadav, J.S. *Synthesis* **2002**, 2254.

[939]Göksu, S.; Soçen, H.; Sütbeyaz, Y. *Synthesis* **2002**, 2373.

[940]In emim, 1-ethyl-3-methylimidazolium: Song, C.E.; Oh, C.R.; Roh, E.J.; Choo, D.J. *Chem. Commun.* **2000**, 1743.

[941]Iranpoor, N.; Kazemi, F. *Synth. Commun.* **1999**, *29*, 561.

[942]Sabitha, G.; Babu, R.S.; Rajkumar, M.; Yadav, J.S. *Org. Lett.* **2002**, *4*, 343.

[943]Kantam, M.L.; Choudary, B.M.; Bharathi, B. *Synth. Commun.* **1999**, *29*, 1121.

[944]Fringuelli, F.; Pizzo, F.; Vaccaro, L. *Tetrahedron Lett.* **2001**, *42*, 1131.

[945]Schneider, C. *Synlett* **2000**, 1840.

Sodium nitrate ($NaNO_2$) reacts with epoxides in the presence of $MgSO_4$ to give the nitro alcohol.[946] The nitro group can also be reduced to give the amine (**19-45**).[947]

Episulfides, which can be generated *in situ* in various ways, react similarly to give β-amino thiols,[948] and aziridines react with amines to give 1,2-diamines (**10-38**). Triphenylphosphine similarly reacts with epoxides to give an intermediate that undergoes elimination to give alkenes (see the Wittig reaction, **16-44**).

OS **X**, 29. See OS **VI**, 652 for a related reaction.

10-36   Formation of Aziridines from Epoxides

Amino-de-alkoxylation

It is possible to prepare aziridines, which are synthetically important molecules, directly from the corresponding epoxide. Reaction of $Ph_3P=NPh$ with an epoxide in the presence of $ZnCl_2$ gives the *N*-phenyl aziridine.[949] Guanidines have also been used to prepare aziridnes from epoxides.[950] Tosylamines react with epoxides to give the *N*-tosylaziridine.[951]

Various methods are available to convert an aminomethyl epoxide to a hydroxy-methyl aziridine, **125**[952]

125

10-37   Amination of Oxetanes

(4)*OC-homoseco*-Amino-de-alkoxylation

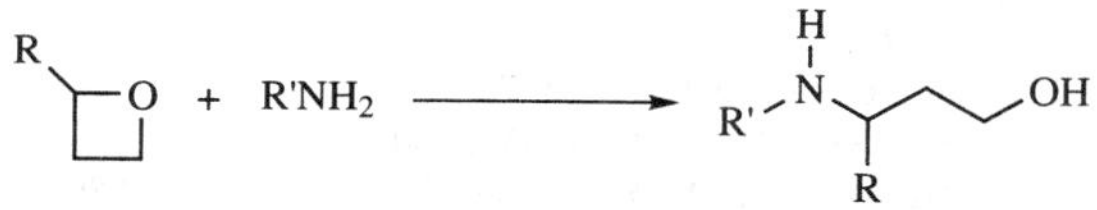

Oxetanes are significantly less reactive with nucleophiles due to diminished ring strain. Under certain conditions, however, amines can open oxetanes to give amino

[946]Kalita, B.; Barua, N.C.; Bezbarua, M.; Bez, G. *Synlett* **2001**, 1411.

[947]Larock, R.C. *Comprehensive Organic Transformations*, 2nd ed., Wiley-VCH, NY, **1999**, p. 821.

[948]Dong, Q.; Fang, X.; Schroeder, J.D.; Garvey, D.S. *Synthesis* **1999**, 1106.

[949]Kühnau, D.; Thomsen, I.; Jørgensen, K.A. *J. Chem. Soc. Perkin Trans. 1*, **1996**, 1167.

[950]Tsuchiya, Y.; Kumamoto, T.; Ishikawa, T. *J. Org. Chem.* **2004**, *69*, 8504.

[951]Albanese, D.; Landini, D.; Penso, M.; Petricci, S. *Tetrahedron* **1999**, *55*, 6387.

[952]Najime, R.; Pilard, S.; Vaultier, M. *Tetrahedron Lett.* **1992**, *33*, 5351; Moulines, J.; Bats, J.-P.; Hautefaye, P.; Nuhrich, A.; Lamidey, A.-M. *Tetrahedron Lett.* **1993**, *34*, 2315; Moulines, J.; Charpentier, P.; Bats, J.-P.; Nuhrich, A.; Lamidey, A.-M. *Tetrahedron Lett.* **1992**, *33*, 487.

alcohols. *tert*-Butylamine reacts with oxetanes in the presence of Yb(OTf)$_3$, for example, to give 3-hydroxy amines.[953] Lithium tetrafluoroborate has also been used for this purpose.[954]

10-38   Reaction of Aziridines With Nitrogen

(3)*NC-seco*-**Amino-de-aminoalkylation**

Just as epoxides can be opened by amines to give hydroxy amines, aziridines can be opened to give diamines.[955] With bicyclic aziridines, the major product is usually the trans diamine. *N*-Aryl or *N*-alkyl aziridines react with amines in the presence of Sn(OTf)$_2$[956] or B(C$_6$F$_5$)$_3$[957] to give the diamine. Amines react with *N*-tosylaziridines, in the presence of various catalysts or additives to give the corresponding diamine.[958] This reaction also takes place on activated silica.[959] The reaction of LiNTf$_2$ and an amine, in the presence of an *N*-alkyl aziridine gives the diamine.[960]

As with epoxides, tosyl-aziridines react with azide to generate azido tosylamines.[961] Reduction of the azide (**19-50**) gives the diamine. Silylazides, such as Me$_3$SiN$_3$, also react with aziridine derivatives to give the azidoamine.[962] This latter reaction can be catalyzed by InCl$_3$.[963]

[953]Crotti, P.; Favero, L.; Macchia, F.; Pineschi, M. *Tetrahedron Lett.* **1994**, *35*, 7089.

[954]Chini, M.; Crotti, P.; Favero, L.; Macchia, F. *Tetrahedron Lett.* **1994**, *35*, 761.

[955]For a review, see Dermer, O.C.; Ham, G.E. *Ethylenimine and Other Aziridines*, Academic Press, NY, **1969**, pp. 262–268. See also, Scheuermann, J.E.W.; Ilyashenko, G.; Griffiths, D.V.; Watkinson, M. *Tetrahedron Asymmetry* **2002**, *13*, 269.

[956]Sekar, G.; Singh, V.K. *J. Org. Chem.* **1999**, *64*, 2537.

[957]Watson, I.D.G.; Yudin, A.K. *J. Org. Chem.* **2003**, *68*, 5160.

[958]Examples include **Yb(OTf)$_3$**: Meguro, M.; Yamamoto, Y. *Heterocycles* **1996**, *43*, 2473. **PBu$_3$**: Fan, R.-H.; Hou, X.-L. *J. Org. Chem.* **2003**, *68*, 726. **Aqueous media with β-cyclodextrin**: Reddy, M.A.; Reddy, L.R.; Bhanamathi, N.; Rao, K.R. *Chem. Lett.* **2001**, 246. **TaCl$_5$/SiO$_2$**: Chandrasekhar, S.; Prakash, S.J.; Shyamsunder, T.; Ramachandar, T. *Synth. Commun.* **2004**, *34*, 3865. **InCl$_3$**: Yadav, J.S.; Reddy, B.V.S.; Abraham, S.; Sabitha, G. *Tetrahedron Lett.* **2002**, *43*, 1565; **InBr$_3$**: Yadav, J.S.; Reddy, B.V.S.; Rao, K.; Raj, K.S.; Prasad, A.R. *Synthesis* **2002**, 1061. **BiCl$_3$**: Swamy, N.R.; Venkateswarlu, Y. *Synth. Commun.* **2003**, *33*, 547. **LiClO$_4$**: Yadav, J.S.; Reddy, B.V.S.; Jyothivmai, B.; Murty, M.S.R. *Synlett* **2002**, 53; Yadav, J.S.; Reddy, B.V.S.; Parimala, G.; Reddy, P.V. *Synthesis* **2002**, 2383.

[959]Anand, R.V.; Pandey, G.; Singh, V.K. *Tetrahedron Lett.* **2002**, *43*, 3975; Kumar, G.D.K.; Baskaran, S. *Synlett* **2004**, 1719.

[960]Cossy, J.; Bellosta, V.; Alauze, V.; Desmurs, J.-R. *Synthesis* **2002**, 2211.

[961]Bisai, A.; Pandey, G.; Pandey, M.K.; Singh, V.K. *Tetrahedron Lett.* **2003**, *44*, 5839.

[962]Chandrasekhar, M.; Sekar, G.; Singh, V.K. *Tetrahedron Lett.* **2000**, *41*, 10079.

[963]Yadav, J.S.; Reddy, B.V.S.; Kumar, G.M.; Murthy, Ch.V.S.R. *Synth. Commun.* **2002**, *32*, 1797.

10-39   Amination of Alkanes

Amino-de-hydrogenation or Amination

$$R_3CH \ + \ NCl_3 \ \xrightarrow[0\,-\,10°C]{AlCl_3} \ R_3CNH_2$$

Alkanes, arylalkanes, and cycloalkanes can be aminated, at tertiary positions only, by treatment with trichloroamine and aluminum chloride at 0–10°C.[964] For example, $p\text{-MeC}_6\text{H}_4\text{CHMe}_2$ gives $p\text{-MeC}_6\text{H}_4\text{CMe}_2\text{NH}_2$, methylcyclopentane gives 1-amino-1-methylcyclopentane, and adamantane gives 1-aminoadamantane, all in good yields. This is a useful reaction, since there are not many other methods for the preparation of *tert*-alkyl amines. The mechanism has been rationalized as an S_N1 process with H^- as the leaving group:[964]

$$NCl_3 \ + \ AlCl_3 \ \longrightarrow \ (Cl_2N\text{-}AlCl_3)^- \ Cl^+$$

$$R_3CH \ \xrightarrow{Cl+} \ R_3C^\oplus \ \xrightarrow{NCl_2^-} \ R_3CNCl_2 \ \xrightarrow[2\,H^+]{-\,2\,Cl^+} \ R_3CNH_2$$

It is noted than under photochemical conditions, ammonia opens cyclopropane derivatives to give the corresponding alkyl amine.[965] See also, **12-12**.

OS **V**, 35.

10-40   Formation of Isocyanides (Isonitriles)

Haloform-isocyanide transformation

$$CHCl_3 \ + \ RNH_2 \ \xrightarrow{-OH} \ R\text{-}\overset{\oplus}{N}\text{≡}C^\ominus$$

Reaction with chloroform under basic conditions is a common test for primary amines, both aliphatic and aromatic, since isocyanides (**126**) have very strong bad odors. The reaction probably proceeds by an S_N1cB mechanism with dichlorocarbene (**127**) as an intermediate.

$$CHCl_3 \ + \ {}^-OH \ \xrightarrow[-Cl^-]{-H^+} \ :CCl_2 \ \xrightarrow{RNH_2} \ \underset{\textbf{126}}{Cl\overset{\textstyle Cl}{\underset{\underset{\textbf{H}}{\overset{\displaystyle\ominus}{\text{C}}}}{\overset{\ominus}{\text{C}}}\overset{\oplus}{\underset{R}{N}}\overset{H}{}} \ \xrightarrow{-2HCl} \ \underset{\textbf{127}}{{}^\ominus C\text{≡}\overset{\oplus}{N}\text{-}R}$$

The reaction can also be used synthetically for the preparation of isocyanides, although yields are generally not high.[966] An improved procedure has been reported.[967] When

[964]Wnuk, T.A.; Chaudhary, S.S.; Kovacic, P. *J. Am. Chem. Soc.* *1976*, *98*, 5678, and references cited therein.
[965]Yasuda, M.; Kojima, R.; Tsutsui, H.; Utsunomiya, D.; Ishii, K.; Jinnouchi, K.; Shiragami, T.; Yamashita, T. *J. Org. Chem.* *2003*, *68*, 7618.
[966]For a review of isocyanides, see Periasamy, M.P.; Walborsky, H.M. *Org. Prep. Proced. Int.* *1979*, *11*, 293.
[967]Weber, W.P.; Gokel, G.W. *Tetrahedron Lett.* *1972*, 1637; Weber, W.P.; Gokel, G.W.; Ugi, I. *Angew. Chem. Int. Ed.* *1972*, *11*, 530.

secondary amines are involved, the adduct **128** cannot lose 2 mol of HCl. Instead it is hydrolyzed to an *N,N*-disubstituted formamide.[968]

$$
\begin{array}{ccc}
\underset{Cl}{\overset{Cl}{\overset{|}{\underset{}{C}}}}\overset{\ominus}{\underset{}{C}}\overset{\oplus}{\underset{\underset{R}{|}}{N}}\overset{H}{\underset{R}{\diagdown}} & \longrightarrow & \underset{Cl}{\overset{H}{\underset{}{\diagup}}}\overset{Cl}{\overset{|}{\underset{}{C}}}\overset{}{\underset{\underset{R}{|}}{N}}\overset{R}{\diagdown} & \overset{H_2O}{\longrightarrow} & \underset{H}{\overset{O}{\underset{\diagup}{\overset{\parallel}{C}}}}NR_2 \\
& & \mathbf{128}
\end{array}
$$

A completely different way of preparing isocyanides involves the reaction of epoxides or oxetanes with trimethylsilyl cyanide and zinc iodide to give the isocyanide **129**.[969]

$$
\underset{\square}{\overset{Me}{\diagdown}}\!\!O \quad \xrightarrow[ZnI_2]{Me_3SiCN} \quad Me_3SiO\diagdown\diagup\overset{Me}{\underset{}{\diagup}}N{\equiv}C \quad \xrightarrow[MeOH]{HCl} \quad Me_3SiO\diagdown\diagup\overset{Me}{\underset{}{\diagup}}NH_2
$$

$$
\qquad\qquad\qquad\qquad\qquad \mathbf{129} \qquad\qquad\qquad\qquad\qquad \mathbf{130}
$$

The products can be hydrolyzed to protected hydroxy-amines, such as **130**.
OS **VI**, 232.

B. Attack by NHCOR

10-41   *N*-Alkylation or *N*-Arylation of Amides and Imides

Acylamino-de-halogenation

$$RX + {}^{\ominus}NHCOR' \longrightarrow RNHCOR'$$

$$ArX + {}^{\ominus}NHCOR' \longrightarrow ArNHCOR'$$

Amides are very weak nucleophiles,[970] far too weak to attack alkyl halides, so they must first be converted to their conjugate bases. By this method, unsubstituted amides can be converted to *N*-substituted, or *N*-substituted to *N,N*-disubstituted, amides.[971] Esters of sulfuric or sulfonic acids can also be substrates. Tertiary substrates give elimination. *O*-Alkylation is at times a side reaction.[972] Both amides and sulfonamides have been alkylated under phase-transfer conditions.[973] Lactams can be alkylated using similar procedures. Ethyl pyroglutamate (5-carboethoxy

[968]Saunders, M.; Murray, R.W. *Tetrahedron* **1959**, *6*, 88; Frankel, M.B.; Feuer, H.; Bank, J. *Tetrahedron Lett.* **1959**, no. 7, 5.

[969]Gassman, P.G.; Haberman, L.M. *Tetrahedron Lett.* **1985**, *26*, 4971, and references cited therein.

[970]Brace, N.O. *J. Org. Chem.* **1993**, *58*, 1804.

[971]For procedures, see Zawadzki, S.; Zwierzak, A. *Synthesis* **1979**, 549; Yamawaki, J.; Ando, T.; Hanafusa, T. *Chem. Lett.* **1981**, 1143; Sukata, K. *Bull. Chem. Soc. Jpn.* **1985**, *58*, 838.

[972]For a review of alkylation of amides, see Challis, B.C.; Challis, J.A., in Zabicky, J. *The Chemistry of Amides*, Wiley, NY, **1970**, pp. 734–754.

[973]Loupy, A.; Sansoulet, J.; Díez-Barra, E.; Carrillo, J.R. *Synth. Commun.* **1992**, *22*, 1661; Salvatore, R.N.; Shin, S.I.; Flanders, V.L.; Jung, K.w. *Tetrahedron Lett.* **2001**, *42*, 1799.

2-pyrrolidinone) and related lactams were converted to *N*-alkyl derivatives via treatment with NaH (short contact time) followed by addition of the halide.[974] 2-Pyrrolidinone derivatives can be alkylated using a similar procedure.[975] Lactams can be reductively alkylated using aldehydes under catalytic hydrogenation conditions (reductive alkylation).[976] *N*-Aryl lactams can be prepared using Ph_3Bi and $Cu(OAc)_2$.[977] *N*-Arylation of sulfonamides has been reported using a palladium catalysis.[978] *N*-Alkenyl amides have been prepared from vinyl iodides and primary amides, using 10% CuI and two equivalents of cesium carbonate.[979] A related palladium-catalyzed vinylation of lactams was repeated using vinyl ethers as a substrate.[980] Oxazolidin-2-ones (a cyclic carbamate) can be *N*-alkylated using an alkyl halide with KF/Al_2O_3.[981]

The *Gabriel synthesis*[982] for converting halides to primary amines is based on this reaction. The halide is treated with potassium phthalimide and the product hydrolyzed (**16-60**):

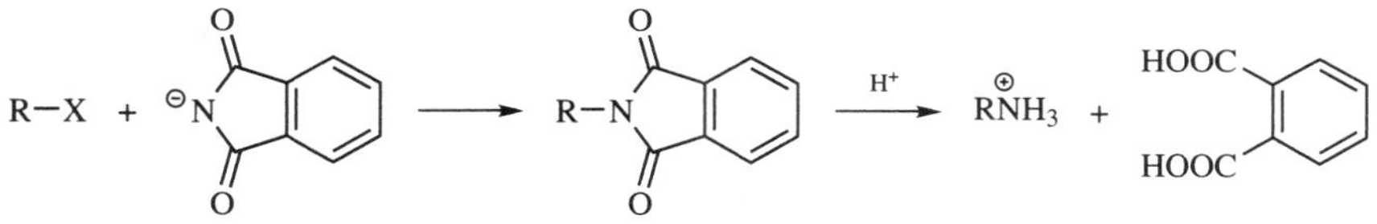

It is obvious that the primary amines formed in this reaction will be uncontaminated by secondary or tertiary amines (unlike **10-31**). The reaction is usually rather slow, but can be conveniently speeded by the use of a dipolar aprotic solvent, such as DMF[983] or with a crown ether.[984] Hydrolysis of the phthalimide, whether acid or base catalyzed (acid catalysis is used far more frequently), is also usually very slow, and better procedures are generally used. A common one is the *Ing–Manske procedure*,[985] in which the phthalimide is heated with hydrazine in an exchange

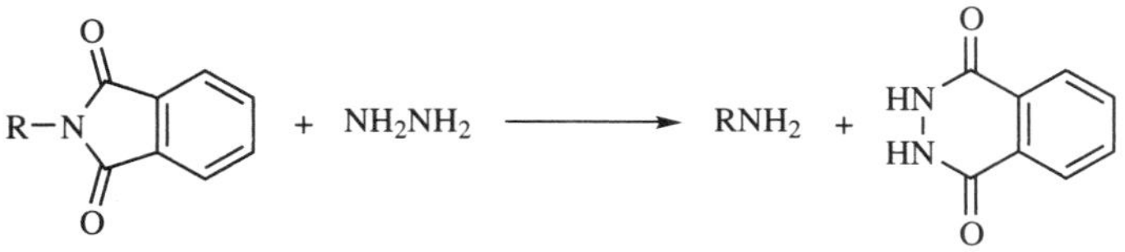

[974]Simandan, T.; Smith, M.B. *Synth. Commun.* **1996**, *26*, 1827; Keusenkothen, P.F.; Smith, M.B. *Synth. Commun.* **1992**, *22*, 2935.

[975]Liu, H.; Ko, S.-B.; Josien, H.; Curran, D.P. *Tetrahedron Lett.* **1995**, *36*, 8917.

[976]Fache, F.; Jacquot, L.; Lemaire, M. *Tetrahedron Lett.* **1994**, *35*, 3313.

[977]Chan, D.M.T. *Tetrahedron Lett.* **1996**, *37*, 9013.

[978]Burton, G.; Cao, P.; Li, G.; Rivero, R. *Org. Lett.* **2003**, *5*, 4373.

[979]Pan, X.; Cai, Q.; Ma, D. *Org. Lett.* **2004**, *6*, 1809.

[980]Brice, J.L.; Meerdink, J.E.; Stahl, S.S. *Org. Lett.* **2004**, *6*, 1845.

[981]Blass, B.E.; Drowns, M.; Harris, C.L.; Liu, S.; Portlock, D.E. *Tetrahedron Lett.* **1999**, *40*, 6545.

[982]For a review, see Gibson, M.S.; Bradshaw, R.W. *Angew. Chem. Int. Ed.* **1968**, *7*, 919.

[983]For example, see Sheehan, J.C.; Bolhofer, W.A. *J. Am. Chem. Soc.* **1950**, *72*, 2786. See also, Landini, D.; Rolla, F. *Synthesis* **1976**, 389.

[984]Soai, K.; Ookawa, A.; Kato, K. *Bull. Chem. Soc. Jpn.* **1982**, *55*, 1671.

[985]Ing, H.R.; Manske, R.H.F. *J. Chem. Soc.* **1926**, 2348.

reaction,[986] but other methods have been introduced, using Na_2S in aqueous THF or acetone,[987] $NaBH_4$-2-propanol followed by acetic acid;[988] and 40% aqueous methylamine.[989] N-aryl imides can be prepared from $ArPb(OAc)_3$ and NaH.[990]

An alternative to the Gabriel synthesis, in which alkyl halides can be converted to primary amines in good yields, involves treatment of the halide with the strong base guanidine followed by alkaline hydrolysis.[991] There are several alternative procedures.[992]

N-Alkyl amides or imides can also be prepared starting from alcohols by treatment of the latter with equimolar amounts of the amide or imide, Ph_3P, and diethyl azodicarboxylate (EtOOCN=NCOOEt) at room temperature (the Mitsunobu reaction, **10-17**).[993] A related reaction treats the alcohol with $ClCH=NMe_2^+Cl^-$, followed by potassium phthalimide and treatment with hydrazine give the amine.[994]

Amides can also be alkylated with diazo compounds, as in **10-34**. Salts of sulfonamides ($ArSO_2NH^-$) can be used to attack alkyl halides to prepare N-alkyl sulfonamides ($ArSO_2NHR$) that can be further alkylated to $ArSO_2NRR'$. Hydrolysis of the latter is a good method for the preparation of secondary amines. Secondary amines can also be made by crown ether assisted alkylation of $F_3CCONHR$ (R = alkyl or aryl) and hydrolysis of the resulting $F_3CCONRR'$.[995]

The reaction of a primary amide and benzaldehyde, in the presence of a silane and trifluoroacetic acid, leads to the corresponding N-benzylamide.[996] This transformation is a reductive alkylation. N-Alkynyl amides have been prepared by the copper-catalyzed reaction of 1-bromoalkynes and secondary amides.[997] 1-Haloalkynes

[986]See Khan, M.N. *J. Org. Chem.* **1995**, *60*, 4536 for the kinetics of hydrazinolysis of phthalimides.

[987]Kukolja, S.; Lammert, S.R. *J. Am. Chem. Soc.* **1975**, *97*, 5582.

[988]Osby, J.O.; Martin, M.G.; Ganem, B. *Tetrahedron Lett.* **1984**, *25*, 2093.

[989]Wolfe, S.; Hasan, S.K. *Can. J. Chem.* **1970**, *48*, 3572.

[990]López-Alvarado, P.; Avendaño, C.; Menéndez, J.C. *Tetrahedron Lett.* **1992**, *33*, 6875.

[991]Hebrard, P.; Olomucki, M. *Bull. Soc. Chim. Fr.* **1970**, 1938.

[992]For other methods, see Mukaiyama, T.; Taguchi, T.; Nishi, M. *Bull. Chem. Soc. Jpn.* **1971**, *44*, 2797; Hendrickson, J.B.; Bergeron R.; Sternbach, D.D. *Tetrahedron* **1975**, *31*, 2517; Clarke, C.T.; Elliott, J.D.; Jones, J.H. *J. Chem. Soc. Perkin Trans 1*, **1978**, 1088; Mukaiyama, T.; Tsuji, T.; Watanabe, Y. *Chem. Lett.* **1978**, 1057; Zwierzak, A.; Pilichowska, S. *Synthesis* **1982**, 922; Calverley, M.J. *Synth. Commun.* **1983**, *13*, 601; Harland, P.A.; Hodge, P.; Maughan, W.; Wildsmith, E. *Synthesis* **1984**, 941; Grehn, L.; Ragnarsson, U. *Synthesis* **1987**, 275; Dalla Croce, P.; La Rosa, C.; Ritieni, A. *J. Chem. Res. (S)* **1988**, 346; Yinglin, H.; Hongwen, H. *Synthesis* **1990**, 122.

[993]Mitsunobu, O.; Wada, M.; Sano, T. *J. Am. Chem. Soc.* **1972**, *94*, 679; Grunewald, G.L.; Paradkar, V.M.; Pazhenchevsky, B.; Pleiss, M.A.; Sall, D.J.; Seibel, W.L.; Reitz, T.J. *J. Org. Chem.* **1983**, *48*, 2321; Ślusarska, E.; Zwierzak, A. *Liebigs Ann. Chem.* **1986**, 402; Kolasa, T.; Miller, M.J. *J. Org. Chem.* **1987**, *52*, 4978; Sammes, P.G.; Thetford, D. *J. Chem. Soc. Perkin Trans. 1* **1989**, 655.

[994]Barrett, A.G.M.; Braddock, D.C.; James, R.A.; Procopiou, P.A. *Chem. Commun.* **1997**, 433.

[995]Nordlander, J.E.; Catalane, D.B.; Eberlein, T.H.; Farkas, L.V.; Howe, R.S.; Stevens, R.M.; Tripoulas, N.A. *Tetrahedron Lett.* **1978**, 4987. For other methods, see Zwierzak, A.; Brylikowska-Piotrowicz, J. *Angew. Chem. Int. Ed.* **1977**, *16*, 107; Briggs, E.M.; Brown, G.W.; Jiricny, J.; Meidine, M.F. *Synthesis* **1980**, 295; Zwierzak, A.; Brylikowska-Piotrowicz, J. *Synthesis* **1982**, 922

[996]Dubé, D.; Scholte, A.A. *Tetrahedron Lett.* **1999**, *40*, 2295.

[997]Zhang, Y.; Hsung, R.P.; Tracey, M.R.; Kurtz, K.C.M.; Vera, E.L. *Org. Lett.* **2004**, *6*, 1151; Frederick, M.O.; Mulder, J.A.; Tracey, M.R.; Hsung, R.P.; Huang, J.; Kurtz, K.C.M.; Shen, L.; Douglas, C.J. *J. Am. Chem. Soc.* **2003**, *125*, 2368.

are typically prepared by base-induced elimination of 1,1-dihaloalkenes[998] or by direct halogenation of an alkyne with sodium or potassium hypohalite, prepared by reaction of the appropriate base with the halogen.[999]

Internal *N*-alkylation has been used to prepare the highly strained compounds α-lactams.[1000]

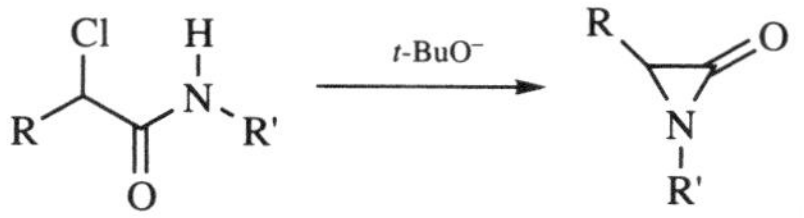

OS **I**, 119, 203, 271; **II**, 25, 83, 208; **III**, 151; **IV**, 810; **V**, 1064; **VI**, 951; **VII**, 501.

C. Other Nitrogen Nucleophiles

10-42   Formation of Nitro Compounds[1001]

Nitro-de-halogenation

$$RX + NO_2^- \longrightarrow RNO_2$$

Sodium nitrite can be used to prepare nitro compounds from primary or secondary alkyl bromides or iodides, but the method is of limited scope. Silver nitrite gives nitro compounds only when RX is a primary bromide or iodide.[1002] Nitrite esters are an important side product in all these cases (**10-22**) and become the major product (by an S_N1 mechanism) when secondary or tertiary halides are treated with silver nitrite. Alkyl nitro compounds can be prepared from the alkyl halide via the corresponding azide, by treatment with HOF in acetonitrile.[1003]

Nitro compounds can be prepared from alcohols using $NaNO_2/AcOH/HCl$.[1004]

OS **I**, 410; **IV**, 368, 454, 724.

10-43   Formation of Azides

Azido-de-halogenation

$$RX + N_3^- \longrightarrow RN_3$$
$$RCOX + N_3^- \longrightarrow RCON_3$$

[998]For an example involving bromine see Bestmann, H.-J.; Frey, H. *Liebigs Ann. Chem.* **1980**, *12*, 2061.
[999]For examples with hypobromite, see Mozūraitis, R.; Būda, V.; Liblikas, I.; Unelius, C.R.; Borg-Karlson, A.-K. *J. Chem. Ecol.* **2002**, *28*, 1191; Barbu, E.; Tsibouklis, J. *Tetrahedron Lett.* **1996**, *37*, 5023; Brandsma, L.; Verkruijsse, H.D *Synthesis* **1990**, 984.
[1000]See Quast, H.; Leybach, H. *Chem. Ber.* **1991**, *124*, 849. For a review of α-lactams, see Lengyel, I.; Sheehan, J.C. *Angew. Chem. Int. Ed.* **1968**, *7*, 25.
[1001]For reviews, see Larson, H.O. in Feuer, H. *The Chemistry of the Nitro and Nitroso Groups*, pt. 1, Wiley, NY, **1969**, pp. 325–339; Kornblum, N. *Org. React.* **1962**, *12*, 101.
[1002]See Ballini, R.; Barboni, L.; Giarlo, G. *J. Org. Chem.* **2004**, *69*, 6907.
[1003]Rozen, S.; Carmeli, M. *J. Am. Chem. Soc.* **2003**, *125*, 8118.
[1004]Baruah, A.; Kalita, B.; Barua, N.C. *Synlett* **2000**, 1064.

Alkyl azides can be prepared by treatment of the appropriate halide with azide ion.[1005] Phase-transfer catalysis,[1006] ultrasound,[1007] and the use of reactive clays[1008] are important variations. Substrates other than alkyl halides have been used,[1009] including OH,[1010] OMs, OTs,[1011] and OAc.[1012] Epoxides react with NaN_3 (**10-35**), $SnCl_2/Mg$ with NaN_3,[1013] $TMSN_3$ and Ph_4SbOH[1014] or SmI_2,[1015] or $(i\text{-}Bu)_2AlHN_3Li$[1016] to give β-azido alcohols; these are easily converted to aziridines, **131**.[1017]

This conversion has been used as a key step in the preparation of optically active aziridines from optically active 1,2-diols (prepared by **15-48**).[1018] Even hydrogen can be the leaving group. Benzylic hydrogens have been replaced by N_3 by treatment with HN_3 in $CHCl_3$ in the presence of DDQ (p. 1710).[1019]

Tertiary alkyl azides can be prepared by stirring tertiary alkyl chlorides with NaN_3 and $ZnCl_2$ in CS_2[1020] or by treating tertiary alcohols with NaN_3 and CF_3COOH[1021] or with HN_3 and $TiCl_4$[1022] or BF_3.[1023] Aryl azides can be prepared from aniline and aniline derivatives.[1024] Acyl azides, which can be used in the Curtius reaction (**18-14**),

[1005]For reviews, see Scriven, E.F.V.; Turnbull, K. *Chem. Rev.* **1988**, *88*, 297; Biffin, M.E.C.; Miller, J.; Paul, D.B., in Patai, S. *The Chemistry of the Azido Group*, Wiley, NY, **1971**, pp. 57–119; Alvarez, S.G.; Alvarez, M.T. *Synthesis* **1997**, 413.

[1006]See Reeves, W.P.; Bahr, M.L. *Synthesis* **1979**, 823; Marti, M.J.; Rico, I.; Ader, J.C.; de Savignac, A.; Lattes, A. *Tetrahedron Lett.* **1989**, *30*, 1245.

[1007]Priebe, H. *Acta Chem. Scand. Ser. B*, **1984**, *38*, 895.

[1008]See, for example, Varma, R.S.; Naicker, K.P.; Aschberger, J. *Synth. Commun.* **1999**, *29*, 2823.

[1009]See, for example, Hojo, K.; Kobayashi, S.; Soai, K.; Ikeda, S.; Mukaiyama, T. *Chem. Lett.* **1977**, 635; Murahashi, T.; Tanigawa, Y.; Imada, Y.; Taniguchi, Y. *Tetrahedron Lett.* **1986**, *27*, 227.

[1010]See, for example, Yu, C.; Liu, B.; Hu, L. *Org. Lett.* **2000**, *2*, 1959.

[1011]Scriven, E.F.V.; Turnbull, K. *Chem. Rev.* **1988**, *88*, 297, see p. 306.

[1012]Murahashi, S.; Taniguchi, Y.; Imada, Y.; Tanigawa, Y. *J. Org. Chem.* **1989**, *54*, 3292.

[1013]Sarangi, C.; Das, N.B.; Nanda, B.; Nayak, A.; Sharma, R.P. *J. Chem. Res. (S)* **1997**, 378.

[1014]Fujiwara, M.; Tanaka, M.; Baba, A.; Ando, H.; Souma, Y. *Tetrahedron Lett.* **1995**, *36*, 4849.

[1015]Van de Weghe, P.; Collin, J. *Tetrahedron Lett.* **1995**, *36*, 1649.

[1016]Youn, Y.S.; Cho, I.S.; Chung, B.Y. *Tetrahedron Lett.* **1998**, *39*, 4337.

[1017]See, for example, Ittah, Y.; Sasson, Y.; Shahak, I.; Tsaroom, S.; Blum, J. *J. Org. Chem.* **1978**, *43*, 4271. For the mechanism of the conversion to aziridines, see Pöchlauer, P.; Müller, E.P.; Peringer, P. *Helv. Chim. Acta* **1984**, *67*, 1238.

[1018]Lohray, B.B.; Gao, Y.; Sharpless, K.B. *Tetrahedron Lett.* **1989**, *30*, 2623.

[1019]Guy, A.; Lemor, A.; Doussot, J.; Lemaire, M. *Synthesis* **1988**, 900.

[1020]Miller, J.A. *Tetrahedron Lett.* **1975**, 2959. See also, Koziara, A.; Zwierzak, A. *Tetrahedron Lett.* **1987**, *28*, 6513.

[1021]Balderman, D.; Kalir, A. *Synthesis* **1978**, 24.

[1022]Hassner, A.; Fibiger, R.; Andisik, D. *J. Org. Chem.* **1984**, *49*, 4237.

[1023]See, for example, Adam, G.; Andrieux, J.; Plat, M. *Tetrahedron* **1985**, *41*, 399.

[1024]Liu, Q.; Tor, Y. *Org. Lett.* **2003**, *5*, 2571.

can be similarly prepared from acyl halides, anhydrides,[1025] esters,[1026] or other acyl derivatives.[1027] Acyl azides can also be prepared form aldehydes using SiCl$_4$/NaN$_3$–MnO$_2$,[1028] TMSN$_3$/CrO$_3$[1029] or the Dess–Martiin periodinane (see p. 1723) with NaN$_3$.[1030]

α-Azido ketones have been prepared from ketones via reaction with [hydroxy (*p*-nitrobenzenesulfonyloxy)iodo]benzene followed by reaction with sodium azide.[1031]

OS **III**, 846; **IV**, 715; **V**, 273, 586; **VI**, 95, 207, 210, 910; **VII**, 433; **VIII**, 116; **IX**, 220; **X**, 378. See also, OS **VII**, 206.

10-44 Formation of Isocyanates and Isothiocyanates

Isocyanato-de-halogenation

Isothiocyanato-de-halogenation

$$RX + NCO^- \longrightarrow RNCO$$
$$RX + NCS^- \longrightarrow RNCS$$

When the reagent is the thiocyanate ion, *S*-alkylation is an important side reaction (**10-30**), but the cyanate ion practically always gives exclusive *N*-alkylation.[478] Primary alkyl halides have been converted to isocyanates by treatment with sodium nitrocyanamide (NaNCNNO$_2$) and *m*-chloroperoxybenzoic acid, followed by heating of the initially produced RN(NO$_2$)CN.[1032] When alkyl halides are treated with NCO$^-$ in the presence of ethanol, carbamates can be prepared directly (see **16-8**).[1033] Acyl halides give the corresponding acyl isocyanates and isothiocyanates.[1034] For the formation of isocyanides, see **10-75**.

OS **III**, 735.

10-45 Formation of Azoxy Compounds

Alkyl-*NNO*-azoxy-de-halogenation

$$R-X \;+\; R'-N=N-O^{\ominus} \;\longrightarrow\; R'-N\overset{\overset{R}{\diagdown}}{\underset{O^{\ominus}}{=}}N^{\oplus}$$

132

[1025]For a review of acyl azides, see Lwowski, W., in Patai, S. *The Chemistry of the Azido Group*, Wiley, NY, *1971*, pp. 503–554.
[1026]Rawal, V.H.; Zhong, H.M. *Tetrahedron Lett.* **1994**, *35*, 4947.
[1027]Affandi, H.; Bayquen, A.V.; Read, R.W. *Tetrahedron Lett.* **1994**, *35*, 2729. For a preparation using triphosgene, see Gumaste, V.K.; Bhawal, B.M.; Deshmukh, A.R.A.S. *Tetrahedron Lett.* **2002**, *43*, 1345.
[1028]Elmorsy, S.S. *Tetrahedron Lett.* **1995**, *36*, 1341.
[1029]Lee, J.G.; Kwak, K.H. *Tetrahedron Lett.* **1992**, *33*, 3165.
[1030]Bose, D.S.; Reddy, A.V.N. *Tetrahedron Lett.* **2003**, *44*, 3543.
[1031]Lee, J.C.; Kim, S.; Shin, W.C. *Synth. Commun.* **2000**, *30*, 4271.
[1032]Manimaran, T.; Wolford, L.T.; Boyer, J.H. *J. Chem. Res. (S)* **1989**, 331.
[1033]Argabright, P.A.; Rider, H.D.; Sieck, R. *J. Org. Chem.* **1965**, *30*, 3317; Effenberger, F.; Drauz, K.; Förster, S.; Müller, W. *Chem. Ber.* **1981**, *114*, 173.
[1034]For reviews of acyl isocyanates, see Tsuge, O., in Patai, S. *The Chemistry of Cyanates and Their Thio Derivatives*, pt. 1, Wiley, NY, *1977*, pp. 445–506; Nuridzhanyan, K.A. *Russ. Chem. Rev.* **1970**, *39*, 130; Lozinskii, M.O.; Pel'kis, P.S. *Russ. Chem. Rev.* **1968**, *37*, 363.

The reaction between alkyl halides and alkanediazotates (**132**) gives azoxyalkanes.[1035] The R and R' groups may be the same or different, but neither may be aryl or tertiary alkyl. The reaction is regioselective; only the isomer shown is obtained.

HALOGEN NUCLEOPHILES[1036]

10-46   Halide Exchange.

Halo-de-halogenation

$$RX + X'^- \rightleftharpoons RX' + X^-$$

Halide exchange, sometimes call the *Finkelstein reaction*, is an equilibrium process, but it is often possible to shift the equilibrium.[1037] The reaction is most often applied to the preparation of iodides and fluorides. Iodides can be prepared from chlorides or bromides by taking advantage of the fact that sodium iodide, but not the bromide or chloride, is soluble in acetone. When an alkyl chloride or bromide is treated with a solution of sodium iodide in acetone, the equilibrium is shifted by the precipitation of sodium chloride or bromide. Since the mechanism is S_N2, the reaction is much more successful for primary halides than for secondary or tertiary halides; sodium iodide in acetone can be used as a test for primary bromides or chlorides. Tertiary chlorides can be converted to iodides by treatment with excess NaI in CS_2, with $ZnCl_2$ as catalyst.[1038] Vinylic bromides give vinylic iodides with retention of configuration when treated with KI and a nickel bromide-zinc catalyst,[1039] or with KI and CuI in hot HMPA.[1040]

Fluorides[1041] are prepared by treatment of other alkyl halides with any of a number of fluorinating agents,[1042] among them anhydrous HF (which is useful only for

---

[1035]For reviews, see Yandovskii, V.N.; Gidaspov, B.V.; Tselinskii, I.V. *Russ. Chem. Rev.* **1980**, *49*, 237; Moss, R.A. *Acc. Chem. Res.* **1974**, *7*, 421.

[1036]For a review of the formation of carbon-halogen bonds, see Hudlický, M.; Hudlicky, T., in Patai, S.; Rappoport, Z. *The Chemistry of Functional Groups, Supplement D*, pt. 2, Wiley, NY, **1983**, pp. 1021–1172.

[1037]For a list of reagents for alkyl halide interconversion, see Larock, R.C. *Comprehensive Organic Transformations*, 2nd ed., Wiley-VCH, NY, **1999**, pp. 667–671.

[1038]Miller, J.A.; Nunn, M.J. *J. Chem. Soc. Perkin Trans. 1* **1976**, 416.

[1039]Takagi, K.; Hayama, N.; Inokawa, S. *Chem. Lett.* **1978**, 1435.

[1040]Suzuki, H.; Aihara, M.; Yamamoto, H.; Takamoto, Y.; Ogawa, T. *Synthesis* **1988**, 236.

[1041]For reviews of the introduction of fluorine into organic compounds, see Mann, J. *Chem. Soc. Rev.* **1987**, *16*, 381; Rozen, S.; Filler, R. *Tetrahedron* **1985**, *41*, 1111; Hudlický, M. *Chemistry of Organic Fluorine Compounds*, pt. 2, Ellis Horwood, Chichester, **1976**, pp. 24–169; Sheppard, W.A.; Sharts, C.M., *Organic Fluorine Chemistry*, W.A. Benjamin, NY, **1969**, pp. 52–184, 409–430.

[1042]For reviews of the use of halogen exchange to prepare alkyl fluorides, see Sharts, C.M.; Sheppard, W.A. *Org. React.* **1974**, *21*, 125; Hudlický, M. *Chemistry of Organic Fluorine Compunds*, pt. 2, Ellis Horwood, Chichester, **1976**, pp. 91–136.

reactive substrates, e.g., benzylic or allylic), AgF, KF,[1043] HgF_2, $Et_3N \cdot 2HF$,[1044] 4-Me-$C_6H_4IF_2$,[1045] and $Me_2SiF_2Ph^+$ $^-NBu_4$.[1046] The equilibria in these cases are shifted because the alkyl fluoride once formed has little tendency to react, owing to the extremely poor leaving-group ability of fluorine. Phase-transfer catalysis of the exchange reaction is a particularly effective way of preparing both fluorides and iodides.[1047]

Primary alkyl chlorides can be converted to bromides with ethyl bromide, N-methyl-2-pyrrolidinone and a catalytic amount of NaBr,[1048] with LiBr under phase-transfer conditions,[1049] and with Bu_4N^+ Br^-.[1050] Primary bromides were converted to chlorides with TMSCl/imidazole in hot DMF.[1051] For secondary and tertiary alkyl chlorides, treatment in CH_2Cl_2 with excess gaseous HBr and an anhydrous $FeBr_3$ catalyst has given high yields[1052] (this procedure is also successful for chloride-to-iodide conversions). Alkyl chlorides or bromides can be prepared from iodides by treatment with HCl or HBr in the presence of HNO_3, making use of the fact that the leaving I^- is oxidized to I_2 by the HNO_3.[1053] Primary iodides give the chlorides when treated with PCl_5 in $POCl_3$.[1054] Alkyl fluorides and chlorides are converted to the bromides and iodides (and alkyl fluorides to the chlorides) by heating with the corresponding HX in excess amounts.[1055]

OS **II**, 476; **IV**, 84, 525; **VIII**, 486; **IX**, 502.

10-47   Formation of Alkyl Halides from Esters of Sulfuric and Sulfonic Acids

Halo-de-sulfonyloxy-substitution, and so on

$$ROSO_2R' + X^- \longrightarrow RX$$

[1043]See Mąkosza, M.; Bujok, R. *Tetrahedron Lett.* **2002**, *43*, 2761.

[1044]Giudicelli, M.B.; Picq, D.; Veyron B. *Tetrahedron Lett.* **1990**, *31*, 6527. For an electrolytic procedure using $Et_3 \cdot n$ HF see Sawaguchi, M.; Ayuba, S.; Nakamura, Y.; Fukuhara, J.; Hara, S.; Yoneda, N. *Synlett* **2000**, 999.

[1045]Sawaguchi, M.; Hara, S.; Nakamura, Y.; Ayuba, S.; Kukuhara, T.; Yoneda, N. *Tetrahedron* **2001**, 57, 3315.

[1046]Kvíala, J.; Mysík, P.; Paleta, O. *Synlett* **2001**, 547.

[1047]For reviews, see Starks, C.M.; Liotta, C. *Phase Transfer Catalysis*, Academic Press, NY, **1978**, pp. 112–125; Weber, W.P.; Gokel, G.W. *Phase Transfer Catalysis in Organic Synthesis*, Springer, NY, **1977**, pp. 117–124. See also, Clark, J.H.; Macquarrie, D.J. *Tetrahedron Lett.* **1987**, *28*, 111; Bram, G.; Loupy, A.; Pigeon, P. *Synth. Commun.* **1988**, *18*, 1661.

[1048]Willy, W.E.; McKean, D.R.; Garcia, B.A. *Bull. Chem. Soc. Jpn.* **1976**, 49, 1989. See also, Babler, J.H.; Spina, K.P. *Synth. Commun.* **1984**, 14, 1313.

[1049]Loupy, A.; Pardo, C. *Synth. Commun.* **1988**, *18*, 1275.

[1050]Bidd, I.; Whiting, M.C. *Tetrahedron Lett.* **1984**, 25, 5949.

[1051]Peyrat, J.-F.; Figadère, B.; Cavé, A. *Synth. Commun.* **1996**, *26*, 4563.

[1052]Yoon, K.B.; Kochi, J.K. *J. Org. Chem.* **1989**, *54*, 3028.

[1053]Svetlakov, N.V.; Moisak, I.E.; Averko-Antonovich, I.G. *J. Org. Chem. USSR* **1969**, *5*, 971.

[1054]Bartley, J.P.; Carman, R.M.; Russell-Maynard, J.K.L. *Aust. J. Chem.* **1985**, *38*, 1879.

[1055]Namavari, M.; Satyamurthy, N.; Phelps, M.E.; Barrio, J.R. *Tetrahedron Lett.* **1990**, *31*, 4973.

Alkyl sulfates, tosylates, and other esters of sulfuric and sulfonic acids can be converted to alkyl halides with any of the four halide ions.[1056] Neopentyl tosylate reacts with Cl⁻, Br⁻, or I⁻ without rearrangement in HMPA.[1057] Similarly, allylic tosylates can be converted to chlorides without allylic rearrangement by reaction with LiCl in the same solvent.[1058] Inorganic esters are intermediates in the conversion of alcohols to alkyl halides with $SOCl_2$, PCl_5, PCl_3, and so on (**10-48**), but are seldom isolated.

OS **I**, 25; **II**, 111, 404; **IV**, 597, 753; **V**, 545.

10-48 Formation of Alkyl Halides from Alcohols

Halo-de-hydroxylation

$$ROH + HX \longrightarrow RX$$
$$ROH + SOCl_2 \longrightarrow RCl$$

Alcohols can be converted to alkyl halides with several reagents,[1059] the most common of which are halogen acids HX and inorganic acid halides, such as $SOCl_2$,[1060] PCl_5, PCl_3, and $POCl_3$.[1061] The reagent HBr is usually used for alkyl bromides[1062] and HI for alkyl iodides. These reagents are often generated *in situ* from the halide ion and an acid such as phosphoric or sulfuric. The use of HI sometimes results in reduction of the alkyl iodide to the alkane (**19–53**) and, if the substrate is unsaturated, can also reduce the double bond.[1063] The reaction can be used to prepare primary, secondary, or tertiary halides, but alcohols of the isobutyl or neopentyl type often give large amounts of rearrangement products.[1064] Tertiary chlorides are easily made with concentrated HCl, but primary and secondary alcohols react with HCl so slowly that a catalyst, usually zinc chloride, is required.[1065] Primary alcohols give good yields of chlorides upon treatment with HCl in

[1056]For a list of reagents, with references, see Larock, R.C. *Comprehensive Organic Transformations*, 2nd ed., Wiley-VCH, NY, *1999*, pp. 697–700.

[1057]Stephenson, B.; Solladié, G.; Mosher, H.S. *J. Am. Chem. Soc.* *1974*, *96*, 3171.

[1058]Stork, G.; Grieco, P.A.; Gregson, M. *Tetrahedron Lett.* *1969*, 1393.

[1059]For a list of reagents, with references, see Larock, R.C. *Comprehensive Organic Transformations*, 2nd ed., Wiley-VCH, NY, *1999*, pp. 689–697.

[1060]For a review of thionyl chloride ($SOCl_2$), see Pizey, J.S. *Synthetic Reagents*, Vol. 1, Wiley, NY, *1974*, pp. 321–357. See Mohanazadeh, F.; Momeni, A.R. *Org. Prep. Proceed. Int.* *1996*, *28*, 492 for the use of $SOCl_2$ on silica gel.

[1061]For a review, see Salomaa, P.; Kankaanperä, A.; Pihlaja, K., in Patai, S. *The Chemistry of the Hydroxyl Group*, pt. 1, Wiley, NY, *1971*, pt. 1, pp. 595–622.

[1062]Mas, J.-M.; Metivier, P. *Synth. Commun.* *1992*, *22*, 2187; Chong, J.M.; Heuft, M.A.; Rabbat, P. *J. Org. Chem.* *2000*, *65*, 5837.

[1063]Jones, R.; Pattison, J.B. *J. Chem. Soc. C 1969*, 1046.

[1064]For a reaction using $CeCl_3\cdot7\ H_2O$ and NaI with neopentyl alcohol to give 2-iodo-2-methylbutane see Di Deo, M.; Marcantoni, E.; Torregiani, E.; Bartoli, G.; Bellucci, M. C.; Bosco, M.; Sambri, L. *J. Org. Chem.* *2000*, *65*, 2830.

[1065]Phase-transfer catalysts have been used instead of $ZnCl_2$; Landini, D.; Montanari, F.; Rolla, F. *Synthesis 1974*, 37.

HMPA.[1066] The inorganic acid chlorides $SOCl_2$,[1067] PCl_3, and so on, give primary, secondary, or tertiary alkyl chlorides with much less rearrangement than is observed with HCl. Iodides have been prepared by simply heating the alcohol with iodine.[1068] Trichloroisocyanuric acid (1,3,5-trichlorohexahydrotriazin-2,4,6-trione) and triphenylphosphine converts primary alcohols to the corresponding chloride.[1069]

Analogous bromides and iodides, especially PBr_3, have also been used, but they are more expensive and used less often than HBr or HI, although some of them may also be generated *in situ* (e.g., PBr_3 from phosphorous and bromine). Bromides have also been prepared with NaBr on doped Montmorillonite K10 clay[1070] and iodides were prepared by using NaI on KSF-clay,[1071] both using with microwave irradiation. Secondary alcohols always gives *some* rearranged bromides if another secondary position is available, even with PBr_3, PBr_5, or $SOBr_2$; thus 3-pentanol gives both 2- and 3-bromopentane. Such rearrangement can be avoided by converting the alcohol to a sulfonate and then using **10-47**,[1072] or by the use of phase transfer catalysis.[1073] Tertiary alcohols can be converted to the bromide with BBr_3 at $0°C$.[1074] HF does not generally convert alcohols to alkyl fluorides.[1075] The most important reagent for this purpose is the commercially available diethylaminosulfur trifluoride Et_2NSF_3 (DAST),[1076] which converts primary, secondary, tertiary, allylic, and benzylic alcohols to fluorides in high yields under mild conditions.[1077] Fluorides have also been prepared from alcohols by treatment with SF_4,[1078] SeF_4,[1079] TsF,[1080] CsI/BF_3,[1081] and indirectly, by conversion to a sulfate or tosylate, and so on (**10-47**). Sodium iodide and Amberlyst-15[1082] or tosic acid and KI with microwave irradiation[1083] converts primary alcohols to the iodide. A mixture of IF_5, NEt_3

[1066]Fuchs, R.; Cole, L.L. *Can. J. Chem.* **1975**, *53*, 3620.

[1067]For a transformation involving a primary benzylic alcohol, thionyl chloride and benzotriazole, see Chaudhari, S.S.; Akamanchi, K.G. *Synlett* **1999**, 1763.

[1068]Joseph, R.; Pallan, P.S.; Sudalai, A.; Ravindranathan, T. *Tetrahedron Lett.* **1995**, *36*, 609.

[1069]Hiegel, G.A.; Rubino, M. *Synth. Commmun.* **2002**, *32*, 2691.

[1070]Kad, G.L.; Singh, V.; Kaur, K.P.; Singh. J. *Tetrahedron Lett.* **1997**, *38*, 1079.

[1071]Kad, G.L.; Kaur, J.; Bansal, P.; Singh, J. *J. Chem. Res. (S)* **1996**, 188.

[1072]Cason, J.; Correia, J.S. *J. Org. Chem.* **1961**, *26*, 3645.

[1073]Dakka, G.; Sasson, Y. *Tetrahedron Lett.* **1987**, *28*, 1223.

[1074]Pelletier, J.D.; Poirier, D. *Tetrahedron Lett.* **1994**, *35*, 1051.

[1075]For an exception, see Hanack, M.; Eggensperger, H.; Hähnle, R. *Liebigs Ann. Chem.* **1962**, *652*, 96; See also, Politanskii, S.F.; Ivanyk, G.D.; Sarancha, V.N.; Shevchuk, V.U. *J. Org. Chem. USSR* **1974**, *10*, 697.

[1076]For a review of this reagent, see Hudlický, M. *Org. React.* **1988**, *35*, 513.

[1077]Middleton, W.J. *J. Org. Chem.* **1975**, *40*, 574.

[1078]For reviews, see Wang, C.J. *Org. React.* **1985**, *34*, 319; Kollonitsch, J. *Isr. J. Chem.* **1978**, *17*, 53; Boswell, Jr., G.A.; Ripka, W.C.; Scribner, R.M.; Tullock, C.W. *Org. React.* **1974**, *21*, 1.

[1079]Olah, G.A.; Nojima, M.; Kerekes, I. *J. Am. Chem. Soc.* **1974**, *96*, 925.

[1080]Shimizu, M.; Nakahara, Y.; Yoshioka, H. *Tetrahedron Lett.* **1985**, *26*, 4207. For another method, see Olah, G.A.; Li, X. *Synlett,* **1990**, 267.

[1081]Hayat, S.; Atta-ur-Rahman, Khan, K.M.; Choudhary, M.I.; Maharvi, G.M.; Zia-Ullah; Bayer, E. *Synth. Commun.* **2003**, *33*, 2531.

[1082]Tajbakhsh, M.; Hosseinzadeh, R.; Lasemi, Z. *Synlett* **2004**, 635.

[1083]Lee, J.C.; Park, J.Y.; Yoo, E.S. *Synth. Commun.* **2004**, *34*, 2095.

and excess KF^{1084} or $(Cl_3CO)_2C=O$, bis(trichloromethyl)carbonate, and KF (which gives COF_2 *in situ*)with 18-crown-6^{1085} also converts primary alcohols to primary fluorides.

Primary, secondary, and tertiary alcohols can be converted to any of the four halides by treatment with the appropriate NaX, KX, or NH_4X in polyhydrogen fluoride–pyridine solution.[1086] This method is even successful for neopentyl halides. Another reagent that converts neopentyl alcohol to neopentyl chloride, in 95% yield, is PPh_3–CCl_3CN.[1087] Ionic liquids can be used for halogenation, and bmim-Cl (1-n-butyl-3-methylimidazolium chloride) generates the chloride directly from the alcohol without any additional reagent.[1088]

Other reagents[1089] have also been used, including $ZrCl_4/NaI$,[1090] 2,4,6-trichloro [1,3,5]triazine (cyanuric acid) and DMF,[1091] Me_3SiCl and $BiCl_3$[1092] or Me_3SiCl and 5% $InCl_3$[1093] or simply Me_3SiCl in DMSO.[1094] Other specialized reagents include $(RO)_3PRX^{1095}$ and $R_3PX_2^{1096}$ (made from R_3P and X_2), which give good yields for primary (including neopentyl), secondary, and tertiary halides without rearrange-

[1084]Yoneda, N. Fukuhara, T. *Chem. Lett.* **2001**, 222.

[1085]Flosser, D.A.; Olofson, R.A. *Tetrahedron Lett.* **2002**, *43*, 4275.

[1086]Olah, G.A.; Welch, J.; Vankar, Y.D.; Nojima, M.; Kerekes, I.; Olah, J.A. *J. Org. Chem.* **1979**, *44*, 3872. See also, Yin, J.; Zarkowsky, D.S.; Thomas, D.W.; Zhao, M.W.; Huffman, M.A. *Org. Lett.* **2004**, *6*, 1465.

[1087]Matveeva, E.D.; Yalovskaya, A.I.; Cherepanov, I.A.; Kurts, A.L.; Bundel', Yu.G. *J. Org. Chem. USSR* **1989**, *25*, 587.

[1088]Ren, R. X.; Wu, J. X. *Org. Lett.* **2001**, *3*, 3727.

[1089]For some other reagents, not listed here, see Echigo, Y.; Mukaiyama, T. *Chem. Lett.* **1978**, 465; Barton, D.H.R.; Stick, R.V.; Subramanian, R. *J. Chem. Soc. Perkin Trans. 1* **1976**, 2112; Savel'yanov, V.P.; Nazarov, V.N.; Savel'yanova, R.T.; Suchkov, V.V. *J. Org. Chem. USSR* **1977**, *13*, 604; Jung, M.E.; Hatfield, G.L. *Tetrahedron Lett.* **1978**, 4483; Sevrin, M.; Krief, A. *J. Chem. Soc., Chem. Commun.* **1980**, 656; Hanessian, S.; Leblanc, Y.; Lavallée, P. *Tetrahedron Lett.* **1982**, *23*, 4411; Cristol, S.J.; Seapy, D.G. *J. Org. Chem.* **1982**, *47*, 132; Richter, R.; Tucker, B. *J. Org. Chem.* **1983**, *48*, 2625; Imamoto, T.; Matsumoto, T.; Kusumoto, T.; Yokoyama, M. *Synthesis* **1983**, 460; Olah, G.A.; Husain, A.; Singh, B.P.; Mehrotra, A.K. *J. Org. Chem.* **1983**, *48*, 3667; Toto, S.D.; Doi, J.T. *J. Org. Chem.* **1987**, *52*, 4999; Camps, F.; Gasol, V.; Guerrero, A. *Synthesis* **1987**, 511; Schmidt, S.P.; Brooks, D.W. *Tetrahedron Lett.* **1987**, *28*, 767; Collingwood, S.P.; Davies, A.P.; Golding, B.T. *Tetrahedron Lett.* **1987**, *28*, 4445; Kozikowski, A.P.; Lee, J. *Tetrahedron Lett.* **1988**, *29*, 3053; Classon, B.; Liu, Z.; Samuelsson, B. *J. Org. Chem.* **1988**, *53*, 6126; Munyemana, F.; Frisque-Hesbain, A.; Devos, A.; Ghosez, L. *Tetrahedron Lett.* **1989**, *30*, 3077; Ernst, B.; Winkler, T. *Tetrahedron Lett.* **1989**, *30*, 3081.

[1090]Firouzabadi, H.; Iranpoor, N.; Jafarpour, M. *Tetrahedron Lett.* **2004**, *45*, 7451.

[1091]De Luca, L.; Giacomelli, G.; Porcheddu, A. *Org. Lett.* **2002**, *4*, 553.

[1092]Labrouillère, M.; LeRoux, C.; Oussaid, A.; Gaspard-Iloughmane, H.; Dubac, J. *Bull. Soc. Chim. Fr.* **1995** *132*, 522.

[1093]Yasuda, M.; Yamasaki, S.; Onishi, Y.; Baba, A. *J. Am. Chem. Soc.* **2004**, *126*, 7186.

[1094]Snyder, D.C. *J. Org. Chem.* **1995**, *60*, 2638.

[1095]Rydon, H.N. *Org. Synth. VI*, 830.

[1096]Wiley, G.A.; Hershkowitz, R.L.; Rein, B.M.; Chung, B.C. *J. Am. Chem. Soc.* **1964**, *86*, 964; Wiley, G.A.; Rein, B.M.; Hershkowitz, R.L. *Tetrahedron Lett.* **1964**, 2509; Schaefer, J.P.; Weinberg, D.S. *J. Org. Chem.* **1965**, *30*, 2635; Kaplan, L. *J. Org. Chem.* **1966**, *31*, 3454; Weiss, R.G.; Snyder, E.I. *J. Org. Chem.* **1971**, *36*, 403; Garegg, P.J.; Johansson, R.; Samuelsson, B. *Synthesis* **1984**, 168; Sandri, J.; Viala, J. *Synth. Commun.* **1992**, *22*, 2945.

ments.[1097] Similarly, Me_2SBr_2[1098] (prepared from Me_2S and Br_2), and a mixture of PPh_3 and CCl_4[1099] (or CBr_4[1100]).

$$ROH + Ph_3P + CCl_4 \longrightarrow RCl + Ph_3PO + HCCl_3$$

The last method converts allylic alcohols[1101] to the corresponding halides without allylic rearrangements[1102] and also cyclopropylcarbinyl alcohols to the halides without ring opening.[1103] A simple method that is specific for benzylic and allylic alcohols (and does not give allylic rearrangement) involves reaction with N-chloro- or N-bromosuccinimide and methyl sulfide.[1104] The specificity of this method is illustrated by the conversion, in 87% yield, of (Z)-$HOCH_2CH_2CMe$=$CHCH_2OH$ to (Z)-$HOCH_2CH_2Me$=$CHCH_2Cl$. Only the allylic OH group was affected. A mixture of NBS, $Cu(OTf)_2$ and diisopropylcarbodiimide converted primary alcohols to the corresponding bromide.[1105] The use of NCS gave the chloride and NIS gave the iodide under identical conditions. Thiols are converted to alkyl bromides by a similar procedure using PPh_3 and NBS.[1106]

Allylic and benzylic alcohols can also be converted to bromides or iodides with NaX-BF_3 etherate,[1107] and to iodides with AlI_3.[1108] A mixture of methanesulfonic acid and NaI also converts benzylic alcohols to benzylic iodides.[1109] Both (chloro-phenylthio-methylene)dimethylammonium chloride[1110] and 2-chloro-1,3-dimethyl-imidazolinium chloride[1111] react with alcohols to give the corresponding chloride.

[1097]For reviews of reactions with these reagents, see Castro, B.R. *Org. React.* **1983**, *29*, 1; Mackie, R.K., in Cadogan, J.I.G. *Organophosphorus Reagents in Organic Synthesis*, Academic Press, NY, **1979**; pp. 433–466.

[1098]Furukawa, N.; Inoue, T.; Aida, T.; Oae, S. *J. Chem. Soc., Chem. Commun.* **1973**, 212.

[1099]For reviews, see Appel, R. *Angew. Chem. Int. Ed.* **1975**, *14*, 801; Appel, R.; Halstenberg, M., in Cadogan, J.I.G. *Organophosphorus Reagents in Organic Synthesis*, Academic Press, NY, **1979**, pp. 387–431. For a discussion of the mechanism, see Slagle, J.D.; Huang, T.T.; Franzus, B. *J. Org. Chem.* **1981**, *46*, 3526. For a similar reaction using hexachloroethane and bis-1,2-diphenylphosphinoethane see Pollastri, M.P.; Sagal, J.F.; Chang, G. *Tetrahedron Lett.* **2001**, *42*, 2459.

[1100]Wagner, A.; Heitz, M.; Mioskowski, C. *Tetrahedron Lett.* **1989**, *30*, 557. See also, Desmaris, L.; Percina, N.; Cottier, L.; Sinou, D. *Tetrahedron Lett.* **2003**, *44*, 7589.

[1101]For a review of the conversion of allylic alcohols to allylic halides, see Magid, R.M. *Tetrahedron* **1980**, *36*, 1901, pp. 1924–1926.

[1102]Snyder, E.I. *J. Org. Chem.* **1972**, *37*, 1466; Axelrod, E.H.; Milne, G.M.; van Tamelen, E.E. *J. Am. Chem. Soc.* **1973**, *92*, 2139.

[1103]Hrubiec, R.T.; Smith, M.B. *Synth. Commun.* **1983**, *13*, 593.

[1104]Corey, E.J.; Kim, C.U.; Takeda, M. *Tetrahedron Lett.* **1972**, 4339.

[1105]Li, Z.; Crosignani, S.; Linclau, B. *Tetrahedron Lett.* **2003**, *44*, 8143; Crosignani, S.; Nadal, B.; Li, Z.; Linclau, B. *Chem. Commun.* **2003**, 260.

[1106]Iranpoor, N.; Firouzabadi, H.; Aghapour, G. *Synlett* **2001**, 1176.

[1107]Vankar, Y.D.; Rao, C.T. *Tetrahedron Lett.* **1985**, *26*, 2717; Mandal, A.K.; Mahajan, S.W. *Tetrahedron Lett.* **1985**, *26*, 3863; Bandgar, B.P.; Sadavarte, V.S.; Uppalla, L.S. *Tetrahedron Lett.* **2001**, *42*, 951.

[1108]Sarmah, P.; Barua, N.C. *Tetrahedron* **1989**, *45*, 3569.

[1109]Kamal, A.; Ramesh, G.; Laxman, N. *Synth. Commun.* **2001**, *31*, 827.

[1110]Gomez, L.; Gellibert, F.; Wagner, A.; Mioskowski, C. *Tetrahedron Lett.* **2000**, *41*, 6049.

[1111]Isobe, T.; Ishikawa, T. *J. Org. Chem.* **1999**, *64*, 5832.

When the reagent is HX, the mechanism is S_N1cA or S_N2cA; that is, the leaving group is not ^-OH, but OH_2 (p. 496). The leaving group is not ^-OH with the other reagents either, since in these cases the alcohol is first converted to an inorganic ester, for example, ROSOCl with $SOCl_2$ (**10-22**). The leaving group is therefore ^-OSOCl or a similar group (**10-47**). These may react by the S_N1 or S_N2 mechanism and, in the case of ROSOCl, by the S_Ni mechanism[1112] (p. 468).

Trialkylsilyl ethers such as $ROSiMe_3$ are converted to the corresponding iodide with SiO_2–Cl/NaI.[1113]

OS **I**, 25, 36, 131, 142, 144, 292, 294, 533; **II**, 91, 136, 159, 246, 308, 322, 358, 399, 476; **III**, 11, 227, 370, 446, 698, 793, 841; **IV**, 106, 169, 323, 333, 576, 681; **V**, 1, 249, 608; **VI**, 75, 628, 634, 638, 781, 830, 835; **VII**, 210, 319, 356; **VIII**, 451. Also see, OS **III**, 818; **IV**, 278, 383, 597.

10-49   Formation of Alkyl Halides from Ethers

Halo-de-alkoxylation

$$ROR' + HI \longrightarrow RI + R'OH$$

Ethers can be cleaved by heating with concentrated HI or HBr.[1114] Hydrogen chloride is seldom successful,[1115] and HBr reacts more slowly than HI, but is often a superior reagent, since it causes fewer side reactions. Phase-transfer catalysis has also been used,[1116] and 47% HBr in ionic liquids has proven effective.[1117] Dialkyl ethers and alkyl aryl ethers can be cleaved. In the latter case the alkyl–oxygen bond is the one broken. As in **10-48**, the actual leaving group is not OR'^-, but OHR'. Although alkyl aryl ethers always cleave so as to give an alkyl halide and a phenol, there is no general rule for dialkyl ethers. Often cleavage occurs from both sides, and a mixture of two alcohols and two alkyl halides is obtained. However, methyl ethers are usually cleaved so that methyl iodide or bromide is a product. An excess of HI or HBr converts the alcohol product into alkyl halide, so that dialkyl ethers (but not alkyl aryl ethers) are converted to 2 equivalents of alkyl halide. This procedure is often carried out so that a mixture of only two products is obtained instead of four. O-Benzyl ethers are readily cleaved to the alcohol and the hydrocarbon via hydrogenolysis, and the most common methods are hydrogenation[1118] or

[1112]Schreiner, P.R.; Schleyer, P.v.R.; Hill, R.K. *J. Org. Chem.* **1993**, *58*, 2822.

[1113]Firouzabadi, H.; Iranpoor, N.; Hazarkhani, H. *Tetrahedron Lett.* **2002**, *43*, 7139.

[1114]For reviews of ether cleavage in general, see Bhatt, M.V.; Kulkarni, S.U. *Synthesis* **1983**, 249; Staude, E.; Patat, F., in Patai, S. *The Chemistry of the Ether Linkage*, Wiley, NY, **1967**, p. 22. For a review of cleavage of aryl alkyl ethers, see Tiecco, M. *Synthesis* **1988**, 749.

[1115]Cleavage with HCl has been accomplished in the presence of surfactants: Juršić, B. *J. Chem. Res. (S)* **1989**, 284.

[1116]Landini, D.; Montanari, F.; Rolla, F. *Synthesis* **1978**, 771.

[1117]In bmim BF₄, 1-*n*-butyl-3-methylimidazolium bromide: Boovanahalli, S.K.; Kim, D.W.; Chi, D.Y. *J. Org. Chem.* **2004**, *69*, 3340.

[1118]Heathcock, C.H.; Ratcliffe, R. *J. Am. Chem. Soc.* **1971**, *93*, 1746.

dissolving metal conditions (Na or K in ammonia).[1119] Heating in anisole with 3% Sc(NTf$_2$)$_3$[1120] or In metal in aqueous ethanol[1121] also cleaves benzyl ethers. Isoprenyl alkyl ethers are cleaved using iodine in dichloromethane,[1122] and allyl alkyl ethers are cleaved with Lewis acids under various conditions.[1123] The OCH$_2$CH=CHPh unit of mixed allyl ethers (O-CH$_2$CH=CH$_2$ and OCH$_2$CH=CHPh) can be cleaved selectively under electrolytic conditions.[1124]

Cyclic ethers (usually tetrahydrofuran derivatives) can be similarly cleaved (see **10-50** for epoxides). Treatment of 2-methyltetrahydrofuran with acetyl chloride and ZnCl$_2$ gave primarily O-acetyl-4-chloro-1-pentanol.[1125] A mixture of Et$_2$NSiMe$_3$/2 MeI cleaved tetrahydrofuran to give the O-trimethylsilyl ether of 4-iodo-1-butanol.[1126] Ethers have also been cleaved with Lewis acids, such as BF$_3$, Ce(OTf)$_4$,[1127] SiCl$_4$/LiI/BF$_3$,[1128] BBr$_3$,[1129] or AlCl$_3$.[1130] In such cases, the departure of the OR is assisted by complex formation with the Lewis acid (see **133**).

$$\begin{array}{c} R \\ \diagdown \\[-2pt] \underset{R'}{\overset{\oplus}{O}} \longrightarrow \overset{\ominus}{BF_3} \\[4pt] \mathbf{133} \end{array}$$

Lewis acids are also used. The reagent NaI—BF$_3$ etherate selectively cleaves ethers in the order benzylic ethers > alkyl methyl ethers > aryl methyl ethers.[1131]

Dialkyl and alkyl aryl ethers are cleaved with iodotrimethylsilane:[1132] ROR' + Me$_3$SiI → RI + Me$_3$SiOR.[1133] A more convenient and less expensive alternative, which gives the same products, is a mixture of chlorotrimethylsilane and

[1119]McCloskey, C.M. *Adv. Carbohydr. Chem.* **1957**, *12*, 137; Reist, E.J.; Bartuska, V.J.; Goodman, L. *J. Org. Chem.* **1964**, *29*, 3725.

[1120]Ishihara, K.; Hiraiwa, Y.; Yamamoto, H. *Synlett* **2000**, 80.

[1121]Moody, C.J.; Pitts, M.R. *Synlett* **1999**, 1575.

[1122]Vatèle, J.-M. *Synlett* **2001**, 1989. For a procedure using DDQ, see Vatèle, J.-M. *Synlett* **2002**, 507

[1123]Examples include SmI$_2$ in the presence of H$_2$O-*i*PrNH$_2$: Dahlen, A.; Sundgren, A.; Lahmann, M.; Oscarson, S.; Hilmersson, G. *Org. Lett.* **2003**, *5*, 4085. CeCl$_3$/NaI: Bartoli, G.; Cupone, G.; Dalpozzo, R.; DeNino, A.; Maiuolo, L.; Marcantoni, E.; Procopio, A. *Synlett* **2001**, 1897. ZnCl$_2$-Pd(PPh$_3$)$_4$: Chandrasekhar, S.; Reddy, Ch.R.; Rao, R.J. *Tetrahedron* **2001**, *57*, 3435. A ruthenium-catalyzed protocol: Tanaka, S.; Saburi, H.; Ishibashi, Y.; Kitamura, M. *Org. Lett.* **2004**, *6*, 1873. See also, Murakami, H.; Minami, T.; Ozawa, F. *J. Org. Chem.* **2004**, *69*, 4482.

[1124]Solis-Oba, A.; Hudlicky, T.; Koroniak, L.; Frey, D. *Tetrahedron Lett.* **2001**, *42*, 1241.

[1125]Mimero, P.; Saluzzo, C.; Amouroux, R. *Tetrahedron Lett.* **1994**, *35*, 1553.

[1126]Ohshita, J.; Iwata, A.; Kanetani, F.; Kunai, A.; Yamamoto, Y.; Matui, C. *J. Org. Chem.* **1999**, *64*, 8024.

[1127]Khalafi-Nezhad, A.; Alamdari, R.F. *Tetrahedron* **2001**, *57*, 6805.

[1128]Zewge, D.; King, A.; Weissman, S.; Tschaen, D. *Tetrahedron Lett.* **2004**, *45*, 3729.

[1129]Press, J.B. *Synth. Commun.* **1979**, *9*, 407; Niwa, H.; Hida, T.; Yamada, K. *Tetrahedron Lett.* **1981**, *22*, 4239.

[1130]For a review, see Johnson, F., in Olah, G.A. *Friedel–Crafts and Related Reactions*, Vol. 4, Wiley, NY, **1965**, pp. 1–109.

[1131]Vankar, Y.D.; Rao, C.T. *J. Chem. Res. (S)* **1985**, 232. See also, Mandal, A.K.; Soni, N.R.; Ratnam, K.R. *Synthesis* **1985**, 274; Ghiaci, M.; Asghari, J. *Synth. Commun.* **1999**, *29*, 973; Sharma, G.V.M.; Reddy, Ch.G.; Krishna, P.R. *J. Org. Chem.* **2003**, *68*, 4574.

[1132]For a review of this reagent, see Olah, G.A.; Prakash, G.K.S.; Krishnamurti, R. *Adv. Silicon Chem.* **1991**, *1*, 1.

[1133]Jung, M.E.; Lyster, M.A. *J. Org. Chem.* **1977**, *42*, 3761; *Org. Synth.* **VI**, 353.

NaI.[1134] Triphenyldibromophosphorane (Ph$_3$PBr$_2$) cleaves dialkyl ethers to give 2 moles of alkyl bromide.[1135] Alkyl aryl ethers can also be cleaved with LiI to give alkyl iodides and salts of phenols[1136] in a reaction similar to **10-51**. Allyl aryl ethers[1137] are efficiently cleaved with NaI/Me$_3$SiCl,[1138] CeCl$_3$/NaI[1139] or ZrCl$_4$/NaBH$_4$.[1140]

A closely related reaction is cleavage of oxonium salts.

$$R_3O^+X^- \longrightarrow RX + R_2O$$

For these substrates, HX is not required, and X can be any of the four halide ions.

tert-Butyldimethylsilyl ethers (ROSiMe$_2$CMe$_3$) can be converted to bromides RBr by treatment with Ph$_3$PBr$_2$,[1141] Ph$_3$P—CBr$_4$,[1142] or BBr$_3$.[1143] Alcohols are often protected by conversion to this kind of silyl ether.[1144]

OS **I**, 150; **II**, 571; **III**, 187, 432, 586, 692, 753, 774, 813; **IV**, 266, 321; **V**, 412; **VI**, 353. See also, OS **VIII**, 161, 556.

10-50   Formation of Halohydrins from Epoxides

(3)*OC-seco*-Halo-de-alkoxylation

This is a special case of **10-49** and is frequently used for the preparation of halohydrins. In contrast to the situation with open-chain ethers and with larger rings, many epoxides react with all four hydrohalic acids, although with HF[1145] the reaction is unsuccessful with simple aliphatic and cycloalkyl epoxides.[1146] Hydrogen fluoride does react with more rigid epoxides, such as those in steroid systems. The reaction can applied to simple epoxides[1147] if polyhydrogen fluoride-pyridine

[1134]Morita, T.; Okamoto, Y.; Sakurai, H. *J. Chem. Soc., Chem. Commun.* **1978**, 874; Olah, G.A.; Narang, S.C.; Gupta, B.G.B.; Malhotra, R. *J. Org. Chem.* **1979**, *44*, 1247; Amouroux, R.; Jatczak, M.; Chastrette, M. *Bull. Soc. Chim. Fr.* **1987**, 505.

[1135]Anderson Jr., A.G.; Freenor, F.J. *J. Org. Chem.* **1972**, *37*, 626.

[1136]Harrison, I.T. *Chem. Commun.* **1969**, 616.

[1137]For cleavage with Pd/C in KOH/MeOH, see Ishizaki, M.; Yamada, M.; Watanabe, S.-i.; Hoshino, O.; Nishitani, K.; Hayashida, M.; Tanaka, A.; Hara, H. *Tetrahedron* **2004**, *60*, 7973.

[1138]Kamal, A.; Laxman, E.; Rao, N.V. *Tetrahedron Lett.* **1999**, *40*, 371.

[1139]Thomas, R.M.; Reddy, G.S.; Iyengar, D.S. *Tetrahedron Lett.* **1999**, *40*, 7293

[1140]Chary, K.P.; Mohan, G.H.; Iyengar, D.S. *Chem. Lett.* **1999**, 1223.

[1141]Aizpurua, J.M.; Cossío, F.P.; Palomo, C. *J. Org. Chem.* **1986**, *51*, 4941.

[1142]Mattes, H.; Benezra, C. *Tetrahedron Lett.* **1987**, *28*, 1697.

[1143]Kim, S.; Park, J.H. *J. Org. Chem.* **1988**, *53*, 3111.

[1144]See Corey, E.J.; Venkateswarlu, A. *J. Am. Chem. Soc.* **1972**, *94*, 6190.

[1145]For a review of reactions HF with epoxides, see Sharts, C.M.; Sheppard, W.A. *Organic Fluorine Chemistry*, W.A. Benjamin, NY, **1969**, pp. 52–184, 409–430. For a related review, see Yoneda, N. *Tetrahedron* **1991**, *47*, 5329.

[1146]Shahak, I.; Manor, S.; Bergmann, E.D. *J. Chem. Soc. C* **1968**, 2129.

[1147]Olah, G.A.; Meidar, D. *Isr. J. Chem.* **1978**, *17*, 148.

is the reagent. The reagent $NEt_3 \cdot 3$ HF converts epoxides to fluorohydrins with microwave irradiation.[1148] The epoxide-to-fluorohydrin conversion has also been carried out with SiF_4 and a tertiary amine.[1149] Chloro-, bromo-, and iodohydrins can also be prepared[1150] by treating epoxides with Ph_3P and X_2,[1151] with $InBr_3/NaBr/H_2O$,[1152] LiBr on Amberlyst-15 resin,[1153] $TiCl_4$-LiCl,[1154] $SiCl_4$,[1155] I_2 with a SmI_2 catalyst,[1156] and LiI on silica gel.[1157] Epoxides can be converted directly to 1,2-dichloro compounds by treatment with $SOCl_2$ and pyridine,[1158] or with Ph_3P and CCl_4.[1159] These are two-step reactions: a halohydrin is formed first and is then converted by the reagents to the dihalide (**10-48**). As expected, inversion is found at both carbons. Meso epoxides were cleaved enantioselectively with the chiral B-halodiisopinocampheylboranes (see **15-16**), where the halogen was Cl, Br, or I.[1160] Diatomic iodine gives an iodohydrin with a 2,6-bis[2-(o-aminophenoxy) methyl]-4-bromo-1-methoxybenzene catalyst.[1161]

Bicyclic epoxides are usually opened to the *trans*-halohydrin. Unsymmetrical epoxides are usually opened to give mixtures of regioisomers. In a typical reaction, the halogen is delivered to the less sterically hindered carbon of the epoxide. In the absence of this structural feature, and in the absence of a directing group, relatively equal mixtures of regioisomeric halohydrins are expected. The phenyl is such as group in 1-phenyl-2-alkyl epoxides, where reaction with $POCl_3/DMAP$ leads to the chlorohydrin with the chlorine on the carbon bearing the phenyl.[1162]

[1148]Inagaki, T.; Fukuhara, T.; Hara, S. *Synthesis* **2003**, 1157.

[1149]Shimizu, M.; Yoshioka, H. *Tetrahedron Lett.* **1988**, *29*, 4101. For other methods, see Muehlbacher, M.; Poulter, C.D. *J. Org. Chem.* **1988**, *53*, 1026; Ichihara, J.; Hanafusa, T. *J. Chem. Soc., Chem. Commun.* **1989**, 1848.

[1150]Einhorn, C.; Luche, J. *J. Chem. Soc., Chem. Commun.* **1986**, 1368; Ciaccio, J.A.; Addess, K.J.; Bell, T.W. *Tetrahedron Lett.* **1986**, *27*, 3697; Spawn, C.; Drtina, G.J.; Wiemer, D.F. *Synthesis* **1986**, 315. For reviews, see Bonini, C.; Righi, G. *Synthesis* **1994**, 225; Chini, M.; Crotti, P.; Gardelli, C.; Macchia, F. *Tetrahedron* **1992**, *48*, 3805.

[1151]Palumbo, G.; Ferreri, C.; Caputo, R. *Tetrahedron Lett.* **1983**, *24*, 1307. See Afonso, C.A.M.; Vieira, N.M.L.; Motherwell, W.B. *Synlett* **2000**, 382.

[1152]Amantini, D.; Fringulli, F.; Pizzo, F.; Vaccaro, L. *J. Org. Chem.* **2001**, *66*, 4463.

[1153]Bonini, C.; Giuliano, C.; Righi, G.; Rossi, L. *Synth. Commun.* **1992**, *22*, 1863.

[1154]Shimizu, M.; Yoshida, A.; Fujisawa, T. *Synlett,* **1992**, 204.

[1155]Denmark, S.E.; Barsanti, P.A.; Wong, K.-T.; Stavenger, R. *J. Org. Chem.* **1998**, *63*, 2428; Tao, B.; Lo, M.M.-C.; Fu, G.C. *J. Am. Chem. Soc.* **2001**, *123*, 353; Reymond, S.; Legrand, O.; Brunel, J.M.; Buono, G. *Eur. J. Org. Chem.* **2001**, 2819.

[1156]Kwon, D.W.; Cho, M.S.; Kim, Y.H. *Synlett* **2003**, 959.

[1157]Kotsuki, H.; Shimanouchi, T. *Tetrahedron Lett.* **1996**, *37*, 1845.

[1158]Campbell, J.R.; Jones, J.K.N.; Wolfe, S. *Can. J. Chem.* **1966**, *44*, 2339.

[1159]Isaacs, N.S.; Kirkpatrick, D. *Tetrahedron Lett.* **1972**, 3869.

[1160]Srebnik, M.; Joshi, N.N.; Brown, H.C. *Isr. J. Chem.* **1989**, *29*, 229.

[1161]Nikam, K.; Nashi, T. *Tetrahedron,* **2002**, *58*, 10259. For an alternative reaction of iodine and a pyridine-containing macrocycle, see Sharghi, H.; Niknam, K.; Pooyan, M. *Tetrahedron* **2001**, *57*, 6057. For the reaction of iodine with a Mn–salen catalyst see Sharghi, H.; Naeimi, H. *Bull. Chem. Soc. Jpn.* **1999**, *72*, 1525.

[1162]Sartillo-Piscil, F.; Quinero, L.; Villegas, C.; Santacruz-Juárez, E.; de Parrodi, C.A. *Tetrahedron Lett.* **2002**, *43*, 15.

When done in an ionic liquid with Me_3SiCl, styrene epoxide gives 2-chloro-2-phenylethanol.[1163] The reaction of thionyl chloride and poly(vinylpyrrolidinone) converts epoxides to the corresponding 2-chloro-1-carbinol.[1164] Bromine with a phenylhydrazine catalyst, however, converts epoxides to the 1-bromo-2-carbinol.[1165] An alkenyl group also leads to a halohydrin with the halogen on the carbon bearing the C=C unit.[1166] Epxoy carboxylic acids are another example. When NaI reacts at pH 4, the major regioisomer is the 2-iodo-3-hydroxy compound, but when $InCl_3$ is added, the major product is the 3-iodo-2-hydroxy carboxylic acid.[1167]

Acyl chlorides react with ethylene oxide in the presence of NaI to give 2-iodoethyl esters.[1168]

Acyl chlorides react with epoxides in the presence of a $Eu(dpm)_3$ catalyst[1169] [dpm = 1,1-bis(diphenylphosphino)methane] or a YCp_2Cl catalyst[1170] to give chloro esters.

A related reaction with epi-sulfides leads to 2-chlorothio-esters.[1171] Aziridines have been opened with $MgBr_2$ to give 2-haloamides in a related reaction.[1172] N-Tosyl aziridines react with KF•2 H_2O to give the 2-fluorotosylamine product.[1173]

OS **I**, 117; **VI**, 424; **IX**, 220.

10-51 Cleavage of Carboxylic Esters With Lithium Iodide

Iodo-de-acyloxy-substitution

$$R'COOR + LiI \xrightarrow[\Delta]{pyridine} RI + R'COOLi$$

[1163]Xu, L.-W.; Li, L.; Xia, C.-G.; Zhao, P.-Q. *Tetrahedron Lett.* **2004**, *45*, 2435.

[1164]Tamami, B.; Ghazi, I.; Mahdavi, H. *Synth. Commun.* **2002**, *32*, 3725.

[1165]Sharghi, H.; Eskandari, M.M. *Synthesis* **2002**, 1519.

[1166]Ha, J.D.; Kim, S.Y.; Lee, S.J.; Kang, S.K.; Ahn, J.H.; Kim, S.S.; Choi, J.-K. *Tetrahedron Lett.* **2004**, *45*, 5969.

[1167]Fringuelli, F.; Pizzo, F.; Vaccaro, L. *J. Org. Chem.* **2001**, *66*, 4719. For a related SmI_2 ring opening of epoxy amides to give the 3-iodo-2-hydroxy compound, see Concellón, J.M.; Bardales, E.; Concellón, C.; García-Granda, S.; Díaz, M.R. *J. Org. Chem.* **2004**, *69*, 6923.

[1168]Belsner, K.; Hoffmann, H.M.R. *Synthesis* **1982**, 239. See also, Roloff, A. *Chimia*, **1985**, *39*, 392; Iqbal, J.; Khan, M.A.; Srivastava, R.R. *Tetrahedron Lett.* **1988**, *29*, 4985.

[1169]Taniguchi, Y.; Tanaka, S.; Kitamura, T.; Fujiwara, Y. *Tetrahedron Lett.* **1998**, *39*, 4559.

[1170]Qian, C.; Zhu, D. *Synth. Commun.* **1994**, *24*, 2203.

[1171]Kameyama, A.; Kiyota, M.; Nishikubo, T. *Tetrahedron Lett.* **1994**, *35*, 4571.

[1172]Righi, G.; D'Achille, R.; Bonini, C. *Tetrahedron Lett.* **1996**, *37*, 6893.

[1173]Fan, R.-H.; Zhou, Y.-G.; Zhang, W.-X.; Hou, X.-L.; Dai, L.-X. *J. Org. Chem.* **2004**, *69*, 335.

Carboxylic esters, where R is methyl or ethyl, can be cleaved by heating with lithium iodide in refluxing pyridine or a higher boiling amine.[1174] The reaction is useful where a molecule is sensitive to acid and base (so that **16-59** cannot be used) or where it is desired to cleave selectively only one ester group in a molecule containing two or more. For example, refluxing *O*-acetyloleanolic acid methyl ester

with LiI in *s*-collidine cleaved only the 17-carbomethoxy group, not the 3-acetyl group.[1175] Esters RCOOR′ and lactones can also be cleaved with a mixture of Me₃SiCl and NaI to give R′I and RCOOH.[1176] The reaction of acetyl chloride and allylic acetate leads to the allylic chloride.[1177]

10-52   Conversion of Diazo Ketones to α-halo Ketones

Hydro, halo-de-diazo-bisubstitution

$$RCOCHN_2 + HBr \longrightarrow RCOCH_2Br$$

When diazo ketones are treated with HBr or HCl, they give the respective α-halo ketones. HI does not give the reaction, since it reduces the product to a methyl ketone (**19-67**). α-Fluoro ketones can be prepared by addition of the diazo ketone to polyhydrogen fluoride–pyridine.[1178] This method is also successful for diazoalkanes.

Diazotization of α-amino acids in the above solvent at room temperature gives α-fluoro carboxylic acids.[1179] If this reaction is run in the presence of excess KCl or KBr, the corresponding α-chloro or α-bromo acid is obtained instead.[1180]

OS **III**, 119.

[1174]Taschner, E.; Liberek, B. *Rocz. Chem.* **1956**, *30*, 323 [*Chem. Abstr.*, **1957**, *51*, 1039]. For a review, see McMurry, J. *Org. React.* **1976**, *24*, 187–224.

[1175]Elsinger, F.; Schreiber, J.; Eschenmoser, A. *Helv. Chim. Acta* **1960**, *43*, 113.

[1176]Olah, G.A.; Narang, S.C.; Gupta, B.G.B.; Malhotra, R. *J. Org. Chem.* **1979**, *44*, 1247. See also, Kolb, M.; Barth, J. *Synth. Commun.* **1981**, *11*, 763.

[1177]Yadav, V.K.; Babu, K.G. *Tetrahedron* **2003**, *59*, 9111.

[1178]Olah, G.A.; Welch, J. *Synthesis* **1974**, 896; Olah, G.A.; Welch, J.; Vankar, Y.D.; Nojima, M.; Kerekes, I.; Olah, J.A. *J. Org. Chem.* **1979**, *44*, 3872.

[1179]Olah, G.A.; Prakash, G.K.S.; Chao, Y.L. *Helv. Chim. Acta* **1981**, *64*, 2528; Faustini, F.; De Munary, S.; Panzeri, A.; Villa, V.; Gandolfi, C.A. *Tetrahedron Lett.* **1981**, *22*, 4533; Barber, J.; Keck, R.; Rétey, J. *Tetrahedron Lett.* **1982**, *23*, 1549.

[1180]Olah, G.A.; Shih, J.; Prakash, G.K.S. *Helv. Chim. Acta* **1983**, *66*, 1028.

10-53   Conversion of Amines to Halides

Halo-de-amination

$$RNH_2 \longrightarrow RNTs_2 \xrightarrow[\text{DMF}]{I^-} RI$$

Primary alkyl amines RNH_2 can be converted[1181] to alkyl halides by (*1*) conversion to $RNTs_2$ (p. 498) and treatment of this with I^- or Br^- in DMF,[385] or to $N(Ts)-NH_2$ derivatives followed by treatment with *N*-bromosuccinimide under photolysis conditions;[1182] (*2*) diazotization with *tert*-butylnitrite and a metal halide such as $TiCl_4$ in DMF;[1183] or (*3*) the Katritzky pyrylium–pyridinium method (p. 498).[1184] Alkyl groups can be cleaved from secondary and tertiary aromatic amines by concentrated HBr in a reaction similar to **10-49**, for example,[1185]

$$ArNR_2 + HBr \longrightarrow RBr + ArNHR$$

Tertiary aliphatic amines are also cleaved by HI, but useful products are seldom obtained. Tertiary amines can be cleaved by reaction with phenyl chloroformate:[1186] $R_3N + ClCOOPh \rightarrow RCl + R_2NCOOPh$. α-Chloroethyl chloroformate behaves similarly.[1187] Alkyl halides may be formed when quaternary ammonium salts are heated: $R_4N^+ \, X^- \rightarrow R_3N + RX$.[1188]

OS **VIII**, 119. See also, OS **I**, 428.

10-54   Conversion of Tertiary Amines to Cyanamides: The von Braun Reaction

Bromo-de-dialkylamino-substitution

$$R_3NH + BrCN \longrightarrow R_2NCN + RBr$$

The *von Braun reaction* involves the cleavage of tertiary amines by cyanogen bromide to give an alkyl bromide and a disubstituted cyanamide, and can be applied to many tertiary amines.[1189] Usually, the R group that cleaves is the one that gives the most reactive halide (e.g., benzyl or allyl). For simple alkyl groups, the smallest

---

[1181]For another method, see Lorenzo, A.; Molina, P.; Vilaplana, M.J. *Synthesis* **1980**, 853.

[1182]Collazo, L.R.; Guziec, Jr., F.S.; Hu, W.-X.; Pankayatselvan, R. *Tetrahedron Lett.* **1994**, *35*, 7911.

[1183]Doyle, M.P.; Bosch, R.J.; Seites, P.G. *J. Org. Chem.* **1978**, *43*, 4120.

[1184]Katritzky, A.R.; Chermprapai, A.; Patel, R.C. *J. Chem. Soc. Perkin Trans. 1* **1980**, 2901.

[1185]Chambers, R.A.; Pearson, D.E. *J. Org. Chem.* **1963**, *28*, 3144.

[1186]Hobson, J.D.; McCluskey, J.G. *J. Chem. Soc. C* **1967**, 2015. For a review, see Cooley, J.H.; Evain, E.J. *Synthesis* **1989**, 1.

[1187]Olofson, R.A.; Martz, J.T.; Senet, J.; Piteau, M.; Malfroot, T. *J. Org. Chem.* **1984**, *49*, 2081; Olofson, R.A.; Abbott, D.E. *J. Org. Chem.* **1984**, *49*, 2795. See also, Campbell, A.L.; Pilipauskas, D.R.; Khanna, I.K.; Rhodes, R.A. *Tetrahedron Lett.* **1987**, *28*, 2331.

[1188]For examples, see Ko, E.C.F.; Leffek, K.T. *Can. J. Chem.* **1970**, *48*, 1865; *1971*, *49*, 129; Deady, L.W.; Korytsky, O.L. *Tetrahedron Lett.* **1979**, 451.

[1189]For a review, see Cooley, J.H.; Evain, E.J. *Synthesis* **1989**, 1.

are the most readily cleaved. One or two of the groups on the amine may be aryl, but they do not cleave. Cyclic amines have been frequently cleaved by this reaction. Secondary amines also give the reaction, but the results are usually poor.[1190]

The mechanism consists of two successive nucleophilic substitutions, with the tertiary amine as the first nucleophile and the liberated bromide ion as the second:

Step 1 $NC\!\!-\!\!Br \; + \; R_3N \quad\longrightarrow\quad NC\!-\!\overset{\oplus}{N}R_3 \; + \; Br^{\ominus}$

Step 2 $R\!-\!\overset{\oplus}{N}R_2CN \; + \; Br^{\ominus} \quad\longrightarrow\quad RBr \; + \; R_2NCN$

The intermediate *N*-cyanoammonium bromide has been trapped, and its structure confirmed by chemical, analytical, and spectral data.[1191] The BrCN in this reaction has been called a *counterattack reagent*; that is, a reagent that accomplishes, in one flask, two transformations designed to give the product.[1192]

OS **III**, 608.

CARBON NUCLEOPHILES

In any heterolytic reaction in which a new carbon–carbon bond is formed,[1193] one carbon atoms attacks as a nucleophile and the other as an electrophile. The classification of a given reaction as nucleophilic or electrophilic is a matter of convention and is usually based on analogy. Although not discussed in this chapter, **11-8–11-25** and **12-16–12-21** are nucleophilic substitutions with respect to one reactant, though, following convention, we classify them with respect to the other. Similarly, all the reactions in this section would be called electrophilic substitution (aromatic or aliphatic) if we were to consider the reagent as the substrate.

In **10-56–10-65** the nucleophile is a "carbanion" part of an organometallic compound, often a Grignard reagent. There is much that is still not known about the mechanisms of these reactions and many of them are not nucleophilic substitutions at all. In those reactions that are nucleophilic substitutions, the attacking carbon brings a pair of electrons with it to the new C—C bond, whether or not free carbanions are actually involved. The connection of two alkyl or aryl groups is called *coupling*. Reactions **10-56–10-65** include both symmetrical and unsymmetrical coupling reactions. The latter are also called *cross-coupling reactions*. Other coupling reactions are considered in later chapters.

[1190]For a detailed discussion of the scope of the reaction and of the ease of cleavage of different groups, see Hageman, H.A. *Org. React.* **1953**, 205.

[1191]Fodor, G.; Abidi, S. *Tetrahedron Lett.* **1971**, 1369; Fodor, G.; Abidi, S.; Carpenter, T.C. *J. Org. Chem.* **1974**, *39*, 1507. See also, Paukstelis, J.V.; Kim, M. *J. Org. Chem.* **1974**, *39*, 1494.

[1192]For a review of counterattack reagents, see Hwu, J.R.; Gilbert, B.A. *Tetrahedron* **1989**, *45*, 1233.

[1193]For a monograph that discusses most of the reactions in this section, see Stowell, J.C. *Carbanions in Organic Synthesis*, Wiley, NY, **1979**. For a review, see Noyori, R., in Alper, H. *Transition Metal Organometallics in Organic Synthesis*, Vol. 1, Academic Press, NY, **1976**, pp. 83–187.

10-55 Coupling With Silanes

De-silylalkyl-coupling

$$R-X + R_3^1Si-CH_2CH{=}CH_2 \longrightarrow R-CH_2CH{=}CH_2$$

Organosilanes $RSiMe_3$ or $RSiMe_2F$ (where R can be vinylic, allylic, or alkynyl) couple with vinylic, allylic, and aryl bromides and iodides R′X, in the presence of certain catalysts, to give RR′ in good yields.[1194] Allylsilanes react with allylic acetates in the presence of iodine.[1195] The transition-metal catalyzed coupling of silanes, particularly allyl silanes, is a mild method for incorporating alkyl fragments into a molecule.[1196] $PhSiMe_2Cl$ couples to give biphenyl in the presence of CuI and Bu_4NF,[1197] and vinyl silanes react with allylic carbonates and a palladium catalyst to give dienes.[1198] Allylsilanes have been coupled to substrates containing a benzotriazole unit, in the presence of BF_3•etherate.[1199] One variation used a silylmethyl-tin derivative in a palladium-catalyzed coupling with aryl iodides.[1200] Homoallyl silanes coupled to Ph_3BiF_2 in the presence of BF_3•OEt_2 to give the phenyl coupling product.[1201]

α-Silyloxy methoxy derivatives, $RCH(OMe)OSiR_3^1$, react with allyltrimethylsilane ($Me_3SiCH_2CH{=}CH_2$) in the presence of TiX_4 derivatives to give displacement of the OMe group and $RCH(OSiR_3^1)CH_2CH{=}CH_2$).[1202] A tertiary silyloxy group was displaced by allyl in the presence of $ZnCl_2$.[1203] Electrolysis with allyltrimethylsilane and $RCH(OMe)SPh$ leads to $RCH(OMe)CH_2CH{=}CH_2$.[1204] Similar reaction with a dithioacetal leads to the allylic silane.[1205] Allylic acetates react with $Me_3SiSiMe_3$ and LiCl with a palladium catalyst to give the allyl silane.[1206] $RSiF_3$ reagents can also be used in coupling reaction with aryl halides.[1207]

[1194]Hatanaka, Y.; Hiyama, T. *J. Org. Chem.* **1988**, *53*, 918; **1989**, *54*, 268; Cho, Y.S.; Kang, S.-H.; Han, J.-S.; Yoo, B.R.; Jung, I.N. *J. Am. Chem. Soc.* **2001**, *123*, 5584.

[1195]Yadav, J.S.; Reddy, B.V.S.; Rao, K.V.; Raj, K.S.; Rao, P.P.; Prasad, A.R.; Gunasekar, D. *Tetrahedron Lett.* **2004**, *45*, 6505.

[1196]For a ruthenium-catalyzed reaction, see Kakiuchi, F.; Tsuchiya, K.; Matsumoto, M.; Mizushima, E.; Chatani, N. *J. Am. Chem. Soc.* **2004**, *126*, 12792. For a Cp_2TiCl_2-catalyzed reaction with allyl phenyl ether and chlorotrialkylsilanes, see Nii, S.; Terao, J.; Kambe, N. *Tetrahedron Lett.* **2004**, *45*, 1699.

[1197]Kang, S.-K.; Kim, T.H.; Pyun, S.-J. *J. Chem. Soc. Perkin Trans. 1* **1997**, 797.

[1198]Matsuhashi, H.; Hatanaka, Y.; Kuroboshi, M.; Hiyama, T. *Tetrahedron Lett.* **1995**, *36*, 1539; Matsuhashi, H.; Asai, S.; Hirabayashi, K.; Hatanaka, Y.; Mori, A.; Hiyama, T. *Bull. Chem. Soc. Jpn.* **1997**, *70*, 1943.

[1199]Katritzky, A.R.; Mehta, S.; He, H.-Y.; Cui, X. *J. Org. Chem.* **2000**, *65*, 4364.

[1200]Itami, K.; Kamei, T.; Yoshida, J.-i. *J. Am. Chem. Soc.* **2001**, *123*, 8773.

[1201]Matano, Y.; Yoshimune, M.; Suzuki, H. *Tetrahedron Lett.* **1995**, *36*, 7475.

[1202]Maeda, K.; Shinokubo, H.; Oshima, K. *J. Org. Chem.* **1997**, *62*, 6429.

[1203]Yokozawa, T.; Furuhashi, K.; Natsume, H. *Tetrahedron Lett.* **1995**, *36*, 5243.

[1204]Yoshida, J.; Sugawara, M.; Kise, N. *Tetrahedron Lett.* **1996**, *37*, 3157.

[1205]Fujiwara, T.; Takamori, M.; Takeda, T. *Chem. Commun.* **1998**, 51.

[1206]Tsuji, Y.; Funato, M.; Ozawa, M.; Ogiyama, H.; Kajita, S.; Kawamura, T. *J. Org. Chem.* **1996**, *61*, 5779.

[1207]Hatanaka, Y.; Goda, K.; Hiyama, T. *Tetrahedron Lett.* **1994**, *35*, 6511; Matsuhashi, H.; Kuroboshi, M.; Hatanaka, Y.; Hiyama, T. *Tetrahedron Lett.* **1994**, *35*, 6507.

Allyl silanes react with epoxides, in the presence of $BF_3 \bullet OEt_2$ to give 2-allyl alcohols.[1208] The reaction of α-bromo lactones and $CH_2{=}CHCH_2Si(SiMe_3)_3$ and AIBN leads to the α-allyl lactone.[1209] On the other hand, silyl epoxides have been prepared from epoxides via reaction with *sec*-butyllithium and chlorotrimethylsilane.[1210] α-Silyl-*N*-Boc-amines were prepared in a similar manner from the *N*-Boc-amine.[1211] Arylsilanes were prepared by reaction of an aryllithium intermediate with $TfOSi(OEt)_3$.[1212] In the presence of $BF_3 \bullet$etherate, allyl silane and α-methoxy *N*-Cbz amines were coupled.[1213] Benzyl silanes coupled with allyl silanes to give $ArCH_2{-}R$ derivatives in the presence of $VO(OEt)Cl_2$[1214] and allyltin compounds couple with allyl silanes in the presence of $SnCl_4$.[1215] Allyl silanes couple to the α-carbon of amines under photolysis conditions.[1216]

The reaction of a vinyl iodide with $(EtO)_3SiH$ with a palladium catalyst generated a good yield of the corresponding vinylsilane.[1217]

OSCV 10, 531.

10-56 Coupling of Alkyl Halides: The Wurtz Reaction

De-halogen-coupling

$$2\,RX + Na \longrightarrow RR$$

The coupling of alkyl halides by treatment with sodium to give a symmetrical product is called the *Wurtz reaction*. Side reactions (elimination and rearrangement) are so common that the reaction is seldom used. Mixed Wurtz reactions of two alkyl halides are even less feasible because of the number of products obtained. A somewhat more useful reaction (though still not very good) takes place when a mixture of an alkyl and an aryl halide is treated with sodium to give an alkylated aromatic compound (the *Wurtz–Fittig reaction*).[1218]

[1208]Burgess, L.E.; Gross, E.K.M.; Jurka, J. *Tetrahedron Lett.* **1996**, *37*, 3255; Prestat, G.; Baylon, C.; Heck, M.-P.; Mioskowski, C. *Tetrahedron Lett.* **2000**, *41*, 3829.
[1209]Chatgilialoglu, C.; Ferreri, C.; Ballestri, M.; Curran, D.P. *Tetrahedron Lett.* **1996**, *37*, 6387; Chatgilialoglu, C.; Alberti, A.; Ballestri, M.; Macciantelli, D.; Curran, D.P. *Tetrahedron Lett.* **1996**, *37*, 6391.
[1210]Hodgson, D.M.; Norsikian, S.L.M. *Org. Lett.* **2001**, *3*, 461.
[1211]Harrison, J.R.; O'Brien, P.; Porter, D.W.; Smith, N.W. *Chem. Commun.* **2001**, 1202.
[1212]Seganish, W.M.; DeShong, P. *J. Org. Chem.* **2004**, *69*, 6790.
[1213]Matos, M.R.P.N.; Afonso, C.A.M.; Batey, R.A. *Tetrahedron Lett.* **2001**, *42*, 7007.
[1214]Hirao, T.; Fujii, T.; Ohshiro, Y. *Tetrahedron Lett.* **1994**, *35*, 8005.
[1215]Takeda, T.; Takagi, Y.; Takano, H.; Fujiwara, T. *Tetrahedron Lett.* **1992**, *33*, 5381.
[1216]Pandey, G.; Rani, K.S.; Lakshimaiah, G. *Tetrahedron Lett.* **1992**, *33*, 5107. See Gelas-Mialhe, Y.; Gramain, J.-C.; Louvet, A.; Remuson, R. *Tetrahedron Lett.* **1992**, *33*, 73 for an internal coupling reaction of an allyl silane and an α-hydoxy lactam.
[1217]Murata, M.; Watanabe, S.; Masuda, Y. *Tetrahedron Lett.* **1999**, *40*, 9255.
[1218]For an example, see Kwa, T.L.; Boelhouwer, C. *Tetrahedron* **1970**, *25*, 5771.

However, the coupling of two aryl halides with sodium is impractical (but see **13-11**). Other metals have also been used to effect Wurtz reactions,[1219] notably silver, zinc,[1220] iron,[1221] activated copper,[1222] In,[1223] La,[1224] and manganese compounds.[1225] Lithium, under the influence of ultrasound, has been used to couple alkyl, aryl, and benzylic halides.[1226] Metallic nickel, prepared by the reduction of nickel halides with Li, dimerizes benzylic halides to give $ArCH_2CH_2Ar$.[1227] The coupling of alkyl halides has also been achieved electrochemically[1228] and photochemically.[1229] In a related reaction, Grignard reagents (**12-38**) have been coupled in the presence of trifluorosulfonic anhydride.[1230]

Tosylates and other sulfonates and sulfates couple with Grignard reagents,[1231] most often those prepared from aryl or benzylic halides.[1232] Alkyl sulfates and sulfonates generally make better substrates in reactions with Grignard reagents than the corresponding halides (**10-57**). The method is useful for primary and secondary R.

One type of Wurtz reaction that is quite useful is the closing of small rings, especially three-membered rings.[1233] For example, 1,3-dibromopropane can be converted to cyclopropane by Zn and NaI.[1234] Two highly strained molecules that

---

[1219]For a list of reagents, including metals and other reagents, with references, see Larock, R.C. *Comprehensive Organic Transformations*, 2nd ed., Wiley-VCH, NY, *1999*, pp. 83–84.

[1220]See, for example, Nosek, J. *Collect. Czech. Chem. Commun. 1964*, 29, 597.

[1221]Nozaki, H.; Noyori, R. *Tetrahedron 1966*, 22, 2163; Onsager, O. *Acta Chem. Scand. Ser. B, 1978*, 32, 15.

[1222]Ginah, F.O.; Donovan, T.A.; Suchan, S.D.; Pfennig, D.R.; Ebert, G.W. *J. Org. Chem. 1990*, 55, 584.

[1223]Ranu, B.C.; Dutta, P.; Sarkar, A. *Tetrahedron Lett. 1998*, 39, 9557.

[1224]Nishino, T.; Watanabe, T.; Okada, M.; Nishiyama, Y.; Sonoda, N. *J. Org. Chem. 2002*, 67, 966.

[1225]Mn/CuCl$_2$: Ma, J.; Chan, T.-H. *Tetrahedron Lett. 1998*, 39, 2499. Mn$_2$(CO)$_{10}$/hv: Gilbert, B.C.; Lindsay, C.I.; McGrail, P.T.; Parsons, A.F.; Whittaker, D.T.E. *Synth. Commun. 1999*, 29, 2711.

[1226]Han, B.H.; Boudjouk, P. *Tetrahedron Lett. 1981*, 22, 2757.

[1227]Inaba, S.; Matsumoto, H.; Rieke, R.D. *J. Org. Chem. 1984*, 49, 2093. For some other reagents that accomplish this, see Sayles, D.C.; Kharasch, M.S. *J. Org. Chem. 1961*, 26, 4210; Cooper, T.A. *J. Am. Chem. Soc. 1973*, 95, 4158; Ho, T.; Olah, G.A. *Synthesis 1977*, 170; Ballatore, A.; Crozet, M.P.; Surzur, J. *Tetrahedron Lett. 1979*, 3073; Yamada, Y.; Momose, D. *Chem. Lett. 1981*, 1277; Iyoda, M.; Sakaitani, M.; Otsuka, H.; Oda, M. *Chem. Lett. 1985*, 127.

[1228]Folest, J.C.; Nédélec, J.Y.; Perichon, J. *J. Chem. Res. (S) 1989*, 394.

[1229]Ouchi, A.; Yabe, A. *Tetrahedron Lett. 1992*, 33, 5359.

[1230]Nishiyama, T.; Seshita, T.; Shodai, H.; Aoki, K.; Kameyama, H.; Komura, K. *Chem. Lett. 1996*, 549.

[1231]For a review, see Kharasch, M.S.; Reinmuth, O. *Grignard Reactions of Nonmetallic Substances*, Prentice-Hall, Englewood Cliffs, NJ, *1954*, pp. 1277–1286.

[1232]For an example involving an allylic rearrangement (conversion of a silylalkyne to a silylallene), see Danheiser, R.L.; Tsai, Y.; Fink, D.M. *Org. Synth.* 66, 1.

[1233]For a review, see Freidlina, R.Kh.; Kamyshova, A.A.; Chukovskaya, E.Ts. *Russ. Chem. Rev. 1982*, 51, 368. For reviews of methods of synthesizing cyclopropane rings, see, in Rappoport *The Chemistry of the Cyclopropyl Group*, pt. 1; Wiley, NY, *1987*, the reviews by Tsuji, T.; Nishida, S. pp. 307–373, and Verhé, R.; De Kimpe, N. pp. 445–564.

[1234]For a discussion of the mechanism, see Applequist, D.E.; Pfohl, W.F. *J. Org. Chem. 1978*, 43, 867.

Tetracyclo[3.3.1.13,7.01,3]decane

have been prepared this way are bicyclobutane[1235] and tetracyclo[3.3.1.1^{3,7}.0^{1,3}]de-cane.[1236] Three- and four-membered rings can also be closed in this manner with certain other reagents,[1237] including benzoyl peroxide,[1238] *t*-BuLi,[1239] and lithium amalgam,[1240] as well as electrochemically.[1241]

134

Vinylic halides can be coupled to give 1,3-butadienes (**134**) by treatment with acti-vated copper powder in a reaction analogous to the Ullmann reaction (**13-11**).[1242] This reaction is stereospecific, with retention of configuration at both carbons. Vinylic halides can also be coupled[1243] with Zn–NiCl$_2$,[1244] and with *n*-BuLi in ether in the presence of MnCl$_2$.[1245] The coupling reaction with vinyltin reagents and vinyl halides occurs with a palladium catalyst.[1246]

[1235]Wiberg, K.B.; Lampman, G.M. *Tetrahedron Lett.* **1963**, 2173; Lampman, G.M.; Aumiller, J.C. *Org. Synth. VI*, 133.

[1236]Pincock, R.E.; Schmidt, J.; Scott, W.B.; Torupka, E.J. *Can. J. Chem.* **1972**, *50*, 3958.

[1237]For a list of reagents, with references, see Larock, R.C. *Comprehensive Organic Transformations*, 2nd ed., Wiley-VCH, NY, **1999**, pp. 175–184.

[1238]Kaplan, L. *J. Am. Chem. Soc.* **1967**, *89*, 1753; *J. Org. Chem.* **1967**, *32*, 4059.

[1239]Bailey, W.F.; Gagnier, R.P. *Tetrahedron Lett.* **1982**, *23*, 5123.

[1240]Connor, D.S.; Wilson, E.R. *Tetrahedron Lett.* **1967**, 4925.

[1241]Rifi, M.R. *J. Am. Chem. Soc.* **1967**, *89*, 4442; *Org. Synth. VI*, 153.

[1242]Cohen, T.; Poeth, T. *J. Am. Chem. Soc.* **1972**, *94*, 4363.

[1243]See Wellmann, J.; Steckhan, E. *Synthesis 1978*, 901; Miyahara, Y.; Shiraishi, T.; Inazu, T.; Yoshino, T. *Bull. Chem. Soc. Jpn.* **1979**, *52*, 953; Grigg, R.; Stevenson, P.; Worakun, T. *J. Chem. Soc., Chem. Commun.* **1985**, 971; Vanderesse, R.; Fort, Y.; Becker, S.; Caubere, P. *Tetrahedron Lett.* **1986**, *27*, 3517.

[1244]Takagi, K.; Mimura, H.; Inokawa, S. *Bull. Chem. Soc. Jpn.* **1984**, *57*, 3517.

[1245]Cahiez, G.; Bernard, D.; Normant, J.F. *J. Organomet. Chem.* **1976**, *113*, 99.

[1246]Paley, R.S.; de Dios, A.; de la Pradilla, R.F. *Tetrahedron Lett.* **1993**, *34*, 2429.

Treatment of conjugated ketones with SmI_2 in HMPA gave the coupled diketone via Wurtz-type coupling.[1247]

It seems likely that the mechanism of the Wurtz reaction consists of two basic steps. The first is halogen-metal exchange to give an organometallic compound $(RX + M \rightarrow RM)$, which in many cases can be isolated (**12-38**). Following this, the organometallic compound reacts with a second molecule of alkyl halide $(RX + RM \rightarrow RR)$. This reaction and its mechanism are considered in the next section (**10-57**).

OS **III**, 157; **V**, 328, 1058; **VI**, 133, 153.

A variation of the Wurtz coupling uses other metals to mediate or facilitate the coupling. In certain cases, such variations can be synthetically useful.

Because of the presence of the 1,5-diene moiety in many naturally occurring compounds, methods that couple[1248] allylic groups[1249] are quite important. In one of these methods, allylic halides, tosylates, and acetates can be symmetrically coupled by treatment with nickel carbonyl[1250] at room temperature in a solvent, such as THF or DMF to give 1,5-dienes.[1251] The order of halide reactivity is I > Br > Cl. With unsymmetrical allylic substrates, coupling nearly always takes place at the less-substituted end. The reaction can be performed intramolecularly; large (11–20 membered) rings can be made in good yields (60–80%) by the use of high dilution.[1252] The mechanism of coupling likely involves reaction of the allylic compound with $Ni(CO)_4$ to give one or more π-allyl complexes, one of which may be the η^3-complex **135**. Loss of CO to give a π-allylnickel bromide (**136**) and ligand transfer leads to coupling and the final product. In some cases, the η^3-complexes **136** can be isolated from the solution and

[1247]Cabrera, A.; Rosas, N.; Sharma, P.; LeLagadec, R.; Velasco, L.; Salmón, M. *Synth. Commun.* **1998**, *28*, 1103.

[1248]For a review of some allylic coupling reactions, see Magid, R.M. *Tetrahedron* **1980**, *36*, 1901, see pp. 1910–1924.

[1249]In this section are discussed methods in which one molecule is a halide. For other allylic coupling reactions, see **10-57**, **10-63**, and **10-60**.

[1250]For a review of the use of organonickel compounds in organic synthesis, see Tamao, K.; Kumada, M., in Hartley, F.R. *The Chemistry of the Metal-Carbon Bond*, Vol. 4, Wiley, NY, **1987**, pp. 819–887.

[1251]For reviews, see Collman, J.P. ; Hegedus, L.; Norton, J.R.; Finke, R. *Principles and Applications of Organotransition Metal Chemsitry*, 2nd ed., University Science Books, Mill Valley, CA, **1987**, pp. 739–748; Billington, D.C. *Chem. Soc. Rev.* **1985**, *14*, 93; Kochi, J.K. *Organometallic Mechanisms and Catalysis*, Academic Press, NY, **1978**, pp. 398–408; Semmelhack, M.F. *Org. React.* **1972**, *19*, 115, see pp. 162–170; Baker, R. *Chem. Rev.* **1973**, *73*, 487, see pp. 512–517; Heimbach, P.; Jolly, P.W.; Wilke, G. *Adv. Organomet. Chem.* **1970**, *8*, 29, see pp. 30–39.

[1252]Corey, E.J.; Wat, E.K.W. *J. Am. Chem. Soc.* **1967**, *89*, 2757. See also, Corey, E.J.; Helquist, P. *Tetrahedron Lett.* **1975**, 4091; Reijnders, P.J.M.; Blankert, J.F.; Buck, H.M. *Recl. Trav. Chim. Pays-Bas* **1978**, *97*, 30.

crystallized as stable solids.

Unsymmetrical coupling can be achieved by treating an alkyl halide directly with **136**, in a polar aprotic solvent, [1253] where coupling occurs at the less substituted end. There is evidence that free radicals are involved in such couplings. [1254] Hydroxy or carbonyl groups in the alkyl halide do not interfere. When **136** reacts with an allylic halide, a mixture of three products is obtained because of halogen–metal interchange. For example, allyl bromide treated with **136** prepared from methallyl bromide gave an approximately statistical mixture of 1,5-hexadiene, 2-methyl-1,5-hexadiene, and 2,5-dimethyl-1,5-hexadiene. [1255] Allylic tosylates can be symmetrically coupled with $Ni(CO)_4$.

Symmetrical coupling of allylic halides can prepared by heating with magnesium in ether, [1256] with a cuprous iodide–dialkylamide complex, [1257] or electrochemically. [1258] The coupling of two different allylic groups has been achieved by treatment of an allylic bromide with an allylic Grignard reagent in THF containing HMPA, [1259] or with an allylic tin reagent. [1260] This type of coupling can be achieved with almost no allylic rearrangement in the substrate (and almost complete allylic rearrangement in the reagent) by treatment of allylic halides with lithium allylic boron ate complexes $(RCH=CHCH_2B^{\ominus} R_3^2 \ Li^+)$. [1261] The reaction between primary and secondary halides and allyltributylstannane provides another method for unsymmetrical coupling

[1253]Corey, E.J.; Semmelhack, M.F. *J. Am. Chem. Soc.* **1967**, *89*, 2755. For a review, see Semmelhack, M.F. *Org. React.* **1972**, *19*, 115, see pp. 147–162. For a discussion of the preparation and handling of π-allylnickel halides, see Semmelhack, M.F. *Org. React.* **1972**, *199*, 115, see pp. 144–146.
[1254]Hegedus, L.S.; Thompson, D.H.P. *J. Am. Chem. Soc.* **1985**, *107*, 5663.
[1255]Corey, E.J.; Semmelhack, M.F.; Hegedus, L.S. *J. Am. Chem. Soc.* **1968**, *90*, 2416.
[1256]Turk, A.; Chanan, H. *Org. Synth. III*, 121.
[1257]Kitagawa, Y.; Oshima, K.; Yamamoto, H.; Nozaki, H. *Tetrahedron Lett.* **1975**, 1859.
[1258]Tokuda, M.; Endate, K.; Suginome, H. *Chem. Lett.* **1988**, 945.
[1259]Stork, G.; Grieco, P.A.; Gregson, M. *Tetrahedron Lett.* **1969**, 1393; Grieco, P.A. *J. Am. Chem. Soc.* **1969**, *91*, 5660.
[1260]Godschalx, J.; Stille, J.K. *Tetrahedron Lett.* **1980**, *21*, 2599; **1983**, *24*, 1905; Hosomi, A.; Imai, T.; Endo, M.; Sakurai, H. *J. Organomet. Chem.* **1985**, *285*, 95. See also, Yanagisawa, A.; Norikate, Y.; Yamamoto, H. *Chem. Lett.* **1988**, 1899.
[1261]Yamamoto, Y.; Yatagai, H.; Maruyama, K. *J. Am. Chem. Soc.* **1981**, *103*, 1969.

$$RX + CH_2=CHCH_2SnBu_3 \rightarrow RCH_2CH=CH_2.^{1262}$$

137

In another method for the coupling of two different allylic groups,[1263] a carbanion derived from a β,γ-unsaturated thioether couples with an allylic halide to give **137**.[1264] The product **137** contains an SPh group that must be removed (with Li in ethylamine) to give the 1,5-diene. Unlike most of the methods previously discussed, this method has the advantage that the coupling preserves the original positions and configurations of the two double bonds; no allylic rearrangements take place.

OS **III**, 121; **IV**, 748; **VI**, 722.

10-57 The Reaction of Alkyl Halides and Sulfonate Esters With Group I and II Organometallic Reagents[1265]

Alkyl-de-halogenation

$$R-Na(K)(Li) + R'X \longrightarrow R-R'$$

A variety of organometallic compounds[1266] have been used to couple with alkyl halides.[1267] Organosodium and organopotassium compounds are more reactive than Grignard reagents and couple even with less reactive halides. Organolithium reagents react with ether solvents, and their half-life in such solvents is known.[1268] The difficulty is in preparing and keeping them long enough for the alkyl halide to be added. Alkenes can be prepared by the coupling of vinylic lithium compounds with primary halides[1269] or of vinylic halides with alkyllithium reagents in the presence of a Pd or

[1262]See Keck, G.E.; Yates, J.B. *J. Am. Chem. Soc.* **1982**, *104*, 5829; Migita, T.; Nagai, K.; Kosugi, M. *Bull. Chem. Soc. Jpn* **1983**, *56*, 2480.

[1263]For other procedures, see Axelrod, E.H.; Milne, G.M.; van Tamelen, E.E. *J. Am. Chem. Soc.* **1970**, *92*, 2139; Morizawa, Y.; Kanemoto, S.; Oshima, K.; Nozaki, H. *Tetrahedron Lett.* **1982**, *23*, 2953.

[1264]Biellmann, J.F.; Ducep, J.B. *Tetrahedron Lett.* **1969**, 3707.

[1265]For a review of the reactions in this section, see Naso, F.; Marchese, G., in Patai, S.; Rappoport, Z. *The Chemstry of Functional Groups, Supplement D*, pt. 2, Wiley, NY, **1983**, pp. 1353–1449.

[1266]For lists of reagents and substrates, with references, see Larock, R.C. *Comprehensive Organic Transformations*, 2nd ed., Wiley-VCH, NY, **1999**, pp. 101–127.

[1267]For a review of the coupling of organic halides with organotin, mercury, and copper compounds catalyzed by palladium complexes, see Beletskaya, I.P. *J. Organomet. Chem.* **1983**, *250*, 551. For a review of palladium-assisted coupling, see Larock, R.C. *Organomercury Compounds in Organic Synthesis*; Springer, NY, **1985**, pp. 249–262.

[1268]Stanetty, P.; Mihovilovic, M.D. *J. Org. Chem.* **1997**, *62*, 1514.

[1269]Millon, J.; Lorne, R.; Linstrumelle, G. *Synthesis* **1975**, 434; Duhamel, L.; Poirier, J. *J. Am. Chem. Soc.* **1977**, *99*, 8356.

Ru catalyst.[1270] Propargyl lithium reagents formed in the presence of mercuric salts couple with halides.[1271] Coupling of organolithium compounds with alkyl halides[1272] or aryl halides[1273] is possible.[1274] Unactivated aryl halides couple with alkyllithium reagents in THF.[1275] The reaction of n-butyllithium–TMEDA with a homoallylic alcohol [$CH_2=C(Me)CH_2CH_2OH$] leads to the allyllithium reagent, and subsequent reaction with an alkyl halide gives the substituted homoallylic alcohol [$CH_2=C(CH_2R)CH_2CH_2OH$].[1276] α-Lithioepoxides can also be formed, and reaction with an alkyl halide gives the substituted epoxide.[1277] Arylsilanes, such as 2-trimethyl-silylpyridine, undergo a deprotonation reaction of a silyl methyl group when treated with $tert$-butyllithium to give the corresponding $ArMe_2SiCH_2Li$ reagent.[1278] Subsequent reaction with an alkyl halide leads to the substituted silane. Organolithium reagents formed by Li—H exchange in the presence of $(-)$-sparteine couple with alkyl halides with high asymmetric induction.[1279] The dianion of $PhC(=Se)NHCH_2Ph$ was generated with n-butyllithium and reaction with bromocyclohexane gave the C-substituted derivative.[1280] Exchange of organotin compounds with organolithium reagents generates a new organolithium, and in one case intramolecular coupling in the presence of $(-)$-sparteine led to chiral pyrrolidine derivatives.[1281] It is noted that 1-lithioalkynes were coupled to alkyl halides in the presence of a palladium catalyst.[1282]

Aryllithium reagents are formed by metal–halogen exchange with aryl halides or H-metal exchange with various aromatic compounds, and they react with alkyl halides. The reaction of **138** with n-butyllithium, for example, generated the

[1270]Murahashi, S.; Yamamura, M.; Yanagisawa, K.; Mita, N.; Kondo, K. *J. Org. Chem.* **1979**, *44*, 2408.
[1271]Ma, S.; Wang, L. *J. Org. Chem.* **1998**, *63*, 3497.
[1272]Snieckus, V.; Rogers-Evans, M.; Beak, P.; Lee, W.K.; Yum, E.K.; Freskos, J. *Tetrahedron Lett.* **1994**, *35*, 4067.
[1273]Dieter, R.K.; Li, S.J. *J. Org. Chem.* **1997**, *62*, 7726; Dieter, R.K.; Dieter, J.W.; Alexander, C.W.; Bhinderwala, N.S. *J. Org. Chem.* **1996**, *61*, 2930. Also see, Beak, P.; Du, H. *J. Am. Chem. Soc.* **1993**, *115*, 2516; Beak, P.; Wu, S.; Yum, E.K.; Jun, Y.M. *J. Org. Chem.* **1994**, *59*, 276.
[1274]For example, see Brimble, M.A.; Gorsuch, S. *Aust. J. Chem.* **1999**, *52*, 965.
[1275]Merrill, R.E.; Negishi, E. *J. Org. Chem.*, **1974**, *39*, 3452. For another method, see Hallberg, A.; Westerlund, C. *Chem. Lett.*, **1982**, 1993.
[1276]Yong, K.H.; Lotoski, J.A.; Chong, J.M. *J. Org. Chem.* **2001**, *66*, 8248.
[1277]Marié, J.-C.; Curillon, C.; Malacria, M. *Synlett* **2002**, 553.
[1278]Itami, K.; Kamei, T.; Mitsudo, K.; Nokami, T.; Yoshida, J.-i. *J. Org. Chem.* **2001**, *66*, 3970.
[1279]Basu, A.; Beak, P. *J. Am. Chem. Soc.* **1996**, *118*, 1575; Wu, S.; Lee, S.; Beak, P. *J. Am. Chem. Soc.* **1996**, *118*, 715; Dieter, R.K.; Sharma, R.R. *Tetrahedron Lett.* **1997**, *38*, 5937.
[1280]Murai, T.; Aso, H.; Kato, S. *Org. Lett.* **2002**, *4*, 1407.
[1281]Serino, C.; Stehle, N.; Park, Y.S.; Florio, S.; Beak, P. *J. Org. Chem.* **1999**, *64*, 1160.
[1282]Yang, L.-M.; Huang, L.-F.; Luh, T.-Y. *Org. Lett.* **2004**, *6*, 1461.

aryllithium (**139**), which reacted with iodomethane to give **140**.[1283] When an aromatic ring has an attached heteroatom or an heteroatom-containing substituent, reaction with a strong base, such as an organolithium reagent, usually leads to an ortho lithiated species.[1284] Subsequent reaction with an electrophilic species gives the ortho substituted product. This phenomenon is known as *directed ortho metalation* (see **13-17**). This selectivity was discovered independently by Gilman and by Wittig in 1939–1940, when anisole was found to give ortho deprotonation in the presence of butyllithium.[1285] Alkylation ortho to a carbonyl is possible, and treatment of the acyl hydrazide $PhC(=O)NHNMe_2$ with *sec*-butyllithium and then iodoethane gave the ortho ethyl derivative.[1286] It is noted that aminonaphthalene derivatives were reacted with *tert*-butyllithium and aryllithium formation occurred on the ring distal to the amino group, and subsequent reaction with iodomethane gave methylation on that ring.[1287]

$$RX \; + \; \overset{3}{LiCH_3} - C \equiv \overset{1}{C} - SiMe_3 \; \longrightarrow \; RCH_2 - C \equiv C - SiMe_3 \; \overset{1.\,Ag^+}{\underset{2.\,CN^-}{\longrightarrow}} \; R - CH_2 - C \equiv C - H$$
$$\mathbf{141}$$

In a method for propargylating an alkyl halide without allylic rearrangement, the halide is treated with lithio-1-trimethylsilylpropyne (**141**), which is a lithium compound protected by an $SiMe_3$ group.[1288] Attack by the ambident nucleophile at its 1 position (which gives an allene) takes place only to a small extent, because of steric blockage by the large $SiMe_3$ group. The $SiMe_3$ group is easily removed by treatment with Ag^+ followed by CN^-. **141** is prepared by treating propynyllithium with Me_3SiCl to give $MeC=CSiMe_3$ from which a proton is removed with BuLi. R may be primary or allylic.[1289] On the other hand, propargylic halides can be alkylated with essentially complete allylic rearrangement, to give allenes, by treatment with Grignard reagents and metallic salts,[1290] or with dialkylcuprates R_2Cu.[1291]

Grignard reagents can be made to couple with alkyl halides in good yields by the use of certain catalysts,[1292] and stereocontrol is possible in these reactions.[1293] Among these are Cu(I) salts (see **10-58**), which permit the coupling of Grignard reagents with

[1283]MacNeil, S.L.; Familoni, O.B.; Snieckus, V. *J. Org. Chem.* **2001**, *66*, 3662.

[1284]For reviews, see Snieckus, V. *Chem. Rev.* **1990**, *90*, 879; Gschwend, H.W.; Rodriguez, H.R. *Org. React.* **1979**, *26*, 1. See also, Green, L.; Chauder, B.; Snieckus, V. *J. Heterocyclic Chem.* **1999**, *36*, 1453; Puterbaugh, W.H.; Hauser, C.R. *J. Org. Chem.* **1964**, *29*, 853;

[1285]Gilman, H.; Bebb, R.L. *J. Am. Chem. Soc.* **1939**, *61*, 109; Wittig, G.; Fuhrman, G. *Chem. Ber.* **1940**, *73*, 1197.

[1286]McCombie, S.W.; Lin, S.-I.; Vice, S.F. *Tetrahedron Lett.* **1999**, *40*, 8767.

[1287]Kraus, G.A.; Kim, J. *J. Org. Chem.* **2002**, *67*, 2358.

[1288]Corey, E.J.; Kirst, H.A.; Katzenellenbogen, J.A. *J. Am. Chem. Soc.* **1970**, *92*, 6314.

[1289]For an alternative procedure, see Ireland, R.E.; Dawson, M.I.; Lipinski, C.A. *Tetrahedron Lett.* **1970**, 2247.

[1290]Pasto, D.J.; Chou, S.; Waterhouse, A.; Shults, R.H.; Hennion, G.F. *J. Org. Chem.* **1978**, *43*, 1385; Jeffery-Luong, T.; Linstrumelle, G. *Tetrahedron Lett.* **1980**, *21*, 5019.

[1291]Pasto, D.J.; Chou, S.; Fritzen, E.; Shults, R.H.; Waterhouse, A.; Hennion, G.F. *J. Org. Chem.* **1978**, *43*, 1389. See also, Tanigawa, Y.; Murahashi, S. *J. Org. Chem.* **1980**, *45*, 4536.

[1292]For reviews, see Erdik, E. *Tetrahedron* **1984**, *40*, 641; Kochi, J.K. *Organometallic Mechanisms and Catalysis*, Academic Press, NY, **1978**, pp. 374–398.

[1293]Bäckvall, J.-E.; Persson, E.S.M.; Bombrun, A. *J. Org. Chem.* **1994**, *59*, 4126.

primary alkyl halides in good yield[1294] (organocopper salts are probably intermediates here). Allylic halides are more reactive than aliphatic alkyl halides, but copper salts have been used to facilitate coupling with alkylmagnesiumhalides.[1295] Iron(III)[1296] or palladium[1297] complexes are also used, and the latter allows the coupling of Grignard reagents and vinylic halides. Vinyl halides[1298] and aryl halides[1299] also couple with alkyl Grignard reagents in the presence of a catalytic amount of Fe(acac)$_3$, where acac = acetylacetonate, as do vinyl triflates with CuI[1300] or vinyl halides with a cobalt catalyst.[1301] Grignard reagents prepared from primary or secondary[1302] alkyl or aryl halides can be coupled with vinylic or aryl halides (see **13-9**) in high yields in the presence of a nickel(II) catalyst.[1303] When a chiral nickel(II) catalyst is used, optically active hydrocarbons can be prepared from achiral reagents.[1304] Neopentyl iodides also couple with aryl Grignard reagents in the presence of a nickel(II) catalyst.[1305]

Aryl halides, even when activated, generally do not couple with Grignard reagents, although certain transition-metal catalysts do effect this reaction in variable yields.[1306] The reaction with Grignard reagents proceeds better when OR can be the leaving group, providing that activating groups are present in the ring. The oxazoline group actives *o*-methoxy and *o*-fluoro groups to reaction with Grignard

---

[1294]Tamura, M.; Kochi, J.K. *J. Am. Chem. Soc.* **1971**, *93*, 1485; Derguini-Boumechal, F.; Linstrumelle, G. *Tetrahedron Lett.* **1976**, 3225; Mirviss, S.B. *J. Org. Chem.* **1989**, *54*, 1948; Terao, J.; Ikumi, A.; Kuniyasu, H.; Kambe, N. *J. Am. Chem. Soc.* **2003**, *125*, 5646.

[1295]Tissot-Croset, K.; Alexakis, A. *Tetrahedron Lett.* **2004**, *45*, 7375; Tissot-Croset, K.; Polet, D.; Alexakis, A. *Angew. Chem. Int. Ed.* **2004**, *43*, 2426.

[1296]Smith, R.S.; Kochi, J.K. *J. Org. Chem.* **1976**, *41*, 502; Walborsky, H.M.; Banks, R.B. *J. Org. Chem.* **1981**, *46*, 5074; Molander, G.A.; Rahn, B.J.; Shubert, D.C.; Bonde, S.E. *Tetrahedron Lett.* **1983**, *24*, 5449. An iron–salen catalyst has been used: see Bedford, R.B.; Bruce, D.W.; Frost, R.M.; Goodby, J.W.; Hird, M. *Chem. Commun.* **2004**, 2822.

[1297]Ratovelomanana, V.; Linstrumelle, G.; Normant, J. *Tetrahedron Lett.* **1985**, *26*, 2575; Minato, A.; Suzuki, K.; Tamao, K. *J. Am. Chem. Soc.* **1987**, *109*, 1257; Frisch, A.C.; Shaikh, N.; Zapf, A.; Beller, M. *Angew. Chem. Int. Ed.* **2002**, *41*, 4056. For other references, see Larock, R.C. *Comprehensive Organic Transformations*, 2nd ed., Wiley-VCH, NY, **1999**, pp. 386–392.

[1298]Cahiez, G.; Avedissian, H. *Synthesis* **1998**, 1199; Nagano, T.; Hayashi, T. *Org. Lett.* **2004**, *6*, 1297.

[1299]Fürstner, A.; Leitner, A. *Angew. Chem. Int. Ed.* **2002**, *41*, 609; Martin, R.; Fürstner, A. *Angew. Chem. Int. Ed.* **2004**, *43*, 3955.

[1300]Karlström, A.S.E.; Rönn, M.; Thorarensen, A. ; Bäckvall, J.-E. *J. Org. Chem.* **1998**, *63*, 2517.

[1301]Cahiez, G.; Avedissian, H. *Tetrahedron Lett.* **1998**, *39*, 6159.

[1302]Hayashi, T.; Konishi, M.; Kobori, Y.; Kumada, M.; Higuchi, T.; Hirotsu, K. *J. Am. Chem. Soc.* **1984**, *106*, 158.

[1303]Corriu, R.J.P.; Masse, J.P. *J. Chem. Soc., Chem. Commun.* **1972**, 144; Böhm, V.P.W.; Gstöttmayr, C.W.K.; Weskamp, T.; Hermann, W.A. *Angew. Chem. Int. Ed.* **2001**, *40*, 3387; Terao, J.; Watanabe, H.; Ikumi, A.; Kuniyasu, H.; Kambe, N. *J. Am. Chem. Soc.* **2002**, *124*, 4222. For a review, see Kumada, M. *Pure Appl. Chem.* **1980**, *52*, 669.

[1304]For a review, see Hayashi, T.; Kumada, M., in Morrison, J.D. *Asymmetic Synthesis*, Vol. 5, Academic Press, NY, **1985**, pp. 147–169. See also, Cross, G.A.; Kellogg, R.M. *J. Chem. Soc., Chem. Commun.* **1987**, 1746; Iida, A.; Yamashita, M. *Bull. Chem. Soc. Jpn.* **1988**, *61*, 2365.

[1305]Yuan, K.; Scott, W.J. *Tetrahedron Lett.* **1991**, *32*, 189.

[1306]See, for example, Sekiya, A.; Ishikawa, N. *J. Organomet. Chem.*, **1976**, *118*, 349; **1977**, *125*, 281; Tiecco, M.; Testaferri, L.; Tingoli, M.; Chianelli, D.; Wenkert, E. *Tetrahedron Lett.*, **1982**, *23*, 4629; Bell, T.W.; Hu, L.; Patel, S.V. *J. Org. Chem.*, **1987**, *52*, 3847; Bumagin, N.A.; Andryukhova, N.L.; Beletskaya, I.P. *Doklad. Chem.*, **1987**, *297*, 524; Ozawa, F.; Kurihara, K.; Fujimori, M.; Hidaka, T.; Toyoshima, T.; Yamamoto, A. *Organometallics* **1989**, *8*, 180.

reagents and organolithiums; the product **142** can be hydrolyzed after coupling[1307] (see **10-74**):

142

gem-Dichlorides have been prepared by coupling alkyl halides to $RCCl_3$ compounds electrochemically, in an undivided cell with a sacrificial anode:[1308]

$$RCCl_3 + R'X + 2\,e^- \longrightarrow RCCl_2R' + Cl^- + X^-$$

R' could also be Cl, in which case the product bears a CCl_3 group.[1309]

Much study has been devoted to the mechanisms of these reactions,[1310] but firm conclusions are still lacking, in part because the mechanisms vary depending on the metal, the R group, the catalyst, if any, and the reaction conditions. Two basic pathways can be envisioned: a nucleophilic substitution process (which might be S_N1 or S_N2) and a free-radical mechanism. This could be an SET pathway, or some other route that provides radicals. In either case the two radicals R• and R'• would be in a solvent cage:

$$RX + R'M \longrightarrow \begin{bmatrix} R\cdot + R'\cdot \\ + MX \end{bmatrix} \longrightarrow RR'$$

Solvent cage

It is necessary to postulate the solvent cage because, if the radicals were completely free, the products would be about 50% RR', 25% RR, and 25% R'R'. This is generally not the case; in most of these reactions RR' is the predominant or exclusive product.[1311] An example where an S_N2 mechanism has been demonstrated (by the finding of inversion of configuration at R) is the reaction between allylic or benzylic lithium reagents with secondary halides.[1312] The fact that in some of these cases the

[1307]For a review of oxazolines in aromatic substitutions, see Reuman, M.; Meyers, A.I. *Tetrahedron*, **1985**, *41*, 837. For the similar use of oxazoles, see Cram, D.J.; Bryant, J.A.; Doxsee, K.M. *Chem. Lett.*, **1987**, 19.

[1308]Nédélec, J.; Aït Haddou Mouloud, H.; Folest, J.; Périchon, J. *J. Am. Chem. Soc.* **1988**, *53*, 4720.

[1309]For the transformation RX→RCF$_3$, see Chen, Q.; Wu, S. *J. Chem. Soc., Chem. Commun.* **1989**, 705.

[1310]For a review, see Beletskaya, I.P.; Artamkina, G.A.; Reutov, O.A. *Russ. Chem. Rev.* **1976**, *45*, 330.

[1311]When a symmetrical distribution of products *is* found, this is evidence for a free-radical mechanism: the solvent cage is not efficient and breaks down.

[1312]Sauer, J.; Braig, W. *Tetrahedron Lett.* **1969**, 4275; Sommer, L.H.; Korte, W.D. *J. Org. Chem.* **1970**, *35*, 22; Korte, W.D.; Kinner, L.; Kaska, W.C. *Tetrahedron Lett.* **1970**, 603. See also, Schlosser, M.; Fouquet, G. *Chem. Ber.* **1974**, *107*, 1162, 1171.

reaction can be successfully applied to aryl and vinylic substrates indicates that a simple S_N process cannot be the only mechanism. One possibility is that the reagents first undergo an exchange reaction: $ArX + RM \rightarrow RX + ArM$, and then a nucleophilic substitution takes place. On the other hand, there is much evidence that many coupling reactions involving organometallic reagents with simple alkyl groups occur by free-radical mechanisms. Among the evidence[1313] is the observation of CIDNP in reactions of alkyl halides with simple organolithium reagents[1314] (see p. 269), the detection of free radicals by esr spectroscopy[1315] (p. 277), and the formation of 2,3-dimethyl-2,3-diphenylbutane when the reaction was carried out in the presence of cumene[1316] (this product is formed when a free-radical abstracts a hydrogen from cumene to give $PhCMe_2$, which dimerizes). Evidence for free-radical mechanisms has also been found for the coupling of alkyl halides with simple organosodium compounds (Wurtz),[1317] with Grignard reagents,[1318] and with lithium dialkylcopper reagents (see **10-58**).[1319] Free radicals have also been implicated in the metal-ion-catalyzed coupling of alkyl and aryl halides with Grignard reagents.[1320]

A much older reaction is the coupling of alkyl halides with Grignard reagents.[1321] Grignard reagents have the advantage that they are usually simpler to prepare than the corresponding R'_2CuLi (see **10-58**), but the reaction is much narrower in scope. Grignard reagents couple only with active halides: allylic (though allylic rearrangements are common) and benzylic. They also couple with tertiary alkyl halides, but generally in low or moderate yields.[1322]

Aryl Grignard reagents usually give better yields in these reactions than alkyl Grignard reagents. Aryl triflates couple with arylmagnesium halides in the presence

[1313]For other evidence, see Muraoka, K.; Nojima, M.; Kusabayashi, S.; Nagase, S. *J. Chem. Soc. Perkin Trans. 2* **1986**, 761.

[1314]Ward, H.R.; Lawler, R.G.; Cooper, R.A. *J. Am. Chem. Soc.* **1969**, *91*, 746; Lepley, A.R.; Landau, R.L. *J. Am. Chem. Soc.* **1969**, *91*, 748; Podoplelov, A.V.; Leshina, T.V.; Sagdeev, R.Z.; Kamkha, M.A.; Shein, S.M. *J. Org. Chem. USSR* **1976**, *12*, 488. For a review, see Ward, H.R.; Lawler, R.G.; Cooper, R.A., in Lepley, A.R.; Closs, G.L. *Chemically Induced Magnetic Polarization*, Wiley, NY, **1973**, pp. 281–322.

[1315]Russell, G.A.; Lamson, D.W. *J. Am. Chem. Soc.* **1969**, *91*, 3967.

[1316]Bryce-Smith, D. *Bull. Soc. Chim. Fr.* **1963**, 1418.

[1317]Garst, J.F.; Cox, R.H. *J. Am. Chem. Soc.* **1970**, *92*, 6389; Kasukhin, L.F.; Gragerov, I.P. *J. Org. Chem. USSR* **1971**, *7*, 2087; Garst, J.F.; Hart, P.W. *J. Chem Soc. Chem. Commun.* **1975**, 215.

[1318]Gough, R.G.; Dixon, J.A. *J. Org. Chem.* **1968**, *33*, 2148; Ward, H.R.; Lawler, R.G.; Marzilli, T.A. *Tetrahedron Lett.* **1970**, 521; Kasukhin, L.F.; Ponomarchuk, M.P.; Buteiko, Zh.F. *J. Org. Chem. USSR* **1972**, *8*, 673; Singh, P.R.; Tayal, S.R.; Nigam, A. *J. Organomet. Chem.* **1972**, *42*, C9.

[1319]Ashby, E.C.; Coleman, D. *J. Org. Chem.* **1987**, *52*, 4554; Bertz, S.H.; Dabbagh, G.; Mujsce, A.M. *J. Am. Chem. Soc.* **1991**, *113*, 631.

[1320]Norman, R.O.C.; Waters, W.A. *J. Chem. Soc.* **1957**, 950; Frey Jr., F.W. *J. Org. Chem.* **1961**, *26*, 5187; Slaugh, L.H. *J. Am. Chem. Soc.* **1961**, *83*, 2734; Davies, D.I.; Done, J.N.; Hey, D.H. *J. Chem. Soc. C* **1969**, 1392, 2021, 2056; Abraham, M.H.; Hogarth, M.J. *J. Organomet. Chem.* **1968**, *12*, 1, 497; Tamura, M.; Kochi, J.K. *J. Am. Chem. Soc.* **1971**, *93*, 1483, 1485, 1487; *J. Organomet. Chem.* **1971**, *31*, 289; *1972*, *42*, 205; Lehr, G.F.; Lawler, R.G. *J. Am. Chem. Soc.* **1986**, *106*, 4048.

[1321]For reviews, see Raston, C.L.; Salem, G., in Hartley, F.R. *The Chemistry of the Metal-Carbon Bond*, Vol. 4, Wiley, NY, **1987**, pp. 161–306, 269–283; Kharasch, M.S.; Reinmuth, O. *Grignard Reactions of Nonmetallic Substances*, Prentice-Hall, Englewood Cliffs, NJ, **1954**, pp. 1046–1165.

[1322]See, for example, Ohno, M.; Shimizu, K.; Ishizaki, K.; Sasaki, T.; Eguchi, S. *J. Org. Chem.* **1988**, *53*, 729.

of a palladium catalyst,[1323] as do vinyl halides with RMgX with a palladium[1324] or nickel catalyst.[1325] It is also possible to couple alkynylmagnesium halides with aryl iodides in the presence of palladium catalysts.[1326] A silica-supported phosphine–palladium (0) medium was used to couple arylmagnesium halides with aryl iodides.[1327] Aryl Grignard reagents couple with alkyl halides, including neopentyl iodide, in the presence of $ZnCl_2$ and a nickel catalyst.[1328]

In some cases, vinyl halides can be coupled. An aryl Grignard reagent was coupled to a vinyl iodide in the presence of an iron catalyst.[1329] Butylmagnesium chloride was coupled to vinyl triflates with $Fe(acac)_3$.[1330] The palladium-catalyzed coupling of arylmagnesium halides and vinyl bromides has also been reported.[1331]

Because Grignard reagents react with the C=O group (**16-24**, **16-82**), they cannot be used to couple with halides containing ketone, COOR, or amide functions. Although the coupling of Grignard reagents with ordinary alkyl halides is usually not useful for synthetic purposes, small amounts of symmetrical coupling product are commonly formed while Grignard reagents are being prepared.

For symmetrical coupling of organometallic reagents (2RM → RR), see **14-24** and **14-25**.

OS **I**, 186; **III**, 121; **IV**, 748; **VI**, 407; **VII**, 77, 172, 326, 485; **VIII**, 226, 396; **IX**, 530; **X**, 332, 396.

10-58 Reaction of Alkyl Halides and Sulfonate Esters with Organocuprates

Alkyl-de-halogenation

$$RX + R_2'CuLi \longrightarrow R-R'$$

The reagents lithium dialkylcopper[1332] (dialkyl cuprates, also called *Gilman reagents*)[1333] react with alkyl bromides, chlorides, and iodides in ether or THF to

[1323]Kamikawa, T.; Hayashi, T. *Synlett*, **1997**, 163.

[1324]Hoffmann, R.W.; Gieson, V.; Fuest, M. *Liebigs Ann. Chem.* **1993**, 629.

[1325]Babudri, F.; Fiandanese, V.; Mazzone, L.; Naso, F. *Tetrahedron Lett.* **1994**, *35*, 8847.

[1326]Negishi, E.; Kotora, M.; Xu, C. *J. Org. Chem.* **1997**, *62*, 8957.

[1327]Cai, M.-Z.; Song, C.-S.; Huang, X. *J. Chem. Res. (S)* **1998**, 264.

[1328]Kondo, S.; Ohira, M.; Kawasoe, S.; Kunisada, H.; Yuki, Y. *J. Org. Chem.* **1993**, *58*, 5003.

[1329]Dohle, W.; Kopp, F.; Cahiez, G.; Knochel, P. *Synlett* **2001**, 1901.

[1330]Scheiper, B.; Bonnekessel, M.; Krause, H.; Fürstner, A. *J. Org. Chem.* **2004**, *69*, 3943.

[1331]Rathore, R.; Deselnicu, M.I.; Burns, C.L. *J. Am. Chem. Soc.* **2002**, *124*, 14832.

[1332]For the structure of Me_2CuLi (a cyclic dimer), see Pearson, R.G.; Gregory, C.D. *J. Am. Chem. Soc.* **1976**, *98*, 4098. See also, Lipshutz, B.H.; Kozlowski, J.A.; Breneman, C.M. *Tetrahedron Lett.* **1985**, *26*, 5911. For a review of the structure and reactions of organocopper compounds, see Collman, J.P.; Hegedus, L.S.; Norton, J.R.; Finke, R.G. *Principles and Applications of Organotransition Metal Chemistry*, 2nd ed., University Science Books, Mill Valley, CA, **1987**, pp. 682–698.

[1333]See Stemmler, T.L.; Barnhart, T.M.; Penner-Hahn, J.E.; Tucker, C.E.; Knochel, P.; Böhme, M.; Frenking, G. *J. Am. Chem. Soc.* **1995**, *117*, 12489 for a discussion concerning the structure of organocuprate reagents. Solution compositions of Gilman reagents have also been studied. See Lipshutz, B.H.; Kayser, F.; Siegmann, K. *Tetrahedron Lett.* **1993**, *34*, 6693.

give good yields of the cross-coupling products.[1334] They are prepared (see **12-36**) by the reaction of an organolithium compound with CuI or CuBr, typically, most other Cu(I) compounds can be used. They are usually generated at temperatures $<0°C$ due to the thermal instability of many dialkyl cuprates. The reaction with alkyl halides is of wide scope[1335] and R in R_2CuLi may be primary alkyl, allylic, benzylic, aryl, vinylic, or allenic, and may contain keto, COOH, COOR, or $CONR_2$ groups.[1336] Inversion of configuration has been shown in the reaction of 2-bromobutane with Ph_2CuLi,[1337] but the same reaction with 2-iodobutane has been reported to proceed with racemization.[1338] The reaction at a vinylic substrate occurs stereospecifically, with retention of configuration.[1339] When the reagent and substrate are both vinylic, yields are low, but the reaction can be pushed to give 1,3-butadienes, stereospecifically and in high yields by the use of $ZnBr_2$ and a Pd(0) complex.[1340] Many *gem*-dihalides do not react, but when the two halogens are on a carbon α to an aromatic ring[1341] or on a cyclopropane ring,[1342] both halogens can be replaced by R, for example, $PhCHCl_2 \rightarrow PhCHMe_2$. However, 1,2-dibromides give exclusive elimination[1337] (**17-22**). Vinylmagnesium halides, upon addition of a catalytic amount of Li_2CuCl_4, couple to alkyl halide.[1343]

Lithium dialkylcopper reagents couple with alkyl tosylates.[1344] High yields are obtained with primary tosylates; secondary tosylates give lower yields.[1345] Aryl tosylates do not react. Vinylic triflates[1346] couple very well to give alkenes[1347] and they

---

[1334]Corey, E.J.; Posner, G.H. *J. Am. Chem. Soc.* **1968**, *90*, 5615; Bergbreiter, D.E.; Whitesides, G.M. *J. Org. Chem.* **1975**, *40*, 779. See Bertz, S.H.; Eriksson, M.; Miao, G.; Snyder, J.P. *J. Am. Chem. Soc.* **1998**, *118*, 10906 for the reactivity of β-silyl organocuprates.

[1335]For reviews see Posner, G.H. *Org. React.* **1975**, *22*, 253; Normant, J.F. *Synthesis* **1972**, 63; Lipshutz, B.H. *Accts. Chem. Res.* **1997**, *30*, 277; Posner, G.H. *An Introduction to Synthesis Using Organocopper Reagents*, Wiley, NY, **1980**. For lists of substrates and reagents, with references, see Larock, R.C. *Comprehensive Organic Transformations*, 2nd ed., Wiley-VCH, NY, **1999**, pp. 392–399, 599–604, 1564.

[1336]For a discussion of the mechansim of S_N2 alkylation with organocuprates see Mori, S.; Nakamura, E.; Morokuma, K. *J. Am. Chem. Soc.* **2000**, *122*, 7294.

[1337]Cahiez, G.; Chaboche, C.; Jézéquel, M. *Tetrahedron* **2000**, *56*, 2733.

[1338]Lipshutz, B.H.; Wilhelm, R.S.; Nugent, S.T.; Little, R.D.; Baizer, M.M. *J. Org. Chem.* **1983**, *48*, 3306.

[1339]Corey, E.J.; Posner, G.H. *J. Am. Chem. Soc.* **1967**, *89*, 3911; Klein, J.; Levene, R. *J. Am. Chem. Soc.* **1972**, *94*, 2520. For a discussion of the mechanism, see Yoshikai, N.; Nakamura, E. *J. Am. Chem. Soc.* **2004**, *126*, 12264.

[1340]Jabri, N.; Alexakis, A.; Normant, J.F. *Tetrahedron Lett.* **1981**, *22*, 959; **1982**, *23*, 1589; *Bull. Soc. Chim. Fr.* **1983**, II-321, II-332.

[1341]Posner, G.H.; Brunelle, D.J. *Tetrahedron Lett.* **1972**, 293.

[1342]See, for example, Kitatani, K.; Hiyama, T.; Nozaki, H. *Bull. Chem. Soc. Jpn.* **1977**, *50*, 1600.

[1343]Posner, G.H.; Ting, J. *Synth. Commun.* **1973**, *3*, 281.

[1344]Johnson, C.R.; Dutra, G.A. *J. Am. Chem. Soc.* **1973**, *95*, 7777, 7783. For examples, see Posner, G.H. *An Introduction to Synthesis Using Organocopper Reagents*, Wiley, NY, **1980**, pp. 85–90.

[1345]Secondary tosylates give higher yields when they contain an O or S atom: Hanessian, S.; Thavonekham, B.; DeHoff, B. *J. Org. Chem.* **1989**, *54*, 5831.

[1346]For a review of coupling reactions of vinylic triflates, see Scott, W.J.; McMurry, J.E. *Acc. Chem. Res.* **1988**, *21*, 47.

[1347]McMurry, J.E.; Scott, W.J. *Tetrahedron Lett.* **1980**, *21*, 4313; Tsushima, K.; Araki, K.; Murai, A. *Chem. Lett.* **1989**, 1313.

also couple with allylic cuprates, to give 1,4-dienes.[1348] Propargylic tosylates couple with vinylic cuprates to give vinylic allenes.[1349]

The R' in R'_2CuLi may be primary alkyl, vinylic, allylic, or aryl. Thus, in the reaction as so far described, the alkyl groups on the organocuprate or the alkyl halide may *not* be secondary or tertiary alkyl. However, secondary and tertiary alkyl coupling can be achieved (on primary RX) by the use of $R'_2CuLi \cdot PBu_3$[1350] (though this procedure introduces problems in the workup) or by the use of PhS(R')CuLi,[1351] which selectively couples a secondary or tertiary R' with a primary iodide RI to give RR'.[1352] It is possible to prepare mixed cuprates, where one ligand is tightly bound to the copper, allowing the other ligand to be transferred in a coupling reaction. A common example is adds a 2-thienyl group to the cuprate to give R(Th)CuLi, where the R group is transferred in lieu of the thienyl unit.[1353] A lithium neopentyl aryl cuprate selectively transferred to aryl group to an allylic halide.[1354]

Coupling to a secondary alkyl halide (R in RX above = secondary) can be achieved in high yield with the reagents $R'_2Cu(CN)Li_2$,[1355] where R' is primary alkyl or vinylic (but not aryl).[1356] This modified reagent is commonly known as a higher order mixed cuprate. The reagents $RCu(PPh_2)Li$, $RCu(NR'_2)Li$, and $RCu(PR'_2)Li$ (R' = cyclohexyl) are more stable than R_2CuLi and can be used at higher temperatures.[1357] However, these reagents are quite reactive. Unactivated aryl triflates[1358] $ArOSO_2CF_3$ react to give ArR in good yields when treated with $R_2Cu(CN)Li_2$,[1359] with R_3Al,[1360] or with R'_3SnR and a Pd complex catalyst.[1361] See section **10-59** for other examples involving Al, Sn and Pd coupling reactions.

[1348]Lipshutz, B.H.; Elworthy, T.R. *J. Org. Chem.* **1990**, *55*, 1695.

[1349]Baudouy, R.; Goré, J. *J. Chem. Res. (S)* **1981**, 278. See also, Elsevier, C.J.; Vermeer, P. *J. Org. Chem.* **1989**, *54*, 3726.

[1350]Whitesides, G.M.; Fischer, Jr., W.F.; San Filippo, Jr., J.; Bashe, R.W.; House, H.O., *J. Am. Chem. Soc.* **1969**, *91*, 4871.

[1351]Prepared as in Ref. 1371 or treatment of PhSCu with RLi: Posner, G.H.; Brunelle, D.J.; Sinoway, L. *Synthesis* **1974**, 662.

[1352]Posner, G.H.; Whitten, C.E.; Sterling, J.J. *J. Am. Chem. Soc.* **1973**, *95*, 7788.

[1353]For an example, see Malmberg, H.; Nilsson, M.; Ullenius, C. *Tetrahedron Lett.* **1982**, *23*, 3823. For an example involving higher order cuprates, see Lipshutz, B.H.; Kozlowski, J.A.; Parker, D.A.; Nguyen, S.L.; McCarthy, K.E. *J. Organomet. Chem.* **1985**, *285*, 437.

[1354]Piazza, C.; Knochel, P. *Angew. Chem. Int. Ed.* **2002**, *41*, 3263.

[1355]For reviews of these and other "higher order" organocuprates, see Lipshutz, B.H.; Wilhelm, R.S.; Kozlowski, J.A. *Tetrahedron* **1984**, *40*, 5005; Lipshutz, B.H. *Synthesis* **1987**, 325; *Synlett* **1990**, 119. See also, Bertz, S.H. *J. Am. Chem. Soc.* **1990**, *112*, 4031; Lipshutz, B.H.; Sharma, S.; Ellsworth, E.L. *J. Am. Chem. Soc.* **1990**, *112*, 4032.

[1356]Lipshutz, B.H.; Wilhelm, R.S.; Floyd, D.M. *J. Am. Chem. Soc.* **1981**, *103*, 7672.

[1357]Bertz, S.H.; Dabbagh, G.; Villacorta, G.M. *J. Am. Chem. Soc.* **1982**, *104*, 5824; Bertz, S.H.; Dabbagh, G. *J. Org. Chem.* **1984**, *49*, 1119.

[1358]For another coupling reaction of aryl triflates, see Aoki, S.; Fujimura, T.; Nakamura, E.; Kuwajima, I. *J. Am. Chem. Soc.*, **1988**, *110*, 3296.

[1359]McMurry, J.E.; Mohanraj, S. *Tetrahedron Lett.*, **1983**, *24*, 2723.

[1360]Hirota, K.; Isobe, Y.; Maki, Y. *J. Chem. Soc., Perkin Trans. 1*, **1989**, 2513.

[1361]Echevarren, E.M.; Stille, J.K. *J. Am. Chem. Soc.*, **1987**, *109*, 5478. For a similar reaction with aryl fluorosulfonates, see Roth, G.P.; Fuller, C.E. *J. Org. Chem.*, **1991**, *56*, 3493.

Both OTf units in $RCH(OTf)_2$ can be replaced with $Me_2(CN)CuLi_2$.[1362] With an allenic substrate, reaction with $R(CN)CuLi$ can give ordinary displacement (with retention of configuration)[1363] or an S_N2' reaction to produce an alkyne.[1364] In the latter case, a chiral allene gave a chiral alkyne. The structures of these "higher order mixed" cuprates has been called into question[1365] by Bertz, who suggested the reagent actually existed as $R_2CuLi•LiCN$ in THF.[1366] This was contradicted by Lipshutz.[1367]

143

The fact that R'_2CuLi do not react with ketones provides a method for the alkylation of ketones via the organocuprate coupling with α-halokeotones, such as **143**[1368] (see also, **10-68** and **10-73**). Note that halogen–metal exchange (**12-39**) is a side reaction and can become the main reaction.[1369] When α,α'-dibromo ketones are treated with Me_2CuLi in ether at $-78°C$ and the mixture quenched with methanol, *mono*methylation takes place[1370] (no dimethylation is observed). It has been suggested that the reaction involves cyclization (**10-56**) to a cyclopropanone followed by nucleophilic attack to give the enolate anion, which is protonated by the methanol. If methyl iodide is added instead of methanol, an α,α'-dimethyl ketone is obtained, presumably from S_N2 attack (**10-68**). Primary, secondary, *and tertiary* monoalkylation can be achieved with a lithium *tert*-butoxy (alkyl)copper reagent[1371] instead of Me_2CuLi, one of the few methods for introducing a tertiary alkyl group a to a carbonyl group.

When dialkylcopperzinc reagents $R_2CuZnCl$ couple with allylic halides, almost complete allylic rearrangement occurs (S_N2'), and the reaction is diastereoselective if the allylic halide contains a δ alkoxy group.[1372] Another type of copper reagent

[1362]Martínez, A.G.; Barcina, J.O.; Díez, B.R.; Subramanian, L.R. *Tetrahedron* **1994**, *50*, 13231.

[1363]Mooiweer, H.H.; Elsevier, C.J.; Wijkens, P.; Vermeer, P. *Tetrahedron Lett.* **1985**, *26*, 65.

[1364]Corey, E.J.; Boaz, N.W. *Tetrahedron Lett.* **1984**, *25*, 3059, 3063. For the reaction of these reagents with haloalkynes, see Yeh, M.C.P.; Knochel, P. *Tetrahedron Lett.* **1989**, *30*, 4799.

[1365]Bertz, S.H.; Miao, G.; Eriksson, M. *Chem. Commun.* **1996**, 815; Snyder, J.P.; Bertz, S.H. *J. Org. Chem.* **1995**, *60*, 4312. Also see, Snyder, J.P.; Tipsword, G.E.; Spangler, D.P. *J. Am. Chem. Soc.* **1992**, *114*, 1507.

[1366]Bertz, S.H. *J. Am. Chem. Soc.* **1990**, *112*, 4031.

[1367]Lipshutz, B.H.; James, B. *J. Org. Chem.* **1994**, *59*, 7585; Lipshutz, B.H.; Sharma, S.; Ellsworth, E.L. *J. Am. Chem. Soc.* **1990**, *112*, 4032.

[1368]Dubois, J.E.; Lion, C.; Moulineau, C. *Tetrahedron Lett.* **1971**, 177; Dubois, J.E.; Fournier, P.; Lion, C. *Bull. Soc. Chim. Fr.* **1976**, 1871.

[1369]See Corey, E.J.; Posner, G.H. *J. Am. Chem. Soc.* **1967**, *89*, 3911; Wakselman, C.; Mondon, M. *Tetrahedron Lett.* **1973**, 4285.

[1370]Posner, G.H.; Sterling, J.J. *J. Am. Chem. Soc.* **1973**, *95*, 3076. See also, Posner, G.H.; Sterling, J.J.; Whitten, C.E.; Lentz, C.M.; Brunelle, D.J. *J. Am. Chem. Soc.* **1975**, *97*, 107; Lion, C.; Dubois, J.E. *Tetrahedron* **1975**, *31*, 1223. The compound Ph_2CuLi behaves similarly: see Lei, X.; Doubleday Jr., C.; Turro, N.J. *Tetrahedron Lett.* **1986**, *27*, 4671.

[1371]Prepared by treating CuI with *t*-BuOLi in THF at $0°C$ and adding RLi to this solution.

[1372]Nakamura, E.; Sekiya, K.; Arai, M.; Aoki, S. *J. Am. Chem. Soc.* **1989**, *111*, 3091.

was prepared from RZnI/CuCN, and this was shown to couple with alkenyl halides,[1373] and diethylzinc in the presence of a catalytic amount of CuBr coupled to allylic chlorides.[1374] When treated with organocopper compounds and Lewis acids (e.g., n-BuCu•BF$_3$), allylic halides give substitution with almost complete allylic rearrangement, irrespective of the degree of substitution at the two ends of the allylic system.[1375]

$$ArI + R_2CuLi \longrightarrow ArR$$

OS **IX**, 502.

10-59   Reaction of Alkyl Halides and Sulfonate Esters With Other Organometallic Reagents

Alkyl-de-halogenation

$$RX + R'-M \longrightarrow R-R'$$

Many other metals and metal complexes can be used to catalyze or mediate coupling reactions. Organoaluminum compounds couple very well with tertiary (to give products containing a quaternary carbon) and benzylic halides at $-78°C$.[1376] This reaction can also be applied to allylic, secondary, and some primary halides, but several days standing at room temperature is required (see also **10-63**). Vinylic aluminum compounds (in the presence of a suitable transition-metal catalyst) couple with allylic halides, acetates, and alcohol derivatives to give 1,4-dienes,[1377] and with vinylic and benzylic halides to give 1,3-dienes and allylic arenes, respectively.[1378] An interesting transformation treated a vinyl nitro compound (PhCH=CHNO$_2$) with Et$_3$Al and a large excess of 2-iodopropane, in the presence of 2 equivalents of dibenzoyl peroxide, to give the coupling product, PhCH=CHi-Pr.[1379] Note that alkylboronic acids are coupled in the presence of Ag$_2$O and a catalytic amount of CrCl$_2$ to give the symmetrical alkyl derivative.[1380]

[1373]Marquais, S.; Cahiez, G.; Knochel, P. *Synlett, 1994*, 849.

[1374]Malda, H.; van Zijl, A.W.; Arnold, L.A.; Feringa, B.L. *Org. Lett. 2001, 3,* 1169.

[1375]Yamamoto, Y.; Yamamoto, S.; Yatagai, H.; Maruyama, K. *J. Am. Chem. Soc. 1980, 102,* 2318. See also, Lipshutz, B.H.; Ellsworth, E.L.; Dimock, S.H. *J. Am. Chem. Soc. 1990, 112,* 5869.

[1376]Miller, D.B. *J. Org. Chem. 1966, 31,* 908; Kennedy, J.P. *J. Org. Chem. 1970, 35,* 532. See also, Kennedy, J.P.; Sivaram, S. *J. Org. Chem. 1973, 38,* 2262; Sato, F.; Kodama, H.; Sato, M. *J. Organomet. Chem. 1978, 157,* C30.

[1377]Lynd, R.A.; Zweifel, G. *Synthesis 1974,* 658; Matsushita, H.; Negishi, E. *J. Am. Chem. Soc. 1981, 103,* 2882; *J. Chem. Soc., Chem. Commun. 1982,* 160. For similar reactions with other metals, see Larock, R.C.; Bernhardt, J.C.; Driggs, R.J. *J. Organomet. Chem. 1978, 156,* 45; Brown, H.C.; Campbell, Jr., J.B. *J. Org. Chem. 1980, 45,* 550; Baeckström, P.; Björkling, F.; Högberg, H.; Norin, T. *Acta Chem. Scand. Ser. B, 1984, 38,* 779.

[1378]Negishi, E.; Takahashi, T.; Baba, S.; Van Horn, D.E.; Okukado, N. *J. Am. Chem. Soc. 1987, 109,* 2393; Negishi, E.; Takahashi, T.; Baba, S. *Org. Synth. 66,* 60.

[1379]Liu, J.-Y.; Liu, J.-T.; Yao, C.-F. *Tetrahedron Lett. 2001, 42,* 3613.

[1380]Falck, J.R.; Mohaptra, S.; Bondlela, M.; Venkataraman, S.K. *Tetrahedron Lett. 2002, 43,* 8149.

Products containing a quaternary carbon can also be obtained by treatment of tertiary halides with dialkyl or diaryl zinc reagents in CH_2Cl_2,[1381] with Me_4Si and $AlCl_3$,[1382] or with alkyltitanium reagents $RTiCl_3$ and R_2TiCl_2.[1383] Dialkylzinc compounds can be coupled to alkyl iodides in the presence of a nickel catalyst,[1384] but with geminal diiodo compounds without a catalyst.[1385] Copper compounds can also be used as catalysts with dialkylzinc reagents.[1386] The reaction of aryl halides with Me_4ZnLi_2, and then $VO(OEt)Cl_2$ leads to the methylated aryl.[1387] Isopropylzinc ($iPrZn$) displaces the iodide in γ-iodo ketones to give the alkyl substitution product, without reaction at the carbonyl.[1388] Reactions of organozinc reagents with a carbonyl compound via acyl addition is presented in **16-31**, the Reformatsky reaction. The titanium method can also be used with secondary halides ($R_2CHCl \rightarrow R_2CHMe$), tertiary ethers ($R_3COR' \rightarrow R_3CMe$), and *gem*-dihalides ($R_2CCl_2 \rightarrow R_2CMe_2$).[1389] Tertiary halides have also been coupled to allyltin reagents in the presence of AIBN.[1390] Alkyl halides can be treated with SmI_2 and then CuBr to give a reactive species that couples with other alkyl halides.[1391] Trialkylindium compounds couple to allylic bromides in the presence of $Cu(OTf)_2\bullet P(OEt)_3$[1392] and vinyl indium compounds are coupled to α-halo esters with a BEt_3 catalyst.[1393] Arylsulfonyl chlorides couple with allyl halides in the presence of bismuth to give allyl-aryls.[1394] Vinyl iodides couple with RMnCl with an iron catalyst[1395] and $Bu_3MnMgBr$ reacted with a geminal dibromocyclopropane to give a dialkylated cyclopropane.[1396] α-Haloketones are coupled with aryl halides using a nickel catalyst.[1397] Allylgallium reagents have been coupled to α-bromo esters in the presence of BEt_3/O_2.[1398]

Arylpalladium salts "ArPdX" prepared from arylmercury compounds and lithium palladium chloride couple with allylic chlorides in moderate yields,

[1381]Reetz, M.T.; Wenderoth, B.; Peter, R.; Steinbach, R.; Westermann, J. *J. Chem. Soc., Chem. Commun.* **1980**, 1202. See also, Klingstedt, T.; Frejd, T. *Organometallics* **1983**, *2*, 598.

[1382]Bolestova, G.I.; Parnes, Z.N.; Latypova, F.M.; Kursanov, D.N. *J. Org. Chem. USSR* **1981**, *17*, 1203.

[1383]Reetz, M.T.; Westermann, J.; Steinbach, R. *Angew. Chem. Int. Ed.* **1980**, *19*, 900, 901.

[1384]Giovannini, R.; Stüdemann, T.; Devasagayaraj, A.; Dussin, G.; Knochel, P. *J. Org. Chem.* **1999**, *64*, 3544; Jensen, A.E.; Knochel, P. *J. Org. Chem.* **2002**, *67*, 79; Zhou, J.; Fu, G.C. *J. Am. Chem. Soc.* **2003**, *125*, 14726; Terao, J.; Todo, H.; Watanabe, H.; Ikumi, A.; Kambe, N. *Angew. Chem. Int. Ed.* **2004**, *43*, 6180.

[1385]Shibli, A.; Varghese, J.P.; Knochel, P.; Marek, I. *Synlett* **2001**, 818.

[1386]Shi, W.J.; Wang, L.-X.; Fu, Y.; Zhu, S.-F.; Zhou, Q.-L. *Tetrahedron Asymmetry* **2003**, *14*, 3867.

[1387]Hu, J.-b.; Zhao, G.; Yang, G.-s.; Ding, Z.-d. *J. Org. Chem.* **2001**, *66*, 303.

[1388]Jensen, A.E.; Knochel, P. *J. Org. Chem.* **2002**, *67*, 79.

[1389]Reetz, M.T.; Steinbach, R.; Wenderoth, B. *Synth. Commun.* **1982**, *11*, 261.

[1390]Kraus, G.A.; Ansersh, B.; Su, Q.; Shi, J. *Tetrahedron Lett.* **1993**, *34*, 1741.

[1391]Berkowitz, W.F.; Wu, Y. *Tetrahedron Lett.* **1997**, *38*, 3171.

[1392]Rodríguez, D.; Sestelo, J.P.; Sarandeses, L.A. *J. Org. Chem.* **2003**, *68*, 2518.

[1393]Takami, K.; Yorimitsu, H.; Oshima, K. *Org. Lett.* **2004**, *6*, 4555.

[1394]Baruah, M.; Boruah, A.; Prajapati, D.; Sandu, J.S. *Synlett*, **1998**, 1083.

[1395]Cahiez, G.; Marquais, S. *Tetrahedron Lett.* **1996**, *37*, 1773.

[1396]Kakiya, H.; Inoue, R.; Shinokubo, H.; Oshima, K. *Tetrahedron* **2000**, *56*, 2131.

[1397]Durandetti, M.; Sibille, S.; Nédélec, J.-Y.; Périchon, J. *Synth. Commun.* **1994**, *24*, 145.

[1398]Usugi, S.-i.; Yorimitsu, H.; Oshima, K. *Tetrahedron Lett.* **2001**, *42*, 4535.

although allylic rearrangements can occur.[1399] The advantage of this procedure is that the aryl group may contain nitro, ester, or aldehyde groups, and so on, which cannot be present in a Grignard reagent. In most cases, a palladium(0) complex is added to the substrate, sometimes in conjunction with another metal, to facilitate coupling. Any arylpalladium species is therefore generated *in situ*. Allylic, benzylic, vinylic, and aryl halides or triflates (trifluoromethylsulfonates) couple with organotin reagents in a reaction catalyzed by palladium complexes.[1400] Such functional groups as COOR, CN, OH, and CHO may be present in either reagent, but the substrate may not bear a β hydrogen on an sp^3 carbon, because that results in elimination. Indium metal has been used to mediate the coupling of an allylic halide and an arylpalladium complex,[1401] and organoindium compounds were coupled to 1-iodonaphthalene with a palladium catalyst.[1402] Dimethylzinc was coupled to aryl halides with a palladium catalyst,[1403] and Reformatsky-type zinc derivatives (**16–28**) have been coupled to aryl halides using a palladium catalyst and microwave irradiation.[1404]

In many cases, the organometallic reagent is prepared from the corresponding organolithium reagent (**10-57**), as in the conversion of an aryllithium to an arylzirconium reagent, which was subsequently coupled to a aryl halide in the presence of a palladium catalyst.[1405] Alkyl or aryl triflates (halides) couple with alkyl or ArZn(halide) reagents in the presence of a palladium catalyst.[1406] This organozinc coupling reaction has been done in ionic liquids.[1407] Vinyl halides coupled with vinyltin reagents in the presence of CuI,[1408] and aryl tin compounds couple with vinyl halides[1409] or vinyl triflates when a palladium catalyst is present.[1410] When the vinyltin reagent is coupled with a vinyl triflate in the presence of a palladium catalyst, the reaction is known as the *Stille reaction* (**12-15**). These latter reactions are obviously related, but the Stille reaction is placed in Chapter 14 for mechanistic reasons related

[1399]Heck, R.F. *J. Am. Chem. Soc.* **1968**, *90*, 5531. See **13-10**. For a review of palladium-assisted coupling, see Heck, R.F. *Palladium Reagents in Organic Syntheses*, Academic Press, NY, **1985**, pp. 208–214, 242–249.

[1400]For a review, see Stille, J.K. *Angew. Chem. Int. Ed.* **1986**, *25*, 508. For a review of the mechanism, see Bumagin, N.A.; Beletskaya, I.P. *Russ. Chem. Rev.* **1990**, *59*, 1174. See also, Stille, J.K.; Simpson, J.H. *J. Am. Chem. Soc.* **1987**, *109*, 2138; Martínez, A.G.; Barcina, J.O.; Heras, Md.R.C.; Cerezo, A.d.F. *Org. Lett.* **2000**, *2*, 1377.

[1401]Lee, P.H.; Sung, S.-y.; Lee, K. *Org. Lett.* **2001**, *3*, 3201.

[1402]Lee, P.H.; Lee, S.W.; Seomoon, D. *Org. Lett.* **2003**, *5*, 4963; Rodríguez, D.; Sestelo, J.P.; Sarandeses, L.A. *J. Org. Chem.* **2004**, *69*, 8136.

[1403]Herbert, J.M. *Tetrahedron Lett.* **2004**, *45*, 817.

[1404]Bentz, E.; Moloney, M.G.; Westaway, S.M. *Tetrahedron Lett.* **2004**, *45*, 7395.

[1405]Frid, M.; Pérez, D.; Peat, A.J.; Buchwald, S.L. *J. Am. Chem. Soc.* **1999**, *121*, 9469. See also, Villiers, P.; Vicart, N.; Ramondenc, Y.; Plé, G. *Tetrahedron Lett.* **1999**, *40*, 8781.

[1406]Piber, M.; Jensen, A.E.; Rottländer, M.; Knochel, P. *Org. Lett.* **1999**, *1*, 1323; Hossain, K.M.; Shibata, T.; Takagi, K. *Synlett* **2000**, 1137; Zhou, J.; Fu, G.C. *J. Am. Chem. Soc.* **2003**, *125*, 12527.

[1407]Sirieix, J.; Oßberger, M.; Betzemeier, B.; Knochel, P. *Synlett* **2000**, 1613.

[1408]Kang, S.-K.; Kim, J.-S.; Choi, S.-C. *J. Org. Chem.* **1997**, *62*, 4208.

[1409]Shen, W.; Wang, L. *J. Org. Chem.* **1999**, *64*, 8873.

[1410]Fouquet, E.; Rodriguez, A.L. *Synlett* **1998**, 1323; Lipshutz, B.H.; Alami, M. *Tetrahedron Lett.* **1993**, *34*, 1433.

to similar palladium-catalyzed coupling reactions. Vinylic triflates, in the presence of $Pd(Ph_3P)_4$ and LiCl, couple with organotin compounds $R'SnMe_3$, where R' can be alkyl, allylic, vinylic, or alkynyl.[1411] The reaction has been performed intramolecularly, to prepare large-ring lactones.[1412] Alkyl halides couple with ArMnCl or RMnCl in the presence of a palladium catalyst.[1413] The coupling of aryl substrates to form biaryls is discussed in **13-9**.

Alkenylboranes ($R_2'C=CHBZ_2$; $Z =$ various groups) couple in high yields with vinylic,[1414] alkynyl, aryl, benzylic, and allylic halides or triflates in the presence of a palladium catalyst and a base to give $R_2'C=CHR$.[1415] 9-Alkyl-9-BBN compounds (**15–16**) also couple with vinylic and aryl halides,[1416] as well as with α-halo ketones, nitriles, and esters.[1417] Another palladium-catalyzed coupling of vinyl halides and alkylboronic acids[1418] gives substituted alkenes, in a reaction that is related to the Suzuki coupling (**13-12**). Arylboronic acids can also be coupled to alkyl halides with a palladium catalyst,[1419] alkylboronic acids can be coupled to aryl halides in a similar manner,[1420] and a cyclopropylboronic acid was coupled to an allylic bromide with silver oxide/KOH and a palladium catalyst.[1421] Vinyl zirconium reagents were coupled to alkyl halides with a palladium catalyst.[1422]

Potassium aryl- and 1-alkenyltrifluoroborates ($ArBF_3K$ and RBF_3K) are easily prepared from organoboronic acids or esters. In general, the trifluoroborates have greater air stability and greater nucleophilicity[1423] when compared to the

[1411]Kwon, H.B.; McKee, B.H.; Stille, J.K. *J. Org. Chem.* **1990**, *55*, 3114. For discussions of the mechanism, see Stang, P.J.; Kowalski, M.H.; Schiavelli, M.D.; Longford, D. *J. Am. Chem. Soc.* **1989**, *111*, 3347; Stang, P.J.; Kowalski, M.H. *J. Am. Chem. Soc.* **1989**, *111*, 3356.

[1412]Stille, J.K.; Tanaka, M. *J. Am. Chem. Soc.* **1987**, *109*, 3785.

[1413]Riquet, E.; Alami, M.; Cahiez, G. *Tetrahedron Lett.* **1997**, *38*, 4397; Cahiez, G.; Marquais, S. *Synlett*, **1993**, 45.

[1414]Occhiato, E.G.; Trabocchi, A.; Guarna, A. *Org. Lett.* **2000**, *2*, 1241.

[1415]Brown, H.C.; Molander, G.A. *J. Org. Chem.* **1981**, *46*, 645; Miyaura, N.; Yamada, K.; Suginome, H.; Suzuki, A. *J. Am. Chem. Soc.* **1985**, *107*, 972; Sato, M.; Miyaura, N.; Suzuki, A. *Chem. Lett.* **1989**, 1405; Rivera, I.; Soderquist, J.A. *Tetrahedron Lett.* **1991**, *32*, 2311; and references cited therein. For a review, see Matteson, D.S. *Tetrahedron* **1989**, *45*, 1859.

[1416]Miyaura, N.; Ishiyama, T.; Sasaki, H.; Ishikawa, M.; Satoh, M.; Suzuki, A. *J. Am. Chem. Soc.* **1989**, *111*, 314. See also, Soderquist, J.A.; Santiago, B. *Tetrahedron Lett.* **1990**, *31*, 5541.

[1417]Ishiyama, T.; Abe, S.; Miyaura, N.; Suzuki, A. *Chem. Lett.* **1992**, 691; Brown, H.C.; Joshi, N.N.; Pyun, C.; Singaram, B. *J. Am. Chem. Soc.* **1989**, *111*, 1754. For another such coupling, see Matteson, D.S.; Tripathy, P.B.; Sarkar, A.; Sadhu, K.M. *J. Am. Chem. Soc.* **1989**, *111*, 4399.

[1418]Bellina, F.; Anselmi, C.; Rossi, R. *Tetrahedron Lett.* **2001**, *42*, 3851. See also, Yoshida, H.; Yamaryo, Y.; Oshita, J.; Kunai, A. *Tetrahedron Lett.* **2003**, *44*, 1541.

[1419]Nobre, S.M.; Monteiro, A.L. *Tetrahedron Lett.* **2004**, *45*, 8225; Liu, X.-x.; Deng, M.-z. *Chem. Commun.* **2002**, 622; Langle, S.; Abarbri, M.; Duchêne, A. *Tetrahedron Lett.* **2003**, *44*, 9255.

[1420]Kondolff, I.; Doucet, H.; Santelli, M. *Tetrahedron* **2004**, *60*, 3813. For a variation involving a borate complex, see Zou, G.; Falck, J.R. *Tetraahedron Lett.* **2001**, *42*, 5817.

[1421]Chen, H.; Deng, M.-Z. *J. Org. Chem.* **2000**, *65*, 4444.

[1422]Wiskur, S.L.; Lorte, A.; Fu, G.C. *J. Am. Chem. Soc.* **2004**, *126*, 82.

[1423]Batey, R.A.; Thadani, A.N.; Smil, D.V.; Lough, A.J. *Synthesis* **2000**, 990; Batey R.A.; Thadani, A.N.; Smil, D.V. *Org. Lett.* **1999**, *1*, 1683; Batey, R.A.; Thadani, A.N.; Smil, D.V. *Tetrahedron Lett.* **1999**, *40*, 4289; Batey, R.A.; MacKay, D.B.; Santhakumar, V. *J. Am. Chem. Soc.* **1999**, *121*, 5075.

corresponding organoboranes and organoboronic acid derivatives. Potassium alkyl-trifluoroborates undergo the palladium-catalyzed coupling reaction with arenediazonium tetrafluoroborates,[1424] diaryliodonium salts,[1425] aryl halides,[1426] as well as with aryl triflates. An example of the latter reaction converted **144** to diphenylmethane via coupling with phenyl triflate.[1427] Alkenyltrifluoroborates can be coupled to aryl halides.[1428]

$$\text{PhCH}_2\text{BF}_3\text{K} + \text{PhOTf} \xrightarrow[\text{9\% PdCl}_2(\text{dppf})\cdot\text{CH}_2\text{Cl}_2]{\text{3 equiv Cs}_2\text{CO}_3, \text{ aq. THF}} \text{PhCH}_2\text{Ph}$$
$$\underset{\textbf{144}}{}$$

OS **VII**, 245; **VIII**, 295; **X**, 391.

10-60   Coupling of Organometallic Reagents With Carboxylic Esters

Alkyl-de-acyloxy-substitution

Several organometallic reagents react with allylic esters and carbonates to give the coupling product. Lithium dialkylcopper reagents couple with allylic acetates to give normal coupling products or those resulting from allylic rearrangement, depending on the substrate.[1429] A mechanism involving a σ-allylic copper(III) complex has been suggested.[1430] Silyl cuprates have also been used, with benzoate esters, to give allyl silanes.[1431] Interestingly, allylic silanes have been coupled to acetates using $B(C_6F_5)_3$[1432] or BF_3.[1433] With propargyl substrates, the products are allenes.[1434]

$$\text{RC}\equiv\text{C}-\text{CR}_2-\text{OAc} + \text{R}_2'\text{CuLi} \longrightarrow \text{RR}'\text{C}=\text{C}=\text{CR}_2$$

[1424]Darses, S.; Michaud, G.; Genêt, J.-P. *Eur. J. Org. Chem.* **1999**, 1875; Darses, S.; Michaud, G.; Genêt, J.-P. *Tetrahedron Lett.* **1998**, *39*, 5045; Darses, S.; Genêt, J.-P.; Brayer, J.-L.; Demoute, J.-P. *Tetrahedron Lett.* **1997**, *38*, 4393.

[1425]Xia, M.; Chen, Z.-C. *Synth. Commun.* **1999**, *29*, 2457.

[1426]Ishikura, M.; Agata, I.; Katagiri, N. *J. Heterocylic Chem.* **1999**, *36*, 873; Molander, G.A.; Biolatto, B. *Org. Lett.* **2002**, *4*, 1867.

[1427]Molander, G.A.; Ito, T. *Org. Lett.* **2001**, *3*, 393.

[1428]Molander, G.A.; Rivero, M.R. *Org. Lett.* **2002**, *4*, 107.

[1429]Rona, P.; Tökes, L.; Tremble, J.; Crabbé, P. *Chem. Commun.* **1969**, 43; Goering, H.L.; Kantner, S.S. *J. Org. Chem.* **1984**, *49*, 422; Purpura, M.; Krause, N. *Eur. J. Org. Chem.* **1999**, 267.

[1430]Goering, H.L.; Kantner, S.S.; Seitz Jr., E.P. *J. Org. Chem.* **1985**, *50*, 5495.

[1431]Fleming, I.; Higgins, D.; Lawrence, N.J.; Thomas, A.P. *J. Chem. Soc. Perkin Trans. 1* **1992**, 3331.

[1432]Rubin, M.; Gevorgyan, V. *Org. Lett.* **2001**, *3*, 2705. For a reaction of a propargyl ester ($-\text{O}_2\text{CCH}_2\text{Cl}$) with an allylic silane and a catalytic amount of $B(C_6F_5)_3$, see Schwier, T.; Rubin, M.; Gevorgyan, V. *Org. Lett.* **2004**, *6*, 1999.

[1433]Smith, D.M.; Tran, M.B.; Woerpel, K.A. *J. Am. Chem. Soc.* **2003**, *125*, 14149; Ayala, L.; Lucero, C.G.; Romero, J.A.C.; Tabacco, S.A.; Woerpel, K.A. *J. Am. Chem. Soc.* **2003**, *125*, 15521.

[1434]Crabbé, P.; Barreiro, E.; Dollat, J.; Luche, J. *J. Chem. Soc., Chem. Commun.* **1976**, 183, and references cited therein.

Allenes are also obtained when propargyl acetates are treated with methylmagnesium iodide.[1435] Lithium dialkylcopper reagents also give normal coupling products with enol acetates of β-dicarbonyl compounds.[1436] It is also possible to carry out the coupling of allylic acetates with Grignard reagents, if catalytic amounts of cuprous salts are present.[1437] With this method yields are better and regioselectivity can be controlled by a choice of cuprous salts.

Allylic, benzylic, and cyclopropylmethyl acetates couple with trialkylaluminums,[1438] and allylic acetates couple with aryl and vinylic tin reagents, in the presence of a palladium catalyst[1439] (see below). Allylic acetates can be symmetrically coupled by treatment with $Ni(CO)_4$ (reaction **10-56**) or with Zn and a palladium-complex catalyst,[1440] or converted to unsymmetrical 1,5-dienes by treatment with an allylic stannane $R_2C=CHCH_2SnR_3$ in the presence of a palladium complex.[1441] Aryl halides can be coupled to allylic acetates with $CoBr_2/Mn/FeBr_2$.[1442] Lactones can be coupled at carbon by an alkylpalladium reagent in the presence of a silane[1443] or by a Grignard reagent with CuBr.[1444]

$$Ph\diagdown\diagup\diagdown OAc \xrightarrow[\substack{BSA,\ KOAc,\ THF,\ 70°C}]{\substack{CH_2(CO_2Et)_2 \\ 5\%\ Pd_2(dba)_3,\ 50\%\ PPh_3}} Ph\diagdown\diagup\diagdown\diagup\overset{\displaystyle CO_2Et}{\underset{\displaystyle CO_2Et}{|}}$$

145   **146**

The most common method now in the literature is the reaction of η^3-π-allyl palladium complexes[1445] (see p. 117) with various nucleophiles,[1446] where the complex is obtained from allylic esters (acetate is the most common) or allylic

[1435]Roumestant, M.; Gore, J. *Bull. Soc. Chim. Fr.* **1972**, 591, 598.

[1436]Casey, C.P.; Marten, D.F. *Synth. Commun.* **1973**, *3*, 321, *Tetrahedron Lett.* **1974**, 925. See also, Posner, G.H.; Brunelle, D.J. *J. Chem. Soc., Chem. Commun.* **1973**, 907; Kobayashi, S.; Takei, H.; Mukaiyama, T. *Chem. Lett.* **1973**, 1097.

[1437]Paisley, S.D.; Schmitter, J.; Lesheski, L.; Goering, H.L. *J. Org. Chem.* **1989**, *54*, 2369; Karlström, A.S.E.; Huerta, F.F.; Muezelaar, G.J.; Bäckvall, J.-E. *Synlett* **2001**, 923; Alexakis, A.; Malan, C.; Lea, L.; Benhaim, C.; Fournioux, X. *Synlett* **2001**, 927.

[1438]Itoh, A.; Oshima, K.; Sasaki, S.; Yamamoto, H.; Hiyama, T.; Nozaki, H. *Tetrahedron Lett.* **1979**, 4751; Gallina, C. *Tetrahedron Lett.* **1985**, *26*, 519; Tolstikov, G.A.; Dzhemilev, U.M. *J. Organomet. Chem.* **1985**, *292*, 133; van Klaveren, M.; Persson, E.S.M.; del Villar, A.; Grove, D.M.; Bäckvall, J.-E.; van Koten, G. *Tetrahedron Lett.* **1995**, *36*, 3059.

[1439]Del Valle, L.; Stille, J.K.; Hegedus, L.S. *J. Org. Chem.* **1990**, *55*, 3019. For another method, see Legros, J.; Fiaud, J. *Tetrahedron Lett.* **1990**, *31*, 7453.

[1440]Sasaoka, S.; Yamamoto, T.; Kinoshita, H.; Inomata, K.; Kotake, H. *Chem. Lett.* **1985**, 315.

[1441]Trost, B.M.; Keinan, E. *Tetrahedron Lett.* **1980**, *21*, 2595.

[1442]Gomes, P.; Gosmini, C.; Périchon, J. *Org. Lett.* **2003**, *5*, 1043.

[1443]Iwata, A.; Ohshita, J.; Tang, H.; Kunai, A.; Yamamoto, Y.; Matui, C. *J. Org. Chem.* **2002**, *67*, 3927.

[1444]Nelson, S.G.; Wan, Z.; Stan, M.A. *J. Org. Chem.* **2002**, *67*, 4680.

[1445]For a review of the use of η^3-allylpalladium complexes to form C–C bonds, see Tsuji, J., in Hartley, F.R.; Patai, S. *The Chemistry of the Metal-Carbon Bond*, Vol. 3, Wiley, NY, **1985**, pp. 163–199.

[1446]For a review related to synthetic applications see Trost, B.M.; Crawley, M.L. *Chem. Rev.* **2003**, *103*, 2921. For a discussion of the mechanism see Tsurugi, K.; Nomura, N.; Aoi, K. *Tetrahedron Lett.* **2002**, *43*, 469.

carbonates.[1447] The mechansim of such π-allyl palladium reactions has been discussed.[1448] A typical transformation is shown for the reaction of **145** with diethyl malonate, BSA and potassium acetate, which gives coupling product **146** in the presence of the palladium catalyst.[1449] This reaction is a variation of the basic transformation reported several years ago by Trost.[1450] Sulfone anions were also used as nucleophiles.[1451] The palladium catalyst used, the reaction conditions, and the nature of the organometallic compounds varies widely. Although two allylic coupling products are possible via the π-allyl intermediate, attack at the less substituted position is generally favored. In most reported cases the R'M species is the anion of an active methylene compound (such as sodium, potassium or lithium dimethylmalonate) or Knoevenagel-type carbanions (**16–38**) or amino acid surrogates.[1452] The use of chiral ligands[1453] or chiral additives that may act as ligands[1454]

[1447]For reviews, see Trost, B.M. *Angew. Chem. Int. Ed.* **1989**, *28*, 1173; *Aldrichimica Acta* **1981**, *14*, 43; *Acc. Chem. Res.* **1980**, *13*, 385; *Tetrahedron* **1977**, *33*, 2615; Tsuji, J.; Minami, I. *Acc. Chem. Res.* **1987**, *20*, 140; Tsuji, J. *Tetrahedron* **1986**, *42*, 4361, *Organic Synthesis with Palladium Compounds*, Springer, Berlin, **1981**, pp. 45, 125; Heck, R.F. *Palladium Reagents in Organic Synthesis*, Academic Press, NY, **1985**, pp. 130–166; Hegedus, L.S., in Buncel, E.; Durst, T. *Comprehensive Carbanion Chemistry*, Vol. 5, pt. B, Elsevier, NY, **1984**, pp. 30–44.

[1448]Trost, B.M.; Strege, P.E.; Weber, L.; Fullerton, T.J.; Dietsche, T.J. *J. Am. Chem. Soc.* **1978**, *100*, 3407; Organ, M.G.; Miller, M.; Konstantinou, Z. *J. Am. Chem. Soc.* **1998**, *120*, 9283; Trost, B.M.; Toste, F.D. *J. Am. Chem. Soc.* **1999**, *121*, 4545

[1449]Poli, G.; Giambastiani, G.; Mordini, A. *J. Org. Chem.* **199**, *64*, 2962.

[1450]Trost, B.M.; Weber, L.; Strege, P.E.; Fullerton, T.J.; Dietsche, T.J. *J. Am. Chem. Soc.* **1978**, *100*, 3416, 3426. These papers include a discussion of the mechanism of this reaction.

[1451]Manchand, P.S.; Wong, H.S.; Blount, J.F. *J. Org. Chem.* **1978**, *43*, 4769.

[1452]Nakoji, M.; Kanayama, T.; Okino, T.; Takemoto, Y. *Org. Lett.* **2001**, *3*, 3329.

[1453]For example see Kodama, H.; Taiji, T.; Ohta, T.; Furkawa, I. *Tetrahedron Asymmetry* **2000**, *11*, 4009; Gong, L.; Chen, G.; Mi, A.; Jiang, Y.; Fu, F.; Cui, X.; Chan, A.S.C. *Tetrahedron Asymmetry* **2000**, *11*, 4297; Mino, T.; Shiotsuki, M.; Yamamoto, N.; Suenaga, T.; Sakamoto, M.; Fujita, T.; Yamashita, M. *J. Org. Chem.* **2001**, *66*, 1795; Zhang, R.; Yu, L.; Xu, L.; Wang, Z.; Ding, K. *Tetrahedron Lett.* **2001**, *42*, 7659; Arena, C.G.; Drommi, D.; Faraone, F. *Tetrahedron Asymmetry* **2000**, *11*, 4753; Kang, J.; Lee, J.H.; Choi, J.S. *Tetrahedron Asymmetry* **2001**, *12*, 33; Mino, T.; Tanaka, Y.; Sakamoto, M.; Fujita, T. *Tetrahedron Asymmetry* **2001**, *12*, 2435; Hu, X.; Dai, H.; Hu, X.; Chen, H.; Wang, J.; Bai, C.; Zheng, Z. *Tetrahedron Asymmetry* **2002**, *13*, 1687; Tollabi, M.; Framery, E.; Goux-Henry, C.; Sinou, D. *Tetrahedron Asymmetry* **2003**, *14*, 3329; Boaz, N.W.; Ponaskik, Jr., J.A.; Large, S.E.; Debenham, S.D. *Tetrahedron Asymmetry* **2004**, *15*, 2151.

[1454]Rabeyrin, C.; Nguefack, C.; Sinou, D. *Tetrahedron Lett.* **2000**, *41*, 7461; Jansat, S.; Gómez, M.; Muller, G.; Diéguez, M.; Aghmiz, A.; Claver, C.; Masdeu-Bultó, A.M.; Flores-Santos, L.; Martin, E.; Maestro, M.A.; Mahía, J. *Tetrahedron Asymmetry* **2001**, *12*, 1469; Fukuda, T.; Takehara, A.; Iwao, M. *Tetrahedron Asymmetry* **2001**, *12*, 2793; Abrunhosa, I.; Gulea, M.; Levillain, J.; Masson, S. *Tetrahedron Asymmetry* **2001**, *12*, 2851; Stranne, R.; Moberg, C. *Eur. J. Org. Chem.* **2001**, 2191; Mancheño, O.G.; Priego, J.; Cabera, S.; Arrayás, R.G.; Llamas, T.; Carretero, J.C. *J. Org. Chem.* **2003**, *68*, 3679; Molander, G.A.; Burke, J.P.; Carroll, P.J. *J. Org. Chem.* **2004**, *69*, 8062; Hou, X.-L.; Sun, N. *Org. Lett.* **2004**, *6*, 4399; Jin, M.-J.; Kim, S.-H.; Lee. S.-J.; Kim, Y.-M. *Tetrahedron Lett.* **2002**, *43*, 7409; Okuyama, Y.; Nakano, H.; Takahashi, K.; Hongo, H.; Kubuto, C. *Chem. Commun.* **2003**, 524; Chen, G.; Li, X.; Zhang, H.; Gong, L.; Mi, A.; Cui, X.; Jiang, Y.; Choi, M.C.K.; Chan, A.S.C. *Tetrahedron Asymmetry* **2002**, *13*, 809; Kloetzing, R.J.; Lotz, M.; Knochel, P. *Tetrahedron Asymmetry* **2003**, *14*, 255; Nakano, H.; Yokayama, J.-i.; Koiyama, Y.; Fujita, R.; Hongo, H. *Tetrahedron Asymmetry* **2003**, *14*, 2361; Mercier, F.; Brebion, F.; Dupont, R.; Mathey, F. *Tetrahedron Asymmetry* **2003**, *14*, 3137.

lead to asymmetric induction in the coupling product.[1455] Enolate anions (see **10–68**) have also been used.[1456] This transformation has been done in ionic liquids[1457] and ionic liquids have been used as additives in catalytic amounts in other solvents.[1458] Palladium nanoparticles have been used to catalyze the reaction.[1459] Other nucleophiles can be used to displace allylic acetates.[1460] S_N2' reactions with allylic acetates have been reported.[1461] Benzoate esters have been used successfully in lieu of the acetate.[1462] Catalyst systems other than palladium have been used for this reaction with allylic acetates.[1463]

            147   **148**

As mentioned above, a common variation is to replace the acetate leaving group with a carbonate ($-OCO_2R$), where methyl carbonate ($-OCO_2Me$) is most common.[1464] A typical reaction is the transformation of **147** to **148**,[1465] where the use of a chiral ligand led to modest asymmetric induction. As with allylic acetates, chiral ligands and chiral additives lad to asymmetric induction.[1466] A variety of active methylene compounds can be used as nucleophiles,[1467] including enolate anions.[1468] Other nucleophiles can be used to

[1455]For a review, see Consiglio, G.; Waymouth, R.M. *Chem. Rev.* **1989**, *89*, 257.

[1456]Braun, M.; Laicher, F.; Meier, T. *Angew. Chem. Int. Ed.* **2000**, *39*, 3494.

[1457]For a reaction in bmim BF_4, 1-butyl-3-methylimidazolium tetrafluoroborate, see Chen, W.; Xu, L.; Chatterton, C.; Xiao, J. *Chem. Commun.* **1999**, 1247.

[1458]Sato, Y.; Yoshino, T.; Mori, M. *Org. Lett.* **2003**, *5*, 31.

[1459]Jansat, S.; Gómez, M.; Philippot, K.; Muller, G.; Guiu, E.; Claver, C.; Castillón, S.; Chaudret, B. *J. Am. Chem. Soc.*, **2004**, *126*, 1592.

[1460]**NaN(CHO)₂:** Wang, Y.; Ding, K. *J. Org. Chem.* **2001**, *66*, 3238. **Indene:** Hayashi, T.; Suzuka, T.; Okada, A.; Kawatsura, M. *Tetrahedron Asymmetry* **2004**, *15*, 545.

[1461]Belelie, J.L.; Chong, J.M. *J. Org. Chem.* **2002**, *67*, 3000.

[1462]Krafft, M.E.; Sugiura, M.; Abboud, K.A. *J. Am. Chem. Soc.* **2001**, *123*, 9174.

[1463]**Pt:** Blacker, A.J.; Clarke, M.L.; Loft, M.S.; Mahon, M.F.; Humphries, M.E.; Williams, J.M.J. *Chem. Eur. J.* **2000**, *6*, 353. **Ir:** Kinoshita, N.; Marx, K.H.; Tanaka, K.; Tsubaki, K.; Kawabata, T.; Yoshikai, N.; Nakamura, E.; Fuji, K. *J. Org. Chem.* **2004**, *69*, 7960; Bartels, B.; García-Yebra, C.; Helmchen, G. *Eur. J. Org. Chem.* **2003**, 1097. **Ru:** Renaud, J.-L.; Bruneau, C.; Demerseman, B. *Synlett* **2003**, 408.

[1464]For a $-OCO_2Ph$ leaving group, see Ito, K.; Kashiwagi, R.; Hayashi, S.; Uchida, T.; Katsuki, T. *Synlett* **2001**, 284.

[1465]Hamada, Y.; Sakaguchi, K.-e.; Hatano, K.; Hara, O. *Tetrahedron Lett.* **2001**, *42*, 1297.

[1466]Kuwano, R.; Kondo, Y.; Matsuyama, Y. *J. Am. Chem. Soc.* **2003**, *125*, 12104; Faller, J.W.; Wilt, J.C. *Tetrahedron Lett.* **2004**, *45*, 7613.

[1467]**Amide esters:** Kazmaier, U.; Zumpe, F.L. *Angew. Chem. Int. Ed.* **1999**, *38*, 1468.

[1468]You, S.-L.; Hou, X.-L.; Dai, L.-X.; Zhu, X.-Z. *Org. Lett.* **2001**, *3*, 149; Evans, P.A.; Leahy, D.K. *J. Am. Chem. Soc.* **2003**, *125*, 7882; Evans, P.A.; Lawler, M.J. *J. Am. Chem. Soc.* **2004**, *126*, 8642. For a reaction of a silyl enol ether, see Muraoka, T.; Matsuda, I.; Itoh, K. *Tetrahedron Lett.* **2000**, *41*, 8807.

displace allylic carbonates,[1469] often in conjunction with chiral ligands to give the product with enantioselectivity. Polymer-supported phosphine ligands have been used successfully.[1470] Catalyst systems other than palladium have been used for this reaction with allylic carbonates.[1471]

Intramolecular cyclization is possible when the active methylene compound and an allylic acetate or carbonate is incorporated into the same molecule.[1472]

Allylic phosphonates have been used as substrates for displacement by higher order cuprates[1473] (see **10-58**) or dialkylzinc reagents.[1474]

10-61 Coupling of Organometallic Reagents With Esters of Sulfates, Sulfoxides, Sulfones, Nitro, and Acetals

Alkyl-de-sulfonyl and de-sulfonyloxy-substitution, and so on; **Alkyl-de-alkoxy-substitution**, and so on; **Alkyl-de-nitration**, and so on.

$$RSO_2X + R'M \longrightarrow R\!-\!R'$$

Leaving groups other than halide, esters or carbonate, or sulfonate esters are sometimes used. Sulfates, sulfonates, and epoxides give the expected products. The reaction of sodium sulfonates and alkyl halides in ionic liquids have been reported.[1475] Acetals can behave as substrates, one OR group being replaced by ZCHZ' in a reaction similar to **10-64**.[1476] Ortho esters behave similarly, but the product loses R'OH to give an enol ether.[1477] The SO$_2$Ph group of allylic sulfones can

[1469]**Aryllithium reagents:** Evans, P.A.; Uraguchi, D. *J. Am. Chem. Soc.* **2003**, *125*, 7158. **Alkoxides:** Evans, P.A.; Leahy, D.K.; Slieker, L.M. *Tetrahedron Asymmetry* **2003**, *14*, 3613. **Phenoxide anions:** Evans, P.A.; Leahy, D.K. *J. Am. Chem. Soc.* **2000**, *122*, 5012; López, F.; Ohmura, T.; Hartwig, J.F. *J. Am. Chem. Soc.* **2003**, *125*, 3426. **Secondary amines:** Matsushima, Y.; Onitsuka, K.; Kondo, T.; Mitsudo, T.-A.; Takahashi, S. *J. Am. Chem. Soc.* **2001**, *123*, 10405. **Primary amines:** Ohmura, T.; Hartwig, J.F. *J. Am. Chem. Soc.* **2002**, *124*, 15164. *N*-**Lithio-sulfonamides**: Evans, P.A.; Robinson, J.E.; Baum, E.W.; Fazal, A.N. *J. Am. Chem. Soc.* **2002**, *124*, 8782. *C*-**Alkylation with an indole**: Bandini, M.; Melloni, A.; Umani-Ronchi, A. *Org. Lett.* **2004**, *6*, 3199. **Michael addition of conjugated esters**: Muraoka, T.; Matsuda, I.; Itoh, K. *J. Am. Chem. Soc.* **2000**, *122*, 9552.

[1470]Uozumi, Y.; Shibatmoi, K. *J. Am. Chem. Soc.* **2001**, *123*, 2919.

[1471]**Ruthenium**: Trost, B.M.; Fraisse, P.L.; Ball, Z.T. *Angew. Chem. Int. Ed.* **2002**, *41*, 1059. **Molybdenum**: Glorius, F.; Pfaltz, A. *Org. Lett.* **1999**, *1*, 141; Malkov, A.V.; Spoor, P.; Vinader, V.; Kočovský, P. *Tetrahedron Lett.* **2001**, *42*, 509. **Iridium**: Alexakis, A.; Polet, D. *Org. Lett.* **2004**, *6*, 3529; Lee, P.H.; Sung, S.-y.; Lee, K.; Chang, S. *Synlett* **2002**, 146.

[1472]Castaño, A.M.; Méndez, M.; Ruano, M.; Echavarren, A.M. *J. Org. Chem.* **2001**, *66*, 589. See also, Zhang, Q.; Lu, X.; Han, X. *J. Org. Chem.* **2001**, *66*, 7676.

[1473]Belelie, J.L.; Chong, J.M. *J. Org. Chem.* **2001**, *66*, 5552.

[1474]Kacprzynski, M.A.; Hoveyda, A.H. *J. Am. Chem. Soc.* **2004**, *126*, 10676.

[1475]In bmim BF$_4$, 1-butyl-3-methylimidazolium tetrafluoroborate: Hu, Y.; Chen, Z.-C.; Le, Z.-G.; Zheng, Q.G. *Synth. Commun.* **2004**, *34*, 4031.

[1476]Yufit, S.S.; Krasnaya, Zh.A.; Levchenko, T.S.; Kucherov, V.F. *Bull. Acad. Sci. USSR Div. Chem. Sci.* **1967**, 123; Aleskerov, M.A.; Yufit, S.S.; Kucherov, V.F. *Bull. Acad. Sci. USSR Div. Chem. Sci.* **1972**, *21*, 2279.

[1477]For a review, see DeWolfe, R.H. *Carboxylic Ortho Acid Derivatives*, Academic Press, NY, **1970**, pp. 231–266.

be a leaving group if a palladium(0) complex is present.[1478] The NR_2 group from Mannich bases, such as $RCOCH_2CH_2NR_2$, can also act as a leaving group in this reaction (elimination–addition mechanism, p. 474). A nitro group can be displaced[1479] from α-nitro esters, ketones, nitriles, and α,α-dinitro compounds,[1480] and even from simple tertiary nitro compounds of the form R_3CNO_2[1481] or $ArR_2C\text{-}NO_2$[1482] by salts of nitroalkanes, for example,

These reactions take place by SET mechanisms.[1483] However, with α-nitro sulfones it is the sulfone group that is displaced, rather than the nitro group.[1484] The SO_2R group of allylic sulfones can be replaced by $CHZZ'$ ($C=CCH_2-SO_2R \rightarrow C=CCH_2-CHZZ'$) if an $Mo(CO)_6$ catalyst is used.[1485]

tert-Butylsulfones react with organolithium reagents, in the presence of a catalytic amount of iron complex, to give coupling.[1486] In this case, the t-$BuSO_2$ unit becomes a "leaving group." A sulfoxide was a "leaving group" in the cyclization of a carboxylic acid contain a sulfoxide unit at C-4. Treatment with phenyliodonium bis(trifluoroacetate) gave the five-membered ring lactone.[1487] Similar displacement of $TolSO_2$ was observed with tolylsulfones and diethylzinc.[1488] Biaryl can be prepared by the reaction of diarylsulfones and arylmagnesium halides, n the presence of a nickel catalyst.[1489]

Phosphonic esters, $ROPO(OR)_2$, react with allylic Grignard reagents to give the coupling product.[1490]

OS **I**, 471; **II**, 47, 360; **VII**, 351; **VIII**, 97, 471.

[1478]Trost, B.M.; Schmuff, N.R.; Miller, M.J. *J. Am. Chem. Soc.* **1980**, *102*, 5979.

[1479]For reviews, see Kornblum, N., in Patai, S. *The Chemistry of Functional Groups, Supplement F*, pt. 1, Wiley, NY, **1982**, pp. 361–393; Kornblum, N. *Angew. Chem. Int. Ed.* **1975**, *14*, 734. For reviews of aliphatic S_N reactions in which NO_2 is a leaving group, see Tamura, R.; Kamimura, A.; Ono, N. *Synthesis* **1991**, 423; Kornblum, N., in Feuer, H.; Nielsen, A.T. *Nitro Compunds: Recent Advances in Synthesis and Chemisry*, VCH, NY, **1990**, pp. 46–85.

[1480]Kornblum, N.; Kelly, W.J.; Kestner, M.M. *J. Org. Chem.* **1985**, *50*, 4720.

[1481]Kornblum, N.; Erickson, A.S. *J. Org. Chem.* **1981**, *46*, 1037.

[1482]Kornblum, N.; Carlson, S.C.; Widmer, J.; Fifolt, M.J.; Newton, B.N.; Smith, R.G. *J. Org. Chem.* **1978**, *43*, 1394.

[1483]For a review of the mechanism, see Beletskaya, I.P.; Drozd, V.N. *Russ. Chem. Rev.* **1979**, *48*, 431. See also, Kornblum, N.; Wade, P.A. *J. Org. Chem.* **1987**, *52*, 5301; Bowman, W.R. *Chem. Soc. Rev.* **1988**, *17*, 283; Ref. 1479.

[1484]Kornblum, N.; Boyd, S.D.; Ono, N. *J. Am. Chem. Soc.* **1974**, *96*, 2580.

[1485]Trost, B.M.; Merlic, C.A. *J. Org. Chem.* **1990**, *55*, 1127.

[1486]Jin, L.; Julia, M.; Verpeaux, J.N. *Synlett* **1994**, 215.

[1487]Casey, M.; Manage, A.C.; Murphy, P.J. *Tetrahedron Lett.* **1992**, *33*, 965.

[1488]Dahmen, S.; Bräse, S. *J. Am. Chem. Soc.* **2002**, *124*, 5940.

[1489]Cho, C.-H.; Yun, H.-S.; Park, K. *J. Org. Chem.* **2003**, *68*, 3017.

[1490]Yanagisawa, A.; Hibino, H.; Nomura, N.; Yamamoto, H. *J. Am. Chem. Soc.* **1993**, *115*, 5879.

10-62   The Bruylants Reaction

Alkyl-de-cyanation

$$\underset{R'}{\overset{NR_2}{\diagdown}}\!\!\!\overset{}{\underset{CN}{|}} \; + \; R^2MgX \; \longrightarrow \; \underset{R'}{\overset{NR_2}{\diagdown}}\!\!\!\overset{}{\underset{R^2}{|}}$$

The *Bruylants reaction* is the reaction of an aminonitrile with a Grignard reagent to give a substituted amine.[1491] This reaction is most often used for the preparation of aliphatic amines via aliphatic Grignard reagents. In a few cases, vinylic Grignard reagents can be used to prepare allylic amines.[1492] The use of AgBF$_4$ to convert amino nitriles to the corresponding iminium ion facilitates the Bruylants reaction with vinylic Grignard reagents.[1493]

Displacement of a cyano group in α-cyanoketones is possible. Treatment of the α-cyanoketone with SmI$_2$ followed by addition of an excess of allyl bromide gave the α-allyl ketone derivative.[1494] α-Cyano amines react with allyl bromide and then zinc metal to give homoallylic amines after treatment with dilute acetic acid in THF.[1495]

10-63   Coupling Involving Alcohols

De-hydroxyl-coupling

$$ROH + R'M \longrightarrow R{-}R'$$

In some cases, it is possible to couple an alcohol with an organometallic compound. Allylic alcohols are coupled with alkylmagnesium bromides in the presence of Ti(OiPr)$_4$, for example.[1496] Allylic alcohols can be coupled with arylboronic acids in ionic liquid solvent and a rhodium catalyst.[1497] The palladium-catalyzed reaction of active methylene compounds with allylic alcohols[1498] or benzylic alcohols[1499] is also known. The coupling of an alcohol to the α-carbon of a ketone

[1491]Bruylants, P. *Bull. Soc. Chem. Belg.* **1924**, *33*, 467.
[1492]Ahlbrecht, H.; Dollinger, H. *Synthesis* **1985**, 743; Trost, B.M.; Spagnol, M.D. *J. Chem. Soc., Perkin Trans. 1* **1995**, 2083.
[1493]Agami, C.; Couty, F.; Evano, G. *Org. Lett.* **2000**, *2*, 2085.
[1494]Zhu, J.-L.; Shia, K.-S.; Liu, H.-J. *Tetrahedron Lett.* **1999**, *40*, 7055.
[1495]Bernardi, L.; Bonini, B.F.; Capitò, E.; Dessole, G.; Fochi, M.; Comes-Franchini, M.; Ricci, A. *Synlett* **2003**, 1778.
[1496]Kulinkovich, O.G.; Epstein, OL.; Isakov, V.E.; Khmel'nitskaya, E.A. *Synlett* **2001**, 49.
[1497]Kabalka, G.W.; Dong, G.; Venkataish, B. *Org. Lett.* **2003**, *5*, 893.
[1498]Manabe, K.; Kobayashi, S. *Org. Lett.* **2003**, *5*, 3241; Kinoshita, H.; Shinokubo, H.; Oshima, K. *Org. Lett.* **2004**, *6*, 4085; Horino, Y.; Naito, M.; Kimura, M. Tanaka, S.; Tamaru, Y. *Tetrahedron Lett.* **2001**, *42*, 3113.
[1499]Bisaro, F.; Prestat, G.; Vitale, M.; Poli, G. *Synlett* **2002**, 1823.

(RCOMe + R'OH) to give a β-substituted alcohol (RCH(OH)CH$_2$R') is possible in the presence of a ruthenium catalyst.[1500]

$$2\,ROH \xrightarrow[-78°C]{MeLi–TiCl_3} RR$$

Allylic or benzylic alcohols can be symmetrically coupled[1501] by treatment with methyllithium and titanium trichloride at −78°C[1502] or by refluxing with TiCl$_3$ and LiAlH$_4$.[1503] When the substrate is an allylic alcohol, the reaction is not regiospecific, but a mixture of normal coupling and allylically rearranged products is found. A free-radical mechanism is involved.[1504] The TiCl$_3$–LiAlH$_4$ reagent can also convert 1,3-diols to cyclopropanes, provided that at least one a phenyl is present.[1505]

Tertiary alcohols react with trimethylaluminum at 80–200°C to give methylation.[1506] The presence of side products from elimination and rearrangement, as well as the lack of stereospecificity,[1507] indicate an

$$R_3COH + Me_3Al \xrightarrow{80-200°C} R_3CMe$$

S$_N$1 mechanism. The reaction can also be applied to primary and secondary alcohols if these contain an aryl group in the a position. Higher trialkylaluminums are far less suitable, because reduction competes with alkylation (see also, reactions of Me$_3$Al with ketones, **16-24**, and with carboxylic acids, **16-82**). The compound Me$_2$TiCl$_2$ also reacts with tertiary alcohols in the same way.[1508] Allylic alcohols couple with a reagent prepared from MeLi, CuI, and R'Li in the presence of (Ph$_3$PNMePh)$^+$ I$^-$ to give alkenes, such as **149**, that are products of allylic rearrangement.[1509]

149

[1500]Cho, C.S.; Kim, B.T.; Kim. T.-J.; Shim. S.C. *J. Org. Chem.* **2001**, *66*, 9020.

[1501]For a review, see Lai, Y. *Org. Prep. Proceed. Int.* **1980**, *12*, 363, pp. 377–388.

[1502]Sharpless, K.B.; Hanzlik, R.P.; van Tamelen, E.E. *J. Am. Chem. Soc.* **1968**, *90*, 209.

[1503]McMurry, J.E.; Silvestri, M.G.; Fleming, M.P.; Hoz, T.; Grayston, M.W. *J. Org. Chem.* **1978**, *43*, 3249. For another method, see Nakanishi, S.; Shundo, T.; Nishibuchi, T.; Otsuji, Y. *Chem. Lett.* **1979**, 955.

[1504]van Tamelen, E.E.; Åkermark, B.; Sharpless, K.B. *J. Am. Chem. Soc.* **1969**, *91*, 1552.

[1505]Baumstark, A.L.; McCloskey, C.J.; Tolson, T.J.; Syriopoulos, G.T. *Tetrahedron Lett.* **1977**, 3003; Walborsky, H.M.; Murati, M.P. *J. Am. Chem. Soc.* **1980**, *102*, 426.

[1506]Meisters, A.; Mole, T. *J. Chem. Soc., Chem. Commun.* **1972**, 595; Harney, D.W.; Meisters, A.; Mole, T. *Aust. J. Chem.* **1974**, *27*, 1639.

[1507]Salomon, R.G.; Kochi, J.K. *J. Org. Chem.* **1973**, *38*, 3715.

[1508]Reetz, M.T.; Westermann, J.; Steinbach, R. *J. Chem. Soc., Chem. Commun.* **1981**, 237.

[1509]Tanigawa, Y.; Ohta, H.; Sonoda, A.; Murahashi, S. *J. Am. Chem. Soc.* **1978**, *100*, 4610; Goering, H.L.; Tseng, C.C. *J. Org. Chem.* **1985**, *50*, 1597. For another procedure, see Yamamoto, Y.; Maruyama, K. *J. Organomet. Chem.* **1978**, *156*, C9.

The reaction gives good yields with primary, secondary, and tertiary alcohols, and with alkyl and aryllithium reagents.[1510] Allylic alcohols also couple with certain Grignard reagents[1511] in the presence of a nickel complex to give both normal products and the products of allylic rearrangement.

Allenic alcohols couple with allyl indium reagents at 140°C to give allylic alcohol products.[1512] Similarly, ω-hydroxy lactones couple with organoindium reagents.[1513] Phenols react with vinyl boronates and a copper catalyst to give aryl vinyl ethers.[1514]

Alcohols react with allylsilanes, in the presence of an $InCl_3$[1515] or $InBr_3$[1516] catalyst to give the corresponding coupling product ($R_2CHOH \rightarrow R_2CH–CH_2CH=CH_2$).

10-64   Coupling of Organometallic Reagents With Compounds Containing the Ether Linkage[1517]

Alkyl-de-alkoxy-substitution

$$R_2C(OR')_2 + R''MgX \longrightarrow R_2CR''(OR') + R'OMgX$$

$$RC(OR')_3 + R''MgX \longrightarrow RCR''(OR')_2 + R'OMgX$$

Acetals,[1518] ketals, and ortho esters[1519] react with Grignard reagents to give, respectively, ethers and acetals (or ketals). The latter can be hydrolyzed to aldehydes or ketones (**10-6**). This procedure is a way of converting a halide $R''X$ (which may be alkyl, aryl, vinylic, or alkynyl) to an aldehyde $R''CHO$, increasing the length of the carbon chain by one carbon (see also, **10-76**). The ketone synthesis generally gives lower yields. Acetals, including allylic acetals, also give this reaction with organocopper compounds and BF_3.[1520] Dihydropyrans react with Grignard reagents in the

[1510]For the allylation of benzylic alcohols, see Cella, J.A. *J. Org. Chem.* **1982**, *47*, 2125.

[1511]Buckwalter, B.L.; Burfitt, I.R.; Felkin, H.; Joly-Goudket, M.; Naemura, K.; Salomon, M.F.; Wenkert, E.; Wovkulich, P.M. *J. Am. Chem. Soc.* **1978**, *100*, 6445; Felkin, H.; Joly-Goudket, M.; Davies, S.G. *Tetrahedron Lett.* **1981**, *22*, 1157; Consiglio, G.; Morandini, F.; Piccolo, O. *J. Am. Chem. Soc.* **1981**, *103*, 1846, and references cited therein. For a review, see Felkin, H.; Swierczewski, G. *Tetrahedron* **1975**, *31*, 2735. For other procedures, see Mukaiyama, T.; Imaoka, M.; Izawa, T. *Chem. Lett.* **1977**, 1257; Fujisawa, T.; Iida, S.; Yukizaki, H.; Sato, T. *Tetrahedron Lett.* **1983**, *24*, 5745.

[1512]Araki, S.; Usui, H.; Kato, M.; Butsugan, Y. *J. Am. Chem. Soc,* **1996**, *118*, 4699.

[1513]Bernardelli, P.; Paquette, L.A. *J. Org. Chem.* **1997**, *62*, 8284.

[1514]McKinley, N.F.; O'Shea, D.F. *J. Org. Chem.* **2004**, *69*, 5087.

[1515]Yasuda, M.; Saito, T.; Ueba, M.; Baba, A. *Angew. Chem. Int. Ed.* **2004**, *43*, 1414.

[1516]Kim, S.H.; Shin, C.; Pae, A.N.; Koh, H.Y.; Chang, M.H.; Chung, B.Y.; Cho, Y.S. *Synthesis* **2004**, 1581.

[1517]For a review, see Trofimov, B.A.; Korostova, S.E. *Russ. Chem. Rev.* **1975**, *44*, 41.

[1518]For a review of coupling reactions of acetals, see Mukaiyama, T.; Murakami, M. *Synthesis* **1987**, 1043. For a discussion of the mechanism, see Abell, A.D.; Massy-Westropp, R.A. *Aust. J. Chem.* **1985**, *38*, 1031. For a list of substrates and reagents, with references, see Larock, R.C. *Comprehensive Organic Transformations*, 2nd ed., Wiley-VCH, NY, **1999**, pp. 934–942.

[1519]For a review of the reaction with ortho esters, see DeWolfe, R.H. *Carboxylic Ortho Acid Derivatives*, Academic Press, NY, **1970**, pp. 44–45, 224–230.

[1520]Normant, J.F.; Alexakis, A.; Ghribi, A.; Mangeney, P. *Tetrahedron* **1989**, *45*, 507; Alexakis, A.; Mangeney, P.; Ghribi, A.; Marek, I.; Sedrani, R.; Guir, C.; Normant, J.F. *Pure Appl. Chem.* **1988**, *60*, 49.

presence of a nickel catalyst.[1521] Acetals also undergo substitution when treated with silyl enol ethers or allylic silanes, with a Lewis acid catalyst,[1522] for example,

$$RH_2C \underset{OR^2}{\overset{OR^2}{<}} + H_2C = C \underset{R^1}{\overset{OSiMe_3}{<}} \xrightarrow{TiCl_4} RH_2C \underset{OR^2 \quad O}{\overset{}{\diagdown\diagup\diagdown\diagup}} R^1$$

ω-Ethoxy lactams react with Grignard regents to give ω-substituted lactams.[1523] Tertiary amines can be prepared by the reaction of amino ethers with Grignard reagents,[1524] ($R_2NCH_2-OR' + R^2MgX \rightarrow R_2NCH_2-R^2$) or with lithium dialkyl-copper reagents.[1525]

Ordinary ethers are not cleaved by Grignard reagents (in fact, diethyl ether and THF are the most common solvents for Grignard reagents), although more active organometallic compounds often do cleave them.[1526] Oxetanes have been opened with organolithium reagents and $BF_3 \cdot OEt_2$[1527] and also with excess lithium metal with a biphenyl catalyst.[1528] Allylic ethers can be cleaved by Grignard reagents in THF if CuBr is present.[1529] The reaction takes place either with or without allylic rearrangement.[1530] Propargylic ethers give allenes.[1531] Vinylic ethers can also be cleaved by Grignard reagents in the presence of a catalyst, in this case, a nickel complex.[1532] Silyl enol ethers $R_2C=CROSiMe_3$ behave similarly.[1533] Bicyclic benzofurans can be opened by dialkylzinc reagents in the presence of a palladium catalyst.[1534]

[1521]Ducoux, J.-P.; LeMénez, P.; Kunesch, N.; Wenkert, E. *J. Org. Chem.* **1993**, *58*, 1290.

[1522]See Mori, I.; Ishihara, K.; Flippen, L.A.; Nozaki, K.; Yamamoto, H.; Bartlett, P.A.; Heathcock, C.H. *J. Org. Chem.* **1990**, *55*, 6107, and references cited therein.

[1523]Wei, Z.Y.; Knaus, E.E. *Org. Prep. Proceed. Int.* **1993**, *25*, 255.

[1524]For example, see Miginiac, L.; Mauzé, B. *Bull. Soc. Chim. Fr.* **1968**, 2544; Eisele, G.; Simchen, G. *Synthesis* **1978**, 757; Kapnang, H.; Charles, G. *Tetrahedron Lett.* **1983**, *24*, 1597; Morimoto, T.; Takahashi, T.; Sekiya, M. *J. Chem. Soc., Chem. Commun.* **1984**, 794; Mesnard, D.; Miginiac, L. *J. Organomet. Chem.* **1989**, *373*, 1. See also, Bourhis, M.; Bosc, J.; Golse, R. *J. Organomet. Chem.* **1983**, *256*, 193.

[1525]Germon, C.; Alexakis, A.; Normant, J.F. *Bull. Soc. Chim. Fr.* **1984**, II-377.

[1526]For a review of the reactions of ethers with Grignard Reagents, see Kharasch, M.S.; Reinmuth, O. *Grignard Reactions of Nonmetallic Substances*, Prentice-Hall, Englewood Cliffs, NJ, **1954**, pp. 1013–1045.

[1527]Bach, T.; Eilers, F. *Eur. J. Org. Chem.* **1998**, 2161.

[1528]Rama, K.; Pasha, M.A. *Tetrahedron Lett.* **2000**, *41*, 1073.

[1529]Commercon, A.; Bourgain, M.; Delaumeny, M.; Normant, J.F.; Villieras, J. *Tetrahedron Lett.* **1975**, 3837; Claesson, A.; Olsson, L. *J. Chem. Soc., Chem. Commun.* **1987**, 621.

[1530]Normant, J.F.; Commercon, A.; Gendreau, Y.; Bourgain, M.; Villieras, J. *Bull. Soc. Chim. Fr.* **1979**, II-309; Gendreau, Y.; Normant, J.F. *Tetrahedron* **1979**, *35*, 1517; Calo, V.; Lopez, L.; Pesce, G. *J. Chem. Soc. Perkin Trans. 1* **1988**, 1301. See also, Valverde, S.; Bernabé, M.; Garcia-Ochoa, S.; Gómez, A.M. *J. Org. Chem.* **1990**, *55*, 2294.

[1531]Alexakis, A.; Marek, I.; Mangeney, P.; Normant, J.F. *Tetrahedron Lett.* **1989**, *30*, 2387; *J. Am. Chem. Soc.* **1990**, *112*, 8042.

[1532]Wenkert, E.; Michelotti, E.L.; Swindell, C.S.; Tingoli, M. *J. Org. Chem.* **1984**, *49*, 4894; Kocieński, P.; Dixon, N.J.; Wadman, S. *Tetrahedron Lett.* **1988**, *29*, 2353.

[1533]Hayashi, T.; Katsuro, Y.; Kumada, M. *Tetrahedron Lett.* **1980**, *21*, 3915.

[1534]Lauens, M.; Renaud, J.-L.; Hiebert, S. *J. Am. Chem. Soc.* **200**, *122*, 1804.

Certain acetals and ketals can be dimerized in a reaction similar to **10-56** by treatment with TiCl$_4$–LiAlH$_4$, for example,[1535]

Ph— with OEt, OEt $\xrightarrow[\text{LiAlH}_4]{\text{TiCl}_4}$ Ph—Ph with OEt, OEt   85%

Also see, **10-65**.
OS **II**, 323; **III**, 701. Also see, OS **V**, 431.

10-65   The Reaction of Organometallic Reagents With Epoxides

3(*OC*)-*seco*-Alkyl-de-alkoxy-substitution

$$-\overset{O}{\underset{\diagdown}{C}}-\overset{}{\underset{}{C}}- \ + \ R-M \ \longrightarrow \ R-\overset{}{\underset{}{C}}-\overset{}{\underset{}{C}}-OM$$

The reaction between Grignard reagents or organolithium reagents and epoxides is very valuable and is often used to increase the length of a carbon chain by two carbons.[1536] The Grignard reagent may be aromatic or aliphatic, although tertiary Grignard reagents give low yields. As expected for an S$_N$2 process, attack is at the less substituted carbon. With allylic Grignard reagents, the addition of a catalytic amount of Yb(OTf)$_3$ facilitated alkylation.[1537] Organolithium reagents,[1538] in the presence of chiral additives lead to the 2-substituted alcohol with good enantioselectivity. Similar reaction with a chiral Schiff base gave the same type of product, with excellent enantioselectivity.[1539]

Lithium dialkylcopper reagents also give the reaction,[1540] as do higher order cuprates,[1541] often producing higher yields. They have the additional advantage that they do not react with ester, ketone, or carboxyl groups so that the epoxide ring of epoxy esters, ketones, and carboxylic acids can be selectively attacked, often

[1535]Ishikawa, H.; Mukaiyama, T. *Bull. Chem. Soc. Jpn.* *1978*, *51*, 2059.

[1536]For a review, see Kharasch, M.S.; Reinmuth, O. *Grignard Reactions of Nonmetallic Substances*, Prentice-Hall, Englewood Cliffs, NJ, *1954*, pp. 961–1012. For a thorough discussion, see Schaap, A.; Arens, J.F. *Recl. Trav. Chim. Pays-Bas 1968*, *87*, 1249. For improved procedures, see Huynh, C.; Derguini-Boumechal, F.; Linstrumelle, G. *Tetrahedron Lett.* *1979*, 1503; Schrumpf, G.; Grätz, W.; Meinecke, A.; Fellenberger, K. *J. Chem. Res. (S) 1982*, 162.

[1537]Likhar, P.R.; Kumar, M.P.; Bandyopadhyay, A.K. *Tetrahedron Lett.* *2002*, *43*, 3333.

[1538]Harder, S.; van Lenthe, J.H.; van Eikkema Hommes, N.J.R.; Schleyer, P.v.R. *J. Am. Chem. Soc.* *1994*, *116*, 2508; Alexakis, A.; Vrancken, E.; Mangeney, P. *Synlett*, *1998*, 1165.; Hodgson, D.M.; Stent, M.A.H.; Štefane, B.; Wilson, F.X. *Org. Biomol. Chem.* *2003*, *1*, 1139; Hodgson, D.M.; Maxwell, C.R.; Miles, T.J.; Paruch, E.; Stent, M.A.H.; Matthews, I.R. Wilson, F.X.; Witherington, J. *Angew. Chem. Int. Ed. 2002*, *41*, 4313.

[1539]Oguni, N.; Miyagi, Y.; Itoh, K. *Tetrahedron Lett.* *1998*, *39*, 9023.

[1540]For examples of the use of this reactions, see Posner, G.H. *An Introduction to Synthesis Using Organocopper Reagents*, Wiley, NY, *1980*, pp. 103–113. See also, Lipshutz, B.H.; Kozlowski, J.; Wilhelm, R.S. *J. Am. Chem. Soc. 1982*, *104*, 2305; Blanchot-Courtois, V.; Hanna, I. *Tetrahedron Lett. 1992*, *33*, 8087.

[1541]Chauret, D.C.; Chong, J.M. *Tetrahedron Lett. 1993*, *34*, 3695.

in a regioselective manner.[1542] The use of BF_3 increases the reactivity of R_2CuLi, enabling it to be used with thermally unstable epoxides.[1543] Lithium diaminocyano cuprates have also been used.[1544]

The reaction has also been performed with other organometallic compounds.[1545] Trialkylaluminum reagents open epoxides with delivery of the alkyl group to carbon.[1546] In the presence of a Lewis acid catalyst, such as BF_3, alkylation can occur at the more substituted carbon.[1547] Friedel–Crafts type alkylation (see **11-11**) is possible when an aromatic compounds reacts with an epoxide and $AlCl_3$.[1548] Epoxides reaction with allyl bromide in the presence of indium metal, with the expected delivery of allyl to the less substituted carbon being the major product.[1549] Other organometallic reagents can be used.[1550] When a substituted epoxide was treated with CO, $BF_3 \cdot OEt_2$ and a cobalt catalyst, carbonylation occurred and the final product was a β-lactone.[1551] Similar β-lactone forming reactions were reported using substituted epoxides, CO and a metal compound–BF_3 complex.[1552] Five-membered ring lactams were also formed from substituted epoxides using $BF_3 \cdot OEt_2$ followed by treatment with KHF_2.[1553] An interesting variation reacted an epoxy acetate (acetoxy at the 3-position relative to the first epoxy carbon) with Cp_2TiCl_2/Zn, and the product was an allylic alcohol where the epoxide ring was opened with loss of the acetoxy group.[1554]

[1542]Johnson, C.R.; Herr, R.W.; Wieland, D.M. *J. Org. Chem.* **1973**, *38*, 4263; Hartman, B.C.; Livinghouse, T.; Rickborn, B. *J. Org. Chem.* **1973**, *38*, 4346; Hudrlik, P.F.; Peterson, D.; Rona, R.J. *J. Org. Chem.* **1975**, *40*, 2263; Chong, J.M.; Sharpless, K.B. *Tetrahedron Lett.* **1985**, *26*, 4683; Chong, J.M.; Cyr, D.R.; Mar, E.K. *Tetrahedron Lett.* **1987**, *28*, 5009; Larchevêque, M.; Petit, Y. *Tetrahedron Lett.* **1987**, *28*, 1993.

[1543]See, for example, Alexakis, A.; Jachiet, D.; Normant, J.F. *Tetrahedron* **1986**, *42*, 5607.

[1544]Yamamoto, Y.; Asao, N.; Meguro, M.; Tsukada, N.; Nemoto, H.; Sadayori, N.; Wilson, J.G.; Nakamura, H. *J. Chem. Soc., Chem. Commun.* **1993**, 1201.

[1545]For lists of organometallic reagents that react with epoxides, see Wardell, J.L.; Paterson, E.S. in Hartley, F.R.; Patai, S. *The Chemistry of the Metal-Carbon Bond*, Vol. 2; Wiley, NY, **1985**, pp. 307–310; Larock, R.C. *Comprehensive Organic Transformations*, 2nd ed., Wiley-VCH, NY, **1999**, pp. 1045–1063.

[1546]Schneider, C.; Brauner, J. *Eur. J. Org. Chem.* **2001**, 4445; Sasaki, M.; Tanino, K.; Miyashita, M. *J. Org. Chem.* **2001**, *66*, 5388; Sasaki, M.; Tanino, K.; Miyashita, M. *Org. Lett.* **2001**, *3*, 1765; Shanmugam, P.; Miyashita, M. *Org. Lett.* **2003**, *5*, 3265 (formation of *O*-silyl ether product). For the reaction in an ionic liquid see Zhou, H.; Campbell, E.J.; Nguyen, S.T. *Org. Lett.* **2001**, *3*, 2229.

[1547]For an example, see Zhao, H.; Pagenkopf, B.L. *Chem. Commun.* **2003**, 2592.

[1548]Lin, J.; Kanazaki, S.; Kashino, S.; Tsuboi, S. *Synlett* **2002**, 899.

[1549]Yadav, J.S.; Anjaneyulu, S.; Ahmed, Md.M.; Subba Reddy, B.V. *Tetrahedron Lett.* **2001**, *42*, 2557; Oh, B.K.; Cha, J.H.; Cho, Y.S.; Choi, K.I.; Koh, H.Y.; Chang, M.H.; Pae, A.N. *Tetrahedron Lett.* **2003**, *44*, 2911; Hirashita, T.; Mitsui, K.; Hayashi, Y.; Araki, S. *Tetrahedron Lett.* **2004**, *45*, 9189. For a reaction using palladium nanoparticles see Jiang, N.; Hu, Q.; Reid, C.S.; Ou, Y.; Li, C.J. *Chem. Commun.* **2003**, 2318.

[1550]**Ba**: Yasue, K.; Yanagisawa, A.; Yamamoto, H. *Bull. Chem. Soc. Jpn.* **1997**, *70*, 493. **Mn**: Tang, J.; Yorimitsu, H.; Kakiya, H.; Inoue, R.; Shinokubo, H.; Oshima, K. *Tetrahedron Lett.* **1997**, *38*, 9019. **Sn**: Yadav, J.S.; Reddy, B.V.S.; Satheesh, G. *Tetrahedron Lett.* **2003**, *44*, 6501. **Zn**: Equey, O.; Vrancken, E.; Alexakis, A. *Eur. J. Org. Chem.* **2004**, 2151.

[1551]Lee, J.T.; Thomas, P.J.; Apler, H. *J. Org. Chem.* **2001**, *66*, 5424.

[1552]Getzler, Y.D.Y.L.; Mahadevan V.; Lobkovsky, E.B.; Coates, G.W. *J. Am. Chem. Soc.* **2002**, *124*, 1174; Schmidt, J.A.R.; Mahadevan, V.; Getzler, Y.D.Y.L.; Coates, G.W. *Org. Lett.* **2004**, *6*, 373.

[1553]Movassaghi, M.; Jacobsen, E.N. *J. Am. Chem. Soc.* **2002**, *124*, 2456.

[1554]Bermejo, F.; Sandoval, C. *J. Org. Chem.* **2004**, *69*, 5275.

In the presence of a scandium catalyst, chiral allylic boranes open epoxides at the less substituted position to generate chiral, homoallylic alcohols.[1555]

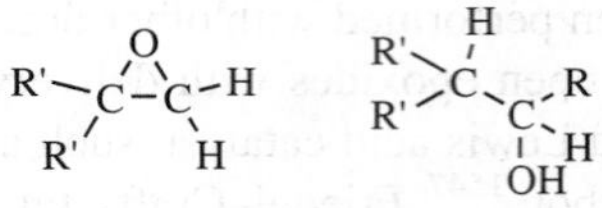

150   **151**

When *gem*-disubstituted epoxides (**150**) are treated with Grignard reagents (and sometimes other epoxides), the product may be **151**, that is, the new alkyl group may appear on the same carbon as the OH. In such cases, the epoxide is isomerized to an aldehyde or a ketone before reacting with the Grignard reagent. Halohydrins are often side products.

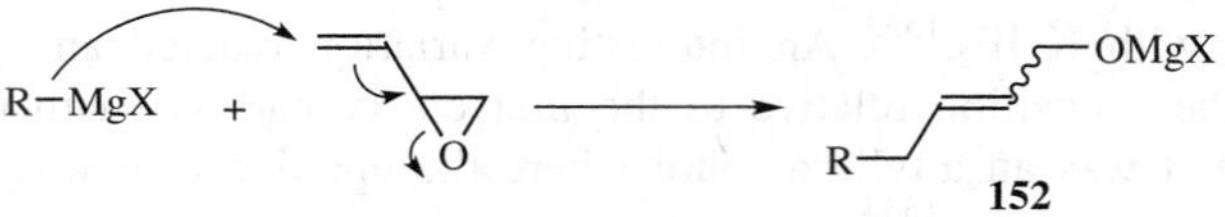

152

When the substrate is a vinylic epoxide,[1556] Grignard reagents generally give a mixture of the normal product and the product of allylic rearrangement (**152**).[1557] Butyllithium reacted with a difluoroalkylidene epoxide ($F_2C=CR$—epoxide) and S_N2' displacement gave alkylation at the difluoro carbon and opened the epoxide.[1558] The latter often predominates. In the case of R_2CuLi,[1559] acyclic substrates give mostly allylic rearrangement (S_N2').[1556] The double bond of the "vinylic" epoxide can be part of an enolate anion. In this case, R_2CuLi give exclusive allylic rearrangement (S_N2') to **153** after hydrolysis, while Grignard and organolithium reagents opened the epoxide directly (S_N2) to give **154** after hydrolysis.[1560]

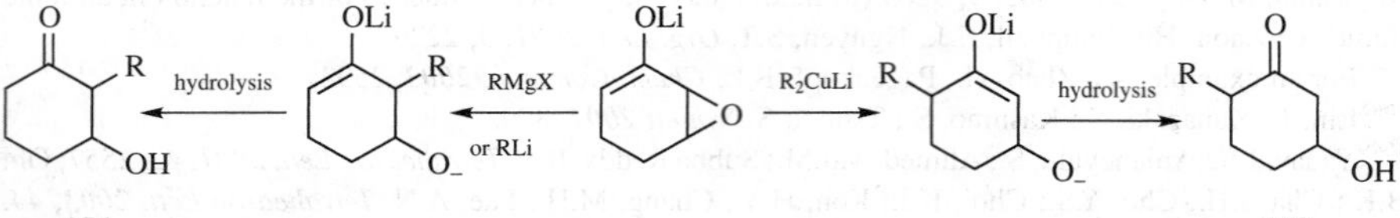

154   **153**

[1555]Lautens, M.; Maddess, M.L.; Sauer, E.L.O.; Oullet, S.G. *Org. Lett.* **2002**, *4*, 83.

[1556]For a list of organometallic reagents that react with vinylic epoxides, with references, see Larock, R.C. *Comprehensive Organic Transformations*, 2nd ed., Wiley-VCH, NY, **1999**, pp. 244–250.

[1557]Anderson, R.J. *J. Am. Chem. Soc.* **1970**, *92*, 4978; Johnson, C.R.; Herr, R.W.; Wieland, D.M. *J. Org. Chem.* **1973**, *38*, 4263; Marshall, J.A.; Trometer, J.D.; Cleary, D.G. *Tetrahedron* **1989**, *45*, 391.

[1558]Ueki, H.; Chiba, T.; Yamazaki, T.; Kitazume, T. *J. Org. Chem.* **2004**, *69*, 7616.

[1559]For a review of the reactions of vinylic epoxides with organocopper reagents, see Marshall, J.A. *Chem. Rev.* **1989**, *89*, 1503.

[1560]Wender, P.A.; Erhardt, J.M.; Letendre, L.J. *J. Am. Chem. Soc.* **1981**, *103*, 2114.

An organometallic equivalent that opens epoxides is a hydrosilane, for example, Me$_3$SiH, and carbon monoxide, catalyzed by dicobalt octacarbonyl:[1561] See **10-55** for other coupling reactions with organosilanes. Silyl enol ethers react with epoxides in a related reaction, but a Lewis acid, such as TiCl$_4$, is required.[1562]

OS **I**, 306; **VII**, 501; **VIII**, 33, 516; **X**, 297.

10-66   Reaction of Organometallics With Aziridines

Aziridines have been opened by organometallic reagents to give amines.[1563] Although less reactive than epoxides, it is also possible to open aziridines[1564] with organometallic reagents particularly when there is a N-sulfonyl group such as tosyl (formally making it a sulfonamide). Grignard reagents react with N-tosyl 2-phenylaziridine to give the corresponding N-tosylamine.[1565] Organocuprates (**10-58**) reaction with N-alkylaziridines to give the corresponding amine.[1566] N-Tosyl aziridines have also been opened with enolate anions, which led to a pyrroline derivative,[1567] and with Me$_2$S=CHCO$_2$Et (see **16-46**) to generate a N-tosyl azetidine.[1568] In a Friedel–Crafts type reaction (**11-11**), aziridines react with benzene, in the presence of In(OTf)$_3$, to give the β-aryl amine.[1569] Allylic alcohols open N-tosylaziridines with KSF–Montmorillonite clay.[1570]

Aziridines react with nucleophiles other than carbon nucleophiles. In the presence of TBAF, trimethylsilyl azide react with N-tosylaziridines to give the azido N-tosylamine.[1571] N-Benzylic aziridines are opened by trimethylsilyl azide in the presence of a chromium catalyst.[1572] Acetic anhydride reacts with N-tosylaziridines, in the presence of PBu$_3$, to give the N-tosylamino acetate.[1573] N-Tosylaziridines react with InCl$_3$ to give the chloro N-tosylamine.[1574]

[1561]Murai, T.; Kato, S.; Murai, T.; Toki, T.; Suzuki, S.; Sonoda, N. *J. Am. Chem. Soc.* **1984**, *106*, 6093.

[1562]Lalić, G.; Petrovski, Ž.; Galonić, D.; Matović, R.; Saičić, R.N. *Tetrahedron* **2001**, *57*, 583.

[1563]See, for example Eis, M.J.; Ganem, B. *Tetrahedron Lett.* **1985**, *26*, 1153; Onistschenko, A.; Buchholz, B.; Stamm, H. *Tetrahedron* **1987**, *43*, 565.

[1564]Crotti, P.; Favero, L.; Gardelli, C.; Macchia, F.; Pineschi, M. *J. Org. Chem.* **1995**, *60*, 2514.

[1565]Toshimitsu, A.; Abe, H.; Hirosawa, C.; Tamao, K. *J. Chem. Soc. Perkin Trans. 1* **1994**, 3465; Müller, P.; Nury, P. *Org. Lett.* **1999**, *1*, 439; Müller, P.; Nury, P. *Helv. Chim. Acta* **2001**, *84*, 662.

[1566]Penkett, C.S.; Simpson, I.D. *Tetrahedron Lett.* **2001**, *42*, 1179.

[1567]Lygo, B. *Synlett*, **1993**, 764.

[1568]Nadir, U.K.; Arora, A. *J. Chem. Soc. Perkin Trans. 1* **1995**, 2605.

[1569]Saidi, M.R.; Azizi, N.; Naimi-Jamal, M.R. *Tetrahedron Lett.* **2001**, *42*, 8111.

[1570]Yadav, J.S.; Reddy, B.V.S.; Balanarsaiah, E.; Raghavendra, S. *Tetrahedron Lett.* **2002**, *43*, 5105.

[1571]Wu, J.; Hou, X.-L.; Dai, L.-X. *J. Org. Chem.* **2000**, *65*, 1344.

[1572]Li, Z.; Fernández, M.; Jacobsen, E.N. *Org. Lett.* **1999**, *1*, 1611.

[1573]Fan, R.-H.; Hou, X.-L. *Tetrahedron Lett.* **2003**, *44*, 4411.

[1574]Yadav, J.S.; Subba Reddy, B.V.; Kumar, G.M. *Synlett* **2001**, 1417.

10-67 Alkylation at a Carbon Bearing an Active Hydrogen

Bis(ethoxycarbonyl)methyl-de-halogenation, and so on.

$$R{-}X \; + \; H{-}\underset{Z}{\overset{Z'}{\underset{|}{\overset{|}{C}}}}{\ominus} \quad \longrightarrow \quad R{-}\underset{Z}{\overset{Z'}{\underset{|}{\overset{|}{C}}}}{-}H$$

The metal-catalyzed displacement of allylic acetates and carbonates (**10-60**) clearly falls in to this category. However, this section will focus on the more general reaction of active methylene compounds with substrates bearing a leaving group, not necessarily allylic substrates or metal catalyzed. When compounds contain two or three strong electron-withdrawing groups on a carbon atom bearing a proton (the so-called α-proton), that proton is more acidic than compounds without such groups (p. 252). Treatment with a suitable base (a base that has a conjugate acid with a pK_a greater than the α-proton) removes the α-proton and generates the corresponding enolate anion (**10-68**). These enolate anions react as carbon nucleophiles and attack alkyl halides, resulting in their alkylation.[1575] Both Z and Z′ may be COOR′, CHO, COR′,[1576] CONR$'_2$, COO$^-$, CN,[1577] NO$_2$, SOR′, SO$_2$R′,[1578] SO$_2$OR′, SO$_2$NR$'_2$ or similar groups.[1579] Some commonly used bases are sodium ethoxide and potassium *tert*-butoxide, each in its respective alcohol as solvent. With particularly acidic compounds (e.g., β-diketones–Z, Z′ = COR′), sodium hydroxide in water or aqueous alcohol or acetone, or even sodium carbonate,[1580] is a strong enough base for the reaction. If at least one Z group is COOR′, saponification is a possible side reaction. In addition to the groups listed above, Z may also be phenyl, but if two phenyl groups are on the same carbon, the acidity is less than in the other cases and a stronger base must be used. However, the reaction can be successfully carried out with diphenylmethane with NaNH$_2$ as the base.[1581] If the solvent used in the reaction is acidic enough to protonate either the enolate anion or the base, an equilibrium will be established leading to only small amounts of the enolate anion (thermodynamic conditions). Such aprotic solvents include

[1575]For dicussions of reactions **10-67** and **10-68**, see House, H.O. *Modern Synthetic Reactions*, 2nd ed., W. A. Benjamin, NY, *1972*, pp. 492–570, 586–595; Carruthers, W. *Some Modern Methods of Organic Synthesis* 3rd ed., Cambridge University Press, Cambridge, *1986*, pp. 1–26.

[1576]For a reaction using *n*-Bu$_4$NF as the base in aq. THF, see Christoffers, J. *Synth. Commun. 1999*, 29, 117.

[1577]For reviews of the reactions of malononitrile CH$_2$(CN)$_2$, see Fatiadi, A.J. *Synthesis 1978*, 165, 241; Freeman, F. *Chem. Rev. 1969*, 69, 591.

[1578]For a review of compounds with two SO$_2$R groups on the same carbon (*gem*-disulfones), see Neplyuev, V.M.; Bazarova, I.M.; Lozinskii, M.O. *Russ. Chem. Rev. 1986*, 55, 883.

[1579]For lists of examples, with references, see Larock, R.C. *Comprehensive Organic Transformations*, 2nd ed., Wiley-VCH, NY, *1999*, pp. 1522–1527 *ff*, 1765–1769.

[1580]See, for example, Fedoryński, M.; Wojciechowski, K.; Matacz, Z.; Makosza, M. *J. Org. Chem. 1978*, 43, 4682.

[1581]Murphy, W.S.; Hamrick, Jr., P.J.; Hauser, C.R. *Org. Synth. V.* 523.

water, alcohols, or amines. In general, solvents that do not contain an acidic proton (aprotic solvents) are used, but protic solvents can be used in some cases. The use of polar aprotic solvents (e.g., DMF or DMSO) markedly increases the rate of alkylation,[1582] but also increases the extent of alkylation at the oxygen rather than the carbon with highly reactive species such as iodomethane (p. 513). In general, enolate anions such as those described here react with alkyl halides via *C*-alkylation, although trialkylsilyl halides and anhydrides tend to react via *O*-alkylation. Phase-transfer catalysis has also been used,[1583] and the use of chiral phase transfer catalysts led to enantioselectivity in the alkylated product.[1584] The reaction is successful for primary and secondary alkyl, allylic (with allylic rearrangement possible), and benzylic RX, but fails for tertiary halides, since these undergo elimination under the reaction conditions (see, however, p. 625). Various functional groups may be present in RX as long as they are not sensitive to base. Side reactions that may cause problems are the above-mentioned competing *O*-alkylation, elimination (if the enolate anion is a strong enough base), and dialkylation.

With substrates, such as ZCH$_2$Z′, it is possible to alkylate twice. Initial removal of the proton with a base followed by alkylation of the resulting enolate anion with RX, can be followed by subsequent removal of the proton from ZCHRZ′ and then alkylation with the same or a different RX. An important example of this reaction is the *malonic ester synthesis*, in which both Z groups are COOEt. The product can be hydrolyzed and decarboxylated (**12-40**) to give a carboxylic acid. An illustration is the preparation of 2-ethylpentanoic acid (**155**) from malonic ester. A variation of this alkylation sequence employs 1,2-dibromoethane as the alkylating agent, and subsequent treatment with DBU leads to incorporation of a vinyl group on the α-carbon.[1585] Another variation involved coupling of a dimalonate with an allylic carbonate (see **10-60**), using a polymer-supported palladium catalyst.[1586]

It is obvious that many carboxylic acids of the formulas RCH$_2$COOH and RR′CHCOOH can be synthesized by this method [for some other ways of preparing

[1582]Zaugg, H.E.; Dunnigan, D.A.; Michaels, R.J.; Swett, L.R.; Wang, T.S.; Sommers, A.H.; DeNet, R.W. *J. Org. Chem.* **1961**, *26*, 644; Johnstone, R.A.W.; Tuli, D.; Rose, M.E. *J. Chem. Res. (S)* **1980**, 283.
[1583]See Sukhanov, N.N.; Trappel', L.N.; Chetverikov, V.P.; Yanovskaya, L.A. *J. Org. Chem. USSR* **1985**, *21*, 2288; Tundo, P.; Venturello, P.; Angeletti, E. *J. Chem. Soc. Perkin Trans. 1* **1987**, 2159.
[1584]Park, E.J.; Kim, M.H.; Kim, D.Y. *J. Org. Chem.* **2004**, *69*, 6897.
[1585]Bunce, R.A.; Burns, S.E. *Org. Prep. Proceed. Int.* **1999**, *31*, 99.
[1586]Akiyama, R.; Kobayashi, S. *J. Am. Chem. Soc.* **2003**, *125*, 3412.

such acids (see **10-70–10-73**)]. Another important example is the *acetoacetic ester synthesis*, in which Z is COOEt and Z' is COCH₃. In this case, the product can be decarboxylated with acid or dilute base (**12-40**) to give a ketone (**156**) or cleaved with concentrated base (**12-43**) to give a carboxylic ester (**157**) and a salt of acetic acid. This reaction has been done in *tert*-butanol in the presence of alumina, *in vacuo*, to give the alkylated keto acid directly from the keto ester.[1587]

Another way of preparing ketones involves alkylation[1588] of β-keto sulfoxides[1589] or sulfones,[1590] to give **158**.

The sulfoxide group in the product (**158**) is easily reduced (desulfurized, see p. $$$) to give the ketone in high yields using aluminum amalgam or by electrolysis.[1591] β-Keto sulfoxides, such as **158** or sulfones (—SO₂—), are easily prepared (**16-86**). When one group attached to the sulfur atom is chiral, the alkylation proceeds to with reasonable enantioselectivity.[1592] Alkylation of α-nitrosulfones was reported, using photochemical conditions, (Me₃Sn)₂ and a secondary iodide.[1593]

Other examples of the reaction are the *cyanoacetic ester synthesis*, in which Z is COOEt and Z' is CN (as in the malonic ester synthesis, the product here can be hydrolyzed and decarboxylated), and the *Sorensen* method of amino acid synthesis, in which

[1587]Bhar, S.; Chaudhuri, S.K.; Sahu, S.G.; Panja, C. *Tetrahedron* **2001**, *57*, 9011.

[1588]For a review of the synthetic uses of β-keto sulfoxides, sulfones, and sulfides, see Trost, B.M. *Chem. Rev.* **1978**, *78*, 363. For a review of asymmetric synthesis with chiral sulfoxides, see Solladié, G. *Synthesis* **1981**, 185.

[1589]Gassman, P.G.; Richmond, G.D. *J. Org. Chem.* **1966**, *31*, 2355. Such sulfoxides can be alkylated on the other side of the C=O group by the use of two moles of base: Kuwajima, I.; Iwasawa, H. *Tetrahedron Lett.* **1974**, 107.

[1590]House, H.O.; Larson, J.K. *J. Org. Chem.* **1968**, *33*, 61; Kurth, M.J.; O'Brien, M.J. *J. Org. Chem.* **1985**, 3846.

[1591]Lamm, B.; Samuelsson, B. *Acta Chem. Scand.* **1969**, *23*, 691.

[1592]Enders, D.; Harnying, W.; Vignola, N. *Eur. J. Org. Chem.* **2003**, 3939.

[1593]Kim, S. Yoon, Y.-y.; Lim, C.J. *Synlett* **2000**, 1151.

the reaction is applied to *N*-acetylaminomalonic ester $(EtOOC)_2CHNHCOCH_3$. Hydrolysis and decarboxylation of the product in this case gives an α-amino acid. The amino group is also frequently protected by conversion to a phthalimido group.

The reaction is not limited to $Z-CH_2-Z'$ compounds. Other compounds have acidic CH hydrogens. Some examples are the methyl hydrogens of α-aminopyridines, the methyl hydrogens of ynamines of the form $CH_3C \equiv CNR_2$[1594] (the product in this case can be hydrolyzed to an amide $RCH_2CH_2CONR_2$), the CH_2 hydrogens of cyclopentadiene and its derivatives (p. 63), hydrogens connected to a triple-bond carbon (**10-74**), and the hydrogen of HCN (**10-75**) can also be removed with a base and the resulting ion alkylated (see also, **10-68** to **10-72**). α-Imino esters have been used since treatment with a strong base with a titanium catalyst followed by an aldehyde leads to hydroxy-amino-esters.[1595]

Alkylation takes place at the most acidic position of a reagent molecule; for example, acetoacetic ester (CH_3COCH_2COOEt) is alkylated at the methylene and not at the methyl group, because the former is more acidic than the latter and hence gives up its proton to the base. However, if 2 equivalents of base are used, then not only is the most acidic proton removed, but also the second most acidic. Alkylation of this doubly charged anion (a dianion) occurs at the less acidic position, in this case the second most acidic position[1596] (see p. 513). The first and second ion pair acidities of β-diketones has been studied.[1597]

When ω,ω′-dihalides are used, ring closures can be effected:[1598]

This method has been used to close rings of from three ($n = 0$) to seven members, although five-membered ring closures proceed in highest yields. Another ring-closing method involves internal alkylation.[1599]

[1594]Corey, E.J.; Cane, D.E. *J. Org. Chem.* ***1970***, *35*, 3405.

[1595]Kanemasa, S.; Mori, T.; Wada, E.; Tatsukawa, A. *Tetrahedron Lett.* ***1993***, *34*, 677. See Kotha, S.; Kuki, A. *Tetrahedron Lett.* ***1992***, *33*, 1565 for a related reaction.

[1596]For a list of references, see Larock, R.C. *Comprehensive Organic Transformations*, 2nd ed., Wiley-VCH, NY, ***1999***, pp. 1540–1541. Also see, Lu, Y.-Q.; Li, C.-J. *Tetrahedron Lett.* ***1996***, *37*, 471.

[1597]Facchetti, A.; Streitwieser, A. *J. Org. Chem.* ***2004***, *69*, 8345.

[1598]Zefirov, N.S.; Kuznetsova, T.S.; Kozhushkov, S.I.; Surmina, L.S.; Rashchupkina, Z.A. *J. Org. Chem. USSR* ***1983***, *19*, 474.

[1599]For example, see Knipe, A.C.; Stirling, C.J.M. *J. Chem. Soc. B* ***1968***, 67; Gosselck, J.; Winkler, A. *Tetrahedron Lett.* ***1970***, 2437; Walborsky, H.M.; Murari, M.P. *Can. J. Chem.* ***1984***, *62*, 2464. For a review of this method as applied to the synthesis of β-lactams, see Bose, A.K.; Manhas, M.S.; Chatterjee, B.G.; Abdulla, R.F. *Synth. Commun.* ***1971***, *1*, 51. For a list of examples, see Larock, R.C. *Comprehensive Organic Transformations*, 2nd ed., Wiley-VCH, NY, ***1999***, pp. 156–157, 165–166.

This method has been shown to be applicable to medium rings (10–14 members) without the use of high-dilution techniques.[1600]

The mechanism of these reactions is usually S_N2 with inversion taking place at a chiral RX, although an SET[1601] mechanism may be involved in certain cases,[1602] especially where the nucleophile is an α-nitro carbanion[1603] and/or the substrate contains a nitro or cyano[1604] group. Tertiary alkyl groups can be introduced by an S_N1 mechanism if the ZCH$_2$Z' compound (not the enolate anion) is treated with a tertiary carbocation generated *in situ* from an alcohol or alkyl halide and BF$_3$ or AlCl$_3$,[1605] or with a tertiary alkyl perchlorate.[1606]

Alkylation α to a nitro group can be achieved with the Katritzky pyrylium–pyridinium reagents.[1607] This reaction probably has a free-radical mechanism.[1608]

OS **I**, 248, 250; **II**, 262, 279, 384, 474; **III**, 213, 219, 397, 405, 495, 705; **IV**, 10, 55, 288, 291, 623, 641, 962; **V**, 76, 187, 514, 523, 559, 743, 767, 785, 848, 1013; **VI**, 223, 320, 361, 482, 503, 587, 781, 991; **VII**, 339, 411; **VIII**, 5, 312, 381. See also, OS **VIII**, 235.

10-68   Alkylation of Ketones, Aldehydes, Nitriles, and Carboxylic Esters

α-Acylalkyl-de-halogenation, and so on

159

Ketones,[1609] nitriles,[1610] and carboxylic esters[1611] can be alkylated in the α position in a reaction similar to **10-67**.[1568] The pK_a of the proton α to the carbonyl or

[1600]Deslongchamps, P.; Lamothe, S.; Lin, H. *Can. J. Chem.* **1984**, *62*, 2395; **1987**, *65*, 1298; Brillon, D.; Deslongchamps, P. *Can. J. Chem.* **1987**, *65*, 43, 56.

[1601]These SET mechanisms are often called $S_{RN}1$ mechanisms. See also, Ref. 96.

[1602]Kornblum, N.; Michel, R.E.; Kerber, R.C. *J. Am. Chem. Soc.* **1966**, *88*, 5660, 5662; Russell, G.A.; Ros, F. *J. Am. Chem. Soc.* **1985**, *107*, 2506; Ashby, E.C.; Argyropoulos, J.N. *J. Org. Chem.* **1985**, *50*, 3274; Bordwell, F.G.; Harrelson, Jr., J.A. *J. Am. Chem. Soc.* **1989**, *111*, 1052.

[1603]For a review of mechanisms with these nucleophiles, see Bowman, W.R. *Chem. Soc. Rev.* **1988**, *17*, 283.

[1604]Kornblum, N.; Fifolt, M. *Tetrahedron* **1989**, *45*, 1311.

[1605]For example, see Boldt, P.; Militzer, H. *Tetrahedron Lett.* **1966**, 3599; Crimmins, T.F.; Hauser, C.R. *J. Org. Chem.* **1967**, *32*, 2615; Boldt, P.; Militzer, H.; Thielecke, W.; Schulz, L. *Liebigs Ann. Chem.* **1968**, *718*, 101.

[1606]Boldt, P.; Ludwieg, A.; Militzer, H. *Chem. Ber.* **1970**, *103*, 1312.

[1607]Katritzky, A.R.; Kashmiri, M.A.; Wittmann, D.K. *Tetrahedron* **1984**, *40*, 1501.

[1608]Katritzky, A.R.; Chen, J.; Marson, C.M.; Maia, A.; Kashmiri, M.A. *Tetrahedron* **1986**, *42*, 101.

[1609]For a review of the alkylation and acylation of ketones and aldehydes, see Caine, D., in Augustine, R.L. *Carbon–Carbon Bond Formation*, Vol. 1, Marcel Dekker, NY, **1979**, pp. 85–352.

[1610]For a review, see Arseniyadis, S.; Kyler, K.S.; Watt, D.S. *Org. React.* **1984**, *31*, 1. For a list of references, see Larock, R.C. *Comprehensive Organic Transformations*, 2nd ed., Wiley-VCH, NY, **1999**, pp. 1801–1808. See Taber, D.F.; Kong, S. *J. Org. Chem.* **1997**, *62*, 8575.

[1611]For a review, see Petragnani, N.; Yonashiro, M. *Synthesis* **1982**, 521. For a list of references, see Larock, R.C. *Comprehensive Organic Transformations*, 2nd ed., Wiley-VCH, NY, **1999**, pp. 1724–1758*ff*.

CN is in the range of 19–25 (see p. 363), and a base that has a conjugate acid with a pK_a greater than that proton must be employed. Note that since only one activating group is present here, compared with two activating groups for the substrates in **10-67**, the pK_a of the α-proton is higher (a weaker acid) and a stronger base is required. Reaction of the α-proton with the base generates they key nucleophilic intermediate, an enolate anion (**159**). The most common bases[1612] are lithium diethylamide (Et$_2$NLi), lithium diisopropylamide[(iPr)$_2$NLi, LDA], t-BuOK, NaNH$_2$, and KH. A combination of lithium hexamethyldisilazide [LiN(SiMe$_3$)$_2$] followed by MnBr$_2$ is also effective for alkylation of ketones.[1613] The base lithium N-isopropyl-N-cyclohexylamide (LICA) is particularly successful for carboxylic esters[1614] and nitriles.[1615] Solid KOH in Me$_2$SO has been used to methylate ketones, in high yields.[1616] Some of these bases are strong enough to convert the ketone, nitrile, or ester completely to its enolate anion conjugate base; others (especially t-BuOK) convert a significant fraction of the molecules. In the latter case, the aldol reaction (**16-34**) or Claisen condensation (**16-85**) may be side reactions, since both the free molecule and its conjugate base are present at the same time. It is therefore important to use a base strong enough to convert the starting compound completely. Both lactones[1617] and lactams are similarly alkylated.[1618] Protic solvents are generally not suitable because they protonate the base (though of course this is not a problem with a conjugate pair, such as t-BuOK in t-BuOH). Some common solvents are 1,2-dimethoxyethane, THF, DMF, and liquid NH$_3$. Phase-transfer catalysis has been used to alkylate many nitriles, as well as some esters and ketones.[1619]

Direct alkylation of aldehydes is difficult when bases, such as KOH and NaOMe, are used due to rapid aldol reaction (**16-34**), but aldehydes bearing only one α hydrogen have been alkylated with allylic and benzylic halides in good yields by the use of the base KH to prepare the potassium enolate,[1620] or in moderate yields, by the use of a phase-transfer catalyst.[1621] Even the use of amide bases such as

[1612]For a list of some bases, with references, see Larock, R.C. *Comprehensive Organic Transformations*, 2nd ed., Wiley-VCH, NY, *1999*, pp. 1476–1479.

[1613]Reetz, M.T.; Haning, H. *Tetrahedron Lett.* *1993*, *34*, 7395.

[1614]Rathke, M.W.; Lindert, A. *J. Am. Chem. Soc.* *1971*, *93*, 2319; Bos, W.; Pabon, H.J.J. *Recl. Trav. Chim. Pays-Bas 1980*, *99*, 141. See also, Cregge, R.J.; Herrmann, J.L.; Lee, C.S.; Richman, J.E.; Schlessinger, R.H. *Tetrahedron Lett.* *1973*, 2425.

[1615]Watt, D.S. *Tetrahedron Lett.* *1974*, 707.

[1616]Langhals, E.; Langhals, H. *Tetrahedron Lett.* *1990*, *31*, 859.

[1617]For a discusson of the stereochemistry of lactone alkylation see Ibrahim-Ouali, M.; Parrain, J.-L.; Santelli, M. *Org. Prep. Proceed. Int.* *1999*, *31*, 467. Enolate anions of β-lactones are subject to ring opening: see Mori, S.; Shindo, M. *Org. Lett.* *2004*, *6*, 3945.

[1618]Matsuo, J.-i.; Kobayashi, S.; Koga, K. *Tetrahedron Lett.* *1998*, *39*, 9723.

[1619]For reviews, see Makosza, M. *Russ. Chem. Rev.* *1977*, *46*, 1151; *Pure Appl. Chem.* *1975*, *43*, 439; Starks, C.M.; Liotta, C. *Phase Transfer Catalysis*, Acaemic Press, NY, *1978*, pp. 170–217; Weber, W.P.; Gokel, G.W. *Phase Transfer Catalysis in Organic Synthesis*, Springer, NY, *1977*, pp. 136–204.

[1620]Groenewegen, F.; Kallenberg, H.; van der Gen, A. *Tetrahedron Lett.* *1978*, 491; Artaud, I.; Torossian, G.; Viout, P. *Tetrahedron 1985*, *41*, 5031.

[1621]Dietl, H.K.; Brannock, K.C. *Tetrahedron Lett.* *1973*, 1273; Purohit, V.G.; Subramanian, R. *Chem. Ind. (London) 1978*, 731; Buschmann, E.; Zeeh, B. *Liebigs Ann. Chem.* *1979*, 1585.

lithium diisopropylamide (LDA), lithium hexamethyldisilazide (LHMDS), or lithium tetramethylpiperidide (LTMP) to generate the enolate anion in an aprotic solvent, such as ether or THF, cannot preclude rapid aldol side reactions.

As in **10-67**, the alkyl halide that reacts with the enolate anion may be primary or secondary. Tertiary halides give elimination. Even primary and secondary halides give predominant elimination if the enolate anion is a strong enough base (e.g., the enolate anion from Me₃CCOMe).[1622] Tertiary alkyl groups, as well as other groups that normally give S_N1 reactions, can be introduced if the reaction is performed on a silyl enol ether[1623] of a ketone, aldehyde, or ester (see **160**) with a Lewis acid catalyst.[1624] Tertiary alkyl fluorides were coupled to silyl enol ethers with BF₃•etherate.[1625] An interesting reaction reacted a methyl ketone, such as acetophenone (1-phenyl-1-ethanone) with tributylamine, in the presence of a ruthenium catalyst at 180°C, and the product resulted from *C*-alkylation (1-phenyl-1-hexanone).[1626] Note that tin enolates (C=C—OSnR₃) react with halides in the presence of a zinc catalyst.[1627] A chiral variation of this latter reaction was reported involving generation of the enolate anion in the presence of Me₃SnCl, a palladium catalyst and a chiral ligand.[1628]

Silyl enol ethers can be converted to the enolate anion, which can then be alkylated in the usual manner. The reaction of silyl enol ether **161** using KOEt followed by LiBr at a catalytic amount of *n*-butyllithium with allyl iodide gave **162**.[1629] Initial conversion of the silyl enol ether to the enolate anion allows the alkylation process to take place.

[1622]Zook, H.D.; Kelly, W.L.; Posey, I.Y. *J. Org. Chem.* **1968**, *33*, 3477.

[1623]For a list of alkylations of silyl enol ethers, see Larock, R.C. *Comprehensive Organic Transformations*, 2nd ed., Wiley-VCH, NY, **1999**, pp. 1494–1505.

[1624]Reetz, M.T.; Sauerwald, M. *J. Organomet. Chem.* **1990**, *382*, 121; Kad, G.L.; Singh, V.; Khurana, A.; Chaudhary, S.; Singh, J. *Synth. Commun.* **1999**, *29*, 3439; Kang, S.-K.; Ryu, H.-C.; Hong, Y.-T. *J. Chem. Soc., Perkin Trans. 1* **2000**, 3350. For a review, see Reetz, M.T. *Angew. Chem. Int. Ed.* **1982**, *21*, 96.

[1625]Hirano, K.; Fujita, K.; Yorimitsu, H.; Shinokubo, H.; Oshima, K. *Tetrahedron Lett.* **2004**, *45*, 2555.

[1626]Cho, C.S.; Kim, B.T.; Lee, M.J.; Kim, T.-J.; Shim, S.C. *Angew. Chem. Int. Ed.* **2001**, *40*, 958.

[1627]Yasuda, M.; Tsuji, S.; Shigeyoshi, Y.; Baba, A. *J. Am. Chem Soc.* **2002**, *124*, 7440.

[1628]Trost, B.M.; Schroeder, G.M. *J. Am. Chem. Soc.* **1999**, *121*, 6759.

[1629]Yu, W.; Jin, Z. *Tetrahedron Lett.* **2001**, *42*, 369.

Enol carbonates react with alkylating agents in the presence of a palladium catalyst. The decarboxylative alkylation of allyl enol carbonates to the corresponding allylcyclohexanone derivatives is known as the *Tsuji alkylation*.[1630] An asymmetric version of this reaction has been reported.[1631] The same reaction can be done using enolate anion and allylic acetates with a palladium catalyst.[1632]

Vinylic and aryl halides can be used to vinylate or arylate carboxylic esters (but not ketones) by the use of $NiBr_2$ as a catalyst.[1633] Ketones have been vinylated by treating their enol acetates with vinylic bromides in the presence of a Pd compound catalyst,[1634] but direct reaction of a ketone, a vinyl halide, sodium *tert*-butoxide and a palladium catalyst also give the α-vinyl ketone.[1635] Also as in **10-67**, this reaction can be used to close rings.[1636] Rings have been closed by treating a dianion of a dialkyl succinate with a 1,ω-dihalide or ditosylate.[1637] This was applied to the synthesis of three-, four-, five-, and six-membered rings. When the attached groups were chiral (e.g., menthyl) the product was formed with >90% ee.[1636]

Efficient enantioselective alkylations are known.[1638] In another method enantioselective alkylation can be achieved by using a chiral base to form the enolate.[1639] Alternatively, a chiral auxiliary can be attached. Many auxiliaries are based on the use of chiral amides[1640] or esters.[1641] Subsequent formation of the enolate anion allows alkylation to proceed with high enantioselectivity. A subsequent step is

[1630]Tsuji, J.; Minami, I. *Acc. Chem. Res.* **1987**, *20*, 140; Tsuji, J.; Minami, I.; Shimizu, I. *Tetrahedron Lett.* **1983**, *24*, 1793; Tsuji, J.; Shimizu, I.; Minami, I.; Ohashi, Y.; Sugiura, T.; Takahashi, K. *J. Org. Chem.* **1985**, *50*, 1523; Tsuji, J.; Minami, I.; Shimizu, I. *Chem. Lett.* **1983**, *12*, 1325. See also, Nicolaou, K.C.; Vassilikogiannakis, G.; Mägerlein, W.; Kranich, R. *Angew. Chem. Int. Ed.* **2001**, *40*, 2482; Herrinton, P.M.; Klotz, K.L.; Hartley, W.M. *J. Org. Chem.* **1993**, *58*, 678.

[1631]Behenna, D.C.; Stoltz, B.M. *J. Am. Chem. Soc.* **2004**, *126*, 15044.

[1632]Trost, B.M.; Schroeder, G.M.; Kristensen, J. *Angew. Chem. Int. Ed.* **2002**, *41*, 3492.

[1633]Millard, A.A.; Rathke, M.W. *J. Am. Chem. Soc.* **1977**, *99*, 4833.

[1634]Kosugi, M.; Hagiwara, I.; Migita, T. *Chem. Lett.* **1983**, 839. For other methods, see Negishi, E.; Akiyoshi, K. *Chem. Lett.* **1987**, 1007; Chang, T.C.T.; Rosenblum, M.; Simms, N. *Org. Synth. 66*, 95.

[1635]Chieffi, A.; Kamikawa, K.; Åhman, J.; Fox, J.M.; Buchwald, S.L *Org. Lett.* **2001**, *3*, 1897.

[1636]For example, see Etheredge, S.J. *J. Org. Chem.* **1966**, *31*, 1990; Wilcox, C.F.; Whitney, G.C. *J. Org. Chem.* **1967**, *32*, 2933; Bird, R.; Stirling, C.J.M. *J. Chem. Soc. B* **1968**, 111; Stork, G.; Boeckman, Jr., R.K. *J. Am. Chem. Soc.* **1973**, *95*, 2016; Stork, G.; Cohen, J.F. *J. Am. Chem. Soc.* **1974**, *96*, 5270. In the last case, the substrate moiety is an epoxide function.

[1637]Misumi, A.; Iwanaga, K.; Furuta, K.; Yamamoto, H. *J. Am. Chem. Soc.* **1985**, *107*, 3343; Furuta, K.; Iwanaga, K.; Yamamoto, H. *Org. Synth. 67*, 76.

[1638]For reviews of stereoselective alkylation of enolates, see Nógrádi, M. *Stereoselective Synthesis*, VCH, NY, **1986**, pp. 236–245; Evans, D.A. in Morrison, J.D. *Asymmetric Synthesis*, Vol. 3, Academic Press, NY, **1984**, pp. 1–110.

[1639]For example, see Murakata, M.; Nakajima, N.; Koga, K. *J. Chem. Soc., Chem. Commun.* **1990**, 1657. For a review, see Cox. P.J.; Simpkins, N.S. *Tetrahedron: Asymmetry* **1991**, *2*, 1, pp. 6–13.

[1640]Chiral oxazolidinones such as the Evan's auxiliaries derived from chiral amino alcohols: Lafontaine, J.A.; Provencal, D.P.; Gardelli, C.; Leahy, J.W. *J. Org. Chem.* **2003**, *68*, 4215; Bull, S.D.; Davies, S.G.; Nicholson, R.L.; Sanganee, H.J.; Smith, A.D. *Tetrahedron Asymmetry* **2000**, *11*, 3475. See Evans, D.A.; Chapman, K.T.; Bisaha, J. *Tetrahedron Lett.* **1984**, *25*, 4071; Evans, D.A. Chapman, K.T.; Bisaha, J. *J. Am. Chem. Soc.* **1984**, *106*, 4261. Oppolzer's sultam: Oppolzer, W.; Chapuis, C.; Dupuis, D.; Guo, M. *Helv. Chim. Acta* **1985**, *68*, 2100. Chiral sulfonamides: Schmierer, R.; Grotemeier, G.; Helmchen, G.; Selim, A. *Angew. Chem. Int. Ed.* **1981**, *20*, 207.

[1641]Oppolzer, W.; Dudfield, P.; Stevenson, T.; Godel, T. *Helv. Chim. Acta* **1985**, *68*, 212.

required to convert the chiral amide or ester to the corresponding carboxylic acid. Chiral additives can also be used.[1642]

When the compound to be alkylated is an unsymmetrical ketone, the question arises as to which side will be alkylated. If a phenyl or a vinylic group is present on one side, alkylation goes predominantly on that side. When only alkyl groups are present, the reaction is generally not regioselective; mixtures are obtained in which sometimes the more alkylated and sometimes the less alkylated side is predominantly alkylated. Which product is found in higher yield depends on the nature of the substrate, the base,[1643] the cation, and the solvent. In any case, di- and trisubstitution are frequent[1644] and it is often difficult to stop with the introduction of just one alkyl group.[1645]

Several methods have been developed for ensuring that alkylation takes place regioselectively on the *desired* side of a ketone.[1646] Among these are

1. Block one side of the ketone by introducing a removable group. Alkylation takes place on the other side; the blocking group is then removed. A common reaction for this purpose is formylation with ethyl formate (**16-86**); this generally blocks the less hindered side. The formyl group is easily removed by alkaline hydrolysis (**12-43**).

2. Introduce an activating group on one side; alkylation then takes place on that side (**10-67**); the activating group is then removed.

3. Prepare the desired one of the two possible enolate anions.[1647] The two ions, for example, **163** and **164** for 2-heptanone, interconvert rapidly only in

the presence of the parent ketone or any stronger acid.[1648] In the absence of such acids, it is possible to prepare either **163** or **164** and thus achieve

[1642]Denmark, S.E.; Stavenger, R.A. *Acc. Chem. Res.* **2000**, *33*, 432; Machajewski, T.D.; Wong, C.-H. *Angew. Chem. Int. Ed.* **2000**, *39*, 1352.

[1643]Sterically hindered bases may greatly favor one enolate over the other. See, for example, Prieto, J.A.; Suarez, J.; Larson, G.L. *Synth. Commun.* **1988**, *18*, 253; Gaudemar, M.; Bellassoued, M. *Tetrahedron Lett.* **1989**, *30*, 2779.

[1644]For a procedure for completely methylating the apositions of a ketone, see Lissel, M.; Neumann, B.; Schmidt, S. *Liebigs Ann. Chem.* **1987**, 263.

[1645]For some methods of reducing dialkylation, see Hooz, J.; Oudenes, J. *Synth. Commun.* **1980**, *10*, 139; Morita, J.; Suzuki, M.; Noyori, R. *J. Org. Chem.* **1989**, *54*, 1785.

[1646]For a review, see House, H.O. *Rec. Chem. Prog.* **1968**, *28*, 99. For a review with respect to cyclohexenones, see Podraza, K.F. *Org. Prep. Proced. Int.* **1991**, *23*, 217.

[1647]For reviews, see d'Angelo, J. *Tetrahedron* **1976**, *32*, 2979; Stork, G. *Pure Appl. Chem.* **1975**, *43*, 553.

[1648]House, H.O.; Trost, B.M. *J. Org. Chem.* **1965**, *30*, 1341.

selective alkylation on either side of the ketone.[1649] The desired enolate anion can be obtained by treatment of the corresponding enol acetate with two equivalents of methyllithium in 1,2-dimethoxyethane. Each enol acetate gives the corresponding enolate, for example,

The enol acetates, in turn, can be prepared by treatment of the parent ketone with an appropriate reagent.[1241] Such treatment generally gives a mixture of the two enol acetates in which one or the other predominates, depending on the reagent. The mixtures are easily separable.[1648] An alternate procedure involves conversion of a silyl enol ether[1650] (see **12-17**) or a dialkylboron enol ether[1651] (an enol borinate, see p. 645) to the corresponding enolate anion. If the less hindered enolate anion is desired (e.g., **126**), it can be prepared directly from the ketone by treatment with LDA in THF or 1,2-dimethoxyethane (DME) at −78°C.[1652]

4. Begin not with the ketone itself, but with an α,β-unsaturated ketone in which the double bond is present on the side where alkylation is desired. Upon treatment with lithium in liquid NH_3, such a ketone is reduced to an enolate anion. When the alkyl halide is added, it must react with the enolate anion on

the side where the double bond was.[1653] Of course, this method is not actually an alkylation of the ketone, but of the α,β-unsaturated ketone, although the

[1649]Whitlock Jr., H.W.; Overman, L.E. *J. Org. Chem.* **1969**, *34*, 1962; House, H.O.; Gall, M.; Olmstead, H.D. *J. Org. Chem.* **1971**, *36*, 2361. For an improved procedure, see Liotta, C.L.; Caruso, T.C. *Tetrahedron Lett.* **1985**, *26*, 1599.
[1650]Stork, G.; Hudrlik, P.F. *J. Am. Chem. Soc.* **1968**, *90*, 4462, 4464. For reviews, see Kuwajima, I.; Nakamura, E. *Acc. Chem. Res.* **1985**, *18*, 181; Fleming, I. *Chimia*, **1980**, *34*, 265; Rasmussen, J.K. *Synthesis* **1977**, 91.
[1651]Pasto, D.J.; Wojtkowski, P.W. *J. Org. Chem.* **1971**, *36*, 1790.
[1652]House, H.O.; Gall, M.; Olmstead, H.D. *J. Org. Chem.* **1971**, *36*, 2361. See also, Corey, E.J.; Gross, A.W. *Tetrahedron Lett.* **1984**, *25*, 495.
[1653]Stork, G.; Rosen, P.; Goldman, N.; Coombs, R.V.; Tsuji, J. *J. Am. Chem. Soc.* **1965**, *87*, 275. For a review, see Caine, D. *Org. React.* **1976**, *23*, 1. For similar approaches, see Coates, R.M.; Sowerby, R.L. *J. Am. Chem. Soc.* **1971**, *93*, 1027; Näf, F.; Decorzant, R. *Helv. Chim. Acta* **1974**, *57*, 1317; Wender, P.A.; Eissenstat, M.A. *J. Am. Chem. Soc.* **1978**, *100*, 292.

product is the same as if the saturated ketone had been alkylated on the desired side.

Both sides of acetone have been alkylated with different alkyl groups, in one operation, by treatment of the N,N-dimethylhydrazone of acetone with n-BuLi, followed by a primary alkyl, benzylic, or allylic bromide or iodide; then another mole of n-BuLi, a second halide, and finally hydrolysis of the hydrazone.[1654] Alkylation of an unsymmetrical ketone at the more substituted position was reported using an alkyl bromide, NaOH, and a calix[n]arene catalyst (see p. 122 for calixarenes).[1655]

Among other methods for the preparation of alkylated ketones are (*1*) Alkylation of silyl enol ethers using various reagents as noted above, (*2*) the Stork enamine reaction (**10-69**), (*3*) the acetoacetic ester synthesis (**10-67**), (*4*) alkylation of β-keto sulfones or sulfoxides (**10-67**), (*5*) acylation of $CH_3SOCH_2^-$ followed by reductive cleavage (**16-86**), (*6*) treatment of α-halo ketones with lithium dialkylcopper reagents (**10-57**), and (*7*) treatment of α-halo ketones with trialkylboranes (**10-73**).

Aldehydes can be indirectly alkylated via an imine derivative of the aldehyde.[1656] The derivative is easily prepared (**16-13**) and the product easily hydrolyzed to the aldehyde (**16-2**). Either or both R groups may be hydrogen, so that

mono-, di-, and trisubstituted acetaldehydes can be prepared by this method. R′ may be primary alkyl, allylic, or benzylic. Imine alkylation can also be applied to the preparation of substituted amine derivatives. An amino acid surrogate, such as $Ph_2C=NCH_2CO_2R$, when treated with KOH and an alkyl halide gives the C-alkylated product.[1657] When a chiral additive is used, good enantioselectivity was observed. This reaction has also been done in the ionic liquid bmim tetrafluoroborate (see p. 415).[1658] It is possible to alkylate α-amino amides directly.[1659]

[1654]Yamashita, M.; Matsuyama, K.; Tanabe, M.; Suemitsu, R. *Bull. Chem. Soc. Jpn.* **1985**, *58*, 407.
[1655]Shimizu, S.; Suzuki, T.; Sasaki, Y.; Hirai, C. *Synlett* **2000**, 1664.
[1656]Cuvigny, T.; Normant, H. *Bull. Soc. Chim. Fr.* **1970**, 3976. For reviews, see Fraser, R.R., in Buncel, E.; Durst, T. *Comprehensive Carbanion Chemistry*, Vol. 5, pt. B, Elsevier, NY, **1984**, pp. 65–105; Whitesell, J.K.; Whitesell, M.A. *Synthesis* **1983**, 517. For a list of references, see Larock, R.C. *Comprehensive Organic Transformations*, 2nd ed., Wiley-VCH, NY, **1999**, pp. 1513–1518. For a method in which the metalated imine is prepared from a nitrile, see Goering, H.L.; Tseng, C.C. *J. Org. Chem.* **1981**, *46*, 5250.
[1657]Park, H.-g.; Jeong, B.-s.; Yoo, M.-s.; Park, M.-k.; Huh, H.; Jew, S.-s. *Tetrahedron Lett.* **2001**, *42*, 4645; Jew, S.-s.; Jeong, B.-s.; Yoo, M.-s.; Huh, H.; Park, H.-g. *Chem. Commun.* **2001**, 1244.
[1658]Pellissier, H.; Santelli, M. *Tetrahedron* **2003**, *59*, 701.
[1659]Myers, A.G.; Schnider, P.; Kwon, S.; Kung, D.W. *J. Org. Chem.* **1999**, *64*, 3322.

Hydrazones and other compounds with C=N bonds can be similarly alkylated.[1639] The use of chiral amines or hydrazines[1660] (followed by hydrolysis **16-2** of the alkylated imine) can lead to chiral alkylated ketones in high optical yields[1661] (for an example, see p. 170).

In α,β-unsaturated ketones, nitriles, and esters (e.g., **165**), the γ hydrogen assumes the acidity normally held by the position α to the carbonyl group, especially when R is not hydrogen and so cannot compete. This principle, called *vinylogy*, operates because the resonance effect is transmitted through the double bond. However, because of the resonance, alkylation at the α position (with allylic rearrangement) competes with alkylation at the γ position and usually predominates.

α-Hydroxynitriles (cyanohydrins), protected by conversion to acetals with ethyl vinyl ether (**15-5**), can be easily alkylated with primary or secondary alkyl or allylic halides.[1662] The R group can be aryl or a saturated or unsaturated alkyl. Since the cyanohydrins[1663] are easily formed from aldehydes (**16-52**) and the product is easily hydrolyzed to a ketone, this is a method for converting an aldehyde

---

[1660]For a review of the alkylation of chiral hydrazones, see Enders, D., in Morrison, J.D. *Asymmetric Synthesis*, Vol. 3, Academic Press, NY, *1984*, pp. 275–339.

[1661]Meyers, A.I.; Williams, D.R.; Erickson, G.W.; White, S.; Druelinger, M. *J. Am. Chem. Soc. 1981, 103*, 3081; Meyers, A.I.; Williams, D.R.; White, S.; Erickson, G.W. *J. Am. Chem. Soc. 1981, 103*, 3088; Enders, D.; Bockstiegel, B. *Synthesis 1989*, 493; Enders, D.; Kipphardt, H.; Fey, P. *Org. Synth.* 65, 183.

[1662]Stork, G.; Maldonado, L. *J. Am. Chem. Soc. 1971, 93*, 5286; Stork, G.; Depezay, J.C.; D'Angelo, J. *Tetrahedron Lett. 1975*, 389. See also, Rasmussen, J.K.; Heilmann, S.M. *Synthesis 1978*, 219; Ahlbrecht, H.; Raab, W.; Vonderheid, C. *Synthesis 1979*, 127; Hünig, S.; Marschner, C.; Peters, K.; von Schnering, H.G. *Chem. Ber. 1989, 122*, 2131, and other papers in this series.

[1663]For a review of **166**, see Albright, J.D. *Tetrahedron 1983, 39*, 3207.

RCHO to a ketone $RCOR'$[1664] (for other methods, see **10-71**, **16-82**, and **18-9**).[1665] In this procedure the normal mode of reaction of a carbonyl carbon is reversed. The C atom of an aldehyde molecule is normally electrophilic and is attacked by nucleophiles (Chapter 16), but by conversion to the protected cyanohydrin this carbon atom has been induced to perform as a nucleophile.[1666] The German word *Umpolung*[1667] is used to describe this kind of reversal (another example is found in **10-71**). Since the ion **166** serves as a substitute for the unavailable $R-^{\ominus}C=O$ anion, it is often called a "masked" $R(^{\ominus}C=O)$ ion. This method fails for formaldehyde ($R = H$), but other masked formaldehydes have proved successful.[1668] In an interesting variation of nitrile alkylation, a quaternary bromide [PhC(Br)(Me)CN] reacted with allyl bromide, in the presence of a Grignard reagent, to give the alkylated product [PhC(CN)(Me)CH$_2$CH=CH$_2$].[1669]

A coupling react of two ketones to form a 1,4-diketone has been reported, using $ZnCl_2/Et_2NH$.[1670]

OS **III**, 44, 219, 221, 223, 397; **IV**, 278, 597, 641, 962; **V**, 187, 514, 559, 848; **VI**, 51, 115, 121, 401, 818, 897, 958, 991; **VII**, 153, 208, 241, 424; **VIII**, 141, 173, 241, 403, 460, 479, 486; **X**, 59, 460; **80**, 31.

10-69   The Stork Enamine Reaction

α-Acylalkyl-de-halogenation[1671]

[1664]For similar methods, see Stetter, H.; Schmitz, P.H.; Schreckenberg, M. *Chem. Ber.* **1977**, *110*, 1971; Hünig, S. *Chimia*, **1982**, *36*, 1.

[1665]For a review of methods of synthesis of aldehydes, ketones and carboxylic acids by coupling reactions, see Martin, S.F. *Synthesis* **1979**, 633.

[1666]For reviews of such reversals of carbonyl group reactivity, see Block, E. *Reactions of Organosulfur Compounds*, Academic Press, NY, **1978**, pp. 56–67; Gröbel, B.; Seebach, D. *Synthesis* **1977**, 357; Lever, Jr., O.W. *Tetrahedron* **1976**, *32*, 1943; Seebach, D.; Kolb, M. *Chem. Ind. (London)* **1974**, 687; Seebach, D. *Angew. Chem. Int. Ed.* **1969**, *8*, 639. For a compilation of references to masked acyl and formyl anions, see Hase, T.A.; Koskimies, J.K. *Aldrichimica Acta* **1981**, *14*, 73. For tables of masked reagents, see Hase, T.A. Umpoled Synthons, Wiley, NY, **1987**, pp. xiii-xiv, 7–18, 219–317. For lists of references, see Larock, R.C. *Comprehensive Organic Transformations*, 2nd ed., Wiley-VCH, NY, **1999**, pp. 1435–1438.

[1667]For a monograph, see Hase, T.A. *Umpoled Synthons*, Wiley, NY, **1987**. For a review, see Seebach, D. *Angew. Chem. Int. Ed.* **1979**, *18*, 239.

[1668]Possel, O.; van Leusen, A.M. *Tetrahedron Lett.* **1977**, 4229; Stork, G.; Ozorio, A.A.; Leong, A.Y.W. *Tetrahedron Lett.* **1978**, 5175.

[1669]Fleming, F.F.; Zhang, Z.; Knochel, P. *Org. Lett.* **2004**, *6*, 501.

[1670]Nevar, N.M.; Kel'in, A.V.; Kulinkovich, O.G. *Synthesis* **2000**, 1259.

[1671]This is the IUPAC name with respect to the halide as substrate.

When enamines are treated with alkyl halides, an alkylation occurs to give an iminium salt via electron transfer from the electron pair on nitrogen, through the C=C to the electrophilic carbon of the alkyl halide.[1672] In effect, an enamine behaves as a "nitrogen enolate" and generally react as carbon nucleophiles.[1673] Hydrolysis of the iminium salt gives a ketone. Since the enamine is normally formed from a ketone (**16-13**), the net result is alkylation of the ketone at the α position. The method, known as the *Stork enamine reaction*,[1674] is an alternative to the ketone alkylation considered in **10-68**, generally giving monoalkylation of the ketone. Alkylation usually takes place on the less substituted side of the original ketone. The most commonly used amines are the cyclic amines piperidine, morpholine, and pyrrolidine.

The method is quite useful for particularly active alkyl halides, such as allylic, benzylic, and propargylic halides, and for α-halo ethers and esters. Other primary and secondary halides can show sluggish reactivity. The react of enamines with benzotriazole derivatives has been reported.[1675] Tertiary halides do not give the reaction at all since, with respect to the halide, this is nucleophilic substitution and elimination predominates. The reaction can also be applied to activated aryl halides (e.g., 2,4-dinitrochlorobenzene; see Chapter 13), to epoxides,[1676] and to activated alkenes, such as acrylonitrile. The latter is a Michael-type reaction (**15–24**) with respect to the alkene.

Acylation[1677] can be accomplished with acyl halides or with anhydrides. Hydrolysis of the resulting iminium salt leads to a 1,3-diketone. A COOEt group can be introduced by treatment of the enamine with ethyl chloroformate ClCOOEt,[1678] a CN group with cyanogen chloride[1679] (not cyanogen bromide or iodide, which leads to halogenation of the enamine), a CHO group with the mixed anhydride of formic and acetic acids[1678] or with DMF and phosgene,[1680] and a C(R)=NR′ group with a nitrilium salt RC≡N⁺R′.[1681] The acylation of the enamine can take

---

[1672]See Adams, J.P. *J. Chem. Soc., Perkin Trans. 1* **2000**, 125.

[1673]For a discussion of structure–nucleophilicity relationships, see Kempf, B.; Hampel, N.; Ofial, A.R.; Mayr, H. *Chem. Eur. J.* **2003**, *9*, 2209.

[1674]Stork, G.; Brizzolara, A.; Landesman, H.; Szmuszkovicz, J.; Terrell, R. *J. Am. Chem. Soc.*, **1963**, *85*, 207. For general reviews of enamines, see Hickmott, P.W. *Tetrahedron*, **1984**, *40*, 2989; **1982**, *38*, 1975, 3363; Granik, V.G. *Russ. Chem. Rev.*, **1984**, *53*, 383. For reviews of this reaction, see, in Cook, A.G. *Enamines*, 2nd ed.; Marcel Dekker, NY, **1988**, the articles by Alt, G.H.; Cook, A.G. pp. 181–246, and Gadamasetti, G.; Kuehne, M.E. pp. 531–689; Whitesell, J.K.; Whitesell, M.A. *Synthesis*, **1983**, 517; Kuehne, M.E. *Synthesis*, **1970**, 510; House, H.O. *Modern Synthetic Reactions*, 2nd ed., W.A. Benjamin, NY, **1972**, pp. 570–582, 766–772; Bláha, K.; Červinka, O. *Adv. Heterocycl. Chem.*, **1966**, *6*, 147, pp. 186.

[1675]Katritzky, A.R.; Fang, Y.; Silina, A. *J. Org. Chem.* **1999**, *64*, 7622; Katritzky, A.R.; Huang, Z.; Fang, Y. *J. Org. Chem.* **1999**, *64*, 7625.

[1676]Britten, A.Z.; Owen, W.S.; Went, C.W. *Tetrahedron* **1969**, *25*, 3157.

[1677]For reviews, see Hickmott, P.W. *Chem. Ind. (London)* **1974**, 731; Hünig, S.; Hoch, H. *Fortschr. Chem. Forsch.* **1970**, *14*, 235.

[1678]Stork, G.; Brizzolara, A.; Landesman, H.; Szmuszkovicz, J.; Terrell, R. *J. Am. Chem. Soc.* **1963**, *85*, 207.

[1679]Kuehne, M.E. *J. Am. Chem. Soc.*, **1959**, *81*, 5400.

[1680]Ziegenbein, W. *Angew. Chem. Int. Ed. Engl.*, **1965**, *4*, 358.

[1681]Baudoux, D.; Fuks, R. *Bull. Soc. Chim. Belg.*, **1984**, *93*, 1009.

place by the same mechanism as alkylation, but another mechanism is also possible, if the acyl halide has an a hydrogen and if a tertiary amine is present, as it often is (it is added to neutralize the HX given off). In this mechanism, the acyl halide is dehydrohalogenated by the tertiary amine, producing a ketene (**17-14**), which adds to the enamine to give a cyclobutanone (**15-63**). This compound can be cleaved in the solution to form the same acylated imine salt (that would form by the more direct mechanism, or it can be isolated (in the case of enamines derived from aldehydes), or it may cleave in other ways.[1682]

N-Alkylation can be a problem, particularly with enamines derived from aldehydes. An alternative method, which gives good yields of alkylation with primary and secondary halides, is alkylation of enamine *salts*, which are prepared by treating an imine with ethylmagnesium bromide in THF:[1683]

The imines are prepared by the reaction of secondary amines with aldehydes or ketones, mainly ketones (**16-13**). The enamine salt method has also been used to give good yields of mono α alkylation of α,β-unsaturated ketones.[1684] Enamines prepared from aldehydes and butylisobutylamine can be alkylated by simple primary alkyl halides in good yields.[1685] *N*-Alkylation in this case is presumably prevented by steric hindrance.

When the nitrogen of the substrate contains a chiral R group, both the Stork enamine synthesis and the enamine salt method can be used to perform enantioselective syntheses.[1686] The use of *S*-proline can generate a chiral enamine *in situ*, thus allowing alkylation to occur, giving alkylated product with good enantioselectivity,. The reaction has been done intramolecularly.[1687]

Conjugate addition (Michael addition) occurs when enamines react with conjugated ketones. This reaction is discussed in Section **15-24**.

Although not formally the enamine synthesis, reaction of an enamine with methyl bromoacetate in the presence of indium metal leads to α-alkylation: $R_2N-CH=CHR \rightarrow R_2N-CH(R')CHR$.[1688]

OS **V**, 533, 869; **VI**, 242, 496, 526; **VII**, 473.

[1682]See Alt, G.H.; Cook, A.G., in Cook, A.G. *Enamines*, 2nd ed., Marcel Dekker, NY, *1988*, pp. 204–215.

[1683]Stork, G.; Dowd, S.R. *J. Am. Chem. Soc.*, *1963*, *85*, 2178.

[1684]Stork, G.; Benaim, J. *J. Am. Chem. Soc.*, *1971*, *93*, 5938.

[1685]Curphey, T.J.; Hung, J.C.; Chu, C.C.C. *J. Org. Chem.*, *1975*, *40*, 607. See also, Ho, T.; Wong, C.M. *Synth. Commun.*, *1974*, *4*, 147.

[1686]For reviews, see Nógrádi, M. *Stereoselective Synthesis*, VCH, NY, *1986*, pp. 248–255; Whitesell, J.K. *Acc. Chem. Res.*, *1985*, *18*, 280; Bergbreiter, D.E.; Newcomb, M., in Morrison, J.D. *Asymmetric Synthesis*, Vol. 2, Academic Press, NY, *1983*, pp. 243–273.

[1687]Vignola, N.; List, B. *J. Am. Chem. Soc. 2004*, *126*, 450.

[1688]Bossard, F.; Dambrin, V.; Lintanf, V.; Beuchet, P.; Mosset, P. *Tetrahedron Lett.*, *1995*, *36*, 6055.

10-70   Alkylation of Carboxylic Acid Salts

α-Carboxyalkyl-de-halogenation

Carboxylic acids can be alkylated in the a position by conversion of their salts to dianions [which have resonance contributors $RCH=C(O^-)_2$[1689]] by treatment with a strong base, such as LDA.[1690] The use of Li^+ as the counterion increases the solubility of the dianionic salt. The reaction has been applied[1691] to primary alkyl, allylic, and benzylic halides, and to carboxylic acids of the form RCH_2COOH and $RR^2CHCOOH$.[1610] Allkylation occurs at carbon, the more nucleophilic site relative to the carboxylate oxygen anion (see p. 513). this procedure is an alternative to the malonic ester synthesis (**10-67**) as a means of preparing carboxylic acids and has the advantage that acids of the form $RR'R^2CCOOH$ can also be prepared. In a related reaction, methylated aromatic acids can be alkylated at the methyl group by a similar procedure.[1692]

OS **V**, 526; **VI**, 517; **VII**, 249. See also, OS **VII**, 164.

10-71   Alkylation at a Position α to a Heteroatom.

2-(2-Alkyl-thio)de-halogenation

The presence of a sulfur atom on a carbon enhances the acidity of a proton on that carbon, and in dithioacetals and dithioketals that proton ($RSCH_2SR$) is even more acidic. 1,3-Dithianes can be alkylated[1693] if a proton is first removed by

[1689]Mladenova, M.; Blagoev, B.; Gaudemar, M.; Dardoize, F.; Lallemand, J.Y. *Tetrahedron* **1981**, *37*, 2153.
[1690]Cregar, P.L. *J. Am. Chem. Soc.* **1967**, *89*, 2500; **1970**, *92*, 1397; Pfeffer, P.E.; Silbert, L.S.; Chirinko, Jr., J.M. *J. Org. Chem.* **1972**, *37*, 451.
[1691]For lists of reagents, with references, see Larock, R.C. *Comprehensive Organic Transformations*, 2nd ed., Wiley-VCH, NY, **1999**, pp. 1717–1720*ff*.
[1692]Cregar, P.L. *J. Am. Chem. Soc.* **1970**, *92*, 1396.
[1693]Seebach, D.; Corey, E.J. *J. Org. Chem.* **1975**, *40*, 231. For reviews, see Page, P.C.B.; van Niel, M.B.; Prodger, J.C. *Tetrahedron* **1989**, *45*, 7643; Ager, D.J., in Hase, T.A. *Umpoled Synthons*, Wiley, NY, **1987**, pp. 19–37; Seebach, D. *Synthesis* **1969**, 17, especially pp. 24–27; Olsen, R.K.; Curriev, Jr., Y.O., in Patai, S. *The Chemistry of the Thiol Group*, pt. 2, Wiley, NY, **1974**, pp. 536–547.

treatment with butyllithium in THF.[1694] Since 1,3-dithianes can be prepared by treatment of an aldehyde or its acetal (see OS **VI**, 556) with 1,3-propanedithiol (**16-11**) and can be hydrolyzed (**10-7**), this is a method for the conversion of an aldehyde to a ketone[1695] (see also, **10-68** and **18-9**):

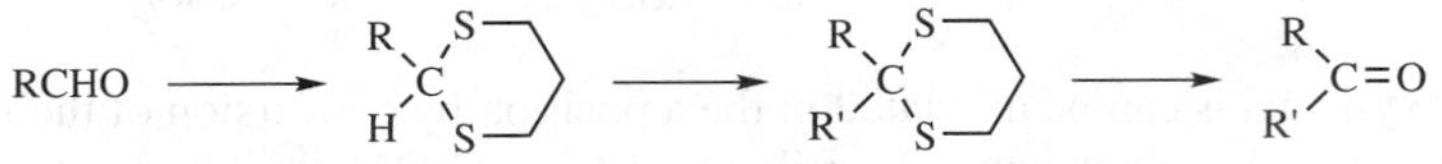

This is another example of Umpolung (see **10-68**);[1664] the normally electrophilic carbon of the aldehyde is made to behave as a nucleophile. The reaction can be applied to the unsubstituted dithiane (R = H) and one or two alkyl groups can be introduced, so a wide variety of aldehydes and ketones can be made starting with formaldehyde.[1696] The R′ group may be primary or secondary alkyl or benzylic. Iodides give the best results. The reaction has been used to close rings.[1697] A similar synthesis of aldehydes can be performed starting with ethyl ethylthiomethyl sulfoxide (EtSOCH$_2$SEt).[1698]

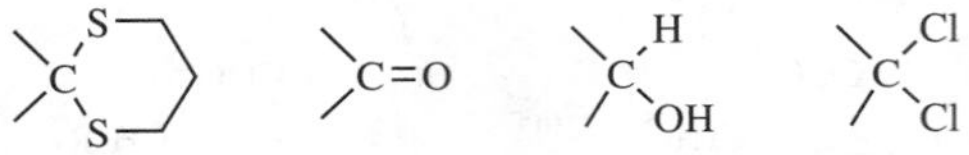

The group **A** may be regarded as a structural equivalent for the carbonyl group **B**, since introduction of **A** into a molecule is actually an indirect means of introducing **B**. It is convenient to have a word for units within molecules; such a word is *synthon*, introduced by Corey,[1699] which is defined as a structural unit within a molecule that can be formed and/or assembled by known or conceivable synthetic operations. There are many other synthons equivalent to **A** and **B**, for example, **C** (by reactions **19-36** and **19-3**) and **D** (by reactions **10-2** and **16-23**).[1700]

Carbanions generated from 1,3-dithianes also react with epoxides[1701] to give the expected products.

[1694]For an improved method of removing the proton, see Lipshutz, B.H.; Garcia, E. *Tetrahedron Lett.* **1990**, *31*, 7261.

[1695]For examples of the use of this reaction, with references, see Larock, R.C. *Comprehensive Organic Transformations*, 2nd ed., Wiley-VCH, NY, **1999**, pp. 1451–1454.

[1696]For a direct conversion of RX to RCHO, see **10-76**.

[1697]For example, see Seebach, D.; Jones, N.R.; Corey, E.J. *J. Org. Chem.* **1968**, *33*, 300; Hylton, T.; Boekelheide, V. *J. Am. Chem. Soc.* **1968**, *90*, 6887; Ogura, K.; Yamashita, M.; Suzuki, M.; Tsuchihashi, G. *Tetrahedron Lett.* **1974**, 3653.

[1698]Richman, J.E.; Herrmann, J.L.; Schlessinger, R.H. *Tetrahedron Lett.* **1973**, 3267. See also, Ogura, K.; Tsuchihashi, G. *Tetrahedron Lett.* **1971**, 3151; Schill, G.; Jones, P.R. *Synthesis* **1974**, 117; Hori, I.; Hayashi, T.; Midorikawa, H. *Synthesis* **1974**, 705.

[1699]Corey, E.J. *Pure Appl. Chem.* **1967**, *14*, 19, pp. 20–23.

[1700]For a long list of synthons for RCO, with references, see Hase, T.A.; Koskimies, J.K. *Aldrichimica Acta* **1982**, *15*, 35.

[1701]For example, see Corey, E.J.; Seebach, D. *J. Org. Chem.* **1975**, *40*, 231; Jones, J.B.; Grayshan, R. *Chem. Commun.* **1970**, 141, 741.

Another useful application of this reaction stems from the fact that dithianes can be desulfurated with Raney nickel (**14-27**). Aldehydes can therefore be converted to chain-extended hydrocarbons:[1702]

$$RCHO \longrightarrow \underset{R'}{\overset{R}{>}}\!\!C\!\!\underset{S}{\overset{S}{<}}\!\!\bigg] \xrightarrow{\text{Raney Ni}} \underset{R'}{\overset{R}{>}}\!\!CH_2$$

Similar reactions have been carried out with other thioacetals, as well as with compounds containing three thioether groups on a carbon.[1703]

If a stabilizing group other than sulfur is attached to the S-CH$_2$ unit of a thioether (RSCH$_2$X, where X is a stabilizing group), formation of the anion and alkylation can be facile. For example, benzylic and allylic thioethers (RSCH$_2$Ar and RSCH$_2$CH=CH$_2$)[1704] and thioethers of the form RSCH$_3$ (R = tetrahydrofuranyl or 2-tetrahydropyranyl)[1705] have been successfully alkylated at the carbon adjacent to the sulfur atom.[1706] Stabilization by one thioether group has also been used in a method for the homologation of primary halides.[1707] Thioanisole is treated with BuLi to give the corresponding anion,[1708] which reacts with the halide to give the thioether, which is then refluxed with a mixture of methyl iodide and sodium iodide in DMF to give the alkyl iodide as the final product (via an intermediate sulfonium salt). By this sequence an alkyl halide RX is converted to its homolog RCH$_2$X by a pathway involving two laboratory steps (see also, **10-64**).

Vinylic sulfides containing an a hydrogen can also be alkylated[1709] by alkyl halides or epoxides. This is a method for converting an alkyl halide RX to an α,β-unsaturated aldehyde, which is the synthetic equivalent of the unknown H$^{\ominus}$C=CH−CHO ion.[1710] Even simple alkyl aryl sulfides (RCH$_2$SAr and RR′CHSAr) have been alkylated to the sulfur.[1711]

[1702]For examples, see Hylton, T.; Boekelheide, V. *J. Am. Chem. Soc.* **1968**, *90*, 6887; Jones, J.B.; Grayshan, R.*Chem. Commun.* **1970**, 141, 741.

[1703]For example, see Seebach, D. *Angew. Chem. Int. Ed.* **1967**, *6*, 442; Olsson, K. *Acta Chem. Scand.* **1968**, *22*, 2390; Mori, K.; Hashimoto, H.; Takenaka, Y.; Takigawa, T. *Synthesis* **1975**, 720; Lissel, M. *Liebigs Ann. Chem.* **1982**, 1589.

[1704]Uemoto, K.; Kawahito, A.; Matsushita, N.; Skamoto, I.; Kaku, H.; Tsunoda, T. *Tetrahedron Lett.* **2001**, *42*, 905.

[1705]Block, E.; Aslam, M. *J. Am. Chem. Soc.* **1985**, *107*, 6729.

[1706]Biellmann, J.F.; Ducep, J.B. *Tetrahedron Lett.* **1968**, 5629; **1969**, 3707; *Tetrahedron* **1971**, *27*, 5861. See also, Narasaka, K.; Hayashi, M.; Mukaiyama, T. *Chem. Lett.* **1972**, 259.

[1707]Corey, E.J.; Jautelat, M. *Tetrahedron Lett.* **1968**, 5787.

[1708]Corey, E.J.; Seebach, D. *J. Org. Chem.* **1966**, *31*, 4097.

[1709]Oshima, K.; Shimoji, K.; Takahashi, H.; Yamamoto, H.; Nozaki, H. *J. Am. Chem. Soc.* **1973**, *95*, 2694.

[1710]For references to other synthetic equivalents of this ion, see Funk, R.L.; Bolton, G.L. *J. Am. Chem. Soc.* **1988**, *110*, 1290.

[1711]Dolak, T.M.; Bryson, T.A. *Tetrahedron Lett.* **1977**, 1961.

Sulfones[1712] and sulfonic esters can also be alkylated in the a position if strong enough bases are used.[1713] Alkylation at the α position of selenoxides allows the formation of alkenes, since selenoxides easily undergo elimination (**17-12**).[1714]

$$\underset{\text{Ph}}{\overset{\text{O}}{\underset{\text{Se}}{\parallel}}}\overset{\ominus}{\underset{\text{H}}{\text{C}}}\text{--CHR'}_2 \xrightarrow{\text{RX}} \underset{\text{Ph}}{\overset{\text{O}}{\underset{\text{Se}}{\parallel}}}\underset{\text{R}}{\overset{}{\text{C}}}\text{--CHR'}_2 \xrightarrow{\textbf{17-12}} \underset{\text{H}}{\overset{\text{R}}{\text{C}}}\text{=CR'}_2$$

Alkylation can also be carried out, in certain compounds, at positions α to other heteroatoms,[1715] for example, at a position α to the nitrogen of tertiary amines.[1716] Alkylation α to the nitrogen of primary or secondary amines is not generally feasible because an NH hydrogen is usually more acidic than a CH

$$\underset{R}{\overset{R^1}{\text{H--C--N}}}\text{--H} \xrightarrow{\textbf{12-50}} \underset{R}{\overset{R^1}{\text{H--C--N}}}\text{--NO} \xrightarrow{(i\text{Pr})_2\text{NLi}} \overset{\ominus}{\underset{R}{\overset{R^1}{\text{C--N}}}}\text{--NO} \xrightarrow{R^2X} \underset{R}{\overset{R^1}{R^2\text{--C--N}}}\text{--NO} \xrightarrow[\text{2. OH}^-]{\text{1. H}^+} \underset{R}{\overset{R^1}{R^2\text{--C--N}}}\text{--H}$$

hydrogen. α-Lithiation of *N*-Boc amines has been accomplished and these react with halides in the presence of a palladium catalyst.[1717] Alkylation α to the nitrogen atom of a carbamate occurs when the carbamate is treated with a Grignard reagent under electrolysis conditions.[1718] α-Methoxy amides also react with allyl halides and zinc metal to give alkylation via replacement of the OMe unit.[1719] It has been accomplished, however, by replacing the NH hydrogens with other (removable) groups.[1720] In one example, a secondary amine is converted to its *N*-nitroso derivative (**12-50**).[1721] The *N*-nitroso product is easily hydrolyzed to the product

---

[1712]For a review, see Magnus, P.D. *Tetrahedron* **1977**, *33*, 2019, 2022–2025. For alkylation of sulfones containing the F_3CSO_2 group, see Hendrickson, J.B.; Sternbach, D.D.; Bair, K.W. *Acc. Chem. Res.* **1977**, *10*, 306.

[1713]For examples, see Truce, W.E.; Hollister, K.R.; Lindy, L.B.; Parr, J.E. *J. Org. Chem.* **1968**, *33*, 43; Julia, M.; Arnould, D. *Bull. Soc. Chim. Fr.* **1973**, 743, 746; Bird, R.; Stirling, C.J.M. *J. Chem. Soc. B* **1968**, 111.

[1714]Reich, H.J.; Shah, S.K. *J. Am. Chem. Soc.* **1975**, *97*, 3250.

[1715]For a review of anions α to a selenium atom on small rings, see Krief, A. *Top. Curr. Chem.* **1987**, *135*, 1. For alkylation α to boron see Pelter, A.; Smith, K.; Brown, H.C. *Borane Reagents*, Academic Press, NY, **1988**, pp. 336–341.

[1716]Lepley, A.R.; Khan, W.A. *J. Org. Chem.* **1966**, *31*, 2061, 2064; *Chem. Commun.* **1967**, 1198; Lepley, A.R.; Giumanini, A.G. *J. Org. Chem.* **1966**, *31*, 2055; Ahlbrecht, H.; Dollinger, H. *Tetrahedron Lett.* **1984**, *25*, 1353.

[1717]Dieter, R.K.; Li, S. *Tetrahedron Lett.* **1995**, *36*, 3613.

[1718]Suga, S.; Okajima, M.; Yoshida, J.-i. *Tetrahedron Lett.* **2001**, *42*, 2173.

[1719]Kise, N.; Yamazaki, H.; Mabuchi, T.; Shono, T. *Tetrahedron Lett.* **1994**, *35*, 1561.

[1720]For a review, see Beak, P.; Zajdel, W.J.; Reitz, D.B. *Chem. Rev.* **1984**, *84*, 471.

[1721]Seebach, D.; Enders, D.; Renger, B. *Chem. Ber.* **1977**, *110*, 1852; Renger, B.; Kalinowski, H.; Seebach, D. *Chem. Ber.* **1977**, *110*, 1866. For a review, see Seebach, D.; Enders, D. *Angew. Chem. Int. Ed.* **1975**, *14*, 15.

amine (**19-51**).[1722] Alkylation of secondary and primary amines has also been accomplished with >10 other protecting groups, involving conversion of amines to amides, carbamates,[1723] formamidines,[1724] and phosphoramides.[1719] In the case of formamidines (**167**), use of a chiral R' leads to a chiral amine, in high ee, even when R is not chiral.[1725]

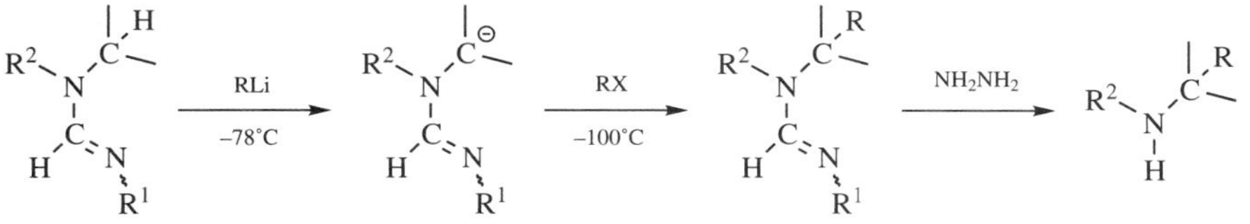

167

A proton can be removed from an allylic ether by treatment with an alkyllithium at about −70°C (at higher temperatures the Wittig rearrangement, **18-22**, takes place) to give the ion **168**, which reacts with alkyl halides to give the two products

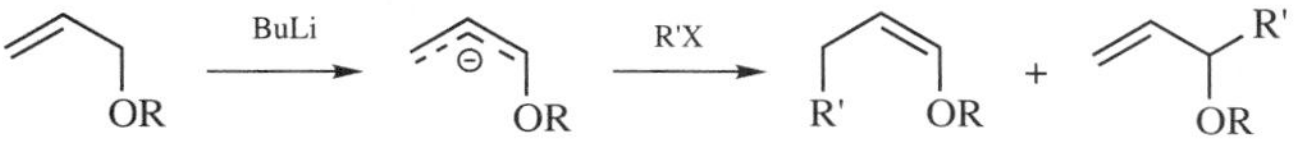

168

shown.[1726] Similar reactions[1727] have been reported for allylic[1728] and vinylic tertiary amines. In the latter case, enamines **169**, treated with a strong base, are converted to anions that are then alkylated, generally at C-3.[1729] (For direct alkylation of enamines at C-2, see **10-69**.)

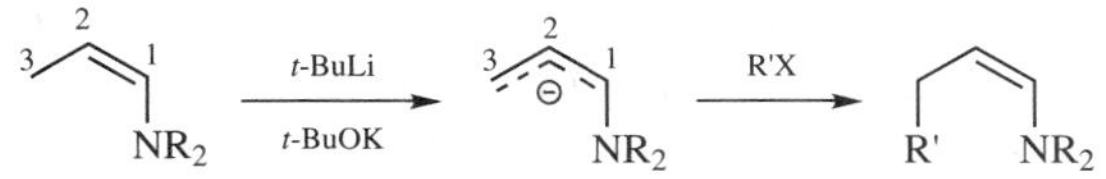

169

[1722]Fridman, A.L.; Mukhametshin, F.M.; Novikov, S.S. *Russ. Chem. Rev.* **1971**, *40*, 34, pp. 41–42.

[1723]For the use of *tert*-butyl carbamates, see Beak, P.; Lee, W. *Tetrahedron Lett.* **1989**, *30*, 1197.

[1724]For a review, see Meyers, A.I. *Aldrichimica Acta* **1985**, *18*, 59.

[1725]Gawley, R.E.; Hart, G.; Goicoechea-Pappas, M.; Smith, A.L. *J. Org. Chem.* **1986**, *51*, 3076; Gawley, R.E. *J. Am. Chem. Soc.* **1987**, *109*, 1265; Meyers, A.I.; Miller, D.B.; White, F. *J. Am. Chem. Soc.* **1988**, *110*, 4778; Gonzalez, M.A.; Meyers, A.I. *Tetrahedron Lett.* **1989**, *30*, 43, 47, and references cited therein.

[1726]Evans, D.A.; Andrews, G.C.; Buckwalter, B. *J. Am. Chem. Soc.* **1974**, *96*, 5560; Still, W.C.; Macdonald, T.L. *J. Am. Chem. Soc.* **1974**, *96*, 5561; Funk, R.L.; Bolton, G.L. *J. Am. Chem. Soc.* **1988**, *110*, 1290. For a similar reaction with triple-bond compounds, see Hommes, H.; Verkruijsse, H.D.; Brandsma, L. *Recl. Trav. Chim. Pays-Bas* **1980**, *99*, 113, and references cited therein.

[1727]For a review of allylic and benzylic carbanions substituted by heteroatoms, see Biellmann, J.F.; Ducep, J. *Org. React.* **1982**, *27*, 1.

[1728]Martin, S.F.; DuPriest, M.T. *Tetrahedron Lett.* **1977**, 3925, and references cited therein.

[1729]For a review, see Ahlbrecht, H. *Chimia* **1977**, *31*, 391.

It is also possible to alkylate a methyl, ethyl, or other primary group of an aryl ester ArCOOR, where Ar is a 2,4,6-trialkylphenyl group.[1730] Since esters can be hydrolyzed to alcohols, this constitutes an indirect alkylation of primary alcohols. Methanol has also been alkylated by converting it to $^{\ominus}CH_2O^{\ominus}$.[1731]

OS **VI**, 316, 364, 542, 704, 869; **VIII**, 573.

10-72   Alkylation of Dihydro-1,3-Oxazine: The Meyers Synthesis of Aldehydes, Ketones, and Carboxylic Acids

A synthesis of aldehydes[1732] developed by Meyers[1733] begins with the commercially available dihydro-1,3-oxazine derivatives **170** (A = H, Ph, or COOEt).[1734] Removal of a proton from the indicated carbon in **170** leads to the resonance stabilized and bidentate anion **172**. Alkylation occurs regioselectively at carbon by a many alkyl bromides and iodides. The R group of RX can be primary or secondary alkyl, allylic, or benzylic and can carry another halogen or a CN group.[1735] The alkylated oxazine **173** is then reduced and hydrolyzed to give an aldehyde containing two more carbons than the starting RX. This method thus complements **10-71**, which converts RX to an aldehyde containing one more carbon. Since A can be H, mono- or disubstituted acetaldehydes can be produced by this method.

The ion **171** also reacts with epoxides, to form γ-hydroxy aldehydes after reduction and hydrolysis,[1736] and with aldehydes and ketones (**16-38**). Similar aldehyde

[1730]Beak, P.; Carter. L.G. *J. Org. Chem.* **1981**, *46*, 2363.

[1731]Seebach, D.; Meyer, N. *Angew. Chem. Int. Ed.* **1976**, *15*, 438.

[1732]For examples of the preparation of aldehydes and ketones by the reactions in this section, see Larock, R.C. *Comprehensive Organic Transformations*, 2nd ed., Wiley-VCH, NY, **1999**, pp. 1461–1465.

[1733]Meyers, A.I.; Nabeya, A.; Adickes, H.W.; Politzer, I.R.; Malone, G.R.; Kovelesky, A.C.; Nolen, R.L.; Portnoy, R.C. *J. Org. Chem.* **1973**, *38*, 36.

[1734]For reviews of the preparation and reactions of **169**, see Schmidt, R.R. *Synthesis* **1972**, 333; Collington, E.W. *Chem. Ind. (London)* **1973**, 987.

[1735]Meyers, A.I.; Malone, G.R.; Adickes, H.W. *Tetrahedron Lett.* **1970**, 3715.

[1736]Adickes, H.W.; Politzer, I.R.; Meyers, A.I. *J. Am. Chem. Soc.* **1969**, *91*, 2155.

synthesis has also been carried out with thiazoles[1737] and thiazolines[1738] (five-membered rings containing N and S in the 1 and 3 positions).

The reaction has been extended to the preparation of ketones:[1739] Treatment of a dihydro-1,3-oxazine (**172**) with iodomethane forms the iminium salt **173** (**10-31**) which, when treated with a Grignard reagent or organolithium compound (**16-31**)

produces **174**, which can be hydrolyzed to a ketone. The R group can be alkyl, cycloalkyl, aryl, benzylic, and so on, and R' of the Grignard reagent can be alkyl, aryl, benzylic, or allylic. Note that the hetereocycles **170**, **172**, or **173** do not react directly with Grignard reagents. In another procedure, 2-oxazolines (**175**)[1740] can be alkylated to give **176**,[1741] which are easily converted directly to the esters **177** by heating in 5–7% ethanolic sulfuric acid.

2-Oxazolines **175** and **176** are thus synthons for carboxylic acids; this is another indirect method for the α alkylation of a carboxylic acid,[1742] representing an alternative to the malonic ester synthesis (**10-67**) and to **10-70** and **10-73**. The method can be adapted to the preparation of optically active carboxylic acids by the use of a chiral reagent.[1743] Note that, unlike **170**, **175** can be alkylated even if R is alkyl. However, the C=N bond of **175** and **176** cannot be effectively reduced, so that aldehyde synthesis is not feasible here.[1744]

OS **VI**, 905.

[1737]Altman, L.J.; Richheimer, S.L. *Tetrahedron Lett.* **1971**, 4709.

[1738]Meyers, A.I.; Durandetta, J.L. *J. Org. Chem.* **1975**, 40, 2021.

[1739]Meyers, A.I.; Smith, E.M. *J. Am. Chem. Soc.* **1970**, 92, 1084; *J. Org. Chem.* **1972**, 37, 4289.

[1740]For a review, see Meyers, A.I.; Mihelich, E.D. *Angew. Chem. Int. Ed.* **1976**, 15, 270.

[1741]Meyers, A.I.; Temple, Jr., D.L.; Nolen, R.L.; Mihelich, E.D. *J. Org. Chem.* **1974**, 39, 2778; Meyers, A.I.; Mihelich, E.D.; Nolen, R.L. *J. Org. Chem.* **1974**, 39, 2783; Meyers, A.I.; Mihelich, E.D.; Kamata, K. *J. Chem. Soc., Chem. Commun.* **1974**, 768.

[1742]For reviews, see Meyers, A.I. *Pure Appl. Chem.* **1979**, 51, 1255; *Acc. Chem. Res.* **1978**, 11, 375. See also, Hoobler, M.A.; Bergbreiter, D.E.; Newcomb, M. *J. Am. Chem. Soc.* **1978**, 100, 8182; Meyers, A.I.; Snyder, E.S.; Ackerman, J.J.H. *J. Am. Chem. Soc.* **1978**, 100, 8186.

[1743]For a review of asymmetric synthesis via chiral oxazolines, see Lutomski, K.A.; Meyers, A.I., in Morrison, J.D. *Asymmetric Synthesis*, Vol. 3, Academic Press, NY, **1984**, pp. 213–274.

[1744]Meyers, A.I.; Temple Jr., D.L. *J. Am. Chem. Soc.* **1970**, 92, 6644, 6646.

10-73 Alkylation with Trialkylboranes

Alkyl-de-halogenation

Trialkylboranes react rapidly and in high yields with α-halo ketones,[1745] α-halo esters,[1746] α-halo nitriles,[1747] and α-halo sulfonyl derivatives (sulfones, sulfonic esters, sulfonamides)[1748] in the presence of a base to give, respectively, alkylated ketones, esters, nitriles, and sulfonyl derivatives.[1749] Potassium *tert*-butoxide is often a suitable base, but potassium 2,6-di-*tert*-butylphenoxide at 0°C in THF gives better results in most cases, possibly because the large bulk of the two *tert*-butyl groups prevents the base from coordinating with the R_3B.[1750] The trialkylboranes are prepared by treatment of 3 equivalents of an alkene with 1 equivalent of BH_3 (**15-16**).[1751] With appropriate boranes, the R group transferred to α-halo ketones, nitriles, and esters can be vinylic,[1752] or (for α-halo ketones and esters) aryl.[1753]

The reaction can be extended to α,α-dihalo esters[1754] and α,α-dihalo nitriles.[1755] It is possible to replace just one halogen or both. In the latter case the two alkyl groups can be the same or different. When dialkylation is applied to dihalo nitriles, the two alkyl groups can be primary or secondary, but with dihalo esters, dialkylation is limited to primary R. Another extension is the reaction of boranes (BR_3) with γ-halo-α,β-unsaturated esters.[1756] Alkylation takes place in the γ position, but the double bond migrates out of conjugation with the COOEt unit [$BrCH_2$ $CH{=}CHCOOEt \rightarrow RCH{=}CHCH_2COOEt$]. In this case, however, double-bond

[1745]Brown, H.C.; Rogić, M.M.; Rathke, M.W. *J. Am. Chem. Soc.* **1968**, *90*, 6218.

[1746]Brown, H.C.; Rogić, M.M.; Rathke, M.W.; Kabalka, G.W. *J. Am. Chem. Soc.* **1968**, *90*, 818.

[1747]Brown, H.C.; Nambu, H.; Rogić, M.M. *J. Am. Chem. Soc.* **1969**, *91*, 6854.

[1748]Truce, W.E.; Mura, L.A.; Smith, P.J.; Young, F. *J. Org. Chem.* **1974**, *39*, 1449.

[1749]For reviews, see Negishi, E.; Idacavage, M.J. *Org. React.* **1985**, *33*, 1, 42–43, 143–150; Weill-Raynal, J. *Synthesis* **1976**, 633; Brown, H.C.; Rogić, M.M. *Organomet. Chem. Synth.* **1972**, *1*, 305; Rogić, M.M. *Intra-Sci. Chem. Rep.* **1973**, *7(2)*, 155; Brown, H.C. *Boranes in Organic Chemistry*, Cornell University Press, Ithaca, NY, **1972**, pp. 372–391, 404–409; Cragg, G.M.L. *Organoboranes in Organic Synthesis*, Marcel Dekker, NY, **1973**, pp. 275–278, 283–287.

[1750]Brown, H.C.; Nambu, H.; Rogić, M.M. *J. Am. Chem. Soc.* **1969**, *91*, 6852, 6854, 6855.

[1751]For an improved procedure, with B-9-BBN (see p. $$$), see Brown, H.C.; Rogić, M.M. *J. Am. Chem. Soc.* **1969**, *91*, 2146; Brown, H.C.; Rogić, M.M.; Nambu, H.; Rathke, M.W. *J. Am. Chem. Soc.* **1969**, *91*, 2147; Katz, J.; Dubois, J.E.; Lion, C. *Bull. Soc. Chim. Fr.* **1977**, 683.

[1752]Brown, H.C.; Bhat, N.G.; Campbell, Jr., J.B. *J. Org. Chem.* **1986**, *51*, 3398.

[1753]Brown, H.C.; Rogić, M.M. *J. Am. Chem. Soc.* **1969**, *91*, 4304.

[1754]Brown, H.C.; Rogić, M.M.; Rathke, M.W.; Kabalka, G.W. *J. Am. Chem. Soc.* **1968**, *90*, 1911.

[1755]Nambu, H.; Brown, H.C. *J. Am. Chem. Soc.* **1970**, *92*, 5790.

[1756]Brown, H.C.; Nambu, H. *J. Am. Chem. Soc.* **1970**, *92*, 1761.

migration is an advantage, because nonconjugated β,γ-unsaturated esters are usually much more difficult to prepare than their α,β-unsaturated isomers.

The alkylation of activated halogen compounds is one of several reactions of trialkylboranes developed by H.C. Brown[1757] (see also, **15-16**, **15-27**, **18-31-18-40**, and so on). These compounds are extremely versatile and can be used for the preparation of many types of compounds. In this reaction, for example, an alkene (through the BR_3 prepared from it) can be coupled to a ketone, a nitrile, a carboxylic ester, or a sulfonyl derivative. Note that this is still another indirect way to alkylate a ketone (see **10-68**) or a carboxylic acid (see **10-70**), and provides an additional alternative to the malonic ester and acetoacetic ester syntheses (**10-67**).

Although superficially this reaction resembles **10-57** it is likely that the mechanism is quite different, involving migration of an R group from boron to carbon (see also, **18-23–18-26**). The mechanism is not known with certainty,[1758] but it may be tentatively shown as (illustrated for an α-halo ketone):

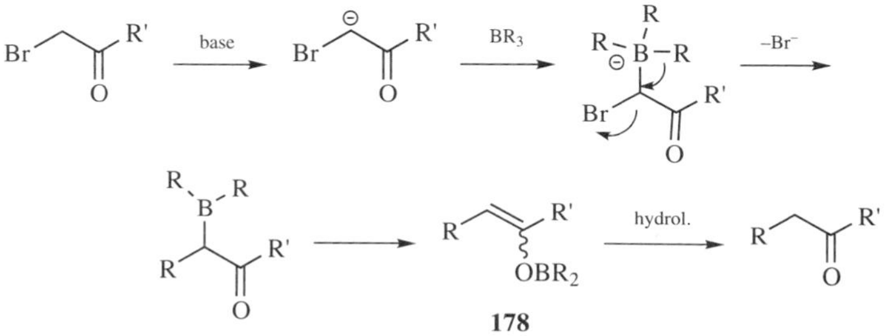

178

The first step is removal of the acidic proton by the base to give an enolate anion that combines with the borane (Lewis acid–base reaction). An R group then migrates, displacing the halogen leaving group.[1759] Another migration follows, this time of BR_2 from carbon to oxygen to give the enol borinate **178**,[1760] which is hydrolyzed. Configuration at R is retained.[1761]

[1757]Brown, H.C. *Organic Syntheses via Boranes*, Wiley, NY, *1975*; *Hydroboration*, W.A. Benjamin, NY, *1962*; *Boranes in Organic Chemistry*, Cornell University Press, Ithaca, NY, *1972*; Pelter, A.; Smith, K.; Brown, H.C. *Borane Reagents*, Academic Press, NY, *1988*.

[1758]See Prager, R.H.; Reece, P.A. *Aust. J. Chem. 1975*, 28, 1775.

[1759]It has been shown that this migration occurs stereospecifically with inversion in the absence of a solvent, but nonstereospecifically in the presence of a solvent, such as THF or dimethyl sulfide: Midland, M.M.; Zolopa, A.R.; Halterman, R.I. *J. Am. Chem. Soc. 1979*, *101*, 248. See also, Midland, M.M.; Preston, S.B. *J. Org. Chem. 1980*, 45, 747.

[1760]Pasto, D.J.; Wojtkowski, P.W. *Tetrahedron Lett. 1970*, 215, Pasto, D.J.; Wojtkowski, P.W. *J. Org. Chem. 1971*, 36, 1790.

[1761]Brown, H.C.; Rogić, M.M.; Rathke, M.W.; Kabalka, G.W. *J. Am. Chem. Soc. 1969*, 91, 2150.

The reaction has also been applied to compounds with other leaving groups. Diazo ketones, diazo esters, diazo nitriles, and diazo aldehydes (**179**)[1762] react with trialkylboranes in a similar manner.

$$ \underset{\textbf{179}}{\overset{\displaystyle O}{\underset{H}{\overset{\|}{C}}} - CHN_2} \xrightarrow[\text{THF} - H_2O]{R_3B} \overset{\displaystyle O}{\underset{H}{\overset{\|}{C}}} - CH_2R $$

The mechanism is probably also similar. In this case a base is not needed, since the carbon already has an available pair of electrons. The reaction with diazo aldehydes[1763] is especially notable, since successful reactions cannot be obtained with α-halo aldehydes.[1764]

OS **VI**, 919; **IX**, 107.

10-74   Alkylation at an Alkynyl Carbon

Alkynyl-de-halogenation

$$ RX + R'C \equiv C^- \longrightarrow RC \equiv CR' $$

The reaction between alkyl halides and acetylide ions is useful but of limited scope.[1765] Only primary halides unbranched in the β-position give good yields, although allylic halides can be used if CuI is present.[1766] If acetylene is the reagent, two different groups can be successively attached. Sulfates, sulfonates, and epoxides[1767] are sometimes used as substrates. The acetylide ion is often prepared by treatment of an alkyne with a strong base such as NaNH$_2$. Magnesium acetylides (ethynyl Grignard reagents; prepared as in **12-22**) are also frequently used, although they react only with active substrates, such as allylic, benzylic, and propargylic halides, and not with primary alkyl halides. Alternatively, the alkyl halide can be treated with a lithium acetylide–ethylenediamine complex.[1768] If 2 equivalents of a very

[1762]Hooz, J.; Gunn, D.M.; Kono, H. *Can. J. Chem.* **1971**, *49*, 2371; Mikhailov, B.M.; Gurskii, M.E. *Bull. Acad. Sci. USSR Div. Chem. Sci.* **1973**, *22*, 2588.

[1763]Hooz, J.; Morrison, G.F. *Can J. Chem.* **1970**, *48*, 868.

[1764]For an improved procedure, see Hooz, J.; Bridson, J.N.; Calzada, J.G.; Brown, H.C.; Midland, M.M.; Levy, A.B. *J. Org. Chem.* **1973**, *38*, 2574.

[1765]For reviews, see Ben-Efraim, D.A., in Patai, S. *The Chemistry of the Carbon–Carbon Triple Bond*, Wiley, NY, *1978*, pp. 790–800; Ziegenbein, W., in Viehe, H.G. *Acetylenes*, Marcel Dekker, NY, *1969*, pp. 185–206, 241–244. For a discussion of the best ways of preparing various types of alkyne, see Bernadou, F.; Mesnard, D.; Miginiac, L. *J. Chem. Res. (S) 1978*, 106; *1979*, 190.

[1766]Bourgain, M.; Normant, J.F. *Bull. Soc. Chim. Fr.* **1973**, 1777; Jeffery, T. *Tetrahedron Lett.* **1989**, *30*, 2225.

[1767]For example, see Fried, J.; Lin, C.; Ford, S.H. *Tetrahedron Lett.* **1969**, 1379; Krause, N.; Seebach, D. *Chem. Ber. 1988*, *121*, 1315.

[1768]Smith, W.N.; Beumel Jr., O.F. *Synthesis 1974*, 441.

strong base are used, alkylation can be effected at a carbon α to a terminal triple bond: $RCH_2C\equiv CH + 2BuLi \rightarrow RCHC\equiv C^- + R'Br \rightarrow RR'CHC\equiv C^-$.[1769] For another method of alkylating at an alkynyl carbon, see **18-26**. An alternative method for generating an alkyne anion treated a trialkylsilyl alkyne with potassium carbonate in methanol, and then methyllithium/LiBr.[1770] In the presence of an alkyl iodide, alkylation at the alkynyl carbon occurred.

Alkynes couple with alkyl halides in the presence of SmI_2/Sm.[1771] Alkynes react with hypervalent iodine compounds[1772] and with reactive alkanes such as adamantane in the presence of AIBN.[1773] The reaction of benzylic amines with terminal alkynes, in the presence of copper triflate and *tert*-butylhydroperoxide leads to incorporation of the alkyne group α to the nitrogen.[1774] A similar reaction occurs at a methyl group of *N,N*-dimethylaniline.[1775] α-Methoxycarbamates (MeO—CHR—NR^1—CO_2R^2) react with terminal alkynes and CuBr to give the alkynylamine.[1776] In the presence of $GaCl_3$, $ClC\equiv CSiMe_3$ reacts with silyl enol ethers to give, after treatment with methanolic acid, an α-ethynyl ketone.[1777]

1-Haloalkynes ($R—C\equiv C—X$) react with $ArSnBu_3$ and CuI to give $R—C\equiv C$ —Ar.[1778] Organozirconium compounds react in a similar manner.[1779] Acetylene reacts with 2 equivalents of iodobenzene, in the presence of a palladium catalyst and CuI, to give 1,2-diphenylethyne.[1780] 1-Trialkylsilyl alkynes react with 1-haloalkynes, in the presence of a CuCl catalyst, to give diynes[1781] and with aryl triflates to give 1-aryl alkynes.[1782]

In a related reaction, terminal alkynes react with silanes (R_3SiH) in the presence of an iridium catalyst to give the 1-trialkylsilyl alkyne.[1783] similar products are obtained when terminal alkynes react with *N*-trialkylsilylamines and $ZnCl_2$.[1784]

[1769]Bhanu, S.; Scheinmann, F. *J. Chem. Soc. Perkin Trans.1*, *1979*, 1218; Quillinan, A.J.; Scheinmann, F. *Org. Synth.* **VI**, 595.
[1770]Fiandanese, V.; Bottalico, D.; Marchese, G.; Punzi, A. *Tetrahedron Lett.* *2003*, *44*, 9087.
[1771]Murakami, M.; Hayashi, M.; Ito, Y. *Synlett*, *1994*, 179.
[1772]Kang, S.-K.; Lim, K.-H.; Ho, P.-S.; Kim, W.-Y. *Synthesis* *1997*, 874.
[1773]Xiang, J.; Jiang, W.; Fuchs, P.L. *Tetrahedron Lett.* *1997*, *38*, 6635.
[1774]Li, Z.; Li, C.-J. *Org. Lett.* *2004*, *6*, 4997.
[1775]Li, Z.; Li, C.-J. *J. Am. Chem. Soc.* *2004*, *126*, 11810.
[1776]Zhang, J.; Wei, C.; Lei, C.-J. *Tetrahedron Lett.* *2002*, *43*, 5731.
[1777]Arisawa, M.; Amemiya, R.; Yamaguchi, M. *Org. Lett.* *2002*, *4*, 2209.
[1778]Kang, S.-K.; Kim, W.-Y.; Jiao, X. *Synthesis* *1998*, 1252.
[1779]Liu, Y.; Xi, C.; Hara, R.; Nakajima, K.; Yamazaki, A.; Kotora, M.; Takahashi, T. *J. Org. Chem.* *2000*, *65*, 6951.
[1780]Pal, M.; Kundu, N.G. *J. Chem. Soc. Perkin Trans 1*, *1996*, 449. Also see, Nguefack, J.-F.; Bolitt, V.; Sinou, D. *Tetrahedron Lett*, *1996*, *37*, 5527.
[1781]Nishihara, Y.; Ikegashira, K.; Mori, A.; Hiyama, T. *Tetrahedron Lett.* *1998*, *39*, 4075.
[1782]Bumagin, N.A.; Sukhmolinova, L.I.; Luzikova, E.V.; Tolstaya, T.P.; Beletskaya, I.P. *Tetrahedron Lett.* *1996*, *37*, 897; Powell, N.A.; Rychnovsky, S.D. *Tetrahedron Lett.* *1996*, *37*, 7901; Nishihara, Y.; Ikegashira, K.; Mori, A.; Hiyama, T. *Chem. Lett.* *1997*, 1233.
[1783]Shimizu, R; Fuchikami, T. *Tetrahedron Lett.* *2000*, *41*, 907.
[1784]Andreev, A.A.; Konshin, V.V.; Komarov, N.V.; Rubin, M.; Brouwer, C.; Gevorgyan, V. *Org. Lett.* *2004*, *6*, 421.

OS **IV**, 117; **VI**, 273, 564, 595; **VIII**, 415; **IX**, 117, 477, 688; **76**, 263. Also see, OS **IV**, 801; **VI**, 925.

10-75 Preparation of Nitriles

Cyano-de-halogenation

$$RX + {}^-CN \longrightarrow RCN$$

The reaction between cyanide ion and alkyl halides is a convenient method for the preparation of nitriles.[1785] Primary, benzylic, and allylic halides give good yields of nitriles; secondary halides give moderate yields. The reaction fails for tertiary halides, which give elimination under these conditions. Many other groups on the molecule do not interfere. A number of solvents have been used, but the high yields and short reaction times observed with DMSO make it a very good solvent for this reaction.[1786] Other ways to obtain high yields under mild conditions are to use a phase-transfer catalyst,[1787] in alternative solvents, such as PEG 400 (a polyethylene glycol),[1788] or with ultrasound.[1789] This is an important way of increasing the length of a carbon chain by one carbon, since nitriles are easily hydrolyzed to carboxylic acids (**16-4**).

The cyanide ion is an ambident nucleophile (it can react via N or via C) and isocyanides (also called isonitriles, R—N≡C) may be side products.[1790] If the preparation of isocyanides is desired, they can be made the main products by the use of reagents with more covalent metal–carbon bonds, such as silver or copper(I) cyanide[1791] (p. 515). However, the use on an excess of LiCN in acetone/THF gave the nitrile as the major product.[1792] Tosyl cyanide (TolSO₂CN) has been used in some cases.[1793]

Vinylic bromides can be converted to vinylic cyanides with CuCN,[1794] with KCN, a crown ether, and a Pd(0) complex,[1795] or with KCN and a Ni(0)

[1785]For reviews, see, in Patai, S.; Rappoport, Z. *The Chemistry of Functional Groups, Supplement C*, pt. 1, Wiley, NY, *1983*, the articles by Fatiadi, A.J. pt. 2, pp. 1057–1303, and Friedrich, K. pt. 2, pp. 1343–1390; Friedrich, K.; Wallenfels, K., in Rappoport, Z. *The Chemistry of the Cyano Group*, Wiley, NY, *1970*, pp. 77–86.

[1786]Smiley, R.A.; Arnold, C. *J. Org. Chem.* *1960*, *25*, 257; Friedman, L.; Shechter, H. *J. Org. Chem.* *1960*, *25*, 877.

[1787]For reviews, see Starks, C.M.; Liotta, C. *Phase Transfer Catalysis*, Acaemic Press, NY, *1978*, pp. 94–112; Weber, W.P.; Gokel, G.W. *Phase Transfer Catalysis in Organic Synthesis*, Springer, NY, *1977*, pp. 96–108. See also, Bram, G.; Loupy, A.; Pedoussaut, M. *Tetrahedron Lett.* *1986*, *27*, 4171; *Bull. Soc. Chim. Fr.* *1986*, 124.

[1788]Cao, Y.-Q.; Che, B.-H.; Pei, B.-G. *Synth. Commun.* *2001*, *31*, 2203.

[1789]Ando, T.; Kawate, T.; Ichihara, J.; Hanafusa, T. *Chem. Lett.* *1984*, 725.

[1790]For a solid-phase synthesis of isonitriles see Luanay, D.; Booth, S.; Clemens, I.; Merritt, A.; Bradley, M. *Tetrahedron Lett.* *2002*, *43*, 7201.

[1791]For an example, see Jackson, H.L.; McKusick, B.C. *Org. Synth.* *IV*, 438.

[1792]Ciaccio, J.A.; Smrtka, M.; Maio, W.A.; Rucando, D. *Tetrahedron Lett.* *2004*, *45*, 7201.

[1793]Kim, S.; Song, H.-J. *Synlett 2002*, 2110.

[1794]For example, see Koelsch, C.F. *J. Am. Chem. Soc.* *1936*, *58*, 1328; Newman, M.S.; Boden, H. *J. Org. Chem.* *1961*, *26*, 2525; Lapouyade, R.; Daney, M.; Lapenue, M.; Bouas-Laurent, H. *Bull. Soc. Chim. Fr.* *1973*, 720.

[1795]Yamamura, K.; Murahashi, S. *Tetrahedron Lett.* *1977*, 4429.

catalyst.[1796] Halides can be converted to the corresponding nitriles by treatment with trimethylsilyl cyanide in the presence of catalytic amounts of $SnCl_4$: $R_3CCl + Me_3 SiCN \rightarrow R_3CCN$.[1797] Primary, secondary, and tertiary alcohols are converted to nitriles in good yields by treatment with NaCN, Me_3SiCl, and a catalytic amount of NaI in DMF–MeCN.[1798] Lewis acids have been used in conjunction with NaCN or KCN.[1799] α,β-Epxoy amides were opened to the β-cyano-α-hydroxyamide with Et_2AlCN.[1800] Cyanohydrins react with alkyl halides in some cases to give the nitrile.[1801]

Substrates that react with cyanide may contain leaving groups other than halides, such as esters of sulfuric and sulfonic acids (sulfates and sulfonates, respectively). Vinylic triflates give vinylic cyanides when treated with LiCN, a crown ether, and a palladium catalyst.[1802] Epoxides give β-hydroxy nitriles. The C-2-selectivity was observed when NaCN and $B(OMe)_3$ were reacted with a disubstituted epoxide.[1803] The use of trimethylsilyl cyanide (Me_3SiCN) and a Lewis acid generates the O-TMS β-hydroxy nitrile, and the use of $YbCl_3$ and a salen complex gave good enantioselectivity.[1804] One alkoxy group of acetals is replaced by CN $[R_2C(OR')_2 \rightarrow R_2C(OR')CN]$ with Me_3SiCN and a catalyst[1805] or with t-BuNC and $TiCl_4$.[1806] Tetrabutylammonium cyanide converted a primary alcohol to the corresponding nitrile in the presence of PPh_3/DDQ.[1807]

Sodium cyanide in HMPA selectively cleaves methyl esters in the presence of ethyl esters:

$$RCOOMe + CN^- \longrightarrow MeCN + RCOO^-. \quad [1808]$$

[1796]Sakakibara, Y.; Yadani, N.; Ibuki, I.; Sakai, M.; Uchino, N. *Chem. Lett.* **1982**, 1565; Procházka, M.; Siroky, M. *Collect. Czech. Chem. Commun.* **1983**, *48*, 1765.
[1797]Reetz, M.T.; Chatziiosifidis, I. *Angew. Chem. Int. Ed.* **1981**, *20*, 1017; Zieger, H.E.; Wo, S. *J. Org. Chem.* **1994**, *59*, 3838. See Tsuji, Y.; Yamada, N.; Tanaka, S. *J. Org. Chem.* **1993**, *58*, 16 for a similar reaction with allylic acetates. See Hayashi, M.; Tamura, M.; Oguni, N. *Synlett,* **1992**, 663 for a similar reaction with epoxides using a titanium catalyst.
[1798]Davis, R.; Untch, K.G. *J. Org. Chem.* **1981**, *46*, 2985. See also, Mizuno, A.; Hamada, Y.; Shioiri, T. *Synthesis* **1980**, 1007; Manna, S.; Falck, J.R.; Mioskowski, C. *Synth. Commun.* **1985**, *15*, 663; Camps, F.; Gasol, V.; Guerrero, A. *Synth. Commun.* **1988**, *18*, 445.
[1799]Ce(OTf)$_4$: Iranpoor, N.; Shekarriz, M. *Synth. Commun.* **1999**, *29*, 2249.
[1800]Ruano, J.L.G.; Fernández-Ibáñez, M.Á.; Castro, A.M.M.; Ramos, J.H.R.; Flamarique, A.C.R. *Tetrahedron Asymmetry* **2002**, *13*, 1321.
[1801]Dowd, P.; Wilk, B.K.; Wlostowski, M. *Synth. Commun.* **1993**, *23*, 2323; Wilk, B.K. *Synth. Commun.* **1993**, *23*, 2481 and see Ohno, H.; Mori, A.; Inoue, S. *Chem. Lett.* **1993**, 975 and Mitchell, D.; Koenig, T.M. *Tetrahedron Lett.* **1992**, *33*, 3281 for similar reactions with epoxides.
[1802]Piers, E.; Fleming, F.F. *J. Chem. Soc., Chem. Commun.* **1989**, 756.
[1803]Sasaki, M.; Tanino, K.; Hirai, A.; Miyashita, M. *Org. Lett.* **2003**, *5*, 1789.
[1804]Schaus, S.E.; Jacobsen, E.N. *Org. Lett.* **2000**, *2*, 1001.
[1805]Torii, S.; Inokuchi, T.; Kobayashi, T. *Chem. Lett.* **1984**, 897; Soga, T.; Takenoshita, H.; Yamada, M.; Mukaiyama, T. *Bull. Chem. Soc. Jpn.* **1990**, *63*, 3122.
[1806]Ito, Y.; Imai, H.; Segoe, K.; Saegusa, T. *Chem. Lett.* **1984**, 937.
[1807]Iranpoor, N.; Firouzabadi, H.; Akhlaghinia, B.; Nowrouzi, N. *J. Org. Chem.* **2004**, *69*, 2562.
[1808]Müller, P.; Siegfried, B. *Helv. Chim. Acta* **1974**, *57*, 987.

OS **I**, 46, 107, 156, 181, 254, 256, 536; **II**, 292, 376; **III**, 174, 372, 557; **IV**, 438, 496, 576; **V**, 578, 614.

10-76 Direct Conversion of Alkyl Halides to Aldehydes and Ketones

Formyl-de-halogenation

$$RX + Na_2Fe(CO)_4 \xrightarrow{PPh_3} \underset{\mathbf{180}}{RCOFe(CO)_3PPh_3^-} \xrightarrow{HOAc} RCHO$$

The direct conversion of alkyl bromides to aldehydes, with an increase in the chain length by one carbon, can be accomplished[1809] by treatment with sodium tetracarbonylferrate(-2)[1810] (*Collman's reagent*) in the presence of triphenylphosphine and subsequent quenching of **180** with acetic acid. The reagent $Na_2Fe(CO)_4$ can be prepared by treatment of iron pentacarbonyl $Fe(CO)_5$ with sodium amalgam in THF. Good yields are obtained from primary alkyl bromides; secondary bromides give lower yields. The reaction is generally not satisfactory for benzylic bromides, but a good yield of the ketone was obtained using benzyl chloride and aryl iodides.[1811] The initial species produced from RX and $Na_2Fe(CO)_4$ is the ion $RFe(CO)_4^-$ (which can be isolated[1812]); it then reacts with Ph_3P to give **180**.[1813]

The synthesis can be extended to the preparation of ketones in six distinct ways.[1814] These include quenching **180** with a second alkyl halide (R'X) rather than acetic acid; omitting PPh_3 with first RX and then adding the second, R'X; treatment with RX in the presence of CO,[1810] followed by treatment with R'X'; treatment with an acyl halide followed by treatment with an alkyl halide or an epoxide, gives an α,β-unsaturated ketone.[1815] The final variations involve reaction of alkyl halides or tosylates with $Na_2Fe(CO)_4$ in the presence of ethylene to give alkyl ethyl ketones;[1816] when 1,4-dihalides are used, five-membered cyclic ketones are prepared.[1817]

[1809]Cooke, Jr., M.P. *J. Am. Chem. Soc.* **1970**, *92*, 6080.

[1810]For a review of this reagent, see Collman, J.P. *Acc. Chem. Res.* **1975**, *8*, 342. For a review of the related tetracarbonylhydridoferrates MHFe(CO)₄, see Brunet, J. *Chem. Rev.* **1990**, *90*, 1041.

[1811]Dolhem, E.; Barhdadi, R.; Folest, J.C.; Nédélec, J.Y.; Troupel, M. *Tetrahedron* **2001**, *57*, 525.

[1812]Siegl, W.O.; Collman, J.P. *J. Am. Chem. Soc.* **1972**, *94*, 2516.

[1813]For the mechanism of the conversion RFe(CO)₄⁻ → **180**, see Collman, J.P.; Finke, R.G.; Cawse, J.N.; Brauman, J.I. *J. Am. Chem. Soc.* **1977**, *99*, 2515; **1978**, *100*, 4766.

[1814]For the first four of these methods, see Collman, J.P.; Winter, S.R.; Clark, D.R. *J. Am. Chem. Soc.* **1972**, *94*, 1788; Collman, J.P.; Hoffman, N.W. *J. Am. Chem. Soc.* **1973**, *95*, 2689.

[1815]Yamashita, M.; Yamamura, S.; Kurimoto, M.; Suemitsu, R. *Chem. Lett.* **1979**, 1067.

[1816]Cooke, Jr., M.P.; Parlman, R.M. *J. Am. Chem. Soc.* **1975**, *97*, 6863. The reaction was not successful for higher alkenes, except that where the double bond and the tosylate group are in the same molecule, five- and six-membered rings can be closed: see McMurry, J.E.; Andrus, A. *Tetrahedron Lett.* **1980**, *21*, 4687, and references cited therein.

[1817]Yamashita, M.; Uchida, M.; Tashika, H.; Suemitsu, R. *Bull. Chem. Soc. Jpn.* **1989**, *62*, 2728.

Yet another approach uses electrolysis conditions with the alkyl chloride, $Fe(CO)_5$ and a nickel catalyst and gives the ketone directly, in one step.[1818] In the first stage of methods 1, 2, and 3, primary bromides, iodides, and tosylates and secondary tosylates can be used. The second stage of the first four methods requires more active substrates, such as primary iodides or tosylates or benzylic halides. Method 5 has been applied to primary and secondary substrates.

Other acyl organometallic reagents are known. An acyl zirconium reagent, such as $RCOZr(Cl)Cp_2$, reacted with allylic bromide in the presence of CuI to give the corresponding ketone, but with allylic rearrangement.[1819]

Symmetrical ketones R_2CO can be prepared by treatment of a primary alkyl or benzylic halide with $Fe(CO)_5$ and a phase transfer catalyst,[1820] or from a halide RX (R = primary alkyl, aryl, allylic, or benzylic) and CO by an electrochemical method involving a nickel complex.[1821] Aryl, benzylic, vinylic, and allylic halides have been converted to aldehydes by treatment with CO and Bu_3SnH, with a Pd(0) catalyst.[1822] Various other groups do not interfere. Several procedures for the preparation of ketones are catalyzed by palladium complexes. Alkyl aryl ketones are formed in good yields by treatment of a mixture of an aryl iodide, an alkyl iodide, and a Zn–Cu couple with CO $(ArI + RI + CO \rightarrow RCOAr)$.[1823] Vinylic halides react with vinylic tin reagents in the presence of CO to give unsymmetrical divinyl ketones.[1824] Aryl, vinylic, and benzylic halides can be converted to methyl ketones $(RX \rightarrow RCOMe)$ by reaction with (α-ethoxyvinyl)tributyltin Bu_3Sn-$C(OEt){=}CH_2$.[1825] In addition, SmI_2 can be used to convert alkyl chloride to ketones, in the presence of 50 atm of CO.[1826] Carbonylation can also be done with Zn/CuI,[1827] Zn, and then $CoBr_2$,[1828] or with AIBN and $(Me_3Si)_3SiH$.[1829]

[1818]Dolhem, E.; Oçafrain, M.; Nédélec, J.Y.; Troupel, M. *Tetrahedron* **1997**, *53*, 17089; Yoshida, K.; Kobayashi, M.; Amano, S. *J. Chem. Soc. Perkin Trans. 1* **1992**, 1127.

[1819]Hanzawa, Y.; Narita, K.; Taguchi, T. *Tetrahedron Lett.* **2000**, *41*, 109.

[1820]Kimura, Y.; Tomita, Y.; Nakanishi, S.; Otsuji, Y. *Chem. Lett.* **1979**, 321; des Abbayes, H.; Clément, J.; Laurent, P.; Tanguy, G.; Thilmont, N. *Organometallics* **1988**, *7*, 2293.

[1821]Garnier, L.; Rollin, Y.; Périchon, J. *J. Organomet. Chem.* **1989**, *367*, 347.

[1822]Baillargeon, V.P.; Stille, J.K. *J. Am. Chem. Soc.* **1986**, *108*, 452. See also, Kasahara, A.; Izumi, T.; Yanai, H. *Chem. Ind. (London)* **1983**, 898; Pri-Bar, I.; Buchman, O. *J. Org. Chem.* **1984**, *49*, 4009; Takeuchi, R.; Tsuji, Y.; Watanabe, Y. *J. Chem. Soc., Chem. Commun.* **1986**, 351; Ben-David, Y.; Portnoy, M.; Milstein, D. *J. Chem. Soc., Chem. Commun.* **1989**, 1816.

[1823]Tamaru, Y.; Ochiai, H.; Yamada, Y.; Yoshida, Z. *Tetrahedron Lett.* **1983**, *24*, 3869.

[1824]Goure, W.F.; Wright, M.E.; Davis, P.D.; Labadie, S.S.; Stille, J.K. *J. Am. Chem. Soc.* **1984**, *106*, 6417. For a similar preparation of diallyl ketones, see Merrifield, J.H.; Godschalx, J.P.; Stille, J.K. *Organometallics* **1984**, *3*, 1108.

[1825]Kosugi, M.; Sumiya, T.; Obara, Y.; Suzuki, M.; Sano, H.; Migita, T. *Bull. Chem. Soc. Jpn.* **1987**, *60*, 767.

[1826]Ogawa, A.; Sumino, Y.; Nanke, T.; Ohya, S.; Sonoda, N.; Hirao, T. *J. Am. Chem. Soc.*, **1997**, *119*, 2745.

[1827]Tsunoi, S.; Ryu, I.; Fukushima, H.; Tanaka, M.; Komatsu, M.; Sonoda, N. *Synlett*, **1995**, 1249.

[1828]Devasagayaraj, A.; Knochel, P. *Tetrahedron Lett.* **1995**, *36*, 8411.

[1829]Ryu, I.; Hasegawa, M.; Kurihara, A.; Ogawa, A.; Tsunoi, S.; Sonoda, N. *Synlett*, **1993**, 143.

The conversion of alkyl halides to aldehydes and ketones can also be accomplished indirectly (**10-71**). See also, **12-33**.

OS **VI**, 807.

10-77   Carbonylation of Alkyl Halides, Alcohols, or Alkanes

Alkoxycarbonyl-de-halogenation

$$RX + CO + R'OH \xrightarrow[-70°C]{SbCl_5-SO_2} RCOOR'$$

A direct method for preparing a carboxylic acid treats an alkyl halide with $NaNO_2$ in acetic acid and DMSO.[1830] Reaction of an alkyl halide with ClCO-CO_2Me and $(Bu_3Sn)_2$ under photochemical conditions leads to the corresponding methyl ester.[1831]

Several methods, all based on carbon monoxide or metal carbonyls, have been developed for converting an alkyl halide to a carboxylic acid or an acid derivative with the chain extended by one carbon.[1832] When an alkyl halide is treated with $SbCl_5-SO_2$ at $-70°C$, it dissociates into the corresponding carbocation (p. 236). If carbon monoxide and an alcohol are present, a carboxylic ester is formed by the following route:[1833]

$$R-X \xrightarrow[-70°C]{SbCl_5-SO_2} R^+ X^- \xrightarrow{CO} \underset{R}{\overset{O \to SbCl_3}{\underset{X}{C}}} \xrightarrow{R'OH} \underset{R}{\overset{O}{C}}\underset{H}{\overset{\oplus}{O}}R' \xrightarrow{-H^+} \underset{R}{\overset{O}{C}}OR'$$

This has also been accomplished with concentrated H_2SO_4 saturated with CO.[1834] Not surprisingly, only tertiary halides perform satisfactorily; secondary halides give mostly rearrangement products. An analogous reaction takes place with alkanes possessing a tertiary hydrogen, using $HF-SbF_5-CO$.[1835]

Carboxylic acids or esters are the products, depending on whether the reaction mixture is solvolyzed with water or an alcohol. Alcohols with more than seven

[1830]Matt, C.; Wagner, A.; Mioskowski, C. *J. Org. Chem.* **1997**, *62*, 234.

[1831]Kim, S.; Jon, S.Y. *Tetrahedron Lett.* **1998**, *39*, 7317.

[1832]For discussions of most of the reactions in this section, see Colquhoun, H.M.; Holton, J.; Thompson, D.J.; Twigg, M.V. *New Pathways for Organic Synthesis*; Plenum, NY, **1984**, pp. 199–204, 212–220, 234–235. For lists of reagents, with references, see Larock, R.C. *Comprehensive Organic Transformations*, 2nd ed., Wiley-VCH, NY, **1999**, pp. 1684–1685, 1694–1698, 1702–1704.

[1833]Yoshimura, M.; Nojima, M.; Tokura, N. *Bull. Chem. Soc. Jpn.* **1973**, *46*, 2164; Puzitskii, K.V.; Pirozhkov, S.D.; Ryabova, K.G.; Myshenkova, T.N.; Éidus, Ya.T. *Bull. Acad. Sci. USSR Div. Chem. Sci.* **1974**, *23*, 192.

[1834]Takahashi, Y.; Yoneda, N. *Synth. Commun.* **1989**, *19*, 1945.

[1835]Paatz, R.; Weisgerber, G. *Chem. Ber.* **1967**, *100*, 984. For a related reaction using $AlBr_3$ see Akhrem, I.; Afanas'eva, L.; Petrovskii, P.; Vitt, S.; Orlinkov, A. *Tetrahedron Lett.* **2000**, *41*, 9903.

carbons are cleaved into smaller fragments by this procedure.[1836] Similarly, tertiary alcohols[1837] react with H_2SO_4 and CO (which is often generated from HCOOH and the H_2SO_4 in the solution) to give trisubstituted acetic acids in a process called the *Koch–Haaf reaction* (see also, **15-35**).[1838] If a primary or secondary alcohol is the substrate, the carbocation initially formed rearranges to a tertiary ion before reacting with the CO. Better results are obtained if trifluoromethanesulfonic acid F_3CSO_2OH is used instead of H_2SO_4.[1839] Iodo alcohols were transformed into lactones under radical conditions (AIBN, allylSnBu$_3$) and 45 atm of CO.[1840]

Another method[1841] for the conversion of alkyl halides to carboxylic esters is treatment of a halide with nickel carbonyl $Ni(CO)_4$ in the presence of an alcohol and its conjugate base.[1842] When R′ is primary, RX may only be a vinylic or an aryl halide; retention of configuration is observed at a vinylic R. Consequently, a carbocation intermediate is not involved here. When R′ is tertiary, R may be primary alkyl as well as vinylic or aryl. This is thus one of the few methods for preparing esters of tertiary alcohols. Alkyl iodides give the best results, then bromides. In the presence of an amine, an amide can be isolated directly, at least in some instances.

$$RX \ + \ Ni(CO)_4 \ \xrightarrow[\text{R'OH}]{\text{R'O}^-} \ RCOOR'$$

Still another method for the conversion of halides to acid derivatives makes use of $Na_2Fe(CO)_4$. As described in **10-76**, primary and secondary alkyl halides and tosylates react with this reagent to give the ion $RFe(CO)_4^-$ or, if CO is present, the ion $RCOFe(CO)_4^-$. Treatment of $RFe(CO)_4^-$ or $RCOFe(CO)_4^-$ with oxygen or sodium hypochlorite gives, after hydrolysis, a carboxylic acid.[1843] Alternatively, $RFe(CO)_4^-$ or $RCOFe(CO)_4^-$ reacts with a halogen (e.g., I_2) in the presence of an

[1836]Yoneda, N.; Takahashi, Y.; Fukuhara, T.; Suzuki, A. *Bull. Chem. Soc. Jpn.* **1986**, *59*, 2819.

[1837]For reviews of other carbonylation reactions of alcohols and other saturated oxygenated compounds, see Bahrmann, H.; Cornils, B., in Falbe, J. *New Syntheses with Carbon Monoxide*, Springer, NY, **1980**, pp. 226–241; Piacenti, F.; Bianchi, M. in Wender, I.; Pino, P. *Organic Syntheses via Metal Carbonyls*, Vol. 2, Wiley, NY, **1977**, pp. 1–42.

[1838]For a review, see Bahrmann, H., in Falbe, J. *New Syntheses with Carbon Monoxide*, Springer, NY, **1980**, pp. 372–413.

[1839]Booth, B.L.; El-Fekky, T.A. *J. Chem. Soc. Perkin Trans. 1* **1979**, 2441.

[1840]Kreimerman, S.; Ryu, I.; Minakata, S.; Komatsu, M. *Org. Lett.* **2000**, *2*, 389.

[1841]For reviews of methods involving transition metals, see Collman, J.P.; Hegedus, L.S.; Norton, J.R.; Finke, R.G. *Principles and Applications of Organotransition Metal Chemistry*, 2nd ed., University Science Books, Mill Valley, CA, **1987**, pp. 749–768; Anderson, G.K.; Davies, J.A., in Hartley, F.R.; Patai, S. *The Chemistry of the Metal–Carbon Bond*, Vol. 3, Wiley, NY, pp. 335–359, pp. 348–356; Heck, R.F. *Adv. Catal.*, **1977**, *26*, 323, see pp. 323; Cassar, L.; Chiusoli, G.P.; Guerrieri, F. *Synthesis* **1973**, 509.

[1842]Corey, E.J.; Hegedus, L.S. *J. Am. Chem. Soc.* **1969**, *91*, 1233. See also, Crandall, J.K.; Michaely, W.J. *J. Organomet. Chem.* **1973**, *51*, 375.

[1843]Collman, J.P.; Winter, S.R.; Komoto, R.G. *J. Am. Chem. Soc.* **1973**, *95*, 249.

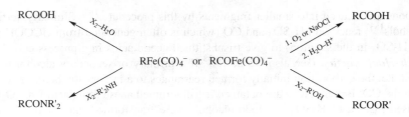

alcohol to give a carboxylic ester,[1844] or in the presence of a secondary amine or water to give, respectively, the corresponding amide or free acid. The compound $RFe(CO)_4^-$ and $RCOFe(CO)_4^-$, what are prepared from primary R, give high yields. With secondary R, the best results are obtained in the solvent THF by the use of $RCOFe(CO)_4^-$ prepared from secondary tosylates. Ester and keto groups may be present in R without being affected. Carboxylic esters RCO_2R' have also been prepared by treating primary alkyl halides RX with alkoxides $R'O^-$ in the presence of $Fe(CO)_5$.[1845] $RCOFe(CO)_4^-$ is presumably an intermediate.

Palladium complexes also catalyze the carbonylation of halides.[1846] Aryl (see **13-15**),[1847] vinylic,[1848] benzylic, and allylic halides (especially iodides) can be converted to carboxylic esters with CO, an alcohol or alkoxide, and a palladium complex.[1849] Similar reactivity was reported with vinyl triflates.[1850] α-Halo ketones are converted to β-keto esters with CO, an alcohol, NBu₃ and a palladium catalyst at 110°C.[1851] Use of an amine instead of the alcohol or alkoxide leads to an amide.[1852]

[1844]Collman, J.P.; Winter, S.R.; Komoto, R.G. *J. Am. Chem. Soc.* **1973**, *95*, 249; Masada, H.; Mizuno, M.; Suga, S.; Watanabe, Y.; Takegami, Y. *Bull. Chem. Soc. Jpn.* **1970**, *43*, 3824.

[1845]Yamashita, M.; Mizushima, K.; Watanabe, Y.; Mitsudo, T.; Takegami,Y. *Chem. Lett.* **1977**, 1355. See also, Tanguy, G.; Weinberger, B.; des Abbayes, H. *Tetrahedron Lett.* **1983**, *24*, 4005.

[1846]For reviews, see Gulevich, Yu.V.; Bumagin, N.A.; Beletskaya, I.P. *Russ. Chem. Rev.* **1988**, *57*, 299, 303–309; Heck, R.F. *Palladium Reagents in Organic Synthesis*, Academic Press, NY, **1985**, pp. 348–356, 366–370.

[1847]For an example, see Bessard, Y; Crettaz, R. *Heterocycles* **1999**, *51*, 2589.

[1848]For conversion of vinylic triflates to carboxylic esters and amides, see Cacchi, S.; Morera, E.; Ortar, G. *Tetrahedron Lett.* **1985**, *26*, 1109.

[1849]Tsuji, J.; Kishi, J.; Imamura, S.; Morikawa, M. *J. Am. Chem. Soc.* **1964**, *86*, 4350; Schoenberg, A.; Bartoletti, I.; Heck, R.F. *J. Org. Chem.* **1974**, *39*, 3318; Adapa, S.R.; Prasad, C.S.N. *J. Chem. Soc. Perkin Trans. 1* **1989**, 1706; Kiji, J.; Okano, T.; Higashimae, Y.; Kukui, Y. *Bull. Chem. Soc. Jpn.* **1996**, *69*, 1029; Okano, T.; Okabe, N.; Kiji, J. *Bull. Chem. Soc. Jpn.* **1992**, *65*, 2589.

[1850]Jutand, A.; Négri, S. *Synlett*, **1997**, 719.

[1851]Lapidus, A.L.; Eliseev, O.L.; Bondarenko, T.N.; Sizan, O.E.; Ostapenko, A.G.; Beletskaya, I.P. *Synthesis* **2002**, 317.

[1852]Schoenberg, A.; Heck, R.F. *J. Org. Chem.* **1974**, *39*, 3327. See also, Lindsay, L.M.; Widdowson, D.A. *J. Chem. Soc. Perkin Trans. 1* **1988**, 569; Cai, M.-Z.; Song, C.-S.; Huang, X. *Synth. Commun.* **1997**, *27*, 361. For a review of some methods of amide formation that involve transition metals, see Screttas, C.G.; Steele, B.R. *Org. Prep. Proceed. Int.* **1990**, *22*, 271, 288–314. See Satoh, T.; Ikeda, M.; Kushino, Y.; Miura, M.; Nomura, M. *J. Org. Chem.* **1997**, *62*, 2662 for the carbonylation of an alcohol to give the corresponding ester by a similar method.

Reaction with an amine, AIBN, CO and a tetraalkyltin catalyst also leads to an amide.[1853] Benzylic and allylic halides were converted to carboxylic acids electrocatalytically, with CO and a cobalt imine complex.[1854] Vinylic halides were similarly converted with CO and nickel cyanide, under phase-transfer conditions.[1855] Allylic O-phosphates were converted to allylic amides with CO and ClTi=NTMS, in the presence of a palladium catalyst.[1856] Terminal alkynes were converted to the alkynyl ester using CO, $PdBr_2$, $CuBr_2$ in methanol and sodium bicarbonate.[1857]

Other organometallic reagents can be used to convert alkyl halides to carboxylic acid derivatives. Benzylic halides were converted to carboxylic esters with CO in the presence of a rhodium complex.[1858] Variations introduce the R' group via an ether $R_2'O$,[1859] a borate ester $B(OR')_3$,[1860] or an Al, Ti, or Zr alkoxide.[1861] The reaction of an alkene, a primary alcohol and CO, in the presence of a rhodium catalyst, led to carbonylation of the alkene and formation of the corresponding ester.[1862] Vinyl triflates were converted to the conjugated carboxylic acid with CO_2 and a nickel catalyst.[1863] Hydrogen peroxide with a catalytic amount of $Na_2WO_4 \cdot 2\ H_2O$ converted benzylic chlorides to the corresponding benzoic acid.[1864] Reaction with an α,ω-diiodide, Bu_4NF and $Mo(CO)_6$ gave the corresponding lactone.[1865]

Reaction of an alkyl halide with $(MeS)_3C$—Li followed by aqueous HBF_4 leads to a thioester.[1866]

A number of double carbonylations have been reported. In these reactions, two molecules of CO are incorporated in the product, leading to α-keto acids or their derivatives.[1867] When the catalyst is a palladium complex, best results are obtained in the formation of α-keto amides.[1868] R is usually aryl or vinylic.[1869] The formation

[1853]Ryu, I.; Nagahara, K.; Kambe, N.; Sonoda, N.; Kreimerman, S.; Komatsu, M. *Chem. Commun.* **1998**, 1953.
[1854]Folest, J.; Duprilot, J.; Perichon, J.; Robin, Y.; Devynck, J. *Tetrahedron Lett.* **1985**, *26*, 2633. See also, Miura, M.; Okuro, K.; Hattori, A.; Nomura, M. *J. Chem. Soc. Perkin Trans. 1* **1989**, 73; Urata, H.; Goto, D.; Fuchikami, T. *Tetrahedron Lett.* **1991**, *32*, 3091; Isse, A.A.; Gennaro, A. *Chem. Commun.* **2002**, 2798.
[1855]Alper, H.; Amer, I.; Vasapollo, G. *Tetrahedron Lett.* **1989**, *30*, 2615. See also, Amer, I.; Alper, H. *J. Am. Chem. Soc.* **1989**, *111*, 927.
[1856]Ueda, K.; Mori, M. *Tetrahedron Lett.* **2004**, *45*, 2907. For an intramolecular carbonylation to generate a cyclic amide, see Trost, B.M.; Ameriks, M.K. *Org. Lett.* **2004**, *6*, 1745.
[1857]Li, J.; Jiang, H.; Chen, M. *Synth. Commun.* **2001**, *31*, 199.
[1858]For an example, see Giroux, A.; Nadeau, C.; Han, Y. *Tetrahedron Lett.* **2000**, *41*, 7601.
[1859]Buchan, C.; Hamel, N.; Woell, J.B.; Alper, H. *Tetrahedron Lett.* **1985**, *26*, 5743.
[1860]Alper, H.; Hamel, N.; Smith, D.J.H.; Woell, J.B. *Tetrahedron Lett.* **1985**, *26*, 2273.
[1861]Woell, J.B.; Fergusson, S.B.; Alper, H. *J. Org. Chem.* **1985**, *50*, 2134.
[1862]Yokoa, K.; Tatamidani, H.; Fukumoto, Y.; Chatani, N. *Org. Lett.* **2003**, *5*, 4329.
[1863]Senboku, H.; Kanaya, H.; Tokuda, M. *Synlett* **2002**, 140.
[1864]Shi, M.; Feng, Y.-S. *J. Org. Chem.* **2001**, *66*, 3235.
[1865]Imbeaux, M.; Mestdagh, H.; Moughamir, K.; Rolando, C. *J. Chem. Soc., Chem. Commun.* **1992**, 1678.
[1866]Barbero, M.; Cadamuro, S.; Degani, I.; Dughera, S.; Fochi, R. *J. Chem. Soc. Perkin Trans. 1* **1993**, 2075.
[1867]For a review, see Collin, J. *Bull. Soc. Chim. Fr.* **1988**, 976.
[1868]Kobayashi, T.; Tanaka, M. *J. Organomet. Chem.* **1982**, *233*, C64; Ozawa, F.; Sugimoto, T.; Yuasa, Y.; Santra, M.; Yamamoto, T.; Yamamoto, A. *Organometallics* **1984**, *3*, 683.
[1869]Son, T.; Yanagihara, H.; Ozawa, F.; Yamamoto, A. *Bull. Chem. Soc. Jpn.* **1988**, *61*, 1251.

of α-keto acids[1870] or esters[1871] requires more severe conditions. α-Hydroxy acids were obtained from aryl iodides when the reaction was carried out in the presence of an alcohol, which functioned as a reducing agent.[1872] Cobalt catalysts have also been used and require lower CO pressures.[1867]

OS **V**, 20, 739.

[1870]Tanaka, M.; Kobayashi, T.; Sakakura, T. *J. Chem. Soc., Chem. Commun.* **1985**, 837.

[1871]See Ozawa, F.; Kawasaki, N.; Okamoto, H.; Yamamoto, T.; Yamamoto, A. *Organometallics* **1987**, *6*, 1640.

[1872]Kobayashi, T.; Sakakura, T.; Tanaka, M. *Tetrahedron Lett.* **1987**, *28*, 2721.

CHAPTER 11

Aromatic Substitution, Electrophilic

Most substitutions at an aliphatic carbon are by nucleophiles. In aromatic systems the situation is reversed, because the high electron density at the aromatic ring leads to its reactivity as a Lewis base or a Brønsted–Lowry base, depending on the positive species. In electrophilic substitutions, a positive ion or the positive end of a dipole or induced dipole is attacked by the aromatic ring. The leaving group (the electrofuge) must necessarily depart without its electron pair. In nucleophilic substitutions, the chief leaving groups are those best able to carry the unshared pair: Br^-, H_2O, OTs^-, and so on., that is, the weakest bases. In electrophilic substitutions the most important leaving groups are those that can best exist without the pair of electrons necessary to fill the outer shell, that is, the weakest Lewis acids.

MECHANISMS

Electrophilic aromatic substitutions are unlike nucleophilic substitutions in that the large majority proceed by just one mechanism with respect to the substrate.[1] In this mechanism, which we call the *arenium ion mechanism*, the electrophile (which can be viewed as a Lewis acid) is attacked by the π-electrons of the aromatic ring (behaving as a Lewis base in most cases) in the first step. This reaction leads to formation of a new C–X bond and a new sp^3 carbon in a positively charged intermediate called an arenium ion, where X is the electrophile. The positively charged intermediate (the arenium ion) is resonance stabilized, but not aromatic. Loss of a proton from the sp^3 carbon that is "adjacent" to the positive carbon in the arenium ion, in what is effectively an E1 process (see p. 1487), is driven by rearomatization of the ring from the arenium ion to give the aromatic substitution product. A proton

[1]For monographs, see Taylor, R. *Electrophilic Aromatic Substitution*, Wiley, NY, ***1990***; Katritzky, A.R.; Taylor, R. *Electrophilic Substitution of Heterocycles: Quantitative Aspects* (Vol. 47 of *Adv. Heterocycl. Chem.*), Academic Press, NY, ***1990***. For a review, see Taylor, R., in Bamford, C.H.; Tipper, C.F.H. *Comprehensive Chemical Kinetics*, Vol. 13, Elsevier, NY, ***1972***, pp. 1–406.

March's Advanced Organic Chemistry: Reactions, Mechanisms, and Structure, Sixth Edition, by Michael B. Smith and Jerry March therefore becomes the leaving group in this overall transformation, where X replaces H. The IUPAC designation for this mechanism is $A_E + D_E$. Another mechanism, much less common, consists of the opposite behavior: a leaving group departs *before* the electrophile arrives. In this case, a substituent (*not* H) is attached to the aromatic ring, and the substituent is lost prior to incorporation of the electrophile. This mechanism, the S_E1 mechanism, corresponds to the S_N1 mechanism of nucleophilic substitution. Simultaneous attack and departure mechanisms (corresponding to S_N2) are not found at all. An addition–elimination mechanism has been postulated in one case (see **11-6**).

The Arenium Ion Mechanism[2]

In the arenium ion mechanism the electrophilic species may be produced in various ways, but when H is replaced by X conversion of the aromatic ring to an arenium ion is basically the same in all cases. For this reason, most attention in the study of this mechanism centers around the identity of the electrophilic entity and how it is produced.

The electrophile may be a positive ion or be a molecule that has a positive dipole. If it is a positive ion, it is attacked by the ring (a pair of electrons from the aromatic sextet is donated to the electrophile) to give a carbocation. This intermediate is a resonance hybrid as shown in **1**, but is often represented as in **2**. For convenience, the H atom to be replaced by X is shown in **1**. Ions of this type are called[3] *Wheland intermediates*, σ *complexes*, or *arenium ions*.[4] The inherent stability associated with aromaticity is no longer present in **1**, but the ion is stabilized by resonance. For this reason, the arenium ion is generally a highly reactive intermediate, although there are cases in which it has been isolated (see p. 661).

Carbocations can react in various ways (see p. 247), but for this type of ion the most likely pathway[5] is loss of either X^+ or H^+. In the second step of the

[2]This mechanism is sometimes called the S_E2 mechanism because it is bimolecular, but in this book we reserve that name for aliphatic substrates (see Chapter 12).
[3]General agreement on what to call these ions has not yet been reached. The term σ complex is a holdover from the time when much less was known about the structure of carbocations and it was thought they might be complexes of the type discussed in Chapter 3. Other names have also been used. We will call them arenium ions, following the suggestion of Olah, G.A. *J. Am. Chem. Soc.* **1971**, *94*, 808.
[4]For reviews of arenium ions formed by addition of a proton to an aromatic ring, see Brouwer, D.M.; Mackor, E.L.; MacLean, C. in Olah, G.A.; Schleyer, P.V.R. *Carbonium Ions*, vol. 2, Wiley, NY, **1970**, pp. 837–897; Perkampus, H. *Adv. Phys. Org. Chem.* **1966**, *4*, 195.
[5]For a discussion of cases in which **1** stabilizes itself in other ways, see de la Mare, P.B.D. *Acc. Chem. Res.* **1974**, *7*, 361.

mechanism, the reaction proceeds with loss of the proton and the aromatic sextet is restored in the final product **3**.

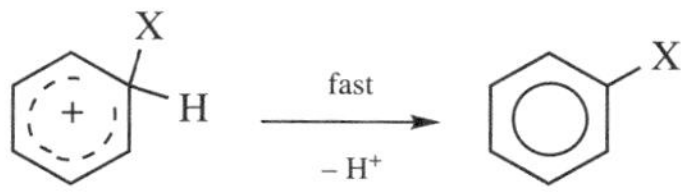

3

The second step is nearly always faster than the first, making the first rate determining, and the reaction is second order. If formation of the attacking species is slower still, the aromatic compound does not take part in the rate expression at all. If X^+ is lost, there is no net reaction, but if H^+ is lost, an aromatic substitution has taken place and a base (generally the counterion of the electrophilic species although solvents can also serve this purpose) is necessary to help remove it.

If the attacking species is not an ion, but a dipole, the product must have a negative charge unless part of the dipole, with its pair of electrons, is broken off somewhere in the process, as in the conversion of **4** to **5**. Note that when the aromatic ring attacks X, Z may be lost directly to give **5**.

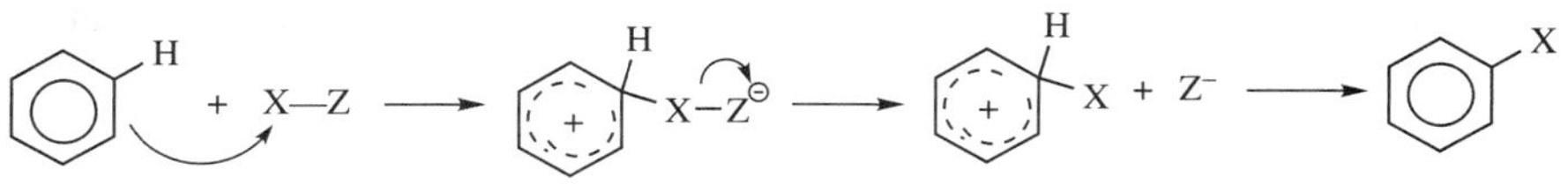

4   **5**

The electrophilic entities and how they are formed are discussed for each reaction in the reactions section of this chapter.

The evidence for the arenium ion mechanism is mainly of two kinds:

1. *Isotope Effects.* If the hydrogen ion departs before the arrival of the electrophile (S_E1 mechanism) or if the arrival and departure are simultaneous, there should be a substantial isotope effect (i.e., deuterated substrates should undergo substitution more slowly than non-deuterated compounds) because, in each case, the C—H bond is broken in the rate-determining step. However, in the arenium ion mechanism, the C—H bond is not broken in the rate-determining step, so no isotope effect should be found. Many such studies have been carried out and, in most cases, especially in the case of nitrations, there is no isotope effect.[6] This result is incompatible with either the S_E1 or the simultaneous mechanism.

However, in many instances, isotope effects have been found. Since the values are generally much lower than expected for either the S_E1 or the simultaneous mechanisms (e.g., 1–3 for k_H/k_D instead of 6–7), we must look elsewhere for

---

[6]The pioneering studies were by Melander, L. *Ark. Kemi* **1950**, *2*, 213; Berglund-Larsson, U.; Melander, L. *Ark. Kemi* **1953**, *6*, 219. See also, Zollinger, H. *Adv. Phys. Org. Chem.* **1964**, *2*, 163.

the explanation. For the case where hydrogen is the leaving group, the arenium ion mechanism can be summarized:

Step 1
$$ArH + Y^+ \underset{k_{-1}}{\overset{k_1}{\rightleftharpoons}} Ar\overset{\oplus}{\underset{Y}{\diagdown}}H$$

Step 2
$$Ar\overset{\oplus}{\underset{Y}{\diagdown}}H \overset{k_2}{\longrightarrow} ArY + H^+$$

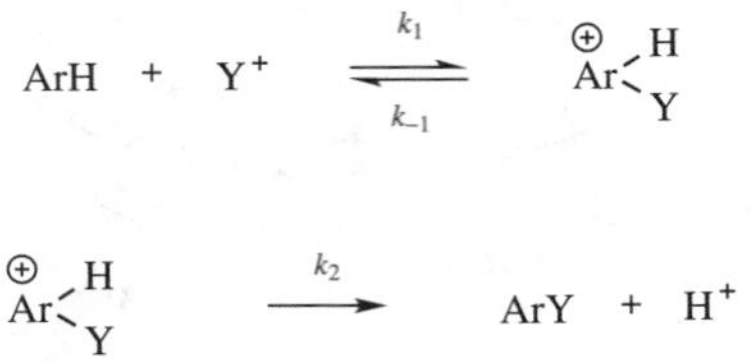

The small isotope effects found most likely arise from the reversibility of step 1 by a *partitioning effect*.[7] The rate at which $ArHY^+$ reverts to ArH should be essentially the same as that at which $ArDY^+$ (or $ArTY^+$) reverts to ArD (or ArT), since the Ar–H bond is not cleaving. However, $ArHY^+$ should go to ArY faster than either $ArDY^+$ or $ArTY^+$, since the Ar–H bond is broken in this step. If $k_2 \gg k_{-1}$, this does not matter; since a large majority of the intermediates go to product, the rate is determined only by the slow step ($k_1[ArH][Y^+]$) and no isotope effect is predicted. However, if $k_2 \leq k_{-1}$, reversion to starting materials is important. If k_2 for $ArDY^+$ (or $ArTY^+$) is $<$ k_2 for $ArHY^+$, but k_{-1} is the same, then a larger proportion of $ArDY^+$ reverts to starting compounds. That is, k_2/k_{-1} (the *partition factor*) for $ArDY^+$ is less than that for $ArHY^+$. Consequently, the reaction is slower for ArD than for ArH and an isotope effect is observed.

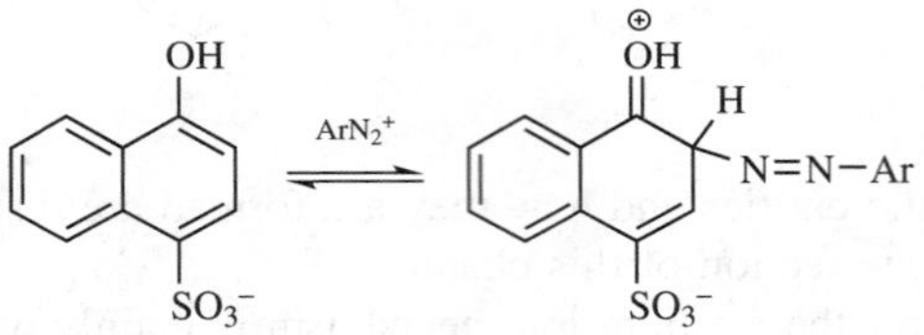

6   7

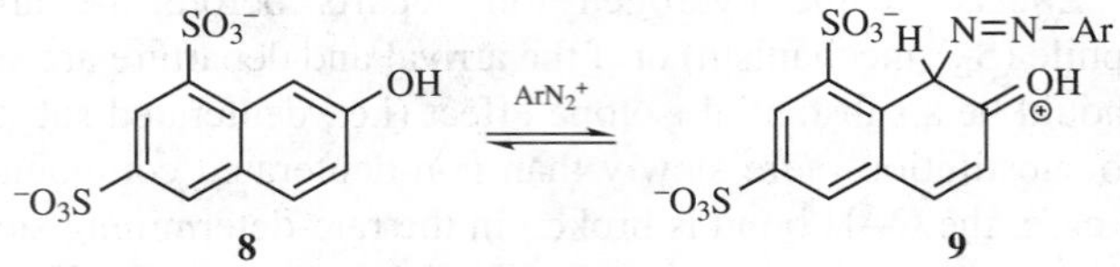

8   9

One circumstance that could affect the k_2/k_{-1} ratio is steric hindrance. Thus, diazonium coupling of **6** gave no isotope effect, while coupling of **8** gave a k_H/k_D ratio of 6.55.[8] For steric reasons, it is much more difficult for **9** to lose a proton (it is harder for a base to approach) than it is for **7**, so k_2 is greater for the latter. Since no base is necessary to remove ArN_2^+, k_{-1} does not depend on steric factors[9] and is about the same for each. Thus the partition factor k_2/k_{-1}

[7]For a discussion, see Hammett, L.P. *Physical Organic Chemistry*, 2nd ed., McGraw-Hill, NY, *1970*, pp. 172–182.

[8]Zollinger, H. *Helv. Chim. Acta* *1955*, *38*, 1597, 1617, 1623.

[9]Snyckers, F.; Zollinger, H. *Helv. Chim. Acta* *1970*, *53*, 1294.

is sufficiently different for **7** and **9** that **8** exhibits a large isotope effect and **6** exhibits none.[10] Base catalysis can also affect the partition factor, since an increase in base concentration increases the rate at which the intermediate goes to product without affecting the rate at which it reverts to starting materials. In some cases, isotope effects can be diminished or eliminated by a sufficiently high concentration of base.

Evidence for the arenium ion mechanism has also been obtained from other kinds of isotope-effect experiments, involving substitutions of the type

$$\text{ArMR}_3 \quad + \quad \text{H}_3\text{O}^+ \quad \longrightarrow \quad \text{ArH} \quad + \quad \text{R}_3\text{MOH}_2{}^+$$

where M is Si, Ge, Sn, or Pb, and R is methyl or ethyl. In these reactions, the proton is the electrophile. If the arenium ion mechanism is operating, then the use of D_3O^+ should give rise to an isotope effect, since the D–O bond would be broken in the rate-determining step. Isotope effects of 1.55–3.05 were obtained,[11] in accord with the arenium ion mechanism.

2. *Isolation of Arenium Ion Intermediates.* Very strong evidence for the arenium ion mechanism comes from the isolation of arenium ions in a number of instances.[12] For example, **7** was isolated as a solid with a

mesitylene   **10**   **11**

melting point of $-15°\text{C}$ from treatment of mesitylene with ethyl fluoride and the catalyst BF_3 at $-80°\text{C}$. When **10** was heated, the normal substitution product **11** was obtained.[13] Even the simplest such ion, the benzenonium ion (**12**), has been prepared in $\text{HF–SbF}_5\text{–SO}_2\text{ClF–SO}_2\text{F}_2$ at $-134°\text{C}$, where it could be studied

[10]For some other examples of isotope effects caused by steric factors, see Helgstrand, E. *Acta Chem. Scand.* **1965**, *19*, 1583; Nilsson, A. *Acta Chem. Scand.* **1967**, *21*, 2423; Baciocchi, E.; Illuminati, G.; Sleiter, G.; Stegel, F. *J. Am. Chem. Soc.* **1967**, *89*, 125; Myhre, P.C.; Beug, M.; James, L.L. *J. Am. Chem. Soc.* **1968**, *90*, 2105; Dubois, J.E.; Uzan, R. *Bull. Soc. Chim. Fr.* **1968**, 3534; Márton, J. *Acta Chem. Scand.* **1969**, *23*, 3321, 3329.

[11]Bott, R.W.; Eaborn, C.; Greasley. P.M. *J. Chem. Soc.* **1964**, 4803.

[12]For reviews, see Koptyug, V.A. *Top. Curr. Chem.* **1984**, *122*, 1; *Bull. Acad. Sci. USSR Div. Chem. Sci.* **1974**, *23*, 1031. For a review of polyfluorinated arenium ions, see Shteingarts, V.D. *Russ. Chem. Rev.* **1981**, *50*, 735. For a review of the protonation of benzene and simple alkylbenzenes, see Fărcaşiu, D. *Acc. Chem. Res.* **1982**, *15*, 46.

[13]Olah, G.A.; Kuhn, S.J. *J. Am. Chem. Soc.* **1958**, *80*, 6541. For some other examples, see Ershov, V.V.; Volod'kin, A.A. *Bull. Acad. Sci. USSR Div. Chem. Sci.* **1962**, 680; Farrell, P.G.; Newton, J.; White, R.F.M. *J. Chem. Soc. B* **1967**, 637; Kamshii, L.P.; Koptyug, V.A. *Bull. Acad. Sci. USSR Div. Chem. Sci.* **1974**, *23*, 232; Olah, G.A.; Spear, R.J.; Messina, G.; Westerman, P.W. *J. Am. Chem. Soc.* **1975**, *97*, 4051; Nambu, N.; Hiraoka, N.; Shigemura, K.; Hamanaka, S.; Ogawa, M. *Bull. Chem. Soc. Jpn.* **1976**, *49*, 3637; Chikinev, A.V.; Bushmelev, V.A.; Shakirov, M.; Shubin, V.G. *J. Org. Chem. USSR* **1986**, *22*, 1311; Knoche, W.; Schoeller, W.W.; Schomäcker, R.; Vogel, S. *J. Am. Chem. Soc.* **1988**, *110*, 7484; Effenberger, F. *Acc. Chem. Res.* **1989**, *22*, 27.

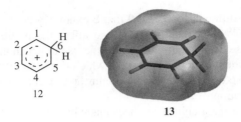

12

13

spectrally.[14] The ^{13}C NMR spectra of the benzenonium ion[15] and the pentamethylbenzenonium ion[16] give graphic evidence for the charge distribution shown in **1** (see the electron density map for the arenium ion, **13**). According to this, the 1, 3, and 5 carbons, each of which bears a positive charge of $+\frac{1}{3}$ [note that C-1,-3,-5 (numbering from **12**) are lighter,indicating less electron density in **13**, whereas C-2,-4 are darker for higher electron density], should have a greater chemical shift in the NMR than the 2 and 4 carbons, which are uncharged. The spectra bear this out. For example, ^{13}C NMR chemical shifts for **12** are C-3: 178.1; C-1 and C-5: 186.6; C-2 and C-4: 136.9, and C-6: 52.2.[15]

In Chapter 3, it was mentioned that positive ions can form addition complexes with π systems. Since the initial step of electrophilic substitution involves attack of a positive ion by an aromatic ring, it has been suggested[17] that such a complex, called a π *complex* (represented as **14**), is formed first, and then is converted to the arenium ion **15**.[18] Stable solutions of arenium ions or π complexes (e.g., with Br_2, I_2,

14   15

picric acid, Ag^+, or HCl) can be formed.[19] For example, π complexes are formed when aromatic hydrocarbons are treated with HCl alone, but the use of HCl plus a

[14]Olah, G.A.; Schlosberg, R.H.; Porter, R.D.; Mo, Y.K.; Kelly, D.P.; Mateescu, G.D. *J. Am. Chem. Soc.* **1972**, *94*, 2034.

[15]Olah, G.A.; Staral, J.S.; Asencio, G.; Liang, G.; Forsyth, D.A.; Mateescu, G.D. *J. Am. Chem. Soc.* **1978**, *100*, 6299.

[16]Lyerla, J.R.; Yannoni, C.S.; Bruck, D.; Fyfe, C.A. *J. Am. Chem. Soc.* **1979**, *101*, 4770.

[17]Dewar, M.J.S. *Electronic Theory of Organic Chemistry;* Clarendon Press: Oxford, **1949**.

[18]For a discussion of both σ- and π-complexes in electrophilic aromatic substitution, see Hubig, S. M.; Kochi, J. K. *J. Org. Chem.* **2000**, *65*, 6807.

[19]For an *ab initio* study involving the interaction of water and hexafluorobenzene, to determine the efficacy of lone-pair binding to a π-system, see Gallivan, J.P.; Dougherty, D.A. *Org. Lett.* **1999**, *1*, 103. For a study concerning preorganization and charge-transfer complexes, see Rosokha, S.V.; Kochi, J.K. *J. Org. Chem.* **2002**, *67*, 1727.

TABLE 11.1. Relative Stabilities of Arenium Ions and π Complexes and Relative Rates of Chlorination and Nitration[a]

Substituents	Relative Arenium Ion Stability[20]	Relative π-Complex Stability[20]	Rate of Chlorination[21]	Rate of Nitration[26]
None (benzene)	0.09	0.61	0.0005	0.51
Me	0.63	0.92	0.157	0.85
p-Me$_2$	1.00	1.00	1.00	1.00
o-Me$_2$	1.1	1.13	2.1	0.89
m-Me$_2$	26	1.26	200	0.84
1,2,4-Me$_3$	63	1.36	340	
1,2,3-Me$_3$	69	1.46	400	
1,2,3,4-Me$_4$	400	1.63	2,000	
1,2,3,5-Me$_4$	16,000	1.67	240,000	
Me$_5$	29,900		360,000	

[a]In each case, p-xylene = 1.00.

Lewis acid (e.g., AlCl$_3$) gives arenium ions. The two types of solution have very different properties. For example, a solution of an arenium ion is colored and conducts electricity (showing positive and negative ions are present), while a π complex formed from HCl and benzene is colorless and does not conduct a current. Furthermore, when DCl is used to form a π complex, no deuterium exchange takes place (because there is no covalent bond between the electrophile and the ring), while formation of an arenium ion with DCl and AlCl$_3$ gives deuterium exchange. The relative stabilities of some methylated arenium ions and π complexes are shown in Table 11.1. The arenium ion stabilities listed were determined by the relative basicity of the substrate toward HF.[20] The π complex stabilities are relative equilibrium constants for the reaction[21] between the aromatic hydrocarbon and HCl. As shown in Table 11.1, the relative stabilities of the two types of species are very different: the π complex stability changes very little with methyl substitution, but the arenium ion stability changes a great deal. It is noted that stable arenium ions have been obtained from large methylene-bridged polycyclic aromatic hydrocarbons.[22]

How can we tell if **14** is present on the reaction path? If it is present, there are two possibilities: (*1*) The formation of **14** is rate determining (the conversion of **14** to **15** is much faster), or (*2*) the formation of **14** is rapid, and the conversion **14** to **15** is rate determining. One way to ascertain which species is formed in the rate-determining step in a given reaction is to use the stability information given in Table 11.1. We measure the relative rates of reaction of a given electrophile with the series of compounds listed in Table 11.1. If the relative rates resemble the arenium ion stabilities, we conclude that the arenium ion is formed in the slow step; but if they

[20]Kilpatrick, M.; Luborsky, F.E. *J. Am. Chem. Soc.* **1953**, *75*, 577.
[21]Brown, H.C.; Brady, J.D. *J. Am. Chem. Soc.* **1952**, *74*, 3570.
[22]Laali, K.K.; Okazaki, T.; Harvey, R.G. *J. Org. Chem.* **2001**, *66*, 3977.

resemble the stabilities of the π complexes, the latter are formed in the slow step.[23] When such experiments are carried out, it is found in most cases that the relative rates are similar to the arenium ion and not to the π complex stabilities. For example, Table 11.1 lists chlorination rates.[21] Similar results were obtained in room-temperature bromination with Br_2 in acetic acid[24] and in acetylation with CH_3CO^+ SbF_6^-.[25] It is clear that in these cases the π complex either does not form at all, or if it does, its formation is not rate determining (unfortunately, it is very difficult to distinguish between these two possibilities).

On the other hand, in nitration with the powerful electrophile NO_2^+ (in the form of NO_2^+ BF_4^-), the relative rates resembled π complex stabilities much more than arenium ion stabilities (Table 11.1).[26] Similar results were obtained for bromination with Br_2 and $FeCl_3$ in nitromethane. These results were taken to mean[27] that in these cases π complex formation is rate determining. However, graphical analysis of the NO_2^+ data showed that a straight line could not be drawn when the nitration rate was plotted against π complex stability,[28] which casts doubt on the rate-determining formation of a π complex in this case.[29] There is other evidence, from positional selectivities (discussed on p. 682), that *some* intermediate is present before the arenium ion is formed, whose formation can be rate determining with powerful electrophiles. Not much is known about this intermediate, which is given the nondescriptive name *encounter complex* and generally depicted as **16**. The arenium complex mechanism is therefore written as[30]

$$1. \ ArH + Y^+ \ \rightleftharpoons \ \overline{Y^+ArH} \qquad 2. \ Y^+ArH \ \rightleftharpoons \ \overset{\oplus}{Ar}\overset{H}{\underset{Y}{\diagup}} \qquad 3. \ \overset{\oplus}{Ar}\overset{H}{\underset{Y}{\diagup}} \ \rightleftharpoons \ ArH + H^+$$

$$\textbf{16}$$

[23]Condon, F.E. *J. Am. Chem. Soc.* **1952**, *74*, 2528.

[24]Brown, H.C.; Stock, L.M. *J. Am. Chem. Soc.* **1957**, *79*, 1421.

[25]Olah, G.A.; Kuhn, S.J.; Flood, S.H.; Hardie, B.A. *J. Am. Chem. Soc.* **1964**, *86*, 2203.

[26]Olah, G.A.; Kuhn, S.J.; Flood, S.H. *J. Am. Chem. Soc.* **1961**, *83*, 4571, 4581.

[27]Olah, G.A.; Kuhn, S.J.; Flood, S.H.; Hardie, B.A. *J. Am. Chem. Soc.* **1964**, *86*, 1039, 1044; Olah, G.A.; Kuhn, S.J.; Flood, S.H. *J. Am. Chem. Soc.* **1961**, *83*, 4571, 4581.

[28]Rys, P.; Skrabal, P.; Zollinger, H. *Angew. Chem. Int. Ed.* **1972**, *11*, 874. See also, DeHaan, F.P.; Covey, W.D.; Delker, G.L.; Baker, N.J.; Feigon, J.F.; Miller, K.D.; Stelter, E.D. *J. Am. Chem. Soc.* **1979**, *101*, 1336; Santiago, C.; Houk, K.N.; Perrin, C.L. *J. Am. Chem. Soc.* **1979**, *101*, 1337.

[29]For other evidence against π complexes, see Tolgyesi, W.S. *Can. J. Chem.* **1965**, *43*, 343; Caille, S.Y.; Corriu, R.J.P. *Tetrahedron* **1969**, *25*, 2005; Coombes, R.G.; Moodie, R.B.; Schofield, K. *J. Chem. Soc. B* **1968**, 800; Hoggett, J.G.; Moodie, R.B.; Schofield, K. *J. Chem. Soc. B* **1969**, 1; Christy, P.F.; Ridd, J.H.; Stears, N.D. *J. Chem. Soc. B* **1970**, 797; Ridd, J.H. *Acc. Chem. Res.* **1971**, *4*, 248; Taylor, R.; Tewson, T.J. *J. Chem. Soc., Chem. Commun.* **1973**, 836; Naidenov, S.V.; Guk, Yu.V.; Golod, E.L. *J. Org. Chem. USSR* **1982**, *18*, 1731. For further support for π complexes, see Olah, G.A. *Acc. Chem. Res.* **1971**, *4*, 240; Olah, G.A.; Lin, H.C. *J. Am. Chem. Soc.* **1974**, *96*, 2892; Koptyug, V.A.; Rogozhnikova, O.Yu.; Detsina, A.N. *J. Org. Chem. USSR* **1983**, *19*, 1007; El-Dsouqui, O.M.E.; Mahmud, K.A.M.; Sulfab, Y. *Tetrahedron Lett.* **1987**, *28*, 2417; Sedaghat-Herati, M.R.; Sharifi, T. *J. Organomet. Chem.* **1989**, *363*, 39. For an excellent discussion of the whole question, see Banthorpe, D.V. *Chem. Rev.* **1970**, *70*, 295, especially Sections VI and IX.

[30]For discussions, see Stock, L.M. *Prog. Phys. Org. Chem.* **1976**, *12*, 21; Ridd, J.H. *Adv. Phys. Org. Chem.* **1978**, *16*, 1.

For the reason given above and for other reasons, it is unlikely that the encounter complex is a π complex, but just what kind of attraction exists between Y^+ and ArH is not known, other than the presumption that they are together within a solvent cage (see also p. 682). There is evidence (from isomerizations occurring in the alkyl group, as well as other observations) that π complexes are present on the pathway from substrate to arenium ion in the gas-phase protonation of alkylbenzenes.[31]

The S_E1 Mechanism

The S_E1 mechanism (*substitution electrophilic unimolecular*) is rare, being found only in certain cases in which carbon is the leaving atom (see **11-33**, **11-35**) or when a very strong base is present (see **11-1**, **11-10**, and **11-39**).[32] It consists of two steps with an intermediate carbanion. The IUPAC designation is $D_E + A_E$.

Reactions **12-41**, **12-45**, and **12-46** also take place by this mechanism when applied to aryl substrates.

ORIENTATION AND REACTIVITY

Orientation and Reactivity in Monosubstituted Benzene Rings[33]

When an electrophilic substitution reaction is performed on a monosubstituted benzene, the new group may be directed primarily to the ortho, meta, or para position and the substitution may be slower or faster than with benzene itself. The group already on the ring determines which position the new group will take and whether the reaction will be slower or faster than with benzene. Groups that increase the reaction rate are called *activating* and those that slow it *deactivating*. Some groups are predominantly meta directing; all of these are deactivating. Others are mostly ortho-para directing; some of these are deactivating too, but most are activating. Groups direct *predominantly*, but usually not *exclusively*. For example, nitration of nitrobenzene gave 93% *m*-dinitrobenzene, 6% of the ortho, and 1% of the para isomer.

The orientation and reactivity effects are explained on the basis of resonance and field effects of each group on the stability of the intermediate arenium ion. To understand why we can use this approach, it is necessary to know that in these reactions

---

[31]Holman, R.W.; Gross, M.L. *J. Am. Chem. Soc.* **1989**, *111*, 3560.

[32]It has also been found with a metal ($SnMe_3$) as electrofuge: Eaborn, C.; Hornfeld, H.L.; Walton, D.R.M. *J. Chem. Soc. B* **1967**, 1036.

[33]For a review of orientation and reactivity in benzene and other aromatic rings, see Hoggett, J.G.; Moodie, R.B.; Penton, J.R.; Schofield, K. *Nitration and Aromatic Reactivity*, Cambridge University Press, Cambridge, **1971**, pp. 122–145, 163–220.

the product is usually kinetically and not thermodynamically controlled (see p. 307). Some of the reactions are irreversible and the others are usually stopped well before equilibrium is reached. *Therefore, which of the three possible intermediates is formed is dependent not on the thermodynamic stability of the products, but on the activation energy necessary to form each of the three intermediates.* It is not easy to predict which of the three activation energies is lowest, but we make the assumption that the free-energy profile resembles either Fig. 6.2(*a* or *b*). In either case, the transition state is closer in energy to the arenium ion intermediate than to the starting compounds. Invoking the Hammond postulate (p. 308), we can then assume that the geometry of the transition state also resembles that of the intermediate and that anything that increases the stability of the intermediate will also lower the activation energy necessary to attain it. Since the intermediate, once formed, is rapidly converted to products, we can use the relative stabilities of the three intermediates as guides to predict which products will predominantly form. Of course, if reversible reactions are allowed to proceed to equilibrium, we may get product ratios that are quite different. For example, the sulfonation of naphthalene at 80°C, where the reaction does not reach equilibrium, gives mostly α-naphthalenesulfonic acid,[34] while at 160°C, where equilibrium is attained, the β isomer predominates[35] (the α isomer is thermodynamically less stable because of steric interaction between the SO_3H group and the hydrogen at the 8 position).

The three possible ions from incorporation of Y at the ortho, meta, and para positions are shown, and each arenium in obviously has a positive charge in the ring.

We can therefore predict that any group Z that has an electron-donating field effect ($+I$, Z will have a − charge or a δ− dipole in most cases) should stabilize all three

[34]Fierz, H.E.; Weissenbach, P. *Helv. Chim. Acta* **1920**, *3*, 312.
[35]Witt, O.N. *Berchti* **1915**, *48*, 743.

ions (relative to **1**), since electron donation to a positive center is stabilizing. On the other hand, electron-withdrawing groups ($-I$, Z will have a + charge or a $\delta+$ dipole in most cases) will increase the positive charge on the ring (like charges repel), and destabilize the arenium ion. Formation of a stabilized ion should be faster than benzene (which generates **1**), or activating, but formation of a destabilized ion should be slower, or deactivating. Such field effects should taper off with distance and are thus strongest at the carbon connected to the group Z (known as the ipso carbon). Of the three arenium ions, only the ortho and para have any positive charge at this carbon. None of the canonical forms of the meta ion has a positive charge at the ipso carbon. Therefore, $+I$ groups should stabilize all three ions but mostly the ortho and para, so they should be not only activating but ortho–para-directing as well. On the other hand, $-I$ groups, by removing electron density, should destabilize all three ions but mostly the ortho and para, and should be not only deactivating but also meta-directing.

These conclusions are correct as far as they go, but they do not lead to the proper results in all cases. In many cases, there is *resonance interaction* between Z and the ring; this also affects the relative stability, in some cases in the same direction as the field effect, in others differently.

Some substituents have a pair of electrons (usually unshared) that may be contributed *toward* the ring. The three arenium ions would then look like this:

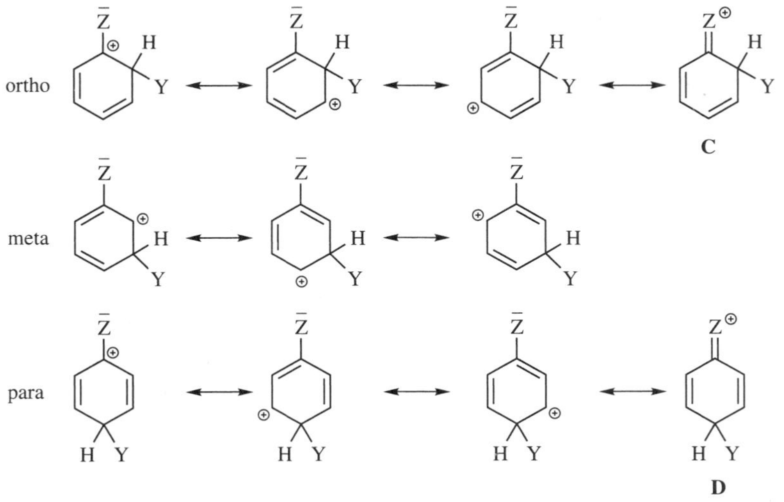

For each ion the same three canonical forms can be drawn as before, but now we can draw an extra form for the ortho and para ions. The stability of these two ions is increased by the extra form not only because it is another canonical form, but because it is more stable than the others and makes a greater contribution to the hybrid. Every atom (except of course hydrogen) in these forms (**C** and **D**) has a complete octet, while all the other forms have one carbon atom with a sextet. No corresponding form can be drawn for the meta isomer. The inclusion of this form in the hybrid lowers the energy not only because of rule 6 (p. 47), but also because it spreads the positive charge over a larger area—out onto the group Z. Groups with a pair of electrons (e.g., as the halogens) to contribute would be expected, then, in the absence of field effects, not only to direct ortho and para, but also to activate these positions for electrophilic attack.

On the basis of these discussions, we can distinguish three types of groups.

1. Groups that contain an unshared pair of electrons on the atom connected to the ring. In this category are O^-, NR_2, NHR, NH_2,[36] OH, OR, NHCOR, OCOR, SR, and the four halogens.[37] The halogens deactivate the aromatic ring to substitution (the rate of reaction is slower than that of benzene), and this effect may arise from the unique energy level of the halogen lone-pair orbital, which is higher than the adjacent π-molecular orbital of benzene (π_1).[38] The widely held explanation for this, however, is that the halogens have a $-I$ effect. The SH group would probably belong here too, except that in the case of thiophenols electrophiles usually attack the sulfur rather than the ring, and ring substitution is not feasible with these substrates. [39] The resonance explanation predicts that all these groups should be ortho–para directing, and they are, though all except O^- are electron withdrawing by the field effect (p. 20). Therefore, for these groups, resonance is more important than the field effect. This is especially true for NR_2, NHR, NH_2, and OH, which are *strongly* activating, as is O^-. The other groups are mildly activating, except for the halogens, which are deactivating. Fluorine is the least deactivating, and fluorobenzenes usually show a reactivity approximating that of benzene itself. The other three halogens deactivate about equally. In order to explain why chlorine, bromine, and iodine deactivate the ring, even though they direct ortho–para, we must assume that the canonical forms **C** and **D** make such great contributions to the respective hybrids that they make the ortho and para arenium ions more stable than the meta, even though the $-I$ effect of the halogen is withdrawing sufficient electron density from the ring to deactivate it. The three halogens make the ortho and para ions more stable than the meta, but less stable than the unsubstituted arenium ion (**1**). For the other groups that contain an unshared pair, the ortho and para ions are more stable than either the meta ion or the unsubstituted ion. For most of

---

[36]It must be remembered that in acid solution amines are converted to their conjugate acids, which for the most part are meta-directing (type 2). Therefore in acid (which is the most common medium for electrophilic substitutions) amino groups may direct meta. However, unless the solution is highly acidic, there will be a small amount of free amine present, and since amino groups are activating and the conjugate acids deactivating, ortho-para direction is often found even under acidic conditions.

[37]For a review of the directing and orienting effects of amino groups, see Chuchani, G., in Patai's. *The Chemistry of the Amino Group*, Wiley, NY, *1968*, pp. 250–265; for ether groups see Kohnstam, G.; Williams, D.L.H., in Patai's. *The Chemistry of the Ether Linkage*, Wiley, NY, *1967*, pp. 132–150.

[38]Tomoda, S.; Takamatsu, K.; Iwaoka, M. *Chem. Lett.* *1998*, 581.

[39]Tarbell, D.S.; Herz, A.H. *J. Am. Chem. Soc.* *1953*, 75, 4657. Ring substitution *is* possible if the SH group is protected. For a method of doing this, see Walker, D. *J. Org. Chem.* *1966*, 31, 835.

the groups in this category, the meta ion is more stable than **1**, so that groups, such as NH_2 and, OH, activate the meta positions too, but not as much as the ortho and para positions (see also the discussion on pp. 677–679).

2. Groups that lack an unshared pair on the atom connected to the ring and that are $-I$. In this category are, in approximate order of decreasing deactivating ability, NR_3^+, NO_2, CF_3,[40] CN, SO_3H, CHO, COR, COOH, COOR, $CONH_2$, CCl_3, and NH_3^+. Also in this category are all other groups with a positive charge on the atom directly connected to the ring[41] (SR_2^+, PR_3^+, etc.) and many groups with positive charges on atoms farther away, since often these are still powerful $-I$ groups. The field-effect explanation predicts that these should all be meta directing and deactivating, and (except for NH_3^+) this is the case. The NH_3^+ group is an anomaly, since this group directs para about as much as or a little more than it directs meta.[42] The NH_2Me^+, $NHMe_2^+$, and NMe_3^+ groups all give more meta than para substitution, the percentage of para product decreasing with the increasing number of methyl groups.[43]

3. Groups that lack an unshared pair on the atom connected to the ring and that are ortho–para directing. In this category are alkyl groups, aryl groups, and the COO^- group,[44] all of which activate the ring. We will discuss them separately. Since aryl groups are $-I$ groups, they might seem to belong to category 2. They are nevertheless ortho–para directing and activating. This can be explained in a similar manner as in category 1, with a pair of electrons from the aromatic sextet playing the part played by the unshared pair, so

that we have forms like **E**. The effect of negatively charged groups like COO^- is easily explained by the field effect (negatively charged groups are of

[40]For the long-range electron-withdrawing effects of this group, see Castagnetti, E.; Schlosser, M. *Chem. Eur. J.* **2002**, *8*, 799.

[41]For discussions, see Gastaminza, A.; Ridd, J.H.; Roy, F. *J. Chem. Soc. B* **1969**, 684; Gilow, H.M.; De Shazo, M.; Van Cleave, W.C. *J. Org. Chem.* **1971**, *36*, 1745; Hoggett, J.G.; Moodie, R.B.; Penton, J.R.; Schofield, K. *Nitration and Aromatic Reactivity*, Cambridge University Press, Cambridge, **1971**, pp. 167–176.

[42]Hartshorn, S.R.; Ridd, J.H. *J. Chem. Soc. B* **1968**, 1063. For a discussion, see Ridd, J.H., in *Aromaticity, Chem. Soc. Spec. Publ., no. 21,* **1967**, 149–162.

[43]Brickman, M.; Utley, J.H.P.; Ridd, J.H. *J. Chem. Soc.* **1965**, 6851.

[44]Spryskov, A.A.; Golubkin, L.N. *J. Gen. Chem. USSR* **1961**, *31*, 833. Since the COO^- group is present only in alkaline solution, where electrophilic substitution is not often done, it is seldom met with.

course electron donating), since there is no resonance interaction between the group and the ring. The effect of alkyl groups can be explained in the same way, but, in addition, we can also draw canonical forms, even though there is no unshared pair. These of course are hyperconjugation forms like **F** (see p. 669). This effect, like the field effect, predicts activation and ortho–para direction, so that it is not possible to say how much each effect contributes to the result. Another way of looking at the effect of alkyl groups (which sums up both field and hyperconjugation effects) is that (for $Z = R$) the ortho and para arenium ions are more stable because each contains a form (**A** and **B**) that is a tertiary carbocation, while all the canonical forms for the meta ion and for **1** are secondary carbocations. In activating ability, alkyl groups usually follow the Baker–Nathan order (p. 96), but not always.[45]

The Ortho/Para Ratio[46]

When an ortho–para-directing group is on a ring, it is usually difficult to predict how much of the product will be the ortho isomer and how much the para isomer. Indeed, these proportions can depend greatly on the reaction conditions. For example, chlorination of toluene gives an ortho/para ratio anywhere from 62:38 to 34:66.[47] Nevertheless, certain points can be made. On a purely statistical basis there would be 67% ortho and 33% para, since there are two ortho positions and only one para. However, the phenonium ion

12

12, which arises from protonation of benzene, has the approximate charge distribution shown[48] (see **13** as well). If we accept this as a model for the arenium ion in aromatic substitution, a para substituent would have a greater stabilizing effect on the adjacent carbon than an ortho substituent. If other effects are absent, this would mean that >33% para and <67% ortho substitution would be found. In hydrogen exchange (reaction **11-1**), where other effects are absent, it has been found for a number of substituents that the average ratio of the logarithms of the partial rate

[45]For examples of situations where the Baker–Nathan order is not followed, see Eaborn, C.; Taylor, R. *J. Chem. Soc.* **1961**, 247; Utley, J.H.P.; Vaughan, T.A. *J. Chem. Soc. B* **1968**, 196; Schubert, W.M.; Gurka, D.F. *J. Am. Chem. Soc.* **1969**, *91*, 1443; Himoe, A.; Stock, L.M. *J. Am. Chem. Soc.* **1969**, *91*, 1452.

[46]For a discussion, see Pearson, D.E.; Buehler, C.A. *Synthesis* **1971**, 455 see pp 455–464. For a discussion of the influence of reaction conditions on the ortho/para ratio, see Effenberger, F.; Maier, A.J. *J. Am. Chem. Soc.* **2001**, *123*, 3429.

[47]Stock, L.M.; Himoe, A. *J. Am. Chem. Soc.* **1961**, *83*, 4605.

[48]Olah, G.A. *Acc. Chem. Res.* **1970**, *4*, 240, p. 248.

factors for these positions (see p. 677 for a definition of partial rate factor) was close to 0.865,[49] which is not far from the value predicted from the ratio of charge densities in **12**. This picture is further supported by the fact that meta-directing groups, which destabilize a positive charge, give ortho/para ratios >67:33[50] (of course the total amount of ortho and para substitution with these groups is small, but the *ratios* are generally >67:33). Another important factor is the steric effect. If either the group on the attacking ring or the group on the electrophile is large, steric hindrance inhibits formation of the ortho product and increases the amount of the para isomer. An example may be seen in the nitration, under the same conditions, of toluene and *tert*-butylbenzene. The former gave 58% of the ortho compound and 37% of the para, while the more bulky *tert*-butyl group gave 16% of the ortho product and 73% of the para.[51] Some groups are so large that they direct almost entirely para.

When the ortho–para-directing group is one with an unshared pair (this of course applies to most of them), there is another effect that increases the amount of para product at the expense of the ortho. A comparison of the intermediates involved (p. 667) shows that **C** is a canonical form with an ortho-quinoid structure, while **D** has a para-quinoid structure. Since we know that para-quinones are more stable than the ortho isomers, it seems reasonable to assume that **D** is more stable than **C**, and therefore contributes more to the hybrid and increases its stability compared to the ortho intermediate.

It has been shown that it is possible to compel regiospecific para substitution by enclosing the substrate molecules in a cavity from which only the para position projects. Anisole was chlorinated in solutions containing a cyclodextrin, a molecule in which the anisole is almost entirely enclosed (see Fig. 3.4). With a high enough concentration of cyclodextrin, it was possible to achieve a para/ortho ratio of 21.6[52] (in the absence of the cyclodextrin the ratio was only 1.48). This behavior is a model for the regioselectivity found in the action of enzymes.

Ipso Attack

We have discussed orientation in the case of monosubstituted benzenes entirely in terms of attachment at the ortho, meta, and para positions, but attachment at the

[49]Ansell, H.V.; Le Guen, J.; Taylor, R. *Tetrahedron Lett.* **1973**, 13.

[50]Hoggett, J.G.; Moodie, R.B.; Penton, J.R.; Schofield, K. *Nitration and Aromatic Reactivity*, Cambridge University Press, Cambridge, **1971**, pp. 176–180.

[51]Nelson, K.L.; Brown, H.C. *J. Am. Chem. Soc.* **1951**, *73*, 5605. For product ratios in the nitration of many monoalkylbenzenes, see Baas, J.M.A.; Wepster, B.M. *Recl. Trav. Chim. Pays-Bas* **1971**, *90*, 1081, 1089;111 **1972**, *91*, 285, 517, 831.

[52]Breslow, R.; Campbell, P. *J. Am. Chem. Soc.* **1969**, *91*, 3085; *Bioorg. Chem.* **1971**, *1*, 140. See also Chen, N.Y.; Kaeding, W.W.; Dwyer, F.G. *J. Am. Chem. Soc.* **1979**, *101*, 6783; Konishi, H.; Yokota, K.; Ichihashi, Y.; Okano, T.; Kiji, J. *Chem. Lett.* **1980**, 1423; Komiyama, M.; Hirai, H. *J. Am. Chem. Soc.* **1983**, *105*, 2018; **1984**, *106*, 174; Chênevert, R.; Ampleman, G. *Can. J. Chem.* **1987**, *65*, 307; Komiyama, M. *Polym. J. (Tokyo)* **1988**, *20*, 439.

position bearing the substituent (called the *ipso position*[53]) can also be important. Ipso attack has mostly been studied for nitration.[54] When attack of NO_2^+ leads to incorporation at the ipso position there are at least five possible fates for the resulting arenium ion (**17**).

Path a. The arenium ion can lose NO_2^+ and revert to the starting compounds. This results in no net reaction and is often undetectable.

Path b. The arenium ion can lose Z^+, in which case this is simply aromatic substitution with a leaving group other than H (see **11-33–11-41**).

Path c. The electrophilic group (in this case NO_2^+) can undergo a 1,2-migration, followed by loss of the proton. The product in this case is the same as that obtained by direct attachment of NO_2^+ at the ortho position of PhZ. It is not always easy to tell how much of the ortho product in any individual case arises from this pathway,[55] though there is evidence that it can be a considerable proportion. Because of this possibility, many of the reported conclusions about the relat'ive reactivity of the ortho, meta, and para positions are cast into doubt, since some of the product may have arisen not from direct attachment at the ortho position, but from attachment at the ipso position followed by rearrangement.[56]

Path d. The ipso substituent (Z) can undergo 1,2-migration, which also produces the ortho product (though the rearrangement would become apparent if there

[53]Perrin, C.L.; Skinner, G.A. *J. Am. Chem. Soc.* **1971**, *93*, 3389. For a review of ipso substitution, see Traynham, J.G. *J. Chem. Educ.* **1983**, *60*, 937.

[54]For a review, see Moodie, R.B.; Schofield, K. *Acc. Chem. Res.* **1976**, *9*, 287. See also, Fischer, A.; Henderson, G.N.; RayMahasay, S. *Can. J. Chem.* **1987**, *65*, 1233, and other papers in this series.

[55]For methods of doing so, see Gibbs, H.W.; Moodie, R.B.; Schofield, K. *J. Chem. Soc. Perkin Trans. 2* **1978**, 1145.

[56]This was first pointed out by Myhre, P.C. *J. Am. Chem. Soc.* **1972**, *94*, 7921.

were other substituents present). The evidence is that this pathway is very minor, at least when the electrophile is NO_2^+.[57]

Path e. Attack of a nucleophile on **17**. In some cases, the products of such an attack (cyclohexadienes) have been isolated[58] (this is 1,4-addition to the aromatic ring), but further reactions are also possible.

Orientation in Benzene Rings With More Than One Substituent[59]

It is often possible in these cases to predict the correct isomer. In many cases, the groups already on the ring reinforce each other. Thus, 1,3-dimethylbenzene is substituted at the 4 position (ortho to one group and para to the other), but not at the 5 position (meta to both). Likewise, the incoming group in *p*-chlorobenzoic acid goes to the position ortho to the chloro and meta to the carboxyl group.

When the groups oppose each other, predictions may be more difficult. In a case such as where two

groups of about equal directing ability are in competing positions, all four products can be expected, and it is not easy to predict the proportions, except that steric hindrance should probably reduce the yield of substitution ortho to the acetamido group, especially for large electrophiles. Mixtures of about equal proportions are frequent in such cases. Nevertheless, even when groups on a ring oppose each other, there are some regularities.

1. If a strong activating group competes with a weaker one or with a deactivating group, the former controls. Thus *o*-cresol gives substitution mainly ortho and para to the *hydroxyl* group and not to the methyl. For this purpose we can arrange the groups in the following order: NH_2, OH, NR_2, O^- > OR, OCOR, NHCOR > R, Ar > halogen > meta-directing groups.

2. All other things being equal, a third group is least likely to enter between two groups in the meta relationship. This is the result of steric hindrance and increases in importance with the size of the groups on the ring and with the size of the attacking species.[60]

[57]For examples of such migration, where Z = Me, see Hartshorn, M.P.; Readman, J.M.; Robinson, W.T.; Sies, C.W.; Wright, G.J. *Aust. J. Chem.* **1988**, *41*, 373.

[58]For examples, see Banwell, T.; Morse, C.S.; Myhre, P.C.; Vollmar, A. *J. Am. Chem. Soc.* **1977**, *99*, 3042; Fischer, A.; Greig, C.C. *Can. J. Chem.* **1978**, *56*, 1063.

[59]For a quantitative discussion, see pp. 677–678.

[60]In some cases, attack at an electrophile preferentially leads to attachment at the position between two groups in the meta relationship. For a list of some of these cases and a theory to explain them, see Kruse, L.I.; Cha, J.K. *J. Chem. Soc., Chem. Commun.* **1982**, 1333.

3. When a meta-directing group is meta to an ortho–para-directing group, the incoming group primarily goes ortho to the meta-directing group rather than para. For example, chlorination of **18** gives mostly **19**.

18	**19**	**20**	**21**

The importance of this effect is underscored by the fact that **20**, which is in violation of the preceding rule, is formed in smaller amounts, but **21** is not formed at all. This is called the *ortho effect*,[61] and many such examples are known.[62] Another is the nitration of *p*-bromotoluene, which gives 2,3-dinitro-4-bromotoluene. In this case, once the first nitro group came in, the second was directed ortho to it rather than para, even though this means that the group has to come in between two groups in the meta position. There is no good explanation yet for the ortho effect, though possibly there is intramolecular assistance from the meta-directing group.

It is interesting that chlorination of **18** illustrates all three rules. Of the four positions open to the electrophile, the 5 position violates rule 1, the 2 position rule 2, and the 4 position rule 3. The principal attachment is therefore at position 6.

Orientation in Other Ring Systems[63]

In fused ring systems, the positions are not equivalent and there is usually a preferred orientation, even in the unsubstituted hydrocarbon. The preferred positions may often be predicted as for benzene rings. Thus it is possible to draw more canonical forms for the arenium ion when attack by naphthalene leads to attachment of the electrophile at the α position than when attack by naphthalene leads to attachment of the electrophile at the β position. Therefore, the α position is the preferred site of attachment,[64] though, as previously mentioned (p. 666), the isomer formed by substitution at the β-position is thermodynamically more stable and is the product if the reaction is reversible and equilibrium is reached. Because of the more extensive delocalization of charges in the corresponding arenium ions, naphthalene is more reactive than benzene and substitution is faster at both positions. Similarly,

[61]This is not the same as the ortho effect mentioned on p. 412.
[62]See Hammond, G.S.; Hawthorne, M.F., in Newman, M.S. *Steric Effects in Organic Chemistry*, Wiley, NY, *1956*, pp. 164–200, 178–182.
[63]For a review of substitution on nonbenzenoid aromatic systems, see Hafner, H.; Moritz, K.L., in Olah, G.A. *Friedel–Crafts and Related Reactions*, Vol. 4, Wiley, NY, *1965*, pp. 127–183. For a review of aromatic substitution on ferrocenes, see Bublitz, D.E.; Rinehart Jr., K.L. *Org. React. 1969*, *17*, 1.
[64]For a discussion on the preferred site of attachment for many ring systems, see de la Mare, P.B.D.; Ridd, J.H. *Aromatic Substitution Nitration and Halogenation*, Academic Press, NY, *1959*, pp. 169–209.

anthracene, phenanthrene, and other fused polycyclic aromatic hydrocarbons are also substituted faster than benzene.

Heterocyclic compounds, too, have nonequivalent positions, and the principles are similar,[65] in terms of mechanism, and rate data is available.[66] Furan, thiophene, and pyrrole are chiefly substituted at the 2 position, and all are substituted faster than benzene.[67] Pyrrole is particularly reactive, with a reactivity approximating that of aniline or the phenoxide ion. For pyridine,[68] it is not the free base that must attack the electrophile, but the conjugate acid (the pyridinium ion),[69] making the reactivity much less than that of benzene, being similar to that of nitrobenzene. The 3 position is most reactive in electrophilic substitution reactions of pyridine. However, groups can be introduced into the 4 position of a pyridine ring indirectly, by performing the reaction on the corresponding pyridine N-oxide.[70] Note that calculations show that the 2-pyridyl and 2-pyrimidyl cations are best represented as *ortho*-hetarynium ions, being more stable than their positional, nonconjugated isomers by as much as 18–28 kcal mol^{-1} (75-11) kJ mol^{-1}.[71]

When fused ring systems contain substituents, successful predictions can often be made by using a combination of the above principles. Thus, ring A of 2-methylnaphthalene (**22**) is activated by the methyl

group; ring B is not (though the presence of a substituent in a fused ring system affects all the rings,[72] the effect is generally greatest on the ring to which it is attached). We therefore expect substitution in ring A. The methyl group activates positions 1 and 3, which are ortho to itself, but not position 4, which is meta to it.

[65]For a monograph, see Katritzky, A.R.; Taylor, R. *Electrophilic Substitution of Heterocycles: Quantitative Aspects* (Vol. 47 of *Adv. Heterocycl. Chem.*), Academic Press, NY, *1990*.

[66]Katritzky, A.R.; Fan, W.-Q. *Heterocycles* *1992*, *34*, 2179.

[67]For a review of electrophilic substitution on five-membered aromatic heterocycles, see Marino, G. *Adv. Heterocycl. Chem.* *1971*, *13*, 235.

[68]For reviews of substitution on pyridines and other six-membered nitrogen-containing aromatic rings, see Comins, D.L.; O'Connor, S. *Adv. Heterocycl. Chem.* *1988*, *44*, 199; Aksel'rod, Zh.I.; Berezovskii, V.M. *Russ. Chem. Rev.* *1970*, *39*, 627; Katritzky, A.R.; Johnson, C.D. *Angew. Chem. Int. Ed.* *1967*, *6*, 608; Abramovitch, R.A.; Saha, J.G. *Adv. Heterocycl. Chem.* *1966*, *6*, 229. For a review of methods of synthesizing 3-substituted pyrroles, see Anderson, H.J.; Loader, C.E. *Synthesis* *1985*, 353.

[69]Olah, G.A.; Olah, J.A.; Overchuk, N.A. *J. Org. Chem.* *1965*, *30*, 3373; Katritzky, A.R.; Kingsland, M. *J. Chem. Soc. B* *1968*, 862.

[70]Jaffé, H.H. *J. Am. Chem. Soc.* *1954*, *76*, 3527.

[71]Gozzo, F.C.; Eberlin, M.N. *J. Org. Chem.* *1999*, *64*, 2188.

[72]See, for example, Ansell, H.V.; Sheppard, P.J.; Simpson, C.F.; Stroud, M.A.; Taylor, R. *J. Chem. Soc. Perkin Trans. 2* *1979*, 381.

However, substitution at the 3 position gives rise to an arenium ion for which it is impossible to write a low-energy canonical form in which ring B has a complete sextet. All we can write are forms like **23**, in which the sextet is no longer intact. In contrast, substitution at the 1 position gives rise to a more stable arenium ion, for which two canonical forms (one of them is **24**) can be written in which ring B is benzenoid. We thus predict predominant substitution at C-1, and that is what is generally found.[73] However, in some cases predictions are much harder to make. For example, chlorination or nitration of **25** gives mainly the 4 derivative, but bromination yields chiefly the 6 compound.[74]

| **25** | Indole | Quinoline |

For fused heterocyclic systems too, we can often make predictions based on the above principles, though many exceptions are known. Thus, indole is chiefly substituted in the pyrrole ring (at position 3) and reacts faster than benzene, while quinoline generally reacts in the benzene ring, at the 5 and 8 positions, and slower than benzene, though faster than pyridine.

| **24** | **26** | **27** |

In alternant hydrocarbons (p. 69), the reactivity at a given position is similar for electrophilic, nucleophilic, and free-radical substitution, because the same kind of resonance can be shown in all three types of intermediate (cf. **24**, **26**, and **27**). Attachment of the electrophile at the position that will best delocalize a positive charge will also best delocalize a negative charge or an unpaired electron. Most results are in accord with these predictions. For example, naphthalene is attacked primarily at the 1 position by NO_2^+, NH_2-, and Ph•, and always more readily than benzene.

[73]For example, see Alcorn, P.G.E.; Wells, P.R. *Aust. J. Chem.* **1965**, *18*, 1377, 1391; Eaborn, C.; Golborn, P.; Spillett, R.E.; Taylor, R. *J. Chem. Soc. B* **1968**, 1112; Kim, J.B.; Chen, C.; Krieger, J.K.; Judd, K.R.; Simpson, C.C.; Berliner, E. *J. Am. Chem. Soc.* **1970**, *92*, 910. For discussions, see Taylor, R. *Chimia* **1968**, *22*, 1; Gore, P.H.; Siddiquei, A.S.; Thorburn, S. *J. Chem. Soc. Perkin Trans. 1* **1972**, 1781.
[74]Bell, F. *J. Chem. Soc.* **1959**, 519.

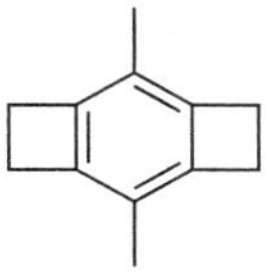

28

When strain due to a ring fused on an aromatic ring deforms that ring out of planarity, the molecule is more reactive to electrophilic aromatic substitution.[75] This has been explained by the presence of a shortened bond for the sp^2 hybridized carbon, increasing the strain at that position, and this is known as the *Mills–Nixon effect*.[76] There is EPR evidence (see p. 267) for 3,6-dimethyl-1,2,4,5-tetrahydrobenzo-bis (cyclobutene) (**28**) that supports the Mills–Nixon effect,[77] and a theoretical study supports this.[78] However, *ab initio* studies of triannelated benzene rings shows *no evidence* for the Mills–Nixon effect, and an new motif for bond-alternating benzenes was proposed.[79] Indeed, it is argued that the Mills–Nixon effect is not real.[80]

Quantitative Treatments of Reactivity in the Substrate

Quantitative rate studies of aromatic substitutions are complicated by the fact that there are usually several hydrogens that can leave, so that measurements of overall rate ratios do not give a complete picture as they do in nucleophilic substitutions, where it is easy to compare substrates that have only one possible leaving group in a molecule. What is needed is not, say, the overall rate ratio for acetylation of toluene versus that for benzene, but the *rate ratio at each position*. These can be calculated from the overall rates and a careful determination of the proportion of isomers formed, provided that the products are kinetically controlled, as is usually the case. We may thus define the *partial rate factor* for a given group and a given reaction as the rate of substitution at a single position relative to a single position in benzene. For example, for acetylation

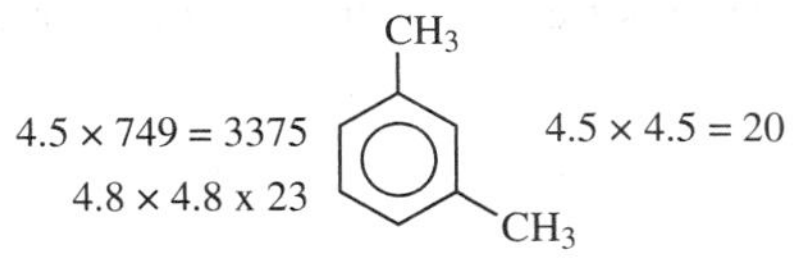

$$4.5 \times 749 = 3375$$
$$4.8 \times 4.8 \times 23$$
$$4.5 \times 4.5 = 20$$

of toluene the partial rate factors are: for the ortho position $o_f^{Me} = 4.5$, for the meta $m_f^{Me} = 4.8$, and for the para $p_f^{Me} = 749$.[81] This means that toluene is acetylated at

[75]Taylor, R. *Electrophilic Aromatic Substitution*, Wiley, Chichester, *1990*, pp. 53.

[76]Mills, W.H.; Nixon, I.G. *J. Chem. Soc.* *1930*, 2510.

[77]Davies, A.G.; Ng, K.M. *J. Chem. Soc. Perkin Trans. 2* *1992*, 1857.

[78]Eckert-Maksić, M.; Maksić, Z.B.; Klessinger, M. *J. Chem. Soc. Perkin Trans. 2* *1994*, 285; Eckert-Maksić, M.; Lesar, A.; Maksić, Z.B. *J. Chem. Soc. Perkin Trans. 2* *1992*, 993.

[79]Baldridge, K.K.; Siegel, J.J. *J. Am. Chem. Soc.* *1992*, *114*, 9583.

[80]Siegel, J.S. *Angew. Chem. Int. Ed.* *1994*, *33*, 1721.

[81]Brown, H.C.; Marino, G.; Stock, L.M. *J. Am. Chem. Soc.* *1959*, *81*, 3310.

the ortho position 4.5 times as fast as a single position in benzene, or 0.75 times as fast as the overall rate of acetylation of benzene. A partial rate factor >1 for a given position indicates that the group in question activates that position for the given reaction. Partial rate factors differ from one reaction to another and are even different, though less so, for the same reaction under different conditions.

Once we know the partial rate factors, we can predict the proportions of isomers to be obtained when two or more groups are present on a ring, *if we make the assumption that the effect of substituents is independent.* For example, if the two methyl groups in *m*-xylene have the same effect as the methyl group in toluene, we can calculate the theoretical partial rate factors at each position by multiplying those from toluene, so they should be as indicated:

TABLE 11.2. Calculated and Experimental Isomer Distributions in the Acetylation of *m*-Xylene[81]

Position	Isomer Distribution, %	
	Calculated	Observed
2	0.30	0
4	9.36	97.5
5	0.34	2.5

From this, it is possible to calculate the overall theoretical rate ratio for acetylation of *m*-xylene relative to benzene, since this is one-sixth the sum of the partial rate factors (in this case 1130), and the isomer distribution if the reaction is kinetically controlled. The overall rate ratio actually is 347[82] and the calculated and observed isomer distributions are listed in Table 11.2.[76] In this case, and in many others, agreement is fairly good, but many cases are known where the effects are not additive (as on p. 671).[83] For example, this treatment predicts that for 1,2,3-trimethylbenzene

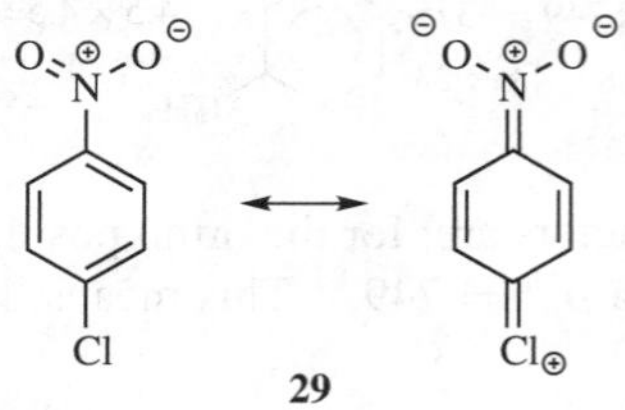

29

[82]Marino, G.; Brown, H.C. *J. Am. Chem. Soc.* **1959**, *81*, 5929.

[83]For some examples where additivity fails, see Fischer, A.; Vaughan, J.; Wright, G.J. *J. Chem. Soc. B* **1967**, 368; Coombes, R.G.; Crout, D.H.G.; Hoggett, J.G.; Moodie, R.B.; Schofield, K. *J. Chem. Soc. B* **1970**, 347; Richards, K.E.; Wilkinson, A.L.; Wright, G.J. *Aust. J. Chem.* **1972**, *25*, 2369; Cook, R.S.; Phillips, R.; Ridd, J.H. *J. Chem. Soc. Perkin Trans. 2* **1974**, 1166. For a theoretical treatment of why additivity fails, see Godfrey, M. *J. Chem. Soc. B* **1971**, 1545.

there should be 35% 5 substitution and 65% 4 substitution, but acetylation gave 79% 5 substitution and 21% of the 4 isomer. The treatment is thrown off by steric effects, such as those mentioned earlier (p. 673), by-products arising from ipso attack (p. 671) and by resonance interaction *between* groups (e.g., **29**), which must make the results deviate from simple additivity of the effects of the groups.

Another approach that avoids the problem created by having competing leaving groups present in the same substrate is the use of substrates that contain only one leaving group. This is most easily accomplished by the use of a leaving group other than hydrogen. By this means overall rate ratios can be measured for specific positions.[84] Results obtained in this way[85] give a reactivity order quite consistent with that for hydrogen as leaving group.

A quantitative scale of reactivity for aromatic substrates (fused, heterocyclic, and substituted rings) has been devised, based on the hard–soft acid–base concept (p. 375).[86] From molecular-orbital theory, a quantity called *activation hardness* can be calculated for each position of an aromatic ring. The smaller the activation hardness, the faster the attachment at that position; hence the treatment predicts the most likely orientations for incoming groups.

A Quantitative Treatment of Reactivity of the Electrophile: The Selectivity Relationship

Not all electrophiles are equally powerful. The nitronium ion attacks not only benzene but also aromatic rings that contain a strongly deactivating group. On the other hand, diazonium ions couple only with rings containing a powerful activating group. Attempts have been made to correlate the influence of substituents with the power of the attacking group. The most obvious way to do this is with the Hammett equation (p. 392):

$$\log \frac{k}{k_0} = \rho\,\sigma$$

For aromatic substitution,[87] k_0 is divided by 6 and, for meta substitution, k is divided by 2, so that comparisons are made for only one position (consequently, k/k_0 for, say, the methyl group at a para position is identical to the partial rate factor p_f^{Me}). It was soon found that, while this approach worked fairly well for electron-withdrawing groups, it failed for those that are electron donating. However, if the equation is modified by the insertion of the Brown σ^+ values instead of the Hammett σ values (because a positive charge develops during the transition state), more satisfactory correlations can be made, even for electron-donating groups (see Table 9.4

[84]For a review of aryl-silicon and Related cleavages, see Eaborn, C. *J. Organomet. Chem.* **1975**, *100*, 43.
[85]See, for example, Deans, F.B.; Eaborn, C. *J. Chem. Soc.* **1959**, 2299; Eaborn, C.; Jackson, P.M. *J. Chem. Soc. B* **1969**, 21.
[86]Zhou, Z.; Parr, R.G. *J. Am. Chem. Soc.* **1990**, *112*, 5720.
[87]See Exner, O.; Böhm, S. *J. Org. Chem.* **2002**, *67*, 6320.

TABLE 11.3. Relative Rates and Product Distributions in Some Electrophilic Substitutions on Toluene and Benzene[89]

Reaction	Relative Rate $k_{toluene}/k_{benzene}$	Product Distribution, %	
		m	p
Bromination	605	0.3	66.8
Chlorination	350	0.5	39.7
Benzoylation	110	1.5	89.3
Nitration	23	2.8	33.9
Mercuration	7.9	9.5	69.5
Isopropylation	1.8	25.9	46.2

for a list of σ^+ values).[88] Groups with a negative value of σ_p^+ or σ_m^+ are activating for that position; groups with a positive value are deactivating. The ρ values correspond to the susceptibility of the reaction to stabilization or destabilization by the Z group and to the reactivity of the electrophile. The ρ values vary not only with the electrophile, but also with conditions. A large negative value of ρ means an electrophile of relatively low reactivity. Of course, this approach is completely useless for ortho substitution, since the Hammett equation does not apply there.

A modification of the Hammett approach, suggested by Brown, called the *selectivity relationship*,[89] is based on the principle that reactivity of a species varies inversely with selectivity. Table 11.3 shows how electrophiles can be arranged in order of selectivity as measured by two indexes: (*1*) their selectivity in attacking toluene rather than benzene, and (*2*) their selectivity between the meta and para positions in toluene.[90] As the table shows, an electrophile more selective in one respect is also more selective in the other. In many cases, electrophiles known to be more stable (hence less reactive) than others show a higher selectivity, as would be expected. For example, the *tert*-butyl cation is more stable and more selective than the isopropyl (p. 236), and Br_2 is more selective than Br^+. However, deviations from the relationship are known.[91] Selectivity depends not only on the nature of the electrophile but also on the temperature. As expected, it normally decreases with increasing temperature.

Brown assumed that a good measurement of selectivity was the ratio of the para and meta partial rate factors in toluene. He defined the selectivity S_f of a reaction as

$$S_f = \log \frac{p_f^{Me}}{m_f^{Me}}$$

[88]For a discussion of the limitations of the Hammett equation approach, see Koptyug, V.A.; Salakhutdinov, N.F.; Detsina, A.N. *J. Org. Chem. USSR* **1984**, *20*, 1039.

[89]Stock, L.M.; Brown, H.C. *Adv. Phys. Org. Chem.* **1963**, *1*, 35.

[90]Stock, L.M.; Brown, H.C. *Adv. Phys. Org. Chem.* **1963**, *1*, 35, see p. 45.

[91]At least some of these may arise from migration of groups already on the ring; see Olah, G.A.; Olah, J.A.; Ohyama, T. *J. Am. Chem. Soc.* **1984**, *106*, 5284.

That is, the more reactive an attacking species, the less preference it has for the para position compared to the meta. If we combine the Hammett–Brown $\sigma^+\rho$ relationship with the linearity between $\log S_f$ and $\log p_f^{Me}$ and between $\log S_f$ and $\log m_f^{Me}$, it is possible to derive the following expressions:

$$\log p_f^{Me} = \frac{\sigma_p^+}{\sigma_p^+ - \sigma_m} S_f$$

$$\log m_f^{Me} = \frac{\sigma_m^+}{\sigma_p^+ - \sigma_m^+} S_f$$

S_f is related to ρ by

$$S_f = \rho(\sigma_p^+ - \sigma_m^+)$$

The general validity of these equations is supported by a great deal of experimental data on aromatic substitution reactions of toluene. Examples of values for some reactions obtained from these equations are given in Table 11.4.[92] For other substituents, the treatment works well with groups that, like methyl, are not very polarizable. For more polarizable groups the correlations are sometimes satisfactory and sometimes not, probably because each electrophile in the transition state makes a different demand on the electrons of the substituent group.

Not only are there substrates for which the treatment is poor, but it also fails with very powerful electrophiles; this is why it is necessary to postulate the encounter complex mentioned on p. 664. For example, relative rates of nitration of *p*-xylene, 1,2,4-trimethylbenzene, and 1,2,3,5-tetramethylbenzene were 1.0, 3.7, and 6.4,[93] though the extra methyl groups should enhance the rates much more (*p*-xylene itself reacted 295 times faster than benzene). The explanation is that with powerful electrophiles the reaction rate is so rapid (reaction taking place at virtually every

TABLE 11.4. Values of m_f^{Me}, p_f^{Me}, S_f, and ρ for Three Reactions of Toluene[92]

Reaction	m_f^{Me}	p_f^{Me}	S_f	ρ
PhMe + EtBr $\xrightarrow[\text{benzene, 25°C}]{\text{GaBr}_3}$	1.56	6.02	0.587	−2.66
PhMe + HNO$_3$ $\xrightarrow[\text{45°C}]{\text{90\% HOAc}}$	2.5	58	1.366	−6.04
PhMe + BR$_2$ $\xrightarrow[\text{25°C}]{\text{85\% HOAc}}$	5.5	2420	2.644	−11.40

[92]Stock, L.M.; Brown, H.C. *J. Am. Chem. Soc.* **1959**, *81*, 3323. Stock, L.M.; Brown, H.C. *Adv. Phys. Org. Chem.* **1963**, *1*, 35 presents many tables of these kinds of data. See also, DeHaan, F.P.; Chan, W.H.; Chang, J.; Ferrara, D.M.; Wainschel, L.A. *J. Org. Chem.* **1986**, *51*, 1591, and other papers in this series.
[93]Olah, G.A.; Lin, H.C. *J. Am. Chem. Soc.* **1974**, *96*, 2892.

encounter[94] between an electrophile and substrate molecule)[95] that the presence of additional activating groups can no longer increase the rate.[96]

Given this behavior (little selectivity in distinguishing between different substrate molecules), the selectivity relationship would predict that positional selectivity should also be very small. However, it is not. For example, under conditions where nitration of p-xylene and 1,2,4-trimethylbenzene takes place at about equal rates, there was no corresponding lack of selectivity at positions *within* the latter.[97] Though

Relative rate ratios

steric effects are about the same at both positions, >10 times as much 5-nitro product was formed as 6-nitro product. It is clear that the selectivity relationship has broken down and it becomes necessary to explain why such an extremely rapid reaction should occur with positional selectivity. The explanation offered is that the rate-determining step is formation of an encounter complex (**12**, p. 664).[98] Since the position of attachment is not determined in the rate-determining step, the 5:6 ratio is not related to the reaction rate. Essentially the same idea was suggested earlier[99] and for the same reason (failure of the selectivity relationship in some cases), but the earlier explanation specifically pictured the complex as a π complex, and we have seen (p. 664) that there is evidence against this.

One interesting proposal[100] is that the encounter pair is a radical pair $\overline{NO_2 \bullet ArH \bullet}^+$ formed by an electron transfer (SET), which would explain why the electrophile, once in the encounter complex, can acquire the selectivity that the free NO_2^+ lacked (it is not proposed that a radical pair is present in all aromatic substitutions; only in those that do not obey the selectivity relationship). The radical

[94]See Coombes, R.G.; Moodie, R.B.; Schofield, K. *J. Chem. Soc. B* **1968**, 800; Moodie, R.B.; Schofield, K.; Thomas, P.N. *J. Chem. Soc. Perkin Trans. 2* **1978**, 318.

[95]For a review of diffusion control in electrophilic aromatic substitution, see Ridd, J.H. *Adv. Phys. Org. Chem.* **1978**, *16*, 1.

[96]Coombes, R.G.; Moodie, R.B.; Schofield, K. *J. Chem. Soc. B* **1968**, 800; Hoggett, J.G.; Moodie, R.B.; Schofield, K. *J. Chem. Soc. B* **1969**, 1; Manglik, A.K.; Moodie, R.B.; Schofield, K.; Dedeoglu, E.; Dutly, A.; Rys, P. *J. Chem. Soc. Perkin Trans. 2* **1981**, 1358.

[97]Barnett, J.W.; Moodie, R.B.; Schofield, K.; Taylor, P.G.; Weston, J.B. *J. Chem. Soc. Perkin Trans. 2* **1979**, 747.

[98]For kinetic evidence in favor of encounter complexes, see Sheats, G.F.; Strachan, A.N. *Can. J. Chem.* **1978**, *56*, 1280. For evidence for such complexes in the gas phase, see Attinà, M.; Cacace, F.; de Petris, G. *Angew. Chem. Int. Ed.* **1987**, *26*, 1177.

[99]Olah, G.A. *Acc. Chem. Res.* **1971**, *4*, 240.

[100]Perrin, C.L. *J. Am. Chem. Soc.* **1977**, *99*, 5516.

pair subsequently collapses to the arenium ion. There is evidence[101] both for and against this proposal.[102]

The Effect of the Leaving Group

30

In the vast majority of aromatic electrophilic substitutions, the leaving group is H^+ as indicated above, and very little work has been done on the relative electrofugal ability of other leaving groups. However, the following orders of leaving-group ability have been suggested:[103] (1) for leaving groups that depart without assistance (S_N1 process with respect to the leaving group), $NO_2^{+\,[104]}$ $< iPr^+ \sim SO_3 < t\text{-}Bu^+ \sim ArN_2^+ < ArCHOH^+ < NO^+ < CO_2$; (2) for leaving groups that depart with assistance from an outside nucleophile (S_N2 process), $Me^+ < Cl^+ < Br^+ < D^+ \sim RCO^+ < H^+ \sim I^+ < Me_3Si^+$. We can use this kind of list to help predict which group, X or Y, will cleave from an arenium ion **30** (see **1**, where $Y = H$) once it has been formed, and so obtain an idea of which electrophilic substitutions are feasible. However, a potential leaving group can also affect a reaction in another way: by influencing the rate at which attack of the original electrophile leads to attachment directly at the ipso position. Partial rate factors for electrophilic attack at a position substituted by a group other than hydrogen are called ipso partial rate factors (i_f^X).[53] Such factors for the nitration of p-haloanisoles are 0.18, 0.08, and 0.06, for p-iodo, p-bromo-, and p-chloroanisole, respectively.[105] This means, for example, that attack at the electrophile in this case leads to attachment at the 4 position of 4-iodoanisole 0.18 times as fast as a single position of benzene. Note that this is far slower than attachment at the 4 position resulting from attack of anisole itself so that the presence of the iodo group greatly slows the reaction at that position. A similar experiment on p-cresol showed that ipso

[101]For evidence in favor of the proposal, see Reents, Jr., W.D.; Freiser, B.S. *J. Am. Chem. Soc.* **1980**, *102*, 271; Morkovnik, A.S.; Dobaeva, N.M.; Panov, V.B.; Okhlobystin, O.Yu. *Doklad. Chem.* **1980**, *251*, 116; Sankararaman, S.; Haney, W.A.; Kochi, J.K. *J. Am. Chem. Soc.* **1987**, *109*, 5235; Keumi, T.; Hamanaka, K.; Hasegawa, K.; Minamide, N.; Inoue, Y.; Kitajima, H. *Chem. Lett.* **1988**, 1285; Johnston, J.F.; Ridd, J.H.; Sandall, J.P.B. *J. Chem. Soc., Chem. Commun.* **1989**, 244. For evidence against it, see Barnes, C.E.; Myhre, P.C. *J. Am. Chem. Soc.* **1978**, *100*, 975; Eberson, L.; Radner, F. *Acc. Chem. Res.* **1987**, *20*, 53; Baciocchi, E.; Mandolini, L. *Tetrahedron* **1987**, *43*, 4035.

[102]For a review, see Morkovnik, A.S. *Russ. Chem. Rev.* **1988**, *57*, 144.

[103]Perrin, C.L. *J. Org. Chem.* **1971**, *36*, 420.

[104]For examples where NO_2^+ is a leaving group (in a migration), see Bullen, J.V.; Ridd, J.H.; Sabek, O. *J. Chem. Soc. Perkin Trans. 2* **1990**, 1681, and other papers in this series.

[105]Perrin, C.L.; Skinner, G.A. *J. Am.Chem. Soc.* **1971**, *93*, 3389. See also, Fischer, P.B.; Zollinger, H. *Helv. Chim. Acta* **1972**, *55*, 2139.

attack at the methyl position was 6.8 times slower than attack of phenol leading to attachment at the para position.[106] Thus, in these cases, both an iodo and a methyl group deactivate the ipso position.[107]

REACTIONS

The reactions in this chapter are classified according to leaving group. Hydrogen replacements are treated first, then rearrangements in which the attacking entity is first cleaved from another part of the molecule (hydrogen is also the leaving group in these cases), and finally replacements of other leaving groups.

Hydrogen as the Leaving Group in Simple Substitution Reactions

A. Hydrogen as the Electrophile

11-1 Hydrogen Exchange

Deuterio-de-hydrogenation or Deuteriation

$$ArH + D^+ \rightleftharpoons ArD + H^+$$

Aromatic compounds can exchange hydrogens when treated with acids. The reaction is used chiefly to study mechanistic questions[108] (including substituent effects), but can also be useful to deuterate (add ^2H) or tritiate (add ^3H) aromatic rings selectively. The usual directive effects apply and, for example, phenol treated with D_2O gives slow exchange on heating, with only ortho and para hydrogens being exchanged.[109] Strong acids, of course, exchange faster with aromatic substrates, and this exchange must be taken into account when studying the mechanism of any aromatic substitution catalyzed by acids. There is a great deal of evidence that exchange takes place by the ordinary arenium ion mechanism. Among the evidence are the orientation effects noted above and the finding that the reaction is general acid catalyzed, which means that a proton is transferred in the slow step[110] (p. 373). Furthermore, many examples have been reported of stable solutions of arenium ions formed by attack of a proton on an aromatic ring.[4] Simple aromatic compounds can be extensively deuterated in a convenient fashion by

[106]Tee, O.; Iyengar, N.R.; Bennett, J.M. *J. Org. Chem.* **1986**, *51*, 2585.

[107]For other work on ipso reactivity, see Baciocchi, E.; Illuminati, G. *J. Am. Chem. Soc.* **1967**, *89*, 4017; Berwin, H.J. *J. Chem. Soc., Chem. Commun.* **1972**, 237; Galley, M.W.; Hahn, R.C. *J. Am. Chem. Soc.* **1974**, *96*, 4337; Clemens, A.H.; Hartshorn, M.P.; Richards, K.E.; Wright, G.J. *Aust. J. Chem.* **1977**, *30*, 103, 113.

[108]For a review, see Taylor, R., in Bamford, C.H.; Tipper, C.F.H. *Comprehensive Chemical Kinetics*, Vol. 13, Elsevier, NY, **1972**, pp. 194–277.

[109]Small, P.A.; Wolfenden, J.H. *J. Chem. Soc.* **1936**, 1811.

[110]For example, see Challis, B.C.; Long, F.A. *J. Am. Chem. Soc.* **1963**, *85*, 2524; Batts, B.D.; Gold, V. *J. Chem. Soc.* **1964**, 4284; Kresge, A.J.; Chiang, Y.; Sato, Y. *J. Am. Chem. Soc.* **1967**, *89*, 4418; Gruen, L.C.; Long, F.A. *J. Am. Chem. Soc.* **1967**, *89*, 1287; Butler, A.B.; Hendry, J.B. *J. Chem. Soc. B* **1970**, 852.

treatment with D_2O and BF_3.[111] It has been shown that tritium exchange takes place readily at the 2 position of **31**, despite the fact that this position is hindered by the bridge. The rates were not very different from the comparison compound 1,3-dimethylnaphthalene.[112]

31

Hydrogen exchange can also be effected with strong bases,[113] such as NH_2-. In these cases, the slow step is the proton transfer:

$$ArH \quad + \quad B \quad \longrightarrow \quad Ar^- \quad + \quad BH^+$$

so the S_E1 mechanism and not the usual arenium ion mechanism is operating.[114] Aromatic rings can also be deuterated by treatment with D_2O and a rhodium(III) chloride[115] or platinum[116] catalyst or with C_6D_6 and an alkylaluminum dichloride catalyst,[117] though rearrangements may take place during the latter procedure. Tritium (^3H, abbreviated T) can be introduced by treatment with T_2O and an alkylaluminum dichloride catalyst.[117] Tritiation at specific sites (e.g., >90% para in toluene) has been achieved with T_2 gas and a microporous aluminophosphate catalyst.[118]

B. Nitrogen Electrophiles

11-2   Nitration or Nitro-de-hydrogenation

$$ArH \quad + \quad HNO_3 \quad \xrightarrow{H_2SO_4} \quad ArNO_2$$

[111]Larsen, J.W.; Chang, L.W. *J. Org. Chem.* **1978**, *43*, 3602.

[112]Laws, A.P.; Neary, A.P.; Taylor, R. *J. Chem. Soc. Perkin Trans. 2* **1987**, 1033.

[113]For a review of base-catalyzed hydrogen exchange on heterocycles, see Elvidge, J.A.; Jones, J.R.; O'Brien, C.; Evans, E.A.; Sheppard, H.C. *Adv. Heterocycl. Chem.* **1974**, *16*, 1.

[114]Shatenshtein, A.I. *Tetrahedron* **1962**, *18*, 95.

[115]Lockley, W.J.S. *Tetrahedron Lett.* **1982**, *23*, 3819; *J. Chem. Res. (S)* **1985**, 178.

[116]See, for example, Leitch, L.C. *Can. J. Chem.* **1954**, *32*, 813; Fraser, R.R.; Renaud, R.N. *J. Am. Chem. Soc.* **1966**, *88*, 4365; Fischer, G.; Puza, M. *Synthesis* **1973**, 218; Blake, M.R.; Garnett, J.L.; Gregor, I.K.; Hannan, W.; Hoa, K.; Long, M.A. *J. Chem. Soc., Chem. Commun.* **1975**, 930. See also, Parshall, G.W. *Acc. Chem. Res.* **1975**, *8*, 113.

[117]Long, M.A.; Garnett, J.L.; West, J.C. *Tetrahedron Lett.* **1978**, 4171.

[118]Garnett, J.L.; Kennedy, E.M.; Long, M.A.; Than, C.; Watson, A.J. *J. Chem. Soc., Chem. Commun.* **1988**, 763.

Most aromatic compounds, whether of high or low reactivity, can be nitrated, because a wide variety of nitrating agents is available.[119] For benzene, the simple alkylbenzenes, and less reactive compounds, the most common reagent is a mixture of concentrated nitric and sulfuric acids,[120] but for active substrates, the reaction can be carried out with nitric acid alone,[121] or in water, acetic acid, acetic anhydride, or chloroform.[122] Nitric acid in acetic anhydride/trifluoroacetic anhydride on zeolite H-β was used to convert toluene to 2,4-dinitrotoluene,[123] and $AcONO_2$ on clay converted ethylbenzene to ortho–para nitro ethylbenzene.[124] In fact, these milder conditions are necessary for active compounds, such as amines, phenols, and pyrroles, since reaction with mixed nitric and sulfuric acids would oxidize these substrates. With active substrates, such as amines and phenols, nitration can be accomplished by nitrosation under oxidizing conditions with a mixture of dilute nitrous and nitric acids.[125] A mixture of $NO_2/O_2/Fe(acac)_3$ can be used for active compounds,[126] as can $NaNO_2$ with trichloroisocyanuric acid on wet silica gel,[127] or N_2O_4 and silica acetate.[128] Trimethoxybenzenes were nitrated easily with ceric ammonium nitrate on silica gel,[129] and mesitylene was nitrated in an

[119]For a discussion of a unified mechansim, see Esteves, P.M.; de M. Carneiro, J.W.; Cardoso, S.P.; Barbosa, A.G.H.; Laali, K.K.; Rasul, G.; Prakash, G.K.S.; Olah, G.A. *J. Am. Chem. Soc.* **2003**, *125*, 4836. For monographs, see Olah, G.A.; Malhotra, R.; Narang, S.C. *Nitration: Methods and Mechanisms*, VCH, NY, *1989*; Schofield, K. *Aromatic Nitration*; Cambridge University Press, Cambridge, *1980*; Hoggett, J.H.; Moodie, R.B.; Penton, J.R.; Schofield, K. *Nitraton and aromatic Reactivity*, Cambridge University Press, Cambridge, *1971*. For reviews, see Weaver, W.M., in Feuer, H. *Chemistry of the Nitro and Nitroso Groups*, pt. 2, Wiley, NY, *1970*, pp. 1–48; de la Mare, P.B.D.; Ridd, J.H. *Aromatic Substitution Nitration and Halogenation*, Academic Press, NY, *1959*, pp. 48–93. See also, Ref. 1. For a review of side reactions, see Suzuki, H. *Synthesis 1977*, 217. Also see, Bosch, E.; Kochi, J.K. *J. Org. Chem.* **1994**, *59*, 3314; Olah, G.A.; Wang, Q.; Li, X.; Bucsi, I. *Synthesis 1992*, 1085; Olah, G.A.; Reddy, V.P.; Prakash, G.K.S. *Synthesis 1992*, 1087.

[120]For the use of sulfuric acid/nitric acid on silica, see Smith, A.C.; Narvaez, L.D.; Akins, B.G.; Langford, M.M.; Gary, T.; Geisler, V.J.; Khan, F.A. *Synth. Commun.* **1999**, *29*, 4187. For a reaction with guanidine–nitric acid with sulfric acid, see Ramana, M.M.V.; Malik, S.S.; Parihar, J.A. *Tetrahedron Lett.* **2004**, *45*, 8681.

[121]For a reaction with nitric acid and a lanthanum salt, see Parac-Vogt, T.N.; Binnesmans, K. *Tetrahedron Lett.* **2004**, *45*, 3137.

[122]Used with $(NH_4)_2SO_4 \bullet NiSO_4 \bullet 6\ H_2O$: Tasneem, Ali, M.M.; Rajanna, K.C.; Saiparakash, P.K. *Synth. Commun.* **2001**, *31*, 1123.

[123]Smith, K.; Gibbons, T.; Millar, R.W.; Claridge, R.P. *J. Chem. Soc., Perkin Trans. 1*, **2000**, 2753.

[124]Rodrigues, J.A.R.; Filho, A.P.O.; Moran, P.J.S. *Synth. Commun.* **1999**, *29*, 2169.

[125]For discussions of the mechanism in this case, see Giffney, J.C.; Ridd, J.H. *J. Chem. Soc. Perkin Trans. 2 1979*, 618; Bazanova, G.V.; Stotskii, A.A. *J. Org. Chem. USSR 1980*, *16*, 2070, 2075; Ross, D.S.; Moran, K.D.; Malhotra, R. *J. Org. Chem. 1983*, *48*, 2118; Dix, L.R.; Moodie, R.B. *J. Chem. Soc. Perkin Trans. 2 1986*, 1097; Leis, J.R.; Peña, M.E.; Ridd, J.H. *Can. J. Chem. 1989*, *67*, 1677. For a review, see Ridd, J.H. *Chem. Soc. Rev. 1991*, *20*, 149.

[126]Suzuki, H.; Yonezawa, S.; Nonoyama, N.; Mori, T. *J. Chem. Soc. Perkin Trans. 1 1996*, 2385.

[127]Zolfigol, M.A.; Madrakian, E.; Ghaemi, E. *Synlett 2003*, 2222.

[128]Iranpoor, N.; Firouzabadi, H.; Heydari, R. *Synth. Commun.* **2003**, *33*, 703.

[129]Khadilkar, B.M.; Madyar, V.R. *Synth. Commun.* **1999**, *29*, 1195.

ionic liquid using nitric acid–acetic anhydride.[130] Phenol can be nitrated in an ionic liquid.[131]

If anhydrous conditions are required, nitration can be effected with N_2O_5[132] in CCl_4 in the presence of P_2O_5, which removes the water formed in the reaction.[133] These reagents can also be used with proton or Lewis acid catalysts. Representative nitrating agents are $NaNO_2$ and trifluoroacetic acid,[134] N_2O_4 (which gives good yields with polycyclic hydrocarbons[135]), N_2O_4/O_2 and a catalytic amount of zeolite Hβ,[136] $Yb(OTf)_3$,[137] and nitronium salts,[138] such as $NO_2^+ BF_4^-$, $NO_2^+ PF_6^-$, and $NO_2^+ CF_3SO_3^-$.[139] A mixture of NO_2 and ozone has also been used.[140] Clays, such as clay-supported cupric nitrate (Claycop),[141,142] or Montmorillonite KSF—$Bi(NO_3)$[143] can be used to nitrate aromatic rings. Nitration of styrene poses a problem since addition occurs to the C=C unit to give a 1-nitroethyl aryl.[144] Heterocycles, such as pyridine, are nitrated with N_2O_5 and SO_2.[145] Deactivated aromatic rings, as in acetophenone, were nitrated with N_2O_5 and $Fe(acac)_2$.[146]

[130]In bmpy NTf₂, 1-butyl-4-methylpyridinium triflimide: Lancaster, N.L.; Llopis-Mestre, V. *Chem. Commun.* **2003**, 2812.

[131]In bbim BF₄, 1,3-dibutylimidazoliiuum tetrafluoroborate: Rajogopal, R.; Srinivasan, K.V. *Synth. Commun.* **2004**, *34*, 961.

[132]For a review of N₂O₅, see Fischer, J.W. in Feuer, H.; Nielsen, A.T. *Nitro Compounds, Recent Advances in synthesis and Chemistry*; VCH, NY, **1990**, pp. 267–365.

[133]For another method, see Olah, G.A.; Krishnamurthy, V.V.; Narang, S.C. *J. Org. Chem.* **1982**, *47*, 596.

[134]Uemura, S.; Toshimitsu, A.; Okano, M. *J. Chem. Soc. Perkin Trans. 1* **1978**, 1076. For a reaction with NaNO₂ and wet silica, see Zolfigol, M.A.; Ghaemi, E.; Madrakian, E. *Synth. Commun.* **2000**, *30* , 1689; Zolfigol, M.A.; Bagherzadeh, M.; Madrakian, E.; Gaemi, E.; Taqian-Nasab, A. *J. Chem. Res. (S)* **2001**, 140.

[135]Radner, F. *Acta Chem. Scand. Ser. B* **1983**, *37*, 65.

[136]Smith, K.; Almeer, S.; Black, S.J. *Chem. Commun.* **2000**, 1571. See also, Smith, K.; Musson, A.; DeBoos, G.A. *J. Org. Chem.* **1998**, *63*, 8448.

[137]Barrett, A.G.M.; Braddock, D.C.; Ducray, R.; McKinnell, R.M.; Waller, F.J. *Synlett* **2000**, 57.

[138]Olah, G.A.; Kuhn, S.J. *J. Am. Chem. Soc.* **1962**, *84*, 3684. These have also been used together with crown ethers: Masci, B. *J. Org. Chem.* **1985**, *50*, 4081; Iranpoor, N.; Firouzabadi, H.; Heydari, R. *Synth. Commun.* **1999**, *29*, 3295. For a review of nitronium salts in organic chemistry, see Guk,Yu. V.; Ilyushin, M.A.; Golod, E.L.; Gidaspov, B.V. *Russ. Chem. Rev.* **1983**, *52*, 284.

[139]This salt gives a very high yield of products at low temperatures, see Coon, C.L.; Blucher, W.G.; Hill, M.E. *J. Org. Chem.* **1973**, *38*, 4243; Effenberger, F.; Geke, J. *Synthesis* **1975**, 40.

[140]Nose, M.; Suzuki, H.; Suzuki, H. *J. Org. Chem.* **2001**, *66*, 4356; Peng, X.; Suzuki, H. *Org. Lett.* **2001**, *3*, 3431; Suzuki, H.; Tomaru, J.-i.; Murashima, T. *J. Chem. Soc. Perkin Trans. 1* **1994**, 2413; Suzuki, H.; Tatsumi, A.; Ishibashi, T.; Mori, T. *J. Chem. Soc. Perkin Trans. 1* **1995**, 339.

[141]For reviews of clay-supported nitrates, see Cornélis, A.; Laszlo, P. *Synthesis* **1985**, 909; Laszlo, P. *Acc. Chem. Res.* **1986**, 121; Laszlo, P.; Cornélis, A. *Aldrichimica Acta* **1988**, *21*, 97.

[142]Cornélis, A.; Delaude, L.; Gerstmans, A.; Laszlo, P. *Tetrahedron Lett.* **1988**, *29*, 5657. See also, Smith, K.; Fry, K.; Butters, M.; Nay, B. *Tetrahedron Lett.* **1989**, *30*, 5333; Cornélis, A.; Laszlo, P.; Pennetreau, P. *Bull. Soc. Chim. Belg.*, **1984**, *93*, 961; Poirier, J.; Vottero, C. *Tetrahedron* **1989**, *45*, 1415. For a method of nitrating phenols in the ortho position, see Pervez, H.; Onyiriuka, S.O.; Rees, L.; Rooney, J.R.; Suckling, C.J. *Tetrahedron* **1988**, *44*, 4555.

[143]Samajdar, S.; Becker, F.F.; Banik, B.K. *Tetrahedron Lett.* **2000**, *41*, 8017.

[144]Lewis, R.J.; Moodie, R.B. *J. Chem. Soc. Perkin Trans. 2* **1997**, 563.

[145]Arnestad, B.; Bakke, J.M.; Hegbom, I.; Ranes, E. *Acta Chem. Scand. B* **1996**, *50*, 556.

[146]Bak, R.R.; Smallridge, A.J. *Tetrahedron Lett.* **2001**, *42*, 6767.

An alternative route for the nitration of activated aromatic compounds, such as anisole, used a nitrate ester ($RONO_2$) with triflic acid in an ionic liquid for ortho-selective nitration.[147] Nitration in alkaline media can be accomplished with esters of nitric acid, such as ethyl nitrate ($EtONO_2$).

When anilines are nitrated under strong acid conditions, meta orientation is generally observed, because the species undergoing nitration is actually the conjugate acid of the amine. If the conditions are less acidic, the free amine is nitrated and the orientation is ortho–para. Although the free base may be present in much smaller amounts than the conjugate acid, it is far more susceptible to aromatic substitution (see also p. 668). Because of these factors and because they are vulnerable to oxidation by nitric acid, primary aromatic amines are often protected before nitration by treatment with acetyl chloride (**16-72**) or acetic anhydride (**16-73**). Nitration of the resulting acetanilide derivative avoids all these problems. There is evidence that when the reaction takes place on the free amine, it is the nitrogen that is attacked to give an *N*-nitro compound $Ar-NH-NO_2$ which rapidly undergoes rearrangement (see **11-28**) to give the product.[148]

Since the nitro group is deactivating, it is usually easy to stop the reaction after one group has entered the ring, but a second and a third group can be introduced if desired, especially when an activating group is also present. Even *m*-dinitrobenzene can be nitrated if vigorous conditions are applied. This has been accomplished with $NO_2^+BF_4^-$ in FSO_3H at 150°C.[149]

With most of the reagents mentioned, the attacking species is the nitronium ion NO_2^+. Among the ways in which this ion is formed are

1. In concentrated sulfuric acid, by an acid–base reaction in which nitric acid is the base:

$$HNO_3 + 2\,H_2SO_4 \rightleftharpoons NO_2^+ + H_3O^+ + 2\,HSO_4^-$$

This ionization is essentially complete.

2. In concentrated nitric acid alone,[150] by a similar acid–base reaction in which one molecule of nitric acid is the acid and another the base:

$$2\,HNO_3 \rightleftharpoons NO_2^+ + NO_3^- + H_2O$$

This equilibrium lies to the left ($\sim$4% ionization), but enough NO_2^+ is formed for nitration to occur.

[147]In emim OTf, 1-ethyl-3-methylimidazolium triflate: Laali, K.K.; Gettwert, V.J. *J. Org. Chem.* **2001**, *66*, 35.

[148]Ridd, J.H.; Scriven, E.F.V. *J. Chem. Soc., Chem. Commun.* **1972**, 641. See also, Helsby, P.; Ridd, J.H. *J. Chem. Soc. Perkin Trans. 2* **1983**, 1191.

[149]Olah, G.A.; Lin, H.C. *Synthesis* **1974**, 444.

[150]See Belson, D.J.; Strachan, A.N. *J. Chem. Soc. Perkin Trans. 2* **1989**, 15.

3. The equilibrium just mentioned occurs to a small extent even in organic solvents.

4. With N_2O_5 in CCl_4, there is spontaneous dissociation:

$$N_2O_5 \rightleftharpoons NO_2^+ + NO_3^-$$

but in this case there is evidence that some nitration also takes place with undissociated N_2O_5 as the electrophile.

5. When nitronium salts are used, NO_2^+ is of course present to begin with. Esters and acyl halides of nitric acid ionize to form NO_2^+. Nitrocyclohexadienones are converted to NO_2^+ and the corresponding phenol.[132]

There is a great deal of evidence that NO_2^+ is present in most nitration reactions and that it is the attacking entity,[151] for example,

1. Nitric acid has a peak in the Raman spectrum. When nitric acid is dissolved in concentrated sulfuric acid, the peak disappears and two new peaks appear, one at 1400 cm^{-1} attributable to NO_2^+ and one at 1050 cm^{-1} due to HSO_4-.[152]

2. On addition of nitric acid, the freezing point of sulfuric acid is lowered about four times the amount expected if no ionization has taken place.[153] This means that the addition of one molecule of nitric acid results in the production of four particles, which is strong evidence for the ionization reaction between nitric and sulfuric acids given above.

3. The fact that nitronium salts in which nitronium ion is known to be present (by X-ray studies) nitrate aromatic compounds shows that this ion does attack the ring.

4. The rate of the reaction with most reagents is proportional to the concentration of NO_2^+, not to that of other species.[154] When the reagent produces this ion in small amounts, the attack is slow and only active substrates can be nitrated. In concentrated and aqueous mineral acids, the kinetics are second order: first order each in aromatic substrate and in nitric acid (unless pure nitric acid is used in which case there are pseudo-first-order kinetics). But in organic solvents such as nitromethane, acetic acid, and CCl_4, the kinetics are first order in nitric acid alone and zero order in aromatic substrate, because the rate-determining step is formation of NO_2^+ and the substrate does not take part in this.

---

[151]For an exhaustive study of this reaction, see Hughes, E.D.; Ingold, C.K.in a series of several papers with several different co-workers, see *J. Chem. Soc.* **1950**, 2400.

[152]Ingold, C.K.; Millen, D.J.; Poole, H.G. *J. Chem. Soc.* **1950**, 2576.

[153]Gillespie, R.J.; Graham, J.; Hughes, E.D.; Ingold, C.K.; Peeling, E.R.A. *J. Chem. Soc.* **1950**, 2504.

[154]This is not always strictly true. See Ross, D.S.; Kuhlmann, K.F.; Malhotra, R. *J. Am. Chem. Soc.* **1983**, *105*, 4299.

An interesting route to nitrobenzene begins with bromobenzene. Reaction with butyllithium gives phenyllithium, which reacts with an excess of N_2O_4 to give nitrobenzene.[155]

In a few cases, depending on the substrate and solvent, there is evidence that the arenium ion is not formed directly, but via the intermediacy of a radical pair (see p. 682) such as **32**.[156]

$$ArH + NO_2^+ \longrightarrow [\,ArH \;\overset{\cdot}{\cdot}\; NO_2{}^{\boldsymbol{\cdot}}\,] \longrightarrow$$

32

Arylboronic acids have been shown to react with ammonium nitrate and trifluoroacetic acid to give the corresponding nitrobenzene.[157]

OS **I**, 372, 396, 408 (see also OS **53**, 129); **II**, 254, 434, 438, 447, 449, 459, 466; **III**, 337, 644, 653, 658, 661, 837; **IV**, 42, 364, 654, 711, 722, 735; **V**, 346, 480, 829, 1029, 1067.

11-3   Nitrosation or Nitroso-de-hydrogenation

Ring nitrosation[158] with nitrous acid is normally carried out only with active substrates, such as amines and phenols. However, primary aromatic amines give diazonium ions (**13-19**) when treated with nitrous acid,[159] and secondary amines tend to give *N*-nitroso rather than *C*-nitroso compounds (**12-50**); hence this reaction is normally limited to phenols and tertiary aromatic amines. Nevertheless, secondary aromatic amines can be *C*-nitrosated in two ways. The *N*-nitroso compound first obtained can be isomerized to a *C*-nitroso compound (**11-29**), or it can be treated with another equivalent of nitrous acid to give an *N,C*-dinitroso compound. Also, a successful nitrosation of anisole has been reported, where the solvent was $CF_3COOH–CH_2Cl_2$.[160]

[155]Tani, K.; Lukin, K.; Eaton, P.E. *J. Am. Chem. Soc.* **1997**, *119*, 1476.

[156]For a review of radical processes in aromatic nitration, see Ridd, J.H. *Chem. Soc. Rev.* **1991**, *20*, 149. For a review of aromatic substitutions involving radical cations, see Kochi, J.K. *Adv. Free Radical Chem. (Greenwich, Conn.)* **1990**, *1*, 53.

[157]Salzbrunn, S.; Simon, J.; Prakash, G.K.S.; Petasis, N.A.; Olah, G.A. *Synlett* **2000**, 1485; Prakash, G.K.S.; Panja, C.; Mathew, T.; Surampudi, V.; Petasis, N.A.; Olah, G.A. *Org. Lett.* **2004**, *6*, 2205.

[158]For a review, see Williams, D.L.H. *Nitrosation*, Cambridge University Press, Cambridge, **1988**, pp. 58–76. Also see Atherton, J.H.; Moodie, R.B.; Noble, D.R.; O'Sullivan, B. *J. Chem. Soc. Perkin Trans. 2* **1997**, 663.

[159]For examples of formation of *C*-nitroso compounds from primary and secondary amines, see Hoefnagel, M.A.; Wepster, B.M. *Recl. Trav. Chim. Pays-Bas* **1989**, *108*, 97.

[160]Radner, F.; Wall, A.; Loncar, M. *Acta Chem. Scand.* **1990**, *44*, 152.

Much less work has been done on the mechanism of this reaction than on **11-2**.[161] In some cases, the attacking entity is NO^+, but in others it is apparently NOCl, NOBr, N_2O_3, and so on, in each of which there is a carrier of NO^+. Both NOCl and NOBr are formed during the normal process of making nitrous acid (the treatment of sodium nitrite with HCl or HBr). Nitrosation requires active substrates because NO^+ is much less reactive than NO_2^+. Kinetic studies have shown that NO^+ is at least 10^{14} times less reactive than NO_2^+.[162] A consequence of the relatively high stability of NO^+ is that this species is easily cleaved from the arenium ion, so that k_{-1} competes with k_2 (p. 660) and isotope effects are found.[163] With phenols, there is evidence that nitrosation may first take place at the OH group, after which the nitrite ester thus formed rearranges to the C-nitroso product.[164] Tertiary aromatic amines substituted in the ortho position generally do not react with HONO, probably because the ortho substituent prevents planarity of the dialkylamino group, without which the ring is no longer activated. This is an example of steric inhibition of resonance (p. 48).

OS **I**, 214, 411, 511; **II**, 223; **IV**, 247.

11-4   Diazonium Coupling

Arylazo-de-hydrogenation

$$ArH + Ar'N_2^+ \longrightarrow Ar-N{=}N-Ar'$$

Aromatic diazonium ions normally couple only with active substrates, such as amines and phenols.[165] Many of the products of this reaction are used as dyes (*azo dyes*).[166] Presumably because of the size of the attacking species, substitution is mostly para to the activating group, unless that position is already occupied, in which case ortho substitution takes place. The pH of the solution is important both for phenols and amines. For amines, the solutions may be mildly acidic or neutral. The fact that amines give ortho and para products shows that even in mildly acidic solution they react in their un-ionized form. If the acidity is too high, the reaction does not occur, because the concentration of free amine becomes too small. Phenols must be coupled in slightly alkaline solution where they are converted to the more reactive phenoxide ions, because phenols themselves are not active enough for the

[161]For a review of nitrosation mechanisms at C and other atoms, see Williams, D.L.H. *Adv. Phys. Org. Chem.* **1983**, *19*, 381. See Williams, D.L.H. *Nitrosation*, Cambridge University Press, Cambridge, **1988**, pp. 58–76; Atherton, J.H.; Moodie, R.B.; Noble, D.R.; O'Sullivan, B. *J. Chem. Soc. Perkin Trans. 2* **1997**, 663.
[162]Challis, B.C.; Higgins, R.J.; Lawson, A.J. *J. Chem. Soc. Perkin Trans. 2* **1972**, 1831; Challis, B.C.; Higgins, R.J. *J. Chem. Soc. Perkin Trans. 2* **1972**, 2365.
[163]Challis, B.C.; Higgins, R.J. *J. Chem. Soc. Perkin Trans. 2* **1973**, 1597.
[164]Gosney, A.P.; Page, M.I. *J. Chem. Soc. Perkin Trans. 2* **1980**, 1783.
[165]For reviews, see Szele, I.; Zollinger, H. *Top. Curr. Chem.* **1983**, *112*, 1; Hegarty, A.F., in Patai's. *The Chemistry of Diazonium and Diazo Groups*, pt. 2, Wiley, NY, **1978**, pp. 545–551.
[166]For reviews of azo dyes, see Zollinger, H. *Color Chemistry*, VCH, NY, **1987**, pp. 85–148; Gordon, P.F.; Gregory, P. *Organic Chemistry in Colour*, Springer, NY, **1983**, pp. 95–162.

reaction. However, neither phenols nor amines react in moderately alkaline solution, because the diazonium ion is converted to a diazo hydroxide Ar—N=N—OH. Primary and secondary amines face competition from attack at the nitrogen.[167] However, the resulting *N*-azo compounds (aryl triazenes) can be isomerized to *C*-azo compounds (**11-30**). In at least some cases, even when the *C*-azo compound is isolated, it is the result of initial *N*-azo compound formation followed by isomerization. It is therefore possible to synthesize the *C*-azo compound directly in one laboratory step.[168] Acylated amines and phenolic ethers and esters are ordinarily not active enough for this reaction, though it is sometimes possible to couple them (as well as such polyalkylated benzenes as mesitylene and pentamethylbenzene) to diazonium ions containing electron-withdrawing groups in the para position, since such groups increase the concentration of the positive charge and thus the electrophilicity of the ArN_2^+. Some coupling reactions which are otherwise very slow (in cases where the coupling site is crowded) are catalyzed by pyridine for reasons discussed on p. 661. Phase transfer catalysis has also been used.[169]

Coupling of a few aliphatic diazonium compounds to aromatic rings has been reported. All the examples reported so far involve cyclopropanediazonium ions and bridgehead diazonium ions, in which loss of N_2 would lead to very unstable carbocations.[170] Azobenzenes have been prepared by Pd-catalyzed coupling of aryl hydrazides with aryl halides, followed by direct oxidation.[171]

The mechanism of Z/E isomerization in Ar-N=NAr systems has been studied.[172]

OS **I**, 49, 374; **II**, 35, 39, 145.

11-5   Direct Introduction of the Diazonium Group

Diazoniation or Diazonio-de-hydrogenation

$$ArH \xrightarrow[\text{HX}]{\text{2 HONO}} ArN_2^+X^-$$

Diazonium salts can be prepared directly by replacement of an aromatic hydrogen without the necessity of going through the amino group.[173] The reaction is essentially limited to active substrates (amines and phenols), since otherwise poor yields are obtained. Since the reagents and the substrate are the same as in reaction **11-3**, the first species formed is the nitroso compound. In the presence of excess nitrous acid, this is converted to the diazonium ion.[174] The reagent

[167]See Penton, J.R.; Zollinger, H. *Helv. Chim. Acta* **1981**, *64*, 1717, 1728.

[168]Kelly, R.P.; Penton, J.R.; Zollinger, H. *Helv. Chim. Acta* **1982**, *65*, 122.

[169]Hashida, Y.; Kubota, K.; Sekiguchi, S. *Bull. Chem. Soc. Jpn.* **1988**, *61*, 905.

[170]See Szele, I.; Zollinger, H. *Top. Curr. Chem.* **1983**, *112*, 1, see pp. 3–6.

[171]Lim, Y.-K.; Lee, K.-S.; Cho, C.-G. *Org. Lett.* **2003**, *5*, 979.

[172]Asano, T.; Furuta, H.; Hofmann, H.-J.; Cimiraglia, R.; Tsuno, Y.; Fujio, M. *J. Org. Chem.* **1993**, *58*, 4418.

[173]Tedder, J.M. *J. Chem. Soc.* **1957**, 4003.

[174]Tedder, J.M.; Theaker, G. *Tetrahedron* **1959**, *5*, 288; Kamalova, F.R.; Nazarova, N.E.; Solodova, K.V. ; Yaskova, M.S. *J. Org. Chem. USSR* **1988**, *24*, 1004.

(azidochloromethylene)dimethylammonium chloride [$Me_2N{=}C(Cl)N_3\ Cl^-$] can also introduce the diazonium group directly into a phenol.[175] A synthesis of solid aryldiazonium chlorides is now available.[176]

11-6   Amination or Amino-de-hydrogenation[177]

$$ArH + HN_3 \xrightarrow{\text{AlCl}_3} ArNH_2$$

Aromatic compounds can be converted to primary aromatic amines, in 10–65% yields, by treatment with hydrazoic acid HN_3 in the presence of $AlCl_3$ or H_2SO_4.[178] Higher yields (>90%) have been reported with trimethylsilyl azide (Me_3SiN_3) and triflic acid F_3CSO_2OH.[179] Treatment of an aromatic compound with tetramethylhydrazonium iodide and then ammonium also give the aryl amine.[180] Tertiary amines have been prepared in ~50–90% yields by treatment of aromatic hydrocarbons with N-chlorodialkylamines; by heating in 96% sulfuric acid; or with $AlCl_3$ or $FeCl_3$ in nitroalkane solvents; or by irradiation.[181] Treatment of an aryl halide with an amine and a palladium catalyst leads to the aniline derivative.[182]

Tertiary (and to a lesser extent, secondary) aromatic amines can also be prepared in moderate to high yields by amination with an N-chlorodialkylamine (or an N-chloroalkylamine) and a metallic-ion catalyst (e.g., Fe^{2+}, Ti^{3+}, Cu^+, Cr^{2+}) in the presence of sulfuric acid.[183] The attacking species in this case is the aminium radical ion $R_2NH\bullet$ formed by[184]

$$R_2\overset{\oplus}{N}HCl \ + \ M^+ \ \longrightarrow \ R_2\overset{\oplus}{N}H\bullet \ + \ M^{2+} \ + \ Cl^-$$

Because attack is by a positive species (even though it is a free radical), orientation is similar to that in other electrophilic substitutions (e.g., phenol and acetanilide give ortho and para substitution, mostly para). When an alkyl group is present, attack at the benzylic position competes with ring substitution. Aromatic rings containing only meta-directing groups do not give the reaction at all. Fused ring systems react well.[185]

---

[175]Kokel, B.; Viehe, H.G. *Angew. Chem. Int. Ed.* **1980**, *19*, 716.

[176]Mohamed, S.K.; Gomaa, M.A.-M.; El-Din, A..M.N. *J. Chem. Res. (S)* **1997**, 166.

[177]For a review, see Kovacic, P., in Olah, G.A. *Friedel–Crafts and Related Reactions*, Vol. 3, Wiley, NY, **1964**, pp. 1493–1506.

[178]Kovacic, P.; Russell, R.L.; Bennett, R.P. *J. Am. Chem. Soc.* **1964**, *86*, 1588.

[179]Olah, G.A.; Ernst, T.D. *J. Org. Chem.* **1989**, *54*, 1203.

[180]Rozhkov, V.V.; Shevelev, S.A.; Chervin, I.T.; Mitchel, A.R.; Schmidt, R.D. *J. Org. Chem.* **2003**, *68*, 2498.

[181]Bock, H.; Kompa, K. *Angew. Chem. Int. Ed.* **1965**, *4*, 783; *Chem. Ber.* **1966**, *99*, 1347, 1357, 1361.

[182]Guram, A.S.; Rennels, R.A.; Buchwald, S.L. *Angew. Chem. Int. Ed. Engl.* **1995**, *34*, 1348.

[183]For reviews, see Minisci, F. *Top. Curr. Chem.* **1976**, *62*, 1, see pp. 6–16, *Synthesis* **1973**, 1, see pp. 2–12, Sosnovsky, G.; Rawlinson, D.J. *Adv. Free-Radical Chem.* **1972**, *4*, 203, see pp. 213–238.

[184]For a review of aminium radical ions, see Chow, Y.L. *React. Intermed. (Plenum)* **1980**, *1*, 151.

[185]The reaction has been extended to the formation of primary aromatic amines, but the scope is narrow: Citterio, A.; Gentile, A.; Minisci, F.; Navarrini, V.; Serravalle, M.; Ventura, S. *J. Org. Chem.* **1984**, *49*, 4479.

Unusual orientation has been reported for amination with haloamines and with NCl_3 in the presence of $AlCl_3$. For example, toluene gave predominately meta amination.[186] It has been suggested that initial attack in this case is by Cl^+ and that a nitrogen nucleophile (whose structure is not known, but is represented here as NH_2^- for simplicity) adds to the resulting arenium ion, so that the initial reaction is addition to a carbon–carbon double bond followed by elimination of HCl from **33**.[187]

According to this suggestion, the electrophilic attack is at the para position (or the ortho, which leads to the same product) and the meta orientation of the amino group arises indirectly. This mechanism is called the σ-*substitution mechanism*.

Diphenylliodonium salts react with amines in the presence of a copper catalyst. Diphenyliodonium tetrafluoroborate, $Ph_2I^+BF_4^-$, reacts with indole in DMF at 150°C with a $Cu(OAc)_2$ catalyst, for example, to give N-phenylindole.[188]

Aromatic compounds that do not contain meta-directing groups can be converted to diarylamines by treatment with aryl azides in the presence of phenol at −60°C: $ArH + Ar'N_3 \rightarrow ArNHAr'$.[189] Diarylamines are also obtained by the reaction of N-arylhydroxylamines with aromatic compounds (benzene, toluene, anisole) in the presence of F_3CCOOH: $ArH + Ar'NHOH \rightarrow ArNHAr'$.[190]

Direct *amidation* can be carried out if an aromatic compound is heated with a hydroxamic acid (**34**) in polyphosphoric acid, but the scope is essentially limited to phenolic ethers.[191] The reaction of an aromatic compound with aniline, Bu_4NF and $KMnO_4$ led to the diarylamine.[192] The formation of hydroindole derivatives was accomplished by reaction of a N-carbamoyl phenylethylamine derivative with phenyliodine (III) diacetate, followed by Bu_4NF.[193] Direct amidation via ipso substitution by nitrogen was accomplished when a N-methoxy arylethylamide (**35**) was

[186] See Strand, J.W.; Kovacic, P. *J. Am. Chem. Soc.* **1973**, *95*, 2977, and references cited therein.

[187] Kovacic, P.; Levisky, J.A. *J. Am. Chem. Soc.* **1966**, *88*, 1000.

[188] Zhou, T.; Chen, Z.-C. *Synth. Commun.* **2002**, *32*, 903.

[189] Nakamura, K.; Ohno, A.; Oka, S. *Synthesis* **1974**, 882. See also, Takeuchi, H.; Takano, K. *J. Chem. Soc. Perkin Trans. 1* **1986**, 611.

[190] Shudo, K.; Ohta, T.; Okamoto, T. *J. Am. Chem. Soc.* **1981**, *103*, 645.

[191] Wassmundt, F.W.; Padegimas, S.J. *J. Am. Chem. Soc.* **1967**, *89*, 7131; March, J.; Engenito Jr., J.S. *J. Org. Chem.* **1981**, *46*, 4304. Also see, Cablewski, T.; Gurr, P.A.; Rander, K.D.; Strauss, C.R. *J. Org. Chem.* **1994**, *59*, 5814.

[192] Huertas, I.; Gallardo, I.; Marquet, J. *Tetrahedron Lett.* **2001**, *42*, 3439.

[193] Pouységu, L.; Avellan, A.-V.; Quideau, S. *J. Org. Chem.* **2002**, *67*, 3425.

treated with [hydroxyl(tosyloxy)iodo]benzene (HTIB) in 2,2,2-trifluoroethanol, giving a *N*-methoxy spirocylcic amide, **36**.[194]

Aromatic compounds add to DEAD (diethyl azodicarboxylate), in the presence of InCl$_3$–SiO$_2$ and microwave irradiation, to give the *N*-aryldiamino compound [ArN(CO$_2$Et)–NHCO$_2$Et].[195]

An interesting variation in the alkylation reaction used five equivalents of aluminum chloride in a reaction of *N*-methyl-*N*-phenylhydrazine and benzene to give *N*-methyl-4-phenylaniline.[196]

Also see **13-5**, **13-16**.

C. Sulfur Electrophiles

11-7   Sulfonation or Sulfo-de-hydrogenation

$$ArH \quad + \quad H_2SO_4 \longrightarrow ArSO_2OH$$

The sulfonation reaction is very broad in scope and many aromatic hydrocarbons (including fused ring systems), aryl halides, ethers, carboxylic acids, amines,[197] acylated amines, ketones, nitro compounds, and sulfonic acids have been sulfonated.[198] Phenols can also be successfully sulfonated, but attack at oxygen may compete.[199] Sulfonation is often accomplished with concentrated sulfuric acid, but it can also be done with fuming sulfuric acid, SO$_3$, ClSO$_2$OH, ClSO$_2$NMe$_2$/In(OTf)$_3$,[200] or other reagents.[201] As with nitration (**11-2**), reagents of a wide variety of activity are available to suit both highly active and highly inactive substrates. Since this is a reversible reaction (see **11-38**), it may be necessary to drive the reaction to completion.

[194]Miyazawa, E.; Sakamoto, T.; Kikugawa, Y. *J. Org. Chem.* **2003**, *68*, 5429.

[195]Yadav, J.S.; Subba Reddy, B.V.; Kumar, G.M.; Madan, C. *Synlett* **2001**, 1781.

[196]Ohwada, A.; Nara, S.; Sakamoto, T.; Kikugawa, Y. *J. Chem. Soc, Perkin Trans. 1* **2001**, 3064.

[197]See Khelevin, R.N. *J. Org. Chem. USSR* **1987**, *23*, 1709; *1988*, *24*, 535, and references cited therein.

[198]For reviews, see Nelson, K.L. in Olah, G.A. *Friedel–Crafts and Related Reactions*, Vol. 3, Wiley, NY, *1964*, pp. 1355–1392; Gilbert, E.E. *Sulfonation and Related Reactions*, Wiley, NY, *1965*, pp. 62–83, 87–124.

[199]See, for example, de Wit, P.; Woldhuis, A.F.; Cerfontain, H. *Recl. Trav. Chim. Pays-Bas* **1988**, *107*, 668.

[200]Frost, C.G.; Hartley, J.P.; Griffin, D. *Synlett* **2002**, 1928.

[201]For a reaction using silica sulfuric acid, see Hajipour, A.R.; Mirjalili, B.B.F.; Zarei, A.; Khazdooz, L.; Ruoho, A.E. *Tetrahedron Lett.* **2004**, *45*, 6607.

However, at low temperatures the reverse reaction is very slow and the forward reaction is practically irreversible.[202] Sulfur trioxide reacts much more rapidly than sulfuric acid with benzene it is nearly instantaneous. Sulfones are often side products. When sulfonation is carried out on a benzene ring containing four or five alkyl and/or halogen groups, rearrangements usually occur (see **11-36**).

A great deal of work has been done on the mechanism,[203] chiefly by Cerfontain and co-workers. Mechanistic study is made difficult by the complicated nature of the solutions. Indications are that the electrophile varies with the reagent, though SO_3 is involved in all cases, either free or combined with a carrier. In aqueous H_2SO_4 solutions, the electrophile is thought to be $H_3SO_4^+$ (or a combination of H_2SO_4 and H_3O^+) at concentrations below ~ 80–85% H_2SO_4, and $H_2S_2O_7$ (or a combination of H_2SO_4 and SO_3) at concentrations higher than this[204] (the change-over point varies with the substrate[205]). Evidence for a change in electrophile is that in the dilute and in the concentrated solutions the rate of the reaction was proportional to the activity of $H_3SO_4^+$ and $H_2S_2O_7$, respectively. Further evidence is that with toluene as substrate the two types of solution gave very different ortho/para ratios. The mechanism is essentially the same for both electrophiles and may be shown as:[204]

The other product of the first step is HSO_4^- or H_2O from $H_2S_2O_7$ or $H_3SO_4^+$, respectively. Path a is the principal route, except at very high H_2SO_4 concentrations, when path b becomes important. With $H_3SO_4^+$ the first step is rate determining under all conditions, but with $H_2S_2O_7$ the first step is the slow step only up to $\sim 96\%$ H_2SO_4, when a subsequent proton transfer becomes partially rate determining.[206] The $H_2S_2O_7$ is more reactive than $H_3SO_4^+$. In fuming sulfuric acid (H_2SO_4 containing excess SO_3), the electrophile is thought to be $H_3S_2O_7^+$ (protonated $H_2S_2O_7$) up to

[202]Spryskov, A.A. *J. Gen. Chem. USSR* **1960**, *30*, 2433.

[203]For a monograph, see Cerfontain, H. *Mechanistic Aspects in Aromatic Sulfonation and Desulfonation*, Wiley, NY, **1968**. For reviews, see Cerfontain, H. *Recl. Trav. Chim. Pays-Bas* **1985**, *104*, 153; Cerfontain, H.; Kort, C.W.F. *Int. J. Sulfur Chem. C* **1971**, *6*, 123; Taylor, R., in Bamford, C.H.; Tipper, C.F.H. *Comprehensive Chemical Kinetics*, Vol. 13, Elsevier, NY, **1972**, pp. 56–77.

[204]Cerfontain, H.; Lambrechts, H.J.A.; Schaasberg-Nienhuis, Z.R.H.; Coombes, R.G.; Hadjigeorgiou, P.; Tucker, G.P. *J. Chem. Soc. Perkin Trans. 2* **1985**, 659, and references cited therein.

[205]See, for example, Kaandorp, A.W.; Cerfontain, H. *Recl. Trav. Chim. Pays-Bas* **1969**, *88*, 725.

[206]Kort, C.W.F.; Cerfontain, H. *Recl. Trav. Chim. Pays-Bas* **1967**, *86*, 865.

$\sim$104% H_2SO_4 and $H_2S_4O_13$ ($H_2SO_4 + 3SO_3$) beyond this concentration.[207] Finally, when pure SO_3 is the reagent in aprotic solvents, SO_3 itself is the actual electrophile.[208] Free SO_3 is the most reactive of all these species, so that attack here is generally fast and a subsequent step is usually rate determining, at least in some solvents.

OS **II**, 42, 97, 482, 539; **III**, 288, 824; **IV**, 364; **VI**, 976.

11-8   Halosulfonation or Halosulfo-de-hydrogenation

$$ArH \quad + \quad ClSO_2OH \xrightarrow{\hspace{3cm}} ArSO_2Cl$$

Aromatic sulfonyl chlorides can be prepared directly, by treatment of aromatic rings with chlorosulfuric acid.[209] Since sulfonic acids can also be prepared by the same reagent (**11-7**), it is likely that they are intermediates, being converted to the halides by excess chlorosulfuric acid.[210] The reaction has also been effected with bromo- and fluorosulfuric acids. Sulfinyl chlorides (ArSOCl) have been prepared by the reaction of thionyl chloride and an aromatic compound on Montmorillonite K10 clay.[211]

OS **I**, 8, 85.

11-9   Sulfonylation

Alkylsulfonylation or **Alkylsulfo-de-hydrogenation**

$$ArH + SOCl_2 \xrightarrow{\text{TfOH}} ArSOAr$$

$$ArH + Ar'SO_2Cl \xrightarrow{\text{AlCl}_3} ArSO_2Ar'$$

Diaryl sulfoxides can be prepared by the reaction of aromatic compounds with thionyl chloride and triflic acid.[212] Diaryl sulfones have also been prepared using thionyl chloride with the ionic liquid [bmim]Cl•AlCl_3.[213] Diaryl sulfones can be formed by treatment of aromatic compounds with aryl sulfonyl chlorides and a Friedel–Crafts catalyst[214] This reaction is analogous to Friedel–Crafts acylation with carboxylic acid halides (**11-17**). In a better procedure, the aromatic compound

---

[207]Koeberg-Telder, A.; Cerfontain, H. *J. Chem. Soc. Perkin Trans. 2* **1973**, 633.

[208]Lammertsma, K.; Cerfontain, H. *J. Chem. Soc. Perkin Trans. 2* **1980**, 28, and references cited therein.

[209]For a review, see Gilbert, E.E. *Sulfonaton and Related Reactions*, Wiley, NY, **1965**, pp. 84–87.

[210]For a discussion of the mechanism with this reagent, see van Albada, M.P.; Cerfontain, H. *J. Chem. Soc. Perkin Trans. 2* **1977**, 1548, 1557.

[211]Karade, N.N.; Kate, S.S.; Adude, R.N. *Synlett* **2001**, 1573.

[212]Olah G.A.; Marinez, E.R.; Prakash, G.K.S. *Synlett* **1999**, 1397.

[213]In [bmim]Cl•AlCl_3, 1-butyl-3-methylimidazolium chloroaluminate: Mohile, S.S.; Potdar, M.K.; Salunkhe, M.M. *Tetrahedron Lett.* **2003**, *44*, 1255.

[214]For reviews, see Taylor, R., in Bamford, C.H.; Tipper, C.F.H *Comprehensive Chemical Kinetics*, Vol. 13, Elsevier, NY, **1972**, pp. 77–83; Jensen, F.R.; Goldman, G. in Olah, G.A. *Friedel–Crafts and Related Reactions*, Vol. 3, Wiley, NY, **1964**, pp. 1319–1347. For a solid-state reaction using Fe^{3+}-Montmorillonite, see Choudary, B.M.; Chowdari, N.S.; Kantam, M.L. *J. Chem. Soc., Perkin Trans. 1*, **2000**, 2689.

is treated with an aryl sulfonic acid and P_2O_5 in polyphosphoric acid.[215] Still another method uses an arylsulfonic trifluoromethanesulfonic anhydride $ArSO_2OSO_2CF_3$ (generated *in situ* from $ArSO_2Br$ and CF_3SO_3Ag) without a catalyst.[216] Indium *tris*(triflate)[217] and indium trichloride[218] give sulfonation with sulfonyl chlorides, and indium bromide was used in indoles.[219] A ferric chloride catalyzed reaction with microwave irradiation has also been reported,[220] as has the use of zinc metal with microwave irradiation.[221]

The reaction can be extended to the preparation of alkyl aryl sulfones by the use of a sulfonyl fluoride.[222]

Direct formation of diaryl sulfones from benzenesulfonic acid and benzene was accomplished using Nafion-H.[223]

OS **X**, 147.

D. Halogen Electrophiles

11-10   Halogenation[224]

Halo-de-hydrogenation

$$ArH + Br_2 \xrightarrow{\text{catalyst}} ArBr$$

1. *Chlorine and Bromine.* Aromatic compounds can be brominated or chlorinated by treatment with bromine or chlorine in the presence of a catalyst. For amines and phenols the reaction is so rapid that it is carried out with a dilute solution of Br_2 or Cl_2 in water at room temperature, or with aqueous HBr in DMSO.[225] Even so, with amines it is not possible to stop the reaction before all the available ortho and para positions are substituted, because the initially formed haloamines are weaker bases than the original amines and are less

[215]Graybill, B.M. *J. Org. Chem.* **1967**, *32*, 2931; Sipe, Jr., H.J.; Clary, D.W.; White, S.B. *Synthesis* **1984**, 283. See also, Ueda, M.; Uchiyama, K.; Kano, T. *Synthesis* **1984**, 323.

[216]Effenberger, F.; Huthmacher, K. *Chem. Ber.* **1976**, *109*, 2315. For similar methods, see Hancock, R.A.; Tyobeka, T.E.; Weigel, H. *J. Chem. Res. (S)* **1980**, 270; Ono, M.; Nakamura, Y.; Sato, S.; Itoh, I. *Chem. Lett.* **1988**, 395.

[217]Frost, C.G.; Hartley, J.P.; Whittle, A.J. *Synlett* **2001**, 830.

[218]Garzya, V.; Forbes, I.T.; Lauru, S.; Maragni, P. *Tetrahedron Lett.* **2004**, *45*, 1499.

[219]Yadav, J.S.; Reddy, B.V.S.; Krishna, A.D.; Swamy, T. *Tetrahedron Lett.* **2003**, *44*, 6055.

[220]Marquié, J.; Laporterie, A.; Dubac, J.; Roques, N.; Desmurs, J.-R. *J. Org. Chem.* **2001**, *66*, 421.

[221]Bandgar, B.P.; Kasture, S.P. *Synth. Commun.* **2001**, *31*, 1065.

[222]Hyatt, J.A.; White, A.W. *Synthesis* **1984**, 214.

[223]Olah, G.A.; Mathew, T.; Prakash, G.K.S. *Chem. Commun.* **2001**, 1696.

[224]For a monograph, see de la Mare, P.B.D. *Electrophilic Halogenation*, Cambridge University Press, Cambridge, **1976**. For reviews, see Buehler, C.A.; Pearson, D.E. *Survey of Organic Synthesis*, Wiley, NY, **1970**, pp. 392–404; Braendlin, H.P.; McBee, E.T., in Olah, G.A. *Friedel–Crafts and Related Reactions*, Vol. 3, Wiley, NY, **1964**, pp. 1517–1593. For a review of the halogenation of heterocyclic compounds, see Eisch, J.J. *Adv. Heterocycl. Chem.* **1966**, *7*, 1. For a list of reagents, with references, see Larock, R.C. *Comprehensive Organic Transformations*, 2nd ed., Wiley-VCH, NY, **1999**, pp. 619–628.

[225]Srivastava, S.K.; Chauhan, P.M.S.; Bhaduri, A.P. *Chem. Commun.* **1996**, 2679.

likely to be protonated by the liberated HX.[226] For this reason, primary amines are often converted to the corresponding anilides if monosubstitution is desired. With phenols it is possible to stop after one group has entered.[227] The rapid room-temperature reaction with amines and phenols is often used as a test for these compounds.

For less activated aromatic rings, iron was commonly used at one time for halogenation, but the real catalyst was shown not to be the iron itself, but rather the ferric bromide or ferric chloride formed in small amounts from the reaction between iron and the reagent. Indeed, ferric chloride and other Lewis acids are typically directly used as catalysts, as is iodine. For active substrates, including amines, phenols, naphthalene, and polyalkylbenzenes,[228] such as mesitylene and isodurene, no catalyst is needed. Many Lewis acids can be used, including thallium(III) acetate, which promotes bromination with high regioselectivity para to an ortho–para-directing group.[229] A mixture of $Mn(OAc)_3$ and acetyl chloride, with ultrasound, chlorinates anisole with high selectivity.[230] Bromination on NaY zeolite occurs with high para selectivity.[231]

Other acids can be used to promote chlorination or bromination. N-Bromosuccinimide and HBF_4 can be used to brominate phenols with high *para*-selectivity,[232] as can pyridinium bromide perbromide,[233] and NBS in acetic acid with ultrasound is effective.[234] The use of NBS with a catalytic amount of HCl has also been reported.[235] Both NCS and NBS with aqueous BF_3 gave the respective chloride or bromide.[236] Note that NBS in an ionic liquid[237] gave the brominated aromatic. Bromine on silica gel gave good yields of the brominated aromatic compound.[238] HBr with hydrogen peroxide

---

[226]Monobromination (para) of aromatic amines has been achieved with tetrabutylammonium tribromide: Berthelot, J.; Guette, C.; Desbène, P.; Basselier, J.; Chaquin, P.; Masure, D. *Can. J. Chem.* **1989**, *67*, 2061. For another procedure, see Onaka, M.; Izumi, Y. *Chem. Lett.* **1984**, 2007.

[227]For a review of the halogenation of phenols, see Brittain, J.M.; de la Mare, P.B.D., in Patai, S.; Rappoport, Z. *The Chemistry of Functional Groups, Supplement D*, pt. 1, Wiley, NY, **1983**, pp. 522–532.

[228]For a review of aromatic substitution on polyalkylbenzenes, see Baciocchi, E.; Illuminati, G. *Prog. Phys. Org. Chem.* **1967**, *5*, 1.

[229]McKillop, A.; Bromley, D.; Taylor, E.C. *J. Org. Chem.* **1972**, *37*, 88.

[230]Prokes, I.; Toma, S.; Luche, J.-L. *J. Chem. Res. (S)* **1996**, 164.

[231]See Smith, K.; Bahzad, D. *Chem. Commun.* **1996**, 467; Smith, K.; Musson, A.; DeBoos, G.A. *J. Org. Chem.* **1998**, *63*, 8448. Also see, Paul, V.; Sudalai, A.; Daniel, T.; Srinivasan, K.V. *Tetrahedron Lett.* **1994**, *35*, 7055.

[232]Oberhauser, T. *J. Org. Chem.* **1997**, *62*, 4504.

[233]Reeves, W.P.; Lu, C.V.; Schulmeier, B.; Jonas, L.; Hatlevik, O. *Synth. Commun.* **1998**, *28*, 499; Reeves, W.P.; King II, R.M. *Synth. Commun.* **1993**, *23*, 855. Also see, Bisarya, S.C.; Rao, R. *Synth. Commun.* **1993**, *23*, 779.

[234]Paul, V.; Sudalai, A.; Daniel, T.; Srinivasan, K.V. *Synth. Commun.* **1995**, *25*, 2401.

[235]Andersh, B.; Murphy, D.L.; Olson, R.J. *Synth. Commun.* **2000**, *30*, 2091.

[236]Prakash, G.K.S.; Mathew, T.; Hoole, D.; Esteves, P.M.; Wang, Q.; Rasul, G.; Olah, G.A. *J. Am. Chem. Soc.* **2004**, *126*, 15770.

[237]In bbim BF_4, 1,3-di-*n*-butylimidazolium tetrafluoroborate: Rajagopal, R.; Jarikote, D.V.; Lahoti, R.J.; Daniel, T.; Srinivasan, K.V. *Tetrahedron Lett.* **2003**, *44*, 1815.

[238]Ghiaci, M.; Asghari, J. *Bull. Chem. Soc. Jpn.* **2001**, *74*, 1151.

converted aniline to 2,4,6-tribromoaniline.[239] Majetich and co-workers reported the use of HBr/DMSO for the remarkably selective bromination of aniline.[240] *para*-Bromination of aniline was reported by mixing aniline with the ionic liquid, bmim Br₃.[241] Similarly, hmim Br₃[242] without another reagent is a brominating agent.

Other reagents have been used for chlorination and bromination, among them HOCl,[243] HOBr, and *N*-chloro and *N*-bromo amides (especially NBS and tetraalkylammonium polyhalides[244]). In all but the last of these cases, the reaction is catalyzed by the addition of acids. Sulfuryl chloride (SO_2Cl_2) in acetic acid effective chlorinates anisole derivatives,[245] and LiBr with ceric ammonium nitrate in acetonitrile brominates.[246] Acetyl chloride with a catalytic amount of ceric ammonium nitrate also converted aromatic compounds to the corresponding chlorinated derivative.[247] A mixture of KCl and Oxone® as chlorinated activated aromatic compounds.[248] Oxone® and KBr gave good para bromination of anisole.[249] Dibromoisocyanuric acid in H_2SO_4 is a very good brominating agent[250] for substrates with strongly deactivating substituents.[251] If the substrate contains alkyl groups, side-chain halogenation (**14-1**) is possible with most of the reagents mentioned, including chlorine and bromine. Since side-chain halogenation is catalyzed by light, the reactions should be run in the absence of light wherever possible. Both NCS in isopropanol[252] and *tert*-butyl hypochlorite[253] chlorinate aniline derivatives, and KBr/NaBO₃•4 H₂O has been used for the bromination of aniline derivatives.[254] Anisole was brominated with para selectivity using HBr, in the presence of *tert*-butyl hydroperoxide and hydrogen peroxide.[255] Potassium bromide (KBr) with a zeolite (HZSM-5), acetic acid and 30% hydrogen peroxide was used to brominate both anisole and aniline derivatives.[256] Conversion of aniline to the *N*-SnMe₃ derivative allowed

[239]Vyas, P.V.; Bhatt, A.K.; Ramachandraiah, G.; Bedekar, A.V. *Tetrahedron Lett.* **2003**, *44*, 4085.

[240]Majetich, G.; Hicks, R.; Reister, S. *J. Org. Chem.* **1997**, *62*, 4321.

[241]1-Butyl-3-methylimidazolium tribromide: Lei, Z.-G.; Chen, Z.-C.; Hu, Y.;. Zheng, Q.-G. *Synthesis* **2004**, 2809.

[242]In hmim, *N*-methylimidazolium: See Chiappe, C.; Leandri, E.; Pieraccini, D. *Chem. Commun.* **2004**, 2536.

[243]For the use of calcium hypochlorite, see Nwaukwa, S.O.; Keehn, P.M. *Synth. Commun.* **1989**, *19*, 799.

[244]See Kajigaeshi, S.; Moriwaki, M.; Tanaka, T.; Fujisaki, S.; Kakinami, T.; Okamoto, T. *J. Chem. Soc. Perkin Trans. 1* **1990**, 897, and other papers in this series.

[245]Yu, G.; Mason, H.J.; Wu, X.; Endo, M.; Douglas, J.; Macor, J.E. *Tetrahedron Lett.* **2001**, *42*, 3247.

[246]Roy, S.C.; Guin, C.; Rana, K.K.; Maiti, G. *Tetrahedron Lett.* **2001**, *42*, 6941.

[247]Roy, S.C.; Rana, K.K.; Guin, C.; Banerjee, B. *Synlett* **2003**, 221.

[248]Narender, N.; Srinivasu, P.; Kulkarni, S.J.; Raghavan, K.V. *Synth. Commun.* **2002**, *32*, 279.

[249]Tamhankar, B.V.; Desai, U.V.; Mane, R.B.; Wadgaonkar, P.P.; Bedekar, A.V. *Synth. Commun.* **2001**, *31*, 2021.

[250]Nitrobenzene is pentabrominated in 1 min with this reagent in 15% oleum at room temperature.

[251]Gottardi, W. *Monatsh. Chem.* **1968**, *99*, 815; **1969**, *100*, 42.

[252]Zanka, A.; Kubota, A. *Synlett* **1999**, 1984.

[253]Lengyel, I.; Cesare, V.; Stephani, R. *Synth. Commun.* **1998**, *28*, 1891.

[254]Roche, D.; Prasad, K.; Repic, O.; Blacklock, T.J. *Tetrahedron Lett.* **2000**, *41*, 2083.

[255]Barhate, N.B.; Gajare, A.S.; Wakharkar, R.D.; Bedekar, A.V. *Tetrahedron* **1999**, *55*, 11127.

[256]Narender, N.; Srinivasu, P.; Kulkarni, S.J.; Raghavan, K.V. *Synth. Commun.* **2000**, *30*, 3669.

in situ bromination with bromine, with high para selectivity after conversion to the free amine with aqueous KF.[257] Pyridinium bromochromate converted phenolic derivatives to brominated phenols.[258]

Chlorine is a more active reagent than bromine. Phenols can be brominated exclusively in the ortho position (disubstitution of phenol gives 2,6-dibromophenol) by treatment with Br_2 at about $-70°C$, in the presence of *tert*-butylamine or triethylenediamine to precipitate out the liberated HBr.[259] Predominant ortho chlorination[260] of phenols has been achieved with chlorinated cyclohexadienes,[261] while para chlorination of phenols, phenolic ethers, and amines can be accomplished with *N*-chloroamines[262] and with *N*-chlorodimethylsulfonium chloride $(Me_2S^+Cl\,Cl^-)$.[263] The last method is also successful for bromination when *N*-bromodimethylsufonium bromide is used. On the other hand, certain alkylated phenols can be brominated in the meta positions with Br_2 in the superacid solution SbF_5—HF.[264] It is likely that the meta orientation is the result of conversion by the super acid of the OH group to the OH_2^+ group, which should be meta directing because of its positive charge. Bromination and the Sandmeyer reaction (**14-20**) can be carried out in one laboratory step to give **37** by treatment of an aromatic primary amine with $CuBr_2$ and *tert*-butyl nitrite, for example[265]

With deactivated aromatic derivatives, such as nitrobenzene, BrF_3 and Br_2 is an effective reagent, gives the *meta*-brominated product.[266] Tetrabutylammonium bromide and P_2O_5 at 100°C has been used to convert 2-hydroxypyridine derivatives to the corresponding 2-bromopyridine.[267] Bromination at C-6 of 2-aminopyridine was accomplished with NBS.[268] An alternative route

[257]Smith, M.B.; Guo, L.; Okeyo, S.; Stenzel, J.; Yanella, J.; La Chapelle, E. *Org. Lett.* **2002**, *4*, 2321.

[258]Patwari, S.B.; Baseer, M.A.; Vibhute, Y.B.; Bhusare, S.R. *Tetrahedron Lett.* **2003**, *44*, 4893.

[259]Pearson, D.E.; Wysong, R.D.; Breder, C.V. *J. Org. Chem.* **1967**, *32*, 2358.

[260]For other methods of regioselective chlorination or bromination, see Kodomari, M.; Takahashi, S.; Yoshitomi, S. *Chem. Lett.* **1987**, 1901; Kamigata, N.; Satoh, T.; Yoshida, M.; Matsuyama, H.; Kameyama, M. *Bull. Chem. Soc. Jpn.* **1988**, *61*, 2226; de la Vega, F.; Sasson, Y. *J. Chem. Soc., Chem. Commun.* **1989**, 653.

[261]Lemaire, M.; Guy, A.; Guette, J. *Bull. Soc. Chim. Fr.* **1985**, 477.

[262]Lindsay Smith, J.R.; McKeer, L.C.; Taylor, J.M. *J. Chem. Soc. Perkin Trans. 2* **1989**, 1529, 1537. See also, Minisci, F.; Vismara, E.; Fontana, F.; Platone, E.; Faraci, G. *J. Chem. Soc. Perkin Trans. 2* **1989**, 123.

[263]Olah, G.A.; Ohannesian, L.; Arvanaghi, M. *Synthesis* **1986**, 868.

[264]Jacquesy, J.; Jouannetaud, M.; Makani, S. *J. Chem. Soc., Chem. Commun.* **1980**, 110.

[265]Doyle, M.P.; Van Lente, M.A.; Mowat, R.; Fobare, W.F. *J. Org. Chem.* **1980**, *45*, 2570.

[266]Rozen, S.; Lerman, O. *J. Org. Chem.* **1993**, *58*, 239.

[267]Kato, Y.; Okada, S.; Tomimoto, K.; Mase, T. *Tetrahedron Lett.* **2001**, *42*, 4849.

[268]Cañibano, V.; Rodríguez, J.F.; Santos, M.; Sanz-Tejedor, A.; Carreño, M.C.; González, G.; García-Ruano, J.L. *Synthesis* **2001**, 2175.

reacted pyridine *N*-oxide was POCl$_3$ and triethylamine to give 2-chloropyridine.[269] Pyridinium dichlorobromate with FeCl$_3$ brominates benzene.[270]

For reactions in the absence of a catalyst, the attacking entity is simply Br$_2$ or Cl$_2$ that has been polarized by the ring.[271]

38

Evidence for molecular chlorine or bromine as the attacking species in these cases is that acids, bases, and other ions, especially chloride ion, accelerate the rate about equally, though if chlorine dissociated into Cl$^+$ and Cl$^-$, the addition of chloride should decrease the rate and the addition of acids should increase it. Intermediate **38** has been detected spectrally in the aqueous bromination of phenol.[272]

When a Lewis acid catalyst is used with chlorine or bromine, the attacking entity may be Cl$^+$ or Br$^+$, formed by FeCl$_3$ + Br$_2$ → FeCl$_3$Br$^-$ + Br$^+$, or it may be Cl$_2$ or Br$_2$, polarized by the catalyst. With other reagents, the attacking entity in brominations may be Br$^+$ or a species, such as H$_2$OBr$^+$ (the conjugate acid of HOBr), in which H$_2$O is a carrier of Br$^+$.[273] With HOCl in water the electrophile may be Cl$_2$O, Cl$_2$, or H$_2$OCl$^+$; in acetic acid it is generally AcOCl. All these species are more reactive than HOCl itself.[274] It is extremely doubtful that Cl$^+$ is a significant electrophile in chlorinations by HOCl.[274] It has been demonstrated in the reaction between *N*-methylaniline and calcium hypochlorite that the chlorine attacking entity attacks the *nitrogen* to give *N*-chloro-*N*-methylaniline, which rearranges (as in **11-31**) to give a mixture of ring-chlorinated *N*-methylanilines in which the ortho isomer predominates.[275] In addition to hypohalous acids and metal hypohalites, organic hypohalites are reactive. An example is *tert*-butylhypobromite (*t*-BuOBr), which brominated toluene in the presence of zeolite HNaX.[276]

[269]Jung, J.-C.; Jung, Y.-J.; Park, O.-S. *Synth. Commun.* **2001**, *31*, 2507.

[270]Muathen, H.A. *Synthesis* **2002**, 169.

[271]For reviews of the mechanism of halogenation, see de la Mare, P.B.D., *Electrophilic Halogenation*, Cambridge University Press, Cambridge, *1976*; de la Mare, P.B.D.; Swedlund, B.E., in Patai. S. *The Chemistry of the Carbon–Halogen Bond*, pt. 1, Wiley, NY, *1973*; pp. 490–536; Taylor, R., in Bamford, C.H.; Tipper, C.F.H *Comprehensive Chemical Kinetics*, Vol. 13, Elsevier, NY, *1972*, pp. 83–139. See also, Schubert, W.M.; Dial, J.L. *J. Am. Chem. Soc.* **1975**, *97*, 3877; Keefer, R.M.; Andrews, L.J. *J. Am. Chem. Soc.* **1977**, *99*, 5693; Tee, O.S.; Paventi, M.; Bennett, J.M. *J. Am. Chem. Soc.* **1989**, *111*, 2233.

[272]Tee, O.S.; Iyengar, N.R.; Paventi, M. *J. Org. Chem.* **1983**, *48*, 759. See also, Tee, O.S.; Iyengar, N.R. *Can. J. Chem.* **1990**, *68*, 1769.

[273]For discussions, see Gilow, H.M.; Ridd, J.H. *J. Chem. Soc. Perkin Trans. 2* **1973**, 1321; Rao, T.S.; Mali, S.I.; Dangat, V.T. *Tetrahedron* **1978**, *34*, 205.

[274]Swain, C.G.; Crist, D.R. *J. Am. Chem. Soc.* **1972**, *94*, 3195.

[275]Gassman, P.G.; Campbell, G.A. *J. Am. Chem. Soc.* **1972**, *94*, 3891; Paul, D.F.; Haberfield, P. *J. Org. Chem.* **1976**, *41*, 3170.

[276]Smith, K.; El-Hiti, G.A.; Hammond, M.E.W.; Bahzad, D.; Li, Z.; Siquet, C. *J. Chem. Soc., Perkin Trans. 1* **2000**, 2745.

When chlorination or bromination is carried out at high temperatures (e.g., 300–400°C), ortho–para-directing groups direct meta and vice versa.[277] A different mechanism operates here, which is not completely understood. It is also possible for bromination to take place by the S_E1 mechanism, for example, in the t-BuOK-catalyzed bromination of 1,3,5-tribromobenzene.[278]

Furan and thiophene are known to polymerize in the presence of strong acid, both Brønsted–Lowry and Lewis. For such highly reactive heteroaromatic systems, alternative halogenating reagents are commonly used. Furan was converted to 2-bromofuran with a bromine•dioxane complex, for example, at $<0°C$.[279] 3-Butylthiophene reacted with NBS/acetic acid to give 2-bromo-3-butylthiophene.[280] N-Methylpyrrole reacted with NBS and a catalytic amount of PBr_3, at $-78°C \rightarrow -10°C$, to give N-methyl-3-bromopyrrole.[281]

2. *Iodine.* Iodine is the least reactive of the halogens in aromatic substitution.[282] Except for active substrates, an oxidizing agent must normally be present to oxidize I_2 to a better electrophile.[283] Examples of such oxidizing agents are HNO_3, HIO_3, SO_3, MnO_2,[284] hypervalent iodine compounds, such as $PhI(OTf)_2$,[285] $NaIO_4$,[286] peroxyacetic acid, H_2O_2[287] peroxydisulfates,[288] and ammonium iodide with Oxone®.[289] The ICl is a better iodinating agent than iodine itself.[290] Among other reagents used have been IF (prepared directly from the elements),[291] and benzyltrialkylammonium dichloroiodate (which iodinates phenols, aromatic amines, and N-acylated aromatic amines,[292] as well

[277]For a review of this type of reaction, see Kooyman, E.C. *Pure. Appl. Chem.* **1963**, *7*, 193.

[278]Mach, M.H.; Bunnett, J.F. *J. Am. Chem. Soc.* **1974**, *96*, 936.

[279]See Baciocchi, E.; Clementi, S.; Sebastiani, G.V. *J. Chem. Soc., Chem. Commun.* **1975**, 875.

[280]Hoffmann, K.J.; Carlsen, P.H.J. *Synth. Commun.* **1999**, *29*, 1607.

[281]Dvornikova, E.; Kamieńska-Trela, K. *Synlett* **2002**, 1152.

[282]For reviews of I_2 as an electrophilic reagent, see Pizey, J.S., in Pizey, J.S. *Synthetic Reagents*, Vol. 3, Wiley, NY, **1977**, pp. 227–276. For a review of aromatic iodination, see Merkushev, E.B. *Synthesis* **1988**, 923.

[283]Butler, A.R. *J. Chem. Educ.* **1971**, *48*, 508.

[284]Luliski, P.; Skulski, L. *Bull. Chem. Soc. Jpn.* **1999**, *72*, 115.

[285]D'Auria, M.; Mauriello, G. *Tetrahedron Lett.* **1995**, *36*, 4883; Togo, H.; Abe, S.; Nogami, G.; Yokoyama, M. *Bull. Chem. Soc. Jpn.* **1999**, *72*, 2351; Panunzi, B.; Rotiroti, L.; Tingoli, M. *Tetrahedron Lett.* **2003**, *44*, 8753.

[286]Luliński, P.; Skulski, L. *Bull. Chem. Soc. Jpn.* **2000**, *73*, 951.

[287]For a discussion, see Makhon'kov, D.I.; Cheprakov, A.V.; Beletskaya, I.P. *J. Org. Chem. USSR* **1989**, *24*, 2029. See Iskra, J.; Stavber, S.; Zupan, M. *Synthesis* **2004**, 1869.

[288]Tajik, H.; Esmaeili, A.A.; Mohammadpoor-Baltork, I.; Ershadi, A.; Tajmehri, H. *Synth. Commun.* **2003**, *33*, 1319.

[289]Mohan, K.V.V.K.; Narender, N.; Kulkarni, S.J. *Tetrahedron Lett.* **2004**, *45*, 8015.

[290]For a review of ICl, see McCleland, C.W., in Pizey, J.S. *Synthetic Reagents*, Vol. 5, Wiley, NY, **1983**, pp. 85–164. For a reaction using ICl, ZnO and an iron catalyst, see Mukaiyama, T.; Kitagawa, H.; Matsuo, J.-i. *Tetrahedron Lett.* **2000**, *41*, 9383.

[291]Rozen, S.; Zamir, D. *J. Org. Chem.* **1990**, *55*, 3552.

[292]See Kajigaeshi, S.; Kakinami, T.; Watanabe, F.; Okamoto, T. *Bull. Chem. Soc. Jpn.* **1989**, *62*, 1349, and references cited therein. For a reaction of anisole with Me_4N ICl_2 to give p-iodoanisole exclusively, see Hajipour, A.R.; Arbabian, M.; Ruoho, A.E. *J. Org. Chem.* **2002**, *67*, 8622.

as unprotected aniline derivatives[293]). Iodination can also be accomplished by treatment of the substrate with NCI and sulfuric acid,[294] NIS and trifluoroacetic acid,[295] KI/KIO$_3$ in aqueous methanol,[296] I$_2$ in the presence of copper salts,[297] Al$_2$O$_3$,[298] and NaI with an iron catalyst.[299] Sodium periodate and iodine was used to iodinate β-carbolines.[300] Sodium iodide in liquid NO$_2$ can be used to iodinate aniline derivatives.[301] A solvent-free iodination was accomplished using NaICl$_2$ and an *N*-bromoammonium salt.[302] Another solvent-free iodination used I$_2$ with Bi(NO$_3$)$_3$ on silica gel.[303] Iodine with Selectfluor also leads to iodination of aromatic compounds.[304]

The actual attacking species is less clear than with bromine or chlorine. Iodine itself is too unreactive, except for active species, such as phenols, where there is good evidence that I$_2$ is the attacking entity.[305] There is evidence that AcOI may be the attacking entity when peroxyacetic acid is the oxidizing agent,[306] and I$_3^+$ when SO$_3$ or HIO$_3$ is the oxidizing agent.[307] The I$^+$ ion has been implicated in several procedures.[308] For an indirect method for accomplishing aromatic iodination, see **12-31**.

Note that conversion of aniline derivatives to the corresponding para aryllithium, followed by reaction with B(OMe)$_3$ and then bromine at −78°C gave *p*-bromoaniline.[309]

3. *Fluorine*. Direct fluorination of aromatic rings with F$_2$ is not feasible at room temperature, because of the extreme reactivity of F$_2$.[310] It has been accomplished at low temperatures (e.g., −70 to −20°C, depending on the

[293]Kosynkin, D.V.; Tour, J.M. *Org. Lett.* **2001**, *3*, 991.

[294]Chaikovskii, V.K.; Shorokhodov, V.I.; Filimonov, V.D. *Russ. J. Org. Chem.* **2001**, *37*, 1503.

[295]Castanet, A.-S.; Colobert, F.; Broutin, P.-E. *Tetrahedron Lett.* **2002**, *43*, 5047.

[296]Adimurthy, S.; Ramachandraiah, G.; Ghosh, P.K.; Bedekar, A.V. *Tetrahedron Lett.* **2003**, *44*, 5099.

[297]Baird Jr., W.C.; Surridge, J.H. *J. Org. Chem.* **1970**, *35*, 3436; Horiuchi, C.A.; Satoh, J.Y. *Bull. Chem. Soc. Jpn.* **1984**, *57*, 2691; Makhon'kov, D.I.; Cheprakov, A.V.; Rodkin, M.A.; Beletskaya, I.P. *J. Org. Chem. USSR* **1986**, *22*, 1003.

[298]Pagni, R.M.; Kabalka, G.W.; Boothe, R.; Gaetano, K.; Stewart, L.J.; Conaway, R.; Dial, C.; Gray, D.; Larson, S.; Luidhart, T. *J. Org. Chem.* **1988**, *53*, 4477.

[299]Firouzabadi, H.; Iranpoor, N.; Shiri, M. *Tetrahedron Lett.* **2003**, *44*, 8781.

[300]Bonesi, S.M.; Erra-Balsells, R. *J. Heterocyclic Chem.* **2001**, *38*, 77.

[301]Suzuki, H.; Nonoyama, N. *Tetrahedron Lett.* **1998**, *39*, 4533.

[302]Hajipour, A.R.; Ruoho, A.E. *Org. Prep. Proceed. Int.* **2002**, *34*, 647.

[303]Alexander, V.M.; Khandekar, A.C.; Samant, S.D. *Synlett* **2003**, 1895.

[304]Stavber, S.; Kralj, P.; Zupan, M. *Synlett* **2002**, 598.

[305]Grovenstein, Jr., E.; Aprahamian, N.S.; Bryan, C.J.; Gnanapragasam, N.S.; Kilby, D.C.; McKelvey Jr., J.M.; Sullivan, R.J. *J. Am. Chem. Soc.* **1973**, *95*, 4261.

[306]Ogata, Y.; Urasaki, I. *J. Chem. Soc. C* **1970**, 1689.

[307]Arotsky, J.; Butler, R.; Darby, A.C. *J. Chem. Soc. C* **1970**, 1480.

[308]Galli, C. *J. Org. Chem.* **1991**, *56*, 3238.

[309]Zhao, J.; Jia, X.; Zhai, H. *Tetrahedron Lett.* **2003**, *44*, 9371.

[310]For a monograph on fluorinating agents, see German, L.; Zemskov, S. *New Fluorinating Agents in Organic Synthesis*, Springer, NY, **1989**. For reviews of F$_2$ in organic synthesis see Purrington, S.T.; Kagen, B.S.; Patrick, T.B. *Chem. Rev.* **1986**, *86*, 997; Grakauskas, V. *Intra-Sci. Chem. Rep.* **1971**, *5*, 85. For a review of fluoroaromatic compounds, see Hewitt, C.D.; Silvester, M.J. *Aldrichimica Acta* **1988**, *21*, 3.

substrate),[311] but the reaction is not yet of preparative significance. Fluorination has also been reported with acetyl hypofluorite CH_3COOF (generated from F_2 and sodium acetate),[312] with XeF_2,[313] and with an N-fluoroperfluoroalkyl sulfonamide, for example $(CF_3SO_2)_2NF$.[314] Pyridine has been converted to 2-fluoropyridine with $F_2/I_2/NEt_3$ in 1,1,2-trichloro-1,2,2-trifluoroethane.[315] However, none of these methods seems likely to displace the Schiemann reaction (**13-23**; heating diazonium tetrafluoroborates) as the most common method for introducing fluorine into aromatic rings.

The overall effectiveness of reagents in aromatic substitution is $Cl_2 > BrCl > Br_2 > ICl > I_2$.

OS **I**, 111, 121, 123, 128, 207, 323; **II**, 95, 97, 100, 173, 196, 343, 347, 349, 357, 592; **III**, 132, 134, 138, 262, 267, 575, 796; **IV**, 114, 166, 256, 545, 547, 872, 947; **V**, 117, 147, 206, 346; **VI**, 181, 700; **VIII**, 167; **IX**, 121, 356. Also see, OS **II**, 128.

E. Carbon Electrophiles

In the reactions in this section, a new carbon–carbon bond is formed. With respect to the aromatic ring, they are electrophilic substitutions, because a positive species attacks the ring. We treat them in this manner because it is customary. However, with respect to the electrophile, most of these reactions are nucleophilic substitutions, and what was said in Chapter 10 is pertinent to them.

11-11   Friedel–Crafts Alkylation

Alkylation or **Alkyl-de-hydrogenation**

$$ArH + RCl \xrightarrow{AlCl_3} ArCl$$

The alkylation of aromatic rings, called *Friedel–Crafts alkylation*, is a reaction of very broad scope.[316] The most important reagents are alkyl halides, alkenes, and

---

[311]Grakauskas, V. *J. Org. Chem.* **1970**, *35*, 723; Cacace, F.; Giacomello, P.; Wolf, A.P. *J. Am. Chem. Soc.* **1980**, *102*, 3511; Stavber, S.; Zupan, M. *J. Org. Chem.* **1983**, *48*, 2223. See also, Purrington, S.T.; Woodard, D.L. *J. Org. Chem.* **1991**, *56*, 142.

[312]See Hebel, D.; Lerman, O.; Rozen, S. *Bull. Soc. Chim. Fr.* **1986**, 861; Visser, G.W.M.; Bakker, C.N.M.; van Halteren, B.W.; Herscheid, J.D.M.; Brinkman, G.A.; Hoekstra, A. *J. Org. Chem.* **1986**, *51*, 1886.

[313]Shaw, M.J.; Hyman, H.H.; Filler, R. **1970**, *92*, 6498; *J. Org. Chem.* **1971**, *36*, 2917; Mackenzie, D.R.; Fajer, J. *J. Am. Chem. Soc.* **1970**, *92*, 4994; Filler, R. *Isr. J. Chem.* **1978**, *17*, 71.

[314]Singh, S.; DesMarteau, D.D.; Zuberi, S.S.; Witz, M.; Huang, H. *J. Am. Chem. Soc.* **1987**, *109*, 7194.

[315]Chambers, R.D.; Parsons, M.; Sandford, G.; Skinner, C.J.; Atherton, M.J.; Moilliet, J.S. *J. Chem. Soc., Perkin Trans. 1* **1999**, 803.

[316]For a monograph, see Roberts, R.M.; Khalaf, A.A. *Friedel–Crafts Alkylation Chemistry*, Marcel Dekker, NY, **1984**. For a treatise on Friedel–Crafts reactions in general, see Olah, G.A. *Friedel–Crafts and Related Reactions*, Wiley, NY, **1963–1965**. Volume 1 covers general aspects, such as catalyst activity, intermediate complexes, and so on. Volume 2 covers alkylation and related reactions. In this volume, the various reagents are treated by the indicated authors as follows: alkenes and alkanes, Patinkin, S.H.; Friedman, B.S. pp. 1–288; dienes and substituted alkenes, Koncos, R.; Friedman, B.S. pp. 289–412; alkynes, Franzen, V. pp. 413–416; alkyl halides, Drahowzal, F.A. pp. 417–475; alcohols and ethers, Schriesheim, A. pp. 477–595; sulfonates and inorganic esters, Drahowzal, F.A. pp. 641–658. For a monograph in which five chapters of the above treatise are reprinted and more recent material added, see Olah, G.A. *Friedel–Crafts Chemistry*, Wiley, NY, **1973**.

alcohols, but other types of reagent have also been employed.[316] Tertiary halides are particularly good substrates since they form relatively stable tertiary carbocations. *tert*-Butyl chloride reacts with phenetole in the presence of a ReBr(CO)$_5$ catalyst, for example, to give the 4-*tert*-butyl isomer as the major product.[317] When alkyl halides are used, the reactivity order is F > Cl > Br > I.[318] This trend can be seen in reactions of dihalo compounds, such as FCH$_2$CH$_2$CH$_2$Cl, which react with benzene to give PhCH$_2$CH$_2$CH$_2$Cl[319] when the catalyst is BCl$_3$. By the use of this catalyst, it is therefore possible to place a haloalkyl group on a ring (see also, **11-14**).[320] Di- and trihalides, when all the halogens are the same, usually react with more than one molecule of an aromatic compound; it is usually not possible to stop the reaction earlier.[321] Thus, benzene with CH$_2$Cl$_2$ gives not PhCH$_2$Cl, but Ph$_2$CH$_2$; benzene with CHCl$_3$ gives Ph$_3$CH. With CCl$_4$, however, the reaction stops when only three rings have been substituted to give Ph$_3$CCl. Functionalized alkyl halides, such as ClCH(SEt)CO$_2$Et, undergo Friedel–Crafts alkylation.[322] Interestingly, benzyl chloride was converted to diphenylmethane in benzene at 130°C with 10 atm of CO,[323] and also with a LiB(C$_6$F$_5$)$_4$ catalyst.[324]

Alkenes are especially good alkylating agents, generally proceeding by formation of an intermediate carbocation that reacts with the electron rich aromatic ring, and the final product (**39**) incorporates a H and Ar from ArH to a C=C double bond. Many variations are possible. This reaction has been accomplished in an ionic liquid, using Sc(OTf)$_3$ as the catalyst.[325] Intramolecular versions lead to polycyclic aromatic compounds.[326] Benzene reacted with 1,2,3,6-tetrahydropyridine in the presence of trifluoromethanesulfonic acid to give 4-phenylpiperidine.[327]

$$\text{Ar–H} + \overset{\diagdown}{\underset{\diagup}{C}}{=}\overset{\diagup}{\underset{\diagdown}{C}} \xrightarrow[\text{H}^+]{\text{AlCl}_3} \text{Ar}-\overset{\diagdown}{\underset{\diagup}{C}}-\overset{\diagup}{\underset{\diagdown}{C}}-\text{H}$$

39

[317]Nishiyama, Y.; Kakushou, F.; Sonoda, N. *Bull. Chem. Soc. Jpn.* **2000**, *73*, 2779.

[318]For example, see Calloway, N.O. *J. Am. Chem. Soc.* **1937**, *59*, 1474; Brown, H.C.; Jungk, H. *J. Am. Chem. Soc.* **1955**, *77*, 5584.

[319]Olah, G.A.; Kuhn, S.J. *J. Org. Chem.* **1964**, *29*, 2317.

[320]For a review of selectivity in this reaction, see Olah, G.A., in Olah, G.A. *Friedel–Crafts and Related Reactions*, Vol. 1, Wiley, NY, **1963**, pp. 881–905. This review also covers the case of alkylation versus acylation.

[321]It has proven possible in some cases. Thus, arenes ArH have been converted to ArCCl$_3$ with CCl$_4$ and excess AlCl$_3$: Raabe, D.; Hörhold, H. *J. Prakt. Chem.* **1987**, *329*, 1131; Belen'kii, L.I.; Brokhovetsky, D.B.; Krayushkin, M.M. *Chem. Scr.*, **1989**, *29*, 81.

[322]For the reaction of anisole using a Yb(OTf)$_3$ catalyst, see Sinha, S.; Mandal, B.; Chandrasekaran, S. *Tetrahedron Lett.* **2000**, *41*, 9109.

[323]Ogoshi, S.; Nakashima, H.; Shimonaka, K.; Kurosawa, H. *J. Am. Chem. Soc.* **2001**, *123*, 8626.

[324]Mukaiyama, T.; Nakano, M.; Kikuchi, W.; Matsuo, J.-i. *Chem. Lett.* **2000**, 1010.

[325]In emim SbF$_6$, 1-ethyl-3-mthylimidazolium: Song, C.E.; Shim, W.H.; Roh, E.J.; Choi, J.H. *Chem. Commun.* **2000**, 1695.

[326]For a RuCl$_3$/AgOTf catalyzed version, see Youn, S.W.; Pastine, S.J.; Sames, D. *Org. Lett.* **2004**, *6*, 581.

[327]Klumpp, D.A.; Beauchamp, P.S.; Sanchez Jr., G.V.; Aguirre, S.; de Leon, S. *Tetrahedron Lett.* **2001**, *42*, 5821.

When 4-methoxyphenol reacted with isobutylene (electrolysis with 3 M LiClO$_4$ in nitromethane and acetic acid, initial reaction with the phenolic oxygen generated an ether moiety and the resulting carbocation was attacked by the aromatic ring to form a benzofuran.[328] Acetylene reacts with 2 mol of aromatic compound to give 1,1-diarylethanes, and phenylacetylene reacted to give 1,1-diarylethenes with a Sc(OTf)$_3$ catalyst.[329] Variations are possible here as well. Phenol reacted with trimethylsilylethyne, in the presence of SnCl$_4$ and 50% BuLi, at 105°C, to give the 2-vinyl phenolic derivative.[330] A palladium-catalyzed reaction of ethyl propiolate and p-xylene, with trifluoroacetic acid, gave the 3-arylalkenyl ester.[331] A ruthenium catalyzed intramolecular reaction with a pendant alkyne unit led to a dihydronapthalene derivative.[332]

Alcohols are more active than alkyl halides, but if a Lewis acid catalyst is used more catalyst is required, since the catalyst complexes with the OH group. However, proton acids, such as H$_2$SO$_4$, are often used to catalyze alkylation with alcohols. An intramolecular cyclization was reported from an allylic alcohol, using P$_2$O$_5$, to give indene derivatives.[333] When carboxylic esters are the reagents, there is competition between alkylation and acylation (**11-17**). This competition can often be controlled by choice of catalyst, and alkylation is usually favored, but carboxylic esters are not often employed in Friedel–Crafts reactions. Other alkylating agents are ethers, thiols, sulfates, sulfonates, alkyl nitro compounds,[334] and even alkanes and cycloalkanes, under conditions where these are converted to carbocations. Notable here are ethylene oxide, which puts the CH$_2$CH$_2$OH group onto the ring,[335] and cyclopropyl[336] units. For all types of reagent the reactivity order is allylic $\sim$ benzylic > tertiary > secondary > primary.

---

[328]Chiba, K.; Fukuda, M.; Kim, S.; Kitano, Y.; Toda, M. *J. Org. Chem.* **1999**, *64*, 7654. For a variation using a seleno ether to form a fused six-membered ring, see Abe, H.; Koshiba, N.; Yamasaki, A.; Harayama, T. *Heterocycles* **1999**, *51* 2301. See also, Shen, Y.; Atobe, M.; Fuchigami, T. *Org. Lett.* **2004**, *6*, 2441.

[329]Tsuchimoto, T.; Maeda, T.; Shirakawa, E.; Kawakami, Y. *Chem. Commun.* **2000**, 1573.

[330]Kobayasshi, K.; Yamaguchi, M. *Org. Lett.* **2001**, *3*, 241.

[331]Jia, C.; Lu, W.; Oyamada, J.; Kitamura, T.; Katsuda, K.; Irie, M.; Fujiwara, Y. *J. Am. Chem. Soc.* **2000**, *122*, 7252.

[332]Chatani, N.; Inoue, H.; Ikeda, T.; Murai, S. *J. Org. Chem.* **2000**, *65*, 4913. For a GaCl$_3$ catalyzed version, see Inoue, H.; Chatani, N.; Murai, S. *J. Org. Chem.* **2002**, *67*, 1414. For a mercuric salt catalyst, see Nishizawa, M.; Takao, H.; Yadav, V.K.; Imagawa, H.; Sugihara, T. *Org. Lett.* **2003**, *5*, 4563. For a BF$_3$ catalyzed version that generates allenes, see Ishikawa, T.; Manabe, S.; Aikawa, T.; Kudo, T.; Saito, S. *Org. Lett.* **2004**, *6*, 2361. See also, Fillion, E.; Carson, R.J.; Trépanier, V.E.; Goll, J.M.; Remorova, A.A. *J. Am. Chem. Soc.* **2004**, *126*, 15354.

[333]Basavaiah, D.; Bakthadoss, M.; Reddy, G.J. *Synthesis* **2001**, 919. For a variation involving a propargylic alcohols with a ruthenium catalyst and ammonium tetrafluoroborate, see Nishibayashi, Y.; Joshikawa, M.; Inada, Y.; Hidai, M.; Uemura, S. *J. Am. Chem. Soc.* **2002**, *124*, 11846.

[334]Bonvino, V.; Casini, G.; Ferappi, M.; Cingolani, G.M.; Pietroni, B.R. *Tetrahedron* **1981**, *37*, 615.

[335]Taylor, S.K.; Dickinson, M.G.; May, S.A.; Pickering, D.A.; Sadek, P.C. *Synthesis* **1998**, 1133. See also, Brandänge, S.; Bäckvall, J.-E.; Leijonmarck, H. *J. Chem. Soc., Perkin Trans. 1* **2001**, 2051.

[336]Patra, P.K.; Patro, B.; Ila, H.; Junjappa, H. *Tetrahedron Lett.* **1993**, *34*, 3951.

Regardless of which reagent is used, a catalyst is nearly always required.[337] Aluminum chloride and boron trifluoride are the most common, but many other Lewis acids have been used, and also proton acids, such as HF and H_2SO_4.[338] For active halides a trace of a less active catalyst, such as $ZnCl_2$, may be enough. For an unreactive halide, such as chloromethane, a more powerful catalyst, such as $AlCl_3$, is needed, and in larger amounts. In some cases, especially with alkenes, a Lewis acid catalyst causes reaction only if a small amount of proton-donating cocatalyst is present. Catalysts have been arranged in the following order of overall reactivity: $AlBr_3 > AlCl_3 > GaCl_3 > FeCl_3 > SbCl_5$[339]$> ZrCl_4, SnCl_4 > BCl_3, BF_3, SbCl_3$;[340] but the reactivity order in each case depends on the substrate, reagent, and conditions.

Alkyl mesylates undergo alkylation reaction with benzene rings in the presence of $Sc(OTf)_3$.[341] Allylic acetates undergo alkylation with $Mo(CO)_6$[342] and allylic chlorides react in the presence of $ZnCl_2/SiO_2$.[343] Montmorillonite clay (K10) is an effective medium for alkylation reactions.[344] Nafion-H, a super acidic perfluorinated resin sulfonic acid, is a very good catalyst for gas phase alkylations with alkyl halides, alcohols,[345] or alkenes.[346]

Friedel–Crafts alkylation is unusual among the principal aromatic substitutions in that the entering group is activating (the product is more reactive than the starting aromatic substrate), and di- and polyalkylation are frequently observed. However, the activating effect of simple alkyl groups (e.g., ethyl, isopropyl) is only ~1.5–3 times as fast as benzene for Friedel–Crafts alkylations,[347] so it is often possible to obtain high yields of monoalkyl product.[348] Actually, the fact that di- and polyalkyl derivatives are frequently obtained is not due to the small difference in reactivity, but to the circumstance that alkylbenzenes are preferentially soluble in the catalyst layer, where the reaction actually takes place.[349] This factor can be removed by the use of a suitable solvent, by high temperatures, or by high–speed stirring.

[337]There are a few exceptions. Certain alkyl and vinylic triflates alkylate aromatic rings without a catalyst, see Gramstad, T.; Haszeldine, R.N. *J. Chem. Soc.* **1957**, 4069; Olah, G.A.; Nishimura, J. *J. Am. Chem. Soc.* **1974**, *96*, 2214; Stang, P.J.; Anderson, A.G. *J. Am. Chem. Soc.* **1978**, *100*, 1520.

[338]For a review of catalysts and solvents in Friedel–Crafts reactions, see Olah, G.A., in Olah, G.A. *Friedel–Crafts and Related Reactions*, Vol. 1, Wiley, NY, **1963**, pp. 201–366, 853–881.

[339]For a review of SbCl₅ as a Friedel–Crafts catalyst, see Yakobson, G.G.; Furin, G.G. *Synthesis* **1980**, 345.

[340]Russell, G.A. *J. Am. Chem. Soc.* **1959**, *81*, 4834.

[341]Kotsuki, H.; Oshisi, T.; Inoue, M.; Kojima, T. *Synthesis* **1999**, 603; Singh, R.P.; Kamble, R.M.; Chandra, K.L.; Saravanani, P.; Singh, V.K. *Tetrahedron* **2001**, *57*, 241.

[342]Shimizu, I.; Sakamoto, T.; Kawaragi, S.; Maruyama, Y.; Yamamoto, A. *Chem. Lett.* **1997**, 137.

[343]Kodomari, M.; Nawa, S.; Miyoshi, T. *J. Chem. Soc. Chem.Commun.* **1995**, 1895.

[344]Sieskind, O.; Albrecht, P. *Tetrahedron Lett.* **1993**, *34*, 1197.

[345]Aleksiuk, O.; Biali, S.E. *Tetrahedron Lett.* **1993**, *34*, 4857.

[346]For a review of Nafion-H in organic synthesis, see Olah, G.A.; Iyer, P.S.; Prakash, G.K.S. *Synthesis* **1986**, 513.

[347]Condon, F.E. *J. Am. Chem. Soc.* **1948**, *70*, 2265; Olah, G.A.; Kuhn, S.J.; Flood, S.H. *J. Am. Chem. Soc.* **1962**, *84*, 1688.

[348]See Davister, M.; Laszlo, P. *Tetrahedron Lett.* **1993**, *34*, 533 for examples of paradoxical selectivity in Friedel–Crafts alkylation.

[349]Francis, A.W. *Chem. Rev.* **1948**, *43*, 257.

It is important to note that the OH, OR, NH$_2$, and so on groups do not facilitate the reaction, since most Lewis acid catalysts coordinate with these basic groups. Although phenols give the usual Friedel–Crafts reactions, orienting ortho and para, the reaction is very poor for aniline derivatives. However, amines can undergo the reaction if alkenes are used as reagents and aluminum anilides as catalysts.[350] In this method, the catalyst is prepared by treating the amine to be alkylated with $\frac{1}{3}$ equivalent of AlCl$_3$. A similar reaction can be performed with phenols, though here the catalyst is Al(OAr)$_3$.[351] Primary aromatic amines (and phenols) can be methylated regioselectively in the ortho position by an indirect method (see **11-23**). For an indirect method for regioselective ortho methylation of phenols (see p. 1247).

Naphthalene and other fused ring compounds are so reactive that they react with the catalyst, and therefore tend to give poor yields in Friedel–Crafts alkylation. Heterocyclic rings are also tend to be poor substrates for the reaction. Although some furans and thiophenes have been alkylated, polymerization is quite common, and a true alkylation of a pyridine or a quinoline has never been described.[352] N-Methylpyrrole reacted with the C=C unit of methacrolein in the presence of a chiral catalyst (a chiral Friedel–Crafts catalyst) to give the 2-alkylated pyrrole, with good enantioselectivity.[353] Alkylation at C-5 of 2-trimethylsilylfuran was accomplished using the carbocation [(p-MeOC$_6$H$_4$)$_2$CH$^+$ OTf] and Proton Sponge (see p. 386).[354] Although mechanistically different, an intramolecular cyclization of an N-allylic pyrrole was accomplished using a rhodium catalyst with 100 atm of CO/H$_2$.[355] Note that alkylation of pyridine and other nitrogen heterocycles can be accomplished by a free radical[356] (**14-19**) and by a nucleophilic method (**13-17**). A variation generates an electrophilic species on the aromatic substrate. The reaction of isoquinoline with ClCO$_2$Ph and AgOTf, followed by reaction with an allylic silane, led to a 2-allylic dihydroisoquinoline.[357]

In most cases, meta-directing groups make the ring too inactive for alkylation. Nitrobenzene cannot be alkylated, and there are only a few reports of successful Friedel–Crafts alkylations when electron-withdrawing groups are present.[358] This is not because the attacking species is not powerful enough; indeed we have

[350]For a review, see Stroh, R.; Ebersberger, J.; Haberland, H.; Hahn, W. *Newer Methods Prep. Org. Chem.* **1963**, *2*, 227. This article also appeared in *Angew. Chem.* **1957**, *69*, 124.

[351]Koshchii, V.A.; Kozlikovskii, Ya.B.; Matyusha, A.A. *J. Org. Chem. USSR* **1988**, *24*, 1358; Laan, J.A.M.; Giesen, F.L.L.; Ward, J.P. *Chem. Ind. (London)* **1989**, 354. For a review, see Stroh, R.; Seydel, R.; Hahn, W. *Newer Methods Prep. Org. Chem.* **1963**, *2*, 337. This article also appeared in *Angew. Chem.* **1957**, *69*, 669.

[352]Drahowzal, F.A., in Olah, G.A., *Friedel–Crafts and Related Reactions*, Vol. 2, Wiley, NY, **1964**, p. 433.

[353]Paras, N.A.; MacMillan, D.W.C. *J. Am. Chem. Soc.* **2001**, *123*, 4370.

[354]Herrlich, M.; Hampel, N.; Mayr, H. *Org. Lett.* **2001**, *3*, 1629.

[355]Settambalo, R.; Caiazzo, A.; Lazzaroni, R. *Tetraehdron Lett.* **2001**, *42*, 4045.

[356]For a silyl-mediated reaction with 2-bromopyridine and 2 equivalents of AIBN, see Núñez, A.; Sánchez, A.; Burgos, C.; Alvarez-Builla, J. *Tetrahedron* **2004**, *60*, 6217.

[357]Yamaguchi, R.; Nakayasu, T.; Hatano, B.; Nagura, T.; Kozima, S.; Fujita, K.-i. *Tetrahedron* **2001**, *57*, 109.

[358]Campbell Jr., B.N.; Spaeth, E.C. *J. Am. Chem. Soc.* **1959**, *81*, 5933; Yoneda, N.; Fukuhara, T.; Takahashi, Y.; Suzuki, A. *Chem. Lett.* **1979**, 1003; Shen, Y.; Liu, H.; Chen,Y. *J. Org. Chem.* **1990**, *55*, 3961.

seen (p. 681) that alkyl cations are among the most powerful of electrophiles. The difficulty is caused by the fact that, with inactive substrates, degradation and poly-merization of the electrophile occurs before it can attack the ring. However, if an activating and a deactivating group are both present on a ring, Friedel–Crafts alky-lation can be accomplished.[359] Aromatic nitro compounds can be methylated by a nucleophilic mechanism (13-17).

The intermediate for Friedel–Crafts alkylation is a carbocation, and rearrange-ment to a more stable cation can be quite facile. Therefore, rearrangement of the alkyl substrate occurs frequently and is an important synthetic limitation of Friedel–Crafts alkylation. For example, benzene treated with n-propyl bromide gives mostly isopropylbenzene (cumene) and much less n-propylbenzene. Rearrangement is usually in the order primary → secondary → tertiary and usually occurs by migra-tion of the smaller group on the adjacent carbon. Therefore, in the absence of spe-cial electronic or resonance influences on the migrating group (such as phenyl), H migrates before methyl, which migrates before ethyl, and so on (see discussion of rearrangement mechanisms in Chapter 18). It is therefore not usually possible to put a primary alkyl group (other than methyl[360] and ethyl) onto an aromatic ring by Friedel–Crafts alkylation. Because of these rearrangements, n-alkylbenzenes are often prepared by *acylation* (11-17), followed by reduction (19-61).

An important use of the Friedel–Crafts alkylation reaction is to effect ring clo-sure.[361] The most common method is to heat with aluminum chloride an aromatic compound having a halogen, hydroxy, or alkene group in the proper position, as, for example, in the preparation of tetralin, **40**.

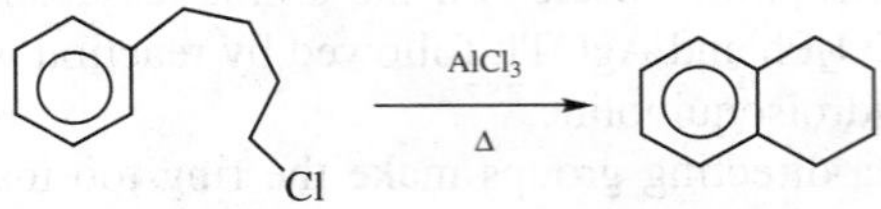

40

Another way of effecting ring closure through Friedel–Crafts alkylation is to use a reagent containing two groups, such as **41**.

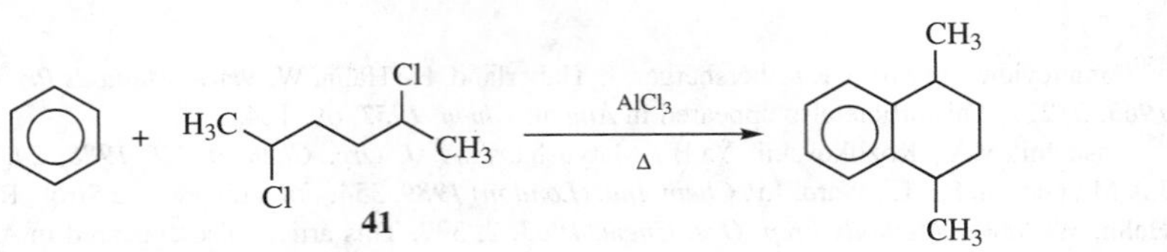

41

These reactions are most successful for the preparation of six-membered rings,[362] though five- and Seven-membered rings have also been closed in this

[359]Olah, G.A. in Olah, G.A. *Friedel–Crafts and Related Reactions*, Vol. 1, Wiley, NY, *1963*, p. 34.
[360]For methylation using a specialized aluminum reagent, with a nickel catalyst, see Gelman, D.; Schumann, H.; Blum, J. *Tetrahedron Lett. 2000, 41*, 7555.
[361]For a review, see Barclay, L.R.C., in Olah, G.A. *Friedel–Crafts and Related Reactions*, Vol. 2, Wiley, NY, *1964*, pp. 785–977.
[362]See Khalaf, A.A.; Roberts, R.M. *J. Org. Chem. 1966, 31*, 89.

manner. For other Friedel–Crafts ring-closure reactions, see **11-15**, **11-13**, and **11-17**. An interesting variation in this reaction showed that *N*-acyl aniline derivatives, upon treatment with $Et_2P(=O)H$ in water and a water soluble initiator (V-501) led to an intramolecular alkylation reaction to give an amide.[363]

As mentioned above, the electrophile in Friedel–Crafts alkylation is a carbocation, at least in most cases.[364] This is in accord with the knowledge that carbocations rearrange in the direction primary $\rightarrow$ secondary $\rightarrow$ tertiary (see Chapter 18). In each case the cation is formed from the attacking reagent and the catalyst. For the three most important types of reagent these reactions are

From alkyl halides: $RCl + AlCl_3 \longrightarrow R^+ + AlCl_4^-$

From alcohols[365] and Lewis acids: $ROH + AlCl_3 \longrightarrow ROAlCl_2 \longrightarrow R^+ + {}^-OAlCl_2$

From alcohols and proton acids: $ROH + H^+ \longrightarrow ROH_2^+ \longrightarrow R^+ + H_2O$

From alkenes (a supply of protons is usually required): $\overset{\diagdown}{\underset{\diagup}{C}}=\overset{\diagup}{\underset{\diagdown}{C}} + H^+ \longrightarrow H-\overset{\diagdown}{\underset{\diagup}{C}}-\overset{\diagup}{\underset{\diagdown}{C}}{}^\oplus$

There is direct evidence, from ir and nmr spectra, that the *tert*-butyl cation is quantitatively formed when *tert*-butyl chloride reacts with $AlCl_3$ in anhydrous liquid HCl.[366] In the case of alkenes, Markovnikov's rule (p. 1019) is followed. Carbocation formation is particularly easy from some reagents, because of the stability of the cations. Triphenylmethyl chloride[367] and 1-chloroadamantane[368] alkylate activated aromatic rings (e.g., phenols, amines) with no catalyst or solvent. Ions as stable as this are less reactive than other carbocations and often attack only active substrates. The tropylium ion, for example, alkylates anisole, but not benzene.[369] It was noted on p. 476 that relatively stable vinylic cations can be generated from certain vinylic compounds. These have been used to introduce vinylic groups into aryl substrates.[370] Lewis acids, such as BF_3[371] or $AlEt_3$,[372] can also be used to alkylation of aromatic rings with alkene units.

[363]Khan, T.A.; Tripoli, R.; Crawford, J.T.; Martin, C.G. Murphy, J.A. *Org. Lett.* **2003**, *5*, 2971.

[364]For a discussion of the mechanism, see Taylor, R. *Electrophilic Aromatic Substitution, Electrophilic Aromatic Substitution*, Wiley, NY, **1990**, pp. 188–213.

[365]See Bijoy, P.; Subba Rao, G.S.R. *Tetrahedron Lett.* **1994**, *35*, 3341 for a double Friedle–Crafts alkylation involving a diol.

[366]Kalchschmid, F.; Mayer, E. *Angew. Chem. Int. Ed.* **1976**, *15*, 773.

[367]See, for example, Hart, H.; Cassis, F.A. *J. Am. Chem. Soc.* **1954**, *76*, 1634; Hickinbottom, W.J. *J. Chem. Soc.* **1934**, 1700; Chuchani, G.; Zabicky, J. *J. Chem. Soc. C* **1966**, 297.

[368]Takaku, M.; Taniguchi, M.; Inamoto, Y. *Synth. Commun.* **1971**, *1*, 141.

[369]Bryce-Smith, D.; Perkins, N.A. *J. Chem. Soc.* **1962**, 5295.

[370]Kitamura, T.; Kobayashi, S.; Taniguchi, H.; Rappoport, Z. *J. Org. Chem.* **1982**, *47*, 5503.

[371]Majetich, G.; Liu, S.; Siesel, D. *Tetrahedron Lett.* **1995**, *36*, 4749; Majetich, G.; Zhang, Y.; Feltman, T.L.; Belfoure, V. *Tetrahedron Lett.* **1993**, *34*, 441; Majetich, G.; Zhang, Y.; Feltman, T.L.; Duncan Jr., S. *Tetrahedron Lett.* **1993**, *34*, 445.

[372]Majetich, G.; Zhang, Y.; Liu, S. *Tetrahedron Lett.* **1994**, *35*, 4887.

There is considerable evidence that many Friedel–Crafts alkylations, especially with primary reagents, do not go through a completely free carbocation. The ion may exist as a tight ion pair with, say, $AlCl_4^-$ as the counterion or as a complex. Among the evidence is that methylation of toluene by methyl bromide and methyl iodide gave different ortho/para/meta ratios,[373] although we would expect the same ratios if the same species attacked in each case. Other evidence is that, in some cases, the reaction kinetics are third order; first order each in aromatic substrate, attacking reagent, and catalyst.[374] In these instances a mechanism in which the carbocation is slowly formed and then rapidly attacked by the aromatic ring is ruled out since, in such a mechanism, the substrate would not appear in the rate expression. Since it is known that free carbocations, once formed, are rapidly attacked by the ring (acting as a nucleophile), there are no free carbocations here. Another possibility (with alkyl halides) is that some alkylations take place by an S_N2 mechanism (with respect to the halide), in which case no carbocations would be involved at all. However, a completely S_N2 mechanism requires inversion of configuration. Most investigations of Friedel–Crafts stereochemistry, even where an S_N2 mechanism might most be expected, have resulted in total racemization, or at best a few percent inversion. A few exceptions have been found,[375] most notably where the reagent was optically active propylene oxide, in which case 100% inversion was reported.[376]

Rearrangement is possible even with a non-carbocation mechanism. The rearrangement could occur *before* the attack on the ring takes place. It has been shown that treatment of $CH_3{}^{14}CH_2Br$ with $AlBr_3$ in the absence of any aromatic compound gave a mixture of the starting material and ${}^{14}CH_3CH_2Br$.[377] Similar results were obtained with $PhCH_2{}^{14}CH_2Br$, in which case the rearrangement was so fast that the rate could be measured only below $-70°C$.[378] Rearrangement could also occur *after* formation of the product, since alkylation is reversible (see **11-33**).[379]

See **14-17** and **14-19** for free-radical alkylation.

A variation of this reaction involves acylation of a β-keto ester, followed by Friedel–Crafts cyclization of the ketone moiety. The product is a coumarin **43**, in what is known as the *Pechmann condensation*.[380] Isolation of esters, such as **42**, is not

[373]Brown, H.C.; Jungk, H. *J. Am. Chem. Soc.* **1956**, *78*, 2182.

[374]For examples see Choi, S.U.; Brown, H.C. *J. Am. Chem. Soc.* **1963**, *85*, 2596.

[375]Some instances of retention of configuration have been reported; a neighboring-group mechanism is likely in these cases: see Masuda, S.; Nakajima, T.; Suga, S. *Bull. Chem. Soc. Jpn.* **1983**, *56*, 1089; Effenberger, F.; Weber, T. *Angew. Chem. Int. Ed.* **1987**, *26*, 142.

[376]Nakajima, T.; Suga, S.; Sugita, T.; Ichikawa, K. *Tetrahedron* **1969**, *25*, 1807. For cases of almost complete inversion, with acyclic reagents, see Piccolo, O.; Azzena, U.; Melloni, G.; Delogu, G.; Valoti, E. *J. Org. Chem.* **1991**, *56*, 183.

[377]Adema, E.H.; Sixma, F.L.J. *Recl. Trav. Chim. Pays-Bas* **1962**, *81*, 323, 336.

[378]For a review of the use of isotopic labeling to study Friedel–Crafts reactions, see Roberts, R.M.; Gibson, T.L. *Isot. Org. Chem.* **1980**, *5*, 103.

[379]For an example, see Lee, C.C.; Hamblin, M.C.; Uthe, J.F. *Can. J. Chem.* **1964**, *42*, 1771.

[380]von Pechmann, H.; Duisberg, C. *Berchti 1883*, *16*, 2119; Sethna, S.; Shah, N.M. *Chem. Rev.* **1945**, *36*, 1 (see p 10); Sethna, S.; Phadke, R. *Org. React.* **1953**, *7*, 1.

always necessary, and protonic acids can be used rather than Lewis acids. The Pechmann condensation is facilitated by the presence of hydroxyl (OH), dimethylamino (NMe$_2$) and alkyl groups meta to the hydroxyl of the phenol.[381] The reaction has been accomplished using microwave irradiation on graphite/Montmorillonite K10.[382] Pechmann condensation in an ionic liquid using ethyl acetate has also been reported.[383]

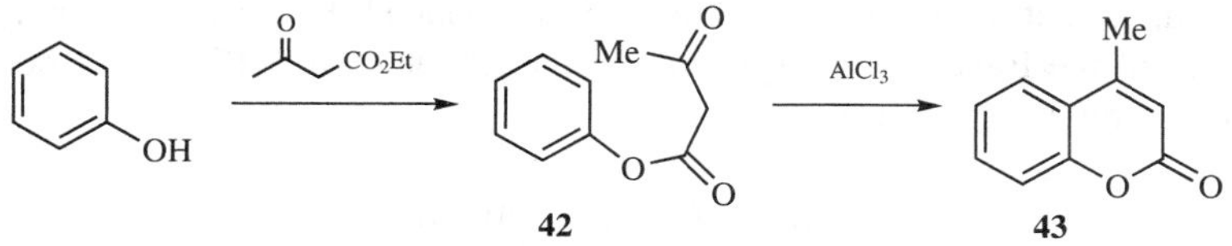

OS **I**, 95, 548; **II**, 151, 229, 232, 236, 248; **III**, 343, 347, 504, 842; **IV**, 47, 520, 620, 665, 702, 898, 960; **V**, 130, 654; **VI**, 109, 744.

11-12 Hydroxyalkylation or Hydroxyalkyl-de-hydrogenation

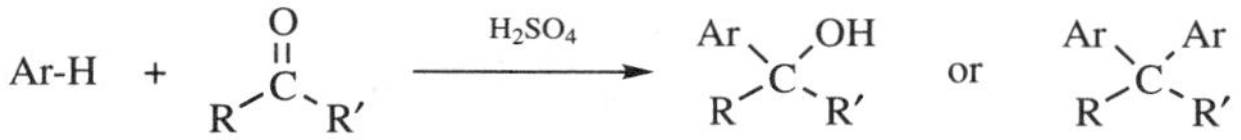

When an aldehyde, ketone, or other carbonyl-containing substrate is treated with a protonic or Lewis acid, an oxygen-stabilized cation is generated. In the presence of an aromatic ring, Friedel–Crafts type alkylation occurs. The condensation of aromatic rings with aldehydes or ketones is called *hydroxyalkylation*.[384] The reaction can be used to prepare alcohols,[385] though more often the alcohol initially produced reacts with another molecule of aromatic compound (**11-11**) to give diarylation. For this the reaction is quite useful, an example being the preparation of DDT, **44**:

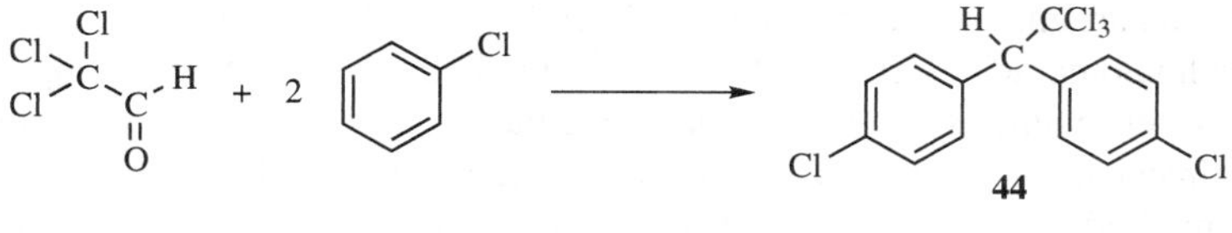

The diarylation reaction is especially common with phenols (the diaryl product here is called a bisphenol). The reaction is normally carried out in alkaline solution on

---

[381]Shah, M.M.; Shah, R.C. *Ber.* **1938**, *71*, 2075; Miyano, M.; Dorn, C.R. *J. Org. Chem.* **1972**, *37*, 259.

[382]Frère, S.; Thiéry, V.; Besson, T. *Tetrahedron Lett.* **2001**, *42*, 2791.

[383]In [bmim]Cl·2AlCl$_3$, 1-butyl-3-methylimidazolium chloroaluminate: Potdar, M.K.; Mohile, S.S.; Salunkhe, M.M. *Tetrahedron Lett.* **2001**, *42*, 9285.

[384]For a review, see Hofmann, J.E.; Schriesheim, A., in Olah, G.A., *Friedel–Crafts and Related Reactions*, Vol. 2, Wiley, NY, **1963**, pp. 597–640.

[385]See, for example, Casiraghi, G.; Casnati, G.; Puglia, G.; Sartori, G. *Synthesis* **1980**, 124.

the phenolate ion.[386] Another variation involved Friedel–Crafts coupling of an aldehyde to an activated aromatic compound (an aniline derivative) to give diaryl carbinols that exhibited atropisomerism (see 146).[387] When the reaction was done with a chiral aluminum complex, modest enantioselectivity was observed.

The hydroxymethylation of phenols with formaldehyde is called the *Lederer–Manasse reaction*. This reaction must be carefully controlled,[388] since it is possible for the para and both ortho positions to be substituted and for each of these to be rearylated, so that a polymeric structure **45** is produced. However, such polymers, which are of the Bakelite type (phenol–formaldehyde resins, **45**), are of considerable commercial importance.

45

The attacking species is the carbocation,

$$R - \overset{\oplus}{\underset{OH}{C}} - R$$

formed from the aldehyde or ketone and the acid catalyst, except when the reaction is carried out in basic solution.

When an aromatic ring is treated with diethyl oxomalonate, $(EtOOC)_2C=O$, the product is an arylmalonic acid derivative $ArC(OH)(COOEt)_2$, which can be converted to an arylmalonic acid, $ArCH(COOEt)_2$.[389] This is therefore a way of applying the malonic ester synthesis (**10-67**) to an aryl group (see also, **13-14**). Of course, the opposite mechanism applies here: The aryl species is the nucleophile.

Two methods, both involving boron-containing reagents, have been devised for the regioselective ortho hydroxymethylation of phenols or aromatic amines.[390]

OS **III**, 326; **V**, 422; **VI**, 471, 856; **VIII**, 75, 77, 80. Also see, OS **I**, 214.

[386]For a review, see Schnell, H.; Krimm, H. *Angew. Chem. Int. Ed.* **1963**, *2*, 373.

[387]Gothelf, A.S.; Hansen, T.; Jørgensen, K.A. *J. Chem. Soc., Perkin Trans. 1* **2001**, 854.

[388]See, for example, Casiraghi, G.; Casnati, G.; Pochini, A.; Puglia, G.; Ungaro, R.; Sartori, G. *Synthesis* **1981**, 143.

[389]Ghosh, S.; Pardo, S.N.; Salomon, R.G. *J. Org. Chem.* **1982**, *47*, 4692.

[390]Sugasawa, T.; Toyoda, T.; Adachi, M.; Sasakura, K. *J. Am. Chem. Soc.* **1978**, *100*, 4842; Nagata, W.; Okada, K.; Aoki, T. *Synthesis* **1979**, 365.

11-13   Cyclodehydration of Carbonyl-Containing Compounds

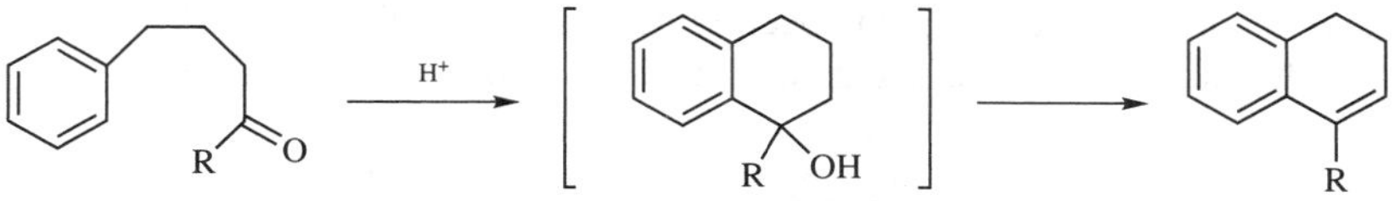

As described in the previous section (**11-12**), the reaction of carbonyl-containing functional groups with protonic or Lewis acids lead to oxygen-stabilized carbocations. When generated in the presence of an aromatic ring, Friedel–Crafts alkylation occurs to give an alcohol or an alkene, if dehydration occurs under the reaction conditions. When an aromatic compound contains an aldehyde or ketone function in a position suitable for closing a suitably sized ring, treatment with acid results in cyclodehydration. The reaction is a special case of **11-12**, but in this case dehydration almost always takes place to give a double bond conjugated with the aromatic ring.[391] The method is very general and is widely used to close both carbocyclic and heterocyclic rings.[392] Polyphosphoric acid is a common reagent, but other acids have also been used. In a variation known as the *Bradsher reaction*,[393] diarylmethanes containing a carbonyl group in the ortho position can be cyclized to anthracene derivatives, **46**. In this case, 1,4-dehydration takes place, at least formally.

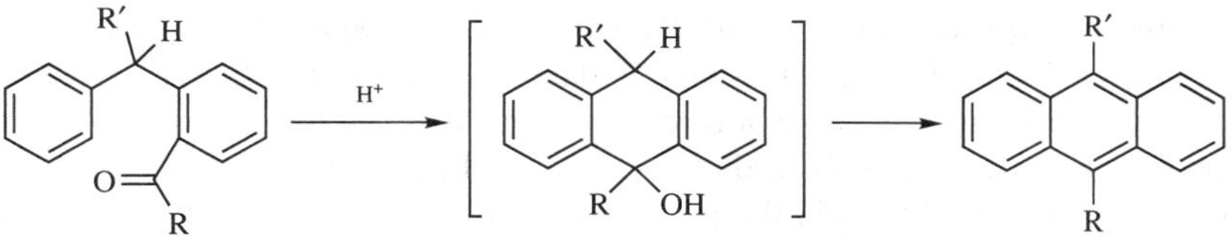

46

An intramolecular cyclization of an aryl ether to the carbonyl of a pendant aryl ketone, on clay with microwave irradiation, led to a benzofuran via Friedel–Crafts cyclization and elimination of water.[394]

The carbonyl unit involved in the cyclization process is not restricted to aldehydes and ketones. The carbonyl of acid derivatives, such as amides can also be utilized. One of the more important cyclodehydration reactions is applied to the formation of heterocyclic systems via cyclization of β-aryl amides, in what is called the *Bischler–Napieralski reaction*.[395] In this reaction amides of the type **47** are

---

[391]For examples where the hydroxy compound was the principal product (with R = CF₃), see Fung, S.; Abraham, N.A.; Bellini, F.; Sestanj, K. *Can. J. Chem.* **1983**, *61*, 368; Bonnet-Delpon, D.; Charpentier-Morize, M.; Jacquot, R. *J. Org. Chem.* **1988**, *53*, 759.

[392]For a review, see Bradsher, C.K. *Chem. Rev.* **1987**, *87*, 1277.

[393]For examples, see Bradsher, C.K. *J. Am. Chem. Soc.* **1940**, *62*, 486; Saraf, S.D.; Vingiello, F.A. *Synthesis* **1970**, 655; Bradsher, C.K. Chem. Rev. **1987**, *87*, 1277, see pp. 1287–1294.

[394]Meshram, H.M.; Sekhar, K.C.; Ganesh, Y.S.S.; Yadav, J.S. *Synlett* **2000**, 1273.

[395]For a review of the mechanism, see Fodor, G.; Nagubandi, S. *Tetrahedron* **1980**, *36*, 1279.

cyclized with phosphorous oxychloride or other reagents, including polyphosphoric acid, sulfuric acid or phosphorus pentoxide, to give a dihydroisoquinoline, **48**. The Bischler–Napieralski reaction has been done in ionic liquids using $POCl_3$.[396] The reaction has also been done using solid-phase (see p. 416) techniques.[397]

If the starting compound contains a hydroxyl group in the α position, an additional dehydration takes place and the product is an isoquinoline.[398] Higher yields can be obtained if the amide is treated with PCl_5 to give an imino chloride $ArCH_2CH_2N{=}CR{-}Cl$, which is isolated and then cyclized by heating.[399] In this latter case, a nitrilium ion $ArCH_2CH_2{}^{\oplus}N{\equiv}CR$ is an intermediate.

Another useful variation is the *Pictet–Spengler isoquinoline synthesis*, also known as the *Pictet–Spengler reaction*.[400] The reactive intermediate is an iminium ion **49** rather than an oxygen-stabilized cation, but attack at the electrophilic carbon of the C=N unit (see **16-31**) leads to an isoquinoline derivative. When a β-arylamine reacts with an aldehyde, the product is an iminium salt, which cyclizes with an aromatic ring to complete the reaction and generate a tetrahydroisoquinoline.[401] A variety of aldehydes can be used, and substitution on the aromatic ring leads to many derivatives. When the reaction is done in the presence of a chiral thiourea catalyst, good enantioselectivity was observed.[402]

Another variation in this basic procedure leads to tetrahydroisoquinolines. When phenethylamine was treated with *N*-hydroxymethylbenzotriazole and then $AlCl_3$ in chloroform, cyclization occurred, and reduction with sodium borohydride gave the 1,2,3,4-tetrahydro-*N*-methylisoquinoline.[403]

---

[396]The reaction was done in bmim PF_6, 1-butyl-3-methylimidazolium hexafluorophosphate: Judeh, Z.M.A.; Ching, C.B.; Bu, J.; McCluskey, A. *Tetrahedron Lett.* **2002**, *43*, 5089.

[397]Chern, M.-S.; Li, W.R. *Tetrahedron Lett.* **2004**, *45*, 8323.

[398]Wang, X.-j.; Tan, J.; Grozinger, K. *Tetrahedron Lett.* **1998**, *39*, 6609.

[399]Fodor, G.; Gal, G.; Phillips, B.A. *Angew. Chem. Int. Ed.* **1972**, *11*, 919.

[400] Pictet, A.; Spengler, T. *Ber.* **1911**, *44*, 2030; Cox, E.D.; Cook, J.M. *Chem. Rev.* **1995**, *95*, 1797. See also Whaley, W.M.; Govindachari, T.R. *Org. React.* **1951**, *6*, 74.

[401]Ong, H.H.; May, E.L. *J. Heterocyclic Chem.* **1971**, *8*, 1007.

[402]Taylor, M.S.; Jacobsen, E.N. *J. Am. Chem. Soc.* **2004**, *126*, 10558.

[403]Locher, C.; Peerzada, N. *J. Chem. Soc., Perkin Trans. 1* **1999**, 179.

OS **I**, 360, 478; **II**, 62, 194; **III**, 281, 300, 329, 568, 580, 581; **IV**, 590; **V**, 550; **VI**, 1. Also see, OS **I**, 54.

11-14   Haloalkylation or Haloalkyl-de-hydrogenation

$$\text{ArH} + \text{HCHO} + \text{HCl} \xrightarrow{\text{ZnCl}_2} \text{ArCH}_2\text{Cl}$$

When certain aromatic compounds are treated with formaldehyde and HCl, the CH_2Cl group is introduced into the ring in a reaction called *chloromethylation*. The reaction has also been carried out with other aldehydes and with HBr and HI. The more general term *haloalkylation* covers these cases.[404] The reaction is successful for benzene, and alkyl-, alkoxy-, and halobenzenes. It is greatly hindered by meta-directing groups, which reduce yields or completely prevent the reactions. Amines and phenols are too reactive and usually give polymers unless deactivating groups are also present, but phenolic ethers and esters successfully undergo the reaction. Compounds of lesser reactivity can often be chloromethylated with chloromethyl methyl ether ($ClCH_2OMe$), or methoxyacetyl chloride $MeOCH_2COCl$.[405] Zinc chloride is the most common catalyst, but other Friedel–Crafts catalysts are also employed. As with reaction **11-12** and for the same reason, an important side product is the diaryl compound Ar_2CH_2 (from formaldehyde).

Apparently, the initial step involves reaction of the aromatic compound with the aldehyde to form the hydroxyalkyl compound, exactly as in **11-12**, and then the HCl converts this to the chloroalkyl compound.[406] The acceleration of the reaction by $ZnCl_2$ has been attributed[407] to the raising of the acidity of the medium, causing an increase in the concentration of $HOCH_2^+$ ions.

OS **III**, 195, 197, 468, 557; **IV**, 980.

11-15   Friedel–Crafts Arylation: The Scholl Reaction

De-hydrogen-coupling

$$2\,\text{ArH} \xrightarrow[\text{H}^+]{\text{AlCl}_3} \text{Ar}{-}\text{Ar} + \text{H}_2$$

The coupling of two aromatic molecules by treatment with a Lewis acid and a proton acid is called the *Scholl reaction*.[408] Yields are low and the synthesis is seldom useful. High temperatures and strong-acid catalysts are required, and the reaction fails for substrates that are destroyed by these conditions. Because the reaction

---

[404]For reviews, see Belen'kii, L.I.; Vol'kenshtein, Yu.B.; Karmanova, I.B. *Russ. Chem. Rev.* **1977**, *46*, 891; Olah, G.A.; Tolgyesi, W.S., in Olah, G.A. *Friedel–Crafts and Related Reactions*, Vol. 2, Wiley, NY, **1963**, pp. 659–784.

[405]McKillop, A.; Madjdabadi, F.A.; Long, D.A. *Tetrahedron Lett.* **1983**, *24*, 1933.

[406]Ziegler, E.; Hontschik, I.; Milowiz, L. *Monatsh. Chem.* **1948**, *79*, 142; Ogata, Y.; Okano, M. *J. Am. Chem. Soc.* **1956**, *78*, 5423. See also, Olah, G.A.; Yu, S.H. *J. Am. Chem. Soc.* **1975**, *97*, 2293.

[407]Lyushin, M.M.; Mekhtiev, S.D.; Guseinova, S.N. *J. Org. Chem. USSR* **1970**, *6*, 1445.

[408]For reviews, see Kovacic, P.; Jones, M.B. *Chem. Rev.* **1987**, *87*, 357; Balaban, A.T.; Nenitzescu, C.D., in Olah, G.A. *Friedel–Crafts and Related Reactions*, Vol. 2, Wiley, NY, **1964**, pp. 979–1047.

becomes important with large fused-ring systems, ordinary Friedel–Crafts reactions (**11-11**) on these systems are rare. For example, naphthalene gives binaphthyl under Friedel–Crafts conditions. Yields can be increased by the addition of a salt, such as $CuCl_2$ or $FeCl_3$, which acts as an oxidant.[409] Rhodium catalysts have also been used.[410]

Intramolecular Scholl reactions, such as formation of **50** from triphenylmethane,

are much more successful than the intermolecular reaction. The mechanism is not clear, but it may involve attack by a proton to give an arenium ion of the type **12** (p. 662), which would be the electrophile that attacks the other ring.[411] Sometimes arylations have been accomplished by treating aromatic substrates with particularly active aryl halides, especially fluorides. For free-radical arylations, see reactions **12-15**, **13-26**, **13-27**, **13-10**, **14-17**, and **14-18**.

OS **IV**, 482; **X**, 359. Also see, OS **V**, 102, 952.

11-16 Arylation of Aromatic Compounds By Metalated Aryls

$$Ar\text{-}M \xrightarrow{\text{Ar}'\text{-}H} Ar\text{—}Ar'$$

Many metalated aryl compounds are known to couple with aromatic compounds. Aniline derivatives react with $ArPb(OAc)_3$, for example, to give the 2-arylaniline.[412] Phenolic anions also react to form biaryls, with modest enantioselectivity in the presence of brucine.[413]

Phenylboronates $[ArB(OR)_2]$ react with electron-deficient aromatic compounds, such as acetophenone, to give the biaryl.[414] Arylboronates also react with π-allyl palladium complexes to form the alkylated aromatic compound.[415]

[409]Kovacic, P.; Koch, Jr., F.W. *J. Org. Chem.* **1965**, *30*, 3176; Kovacic, P.; Wu, C. *J. Org. Chem.* **1961**, *26*, 759, 762. For examples with references, see Larock, R.C. *Comprehensive Organic Transformations*, 2nd ed., Wiley-VCH, NY, **1999**, pp. 77–84; Sartori, G.; Maggi, R.; Bigi, F.; Grandi, M. *J. Org. Chem.* **1993**, *58*, 7271

[410]Barrett, A.G.M.; Itoh, T.; Wallace, E.M. *Tetrahedron Lett.* **1993**, *34*, 2233.

[411]For a discussion, see Clowes, G.A. *J. Chem. Soc., C* **1968**, 2519.

[412]Saito, S.; Kano, T.; Ohyabu, Y.; Yamamoto, H. *Synlett* **2000**, 1676.

[413]Kano, T.; Ohyabu, Y.; Saito, S.; Yamamoto, H. *J. Am. Chem. Soc.* **2002**, *124*, 5365.

[414]Kakiuchi, F.; Kan, S.; Igi, K.; Chatani, N.; Murai, S. *J. Am. Chem. Soc.* **2003**, *125*, 1698.

[415]Ortar, G. *Tetrahedron Lett.* **2003**, *44*, 4311.

11-17   Friedel–Crafts Acylation

Acylation or Acyl-de-hydrogenation

$$\text{ArH} + \text{RCOCl} \xrightarrow{\text{AlCl}_3} \text{ArCOR}$$

The most important method for the preparation of aryl ketones is known as *Friedel–Crafts acylation.*[416] The reaction is of wide scope. Reagents other than acyl halides can be used,[417] including carboxylic acids,[418] anhydrides, and ketenes. Oxalyl chloride has been used to give diaryl 1,2-diketones.[419] Carboxylic esters usually give alkylation as the predominant product (see **11-11**).[420] *N*-Carbamoyl β-lactams reacted with naphthalene in the presence of trifluoromethanesulfonic acid to give the keto-amide.[421]

The alkyl group (R in RCOCl) may be aryl as well as alkyl. The major disadvantages of Friedel–Crafts alkylation, polyalkylation, and rearrangement of the intermediate carbocation, are not a problem in Friedel–Crafts acylation. Rearrangement of the alkyl group (R in RCOCl) is never found because the intermediate is an acylium ion (an acyl cation, $\text{RC} \equiv \text{O}^+$, see below). Because the RCO group is deactivating, the reaction stops cleanly after one group is introduced. All four acyl halides can be used, though chlorides are most commonly employed. The order of activity is usually, but not always, $\text{I} > \text{Br} > \text{Cl} > \text{F}$.[422] Catalysts are Lewis acids,[423] similar to those in reaction **11-11**, but in acylation a little $>$ than 1 equivalent of catalyst is required per mole of reagent, because the first mole coordinates

[416]For reviews of Friedel–Crafts acylation, see Olah, G.A. *Friedel–Crafts and Related Reactions*, Wiley, NY, *1963–1964*, as follows: Vol. 1, Olah, G.A. pp. 91–115; Vol. 3, Gore, P.H. pp. 1–381; Peto, A.G. pp. 535–910; Sethna, S. pp. 911–1002; Jensen, F.R.; Goldman, G. pp. 1003–1032. For another review, see Gore, P.H. *Chem. Ind. (London) 1974*, 727.

[417]For a list of reagents, with references, see Larock, R.C. *Comprehensive Organic Transformations*, 2nd ed., Wiley-VCH, NY, *1999*, pp. 1423–1426.

[418]Ranu, B.C.; Ghosh, K.; Jana, U. *J. Org. Chem. 1996*, *61*, 9546; Kawamura, M.; Cui, D.-M.; Hayashi, T.; Shimada, S. *Tetrahedron Lett. 2003*, *44*, 7715. For an example of acylation by heating with octanoic acid, without a catalyst, see Kaur, J.; Kozhevnikov, I.V. *Chem.Commun. 2002*, 2508.

[419]Mohr, B.; Enkelmann, V.; Wegner, G. *J. Org. Chem. 1994*, *59*, 635; Taber, D.F.; Sethuraman, M.R. *J. Org. Chem. 2000*, *65*, 254.

[420]For a reaction involving the Friedel–Crafts acylation using an ester, see Hwang, J.P.; Prakash, G.K.S.; Olah, G.A. *Tetrahedron 2000*, *56*, 7199.

[421]Anderson, K.W.; Tepe, J. *Org. Lett. 2002*, *4*, 459.

[422]Yamase, Y. *Bull. Chem. Soc. Jpn. 1961*, *34*, 480; Corriu, R. *Bull. Soc. Chim. Fr. 1965*, 821.

[423]The usual Lewis acids can be used, as described in **11-11**, and ferric chloride, iodine, zinc chloride, and iron are probably the most common catalysts. For a review, see Pearson, D.E.; Buehler, C.A. *Synthesis 1972*, 533. Recently employed catalysts include, **Ga(ONf)$_3$**, where Nf = nonafluorobutanesulfonate: Matsu, J.-i.; Odashima, K.; Kobayashi, S. *Synlett 2000*, 403. **In(OTf)$_3$** with LiClO$_4$: Chapman, C.J.; Frost, C.G.; Hartley, J.P.; Whittle, A.J. *Tetrahedron Lett. 2001*, *42*, 773. **InCl$_3$**: Choudhary, V.R.; Jana, S.K.; Patil, N.S. *Tetrahedron Lett. 2002*, *43*, 1105. **Sc(OTf)$_3$**: Kawada, A.; Mitamura, S.; Matsuo, J-i.; Tsuchiya, T.; Kobayashi, S. *Bull. Chem. Soc. Jpn. 2000*, *73*, 2325. **Yb[C(SO$_2$C$_4$F$_4$)$_3$]$_3$**: Barrett, A.G.M.; Bouloc, N.; Braddock, D.C.; Chadwick, D.; Henderson, D.A. *Synlett 2002*, 1653. **BiOCl$_3$**: Répichet, S.; Le Roux, C.; Roques, N.; Dubac, J. *Tetrahedron Lett. 2003*, *44*, 2037. **ZnO**: Sarvari, M.H.; Sharghi, H. *J. Org. Chem. 2004*, *69*, 6953.

with the oxygen of the reagent [as in $R(Cl)C=O^+$ $^-AlCl_3$].[424] A reusable catalyst [$Ln(OTf)_3$–$LiClO_4$] has been developed as well.[425] HY-Zeolite has also been used to facilitate the reaction with acetic anhydride.[426] A platinum catalyst was used with acetic anhydride,[427] $TiCl_4$ with acetyl chloride[428] or acetyl chloride and zinc powder with microwave irradiation.[429] Friedel–Crafts acylation using a carboxylic acid with a catalyst called Envirocat-EPIC (an acid-treated clay-based material was reported.[430] Friedel–Crafts acylation was reported in an ionic liquid.[431] An interesting acylation reaction was reported that coupled trichlorophenylmethane to benzene, giving benzophenone in the presence of the ionic liquid $AlCl_3$-n-BPC.[432] Acylation has been accomplished in carbon disulfide.[433]

Proton acids can be used as catalysts when the reagent is a carboxylic acid. The mixed carboxylic sulfonic anhydrides $RCOOSO_2CF_3$ are extremely reactive acylating agents and can smoothly acylate benzene without a catalyst.[434] With active substrates (e.g., aryl ethers, fused-ring systems, thiophenes), Friedel–Crafts acylation can be carried out with very small amounts of catalyst, often just a trace, or even sometimes with no catalyst at all.

The reaction is quite successful for many types of substrate, including fused ring systems, which give poor results in **11-11**. Compounds containing ortho–para-directing groups, including alkyl, hydroxy, alkoxy, halogen, and acetamido groups, are easily acylated and give mainly or exclusively the para products, because of the relatively large size of the acyl group. However, aromatic amines give poor results. With amines and phenols there may be competition from *N*- or *O*-acylation; however, *O*-acylated phenols can be converted to *C*-acylated phenols by the Fries rearrangement (**11-27**). Friedel–Crafts acylation is usually prevented by meta-directing groups. Indeed, nitrobenzene is often used as a solvent for the reaction. Many heterocyclic systems, including furans, thiophenes, pyrans, and pyrroles[435]

[424]The crystal structures of several of these complexes have been reported: Rasmussen, S.E.; Broch, N.C. *Acta Chem. Scand.* **1966**, *20*, 1351; Chevrier, B.; Le Carpentier, J.; Weiss, R. *J. Am. Chem. Soc.* **1972**, *94*, 5718. For a review of these complexes, see Chevrier, B.; Weiss, R. *Angew. Chem. Int. Ed.* **1974**, *13*, 1.

[425]Kawada, A.; Mitamura, S.; Kobayashi, S. *Chem. Commun.* **1996**, 183. See Kawada, A.; Mitamura, S.; Kobayashi, S. *SynLett*, **1994**, 545 for the use of Sc(OTf)$_3$ with acetic anhydride and Hachiya, I.; Moriwaki, M.; Kobayashi, S. *Tetrahedron Lett.* **1995**, *36*, 409 for the use of Hf(OTf)$_4$.

[426]Sreekumar, R.; Padmukumar, R. *Synth. Commun.* **1997**, *27*, 777. See Paul, V.; Sudalai, A.; Daniel, T.; Srinivasan, K.V. *Tetrahedron Lett.* **1994**, *35*, 2601 for the use of an acidic zeolite.

[427]Fürstner, A.; Voigtländer, D.; Schrader, W.; Giebel, D.; Reetz, M.T. *Org. Lett.* **2001**, *3*, 417.

[428]Bensari, A.; Zaveri, N.T. *Synthesis* **2003**, 267.

[429]Paul, S.; Nanda, P.; Gupta, R.; Loupy, A. *Synthesis* **2003**, 2877.

[430]Bandgari, B.P.; Sadavarte, V.S. *Synth. Commun.* **1999**, *29*, 2587.

[431]The reaction was catalyzed by Br$_2$O$_3$ in bmim NTf$_2$, 1-butyl-3-methylimidazolium triflimide: Gmouth, S.; Yang, H.; Vaultier, M. *Org. Lett.* **2003**, *5*, 2219.

[432]This catalyst is *n*-butylpyridinium chloroaluminate, see Rebeiro, G.L.; Khadilkar, B.M. *Synth. Commun.* **2000**, *30*, 1605.

[433]Georgakilas, V.; Perdikomatis, G.P.; Triantafyllou, A.S.; Siskos, M.G.; Zarkadis, A.K. *Tetrahedron* **2002**, *58*, 2441.

[434]Effenberger, F.; Sohn, E.; Epple, G. *Chem. Ber.* **1983**, *116*, 1195. See also, Keumi, T.; Yoshimura, K.; Shimada, M.; Kitajima, H. *Bull. Chem. Soc. Jpn.* **1988**, *44*, 455.

[435]Yadav, J.S.; Reddy, B.V.S.; Kondaji, G.; Rao, R.S.; Kumar, S.P. *Tetrahedron Lett.* **2002**, *43*, 8133.

but not pyridines or quinolines, can be acylated in good yield. Initial reaction of indole with Et_2AlCl[436] or $SnCl_4$,[437] followed by acetyl chloride leads to 3-acetylindole. By comparison, the reaction of *N*-acetylindole with acetic anhydride and $AlCl_3$ gave *N*,6-diacetylindole.[438] Acetylation at C-3 was also accomplished with acetyl chloride in the ionic liquid emimcl-$AlCl_3$.[439] Gore, in Ref. 417 (pp. 36–100; with tables, pp. 105–321), presents an extensive summary of the substrates to which this reaction has been applied. Pyridines and quinolines can be also be acylated by a free-radical mechanism (reaction **14-19**).

When a mixed-anhydride $RCOOCOR'$ is the reagent, two products are possible: ArCOR and $ArCOR'$. Which product predominates depends on two factors. If R contains electron-withdrawing groups, then $ArCOR'$ is chiefly formed, but if this factor is approximately constant in R and R', the ketone with the larger R group predominantly forms.[440] This means that *formylations* of the ring do not occur with mixed anhydrides of formic acid HCOOCOR.

An important use of the Friedel–Crafts acylation is to effect ring closure.[441] This can be done if an acyl halide, anhydride, or carboxylic acid[442] group is in the proper position. An example is the conversion of **51** to **52**.

The reaction is used mostly to close six-membered rings, but has also been done for five- and seven-membered rings, which close less readily. Even larger rings can be closed by high-dilution techniques.[443] Tricyclic and larger systems are often made by using substrates containing one of the acyl groups on a ring. Many fused-ring systems are made in this manner. If the bridging group is CO, the product is a quinone.[444] One of the most common catalysts for intramolecular Friedel–Crafts

[436]Okauchi, T.; Itonaga, M.; Minami, T.; Owa, T.; Kitoh, K.; Yoshino, H. *Org. Lett.* **2000**, *2*, 1485; Zhang, Z; Yang, Z.; Wong, H.; Zhu, J.; Meanwell, N.A.; Kadow, J.F.; Wang, T. *J. Org. Chem.* **2002**, *67*, 6226.

[437]Ottoni, O.; de V.F. Neder, A.; Dias, A.K.B.; Cruz, R.P.A.; Aquino, L.B. *Org. Lett.* **2001**, *3*, 1005.

[438]Cruz, R.P.A.; Ottoni, O.; Abella, C.A.M.; Aquino, L.B. *Tetrahedron Lett.* **2001**, *42*, 1467. 3-Methylindole was converted to 2-acetyl-3-methylindole with acetyl chloride and zinc(II) chloride: see Pal, M.; Dakarapu, R.; Padakanti, S. *J. Org. Chem.* **2004**, *69*, 2913.

[439]The ionic liquid emimcl-$AlCl_3$ is 1-ethyl-3-methylimidazolium chloroaluminate, see Yeung, K.-S.; Farkas, M.E.; Qiu, Z.; Yang, Z. *Tetrahedron lett.* **2002**, *43*, 5793.

[440]Edwards, Jr., W.R.; Sibelle, E.C. *J. Org. Chem.* **1963**, *28*, 674.

[441]For a review, see Sethna, S., in Olah, G.A. *Friedel–Crafts and Related Reactions*, Vol. 3, Wiley, NY, **1964**, pp. 911–1002;. For examples with references, see Larock, R.C. *Comprehensive Organic Transformations*, 2nd ed., Wiley-VCH, NY, **1999**, pp. 1427–1431.

[442]For an example using $Tb(OTf)_3$, see Cui, D.-M.; Zhang, C.; Kawamura, M.; Shimada, S. *Tetrahedron Lett.* **2004**, *45*, 1741.

[443]For example, see Schubert, W.M.; Sweeney, W.A.; Latourette, H.K. *J. Am. Chem. Soc.* **1954**, *76*, 5462.

[444]For discussions, see Naruta, Y.; Maruyama, K., in Patai, S.; Rappoport, Z. *The Chemistry of the Quinonoid Compounds*, Vol. 2, pt. 1, Wiley, NY, **1988**, pp. 325–332; Thomson, R.H., in Patai, S. *The Chemistry of the Quinonoid Compounds*, Vol. 1, pt. 1, Wiley, NY, **1974**; pp. 136–139.

acylation is polyphosphoric acid[445] (because of its high potency), but $AlCl_3$, H_2SO_4, and other Lewis and proton acids are also used, though acylations with acyl halides are not generally catalyzed by proton acids.

Friedel–Crafts acylation can be carried out with cyclic anhydrides,[446] in which case the product contains a carboxyl group in the side chain (**53**). When succinic anhydride is used, the product is $ArCOCH_2CH_2COOH$. This can be reduced (**19-61**) to $ArCH_2CH_2CH_2COOH$, which can then be cyclized by an internal Friedel–Crafts acylation to give **54**. The total process is called the *Haworth reaction*:[447]

The mechanism of Friedel–Crafts acylation is not completely understood,[448] but at least two mechanisms probably operate, depending on conditions.[449] In most cases the attacking species is the acyl cation, either free or as an ion pair, formed by[450]

$$RCOCl + AlCl_3 \longrightarrow RCO^+ + AlCl_4^-$$

If R is tertiary, RCO^+ may lose CO to give R^+, so that the alkyl arene ArR is often a side product or even the main product. This kind of cleavage is much more likely with relatively unreactive substrates, where the acylium ion has time to break down. For example, pivaloyl chloride Me_3CCOCl gives the normal acyl product with anisole, but the alkyl product Me_3CPh with benzene. In the other mechanism, an acyl cation is not involved, but the 1:1 complex (**55**) attacks directly.[451]

[445]For a review of polyphosphoric acid, see Rowlands, D.A., in Pizey, J.S. *Synthetic Reagents*, Vol. 6, Wiley, NY, *1985*, pp. 156–414.

[446]For a review see Peto, A.G., in Olah, G.A. *Friedel–Crafts and Related Reactions*, Vol. 3, Wiley, NY, *1964*, p. 535.

[447]See Agranat, I.; Shih, Y. *J. Chem. Educ. 1976, 53*, 488.

[448]See Effenberger, F.; Eberhard, J.K.; Maier, A.H. *J. Am. Chem. Soc. 1996, 118*, 12572 for first evidence of the reacting electrophile.

[449]For a review of the mechanism, see Taylor, R. *Electrophilic Aromatic Substitution*, Wiley, NY, *1990*, pp. 222–237.

[450]After 2 min, exchange between PhCOCl and $Al(^{36}Cl)_3$ is complete: Oulevey, G.; Susz, P.B. *Helv. Chim. Acta 1964, 47*, 1828.

[451]For example, see Corriu, R.; Dore, M.; Thomassin, R. *Tetrahedron 1971, 27*, 5601, 5819; Tan, L.K.; Brownstein, S. *J. Org. Chem. 1983, 48*, 302.

Free-ion attack is more likely for sterically hindered R.[452] The ion CH_3CO^+ has been detected (by IR spectroscopy) in the liquid complex between acetyl chloride and aluminum chloride, and in polar solvents, such as nitrobenzene; but in nonpolar solvents, such as chloroform, only the complex and not the free ion is present.[453] In any event, 1 equivalent of catalyst certainly remains complexed to the product at the end of the reaction. When the reaction is performed with $RCO^+SbF_6^-$, no catalyst is required and the free ion[454] (or ion pair) is undoubtedly the attacking entity.[455] The use of $LiClO_4$ on the metal triflate-catalyzed Friedel–Crafts acylation of methoxy-naphthalene derivatives has been examined, and the presence of the lithium salt leads to acylation in the ring containing the methoxy unit, whereas reaction occurs in the other ring in the absence of lithium salts.[456] Note that lithium perchlorate forms a complex with acetic anhydride, which can be used for the Friedel–Crafts acetylation of activated aromatic compounds.[457]

OS **I**, 109, 353, 476, 517; **II**, 3, 8, 15, 81, 156, 169, 304, 520, 569; **III**, 6, 14, 23, 53, 109, 183, 248, 272, 593, 637, 761, 798; **IV**, 8, 34, 88, 898, 900; **V**, 111; **VI**, 34, 618, 625 **X**, 125.

Reaction **11-18** is a direct formylation of the ring.[458] Reaction **11-17** has not been used for formylation, since neither formic anhydride nor formyl chloride is stable at ordinary temperatures. Formyl chloride has been shown to be stable in chloroform solution for 1 h at $-60°C$,[459] but it is not useful for formylating aromatic rings under these conditions. Formic anhydride has been prepared in solution, but has not been isolated.[460] Mixed anhydrides of formic and other acids are known[461] and can be used to formylate amines (see **16-73**) and alcohols, but no formylation takes place when they are applied to aromatic rings. See **13-17** for a nucleophilic method for the formylation of aromatic rings.

A related reaction involves a biaryl, where one ring is a phenol. Treatment with BCl_3 and an $AlCl_3$ catalyst, followed by reaction with CO and $Pd(OAc)_2$, led to

[452]Yamase, Y. *Bull. Chem. Soc. Jpn.* **1961**, *34*, 484; Gore, P.H. *Bull. Chem. Soc. Jpn.* **1962**, *35*, 1627; Satchell, D.P.N. *J. Chem. Soc.* **1961**, 5404.

[453]Cook, D. *Can. J. Chem.* **1959**, *37*, 48; Cassimatis, D.; Bonnin, J.P.; Theophanides, T. *Can. J. Chem.* **1970**, *48*, 3860.

[454]Crystal structures of solid RCO^+ SbF_6^- salts have been reported: Boer, F.P. *J. Am. Chem. Soc.* **1968**, *90*, 6706; Chevrier, B.; Le Carpentier, J.; Weiss, R. *Acta Crystallogr., Sect. B,* **1972**, *28*, 2673; *J. Am. Chem. Soc.* **1972**, *94*, 5718.

[455]Olah, G.A.; Lin, H.C.; Germain, A. *Synthesis* **1974**, 895. For a review of acylium salts in organic synthesis, see Al-Talib, M.; Tashtoush, H. *Org. Prep. Proced. Int.* **1990**, *22*, 1.

[456]Kobayashi, S.; Komoto, I. *Tetrahedron* **2000**, *56*, 6463.

[457]Bartoli, G.; Bosco, M.; Marcantoni, E.; Massaccesi, M.; Rinalde, S.; Sambri, L. *Tetrahedron Lett.* **2002**, *43*, 6331.

[458]For a review, see Olah, G.A.; Kuhn, S.J. Olah, G.A. *Friedel–Crafts and Related Reactions*, Vol. 3, Wiley, NY, **1964**, pp. 1153–1256. For a review of formylating agents, see Olah, G.A.; Ohannesian, L.; Arvanaghi, M. *Chem. Rev.* **1987**, *87*, 671. For a list of reagents, with references, see Larock, R.C. *Comprehensive Organic Transformations*, 2nd ed., Wiley-VCH, NY, **1999**, pp. 1423–1426.

[459]Staab, H.A.; Datta, A.P. *Angew. Chem. Int. Ed.* **1964**, *3*, 132.

[460]Olah, G.A.; Vankar, Y.D.; Arvanaghi, M.; Sommer, J. *Angew. Chem. Int. Ed.* **1979**, *18*, 614; Schijf, R.; Scheeren, J.W.; van Es, A.; Stevens, W. *Recl. Trav. Chim. Pays-Bas* **1965**, *84*, 594.

[461]Stevens, W.; van Es, A. *Recl. Trav. Chim. Pays-Bas* **1964**, *83*, 863.

carbonylation and acylation to give the corresponding lactone.[462] Carbonylation of aromatic compounds can lead to aryl ketones. Heating an aromatic compound with $Ru(CO)_{12}$, ethylene and 20 atm of CO gave the corresponding aryl ethyl ketone.[463]

11-18 Formylation

Formylation or Formyl-de-hydrogenation

$$Ar-H \longrightarrow Ar-CHO$$

The reaction with disubstituted formamides R_2N-CHO and phosphorus oxychloride, called the *Vilsmeier* or the *Vilsmeier–Haack reaction*,[464] is the most common method for the formylation of aromatic rings.[465] However, it is applicable only to active substrates, such as amines and phenols. An intramolecular version is also known.[466] Aromatic hydrocarbons and heterocycles can also be formylated, but only if they are much more active than benzene (e.g., azulenes, ferrocenes). Although *N*-phenyl-*N*-methylformamide is a common reagent, other arylalkyl amides and dialkyl amides are also used.[467] Phosgene ($COCl_2$) has been used in place of $POCl_3$. The reaction has also been carried out with other amides to give ketones (actually an example of **11-17**), but not often. The attacking species[468] is **56**,[469] and the mechanism is probably that shown to give **57**, which is unstable and easily hydrolyzes to the product. Either formation of **56** or the reaction of **56** with the substrate can be rate determining, depending on the reactivity of the substrate.[470]

[462]Zhou, Q.J.; Worm, K.; Dolle, R.E. *J.Org. Chem.* **2004**, *69*, 5147.

[463]Ie, Y.; Chatani, N.; Ogo, T.; Marshall, D.R.; Fukuyama, T.; Kakiuchi, F.; Murai, S. *J. Org. Chem.* **2000**, *65*, 1475.

[464]See Blaser, D.; Calmes, M.; Daunis, J.; Natt, F.; Tardy-Delassus, A.; Jacquier, R. *Org. Prep. Proceed. Int.* **1993**, *25*, 338 for improvements in this reaction.

[465]For a review, see Jutz, C. *Adv. Org. Chem.* **1976**, *9*, pt. 1, 225.

[466]Meth-Cohn, O.; Goon, S. *J. Chem. Soc. Perkin Trans. 1* **1997**, 85.

[467]For a review of dimethylformamide, see Pizey, J.S. *Synthetic Reagents*, Vol. 1, Wiley, NY, **1974**, pp. 1–99.

[468]For a review of such species, see Kantlehner, W. *Adv. Org. Chem.* **1979**, *9*, pt. 2, 5.

[469]See Arnold, Z.; Holy, A. *Collect. Czech. Chem. Commun.* **1962**, *27*, 2886; Fritz, H.; Oehl, R. *Liebigs Ann. Chem.* **1971**, *749*, 159; Jugie, G.; Smith, J.A.S.; Martin, G.J. *J. Chem. Soc. Perkin Trans. 2* **1975**, 925.

[470]Alunni, S.; Linda, P.; Marino, G.; Santini, S.; Savelli, G. *J. Chem. Soc. Perkin Trans. 2* **1972**, 2070.

When $(CF_3SO_2)_2O$ was used instead of $POCl_3$, the reaction was extended to some less-active compounds, including naphthalene and phenanthrene.[471] In a related reaction, paraformaldehyde can be used, with $MgCl_2$–NEt_3, to convert phenol to phenol 2-carboxaldehyde.[472] Another variation treated acetanilide with $POCl_3$–DMF and generated 2-chloroquinoline-3-carboxaldehyde.[473] Used in conjunction with conjugated hydroxylamines, a tandem Vilsmeier–Beckman reaction (see **18-17** for the Beckman rearrangement) leads to pyridines (2-chloro-3-carboxaldehyde).[474] A chain-extension variation has been reported in which an aryl alkyl ketone is treated with $POCl_3$/DMF on silica with microwave irradiation to give a conjugated aldehyde, $ArC(=O)R \rightarrow ArC(Cl)=CHCHO$.[475]

OS **I**, 217; **III**, 98, **IV**, 331, 539, 831, 915.

$$ArH + Zn(CN)_2 \xrightarrow{\ HCl\ } ArCH=NH_2{}^+\ Cl^- \xrightarrow{\ H_2O\ } ArCHO$$

Formylation with $Zn(CN)_2$ and HCl is called the *Gatterman reaction*.[476] It can be applied to alkylbenzenes, phenols and their ethers, and many heterocyclic compounds. However, it cannot be applied to aromatic amines. In the original version of this reaction the substrate was treated with HCN, HCl, and $ZnCl_2$, but the use of $Zn(CN)_2$ and HCl (HCN and $ZnCl_2$ are generated *in situ*) makes the reaction more convenient to carry out and yields are not diminished. The mechanism of the Gatterman reaction has not been investigated very much, but it is known that an initially formed but not isolated nitrogen-containing product is hydrolyzed to aldehyde. This product is presumed to be $ArCH=NH_2{}^+Cl^-$, as shown. When benzene was treated with NaCN under superacid conditions (F_3CSO_2OH–SbF_5, see p. 236), a good yield of product was obtained, leading to the conclusion that the electrophile in this case was ${}^+C(H)=N^+H_2$.[477] The Gatterman reaction may be regarded as a special case of **11-24**.

Another method, formylation with CO and HCl in the presence of $AlCl_3$ and $CuCl$[478] (the *Gatterman–Koch reaction*), is limited to benzene and alkylbenzenes.[479]

[471]Martínez, A.G.; Alvarez, R.M.; Barcina, J.O.; Cerero, S. de la M.; Vilar, E.T.; Fraile, A.G.; Hanack, M.; Subramanian, L.R. *J. Chem. Soc., Chem. Commun.* **1990**, 1571.

[472]Hofsløkken, N.U.; Skattebøl, L. *Acta Chem. Scand.* **1999**, *53*, 258.

[473]Ali, M.M.; Tasneem, Rajanna, K.C.; Prakash, P.K.S. *Synlett* **2001**, 251. For another variation to generate 4-chloro-2-phenyl-*N*-formyldihydroquinoline derivatives, see Akila, S.; Selvi, S.; Balasubramanian, K. *Tetrahedron* **2001**, *57*, 3465.

[474]Amaresh, R.R.; Perumal, P.T. *Synth. Commun.* **2000**, *30*, 2269.

[475]Paul, S.; Gupta, M.; Gupta, R. *Synlett* **2000**, 1115.

[476]For a review, see Truce, W.E. *Org. React.* **1957**, *9*, 37. See Tanaka, M.; Fujiwara, M.; Ando, H. *J. Org. Chem.* **1995**, *60*, 2106 for rate studies.

[477]Yato, M.; Ohwada, T.; Shudo, K. *J. Am. Chem. Soc.* **1991**, *113*, 691.

[478]The CuCl is not always necessary: see Toniolo, L.; Graziani, M. *J. Organomet. Chem.* **1980**, *194*, 221.

[479]For a review, see Crounse, N.N. *Org. React.* **1949**, *5*, 290.

OS **II**, 583; **III**, 549.

In the *Reimer–Tiemann reaction*, aromatic rings are formylated by reaction with chloroform and hydroxide ion.[480] The method is useful only for phenols and certain heterocyclic compounds such as pyrroles and indoles. Unlike the previous formylation methods (**11-18**), this one is conducted in basic solution. Yields are generally low, seldom rising above 50%.[481] The incoming group is directed ortho, unless both ortho positions are filled, in which case the attack is para.[482] Certain substrates have been shown to give abnormal products instead of or in addition to the normal ones. For example, **58** and **60** gave, respectively, **59** and **61** as well as the normal aldehyde products. From the nature of the reagents and

from the kind of abnormal products obtained, it is clear that the reactive entity in this reaction is dichlorocarbene CCl$_2$.[483] This is known to be produced by treatment of chloroform with bases (p. 521); it is an electrophilic reagent and is known to give ring expansion of aromatic rings (see **15-64**), accounting for products like **58**. The mechanism of the normal reaction is thus something like this.[484]

[480]For a review, see Wynberg, H.; Meijer, E.W. *Org. React.* **1982**, *28*, 1.

[481]For improved procedures, see Thoer, A.; Denis, G.; Delmas, M.; Gaset, A. *Synth. Commun.* **1988**, *18*, 2095; Cochran, J.C.; Melville, M.G. *Synth. Commun.* **1990**, *20*, 609.

[482]Increased para selectivity has been achieved by the use of polyethylene glycol: Neumann, R.; Sasson, Y. *Synthesis* **1986**, 569.

[483]For a review of carbene methods for introducing formyl and acyl groups into organic molecules see Kulinkovich, O.G. *Russ. Chem. Rev.* **1989**, *58*, 711.

[484]Robinson, E.A. *J. Chem. Soc.* **1961**, 1663; Hine, J.; van der Veen, J.M. *J. Am. Chem. Soc.* **1959**, *81*, 6446. See also, Langlois, B.R. *Tetrahedron Lett.* **1991**, *32*, 3691.

$$\text{(phenolate)} + :CCl_2 \longrightarrow \cdots \longrightarrow \left[\cdots \longleftrightarrow \cdots \right] \longrightarrow \text{Hydrolysis}$$

The formation of **61** in the case of **60** can be explained by attack of some of the CCl_2 ipso to the CH_3 group. Since this position does not contain a hydrogen, normal proton loss cannot take place and the reaction ends when the CCl_2^- moiety acquires a proton.

A method closely related to the Reimer–Tiemann reaction is the *Duff reaction*, in which hexamethylenetetramine $(CH_2)_6N_4$ is used instead of chloroform. This reaction can be applied only to phenols and amines; ortho substitution is generally observed and yields are low. A mechanism[485] has been proposed that involves initial aminoalkylation (**11-22**) to give $ArCH_2NH_2$, followed by dehydrogenation to $ArCH=NH$ and hydrolysis of this to the aldehyde product. When $(CH_2)_6N_4$ is used in conjunction with F_3CCOOH, the reaction can be applied to simple alkylbenzenes; yields are much higher and a high degree of regioselectively para substitution is found.[486] In this case too an imine seems to be an intermediate.

OS **III**, 463; **IV**, 866

$$ArH + Cl_2CHOMe \xrightarrow{\;AlCl_3\;} ArCHO$$

Besides **11-18**, several other formylation methods are known.[487] In one of these, dichloromethyl methyl ether formylates aromatic rings with Friedel–Crafts catalysts.[488] The ArCHClOMe compound is probably an intermediate. Orthoformates have also been used.[489] In another method, aromatic rings are formylated with formyl fluoride HCOF and BF_3.[490] Unlike formyl chloride, formyl fluoride is stable enough for this purpose. This reaction was successful for benzene, alkylbenzenes, PhCl, PhBr, and naphthalene. Phenols can be regioselectively formylated in the ortho position in high yields by treatment with 2 equivalents of paraformaldehyde in aprotic solvents in the presence of $SnCl_4$ and a tertiary amine.[491] Phenols have also been formylated indirectly by conversion to the aryllithium reagent followed by treatment with *N*-formyl piperidine.[492] See also the indirect method mentioned at **11-23**.

[485]Ogata, Y.; Kawasaki, A.; Sugiura, F. *Tetrahedron* **1968**, *24*, 5001.
[486]Smith, W.E. *J. Org. Chem.* **1972**, *37*, 3972.
[487]For methods other than those described here, see Smith, R.A.J.; Manas, A.R.B. *Synthesis* **1984**, 166; Olah, G.A.; Laali, K.; Farooq, O. *J. Org. Chem.* **1985**, *50*, 1483; Nishino, H.; Tsunoda, K.; Kurosawa, K. *Bull. Chem. Soc. Jpn.* **1989**, *62*, 545.
[488]Rieche, A.; Gross, H.; Höft, E. *Chem. Ber.* **1960**, *93*, 88; Lewin, A.H.; Parker, S.R.; Fleming, N.B.; Carroll, F.I. *Org. Prep. Proceed. Int.* **1978**, *10*, 201.
[489]Gross, H.; Rieche, A.; Matthey, G. *Chem. Ber.* **1963**, *96*, 308.
[490]Olah, G.A.; Kuhn, S.J. *J. Am. Chem. Soc.* **1960**, *82*, 2380.
[491]Casiraghi, G.; Casnati, G.; Puglia, G.; Sartori, G.; Terenghi, G. *J. Chem. Soc. Perkin Trans. 1* **1980**, 1862.
[492]Hardcastle, I.R.; Quayle, P.; Ward, E.L.M. *Tetrahedron Lett.* **1994**, *35*, 1747.

OS **V**, 49; **VII**, 162.

Reactions **11-19** and **11-20** are direct carboxylations[493] of aromatic rings.[494]

11-19 Carboxylation With Carbonyl Halides

Carboxylation or **Carboxy-de-hydrogenation**

$$\text{ArH} + \text{COCl}_2 \xrightarrow{\text{AlCl}_3} \text{ArCOOH}$$

Phosgene, in the presence of Friedel–Crafts catalysts, can carboxylate the ring. This process is analogous to **11-17**, but the ArCOCl initially produced hydrolyzes to the carboxylic acid. However, in most cases the reaction does not take this course, but instead the ArCOCl attacks another ring to give a ketone ArCOAr. A number of other reagents have been used to get around this difficulty, among them oxalyl chloride, urea hydrochloride, chloral Cl_3CCHO,[495] carbamoyl chloride H_2NCOCl, and N,N-diethylcarbamoyl chloride.[496] With carbamoyl chloride the reaction is called the *Gatterman amide synthesis* and the product is an amide. Among compounds carboxylated by one or another of these reagents are benzene, alkylbenzenes, and fused ring systems.[497]

Although mechanistically different, other methods are available to convert aromatic compounds to aromatic carboxylic acids. The palladium-catalyzed reaction of aromatic compounds and formic acid leads to benzoic acid derivatives.[498] Diphenyliodonium tetrafluoroborate, $Ph_2I^+BF_4^-$ reacts with CO and In in DMF, with a palladium catalyst, to give benzophenone.[499]

OS **V**, 706; **VII**, 420.

11-20 Carboxylation With Carbon Dioxide: The Kolbe–Schmitt Reaction

Carboxylation or **Carboxy-de-hydrogenation**

[493]For other carboxylation methods, one of which leads to the anhydride, see Sakakibara, T.; Odaira, M. *J. Org. Chem.* **1976**, *41*, 2049; Fujiwara, Y.; Kawata, I.; Kawauchi, T.; Taniguchi, H. *J. Chem. Soc., Chem. Commun.* **1982**, 132.

[494]For a review, see Olah, G.A.; Olah, J.A., in Olah, G.A. *Friedel–Crafts and Related Reactions*, Vol. 3, Wiley, NY, **1964**, pp. 1257–1273.

[495]Menegheli, P.; Rezende, M.C.; Zucco, C. *Synth. Commun.* **1987**, *17*, 457.

[496]Naumov, Yu.A.; Isakova, A.P.; Kost, A.N.; Zakharov, V.P.; Zvolinskii, V.P.; Moiseikina, N.F.; Nikeryasova, S.V. *J. Org. Chem. USSR* **1975**, *11*, 362.

[497]For the use of phosgene to carboxylate phenols, see Sartori, G.; Casnati, G.; Bigi, F.; Bonini, G. *Synthesis* **1988**, 763.

[498]Shibahara, F.; Kinoshita, S.; Nozaki, K. *Org. Lett.* **2004**, *6*, 2437.

[499]Zhou, T.; Chen, Z.-C. *Synth. Commun.* **2002**, *32*, 3431.

Sodium phenoxides can be carboxylated, mostly in the ortho position, by carbon dioxide (the *Kolbe–Schmitt reaction*). The mechanism is not clearly understood, but apparently some kind of a complex is formed between the reactants,[500] making the carbon of the CO_2 more positive and putting it in a good

position to attack the ring. Potassium phenoxide, which is less likely to form such a complex,[501] is chiefly attacked in the para position.[502] Carbon tetrachloride can be used instead of CO_2 under Reimer–Tiemann (**11-18**) conditions.

Sodium or potassium phenoxide can be carboxylated regioselectively in the para position in high yield by treatment with sodium or potassium carbonate and carbon monoxide.[503] [14]C Labeling showed that it is the carbonate carbon that appears in the *p*-hydroxybenzoic acid product.[504] The CO is converted to sodium or potassium formate. Carbon monoxide has also been used to carboxylate aromatic rings with palladium compounds as catalysts.[505] In addition, a palladium-catalyzed reaction has been used directly to prepare acyl fluorides ArH → ArCOF.[506]

An enzymatic carboxylation was reported, in supercritical CO_2 (see p. $$$), in which exposure of pyrrole to *Bacillus megaterium* PYR2910 and $KHCO_3$ gave the potassium salt of pyrrole-2-carboxylic acid.[507]

OS **II**, 557.

11-21 Amidation

N-**Alkylcarbamoyl-de-hydrogenation**

$$\text{ArH + RNCO} \xrightarrow{\text{AlCl}_3} \text{ArCONHR}$$

[500]Hales J.L.; Jones, J.I.; Lindsey, A.S. *J. Chem. Soc.* **1954**, 3145.

[501]There is evidence that, in the complex formed from potassium salts, the bonding is between the aromatic compound and the carbon atom of CO_2: Hirao, I.; Kito, T. *Bull. Chem. Soc. Jpn.* **1973**, *46*, 3470.

[502]Actually, the reaction seems to be more complicated than this. At least part of the potassium *p*-hydroxybenzoate that forms comes from a rearrangement of initially formed potassium salicylate. Sodium salicylate does not rearrange. See Shine, H.J. *Aromatic Rearrangements*, Elsevier, NY, **1967**, pp. 344–348. See also, Ota, K. *Bull. Chem. Soc. Jpn.* **1974**, *47*, 2343.

[503]Yasuhara, Y.; Nogi, T. *J. Org. Chem.* **1968**, *33*, 4512, *Chem. Ind. (London)* **1969**, 77.

[504]Yasuhara, Y.; Nogi, T.; Saishō *Bull. Chem. Soc. Jpn.* **1969**, *42*, 2070.

[505]See Sakakibara, T.; Odaira, Y. *J. Org. Chem.* **1976**, *41*, 2049; Jintoku, T.; Taniguchi, H.; Fujiwara, Y. *Chem. Lett.* **1987**, 1159; Ugo, R.; Chiesa, A. *J. Chem. Soc. Perkin Trans. 1* **1987**, 2625.

[506]Sakakura, T.; Chaisupakitsin, M.; Hayashi, T.; Tanaka, M. *J. Organomet. Chem.* **1987**, *334*, 205.

[507]Matsuda, T.; Ohashi, Y.; Harada, T.; Yanagihara, R.; Nagasawa, T.; Nakamura, K. *Chem. Commun.* **2001**, 2194.

N-Substituted amides can be prepared by direct attack of isocyanates on aromatic rings.[508] The R group may be alkyl or aryl, but if the latter, dimers and trimers are also obtained. Isothiocyanates similarly give thioamides.[509] The reaction has been carried out intramolecularly both with aralkyl isothiocyanates and acyl isothiocyanates.[510] In the latter case, the product is easily hydrolyzable to a dicarboxylic acid; this is a way

of putting a carboxyl group on a ring ortho to one already there (**62** is prepared by treatment of the acyl halide with lead thiocyanate). The reaction gives better yields with substrates of the type ArCH$_2$CONCS, where six-membered rings are formed.

There are interesting transition metal-catalyzed-reactions that lead to aryl amides. The use of POCl$_3$ and DMF, with a palladium catalyst, converts aryl iodides to benzamides.[511] A palladium-catalyzed reaction of aryl halides and formamide leads to benzamide derivatives.[512] Carbonylation is another method that generates amides. When an aryl iodide was treated with a secondary amine and Mo(CO)$_6$, in the presence of 3 equivalents of DBU, 10% Pd(OAc)$_2$, with microwave irradiation at 100°C, the corresponding benzamide was obtained.[513]

OS **V**, 1051; **VI**, 465.

Reactions **11-12–11-23** involve the introduction of a CH$_2$Z group, where Z is halogen, hydroxyl, amino, or alkylthio. They are all Friedel–Crafts reactions of aldehydes and ketones and, with respect to the carbonyl compound, additions to the C=O double bond. They follow mechanisms discussed in Chapter 16.

11-22   Aminoalkylation and Amidoalkylation

Dialkylaminoalkylation or Dialkylamino-de-hydrogenation

[508]Effenberger, F.; Gleiter, R.; Heider, L.; Niess, R. *Chem. Ber.* **1968**, *101*, 502; Piccolo, O.; Filippini, L.; Tinucci, L.; Valoti, E.; Citterio, A. *Tetrahedron* **1986**, *42*, 885.

[509]Jagodziński, T. *Synthesis* **1988**, 717.

[510]Smith, P.A.S.; Kan, R.O. *J. Org. Chem.* **1964**, *29*, 2261.

[511]Hosoi, K.; Nozaki, K.; Hiyama, T. *Org. Lett.* **2002**, *4*, 2849.

[512]Schnyder, A.; Beller, M.; Mehltretter, G.; Nsenda, T.; Studer, M.; Indolese, A.F. *J. Org. Chem.* **2001**, *66*, 4311. See also, Schnyder, A.; Indolese, A.F. *J. Org. Chem.* **2002**, *67*, 594.

[513]Wannberg, J.; Larhed, M. *J. Org. Chem.* **2003**, *68*, 5750.

Phenols, secondary and tertiary aromatic amines,[514] pyrroles, and indoles can be aminomethylated by treatment with formaldehyde and a secondary amine. Other aldehydes have sometimes been employed. Aminoalkylation is a special case of the Mannich reaction (**16-19**). When phenols and other activated aromatic compounds are treated with *N*-hydroxymethylchloroacetamide, *amidomethylation* takes place[515] to

give **63**, which is often hydrolyzed *in situ* to the aminoalkylated product. Other *N*-hydroxyalkyl and *N*-chlorinated compounds have also been used.[374]

OS **I**, 381; **IV**, 626; **V**, 434; **VI**, 965; **VII**, 162.

11-23   Thioalkylation

Alkylthioalkylation or **Alkylthioalkyl-de-hydrogenation**

A methylthiomethyl group can be inserted into the ortho position of phenols by heating with dimethyl sulfoxide and dicyclohexylcarbodiimide (DCC).[516] Other reagents can be used instead of DCC, among them $SOCl_2$,[517] and acetic anhydride.[518] Alternatively, the phenol can be treated with dimethyl sulfide and *N*-chlorosuccinimide, followed by triethylamine.[519] The reaction can be applied to amines (to give *o*-$NH_2C_6H_4CH_2SMe$) by treatment with *t*-BuOCl, Me_2S, and NaOMe in CH_2Cl_2.[520] Aromatic hydrocarbons have been thioalkylated with ethyl α-(chloromethylthio)-acetate $ClCH_2SCH_2COOEt$ (to give $ArCH_2SCH_2CO$-OEt)[521] and with methyl methylsulfinylmethyl sulfide $MeSCH_2SOMe$ or methylthiomethyl *p*-tolyl sulfone $MeSCH_2$-$SO_2C_6H_4Me$ (to give $ArCH_2SMe$),[522] in each case with a Lewis acid catalyst.

OS **VI**, 581, 601.

[514]Miocque, M.; Vierfond, J. *Bull. Soc. Chim. Fr. 1970*, 1896, 1901, 1907.

[515]For a review, see Zaugg, H.E. *Synthesis 1984*, 85.

[516]Burdon, M.G.; Moffatt, J.G. *J. Am. Chem. Soc. 1966, 88*, 5855, *1967, 89*, 4725; Olofson, R.A.; Marino, J.P. *Tetrahedron 1971, 27*, 4195.

[517]Sato, K.; Inoue, S.; Ozawa, K.; Tazaki, M. *J. Chem. Soc. Perkin Trans. 1 1984*, 2715.

[518]Hayashi, Y.; Oda, R. *J. Org. Chem. 1967, 32*, 457; Pettit, G.H.; Brown, T.H. *Can. J. Chem. 1967, 45*, 1306; Claus, P. *Monatsh. Chem. 1968, 99*, 1034.

[519]Gassman, P.G.; Amick, D.R. *J. Am. Chem. Soc. 1978, 100*, 7611.

[520]Gassman, P.G.; Gruetzmacher, G. *J. Am. Chem. Soc. 1973, 95*, 588; Gassman, P.G.; van Bergen, T.J. *J. Am. Chem. Soc. 1973, 95*, 590, 591.

[521]Tamura, Y.; Tsugoshi, T.; Annoura, H.; Ishibashi, H. *Synthesis 1984*, 326.

[522]Torisawa, Y.; Satoh, A.; Ikegami, S. *Tetrahedron Lett. 1988, 29*, 1729.

11-24 Acylation with Nitriles: The Hoesch Reaction

Acylation or **Acyl-de-hydrogenation**

$$ArH + RCN \xrightarrow[\text{ZnCl}_2]{\text{HCl}} ArCOR$$

Friedel–Crafts acylation with nitriles and HCl is called the *Hoesch* or the *Houben–Hoesch reaction*.[523] In most cases, a Lewis acid is necessary; zinc chloride is the most common. The reaction is generally useful only with phenols, phenolic ethers, and some reactive heterocyclic compounds such as pyrrole, but it can be extended to aromatic amines by the use of BCl_3.[524] Acylation in the case of aniline derivatives is regioselectively ortho. Monohydric phenols, however, generally do not give ketones[525] but are attacked at the oxygen to

$$Ar\overset{\displaystyle O}{\diagdown}\underset{\underset{\displaystyle \overset{\oplus}{N}H_2 \ \ \overset{\ominus}{Cl}}{\|}}{C}\diagup R$$

An imino ester

produce imino esters. Many nitriles have been used. Even aryl nitriles give good yields if they are first treated with HCl and $ZnCl_2$ and then the substrate added at 0°C.[526] In fact, this procedure increases yields with any nitrile. If thiocyanates RSCN are used, thiol esters ArCOSR can be obtained. The Gatterman reaction (**11-18**) is a special case of the Hoesch synthesis.

The reaction mechanism is complex and not completely settled.[527] The first stage consists of an attack on the substrate by a species containing the nitrile and HCl (and the Lewis acid, if present) to give an imine salt (**66**). Among the possible attacking species are **64** and **65**. In the second stage, the salts are hydrolyzed to the products, first the iminium salt, and then the ketone. Ketones can also be obtained by treating phenols or phenolic ethers with a nitrile in the presence of F_3CSO_2OH.[528] The mechanism in this case is different.

$$Ar\text{-}H \ + \ R\text{-}C\!\!=\!\!\overset{\oplus}{N}H \quad \overset{\ominus}{Cl} \xrightarrow{\quad H^+ \quad}$$
64

$$Ar\text{-}H \ + \ ZnCl_2(RCN)_2 \ + \ HCl$$
65

$$Ar\diagdown\underset{\underset{\displaystyle \overset{\oplus}{N}H_2 \ \ \overset{\ominus}{Cl}}{\|}}{C}\diagup R \longrightarrow Ar\diagdown\underset{\underset{\displaystyle O}{\|}}{C}\diagup R$$
66

OS **II**, 522.

[523]For a review, see Ruske, W., in Olah, G.A. *Friedel–Crafts and Related Reactions*, Vol. 3, Wiley, NY, *1964*, pp. 383–497.

[524]Sugasawa, T.; Toyoda, T.; Adachi, M.; Sasakura, K. *J. Am. Chem. Soc.* *1978*, *100*, 4842; Sugasawa, T.; Adachi, M.; Sasakura, K.; Kitagawa, A. *J. Org. Chem.* *1979*, *44*, 578.

[525]For an exception, see Toyoda, T.; Sasakura, K.; Sugasawa, T. *J. Org. Chem.* *1981*, *46*, 189.

[526]Zil'berman, E.N.; Rybakova, N.A. *J. Gen. Chem. USSR 1960*, *30*, 1972.

[527]For discussions, see Ruske, W., in Olah, G.A. *Friedel–Crafts and Related Reactions*, Vol. 3, Wiley, NY, *1964*, p. 383; Jeffery, E.A.; Satchell, D.P.N. *J. Chem. Soc. B 1966*, 579.

[528]Amer, M.I.; Booth, B.L.; Noori, G.F.M.; Proença, M.F.J.R.P. *J. Chem. Soc. Perkin Trans. 1 1983*, 1075.

11-25   Cyanation or Cyano-de-hydrogenation

$$\text{ArH} + \text{Cl}_3\text{CCN} \xrightarrow{\text{HCl}} \underset{\overset{\oplus}{\text{NH}_2} \; \overset{\ominus}{\text{Cl}}}{\text{Ar}\diagdown\underset{\|}{\text{C}}\diagup\text{CCl}_3} \xrightarrow{\text{NaOH}} \text{ArCN}$$

Aromatic hydrocarbons (including benzene), phenols, and phenolic ethers can be cyanated with trichloroacetonitrile, BrCN, or mercury fulminate Hg(ONC)_2.[529] In the case of Cl_3CCN, the actual attacking entity is probably $\text{Cl}_3\text{C}-\text{C}=\text{NH}$, formed by addition of a proton to the cyano nitrogen. Secondary aromatic amines ArNHR, as well as phenols, can be cyanated in the ortho position with Cl_3CCN and BCl_3.[530]

It is noted that aryl triflates are converted to the aryl nitrile by treatment with Zn(CN)_2 and a palladium catalyst.[531]

OS **III**, 293.

F. Oxygen Electrophiles

Oxygen electrophiles are very uncommon, since oxygen does not bear a positive charge very well. However, there is one reaction that can be mentioned.

11-26   Hydroxylation or Hydroxy-de-hydrogenation

$$\text{Ar-H} \; + \; \underset{\text{F}_3\text{C}}{\overset{\overset{\text{O}}{\|}}{\text{C}}}\diagdown\underset{\text{O}}{}\diagup\text{OH} \xrightarrow{\text{BF}_3} \text{Ar-OH}$$

There have been only a few reports of direct hydroxylation[532] by an electrophilic process (see, however, **14-5**).[533] In general, poor results are obtained, partly because the introduction of an OH group activates the ring to further attack. Quinone formation is common. However, alkyl-substituted benzenes, such as mesitylene or durene can be hydroxylated in good yield with trifluoroperacetic acid and boron trifluoride.[534] In the case of mesitylene, the product (**67**) is not subject to further attack.

67

[529]Olah, G.A., in Olah, G.A. *Friedel–Crafts and Related Reactions*, Vol. 1, Wiley, NY, *1963*, pp. 119–120.
[530]Adachi, M.; Sugasawa, T. *Synth. Commun.* *1990*, 20, 71.
[531]Kubota, H.; Rice, K.C. *Tetrahedron Lett.* *1998*, 39, 2907.
[532]For a list of hydroxylation reagents, with references, see Larock, R.C. *Comprehensive Organic Transformations*, 2nd ed., Wiley-VCH, NY, *1999*, pp. 977–978.
[533]For reviews of electrophilic hydroxylation, see Jacquesy, J.; Gesson, J.; Jouannetaud, M. *Rev. Chem. Intermed.* *1988*, 9, 1, see pp. 5–10; Haines, A.H. *Methods for the Oxidation of Organic Compounds*, Academic Press, NY, *1985*, pp. 173–176, 347–350.
[534]Hart, H.; Buehler, C.A. *J. Org. Chem.* *1964*, 29, 2397. See also, Hart, H. *Acc. Chem. Res.* *1971*, 4, 337.

In a related procedure, even benzene and substituted benzenes (e.g., PhMe, PhCl, xylenes) can be converted to phenols in good yields with sodium perborate–F_3CSO_2OH.[535] Aromatic amines, N-acyl amines, and phenols were hydroxylated with H_2O_2 in SbF_5—HF.[536] Pyridine and quinoline were converted to their 2-acetoxy derivatives in high yields with acetyl hypofluorite AcOF at $-75°C$.[537]

Another hydroxylation reaction is the *Elbs reaction*.[538] In this method phenols can be oxidized to *p*-diphenols with $K_2S_2O_8$ in alkaline solution.[539] Primary, secondary, or tertiary aromatic amines give predominant or exclusive ortho substitution unless both ortho positions are blocked, in which case para substitution is found. The reaction with amines is called the *Boyland–Sims oxidation*. Yields are low with either phenols or amines, generally <50%. The mechanisms are not clear,[540] but for the Boyland–Sims oxidation there is evidence that the $S_2O_8^{2-}$ ion attacks at the ipso position, and then a migration follows.[541]

Electrolysis of benzene, in the presence of trifluoroacetic acid and triethylamine, leads to a 73% yield of phenol.[542]

G. Metal Electrophiles

Reactions in which a metal replaces the hydrogen of an aromatic ring are considered along with their aliphatic counterparts in Chapter 12 (**12-22** and **12-23**).

HYDROGEN AS THE LEAVING GROUP
IN REARRANGEMENT REACTIONS

In these reactions, a group is detached from a *side chain* and then attacks the ring, but in other aspects they resemble the reactions already treated in this chapter.[543] Since a group moves from one position to another in a molecule, these are rearrangements. In all these reactions, the question arises as to whether the group that cleaves from a given molecule attacks the same molecule or another one, that is is the reaction intramolecular or intermolecular? For intermolecular reactions the mechanism is the same as ordinary aromatic substitution, but for intramolecular

[535]Prakash, G.K.S.; Krass, N.; Wang, Q.; Olah, G.A. *Synlett* **1991**, 39.

[536]Berrier, C.; Carreyre, H.; Jacquesy, J.; Joannetaud, M. *New J. Chem.* **1990**, *14*, 283, and cited references.

[537]Rozen, S.; Hebel, D.; Zamir, D. *J. Am. Chem. Soc.* **1987**, *109*, 3789.

[538]For a review of the Elbs and Boyland–Sims reactions, see Behrman, E.J. *Org. React.* **1988**, *35*, 421.

[539]For a method for the ortho hydroxylation of phenols, see Capdevielle, P.; Maumy, M. *Tetrahedron Lett.* **1982**, *23*, 1573, 1577.

[540]Behrman, E.J. *J. Am. Chem. Soc.* **1967**, *89*, 2424; Ogata, Y.; Akada, T. *Tetrahedron* **1970**, *26*, 5945; Walling, C.; Camaioni, D.M.; Kim, S.S. *J. Am. Chem. Soc.* **1978**, *100*, 4814.

[541]Srinivasan, C.; Perumal, S.; Arumugam, N. *J. Chem. Soc. Perkin Trans. 2* **1985**, 1855.

[542]Fujimoto, K.; Tokuda, Y.; Maekawa, H.; Matsubara, Y.; Mizuno, T.; Nishiguchi, I. *Tetrahedron* **1996**, *52*, 3889; Fujimoto, K.; Maekawa, H.; Tokuda, Y.; Matsubara, Y.; Mizuno, T.; Nishiguchi, I. *SynLett*, **1995**, 661.

[543]For a monograph, see Shine, H.J. *Aromatic Rearrangements*, Elsevier, NY, **1967**. For reviews, see Williams, D.L.H.; Buncel, I.M. *Isot. Org. Chem.* **1980**, *5*, 147; Williams, D.L.H., in Bamford, C.H.; Tipper, C.F.H. *Comprehensive Chemical Kinetics*, Vol. 13, Elsevier, NY, **1972**, pp. 433–486.

cases the migrating group could never be completely free, or else it would be able to attack another molecule. Since the migrating species in intramolecular rearrangements is thus likely to remain near the atom from which it cleaved, it has been suggested that intramolecular reactions are more likely to lead to ortho products than are the intermolecular type. This characteristic has been used, among others, to help decide whether a given rearrangement is inter- or intramolecular, though there is evidence that at least in some cases, an intermolecular mechanism can still result in a high degree of ortho migration.[544]

The Claisen (**18-33**) and benzidine (**18-36**) rearrangements, which superficially resemble those in this section, have different mechanisms and are treated in Chapter 18.

A. Groups Cleaving from Oxygen

11-27   The Fries Rearrangement

1/*C*-Hydro,5/*O*-acyl-interchange[545]

Phenolic esters can be rearranged by heating with Friedel–Crafts catalysts in a synthetically useful reaction known as the *Fries rearrangement.*[546] Both *o*- and *p*-acylphenols can be produced, and it is often possible to select conditions so that either one predominates. The ortho/para ratio is dependent on the temperature, solvent, and amount of catalyst used. Exceptions are known, but low temperatures generally favor the para product and high temperatures the ortho product. The R group may be aliphatic or aromatic. Any meta-directing substituent on the ring interferes with the reactions, as might be expected for a Friedel–Crafts process. In the case of aryl benzoates treated with F_3CSO_2OH, the Fries rearrangement was shown to be reversible and an equilibrium was established.[547] Transition-metal-catalyzed Fries rearrangements have been reported.[548]

[544]See Dawson, I.M.; Hart, L.S.; Littler, J.S. *J. Chem. Soc. Perkin Trans. 2* **1985**, 1601.

[545]This is the name for the para migration. For the ortho migration, the name is 1/*C*-hydro,3/*O*-acyl-interchange.

[546]For reviews, see Shine, H.J. *Aromatic Rearrangements*, Elsevier, NY, **1967**, pp. 72–82, 365–368; Gerecs, A., in Olah, G.A. *Friedel–Crafts and Related Reactions*, Vol. 3, Wiley, NY, **1964**, pp. 499–533. For a list of references, see Larock, R.C. *Comprehensive Organic Transformations*, 2nd ed., Wiley-VCH, NY, **1999**, p. 1310.

[547]Effenberger, F.; Gutmann, R. *Chem. Ber.* **1982**, *115*, 1089.

[548]With **Sc(OTf)₃**, see Kobayashi, S.; Moriwaki, M.; Hachiya, I. *Tetrahedron Lett.* **1996**, *37*, 4183; with ZrCl₄ see Harrowven, D.C.; Dainty, R.F. *Tetrahedron Lett.* **1996**, *37*, 7659; with **Hf(OTf)₄** see Kobayashi, S.; Moriwaki, M.; Hachiya, I. *Tetrahedron Lett.* **1996**, *37*, 2053. Also see Kobayashi, S.; Moriwaki, M.; Hachiya, I. *J. Chem. Soc., Chem. Commun.* **1995**, 1527.

The exact mechanism has still not been completely worked out.[549] Opinions have been expressed that it is completely intermolecular,[550] completely intramolecular,[551] and partially inter- and intramolecular.[552] One way to decide between inter- and intramolecular processes is to run the reaction of the phenolic ester in the presence of another aromatic compound, say, toluene. If some of the toluene is acylated, the reaction must be, at least in part, intermolecular. If the toluene is not acylated, the presumption is that the reaction is intramolecular, though this is not certain, for it may be that the toluene is not attacked because it is less active than the other. A number of such experiments (called *crossover experiments*) have been carried out; sometimes crossover products have been found and sometimes not. As in **11-17**, an initial complex (**68**) is formed between the substrate and the catalyst, so that a catalyst/substrate molar ratio of at least 1:1 is required. In the presence of aluminum chloride, the Fries rearrangement can be induced with microwave irradiation.[553] Simply heating phenyl acetate with microwave irradiation gives the Fries rearrangement.[554] The Fries rearrangement has been carried out in ionic melts.[555]

$$\underset{\text{ArO}}{\overset{\overset{\oplus}{O}\,\cdot\,\overset{\ominus}{AlCl_3}}{\underset{\underset{\textbf{68}}{}}{\overset{\|}{C}}}}R$$

The Fries rearrangement can also be carried out with UV light, in the absence of a catalyst.[556] This reaction, called the *photo-Fries rearrangement*,[557] is predominantly an intramolecular free-radical process. Both ortho and para migration are observed.[558] Unlike the Lewis acid-catalyzed Fries rearrangement, the photo-Fries reaction can be accomplished, though often in low yields, when meta-directing groups are on the ring. The available evidence strongly suggests the following

[549]For the mechanism in polyphosphoric acid, see Sharghi, H.; Eshghi, H. *Bull. Chem. Soc. Jpn.* *1993*, *66*, 135.

[550]Martin, R.; Gavard, J.; Delfly, M.; Demerseman, P.; Tromelin, A. *Bull. Soc. Chim. Fr.* *1986*, 659 and cited references.

[551]Ogata, Y.; Tabuchi, H. *Tetrahedron* *1964*, *20*, 1661.

[552]Munavilli, S. *Chem. Ind. (London)* *1972*, 293; Warshawsky, A.; Kalir, R.; Patchornik, A. *J. Am. Chem. Soc.* *1978*, *100*, 4544; Dawson, I.M.; Hart, L.S.; Littler, J.S. *J. Chem. Soc. Perkin Trans. 2* *1985*, 1601.

[553]Khadilkar, B.M.; Madyar, V.R. *Synth. Commun.* *1999*, *29*, 1195.

[554]Paul, S.; Gupta, M. *Synthesis* *2004*, 1789.

[555]Harjani, J.R.; Nara, S.J.; Salunkhe, M.M. *Tetrahedron Lett.* *2001*, *42*, 1979.

[556]Kobsa, H. *J. Org. Chem.* *1962*, 27, 2293; Anderson, J.C.; Reese, C.B. *J. Chem. Soc.* *1963*, 1781; Finnegan, R.A.; Matice, J.J. *Tetrahedron* *1965*, *21*, 1015.

[557]For reviews, see Belluš, D. *Adv. Photochem.* *1971*, *8*, 109; Belluš, D.; Hrdlovič, P. *Chem. Rev.* *1967*, *67*, 599; Stenberg, V.I. *Org. Photochem.* *1967*, *1*, 127. See Cui, C.; Wang, X.; Weiss, R.G. *J. Org. Chem.* *1996*, *61*, 1962.

[558]The migration can be made almost entirely ortho by cyclodextrin encapsulation (see p. 129): Syamala, M.S.; Rao, B.N.; Ramamurthy, V. *Tetrahedron* *1988*, *44*, 7234. See also, Veglia, A.V.; Sanchez, A.M.; de Rossi, R.H. *J. Org. Chem.* *1990*, *55*, 4083.

mechanism involving formation of the excited state ester followed by dissociation to a radical pair[559] for the photo-Fries rearrangement[560] (illustrated for para attack).

The phenol ArOH is always a side product, resulting from some ArO• that leaks from the solvent cage and abstracts a hydrogen atom from a neighboring molecule. When the reaction was performed on phenyl acetate in the gas phase, where there are no solvent molecules to form a cage (but in the presence of isobutane as a source of abstractable hydrogens), phenol was the chief product and virtually no o- or p-hydroxyacetophenone was found.[561] Other evidence[562] for the mechanism is that CIDNP has been observed during the course of the reaction[563] and that the ArO•radical has been detected by flash photolysis[564] and by nanosecond time-resolved Raman spectroscopy.[565]

Treatment of O-arylsulfonate esters with $AlCl_3-ZnCl_2$, on silica with microwave irradiation, leads to 2-sulfonyl phenols in a thia-Fries rearrangement.[566] A similar reaction was reported with O-arylsulfonamides.[567]

OS **II**, 543; **III**, 280, 282.

B. Groups Cleaving from Nitrogen[568]

It has been shown that $PhNH_2D$ rearranges to o- and p-deuterioaniline.[569] The migration of OH, formally similar to reactions **11-28–11-32**, is a nucleophilic substitution and is treated in Chapter 13 (**13-32**).

[559]Proposed by Kobsa, H. *J. Org. Chem.* **1962**, *27*, 2293.
[560]It has been suggested that a second mechanism, involving a four-center transition state, is also possible: Bellus, D.; Schaffner, K.; Hoigné, J. *Helv. Chim. Acta* **1968**, *51*, 1980; Sander, M.R.; Hedaya, E.; Trecker, D.J. *J. Am. Chem. Soc.* **1968**, *90*, 7249; Belluš, D. *Adv. Photochem.* **1971**, *8*, 109.
[561]Meyer, J.W.; Hammond, G.S. *J. Am. Chem. Soc.* **1972**, *94*, 2219.
[562]For evidence from isotope effect studies, see Shine, H.J.; Subotkowski, W. *J. Org. Chem.* **1987**, *52*, 3815.
[563]Adam, W.; Arce de Sanabia, J.; Fischer, H. *J. Org. Chem.* **1973**, *38*, 2571; Adam, W. *J. Chem. Soc., Chem. Commun.* **1974**, 289.
[564]Kalmus, C.E.; Hercules D.M. *J. Am. Chem. Soc.* **1974**, *96*, 449.
[565]Beck, S.M.; Brus, L.E. *J. Am. Chem. Soc.* **1982**, *104*, 1805.
[566]Moghaddam, F.M.; Dakamin, M.G. *Tetrahedron Lett.* **2000**, *41*, 3479.
[567]Benson, G.A.; Maughan, P.J.; Shelly, D.P.; Spillane, W.J. *Tetrahedron Lett.* **2001**, *42*, 8729.
[568]For a review, see Stevens, T.S.; Watts, W.E. *Selected Molecular Rearrangements*, Van Nostrand-Reinhold, Princeton, NJ **1973**, pp. 192–199.
[569]Okazaki, N.; Okumura, A. *Bull. Chem. Soc. Jpn.* **1961**, *34*, 989.

11-28 Migration of the Nitro Group

1/C-Hydro,3/N-nitro-interchange

N-Nitro aromatic amines rearrange on treatment with acids to *o*- and *p*-nitroamines with the ortho compounds predominating.[570] Aside from this indication of an intramolecular process, there is also the fact that virtually no meta isomer is produced in this reaction,[571] although direct nitration of an aromatic amine generally gives a fair amount of meta product. Thus a mechanism in which NO_2^+ is dissociated from the ring, and then attacks another molecule must be ruled out. Further results indicating an intramolecular process include the observation that rearrangement of several substrates in the presence of $K^{15}NO_3$ gave products containing no ^{15}N,[572] and that rearrangement of a mixture of $PhNH^{15}NO_2$ and unlabeled *p*-$MeC_6H_4NHNO_2$ gave 2-nitro-4-methylaniline containing no ^{15}N.[573] On the other hand, rearrangement of **69** in the presence of

unlabeled $PhNMeNO_2$ gave labeled **70**, which did not arise by displacement of F.[574] The R group may be hydrogen or alkyl. Two principal mechanisms have been suggested, one involving cyclic attack by the oxygen of the nitro group at the ortho position before the group cleaves,[575] and the other involving a cleavage into a

[570]For reviews, see Williams, D.L.H., in Patai, S. *The Chemistry of Functional Groups, Supplement F*, pt. 1, Wiley, NY, *1982*, pp. 127–153; White, W.N. *Mech. Mol. Migr. 1971*, *3*, 109–143; Shine, H.J. *Aromatic Rearrangements*, Elsevier, NY, *1967*, pp. 235–249.

[571]Hughes, E.D.; Jones, G.T. *J. Chem. Soc. 1950*, 2678.

[572]Brownstein, S.; Bunton, C.A.; Hughes, E.D. *J. Chem. Soc. 1958*, 4354; Banthorpe, D.V.; Thomas, J.A.; Williams, D.L.H. *J. Chem. Soc. 1965*, 6135.

[573]Geller, B.A.; Dubrova, L.N. *J. Gen. Chem. USSR 1960*, *30*, 2627.

[574]White, W.N.; Golden, J.T. *J. Org. Chem. 1970*, *35*, 2759.

[575]Banthorpe, D.V.; Thomas, J.A. *J. Chem. Soc. 1965*, 7149, 7158. Also see, Brownstein, S.; Bunton, C.A.; Hughes, E.D. *J. Chem. Soc. 1958*, 4354; Banthorpe, D.V.; Thomas, J.A.; Williams, D.L.H. *J. Chem. Soc. 1965*, 6135.

radical and a radical ion held together in a solvent cage.[576] Among the evidence for the latter view[577] are

Solvent cage

the effects of substituents on the rate of the reaction,[578] ^{15}N and ^{14}C kinetic isotope effects that show non-concertedness,[579] and the fact that both *N*-methylaniline and nitrous acid are produced in sizable and comparable amounts in addition to the normal products *o*- and *p*-nitro-*N*-methylaniline.[580] These side products are formed when the radicals escape from the solvent cage.

11-29   Migration of the Nitroso Group: The Fischer–Hepp Rearrangement

1/*C*-Hydro-5/*N*-nitroso-interchange

The migration of a nitroso group, formally similar to **11-28**, is important because *p*-nitroso secondary aromatic amines cannot generally be prepared by direct *C*-nitrosation of secondary aromatic amines (see **12-50**). The reaction, known as the *Fischer–Hepp rearrangement*,[581] is brought about by treatment of *N*-nitroso secondary aromatic amines with HCl. Other acids give poor or no results. In benzene systems the para product is usually formed exclusively.[582] The mechanism of the rearrangement is not completely understood. The fact that the reaction takes place in a large excess of urea[583] shows that it is intramolecular[584] since, if NO^+, NOCl,

[576]White, W.N.; White, H.S.; Fentiman, A. *J. Org. Chem.* **1976**, *41*, 3166.

[577]For additional evidence, see White, W.N.; Klink, J.R. *J. Org. Chem.* **1977**, *42*, 166; Ridd, J.H.; Sandall, J.P.B. *J. Chem. Soc., Chem. Commun.* **1982**, 261.

[578]White, W.N.; Klink, J.R. *J. Org. Chem.* **1970**, *35*, 965.

[579]Shine, H.J.; Zygmunt, J.; Brownawell, M.L.; San Filippo, Jr., J. *J. Am. Chem. Soc.* **1984**, *106*, 3610.

[580]White, W.N.; White, H.S. *J. Org. Chem.* **1970**, *35*, 1803.

[581]For reviews, see Williams, D.L.H. *Nitrosation*, Cambridge University Press, Cambridge, **1988**, pp. 113–128; Williams, D.L.H., in Patai, S. *The Chemistry of Functional Groups, Supplement F*, pt. 1, Wiley, NY, **1982**, Shine, H.J. *Aromatic Rearrangements*, Elsevier, NY, **1967**, pp. 231–235.

[582]For a report of formation of 15% ortho product in the case of *N,N*-diaryl-*N*-nitroso amides, see Titova, S.P.; Arinich, A.K.; Gorelik, M.V. *J. Org. Chem. USSR* **1986**, *22*, 1407.

[583]Aslapovskaya, T.I.; Belyaev, E.Yu.; Kumarev, V.P.; Porai-Koshits, B.A. *Org. React. USSR* **1968**, *5*, 189; Morgan, T.D.B.; Williams, D.L.H. *J. Chem. Soc. Perkin Trans. 2* **1972**, 74.

[584]See also, Belyaev, E.Yu.; Nikulicheva, T.I. *Org. React. USSR* **1971**, *7*, 165; Williams, D.L.H. *Tetrahedron* **1975**, *31*, 1343; *J. Chem. Soc. Perkin Trans. 2* **1982**, 801.

or some similar species were free in the solution, it would be captured by the urea, preventing the rearrangement.

11-30 Migration of an Arylazo Group

1/C-Hydro-5/N-arylazo-interchange

Rearrangement of aryl triazenes can be used to prepare azo derivatives of primary and secondary aromatic amines.[585] These are first diazotized at the amino group (see **11-4**) to give triazenes, which are then rearranged by treatment with acid. The rearrangement always gives the para isomer, unless that position is occupied.

11-31 Migration of Halogen: The Orton Rearrangement

1/C-Hydro-5/N-halo-interchange

Migration of a halogen from a nitrogen side chain to the ring by treatment with HCl is called the *Orton rearrangement*.[586] The main product is the para isomer, though some ortho product may also be formed. The reaction has been carried out with N-chloro- and N-bromoamines and less often with N-iodo compounds. The amine must be acylated, except that $PhNCl_2$ gives 2,4-dichloroaniline. The reaction is usually performed in water or acetic acid. There is considerable evidence (cross-halogenation, labeling, etc.) that this is an intermolecular process.[587] First, the HCl reacts with the starting material to give $ArNHCOCH_3$ and Cl_2; then the chlorine halogenates the ring as in **11-10**. Among the evidence is that chlorine has been isolated from the reaction mixture. The Orton rearrangement can also

[585]For a review, see Shine, H.J. *Aromatic Rearrangements*, Elsevier, NY, *1967*, pp. 212–221.
[586]For reviews, see Shine, H.J. *Aromatic Rearrangements*, Elsevier, NY, *1967*, pp. 221–230, 362–364: Bieron, J.F.; Dinan, F.J., in Zabicky, J. *The Chemistry of Amides*, Wiley, NY, *1970*, pp. 263–269.
[587]The reaction has been found to be intramolecular in aprotic solvents: Golding, P.D.; Reddy, S.; Scott, J.M.W.; White, V.A.; Winter, J.G. *Can. J. Chem. 1981*, *59*, 839.

be brought about photochemically[588] and by heating in the presence of benzoyl peroxide.[589] These are free-radical processes.

11-32 Migration of an Alkyl Group[590]

1/*C*-Hydro-5/*N*-alkyl-interchange

When HCl salts of arylalkylamines are heated at ~200–300°C, migration occurs in what is called the *Hofmann–Martius reaction*. It is an intermolecular reaction, since crossing is found. For example, methylanilinium bromide gave not only the normal products *o*- and *p*-toluidine but also aniline and di- and trimethylanilines.[591] As would be expected for an intermolecular process, there is isomerization when R is primary.

With primary R, the reaction probably goes through the alkyl halide formed initially in an S_N2 reaction:

$$\overset{\oplus}{R}NH_2Ar + Cl^- \longrightarrow RCl + ArNH_2$$

Evidence for this view is that alkyl halides have been isolated from the reaction mixture and that Br^-, Cl^-, and I^- gave different ortho/para ratios, which indicates that the halogen is involved in the reaction.[591] Further evidence is that the alkyl halides isolated are not rearranged (as would be expected if they are formed by an S_N2 mechanism), even though the alkyl groups in the ring are rearranged. Once the alkyl halide is formed, it reacts with the substrate by a normal Friedel–Crafts alkylation process (**11-11**), accounting for the rearrangement. When R is secondary or tertiary, carbocations may be directly formed so that the reaction does not go through the alkyl halides.[592]

It is also possible to carry out the reaction by heating the amine (not the salt) at a temperature between 200 and 350°C with a metal halide, such as $CoCl_2$, $CdCl_2$, or $ZnCl_2$. When this is done, the reaction is called the *Reilly–Hickinbottom rearrangement*. Primary R groups larger than ethyl give both rearranged and unrearranged products.[593] The reaction is not generally useful for secondary and tertiary R groups, which are usually cleaved to alkenes under these conditions.

---

[588]For example, see Hodges, F.W. *J. Chem. Soc.* **1933**, 240.

[589]For example, Ayad, K.N.; Beard, C.; Garwood, R.F.; Hickinbottom, W.J. *J. Chem. Soc.* **1957**, 2981; Coulson, J.; Williams, G.H.; Johnston, K.M. *J. Chem. Soc. B* **1967**, 174.

[590]For reviews, see Grillot, G.F. *Mech. Mol. Migr.* **1971**, *3* 237; Shine, H.J. *Aromatic Rearrangements*, Elsevier, NY, **1967**, pp. 249–257.

[591]Ogata, Y.; Tabuchi, H.; Yoshida, K. *Tetrahedron* **1964**, *20*, 2717.

[592]Hart, H.; Kosak, J.R. *J. Org. Chem.* **1962**, *27*, 116.

[593]For example, see Birchal, J.M.; Clark, M.T.; Goldwhite, H.; Thorpe, D.H. *J. Chem. Soc. Perkin Trans. 1* **1972**, 2579.

When acylated arylamines are photolyzed, migration of an acyl group takes place[594] in a process that resembles the photo-Fries reaction (**11-27**).

OTHER LEAVING GROUPS

Three types of reactions are considered in this section.

1. Reactions in which hydrogen replaces another leaving group:

$$ArX + H^+ \longrightarrow ArH$$

2. Reactions in which an electrophile other than hydrogen replaces another leaving group:

$$ArX + Y^+ \longrightarrow ArY$$

3. Reactions in which a group (other than hydrogen) migrates from one position in a ring to another. Such migrations can be either inter- or intramolecular:

The three types are not treated separately, but reactions are classified by leaving group.

A. Carbon Leaving Groups

11-33 Reversal of Friedel–Crafts Alkylation

Hydro-de-alkylation or **Dealkylation**

$$ArR + H^+ \xrightarrow{AlCl_3} ArH$$

Alkyl groups can be cleaved from aromatic rings by treatment with proton and/ or Lewis acids. Tertiary R groups are the most easily cleaved; because this is true, the *tert*-butyl group is occasionally introduced into a ring, used to direct another

[594]For examples see Elad, D.; Rao, D.V.; Stenberg, V.I. *J. Org. Chem.* **1965**, *30*, 3252; Shizuka, H.; Tanaka, I. *Bull. Chem. Soc. Jpn.* **1968**, *41*, 2343; **1969**, *42*, 909; Fischer, M. *Tetrahedron Lett.* **1968**, 4295; Hageman, H.J. *Recl. Trav. Chim. Pays-Bas* **1972**, *91*, 1447; Chênevert, R.; Plante, R. *Can. J. Chem.* **1983**, *61*, 1092; Abdel-Malik, M.M.; de Mayo, P. *Can. J. Chem.* **1984**, *62*, 1275; Nassetta, M.; de Rossi, R.H.; Cosa, J.J. *Can. J. Chem.* **1988**, *66*, 2794.

group, and then removed.[595] For example, 4-*tert*-butyltoluene (**71**) reacted with benzoyl chloride and $AlCl_3$ to give the acylated product, and subsequent treatment with $AlCl_3$ led to loss of the *tert*-butyl group to give **72**.[596]

Secondary R groups are harder to cleave, and primary R harder still. Because of this reaction, care must be taken when using Friedel–Crafts catalysts (Lewis or proton acids) on aromatic compounds containing alkyl groups. True cleavage, in which the R becomes an alkene, occurs only at high temperatures, >400°C.[597] At ordinary temperatures, the R group attacks another ring, so that the bulk of the product may be dealkylated, but there is a residue of heavily alkylated material. The isomerization reaction, in which a group migrates from one position in a ring to another or to a different ring, is therefore more important than true cleavage. In these reactions, the meta isomer is generally the most favored product among the dialkylbenzenes; and the 1,3,5 product the most favored among the trialkylbenzenes, because they have the highest thermodynamic stabilities. Alkyl migrations can be inter- or intramolecular, depending on the conditions and on the R group. The following experiments can be cited: Ethylbenzene treated with HF and BF_3 gave, almost completely, benzene and diethylbenzenes[598] (entirely intermolecular); propylbenzene labeled in the β position gave benzene, propylbenzene, and di- and tripropylbenzenes, but the propylbenzene recovered was partly labeled in the a position and not at all in the γ position[599] (both intra- and intermolecular); *o*-xylene treated with HBr and $AlBr_3$ gave a mixture of *o*- and *m*-, but no *p*-xylene, while *p*-xylene gave *p*- and *m*-, but no *o*-xylene, and no trimethyl compounds could be isolated in these experiments[600] (exclusively intramolecular rearrangement). Apparently, methyl groups migrate only intramolecularly, while other groups may follow either path.[601]

[595]For reviews of such reactions, where the blocking group is *tert*-butyl, benzyl, or a halogen, see Tashiro, M. *Synthesis* **1979**, 921; Tashiro, M.; Fukata, G. *Org. Prep. Proced. Int.* **1976**, *8*, 51.
[596]Hofman, P.S.; Reiding, D.J.; Nauta, W.T. *Recl. Trav. Chim. Pays-Bas* **1960**, *79*, 790.
[597]Olah, G.A., in Olah, G.A. *Friedel–Crafts and Related Reactions*, Vol. 1, Wiley, NY, **1963**, pp. 36–38.
[598]McCaulay, D.A.; Lien, A.P. *J. Am. Chem. Soc.* **1953**, *75*, 2407. For similar results, see Roberts, R.M.; Roengsumran, S. *J. Org. Chem.* **1981**, *46*, 3689; Bakoss, H.J.; Roberts, R.M.G.; Sadri, A.R. *J. Org. Chem.* **1982**, *47*, 4053.
[599]Roberts, R.M.G.; Douglass, J.E. *J. Org. Chem.* **1963**, *28*, 1225.
[600]Brown, H.C.; Jungk, H. *J. Am. Chem. Soc.* **1955**, *77*, 5579; Allen, R.H.; Yats, L.D. *J. Am. Chem. Soc.* **1959**, *81*, 5289.
[601]Allen, R.H. *J. Am. Chem. Soc.* **1960**, *82*, 4856.

The mechanism[602] of intermolecular rearrangement can involve free alkyl cations, but there is much evidence to show that this is not necessarily the case. For example, many of them occur without rearrangement within the alkyl group. The following mechanism has been proposed for intermolecular rearrangement without the involvement of carbocations that are separated from the ring.[603]

Evidence for this mechanism is that optically active $PhCHDCH_3$ labeled in the ring with ^{14}C and treated with $GaBr_3$ in the presence of benzene gave ethylbenzene containing no deuterium and two deuterium atoms and that the rate of loss of radioactivity was about equal to the rate of loss of optical activity.[603] The mechanism of intramolecular rearrangement is not very clear. 1,2-shifts of this kind have been proposed:[604]

There is evidence from ^{14}C labeling that intramolecular migration occurs only through 1,2-shifts.[605] Any 1,3 or 1,4 migration takes place by a series of two or more 1,2-shifts.

Phenyl groups have also been found to migrate. Thus o-terphenyl, heated with $AlCl_3-H_2O$, gave a mixture containing 7% o-, 70% m-, and 23% p-terphenyl.[606] Alkyl groups have also been replaced by groups other than hydrogen (e.g., nitro groups).

Unlike alkylation, *Friedel–Crafts acylation* has been generally considered to be irreversible, but a number of instances of electrofugal acyl groups have been reported,[607] especially where there are two ortho substituents, for example the

---

[602]For a review of the mechanism of this and closely related reactions, see Shine, H.J. *Aromatic Rearrangements*, Elsevier, NY, *1967*, pp. 1–55.

[603]Streitwieser, Jr., A.; Reif, L. *J. Am. Chem. Soc.* *1964*, 86, 1988.

[604]Olah, G.A.; Meyer, M.W.; Overchuk, N.A. *J. Org. Chem.* *1964*, 29, 2313.

[605]See, for example, Steinberg, H.; Sixma, F.L.J. *Recl. Trav. Chim. Pays-Bas* *1962*, 81, 185; Koptyug, V.A.; Isaev, I.S.; Vorozhtsov, Jr., N.N. *Doklad. Akad. Nauk SSSR*, *1963*, 149, 100.

[606]Olah, G.A.; Meyer, M.W. *J. Org. Chem.* *1962*, 27, 3682.

[607]For some other examples see Agranat, I.; Bentor, Y.; Shih, Y. *J. Am. Chem. Soc.* *1977*, 99, 7068; Bokova, A.I.; Buchina, I.K. *J. Org. Chem. USSR* *1984*, 20, 1199; Benedikt, G.M.; Traynor, L. *Tetrahedron Lett.* *1987*, 28, 763; Gore, P.H.; Moonga, B.S.; Short, E.L. *J. Chem. Soc. Perkin Trans. 2* *1988*, 485; Keumi, T.; Morita, T.; Ozawa, Y.; Kitajima, H. *Bull. Chem. Soc. Jpn.* *1989*, 62, 599; Giordano, C.; Villa, M.; Annunziata, R. *Synth. Commun.* *1990*, 20, 383.

hydro-de-benzoylation of **73**.[608]

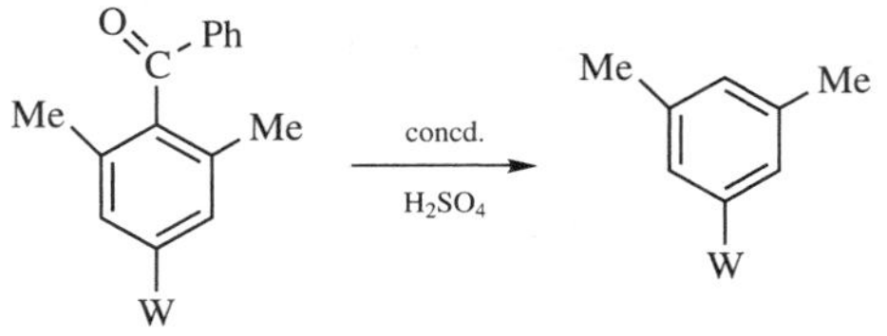

73

OS **V**, 332. Also see OS **III**, 282, 653; **V**, 598.

11-34 Decarbonylation of Aromatic Aldehydes

Hydro-de-formylation or **Deformylation**

$$ArCHO \xrightarrow{H_2SO_4} ArH + CO$$

The decarbonylation of aromatic aldehydes with sulfuric acid[609] is the reverse of the *Gatterman–Koch reaction* (**11-18**). It has been carried out with trialkyl- and trialkoxybenzaldehydes. The reaction takes place by the ordinary arenium ion mechanism: the attacking species is H^+ and the leaving group is HCO^+, which can lose a proton to give CO or combine with OH^- from the water solvent to give formic acid.[610] Aromatic aldehydes have also been decarbonylated with basic catalysts.[611] When basic catalysts are used, the mechanism is probably similar to the S_E1 process of **11-35** (see also **14-32**).

11-35 Decarboxylation of Aromatic Acids

Hydro-de-carboxylation or **Decarboxylation**

$$ArCOOH \xrightarrow[\text{quinoline}]{Cu} ArH + CO_2$$

The decarboxylation of aromatic acids is most often carried out by heating with copper and quinoline. However, two other methods can be used with certain substrates. In one method the salt of the acid ($ArCOO^-$) is heated, and in the other the carboxylic acid is heated with a strong acid, often sulfuric. The latter method is accelerated by the presence of electron-donating groups in ortho and para positions

---

[608]Al-Ka'bi, J.; Farooqi, J.A.; Gore, P.H.; Moonga, B.S.; Waters, D.N. *J. Chem. Res. (S)* **1989**, 80.

[609]For reviews of the mechanism, see Taylor, R. in Bamford, C.H.; Tipper, C.F.H. *Comprehensive Chemical Kinetics*, Vol. 13, Elsevier, NY, **1972**, pp. 316–323; Schubert, W.M.; Kintner, R.R., in Patai, S. *The Chemistry of the Carbonyl Group*, Vol. 1, Wiley, NY, **1966**, pp. 695–760.

[610]Burkett, H.; Schubert, W.M.; Schultz, F.; Murphy, R.B.; Talbott, R. *J. Am. Chem. Soc.* **1959**, *81*, 3923.

[611]Bunnett, J.F.; Miles J.H.; Nahabedian, K.V. *J. Am. Chem. Soc.* **1961**, *83*, 2512; Forbes, E.J.; Gregory, M.J. *J. Chem. Soc. B* **1968**, 205.

and by the steric effect of groups in the ortho positions; in benzene systems it is generally limited to substrates that contain such groups. In this method, decarboxylation takes place by the arenium ion mechanism,[612] with

$$ArCOOH \xrightarrow{H^+} \underset{H}{Ar^{\oplus}{\diagdown}}{}^{COOH} \xrightarrow{-H^+} \underset{H}{Ar^{\oplus}{\diagdown}}{}^{COO^-} \xrightarrow{-CO_2} ArH + CO_2$$

H^+ as the electrophile and CO_2 as the leaving group.[613] Evidently, the order of electrofugal ability is $CO_2 > H^+ > COOH^+$, so that it is necessary, at least in most cases, for the COOH to lose a proton before it can cleave.

When carboxylate *ions* are decarboxylated, the mechanism is entirely different, being of the S_E1 type. Evidence for this mechanism is that the reaction is first order and that electron-withdrawing groups, which would stabilize a carbanion, facilitate the reaction.[614]

Step 1

Step 2

Despite its synthetic importance, the mechanism of the copper–quinoline method has been studied very little, but it has been shown that the actual catalyst is cuprous ion.[615] In fact, the reaction proceeds much faster if the acid is heated in quinoline with cuprous oxide instead of copper, provided that atmospheric oxygen is rigorously excluded. A mechanism has been suggested in which it is the cuprous salt of the acid that actually undergoes the decarboxylation.[615] It has been shown that cuprous salts of aromatic acids are easily decarboxylated by heating in quinoline[616] and that arylcopper compounds are intermediates that can be isolated in some cases.[617] Metallic silver has been used in place of copper, with higher yields.[618]

[612]For a review, see Taylor, R., in Bamford, C.H.; Tipper, C.F.H. *Comprehensive Chemical Kinetics*, Vol. 13, Elsevier, NY, *1972*, pp. 303–316. For a review of isotope effect studies of this reaction, see Willi, A.V. *Isot. Org. Chem.* *1977*, *3*, 257.

[613]See, for example, Los, J.M.; Rekker, R.F.; Tonsbeeck, C.H.T. *Recl. Trav. Chim. Pays-Bas* *1967*, *86*, 622; Huang, H.H.; Long, F.A. *J. Am. Chem. Soc.* *1969*, *91*, 2872; Willi, A.V.; Cho, M.H.; Won, C.M. *Helv. Chim. Acta* *1970*, *53*, 663.

[614]See, for example, Segura, P.; Bunnett, J.F.; Villanova, L. *J. Org. Chem.* *1985*, *50*, 1041.

[615]Cohen, T.; Schambach, R.A. *J. Am. Chem. Soc.* *1970*, *92*, 3189. See also, Aalten, H.L.; van Koten, G.; Tromp, J.; Stam, C.H.; Goubitz, K.; Mak, A.N.S. *Recl. Trav. Chim. Pays-Bas* *1989*, *108*, 295.

[616]Cairncross, A.; Roland, J.R.; Henderson, R.M.; Sheppard, W.A. *J. Am. Chem. Soc.* *1970*, *92*, 3187; Cohen, T.; Berninger, R.W.; Wood, J.T. *J. Org. Chem.* *1978*, *43*, 37.

[617]For example, see Ibne-Rasa, K.M. *J. Am. Chem. Soc.* *1962*, *84*, 4962; Tedder, J.M.; Theaker, G. *J. Chem. Soc.* *1959*, 257.

[618]Chodowska-Palicka, J.; Nilsson, M. *Acta Chem. Scand.* *1970*, *24*, 3353.

In certain cases, the carboxyl group can be replaced by electrophiles other than hydrogen, for example NO,[618] I,[619] Br,[620] or Hg.[621]

Rearrangements are also known to take place. For example, when the phthalate ion is heated with a catalytic amount of cadmium, the terphthalate ion (**74**) is produced:[622]

Phthalate ion   **74**

In a similar process, potassium benzoate heated with cadmium salts disproportionates to benzene and **74**. The term *Henkel reaction* (named for the company that patented the process) is used for these rearrangements.[623] An S_E1 mechanism has been suggested.[624] The terphthalate is the main product because it crystallizes from the reaction mixture, driving the equilibrium in that direction.[625]

For aliphatic decarboxylation, see **12-40**.

OS **I**, 274, 455, 541; **II**, 100, 214, 217, 341; **III**, 267, 272, 471, 637; **IV**, 590, 628; **V**, 635, 813, 982, 985. Also see, OS **I**, 56.

11-36   The Jacobsen Reaction

When polyalkyl- or polyhalobenzenes are treated with sulfuric acid, the ring is sulfonated, but rearrangement also takes place. The reaction, known as the *Jacobsen reaction*, is limited to benzene rings that have at least four substituents, which can be any combination of alkyl and halogen groups, where the alkyl groups can be

[619]Singh, R.; Just, G. *Synth. Commun.* **1988**, *18*, 1327.

[620]For example, see Grovenstein, Jr., E.; Ropp, G.A. *J. Am. Chem. Soc.* **1956**, *78*, 2560.

[621]For a review, see Larock, R.C. *Organomercury Compounds in Organic Synthesis*, Springer, NY, **1985**, pp. 101–105.

[622]Raecke, B. *Angew. Chem.* **1958**, *70*, 1; Riedel, O.; Kienitz, H. *Angew. Chem.* **1960**, *72*, 738; McNelis, E. *J. Org. Chem.* **1965**, *30*, 1209; Ogata, Y.; Nakajima, K. *Tetrahedron* **1965**, *21*, 2393; Ratusky, J.; Sorm, F. *Chem. Ind. (London)*, **1966**, 1798.

[623]For a review, see Ratusky, J., in Patai, S. *The Chemistry of Acid Derivatives*, pt. 1, Wiley, NY, **1979**, pp. 915–944.

[624]See Ratusky, J. *Collect. Czech. Chem. Commun.* **1973**, *38*, 74, 87, and references cited therein.

[625]Ratusky, J. *Collect. Czech. Chem. Commun.* **1968**, *33*, 2346.

ethyl or methyl and the halogen iodo, chloro, or bromo. When isopropyl or *tert*-butyl groups are on the ring, these groups are cleaved to give alkenes. Since a sulfo group can later be removed (**11-38**), the Jacobsen reaction can be used as a means of rearranging polyalkylbenzenes. The rearrangement always brings the alkyl or halo groups closer together than they were originally. Side products in the case illustrated above are pentamethylbenzenesulfonic acid, 2,4,5-trimethylbenzenesulfonic acid, and so on, indicating an intermolecular process, at least partially.

The mechanism of the Jacobsen reaction is not established,[626] but there is evidence, at least for polymethylbenzenes, that the rearrangement is intermolecular, and that the species to which the methyl group migrates is a polymethylbenzene, not a sulfonic acid. Sulfonation takes place after the migration.[627] It has been shown by labeling that ethyl groups migrate without internal rearrangement.[628]

Isomerization of alkyl groups in substituted biphenyls has been observed[629] when the medium is a superacid (see p. 236).

B. Oxygen Leaving Groups

11-37   Deoxygenation

$$\text{ArOR} \longrightarrow \text{ArH}$$

In a few cases, it is possible to remove an oxygen substituent directly from the aromatic ring. Treatment of an aryl mesylate (ArOMs) with a nickel catalyst in DMF, for example, leads to the deoxygenated product, Ar—H.[630]

C. Sulfur Leaving Groups

11-38   Desulfonation or Hydro-de-sulfonation

$$\text{ArSO}_3\text{H} \xrightarrow[\text{dil. H}_2\text{SO}_4]{135-200°\text{C}} \text{ArH} + \text{H}_2\text{SO}_4$$

The cleavage of sulfo groups from aromatic rings is the reverse of **11-7**.[631] By the principle of microscopic reversibility, the mechanism is also the reverse.[632] Dilute H_2SO_4 is generally used, as the reversibility of sulfonation decreases with

[626]For discussions, see Suzuki, H. *Bull. Chem. Soc. Jpn.* **1963**, *36*, 1642; Koeberg-Telder, A.; Cerfontain, H. *J. Chem. Soc. Perkin Trans. 2* **1977**, 717; Cerfontain, H. *Mechanistic Aspects in Aromatic Sulfonation and Desulfonation*, Wiley, NY, **1968**, pp. 214–226; Taylor, R., in Bamford, C.H.; Tipper, C.F.H. *Comprehensive Chemical Kinetics*, Vol. 13, Elsevier, NY, **1972**, pp. 22–32, 48–55.

[627]Koeberg-Telder, A.; Cerfontain, H. *Recl. Trav. Chim. Pays-Bas* **1987**, *106*, 85; Cerfontain, H.; Koeberg-Telder, A. *Can. J. Chem.* **1988**, *66*, 162.

[628]Marvell, E.N.; Webb, D. *J. Org. Chem.* **1962**, *27*, 4408.

[629]Sherman, S. C.; Iretskii, A. V.; White, M. G.; Gumienny, C.; Tolbert, L. M.; Schiraldi, D. A. *J. Org. Chem.* **2002**, *67*, 2034.

[630]Sasaki, K.; Kubo, T.; Sakai, M.; Kuroda, Y. *Chem. Lett.* **1997**, 617.

[631]For reviews, see Cerfontain, H. *Mechanistic Aspects in Aromatic Sulfonation and Desulfonation*, Wiley, NY, **1968**, pp. 185–214; Taylor, R., in Bamford, C.H.; Tipper, C.F.H. *Comprehensive Chemical Kinetics*, Vol. 13, Elsevier, NY, **1972**, pp. 349–355; Gilbert, E.E. *Sulfonation and Related Reactions*, Wiley, NY, **1965**, pp. 427–442. See also, Krylov, E.N. *J. Org. Chem. USSR* **1988**, *24*, 709.

[632]For a discussion, see Kozlov, V.A.; Bagrovskaya, N.A. *J. Org. Chem. USSR* **1989**, *25*, 1152.

increasing H_2SO_4 concentration. The reaction permits the sulfo group to be used as a blocking group to direct meta and then to be removed. The sulfo group has also been replaced by nitro and halogen groups. Sulfo groups have also been removed from the ring by heating with an alkaline solution of Raney nickel.[633] In another catalytic process, aromatic sulfonyl bromides or chlorides are converted to aryl bromides or chlorides, respectively, on heating with chlorotris(triphenylphosphine) rhodium(I).[634] This reaction is similar to the decarbonylation of aromatic acyl halides mentioned in **14-32**.

$$\text{ArSO}_2\text{Br} \xrightarrow{\text{RhCl(PPh}_3)_3} \text{ArBr}$$

OS **I**, 388; **II**, 97; **III**, 262; **IV**, 364. Also see OS **I**, 519; **II**, 128; **V**, 1070.

D. Halogen Leaving groups

11-39   Dehalogenation or Hydro-de-halogenation

$$\text{ArX} \xrightarrow{\text{AlCl}_3} \text{ArH}$$

Aryl halides can be dehalogenated by Friedel–Crafts catalysts. Iodine is the most easily cleaved. Dechlorination is seldom performed and defluorination apparently never. The reaction is most successful when a reducing agent, say, Br^- or I^- is present to combine with the I^+ or Br^+ coming off.[635] Except for deiodination, the reaction is seldom used for preparative purposes. Migration of halogen is also found,[636] both intramolecular[637] and intermolecular.[638] The mechanism is probably the reverse of that of **11-10**.[639] Debromination of aromatic rings having two attached amino groups was accomplished by refluxing in aniline containing acetic acid/HBr.[640]

Rearrangement of polyhalobenzenes can also be catalyzed by very strong bases; for example 1,2,4-tribromobenzene is converted to 1,3,5-tribromobenzene by treatment with PhNHK.[641] This reaction, which involves aryl carbanion intermediates (S_E1 mechanism), has been called the *halogen dance*.[642]

[633]Feigl, F. *Angew. Chem.* **1961**, *73*, 113.
[634]Blum, J.; Scharf, G. *J. Org. Chem.* **1970**, *35*, 1895.
[635]Pettit, G.R.; Piatak, D.M. *J. Org. Chem.* **1960**, *25*, 721.
[636]Olah, G.A.; Tolgyesi, W.S.; Dear, R.E.A. *J. Org. Chem.* **1962**, *27*, 3441, 3449, 3455; De Valois, P.J.; Van Albada, M.P.; Veenland, J.U. *Tetrahedron* **1968**, *24*, 1835; Olah, G.A.; Meidar, D.; Olah, J.A. *Nouv. J. Chim.*, **1979**, *3*, 275.
[637]Koptyug, V.A.; Isaev, I.S.; Gershtein, N.A.; Berezovskii, G.A. *J. Gen. Chem. USSR* **1964**, *34*, 3830; Erykalov, Yu.G.; Becker, H.; Belokurova, A.P. *J. Org. Chem. USSR* **1968**, *4*, 2054; Jacquesy, J.; Jouannetaud, M. *Tetrahedron Lett.* **1982**, *23*, 1673.
[638]Augustijn, G.J.P.; Kooyman, E.C.; Louw, R. *Recl. Trav. Chim. Pays-Bas* **1963**, *82*, 965.
[639]Choguill, H.S.; Ridd, J.H. *J. Chem. Soc.* **1961**, 822; Shine, H.J. *Aromatic Rearrangements, Elsevier, NY*, **1967**, p. 1; Ref. 636.
[640]Choi, H.; Chi, D.Y. *J. Am. Chem. Soc.* **2001**, *123*, 9202.
[641]Moyer, Jr., C.E.; Bunnett, J.F. *J. Am. Chem. Soc.* **1963**, *85*, 1891.
[642]Bunnett, J.F. *Acc. Chem. Res.* **1972**, *5*, 139; Mach, M.H.; Bunnett, J.F. *J. Org. Chem.* **1980**, *45*, 4660; Sauter, F.; Fröhlich, H.; Kalt, W. *Synthesis* **1989**, 771.

Removal of halogen from aromatic rings can also be accomplished by various reducing agents, among them Bu_3SnH,[643] catalytic hydrogenolysis,[644] catalytic transfer hydrogenolysis,[645] $Fe(CO)_5$,[646] Na–Hg in liquid NH_3,[647] $LiAlH_4$,[648] $LiAlH_4$ and a $NbCl_5$ catalyst,[649] $NaBH_4$ and a catalyst,[650] Ni/C with $Me_2NH\cdot BH_3$,[651] NaH,[652] $HCOOH$[653] or aqueous $HCOO^-$[654] with Pd/C, and Raney nickel in alkaline solution,[655] the last method being effective for fluorine, as well as for the other halogens. Carbon monoxide, with potassium tetracarbonylhydridoferrate $KHFe(CO)_4$ as a catalyst, specifically reduces aryl iodides.[656] Polymethylhydrosiloxane (PHMS) and KF, with a palladium catalyst, also reduces aryl iodides.[657] Not all these reagents operate by electrophilic substitution mechanisms. Some are nucleophilic substitutions and some are free-radical processes. Photochemical[658] and electrochemical[659] reduction are also known. Halogen can also be removed from aromatic rings indirectly by conversion to Grignard reagents (**12-38**) followed by hydrolysis (**11-41**).

OS **III**, 132, 475, 519; **V**, 149, 346, 998; **VI**, 82, 821.

11-40 Formation of Organometallic Compounds

$$ArBr + M \longrightarrow ArM$$
$$ArBr + RM \longrightarrow ArM + RBr$$

[643]Maitra, U.; Sarma, K.D. *Tetrahedron Lett.* **1994**, *35*, 7861.

[644]For example, see Subba Rao, Y.V.; Mukkanti, K.; Choudary, B.M. *J. Organomet. Chem.* **1989**, *367*, C29. See also, Sajiki, H.; Kume, A.; Hattori, K.; Hirota, K. *Tetrahedron Lett.* **2002**, *43*, 7247.

[645]Anwer, M.K.; Spatola, A.F. *Tetrahedron Lett.* **1985**, *26*, 1381.

[646]Brunet, J.-J.; El Zaizi, A. *Bull. Soc. Chim. Fr.* **1996**, *133*, 75.

[647]Austin, E.; Alonso, R.A.; Rossi, R.A. *J. Chem. Res. (S)* **1990**, 190.

[648]Karabatsos, G.J.; Shone, R.L. *J. Org. Chem.* **1968**, *33*, 619; Brown, H.C.; Chung, S.; Chung, F. *Tetrahedron Lett.* **1979**, 2473. Evidence for a free-radical mechanism has been found in this reaction; see Chung, F.; Filmore, K.L. *J. Chem. Soc., Chem. Commun.* **1983**, 358; Beckwith, A.L.J.; Goh, S.H. *J. Chem. Soc., Chem. Commun.* **1983**, 905. See also, Beckwith, A.L.J.; Goh, S.H. *J. Chem. Soc., Chem. Commun.* **1983**, 907; Han, B.H.; Baudjouk, P. *Tetrahedron Lett.* **1982**, *23*, 1643.

[649]Fuchibe, K.; Akiyama, T. *Synlett* **2004**, 1282.

[650]Egli, R.A. *Helv. Chim. Acta* **1968**, *51*, 2090; Lin, S.; Roth, J.A. *J. Org. Chem.* **1979**, *44*, 309; Narisada, M.; Horibe, I.; Watanabe, F.; Takeda, K. *J. Org. Chem.* **1989**, *54*, 5308.

[651]Lipshutz, B.H.; Tomioka, T.; Sato, K. *Synlett* **2001**, 970; Lipshutz, B.H.; Tomioka, T.; Pfeiffer, S.S. *Tetrahedron Lett.* **2001**, *42*, 7737.

[652]Nelson, R.B.; Gribble, G.W. *J. Org. Chem.* **1974**, *39*, 1425.

[653]Barren, J.P.; Baghel, S.S.; McCloskey, P.J. *Synth. Commun.* **1993**, *23*, 1601.

[654]Arcadi, A.; Cerichelli, G.; Chiarini, M.; Vico, R.; Zorzan, D. *Eur. J. Org. Chem.* **2004**, 3404.

[655]Buu-Hoï, N.P.; Xuong, N.D.; van Bac, N. *Bull. Soc. Chim. Fr.* **1963**, 2442; de Koning, A.J. *Org. Prep. Proced. Int.* **1975**, *7*, 31.

[656]Brunet, J.; Taillefer, M. *J. Organomet. Chem.* **1988**, *348*, C5.

[657]Maleczka, Jr., R.E.; Rahaim, Jr., R.J.; Teixeira, R.R. *Tetrahedron Lett.* **2002**, *43*, 7087.

[658]See, for example, Pinhey, J.T.; Rigby, R.D.G. *Tetrahedron Lett.* **1969**, 1267, 1271; Barltrop, J.A.; Bradbury, D. *J. Am. Chem. Soc.* **1973**, *95*, 5085.

[659]See Fry, A.J. *Synthetic Organic Electrochemistry*, 2nd ed., Wiley, NY, **1989**, pp. 142–143. Also see, Bhuvaneswari, N.; Venkatachalam, C.S.; Balasubramanian, K.K. *Tetrahedron Lett.* **1992**, *33*, 1499.

These reactions are considered along with their aliphatic counterparts at reactions **12-38** and **12-39**.

E. Metal Leaving Groups

11-41   Hydrolysis of Organometallic Compounds

Hydro-de-metallation or **Demetallation**

$$\text{ArM} + \text{H}^+ \longrightarrow \text{ArH} + \text{M}^+$$

Organometallic compounds can be hydrolyzed by acid treatment. For active metals, such as Mg, Li, and so on water is sufficiently acidic. The most important example of this reaction is hydrolysis of Grignard reagents, but M may be many other metals or metalloids. Examples are SiR_3, HgR, Na, and $B(OH)_2$. Since aryl Grignard and aryllithium compounds are fairly easy to prepare, they are often used to prepare salts of weak acids, such as alkynes.

$$\text{PhMgBr} + \text{H}-\text{C}{\equiv}\text{C}-\text{H} \longrightarrow \text{H}-\text{C}{\equiv}\text{C:}^{-\ +}\text{MgBr} + \text{PhH}$$

Where the bond between the metal and the ring is covalent, the usual arenium ion mechanism operates.[660] Where the bonding is essentially ionic, this is a simple acid–base reaction. For the aliphatic counterpart of this reaction, see reaction **12-24**.

Other reactions of aryl organometallic compounds are treated with their aliphatic analog: reactions **12-25–12-27** and **12-30–12-37**.

[660]For a discussion of the mechanism, see Taylor, R., in Bamford, C.H.; Tipper, C.F.H. *Comprehensive Chemical Kinetics*, Vol. 13, Elsevier, NY, *1972*, pp. 278–303, 324–349.

CHAPTER 12

Aliphatic, Alkenyl, and Alkynyl Substitution, Electrophilic and Organometallic

In Chapter 11, it was pointed out that the most important leaving groups in electrophilic substitution are those that can best exist with an outer shell that is deficient in a pair of electrons. For aromatic systems, the most common leaving group is the proton. The proton is also a leaving group in aliphatic systems, but the reactivity depends on the acidity. Protons in saturated alkanes are very unreactive, but electrophilic substitutions are often easily carried out at more acidic positions, for example, α to a carbonyl group, or at an alkynyl position (RC≡CH). Since metallic ions are easily able to bear positive charges, we might expect that organometallic compounds would be especially susceptible to electrophilic substitution, and this is indeed the case.[1] Another important type of electrophilic substitution, known as *anionic cleavage*, involves the breaking of C–C bonds; in these reactions there are carbon leaving groups (**12-40–12-46**). A number of electrophilic substitutions at a nitrogen atom are treated at the end of the chapter.

Since a carbanion is what remains when a positive species is removed from a carbon atom, the subject of carbanion structure and stability (Chapter 5) is inevitably related to the material in this chapter. So is the subject of very weak acids and very strong bases (Chapter 8), because the weakest acids are those in which the hydrogen is bonded to carbon.

[1]For books on the preparation and reactions of organometallic compounds, see Hartley, F.R.; Patai, S. *The Chemistry of the Metal-Carbon Bond*, 5 vols., Wiley, NY, ***1984–1990***; Haiduc, I.; Zuckerman, J.J. *Basic Organometallic Chemistry*, Walter de Gruyter, NY, ***1985***; Negishi, E. *Organometallics in Organic Synthesis*, Wiley, NY, ***1980***; Aylett, B.J. *Organometallic Compounds*, 4th ed., Vol. 1, pt. 2, Chapman and Hall, NY, ***1979***; Coates, G.E.; Green, M.L.H.; Wade, K. *Organometallic Compounds*, 3rd ed., 2 vols., Methuen, London, ***1967–1968***; Eisch, J.J. *The Chemistry of Organometallic Compounds*, Macmillan, NY, ***1967***. For reviews, see Maslowsky, Jr., E. *Chem. Soc. Rev.* ***1980***, *9*, 25, and in Tsutsui, M. *Characterization of Organometallic Compounds*, Wiley, NY, ***1969–1971***, the articles by Cartledge, F.K.; Gilman, H. pt. 1, pp. 1–33, and by Reichle, W.T. pt. 2, pp. 653–826.

March's Advanced Organic Chemistry: Reactions, Mechanisms, and Structure, Sixth Edition, by Michael B. Smith and Jerry March

MECHANISMS

For aliphatic electrophilic substitution, we can distinguish at least four possible major mechanisms,[2] which we call S_E1, S_E2 (front), S_E2 (back), and S_Ei. The S_E1 is unimolecular; the other three are bimolecular. It is noted that the term "S_EAr" has been proposed to represent electrophilic aromatic substitution, so that the term "S_E2" refers exclusively to electrophilic substitutions where a steric course is possible.[3] To describe the steric course of an aliphatic substitution reaction, the suffixes "ret" and "inv" were proposed, referring to retention and inversion of configuration, respectively.

BIMOLECULAR MECHANISMS. S_E2 AND S_Ei

The bimolecular mechanisms for electrophilic aliphatic substitution are analogous to the S_N2 mechanism in that the new bond forms as the old one breaks. However, in the S_N2 mechanism the incoming group brings with it a pair of electrons, and this orbital can overlap with the central carbon only to the extent that the leaving group takes away its electrons; otherwise the carbon would have more than eight electrons at once in its outer shell. Since electron clouds repel, this means also that the incoming group attacks backside, at a position 180° from the leaving group, resulting in inversion of configuration. When the nucleophilic species attacks (donates electrons to) an electrophile, which brings to the substrate only a vacant orbital, predicting the direction the attack is not as straightforward. We can imagine two main possibilities: delivery of the electrophile to the front, which we call S_E2 (front), and delivery of the electrophile to the rear, which we call S_E2 (back). The possibilities can be pictured (charges not shown):

Both the S_E2 (front) and S_E2 (back) mechanisms are designated D_EA_E in the IUPAC system. With substrates in which we can distinguish the possibility, the former

[2]For monographs, see Abraham, M.H. *Comprehensive Chemical Kinetics*, Bamford, C.H.; Tipper, C.F.H. Eds., Vol. 12, Elsevier, NY, *1973*; Jensen, F.R.; Rickborn, B. *Electrophilic Substitution of Organomercurials*, McGraw-Hill, NY, *1968*; Reutov, O.A.; Beletskaya, I.P. *Reaction Mechanisms of Organometallic Compounds*, North-Holland Publishing Company, Amsterdam, The Netherlands, *1968*. For reviews, see Abraham, M.H.; Grellier, P.L., in Hartley, F.R.; Patai, S. *The Chemistry of the Metal-Carbon Bond*, Vol. 2, Wiley, NY, pp. 25–149; Beletskaya, I.P. *Sov. Sci. Rev. Sect. B 1979*, *1*, 119; Reutov, O.A. *Pure Appl. Chem.* *1978*, *50*, 717; *1968*, *17*, 79; *Tetrahedron 1978*, *34*, 2827; *J. Organomet. Chem. 1975*, *100*, 219; *Russ. Chem. Rev. 1967*, *36*, 163; *Fortschr. Chem. Forsch. 1967*, *8*, 61; Matteson, D.S. *Organomet. Chem. Rev. Sect. A 1969*, *4*, 263; Dessy, R.E.; Kitching, W. *Adv. Organomet. Chem. 1966*, *4*, 267.
[3]Gawley, R.E. *Tetrahedron Lett. 1999*, *40*, 4297.

mechanism should result in retention of configuration and the latter in inversion. The reaction of allylsilanes with adamantyl chloride and $TiCl_4$, for example, gives primarily the antiproduct via a S_E2' reaction.[4] When the electrophile reacts from the front, there is a third possibility. A portion of the electrophile may assist in the removal of the leaving group, forming a bond with it at the same time that the new C–Y bond is formed:

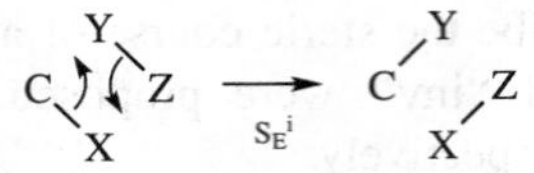

This mechanism, which we call the S_Ei mechanism[5] (IUPAC designation: *cyclo*-$D_EA_ED_nA_n$), also results in retention of configuration.[6] Plainly, where a second-order mechanism involves this kind of internal assistance, backside attack is impossible.

It is evident that these three mechanisms are not easy to distinguish. All three give second-order kinetics, and two result in retention of configuration.[7] In fact, although much work has been done on this question, there are few cases in which we can unequivocally say that one of these three and not another is actually

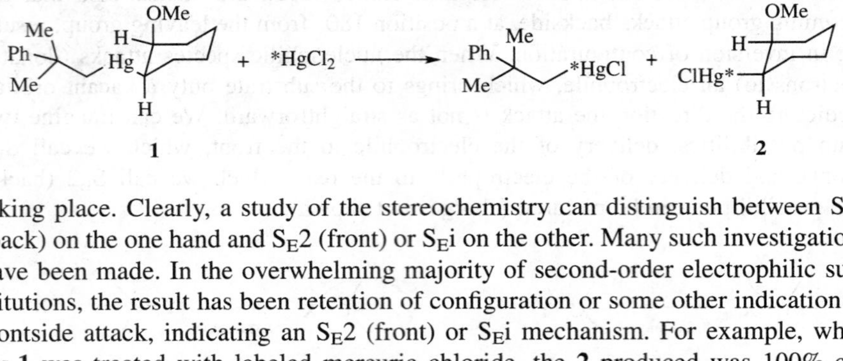

taking place. Clearly, a study of the stereochemistry can distinguish between S_E2 (back) on the one hand and S_E2 (front) or S_Ei on the other. Many such investigations have been made. In the overwhelming majority of second-order electrophilic substitutions, the result has been retention of configuration or some other indication of frontside attack, indicating an S_E2 (front) or S_Ei mechanism. For example, when *cis*-**1** was treated with labeled mercuric chloride, the **2** produced was 100% *cis*. The bond between the mercury and the ring must have been broken (as well as the other Hg–C bond), since each of the products contained about half of the labeled mercury.[8] Another indication of frontside attack is that second-order

[4]Buckle, M.J.C.; Fleming, I.; Gil, S. *Tetrahedron Lett.* **1992**, *33*, 4479.

[5]The names for these mechanisms vary throughout the literature. For example, the S_Ei mechanism has also been called the S_F2, the S_E2 (closed), and the S_E2 (cyclic) mechanism. The original designations, S_E1, S_E2, and so on, were devised by the Hughes–Ingold school.

[6]It has been contended that the S_Ei mechanism violates the principle of conservation of orbital symmetry (p. 1208), and that the S_E2 (back) mechanism partially violates it: Slack, D.A.; Baird, M.C. *J. Am. Chem. Soc.* **1976**, *98*, 5539.

[7]For a review of the stereochemistry of reactions in which a carbon-transition-metal σ bond is formed or broken, see Flood, T.C. *Top. Stereochem.* **1981**, *12*, 37. See also Jensen, F.R.; Davis, D.D. *J. Am. Chem. Soc.* **1971**, *93*, 4048.

[8]Winstein, S.; Traylor, T.G.; Garner, C.S. *J. Am. Chem. Soc.* **1955**, *77*, 3741.

electrophilic substitutions proceed very easily at *bridgehead* carbons (see p. 429).[9] Still another indication is the behavior of neopentyl as a substrate. S$_N$2 reactions at neopentyl are extremely slow (p. 479), because attack from the rear is blocked and the transition state for the reaction lies very high in energy. The fact that neopentyl systems undergo electrophilic substitution only slightly more slowly than ethyl[10] is further evidence for frontside attack. One final elegant experiment may be noted.

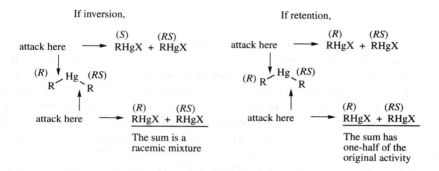

The compound di-*sec*-butylmercury was prepared with one *sec*-butyl group optically active and the other racemic.[11] This was accomplished by treatment of optically active *sec*-butylmercuric bromide with racemic *sec*-butylmagnesium bromide. The di-*sec*-butyl compound was then treated with mercuric bromide to give 2 eqiuivalents of *sec*-butylmercuric bromide. The steric course of the reaction could then be predicted by the following analysis, assuming that the bonds between the mercury and each carbon have a 50% chance of breaking. The original activity referred to is the activity of the optically active *sec*-butylmercuric bromide used to make the dialkyl compound. The actual result was that, under several different sets of conditions, the product had one-half of the original activity, demonstrating retention of configuration.

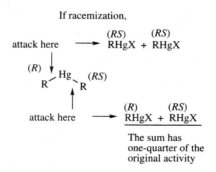

[9]Winstein, S.; Traylor, T.G. *J. Am. Chem. Soc.* **1956**, *78*, 2597; Schöllkopf, U. *Angew. Chem.* **1960**, *72*, 147. For a discussion, see Fort Jr., R.C.; Schleyer, P.v.R. *Adv. Alicyclic Chem.* **1966**, *1*, 283, pp. 353–370.
[10]Hughes, E.D.; Volger, H.C. *J. Chem. Soc.* **1961**, 2359.
[11]Jensen, F.R. *J. Am. Chem. Soc.* **1960**, *82*, 2469; Ingold, C.K. *Helv. Chim. Acta* **1964**, *47*, 1191.

However, inversion of configuration has been found in certain cases, demonstrating that the S_E2 (back) mechanism can take place. For example, the reaction of optically active *sec*-butyltrineopentyltin with bromine (**12-40**) gives inverted *sec*-butyl bromide.[12] A number of other organometallic compounds have also been shown to give inversion when treated with halogens,[13] although others do not.[14] So far, no inversion has been found with an organomercury substrate. It may be that still other examples of backside

$$sec\text{-BuSnR}_3 \quad + \quad \text{Br}_2 \xrightarrow{\hspace{3cm}} sec\text{-BuBr} \quad \text{R} = \text{neopentyl}$$

attack exist,[15] but have escaped detection because of the difficulty in preparing compounds with a configurationally stable carbon–metal bond. Compounds that are chiral because of a stereogenic carbon at which a carbon–metal bond is located[16] are often difficult to resolve and once resolved are often easily racemized. The resolution has been accomplished most often with organomercury compounds,[17] and most stereochemical investigations have therefore been made with these substrates. Only a few optically active Grignard reagents have been prepared[18] (i.e., in which the only stereogenic center is the carbon bonded to the magnesium). Because of this, the steric course of electrophilic substitutions at the C–Mg bond has not often been determined. However, in one such case, the reaction of both the exo and endo isomers of the 2-norbornyl Grignard reagent with HgBr_2 (to give 2-norbornylmercuric bromide) has been shown to proceed with retention of configuration.[19] It is likely that inversion takes place only when steric hindrance

[12]Jensen, F.R.; Davis, D.D. *J. Am. Chem. Soc.* **1971**, *93*, 4048. For a review of the stereochemistry of S_E2 reactions with organotin substrates, see Fukuto, J.M.; Jensen, F.R. *Acc. Chem. Res.* **1983**, *16*, 177.

[13]For example, See Applequist, D.E.; Chmurny, G.N. *J. Am. Chem. Soc.* **1967**, *89*, 875; Glaze, W.H.; Selman, C.M.; Ball Jr., A.L.; Bray, L.E. *J. Org. Chem.* **1969**, *34*, 641; Brown, H.C.; Lane, C.F. *Chem. Commun.* **1971**, 521; Jensen, F.R.; Madan, V.; Buchanan, D.H. *J. Am. Chem. Soc.* **1971**, *93*, 5283; Espenson, J.H.; Williams, D.A. *J. Am. Chem. Soc.* **1974**, *96*, 1008; Bock, P.L.; Boschetto, D.J.; Rasmussen, J.R.; Demers, J.P.; Whitesides, G.M. *J. Am. Chem. Soc.* **1974**, *96*, 2814; Magnuso, R.H.; Halpern, J.; Levitin, I.Ya.; Vol'pin, M.E. *J. Chem. Soc. Chem. Commun.* **1978**, 44.

[14]See, for example, Rahm, A.; Pereyre, M. *J. Am. Chem. Soc.* **1977**, *99*, 1672; McGahey, L.F.; Jensen, F.R. *J. Am. Chem. Soc.* **1979**, *101*, 4397. Electrophilic bromination of certain organotin compounds was found to proceed with inversion favored for equatorial and retention for axial C–Sn bonds: Olszowy, H.A.; Kitching, W. *Organometallics* **1984**, *3*, 1676. For a similar result, see Rahm, A.; Grimeau, J.; Pereyre, M. *J. Organomet. Chem.* **1985**, *286*, 305.

[15]Cases of inversion involving replacement of a metal by a metal have been reported. See Tada, M.; Ogawa, H. *Tetrahedron Lett.* **1973**, 2639; Fritz, H.L.; Espenson, J.H.; Williams, D.A.; Molander, G.A. *J. Am. Chem. Soc.* **1974**, *96*, 2378; Gielen, M.; Fosty, R. *Bull. Soc. Chim. Belg.* **1974**, *83*, 333; Bergbreiter, D.E.; Rainville, D.P. *J. Organomet. Chem.* **1976**, *121*, 19.

[16]For a monograph, see Sokolov, V.I. *Chirality and Optical Activity in Organometallic Compounds*, Gordon and Breach, NY, *1990*.

[17]Organomercury compounds were first resolved by three groups: Jensen, F.R.; Whipple, L.D.; Wedegaertner, D.K.; Landgrebe, J.A. *J. Am. Chem. Soc.* **1959**, *81*, 1262; Charman, H.B.; Hughes, E.D.; Ingold, C.K. *J. Chem. Soc.* **1959**, 2523, 2530; Reutov, O.A.; Uglova, E.V. *Bull. Acad. Sci. USSR Div. Chem. Sci.* **1959**, 735.

[18]This was done first by Walborsky, H.M.; Young, A.E. *J. Am. Chem. Soc.* **1964**, *86*, 3288.

[19]Jensen, F.R.; Nakamaye, K.L. *J. Am. Chem. Soc.* **1966**, *88*, 3437.

prevents reaction on the frontside and when the electrophile does not carry a Z group (p. 754).

The S_E2 (back) mechanism can therefore be identified in certain cases (if inversion of configuration is found), but it is plain that stereochemical investigations cannot distinguish between the S_E2 (front) and the S_Ei mechanisms and that, in the many cases where configurationally stable substrates cannot be prepared, such investigations are of no help at all in distinguishing among all three of the second-order mechanisms. Unfortunately, there are not many other methods that lead to unequivocal conclusions. One method that has been used in an attempt to distinguish between the S_Ei mechanism on the one hand and the S_E2 pathways on the other involves the study of salt effects on the rate. It may be recalled (p. 501) that reactions in which neutral starting molecules acquire charges in the transition state are aided by an increasing concentration of added ions. Thus the S_Ei mechanism would be less influenced by salt effects than would either of the S_E2 mechanisms. On this basis, Abraham and co-workers[20] concluded that the reactions $R_4Sn + HgX_2 \rightarrow RHgX + R_3SnX$ (X = Cl or I) take place by S_E2 and not by S_Ei mechanisms. Similar investigations involve changes in solvent polarity (see also, p. 765).[21] In the case of the reaction

$$sec\text{-}BuSnR_2R' + Br_2 \longrightarrow sec\text{-}BuBr$$

(where R = R′ = iPr and R = iPr, R′ = neopentyl), the use of polar solvents gave predominant inversion, while nonpolar solvents gave predominant retention.[22]

On the basis of evidence from reactivity studies, it has been suggested[23] that a variation of the S_Ei mechanism is possible in which the group Z becomes attached to X before the latter becomes detached:

This process has been called the S_EC[22] or S_E2 (co-ord)[24] mechanism (IUPAC designation $A_n + cyclo\text{-}D_EA_ED_n$).

It has been shown that in certain cases (e.g., $Me_4Sn + I_2$) the reactants in an S_E2 reaction, when mixed, give rise to an immediate charge-transfer spectrum (p. 115), showing that an electron donor–acceptor (EDA) complex has been formed.[25] In these cases it is likely that the EDA complex is an intermediate in the reaction.

[20]Abraham, M.H.; Johnston, G.F. *J. Chem. Soc. A*, *1970*, 188.

[21]See, for example, Abraham, M.H.; Dorrell, F.J. *J. Chem. Soc. Perkin Trans. 2 1973*, 444.

[22]Fukuto, J.M.; Newman, D.A.; Jensen, F.R. *Organometallics 1987*, 6, 415.

[23]Abraham, M.H.; Hill, J.A. *J. Organomet. Chem. 1967*, 7, 11.

[24]Abraham, M.H. *Comprehensive Chemical Kinetics*, Bamford, C.H.; Tipper, C.F.H. Eds., Vol. 12, Elsevier, NY, *1973*, p. 15.

[25]Fukuzumi, S.; Kochi, J.K. *J. Am. Chem. Soc. 1980*, 102, 2141, 7290.

THE S_E1 MECHANISM

The S_E1 mechanism is analogous to the S_N1. It involves two steps: a slow ionization and a fast combination.

Step 1 $R\text{-}X \xrightarrow{\text{slow}} R^- + X^+$

Step 2 $R^- + Y^+ \longrightarrow R\text{—}Y$

The IUPAC designation is $D_E + A_E$. First-order kinetics are predicted and many such examples have been found. Other evidence for the S_E1 mechanism was obtained in a study of base-catalyzed tautomerization. In the reaction

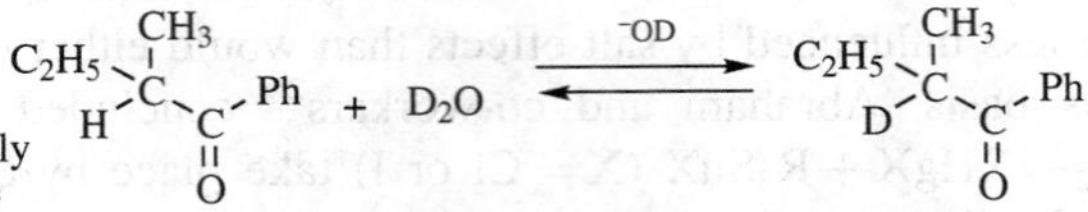

the rate of deuterium exchange was the same as the rate of racemization[26] and there was an isotope effect.[27]

The S_N1 reactions do not proceed at strained bridgehead carbons (e.g., in [2.2.1] bicyclic systems, p. 435) because planar carbocations cannot form at these carbons. However, carbanions not stabilized by resonance are probably not planar, and S_E1 reactions readily occur with this type of substrate. Indeed, the question of carbanion structure is intimately tied into the problem of the stereochemistry of the S_E1 reaction. If a carbanion is planar, racemization should occur. If it is pyramidal and *can hold its structure*, the result should be retention of configuration. On the other hand, even a pyramidal carbanion will give racemization if it cannot hold its structure, that is, if there is pyramidal inversion as with amines (p. 142). Unfortunately, the only carbanions that can be studied easily are those stabilized by resonance, which makes them planar, as expected (p. 258). For simple alkyl carbanions, the main approach to determining structure has been to study the stereochemistry of S_E1 reactions rather than the other way around. Racemization is almost always observed, but whether this is caused by planar carbanions or by oscillating pyramidal carbanions is not known. In either, case racemization occurs whenever a carbanion is completely free or is symmetrically solvated.

However, even planar carbanions need not give racemization. Cram found that retention and even inversion can occur in the alkoxide (see **3**) cleavage reaction (**12-41**):

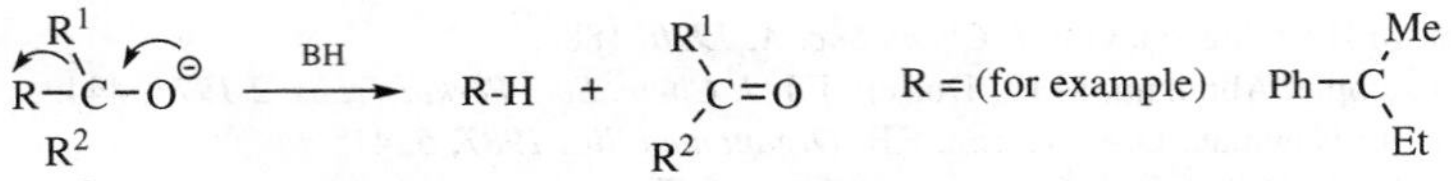

[26]Hsu, S.K.; Ingold, C.K.; Wilson, C.L. *J. Chem. Soc.* **1938**, 78.
[27]Wilson, C.L. *J. Chem. Soc.* **1936**, 1550.

which is a first-order S$_E$1 reaction involving resonance-stabilized planar carbanions (here designated R⁻).[28] By changing the solvent Cram was able to produce products ranging from 99% retention to 60% inversion and including complete racemization. These results are explained by a carbanion that is not completely free but is solvated. In nondissociating, nonpolar solvents, such as benzene or dioxane, the alkoxide ion exists as an ion pair, solvated by the solvent BH:

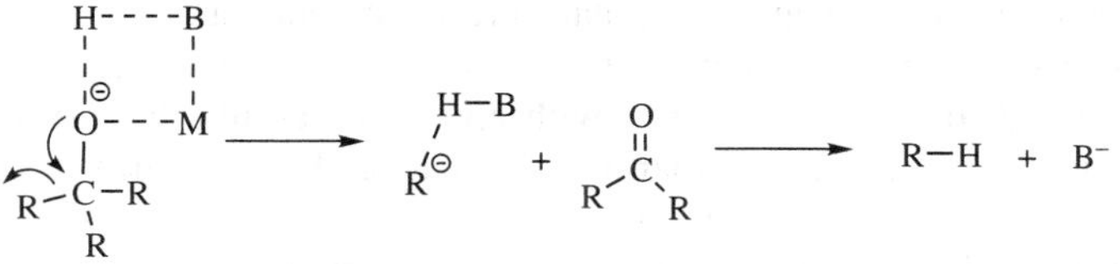

In the course of the cleavage, the proton of the solvent moves in to solvate the newly forming carbanion. As is easily seen, this solvation is asymmetrical since the solvent molecule is already on the front side of the carbanion. When the carbanion actually bonds with the proton, the result is retention of the original configuration. In protic solvents, such as diethylene glycol, a good deal of inversion is found. In these solvents, the *leaving group* solvates the carbanion, so the solvent can solvate it only from the opposite side:

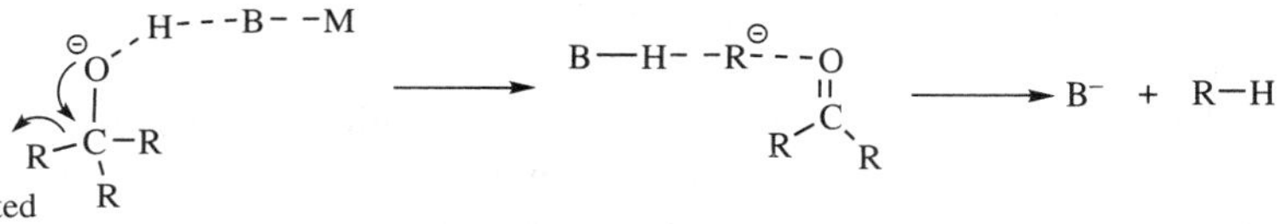

When C–H bond formation occurs, the result is inversion. Racemization results in polar aprotic solvents, such as DMSO. In these solvents, the carbanions are relatively long lived (because the solvent has no proton to donate) and symmetrically solvated.

Similar behavior was found for carbanions generated by base-catalyzed hydrogen exchange (reaction **12-1**):[29]

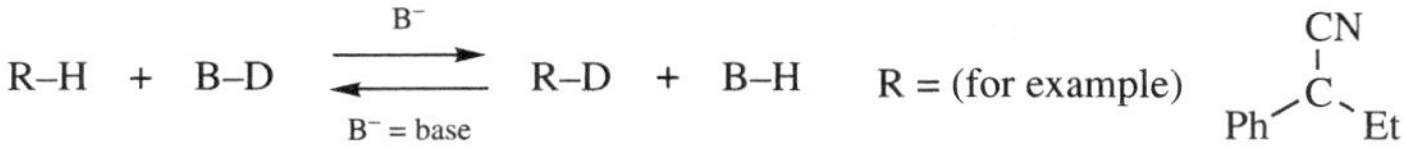

---

[28]See Cram, D.J.; Langemann, A.; Allinger, J.; Kopecky, K.R. *J. Am. Chem. Soc.* **1959**, *81*, 5740; Hoffman, T.D.; Cram, D.J. *J. Am. Chem. Soc.* **1969**, *91*, 1009. For a discussion, see Cram, D.J. *Fundamentals of Carbanion Chemistry*, Academic Press, NY, **1965**, pp. 138–158.
[29]See Roitman, J.N.; Cram, D.J. *J. Am. Chem. Soc.* **1971**, *93*, 2225, 2231 and references cited therein; Cram, J.M.; Cram, D.J. *Intra-Sci. Chem. Rep.* **1973**, *7(3)*, 1. For a discussion, see Cram, D.J. *Fundamentals of Carbanion Chemistry*, Academic Press, NY, **1965**, pp. 85–105.

In this case, information was obtained from measurement of the ratio of k_e (rate constant for isotopic exchange) to k_a (rate constant for racemization). A k_e/k_a ratio substantially >1 means retention of configuration, since many individual isotopic exchanges are not producing a change in configuration. A k_e/k_a ratio of ~1 indicates racemization and a ratio of $\frac{1}{2}$ corresponds to inversion (see p. 430). All three types of steric behavior were found, depending on R, the base, and the solvent. As with the alkoxide cleavage reaction, retention was generally found in solvents of low dielectric constant, racemization in polar aprotic solvents, and inversion in protic solvents. However, in the proton-exchange reactions, a fourth type of behavior was encountered. In aprotic solvents, with aprotic bases like tertiary amines, the k_e/k_a ratio was found to be *less* than 0.5, indicating that racemization took place *faster* than isotopic exchange (this process is known as *isoracemization*). Under these conditions, the conjugate acid of the amine remains associated with the carbanion as an ion pair. Occasionally, the ion pair dissociates long enough for the carbanion to turn over and recapture the proton:

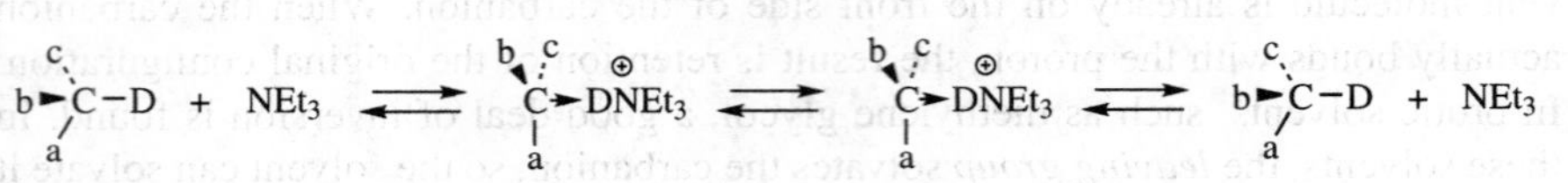

Thus, inversion (and hence racemization, which is produced by repeated acts of inversion) occurs without exchange. A single act of inversion without exchange is called *isoinversion*.

The isoinversion process can take place by a pathway in which a positive species migrates in a stepwise fashion around a molecule from one nucleophilic position to another. For example, in the exchange reaction of 3-carboxamido-9-methylfluorene (**4**) with Pr_3N in *t*-BuOH, it has been proposed that the amine removes

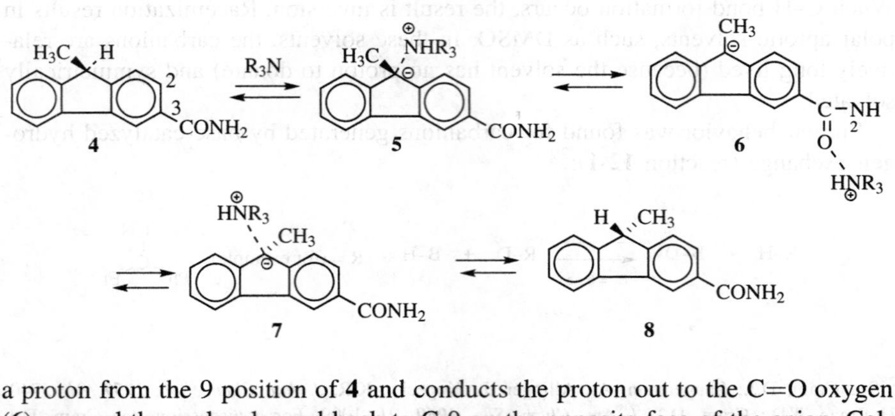

a proton from the 9 position of **4** and conducts the proton out to the C=O oxygen (**6**), around the molecule, and back to C-9 on the opposite face of the anion. Collapse of **7** gives the inverted product **8**. Of course, **6** could also go back to **4**, but a molecule that undergoes the total process **4** → **5** → **6** → **7** → **8** has experienced an inversion without an exchange. Evidence for this pathway, called the *conducted* *tour mechanism,*[30] is that the 12-carboxamido isomer of **4** does not give isoracemization. In this case, the negative charge on the oxygen atom in the anion corresponding to **6** is less, because a canonical form in which oxygen acquires a full negative charge (**9**) results in disruption of the aromatic sextet in both

| 9 | 10 |

benzene rings (cf. **10** where one benzene ring is intact). Whether the isoracemization process takes place by the conducted tour mechanism or a simple nonstructured contact ion-pair mechanism depends on the nature of the substrate (e.g., a proper functional group is necessary for the conducted tour mechanism) and of the base.[31]

It is known that vinylic carbanions *can* maintain configuration, so that S$_E$1 mechanisms should produce retention there. This has been found to be the case. For example, *trans*-2-bromo-2-butene was converted to 64–74% angelic acid:[32]

Only ~5% of the cis isomer, tiglic acid, was produced. In addition, certain carbanions in which the negative charge is stabilized by *d*-orbital overlap can maintain configuration (p. 258) and S$_E$1 reactions involving them proceed with retention of configuration.

Electrophilic Substitution Accompanied by Double-Bond Shifts

| 11 | 12 |

When electrophilic substitution is carried out at an allylic substrate, the product may be rearranged (**11** → **12**). This type of process is analogous to the nucleophilic

[30]Cram, D.J.; Ford, W.T.; Gosser, L. *J. Am. Chem. Soc.* **1968**, *90*, 2598; Ford, W.T.; Cram, D.J. *J. Am. Chem. Soc.* **1968**, *90*, 2606, 2612. See also Wong, S.M.; Fischer, H.P.; Cram, D.J. *J. Am. Chem. Soc.* **1971**, *93*, 2235; Buchholz, S.; Harms, K.; Massa, W.; Boche, G. *Angew. Chem. Int. Ed.* **1989**, *28*, 73.

[31]Almy, J.; Hoffman, D.H.; Chu, K.C.; Cram, D.J. *J. Am. Chem. Soc.* **1973**, *95*, 1185.

[32]Dreiding, A.S.; Pratt, R.J. *J. Am. Chem. Soc.* **1954**, *76*, 1902. See also Walborsky, H.M.; Turner, L.M. *J. Am. Chem. Soc.* **1972**, *94*, 2273.

allylic rearrangements discussed in Chapter 10 (p. 468). There are two principal pathways. The first of these is analogous to the S_E1 mechanism in that the leaving group is first removed, giving a resonance-stabilized allylic carbanion, which then attacks the electrophile.

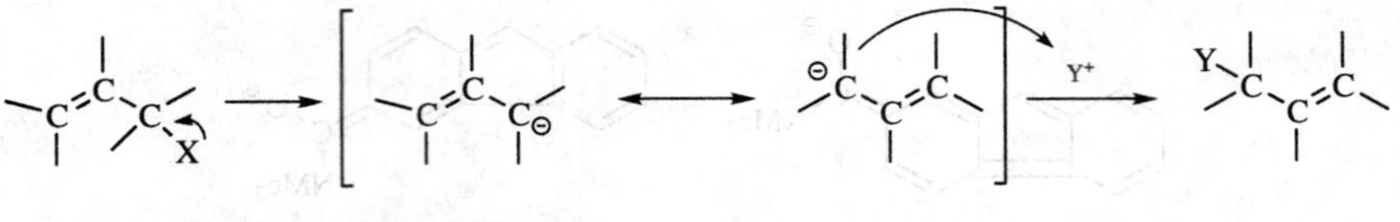

In the other pathway, the Y group is first attacked by the π-bond, giving a carbocation, which then loses X with formation of the alkene unit.

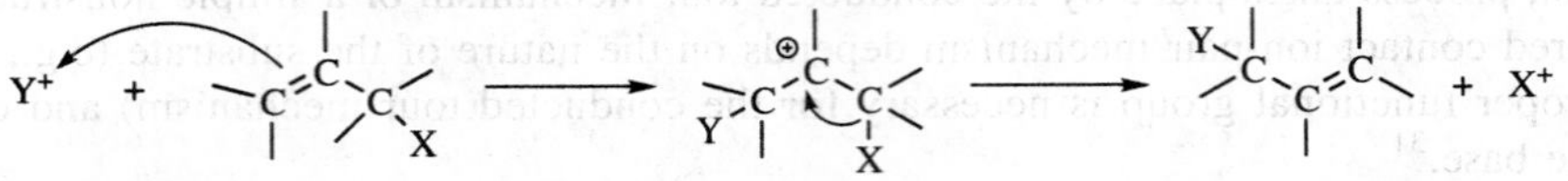

These mechanisms are more fully discussed under reaction **12-2**.

Most electrophilic allylic rearrangements involve loss of hydrogen, but they have also been observed with metallic leaving groups.[33] Sleezer, Winstein, and Young found that crotylmercuric bromide reacted with HCl ~10^7 times faster than *n*-butyl-mercuric bromide and the product was >99% 1-butene.[34] These facts point to an S_Ei' mechanism (IUPAC designation *cyclo*-1/3/$D_EA_ED_nA_n$):

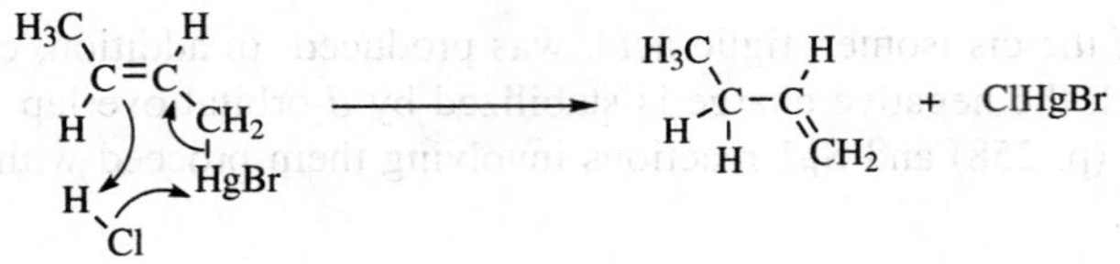

The reaction of the same compound with acetic acid-perchloric acid seems to proceed by an S_E2' mechanism (IUPAC designation 1/3/D_EA_E):[34]

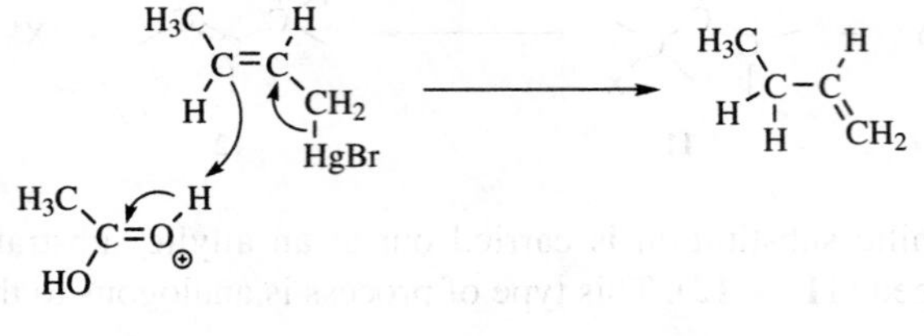

---

[33]For a review of reactions of allylic organometallic compounds, see Courtois, G.; Miginiac, L. *J. Organomet. Chem.* **1974**, *69*, 1.

[34]Sleezer, P.D.; Winstein, S.; Young, W.G. *J. Am. Chem. Soc.* **1963**, *85*, 1890. See also, Cunningham, I.M.; Overton, K.H. *J. Chem. Soc. Perkin Trans. 1* **1975**, 2140; Kashin, A.N.; Bakunin, V.N.; Khutoryanskii, V.A.; Beletskaya, I.P.; Reutov, O.A. *J. Org. Chem. USSR* **1979**, *15*, 12; *J. Organomet. Chem.* **1979**, *171*, 309.

The geometry of electrophilic allylic rearrangement has not been studied very much (cf. the nucleophilic case, p. 471), but in most cases the rearrangement takes place with anti stereoselectivity,[35] although syn stereoselectivity has also been demonstrated.[36] In one case, use of the electrophile H^+ and the leaving group $SnMe_3$ gave both syn and anti stereoselectivity, depending on whether the substrate was cis or trans.[37]

Other Mechanisms

Addition–elimination (**12-16**) and cyclic mechanisms (**12-40**) are also known.

Much less work has been done on electrophilic aliphatic substitution mechanisms than on nucleophilic substitutions, and the exact mechanisms of many of the reactions in this chapter are in doubt. For many of them, not enough work has been done to permit us to decide which of the mechanisms described in this chapter is operating, if indeed any is. There may be other electrophilic substitution mechanisms, and some of the reactions in this chapter may not even be electrophilic substitutions at all.

REACTIVITY

Only a small amount of work has been done in this area, compared to the vast amount done for aliphatic nucleophilic substitution and aromatic electrophilic substitution. Only a few conclusions, most of them sketchy or tentative, can be drawn.[38]

1. *Effect of Substrate.* For S_E1 reactions electron-donating groups decrease rates and electron-withdrawing groups increase them. This is as would be expected from a reaction in which the rate-determining step is analogous to the cleavage of a proton from an acid. For the S_E2 (back) mechanism, Jensen and Davis[12] showed that the reactivity of alkyl groups is similar to that for the S_N2 mechanism (i.e., Me > Et > Pr > iPr > neopentyl), as would be expected, since both involve backside attack and both are equally affected by steric hindrance. In fact, this pattern of reactivity can be regarded as evidence for the occurrence of the S_E2 (back) mechanism in cases where

[35]Hayashi, T.; Ito, H.; Kumada, M. *Tetrahedron Lett.* *1982*, *23*, 4605; Wetter, H.; Scherer, P. *Helv. Chim. Acta* *1983*, *66*, 118; Wickham, G.; Kitching, W. *J. Org. Chem.* *1983*, *48*, 612; Fleming, I.; Kindon, N.D.; Sarkar, A.K. *Tetrahedron Lett.* *1987*, *28*, 5921; Hayashi, T.; Matsumoto, Y.; Ito, Y. *Chem. Lett.* *1987*, 2037, *Organometallics* *1987*, *6*, 885; Matassa, V.G.; Jenkins, P.R.; Kümin, A.; Damm, L.; Schreiber, J.; Felix, D.; Zass, E.; Eschenmoser, A. *Isr. J. Chem.* *1989*, *29*, 321.
[36]Wetter, H.; Scherer, P.; Schweizer, W.B. *Helv. Chim. Acta* *1979*, *62*, 1985; Young, D.; Kitching, W. *J. Org. Chem.* *1983*, *48*, 614; *Tetrahedron Lett.* *1983*, *24*, 5793.
[37]Kashin, A.N.; Bakunin, V.N.; Beletskaya, I.P.; Reutov, O.A. *J. Org. Chem. USSR* *1982*, *18*, 1973. See also, Wickham, G.; Young, D.; Kitching, W. *Organometallics* *1988*, *7*, 1187.
[38]For a discussion, see Abraham, M.H. *Comprehensive Chemical Kinetics*, Bamford, C.H.; Tipper, C.F.H., Eds., Vol. 12; Elsevier, NY, *1973*, pp. 211–241.

TABLE 12.1. Relative Rates of the Reaction of RHgBr with Br$_2$ and Br$^-$ [41]

R	Relative Rate	R	Relative Rate
Me	1	Et	10.8
Et	10.8	iBu	1.24
iPr	780	Neopentyl	0.173
t-Bu	3370		

stereochemical investigation is not feasible.[39] For S$_E$2 reactions that proceed with retention, several studies have been made with varying results, depending on the reaction.[40] One such study, which examined the reaction RHgBr + Br$_2$ → RBr catalyzed by Br$^-$, gave the results shown in Table 12.1.[41] As can be seen, a branching increased the rates, while β branching decreased them. Sayre and Jensen attributed the decreased rates to steric hindrance, although attack here was definitely frontside, and the increased rates to the electron-donating effect of the alkyl groups, which stabilized the electron-deficient transition state.[42] Of course, steric hindrance should also be present with the a branched groups, so these workers concluded that if it were not, the rates would be even greater. The Br electrophile is a rather large one and it is likely that smaller steric effects are present with smaller electrophiles. The rates of certain second-order substitutions of organotin compounds have been found to increase with increasing electron withdrawal by substituents. This behavior has been ascribed[43] to an S$_E$2 mechanism involving ion pairs, analogous to Sneen's ion-pair mechanism for nucleophilic substitution (p. 441). Solvolysis of 2-bromo-1,1,1-trifluoro-2-(p-methoxyphenyl)ethane in water proceeds via a free carbocation intermediate, but ion pairing influences the reaction in the presence of bromide ion.[44]

2. *Effect of Leaving Group.* For both S$_E$1 and second-order mechanisms, the more polar the C–X bond, the easier it is for the electrofuge to cleave. For metallic leaving groups in which the metal has a valence >1, the nature of the other group or groups attached to the metal thus has an effect on the reaction.

---

[39]Another method involves measurement of the susceptibility of the rate to increased pressure: See Isaacs, N.S.; Javaid, K. *Tetrahedron Lett.* **1977**, 3073; Isaacs, N.S.; Laila, A.H. *Tetrahedron Lett.* **1984**, 25, 2407.
[40]For some of these, see Abraham, M.H.; Grellier, P.L. *J. Chem. Soc. Perkin Trans. 2* **1973**, 1132; Dessy, R.E.; Reynolds, G.F.; Kim, J. *J. Am. Chem. Soc.* **1959**, 81, 2683; Minato, H.; Ware, J.C.; Traylor, T.G. *J. Am. Chem. Soc.* **1963**, 85, 3024; Boué, S.; Gielen, M.; Nasielski, J. *J. Organomet. Chem.* **1967**, 9, 443; Abraham, M.H.; Broadhurst, A.T.; Clark, I.D.; Koenigsberger, R.U.; Dadjour, D.F. *J. Organomet. Chem.* **1981**, 209, 37.
[41]Sayre, L.M.; Jensen, F.R. *J. Am. Chem. Soc.* **1979**, 101, 6001.
[42]A similar conclusion, that steric and electronic effects are both present, was reached for a different system by Nugent, W.A.; Kochi, J.K. *J. Am. Chem. Soc.* **1976**, 98, 5979.
[43]Reutov, O.A. *J. Organomet. Chem.* **1983**, 250, 145. See also, Butin, K.P.; Magdesieva, T.V. *J. Organomet. Chem.* **1985**, 292, 47; Beletskaya, I.P. *Sov. Sci. Rev. Sect. B* **1979**, 1, 119.
[44]Richard, J.P. *J. Org. Chem.* **1992**, 57, 625.

For example, consider a series of organomercurials RHgW. Because a more electronegative W decreases the polarity of the C–Hg bond and furthermore results in a less stable HgW$^+$, the electrofugal ability of HgW decreases with increasing electronegativity of W. Thus, HgR′ (from RHgR′) is a better leaving group than HgCl (from RHgCl). Also in accord with this is the leaving-group order Hg-*t*-Bu > Hg-*i*Pr > HgEt > HgMe, reported for acetolysis of R$_2$Hg,[42] since the more highly branched alkyl groups better help to spread the positive charge. It might be expected that, when metals are the leaving groups, S$_E$1 mechanisms would be favored, while with carbon leaving groups, second-order mechanisms would be found. However, the results so far reported have been just about the reverse of this. For carbon leaving groups the mechanism is usually S$_E$1, while for metallic leaving groups the mechanism is almost always S$_E$2 or S$_E$i. A number of reports of S$_E$1 reactions with metallic leaving groups have appeared,[45] but the mechanism is not easy to prove and many of these reports have been challenged.[46] Reutov and co-workers[45] have expressed the view that in such reactions a nucleophile (which may be the solvent) must assist in the removal of the electrofuge and refer to such processes as S$_E$1(*N*) reactions.

3. *Effect of Solvent.*[47] In addition to the solvent effects on certain S$_E$1 reactions, mentioned earlier (p. 758), solvents can influence the mechanism that is preferred. As with nucleophilic substitution (p. 501), an increase in solvent polarity increases the possibility of an ionizing mechanism, in this case S$_E$1, in comparison with the second-order mechanisms, which do not involve ions. As previously mentioned (p. 758), the solvent can also exert an influence between the S$_E$2 (front or back) and S$_E$i mechanisms in that the rates of S$_E$2 mechanisms should be increased by an increase in solvent polarity, while S$_E$i mechanisms are much less affected.

REACTIONS

The reactions in this chapter are arranged in order of leaving group: hydrogen, metals, halogen, and carbon. Electrophilic substitutions at a nitrogen atom are treated last.

Hydrogen as Leaving Group

A. Hydrogen as the Electrophile

[45]For discussions, see Reutov, O.A. *Bull. Acad. Sci. USSR Div. Chem. Sci.* **1980**, *29*, 1461; Beletskaya, I.P.; Butin, K.P.; Reutov, O.A. *Organomet. Chem. Rev. Sect. A* **1971**, *7*, 51. See also, Deacon, G.B.; Smith, R.N.M. *J. Org. Chem. USSR* **1982**, *18*, 1584; Dembech, P.; Eaborn, C.; Seconi, G. *J. Chem. Soc. Chem. Commun.* **1985**, 1289.
[46]For a discussion, see Kitching, W. *Rev. Pure Appl. Chem.* **1969**, *19*, 1.
[47]For a discussion of solvent effects on organotin alkyl exchange reactions, see Petrosyan, V.S. *J. Organomet. Chem.* **1983**, *250*, 157.

12-1   Hydrogen Exchange

Deuterio-de-hydrogenation or Deuteriation

$$R-H + D^+ \rightleftharpoons R-D + H^+$$

Hydrogen exchange can be accomplished by treatment with acids or bases. As with **11-1**, the exchange reaction is mostly used to study mechanistic questions, such as relative acidities, but it can be used synthetically to prepare deuterated or tritiated molecules. When ordinary strong acids, such as H_2SO_4 are used, only fairly acidic protons n carbon exchange, for example, acetylenic and allylic. However, primary, secondary, and tertiary hydrogens of alkanes can be exchanged by treatment with superacids (p. 236).[48] The order of hydrogen reactivity is tertiary > secondary > primary. Where C–C bonds are present, they may be cleaved also (**12-47**). The mechanism of the exchange (illustrated for methane) has been formulated as involving attack of H^+ on the C–H bond to give the pentavalent methanonium ion that loses H_2 to give a tervalent

$$H_3C-H + H^+ \rightleftharpoons \left[H_3C \cdots \begin{matrix} H \\ H \end{matrix} \right]^+ \rightleftharpoons CH_3^+ + H_2$$

Methanonium ion

carbocation.[49] The methanonium ion CH_5^+ has a three-center, two-electron bond.[50] It is not known whether the methanonium ion is a transition state or a true intermediate, but an ion CH_5^+ has been detected in the mass spectrum.[51] The IR spectrum of the ethanonium ion $C_2H_7^+$ has been measured in the gas phase.[52] Note that the two electrons in the three-center, two-electron bond can move in three directions, in accord with the threefold symmetry of such a structure. The electrons can move to unite the two hydrogens, leaving the CH_3^+ free (the forward reaction), or they can unite the CH_3 with either of the two hydrogens, leaving the other hydrogen as a free H^+ ion (the reverse reaction). Actually, the methyl cation is not stable under these conditions. It can go back to CH_4 by the route shown (leading to H^+ exchange) or it can react with additional CH_4 molecules (**12-20**) to eventually yield the *tert*-butyl cation, which is stable in these superacid solutions. Hydride ion can also be removed from alkanes (producing tervalent carbocations) by treatment with pure SbF_5 in the absence of any source of H^+.[53] Complete or almost complete perdeuteration of cyclic alkenes has been achieved by treatment with dilute DCl/D_2O in sealed Pyrex tubes at 165–280°C.[54]

[48]For reviews, see Olah, G.A.; Prakash, G.K.S.; Sommer, J. *Superacids*, Wiley, NY, *1985*, pp. 244–249; Olah, G.A. *Angew. Chem. Int. Ed. 1973*, *12*, 173; Brouwer, D.M.; Hogeveen, H. *Prog. Phys. Org. Chem. 1972*, *9*, 179, 180–203.

[49]The mechanism may not be this simple in all cases. For discussions, see McMurry, J.E.; Lectka, T. *J. Am. Chem. Soc. 1990*, *112*, 869; Culmann, J.; Sommer, J. *J. Am. Chem. Soc. 1990*, *112*, 4057.

[50]For a monograph on this type of species, see Olah, G.A.; Prakash, G.K.S.; Williams, R.E.; Field, L.D.; Wade, K. *Hypercarbon Chemistry*; Wiley, NY, *1987*.

[51]See, for example, Sefcik, M.D.; Henis, J.M.S.; Gaspar, P.P. *J. Chem. Phys. 1974*, *61*, 4321.

[52]Yeh, L.I.; Pric, J.M.; Lee, Y.T. *J. Am. Chem. Soc. 1989*, *111*, 5597.

[53]Lukas, J.; Kramer, P.A.; Kouwenhoven, A.P. *Recl. Trav. Chim. Pays-Bas 1973*, *92*, 44.

[54]Werstiuk, N.H.; Timmins, G. *Can. J. Chem. 1985*, *63*, 530; *1986*, *64*, 1564.

Exchange with bases involves an S_E1 mechanism.

Step 1   RH + B$^-$ $\longrightarrow$ R$^-$ + BH

Step 2   R$^-$ + BD $\longrightarrow$ RD + B$^-$

Of course, such exchange is most successful for relatively acidic protons, such as those a to a carbonyl group, but even weakly acidic protons can exchange with bases if the bases are strong enough (see p. 251).

Alkanes and cycloalkanes, of both low and high molecular weight, can be fully perdeuterated treatment with D_2 gas and a catalyst, such as Rh, Pt, or Pd.[55]

OS **VI**, 432.

12-2   Migration of Double Bonds

3/Hydro-de-hydrogenation

$$C_5H_{11}-CH_2-CH=CH_2 \xrightarrow[\text{Me}_2\text{SO}]{\text{KNH}_2} C_5H_{11}-CH=CH-CH_3$$

The double bonds of many unsaturated compounds are shifted[56] on treatment with strong bases.[57] In many cases, equilibrium mixtures are obtained and the thermodynamically most stable isomer predominates.[58] Thus, if the new double bond can be in conjugation with one already present or with an aromatic ring, the migration favors the conjugated compound.[59] If the choice is between an exocyclic and an endocyclic double bond (particularly with six-membered rings), it generally chooses the latter. In the absence of considerations like these, Zaitsev's rule (p. 1497) applies and the double bond goes to the carbon with the fewest hydrogens. All these considerations lead us to predict that terminal alkenes can be isomerized to internal ones, nonconjugated alkenes to conjugated, exo six-membered-ring alkenes to endo, and so on, and not the other way around. This is indeed usually the case.

[55]See, for example, Atkinson, J.G.; Luke, M.O.; Stuart, R.S. *Can. J. Chem.* **1967**, *45*, 1511.

[56]For a list of methods used to shift double and triple bonds, with references, see Larock, R.C. *Comprehensive Organic Transformations*, 2nd ed., Wiley-VCH, NY, **1999**, pp. 220–226, 567–568.

[57]For reviews of double-bond migrations, see Pines, H.; Stalick, W.M. *Base-Catalyzed Reactions of Hydrocarbons and Related Compounds*, Academic Press, NY, **1977**, pp. 25–123; DeWolfe, R.H., in Bamford, C.H.; Tipper, C.F.H. *Comprehensive Chemical Kinetics*, Vol. 9, Elsevier, NY, **1973**, pp. 437–449; Yanovskaya, L.A.; Shakhidayatov, Kh. *Russ. Chem. Rev.* **1970**, *39*, 859; Hubert, A.J.; Reimlinger, H. *Synthesis* **1969**, 97; **1970**, 405; Mackenzie, K., in *The Chemistry of Alkenes*, Vol. 1, Patai, S. pp. 416–436, vol. 2, Zabicky, J. pp. 132–148; Wiley, NY, 1964, **1970**; Broaddus, C.D. *Acc. Chem. Res.* **1968**, *1*, 231; Cram, D.J. *Fundamentals of Carbanion Chemistry*, Academic Press, NY, **1965**, pp. 175–210.

[58]For lists of which double bonds are more stable in conversions of $XCH_2CH=CHY$ to $XCH=CHCH_2Y$, see Hine, J.; Skoglund, M.J. *J. Org. Chem.* **1982**, *47*, 4766. See also, Hine, J.; Linden, S. *J. Org. Chem.* **1983**, *48*, 584.

[59]For a review of conversions of β,γ enones to α,β enones, see Pollack, R.M.; Bounds, P.L.; Bevins, C.L., in Patai, S.; Rappoport, Z. *The Chemistry of Enones*, pt. 1, Wiley, NY, **1989**, pp. 559–597.

This reaction, for which the term *prototropic rearrangement* is sometimes used, is an example of electrophilic substitution with accompanying allylic rearrangement. The mechanism involves abstraction by a base to give a resonance-stabilized carbanion, which then combines with a proton at the position that will give the more stable alkene:[60]

Step 1   R—CH=CH—CH₃ + B ⟶ [R—CH̄—CH=CH₂ ⟷ R—CH=CH—CH̄₂] + HB⁺

Step 2   [R—CH̄—CH=CH₂ ⟷ R—CH=CH—CH̄₂] —BH⁺→ R—CH=CH—CH₃ + B

This mechanism is exactly analogous to the allylic-rearrangement mechanism for nucleophilic substitution (p. 468). UV spectra of allylbenzene and 1-propenylbenzene in solutions containing NH_2^- are identical, which shows that the same carbanion is present in both cases, as required by this mechanism.[61] The acid BH^+ protonates the position that will give the more stable product, although the ratio of the two possible products can vary with the identity of BH^+.[62] It has been shown that base-catalyzed double-bond shifts are partially intramolecular, at least in some cases.[63] The intramolecular nature has been ascribed to a *conducted tour mechanism* (p. 761) in which the base leads the proton from one carbanionic site to the other (**13** → **14**).[64]

R—CH=CH—CH₂ + B ⇌ R—[13]— H—B ⇌ R—[14]— B—H ⇌ R—CH=CH—CH₃ + B

13   **14**

Triple bonds can also migrate in the presence of bases,[65] but through the allene intermediate:[66]

R—CH₂—C≡CH ⇌ R—CH=C=CH₂ ⇌ R—C≡C—CH₃

---

[60]See, for example, Hassan, M.; Nour, A.R.O.A.; Satti, A.M.; Kirollos, K.S. *Int. J. Chem. Kinet.* **1982**, *14*, 351; Pollack, R.M.; Mack, J.P.G.; Eldin, S. *J. Am. Chem. Soc.* **1987**, *109*, 5048.

[61]Rabinovich, E.A.; Astaf'ev, I.V.; Shatenshtein, A.I. *J. Gen. Chem. USSR* **1962**, *32*, 746.

[62]Hünig, S.; Klaunzer, N.; Schlund, R. *Angew. Chem. Int. Ed.* **1987**, *26*, 1281.

[63]See, for example, Cram, D.J.; Uyeda, R.T. *J. Am. Chem. Soc.* **1964**, *86*, 5466; Bank, S.; Rowe, Jr., C.A.; Schriesheim, A. *J. Am. Chem. Soc.* **1963**, *85*, 2115; Doering, W. von E.; Gaspar, P.P. *J. Am. Chem. Soc.* **1963**, *85*, 3043; Ohlsson, L.; Wold, S.; Bergson, G. *Ark. Kemi.*, **1968**, *29*, 351.

[64]Almy, J.; Cram, D.J. *J. Am. Chem. Soc.* **1969**, *91*, 4459; Hussénius, A.; Matsson, O.; Bergson, G. *J. Chem. Soc. Perkin Trans. 2* **1989**, 851.

[65]For reviews, see Pines, H.; Stalick, W.M. *Base-Catalyzed Reactions of Hydrocarbons and Related Compounds*, Academic Press, NY, **1977**, pp. 124–204; Théron F.; Verny, M.; Vessière, R. in Patai, S. *The Chemistry of Carbon–Carbon Triple Bond*, pt. 1, Wiley, NY, **1978**, pp. 381–445; Bushby, R.J. *Q. Rev. Chem. Soc.* **1970**, *24*, 585; Iwai, I. *Mech. Mol. Migr.* **1969**, *2*, 73; Wotiz, J.H., in Viehe, H.G. *Acetylenes*, Marcel Dekker, NY, **1969**, pp. 365–424; Vartanyan, S.A.; Babanyan, Sh.O. *Russ. Chem. Rev.* **1967**, *36*, 670.

[66]For a review of rearrangements involving allenes, see Huntsman, W.D., in Patai, S. *The Chemistry of Ketenes, Allenes, and Related Compounds*, pt. 2; Wiley, NY, **1980**, pp. 521–667.

In general, strong bases, for example, NaNH$_2$, convert internal alkynes to terminal alkynes (a particularly good base for this purpose is potassium 3-aminopropylamide NH$_2$CH$_2$CH$_2$CH$_2$NHK[67]), because the equilibrium is shifted by formation of the acetylid ion. With weaker bases such as NaOH (which are not strong enough to remove the acetylenic proton), the internal alkynes are favored because of their greater thermodynamic stability. In some cases the reaction can be stopped at the allene stage.[68] The reaction then becomes a method for the preparation of allenes.[69] The reaction of propargylic alcohols with tosylhydrazine, PPh$_3$, and DEAD also generates allenes.[70]

Double-bond rearrangements can also take place on treatment with acids. Both proton and Lewis[71] acids can be used. The mechanism in the case of proton acids is the reverse of the previous one; first a proton is gained, giving a carbocation, and then another is lost:

Step 1 CH_3—CH_2—CH=CH_2 + H^+ $\longrightarrow$ CH_3—CH_2—$\overset{\oplus}{CH}$—CH_3

Step 2 CH_3—CH_2—$\overset{\oplus}{CH}$—CH_3 $\longrightarrow$ CH_3—CH=CH—CH_3 + H^+

As in the case of the base-catalyzed reaction, the thermodynamically most stable alkene is the one predominantly formed. However, the acid-catalyzed reaction is much less synthetically useful because carbocations give rise to many side products. If the substrate has several possible locations for a double bond, mixtures of all possible isomers are usually obtained. Isomerization of 1-decene, for example, gives a mixture that contains not only 1-decene and *cis-* and *trans*-2-decene, but also the cis and trans isomers of 3-, 4-, and 5-decene as well as branched alkenes resulting from rearrangement of carbocations. It is true that the most stable alkenes predominate, but many of them have stabilities that are close together. Acid-catalyzed migration of triple bonds (with allene intermediates) can be accomplished if very strong acids (e.g., HF–PF$_5$) are used.[72] If the mechanism is the same as that for double bonds, vinyl cations are intermediates.

Double-bond isomerization can also take place in other ways. Nucleophilic allylic rearrangements were discussed in Chapter 10 (p. 468). Electrocyclic and sigmatropic rearrangements are treated at **18-27–18-35**. Double-bond migrations have also been accomplished photochemically,[73] and by means of metallic ion (most

[67]Brown, C.A.; Yamashita, A. *J. Am. Chem. Soc.* **1975**, *97*, 891; Macaulay, S.R. *J. Org. Chem.* **1980**, *45*, 734; Abrams, S.R. *Can. J. Chem.* **1984**, *62*, 1333.

[68]For an example, see Oku, M.; Arai, S.; Katayama, K.; Shioiri, T. *Synlett* **2000**, 493.

[69]See Enomoto, M.; Katsuki, T.; Yamaguchi, M. *Tetrahedron Lett.* **1986**, *27*, 4599; Cunico, R.F.; Zaporowski, L.F.; Rogers, M. *J. Org. Chem.* **1999**, *64*, 9307.

[70]Myers, A.G.; Zheng, B. *J. Am. Chem. Soc.* **1996**, *118*, 4492. See Moghaddam, F.M.; Emami, R. *Synth. Commun.* **1997**, *27*, 4073 for the formation of alkoxy allenes from propargyl ethers.

[71]For an example of a Lewis acid catalyzed rearrangement, see Cameron G.S.; Stimson, V.R. *Aust. J. Chem.* **1977**, *30*, 923.

[72]Barry, B.J.; Beale, W.J.; Carr, M.D.; Hei, S.; Reid, I. *J. Chem. Soc. Chem. Commun.* **1973**, 177.

[73]Schönberg, A. *Preparative Organic Photochemistry*, Springer, NY, **1968**, pp. 22–24.

often complex ions containing Pt, Rh, or Ru) or metal carbonyl catalysts.[74] In the latter case, there are at least two possible mechanisms. One of these, which requires external hydrogen, is called the *metal hydride addition–elimination mechanism*:

$$R\diagdown\diagup\diagdown \underset{}{\overset{MH}{\rightleftharpoons}} R\diagdown\diagup\overset{}{\underset{CH_3}{\diagdown}} \underset{M}{\overset{-MH}{\rightleftharpoons}} R\diagdown\diagdown\diagup\diagup CH_3$$

The other mechanism, called the *π-allyl complex mechanism*, does not require external hydrogen and proceeds by hydrogen abstraction to form the η^3-π-allyl complex **15** (see p. 117 and **10-60**).

$$R\diagdown\diagup\diagdown \overset{M}{\rightleftharpoons} R\overset{M}{\diagdown\diagup\diagdown} \rightleftharpoons R\overset{H}{\underset{M}{\diagdown\diagdown\diagup}} \rightleftharpoons R\diagdown\diagup\overset{M}{\diagdown\diagup}CH_3 \overset{-M}{\rightleftharpoons} R\diagdown\diagup\diagdown\diagup CH_3$$

15

Another difference between the two mechanisms is that the former involves 1,2- and the latter 1,3-shifts. The isomerization of 1-butene by rhodium(I) is an example of a reaction that takes place by the metal hydride mechanism,[75] while an example of the π-allyl complex mechanism is found in the $Fe_3(CO)_{12}$-catalyzed isomerization of 3-ethyl-1-pentene.[76] A palladium catalyst was used to convert alkynones $RCOC\equiv CCH_2CH_2R'$ to 2,4-alkadien-1-ones, $RCOCH=CHCH=CHCHR'$.[77] The reaction of an en-yne with $HSiCl_3$ and a palladium catalyst generated an allene with moderate enantioselectivity (see p 148 for chiral allenes).[78]

The metal catalysis method has been used for the preparation of simple enols, by isomerization of allylic alcohols, for example,[79] these enols are stable enough for isolation (see p. 231), but slowly tautomerize to the aldehyde or ketone, with half-lives ranging from 40 to 50 min to several days.[79]

---

[74]For reviews, see Rodriguez, J.; Brun, P.; Waegell, B. *Bull. Soc. Chim. Fr.* **1989**, 799–823; Jardine, F.R., in Hartley, F.R.; Patai, S. *The Chemistry of the Metal-Carbon Bond*, Vol. 4, Wiley, NY, pp. 733–818, 736–740; Otsuka, S.; Tani, K., in Morrison, J.D. *Asymmetric Synthesis*, Vol. 5, Academic Press, NY, **1985**, pp. 171–191 (enantioselective); Colquhoun, H.M.; Holton, J.; Thompson, D.J.; Twigg, M.V. *New Pathways for Organic Synthesis*, Plenum, NY, **1984**, pp. 173–193; Khan, M.M.T.; Martell, A.E. *Homogeneous Catalysis by Metal Complexes*, Academic Press, NY, **1974**, pp. 9–37; Heck, R.F. *Organotransition Metal Chemistry*, Academic Press, NY, **1974**, pp. 76–82; Jira, R.; Freiesleben, W. *Organomet. React.* **1972**, *3*, 1, 133–149; Biellmann, J.F.; Hemmer, H.; Levisalles, J., in Hartley, F.R.; Patai, S. *The Chemistry of the Metal-Carbon Bond*, Vol. 2, Wiley, NY, pp. 224–230; Bird, C.W. *Transition Metal Intermediates in Organic Synthesis*, Academic Press, NY, **1967**, pp. 69–87; Davies, N.R. *Rev. Pure Appl. Chem.* **1967**, *17*, 83; Orchin, M. *Adv. Catal.* **1966**, *16*, 1.
[75]Cramer, R. *J. Am. Chem. Soc.* **1966**, *88*, 2272.
[76]Casey, C.P.; Cyr, C.R. *J. Am. Chem. Soc.* **1973**, *95*, 2248.
[77]Trost, B.M.; Schmidt, T. *J. Am. Chem. Soc.* **1988**, *110*, 2301.
[78]Han, J.W.; Tokunaga, N.; Hayashi, T. *J. Am. Chem. Soc.* **2001**, *123*, 12915.
[79]Bergens, S.H.; Bosnich, B. *J. Am. Chem. Soc.* **1991**, *113*, 958.

No matter which of the electrophilic methods of double-bond shifting is employed, the thermodynamically most stable alkene is usually formed in the largest amount in most cases, although a few anomalies are known. However, an indirect method of double-bond isomerization us known, leading to migration in the other direction. This involves conversion of the alkene to a borane (**15-16**), rearrangement of the borane (**18-11**), oxidation and hydrolysis of the newly formed borane to the alcohol **17** (see **12-31**), and dehydration of the alcohol (**17-1**) to the alkene. The reaction is driven by the fact that with heating the addition of borane is reversible, and the equilibrium favors formation of the less sterically hindered borane, which is **16** in this case.

Since the migration reaction is always toward the end of a chain, terminal alkenes can be produced from internal ones, so the migration is often opposite to that with the other methods. Alternatively, the rearranged borane can be converted directly to the alkene by heating with an alkene of molecular weight higher than that of the product (**17-15**). Photochemical isomerization can also lead to the thermodynamically less stable isomer.[80]

If a hydroxy group is present in the chain, *it* may lose a proton, so that a ketone is the product, for example,[81]

$$R_2C=CHCH_2CH_2CHOHCH_3 \xrightarrow[\text{acid}]{\text{polyphosphoric}} R_2CHCH_2CH_2CH_2COCH_3$$

Similarly, α-hydroxy triple-bond compounds have given α,β-unsaturated ketones.[82]

---

[80]For example, see Kropp, P.J.; Krauss, H.J. *J. Am. Chem. Soc.* **1967**, *89*, 5199; Reardon, Jr., E.J.; Krauss, H. *J. Am. Chem. Soc.* **1971**, *93*, 5593; Duhaime, R.M.; Lombardo, D.A.; Skinner, I.A.; Weedon, A.C. *J. Org. Chem.* **1985**, *50*, 873.
[81]Colonge, J.; Brunie, J. *Bull. Soc. Chim. Fr.* **1963**, 1799. For an example with basic catalysis, see Hoffmann, H.M.R.; Köver, A.; Pauluth, D. *J. Chem. Soc. Chem. Commun.* **1985**, 812. For an example with a ruthenium complex catalyst, see Trost, B.M.; Kulawiec, R.J. *Tetrahedron Lett.* **1991**, *32*, 3039.
[82]For example, see Chabardes, P. *Tetrahedron Lett.* **1988**, *29*, 6253.

See **15-1** for related reactions in which double bonds migrate or isomerize.

OS **II**, 140; **III**, 207; **IV**, 189, 192, 195, 234, 398, 683; **VI**, 68, 87, 815, 925; **VII**, 249; **VIII**, 146, 196, 251, 396, 553; **X**, 156, 165; **81**, 147

12-3 Keto–Enol Tautomerization

3/*O*-Hydro-de-hydrogenation

The tautomeric equilibrium between enols and ketones or aldehydes (keto–enol tautomerism) is a form of prototropy,[83] but is not normally a preparative reaction. For some ketones, however, both forms can be prepared (see p. 101 for a discussion of this and other aspects of tautomerism). Keto–enol tautomerism occurs in systems containing one or more carbonyl groups linked to sp^3 carbons bearing one or more hydrogen atoms. The keto tautomer is generally more stable than the enol tautomer for neutral systems, and for most ketones and aldehydes only the keto form is detectable under ordinary conditions. The availability of additional intramolecular stabilization through hydrogen bonding or complete electron delocalization (as in phenol), may cause the enol tautomer to be favored.

Keto–enol tautomerism cannot take place without at least a trace of acid or base,[84] since the acidic or basic center or both in the tautomeric substance is too weak.[85] In this equilibrium, the heteroatom is the basic site the proton is the acidic site. For tautomerism in general (see p 98),[86] the presence of an acid or a base is not necessary to initiate the isomerization since each tautomeric substance possesses amphiprotic properties.[85] Keto-enol tautomerism is therefore the exception.

[83]Patai, S. *The Chemistry of the Carbonyl Group*, Wiley, London, *1966*; Rappoport, Z. *The Chemistry of Enols*, Wiley, NY, *1990*; Kresge, A.J. *Chem. Soc. Rev. 1996*, 25, 275; Karelson, M.; Maran, U.; Katritzky, A.R. *Tetrahedron 1996*, 52, 11325; Rappoport, Z.; Frey, J.; Sigalov, M.; Rochlin, E. *Pure Appl. Chem. 1997*, 69, 1933; Fontana, A.; De Maria, P.; Siani, G.; Pierini, M.; Cerritelli, S.; Ballini, R. *Eur. J. Org. Chem. 2000*, 1641; Iglesias, E. *Curr. Org. Chem. 2004*, 8, 1.

[84]Bell, R.P. *Acid–Base Catalysis*, Oxford University Press, Oxford, *1941*; Jones, J.R. *The Ionisation of Carbon Acids*, Academic Press, London, *1973*; Pederson, K.J. *J. Phys. Chem. 1934*, 38, 581; Lienhard, G.E.; Wang, T. C. *J. Am. Chem. Soc. 1969*, 91, 1146; Toullec, J. *Adv. Phys. Org. Chem. 1982*, 18, 1. See also, Chiang, Y.; Kresge, A.J.; Santaballa, J.A.; Wirz, J. *J. Am. Chem. Soc. 1988*, 110, 5506.

[85]Raczynska, E. D.; Kosinska, W.; Osmialowski, B.; Gawinecki, R. *Chem. Rev. 2005*, 105, 3561.

[86]See Patai, S. *The Chemistry of the Carbonyl Group*, Wiley, London, *1966*; Rappoport, Z. *The Chemistry of Enols*, Wiley, NY, *1990*; Patai, S. *The Chemistry of the Thiol Group*, Wiley, London, *1974*; Zabicky, J. *The Chemistry of Amides*, Wiley, London, *1970*; Boyer, J. H. *The Chemistry of the Nitro and Nitroso Groups*, Interscience Publishers, NY, *1969*; Patai, S. *The Chemistry of Amino, Nitroso, Nitro Compounds and their Derivatives*, Wiley, NY, *1982*; Patai, S. *The Chemistry of Amino, Nitroso, Nitro and Related Groups, Supplement F2*, Wiley, Chichester, *1996*; Cook, A. G. *Enamines*, 2nd ed., Marcel Dekker, NY, *1998*.

Polar protic solvents, such as water or alcohol, may participate in the proton transfer by forming a cyclic or a linear complex with the tautomers.[87] Whether the complex formed is cyclic or linear depends on the conformation and configuration of the tautomers. In a strongly polar aprotic solvent and in the presence of an acid or a base, the tautomeric molecule may lose or gain a proton and form the corresponding mesomeric anion or cation, which, in turn, may gain or lose a proton, respectively, and yield a new tautomeric form.[88] The structural features of the carbonyl compound influences the equilibrium.[89] There is a rate acceleration when $LiN(SiMe_3)_2$–NEt_3 is used.[90] It has been shown that ring strain plays no significant role on the rate of base-catalyzed enolization.[91] Differing conjugative stabilization by CH-π orbital overlap does not directly influence stereoselectivity, and steric effects are generally not large enough to cause the several kcal/mol energy difference seen between transition structures unless there is exceptional crowding.[92] It is noted that sterically stabilized enols are known,[93] including arylacetaldehydes.[94] Torsional strain involving vicinal bonds does contribute significantly to stereoselectivity in enolate formation.[92]

The acid and base catalyzed mechanisms are identical to those in **12-2**.[95]

Acid-catalyzed

[87]Lledós, A.; Bertran, J. *Tetrahedron Lett.* **1981**, *22*, 775; Zielinski, T.J.; Poirier, R.A.; Peterson, M.R.; Csizmadia, I.G. *J. Comput. Chem.* **1983**, *4*, 419; Yamabe, T.; Yamashita, K.; Kaminoyama, M.; Koizumi, M.; Tachibana, A.; Fukui, K. *J. Phys. Chem.* **1984**, *88*, 1459; Chen, Y.; Gai, F.; Petrich, J.W. *J. Am. Chem. Soc.* **1993**, *115*, 10158; Herbich, J.; Dobkowski, J.; Thummel, R.P.; Hegde, V.; Waluk, J. *J. Phys. Chem. A* **1997**, *101*, 5839; Gorb, L.; Leszczynski, J. *J. Am. Chem. Soc.* **1998**, *120*, 5024; Guo, J. X.; Ho, J. J. *J. Phys. Chem. A* **1999**, *103*, 6433.

[88]Watson, H.B. *Trans. Faraday Soc.* **1941**, *37*, 713; Kabachnik, M.I. *Dokl. Akad. Nauk SSSR* **1952**, *83*, 407; Perez Ossorio, R.; Hughes, E.D. *J. Chem. Soc.* **1952**, 426; Briegleb, G.; Strohmeier, W. *Angew. Chem.* **1952**, *64*, 409; Baddar, F.G.; Iskander, Z. *J. Chem. Soc.* **1954**, 203.

[89]Hegarty, A.F.; Dowling, J.P.; Eustace, S.J.; McGarraghy, M. *J. Am. Chem. Soc.* **1998**, *120*, 2290.

[90]Zhao, P.; Collum, D.B. *J. Am. Chem. Soc.* **2003**, *125*, 4008.

[91]Cantlin, R.J.; Drake, J.; Nagorski, R.W. *Org. Lett.* **2002**, *4*, 2433.

[92]Behnam, S.M.; Behnam, S.E.; Ando, K.; Green, N.S.; Houk, K.N. *J. Org. Chem.* **2000**, *65*, 8970.

[93]Miller, A.R. *J. Org. Chem.*, **1976**, *41*, 3599.

[94]Fuson, R.C.; Southwick, P.L.; Rowland, Jr., S.P. *J. Am. Chem. Soc.* **1944**, *66*, 1109; Fuson, R.C.; Tan, T.-L. *J. Am. Chem. Soc.* **1948**, *70*, 602.

[95]For reviews of the mechanism, see Keeffe, J.R.; Kresge, A.J., in Rappoport, Z. *The Chemistry of Enols*, Wiley, NY, **1990**, pp. 399–480; Toullec, J. *Adv. Phys. Org. Chem.* **1982**, *18*, 1; Lamaty, G. *Isot. Org. Chem.* **1976**, *2*, 33. For discussions, see Ingold, C.K. *Structure and Mechanism in Organic Chemistry*, 2nd ed., Cornell University Press, Ithaca, NY, **1969**, pp. 794–837; Bell, R.P. *The Proton in Chemistry*, 2nd ed., Cornell University Press, Ithaca, NY, **1973**, pp. 171–181; Bruice, P.Y.; Bruice, T.C. *J. Am. Chem. Soc.* **1976**, *98*, 844; Shelly, K.P.; Venimadhavan, S.; Nagarajan, K.; Stewart, R. *Can. J. Chem.* **1989**, *67*, 1274. For a review of stereoelectronic control in this mechanism, see Pollack, R.M. *Tetrahedron* **1989**, *45*, 4913.

Base-catalyzed[96]

18

For each catalyst, the mechanism for one direction is the exact reverse of the other, by the principle of microscopic reversibility.[97] As expected from mechanisms in which the C–H bond is broken in the rate-determining step, substrates of the type RCD_2COR show deuterium isotope effects (of ~5) in both the basic-[98] and the acid[99]-catalyzed processes. The keto–enol/enolate anion equilibrium has been studied in terms of the influence of β-oxygen[100] or β-nitrogen[101] substituents.

Although the conversion of an aldehyde or a ketone to its enol tautomer is not generally a preparative procedure, the reactions do have their preparative aspects. If a full equivalent of base per equivalent of ketone is used, the enolate ion (**18**) is formed and can be isolated[102] (see, e.g., the alkylation reaction in **10-68**).[103] When enol ethers or esters are hydrolyzed, the enols initially formed immediately tautomerize to the aldehydes or ketones. In addition, the overall processes (forward plus reverse reactions) are often used for equilibration purposes. When an optically active compound in which the chirality is due to an stereogenic carbon α to a carbonyl group (as in **19**) is treated with acid or base, racemization results.[104]

---

[96]Another mechanism for base-catalyzed enolization has been reported when the base is a tertiary amine: See Bruice, P.Y. *J. Am. Chem. Soc.* **1983**, *105*, 4982; **1989**, *111*, 962; **1990**, *112*, 7361.

[97]It has been proposed that the acid-catalyzed ketonization of simple enols is concerted; that is, both of the processes shown in the equation take place simultaneously. This would mean that in these cases the forward reaction is also concerted. For evidence in favor of this proposal, see Capon, B.; Siddhanta, A.K.; Zucco, C. *J. Org. Chem.* **1985**, *50*, 3580. For evidence against it, see Chiang, Y.; Hojatti, M.; Keeffe, J.R.; Kresge, A.J.; Schepp, N.P.; Wirz, J. **1987**, *109*, 4000 and references cited therein.

[98]Riley, T.; Long, F.A. *J. Am. Chem. Soc.* **1962**, *84*, 522; Xie, L.; Saunders, Jr., W.H. *J. Am. Chem. Soc.* **1991**, *113*, 3123.

[99]Swain, C.G.; Stivers, E.C.; Reuwer Jr., J.F.; Schaad, L.J. *J. Am. Chem. Soc.* **1958**, *80*, 5885; Lienhard, G.E.; Wang, T. *J. Am. Chem. Soc.* **1969**, *91*, 1146. See also Toullec, J.; Dubois, J.E. *J. Am. Chem. Soc.* **1974**, *96*, 3524.

[100]Chiang, Y.; Kresge, A.J.; Meng, Q.; More, O'Farrall, R.A.; Zhu, Y. *J. Am. Chem. Soc.* **2001**, *123*, 11562.

[101]Chiang, Y.; Griesbeck, A. G.; Heckroth, H.; Hellrung, B.; Kresge, A. J.; Meng, Q.; O'Donoghue, A. C.; Richard, J. P.; Wirz, J. *J. Am. Chem. Soc.* **2001**, *123*, 8979.

[102]For nmr studies of the Li enolate of acetaldehyde in solution, see Wen, J.Q.; Grutzner, J.B. *J. Org. Chem.* **1986**, *51*, 4220.

[103]For a review of the preparation and uses of enolates, see d'Angelo, J. *Tetrahedron* **1976**, *32*, 2979. For a discussion of solid state enolate chemistry, see Fruchart, J.-S.; Lippens, G.; Kuhn, C.; Gran-Masse, H.; Melnyk, O. *J. Org. Chem.* **2002**, *67*, 526.

[104]For an exception, see Guthrie, R.D.; Nicolas, E.C. *J. Am. Chem. Soc.* **1981**, *103*, 4637.

If there is another

19   *cis*-Decalone   *trans*-Decalone

stereogenic center in the molecule, the less stable epimer can be converted to the more stable one in this manner, and this is often done. For example, *cis*-decalone can be equilibrated to the trans isomer. Isotopic exchange can also be accomplished at the a position of an aldehyde or ketone in a similar manner. The role of additives, such as $ZnCl_2$ on the stereogenic enolization reactions using chiral cases has been discussed.[105] Enantioselective enolate anion protonation reactions have been studied.[106] For the acid-catalyzed process, exchange or equilibration is accomplished only if the carbonyl compound is completely converted to the enol and then back, but in the base-catalyzed process exchange or equilibration can take place if only the first step (conversion to the enolate ion) takes place. The difference is usually academic. In cyclic compounds, cis- to trans-isomerization can occur via the enol.[107]

20   **21**   **22**   **20**
Racemic   Optically active

In the case of the ketone **20**, a racemic mixture was converted to an optically active mixture (optical yield 46%) by treatment with the chiral base **21**.[108] This happened because **21** reacted with one enantiomer of **20** faster than with the other (an example of kinetic resolution). The enolate **22** must remain coordinated with the chiral amine, and it is the amine that reprotonate **22**, not an added proton donor.

Enolizable hydrogens can be replaced by deuterium (and ^{16}O by ^{18}O) by passage of a sample through a deuterated (or ^{18}O-containing) gas-chromatography column.[109]

[105]Coggins, P.; Gaur, S.; Simpkins, N.S. *Tetrahedron Lett.* **1995**, *36*, 1545.

[106]Vedejs, E.; Kruger, A.W.; Suna, E. *J. Org. Chem.* **1999**, *64*, 7863.

[107]Dechoux, L.; Doris, E. *Tetrahedron Lett.* **1994**, *35*, 2017.

[108]Eleveld, M.B.; Hogeveen, H. *Tetrahedron Lett.* **1986**, *27*, 631. See also, Shirai, R.; Tanaka, M.; Koga, K. *J. Am. Chem. Soc.* **1986**, *108*, 543; Cain, C.M.; Cousins, R.P.C.; Coumbarides, G.; Simpkins, N.S. *Tetrahedron* **1990**, *46*, 523.

[109]Senn, M.; Richter, W.J.; Burlingame, A.L. *J. Am. Chem. Soc.* **1965**, *87*, 680; Richter, W.J.; Senn, M.; Burlingame, A.L. *Tetrahedron Lett.* **1965**, 1235.

There are many enol-keto interconversions and acidification reactions of enolate ions to the keto forms listed in *Organic Syntheses*. No attempt is made to list them here.

B. Halogen Electrophiles

Halogenation of unactivated hydrocarbons is discussed in **14-1**.

12-4 Halogenation of Aldehydes and Ketones

Halogenation or **Halo-de-hydrogenation**

Aldehydes and ketones can be halogenated in the a position with bromine, chlorine, or iodine.[110] The reaction is not successful with fluorine.[111] Sulfuryl chloride,[112] $NaClO_2/Mn(acac)_3$,[113] $Me_3SiCl–Me_2SO$,[114] $Me_3SiCl–MnO_2$,[115] and cupric chloride[116] have been used as reagents for chlorination, and N-bromosuccinimide (see **14-3**),[117] t-BuBr–DMSO,[118] $Me_3SiBr–DMSO$,[119] tetrabutylammonium tribromide,[120] and bromine • dioxane on silica with microwave irradiation[121] for bromination. Bromination of methyl ketones was done using PhI(OH)OTs with microwave irradiation, followed by treatment with $MgBr_2$ and microwave irradiation.[122] α-Chloro aldehydes are formed with Cl_2 and a catalytic amount of tetraethylammonium chloride.[123] Chlorination of aldehydes with good enantioselectivity was

[110]For a review, see House, H.O. *Modern Synthetic Reactions*, 2nd ed., W.A. Benjamin, NY, *1972*, pp. 459–478. For lists of reagents, with references, see Larock, R.C. *Comprehensive Organic Transformations*, 2nd ed., Wiley-VCH, NY, *1999*, pp.709–719. For a monograph, see De Kimpe, N.; Verhé, R. *The Chemistry of a Haloketones, α-Haloaldehydes, and α-Haloimines,* Wiley, NY, 1988.

[111]For a review of the preparation of α-fluoro carbonyl compounds, see Rozen, S.; Filler, R. *Tetrahedron* *1985*, *41*, 1111. For a monograph, see German, L.; Zemskov, S. *New Fluorinating Agents in Organic Chemistry,* Springer, NY, *1989*.

[112]For a review of sulfuryl chloride, see Tabushi, I.; Kitaguchi, H. in Pizey, J.S. *Synthetic Reagents*, Vol. 4; Wiley, NY, *1981*, pp. 336–396.

[113]Yakabe, S.; Hirano, M.; Morimoto, T. *Synth. Commun.* *1998*, *28*, 131.

[114]Bellesia, F.; Ghelfi, F.; Grandi, R.; Pagnoni, U.M. *J. Chem. Res. (S) 1986*, 426; Fraser, R.R.; Kong, F. *Synth. Commun.* *1988*, *18*, 1071.

[115]Bellesia, F.; Ghelfi, F.; Pagnoni, U.M.; Pinetti, A. *J. Chem. Res. (S) 1990*, 188.

[116]For a review, see Nigh, W.G., in Trahanovsky, W.S. *Oxidation in Organic Chemistry*, pt. B, Academic Press, NY, *1973*, pp. 67–81. Cupric chloride has been used to chlorinate α,β-unsaturated aldehydes and ketones in the γ position: Dietl, H.K.; Normark, J.R.; Payne, D.A.; Thweatt, J.G.; Young, D.A. *Tetrahedron Lett.* *1973*, 1719.

[117]For an example, see Tanemura, K.; Suzuki, T.; Nishida, Y.; Satsumabayashi, K.; Horaguchi, T. *Chem. Commun.* *2004*, 470.

[118]Armani, E.; Dossena, A.; Marchelli, R.; Casnati, G. *Tetrahedron 1984*, *40*, 2035.

[119]Bellesia, F.; Ghelfi, F.; Grandi, R.; Pagnoni, U.M. *J. Chem. Res. (S) 1986*, 428.

[120]Kajigaeshi, S.; Kakinami, T.; Okamoto, T.; Fujisaki, S. *Bull. Chem. Soc. Jpn. 1987*, *60*, 1159.

[121]Paul, S.; Gupta, V.; Gupta, R.; Loupy, A. *Tetrahedron Lett. 2003*, *44*, 439.

[122]Lee, J.C.; Park, J.Y.; Yoon, S.Y.; Bae, Y.H.; Lee, S.J. *Tetrahedron Lett. 2004*, *45*, 191.

[123]Bellesia, F.; DeBuyck, L.; Ghelfi, F.; Pagnoni, U.M.; Parson, A.F.; Pinetti, A. *Synthesis 2003*, 2173.

reported using a chlorinated quinone and L-proline, with the reaction proceeding via the chiral enamine.[124] Iodination has been accomplished with I_2–$HgCl_2$,[125] with I_2-cerium(IV) ammonium nitrate,[126] and with iodine using 1-chloromethyl-4-fluoro-1,4-diazoniabicyclo[2.2.2]octane bis(tetrafluoroborate), known as Selectfluor F–TEDA–BF_4, in methanol.[127] Treatment of a ketone with (hydroxy-*p*-nitrobenzene-sulfonyloxy)benzene followed by SmI_2 give the α-iodo ketone.[128] Methyl ketones react with N-iodosuccinimide (NIS) and tosic acid with microwave irradiation without solvent to give the α-iodo ketone.[129] Several methods have been reported for the preparation of α-fluoro aldehydes and ketones.[130] Another Selectfluor, 1-Fluoro-4-hydroxy-1,4-diazoniabicyclo[2.2.2]octane bis(tetrafluoroborate) has been used for the monofluorination of ketones,[131] as has a mixture of KI–KIO_3–H_2SO_4.[132] Active compounds, such as β-keto esters and β-diketones, have been fluorinated with an N-fluoro-N-alkylsulfonamide[133] (this can result in enantioselective fluorination, if an optically active N-fluorosulfonamide is used[134]), with F_2/N_2–HCOOH,[135] with NF_3O/Bu_4NOH,[136] and with acetyl hypofluorite.[137] The last reagent also fluorinates simple ketones in the form of their lithium enolates.[138]

For unsymmetrical ketones, the preferred position of halogenation is usually the more substituted: a CH group, then a CH_2 group, and then CH_3;[139] however, mixtures are frequent. With aldehydes the aldehydic hydrogen is sometimes replaced (see **14-4**). It is also possible to prepare di- and polyhalides. When basic catalysts are used, one a position of a ketone is completely halogenated before the other is

[124]Brochu, M.P.; Brown, S.P.; MacMillan, D.W.C. *J. Am. Chem. Soc.* **2004**, *126*, 4108. For this chlorination using a chiral pyrrolidine derivative with NCS, see Halland, N.; Braunton, A.; Bachmann, S.; Marigo, M.; Jorgensen, K.A. *J. Am. Chem. Soc.* **2004**, *126*, 4790. See Wack, H.; Taggi, A.E.; Hafez, A.M.; Drury III, W. J.; Lectka, T. *J. Am. Chem. Soc.* **2001**, *123*, 1531; Hafez, A.M.; Taggi, A.E.; Wack, H.; Esterbrook III, J.; Lectka, T. *Org. Lett.* **2001**, *3*, 2049.

[125]Barluenga, J.; Martinez-Gallo, J.M.; Najera, C.; Yus, M. *Synthesis* **1986**, 678.

[126]Horiuchi, C.A.; Kiji, S. *Bull. Chem. Soc. Jpn.* **1997**, *70*, 421. For another reagent, see Sket, B.; Zupet, P.; Zupan, M.; Dolenc, D. *Bull. Chem. Soc. Jpn.* **1989**, *62*, 3406.

[127]Jereb, M.; Stavber, S.; Zupan, M. *Tetrahedron* **2003**, *59*, 5935.

[128]Lee, J.C.; Jin, Y.S. *Synth. Commun.* **1999**, *29*, 2769.

[129]Lee, J.C.; Bae, Y.H. *Synlett* **2003**, 507.

[130]Davis, F.A.; Kasu, P.V.N. *Org. Prep. Proceed. Int.* **1999**, *31*, 125.

[131]Stavber, S.; Zupan, M. *Tetrahedron Lett.* **1996**, *37*, 3591.

[132]Okamoto, T.; Kakinami, T.; Nishimura, T.; Hermawan, I.; Kajigaeshi, S. *Bull. Chem. Soc. Jpn.* **1992**, *65*, 1731.

[133]Barnette, W.E. *J. Am. Chem. Soc.* **1984**, *106*, 452; Ma, J.-A. For an example using a chiral copper catalyst for asymmetric induction, see Cahard, D. *Tetrahedron Asymm* **2004**, *15*, 1007.

[134]Differding, E.; Lang, R.W. *Tetrahedron* **1988**, *29*, 6087.

[135]Chambers, R.D.; Greenhall, M.P.; Hutchinson, J. *J. Chem. Soc. Chem. Commun.* **1995**, 21.

[136]Gupta, O.D.; Shreeve, J.M. *Tetrahedron Lett.* **2003**, *44*, 2799.

[137]Lerman, O.; Rozen, S. *J. Org. Chem.* **1983**, *48*, 724. See also Purrington, S.T.; Jones, W.A. *J. Org. Chem.* **1983**, *48*, 761.

[138]Rozen, S.; Brand, M. *Synthesis* **1985**, 665. For another reagent, see Davis, F.A.; Han, W. *Tetrahedron Lett.* **1991**, *32*, 1631.

[139]For chlorination this is reversed if the solvent is methanol: Gallucci, R.R.; Going, R. *J. Org. Chem.* **1981**, *46*, 2532.

attacked, and the reaction cannot be stopped until all the hydrogens of the first carbon have been replaced (see below). If one of the groups is methyl, the haloform reaction (**12-44**) takes place. With acid catalysts, it is easy to stop the reaction after only one halogen has entered, although a second halogen can be introduced by the use of excess reagent. In chlorination the second halogen generally appears on the same side as the first,[140] while in bromination the α,α'-dibromo product is found.[141] Actually, with both halogens it is the α,α-dihalo ketone that is formed first, but in the case of bromination this compound isomerizes under the reaction conditions to the α,α' isomer.[140] α,α'-Dichloro ketones are formed by reaction of a methyl ketone with an excess of $CuCl_2$ and LiCl in DMF[142] or with HCl and H_2O_2 in methanol.[143] Aryl methyl ketones can be dibrominated ($ArCOCH_3 \rightarrow ArCOCHBr_2$) in high yields with benzyltrimethylammonium tribromide.[144] Active methylene compounds are chlorinated with NCS and $Mg(ClO_4)_2$.[145] Similar chlorination in the presence of a chiral copper catalyst led to α-chlorination with modest enantioselectivity.[146]

It is not the aldehyde or ketone itself that is halogenated, but the corresponding enol or enolate ion. The purpose of the catalyst is to provide a small amount of enol or enolate. The reaction is often done without addition of acid or base, but traces of acid or base are always present, and these are enough to catalyze formation of the enol or enolate. With acid catalysis the mechanism is

[140]Rappe, C. *Ark. Kemi* **1965**, *24*, 321. But see also Teo, K.E.; Warnhoff, E.W. *J. Am. Chem. Soc.* **1973**, *95*, 2728.
[141]Rappe, C.; Schotte, L. *Acta Chem. Scand.* **1962**, *16*, 2060; Rappe, C. *Ark. Kemi* **1964**, *21*, 503; Garbisch, Jr., E.W. *J. Org. Chem.* **1965**, *30*, 2109.
[142]Nobrega, J.A.; Gonalves, S.M.C.; Reppe, C. *Synth. Commun.* **2002**, *32*, 3711.
[143]Terent'ev, A.O.; Khodykin, S.V.; Troitskii, N.A.; Ogibin, Y.N.; Nikishin, G.I. *Synthesis* **2004**, 2845.
[144]Kajigaeshi, S.; Kakinami, T.; Tokiyama, H.; Hirakawa, T.; Okamoto, T. *Bull. Chem. Soc. Jpn.* **1987**, *60*, 2667.
[145]Yang, D.; Yan, Y.-L.; Lui, B. *J. Org. Chem.* **2002**, *67*, 7429.
[146]Marigo, M.; Kumaragurubaran, N.; Jørgensen, K.A. *Chem. Eur. J.* **2004**, *10*, 2133.

The first step, as we have already seen (**12-3**), actually consists of two steps. The second step is very similar to the first step in electrophilic addition to double bonds (p. 999). There is a great deal of evidence for this mechanism: (*1*) the rate is first order in substrate; (*2*) bromine does not appear in the rate expression at all,[147] a fact consistent with a rate-determining first step;[148] (*3*) the reaction rate is the same for bromination, chlorination, and iodination under the same conditions;[149] (*4*) the reaction shows an isotope effect; and (*5*) the rate of the step 2–step 3 sequence has been independently measured (by starting with the enol) and found to be very fast.[150]

With basic catalysts the mechanism may be the same as that given above (since bases also catalyze formation of the enol), or the reaction may go directly through the enolate ion without formation of the enol:

Step 1   (reaction scheme)

Step 2   (reaction scheme)

It is difficult to distinguish the two possibilities. It was mentioned above that in the base-catalyzed reaction, if the substrate has two or three a halogens on the same side of the C=O group, it is not possible to stop the reaction after just one halogen atom has entered. The reason is that the electron-withdrawing field effect of the first halogen increases the acidity of the remaining hydrogens, that is, a CHX group is more acidic than a CH_2 group, so that initially formed halo ketone is converted to enolate ion (and hence halogenated) more rapidly than the original substrate. Other halogenating agents can be used in this reaction. Reaction of a lithium enolate anion with tosyl chloride gave the corresponding α-chloro ketone.[151] When an aldehyde was treated with a catalytic amount of 2,5-lutidine to generate the enolate anion, reaction with 35% HCl in dichloromethane gave the α,α-dichloroaldehyde.[152]

[147]When the halogenating species is at low concentration or has a low reactivity, it can appear in the rate expression. The reaction becomes first order in the halogenating species. See, for example, Tapuhi, E.; Jencks, W.P. *J. Am. Chem. Soc.* **1982**, *104*, 5758. For a case in which the reaction is first order in bromine, even at relatively high Br_2 contentration, see Pinkus, A.G.; Gopalan, R. *J. Am. Chem. Soc.* **1984**, *106*, 2630. For a study of the kinetics of iodination, see Pinkus, A.G.; Gopalan, R. *Tetrahedron* **1986**, *42*, 3411.
[148]Under some conditions it is possible for step 2 to be rate-determining: Deno, N.C.; Fishbein, R. *J. Am. Chem. Soc.* **1973**, *95*, 7445.
[149]Bell, R.P.; Yates, K. *J. Chem. Soc.* **1962**, 1927.
[150]Hochstrasser, R.; Kresge, A.J.; Schepp, N.P.; Wirz, J. *J. Am. Chem. Soc.* **1988**, *110*, 7875.
[151]Brummond, K.M.; Gesenberg, K.D. *Tetrahedron Lett.* **1999**, *40*, 2231.
[152]Bellesia, F.; DeBuyck, L.; Ghelfi, F.; Libertini, E.; Pagnoni, U.M.; Roncaglia, F. *Tetrahedron* **2000**, *56*, 7507.

Regioselectivity in the halogenation of unsymmetrical ketones can be attained by treatment of the appropriate enol borinate of the ketone with N-bromo- or N-chlorosuccinimide.[153] The desired halo

ketone is formed in high yield. Another method for achieving the same result involves bromination of the appropriate lithium enolate at a low temperature[154] (see p. 630 for the regioselective formation of enolate ions). In a similar process, α-halo aldehydes have been prepared in good yield by treatment of silyl enol ethers $R_2C=CHOSiMe_3$ with Br_2 or Cl_2,[155] with sulfuryl chloride SO_2Cl_2;[156] or with I_2 and silver acetate.[157] Other chlorinating agents can be used with a variety of silyl enol ethers to generate α-chloroketones with good enantioselectivity, including $ZrCl_4$ in conjunction with an α,α-dichloromalonate ester.[158] Silyl enol ethers can also be fluorinated, with XeF_2[159] or with 5% F_2 in N_2 at $-78°C$ in $FCCl_3$.[160] Enol acetates have been regioselectively iodinated with I_2 and either thallium(I) acetate[161] or copper(II) acetate.[162]

α,β-Unsaturated ketones can be converted to α-halo-α,β-unsaturated ketones by treatment with phenylselenium bromide or chloride,[163] and to α-halo-β,γ-unsaturated ketones by two-phase treatment with HOCl.[164] Conjugated ketones were converted to the α-bromo conjugated ketone (a vinyl bromide) using the Dess–Martin periodinane (see p. 1723) and tetraethylammonium bromide.[165]

OS **I**, 127; **II**, 87, 88, 244, 480; **III**, 188, 343, 538; **IV**, 110, 162, 590; **V**, 514; **VI**, 175, 193, 368, 401, 512, 520, 711, 991; **VII**, 271; **VIII**, 286. See also, OS **VI**, 1033; **VIII**, 192.

[153]Hooz, J.; Bridson, J.N. *Can. J. Chem.* **1972**, *50*, 2387.

[154]Stotter, P.L.; Hill, K.A. *J. Org. Chem.* **1973**, *38*, 2576.

[155]Reuss, R.H.; Hassner, A. *J. Org. Chem.* **1974**, *39*, 1785; Blanco, L.; Amice, P.; Conia, J.M. *Synthesis* **1976**, 194.

[156]Olah, G.A.; Ohannesian, L.; Arvanaghi, M.; Prakash, G.K.S. *J. Org. Chem.* **1984**, *49*, 2032.

[157]Rubottom, G.M.; Mott, R.C. *J. Org. Chem.* **1979**, *44*, 1731.

[158]Zhang, Y.; Shibatomi, K.; Yamamoto, H. *J. Am. Chem. Soc.* **2004**, *126*, 15038.

[159]Tsushima, T.; Kawada, K.; Tsuji, T. *Tetrahedron Lett.* **1982**, *23*, 1165.

[160]Purrington, S.T.; Bumgardner, C.L.; Lazaridis, N.V.; Singh, P. *J. Org. Chem.* **1987**, *52*, 4307.

[161]Cambie, R.C.; Hayward, R.C.; Jurlina, J.L.; Rutledge, P.S.; Woodgate, P.D. *J. Chem. Soc. Perkin Trans. 1* **1978**, 126.

[162]Horiuchi, C.A.; Satoh, J.Y. *Synthesis* **1981**, 312.

[163]Ley, S.V.; Whittle, A.J. *Tetrahedron Lett.* **1981**, *22*, 3301.

[164]Hegde, S.G.; Wolinsky, J. *Tetrahedron Lett.* **1981**, *22*, 5019.

[165]Fache, F.; Piva, O. *Synlett* **2002**, 2035.

12-5   Halogenation of Carboxylic Acids and Acyl Halides

Halogenation or **Halo-de-hydrogenation**

$$ R\overset{}{\frown}COOH \;+\; Br_2 \;\xrightarrow{\;PBr_3\;}\; \underset{R\overset{}{\frown}COOH}{\overset{Br}{|}} $$

Using a phosphorus halide as catalyst, the α hydrogens of carboxylic acids can be replaced by bromine or chlorine.[166] The reaction, known as the *Hell–Volhard–Zelinskii reaction*, is not applicable to iodine or fluorine. When there are two α hydrogens, one or both may be replaced, although it is often hard to stop with just one. The reaction actually takes place on the acyl halide formed from the carboxylic acid and the catalyst. The acids alone are inactive, except for those with relatively high enol content, such as malonic acid. Less than one full mole of catalyst (per mole of substrate) is required, because of the exchange reaction between carboxylic acids and acyl halides (see **16-79**). Each molecule of acid is α halogenated while it is in the acyl halide stage. The halogen from the catalyst does not enter the α position. For example, the use of Cl_2 and PBr_3 results in α chlorination, not bromination. As expected from the foregoing, acyl halides undergo a halogenation without a catalyst. An enantioselective α-halogenation was reported yielding via an alkaloid catalyzed reaction of acyl halides with perhaloquinone-derived reagents to give to chiral α-haloesters.[167] So do anhydrides and many compounds that enolize easily (e.g., malonic ester and aliphatic nitro compounds). The mechanism is usually regarded as proceeding through the enol as in **12-4**.[168] If chlorosulfuric acid $ClSO_2OH$ is used as a catalyst, carboxylic acids can be α-iodinated,[169] as well as chlorinated or brominated.[170] *N*-Bromosuccinimide in a mixture of sulfuric acid–trifluoroacetic acid can mono-brominate simple carboxylic acids.[171]

A number of other methods exist for the a halogenation of carboxylic acids or their derivatives.[172] Under electrolytic conditions with NaCl, malonates are converted to 2-chloro malonates.[173] Acyl halides can be a brominated or chlorinated by use of *N*-bromo- or *N*-chlorosuccinimide and HBr or HCl.[174] The latter is an ionic, not a free-radical halogenation (see **14-3**). Direct iodination of carboxylic acids has been achieved with I_2–Cu(II) acetate in HOAc.[175] Acyl chlorides can

[166]For a review, see Harwood, H.J. *Chem. Rev.* **1962**, *62*, 99, pp. 102-103.

[167]Wack, H.; Taggi, A.E.; Hafez, A.M.; Drury III, W.J.; Lectka, T. *J. Am. Chem. Soc.* **2001**, *123*, 1531. See also, France, S.; Wack, H.; Taggi, A.E.; Hafez, A.M.; Wagerle, Ty.R.; Shah, M.H.; Dusich, C.L.; Lectka, T. *J. Am. Chem. Soc.* **2004**, *126*, 4245.

[168]See, however, Kwart, H.; Scalzi, F.V. *J. Am. Chem. Soc.* **1964**, *86*, 5496.

[169]Ogata, Y.; Watanabe, S. *J. Org. Chem.* **1979**, 44, 2768; **1980**, *45*, 2831.

[170]Ogata, Y.; Adachi, K. *J. Org. Chem.* **1982**, *47*, 1182.

[171]Zhang, L.H.; Duan, J.; Xu, Y.; Dolbier, Jr., W.R. *Tetrahedron Lett.* **1998**, *39*, 9621.

[172]For a list of reagents, with references, see Larock, R.C. *Comprehensive Organic Transformations*, 2nd ed., Wiley-VCH, NY, **1999**, pp. 730–738.

[173]Okimoto, M.; Takahashi, Y. *Synthesis* **2002**, 2215.

[174]Harpp, D.N.; Bao, L.Q.; Black, C.J.; Gleason, J.G.; Smith, R.A. *J. Org. Chem.* **1975**, *40*, 3420.

[175]Horiuchi, C.A.; Satoh, J.Y. *Chem. Lett.* **1984**, 1509.

be a iodinated with I_2 and a trace of HI.[176] Carboxylic esters can be a halogenated by conversion to their enolate ions with lithium N-isopropylcyclohexylamide in THF and treatment of this solution at $-78°$ with I_2[176] or with a carbon tetrahalide.[177] Carboxylic acids, esters, and amides have been α-fluorinated at $-78°C$ with F_2 diluted in N_2.[178] Amides have been α-iodinated using iodine and s-collidine.[179]

OS **I**, 115, 245; **II**, 74, 93; **III**, 347, 381, 495, 523, 623, 705, 848; **IV**, 254, 348, 398, 608, 616; **V**, 255; **VI**, 90, 190, 403; **IX**, 526. Also see, OS **IV**, 877; **VI**, 427.

12-6 Halogenation of Sulfoxides and Sulfones

Halogenation or Halo-de-hydrogenation

Sulfoxides can be chlorinated in the α position[180] by treatment with Cl_2[181] or N-chlorosuccinimide,[182] in the presence of pyridine. These methods involve basic conditions. The reaction can also be accomplished in the absence of base with SO_2Cl_2 in CH_2Cl_2,[183] or with $TsNCl_2$.[184] The bromination of sulfoxides with bromine[185] and with NBS-bromine[186] have also been reported. Sulfones have been chlorinated by treatment of their conjugate bases $RSO_2C^{\ominus}HR'$ with various reagents, among them SO_2Cl_2, CCl_4,[187] N-chlorosuccinimide,[188] and hexachloroethane.[189] The α fluorination of sulfoxides has accomplished in a two-step procedure. Treatment with diethylaminosulfur trifluoride Et_2NSF_3 (DAST) produces an

[176]Rathke, M.W.; Lindert, A. *Tetrahedron Lett.* **1971**, 3995.

[177]Arnold, R.T.; Kulenovic, S.T. *J. Org. Chem.* **1978**, *43*, 3687.

[178]Purrington, S.T.; Woodard, D.L. *J. Org. Chem.* **1990**, *55*, 3423.

[179]Kitagawa, O.; Hanano, T.; Hirata, T.; Inoue, T.; Taguchi, T. *Tetrahedron Lett.* **1992**, *33*, 1299.

[180]For a review, see Venier, C.G.; Barager III, H.J. *Org. Prep. Proced. Int.* **1974**, *6*, 77, pp. 81–84.

[181]Tsuchihashi, G.; Iriuchijima, S. *Bull. Chem. Soc. Jpn.* **1970**, *43*, 2271.

[182]Ogura, K.; Imaizumi, J.; Iida, H.; Tsuchihashi, G. *Chem. Lett.* **1980**, 1587.

[183]Tin, K.; Durst, T. *Tetrahedron Lett.* **1970**, 4643.

[184]Kim, Y.H.; Lim, S.C.; Kim, H.R.; Yoon, D.C. *Chem. Lett.* **1990**, 79.

[185]Cinquini, M.; Colonna, S. *J. Chem. Soc. Perkin Trans. 1* 1972, 1883. See also, Cinquini, M.; Colonna, S. *Synthesis* **1972**, 259.

[186]Iriuchijima, S.; Tsuchihashi, G. *Synthesis* **1970**, 588.

[187]Regis, R.R.; Doweyko, A.M. *Tetrahedron Lett.* **1982**, *23*, 2539.

[188]Paquette, L.A.; Houser, R.W. *J. Org. Chem.* **1971**, *36*, 1015.

[189]Kattenberg, J.; de Waard, E.R.; Huisman, H.O. *Tetrahedron* **1973**, *29*, 4149; **1974**, *30*, 463.

α-fluoro thioether, usually in high yield. Oxidation of this compound with *m*-chloroperoxybenzoic acid gives the sulfoxide.[190]

C. Nitrogen Electrophiles

12-7   Aliphatic Diazonium Coupling

Arylhydrazono-de-dihydro-bisubstitution

$$Z \diagdown Z' \ + \ ArN_2{}^+ \ \xrightarrow{\ OAc^-\ } \ \underset{Z}{\overset{Z'}{\diagup}}\!\!=\!N\diagup\!\!\sim NHAr$$

If a C–H bond is acidic enough, it couples with diazonium salts in the presence of a base, most often aqueous sodium acetate.[191] The reaction is commonly carried out on compounds of the form Z–CH$_2$–Z′, where Z and Z′ are as defined on p. 1358, for example, β-keto esters, β-keto amides, malonic ester.

The mechanism is probably of the simple S_E1 type:

$$Z \diagdown Z' \ \xrightarrow{\ B\ } \ Z \diagdown Z'^{\ominus} \ + \ ArN_2{}^+ \ \longrightarrow \ \underset{Z}{\overset{Z'}{\diagup}}\!\!-\!N\!\!=\!\!N\!-\!Ar \ \longrightarrow \ \underset{Z}{\overset{Z'}{\diagup}}\!\!=\!N\diagup\!\!\sim NH\!-\!Ar$$

$$\textbf{23}$$

Aliphatic azo compounds in which the carbon containing the azo group is attached to a hydrogen are unstable and tautomerize to the isomeric hydrazones (**23**), which are therefore the products of the reaction.

When the reaction is carried out on a compound of the form Z–CHR–Z′, so that the azo compound does not have a hydrogen that can undergo tautomerism, if at least one Z is acyl or carboxyl, this group usually cleaves:

$$H_3C\overset{O}{\underset{}{\diagdown}}\!\!\underset{\underset{Z'}{R}}{\diagup}\!\!-\!N\!\!=\!\!N\diagup Ar \ \xrightarrow{\ B^-\ } \ \underset{Z'}{\overset{R}{\diagup}}\!\!=\!N\!-\!N\underset{\ominus}{\diagup}Ar \ \longrightarrow \ \underset{Z'}{\overset{R}{\diagup}}\!\!=\!N\!-\!NH\diagup Ar$$

$$\textbf{24}$$

so the product in this case is also the hydrazone, and not the azo compound. In fact, compounds of the type **24** are seldom isolable from the reaction, although this has been accomplished.[192] The cleavage step shown is an example of **12-43** and, when a carboxyl group cleaves, of **12-40**. The overall process in this case is called the *Japp–Klingemann reaction*[193] and involves conversion of a ketone (**25**) or a

---

[190]McCarthy, J.R.; Pee, N.P.; LeTourneau, M.E.; Inbasekaran, M. *J. Am. Chem. Soc.* **1985**, *107*, 735. See also, Umemoto, T.; Tomizawa, G. *Bull. Chem. Soc. Jpn.* **1986**, *59*, 3625.

[191]For a review, see Parmerter, S.M. *Org. React.* **1959**, *10*, 1.

[192]See, for example, Yao, H.C.; Resnick, P. *J. Am. Chem. Soc.* **1962**, *84*, 3514.

[193]For a review, see Phillips, R.R. *Org. React.* **1959**, *10*, 143.

carboxylic acid (**26**)

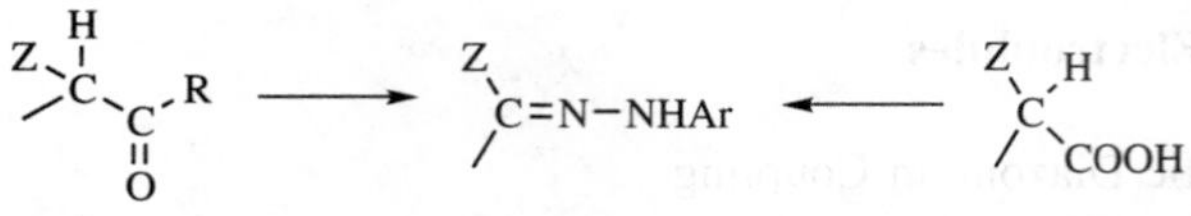

$$
\underset{\textbf{25}}{\overset{\displaystyle Z\diagdown \overset{\displaystyle \overset{H}{|}}{\underset{\displaystyle \underset{\displaystyle \overset{||}{O}}{C}}{C}}{\diagup} \diagup^{R}}{}
\quad \longrightarrow \quad
\underset{\textbf{26}}{\overset{\displaystyle Z\diagdown}{\underset{\displaystyle \diagup}{C}}{=}N{-}NHAr}
\quad \longleftarrow \quad
\underset{\textbf{27}}{\overset{\displaystyle Z\diagdown \overset{H}{\diagup}}{\underset{\displaystyle \diagup}{C}}{\diagdown COOH}}
$$

to a hydrazone (**27**). When an acyl and a carboxyl group are both present, the leaving group order has been reported to be MeCO > COOH > PhCO.[194] When there is no acyl or carboxyl group present, the aliphatic azo compound is stable.

OS **III**, 660; **IV**, 633.

12-8 Nitrosation at a Carbon Bearing an Active Hydrogen

Hydroxyimino-de-dihydro-bisubstitution

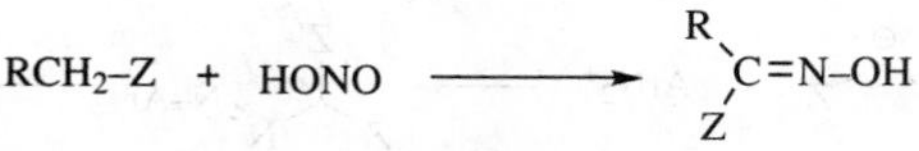

$$
RCH_2{-}Z + HONO \longrightarrow \overset{\displaystyle R\diagdown}{\underset{\displaystyle Z\diagup}{C}}{=}N{-}OH
$$

Nitrosation or **Nitroso-de-hydrogenation**

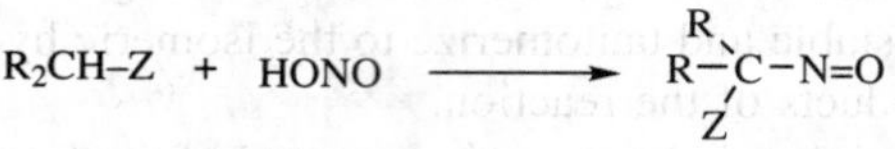

$$
R_2CH{-}Z + HONO \longrightarrow \overset{\displaystyle R}{\underset{\displaystyle Z}{R{-}\overset{|}{\underset{|}{C}}{-}N{=}O}}
$$

Carbons adjacent to a Z group (as defined on p. 622) can be nitrosated with nitrous acid or alkyl nitrites.[195] The initial product is the *C*-nitroso compound, but these are stable only when there is no hydrogen that can undergo tautomerism. When there is, the product is the more stable oxime. The situation is analogous to that with azo compounds and hydrazones (**12-7**). The mechanism is similar to that in **12-7**:[196] R–H → R⁻ + ⁺N=O → R–N=O. The attacking species is either NO⁺ or a carrier of it. When the substrate is a simple ketone, the mechanism goes through the enol (as in halogenation **12-4**):

Evidence is that the reaction, in the presence of X⁻ (Br⁻, Cl⁻, or SCN⁻), was first order in ketone and in H⁺, but zero order in HNO_2 and X⁻.[197] Furthermore, the rate of the nitrosation was about the same as that for enolization of the same ketones. The species NOX is formed by $HONO + X^- + H^+ \rightarrow HOX + H_2O$. In

[194]Neplyuev, V.M.; Bazarova, I.M.; Lozinskii, M.O. *J. Org. Chem. USSR* **1989**, *25*, 2011. This paper also includes a sequence of leaving group ability for other Z groups.
[195]For a review, see Williams, D.L.H. *Nitrosation*, Cambridge University Press, Cambridge, **1988**, pp. 1–45.
[196]For a review, see Williams, D.L.H. *Adv. Phys. Org. Chem.* **1983**, *19*, 381. See also Williams, D.L.H. *Nitrosation*, Cambridge Univ. Press, Cambridge, **1988**.
[197]Leis, J.R.; Peña, M.E.; Williams, D.L.H.; Mawson, S.D. *J. Chem. Soc. Perkin Trans. 2* **1988**, 157.

the cases of $F_3CCOCH_2COCF_3$ and malononitrile the nitrosation went entirely through the enolate ion rather than the enol.[198]

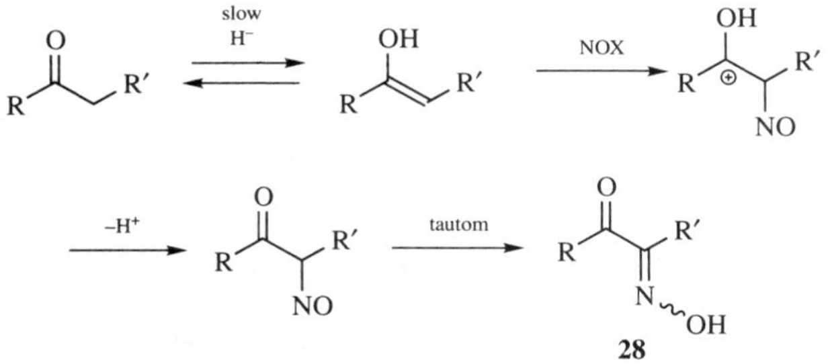

As in the Japp–Klingemann reaction, when Z is an acyl or carboxyl group (in the case of $R_2CH–Z$), it can be cleaved. Since oximes and nitroso compounds can be reduced to primary amines, this reaction often provides a route to amino acids. As in the case of **12-4**, the silyl enol ether of a ketone can be used instead of the ketone itself.[199] Good yields of α-oximinoketones (**28**) can be obtained by treating ketones with *tert*-butyl thionitrate.[200]

Imines can be prepared in a similar manner by treatment of an active hydrogen compound with a nitroso compound:

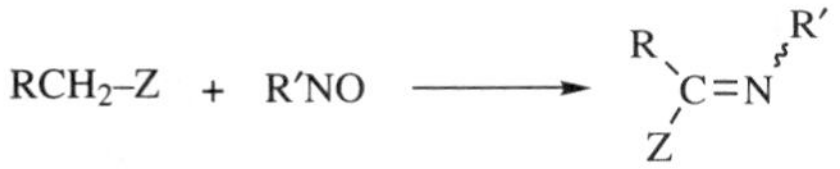

Alkanes can be nitrosated photochemically, by treatment with NOCl and UV light.[201] For nitration at an activated carbon, see **12-9**. Trialkyltin enol ethers ($C=C–O–SnR_3$) react with PhNO to give α-(N-hydroxylamino)ketones.[202]

OS **II**, 202, 204, 223, 363; **III**, 191, 513; **V**, 32, 373; **VI**, 199, 840. Also see, OS **V**, 650.

12-9   Nitration of Alkanes

Nitration or Nitro-de-hydrogenation

$$RH \quad + \quad HNO_3 \quad \xrightarrow{400°C} \quad RNO_2$$

---

[198]Iglesias, E.; Williams, D.L.H. *J. Chem. Soc. Perkin Trans. 2* **1989**, 343; Crookes, M.J.; Roy, P.; Williams, D.L.H. *J. Chem. Soc. Perkin Trans. 2* **1989**, 1015. See also Graham, A.; Williams, D.L.H. *J. Chem. Soc. Chem. Commun.* **1991**, 407.

[199]Rasmussen, J.K.; Hassner, A. *J. Org. Chem.* **1974**, *39*, 2558.

[200]Kim, Y.H.; Park, Y.J.; Kim, K. *Tetrahedron Lett.* **1989**, *30*, 2833.

[201]For a review, see Pape, M. *Fortschr. Chem. Forsch.* **1967**, *7*, 559.

[202]Momiyama, N.; Yamamoto, H. *Org. Lett.* **2002**, *4*, 3579.

Nitration of alkanes[203] can be carried out in the gas phase at $\sim 400°C$ or in the liquid phase. The reaction is not practical for the production of pure products for any alkane except methane. For other alkanes, not only does the reaction produce mixtures of the mono-, di-, and polynitrated alkanes at every combination of positions, but extensive chain cleavage occurs.[204] A free-radical mechanism is involved.[205]

$$-\overset{|}{\underset{|}{C}}{}^{\ominus} \quad + \quad MeONO_2 \quad \longrightarrow \quad -\overset{|}{\underset{|}{C}}-NO_2 \quad + \quad {}^-OMe$$

Activated positions (e.g., ZCH_2Z' compounds) can be nitrated by fuming nitric acid in acetic acid, by acetyl nitrate and an acid catalyst,[206] or by alkyl nitrates under alkaline conditions.[207] In the latter case, it is the carbanionic form of the substrate that is actually nitrated. What is isolated under these alkaline conditions is the conjugate base of the nitro compound. Yields are not high. Of course, the mechanism in this case is not of the free-radical type, but is electrophilic substitution with respect to the carbon (similar to the mechanisms of **12-7** and **12-8**). Positions activated by only one electron-withdrawing group, for example, a positions of simple ketones, nitriles, sulfones, or *N,N*-dialkyl amides, can be nitrated with alkyl nitrates if a very strong base, for example, *t*-BuOK or $NaNH_2$, is present to convert the substrate to the carbanionic form.[208]

Electrophilic nitration of alkanes has been performed with nitronium salts, for example, NO_2^+ PF_6^- and with HNO_3–H_2SO_4 mixtures, but mixtures of nitration and cleavage products are obtained and yields are generally low.[209] The reaction of alkanes with nitric acid and *N*-hydroxysuccinimide (NHS), however, gave moderate-to-good yields of the corresponding nitroalkane.[210] Similar nitration was accomplished with NO_2, NHS and air.[211]

Aliphatic nitro compounds can be a nitrated $[R_2C^{\ominus}NO_2 \rightarrow R_2C(NO_2)_2]$ by treatment of their conjugate bases $RCNO_2$ with NO_2^- and $K_3Fe(CN)_6$.[212]

[203]For reviews, see Olah, G.A.; Malhotra, R.; Narang, S.C. *Nitration*, VCH, NY, *1989*, pp. 219–295; Ogata, Y. in Trahanovsky, W.S. *Oxidation in Organic Chemisry*, part C, Academic Press, NY, *1978*, pp. 295–342; Ballod, A.P.; Shtern, V.Ya. *Russ. Chem. Rev. 1976*, *45*, 721.

[204]For a discussion of the mechanism of this cleavage, see Matasa, C.; Hass, H.B. *Can. J. Chem. 1971*, *49*, 1284.

[205]Titov, A.I. *Tetrahedron 1963*, *19*, 557.

[206]Sifniades, S. *J. Org. Chem. 1975*, *40*, 3562.

[207]For a review, see Larson, H.O., in Feuer, H. *The Chemistry of the Nitro and Nitroso Groups*, Vol. 1, Wiley, NY, *1969*, pp. 310–316.

[208]For examples, see Truce, W.E.; Christensen, L.W. *Tetrahedron 1969*, *25*, 181; Pfeffer, P.E.; Silbert, L.S. *Tetrahedron Lett. 1970*, 699; Feuer, H.; Spinicelli, L.F. *J. Org. Chem. 1976*, *41*, 2981; Feuer, H.; Van Buren II, W.D.; Grutzner, J.B. *J. Org. Chem. 1978*, *43*, 4676.

[209]Olah, G.A.; Lin, H.C. *J. Am. Chem. Soc. 1973*, *93*, 1259. See also, Bach, R.D.; Holubka, J.W.; Badger, R.C.; Rajan, S. *J. Am. Chem. Soc. 1979*, *101*, 4416.

[210]Isozaki, S.; Nishiwaki, Y.; Sakaguchi, S.; Ishii, Y. *Chem. Commun. 2001*, 1352.

[211]Sakaguchi, S.; Nishiwaki, Y.; Kitamura, T.; Ishii, Y. *Angew. Chem. Int. Ed. 2001*, *40*, 222; Nishiwaki, Y.; Sakaguchi, S.; Ishii, Y. *J. Org. Chem. 2002*, *67*, 5663.

[212]Matacz, Z.; Piotrowska, H.; Urbanski, T. *Pol. J. Chem. 1979*, *53*, 187; Kornblum, N.; Singh, H.K.; Kelly, W.J. *J. Org. Chem. 1983*, *48*, 332; Garver, L.C.; Grakauskas, V.; Baum, K. *J. Org. Chem. 1985*, *50*, 1699.

A novel reaction converted a vinyl methyl moiety to a vinyl nitro. The reaction of $MeCH=C(Ph)CN$ with NO_x and iodine gave $O_2NCH=C(Ph)CN$.[213]

OS **I**, 390; **II**, 440, 512.

12-10   Direct Formation of Diazo Compounds

Diazo-de-dihydro-bisubstitution

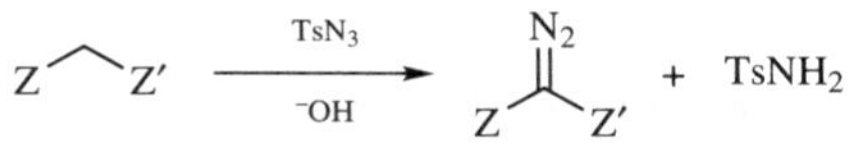

Compounds containing a CH_2 bonded to two Z groups (active methylene compounds, with Z as defined on p. 622) can be converted to diazo compounds on treatment with tosyl azide in the presence of a base.[214] The use of phase-transfer catalysis increases the convenience of the method.[215] p-Dodecylbenzenesulfonyl azide,[216] methanesulfonyl azide,[217] and p-acetamidobenzenesulfonyl azide[218] also give the reaction. The reaction, which is called the *diazo-transfer reaction*, can also be applied to other reactive positions (e.g., the 5 position of cyclopentadiene).[219] The mechanism is probably as follows:

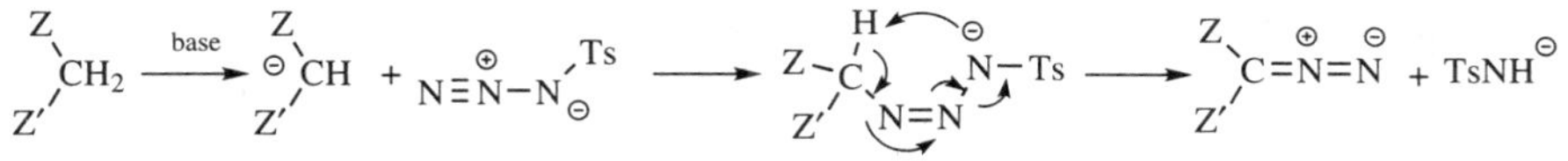

A diazo group can be introduced adjacent to a single carbonyl group indirectly by first converting the ketone to an α-formyl ketone (**16-85**) and then treating it with tosyl azide. As in the similar cases of

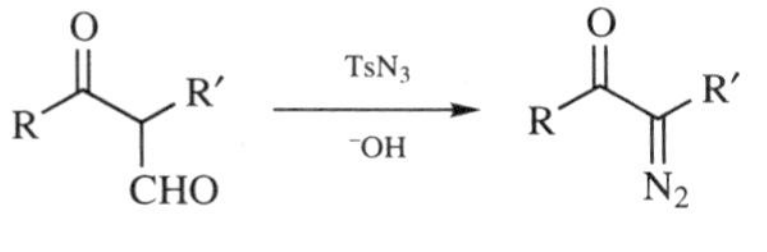

12-7 and **12-8**, the formyl group is cleaved during the reaction.[220]

OS **V**, 179; **VI**, 389, 414.

[213]Navarro-Ocaña, A.; Barzana, E.; López-González, D.; Jiménez-Estrada, M. *Org. Prep. Proceed. Int.* **1999**, *31*, 117.

[214]For reviews, see Regitz, M.; Maas, G. *Diazo Compounds*, Academic Press, NY, **1986**, pp. 326–435; Regitz, M. *Synthesis* **1972**, 351; *Angew. Chem. Int. Ed.* **1967**, *6*, 733; *Newer Methods Prep. Org. Chem.* **1971**, *6*, 81. See also, Hünig, S. *Angew. Chem. Int. Ed.* **1968**, *7*, 335; Koskinen, A.M.P.; Muñoz, L. *J. Chem. Soc. Chem. Commun.* **1990**, 652.

[215]Ledon, H. *Synthesis* **1974**, 347, *Org. Synth.* **VI**, 414. For another convenient method, see Ghosh, S.; Datta, I. *Synth. Commun.* **1991**, *21*, 191.

[216]Hazen, G.G.; Weinstock, L.M.; Connell, R.; Bollinger, F.W. *Synth. Commun.* **1981**, *11*, 947.

[217]Taber, D.F.; Ruckle Jr., R.E.; Hennessy, M.J. *J. Org. Chem.* **1986**, *51*, 4077.

[218]Baum, J.S.; Shook, D.A.; Davies, H.M.L.; Smith, H.D. *Synth. Commun.* **1987**, *17*, 1709.

[219]Doering, W. von E.; DePuy, C.H. *J. Am. Chem. Soc.* **1953**, *75*, 5955.

[220]For a similar approach, see Danheiser, R.L.; Miller, R.F.; Brisbois, R.G.; Park, S.Z. *J. Org. Chem.* **1990**, *55*, 1959.

12-11 Conversion of Amides to α-Azido Amides

Azidation or **Azido-de-hydrogenation**

In reaction **12-10**, treatment of Z–CH$_2$–Z′ with tosyl azide gave the α-diazo compound via diazo transfer. When this reaction is performed on a compound with a single Z group such as an amide, formation of the azide becomes a competing process via the enolate anion.[221] Factors favoring azide formation rather than diazo transfer include K$^+$ as the enolate counterion rather than Na$^+$ or Li$^+$ and the use of 2,4,6-triisopropylbenzenesulfonyl azide rather than TsN$_3$. When the reaction was applied to amides with a chiral R′, such as the oxazolidinone derivative **29**, it was highly stereoselective, and the product could be converted to an optically active amino acid.[221]

12-12 Direct Amination at an Activated Position

Alkyamino-de-hydrogenation, and so on

Alkenes can be aminated[222] in the allylic position by treatment with solutions of imido selenium compounds R–N=Se=N–R.[223] The reaction, which is similar to the allylic oxidation of alkenes with SeO$_2$ (see **19-14**), has been performed with R = t-Bu and R = Ts. The imido sulfur compound TsN=S=NTs has also been used,[224] as well

[221]Evans, D.A.; Britton, T.C. *J. Am. Chem. Soc.* **1987**, *109*, 6881, and references cited therein.

[222]For a review of direct aminations, see Sheradsky, T., in Patai, S. *The Chemistry of Functional Groups, Supplement F*, pt. 1, Wiley, NY, **1982**, pp. 395–416.

[223]Sharpless, K.B.; Hori, T.; Truesdale, L.K.; Dietrich, C.O. *J. Am. Chem. Soc.* **1976**, *98*, 269. For another method, see Kresze, G.; Münsterer, H. *J. Org. Chem.* **1983**, *48*, 3561. For a review, see Cheikh, R.B.; Chaabouni, R.; Laurent, A.; Mison, P.; Nafti, A. *Synthesis* **1983**, 685, pp. 691–696.

[224]Sharpless, K.B.; Hori, T. *J. Org. Chem.* **1979**, *41*, 176; Singer, S.P.; Sharpless, K.B. *J. Org. Chem.* **1978**, *43*, 1448. For other reagents, see Mahy, J.P.; Bedi, G.; Battioni, P.; Mansuy, D. *Tetrahedron Lett.* **1988**, *29*, 1927; Tsushima, S.; Yamada, Y.; Onami, T.; Oshima, K.; Chaney, M.O.; Jones, N.D.; Swartzendruber, J.K. *Bull. Chem. Soc. Jpn.* **1989**, *62*, 1167.

as PhNHOH–FeCl$_2$/FeCl$_3$.[225] Benzylic positions can be aminated with *t*-BuOO-CONHTs in the presence of a catalytic amount of Cu(OTf)$_2$.[226] In another reaction, compounds containing an active hydrogen can be converted to primary amines (**30**) in moderate yields by treatment with *O*-(2,4-dinitrophenyl)hydroxylamine.[227]

Tertiary alkyl hydrogen can be replaced in some cases via C–H nitrogen insertion. The reaction of sulfamate ester **31** with PhI(OAc)$_2$, MgO and a dinuclear Rh carboxylate catalyst, for example, generated oxathiazinane **32**.[228] This transformation is a formal oxidation, and primary carbamates have been similarly converted to oxazolidin-2-ones.[229]

In an indirect amination process, acyl halides are converted to amino acids.[230] Reaction of the acyl halide with a chiral oxazolidinone leads to a chiral amide, which reacts with the N=N unit of a dialkyl azodicarboxylate[R^{2}O$_2$C–N=N–CO$_2$R′]. Hydrolysis and catalytic hydrogenation leads to an amino acid with good enantioselectivity.[226]

See also, **10-39**.

12-13   Insertion by Nitrenes

CH-[Acylimino]-insertion, and so on

---

[225]Srivastava, R.S.; Nicholas, K.M. *Tetrahedron Lett.* **1994**, *35*, 8739.

[226]Kohmura, Y.; Kawasaki, K.; Katsuki, T. *Synlett*, **1997**, 1456.

[227]Sheradsky, T.; Salemnick, G.; Nir, Z. *Tetrahedron* **1972**, *28*, 3833; Radhakrishna, A.; Loudon, G.M.; Miller, M.J. *J. Org. Chem.* **1979**, *44*, 4836.

[228]Espino, C. G.; Wehn, P. M.; Chow, J.; Du Bois, J. *J. Am. Chem. Soc.* **2001**, *123*, 6935.

[229]Espino, C.G.; Du Bois, J. *Angew. Chem. Int. Ed.* **2001**, *40*, 598.

[230]Trimble, L.A.; Vederas, J.C. *J. Am. Chem. Soc.* **1986**, *108*, 6397; Evans, D.A.; Britton, T.C.; Dorow, R.L.; Dellaria, J.F. *Tetrahedron* **1988**, *44*, 5525; Gennari, C.; Colombo, L.; Bertolini, G. *J. Am. Chem. Soc.* **1986**, *108*, 6394; Oppolzer, W.; Moretti, R. *Helv. Chim. Acta* **1986**, *69*, 1923; *Tetrahedron* **1988**, *44*, 5541; Guanti, G.; Banfi, L.; Narisano, E. *Tetrahedron* **1988**, *44*, 5523.

Carbonylnitrenes: NCOW (W = R′, Ar, or OR′) are very reactive species (p. 293) and insert into the C–H bonds of alkanes to give amides (W = R′ or Ar) or carbamates (W = OR′).[231] The nitrenes are generated as discussed on p. 293. The order of reactivity among alkane C–H bonds is tertiary > secondary > primary.[232] Indications are that in general it is only singlet and not triplet nitrenes that insert.[233] Retention of configuration is found at a chiral carbon.[234] The mechanism is presumably similar to the simple one-step mechanism for insertion of carbenes (**12-21**). Other nitrenes [e.g., cyanonitrene (NCN)[235] and arylnitrenes (NAr)[236]] can also insert into C–H bonds, but alkylnitrenes usually undergo rearrangement before they can react with the alkane. N-Carbamoyl nitrenes undergo insertion reactions that often lead to mixtures of products, but exceptions are known,[237] chiefly in cyclizations.[238] For example, heating of 2-(2-methylbutyl)phenyl azide gave ~60% 2-ethyl-2-methylindoline.[234] Enantioselective nitrene insertion reactions are known.[239]

D. Sulfur Electrophiles

12-14 Sulfenylation, Sulfonation, and Selenylation of Ketones and Carboxylic Esters

Alkylthio-de-hydrogenation, and so on

[231]For a review, see Lwowski, W., in Lwowski, W. *Nitrenes*, Wiley, NY, *1970*, pp. 199–207.

[232]For example, see Maslak, P. *J. Am. Chem. Soc.* *1989*, *111*, 8201. Nitrenes are much more selective (and less reactive) in this reaction than carbenes (**12-17**). For a discussion, see Alewood, P.F.; Kazmaier, P.M.; Rauk, A. *J. Am. Chem. Soc.* *1973*, *95*, 5466.

[233]For example, see Simson, J.M.; Lwowski, W. *J. Am. Chem. Soc.* *1969*, *91*, 5107; Inagaki, M.; Shingaki, T.; Nagai, T. *Chem. Lett.* *1981*, 1419.

[234]Smolinsky, G.; Feuer, B.I. *J. Am. Chem. Soc.* *1964*, *86*, 3085.

[235]For a review of cyanonitrenes, see Anastassiou, A.G.; Shepelavy, J.N.; Simmons, H.E.; Marsh, F.D., in Lwowski, W. *Nitrenes*, Wiley, NY, *1970*, pp. 305–344.

[236]For a review of arylnitrenes, see Scriven, E.F.V. *Azides and Nitrenes*, Academic Press, NY, *1984*, pp. 95–204.

[237]For a synthetically useful noncyclization example, see Meinwald, J.; Aue, D.H. *Tetrahedron Lett.* *1967*, 2317.

[238]For a list of examples, with references, see Larock, R.C. *Comprehensive Organic Transformations*, 2nd ed., Wiley-VCH, NY, *1999*, pp. 1148–1149.

[239]For a review, see Müller, P.; Fruit, C. *Chem. Rev.* *2003*, *103*, 2905.

Sulfonation or Sulfo-de-hydrogenation

$$R\overset{\overset{\displaystyle}{}}{\underset{\underset{\displaystyle O}{}}{\diagdown}}\,R' \quad\xrightarrow{\text{SO}_3}\quad R\overset{\overset{\displaystyle SO_2H}{}}{\underset{\underset{\displaystyle O}{}}{\diagdown}}\,R'$$

Ketones, carboxylic esters (including lactones),[240] and amides (including lactams)[241] can be sulfenylated[242] in the α position by conversion to the enolate ion with a base, such as lithium N-isopropylcyclohexylamide and subsequent treatment with a disulfide.[243] The reaction, shown above for ketones, involves nucleophilic substitution at sulfur. Analogously, α-phenylseleno ketones RCH(SePh)COR' and α-phenylseleno esters RCH(SePh)COOR' can be prepared[244] by treatment of the corresponding enolate anions with PhSeBr,[245] PhSeSePh,[246] or benzeneseleninic anhydride PhSe(O)OSe(O)Ph.[247] Another method for the introduction of a phenylseleno group into the α position of a ketone involves simple treatment of an ethyl acetate solution of the ketone with PhSeCl (but not PhSeBr) at room temperature.[248] This procedure is also successful for aldehydes, but not for carboxylic esters. N-Phenylselenophthalimide has been used to convert ketones[249] and aldehydes[250] to the α- PhSe derivative. In another method that avoids the use of PhSeX reagents, a ketone enolate is treated with selenium to give an R'COCHRSe– ion, which is treated with MeI, producing the α-methylseleno ketone R'COCHRSeMe.[251] This method has also been applied to carboxylic esters.

[240]Trost, B.M.; Salzmann, T.N. *J. Am. Chem. Soc.* **1973**, *95*, 6840; Seebach, D.; Teschner, M. *Tetrahedron Lett.* **1973**, 5113. For discussions, see Trost, B.M. *Pure Appl. Chem.* **1975**, *43*, 563, pp. 572–578; Caine, D., in Augustine, R.L. *Carbon–Carbon Bond Formation*, Vol. 1, Marcel Dekker, NY, **1979**, pp. 278–282.

[241]Zoretic, P.A.; Soja, P. *J. Org. Chem.* **1976**, *41*, 3587; Gassman, P.G.; Balchunis, R.J. *J. Org. Chem.* **1977**, *42*, 3236.

[242]For a discussion of the synthesis of sulfenates, see Sandrinelli, F.; Fontaine, G.; Perrio, S.; Beslin, P. *J. Org. Chem.* **2004**, *69*, 6916.

[243]For another reagent, see Scholz, D. *Synthesis* **1983**, 944.

[244]For reviews of selenylations, see Back, T.G., in Liotta, D.C. *Organoselenium Chemistry*, Wiley, NY, **1987**, pp. 1–125; Paulmier, C. *Selenium Reagents and Intermediates in Organic Synthesis*, Pergamon, Elmsford, NY, **1986**, pp. 95–98.

[245]Reich, H.J.; Reich, I.J.; Renga, J.M. *J. Am. Chem. Soc.* **1973**, *95*, 5813; Clive, D.L.J. *J. Chem. Soc. Chem. Commun.* **1973**, 695; Brocksom, T.J.; Petragnani, N.; Rodrigues, R. *J. Org. Chem.* **1974**, *39*, 2114; Schwartz, J.; Hayasi, Y. *Tetrahedron Lett.* **1980**, *21*, 1497. See also Liotta, D. *Acc. Chem. Res.* **1984**, *17*, 28.

[246]Grieco, P.A.; Miyashita, M. *J. Org. Chem.* **1974**, *39*, 120. α-Phenylselenation can also be accomplished with PhSeSePh, SeO$_2$, and an acid catalyst: Miyoshi, N.; Yamamoto, T.; Kambe, N.; Murai, S.; Sonoda, N. *Tetrahedron Lett.* **1982**, *23*, 4813.

[247]Barton, D.H.R.; Morzycki, J.W.; Motherwell, W.B.; Ley, S.V. *J. Chem. Soc. Chem. Commun.* **1981**, 1044.

[248]Sharpless, K.B.; Lauer, R.F.; Teranishi, A.Y. *J. Am. Chem. Soc.* **1973**, *95*, 6137.

[249]Cossy, J.; Furet, N. *Tetrahedron Lett.* **1993**, *34*, 7755.

[250]Wang, W.; Wang, K.; Li, H. *Org. Lett.* **2004**, *6*, 2817.

[251]Saindane, M.; Barnum, C.; Ensley, H.; Balakrishnan, P. *Tetrahedron Lett.* **1981**, *22*, 3043; Liotta, D. *Acc. Chem. Res.* **1984**, *17*, 28.

Silyl enol ethers are converted to α-thioalkyl and α-thioaryl ketones via a sulfenylation method, driven by aromatization of an added quinone mono-O,S-acetal in the presence of Me$_3$SiOTf.[252]

The α-seleno and α-sulfenyl carbonyl compounds prepared by this reaction can be converted to α,β-unsaturated carbonyl compounds (**17-12**). The sulfenylation reaction has also been used[253] as a key step in a sequence for moving the position of a carbonyl group to an adjacent carbon.[254]

OS VI, 23, 109; 68, 8.

Aldehydes, ketones, and carboxylic acids containing α hydrogens can be sulfonated with sulfur trioxide.[255] The mechanism is presumably similar to that of **12-4**. Sulfonation has also been accomplished at vinylic hydrogen.

OS **VI**, 23, 109; **VIII**, 550. OS **IV**, 846, 862.

E. Carbon Reagents

12-15 Arylation and Alkylation of Alkenes

Alkylation or **Alkyl-de-oxysulfonation (de-halogenation)**, **Arylation** or **Aryl-de-oxysulfonation (de-halogenation)**, and so on

Vinyl triflates ($C{=}C{-}OSO_2CF_3$) react with vinyl tin derivatives in the presence of palladium catalysts to form dienes, in what is known as the *Stille coupling*.[256] Vinyl triflates can be prepared from the enolate by reaction with N-phenyl triflimide.[257] Vinyltin compounds are generally prepared by the reaction of an alkyne with an trialkyltin halide (see **15-17** and **15-21**).[258] Still cross-coupling reactions are quite important.[259] Stille reactions are compatible with many functional groups,

---

[252]Matsugi, M.; Murata, K.; Gotanda, K.; Nambu, H.; Anilkumar, G.; Matsumoto, K.; Kita, Y. *J. Org. Chem.*, **2001**, *66*, 2434.

[253]Trost, B.M.; Hiroi, K.; Kurozumi, S. *J. Am. Chem. Soc.* **1975**, *97*, 438.

[254]There are numerous other ways of achieving this conversion. For reviews, see Morris, D.G. *Chem. Soc. Rev.* **1982**, *11*, 397; Kane, V.V.; Singh, V.; Martin, A.; Doyle, D.L. *Tetrahedron* **1983**, *39*, 345.

[255]For a review, see Gilbert, E.E. *Sulfonation and Related Reactions*, Wiley, NY, **1965**, pp. 33–61.

[256]Scott, W.J.; Crisp, G.T.; Stille, J.K. *J. Am. Chem. Soc.* **1984**, *106*, 4630. See Roth, G.P.; Farina, V.; Liebeskind, L.S.; Peña-Cabrera, E. *Tetrahedron Lett.* **1995**, *36*, 2191 for an optimized version of this reaction.

[257]McMurry, J.E.; Scott, W.J. *Tetrahedron Lett.* **1983**, *24*, 979.

[258]For an example, see Maleczka Jr., R.E.; Lavis, J.M.; Clark, D.H.; Gallagher, W.P. *Org. Lett.* **2000**, *2*, 3655.

[259]Stille, J.K. *Angew. Chem. Int. Ed.* **1986**, *25*, 508; Stille, J.K.; Groh, B.L. *J. Am. Chem. Soc.* **1987**, *109*, 813; Farina, V.; Krishnamurthy, V.; Scott, W.J. *Org. React.* **1997**, *50*, 1.

proceed with a retention of geometry of the C=C units, and are usually regiospe-cific with respect to the newly formed C–C σ-bond. Vinyl halides can be used,[260] and allenic tin compounds have been used.[261] Intamolecualr reactions are possi-ble.[262] Stille coupling has been done using microwave irradiation,[263] in fluorous solvents,[264] and in supercritical carbon dioxide (see p. 415).[265] One-pot hydrostan-nylation/Stille coupling has been reported using catalytic amounts of tin with alkyne substrates reacting with vinyl halides.[266]

This reaction is highly stereoselective. Cine substitution is known with this reac-tion, and its mechanism has been studied.[267] Using ArSnCl$_3$ derivatives, Stille cou-pling can be done in aq. KOH.[268] A related reaction couples reagents with C=C–I$^+$Ph reagents, in the presence of a palladium catalyst.[269] Aryl halides[270] and het-eroaryl halides[271] can be coupled to vinyltin reagents[272] using a palladium catalyst. Vinylation of heteroaryl triflates[273] also possible. Vinyl halides can be coupled to alkenes to form dienes.[274] The reaction of dihydrofurans with vinyl triflates and a palladium catalyst leads to a nonconjugated diene, **33**.[275] This example illustrates that the product is formed by an elimination step, as with the Heck reaction (**13-10**), and double bond migration can occur resulting in allylic rearrangement.

[260]Johnson, C.R.; Adams, J.P.; Braun, M.P.; Senanayake, C.B.W. *Tetrahedron Lett.* **1992**, *33*, 919.

[261]Badone, D.; Cardamone, R.; Guzzi, U. *Tetrahedron Lett.* **1994**, *35*, 5477.

[262]Segorbe, M.M.; Adrio, J.; Carretero, J.C. *Tetrahedron Lett.* **2000**, *41*, 1983.

[263]Larhed, M.; Hoshino, M.; Hadida, S.; Curran, D.P.; Hallberg, A. *J. Org. Chem.* **1997**, *62*, 5583; Olofsson, K.; Kim, S.-Y.; Larhed, M.; Curran, D.P.; Hallberg, A. *J. Org. Chem.* **1999**, *64*, 4539.

[264]Olofsson, K.; Kim, S.-Y.; Larhed, M.; Curran, D.P.; Hallberg, A. *J. Org. Chem.* **1999**, *64*, 4539; Hoshino, M.; Degenkolb, P.; Curran, D.P. *J. Org. Chem.* **1997**, *62*, 8341; Curran, D.P.; Hadida, S. *J. Am. Chem. Soc.* **1996**, *118*, 2531.

[265]Jessop, P. G.; Ikariya, T.; Noyori, R. *Chem. Rev.* **1999**, *99*, 475.

[266]Maleczka Jr., R.E.; Gallagher, W.P.; Terstiege, I. *J. Am. Chem. Soc.* **2000**, *122*, 384; Gallagher, W.P.; Terstiege, I.; Maleczka Jr., R.E. *J. Am. Chem. Soc.* **2001**, *123*, 3194.

[267]Farina, V.; Hossain, M.A. *Tetrahedron Lett.* **1996**, *37*, 6997.

[268]Rai, R.; Aubrecht, K.B.; Collum, D.B. *Tetrahedron Lett.* **1995**, *36*, 3111.

[269]Moriarty, R.M.; Epa, W.R. *Tetrahedron Lett.* **1992**, *33*, 4095.

[270]Corriu, R.J.P.; Geng, B.; Moreau, J.J.E. *J. Org. Chem.* **1993**, *58*, 1443; Levin, J.I. *Tetrahedron Lett.* **1993**, *34*, 6211; Littke, A.F.; Fu, G.C. *Angew. Chem. Int. Ed.* **1999**, *38*, 2411.

[271]Barchín, B.M.; Valenciano, J.; Cuadro, A.M.; Builla-Alvarez, J.; Vaquero, J.J. *Org. Lett.* **1999**, *1*, 545; Clapham, B.; Sutherland, A.J. *J. Org. Chem.* **2001**, *66*, 9033.

[272]For a coupling reaction using a butenolide-vinyltin reagent, see Rousset, S.; Abarbri, M.; Thibonnet, J.; Duchêne, A.; Parrain, J.-L. *Org. Lett.* **1999**, *1*, 701. For a vinyltin reagent with a nitrogen substituent (a tinylated enamide), see Minière, S.; Cintrat, J.-C. *J. Org. Chem.* **2001**, *66*, 7385.

[273]Bernabé, P.; Rutjes, P.J.T.; Hiemstra, H.; Speckamp, W.N. *Tetrahedron Lett.* **1996**, *37*, 3561; Schaus, J.V.; Panek, J.S. *Org. Lett.* **2000**, *2*, 469.

[274]Voigt, K.; Schick, U.; Meyer, F.E.; de Meijere, A. *Synlett* **1994**, 189.

[275]Gilbertson, S.R.; Fu, Z.; Xie, D. *Tetrahedron Lett.* **2001**, *42*, 365.

The accepted mechanism for the Stille reaction involves a catalytic cycle[276] in which an oxidative addition[277] and a reductive elimination step[278] are fast, relative to Sn/Pd transmetallation (the rate-determining step).[279] It appears that the more coordinatively unsaturated species, probably with a coordinated solvent molecule, is involved in the electrophilic substitution at tin. Another mechanism has been proposed, in which oxidative addition of the vinyl triflate to the ligated palladium gives a *cis*-palladium complex that isomerizes rapidly to *trans*-palladium complex, which then reacts with the organotin compound following a S_E2 (cyclic) mechanism, with release of a ligand.[280] This pathway gives a bridged intermediate, and subsequent elimination of $XSnBu_3$ yields a three-coordinate species cis-palladium complex, which readily gives the coupling product.[280]

Cyclopropylboronic acids (**12-28**) couple with vinylic halides[281] or vinyl triflates[282] to give vinylcyclopropanes, using a palladium catalyst. Vinyl borates (**12-28**) were coupled to vinyl triflates using a palladium catalyst.[283] In a variation, phenylboronic acid reacted with a symmetrical internal alkyne and a nickel catalyst to give a conjugated diene bearing a phenyl group.[284] Stille coupling to enols has been reported.[285] A variation of this latter reaction coupled vinyl triflates to vinyl ethers, without a palladium catalyst, but using microwave irradiation.[286] The

[276]Stanforth, S.P. *Tetrahedron* **1998**, *54*, 263; Farina, V.; Roth, G.P. *Adv. Metalorg. Chem.* **1996**, *5*, 1; Curran, D.P.; Hoshino, M. *J. Org. Chem.* **1996**, *61*, 6480; Mateo, C.; Cárdenas, D.J.; Fernández-Rivas, C.; Echavarren, A.M. *Chem. Eur. J.* **1996**, *2*, 1596; Roth, G.P.; Farina, V.; Liebeskind, L.S.; Peña-Cabrera, E. *Tetrahedron Lett.* **1995**, *36*, 2191; Mitchell, T.N. *Synthesis* **1992**, 803; Scott, W.J.; Stille, J.K. *J. Am. Chem. Soc.* **1986**, *108*, 3033; Stille, J.K. *Angew. Chem., Int. Ed.* **1986**, *25*, 508; Beletskaya, I.P. *J. Organomet. Chem.* **1983**, *250*, 551; Farina, V., in, Abel, E. W., Stone, F. G. A., Wilkinson, G. *Comprehensive Organometallic Chemistry II*, Vol. 12, Pergamon, Oxford, U.K., **1995**, Chapter 3.4.; Brown, J.M.; Cooley, N.A. *Chem. Rev.* **1988**, *88*, 1031.

[277]Amatore, C.; Jutand, A.; Suarez, A. *J. Am. Chem. Soc.* **1993**, *115*, 9531; Amatore, C.; Pflüger, F. *Organometallics* **1990**, *9*, 2276, and references cited therein.

[278]Ozawa, F.; Fujimori, M.; Yamamoto, T.; Yamamoto, A. *Organometallics* **1986**, *5*, 2144; Tatsumi, K.; Hoffmann, R.; Yamamoto, A.; Stille, J.K. *Bull. Chem. Soc. Jpn.* **1981**, *54*, 1857; Ozawa, F.; Ito, T.; Nakamura, Y.; Yamamoto, A. *Bull. Chem. Soc. Jpn.* **1981**, *54*, 1868; Moravsikiy, A.; Stille, J.K. *J. Am. Chem. Soc.* **1981**, *103*, 4182; Loar, M.K.; Stille, J.K. *J. Am. Chem. Soc.* **1981**, *103*, 4174; Ozawa, F.; Ito, T.; Yamamoto, A. *J. Am. Chem. Soc.* **1980**, *102*, 6457; Gillie, A.; Stille, J.K. *J. Am. Chem. Soc.* **1980**, *102*, 4933; Komiya, S.; Albright, T.A.; Hoffmann, R.; Kochi, J.K. *J. Am. Chem. Soc.* **1976**, *98*, 7255.

[279]Labadie, J.W.; Stille, J.K. *J. Am. Chem. Soc.* **1983**, *105*, 6129; Eaborn, C.; Odell, K.J.; Pidcock, A. *J. Chem. Soc., Dalton Trans.* **1978**, 357; Eaborn, C.; Odell, K.J.; Pidcock, A. *J. Chem. Soc., Dalton Trans.* **1979**, 758; Deacon, G.B.; Gatehouse, B.M.; Nelson-Reed, K.T. *J. Organomet. Chem.* **1989**, *359*, 267.

[280]Casado, A.L.; Espinet, P.; Gallego, A.M. *J. Am. Chem. Soc.* **2000**, *122*, 11771; Casado, A.L.; Espinet, P. *J. Am. Chem. Soc.* **1998**, *120*, 8978.

[281]Zhou, S.-m.; Deng, M.-z. *Tetrahedron Lett.* **2000**, *41*, 3951.

[282]Yao, M.-L.; Deng, M.-Z. *J. Org. Chem.* **2000**, *65*, 5034; Yao, M.-L.; Deng, M.-Z. *Tetrahedron Lett.* **2000**, *41*, 9083.

[283]Occhiato, E.G.; Trabocchi, A.; Guarna, A. *J. Org. Chem.* **2001**, *66*, 2459.

[284]Shirakawa, E.; Takahashi, G.; Tsuchimoto, T.; Kawakami, Y. *Chem. Commun.* **2001**, 2688.

[285]See Fu, X.; Zhang, S.; Yin, J.; McAllister, T.L.; Jiang, S.A.; Tann, C.-H.; Thiruvengadam, T.K.; Zhang, F. *Tetrahedron Lett.* **2002**, *43*, 573.

[286]Vallin, K.S.A.; Larhed, M.; Johansson, K.; Hallberg, A. *J. Org. Chem.* **2000**, *65*, 4537.

coupling of vinyl silanes to give the symmetrically conjugated diene using CuCl and air was reported.[287] Vinyl zinc halides were coupled to 1-halo enol ether to give a conjugated diene bearing a vinyl ether unit, using a palladium catalyst.[288] Tertiary propargyl alcohols (R-C≡C-CMe₂OH) are coupled to conjugated alkenes in a Heck-like process using a palladium catalyst and oxygen to give the conjugated ene-yne.[289]

Coupling is not restricted to two vinyl units or an aryl with a vinyl. 1-Lithioalkynes were coupled to vinyl tellurium compounds (C=C–TeBu) using a nickel catalyst[290] or a palladium catalyst[291] to give a conjugated en-yne. 2-Alkynes (R-C≡C-Me) react with HgCl₂, n-butyllithium, and ZnBr₂, sequentially, and then with vinyl iodides and a palladium catalyst to give the nonconjugated en-yne.[292] Alkynyl groups can be coupled to vinyl groups to give ene-ynes, via reaction of silver alkynes (Ag-C≡C-R) with vinyl triflates and a palladium catalyst.[293] In the presence of CuI and a palladium catalyst, vinyl triflates[294] or vinyl halides[295] couple to terminal alkynes. Alkynyl zinc reagents (R-C≡C-ZnBr) can be coupled to vinyl halides with a palladium catalyst to give the conjugate ene-yne.[296]

Alkyl groups can be coupled to a vinyl unit to give substituted alkenes. The reaction of vinyl iodides and EtZnBr, with a palladium catalyst, gave the ethylated alkene (C=C–Et).[297] A similar coupling reaction was observed with RZnI reagents and vinyl nitro compounds (C=C–NO₂), which gave the alkyne (C≡C–R) with microwave irradiation.[298] Aliphatic alkyl bromides reacted with vinyltin compounds to give the alkylated alkene using a palladium catalyst.[299] Allylic tosylates were coupled to conjugated alkenes to give a non-conjugated diene using a palladium catalyst.[300] An internal coupling reaction was reported in which an alkenyl enamide (**34**) reacted with Ag₃PO₄ and a chiral palladium catalyst to give **35** enantioselectively.[301]

[287]Nishihara, Y.; Ikegashira, K.; Toriyama, F.; Mori, A.; Hiyama, T. *Bull. Chem. Soc. Jpn.* **2000**, *73*, 985.

[288]Su, M.; Kang, Y.; Yu, W.; Hua, Z.; Jin, Z. *Org. Lett.* **2002**, *4*, 691.

[289]Nishimura, T.; Araki, H.; Maeda, Y.; Uemura, S. *Org. Lett.* **2003**, *5*, 2997.

[290]Raminelli, C.; Gargalak, Jr., J.; Silveira, C.C.; Comasseto, J.V. *Tetrahedron Lett.* **2004**, *45*, 4927; Silveira, C.C.; Braga, A.L.; Vieira, A.S.; Zeni, G. *J. Org. Chem.* **2003**, *68*, 662.

[291]Zeni, G.; Comasseto, J.V. *Tetrahedron Lett.* **1999**, *40*, 4619.

[292]Ma, S.; Zhang, A.; Yu, Y.; Xia, W. *J. Org. Chem.* **2000**, *65*, 2287.

[293]Dillinger, S.; Bertus, P.; Pale, P. *Org. Lett.* **2001**, *3*, 1661. See Halbes, U.; Bertus, P.; Pale, P. *Tetrahedron Lett.* **2001**, *42*, 8641; Bertus, P.; Halbes, U.; Pale, P. *Eur. J. Org. Chem.* **2001**, 4391.

[294]Braga, A.L.; Emmerich, D.J.; Silveira, C.C.; Martins, T.L.C.; Rodrigues, O.E.D. *Synlett* **2001**, 369.

[295]Lee, J.-H.; Park, J.-S.; Cho, C.-G. *Org. Lett.* **2002**, *4*, 1171. For an example using another copper catalyst, see Bates, C.G.; Saejueng, P.; Venkataraman, D. *Org. Lett.* **2004**, *6*, 1441.

[296]Negishi, E.; Qian, M.; Zeng, F.; Anastasia, L.; Babinski, D. *Org. Lett.* **2003**, *5*, 1597.

[297]Abarbri, M.; Parrain, J.-L.; Kitamura, M.; Noyori, R.; Duchêne, A. *J. Org. Chem.* **2000**, *65*,7475.

[298]Hu, Y.; Yu, J.; Yang, S.; Wang, J.-X.; Yin, Y. *Synth. Commun.* **1999**, *29*, 1157.

[299]Menzel, K.; Fu, G.C. *J. Am. Chem. Soc.* **2003**, *125*, 3718.

[300]Tsukada, N.; Sato, T.; Inoue, Y. *Chem. Commun.* **2003**, 2404.

[301]Kiewel, K.; Tallant, M.; Sulikowski, G.A. *Tetrahedron Lett.* **2001**, *42*, 6621.

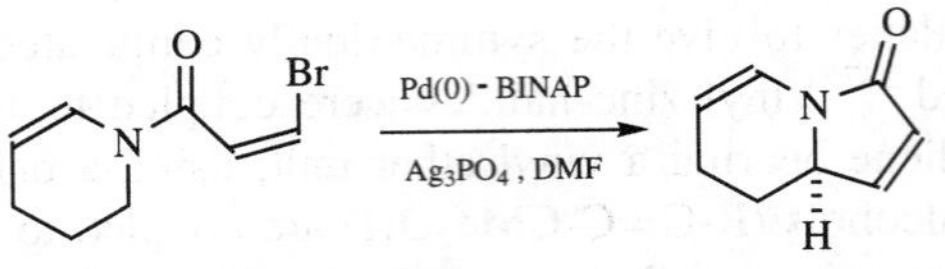

34   **35**

12-16   Acylation at an Aliphatic Carbon

Acylation or Acyl-de-hydrogenation

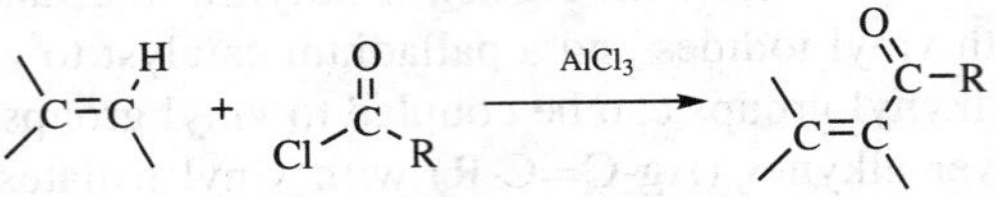

Alkenes can be acylated with an acyl halide and a Lewis acid catalyst in what is essentially a Friedel–Crafts reaction at an aliphatic carbon.[302] The product can arise by two paths. The initial attack is by the π-bond of the alkene unit on the acyl cation (RCO^+; or on the acyl halide free or complexed; see **11-17**) to give a carbocation, **36**.

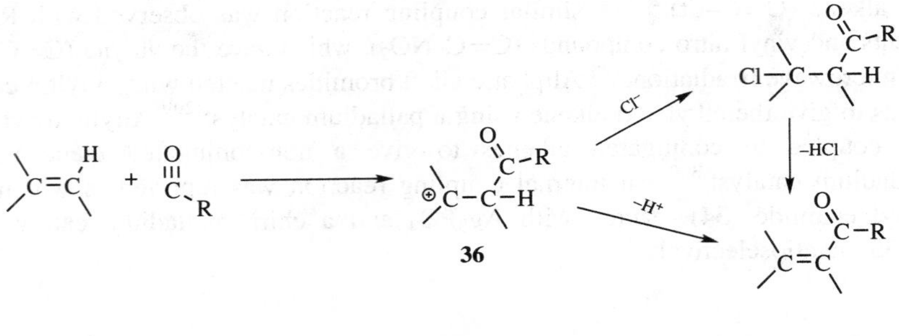

36

Ion **36** can either lose a proton or combine with chloride ion. If it loses a proton, the product is an unsaturated ketone; the mechanism is similar to the tetrahedral mechanism of Chapter 10, but with the charges reversed. If it combines with chloride, the product is a β-halo ketone, which can be isolated, so that the result is addition to the double bond (see **15-47**). On the other hand, the β-halo ketone may, under the conditions of the reaction, lose HCl to give the unsaturated ketone, this time by an addition–elimination mechanism. In the case of unsymmetrical alkenes, the more stable alkene is formed (the more highly substituted and/or conjugated alkene, following Markovnikov's rule, see p. 1019). Anhydrides and carboxylic acids (the latter with a proton acid such as anhydrous HF, H_2SO_4, or polyphosphoric acid as a catalyst) are sometimes used instead of acyl halides. With some sub-

---

[302]For reviews, see Groves, E.E. *Chem. Soc. Rev. 1972, 1,* 73; Satchell, D.P.N.; Satchell, R.S., in Patai, S. *The Chemistry of the Carbonyl Group,* Vol. 1, Wiley, NY, *1966,* pp. 259–266, 270–273; Nenitzescu, C.D.; Balaban, A.T., in Olah A, G.A. *Friedel-Crafts and Related Reactions,* Vol. 3, Wiley, NY, *1964,* pp. 1033–1152.

strates and catalysts double-bond migrations are occasionally encountered so that, for example, when 1-methylcyclohexene was acylated with acetic anhydride and zinc chloride, the major product was 6-acetyl-1-methylcyclohexene.[303]

Conjugated dienes can be acylated by treatment with acyl- or alkylcobalt tetracarbonyls, followed by base-catalyzed cleavage of the resulting π-allyl carbonyl derivatives[304] (π-allyl metal complexes were discussed on p. 117. The reaction is very general. With unsymmetrical dienes, the acyl group generally substitutes most readily at a cis double bond, next at a terminal alkenyl group, and least readily at a trans double bond. The most useful bases are strongly basic, hindered amines, such as dicyclohexylethylamine. The use of an alkylcobalt tetracarbonyl $RCo(CO)_4$ gives the same product as that shown above. Acylation of vinylic ethers has been accomplished with aromatic acyl chlorides, a base, and a palladium catalyst: $ROCH=CH_2 \rightarrow ROCH=CHCOAr$.[305]

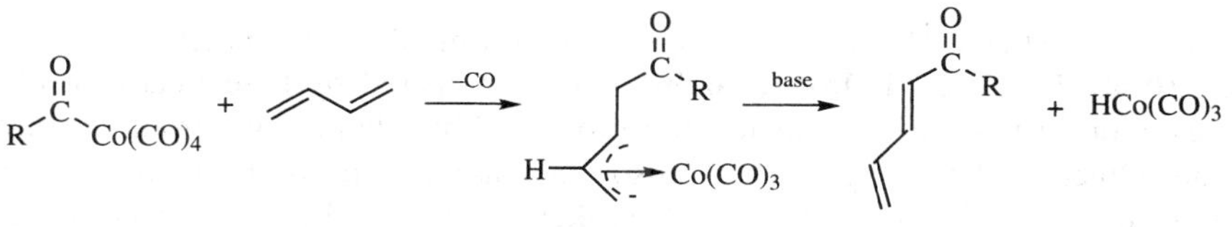

Formylation of alkenes can be accomplished with *N*-disubstituted formamides and $POCl_3$.[306] This is an aliphatic Vilsmeier reaction (see **11-18**). Vilsmeier formylation can also be performed on the α position of acetals and ketals, so that hydrolysis of the products gives keto aldehydes or dialdehydes:[307] A variation of this reaction heated a 1,1-dibromoalkene with a secondary amine in aq. DMF to give the corresponding amide.[308]

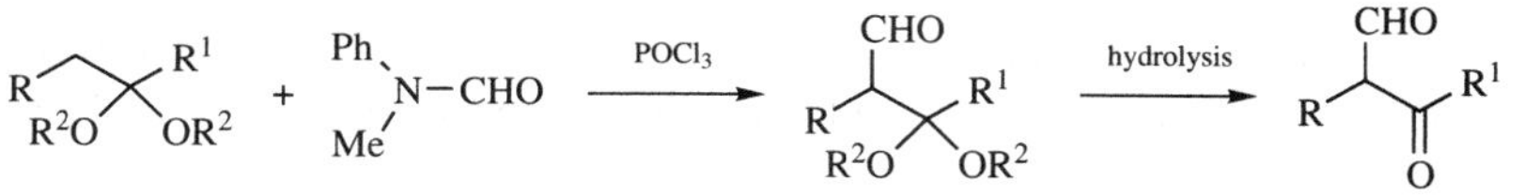

Acetylation of acetals or ketals can be accomplished with acetic anhydride and BF_3-etherate.[309] The mechanism with acetals or ketals also involves attack at an

[303]Deno, N.C.; Chafetz, H. *J. Am. Chem. Soc.* **1952**, *74*, 3940. For other examples, see Beak, P.; Berger, K.R. *J. Am. Chem. Soc.* **1980**, *102*, 3848; Dubois, J.E.; Saumtally, I.; Lion, C. *Bull. Soc. Chim. Fr.* **1984**, II-133; Grignon-Dubois, M.; Cazaux, M. *Bull. Soc. Chim. Fr.* **1986**, 332.

[304]For a review, see Heck, R.F., in Wender, I.; Pino, P. *Organic Syntheses via Metal Carbonyls*, Vol. 1, Wiley, NY, **1968**, pp. 388–397.

[305]Andersson, C.; Hallberg, A. *J. Org. Chem.* **1988**, *53*, 4257.

[306]For reviews, see Burn, D. *Chem. Ind. (London)* **1973**, 870; Satchell, D.P.N.; Satchell, R.S., in Patai, S. *The Chemistry of the Carbonyl Group*, Vol. 1, Wiley, NY, **1966**, pp. 281–282.

[307]Youssefyeh, R.D. *Tetrahedron Lett.* **1964**, 2161.

[308]Shen, W.; Kunzer, A. *Org. Lett.* **2002**, *4*, 1315.

[309]Youssefyeh, R.D. *J. Am. Chem. Soc.* **1963**, *85*, 3901.

alkenyl carbon, since enol ethers are intermediates.[309] Ketones can be formylated in the α position by treatment with CO and a strong base.[310]

OS **IV**, 555, 560; **VI**, 744. Also see OS **VI**, 28.

12-17 Conversion Of Enolates to Silyl Enol Ethers, Silyl Enol Esters, and Silyl Enol Sulfonate Esters

3/*O*-Trimethylsilyl-de-hydrogenation

Silyl enol ethers,[311] important reagents with a number of synthetic uses (see, e.g., **10-68**, **12-4**, **15-24**, **15-64**, **16-36**), can be prepared by base treatment of a ketone (converting it to its enolate anion) followed by addition of a trialkylchlorosilane. Other silylating agents have also been used.[312] Both strong bases, e.g., lithium diisopropylamide (LDA), and weaker bases (e.g. Et$_3$N) have been used for this purpose.

In some cases, the base and the silylating agent can be present at the same time.[313] Enolates prepared in other ways (e.g., as shown on p. 603) also give the reaction.[314] The reaction can be applied to aldehydes by the use of the base KH in 1,2-dimethoxyethane.[315] A particularly mild method for conversion of ketones

[310]See, for example, van der Zeeuw, A.J.; Gersmann, H.R. *Recl. Trav. Chim. Pays-Bas* **1965**, *84*, 1535.

[311]For reviews of these compounds, see Poirier, J. *Org. Prep. Proced. Int.* **1988**, *20*, 319; Brownbridge, P. *Synthesis* **1983**, 1, 85; Rasmussen, J.K. *Synthesis* **1977**, 91. For monographs on silicon reagents in organic synthesis, see Colvin, E.W. *Silicon Reagents in Organic Synthesis*, Academic Press, NY, **1988**. For reviews, see Colvin, E.W. in Hartley, C.R.; Patai, S. *The Chemistry of the Metal-Carbon Bond*, Vol. 4, Wiley, NY, pp. 539–621; Ager, D.J. *Chem. Soc. Rev.* **1982**, *11*, 493; Colvin, E.W. *Chem. Soc. Rev.* **1978**, *7*, 15, pp. 43–50.

[312]For a review of silylating agents, see Mizhiritskii, M.D.; Yuzhelevskii, Yu.A. *Russ. Chem. Rev.* **1987**, *56*, 355. For a list, with references, see Larock, R.C. *Comprehensive Organic Transformations*, 2nd ed., Wiley-VCH, NY, **1999**, pp. 1488–1491.

[313]Corey, E.J.; Gross, A.W. *Tetrahedron Lett.* **1984**, *25*, 495. See Lipshutz, B.H.; Wood, M.R.; Lindsley, C.W. *Tetrahedron Lett.* **1995**, *36*, 4385 for a discussion of the role of Me$_3$SiCl in deprotonations with LiNR$_2$.

[314]See Cahiez, G.; Figadère, B.; Cléry, P. *Tetrahedron Lett.* **1994**, *35*, 6295.

[315]Ladjama, D.; Riehl, J.J. *Synthesis* **1979**, 504. This base has also been used for ketones: See Orban, J.; Turner, J.V.; Twitchin, B. *Tetrahedron Lett.* **1984**, *25*, 5099.

or aldehydes to silyl enol ethers uses Me_3SiI and the base hexamethyldisilazane, $(Me_3Si)_2NH$.[316] Cyclic ketones can be converted to silyl enol ethers in the presence of acyclic ketones, by treatment with Me_3SiBr, tetraphenylstibonium bromide, Ph_4SbBr, and an aziridine.[317] bis(Trimethylsilyl)acetamide is an effective reagent for the conversion of ketones to the silyl enol ether, typically giving the thermodynamic product (see below).[318] Silyl enol ethers have also been prepared by the direct reaction of a ketone and a silane (R_3SiH) with a platinum complex catalyst.[319]

Unsymmetrical ketones can give the more substituted (thermodynamic) silyl enol ether or the less substituted (kinetic) product, depending on the use of thermodynamic conditions (protic solvents, e.g., ethanol, water, or ammonia; a base generating a conjugate acid stronger than the starting ketone; more ionic counterions, e.g., K or Na; higher temperatures and longer reaction times) or kinetic conditions (aprotic solvents, such as ether or THF; a base generating a conjugate acid weaker than the starting ketone; more covalent counterions, e.g., Li; lower temperatures and relatively short reaction times). Other reaction conditions have been developed to control or influence the relative amounts of kinetic or thermodynamic silyl enol ether. Magnesium diisopropyl amide has been used to prepare kinetic silyl enol ethers in virtual quantitative yield.[320] Reaction with Me_3SiCl/KI in DMF gives primarily the thermodynamic silyl enol ether.[321] The reaction of an unsymmetrical ketone with Mg and TMSCl in DMF gives a roughly 2:1 mixture of thermodynamic: kinetic silyl enol ether.[322]

An interesting synthesis of silyl enol ethers involves chain extension of an aldehyde. Aldehydes are converted to the silyl enol ether of a ketone upon reaction with lithium (trimethylsilyl)diazomethane and then a dirhodium catalyst.[323] Initial reaction of lithium(trimethylsilyl)diazomethane [LTMSD, prepared *in situ* by reaction of butyllithium with (trimethylsilyl)diazomethane] to the aldehyde (e.g., **37**) gave the alkoxide addition product. Protonation, and then capture by a transition-metal catalyst, and a 1,2-hydride migration gave the silyl enol ether, **38**.

[316]Miller, R.D.; McKean, D.R. *Synthesis* **1979**, 730; *Synth. Commun.* **1982**, *12*, 319. See also, Cazeau, P.; Duboudin, F.; Moulines, F.; Babot, O.; Dunogues, J. *Tetrahedron* **1987**, *43*, 2075, 2089; Ahmad, S.; Khan, M.A.; Iqbal, J. *Synth. Commun.* **1988**, *18*, 1679.
[317]Fujiwara, M.; Baba, A.; Matsuda, H. *Chem. Lett.* **1989**, 1247.
[318]Smietana, M.; Mioskowski, C. *Org. Lett.* **2001**, *3*, 1037. See also, Tanabe, Y.; Misaki, T.; Kurihara, M.; Iida, A.; Nishii, Y. *Chem. Commun.* **2002**, 1628.
[319]Ozawa, F.; Yamamoto, S.; Kayagishi, S.; Hiraoka, M.; Ideda, S.;Minami, T.; Ito, S.; Yoshifuji, M. *Chem. Lett.* **2001**, 972. For the conversion of a conjugated ketone to a silyl enol ether with R_3SiH and a triarylborane catalyst, see Blackwell, J.M.; Morrison, D.J.; Piers, W.E. *Tetahedron* **2002**, *58*, 8247. For the conversion of a conjugated ketone to a silyl enol ether with R_3SiH and a rhodium catalyst, see Mori, A.; Kato, T. *Synlett* **2002**, 1167.
[320]Lessène, G.; Tripoli, R.; Cazeau, P.; Biran, C.; Bordeau, M. *Tetrahedron Lett.* **1999**, *40*, 4037.
[321]Lin, J.-M.; Liu, B.-S. *Synth. Commun.* **1997**, *27*, 739.
[322]Patonay, T.; Hajdu, C.; Jekö, J.; Lévai, A.; Micskei, K.; Zucchi, C. *Tetrahedron Lett.* **1999**, *40*, 1373.
[323]Aggarwal, V. K.; Sheldon, C. G.; Macdonald, G. J.; Martin, W. P. *J. Am. Chem. Soc.* **2002**, *124*, 10300.

37   **38**

Enol acetates are generally prepared by the reaction of an enolate anion with a suitable acylating reagent.[324] Enolate anions react with acyl halides and with anhydrides to give the acylated product. Both C-acylation and O-acylation are possible, but in general O-acylation predominates.[325] Note that the extent of O- versus C-acylation is very dependent on the local environment and electronic effects within the enolate anion.[326] Silyl sulfonate esters can be prepared by similar methods, using sulfonic acid anhydrides rather than carboxylic anhydrides. A polymer-supported triflating agent was used to prepare silyl enol triflate from ketones, in the presence of diisopropylethylamine.[327]

When a silyl enol ether is the trimethylsilyl derivative ($Me_3Si–O–C=C$), treatment with methyllithium will regenerate the lithium enolate anion and the volatile trimethylsilane (Me_3SiH).[328] The enolate anion can be used in the usual reactions. In a similar reaction, a trimethylsilyl enol ether was treated with Cp_2Zr (from Cp_2ZrCl_2/2 BuLi/THF/–78°C), and subsequent quenching with D_2O led to incorporation of deuterium at the vinyl carbon ($C=C–D$).[329]

OS **VI**, 327, 445; **VII**, 282, 312, 424, 512; **VIII**, 1, 286, 460; **IX**, 573. See also OS **VII**, 66, 266. For the conversion of ketones to vinylic triflates,[330] see OS **VIII**, 97, 126.

12-18 Conversion of Aldehydes to β-Keto Esters or Ketones

Alkoxycarbonylalkylation or **Alkoxycarbonylalkyl-de-hydrogenation**

β-Keto esters have been prepared in moderate to high yields by treatment of aldehydes with diethyl diazoacetate in the presence of a catalytic amount of a Lewis acid, such as $SnCl_2$, BF_3, or $GeCl_2$.[331] The reaction was successful for both aliphatic and aromatic aldehydes, but the former react more rapidly than the latter, and the

---

[324]For the synthesis of enol acetates, see Larock, R.C. *Comprehensive Organic Transformations*, 2nd ed., Wiley-VCH, NY, *1999*, 1484–1485.

[325]See Krapcho, A.P.; Diamanti, J.; Cayen, C.; Bingham, R. *Org. Synth. Coll. Vol. V 1973*, 198.

[326]For example, see Honda, T.; Namiki, H.; Kudoh, M.; Watanabe, N.; Nagase, H.; Mizutani, H. *Tetrahedron Lett.* *2000*, *41*, 5927.

[327]Wentworth, A.D.; Wentworth, Jr., P.; Mansoor, U.F.; Janda, K.D. *Org. Lett.* *2000*, *2*, 477.

[328]House, H.O.; Czuba, L.J.; Gall, M.; Olmstead, H.D. *J. Org. Chem.* *1969*, *34*, 2324.

[329]Ganchegui, B.; Bertus, P.; Szymoniak, J. *Synlett 2001*, 123.

[330]Comins, D.L.; Dehghani, A. *Tetrahedron Lett.* *1992*, *33*, 6299.

[331]Holmquist, C.R.; Roskamp, E.J. *J. Org. Chem.* *1989*, *54*, 3258.

difference is great enough to allow selective reactivity. In a similar process, aldehydes react with certain carbanions stabilized by boron, in the presence of $(F_3CCO)_2O$ or NCS, to give ketones.[332]

Ketones can be prepared from aryl aldehydes (ArCHO) by treatment with a rhodium complex $(Ph_3P)_2Rh(CO)Ar'$, whereby the Ar group is transferred to the aldehyde, producing the ketone, Ar–CO–Ar'.[333] In a rhodium catalyzed reaction, aryl aldehydes (ArCHO) react with Me_3SnAr' to give the diaryl ketone Ar–CO–Ar'.[334]

12-19   Cyanation or Cyano-de-hydrogenation

There are several reactions in which a C–H unit is replaced by C–CN. In virtually all cases, the hydrogen being replaced is on a carbon α to a heteroatom or functional group. There are several examples.

Introduction of a cyano group α to the carbonyl group of a ketone can be accomplished by prior formation of the enolate anion with LDA in THF and addition of this solution to p-TsCN at −78°C.[335] The products are formed in moderate to high yields but the reaction is not applicable to methyl ketones. Treatment of $TMSCH_2N(Me)C=Nt$-Bu with sec-butyllithium and $R_2C=O$, followed by iodomethane and NaOMe leads to the nitrile, $R_2CH–CN$.[336]

Cyanation has been shown to occur α to a nitrogen, specifically in N,N-dimethylaniline derivatives. Treatment with a catalytic amount of $RuCl_3$ in the presence of oxygen and NaCN leads to the corresponding cyanomethylamine.[337]

[332]Pelter, A.; Smith, K.; Elgendy, S.; Rowlands, M. *Tetrahedron Lett.* **1989**, *30*, 5643.

[333]Krug, C.; Hartwig, J. F *J. Am. Chem. Soc.* **2002**, *124*, 1674.

[334]Pucheault, M.; Darses, S.; Genet, J.-P. *J. Am. Chem. Soc.* **2004**, *126*, 15356.

[335]Kahne, D.; Collum, D.B. *Tetrahedron Lett.* **1981**, *22*, 5011.

[336]Santiago, B.; Meyers, A.I. *Tetrahedron Lett.* **1993**, *34*, 5839.

[337]Murahashi, S.-I.; Komiya, N.; Terai, H.; Nakae, T. *J. Am. Chem. Soc.* **2003**, *125*, 15312; North, M. *Angew. Chem. Int. Ed.* **2004**, *43*, 4126.

In a different kind of reaction, nitro compounds are α-cyanated by treatment with $^-$CN and $K_3Fe(CN)_6$.[338] The mechanism probably involves ion radicals. In still another reaction, secondary amines are converted to α-cyanoamines by treatment with phenylseleninic anhydride and NaCN or Me_3SiCN.[339] The compound Me_3SiCN has also been used in a reaction that cyanates benzylic positions.[340]

12-20 Alkylation of Alkanes

Alkylation or Alkyl-de-hydrogenation

$$RH \ + \ R'^+ \longrightarrow R{-}R' \ + \ H^+$$

Alkanes can be alkylated by treatment with solutions of stable carbocations[341] (p. 235), but the availability of such carbocations is limited and mixtures are usually obtained. In a typical experiment, the treatment of propane with isopropyl fluoroantimonate (Me_2C^+ $SbF_6{-}$) gave 26% 2,3-dimethylbutane, 28% 2-methylpentane, 14% 3-methylpentane, and 32% *n*-hexane, as well as some butanes, pentanes (formed by **12-47**), and higher alkanes. Mixtures arise in part because intermolecular hydrogen exchange ($RH + R'^+ \ R^+ + R'H$) is much faster than alkylation, so that alkylation products are also derived from the new alkanes and carbocations formed in the exchange reaction. Furthermore, the carbocations present are subject to rearrangement (Chapter 18), giving rise to new carbocations. Products result from all the hydrocarbons and carbocations present in the system. As expected from their relative stabilities, secondary alkyl cations alkylate alkanes more readily than tertiary alkyl cations (the *tert*-butyl cation does not alkylate methane or ethane). Stable primary alkyl cations are not available, but alkylation has been achieved with complexes formed between CH_3F or C_2H_5F and SbF_5.[342] The mechanism of alkylation can be formulated (similar to that shown in hydrogen exchange with superacids, **12-1**) as

$$R{-}H \ + \ R'^+ \longrightarrow \left[R \overset{\overset{\textstyle H}{\cdot}}{\underset{\underset{\textstyle R'}{\cdot}}{}} \right]^+ \xrightarrow{-H^+} R{-}R'$$

[338]Matacz, Z.; Piotrowska, H.; Urbanski, T. *Pol. J. Chem.* **1979**, *53*, 187; Kornblum, N.; Singh, N.K.; Kelly, W.J. *J. Org. Chem.* **1983**, *48*, 332.

[339]Barton, D.H.R.; Billion, A.; Boivin, J. *Tetrahedron Lett.* **1985**, *26*, 1229.

[340]Lemaire, M.; Doussot, J.; Guy, A. *Chem. Lett.* **1988**, 1581. See also, Hayashi, Y.; Mukaiyama, T. *Chem. Lett.* **1987**, 1811.

[341]Olah, G.A.; Mo, Y.K.; Olah, J.A. *J. Am. Chem. Soc.* **1973**, *95*, 4939. For reviews, see Olah, G.A.; Farooq, O.; Prakash, G.K.S., in Hill, C.L. *Activation and Functionalization of Alkanes*, Wiley, NY, **1989**, pp. 27–78; Ref. 48. For a review of the thermodynamic behavior of alkanes in superacid media, see Fabre, P.; Devynck, J.; Trémillon, B. *Chem. Rev.* **1982**, *82*, 591. See also, Olah, G.A.; Prakash, G.K.S.; Williams, R.E.; Field, L.D.; Wade, K. *Hypercarbon Chemistry*, Wiley, NY, **1987**.

[342]Olah, G.A.; DeMember, J.R.; Shen, J. *J. Am. Chem. Soc.* **1973**, *95*, 4952. See also, Sommer, J.; Muller, M.; Laali, K. *Nouv. J. Chem.* **1982**, *6*, 3.

It is by means of successive reactions of this sort that simple alkanes like methane and ethane give *tert*-butyl cations in superacid solutions (p. 236).[343]

Intramolecular insertion has been reported. The positively charged carbon of the carbocation **40**, generated from the diazonium salt of the triptycene compound **39**, reacted with the CH_3 group in close proximity with it.[344]

12-21   Insertion by Carbenes

CH-Methylene-insertion

$$RH \quad + \quad :CH_2 \quad \longrightarrow \quad RCH_3$$

The highly reactive species methylene ($:CH_2$) inserts into C–H bonds,[345] both aliphatic and aromatic,[346] although with aromatic compounds subsequent ring expansion is also possible (see **15-64**). This is effectively a homologation reaction.[347] The methylene insertion reaction has limited utility because of its nonselectivity (see p. 284). The insertion reaction of carbenes has been used for synthetic purposes.[348] The carbenes can be generated in any of the ways mentioned in Chapter 5 (p. 287). Alkylcarbenes usually rearrange rather than give

---

[343]For example, see Hogeveen, H.; Roobeek, C.F. *Recl. Trav. Chim. Pays-Bas* **1972**, *91*, 137.

[344]Yamamoto, G.; Ō ki, M. *Chem. Lett.* **1987**, 1163.

[345]First reported by Meerwein, H.; Rathjen, H.; Werner, H. *Berchtt.* **1942**, *75*, 1610. For reviews, see Bethell, D., in McManus, S.P. *Organic Reactive Intermediates*, Academic Press, NY, **1973**, pp. 92–101; Kirmse, W. *Carbene Chemistry*, 2nd ed., Academic Press, NY, **1971**, pp. 209–266.

[346]Terao, T.; Shida, S. *Bull. Chem. Soc. Jpn.* **1964**, *37*, 687. See also, Moss, R.A.; Fedé, J.-M.; Yan, S. *J. Am. Chem. Soc.* **2000**, *122*, 9878.

[347]For a discussion of organozinc carbenoid homologation reactions, see Marek, I. *Tetrahedron* **2002**, *58*, 9463.

[348]For some examples of intramolecular carbene insertions used synthetically, see Gilbert, J.C.; Giamalva, D.H.; Weerasooriya, U. *J. Org. Chem.* **1983**, *48*, 5251; Taber, D.F.; Ruckle, Jr., R.E. *J. Am. Chem. Soc.* **1986**, *108*, 7686; Paquette, L.A.; Kobayashi, T.; Gallucci, J.C. *J. Am. Chem. Soc.* **1988**, *110*, 1305; Adams, J.; Poupart, M.; Grenier, L.; Schaller, C.; Ouimet, N.; Frenette, R. *Tetrahedron Lett.* **1989**, *30*, 1749; Doyle, M.P.; Bagheri, V.; Pearson, M.M.; Edwards, J.D. *Tetrahedron Lett.* **1989**, *30*, 7001.

insertion (p. 291), but, when this is impossible, *intramolecular* insertion[349] is found rather than intermolecular.[350] Methylene (:CH$_2$) generated by photolysis of diazomethane (CH$_2$N$_2$) in the liquid phase is indiscriminate (totally nonselective) in its reactivity (p. 288). Methylene (:CH$_2$) generated in other ways and mono-alkyl and dialkyl carbenes are less reactive and insert in the order tertiary > secondary > primary.[351] Carbene insertion with certain allylic systems can proceed with rearrangement of the double bond.[352] Carbenes have been generated in the presence of ultrasound.[353] Halocarbenes (:CCl$_2$, :CBr$_2$, etc.) insert much less readily, although a number of instances have been reported.[354] Insertion into the O–H bond of alcohols, to produce ethers, has been reported using a diazocarbonyl compound and an In(OTf)$_3$ catalyst.[355]

For the similar insertion reaction of nitrenes, see **12-13**.

The metal carbene insertion reaction, in contrast to the methylene insertion reaction, can be highly selective,[356] is very useful in synthesis,[357] and there are numerous examples, usually requiring a catalyst.[358] The catalyst typically convert a diazoalkane or diazocarbonyl compound to the metal carbene *in situ*, allowing the subsequent insertion reaction. Intermolecular reactions are known, including diazoalkane insertion reaction with a dirhodium catalyst.[359] When chiral ligands are present good enantioselectivity is observed in the insertion product.[360] Insertion at an allylic carbon of alkenes has been reported.[361] Insertion into a 2-pyrrolidinone derivative using Me$_3$SiCH$_2$N$_2$ followed by AgCO$_2$Ph with ultrasound gave a

[349]Kirmse, W.; Doering, W. von E. *Tetrahedron* **1960**, *11*, 266; Friedman, L.; Berger, J.G. *J. Am. Chem. Soc.* **1961**, *83*, 492, 500. See Padwa, A.; Krumpe, K.E. *Tetrahedron* **1992**, *48*, 5385.

[350]For a review of the intramolecular insertions of carbenes or carbenoids generated from diazocarbonyl compounds, see Burke, S.D.; Grieco, P.A. *Org. React.* **1979**, *26*, 361.

[351]Doering, W. von E.; Knox, L.H. *J. Am. Chem. Soc.* **1961**, *83*, 1989.

[352]Carter, D.S.; Van Vranken, D.L. *Org. Lett.* **2000**, *2*, 1303; Kirmse, W.; Kapps, M. *Chem. Ber.* **1968**, *101*, 994; Doyle, M.P.; Griffin, J.H.; Chinn, M.S.; van Leusen, D. *J. Org. Chem.* **1984**, *49*, 1917; Doyle, M.P.; McKervey, M.A.; Ye, T. *Modern Catalytic Methods for Organic Synthesis with Diazo Compounds: From Cyclopropanes to Ylides*, Wiley,NY, **1998**; Meyer, O.; Cagle, P.C.; Weickhardt, K.; Vichard, D.; Gladysz, J.A. *Pure Appl. Chem.* **1996**, *68*, 79.

[353]Bertram, A.K.; Liu, M.T.H. *J. Chem. Soc. Chem. Commun.* **1993**, 467.

[354]For example, see Parham, W.E.; Koncos, R. *J. Am. Chem. Soc.* **1961**, *83*, 4034; Fields, E.K. *J. Am. Chem. Soc.* **1962**, *82*, 1744; Anderson, J.C.; Lindsay, D.G.; Reese, C.B. *J. Chem. Soc.* **1964**, 4874; Seyferth, D.; Cheng, Y.M. *J. Am. Chem. Soc.* **1973**, *95*, 6763; *Synthesis* **1974**, 114; Steinbeck, K. *Tetrahedron Lett.* **1978**, 1103; Boev, V.I. *J. Org. Chem. USSR* **1981**, *17*, 1190.

[355]Matusamy, S.; Arulananda, S.; Babu, A.; Gunanathan, C. *Tetrahedron Lett.* **2002** *43*, 3133.

[356]Particularly the C–H insertion reaction, see Sulikowski, G.A.; Cha, K.L.; Sulikowski, M.M. *Tetrahedron Asymmetry*, **1998**, *9*, 3145; Taber, D.F.; Meagley, R.P. *Tetrahedron Lett.* **1994**, *35*, 7909.

[357]Ye, T.; McKervey, M.A. *Chem. Rev.* **1994**, *94*, 1091.

[358]Doyle, M.P. *Pure Appl. Chem.* **1998**, *70*, 1123. See Taber, D.F.; Malcolm, S.C. *J. Org. Chem.* **1998**, *63*, 3717 for a discussion of transition state geometry in rhodium mediated C—H insertion.

[359]Davies, H.M.; Hansen, T.; Churchill, M.R. *J. Am. Chem. Soc.* **2000**, *122*, 3063; Davies, H.M.L.; Jin, Q.; Ren, P.; Kovalensky, A.Yu. *J. Org. Chem.* **2002**, *67*, 4165; Davies, H.M.L.; Beckwith, R.E.J.; Antoulinakis, E.G.; Jin, Q. *J. Org. Chem.* **2003**, *68*, 6126; Davies, H.M.L.; Jin, Q. *Org. Lett.* **2004**, *6*, 1769. For a review, see Davies, H.M.L.; Loe, Ø. *Synthesis* **2004**, 2595.

[360]For a review, see Davies, H.M.L.; Beckwith, R.E.J. *Chem. Rev.* **2003**, *103*, 2861.

[361]Davies, H.M.L.; Ren, P.; Jin, Q. *Org. Lett.* **2001**, *3*, 3587.

2-piperidone derivative.[362] The copper-catalyzed insertion of a diazo ester into an oxetane gives the ring-expanded tetrahydrofuran derivative.[363] Dirhodium catalyzed insertion into H–C^{sp2} bonds is also known,[364] and also H–C^{sp} bonds.[365] Insertion of diazoalkane and diazocarbonyl compounds can be catalyzed by copper compounds[366] and silver compounds[367] as well. Intramolecular insertion reactions are well known, and tolerate a variety of functional groups.[368] Intramolecular insertion at the α-carbon of a ketone by a diazoketone, using TiCl$_4$, gives a bicyclic 1,3-diketone.[369] A typical example is the insertion of the diazocarbonyl unit into the C–H bond to give the lactam.[370] Similar insertion at the α-carbon of an ether leads to cyclic ethers, with high enantioselectivity when a chiral ligand is used with a rhodium catalyst.[371] Similar insertion at the α-carbon of silyl ethers has been reported.[372] Aryl ketenes react with Me$_3$SiCHN$_2$ and then silica to give 2-indanone derivatives.[373]

The mechanism[374] of the insertion reaction is not known with certainty, but there seem to be at least two possible pathways.

[362]Coutts, I.G.C.; Saint, R.E.; Saint, S.L.; Chambers-Asman, D.M. *Synthesis* **2001**, 247.

[363]Lo, M.M.-C.; Fu, G.C. *Tetrahedron* **2001**, *57*, 2621.

[364]Gibe, R.; Kerr, M.A. *J. Org. Chem.* **2002**, *67*, 6247.

[365]Arduengo III, A.J.; Calabrese, J.C.; Davidson, F.; Dias, H.V.R.; Goerlich, J.R.; Krafczyk, R.; Marshall, W.J.; Tamm, M.; Schmutzler, R. *Helv. Chim. Acta.* **1999**, *82*, 2348.

[366]See Caballero, A.; Díaz-Requejo, M.M.; Belderraín, T.R.; Nicasio, M.C.; Trofimenko, S.; Pérez, P. J. *J. Am. Chem. Soc.* **2003**, *125*, 1446.

[367]Dias, H.V.R.; Browning, R.G.; Polach, S.A.; Diyabalanage, H.V.K.; Lovely, C.J. *J. Am. Chem. Soc.* **2003**, *125*, 9270.

[368]For examples, see Marmsäter, F.P.; Murphy, G.K.; West, F.G. *J. Am. Chem. Soc.* **2003**, *125*, 14724; Müller, P.; Polleux, P. *Helv. Chim. Acta* **1994**, *77*, 645; Doyle, M.P.; Kalinin, A.V. *Synlett*, **1995**, 1075; Watanabe, N.; Ohtake, Y.; Hashimoto, S.; Shiro, M.; Ikegami, S. *Tetrahedron Lett.* **1995**, *36*, 1491; Maruoka, K.; Concepcion, A.B.; Yamamoto, H. *J. Org. Chem.* **1994**, *59*, 4725; Spero, D.M.; Adams, J. *Tetrahedron Lett.* **1992**, *33*, 1143.

[369]Muthusamy, S.; Babu, S.A.; Gunanathan, C. *Synth. Commun.* **2001**, *31*, 1205.

[370]Doyle, M.P.; Protopopova, M.N.; Winchester, W.R.; Daniel, K.L. *Tetrahedron Lett.* **1992**, *33*, 7819. See also, Wang, J.; Hou, Y.; Wu, P. *J. Chem. Soc., Perkin Trans. 1* **1999**, 2277; Clark, J.S.; Hodgson, P.B.; Goldsmith, M.D.; Street, L.J. *J. Chem. Soc., Perkin Trans. 1* **2001**, 3312. For a related reaction, see Yang, H.; Jurkauskas, V.; Mackintosh, N.; Mogren, T.; Stephenson, C.R.J.; Foster, K.; Brown, W.; Roberts, E. *Can. J. Chem.* **2000**, *78*, 800.

[371]Davies, H.M.L.; Grazini, M.V.A.; Aouad, E. *Org. Lett.* **2001**, *3*, 1475.

[372]Yoon, C.H.; Zaworotko, M.J.; Moulton, B.; Jung, K.W. *Org. Lett.* **2001**, *3*, 3539.

[373]Dalton, A.M.; Zhang, Y.; Davie, C.P.; Danheiser, R.L. *Org. Lett.* **2002**, *4*, 2465.

[374]For a discussion, see Bethell, D. *Adv. Phys. Org. Chem.* **1969**, *7*, 153, pp. 190–194.

1. A simple one-step process involving a three-center cyclic transition state:

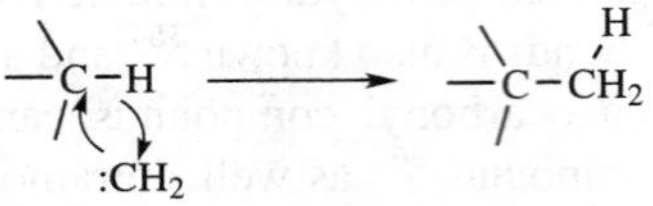

The most convincing evidence for this mechanism is that in the reaction between isobutene-1-^{14}C and carbene the product 2-methyl-1-butene was labeled only in the 1 position.[375] This rules out a free radical or a carbocation or carbanion intermediate. If **41** (or a corresponding ion) were an intermediate, resonance would ensure that some carbene attacked at the 1 position:

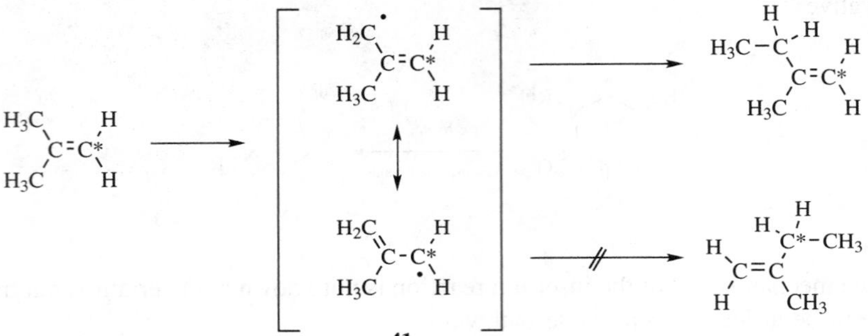

Other evidence is that retention of configuration, which is predicted by this mechanism, has been found in a number of instances.[376] An ylid intermediate was trapped in the reaction of :CH$_2$ with allyl alcohol.[377]

2. A free-radical process in which the carbene directly abstracts a hydrogen from the substrate to generate a pair of free radicals:

$$RH \ + \ CH_2 \longrightarrow R\bullet \ + \ \bullet CH_3$$
$$R\bullet \ + \ \bullet CH_3 \longrightarrow RCH_3$$

One fact supporting this mechanism is that among the products obtained (beside butane and isobutane) on treatment of propane with CH$_2$ (generated by photolysis of diazomethane and ketene) were propene and ethane,[378] which could arise, respectively, by

$$2 \ CH_3CH_2CH_2\bullet \longrightarrow CH_3CH=CH_2 \ + \ CH_3CH_2CH_3 \ \text{(disproportionation)}$$

[375]Doering, W. von E.; Prinzbach, H. *Tetrahedron* **1959**, *6*, 24.
[376]See, for example, Kirmse, W.; Buschhoff, M. *Chem. Ber.* **1969**, *102*, 1098; Seyferth, D.; Cheng, Y.M. *J. Am. Chem. Soc.* **1971**, *93*, 4072.
[377]Sobery, W.; DeLucca, J.P. *Tetrahedron Lett.* **1995**, *36*, 3315.
[378]Frey, H.M. *Proc. Chem. Soc.* **1959**, 318.

and

$$CH_3CH_2CH_3 + :CH_2 \longrightarrow CH_3CH_2CH_2\bullet + \bullet CH_3$$
$$2 \bullet CH_3 \longrightarrow CH_3CH_3$$

That this mechanism can take place under suitable conditions has been demonstrated by isotopic labeling[379] and by other means.[380] However, the formation of disproportionation and dimerization products does not always mean that the free-radical abstraction process takes place. In some cases, these products arise in a different manner.[381] We have seen that the product of the reaction between a carbene and a molecule may have excess energy (p. 288). Therefore it is possible for the substrate and the carbene to react by mechanism 1 (the direct-insertion process) and for the excess energy to cause the compound thus formed to cleave to free radicals. When this pathway is in operation, the free radicals are formed *after* the actual insertion reaction.

The mechanism of cyclopropylcarbene reactions has also been discussed.[382]

It has been suggested[383] that singlet carbenes insert by the one-step direct-insertion process and triplets (which, being free radicals, are more likely to abstract hydrogen) by the free-radical process. In support of this suggestion is that CIDNP signals[384] (p. 269) were observed in the ethylbenzene produced from toluene and triplet CH_2, but not from the same reaction with singlet CH_2.[385] Carbenoids (e.g., compounds of the form R_2CMCl, see **12-39**) can insert into a C–H bond by a different mechanism, similar to pathway 2, but involving abstraction of a hydride ion rather than a hydrogen atom.[386]

An interesting insertion reaction involves $EtZnCH_2I$ and β-keto carbonyl compounds. The reaction of this reagent with *N,N*-dibutyl-3-oxobutanamide, for example, gives the methylene insertion product *N,N*-dibutyl 4-oxopentanamide.[387]

The reaction in which aldehydes are converted to methyl ketones, $RCHO + CH_2N_2 \rightarrow RCOCH_3$, while apparently similar, does not involve a free carbene intermediate. It is considered in Chapter 18 (**18-9**).

OS **VII**, 200.

[379]Halberstadt, M.L.; McNesby, J.R. *J. Chem. Phys.* **1966**, *45*, 1666; McNesby, J.R.; Kelly, R.V. *Int. J. Chem. Kinet.*, **1971**, *3*, 293.

[380]Ring, D.F.; Rabinovitch, B.S. *J. Am. Chem. Soc.* **1966**, *88*, 4285; *Can J. Chem.* **1968**, *46*, 2435.

[381]Bell, J.A. *Prog. Phys. Org. Chem.* **1964**, *2*, 1, pp. 30–43.

[382]Cummins, J.M.; Porter, T.A.; Jones Jr., M. *J. Am. Chem. Soc.* **1998**, *120*, 6473.

[383]Richardson, D.B.; Simmons, M.C.; Dvoretzky, I. *J. Am. Chem. Soc.* **1961**, *83*, 1934.

[384]For a review of the use of CIDNP to study carbene mechanisms, see Roth, H.D. *Acc. Chem. Res.* **1977**, *10*, 85.

[385]Roth, H.D. *J. Am. Chem. Soc.* **1972**, *94*, 1761. See also Closs, G.L.; Closs, L.E. *J. Am. Chem. Soc.* **1969**, *91*, 4549; Bethell, D.; McDonald, K. *J. Chem. Soc. Perkin Trans. 2* **1977**, 671.

[386]See Oku, A.; Yamaura, Y.; Harada, T. *J. Org. Chem.* **1986**, *51*, 3730; Ritter, R.H.; Cohen, T. *J. Am. Chem. Soc.* **1986**, *108*, 3718.

[387]Hilgenkamp, R.; Zercher, C.K. *Tetrahedron* **2001**, *57*, 8793.

F. Metal Electrophiles

12-22 Metalation With Organometallic Compounds

Metalation or Metalo-de-hydrogenation

$$RH + R'M \longrightarrow RM + R'M$$

Many organic compounds can be metalated by treatment with an organometallic compound.[388] Since the reaction involves a proton transfer, the equilibrium lies on the side of the weaker acid.[389] For example, fluorene reacts with butyllithium to give butane and 9-fluoryllithium. Since aromatic hydrocarbons are usually stronger acids than aliphatic ones, R is most often aryl. The most common reagent is butyl-lithium.[390] Normally, only active aromatic rings react with butyllithium. Benzene itself reacts very slowly and in low yield, although benzene can be metalated by butyllithium either in the presence of t-BuOK[391] or by n-butyllithium that is coordinated with various diamines.[392] Metalation of aliphatic RH is most successful when the carbanions are stabilized by resonance (allylic, benzylic, propargylic,[393] etc.) or when the negative charge is at an sp carbon (at triple bonds). Very good reagents for allylic metalation are trimethylsilylmethyl potassium Me_3SiCH_2K[394] and a combination of an organolithium compound with a bulky alkoxide (LICKOR superbase).[395] The former is also useful for benzylic positions. A combination of BuLi, t-BuOK, and tetramethylethylenediamine has been used to convert ethylene to vinylpotassium.[396] In certain cases, *gem*-dialkali metal or 1,1,1-trialkali metal compounds can be prepared.[397] Examples are the conversion of phenylacetonitrile

[388]For reviews, see Wardell, J.L., in Zuckerman,J.J. *Inorganic Reactions and Methods*, Vol. 11, VCH, NY, *1988*, pp. 44–107; Wardell, J.L., in Hartley, F.R.; Patai, S. *The Chemistry of the Metal-Carbon Bond*, Vol. 4, Wiley, NY, pp. 1–157, 27–71; Narasimhan, M.S.; Mali, R.S. *Synthesis 1983*, 957; Biellmann, J.F.; Ducep, J. *Org. React. 1982*, 27, 1; Gschwend, H.W.; Rodriguez, H.R. *Org. React. 1979*, 26, 1; Mallan, J.M.; Bebb, R.L. *Chem. Rev. 1969*, 69, 693.

[389]See Saá, J.M.; Martorell, G.; Frontera, A. *J. Org. Chem. 1996*, 61, 5194 for a discussion of the mechanism of lithiation of aromatic species.

[390]For a review, see Durst, T., in Buncel, E.; Durst, T. *Comprehensive Carbanion Chemistry*, Vol. 5, pt. B, Elsevier, NY, *1984*, pp. 239–291, 265–279. For an article on the safe handling of RLi compounds, see Anderson, R. *Chem. Ind. (London) 1984*, 205.

[391]Schlosser, M. *J. Organomet. Chem. 1967*, 8, 9. See also, Schlosser, M.; Katsoulos, G.; Takagishi, S. *Synlett, 1990*, 747.

[392]Eberhardt, G.G.; Butte, W.A. *J. Org. Chem. 1964*, 29, 2928; Langer, Jr., A.W. *Trans. N.Y. Acad. Sci. 1965*, 27, 741; Eastham, J.F.; Screttas, C.G. *J. Am. Chem. Soc. 1965*, 87, 3276; Rausch, M.D.; Ciappenelli, D.J. *J. Organomet. Chem. 1967*, 10, 127.

[393]For a review of directive effects in allylic and benzylic metallation, see Klein, J. *Tetrahedron 1983*, 39, 2733. For a review of propargylic metallation, see Klein, J., in Patai, S. *The Chemistry of the Carbon-Carbon Triple Bond*, pt. 1, Wiley, NY, *1978*, pp. 343–379.

[394]Hartmann, J.; Schlosser, M. *Helv. Chim. Acta 1976*, 59, 453.

[395]Schlosser, M. *Pure Appl. Chem. 1988*, 60, 1627. For sodium analogs, see Schlosser, M.; Hartmann, J.; Stähle, M.; Kramǎr, J.; Walde, A.; Mordini, A. *Chimia, 1986*, 40, 306.

[396]Brandsma, L.; Verkruijsse, H.D.; Schade, C.; Schleyer, P.v.R. *J. Chem. Soc. Chem. Commun. 1986*, 260.

[397]For a review of di- and polylithium compounds, see Maercker, A.; Theis, M. *Top. Curr. Chem. 1987*, 138, 1.

to 1,1-dilithiophenylacetonitrile (PhCLi$_2$CN)[398] and propyne to tetralithiopropyne (Li$_3$CC≡CLi)[399] in each case by treatment with excess butyllithium. The reaction can be used to determine relative acidities of very weak acids by allowing two R–H compounds to compete for the same R'M and to determine which proton in a molecule is the most acidic.[400]

In general, the reaction can be performed only with organometallics of active metals such as lithium, sodium, and potassium, but Grignard reagents abstract protons from a sufficiently acidic C–H bond, as in R–C≡C–H → R–C≡C–MgX. This is the best method for the preparation of alkynyl Grignard reagents.[401]

When a heteroatom, such as N, O, S,[402] or a halogen,[403] is present in a molecule containing an aromatic ring or a double bond, lithiation is usually quite regioselective.[404] The lithium usually bonds with the sp^2 carbon closest to the heteroatom, probably because the attacking species coordinates with the heteroatom.[405] Such reactions with compounds such as anisole are often called directed metalations.[406] In the case of aromatic rings this means attack at the ortho position,[407] but this is considered in **13-17**.

Ref. 408

[398]Kaiser, E.M.; Solter, L.E.; Schwartz, R.A.; Beard, R.D.; Hauser, C.R. *J. Am. Chem. Soc.* **1971**, *93*, 4237. See also, Kowalski, C.J.; O'Dowd, M.L.; Burke, M.C.; Fields, K.W. *J. Am. Chem. Soc.* **1980**, *102*, 5411.

[399]Priester, W.; West, R. *J. Am. Chem. Soc.* **1976**, *98*, 8421, 8426, and references cited therein.

[400]For examples, see Broaddus, C.D.; Logan, T.J.; Flautt, T.J. *J. Org. Chem.* **1963**, *28*, 1174; Finnegan, R.A.; McNees, R.S. *J. Org. Chem.* **1964**, *29*, 3234; Shirley, D.A.; Hendrix, J.P. *J. Organomet. Chem.* **1968**, *11*, 217.

[401]For a review of the synthetic applications of metallation by Grignard reagents at positions other than at triple bonds, see Blagoev, B.; Ivanov, D. *Synthesis* **1970**, 615.

[402]For example, see Figuly, G.D.; Loop, C.K.; Martin, J.C. *J. Am. Chem. Soc.* **1989**, *111*, 654; Block, E.; Eswarakrishnan, V.; Gernon, M.; Ofori-Okai, G.; Saha, C.; Tang, K.; Zubieta, J. *J. Am. Chem. Soc.* **1989**, *111*, 658; Smith, K.; Lindsay, C.M.; Pritchard, G.J. *J. Am. Chem. Soc.* **1989**, *111*, 665.

[403]Fluorine is an especially powerful ortho director in lithiation of aromatic systems: Gilday, J.P.; Negri, J.T.; Widdowson, D.A. *Tetrahedron* **1989**, *45*, 4605.

[404]For a review of regioselective lithiation of heterocycles, see Katritzky, A.R.; Lam, J.N.; Sengupta, S. *Prog. Heterocycl. Chem.* **1989**, *1*, 1.

[405]For many examples with references, see Ref. 388; Beak, P.; Meyers, A.I. *Acc. Chem. Res.* **1986**, *19*, 356; Beak, P.; Snieckus, V. *Acc. Chem. Res.* **1982**, *15*, 306; Snieckus, V. *Bull. Soc. Chim. Fr.* **1988**, 67; Narasimhan, N.S.; Mali, R.S. *Top. Curr. Chem.* **1987**, *138*, 63; Reuman, M.; Meyers, A.I. *Tetrahedron* **1985**, *41*, 837; and the papers in *Tetrahedron* **1983**, *39*, 1955.

[406]Slocum, D.W.; Moon, R.; Thompson, J.; Coffey, D.S.; Li, J.D.; Slocum, M.G.; Siegel, A.; Gayton-Garcia, R. *Tetrahedron Lett.* **1994**, *35*, 385; Slocum, D.W.; Coffey, D.S.; Siegel, A.; Grimes, P. *Tetrahedron Lett.* **1994**, *35*, 389.

[407]For reviews of ortho metallation, see Snieckus, V. *Chem. Rev.* **1990**, *90*, 879; *Pure Appl. Chem.* **1990**, *62*, 2047. For a discussion of the mechanism, see Bauer, W.; Schleyer, P. v.R. *J. Am. Chem. Soc.* **1989**, *111*, 7191.

[408]Baldwin, J.E.; Höfle, G.A.; Lever, Jr., O.W. *J. Am. Chem. Soc.* **1974**, *96*, 7125.

In the case of γ,δ-unsaturated disubstituted amides (**42**),the lithium does not go to the closest position, but in this case too the regiochemistry is controlled

42

by coordination to the oxygen.[409]

The mechanism involves an attack by $R'-$ (or a polar R') on the *hydrogen*[410] (an acid–base reaction) Evidence is that resonance effects of substituents in R seem to make little difference. When R is aryl, OMe and CF_3 *both* direct ortho, while isopropyl directs meta and para (mostly meta).[411] These results are exactly what would be expected from pure field effects, with no contribution from resonance effects, which implies that attack occurs at the hydrogen and not at R. Other evidence for the involvement of H in the rate-determining step is that there are large isotope effects.[412] The nature of R' also has an effect on the rate. In the reaction between triphenylmethane and $R'Li$, the rate decreased in the order $R' = $ allyl $>$ Bu $>$ Ph $>$ vinyl $>$ Me, although this order changed with changing concentration of the $R'Li$, because of varying degrees of aggregation of the $R'Li$.

With respect to the reagent, this reaction is a special case of **12-24**.

A closely related reaction is formation of nitrogen ylids[414] from quaternary ammonium salts (see **17-8**):

Phosphonium salts undergo a similar reaction (see **16-44**).

OS **II**, 198; **III**, 413, 757; **IV**, 792; **V**, 751; **VI**, 436, 478, 737, 979; **VII**, 172, 334, 456, 524; **VIII**, 19, 391, 396, 606.

[409]Beak, P.; Hunter, J.E.; Jun, Y.M.; Wallin, A.P. *J. Am. Chem. Soc.* **1987**, *109*, 5403. See also, Stork, G.; Polt, R.L.; Li, Y.; Houk, K.N. *J. Am. Chem. Soc.* **1988**, *110*, 8360; Barluenga, J.; Foubelo, F.; Fañanas, F.J.; Yus, M. *J. Chem. Res. (S)* **1989**, 200.

[410]Benkeser, R.A.; Trevillyan, E.A.; Hooz, J. *J. Am. Chem. Soc.* **1962**, *84*, 4971.

[411]Bryce-Smith, D. *J. Chem. Soc.* **1963**, 5983; Benkeser, R.A.; Hooz, J.; Liston, T.V.; Trevillyan, E.A. *J. Am. Chem. Soc.* **1963**, *85*, 3984.

[412]Bryce-Smith, D.; Gold, V.; Satchell, D.P.N. *J. Chem. Soc.* **1954**, 2743; Pocker, Y.; Exner, J.H. *J. Am. Chem. Soc.* **1968**, *90*, 6764.

West, P.; Waack, R.; Purmort, J.I. *J. Am. Chem. Soc.* **1970**, *92*, 840.

[414]Zugravescu, I.; Petrovanu, M. *Nitrogen-Ylid Chemistry*, McGraw-Hill, NY, **1976**, pp 251–283; Kröhnke, F. *Berchtt* **1935**, *68*, 1177; Wittig, G.; Wetterling, M. *Ann.* **1947**, *557*, 193; Wittig, G.; Rieber, M. *Ann.* **1949**, *562*, 177; Wittig, G.; Polster, R. *Ann.* **1956**, *599*, 1.

12-23   Metalation With Metals and Strong Bases

Metalation or **Metalo-de-hydrogenation**

$$2 \ RH \quad + \quad M \quad \longrightarrow \quad 2 \ RM \quad + \quad H_2$$

Organic compounds can be metalated at suitably acidic positions by active metals and by strong bases.[415] The reaction has been used to study the acidities of very weak acids (see p. 250). The conversion of terminal alkynes to acetylid ions is one important application.[416] Synthetically, an important use of the method is to convert aldehydes and ketones,[417] carboxylic esters, and similar compounds to their enolate forms,[418] for example,

for use in nucleophilic substitutions (**10-67**, **10-68**, and **13-14**) and in additions to multiple bonds (**15-24** and **16-53**). It has been shown that lithiation with lithium amides can also be regioselective (see **12-22**).[419] Lithium enolates exist as aggregates in solution.[420] For very weak acids, the most common reagents for synthetic purposes are lithium amides, especially LDA, which has the structure $(iPr)_2NLi$.[421] The mechanism for this deprotonation reaction has been studied,[422] as has the rate of deprotonation.[423]

OS **I**, 70, 161, 490; **IV**, 473; **VI**, 468, 542, 611, 683, 709; **VII**, 229, 339. Conversions of ketones or esters to enolates are not listed.

---

[415]For a review, see Durst, T., in Buncel, E.; Durst, T. *Comprehensive Carbanion Chemistry*, Vol. 5, pt. B, Elsevier, NY, *1984*, pp. 239–291. For reviews with respect to lithium, see Wardell, J.L. Ref. 388; Wakefield, B.J. *Organolithium Methods*, Academic Press, NY, *1988*, pp. 32–44.

[416]For a review, see Ziegenbein, W., in Viehe, H.G. *Acetylenes*, Marcel Dekker, NY, *1969*, pp. 170–185. For an improved method, see Fisch, A.; Coisne, J.M.; Figeys, H.P. *Synthesis 1982*, 211.

[417]Hegarty, A.F.; Dowling, J.P.; Eustace, S.J.; McGarraghy, M. *J. Am. Chem. Soc. 1998, 120*, 2290.

[418]For a review, see Caine, D. in Augustine, R.L. *Carbon–Carbon Bond Formation*, Vol. 1, Marcel Dekker, NY,*1979*, pp. 95–145, 284–291.

[419]For example, see Comins, D.L.; Killpack, M.O. *J. Org. Chem. 1987, 52*, 104. See Xie, L.; Isenberger, K.M.; Held, G.; Dahl, M. *J. Org. Chem. 1997, 62*, 7516 for steric versus electronic effects in kinetic enolate formation.

[420]Abu-Hasanayn, F.; Stratakis, M.; Streitwieser, A. *J. Org. Chem. 1995, 60*, 4688; Jackman, L.M.; Szeverenyi, N.M. *J. Am. Chem. Soc. 1977, 99*, 4954; Jackman, L.M.; Lange, B.C. *J. Am. Chem. Soc. 1981, 103*, 4494; House, H.O.; Gall, M.; Olmstead, H.D. *J. Org. Chem. 1971, 36*, 2361; Zook, H.D.; Kelly, W.L.; Posey, I.Y. *J. Org. Chem. 1968, 33*, 3477; Stork, G.; Hudrlik, P.F. *J. Am. Chem. Soc. 1968, 90*, 4464.

[421]The alkali metal hydrides, LiH, NaH, and KH, when prepared in a special way, are very rapid metallation agents: Klusener, P.A.A.; Brandsma, L.; Verkruijsse, H.D.; Schleyer, P.v.R.; Friedl, T.; Pi, R. *Angew. Chem. Int. Ed. 1986, 25*, 465.

[422]Romesberg, F.E.; Collum, D.B. *J. Am. Chem. Soc. 1995, 117*, 2166; Sun, X.; Kenkre, S.L.; Remenar, J.F.; Gilchrist, J.H. *J. Am. Chem. Soc. 1997, 119*, 4765.

[423]Majewski, M.; Nowak, P. *Tetrahedron Lett. 1998, 39*, 1661.

METALS AS LEAVING GROUPS

A. Hydrogen as the Electrophile

12-24   Replacement of Metals by Hydrogen

Hydro-de-metallation or **Demetallation**

$$RM + HA \longrightarrow RH + MA$$

Organometallic compounds, including enolate anions, react with acids in reactions in which the metal is replaced by hydrogen.[424] The R group may be aryl (see **11-41**). The reaction is often used to introduce deuterium or tritium into susceptible positions. For Grignard reagents, water is usually a strong enough acid, but stronger acids are also used. An important method for the reduction of alkyl halides consists of the process RX → RMgX → RH.

Other organometallic compounds that are hydrolyzed by water are those of sodium, potassium, lithium, zinc, and so on, the ones high in the electromotive series. Enantioselective protonation of lithium enolates[425] and cyclopropyllithium compounds[426] have been reported. When the metal is less active, stronger acids are required. For example, R_2Zn compounds react explosively with water, R_2Cd slowly, and R_2Hg not at all, although the latter can be cleaved with concentrated HCl. However, this general statement has many exceptions, some hard to explain. For example, BR_3 compounds are completely inert to water, and GaR_3 at room temperature cleave just one R group, but AlR_3 react violently with water. However, BR_3 can be converted to RH with carboxylic acids.[427] For less active metals it is often possible to cleave just one R group from a multivalent metal. For example,

$$R_2Hg + HCl \longrightarrow RH + RHgCl$$

Organometallic compounds of less active metals and metalloids (e.g., silicon,[428] antimony, and bismuth, are quite inert to water. Organomercury compounds (RHgX or R_2Hg) can be reduced to RH by H_2, $NaBH_4$, or other reducing agents.[429] The reduction with $NaBH_4$ takes place by a free-radical mechanism.[430] Alkyl–Si

---

[424]For reviews, see Abraham, M.H.; Grellier, P.L., in Hartley, FR.; Patai, S. *The Chemistry of the Metal–Carbon Bond*, Vol. 2, Wiley, NY, pp. 25–149, 105–136; Abraham, M.H. *Comprehensive Chemical Kinetics*, Bamford, C.H.; Tipper, C.F.H., eds., Vol. 12, Elsevier, NY, *1973*, pp. 107–134; Jensen, F.R.; Rickborn, B. *Electrophilic Substitution of Organomercurials*, McGaw-Hill, NY, *1968*, pp. 45–74; Schlosser, M. *Angew. Chem. Int. Ed. 1964*, *3*, 287, 362; *Newer Methods Prep. Org. Chem. 1968*, *5*, 238.

[425]Fehr, C. *Angew. Chem. Int. Ed. 1996*, *35*, 2567.

[426]Walborsky, H.M.; Ollman, J.; Hamdouchi, C.; Topolski, M. *Tetrahedron Lett. 1992*, *33*, 761.

[427]Brown, H.C.; Murray, K.J. *Tetrahedron 1986*, *42*, 5497; Pelter, A.; Smith, K.; Brown, H.C. *Borane Reagents*, Academic Press, NY, *1988*, pp. 242–244.

[428]For a review of hydro-de-silylation of allylic and vinylic silanes, see Fleming, I.; Dunoguès, J.; Smithers, R. *Org. React. 1989*, *37*, 57, see pp. 89–97, 194–243. Also see, **10-12**

[429]For a review, see Makarova, L.G. *Organomet. React. 1970*, *1*, 119, see pp. 251–270, 275–300.

[430]For a review of this and other free-radical reactions of organomercury compounds, see Barluenga, J.; Yus, M. *Chem. Rev. 1988*, *88*, 487.

bonds can be cleaved by H_2SO_4, for example, $HOOCCH_2CH_2SiMe_3 \rightarrow 2\ CH_4 +$ $(HOOCCH_2CH_2SiMe_2)_2O$.[431]

When the hydrogen of the HA is attached to carbon, this reaction is the same as **12-22**.

We do not list the many hydrolyses of sodium or potassium enolates, and so on found in *Organic Syntheses*. The hydrolysis of a Grignard reagent to give an alkane is found at OS **II**, 478; the reduction of a vinylic tin compound at OS **VIII**, 381; and the reduction of an alkynylsilane at OS **VIII**, 281.

B. Oxygen Electrophiles

12-25   The Reaction between Organometallic Reagents and Oxygen[432]

Hydroperoxy-de-metalation; Hydroxy-de-metalation

Oxygen reacts with Grignard reagents to give either hydroperoxides[433] or alcohols. The reaction can be used to convert alkyl halides to alcohols without side reactions. With aryl Grignard reagents yields are lower and only phenols are obtained, not hydroperoxides. Because of this reaction, oxygen should be excluded when Grignard reagents are prepared and used in various reactions.

Most other organometallic compounds also react with oxygen. Trialkylboranes and alkyldichloroboranes $RBCl_2$ can be conveniently converted to hydroperoxides by treatment with oxygen followed by hydrolysis.[434] Dilithiated carboxylic acids (see **10-70**) react with oxygen to give (after hydrolysis) α-hydroxy carboxylic acids.[435] There is evidence that the reaction between Grignard reagents and oxygen involves a free-radical mechanism.[436]

The 1,1-dimetallic compounds $R_2C(SnMe_3)ZnBr$ were oxidized by dry air at -10 to $0°C$ in the presence of Me_3SiCl to give aldehydes or ketones $R_2C{=}O$.[437]

OS **V**, 918. See also, OS **VIII**, 315.

---

[431]Sommer, L.H.; Marans, N.S.; Goldberg, G.M.; Rockett, J.; Pioch, R.P. *J. Am. Chem. Soc.* *1951*, *73*, 882. See also, Abraham, M.H.; Grellier, P.L., in Hartley, F.R.; Patai, S. *The Chemistry of the Metal–Carbon Bond*, Vol. 2, Wiley, NY, p. 117.

[432]For a monograph, see Brilkina, T.G.; Shushunov, V.A. *Reactions of Organometallic Compounds with Oxygen and Peroxides*, CRC Press, Boca Raton, FL, *1969*. For a review, see Wardell, J.L.; Paterson, E.S., in Hartley, F.R.; Patai, S. *The Chemistry of the Metal–Carbon Bond*, Vol. 2, Wiley, NY, *1985*, pp. 219–338, see pp. 311–316.

[433]For the preparation of propargyl hydroperoxides, see Harada, T.; Kutsuwa, E. *J. Org. Chem.* *2003*, *68*, 6716.

[434]Brown, H.C.; Midland, M.M. *Tetrahedron* *1987*, *43*, 4059.

[435]Moersch, G.W.; Zwiesler, M.L. *Synthesis* *1971*, 647; Adam, W.; Cueto, O. *J. Org. Chem.* *1977*, *42*, 38.

[436]Davies, A.G.; Roberts, B.P. *J. Chem. Soc. B*, *1969*, 317; Walling, C.; Cioffari, A. *J. Am. Chem. Soc.* *1970*, *92*, 6609; Garst, J.F.; Smith, C.D.; Farrar, A.C. *J. Am. Chem. Soc.* *1972*, *94*, 7707. For a review, see Davies, A.G. *J. Organomet. Chem.* *1980*, *200*, 87.

[437]Knochel, P.; Xiao, C.; Yeh, M.C.P. *Tetrahedron Lett.* *1988*, *29*, 6697.

12-26 Reaction between Organometallic Reagents and Peroxides

tert-But**oxy-de-metalation**

$$R\text{-}MgX \ + \ \underset{t\text{-}Bu}{}O\text{-}O\overset{O}{\underset{\|}{C}}R' \ \longrightarrow \ R\text{-}O\text{-}t\text{-}Bu \ + \ R'\overset{O}{\underset{\|}{C}}OMgX$$

A convenient method of preparation of *tert*-butyl ethers consists of treating Grignard reagents with *tert*-butyl acyl peroxides.[438] Both alkyl and aryl Grignard reagents can be used. The application of this reaction to Grignard reagents prepared from cyclopropyl halides permits cyclopropyl halides to be converted to *tert*-butyl ethers of cyclopropanols,[439] which can then be easily hydrolyzed to the cyclopropanols. The direct conversion of cyclopropyl halides to cyclopropanols by **10-1** is not generally feasible, because cyclopropyl halides do not generally undergo nucleophilic substitutions without ring opening.

Vinylic lithium reagents (**43**) react with silyl peroxides to give high yields of silyl enol ethers with retention of configuration.[440] Since the preparation of **43** from vinylic halides

$$\underset{R^1}{\overset{R^2}{}}C=C\underset{Li}{\overset{R^3}{}} \ + \ Me_3Si\text{-}O\text{-}O\text{-}SiMe_3 \ \longrightarrow \ \underset{R^1}{\overset{R^2}{}}C=C\underset{OSiMe_3}{\overset{R^3}{}}$$

43

(**12-39**) also proceeds with retention, the overall procedure is a method for the stereospecific conversion of a vinylic halide to a silyl enol ether. In a related reaction, alkynyl esters can be prepared from lithium acetylides and phenyliodine(III) dicarboxylates.[441]

$$R\text{-}C\equiv C\text{-}Li \ + \ Ph\text{-}I\underset{O_2CR'}{\overset{O_2CR'}{}} \ \longrightarrow \ R\text{-}C\equiv C\text{-}O\overset{O}{\underset{\|}{C}}R'$$

OS **V**, 642, 924.

12-27 Oxidation of Trialkylboranes to Borates

$$R_3B \ \xrightarrow[NaOH]{H_2O_2} \ (RO)_3B \ \longrightarrow \ 3\ ROH \ + \ B(OH)_3$$

[438]Lawesson, S.; Frisell, C.; Denney, D.B.; Denney, D.Z. *Tetrahedron* **1963**, *19*, 1229. For a monograph on the reactions of organometallic compounds with peroxides, see Brilkina, T.G.; Shushunov, V.A. *Reactions of Organometallic Compounds with Oxygen and Peroxides*, CRC Press, Boca Raton, FL, *1969*. For a review, see Razuvaev, G.A.; Shushunov, V.A.; Dodonov, V.A.; Brilkina, T.G., in Swern, D. *Organic Peroxides*, Vol. 3, Wiley, NY, *1972*, pp. 141–270.

[439]Longone, D.T.; Miller, A.H. *Tetrahedron Lett.* **1967**, 4941.

[440]Davis, F.A.; Lal, G.S.; Wei, J. *Tetrahedron Lett.* **1988**, *29*, 4269.

[441]Stang, P.J.; Boehshar, M.; Wingert, H.; Kitamura, T. *J. Am. Chem. Soc.* **1988**, *110*, 3272.

The reaction of alkenes with borane, monoalkyl and dialkylboranes leads to a new organoborane (see **15-16**). Treatment of organoboranes with alkaline H_2O_2 oxidizes trialkylboranes to esters of boric acid.[442] This reaction does not affect double or triple bonds, aldehydes, ketones, halides, or nitriles that may be present elsewhere in the molecule. There is no rearrangement of the R group itself, and this reaction is a step in the hydroboration method of converting alkenes to alcohols (**15-16**). The mechanism has been formulated as involving initial formation of an ate complex when the hydroperoxide anion attacks the electrophilic boron atom. Subsequent rearrangement from boron to oxygen,[442] as shown, leads to the B–O–R unit.

Similar migration of the other two R groups and hydrolysis of the B–O bonds leads to the alcohol and boric acid. Retention of configuration is observed in R. Boranes can also be oxidized to borates in good yields with oxygen,[443] with sodium perborate $NaBO_3$,[444] and with trimethylamine oxide, either anhydrous[445] or in the form of the dihydrate.[446] The reaction with oxygen is free radical in nature.[447]

OS **V**, 918; **VI**, 719, 852, 919.

12-28 Preparation of Borates and Boronic Acids

$$R\text{-}M \longrightarrow R\text{-}B(OH)_2$$

$$Ar\text{-}M \longrightarrow Ar\text{-}B(OH)_2$$

$$R\text{-}OH + BX_3 \text{ or } B(OH)_3 \longrightarrow B(OR)_3$$

Alkylboronic acids and arylboronic acids, $RB(OH)_2$, and $ArB(OH)_2$, respectively, are increasingly important in organic chemistry. The palladium catalyzed coupling reaction of aryl halides and aryl triflates with arylboronic acids (the Suzuki–Miyaura

[442]For reviews, see Pelter, A.; Smith, K.; Brown, H.C. *Borane Reagents*, Aademic Press, NY, *1988*, pp. 244–249; Brown, H.C. *Boranes in Organic Chemistry*; Cornell University Press, Ithaca, NY, *1972*, pp. 321–325; Matteson, D.S., in Hartley, F.R.; Patai, S. *The Chemistry of the Metal–Carbon Bond*, Vol. 4, Wiley, NY, pp. 307–409, 337–340. See also, Brown, H.C.; Snyder, C.; Subba Rao, B.C.; Zweifel, G. *Tetrahedron 1986*, *42*, 5505.

[443]Brown, H.C.; Midland, M.M.; Kabalka, G.W. *J. Am. Chem. Soc. 1971*, *93*, 1024; *Tetrahedron 1986*, *42*, 5523.

[444]Kabalka, G.W.; Shoup, T.M.; Goudgaon, N.M. *J. Org. Chem. 1989*, *54*, 5930.

[445]Köster, R.; Arora, S.; Binger, P. *Angew. Chem. Int. Ed. 1969*, *8*, 205.

[446]Kabalka, G.W.; Hedgecock, Jr., H.C. *J. Chem. Educ. 1975*, *52*, 745; Kabalka, G.W.; Slayden, S.W. *J. Organomet. Chem. 1977*, *125*, 273.

[447]Mirviss, S.B. *J. Am. Chem. Soc. 1961*, *83*, 3051; *J. Org. Chem. 1967*, *32*, 1713; Davies, A.G.; Roberts, B.P. *Chem. Commun. 1966*, 298; Midland, M.M.; Brown, H.C. *J. Am. Chem. Soc. 1971*, *93*, 1506.

reaction, **13-12**) is probably the most notable example. A simple synthesis involve the reaction of a Grignard reagent, such as phenylmagnesium bromide with an alkyl borate to give phenylboronic acid.[448] Alkylboronic acids are similarly prepared.[449] Note that boronic acids are subject to cyclic trimerization with loss of water to form boroxines. Trimethylborate, $B(OMe)_3$, can be used in place of tri-n-butyl borate.[450] Newer methods involve the palladium-mediated borylation of alcohols with bis(pinacolato)diboron[451] or pinacolborane,[452] but deprotection of the boronate esters can be a problem. Diolboranes, such as catecholborane **44**,[453] are prepared by the reaction of a diol with borane. Cedranediolborane (**45**, prepared from the cedrane-8,9-diol[454] by treatment with borane•dimethyl sulfide) can be coupled to aryl iodides with a palladium catalyst, and generates the free boronic acid by treatment with diethanolamine and then aqueous acid.[455] Boronate esters are often prepared as a means to purify the organoboron species, but some of these esters are hydrolytically unstable and difficult to deal with upon completion of the reaction.[456]

Alkeneboronic esters and acids are also readily available, as in the addition of vinyl-magnesium chloride[457] to trimethyl borate below $-50°C$, followed by hydrolysis.[458]

[448]Bean, F.R.; Johnson, J.R. *J. Am. Chem Soc.* **1932**, *54*, 4415. For a review, see Lappert, M.F. *Chem. Rev.* **1956**, *56*, 959.

[449]Khotinsky, E.; Melamed, M. *Chem. Ber.* **1909**, *42*, 3090.

[450]Soloway, A.H. *J. Am. Chem. Soc.* **1959**, *81*, 3017.

[451]Ishiyama, T.; Murata, M.; Miyaura, N. *J. Org. Chem.* **1995**, *60*, 7508.

[452]Murata, M.; Oyama, T.; Watanabe, S.; Masuda, Y. *J. Org. Chem.* **2000**, *65*, 164; Song, Y.L. *Synlett* **2000**, 1210.

[453]Brown, H.C.; Gupta, S.K. *J. Am. Chem. Soc.* **1972**, *94*, 4370; Kanth, J. V. B.; Periasamy, M.; Brown, H.C. *Org. Process Res. Dev.* **2000**, *4*, 550.

[454]Narula, A.S.; Trifilieff, E.; Bang, L.; Ourisson, G. *Tetrahedron Lett.* **1977**, *18*, 3959; Song, Y.; Ding, Z.; Wang, Q.; Tao, F. *Synth. Commun.* **1998**, *28*, 3757.

[455]Song, Y.-L.; Morin, C. *Synlett* **2001**, 266.

[456]Lightfoot, A.P.; Maw, G.; Thirsk, C.; Twiddle, S.J.R.; Whiting, A. *Tetrahedron Lett.* **2003**, *44*, 7645.

[457]Ramsden, H.E.; Leebrick, J.R.; Rosenberg, S.D.; Miller, E.H.; Walburn, J.J.; Balint, A.E.; Cserr, R. *J. Org, Chem.*, **1957**, 22, 1602.

[458]D.S. Matteson *J. Am. Chem. Soc.* **1960**, 82, 4228; Matteson, D.S. *Acc. Chem. Res.* **1970**, *3*, 186; Matteson, D.S. *Progr. Boron Chem.* **1970**, *3*, 117.

A nonaqueous workup procedure has been reproted for the preparation of arylboronic esters [ArB(OR'$_2$)].[459] Uncontrollable polymerization or oxidation of much of the boronic acid occurred during the final stages of the isolation procedure, but could be avoided by *in situ* conversion to the dibutyl ester by adding the crude product to 1-butanol. The samarium(III)-catalyzed hydroboration of olefins with catecholborane is a good synthesis of boronate esters.[460]

Trialkyl borates (called orthoborates) can be prepared by heating the appropriate alcohol with boron trichloride in a sealed tube, but the procedure works well only for relatively simple alkyl groups.[461] Heating alcohols with boron trioxide (B$_2$O$_3$) in an autoclave at 110–170°C give the trialkyl borate.[462] Boric acid can be used for the preparation of orthoborates[463] by heating with alcohols in the presence of either hydrogen chloride or concentrated sulfuric acid. Removal of water as an azeotrope with excess alcohol improves the yield,[464] and good yields can be obtained for trialkyl borates[465] and even for triphenyl borate.[466] This method is unsuccessful for those borates whose parent alcohols do not form azeotropes with water and for the tertiary alkyl borates,[467] impure samples are usually obtained.[468]

Potassium organotrifluoroborates (RBF$_3$K) are readily prepared by the addition of inexpensive KHF$_2$ to a variety of organoboron intermediates.[469] They are monomeric, crystalline solids that are readily isolated and indefinitely stable in the air. These reagents can be used in several of the applications where boronic acids or esters are used (**13-10–13-13**).[470] Note that vinylboronic acid and even vinylboronate esters are unstable to polymerization,[471] whereas the analogous vinyltrifluoroborate is readily synthesized and completely stable.[472]

O.S. 13, 16; **81,** 134.

[459]Wong, K.-T.; Chien, Y.-Y.; Liao, Y.-L.; Lin, C.-C.; Chou, M.-Y.; Leung, M.-K. *J. Org. Chem.* **2002,** *67,* 1041.

[460]Evans, D.A.; Muci, A.R.; Stuermer, R. *J. Org. Chem.,* **1993,** *58,* 5307.

[461]Councler, C. *Ber.* **1876,** *9,* 485; **1877,** *10,* 1655; **1878,** *11,* 1106.

[462]Schiff, H. *Ann. Suppl.* **1867,** *6,* 158; Councler, C. *J. Prakt. Chem.* **1871,** *16,* 371.

[463]Cohn, G. *Pharm. Zentr.* **1911,** *62,* 479.

[464]Bannister, W.J. U.S. Patent 1,668,797 (*Chem. Abstr.* **1928,** 22:2172).

[465]Ballard, S.A, U.S. Patent 2,431,224 (*Chem. Abstr.* **1948,** 42:1960); Haider, S.Z.; Khundhar, M.H.; Siddiqulah, Md. *J. Appl. Chem.* **1954,** *4,* 93; Scattergood, A.; Miller, W.H.; Gammon, J. *J. Am. Chem. Soc.* **1945,** *67,* 2150; Wuyts, H.; Duquesne, A. *Bull. Soc. Chim. Belg.* **1939,** *48,* 77.

[466]Colclough, T.; Gerrard, W.; Lappert, M.F. *J. Chem. Soc.* **1955,** 907.

[467]Haider, S.Z.; Khundhar, M.H.; Siddiqullah, Md. *J. Appl. Chem.* **1954,** *4,* 93; Scattergood, A., Miller, W.H.; Gammon, J. *J. Am. Chem. Soc.* **1945,** *67,* 2150.

[468]Ahmad, T.; Khundkar, M.H. *Chem. Ind.* **1954,** 248.

[469]Vedejs, E.; Chapman, R.W.; Fields, S.C.; Lin, S.; Schrimpf, M.R. *J. Org. Chem.* **1995,** *60,* 3020; Vedejs, E.; Fields, S.C.; Hayashi, R.; Hitchcock, S.R.; Powell, D.R.; Schrimpf, M.R. *J. Am. Chem. Soc.* **1999,** *121,* 2460.

[470]Molander, G.A.; Ito, T. *Org. Lett.* **2001,** *3,* 393; Molander, G.A.; Biolatto, B. *Org. Lett.* **2002,** *4,* 1867; Molander, G.A.; Biolatto, B. *J. Org. Chem.* **2003,** *68,* 4302; Molander, G.A.; Katona, B.W.; Machrouhi, F. *J. Org. Chem.* **2002,** *67,* 8416; Molander, G.A.; Yun, C.; Ribagorda, M.; Biolatto, B. *J. Org. Chem.* **2003,** *68,* 5534; Molander, G.A.; Ribagorda, M. *J. Am. Chem. Soc.* **2003,** *125,* 11148.

[471]Matteson, D.S. *J. Am. Chem. Soc.* **1960,** *82,* 4228.

[472]Molander, G.A.; Felix, L.A. *J. Org. Chem.* **2005,** *70,* 3950.

12-29 Oxygenation of Organometallic Reagents and Other Substrates to O-Esters and Related Compounds

$$R-M \longrightarrow R-OOCR'$$

$$R-Y \longrightarrow R-OOCR'$$

In some cases, it is possible to oxygenate a nonaromatic carbon atom using various reagents, where the product is an O- ester rather than an alcohol. In one example, a vinyl iodonium salt was heated with DMF to product the corresponding formate ester.[473]

C. Sulfur Electrophiles

12-30 Conversion of Organometallic Reagents to Sulfur Compounds

Thiols and sulfides are occasionally prepared by treatment of Grignard reagents with sulfur.[474] Analogous reactions are known for selenium and tellurium compounds. Grignard reagents and other organometallic

$$RMgX + SO_2Cl_2 \longrightarrow RSO_2Cl$$
$$RMgX + R^1SO-OR^2 \longrightarrow RSOR^1$$
$$RMgX + R^1SSR^1 \longrightarrow RSR^1$$
$$RMgX + SO_2 \longrightarrow RSO-OMgX$$

with a branch to RSO_2H (via H^+) and RSO_2X (via X_2)

compounds[475] react with sulfuryl chloride to give sulfonyl chlorides,[476] with esters of sulfinic acids to give (stereospecifically) sulfoxides,[477] with disulfides to give

[473]Ochiai, M.; Yamamoto, S.; Sato, K. *Chem. Commun.* **1999**, 1363.

[474]For reviews of the reactions in this section, see Wardell, J.L.; Paterson, E.S., in Hartley, F.R.; Patai, S. *The Chemistry of the Metal-Carbon Bond*, Vol. 2, Wiley, NY, **1985**, pp. 316–323; Wardell, J.L., in Patai, S. *The Chemistry of the Thiol Group*, pt. 1, Wiley, NY, **1974**, pp. 211–215; Wakefield, B.J. *Organolithium Methods*, Academic Press, NY, **1988**, pp. 135–142.

[475]For a discussion of conversions of organomercury compounds to sulfur-containing compounds, see Larock, R.C. *Organomercury Compounds in Organic Synthesis*, Springer, NY, **1985**, pp. 210–216.

[476]Bhattacharya, S.N.; Eaborn, C.; Walton, D.R.M. *J. Chem. Soc. C* **1968**, 1265. For similar reactions with organolithiums, see Quast, H.; Kees, F. *Synthesis* **1974**, 489; Hamada, T.; Yonemitsu, O. *Synthesis* **1986**, 852.

[477]Harpp, D.N.; Vines, S.M.; Montillier, J.P.; Chan, T.H. *J. Org. Chem.* **1976**, *41*, 3987.

sulfides,[478] and with SO_2 to give sulfinic acid salts[479] which can be hydrolyzed to sulfinic acids or treated with halogens to give sulfonyl halides.[480]

OS **III**, 771; **IV**, 667; **VI**, 533, 979.

D. Halogen Electrophiles

12-31   Halo-de-metalation

$$RMgX \quad + \quad I_2 \quad \longrightarrow \quad RI \quad + \quad MgIX$$

Grignard reagents react with halogens to give alkyl halides. The reaction is useful for the preparation of iodo compounds from the corresponding chloro or bromo compounds. The reaction is not useful for preparing chlorides, since the reagents RMgBr and RMgI react with Cl_2 to give mostly RBr and RI, respectively.[481]

Most organometallic compounds, both alkyl and aryl, also react with halogens to give alkyl or aryl halides.[482] The reaction can be used to convert acetylide ions to 1-haloalkynes.[483] Since acetylide ions are easily prepared from alkynes (**12-23**), this provides a means of accomplishing the conversion $RC{\equiv}CH \rightarrow RC{\equiv}CX$. Vinyliodonium tetrafluoroborates were converted to vinyl fluorides by heating.[484] Similarly, vinyl trifluoroborates were converted to the vinyl iodide with NaI and chloramine-T in aq. THF.[485] The reaction of an alkene with CuO•BF4, iodine and triethylsilane gave the 2-iodo alkane.[486]

Trialkylboranes react rapidly with I_2[487] or Br_2[488] in the presence of NaOMe in methanol, or with $FeCl_3$ or other reagents[489] to give alkyl iodides, bromides, or chlorides, respectively. Combined with the hydroboration reaction (**15-16**), this is an indirect way of adding HBr, HI, or HCl to a double bond to give products with an

---

[478]For a discussion, see Negishi, E. *Organometallics in Organic Synthesis*, Wiley, NY, *1980*, pp. 243–247.

[479]For a review of the reactions of organometallic compounds with SO_2, see Kitching, W.; Fong, C.W. *Organomet. Chem. Rev. Sect. A* *1970*, *5*, 281.

[480]Asinger, F.; Laue, P.; Fell, B.; Gubelt, C. *Chem. Ber.* *1967*, *100*, 1696.

[481]Zakharkin, L.I.; Gavrilenko, V.V.; Paley, B.A. *J. Organomet. Chem.* *1970*, *21*, 269.

[482]For a review, see Abraham, M.H.; Grellier, P.L., in Hartley, F.R.; Patai, S. *The Chemistry of the Metal–Carbon Bond*, Vol. 2, Wiley, NY, pp. 72–105. For reviews with respect to organomercury compounds, see Larock, R.C. *Organomercury Compounds in Organic Synthesis*, Springer, NY, *1985*, pp. 158–178; Makarova, L.G. *Organomet. React.* *1970*, *1*, 119, pp. 325–348.

[483]For a review, see Delavarenne, S.Y.; Viehe, H.G., in Viehe, H.G. *Acetylenes*, Marcel Dekker, NY, *1969*, pp. 665–688. For a list of reagents, with references, see Larock, R.C. *Comprehensive Organic Transformations*, 2nd ed., Wiley-VCH, NY, *1999*, pp. 655–656. For an improved procedure, see Brandsma, L.; Verkruijsse, H.D. *Synthesis 1990*, 984.

[484]Okuyama, T.; Fujita, M.; Gronheid, R.; Lodder, G. *Tetrahedron Lett.* *2000*, *41*. 5125.

[485]Kabalka, G.W.; Mereddy, A.R. *Tetrahedron Lett.* *2004*, *45*, 1417.

[486]Campos, P.J.; García, B.; Rodríguez, M.A. *Tetrahedron Lett.* *2002*, *43*, 6111.

[487]Brown, H.C.; Rathke, M.W.; Rogić, M.M.; De Lue, N.R. *Tetrahedron 1988*, *44*, 2751.

[488]Brown, H.C.; Lane, C.F. *Tetrahedron 1988*, *44*, 2763; Brown, H.C.; Lane, C.F.; De Lue, N.R. *Tetrahedron 1988*, *44*, 2273. For another reagent, see Nelson, D.J.; Soundararajan, R. *J. Org. Chem.* *1989*, *54*, 340.

[489]Nelson, D.J.; Soundararajan, R. *J. Org. Chem.* *1988*, *53*, 5664. For other reagents, see Jigajinni, V.B.; Paget, W.E.; Smith, K. *J. Chem. Res. (S) 1981*, 376; Brown, H.C.; De Lue, N.R. *Tetrahedron 1988*, *44*, 2785.

anti-Markovnikov orientation (see **15-1**). Trialkylboranes can also be converted to alkyl iodides by treatment with allyl iodide and air in a free-radical process.[490] *trans*-1-Alkenylboronic acids **47**, prepared by hydroboration of terminal alkynes with catecholborane to give **46**[491] (**15-16**), followed by hydrolysis, react with I_2 in the presence of NaOH at $0°C$ in ethereal solvents to give trans vinylic iodides.[492] Treatment with ICl also gives the vinyl iodide.[493] This is an indirect way of accomplishing the anti-Markovnikov addition of HI to a

Catecholborane   **46**   **47**

terminal triple bond. The reaction cannot be applied to alkenylboronic acids prepared from internal alkynes. However, alkenylboronic acids prepared from both internal and terminal alkynes react with Br_2 (2 equivalents of Br_2 must be used) followed by base to give the corresponding vinylic bromide, but in this case with *inversion* of configuration; so the product is the cis vinylic bromide.[494] Alkenylboronic acids also give vinylic bromides and iodides when treated with a mild oxidizing agent and NaBr or NaI, respectively.[495] Treatment of **47** (prepared from terminal alkynes) with Cl_2 gave vinylic chlorides with inversion.[496] Vinylic boranes can be converted to the corresponding vinylic halide by treatment with NCS or NBS.[497] Vinylic halides can also be prepared from vinylic silanes[498] and from vinylic copper reagents. The latter react with I_2 to give iodides,[499] and with NCS or NBS at $-45°C$ to give chlorides or bromides.[500] T

For the reaction of lithium enolate anions of esters with I_2 or CX_4, see **12-5**.

The conversion of terminal alkynes to 1-iodo-1-alkynes was reported using NaI under electrochemical conditions.[501] The reaction of an aryl alkyne with $HInCl_2/BEt_3$,

[490]Suzuki, A.; Nozawa, S.; Harada, M.; Itoh, M.; Brown, H.C.; Midland, M.M. *J. Am. Chem. Soc.* **1971**, *93*, 1508. For reviews, see Brown, H.C.; Midland, M.M. *Angew. Chem. Int. Ed.* **1972**, *11*, 692, pp. 699–700; Brown, H.C. *Boranes in Organic Chemistry*, Cornell Univ. Press, Ithica, NY, **1972**, pp. 442–446.

[491]For a review of this reagent, see Kabalka, G.W. *Org. Prep. Proced. Int.* **1977**, *9*, 131.

[492]Brown, H.C.; Hamaoka, T.; Ravindran, N.; Subrahmanyam, C.; Somayaji, V.; Bhat, N.G. *J. Org. Chem.* **1989**, *54*, 6075. See also, Kabalka, G.W.; Gooch, E.E.; Hsu, H.C. *Synth. Commun.* **1981**, *11*, 247.

[493]Stewart, S.K.; Whiting, A. *Tetrahedron Lett.* **1995**, *36*, 3929.

[494]Brown, H.C.; Hamaoka, T.; Ravindran, N. *J. Am. Chem. Soc.* **1973**, *95*, 6456. See also, Brown, H.C.; Bhat, N.G. *Tetrahedron Lett.* **1988**, *29*, 21.

[495]See Kabalka, G.W.; Sastry, K.A.R.; Knapp, F.F.; Srivastava, P.C. *Synth. Commun.* **1983**, *13*, 1027.

[496]Kunda, S.A.; Smith, T.L.; Hylarides, M.D.; Kabalka, G.W. *Tetrahedron Lett.* **1985**, *26*, 279.

[497]Hoshi, M.; Shirakawa, K. *Tetrahedron Lett.* **2000**, *41*, 2595.

[498]See, for example, Chou, S.P.; Kuo, H.; Wang, C.; Tsai, C.; Sun, C. *J. Org. Chem.* **1989**, *54*, 868.

[499]Normant, J.F.; Chaiez, G.; Chuit, C.; Villieras, J. *J. Organomet. Chem.* **1974**, *77*, 269; *Synthesis* **1974**, 803.

[500]Westmijze, H.; Meijer, J.; Vermeer, P. *Recl. Trav. Chim. Pays-Bas* **1977**, *96*, 168; Levy, A.B.; Talley, P.; Dunford, J.A. *Tetrahedron Lett.* **1977**, 3545.

[501]Nishiguchi, I.; Kanbe, O.; Itoh, K.; Maekawa, H. *Synlett* **2000**, 89.

and then iodine leads to a Z-vinyl iodide with respect to the aryl group and the iodine atom.[502] 1-Bromo-1-alkynes were converted to the 1-iodo-1-alkyne with CuI.[503]

It is unlikely that a single mechanism suffices to cover all conversions of organometallic compounds to alkyl halides.[504] In a number of cases, the reaction has been shown to involve inversion of configuration (see p. 757), indicating an S_E2 (back) mechanism, while in other cases retention of configuration has been shown,[505] implicating an S_E2 (front) or S_Ei mechanism. In still other cases, complete loss of configuration as well as other evidence have demonstrated the presence of a free-radical mechanism.[505,506]

OS **I**, 125, 325, 326; **III**, 774, 813; **V**, 921; **VI**, 709; **VII**, 290; **VIII**, 586; **IX**, 573. Also see, OS **II**, 150.

E. Nitrogen Electrophiles

12-32   The Conversion of Organometallic Compounds to Amines

Amino-de-metalation

$$RLi \xrightarrow[\text{MeLi}]{CH_3ONH_2} RNH_2$$

There are several methods for conversion of alkyl- or aryllithium compounds to primary amines.[507] The two most important are treatment with hydroxylamine derivatives and with certain azides.[508] In the first of these methods, treatment of RLi with methoxyamine and MeLi in ether at $-78°C$ gives RNH_2.[509] Grignard reagents from aliphatic halides give lower yields. The reaction can be extended to give secondary amines by the use of N-substituted methoxyamines (CH_3ONHR').[510] There is evidence[511] that the mechanism involves the direct displacement of OCH_3 by R

[502]Takami, K.; Yorimitsu, H.; Oshima, K. *Org. Lett.* **2002**, *4*, 2993.

[503]Abe, H.; Suzuki, H. *Bull. Chem. Soc. Jpn.* **1999**, *72*, 787.

[504]For reviews of the mechanisms, see Abraham, M.H.; Grellier, P.L., in Hartley, F.R.; Patai, S. *The Chemistry of the Carbon–Metal Bond*, Vol. 2, Wiley, NY, p. 72; Abraham, M.H. *Comprehensive Chemical Kinetics*, Bamford, C.H.; Tipper, C.F.H., Eds., Vol. 12; Elsevier, NY, *1973*, pp. 135–177; Jensen, F.R.; Rickborn, B. *Electrophilic Substitution of Organomercurials*, McGraw-Hil, NY, *1968*, pp. 75–97.

[505]For example, see Jensen, F.R.; Gale, L.H. *J. Am. Chem. Soc.* **1960**, *82*, 148.

[506]See, for example, Beletskaya, I.P.; Reutov, O.A.; Gur'yanova, T.P. *Bull. Acad. Sci. USSR Div. Chem. Sci.* **1961**, 1483; Beletskaya, I.P.; Ermanson, A.V.; Reutov, O.A. *Bull. Acad. Sci. USSR Div. Chem. Sci.* **1965**, 218; de Ryck, P.H.; Verdonck, L.; Van der Kelen, G.P. *Bull. Soc. Chim. Belg.*, *1985*, *94*, 621.

[507]For a review of methods for achieving the conversion RM → RNH_2, see Erdik, E.; Ay, M. *Chem. Rev.* **1989**, *89*, 1947.

[508]For some other methods of converting organolithium or Grignard reagents to primary amines, see Alvernhe, G.; Laurent, A. *Tetrahedron Lett.* **1972**, 1007; Hagopian, R.A.; Therien, M.J.; Murdoch, J.R. *J. Am. Chem. Soc.* **1984**, *106*, 5753; Genet, J.P.; Mallart, S.; Greck, C.; Piveteau, E. *Tetrahedron Lett.* **1991**, *32*, 2359.

[509]Beak, P.; Kokko, B.J. *J. Org. Chem.* **1982**, *47*, 2822. For other hydroxylamine derivatives, see Colvin, E.W.; Kirby, G.W.; Wilson, A.C. *Tetrahedron Lett.* **1982**, *23*, 3835; Boche, G.; Bernheim, M.; Schrott, W. *Tetrahedron Lett.* **1982**, *23*, 5399; Boche, G.; Schrott, W. *Tetrahedron Lett.* **1982**, *23*, 5403.

[510]Kokko, B.J.; Beak, P. *Tetrahedron Lett.* **1983**, *24*, 561.

[511]Beak, P.; Basha, A.; Kokko, B.; Loo, D. *J. Am. Chem. Soc.* **1986**, *108*, 6016.

on an intermediate $CH_2ONR'^-$ ($CH_3ONR'^-$ $Li^+ + RLi \rightarrow CH_3OLi + RNR'^-$ Li^+). The most useful azide is tosyl azide TsN_3.[512] The initial product is usually RN_3, but this is easily reducible to the amine (**19-51**). With some azides, such as azidomethyl phenyl sulfide ($PhSCH_2N_3$), the group attached to the N_3 is a poor leaving group, so the initial product is a triazene (in this case $ArNHN=NCH_2SPh$ from ArMgX), which can be hydrolyzed to the amine.[513]

$$R_3B \xrightarrow{NH_3-NaOCl} 2\,RNH_2 + RB(OH)_2$$

Organoboranes react with a mixture of aqueous NH_3 and NaOCl to produce primary amines.[514] It is likely that the actual reagent is chloramine (NH_2Cl). Chloramine itself,[515] hydroxylamine-O-sulfonic acid in diglyme,[516] and trimethylsilyl azide[517] also give the reaction. Since the boranes can be prepared by the hydroboration of alkenes (**15-16**), this is an indirect method for the addition of NH_3 to a double bond with anti-Markovnikov orientation. Secondary amines can be prepared[518] by the treatment of alkyl- or aryldichloroboranes or dialkylchloroboranes with alkyl or aryl azides.

$$RBCl_2 + R'N_3 \longrightarrow RR'NBCl_2 \xrightarrow[OH^-]{H_2O} RNHR'$$

$$R_2BCl + R'N_3 \xrightarrow[2.H_2O]{1.Et_2O} RNHR'$$

The use of an optically active R^*BCl_2 gave secondary amines of essentially 100% optical purity.[519] Aryllead triacetates, $ArPb(OAc)_3$, give secondary amines ($ArNHAr'$) when treated with primary aromatic amines $Ar'NH_2$ and $Cu(OAc)_2$.[520]

Secondary amines have been converted to tertiary amines by treatment with lithium dialkylcuprate reagents: $R_2CuLi + NHR \rightarrow RNR_2'$.[521] The reaction was also used to convert primary amines to secondary, but yields were lower.[522]

[512]See, for example, Spagnolo, P.; Zanirato, P.; Gronowitz, S. *J. Org. Chem.* **1982**, *47*, 3177; Reed, J.N.; Snieckus, V. *Tetrahedron Lett.* **1983**, *24*, 3795. For other azides, see Hassner, A.; Munger, P.; Belinka Jr., B.A. *Tetrahedron Lett.* **1982**, *23*, 699; Mori, S.; Aoyama, T.; Shioiri, T. *Tetrahedron Lett.* **1984**, *25*, 429.

[513]Trost, B.M.; Pearson, W.H. *J. Am. Chem. Soc.* **1981**, *103*, 2483; **1983**, *105*, 1054.

[514]Kabalka, G.W.; Wang, Z.; Goudgaon, N.M. *Synth. Commun.* **1989**, *19*, 2409. For the extension of this reaction to the preparation of secondary amines, see Kabalka, G.W.; Wang, Z. *Organometallics* **1989**, *8*, 1093; *Synth. Commun.* **1990**, *20*, 231.

[515]Brown, H.C.; Heydkamp, W.R.; Breuer, E.; Murphy, W.S. *J. Am. Chem. Soc.* **1964**, *86*, 3565.

[516]Brown, H.C.; Kim, K.; Srebnik, M.; Singaram, B. *Tetrahedron* **1987**, *43*, 4071. For a method of using this reaction to prepare optically pure chiral amines, see Brown, H.C.; Kim, K.; Cole, T.E.; Singaram, B. *J. Am. Chem. Soc.* **1986**, *106*, 6761.

[517]Kabalka, G.W.; Goudgaon, N.M.; Liang, Y. *Synth. Commun.* **1988**, *18*, 1363.

[518]Brown, H.C.; Midland, M.M.; Levy, A.B.; Suzuki, A.; Sono, S.; Itoh, M. *Tetrahedron* **1987**, *43*, 4079; Carboni, B.; Vaultier, M.; Courgeon, T.; Carrié, R. *Bull. Soc. Chim. Fr.* **1989**, 844.

[519]Brown, H.C.; Salunkhe, A.M.; Singaram, B. *J. Org. Chem.* **1991**, *56*, 1170.

[520]Barton, D.H.R.; Donnelly, D.M.X.; Finet, J.; Guiry, P.J. *Tetrahedron Lett.* **1989**, *30*, 1377.

[521]Yamamoto, H.; Maruoka, K. *J. Org. Chem.* **1980**, *45*, 2739.

[522]Merkushev, E.B. *Synthesis* **1988**, 923

In the presence of a CuI catalyst, acetamide reacted with vinyl iodides to give the corresponding enamide, where the nitrogen of the amide replaced the iodine atom.[523]

Terminal alkynes reacted with chlorodiphenylphosphine (Ph$_2$PCl) and a nickel catalyst to give the 1-diphenylphosphino alkyne (R-C≡C-PPh$_2$).[524] Alkynyl halides can be used for a similar reaction. Treatment of methyl carbamates with KHMDS and CuI, followed by two equivalents of 1-bromo phenylacetylene gave the *N*-substituted alkyne, Ph–C≡C–N(CO$_2$Me)R.[525]

OS **VI**, 943.

F. Carbon Electrophiles

12-33   The Conversion of Organometallic Compounds to Ketones, Aldehydes, Carboxylic Esters, or Amides

Acyl-de-metalation, and so on

$$R\text{-HgX} \quad + \quad Co_2(CO)_8 \quad \xrightarrow{THF} \quad \underset{R}{\overset{O}{\underset{}{\|}}}\overset{}{C}\underset{R}{}$$

Symmetrical ketones[526] can be prepared in good yields by the reaction of organomercuric halides[527] with dicobalt octacarbonyl in THF,[528] or with nickel carbonyl in DMF or certain other solvents.[529] The R group may be aryl or alkyl. However, when R is alkyl, rearrangements may intervene in the $CO_2(CO)_8$ reaction, although the $Ni(CO)_4$ reaction seems to be free from such rearrangements.[530] Divinylic ketones (useful in the Nazarov cyclization, **15-20**) have been prepared in high yields by treatment of vinylic mercuric halides with CO and a rhodium catalyst.[530] In a more general synthesis of unsymmetrical ketones, tetraalkyltin compounds (R$_4$Sn) are treated with a halide R′X (R′ = aryl, vinylic, benzylic), CO, and a Pd complex catalyst.[531] Similar reactions use Grignard reagents, Fe(CO)$_5$, and an alkyl halide.[532] Cyclobutanone derivatives were prepared by carbonylation (treatment with CO) of a cyclic titanium compound.[533]

Grignard reagents react with formic acid to give good yields of aldehydes. Two equivalents of RMgX are used; the first converts HCOOH to HCOO–, which reacts

[523]Jiang, L.; Job, G.E.; Klapars, A.; Buchwald, S.L. *Org. Lett.* **2003**, *5*, 3667.
[524]Beletskaya, I.P.; Affanasiev, V.V.; Kazankova, M.A.; Efimova, I.V. *Org. Lett.* **2003**, *5*, 4309.
[525]Dunetz, J.R.; Danheiser, R.L. *Org. Lett.* **2003**, *5*, 4011.
[526]For reviews of the reactions in this section, and related reactions, see Narayana, C.; Periasamy, M. *Synthesis* **1985**, 253; Gulevich, Yu.V.; Bumagin, N.A.; Beletskaya, I.P. *Russ. Chem. Rev.* **1988**, *57*, 299.
[527]For a monograph on the synthetic uses of organomercury compounds, see Larock, R.C. *Organomercury Compounds in Organic Synthesis*, Springer, NY, **1985**. For reviews, see Larock, R.C. *Tetrahedron* **1982**, *38*, 1713; *Angew. Chem. Int. Ed.* **1978**, *17*, 27.
[528]Seyferth, D.; Spohn, R.J. *J. Am. Chem. Soc.* **1969**, *91*, 3037.
[529]Ryu, I.; Ryang, M.; Rhee, I.; Omura, H.; Murai, S.; Sonoda, N. *Synth. Commun.* **1984**, *14*, 1175 and references cited therein. For another method, see Hatanaka, Y.; Hiyama, T. *Chem. Lett.* **1989**, 2049.
[530]Larock, R.C.; Hershberger, S.S. *J. Org. Chem.* **1980**, *45*, 3840.
[531]Tanaka, M. *Tetrahedron Lett.* **1979**, 2601.
[532]Yamashita, M.; Suemitsu, R. *Tetrahedron Lett.* **1978**, 761. See also, Vitale, A.A.; Doctorovich, F.; Nudelman, N.S. *J. Organomet. Chem.* **1987**, *332*, 9.
[533]Carter, C.A.G.; Greidanus, G.; Chen, J.-x.; Stryker, J.M. *J. Am. Chem. Soc.* **2001**, *123*, 8872.

with the second equivalent to give RCHO.[534] Alkyllithium reagents and Grignard reagents react with CO to give symmetrical ketones.[535] An interesting variation reacts CO_2 with an organolithium, which is then treated with a different organolithium reagent to give the unsymmetrical ketone.[536] α,β-Unsaturated aldehydes can be prepared by treatment of vinylic silanes with dichloromethyl methyl ether and $TiCl_4$ at $-90°C$.[537] α,β-Unsaturated esters can be prepared by treating boronic esters **27** with CO, $PdCl_2$, and NaOAc in MeOH.[538] The synthesis of α,β-unsaturated esters has also been accomplished by treatment of vinylic mercuric chlorides with CO at atmospheric pressure and a Pd catalyst in an alcohol as solvent, for example,[539]

Alkyl and aryl Grignard reagents can be converted to carboxylic esters with $Fe(CO)_5$ instead of CO.[540]

Amides have been prepared by the treatment of trialkyl or triarylboranes with CO and an imine, in the presence of catalytic amounts of cobalt carbonyl:[541]

In another method for the conversion RM $\rightarrow$ RCONR, Grignard reagents, and organolithium compounds are treated with a formamide ($HCONR_2'$) to give the intermediate $RCH(OM)NR_2'$, which is not isolated, but treated with PhCHO or Ph_2CO to give the product $RCONR_2'$.[542]

Direct conversion of a hydrocarbon to an aldehyde (R–H $\rightarrow$ R–CHO) was reported by treatment of the hydrocarbon with $GaCl_3$ and CO.[543]

See also, reactions **10-76**, **15-32**, and **18-23–18-24**.

OS **VIII**, 97.

[534]Sato, F.; Oguro, K.; Watanabe, H.; Sato, M. *Tetrahedron Lett.* **1980**, *21*, 2869. For another method of converting RMgX to RCHO, see Meyers, A.I.; Comins, D.L. *Tetrahedron Lett.* **1978**, 5179; Comins, D.L.; Meyers, A.I. *Synthesis* **1978**, 403; Amaratunga, W.; Fréchet, J.M.J. *Tetrahedron Lett.* **1983**, *24*, 1143.

[535]Ryang, M.; Sawa, Y.; Hasimoto, T.; Tsutsumi, S. *Bull. Chem. Soc. Jpn.* **1964**, *37*, 1704; Trzupek, L.S.; Newirth, T.L.; Kelly, E.G.; Sbarbati, N.E.; Whitesides, G.M. *J. Am. Chem. Soc.* **1973**, *95*, 8118.

[536]Zadel, G.; Breitmaier, E. *Angew. Chem. Int. Ed.* **1992**, *31*, 1035.

[537]Yamamoto, K.; Yohitake, J.; Qui, N.T.; Tsuji, J. *Chem. Lett.* **1978**, 859.

[538]Miyaura, N.; Suzuki, A. *Chem. Lett.* **1981**, 879. See also Yamashina, N.; Hyuga, S.; Hara, S.; Suzuki, A. *Tetrahedron Lett.* **1989**, *30*, 6555.

[539]Larock, R.C. *J. Org. Chem.* **1975**, *40*, 3237.

[540]Yamashita, M.; Suemitsu, R. *Tetrahedron Lett.* **1978**, 1477.

[541]Alper, H.; Amaratunga, S. *J. Org. Chem.* **1982**, *47*, 3593.

[542]Screttas, C.G.; Steele, B.R. *J. Org. Chem.* **1988**, *53*, 5151.

[543]Oshita, M.; Chatani, N. *Org. Lett.* **2004**, *6*, 4323.

12-34 Cyano-de-metalation

$$R\text{-}M \quad + \quad CuCN \longrightarrow R\text{-}CN$$

Vinylic copper reagents react with ClCN to give vinyl cyanides, although BrCN and ICN give the vinylic halide instead.[544] Vinylic cyanides have also been prepared by the reaction between vinylic lithium compounds and phenyl cyanate (PhOCN).[545] Alkyl nitriles (RCN) have been prepared, in varying yields, by treatment of sodium trialkylcyanoborates with NaCN and lead tetraacetate.[546] Vinyl bromides reacted with KCN, in the presence of a nickel complex and zinc metal to give the vinyl nitrile.[547] Vinyl triflates react with LiCN, in the presence of a palladium catalyst, to give the vinyl nitrile.[548]

For other electrophilic substitutions of the type RM → RC, which are discussed under nucleophilic substitutions in Chapter 10. See also, **16-81–16-85** and **16-99**.

OS **IX**, 548

G. Metal Electrophiles

12-35 Transmetallation With a Metal

Metalo-de-metalation

$$RM + M' \rightleftharpoons RM' + M$$

Many organometallic compounds are best prepared by this reaction, which involves replacement of a metal in an organometallic compound by another metal. The RM' compound can be successfully prepared only when M' is above M in the electromotive series, unless some other way is found to shift the equilibrium. That is, RM is usually an unreactive compound and M' is a metal more active than M. Most often, RM is R_2Hg, since mercury alkyls[527] are easy to prepare and mercury is far down in the electromotive series.[549] Alkyls of Li, Na, K, Be, Mg, Al, Ga, Zn, Cd, Te, Sn, and so on have been prepared this way. An important advantage of this method over **12-38** is that it ensures that the organometallic compound will be prepared free of any possible halide. This method can be used for the isolation of solid sodium and potassium alkyls.[550] If the metals lie too close together in the series, it may not be possible to shift the equilibrium. For example, alkylbismuth compounds cannot be prepared in this way from alkylmercury compounds.

OS **V**, 1116.

[544]Westmijze, H.; Vermeer, P. *Synthesis* **1977**, 784.

[545]Murray, R.E.; Zweifel, G. *Synthesis* **1980**, 150.

[546]Masuda, Y.; Hoshi, M.; Yamada, T.; Arase, A. *J. Chem. Soc. Chem. Commun.* **1984**, 398.

[547]Sakakibara, Y.; Enami, H.; Ogawa, H.; Fujimoto, S.; Kato, H.; Kunitake, K.; Sasaki, K.; Sakai, M. *Bull. Chem. Soc. Jpn.* **1995**, *68*, 3137.

[548]Piers, E.; Fleming, F.F. *Can. J. Chem.* **1993**, *71*, 1867.

[549]For a review of the reaction when M is Hg, see Makarova, L.G. *Organomet. React.* **1970**, *1*, 119, pp. 190–226. For a review where M' is Li, see Wardell, J.L., in Zuckerman, J.J. *Inorganic Reactions and Methods*, Vol. 11, VCH, NY, **1988**, pp. 31–44.

[550]BuNa and BuK have also been prepared by exchange of BuLi with *t*-BuONa or *t*-AmOK: Pi, R.; Bauer, W.; Brix, B.; Schade, C.; Schleyer, P.v.R. *J. Organomet. Chem.* **1986**, *306*, C1.

12-36   Transmetallation With a Metal Halide

Metalo-de-metalation

$$RM + M'X \rightleftharpoons RM' + MX$$

In contrast to **12-35**, the reaction between an organometallic compound and a metal *halide* is successful only when M' is *below* M in the electromotive series.[551] The two reactions considered together therefore constitute a powerful tool for preparing all kinds of organometallic compounds. In this reaction, the most common substrates are Grignard reagents and organolithium compounds.[552]

The MgX of Grignard reagents[553] can migrate to terminal positions in the presence of small amounts of $TiCl_4$.[554] The proposed mechanism consists of metal exchange (**12-36**), elimination–addition, and metal exchange:

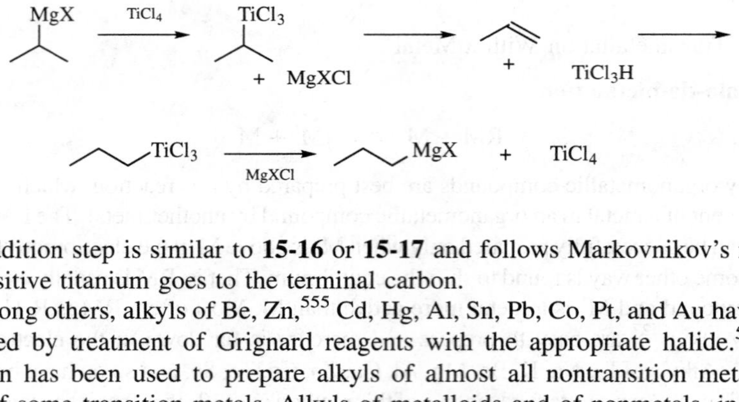

The addition step is similar to **15-16** or **15-17** and follows Markovnikov's rule, so the positive titanium goes to the terminal carbon.

Among others, alkyls of Be, Zn,[555] Cd, Hg, Al, Sn, Pb, Co, Pt, and Au have been prepared by treatment of Grignard reagents with the appropriate halide.[556] The reaction has been used to prepare alkyls of almost all nontransition metals and even of some transition metals. Alkyls of metalloids and of nonmetals, including

[551]For reviews of the mechanism, see Abraham, M.H.; Grellier, P.L. in Hartley, F.R.; Patai, S. *The Chemistry of the Carbon–Metal Bond*, Vol. 2, Wiley, NY, pp. 25–149; Abraham, M.H. *Comprehensive Chemical Kinetics*, Bamford, C.H.; Tipper, C.F.H., Eds., Vol. 12; Elsevier, NY, *1973*, pp. 39–106; Jensen, F.R.; Rickborn, B. *Electrophilic Substituton of Organomercurials*, McGraw-Hill, NY, *1968*, pp. 100–192. Also see Schlosser, M. *Angew. Chem. Int. Ed. 1964*, *3*, 287, 362; *Newer Methods Prep. Org. Chem. 1968*, *5*, 238.

[552]For monographs on organolithium compounds, see Wakefield, B.J. *Organolithium Methods*, Academic Press, NY, *1988*; Wakefield, B.J. *The Chemistry of Organolithium Compounds*, Pergamon: Elmsford, NY, *1974*.

[553]For reviews of rearrangements in organomagnesium chemistry, see Hill, E.A. *Adv. Organomet. Chem. 1977*, *16*, 131; *J. Organomet. Chem. 1975*, *91*, 123.

[554]Cooper, G.D.; Finkbeiner, H.L. *J. Org. Chem. 1962*, *27*, 1493; Fell, B.; Asinger, F.; Sulzbach, R.A. *Chem. Ber. 1970*, *103*, 3830. See also, Ashby, E.C.; Ainslie, R.D. *J. Organomet. Chem. 1983*, *250*, 1.

[555]For a review of the use of activated zinc, see Erdik, E. *Tetrahedron 1987*, *43*, 2203.

[556]For a review, see Noltes, J.G. *Bull. Soc. Chim. Fr. 1972*, 2151.

Si, B,[557] Ge, P, As, Sb, and Bi, can also be prepared in this manner.[558] Except for alkali-metal alkyls and Grignard reagents, the reaction between RM and M'X is the most common method for the preparation of organometallic compounds.[559]

Lithium dialkylcopper reagents can be prepared by mixing 2 equivalents of RLi with 1 equivalent of a cuprous halide in ether at low temperatures:[560]

$$2 \text{ RLi} \quad + \quad \text{CuX} \quad \longrightarrow \quad \text{R}_2\text{CuLi} \quad + \quad \text{LiX}$$

Another way is to dissolve an alkylcopper compound in an alkyllithium solution. Higher order cuprates can also be prepared, as well as "non-ate" copper reagents.[561]

Metallocenes (**48**, see p. 66) are usually made by this method:

48

Among others, metallocenes of Sc, Ti, V, Cr, Mn, Fe, Co, and Ni have been prepared in this manner.[562]

Metal nitrates are sometimes used instead of halides.

In a related reaction sulfurated boranes (R_2B–$SSiR'_2$) react with Grignard reagents, such as methylmagneisum bromide to give the B-alkyl borane (e.g., R_2B–Me) upon heating *in vacuo*.[563]

OS **I**, 231, 550; **III**, 601; **IV**, 258, 473, 881; **V**, 211, 496, 727, 918, 1001; **VI**, 776, 875, 1033; **VII**, 236, 290, 524; **VIII**, 23, 57, 268, 474, 586, 606, 609. Also see, OS **IV**, 476

---

[557]For a method of preparing organoboranes from RMgX and BF_3, where the RMgX is present only *in situ*, see Brown, H.C.; Racherla, U.S. *Tetrahedron Lett.* **1985**, *26*, 4311.

[558]For reviews as applied to Si, B, and P, see Wakefield, B.J. *Organolithium Methods*, Academic Press, NY, *1988*, pp. 149–158; Kharasch, M.S.; Reinmuth, O. *Grignard Reactions of Nonmetallic Substances*; Prentice-Hall: Englewood Cliffs, NJ, *1954*, pp. 1306–1345.

[559]For a review with respect to Al, see Mole, T. *Organomet. React.* **1970**, *1*, 1, pp. 31–43; to Hg, see Larock, R.C. *Organomercury Compounds in Organic Synthesis*, Springer, NY, *1985*, pp. 9–26; Makarova, L.G. *Organomet. React.* **1970**, *1*, 119, pp. 129–178, 227–240; to Cu, Ag, or Au, see van Koten, G., in Zuckerman, J.J. *Inorganic Reactions and Methods*, Vol. 11, VCH, NY, *1988*, pp. 219–232; to Zn, Cd, or Hg, see Wardell, J.L. in Zuckerman, J.J. *Inorganic Reactions and Methods*, Vol. 11, VCH, NY, *1988*, pp. 248–270.

[560]House, H.O.; Chu, C.; Wilkins, J.M.; Umen, M.J. *J. Org. Chem.* **1975**, *40*, 1460. But see also, Lipshutz, B.H.; Whitney, S.; Kozlowski, J.A.; Breneman, C.M. *Tetrahedron Lett.* **1986**, *27*, 4273; Bertz, S.H.; Dabbagh, G. *Tetrahedron* **1989**, *45*, 425.

[561]Stack, D.E.; Klein, W.R.; Rieke, R.D. *Tetrahedron Lett.* **1993**, *34*, 3063.

[562]For reviews of the preparation of metallocenes, see Bublitz, D.E.; Rinehart, Jr., K.L. *Org. React.* **1969**, *17*, 1; Birmingham, J.M. *Adv. Organomet. Chem.* **1965**, *2*, 365, p. 375.

[563]Soderquist, J.A.; DePomar, J.C.J. *Tetrahedron Lett.* **2000**, *41*, 3537.

12-37   Transmetalation With an Organometallic Compound

Metalo-de-metalation

$$RM \ + \ R'M' \ \rightleftharpoons \ RM' \ + \ R'M$$

This type of metallic exchange is used much less often than **12-35** and **12-36**. It is an equilibrium reaction and is useful only if the equilibrium lies in the desired direction. Usually the goal is to prepare a lithium compound that is not prepared easily in other ways,[564] for example, a vinylic or an allylic lithium, most commonly from an organotin substrate. Examples are the preparation of vinyllithium from phenyllithium and tetravinyltin and the formation of α-dialkylamino organolithium compounds from the corresponding organotin compounds[565]

$$RR'NCH_2SnBu_3 \ + \ BuLi \ \xrightarrow{0°C} \ RR'NCH_2Li \ + \ Bu_4Sn$$

The reaction has also been used to prepare 1,3-dilithiopropanes[566] and 1,1-dilithiomethylenecyclohexane[567] from the corresponding mercury compounds. In general, the equilibrium lies in the direction in which the more electropositive metal is bonded to that alkyl or aryl group that is the more stable carbanion (p. 250). The reaction proceeds with retention of configuration;[568] an S_Ei mechanism is likely.[569]

"Higher order" cuprates[570] (see **10-58**) have been produced by this reaction starting with a vinylic tin compound:[571]

$$RSnR'_3 + Me_2Cu(CN)Li_2 \longrightarrow RCuMe(CN)Li_2 + MeSnR'_{3-} \quad R = \text{a vinylic group}$$

[564]For reviews, see Wardell, J.L. in Hartley, F.R; Patai, S. *The Chemistry of the Carbon-Metal Bond*, Vol. 4, Wiley, NY, pp. 1–157, see pp. 81–89; Kauffmann, T. *Top. Curr. Chem.* **1980**, *92*, 109, p. 130.

[565]Peterson, D.J.; Ward, J.F. *J. Organomet. Chem.* **1974**, *66*, 209; Pearson, W.H.; Lindbeck, A.C. *J. Org. Chem.* **1989**, *54*, 5651.

[566]Seetz, J.W.F.L.; Schat, G.; Akkerman, O.S.; Bickelhaupt, F. *J. Am. Chem. Soc.* **1982**, *104*, 6848.

[567]Maercker, A.; Dujardin, R. *Angew. Chem. Int. Ed.* **1984**, *23*, 224.

[568]Seyferth, D.; Vaughan, L.G. *J. Am. Chem. Soc.* **1964**, *86*, 883; Sawyer, J.S.; Kucerovy, A.; Macdonald, T.L.; McGarvey, G.J. *J. Am. Chem. Soc.* **1988**, *110*, 842.

[569]Dessy, R.E.; Kaplan, F.; Coe, G.R.; Salinger, R.M. *J. Am. Chem. Soc.* **1963**, *85*, 1191.

[570]For reviews of these and other "higher order" organocuprates, see Lipshutz, B.H.; Wilhelm, R.S.; Kozlowski, J.A. *Tetrahedron* **1984**, *40*, 5005; Lipshutz, B.H. *Synthesis* **1987**, 325; *Synlett*, **1990**, 119. See also, Bertz, S.H. *J. Am. Chem. Soc.* **1990**, *112*, 4031; Lipshutz, B.H.; Sharma, S.; Ellsworth, E.L. *J. Am. Chem. Soc.* **1990**, *112*, 4032.

[571]Behling, J.R.; Babiak, K.A.; Ng, J.S.; Campbell, A.L.; Moretti, R.; Koerner, M.; Lipshutz, B.H. *J. Am. Chem. Soc.* **1988**, *110*, 2641.

These compounds are not isolated, but used directly *in situ* for conjugate addition reactions (**15-25**). Another method for the preparation of such reagents (but with Zn instead of Li) allows them to be made from α-acetoxy halides:[572]

OS **V**, 452; **VI**, 815; **VIII**, 97.

HALOGEN AS LEAVING GROUP

The reduction of alkyl halides can proceed by an electrophilic substitution mechanism, but it is considered in Chapter 19 (**19-53**).

12-38   Metalo-de-halogenation

$$RX \quad + \quad M \quad \longrightarrow \quad RM$$

Alkyl halides react directly with certain metals to give organometallic compounds.[573] The most common metal is magnesium, and of course this is by far the most common method for the preparation of Grignard reagents.[574] The order of halide activity is I > Br > Cl. The reaction can be applied to many alkyl halides primary, secondary, and tertiary and to aryl halides, although aryl *chlorides* require the use of THF or another higher boiling solvent instead of the usual ether, or special entrainment methods.[575] Aryl iodides and bromides can be treated in the usual manner. Allylic Grignard reagents can also be prepared in the usual manner (or in THF),[576] although in the presence of excess halide these may give Wurtz-type coupling products (see **10-56**).[577] Like aryl chlorides, vinylic halides require higher boiling solvents (see OS **IV**, 258). A good procedure for benzylic and allylic halides is to use magnesium anthracene (prepared from Mg and anthracene in THF)[578]

---

[572]Chou, T.; Knochel, P. *J. Org. Chem.* **1990**, *55*, 4791.

[573]For reviews, see Massey, A.G.; Humphries, R.E. *Aldrichimica Acta* **1989**, *22*, 31; Negishi, E. *Organometallics in Organic Synthesis*, Wiley, NY, **1980**, pp. 30–37; Rochow, E.G. *J. Chem. Educ.* **1966**, *43*, 58.

[574]For reviews, see Raston, C.L.; Salem, G., in Hartley, F.R.; Patai, S. *The Chemistry of the Carbon–Metal Bond*, Vol. 4, Wiley, NY, pp. 159–306, 162–175; Kharasch, M.S.; Reinmuth, O. *Grignard Reactions of Monmetallic Substances*, Prentice-Hall, Englewood Cliffs, NJ, **1954**, pp. 5–91.

[575]Pearson, D.E.; Cowan, D.; Beckler, J.D. *J. Org. Chem.* **1959**, *24*, 504.

[576]For a review of allyl and crotyl Grignard reagents, see Benkeser, R.A. *Synthesis* **1971**, 347.

[577]For a method of reducing coupling in the formation of allylic Grignard reagents, see Oppolzer, W.; Schneider, P. *Tetrahedron Lett.* **1984**, *25*, 3305.

[578]Freeman, P.K.; Hutchinson, L.L. *J. Org. Chem.* **1983**, *48*, 879; Bogdanović, B.; Janke, N.; Kinzelmann, H. *Chem. Ber.* **1990**, *123*, 1507, and other papers in this series.

instead of ordinary magnesium,[579] although activated magnesium turnings have also been used.[580] Alkynyl Grignard reagents are not generally prepared by this method at all. For these, **12-22** is used. Grignard reagents can also be formed from an alkyl halide and 1,2-dibromoethane with iodine as an initiator.[581]

Dihalides[582] can be converted to Grignard reagents if the halogens are different and are at least three carbons apart. If the halogens are the same, it is possible to obtain dimagnesium compounds (e.g., $BrMg(CH_2)_4MgBr$).[583] 1,2-Dihalides give elimination[584] instead of Grignard reagent formation (**17-22**), and the reaction is seldom successful with 1,1-dihalides, although the preparation of *gem*-disubstituted compounds, such as $CH_2(MgBr)_2$, has been accomplished with these substrates.[585] α-halo Grignard reagents and α-halolithium reagents can be prepared by the method given in **12-39**.[586] Alkylmagnesium fluorides can be prepared by refluxing alkyl fluorides with Mg in the presence of appropriate catalysts (e.g., I_2 or EtBr) in THF for several days.[587] Nitrogen-containing Grignard reagents have been prepared.[588]

The presence of other functional groups in the halide usually affects the preparation of the Grignard reagent. Groups that contain active hydrogen (defined as any hydrogen that will react with a Grignard reagent), such as OH, NH_2, and COOH, can be present in the molecule, but only if they are converted to the salt form (O^-, NH^-, COO^-, respectively). Groups that react with Grignard reagents, such as C=O, C≡N, NO_2, COOR, inhibit Grignard formation entirely. In general, the only functional groups that may be present in the halide molecule without any interference at all are double and triple bonds (except terminal triple bonds) and OR and NR_2 groups. However, β-halo ethers generally give β elimination when treated with

[579]Gallagher, M.J.; Harvey, S.; Raston, C.L.; Sue, R.E. *J. Chem. Soc. Chem. Commun.* **1988**, 289.

[580]Baker, K.V.; Brown, J.M.; Hughes, N.; Skarnulis, A.J.; Sexton, A. *J. Org. Chem.* **1991**, *56*, 698. For a review of the use of activated magnesium, see Lai, Y. *Synthesis* **1981**, 585.

[581]Li, J.; Liao, X.; Liu, H.; Xie, Q.; Liu, Z.; He, X. *Synth. Commun.* **1999**, *29*, 1037.

[582]For reviews of the preparation of Grignard reagents from dihalides, see Raston, C.L.; Salem, G. in Hartley, F.R.; Patai, S. *The Chemistry of the Carbon–Metal Bond*, Vol. 4, Wiley, NY, pp. 187–193; Heaney, H. *Organomet. Chem. Rev.* **1966**, *1*, 27. For a review of di-Grignard reagents, see Bickelhaupt, F. *Angew. Chem. Int. Ed.* **1987**, *26*, 990.

[583]For example, see Denise, B.; Ducom, J.; Fauvarque, J. *Bull. Soc. Chim. Fr.* **1972**, 990; Seetz, J.W.F.L.; Hartog, F.A.; Böhm, H.P.; Blomberg, C.; Akkerman, O.S.; Bickelhaupt, F. *Tetrahedron Lett.* **1982**, *23*, 1497.

[584]For formation of 1,2-dilithio compounds and 1,2-di-Grignard reagents, but not by this method, see van Eikkema Hommes, N.J.R.; Bickelhaupt, F.; Klumpp, G.W. *Recl. Trav. Chim. Pays-Bas* **1988**, *107*, 393; *Angew. Chem. Int. Ed.* **1988**, *27*, 1083.

[585]For example, see Bertini, F.; Grasselli, P.; Zubiani, G.; Cainelli, G. *Tetrahedron* **1970**, *26*, 1281; Bruin, J.W.; Schat, G.; Akkerman, O.S.; Bickelhaupt, F. *J. Organomet. Chem.* **1985**, *288*, 13. For the synthesis of *gem*-dilithio and 1,1,1-trilithio compounds, see Baran, Jr., J.R.; Lagow, R. *J. Am. Chem. Soc.* **1990**, *112*, 9415.

[586]For a review of compounds containing both carbon–halogen and carbon-metal bonds, see Chivers, T. *Organomet. Chem. Rev. Sect. A* **1970**, *6*, 1.

[587]Yu, S.H.; Ashby, E.C. *J. Org. Chem.* **1971**, *36*, 2123.

[588]Sugimoto, O.; Yamada, S.; Tanji, K. *J. Org. Chem.* **2003**, *68*, 2054.

magnesium (see **17-24**), and Grignard reagents from α-halo ethers[589] can only be formed in THF or dimethoxymethane at a low temperature, for example,[590]

$$\text{EtOCH}_2\text{Cl} + \text{Mg} \xrightarrow[-30^\circ\text{C}]{\text{THF or CH}_2(\text{OMe})_2} \text{EtOCH}_2\text{MgCl}$$

because such reagents immediately undergo a elimination (see **12-39**) at room temperature in ether solution.

Because Grignard reagents react with water (**12-24**) and with oxygen (**12-25**), it is generally best to prepare them in an anhydrous nitrogen atmosphere. Grignard reagents are generally neither isolated nor stored; solutions of Grignard reagents are used directly for the required synthesis. Grignard reagents can also be prepared in benzene or toluene, if a tertiary amine is added to complex with the RMgX.[591] This method eliminates the need for an ether solvent. With certain primary alkyl halides it is even possible to prepare alkylmagnesium compounds in hydrocarbon solvents in the absence of an organic base.[592] It is also possible to obtain Grignard reagents in powdered form, by complexing them with the chelating agent tris(3,6-dioxaheptyl)amine, $N(\text{CH}_2\text{CH}_2\text{OCH}_2\text{CH}_2\text{OCH}_3)_3$.[593]

Next to the formation of Grignard reagents, the most important application of this reaction is the conversion of alkyl and aryl halides to organolithium compounds,[594] but it has also been carried out with many other metals (e.g., Na, Be, Zn, Hg, As, Sb, and Sn). With sodium, the Wurtz reaction (**10-56**) is an important side reaction. In some cases where the reaction between a halide and a metal is too slow, an alloy of the metal with potassium or sodium can be used instead. The most important example is the preparation of tetraethyl lead from ethyl bromide and a Pb–Na alloy.

The efficiency of the reaction can often be improved by use of the metal in its powdered[595] or vapor[596] form. These techniques have permitted the preparation of some organometallic compounds that cannot be prepared by the standard

---

[589]For a review of organometallic compounds containing a hetero atom (N, O, P, S, or Si), see Peterson, D.J. *Organomet. Chem. Rev. Sect. A* **1972**, *7*, 295.

[590]For example, see Normant, H.; Castro, B. *C. R. Acad. Sci.* **1963**, *257*, 2115; **1964**, *259*, 830; Castro, B. *Bull. Soc. Chim. Fr.* **1967**, 1533, 1540, 1547; Taeger, E.; Kahlert, E.; Walter, H. *J. Prakt. Chem.* **1965**, *[4] 28*, 13.

[591]Ashby, E.C.; Reed, R. *J. Org. Chem.* **1966**, *31*, 971; Gitlitz, M.H.; Considine, W.J. *J. Organomet. Chem.* **1970**, *23*, 291.

[592]Smith Jr., W.N. *J. Organomet. Chem.* **1974**, *64*, 25.

[593]Boudin, A.; Cerveau, G.; Chuit, C.; Corriu, R.J.P.; Reye, C. *Tetrahedron* **1989**, *45*, 171.

[594]For reviews, see Wakefield, B.J. *Organolithium Methods*, Academic Press, NY, **1988**, pp. 21–32; Wardell, J.L., in Hartley, F.R.; Patai, S. Vol. 4, pp. 1–157, 5–27; Newcomb, M.E., in Zuckerman, J.J. *Inorganic Reactions and Methods*, Vol. 11, VCH, NY, **1988**, pp. 3–14.

[595]For a review, see Rieke, R.D. *Science* **1989**, *246*, 1260.

[596]For reviews, see Klabunde, K.J. *React. Intermed. (Plenum)* **1980**, *1*, 37; *Acc. Chem. Res.*; **1975**, *8*, 393; Skell, P.S. Havel, J.J.; McGlinchey, M.J. *Acc. Chem. Res.* **1973**, *6*, 97; Timms, P.L. *Adv. Inorg. Radiochem.* **1972**, *14*, 121.

procedures. Among the metals produced in an activated form are Mg,[597] Ca,[598] Zn,[599] Al, Sn, Cd,[600] Ni, Fe, Ti, Cu,[601] Pd, and Pt.[602]

The mechanism of Grignard reagent formation involves free radicals,[603] and there is much evidence for this, from CIDNP[604] (p. 269) and from stereochemical, rate, and product studies.[605] Further evidence is that free radicals have been trapped,[606] and that experiments that studied the intrinsic reactivity of MeBr on a magnesium single-crystal surface showed that Grignard reagent formation does not take place by a single-step insertion mechanism.[607] The following SET mechanism has been proposed:[604]

$$R-X + \overline{Mg} \longrightarrow R-X^{\cdot -} + Mg_s^{\cdot +}$$

$$R-X^{\cdot -} \longrightarrow R^{\cdot} + X^{-}$$

$$X^{-} + Mg_s^{\cdot +} \longrightarrow XMg_s^{\cdot}$$

$$R^{\cdot} + XMg_s^{\cdot} \longrightarrow RMgX$$

Other evidence has been offered to support a SET-initiated radical process for the second step of this mechanism.[608] The species $R-X^{\cdot -}$ and $Mg^{\cdot +}$ are radical ions.[609] The subscript "s" is meant to indicate that the species so marked are bound to the surface of the magnesium. It is known that this is a surface reaction.[610] It has been suggested that some of the $R^{\cdot}$ radicals diffuse from the magnesium surface into the solution and then return to the surface to react with the $XMg^{\cdot}$. There is evidence

[597]Ebert, G.W.; Rieke, R.D. *J. Org. Chem.* **1988**, *53*, 4482. See also, Baker, K.V.; Brown, J.M.; Hughes, N.; Skarnulis, A.J.; Sexton, A. *J. Org. Chem.* **1991**, *56*, 698.

[598]Wu, T.; Xiong, H.; Rieke, R.D. *J. Org. Chem.* **1990**, *55*, 5045.

[599]Rieke, R.D.; Li, P.T.; Burns, T.P.; Uhm, S.T. *J. Org. Chem.* **1981**, *46*, 4323. See also, Grondin, J.; Sebban, M.; Vottero, G.P.; Blancou, H.; Commeyras, A. *J. Organomet. Chem.* **1989**, *362*, 237; Berk, S.C.; Yeh, M.C.P.; Jeong, N.; Knochel, P. *Organometallics* **1990**, *9*, 3053; Zhu, L.; Wehmeyer, R.M.; Rieke, R.D. *J. Org. Chem.* **1991**, *56*, 1445.

[600]Burkhardt, E.R.; Rieke, R.D. *J. Org. Chem.* **1985**, *50*, 416.

[601]Stack, D.E.; Dawson, B.T.; Rieke, R.D. *J. Am. Chem. Soc.* **1991**, *113*, 4672, and references cited therein.

[602]For reviews, see Lai, Y. *Synthesis* **1981**, 585; Rieke, R.D. *Acc. Chem. Res.* **1977**, *10*, 301; *Top. Curr. Chem.* **1975**, *59*, 1.

[603]For a review, see Blomberg, C. *Bull. Soc. Chim. Fr.* **1972**, 2143.

[604]Bodewitz, H.W.H.J.; Blomberg, C.; Bickelhaupt, F. *Tetrahedron Lett.* **1975**, 2003; *Tetrahedron* **1975**, *31*, 1053. See also, Lawler, R.G.; Livant, P. *J. Am. Chem. Soc.* **1976**, *98*, 3710; Schaart, B.J.; Blomberg, C.; Akkerman, O.S.; Bickelhaupt, F. *Can. J. Chem.* **1980**, *58*, 932.

[605]See, for example, Walborsky, H.M.; Aronoff, M.S. *J. Organomet. Chem.* **1973**, *51*, 31; Czernecki, S.; Georgoulis, C.; Gross, B.; Prevost, C. *Bull. Soc. Chim. Fr.* **1968**, 3720; Rogers, H.R.; Hill, C.L.; Fujiwara, Y.; Rogers, R.J.; Mitchell, H.L.; Whitesides, G.M. *J. Am. Chem. Soc.* **1980**, *102*, 217; Barber, J.J.; Whitesides, G.M. *J. Am. Chem. Soc.* **1980**, *102*, 239.

[606]Root, K.S.; Hill, C.L.; Lawrence, L.M.; Whitesides, G.M. *J. Am. Chem. Soc.* **1989**, *111*, 5405.

[607]Nuzzo, R.G.; Dubois, L.H. *J. Am. Chem. Soc.* **1986**, *108*, 2881.

[608]Hoffmann, R. W.; Brönstrup, M.; Müller, M. *Org. Lett.* **2003**, *5*, 313.

[609]For additional evidence for this mechanism, see Vogler, E.A.; Stein, R.L.; Hayes, J.M. *J. Am. Chem. Soc.* **1978**, *100*, 3163; Sergeev, G.B.; Zagorsky, V.V.; Badaev, F.Z. *J. Organomet. Chem.* **1983**, *243*, 123. However, there is evidence that the mechanism may be more complicated: de Souza-Barboza, J.C.; Luche, J.; Pétrier, C. *Tetrahedron Lett.* **1987**, *28*, 2013.

[610]Walborsky, H.M.; Topolski, M. *J. Am. Chem. Soc.* **1992**, *114*, 3455; Walborsky, H.M.; Zimmermann, C. *J. Am. Chem. Soc.* **1992**, *114*, 4996; Walborsky, H.M. *Accts. Chem. Res.* **1990**, *23*, 286.

both for[611] and against[612] this suggestion. Another proposal is that the fourth step is not the one shown here, but that the R• is reduced by Mg^+ to the carbanion R^-, which combines with MgX^+ to give $RMgX$.[613]

There are too many preparations of Grignard reagents in *Organic Syntheses* for us to list here. Chiral Grignard reagents are rare, since they are configurationally unstable in most cases. However, a few chiral Grignard reagents are known.[614] Use of the reaction to prepare other organometallic compounds can be found in OS **I**, 228; **II**, 184, 517, 607; **III**, 413, 757; **VI**, 240; **VII**, 346; **VIII**, 505. The preparation of unsolvated butylmagnesium bromide is described at OS **V**, 1141. The preparation of highly reactive (powdered) magnesium is given at OS **VI**, 845.

12-39   Replacement of a Halogen by a Metal from an Organometallic Compound

Metalo-de-halogenation

$$RX + R'M \longrightarrow RM + R'X$$

The exchange reaction between halides and organometallic compounds occurs most readily when M is lithium and X is bromide or iodide,[615] although it has been shown to occur with magnesium.[616] The R' group is usually, although not always, alkyl, and often butyl; R is usually aromatic.[617] Alkyl halides are generally not reactive enough, while allylic and benzylic halides usually give Wurtz coupling. Of course, the R that becomes bonded to the halogen is the one for which RH is the weaker acid. Despite the preponderance of reactions with bromides and iodides, it is noted that the reaction of 1-fluorooctane with 4–10 equivalents of lithium powder and 2–4 equivalents of DTBB (4,4′-di-*tert*-butylbiphenyl) in THP at 0°C for 5 min, was shown to give a solution of the corresponding 1-octyllithium.[618] Vinylic halides react with retention of configuration.[619] The

[611]Garst, J.F.; Deutch, J.E.; Whitesides, G.M. *J. Am. Chem. Soc.* **1986**, *108*, 2490; Ashby, E.C.; Oswald, J. *J. Org. Chem.* **1988**, *53*, 6068; Garst, J.F. *Acc. Chem. Res.* **1991**, *24*, 95; Garst, J.F.; Ungváry, F.; Batlaw, R.; Lawrence, K.E. *J. Am. Chem. Soc.* **1991**, *113*, 5392.

[612]Walborsky, H.M.; Rachon, J. *J. Am. Chem. Soc.* **1989**, *111*, 1896; Rachon, J.; Walborsky, H.M. *Tetrahedron Lett.* **1989**, *30*, 7345; Walborsky, H.M. *Acc. Chem. Res.* **1990**, *23*, 286.

[613]de Boer, H.J.R.; Akkerman, O.S.; Bickelhaupt, F. *Angew. Chem. Int. Ed.* **1988**, *27*, 687.

[614]See Hölzer, B.; Hoffmann, R.W. *Chem. Commun.* **2003**, 732; Walborsky, H.M.; Impastato, F.J.; Young, A.E. *J. Am. Chem. Soc.* **1964**, *86*, 3283; Tanaka, M.; Ogata, I. *Bull. Chem. Soc. Jpn.* **1975**, *48*, 1094; Schumann, H.; Wassermann, B.C.; Hahn, F.E. *Organometallics* **1992**, *11*, 2803; Dakternieks, D.; Dunn, K.; Henry, D.J.; Schiesser, C.H.; Tiekink, E.R. *Organometallics* **1999**, *18*, 3342.

[615]For reviews, see Wardell, J.L., in Zuckerman, J.J. *Inorganic Reactions and Methods*, Vol. 11, VCH, NY, **1988**, pp. 107–129; Parham, W.E.; Bradsher, C.K. *Acc. Chem. Res.* **1982**, *15*, 300.

[616]See, for example, Zakharkin, L.I.; Okhlobystin, O.Yu.; Bilevitch, K.A. *J. Organomet. Chem.* **1964**, *2*, 309; Tamborski, C.; Moore, G.J. *J. Organomet. Chem.* **1971**, *26*, 153.

[617]For the preparation of primary alkyllithiums by this reaction, see Bailey, W.F.; Punzalan, E.R. *J. Org. Chem.* **1990**, *55*, 5404; Negishi, E.; Swanson, D.R.; Rousset, C.J. *J. Org. Chem.* **1990**, *55*, 5406.

[618]Yus, M.; Herrera, R.P.; Guijarro, A. *Tetrahedron Lett.*, **2003**, *44*, 5025.

[619]For examples of exchange where R = vinylic, see Neumann, H.; Seebach, D. *Chem. Ber.* **1978**, *111*, 2785; Miller, R.B.; McGarvey, G. *Synth. Commun.* **1979**, *9*, 831; Sugita, T.; Sakabe, Y.; Sasahara, T.; Tsukuda, M.; Ichikawa, K. *Bull. Chem. Soc. Jpn.* **1984**, *57*, 2319.

reaction can be used to prepare α-halo organolithium and α-halo organomagnesium compounds,[620] for example,[621]

$$CCl_4 + BuLi \xrightarrow[-105°C]{THF} Cl_3C-Li$$

Such compounds can also be prepared by hydrogen–metal exchange, for example,[622]

$$Br_3CH + iPrMgCl \xrightarrow[-95°C]{THF-HMPA} Br_3C-MgCl + C_3H_8$$

This is an example of **12-22**. However, these α-halo organometallic compounds are stable (and configurationally stable as well[623]) only at low temperatures (ca. −100°C) and only in THF or mixtures of THF and other solvents (e.g., HMPA). At ordinary temperatures they lose MX (α elimination) to give carbenes (which then react further) or carbenoid reactions. The α-chloro-α-magnesio sulfones $ArSO_2CH(Cl)MgBr$ are exceptions, being stable in solution at room temperature and even under reflux.[624] Compounds in which a halogen and a transition metal are on the same carbon can be more stable than the ones with lithium.[625]

There is evidence that the mechanism[626] of the reaction of alkyllithium compounds with alkyl and aryl iodides involves free radicals.[627]

$$RX + R'M \rightleftharpoons \frac{[R\cdot, X, M, R'\cdot]}{\text{Solvent cage}} \rightleftharpoons RM + R'X$$

Among the evidence is the fact that coupling and disproportionation products are obtained from R• and R'• and the observation of CIDNP.[627,628] However, in the degenerate exchange between PhI and PhLi the ate complex $Ph_2I^- Li^+$ has been

[620]For reviews of such compounds, see Siegel, H. *Top. Curr. Chem.* **1982**, *106*, 55; Negishi, E. *Organometallics in Organic Synthesis*, Wiley, NY, **1980**, pp. 136–151; Köbrich, G. *Angew. Chem. Int. Ed.* **1972**, *11*, 473; **1967**, *6*, 41; *Bull. Soc. Chim. Fr.* **1969**, 2712; Villieras, J. *Organomet. Chem. Rev. Sect. A* **1971**, *7*, 81. For related reviews, see Krief, A. *Tetrahedron* **1980**, *36*, 2531; Normant, H. *J. Organomet. Chem.* **1975**, *100*, 189; Zhil'tsov, S.F.; Druzhkov, O.N. *Russ. Chem. Rev.* **1971**, *40*, 126.

[621]Hoeg, D.F.; Lusk, D.I.; Crumbliss, A.L. *J. Am. Chem. Soc.* **1965**, *87*, 4147. See also, Villieras, J.; Tarhouni, R.; Kirschleger, B.; Rambaud, M. *Bull. Soc. Chim. Fr.* **1985**, 825.

[622]Villieras, J. *Bull. Soc. Chim. Fr.* **1967**, 1520.

[623]Schmidt, A.; Köbrich, G.; Hoffmann, R.W. *Chem. Ber.* **1991**, *124*, 1253; Hoffmann, R.W.; Bewersdorf, M. *Chem. Ber.* **1991**, *124*, 1259.

[624]Stetter, H.; Steinbeck, K. *Liebigs Ann. Chem.* **1972**, *766*, 89.

[625]Kauffmann, T.; Fobker, R.; Wensing, M. *Angew. Chem. Int. Ed.* **1988**, *27*, 943.

[626]For reviews of the mechanism, see Bailey, W.F.; Patricia, J.J. *J. Organomet. Chem.* **1988**, *352*, 1; Beletskaya, I.P.; Artamkina, G.A.; Reutov, O.A. *Russ. Chem. Rev.* **1976**, *45*, 330.

[627]Ward, H.R.; Lawler, R.G.; Cooper, R.A. *J. Am. Chem. Soc.* **1969**, *91*, 746; Lepley, A.R.; Landau, R.L. *J. Am. Chem. Soc.* **1969**, *91*, 748; Ashby, E.C.; Pham, T.N. *J. Org. Chem.* **1987**, *52*, 1291. See also, Bailey, W.F.; Patricia, J.J.; Nurmi, T.T.; Wang, W. *Tetrahedron Lett.* **1986**, *27*, 1861.

[628]Ward, H.R.; Lawler, R.G.; Loken, H.Y. *J. Am. Chem. Soc.* **1968**, *90*, 7359.

shown to be an intermediate,[629] and there is other evidence that radicals are not involved in all instances of this reaction.[630]

In a completely different kind of process, alkyl halides can be converted to certain organometallic compounds by treatment with organometalate ions, for example,

$$RX + R'_3SnLi \longrightarrow RSnR'_3 + LiX$$

Most of the evidence is in accord with a free-radical mechanism involving electron transfer, although an S_N2 mechanism can compete under some conditions.[631]

OS **VI**, 82; **VII**, 271, 326, 495; **VIII**, 430. See also, OS **VII**, 512; **VIII**, 479.

CARBON LEAVING GROUPS

In these reactions (**12-40–12-48**), a carbon–carbon bond cleaves. We regard as the substrate the side that retains the electron pair; hence the reactions are considered electrophilic substitutions. The incoming group is hydrogen in all but one (**12-42**) of the cases. The reactions in groups A and B are sometimes called *anionic cleavages*,[632] although they do not always occur by mechanisms involving free carbanions (S_E1). When they do, the reactions are facilitated by increasing stability of the carbanion.

A. Carbonyl-Forming Cleavages

These reactions follow the pattern

The leaving group is stabilized because the electron deficiency at its carbon is satisfied by a pair of electrons from the oxygen. With respect to the leaving group the reaction is elimination to form a C=O bond. Retrograde aldol reactions (**16-34**) and cleavage of cyanohydrins (**16-52**) belong to this classification but are treated in Chapter 16 under their more important reverse reactions. Other eliminations to form C=O bonds are discussed in Chapter 17 (**17-32**).

12-40   Decarboxylation of Aliphatic Acids

Hydro-de-carboxylation

$$RCOOH \longrightarrow RH + CO_2$$

[629]See Farnham, W.B.; Calabrese, J.C. *J. Am. Chem. Soc.* **1986**, *108*, 2449; Reich, H.J.; Green, D.P.; Phillips, N.H. *J. Am. Chem. Soc.* **1989**, *111*, 3444.

[630]Rogers, H.R.; Houk, J. *J. Am. Chem. Soc.* **1982**, *104*, 522; Beak, P.; Allen, D.J.; Lee, W.K. *J. Am. Chem. Soc.* **1990**, *112*, 1629.

[631]See San Filippo, Jr., J.; Silbermann, J. *J. Am. Chem. Soc.* **1982**, *104*, 2831; Ashby, E.C.; Su, W.; Pham, T.N. *Organometallics* **1985**, *4*, 1493; Alnajjar, M.S.; Kuivila, H.G. *J. Am. Chem. Soc.* **1985**, *107*, 416.

[632]For a review, see Artamkina, G.A.; Beletskaya, I.P. *Russ. Chem. Rev.* **1987**, *56*, 983.

TABLE 12.2. Some Acids that Undergo Decarboxylation Fairly Readily[a]

Acid Type		Decarboxylation Product
Malonic	$HOOC-\overset{\diagup}{\underset{\diagdown}{C}}-COOH$	$HOOC-\overset{\diagup}{\underset{\diagdown}{C}}-H$
α-Cyano	$HOOC-\overset{\diagup}{\underset{\diagdown}{C}}-CN$	$H-\overset{\diagup}{\underset{\diagdown}{C}}-CN$ or $HOOC-\overset{\diagup}{\underset{\diagdown}{C}}-H$
α-Nitro	$HOOC-\overset{\diagup}{\underset{\diagdown}{C}}-NO_2$	$O_2N-\overset{\diagup}{\underset{\diagdown}{C}}-H$
α-Aryl	$HOOC-\overset{\diagup}{\underset{\diagdown}{C}}-Ar$	$Ar-\overset{\diagup}{\underset{\diagdown}{C}}-H$
α,α,α-Trihalo	$X_3C-COOH$	X_3C-H
β-Keto	$\overset{\diagup}{\underset{\diagdown}{C}}-\overset{O}{\overset{\|}{C}}-\overset{\diagup}{\underset{\diagdown}{C}}-COOH$	$\overset{\diagup}{\underset{\diagdown}{C}}-\overset{O}{\overset{\|}{C}}-\overset{\diagup}{\underset{\diagdown}{C}}-H$
β,γ-Unsaturated	$\overset{\diagup}{\underset{\diagdown}{C}}=\overset{\|}{C}-\overset{\diagup}{\underset{\diagdown}{C}}-COOH$	$\overset{\diagup}{\underset{\diagdown}{C}}=\overset{\|}{C}-\overset{\diagup}{\underset{\diagdown}{C}}-H$

[a]Others are described in the text.

Many carboxylic acids can be successfully decarboxylated, either as the free acid or in the salt form, but not simple fatty acids.[633] An exception is acetic acid, which as the acetate, heated with base, gives good yields of methane. Malonic acid derivatives are the most common substrates for decarboxylation, giving the corresponding monocarboxylic acid. Decarboxylation of 2-substituted malonic acids has been reported using microwave irradiation.[634] Aliphatic acids that do undergo successful decarboxylation have certain functional groups or double or triple bonds in the α or β position. Some of these are shown in Table 12.2. For decarboxylation of aromatic acids, see **11-35**. Decarboxylation of an α-cyano acid can give a nitrile or a carboxylic acid, since the cyano group may or may not be hydrolyzed in the course of the reaction. In addition to the compounds listed in Table 12.2, decarboxylation can also be carried out on α,β-unsaturated and α,β-acetylenic acids. α,β-Unsaturated acids can also be decarboxylated[635] with copper

[633]March, J. *J. Chem. Educ.* **1963**, *40*, 212.
[634]Zara, C.L.; Jin, T.; Giguere, R.J. *Synth. Commun.* **2000**, *30*, 2099.
[635]For an example involving the conversion of C=C-COOH to C=C-Br with LiBr and ceric ammonium nitrate in aqueous acetonitrile, see Roy, S.C.; Guin, C.; Maiti, G. *Tetrahedron Lett.* **2001**, *42*, 9253.

and quinoline in a manner similar to that discussed in **11-35**. Glycidic acids give aldehydes on decarboxylation. The following mechanism has been suggested:[636]

The direct product is an enol that tautomerizes to the aldehyde.[637] This is the usual last step in the Darzens reaction (**16-40**).

Decarboxylations can be regarded as reversals of the addition of carbanions to carbon dioxide (**16-82**), but free carbanions are not always involved.[638] When the carboxylate *ion* is decarboxylated, the mechanism can be either S_E1 or S_E2. In the case of the S_E1 mechanism, the reaction is of course aided by the presence of electron-withdrawing groups, which stabilize the carbanion.[639] Decarboxylations of carboxylate ions can be accelerated by the addition of a suitable crown ether, which in effect removes the metallic ion.[640] The reaction without the metallic ion has also been performed in the gas phase.[641] But some acids can also be decarboxylated directly and, in most of these cases, there is a cyclic, six-center mechanism:

$$+ \ O{=}C{=}O$$

Here too there is an enol that tautomerizes to the product. The mechanism is illustrated for the case of β-keto acids,[642] but it is likely that malonic acids, α-cyano acids, α-nitro acids, and β,γ-unsaturated acids[643] behave similarly,

[636]Singh, S.P.; Kagan, J. *J. Org. Chem.* **1970**, *35*, 2203.

[637]Shiner, Jr., V.J.; Martin, B. *J. Am. Chem. Soc.* **1962**, *84*, 4824.

[638]For reviews of the mechanism, see Richardson, W.H.; O'Neal, H.E., in Bamford, C.H.; Tipper, C.F.H. *Comprehensive Chemical Kinetics*, Vol. 5, Elsevier, NY, **1972**, pp. 447–482; Clark, L.W., in Patai, S. *The Chemistry of Carboxylic Acids and Esters*; Wiley, NY, **1969**, pp. 589–622. For a review of carbon isotope effect studies, see Dunn, G.E. *Isot. Org. Chem.* **1977**, *3*, 1.

[639]See, for example, Oae, S.; Tagaki, W.; Uneyama, K.; Minamida, I. *Tetrahedron* **1968**, *24*, 5283; Buncel, E.; Venkatachalam, T.K.; Menon, B.C. *J. Org. Chem.* **1984**, *49*, 413.

[640]Hunter, D.H.; Patel, V.; Perry, R.A. *Can. J. Chem.* **1980**, *58*, 2271, and references cited therein.

[641]Graul, S.T.; Squires, R.R. *J. Am. Chem. Soc.* **1988**, *110*, 607.

[642]For a review of the mechanism of the decarboxylation of β-keto acids, see Jencks, W.P. *Catalysis in Chemistry and Enzmology*; McGraw-Hill, NY, **1969**, pp. 116–120.

[643]Bigley, D.B.; Clarke, M.J. *J. Chem. Soc. Perkin Trans. 2* **1982**, 1, and references cited therein. For a review, see Smith, G.G.; Kelly, F.W. *Prog. Phys. Org. Chem.* **1971**, *8*, 75, pp. 150–153.

since similar six-membered transition states can be written for them. Some α,β-unsaturated acids are also decarboxylated by this mechanism by isomerizing to the β,γ-isomers before they

49

actually decarboxylate.[644] Evidence is that **49** and similar bicyclic β-keto acids resist decarboxylation.[645] In such compounds, the six-membered cyclic transition state cannot form for steric reasons, and if it could, formation of the intermediate enol would violate Bredt's rule (p. 229).[646] Some carboxylic acids that cannot form a six-membered transition state can still be decarboxylated, and these presumably react through an S_E1 or S_E2 mechanism.[647] Further evidence for the cyclic mechanism is that the reaction rate varies very little with a change from a nonpolar to a polar solvent (even from benzene to water[648]), and is not subject to acid catalysis.[649] The rate of decarboxylation of a β,γ-unsaturated acid was increased $\sim10^5$-10^6 times by introduction of a β-methoxy group, indicating that the cyclic transition state has dipolar character.[650]

β-Keto acids[651] are easily decarboxylated, but such acids are usually prepared from β-keto esters, and the esters are easily decarboxylated themselves on hydrolysis without isolation of the acids.[652] This decarboxylation of β-keto esters

[644]Bigley, D.B. *J. Chem. Soc.* **1964**, 3897.

[645]Wasserman, H.H., in Newman *Steric Effects in Organic Chemistry*, Wiley, NY, **1956**, p. 352. See also, Buchanan, G.L.; Kean, N.B.; Taylor, R. *Tetrahedron* **1975**, *31*, 1583.

[646]Sterically hindered β-keto acids decarboxylate more slowly: Meier, H.; Wengenroth, H.; Lauer, W.; Krause, V. *Tetrahedron Lett.* **1989**, *30*, 5253.

[647]For example, see Ferris, J.P.; Miller, N.C. *J. Am. Chem. Soc.* **1966**, *88*, 3522.

[648]Westheimer, F.H.; Jones, W.A. *J. Am. Chem. Soc.* **1941**, *63*, 3283; Swain, C.G.; Bader, R.F.W.; Esteve Jr., R.M.; Griffin, R.N. *J. Am. Chem. Soc.* **1961**, *83*, 1951.

[649]Pedersen, K.J. *Acta Chem. Scand.* **1961**, *15*, 1718; Noyce, D.S.; Metesich, M.A. *J. Org. Chem.* **1967**, *32*, 3243.

[650]Bigley, D.B.; Al-Borno, A. *J. Chem. Soc. Perkin Trans. 2* **1982**, 15.

[651]For a review of β-keto acids, see Oshry, L.; Rosenfeld, S.M. *Org. Prep. Proced. Int.* **1982**, *14*, 249.

[652]For a list examples, with references, see Larock, R.C. *Comprehensive Organic Transformations*, 2nd ed., Wiley-VCH, NY, **1999**, pp. 1542–1543. For an example of decarboxylation of the β-keto ester with Cp2TiCl2 and *i*-PrMgBr, followed by treatment with 2*N* HCl, see Yu, Y.; Zhang, Y. *Synth. Commun.* **1999**, *29*, 243.

involving cleavage on the carboxyl side of the substituted methylene group (arrow) is carried out under acidic, neutral, or slightly basic conditions to yield a ketone. When strongly basic conditions are used, cleavage occurs on the other side of the CR_2 group (**12-43**). β-Keto esters can be decarbalkoxylated without passing through the free-acid stage by treatment with boric anhydride (B_2O_3) at 150°C.[653] The alkyl portion of the ester (R') is converted to an alkene or, if it lacks a β hydrogen, to an ether R'OR'. Another method for the decarbalkoxylation of β-keto esters, malonic esters, and α-cyano esters consists of heating the substrate in wet DMSO containing NaCl, Na_3PO_4, or some other simple salt.[654] In this method too, the free acid is probably not an intermediate, but here the alkyl portion of the substrate is converted to the corresponding alcohol. Ordinary carboxylic acids, containing no activating groups, can be decarboxylated by conversion to esters of *N*-hydroxypyridine-2-thione and treatment of these with Bu_3SnH.[655] A free-radical mechanism is likely. α-Amino acids have been decarboxylated by treatment with a catalytic amount of 2-cyclohexenone.[656] Amino acids are decarboxylated by sequential treatment with NBS at pH 5 followed by $NaBH_4$ and $NiCl_2$.[657] Certain decarboxylations can also be accomplished photochemically.[658] See also, the decarbonylation of acyl halides, mentioned in **14-32**. In some cases, decarboxylations can give organometallic compounds: $RCOOM \rightarrow RM + CO_2$.[659]

Some of the decarboxylations listed in *Organic Syntheses* are performed with concomitant ester or nitrile hydrolysis and others are simple decarboxylations.

With ester or nitrile hydrolysis: OS **I**, 290, 451, 523; **II**, 200, 391; **III**, 281, 286, 313, 326, 510, 513, 591; **IV**, 55, 93, 176, 441, 664, 708, 790, 804; **V**, 76, 288, 572, 687, 989; **VI**, 615, 781, 873, 932; **VII**, 50, 210, 319; **VIII**, 263.

Simple decarboxylations: OS **I**, 351, 401, 440, 473, 475; **II**, 21, 61, 93, 229, 302, 333, 368, 416, 474, 512, 523; **III**, 213, 425, 495, 705, 733, 783; **IV**, 234, 254, 278, 337, 555, 560, 597, 630, 731, 857; **V**, 251, 585; **VI**, 271, 965; **VII**, 249, 359; **VIII**, 235, 444, 536; **75**, 195. Also see, OS **IV**, 633.

---

[653]Lalancette, J.M.; Lachance, A. *Tetrahedron Lett.* **1970**, 3903.

[654]For a review of the synthetic applications of this method, see Krapcho, A.P. *Synthesis* **1982**, 805, 893. For other methods, see Aneja, R.; Hollis, W.M.; Davies, A.P.; Eaton, G. *Tetrahedron Lett.* **1983**, *24*, 4641; Brown, R.T.; Jones, M.F. *J. Chem. Res. (S)* **1984**, 332; Dehmlow, E.V.; Kunesch, E. *Synthesis* **1985**, 320; Taber, D.F.; Amedio, Jr., J.C.; Gulino, F. *J. Org. Chem.* **1989**, *54*, 3474.

[655]Barton, D.H.R.; Crich, D.; Motherwell, W.B. *Tetrahedron* **1985**, *41*, 3901; Della, E.W.; Tsanaktsidis, J. *Aust. J. Chem.* **1987**, *39*, 2061. For another method of more limited scope, see Maier, W.F.; Roth, W.; Thies, I.; Schleyer, P.v.R. *Chem. Ber.* **1982**, *115*, 808.

[656]Hashimoto, M.; Eda, Y.; Osanai, Y.; Iwai, T.; Aoki, S. *Chem. Lett.* **1986**, 893.

[657]Laval, G.; Golding, B.T. *Synlett* **2003**, 542.

[658]See Davidson, R.S.; Steiner, P.R. *J. Chem. Soc. Perkin Trans. 2* **1972**, 1357; Kraeutler, B.; Bard, A.J. *J. Am. Chem. Soc.* **1978**, *100*, 5985; Hasebe, M.; Tsuchiya, T. *Tetrahedron Lett.* **1987**, *28*, 6207; Okada, K.; Okubo, K.; Oda, M. *Tetrahedron Lett.* **1989**, *30*, 6733.

[659]For reviews, see Deacon, G.B. *Organomet. Chem. Rev. A* **1970**, 355; Deacon, G.B.; Faulks, S.J.; Pain, G.N. *Adv. Organomet. Chem.* **1986**, *25*, 237.

12-41   Cleavage of Alkoxides

Hydro-de-(α-oxidoalkyl)-substitution

$$
\overset{R^1}{\underset{R^2}{R-C-O^{\ominus}}} \quad \xrightarrow[\text{2. HA}]{\text{1. } \Delta} \quad R-H \; + \; \overset{R^1}{\underset{R^2}{C=O}}
$$

Alkoxides of tertiary alcohols can be cleaved in a reaction that is essentially the reverse of addition of carbanions to ketones (**16-24**).[660] The reaction is unsuccessful when the R groups are simple unbranched alkyl groups, for example, the alkoxide of triethylcarbinol. Cleavage is accomplished with branched alkoxides, such as the alkoxides of diisopropylneopentylcarbinol or tri-*tert*-butylcarbinol.[661] Allylic,[662] benzylic,[663] and aryl groups also cleave; for example, the alkoxide of triphenylcarbinol gives benzene and benzophenone. Studies in the gas phase show that the cleavage is a simple one, giving the carbanion and ketone directly in one step.[664] However, with some substrates in solution, substantial amounts of dimer R–R have been found, indicating a radical pathway.[665] Hindered alcohols (not the alkoxides) also lose one R group by cleavage, also by a radical pathway.[666]

The reaction has been used for extensive mechanistic studies (see p. 758).

OS **VI**, 268.

12-42   Replacement of a Carboxyl Group by an Acyl Group

Acyl-de-carboxylation

$$
\overset{NH_2}{\underset{R}{\overset{|}{\diagdown}}COOH} \; + \; \overset{O}{\underset{R'}{\overset{\|}{\diagup}}}O\overset{O}{\underset{R'}{\overset{\|}{\diagdown}}} \xrightarrow{\text{pyridine}} \quad R'\overset{O}{\overset{\|}{C}}\underset{\underset{R}{\overset{|}{C}}\underset{\overset{\|}{O}}{R'}}{N}\overset{H}{} \; + \; CO_2
$$

[660]Zook, H.D.; March, J.; Smith, D.F. *J. Am. Chem. Soc.* **1959**, *81*, 1617; Barbot, F.; Miginiac, P. *J. Organomet. Chem.* **1977**, *132*, 445; Benkeser, R.A.; Siklosi, M.P.; Mozdzen, E.C. *J. Am. Chem. Soc.* **1978**, *100*, 2134.

[661]Arnett, E.M.; Small, L.E.; McIver Jr., R.T.; Miller, J.S. *J. Org. Chem.* **1978**, *43*, 815. See also Lomas, J.S.; Dubois, J.E. *J. Org. Chem.* **1984**, *49*, 2067.

[662]See Snowden, R.L.; Linder, S.M.; Muller, B.L.; Schulte-Elte, K.H. *Helv. Chim. Acta* **1987**, *70*, 1858, 1879.

[663]Partington, S.M.; Watt, C.I.F. *J. Chem. Soc. Perkin Trans. 2* **1988**, 983.

[664]Tumas, W.; Foster, R.F.; Brauman, J.I. *J. Am. Chem. Soc.* **1988**, *110*, 2714; Ibrahim, S.; Watt, C.I.F.; Wilson, J.M.; Moore, C. *J. Chem. Soc. Chem. Commun.* **1989**, 161.

[665]Paquette, L.A.; Gilday, J.P.; Maynard, G.D. *J. Org. Chem.* **1989**, *54*, 5044; Paquette, L.A.; Maynard, G.D. *J. Org. Chem.* **1989**, *54*, 5054.

[666]See Lomas, J.S.; Fain, D.; Briand, S. *J. Org. Chem.* **1990**, *55*, 1052, and references cited therein.

When an α-amino acid is treated with an anhydride in the presence of pyridine, the carboxyl group is replaced by an acyl group and the NH_2 becomes acylated. This is called the *Dakin–West reaction*.[667] The mechanism involves formation of an oxazolone.[668] The reaction sometimes takes place on carboxylic acids even when an a amino group is not present. A number of *N*-substituted amino acids, RCH(NHR′)COOH, give the corresponding *N*-alkylated products.

OS **IV**, 5; **V**, 27.

B. Acyl Cleavages

In these reactions (**12-43–12-46**), a carbonyl group is attacked by a hydroxide ion (or amide ion), giving an intermediate that undergoes cleavage to a carboxylic acid (or an amide). With respect to the leaving group, this is nucleophilic substitution at a carbonyl group and the mechanism is the tetrahedral one discussed in Chapter 10.

With respect to R this is of course electrophilic substitution. The mechanism is usually S_E1.

12-43   Basic Cleavage of β-Keto Esters and β-Diketones

Hydro-de-acylation

When β-keto esters are treated with concentrated base, cleavage occurs, but is on the keto side of the CR_2 group (arrow) in contrast to the acid cleavage mentioned on page 838. The products are a carboxylic ester and the salt of an acid. However, the utility of the reaction is somewhat limited by the fact that decarboxylation is a side reaction, even under basic conditions. β-Diketones behave similarly to give a ketone and the salt of a carboxylic acid. With both β-keto esters and β-diketones, ⁻OEt can be used instead of ⁻OH, in which case the ethyl esters of the corresponding acids are obtained instead of the salts. In the case of β-keto esters, this is the reverse of Claisen condensation (**16-85**). The similar cleavage of cyclic α-cyano

[667]For a review, see Buchanan, G.L. *Chem. Soc. Rev.* **1988**, *17*, 91.

[668]Allinger, N.L.; Wang, G.L.; Dewhurst, B.B. *J. Org. Chem.* **1974**, *39*, 1730.

ketones, in an intramolecular fashion, has been used to effect a synthesis of macro-cyclic lactones such as **50**.[669]

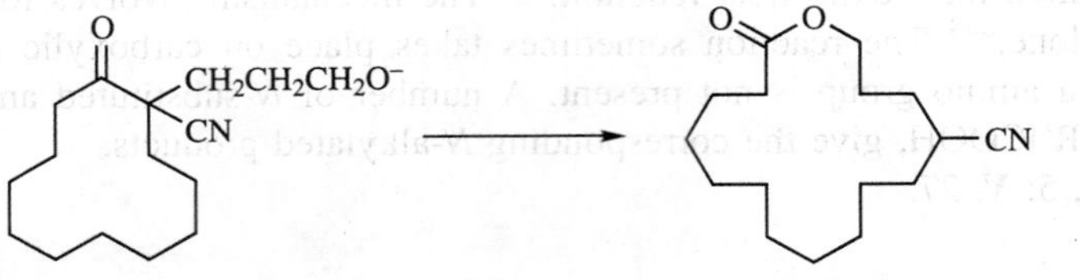

50

Activated F⁻ (from KF and a crown ether) has been used as the base to cleave an α-cyano ketone.[670]

OS **II**, 266, 531; **III**, 379; **IV**, 415, 957; **V**, 179, 187, 277, 533, 747, 767.

12-44   Haloform Reaction

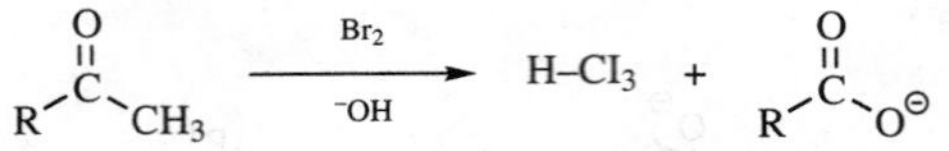

In the *haloform reaction*, methyl ketones (and the only methyl aldehyde, acet-aldehyde) are cleaved with halogen and a base.[671] The halogen can be bromine, chlorine, or iodine. What takes place is actually a combination of two reactions. The first is an example of **12-4**, in which, under the basic conditions employed, the methyl group is trihalogenated. Then the resulting trihalo ketone is attacked by hydroxide ion to give tetrahedral intermediate **51**.[672] The X_3C^- group is a sufficiently good leaving group (not HX_2C^- or H_2XC^-) that a carboxylic acid is formed, with quickly reacts with the carbanion to give the final products. Primary or secondary methylcarbinols also give the reaction, because they are oxidized to the carbonyl compounds under the conditions employed.

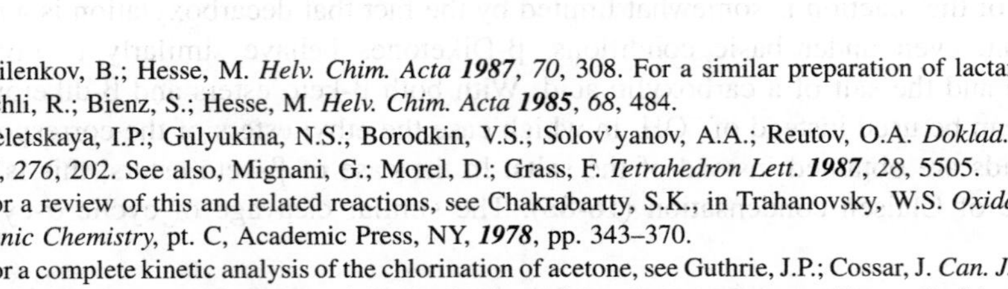

51

[669]Milenkov, B.; Hesse, M. *Helv. Chim. Acta* **1987**, *70*, 308. For a similar preparation of lactams, see Wälchli, R.; Bienz, S.; Hesse, M. *Helv. Chim. Acta* **1985**, *68*, 484.

[670]Beletskaya, I.P.; Gulyukina, N.S.; Borodkin, V.S.; Solov'yanov, A.A.; Reutov, O.A. *Doklad. Chem.* **1984**, *276*, 202. See also, Mignani, G.; Morel, D.; Grass, F. *Tetrahedron Lett.* **1987**, *28*, 5505.

[671]For a review of this and related reactions, see Chakrabartty, S.K., in Trahanovsky, W.S. *Oxidation in Organic Chemistry*, pt. C, Academic Press, NY, **1978**, pp. 343–370.

[672]For a complete kinetic analysis of the chlorination of acetone, see Guthrie, J.P.; Cossar, J. *Can. J. Chem.* **1986**, *64*, 1250. For a discussion of the mechanism of the cleavage step, see Zucco, C.; Lima, C.F.; Rezende, M.C.; Vianna, J.F.; Nome, F. *J. Org. Chem.* **1987**, *52*, 5356.

As with **12-4**, the rate-determining step is the preliminary enolization of the methyl ketone.[673] A side reaction is α halogenation of the non-methyl R group. Sometimes these groups are also cleaved.[674] The reaction cannot be applied to F_2, but ketones of the form $RCOCF_3$ (R = alkyl or aryl) give fluoroform and $RCOO^-$ when treated with base.[675] Rate constants for cleavage of X_3CCOPh (X = F, Cl, Br) were found to be in the ratio $1 : 5.3 \times 10^{10} : 2.2 \times 10^{13}$, showing that an F_3C^- group cleaves much more slowly than the others.[676] The haloform reaction is often used as a test for methylcarbinols and methyl ketones. Iodine is most often used as the test reagent, since iodoform (HCI_3) is an easily identifiable yellow solid. The reaction is also frequently used for synthetic purposes. Methyl ketones $RCOCH_3$ can be converted directly to methyl esters $RCOOCH_3$ by an electrochemical reaction.[677] Trifluoromethyl ketones have been converted to ethyl esters via treatment with NaH in aqueous DMF followed by reaction with bromoethane.[678]

OS **I**, 526; **II**, 428; **III**, 302; **IV**, 345; **V**, 8. Also see, OS **VI**, 618.

12-45   Cleavage of Nonenolizable Ketones

Hydro-de-acylation

Ordinary ketones are generally much more difficult to cleave than trihalo ketones or β-diketones, because the carbanion intermediates in these cases are more stable than simple carbanions. However, nonenolizable ketones can be cleaved by treatment with a 10:3 mixture of t-BuOK–H_2O in an aprotic solvent, such as ether, DMSO, 1,2-dimethoxyethane (glyme),[679] or with solid t-BuOK in the absence of a solvent.[680] When the reaction is applied to monosubstituted diaryl ketones, that aryl group preferentially cleaves that comes off as the more stable carbanion, except that aryl groups substituted in the ortho position are more readily cleaved than otherwise because of the steric effect (relief of strain).[680,681] In certain cases, cyclic ketones can be cleaved by base treatment, even if they are enolizable.[682]

OS **VI**, 625. See also, OS **VII**, 297.

[673]Pocker, Y. *Chem. Ind.* (*London*) *1959*, 1383.

[674]Levine, R.; Stephens, J.R. *J. Am. Chem. Soc. 1950*, 72, 1642.

[675]See Hudlicky, M. *Chemistry of Organic Fluorine Compounds*, 2nd ed.; Ellis Horwood: Chichester, *1976*, pp. 276–278.

[676]Guthrie, J.P.; Cossar, J. *Can. J. Chem. 1990*, 68, 1640.

[677]Nikishin, G.I.; Elinson, M.N.; Makhova, I.V. *Tetrahedron 1991*, 47, 895.

[678]Delgado, A.; Clardy, J. *Tetrahedron Lett. 1992*, 33, 2789.

[679]Swan, G.A. *J. Chem. Soc. 1948*, 1408; Gassman, P.G.; Lumb, J.T.; Zalar, F.V. *J. Am. Chem. Soc. 1967*, 89, 946.

[680]March, J.; Plankl, W. *J. Chem. Soc. Perkin Trans. 1 1977*, 460.

[681]Davies, D.G.; Derenberg, M.; Hodge, P. *J. Chem. Soc. C 1971*, 455.

[682]For example, see Swaminathan, S.; Newman, M.S. *Tetrahedron 1958*, 2, 88; Hoffman, T.D.; Cram, D.J. *J. Am. Chem. Soc. 1969*, 91, 1009.

12-46   The Haller–Bauer Reaction

Hydro-de-acylation

$$R \overset{O}{\underset{}{\overset{\|}{C}}} R' \xrightarrow{\ ^-NH_2\ } R\text{-H} \ + \ \underset{HN}{\overset{O}{\underset{}{\overset{\ominus\ \|}{C}}}} R'$$

Cleavage of ketones with sodium amide is called the *Haller–Bauer reaction*.[683] As with **12-45**, which is exactly analogous, the reaction is usually applied only to non-enolizable ketones, most often to ketones of the form $ArCOCR_3$, where the products R_3CCONH_2 are not easily attainable by other methods. However, many other ketones have been used, although benzophenone is virtually unaffected. It has been shown that the configuration of optically active alkyl groups (R) is retained.[684] The NH_2 loses its proton from the tetrahedral intermediate **52** before the R group is cleaved.[685]

$$R \overset{O}{\underset{}{\overset{\|}{C}}} R' \ + \ ^-NH_2 \ \longrightarrow \ \underset{NH_2}{R\overset{O^\ominus}{\underset{}{\overset{|}{C}}} R'} \ \longrightarrow \ \underset{\ominus NH}{R\overset{O^\ominus}{\underset{}{\overset{|}{C}}} R'} \xrightarrow{\ HA\ } R\text{-H} \ + \ \underset{HN}{\overset{O}{\underset{}{\overset{\ominus\ \|}{C}}}} R'$$

$$\mathbf{52}$$

An extension of this cleavage process involves the reaction of α-nitro ketones $(O{=}C\text{-}CHRNO_2)$ with a primary amine, neat, to give the corresponding amide $(O{=}C\text{-}NHR')$.[686]

OS **V**, 384, 1074.

C. Other Cleavages

12-47   The Cleavage of Alkanes

Hydro-de-*tert*-butylation, and so on

$$(CH_3)_4C \xrightarrow{\ FSO_3H\text{-}SbF_5\ } CH_4 \ + \ (CH_3)_3C +$$

The C–C bonds of alkanes can be cleaved by treatment with superacids[48] (p. 236). For example, neopentane in $FSO_3H–SbF_5$ can cleave to give methane and the *tert*-butyl cation. The C–H cleavage (see **12-1**) is a competing reaction and, for example, neopentane can give H_2 and the *tert*-pentyl cation (formed by rearrangement of the initially formed neopentyl cation) by this pathway. In general, the order of reactivity is tertiary C–H > C–C > secondary C–H ≫ primary C–H,

[683]For a review, see Gilday, J.P.; Paquette, L.A. *Org. Prep. Proced. Int.* **1990**, *22*, 167. For an improved procedure, see Kaiser, E.M.; Warner, C.D. *Synthesis* **1975**, 395.

[684]Impastato, F.J.; Walborsky, H.M. *J. Am. Chem. Soc.* **1962**, *84*, 4838; Paquette, L.A.; Gilday, J.P. *J. Org. Chem.* **1988**, *53*, 4972; Paquette, L.A.; Ra, C.S. *J. Org. Chem.* **1988**, *53*, 4978.

[685]Bunnett, J.F.; Hrutfiord, B.F. *J. Org. Chem.* **1962**, *27*, 4152.

[686]Ballini, R.; Bosica, G.; Fiorini, D. *Tetrahedron* **2003**, *59*, 1143.

although steric factors cause a shift in favor of C–C cleavage in such a hindered compound as tri-*tert*-butylmethane. The mechanism is similar to that shown in **12-1** and **12-20** and involves attack by H^+ on the C–C bond to give a pentavalent cation.

Catalytic hydrogenation seldom breaks unactivated C–C bonds (i.e., $R-R' + H_2 \rightarrow RH + R'H$), but methyl and ethyl groups have been cleaved from substituted adamantanes by hydrogenation with a $Ni-Al_2O_3$ catalyst at about 250°C.[687] Certain C–C bonds have been cleaved by alkali metals.[688]

The C–C bond of 2-allyl-2-arylmalonate derivatives was cleaved, with loss of the allylic group to give the 2-arylmalonate, by treatment with a nickel catalyst.[689]

12-48   Decyanation or Hydro-de-cyanation

$$ RCN \xrightarrow{\text{Na-NH}_3 \text{ or } \text{Na-Fe(acac)}_3} RH $$

The cyano group of alkyl nitriles can be removed[690] by treatment with metallic sodium, either in liquid ammonia,[691] or together with tris(acetylacetonato)iron(III) [Fe(acac)$_3$][692] or, with lower yields, titanocene. The two procedures are complementary. Although both can be used to decyanate many kinds of nitriles, the Na–NH$_3$ method gives high yields with R groups, such as trityl, benzyl, phenyl, and tertiary alkyl, but lower yields ($\sim$35–50%) when R = primary or secondary alkyl. On the other hand, primary and secondary alkyl nitriles are decyanated in high yields by the Na–Fe(acac)$_3$ procedure. Sodium in liquid ammonia is known to be a source of solvated electrons, and the reaction may proceed through the free radical R• that would then be reduced to the carbanion R^-, which by abstraction of a proton from the solvent, would give RH. The mechanism with Fe(acac)$_3$ is presumably different. Another procedure,[693] which is successful for R = primary, secondary, or tertiary, involves the use of potassium metal and the crown ether dicyclohexano-18-crown-6 in toluene.[694]

---

[687]Grubmüller, P.; Schleyer, P.v.R.; McKervey, M.A. *Tetrahedron Lett.* **1979**, 181.

[688]For examples and references, see Grovenstein, Jr., E.; Bhatti, A.M.; Quest, D.E.; Sengupta, D.; VanDerveer, D. *J. Am. Chem. Soc.* **1983**, *105*, 6290.

[689]Nečas, D.; Turský, M.; Kotora, M. *J. Am. Chem. Soc.* **2004**, *126*, 10222.

[690]For a list of procedures, with references, see Larock, R.C. *Comprehensive Organic Transformations*, 2nd ed., Wiley-VCH, NY, **1999**, p. 75.

[691]Büchner, W.; Dufaux, R. *Helv. Chim. Acta* **1966**, *49*, 1145; Arapakos, P.G.; Scott, M.K.; Huber, Jr., F.E. *J. Am. Chem. Soc.* **1969**, *91*, 2059; Birch, A.J.; Hutchinson, E.G. *J. Chem. Soc. Perkin Trans. 1* **1972**, 1546; Yamada, S.; Tomioka, K.; Koga, K. *Tetrahedron Lett.* **1976**, 61.

[692]Van Tamelen, E.E.; Rudler, H.; Bjorklund, C. *J. Am. Chem. Soc.* **1971**, *93*, 7113.

[693]For other procedures, see Cuvigny, T.; Larcheveque, M.; Normant, H. *Bull. Soc. Chim. Fr.* **1973**, 1174; Berkoff, C.E.; Rivard, D.E.; Kirkpatrick, D.; Ives, J.L. *Synth. Commun.* **1980**, *10*, 939; Savoia, D.; Tagliavini, E.; Trombini, C.; Umani-Ronchi, A. *J. Org. Chem.* **1980**, *45*, 3227; Ozawa, F.; Iri, K.; Yamamoto, A. *Chem. Lett.* **1982**, 1707.

[694]Ohsawa, T.; Kobayashi, T.; Mizuguchi, Y.; Saitoh, T.; Oishi, T. *Tetrahedron Lett.* **1985**, *26*, 6103.

α-Amino and α-amido nitriles RCH(CN)NR'$_2$ and RCH(CN)NHCOR' can be decyanated in high yield by treatment with NaBH$_4$.[695]

ELECTROPHILIC SUBSTITUTION AT NITROGEN

In most of the reactions in this section, an electrophile bonds with the unshared pair of a nitrogen atom. The electrophile may be a free positive ion or a positive species attached to a carrier that breaks off in the course of the attack or shortly after:

$$-\underset{\diagdown}{\overset{\diagup}{N}} \ + \ Y-Z \ \longrightarrow \ -\underset{\diagup}{\overset{\diagdown}{N}}\overset{\oplus}{-}Y \ + \ Z$$
$$\mathbf{53}$$

Further reaction of **53** depends on the nature of Y and of the other groups attached to the nitrogen.

12-49   The Conversion of Hydrazines to Azides

Hydrazine–azide transformation

$$RNHNH_2 \ + \ HONO \ \longrightarrow \ R-\overset{\oplus}{N}=N\overset{\ominus}{=}N$$

Monosubstituted hydrazines treated with nitrous acid give azides in a reaction exactly analogous to the formation of aliphatic diazo compounds mentioned in **13-19**. Among other reagents used for this conversion have been N$_2$O$_4$[696] and nitrosyl tetrafluoroborate (NOBF$_4$).[697]

OS **III**, 710; **IV**, 819; **V**, 157.

12-50   N-Nitrosation

N-Nitroso-de-hydrogenation

$$R_2NH \ + \ HONO \ \longrightarrow \ R_2N-NO$$

When secondary amines are treated with nitrous acid (typically formed from sodium nitrite and a mineral acid),[698] N-nitroso compounds (also called

[695]Yamada, S.; Akimoto, H. *Tetrahedron Lett.* **1969**, 3105; Fabre, C.; Hadj Ali Salem, M.; Welvart, Z. *Bull. Soc. Chim. Fr.* **1975**, 178. See also Ogura, K.; Shimamura, Y.; Fujita, M. *J. Org. Chem.* **1991**, *56*, 2920.

[696]Kim, Y.H.; Kim, K.; Shim, S.B. *Tetrahedron Lett.* **1986**, *27*, 4749.

[697]Pozsgay, V.; Jennings, H.J. *Tetrahedron Lett.* **1987**, *28*, 5091.

[698]From NaNO$_2$/oxalic acid: Zolfigol, M.A. *Synth. Commun.* **1999**, *29*, 905. From NaNO$_2$ on wet silica: Zolfigol, M.A.; Ghaemi, E.; Madrikian, E.; Kiany-Burazjani, M. *Synth. Commun.* **2000**, *30*, 2057.

nitrosamines) are formed.[699] The reaction can be accomplished with dialkyl-, diaryl-, or alkylarylamines, and even with mono-*N*-substituted amides: RCONHR' + HONO → RCON(NO)R'.[700] Tertiary amines have also been *N*-nitrosated, but in these cases one group cleaves, so that the product is the nitroso derivative of a secondary amine.[701] The group that cleaves appears as an aldehyde or ketone. Other reagents have also been used, for example, NOCl, which is useful for amines or amides that are not soluble in an acidic aqueous solution or where the *N*-nitroso compounds are highly reactive. *N*-Nitroso compounds can be prepared in basic solution by treatment of secondary amines with gaseous N_2O_3, N_2O_4,[702] or alkyl nitrites,[703] and, in aqueous or organic solvents, by treatment with $BrCH_2NO_2$.[704] Secondary amines are converted to the *N*-nitroso compound with H_5IO_6 on wet silica.[705]

$$Ar \diagdown \underset{R'}{\overset{}{N}} - N = O$$
$$\textbf{54}$$

The mechanism of nitrosation is essentially the same as in **13-19** up to the point where **54** is formed. Since this species cannot lose a proton, it is stable and the reaction ends there. The attacking entity can be any of those mentioned in **13-19**. The following has been suggested as the mechanism for the reaction with tertiary amines:[706]

[699]For reviews, see Williams, D.L.H. Williams, D.L.H. *Nitrosation*; Cambridge University Press, Cambridge, *1988*, pp. 95–109; Kostyukovskii, Ya.L.; Melamed, D.B. *Russ. Chem. Rev. 1988*, *57*, 350; Saavedra, J.E. *Org. Prep. Proced. Int. 1987*, *19*, 83; Williams, D.L.H. *Adv. Phys. Org. Chem. 1983*, *19*, 381; Challis, B.C.; Challis, J.A. in Patai, S.; Rappoport, Z. *The Chemistry of the Functional Groups Supplement F*, pt. 2, Wiley, NY, *1982*, pp. 1151–1223; Ridd, J.H. *Q. Rev. Chem. Soc. 1961*, *15*, 418. For a review of the chemistry of aliphatic *N*-nitroso compounds, including methods of synthesis see Fridman, A.L.; Mukhametshin, F.M.; Novikov, S.S. *Russ. Chem. Rev. 1971*, *40*, 34. For a discussion of encapsulated reagents used for nitrosation, see Zyranov, G.V.; Rudkevich, D.M. *Org. Lett. 2003*, *5*, 1253.
[700]For a discussion of the mechanism with amides, see Castro, A.; Iglesias, E.; Leis, J.R.; Peña, M.E.; Tato, J.V. *J. Chem. Soc. Perkin Trans. 2 1986*, 1725.
[701]Hein, G.E. *J. Chem. Educ. 1963*, *40*, 181. See also, Verardo, G.; Giumanini, A.G.; Strazzolini, P. *Tetrahedron 1990*, *46*, 4303.
[702]Challis, B.C.; Kyrtopoulos, S.A. *J. Chem. Soc. Perkin Trans. 1 1979*, 299.
[703]Casado, J.; Castro, A.; Lorenzo, F.M.; Meijide, F. *Monatsh. Chem. 1986*, *117*, 335.
[704]Challis, B.C.; Yousaf, T.I. *J. Chem. Soc. Chem. Commun. 1990*, 1598.
[705]Zolfigol, M.A.; Choghamarani, A.G.; Shivini, F.; Keypour, H.; Salehzadeh, S. *Synth. Commun. 2001*, *31*, 359. Also with $KHSO_5$ on wet silica, see Zolfigol, M.A.; Bagherzadeh, M.; Choghamarani, A.G.; Keypour, H.; Salehzadeh, S. *Synth. Commun. 2001*, *31*, 1161.
[706]Smith, P.A.S.; Loeppky, R.N. *J. Am. Chem. Soc. 1967*, *89*, 1147; Smith, P.A.S.; Pars, H.G. *J. Org. Chem. 1959*, *24*, 1324; Gowenlock, B.G.; Hutchison, R.J.; Little, J.; Pfab, J. *J. Chem. Soc. Perkin Trans. 2 1979*, 1110. See also, Loeppky, R.N.; Outram, J.R.; Tomasik, W.; Faulconer, J.M. *Tetrahedron Lett. 1983*, *24*, 4271.

The evidence for this mechanism includes the facts that nitrous oxide is a product (formed by $2\ HNO \rightarrow H_2O + N_2O$) and that quinuclidine, where the nitrogen is at a bridgehead, and therefore cannot give elimination, does not react. Tertiary amines have also been converted to nitrosamines with nitric acid in Ac_2O[707] and with N_2O_4.[708]

Amines and amides can be N-nitrated[709] with nitric acid,[710] or NO_2^+,[711] and aromatic amines can be converted to triazenes with diazonium salts. Aliphatic primary amines can also be converted to triazenes if the diazonium salts contain electron-withdrawing groups.[712] C-Nitrosation is discussed at **11-3** and **12-8**.

OS **I**, 177, 399, 417; **II**, 163, 211, 290, 460, 461, 462, 464 (also see **V**, 842); **III**, 106, 244; **IV**, 718, 780, 943; **V**, 336, 650, 797, 839, 962; **VI**, 542, 981. Also see, OS **III**, 711.

12-51 Conversion of Nitroso Compounds to Azoxy Compounds

In a reaction similar to **13-24**, azoxy compounds can be prepared by the condensation of a nitroso compound with a hydroxylamine.[713] The position of the oxygen in the final product is determined by the nature of the R groups, not by which R groups came from which starting compound. Both R and R' can be alkyl or aryl, but when two different aryl groups are involved, mixtures of azoxy compounds

[707]Boyer, J.H.; Pillai, T.P.; Ramakrishnan, V.T. *Synthesis* **1985**, 677.

[708]Boyer, J.H.; Kumar, G.; Pillai, T.P. *J. Chem. Soc. Perkin Trans. 1* **1986**, 1751.

[709]For other reagents, see Mayants, A.G.; Pyreseva, K.G.; Gordeichuk, S.S. *J. Org. Chem. USSR* **1986**, *22*, 1900; Bottaro, J.C.; Schmitt, R.J.; Bedford, C.D. *J. Org. Chem.* **1987**, *52*, 2292; Suri, S.C.; Chapman, R.D. *Synthesis* **1988**, 743; Carvalho, E.; Iley, J.; Norberto, F.; Rosa, E. *J. Chem. Res. (S)* **1989**, 260.

[710]Cherednichenko, L.V.; Dmitrieva, L.G.; Kuznetsov, L.L.; Gidaspov, B.V. *J. Org. Chem. USSR* **1976**, *12*, 2101, 2105.

[711]Ilyushin, M.A.; Golod, E.L.; Gidaspov, B.V. *J. Org. Chem. USSR* **1977**, *13*, 8; Andreev, S.A.; Lededev, B.A.; Tselinskii, I.V. *J. Org. Chem. USSR* **1980**, *16*, 1166, 1170, 1175, 1179.

[712]For a review of alkyl traizenes, see Vaughan, K.; Stevens, M.F.G. *Chem. Soc. Rev.* **1978**, *7*, 377.

[713]Boyer, J.H., in Feuer, H. *The Chemistry of the Nitro and Nitroso Groups*, pt. 1, Wiley, NY, **1969**, pp. 278–283.

(ArNONAr, ArNONAr′, and Ar′NONAr′) are obtained[714] and the unsymmetrical product (ArNONAr′) is likely to be formed in the smallest amount. This behavior is probably caused by an equilibration between the starting compounds prior to the actual reaction (ArNO + Ar′NHOH → Ar′NO + ArNHOH).[715] The mechanism[716] has been investigated in the presence of base. Under these conditions both reactants are converted to radical anions, which couple:

$$R\text{-}N{=}O \;+\; R'NHOH \;\longrightarrow\; 2\;Ar{-}\overset{\cdot}{N}{-}O^{\ominus} \;\longrightarrow\; Ar{\overset{O^{\ominus}}{\underset{\underset{O^{\ominus}}{|}}{\overset{|}{N}}}}\overset{Ar}{\underset{N}{\diagdown}} \;\xrightarrow[\text{H}_2\text{O}]{-2\text{ }^{-}\text{OH}}\; Ar{\overset{N}{\underset{\underset{O^{\ominus}}{|}}{\diagup}}}\overset{\oplus}{\underset{N}{}}{\diagdown}Ar$$

These radical anions have been detected by esr.[717] This mechanism is consistent with the following result: when nitrosobenzene and phenylhydroxylamine are coupled, [18]O and ^{15}N labeling show that the two nitrogens and the two oxygens become equivalent.[718] Unsymmetrical azoxy compounds can be prepared[719] by combination of a nitroso compound with an *N,N*-dibromoamine. Symmetrical and unsymmetrical azo and azoxy compounds are produced when aromatic nitro compounds react with aryliminodimagnesium reagents ArN(MgBr)$_2$.[720]

12-52 *N*-Halogenation

N-Halo-de-hydrogenation

$$RNH_2 \;+\; NaOCl \;\longrightarrow\; RNHCl$$

Treatment with sodium hypochlorite or hypobromite converts primary amines into *N*-halo- or *N,N*-dihaloamines. Secondary amines can be converted to *N*-halo secondary amines. Similar reactions can be carried out on unsubstituted and *N*-substituted amides and on sulfonamides. With unsubstituted amides the *N*-halogen product is seldom isolated but usually rearranges (see **18-13**); however, *N*-halo-*N*-alkyl amides and *N*-halo imides are quite stable. The important reagents NBS and NCS are made in this manner. *N*-Halogenation has also been accomplished with other

[714]See, for example, Ogata, Y.; Tsuchida, M.; Takagi, Y. *J. Am. Chem. Soc.* **1957**, *79*, 3397.

[715]Knight, G.T.; Saville, B. *J. Chem. Soc. Perkin Trans. 2* **1973**, 1550.

[716]For discussions of the mechanism in the absence of base, see Darchen, A.; Moinet, C. *Bull. Soc. Chim. Fr.* **1976**, 812; Becker, A.R.; Sternson, L.A. *J. Org. Chem.* **1980**, *45*, 1708. See also, Pizzolatti, M.G.; Yunes, R.A. *J. Chem. Soc. Perkin Trans. 1* **1990**, 759.

[717]Russell, G.A.; Geels, E.J.; Smentowski, F.J.; Chang, K.; Reynolds, J.; Kaupp, G. *J. Am. Chem. Soc.* **1967**, *89*, 3821.

[718]Shemyakin, M.M.; Maimind, V.I.; Vaichunaite, B.K. *Izv. Akad. Nauk SSSR, Ser. Khim.* **1957**, 1260; Oae, S.; Fukumoto, T.; Yamagami, M. *Bull. Chem. Soc. Jpn.* **1963**, *36*, 728.

[719]Zawalski, R.C.; Kovacic, P. *J. Org. Chem.* **1979**, *44*, 2130. For another method, see Moriarty, R.M.; Hopkins, T.E.; Prakash, I.; Vaid, B.K.; Vaid, R.K. *Synth. Commun.* **1990**, *20*, 2353.

[720]O kubo, M.; Matsuo, K.; Yamauchi, A. *Bull. Chem. Soc. Jpn.* **1989**, *62*, 915, and other papers in this series.

reagents (e.g., sodium bromite $NaBrO_2$),[721] benzyltrimethylammonium tribromide $(PhCH_2NMe_3{}^+ Br_3-)$,[722] NaCl with Oxone®,[723] and N-chlorosuccinimide.[724] The mechanisms of these reactions[725] involve attack by a positive halogen and are probably similar to those of **13-19** and **12-50**.[726] N-Fluorination can be accomplished by direct treatment of amines[727] or amides[728] with F_2. Fluorination of N-alkyl-N-fluoro amides (RRN(F)COR') results in cleavage to N,N-difluoroamines (RNF_2).[728,729] Trichloroisocyanuric acid converts primary amines to the N,N-dichloroamine.[730]

OS **III**, 159; **IV**, 104, 157; **V**, 208, 663, 909; **VI**, 968; **VII**, 223; **VIII**, 167, 427.

12-53 The Reaction of Amines With Carbon Monoxide or Carbon Dioxide

N-**Formylation** or *N*-**Formyl-de-hydrogenation**, and so on

$$RNH_2 + CO \xrightarrow{\text{catalyst}} \underset{H}{\overset{O}{\underset{\|}{C}}}\text{-NHR} \quad \text{or} \quad \underset{RHN}{\overset{O}{\underset{\|}{C}}}\text{-NHR} \quad \text{or} \quad R\text{-}N\text{=}C\text{=}O$$

Three types of product can be obtained from the reaction of amines with carbon monoxide, depending on the catalyst. (*1*) Both primary and secondary amines react with CO in the presence of various catalysts [e.g., $Cu(CN)_2$, Me_3N–H_2Se, rhodium or ruthenium complexes] to give N-substituted and N,N-disubstituted formamides, respectively.[731] Primary aromatic amines react with ammonium formate to give the formamide.[732] Tertiary amines react with CO and a palladium catalyst to give an amide.[733] (*2*) Symmetrically substituted ureas can be prepared by treatment of a primary amine (or ammonia) with CO[734] in the presence of selenium[735] or

[721]Kajigaeshi, S.; Nakagawa, T.; Fujisaki, S. *Chem. Lett.* **1984**, 2045.

[722]Kajigaeshi, S.; Murakawa, K.; Asano, K.; Fujisaki, S.; Kakinami, T. *J. Chem. Soc. Perkin Trans. 1* **1989**, 1702.

[723]Curini, M.; Epifano, F.; Marcotullio, M.C.; Rosati, O.; Tsadjout, A. *Synlett* **2000**, 813.

[724]See Deno, N.C.; Fishbein, R.; Wyckoff, J.C. *J. Am. Chem. Soc.* **1971**, 93, 2065; Guillemin, J.; Denis, J.N. *Synthesis* **1985**, 1131.

[725]For a study of the mechanism, see Matte, D.; Solastiouk, B.; Merlin, A.; Deglise, X. *Can. J. Chem.* **1989**, 67, 786.

[726]For studies of reactivity in this reaction, see Thomm, E.W.C.W.; Wayman, M. *Can. J. Chem.* **1969**, 47, 3289; Higuchi, T.; Hussain, A.; Pitman, I.H. *J. Chem. Soc. B*, **1969**, 626.

[727]Sharts, C.M. *J. Org. Chem.* **1968**, 33, 1008.

[728]Grakauskas, V.; Baum, K. *J. Org. Chem.* **1969**, 34, 2840; **1970**, 35, 1545.

[729]See Barton, D.H.R.; Hesse, R.H.; Klose, T.R.; Pechet, M.M. *J. Chem. Soc. Chem. Commun.* **1975**, 97.

[730]DeLuca, L.; Giacomelli, G. *Synlett* **2004**, 2180.

[731]See Saegusa, T.; Kobayashi, S.; Hirota, K.; Ito, Y. *Bull. Chem. Soc. Jpn.* **1969**, 42, 2610; Nefedov, B.K.; Sergeeva, N.S.; Éidus, Ya.T. *Bull. Acad. Sci. USSR Div. Chem. Sci.* **1973**, 22, 784; Yoshida, Y.; Asano, S.; Inoue, S. *Chem. Lett.* **1984**, 1073; Bitsi, G.; Jenner, G. *J. Organomet. Chem.* **1987**, 330, 429.

[732]Reddy, P.G.; Kumar,. D.K.; Baskaran, S. *Tetrahedron Lett.* **2000**, 41, 9149.

[733]Murahashi, S.-I.; Imada, Y.; Nishimura, K. *Tetrahedron*, **1994**, 50, 453.

[734]For a synthesis involving a palladium catalyst, see Gabriele, B.; Salerno, G.; Mancuso, R.; Costa, M. *J. Org. Chem.* **2004**, 69, 4741.

[735]Sonoda, N.; Yasuhara, T.; Kondo, K.; Ikeda, T.; Tsutsumi, S. *J. Am. Chem. Soc.* **1971**, 93, 6344.

sulfur.[736] R can be alkyl or aryl. The same thing can be done with secondary amines, by using $Pd(OAc)_2$–I_2–K_2CO_3.[737] Primary aromatic amines react with β-keto esters and a Mo–ZrO_2 catalyst to give the symmetrical urea.[738] Treatment of a secondary amine with nitrobenzene, selenium, and carbon monoxide leads to the unsymmetrical urea.[739] (*3*) When $PdCl_2$ is the catalyst, primary amines yield isocyanates.[740] Isocyanates can also be obtained by treatment of CO with azides: $RN_3 + CO \rightarrow RNCO$,[741] or with an aromatic nitroso or nitro compound and a rhodium complex catalyst.[742] Primary amines react with di-*tert*-butyltricarbonate to give the isocyanate.[743] Lactams are converted to the corresponding *N*-chloro lactam with $Ca(OCl)_2$ with moist alumina in dichloromethane.[744]

A fourth type of product, a carbamate RNHCOOR', can be obtained from primary or secondary amines, if these are treated with CO, O_2, and an alcohol R'OH in the presence of a catalyst.[745] Primary amines react with dimethyl carbonate in supercritical CO_2 (see p. 414) to give a carbamate.[746] Carbamates can also be obtained from nitroso compounds, by treatment with CO, R'OH, $Pd(OAc)_2$, and $Cu(OAc)_2$,[747] and from nitro compounds.[748] When allylic amines ($R_2C=CHRCHRNR'_2$) are treated with CO and a palladium–phosphine catalyst, the CO inserts to produce the β,γ-unsaturated amides ($R_2C=CHRCHRCONR'_2$) in good yields.[749] Ring-expanded lactams are obtained from cyclic amines via a similar reaction[750] (see also, **16-22**). Silyloxy carbamates ($RNHCO_2SiR'_3$) can be prepared by the reaction of a primary amine with carbon dioxide and triethylamine, followed by reaction with triisopropylsilyl triflate and tetrabutylammonium fluoride.[751]

Carbon dioxide reacts with amines ($ArNH_2$) and alkyl halides, under electrolysis conditions, to give the corresponding carbamate ($ArNHCO_2Et$).[752] Secondary

[736]Franz, R.A.; Applegath, F.; Morriss, F.V.; Baiocchi, F.; Bolze, C. *J. Org. Chem.* **1961**, *26*, 3309.

[737]Pri-Bar, I.; Alper, H. *Can. J. Chem.* **1990**, *68*, 1544.

[738]Reddy, B.M.; Reddy, V.R. *Synth. Commun.* **1999**, *29*, 2789.

[739]Yang, Y.; Lu, S. *Tetrahedron Lett.* **1999**, *40*, 4845.

[740]Stern, E.W.; Spector, M.L. *J. Org. Chem.* **1966**, *31*, 596.

[741]Bennett, R.P.; Hardy, W.B. *J. Am. Chem. Soc.* **1968**, *90*, 3295.

[742]Unverferth, K.; Rüger, C.; Schwetlick, K. *J. Prakt. Chem.* **1977**, *319*, 841; Unverferth, K.; Tietz, H.; Schwetlick, K. *J. Prakt. Chem.* **1985**, *327*, 932. See also, Braunstein, P.; Bender, R.; Kervennal, J. *Organometallics* **1982**, *1*, 1236; Kunin, A.J.; Noirot, M.D.; Gladfelter, W.L. *J. Am. Chem. Soc.* **1989**, *111*, 2739.

[743]Peerlings, H.W.I.; Meijer, E.W. *Tetrahedron Lett.* **1999**, *40*, 1021.

[744]Larionov, O.V.; Kozhushkov, S.I.; de Meijere, A. *Synthesis* **2003**, 1916.

[745]Fukuoka, S.; Chono, M.; Kohno, M. *J. Org. Chem.* **1984**, *49*, 1458; *J. Chem. Soc. Chem. Commun.* **1984**, 399; Feroci, M.; Inesi, A.; Rossi, L. *Tetrahedron Lett.* **2000**, *41*, 963.

[746]Selva, M.; Tundo, P.; Perosa, A. *Tetrahedron Lett.* **2002**, *43*, 1217.

[747]Alper, H.; Vasapollo, G. *Tetrahedron Lett.* **1987**, *28*, 6411.

[748]Cenini, S.; Crotti, C.; Pizzotti, M.; Porta, F. *J. Org. Chem.* **1988**, *53*, 1243; Reddy, N.P.; Masdeu, A.M.; El Ali, B.; Alper, H. *J. Chem. Soc. Chem. Commun.* **1994**, 863.

[749]Murahashi, S.; Imada, Y.; Nishimura, K. *J. Chem. Soc. Chem. Commun.* **1988**, 1578.

[750]Wang, M.D.; Alper, H. *J. Am. Chem. Soc.* **1992**, *114*, 7018.

[751]Lipshutz, B.H.; Papa, P.; Keith, J.M. *J. Org. Chem.* **1999**, *64*, 3 792.

[752]Casadei, M.A.; Inesi, A.; Moracci, F.M.; Rossi, L. *Chem. Commun.* **1996**, 2575; Feroci, M.; Casadei, M.A.; Orsini, M.; Palombi, L.; Inesi, A. *J. Org. Chem.* **2003**, *68*, 1548.

amines react with all halides and an onium salt in supercritical CO_2 (see p. 414) to give the carbamate.[753] N-phenylthioamines react with CO and a palladium catalyst to give a thiocarbamate $(ArSCO_2NR'_2)$.[754] Urea derivatives were obtained from amines, CO_2, and an antimony catalyst.[755]

Aziridines can be converted to cyclic carbamates (oxazolidinones) by heating with carbon dioxide and a chromium–salen catalyst.[756] The reaction of aziridines with LiI, and then CO_2 also generates oxazolidinones.[757]

[753]Yoshida, M.; Hara, N.; Okuyama, S. *Chem. Commun.* **2000**, 151.
[754]Kuniyasu, H.; Hiraike, H.; Morita, M.; Tanaka, A.; Sugoh, K.; Kurosawa, H. *J. Org. Chem.* **1999**, *64*, 7305.
[755]Nomura, R.; Hasegawa, Y.; Ishimoto, M.; Toyosaki, T.; Matsuda, H. *J. Org. Chem.* **1992**, *57*, 7339.
[756]Miller, A.W.; Nguyen, S.T. *Org. Lett.* **2004**, *6*, 2301.
[757]Hancock, M.T.; Pinhas, A.R. *Tetrahedron Lett.* **2003**, *44*, 5457.

CHAPTER 13

Aromatic Substitution, Nucleophilic and Organometallic

On p. 481, it was pointed out that nucleophilic substitutions proceed so slowly at an aromatic carbon that the reactions of Chapter 10 are not feasible for aromatic substrates. There are, however, exceptions to this statement, and it is these exceptions that form the subject of this chapter.[1] Reactions that *are* successful at an aromatic substrate are largely of four kinds: (*1*) reactions activated by electron-withdrawing groups ortho and para to the leaving group; (*2*) reactions catalyzed by very strong bases and proceeding through aryne intermediates; (*3*) reactions initiated by electron donors; and (*4*) reactions in which the nitrogen of a diazonium salt is replaced by a nucleophile. It is noted that solvent effects can be important.[2] Also, not all the reactions discussed in this chapter fit into these categories, and certain transition-metal catalyzed coupling reaction are included because they involve replacement of a leaving group on an aromatic ring.

MECHANISMS

There are four principal mechanisms for aromatic nucleophilic substitution.[3] Each of the four is similar to one of the aliphatic nucleophilic substitution mechanisms discussed in Chapter 10.

---

[1]For a review of aromatic nucleophilic substitution, see Zoltewicz, J.A. *Top. Curr. Chem.* **1975**, *59*, 33.

[2]Acevedo, O.; Jorgensen, W.L. *Org. Lett.* **2004**, *6*, 2881.

[3]For a monograph on aromatic nucleophilic substitution mechanisms, see Miller, J. *Aromatic Nucleophilic Substitution*, Elsevier, NY, **1968**. For reviews, see Bernasconi, C.F. *Chimia* **1980**, *34*, 1; *Acc. Chem. Res.* **1978**, *11*, 147; Bunnett, J.F. *J. Chem. Educ.* **1974**, *51*, 312; Ross, S.D., in Bamford, C.H.; Tipper, C.F.H. *Comprehensive Chemical Kinetics*, Vol. 13; Elsevier, NY, **1972**, pp. 407–431; Buck, P. *Angew. Chem, Int. Ed.* **1969**, *8*, 120; Buncel, E.; Norris, A.R.; Russell, K.E. *Q. Rev. Chem. Soc.* **1968**, *22*, 123; Bunnett, J.F. *Tetrahedron* **1993**, *49*, 4477; Zoltewicz, J.A. *Top. Curr. Chem.* **1975**, *59*, 33.

March's Advanced Organic Chemistry: Reactions, Mechanisms, and Structure, Sixth Edition, by Michael B. Smith and Jerry March
Copyright © 2007 John Wiley & Sons, Inc.

The S_NAr Mechanism[4]

By far the most important mechanism for nucleophilic aromatic substitution consists of two steps, attack of the nucleophilic species at the ipso carbon of the aromatic ring (the carbon bearing the leaving group in this case), followed by elimination of the leaving group and regeneration of the aromatic ring.

The first step is usually, but not always, rate determining. It can be seen that this mechanism greatly resembles the tetrahedral mechanism discussed in Chapter 16 and, in another way, the arenium ion mechanism of electrophilic aromatic substitution discussed in Chapter 11. In all three cases, the attacking species forms a bond with the substrate, giving an intermediate, such as **1**, and then the leaving group departs. We refer to this mechanism as the S_NAr mechanism.[5] The IUPAC designation is $A_N + D_N$ (the same as for the tetrahedral mechanism; compare the designation $A_E + D_E$ for the arenium ion mechanism). This mechanism is generally found where activating groups are present on the ring (see p. 864).

There is a great deal of evidence for the mechanism; we shall discuss only some of it.[3] Probably the most convincing evidence was the isolation, as long ago as 1902, of the intermediate **2** in the reaction between 2,4,6-trinitrophenetole and methoxide ion.[6] Intermediates of this type are stable salts, called *Meisenheimer* or *Meisenheimer–Jackson salts*,[7] and many more have been isolated.[8] The structures

[4]High pressure S_NAr reactions are known. see Barrett, I.C.; Kerr, M.A. *Tetrahedron Lett.* **1999**, *40*, 2439.
[5]The mechanism has also been called by other names, including the S_N2Ar, the addition–elimination, and the intermediate complex mechanism. See Wu, Z.; Glaser, R. *J. Am. Chem. Soc.* **2004**, *126*, 10632. See also, Terrier, F.; Mokhtari, M.; Goumont, T.; Hallé, J.-C.; Buncel, E. *Org. Biomol. Chem.* **2003**, *1*, 1757.
[6]Meisenheimer, J. *Liebigs Ann. Chem.* **1902**, *323*, 205. Similar salts were isolated even earlier by Jackson, C.L.; see Jackson, C.L.; Gazzolo, F.H. *Am. Chem. J.* **1900**, *23*, 376; Jackson, C.L.; Earle, R.B. *Am. Chem. J.*, **1903**, *29*, 89.
[7]Nucleophilic aromatic substitution for heteroatom nucleophiles through electrochemical oxidation of intermediate σ-complexes (Meisenheimer complexes) in simple nitroaromatic compounds has been reported, see Gallardo, I.; Guirado, G.; Marquet, J. *J. Org. Chem.* **2002**, *67*, 2548.
[8]For a monograph on Meisenheimer salts and on this mechanism, see Buncel, E.; Crampton, M.R.; Strauss, M.J.; Terrier, F. *Electron Deficient Aromatic- and Heteroaromatic-Base Interactions*, Elsevier, NY, **1984**. For reviews of structural and other studies, see Illuminati, G.; Stegel, F. *Adv. Heterocycl. Chem.* **1983**, *34*, 305; Artamkina, G. A.; Egorov, M.P.; Beletskaya, I.P. *Chem. Rev.* **1982**, *82*, 427; Terrier, F. *Chem. Rev.* **1982**, *82*, 77; Strauss, M.J. *Chem. Rev.* **1970**, *70*, 667; *Acc. Chem. Res.* **1974**, *7*, 181; Hall, T.N.; Poranski, Jr., C.F., in Feuer, H. *The Chemistry of the Nitro and Nitroso Groups*, pt. 2; Wiley, NY, **1970**, pp. 329–384; Crampton, M.R. *Adv. Phys. Org. Chem.* **1969**, *7*, 211; Foster R.; Fyfe, C.A. *Rev. Pure Appl. Chem.* **1966**, *16*, 61.

of several of these intermediates

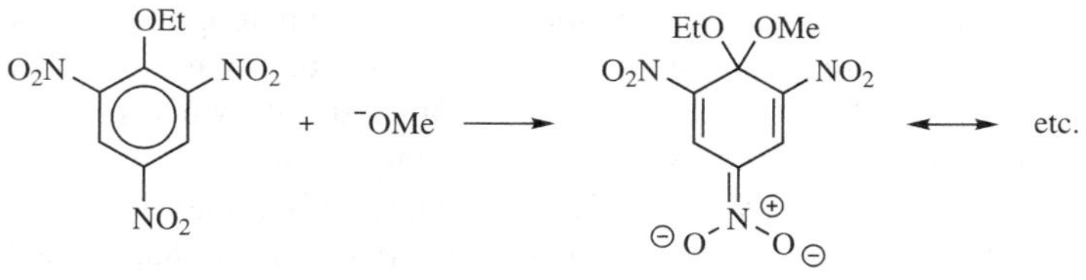

2

have been proved by NMR[9] and by X-ray crystallography.[10] Further evidence comes from studies of the effect of the leaving group on the reaction. If the mechanism were similar to either the S_N1 or S_N2 mechanisms described in Chapter 10, the Ar–X bond would be broken in the rate-determining step. In the S_NAr mechanism, this bond is not broken until after the rate-determining step (i.e. if step 1 is rate determining). There is some evidence that electron transfer may be operative during this process.[11] We would predict from this that if the S_NAr mechanism is operating, a change in leaving group should not have much effect on the reaction rate. In the reaction of dinitro compound **3** with piperidine,

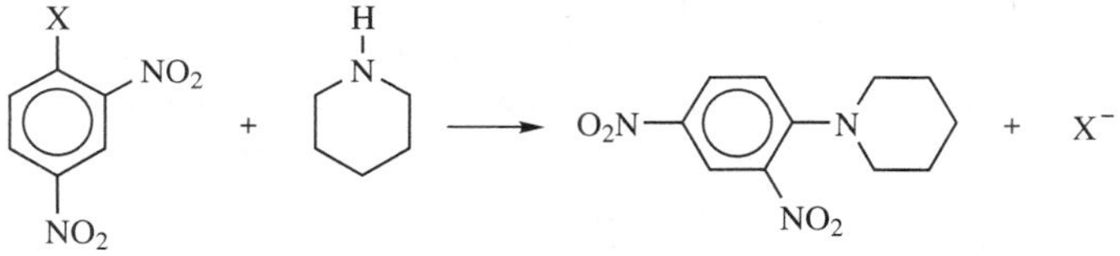

3

when X was Cl, Br, I, SOPh, SO$_2$Ph, or *p*-nitrophenoxy, the rates differed only by a factor of ~5.[12] This behavior would not be expected in a reaction in which the Ar–X bond is broken in the rate-determining step. We do not expect the rates to be *identical*, because the nature of X affects the rate at which Y attacks. An increase in the electronegativity of X causes a decrease in the electron density at the site of attack, resulting in a faster attack by a nucleophile. Thus, in the reaction just mentioned, when X = F, the relative rate was 3300 (compared with I = 1). The very fact that fluoro is the best leaving group among the halogens in most aromatic nucleophilic substitutions is good evidence that the mechanism is different from the S_N1

[9]First done by Crampton, M.R.; Gold, V. *J. Chem. Soc. B 1966*, 893. A good review of spectral studies is found, in Buncel, E.; Crampton, M.R.; Strauss, M.J.; Terrier, F. *Electron Deficient Aromatic- and Heteroaromatic-Base Interactions*, Elsevier, NY, *1984*, pp. 15–133.

[10]Destro, R.; Gramaccioli, C.M.; Simonetta, M. *Acta Crystallogr. 1968*, *24*, 1369; Ueda, H.; Sakabe, M.; Tanaka, J.; Furusaki, A. *Bull. Chem. Soc. Jpn. 1968*, *41*, 2866; Messmer, G.G.; Palenik, G.J. *Chem. Commun. 1969*, 470.

[11]Grossi, L. *Tetrahedron Lett. 1992*, *33*, 5645.

[12]Bunnett, J.F.; Garbisch Jr., E.W.; Pruitt, K.M. *J. Am. Chem. Soc. 1957*, *79*, 385. See Gandler, J.R.; Setiarahardjo, I.U.; Tufon, C.; Chen, C. *J. Org. Chem. 1992*, *57*, 4169 for a more recent example.

and the S_N2 mechanisms, where fluoro is by far the poorest leaving group of the halogens. This is an example of the element effect (p. 475).

The pattern of base catalysis of reactions with amine nucleophiles provides additional evidence. These reactions are catalyzed by bases only when a relatively poor leaving group (e.g., OR) is present (not Cl or Br) and only when relatively bulky amines are nucleophiles.[13] Bases could not catalyze step 1, but if amines are nucleophiles, bases can catalyze step 2. Base catalysis is found precisely in those cases where the amine moiety cleaves easily but X does not, so that k_{-1} is large and step 2 is rate determining. This is evidence for the S_NAr mechanism because it implies two steps. Furthermore, in cases where bases *are* catalysts, they catalyze only at

low base concentrations: a plot of the rate against the base concentration shows that small increments of base rapidly increase the rate until a certain concentration of base is reached, after which further base addition no longer greatly affects the rate. This behavior, based on a partitioning effect (see p. 660), is also evidence for the S_NAr mechanism. At low base concentration, each increment of base, by increasing the rate of step 2, increases the fraction of intermediate that goes to product rather than reverting to reactants. At high base concentration the process is virtually complete: there is very little reversion to reactants and the rate becomes dependent on step 1. Just how bases catalyze step 2 has been investigated. For protic solvents two proposals have been presented. One is that step 2 consists of two steps: rate-determining deprotonation of **4** followed by rapid loss of X,

and that bases catalyze the reaction by increasing the rate of the deprotonation step.[14] According to the other proposal, loss of X assisted by BH^+ is rate determining.[15] Two mechanisms, both based on kinetic evidence, have been proposed for aprotic solvents, such as benzene. In both proposals the ordinary S_NAr mechanism

[13]Kirby, A.J.; Jencks, W.P. *J. Am. Chem. Soc.* **1965**, *87*, 3217; Bunnett, J.F.; Bernasconi, C.F. *J. Org. Chem.* **1970**, *35*, 70; Bernasconi, C.F.; Schmid, P. *J. Org. Chem.* **1967**, *32*, 2953; Bernasconi, C.F.; Zollinger, H. *Helv. Chim. Acta* **1966**, *49*, 103; **1967**, *50*, 1; Pietra, F.; Vitali, D. *J. Chem. Soc. B* **1968**, 1200; Chiacchiera, S.M.; Singh, J.O.; Anunziata, J.D.; Silber, J.J. *J. Chem. Soc. Perkin Trans. 2* **1987**, 987.

[14]Bernasconi, C.F.; de Rossi, R.H.; Schmid, P. *J. Am. Chem. Soc.* **1977**, *99*, 4090, and references cited therein.

[15]Bunnett, J.F.; Sekiguchi, S.; Smith, L.A. *J. Am. Chem. Soc.* **1981**, *103*, 4865, and references cited therein.

operates, but in one the attacking species involves two molecules of the amine (the *dimer mechanism*),[16] while in the other there is a cyclic transition state.[17] Further evidence for the S$_N$Ar mechanism has been obtained from ^{18}O/^{16}O and ^{15}N/^{14}N isotope effects.[18]

Step 1 of the S$_N$Ar mechanism has been studied for the reaction between picryl chloride (as well as other substrates) and $^-$OH ions (**13-1**), and spectral evidence has been reported[19] for two intermediates, one a π complex (p. 662), and the other a radical ion–radical pair:

| Picryl chloride | π Complex | Radical ion–radical pair |

As with the tetrahedral mechanism at an acyl carbon, nucleophilic catalysis (p. 1259) has been demonstrated with an aryl substrate, in certain cases.[20] There is also evidence of an interaction of anions with the π-cloud of aromatic compounds.[21]

The S$_N$1 Mechanism

For aryl halides and sulfonates, even active ones, a unimolecular S$_N$1 mechanism (IUPAC: D$_N$ + A$_N$) is very rare; it has only been observed for aryl triflates in which both ortho positions contain bulky groups (*tert*-butyl or SiR$_3$).[22] It is in reactions with diazonium salts[23] that this mechanism is important:[24]

---

[16]For a review of this mechanism, see Nudelman, N.S. *J. Phys. Org. Chem.* **1989**, *2*, 1. See also Nudelman, N.S.; Montserrat, J.M. *J. Chem. Soc. Perkin Trans. 2* **1990**, 1073.

[17]Banjoko, O.; Bayeroju, I.A. *J. Chem. Soc. Perkin Trans. 2* **1988**, 1853; Jain, A.K.; Gupta, V.K.; Kumar, A. *J. Chem. Soc. Perkin Trans. 2* **1990**, 11.

[18]Hart, C.R.; Bourns, A.N. *Tetrahedron Lett.* **1966**, 2995; Ayrey, G.; Wylie, W.A. *J. Chem. Soc. B* **1970**, 738.

[19]Bacaloglu, R.; Blaskó, A.; Bunton, C.A.; Dorwin, E.; Ortega, F.; Zucco, C. *J. Am. Chem. Soc.* **1991**, *113*, 238, and references cited therein. For earlier reports, based on kinetic data, of complexes with amine nucleophiles, see Forlani, L. *J. Chem. Res. (S)* **1984**, 260; Hayami, J.; Otani, S.; Yamaguchi, F.; Nishikawa, Y. *Chem. Lett.* **1987**, 739; Crampton, M.R.; Davis, A.B.; Greenhalgh, C.; Stevens, J.A. *J. Chem. Soc. Perkin Trans. 2* **1989**, 675.

[20]See Muscio, Jr., O.J.; Rutherford, D.R. *J. Org. Chem.* **1987**, *52*, 5194.

[21]Quiñonero, D.; Garau, C.; Rotger, C.; Frontera, A.; Ballester, P.; Costa, A.; Deyà. P.M. *Angew. Chem. Int. Ed.* **2002**, *41*, 3389, and references cited therein.

[22]Himeshima, Y.; Kobayashi, H.; Sonoda, T. *J. Am. Chem. Soc.* **1985**, *107*, 5286.

[23]See Glaser, R.; Horan, C.J.; Nelson, E.D.; Hall, M.K. *J. Org. Chem.* **1992**, *57*, 215 for the influence of neighboring group interactions on the electronic structure of diazonium ions.

[24]Aryl iodonium salts Ar$_2$I$^+$ also undergo substitutions by this mechanism (and by a free-radical mechanism).

Step 1

Step 2

Among the evidence for the S_N1 mechanism[25] with aryl cations as intermediates,[26,27] is the following:[28]

1. The reaction rate is first order in diazonium salt and independent of the concentration of Y.

2. When high concentrations of halide salts are added, the product is an aryl halide but the rate is independent of the concentration of the added salts.

3. The effects of ring substituents on the rate are consistent with a unimolecular rate-determining cleavage.[29]

4. When reactions were run with substrate deuterated in the ortho position, isotope effects of ~1.22 were obtained.[30] It is difficult to account for such high secondary isotope effects in any other way except that an incipient phenyl cation is stabilized by hyperconjugation,[31] which is reduced when hydrogen is replaced by deuterium.

5. That the first step is reversible cleavage[32] was demonstrated by the observation that when $Ar^{15}\overset{H}{N}{\equiv}N$ was the reaction species, recovered starting

[25]For additional evidence, see Lorand, J.P. *Tetrahedron Lett.* **1989**, *30*, 7337.

[26]For a review of aryl cations, see Ambroz, H.B.; Kemp, T.J. *Chem. Soc. Rev.* **1979**, *8*, 353.

[27]For a monograph, see Stang, P.J.; Rappoport, Z.; Hanack, M.; Subramanian, L.R. *Vinyl Cations*, Academic Press, NY, 1979. For reviews of aryl and/or vinyl cations, see Hanack, M. *Pure Appl. Chem.* **1984**, *56*, 1819, *Angew. Chem. Int. Ed.* **1978**, *17*, 333; *Acc. Chem. Res.* **1976**, *9*, 364; Rappoport, Z. *Reactiv. Intermed.* (Plenum) **1983**, *3*, 427; Ambroz, H.B.; Kemp, T.J. *Chem. Soc. Rev.* **1979**, *8*, 353; Modena, G.; Tonellato, U. *Adv. Phys. Org. Chem.* **1971**, *9*, 185; Stang, P.J. *Prog. Org. Chem.* **1973**, *10*, 205. See also, Charton, M. *Mol. Struct. Energ.* **1987**, *4*, 271. For a computational study, see Glaser, R.; Horan, C.J.; Lewis, M.; Zollinger, H. *J. Org. Chem.* **1999**, *64*, 902.

[28]For a review, see Zollinger, H. *Angew. Chem, Int. Ed.* **1978**, *17*, 141. For discussions, see Swain, C.G.; Sheats, J.E.; Harbison, K.G. *J. Am. Chem. Soc.* **1975**, *97*, 783, 796; Burri, P.; Wahl, Jr., G.H.; Zollinger, H. *Helv. Chim. Acta* **1974**, *57*, 2099; Richey Jr., H.G.; Richey, J.M., in Olah, G.A.; Schleyer, P.v.R. *Carbonium Ions*, Vol. 2, Wiley, NY, **1970**, pp. 922–931; Zollinger, H. *Azo and Diazo Chemistry*, Wiley, NY, **1961**, pp. 138–142; Miller, J. *Aromatic Nucleophilic Substitution*, Elsevier, NY, **1968**, pp. 29–40.

[29]Lewis, E.S.; Miller, E.B. *J. Am. Chem. Soc.* **1953**, *75*, 429.

[30]Swain, C.G.; Sheats, J.E.; Gorenstein, D.G.; Harbison, K.G. *J. Am. Chem. Soc.* **1975**, *97*, 791.

[31]See Apeloig, Y.; Arad, D. *J. Am. Chem. Soc.* **1985**, *107*, 5285.

[32]For discussions, see Williams, D.L.H.; Buncel, E. *Isot. Org. Chem.* Vol. 5, Elsevier, Amsterdam, The Netherlands, **1980**, 147, 212; Zollinger, H. *Pure Appl. Chem.* **1983**, *55*, 401.

material contained not only $Ar^{15}\overset{H}{N}{\equiv}N$, but also $Ar\overset{H}{N}{\equiv}\overset{15}{N}$.[33,34] This could arise only If the nitrogen breaks away from the ring and then returns. Additional evidence was obtained by treating $Ph\overset{H}{N}{\equiv}\overset{15}{N}$ with unlabeled N_2 at various pressures. At 300 atm, the recovered product had lost ~3% of the labeled nitrogen, indicating that PhN_2^+ was exchanging with atmospheric N_2.[34]

There is kinetic and other evidence[35] that step 1 is more complicated and involves two steps, both reversible:

$$ArN_2^+ \rightleftarrows \underset{5}{[Ar^+ N_2]} \rightleftarrows Ar^+ + N_2$$

Intermediate **5**, which is probably some kind of a tight ion–molecule pair, has been trapped with carbon monoxide.[36]

The Benzyne Mechanism[37]

Some aromatic nucleophilic substitutions are clearly different in character from those that occur by the S_NAr mechanism (or the S_N1 mechanism). These substitutions occur on aryl halides that have no activating groups; bases are required that are stronger than those normally used; and most interesting of all, the incoming group does not always take the position vacated by the leaving group. That the latter statement is true was elegantly demonstrated by the reaction of 1-^{14}C-chlorobenzene with potassium amide:

The product consisted of almost equal amounts of aniline labeled in the 1 position and in the 2 position.[38]

[33]Lewis, E.S.; Kotcher, P.G. *Tetrahedron* **1969**, *25*, 4873; Lewis, E.S.; Holliday, R.E. *J. Am. Chem. Soc.* **1969**, *91*, 426; Tröndlin, F.; Medina, R.; Rüchardt, C. *Chem. Ber.* **1979**, *112*, 1835.
[34]Bergstrom, R.G.; Landell, R.G.M.; Wahl Jr., G.H.; Zollinger, H. *J. Am. Chem. Soc.* **1976**, *98*, 3301.
[35]Szele, I.; Zollinger, H. *Helv. Chim. Acta* **1981**, *64*, 2728.
[36]Ravenscroft, M.D.; Skrabal, P.; Weiss, B.; Zollinger, H. *Helv. Chim. Acta* **1988**, *71*, 515.
[37]For a monograph, see Hoffmann, R.W. *Dehydrobenzene and Cycloalkynes*, Academic Press, NY, **1967**. For reviews, see Gilchrist, T.L., in Patai, S.; Rappoport, Z. *The Chemistry of Functional Groups, Supplement C* pt. 1, Wiley, NY, **1983**, pp. 383–419; Bryce, M.R.; Vernon, J.M. *Adv. Heterocycl. Chem.* **1981**, *28*, 183; Levin R.H. *React. Intermed. (Wiley)* **1985**, *3*, 1; **1981**, *2*, 1; **1978**, *1*, 1; Nefedov, O.M.; D'yachenko, A.I.; Prokof'ev, A.K. *Russ. Chem. Rev.* **1977**, *46*, 941; Fields, E.K., in McManus, S.P. *Organic Reactive Intermediates*, Academic Press, NY, **1973**, pp. 449–508; Heaney, H. *Fortschr. Chem. Forsch.* **1970**, *16*, 35; *Essays Chem.* **1970**, *1*, 95; Hoffmann, R.W., in Viehe, H.G. *Acetylenes*, Marcel Dekker, NY, **1969**, pp. 1063–1148; Fields, E.K.; Meyerson, S. *Adv. Phys. Org. Chem.* **1968**, *6*, 1; Witting, G. *Angew. Chem. Int. Ed.* **1965**, *4*, 731.
[38]Roberts, J.D.; Semenow, D.A.; Simmons, H.E.; Carlsmith, L.A. *J. Am. Chem. Soc.* **1965**, *78*, 601.

A mechanism that can explain all these facts involves elimination followed by addition. In step 1, a suitable base removes the ortho hydrogen, with subsequent (or concomitant) loss of the chlorine (leaving group) to

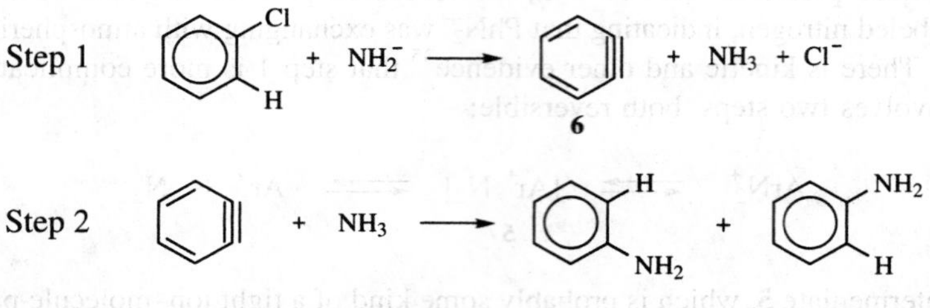

Step 1

generate symmetrical intermediate **6**[39] is called benzyne (see below).[40] In step 2, benzyne is attacked by the NH_3 at either of two positions, which explains why about half of the aniline produced from the radioactive chlorobenzene was labeled at the 2 position. The fact that the 1 and 2 positions were not labeled equally is the result of a small isotope effect. Other evidence for this mechanism is the following:

1. If the aryl halide contains two ortho substituents, the reaction should not be able to occur. This is indeed the case.[36]

2. It had been known many years earlier that aromatic nucleophilic substitution occasionally results in substitution at a different position. This is called *cine substitution*[41] and can be illustrated by the conversion of *o*-bromoanisole to *m*-aminoanisole.[42] In this particular case, only the meta isomer is

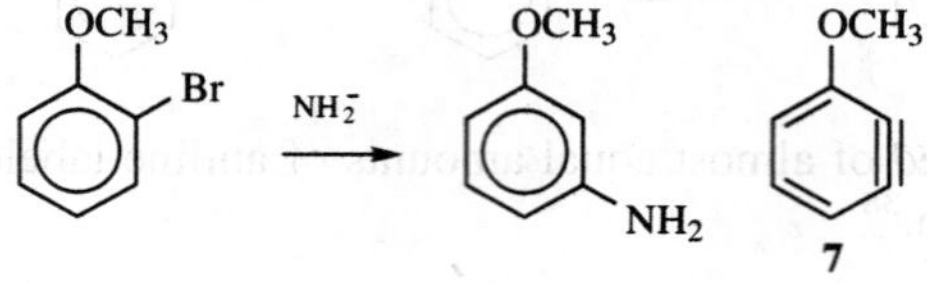

formed. The reason a 1:1 mixture is not formed is that the intermediate **7** is not symmetrical and the methoxy group directs the incoming group meta, but not ortho (see p. 867). However, not all cine substitutions proceed by this kind of mechanism (see **13-30**).

3. The fact that the order of halide reactivity is $Br > I > Cl > F$ (when the reaction is performed with KNH_2 in liquid NH_3) shows that the S_NAr mechanism is not operating here.[38]

---

[39]For a discussion of the structure of *m*- and *p*-benzynes, see Hess, Jr., B.A. *Eur. J. Org. Chem.* **2001**, 2185.
[40]For other methods to generate benzyne, see Kitamura, T.; Meng, Z.; Fujiwara, Y. *Tetrahedron Lett.* **2000**, *41*, 6611, and references cited therein; Kawabata, H.; Nishino, T.; Nishiyama, Y.; Sonoda, N. *Tetrahedron Lett.* **2002**, *43*, 4911, and references cited therein.
[41]For a review, see Suwiński, J.; wierczek, K. *Tetrahedron* **2001**, *57*, 1639.
[42]This example is from Gilman, H.; Avakian, S. *J. Am. Chem. Soc.* **1945**, *67*, 349. For a table of many such examples, see Bunnett, J.F.; Zahler, R.E. *Chem. Rev.* **1951**, *49*, 273, p. 385.

In the conversion of the substrate to **7**, either proton removal or subsequent loss of halide ion can be rate determining. In fact, the unusual leaving-group order just mentioned ($Br > I > Cl$) stems from a change in the rate-determining step. When the leaving group is Br or I, proton removal is rate-determining and the rate order for this step is $F > Cl > Br > I$. When Cl or F is the leaving group, cleavage of the C–X bond is rate determining and the order for this step is $I > Br > Cl > F$. Confirmation of the latter order was found in a direct competitive study. *meta*-Dihalobenzenes in which the two halogens are different were treated with $^-NH_2$.[43] In such compounds, the most acidic hydrogen is the one between the two halogens; when it leaves, the remaining anion can lose either halogen. Therefore, a study of which halogen is preferentially lost provides a direct measure of leaving-group ability. The order was found to be $I > Br > Cl$.[43,44]

Species, such as **6** and **7**, are called *benzynes* (sometimes *dehydrobenzenes*), or more generally, *arynes*,[45] and the mechanism is known as the *benzyne mechanism*. Benzynes are very reactive. Neither benzyne nor any other aryne has yet been isolated under ordinary conditions,[46] but benzyne has been isolated in an argon matrix at 8 K,[47] where its IR spectrum could be observed. In addition, benzynes can be trapped; for example, they undergo the Diels–Alder reaction (see **15-60**). Note that the extra pair of electrons does not affect the

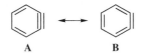

aromaticity. However, evaluation by a series of aromaticity indicators, including magnetic susceptibility anisotropies and exaltations, nucleus-independent chemical shifts (NICS), and aromatic stabilization energies, and valence-bond Pauling resonance energies point to the *o*-benzyne > *m*-benzyne > *p*-benzyne aromaticity order.[48] The relative order with respect to benzene depends on the aromaticity criterion.[48] The aromtic sextet from the aromatic precursor functions as a closed ring, and the two additional electrons are merely located in a π orbital that covers only two carbons. Benzynes do not have a formal triple bond, since two canonical forms (**A** and **B**) contribute to the hybrid. The IR spectrum, mentioned above, indicates that **A** contributes more than **B**. Not only benzene rings, but other aromatic

[43]Bunnett, J.F.; Kearley, Jr., F.J. *J. Org. Chem.* **1971**, *36*, 184.

[44]For a discussion of the diminished reactivity of ortho-substituted bromides, see Kalendra, D.M.; Sickles, B.R. *J. Org. Chem.* **2003**, *68*, 1594.

[45]For the use of arynes in organic synthesis see Pellissier, H.; Santelli, M. *Tetrahedron* **2003**, *59*, 701.

[46]For the measurement of aryne lifetimes in solution, see Gaviña, F.; Luis, S.V.; Costero, A.M.; Gil, P. *Tetrahedron* **1986**, *42*, 155.

[47]Chapman, O.L.; Mattes, K.; McIntosh, C.L.; Pacansky, J.; Calder, G.V.; Orr, G. *J. Am. Chem. Soc.* **1973**, *95*, 6134. For the ir spectrum of pyridyne trapped in a matrix, see Nam, H.; Leroi, G.E. *J. Am. Chem. Soc.* **1988**, *110*, 4096. For spectra of transient arynes, see Berry, R.S.; Spokes, G.N.; Stiles, M. *J. Am. Chem. Soc.* **1962**, *84*, 3570; Brown, R.D.; Godfrey, P.D.; Rodler, M. *J. Am. Chem. Soc.* **1986**, *108*, 1296.

[48]DeProft, F.; Schleyer, P.v.R.; van Lenthe, J.H.; Stahl, F.; Geerlings, P. *Chem. Eur. J.* **2002**, *8*, 3402.

rings[49] and even nonaromatic rings (p. 475) can react through this kind of intermediate. Of course, the non-aromatic rings do have a formal triple bond. When a benzyne unit is fused to a small ring, strain induced regioselectivity observed in its reactions.[50]

The $S_{RN}1$ Mechanism

When 5-iodo-1,2,4-trimethylbenzene **7** was treated with KNH_2 in NH_3, **8** and **10** were formed in the ratio 0.63:1. From what we have already seen, the presence of an unactivated substrate, a strong base, and the occurrence of cine substitution along with normal substitution would be strong indications of a benzyne mechanism. Yet if that were so, the 6-iodo isomer of **8** should have given **9** and **10** in the same ratio (because the same aryne intermediate would be formed in both cases), but in this case the ratio of **9–10** was 5.9:1 (the chloro and bromo analogs did give the same ratio, 1.46:1, showing that the benzyne mechanism may be taking place there).

To explain the iodo result, it has been proposed[51] that besides the benzyne mechanism, this free-radical mechanism is also operating here:

$$ArI \xrightarrow[\text{donor}]{\text{electron}} ArI^{\cdot-} \longrightarrow Ar^{\cdot} + I^-$$

$$Ar^{\cdot} + NH_2^- \longrightarrow ArNH_2^{\cdot-} + ArI \longrightarrow ArNH_2 + ArI^{\cdot-}$$

followed by terminations steps

This is called the $S_{RN}1$ mechanism,[52] and many other examples are known (see **13-3**, **13-4**, **13-6**, **13-14**). The IUPAC designation is $T + D_N + A_N$.[53] Note that the

---

[49]For reviews of *hetarynes* (benzyne intermediates in heterocyclic rings), see van der Plas, H.C.; Roeterdink, F., in Patai, S.; Rappoport, Z. *The Chemistry of Functional Groups, Supplement C*, pt. 1, Wiley, NY, *1983*, pp. 421–511; Reinecke, M.G. *React. Intermed. (Plenum) 1982*, *2*, 367; *Tetrahedron 1982*, *38*, 427; den Hertog, H.J.; van der Plas, H.C., in Viehe, H.G. *Acetylenes*, Marcel Dekker, NY, *1969*, pp. 1149–1197, *Adv. Heterocycl. Chem. 1971*, *40*, 121; Kauffmann, T.; Wirthwein, R. *Angew. Chem, Int. Ed. 1971*, *10*, 20; Kauffmann, T. *Angew. Chem, Int. Ed. 1965*, *4*, 543; Hoffmann, R.W. *Dehydrobenzene and Cycloalkynes*, Academic Press, NY, *1967*, pp. 275–309.

[50]Hamura, T.; Ibusuki, Y.; Sato, K.; Matsumoto, T.; Osamura, Y.; Suzuki, K. *Org. Lett. 2003*, *5*, 3551.

[51]Kim, J.K.; Bunnett, J.F. *J. Am. Chem. Soc. 1970*, *92*, 7463, 7464.

[52]For a monograph, see Rossi, R.A.; de Rossi, R.H. *Aromatic Substitution by the $S_{RN}1$ Mechanism*, American Chemical Society, Washington, 1983. For reviews, see Savéant, J. *Adv. Phys. Org. Chem. 1990*, *26*, 1; Russell, G.A. *Adv. Phys. Org. Chem. 1987*, *23*, 271; Norris, R.K. in Patai, S. Rappoport, Z. *The Chemistry of Functional Groups, Supplement D*, pt. 1, Wiley, NY, *1983*, pp. 681–701; Chanon, M.; Tobe, M.L. *Angew. Chem, Int. Ed. 1982*, *21*, 1; Rossi, R.A. *Acc. Chem. Res. 1982*, *15*, 164; Beletskaya, I.P.; Drozd, V.N. *Russ. Chem. Rev. 1979*, *48*, 431; Bunnett, J.F. *Acc. Chem. Res. 1978*, *11*, 413; Wolfe, J.F.; Carver, D.R. *Org. Prep. Proced. Int. 1978*, *10*, 225. For a review of this mechanism with aliphatic substrates, see Rossi, R.A.; Pierini, A.B.; Palacios, S.M. *Adv. Free Radical Chem.* (*Greenwich, Conn.*) *1990*, *1*, 193. For 'thermal' $S_{RN}1$ reactions, see Costentin, C.; Hapiot, P.; Médebielle, M.; Savéant, J.-M. *J. Am. Chem. Soc. 1999*, *121*, 4451.

[53]The symbol T is used for electron transfer.

last step of the mechanism produces $ArI^{•-}$ radical ions, so the process is a chain mechanism (see p. 936).[54] An electron donor is required to initiate the reaction. In the case above it was solvated electrons from KNH_2 in NH_3. Evidence was that the addition of potassium metal (a good producer of solvated electrons in ammonia) completely suppressed the cine substitution. Further evidence for the $S_{RN}1$ mechanism was that addition of radical scavengers (which would suppress a free-radical mechanism) led to **9:10** ratios much closer to 1.46:1. Numerous other observations of $S_{RN}1$ mechanisms that were stimulated by solvated electrons and inhibited by radical scavengers have also been recorded.[55] Further evidence for the $S_{RN}1$ mechanism in the case above was that some 1,2,4-trimethylbenzene was found among the products. This could easily be formed by abstraction by $Ar^•$ of H from the solvent NH_3. Besides initiation by solvated electrons,[56] $S_{RN}1$ reactions have been initiated photochemically,[57] electrochemically,[58] and even thermally.[59]

The $S_{RN}1$ reactions have a fairly wide scope. The efficiency of the reaction has been traced to the energy level of the radical anion of the substitution product.[60] There is no requirement for activating groups or strong bases, but in DMSO haloarenes are less reactive as the stability of the anion increases.[61] The reaction has also been done in liquid ammonia, promoted by ultrasound (p. 349),[62] and ferrous ion has been used as a catalyst.[63] Alkyl, alkoxy, aryl, and COO^- groups do not interfere, although Me_2N, O^-, and NO_2 groups do interfere. Cine substitution is not found.

Other Mechanisms

There is no clear-cut proof that a one-step S_N2 mechanism, so important at a saturated carbon, ever actually occurs with an aromatic substrate. The hypothetical aromatic S_N2 process is sometimes called the *one-stage* mechanism to distinguish it from the *two-stage* S_NAr mechanism. A "clean" example of a $S_{RN}2$ reaction has been reported, the conversion of **11** to **12** in methanol.[64] Both the $S_{RN}1$ and S_{RN}^2 reactions have been reviewed.[65]

[54]For a discussion, see Amatore, C.; Pinson, J.; Savéant, J.; Thiébault, A. *J. Am. Chem. Soc.* ***1981***, *103*, 6930.

[55]Bunnett, J.F. *Acc. Chem. Res.* ***1978***, *11*, 413.

[56]Savéant, J.-M. *Tetrahedron* ***1994***, *50*, 10117.

[57]For reviews of photochemical aromatic nucleophilic substitutions, see Cornelisse, J.; de Gunst, G.P.; Havinga, E. *Adv. Phys. Org. Chem.* ***1975***, *11*, 225; Cornelisse, J. *Pure Appl. Chem.* ***1975***, *41*, 433; Pietra, F. *Q. Rev. Chem. Soc.* ***1969***, *23*, 504, p. 519.

[58]For a review, see Savéant, J. *Acc. Chem. Res.* ***1980***, *13*, 323. See also, Alam, N.; Amatore, C.; Combellas, C.; Thiébault, A.; Verpeaux, J.N. *J. Org. Chem.* ***1990***, *55*, 6347.

[59]Swartz, J.E.; Bunnett, J.F. *J. Org. Chem.* ***1979***, *44*, 340, and references cited therein.

[60]Galli, C.; Gentili, P.; Guarnieri, A. *Gazz. Chim. Ital.*, ***1995***, *125*, 409.

[61]Borosky, G.L.; Pierini, A.B.; Rossi, R.A. *J. Org. Chem.* ***1992***, *57*, 247.

[62]Manzo, P.G.; Palacios, S.M.; Alonso, R.A. *Tetrahedron Lett.* ***1994***, *35*, 677.

[63]Galli, C.; Gentili, P.; *J. Chem. Soc. Perkin Trans. 2* ***1993***, 1135.

[64]Marquet, J.; Jiang, Z.; Gallardo, I.; Batlle, A.; Cayón, E. *Tetrahedron Lett.* ***1993***, *34*, 2801. Also see, Keegstra, M.A. *Tetrahedron* ***1992***, *48*, 2681.

[65]Rossi, R.A.; Palacios, S.M. *Tetrahedron* ***1993***, *49*, 4485.

Some of the reactions in this chapter operate by still other mechanisms, among them an addition–elimination mechanism (see **13-17**). A new mechanism has been reported in aromatic chemistry, a reductively activated 'polar' nucleophilic aromatic substitution.[66] The reaction of phenoxide with *p*-dinitrobenzene in DMF shows radical features that cannot be attributed to a radical anion, and it is not $S_{RN}2$. The new designation was proposed to account for these results.

REACTIVITY

The Effect of Substrate Structure

In the discussion of electrophilic aromatic substitution (Chapter 11) equal attention was paid to the effect of substrate structure on reactivity (activation or deactivation) and on orientation. The question of orientation was important because in a typical substitution there are four or five hydrogens that could serve as leaving groups. This type of question is much less important for aromatic nucleophilic substitution, since in most cases there is only one potential leaving group in a molecule. Therefore attention is largely focused on the reactivity of one molecule compared with another and not on the comparison of the reactivity of different positions within the same molecule.

S_NAr Mechanism. These substitutions are accelerated by electron-withdrawing groups, especially in positions ortho and para to the leaving group[67] and hindered by electron-attracting groups. This is, of course, opposite to the effects of these groups on electrophilic substitutions, and the reasons are similar to those discussed in Chapter 11 (p. 660). Table 13.1 contains a list of groups arranged approximately in order of activating or deactivating ability.[68] Nitrogen atoms are also strongly activating (especially to the α and γ positions) and are even more so when quaternized.[69] Both 2- and 4-chloropyridine, for example, are often used as substrates. Heteroaromatic amine *N*-oxides are readily attacked by nucleophiles in the 2 and 4 positions, but the oxygen is generally lost in these reactions.[70]

[66]Marquet, J.; Casado, F.; Cervera, M.; Espín, M.; Gallardo, I.; Mir, M.; Niat, M. *Pure Appl. Chem.* **1995**, *67*, 703.

[67]The effect of meta substituents has been studied much less, but it has been reported that here too, electron-withdrawing groups increase the rate: See Nurgatin, V.V.; Sharnin, G.P.; Ginzburg, B.M. *J. Org. Chem, USSR* **1983**, *19*, 343.

[68]For additional tables of this kind, see Miller, J. *Aromatic Nucleophilic Substitution*, Elsevier, NY, *1968*, pp. 61–136.

[69]Miller, J.; Parker, A.J. *Aust. J. Chem.* **1958**, *11*, 302.

[70]Berliner, E.; Monack, L.C. *J. Am. Chem. Soc.* **1952**, *74*, 1574.

TABLE 13.1. Groups Listed in Approximate Descending Order of Activating Ability in the S_NAr Mechanism[68]

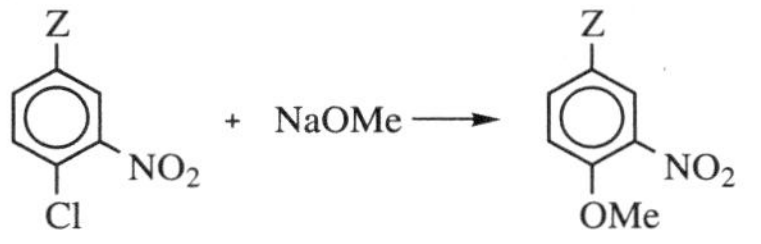

at $0°C^{71}$ $(a)^a$

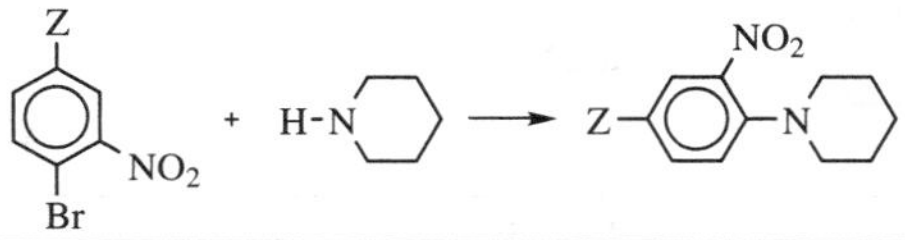

at $25°C^{72}$ $(a)^a$

Comments[b]	Group Z	Relative Rate of Reaction	
		(a) H $= 1^{69}$	(b) NH$_2 = 1^{70}$
Activates halide exchange at room temperature	N_2^+		
Activates reaction with strong nucleophiles at room temperature	$\overset{\oplus}{\underset{/\!/}{N}}$–R (heterocyclic)		
Activate reactions with strong nucleophiles at 80–100°C	NO	5.22×10^6	
	NO$_2$	6.73×10^5	Very fast
	$\underset{/\!/}{N}$ (heterocyclic)		
With nitro also present, activate reactions with strong nucleophiles at room temperature	SO$_2$Me		
	NMe$_3^+$		
	CF$_3$		
	CN	3.81×10^4	
	CHO	2.02×10^4	
With nitro also present, activate reactions with strong nucleophiles at 40–60°C	COR		
	COOH		
	SO$_3^-$		
	Br		6.31×10^4
	Cl		4.50×10^4
	I		4.36×10^4
	COO$^-$		2.02×10^4
	H		8.06×10^3
	F		2.10×10^3
	CMe$_3$		1.37×10^3
	Me		1.17×10^3

(*continued*)

[71]For reviews of reactivity of nitrogen-containing heterocycles, see Illuminati, G. *Adv. Heterocycl. Chem.* **1964**, *3*, 285; Shepherd, R.G.; Fedrick, J.L. *Adv. Heterocycl. Chem.* **1965**, *4*, 145.

[72]For reviews, see Albini, A.; Pietra, S. *Heterocyclic N-Oxides*; CRC Press: Boca Raton, FL, **1991**, pp. 142–180; Katritzky, A.R.; Lagowski, J.M. *Chemistry of the Heterocyclic N-Oxides*, Academic Press, NY, **1971**, pp. 258–319, 550–553.

TABLE 13.1. (*Continued*)

		Relative Rate of Reaction	
Comments[b]	Group Z	(a) H = 1[69]	(b) NH$_2$ = 1[70]
	OMe		145
	NMe$_2$		9.77
	OH		4.70
	NH$_2$		1

[a]For reaction (a) the rates are relative to **H**; *for (b)* they are relative to NH$_2$.
[b]The comments on the left column are from Ref. 73.

The most highly activating group, N$_2^+$, is seldom deliberately used to activate a reaction, but it sometimes happens that in the diazotization of a compound, such as *p*-nitroaniline or *p*-chloroaniline, the group para to the diazonium group is replaced by OH from the solvent or by X from ArN$_2^+$ X$^-$, to the surprise and chagrin of the investigator, who was trying only to replace the diazonium group and to leave the para group untouched. By far, the most common activating group is the nitro group and the most common substrates are 2,4-dinitrophenyl halides and 2,4,6-trinitrophenyl halides (also called picryl halides).[74] Polyfluorobenzenes[75] (e.g., C$_6$F$_6$), also undergo aromatic nucleophilic substitution quite well.[76] Benzene rings that lack activating substituents are generally not useful substrates for the S$_N$Ar mechanism, because the two extra electrons in **1** are in an antibonding orbital (p. 34). Activating groups, by withdrawing electron density, are able to stabilize the intermediates and the

[73]Bunnett, J.F.; Zahler, R.E. *Chem. Rev.* **1951**, *49*, 273, p. 308.
[74]For a review of the activating effect of nitro groups, see de Boer, T.J.; Dirkx, I.P., in Feuer, H. *The Chemistry of the Nitro and Nitroso Groups*, pt. 1, Wiley, NY, **1970**, pp. 487–612.
[75]Fluorine significantly activates ortho and meta positions, and slightly deactivates (see Table 13.1) para positions: Chambers, R.D.; Seabury, N.J.; Williams, D.L.H.; Hughes, N. *J. Chem. Soc. Perkin Trans. 1* **1988**, 255.
[76]For reviews, see Yakobson, G.G.; Vlasov, V.M. *Synthesis* **1976**, 652; Kobrina, L.S. *Fluorine Chem. Rev.* **1974**, *7*, 1.

transition states leading to them. Reactions taking place by the S_NAr mechanism are also accelerated when the aromatic ring is coordinated with a transition metal.[77]

Just as electrophilic aromatic substitutions were found more or less to follow the Hammett relationship (with σ^+ instead of σ; see p. 402), so do nucleophilic substitutions, with σ^- instead of σ for electron-withdrawing groups.[78]

Benzyne Mechanism. Two factors affect the position of the incoming group, the first being the direction in which the aryne forms.[79] When there are groups ortho or para to the leaving group, there is no choice:

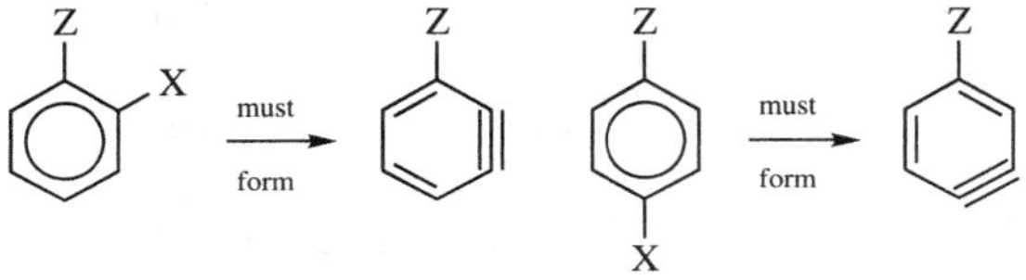

but when a meta group is present, the aryne can form in two different ways:

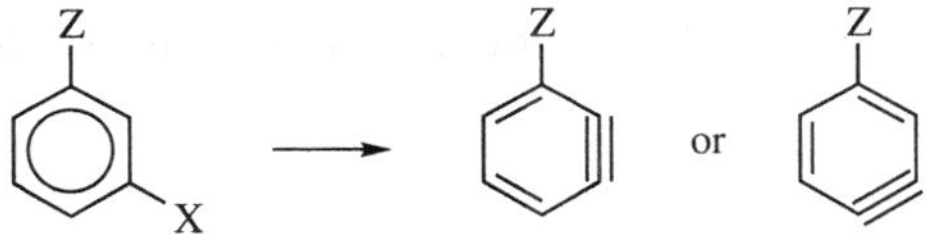

In such cases, the more acidic hydrogen is removed. Since acidity is related to the field effect of Z, it can be stated that an electron-attracting Z favors removal of the ortho hydrogen while an electron-donating Z favors removal of the para hydrogen. The second factor is that the aryne, once formed, can be attacked at two positions. The favored position for nucleophilic attack is the one that leads to the more stable carbanion intermediate, and this in turn also depends on the field effect of Z. For *-I* groups, the more stable carbanion is the one in which the negative charge is closer to the substituent. These principles are illustrated by the reaction of the three dichlorobenzenes (**13-15**) with alkali-metal

[77]For a review, see Balas, L.; Jhurry, D.; Latxague, L.; Grelier, S.; Morel, Y.; Hamdani, M.; Ardoin, N.; Astruc, D. *Bull. Soc. Chim. Fr.* **1990**, 401. For a discussion of iron assisted nucleophilic aromatic substitution on the solid phase, see Ruhland, T.; Bang, K.S.; Andersen, K. *J. Org. Chem.* **2002**, *67*, 5257.

[78]For a discussion of linear free-energy relationships in this reaction, see Bartoli, G.; Todesco, P.E. *Acc. Chem. Res.* **1977**, *10*, 125. For a list of σ^- values, see Table 9.4 on p. 404.

[79]This analysis is from Roberts, J.D.; Vaughan, C.W.; Carlsmith, L.A.; Semenow, D.A. *J. Am. Chem. Soc.* **1956**, *78*, 611. For a discussion, see Hoffmann, R.W. *Dehydrobenzene and Cycloalkynes*, Academic Press, NY, **1973**, pp. 134–150.

amides to give the predicted products shown.

13: 1,2-dichlorobenzene —(must give)→ 3-chlorobenzyne —(should give)→ *m*-chloroaniline (Cl, NH$_2$)

14: 1,4-dichlorobenzene —(must give)→ benzyne —(should give)→ *p*-chloroaniline (Cl, NH$_2$)

15: 1,3-dichlorobenzene —(should give)→ chlorobenzyne —(should give)→ *m*-chloroaniline (Cl, NH$_2$)

In each case, the predicted product was the one chiefly formed.[80] The observation that *m*-aminoanisole is obtained, mentioned on p. 860, is also in accord with these predictions.

The Effect of the Leaving Group[81]

The common leaving groups in aliphatic nucleophilic substitution (halide, sulfate, sulfonate, NR_3^+, etc.) are also common leaving groups in aromatic nucleophilic substitutions, but the groups NO_2, OR, OAr, SO_2R,[82] and SR, which are not generally lost in aliphatic systems, *are* leaving groups when attached to aromatic rings. Surprisingly, NO_2 is a particularly good leaving group.[83] An approximate order of leaving-group ability is[84] $F > NO_2 > OTs > SOPh > Cl$, Br, $I > N_3 > NR_3^+ >$ OAr, OR, SR, NH_2. However, this depends greatly on the nature of the nucleophile, as illustrated by the fact that $C_6Cl_5OCH_3$ treated with NH_2^- gives mostly $C_6Cl_5NH_2$; that is, one methoxy group is replaced in preference to five chlorines.[85] As usual, OH can be a leaving group if it is converted to an inorganic ester. Among

---

[80]Wotiz, J.H.; Huba, F. *J. Org. Chem.* **1959**, *24*, 595. Eighteen other reactions also gave products predicted by these principles. See also, Caubere, P.; Lalloz, L. *Bull. Soc. Chim. Fr.* **1974**, 1983, 1989, 1996; Biehl, E.R.; Razzuk, A.; Jovanovic, M.V.; Khanapure, S.P. *J. Org. Chem.* **1986**, *51*, 5157.

[81]For a review, see Miller, J. *Aromatic Nucleophilic Substitution*, Elsevier, NY, **1968**, pp. 137–179.

[82]See, for example, Furukawa, N.; Ogawa, S.; Kawai, T.; Oae, S. *J. Chem. Soc. Perkin Trans. 1* **1984**, 1839.

[83]For a review, see Beck, J.R. *Tetrahedron* **1978**, *34*, 2057. See also, Effenberger, F.; Koch, M.; Streicher, W. *Chem. Ber.* **1991**, *24*, 163.

[84]Loudon, J.D.; Shulman, N. *J. Chem. Soc.* **1941**, 772; Suhr, H. *Chem. Ber.* **1963**, *97*, 3268.

[85]Kobrina, L.S.; Yakobson, G.G. *J. Gen. Chem. USSR* **1963**, *33*, 3238.

the halogens, fluoro is generally a much better leaving group than the other halogens, which have reactivities fairly close together. The order is usually $Cl > Br > I$, but not always.[86] The leaving-group order is quite different from that for the S_N1 or S_N2 mechanisms. The most likely explanation is that the first step of the S_NAr mechanism is usually rate determining, and this step is promoted by groups with strong $-I$ effects. This would explain why fluoro and nitro are such good leaving groups when this mechanism is operating. Fluoro is the poorest leaving group of the halogens when the second step of the S_NAr mechanism is rate determining or when the benzyne mechanism is operating. The four halogens, as well as SPh, NMe_3^+, and $OPO(OEt)_2$, have been shown to be leaving groups in the $S_{RN}1$ mechanism.[55] The only important leaving group in the S_N1 mechanism is N_2^+.

The Effect of the Attacking Nucleophile[87]

It is not possible to construct an invariant nucleophilicity order because different substrates and different conditions lead to different orders of nucleophilicity, but an overall approximate order is $^-NH_2 > Ph_3C^- > PhNH^-$ (aryne mechanism) $> ArS^- > RO^{\neq} > R_2NH > ArO^- > {}^-OH > ArNH_2 > NH_3 > I^- > Br^- > Cl^- > H_2O > ROH$.[88] As with aliphatic nucleophilic substitution, nucleophilicity is generally dependent on base strength and nucleophilicity increases as the attacking atom moves down a column of the periodic table, but there are some surprising exceptions, for example, ^-OH, a stronger base than ArO^-, is a poorer nucleophile.[89] In a series of similar nucleophiles, such as substituted anilines, nucleophilicity *is* correlated with base strength. Oddly, the cyanide ion is not a nucleophile for aromatic systems, except for sulfonic acid salts and in the von Richter (**13-30**) and Rosenmund-von Braun (**13-8**) reactions, which are special cases.

REACTIONS

In the first part of this section, reactions are classified according to attacking species, with all leaving groups considered together, except for hydrogen and N_2^+, which are treated subsequently. Finally, a few rearrangement reactions are discussed.

[86]Reinheimer, J.D.; Taylor, R.C.; Rohrbaugh, P.E. *J. Am. Chem. Soc.* **1961**, *83*, 835; Ross, S.D. *J. Am. Chem. Soc.* **1959**, *81*, 2113; Bunnett, J.F.; Garbisch Jr., E.W.; Pruitt, K.M. *J. Am. Chem. Soc.* **1957**, *79*, 385; Parker, R.E.; Read, T.O. *J. Chem. Soc.* **1962**, 9, 3149; Litvinenko, L.M.; Shpan'ko, L.V.; Korostylev, A.P. *Doklad. Chem.* **1982**, *266*, 309.
[87]For a review, see Miller, J. *Aromatic Nucleophilic Substitution*, Elsevier, NY, **1968**, pp. 180–233.
[88]This list is compiled from data, in Bunnett, J.F. ; Zahler, R.E. *Chem. Rev.* **1951**, *49*, 273, p. 340; Bunnett, J.F. *Q. Rev. Chem. Soc.* **1958**, *12*, 1, p. 13; Sauer, J.; Huisgen, R. *Angew. Chem.* **1960**, *72*, 294, p. 311; Bunnett, J.F. *Annu. Rev. Phys. Chem.* **1963**, *14*, 271.
[89]For studies of nucleophilicity in the $S_{RN}1$ mechanism, see Amatore, C.; Combellas, C.; Robveille, S.; Savéant, J.; Thiébault, A. *J. Am. Chem. Soc.* **1986**, *108*, 4754, and references cited therein.

ALL LEAVING GROUPS EXCEPT HYDROGEN AND N_2^+

A. Oxygen Nucleophiles

13-1 Hydroxylation of Aromatic Compounds

Hydroxy-de-halogenation

$$ArBr + OH^- \longrightarrow ArOH$$

Aryl halides are converted to phenols if activating groups are present or if exceedingly strenuous conditions are employed.[90] When the reaction is carried out at high temperatures, cine substitution is observed, indicating a benzyne mechanism.[91] The reaction has been done using NaOH on Montmorillonite K10 and $AgNO_3$ with microwave irradiation.[92]

A slightly related reaction involves the amino group of naphthylamines can be replaced by a hydroxyl group by treatment with aqueous bisulfite.[93] The scope is greatly limited; the amino group (which may be NH_2 or NHR) must be on a naphthalene ring, with very few exceptions. The reaction is reversible (see **13-6**), and both the forward and reverse reactions are called the *Bucherer reaction*.

$$ArMgX \xrightarrow{\text{B(OMe)}_3} ArB(OMe)_2 \xrightarrow[\text{H}_2\text{O}_2]{\text{H}^+} ArOH$$

An indirect method for conversion of an aryl halide to a phenol involves initial conversion to an organometallic, followed by oxidation to the phenol. For the conversion of aryl Grignard reagents to phenols, a good procedure is the use of trimethyl borate followed by oxidation with H_2O_2 in acetic acid[94] (see **12-31**). Phenols have been obtained from unactivated aryl halides by treatment with borane and a metal such as lithium, followed by oxidation with alkaline H_2O_2.[95] Arylboronic acids, $ArB(OH)_2$, are oxidized by aqueous hydrogen peroxide to give the corresponding phenol.[96] The reaction of an aromatic compound with a borane in the

[90]For a review of ⁻OH and ⁻OR as nucleophiles in aromatic substitution, see Fyfe, C.A., in Patai, S. *The Chemistry of the Hydroxyl Group*, pt. 1, Wiley, NY, *1971*, pp. 83–124.

[91]The benzyne mechanism for this reaction is also supported by ^{14}C labeling experiments: Bottini, A.T.; Roberts, J.D. *J. Am. Chem. Soc. 1957*, *79*, 1458; Dalman, G.W.; Neumann, F.W. *J. Am. Chem. Soc. 1968*, *90*, 1601.

[92]Hashemi, M.M.; Akhbari, M. *Synth. Commun. 2004*, *34*, 2783.

[93]For reviews, see Seeboth, H. *Angew. Chem, Int. Ed. 1967*, *6*, 307; Gilbert, E.E. *Sulfonation and Related Reactions*; Wiley, NY, *1965*, pp. 166–169.

[94]Hawthorne, M.F. *J. Org. Chem. 1957*, *22*, 1001. For other procedures, see Lewis, N.J.; Gabhe, S.Y. *Aust. J. Chem. 1978*, *31*, 2091; Hoffmann, R.W.; Ditrich, K. *Synthesis 1983*, 107.

[95]Pickles, G.M.; Thorpe, F.G. *J. Organomet. Chem. 1974*, *76*, C23.

[96]Simon, J.; Salzbrunn, S.; Prakash, G.K.S.; Petasis, N.A.; Olah, G.A. *J. Org. Chem. 2001*, *66*, 633.

presence of an iridium catalyst, followed by oxidation with aqueous Oxone® gave the corresponding phenol.[97] Aryllithium reagents have been converted to phenols by treatment with oxygen.[98] In a related indirect method, arylthallium bis(trifluoroacetates) (prepared by **12-23**) can be converted to phenols by treatment with lead tetraacetate followed by triphenylphosphine and then dilute NaOH.[99] Diarylthallium trifluoroacetates undergo the same reaction.[100]

OS **I**, 455; **II**, 451; **V**, 632. Also see, OS **V**, 918.

13-2 Alkali Fusion of Sulfonate Salts

Oxido-de-sulfonato-substitution

$$ArSO_3^- \xrightarrow[\text{300--320°C}]{\text{NaOH fusion}} ArO^-$$

Aryl sulfonic acids can be converted, through their salts, to phenols, by alkali fusion. In spite of the extreme conditions, the reaction gives fairly good yields, except when the substrate contains other groups that are attacked by alkali at the fusion temperatures. Milder conditions can be used when the substrate contains activating groups, but the presence of deactivating groups hinders the reaction. The mechanism is obscure, but a benzyne intermediate has been ruled out by the finding that cine substitution does not occur.[101]

OS **I**, 175; **III**, 288.

13-3 Replacement by OR or OAr

Alkoxy-de-halogenation

$$ArBr \;+\; OR^- \longrightarrow ArOR$$

This reaction is similar to **13-1** and, like that one, generally requires activated substrates.[90,102] With unactivated substrates, side reactions predominate, though aryl methyl ethers have been prepared from unactivated chlorides by treatment with MeO⁻ in HMPA.[103] This reaction gives better yields than **13-1** and is

---

[97]Maleczka Jr., R.E.; Shi, F.; Holmes, D.; Smith III, M.R. *J. Am. Chem. Soc.* **2003**, *125*, 7792.

[98]Parker, K.A.; Koziski, K.A. *J. Org. Chem.* **1987**, *52*, 674. For other reagents, see Taddei, M.; Ricci, A. *Synthesis* **1986**, 633; Einhorn, J.; Luche, J.; Demerseman, P. *J. Chem. Soc. Chem. Commun.* **1988**, 1350.

[99]Taylor, E.C.; Altland, H.W.; Danforth, R.H.; McGillivray, G.; McKillop, A. *J. Am. Chem. Soc.* **1970**, *92*, 3520.

[100]Taylor, E.C.; Altland, H.W.; McKillop, A. *J. Org. Chem.* **1975**, *40*, 2351.

[101]Buzbee, L.R. *J. Org. Chem.* **1966**, *31*, 3289; Oae, S.; Furukawa, N.; Kise, M.; Kawanishi, M. *Bull. Chem. Soc. Jpn.* **1966**, *39*, 1212.

[102]See Gujadhur, R.; Venkataraman, D. *Synth. Commun.* **2001**, *31*, 2865.

[103]Shaw, J.E.; Kunerth, D.C.; Swanson, S.B. *J. Org. Chem.* **1976**, *41*, 732; Testaferri, L.; Tiecco, M.; Tingoli, M.; Chianelli, D.; Montanucci, M. *Tetrahedron* **1983**, *39*, 193.

used more often. A good solvent is liquid ammonia. Aryl chlorides react with phenol and KOH with microwave irradiation to give the diaryl ether.[104] Potassium phenoxide reacts with iodobenzene in an ionic solvent at 100°C with CuCl.[105] The NaOMe reacted with o- and p-fluoronitrobenzenes $\sim 10^9$ times faster in NH_3 at -70°C than in MeOH.[106] Phase-transfer catalysis has also been used.[107] Phenols reacted with aryl fluorides with K_2CO_3/DMSO[108] or aryl chlorides with KOH,[109] with microwave irradiation, to give the diaryl ether. Aryl carbonates react with aryl oxides.[110] Phenolic compounds react with aryl fluorides in the presence of LiOH in DMF to give the diaryl ether.[111] Aryl iodides react with phenols in the presence of K_2CO_3, CuI and Raney nickel alloy.[112]

In addition to halides, leaving groups can be other OR, and so on, even OH.[113] Acid salts, $RCOO^-$, are sometimes used as nucleophiles. Good yields of aryl benzoates can be obtained by the treatment of aryl halides with cuprous benzoate in diglyme or xylene at 140–160°C.[114] Unactivated substrates have been converted to carboxylic esters in low-to-moderate yields under oxidizing conditions.[115] The following chain mechanism, called the $S_{ON}2$ mechanism,[116] has been suggested:[115]

For aroxide nucleophiles, the reaction is promoted by copper salts,[117] and when these are used, activating groups need not be present. Indeed, unactivated aryl

---

[104]Rebeiro, G.L.; Khadilkar, B.M. *Synth. Commun.* ***2003***, *33*, 1405.

[105]In bmim BF_4, 1-butyl-3-methylimidazolium tetrafluoroborate: Chauhan, S.M.S.; Jain, N.; Kumar, A.; Srinivas, K.A. *Synth. Commun.* ***2003***, *33*, 3607.

[106]Kizner, T.A.; Shteingarts, V.D. *J. Org. Chem, USSR* ***1984***, *20*, 991.

[107]Artamanova, N.N.; Seregina, V.F.; Shner, V.F.; Salov, B.V.; Kokhlova, V.M.; Zhdamarova, V.N. *J. Org. Chem, USSR* ***1989***, *25*, 554.

[108]Li, F.; Wang, Q.; Ding, Z.; Tao, F. *Org. Lett.* ***2003***, *5*, 2169.

[109]Chaouchi, M.; Loupy, A.; Marque, S.; Petit, A. *Eur. J. Org. Chem.* ***2002***, 1278.

[110]Castro, E.A.; Pavez, P.; Santos, J.G. *J. Org. Chem.* ***2001***, *66*, 3129.

[111]Ankala, S.V.; Fenteany, G. *Synlett* ***2003***, 825.

[112]Xu, L.-W.; Xia, C.-G.; Li, J.-W.; Hu, X.-X. *Synlett* ***2003***, 2071.

[113]Oae, S.; Kiritani, R. *Bull. Chem. Soc. Jpn.* ***1964***, *37*, 770; ***1966***, *39*, 611.

[114]Cohen, T.; Wood, J.; Dietz Jr., A.G. *Tetrahedron Lett.* ***1974***, 3555.

[115]Jönsson, L.; Wistrand, L. *J. Org. Chem.* ***1984***, *49*, 3340.

[116]First proposed by Alder, R.W. *J. Chem. Soc. Chem. Commun.* ***1980***, 1184.

[117]For a review of copper-assisted aromatic nucleophilic substitution, see Lindley, J. *Tetrahedron* ***1984***, *40*, 1433. For other examples, see Marcoux, J.-F.; Doye, S.; Buchwald, S.L. *J. Am. Chem. Soc.* ***1997***, *119*, 10539; Ma, D.; Cai, Q. *Org. Lett.* ***2003***, *5*, 3799.

halides, such as 4-iodoanisole, were coupled to allylic alcohols using a CuI catalyst in the presence of 2% 1,10-phenanthroline and cesium carbonate.[118] This method of preparation of diaryl ethers is called the *Ullmann ether synthesis*[119] and should not be confused with the Ullmann biaryl synthesis (**13-11**). The reactivity order is typical of nucleophilic substitutions, despite the presence of the copper salts.[120] Because aryloxycopper(I) reagents ArOCu react with aryl halides to give ethers, it has been suggested that they are intermediates in the Ullmann ether synthesis.[121] Indeed, high yields of ethers can be obtained by reaction of ROCu or ArOCu with aryl halides.[122] Alcohols, via the alkoxide, displace the halogen group in aryl halides to give aryl ethers in the presence of a palladium catalyst.[123] A palladium catalyzed, intramolecular displacement of an aryl halide with a pendant alkoxide unit leads to dihydrobenzofurans.[124] Nickel catalysts have also been used.[125] The reaction has been done by heating aryl iodides and phenols in an ionic liquid.[126]

Unactivated substrates also react with phenoxide ion with electrochemical catalysis in liquid NH_3–Me_2SO, to give diaryl ethers, presumably by the $S_{RN}1$ mechanism.[127] Diaryl ethers can be prepared from activated aryl halides by treatment with a triaryl phosphate, $(ArO)_3PO$.[128]

OS **I**, 219; **II**, 445; **III**, 293, 566; **V**, 926; **VI**, 150; **X**, 418.

B. Sulfur Nucleophiles

13-4 Replacement by SH or SR

Mercapto-de-halogenation $ArBr + SH^- \longrightarrow ArSH$

Alkylthio-de-halogenation $ArBr + SR^- \longrightarrow ArSR$

[118]Wolter, M.; Nordmann, G.; Job, G.E.; Buchwald, S.L. *Org. Lett.* **2002**, *4*, 973.

[119]For reviews of the Ullmann ether synthesis see Moroz, A.A.; Shvartsberg, M.S. *Russ. Chem. Rev.* **1974**, *43*, 679; Kunz, K.; Scholz, U.; Ganzer, D. *Synlett* **2003**, 2428.

[120]Weingarten, H. *J. Org. Chem.* **1964**, *29*, 977, 3624.

[121]Kawaki, T.; Hashimoto, H. *Bull. Chem. Soc. Jpn.* **1972**, *45*, 1499.

[122]Whitesides, G.M.; Sadowski, J.S.; Lilburn, J. *J. Am. Chem. Soc.* **1974**, *96*, 2829.

[123]Parrish, C.A.; Buchwald, S.L. *J. Org. Chem.* **2001**, *66*, 2498; Torraca, K.E.; Huang, X.; Parrish, C.A.; Buchwald, S.L. *J. Am. Chem. Soc.* **2001**, *123*, 10770.

[124]Kuwabe, S.-i.; Torraca, K.E.; Buchwald, S.L. *J. Am. Chem. Soc.* **2001**, *123*, 12202.

[125]Mann, G.; Hartwig, J.F. *J. Org. Chem.* **1997**, *62*, 5413.

[126]In bmiI, 1-butyl-3-methylimidazolium iodide: Luo, Y.; Wu, J.X.; Ren, R.X. *Synlett* **2003**, 1734.

[127]Alam, N.; Amatore, C.; Combellas, C.; Pinson, J.; Savéant, J.; Thiébault, A.; Verpeaux, J. *J. Org. Chem.* **1988**, *53*, 1496.

[128]Ohta, A.; Iwasaki, Y.; Akita, Y. *Synthesis* **1982**, 828. For other procedures, see Bates, R.B.; Janda, K.D. *J. Org. Chem.* **1982**, *47*, 4374; Sammes, P.G.; Thetford, D.; Voyle, M. *J. Chem. Soc. Perkin Trans. 1* **1988**, 3229.

Aryl thiols and thioethers can be prepared by reactions that are similar to **13-1** and **13-3**.[129] Activated aryl halides generally give good results, but side reactions are occasionally important. Some reagents give the thiol directly. 4-bromonitrobenzene reacts with Na_3SPO_3, in refluxing methanol, to give 4-nitrothiophenol.[130] Diaryl sulfides can be prepared by the use of ^-SAr.[131] Even unactivated aryl halides react with ^-SAr if polar aprotic solvents, for example, DMF,[132] DMSO[133] 1-methyl-2-pyrrolidinone,[134] or HMPA,[135] are used, though the mechanisms are still mostly or entirely nucleophilic substitution. 2-Iodothiophene reacts directly with thiophenol to give 2-phenylthiothiophene.[136] Unactivated aryl halides also give good yields of sulfides on treatment with ArS^- or RS^- (generated *in situ* from the corresponding thiol) in the presence of a palladium catalyst.[137] Copper catalysts have also been used.[138] Thiophenols were coupled to indoles in the presence of a vanadium catalyst.[139] Aryl iodides react with dialkyl disulfides and a nickel catalyst to give aryl alkyl sulfides.[140] Diaryl sulfides can also be prepared (in high yields) by treatment of unactivated aryl iodides with ArS^- in liquid ammonia under irradiation.[141] The mechanism in this case is probably $S_{RN}1$. The reaction (with unactivated halides) has also been carried out electrolytically, with a nickel complex catalyst.[142]

Arylboronic acids, $(ArB(OH)_2$, react with thiols and copper(II) acetate to give the corresponding alkyl aryl sulfide.[143] Arylboronic acids also react with *N*-methylthiosuccinimide, with a copper catalyst, to give the aryl methyl sulfide.[144] In the presence of a palladium catalyst, thiophenols react with diaryliodonium salts, Ar_2I^{+-} BF_4, to give the unsymmetrical diaryl sulfide.[145]

[129]For a review of sulfur nucleophiles in aromatic substitution, see Peach, M.E., in Patai, S. *The Chemistry of the Thiol Group*, pt. 2, Wiley, NY, *1974*, pp. 735–744.

[130]Bieniarz, C.; Cornwell, M.J. *Tetrahedron Lett.* *1993*, *34*, 939.

[131]For generation of ArS^- with a phosphazine base and the copper-catalyzed displacement of Ar'I, see Palomo, C.; Oiarbide, M.; López, R.; Gómez-Bengoa, E. *Tetrahedron Lett.* *2000*, *41*, 1283.

[132]Campbell, J.R. *J. Org. Chem.* *1964*, *29*, 1830; Testaferri, L.; Tiecco, M.; Tingoli, M.; Chianelli, D.; Montanucci, M. *Synthesis 1983*, 751. For the extension of this to selenides, see Tiecco, M.; Testaferri, L.; Tingoli, M.; Chianelli, D.; Montanucci, M. *J. Org. Chem.* *1983*, *48*, 4289.

[133]Bradshaw, J.S.; South, J.A.; Hales, R.H. *J. Org. Chem.* *1972*, *37*, 2381.

[134]Caruso, A.J.; Colley, A.M.; Bryant, G.L. *J. Org. Chem. 1991*, *56*, 862; Shaw, J.E. *J. Org. Chem. 1991*, *56*, 3728.

[135]Cogolli, P.; Maiolo, F.; Testaferri, L.; Tingoli, M.; Tiecco, M. *J. Org. Chem. 1979*, *44*, 2642. See also Testaferri, L.; Tingoli, M.; Tiecco, M. *Tetrahedron Lett. 1980*, *21*, 3099; Suzuki, H.; Abe, H.; Osuka, A. *Chem. Lett. 1980*, 1363.

[136]Lee, S.B.; Hong, J.-I. *Tetrahedron Lett. 1995*, *36*, 8439.

[137]Itoh, T.; Mase, T. *Org. Lett. 2004*, *6*, 4587.

[138]Kwong, F.Y.; Buchwald, S.L. *Org. Lett. 2002*, *4*, 3517; Wu, Y.-J.; He, H. *Synlett 2003*, 1789; Deng, W.; Zou, Y.; Wang, Y.-F.;Liu, L.; Guo, Q.-X. *Synlett 2004*, 1254.

[139]Maeda, Y.; Koyabu, M.; Nishimura, T.; Uemura, S. *J. Org. Chem. 2004*, *69*, 7688.

[140]Tankguchi, N. *J. Org. Chem. 204*, *69*, 6904.

[141]Bunnett, J.F.; Creary, X. *J. Org. Chem. 1974*, *39*, 3173, 3611.

[142]Meyer, G.; Troupel, M. *J. Organomet. Chem. 1988*, *354*, 249.

[143]Herradua, P.S.; Pendola, K.A.; Guy, R.K. *Org. Lett. 2000*, 2, 2019.

[144]Savarin, C.; Srogl, J.; Liebeskind, L.S. *Org. Lett. 2002*, *4*, 4309.

[145]Wang, L.; Chen, Z.-C. *Synth. Commun. 2001*, *31*, 1227.

Other sulfur nucleophiles also react with activated aryl halides:

$$2\,ArX + S_2^{2-} \longrightarrow Ar-S-S-Ar \qquad ArX + SCN^- \longrightarrow ArSCN$$
$$ArX + SO_3^{2-} \longrightarrow Ar-SO_3^- \qquad ArX + RSO_2^- \longrightarrow Ar-SO_2-R$$

Aryl sulfones have been prepared from sulfinic acid salts, aryl iodides and CuI.[146] Formation of thiocyanates from unactivated aryl halides has been accomplished with charcoal supported copper(I) thiocyanate.[147] The copper catalyzed reaction of NaO$_2$SMe and aryl iodides give the aryl methyl sulfone.[148] A similar synthesis of diaryl sulfones has been reported using a palladium catalyst.[149]

An indirect method for the synthesis of aryl alkyl sulfides involves treatment of an aryl halide with butyllithium and then elemental sulfur. The resulting thiophenoxide anion reacts with an alkyl halide to give the targeted sulfide.[150]

Aryl selenides (ArSeAr and ArSeAr') can be prepared by similar methodology. Symmetrical diaryl selenides were prepared by the reaction of iodobenzene with diphenyl diselenide (PhSeSePh), in the presence of Mg and a copper catalyst.[151] Aryl halides react with tin selenides (ArSeSnR$_3$), with a copper catalyst, to give the diaryl selenide.[152]

OS **I**, 220; **III**, 86, 239, 667; **V**, 107, 474; **VI**, 558, 824. Also see, OS **V**, 977.

C. Nitrogen Nucleophiles

13-5   Replacement by NH$_2$, NHR, or NR$_2$

Amino-de-halogenation

Amido-de-halogenation

$$R_3N \quad + \quad Ar-X \xrightarrow[\;R = H,\ alkyl\ (1\ and\ 2°)\;]{} R_2N-Ar$$

Activated aryl halides react quite well with ammonia and with primary and secondary amines to give the corresponding arylamines. Primary and secondary amines usually give better results than ammonia, with piperidine especially reactive. Picryl chloride (2,4,6-trinitrochlorobenzene) is often used to form amine derivatives. 2,4-Dinitrofluorobenzene is used to tag the amino end of a peptide or protein chain. Other leaving groups in this reaction may be NO$_2$,[153] N$_3$, OSO$_2$R, OR, SR, N=NAr (where Ar contains electron-withdrawing groups)[154] and even NR$_2$.[155]

[146]Suzuki, H.; Abe, H. *Tetrahedron Lett.* **1995**, *36*, 6239.

[147]Clark, J.H.; Jones, C.W.; Duke, C.V.A.; Miller, J.M. *J. Chem. Soc. Chem. Commun.* **1989**, 81. See also, Yadav, J.S.; Reddy, B.V.S.; Shubashree, S.; Sadashiv, K. *Tetrahedron Lett.* **2004**, *45*, 2951.

[148]Baskin, J.M.; Wang, Z. *Org. Lett.* **2002**, *4*, 4423.

[149]Cacchi, S.; Fabrizi, G.; Goggiamani, A.; Parisi, L.M. *Org. Lett.* **2002**, *4*, 4719; Cacchi, S.; Fabrizi, G.; Goggiamani, A.; Parisi, L.M.; Bernini, R. *J. Org. Chem.* **2004**, *69*, 5608.

[150]Ham, J.; Yang, I.; Kang, H. *J. Org. Chem.* **2004**, *69*, 3236.

[151]Taniguchi, N.; Onami, T. *J. Org. Chem.* **2004**, 69, 915; Taniguchi, N.; Onami, T. *Synlett* **2003**, 829.

[152]Beletskaya, I.P.; Sigeev, A.S.; Peregudov, A.S.; Petrovlskii, P.V. *Tetrahedron Lett.* **2003**, *44*, 7039.

[153]For a reaction with an aryllithium reagent, see Yang, T.; Cho, B.P. *Tetrahedron Lett.* **2003**, *44*, 7549.

[154]Kazankov, M.V.; Ginodman, L.G. *J. Org. Chem, USSR* **1975**, *11*, 451.

[155]Sekiguchi, S.; Horie, T.; Suzuki, T. *J. Chem. Soc. Chem. Commun.* **1988**, 698.

Aryl triflates were shown to react directly with secondary amines in N-methylpyrrolidine solvent using microwave irradiation.[156] Activated halides can be converted to diethylamino compounds $ArX \rightarrow ArNMe_2$ by treatment with HMPA.[157] Aniline derivatives react with activated aromatic rings, in the presence of tetrabutylammonium fluoride and under photolysis conditions, to give a N,N-diarylamine.[158] Arylation of amines with aryl halides has also been done in ionic liquids.[159]

Unactivated aryl halides can be converted to amines by the use of $NaNH_2$, $NaNHR$, or $NaNR_2$.[160] Lithium dialkylamides also react with aryl halides to give the N-arylamine.[161] With these reagents, the benzyne mechanism generally operates, so cine substitution is often found. The reaction of an amine, an aryl halide, and potassium $tert$-butoxide generates the N-aryl amine.[162] N-Arylation was accomplished with butyllithium and a secondary amine using Ni/C-diphenylphosphinoferrocene (dppf).[163] Ring closure has been effected by this type of reaction,[164] as in the conversion of **16** to the tetrahydroquinoline.

Larger rings can be prepared using this approach: 8 and even 12 membered. Triarylamines have been prepared in a similar manner from ArI and Ar_2' NLi, even with unactivated ArI.[165] In the *Goldberg reaction*, an aryl bromide reacts with an acetanilide in the presence of K_2CO_3 and CuI to give an N-acetyldiarylamine, which can be hydrolyzed to a diarylamine: $ArBr + Ar'NHAc \rightarrow ArAr'NAc$.[166] Aryl fluorides react in the presence of KF-alumina and 18-crown-6 in DMSO.[167] Lithium amides have been shown to react directly with aryl halides.[168] Aryl fluorides react

[156]Xu, G.; Wang, Y.-G. *Org. Lett.* **2004**, *6*, 985.

[157]See, for example, Gupton, J.T.; Idoux, J.P.; Baker, G.; Colon, C.; Crews, A.D.; Jurss, C.D.; Rampi, R.C. *J. Org. Chem.* **1983**, *48*, 2933.

[158]Hertas, I.; Gallardo, I.; Marquet, J. *Tetrahedron Lett.* **2000**, *41*, 279.

[159]In bmim PF$_6$, 1-butyl-3-methylimidazolium hexafluorophosphate: See Yadav, J.S.; Reddy, B.V.S.; Basak, A.K.; Narsaiah, A.V. *Tetrahedron Lett.* **2003**, *44*, 2217.

[160]For a review, see Heaney, H. *Chem Rev.* **1962**, *62*, 81, see p. 83.

[161]Tripathy, S.; Le Blanc, R.; Durst, T. *Org. Lett.* **1999**, *1*, 1973.

[162]Beller, M.; Breindl, C.; Riermeier, T.H.; Tillack, A. *J. Org. Chem.* **2001**, *66*, 1403; Shi, L.; Wang, M.; Fan, C.-A.; Zhang, F.-M.; Tu, Y.-Q. *Org. Lett.* **2003**, *5*, 3515.

[163]Tasler, S.; Lipshutz, B.H. *J. Org. Chem.* **2003**, *68*, 1190.

[164]Huisgen, R.; König, H.; Lepley, A.R. *Chem. Ber.* **1960**, *93*, 1496; Bunnett, J.F.; Hrutfiord, B.F. *J. Am. Chem. Soc.* **1961**, *83*, 1691. For a review of ring closures by the benzyne mechanism, see Hoffmann, R.W. *Dehydrobenzene and Cycloalkynes*, Academic Press, NY, **1973**, pp. 150–164.

[165]Neunhoeffer, O.; Heitmann, P. *Chem. Ber.* **1961**, *94*, 2511.

[166]See Freeman, H.S.; Butler, J.R.; Freedman, L.D. *J. Org. Chem.* **1978**, *43*, 4975; Renger, B. *Synthesis* **1985**, 856.

[167]Smith III, W.J.; Sawyer, J.S. *Tetrahedron Lett.* **1996**, *37*, 299.

[168]Kanth, J.V.B.; Periasamy, M. *J. Org. Chem.* **1993**, *58*, 3156.

with amines in the presence of potassium carbonate/DMSO and ultrasound,[169] and aryl chlorides react on basic alumina with microwave irradiation.[170] 2-Chloronitrobenzene also reacts with aniline derivatives directly with microwave irradiation.[171] 2-Fluoropyridine reacts with R_2NBH_3Li to give the 2-aminoalkylpyridine.[172]

The reaction of amines with unactivated aryl halides requires a catalyst in most cases to initiate the reaction. There are several approaches that result in *N*-aryl amines, but recent work with aryl halides, amines, and palladium catalysts has proven quite useful.[173] Aryl halides react with amines (including aniline derivatives) in the presence of palladium catalysts to give the *N*-aryl amine.[174] Palladium catalysts have been used with aniline and or triflates[175] to give the secondary amine. Palladium catalysts have been used in conjunction with aryl halides and aliphatic amines–amide bases.[176] A considerable amount of work[177] has been done to vary the nature of the ligand and the palladium catalyst, as well as the base.[178] Aryl halides also react with aliphatic amines,[179] including cyclopropylamines,[180] and an intramolecular version of this reaction generates bicyclic amines (hydroindole derivatives).[181] Primary aliphatic amines can be converted to tertiary *N,N*-diarylalkylamines in a two-step procedure using palladium catalysts.[182] Aryl halides are

[169]Magdolen, P.; Mečiarová, M.; Toma, Š. *Tetrahedron* **2001**, *57*, 4781.

[170]Kidwai, M.; Sapra, P.; Dave, B. *Synth. Commun.* **2000**, *30*, 4479.

[171]Xu, Z.-B.; Lu, Y.; Guo, Z.-R. *Synlett* **2003**, 564. See Li, W.; Yun, L.; Wang, H. *Synth. Commun.* **2002**, *32*, 2657.

[172]Thomas, S.; Roberts, S.; Pasumansky, L.; Gamsey, S.; Singaram, B. *Org. Lett.* **2003**, *5*, 3867.

[173]For a discussion of the mechanism of the palladium-catalyzed amination of aryl chlorides, see Alcazar-Roman, L.M.; Hartwig, J.F. *J. Am.Chem. Soc.* **2001**, *123*, 12905.

[174]Driver, M.S.; Hartwig, J.F. *J. Am. Chem. Soc.* **1996**, *118*, 7217; Reddy, N.P.; Tanaka, M. *Tetrahedron Lett.* **1997**, *38*, 4807; Wolfe, J.P.; Buchwald, S.L. *J. Org. Chem.* **1997**, *62*, 6066; Marcoux, J.-F.; Wagaw, S.; Buchwald, S.L. *J. Org. Chem.* **1997**, *62*, 1568; Maes, B.U.W.; Loones, K.T.J.; Lemière, G.L.F.; Dommisse, R.A. *Synlett* **2003**, 1822; Wan, Y.; Alterman, M.; Hallberg, A. *Synthesis* **2002**, 1597.

[175]Louie, J.; Driver, M.S.; Hamann, B.C.; Hartwig, J.F. *J. Org. Chem.* **1997**, *62*, 1268; Wolfe, J.P.; Buchwald, S.L. *J. Org. Chem.* **1997**, *62*, 1264.

[176]Harris, M.C.; Huang, X.; Buchwald, S.L. *Org. Lett.* **2003**, *4*, 2885.

[177]Bei, X.; Guram, A.S.; Turner, H.W.; Weinberg, W.H. *Tetrahedron Lett.* **1999**, *40*, 1237; Guari, Y.; van Es, D.S.; Reek, J.N.H.; Kamer, P.C.J.; van Leeuwen, P.W.N.M. *Tetrahedron Lett.* **1999**, *40*, 3789; Urgaonkar, S.; Xu, J.-H.; Verkade, J.G. *J. Org. Chem.* **2003**, *68*, 8416; Viciu, M.S.; Kissling, R.M.; Stevens, E.D.; Nolan, S.P. *Org. Lett.* **2002**, *4*, 2229; Gajare, A.S.; Toyota, K.; Yoshifuji, M.; Ozawa, F. *J. Org. Chem.* **2004**, *69*, 6504; Huang, X.; Anderson, K.W.; Zim, D.; Jiang, L.; Klapars, A.; Buchwald, S.L. *J. Am. Chem. Soc.* **2004**, *125*, 6653; Singer, R.A.; Tom, N.J.; Frost, H.N.; Simon, W.M. *Tetrahedron Lett.* **2004**, *45*, 4715; Smith, C.J.; Early, T.R.; Holmes, A.B.; Shute, R.E. *Chem. Commun.* **2004**, 1976.

[178]For a study of the mechanism of the palladium catalyzed amination of aryl halides, see Singh, U.K.; Strieter, E.R.; Blackmond, D.G.; Buchwald, S.L. *J. Am. Chem. Soc.* **2002**, *124*, 14104. For a study of rate enhancement by the added base, see Meyers, C.; Maes, B.U.W.; Loones, K.T.J.; Bal, G.; Lemière, G.L.F.; Dommisse, R.A. *J. Org. Chem.* **2004**, *69*, 6010.

[179]Ali, M.H.; Buchwald, S.L. *J. Org. Chem.* **2001**, *66*, 2560; Cheng, J.; Trudell, M.L. *Org. Lett.* **2001**, *3*, 1371; Kuwano, R.; Utsunomiya, M.; Hartwig, J.F. *J. Org. Chem.* **2002**, *67*, 6479; Urgaonkar, S.; Nagarajan, M.; Verkade, J.G. *J. Org.Chem.* **2003**, *68*, 452; Prashad, M.; Mak, X.Y.;Lium Y.; Repi, O. *J. Org. Chem.* **2003**, *68*, 1163.

[180]Cui, W.; Loeppky, R.N. *Tetrahedron* **2001**, *57*, 2953.

[181]Johnston, J.N.; Plotkin, M.A.; Viswanathan, R.; Prabhakaran, E.N. *Org. Lett.* **2001**, *3*, 1009.

[182]Harris, M.C.; Geis, O.; Buchwald, S.L. *J. Org. Chem.* **1999**, *64*, 6019.

converted to *N,N*-diaryl tertiary amines by reaction with *N*-alkylaniline derivatives and a palladium catalyst.[183] Beginning with a primary aromatic amine and two different aryl halides (ArBr and Ar'Cl), a triarylamine with three different aryl groups can be prepared using a palladium catalyst.[184] Polymer-bound phosphine ligands have been used in conjunction with a palladium catalyst,[185] and polymer-bound amines have been *N*-arylated with a palladium catalyst followed by treatment with trifluoroacetic acid to release the aniline derivative.[186] Palladium-catalyzed aminoalkylation of aryl halides has been reported using microwave irradiation.[187] Aryl halides (Ar–X) have also been converted to the aniline derivative (Ar–NH$_2$) by reaction of the halide with an imine and a palladium catalyst, followed by hydrolysis.[188] Similarly, aniline derivatives have been prepared by the reaction of aryl chlorides with silylamines (Ph$_3$SiNH$_2$) using lithium hexamethyldisilazide and a palladium catalyst.[189] Amines react with Ph$_2$I$^+$BF$_4^-$, in the presence of palladium catalysts,[190] or a CuI catalyst[191] to give the *N*-phenyl amine. These reactions have been done in ionic liquids using a palladium catalyst.[192] Arylation of the amine unit of primary enamino ketones was accomplished using a palladium catalyst.[193] Mono-arylation of a 1,2-diamine is possible.[194] Aminoalkylation of heteroaromatic rings is possible, as in the reaction of 3-bromothiophene with a primary amine and a palladium catalyst.[195] 2-Halopyridines react to give the 2-aminoalkyl pyridine.[196] Carbazole derivatives were prepared from 2-iodoaniline and 2-trimethylsilylphenol *O*-triflates, using cesium fluoride and then a palladium catalyst.[197]

Nickel catalysts have been used in the reaction of aryl halides with *N*-alkyl aniline derivatives.[198] Nickel catalyst also allow the conversion of aryl halides to *N*-arylamines via reaction with aliphatic amines.[199] An intramolecular reaction of a

[183]Wolfe, J.P.; Buchwald, S.L. *J. Org. Chem.* **2000**, *65*, 1144; Wolfe, J.P.; Tomori, H.; Sadighi, J.P.; Yin, J.; Buchwald, S.L. *J. Org. Chem.* **2000**, *65*, 1158.

[184]Harris, M.C.; Buchwald, S.L. *J. Org. Chem.* **2000**, *65*, 5327.

[185]Parrish, C.A.; Buchwald, S.L. *J. Org. Chem.* **2001**, *66*, 3820.

[186]Weigand, K.; Pelka, S. *Org. Lett.* **2002**, *4*, 4689.

[187]Wang, T.; Magnin, D.R.; Hamann, L.G. *Org. Lett.* **2003**, *5*, 897; Jensen, T.A.; Liang, X.; Tanner, D.; Skjaerbaek, N. *J. Org. Chem.* **2004**, *69*, 4936.; Maes, B.U.W.; Loones, K.T.J.; Hostyn, S.; Diels, G.; Rombouts, G. *Tetrahedron* **2004**, *60*, 11559.

[188]Wolfe, J.P.; Åhman, J.; Sadighi, J.P.; Singer, R.A.; Buchwald, S.L. *Tetrahedron Lett.* **1997**, *38*, 6367. For a variation, see Erdik, E.; Daşkapan, T. *Tetrahedron Lett.* **2002**, *43*, 6237.

[189]Huang, X.; Buchwald, S.L. *Org. Lett.* **2001**, *3*, 3417.

[190]Kang, S.-K.; Lee, H.-W.; Choi, W.-K.; Hong, R.-K.; Kim, J.-S. *Synth. Commun.* **1996**, *26*, 4219.

[191]Kang, S.-K.; Lee, S.-H.; Lee, D. *Synlett* **2000**, 1022.

[192]In diarylimidazolium salts: Grasa, G.A.; Viciu, M.S.; Huang, J.; Nolan, S.P. *J. Org. Chem.* **2001**, *66*, 7729.

[193]Edmondson, S.D.; Mastracchio, A.; Parmee, E.R. *Org. Lett.* **2000**, *2*, 1109 and references cited therein.

[194]Frost, C.G.; Mendonça, P. *Tetahedron Asymmetry* **1999**, *10*, 1831.

[195]Ogawa, K.; Radke, K.R.; Rothstein, S.D.; Rasmussen, S.C. *J. Org. Chem.* **2001**, *66*, 9067.

[196]Junckers, T.H.M.; Maes, B.U.W.; Lemière, G.L.F.; Dommisse, R. *Tetrahedron* **2001**, *57*, 7027; Basu, B.; Jha, S.; Mridha, N.K.; Bhuiyan, Md.M.H. *Tetrahedron Lett.* **2002**, *43*, 7967.

[197]Liu, Z.; Larock, R.C. *Org. Lett.* **2004**, *6*, 3739.

[198]Wolfe, J.P.; Buchwald, S.L. *J. Am. Chem. Soc.* **1997**, *119*, 6054; Lipshutz, B.H.; Ueda, H. *Angew. Chem. Int. Ed.* **2000**, *39*, 4492.; Brenner, E.; Schneider, R.; Fort, Y. *Tetrahedron* **2002**, *58*, 6913.

[199]Desmarets, C.; Schneider, R.; Fort, Y. *Tetrahedron Lett.* **2001**, *42*, 247.

pendant aminoalkyl unit with an aryl chloride moiety, catalyzed by nickel(0) gave a dihydroindole.[200] Copper catalysts allow the reaction of diarylamines and aryl halides to give the corresponding triarylamine,[201] or with aliphatic amines to give the N-arylamine.[202] Aniline reacts with aryl iodides an a copper catalyst and potassium *tert*-butoxide to give triphenylamine.[203] A polymer-bound copper cata-lyst was used in conjunction with aliphatic amines and arylboronic acids.[204] Amino alcohols react with aryl iodides and a copper catalyst to give the N-arylamino alco-hol.[205] Treatment of alkylamines with arylboronic acids ArB(OH)$_2$ and Cu(OAc)$_2$ gave the N-aryl amine in 63% yield.[206] Similar reaction with arylamines, such as aniline, gave the diarylamine.[207] Arylboronic acids convert aziridines to N-arylazir-idines,[208] and amino esters to N-arylated amino esters,[209] both reactions using a copper catalyst. An arylbismuth reagent reacts with aliphatic amines, in the pre-sence of copper(II) acetate, to give an N-arylamine.[210] N-Arylation of pyrroles was accomplished by the reaction of an arylboronic acid and a copper catalyst.[211] N-Arylindoles[212] and N-arylimidazoles[213] were prepared from aryl halide using a copper catalyst. Diarylzinc reagents react with N-(OBz) amine derivatives, with a copper catalyst, to give the N-aryl amine.[214]

In a related reaction, trifluoroarylboronates react with copper(II) acetate and then an aliphatic amine to give the N-phenylamine.[215]

The metal catalyzed reaction with ammonia or amines likely proceeds by the S$_N$Ar mechanism.[216] This reaction, with phase-transfer catalysis, has been used to synthesize triarylamines.[217] Copper ion catalysts (especially cuprous oxide or iodide) also permit the Gabriel synthesis (**10-41**) to be applied to aromatic sub-strates. Aryl bromides or iodides are refluxed with potassium phthalimide and

[200]Omar-Amrani, R.; Thomas, A.; Brenner, E.; Schneider, R.; Fort, Y. *Org. Lett.* **2003**, *5*, 2311.

[201]Gujadhur, R.K.; Bates, C.G.; Venkataraman, D. *Org. Lett.* **2001**, *3*, 4315.; Klapars, A.; Antilla, J.C.; Huang, X.; Buchwald, S.L. *J. Am. Chem. Soc.* **2001**, *123*, 7727.

[202]Kwong, F.Y.; Buchwald, S.L. *Org. Lett.* **2003**, *5*, 793; Ma, D.; Cai, Q.; Zhang, H. *Org. Lett.* **2003**, *5*, 2453; Okano, K.; Tokuyama, H.; Fukuyama, T. *Org. Lett.* **2003**, *5*, 4987; Lu, Z.; Twieg, R.J.; Huang, S.D. *Tetrahedron Lett.* **2003**, *44*, 6289.

[203]Kelkar, A.A.; Patil, N.M.; Chaudhari, R.V. *Tetrahedron Lett.* **2002**, *43*, 7143.

[204]Chiang, G.C.H.; Olsson, T. *Org.Lett.* **2004**, *6*, 3079.

[205]Job, G.E.; Buchwald, S.L. *Org. Lett.* **2002**, *4*, 3703.

[206]Lan, J.-B.; Zhang, G.-L.; Yu, X.-Q.; You, J.-S.; Chen, L.; Yan, M.; Xie, R.-G. *Synlett* **2004**, 1095.

[207]Antilla, J.C.; Buchwald, S.L. *Org. Lett.* **2001**, *3*, 2077.

[208]Sasaki, M.; Dalili, S.; Yudin, A.K. *J. Org. Chem.* **2003**, *68*, 2045.

[209]Lam, P.Y.S.; Bonne, D.; Vincent, G.; Clark, C.G.; Combs, A.P. *Tetrahedron Lett.* **2003**, *44*, 1691.

[210]Fedorov, A.Yu.; Finet, J.-P. *J. Chem. Soc. Perkin Trans. 1* **2000**, 3775.

[211]Yu, S.; Saenz, J.; Srirangam, J.K. *J. Org. Chem.* **2002**, *67*, 1699.

[212]Antilla, J.C.; Klapars, A.; Buchwald, S.L. *J. Am. Chem. Soc.* **2002**, *124*, 11684.

[213]Wu, Y.-J.; He, H.; L'Hereux, A. *Tetrahedron Lett.* **2003**, *44*, 4217.

[214]Berman, A.M.; Johnson, J.S. *J. Am. Chem. Soc.* **2004**, *126*, 5680.

[215]Quach, T.D.; Batey, R.A. *Org. Lett.* **2003**, *5*, 4397.

[216]For discussions of the mechanism, see Bethell, D.; Jenkins, I.L.; Quan, P.M. *J. Chem. Soc. Perkin Trans. 1* **1985**, 1789; Tuong, T.D.; Hida, M. *J. Chem. Soc. Perkin Trans. 2* **1974**, 676; Kondratov, S.A.; Shein, S.M. *J. Org. Chem, USSR* **1979**, *15*, 2160; Paine, A.J. *J. Am. Chem. Soc.* **1987**, *109*, 1496.

[217]Gauthier, S.; Fréchet, J.M.J. *Synthesis* **1987**, 383.

Cu_2O or CuI in dimethylacetamide to give *N*-aryl phthalimides, which can be hydrolyzed to primary aryl amines.[218]

In certain cases, the $S_{RN}1$ mechanism has been found (p. 550). When the substrate is a heterocyclic aromatic nitrogen compound, still a different mechanism [the S_N(ANRORC) mechanism], involving opening and reclosing of the aromatic ring, has been shown to take place.[219]

There are a number of indirect approaches for the preparation of aryl amines. Activated aromatic compounds can be directly converted to the *N*-aryl amine with hydroxylamine in the presence of strong bases.[220] Conditions are mild and yields are high. Aryl halides can be converted to the corresponding Grignard reagent (**12-38**). Subsequent reaction of arylmagnesium halides with allyl azide ($CH_2=CHCH_2N_3$) followed by hydrolysis leads to the corresponding aniline derivative.[221] Aryl halides can be converted to the aryllithium via halogen–lithium exchange or hydrogen–lithium exchange (**12-38, 12-39**). Molecular nitrogen (N_2) reacts with aryllithium compounds in the presence of compounds of such transition metals as titanium (e.g., $TiCl_4$), chromium, molybdenum, or vanadium to give (after hydrolysis) primary aromatic amines ($ArLi + N_2 +$ transition metal salts $\rightarrow ArNH_2$, after hydrolysis).[222] Primary aromatic amines $ArNH_2$ were converted to diaryl amines ArNHPh by treatment with $Ph_3Bi(OAc)_2$[223] and a copper powder catalyst.[224] Aryl Grignard reagents react with nitroaryl compounds to give, after reduction with $FeCl_3/NaBH_4$, a diaryl amine.[225]

$$R^1-\overset{\overset{O}{\|}}{C}-NHR \quad \xrightarrow[\text{catalyst}]{Ar-X} \quad R^1-\overset{\overset{O}{\|}}{C}-\underset{\underset{Ar}{|}}{N}-R \quad (R = H, \text{alkyl, aryl})$$

The use of transition-metal catalysts allows aryl halides to react with the nitrogen of amides or carbamates, as well as amines, to give the corresponding *N*-aryl amide or *N*-aryl carbamate. Amides react with aryl halides in the presence of a palladium catalyst[226] or a copper catalyst.[227] *N*-Aryl lactams are prepared by the reaction of a lactam with an aryl halide in the presence of a palladium catalyst.[228]

[218]Bacon, R.G.R.; Karim, A. *J. Chem. Soc. Perkin Trans. 1* **1973**, 272, 278; Sato, M.; Ebine, S.; Akabori, S. *Synthesis* **1981**, 472. See also Yamamoto, T.; Kurata, Y. *Can. J. Chem.* **1983**, *61*, 86.

[219]For reviews, see van der Plas, H.C. *Tetrahedron* **1985**, *41*, 237; *Acc. Chem. Res.* **1978**, *11*, 462.

[220]See Chupakhin, O.N.; Postovskii, I.Ya. *Russ. Chem. Rev.* **1976**, *45*, 454, p. 456.

[221]Kabalka, G.W.; Li, G. *Tetrahedron Lett.* **1997**, *38*, 5777.

[222]Vol'pin, M.E. *Pure Appl. Chem.* **1972**, *30*, 607.

[223]For a review of arylations with bismuth reagents, see Finet, J. *Chem. Rev.* **1989**, *89*, 1487.

[224]Dodonov, V.A.; Gushchin, A.V.; Brilkina, T.G. *Zh. Obshch. Khim.*, **1985**, *55*, 466 [*Chem. Abstr., 103*, 22218z]; Barton, D.H.R.; Yadav-Bhatnagar, N.; Finet, J.; Khamsi, J. *Tetrahedron Lett.* **1987**, *28*, 3111.

[225]Sapountzis, I.; Knochel, P. *J. Am.Chem. Soc.* **2002**, *124*, 9390.

[226]Yin, J.; Buchwald, S.L. *J. Am. Chem. Soc.* **2002**, *124*, 6043. For an intramolecular reaction, see Yang, B.H.; Buchwald, S.L. *Org. Lett.* **1999**, *1*, 35.

[227]Hosseinzadeh, R.; Tajbakhsh, M.; Mohadjerani, M.; Mehdinejad, H. *Synlett* **2004**, 1517.

[228]Browning, R.G.; Badaringarayana, V.; Mahmud, H.; Lovely, C.J. *Tetrahedron* **2004**, *60*, 359; Deng, W.; Wang, Y.-F.; Zou, Y.; Liu, L.; Guo, Q.-X. *Tetrahedron Lett.* **2004**, *45*, 2311; Shakespeare, W.C. *Tetrahedron Lett.* **1999**, *40*, 2035. For the synthesis of *N*-(2-thiophene)-2-pyrrolidinone by coupling 2-iodothiophene and 2-pyrrolidinone, see Klapars, A.; Huang, X.; Buchwald, S.L. *J. Am. Chem. Soc.* **2002**, *124*, 7421. See also, Ferraccioli, R.; Carenzi, D.; Rombolà, O.; Catellani, M. *Org. Lett.* **2004**, *6*, 4759.

β-Lactams also react.[229] The reaction of 2-oxazolidinones with aryl halides in the presence of a palladium catalyst gave the *N*-aryl-2-oxazolidinone.[230] Amides react with PhSi(OMe)$_3$/Cu(OAc)$_2$/Bu$_4$NF to give the *N*-aryl amide.[231] *N*-Boc hydrazine derivatives (BocNHNH$_2$) gave the *N*-phenyl derivative BocN(Ph)NH$_2$ when reacted with iodobenzene and a catalytic amount of CuI and 10% of 1,10-phenanthroline.[232] 3-Bromothiophene was converted to the 3-amido derivative with an amide and CuI-dimethylethylenediamine,[233] and *N*-(2-thiophene)-2-pyrrolidinone was similarly prepared from 2-iodothiophene, the lactam and a copper catalyst.[234] *N*-Arylation of urea is also possible using a copper catalyst.[235] The reaction of a vinyl triflate with benzamide and a palladium catalyst gave the corresponding enamide (C=C−NHC=O).[236]

The transition-metal catalyzed couplings of primary or secondary phosphines with aryl halides or sulfonate esters to give arylphosphines is known.[237] Palladium catalyzed conversion of aryl halides to aryl phosphines using (trimethylsilyl)diphenylphosphine is known, and tolerates many functional groups (not those that are easily reducible, such as aldehydes because zinc metal[238] is often used as a coreagent), but it is mainly limited to aryl iodides.[239] Diphenylphosphine reacts with aryl iodides and a copper catalyst to give the triarylphosphine.[240] Aryl iodides also react with secondary phosphine and 5% Pd/C to give the *P*-arylphosphine.[241] Tertiary phosphines can also be used via aryl–aryl exchange, as in the reaction of an aryl triflate and triphenylphosphine and a palladium catalyst, for example, gave the arylphosphine (ArPPh$_2$).[242]

Arylsulfonic acid chlorides (ArSO$_2$Cl) have been shown to react with arylboronic acids, Ar'B(OH)$_2$, in the presence of a palladium catalyst, to give the corresponding biaryl (Ar−Ar').[243]

[229]For a variation of this reaction, see Klapars, A.; Parris, S.; Anderson, K.W.; Buchwald, S.L. *J. Am. Chem. Soc.* **2004**, *126*, 3529.

[230]Cacchi, S.; Fabrizi, G.; Goggiamani, A.; Zappia, G. *Org. Lett.* **2001**, *3*, 2539.

[231]Lam, P.Y.S.; Deudon, S.; Hauptman, E.; Clark, C.G. *Tetrahedron Lett.* **2001**, *42*, 2427.

[232]Wolter, M.; Klapars, A.; Buchwald, S.L. *Org. Lett.* **2001**, *3*, 3803.

[233]Padwa, A.; Crawford, K.R.; Rashatasakhon, P.; Rose, M. *J. Org. Chem.* **2003**, *68*, 2609.

[234]Kang, S.-K.; Kim, D.-H.; Park, J.-N. *Synlett* **2002**, 427.

[235]Nandakumar, M.V. *Tetrahedron Lett.* **2004**, *45*, 1989.

[236]Wallace, D.J.; Klauber, D.J.; Chen, C.-y.; Volante, R.P. *Org. Lett.* **2003**, *5*, 4749.

[237]Cai, D.; Payack, J.F.; Bender, D.R.; Hughes, D.L.; Verhoeven, T.R.; Reider, P.J. *J. Org. Chem.* **1994**, *59*, 7180; Herd, O.; Heßler, A.; Machnitzki, P.; Tepper, M.; Stelzer, O. *Catalysis Today* **1998**, *42*, 413; Gelpke, A.E.S.; Kooijman, H.; Spek, A.L.; Hiemstra, H. *Chem. Eur. J.* **1999**, *5*, 2472; Ding, K.; Wang, Y.; Yun, H.; Liu, J.; Wu, Y.; Terada, M.; Okubo, Y.; Mikami, K. *Chem. Eur. J.* **1999**, *5*, 1734; Vyskocil, S.; Smrcina, M.; Hanus, V.; Polasek, M.; Kocovsky, P. *J. Org. Chem.* **1998**, *63*, 7738; Martorell, G.; Garcias, X.; Janura, M.; Saá, J.M. *J. Org. Chem.* **1998**, *63*, 3463; Bringmann, G.; Wuzik, A.; Vedder, C.; Pfeiffer, M.; Stalke, D. *Chem. Commun.* **1998**, 1211; Lipshutz, B.H.; Buzard, D.H.; Yun, C.S. *Tetrahedron Lett.* **1999**, *40*, 201.

[238]Ager, D.J.; Laneman, S. *Chem. Commun.* **1997**, 2359.

[239]Tunney, B.H.; Stille, J.K. *J. Org. Chem.* **1987**, *52*, 748.

[240]Van Allen, D.; Venkataraman, D. *J. Org. Chem.* **2003**, *68*, 4590.

[241]Stadler, A.; Kappe, C.O. *Org. Lett.* **2002**, *4*, 3541.

[242]Kwong, F.Y.; Lai, C.W.; Tian, Y.; Chan, K.S. *Tetrahedron Lett.* **2000**, *41*, 10285; Kwong, F.Y.; Lai, C.W.; Chan, K.S. *Tetrahedron Lett.* **2002**, *43*, 3537.

[243]Dubbaka, S.R.; Vogel, P. *Org. Lett.* **2004**, *6*, 95.

OS **I**, 544; **II**, 15, 221, 228; **III**, 53, 307, 573; **IV**, 336, 364; **V**, 816, 1067; **VII**, 15. OS **III**, 664.

OS X, 423.

13-6 Replacement of a Hydroxy Group by an Amino Group
Amino-de-hydroxylation

The reaction of naphthols with ammonia and sodium bisulfite[81] is called the *Bucherer reaction*. Primary amines can be used instead of ammonia, in which case *N*-substituted naphthylamines are obtained. In addition, primary naphthylamines can be converted to secondary ($ArNH_2 + RNH_2 + NaSO_3 \rightarrow ArNHR$), by a transamination reaction. The mechanism of the Bucherer reaction amounts to a kind of overall addition–elimination, via **18** and **19**.[244]

The first step in either direction consists of addition of $NaHSO_3$ to one of the double bonds of the ring, which gives an enol from **17** (or enamine from **20**) that tautomerizes to the keto form **18** (or imine form, **19**). The conversion of **18** to **19** (or vice versa) is an example of **16-13** (or **16-2**). Evidence for this mechanism was the isolation of **18**[245] and the demonstration that for β-naphthol treated with ammonia and HSO_3^-, the rate of the reaction depends only on the substrate and on

[244]Rieche, A.; Seeboth, H. *Liebigs Ann. Chem.* **1960**, *638*, 66.
[245]Rieche, A.; Seeboth, H. *Liebigs Ann. Chem.* **1960**, *638*, 43, 57.

HSO_3^-, indicating that ammonia is not involved in the rate-determining step.[246] If the starting compound is a β-naphthol, the intermediate is a 2-keto-4-sulfonic acid compound, so the sulfur of the bisulfite in either case attacks meta to the OH or NH_2.[247]

Hydroxy groups on benzene rings can be replaced by NH_2 groups if they are first converted to aryl diethyl phosphates. Treatment of these with KNH_2 and potassium metal in liquid ammonia gives the corresponding primary aromatic amines.[248] The mechanism of the second step is $S_{RN}1$.[249]

OS **III**, 78.

D. Halogen Nucleophiles

13-7   The Introduction of Halogens

Halo-de-halogenation, and so on.

$$Ar-X + X'^- \rightleftharpoons Ar-X' + X^-$$

It is possible to replace a halogen on a ring by another halogen[250] if the ring is activated. In such cases there is an equilibrium, but it is usually possible to shift this in the desired direction by the use of an excess of added halide ion.[251] A phenolic hydroxy group can be replaced by chloro with PCl_5 or $POCl_3$, but only if activated. Unactivated phenols give phosphates when treated with $POCl_3$: 3 $ArOH + POCl_3 \rightarrow$ $(ArO)_3PO$. Phenols, even unactivated ones, can be converted to aryl bromides by treatment with Ph_3PBr_2[252] (see **10-47**) and to aryl chlorides by treatment with $PhPCl_4$.[253]

Halide exchange is particularly useful for putting fluorine into a ring, since there are fewer alternate ways of doing this than for the other halogens. Activated aryl chlorides give fluorides when treated with KF in DMF, DMSO, or dimethyl sulfone.[254] Reaction of aryl halides with Bu_4PF/HF is also effective for exchanging a halogen with fluorine.[255] Halide exchange can also be accomplished with copper halides. Since the leaving-group order in this case is $I > Br > Cl \gg F$ (which means that iodides cannot normally be made by this method), the S_NAr mechanism is

[246]Kozlov, V.V.; Veselovskaia, I.K. *J. Gen. Chem. USSR* **1958**, *28*, 3359.

[247]Rieche, A.; Seeboth, H. *Liebigs Ann. Chem.* **1960**, *638*, 76.

[248]Rossi, R.A.; Bunnett, J.F. *J. Org. Chem.* **1972**, *37*, 3570.

[249]For another method of converting phenols to amines, see Scherrer, R.A.; Beatty, H.R. *J. Org. Chem.* **1972**, *37*, 1681.

[250]For a list of reagents, with references, see Larock, R.C. *Comprehensive Organic Transformations*, 2nd ed., Wiley-VCH, NY, **1999**, pp. 671–672.

[251]Sauer, J.; Huisgen, R. *Angew. Chem.* **1960**, *72*, 294, p. 297.

[252]Wiley, G.A.; Hershkowitz, R.L.; Rein, B.M.; Chung, B.C. *J. Am. Chem. Soc.* **1964**, *86*, 964; Wiley, G.A.; Rein, B.M.; Hershkowitz, R.L. *Tetrahedron Lett.* **1964**, 2509; Schaefer, J.P.; Higgins, J. *J. Org. Chem.* **1967**, *32*, 1607.

[253]Bay, E.; Bak, D.A.; Timony, P.E.; Leone-Bay, A. *J. Org. Chem.* **1990**, *55*, 3415.

[254]Kimura, Y.; Suzuki, H. *Tetrahedron Lett.* **1989**, *30*, 1271. For the use of phase-transfer catalysis in this reaction, see Yoshida, Y.; Kimura, Y. *Chem. Lett.* **1988**, 1355. For a review of the preparation of aryl fluorides by halogen exchange, see Dolby-Glover, L. *Chem. Ind.* (*London*) **1986**, 518.

[255]Uchibori, Y.; Umeno, M.; Seto, H.; Qian, Z.; Yoshioka, H. *Synlett* **1992**, 345.

probably not operating.[256] However, aryl iodides have been prepared from bromides, by the use of Cu supported on charcoal or Al_2O_3,[257] with an excess of NaI and a copper catalyst,[258] and by treatment with excess KI and a nickel catalyst.[259] Interestingly, aryl chlorides have been prepared from aryl iodides using 2 equivalents of $NiCl_2$ in DMF, with microwave irradiation.[260]

An indirect halogen exchange treated aryl bromides with n-butyllithium and the 5-(iodomethyl)-γ-butyrolactone, giving the aryl iodide and the lithium salt of 4-pentenoic acid.[261] Aryl iodides[262] and fluorides can be prepared from arylthallium bis(trifluoroacetates) (see **12-23**), indirectly achieving the conversions ArH → ArI and ArH → ArF. The bis(trifluoroacetates) react with KI to give ArI in high yields.[263] Aryllead triacetates $ArPb(OAc)_3$ can be converted to aryl fluorides by treatment with BF_3–etherate.[264] Treatment of $PhB(OH)_2$ with N-iodosuccinimide gives iodobenzene.[265] Arylboronic acids (**12-28**) can be converted to the corresponding aryl bromides by reaction with 1,3-dibromo-5,5-dimethylhydantoin and 5 mol % NaOMe.[266] Other aryl halides can be prepared using 1,3-dihalo-5,5-dimethylhydantoins.

OS **III**, 194, 272, 475; **V**, 142, 478; **VIII**, 57; **81**, 98.

The reduction of phenols and phenolic esters and ethers is discussed in Chapter 19 (see **19-38** and **19-35**). The reaction ArX → ArH is treated in Chapter 11 (reaction **11-39**), although, depending on reagent and conditions, it can be nucleophilic or free-radical substitution, as well as electrophilic.

E. Carbon Nucleophiles[267]

Some formations of new aryl–carbon bonds formed from aryl substrates have been considered in Chapter 10 (see **10-57, 10-68, 10-76, 10-77**).

[256]Bacon, R.G.R.; Hill, H.A.O. *J. Chem. Soc.* **1964**, 1097, 1108. See also Nefedov, V.A.; Tarygina, L.K.; Kryuchkova, L.V.; Ryabokobylko, Yu.S. *J. Org. Chem, USSR* **1981**, *17*, 487; Suzuki, H.; Kondo, A.; Ogawa, T. *Chem. Lett.* **1985**, 411; Liedholm, B.; Nilsson, M. *Acta Chem. Scand. Ser. B* **1988**, *42*, 289; Clark, J.H.; Jones, C.W.; Duke, C.V.A.; Miller, J.M. *J. Chem. Res. (S)* **1989**, 238.

[257]Clark, J.H.; Jones, C.W. *J. Chem. Soc. Chem. Commun.* **1987**, 1409.

[258]Klapars, A.; Buchwald, S.L. *J. Am. Chem. Soc.* **2002**, *124*, 14844.

[259]Yang, S.H.; Li, C.S.; Cheng, C.H. *J. Org. Chem.* **1987**, *52*, 691.

[260]Arvela, R.K.; Leadbeater, N.E. *Synlett* **2003**, 1145.

[261]Harrowven, D.C.; Nunn, M.I.T.; Fenwick, D.R. *Tetrahedron Lett.* **2001**, *42*, 7501.

[262]For reviews of the synthesis of aryl iodides, see Merkushev, E.B. *Synthesis* **1988**, 923; *Russ. Chem. Rev.* **1984**, *53*, 343.

[263]Taylor, E.C.; Kienzle, F.; McKillop, A. *Org. Synth.* **VI**, 826; Taylor, E.C.; Katz, A.H.; Alvarado, S.I.; McKillop, A. *J. Organomet. Chem.* **1985**, *285*, C9. For reviews, see Usyatinskii, A.Ya.; Bregadze, V.I. *Russ. Chem. Rev.* **1988**, *57*, 1054; Uemura, S., in Hartley, F. R.; Patai, S. *The Chemistry of the Metal–Carbon Bond*, Vol. 4, Wiley, NY, pp. 473–538. See also, Ishikawa, N.; Sekiya, A. *Bull. Chem. Soc. Jpn.* **1974**, *47*, 1680; Taylor, E.C.; Altland, H.W.; McKillop, A. *J. Org. Chem.* **1975**, *40*, 2351.

[264]De Meio, G.V.; Pinhey, J.T. *J. Chem. Soc. Chem. Commun.* **1990**, 1065.

[265]Thiebes, C.; Prakash, G.K.S.; Petasis N.A.; Olah, G.A. *Synlett* **1998**, 141.

[266]Szumigala, Jr., R.H.; Devine, P.N.; Gauthier Jr., D.R.; Volante, R.P. *J. Org. Chem.* **2004**, *69*, 566.

[267]For a review of many of these reactions, see Artamkina, G.A.; Kovalenko, S.V.; Beletskaya, I.P.; Reutov, O.A. *Russ. Chem. Rev.* **1990**, *59*, 750.

13-8   Cyanation of Aromatic Rings

Cyano-de-halogenation
Cyano-de-metalation

Ar-X   ⟶   Ar-CN

The reaction between aryl halides and cuprous cyanide is called the *Rosenmund-von Braun reaction*.[268] Reactivity is in the order I > Br > Cl > F, indicating that the S_NAr mechanism does not apply.[269] Other cyanides (e.g., KCN and NaCN) do not react with aryl halides, even activated ones. This reaction has been done in ionic liquids using CuCN.[270] The reaction has also been done in water using CuCN, a phase transfer catalyst, and microwave irradiation.[271]

Aryl halides reaction with metal cyanides, often with another transition metal catalyst, to give aryl nitriles (aryl cyanides). Aryl halides react with $Zn(CN)_2$ and a palladium catalyst, for example, to give the aryl nitrile.[272] Similarly, aryl iodides react with CuCN and a palladium catalyst to give the aryl nitrile.[273] Potassium cyanide (KCN) reacts in a similar manner with a palladium catalyst.[274] Sodium cyanide has been used with a copper catalyst and 20% KI.[275] The reaction of aryl iodides and sodium cyanoborohydride/catechol, with a palladium catalyst, generates the aryl nitrile.[276] Aryl bromides react with $Ni(CN)_2$ with microwave irradiation to give ArCN.[277] In general, alkali cyanides do convert aryl halides to nitriles[278] in dipolar aprotic solvents in the presence of Pd(II) salts[279] or copper[280] or nickel[281]

---

[268]For a review of cyano-de-halogenation, see Ellis, G.P.; Romney-Alexander, T.M. *Chem. Rev.* **1987**, *87*, 779.

[269]For discussions of the mechanism, see Couture, C.; Paine, A.J. *Can. J. Chem.* **1985**, *63*, 111; Connor, J.A.; Leeming, S.W.; Price, R. *J. Chem. Soc. Perkin Trans. 1* **1990**, 1127.

[270]In bmiI, 1-*n*-butyl-3-methylimidazolium iodide: Wu, J.X.; Beck, B.; Ren, R.X. *Tetrahedron Lett.* **2002**, *43*, 387.

[271]Arvela, R.K.; Leadbeater, N.W.; Torenius, H.M.; Tye, H. *Org. Biomol. Chem.* **2003**, *1*, 1119.

[272]Jin, F.; Confalone, P.N. *Tetrahedron Lett.* **2000**, *41*, 3271; Zhang, A.; Neumeyer, J.L. *Org. Lett.* **2003**, *5*, 201; Marcantonio, K.M.; Frey, L.F.; Liu, Y.; Chen, Y.; Strine, J.; Phenix, B.; Wallace, D.J.; Chen, C.-y. *Org. Lett.* **2004**, *6*, 3723; Ramnauth, J.; Bhardwaj, N.; Renton, P.; Rakhit, S.; Maddaford, S.P. *Synlett* **2003**, 2237. See Erker, T.; Nemec, S. *Synthesis* **2004**, 23.

[273]Sakamoto, T.; Ohsawa, K. *J. Chem. Soc. Perkin Trans. 1* **1999**, 2323.

[274]Sundermeier, M.; Zapf, A.; Beller, M.; Sans, J. *Tetrahedron Lett.* **2001**, *42*, 6707; Yang, C.; Williams, J.M. *Org. Lett.* **2004**, *6*, 2837 (this reaction used a catalytic amount of tributyltin chloride as well).

[275]Zanon, J.; Klapers, A.; Buchwald, S.L. *J. Am. Chem. Soc.* **2003**, *125*, 2890.

[276]Jiang, B.; Kan, Y.; Zhang, A. *Tetrahedron* **2001**, *57*, 1581.

[277]Arvela, R.K.; Leadbeater, N.E. *J. Org. Chem.* **2003**, *68*, 9122.

[278]For a list of reagents that convert aryl halides to cyanides, with references, see Larock, R.C. *Comprehensive Organic Transformations*, 2nd ed., Wiley-VCH, NY, **1999**, pp. 1705–1709.

[279]Takagi, K.; Okamoto, T.; Sakakibara, Y.; Ohno, A.; Oka, S.; Hayama, N. *Bull. Chem. Soc. Jpn,.* **1975**, *48*, 3298; *1976*, *49*, 3177. See also Sekiya, A.; Ishikawa, N. *Chem. Lett.* **1975**, 277; Takagi, K.; Sasaki, K.; Sakakibara, Y. *Bull. Chem. Soc. Jpn.* **1991**, *64*, 1118.

[280]Connor, J.A.; Gibson, D.; Price, R. *J. Chem. Soc. Perkin Trans. 1* **1987**, 619.

[281]Cassar, L.; Foà, M.; Montanari, F.; Marinelli, G.P. *J. Organomet. Chem.* **1979**, *173*, 335; Sakakibara, Y.; Okuda, F.; Shimobayashi, A.; Kirino, K.; Sakai, M.; Uchino, N.; Takagi, K. *Bull. Chem. Soc. Jpn.* **1988**, *61*, 1985.

complexes. A nickel complex also catalyzes the reaction between aryl triflates and KCN to give aryl nitriles.[282]

Arylthallium bis(trifluoroacetates) (see **12-23**) can be converted to aryl nitriles by treatment with copper(I) cyanide in acetonitrile.[283] Another procedure uses excess aqueous KCN followed by photolysis of the resulting complex ion $ArTl(CN)_3^-$ in the presence of excess KCN.[284] Alternatively, arylthallium acetates react with $Cu(CN)_2$ or CuCN to give aryl nitriles.[285] Yields from this procedure are variable, ranging from almost nothing to 90 or 100%. Aromatic ethers ArOR[286] have been photochemically converted to ArCN.

An indirect method involves the reaction of an aromatic ring with *tert*-butyllithium, particularly when there is a directing group (see **13-17**), followed by reaction with PhOCN (phenyl cyanate) to give the aryl nitrile.[287] another indirect method involve the palladium catalyzed reaction of aryl bromides with the cyanohydrin of acetone [$Me_2C(OH)CN$] to give ArCN.[288]

OS **III**, 212, 631.

13-9 Coupling of Aryl and Alkyl Organometallic Compounds with Aryl Halides, Ethers, and Carboxylic Esters

Aryl-de-halogenation, and so on

$$Ar-X \ + \ Ar'-M \longrightarrow Ar-Ar'$$

$$Ar-X \ + \ R-M \longrightarrow Ar-R$$

A number of methods involving transition metals have been used to prepare unsymmetrical biaryls (see also, **13-11**). The uncatalyzed coupling of aryl halides and metalated aryls (particularly aryllithium reagents) is also known, including cyclization of organolithium reagents to aromatic rings.[289] Noncatalyzed coupling reactions of aryllithium reagents and haloarenes can proceed via the well-known aryne route but in some cases, a novel addition–elimination pathway is possible when substituents facilitate a chelation-driven nucleophilic substitution pathway.[290] Such noncatalyzed coupling reactions often proceed with high regioselectivity and high yield.[290] Several noncatalyzed alternative routes are available. 2-Bromopyridine reacts with pyrrolidine, at 130°C with microwave irradiation, to give 2-(2-pyrrolidino)pyridine.[291] Aryl iodides undergo homo-coupling to give the biaryl by

[282]Chambers, M.R.I.; Widdowson, D.A. *J. Chem. Soc. Perkin Trans. 1* **1989**, 1365; Takagi, K.; Sakakibara, Y. *Chem. Lett.* **1989**, 1957.

[283]Taylor, E.C.; Katz, A.H.; McKillop, A. *Tetrahedron Lett.* **1984**, 25, 5473.

[284]Taylor, E.C.; Altland, H.W.; McKillop, A. *J. Org. Chem.* **1975**, 40, 2351.

[285]Uemura, S.; Ikeda, Y.; Ichikawa, K. *Tetrahedron* **1972**, 28, 3025.

[286]Letsinger. R.L.; Colb, A.L. *J. Am. Chem. Soc.* **1972**, 94, 3665.

[287]Sato, N. *Tetrahedron Lett.* **2002**, 43, 6403.

[288]Sundermeier, M.; Zapf, A.; Beller, M. *Angew. Chem. Int. Ed.* **2003**, 42, 1661.

[289]For a review of cyclization of organolithium reagents, see Clayden, J.; Kenworthy, M.N. *Synthesis* **2004**, 1721.

[290]See Becht, J.-M.; Gissot, A.; Wagner, A.; Mioskowski, C. *Chem. Eur. J.* **2003**, 9, 3209.

[291]Narayan, S.; Seelhammer, T.; Gawley, R.E. *Tetrahedron Lett.* **2004**, 45, 757.

heating with triethylamine in an ionic liquid.[292] Arylsiloxanes react with aryl halides, for example, to give the biaryl derivative.[293] The reaction of $NaBPh_4$ (sodium tetraphenylborate) and a silyl dichloride (Ph_2SiCl_2) gives biphenyl.[294]

There are many catalytic methods. A homo-coupling type reaction was reported in which $PhSnBu_3$ was treated with 10% $CuCl_2$, 0.5 equivalents of iodine and heated in DMF to give biphenyl.[295] Arylsulfonyl chlorides also react with $ArSnBu_3$ with palladium and copper catalysts to give the biaryl.[296] Aryl halides undergo homo-coupling to give the biaryl with a palladium catalyst[297] or a nickel catalyst.[298] In general, aryl tin compounds couple with aryl halides.[299] An aryltin–aryl halide coupling has been done in ionic liquids.[300] Aryl iodides have been coupled to form symmetric biphenyls using $Pd(OAc)_2$[301] and self-coupling occurs with aryl triflates under electrolysis conditions with a palladium catalyst.[302] A "double-coupling" reaction involving 2-trimethysilylphenol O-triflate, allyltributyltin and allyl chloride, with CsF and a palladium catalyst, gave 1,2-diallylbenzene.[303] Another homo-coupling reaction of pyridyl bromides was reported using $NiBr_2$ under electrolytic conditions.[304] Thiophene derivatives,[305] pyrrole,[306] azoles,[307] quinoline,[308] and indolizine[309] have been coupled to aryl halides using a palladium catalyst.

Grignard reagents couple with aryl halides without a palladium catalyst, by the benzyne mechanism,[310] but an iron catalyzed coupling reaction was reported,[311] as

[292]In bmim PF_6, 1-butyl-3-methylimidazolium hexafluorophosphate: Park, S.B.; Alper, H. *Tetrahedron Lett.* **2004**, *45*, 5515.

[293]Mori, A.; Suguro, M. *Synlett* **2001**, 845; Murata, M.; Shimazaki, R.; Watanabe, S.; Masuda, Y. *Synthesis* **2001**, 2231.

[294]Sakurai, H.; Morimoto, C.; Hirao, T. *Chem. Lett.* **2001**, 1084. See also, Powell, D.A.; Fu, G.C. *J. Am. Chem. Soc.* **2004**, *126*, 7788.

[295]Kang, S.-K.; Baik, T.-G.; Jiao, X.H.; Lee, Y.-T. *Tetrahedron Lett.* **1999**, *40*, 2383.

[296]Dubbaka, S.R.; Vogel, P. *J. Am. Chem. Soc.* **2003**, *125*, 15292.

[297]Silveira, P.B.; Lando, V.R.; Dupont, J.; Monteiro, A.L. *Tetrahedron Lett.* **2002**, *43*, 2327; Kuroboshi, M.; Waki, Y.; Tanaka, H. *Synlett* **2002**, 637. See also, Venkatraman, S.; Li, C.-J. *Org. Lett.* **1999**, *1*, 1133.

[298]Leadbeater, N.E.; Resouly, S.M. *Tetrahedron Lett.* **1999**, *40*, 4243.

[299]Wang, J.; Scott, A.I. *Tetrahedron Lett.* **1996**, *37*, 3247; Saá, J.M.; Martorell, G.; García-Raso, A. *J. Org. Chem.* **1992**, *57*, 678; Littke, A.F.; Schwarz, L.; Fu, G.C. *J. Am. Chem. Soc.* **2002**, *124*, 6343; Kim, Y.M.; Yu, S. *J. Am. Chem. Soc.* **2003**, *125*, 1696.

[300]Grasa, G.A.; Nolan, S.P. *Org. Lett.* **2001**, *3*, 119.

[301]Penalva, V.; Hassan, J.; Lavenot, L.; Gozzi, C.; Lemaire, M. *Tetrahedron Lett.* **1998**, *39*, 2559.

[302]Jutand, A.; Négri, S.; Mosleh, A. *J. Chem. Soc, Chem. Commun.* **1992**, 1729.

[303]Yoshikawa, E.; Radhakrishnan, K.V.; Yamamoto, Y. *Tetrahedron Lett.* **2000**, *41*, 729.

[304]de Franç a, K.W.R.; Navarro, M.; Léonel, É; Durandetti, M.; Nédélec, J.-Y. *J. Org. Chem.* **2002**, *67*, 1838.

[305]Glover, B.; Harvey, K.A.; Liu, B.; Sharp, M.J.; Tymoschenko, M.F. *Org. Lett.* **2003**, *5*, 301.

[306]With $ZnCl_2$ as an additive, see Rieth, R.D.; Mankand, N.P.; Calimano, E.; Sadighi, J.P. *Org. Lett.* **2004**, *6*, 3981.

[307]Sezen, B.; Sames, D. *Org. Lett.* **2003**, *5*, 3607.

[308]Quintin, J.; Franck, X.; Hocquemiller, R.; Figadère, B. *Tetrahedron Lett.* **2002**, *43*, 3547.

[309]Park, C.-H.; Ryabova, V.; Seregin, I. V.; Sromek, A. W.; Gevorgyan, V. *Org. Lett.* **2004**, *6*, 1159.

[310]Du, C.F.; Hart, H.; Ng, K.D. *J. Org. Chem.* **1986**, *51*, 3162.

[311]Fürstner, A.; Leitner, A.; Méndez, M.; Krause, H. *J. Am. Chem. Soc.* **2002**, *124*, 13856.

well as a nickel-[312] and a cobalt-catalyzed reaction.[313] The coupling reaction of an excess of a Grignard reagent (RMgX) with methoxy aromatic compounds, when the aromatic ring contains multiple alkoxy groups, proceeds with replacement of the OMe group by R.[314] Aryl Grignard reagents coupled with phenyl allyl sulfone, in the presence of an iron catalyst, to give $ArCH_2CH=CH_2$.[315] In a similar manner, aryl sulfone coupled with aryl Grignard reagents in the presence of a nickel catalyst.[316] Arylmagnesium compounds couple to give the symmetrical biaryl in the presence of $TiCl_4$.[317] Arylmagnesium halides couple with aryl tosylates in the presence of a palladium catalyst to give unsymmetrical biaryls,[318] and to halopyridines to give the arylated pyridine.[319] Aryl Grignard reagents can be coupled to aryliodonium salts, with $ZnCl_2$ and a palladium catalyst, to give the biaryl.[320] Specialized aryl bismuth compounds have been used with a palladium catalyst to convert aryl chlorides to biaryls,[321] and specialized alkyl indium complexes have been used with a palladium catalyst to give arenes.[322] α-Lithio lactams are coupled to aryl bromides using a palladium catalyst, giving α-aryl lactams.[323]

The homo-coupling of arylzinc iodides with a palladium catalyst has been reported.[324] Vinyl halides, in the presence of an arylmagnesium halides, $ZnCl_2$ and a palladium catalyst, give the styrene compound.[325] Aryl triflates (halides) couple with ArZn(halide) reagents in the presence of a nickel catalyst.[326] Aryl triflates were coupled to triphenylbismuth using a palladium catalyst.[327] Homo-coupling of triphenylbismuth is known,[328] as well as the coupling of arylbismuth reagents to aryliodonium salts[329] and to aryltin compounds[330] with palladium chloride. Similar coupling was accomplished with aryltellurium compounds.[331] Aryl iodides undergo

[312]Dankwardt, J.W. *Angew. Chem. Int. Ed.* **2004**, *43*, 2428; Mongin, F.; Mojovic, L.; Guillamet, B.; Trécourt, F.; Quéguiner, G. *J. Org. Chem.* **2002**, *67*, 8991.

[313]Korn, T.J.; Cahiez, G.; Knochel, P. *Synlett* **2003**, 1892.

[314]Kojima, T.; Ohishi, T.; Yamamoto, I.; Matsuoka, T.; Kotsuki, H. *Tetrahedron Lett.* **2001**, *42*, 1709.

[315]Gai, Y.; Julia, M.; Verpeaux, J.-N. *Bull. Soc. Chim. Fr.* **1996**, *133*, 805.

[316]Clayden, J.; Cooney, J.J.A.; Julia, M. *J. Chem. Soc. Perkin Trans. 1* **1995**, 7.

[317]Inoue, A.; Kitagawa, K.; Shinokubo, H.; Oshima, K. *Tetrahedron* **2000**, *56*, 9601.

[318]Roy, A.H.; Hartwig, J.F. *J. Am. Chem. Soc.* **2003**, *125*, 8704.

[319]Bonnet, V.; Mongin, F.; Trècourt, F.; Quèguiner, G.; Knochel, P. *Tetrahedron Lett.* **2001**, *42*, 5717.

[320]Wang, L.; Chen, Z.-C. *Synth. Commun.* **2000**, *30*, 3607.

[321]Yamazaki, O.; Tanaka, T.; Shimada, S.; Suzuki, Y.; Tanaka, M. *Synlett* **2004**, 1921.

[322]Shenglof, M.; Gelman, D.; Heymer, B.; Schumann, H.; Molander, G.A.; Blum, J. *Synthesis* **2003**, 302.

[323]Cossy, J.; de Filippis, A.; Pardo, D.G. *Synlett* **2003**, 2171.

[324]With NCS, Hossain, K.M.; Kameyama, T.; Shibata, T.; Takagi, K. *Bull. Chem. Soc. Jpn.* **2001**, *74*, 2415. See also, Venkatraman, S.; Li, C.-J. *Tetrahedron Lett.* **2000**, *41*, 4831; Albanese, D.; Landini, D.; Penso, M.; Petricci, S. *Synlett* **1999**, 199.

[325]Peyrat, J.-F.; Thomas, E.; L'Hermite, N.; Alami, M.; Brion, J.-D. *Tetrahedron Lett.* **2003**, *44*, 6703.

[326]Quesnelle, C.A.; Familoni, O.B.; Snieckus, V. *Synlett* **1994**, 349. For the use of $NiCl_2/CrCl_2/Mn$, see Chen, C. *Synlett* **2000**, 1491. For a reaction done with microwave irradiation, see Walla, P.; Kappe, C.O. *Chem. Commun.* **2004**, 564.

[327]Rao, M.L.N.; Yamazaki, O.; Shimada, S.; Tanaka, T.; Suzuki, Y.; Tanaka, M. *Org. Lett.* **2001**, *3*, 4103.

[328]Ohe, T.; Tanaka, T.; Kuroda, M.; Cho, C.S.; Ohe, K.; Uemura, S. *Bull. Chem. Soc. Jpn.* **1999**, *72*, 1851.

[329]Kang, S.-K.; Ryu, H.-C.; Kim, J.-W. *Synth. Commun.* **2001**, *31*, 1021.

[330]Kang, S.-K.; Ryu, H.-C.; Lee, S.-W. *Synth. Commun.* **2001**, *31*, 1027.

[331]Kang, S.-K.; Lee, S.-W.; Kim, M.-S.; Kwon, H.S. *Synth. Commun.* **2001**, *31*, 1721.

a homo-coupling in the presence of hydroquinone and a palladium catalyst.[332] Arylgermanium compounds are coupled with aryl iodides using tetrabutylammonium fluoride and a palladium catalyst.[333] Both alkylmanganese compounds $(RMnCl)$[334] and Ph_3In[335] react with aryl halides or aryl triflates to give the arene, as do arylbismuth regents with aryl triflates.[336] Aryl halides couple to vinyl acetates, with a cobalt catalyst, to give the styrene derivative.[337] Aryl halides react with cyclopentadiene and Cp_2ZrCl_2 and a palladium catalyst to give pentaphenylcyclopentadiene.[338] Aryl halides also react with phenols to form biaryls using a rhodium catalyst.[339] Diaryliodonium salts react with $PhPb(OAc)_3$ and a palladium catalyst to give the biaryl.[340] Arylsilanes can be coupled to aryl iodides using a palladium catalyst.[341] Aryl halides reacts with acrolein diethyl acetal under electrolysis conditions and a nickel catalyst to give the allyl arene $(Ar-CH_2CH=CHOEt)$.[342]

Unsymmetrical binaphthyls were synthesized by photochemically stimulated reaction of naphthyl iodides with naphthoxide ions in an $S_{RN}1$ reaction.[343] Methyl chloroacetate coupled with aryl iodides under electrolysis conditions, using a nickel catalyst.[344] Unsymmetrical biaryls were prepared from two aryl iodides using a CuI catalyst and microwave irradiation.[345]

Alkylboronic acids are coupled to aryl halides using a palladium catalyst,[346] analogous to the Suzuki reaction in **13-12**. Conversely, arylboronic acids can be coupled to aliphatic halides.[347] Arylboronic acids can be coupled to allylic alcohols as well.[348] Arylboronic acids (**12-28**) were shown to react directly with benzene in the presence of $Mn(OAc)_3$.[349] Arylboronic acids also couple with alkyl halides in

[332]Hennings, D.D.; Iwama, T.; Rawal, V.H. *Org. Lett.* **1999**, *1*, 1205.

[333]Nakamura, T.; Kinoshita, H.; Shinokubo, H.; Oshima, K. *Org. Lett.* **2002**, *4*, 3165.

[334]Cahiez, G.; Luart, D.; Lecomte, F. *Org. Lett.* **2004**, *6*, 4395.

[335]Pérez, I.; Sestelo, J.P.; Sarandeses, L.A. *Org. Lett.* **1999**, *1*, 1267; *J. Am. Chem. Soc.* **2001**, *123*, 4155.

[336]Rao, M.L.N.; Shimada, S.; Tanaka, M. *Org. Lett.* **1999**, *1*, 1271.

[337]Gomes, P.; Gosmini, C.; Périchon, J. *Tetrahedron* **2003**, *59*, 2999.

[338]Dyker, G.; Heiermann, J.; Miura, M.; Inoh, J.-I.; Pivsa-Ast, S.; Satoh, T.; Nomura, M. *Chem. Eur. J.* **2000**, *6*, 3426.

[339]Bedford, R.B.; Limmert, M.E. *J. Org. Chem.* **2003**, *68*, 8669.

[340]Kang, S.-K.; Choi, S.-C.; Baik, T.-G. *Synth. Commun.* **1999**, *29*, 2493.

[341]Denmark, S.E.; Wu, Z. *Org. Lett.* **1999**, *1*, 1495; Lee, H.M.; Nolan, S.P. *Org. Lett.* **2000**, *2*, 2053.

[342]Condon, S.; Dupré, D.; Nédélec, J.Y. *Org. Lett.* **2003**, *5*, 4701.

[343]Beugelmans, R.; Bois-Choussy, M.; Tang, Q. *Tetrahedron Lett.* **1988**, *29*, 1705. For other preparations of biaryls via $S_{RN}1$ processes, see Alam, N.; Amatore, C.; Combellas, C.; Thiébault, A.; Verpeaux, J.N. *Tetrahedron Lett.* **1987**, *28*, 6171; Pierini, A.B.; Baumgartner, M.T.; Rossi, R.A. *Tetrahedron Lett.* **1988**, *29*, 3429.

[344]Durandetti, M.; Nédélec, J.-Y.; Périchon, J. *J. Org. Chem.* **1996**, *61*, 1748.

[345]He, H.; Wu, Y.-J. *Tetrahedron Lett.* **2003**, *44*, 3445.

[346]Zou, G.; Reddy, Y.K.; Falck, J.R. *Tetrahedron Lett.* **2001**, *42*, 7217; Molander, G.A.; Yun, C.-S. *Tetrahedron* **2002**, *58*, 1465.

[347]Duan, Y.-Z.; Deng, M.-Z. *Tetrahedron Lett.* **2003**, *44*, 3423; Bandgar, B.P.; Bettigeri, s.V.; Phopase, J. *Tetrahedron Lett.* **2004**, *45*, 6959.

[348]Tsukamoto, H.; Sato, M.; Kondo, Y. *Chem. Commun.* **2004**, 1200; Kayaki, Y.; Koda, T.; Ikariya, T. *Eur. J. Org. Chem.* **2004**, 4989.

[349]Demir, A.S.; Reis, Ö; Emrullahoglu, M. *J. Org. Chem.* **2003**, *68*, 578.

the presence of palladium(II) acetate[350] or a nickel catalyst.[351] Vinylboronic acids coupled to aryl halides to give the vinyl coupling product.[352] Vinylboronic acids have been coupled to aryldiazonium salts (**13-25**) without added base, using a palladium catalyst with an imidazolium ligand.[353]

Alkyltrifluoroborates (RBF$_3$K, see **12-28**) react with aryl triflates[354] or aryl halides,[355] or aryliodonium salts[356] with a palladium catalyst, to give the arene. The reaction is compatible with sensitive functionality, such as an epoxide unit.

It is possible to couple metalated alkyl compounds to aryl compounds. The lithium enolate anion of an ester was coupled to an aryl halide, for example, using a palladium catalyst.[357]

Chiral vinyl sulfoxides have been coupled to aryl iodides to give a chiral allylic aryl compounds (C=C-CH$_2$-Ar), in a three-step procedure with good enantioselectivity.[358]

The reaction of a cyclic zirconium–diene complex and an aryl diiodide, with CuCl, leads to highly substituted naphthalene derivatives.[359]

OS **VI**, 916; **VIII**, 430, 586; **X**, 9, 448.

13-10 Arylation and Alkylation of Alkenes

Alkylation or **Alkyl-de-hydrogenation**, and so on

$$R_2C=CH_2 \ + \ Ar\text{-}X \ \xrightarrow{\text{Pd(0)}} \ R_2C=CH\text{-}Ar$$

Arylation of alkenes can also be achieved[360] by treatment with an "arylpalladium" reagent, typical generated *in situ* from an aryl halide or other suitably functionalized aromatic compound and a palladium(0) catalyst.[361] Other methods

[350]Kirchhoff, J.H.; Netherton, M.R. Hills, I.D.; Fu, G.C. *J. Am. Chem. Soc.* **2002**, *124*, 13662.

[351]Zhou, J.; Fu, G.C. *J. Am. Chem. Soc.* **2004**, *126*, 1340.

[352]Collet, S.; Danion-Bougot, R.; Danion, D. *Synth. Commun.* **2001**, *31*, 249.

[353]Andrus, M.B.; Song, C. *Org. Lett.* **2001**, *3*, 3761; Andrus, M.B.; Song, C.; Zhang, J. *Org. Lett.* **2002**, *4*, 2079.

[354]Molander, G.A.; Yun, C.-S.; Ribagorda, M.; Biolatto, B. *J. Org. Chem.* **2003**, *68*, 5534.

[355]Molander, G.A.; Ribagorda, M. *J. Am. Chem. Soc.* **2003**, *125*, 11148.

[356]Xia, M.; Chen, Z.-C. *Synth. Commun.* **1999**, *29*, 2457.

[357]Moradi, W.A.; Buchwald, S.L. *J. Am. Chem. Soc.* **2001**, *123*, 7996.

[358]de la Rosa, J.C.; Díaz, N.; Carretero, J.C. *Tetrahedron Lett.* **2000**, *41*, 4107.

[359]Zhou, X.; Li, Z.; Wang, H.; Kitamura, M.; Kanno, K.-i.; Nakajima, K.; Takahashi, T. *J. Org. Chem.* **2004**, *69*, 4559.

[360]For reviews of this and related reactions, see Heck, R.F. *Palladium Reagents in Organic Syntheses*, Academic Press, NY, *1985*, pp. 179–321; Ryabov, A.D. *Synthesis 1985*, 233; Heck, R.F. *Org. React. 1982*, 27, 345; Moritani, I.; Fujiwara, Y. *Synthesis 1973*, 524. See Cabri, W.; Candiani, I. *Acc. Chem. Res. 1995*, 28, 2.

[361]For reviews, see Heck, R.F. *Acc. Chem. Res. 1979*, 12, 146; *Pure Appl. Chem. 1978*, *50*, 691; Kozhevnikov, I.V. *Russ. Chem. Rev. 1983*, *52*, 138. See also Bender, D.D.; Stakem, F.G.; Heck, R.F. *J. Org. Chem. 1982*, 47, 1278; Spencer, A. *J. Organomet. Chem. 1983*, *258*, 101. See also Bozell, J.J.; Vogt, C.E. *J. Am. Chem. Soc. 1988*, *110*, 2655; Andersson, C.; Karabelas, K.; Hallberg, A.; Andersson, C. *J. Org. Chem. 1985*, *50*, 3891; Merlic, C.A.; Semmelhack, M.F. *J. Organomet. Chem. 1990*, *391*, C23; Larock, R.C.; Johnson, P.L. *J. Chem. Soc. Chem. Commun. 1989*, 1368.

are available for this arylation reaction.[362] Treatment of an arylmercury compound (either Ar₂Hg or ArHgX) with LiPdCl₃ (ArHgX → "ArPdX") can generate the appropriate intermediate,[363] and in some cases other noble metal salts have been used. The palladium catalyzed aryl–alkene coupling reaction is known as *the Heck reaction*. The reaction works best with aryl iodides, although conditions have been developed for aryl bromides and aryl chlorides.[364] Aryldiazonium salts (**13-25**), rather than aryl halides, have also been used in the Heck reaction.[365] When 2,3,4,5,6-pentafluorobromobezene was used as a substrate, coupling occurred via the bromine, giving the pentafluorophenyl alkene.[366] Aryl halides bearing ortho-substituents also under the coupling reaction.[367] Heteroaryl halides can be used in the couple reaction.[368] Note that acetanilide derivatives reacted with conjugated esters to give the Heck product in acetic acid using a palladium catalyst.[369] Other activated aromatic compounds couple in a similar manner using palladium catalysts[370] unactivated aromatic compounds using special reaction conditions.[371]

Unlike **13-26**, the Heck reaction is not limited to activated substrates. The substrate can be a simple alkene, or it can contain a variety of functional groups, such as ester, ether,[372,373] carboxyl, phenolic, or cyano groups.[374] Coupling with vinyl ethers has been reported, C=C—OR → C=C(Ar)OR.[375] The Heck reaction can be done with heterocyclic compounds,[376] and the C—C unit of compounds, such as indene, react with aryl iodides and palladium catalyst without the need for

---

[362]For other methods, see Tsuji, J.; Nagashima, H. *Tetrahedron* **1984**, *40*, 2699; Kikukawa, K.; Naritomi, M.; He, G.; Wada, F.; Matsuda, T. *J. Org. Chem.* **1985**, *50*, 299; Chen, Q.; Yang, Z. *Tetrahedron Lett.* **1986**, *27*, 1171; Kasahara, A.; Izumi, T.; Miyamoto, K.; Sakai, T. *Chem. Ind.* (*London*) **1989**, 192; Miura, M.; Hashimoto, H.; Itoh, K.; Nomura, M. *Tetrahedron Lett.* **1989**, *30*, 975.

[363]Heck, R.F. *J. Am. Chem. Soc.* **1968**, *90*, 5518, 5526, 5535. For a review, see Larock, R.C. *Organomercury Compounds in Organic Synthesis*, Springer, NY, **1985**, pp. 273–292.

[364]For reviews, see Whitcombe, N.J.; Hii, K.K.; Gibson, S.E. *Tetrahedron* **2001**, *57*, 7449; Littke, A.F.; Fu, G.C. *Angew. Chem. Int. Ed.* **2002**, *41*, 4176.

[365]Sengupta, S.; Bhattacharyya, S. *Tetrahedron Lett.* **2001**, *42*, 2035; Masllorens, J.; Moreno-Mañas, M.; Pla-Quintana, A.; Roglans, A. *Org. Lett.* **2003**, *5*, 1559; Dai, M.; Liang, B.; Wang, C.; Chen, J.; Yang, Z. *Org. Lett.* **2004**, *6*, 221.

[366]Albéniz, A.C.; Espinet, P.; Martín-Ruiz, B.; Milstein, D. *J. Am. Chem. Soc.* **2001**, *123*, 11504.

[367]Littke, A.F.; Fu, G.C. *J. Am. Chem. Soc.* **2001**, *123*, 6989; Feuerstein, M.; Doucet, H.; Santelli, M. *Synlett* **2001**, 1980.

[368]See Park, S.B.; Alper, H. *Org. Lett.* **2003**, *5*, 3209. See also, Zeni, G.; Larock, R.C. *Chem. Rev.* **2004**, *104*, 2285.

[369]Boele, M. D. K.; van Strijdonck, G. P. F.; de Vries, A. H. M.; Kamer, P. C. J.; de Vries, J. G.; van Leeuwen, P. W. N. M. *J. Am. Chem. Soc.* **2002**, *124*, 1586.

[370]Myers, A.G.; Tanaka, D.; Mannion, M.R. *J. Am. Chem. Soc.* **2002**, *124*, 11250.

[371]Yokota, T.; Tani, M.; Sakaguchi, S.; Ishii, Y. *J. Am. Chem. Soc.* **2003**, *125*, 1476.

[372]For a review pertaining to enol ethers, see Daves, Jr., G.D. *Adv. Met.- Org. Chem.* **1991**, *2*, 59.

[373]Larhed, M.; Hallberg, A. *J. Org. Chem.* **1996**, *61*, 9582.

[374]For a review of cases where the alkene contains an heteroatom, see Daves, Jr., G.D.; Hallberg, A. *Chem. Rev.* **1989**, *89*, 1433.

[375]Andappan, M.M.S.; Nilsson, P.; von Schenck, H.; Larhed, M. *J. Org. Chem.* **2004**, *69*, 5212.

[376]Pyridines: Draper, T.L.; Bailey, T.R. *Synlett* **1995**, 157.

preparing the halide.[377] The Heck reaction has also been performed intramolecularly.[378] Asymmetric Heck reactions are known[379] and the effects of high pressure have been studied.[380]

Ethylene is the most reactive alkene. Increasing substitution lowers the reactivity. Substitution therefore takes place at the less highly substituted side of the double bond.[381] The aryl halide or aryl triflate can be coupled to dienes,[382] allenes,[383] allylic silanes,[384] allylic amines,[385] vinyl phosphonate esters,[386] and with terminal alkynes.[387] Alkylation can also be accomplished, but only if the alkyl group lacks a β-hydrogen, for example, the reaction is successful for the introduction of methyl, benzyl, and neopentyl groups.[388] However, vinylic groups, even those possessing β-hydrogens, have been successfully introduced (to give 1,3-dienes) by the reaction of the alkene with a vinylic halide in the presence of a trialkylamine and a palladium(0) catalyst.[389] Aryl iodides can be coupled to 1-methyl-1-vinyl- and 1-methyl-1-(prop-2-enyl)silacyclobutane with desilation, using a palladium catalyst and Bu$_4$NF, to give the corresponding styrene derivative.[390] Indene reacts with iodobenzene with a palladium catalyst to give the phenylindene (80:20 C3/C2).[391]

Control of regiochemistry is a serious problem in the addition to unsymmetrical alkenes. Some regioselectivity can be obtained by the use of alkenes attached to an

[377]Nifant'ev, I.E.; Sitnikov, A.A.; Andriukhova, N.V.; Laishevtsev, I.P.; Luzikov, Y.N. *Tetrahedron Lett.* **2002**, *43*, 3213.

[378]See, for example, Negishi, E.; Zhang, Y.; O'Connor, B. *Tetrahedron Lett.* **1988**, *29*, 2915; Larock, R.C.; Song, H.; Baker, B.E.; Gong, W.H. *Tetrahedron Lett.* **1988**, *29*, 2919; Dounay, A.B.; Hatanaka, K.; Kodanko, J.J.; Oestreich, M.; Overman, L.E.; Pfeifer, L.A.; Weiss, M.M. *J. Am. Chem. Soc.* **2003**, *125*, 6261. For a review of the asymmetric intramolecular Heck reaction, see Dounay, A.B.; Overman, L.E. *Chem. Rev.* **2003**, *103*, 2945. Also see Lee, S.W.; Fuchs, P.L. *Tetrahedron Lett.* **1993**, *34*, 5209; Echavarren, A.M.; Gómez-Lor, B.; González, J.J.; de Frutos, Ó. *Synlett* **2003**, 585.

[379]Shibasaki, M.; Boden, D.J.; Kojima, A. *Tetrahedron* **1997**, *53*, 737.

[380]Sugihara, T.; Yakebayashi, M.; Kaneko, C. *Tetrahedron Lett.* **1995**, *36*, 5547; Buback, M.; Perković, T.; Redlich, S.; de Meijere, A. *Eur. J. Org. Chem.* **2003**, 2375.

[381]Heck, R.F. *J. Am. Chem. Soc.* **1969**, *91*, 6707; **1971**, *93*, 6896.

[382]Jeffery, T. *Tetrahedron Lett.* **1992**, *33*, 1989.

[383]Chang, H.-M.; Cheng, C.-H. *J. Org. Chem.* **2000**, *65*, 1767.

[384]Jeffery, T. *Tetrahedron Lett.* **2000**, *41*, 8445.

[385]Olofsson, K.; Larhed, M.; Hallberg, A. *J. Org. Chem.* **2000**, *65*, 7235; Wu, J.; Marcoux, J.-F. Davies, I.W.; Reider, P.J. *Tetrahedron Lett.* **2001**, *42*, 159.

[386]Kabalka, G.W.; Guchhait, S.K.; Naravane, A. *Tetrahedron Lett.* **2004**, *45*, 4685.

[387]Cassar, L. *J. Organomet. Chem.* **1975**, *93*, 253; Dieck, H.A.; Heck, R.F. *J. Organomet. Chem.* **1975**, *93*, 259; Kundu, N.G.; Pal, M.; Mahanty, J.S.; Dasgupta, S.K. *J. Chem. Soc. Chem. Commun.* **1992**, 41. See also, Heck, R.F. *Palladium Reagents in Organic Syntheses*, Academic Press, NY, **1985**, pp. 299–306.

[388]Heck, R.F. *J. Organomet. Chem.* **1972**, *37*, 389; Heck, R.F.; Nolley Jr., J.P. *J. Org. Chem.* **1972**, 3720.

[389]Kim, J.I.; Patel, B.A.; Heck, R.F. *J. Org. Chem.* **1981**, *46*, 1067; Heck, R.F. *Pure Appl. Chem.* **1981**, *53*, 2323. See also Luong-Thi, N.; Riviere, H. *Tetrahedron Lett.* **1979**, 4657; Jeffery, T. *J. Chem. Soc. Chem. Commun.* **1991**, 324; Scott, W.J.; Peña, M.R.; Swärd, K.; Stoessel, S.J.; Stille, J.K. *J. Org. Chem.* **1985**, *50*, 2302; Larock, R.C.; Gong, W.H. *J. Org. Chem.* **1989**, *54*, 2047. For a new palladium catalyst on intercalated clay, see Varma, R.S.; Naicker, K.P.; Liesen, P.J. *Tetrahedron Lett.* **1999**, *40*, 2075.

[390]Denmark, S.E.; Wang, Z. *Synthesis* **2000**, 999.

[391]Nifant'ev, I.E.; Sitnikov, A.A.; Andriukhova, N.V.; Laishevtsev, I.P.; Luzikov, Y.N. *Tetrahedron Lett.* **2002**, *43*, 3213.

auxiliary coordinating group,[392] the use of special ligands and acrylate or styrene as substrates.[393] Steric effects are thought to control regioselectivity,[394] but electronic influences have also been proposed.[395] It has been shown that the presence of steric effects generally improve 1,2-selectivity, and that electronic effects can be used to favor 1,2- or 2,1-selectivity.[396]

Phosphine free catalysts[397] and halogen-free reactions[398] are known for the Heck reaction. Improvements on the palladium catalyst system are constantly being reported,[399] including polymer-supported catalysts.[400] The influence of the ligand has been examined.[401] Efforts have been made to produce a homogeneous catalyst for the Heck reaction.[402] The Heck reaction can be done in aq. media,[403] in perfluorinated solvents,[404] in polyethylene glycol,[405] in neat tricaprylmethylammonium

[392]Nilsson, P.; Larhed, M.; Hallberg, A. *J. Am. Chem. Soc.* **2001**, *123*, 8217 and earlier references.

[393]Ludwig, M.; Strömberg, S.; Svensson, M.; Åkermark, B. *Organometallics* **1999**, *18*, 970; Brown, J.M.; Hii, K.K. *Angew. Chem. Int. Ed.* **1996**, *35*, 657.

[394]Collman, J.P.; Hegedus, L.S.; Norton, J.R.; Finke, R.G., *Principles and Applications of Organotransition Metal Chemistry*, 2nd ed., University Science Books, Mill Valley, CA, **1987**; Cornils, B., Herrmann, A.W., Eds., *Applied Homogeneous Catalysis with Organometallic Compounds*, Wiley, NY, **1996**; Vol. 2; Heck, R.F. *Acc. Chem. Res.* **1979**, *12*, 146.

[395]Cabri, W.; Candiani, I. *Acc. Chem. Res.* **1995**, *28*, 2.

[396]von Schenck, H.; Akermark, B.; Svensson, M. *J. Am. Chem. Soc.* **2003**, *125*, 3503.

[397]Reetz, M.T.; Westermann, E.; Lohmer, R.; Lohmer, G. *Tetrahedron Lett.* **1998**, *39*, 8449; Gruber, A.S.; Pozebon, D.; Monteiro, A.L.; Dupont, J. *Tetrahedron Lett.* **2001**, *42*, 7345.

[398]Hirabayashi, K.; Nishihara, Y.; Mori, A.; Hiyama, T. *Tetrahedron Lett.* **1998**, *39*, 7893.

[399]Miyazaki, F.; Yamaguchi, K.; Shibasaki, M. *Tetrahedron Lett.* **1999** *40*, 7379; Calò, V.; Nacci, A.; Lopez, L.; Mannarini, N. *Tetrahedron Lett.* **2000**, *41*, 8973; Iyer, S.; Ramesh, C. *Tetrahedron Lett.* **2000**, *41*, 8981; Rosner, T.; Le Bars, J.; Pfaltz, A.; Blackmond, D.G. *J. Am. Chem. Soc.* **2001**, *123*, 1848; Iyer, S.; Jayanthi, A. *Tetrahedron Lett.* **2001**, *42*, 7877; Selvakumar, K.; Zapf, A.; Beller, M. *Org. Lett.* **2002**, *4*, 3031; Feuerstein, M.; Doucet, H.; Santelli, M. *Tetrahedron Lett.* **2002**, *43*, 2191. For a recyclable catalyst, see Bergbreiter, D.E.; Osburn, P.L.; Liu, Y.-S. *J. Am. Chem. Soc.* **1999**, *121*, 9531. For the use of palladium nanoparticles, see Calò, V.; Nacci, A.; Monopoli, A.; Laera, S.; Cioffi, N. *J. Org. Chem.* **2003**, *68*, 2929. For a heterogeneous catalyst, see Srivastava, R.; Venkatathri, N.; Srinivas, D.; Ratnasamy, P. *Tetrahedron Lett.* **2003**, *44*, 3649.

[400]Leese, M.P.; Williams, J.M.J. *Synlett* **1999**, 1645; Lin, C.-A.; Luo, F.-T. *Tetrahedron Lett.* **2003**, *44*, 7565. For a zeolite-supported palladium catalyst, see Djakovitch, L.; Koehler, K. *J. Am. Chem. Soc.* **2001**, *123*, 5990.

[401]For a review, see Qadir, M.; Möchel, T.; Hii, K.K. *Tetrahedron* **2000**, *56*, 7975. Feuerstein, M.; Doucet, H.; Santelli, M. *J. Org. Chem.* **2001**, *66*, 5923; Yang, C.; Lee, H.M.; Nolan, S.P. *Org. Lett.* **2001**, *3*, 1511; Tani, M.; Sakaguchi, S.; Ishii, Y. *J. Org. Chem.* **2004**, *69*, 1221; Yang, D.; Chen, Y.-C.; Zhu, N.-Y. *Org. Lett.* **2004**, *6*, 1577; Eberhard, M.R. *Org. Lett.* **2004**, *6*, 2125; Berthiol, F.; Doucet, H.; Santelli, M. *Tetrahedron Lett.* **2003**, *44*, 1221; Liu, J.; Zhao, Y.; Zhou, Y.; Li, L.; Zhang, T.Y.; Zhang, H. *Org. Biomol. Chem.* **2003**, *1*, 3227. A reaction was reported using palladium acetate in dimethylacetamide and no added ligand, see Yao, Q.; Kinney, E.P.; Yang, Z. *J. Org. Chem.* **2003**, *68*, 7528. For a phosphine-free reaction see Consorti, C.S.; Zanini, M.L.; Leal, S.; Ebeling, G.; Dupont, J. *Org. Lett.* **2003**, *5*, 983.

[402]Nair, D.; Scarpello, J.T.; White, L.S.; dos Santos, L.M.F.; Vankelecom, I.F.J.; Livingston, A.G. *Tetrahedron Lett.* **2001**, *42*, 8219.

[403]Jeffery, T. *Tetrahedron Lett.* **1994**, *35*, 3051; Gron, L.U.; Tinsley, A.S. *Tetrahedron Lett.* **1999**, *40*, 227.

[404]Moineau, J.; Pozzi, G.; Quici, S.; Sinou, D. *Tetrahedron Lett.* **1999** *40*, 7683.

[405]Chandrasekhar, S.; Narsihmulu, Ch.; Sultana, S.S.; Reddy, N.R. *Org. Lett.* **2002**, *4*, 4399.

chloride,[406] and in supercritical CO_2 (see p. 414).[407] A noncatalytic reaction was reported using supercritical water.[408] The reaction has been done on solid support,[409] including Montmorillonite clay,[410] glass beads,[411] on a reverse-phase silica support,[412] and using microwave irradiation.[413] A microwave irradiated Heck coupling was done in water using a palladium catalyst.[414] The Heck reaction has also been in ionic liquids,[415] and it is known that the nature of the halide is important in such reactions.[416]

The evidence is in accord with an addition–elimination mechanism (addition of ArPdX followed by elimination of HPdX) in most cases.[417] In the conventionally accepted reaction mechanism,[418] a four-coordinate aryl–Pd(II) intermediate is formed by oxidative addition of the aryl halide to a Pd(0) complex prior to olefin addition. This suggests that cleavage of the dimeric precursor complex, reduction of Pd^{2+}, and ligand dissociation combine to give a viable catalytic species.[419] If these processes occur on a time scale comparable to that of the catalytic reaction, non-steady-state catalysis could occur while the active catalyst is forming, and an

[406]Perosa, A.; Tundo, P.; Selva, M.; Zinovyev, S.; Testa, A. *Org. Biomol. Chem.* **2004**, *2*, 2249.

[407]Shezad, N.; Oakes, R.S.; Clifford, A.A.; Rayner, C.M. *Tetrahedron Lett.* **1999** *40*, 2221; Bhanage, B.M.; Ikushima, Y.; Shirai, M.; Arai, M. *Tetrahedron Lett.* **1999** *40*, 6427; Cacchi, S.; Fabrizi, G.; Gasparrini, F.; Villani, C. *Synlett* **1999**, 345; Early, T.R.; Gordon, R.S.; Carroll, M.A.; Holmes, A.B.; Shute, R.E.; McConvey, I.F. *Chem. Commun.* **2001**, 1966. For a discussion of selectivity in scCO2, see Kayaki, Y.; Noguchi, Y.; Ikariya, T. *Chem. Commun.* **2000**, 2245.

[408]Zhang, R.; Sato, O.; Zhao, F.; Sato, M.; Ikushima, Y. *Chem. Eur. J.* **2004**, *10*, 1501.

[409]Franzén, R. *Can. J. Chem.* **2000**, *78*, 957.

[410]Ramchandani, R.K.; Uphade, B.S.; Vinod, M.P.; Wakharkar, R.D.; Choudhary, V.R.; Sudalai, A. *Chem. Commun.* **1997**, 2071.

[411]Tonks, L.; Anson, M.S.; Hellgardt, K.; Mirza, A.R.; Thompson, D.F.; Williams, J.M.J. *Tetrahedron Lett.* **1997**, *38*, 4319.

[412]Anson, M.S.; Mirza, A.R.; Tonks, L.; Williams, J.M.J. *Tetrahedron Lett.* **1999** *40*, 7147.

[413]Li, J.; Mau, A.W.-H.; Struass, C.R. *Chem. Commun.* **1997**, 1275; Díaz-Ortiz, Á.; Prieto, P.; Vázquez, E. *Synlett* **1997**, 269; Xie, X.; Lu, J.; Chen, B.; Han, J.; She, X.; Pan, X. *Tetrahedron Lett.* **2004**, *45*, 809. For microwave assisted, enantioselective Heck reactions, see Nilsson, P.; Gold, H.; Larhed, M.; Hallberg, A. *Synthesis* **2002**, 1611.

[414]Wang, J.-X.; Liu, Z.; Hu, Y.; Wei, B.; Bai, L. *Synth. Commun.* **2002**, *32*, 1607.

[415]In **bmim PF₆**, 1-butyl-3-methylimidazolium hexafluorophosphate: Carmichael, A.J.; Earle, M.J.; Holbrey, J.D.; McCormac, P.B.; Seddon, K.R. *Org. Lett.* **1999**, *1*, 997; Hagiwara, H.; Shimizu, Y.; Hoshi, T.; Suzuki, T.; Ando, M.; Ohkubo, K.; Yokoyama, C. *Tetrahedron Lett.* **2001**, *42*, 4349. In bmim PF₆ with microwave irradiation, see Vallin, K.S.A.; Emilsson, P.; Larhed, M.; Hallberg, A. *J. Org. Chem.* **2002**, *67*, 6243. In **bmim BF₄**, 1-butyl-3-methylimidazolium tetrafluoroborate: Handy, S.T.; Zhang, X. *Org. Lett.* **2001**, *3*, 233. In **bbim Br**, 1,3-di-*n*-butylimidazolium bromide, Deshmukh, R.R.; Rajagopal, R.; Srinivasan, K.V. *Chem. Commun.* **2001**, 1544. In a **tetrabutylammonium bromide melt**, Calò, V.; Nacci, A.; Monopoli, A.; Lopez, L.; di Cosmo, A. *Tetrahedron* **2001**, *57*, 6071. See Hagiwara, H.; Sugawara, Y.; Isobe, K.; Hoshi, T.; Suzuki, T. *Org. Lett.* **2004**, *6*, 2325; Okubo, K.; Shirai, M.; Yokoyama, C. *Tetrahedron Lett.* **2002**, *43*, 7115.

[416]Handy, S.T.; Okello, M. *Tetrahedron Lett.* **2003**, *44*, 8395.

[417]Heck, R.F. *J. Am. Chem. Soc.* **1969**, *91*, 6707; Shue, R.S. *J. Am. Chem. Soc.* **1971**, *93*, 7116; Heck, R.F.; Nolley Jr., J.P. *J. Org. Chem.* **1972**, 3720.

[418]Heck, R. F. *Comprehensive Organic Synthesis*, Vol. 4, Trost, B.M., Fleming, I., Eds., Pergamon, Oxford, NY, **1991**, p 833; de Meijere, A.; Meyer, F.E. *Angew. Chem. Int. Ed.* **1994**, *33*, 2379; Cabri, W.; Candiani, I. *Acc. Chem. Res.* **1995**, *28*, 2; Crisp, G.T. *Chem. Soc. Rev.* **1998**, *27*, 427.

[419]Rosner, T.; Pfaltz, A.; Blackmond, D. G. *J. Am. Chem. Soc.* **2001**, *123*, 4621.

analysis of reaction kinetics under dry conditions was reported.[419] In this study, the mechanism requires a first-order dependence on olefin concentration, and anomalous kinetics may be observed when the rate-limiting step is not directly on the catalytic cycle.[419]

The reactions are stereospecific, yielding products expected from syn addition followed by syn elimination.[420] Because the product is formed by an elimination step, with suitable substrates double bond migration can occur, resulting in allylic rearrangement (as in the reaction of cyclopentene and iodobenzene to give **21**).[421] Primary and secondary allylic alcohols (and even non-allylic unsaturated alcohols[422]) give aldehydes, such as **22** or ketones that are products of double-bond migration.[423] Similarly, dihydrofurans react with aryl triflates and a palladium catalyst that includes a chiral ligand, to give the 5-phenyl-3,4-dihydrofuran with good enantioselectivity.[424] A similar reaction was reported for an N-carbamoyl dihydropyrrole.[425] It has been reported that double bond isomerization can be suppressed in intramolecular Heck reactions done in supercritical CO_2 (see p. 414).[426] The mechanistic implications of asymmetric Heck reactions has been examined.[427]

There are a number of variations of this reaction, including the use of transition metal catalyst other than palladium. A silane-tethered, intramolecular Heck reaction has been reported.[428] Arylphosphonic acids, $ArP(=O)(OH)_2$, couple to aryl alkenes in the presence of a palladium catalyst.[429] Aryl halides couple with vinyl

[420]Heck, R.F. *J. Am. Chem. Soc.* **1969**, *91*, 6707; Moritani, I.; Danno, S.; Fujiwara, Y.; Teranishi, S. *Bull. Chem. Soc. Jpn.* **1971**, *44*, 578. See Masllorens, J.; Moreno-Mañas, M.; Pla-Quintana, A.; Plexats, R.; Roglans, A. *Synthesis* **2002**, 1903.

[421]Larock, R.C.; Baker, B.E. *Tetrahedron Lett.* **1988**, *29*, 905. Also see, Larock, R.C.; Gong, W.H.; Baker, B.E. *Tetrahedron Lett.* **1989**, *30*, 2603.

[422]Larock, R.C.; Leung, W.; Stolz-Dunn, S. *Tetrahedron Lett.* **1989**, *30*, 6629.

[423]See, for example, Melpolder, J.P.; Heck, R.F. *J. Org. Chem.* **1976**, *41*, 265; Chalk, A.J.; Magennis, S.A. *J. Org. Chem.* **1976**, *41*, 273, 1206.

[424]Tietze, L.F.; Thede, K.; Sannicolò, F. *Chem. Commun.* **1999**, 1811; Hashimoto, Y.; Horie, Y.; Hayashi, M.; Saigo, K. *Tetahedron Asymmetry* **2000**, *11*, 2205; Gilbertson, S.R.; Xie, D.; Fu, Z. *J. Org. Chem.* **2001**, *66*, 7240; Gilbertson, S.R.; Fu, Z. *Org. Lett.* **2001**, *3*, 161; Hennessy, A.J.; Connolly, D.J.; Malone, Y.M.; Buiry, P.J. *Tetrahedron Lett.* **2000**, *41*, 7757.

[425]Servino, E.A.; Correia, C.R.D. *Org. Lett.* **2000**, *2*, 3039.

[426]Shezad, N.; Clifford, A.A.; Rayner, C.M. *Tetrahedron Lett.* **2001**, *42*, 323.

[427]Hii, K.K.; Claridge, T.D.W.; Brown, J.M.; Smith, A.; Deeth, R.J. *Helv. Chim. Acta* **2001**, *84*, 3043.

[428]Mayasundari, A.; Young, D.G.J. *Tetrahedron Lett.* **2001**, *42*, 203.

[429]Inoue, A.; Shinokubo, H.; Oshima, K. *J. Am. Chem. Soc.* **2003**, *125*, 1484.

tin reagents to form styrene derivatives in the presence of a nickel catalyst.[430] Aryl chlorides were coupled to conjugated esters using a $RuCl_3 \cdot 3\ H_2O$, in an atmosphere of O_2 and CO.[431] Alkenyl organometallic compounds have been coupled to aryl halides, including allenyltin compounds $(C=C=C-SnR_3)$.[432] Divinylindium chloride, $(CH_2=CH)_2InCl$, reacted with an aryl iodide in aq. THF with a palladium catalyst to give the styrene derivative.[433] Trialkenylindium reagents reacted similarly with aryl halides and a palladium catalyst.[434] Arylzinc chlorides (ArZncl) were coupled to vinyl chlorides using a palladium catalyst,[435] and vinyl zinc compounds were coupled to aryl iodides.[436] Aryliodonium salts can be coupled to conjugated alkenes in a Heck-like manner using a palladium catalyst.[437] In the presence of trimethylsilylmagnesium chloride, primary alkyl halides coupled to aryl alkenes to give the substituted alkene $(R'—CH=CHAr)$, using a cobalt catalyst.[438]

Arylboronic acids (**12-28**) have been coupled to conjugated alkenes to give the aryl–alkene coupling product using a palladium catalyst,[439] a ruthenium catalyst with copper(II) acetate,[440] or a rhodium catalyst.[441] Arylboronic acids have also been coupled to vinyl halides[442] or vinyl tosylates[443] using a palladium catalyst. Note that the reaction of an arylboronic acid and 1,2-dibromoethane, with KOH and a palladium catalyst leads to the styrene derivative.[444] vinylboronic acids have been coupled to aryl halides using a palladium catalyst.[445] Styrene derivatives have been prepared by the reaction of aryl halides and 2,4,6-trivinylcyclotriboroxane, with a palladium catalyst.[446] Conjugated esters can be coupled to benzene using a palladium acetate/benzoquinone catalyst, *tert*-butyl hydroperoxide in acetic acid–acetic anhydride, at 90°C in a sealed tube.[447] Vinyl silanes were converted to styrene derivatives upon treatment with Bu_4NF, and aryl iodide and a palladium

[430]Shirakawa, E.; Yamasaki, K.; Hiyama, T. *Synthesis* **1998**, 1544; Chen, C.; Wilcoxen, K.; Zhu, Y.-F.; Kim, K.-i.; McCarthy, J.R. *J. Org. Chem.* **1999**, *64*, 3476.

[431]Weissman, H.; Song, X.; Milstein, D. *J. Am. Chem. Soc.* **2001**, *123*, 337.

[432]Huang, C.-W.; Shanmugasundarm, M.; Chang, H.-M.; Cheng, C.-H. *Tetrahedron* **2003**, *59*, 3635.

[433]Takami, K.; Yorimitsu, H.; Shinokubo, H.; Matsubara, S.; Oshima, K. *Org. Lett.* **2001**, *3*, 1997.

[434]Lehmann, U.; Awasthi, S.; Minehan, T. *Org. Lett.* **2003**, *5*, 2405.

[435]Dai, C.; Fu, G.C. *J. Am. Chem. Soc.* **2001**, *123*, 2719.

[436]Jalil, A.A.; Kurono, N.; Tokuda, M. *Synlett* **2001**, 1944.

[437]Xia, M.; Chen, Z.C. *Synth. Commun.* **2000**, *30*, 1281.

[438]Ikeda, Y.; Nakamura, T.; Yorimitsu, H.; Oshima, K. *J. Am. Chem. Soc.* **2002**, *124*, 6514.

[439]Du, X.; Suguro, M.; Hirabayashi, K.; Mori, A.; Nishikata, T.; Hagiwara, N.; Kawata, K.; Okeda, T.; Wang, H.-F.; Fugami, K.; Kosugi, M. *Org. Lett.* **2001**, *3*, 3313; Jung, Y.C.; Mishra, R.K.; Yoon, C.H.; Jung, K.W. *Org. Lett.* **2003**, *5*, 2231.

[440]Farrington, E.J.; Brown, J.M.; Barnard, C.F.J.; Rowsell, E. *Angew. Chem. Int. Ed.* **2002**, *41*, 169.

[441]Lautens, M.; Roy, A.; Fukuoka, K.; Fagnou, K.; Martín-Matute, B. *J. Am. Chem. Soc.* **2001**, *123*, 5358.

[442]Bauer, A.; Miller, M.W.; Vice, S.F.; McCombie, S.W. *Synlett* **2001**, 254; Poondra, R.R.; Fischer, P.M.; Turner, N.J. *J. Org. Chem.* **2004**, *69*, 6920.

[443]Wu, J.; Zhu, Q.; Wang, L.; Fathi, R.; Yang, Z. *J. Org. Chem.* **2003**, *68*, 670.

[444]Lando, V.R.; Monteiro, A.L. *Org. Lett.* **2003**, *5*, 2891.

[445]Peyroux, E.; Berthiol, F.; Doucet, H.; Santelli, M. *Eur. J. Org. Chem.* **2004**, 1075.

[446]Kerins, F.; O'Shea, D. F. *J. Org. Chem.* **2002**, *67*, 4968.

[447]Jia, C.; Lu, W.; Kitamura, T.; Fujiwara, Y. *Org. Lett.* **1999**, *1*, 2097.

catalyst.[448] Arylsilanes were coupled to alkenes to give the styrene derivative using palladium acetate and an oxygen atmosphere,[449] for Bu_4NF and an iridium catalyst.[450]

In a related reaction, vinyltrifluoroborates $C=C-BF_3^+ X^-$ (**12-28**), are coupled to aryl halides with a palladium catalyst to give the styrene derivative.[451]

In an unusual variation, an aryl compound bearing a tertiary alcohol substituent ($ArCMe_2OH$) reacted with aryl halides and a palladium catalyst to give the biaryl.[452] Benzoyl chloride was coupled to styrene to form $PhCH=CHPh$ using a rhodium catalyst.[453] Benzoic acid was coupled to styrene to give the same type of product using a palladium catalyst and a diacyl peroxide.[454]

OS **VI**, 815; **VII**, 361; **81**, 42, 54, 63, 263

13-11 Homo-Coupling of Aryl Halides: The Ullmann Reaction

De-halogen-coupling

$$2 \; ArI \quad \xrightarrow[\Delta]{Cu} \quad Ar-Ar$$

The coupling of aryl halides with copper is called the *Ullmann reaction*.[455] The reaction is clearly related to **13-9**, but involves aryl copper intermediates. The reaction is of broad scope and has been used to prepare many symmetrical and unsymmetrical biaryls.[456] When a mixture of two different aryl halides is used, there are three possible products, but often only one is obtained. For example, picryl chloride and iodobenzene gave only 2,4,6-trinitrobiphenyl.[457] The best leaving group is iodo, and the reaction is most often done on aryl iodides, but bromides, chlorides, and even thiocyanates have been used.

The effects of other groups on the ring are unusual. The nitro group is strongly activating, but only in the ortho (not meta or para) position.[458] Both R and OR groups activate in all positions. Not only do OH, NH_2, NHR, and NHCOR inhibit

---

[448]Denmark, S.E.; Yans, S.-M. *Org. Lett.* **2001**, *3*, 1749; Itami, K.; Nokami, T.; Yoshida, J.-I. *J. Am. Chem. Soc.* **2001**, *123*, 5600; Hanamoto, T.; Kobayashi, T.; Kondo, M. *Synlett* **2001**, 281; Hanamoto, T.; Kobayashi, T. *J. Org. Chem.* **2003**, *68*, 6354. See also Taguchi, H.; Ghoroku, K.; Tadaki, M.; Tsubouchi, A.; Takeda, T. *J. Org. Chem.* **2002**, *67*, 8450.

[449]Parrish, J.P.; Jung, Y.C.; Shin, S.I.; Jung, K.W. *J. Org. Chem.* **2002**, *67*, 7127; Hirabayashi, K.; Ando, J.-i.; Kawashima, J.; Nishihara, Y.; Mori, A.; Hiyama, T. *Bull. Chem. Soc. Jpn.* **2000**, *73*, 1409.

[450]Koike, T.; Du, X.; Sanada, T.; Danda, Y.; Mori, A. *Angew. Chem. Int. Ed.* **2003**, *42*, 89.

[451]Molander, G. A.; Bernardi, C.R. *J. Org. Chem.* **2002**, *67*, 8424.

[452]Terao, Y.; Wakui, H.; Satoh, T.; Miura, M.; Nomura, M. *J. Am. Chem. Soc.* **2001**, *123*, 10407.

[453]Sugihara, T.; Satoh, T.; Miura, M.; Nomura, M. *Angew. Chem. Int. Ed.* **2003**, *42*, 4672.

[454]Gooßen, L.J.; Paetzold, J.; Winkel, L. *Synlett* **2002**, 1721.

[455]For reviews, see Fanta, P.E. *Synthesis* **1974**, 9; Goshaev, M.; Otroshchenko, O.S.; Sadykov, A.S. *Russ. Chem. Rev.* **1972**, *41*, 1046.

[456]For reviews of methods of aryl–aryl bond formation, see Bringmann, G.; Walter, R.; Weirich, R. *Angew. Chem. Int. Ed.* **1990**, *29*, 977; Sainsbury, M. *Tetrahedron* **1980**, *36*, 3327. Also see, Meyers, A.I.; Price, A. *J. Org. Chem.* **1998**, *63*, 412.

[457]Rule, H.G.; Smith, F.R. *J. Chem. Soc.* **1937**, 1096.

[458]Forrest, J. *J. Chem. Soc.* **1960**, 592.

the reaction, as would be expected for aromatic nucleophilic substitution, but so do COOH (but not COOR), SO_2NH_2, and similar groups for which the reaction fails completely. These groups inhibit the coupling reaction by causing side reactions.

The mechanism is not known with certainty. It seems likely that it is basically a two-step process, similar to that of the Wurtz reaction (**10-56**), which can be represented schematically by

Step 1   ArI + Cu $\longrightarrow$ ArCu

Step 2   ArCu + ArI $\longrightarrow$ Ar—Ar

Organocopper compounds have been trapped by coordination with organic bases.[459] In addition, aryl copper compounds (ArCu) have been independently prepared and shown to give biaryls (Ar—Ar') when treated with aryl iodides Ar'I.[460] A similar reaction has been used for ring closure:[461]

An important alternative to the Ullmann method is the use of certain nickel complexes.[462] This method has also been used intramolecularly.[463] Aryl halides ArX can also be converted to Ar—Ar[464] by treatment with activated Ni metal,[465] with Zn and nickel complexes,[466] with aqueous alkaline sodium formate, Pd—C, and a phase-transfer catalyst,[467] and in an electrochemical process catalyzed by a nickel complex.[468]

An asymmetric Ullmann reaction has also been reported.[469]

OS **III**, 339; **V**, 1120.

[459]Lewin, A.H.; Cohen, T. *Tetrahedron Lett.* **1965**, 4531.

[460]For examples, see Nilsson, M. *Tetrahedron Lett.* **1966**, 675; Cairncross, A.; Sheppard, W.A. *J. Am. Chem. Soc.* **1968**, *90*, 2186; Ullenius, C. *Acta Chem. Scand.* **1972**, *26*, 3383; Mack, A.G.; Suschitzky, H.; Wakefield, B.J. *J. Chem. Soc. Perkin Trans. 1* **1980**, 1682.

[461]Salfeld, J.C.; Baume, E. *Tetrahedron Lett.* **1966**, 3365; Lothrop, W.C. *J. Am. Chem. Soc.* **1941**, *63*, 1187.

[462]See, for example Semmelhack, M.F.; Helquist, P.M.; Jones, L.D. *J. Am. Chem. Soc.* **1971**, *93*, 5908; Clark, F.R.S.; Norman, R.O.C.; Thomas, C.B. *J. Chem. Soc. Perkin Trans. 1* **1975**, 121; Tsou, T.T.; Kochi, J.K. *J. Am. Chem. Soc.* **1979**, *101*, 7547; Colon, I.; Kelsey, D.R. *J. Org. Chem.* **1986**, *51*, 2627; Lourak, M.; Vanderesse, R.; Fort, Y.; Caubere, P. *J. Org. Chem.* **1989**, *54*, 4840, 4844; Iyoda, M.; Otsuka, H.; Sato, K.; Nisato, N.; Oda, M. *Bull. Chem. Soc. Jpn.* **1990**, *63*, 80. For a review of the mechanism, see Amatore, C.; Jutand, A. *Acta Chem. Scand.* **1990**, *44*, 755.

[463]See, for example, Karimipour, M.; Semones, A.M.; Asleson, G.L.; Heldrich, F.J. *Synlett*, **1990**, 525.

[464]For a list of reagents, with references, see Larock, R.C. *Comprehensive Organic Transformations*, 2nd ed., Wiley-VCH, NY, **1999**, pp. 82–84.

[465]Inaba, S.; Matsumoto, H.; Rieke, R.D. *Tetrahedron Lett.* **1982**, *23*, 4215; Matsumoto, H.; Inaba, S.; Rieke, R.D. *J. Org. Chem.* **1983**, *48*, 840; Chao, C.S.; Cheng, C.H.; Chang, C.T. *J. Org. Chem.* **1983**, *48*, 4904.

[466]Takagi, K.; Hayama, N.; Sasaki, K. *Bull. Chem. Soc. Jpn.* **1984**, *57*, 1887.

[467]Bamfield, P.; Quan, P.M. *Synthesis* **1978**, 537.

[468]Meyer, G.; Rollin, Y.; Perichon, J. *J. Organomet. Chem.* **1987**, *333*, 263.

[469]Nelson, T.D.; Meyers, A.I. *J. Org. Chem.* **1994**, *59*, 2655; Nelson, T.D.; Meyers, A.I. *Tetrahedron Lett.* **1994**, *35*, 3259.

The palladium-catalyzed coupling of aryl halides and other aryl substrates with aromatic rings containing a suitable leaving group is now well established. Other nucleophiles can be coupled to aryl halides.[470] The reaction has become so significant in organic chemistry that the transformations have been categorized as named reactions, and are discussed in Sections **13-14** and **13-15**.

13-12   Coupling of Aryl Compounds With Arylboronic acid Derivatives

Aryl-de-halogenation, and so on **Aryl-de-boronylation**, and so on

$$Ar\!-\!Br \quad + \quad Ar'B(OH)_2 \quad \xrightarrow{\;\;PdL_4\;\;} \quad Ar\!-\!Ar'$$

Aryl triflates react with arylboronic acids, $ArB(OH)_2$ (**12-28**),[471] or with organoboranes,[472] in the presence of a palladium catalyst,[473] to give the arene in what is called *Suzuki coupling* (or *Suzuki–Miyaura coupling*).[474] Aryl halides are commonly used, and aryl sulfonates have been used.[475] Even hindered boronic acids give good yields of the coupled product.[476] Homo-coupling of arylboronic acids has been reported.[477] Coupling of the alkynes to form a diyne (see **14-16**) can be a problem is some cases, although the aryl–alkyne coupling usually predominates.[478] Some aromatic compounds are so reactive that a catalyst may not be required. Using tetrabutylammonium bromide, phenylboronic acid was coupled to 2-bromofuran without a catalyst.[479]

[470]For a review, see Prim, D.; Campagne, J.-M.; Joseph, D.; Andrioletti, B. *Tetrahedron* **2002**, *58*, 2041.

[471]Miyaura, N.; Yanagi, T.; Suzuki, A. *Synth. Commun.* **1981**, *11*, 513; Cheng, W.; Snieckus, V. *Tetrahedron Lett.* **1987**, *28*, 5097; Badone, D.; Baroni, M.; Cardomone, R.; Ielmini, A.; Guzzi, U. *J. Org. Chem.* **1997**, *62*, 7170. For a review of the synthesis and applications of heterocyclic boronic acids, see Torrell, E.; Brookes, P. *Synthesis* **2003**, 469.

[472]Fürstner, A.; Seidel, G. *Synlett*, **1998**, 161.

[473]For new palladium catalysts, see Wolfe, J.P.; Singer, R.A.; Yang, B.H.; Buchwald, S.L. *J. Am. Chem. Soc.* **1999**, *121*, 9550; Bedford, R.B.; Cazin, C.S.J. *Chem. Commun.* **2001**, 1540. For a review, see Bellina, F.; Carpita, A.; Rossi, R. *Synthesis* **2004**, 2419.

[474]Miyaura, N.; Suzuki, A. *Chem. Rev.* **1995**, *95*, 2457. For a review of the Suzuki couping in synthesis see Kotha, S.; Lahiri, K.; Kashinath, D. *Tetrahedron* **2002**, *58*, 9633.

[475]Zim, D.; Lando, V.R.; Dupont, J.; Monteiro, A.L. *Org. Lett.* **2001**, *3*, 3049; Zhang, W.; Chen, C.H.-T.; Lu, Y.; Nagashima, T. *Org. Lett.* **2004**, *6*, 1473.

[476]Watanabe, T.; Miyaura, N.; Suzuki, A. *Synlett* **1992**, 207.

[477]Lei, A.; Zhang, X. *Tetrahedron Lett.* **2002**, *43*, 2525; Parrish, J.P.; Jung, Y.C.; Floyd, R.J.; Jung, K.W. *Tetrahedron Lett.* **2002**, *43*, 7899.

[478]See, for example, Chow, H.-F.; Wan, C.-W.; Low, K.-H.; Yeung, Y.-Y. *J. Org. Chem.* **2001**, *66*, 1910.

[479]Bussolari, J.C.; Rehborn, D.C. *Org. Lett.* **1999**, *1*, 965.

Different conditions (including additives and solvent) for the reaction have been reported,[480] often focusing on the palladium catalyst itself,[481] or the ligand.[482] Catalysts have been developed for deactivated aryl chlorides,[483] and nickel catalysts have been used.[484] Modifications to the basic procedure include tethering the aryl triflate[485] or the boronic acid[486] to a polymer, allowing a polymer-supported Suzuki reaction. Polymer-bound palladium complexes have also been used.[487,488] The reaction has been done neat on alumina,[489] and on alumina with microwave irradiation.[490] Suzuki coupling has also been done in ionic liquids,[491] in supercritical

[480]Littke, A.F.; Dai, C.; Fu, G.C. *J. Am. Chem. Soc.* **2000**, *122*, 4020; Grasa, G.A.; Hillier, A.C.; Nolan, S.P. *Org. Lett.* **2001**, *3*, 1077; Le Blond, C.R.; Andrews, A.T.; Sun, Y.; Sowa, Jr., J.R. *Org. Lett.* **2001**, *3*, 1555; Savarin, C.; Liebeskind, L.S. *Org. Lett.* **2001**, *3*, 2149; Liu, S.-Y.; Choi, M.J.; Fu, G.C. *Chem. Commun.* **2001**, 2408; Li, G.-Y. *J. Org. Chem.* **2002**, *67*, 3643; Fairlamb, I.J.S.; Kapdi, A.R.; Lee, A.F. *Org. Lett.* **2004**, *6*, 4435; Oh, C.H.; Lim, Y.M.; You, C.H. *Tetrahedron Lett.* **2002**, *43*, 4645; Tao, B.; Boykin, D.W. *Tetrahedron Lett.* **2002**, *43*, 4955; Arentsen, K.; Caddick, S.; Cloke, G.N.; Herring, A.P.; Hitchcock, P.B. *Tetrahedron Lett.* **2004**, *45*, 3511; Artok, L.; Bulat, H. *Tetrahedron Lett.* **2004**, *45*, 3881; Arcadi, A.; Cerichelli, G.; Chiarini, M.; Correa, M.; Zorzan, D. *Eur. J. Org. Chem.* **2003**, 4080.
[481]Pd/C has been reported as a reusable catalyst, see Sakurai, H.; Tsukuda, T.; Hirao, T. *J. Org. Chem.* **2002**, *67*, 2721. For other recoverable or recyclable catalysts, see Nobre, S.M.; Wolke, S.I.; da Rosa, R.G.; Monteiro, A.L. *Tetrahedron Lett.* **2004**, *45*, 6527; Blanco, B.; Mehdi, A.; Moreno-Mañas, M.; Pleixats, R.; Reyé, C. *Tetrahedron Lett.* **2004**, *45*, 8789. A palladium catalyst was developed on nanoparticles, see Kogan V.; Aizenshtat, Z.; Popovitz-Biro, R.; Neumann, R. *Org. Lett.* **2002**, *4*, 3529. For a colloid-metal catalyst, see Thathagar, M.B.; Beckers, J.; Rothenberg, G. *J. Am. Chem Soc.* **2002**, *124*, 11858. Palladium (II)-sepiolite has been used, see Shimizu, K.-i.; Kan-no, T.; Kodama, T.; Hagiwara, H.; Kitayama, Y. *Tetrahedron Lett.* **2002**, *43*, 5653. See Zhao, Y.; Zhou, Y.; Ma, D.; Liu, J.; Li, L.; Zhang, T.Y.; Zhang, H. *Org. Biomol. Chem.* **2003**, *1*, 1643.
[482]Kataoka, N.; Shelby, Q.; Stambuli, J.P.; Hartwig, J.F. *J. Org. Chem.* **2002**, *67*, 5553. Ligand-free catalyst systems have been developed, see Klingensmith, L.M.; Leadbeater, N.E. *Tetrahedron Lett.* **2003**, *44*, 765; Deng, Y.; Gong, L.; Mi, A.; Li, H.; Jiang, Y. *Synthesis* **2003**, 337. For a phosphine-free reaction, see Mino, T.; Shirae, Y.; Sakamoto, M.; Fujita, T. *Synlett* **2003**, 882.
[483]Zapf, A.; Ehrentraut, A.; Beller, M. *Angew. Chem. Int. Ed.* **2000**, *39*, 4153.
[484]Zim, D.; Monteiro, A.L. *Tetrahedron Lett.* **2002**, *43*, 4009; Percec, V.; Golding, G.M.; Smidrkal, J.; Weichold, O. *J. Org. Chem.* **2004**, *69*, 3447.
[485]Blettner, C.G.; König, W.A.; Stenzel, W.; Schotten, T. *J. Org. Chem.* **1999**, *64*, 3885. For other reactions on solid support, see Franzén, R. *Can. J. Chem.* **2000**, *78*, 957.
[486]Hebel, A.; Haag, R. *J. Org. Chem.* **2002**, *67*, 9452.
[487]Inada, K.; Miyaura, N. *Tetrahedron* **2000**, *56*, 8661, 8657; Uozumi, Y.; Nakai, Y. *Org. Lett.* **2002**, *4*, 2997; Okamoto, K.; Akiyama, R.; Kobayashi, S. *Org. Lett.* **2004**, *6*, 1987; Lin, C.-A.; Luo, F.-T. *Tetrahedron Lett.* **2003**, *44*, 7565; Shieh, W.-C.; Shekhar, R.; Blacklock, T.; Tedesco, A. *Synth. Commun.* **2002**, *32*, 1059.
[488]Wang, Y; Sauer, D.R. *Org. Lett.* **2004**, *6*, 2793.
[489]Kabalka, G.W.; Pagni, R.M.; Hair, C.M. *Org. Lett.* **1999**, *1*, 1423. The reaction has also been done on KF-alumina with microwave irradiation, see Basu, B.; Das, P.; Bhuiyan, Md.M.H.; Jha, S. *Tetrahedron Lett.* **2003**, *44*, 3817.
[490]Villemin, D.; Caillot, F. *Tetrahdron Lett.* **2001**, *42*, 639.
[491]In bmim PF$_6$, 1-butyl-3-methylimidazolium hexafluorophosphate, with a nickel catalyst: Howarth, J.; James, P.; Dai, J. *Tetrahedron Lett.* **2000**, *41*, 10319. In bbim BF$_4$, 1,3-di-*n*-butylimidazolium tetrafluoroborate, with ultrasound: Rajagopal, R.; Jarikote, D.V.; Srinivasan, K.V. *Chem. Commun.* **2002**, 616. In dodecyltrihexylphosphonium chloride: McNulty, J.; Capretta, Wilson, J.; Dyck, J.; Adjabeny, G.; Robertson, A. *Chem. Commun.* **2002**, 1986.

CO_2[492] (see p. 414), and in water with microwave irradiation[493] or in water with a palladium catalyst, air, and tetrabutylammonium fluoride.[494] A solvent free (neat) Suzuki reactions have been reported.[495] A variety of functional groups are compatible with Suzuki coupling, including $Ar_2P=O$,[496] CHO,[497] $C=O$ of a ketone,[498] CO_2R,[499] cyclopropyl,[500] NO_2,[501] CN,[488] and halogen substituents.[502]

There are many structural variations of the reaction that give it enormous synthetic potential. Halogenated heteroaromatic compounds react. 2-Halopyridines react with arylboronic acids and a palladium catalyst to give 2-arylpyridines.[503] Other heterocycles have been similarly arylated.[504] 4-Pyridylboronic acids have been used.[505] The reaction of phenylboronic acid and a diallyl amide which contained a vinyl bromide, led to ring closure as well as incorporation of the phenyl group, give an *N*-tosylpyrrolidine with an exocyclic methylene unit.[506] Vinyl halides react with arylboronic acids to give alkenyl derivatives (vinyl arenes, $C=C-Ar$).[507] Alkylation can accompany arylation if alkyl halides are added, as in the conversion of iodobenzene to 2,6-dibutylbiphenyl.[508]

[492]Early, T.R.; Gordon, R.S.; Carroll, M.A.; Holmes, A.B.; Shute, R.E.; McConvey, I.F. *Chem. Commun.* **2001**, 1966.

[493]Leadbeater, N.E.; Marco, M. *J. Org. Chem.* **2003**, *68*, 888; Leadbeater, N.E.; Marco, M. *J. Org. Chem.* **2003**, *68*, 5660; Leadbeater, N.E.; Marco, M. *Org. Lett.* **2002**, *4*, 2973.

[494]Punna, S.; Díaz, D.D.; Finn, M.G. *Synlett* **2004**, 2351.

[495]Nielsen, S.F.; Peters, D.; Axelsson, O. *Synth. Commun.* **2000**, *30*, 3501; Kabalka, G.W.; Wang, L.; Pagni, R.M.; Hair, C.M.; Namboodiri, V. *Synthesis* **2003**, 217.

[496]Baillie, C.; Chen, W.; Xiao, J. *Tetrahedron Lett.* **2001**, *42*, 9085.

[497]Hesse, S.; Kirsch, G. *Synthesis* **2001**, 755; Phan, N.T.S.; Brown, D.H.; Styring, P. *Tetrahedron Lett.* **2004**, *45*, 7915.

[498]Bedford, R.B.; Welch, S.L. *Chem. Commun.* **2001**, 129; Baille, C.; Zhang, L.; Xiao, J. *J. Org. Chem.* **2004**, *69*, 7779.

[499]Mutule, I.; Suna, E. *Tetrahedron Lett.* **2004**, *45*, 3909.

[500]Ma, H.-r.; Wang, X.-L.; Deng, M.-z. *Synth. Commun.* **1999**, *29*, 2477.

[501]Tao, B.; Boykin, D.W. *J. Org. Chem.* **2004**, *69*, 4330; Li, J.-H.; Liu, W.-J. *Org. Lett.* **2004**, *6*, 2809; Widdowson, D.A.; Wilhelm, R. *Chem. Commun.* **2003**, 578.

[502]Colacot, T.J.; Shea, H.A. *Org. Lett.* **2004**, *6*, 3731; DeVasher, R.B.; Moore, L.R.; Shaughnessy, K.H. *J. Org. Chem.* **2004**, *69*, 7919.

[503]Lohse, O.; Thevenin, P.; Waldvogel, E. *Synlett* **1999**, 45; Gong, Y.; Pauls, H.W. *Synlett* **2000**, 829; Navaro, O.; Kaur, H.; Mahjoor, P.; Nolan, S.P. *J. Org. Chem.* **2004**, *69*, 3173. For a variation using hexamethyl ditin as an additive, see Zhang, N.; Thomas, L.; Wu, B. *J. Org. Chem.* **2001**, *66*, 1500.

[504]Indole derivatives: at C2, Lane, B.S.; Sames, D. *Org. Lett.* **2004**, *6*, 2897; Denmark, S.E.; Baird, J.D. *Org. Lett.* **2004**, *6*, 3649. at C3, Liu, Y.; Gribble, G.W. *Tetrahedron Lett.* **2000**, *41*, 8717. At C6, Allegretti, M.; Arcadi, A.; Marinelli, F.; Nicolini, L. *Synlett* **2001**, 609. See also, Prieto, M.; Zurita, E.; Rosa, E.; Muñoz, L.; Lloyd-Williams, P.; Giralt, E. *J. Org. Chem.* **2004**, *69*, 6812. Pyrimidines: Cooke, G.; de Cremiers, V.A.; Rotello, V.M.; Tarbit, B.; Vanderstraeten, P.E. *Tetrahedron* **2001**, *57*, 2787. Furan was coupled to pyridine: Gauthier, Jr., D.R.; Szumigala Jr., R.H.; Dormer, P.G.; Armstrong III, J.D.; Volante, R.P.; Reider, P.J. *Org. Lett.* **2002**, *4*, 375. Pyrimidines and Pyridines: Liebeskind, L.S.; Srogl, J. *Org. Lett.* **2002**, *4*, 979; Quinolines: Wolf, C.; Lerebours, R. *J. Org. Chem.* **2003**, *68*, 7551.

[505]Morris, G.A.; Nguyen, S.T. *Tetrahedron Lett.* **2000**, *42*, 2093.

[506]Lee, C.-W.; Oh, K.S.; Kim, K.S. ; Ahn, K.H. *Org. Lett.* **2000**, *2*, 1213.

[507]Shen, W. *Synlett* **2000**, 737.

[508]Catellani, M.; Motti, E.; Minari, M. *Chem. Commun.* **2000** 157. For a different approach using aryl halides having a phosphonate ester group, see Yin, J.; Buchwald, S.L. *J. Am. Chem. Soc.* **2000**, *122*, 12051.

Since many biaryls are chiral due to atropisomerism (see p. 147), the use of a chiral catalyst, and/or a chiral ligand can lead to enantioselectivity in the Suzuki coupling.[509]

Arylsulfonates can be coupled to aryl triflates using a palladium catalyst,[510] and arylboronic acids couple with aryl sulfonate esters.[511] Aryl boronic acids are coupled with aryl ammonium salts to give the biaryl, with a nickel catalyst.[512] Allylic acetates have been coupled to arylboronic acids using nickel bis(acetylacetonate) and diisobutylaluminum hydride.[513] Aryl halides couple with ArB(IR'$_2$) species with a palladium catalyst.[514] Arylboronic acids couple with the phenyl group of Ph$_2$TeCl$_2$ with a palladium catalyst.[515] 3-Iodopyridine reacted with NaBPh and palladium acetate, with microwave irradiation, to give 3-phenylpyridine.[516] Tributyltinaryl compounds were coupled to the aryl group of Ar$_2$I$^+$BF$_4^-$ with a nickel catalyst.[517] Organoboranes are coupled to aryl halides with a palladium catalyst.[518] Aryl silanes can be coupled to aryl iodides using Ag$_2$O and a palladium catalyst,[519] and arylsiloxanes ArSi(OR)$_3$, are coupled to aryl halides with Bu$_4$NF and a palladium catalyst.[520]

Arylborates (**12-28**), ArB(OR)$_2$, can be used in place of the boronic acid. The coupling reaction of aryl iodide **23** with boronate **24**, for example, gave the biaryl.[521] Aryl and heteroarylboroxines (**25**) can be coupled to aryl halides using a palladium catalyst.[522]

For a mechanistic viewpoint,[523] the Suzuki coupling proceeds via oxidative addition of areneboronic acids to give a Pd(0) species, followed by 1,2 arene migration to an electron-deficient palladium atom, eventually leading to very fast reductive

[509]Nishimura, T.; Araki, H.; Maeda, Y.; Uemura, S. *Org. Lett.* **2003**, *5*, 2997; Navarro, O.; Kelly III, R.A.; Nolan, S.P. *J. Am. Chem. Soc.* **2003**, *125*, 16194.

[510]Riggleman, S.; DeShong, P. *J. Org. Chem.* **2003**, *68*, 8106.

[511]Using a nickel catalyst, Tang, Z.-Y.; Hu, Q.-S. *J. Am. Chem. Soc.* **2004**, *126*, 3058.

[512]Blakey, S.B.; MacMillan, D.W.C. *J. Am. Chem. Soc.* **2003**, *125*, 6046.

[513]Chung, K.-G.; Miyake, Y.; Uemura, S. *J. Chem. Soc. Perkin Trans. 1* **2000**, 15.

[514]Bumagin, N.A.; Tsarev, D.A. *Tetrahedron Lett.* **1998**, *39*, 8155; Shen, W. *Tetrahedron Lett.* **1997**, *38*, 5575.

[515]Kang, S.-K.; Hong, Y.-T.; Kim, D.-H.; Lee, S.-H. *J. Chem. Res. (S)* **2001**, 283.

[516]Villemin, D.; Gómez-Escalonilla, M.J.; Saint-Clair, J.-F. *Tetrahedron Lett.* **2001**, *42*, 635.

[517]Kang, S.-K.; Ryu, H.-C.; Lee, S.-W. *J. Chem. Soc. Perkin Trans. 1* **1999**, 2661.

[518]Iglesias, B.; Alvarez, R.; de Lera, A.R. *Tetrahedron* **2001**, *57*, 3125.

[519]Hirabayashi, K.; Kawashima, J.; Nishihara, Y.; Mori, A.; Hiyama, T. *Org. Lett.* **1999**, *1*, 299.

[520]Mowery, M.E.; DeShong, P. *Org. Lett.* **1999**, *1*, 2137.

[521]Chaumeil, H.; Signorella, S.; Le Drian, C. *Tetrahedron* **2000**, *56*, 9655.

[522]Cioffi, C.L.; Spencer, W.T.; Richards, J.J.; Herr, R.J. *J. Org.Chem.* **2004**, *69*, 2210.

[523]For a review, see Esponet, P.; Echavarren, A.M. *Angew. Chem. Int. Ed.* **2004**, *43*, 4704.

elimination to afford biaryls.[524] Several intermediates of the oxidative coupling process have been identified by electrospray ionization mass spectrometry.[525]

25

A Suzuki-type coupling reaction has been reported involving acyl halides. When arylboronic acids were reacted with benzoyl chloride and $PdCl_2$, the product was the diaryl ketone.[526] This coupling reaction was also accomplished using a palladium(0) catalyst.[527] Cyclopropylboronic acids couple with benzoyl chloride, in the presence of Ag_2O and a palladium catalyst, to give the cyclopropyl ketone.[528] A nickel catalyst has been used,[529] and $Ph_3P/Ni/C$—BuLi has also been used.[530] Arylboronic acids have also been coupled to anhydrides,[531] and the methoxy group of anisole derivatives has been replaced with phenyl using phenylboronic acid and a ruthenium catalyst.[532]

In a related reaction, aryltrifluoroborates $PhBF_3^+$ X^- (**12-28**), are coupled to aryl halides with a palladium catalyst to give the biaryl.[533]

OS **75**, 53, 61

The coupling reactions of alkylboronic acids are covered in **13-17**.

OS X, 102, 467; **81**, 89.

13-13   Aryl–Alkyne Coupling Reactions

Alkynyl-de-halogenation, and so on

$$ArI + RC{\equiv}CCu \longrightarrow ArC{\equiv}CR$$

When aryl halides react with copper acetylides to give 1-aryl alkynes, the reaction is known as *Stephens–Castro coupling.*[534] Both aliphatic and aromatic

---

[524]Moreno-Mañas, M.; Pérez, M.; Pleixats, R. *J. Org. Chem.* **1996**, *61*, 2346.

[525]Aramendia, M.A.; Lafont, F.; Moreno-Mañas, M.; Pleixats, R.; Roglans, A. *J. Org. Chem.* **1999**, *64*, 3592.

[526]Bumagin, N.A.; Korolev, D.N. *Tetrahedron Lett.* **1999**, *40*, 3057.

[527]Haddach, M.; McCarthy, J.R. *Tetrahedron Lett.* **1999**, *40* , 3109.

[528]Chen, H.; Deng, M.-Z. *Org. Lett.* **2000**, *2*, 1649.

[529]Leadbeater, N.E.; Resouly, S.M. *Tetrahedron* **1999**, *55*, 11889.

[530]Lipshutz, B.H.; Sclafani, J.A.; Blomgren, P.A. *Tetrahedron* **2000**, *56*, 2139.

[531]Gooßen, L.J.; Ghosh, K. *Angew. Chem. Int. Ed.* **2001**, *40*, 3458.

[532]Kakiuchi, F.; Usai, M.; Ueno, S.; Chatani, N.; Murai, S. *J. Am. Chem. Soc.* **2004**, *126*, 2706.

[533]Batey, R.A.; Quach, T.D. *Tetrahedron Lett.* **2001**, *42*, 9099; Barder, T.E.; Buchwald, S.L. *Org. Lett.* **2004**, *6*, 2649; Molander, G. A.; Biolatto, B. *J. Org. Chem.* **2003**, *68*, 4302. See Ito, T.; Iwai, T.; Mizuno, T.; Ishino, Y. *Synlett* **2003**, 1435.

[534]Castro, C.E.; Stephens, R.D. *J. Org. Chem.* **1963**, *28*, 2163; Stephens, R.D.; Castro, C.E. *J. Org. Chem.* **1963**, *28*, 3313; Sladkov, A.M.; Ukhin, L.Yu.; Korshak, V.V. *Bull. Acad. Sci. USSR., Div. Chem. Sci.* **1963**, 2043. For a review, see Sladkov, A.M.; Gol'ding, I.R. *Russ. Chem. Rev.* **1979**, *48*, 868. For an improved procedure, see Bumagin, N.A.; Kalinovskii, I.O.; Ponomarov, A.B.; Beletskaya, I.P. *Doklad. Chem.* **1982**, *265*, 262.

substituents can be attached to the alkyne unit, and a variety of aryl iodides has been used. Benzonitrile was shown to react with alkynyl zinc bromides, with a nickel catalyst and after electrolysis to give the diarylalkyne, where the cyano unit was replaced with an alkyne unit.[535]

$$\text{Ar-X} + \text{RC}\equiv\text{CH} \xrightarrow{\text{Pd(0)}} \text{Ar-C}\equiv\text{CR}$$

A palladium–catalyzed variation is also known in which an aryl halide reacts with a terminal alkyne to give 1-aryl alkynes is called the *Sonogashira coupling*.[536] Terminal aryl alkynes react with aryl iodides and palladium(0)[537] to give the corresponding diaryl alkyne.[538] As with all of the metal-catalyzed reactions in this chapter, work has been done to vary reaction conditions, including the catalyst,[539] the ligand, the solvent,[540] and additives.[541] copper-free palladium/DABCO catalysts have been used.[542] Aryl iodides are more reactive than aryl fluorides.[543] Alkynes can be coupled to heteroaromatic compounds via the heteroaryl halide.[544] The coupling reaction has been done neat, with microwave irradiation on KF-alumina,[545] and in aqueous polyethylene glycol.[546] The aryl–alkyne coupling has also been done in solution with microwave irradiation.[547] Sonogashira coupling

[535]Penney, J.M.; Miller, J.A. *Tetrahedron Lett.* **2004**, *45*, 4989.

[536]Sonogashira, K.; Tohda, Y.; Hagihara, N. *Tetrahedron Lett.* **1975**, 4467; Sonogashira, K., in Trost, B.M.; Fleming, I.*Comprehensive Organic Synthesis*, Pergamon Press, NY, **1991**, Vol. 3, Chapter 2.4; Rossi, R.; Carpita, A.; Bellina, F. *Org. Prep. Proceed. Int.* **1995**, *27*, 127; Sonogashira, K., in Diederich, F.; Stang, P.J. *Metal–Catalyzed Cross–Coupling Reactions*, Wiley–VCH, NY, **1998**, Chapter 5.

[537]Pd/C has also been used as a catalyst, see Novák, Z.; Szabó, A.; Répási, J.; Kotschy, A. *J. Org. Chem.* **2003**, *68*, 3327.

[538]Böhm, V.P.W.; Herrmann, W.A. *Eur. J. Org. Chem.* **2000**, 3679. For an example with a CuI catalyst, see Nakamura, K.; Okubo, H.; Yamaguchi, M. *Synlett* **1999**, 549; Mori, A.; Shimada, T.; Kondo, T.; Sekiguchi, A. *Synlett* **2001**, 649.

[539]Köllhofer, A.; Pullmann, T.; Plenio, H. *Angew. Chem. Int. Ed.* **2003**, *42*, 1056; Feuerstein, M.; Berthiol, F.; Doucet, H.; Santelli, M. *Org. Biomol. Chem.* **2003**, *1*, 2235. For a reaction with a nickel catalyst, see Wang, L.; Li, P.; Zhang, Y. *Chem. Commun.* **2004**, 514; Hundertmark, T.; Littke, A.F.; Buchwald, S.L.; Fu, G.C. *Org. Lett.* **2000**, *2*, 1729.

[540]The reaction has been done in aqueous media, see Bhattacharya, S.; Sengupta, S. *Tetrahedron Lett.* **2004**, *45*, 8733.

[541]See Soheili, A.; Albaneze-Walker, J.; Murry, J.A.; Dormer, P.G.; Hughes, D.L. *Org. Lett.* **2003**, *5*, 4191; Sakai, N.; Annaka, K.; Konakahara, T. *Org. Lett.* **2004**, *6*, 1527; Leadbeater, N.E.; Tominack, B.J. *Tetrahedron Lett.* **2003**, *44*, 8653; Djakovitch, L.; Rollet, P. *Tetrahedron Lett.* **2004**, *45*, 1367; Hierso, J.-C.; Fihri, A.; Amardeil, R.; Meunier, P.; Doucet, H.; Santelli, M.; Ivanov, V. V. *Org. Lett.* **2004**, *6*, 3473.

[542]See Li, J.-H.; Zhang, X.-D.; Xie, Y.-X. *Synthesis* **2005**, 804.

[543]See, for example, Mio, M.J.; Kopel, L.C.; Braun, J.B.; Gadzikwa, T.L.; Hull, K..; Brisbois, R.G.; Markworth, C.J.; Grieco, P.A. *Org. Lett.* **2002**, *4*, 3199.

[544]Elangovan, A.; Wang, Y.-H.; Ho, T.-I. *Org. Lett.* **2003**, *5*, 1841; García, D.; Cuadro, A.M.; Alvarez-Builla, J.; Vaquero, J.J. *Org. Lett.* **2004**, *6*, 4175; Wolf, C.; Lerebours, R. *Org. Biomol. Chem.* **2004**, *2*, 2161.

[545]Kabalka, G.W.; Wang, L.; Namboodiri, V.; Pagni, R.M. *Tetrahedron Lett.* **2000**, *41*, 5151.

[546]Leadbeater, N.E.; Marco, M.; Tominack, B.J. *Org. Lett.* **2003**, *5*, 3919.

[547]Erdélyi, M.; Gogoll, A. *J. Org. Chem.* **2001**, *66*, 4165; Appukkuttan, P.; Dehaen, W.; van der Eyken, E. *Eur. J. Org. Chem.* **2003**, 4713.

was reported on microbeads,[548] with nanoparticulate nickel powder,[549] and the aryl iodide was tethered to a polymer for a solid-state reaction that included the use of microwave irradiation, and cleavage from the polymer using trifluoroacetic acid.[550] Polymer supported catalysts are known.[551] Conversion of 1-lithioalkynes to the corresponding alkynyl zinc reagent allows coupling with aryl iodides when a palladium catalyst is used.[552] Coupling with alkynyl tin compounds is also known.[553] The 1-lithioalkyne was directly coupled to aryl bromides in the presence of B(O*i*Pr)$_3$ and a palladium catalyst,[554] where an alkynylboronic acid was generated *in situ*.

A variation was reported with environmental importance, where the triphenylphosphine by-product was scavenged by addition of Merrifield resin.[555] A copper-free Sonogashira coupling has been reported, in triethylamine[556] and in an ionic liquid.[557] A copper and amine-free reaction was reported in normal solvents, such as THF.[558] An interesting example of the versatility of the coupling reaction is the coupling of propargyl bromide and an aryl iodide, in the presence of an amine, giving the aryl aminomethylalkyne.[559] The coupling of 4-chloroacetophenone with 1-phenylethyne shows that the carbonyl group is compatible with this reaction.[560]

Diaryliodonium salts react with terminal alkynes to give the phenyl alkyne.[561] A variation couples the phenyl group of Ph$_2$I$^+$OTf$^-$ with an en-yne using a palladium catalyst.[562] Aryl sulfonate esters can be coupled to terminal alkynes using a palladium catalyst in polymethylhydrosiloxane.[563] Aryl halides are coupled to alkynyltrifluoroborates (R—C≡C—BF$_3$K, **12-28**) using a palladium catalyst.[564] The boron trifluoride induced palladium-catalyzed cross-coupling reaction of 1-aryltriazenes with areneboronic acids has been reported.[565]

A variation of this aryl–alkyne coupling reaction reacted methylthioalkynes (R—C≡C—SMe) with arylboronic acids and a palladium catalyst to give the aryl alkyne (R—C≡C—Ar).[566] 1-Trialkylsilylalkynes (R$_3$Si—C≡C—R′) were coupled

[548]Liao, Y.; Fathi, R.; Reitman, M.; Zhang, Y.; Yang, Z. *Tetrahedron Lett.* **2001**, *42*, 1815; Gonthier, E.; Breinbauer, R. *Synlett* **2003**, 1049.

[549]Wang, M.; Li, P.; Wang, L. *Synth. Commun.* **2004**, *34*, 2803.

[550]Erdélyi, M.; Gogoll, A. *J. Org. Chem.* **2003**, *68*, 6431.

[551]Lin, C.-A.; Luo, F.-T. *Tetrahedron Lett.* **2003**, *44*, 7565.

[552]Anastasia, L.; Negishi, E. *Org. Lett.* **2001**, *3*, 3111.

[553]See Jeganmohan, M.; Cheng, C.-H. *Org. Lett.* **2004**, *6*, 2821.

[554]Castanet, A.-S.; Colobert, F.; Schlama, T. *Org. Lett.* **2000**, *2*, 3559.

[555]Lipshutz, B.H.; Blomgren, P.A. *Org. Lett.* **2001**, *3*, 1869.

[556]Méry, D.; Heuzé, K.; Astruc, D. *Chem. Commun.* **2003**, 1934.

[557]In bmim PF$_6$, 1-butyl-3-methylimidazolium hexafluorophosphate: Fukuyama, T.; Shinmen, M.; Nishitani, S.; Sato, M.; Ryu, I. *Org. Lett.* **2002**, *4*, 1691; Park, S.B.; Alper, H. *Chem. Commun.* **2004**, 1306.

[558]Cheng, J.; Sun, Y.; Wang, F.; Guo, M.; Xu, J.-H.; Pan, Y.; Zhang, Z. *J. Org. Chem.* **2004**, *69*, 5428; Urgaonkar, S.; Verkade, J.G. *J. Org. Chem.* **2004**, *69*, 5752.

[559]Olivi, N.; Spruyt, P.; Peyrat, J.-F.; Alami, M.; Brion, J.-D. *Tetrahedron Lett.* **2004**, *45*, 2607.

[560]Feuerstein, M.; Doucet, H.; Santelli, M. *Tetrahedron Lett.* **2004**, *45*, 8443.

[561]Kang, S.-K.; Yoon, S.-K.; Kim, Y.-M. *Org. Lett.* **2001**, *3*, 2697.

[562]Radhakrishnan, U.; Stang, P.J. *Org. Lett.* **2001**, *3*, 859.

[563]Gallagher, W.P.; Maleczka, Jr., R.E. *J. Org. Chem.* **2003**, *68*, 6775.

[564]Molander, G.A.; Katona, B.W.; Machrouhi, F. *J. Org. Chem.* **2002**, *67*, 8416.

[565]Saeki, T.; Son, E.-C.; Tamao, K. *Org.Lett.* **2004**, *6*, 617.

[566]Savarin, C.; Srogl, J.; Liebeskind, L.S. *Org. Lett.* **2001**, *3*, 91.

to aryl iodides using a palladium catalyst.[567] A triphenylstibine, $Ph_3Sb(OAc)_2$, was used to transfer a phenyl group to the alkyne carbon of $PhC{\equiv}CSiMe_3$, using palladium and CuI catalysts.[568] Aryl iodides were also coupled to lithium alkynyl borate complexes, $Li[R-C{\equiv}C-B(OR')_3$, to give the aryl alkyne.[569] Note that diphenylethyne was prepared from bromobenzene and 2-chloro-1-bromoethane using KOH, 18-crown-6 and a palladium catalyst.[570]

13-14 Arylation at a Carbon Containing an Active Hydrogen

Bis(ethoxycarbonyl)methyl-de-halogenation, and so on

$$Ar-Br \quad + \quad Z \overset{\ominus}{\frown} Z' \quad \longrightarrow \quad \underset{Z \quad Z'}{\overset{Ar}{\diagdown}}$$

The arylation of compounds of the form ZCH_2Z' is analogous to **10-67**, where Z is as defined as an electron withdrawing group (ester, cyano, sulfonyl, etc.). Activated aryl halides generally give good results.[571] Even unactivated aryl halides can be employed if the reaction is carried out in the presence of a strong base, such as $NaNH_2$[572] or LDA. Compounds of the form ZCH_2Z', even simple ketones[573] and carboxylic esters have been arylated in this manner. The reaction with unactivated halides proceeds by the benzyne mechanism and represents a method for extending the malonic ester (and similar) syntheses to aromatic compounds. The base performs two functions: it removes a proton from ZCH_2Z' and catalyzes the benzyne mechanism. The reaction has been used for ring closure, as in the formation of **26**.[574]

[567]Chang, S.; Yang, S.H.; Lee, P.H. *Tetahedron Lett.* **2001**, *42*, 4833; Kabalka, G.W.; Wang, L.; Pagni, R.M. *Tetrahedron* **2001**, *57*, 8017; Denmark, S.E.; Tymonko, S.A. *J. Org. Chem.* **2003**, *68*, 9151.
[568]Kang, S.-K.; Ryu, H.-C.; Hong, Y-T. *J. Chem. Soc. Perkin Trans. 1* **2001**, 736.
[569]Oh, C.H.; Jung, S.H. *Tetrahedron Lett.* **2000**, *41*, 8513.
[570]Abele, E.; Abele, R.; Arsenyan, P.; Lukevics, E. *Tetrahedron Lett.* **2003**, *44*, 3911.
[571]There is evidence for both the S_NAr mechanism (see Leffek, K.T.; Matinopoulos-Scordou, A.E. *Can. J. Chem.* **1977**, *55*, 2656, 2664) and the $S_{RN}1$ mechanism (see Zhang, X.; Yang, D.; Liu, Y.; Chen, W.; Cheng, J. *Res. Chem. Intermed.* **1989**, *11*, 281).
[572]Leake, W.W.; Levine, R. *J. Am. Chem. Soc.* **1959**, *81*, 1169, 1627.
[573]For example, see Caubere, P.; Guillaumet, G. *Bull. Soc. Chim. Fr.* **1972**, 4643, 4649.
[574]Bunnett, J.F.; Kato, T.; Flynn, R.; Skorcz, J.A. *J. Org. Chem.* **1963**, *28*, 1. For reviews, see Biehl, E.R.; Khanapure, S.P. *Acc. Chem. Res.* **1989**, *22*, 275; Hoffmann, R.W. *Dehydrobenzene and Cycloalkynes*, Academic Press, NY, *1967*, pp. 150–164. See also, Kessar, S.V. *Acc. Chem. Res.* **1978**, *11*, 283.

The coupling of active methylene compounds and unactivated aryl halides can also be done with copper halide catalysts[93] (the *Hurtley reaction*).[575] A palladium catalyst can be used for the coupling of malonate esters with unactivated aryl halides.[576] Bis(sulfones), $CH_2(SO_2Ar)_2$, react with aryl halides in the presence of a palladium catalyst.[577] Similar coupling was accomplished with $CH_2(CN)_2$ and a nickel catalyst.[578] Malonic and β-keto esters can be arylated at the α-carbon in high yields by treatment with aryllead tricarboxylates [$ArPb(OAc)_3$],[579] and with triphenylbismuth carbonate (Ph_3BiCO_3)[580] and other bismuth reagents.[581] In a related process, manganese(III) acetate was used to convert a mixture of ArH and ZCH_2Z' to $ArCHZZ'$.[582]

The reaction of the enolate anions ketones and aldehydes, generated *in situ* by addition of a suitable base, with aryl halides can be accomplished by treatment with a palladium catalyst.[583] Formation of an enolate anion of a conjugated ketone (cyclohexenone) via reaction with LDA (see p. 389), in the presence of Ph_3BiCl_2, leads to the α-phenyl conjugated ketone (6-phenylcyclohex-2-enone).[584] An ester reacted with $TiCl_4$ and *N,N*-dimethylanline to give the para-substitution product. ($Me_2N—Ar—CHRCO_2Et$).[585] The enolate anion of lactams will react with aryl halides in the presence of a palladium catalyst go via the 3-aryl lactam.[586] When the enolate anion of a ketone is generated in the presence of a palladium catalyst and a chiral phosphine ligand, the α-aryl ketone is formed with good enantioselectivity.[587]

Compounds of the form CH_3Z can be arylated by treatment with an aryl halide in liquid ammonia containing Na or K, as in the formation of **27** and **28**.[588]

---

[575]See Bruggink, A.; McKillop, A. *Tetrahedron* **1975**, *31*, 2607; McKillop, A.; Rao, D.P. *Synthesis* **1977**, 759; Osuka, A.; Kobayashi, T.; Suzuki, H. *Synthesis* **1983**, 67; Hennessy, E.J.; Buchwald, S.L. *Org. Lett.*, **2002**, 4, 269.

[576]Aramendía, M.A.; Borau, V.; Jiménez, C.; Marinas, J.M.; Ruiz, J.R.; Urbano, F.J. *Tetrahedron Lett.* **2002**, *43*, 2847.

[577]Kashin, A.N.; Mitin, A.V.; Beletskaya, I.P.; Wife, R. *Tetrahedron Lett.* **2002**, *43*, 2539.

[578]Cristau, H.J.; Vogel, R.; Taillefer, M.; Gadras, A. *Tetrahedron Lett.* **2000**, *41*, 8457.

[579]Elliott, G.I.; Konopelski, J.P.; Olmstead, M.M. *Org. Lett.* **1999**, *1*, 1867, and Refs. 3–7 therein.

[580]For a review of the aryllead and arylbismuth, and related reactions, see Elliott, G.I.; Konopelski, J.P. *Tetrahedron* **2001**, *57*, 5683; Abramovitch, R.A.; Barton, D.H.R.; Finet, J. *Tetrahedron* **1988**, *44*, 3039.

[581]Barton, D.H.R.; Blazejewski, J.; Charpiot, B.; Finet, J.; Motherwell, W.B.; Papoula, M.T.B.; Stanforth, S.P. *J. Chem. Soc. Perkin Trans. 1* **1985**, 2667; O'Donnell, M.J.; Bennett, W.D.; Jacobsen, W.N.; Ma, Y. *Tetrahedron Lett.* **1989**, *30*, 3913.

[582]Citterio, A.; Santi, R.; Fiorani, T.; Strologo, S. *J. Org. Chem.* **1989**, *54*, 2703; Citterio, A.; Fancelli, D.; Finzi, C.; Pesce, L.; Santi, R. *J. Org. Chem.* **1989**, *54*, 2713.

[583]Uno, M.; Seto, K.; Ueda, W.; Masuda, M.; Takahashi, S. *Synthesis* **1985**, 506; Kawatsura, M.; Hartwig, J.F. *J. Am. Chem. Soc.* **1999**, *121*, 1473; Jørgensen, M.; Lee, S.; Liu, X.; Wolkowski, J.P.; Hartwig, J.F. *J. Am. Chem. Soc.* **2002**, *124*, 12557; Fox, J.M.; Huang, X.; Chieffi, A.; Buchwald, S.L. *J. Am. Chem. Soc.* **2000**, *122*, 1360. For a review, see Culkin, D.A.; Hartwig, J.F. *Acc. Chem. Res.* **2003**, *36*, 234.

[584]Arnauld, T.; Barton, D.H.R.; Normat, J.-F.; Doris, E. *J. Org. Chem.* **1999**, *64*, 6915.

[585]Periasamy, M.; KishoreBabu N.; Jayakumar, K.N. *Tetrahedron Lett.* **2003**, *44*, 8939.

[586]Cossy, J.; de Filippis, A.; Pardo, D.G. *Org. Lett.* **2003**, *5*, 3037.

[587]Hamada, T.; Chieffi, A.; Åhman, J.; Buchwald, S.L. *J. Am. Chem. Soc.* **2002**, *124*, 1261.

[588]Rossi, R.A.; Bunnett, J.F. *J. Org. Chem.* **1973**, *38*, 3020; Bunnett, J.F.; Gloor, B.F. *J. Org. Chem.* **1973**, *38*, 4156; **1974**, *39*, 382.

$$\text{Ar–X} + \underset{\text{Me}}{\overset{\text{O}}{\underset{\|}{\text{C}}}}\text{Me} \xrightarrow[\text{NH}_3]{\text{K}} \underset{\text{Me}}{\overset{\text{O}}{\underset{\|}{\text{C}}}}\text{Ar} + \underset{\text{Me}}{\overset{\text{OH}}{\text{C}}}\text{Ar}$$

27   **28**

When the solution is irradiated with near-UV light, but Na or K is omitted, the same products are obtained (though in different proportions).[589] In either case, other leaving groups can be used instead of halogens (e.g., NR_3^+, SAr) and the mechanism is the $S_{RN}1$ mechanism. Iron(II) salts have also been used to initiate this reaction.[590] The reaction can also take place without an added initiator. The reaction of 2-fluoroanisole and KHMDS, and 4 equivalents of 2-cyanopropane, leads to substitution of the fluorine atom by CMe_2CN.[591] A similar reaction as reported using a palladium catalyst.[592] Nitroethane was converted to 2-phenylnitroethane using bromobenzene and a palladium catalyst.[593]

Enolate ions of ketones react with PhI in the dark.[594] In this case, it has been suggested[595] that initiation takes place by formation of a radical, such as **29**.

$$\underset{R}{\overset{R}{\text{C}}}=\underset{R}{\overset{R}{\text{C}}}\text{–O}^{\ominus} + \text{Ar–I} \longrightarrow \underset{R}{\overset{R}{\text{C}}}-\underset{R}{\overset{\text{O}}{\underset{\|}{\text{C}}}}R + \text{Ar–I}^{\overset{\cdot}{-}}$$

29

This is an SET mechanism (see p. 444). The photostimulated reaction has also been used for ring closure.[596] In certain instances of the intermolecular reaction there is evidence that the leaving group exerts an influence on the product ratios, even when it has already departed at the time that product selection takes place.[597]

OS **V**, 12, 263; **VI**, 36, 873, 928; **VII**, 229.

13-15 Conversion of Aryl Substrates to Carboxylic Acids, Their Derivatives, Aldehydes, and Ketones[598]

[589]Hay, J.V.; Hudlicky, T.; Wolfe, J.F. *J. Am. Chem. Soc.* **1975**, *97*, 374; Bunnett, J.F.; Sundberg, J.E. *J. Org. Chem.* **1976**, *41*, 1702; Rajan, S.; Muralimohan, K. *Tetrahedron Lett.* **1978**, 483; Rossi, R.A.; de Rossi, R.H.; Pierini, A.B. *J. Org. Chem.* **1979**, *44*, 2662; Rossi, R.A.; Alonso, R.A. *J. Org. Chem.* **1980**, *45*, 1239; Beugelmans, R. *Bull. Soc. Chim. Belg.* **1984**, *93*, 547.
[590]Galli, C.; Bunnett, J.F. *J. Org. Chem.* **1984**, *49*, 3041.
[591]Caron, S.; Vasquez, E.; Wojcik, J.M. *J. Am. Chem. Soc.* **2000**, *122*, 712.
[592]You, J.; Verkade, J.G. *Angew. Chem. Int. Ed.* **2003**, *42*, 5051.
[593]Vogl, E.M.; Buchwald, S.L. *J. Org. Chem.* **2002**, *67*, 106.
[594]Scamehorn, R.G.; Hardacre, J.M.; Lukanich, J.M.; Sharpe, L.R. *J. Org. Chem.* **1984**, *49*, 4881.
[595]Aoki, S.; Fujimura, T.; Nakamura, E.; Kuwjima, I. *J. Am. Chem. Soc.* **1988**, *110*, 3296.
[596]See Semmelhack, M.F.; Bargar, T. *J. Am. Chem. Soc.* **1980**, *102*, 7765; Bard, R.R.; Bunnett, J.F. *J. Org. Chem.* **1980**, *45*, 1546.
[597]Bard, R.R.; Bunnett, J.F.; Creary, X.; Tremelling, M.J. *J. Am. Chem. Soc.* **1980**, *102*, 2852; Tremelling, M.J.; Bunnett, J.F. *J. Am. Chem. Soc.* **1980**, *102*, 7375.
[598]For a review, see Weil, T.A.; Cassar, L.; Foà, M., in Wender, I.; Pino, P. *Organic Synthesis Via Metal Carbonyls*, Vol. 2, Wiley, NY, **1977**, pp. 517–543.

Alkoxycarbonyl-de-halogenation, and so on

$$\text{ArX} \quad + \quad \text{CO} \quad + \quad \text{ROH} \quad \xrightarrow[\text{Pd complex}]{\text{base}} \quad \text{ArCOOR}$$

Carbonylation of aryl bromides and iodides with carbon monoxide, an alcohol, a base, and a palladium catalyst, give carboxylic esters. Even very sterically hindered alkoxides can be used to produce the corresponding ester.[599] The use of H_2O, RNH_2, or an alkali metal or calcium carboxylate[600] instead of ROH, gives the carboxylic acid,[601] amide,[602] or mixed anhydride, respectively.[603] Heating an aryl iodide, CO in ethanol and DBU, with a palladium catalyst, gave the ethyl ester of the aryl carboxylic acid.[604] A similar result was obtained when an aryl iodide was heated in ethanol with triethylamine, CO and Pd/C.[605] Ester formation via carbonylation was done is supercritical CO_2 (see p. 414).[606] With certain palladium catalysts, aryl chlorides[607] and aryl triflates[608] can also be substrates. Aryl carboxylic acids were also prepared from aryl iodides by heating in DMF with lithium formate, LiCl, acetic anhydride and a palladium catalyst.[609] A silica-supported palladium reagent has been used to convert iodobenzene to butyl benzoate, in the presence of CO and butanol.[610] 2-Chloropyridine was converted the butyl pyridine 2-carboxylate with this procedure.[611] Halogenated biaryls can be converted to the tricyclic ketone, 9-fluorenone, by an intramolecular carbonylation reaction with CO and a palladium catalyst.[612] A surrogate reagent used instead of CO is dicobalt octacarbonyl $CO_2(CO)_8$.[613] Aryl chlorides have been converted to carboxylic acids by an electrochemical synthesis,[614] and aryl iodides to aldehydes by treatment with

[599]Kubota, Y.; Hanaoka, T.-a.; Takeuchi, K.; Sugi, Y. *J. Chem. Soc. Chem. Commun.* **1994**, 1553; Antebi, S.; Arya, P.; Manzer, L.E.; Alper, H. *J. Org. Chem.* **2002**, *67*, 6623. For an interesting variation that generated a lactone ring, see Cho, C.S.; Baek, D.Y.; Shim, S.C. *J. Heterocyclic Chem.* **1999**, *36*, 289.

[600]Pri-Bar, I.; Alper, H. *J. Org. Chem.* **1989**, *54*, 36.

[601]For example, see Bumagin, N.A.; Nikitin, K.V.; Beletskaya, I.P. *Doklad. Chem.* **1990**, *312*, 149.

[602]Lin, Y.-S.; Alper, H. *Angew. Chem. Int. Ed.* **2001**, *40*, 779; Wan, Y.; Alterman, M.; Larhed, M.; Hallberg, A. *J. Org. Chem.* **2002**, *67*, 6232. For another reagent that also gives amides, see Bumagin, N.A.; Gulevich, Yu.V.; Beletskaya, I.P. *J. Organomet. Chem.* **1985**, *285*, 415.

[603]For a review, see Heck, R.F. *Palladium Reagents in Organic Synthesis*, Academic Press, NY, **1985**, pp. 348–358.

[604]Ramesh, C.; Kubota, Y.; Miwa, M.; Sugi, Y. *Synthesis* **2002**, 2171.

[605]Ramesh, C.; Nakamura, R.; Kubota, Y.; Miwa, M.; Sugi, Y. *Synthesis* **2003**, 501.

[606]Albaneze-Walker, J.; Bazaral, C.; Leavey, T.; Dormer, P.G.; Murry, J.A. *Org. Lett.* **2004**, *6*, 2097.

[607]Ben-David, Y.; Portnoy, M.; Milstein, D. *J. Am. Chem. Soc.* **1989**, *111*, 8742.

[608]Cacchi, S.; Ciattini, P.G.; Morera, E.; Ortar, G. *Tetrahedron Lett.* **1986**, *27*, 3931; Kubota, Y.; Nakada, S.; Sugi, Y. *Synlett*, **1998**, 183; Garrido, F.; Raeppel, S.; Mann, A.; Lautens, M. *Tetrahedron Lett.* **2001**, *42*, 265.

[609]Cacchi, S.; Babrizi, G.; Goggiamani, A. *Org. Lett.* **2003**, *5*, 4269.

[610]Cai, M.-Z.; Song, C.-S.; Huang, X. *J. Chem. Soc. Perkin Trans. 1*, **1997**, 2273.

[611]Beller, M.; Mägerlein, W.; Indolese, A.F.; Fischer, C. *Synthesis* **2001**, 1098.

[612]Campo, M.A.; Larock, R.C. *Org. Lett.* **2000**, *2*, 3675.

[613]Brunet, J.; Sidot, C.; Caubere, P. *J. Org. Chem.* **1983**, *48*, 1166. See also, Foà, M.; Francalanci, F.; Bencini, E.; Gardano, A. *J. Organomet. Chem.* **1985**, *285*, 293; Kudo, K.; Shibata, T.; Kashimura, T.; Mori, S.; Sugita, N. *Chem. Lett.* **1987**, 577.

[614]Heintz, M.; Sock, O.; Saboureau, C.; Périchon, J. *Tetrahedron* **1988**, *44*, 1631.

CO, Bu$_3$SnH, and NCCMe$_2$N=NCMe$_2$CN (AIBN).[615] Aryl ketones can be prepared from aryltrimethylsilanes ArSiMe$_3$ and acyl chlorides in the presence of AlCl$_3$.[616] Aryllithium and Grignard reagents react with iron pentacarbonyl to give aldehydes ArCHO.[617] The reaction of CO with aryllithium may occur by electron transfer.[618]

Aryl iodides are converted to unsymmetrical diaryl ketones on treatment with arylmercury halides and nickel carbonyl: ArI + Ar′HgX + Ni(CO)$_4$ → ArCOAr′.[619] Aryl iodides are carbonylated to give the aryl alkyl ketone with CO and R$_3$In.[620] Arylthallium bis(trifluoroacetates), ArTl(O$_2$CCF$_3$)$_2$ (see **12-23**), can be carbonylated with CO, an alcohol, and a PdCl$_2$ catalyst to give esters.[621] Organomercury compounds undergo a similar reaction.[622] The aryllead reagent PhPb(OAc)$_3$, was converted to benzophenone using NaOMe, CO and a palladium catalyst.[623] Aryl iodides containing an ortho substituent with a β-cyano group that served as the source of a carbonyl group, was converted to a bicyclic ketone with a palladium catalyst at 130°C in aqueous DMF.[624]

Diaryl ketones can also be prepared by coupling aryl iodides with phenylboronic acid (**12-28**), in the presence of CO and a palladium catalyst.[625] This reaction has been extended to heteroaromatic systems, with the preparation of phenyl 4-pyridyl ketone from phenylboronic acid and 4-iodopyridine.[626] 2-Bromopyridine as coupled with phenylboronic acid, CO and a palladium catalyst to give phenyl 2-pyridyl ketone.[627] An interesting reaction treated a titanocycle (**30**) with CO to give the cyclobutanone.[628] Carbonylation of an alkyne and an aryl halide, with CO

[615]Ryu, I.; Kusano, K.; Masumi, N.; Yamazaki, H.; Ogawa, A.; Sonoda, N. *Tetrahedron Lett.* **1990**, *31*, 6887.

[616]Dey, K.; Eaborn, C.; Walton, D.R.M. *Organomet. Chem. Synth.* **1971**, *1*, 151.

[617]Ryang, M.; Rhee, I.; Tsutsumi, S. *Bull. Chem. Soc. Jpn.* **1964**, *37*, 341; Giam, C.; Ueno, K. *J. Am. Chem. Soc.* **1977**, *99*, 3166; Yamashita, M.; Miyoshi, K.; Nakazono, Y.; Suemitsu, R. *Bull. Chem. Soc. Jpn.* **1982**, *55*, 1663. For another method, see Gupton, J.T.; Polk, D.E. *Synth. Commun.* **1981**, *11*, 571.

[618]Nudelman, N.S.; Doctorovich, F. *Tetrahedron* **1994**, *50*, 4651.

[619]Rhee, I.; Ryang, M.; Watanabe, T.; Omura, H.; Murai, S.; Sonoda, N. *Synthesis* **1977**, 776; Ryu, I.; Ryang, M.; Rhee, I.; Omura, H.; Murai, S.; Sonoda, N. *Synth. Commun.* **1984**, *14*, 1175 and cited references. For other acylation reactions, see Tanaka, M. *Synthesis* **1981**, 47; *Bull. Chem. Soc. Jpn.* **1981**, *54*, 637; Bumagin, N.A.; Ponomarov, A.B.; Beletskaya, I.P. *Tetrahedron Lett.* **1985**, *26*, 4819; Koga, T.; Makinouchi, S.; Okukado, N. *Chem. Lett.* **1988**, 1141; Echavarren, A.M.; Stille, J.K. *J. Am. Chem. Soc.* **1988**, *110*, 1557; Hatanaka, Y.; Hiyama, T. *Chem. Lett.* **1989**, 2049.

[620]Lee, P.H.; Lee, S.W.; Lee, K. *Org. Lett.* **2003**, *5*, 1103. With a palladium catalyst, see Pena, M.A.; Sestelo, J.P.; Sarandeses, L.A. *Synthesis* **2003**, 780. For a reaction that used R$_4$InLi, see Lee, S.W.; Kee, K.; Seomoon, D.; Kim, S.; Kim, H.; Kim, H.; Shim, E.; Lee, M.; Lee, S.; Kim, S.; Lee, P.H. *J. Org. Chem.* **2004**, *69*, 4852.

[621]Larock, R.C.; Fellows, C.A. *J. Am. Chem. Soc.* **1982**, *104*, 1900.

[622]Baird Jr., W.C.; Hartgerink, R.L.; Surridge, J.H. *J. Org. Chem.* **1985**, *50*, 4601.

[623]Kang, S.-K.; Ryu, H.-C.; Choi, S.-C. *Synth. Commun.* **2001**, *31*, 1035.

[624]Pletnev, A.A.; Larock, R.C. *J. Org. Chem.* **2002**, *67*, 9428.

[625]Ishiyama, T.; Kizaki, H.; Miyaura, N.; Suzuki, A. *Tetrahedron Lett.* **1993**, *34*, 7595.

[626]Couve-Bonnaire, S.; Caprentier, J.-F.; Mortreux, A.; Castanet, Y. *Tetrahedron Lett.* **2001**, *42*, 3689.

[627]Maerten, E.; Hassouna, F.; Couve-Bonnaire, S.; Mortreux, A.; Carpentiere, J.-F.; Castanet, Y. *Synlett* **2003**, 1874.

[628]Carter, C.A.G.; Greidanus, G.; Chen, J.-X.; Stryker, J.M. *J. Am. Chem. Soc.* **2001**, *123*, 8872.

and palladium and copper catalysts, gave the alkynyl ketone RC≡C(C=O)Ar.[629]

$$\underset{\textbf{30}}{C_5Me_5\diagdown}\overset{C_5Me_5\diagup}{Ti}\diagdown\diagup\diagdown \quad \xrightarrow[\text{pentane}]{\text{CO, 45°C}} \quad O=\diagdown\diagdown\diagup$$

Note that seleno esters (ArCOSeAr) were prepared from aryl iodides, CO, PhSeSnBu$_3$, and a palladium catalyst.[630]

13-16   Arylation of Silanes

Silyl and Silyloxy-de-halogenation, and so on

$$Ar–X \quad + \quad Ar'SiR_2 \quad \xrightarrow{\hspace{4cm}} \quad Ar–SiR_3$$

In the presence of transition-metal catalysts, such as palladium, trialkoxysilanes [HSi(OR)$_3$] react with aryl halides to give the corresponding arylsilane.[631] This transformation is an alternative to the Suzuki coupling (**13-10**).[632] A similar reaction was reported using a rhodium catalyst.[633] Arylsilanes can be coupled to aryl iodides in aqueous media.[634] Arylsilanes react with alkyl halides to give the corresponding arene, in the presence of a palladium catalyst.[635] Suzuki-type coupling using Me$_3$SiSiMe$_3$ leads to aryl silanes.[636]

An alternative approach reacts aryllithium reagents with siloxanes [Si(OR)$_4$], to give the aryl derivative ArSi(OR)$_3$.[637]

HYDROGEN AS LEAVING GROUP[638]

13-17   Alkylation and Arylation

Alkylation or Alkyl-de-hydrogenation, and so on

$$\underset{\textbf{31}}{\overset{NMe_2}{\diagup}\diagdown OMe} \quad \xrightarrow{n\text{-BuLi}} \quad \underset{\textbf{32}}{\overset{NMe_2}{\diagup}\diagdown\overset{Li}{\diagup} OMe}$$

[629]Ahmed, M.S.M.; Mori, A. *Org. Lett.* **2003**, *5*, 3057.

[630]Nishiyama, Y.; Tokunaga, K.; Kawamatsu, H.; Sonoda, N. *Tetrahedron Lett.* **2002**, *43*, 1507.

[631]Manoso, A.S.; DeShong, P. *J. Org. Chem.* **2001**, *66*, 7449.

[632]DeShong, P.; Handy, C.J.; Mowery, M.W. *Pure Appl. Chem.* **2000**, *72*, 1655; Seganish, W.M.; DeShong, P. *Org. Lett.* **2004**, *6*, 4379.

[633]Murata, M.; Ishikura, M.; Nagata, M.; Watanabe, S.; Masuda, Y. *Org. Lett.* **2002**, *4*, 1843.

[634]Denmark, S.E.; Ober, M.H. *Org. Lett.* **2003**, *5*, 1357.

[635]Mowery, M.E.; DeShong, P. *J. Org. Chem.* **1999**, *64*, 3266; Lee, J.-y.; Fu, G.C. *J. Am. Chem. Soc.* **2003**, *125*, 5616.

[636]Gooßen, L.J.; Ferwanah, A.-R.S. *Synlett* **2000**, 1801.

[637]Manoso, A.S.; Ahn, C.; Soheili, A.; Handy, C.J.; Correia, R.; Seganish, W.M.; DeShong, P. *J. Org. Chem.* **2004**, *69*, 8305.

[638]For reviews, see Chupakhin, O.N.; Postovskii, I.Ya. *Russ. Chem. Rev.* **1976**, *45*, 454. For a review of reactivity and mechanism in these cases, see Chupakhin, O.N.; Charushin, V.N.; van der Plas, H.C. *Tetrahedron* **1988**, *44*, 1.

The alkylation of aromatic rings was introduced, in part, in section **10-57**. The reaction of an aromatic ring with an organolithium reagent can give H—Li exchange to form an aryllithium. This reaction tends to be slow in the absence of diamine additives or if there are activating substituents on the aryl halide.[639] When heteroatom substituents are present as in **31**, however, the reaction is facile and the lithium goes into the 2 position (as in **32**).[640] This regioselectivity can be quite valuable synthetically, and is now known as *directed ortho metalation*[641] (see **10-57**). Lithiation reactions do not necessarily rely on a complex-induced proximity effect.[642] With TMEDA/*n*-butyllithium-mediate arene lithiation reactions, the viability of directive effects (complex-induced proximate effects) has been questioned,[643] although it is not clear if this extends to other systems (particularly when there is a strong coordinating group, such as carbamate).[644] The 2 position is much more acidic than the 3 position (see Table 8.1), but a negative charge at C-3 is in a more favorable position to be stabilized by the Li^+. Formation of the ortho arylmagnesium compound has been accomplished with bases of the form $(R_2N)_2Mg$.[645] Note that H—Li exchange can be faster than Cl—Li exchange. Treatment of 2-chloro-5-phenylpyridine with *tert*-butyllithium leads to lithiation on the phenyl ring rather than Li—Cl exchange, and subsequent treatment with dimethyl sulfate gave 2-chloro-5-(2-methylphenyl)pyridine.[646] Heteroaromatic rings do react, however. The reaction of 2-chloropyridine with 3 equivalents of butyllithium-$Me_2NCH_2CH_2OLi$ and then iodomethane gave 2-chloro-6-methylpyridine.[647] The reaction of *N*-triisopropylsilyl indole with *tert*-butyllithium and then iodomethane gave the 3-methyl derivative.[648] Furfural (furan 2-carboxaldehyde) reacts with aryl iodides in the presence of a palladium catalyst to give the 5-arylfuran 2-carboxaldehyde.[649]

Benzene, naphthalene, and phenanthrene have been alkylated with alkyllithium reagents, though the usual reaction with these reagents is **12-22**,[650] and Grignard reagents have been used to alkylate naphthalene.[651] The addition–elimination

[639]See, for example, Becht, J.-M.; Gissot, A.; Wagner, A.; Misokowski, C. *Tetrahedron Lett.* **2004**, *45*, 9331.

[640]Slocum, D.W.; Jennings, C.A. *J. Org. Chem.* **1976**, *41*, 3653. However, the regioselectivity can depend on reaction conditions: See Meyers, A.I.; Avila, W.B. *Tetrahedron Lett.* **1980**, 3335.

[641]For a reviews of directed ortho metallation, see Snieckus, V. *Chem. Rev.* **1990**, *90*, 879; Gschwend, H.W.; Rodriguez, H.R. *Org. React.* **1979**, *26*, 1. See Green, L.; Chauder, B.; Snieckus, V. *J. Heterocylic Chem.* **1999**, *36*, 1453. Also see Green, L.; Chauder, B.; Snieckus, V. *J. Heterocyclic Chem.* **1999**, *36*, 1453. See Slocum, D.W.; Dietzel, P. *Tetrahedron Lett.* **1999**, *40*, 1823.

[642]Chadwick, S.T.; Rennels, R.A.; Rutherford, J.L.; Collum, D.B. *J. Am. Chem. Soc.* **2000**, *122*, 8640; Collum, D.B. *Acc. Chem. Res.* **1992**, *25*, 448.

[643]Chadwick, S.T.; Rennels, R.A.; Rutherford, J.L.; Collum, D.B. *J. Am. Chem. Soc.* **2000**, *122*, 8640.

[644]Hay, D. R.; Song, Z.; Smith, S.G.; Beak, P. *J. Am. Chem. Soc.* **1988**, *110*, 8145.

[645]Eaton, P.E.; Lee, C.; Xiong, Y. *J. Am. Chem. Soc.* **1989**, *111*, 8016.

[646]Fort, Y. Rodriguez, A.L. *J. Org. Chem.* **2003**, *68*, 4918.

[647]Choppin, S. Gros, P.; Fort, Y. *Org. Lett.* **2000**, *2*, 803.

[648]Matsuzono, M.; Fukuda, T.; Iwao, M. *Tetrahedron Lett.* **2001**, *42*, 7621.

[649]McClure, M.S.; Glover, B.; McSorley, E.; Millar, A.; Osterhout, M.H.; Roschangar, F. *Org. Lett.* **2001**, *3*, 1677.

[650]Eppley, R.L.; Dixon, J.A. *J. Am. Chem. Soc.* **1968**, *90*, 1606.

[651]Bryce-Smith, D.; Wakefield, B.J. *Tetrahedron Lett.* **1964**, 3295.

mechanism apparently applies in these cases too. A protected form of benzaldehyde (protected as the benzyl imine) has been similarly alkylated at the *ortho*-position with butyllithium.[652]

The alkylation of heterocyclic nitrogen compounds[653] with alkyllithium reagents is called *Ziegler alkylation*. Aryllithium reagents give arylation. The reaction occurs by an addition–elimination mechanism and the adduct can be isolated.[654] Upon heating of the adduct, elimination of LiH occurs and an alkylated product is obtained. With respect to the 2-carbon the first step is the same as that of the S_NAr mechanism. The difference is that the unshared pair of electrons on the nitrogen combines with the lithium, so the extra pair of ring electrons has a place to go: it becomes the new unshared pair on the nitrogen. Heteroaromatic compounds can be alkylated. Pyrrole, for example, reacts with an allylic halide and zinc to give primarily the 3-substituted pyrrole.[655]

Mercuration of aromatic compounds[656] can be accomplished with mercuric salts, most often $Hg(OAc)_2$[657] to give ArHgOAc. This is ordinary electrophilic aromatic substitution and takes place by the arenium ion mechanism (p. 657).[658] Aromatic compounds can also be converted to arylthallium bis(trifluoroacetates) $ArTl(OOCCF_3)_2$ by treatment with thallium(III) trifluoroacetate[659] in trifluoroacetic acid.[660] These arylthallium compounds can be converted to phenols, aryl iodides or fluorides (**12-31**), aryl cyanides (**12-34**), aryl nitro compounds,[661] or aryl esters

---

[652]Flippin, L.A.; Carter, D.S.; Dubree, N.J.P. *Tetrahedron Lett.* **1993**, *34*, 3255.

[653]For a review of substitution by carbon groups on a nitrogen heterocycle, see Vorbrüggen, H.; Maas, M. *Heterocycles*, **1988**, *27*, 2659. For a related review, see Comins, D.L.; O'Connor, S. *Adv. Heterocycl. Chem.* **1988**, *44*, 199.

[654]See, for example, Armstrong, D.R.; Mulvey, R.E.; Barr, D.; Snaith, R.; Reed, D. *J. Organomet. Chem.* **1988**, *350*, 191.

[655]Yadav, J.S.; Reddy, B.V.S.; Reddy, P.M.; Srinivas, Ch. *Tetrahedron Lett.* **2002**, *43*, 5185.

[656]For reviews, see Larock, R.C. *Organomercury Compounds in Organic Synthesis*, Springer, NY, **1985**, pp. 60–97; Wardell, J.L., in Zuckerman, J.J. *Inorganic Reactions and Methods*, Vol. 11, VCH, NY, **1988**, pp. 308–318.

[657]For a review of mercuric acetate, see Butler, R.N., in Pizey, J.S. *Synthetic Reagents*, Vol. 4, Wiley, NY, **1981**, pp. 1–145.

[658]For a review, see Taylor, R., in Bamford, C.H.; Tipper, C.F.H. *Comprehensive Chemical Kinetics*, vol. 13, Elsevier, NY, **1972**, pp. 186–194. An alternative mechanism, involving radial cations, has been reported: Courtneidge, J.L.; Davies, A.G.; McGuchan, D.C.; Yazdi, S.N. *J. Organomet. Chem.* **1988**, *341*, 63.

[659]For a review of this reagent, see Uemura, S., in Pizey, J.S. *Synthetic Reagents*, Vol. 5, Wiley, NY, **1983**, pp. 165–241.

[660]Taylor, E.C.; Kienzle, F.; McKillop, A. *Org. Synth.* *VI*, 826; Taylor, E.C.; Katz, A.H.; Alvarado, S.I.; McKillop, A. *J. Organomet. Chem.* **1985**, *285*, C9. For reviews, see Usyatinskii, A.Ya.; Bregadze, V.I. *Russ. Chem. Rev.* **1988**, *57*, 1054; Uemura, S., in Hartley, F.R.; Patai, S. *The Chemistry of the Metal-Carbon Bond*, Vol. 4, Wiley, NY, pp. 473–538.

[661]Uemura, S.; Toshimitsu, A.; Okano, M. *Bull. Chem. Soc. Jpn.* **1976**, *49*, 2582.

(**12-33**). The mechanism of thallation appears to be complex, with electrophilic and electron-transfer mechanisms both taking place.[662] Transient metalated aryl complexes can be formed that react with another aromatic compound. Aryl iodides reacted with benzene to form a biaryl in the presence of an iridium catalyst.[663] Aniline derivatives reacted with $TiCl_4$ to give the para-homo coupling product $(R_2N-Ar-Ar-NR_2)$.[664]

Aromatic nitro compounds can be methylated with dimethyloxosulfonium methylid[665] or the methylsulfinyl carbanion (obtained by treatment of DMSO with a strong base):[666]

The latter reagent also methylates certain heterocyclic compounds (e.g., quinoline) and certain fused aromatic compounds (e.g., anthracene, phenanthrene).[666,667] The reactions with the sulfur carbanions are especially useful, since none of these substrates can be methylated by the Friedel–Crafts procedure (**11-10**). It has been reported[668] that aromatic nitro compounds can also be alkylated, not only with methyl but with other alkyl and substituted alkyl groups as well, in ortho and para positions, by treatment with an alkyllithium compound (or, with lower yields, a Grignard reagent), followed by an oxidizing agent, such as Br_2 or DDQ (p. 1710).

A different kind of alkylation of nitro compounds uses carbanion nucleophiles that have a chlorine at the carbanionic carbon. The following process takes place:[669]

[662]Lau, W.; Kochi, J.K. *J. Am. Chem. Soc.* **1984**, *106*, 7100; **1986**, *108*, 6720.

[663]Fujita, K.-i.; Nonogawa, M.; Yamaguchi, R. *Chem. Commun.* **2004**, 1926.

[664]Periasamy, M.; Jayakumar, K.N.; Bharathi, P. *J. Org. Chem.* **2000**, *65*, 3548.

[665]Traynelis, V.J.; McSweeney, J.V. *J. Org. Chem.* **1966**, *31*, 243.

[666]Russell, G.A.; Weiner, S.A. *J. Org. Chem.* **1966**, *31*, 248.

[667]Argabright, P.A.; Hofmann, J.E.; Schriesheim, A. *J. Org. Chem.* **1965**, *30*, 3233; Trost, B.M. *Tetrahedron Lett.* **1966**, 5761; Yamamoto, Y.; Nisimura, T.; Nozaki, H. *Bull. Chem. Soc. Jpn.* **1971**, *44*, 541.

[668]Kienzle, F. *Helv. Chim. Acta* **1978**, *61*, 449.

[669]In some cases, the intermediate bearing the CHCl(Z) unit has been isolated: Stahly, G.P.; Stahly, B.C.; Maloney, J.R. *J. Org. Chem.* **1988**, *53*, 690.

This type of process is called *vicarious nucleophilic substitution of hydrogen.*[670] The Z group is electron -withdrawing (e.g., SO_2R, SO_2OR, SO_2NR_2, COOR, or CN); it stabilizes the negative charge. The carbanion attacks the activated ring ortho or para to the nitro group.[671] Hydride ion H^- is not normally a leaving group, but in this case the presence of the adjacent Cl allows the hydrogen to be replaced. Hence, Cl is a "vicarious" leaving group. Other leaving groups have been used (e.g., OMe, SPh), but Cl is generally the best. Many groups W in ortho, meta, or para positions do not interfere. The reaction is also successful for di- and trinitro compounds, for nitronaphthalenes,[672] and for many nitro heterocycles. $Z—^{\ominus}CR–Cl$ may also be used.[673] When Br_3C^- or Cl_3C^- is the nucleophile the product is $ArCHX_2$, which can easily be hydrolyzed to ArCHO.[674] This is therefore an indirect way of formylating an aromatic ring containing one or more NO_2 groups, which cannot be done by any of the formylations mentioned in Chapter 11 (**11-1–11-18**).

Replacement of an amino group is possible. When aniline derivatives were treated with allyl bromide and *tert*-butyl nitrite (*t*-BuONO), the aryl–allyl coupling product was formed ($Ar–NH_2 \rightarrow Ar–CH_2CH=CH_2$).[675]

For the introduction of CH_2SR groups into phenols, see **11-23**. See also **14-19**. OS **II**, 517.

13-18   Amination of Nitrogen Heterocycles

Amination or **Amino-de-hydrogenation**

Pyridine and other heterocyclic nitrogen compounds can be aminated with alkali-metal amides in a process called the *Chichibabin reaction.*[676] The attack is always in the 2 position unless both such positions are filled, in which case the 4 position is attacked. Substituted alkali-metal amides (e.g., RNH^- and R_2N^-) have also been used. The mechanism is probably similar to that of **13-17** The existence of intermediate ions, such as **33**

[670]Goliński, J.; Mąkosza, M. *Tetrahedron Lett.* *1978*, 3495. For reviews, see Mąkosza, M. *Synthesis* *1991*, 103; *Russ. Chem. Rev.* *1989*, *58*, 747; Mąkosza, M.; Winiarski, J. *Acc. Chem. Res.* *1987*, *20*, 282.

[671]For a discussion of the mechanism, of vicarious nucleophilic aromatic substitution, see Mąkosza, M.; Lemek, T.; Kwast, A.; Terrier, F. *J. Org. Chem.* *2002*, *67*, 394.

[672]Mąkosza, M.; Danikiewicz, W.; Wojciechowski, K. *Liebigs Ann. Chem.* *1987*, 711.

[673]See Mudryk, B.; Mąkosza, M. *Tetrahedron* *1988*, *44*, 209.

[674]Mąkosza, M.; Owczarczyk, Z. *J. Org. Chem.* *1989*, *54*, 5094. See also, Mąkosza, M.; Winiarski, J. *Chem. Lett.* *1984*, 1623.

[675]Ek, F.; Axelsson, O.; Wistrand, L.-G.; Frejd, T. *J. Org. Chem.* *2002*, *67*, 6376.

[676]For reviews, see Vorbrüggen, H. *Adv. Heterocycl. Chem.* *1990*, *49*, 117; McGill, C.K.; Rappa, A. *Adv. Heterocycl. Chem.* *1988*, *44*, 1; Pozharskii, A.F.; Simonov, A.M.; Doron'kin, V.N. *Russ. Chem. Rev.* *1978*, *47*, 1042.

33

(from quinoline) has been demonstrated by NMR spectra.[677] A pyridyne type of intermediate was ruled out by several observations including the facts that 3-ethylpyridine gave 2-amino-3-ethylpyridine[678] and that certain heterocycles that cannot form an aryne could nevertheless be successfully aminated. Nitro compounds do not give this reaction,[679] but they have been aminated (ArH → ArNH$_2$ or ArNHR) via the vicarious substitution principle (see **13-17**), using 4-amino- or 4-alkylamino-1,2,4-triazoles as nucleophiles.[680] The vicarious leaving group in this case is the triazole ring. Note, however, that 3-nitropyridine was converted to 6-amino-3-nitropyridine by reaction with KOH, hydroxylamine and ZnCl$_2$.[681]

Analogous reactions have been carried out with hydrazide ions, R_2NNH^-.[682] A mixture of NO$_2$ and O$_3$, with excess NaHSO$_3$, converted pyridine to 3-aminopyridine.[683] For other methods of aminating aromatic rings, see **11-6**.

There are no *Organic Syntheses* references, but see OS **V**, 977, for a related reaction.

NITROGEN AS LEAVING GROUP

The diazonium group can be replaced by a number of groups.[684] Some of these are nucleophilic substitutions, with S$_N$1 mechanisms (p. 432), but others are free-radical reactions and are treated in Chapter 14. The solvent in all these reactions is usually water. With other solvents it has been shown that the S$_N$1 mechanism is favored by solvents of low nucleophilicity, while those of high nucleophilicity favor free-radical mechanisms.[685] The N_2^+ group[686] can be replaced by Cl$^-$, Br$^-$, and CN$^-$, by a nucleophilic mechanism (see OS **IV**, 182), but the Sandmeyer reaction is much more useful (**14-20**). Transition metal catalyzed reactions are known involving aryl-diazonium salts, and diazonium variants of the Heck reaction (**13-10**) and Suzuki coupling (**13-12**) were discussed previously. As mentioned on p. 866 it must be

[677]Zoltewicz, J.A.; Helmick, L.S.; Oestreich, T.M.; King, R.W.; Kandetzki, P.E. *J. Org. Chem.* **1973**, *38*, 1947; Woźniak, M.; Baránski, A.; Nowak, K.; van der Plas, H.C. *J. Org. Chem.* **1987**, *52*, 5643.

[678]Ban, Y.; Wakamatsu, T. *Chem. Ind.* (*London*) **1964**, 710.

[679]See, for example, Levitt, L.S.; Levitt, B.W. *Chem. Ind.* (*London*) **1975**, 520.

[680]Katritzky, A.R.; Laurenzo, K.S. *J. Org. Chem.* **1986**, *51*, 5039; **1988**, *53*, 3978.

[681]Bakke, J.M.; Svensen, H.; Trevisan, R. *J. Chem. Soc. Perkin Trans. 1* **2001**, 376.

[682]Kauffmann, T.; Hansen, J.; Kosel, C.; Schoeneck, W. *Liebigs Ann. Chem.* **1962**, *656*, 103.

[683]Suzuki, H.; Iwaya, M.; Mori, T. *Tetrahedron Lett.* **1997**, *38*, 5647.

[684]For a review of such reactions, see Wulfman, D.S., in Patai, S. *The Chemistry of Diazonium and Diazo Groups*, pt. 1, Wiley, NY, **1978**, pp. 286–297.

[685]Szele, I.; Zollinger, H. *Helv. Chim. Acta* **1978**, *61*, 1721.

[686]For a discussion of the global and local electrophilicity patterns of diazonium ions, see Pérez, P. *J. Org. Chem.* **2003**, *68*, 5886.

kept in mind that the N_2^+ group can activate the removal of another group on the ring. In a few cases, nitrogen groups, such as nitro or ammonium can be replaced.

13-19 Diazotization

$$Ar\!-\!NH_2 \;+\; HONO \longrightarrow Ar\!-\!\overset{\oplus}{N}\!\equiv\!N$$

When primary aromatic amines are treated with nitrous acid, diazonium salts are formed.[687] The reaction also occurs with aliphatic primary amines, but aliphatic diazonium ions are extremely unstable, even in solution (see p. 500). Aromatic diazonium ions are more stable, because of the resonance interaction between the nitrogens and the ring:

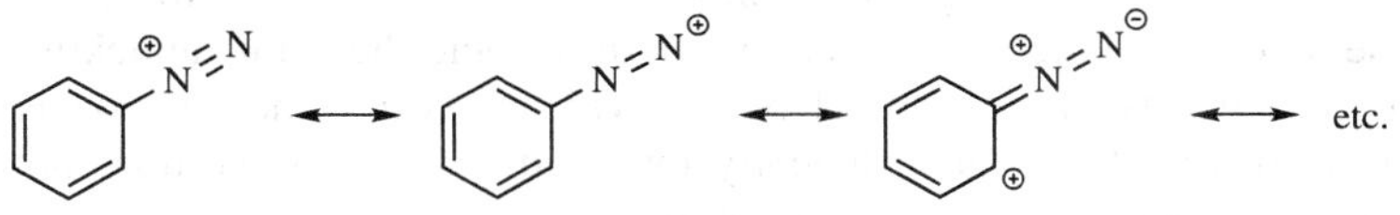

34   **35**

Incidentally, **34** contributes more to the hybrid than **35**, as shown by bond-distance measurements.[688] In benzenediazonium chloride, the C–N distance is ~1.42 Å, and the N–N distance ~1.08 Å,[689] which values fit more closely to a single and a triple bond than to two double bonds (see Table 1.5). Even aromatic diazonium salts are stable only at low temperatures, usually only $< 5°C$, although more stable ones, such as the diazonium salt obtained from sulfanilic acid, are stable up to 10 or 15°C. Diazonium salts are usually prepared in aqueous solution and used without isolation,[690] although it is possible to prepare solid diazonium salts if desired (see **13-23**). The stability of aryl diazonium salts can be increased by crown ether complexion.[691]

For aromatic amines, the reaction is very general. Halogen, nitro, alkyl, aldehyde, sulfonic acid, and so on, groups do not interfere. Since aliphatic amines do not react with

[687]For reviews, see, in Patai, S. *The Chemistry of Diazonium and Diazo Groups*, Wiley, NY, *1978*, the articles by Hegarty, A.F. pt. 2, pp. 511–591, and Schank, K. pt. 2, pp. 645–657; Godovikova, T.I.; Rakitin, O.A.; Khmel'nitskii, L.I. *Russ. Chem. Rev. 1983*, 52, 440; Challis, B.C.; Butler, A.R., in Patai, S. *The Chemistry of the Amino Group*; Wiley, NY, *1968*, pp. 305–320. For a review with respect to heterocyclic amines, see Butler, A.R. *Chem. Rev. 1975*, 75, 241.
[688]For a review of diazonium salt structures, see Sorriso, S., in Patai, S. *The Chemistry of Diazonium and Diazo Groups*, pt. 1, Wiley, NY, *1978*, pp. 95–105.
[689]Rømming, C. *Acta Chem. Scand. 1959*, 13, 1260; *1963*, 17, 1444; Sorriso, S., in Patai, S. *The Chemistry of Diazonium and Diazo Groups*, pt. 1, Wiley, NY, *1978*, p. 98; Ball, R.G.; Elofson, R.M. *Can. J. Chem. 1985*, 63, 332.
[690]For a review of reactions of diazonium salts, see Wulfman, D.S., in Patai, S. *The Chemistry of Diazonium and Diazo Groups*, pt. 1, Wiley, NY, *1978*, pp. 247–339.
[691]Korzeniowski, S.H.; Leopold, A.; Beadle, J.R.; Ahern, M.F.; Sheppard, W.A.; Khanna, R.K.; Gokel, G.W. *J. Org. Chem. 1981*, 46, 2153, and references cited therein. For reviews, see Bartsch, R.A., in Patai, S.; Rappoport, Z. *The Chemistry of Functional Groups, Supplement C* pt. 1, Wiley, NY, *1983*, pp. 889–915; Bartsch, R.A. *Prog. Macrocyclic Chem. 1981*, 2, 1.

nitrous acid below a pH $\sim$ 3, it is even possible, by working at a pH $\sim$ 1, to diazotize an aromatic amine without disturbing an aliphatic amino group in the same molecule.[692]

$$\text{EtOOC—CH}_2\text{—NH}_2 \quad + \quad \text{HONO} \quad \longrightarrow \quad \text{EtOOC—CH=}\overset{\oplus}{\text{N}}\text{=N}$$

If an aliphatic amino group is α to a COOR, CN, CHO, COR, and so on, and has an α hydrogen, treatment with nitrous acid gives not a diazonium salt, but a *diazo compound*.[693] Such diazo compounds can also be prepared, often more conveniently, by treatment of the substrate with isoamyl nitrite and a small amount of acid.[694] Certain heterocyclic amines also give diazo compounds rather than diazonium salts.[695]

Despite the fact that diazotization takes place in acid solution, the actual species attacked is not the salt of the amine, but the small amount of free amine present.[696] It is because aliphatic amines are stronger bases than aromatic ones that at pH values < 3 there is not enough free amine present for the former to be diazotized, while the latter still undergo the reaction. In dilute acid the actual attacking species is N_2O_3, which acts as a carrier of NO^+. Evidence is that the reaction is second order in nitrous acid and, at sufficiently low acidities, the amine does not appear in the rate expression.[697] Under these conditions the mechanism is

Step 1 $2 \text{ HONO} \xrightarrow{\text{slow}} N_2O_3 + H_2O$

Step 2 $\text{Ar}\overline{\text{N}}\text{H}_2 + N_2O_3 \longrightarrow \text{Ar}-\overset{\overset{\text{H}}{|}}{\underset{|}{\overset{\oplus}{\text{N}}}}-\text{N=O} \quad + \quad NO_2^-$

Step 3 $\text{Ar}-\overset{\overset{\text{H}}{|}}{\underset{|}{\overset{\oplus}{\text{N}}}}-\text{N=O} \xrightarrow{-H^+} \text{Ar}-\overset{|}{\underset{\text{H}}{\text{N}}}-\text{N=O}$

33

Step 4 $\text{Ar}-\overset{|}{\underset{\text{H}}{\text{N}}}-\text{N=O} \overset{\text{tautom.}}{\rightleftharpoons} \text{Ar}-\text{N=N}-\text{O}-\text{H}$

Step 5 $\text{Ar}-\text{N=N}-\text{O}-\text{H} \xrightarrow{H^+} \text{Ar}-\text{N}\overset{\oplus}{\equiv}\text{N} \quad + \quad H_2O$

[692]Kornblum, N.; Iffland, D.C. *J. Am. Chem. Soc.* **1949**, *71*, 2137.

[693]For a monograph on diazo compounds, see Regitz, M.; Maas, G. *Diazo Compounds*, Academic Press, NY, **1986**. For reviews, see, in Patai, S. *The Chemistry of Diazonium and Diazo Groups*, pt. 1, Wiley, NY, **1978**, the articles by Regitz, M. pt. 2, pp. 659–708, 751–820, and Wulfman, D.S.; Linstrumelle, G.; Cooper, C.F. pt. 2, pp. 821–976.

[694]Takamura, N.; Mizoguchi, T.; Koga, K.; Yamada, S. *Tetrahedron* **1975**, *31*, 227.

[695]Butler, R.N., in Patai, S. *The Chemistry of the Amino Group*, Wiley, NY, **1968**, p. 305.

[696]Challis, B.C.; Ridd, J.H. *J. Chem. Soc.* **1962**, 5197, 5208; Challis, B.C.; Larkworthy, L.F.; Ridd, J.H. *J. Chem. Soc.* **1962**, 5203.

[697]Hughes, E.D.; Ingold, C.K.; Ridd, J.H. *J. Chem. Soc.* **1958**, 58, 65, 77, 88; Hughes, E.D.; Ridd, J.H. *J. Chem. Soc.* **1958**, 70, 82.

There exists other evidence for this mechanism.[698] Other attacking species can be NOCl, $H_2NO_2^+$, and at high acidities even NO^+. Nucleophiles (e.g., Cl^-, SCN^-, thiourea) catalyze the reaction by converting the HONO to a better electrophile (e.g., $HNO_2 + Cl^- + H^+ \rightarrow NOCl + H_2O$).[699]

N-Aryl ureas are converted to the aryldiazonium nitrate upon treatment with $NaNO_2$ and H_2SO_4 in dioxane[700] or with $DMF-NO_2$ in DMF.[701]

There are many preparations of diazonium salts listed in *Organic Syntheses*, but they are always prepared for use in other reactions. We do not list them here, but under reactions in which they are used. The preparation of aliphatic diazo compounds can be found in OS **III**, 392; **IV**, 424. See also, OS **VI**, 840.

13-20 Hydroxylation of Aryldiazonium Salts

Hydroxy-de-diazoniation

$$ArN_2^+ \ + \ H_2O \ \longrightarrow \ ArOH$$

This reaction is formally analogous to **13-1**, but with a N_2^+ leaving group rather than a halide. Water is usually present whenever diazonium salts are made, but at these temperatures (0–5°C) the reaction proceeds very slowly. When it is *desired* to have OH replace the diazonium group, the excess nitrous acid is destroyed and the solution is usually boiled. Some diazonium salts require even more vigorous treatment, for example, boiling with aqueous sulfuric acid or with trifluoroacetic acid containing potassium trifluoroacetate.[702] The reaction can be performed on solutions of any diazonium salts, but hydrogen sulfates are preferred to chlorides or nitrates, since in these cases there is competition from the nucleophiles Cl^- or NO_3^-. A better method, which is faster, avoids side reactions, takes place at room temperature, and gives higher yields consists of adding Cu_2O to a dilute solution of the diazonium salt dissolved in a solution containing a large excess of $Cu(NO_3)_2$.[703] Aryl radicals are intermediates when this method is used. It has been shown that aryl radicals are at least partly involved when ordinary hydroxy-de-diazoniation is carried out in weakly alkaline aqueous solution.[704] Decomposition of arenediazonium tetrafluoroborates in F_3CSO_2OH gives aryl triflates directly, in high yields.[705]

OS **I**, 404; **III**, 130, 453, 564; **V**, 1130.

---

[698]For discussions, see Williams, D.L.H. *Nitrosation*, Cambridge University Press, Cambridge, *1988*, pp. 95–109; Ridd, J.H. *Q. Rev. Chem. Soc.* *1961*, *15*, 418, p. 422.

[699]Williams, D.L.H. *Nitrosation*; Cambridge University Press, Cambridge, *1988*, pp. 84–93.

[700]Zhang, Z.; Zhang, Q.; Zhang, S.; Liu,, X.; Zhao, G. *Synth. Commun.* *2001*, *31*, 329.

[701]Zhang, O.Z.; Zhang, S.; Zhang, J. *Synth. Commun.* *2001*, *31*, 1243.

[702]Horning, D.E.; Ross, D.A.; Muchowski, J.M. *Can. J. Chem.* *1973*, *51*, 2347.

[703]Cohen, T.; Dietz, Jr., A.G.; Miser, J.R. *J. Org. Chem.* *1977*, *42*, 2053.

[704]Dreher, E.; Niederer, P.; Rieker, A.; Schwarz, W.; Zollinger, H. *Helv. Chim. Acta* *1981*, *64*, 488.

[705]Yoneda, N.; Fukuhara, T.; Mizokami, T.; Suzuki, A. *Chem. Lett.* *1991*, 459.

13-21   Replacement by Sulfur-Containing Groups

Mercapto-de-diazoniation, and so on

$$ArN_2^+ \;+\; HS^- \longrightarrow ArSH$$

$$ArN_2^+ \;+\; S^{2-} \longrightarrow ArSAr$$

$$ArN_2^+ \;+\; RS^- \longrightarrow ArSR$$

$$ArN_2^+ \;+\; SCN^- \longrightarrow ArSCN \;+\; ArNCS$$

These reactions are convenient methods for incorporating a sulfur-containing group onto an aromatic ring. With $Ar'S^-$, diazosulfides $Ar-N=N-S-Ar'$ are intermediates,[706] which can in some cases be isolated.[707] Thiophenols can be made as shown above, but more often the diazonium ion is treated with $EtO-CSS^-$ or S_2^{2-}, which give the expected products, and these are easily convertible to thiophenols. Aryldiazonium salts are prepared by the reaction of an aniline derivative with an alkyl nitrite (RONO), and when formed in the presence of dimethyl disulfide (MeS–SMe), the product is the thioether, Ar–S–Me.[708] Aryl triflates have been converted to the aryl thiol using NaST(P5) and a palladium catalyst, followed by treatment with tetrabutylammonium fluoride[709] (see also, **14-22**).
OS **II**, 580; **III**, 809 (but see OS **V**, 1050). Also see, OS **II**, 238.

13-22   Replacement by Iodine

Iodo-de-diazoniation

$$ArN_2^+ \;+\; I^- \longrightarrow ArI$$

One of the best methods for the introduction of iodine into aromatic rings (see **13-7**) is the reaction of diazonium salts with iodide ions. Analogous reactions with chloride, bromide, and fluoride ions give poorer results, and **14-20** and **13-23** are preferred for the preparation of aryl chlorides, bromides, and fluorides. However, when other diazonium reactions are carried out in the presence of these ions, halides are usually side products. Aniline has also been converted to fluorobenzene by treatment with t-BuONO and SiF_4 followed by heating.[710] A related reaction between $PhN=N-NC_4H_8$ and iodine gave iodobenzene.[711]
The actual attacking species is probably not only I^- if it is I^- at all. The iodide ion is oxidized (by the diazonium ion, nitrous acid, or some other oxidizing agent)

[706]Abeywickrema, A.N.; Beckwith, A.L.J. *J. Am. Chem. Soc.* **1986**, *108*, 8227, and references cited therein.
[707]See, for example, Price, C.C.; Tsunawaki, S. *J. Org. Chem.* **1963**, *28*, 1867.
[708]Allaire, F.S.; Lyga, J.W. *Synth. Commun.* **2001**, *31*, 1857.
[709]Arnould, J.C.; Didelot, M.; Cadilhac, C.; Pasquet, M.J. *Tetrahedron Lett.* **1996**, *37*, 4523.
[710]Tamura, M.; Shibakami, M.; Sekiya, A. *Eur. J. Org. Chem.* **1998**, 725.
[711]Wu, Z.; Moore, J.S. *Tetrahedron Lett.* **1994**, *35*, 5539.

to iodine, which in a solution containing iodide ions is converted to I_3^-; this is the actual attacking species, at least partly. This was shown by isolation of $ArN_2^+ \; I_3^-$ salts, which, on standing, gave ArI.[712] From this, it can be inferred that the reason the other halide ions give poor results is not that they are poor nucleophiles but that they are poor reducing agents (compared with iodide). There is also evidence for a free-radical mechanism.[713]

The hydroxyl group of a phenol can be replaced with iodine. The reaction of phenol with a boronic ester and a palladium catalyst, followed by reaction with NaI and chloramine-T converts phenol to iodobenzene.[714]

OS **II**, 351, 355, 604; **V**, 1120.

13-23 The Schiemann Reaction

Fluoro-de-diazoniation (overall transformation)

$$ArN_2^+ \; BF_4^- \quad \xrightarrow{\Delta} \quad ArF \;+\; N_2 \quad BF_3$$

Heating of diazonium fluoroborates (the *Schiemann* or *Balz–Schiemann reaction*) is by far the best way of introducing fluorine into an aromatic ring.[715] In the most common procedure, the fluoroborate salts are prepared by diazotizing as usual with nitrous acid and HCl and then adding a cold aqueous solution of $NaBF_4$, HBF_4, or NH_4BF_4. A precipitate forms, which is dried, and the salt is heated in the dry state. These salts are unusually stable for diazonium salts, and the reaction is usually successful. In general, any aromatic amine that can be diazotized will form a BF_4^- salt, usually with high yields. The diazonium fluoroborates can be formed directly from primary aromatic amines with *tert*-butyl nitrite and BF_3–etherate.[716] The reaction has also been carried out on $ArN_2^+ \; PF_6^-$, $ArN_2^+ \; SbF_6^-$, and ArN_2^+ AsF_6^- salts, in many cases with better yields.[717] Aryl chlorides and bromides are more commonly prepared by the Sandmeyer reaction (**14-20**). In an alternative procedure, aryl fluorides have been prepared by treatment of aryltriazenes Ar—N=N—NR$_2$ with 70% HF in pyridine.[718]

The mechanism is of the S_N1 type. That aryl cations are intermediates was shown by the following experiments:[719] Aryl diazonium chlorides are known to

[712]Carey, J.G.; Millar, I.T. *Chem. Ind. (London)* **1960**, 97.

[713]Singh, P.R.; Kumar, R. *Aust. J. Chem.* **1972**, 25, 2133; Kumar, R.; Singh, P.R. *Tetrahedron Lett.* **1972**, 613; Meyer, G.; Rössler, K.; Stöcklin, G. *J. Am. Chem. Soc.* **1979**, 101, 3121; Packer, J.E.; Taylor, R.E.R. *Aust. J. Chem.* **1985**, 38, 991; Abeywickrema, A.N.; Beckwith, A.L.J. *J. Org. Chem.* **1987**, 52, 2568.

[714]Thompson, A.L.S.; Kabalka, G.W.; Akula, M.R.; Huffman, J.W. *Synthesis* **2005**, 547.

[715]For a review, see Suschitzky, H. *Adv. Fluorine Chem.* **1965**, 4, 1.

[716]Doyle, M.P.; Bryker, W.J. *J. Org. Chem.* **1979**, 44, 1572.

[717]Rutherford, K.G.; Redmond, W.; Rigamonti, J. *J. Org. Chem.* **1961**, 26, 5149; Sellers, C.; Suschitzky, H. *J. Chem. Soc. C* **1968**, 2317.

[718]Rosenfeld, M.N.; Widdowson, D.A. *J. Chem. Soc. Chem. Commun.* **1979**, 914. For another alternative procedure, see Yoneda, N.; Fukuhara, T.; Kikuchi, T.; Suzuki, A. *Synth. Commun.* **1989**, 19, 865.

[719]See also, Swain, C.G.; Sheats, J.E.; Harbison, K.G. *J. Am. Chem. Soc.* **1975**, 97, 783, 796; Becker, H.G.O.; Israel, G. *J. Prakt. Chem.* **1979**, 321, 579.

arylate other aromatic rings by a free-radical mechanism (see **13-27**). In radical arylation it does not matter whether the other ring contains electron-withdrawing or electron-donating groups; in either case a mixture of isomers is obtained, since the attack is not by a charged species. If an aryl radical were an intermediate in the Schiemann reaction and the reaction were run in the presence of other rings, it should not matter what kinds of groups were on these other rings: Mixtures of biaryls should be obtained in all cases. But if an aryl cation is an intermediate in the Schiemann reaction, compounds containing meta-directing groups, that is, meta directing for *electrophilic* substitutions, should be meta-arylated and those containing ortho–para-directing groups should be ortho– and para arylated, since an aryl cation should behave in this respect like any electrophile (see Chapter 11). Experiments have shown[720] that such orientation is observed, demonstrating that the Schiemann reaction has a positively charged intermediate. The attacking species, in at least some instances, is not F^- but BF_4^-.[721]

OS **II**, 188, 295, 299; **V**, 133.

13-24   Conversion of Amines to Azo Compounds

N-Arylimino-de-dihydro-bisubstitution

$$\text{ArNH}_2 \ + \ \text{Ar'NO} \ \xrightarrow{\text{HOAc}} \ \text{Ar}-\text{N}=\text{N}-\text{Ar'}$$

Aromatic nitroso compounds combine with primary arylamines in glacial acetic acid to give symmetrical or unsymmetrical azo compounds (the *Mills reaction*).[722] A wide variety of substituents may be present in both aryl groups. Unsymmetrical azo compounds have also been prepared by the reaction between aromatic nitro compounds ArNO_2 and *N*-acyl aromatic amines Ar'NHAc.[723] The use of phase-transfer catalysis increased the yields.

13-25   Methylation, Vinylation, and Arylation of Diazonium Salts

Methyl-de-diazoniation, and so on

$$\text{ArN}_2^+ \ + \ \text{Me}_4\text{Sn} \ \xrightarrow[\text{MeCN}]{\text{Pd(OAc)}_2} \ \text{ArMe}$$

A methyl group can be introduced into an aromatic ring by treatment of diazonium salts with tetramethyltin and a palladium acetate catalyst.[724] The reaction has been performed with Me, Cl, Br, and NO_2 groups on the ring. A vinylic group can

[720]Makarova, L.G.; Matveeva, M.K. *Bull. Acad. Sci. USSR Div. Chem. Sci.* **1958**, 548; Makarova, L.G.; Matveeva, M.K.; Gribchenko, E.A. *Bull. Acad. Sci. USSR Div. Chem. Sci.* **1958**, 1399.

[721]Swain, C.G.; Rogers, R.J. *J. Am. Chem. Soc.* **1975**, *97*, 799.

[722]For a review, see Boyer, J.H., in Feuer, H. *The Chemistry of the Nitro and Nitroso Groups*, pt. 1, Wiley, NY, **1969**, pp. 278–283.

[723]Ayyangar, N.R.; Naik, S.N.; Srinivasan, K.V. *Tetrahedron Lett.* **1989**, *30*, 7253.

[724]Kikukawa, K.; Kono, K.; Wada, F.; Matsuda, T. *J. Org. Chem.* **1983**, *48*, 1333.

be introduced with $CH_2=CHSnBu_3$. When an aryl amine is treated with *tert*-butyl hyponitrite (*t*-BuONO) and allyl bromide, the nitrogen is displaced to the give allyl–aryl compound.[725]

Aryl diazonium salts can be used coupled with alkenes in a Heck-like reaction (**12-15**).[726] Other reactive aryl species also couple with aryldiazonium salts in the presence of a palladium catalyst.[727] A Suzuki type coupling (**13-9**) has also been reported using arylboronic acids, aryldiazonium salts and a palladium catalyst.[728]

Aryltrifluoroborates (**12-28**) react with aryldiazonium salts in the presence of a palladium catalyst to give the corresponding biaryl.[729] Arylborate esters also reacat using a palladium catlsyt, and the aryl dizaonium unit reacts faster than an aryl halide.[730]

13-26   Arylation of Activated Alkenes by Diazonium Salts: Meerwein Arylation

Arylation or **Aryl-de-hydrogenation**

Alkenes activated by an electron-withdrawing group (Z may be C=C, halogen, C=O, Ar, CN, etc.) can be arylated by treatment with a diazonium salt and a cupric chloride[731] catalyst. This is called the *Meerwein*

arylation reaction.[732] Addition of ArCl to the double bond (to give) is a side reaction (**15-46**). In an improved procedure, an arylamine is treated with an alkyl nitrite (generating ArN_2^+ *in situ*) and a copper(II) halide in the presence of the alkene.[733]

The mechanism is probably of the free-radical type, with AR• (**36**) forming as in **14-20**, and then halogen transfer to give **37** or elimination to give **38**.[734]

[725]Ek, F.; Wistrand, L.-G.; Frejd, T. *J. Org. Chem.* **2003**, *68*, 1911.

[726]Sengupta, S.; Bhattacharya, S. *J. Chem. Soc. Perkin Trans. 1* **1993**, 1943.

[727]Darses, S.; Genêt, J.-P.; Brayer, J.-L.; Demoute, J.-P. *Tetrahedron Lett.* **1997**, *38*, 4393.

[728]Darses, S.; Jeffery, T.; Genêt, J.-P.; Brayer, J.-L.; Demoute, J.-P. *Tetrahedron Lett.* **1996**, *37*, 3857.

[729]Darses, S.; Michaud, G.; Genêt, J.-P. *Eur. J. Org. Chem.* **1999**, 1875.

[730]Willis, D.M.; Strongin, R.M. *Tetahedron Lett.* **2000**, *41*, 6271.

[731]FeCl$_2$ is also effective: Ganushchak, N.I.; Obushak, N.D.; Luka, G.Ya. *J. Org. Chem. USSR* **1981**, *17*, 765.

[732]For reviews, see Dombrovskii, A.V. *Russ. Chem. Rev.*, **1984**, *53*, 943; Rondestvedt, Jr., C.S. *Org. React.*, **1976**, *24*, 225.

[733]Doyle, M.P.; Siegfried, B.; Elliott, R.C.; Dellaria Jr., J.F. *J. Org. Chem.* **1977**, *42*, 2431.

[734]Dickerman, S.C.; Vermont, G.B. *J. Am. Chem. Soc.* **1962**, *84*, 4150; Morrison, R.T.; Cazes, J.; Samkoff, N.; Howe, C.A. *J. Am. Chem. Soc.* **1962**, *84*, 4152.

The radical **36** can react with cupric chloride by two pathways, one of which leads to addition and the other to substitution. Even when the addition pathway is taken, however, the substitution product may still be formed by subsequent elimination of HCl. Note that radical reactions are presented in Chapter 14, but the coupling of an alkene with an aromatic compound containing a leaving group prompted its placement here. Note also the similarity to the Heck reaction in **13-10**.

A variation of this reaction uses a palladium–copper catalyst on Montmorillonite clay. When aniline reacted with methyl acrylate in acetic acid and the Pd–Cu–Montmorillonite K10, PhCH=CHCO₂Me was obtained.[735]

OS **IV**, 15.

13-27 Arylation of Aromatic Compounds by Diazonium Salts

Arylation or **Aryl-de-hydrogenation**

$$ ArH \ + \ Ar'N_2^+ \ X^- \ \xrightarrow{\ OH^-\ } \ Ar\!-\!Ar' $$

When the normally acidic solution of a diazonium salt is made alkaline, the aryl portion of the diazonium salt can couple with another aromatic ring. Known as the *Gomberg* or *Gomberg–Bachmann reaction*,[736] it has been performed on several types of aromatic rings and on quinones. Yields are not high (usually <40%) because of the many side reactions undergone by diazonium salts, though higher yields have been obtained under phase-transfer conditions.[737] The conditions of the Meerwein reaction (**13-26**), treatment of the solution with a copper-ion catalyst, have also been used, as has the addition of sodium nitrite in Me₂SO (to benzene diazonium fluoroborate in DMSO).[738]

39

[735]Waterlot, C.; Couturier, D.; Rigo, B. *Tetrahedron Lett.* **2000** *41*, 317.

[736]For reviews, see Bolton, R.; Williams, G.H. *Chem. Soc. Rev.*, **1986**, *15*, 261; Hey, D.H. *Adv. Free-Radical Chem.* **1966**, *2*, 47. For a review applied to heterocyclic substrates, see Vernin, G.; Dou, H.J.; Metzger, J. *Bull. Soc. Chim. Fr.* **1972**, 1173.

[737]Beadle, J.R.; Korzeniowski, S.H.; Rosenberg, D.E.; Garcia-Slanga, B.J.; Gokel, G.W. *J. Org. Chem.* **1984**, *49*, 1594.

[738]Kamigata, N.; Kurihara, T.; Minato, H.; Kobayashi, M. *Bull. Chem. Soc. Jpn.* **1971**, *44*, 3152.

When the Gomberg–Bachmann reaction is performed intramolecularly as in the formation of **39**, either by the alkaline solution or by the copper-ion procedure, it is called the *Pschorr ring closure*[739] and yields are usually somewhat higher. Still higher yields have been obtained by carrying out the Pschorr reaction electrochemically.[740] The Pschorr reaction has been carried out for Z = CH=CH, CH_2CH_2, NH, C=O, CH_2, and quite a few others. A rapid and convenient way to carry out the Pschorr synthesis is to diazotize the amine substrate with isopropyl nitrite in the presence of sodium iodide, in which case the ring-closed product is formed in one step.[741]

Other compounds with nitrogen–nitrogen bonds have been used instead of diazonium salts. Among these are *N*-nitroso amides [ArN(NO)COR], triazenes,[742] and azo compounds. Still another method involves treatment of an aromatic primary amine directly with an alkyl nitrite in an aromatic substrate as solvent.[743]

In each case, the mechanism involves generation of an aryl radical from a covalent azo compound. In acid solution, diazonium salts are ionic and their reactions are polar. When they cleave, the product is an aryl cation (see p. 856). However, in neutral or basic solution, diazonium ions are converted to covalent compounds, and these cleave to give free radicals (Ar• and Z•). Note that radical reactions are presented in Chapter 14, but the coupling of an aromatic ring with an aromatic compound containing a leaving group prompted its placement here. Note the similarity to the Suzuki reaction in **13-12**.

$$\text{Ar—N=N—Z} \longrightarrow \text{Ar•} + \text{N≡N} + \text{Z•}$$

Under Gomberg–Bachmann conditions, the species that cleaves is the anhydride, **40**.[744]

$$\text{Ar—N=N—O—N=N—Ar} \longrightarrow \text{Ar•} + N_2 + \text{•O—N=N—Ar}$$
40   **41**

42   **43**

[739]For a review, see Abramovitch, R.A. *Adv. Free-Radical Chem.* **1966**, *2*, 87.
[740]Elofson, R.M.; Gadallah, F.F. *J. Org. Chem.* **1971**, *36*, 1769.
[741]Chauncy, B.; Gellert, E. *Aust. J. Chem.* **1969**, *22*, 993. See also, Duclos, Jr., R.I.; Tung, J.S.; Rappoport, H. *J. Org. Chem.* **1984**, *49*, 5243.
[742]See, for example, Patrick, T.B.; Willaredt, R.P.; DeGonia, D.J. *J. Org. Chem.* **1985**, *50*, 2232; Butler, R.N.; O'Shea, P.D.; Shelly, D.P. *J. Chem. Soc. Perkin Trans. 1*, **1987**, 1039.
[743]Cadogan, J.I.G. *J. Chem. Soc.* **1962**, 4257; Fillipi, G.; Vernin, G.; Dou, H.J.; Metzger, J.; Perkins, M.J. *Bull. Soc. Chim. Fr.* **1974**, 1075.
[744]Rüchardt, C.; Merz, E. *Tetrahedron Lett.* **1964**, 2431; Eliel, E.L.; Saha, J.G.; Meyerson, S. *J. Org. Chem.* **1965**, *30*, 2451.

The aryl radical thus formed attacks the substrate to give the intermediate **42** (see p. 940), from which the radical **41** abstracts hydrogen to give the product, **43**. *N*-Nitroso amides probably rearrange to *N*-acyloxy compounds (**44**), which cleave to give aryl radicals.[745] There is evidence that the reaction with alkyl nitrites also involves attack by aryl radicals.[746]

$$
2 \quad \underset{\text{Ar}}{\overset{\text{N}^{\nearrow \text{O}}}{\text{N}}}\text{-}\underset{\overset{\|}{\text{O}}}{\overset{\text{O}}{\text{C}}}\text{-R} \longrightarrow 2 \quad \underset{\text{Ar-N=N-O}}{\overset{\overset{\text{O}}{\|}}{\text{C-R}}} \longrightarrow \text{Ar}\bullet + \text{Ar-N=N-O}^{\bullet}
$$
$$
\mathbf{44}
$$
$$
+ \; N_2 \; + \; \underset{R}{\overset{O}{\|}}\text{C}\underset{O}{\text{-}}\underset{R}{\overset{O}{\|}}\text{C}
$$

The Pschorr reaction can take place by two different mechanisms, depending on conditions: (*1*) attack by an aryl radical (as in the Gomberg–Bachmann reaction) or (*2*) attack by an aryl cation (similar to the S_N1 mechanism discussed on p. 857).[747] Under certain conditions the ordinary Gomberg–Bachmann reaction can also involve attack by aryl cations.[748]

OS **I**, 113; **IV**, 718.

13-28 Aryl Dimerization With Diazonium Salts

De-diazonio-coupling; Arylazo-de-diazonio-substitution

$$
2\,\text{ArN}_2^+ \xrightarrow[\text{or} \;\; \text{Cu} + \text{H}^+]{\text{Cu}^+} \text{Ar-Ar} + 2\,N_2 \quad \text{or} \quad \text{Ar-N=N-Ar} + N_2
$$

When diazonium salts are treated with cuprous ion (or with copper and acid, in which case it is called the *Gatterman method*), two products are possible. If the ring contains electron-withdrawing groups, the main product is the biaryl, but the presence of electron-donating groups leads mainly to the azo compound. This reaction is different from **13-27** (and from **19-14**) in that *both* aryl groups in the product originate from ArN_2^+, that is, hydrogen is not a leaving group in this reaction. The mechanism probably involves free radicals.[749]

OS **I**, 222; **IV**, 872. Also see, OS **IV**, 273.

[745]Cadogan, J.I.G.; Murray, C.D.; Sharp, J.T. *J. Chem. Soc. Perkin Trans. 2*, *1976*, 583, and references cited therein.

[746]Gragerov, I.P.; Levit, A.F. *J. Org. Chem. USSR 1968*, 4, 7.

[747]For an alternative to the second mechanism, see Gadallah, F.F.; Cantu, A.A.; Elofson, R.M. *J. Org. Chem. 1973*, 38, 2386.

[748]For examples; see Kobori, N.; Kobayashi, M.; Minato, H. *Bull. Chem. Soc. Jpn. 1970*, 43, 223; Cooper, R.M.; Perkins, M.J. *Tetrahedron Lett. 1969*, 2477; Burri, P.; Zollinger, H. *Helv. Chim. Acta 1973*, 56, 2204; Eustathopoulos, H.; Rinaudo, J.; Bonnier, J.M. *Bull. Soc. Chim. Fr. 1974*, 2911. For a discussion, see Zollinger, H. *Acc. Chem. Res. 1973*, 6, 335, 338.

[749]See Cohen, T.; Lewarchik, R.J.; Tarino, J.Z. *J. Am. Chem. Soc. 1974*, 96, 7753.

13-29   Replacement of Nitro

Alkyl-de-nitration, Hydroxy and alkoxy-de-nitration, Halo-de-nitration

$$Ar\text{-}NO_2 \longrightarrow Ar\text{-}R$$

In some cases, the nitrogen group of an aromatic nitro compound can be replaced with an alkyl group. The reaction of 1,4-dinitrobenzene with potassium *tert*-butoxide in the presence of BEt$_3$, for example, gave 4-ethylnitrobenzene.[750] Other nucleophiles can replace a nitrogen-containing group. The reaction of hydroxide with Ar–Y, where Y = nitro,[751] azide, NR$_3^+$, and so on gives the corresponding phenol. This latter reaction works with alkoxide nucleophiles to give the corresponding aryl ether. The nitro can be replaced with chloro by use of NH$_4$Cl, PCl$_5$, SOCl$_2$, HCl, Cl$_2$, or CCl$_4$. Some of these reagents operate only at high temperatures and the mechanism is not always nucleophilic substitution. Activated aromatic nitro compounds can be converted to fluorides with fluoride ion.[752]

The reaction of vinyl nitro compounds (C=C–NO$_2$) and aryl iodide to give the styrene compound (C=C–Ar) was reported using BEt$_3$ and exposure to air.[753]

REARRANGEMENTS

13-30   The von Richter Rearrangement

Hydro-de-nitro-*cine*-substitution

When aromatic nitro compounds are treated with cyanide ion, the nitro group is displaced and a carboxyl group enters with cine substitution (p. 860), always ortho to the displaced group, never meta or para. The scope of this reaction, called the *von Richter rearrangement*, is variable.[754] As with other nucleophilic aromatic substitutions, the reaction gives best results when electron-withdrawing groups are in ortho and para positions, but yields are low, usually <20% and never >50%.

[750]Palani, N.; Jayaprakash, K.; Hoz, S. *J. Org. Chem.* **2003**, *68*, 4388.

[751]For a convenient way of achieving this conversion, see Knudsen, R.D.; Snyder, H.R. *J. Org. Chem.* **1974**, *39*, 3343.

[752]Attiná, M.; Cacace, F.; Wolf, A.P. *J. Chem. Soc. Chem. Commun.* **1983**, 108; Clark, J.H.; Smith, D.K. *Tetrahedron Lett.* **1985**, *26*, 2233; Suzuki, H.; Yazawa, N.; Yoshida, Y.; Furusawa,O.; Kimura, O. *Bull. Chem. Soc. Jpn.* **1990**, *63*, 2010; Effenberger, F.; Streicher, W. *Chem. Ber.* **1991**, *124*, 157.

[753]Liu, J.-T.; Jang, Y.-J.; Shih, Y.-K.; Hu, S.-R.; Chu, C.-M.; Yao, C.-F. *J. Org. Chem.* **2001**, *66*, 6021.

[754]For a review, see Shine, H.J. *Aromatic Rearrangements*, Elsevier, NY, **1967**, pp. 326–335.

At one time it was believed that a nitrile, ArCN, was an intermediate, since cyanide is the reagent and nitriles are hydrolyzable to carboxylic acids under the reaction conditions (**16-4**). However, a remarkable series of results proved this belief to be in error. Bunnett and Rauhut demonstrated[755] that α-naphthyl cyanide is *not* hydrolyzable to α-naphthoic acid under conditions at which β-nitronaphthalene undergoes the von Richter rearrangement to give α-naphthoic acid. This proved that the nitrile cannot be an intermediate. It was subsequently demonstrated that N_2 is a major product of the reaction.[756] It had previously been assumed that all the nitrogen in the reaction was converted to ammonia, which would be compatible with a nitrile intermediate, since ammonia is a hydrolysis product of nitriles. At the same time it was shown that NO_2^- is not a major product. The discovery of nitrogen indicated that a nitrogen–nitrogen bond must be formed during the course of the reaction. A mechanism in accord with all the facts was proposed by Rosenblum:[756]

Note that **46** is a stable compound; hence it should be possible to prepare it independently and to subject it to the conditions of the von Richter rearrangement. This was done and the correct products are obtained.[757] Further evidence is that when **45** (Z = Cl or Br) was treated with cyanide in $H_2{}^{18}O$, half the oxygen in the product was labeled, showing that one of the oxygens of the carboxyl group came from the nitro group and one from the solvent, as required by this mechanism.[758]

[755]Bunnett, J.F.; Rauhut, M.M. *J. Org. Chem.* **1956**, *21*, 934, 944.
[756]Rosenblum, M. *J. Am. Chem. Soc.* **1960**, *82*, 3796.
[757]Ibne-Rasa, K.M.; Koubek, E. *J. Org. Chem.* **1963**, *28*, 3240.
[758]Samuel, D. *J. Chem. Soc.* **1960**, 1318. For other evidence, see Cullen, E.; L'Ecuyer, P. *Can. J. Chem.* **1961**, *39*, 144, 155, 382; Ullman, E.F.; Bartkus, E.A. *Chem. Ind.* (*London*) **1962**, 93.

13-31   The Sommelet–Hauser Rearrangement

Benzylic quaternary ammonium salts, when treated with alkali-metal amides, undergo a rearrangement called the *Sommelet–Hauser rearrangement*.[759] Since the product is a benzylic tertiary amine, it can be further alkylated and the product again subjected to the rearrangement. This process can be continued around the ring until an ortho position is blocked.[760]

The rearrangement occurs with high yields and can be performed with various groups present in the ring.[761] The reaction is most often carried out with three methyl groups on the nitrogen, but other groups can also be used, though if a β-hydrogen is present, Hofmann elimination (**17-7**) often competes. The *Stevens rearrangement* (**18-21**) is also a competing process.[762] When both rearrangements are possible, the Stevens is favored at high temperatures and the Sommelet–Hauser at low temperatures.[763] The mechanism is

The benzylic hydrogen is most acidic and is the one that first loses a proton to give the ylid **47**. However, **48**, which is present in smaller amount, is the species

[759]For reviews, see Pine, S.H. *Org. React.*, *1970*, *18*, 403; Lepley, A.R.; Giumanini, A.G. *Mech. Mol. Migr.* *1971*, *3*, 297; Wittig, G. *Bull. Soc. Chim. Fr.* *1971*, 1921; Stevens, T.S.; Watts, W.E. *Selected Molecular Rearrangements*, Van Nostrand-Reinhold, Princeton, *1973*, pp. 81–88; Shine, H.J.*Aromatic Rearrangements*, Elsevier, NY, *1967*, pp. 316–326. Also see, Klunder, J.M. *J. Heterocyclic Chem. 1995*, *32*, 1687.

[760]Beard, W.Q.; Hauser, C.R. *J. Org. Chem. 1960*, *25*, 334.

[761]Jones, G.C.; Beard, W.Q.; Hauser, C.R. *J. Org. Chem. 1963*, *28*, 199.

[762]For a method that uses nonbasic conditions, and gives high yields of the Sommelet–Hauser product, with little or no Stevens rearrangement, see Nakano, M.; Sato, Y. *J. Org. Chem. 1987*, *52*, 1844; Shirai, N.; Sato, Y. *J. Org. Chem. 1988*, *53*, 194.

[763]Wittig, G.; Streib, H. *Liebigs Ann. Chem. 1953*, *584*, 1.

that undergoes the rearrangement, shifting the equilibrium in its favor. This mechanism is an example of a [2,3] sigmatropic rearrangement (see **18-35**). Another mechanism that might be proposed is one in which a methyl group actually breaks away (in some form) from the nitrogen and then attaches itself to the ring. That this is not so was shown by a product study.[764] If the second mechanism were true, **49** should give **50**, but the first mechanism predicts the formation of **51**, which is what was actually obtained.[765]

The mechanism as we have pictured it can lead only to an ortho product. However, a small amount of para product has been obtained in some cases.[766] A mechanism[767] in which there is a dissociation of the ArC—N bond (similar to the ion-pair mechanism of the Stevens rearrangement, p. 1622) has been invoked to explain the para products that are observed.

Sulfur ylids containing a benzylic group (analogous to **48**) undergo an analogous rearrangement.[768]

OS **IV**, 585.

13-32 Rearrangement of Aryl Hydroxylamines

1/C-Hydro-5/N-hydroxy-interchange

Aryl hydroxylamines treated with acids rearrange to aminophenols.[769] Although this reaction (known as the *Bamberger rearrangement*) is similar in appearance to

[764]For other evidence for the mechanism given, see Hauser, C.R.; Van Eenam, D.N. *J. Am. Chem. Soc.* **1957**, *79*, 5512; Jones, F.N.; Hauser, C.R. *J. Org. Chem.* **1961**, *26*, 2979; Puterbaugh, W.H.; Hauser, C.R. *J. Am. Chem. Soc.* **1964**, *86*, 1105; Pine, S.H.; Sanchez, B.L. *Tetrahedron Lett.* **1969**, 1319; Shirai, N.; Watanabe, Y.; Sato, Y. *J. Org. Chem.* **1990**, *55*, 2767.

[765]Kantor, S.W.; Hauser, C.R. *J. Am. Chem. Soc.* **1951**, *73*, 4122.

[766]Pine, S.H. *Tetrahedron Lett.* **1967**, 3393; Pine, S.H. *Org. React.* **1970**, *18*, 403, p. 418.

[767]Bumgardner, C.L. *J. Am. Chem. Soc.* **1963**, *85*, 73.

[768]See Block, E. *Reactions of Organosulfur Compounds*, Academic Press, NY, **1978**, pp. 118–124.

[769]For a review, see Shine, H.J. *Aromatic Rearrangements*, Elsevier, NY, **1967**, pp. 182–190.

11-28–11-32, the attack on the ring is not electrophilic but nucleophilic. The rearrangement is intermolecular, with the following mechanism:

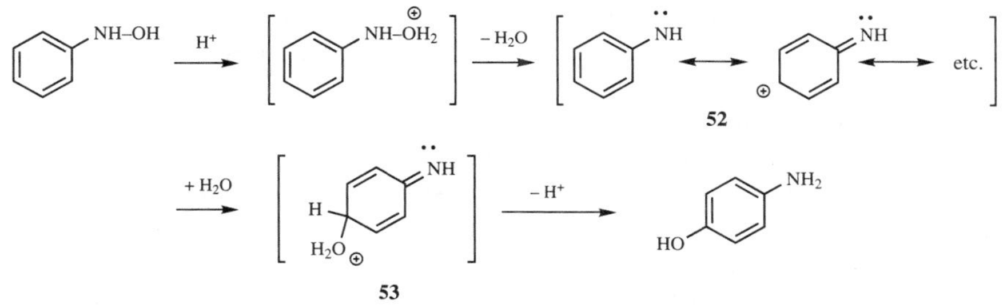

Among the evidence[770] for this mechanism are the facts that other products are obtained when the reaction is run in the presence of competing nucleophiles, for example, p-ethoxyaniline when ethanol is present, and that when the para position is blocked, compounds similar to **53** are isolated. In the case of 2,6-dimethylphenylhydroxylamine, the intermediate nitrenium ion **52** was trapped, and its lifetime in solution was measured.[771] The reaction of **52** with water was found to be diffusion controlled.[288]

OS **IV**, 148.

13-33   The Smiles Rearrangement

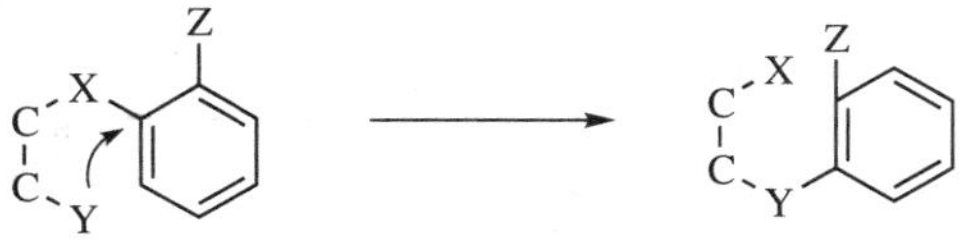

The *Smiles rearrangement* actually comprises a group of rearrangements that follow the pattern given above.[772] A specific example is the reaction of **54** with hydroxide to give **55**.

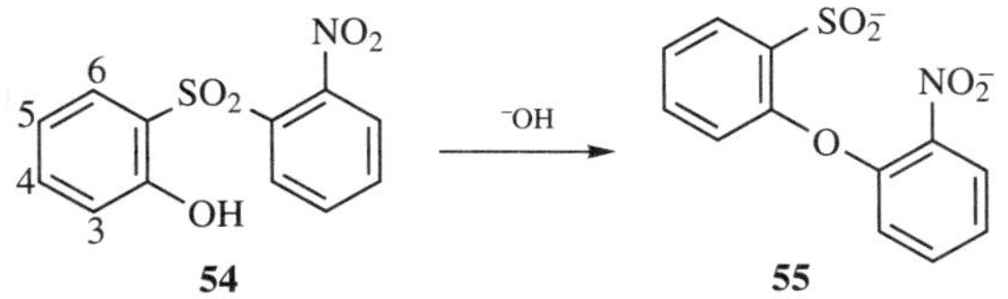

Smiles rearrangements are simply intramolecular nucleophilic substitutions. In the example given, SO_2Ar is the leaving group and ArO^- the nucleophile, and the nitro group serves to activate its ortho position. Halogens also serve as activating

---

[770]For additional evidence, see Sone, T.; Hamamoto, K.; Seiji, Y.; Shinkai, S.; Manabe, O. *J. Chem. Soc. Perkin Trans. 2* **1981**, 1596; Kohnstam, G.; Petch, W.A.; Williams, D.L.H. *J. Chem. Soc. Perkin Trans. 2* **1984**, 423; Sternson, L.A.; Chandrasakar, R. *J. Org. Chem. 1984, 49*, 4295, and references cited therein.
[771]Fishbein, J.C.; McClelland, R.A. *J. Am. Chem. Soc. 1987, 109*, 2824.
[772]For reviews, see Truce, W.E.; Kreider, E.M.; Brand, W.W. *Org. React., 1971, 18*, 99; Shine, H.J. *Aromatic Rearrangements*, Elsevier, NY, **1967**, pp. 307–316; Stevens, T.S.; Watts, W.E. *Selected Molecular Rearrangements*, Van Nostrand-Reinhold, Princeton, NJ, **1973**, pp. 120–126.

groups.[773] The ring at which the substitution takes place is nearly always activated, usually by ortho or para nitro groups. Here X is usually S, SO, SO_2,[774] O, or COO, and Y is usually the conjugate base of OH, NH_2, NHR, or SH. The reaction has even been carried out with $Y = CH_2^-$ (phenyllithium was the base here).[775]

The reaction rate is greatly enhanced by substitution in the 6 position of the attacking ring, for steric reasons. For example, a methyl, chloro, or bromo group in the 6 position of **54** caused the rate to be $\sim 10^5$ times faster than when the same groups were in the 4 position,[776] though electrical effects should be similar at these positions. The enhanced rate comes about because the most favorable conformation the molecule can adopt to suit the bulk of the 6-substituent is also the conformation required for the rearrangement. Thus, less entropy of activation is required.

Although the Smiles rearrangement is usually carried out on compounds containing two rings, this need not be the case, as in the formation of **56**.[777]

In this case, the sulfenic acid (**56**) is unstable[778] and the actual products isolated were the corresponding sulfinic acid (RSO_2H) and disulfide (R_2S_2).

[773]Bonvicino, G.E.; Yogodzinski, L.H.; Hardy Jr., R.A. *J. Org. Chem.* **1962**, *27*, 4272; Nodiff, E.H.; Hausman, M. *J. Org. Chem.* **1964**, *29*, 2453; Grundon, M.F.; Matier, W.L. *J. Chem. Soc., B*, **1966**, 266; Schmidt, D.M.; Bonvicino, G.E. *J. Org. Chem.* **1984**, *49*, 1664.

[774]For a review for the case of $X = SO_2$, see Cerfontain, H. *Mechanistic Aspects in Aromatic Sulfonation and Desulfonation*; Wiley, NY, **1968**, pp. 262–274.

[775]Truce, W.E.; Robbins, C.R.; Kreider, E.M. *J. Am. Chem. Soc.* **1966**, *88*, 4027; Drozd, V.N.; Nikonova, L.A. *J. Org. Chem, USSR* **1969**, *5*, 313.

[776]Bunnett, J.F.; Okamoto, T. *J. Am. Chem. Soc.* **1956**, *78*, 5363.

[777]Kent, B.A.; Smiles, S. *J. Chem. Soc.* **1934**, 422.

[778]For a stable sulfenic acid, see Nakamura, N. *J. Am. Chem. Soc.* **1983**, *105*, 7172.

In the Smiles rearrangement, the nucleophile Y is most often the conjugate base of SH, SO$_2$NHR, SO$_2$NH$_2$, NH$_2$, NHR, OH, OR. There are few examples where Y is a carbanion, and the most common example is probably the *Truce–Smiles rearrangement*, where L—YH is an *o*-tolyl group.[779] The prototypical Truce–Smiles rearrangement requires use of a strong base to form the benzylic carbanion that undergoes the rearrangement. When sulfone **57** was treated with butyllithium, for example, deprotonation led to the benzylic lithium compound **58**. Truce–Smiles rearrangement led to **59**, and hydrolysis gave the sulfinic acid, **60**.[779] Truce–Smiles rearrangements with stabilized benzylic carbanions are known,[780] and rearrangements of carbanions in general fall under this category.[781] Relatively few examples have been reported, however.[782] Truce–Smiles rearrangements of sulfones that proceed through a six-membered transition state have been reported.[783] In another example, displacement of an activated aryl fluoride with *o*-hydroxyacetophenone gave a product that was *C*-arylated adjacent to the ketone.[784]

[779]Truce, W.E.; Ray Jr., W.J.; Norman, O.L.; Eickemeyer, D.B. *J. Am. Chem. Soc.* **1958**, *80*, 3625.
[780]Erickson, W.R.; McKennon, M.J. *Tetrahedron Lett.* **2000**, *41*, 4541.
[781]Fukazawa, Y.; Kato, N.; Itô, S. *Tetrahedron Lett.* **1982**, *23*, 437.
[782]Hirota, T.; Tomita, K.; Sasaki, K.; Okuda, K.; Yoshida, M.; Kashino, S. *Heterocycles* **2001**, *55*, 741; Bayne, D.W.; Nicol, A.J.; Tennant, G.*J. Chem. Soc. Chem. Commun.* **1975**, 782; Hoffman, R.V.; Jankowski, B.C.; Carr, C.S.; Duesler, E.N. *J. Org. Chem.* **1986**, *51*, 130.
[783]Truce, W.E.; Hampton, D.C. *J. Org. Chem.* **1963**, *28*, 2276.
[784]Mitchell, L.H.; Barvian, N.C. *Tetrahedron Lett.* **2004**, *45*, 5669.

CHAPTER 14

Substitution Reactions: Free Radicals

MECHANISMS

Free-Radical Mechanisms in General[1]

A free-radical process consists of at least two steps. The first step involves the *formation* of free radicals, usually by homolytic cleavage of bond, that is, a cleavage in which each fragment retains one electron:

$$A{-}B \longrightarrow A\cdot \; + \; B\cdot$$

This is called an *initiation* step. It may happen spontaneously or may be induced by heat[2] or light (see the discussion on p. 279), depending on the type of bond.[3] Peroxides, including hydrogen peroxide, dialkyl, diacyl, and alkyl acyl peroxides, and peroxyacids are the most common source of free radicals induced spontaneously or by heat, but other organic compounds with low-energy bonds, such as azo compounds, are also used. Molecules that are cleaved by light are most often chlorine, bromine, and various ketones (see Chapter 7). Radicals can also be formed

---

[1]For books on free-radical mechanisms, see Nonhebel, D.C.; Tedder, J.M.; Walton, J.C. *Radicals*, Cambridge University Press, Cambridge, *1979*; Nonhebel, D.C.; Walton. J.C. *Free-Radical Chemistry*, Cambridge University Press, London, *1974*; Huyser, E.S. *Free-Radical Chain Reactions*, Wiley, NY, *1970*; Pryor, W.A. *Free Radicals*, McGraw-Hill, NY, *1966*; For reviews, see Huyser, E.S., in McManus, S.P. *Organic Reactive Intermediates*, Academic Press, NY, *1973*, pp. 1–59. For monographs on the use of free-radical reactions in synthesis see Giese, B. *Radicals in Organic Synthesis, Formation of Carbon-Carbon Bonds*, Pergamon, Elmsford, NY, *1986*; Davies, D.I.; Parrott, M.J. *Free Radicals in Organic Synthesis*, Springer, NY, *1978*. For reviews, see Curran, D.P. *Synthesis 1988*, 417, 489; Ramaiah, M. *Tetrahedron 1987*, *43*, 3541.

[2]For a study of the thermolysis of free-radical initiators, see Engel, P.S.; Pan, L.; Ying, Y.; Alemany, L.B. *J. Am. Chem. Soc. 2001*, *123*, 3706.

[3]See Fokin, A.A.; Schreiner, P.R. *Chem. Rev. 2002*, *102*, 1551.

March's Advanced Organic Chemistry: Reactions, Mechanisms, and Structure, Sixth Edition, by Michael B. Smith and Jerry March
Copyright © 2007 John Wiley & Sons, Inc.

in another way, by a one-electron transfer (loss or gain), for example, $A^+ + e^- \rightarrow A\cdot$. One-electron transfers usually involve inorganic ions or electrochemical processes.

Dialkyl peroxides (ROOR) or alkyl hydroperoxides (ROOH) decompose to hydroxy radicals (HO•) or alkoxy radicals (RO•) when heated.[4] Cumene hydroperoxide (PhCMe$_2$OOH), bi-*tert*-butylperoxide (Me$_3$COOCMe$_3$),[5] and benzoyl peroxide [(PhCO)O$_2$] undergo homolytic cleavage at temperatures compatible with many organic reactions, allowing some control of the reaction, and they are reasonably soluble in organic solvents. In general, when a peroxide decomposes, the oxygen radical remains in a "cage" for $\sim 10^{-11}$ s before diffusing away. The radical can recombine (dimerize), or react with other molecules. Azo compounds, characterized by a —N=N— bond, are free-radical precursors that liberate nitrogen gas (N≡N) upon decomposition. azobis(isobutyronitrile) (AIBN, **1**) is a well-known example, which decomposes to give nitrogen and the cyano stabilized radical, **2**.[6] Homolytic dissociation of symmetrical diazo compounds may be stepwise.[7] A derivative has been developed that decomposes to initiate radical reactions at room temperature, 2,2′-azobis(2,4-dimethyl-4-methoxyvaleronitrile), **3**.[8] Water soluble azo compounds are known, and can be used as radical initiators.[9] Other sources of useful radicals are available. Alkyl hypochlorites (R—O—Cl) generate chlorine radicals (Cl•) and alkoxy radicals (RO•) when heated.[10] Heating N-alkoxydithiocarbamates is another useful source of alkoxy radicals, RO•.[11]

[4]For a table of approximate decomposition temperatures for several common peroxides, see Lazár, M.; Rychlý, J.; Klimo, V.; Pelikán, P.; Valko, L. *Free Radicals in Chemistry and Biology*, CRC Press, Washington, DC, *1989*, p. 12.

[5]Lazár, M.; Rychlý, J.; Klimo, V.; Pelikán, P.; Valko, L. *Free Radicals in Chemistry and Biology*, CRC Press, Washington, DC, *1989*, p. 13.

[6]Yoshino, K.; Ohkatsu, J.; Tsuruta, T. *Polym. J.* *1977*, 9, 275; von J. Hinz, A.; Oberlinner, A.; Rüchardt, C. *Tetrahedron Lett.* *1973*, 1975.

[7]Dannenberg, J.J.; Rocklin, D. *J. Org. Chem.* *1982*, 47, 4529. See also, Newman, Jr, R.C.; Lockyer Jr, G.D. *J. Am. Chem Soc.* *1983*, 105, 3982.

[8]Kita, Y.; Sano, A.; Yamaguchi, T.; Oka, M.; Gotanda, K.; Matsugi, M. *Tetrahedron Lett.* *1997*, 38, 3549.

[9]Yorimitsu, H.; Wakabayashi, K.; Shinokubo, H; Oshima, K. *Tetrahedron Lett.* *1999*, 40 , 519.

[10]Davies, D.I.; Parrott, M.J. *Free Radicals in Organic Synthesis*, Springer–Verlag, Berlin, *1978*, p. 9; Chattaway, F.D.; Baekeberg, O.G. *J. Chem. Soc. 1923*, 123, 2999.

[11]Kim, S.; Lim, C.J.; Song, S.-E.; Kang, H.-Y. *Synlett 2001*, 688.

Note that aldehydes can also be a source of acyl radicals (•C=O) via reaction with transition metal salts such as Mn(III) acetate or Fe(II) compounds.[12] Another useful variation employs imidoyl radicals as synthons for unstable aryl radicals.[13]

The second step involves the *destruction* of free radicals. This usually happens by a process opposite to the first, namely, a combination of two like or unlike radicals to form a new bond:[14]

$$A\bullet \ + \ B\bullet \ \longrightarrow \ A{-}B$$

This type of step is called *termination*, and it ends the reaction as far as these particular radicals are concerned.[15] However, it is not often that termination follows *directly* upon initiation. The reason is that most radicals are very reactive and will react with the first available species with which they come in contact. In the usual situation, in which the concentration of radicals is low, this is much more likely to be a molecule than another radical. When a radical (which has an odd number of electrons) reacts with a molecule (which has an even number), the total number of electrons in the products must be odd. The product in a particular step of this kind may be one particle, as in the addition of a radical to a π-bond, which in this case is

$$R\bullet \ + \ \underset{}{}C{=}C \ \longrightarrow \ \underset{}{}\overset{R}{\underset{}{-C-C\bullet}}$$

4

another free radical, **4**; or abstraction of an atom such as hydrogen to give two particles, R—H and the new radical R'•.

$$R\bullet \ + \ R'H \ \longrightarrow \ RH \ + \ R'\bullet$$

In this latter case, one particle must be a neutral molecule and one a free radical. In both of these examples, a *new radical is generated*. This type of step is called *propagation*, since the newly formed radical can now react with another molecule and produce another radical, and so on, until two radicals do meet each other and terminate the sequence. The process just described is called a *chain reaction*,[16] and there may be hundreds or thousands of propagation steps between an initiation and a termination. Two other types of propagation reactions do not involve a

[12]Davies, D.I.; Parrott, M.J. *Free Radicals in Organic Synthesis* Springer–Verlag, Berlin, *1978*, p. 69; Sosnovsky, G. *Free Radical Reactions in Preparative Organic Chemistry*, MacMillan, New York, *1964*; Vinogradov, M.G.; Nikishin, G.I. *Usp. Khim*, *1971*, *40*, 1960; Nikishin, G.I.; Vinogradov, M.G.; Il'ina, G.P. *Synthesis 1972*, 376; Nikishin, G.I.; Vinogradov, M.G.; Verenchikov, S.P.; Kostyukov, I.N.; Kereselidze, R.V. *J. Org. Chem, USSR 1972*, *8*, 539 (Engl, p. 544).

[13]Fujiwara, S.-i.; Matsuya, T.; Maeda, H.; Shin-ike, T.; Kambe, N.; Sonoda, N. *J. Org. Chem. 2001*, *66*, 2183.

[14]For a review of the stereochemistry of this type of combination reaction, see Porter, N.A.; Krebs, P.J. *Top. Stereochem. 1988*, *18*, 97.

[15]Another type of termination is disproportionation (see p. 280).

[16]For a discussion of radical chain reactions from a synthetic point of view, see Walling, C. *Tetrahedron 1985*, *41*, 3887.

molecule at all. These are (1) cleavage of a radical into, necessarily, a radical and a molecule and (2) rearrangement of one radical to another (see Chapter 18). When radicals are highly reactive, for example, alkyl radicals, chains are long, since reactions occur with many molecules; but with radicals of low reactivity, for example, aryl radicals, the radical may be unable to react with anything until it meets another radical, so that chains are short, or the reaction may be a nonchain process. In any particular chain process, there is usually a wide variety of propagation and termination steps. Because of this, these reactions lead to many products and are often difficult to treat kinetically.[17]

$$\text{R-CH}_2\bullet + \textit{n}\text{-Bu}_3\text{Sn-H} \longrightarrow \text{R-CH}_2\text{-H} + \textit{n}\text{-Bu}_3\text{Sn}\bullet$$

$$\textit{n}\text{-Bu}_3\text{Sn}\bullet + \textit{n}\text{-Bu}_3\text{Sn}\bullet \longrightarrow \textit{n}\text{-Bu}_3\text{Sn-Sn}\textit{n}\text{-Bu}_3$$

A useful variation of propagation and termination combines the two processes. When a carbon radical (R•) is generated in the presence of tributyltin hydride ($\textit{n}$-Bu$_3$SnH), a hydrogen atom is transferred to the radical to give R—H and a new radical, $\textit{n}$-Bu$_3$Sn•. The tin radical reacts with a second tin radical to give $\textit{n}$-Bu$_3$ Sn—Sn—$\textit{n}$-Bu$_3$. The net result is that the carbon radical is reduced to give the desired product and the tin dimer can be removed from the reaction. Tin hydride transfers a hydrogen atom in a chain propagation sequence that produces a new radical, but terminates the carbon radical sequence. Dimerization of the tin radical then terminates that radical process. Silanes, such as triethylsilane (Et$_3$SiH), has also been used as an effective radical reducing agent.[18] The rate constants for the reaction of both tributytin hydride and (Me$_3$Si)$_3$Si—H with acyl radical has been measured and the silane quenches the radical faster than the tin hydride.[19] bis(Tri-$\textit{n}$-butylstannyl)benzopinacolate has also been used as a thermal source of $\textit{n}$-Bu$_3$Sn•, used to mediate radical reactions.[20]

The following are some general characteristics of free-radical reactions:[21]

1. Reactions are fairly similar whether they are occurring in the vapor or liquid phase, though solvation of free radicals in solution does cause some differences.[22]

2. They are largely unaffected by the presence of acids or bases or by changes in the polarity of solvents, except that nonpolar solvents may suppress competing ionic reactions.

[17]For a discussion of the kinetic aspects of radical chain reactions, see Huyser, E.S. *Free-Radical Chain Reactions*, Wiley, NY, *1970*, pp. 39–65.

[18]Chatgilialoglu, C.; Ferreri, C.; Lucarini, M. *J. Org. Chem.* *1993*, *58*, 249.

[19]Chatgilialoglu, C.; Lucarini, M. *Tetrahedron Lett.* *1995*, *36*, 1299.

[20]Hart, D.J.; Krishnamurthy, R.; Pook, L.M.; Seely, F.L. *Tetrahedron Lett.* *1993*, *34*, 7819.

[21]See Beckwith, A.L.J. *Chem. Soc. Rev.* *1993*, *22*, 143 for a discussion of selectivity in radical reactions.

[22]For a discussion, see Mayo, F.R. *J. Am. Chem. Soc.* *1967*, *89*, 2654.

3. They are initiated or accelerated by typical free-radical sources, such as the peroxides, referred to, or by light. In the latter case, the concept of quantum yield applies (p. 349). Quantum yields can be quite high, for example, 1000, if each quantum generates a long chain, or low, in the case of nonchain processes.

4. Their rates are decreased or the reactions are suppressed entirely by substances that scavenge free radicals, for example, nitric oxide, molecular oxygen, or benzoquinone. These substances are called *inhibitors*.[23]

This chapter discusses free-radical substitution reactions. Free-radical additions to unsaturated compounds and rearrangements are discussed in Chapters 15 and 18, respectively. Fragmentation reactions are covered, in part, in Chapter 17. In addition, many of the oxidation–reduction reactions considered in Chapter 19 involve free-radical mechanisms. Several important types of free-radical reactions do not usually lead to reasonable yields of pure products and are not generally treated in this book. Among these are polymerizations and high-temperature pyrolyses.

Free-Radical Substitution Mechanisms[24]

In a free-radical substitution reaction

$$R–X \longrightarrow R–Y$$

there must first be a cleavage of the substrate RX so that R• radicals are produced. This can happen by a spontaneous cleavage

$$R–X \longrightarrow R• + X•$$

or it can be caused by light or heat, or, more often, there is no actual cleavage, but R• is produced by an *abstraction* of another atom, X but the radical W•.

$$R–X + W• \longrightarrow R• + W–X$$

The radical W• is produced by adding a compound, such as a peroxide, that spontaneously forms free radicals. Such a compound is called an *initiator* (see above). Once R• is formed, it can go to product in two ways, by another atom abstraction, such as the reaction with A—B to form R—A and a new radical B•.

$$R• + A–B \longrightarrow R–A + B•$$

Another reaction is coupling with another radical to form the neutral product R—Y.

$$R• + Y• \longrightarrow R–Y$$

---

[23]For a review of the action of inhibitors, see Denisov, E.T.; Khudyakov, I.V. *Chem. Rev.* **1987**, *87*, 1313.
[24]For a review, see Poutsma, M.L., in Kochi, J.K. *Free Radicals*, Vol. 2, Wiley, NY, **1973**, pp. 113–158.

In a reaction with a moderately long chain, much more of the product will be produced by abstraction (4) than by coupling (5). Cleavage steps like (2) have been called S_H1 (H for homolytic), and abstraction steps like (3) and (4) have been called S_H2; reactions can be classified as S_H1 or SH_2 on the basis of whether RX is converted to R by (2) or (3).[25] Most chain substitution mechanisms follow the pattern (3), (4), (3), (4)••• Chains are long and reactions go well where both (3) and (4) are energetically favored (no worse that slightly endothermic (see pp. 944, 959). The IUPAC designation of a chain reaction that follows the pattern (3),(4)••• is $A_rD_R + A_RD_r$ (R stands for radical).

With certain radicals the transition state in an abstraction reaction has some polar character. For example, consider the abstraction of hydrogen from the methyl group of toluene by a bromine atom. Since bromine is more electronegative than carbon, it is reasonable to assume that in the transition state there is a separation of charge, with a partial negative charge on the halogen and a partial positive charge on the carbon:

$$\overset{\delta+}{PhCH_2}\bullet\bullet\bullet\bullet\bullet\bullet\bullet\bullet\bullet\bullet\bullet\bullet H\bullet\bullet\bullet\bullet\bullet\bullet\bullet\bullet\bullet\bullet\bullet\bullet\bullet\bullet\overset{\delta-}{Br}$$

Evidence for the polar character of the transition state is that electron-withdrawing groups in the para position of toluene (which would destabilize a positive charge) decrease the rate of hydrogen abstraction by bromine while electron-donating groups increase it.[26] However, substituents have a smaller effect here ($\rho \sim -1.4$) than they do in reactions where a completely ionic intermediate is involved, for example, the S_N1 mechanism (see p. 487). Other evidence for polar transition states in radical abstraction reactions is mentioned on p. 948. For abstraction by radicals such as methyl or phenyl, polar effects are very small or completely absent. For example, rates of hydrogen abstraction from ring-substituted toluenes by the methyl radical were relatively unaffected by the presence of electron-donating or electron-withdrawing substituents.[27] Those radicals (e.g., Br•) that have a tendency to abstract electron-rich hydrogen atoms are called *electrophilic radicals*.

When the reaction step R—X → R• takes place at a chiral carbon, racemization is almost always observed because free radicals do not retain configuration. Exceptions to this rule are found at cyclopropyl substrates, where both inversion[28] and retention[29] of configuration have been reported, and in the reactions mentioned on p. 942. Enantioselective radical processes have been reviewed.[30]

[25]Eliel, E.L., in Newman, M.S. *Steric Effects in Organic Chemistry*, Wiley, NY, *1956*, pp. 142–143.

[26]For example, see Pearson, R.; Martin, J.C. *J. Am. Chem. Soc. 1963, 85*, 354, 3142; Kim, S.S.; Choi, S.Y.; Kang, C.H. *J. Am. Chem. Soc. 1985, 107*, 4234.

[27]For example, see Kalatzis, E.; Williams, G.H. *J. Chem. Soc. B 1966*, 1112; Pryor, W.A.; Tonellato, U.; Fuller, D.L.; Jumonville, S. *J. Org. Chem. 1969, 34*, 2018.

[28]Altman, L.J.; Nelson, B.W. *J. Am. Chem. Soc. 1969, 91*, 5163.

[29]Jacobus, J.; Pensak, D. *Chem. Commun. 1969*, 400.

[30]Sibi, M.P.; Manyem, S.; Zimmerman, J. *Chem. Rev. 2003, 103*, 3263.

Mechanisms at an Aromatic Substrate[31]

When R in reaction (1) is aromatic, the simple abstraction mechanism just discussed may be operating, especially in gas-phase reactions. However, mechanisms of this type cannot account for all reactions of aromatic substrates. In processes, such as the following (see **13-27**, **14-17**, and **14-18**):

$$Ar\bullet \ + \ ArH \longrightarrow Ar-Ar$$

which occur in solution, the coupling of two rings cannot be explained on the basis of a simple abstraction

$$Ar\bullet \ + \ ArH \longrightarrow Ar-Ar \ + \ H\bullet$$

since, as discussed on p. 944, abstraction of an entire group, such as phenyl, by a free radical is very unlikely. The products can be explained by a mechanism similar to that of electrophilic and nucleophilic aromatic substitution. In the first step, the radical attacks the ring in much the same way as would an electrophile or a nucleophile:

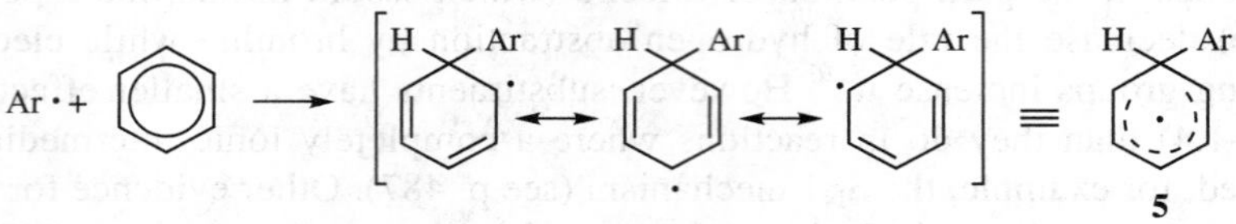

The intermediate radical **5** is relatively stable because of the resonance. The reaction can terminate in three ways: by simple coupling to give **6**, by disproportionation to give **7**,

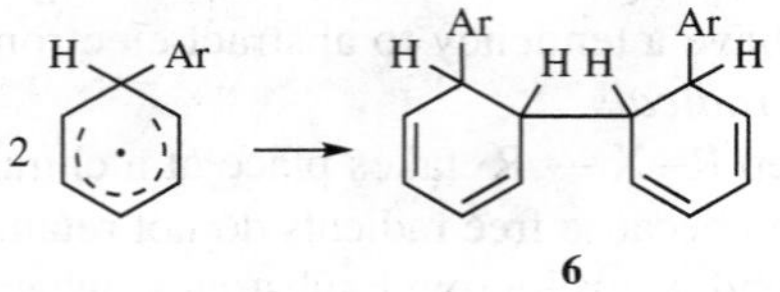

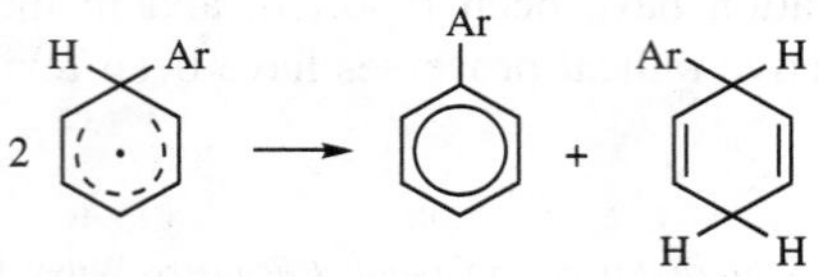

[31]For reviews, see Kobrina, L.S. *Russ. Chem. Rev.* **1977**, *46*, 348; Perkins, M.J., in Kochi, J.K. *Free Radicals*, Vol. 2, Wiley, NY, **1973**, pp. 231–271; Bolton, R.; Williams, G.H. *Adv. Free-Radical Chem.* **1975**, *5*, 1; Nonhebel, D.C.; Walton, J.C. *Free-Radical Chemistry*, Cambridge University Press, London, **1974**, pp. 417–469; Minisci, F.; Porta, O. *Adv. Heterocycl. Chem.* **1974**, *16*, 123; Bass, K.C.; Nababsing, P. *Adv. Free-Radical Chem.* **1972**, *4*, 1; Hey, D.H. *Bull. Soc. Chim. Fr.* **1968**, 1591.

or, if a species (R′•) is present that abstracts hydrogen, by abstraction to give **8**.[32]

8

Coupling product **6** is a partially hydrogenated quaterphenyl. Of course, the coupling need not be ortho–ortho, and other isomers can also be formed. Among the evidence for steps (9) and (10) was isolation of compounds of types **6** and **7**,[33] though normally under the reaction conditions dihydrobiphenyls like **7** are oxidized to the corresponding biphenyls. Other evidence for this mechanism is the detection of the intermediate **5** by CIDNP[34] and the absence of isotope effects, which would be expected if the rate-determining step were (7), which involves cleavage of the Ar—H bond. In the mechanism just given, the rate-determining step (8) does not involve loss of hydrogen. The reaction between aromatic rings and the HO• radical takes place by the same mechanism. Intramolecular hydrogen-transfer reactions of aryl radicals are known.[35] A similar mechanism has been shown for substitution at some vinylic[36] and acetylenic substrates, giving the substituted alkene **9**.[37] The kinetics of radical heterolysis reactions that form alkene radical cations has been studied.[38]

9

This is reminiscent of the nucleophilic tetrahedral mechanism at a vinylic carbon (p. 477).

There are a number of transition-metal mediated coupling reaction of aromatic substrates that probably proceed by radical coupling. It is also likely that many of these reactions do not proceed by free radicals, but rather by metal-mediated radicals or by ligand transfer on the metal. Reactions in these categories were presented

---

[32]Compound **5** can also be oxidized to the arene ArPh by atmospheric O_2. For a discussion of the mechanism of this oxidation, see Narita, N.; Tezuka, T. *J. Am. Chem. Soc.* **1982**, *104*, 7316.

[33]De Tar, D.F.; Long, R.A.J. *J. Am. Chem. Soc.* **1958**, *80*, 4742. See also, DeTar, D.F.; Long, R.A.J.; Rendleman, J.; Bradley, J.; Duncan, P. *J. Am. Chem. Soc.* **1967**, *89*, 4051; DeTar, D.F. *J. Am. Chem. Soc.* **1967**, *89*, 4058. See also, Jandu, K.S.; Nicolopoulou, M.; Perkins, M.J. *J. Chem. Res. (S)* **1985**, 88.

[34]Fahrenholtz, S.R.; Trozzolo, A.M. *J. Am. Chem. Soc.* **1972**, *94*, 282.

[35]Curran, D.P.; Fairweather, N. *J. Org. Chem.* **2003**, *68*, 2972.

[36]The reaction of vinyl chloride with Cl^- favors the σ-route (nucleophilic attack at the σ-bond) over the π-route (nucleophilic attack at the π-bond), but vinyl chloride is not an experimentally viable substrate and cannot be considered as representative for the vinyl S_N2 reaction. The π-route is anticipated in substituted vinylic halide reactions, where electron-withdrawing groups are attached to the vinylic carbon. See Bach, R. D.; Baboul, A. G.; Schlegel, H. B. *J. Am. Chem. Soc,* **2001**, *123*, 5787.

[37]Russell, G.A.; Ngoviwatchai, P. *Tetrahedron Lett.* **1986**, *27*, 3479, and references cited therein.

[38]Horner, J.H.; Bagnol, L.; Newcomb, M. *J. Am. Chem. Soc.* **2004**, *126*, 14979.

in Chapter 13 for convenient correlation with other displacement reactions of aryl halides, aryl diazonium salts, and so on.

Neighboring-Group Assistance in Free-Radical Reactions

In a few cases, it has been shown that cleavage steps (2) and abstraction steps (3) have been accelerated by the presence of neighboring groups. Photolytic halogenation (**14-1**) is a process that normally leads to mixtures of many products. However, bromination of carbon chains containing a bromine atom occurs with high regioselectivity. Bromination of alkyl bromides gave 84–94% substitution at the carbon adjacent to the bromine already in the molecule.[39] This result is especially surprising because, as we will see (p. 947), positions close to a polar group, such as bromine, should actually be *deactivated* by the electron-withdrawing field effect of the bromine. The unusual regioselectivity is explained by a mechanism in which abstraction (3) is assisted by a neighboring bromine atom, as in **10**.[40]

In the normal mechanism, Br• abstracts a hydrogen from RH, leaving R•. When a bromine is present in the proper position, it assists this process, giving a cyclic intermediate (a *bridged free radical*, **11**).[41] In the final step (very similar to R• + Br$_2$ → RBr + Br•), the ring is broken. If this mechanism is correct, the configuration at the substituted carbon (marked *) should be retained. This has been shown to be the case: optically active 1-bromo-2-methylbutane gave 1,2-dibromo-2-methylbutane with retention of configuration.[40] Furthermore, when this reaction was carried out in the presence of DBr, the "recovered" 1-bromo-2-methylbutane was found to be deuterated in the 2 position, and its configuration was retained.[42] This is just what would be predicted if some of the **11** present abstracted D from DBr. There is evidence that Cl can form bridged radicals,[43]

[39]Thaler, W.A. *J. Am. Chem. Soc.* **1963**, *85*, 2607. See also, Traynham, J.G.; Hines, W.G. *J. Am. Chem. Soc.* **1968**, *90*, 5208; Ucciani, E.; Pierri, F.; Naudet, M. *Bull. Soc. Chim. Fr.* **1970**, 791; Hargis, J.H. *J. Org. Chem.* **1973**, *38*, 346.

[40]Skell, P.S.; Tuleen, D.L.; Readio, P.D. *J. Am. Chem. Soc.* **1963**, *85*, 2849. For other stereochemical evidence, see Huyser, E.S.; Feng, R.H.C. *J. Org. Chem.* **1971**, *36*, 731. For another explanation, see Lloyd, R.V.; Wood, D.E. *J. Am. Chem. Soc.* **1975**, *97*, 5986. Also see Cope, A.C.; Fenton, S.W. *J. Am. Chem. Soc.* **1951**, *73*, 1668.

[41]For a monograph, see Kaplan, L. *Bridged Free Radicals*, Marcel Dekker, NY, *1972*. For reviews, see Skell, P.S.; Traynham, J.G. *Acc. Chem. Res.* **1984**, *17*, 160; Skell, P.S.; Shea, K.J. in Kochi, J.K. *Free Radicals*, Vol. 2, Wiley, NY, *1973*, pp. 809–852.

[42]Shea, K.J.; Skell, P.S. *J. Am. Chem. Soc.* **1973**, *95*, 283.

[43]Everly, C.R.; Schweinsberg, F.; Traynham, J.G. *J. Am. Chem. Soc.* **1978**, *100*, 1200; Wells, P.R.; Franke, F.P. *Tetrahedron Lett.* **1979**, 4681.

though ESR spectra show that the bridging is not necessarily symmetrical.[44] Still more evidence for bridging by Br has been found in isotope effect and other studies.[45] However, evidence from CIDNP shows that the methylene protons of the β-bromoethyl radical are not equivalent, at least while the radical is present in the radical pair [PhCOO··CH$_2$CH$_2$Br] within a solvent cage.[46] This evidence indicates that under these conditions BrCH$_2$CH$_2$• is not a symmetrically bridged radical, but it could be unsymmetrically bridged. A bridged intermediate has also been invoked, when a bromo group is in the proper position, in the Hunsdiecker reaction[47] (**14-30**), and in abstraction of iodine atoms by the phenyl radical.[48] Participation by other neighboring groups (e.g. SR, SiR$_3$, SnR$_3$) has also been reported.[49]

REACTIVITY

Reactivity for Aliphatic Substrates[50]

In a chain reaction, the step that determines what the product will be is most often an abstraction step. What is abstracted by a free radical is almost never a tetra-[51] or tervalent atom[52] (except in strained systems, see p. 1027)[53] and seldom a divalent one.[54] Nearly always it is univalent, and so, for organic compounds, it is hydrogen or halogen. For example, a reaction between a chlorine atom and ethane gives an

[44]Bowles, A.J.; Hudson, A.; Jackson, R.A. *Chem. Phys. Lett.* **1970**, *5*, 552; Cooper, J.; Hudson, A.; Jackson, R.A. *Tetrahedron Lett.* **1973**, 831; Chen, K.S.; Elson, I.H.; Kochi, J.K. *J. Am. Chem. Soc.* **1973**, *95*, 5341.
[45]Skell, P.S.; Pavlis, R.R.; Lewis, D.C.; Shea, K.J. *J. Am. Chem. Soc.* **1973**, *95*, 6735; Juneja, P.S.; Hodnett, E.M. *J. Am. Chem. Soc.* **1967**, *89*, 5685; Lewis, E.S.; Kozuka, S. *J. Am. Chem. Soc.* **1973**, *95*, 282; Cain, E.N.; Solly, R.K. *J. Chem. Soc., Chem. Commun.* **1974**, 148; Chenier, J.H.B.; Tremblay, J.P.; Howard, J.A. *J. Am. Chem. Soc.* **1975**, *97*, 1618; Howard, J.A.; Chenier, J.H.B.; Holden, D.A. *Can. J. Chem.* **1977**, *55*, 1463. See, however, Tanner, D.D.; Blackburn, E.V.; Kosugi, Y.; Ruo, T.C.S. *J. Am. Chem. Soc.* **1977**, *99*, 2714.
[46]Hargis, J.H.; Shevlin, P.B. *J. Chem. Soc., Chem. Commun.* **1973**, 179.
[47]Applequist, D.E.; Werner, N.D. *J. Org. Chem.* **1963**, *28*, 48.
[48]Danen, W.C.; Winter, R.L. *J. Am. Chem. Soc.* **1971**, *93*, 716.
[49]Tuleen, D.L.; Bentrude, W.G.; Martin, J.C. *J. Am. Chem. Soc.* **1963**, *85*, 1938; Fisher, T.H.; Martin, J.C. *J. Am. Chem. Soc.* **1966**, *88*, 3382; Jackson, R.A.; Ingold, K.U.; Griller, D.; Nazran, A.S. *J. Am. Chem. Soc.* **1985**, *107*, 208. For a review of neighboring-group participation in cleavage reactions, especially those involving SiR$_3$ as a neighboring group, see Reetz, M.T. *Angew. Chem. Int. Ed.* **1979**, *18*, 173.
[50]For a review of the factors involved in reactivity and regioselectivity in free-radical substitutions and additions, see Tedder, J.M. *Angew. Chem. Int. Ed.* **1982**, *21*, 401.
[51]Abstraction of a tetravalent carbon has been seen in the gas phase in abstraction by F• of R from RCl: Firouzbakht, M.L.; Ferrieri, R.A.; Wolf, A.P.; Rack, E.P. *J. Am. Chem. Soc.* **1987**, *109*, 2213.
[52]See, for example, Back, R.A. *Can. J. Chem.* **1983**, *61*, 916.
[53]For an example of an abstraction occurring to a small extent at an unstrained carbon atom, see Jackson, R.A.; Townson, M. *J. Chem. Soc. Perkin Trans. 2* **1980**, 1452. See also, Johnson, M.D. *Acc. Chem. Res.* **1983**, *16*, 343.
[54]For a monograph on abstractions of divalent and higher valent atoms, see Ingold, K.U.; Roberts, B.P. *Free-Radical Substitution Reactions*, Wiley, NY, **1971**.

ethyl radical, not a hydrogen atom:

$$H-Cl + CH_3CH_2 \cdot \qquad \Delta H = -3 \text{ kcal mol}^{-1},\ -13 \text{ kJ mol}^{-1}$$

$$CH_3CH_3 + Cl \cdot$$

$$CH_3CH_2-Cl + H \cdot \qquad \Delta H = +18 \text{ kcal mol}^{-1},\ +76 \text{ kJ mol}^{-1}$$

The principal reason for this is steric. A univalent atom is much more exposed to attack by the incoming radical than an atom with a higher valence. Another reason is that in many cases abstraction of a univalent atom is energetically more favored. For example, in the reaction given above, a C_2H_5—H bond is broken ($D = 100 \text{ kcal mol}^{-1}$, 419 kJ mol^{-1}, from Table 5.3) whichever pathway is taken, but in the former case an H—Cl bond is formed ($D = 103 \text{ kcal mol}^{-1}$, 432 kJ mol^{-1}) while in the latter case it is a C_2H_5—Cl bond ($D = 82 \text{ kcal mol}^{-1}$, 343 kJ mol^{-1}). Thus the first reaction is favored because it is exothermic by 3 kcal mol^{-1} (100–103) [13 kJ mol^{-1} (419–432)], while the latter is endothermic by 18 kcal mol^{-1} (100–82) [76 kJ mol^{-1} (419–343)].[55] However, the steric reason is clearly more important, because even in cases where ΔH is not very different for the two possibilities, the univalent atom is chosen.[56] *Ab initio* studies have probed the transition structures for radical hydrogen abstractions.[57]

Most studies of aliphatic reactivity have been made with hydrogen as the leaving atom and chlorine atoms as the abstracting species.[58] In these reactions, every hydrogen in the substrate is potentially replaceable and mixtures are usually obtained. However, the abstracting radical is not totally unselective, and some positions on a molecule lose hydrogen more easily than others. *Ab initio* studies have studied the factors controlling hydrogen abstraction by radicals.[59] For hydrogen abstraction by the *tert*-butoxy radical (*t*-Bu—O•) the factors that influence rate in their order of importance are structure of the radical > substituent effects[60] > solvent effects.[61] We discuss the position of attack under several headings:[62]

---

[55]The parameter ΔH for a free-radical abstraction reaction can be regarded simply as the difference in D values for the bond being broken and the one formed.

[56]Giese, B.; Hartung, J. *Chem. Ber.* **1992**, *125*, 1777.

[57]Eksterowicz, J.E.; Houk, K.N. *Tetrahedron Lett.* **1993**, *34*, 427; Damm, W.; Dickhaut, J.; Wetterich, F.; Giese, B. *Tetrahedron Lett.* **1993**, *34*, 431.

[58]For a review that lists many rate constants for abstraction of hydrogen at various positions of many molecules, see Hendry, D.G.; Mill, T.; Piszkiewicz, L.; Howard, J.A.; Eigenmann, H.K. *J. Phys. Chem. Ref. Data* **1974**, *3*, 937; Roberts, B.P.; Steel, A.J. *Tetrahedron Lett.* **1993**, *34*, 5167. See Tanko, J.M.; Blackert, J.F. *J. Chem. Soc. Perkin Trans. 2* **1996**, 1775 for the absolute rate constants for abstraction of chlorine by alkyl radicals.

[59]Zavitsas, A.A. *J. Chem. Soc. Perkin Trans. 2* **1998**, 499; Roberts, B.P. *J. Chem. Soc. Perkin Trans. 2* **1996**, 2719.

[60]See Wen, Z.; Li, Z.; Shang, Z.; Cheng, J.-P. *J. Org. Chem.* **2001**, *66*, 1466.

[61]Kim, S.S.; Kim, S.Y.; Ryou, S.S.; Lee, C.S.; Yoo, K.H. *J. Org. Chem.* **1993**, *58*, 192.

[62]For reviews, see Tedder, J.M. *Tetrahedron* **1982**, *38*, 313; Kerr, J.A., in Bamford, C.H.; Tipper, C.F.H. *Comprehensive Chemical Kinetics*, Vol. 18, Elsevier, NY, **1976**, pp. 39–109; Russell, G.A., in Kochi, J.K. *Free Radicals*, Vol. 2, Wiley, NY, **1973**, pp. 275–331; Rüchardt, C. *Angew. Chem. Int. Ed.* **1970**, *9*, 830; Poutsma, M.L. *Methods Free-Radical Chem.* **1969**, *1*, 79; Davidson, R.S. *Q. Rev. Chem. Soc.* **1967**, *21*, 249; Pryor, W.A.; Fuller, D.L.; Stanley, J.P. *J. Am. Chem. Soc.* **1972**, *94*, 1632.

**TABLE 14.1. Relative Susceptibility to Attack by Cl•
of Primary, Secondary, and Tertiary Positions at 100 and
600°C in the Gas Phase[63]**

Temperature, °C	Primary	Secondary	Tertiary
100	1	4.3	7.0
600	1	2.1	2.6

1. *Alkanes.* The tertiary hydrogens of an alkane are the ones preferentially abstracted by almost any radical, with secondary hydrogens being next preferred. This is in the same order as D values for these types of C—H bonds (Table 5.3). The extent of the preference depends on the selectivity of the abstracting radical and on the temperature. Table 14.1 shows[63] that at high temperatures selectivity decreases, as might be expected.[64] An example of the effect of radical selectivity may be noted in a comparison of fluorine atoms with bromine atoms. For the former, the ratio of primary to tertiary abstraction (of hydrogen) is 1:1.4, while for the less reactive bromine atom this ratio is 1:1600. With certain large radicals there is a steric factor that may change the selectivity pattern. For example, in the photochemical chlorination of isopentane in H_2SO_4 with *N*-chloro-di-*tert*-butylamine and *N*-chloro-*tert*-butyl-*tert*-pentylamine, the primary hydrogens are abstracted 1.7 times *faster* than the tertiary hydrogen.[65] In this case, the attacking radicals (the radical ions $R_2NH•^+$, see p. 958) are bulky enough for steric hindrance to become a major factor.

12

Cyclopropylcarbinyl radicals (**12**) are alkyl radicals, but they undergo rapid ring opening to give butenyl radicals.[66] The rate constant for this process has been measured by picosecond radical kinetic techniques to be in the range of $10^7\,M^{-1}\,s^{-1}$ for the parent[67] to $10^{10}\,M^{-1}\,s^{-1}$ for substituted derivatives.[68] Cyclobutylcarbinyl radicals undergo the cyclobutylcarbinyl to

[63]Hass, H.B.; McBee, E.T.; Weber, P. *Ind. Eng. Chem.* **1936**, *28*, 333.

[64]For a similar result with phenyl radicals, see Kopinke, F.; Zimmermann, G.; Anders, K. *J. Org. Chem.* **1989**, *54*, 3571.

[65]Deno, N.C.; Fishbein, R.; Wyckoff, J.C. *J. Am. Chem. Soc.* **1971**, *93*, 2065. Similar steric effects, though not a reversal of primary-tertiary reactivity, were found by Dneprovskii, A.N.; Mil'tsov, S.A. *J. Org. Chem. USSR* **1988**, *24*, 1836.

[66]Nonhebel, D.C. *Chem. Soc. Rev.* **1993**, *22*, 347.

[67]Engel, P.S.; He, S.-L.; Banks, J.T.; Ingold, K.U.; Lusztyk, J. *J. Org. Chem.* **1997**, *62*, 1210.

[68]Choi, S.-Y.; Newcomb, M. *Tetrahedron* **1995**, *51*, 657; Choi, S.-Y.; Toy, P.H.; Newcomb, M. *J. Org. Chem.* **1998**, *63*, 8609. See Martinez, F.N.; Schlegel, H.B.; Newcomb, M. *J. Org. Chem.* **1996**, *61*, 8547; **1998**, *63*, 3618 for *ab initio* studies to determine rate constants.

4-pentenyl radical process,[69] but examples are generally limited to the parent system and phenyl-substituted derivatives.[70] Cyclization of the 4-pentenyl radical is usually limited to systems where a stabilized radical can be formed.[71] The effect of substituents has been studied.[72] This process has been observed in bicyclo[4.1.0]heptan-4-ones.[73]

The rate of the ring-opening reaction of **5**,[74] and other substrates have been determined using an indirect method for the calibration[75] of fast radical reactions, applicable for radicals with lifetimes as short as 1 ps.[76] This 'radical clock'[77] method is based on the use of Barton's use of pyridine-2-thione-N-oxycarbonyl esters as radical precursors and radical trapping by the highly reactive thiophenol and benzeneselenol.[78] A number of radical clock substrates are known.[79] Other radical clock processes include: racemization of radicals with chiral conformations,[80] one-carbon ring expansion in cyclopentanones,[81] norcarane and spiro[2,5]octane,[82] α- and β-thujone radical rearrangements,[83] and cyclopropylcarbinyl radicals or alkoxycarbonyl radicals containing stabilizing substituents.[84]

[69]For a triplet radical in electron transfer cycloreversion of a cyclobutane, see Miranda, M.A.; Izquierdo, M.A.; Galindo, F. *J. Org. Chem.* **2002**, *67*, 4138.

[70]Beckwith, A.L.J.; Moad, G. *J. Chem. Soc, Perkin Trans. 2* **1980**, 1083; Ingold, K.U.; Maillard, B.; Walton, J.C. *J. Chem. Soc, Perkin Trans. 2* **1981**, 970; Walton, J.C. *J. Chem. Soc, Perkin Trans. 2* **1989**, 173; Choi, S.-Y.; Horner, J.H.; Newcomb, M. *J. Org. Chem.* **2000**, *65*, 4447; Newcomb, M.; Horner, J.H.; Emanuel, C.J. *J. Am. Chem. Soc.* **1997**, *119*, 7147.

[71]Clark, A.J.; Peacock, J.L. *Tetrahedron Lett.* **1998**, *39*, 1265; Cerreti, A.; D'Annibale, A.; Trogolo, C.; Umani, F. *Tetrahedron Lett.* **2000**, *41*, 3261; Ishibashi, H.; Higuchi, M.; Ohba, M.; Ikeda, M. *Tetrahedron Lett.* **1998**, *39*, 75; Ishibashi, H.; Nakamura, N.; Sato, S.; Takeuchi, M.; Ikeda, M. *Tetrahedron Lett.* **1991**, *32*, 1725; Ogura, K.; Sumitani, N.; Kayano, A.; Iguchi, H.; Fujita, M. *Chem. Lett.* **1992**, 1487.

[72]Baker, J.M.; Dolbier Jr, W.R. *J. Org. Chem.* **2001**, *66*, 2662.

[73]Kirschberg, T.; Mattay, J. *Tetrahedron Lett.* **1994**, *35*, 7217.

[74]Mathew, L.; Warkentin, J. *J. Am. Chem. Soc.* **1986**, *108*, 7981; For an article clocking tertiary cyclopropylcarbinyl radical rearrangements, see Engel, P.S.; He, S.-L.; Banks, J.T.; Ingold, K.U.; Lusztyk, J. *J. Org. Chem.* **1997**, *62*, 1212, 5656.

[75]See Hollis, R.; Hughes, L.; Bowry, V.W.; Ingold, K.U. *J. Org. Chem.* **1992**, *57*, 4284.

[76]Newcomb, M.; Toy, P.H. *Acc. Chem. Res.* **2000**, *33*, 449. See Horn, A.H.C.; Clark, T. *J. Am. Chem. Soc.* **2003**, *125*, 2809.

[77]For a review, see Griller, D.; Ingold, K.U. *Acc. Chem. Res.* **1980**, *13*, 317.

[78]Newcomb, M.; Park, S.-U. *J. Am. Chem. Soc.* **1986**, *108*, 4132; Newcomb, M.; Glenn, A.G. *J. Am. Chem. Soc.* **1989**, *111*, 275; Newcomb, M.; Johnson, C.C.; Manek, M.B.; Varick, T.R. *J. Am. Chem. Soc.* **1992**, *114*, 10915; Newcomb, M.; Varick, T.R.; Ha, C.; Manek, M.B.; Yue, X. *J. Am. Chem. Soc.* **1992**, *114*, 8158.

[79]See Kumar, D.; de Visser, S.P.; Sharma, P.K.; Cohen, S.; Shaik, S. *J. Am. Chem. Soc.* **2004**, *126*, 1907.

[80]Buckmelter, A.J.; Kim, A.I.; Rychnovsky, S.D. *J. Am. Chem. Soc.* **2000**, *122*, 9386; Rychnovsky, S.D.; Hata, T.; Kim, A.I.; Buckmelter, A.J. *Org. Lett.* **2001**, *3*, 807.

[81]Chatgilialoglu, C.; Timokhin, V. I.; Ballestri, M. *J. Org. Chem.* **1998**, *63*, 1327.

[82]For an application and leading references, see Auclair, K.; Hu, Z.; Little, D. M.; Ortiz de Montellano, P. R.; Groves, J. T. *J. Am. Chem. Soc.* **2002**, *124*, 6020.

[83]He, X.; Ortiz de Montellano, P. R. *J. Org. Chem.* **2004**, *69*, 5684.

[84]Beckwith, A.L.J.; Bowry, V.W. *J. Am. Chem. Soc.* **1994**, *116*, 2710. See Cooksy, A.L.; King, H.F.; Richardson, W.H. *J. Org. Chem.* **2003**, *68*, 9441.

2. *Alkenes.* When the substrate molecule contains a double bond, treatment with chlorine or bromine usually leads to addition rather than substitution. However, for other radicals (and even for chlorine or bromine atoms when they do abstract a hydrogen) the position of attack is perfectly clear. Vinylic hydrogens are practically never abstracted, and allylic hydrogens are greatly preferred to other positions of the molecule. Allylic hydrogen abstraction from a cyclic alkenes is usually faster than abstraction from an acyclic alkene.[85] This is generally attributed[86] to resonance stabilization of the allylic radical, **13**. As might be expected, allylic rearrangements (see p. 469) are common in these cases.[87]

3. *Alkyl Side Chains of Aromatic Rings.* The preferential position of attack on a side chain is usually the one to the ring. Both for active radicals, such as chlorine and phenyl, and for more selective ones, such as bromine, such attack is faster than that at a primary carbon, but for the active radicals benzylic attack is slower than for tertiary positions, while for the selective ones it is faster. Two or three aryl groups on a carbon activate its hydrogens even more, as would be expected from the resonance involved. These statements can be illustrated by the following abstraction ratios:[88]

	Me–H	MeCH$_2$–H	Me$_2$CH–H	Me$_3$C–H	PhCH$_2$–H	Ph$_2$CH–H	Ph$_3$C–H
Br	0.0007	1	220	19,400	64,000	1.1×10^6	6.4×10^6
Cl	0.004	1	4.3	6.0	1.3	2.6	9.5

However, many anomalous results have been reported for these substrates. The benzylic position is not always the most favored. One thing certain is that *aromatic* hydrogens are seldom abstracted if there are aliphatic ones to compete (note from Table 5.3, that D for Ph–H is higher than that for any alkyl H bond). Several $\sigma \bullet$ scales (similar to the σ, σ^+, and σ^- scales discussed in Chapter 9) have been developed for benzylic radicals.[89]

[85]Rothenberg, G.; Sasson, Y. *Tetrahedron* **1998**, *54*, 5417.
[86]See however Kwart, H.; Brechbiel, M.; Miles, W.; Kwart, L.D. *J. Org. Chem.* **1982**, *47*, 4524.
[87]For reviews, see Wilt, J.W., in Kochi, J.K. *Free Radicals*, Vol. 1, Wiley, NY, *1973*, pp. 458–466.
[88]Russell, G.A., in Kochi, J.K. *Free Radicals*, Vol. 2, Wiley, NY, *1973*, p. 289.
[89]See, for example, Dinçtürk, S.; Jackson, R.A. *J. Chem. Soc. Perkin Trans. 2 1981*, 1127; Dust, J.M.; Arnold, D.R. *J. Am. Chem. Soc.* **1983**, *105*, 1221, 6531; Creary, X.; Mehrsheikh-Mohammadi, M.E.; McDonald, S. *J. Org. Chem.* **1987**, *52*, 3254; *1989*, *54*, 2904; Fisher, T.H.; Dershem, S.M.; Prewitt, M.L. *J. Org. Chem.* **1990**, *55*, 1040.

4. *Compounds Containing Electron-Withdrawing Substituents.* In halogenations, electron-withdrawing groups greatly deactivate adjacent positions. Compounds of the type Z–CH$_2$–CH$_3$ are attacked predominantly or exclusively at the β position when Z is COOH, COCl, COOR, SO$_2$Cl, or CX$_3$. Such compounds as acetic acid and acetyl chloride are not attacked at all. This is in sharp contrast to electrophilic halogenations (**12-4–12-6**), where *only* the α position is substituted. This deactivation of α positions is also at variance with the expected stability of the resulting radicals, since they would be expected to be stabilized by resonance similar to that for allylic and benzylic radicals. This behavior is a result of the polar transition states discussed on p. 939. Halogen atoms are electrophilic radicals and look for positions of high electron density. Hydrogens on carbon atoms next to electron-withdrawing groups have low electron densities (because of the field effect of Z) and are therefore shunned. Radicals that are not electrophilic do not display this behavior. For example, the methyl radical is essentially nonpolar and does not avoid positions next to electron-withdrawing groups; relative rates of abstraction at the α and β carbons of propionic acid are:[90]

	CH$_3$–CH$_2$–COOH	
Me•	1	7.8
Cl•	1	0.02

It is possible to generate radicals adjacent to electron-withdrawing groups. Radical **14** can be generated and it undergoes coupling reactions with little selectivity. When **15** is generated, however, it rapidly disproportionates rather than couples, giving the corresponding alkene and alkane.[91] Such radicals have also been shown to have a conformational preference for orientation of the orbital containing the single electron. In such cases, hydrogen abstraction proceeds with good stereoselectivity.[92]

| **14** | **15** |

Some radicals, for example, *tert*-butyl,[93] benzyl,[94] and cyclopropyl,[95] are *nucleophilic* (they tend to abstract electron-poor hydrogen atoms). The

[90]Russell, G.A., in Kochi, J.K. *Free Radicals*, Vol. 2, Wiley, NY, *1973*, p. 311.

[91]Porter, N.A.; Rosenstein, I.J. *Tetrahedron Lett. 1993*, *34*, 7865.

[92]Giese, B.; Damm, W.; Wetterich, F.; Zeitz, H.-G. *Tetrahedron Lett. 1992*, *33*, 1863.

[93]Pryor, W.A.; Tang, F.Y.; Tang, R.H.; Church, D.F. *J. Am. Chem. Soc. 1982*, *104*, 2885; Dütsch, H.R.; Fischer, H. *Int. J. Chem. Kinet. 1982*, *14*, 195.

[94]Clerici, A.; Minisci, F.; Porta, O. *Tetrahedron 1973*, *29*, 2775.

[95]Stefani, A.; Chuang, L.; Todd, H.E. *J. Am. Chem. Soc. 1970*, *92*, 4168.

phenyl radical appears to have a very small degree of nucleophilic character.[96] For longer chains, the field effect continues, and the β position is also deactivated to attack by halogen, though much less so than the α position. We have already mentioned (p. 939) that abstraction of an α hydrogen atom from ring-substituted toluenes can be correlated by the Hammett equation.

5. *Stereoelectronic Effects.* On p. 1258, we will see an example of a stereoelectronic effect. It has been shown that such effects are important where a hydrogen is abstracted from a carbon adjacent to a C—O or C—N bond. In such cases, hydrogen is abstracted from C—H bonds that have a relatively small dihedral angle (~30°) with the unshared orbitals of the O or N much more easily than from those with a large angle (~90°). For example, the starred hydrogen of **16** was abstracted ~8 times faster than the starred hydrogen of **17**.[97]

16   **17**

The presence of an OR or SiR_3 substituent β- to the carbon bearing the radical accelerates the rate of halogen abstraction.[98]

Abstraction of a halogen has been studied much less,[99] but the order of reactivity is $RI > RBr > RCl \gg RF$.

There are now many cases where free-radical reactions are promoted by transition metals.[100]

Reactivity at a Bridgehead[101]

Many free-radical reactions have been observed at bridgehead carbons, as in formation of bromide **18** (see **14-30**),[102] demonstrating that the free radical need not be planar. However, treatment of norbornane with sulfuryl chloride and benzoyl

---

[96]Suehiro, T.; Suzuki, A.; Tsuchida, Y.; Yamazaki, J. *Bull. Chem. Soc. Jpn.* **1977**, *50*, 3324.

[97]Hayday, K.; McKelvey, R.D. *J. Org. Chem.* **1976**, *41*, 2222. For additional examples, see Malatesta, V.; Ingold, K.U. *J. Am. Chem. Soc.* **1981**, *103*, 609; Beckwith, A.L.J.; Easton, C.J. *J. Am. Chem. Soc.* **1981**, *103*, 615; Beckwith, A.L.J.; Westwood, S.W. *Aust. J. Chem.* **1983**, *36*, 2123; Griller, D.; Howard, J.A.; Marriott, P.R.; Scaiano, J.C. *J. Am. Chem. Soc.* **1981**, *103*, 619. For a stereoselective abstraction step, see Dneprovskii, A.S.; Pertsikov, B.Z.; Temnikova, T.I. *J. Org. Chem. USSR* **1982**, *18*, 1951. See also, Bunce, N.J.; Cheung, H.K.Y.; Langshaw, J. *J. Org. Chem.* **1986**, *51*, 5421.

[98]Roberts, B.P.; Steel, A.J. *J. Chem. Soc. Perkin Trans. 2* **1994**, 2411.

[99]For a review, see Danen, W.C. *Methods Free-Radical Chem.* **1974**, *5*, 1.

[100]Iqbal, J.; Bhatia, B.; Nayyar, N.K. *Chem. Rev.* **1994**, *94*, 519. See Hasegawa, E.; Curran, D.P. *Tetrahedron Lett.* **1993**, *34*, 1717 for the rate of reaction for a primary akyl radical in the presence of SmI_2.

[101]For reviews, see Bingham, R.C.; Schleyer, P.v.R. *Fortschr. Chem. Forsch.* **1971**, *18*, 1, see pp. 79–81; Fort, Jr, R.C.; Schleyer, P.v.R. *Adv. Alicyclic Chem.* **1966**, *1*, 283, see p. 337.

[102]Grob, C.A.; Ohta, M.; Renk, E.; Weiss, A. *Helv. Chim. Acta* **1958**, *41*, 1191.

peroxide gave mostly 2-chloronorbornane, though the bridgehead position is tertiary.[103] So, while bridgehead free-radical substitution is possible, it is not preferred, presumably because of the strain involved.[104]

Reactivity in Aromatic Substrates

Free-radical substitution at an aromatic carbon seldom takes place by a mechanism in which a hydrogen is abstracted to give an aryl radical. Reactivity considerations here are similar to those in Chapters 11 and 13; that is, we need to know which position on the ring will be attacked to give the intermediate, **19**.

19

The obvious way to obtain this information is to carry out reactions with various Z groups and to analyze the products for percent ortho, meta, and para isomers, as has so often been done for electrophilic substitution. However, this procedure is much less accurate in the case of free-radical substitutions because of the many side reactions. It may be, for example, that in a given case the ortho position is more reactive than the para, but the intermediate from the para attack may go on to product while that from ortho attack gives a side reaction. In such a case, analysis of the three products does not give a true picture of which position is most susceptible to attack. The following generalizations can nevertheless be drawn, though there has been much controversy over just how meaningful such conclusions are[105]

1. All substituents increase reactivity at ortho and para positions over that of benzene. There is no great difference between electron-donating and electron-withdrawing groups.

2. Reactivity at meta positions is usually similar to that of benzene, perhaps slightly higher or lower. This fact, coupled with the preceding one, means that all substituents are activating and ortho–para directing; none are deactivating or (chiefly) meta directing.

[103]Roberts, J.D.; Urbanek, L.; Armstrong, R. *J. Am. Chem. Soc.* **1949**, *71*, 3049. See also, Kooyman, E.C.; Vegter, G.C. *Tetrahedron* **1958**, *4*, 382; Walling, C.; Mayahi, M.F. *J. Am. Chem. Soc.* **1959**, *81*, 1485.
[104]See, for example, Koch, V.R.; Gleicher, G.J. *J. Am. Chem. Soc.* **1971**, *93*, 1657.
[105]De Tar, D.F. *J. Am. Chem. Soc.* **1961**, *83*, 1014 (book review); Dickerman, S.C.; Vermont, G.B. *J. Am. Chem. Soc.* **1962**, *84*, 4150; Morrison, R.T.; Cazes, J.; Samkoff, N.; Howe, C.A. *J. Am. Chem. Soc.* **1962**, *84*, 4152; Ohta, H.; Tokumaru, K. *Bull. Chem. Soc. Jpn.* **1971**, *44*, 3218; Vidal, S.; Court, J.; Bonnier, J. *J. Chem. Soc. Perkin Trans. 2* **1973**, 2071; Tezuka, T.; Ichikawa, K.; Marusawa, H.; Narita, N. *Chem. Lett.* **1983**, 1013.

TABLE 14.2. Partial Rate Factors for Attack of Substituted Benzenes by Phenyl Radicals Generated from Bz_2O_2[108]

Z	Partial Rate Factor		
	o	*m*	*p*
H	1	1	1
NO_2	5.50	0.86	4.90
CH_3	4.70	1.24	3.55
CMe_3	0.70	1.64	1.81
Cl	3.90	1.65	2.12
Br	3.05	1.70	1.92
MeO	5.6	1.23	2.31

3. Reactivity at ortho positions is usually somewhat greater than at para positions, except where a large group decreases ortho reactivity for steric reasons.

4. In direct competition, electron-withdrawing groups exert a somewhat greater influence than electron-donating groups. Arylation of para-disubstituted compounds XC_6H_4Y showed that substitution ortho to the group X became increasingly preferred as the electron-withdrawing character of X increases (with Y held constant).[106] The increase could be correlated with the Hammett σ_p values for X.

5. Substituents have a much smaller effect than in electrophilic or nucleophilic substitution; hence the partial rate factors (see p. 677) are not great.[107] Partial rate factors for a few groups are given in Table 14.2.[108]

6. Although hydrogen is the leaving group in most free-radical aromatic substitutions, ipso attack (p. 671) and ipso substitution (e.g., with Br, NO_2, or CH_3CO as the leaving group) have been found in certain cases.[109]

Reactivity in the Attacking Radical[110]

We have already seen that some radicals are much more selective than others (p. 944). The bromine atom is so selective that when only primary hydrogens are available, as in neopentane or *tert*-butylbenzene, the reaction is slow or nonexistent; and isobutane can be selectively brominated to give *tert*-butyl bromide in high yields.

[106]Davies, D.I.; Hey, D.H.; Summers, B. *J. Chem. Soc. C* **1970**, 2653.

[107]For a quantitative treatment, see Charton, M.; Charton, B. *Bull. Soc. Chim. Fr.* **1988**, 199.

[108]Davies, D.I.; Hey, D.H.; Summers, B. *J. Chem. Soc. C* **1971**, 2681.

[109]For reviews, see Traynham, J.G. *J. Chem. Educ.* **1983**, *60*, 937; *Chem. Rev.* **1979**, *79*, 323; Tiecco, M. *Acc. Chem. Res.* **1980**, *13*, 51; *Pure Appl. Chem.* **1981**, *53*, 239.

[110]For reviews with respect to $CH_3\bullet$ and $CF_3\bullet$, see Trotman-Dickenson, A.F. *Adv. Free-Radical Chem.* **1965**, *1*, 1; Spirin, Yu.L. *Russ. Chem. Rev.* **1969**, *38*, 529; Gray, P.; Herod, A.A.; Jones, A. *Chem. Rev.* **1971**, *71*, 247.

TABLE 14.3. Some Common Free Radicals in Decreasing Order of Activity[a]

Radical	E		Radical	E	
	kcal mol^{-1}e	kJ mol^{-1}e		kcal mol^{-1}	kJ mol^{-1}
F•	0.3	1.3	H•	9.0	38
Cl•	1.0	4.2	Me•	11.8	49.4
MeO•	7.1	30	Br•	13.2	55.2
CF$_3$•	7.5	31			

[a]The E values represent activation energies for the reaction
$$X\bullet + C_2H_6 \longrightarrow X-H + C_2H_5\bullet \qquad \text{(Ref. 112)}$$
i-Pr• is less active than Me• and t-Bu• still less so.[113]

However, toluene reacts with bromine atoms instantly. Bromination of other alkyl-benzenes, for example, ethylbenzene and cumene, takes place exclusively at the a position,[111] emphasizing the selectivity of Br•. The dissociation energy D of the C—H bond is more important for radicals of low reactivity than for highly reactive radicals, since bond breaking in the transition state is greater. Thus, bromine shows a greater tendency than chlorine to attack α to an electron-withdrawing group because the energy of the C—H bond there is lower than in other places in the molecule.

Some radicals, for example, triphenylmethyl, are so unreactive that they abstract hydrogens very poorly if at all. Table 14.3 lists some common free radicals in approximate order of reactivity.[112]

It has been mentioned that some free radicals (e.g., chloro) are electro-philic and some (e.g., *tert*-butyl) are nucleophilic. It must be borne in mind that these tendencies are relatively slight compared with the electrophilicity of a positive ion or the nucleophilicity of a negative ion. The predominant character of a free radical is neutral, whether it has slight electrophilic or nucleophilic tendencies.

The Effect of Solvent on Reactivity[114]

As noted earlier, the solvent usually has little effect on free-radical substitutions in contrast to ionic ones: indeed, reactions in solution are often quite similar in character to those in the gas phase, where there is no solvent at all. However, in certain cases the solvent *can* make an appreciable difference. Chlorination of 2,3-dimethylbutane in aliphatic solvents gave about 60% $(CH_3)_2CHCH(CH_3)CH_2Cl$

[111]Huyser, E.S. *Free-Radical Chain Reactions*, Wiley, NY, *1970*, p. 97.
[112]Trotman-Dickenson, A.F. *Adv. Free-Radical Chem.* *1965*, *1*, 1.
[113]Kharasch, M.S.; Hambling, J.K.; Rudy, T.P. *J. Org. Chem.* *1959*, *24*, 303.
[114]For reviews, see Reichardt, C. *Solvent Effects in Organic Chemistry*; Verlag Chemie: Deerfield Beach, FL, *1979*, pp. 110–123; Martin, J.C., in Kochi, J.K. *Free Radicals*, Vol. 2, Wiley, NY, *1973*, pp. 493–524; Huyser, E.S. *Adv. Free-Radical Chem.* *1965*, *1*, 77.

and 40% $(CH_3)_2CHCCl(CH_3)_2$, while in aromatic solvents the ratio became ~10:90.[115] This result is attributed to complex formation between the aromatic solvent and the

20

chlorine atom that makes the chlorine more selective.[116] This type of effect is not found in cases where the differences in ability to abstract the atom are caused by field effects of electron-withdrawing groups (p. 948). In such cases, aromatic solvents make little difference.[117] The complex **20** has been detected[118] as a very short-lived species by observation of its visible spectrum in the pulse radiolysis of a solution of benzene in CCl_4.[119] Differences caused by solvents have also been reported in reactions of other radicals.[120] Some of the anomalous results obtained in the chlorination of aromatic side chains (p. 947) can also be explained by this type of complexing, in this case not with the solvent but with the reacting species.[121] Much smaller, though real, differences in selectivity have been found when the solvent in the chlorination of 2,3-dimethylbutane is changed from an alkane to CCl_4.[122] However, these differences are not caused by formation of a complex between Cl• and the solvent. There are cases,

[115]Russell, G.A. *J. Am. Chem. Soc.* **1958**, *80*, 4987, 4997, 5002; *J. Org. Chem.* **1959**, *24*, 300.

[116]See also, Soumillion, J.P.; Bruylants, A. *Bull. Soc. Chim. Belg.* **1969**, *78*, 425; Potter, A.; Tedder, J.M. *J. Chem. Soc. Perkin Trans. 2* **1982**, 1689; Aver'yanov, V.A.; Ruban, S.G.; Shvets,V.F. *J. Org. Chem. USSR* **1987**, *23*, 782; Aver'yanov, V.A.; Ruban, S.G. *J. Org. Chem. USSR* **1987**, *23*, 1119; Raner, K.D.; Lusztyk, J.; Ingold, K.U. *J. Am. Chem. Soc.* **1989**, *111*, 3652; Ingold, K.U.; Lusztyk, J.; Raner, K.D. *Acc. Chem. Res.* **1990**, *23*, 219.

[117]Russell, G.A. *Tetrahedron* **1960**, *8*, 101; Nagai, T.; Horikawa, Y.; Ryang, H.S.; Tokura, N. *Bull. Chem. Soc. Jpn.* **1971**, *44*, 2771.

[118]It has been contended that another species, a chlorocyclohexadienyl radical (the structure of which is the same as **5**, except that Cl replaces Ar), can also be attacking when the solvent is benzene: Skell, P.S.; Baxter III, H.N.; Taylor, C.K. *J. Am. Chem. Soc.* **1983**, *105*, 120; Skell, P.S.; Baxter III, H.N.; Tanko, J.M.; Chebolu, V. *J. Am. Chem. Soc.* **1986**, *108*, 6300. For arguments against this proposal, see Bunce, N.J.; Ingold, K.U.; Landers, J.P.; Lusztyk, J.; Scaiano, J.C. *J. Am. Chem. Soc.* **1985**, *107*, 5464; Walling, C. *J. Org. Chem.* **1988**, *53*, 305; Aver'yanov, V.A.; Shvets, V.F.; Semenov, A.O. *J. Org. Chem. USSR* **1990**, *26*, 1261.

[119]Bühler, R.E. *Helv. Chim. Acta* **1968**, *51*, 1558. For other spectral observations, see Raner, K.D.; Lusztyk, J.; Ingold, K.U. *J. Phys. Chem.* **1989**, *93*, 564.

[120]Walling, C.; Azar, J.C. *J. Org. Chem.* **1968**, *33*, 3885; Ito, O.; Matsuda, M. *J. Am. Chem. Soc.* **1982**, *104*, 568; Minisci, F.; Vismara, E.; Fontana, F.; Morini, G.; Serravalle, M.; Giordano, C. *J. Org. Chem.* **1987**, *52*, 730.

[121]Russell, G.A.; Ito, O.; Hendry, D.G. *J. Am. Chem. Soc.* **1963**, *85*, 2976; Corbiau, J.L.; Bruylants, A. *Bull. Soc. Chim. Belg.* **1970**, *79*, 203, 211; Newkirk, D.D.; Gleicher, G.J. *J. Am. Chem. Soc.* **1974**, *96*, 3543.

[122]See Raner, K.D.; Lusztyk, J.; Ingold, K.U. *J. Org. Chem.* **1988**, *53*, 5220.

however, where the rate of reaction for trapping a radical depends on the polarity of the solvent, particularly in water.[123]

REACTIONS

The reactions in this chapter are classified according to leaving group. The most common leaving groups are hydrogen and nitrogen (from the diazonium ion); these are considered first.

HYDROGEN AS LEAVING GROUP

A. Substitution by Halogen

14-1   Halogenation at an Alkyl Carbon[124]

Halogenation or **Halo-de-hydrogenation**

$$R-H \ + \ Cl_2 \ \xrightarrow{hv} \ R-Cl$$

Alkanes can be chlorinated or brominated by treatment with chlorine or bromine in the presence of visible or UV light.[125] These reactions require a radical chain initiator, light, or higher temperatures.[126] The reaction can also be applied to alkyl chains containing many functional groups. The chlorination reaction is usually not useful for preparative purposes precisely because it is so general: Not only does substitution take place at virtually every alkyl carbon in the molecule, but di- and polychloro substitution almost invariably occur even if there is a large molar ratio of substrate to halogen.

When functional groups are present, the principles are those outlined on p. 945; favored positions are those α to aromatic rings, while positions α to electron-withdrawing groups are least likely to be substituted. Tertiary carbons are most likely to be attacked and primary least. Positions α to an OR group are very readily attacked. Nevertheless, mixtures are nearly always obtained. This can be contrasted to the regioselectivity of electrophilic halogenation (**12-4–12-6**), which always takes place α to a carbonyl group (except when the reaction is catalyzed by $AgSbF_6$; see following). Of course, if a *mixture* of chlorides is wanted, the reaction is usually

[123]Tronche, C.; Martinez, F.N.; Horner, J.H.; Newcomb, M.; Senn, M.; Giese, B. *Tetrahedron Lett.* **1996**, *37*, 5845.

[124]For lists of reagents, with references, see Larock, R.C. *Comprehensive Organic Transformations*, 2nd ed, Wiley-VCH, NY, **1999**, pp. 611–617.

[125]For reviews, see Poutsma, M.L., in Kochi, J.K. *Free Radicals*, Vol. 2, Wiley, NY, **1973**, pp. 159–229; Huyser, E.S., in Patai, S. *The Chemistry of the Carbon-Halogen Bond*, pt. 1, Wiley, NY, **1973**, pp. 549–607; Poutsma, M.L. *Methods Free-Radical Chem.* **1969**, *1*, 79 (chlorination); Thaler, W.A. *Methods Free-Radical Chem.* **1969**, *2*, 121 (bromination).

[126]Hill, C.L. *Activation and Functionalization of Alkanes*, Wiley, NY, **1989**.

quite satisfactory. For obtaining pure compounds, the chlorination reaction is essentially limited to substrates with only one type of replaceable hydrogen (e.g., ethane, cyclohexane, neopentane). The most common are methylbenzenes and other substrates with methyl groups on aromatic rings, since few cases are known where halogen atoms substitute at an aromatic position.[127] Of course, ring substitution *does* take place in the presence of a positive-ion-forming catalyst (**11-10**). In addition to mixtures of various alkyl halides, traces of other products are obtained. These include H_2, alkenes, higher alkanes, lower alkanes, and halogen derivatives of these compounds. Solvent plays an important role in this process.[128]

The bromine atom is much more selective than the chlorine atom. As indicated on p. 952, it is often possible to brominate tertiary and benzylic positions selectively. High regioselectivity can also be obtained where the neighboring-group mechanism (p. 942) can operate.

As already mentioned, halogenation can be performed with chlorine or bromine. Fluorine has also been used,[129] but seldom, because it is too reactive and hard to control.[130] It often breaks carbon chains down into smaller units, a side reaction that sometimes becomes troublesome in chlorinations too. Fluorination[131] has been achieved by the use of chlorine trifluoride ClF_3 at $-75°C$.[132] For example, cyclohexane gave 41% fluorocyclohexane and methylcyclohexane gave 47% 1-fluoro-1-methylcyclohexane. Fluoroxytrifluoromethane CF_3OF fluorinates tertiary positions of certain molecules in good yields with high regioselectivity.[133] For example, adamantane gave 75% 1-fluoroadamantane. Fluorine at $-70°C$, diluted with N_2,[134] and bromine trifluoride at $25–35°C$[135] are also highly regioselective for

[127]Dermer, O.C.; Edmison, M.T. *Chem. Rev.* **1957**, *57*, 77, pp. 110–112. An example of free-radical ring halogenation can be found in Engelsma, J.W.; Kooyman, E.C. *Revl. Trav. Chim. Pays-Bas*, **1961**, *80*, 526, 537. For a review of aromatic halogenation in the gas phase, see Kooyman, E.C. *Adv. Free-Radical Chem.* **1965**, *1*, 137.

[128]Dneprovskii, A.S.; Kuznetsov, D.V.; Eliseenkov, E.V.; Fletcher, B.; Tanko, J.M. *J. Org. Chem.* **1998**, *63*, 8860.

[129]Rozen, S. *Acc. Chem. Res.* **1988**, *21*, 307; Purrington, S.T.; Kagen, B.S.; Patrick, T.B. *Chem. Rev.* **1986**, *86*, 997, pp. 1003–1005; Gerstenberger, M.R.C.; Haas, A. *Angew. Chem. Int. Ed.* **1981**, *20*, 647; Hudlicky, M. *The Chemistry of Organic Fluorine Compounds*, 2nd ed., Ellis Horwood, Chichester, **1976**; pp. 67–91. For descriptions of the apparatus necessary for handling F_2, see Vypel, H. *Chimia*, **1985**, *39*, 305.

[130]However, there are several methods by which all the C—H bonds in a molecule can be converted to C—F bonds. For reviews, see Rozhkov, I.N., in Baizer, M.M.; Lund, H. *Organic Electrochemistry*, Marcel Dekker, NY, **1983**, pp. 805–825; Lagow, R.J.; Margrave, J.L. *Prog. Inorg. Chem.* **1979**, *26*, 161. See also, Adcock, J.L.; Horita, K.; Renk, E. *J. Am. Chem. Soc.* **1981**, *103*, 6937; Adcock, J.L.; Evans, W.D. *J. Org. Chem.* **1984**, *49*, 2719; Huang, H.; Lagow, R.J. *Bull. Soc. Chim. Fr.* **1986**, 993.

[131]For a monograph on fluorinating agents, see German, L.; Zemskov, S. *New Fluorinating Agents in Organic Synthesis*, Springer, NY, **1989**.

[132]Brower, K.R. *J. Org. Chem.* **1987**, *52*, 798.

[133]Alker, D.; Barton, D.H.R.; Hesse, R.H.; Lister-James, J.; Markwell, R.E.; Pechet, M.M.; Rozen, S.; Takeshita, T.; Toh, H.T. *Nouv. J. Chem.* **1980**, *4*, 239.

[134]Rozen, S.; Ben-Shushan, G. *J. Org. Chem.* **1986**, *51*, 3522; Rozen, S.; Gal, C. *J. Org. Chem.* **1987**, *52*, 4928; **1988**, *53*, 2803; Alker, D.; Barton, D.H.R.; Hesse, R.H.; Lister-James, J.; Markwell, R.E.; Pechet, M.M.; Rozen, S.; Takeshita, T.; Toh, H.T. *Nouv. J. Chem.* **1980**, *4*, 239.

[135]Boguslavskaya, L.S.; Kartashov, A.V.; Chuvatkin, N.N. *J. Org. Chem. USSR* **1989**, *25*, 1835.

tertiary positions. These reactions probably have electrophilic,[136] not free-radical mechanisms. In fact, the success of the F_2 reactions depends on the suppression of free radical pathways, by dilution with an inert gas, by working at low temperatures, and/or by the use of radical scavengers.

Iodine can be used if the activating light has a wavelength of 184.9 nm,[137] but iodinations using I_2 alone are seldom attempted, largely because the HI formed reduces the alkyl iodide. The direct free-radical halogenation of aliphatic hydrocarbons with iodine is significantly endothermic relative to the other halogens, and the requisite chain reaction does not occur.[138] On the other hand, when iodine, $CCl_4 \cdot 2$ AlI_3 react with an alkane in dibromomethane at $-20°C$, good yields of the iodoalkane are obtained.[139] The reaction of an alkane with *tert*-butylhypoiodite (*t*-BuOI) at 40°C gave the iodoalkane in good yield.[140] The reaction of alkanes with iodine and $PhI(OAc)_2$ generates the iodoalkane.[141] A radical protocol was developed using CI_4 with base. Cyclohexane could be iodinated, for example, with CI_4 in the presence of powdered NaOH.[142] The reaction led to the use of iodoform on solid NaOH as the iodination reagent of choice. α-Iodo ethers and α-iodolactones have been prepared from the parent ether or lactone via treatment with $Et_4N \cdot 4$ HF under electrolytic conditions.[143]

Many other halogenation agents have been employed, the most common of which is sulfuryl chloride SO_2Cl_2.[144] A mixture of Br_2 and HgO is a more active brominating agent than bromine alone.[145] The actual brominating agent in this case is believed to be bromine monoxide Br_2O. Among other agents used have been *N*-bromosuccinimide (NBS, see **14-3**), CCl_4,[146] $BrCCl_3$,[147] PCl_5,[148] and *N*-haloamines and sulfuric acid.[149] In all these cases, a chain-initiating catalyst is required, usually peroxides or UV light.

[136]See, for example, Rozen, S.; Gal, C. *J. Org. Chem.* **1987**, *52*, 2769.

[137]Gover, T.A.; Willard, J.E. *J. Am. Chem. Soc.* **1960**, *82*, 3816.

[138]Liguori, L.; Bjørsvik, H.-R.; Bravo, A.; Fontana, R.; Minisci, F. *Chem. Commun.* **1997**, 1501; Tanner, D.D.; Gidley, G.C. *J. Am. Chem. Soc.* **1968**, *90*, 808; Tanner, D.D.; Rowe, J.R.; Potter, A. *J. Org. Chem.* **1986**, *51*, 457.

[139]Akhrem, I.; Orlinkov, A.; Vitt, S.; Chistyakov, A. *Tetrahedron Lett.* **2002**, *43*, 1333.

[140]Montoro, R.; Wirth, T. *Org. Lett.* **2003**, *5*, 4729.

[141]Barluenga, J.; González-Bobes, F.; González, J.M. *Angew. Chem. Int. Ed.* **2002**, *41*, 2556.

[142]Schreiner, P.R.; Lauenstein, O.; Butova, E.D.; Fokin, A.A. *Angew. Chem. Int. Ed.* **1999**, *38*, 2786.

[143]Hasegawa, M.; Ishii, H.; Fuchigami, T. *Tetrahedron Lett.* **2002**, *43*, 1503.

[144]For a review of this reagent, see Tabushi, I.; Kitaguchi, H., in Pizey, J.S. *Synthetic Reagents*, Vol. 4, Wiley, NY, *1981*, pp. 336–396.

[145]Bunce, N.J. *Can. J. Chem.* **1972**, *50*, 3109.

[146]For a discussion of the mechanism with this reagent, see Hawari, J.A.; Davis, S.; Engel, P.S.; Gilbert, B.C.; Griller, D. *J. Am. Chem. Soc.* **1985**, *107*, 4721.

[147]Huyser, E.S. *J. Am. Chem. Soc.* **1960**, *82*, 391; Baldwin, S.W.; O'Neill, T.H. *Synth. Commun.* **1976**, *6*, 109.

[148]Wyman, D.P.; Wang, J.Y.C.; Freeman, W.R. *J. Org. Chem.* **1963**, *28*, 3173.

[149]For reviews, see Minisci, F. *Synthesis* **1973**, 1; Deno, N.C. *Methods Free-Radical Chem.* **1972**, *3*, 135; Sosnovsky, G.; Rawlinson, D.J. *Adv. Free-Radical Chem.* **1972**, *4*, 203.

A base-induced bromination has been reported. 2-Methyl butane reacts with 50% aq. NaOH and CBr$_4$, in a phase-transfer catalyst, to give a modest yields of 2-bromo-2-methylbutane.[150]

When chlorination is carried out with N-haloamines and sulfuric acid (catalyzed by either uv light or metal ions), selectivity is much greater than with other reagents.[149] In particular, alkyl chains are chlorinated with high regioselectivity at the position next to the end of the chain (the ω - 1 position).[151] Some typical selectivity values are[152]

$$CH_3{-}CH_2{-}CH_2{-}CH_2{-}CH_2{-}CH_2{-}CH_3 \qquad\qquad \text{Ref. 153}$$
1   56   29   14

$$CH_3{-}CH_2{-}CH_2{-}CH_2{-}CH_2{-}CH_2{-}CH_2{-}OH \qquad \text{Ref. 154}$$
1   92   3   1   1   2   0   0

$$CH_3{-}CH_2{-}CH_2{-}CH_2{-}CH_2{-}CH_2{-}COOMe \qquad\qquad \text{Ref. 155}$$
3   72   20   4   1   0

Furthermore, di- and polychlorination are much less prevalent. Dicarboxylic acids are predominantly chlorinated in the middle of the chain,[156] and adamantane and bicyclo[2.2.2]octane at the bridgeheads[157] by this procedure. The reasons for the high ω - 1 specificity are not clearly understood.[158] Alkyl bromides can be regioselectively chlorinated one carbon away from the bromine (to give *vic*-bromochlorides) by treatment with PCl$_5$.[159] Alkyl chlorides can be converted to *vic*-dichlorides by treatment with MoCl$_5$.[160] Enhanced selectivity at a terminal position of *n*-alkanes has been achieved by absorbing the substrate onto a pentasil zeolite.[161] In another regioselective chlorination, alkanesulfonamides

[150]Schreiner, P.R.; Lauentstein, O.; Kolomitsyn, I.V.; Nadi, S.; Kokin, A.A. *Angew. Chem. Int. Ed.* **1998**, *37*, 1895.

[151]The ω - 1 regioselectivity diminishes when the chains are >10 carbons; see Deno, N.C.; Jedziniak, E.J. *Tetrahedron Lett.* **1976**, 1259; Konen, D.A.; Maxwell, R.J.; Silbert, L.S. *J. Org. Chem.* **1979**, *44*, 3594.

[152]The ω − 1 selectivity values shown here may actually be lower than the true values because of selective solvolysis of the ω − 1 chlorides in concentrated H$_2$SO$_4$: see Deno, N.C.; Pohl, D.G. *J. Org. Chem.* **1975**, *40*, 380.

[153]Bernardi, R.; Galli, R.; Minisci, F. *J. Chem. Soc. B* **1968**, 324. See also, Deno, N.C.; Gladfelter, E.J.; Pohl, D.G. *J. Org. Chem.* **1979**, *44*, 3728; Fuller, S.E.; Lindsay Smith, J.R.; Norman, R.O.C.; Higgins, R. *J. Chem. Soc. Perkin Trans. 2* **1981**, 545.

[154]Deno, N.C.; Billups, W.E.; Fishbein, R.; Pierson, C.; Whalen, R.; Wyckoff, J.C. *J. Am. Chem. Soc.* **1971**, *93*, 438.

[155]Minisci, F.; Gardini, G.P.; Bertini, F. *Can. J. Chem.* **1970**, *48*, 544.

[156]Kämper, F.; Schäfer, H.J.; Luftmann, H. *Angew. Chem. Int. Ed.* **1976**, *15*, 306.

[157]Smith, C.V.; Billups, W.E. *J. Am. Chem. Soc.* **1974**, *96*, 4307.

[158]It has been reported that the selectivity in one case is in accord with a pure electrostatic (field effect) explanation: Dneprovskii, A.S.; Mil'tsov, S.A.; Arbuzov, P.V. *J. Org. Chem. USSR* **1988**, *24*, 1826. See also, Tanner, D.D.; Arhart, R.; Meintzer, C.P. *Tetrahedron* **1985**, *41*, 4261; Deno, N.C.; Pohl, D.G. *J. Org. Chem.* **1975**, *40*, 380.

[159]Luche, J.L.; Bertin, J.; Kagan, H.B. *Tetrahedron Lett.* **1974**, 759.

[160]San Filippo Jr, J.; Sowinski, A.F.; Romano, L.J. *J. Org. Chem.* **1975**, *40*, 3463.

[161]Turro, N.J.; Fehlner, J.R.; Hessler, D.P.; Welsh, K.M.; Ruderman, W.; Firnberg, D.; Braun, A.M. *J. Org. Chem.* **1988**, *53*, 3731.

$RCH_2-CH_2CH_2SO_2NHR'$ are converted primarily to $RCHClCH_2CH_2SO_2NHR'$ by sodium peroxydisulfate $Na_2S_2O_8$ and $CuCl_2$.[162] For regioselective chlorination at certain positions of the steroid nucleus, see **19-2**.

In almost all cases, the mechanism involves a free-radical chain:

$$\text{Initiation} \qquad X_2 \xrightarrow{hv} 2\,X\cdot$$

$$RH + X\cdot \longrightarrow R\cdot + XH$$

$$\text{Propagation} \qquad R\cdot + X_2 \longrightarrow RX + X\cdot$$

$$\text{Termination} \qquad R\cdot + X\cdot \longrightarrow RX$$

When the reagent is halogen, initiation occurs as shown above.[163] When it is another reagent, a similar cleavage occurs (catalyzed by light or, more commonly, peroxides), followed by propagation steps that do not necessarily involve abstraction by halogen. For example, the propagation steps for chlorination by *tert*-butyl hypochlorite (*t*-BuOCl) have been formulated as[164]

$$RH + t\text{-BuO}\cdot \longrightarrow R\cdot + t\text{-BuOH}$$

$$R\cdot + t\text{-BuOCl} \longrightarrow RCl + t\text{-BuO}\cdot$$

and the abstracting radicals in the case of *N*-haloamines are the aminium radical cations $R_2NH\cdot^+$ (p. 693), with the following mechanism (in the case of initiation by Fe^{2+}):[149]

$$\text{Initiation} \qquad R_2NCl \xrightarrow{H^+} R_2\overset{\oplus}{N}HCl \xrightarrow{Fe^{2+}} R_2\overset{+}{N}H\cdot + \overset{\oplus}{F}eCl$$

$$R_2\overset{+}{N}H\cdot + RH \longrightarrow R_2NH_2 + R\cdot$$

$$\text{Propagation} \qquad R\cdot + R_2\overset{}{N}HCl \longrightarrow RCl + R_2\overset{+}{N}H\cdot$$

This mechanism is similar to that of the Hofmann–Löffler reaction (**18-40**).

The two propagation steps shown above for X_2 are those that lead directly to the principal products (RX and HX), but many other propagation steps are possible and many occur. Similarly, the only termination step shown is the one that leads to RX, but any two radicals may combine ($\cdot$H, $\cdot CH_3$, $\cdot$Cl, $\cdot CH_2CH_3$ in all combinations).

---

[162]Nikishin, G.I.; Troyansky, E.I.; Lazareva, M.I. *Tetrahedron Lett.* **1985**, *26*, 3743.

[163]There is evidence (unusually high amounts of multiply chlorinated products) that under certain conditions in the reaction of RH with Cl_2, the products of the second propagation step (RX + X$\cdot$) are enclosed within a solvent cage. See Skell, P.S.; Baxter III, H.N. *J. Am. Chem. Soc.* **1985**, *107*, 2823; Raner, K.D.; Lusztyk, J.; Ingold, K.U. *J. Am. Chem. Soc.* **1988**, *110*, 3519; Tanko, J.M.; Anderson III, F.E. *J. Am. Chem. Soc.* **1988**, *110*, 3525.

[164]Carlsson, D.J.; Ingold, K.U. *J. Am. Chem. Soc.* **1967**, *89*, 4885, 4891; Walling, C.; McGuiness, J.A. *J. Am. Chem. Soc.* **1969**, *91*, 2053. See also, Zhulin, V.M.; Rubinshtein, B.I. *Bull. Acad. Sci. USSR Div. Chem. Sci*, **1977**, *26*, 2082.

Thus, products like H_2, higher alkanes, and higher alkyl halides can be accounted for. When methane is the substrate, the rate-determining step is

$$CH_4 + Cl\cdot \longrightarrow \cdot CH_3 + HCl$$

since an isotope effect of 12.1 was observed at $0°C$.[165] For chlorinations, chains are very long, typically 10^4–10^6 propagations before a termination step takes place.

The order of reactivity of the halogens can be explained by energy considerations. For the substrate methane, ΔH values for the two principal propagation steps are

	kcal mol^{-1}				kJ mol^{-1}			
Reaction	F_2	Cl_2	Br_2	I_2	F_2	Cl_2	Br_2	I_2
$CH_4 + X\cdot \to CH_3\cdot + HX$	-31	$+2$	$+17$	$+34$	-132	$+6$	$+72$	$+140$
$CH_4 + X_2 \to CH_3X + X\cdot$	-70	-26	-24	-21	-293	-113	-100	-87

In each case, D for CH_3—H is 105 kcal mol^{-1} (438 kJ mol^{-1}), while D values for the other bonds involved are given in Table 14.4.[166] Fluorine is so reactive[167] that neither uv light nor any other initiation is needed (total $\Delta H = -101$ kcal mol^{-1}; -425 kJ mol^{-1});[168] while Br_2 and I_2 essentially do not react with methane. The second step is exothermic in all four cases, but it cannot take place before the first, and it is this step that is very unfavorable for Br_2 and I_2. It is apparent that the most important single factor causing the order of halogen reactivity to be $F_2 > Cl_2 > Br_2 > I_2$ is the decreasing strength of the HX bond in the order HF > HCl > HBr > HI. The increased reactivity of secondary and tertiary positions is in accord with the decrease in D values for R—H in the order primary > secondary > tertiary (Table 5.3). (Note that for chlorination step 1 is exothermic for practically all substrates other than CH_4, since most other aliphatic C—H bonds are weaker than those in CH_4.)

Bromination and chlorination of alkanes and cycloalkanes can also take place by an electrophilic mechanism if the reaction is catalyzed by $AgSbF_6$.[169] Direct

[165]Wiberg, K.B.; Motell, E.L. *Tetrahedron* **1963**, *19*, 2009.

[166]Kerr, J.A., in Weast, R.C. *Handbook of Chemistry and Physics*, 69th ed., CRC Press, Boca Raton, FL, **1988**, pp. F174–F189.

[167]It has been reported that the reaction of F atoms with CH_4 at 25 K takes place with practically zero activation energy: Johnson, G.L.; Andrews, L. *J. Am. Chem. Soc.* **1980**, *102*, 5736.

[168]For F_2, the following initiation step is possible: $F_2 + RH \to R\cdot + F\cdot + HF$ [first demonstrated by Miller, Jr, W.T.; Koch, Jr, S.D.; McLafferty, F.W. *J. Am. Chem. Soc.* **1956**, *78*, 4992]. ΔH for this reaction is equal to the small positive value of 5 kcal mol^{-1} (21 kJ mol^{-1}). The possibility of this reaction (which does not require an initiator) explains why fluorination can take place without UV light [which would otherwise be needed to furnish the 38 kcal mol^{-1} (159 kJ mol^{-1}) necessary to break the F—F bond]. Once the reaction has been initiated, the large amount of energy given off by the propagation steps is ample to cleave additional F_2 molecules. Indeed, it is the magnitude of this energy that is responsible for the cleavage of carbon chains by F_2.

[169]Olah, G.A.; Renner, R.; Schilling, P.; Mo, Y.K. *J. Am. Chem. Soc.* **1973**, *95*, 7686. See also, Olah, G.A.; Wu, A.; Farooq, O. *J. Org. Chem.* **1989**, *54*, 1463.

TABLE 14.4. Some D Values[166]

Bond	D	
	kcal mol^{-1}	kJ mol^{-1}
H–F	136	570
H–Cl	103	432
H–Br	88	366
H–I	71	298
F–F	38	159
Cl–Cl	59	243
Br–Br	46	193
I–I	36	151
CH$_3$–F	108	452
CH$_3$–Cl	85	356
CH$_3$–Br	70	293
CH$_3$–I	57	238

chlorination at a vinylic position by an electrophilic mechanism has been achieved with benzeneseleninyl chloride PhSe(O)Cl and AlCl$_3$ or AlBr$_3$.[170] However, while some substituted alkenes give high yields of chloro substitution products, others (e.g., styrene) undergo addition of Cl$_2$ to the double bond (**15-39**).[131] Electrophilic fluorination has already been mentioned (p. 956).

OS **II**, 89, 133, 443, 549; **III**, 737, 788; **IV**, 807, 921, 984; **V**, 145, 221, 328, 504, 635, 825; **VI**, 271, 404, 715; **VII**, 491; **VIII**, 161.

14-2 Halogenation at Silicon

Halogenation or Halo-de-hydrogenation

$$R_3Si–H \longrightarrow R_3Si–X$$

Just as free-radical halogenation occurs at the carbon of an alkane, via hydrogen abstraction to form the radical, a similar reaction occurs at silicon. When triisopropylsilane (iPr$_3$Si–H) reacts with *tert*-butyl hypochlorite at $-10°C$, the product is triisopropylchlorosilane (iPr$_3$Si–Cl).[171]

14-3 Allylic and Benzylic Halogenation

Halogenation or Halo-de-hydrogenation

[170]Kamigata, N.; Satoh, T.; Yoshida, M. *Bull. Chem. Soc. Jpn.* **1988**, *44*, 449.
[171]Chawla, R.; Larson, G.L. *Synth. Commun.* **1999**, *29*, 3499.

This reaction is a special case of **14-1**, but is important enough to be treated separately.[172] Alkenes can be halogenated in the allylic position and also a benzylic position by a number of reagents, of which NBS[173] is by far the most common. When this reagent is used, the reaction is known as *Wohl–Ziegler bromination*. A nonpolar solvent is used, most often CCl_4, but the reaction has been done in an ionic liquid.[174] A variation in the reaction used NBS with 5% $Yb(OTf)_3$ and 5% $ClSiMe_3$.[175] Other *N*-bromo amides have also been used. Allylic chlorination has been carried out, with *N*-chlorosuccinimide, *tert*-butyl hypochlorite,[176] or with $NaClO/CeCl_3 \cdot 7\ H_2O$.[177] With any reagent an initiator is needed; this is usually AIBN (**1**), a peroxide, such as di-*tert*-butyl peroxide or benzoyl peroxide or, less often, UV light.

The reaction is usually quite specific at an allylic or benzylic position and good yields are obtained. However, when the allylic radical intermediate is unsymmetrical, allylic rearrangements can take place, so that mixtures of both possible products are obtained, **21** and **22**.

$$ \text{alkene} + \text{NBS} \longrightarrow \underset{\textbf{21}}{\text{(Br-substituted alkene)}} + \underset{\textbf{22}}{\text{(allylic Br product)}} $$

When a double bond has two different allylic positions (e.g., $CH_3CH{=}CHCH_2CH_3$), a secondary position is substituted more readily than a primary. The relative reactivity of tertiary hydrogen is not clear, though many substitutions at allylic tertiary positions have been performed.[178] It is possible to brominate both sides of the double bond.[179] Because of the electron-withdrawing nature of bromine, the second bromine substitutes on the other side of the double bond rather than α to the first bromine. Molecules with a benzylic hydrogen, such as toluene, react rapidly to give α-bromomethyl benzene (e.g., $PhCH_3 \rightarrow PhCH_2Br$).

N-Bromosuccinimide is also a highly regioselective brominating agent at other positions, including positions α to a carbonyl group, to a C≡C triple bond, and to an aromatic ring (benzylic position). When both a double and a triple bond are in the same molecule, the preferred position is α to the triple bond.[180]

Dauben and McCoy demonstrated that the mechanism of allylic bromination is of the free-radical type,[181] showing that the reaction is very sensitive to free-radical

[172]For a review, see Nechvatal, A. *Adv. Free-Radical Chem.* **1972**, *4*, 175.

[173]For a review of this reagent, see Pizey, J.S. *Synthetic Reagents*, Vol. 2, Wiley, NY, **1974**, pp. 1–63.

[174]In bmim PF_6, 1-butyl-3-methylimidazolium hexafluoorophosphate: Togo, H.; Hirai, T. *Synlett* **2003**, 702.

[175]Yamanaka, M.; Arisawa, M.; Nishida, A.; Nakagawa, M. *Tetrahedron Lett.* **2002**, *43*, 2403.

[176]Walling, C.; Thaler, W.A. *J. Am. Chem. Soc.* **1961**, *83*, 3877.

[177]Moreno-Dorado, F.J.; Guerra, F.M.; Manzano, F.L.; Aladro, F.J.; Jorge, Z.S.; Massanet, G.M. *Tetrahedron Lett.* **2003**, *44*, 6691.

[178]Dauben, Jr, H.J.; McCoy, L.L. *J. Org. Chem.* **1959**, *24*, 1577.

[179]Ucciani, E.; Naudet, M. *Bull. Soc. Chim. Fr.* **1962**, 871.

[180]Peiffer, G. *Bull. Soc. Chim. Fr.* **1963**, 537.

[181]Dauben, Jr, H.J.; McCoy, L.L. *J. Am. Chem. Soc.* **1959**, *81*, 4863.

initiators and inhibitors and indeed does not proceed at all unless at least a trace of initiator is present. Subsequent work indicated that the species that actually abstracts hydrogen from the substrate is the bromine atom. The reaction is initiated by small amounts of Br•. Once it is formed, the main propagation steps are

Step 1 $Br• + RH \longrightarrow R• + HBr$

Step 2 $R• + Br_2 \longrightarrow RBr + Br•$

The source of the Br_2 is a fast ionic reaction between NBS and the HBr liberated in step 1:

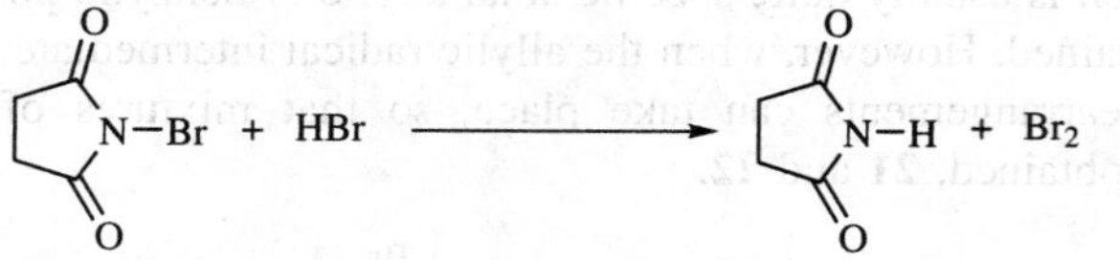

The function of the NBS is therefore to provide a source of Br_2 in a low, steady-state concentration and to use up the HBr liberated in step 1.[182] The main evidence for this mechanism is that NBS and Br_2 show similar selectivity[183] and that the various N-bromo amides also show similar selectivity,[184] which is consistent with the hypothesis that the same species is abstracting in each case.[185]

It may be asked why, if Br_2 is the reacting species, it does not add to the double bond, either by an ionic or by a free-radical mechanism (see **15-39**). Apparently the concentration is too low. In bromination of a double bond, only one atom of an attacking bromine molecule becomes attached to the substrate, whether the addition is electrophilic or free radical:

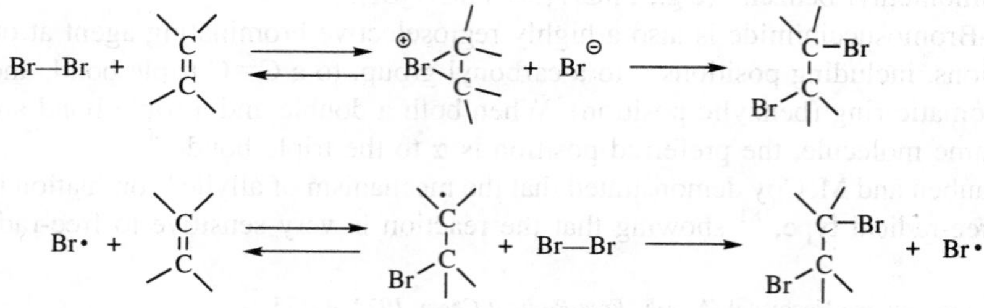

---

[182]This mechanism was originally suggested by Adam, J.; Gosselain, P.A.; Goldfinger, P. *Nature (London)*, **1953**, *171*, 704; *Bull. Soc. Chim. Belg.* **1956**, *65*, 533.

[183]Walling, C.; Rieger, A.L.; Tanner, D.D. *J. Am. Chem. Soc.* **1963**, *85*, 3129; Russell, G.A.; Desmond, K.M. *J. Am. Chem. Soc.* **1963**, *85*, 3139; Russell, G.A.; DeBoer, C.D.; Desmond, K.M. *J. Am. Chem. Soc.* **1963**, *85*, 365; Pearson, R.; Martin, J.C. *J. Am. Chem. Soc.* **1963**, *85*, 3142; Skell, P.S.; Tuleen, D.L.; Readio, P.D. *J. Am. Chem. Soc.* **1963**, *85*, 2850.

[184]Walling, C.; Rieger, A.L. *J. Am. Chem. Soc.* **1963**, *85*, 3134; Pearson, R.; Martin, J.C. *J. Am. Chem. Soc.* **1963**, *85*, 3142; Incremona, J.H.; Martin, J.C. *J. Am. Chem. Soc.* **1970**, *92*, 627.

[185]For other evidence, see Day, J.C.; Lindstrom, M.J.; Skell, P.S. *J. Am. Chem. Soc.* **1974**, *96*, 5616.

The other bromine atom comes from another bromine-containing molecule or ion. This is clearly not a problem in reactions with benzylic species since the benzene ring is not prone to such addition reactions. If the concentration is sufficiently low, there is a low probability that the proper species will be in the vicinity once the intermediate forms. The intermediate in either case reverts to the initial species and the allylic substitution competes successfully. If this is true, it should be possible to brominate an alkene in the allylic position without competition from addition, even in the absence of NBS or a similar compound, if a very low concentration of bromine is used and if the HBr is removed as it is formed so that it is not available to complete the addition step. This has indeed been demonstrated.[186]

$$O \underset{}{=} \overset{\bullet}{\underset{}{N}} \underset{}{=} O$$

23

When NBS is used to brominate non-alkenyl substrates, such as alkanes, another mechanism, involving abstraction of the hydrogen of the substrate by the succinimidyl radical[187] **23** can operate.[188] This mechanism is facilitated by solvents (e.g., CH_2Cl_2, $CHCl_3$, or MeCN) in which NBS is more soluble, and by the presence of small amounts of an alkene that lacks an allylic hydrogen (e.g., ethene). The alkene serves to scavenge any Br• that forms from the reagent. Among the evidence for the mechanism involving **23** are abstraction selectivities similar to those of Cl• atoms and the isolation of β-bromopropionyl isocyanate ($BrCH_2CH_2CONCO$) which is formed by ring opening of **23**.

Allylic chlorination has also been carried out[189] with N-chlorosuccinimide (NCS) and either arylselenyl chlorides (ArSeCl), aryl diselenides (ArSeSeAr), or TsNSO as catalysts. Use of the selenium catalysts produces almost entirely the allylically rearranged chlorides in high yields. With TsNSO the products are the unrearranged chlorides in lower yields. Dichlorine monoxide Cl_2O, with no catalyst, also gives allylically rearranged chlorides in high yields.[190] A free-radical mechanism is unlikely in these reactions.

Allyl silanes react with transition metals bearing chlorine ligands to give allyl chlorides, where a chlorine replaces a Me_3Si unit.[191]

OS **IV**, 108; **V**, 825; **VI**, 462; **IX**, 191.

[186]McGrath, B.P.; Tedder, J.M. *Proc. Chem. Soc.* **1961**, 80.

[187]For a review of this radical, see Chow, Y.L.; Naguib, Y.M.A. *Rev. Chem. Intermed.* **1984**, *5*, 325.

[188]Skell, P.S.; Day, J.C. *Acc. Chem. Res.* **1978**, *11*, 381; Tanner, D.D.; Reed, D.W.; Tan, S.L.; Meintzer, C.P.; Walling, C.; Sopchik, A. *J. Am. Chem. Soc.* **1985**, *107*, 6576; Lüning, U.; Seshadri, S.; Skell, P.S. *J. Org. Chem.* **1986**, *51*, 2071; Zhang, Y.; Dong, M.; Jiang, X.; Chow, Y.L. *Can. J. Chem.* **1990**, *68*, 1668.

[189]Hori, T.; Sharpless, K.B. *J. Org. Chem.* **1979**, *44*, 4204.

[190]Torii, S.; Tanaka, H.; Tada, N.; Nagao, S.; Sasaoka, M. *Chem. Lett.* **1984**, 877.

[191]Fujii, T.; Hirao, Y.; Ohshiro, Y. *Tetrahedron Lett.* **1993**, *34*, 5601.

14-4   Halogenation of Aldehydes

Halogenation or Halo-de-hydrogenation

$$RCHO + Cl_2 \longrightarrow RCOCl$$

The α-halogenation reaction of carbonyl compounds was mentioned in Section **14-2**. A different halogenation reaction is possible in which aldehydes can be directly converted to acyl chlorides by treatment with chlorine, but the reaction operates only when the aldehyde does not contain an α hydrogen and even then it is not very useful. When there is an α hydrogen, α halogenation (**14-2, 12-4**) occurs instead. Other sources of chlorine have also been used, among them SO_2Cl_2[192] and t-BOCl.[193] The mechanisms are probably of the free-radical type. N-Bromosuccinimide, with AIBN (p. 935) as a catalyst, has been used to convert aldehydes to acyl bromides.[194]

OS **I**, 155.

B. Substitution by Oxygen

14-5   Hydroxylation at an Aromatic Carbon[195]

Hydroxylation or Hydroxy-de-hydrogenation

$$ArH + H_2O_2 + FeSO_4 \longrightarrow ArOH$$

A mixture of hydrogen peroxide and ferrous sulfate,[196] called *Fenton's reagent*,[197] can be used to hydroxylate aromatic rings, though yields are usually not high.[198] Biaryls are usually side products.[199] Among other reagents used have been H_2O_2 and titanous ion; O_2 and Cu(I)[200] or Fe(III),[201] a mixture of ferrous

[192]Arai, M. *Bull. Chem. Soc. Jpn.* **1964**, *37*, 1280; **1965**, *38*, 252.

[193]Walling, C.; Mintz, M.J. *J. Am. Chem. Soc.* **1967**, *89*, 1515.

[194]Markó, I.E.; Mekhalfia, A. *Tetrahedron Lett.* **1990**, *31*, 7237. For a related procedure, see Cheung, Y. *Tetrahedron Lett.* **1979**, 3809.

[195]For reviews, see Vysotskaya, N.A. *Russ. Chem. Rev.* **1973**, *42*, 851; Sangster, D.F., in Patai, S. *The Chemistry of the Hydroxyl Group*, pt. 1, Wiley, NY, **1971**, pp. 133–191; Metelitsa, D.I. *Russ. Chem. Rev.* **1971**, *40*, 563; Enisov, E.T.; Metelitsa, D.I. *Russ. Chem. Rev.* **1968**, *37*, 656; Loudon, J.D. *Prog. Org. Chem.* **1961**, *5*, 47.

[196]For a review of reactions of H_2O_2 and metal ions with all kinds of organic compounds, including aromatic rings, see Sosnovsky, G.; Rawlinson, D.J., in Swern, D. *Organic Peroxides*, Vol. 2, Wiley, NY, **1970**, pp. 269–336. See also, Sheldon, R.A.; Kochi, J.K. *Metal-Catalyzed Oxidations of Organic Compounds*, Academic Press, NY, **1981**.

[197]For a discussion of Fenton's reagent, see Walling, C. *Acc. Chem. Res.* **1975**, *8*, 125.

[198]Yields can be improved with phase transfer catalysis: Karakhanov, E.A.; Narin, S.Yu.; Filippova, T.Yu.; Dedov, A.G. *Doklad. Chem.* **1987**, *292*, 81.

[199]See the discussion of the aromatic free-radical substitution mechanism on pp. $$$–$$$.

[200]See Karlin, K.D.; Hayes, J.C.; Gultneh, Y.; Cruse, R.W.; McKown, J.W.; Hutchinson, J.P.; Zubieta, J. *J. Am. Chem. Soc.* **1984**, *106*, 2121; Cruse, R.W.; Kaderli, S.; Meyer, C.J.; Zuberbühler, A.D.; Karlin, K.D. *J. Am. Chem. Soc.* **1988**, *110*, 5020; Ito, S.; Kunai, A.; Okada, H.; Sasaki, K. *J. Org. Chem.* **1988**, *53*, 296.

[201]Funabiki, T.; Tsujimoto, M.; Ozawa, S.; Yoshida, S. *Chem. Lett.* **1989**, 1267.

ion, oxygen, ascorbic acid, and ethylenetetraaminetetraacetic acid (*Udenfriend's reagent*);[202] O_2 and KOH in liquid NH_3;[203] and peroxyacids such as peroxynitrous and trifluoroperoxyacetic acids.

Much work has been done on the mechanism of the reaction with Fenton's reagent, and it is known that free aryl radicals (formed by a process, e.g., $HO\bullet + ArH \rightarrow AR\bullet + H_2O$) are not intermediates. The mechanism is essentially that outlined on p. $$$, with $HO\bullet$ as the attacking species,[204] formed by

$$Fe^{2+} + H_2O_2 \longrightarrow Fe^{3+} + OH^- + HO\bullet$$

The rate-determining step is formation of $HO\bullet$ and not its reaction with the aromatic substrate.

An alternative oxidation of arene to phenol was reported using $Cu(NO_3)\bullet 3\ H_2O$, 30% hydrogen peroxide and a phosphate buffer.[205]

See also, **11-26**.

14-6 Formation of Cyclic Ethers

(5)*OC-cyclo*-Alkoxy-de-hydro-substitution

Alcohols with a hydrogen in the δ position can be cyclized with lead tetraacetate.[206] The reaction is usually carried out at ~80°C (most often in refluxing benzene), but can also be done at room temperature if the reaction mixture is irradiated with uv light. Tetrahydrofurans are formed in high yields. Little or no four- and six-membered cyclic ethers (oxetanes and tetrahydropyrans, respectively) are obtained even when γ and ε hydrogens are present. The reaction has also been carried out with a mixture of halogen (Br_2 or I_2) and a salt or oxide of silver or mercury (especially HgO or AgOAc),[207] with iodosobenzene diacetate and I_2,[208] and with ceric

[202]Udenfriend, S.; Clark, C.T.; Axelrod, J.; Brodie, B.B. *J. Biol. Chem.* **1954**, *208*, 731; Brodie, B.B.; Shore, P.A.; Udenfriend, S. *J. Biol. Chem.* **1954**, *208*, 741. See also, Tamagaki, S.; Suzuki, K.; Tagaki, W. *Bull. Chem. Soc. Jpn.* **1989**, *62*, 148, 153, 159.

[203]Malykhin, E.V.; Kolesnichenko, G.A.; Shteingarts, V.D. *J. Org. Chem. USSR* **1986**, *22*, 720.

[204]Jefcoate, C.R.E.; Lindsay Smith, J.R.; Norman, R.O.C. *J. Chem. Soc. B* **1969**, 1013; Brook, M.A.; Castle, L.; Lindsay Smith, J.R.; Higgins, R.; Morris, K.P. *J. Chem. Soc. Perkin Trans. 2* **1982**, 687; Lai, C.; Piette, L.H. *Tetrahedron Lett.* **1979**, 775; Kunai, A.; Hata, S.; Ito, S.; Sasaki, K. *J. Am. Chem. Soc.* **1986**, *108*, 6012.

[205]Nasreen, A.; Adapa, S.R. *Org. Prep. Proceed. Int.* **2000**, *32*, 373.

[206]For reviews, see Mihailović, M.Lj.; Partch, R. *Sel. Org. Transform.* **1972**, *2*, 97; Milhailović, M.Lj.; Čeković, Z. *Synthesis* **1970**, 209. For a review of the chemistry of lead tetraacetate, see Butler, R.N., in Pizey, J.S. *Synthetic Reagents*, Vol. 3, Wiley, NY, **1977**, pp. 277–419.

[207]Akhtar, M.; Barton, D.H.R. *J. Am. Chem. Soc.* **1964**, *86*, 1528; Sneen, R.A.; Matheny, N.P. *J. Am. Chem. Soc.* **1964**, *86*, 3905, 5503; Roscher, N.M.; Shaffer, D.K. *Tetrahedron* **1984**, *40*, 2643. For a review, see Kalvoda, J.; Heusler, K. *Synthesis* **1971**, 501. For a list of references, see Larock, R.C. *Comprehensive Organic Transformations*, 2nd ed, Wiley-VCH, NY, **1999**, pp. 889–890.

[208]Concepción, J.I.; Francisco, C.G.; Hernández, R.; Salazar, J.A.; Suárez, E. *Tetrahedron Lett.* **1984**, *25*, 1953; Furuta, K.; Nagata, T.; Yamamoto, H. *Tetrahedron Lett.* **1988**, *29*, 2215.

ammonium nitrate (CAN).[209] The following mechanism is likely for the lead tetra-acetate reaction:[210]

though **24** has never been isolated. The step marked **A** is a 1,5 internal hydrogen abstraction. Such abstractions are well known (see **18-40**) and are greatly favored over 1,4 or 1,6 abstractions (the small amounts of tetrahydropyran formed result from 1,6 abstractions).[211]

Reactions that sometimes compete are oxidation to the aldehyde or acid (**19-3** and **19-22**) and fragmentation of the substrate. When the OH group is on a ring of at least seven members, a transannular product can be formed, as in the cyclization reaction of 1-octanol to **25**.[212]

β-Hydroxy ethers can give cyclic acetals, such as **26**.[213]

There are no references in *Organic Syntheses*, but see OS **V**, 692; **VI**, 958, for related reactions.

[209]See, for example, Trahanovsky, W.S.; Young, M.G.; Nave, P.M. *Tetrahedron Lett.* **1969**, 2501; Doyle, M.P.; Zuidema, L.J.; Bade, T.R. *J. Org. Chem.* **1975**, *40*, 1454.
[210]Mihailović, M.Lj.; Čeković, Z.; Maksimović, Z.; Jeremić, D.; Lorenc, Lj.; Mamuzi, R.I. *Tetrahedron* **1965**, *21*, 2799.
[211]Mihailović, M.Lj.; Čeković, Z.; Jeremić, D. *Tetrahedron* **1965**, *21*, 2813.
[212]Cope, A.C.; Gordon, M.; Moon, S.; Park, C.H. *J. Am. Chem. Soc.* **1965**, *87*, 3119; Moriarty, R.M.; Walsh, H.G. *Tetrahedron Lett.* **1965**, 465; Mihailović, M.Lj.; Čeković, Z.; Andrejević, V.; Matić, R.; Jeremić, D. *Tetrahedron* **1968**, *24*, 4947.
[213]Furuta, K.; Nagata, T.; Yamamoto, H. *Tetrahedron Lett.* **1988**, *29*, 2215.

14-7   Formation of Hydroperoxides

Hydroperoxy-de-hydrogenation

$$RH \ + \ O_2 \ \longrightarrow \ R-O-O-H$$

The slow atmospheric oxidation (*slow* meaning without combustion) of C—H to C—O—O—H is called *autoxidation*.[214] The reaction occurs when compounds are allowed to stand in air and is catalyzed by light, so unwanted autoxidations can be greatly slowed by keeping the compounds in dark places. Most autoxidations proceed by free-radical chain processes that involve peroxyl radicals.[215] To suppress autoxidation, an antioxidant can be added that will prevent or retard the reaction with atmospheric oxygen.[216] Although some lactone compounds are sold as antioxidants, many radicals derived from lactones show poor or no reactivity toward oxygen.[216] The hydroperoxides produced often react further to give alcohols, ketones, and more complicated products, so the reaction is not often used for preparative purposes, although in some cases hydroperoxides have been prepared in good yield.[217] It is because of autoxidation that foods, rubber, paint, lubricating oils, and so on deteriorate on exposure to the atmosphere over periods of time. On the other hand, a useful application of autoxidation is the atmospheric drying of paints and varnishes. As with other free-radical reactions of C—H bonds, some bonds are attacked more readily than others,[218] and these are the ones we have seen before (pp. 943–949), though the selectivity is very low at high temperatures and in the gas phase. The reaction can be carried out successfully at tertiary (to a lesser extent, secondary), benzylic,[219] and allylic (though allylic rearrangements are common) R.[220] 2-Phenylpropane reacted with oxygen to give $PhMe_2C-OOH$, for example. Another susceptible position is aldehydic C—H,

---

[214]The term autoxidation actually applies to any slow oxidation with atmospheric oxygen. See Goosen, A.; Morgan, D.H. *J. Chem. Soc. Perkin Trans. 2* **1994**, 557. For reviews, see Sheldon, R.A.; Kochi, J.K. *Adv. Catal.*, **1976**, *25*, 272; Howard, W.G., in Kochi, J.K. *Free Radicals*, Vol. 2, Wiley, NY, **1973**, pp. 3–62; Lloyd, W.G. *Methods Free-Radical Chem.* **1973**, *4*, 1; Betts, J. *Q. Rev. Chem. Soc.* **1971**, *25*, 265; Huyser, E.S. *Free-Radical Chain Reactions*, Wiley, NY, **1970**, pp. 306–312; Chinn, L.J. *Selection of Oxidants in Synthesis* Marcel Dekker, NY, **1971**, pp. 29–39; Ingold, K.U. *Acc. Chem. Res.* **1969**, *2*, 1; Mayo, F.R. *Acc. Chem. Res.* **1968**, *1*, 193. For monographs on these and similar reactions, see Bamford, C.H.; Tipper, C.F.H. *Comprehensive Chemical Kinetics*, Vol. 16, Elsevier, NY, **1980**; Sheldon, R.A.; Kochi, J.K. *Metal-Catalyzed Oxidations of Organic Compounds*, Academic Press, NY, **1981**.

[215]Ingold, K.U. *Acc. Chem. Res.* **1969**, *2*, 1.

[216]Bejan, E.V.; Font-Sanchis, E.; Scaiano, J.C. *Org. Lett*, **2001**, *3*, 4059; Scaiano, J.C.; Martin, A.; Yap, G.P.A.; Ingold, K.U. *Org. Lett.* **2000**, *2*, 899.

[217]For a review of the synthesis of alkyl peroxides and hydroperoxides, see Sheldon, R.A., in Patai, S. *The Chemistry of Peroxides*, Wiley, NY, **1983**, pp. 161–200.

[218]For a discussion, see Korcek, S.; Chenier, J.H.B.; Howard, J.A.; Ingold, K.U. *Can. J. Chem.* **1972**, *50*, 2285, and other papers in this series.

[219]For a method that gives good yields at benzylic positions, see Santamaria, J.; Jroundi, R.; Rigaudy, J. *Tetrahedron Lett.* **1989**, *30*, 4677.

[220]For a review of autoxidation at allylic and benzylic positions, see Voronenkov, V.V.; Vinogradov, A.N.; Belyaev, V.A. *Russ. Chem. Rev.* **1970**, *39*, 944.

but the peroxyacids so produced are not easily isolated[221] since they are converted to the corresponding carboxylic acids (**19-23**). The α positions of ethers are also easily attacked by oxygen [RO—C—H → RO—C—OOH], but the resulting hydroperoxides are seldom isolated. However, this reaction constitutes a hazard in the storage of ethers since solutions of these hydroperoxides and their rearrangement products in ethers are potential spontaneous explosives.[222]

Oxygen itself (a diradical) is not reactive enough to be the species that actually abstracts the hydrogen. But if a trace of free radical (say R′•) is produced by some initiating process, *it* reacts with oxygen[223] to give R′—O—O•; since this type of radical *does* abstract hydrogen, the chain is

$$R'OO\bullet \; + \; RH \longrightarrow \; R\bullet \; + \; R'OOH$$
$$R\bullet \; + \; O_2 \longrightarrow \; R-O-O\bullet$$
$$etc.$$

In at least some cases (in alkaline media)[224] the radical R• can be produced by formation of a carbanion and its oxidation (by O_2) to a radical, such as allylic radical **27**.[225]

Autoxidations in alkaline media can also proceed by a different mechanism: R—H + base → R⁻ + O_2 → ROO⁻.[226]

When alkenes are treated with oxygen that has been photosensitized (p. 341), they are substituted by OOH in the allylic position in a synthetically useful reaction.[227] Although superficially similar to autoxidation, this reaction is clearly different because 100% allylic rearrangement always takes place. The reagent here is not

[221]Swern D. *Organic Peroxides*, Vol. 1, Wiley, NY, *1970*, p. 313.

[222]For methods of detection and removal of peroxides from ether solvents, see Gordon, A.J.; Ford, R.A. *The Chemist's Companion*, Wiley, NY, *1972*, p. 437; Burfield, D.R. *J. Org. Chem.* *1982*, *47*, 3821.

[223]See, for example, Schwetlick, K. *J. Chem. Soc. Perkin Trans. 2* *1988*, 2007.

[224]For a review of base-catalyzed autoxidations in general, see Sosnovsky, G.; Zaret, E.H., in Swern, D. *Organic Peroxides*, Vol. 1, Wiley, NY, *1970*, pp. 517–560.

[225]Barton, D.H.R.; Jones, D.W. *J. Chem. Soc.* *1965*, 3563; Russell, G.A.; Bemis, A.G. *J. Am. Chem. Soc.* *1966*, *88*, 5491.

[226]Gersmann, H.R.; Bickel, A.F. *J. Chem. Soc. B* *1971*, 2230.

[227]For reviews, see Frimer, A.A.; Stephenson, L.M. in Frimer, A.A. *Singlet O₂*, Vol. 2, CRC Press, Boca Raton, FL, *1985*, pp. 67–91; Wasserman, H.H.; Ives, J.L. *Tetrahedron* *1981*, *37*, 1825; Gollnick, K.; Kuhn, H.J., in Wasserman, H.H.; Murray, R.W. *Singlet Oxygen*, Academic Press, NY, *1979*, pp. 287–427; Denny, R.W.; Nickon, A. *Org. React.* *1973*, *20*, 133; Adams, W.R., in Augustine, R.L. *Oxidation*, Vol. 2, Marcel Dekker, NY, *1969*, pp. 65–112.

the ground-state oxygen (a triplet), but an excited singlet state[228] (in which all electrons are paired), and the function of the photosensitization is to promote the oxygen to this singlet state. Singlet oxygen can also be produced by nonphotochemical means,[229] for example, by the reaction between H_2O_2 and $NaOCl$[230] or sodium molybdate,[231] or between ozone and triphenyl phosphite.[232] Calcium peroxide diperoxohydrate ($CaO_2, 2 H_2O_2$) has been reported as a storable compound used for the chemical generation of singlet oxygen.[233] The oxygen generated by either photochemical or nonphotochemical methods reacts with alkenes in the same way;[234] this is evidence that singlet oxygen is the reacting species in the photochemical reaction and not some hypothetical complex between triplet oxygen and the photosensitizer, as had previously been suggested. The fact that 100% allylic rearrangement always takes place is incompatible with a free-radical mechanism, and

28   **29**   **30**

further evidence that free radicals are not involved comes from the treatment of optically active limonene (**28**) with singlet oxygen. Among other products is the optically active hydroperoxide **29**, though if **30** were an intermediate, it could not give an optically active product since it possesses a plane of symmetry.[235] In contrast, autoxidation of **28** gave optically inactive **29** (a mixture of four diastereomers in which the two pairs of enantiomers are present as racemic mixtures). As this example shows, singlet oxygen reacts faster with more-highly substituted than with less-highly substituted alkenes. The order of alkene reactivity is

---

[228]For books on singlet oxygen, see Frimer, A.A. *Singlet O₂*, 4 vols., CRC Press, Boca Raton, FL, *1985*; Wasserman, H.H.; Murray, R.W. *Singlet Oxygen*, Academic Press, NY, *1979*. For reviews, see Frimer, A.A., in Patai, S. *The Chemistry of Peroxides*, Wiley, NY, *1983*, pp. 201–234; Gorman, A.A.; Rodgers, M.A.J. *Chem. Soc. Rev.* **1981**, *10*, 205; Shinkarenko, N.V.; Aleskovskii, V.B. *Russ. Chem. Rev.* **1981**, *50*, 220; Shlyapintokh, V.Ya.; Ivanov, V.B. *Russ. Chem. Rev.* **1976**, *45*, 99; Ohloff, G. *Pure Appl. Chem.* **1975**, *43*, 481; Kearns, D.R. *Chem. Rev.* **1971**, *71*, 395; Wayne, R.P. *Adv. Photochem.* **1969**, *7*, 311.

[229]For reviews, see Turro, N.J.; Ramamurthy, V., in de Mayo, P. *Rearrangements in Ground and Excited States*, Vol. 3, Academic Press, NY, *1980*, pp. 1–23; Murray, R.W., in Wasserman, H.H.; Murray, R.W. *Singlet Oxygen*, Academic Press, NY, *1979*, pp. 59–114. For a general monograph, see Adam, W.; Cilento, G. *Chemical and Biological Generation of Excited States*; Academic Press, NY, *1982*.

[230]Foote, C.S.; Wexler, S. *J. Am. Chem. Soc.* **1964**, *86*, 3879.

[231]Aubry, J.M.; Cazin, B.; Duprat, F. *J. Org. Chem.* **1989**, *54*, 726.

[232]Murray, R.W.; Kaplan, M.L. *J. Am. Chem. Soc.* **1969**, *91*, 5358; Bartlett, P.D.; Mendenhall, G.D.; Durham, D.L. *J. Org. Chem.* **1980**, *45*, 4269.

[233]Pierlot, C.; Nardello, V.; Schrive, J.; Mabille, C.; Barbillat, J.; Sombret, B.; Aubry, J.-M. *J. Org. Chem*, **2002**, *67*, 2418.

[234]Foote, C.S.; Wexler, S.; Ando, W.; Higgins, R. *J. Am. Chem. Soc.* **1968**, *90*, 975. See also, McKeown, E.; Waters, W.A. *J. Chem. Soc. B* **1966**, 1040.

[235]Schenck, G.O.; Gollnick, K.; Buchwald, G.; Schroeter, S.; Ohloff, G. *Liebigs Ann. Chem.* **1964**, *674*, 93; Schenck, G.O.; Neumüller, O.; Ohloff, G.; Schroeter, S. *Liebigs Ann. Chem.* **1965**, *687*, 26.

tetrasubstituted > trisubstituted > disubstituted. Electron-withdrawing substituents deactivate the alkene.[236] In simple trisubstituted alkenes, there is a general preference for the hydrogen to be removed from the more highly congested side of the double bond.[237] With *cis*-alkenes of the form RCH=CHR', the hydrogen is removed from the larger R group.[238] Many functional groups in an allylic position cause the hydrogen to be removed from that side rather than the other (geminal selectivity).[239] Also, in alkyl-substituted alkenes, the hydrogen that is preferentially removed is the one geminal to the larger substituent on the double bond.[240]

Several mechanisms have been proposed for the reaction with singlet oxygen.[241] One of these is a pericyclic mechanism, similar to that of the ene synthesis (**15-23**) and to the first step of the reaction between alkenes and SeO$_2$ (**19-14**). However, there is strong evidence against this mechanism,[242] and a more likely mechanism involves addition of singlet oxygen to the double bond to give a perepoxide (**31**),[243] followed by internal proton transfer.[244]

31

Still other proposed mechanisms involve diradicals or dipolar intermediates.[245]
OS **IV**, 895.

[236]For example, see Foote, C.S.; Denny, R.W. *J. Am. Chem. Soc.* **1971**, *93*, 5162.

[237]Orfanopoulos, M.; Bellamine, M.; Grdina, M.J.; Stephenson, L.M. *J. Am. Chem. Soc.* **1979**, *101*, 275; Rautenstrauch, V.; Thommen, W.; Schulte-Elte, K.H. *Helv. Chim. Acta* **1986**, *69*, 1638 and references cited therein.

[238]Orfanopoulos, M.; Stratakis, M.; Elemes, Y. *Tetrahedron Lett.* **1989**, *30*, 4875.

[239]Clennan, E.L.; Chen, X.; Koola, J.J. *J. Am. Chem. Soc.* **1990**, *112*, 5193, and references cited therein.

[240]Orfanopoulos, M.; Stratakis, M.; Elemes, Y. *J. Am. Chem. Soc.* **1990**, *112*, 6417.

[241]For reviews of the mechanism, see Frimer, A.A.; Stephenson, L.M., in Frimer, A.A. *Singlet O$_2$*, Vol. 2, CRC Press, Boca Raton, FL, **1985**, pp. 80–87; Stephenson, L.M.; Grdina, M.J.; Orfanopoulos, M. *Acc. Chem. Res.* **1980**, *13*, 419; Gollnick, K.; Kuhn, H.J. Wasserman, H.H.; Murray, R.W. *Singlet Oxygen*, Academic Press, NY, **1979**, pp. 288–341; Frimer, A.A. *Chem. Rev.* **1979**, *79*, 359; Foote, C.S. *Acc. Chem. Res.* **1968**, *1*, 104; *Pure Appl. Chem.* **1971**, *27*, 635; Gollnick, K. *Adv. Photochem.* **1968**, *6*, 1; Kearns, D.R. *Chem. Rev.* **1971**, *71*, 395.

[242]Asveld, E.W.H.; Kellogg, R.M. *J. Org. Chem.* **1982**, *47*, 1250.

[243]For a review of perepoxides as intermediates in organic reactions, see Mitchell, J.C. *Chem. Soc. Rev.* **1985**, *14*, 399, p. 401.

[244]For evidence in favor of this mechanism, at least with some kinds of substrates, see Jefford, C.W.; Rimbault, C.G. *J. Am. Chem. Soc.* **1978**, *100*, 6437; Okada, K.; Mukai, T. *J. Am. Chem. Soc.* **1979**, *100*, 6509; Paquette, L.A.; Hertel, L.W.; Gleiter, R.; Böhm, M. *J. Am. Chem. Soc.* **1978**, *100*, 6510; Wilson, S.L.; Schuster, G.B. *J. Org. Chem.* **1986**, *51*, 2056; Davies, A.G.; Schiesser, C.H. *Tetrahedron Lett.* **1989**, *30*, 7099; Orfanopoulos, M.; Smonou, I.; Foote, C.S. *J. Am. Chem. Soc.* **1990**, *112*, 3607.

[245]See, for example, Jefford, C.W. *Helv. Chim. Acta* **1981**, *64*, 2534.

14-8   Formation of Peroxides

Alkyldioxy-de-hydrogenation

$$RH \quad + \quad R'OOH \quad \xrightarrow{\text{CuCl}} \quad ROOR'$$

Peroxy groups (ROO) can be introduced into susceptible organic molecules by treatment with a hydroperoxide in the presence of cuprous chloride or other catalysts, for example, cobalt and manganese salts.[246] Very high yields can be obtained. The type of hydrogen replaced is similar to that with NBS **(14-3)**, that is, mainly benzylic, allylic, and tertiary. The mechanism is therefore of the free-radical type, involving ROO• formed from ROOH and the metal ion. The reaction can be used to demethylate tertiary amines of the form R_2NCH_3, since the product R_2NHCH_2OOR' can easily be hydrolyzed by acid **(10-6)** to give R_2NH.[247]

14-9   Acyloxylation

Acyloxylation or **Acyloxy-de-hydrogenation**

Susceptible positions of organic compounds can be directly acyloxylated[248] by *tert*-butyl peroxyesters, the most frequently used being acetic and benzoic (R' = Me or Ph).[249] The reaction requires a catalyst (cuprous ion is the actual catalyst, but a trace is all that is necessary, and such traces are usually present in cupric compounds, so that these are often used) and without it is not selective. Susceptible positions are similar to those in **14-6**: benzylic, allylic, and the a position of ethers and sulfides. Terminal alkenes are substituted almost entirely in the 3 position, that is, with only a small amount of allylic rearrangement, but internal alkenes generally give mixtures containing a large amount of allylic-shift product. If the reaction with alkenes is carried out in an excess of

[246]For a review, see Sosnovsky, G.; Rawlinson, D.J., in Swern, D. *Organic Peroxides*, Vol. 2, Wiley, NY, *1970*, pp. 153–268. See also, Murahashi, S.; Naota, T.; Kuwabara, T.; Saito, T.; Kumobayashi, H.; Akutagawa, S. *J. Am. Chem. Soc. 1990, 112*, 7820; Sheldon, R.A., in Patai, S. *The Chemistry of Peroxides*, Wiley, NY, *1983*, p. 161.

[247]See Murahashi, S.; Naota, T.; Yonemura, K. *J. Am. Chem. Soc. 1988, 110*, 8256.

[248]For a list of reagents, with references, see Larock, R.C. *Comprehensive Organic Transformations*, 2nd ed, Wiley-VCH, NY, *1999*, pp. 1625–1630 *ff*, 1661–1663.

[249]For reviews, see Rawlinson, D.J.; Sosnovsky, G. *Synthesis 1972*, 1; Sosnovsky, G.; Rawlinson, D.J., in Swern, D. *Organic Peroxides*, Vol. 1, Wiley, NY, *1970*, pp. 585–608; Doumaux, Jr, A.R. in Augustine, R.L. *Oxidation*, Vol. 2, Marcel Dekker, NY, *1971*, pp. 141–185.

another acid R″COOH, the ester produced is of *that* acid ROCOR″. Aldehydes give anhydrides:

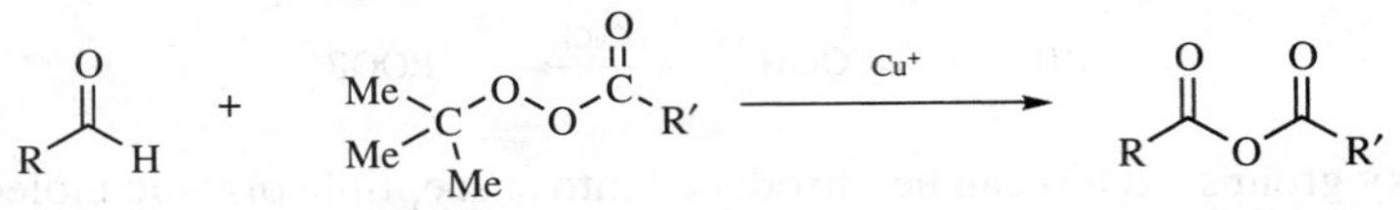

Acyloxylation has also been achieved with metallic acetates, such as lead tetraacetate,[250] mercuric acetate,[251] and palladium(II) acetate.[252] In the case of the lead and mercuric acetates, not only does the reaction take place at allylic and benzylic positions and at those α to an OR or SR group, but also at positions α to the carbonyl groups of aldehydes, ketones, or esters and at those a to two carbonyl groups (ZCH$_2$Z′). It is likely that in the latter cases it is the enol forms that react. Ketones can be α-acyloxylated indirectly by treatment of various enol derivatives with metallic acetates, for example, silyl enol ethers with silver carboxylates-iodine,[253] enol thioethers with lead tetraacetate,[254] and enamines[255] with lead tetraacetate[256] or thallium triacetate.[257] α,β-Unsaturated ketones can be acyloxylated in good yields in the α′ position with manganese triacetate.[258] Palladium acetate converts alkenes to vinylic and/or allylic acetates.[259] Lead tetraacetate even acyloxylates alkanes, in a slow reaction (10 days to 2 weeks), with tertiary and secondary positions greatly favored over primary ones.[260] Yields are as high as 50%. Acyloxylation of certain alkanes has also been reported with palladium(II) acetate.[261]

[250]For a review of lead tetraacetate, see Butler, R.N., in Pizey, J.S. *Synthetic Reagents*, Vol. 3, Wiley, NY, p. 277.
[251]For reviews, see Larock, R.C. *Organomercury Compounds in Organic Synthesis*, Springer, NY, *1985*, pp. 190–208; Rawlinson, D.J.; Sosnovsky, G. *Synthesis 1973*, 567.
[252]Hansson, S.; Heumann, A.; Rein, T.; Åkermark, B. *J. Org. Chem. 1990*, *55*, 975; Byström, S.E.; Larsson, E.M.; Åkermark, B. *J. Org. Chem. 1990*, *55*, 5674.
[253]Rubottom, G.M.; Mott, R.C.; Juve Jr, H.D. *J. Org. Chem. 1981*, *46*, 2717.
[254]Trost, B.M.; Tanigawa, Y. *J. Am. Chem. Soc. 1979*, *101*, 4413.
[255]For a review, see Cook, A.G., in Cook, A.G. *Enamines*, 2nd ed., Marcel Dekker, NY, *1988*, pp. 251–258.
[256]See Butler, R.N. *Chem. Ind. (London) 1976*, 499.
[257]Kuehne, M.E.; Giacobbe, T.J. *J. Org. Chem. 1968*, *33*, 3359.
[258]Demir, A.S.; Sayrac, T.; Watt, D.S. *Synthesis 1990*, 1119.
[259]For reviews, see Rylander, P.N. *Organic Synthesis with Noble Metal Catalysts*, Academic Press, NY, *1973*, pp. 80–87; Jira, R.; Freiesleben, W. *Organomet. React. 1972*, *3*, 1, pp. 44–84; Heck, R.F. *Fortschr. Chem. Forsch. 1971*, *16*, 221, pp. 231–237; Tsuji, J. *Adv. Org. Chem. 1969*, *6*, 109, pp. 132–143.
[260]Bestre, R.D.; Cole, E.R.; Crank, G. *Tetrahedron Lett. 1983*, *24*, 3891; Mosher, M.W.; Cox, J.L. *Tetrahedron Lett. 1985*, *26*, 3753.
[261]This was done in trifluoroacetic acid, and the products were trifluoroacetates: Sen, A.; Gretz, E.; Oliver, T.F.; Jiang, Z. *New J. Chem. 1989*, *13*, 755.

Studies of the mechanism of the cuprous-catalyzed reaction show that the most common mechanism is the following:[262]

Step 1 R'–C(=O)–O–O–t-Bu + Cu⁺ ⟶ R'–C(=O)–O–O–Cu⁺ (II) + t-BuO•

Step 2 R-H + t-BuO• ⟶ R• + t-BuO•

Step 3 R• + R'–C(=O)–O–O–Cu⁺ (II) ⟶ R'–C(=O)–O–R + Cu⁺

32

This mechanism, involving a free radical R•, is compatible with the allylic rearrangements found.[263] The finding that *tert*-butyl peroxyesters labeled with ^{18}O in the carbonyl oxygen gave ester with 50% of the label in each oxygen[264] is in accord with a combination of R• with the intermediate **32**, in which the copper is ionically bound, so that the oxygens are essentially equivalent. Other evidence is that *tert*-butoxy radicals have been trapped with dienes.[265] Much less is known about the mechanisms of the reactions with metal acetates.[266]

Free-radical acyloxylation of aromatic substrates[267] has been accomplished with a number of reagents including copper(II) acetate,[268] benzoyl peroxide-iodine,[269] silver(II) complexes,[270] and cobalt(III) trifluoroacetate.[271]

OS **III**, 3; **V**, 70, 151; **VIII**, 137.

C. Substitution by Sulfur

14-10 Chlorosulfonation or Chlorosulfo-de-hydrogenation

$$RH + SO_2 + Cl_2 \xrightarrow{h\nu} RSO_2Cl$$

[262]Kharasch, M.S.; Sosnovsky, G.; Yang, N.C. *J. Am. Chem. Soc.* **1959**, *81*, 5819; Kochi, J.K.; Mains, H.E. *J. Org. Chem.* **1965**, *30*, 1862. See also, Beckwith, A.L.J.; Zavitsas, A.A. *J. Am. Chem. Soc.* **1986**, *108*, 8230.
[263]Goering, H.L.; Mayer, U. *J. Am. Chem. Soc.* **1964**, *86*, 3753; Denney, D.B.; Appelbaum, A.; Denney, D.Z. *J. Am. Chem. Soc.* **1962**, *84*, 4969.
[264]Denney, D.B.; Denney, D.Z.; Feig, G. *Tetrahedron Lett.* **1959**, no. 15, p. 19.
[265]Kochi, J.K. *J. Am. Chem. Soc.* **1962**, *84*, 2785, 3271; Story, P.R. *Tetrahedron Lett.* **1962**, 401.
[266]See, for example, Jones, S.R.; Mellor, J.H. *J. Chem. Soc. Perkin Trans. 2* **1977**, 511.
[267]For a review, see Haines, A.H. *Methods for the Oxidation of Organic Compounds*, Academic Press, NY, **1985**, pp. 177–180, 351–355.
[268]Takizawa, Y.; Tateishi, A.; Sugiyama, J.; Yoshida, H.; Yoshihara, N. *J. Chem. Soc., Chem. Commun.* **1991**, 104. See also Kaeding, W.W.; Kerlinger, H.O.; Collins, G.R. *J. Org. Chem.* **1965**, *30*, 3754.
[269]For example, see Kovacic, P.; Reid, C.G.; Brittain, T.J. *J. Org. Chem.* **1970**, *35*, 2152.
[270]Nyberg, K.; Wistrand, L.G. *J. Org. Chem.* **1978**, *43*, 2613.
[271]Kochi, J.K.; Tank, R.T.; Bernath, T. *J. Am. Chem. Soc.* **1973**, *95*, 7114; DiCosimo, R.; Szabo, H. *J. Org. Chem.* **1986**, *51*, 1365.

The chlorosulfonation of organic molecules with chlorine and sulfur dioxide is called the *Reed reaction*.[272] In scope and range of products obtained, the reaction is similar to **14-1**. The mechanism is also similar, except that there are two additional main propagation steps:

$$R\bullet + SO_2 \longrightarrow R{-}SO_2\bullet$$
$$R{-}SO_2\bullet + Cl_2 \longrightarrow R{-}SO_2Cl + Cl\bullet$$

Chlorosulfenation[273] can be accomplished by treatment with SCl_2 and UV light: $RH + SCl_2 \xrightarrow{h\nu} RSCl$.

D. Substitution by Nitrogen

14-11   The Direct Conversion of Aldehydes to Amides

Amination or **Amino-de-hydrogenation**

$$ArCHO \xrightarrow[\text{NBS-AIBN}]{\text{NH}_3} ArCONH_2$$

Aliphatic and aromatic aldehydes have been converted to the corresponding amides with ammonia or a primary or secondary amine, NBS, and a catalytic amount of AIBN (p. 935).[274] In a reaction of more limited scope, amides are obtained from aromatic and α,β-unsaturated aldehydes by treatment with dry ammonia gas and nickel peroxide.[275] Best yields (80–90%) are obtained at -25 to $-20°C$. In the nickel peroxide reaction the corresponding alcohols ($ArCH_2OH$) have also been used as substrates.

The reaction has also been performed with MnO_2 and $NaCN$ along with ammonia or an amine at $0°C$ in isopropyl alcohol.[276] Aldehydes were also shown to react with hydroxylamine hydrochloride at $140°C$ in the presence of aluminum oxide and methanesulfonic acid.[277] Treatment of a aldehyde with iodine in aqueous ammonia, followed by oxidation with aqueous hydrogen peroxide generates a primary amide.[278] Secondary amines react with aldehydes to the an amide in using a palladium catalyst[279] or a rhodium catalyst.[280] For an indirect way of converting aldehydes to amides, see **12-32**. Thioamides $RCSNR'_2$ have been prepared in good yield

[272]For a review, see Gilbert, E.E. *Sulfonation and Related Reactions*, Wiley, NY, *1965*, pp. 126–131.

[273]Müller, E.; Schmidt, E.W. *Chem. Ber. 1963*, 96, 3050; *1964*, 97, 2614. For a review of the formation and reactions of sulfenyl halides, see Kühle, E. *Synthesis 1970*, 561; *1971*, 563, 617.

[274]Markó, I.E.; Mekhalfia, A. *Tetrahedron Lett. 1990*, 31, 7237.

[275]Nakagawa, K.; Onoue, H.; Minami, K. *Chem. Commun. 1966*, 17.

[276]Gilman, N.W. *Chem. Commun. 1971*, 733.

[277]Sharghi, H.; Sarvari, M.H. *J. Chem. Res. (S) 2001*, 446.

[278]Shie, J.-J.; Fang, J.-M. *J. Org. Chem. 2003*, 68, 1158.

[279]Tamaru, Y.; Yamada, Y.; Yoshida, Z. *Synthesis 1983*, 474.

[280]Tillack, A.; Rudloff, I.; Beller, M. *Eur. J. Org. Chem. 2001*, 523.

from thioaldehydes (produced *in situ* from phosphoranes and sulfur) and secondary amines.[281]

14-12   Amidation and Amination at an Alkyl Carbon

Acylamino-de-hydrogenation

$$R_3C\text{-}H \quad + \quad CH_3CN \quad \xrightarrow[\text{H}_3(\text{PW}_{12}\text{O}_{40})\cdot\text{H}_2\text{O}]{h\nu} \quad R_3C\!-\!\underset{\underset{H}{|}}{N}\!-\!\overset{\overset{O}{\|}}{C}\!-\!CH_3 \quad + \quad H_2$$

When alkanes bearing a tertiary hydrogen are exposed to UV light in acetonitrile containing a heteropolytungstic acid, they are amidated.[282] The oxygen in the product comes from the tungstic acid. When the substrate bears two adjacent tertiary hydrogens, alkenes are formed (by loss of two hydrogens), rather than amides (**19-2**). Amidyl radicals can be generated by other means.[283]

An electrochemical method for amination has been reported by Shono and co-workers.[284] Derivatives of malonic esters containing an *N*-tosyl group were cyclized in high yields by anodic oxidation:

Three-, four-, and five-membered rings were synthesized by this procedure.

14-13   Substitution by Nitro

Nitro-de-carboxylation

In a reaction termed a "nitro-Hunsdiecker" (see **14-30**), vinyl carboxylic acids (conjugated acids) are treated with nitric acid and a catalytic amount of AIBN (p. 935). The product is the vinyl nitro compound, generated via decarboxylation of a radical intermediate.[285]

[281]Okuma, K.; Komiya, Y.; Ohta, H. *Chem. Lett.* **1988**, 1145.
[282]Renneke, R.F.; Hill, C.L. *J. Am. Chem. Soc.* **1986**, *108*, 3528.
[283]Moutrille, C.; Zard, S.Z. *Chem. Commun.* **2004**, 1848.
[284]Shono, T.; Matsumura, Y.; Katoh, S.; Ohshita, J. *Chem. Lett.* **1988**, 1065.
[285]Das, J.P.; Sinha, P.; Roy, S. *Org. Lett.* **2002**, *4*, 3055.

E. Substitution by Carbon

In these reactions, a new carbon–carbon bond is formed and they may be given the collective title *coupling reactions*. In each case, an alkyl or aryl radical is generated and then combines with another radical (a termination process) or attacks an aromatic ring or alkene to give the coupling product.[286]

14-14   Simple Coupling at a Susceptible Position

De-hydrogen-coupling

$$2 \text{ RH} \longrightarrow \text{R—R}$$

Alkane and alkyl substrates RH are treated with peroxides, which decompose to give a radical that abstracts a hydrogen from RH to give R•, which dimerizes. Dialkyl and diacyl peroxides have been used, as well as Fenton's reagent (p. 964). This reaction is far from general, though in certain cases respectable yields have been obtained. Among susceptible positions are those at a tertiary carbon,[287] as well as those α to a phenyl group (especially if there is also an α-alkyl or α-chloro group),[288] an ether group,[289] a carbonyl group,[290] a cyano group,[291] a dialkylamino group,[292] or a carboxylic ester group, either the acid or alcohol side.[293] Cross-coupling is possible in some cases. When toluene was heated with allyl bromide, in the presence of di-*tert*-butyl peroxide, 4-phenyl-1-butene was formed quantitatively.[294]

Conjugated amide were coupled via the γ-carbon to give good yields of the dimeric diamide, with an excess of samarium (II) iodide, and with modest enantioselectivity using a chiral additive.[295]

$$2 \text{ RH} \xrightarrow[\text{Hg}]{h\nu} \text{R—R} + \text{H}_2$$

[286]For a monograph on the formation of C—C bonds by radical reactions, see Giese, B. *Radicals in Organic Synthesis: Formation of Carbon–Carbon Bonds*, Pergamon, Elsmsford, NY, *1986*. For a review of arylation at carbon, see Abramovitch, R.A.; Barton, D.H.R.; Finet, J. *Tetrahedron* *1988*, *44*, 3039. For a review of aryl–aryl coupling, see Sainsbury, M. *Tetrahedron* *1980*, *36*, 3327.

[287]Meshcheryakov, A.P.; Érzyutova, E.I. *Bull. Acad. Sci. USSR Div. Chem. Sci*, *1966*, 94.

[288]McBay, H.C.; Tucker, O.; Groves, P.T. *J. Org. Chem.* *1959*, *24*, 536; Johnston, K.M.; Williams, G.H. *J. Chem. Soc.* *1960*, 1168.

[289]Pfordte, K.; Leuschner, G. *Liebigs Ann. Chem.* *1961*, *643*, 1.

[290]Kharasch, M.S.; McBay, H.C.; Urry, W.H. *J. Am. Chem. Soc.* *1948*, *70*, 1269; Leffingwell, J.C. *Chem. Commun.* *1970*, 357; Hawkins, E.G.E.; Large, R. *J. Chem. Soc. Perkin Trans. 1 1974*, 280.

[291]Kharasch, M.S.; Sosnovsky, G. *Tetrahedron* *1958*, *3*, 97.

[292]Schwetlick, K.; Jentzsch, J.; Karl, R.; Wolter, D. *J. Prakt. Chem.* *1964*, *[4] 25*, 95.

[293]Boguslavskaya, L.S.; Razuvaev, G.A. *J. Gen. Chem. USSR 1963*, *33*, 1967.

[294]Tanko, J.M.; Sadeghipour, M. *Angew. Chem. Int. Ed. 1999*, *38*, 159.

[295]Kikukawa, T.; Hanamoto, T.; Inanaga, J. *Tetrahedron Lett. 1999*, *40*, 7497.

Alkanes can be dimerized by vapor-phase mercury photosensitization[296] in a synthetically useful process. Best results are obtained for coupling at tertiary positions, but compounds lacking tertiary hydrogens (e.g., cyclohexane) also give good yields. Dimerization of *n*-alkanes gives secondary-secondary coupling in a nearly statistical distribution, with primary positions essentially unaffected. Alcohols and ethers dimerize at the position α to the oxygen [e.g., 2 EtOH → MeCH(OH)CH(OH)Me].

When a mixture of compounds is treated, cross-dimerization (to give **33**) and homodimerization take place statistically.

Even with the limitation on yield implied by the statistical process, cross-dimerization is still useful when one of the reactants is an alkane, because the products are easy to separate, and because of the few other ways to functionalize an alkane. The cross-coupling of an alkane with trioxane is especially valuable, because hydrolysis of the product (**10-6**) gives an aldehyde, thus achieving the conversion RH → RCHO. The mechanism probably involves abstraction of H by the excited Hg atom, and coupling of the resulting radicals.

The reaction has been extended to ketones, carboxylic acids and esters (all of which couple a to the C=O group), and amides (which couple α to the nitrogen) by running it in the presence of H_2.[297] Under these conditions it is likely that the excited Hg abstracts H• from H_2, and that the remaining H• abstracts H from the substrate. Radicals have also been generated at benzylic positions and shown to couple with epoxides, forming an alcohol.[298]

OS **IV**, 367; **V**, 1026; **VII**, 482.

14-15   Coupling at a Susceptible Position Via Silanes

De-silyl-coupling

Under electrochemical conditions it is possible to couple two silanes. The reaction of **34** and allyltrimethylsilane, for example, gave the corresponding homoallylic ether.[299]

[296]Brown, S.H.; Crabtree, R.H. *J. Am. Chem. Soc.* **1989**, *111*, 2935, 2946; *J. Chem. Educ.* **1988**, *65*, 290.
[297]Boojamra, C.G.; Crabtree, R.H.; Ferguson, R.R.; Muedas, C.A. *Tetrahedron Lett.* **1989**, *30*, 5583.
[298]Rawal, V.H.; Krishnamurthy, V.; Fabre, A. *Tetrahedron Lett.* **1993**, *34*, 2899.
[299]Suga, S.; Suzuki, S.; Yamamoto, A.; Yoshida, J.-i. *J. Am. Chem. Soc.* **2000**, *122*, 10244.

14-16 Coupling of Alkynes[300]

De-hydrogen-coupling

$$2 \ R-C\equiv C-H \xrightarrow[\text{pyridine}]{\text{CuX}_2} R-C\equiv C-C\equiv C-R$$

Terminal alkynes can be coupled by heating with stoichiometric amounts of cupric salts in pyridine or a similar base. This reaction, which produces symmetrical diynes in high yields, is called the *Eglinton reaction.*[301] The large-ring annulenes of Sondheimer et al. (see p. 71) were prepared by rearrangement and hydrogenation of cyclic polyynes,[302] prepared by the Eglinton reaction with terminal diynes to give **35**, a cyclic trimer of 1,5-hexadiyne.[303] The corresponding tetramers (C_{24}), pentamers (C_{30}), and hexamers (C_{36}) were also formed. The Eglinton reaction is of wide scope. Many functional groups can be present on the alkyne. The oxidation is usually quite specific for triple-bond hydrogen.

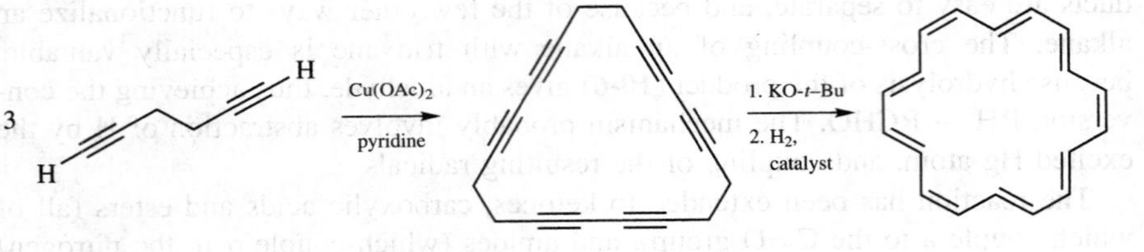

35

Another common procedure is the use of catalytic amounts of cuprous salts in the presence of ammonia or ammonium chloride (this method is called the *Glaser reaction*). Atmospheric oxygen or some other oxidizing agent, such as permanganate or hydrogen peroxide is required in the latter procedure. This method is not satisfactory for cyclic coupling. Hydrogen peroxide, potassium permanganate, potassium ferricyanide, iodine or Cu(II) can be used instead of oxygen as oxidants.[304] Isolation of copper acetylide during the reaction can be avoided by doing the reaction in pyridine or cyclohexylamine, in the presence of catalytic amount of

---

[300]For a review, see Siemsen, P.; Livingston, R.C.; Diederich, F. *Angew. Chem. Int. Ed.* **2000**, *39*, 2632.

[301]For reviews, see Simándi, L.I., in Patai, S.; Rappoport, Z. *The Chemistry of Functional Groups, Supplement C* pt. 1, Wiley, NY, **1983**, pp. 529–534; Nigh, W.G., in Trahanovsky, W.S. *Oxidation in Organic Chemistry*, pt. B, Academic Press, NY, **1973**, pp. 11–31; Cadiot, P.; Chodkiewicz, W., in Viehe, H.G. *Acetylenes*; Marcel Dekker, NY; **1969**, pp. 597–647.

[302]For a review of cyclic alkynes, see Nakagawa, M., in Patai, S. *The Chemistry of the Carbon-Carbon Triple Bond*, pt. 2, Wiley, NY, **1978**, pp. 635–712.

[303]Sondheimer, F.; Wolovsky, R. *J. Am. Chem. Soc.* **1962**, *84*, 260; Sondheimer, F.; Wolovsky, R.; Amiel, Y. *J. Am. Chem. Soc.* **1962**, *84*, 274.

[304]Gunter, H.V. *Chemistry of Acetylenes*, Marcel Dekker, NY, **1969**, pp. 597–647 and references cited therein.

$CuCl_3$.[305] If the Glaser reaction is done with a N,N,N',N'-tetramethylethylenediamine–CuCl complex, the reaction proceeds in good yield in virtually any organic solvent.[306] When molecular oxygen is the oxidant, this modification of Glaser condensation is known as the *Hay reaction*. A variation couples terminal alkynes using $CuCl_2$ in supercritical CO_2 (see p. 414),[307] and in ionic liquids.[308] Coupling was also achieved using $CuCl_2$ on KF–Al_2O_3 with microwave irradiation.[309] Homocoupling of alkynyl amines R_2N–C≡CH to give the diyne R_2N–C≡C–C≡C-NR_2 was reported in aerated acetone with 10% CuI and 20% TMEDA.[310]

Unsymmetrical diynes can be prepared by *Cadiot–Chodkiewicz* coupling:[311]

$$R-C{\equiv}C-H \quad + \quad R'-C{\equiv}C-Br \quad \xrightarrow{\text{Cu}^+} \quad R-C{\equiv}C-C{\equiv}C-R' \quad + \quad HBr$$

This may be regarded as a variation of **10-74**, but it must have a different mechanism since acetylenic halides give the reaction but ordinary alkyl halides do not, which is hardly compatible with a nucleophilic mechanism. However, the mechanism is not fully understood. One version of this reaction binds the alkynyl bromide unit to a polymer, and the di-yne is released from the polymer after the solid state transformation.[312] Alkynes have also been coupled using CuI and a palladium catalyst.[313] Propargyl halides also give the reaction,[314] as do 1-bromo propargylic alcohols (Br–C≡C–CH_2OH).[315] A variation of the Cadiot–Chodkiewicz method consists of treating a haloalkyne (R'C≡CX) with a copper acetylide (RC≡CCu).[316] The Cadiot–Chodkiewicz procedure can be adapted to the preparation of diynes in which R' = H by the use of BrC≡C$SiEt_3$ and subsequent cleavage of the $SiEt_3$ group.[317] This protecting group can also be used in the Eglinton or Glaser methods.[318]

The mechanism of the Eglinton and Glaser reactions probably begins with loss of a proton

$$R-C{\equiv}C-H \quad \xrightarrow{\text{base}} \quad R-C{\equiv}C^-$$

[305]Stansbury, H A.; Proops, W.R. *J. Org. Chem.* **1962**, *27*, 320.

[306]Hay, A.S. *J. Org. Chem.* **1960**, *25*, 1275; Hay, A S. *J. Org. Chem.* **1962**, *27*, 3320.

[307]Li, J.; Jiang, H. *Chem. Commun.* **1999**, 2369.

[308]In bmim PF$_6$, 1-butyl-3-methylimidazolium hexafluorophosphate: Yadav, J.S.; Reddy, B.V.S.; Reddy, K.B.; Gayathri, K.U.; Prasad, A.R. *Tetrahedron Lett.* **2003**, *44*, 6493.

[309]Kabalka, G.W.; Wang, L.; Pagni, R.M. *Synlett* **2001**, 108.

[310]Rodríguez, D.; Castedo, L.; Saá, C. *Synlett* **2004**, 377.

[311]Chodkiewicz, W. *Ann. Chim. (Paris)* **1957**, *[13] 2*, 819.

[312]Montierth, J.M.; DeMario, D.R.; Kurth, M.J.; Schore, N.E. *Tetrahedron* **1998**, *54*, 11741.

[313]Liu, Q.; Burton, D.J. *Tetrahedron Lett.* **1997**, *38*, 4371.

[314]Sevin, A.; Chodkiewicz, W.; Cadiot, P. *Bull. Soc. Chim. Fr.* **1974**, 913.

[315]Marino, J.P.; Nguyen, H.N. *J. Org. Chem.* **2002**, *67*, 6841.

[316]Curtis, R.F.; Taylor, J.A. *J. Chem. Soc. C* **1971**, 186.

[317]Eastmond, R.; Walton, D.R.M. *Tetrahedron* **1972**, *28*, 4591; Ghose, B.N.; Walton, D.R.M. *Synthesis* **1974**, 890.

[318]Johnson, T.R.; Walton, D.R.M. *Tetrahedron* **1972**, *28*, 5221.

since there is a base present and acetylenic protons are acidic. It is known, of course, that cuprous ion can form complexes with triple bonds. The last step is probably the coupling of two radicals:

$$R-C\equiv C\cdot \longrightarrow R-C\equiv C-C\equiv C-R$$

but just how the carbanion becomes oxidized to the radical and what part the cuprous ion plays (other than forming the acetylide salt) are matters of considerable speculation,[319] and depend on the oxidizing agent. One proposed mechanism postulated Cu(II) as the oxidant.[320] It has been shown that molecular oxygen forms adducts with Cu(I) supported by tertiary amines, which might be the intermediates in the Glaser reaction where molecular oxygen is the oxidant.[321] For the Hay reaction, the mechanism involves a Cu^I/Cu^{III}/Cu^{II}/Cu^I catalytic cycle, and the key step for this reaction is the dioxygen activation during complexation of two molecules of acetylide with molecular oxygen, giving a Cu(III) complex.[322] This mechanism is supported by isolation and characterization of Cu(III) complexes formed under the conditions of the Glaser coupling.

Terminal alkynes are not the only reaction partners. 1-Trimethylsilyl alkynes ($R-C\equiv C-SiMe_3$) give the diyne $R-C\equiv C-C\equiv C-R$) upon reaction with CuCl[323] or Cu(OAc)$_2$/Bu$_4$NF.[324]

OS **V**, 517; **VI**, 68, 925; **VIII**, 63.

14-17   Alkylation and Arylation of Aromatic Compounds by Peroxides

Alkylation or **Alkyl-de-hydrogenation**

$$Ar-H \quad + \quad \underset{\overset{\|}{O}}{R-\overset{\overset{O}{\|}}{C}}-O-O-\underset{\overset{\|}{O}}{\overset{}{C}}-R \quad \longrightarrow \quad Ar-R$$

This reaction is most often carried out with R = aryl, so the net result is the same as in **13-27**, though the reagent is different.[325] It is used less often than **13-27**, but

[319]See the discussions, in Nigh, W.G., in Trahanovsky, W.S. *Oxidation in Organic Chemistry*, pt. B, Academic Press, NY, *1973*, pp. 27–31; Fedenok, L.G.; Berdnikov, V.M.; Shvartsberg, M.S. *J. Org. Chem. USSR 1973*, *9*, 1806; Clifford, A.A.; Waters, W.A. *J. Chem. Soc. 1963*, 3056.

[320]Bohlmann, F.; Schönowsky, H.; Inhoffen, E.; Grau, G. *Chem. Ber. 1964*, *97*, 794.

[321]Wieghardt, K.; Chaudhuri, P. *Prog. Inorg. Chem. 1987*, *37*, 329.

[322]Fomina, L.; Vazquez, B.; Tkatchouk, E.; Fomine, S. *Tetrahedron 2002*, *58*, 6741.

[323]Nishihara, Y.; Ikegashira, K.; Hirabayashi, K.; Ando, J.-i.; Mori, A.; Hiyama, T. *J. Org. Chem. 2000*, *65*, 1780.

[324]Heuft, M.A.; Collins, S.K.; Yap, G.P.A.; Fallis, A.E. *Org. Lett. 2001*, *3*, 2883.

[325]For reviews, see Bolton, R.; Williams, G.H. *Chem. Soc. Rev. 1986*, *15*, 261; Hey, D.H. *Adv. Free-Radical Chem. 1966*, *2*, 47.

the scope is similar. When R = alkyl, the scope is more limited.[326] Only certain aromatic compounds, particularly benzene rings with two or more nitro groups, and fused ring systems, can be alkylated by this procedure. 1,4-Quinones can be alkylated with diacyl peroxides or with lead tetraacetate (methylation occurs with this reagent).

The mechanism is as shown on p. 940 (CIDNP has been observed[327]); the radicals are produced by

Since no relatively stable free radical is present (such as •O—N=N—Ar in **13-27**), most of the product arises from dimerization and disproportionation.[328] The addition of a small amount of nitrobenzene increases the yield of arylation product because the nitrobenzene is converted to diphenyl nitroxide, which abstracts the hydrogen from **5** and reduces the extent of side reactions.[329]

$$ArH \ + \ Ar'Pb(OAc)_3 \ \longrightarrow \ ArAr'$$

Aromatic compounds can also be arylated by aryllead tricarboxylates.[330] Best yields (~70–85%) are obtained when the substrate contains alkyl groups; an electrophilic mechanism is likely. Phenols are phenylated ortho to the OH group (and enols are a phenylated) by triphenylbismuth dichloride or by certain other Bi(V) reagents.[331] O-Phenylation is a possible side reaction. As with the aryllead tricarboxylate reactions, a free-radical mechanism is unlikely.[332]

OS **V**, 51. See also, OS **V**, 952; **VI**, 890.

14-18 Photochemical Arylation of Aromatic Compounds

Arylation or Aryl-de-hydrogenation

$$ArH \ + \ Ar'I \ \xrightarrow{\textit{hv}} \ ArAr'$$

[326]For reviews of the free-radical alkylation of aromatic compounds, see Tiecco, M.; Testaferri, L. *React. Intermed. (Plenum)* **1983**, *3*, 61; Dou, H.J.; Vernin, G.; Metzger, J. *Bull. Soc. Chim. Fr.* **1971**, 4593.

[327]Kaptein, R.; Freeman, R.; Hill, H.D.W.; Bargon, J. *J. Chem. Soc., Chem. Commun.* **1973**, 953.

[328]We have given the main steps that lead to biphenyls. The mechanism is actually more complicated than this and includes >100 elementary steps resulting in many side products, including those mentioned on p. $$$: DeTar, D.F.; Long, R.A.J.; Rendleman, J.; Bradley, J.; Duncan, P. *J. Am. Chem. Soc.* **1967**, *89*, 4051; DeTar, D.F. *J. Am. Chem. Soc.* **1967**, *89*, 4058. See also, Jandu, K.S.; Nicolopoulou, M.; Perkins, M.J. *J. Chem. Res. (S)* **1985**, 88.

[329]Chalfont, G.R.; Hey, D.H.; Liang, K.S.Y.; Perkins, M.J. *J. Chem. Soc. B* **1971**, 233.

[330]Bell, H.C.; Kalman, J.R.; May, G.L.; Pinhey, J.T.; Sternhell, S. *Aust. J. Chem.* **1979**, *32*, 1531.

[331]For a review, see Abramovitch, R.A.; Barton, D.H.R.; Finet, J. *Tetrahedron* **1988**, *44*, 3039, pp. 3040–3047.

[332]Barton, D.H.R.; Finet, J.; Giannotti, C.; Halley, F. *J. Chem. Soc. Perkin Trans. 1* **1987**, 241.

Another free-radical arylation method consists of the photolysis of aryl iodides in an aromatic solvent.[333] Yields are generally higher than in **13-27** or **14-17**. The aryl iodide may contain OH or COOH groups. The coupling reaction of iodobenzene and azulene to give a phenylazulene was reported (41% conversion and 85% yield).[334] The mechanism is similar to that of **13-27**. The aryl radicals are generated by the photolytic cleavage $ArI \rightarrow AR\cdot + I\cdot$. The reaction has been applied to intramolecular arylation (analogous to the Pschorr reaction).[335] A similar reaction is photolysis of an arylthallium bis(trifluoroacetate) (**12-23**) in an aromatic solvent. Here too, an unsymmetrical biaryl is produced in good yields.[336] In this case, it is the C—Tl bond that is cleaved to give aryl radicals.

$$Ar'Tl(OCOCF_3)_2 \quad \xrightarrow[ArH]{hv} \quad ArAr'$$

14-19 Alkylation, Acylation, and Carbalkoxylation of Nitrogen Heterocycles[337]

Alkylation or **Alkyl-de-hydrogenation**, and so on

$$\text{pyridine} + RCOOH \quad \xrightarrow[\text{2. } (NH_4)_2S_2O_8]{\text{1. } AgNO_3 - H_2SO_4 - H_2O} \quad \text{4-R-pyridine} + \text{2-R-pyridine}$$

Alkylation of protonated nitrogen heterocycles (e.g., pyridines, quinolines) can be accomplished by treatment with a carboxylic acid, silver nitrate, sulfuric acid, and ammonium peroxydisulfate.[338] The R group can be primary, secondary, or tertiary. The attacking species is R•, formed by[339]

$$2\,Ag^+ + S_2O_8^{2-} \quad \longrightarrow \quad 2\,Ag^{2+} + 2\,SO_4^{2-}$$

$$RCOOH + Ag^{2+} \quad \longrightarrow \quad RCOO\cdot + H^+ + Ag^+$$

$$RCOO\cdot \quad \longrightarrow \quad R\cdot + CO_2$$

[333]Wolf, W.; Kharasch, N. *J. Org. Chem.* **1965**, *30*, 2493. For a review, see Sharma, R.K.; Kharasch, N. *Angew. Chem. Int. Ed.* **1968**, *7*, 36.

[334]Ho, T.-I.; Ku, C.-K.; Liu, R.S.H. *Tetrahedron Lett.* **2001**, *42*, 715.

[335]See, for example, Kupchan, S.M.; Wormser, H.C. *J. Org. Chem.* **1965**, *30*, 3792; Jeffs, P.W.; Hansen, J.F. *J. Am. Chem. Soc.* **1967**, *89*, 2798; Thyagarajan, B.S.; Kharasch, N.; Lewis, H.B.; Wolf, W. *Chem. Commun.* **1967**, 614.

[336]Taylor, E.C.; Kienzle, F.; McKillop, A. *J. Am. Chem. Soc.* **1970**, *92*, 6088.

[337]For reviews; see Heinisch, G. *Heterocycles* **1987**, *26*, 481; Minisci, F.; Vismara, E.; Fontana, F. *Heterocycles* **1989**, *28*, 489; Minisci, F. *Top. Curr. Chem.* **1976**, *62*, 1, pp. 17; *Synthesis* **1973**, 1, pp. 12–19. For a review of substitution of carbon groups on nitrogen heterocycles see Vorbrüggen, H.; Maas, M. *Heterocycles* **1988**, *27*, 2659.

[338]Fontana, F.; Minisci, F.; Barbosa, M.C.N.; Vismara, E. *Tetrahedron* **1990**, *46*, 2525.

[339]Anderson, J.M.; Kochi, J.K. *J. Am. Chem. Soc.* **1970**, *92*, 1651.

A hydroxymethyl group can be introduced (ArH $\rightarrow$ ArCH$_2$OH) by several varia-tions of this method.[340] Alkylation of these substrates can also be accomplished by generating the alkyl radicals in other ways: from hydroperoxides and FeSO$_4$,[341] from alkyl iodides and H$_2$O$_2$–Fe(II),[342] from carboxylic acids and lead tetraacetate, or from the photochemically induced decarboxylation of carboxylic acids by iodosobenzene diacetate.[343]

Protonated nitrogen heterocycles, such as quinoxaline (**36**), can be acylated by treatment with an aldehyde, *tert*-butyl hydroperoxide, sulfuric acid, and ferrous sul-fate, in this case giving **37**.[344]

Photochemical alkylation of protonated quinoline occurred with Ph$_2$Se(O$_2$Cc-C$_6$H$_{11}$)$_2$.[345]

Other positively charged heterocycles react as well. When *N*-fluoropyridinium triflate was treated with the enolate anion of acetone, 2-(2-oxopropyl)pyridine was formed in modest yield.[346]

These alkylation and acylation reactions are important because Friedel–Crafts alkylation and acylation (**11-11, 11-17**) cannot be applied to most nitrogen hetero-cycles (see also **13-17**).

Protonated nitrogen heterocycles can be carbalkoxylated[347] by treatment with esters of α-keto acids and Fenton's reagent. Pyridine is carbalkoxylated at C-2 and C-4, for example. The attack is by •COOR radicals generated from the esters via a hydroperoxide (**38**).

[340]See Citterio, A.; Gentile, A.; Minisci, F.; Serravalle, M.; Ventura, S. *Tetrahedron* **1985**, *41*, 617; Katz, R.B.; Mistry, J.; Mitchell, M.B. *Synth. Commun.* **1989**, *19*, 317.

[341]Minisci, F.; Selva, A.; Porta, O.; Barilli, P.; Gardini, G.P. *Tetrahedron* **1972**, *28*, 2415.

[342]Fontana, F.; Minisci, F.; Barbosa, M.C.N.; Vismara, E. *Acta Chem. Scand*, **1989**, *43*, 995.

[343]Minisci, F.; Vismara, E.; Fontana, F.; Barbosa, M.C.N. *Tetrahedron Lett.* **1989**, *30*, 4569.

[344]Caronna, T.; Gardini, G.P.; Minisci, F. *Chem. Commun.* **1969**, 201; Arnoldi, A.; Bellatti, M.; Caronna, T.; Citterio, A.; Minisci, F.; Porta, O.; Sesana, G. *Gazz. Chim. Ital.* **1977**, *107*, 491.

[345]Togo, H.; Miyagawa, N.; Yokoyama, M. *Chem. Lett.* **1992**, 1677.

[346]Kiselyov, A.S.; Strekowski, L. *J. Org. Chem.* **1993**, *58*, 4476.

[347]Bernardi, R.; Caronna, T.; Galli, R.; Minisci, F.; Perchinunno, M. *Tetrahedron Lett.* **1973**, 645; Heinisch, G.; Lötsch, G. *Angew. Chem. Int. Ed.* **1985**, *24*, 692.

Similarly, a carbamoyl group can be introduced[348] by the use of the radicals

$$H_2N-\overset{\underset{\|}{O}}{C}\,\bullet \qquad Me_2N-\overset{\underset{\|}{O}}{C}\,\bullet$$

H₂N—C• · · · Me₂N—C• generated from formamide or DMF and H_2SO_4, H_2O_2, and $FeSO_4$ or other oxidants.

N₂ AS LEAVING GROUP[349]

In these reactions diazonium salts are cleaved to aryl radicals,[350] in most cases with the assistance of copper salts. Reactions **13-27** and **13-26** may also be regarded as belonging to this category with respect to the attacking compound. For nucleophilic substitutions of diazonium salts (see **13-20–13-23**). Removal of nitrogen and replacement with a hydrogen atom is a reduction, found in Chapter 19.

14-20   Replacement of the Diazonium Group by Chlorine or Bromine

Chloro-de-diazoniation, and so on

$$ArN_2^+ + CuCl \longrightarrow ArCl$$

Treatment of diazonium salts with cuprous chloride or bromide leads to aryl chlorides or bromides, respectively. In either case, the reaction is called the *Sandmeyer reaction*.[351] The reaction can also be carried out with copper and HBr or HCl, in which case it is called the *Gatterman reaction* (not to be confused with **11-18**). The Sandmeyer reaction is not useful for the preparation of fluorides or iodides, but for bromides and chlorides it is of wide scope and is probably the best way of introducing bromine or chlorine into an aromatic ring. The yields are usually high.

The mechanism is not known with certainty, but is believed to take the following course:[352]

$$ArN_2^+ X^- + CuX \longrightarrow Ar\bullet + N_2 + CuX_2$$
$$Ar\bullet + CuX_2 \longrightarrow ArX + CuX$$

---

[348]Minisci, F.; Citterio, A.; Vismara, E.; Giordano, C. *Tetrahedron* **1985**, *41*, 4157.

[349]For a review, see Wulfman, D.S., in Patai, S. *The Chemistry of Diazonium and Diazo Groups*, pt. 1, Wiley, NY, **1978**, pp. 286–297.

[350]For reviews, see Galli, C. *Chem. Rev.* **1988**, *88*, 765; Zollinger, H. *Acc. Chem. Res.* **1973**, *6*, 355, pp. 339–341.

[351]Rate constants for this reaction have been determined. See Hanson, P.; Hammond, R.C.; Goodacre, P.R.; Purcell, J.; Timms, A.W. *J. Chem. Soc. Perkin Trans. 2* **1994**, 691.

[352]Dickerman, S.C.; Weiss, K.; Ingberman, A.K. *J. Am. Chem. Soc.* **1958**, *80*, 1904; Kochi, J.K. *J. Am. Chem. Soc.* **1957**, *79*, 2942; Dickerman, S.C.; DeSouza, D.J.; Jacobson, N. *J. Org. Chem.* **1969**, *34*, 710; Galli, C. *J. Chem. Soc. Perkin Trans. 2* **1981**, 1459; **1982**, 1139; **1984**, 897. See also, Hanson, P.; Jones, J.R.; Gilbert, B.C.; Timms, A.W. *J. Chem. Soc. Perkin Trans. 2* **1991**, 1009.

The first step involves a reduction of the diazonium ion by the cuprous ion, which results in the formation of an aryl radical. In the second step, the aryl radical abstracts halogen from cupric chloride, reducing it. The CuX is regenerated and is thus a true catalyst.

Aryl bromides and chlorides can be prepared from primary aromatic amines in one step by several procedures,[353] including treatment of the amine (*1*) with *tert*-butyl nitrite and anhydrous $CuCl_2$ or $CuBr_2$ at 65°C,[354] and (*2*) with *tert*-butyl thionitrite or *tert*-butyl thionitrate and $CuCl_2$ or $CuBr_2$ at room temperature.[355] These procedures are, in effect, a combination of **13-19** and the Sandmeyer reaction. A further advantage is that cooling to 0°C is not needed. A mixture of Me_3SiCl and $NaNO_2$ was used to convert aniline to chlorobenzene in a related reaction.[356]

For the preparation of fluorides and iodides from diazonium salts (see **13-32** and **13-31**).

$$ArN_2^+ + CuCN \longrightarrow ArCN$$

It is noted that the reaction of aryl diazonium salts with CuCN to give benzonitrile derivatives is also called the *Sandmeyer reaction*. It is usually conducted in neutral solution to avoid liberation of HCN.

OS **I**, 135, 136, 162, 170; **II**, 130; **III**, 185; **IV**, 160. Also see, OS **III**, 136; **IV**, 182. For the reaction with CuCN, see OS **I**, 514.

14-21   Replacement of the Diazonium Group by Nitro

Nitro-de-diazoniation

$$ArN_2^+ + NaNO_2 \xrightarrow{Cu^+} ArNO_2$$

Nitro compounds can be formed in good yields by treatment of diazonium salts with sodium nitrite in the presence of cuprous ion. The reaction occurs only in neutral or alkaline solution. This is not usually called the Sandmeyer reaction, although, like **14-20**, it was discovered by Sandmeyer. Tetrafluoroborate (BF_4^-) is often used as the negative ion since the diminished nucleophilicity avoids competition from the chloride ion. The mechanism is probably like that of **14-20**.[357] If electron-withdrawing groups are present, the catalyst is not needed; $NaNO_2$ alone gives nitro compounds in high yields.[358]

[353]For other procedures, see Brackman,W.; Smit, P.J. *Recl. Trav. Chim. Pays-Bas,* **1966**, *85*, 857; Cadogan, J.I.G.; Roy, D.A.; Smith, D.M. *J. Chem. Soc. C* **1966**, 1249.
[354]Doyle, M.P.; Siegfried, B.; Dellaria, Jr, J.F. *J. Org. Chem.* **1977**, *42*, 2426.
[355]Oae, S.; Shinhama, K.; Kim, Y.H. *Bull. Chem. Soc. Jpn.* **1980**, *53*, 1065.
[356]Lee, J.G.; Cha, H.T. *Tetrahedron Lett.* **1992**, *33*, 3167.
[357]For discussions, see Opgenorth, H.; Rüchardt, C. *Liebigs Ann. Chem.* **1974**, 1333; Singh, P.R.; Kumar, R.; Khanna, R.K. *Tetrahedron Lett.* **1982**, *23*, 5191.
[358]Bagal, L.I.; Pevzner, M.S.; Frolov, A.N. *J. Org. Chem. USSR* **1969**, *5*, 1767.

An alternative procedure used electrolysis, in 60% HNO_3 to convert 1-amino-naphthalene to naphthalene.[359]

OS **II**, 225; **III**, 341.

14-22   Replacement of the Diazonium Group by Sulfur-Containing Groups

Chlorosulfo-de-diazoniation

$$ArN_2{}^+ \quad + \quad SO_2 \quad \xrightarrow[\text{HCl}]{\text{CuCl}_2} \quad ArSO_2Cl$$

Diazonium salts can be converted to sulfonyl chlorides by treatment with sulfur dioxide in the presence of cupric chloride.[360] The use of $FeSO_4$ and copper metal instead of $CuCl_2$ gives sulfinic acids $(ArSO_2H)$[361] (see also, **13-21**).

OS **V**, 60; **VII**, 508.

14-23   Conversion of Diazonium Salts to Aldehydes, Ketones, or Carboxylic Acids

Acyl-de-diazoniation, and so on

Diazonium salts react with oximes to give aryl oximes, which are easily hydro-lyzed to aldehydes $(R = H)$ or ketones.[362] A copper sulfate-sodium sulfite catalyst is essential. In most cases higher yields (40–60%) are obtained when the reaction is used for aldehydes than for ketones. In another method[363] for achieving the conver-sion $ArN_2^+ \rightarrow ArCOR$, diazonium salts are treated with R_4Sn and CO with palla-dium acetate as catalyst.[364] In a different kind of reaction, silyl enol ethers of aryl ketones $Ar'C(OSiMe_3){=}CHR$ react with solid diazonium fluoroborates $(ArN_2^+$ $BF_4^-)$ to give ketones $(ArCHRCOAr')$.[365] This is, in effect, an arylation of the aryl ketone.

Carboxylic acids can be prepared in moderate-to-high yields by treatment of diazonium fluoroborates with carbon monoxide and palladium acetate[366] or

[359]Torii, S.; Okumoto, H.; Satoh, H.; Minoshima, T.; Kurozumi, S. *SynLett,* **1995**, 439.

[360]Gilbert, E.E. *Synthesis* **1969**, 1, p. 6.

[361]Wittig, G.; Hoffmann, R.W. *Org. Synth.* **V**, 60.

[362]Beech, W.F. *J. Chem. Soc.* **1954**, 1297.

[363]For still another method, see Citterio, A.; Serravalle, M.; Vimara, E. *Tetrahedron Lett.* **1982**, *23*, 1831.

[364]Kikukawa, K.; Idemoto, T.; Katayama, A.; Kono, K.; Wada, F.; Matsuda, T. *J. Chem. Soc. Perkin Trans. 1* **1987**, 1511.

[365]Sakakura, T.; Hara, M.; Tanaka, M. *J. Chem. Soc., Chem. Commun.* **1985**, 1545.

[366]Nagira, K.; Kikukawa, K.; Wada, F.; Matsuda, T. *J. Org. Chem.* **1980**, *45*, 2365.

copper(II) chloride.[367] The mixed anhydride ArCOOCOMe is an intermediate that can be isolated. Other mixed anhydrides can be prepared by the use of other salts instead of sodium acetate.[368] An arylpalladium compound is probably an intermediate.[368]

OS **V**, 139.

METALS AS LEAVING GROUPS

14-24   Coupling of Grignard Reagents

De-metallo-coupling

$$2\,RMgX \xrightarrow{\text{TlBr}} RR$$

This organometallic coupling reaction is clearly related to the Wurtz coupling, discussed in **10-56**, and the coupling of other organometallic compounds is discussed in **14-25**. Grignard reagents can be coupled to give symmetrical dimers[369] by treatment with either thallium(I) bromide[370] or with a transition-metal halide, such as $CrCl_2$, $CrCl_3$, $CoCl_2$, $CoBr_2$, or $CuCl_2$.[371] The metallic halide is an oxidizing agent and becomes reduced. Both aryl and alkyl Grignard reagents can be dimerized by either procedure, though the TlBr method cannot be applied to R = primary alkyl or to aryl groups with ortho substituents. Aryl Grignard reagents can also be dimerized by treatment with 1,4-dichloro-2-butene, 1,4-dichloro-2-butyne, or 2,3-dichloropropene.[372] Vinylic and alkynyl Grignard reagents can be coupled (to give 1,3-dienes and 1,3-diynes, respectively) by treatment with thionyl chloride.[373] Primary alkyl, vinylic, aryl, and benzylic Grignard reagents give symmetrical dimers in high yield (~90%) when treated with a silver(I) salt (e.g., $AgNO_3$, AgBr, $AgClO_4$) in the presence of a nitrogen-containing oxidizing agent, such as lithium nitrate, methyl nitrate, or NO_2.[374] This method has been used to close rings of four, five, and six members.[375]

---

[367]Olah, G.A.; Wu, A.; Bagno, A.; Prakash, G.K.S. *Synlett*, **1990**, 596.

[368]Kikukawa, K.; Kono, K.; Nagira, K.; Wada, F.; Matsuda, T. *J. Org. Chem.* **1981**, *46*, 4413.

[369]For a list of reagents, with references, see Larock, R.C. *Comprehensive Organic Transformations*, 2nd ed, Wiley-VCH, NY, **1999**, pp. 85–88.

[370]McKillop, A.; Elsom, L.F.; Taylor, E.C. *Tetrahedron* **1970**, *26*, 4041.

[371]For reviews, see Kauffmann, T. *Angew. Chem. Int. Ed.* **1974**, *13*, 291; Elsom, L.F.; Hunt, J.D.; McKillop, A. *Organomet. Chem. Rev. Sect. A* **1972**, *8*, 135; Nigh, W.G., in Trahanovsky, W.S. *Oxidation in Organic Chemistry*, pt. B, Academic Press, NY, **1973**, pp. 85–91.

[372]Taylor, S.K.; Bennett, S.G.; Heinz, K.J.; Lashley, L.K. *J. Org. Chem.* **1981**, *46*, 2194; Cheng, J.; Luo, F. *Tetrahedron Lett.* **1988**, *29*, 1293.

[373]Uchida, A.; Nakazawa, T.; Kondo, I.; Iwata, N.; Matsuda, S. *J. Org. Chem.* **1972**, *37*, 3749.

[374]Tamura, M.; Kochi, J.K. *Bull. Chem. Soc. Jpn.* **1972**, *45*, 1120.

[375]Whitesides, G.M.; Gutowski, F.D. *J. Org. Chem.* **1976**, *41*, 2882.

The mechanisms of the reactions with metal halides, at least in some cases, probably begin with conversion of RMgX to the corresponding RM (**12-36**), followed by its decomposition to free radicals.[376]

OS **VI**, 488.

14-25   Coupling of Other Organometallic Reagents[332]

De-metallo-coupling

$$R_2CuLi \xrightarrow[-78°C, THF]{O_2} RR$$

Lithium dialkylcopper reagents can be oxidized to symmetrical dimers by O_2 at $-78°C$ in THF.[377] The reaction is successful for R = primary and secondary alkyl, vinylic, or aryl. Other oxidizing agents, for example, nitrobenzene, can be used instead of O_2. Vinylic copper reagents dimerize on treatment with oxygen, or simply on standing at $0°C$ for several days or at $25°C$ for several hours, to yield 1,3-dienes.[378] The finding of retention of configuration for this reaction demonstrates that free-radical intermediates are not involved.

The coupling reaction of Grignard reagents was discussed in **14-24**. Lithium organoaluminates ($LiAlR_4$) are dimerized to RR by treatment with $Cu(OAc)_2$.[379] Terminal vinylic alanes (prepared by **15-17**) can be dimerized to 1,3-dienes with CuCl in THF.[380] Symmetrical 1,3-dienes can also be prepared in high yields by treatment of vinylic mercury chlorides[381] with LiCl and a rhodium catalyst[382] and by treatment of vinylic tin compounds with a palladium catalyst.[383] Arylmercuric salts are converted to biaryls by treatment with copper and a catalytic amount of $PdCl_2$.[384] Vinylic, alkynyl, and aryl tin compounds were dimerized with

[376]For a review of the mechanism, see Kashin, A.N.; Beletskaya, I.P. *Russ. Chem. Rev.* **1982**, *51*, 503.

[377]Whitesides, G.M.; San Filippo, Jr, J.; Casey, C.P.; Panek, E.J. *J. Am. Chem. Soc.* **1967**, *89*, 5302. See also, Kauffmann, T.; Kuhlmann, D.; Sahm, W.; Schrecken, H. *Angew. Chem. Int. Ed.* **1968**, *7*, 541; Bertz, S.H.; Gibson, C.P. *J. Am. Chem. Soc.* **1986**, *108*, 8286.

[378]Whitesides, G.M.; Casey, C.P.; Krieger, J.K. *J. Am. Chem. Soc.* **1971**, *93*, 1379; Walborsky, H.M.; Banks, R.B.; Banks, M.L.A.; Duraisamy, M. *Organometallics* **1982**, *1*, 667; Rao, S.A.; Periasamy, M. *J. Chem. Soc., Chem. Commun.* **1987**, 495. See also, Lambert, G.J.; Duffley, R.P.; Dalzell, H.C.; Razdan, R.K. *J. Org. Chem.* **1982**, *47*, 3350.

[379]Sato, F.; Mori, Y.; Sato, M. *Chem. Lett.* **1978**, 1337.

[380]Zweifel, G.; Miller, R.L. *J. Am. Chem. Soc.* **1970**, *92*, 6678.

[381]For reviews of coupling with organomercury compounds, see Russell, G.A. *Acc. Chem. Res.* **1989**, *22*, 1; Larock, R.C. *Organomercury Compounds in Organic Synthesis*, Springer, NY, **1985**, pp. 240–248.

[382]Larock, R.C.; Bernhardt, J.C. *J. Org. Chem.* **1977**, *42*, 1680. For extension to unsymmetrical 1,3-dienes, see Larock, R.C.; Riefling, B. *J. Org. Chem.* **1978**, *43*, 1468.

[383]Tolstikov, G.A.; Miftakhov, M.S.; Danilova, N.A.; Vel'der, Ya.L.; Spirikhin, L.V. *Synthesis* **1989**, 633.

[384]Kretchmer, R.A.; Glowinski, R. *J. Org. Chem.* **1976**, *41*, 2661. See also, Bumagin, N.A.; Kalinovskii, I.O.; Beletskaya, I.P. *J. Org. Chem. USSR* **1982**, *18*, 1151; Larock, R.C.; Bernhardt, J.C. *J. Org. Chem.* **1977**, *42*, 1680.

$Cu(NO_3)_2$.[385] Alkyl- and aryllithium compounds can be dimerized by transition-metal halides in a reaction similar to **14-24**.[386] Triarylbismuth compounds Ar_3Bi react with palladium(0) complexes to give biaryls ArAr.[387] Diethylzinc reacted with Ph_2I^+ BF_4^- in the presence of palladium acetate, to give biphenyl.[388]

Unsymmetrical coupling of vinylic, alkynyl, and arylmercury compounds was achieved in moderate-to-good yields by treatment with alkyl and vinylic dialkylcopper reagents, for example, $PhCH=CHHgCl + Me_2CuLi \rightarrow PhCH=CHMe$.[389] Unsymmetrical biaryls were prepared by treating a cyanocuprate (ArCu(CN)Li, prepared from ArLi and CuCN) with an aryllithium (Ar'Li).[390]

A radical coupling reaction has been reported, in which an aryl halide reacted with Bu_3SnH, AIBN, and benzene, followed by treatment with methyllithium to give the biaryl.[391]

14-26   Coupling of Boranes

Alkyl-de-dialkylboration

Alkylboranes can be coupled by treatment with silver nitrate and base.[392] Since alkylboranes are easily prepared from alkenes (**15-16**), this is essentially a way of coupling and reducing alkenes; in fact, alkenes can be hydroborated and coupled in the same flask. For symmetrical coupling (R = R') yields range from 60 to 80% for terminal alkenes and from 35 to 50% for internal ones. Unsymmetrical coupling has also been carried out,[393] but with lower yields. Arylboranes react similarly, yielding biaryls.[394] The mechanism is probably of the free-radical type.

Dimerization of two vinylborane units to give a conjugated diene can be achieved by treatment of divinylchloroboranes (prepared by addition of BH_2Cl to alkynes; see **15-16**) with methylcopper. (*E,E*)-1,3-Dienes are prepared in high

[385]Ghosal, S.; Luke, G.P.; Kyler, K.S. *J. Org. Chem.* **1987**, *52*, 4296.

[386]Morizur, J. *Bull. Soc. Chim. Fr.* **1964**, 1331.

[387]Barton, D.H.R.; Ozbalik, N.; Ramesh, M. *Tetrahedron* **1988**, *44*, 5661.

[388]Kang, S.-K.; Hong, R.-K.; Kim, T.-H.; Pyun, S.-J. *Synth. Commun.* **1997**, *27*, 2351.

[389]Larock, R.C.; Leach, D.R. *Tetrahedron Lett.* **1981**, *22*, 3435; *Organometallics* **1982**, *1*, 74. For another method, see Larock, R.C.; Hershberger, S.S. *Tetrahedron Lett.* **1981**, *22*, 2443.

[390]Lipshutz, B.H.; Siegmann, K.; Garcia, E. *J. Am. Chem. Soc.* **1991**, *113*, 8161.

[391]Studer, A. ; Bossart, M.; Vasella, T. *Org. Lett.* **2000**, *2*, 985.

[392]Pelter, A.; Smith, K.; Brown, H.C. *Borane Reagents*, Academic Press, NY, **1988**, pp. 306–308.

[393]Brown, H.C.; Verbrugge, C.; Snyder, C.H. *J. Am. Chem. Soc.* **1961**, *83*, 1001.

[394]Breuer, S.W.; Broster, F.A. *Tetrahedron Lett.* **1972**, 2193.

yields.[395]

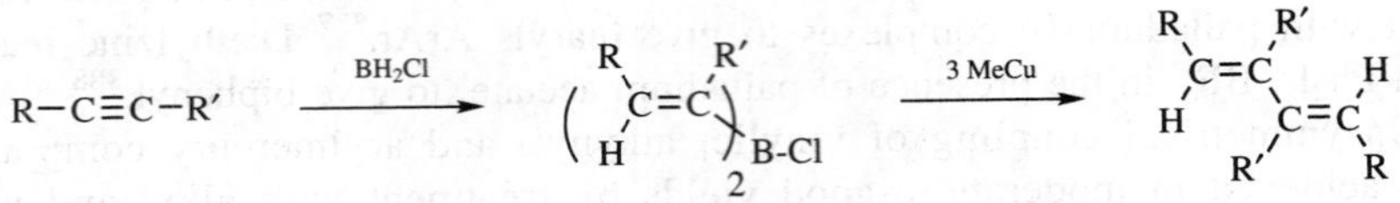

In a similar reaction, symmetrical conjugated diynes RC≡C–C≡CR can be prepared by reaction of lithium dialkyldialkynylborates, Li^+ $[R'_2B(C≡CR)_2]^-$, with iodine.[396]

HALOGEN AS LEAVING GROUP

The conversion of RX to RH can occur by a free-radical mechanism but is treated at **19-53**.

SULFUR AS LEAVING GROUP

14-27 Desulfurization

Hydro-de-thio-substitution, and so on

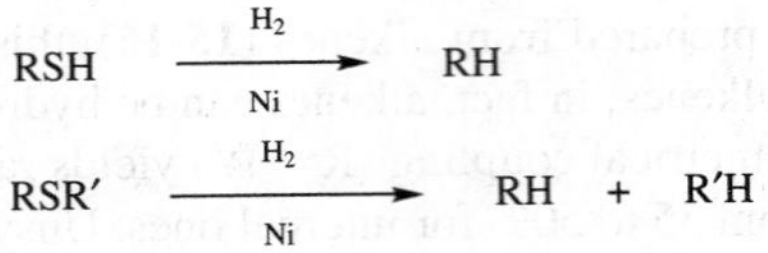

$$RSH \xrightarrow[Ni]{H_2} RH$$

$$RSR' \xrightarrow[Ni]{H_2} RH + R'H$$

Thiols and thioethers,[397] both alkyl and aryl, can be desulfurized by hydrogenolysis with Raney nickel.[398] The hydrogen is usually not applied externally, since Raney nickel already contains enough hydrogen for the reaction. Other sulfur compounds can be similarly desulfurized, among them disulfides (RSSR),

---

[395]Yamamoto, Y.; Yatagai, H.; Maruyama, K.; Sonoda, A.; Murahashi, S. *J. Am. Chem. Soc.* **1977**, *99*, 5652; *Bull. Chem. Soc. Jpn.* **1977**, *50*, 3427. For other methods of dimerizing vinylic boron compounds, see Rao, V.V.R.; Kumar, C.V.; Devaprabhakara, D. *J. Organomet. Chem.* **1979**, *179*, C7; Campbell, Jr, J.B.; Brown, H.C. *J. Org. Chem.* **1980**, *45*, 549.

[396]Pelter, A.; Smith, K.; Tabata, M. *J. Chem. Soc., Chem. Commun.* **1975**, 857. For extensions to unsymmetrical conjugated diynes, see Pelter, A.; Hughes, R.; Smith, K.; Tabata, M. *Tetrahedron Lett.* **1976**, 4385; Sinclair, J.A.; Brown, H.C. *J. Org. Chem.* **1976**, *41*, 1078.

[397]For a review of the reduction of thioethers, see Block, E., in Patai, S. *The Chemistry of Functional Groups, Supplement E*, pt. 1, Wiley, NY, **1980**, pp. 585–600.

[398]For reviews, see Belen'kii, L.I., in Belen'kii, L.I. *Chemistry of Organosulfur Compounds*, Ellis Horwood, Chichester, **1990**, pp. 193–228; Pettit, G.R.; van Tamelen, E.E. *Org. React.* **1962**, *12*, 356; Hauptmann, H.; Walter, W.F. *Chem. Rev.* **1962**, *62*, 347.

thiono esters (RCSOR′),[399] thioamides (RCDNHR′), sulfoxides, and dithioacetals. The last reaction, which is an indirect way of accomplishing reduction of a carbonyl to a methylene group (see **19-61**), can also give the alkene if an α hydrogen is present.[400] In most of the examples given, R can also be aryl. Other reagents[401] have also been used,[402] including samarium in acetic acid for desulfurization of vinyl sulfones.[403]

An important special case of RSR reduction is desulfurization of thiophene derivatives. This proceeds with concomitant reduction of the double bonds. Many compounds have been made by alkylation of thiophene (see **39**), followed by reduction to the corresponding alkane.

Thiophenes can also be desulfurized to alkenes ($RCH_2CH{=}CHCH_2R'$ from **39**) with a nickel boride catalyst prepared from nickel(II) chloride and $NaBH_4$ in methanol.[404] It is possible to reduce just one SR group of a dithioacetal by treatment with borane–pyridine in trifluoroacetic acid or in CH_2Cl_2 in the presence of $AlCl_3$.[405] Phenyl selenides RSePh can be reduced to RH with Ph_3SnH[406] and with nickel boride.[407]

The exact mechanisms of the Raney nickel reactions are still in doubt, though they are probably of the free-radical type.[408] It has been shown that reduction of thiophene proceeds through butadiene and butene, not through 1-butanethiol or other sulfur compounds, that is, the sulfur is removed before the double bonds

[399]See Baxter, S.L.; Bradshaw, J.S. *J. Org. Chem.* **1981**, *46*, 831.

[400]Fishman, J.; Torigoe, M.; Guzik, H. *J. Org. Chem.* **1963**, *28*, 1443.

[401]For lists of reagents, with references, see Larock, R.C. *Comprehensive Organic Transformations*, 2nd ed, Wiley-VCH, NY, *1999*, pp. 53–60. For a review with respect to transition-metal reagents, see Luh, T.; Ni, Z. *Synthesis* **1990**, 89. For some very efficient nickel-containing reagents, see Becker, S.; Fort, Y.; Vanderesse, R.; Caubère, P. *J. Org. Chem.* **1989**, *54*, 4848.

[402]For example, diphosphorus tetraiodide by Suzuki, H.; Tani, H.; Takeuchi, S. *Bull. Chem. Soc. Jpn.* **1985**, *58*, 2421; Shigemasa, Y.; Ogawa, M.; Sashiwa, H.; Saimoto, H. *Tetrahedron Lett.* **1989**, *30*, 1277; NiBr₂–Ph₃P–LiAlH₄ by Ho, K.M.; Lam, C.H.; Luh, T. *J. Org. Chem.* **1989**, *54*, 4474.

[403]Liu, Y.; Zhang, Y. *Org. Prep. Proceed. Int.* **2001**, *33*, 376.

[404]Schut, J.; Engberts, J.B.F.N.; Wynberg, H. *Synth. Commun.* **1972**, *2*, 415.

[405]Kikugawa, Y. *J. Chem. Soc. Perkin Trans. 1* **1984**, 609.

[406]Clive, D.L.J.; Chittattu, G.; Wong, C.K. *J. Chem. Soc., Chem. Commun.* **1978**, 41.

[407]Back, T.G. *J. Chem. Soc., Chem. Commun.* **1984**, 1417.

[408]For a review, see Bonner, W.A.; Grimm, R.A., in Kharasch, N.; Meyers, C.Y. *The Chemistry of Organic Sulfur Compounds*, Vol. 2, Pergamon, NY, *1966*, pp. 35–71, 410–413. For a review of the mechanism of desulfurization on molybdenum surfaces, see Friend, C.M.; Roberts, J.T. *Acc. Chem. Res.* **1988**, *21*, 394.

are reduced. This was demonstrated by isolation of the olefins and the failure to isolate any potential sulfur-containing intermediates.[409]

OS **IV**, 638; **V**, 419; **VI**, 109, 581, 601. See also OS **VII**, 124, 476.

14-28 Conversion of Sulfides to Organolithium Compounds

Lithio-de-phenylthio-substitution

$$RSPh \xrightarrow[\text{THF}]{\text{Li naphthalenide}} RLi$$

Sulfides can be cleaved, with a phenylthio group replaced by a lithium,[410] by treatment with lithium or lithium naphthalenide in THF.[411] Good yields have been obtained with R = primary, secondary, or tertiary alkyl, or allylic,[412] and containing groups, such as double bonds or halogens. Dilithio compounds can be made from compounds containing two separated SPh groups, but it is also possible to replace just one SPh from a compound with two such groups on a single carbon, to give an α-lithio sulfide.[413] The reaction has also been used to prepare α-lithio ethers and α-lithio organosilanes.[410] For some of these compounds lithium 1-(dimethylamino)naphthalenide is a better reagent than either Li or lithium naphthalenide.[414] The mechanism is presumably of the free-radical type.

CARBON AS LEAVING GROUP

14-29 Decarboxylative Dimerization: The Kolbe Reaction

De-carboxylic-coupling

$$2\,RCOO^- \xrightarrow{\text{electrolysis}} R{-}R$$

Electrolysis of carboxylate ions, results in decarboxylation and combination of the resulting radicals to give the coupling product R—R. This coupling

[409]Owens, P.J.; Ahmberg, C.H. *Can. J. Chem.* *1962*, *40*, 941.

[410]For a review, see Cohen, T.; Bhupathy, M. *Acc. Chem. Res.* *1989*, *22*, 152.

[411]Screttas, C.G.; Micha-Screttas, M. *J. Org. Chem.* *1978*, *43*, 1064; *1979*, *44*, 713.

[412]See Cohen, T.; Guo, B. *Tetrahedron* *1986*, *42*, 2803.

[413]See, for example, Cohen, T.; Sherbine, J.P.; Matz, J.R.; Hutchins, R.R.; McHenry, B.M.; Willey, P.R. *J. Am. Chem. Soc.* *1984*, *106*, 3245; Ager, D.J. *J. Chem. Soc. Perkin Trans. 1* *1986*, 183; Screttas, C.G.; Micha-Screttas, M. *J. Org. Chem.* *1978*, *43*, 1064; *1979*, *44*, 713.

[414]See Cohen, T.; Matz, J.R. *Synth. Commun.* *1980*, *10*, 311.

reaction is called the *Kolbe reaction* or the *Kolbe electrosynthesis*.[415] It is used to prepare symmetrical R—R, where R is straight chained, since little or no yield is obtained when there is a branching. The reaction is not successful for R = aryl. Many functional groups may be present, though many others inhibit the reaction.[415] Unsymmetrical RR′ have been made by coupling mixtures of acid salts.

A free-radical mechanism is involved:

$$RCOO^- \xrightarrow[\text{oxidation}]{\text{electrolytic}} RCOO\bullet \xrightarrow{-CO_2} R\bullet \longrightarrow R\text{--}R$$

There is much evidence[416] for this mechanism, including side products (RH, alkenes) characteristic of free-radical intermediates and the fact that electrolysis of acetate ion in the presence of styrene caused some of the styrene to polymerize to polystyrene (such polymerizations can be initiated by free radicals, see p. 1015). Other side products (ROH, RCOOR) are sometimes found, stemming from further oxidation of the radical R• to a carbocation R^+.[417]

When the reaction is conducted in the presence of 1,3-dienes, additive dimerization can occur:[418]

$$2\,RCOO^- + CH_2=CH\text{--}CH=CH_2 \longrightarrow RCH_2CH=CHCH_2CH_2CH=CHCH_2R$$

The radical R• adds to the conjugated system to give $RCH_2CH=CHCH_2\bullet$, which dimerizes. Another possible product is $RCH_2CH=CHCH_2R$, from coupling of the two kinds of radicals.[419]

In a nonelectrolytic reaction, which is limited to R = primary alkyl, the thiohydroxamic esters **40** give dimers when irradiated at −64°C in an argon

[415]For reviews, see Nuding, G.; Vögtle, F.; Danielmeier, K.; Steckhan, E. *Synthesis* **1996**, 71; Schäfer, H.J. *Top. Curr. Chem.* **1990**, *152*, 91; *Angew. Chem. Int. Ed.* **1981**, *20*, 911; Fry, A.J. *Synthetic Organic Electrochemistry*, 2nd ed, Wiley, NY, **1989**, pp. 238–253; Eberson, L.; Utley, J.H.P., in Baizer, M.M.; Lund, H. *Organic Electrochemistry*, Marcel Dekker, NY, **1983**, pp. 435–462; Gilde, H. *Methods Free-Radical Chem.* **1972**, *3*, 1; Eberson, L., in Patai, S. *The Chemistry of Carboxylic Acids and Esters*, Wiley, NY, **1969**, pp. 53–101; Vijh, A.K.; Conway, B.E. *Chem. Rev.* **1967**, *67*, 623.

[416]For other evidence, see Kreautler, B.; Jaeger, C.D.; Bard, A.J. *J. Am. Chem. Soc.* **1978**, *100*, 4903.

[417]See Corey, E.J.; Bauld, N.L.; La Londe, R.T.; Casanova, Jr, J.; Kaiser, E.T. *J. Am. Chem. Soc.* **1960**, *82*, 2645.

[418]Lindsey, Jr, R.V.; Peterson, M.L. *J. Am. Chem. Soc.* **1959**, *81*, 2073; Khrizolitova, M.A.; Mirkind, L.A.; Fioshin, M.Ya. *J. Org. Chem. USSR* **1968**, *4*, 1640; Bruno, F.; Dubois, J.E. *Bull. Soc. Chim. Fr.* **1973**, 2270.

[419]Smith, W.B.; Gilde, H. *J. Am. Chem. Soc.* **1959**, *81*, 5325; **1961**, *83*, 1355; Schäfer, H.; Pistorius, R. *Angew. Chem. Int. Ed.* **1972**, *11*, 841.

atmosphere:[420]

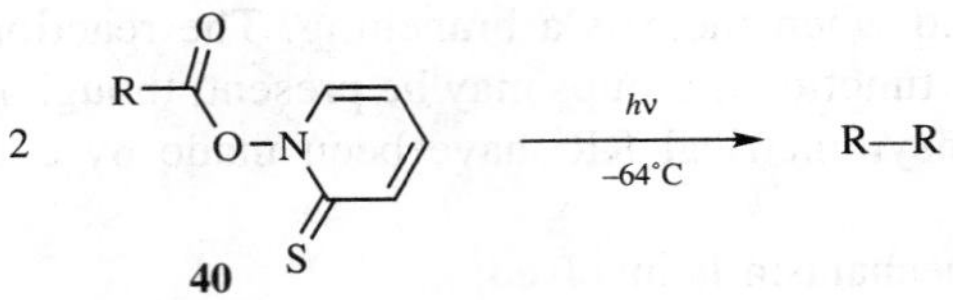

In another nonelectrolytic process, aryl acetic acids are converted to *vic*-diaryl compounds $2ArCR_2COOH \rightarrow ArCR_2CR_2Ar$ by treatment with sodium persulfate $Na_2S_2O_8$ and a catalytic amount of $AgNO_3$.[421] Photolysis of carboxylic acids in the presence of Hg_2F_2 leads to the dimeric alkane via decarboxylation.[422] Both of these reactions involve dimerization of free radicals. In still another process, electron-deficient aromatic acyl chlorides are dimerized to biaryls $(2\ ArCOCl \rightarrow Ar-Ar)$ by treatment with a disilane R_3SiSiR_3 and a palladium catalyst.[423]

OS **III**, 401; **V**, 445, 463; **VII**, 181.

14-30 The Hunsdiecker Reaction

Bromo-de-carboxylation

$$RCOOAg + Br_2 \longrightarrow RBr + CO_2 + AgBr$$

Reaction of a silver salt of a carboxylic acid with bromine is called the *Hunsdiecker reaction*[424] and is a way of decreasing the length of a carbon chain by one unit.[425] The reaction is of wide scope, giving good results for *n*-alkyl R from 2 to 18 carbons and for many branched R too, producing primary, secondary, and tertiary bromides. Many functional groups may be present as long as they are not a substituted. The group R may also be aryl. However, if R contains unsaturation, the reaction seldom gives good results. Although bromine is the most often used halogen, chlorine and iodine have also been used. Catalytic Hunsdiecker reactions are known.[426]

When iodine is the reagent, the ratio between the reactants is very important and determines the products. A 1:1 ratio of salt/iodine gives the alkyl halide, as above.

[420]Barton, D.H.R.; Bridon, D.; Fernandez-Picot, I.; Zard, S.Z. *Tetrahedron* **1987**, *43*, 2733.

[421]Fristad, W.E.; Klang, J.A. *Tetrahedron Lett.* **1983**, *24*, 2219.

[422]Habibi, M.H.; Farhadi, S. *Tetrahedron Lett.* **1999**, *40*, 2821.

[423]Krafft, T.E.; Rich, J.D.; McDermott, P.J. *J. Org. Chem.* **1990**, *55*, 5430.

[424]This reaction was first reported by the Russian composer–chemist Alexander Borodin: *Liebigs Ann. Chem.* **1861**, *119*, 121.

[425]For reviews, see Wilson, C.V. *Org. React.* **1957**, *9*, 332; Johnson, R.G.; Ingham, R.K. *Chem. Rev.* **1956**, *56*, 219. Also see, Naskar, D.; Chowdhury, S.; Roy, S. *Tetrahedron Lett.* **1998**, *39*, 699.

[426]Das, J.P.; Roy, S. *J. Org. Chem.* **2002**, *67*, 7861.

A 2:1 ratio, however, gives the ester RCOOR. This is called the *Simonini reaction* and is sometimes used to prepare carboxylic esters. The Simonini reaction can also be carried out with lead salts of acids.[427] A more convenient way to perform the Hunsdiecker reaction is by use of a mixture of the acid and mercuric oxide instead of the salt, since the silver salt must be very pure and dry and such pure silver salts are often not easy to prepare.[428]

Other methods for accomplishing the conversion RCOOH → RX are[429] (*1*) treatment of thallium(I) carboxylates[430] with bromine;[431] (*2*) treatment of carboxylic acids with lead tetraacetate and halide *ions* (Cl^-, Br^-, or I^-);[432] (*3*) reaction of the acids with lead tetraacetate and NCS, which gives tertiary and secondary chlorides in good yields, but is not good for R = primary alkyl or phenyl;[433] (*4*) treatment of thiohydroxamic esters with CCl_4, $BrCCl_3$ (which gives bromination), CHI_3, or CH_2I_2 in the presence of a radical initiator;[434] (*5*) photolysis of benzophenone oxime esters of carboxylic acids in CCl_4 ($RCON=CPh_2 \rightarrow RCl$).[435] Alkyl fluorides can be prepared in moderate to good yields by treating carboxylic acids RCOOH with XeF_2.[436] This method works best for R = primary and tertiary alkyl, and benzylic. Aromatic and vinylic acids do not react.

The mechanism of the Hunsdiecker reaction is believed to be as follows:

[427]Bachman, G.B.; Kite, G.F.; Tuccarbasu, S.; Tullman, G.M. *J. Org. Chem.* **1970**, *35*, 3167.

[428]Cristol, S.J.; Firth, W.C. *J. Org. Chem.* **1961**, *26*, 280. See also, Meyers, A.I.; Fleming, M.P. *J. Org. Chem.* **1979**, *44*, 3405, and references cited therein.

[429]For a list of reagents, with references, see Larock, R.C. *Comprehensive Organic Transformations*, 2nd ed, Wiley-VCH, NY, **1999**, pp. 741–744.

[430]These salts are easy to prepare and purify; see Ref. 501.

[431]McKillop, A.; Bromley, D.; Taylor, E.C. *J. Org. Chem.* **1969**, *34*, 1172; Cambie, R.C.; Hayward, R.C.; Jurlina, J.L.; Rutledge, P.S.; Woodgate, P.D. *J. Chem. Soc. Perkin Trans. 1* **1981**, 2608.

[432]Kochi, J.K. *J. Am. Chem. Soc.* **1965**, *87*, 2500; *J. Org. Chem.* **1965**, *30*, 3265. For a review, see Sheldon, R.A.; Kochi, J.K. *Org. React.* **1972**, *19*, 279, pp. 326–334, 390–399.

[433]Becker, K.B.; Geisel, M.; Grob, C.A.; Kuhnen, F. *Synthesis* **1973**, 493.

[434]Barton, D.H.R.; Lacher, B.; Zard, S.Z. *Tetrahedron* **1987**, *43*, 4321; Stofer, E.; Lion, C. *Bull. Soc. Chim. Belg.* **1987**, *96*, 623; Della, E.W.; Tsanaktsidis, J. *Aust. J. Chem.* **1989**, *42*, 61.

[435]Hasebe, M.; Tsuchiya, T. *Tetrahedron Lett.* **1988**, *29*, 6287.

[436]Patrick, T.B.; Johri, K.K.; White, D.H.; Bertrand, W.S.; Mokhtar, R.; Kilbourn, M.R.; Welch, M.J. *Can. J. Chem.* **1986**, *64*, 138. For another method, see Grakauskas, V. *J. Org. Chem.* **1969**, *34*, 2446.

The first step is not a free-radical process, and its actual mechanism is not known.[437] Compound **41** is an acyl hypohalite and is presumed to be an intermediate, though it has never been isolated from the reaction mixture. Among the evidence for the mechanism is that optical activity at R is lost (except when a neighboring bromine atom is present, see p. 942); if R is neopentyl, there is no rearrangement, which would certainly happen with a carbocation; and the side products, notably RR, are consistent with a free-radical mechanism. There is evidence that the Simonini reaction involves the same mechanism as the Hunsdiecker reaction, but that the alkyl halide formed then reacts with excess RCOOAg (**10-17**) to give the ester[438] (see also **19-12**).

Vinyl carboxylic acids (conjugated acids) were shown to react with NBS and lithium acetate in aqueous acetonitrile, to give the corresponding vinyl bromide ($C=C-COOH \rightarrow C=C-Br$), using microwave irradiation.[439] A similar reaction was reported using Na_2MoO_4, KBr and aqueous hydrogen peroxide.[440]

A related reaction reacts the sodium salt of an alkylsulfonic acid with thionyl chloride at 100°C, to give the alkyl chloride.[441]

OS **III**, 578; **V**, 126; **VI**, 179; **75**, 124; **X**, 237. See also OS **VI**, 403.

14-31   Decarboxylative Allylation

Allyl-de-carboxylation

The COOH group of a β-keto acid is replaced by an allylic group when the acid is treated with an allylic acetate and a palladium catalyst at room temperature.[442] The reaction is successful for various substituted allylic groups. The less highly substituted end of the allylic group forms the new bond. Thus, both $CH_2=CHCHMeOAc$ and $MeCH=CHCH_2OAc$ gave $O=C(R)-\overset{|}{\underset{|}{C}}-CH_2CH=CHMe$ as the product.

[437]When Br_2 reacts with aryl R, at low temperature in inert solvents, it is possible to isolate a complex containing both Br_2 and the silver carboxylate: see Bryce-Smith, D.; Isaacs, N.S.; Tumi, S.O. *Chem. Lett.* **1984**, 1471.

[438]Oae, S.; Kashiwagi, T.; Kozuka, S. *Bull. Chem. Soc. Jpn.* **1966**, *39*, 2441; Bunce, N.J.; Murray, N.G. *Tetrahedron* **1971**, *27*, 5323.

[439]Kuang, C.; Senboku, H.; Tokuda, M. *Synlett* **2000**, 1439.

[440]Sinha, J.; Layek, S.; Bhattacharjee, M.; Mandal, G.C. *Chem. Commun.* **2001**, 1916.

[441]Carlsen, P.H.J.; Rist, Ø.; Lund, T.; Helland, I. *Acta Chem. Scand. B* **1995**, *49*, 701.

[442]Tsuda, T.; Okada, M.; Nishi, S.; Saegusa, T. *J. Org. Chem.* **1986**, *51*, 421.

14-32   Decarbonylation of Aldehydes and Acyl Halides

Carbonyl-Extrusion

$$RCHO \xrightarrow{\text{RhCl(Ph}_3\text{P)}_3} RH$$

Aldehydes, both aliphatic and aromatic, can be decarbonylated[443] by heating with a rhodium catalyst[444] or other catalysts, such as palladium.[445] $RhCl(Ph_3P)_3$ is often called *Wilkinson's catalyst*.[446] In an older reaction, aliphatic (but not aromatic) aldehydes are decarbonylated by heating with di-*tert*-butyl peroxide or other peroxides,[447] usually in a solution containing a hydrogen donor, such as a thiol. The reaction has also been initiated with light, and thermally (without an initiator) by heating at ~500°C.

Wilkinson's catalyst has also been reported to decarbonylate aromatic acyl halides at 180°C (ArCOX → ArX).[448] This reaction has been carried out with acyl iodides,[449] bromides, and chlorides. Aliphatic acyl halides that lack an a hydrogen also give this reaction,[450] but if an α hydrogen is present, elimination takes place instead (**17-17**). Aromatic acyl cyanides give aryl cyanides (ArCOCN → ArCN).[451] Aromatic acyl chlorides and cyanides can also be decarbonylated with palladium catalysts.[452]

It is possible to decarbonylate acyl halides in another way, to give alkanes (RCOCl → RH). This is done by heating the substrate with tripropylsilane Pr_3SiH

[443]For reviews, see Collman, J.P.; Hegedus, L.S.; Norton, J.R.; Finke, R.G. *Principles and Applications of Organotransition Metal Chemistry*, University Science Books, Mill Valley, CA **1987**, pp. 768–775; Baird, M.C., in Patai, S. *The Chemistry of Functional Groups, Supplement B* pt. 2, Wiley, NY, **1979**, pp. 825–857; Tsuji, J., in Wender, I.; Pino, P. *Organic Syntheses Via Metal Carbonyls*, Vol. 2, Wiley, NY, **1977**, pp. 595–654; Tsuji, J.; Ohno, K. *Synthesis* **1969**, 157; Bird, C.W. *Transition Metal Intermediates in Organic Synthesis*, Academic Press, NY, **1967**, pp. 239–247.
[444]Ohno, K.; Tsuji, J. *J. Am. Chem. Soc.* **1968**, *90*, 99; Baird, C.W.; Nyman, C.J.; Wilkinson, G. *J. Chem. Soc. A* **1968**, 348.
[445]For a review, see Rylander, P.N. *Organic Synthesis with Noble Metal Catalysts*, Academic Press, NY, **1973**, pp. 260–267.
[446]For a review of this catalyst, see Jardine, F.H. *Prog. Inorg. Chem.* **1981**, *28*, 63.
[447]For reviews of free-radical aldehyde decarbonylations, see Vinogradov, M.G.; Nikishin, G.I. *Russ. Chem. Rev.* **1971**, *40*, 916; Schubert, W.M.; Kintner, R.R., in Patai, S. *The Chemistry of the Carbonyl Group*, Vol. 1, Wiley, NY, **1966**, pp. 711–735.
[448]Kampmeier, J.A.; Rodehorst, R.; Philip, Jr, J.B. *J. Am. Chem. Soc.* **1981**, *103*, 1847; Blum, J.; Oppenheimer, E.; Bergmann, E.D. *J. Am. Chem. Soc.* **1967**, *89*, 2338.
[449]Blum, J.; Rosenman, H.; Bergmann, E.D. *J. Org. Chem.* **1968**, *33*, 1928.
[450]Tsuji, J.; Ohno, K. *Tetrahedron Lett.* **1966**, 4713; *J. Am. Chem. Soc.* **1966**, *88*, 3452.
[451]Blum, J.; Oppenheimer, E.; Bergmann, E.D. *J. Am. Chem. Soc.* **1967**, *89*, 2338.
[452]Verbicky, Jr, J.W.; Dellacoletta, B.A.; Williams, L. *Tetrahedron Lett.* **1982**, *23*, 371; Murahashi, S.; Naota, T.; Nakajima, N. *J. Org. Chem.* **1986**, *51*, 898.

in the presence of *tert*-butyl peroxide.[453] Yields are good for R = primary or secondary alkyl and poor for R = tertiary alkyl or benzylic. There is no reaction when R = aryl. (See also the decarbonylation ArCOCl → ArAr mentioned in **14-29**.)

The mechanism of the peroxide- or light-induced reaction seems to be as follows (in the presence of thiols):[454]

$$
\underset{R}{\overset{O}{\underset{}{\parallel}}}\!\!C\!-\!H \quad \xrightarrow[\text{source}]{\text{radical}} \quad \underset{R}{\overset{O}{\underset{}{\parallel}}}\!\!C\cdot \quad \longrightarrow \quad R\cdot \;+\; C\!\equiv\!O
$$

$$
R\cdot \;+\; R'\text{-SH} \quad \longrightarrow \quad R\text{-H} \;+\; R'S\cdot
$$

$$
\underset{R}{\overset{O}{\underset{}{\parallel}}}\!\!C\!-\!H \;+\; R'S\cdot \quad \longrightarrow \quad \underset{R}{\overset{O}{\underset{}{\parallel}}}\!\!C\cdot \;+\; R\text{-SH} \quad \text{etc.}
$$

The reaction of aldehydes with Wilkinson's catalyst goes through complexes of the form **42** and **43**, which have been trapped.[455] The reaction has been shown to give retention of configuration at a chiral R;[456] and deuterium labeling demonstrates that the reaction is intramolecular: RCOD give RD.[457] Free radicals are not involved.[458] The mechanism with acyl halides appears to be more complicated.[459]

42   **43**

For aldehyde decarbonylation by an electrophilic mechanism (see **11-34**).

[453]Billingham, N.C.; Jackson, R.A.; Malek, F. *J. Chem. Soc. Perkin Trans. 1* **1979**, 1137.

[454]Slaugh, L.H. *J. Am. Chem. Soc.* **1959**, *81*, 2262; Berman, J.D.; Stanley, J.H.; Sherman, V.W.; Cohen, S.G. *J. Am. Chem. Soc.* **1963**, *85*, 4010.

[455]Suggs, J.W. *J. Am. Chem. Soc.* **1978**, *100*, 640; Kampmeier, J.A.; Harris, S.H.; Mergelsberg, I. *J. Org. Chem.* **1984**, *49*, 621.

[456]Walborsky, H.M.; Allen, L.E. *J. Am. Chem. Soc.* **1971**, *93*, 5465. See also, Tsuji, J.; Ohno, K. *Tetrahedron Lett.* **1967**, 2173.

[457]Prince, R.H.; Raspin, K.A. *J. Chem. Soc. A* **1969**, 612; Walborsky, H.M.; Allen, L.E. *J. Am Chem. Soc.* **1971**, *93*, 5465. See, however, Baldwin, J.E.; Bardenm, T.C.; Pugh, R.L.; Widdison, W.C. *J. Org. Chem.* **1987**, *52*, 3303.

[458]Kampmeier, J.A.; Harris, S.H.; Wedegaertner, D.K. *J. Org. Chem.* **1980**, *45*, 315.

[459]Kampmeier, J.A.; Liu, T. *Organometallics* **1989**, *8*, 2742.

CHAPTER 15

Addition to Carbon–Carbon Multiple Bonds

There are four fundamental ways in which addition to a double or triple bond can take place. Three of these are two-step processes, with initial attack by a nucleophile, or attack upon an electrophile or a free radical. The second step consists of combination of the resulting intermediate with, respectively, a positive species, a negative species, or a neutral entity. In the fourth type of mechanism, attack at the two carbon atoms of the double or triple bond is simultaneous (concerted). Which of the four mechanisms is operating in any given case is determined by the nature of the substrate, the reagent, and the reaction conditions. Some of the reactions in this chapter can take place by all four mechanistic types.

MECHANISMS

Electrophilic Addition[1]

In this mechanism, a positive species approaches the double or triple bond and in the first step forms a bond by donation of the π pair of electrons[2] to the electrophilic

---

[1]For a monograph, see de la Mare, P.B.D.; Bolton, R. *Electrophilic Additions to Unsaturated Systems*, 2nd ed.; Elsevier, NY, *1982*. For reviews, see Schmid, G.H., in Patai, S. *Supplement A: The Chemistry of Double-bonded Functional Groups*, Vol. 2, pt. 1, Wiley, NY, *1989*, pp. 679–731; Smit, W.A. *Sov. Sci. Rev. Sect. B 1985*, 7, 155; V'yunov, K.A.; Ginak, A.I. *Russ. Chem. Rev. 1981*, 50, 151; Schmid, G.H.; Garratt, D.G., in Patai, S. *Supplement A: The Chemistry of Double-bonded Functional Groups*, Vol. 1, pt. 2, Wiley, NY, *1977*, pp. 725–912; Freeman, F. *Chem. Rev. 1975*, 75, 439; Bolton, R., in Bamford, C.H.; Tipper, C.F.H. *Comprehensive Chemical Kinetics*, Vol. 9, Elsevier, NY, *1973*, pp. 1–86; Dolbier, Jr., W.R. *J. Chem. Educ. 1969*, 46, 342.

[2]For a review of the π-nucleophilicity in carbon–carbon bond-forming reactions, see Mayr, H.; Kempf, B.; Ofial, A.R. *Acc. Chem. Res. 2003*, 36, 66.

March's Advanced Organic Chemistry: Reactions, Mechanisms, and Structure, Sixth Edition, by Michael B. Smith and Jerry March
Copyright © 2007 John Wiley & Sons, Inc.

species to form a σ pair:

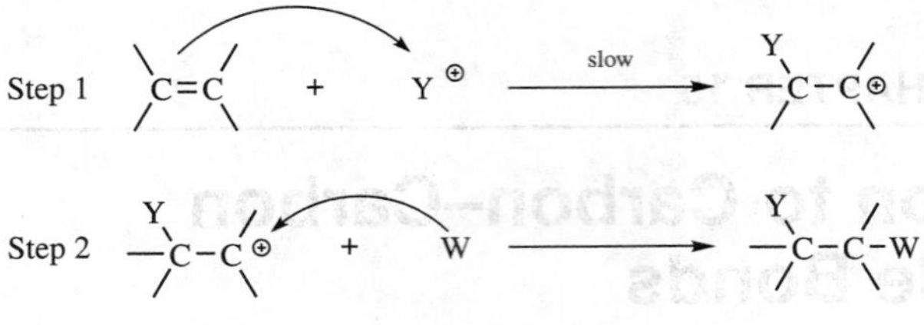

Step 1 C=C + $Y^{\oplus}$ $\xrightarrow{\text{slow}}$ $-\overset{Y}{\underset{}{C}}-\overset{}{\underset{}{C}}{}^{\oplus}$

Step 2 $-\overset{Y}{\underset{}{C}}-\overset{}{\underset{}{C}}{}^{\oplus}$ +   W $\xrightarrow{\hspace{1.5cm}}$ $-\overset{Y}{\underset{}{C}}-\overset{}{\underset{}{C}}-W$

1

The IUPAC designation for this mechanism is $A_E + A_N$ (or $A_H + A_N$ if $Y^+ = H^+$). As in electrophilic substitution (p. 658), Y need not actually be a positive ion but can be the positive end of a dipole or an induced dipole, with the negative part breaking off either during the first step or shortly after. The second step is a combination of **1** with a species carrying an electron pair and often bearing a negative charge. This step is the same as the second step of the S_N1 mechanism. Not all electrophilic additions follow the simple mechanism given above. In many brominations it is fairly certain that **1**, if formed at all, very rapidly cyclizes to a bromonium ion (**2**):

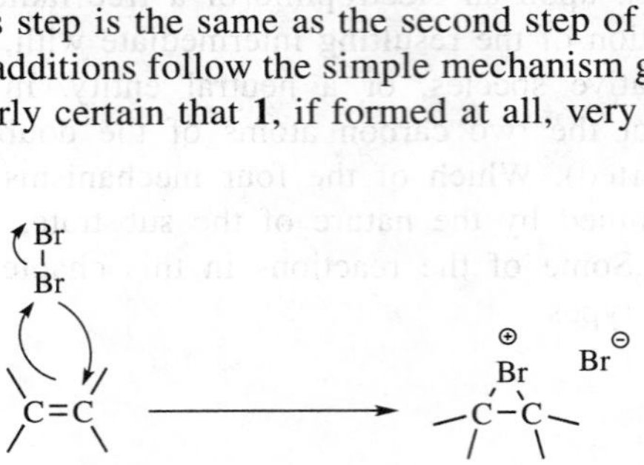

C=C $\xrightarrow{\hspace{2cm}}$ $-\overset{\overset{\oplus}{Br}}{\underset{}{C}}-\overset{}{\underset{}{C}}-$ $Br^{\ominus}$

2

This intermediate is similar to those encountered in the neighboring-group mechanism of nucleophilic substitution (see p. 446). The attack of $\overline{w}$ on an intermediate like **2** is an S_N2 step. Whether the intermediate is **1** or **2**, the mechanism is called Ad_E2 (electrophilic addition, bimolecular).

In investigating the mechanism of addition to a double bond, perhaps the most useful type of information is the stereochemistry of the reaction.[3] The two carbons of the double bond and the four atoms immediately attached to them are all in a plane (p. 9); there are thus three possibilities. Both Y and W may enter from the same side of the plane, in which case the addition is stereospecific and syn; they may enter from opposite sides for stereospecific anti addition; or the reaction may be nonstereospecific. In order to determine which of these possibilities is occurring in a given reaction, the following type of experiment is often done: YW is added to the cis and trans isomers of an alkene of the form ABC=CBA. We may use the cis alkene as an example. If the addition is syn, the product

---

[3]For a review of the stereochemistry of electrophilic additions to double and triple bonds, see Fahey, R.C. *Top. Stereochem.* **1968**, *3*, 237. For a review of the synthetic uses of stereoselective additions, see Bartlett, P.A. *Tetrahedron* **1980**, *36*, 2, pp. 3–15.

will be the erythro *dl* pair, because each carbon has a 50% chance of being attacked by Y:

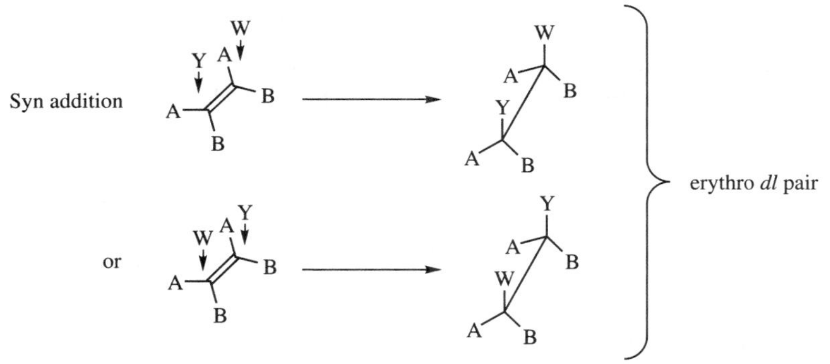

Syn addition

or

erythro *dl* pair

On the other hand, if the addition is anti, the threo *dl* pair will be formed:

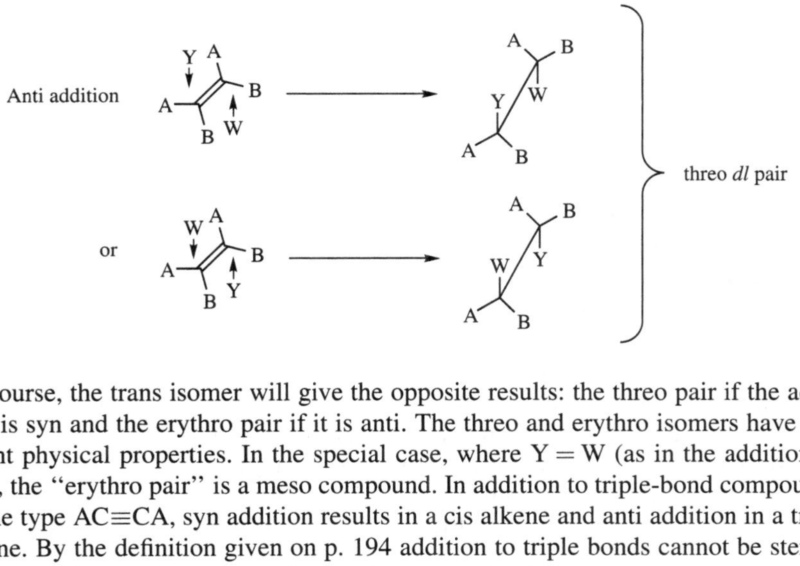

Anti addition

or

threo *dl* pair

Of course, the trans isomer will give the opposite results: the threo pair if the addition is syn and the erythro pair if it is anti. The threo and erythro isomers have different physical properties. In the special case, where $Y = W$ (as in the addition of Br_2), the "erythro pair" is a meso compound. In addition to triple-bond compounds of the type $AC \equiv CA$, syn addition results in a cis alkene and anti addition in a trans alkene. By the definition given on p. 194 addition to triple bonds cannot be stereospecific, although it can be, and often is, stereoselective.

It is easily seen that in reactions involving cyclic intermediates like **2**, addition must be anti, since the second step is an S_N2 step and must occur from the back side. It is not so easy to predict the stereochemistry for reactions involving **1**. If **1** has a relatively long life, the addition should be nonstereospecific, since there will be free rotation about the single bond. On the other hand, there may be some factor that maintains the configuration, in which case W may come in from the same side or the opposite side, depending on the circumstances. For example, the positive charge might be stabilized by an attraction for Y that does not involve a full bond (see **3**).

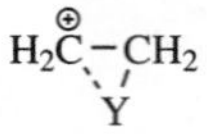

$$H_2\overset{\oplus}{C} - CH_2$$
$$|$$
$$Y$$
$$\mathbf{3}$$

The second group would then come in anti. A circumstance that would favor syn addition would be the formation of an ion pair after the addition of Y:[4]

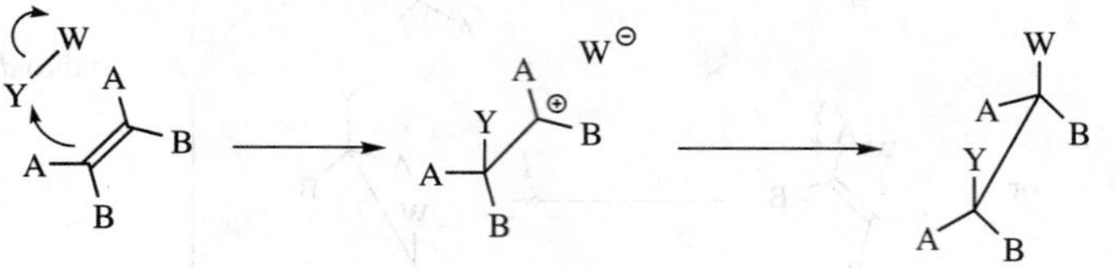

Since W is already on the same side of the plane as Y, collapse of the ion pair leads to syn addition.

Another possibility is that anti addition might, at least in some cases, be caused by the operation of a mechanism in which attack by W and Y are essentially simultaneous but from opposite sides:

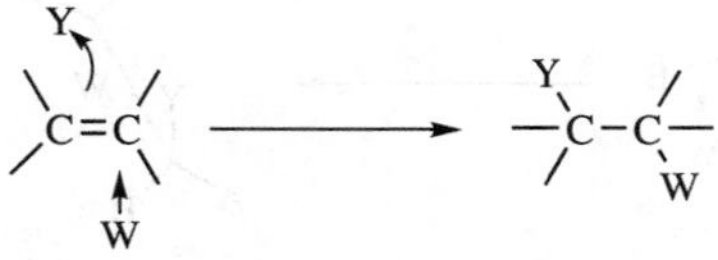

This mechanism, called the Ad_E3 mechanism (*termolecular addition*, IUPAC A_NA_E),[5] has the disadvantage that three molecules must come together in the transition state. However, it is the reverse of the E2 mechanism for elimination, for which the transition state is known to possess this geometry (p. 1478).

There is much evidence that when the attack is on Br^+ (or a carrier of it), the bromonium ion **2** is often an intermediate and the addition is anti. As long ago as 1911, McKenzie and Fischer independently showed that treatment of maleic acid with bromine gave the *dl* pair of 2,3-dibromosuccinic acid, while fumaric acid (the trans isomer) gave the meso compound.[6] Many similar experiments have been performed since with similar results. For triple bonds, stereoselective anti addition was shown even earlier. Bromination of dicarboxyacetylene gave 70%

[4]Dewar, M.J.S. *Angew. Chem. Int. Ed.* **1964**, *3*, 245; Heasley, G.E.; Bower, T.R.; Dougharty, K.W.; Easdon, J.C.; Heasley, V.L.; Arnold, S.; Carter, T.L.; Yaeger, D.B.; Gipe, B.T.; Shellhamer, D.F. *J. Org. Chem.* **1980**, *45*, 5150.

[5]For evidence for this mechanism, see, for example, Hammond, G.S.; Nevitt, T.D. *J. Am. Chem. Soc.* **1954**, *76*, 4121; Bell, R.P.; Pring, M. *J. Chem. Soc. B* **1966**, 1119; Pincock, J.A.; Yates, K. *J. Am. Chem. Soc.* **1968**, *90*, 5643; Fahey, R.C.; Payne, M.T.; Lee, D. *J. Org. Chem.* **1974**, *39*, 1124; Roberts, R.M.G. *J. Chem. Soc. Perkin Trans. 2*, **1976**, 1374; Pasto, D.J.; Gadberry, J.F. *J. Am. Chem. Soc.* **1978**, *100*, 1469; Naab, P.; Staab, H.A. *Chem. Ber.* **1978**, *111*, 2982.

[6]This was done by Fischer, E. *Liebigs Ann. Chem.* **1911**, *386*, 374; McKenzie, A. *Proc. Chem. Soc.* **1911**, 150; *J. Chem. Soc.* **1912**, *101*, 1196.

of the trans isomer.[7]

$$HOOC-C \equiv C-COOH \; + \; Br_2 \longrightarrow \underset{Br}{\overset{HOOC}{>}} C = C \overset{Br}{\underset{COOH}{<}} \quad 70\% \; trans$$

There is other evidence for mechanisms involving **2**. We have already mentioned (p. 449) that bromonium ions have been isolated in stable solutions in nucleophilic substitution reactions involving bromine as a neighboring group. Such ions have also been isolated in reactions involving addition of a Br^+ species to a double bond.[8] The following is further evidence. If the two bromines approach the double bond from opposite sides, it is very unlikely that they could come from the same bromine molecule. This means that if the reaction is performed in the presence of nucleophiles, some of these will compete in the second step with the bromide liberated from the bromine. It has been found, indeed, that treatment of ethylene with bromine in the presence of chloride ions gives some 1-chloro-2-bromoethane along with the dibromoethane.[9] Similar results are found when the reaction is carried out in the presence of water (**15-40**) or of other nucleophiles.[10] *Ab initio* molecular orbital studies show that **2** is more stable than its open isomer **1** (Y = Br).[11] There is evidence that formation of **2** is reversible.[12]

However, a number of examples have been found where addition of bromine is not stereospecifically anti. For example, the addition of Br_2 to *cis*- and *trans*-1-phenylpropenes in CCl_4 was nonstereospecific.[13] Furthermore, the stereospecificity of bromine addition to stilbene depends on the dielectric constant of the solvent. In solvents of low dielectric constant, the addition was 90–100% anti, but with an increase in dielectric constant, the reaction became less stereospecific, until, at a dielectric constant of ~35, the addition was completely nonstereospecific.[14] Likewise in the case of triple bonds, stereoselective anti addition was found in bromi-

[7]Michael, A. *J. Prakt. Chem.* **1892**, *46*, 209.

[8]Strating, J.; Wieringa, J.H.; Wynberg, H. *Chem. Commun.* **1969**, 907; Olah, G.A. *Angew. Chem. Int. Ed.* **1973**, *12*, 173, p. 207; Slebocka-Tilk, H.; Ball, R.G.; Brown, R.S. *J. Am. Chem. Soc.* **1985**, *107*, 4504.

[9]Francis, A.W. *J. Am. Chem. Soc.* **1925**, *47*, 2340.

[10]See, for example, Zefirov, N.S.; Koz'min, A.S.; Dan'kov, Yu.V.; Zhdankin, V.V.; Kirin, V.N. *J. Org. Chem. USSR* **1984**, *20*, 205.

[11]Hamilton, T.P.; Schaefer III, H.F. *J. Am. Chem. Soc.* **1990**, *112*, 8260.

[12]Brown, R.S.; Gedye, R.; Slebocka-Tilk, H.; Buschek, J.M.; Kopecky, K.R. *J. Am. Chem. Soc.* **1984**, *106*, 4515; Ruasse, M.; Motallebi, S.; Galland, B. *J. Am. Chem. Soc.* **1991**, *113*, 3440; Bellucci, G.; Bianchini, R.; Chiappe, C.; Brown, R.S.; Slebocka-Tilk, H. *J. Am. Chem. Soc.* **1991**, *113*, 8012; Bennet, A.J.; Brown, R.S.; McClung, R.E.D.; Klobukowski, M.; Aarts, G.H.M.; Santarsiero, B.D.; Bellucci, G.; Bianchini, R. *J. Am. Chem. Soc.* **1991**, *113*, 8532.

[13]Fahey, R.C.; Schneider, H. *J. Am. Chem. Soc.* **1968**, *90*, 4429. See also, Rolston, J.H.; Yates, K. *J. Am. Chem. Soc.* **1969**, *91*, 1469, 1477, 1483.

[14]Heublein, G. *J. Prakt. Chem.* **1966**, *[4] 31*, 84. See also, Buckles, R.E.; Miller, J.L.; Thurmaier, R.J. *J. Org. Chem.* **1967**, *32*, 888; Heublein, G.; Lauterbach, H. *J. Prakt. Chem.* **1969**, *311*, 91; Ruasse, M.; Dubois, J.E. *J. Am. Chem. Soc.* **1975**, *97*, 1977. For the dependence of stereospecificity in this reaction on the solvent concentration, see Bellucci, G.; Bianchini, R.; Chiappe, C.; Marioni, F. *J. Org. Chem.* **1990**, *55*, 4094.

nation of 3-hexyne, but both cis and trans products were obtained in bromination of phenylacetylene.[15] These results indicate that a bromonium ion is not formed where the open cation can be stabilized in other ways (e.g., addition of Br^+ to 1-phenylpropene gives the ion $PhC^\oplus HCHBrCH_3$, which is a relatively stable benzylic cation) and that there is probably a spectrum of mechanisms between complete bromonium ion (**2**, no rotation) formation and completely open-cation (**1**, free rotation) formation, with partially bridged bromonium ions (**3**, restricted rotation) in between.[16] We have previously seen cases (e.g., p. 461) where cations require more stabilization from outside sources as they become intrinsically less stable themselves.[17] Further evidence for the open cation mechanism where aryl stabilization is present was reported in an isotope effect study of addition of Br_2 to $ArCH=CHCHAr'$ ($Ar = p$-nitrophenyl, $Ar' = p$-tolyl). The ^{14}C isotope effect for one of the double-bond carbons (the one closer to the NO_2 group) was considerably larger than for the other one.[18]

When the π-bond of an alkene attacks Cl^+,[19] I^+,[20] and RS^+,[21] the result is similar to that when the electrophile is Br^+; there is a spectrum of mechanisms between cyclic intermediates and open cations. As might be expected from our discussion in Chapter 10 (p. 446), iodonium ions compete with open carbocations more effectively than bromonium ions, while chloronium ions compete less effectively. There is kinetic and spectral evidence that at least in some cases, for example, in the addition of Br_2 or ICl, the electrophile forms a π complex with the alkene before a covalent bond is formed.[22]

[15]Pincock, J.A.; Yates, K. *Can. J. Chem.* **1970**, *48*, 3332.

[16]For other evidence for this concept, see Pincock, J.A.; Yates, K. *Can. J. Chem.* **1970**, *48*, 2944; Heasley, V.L.; Chamberlain, P.H. *J. Org. Chem.* **1970**, *35*, 539; Dubois, J.E.; Toullec, J.; Barbier, G. *Tetrahedron Lett.* **1970**, 4485; Dalton, D.R.; Davis, R.M. *Tetrahedron Lett.* **1972**, 1057; Wilkins, C.L.; Regulski, T.W. *J. Am. Chem. Soc.* **1972**, *94*, 6016; Sisti, A.J.; Meyers, M. *J. Org. Chem.* **1973**, *38*, 4431; McManus, S.P.; Peterson, P.E. *Tetrahedron Lett.* **1975**, 2753; Abraham, R.J.; Monasterios, J.R. *J. Chem. Soc. Perkin Trans. 1*, **1973**, 1446; Schmid, G.H.; Modro, A.; Yates, K. *J. Org. Chem.* **1980**, *45*, 665; Ruasse, M.; Argile, A. *J. Org. Chem.* **1983**, *48*, 202; Cadogan, J.I.G.; Cameron D.K.; Gosney, I.; Highcock, R.M.; Newlands, S.F. *J. Chem. Soc., Chem. Commun.* **1985**, 1751. For a review, see Ruasse, M. *Acc. Chem. Res.* **1990**, *23*, 87.

[17]In a few special cases, stereospecific syn addition of Br_2 has been found, probably caused by an ion pair mechanism as shown on p. 1002: Naae, D.G. *J. Org. Chem.* **1980**, *45*, 1394.

[18]Kokil, P.B.; Fry, A. *Tetrahedron Lett.* **1986**, *27*, 5051.

[19]Fahey, R.C. *Top. Stereochem.* **1968**, *3*, 237, pp. 273–277.

[20]Hassner, A.; Boerwinkle, F.; Levy, A.B. *J. Am. Chem. Soc.* **1970**, *92*, 4879.

[21]For reviews of thiiranium and/or thiirenium ions, see Capozzi, G.; Modena, G., in Bernardi, F.; Csizmadia, I.G.; Mangini, A. *Organic Sulfur Chemistry*, Elsevier, NY, **1985**, pp. 246–298; Smit, W.A. *Sov. Sci. Rev. Sect. B* **1985**, *7*, 155, see pp. 180–202; Dittmer, D.C.; Patwardhan, B.H., in Stirling, C.J.M. *The Chemistry of the Sulphonium Group*, pt. 1, Wiley, NY, **1981**, pp. 387–412; Capozzi, G.; Lucchini, V.; Modena, G.; *Rev. Chem. Intermed.* **1979**, *2*, 347; Schmid, G.H. *Top. Sulfur Chem.* **1977**, *3*, 102; Mueller, W.H. *Angew. Chem. Int. Ed.* **1969**, *8*, 482. The specific nature of the three-membered sulfur-containing ring is in dispute; see Smit, W.A.; Zefirov, N.S.; Bodrikov, I.V.; Krimer, M.Z. *Acc. Chem. Res.* **1979**, *12*, 282; Bodrikov, I.V.; Borisov, A.V.; Chumakov, L.V.; Zefirov, N.S.; Smit, W.A. *Tetrahedron Lett.* **1980**, *21*, 115; Schmid, G.H.; Garratt, D.G.; Dean, C.L. *Can. J. Chem.* **1987**, *65*, 1172; Schmid, G.H.; Strukelj, M.; Dalipi, S. *Can. J. Chem.* **1987**, *65*, 1945.

[22]See Nordlander, J.E.; Haky, J.E.; Landino, J.P. *J. Am. Chem. Soc.* **1980**, *102*, 7487; Fukuzumi, S.; Kochi, J.K. *Int. J. Chem. Kinet.* **1983**, *15*, 249; Schmid, G.H.; Gordon, J.W. *Can. J. Chem.* **1984**, *62*, 2526; **1986**, *64*, 2171; Bellucci, G.; Bianchini, R.; Chiappe, C.; Marioni, F.; Ambrosetti, R.; Brown, R.S.; Slebocka-Tilk, H. *J. Am. Chem. Soc.* **1989**, *111*, 2640.

When the electrophile is a proton,[23] a cyclic intermediate is not possible, and the mechanism is the simple $A_H + A_N$ process shown before

This is an A-S_E2 mechanism (p. 525). There is a great deal of evidence[24] for it, including:

1. The reaction is general-acid, not specific-acid-catalyzed, implying rate-determining proton transfer from the acid to the double bond.[25]

2. The existence of open carbocation intermediates is supported by the contrast in the pattern of alkyl substituent effects[26] with that found in brominations, where cyclic intermediates are involved. In the latter case, substitution of alkyl groups on $H_2C=CH_2$ causes a cumulative rate acceleration

until all four hydrogens have been replaced by alkyl groups, because each group helps to stabilize the positive charge.[27] In addition of HX, the effect is not cumulative. Replacement of the two hydrogens on one carbon causes great rate increases (primary → secondary → tertiary carbocation), but additional substitution on the other carbon produces little or no acceleration.[28] This is evidence for open cations when a proton is the electrophile.[29]

---

[23]For a review of the addition of HCl, see Sergeev, G.B.; Smirnov, V.V.; Rostovshchikova, T.N. *Russ. Chem. Rev.* **1983**, *52*, 259.

[24]For other evidence, see Baliga, B.T.; Whalley, E. *Can. J. Chem.* **1964**, *42*, 1019; **1965**, *43*, 2453; Gold, V.; Kessick, M.A. *J. Chem. Soc.* **1965**, 6718; Corriu, R.; Guenzet, J. *Tetrahedron* **1970**, *26*, 671; Simandoux, J.; Torck, B.; Hellin, M.; Coussemant, F. *Bull. Soc. Chim. Fr.* **1972**, 4402, 4410; Bernasconi, C.F.; Boyle, Jr., W.J. *J. Am. Chem. Soc.* **1974**, *96*, 6070; Hampel, M.; Just, G.; Pisanenko, D.A.; Pritzkow, W. *J. Prakt. Chem.* **1976**, *318*, 930; Allen, A.D.; Tidwell, T.T. *J. Am. Chem. Soc.* **1983**, *104*, 3145.

[25]Loudon, G.M.; Noyce, D.S. *J. Am. Chem. Soc.* **1969**, *91*, 1433; Schubert, W.M.; Keeffe, J.R. *J. Am. Chem. Soc.* **1972**, *94*, 559; Chiang, Y.; Kresge, A.J. *J. Am. Chem. Soc.* **1985**, *107*, 6363.

[26]Bartlett, P.D.; Sargent, G.D. *J. Am. Chem. Soc.* **1965**, *87*, 1297; Schmid, G.H.; Garratt, D.G. *Can. J. Chem.* **1973**, *51*, 2463.

[27]See, for example, Anantakrishnan, S.V.; Ingold, C.K. *J. Chem. Soc.* **1935**, 1396; Swern, D. in Swern *Organic Peroxides*, Vol. 2, Wiley, NY, **1971**, pp. 451–454; Nowlan, V.J.; Tidwell, T.T. *Acc. Chem. Res.* **1977**, *10*, 252.

[28]Bartlett, P.D.; Sargent, G.D. *J. Am. Chem. Soc.* **1965**, *87*, 1297; Riesz, P.; Taft, R.W.; Boyd, R.H. *J. Am. Chem. Soc.* **1957**, *79*, 3724.

[29]A similar result (open cations) was obtained with carbocations Ar_2CH^+ as electrophiles: Mayr, H.; Pock, R. *Chem. Ber.* **1986**, *119*, 2473.

3. Open carbocations are prone to rearrange (Chapter 18). Many rearrangements have been found to accompany additions of HX and H_2O.[30]

It may also be recalled that vinylic ethers react with proton donors in a similar manner (see **10-6**).

The stereochemistry of HX addition is varied. Examples are known of predominant syn, anti, and nonstereoselective addition. It was found that treatment of 1,2-dimethylcyclohexene (**4**) with HBr gave predominant anti addition,[31] while addition of water to **4** gave equal amounts of the cis and trans alcohols:[32]

On the other hand, addition of DBr to acenaphthylene (**5**) and to indene and 1-phenylpropene gave predominant syn addition.[33]

In fact, it has been shown that the stereoselectivity of HCl addition can be controlled by changing the reaction conditions. Addition of HCl to **4** in CH_2Cl_2 at $-98°C$ gave predominantly syn addition, while in ethyl ether at $0°C$, the addition was mostly anti.[34]

---

[30]For example, see Whitmore, F.C.; Johnston, F. *J. Am. Chem. Soc.* **1933**, *55*, 5020; Fahey, R.C.; McPherson, C.A. *J. Am. Chem. Soc.* **1969**, *91*, 3865; Bundel, Yu.G.; Ryabstev, M.N.; Sorokin, V.I.; Reutov, O.A. *Bull. Acad. Sci. USSR Div. Chem. Sci.* **1969**, 1311; Pocker, Y.; Stevens, K.D. *J. Am. Chem. Soc.* **1969**, *91*, 4205; Staab, H.A.; Wittig, C.M.; Naab, P. *Chem. Ber.* **1978**, *111*, 2965; Stammann, G.; Griesbaum, K. *Chem. Ber.* **1980**, *113*, 598.

[31]Hammond, G.S.; Nevitt, T.D. *J. Am. Chem. Soc.* **1954**, *76*, 4121; See also, Fahey, R.C.; Monahan, M.W. *J. Am. Chem. Soc.* **1970**, *92*, 2816; Pasto, D.J.; Meyer, G.R.; Lepeska, B. *J. Am. Chem. Soc.* **1974**, *96*, 1858.

[32]Collins, C.H.; Hammond, G.S. *J. Org. Chem.* **1960**, *25*, 911.

[33]Dewar, M.J.S.; Fahey, R.C. *J. Am. Chem. Soc.* **1963**, *85*, 2245, 2248. For a review of syn addition of HX, see Dewar, M.J.S. *Angew. Chem. Int. Ed.* **1964**, *3*, 245; Heasley, G.E.; Bower, T.R.; Dougharty, K.W.; Easdon, J.C.; Heasley, V.L.; Arnold, S.; Carter, T.L.; Yaeger, D.B.; Gipe, B.T.; Shellhamer, D.F. *J. Org. Chem.* **1980**, *45*, 5150.

[34]Becker, K.B.; Grob, C.A. *Synthesis* **1973**, 789. See also, Marcuzzi, F.; Melloni, G.; Modena, G. *Tetrahedron Lett.* **1974**, 413; Naab, P.; Staab, H.A. *Chem. Ber.* **1978**, *111*, 2982.

Addition of HX to triple bonds has the same mechanism, although the intermediate in this case is a vinylic cation, **6**.[35]

$$-C\equiv C- \quad \xrightarrow{\quad H^+ \quad} \quad \overset{\oplus}{\underset{/}{C}}=\overset{H}{\underset{\diagdown}{C}}$$
6

In all these cases (except for the Ad_E3 mechanism), we assumed that formation of the intermediate (**1**, **2**, or **3**) is the slow step and attack by the nucleophile on the intermediate is rapid, and this is probably true in most cases. However, some additions have been found in which the second step is rate determining.[36]

Nucleophilic Addition[37]

In the first step of nucleophilic addition, a nucleophile brings its pair of electrons to one carbon atom of the double or triple bond, creating a carbanion. The second step is combination of this carbanion with a positive species:

$$\text{Step 1} \quad \underset{/}{\overset{\diagdown}{C}}=\underset{\diagdown}{\overset{/}{C}} \quad + \quad Y^{\ominus} \quad \longrightarrow \quad -\underset{/}{\overset{\diagdown}{C}}-\underset{\diagdown}{\overset{/}{C}}^{\ominus}$$

$$\text{Step 2} \quad -\underset{/}{\overset{Y}{\overset{\diagdown}{C}}}-\underset{\diagdown}{\overset{/}{C}}^{\ominus} \quad + \quad W^{\oplus} \quad \longrightarrow \quad -\underset{/}{\overset{Y}{\overset{\diagdown}{C}}}-\underset{\diagdown}{\overset{/}{C}}-W$$

This mechanism is the same as the simple electrophilic one shown on p. 999 except that the charges are reversed (IUPAC $A_N + A_E$ or $A_N + A_H$). When the alkene contains a good leaving group (as defined for nucleophilic substitution), substitution is a side reaction (this is nucleophilic substitution at a vinylic substrate, see p. $$$).

In the special case of addition of HY to a substrate of the form $-C=C-Z$, where $Z = CHO$, COR^{38} (including quinones[39]), COOR, CONH$_2$, CN, NO$_2$, SOR,

[35]For reviews of electrophilic addition to alkynes, including much evidence, see Rappoport, Z. *React. Intermed. (Plenum)* **1983**, *3*, 427, pp. 428–440; Stang, P.J.; Rappoport, Z.; Hanack, M.; Subramanian, L.R. *Vinyl Cations*; Academic Press, NY, **1979**, pp. 24–151; Stang, P.J. *Prog. Phys. Org. Chem.* **1973**, *10*, 205; Modena, G.; Tonellato, U. *Adv. Phys. Org. Chem.* **1971**, *9*, 185, pp. 187–231; Richey, Jr., H.G.; Richey, J.M., in Olah, G.A.; Schleyer, P.V.R. *Carbonium Ions*, Vol. 2, Wiley, NY, **1970**, pp. 906–922.

[36]See, for example, Rau, M.; Alcais, P.; Dubois, J.E. *Bull. Soc. Chim. Fr.* **1972**, 3336; Bellucci, G.; Berti, G.; Ingrosso, G.; Mastrorilli, E. *Tetrahedron Lett.* **1973**, 3911.

[37]For a review, see Patai, S.; Rappoport, Z., in Patai, S. *The Chemistry of Alkenes*, Vol. 1, Wiley, NY, **1964**, pp. 469–584.

[38]For reviews of reactions of C=C—C=O compounds, see, in Patai, S.; Rappoport, Z. *The Chemistry of Enones*, pt. 1, Wiley, NY, **1989**, the articles by Boyd, G.V. pp. 281–315; Duval, D.; Géribaldi, S. pp. 355–469.

[39]For reviews of addition reactions of quinones, see Kutyrev, A.A.; Moskva, V.V. *Russ. Chem. Rev.* **1991**, *60*, 72; Finley, K.T., in Patai, S.; Rappoport, Z. *The Chemistry of the Quinonoid Compounds*, Vol. 2, pt. 1, Wiley, NY, **1988**, pp. 537–717, see pp. 539–589; Finley, K.T., in Patai, S. *The Chemistry of the Quinonoid Compounds*, pt. 2, Wiley, NY, **1974**, pp. 877–1144.

SO_2R,[40] and so on, addition nearly always follows a nucleophilic mechanism,[41] with Y^- bonding with the carbon *away* from the Z group, for example,

Protonation of the enolate ion is chiefly at the oxygen, which is more negative than the carbon, but this produces the enol, which tautomerizes (see p. 102). So although the net result of the reaction is addition to a carbon–carbon double bond, the *mechanism* is 1,4-nucleophilic addition to the C=C—C=O (or similar) system and is thus very similar to the mechanism of addition to carbon–oxygen double and similar bonds (see Chapter 16). When Z is CN or a C=O group, it is also possible for Y^- to attack at *this* carbon, and this reaction sometimes competes. When it happens, it is called 1,2-addition. 1,4-Addition to these substrates is also known as *conjugate addition*. The Y^- ion almost never attacks at the 3 position, since the resulting carbanion would have no resonance stabilization:[42]

An important substrate of this type is acrylonitrile, and 1,4-addition to it is called *cyanoethylation* because the Y is cyanoethylated:

$$H_3C=CH-CN \quad + \quad H-Y \quad \longrightarrow \quad Y-CH_2-CH_2-CV$$

With any substrate, when Y is an ion of the type $Z-C^{\ominus}R_2$ (Z is as defined above; R may be alkyl, aryl, hydrogen, or another Z), the reaction is called the *Michael reaction* (see **15-24**). In this book we will call all other reactions that follow this mechanism *Michael-type additions*. Systems of the type C=C—C=C—Z can give

[40]For a review of vinylic sulfones, see Simpkins, N.S. *Tetrahedron* **1990**, *46*, 6951. For a review of conjugate addition to cycloalkenyl sulfones, see Fuchs, P.L.; Braish, T.F. *Chem. Rev.* **1986**, *86*, 903.

[41]For a review of the mechanism with these substrates, see Bernasconi, C.F. *Tetrahedron* **1989**, *45*, 4017.

[42]For 1,8-addition to a trienone, see Barbot, F.; Kadib-Elban, A.; Miginiac, P. *J. Organomet. Chem.* **1988**, *345*, 239.

1,2-1,4- or 1,6-addition.[43] Michael-type reactions are reversible, and compounds of the type YCH_2CH_2Z can often be decomposed to YH and $CH_2{=}CHZ$ by heating, either with or without alkali.

If the mechanism for nucleophilic addition is the simple carbanion mechanism outlined on p. 1007, the addition should be nonstereospecific, although it might well be stereoselective (see p. 194 for the distinction). For example, the (E) and (Z) forms of an alkene ABC=CDE would give **7** and **8**.

If the carbanion has even a short lifetime, **7** and **8** will assume the most favorable conformation before the attack of W. This is of course the same for both, and when W attacks, the same product will result from each. This will be one of two possible diastereomers, so the reaction will be stereoselective; but since the cis and trans isomers do not give rise to different isomers, it will not be stereospecific. Unfortunately, this prediction has not been tested on open-chain alkenes. Except for Michael-type substrates, the stereochemistry of nucleophilic addition to double bonds has been studied only in cyclic systems, where only the cis isomer exists. In these cases the reaction has been shown to be stereoselective, with syn addition reported in some cases[44] and anti addition in others.[45] When the reaction is performed on a Michael-type substrate, C=C–Z, the hydrogen does not arrive at the carbon directly but only through a tautomeric equilibrium. The product naturally assumes the most thermodynamically stable configuration, without relation to the direction of original attack of Y. In one such case (the addition of EtOD and of Me_3CSD to *trans*-MeCH=CHCOOEt) predominant anti addition was found; there is evidence that the stereoselectivity here results from the final protonation of the enolate, and not from the initial attack.[46] For obvious reasons, additions to triple bonds cannot be stereospecific. As with electrophilic additions, nucleophilic additions to triple bonds are usually stereoselective and

[43]However, attack at the 3 position has been reported when the 4 position contains one or two carbanion-stabilizing groups such as $SiMe_3$: Klumpp, G.W.; Mierop, A.J.C.; Vrielink, J.J.; Brugman, A.; Schakel, M. *J. Am. Chem. Soc.* **1985**, *107*, 6740.

[44]For example, Truce, W.E.; Levy, A.J. *J. Org. Chem.* **1963**, *28*, 679.

[45]For example, Truce, W.E.; Levy, A.J. *J. Am. Chem. Soc.* **1961**, *83*, 4641; Zefirov, N.S.; Yur'ev, Yu.K.; Prikazchikova, L.P.; Bykhovskaya, M.Sh. *J. Gen. Chem. USSR* **1963**, *33*, 2100.

[46]Mohrig, J.R.; Fu, S.S.; King, R.W.; Warnet, R.; Gustafson, G. *J. Am. Chem. Soc.* **1990**, *112*, 3665.

anti,[47] although syn addition[48] and nonstereoselective addition[49] have also been reported.

Free-Radical Addition

The mechanism of free-radical addition[50] follows the pattern discussed in Chapter 14 (pp. 934–939). The method of principal component analysis has been used to analyze polar and enthalpic effect in radical addition reactions.[51] A radical is generated by

$$YW \xrightarrow[\text{dissociation}]{h\nu \text{ or spontaneous}} Y\cdot \ + \ W\cdot$$

or

$$R\cdot \text{ (from some other source)} \ + \ YW \longrightarrow RW \ + \ Y\cdot$$

Propagation then occurs by

Step 1

Step 2

[47]Truce, W.E.; Simms, J.A. *J. Am. Chem. Soc.* **1956**, *78*, 2756; Shostakovskii, M.F.; Chekulaeva, I.A.; Kondrat'eva, L.V.; Lopatin, B.V. *Bull. Acad. Sci. USSR Div. Chem. Sci.* **1962**, 2118; Théron F.; Vessière, R. *Bull. Soc. Chim. Fr.* **1968**, 2994; Bowden, K.; Price, M.J. *J. Chem. Soc. B* **1970**, 1466, 1472; Raunio, E.K.; Frey, T.G. *J. Org. Chem.* **1971**, *36*, 345; Truce, W.E.; Tichenor, G.J.W. *J. Org. Chem.* **1972**, *37*, 2391.

[48]Truce, W.E.; Goldhamer, D.M.; Kruse, R.B. *J. Am. Chem. Soc.* **1959**, *81*, 4931; Dolfini, J.E. *J. Org. Chem.* **1965**, *30*, 1298; Winterfeldt, E.; Preuss, H. *Chem. Ber.* **1966**, *99*, 450; Hayakawa, K.; Kamikawaji, Y.; Wakita, A.; Kanematsu, K. *J. Org. Chem.* **1984**, *49*, 1985.

[49]Gracheva, E.P.; Laba, V.I.; Kul'bovskaya, N.K.; Shostakovskii, M.F. *J. Gen. Chem. USSR* **1963**, *33*, 2431; Truce, W.E.; Brady, D.G. *J. Org. Chem.* **1966**, *31*, 3543; Prilezhaeva, E.N.; Vasil'ev, G.S.; Mikhaleshvili, I.L.; Bogdanov, V.S. *Bull. Acad. Sci. USSR Div. Chem. Sci.* **1970**, 1820.

[50]For a monograph on this subject, see Huyser, E.S. *Free-Radical Chain Reactions*, Wiley, NY, **1970**. Other books with much of interest in this field are Nonhebel, D.C.; Walton, J.C. *Free-Radical Chemistry*; Cambridge University Press: London, *1974*; Pyor, W.A. *Free Radicals*; McGraw-Hill, NY, **1965**. For reviews, see Giese, B. *Rev. Chem. Intermed.* **1986**, *7*, 3; *Angew. Chem. Int. Ed.* **1983**, *22*, 753; Amiel, Y., in Patai, S.; Rappoport, Z. *The Chemistry of Functional Groups, Supplement C* pt. 1, Wiley, NY, **1983**, pp. 341–382; Abell, P.I., in Bamford, C.H.; Tipper, C.F.H. *Comprehensive Chemical Kinetics*, Vol. 18; Elsevier, NY, *1976*, pp. 111–165; Abell, P.I. in Kochi, J.K. *Free Radicals*, Vol. 2, Wiley, NY, *1973*, pp. 63–112; Minisci, F. *Acc. Chem. Res.* **1975**, *8*, 165; Julia, M., in Viehe, H.G. *Acetylenes*; Marcel Dekker, NY, *1969*, pp. 335–354; Elad, D. *Org. Photochem.* **1969**, *2*, 168; Schönberg, A. *Preparative Organic Photochemistry*, Springer, NY, *1968*, pp. 155–181; Cadogan, J.I.G.; Perkins, M.J., in Patai, S. *The Chemistry of Alkenes*, Vol. 1, Wiley, NY, *1964*, pp. 585–632.

[51]Héberger, K.; Lopata, A. *J. Chem. Soc. Perkin Trans. 2*, *1995*, 91.

Step 2 is an abstraction (an atom transfer), so W is nearly always univalent, either hydrogen or halogen (p. 943). Termination of the chain can occur in any of the ways discussed in Chapter 14. If **9** adds to another alkene molecule,

$$\underset{\mathbf{9}}{\overset{Y}{\underset{/}{\overset{|}{-}}}\overset{}{\underset{\backslash}{C}}\overset{/}{\underset{\backslash}{C}}\cdot} \;+\; \overset{\backslash}{\underset{/}{C}}=\overset{/}{\underset{\backslash}{C}} \;\longrightarrow\; \overset{Y}{\underset{/}{-}}\overset{}{\underset{\backslash}{C}}\overset{/}{\underset{\backslash}{C}}\!\!-\!\!\overset{\backslash}{\underset{/}{C}}\overset{/}{\underset{\backslash}{-}}C\overset{/}{\underset{\backslash}{C}}\cdot$$

a dimer is formed. This can add to still another, and chains, long or short, may be built up. This is the mechanism of free-radical polymerization. Short polymeric molecules (called *telomers*), formed in this manner, are often troublesome side products in free-radical addition reactions.

When free radicals are added to 1,5- or 1,6-dienes, the initially formed radical (**10**) can add intramolecularly to the other bond, leading to a cyclic product (**11**).[52] When the radical is generated from an precursor that gives vinyl radical **12**, however, cyclization leads to **13**, which is in equilibrium with cyclopropylcarbinyl radical (**14**) via a 5-exo-trig reaction.[53] A 6-endo-trig reaction leads to **15**, but unless there are perturbing substituent effects, however, cyclopropanation should be the major process.

Radicals of the type **10**, generated in other ways, also undergo these cyclizations. Both five- and six-membered rings can be formed in these reactions (see p. 1021).

The free-radical addition mechanism just outlined predicts that the addition should be non-stereospecific, at least if **9** has any, but an extremely short lifetime. However, the reactions may be stereoselective, for reasons similar to those discussed for nucleophilic addition on p. 1007. Not all free-radical additions have been found to be selective, but many are. For example, addition of HBr to 1-bromocyclohexene is regioselective in that it gave only *cis*-1,2-dibromocyclohexane

[52]For reviews of these and other free-radical cyclization reactions, see RajanBabu, T.V. *Acc. Chem. Res.* **1991**, *24*, 139; Beckwith, A.L.J. *Rev. Chem. Intermed.* **1986**, *7*, 143; Giese, B. *Radicals in Organic Synthesis: Formation of Carbon-Carbon Bonds*, Pergamon, Elmsford, NY, **1986**, pp. 141–209; Surzur, J. *React. Intermed. (Plenum)* **1982**, *2*, 121–295; Julia, M. *Acc. Chem. Res.* **1972**, *4*, 386; *Pure Appl. Chem.* **1974**, *40*, 553; **1967**, *15*, 167–183; Nonhebel, D.C.; Walton, J.C. *Free-Radical Chemistry*, Cambridge University Press, London, **1974**, pp. 533–544; Wilt, J.W., in Kochi, J.K. *Free Radicals*, Vol. 1, Wiley, NY, **1973**, pp. 418–446. For a review of cyclizations in general, see Thebtaranonth, C.; Thebtaranonth, Y. *Tetrahedron* **1990**, *46*, 1385.

[53]Denis, R.C.; Rancourt, J.; Ghiro, E.; Boutonnet, F.; Gravel, D. *Tetrahedron Lett.* **1993**, *34*, 2091.

and none of the trans isomer (anti addition),[54] and propyne (at -78 to $-60°C$) gave only *cis*-1-bromopropene (anti addition), making it stereoselective.[55] However, stereospecificity has been found only in a few cases. Selectivity was observed in radical cyclization reactions of functionalized alkenes, which proceeded via a trans-ring closure.[56] The most important case is probably addition of HBr to 2- bromo-2-butene under free-radical conditions at $-80°C$. Under these conditions, the cis isomer gave 92% of the meso product, while the trans isomer gave mostly the *dl* pair.[57] This stereospecificity disappeared at room temperature, where both alkenes gave the same mixture of products ($\sim$78% of the *dl* pair and 22% of the meso compound), so the addition was still stereoselective but no longer stereospecific. The stereospecificity at low temperatures is probably caused by a stabilization of the intermediate radical through the formation of a bridged bromine radical, of the type mentioned on p. 942:

$$\begin{array}{c} \overset{Br}{\underset{|}{}} \\ -\overset{|}{C}-\overset{|}{C}\cdot \end{array} \longrightarrow \begin{array}{c} \overset{\cdot Br}{\underset{|}{}} \\ -\overset{|}{C}-\overset{|}{C}- \end{array}$$

This species is similar to the bromonium ion that is responsible for stereospecific anti addition in the electrophilic mechanism. Further evidence for the existence of such bridged radicals was obtained by addition of Br• to alkenes at 77 K. The ESR spectra of the resulting species were consistent with bridged structures.[58]

For many radicals, step 1 ($C=C + Y• \rightarrow •C-C-Y$) is reversible. In such cases, free radicals can cause cis $\rightarrow$ trans isomerization of a double bond by the pathway[59]

$$\underset{R^3}{\overset{R^1}{\diagdown}}C=C\underset{R^4}{\overset{R^2}{\diagup}} \;\;\overset{Y}{\underset{}{\rightleftarrows}}\;\; Y-\underset{R^3}{\overset{R^1}{\underset{|}{C}}}-\underset{R^4}{\overset{R^2}{\underset{|}{C}}}\cdot \;\;\overset{rotation}{\rightleftarrows}\;\; Y-\underset{R^3}{\overset{R^1}{\underset{|}{C}}}-\underset{R^2}{\overset{R^4}{\underset{|}{C}}}\cdot \;\;\overset{-Y}{\rightleftarrows}\;\; \underset{R^3}{\overset{R^1}{\diagdown}}C=C\underset{R^2}{\overset{R^4}{\diagup}}$$

Cyclic Mechanisms

There are some addition reactions where the initial attack is not at one carbon of the double bond, but both carbons are attacked simultaneously. Some of these are

[54]Goering, H.L.; Abell, P.I.; Aycock, B.F. *J. Am. Chem. Soc.* **1952**, *74*, 3588. See also, LeBel, N.A.; Czaja, R.F.; DeBoer, A. *J. Org. Chem.* **1969**, *34*, 3112.

[55]Skell, P.S.; Allen, R.G. *J. Am. Chem. Soc.* **1958**, *80*, 5997.

[56]Ogura, K.; Kayano, A.; Fujino, T.; Sumitani, N.; Fujita, M. *Tetrahedron Lett.* **1993**, *34*, 8313.

[57]Goering, H.L.; Larsen, D.W. *J. Am. Chem. Soc.* **1957**, *79*, 2653; **1959**, *81*, 5937. Also see, Skell, P.S.; Freeman, P.K. *J. Org. Chem.* **1964**, *29*, 2524.

[58]Abell, P.I.; Piette, L.H. *J. Am. Chem. Soc.* **1962**, *84*, 916. See also, Leggett, T.L.; Kennerly, R.E.; Kohl, D.A. *J. Chem. Phys.* **1974**, *60*, 3264.

[59]Benson, S.W.; Egger, K.W.; Golden, D.M. *J. Am. Chem. Soc.* **1965**, *87*, 468; Golden, D.M.; Furuyama, S.; Benson, S.W. *Int. J. Chem. Kinet.* **1969**, *1*, 57.

four-center mechanisms, which follow this pattern:

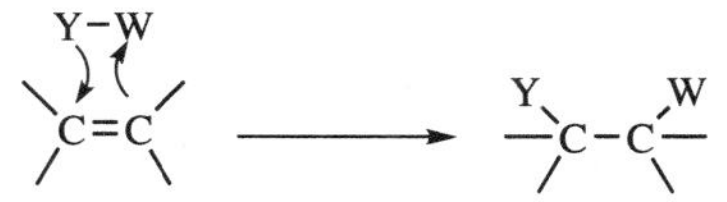

$$\underset{/}{\overset{\backslash}{C}}=\underset{\backslash}{\overset{/}{C}} \longrightarrow -\underset{/}{\overset{Y}{C}}-\underset{\backslash}{\overset{W}{C}}-$$

In others, there is a five- or a six-membered transition state. In these cases the addition to the double or triple bond must be syn. The most important reaction of this type is the Diels–Alder reaction (**15-60**).

Addition to Conjugated Systems

When electrophilic addition is carried out on a compound with two double bonds in conjugation, a 1,2-addition product (**16**) is often obtained, but in most cases there is also a 1,4-addition product (**17**), often in larger yield:[60]

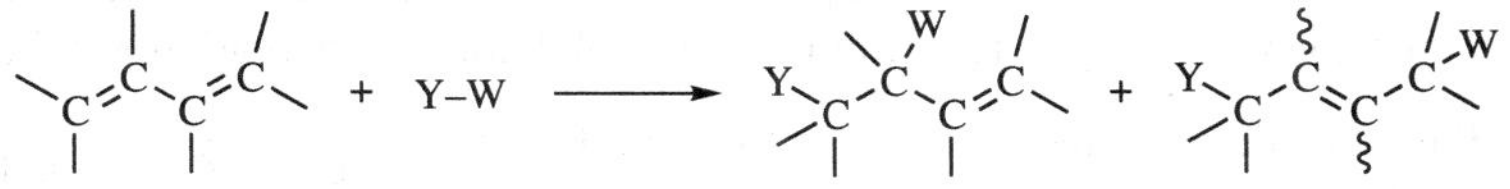

If the diene is unsymmetrical, there may be two 1,2-addition products. The competition between two types of addition product comes about because the carbocation resulting from attack on Y^+ is a resonance hybrid, with partial positive charges at the 2 and 4 positions:

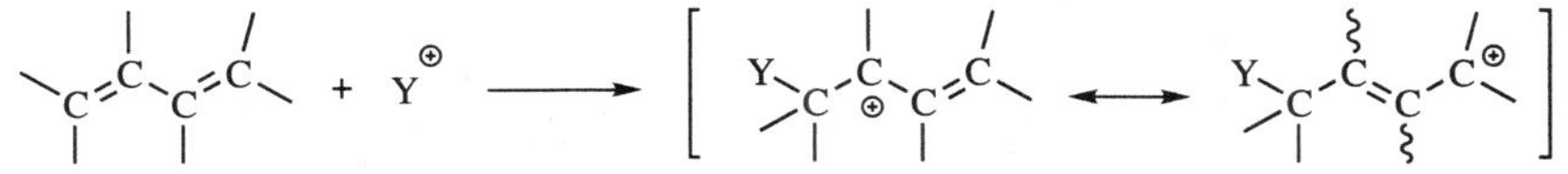

W^- may then attack either position. The original attack of Y^+ is always at the end of the conjugated system because an attack at a middle carbon would give a cation unstabilized by resonance:

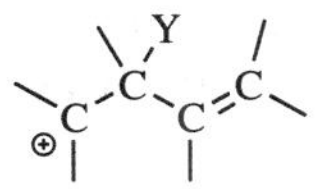

In the case of electrophiles like Br^+, which can form cyclic intermediates, both 1,2- and 1,4-addition products can be rationalized as stemming from an intermediate like **18**. Direct nucleophilic attack by W^- would give the 1,2-product, while the 1,4-product could be formed by attack at the 4 position, by an S_N2'-type mechanism (see p. 470). Intermediates like **19** have been postulated, but ruled out for Br and Cl

---

[60]For a review of electrophilic addition to conjugated dienes, see Khristov, V.Kh.; Angelov, Kh.M.; Petrov, A.A. *Russ. Chem. Rev.* **1991**, *60*, 39.

by the observation that chlorination

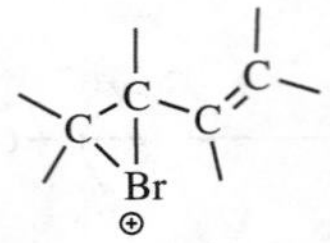

18 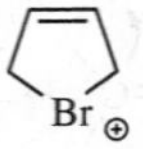   **19**

or bromination of butadiene gives trans 1,4-products.[61] If an ion like **19** were the intermediate, the 1,4-products would have to have the cis configuration.

In most cases, more 1,4- than 1,2-addition product is obtained. This may be a consequence of thermodynamic control of products, as against kinetic. In most cases, under the reaction conditions, **16** is converted to a mixture of **16** and **17** which is richer in **17**. That is, either isomer gives the same mixture of both, which contains more **17**. It was found that at low temperatures, butadiene and HCl gave only 20–25% 1,4-adduct, while at high temperatures, where attainment of equilibrium is more likely, the mixture contained 75% 1,4-product.[62] 1,2-Addition predominated over 1,4- in the reaction between DCl and 1,3-pentadiene, where the intermediate was the symmetrical (except for the D label) $H_3CHC\overset{\oplus}{-}\overline{CH-CHCH_2D}$[63]

Ion pairs were invoked to explain this result, since a free ion would be expected to be attacked by Cl⁻ equally well at both positions, except for the very small isotope effect.

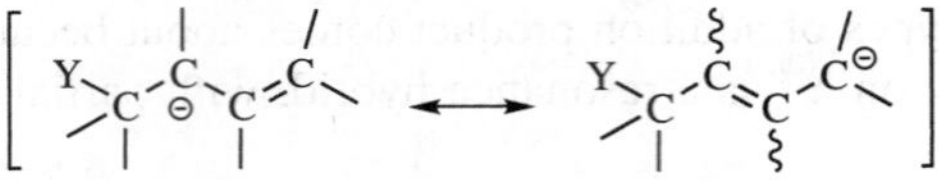

Addition to conjugated systems can also be accomplished by any of the other three mechanisms. In each case, there is competition between 1,2- and 1,4-addition. In the case of nucleophilic or free-radical attack,[64] the intermediates are resonance hybrids and behave like the intermediate from electrophilic attack. Dienes can give 1,4-addition by a cyclic mechanism in this way:

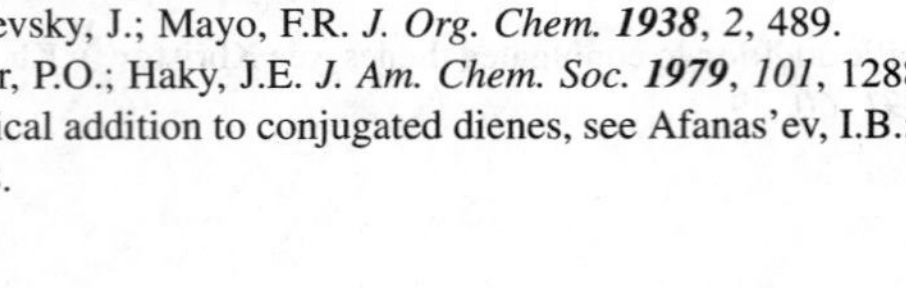

[61]Mislow, K. *J. Am. Chem. Soc.* **1953**, *75*, 2512.
[62]Kharasch, M.S.; Kritchevsky, J.; Mayo, F.R. *J. Org. Chem.* **1938**, *2*, 489.
[63]Nordlander, J.E.; Owuor, P.O.; Haky, J.E. *J. Am. Chem. Soc.* **1979**, *101*, 1288.
[64]For a review of free-radical addition to conjugated dienes, see Afanas'ev, I.B.; Samokhvalov, G.I. *Russ. Chem. Rev.* **1969**, *38*, 318.

Other conjugated systems, including trienes, enynes, diynes, and so on, have been studied much less, but behave similarly. 1,4-Addition to enynes is an important way of making allenes:

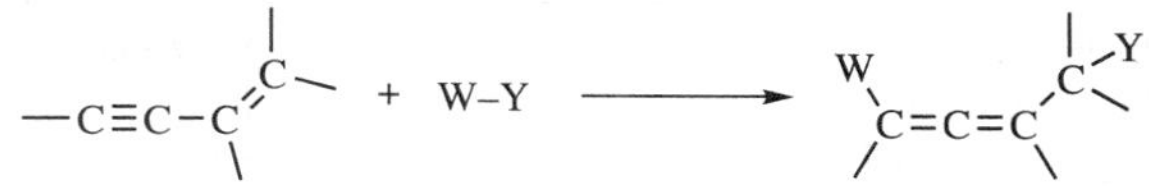

Radical addition to conjugated systems is an important part of chain propagation reactions. The rate constants for addition of cyclohexyl radical to conjugated amides have been measured, and shown to be faster than addition to styrene.[65] In additions to $RCH=C(CN)_2$ systems, where the R group has a chiral center, the Felkin–Ahn rule (p. 169) is followed and the reaction proceeds with high selectivity.[66] Addition of some radicals, such as $(Me_3Si)_3Si\bullet$, is reversible and this can lead to poor selectivity or isomerization.[67]

ORIENTATION AND REACTIVITY

Reactivity

As with electrophilic aromatic substitution (Chapter 11), electron-donating groups increase the reactivity of a double bond toward electrophilic addition and electron-with-drawing groups decrease it. This is illustrated in Tables 15.1 and 15.2.[68] As a further illustration it may be mentioned that the reactivity toward electrophilic addition of a group of alkenes increased in the order $CCl_3CH=CH_2 < Cl_2CHCH=CH_2 < ClCH_2CH=CH_2 < CH_3CH_2=CH_2$.[69] For nucleophilic addition the situation is reversed. These reactions are best carried out on substrates containing three or four electron-withdrawing groups, two of the most common being $F_2C=CF_2$[70] and $(NC)_2C=C(CN)_2$.[71] The effect of substituents is so great that it is possible to make the statement that *simple alkenes do not react by the nucleophilic mechanism, and poly-halo or polycyano alkenes do not generally react by the electrophilic mechanism.*[72]

[65]Curran, D.P.; Qi, H.; Porter, N.A.; Su, Q.; Wu, W.-X. *Tetrahedron Lett.* *1993*, *34*, 4489.

[66]Giese, B.; Damm, W.; Roth, M.; Zehnder, M. *Synlett* *1992*, 441.

[67]Ferreri, C.; Ballestri, M.; Chatgilialoglu, C. *Tetrahedron Lett.* *1993*, *34*, 5147.

[68]Table 15.1 is from de la Mare, P.B.D. *Q. Rev. Chem. Soc.* *1949*, *3*, 126, p. 145. Table 15.2 is from Dubois, J.E.; Mouvier, G. *Tetrahedron Lett.* *1963*, 1325. See also, Dubois, J.E.; Mouvier, G. *Bull. Soc. Chim. Fr.* *1968*, 1426; Grosjean, D.; Mouvier, G.; Dubois, J.E. *J. Org. Chem.* *1976*, *41*, 3869, 3872.

[69]Shelton, J.R.; Lee, L. *J. Org. Chem.* *1960*, *25*, 428.

[70]For a review of additions to $F_2C=CF_2$ and other fluoroalkenes, see Chambers, R.D.; Mobbs, R.H. *Adv. Fluorine Chem.* *1965*, *4*, 51.

[71]For reviews of additions to tetracyanoethylene, see Fatiadi, A.J. *Synthesis* *1987*, 249, 749; Dhar, D.N. *Chem. Rev.* *1967*, *67*, 611.

[72]Such reactions can take place under severe conditions. For example, electrophilic addition could be accomplished with $F_2C=CHF$ in super acid solutions [Olah, G.A.; Mo, Y.K. *J. Org. Chem.* *1972*, *37*, 1028] although $F_2C=CF_2$ did not react under these conditions. For reviews of electrophilic additions to fluoroalkenes, see Belen'kii, G.G.; German, L.S. *Sov. Sci. Rev. Sect. B* *1984*, *5*, 183; Dyatkin, B.L.; Mochalina, E.P.; Knunyants, I.L. *Russ. Chem. Rev.* *1966*, *35*, 417; *Fluorine Chem. Rev.* *1969*, *3*, 45; Chambers, R.D.; Mobbs, R.H. *Adv. Fluorine Chem.* *1965*, *4*, 51, pp. 77–81.

TABLE 15.1. Relative Reactivity of Some Alkenes Toward Bromine in Acetic Acid at 24°C[68]

Alkene	Relative Rate
$PhCH{=}CH_2$	Very fast
$PhCH{=}CHPh$	18
$CH_2{=}CHCH_2Cl$	1.6
$CH_2{=}CHCH_2Br$	1.0
$PhCH{=}CHBr$	0.11
$CH_2{=}CHBr$	0.0011

TABLE 15.2. Relative Reactivity of Some Alkenes Toward Bromine in Methanol[68]

Alkene	Relative Rate
$CH_2{=}CH_2$	3.0×10^1
$CH_3CH_2CH{=}CH_2$	$2/9 \times 10^3$
$cis\text{-}CH_3CH_2CH{=}CHCH_3$	1.3×10^5
$(CH_3)_2C{=}C(CH_3)_2$	2.8×10^7

There are some reagents that attack only as nucleophiles, for example, ammonia, and these add only to substrates susceptible to nucleophilic attack. Other reagents attack only as electrophiles, and, for example, $F_2C{=}CF_2$ does not react with these. In still other cases, the same reagent reacts with a simple alkene by the electrophilic mechanism and with a polyhalo alkene by a nucleophilic mechanism. For example, Cl_2 and HF are normally electrophilic reagents, but it has been shown that Cl_2 adds to $(N{\equiv}C)_2C{=}CHC{\equiv}N$ with initial attack by Cl^{-}[73] and that HF adds to $F_2C{=}CClF$ with initial attack by F^{-}.[74] Compounds that have a double bond conjugated with a Z group (as defined on p. 1007) nearly always react by a nucleophilic mechanism.[75] These are actually 1,4-additions, as discussed on p. 1008. A number of studies have been made of the relative activating abilities of various Z groups.[76] On the basis of these studies, the following order of decreasing activating ability has been suggested: $Z = NO_2$, COAr, CHO, COR, SO_2Ar, CN, COOR, SOAr, $CONH_2$, CONHR.[77]

It seems obvious that electron-withdrawing groups enhance nucleophilic addition and inhibit electrophilic addition because they lower the electron density of

[73]Dickinson, C.L., Wiley, D.W.; McKusick, B.C. *J. Am. Chem. Soc.* **1960**, *82*, 6132. For another example, see Atkinson, R.C.; de la Mare, P.B.D.; Larsen, D.S. *J. Chem. Soc. Perkin Trans. 2*, **1983**, 271.

[74]Miller, Jr., W.T.; Fried, J.H.; Goldwhite, H. *J. Am. Chem. Soc.* **1960**, *82*, 3091.

[75]For a review of electrophilic reactions of such compounds, see Müllen, K.; Wolf, P., in Patai, S.; Rappoport, Z. *The Chemistry of Enones*, pt. 1, Wiley, NY, **1989**, pp. 513–558.

[76]See, for example, Friedman, M.; Wall, J.S. *J. Org. Chem.* **1966**, *31*, 2888; Ring, R.N.; Tesoro, G.C.; Moore, D.R. *J. Org. Chem.* **1967**, *32*, 1091.

[77]Shenhav, H.; Rappoport, Z.; Patai, S. *J. Chem. Soc. B* **1970**, 469.

the double bond. Addition of electrophilic radicals to electron rich alkenes has been reported,[78] so the reaction is possible in some cases. This is probably true, and yet similar reasoning does not always apply to a comparison between double and triple bonds.[79] There is a higher concentration of electrons between the carbons of a triple bond than in a double bond, and yet triple bonds are *less* subject to attack at an electrophilic site and *more* subject to nucleophilic attack than double bonds.[80] This statement is not universally true, but it does hold in most cases. In compounds containing both double and triple bonds (nonconjugated), bromine, an electrophilic reagent, always adds to the double bond.[81] In fact, all reagents that form bridged intermediates like **2** react faster with double than with triple bonds. On the other hand, addition of electrophilic H^+ (acid-catalyzed hydration, **15-3**; addition of hydrogen halides, **15-2**) takes place at about the same rates for alkenes as for corresponding alkynes.[82] Furthermore, the presence of electron-withdrawing groups lowers the alkene/alkyne rate ratio. For example, while styrene $PhCH=CH_2$ was brominated 3000 times faster than $PhC\equiv CH$, the addition of a second phenyl group ($PhCH=CHPh$ versus $PhC\equiv CPh$) lowered the rate ratio to about 250.[83] In the case of *trans*-$MeOOCCH=CHCOOMe$ versus $MeOOCC\equiv CCOOMe$, the triple bond compound was actually brominated faster.[84]

As mentioned, it is true that in general triple bonds are more susceptible to nucleophilic and less to attack on an electrophilic site than double bonds, in spite of their higher electron density. One explanation is that the electrons in the triple bond are held more tightly because of the smaller carbon–carbon distance; it is thus harder for an attacking electrophile to pull out a pair. There is evidence from far-UV spectra to support this conclusion.[85] Another possible explanation has to do with the availability of the unfilled orbital in the alkyne. It has been shown that a π^* orbital of bent alkynes (e.g., cyclooctyne) has a lower energy than the π^* orbital of alkenes, and it has been suggested[86] that linear alkynes can achieve a bent structure in their transition states when reacting with an electrophile. Where electrophilic addition involves bridged-ion intermediates, those arising from triple bonds (**20**) are more strained than the corresponding **21** and furthermore are antiaromatic systems

[78]Curran, D.P.; Ko, S.-B. *Tetrahedron Lett.* **1998**, *39*, 6629.

[79]For reviews of ionic additions to triple bonds, see, in Patai, S. *The Chemistry of the Carbon–Carbon Triple Bond*, Wiley, NY, **1978**, the articles by Schmid, G.H. pt. 1, pp. 275–341, and by Dickstein, J.I.; Miller, S.I. pt. 2, pp. 813–955; Miller, S.I.; Tanaka, R. *Sel. Org. Transform.* **1970**, *1*, 143; Winterfeldt, E., in Viehe, H.G. *Acetylenes*, Marcel Dekker, NY, **1969**, pp. 267–334. For comparisons of double and triple bond reactivity, see Melloni, G.; Modena, G.; Tonellato, U. *Acc. Chem. Res.* **1981**, *14*, 227; Allen, A.D.; Chiang, Y.; Kresge, A.J.; Tidwell, T.T. *J. Org. Chem.* **1982**, *47*, 775.

[80]For discussions, see Daniels, R.; Bauer, L. *J. Chem. Educ.* **1958**, *35*, 444; DeYoung, S.; Ehrlich, S.; Berliner, E. *J. Am. Chem. Soc.* **1977**, *99*, 290; Strozier, R.W.; Caramella, P.; Houk, K.N. *J. Am. Chem. Soc.* **1979**, *101*, 1340.

[81]Petrov, A.A. *Russ. Chem. Rev.* **1960**, *29*, 489.

[82]Melloni, G.; Modena, G.; Tonellato, U. *Acc. Chem. Res.* **1981**, *14*, 227, p. 228.

[83]Robertson, P.W.; Dasent, W.E.; Milburn, R.M.; Oliver, W.H. *J. Chem. Soc.* **1950**, 1628.

[84]Wolf, S.A.; Ganguly, S.; Berliner, E. *J. Am. Chem. Soc.* **1985**, *50*, 1053.

[85]Walsh, A.D. *Q. Rev. Chem. Soc.* **1948**, *2*, 73.

[86]Ng, L.; Jordan, K.D.; Krebs, A.; Rüger, W. *J. Am. Chem. Soc.* **1982**, *104*, 7414.

(see p. 73), which **21** are not. This may be a reason why electrophilic addition by such electrophiles as Br, I, SR, and so on, is slower for triple than for double bonds.[87] As might be expected, triple bonds connected to a Z group (C≡C–Z) undergo nucleophilic addition especially well.[88]

$$
\underset{\textbf{20}}{\overset{\overset{\oplus}{X}}{\triangle}} \qquad \underset{\textbf{21}}{\overset{\overset{\oplus}{X}}{\triangle}}
$$

Although alkyl groups in general increase the rates of electrophilic addition, we have already mentioned (p. 1005) that there is a different pattern depending on whether the intermediate is a bridged ion or an open carbocation. For brominations and other electrophilic additions in which the first step of the mechanism is rate determining, the rates for substituted alkenes correlate well with the ionization potentials of the alkenes, which means that steric effects are not important.[89] Where the second step is rate determining [e.g., oxymercuration (**15-3**), hydroboration (**15-17**)], steric effects are important.[88]

Free-radical additions can occur with any type of substrate. The determining factor is the presence of a free-radical attacking species. Some reagents (e.g., HBr, RSH) attack by ionic mechanisms if no initiator is present, but in the presence of a free-radical initiator, the mechanism changes and the addition is of the free-radical type. Nucleophilic radicals (see p. 938) behave like nucleophiles in that the rate is increased by the presence of electron-withdrawing groups in the substrate. The reverse is true for electrophilic radicals.[90] However, nucleophilic radicals react with alkynes more slowly than with the corresponding alkenes,[91] which is contrary to what might have been expected.[92]

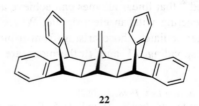

22

[87]Nevertheless, bridged ions **15** have been implicated in some additions to triple bonds. See, for example, Pincock, J.A.; Yates, K. *Can. J. Chem.* **1970**, *48*, 3332; Mauger, E.; Berliner, E. *J. Am. Chem. Soc.* **1972**, *94*, 194; Bassi, P.; Tonellato, U. *J. Chem. Soc. Perkin Trans. 1*, **1973**, 669; Schmid, G.H.; Modro, A.; Lenz, F.; Garratt, D.G.; Yates, K. *J. Org. Chem.* **1976**, *41*, 2331.

[88]For a review of additions to these substrates, see Winterfeldt, E. *Angew. Chem. Int. Ed.* **1967**, *6*, 423; *Newer Methods Prep. Org. Chem.* **1971**, *6*, 243.

[89]Nelson, D.J.; Cooper, P.J.; Soundararajan, R. *J. Am. Chem. Soc.* **1989**, *111*, 1414; Nelson, D.J.; Soundararajan, R. *Tetrahedron Lett.* **1988**, *29*, 6207.

[90]For reviews of reactivity in free-radical additions, see Tedder, J.M. *Angew. Chem. Int. Ed.* **1982**, *21*, 401; Tedder, J.M.; Walton, J.C. *Tetrahedron* **1980**, *36*, 701.

[91]Giese, B.; Lachhein, S. *Angew. Chem. Int. Ed.* **1982**, *21*, 768.

[92]For a discussion of reactivity and orientation of polar radicals, see Volovik, S.V.; Dyadyusha, G.G.; Staninets, V.I. *J. Org. Chem. USSR* **1986**, *22*, 1224.

Steric influences are important in some cases. In catalytic hydrogenation, where the substrate must be adsorbed onto the catalyst surface, the reaction becomes more difficult with increasing substitution. The hydrocarbon **22**, in which the double bond is entombed between the benzene rings, does not react with Br_2, H_2SO_4, O_3, BH_3, :CBr_2, or other reagents that react with most double bonds.[93] A similarly inactive compound is tetra-*tert*-butylallene (t-Bu)$_2$C=C=C(t-Bu)$_2$, which is inert to Br_2, Cl_2, O_3, and catalytic hydrogenation.[94]

Orientation

When an unsymmetrical reagent is added to an unsymmetrical substrate, the question arises: Which side of the reagent goes to which side of the double or triple bond? The terms side and face are arbitrary, and a simple guide is shown to help understand the arguments used here. For electrophilic attack, the answer is given by

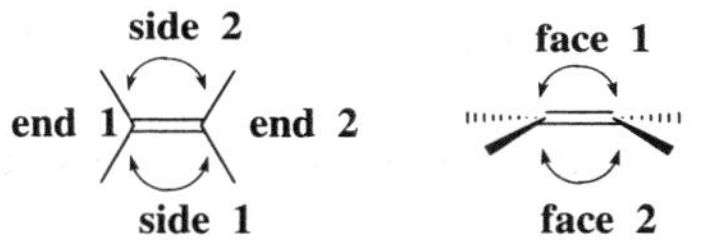

Markovnikov's rule: The positive portion of the reagent goes to the side of the double or triple bond that has more hydrogens.[95] A number of explanations have been suggested for this regioselectivity, but the most probable is that Y^+ adds to that side that will give the more stable carbocation. This premise has been examined by core electron spectroscopy and by theoretical analysis.[96] Thus, when an alkyl group is present, secondary carbocations are more stable than primary:

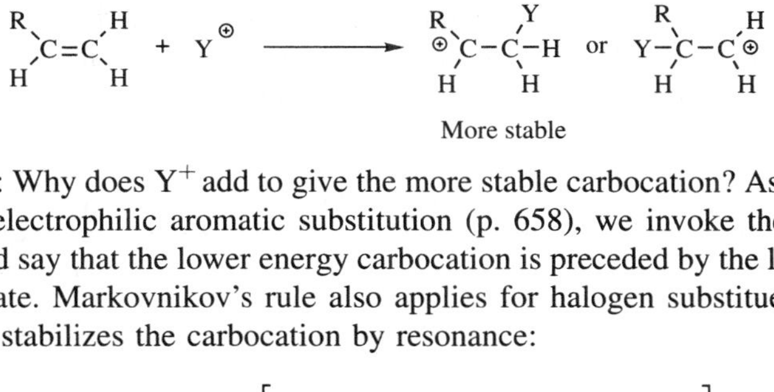

We may ask: Why does Y^+ add to give the more stable carbocation? As in the similar case of electrophilic aromatic substitution (p. 658), we invoke the Hammond postulate and say that the lower energy carbocation is preceded by the lower energy transition state. Markovnikov's rule also applies for halogen substituents because the halogen stabilizes the carbocation by resonance:

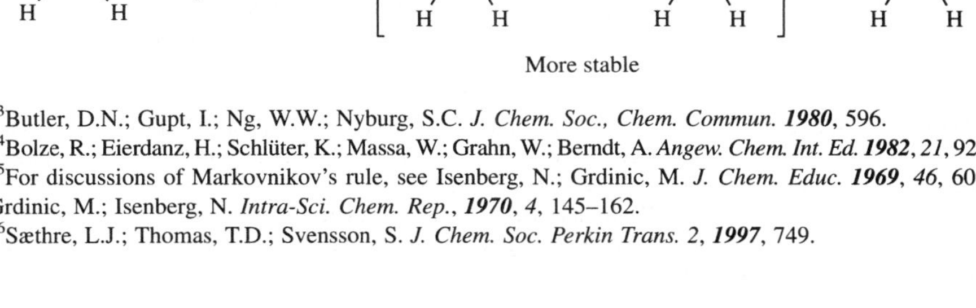

[93]Butler, D.N.; Gupt, I.; Ng, W.W.; Nyburg, S.C. *J. Chem. Soc., Chem. Commun.* **1980**, 596.
[94]Bolze, R.; Eierdanz, H.; Schlüter, K.; Massa, W.; Grahn, W.; Berndt, A. *Angew. Chem. Int. Ed.* **1982**, *21*, 924.
[95]For discussions of Markovnikov's rule, see Isenberg, N.; Grdinic, M. *J. Chem. Educ.* **1969**, *46*, 601; Grdinic, M.; Isenberg, N. *Intra-Sci. Chem. Rep.*, **1970**, *4*, 145–162.
[96]Sæthre, L.J.; Thomas, T.D.; Svensson, S. *J. Chem. Soc. Perkin Trans. 2*, **1997**, 749.

Markovnikov's rule is also usually followed where bromonium ions or other three-membered rings are intermediates.[97] This means that in these cases attack by W must resemble the S_N1 rather than the S_N2 mechanism (see p. 517), although the overall stereospecific anti addition in these reactions means that the nucleophilic substitution step is taking place with inversion of configuration.

$$H-\overset{\displaystyle H}{\underset{\displaystyle H}{C}}\overset{\overset{\displaystyle Y \oplus}{\diagup\ \diagdown}}{-}\overset{}{C}-R \quad\xrightarrow{\ W\ }\quad H-\overset{\displaystyle Y}{\underset{\displaystyle H}{C}}-\overset{\displaystyle W}{\underset{\displaystyle H}{C}}-R$$

Alkenes containing strong electron-withdrawing groups may violate Markovnikov's rule. For example, attack at the Markovnikov position of $Me_3N^+-CH=CH_2$ would give an ion with positive charges on adjacent atoms. The compound $CF_3CH=CH_2$ has been reported to give electrophilic addition with acids in an anti-Markovnikov direction, but it has been shown[98] that, when treated with acids, this compound does not give simple electrophilic addition at all; the apparently anti-Markovnikov products are formed by other pathways. Molecular electrostatic potentials for the π-region of substituted alkenes were studied, with electron donating and withdrawing substituents (based on the increase or decrease in the negative character of V_{min}-most negative-valued point), and plots of V_{min} shows a good linear correlation with the Hammett $\sigma\rho$ constants, suggesting similar substituent electronic effects for substituted ethylenes and substituted benzenes.[99]

For nucleophilic addition the direction of attack has been studied very little, except for Michael-type addition, with compounds of the type C=C–Z. Here the negative part of the reagent almost always attacks regioselectively at the carbon that does not carry the Z (see p. 1008).

In free-radical addition[100] the main effect seems to be steric.[101] All substrates $CH_2=CHX$ preferentially react at the CH_2, regardless of the identity of X or of the radical. With a reagent such as HBr, this means that the addition is anti-Markovnikov:

$$\underset{H}{\overset{R}{\diagdown}}C=C\underset{H}{\overset{H}{\diagup}} + Br\cdot \longrightarrow \cdot\underset{H}{\overset{R}{|}}C-\underset{H}{\overset{Br}{|}}C-H \longrightarrow R-\underset{H}{\overset{H}{|}}C-\underset{H}{\overset{Br}{|}}C-H$$

Preferentially   Product
formed
intermediate

[97]This has been graphically demonstrated by direct treatment of stabilized bromonium ions by nucleophiles: Dubois, J.E.; Chrétien, J.R. *J. Am. Chem. Soc.* **1978**, *100*, 3506.

[98]Myhre, P.C.; Andrews, G.D. *J. Am. Chem. Soc.* **1970**, *92*, 7595, 7596. See also, Newton, T.A. *J. Chem. Educ.* **1987**, *64*, 531.

[99]Suresh, C.H.; Koga, N.; Gadre, S.R. *J. Org. Chem.* **2001**, *66*, 6883.

[100]For reviews of orientation in free-radical additions, see Tedder, J.M.; Walton, J.C. *Tetrahedron* **1980**, *36*, 701; *Adv. Phys. Org. Chem.* **1978**, *16*, 51; *Acc. Chem. Res.* **1976**, *9*, 183. See also, Giese, B. *Rev. Chem. Intermed.* **1986**, *7*, 3; Tedder, J.M. *J. Chem. Educ.* **1984**, *61*, 237.

[101]See, however, Riemenschneider, K.; Bartels, H.M.; Dornow, R.; Drechsel-Grau, E.; Eichel, W.; Luthe, H.; Matter, Y.M.; Michaelis, W.; Boldt, P. *J. Org. Chem.* **1987**, *52*, 205; Gleicher, G.J.; Mahiou, B.; Aretakis, A.J. *J. Org. Chem.* **1989**, *54*, 308.

Thus the observed orientation in both kinds of HBr addition (Markovnikov electrophilic and anti-Markovnikov free radical) is caused by formation of the secondary intermediate. In the electrophilic case it forms because it is more stable than the primary; in the free-radical case because it is sterically preferred. The stability order of the free-radical intermediates is also usually in the same direction: $3° > 2° > 1°$ (p. 272), but this factor is apparently less important than the steric factor. Internal alkenes with no groups present to stabilize the radical usually give an ~1:1 mixture via 5-exo–trig and 6-endo–trig (see Badwin's rules p. $$$) reactions.

Favored

In *intramolecular* additions of radicals containing a 5,6 double bond,[52] both five- and six-membered rings can be formed, but in most cases[102] the five-membered rings are greatly preferred kinetically, even (as in the case shown) where five-membered ring closure means generating a primary radical and six-membered ring closure a secondary radical. This phenomenon may be caused by more favorable entropy factors leading to a five-membered ring, as well as by stereoelectronic factors, but other explanations have also been offered.[103] Similar behavior is found when the double bond is in other positions (from the 3,4 to the 7,8 position). In each case, the smaller ring (exo–trig addition) is preferred to the larger (endo–trig addition)[104] (see the Baldwin rules, p. 305). However, when a radical that is unsaturated in the 5,6 position contains an alkyl group in the 5 position, formation of the six-membered ring is generally favored.[105]

For conjugated dienes, attack by a positive ion, a negative ion, or a free radical is almost always at the *end* of the conjugated system, since in each case this gives an intermediate stabilized by resonance. In the case of an unsymmetrical diene, the more stable ion is formed. For example, isoprene (CH_2=$CMeCH$=CH_2), treated with HCl gives only Me_2CClCH=CH_2 and Me_2C=$CHCH_2Cl$, with none of the product arising from attack at the other end. $PhCH$=$CHCH$=CH_2 gives only $PhCH$=$CHCHClCH_3$ since it is the only one of the eight possible products that has a double bond in conjugation with the ring and that results from attack by H^+ at an end of the conjugated system.

---

[102]For an exception, see Wilt, J.W. *Tetrahedron* **1985**, *41*, 3979.

[103]For discussions, see Beckwith, A.L.J. *Tetrahedron* **1981**, *37*, 3073; Verhoeven, J.W. *Revl. Trav. Chim. Pays-Bas* **1980**, *99*, 143. For molecular mechanics force-field approaches to this problem, see Beckwith, A.L.J.; Schiesser, C.H. *Tetrahedron* **1985**, *41*, 3925; Spellmeyer, D.C.; Houk, K.N. *J. Org. Chem.* **1987**, *52*, 959.

[104]See Beckwith, A.L.J.; Easton, C.J.; Serelis, A.K. *J. Chem. Soc., Chem. Commun.* **1980**, 482.

[105]See Chuang, C.; Gallucci, J.C.; Hart, D.J.; Hoffman, C. *J. Org. Chem.* **1988**, *53*, 3218, and references cited therein.

When allenes attack electrophilic reagents,[106] Markovnikov's rule would predict that the formation of the new bond should be at the end of the system, since there are no hydrogens in the middle. Reaction at the center gives a carbocation stabilized by resonance, but not immediately. In order for such stabilization to be in effect the three p orbitals must be parallel, and it requires a rotation about the C–C bond for this to happen.[107] Therefore, the stability of the allylic cation has no effect on the transition state, which still has a geometry similar to that of the original allene (p. 148). Probably because of this, attack on the unsubstituted $CH_2=C=CH_2$ is most often at the end carbon, to give a vinylic cation, although center attack has also been reported. However, as alkyl or aryl groups are substituted on the allene carbons, attack at the middle carbon becomes more favorable because the resulting cation is stabilized by the alkyl or aryl groups (it is now a secondary, tertiary, or benzylic cation). For example, allenes of the form $RCH=C=CH_2$ are still attacked most often at the end, but with $RCH=C=CHR'$ center attack is more prevalent. Tetramethylallene is also attacked predominantly at the center carbon.[108] Free radicals[109] attack allenes most often at the end,[110] although attack at the middle has also been reported.[111] As with electrophilic attack and for the same reason, the stability of the allylic radical has no effect on the transition state of the reaction between a free radical and an allene. Again, as with electrophilic attack, the presence of alkyl groups increases the extent of attack by a radical at the middle carbon.[112]

Stereochemical Orientation

It has already been pointed out that some additions are syn, with both groups, approaching from the same side, and that others are anti, with the groups approaching from opposite sides of the double or triple bond. For cyclic compounds steric orientation must be considered. In syn addition to an unsymmetrical cyclic alkene,

---

[106]For a monograph on addition to allenes, see Schuster, H.F.; Coppola, G.M. *Allenes in Organic Synthesis* Wiley, NY, *1984*. For reviews, see Pasto, D.J. *Tetrahedron 1984*, *40*, 2805; Smadja, W. *Chem. Rev. 1983*, *83*, 263; in Landor, S.R. *The Chemistry of Allenes*, Vol. 2; Academic Press, NY, *1982*, articles by Landor, S.R., Jacobs, T.L.; Hopf, H. pp. 351–577; Stang, P.J.; Rappoport, Z.; Hanack, M.; Subramanian, L.R. *Vinyl Cations*, Academic Press, NY, *1979*, pp. 152–167; Blake, P., in Patai, S. *The Chemistry of Ketenes, Allenes and Related Compounds*, pt. 1, Wiley, NY, *1980*; pp. 342–357; Modena, G.; Tonellato, U. *Adv. Phys. Org. Chem. 1971*, *9*, 185, pp. 215–231; Richey, Jr., H.G.; Richey, J.M., in Olah, G.A.; Schleyer, P.V.R. *Carbonium Ions*, Vol. 2, Wiley, NY, *1970*, pp. 917–922; Caserio, M.C. *Sel. Org. Transform.*, *1970*, *1*, 239; Taylor, D.R. *Chem. Rev. 1967*, *67*, 317, 338–346; Mavrov, M.V.; Kucherov, V.F. *Russ. Chem. Rev. 1967*, *36*, 233; Griesbaum, K. *Angew. Chem. Int. Ed. 1966*, *5*, 933.
[107]For evidence that this is so, see Okuyama, T.; Izawa, K.; Fueno, T. *J. Am. Chem. Soc. 1973*, *95*, 6749.
[108]For example, see Bianchini, J.; Guillemonat, A. *Bull. Soc. Chim. Fr. 1968*, 2120; Pittman Jr., C.U. *Chem. Commun. 1969*, 122; Poutsma, M.L.; Ibarbia, P.A. *J. Am. Chem. Soc. 1971*, *93*, 440.
[109]For a review, see Jacobs, T.L., in Landor, S.R. *The Chemistry of Allenes*, Vol. 2, Academic Press, NY, *1982*, pp. 399–415.
[110]Griesbaum, K.; Oswald, A.A.; Quiram, E.R.; Naegele, W. *J. Org. Chem. 1963*, *28*, 1952.
[111]See, for example, Pasto, D.J.; L'Hermine, G. *J. Org. Chem. 1990*, *55*, 685.
[112]For example, see Byrd, L.R.; Caserio, M.C. *J. Org. Chem. 1972*, *37*, 3881; Pasto, D.J.; Warren, S.E.; Morrison, M.A. *J. Org. Chem. 1981*, *46*, 2837. See, however, Bartels, H.M.; Boldt, P. *Liebigs Ann. Chem. 1981*, 40.

the two groups can come in from the more- or from the less-hindered face of the double bond. The rule is that syn addition is usually, although not always, from the less-hindered face. For example, epoxidation of 4-methylcyclopentene gave 76% addition from the less-hindered and 24% from the more-hindered face.[113]

In anti addition to a cyclic substrate, the initial attack on the electrophile is also from the less-hindered face. However, many (although not all) electrophilic additions to norbornene and similar strained bicycloalkenes are syn additions.[114] In these cases reaction is always from the exo side, as in formation of **23**,[115]

unless the exo side is blocked by substituents in the 7 position, in which case endo attack may predominate; for example, 7,7-dimethylnorbornene undergoes syn–endo epoxidation (**15-50**) and hydroboration[116] (**15-16**). However, addition of DCl and F_3CCOOD to, and oxymercuration (**15-2**) of, 7,7-dimethylnorbornene proceeds syn–exo in spite of the methyl groups in the 7 position.[117] Similarly, free-radical additions to norbornene and similar molecules are often syn–exo, although anti additions and endo attacks are also known.[118]

Electronic effects can also play a part in determining which face reacts preferentially with the electrophilic species. In the adamantane derivative **24**, steric

---

[113]Henbest, H.B.; McCullough, J.J. *Proc. Chem. Soc.* **1962**, 74.

[114]For a discussion, see Traylor, T.G. *Acc. Chem. Res.* **1969**, *2*, 152.

[115]Cristol, S.J.; Morrill, T.C.; Sanchez, R.A. *J. Org. Chem.* **1966**, *31*, 2719; Brown, H.C.; Kawakami, J.H.; Liu, K. *J. Am. Chem. Soc.* **1970**, *92*, 5536; Alvernhe, G.; Anker, D.; Laurent, A.; Haufe, G.; Beguin, C. *Tetrahedron* **1988**, *44*, 3551; Koga, N.; Ozawa, T.; Morokuma, K. *J. Phys. Org. Chem.* **1990**, *3*, 519.

[116]Brown, H.C.; Kawakami, J.H.; Liu, K. *J. Am. Chem. Soc.* **1973**, *95*, 2209.

[117]Brown, H.C.; Liu, K. *J. Am. Chem. Soc.* **1975**, *97*, 600, 2469; Tidwell, T.T.; Traylor, T.G. *J. Org. Chem.* **1968**, *33*, 2614.

[118]For a review of free-radical addition to these systems, see Azovskaya, V.A.; Prilezhaeva, E.N. *Russ. Chem. Rev.* **1972**, *41*, 516.

effects are about the same for each face of the double bond. Yet epoxidation, dibromocarbene reactions (**15-64**), and hydroboration (**15-16**) all predominantly take place from the face that is syn to the electron-withdrawing fluorine.[119] In the case shown, about twice as much **25** was formed, compared to **26**. Similar results have been obtained on other substrates:[120] groups that are electron withdrawing by the field effect $(-I)$ direct attack from the syn face; $+I$ groups from the anti face, for both electrophilic and nucleophilic attack. These results are attributed[121] to hyperconjugation: For the adamantane case, there is overlap between the σ^* orbital of the newly forming bond (between the attacking species and C-2 in **24**) and the filled σ orbitals of the C_α–C_β bonds on the opposite side. This is called the *Cieplak effect*. The LiAlH$_4$ reduction of 2-axial methyl or methoxy cyclohexanones supports Cieplak's proposal.[122] In addition reactions of methanol to norbornanones, however, little evidence was found to support the Cieplak effect.[123] The four possible bonds are C-3–C-4 and C-1–C-9 on the syn side and C-3–C-10 and C-1–C-8 on the anti side. The preferred pathway is the one where the incoming group has the more electron-rich bonds on the side *opposite* to it (these are the ones it overlaps with). Since the electron-withdrawing F has its greatest effect on the bonds closest to it, the C-1–C-8 and C-3–C-10 bonds are more electron rich, and the group comes in on the face syn to the F.

It has been mentioned that additions of Br$_2$ and HOBr are often anti because of formation of bromonium ions and that free-radical addition of HBr is also anti. When the substrate in any of these additions is a cyclohexene, the addition is not only anti but the initially formed product is conformationally specific too, being mostly diaxial.[124] This is so because diaxial opening of the three-membered ring preserves a maximum coplanarity of the participating centers in the transition state; indeed, on opening, epoxides also give diaxial products.[125] However, the initial diaxial product may then pass over to the diequatorial conformer unless other groups on the ring render the latter less stable than the former. In free-radical additions to cyclohexenes in which cyclic intermediates are not involved, the initial reaction with the radical is also usually from the axial direction,[126] resulting in a diaxial initial product if the overall addition is anti. The direction from which unsymmetrical radicals react has also been studied.[127] For example, when the radical **27** adds

[119]Srivastava, S.; le Noble, W.J. *J. Am. Chem. Soc.* **1987**, *109*, 5874. See also, Bodepudi, V.R.; le Noble, W.J. *J. Org. Chem.* **1991**, *56*, 2001.

[120]Cieplak, A.S.; Tait, B.D.; Johnson, C.R. *J. Am. Chem. Soc.* **1989**, *111*, 8447.

[121]Cieplak, A.S. *J. Am. Chem. Soc.* **1981**, *103*, 4540. See also, Jorgensen, W.L. *Chemtracts: Org. Chem.* **1988**, *1*, 71.

[122]Senda, Y.; Nakano, S.; Kunii, H.; Itoh, H. *J. Chem. Soc. Perkin Trans. 2*, **1993**, 1009.

[123]Coxon, J.M.; McDonald, D.Q. *Tetrahedron* **1992**, *48*, 3353.

[124]Barton, D.H.R., in *Theoretical Organic Chemistry The Kekulé Symposium*, Butterworth: London, **1959**, pp. 127–143; Goering, H.L.; Sims, L.L. *J. Am. Chem. Soc.* **1955**, *77*, 3465; Shoppee, C.W.; Akhtar, M.I.; Lack, R.E. *J. Chem. Soc.* **1964**, 877; Readio, P.D.; Skell, P.S. *J. Org. Chem.* **1966**, *31*, 753, 759.

[125]For example, see Anselmi, C.; Berti, G.; Catelani, G.; Lecce, L.; Monti, L. *Tetrahedron* **1977**, *33*, 2771.

[126]Huyser, E.S.; Benson, H.; Sinnige, H.J. *J. Org. Chem.* **1967**, *32*, 622; LeBel, N.A.; Czaja, R.F.; DeBoer, A. *J. Org. Chem.* **1969**, *34*, 3112

[127]For a review, see Giese, B. *Angew. Chem. Int. Ed.* **1989**, *28*, 969.

to a double bond it preferentially does so anti to the OH group, leading to a diaxial trans product.[125]

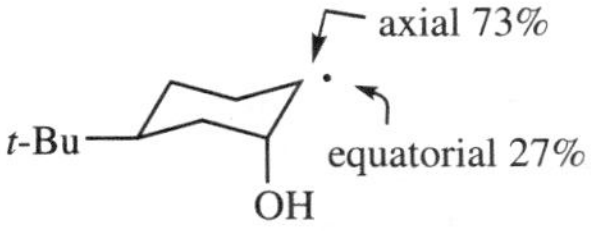

axial 73%

t-Bu — [structure] — equatorial 27%

OH

27

Addition to Cyclopropane Rings[128]

We have previously seen (p. 218) that in some respects, cyclopropane rings resemble double bonds.[129] It is not surprising, therefore, that cyclopropanes undergo addition reactions analogous to those undergone by double-bond compounds, resulting in the opening of the three-membered rings, as in the two examples shown where reaction numbers relating the reaction to alkene chemistry are in parentheses.

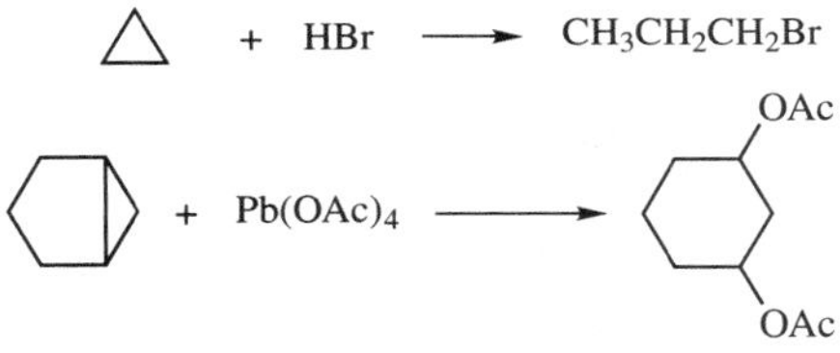

$$\triangle \quad + \quad HBr \quad \longrightarrow \quad CH_3CH_2CH_2Br \qquad \textbf{(15-2)}$$

[cyclopropane-fused ring] + Pb(OAc)$_4$ $\longrightarrow$ [cyclohexane with OAc, OAc]   **(15-47)**   Ref. [130]

Other examples are discussed at **15-3**, **15-15**, and **15-63**.

Additions to cyclopropanes can take place by any of the four mechanisms already discussed in this chapter, but the most important type involves attack on an electrophile.[131] For substituted cyclopropanes, these reactions usually follow Markovnikov's rule, although exceptions are known and the degree of regioselectivity is often small. The application of Markovnikov's rule to these substrates can be illustrated by the reaction of 1,1,2-trimethylcyclopropane with HX.[132] The rule predicts that the electrophile (in this case H$^+$) goes to the carbon

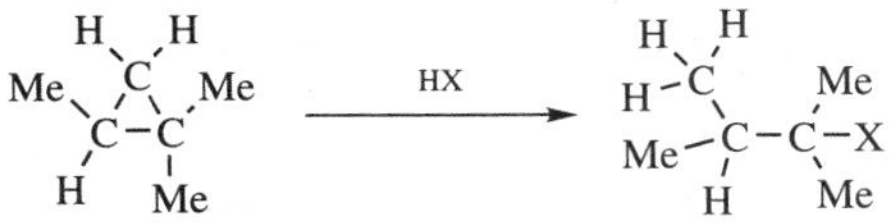

[structure of 1,1,2-trimethylcyclopropane] $\xrightarrow{\text{HX}}$ [product structure]

---

[128]For a review, see Charton, M., in Zabicky, J. *The Chemistry of Alkenes*, Vol 2., Wiley, NY, *1970*, pp. 569–592. For reviews of the use of cyclopropanes in organic synthesis see Reissig, H. *Top. Curr. Chem.* *1988*, *144*, 73; Wong, H.N.C.; Hon, M.; Tse, C.; Yip, Y.; Tanko, J.; Hudlicky, T. *Chem. Rev.* *1989*, *89*, 165.

[129]The analogies are by no means complete: see Gordon, A.J. *J. Chem. Educ.* *1967*, *44*, 461.

[130]Moon, S. *J. Org. Chem.* *1964*, *39*, 3456.

[131]For a review, see DePuy, C.H. *Top. Curr. Chem.* *1973*, *40*, 73–101. For a list of references to pertinent mechanistic studies, see Wiberg, K.B.; Kass, S.R. *J. Am. Chem. Soc.* *1985*, *107*, 988.

[132]Kramer, G.M. *J. Am. Chem. Soc.* *1970*, *92*, 4344.

with the most hydrogens and the nucleophile goes to the carbon that can best stabilize a positive charge (in this case the tertiary rather than the secondary carbon). The stereochemistry of the reaction can be investigated at two positions the one that becomes connected to the electrophile and the one that becomes connected to the nucleophile. The results at the former position are mixed. Additions have been found to take place with 100% retention,[133] 100% inversion,[134] and with mixtures of retention and inversion.[135] At the carbon that becomes connected to the nucleophile the result is usually inversion, although retention has also been found,[136] and elimination, rearrangement, and racemization processes often compete, indicating that in many cases a positively charged carbon is generated at this position.

At least three mechanisms have been proposed for electrophilic addition (these mechanisms are shown for attack by HX, but analogous mechanisms can be written for other electrophiles).

Mechanism *a*

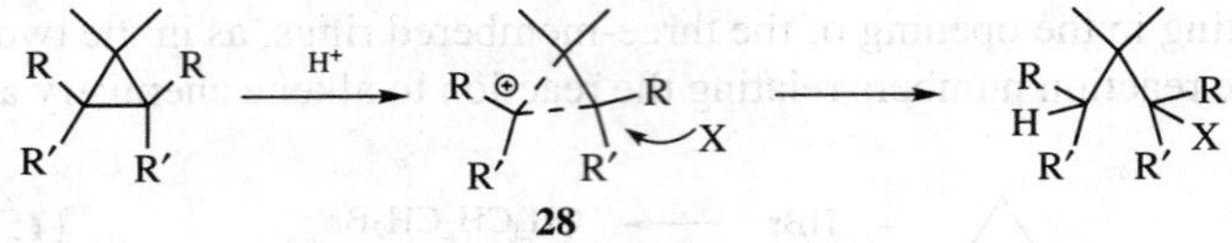

28

Mechanism *b*

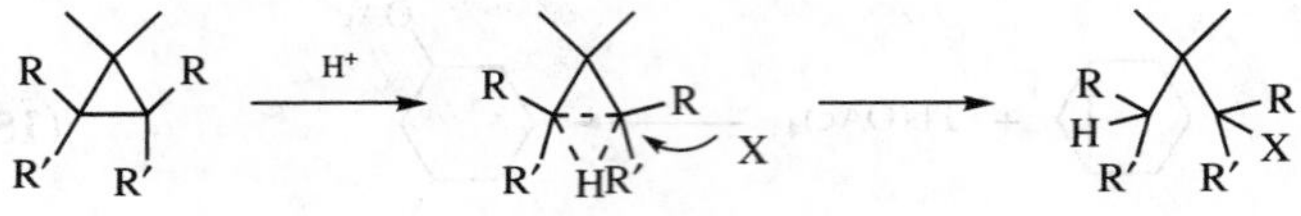

29

Mechanism *c*

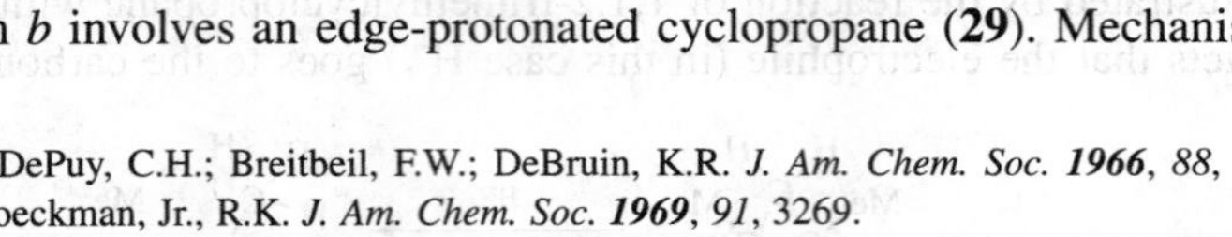

30

Mechanism *a* involves a corner-protonated cyclopropane[137] (**28**); we have already seen examples of such ions in the 2-norbornyl and 7-norbornenyl cations (pp. 453, 460). Mechanism *b* involves an edge-protonated cyclopropane (**29**). Mechanism *c*

---

[133]For example, see DePuy, C.H.; Breitbeil, F.W.; DeBruin, K.R. *J. Am. Chem. Soc.* **1966**, *88*, 3347; Hendrickson, J.B.; Boeckman, Jr., R.K. *J. Am. Chem. Soc.* **1969**, *91*, 3269.

[134]For example, see LaLonde, R.T.; Ding, J.; Tobias, M.A. *J. Am. Chem. Soc.* **1967**, *89*, 6651; Warnet, R.J.; Wheeler, D.M.S. *Chem. Commun.* **1971**, 547; Hogeveen, H.; Roobeek, C.F.; Volger, H.C. *Tetrahedron Lett.* **1972**, 221; Battiste, M.A.; Mackiernan, J. *Tetrahedron Lett.* **1972**, 4095. See also, Jensen, F.R.; Patterson, D.B.; Dinizo, S.E. *Tetrahedron Lett.* **1974**, 1315; Coxon, J.M.; Steel, P.J.; Whittington, B.I. *J. Org. Chem.* **1990**, *55*, 4136.

[135]Nickon, A.; Hammons, J.H. *J. Am. Chem. Soc.* **1964**, *86*, 3322; Hammons, J.H.; Probasco, E.K.; Sanders, L.A.; Whalen, E.J. *J. Org. Chem.* **1968**, *33*, 4493; DePuy, C.H.; Fünfschilling, P.C.; Andrist, A.H.; Olson, J.M. *J. Am. Chem. Soc.* **1977**, *99*, 6297.

[136]Cristol, S.J.; Lim, W.Y.; Dahl, A.R. *J. Am. Chem. Soc.* **1970**, *92*, 4013; Hendrickson, J.B.; Boeckman, Jr., R.K. *J. Am. Chem. Soc.* **1971**, *93*, 4491.

[137]For reviews of protonated cyclopropanes, see Collins, C.J. *Chem. Rev.* **1969**, *69*, 543; Lee, C.C. *Prog. Phys. Org. Chem.* **1970**, *7*, 129.

consists of a one-step S_E2-type attack on H^+ to give the classical cation **30**, which then reacts with the nucleophile. Although the three mechanisms as we have drawn them show retention of configuration at the carbon that becomes attached to the proton, mechanisms *a* and *c* at least can also result in inversion at this carbon. Unfortunately, the evidence on hand at present does not allow us unequivocally to select any of these as the exclusive mechanism in all cases. Matters are complicated by the possibility that more than one edge-protonated cyclopropane is involved, at least in some cases. There is strong evidence for mechanism *b* with the electrophiles Br^+ and Cl^+;[138] and for mechanism *a* with D^+ and Hg^{2+}.[139] *Ab initio* studies show that the corner-protonated **28** is slightly more stable ($\sim$1.4 kcal mol^{-1}, 6 kJ mol^{-1}) than the edge-protonated **29**.[140] There is some evidence against mechanism *c*.[141]

Free-radical additions to cyclopropanes have been studied much less, but it is known that Br_2 and Cl_2 add to cyclopropanes by a free-radical mechanism in the presence of UV light. The addition follows Markovnikov's rule, with the initial radical reacting at the least-substituted carbon and the second group going to the most-substituted position. Several investigations have shown that the reaction is stereospecific at one carbon, taking place with inversion there, but nonstereospecific at the other carbon.[142] A mechanism that accounts for this behavior is[143]

In some cases, conjugate addition has been performed on systems where a double bond is "conjugated" with a cyclopropyl ring. An example is the formation of **31**.[144]

(15-6)

31

[138]Coxon, J.M.; Steel, P.J.; Whittington, B.I.; Battiste, M.A. *J. Org. Chem.* **1989**, *54*, 1383; Coxon, J.M.; Steel, P.J.; Whittington, B.I. *J. Org. Chem.* **1989**, *54*, 3702.

[139]Lambert, J.B.; Chelius, E.C.; Bible, Jr., R.H.; Hadju, E. *J. Am. Chem. Soc.* **1991**, *113*, 1331.

[140]Koch, W.; Liu, B.; Schleyer, P.v.R. *J. Am. Chem. Soc.* **1989**, *111*, 3479, and references cited therein.

[141]Wiberg, K.B.; Kass, S.R. *J. Am. Chem. Soc.* **1985**, *107*, 988.

[142]Maynes, G.G.; Applequist, D.E. *J. Am. Chem. Soc.* **1973**, *95*, 856; Incremona, J.H.; Upton, C.J. *J. Am. Chem. Soc.* **1972**, *94*, 301; Shea, K.J.; Skell, P.S. *J. Am. Chem. Soc.* **1973**, *95*, 6728; Poutsma, M.L. *J. Am. Chem. Soc.* **1965**, *87*, 4293; Jarvis, B.B. *J. Org. Chem.* **1970**, *35*, 924; Upton, C.J.; Incremona, J.H. *J. Org. Chem.* **1976**, *41*, 523.

[143]For free-radical addition to [1.1.1]propellane and bicyclo[1.1.0]butane, see Wiberg, K.B.; Waddell, S.T.; Laidig, K. *Tetrahedron Lett.* **1986**, *27*, 1553.

[144]Sarel, S.; Ben-Shoshan, B. *Tetrahedron Lett.* **1965**, 1053. See also, Danishefsky, S. *Acc. Chem. Res.* **1979**, *12*, 66.

REACTIONS

Reactions are classified by type of reagent. Isomerization of double and triple bonds is followed by examination of all reactions, where hydrogen adds to one side of the double or triple bond.

ISOMERIZATION OF DOUBLE AND TRIPLE BONDS

15-1 Isomerization

There are several reagents that lead to isomerization of a double bond to form a new alkene. In general, there is an energetic preference of an α,β- versus. β,γ- double bond.[145] Transition metals have been used to induce isomerization of alkenes. Allylic arenes (Ar-CH$_2$CH=CH$_2$) have been converted to the corresponding (Z-)1-propenyl arene (Ar-CH=CHMe using a ruthenium catalyst[146] or a polymer-supported iridium catalyst.[147] Allyl decyl ether (CH$_2$=CHCH$_2$OC$_{10}$H$_{21}$) was isomerized to 1-decyloxy-1-propene (CH$_3$CH=CHOC$_{10}$H$_{21}$) by treatment with NaHFe(CO)$_4$.[148] Double-bond migration has been observed in sulfide photo-irradiation, induced by singlet oxygen.[149] N-Acyl allylamine can be isomerized to the N-acyl enamine by heating with a ruthenium catalyst.[150] Many of these reactions were discussed in **12-2**.

For conjugated carbonyl compounds that have a hydrogen atom at the γ-position (C-4), it is possible to move a double bond *out* of conjugation. Photolysis of conjugated esters, at $-40°C$ in the presence of N,N- dimethylaminoethanol, gave the nonconjugated ester.[151] Heating an N-allylic amide (N-C—C=C) with Fe(CO)$_5$, neat, gave the enamide (N-C=C—C).[152]

Isomerization of (E/Z) isomers is another important transformation.[153] Isomerization of (E)- and (Z)-conjugated amides is effected photochemically[154]

[145]Lee, P.S.; Du, W.; Boger, D.L.; Jorgensen, W.L. *J. Org. Chem.* **2004**, *69*, 5448.

[146]Sato, T.; Komine, N.; Hirano, M.; Komiya, S. *Chem. Lett.* **1999**, 441.

[147]Baxendale, I.R.; Lee, A.-L.; Ley, S.V. *Synlett* **2002**, 516.

[148]Crivello, J.V.; Kong, S. *J. Org. Chem.* **1998**, *63*, 6745.

[149]Clennan, E.L.; Aebisher, D. *J. Org. Chem.* **2002**, *67*, 1036.

[150]Krompiec, S.; Pigulla, M.; Krompiec, M.; Baj, S.; Mrowiec-Bialon, J.; Kasperczyk, J. *Tetrahedron Lett.* **2004**, *45*, 5257.

[151]Bargiggia, F.; Piva, O. *Tetrahedron Asymmetry* **2001**, *12*, 1389.

[152]Sergeyev, S.; Hesse, M. *Synlett* **2002**, 1313.

[153]For a review, see Dugave, C.; Demange, L. *Chem. Rev.* **2003**, *103*, 2475.

[154]Kinbara, K.; Saigo, K. *Bull. Chem. Soc. Jpn.* **1996**, *69*, 779; Wada, T.; Shikimi, M.; Inoue, Y.; Lem, G.; Turro, N.J. *Chem. Commun.* **2001**, 1864.

(photoisomerization[155]). There is a rather high energy barrier for the excited state required for (*E/Z*) isomerization.[156] Isomerization of the C=C units in dienes is also induced photochemically.[157] Isomerization of cyclic alkenes is more difficult but cyclooctene is isomerized photochemically.[158] The photosensitized cis–trans isomerization of 1,2-dichloroethylenes have been reported,[159] and also the photo-isomerization of cis/trans cyclooctene.[160] Radical-induced (*E/Z*) isomerization is known.[161]

Conjugated aldehydes have been isomerized using thiourea in DMF.[162] A 1:1 mixture of cis/trans styrene derivatives was isomerized to a 90% yield of the trans styrene derivatives was reported using a palladium catalyst.[163] Thermal cis–trans isomerization of 1,3-diphenyltriazenes has been reported, in aqueous solution.[164]

REACTIONS IN WHICH HYDROGEN ADDS TO ONE SIDE

A. Halogen on the Other Side

15-2 Addition of Hydrogen Halides

Hydro-halo-addition

$$\backslash\!C=\!C\!/ + \text{H-X} \longrightarrow -\overset{\text{H}}{\underset{|}{C}}-\overset{\text{X}}{\underset{|}{C}}-$$

Any of the four hydrogen halides can be added to double bonds.[165] HI, HBr, and HF[166] add at room temperature. The addition of HCl is more difficult and usually requires heat,[23] although HCl adds easily in the presence of silica gel.[167] The reaction has been carried out with a large variety of double-bond compounds, including

[155]Inoue, Y.; Yamasaki, N.; Yokoyama, T.; Tai, A. *J. Org. Chem.* **1992**, *57*, 1332.

[156]Arai, T.; Takahashi, O. *J. Chem. Soc., Chem. Commun.* **1995**, 1837.

[157]Wakamatsu, K.; Takahashi, Y.; Kikuchi, K.; Miyashi, T. *J. Chem. Soc. Perkin Trans. 2*, **1996**, 2105.

[158]Tsuneishi, H.; Hakushi, T.; Inoue, Y. *J. Chem. Soc. Perkin Trans. 2*, **1996**, 1601; Inoue, Y.; Tsuneishi, H.; Hakushi, T.; Yagi, K.; Awazu, K.; Onuki, H. *Chem. Commun.* **1996**, 2627; Tsuneishi, H.; Hakushi, T.; Tai, A.; Inoue, Y. *J. Chem. Soc. Perkin Trans. 2*, **1995**, 2057.

[159]Kokubo, K.; Kakimoto, H.; Oshima, T. *J. Am. Chem. Soc.* **2002**, *124*, 6548.

[160]Wada, T.; Sugahara, N.; Kawano, M.; Inoue, Y. *Chem. Lett.* **2000**, 1174.

[161]Baag, Md.M.; Kar, A.; Argade, N.P. *Tetrahedron* **2003**, *59*, 6489.

[162]Phillips, O.A.; Eby, P.; Maiti, S.N. *Synth. Commun.* **1995**, *25*, 87.

[163]Yu, J.; Gaunt, M.J.; Spencer, J.B. *J. Org. Chem.* **2002**, *67*, 4627.

[164]Chen, N.; Barra, M.; Lee, I.; Chahal, N. *J. Org. Chem.* **2002**, *67*, 2271.

[165]For a list of references, see Larock, R.C. *Comprehensive Organic Transformations*, 2nd ed., Wiley-VCH, NY, **1999**, pp. 633–636.

[166]For reviews of addition of HF, see Sharts, C.M.; Sheppard, W.A. *Org. React.* **1974**, *21*, 125, 192–198, 212–214; Hudlický, M. *The Chemistry of Organic Fluorine Compounds*, 2nd ed., Ellis Horwood, Chichester, **1976**, pp. 36–41.

[167]Kropp, P.J.; Daus, K.A.; Tubergen, M.W.; Kepler, K.D.; Wilson, V.P.; Craig, S.L.; Baillargeon, M.M.; Breton, G.W. *J. Am. Chem. Soc.* **1993**, *115*, 3071.

conjugated systems, where both 1,2- and 1,4-addition are possible. A convenient method for the addition of HF involves the use of a polyhydrogen fluoride-pyridine solution.[168] When the substrate is mixed with this solution in a solvent, such as THF at 0°C, alkyl fluorides are obtained in moderate-to-high yields.

The addition of hydrogen halides to simple alkenes, in the absence of peroxides, takes place by an electrophilic mechanism, and the orientation is in accord with Markovnikov's rule.[169] The addition follows second order kinetics.[170] When peroxides are added, the addition of HBr occurs by a free-radical mechanism and the orientation is anti-Markovnikov (p. 1021).[171] It must be emphasized that this is true *only* for HBr. Free-radical addition of HF and HI has never been observed, even in the presence of peroxides, and of HCl only rarely. In the rare cases where free-radical addition of HCl was noted, the orientation was still Markovnikov, presumably because the more stable *product* was formed.[172] Free-radical addition of HF, HI, and HCl is energetically unfavorable (see the discussions on pp. 943, 959). It has often been found that anti-Markovnikov addition of HBr takes place even when peroxides have not been added. This happens because the substrate alkenes absorb oxygen from the air, forming small amounts of peroxides (**14-7**). Markovnikov addition can be ensured by rigorous purification of the substrate, but in practice this is not easy to achieve, and it is more common to add inhibitors, for example, phenols or quinones, which suppress the free-radical pathway. The presence of free-radical precursors, such as peroxides does not inhibit the ionic mechanism, but the radical reaction, being a chain process, is much more rapid than the electrophilic reaction. In most cases, it is possible to control the mechanism (and hence the orientation) by adding peroxides to achieve complete free-radical addition, or inhibitors to achieve complete electrophilic addition, although there are some cases where the ionic mechanism is fast enough to compete with the free-radical mechanism and complete control cannot be attained. Markovnikov addition of HBr, HCl, and HI has also been accomplished, in high yields, by the use of phase-transfer catalysis.[173] For alternative methods of adding HBr (or HI) with anti-Markovnikov orientation, see **12-31**.

It is also possible to add 1[174] or 2 equivalents of any of the four hydrogen halides to triple bonds. Markovnikov's rule ensures that *gem*-dihalides and not *vic*-dihalides

[168]Olah, G.A.; Welch, J.T.; Vankar, Y.D.; Nojima, M.; Kerekes, I.; Olah, J.A. *J. Org. Chem.* **1979**, *44*, 3872. For related methods, see Yoneda, N.; Abe, T.; Fukuhara, T.; Suzuki, A. *Chem. Lett.* **1983**, 1135; Olah, G.A.; Li, X. *Synlett* **1990**, 267.

[169]For reviews of electrophilic addition of HX, see Sergeev, G.B.; Smirnov, V.V.; Rostovshchikova, T.N.; *Russ. Chem. Rev.* **1983**, *52*, 259, and Dewar, M.J.S. *Angew. Chem. Int. Ed.* **1964**, *3*, 245.

[170]Boregeaud, R.; Newman, H.; Schelpe, A.; Vasco, V.; Hughes, D.E.P. *J. Chem. Soc., Perkin Trans. 2*, **2002**, 810.

[171]For reviews of free-radical addition of HX, see Thaler, W.A. *Methods Free-Radical Chem.* **1969**, *2*, 121, see pp. 182–195.

[172]Mayo, F.R. *J. Am. Chem. Soc.* **1962**, *84*, 3964.

[173]Landini, D.; Rolla, F. *J. Org. Chem.* **1980**, *45*, 3527.

[174]For a convenient method of adding one mole of HCl or HBr to a triple bond, see Cousseau, J.; Gouin, L. *J. Chem. Soc. Perkin Trans. 1*, **1977**, 1797; Cousseau, J. *Synthesis* **1980**, 805. For the addition of one mole of HI, see Kamiya, N.; Chikami, Y.; Ishii, Y. *Synlett* **1990**, 675.

are the products of the addition of two equivalents.

$$—C\ C— \xrightarrow{\text{HX}} —CH{=}CX— \xrightarrow{\text{HX}} —CH_2—CX_2—$$

Chlorotrimethylsilane can be added to alkenes to give alkyl chlorides. 1-Hexene reacts with Me$_3$SiCl in water to give 2-chlorohexane.[175] Treatment of an alkene with KHF$_2$ and SiF$_4$ leads to the alkyl fluoride,[176] and bromotrimethylsilane adds to alkynes to give the vinyl bromide.[177] Trichloroisocyanuric acid reacts with terminal alkenes in water to give the 1-chloro alkane.[178]

HX are electrophilic reagents, and many polyhalo and polycyano alkenes, for example, Cl$_2$C=CHCl, do not react with them at all in the absence of free-radical conditions. Vinylcyclopropanes, however, react with opening of the cyclopropane ring to give a homoallylic chloride.[179] When such reactions do occur, however, they take place by a nucleophilic addition mechanism, that is, initial attack is by X$^-$. This type of mechanism also occurs with Michael-type substrates C=C–Z,[180] There the orientation is always such that the halogen goes to the carbon that does not bear the Z, so the product is of the form X–C–CH–Z, even in the presence of free-radical initiators. Hydrogen iodine adds 1,4 to conjugated dienes in the gas phase by a pericyclic mechanism:[181]

HX can be added to ketenes[182] to give acyl halides:

OS **I**, 166; **II**, 137, 336; **III**, 576; **IV**, 238, 543; **VI**, 273; **VII**, 59; **80**, 129.

[175]Boudjouk, P.; Kim, B.-K.; Han, B.-H. *Synth. Commun.* **1996**, *26*, 3479.
[176]Tamura, M.; Shibakami, M.; Kurosawa, S.; Arimura, T.; Sekiya, A. *J. Chem. Soc., Chem. Commun.* **1995**, 1891.
[177]Su, M.; Yu, W.; Jin, Z. *Tetrahedron Lett.* **2001**, *42*, 3771.
[178]Mendonça, G.F.; Sanseverino, A.M.; de Mattos, M.C.S. *Synthesis* **2003**, 45.
[179]Siriwardana, A.I.; Nakamura, I.; Yamamoto, Y. *Tetrahedron Lett.* **2003**, *44*, 985.
[180]For an example, see Marx, J.N. *Tetrahedron* **1983**, *39*, 1529.
[181]Gorton, P.J.; Walsh, R. *J. Chem. Soc., Chem. Commun.* **1972**, 782. For evidence that a pericyclic mechanism may be possible, even for an isolated double bond, see Sergeev, G.B.; Stepanov, N.F.; Leenson, I.A.; Smirnov, V.V.; Pupyshev, V.I.; Tyurina, L.A.; Mashyanov, M.N. *Tetrahedron* **1982**, *38*, 2585.
[182]For reviews of additions to ketenes, and their mechanisms, see Tidwell, T.T. *Acc. Chem. Res.* **1990**, *23*, 273; Seikaly, H.R.; Tidwell, T.T. *Tetrahedron* **1986**, *42*, 2587; Satchell, D.P.N.; Satchell, R.S. *Chem. Soc. Rev.* **1975**, *4*, 231.

B. Oxygen on the Other Side

15-3   Hydration of Double bonds

Hydro-hydroxy-addition

$$\text{\Large $\underset{/}{\overset{\backslash}{C}}=\underset{\backslash}{\overset{/}{C}}$} \longrightarrow \text{\Large $-\underset{/}{\overset{H}{C}}-\underset{\backslash}{\overset{OH}{C}}-$}$$

Double bonds can be hydrated by treatment with water and an acid catalyst. The most common catalyst is sulfuric acid, but other acids that have relatively non-nucleophilic counterions, such as nitric or perchloric can also be used. The mechanism is electrophilic and begins with attack of the π-bond on an acidic proton (see p. 1005). The resulting carbocation is then attacked by negative species, such as HSO_4^- (or similar counterion in the case of other acids), to give the initial product **32**, which can be isolated in some cases, but under the conditions of the

$$-\underset{/}{\overset{H}{C}}-\underset{\backslash}{\overset{OSO_2OH}{C}}-$$

$$\textbf{32}$$

reaction, is usually hydrolyzed to the alcohol (**10-4**). However, the conjugate base of the acid is not the only possible species that attacks the initial carbocation. The attack can also be by water to form **33**.

$$\textbf{33}$$

When the reaction proceeds by this pathway, **32** and similar intermediates are not involved and the mechanism is exactly (by the principle of microscopic reversibility) the reverse of El elimination of alcohols (**17-1**).[183] It is likely that the mechanism involves both pathways. *The initial carbocation occasionally rearranges to a more stable one.* For example, hydration of $CH_2{=}CHCH(CH_3)_2$ gives $CH_3CH_2COH(CH_3)_2$. With ordinary alkenes the addition predominantly follows Markovnikov's rule. Another method for Markovnikov addition of water consists of simultaneously adding an oxidizing agent (O_2) and a reducing agent (either Et_3SiH[184] or a secondary alcohol, e.g., 2-propanol[185]) to the alkene in the presence of a cobalt-complex catalyst. No rearrangement is observed with this

[183]For discussions of the mechanism, see Vinnik, M.I.; Obraztsov, P.A. *Russ. Chem. Rev.* **1990**, *59*, 63; Liler, M. *Reaction Mechanisms in Sulphuric Acid*, Academic Press, NY, **1971**, pp. 210–225.
[184]Isayama, S.; Mukaiyama, T. *Chem. Lett.* **1989**, 569.
[185]Inoki, S.; Kato, K.; Takai, T.; Isayama, S.; Yamada, T.; Mukaiyama, T. *Chem. Lett.* **1989**, 515.

method. The corresponding alkane and ketone are usually side products.

Alkenes can be hydrated quickly under mild conditions in high yields without rearrangement products by the use of *oxymercuration*[186] (addition of oxygen and mercury) followed by *in situ* treatment with sodium borohydride[187] (**12-24**). For example, 2-methyl-1-butene treated with mercuric acetate,[188] followed by NaBH$_4$, gave 2-methyl-2-butanol.

This method, which is applicable to mono-, di-, tri-, and tetraalkyl as well as phenyl-substituted alkenes, gives almost complete Markovnikov addition. Hydroxy, methoxy, acetoxy, halo, and other groups may be present in the substrate without, in general, causing difficulties.[189] When two double bonds are present in the same molecule, the use of ultrasound allows oxymercuration of the less-substituted one without affecting the other.[190] A related reaction treats an alkene with zinc borohydride on silica gel to give a 35:65 mixture of secondary:primary alcohols.[191]

Water can be added indirectly, with anti-Markovnikov orientation, by treatment of the alkene with a 1:1 mixture of PhCH$_2$NEt$_3$$^+$ BH$_4$$^-$ and Me$_3$SiCl, followed by addition of an aqueous solution of K$_2$CO$_3$.[192] Reaction of alkenes with Ti(BH$_4$)$_3$, and then aqueous K$_2$CO$_3$ also leads to the anti-Markovnikov alcohol.[193] Reaction of

---

[186]For a monograph, see Larock, R.C. *Solvation/Demercuration Reactions in Organic Synthesis*, Springer, NY, *1986*. For reviews of this and other oxymetallation reactions, see Kitching, W. *Organomet. React. 1972*, *3*, 319; *Organomet. Chem. Rev. 1968*, *3*, 61; Oullette, R.J., in Trahanovsky, W.S. *Oxidation in Organic Chemistry*, pt. B; Academic Press, NY, *1973*, pp. 140–166; House, H.O. *Modern Synthetic Reactions*, 2nd ed., W.A. Benjamin, NY, *1972*, pp. 387–396; Zefirov, N.S. *Russ. Chem. Rev. 1965*, *34*, 527.

[187]Brown, H.C.; Geoghegan, Jr., P.J. *J. Org. Chem. 1972*, *37*, 1937; Brown, H.C.; Geoghegan, Jr., P.J.; Lynch, G.J.; Kurek, J.T. *J. Org. Chem. 1972*, *37*, 1941; Moon, S.; Takakis, I.M.; Waxman, B.H. *J. Org. Chem. 1969*, *34*, 2951; Moon, S.; Ganz, C.; Waxman, B.H. *Chem. Commun. 1969*, 866; Johnson, M.R.; Rickborn, B. *Chem. Commun. 1968*, 1073; Klein, J.; Levene, R. *Tetrahedron Lett. 1969*, 4833; Chamberlain, P.; Whitham, G.H. *J. Chem. Soc. B 1970*, 1382; Barrelle, M.; Apparu, M. *Bull. Soc. Chim. Fr. 1972*, 2016.

[188]For a review of this reagent, see Butler, R.N., in Pizey, J.S. *Synthetic Reagents*, Vol. 4, Wiley, NY, *1981*, pp. 1–145.

[189]See the extensive tables, in Larock, R.C. *Solvation/Demercuration Reactions in Organic Synthesis*, Springer, NY, *1986*, pp. 4–71.

[190]Einhorn, J.; Einhorn, C.; Luche, J.L. *J. Org. Chem. 1989*, *54*, 4479.

[191]Ranu, B.C.; Sarkar, A.; Saha, M.; Chakraborty, R. *Tetrahedron 1994*, *50*, 6579; Campelo, J.M.; Chakraborty, R.; Marinas, J.M. *Synth. Commun. 1996*, *26*, 1639; Ranu, B.C.; Chakraborty, R.; Saha, M. *Tetrahedron Lett. 1993*, *34*, 4659.

[192]Baskaran, S.; Gupta, V.; Chidambaram, N.; Chandrasekaran, S. *J. Chem. Soc., Chem. Commun. 1989*, 903.

[193]Kumar, K.S.R.; Baskaran, S.; Chandrasekaran, S. *Tetrahedron Lett. 1993*, *34*, 171.

terminal alkynes with water and ruthenium catalyst, followed by sequential treatment with long chain sulfates and then ammonium salts gave the aldehyde via anti-Markovnikov addition of water.[194] With substrates of the type C=C–Z (Z is as defined on p. 1007) the product is almost always HO–C–CH–Z and the mechanism is usually nucleophilic,[195] although electrophilic addition gives the same product[196] since a cation CH–C–Z would be destabilized by the positive charges (full or partial) on two adjacent atoms. However, the α-hydroxy compound HC–CH(OH)Z, was obtained by treatment of the substrate with O_2, PhSiH$_3$, and a manganese- complex catalyst.[197] When the substrate is of the type RCH=CZZ', CZZ', addition of water may result in cleavage of the adduct, to give an aldehyde and CH_2ZZ', **34**.[198] The cleavage step is an example of **12-41**

For another method of anti-Markovnikov hydration, see hydroboration (**15-16**).

Alkenes react with PhO$_2$BH and a niobium catalyst, followed by oxidation with NaOO$^-$, to give the alcohol,[199] and Cp$_2$TiCl$_4$ can also be used.[200] Reaction with HSiCl$_3$ and a chiral palladium catalyst, followed by reaction with KF and hydrogen peroxide, leads to the alcohol with high asymmetric induction.[201] Conjugated alkenes also react with PhSiH$_2$ and oxygen, with a manganese catalyst, to give an α-hydroxy ketone.[202] Alkenes react with molecular oxygen in the presence of a cobalt porphyrin catalyst, and reduction with P(OMe)$_3$ leads to the secondary alcohol.[203] This procedure has also been used to hydrate conjugated dienes,[204] although conjugated dienes are seldom hydrated.

[194]Alvarez, P.; Basetti, M.; Gimeno, J.; Mancini, G. *Tetrahedron Lett.* **2001**, *42*, 8467.

[195]For example, see Fedor, L.R.; De, N.C.; Gurwara, S.K. *J. Am. Chem. Soc.* **1973**, *95*, 2905; Jensen, J.L.; Hashtroudi, H. *J. Org. Chem.* **1976**, *41*, 3299; Bernasconi, C.F.; Leonarduzzi, G.D. *J. Am. Chem. Soc.* **1982**, *104*, 5133, 5143.

[196]For example, see Noyce, D.S.; DeBruin, K.E. *J. Am. Chem. Soc.* **1968**, *90*, 372.

[197]Inoki, S.; Kato, K.; Isayama, S.; Mukaiyama, T. *Chem. Lett.* **1990**, 1869; Magnus, P.; Scott, D.A.; Fielding, M.R. *Tetrahedron Lett.* **2001**, *42*, 4127.

[198]Bernasconi, C.F.; Fox, J.P.; Kanavarioti, A.; Panda, M. *J. Am. Chem. Soc.* **1986**, *108*, 2372; Bernasconi, C.F.; Paschalis, P. *J. Am. Chem. Soc.* **1989**, *111*, 5893, and other papers in this series.

[199]Burgess, K.; Jaspars, M. *Tetrahedron Lett.* **1993**, *34*, 6813.

[200]Burgess, K.; van der Donk, W.A. *Tetrahedron Lett.* **1993**, *34*, 6817.

[201]Uozumi, Y.; Hayashi, T. *Tetrahedron Lett.* **1993**, *34*, 2335.

[202]Magnus, P.; Payne, A.H.; Waring, M.J.; Scott, D.A.; Lynch, V. *Tetrahedron Lett.* **2000**, *41*, 9725.

[203]Matsushita, Y.; Sugamoto, K.; Matsui, T. *Chem. Lett.* **1993**, 925.

[204]Matshshita, Y.; Sugamoto, K.; Nakama, T.; Sakamoto, T.; Matsui, T.; Nakayama, M. *Tetrahedron Lett.* **1995**, *36*, 1879.

The addition of water to enol ethers causes hydrolysis to aldehydes or ketones (**10-6**). Ketenes add water to give carboxylic acids ($R_2C{=}C{=}O \rightarrow R_2COOH$) in a reaction catalyzed by acids:[205]

OS **IV**, 555, 560; **VI**, 766. Also see, OS **V**, 818.

15-4 Hydration of Triple Bonds

Dihydro-oxo-biaddition

The hydration of triple bonds is generally carried out with mercuric ion salts (often the sulfate or acetate) as catalysts.[206] Mercuric oxide in the presence of an acid is also a common reagent. Since the addition follows Markovnikov's rule, only acetylene gives an aldehyde. All other triple-bond compounds give ketones (for a method of reversing the orientation for terminal alkynes, see **15-16**). With alkynes of the form $RC{\equiv}CH$ methyl ketones are formed almost exclusively, but with $RC{\equiv}CR'$ both possible products are usually obtained. The reaction can be conveniently carried out with a catalyst prepared by impregnating mercuric oxide onto Nafion-H (a superacidic perfluorinated resinsulfonic acid, see p. 236).[207] Terminal alkynes react with water at 200°C with microwave irradiation to give the corresponding methyl ketone.[208] A gold catalyst was used in aqueous methanol with 50% sulfuric acid to convert terminal alkynes to the ketone.[209] Conversion of phenyl acetylene to acetophenone was accomplished in water at 100°C with a catalytic amount of Tf_2NH (trifluoromethanesulfonimide).[210] In a modified reaction, internal alkynes were treated with 2-aminophenol in refluxing dioxane using a palladium catalyst to produce the corresponding ketone.[211]

Hydration of terminal alkynes can proceed with anti-Markovnikov addition. When 1-octyne was heated with water, isopropanol and a ruthenium catalyst, for example, the product was octanal.[212] A similar reaction was reported in aqueous acetone using a ruthenium catalyst.[213] The presence of certain functionality can

---

[205]For discussions of the mechanism, see Poon, N.L.; Satchell, D.P.N. *J. Chem. Soc. Perkin Trans. 2*, *1983*, 1381; *1986*, 1485; Tidwell, T.T. *Acc. Chem. Res.* *1990*, 23, 273; Seikaly, H.R.; Tidwell, T.T. *Tetrahedron* *1986*, 42, 2587; Satchell, D.P.N.; Satchell, R.S. *Chem. Soc. Rev.* *1975*, 4, 231.

[206]For reviews, see Larock, R.C. *Solvation/Demercuration Reactions in Organic Synthesis*, Springer, NY, *1986*, pp. 123–148; Khan, M.M.T.; Martell, A.E. *Homogeneous Catalysis by Metal Complexes*, Vol. 2, Academic Press, NY, *1974*, pp. 91–95. For a list of reagents, with references, see Larock, R.C. *Comprehensive Organic Transformations*, 2nd ed., Wiley-VCH, NY, *1999*, pp. 1217–1219.

[207]Olah, G.A.; Meidar, D. *Synthesis 1978*, 671.

[208]Vasudevan, A.; Verzas, M.K. *Synlett 2004*, 631.

[209]Mizushima, E.; Sato, K.; Hayashi, T.; Tanaka, M. *Angew. Chem. Int. Ed. 2002*, 41, 4563.

[210]Tsuchimoto, T.; Joya, T.; Shirakawa, E.; Kawakami, Y. *Synlett 2000*, 1777.

[211]Shimada, T.; Yamamoto, Y. *J. Am. Chem. Soc. 2002*, 124, 12670.

[212]Suzuki, T.; Tokunaga, M.; Wakatsuki, Y. *Org. Lett. 2001*, 3, 735.

[213]Grotjahn, D.B.; Lev, D.A. *J. Am. Chem. Soc. 2004*, 126, 12232.

influence the regioselectivity of hydration. 1-Seleno alkynes, such as PhSe-C≡C-Ph, react with tosic acid in dichloromethane to give a seleno ester PhSeC(=O)SH$_2$Ph after treatment with water.[214]

The first step of the mechanism is formation of a complex (**35**) (ions like Hg^{2+} form complexes with alkynes, p. 115). Water then attacks in an S$_N$2-type process to give the intermediate **36**,

which loses a proton to give **37**. Hydrolysis of **37** (an example of **12-34**) gives the enol, which tautomerizes to the product. A spectrum of the enol was detected by flash photolysis when phenylacetylene was hydrated photolytically.[215]

Carboxylic esters, thiol esters, and amides can be made, respectively, by acid-catalyzed hydration of acetylenic ethers, thioethers,[216] and ynamines, without a mercuric catalyst:[217]

This is ordinary electrophilic addition, with rate-determining protonation as the first step.[218] Certain other alkynes have also been hydrated to ketones with strong acids in the absence of mercuric salts.[219] Simple alkynes can also be converted to ketones by heating with formic acid, without a catalyst.[220] Lactones have been prepared from trimethylsilyl alkenes containing an hydroxyl unit elsewhere in the molecule, when reacted with molecular oxygen, CuCl$_2$, and a palladium catalyst.[221]

[214]Sheng, S.; Liu, X. *Org. Prep. Proceed. Int.* **2002**, *34*, 499.

[215]Chiang, Y.; Kresge, A.J.; Capponi, M.; Wirz, J. *Helv. Chim. Acta* **1986**, *69*, 1331.

[216]Braga, A.L.; Martins, T.L.C.; Silveira, C.C.; Rodrigues, O.E.D. *Tetrahedron* **2001**, *57*, 3297. For a review of acetylenic ethers and thioethers, see Brandsma, L.; Bos, H.J.T.; Arens, J.F., in Viehe, H.G. *Acetylenes*, Marcel Dekker, NY, **1969**, pp. 751–860.

[217]Arens, J.F. *Adv. Org. Chem.* **1960**, *2*, 163; Brandsma, L.; Bos, H.J.T.; Arens, J.F., in Viehe, H.G. *Acetylenes*, Marcel Dekker, NY, **1969**, pp. 774–775.

[218]Hogeveen, H.; Drenth, W. *Recl. Trav. Chim. Pays-Bas* **1963**, *82*, 375, 410; Verhelst, W.F.; Drenth, W. *J. Am. Chem. Soc.* **1974**, *96*, 6692; Banait, N.; Hojatti, M.; Findlay, P.; Kresge, A.J. *Can. J. Chem.* **1987**, *65*, 441.

[219]See, for example, Noyce, D.S.; Schiavelli, M.D. *J. Org. Chem.* **1968**, *33*, 845; *J. Am. Chem. Soc.* **1968**, *90*, 1020, 1023.

[220]Menashe, N.; Reshef, D.; Shvo, Y. *J. Org. Chem.* **1991**, *56*, 2912.

[221]Compain, P.; Goré, J.; Vatèle, J.-M. *Tetrahedron* **1996**, *52*, 10405.

Allenes can also be hydrolyzed to ketones, with an acid catalyst.[222]

$$\text{(allene structure)} \quad \xrightarrow[\text{H}_2\text{O}]{\text{H}^+} \quad \text{(enol structure)} \quad \xrightarrow{\text{tautom.}} \quad \text{(ketone structure)}$$

OS **III**, 22; **IV**, 13; **V**, 1024.

15-5   Addition of Alcohols and Phenols

Hydro-alkoxy-addition

$$\text{C=C} + \text{ROH} \longrightarrow \text{H–C–C–OR}$$

The addition of alcohols and phenols to double bonds is catalyzed by acids or bases. When the reactions are acid catalyzed, the mechanism is electrophilic, with H^+ as the species attacked by the π-bond. The resulting carbocation combines with a molecule of alcohol to give an oxonium ion, **38**.

$$\text{C=C} + \text{H}^+ \longrightarrow \text{H–C–C}^{\oplus} + \text{ROH} \longrightarrow \text{H–C–C–O}^{\oplus}_{R}\text{H} \xrightarrow{-\text{H}^+} \text{H–C–C–OR}$$

38

The addition, therefore, follows Markovnikov's rule. Primary alcohols give better results than secondary, and tertiary alcohols are very inactive. This is a convenient method for the preparation of tertiary ethers by the use of a suitable alkene, such as $Me_2C{=}CH_2$. Addition of alcohols to allylic systems can proceed with rearrangement, and the use of chiral additive can lead to asymmetric induction.[223]

Alcohols add intramolecularly to alkenes to generate cyclic ethers, often bearing a hydroxyl unit,[224] but not always.[225] Furan derivatives are available for alkeneketones using $CuCl_2$ and a palladium catalyst,[226] but chromium catalysts have been used for a similar purpose.[227] A gold catalyst was used with conjugated ketones bearing an alkyne substituent to give fused-ring furans.[228] Pyrone derivatives are available by the coupling of conjugated ketones bearing an alcohol unit, via an addition

[222]For example, see Fedorova, A.V.; Petrov, A.A. *J. Gen. Chem. USSR* **1962**, *32*, 1740; Mühlstadt, M.; Graefe, J. *Chem. Ber.* **1967**, *100*, 223; Cramer, P.; Tidwell, T.T. *J. Org. Chem.* **1981**, *46*, 2683.

[223]See Nakamura, H.; Ishihara, K.; Yamamoto, H. *J. Org. Chem.* **2002**, *67*, 5124.

[224]Bhaumik, A.; Tatsumi, T. *Chem. Commun.* **1998**, 463; Gruttadauria, M.; Aprile, C.; Riela, S.; Noto, R. *Tetrahedron Lett.* **2001**, *42*, 2213.

[225]Miura, K.; Hondo, T.; Okajima, S.; Nakagawa, T.; Takahashi, T.; Hosomi, A. *J. Org. Chem.* **2002**, *67*, 6082; Marotta, E.; Foresti, E.; Marcelli, T.; Peri, F.; Righi, P.; Scardovi, N.; Rosini, G. *Org. Lett.* **2002**, *4*, 4451.

[226]Han, X.; Widenhoefer, R.A. *J. Org. Chem.* **2004**, *69*, 1738.

[227]Miki, K.; Nishino, F.; Ohe, K.; Uemura, S. *J. Am. Chem. Soc.* **2002**, *124*, 5260.

[228]Yao, T.; Zhang, X.; Larock, R.C. *J. Am. Chem. Soc.* **2004**, *126*, 11164.

elimination process mediated by a palladium catalyst.[229] Intramolecular addition of alcohols to alkenes can be promoted by a palladium catalyst, with migration of the double bond in the final product.[230] Rhenium compounds,[231] titanium compounds,[232] or platinum compounds[233] facilitate this cyclization reaction to form functionalized tetrahydrofurans or tetrahydrofurans. Allylic alcohols have been converted to 2-bromo oxetanes using $Br(collidine)_2^+ PF_6^-$.[234] It is noted that the reaction of an alkene–alcohol and N-iodosuccinimide with a chiral titanium catalyst leads to a tetrahydrofuran with a pendant iodoalkyl group, with modest enantioselectivity.[235]

Allenes react with alcohols and allenic alcohols have been converted to tetrahydrofuran derivatives bearing a vinyl group at the α-position, using diphenyliodonium salts.[236] In the presence of allylic bromide and a palladium catalyst, allenic alcohols lead to allylically substituted dihydrofurans.[237] Intramolecular addition of alcohols to allenes leads to cyclic vinyl ethers.[238] Alcohols add intramolecularly to a vinylidene dithiane under electrolytic conditions to form a tetrahydrofuran derivative with a pendant dithiane group.[239]

In the presence of other reagents, functionalized ethers can be formed. In methanol with an R—Se—Br reagent, alkenes are converted to selenoalkyl ethers (MeO—C—C—SeR).[240]

An interesting "double" addition was reported in which 2-(hydroxymethyl)phenol reacted with 2,3-dimethyl-2-butene in the presence of lithium perchlorate and Montmorillonite clay/water to give benzopyrans, but the reaction proceeded via an O-quinomethane generated in situ.[241]

Alcohols add to alkynes under certain conditions to give vinyl ethers. In an excess of alcohol, and in the presence of a platinum catalyst, internal alkynes are converted to ketals.[242] The alcohol to alkyne addition reaction is quite useful for the preparation of heterocycles. Dihydrofurans,[243] furans,[244] benzofurans,[245] and pyran

[229]Reiter, M.; Ropp, S.; Gouverneur, V. *Org. Lett.* **2004**, *6*, 91

[230]Rönn, M.; Bäckvall, J.-E.; Andersson, P.G. *Tetrahedron Lett.* **1995**, *36*, 7749; Semmelhack, M.F.; Epa, W.R. *Tetrahedron Lett.* **1993**, *34*, 7205. See Tiecco, M.; Testaferri, L.; Santi, C. *Eur. J. Org. Chem.* **1999**, 797.

[231]Kennedy, R.M.; Tang, S. *Tetrahedron Lett.* **1992**, *33*, 3729; McDonald, F.E.; Towne, T.B. *J. Org. Chem.* **1995**, *60*, 5750.

[232]Lattanzi, A.; Della Sala, G.G.D.; Russo, M.; Screttri, A. *Synlett* **2001**, 1479.

[233]Qian, H.; Han, X.; Widenhoefer, R.A. *J. Am. Chem. Soc.* **2004**, *126*, 9536.

[234]Albert, S.; Robin, S.; Rousseau, G. *Tetrahedron Lett.* **2001**, *42*, 2477.

[235]Kang, S.H.; Park, C.M.; Lee, S.B.; Kim, M. *Synlett* **2004**, 1279.

[236]In this case, the phenyl group also added to the allene. Kang, S.-K.; Baik, T.-G.; Kulak, A.N. *Synlett* **1999**, 324.

[237]Ma, S.; Gao, W. *J. Org. Chem.* **2002**, *67*, 6104.

[238]Mukai, C.; Ohta, M.; Yamashita, H.; Kitagaki, S. *J. Org. Chem.* **2004**, *69*, 6867.

[239]Sun, Y.; Liu, B.; Kao, J.; Andred'Avignon, D.; Moeller, K.D. *Org. Lett.* **2001**, *3*, 1729. See also, Mukai, C.; Yamashita, H.; Hanaoka, M. *Org. Lett.* **2001**, *3*, 3385.

[240]Back, T.G.; Moussa, Z.; Parvez, M. *J. Org. Chem.* **2002**, *67*, 499.

[241]Chiba, K.; Hirano, T.; Kitano, Y.; Tada, M. *Chem. Commun.* **1999**, 691.

[242]Hartman, J.W.; Sperry, L. *Tetrahedron Lett.* **2004**, *45*, 3787.

[243]Gabriele, B.; Salerno, G.; Lauria, E. *J. Org. Chem.* **1999**, *64*, 7687.

[244]Qing, F.L.; Gao, W.-Z.; Ying, J. *J. Org. Chem.* **2000**, *65*, 2003. See Kel'in, A.V.; Gevorgyan, V. *J. Org. Chem.* **2002**, *67*, 95.

[245]Nan, Y.; Miao, H.; Yang, Z. *Org. Lett.* **2000**, *2*, 297. See also, Arcadi, A.; Cacchi, S.; DiGiuseppe, S.; Fabrizi, G.; Marinelli, F. *Synlett* **2002**, 453.

derivatives[246] have been prepared using this approach. Tetrahydrofurans bearing an exocyclic double bond (vinylidene tetrahydrofurans) were prepared from alkynyl alcohols and a silver carbonate catalyst.[247]

For those substrates, more susceptible to nucleophilic attack, for example, poly-halo alkenes and alkenes of the type C=C–Z, it is better to carry out the reaction in basic solution, where the attacking species is RO$^-$.[248] The reactions with C=C–Z are of the Michael type, and OR goes to the side away from the Z.[249] Since triple bonds are more susceptible to nucleophilic attack than double bonds, it might be expected that bases would catalyze addition to triple bonds particularly well. This is the case, and enol ethers and acetals can be produced by this reaction.[250] Because enol ethers are more susceptible than triple bonds to electrophilic attack, the addition of alcohols to enol ethers can also be catalyzed by acids.[251] One utilization of this reaction involves the compound dihydropyran

39   **40**

(**39**), which is often used to protect the OH groups of primary and secondary[252] alcohols and phenols.[253] The tetrahydropyranyl acetal formed by this reaction (**40**) is stable to bases, Grignard reagents, LiAlH$_4$, and oxidizing agents, any of which can be used to react with functional groups located within the R group. When the reactions are completed, **40** is easily cleaved by treatment with dilute acids (**10-6**). The addition of alcohols to enol ethers is also catalyzed by CoCl$_2$.[254]

Conjugate addition of alcohols to conjugated esters, using ceric ammonium nitrate and LiBr, gave the corresponding α-bromo-β-alkoxy ester.[255]

In base-catalyzed addition to triple bonds, the rate falls in going from a primary to a tertiary alcohol, and phenols require more severe conditions. Other catalysts, namely, BF$_3$ and mercuric salts, have also been used in addition of ROH to triple bonds.

[246]Davidson, M.H.; McDonald, F.E. *Org. Lett.* **2004**, *6*, 1601.

[247]Pale, P.; Chuche, J. *Eur. J. Org. Chem.* **2000**, 1019.

[248]For a review with respect to fluoroalkenes, see Chambers, R.D.; Mobbs, R.H. *Adv. Fluorine Chem.* **1965**, *4*, 51, pp. 53–61.

[249]For an example using a rhodium catalyst, See Farnsworth, M.V.; Cross, M.J.; Louie, J. *Tetrahedron Lett.* **2004**, *45*, 7441.

[250]For a review, see Shostakovskii, M.F.; Trofimov, B.A.; Atavin, A.S.; Lavrov, V.I. *Russ. Chem. Rev.* **1968**, *37*, 907.

[251]For discussions of the mechanism, see Toullec, J.; El-Alaoui, M.; Bertrand, R. *J. Chem. Soc. Perkin Trans. 2*, **1987**, 1517; Kresge, A.J.; Yin, Y. *J. Phys. Org. Chem.* **1989**, *2*, 43.

[252]Tertiary alcohols can also be protected in this way if triphenylphosphine hydrobromide is used as a catalyst: Bolitt, V.; Mioskowski, C.; Shin, D.; Falck, J.R. *Tetrahedron Lett.* **1988**, *29*, 4583.

[253]For useful catalysts for this reaction, some of which are also applicable to tertiary alcohols, see Miyashita, M.; Yoshikoshi, A.; Grieco, P.A. *J. Org. Chem.* **1977**, *42*, 3772; Olah, G.A.; Husain, A.; Singh, B.P. *Synthesis* **1985**, 703; Johnston, R.D.; Marston, C.R.; Krieger, P.E.; Goem G.L. *Synthesis* **1988**, 393.

[254]Iqbal, J.; Srivastava, R.R.; Gupta, K.B.; Khan, M.A. *Synth. Commun.* **1989**, *19*, 901.

[255]Roy, S.C.; Guin, C.; Rana, K.K.; Maiti, G. *Synlett* **2001**, 226.

Alcohols can be added to certain double-bond compounds (cyclohexenes, cyclo-heptenes) photochemically[256] in the presence of a photosensitizer such as benzene. The mechanism is electrophilic and Markovnikov orientation is found. The alkenes react in their first excited triplet states.[257]

The oxymercuration–demercuration procedure mentioned in **15-3** can be adapted to the preparation of ethers (Markovnikov orientation) if the oxymercuration is carried out in an alcohol ROH as solvent,[258] for example 2-methyl-1-butene in ethanol gives $EtMe_2COEt$.[259] Primary alcohols give good yields when mercuric acetate is used, but for secondary and tertiary alcohols it is necessary to use mercuric tri-fluoroacetate.[260] However, even this reagent fails where the product would be a ditertiary ether. It is possible to combine the alcohol reactant with another reagent. The reaction of an alkene with iodine and allyl alcohol, in the presence of HgO, gave the *vic*-iodo ether.[261] Alkene-alcohols react with mercuric trifluoroacetate and the aq. KBr (with $LiBH_4/BEt_3$) to give a derivative bearing an iodoalkyl substituent, $-O-C-CH(I)R$.[262] Alkynes generally give acetals. If the oxymercuration is carried out in the presence of a hydroperoxide instead of an alcohol, the product (after demercuration with $NaBH_4$) is an alkyl peroxide (peroxy-mercuration).[263] This can be done intramolecularly.[264]

Both alcohols and phenols add to ketenes to give carboxylic esters $[R_2C=C=O + ROH \rightarrow R_2CHCO_2R]$.[265] This has been done intramolecularly (with the ketene end of the molecule generated and used *in situ*) to form medium- and large-ring lactones.[266] In the presence of a strong acid, ketene reacts with aldehydes or ketones (in their enol forms) to give enol acetates. 1,4-Asymmetric induction is possible when chiral alcohols add to ketenes.[267]

[256]For a review of the photochemical protonation of double and triple bonds, see Wan, P.; Yates, K. *Rev. Chem. Intermed.* **1984**, *5*, 157.

[257]Marshall, J.A. *Acc. Chem. Res.* **1969**, *2*, 33.

[258]For a review, with tables of many examples, see Larock, R.C. *Solvation/Demercuration Reactions in Organic Synthesis*, Springer, NY, **1986**, pp. 162–345.

[259]Brown, H.C.; Rei, M. *J. Am. Chem. Soc.* **1969**, *91*, 5646.

[260]Brown, H.C.; Kurek, J.T.; Rei, M.; Thompson, K.L. *J. Org. Chem.* **1984**, *49*, 2551; **1985**, *50*, 1171.

[261]Talybov, G.M.; Mekhtieva, V.Z.; Karaev, S.F. *Russ. J. Org. Chem.* **2001**, *37*, 600.

[262]Kang, S.H.; Kim, M. *J. Am. Chem. Soc.* **2003**, *125*, 4684. For an enantioselective example, see Kang, S.H.; Lee, S.B.; Park, C.M. *J. Am. Chem. Soc.* **2003**, *125*, 15748.

[263]Ballard, D.H.; Bloodworth, A.J. *J. Chem. Soc. C* **1971**, 945; Sokolov, V.I.; Reutov, O.A. *J. Org. Chem. USSR* **1969**, *5*, 168. For a review, see Larock, R.C. *Solvation/Demercuration Reactions in Organic Synthesis*, Springer, NY, **1986**, pp. 346–366.

[264]Garavelas, A.; Mavropoulos, I.; Perlmutter, P.; Westman, F. *Tetrahedron Lett.* **1995**, *36*, 463.

[265]Quadbeck, G. *Newer Methods Prep. Org. Chem.* **1963**, *2*, 133–161. See also, Chihara, T.; Teratini, S.; Ogawa, H. *J. Chem. Soc., Chem. Commun.* **1981**, 1120. For discussions of the mechanism, see Tille, A.; Pracejus, H. *Chem. Ber.* **1967**, *100*, 196–210; Brady, W.T.; Vaughn, W.L.; Hoff, E.F. *J. Org. Chem.* **1969**, *34*, 843; Tidwell, T.T. *Acc. Chem. Res.* **1990**, *23*, 273; Seikaly, H.R.; Tidwell, T.T. *Tetrahedron* **1986**, *42*, 2587; Satchell, D.P.N.; Satchell, R.S. *Chem. Soc. Rev.* **1975**, *4*, 231.; Jähme, J.; Rüchardt, C. *Tetrahedron Lett.* **1982**, *23*, 4011; Poon, N.L.; Satchell, D.P.N. *J. Chem. Soc. Perkin Trans. 2*, **1984**, 1083; **1985**, 1551.

[266]Boeckman, Jr., R.K.; Pruitt, J.R. *J. Am. Chem. Soc.* **1989**, *111*, 8286.

[267]Cannizzaro, C.E.; Strassner, T.; Houk, K.N. *J. Am. Chem. Soc.* **2001**, *123*, 2668.

Alcohols can also add to alkenes via the α-carbon (see **15-33**).

OS **III**, 371, 774, 813; **IV**, 184, 558; **VI**, 916; **VII**, 66, 160, 304, 334, 381; **VIII**, 204, 254; **IX**, 472.

15-6 Addition of Carboxylic Acids to Form Esters

Hydro-acyloxy-addition

Carboxylic esters are produced by the addition of carboxylic acids to alkenes, a reaction that is usually acid-catalyzed (by proton or Lewis acids[268]) and similar in mechanism to **15-5**. Since Markovnikov's rule is followed, hard-to-get esters of tertiary alcohols can be prepared from alkenes of the form $R_2C=CHR$.[269] A combination of V_2O_5 and trifluoroacetic acid converts alkenes to trifluoroacetate esters.[270] When a carboxylic acid that contains a double bond in the chain is treated with a strong acid, the addition occurs internally and the product is a γ- and/or a δ-lactone, regardless of the original position of the double bond in the chain, since strong acids catalyze double-bond shifts (**12-2**).[271] The double bond always migrates (also see, **15-1**) to a position favorable for the reaction, whether this has to be toward or away from the carboxyl group. The use of a chiral *Cinchonidine* alkaloid additive leads to lactone formation with modest enantioselectivity.[272] In the presence of diphenyl diselenide and DDQ, alkene carboxylic acids react of form the lactone with a phenylselenomethyl group (PhSeCH$_2$–) at C-5.[273] Carboxylic esters have also been prepared by the acyloxymercuration-demercuration of alkenes (similar to the procedures mentioned in **15-3** and **15-4**).[274] Conjugated esters has been converted to β-lactones with photolysis and added tributyltin hydride, radical cyclization conditions (**15-30**).[275] Addition of carboxylic acids to alkenes to form esters or lactones is catalyzed by palladium compounds.[276] Thallium acetate also promotes this cyclization reaction.[277]

---

[268]See, for example, Guenzet, J.; Camps, M. *Tetrahedron* **1974**, *30*, 849; Ballantine, J.A.; Davies, M.; Purnell, H.; Rayanakorn, M.; Thomas, J.M.; Williams, K.J. *J. Chem. Soc., Chem. Commun.* **1981**, 8.

[269]See, for example, Peterson, P.E.; Tao, E.V.P. *J. Org. Chem.* **1964**, *29*, 2322.

[270]Choudary, B.M.; Reddy, P.N. *J. Chem. Soc., Chem. Commun.* **1993**, 405.

[271]For a review of such lactonizations, see Ansell, M.F.; Palmer, M.H. *Q. Rev. Chem. Soc.* **1964**, *18*, 211.

[272]Wang, M.; Gao, L.X.; Mai, W.P.; Xia, A.X.; Wang, F.; Zhang, S.B. *J. Org. Chem.* **2004**, *69*, 2874.

[273]Tiecco, M.; Testaferri, L.; Temperini, A.; Bagnoli, L.; Marini, F.; Santi, C. *Synlett* **2001**, 1767.

[274]For a review, see Larock, R.C. *Solvation/Demercuration Reactions in Organic Synthesis*, Springer, NY, **1986**, pp. 367–442.

[275]Castle, K.; Hau, C.-S.; Sweeney, J.B.; Tindall, C. *Org. Lett.* **2003**, *5*, 757.

[276]Larock, R.C.; Hightower, T.R. *J. Org. Chem.* **1993**, *58*, 5298; Annby, U.; Stenkula, M.; Andersson, C.-M. *Tetrahedron Lett.* **1993**, *34*, 8545.

[277]Ferraz, H.M.C.; Ribeiro, C.M.R. *Synth. Commun.* **1992**, *22*, 399.

Triple bonds can give enol esters[278] or acylals when treated with carboxylic acids. Mercuric salts are usually catalysts,[279] and vinylic mercury compounds

$$-\overset{\displaystyle |}{C}=\overset{\displaystyle |}{\underset{\displaystyle HgX}{C}}-OCOR$$ are intermediates.[280] Terminal alkynes $RC\equiv CH$ react with CO_2,

a secondary amine R'_2 NH, and a ruthenium complex catalyst, to give enol carbamates $RCH=CHOC(=O)NR$. [281] This reaction has also been performed intramolecularly, to produce unsaturated lactones.[282] Cyclic unsaturated lactones (internal vinyl esters) have been generated from alkyne-carboxylic acids using a palladium catalyst[283] or a ruthenium catalyst.[284] Carboxylic esters can also be obtained by the addition to alkenes of diacyl peroxides.[285] These reactions are catalyzed by copper and are free-radical processes.

Allene carboxylic acids have been cyclized to butenolides with copper(II) chloride.[286] Allene esters were converted to butenolides by treatment with acetic acid and LiBr.[287] Cyclic carbonates can be prepared from allene alcohols using carbon dioxide and a palladium catalyst, and the reaction was accompanied by arylation when iodobenzene was added.[288] Diene carboxylic acids have been cyclized using acetic acid and a palladium catalyst to form lactones that have an allylic acetate elsewhere in the molecule.[289] With ketenes, carboxylic acids give anhydrides[290] and acetic anhydride is prepared industrially in this manner $[CH_2=C=O + MeCO_2H \rightarrow (MeC=O)_2O]$.

[278]Goossen, L.J.; Paetzold, J.; Koley, D. *Chem. Commun.* **2003**, 706. For a rhenium catalyzed example, see Hua, R.; Tian, X. *J. Org. Chem.* **2004**, *69*, 5782.

[279]For the use of rhodium complex catalysts, see Bianchini, C.; Meli, A.; Peruzzini, M.; Zanobini, F.; Bruneau, C.; Dixneuf, P.H. *Organometallics* **1990**, *9*, 1155.

[280]See for example, Bach, R.D.; Woodard, R.A.; Anderson, T.J.; Glick, M.D. *J. Org. Chem.* **1982**, *47*, 3707; Bassetti, M.; Floris, B. *J. Chem. Soc. Perkin Trans. 2*, **1988**, 227; Grishin, Yu.K.; Bazhenov, D.V.; Ustynyuk, Yu.A.; Zefirov, N.S.; Kartashov, V.R.; Sokolova, T.N.; Skorobogatova, E.V.; Chernov, A.N. *Tetrahedron Lett.* **1988**, *29*, 4631. Ruthenium complexes have also been used as catalysts. See Rotem, M.; Shvo, Y. *Organometallics* **1983**, *2*, 1689; Mitsudo, T.; Hori, Y.; Yamakawa, Y.; Watanabe, Y. *J. Org. Chem.* **1987**, *52*, 2230.

[281]Mitsudo, T.; Hori, Y.; Yamakawa, Y.; Watanabe, Y. *Tetrahedron Lett.* **1987**, *28*, 4417; Mahé, R.; Sasaki, Y.; Bruneau, C.; Dixneuf, P.H. *J. Org. Chem.* **1989**, *54*, 1518.

[282]See, for example, Sofia, M.J.; Katzenellenbogen, J.A. *J. Org. Chem.* **1985**, *50*, 2331. For a list of other examples, see Larock, R.C. *Comprehensive Organic Transformations*, 2nd ed., Wiley-VCH, NY, **1999**, p. 1895.

[283]Liao, H.-Y.; Cheng, C.-H. *J. Org. Chem.* **1995**, *60*, 3711

[284]Jiménez-Tenorio, M.; Puerta, M.C.; Valerga, P.; Moreno-Dorado, F.J.; Guerra, F.M.; Massanet, G.M. *Chem. Commun.* **2001**, 2324.

[285]Kharasch, M.S.; Fono, A. *J. Org. Chem.* **1959**, *24*, 606; Kochi, J.K. *J. Am. Chem. Soc.* **1962**, *84*, 1572.

[286]Ma, S.; Wu, S. *J. Org. Chem.* **1999**, *64*, 9314.

[287]Ma, S.; Li, L.; Wei, Q.; Xie, H.; Wang, G.; Shi, Z.; Zhang, J. *Pure. Appl. Chem.* **2000**, *72*, 1739.

[288]Uemura, K.; Shiraishi, D.; Noziri, M.; Inoue, Y. *Bull. Chem. Soc. Jpn.* **1999**, *72*, 1063.

[289]Verboom, R.C.; Persson, B.A.; Bäckvall, J.-E. *J. Org. Chem.* **2004**, *69*, 3102.

[290]For discussions of the mechanism, see Briody, J.M.; Lillford, P.J.; Satchell, D.P.N. *J. Chem. Soc. B* **1968**, 885; Corriu, R.; Guenzet, J.; Camps, M.; Reye, C. *Bull. Soc. Chim. Fr.* **1970**, 3679; Blake, P.G.; Vayjooee, M.H.B. *J. Chem. Soc. Perkin Trans. 2*, **1976**, 1533.

Sulfonic acids add to alkenes and alkynes. The reaction of an alkyne with *para*-toluenesulfonic acid and treatment with silica gives the vinyl sulfonate (C=C–OSO$_2$Tol).[291] Cyclic sulfonates can be generated by the reaction of an allylic sulfonate salt (C=C–C–OSO$_3$$^-$) with silver nitrate in acetonitrile containing an excess of bromine and a catalytic amount of water.[292] Sultones are formed when alkenes react with PhIO and two equivalents of Me$_2$SiSO$_3$Cl.[293]

OS **III**, 853; **IV**, 261, 417, 444; **V**, 852, 863; **VII**, 30, 411. Also see, OS **I**, 317.

C. Sulfur on the Other Side

15-7   Addition of H$_2$S and Thiols

Hydro-alkylthio-addition

Hydrogen sulfide (H$_2$S) and thiols add to alkenes to give alkyl thiols or sulfides by electrophilic, nucleophilic, or free-radical mechanisms.[294] In the absence of initiators, the addition to simple alkenes is by an electrophilic mechanism, similar to that in **15-5**, and Markovnikov's rule is followed. However, this reaction is usually very slow and often cannot be done or requires very severe conditions unless a proton or Lewis acid catalyst is used. For example, the reaction can be performed in concentrated H$_2$SO$_4$[295] or together with AlCl$_3$.[296] In the presence of free-radical initiators, H$_2$S and thiols add to double and triple bonds by a free-radical mechanism and the orientation is anti-Markovnikov.[297] The addition of thiophenol to an alkene with a zeolite also leads to the anti-Markovnikov sulfide.[298] Additives can influence the regioselectivity. Styrene reacts with thiophenol to give primarily the anti-Markovnikov product, whereas addition of thiophenol in the presence of Montmorillonite K10 clay gives primarily the Markovnikov addition product.[299] In fact, the orientation can be used as a diagnostic tool to indicate which mechanism is operating. Free-radical addition can be done with H$_2$S, RSH (R may be primary,

---

[291]Braga, A.L.; Emmerich, D.J.; Silveira, C.C.; Martins, T.L.C.; Rodrigues, O.E.D. *Synlett* **2001**, 371.

[292]Steinmann, J.E.; Phillips, J.H.; Sanders, W.J.; Kiessling, L.L. *Org. Lett.* **2001**, *3*, 3557.

[293]Bassindale, A.R.; Katampe, I.; Maesano, M.G.; Patel, P.; Taylor, P.G. *Tetrahedron Lett.* **199**, *40*, 7417.

[294]For a review, see Wardell, J.L., in Patai, S. *The Chemistry of the Thiol Group*, pt. 1, Wiley, NY, **1974**, pp. 69–178.

[295]Shostakovskii, M.F.; Kul'bovskaya, N.K.; Gracheva, E.P.; Laba, V.I.; Yakushina, L.M. *J. Gen. Chem. USSR* **1962**, *32*, 707.

[296]Belley, M.; Zamboni, R. *J. Org. Chem.* **1989**, *54*, 1230.

[297]For reviews of free-radical addition of H$_2$S and RSH, see Voronkov, M.G.; Martynov, A.V.; Mirskova, A.N. *Sulfur Rep.*, **1986**, *6*, 77; Griesbaum, K. *Angew. Chem. Int. Ed.* **1970**, *9*, 273; Oswald, A.A.; Griesbaum, K., in Kharasch, N.; Meyers, C.Y. *Organic Sulfur Compounds*, Vol. 2, Pergamon, Elmsford, NY, **1966**, pp. 233–256; Stacey, F.W.; Harris Jr., J.F. *Org. React.* **1963**, *13*, 150, pp. 165–196, 247–324.

[298]Kumar, P.; Pandey, R.K.; Hegde, V.R. *Synlett* **1999**, 1921.

[299]Kanagasabapathy, S.; Sudalai, A.; Benicewicz, B.C. *Tetrahedron Lett.* **2001**, *42*, 3791.

secondary, or tertiary), ArSH, or RCOSH.[300] The R group may contain various functional groups. The alkenes may be terminal, internal, contain branching, be cyclic, and have various functional groups including OH, COOH, COOR, NO$_2$, RSO$_2$, and so on. Addition of Ph$_3$SiSH to terminal alkenes under radical conditions also leads to the primary thiol.[301]

Alkynes react with thiols to give vinyl sulfides. With alkynes it is possible to add 1 or 2 equivalents of RSH, giving a vinyl sulfide[302] or a dithioketal, respectively. Alternative preparations are available, as in the reaction of a terminal alkyne with Cp$_2$Zr(H)Cl followed by PhSCl to give the vinyl sulfide with the SPh unit at the less substituted position (PhCH=CHSPh).[303] The intramolecular addition of a thiol to an ene-yne, with a palladium catalyst, leads to substituted thiophene derivatives.[304]

The fundamental addition reaction can be modified by the use of transition metals and different reagents. Alkenes react with diphenyl disulfide in the presence of GaCl$_3$ to give the product with two phenylthio units, PhS–C–C–SPh).[305] The reaction of an alkyne with diphenyl disulfide and a palladium catalyst leads to the bis-vinyl sulfide, PhS–C=C–SPh.[306]

When thiols are added to substrates susceptible to nucleophilic attack, bases catalyze the reaction and the mechanism is nucleophilic. These substrates may be of the Michael type[307] or may be polyhalo alkenes or alkynes.[250] As with the free-radical mechanism, alkynes can give either vinylic thioethers or dithioacetals:

$$
-C\equiv C- \;+\; RSH \;\xrightarrow{\;OH^-\;}\; \underset{\diagup}{\overset{H}{\diagdown}}C{=}C\underset{\diagdown}{\overset{SR}{\diagup}} \;+\; RSH \;\xrightarrow{\;OH^-\;}\; -\underset{H}{\overset{H}{C}}-\underset{SR}{\overset{SR}{C}}-
$$

Thiols add to alkenes under photochemical conditions to form thioethers, and the reaction can be done intramolecularly to give cyclic thioethers.[308] Thiols also add to alkynes and with a palladium catalyst, vinyl sulfides can be formed.[309] Thiocarbonates function as thiol surrogates, converting alkenes to alkyl thiol in the presence of TiCl$_4$; and CuO.[310]

By any mechanism, the initial product of addition of H$_2$S to a double bond is a thiol, which is capable of adding to a second molecule of alkene, so that sulfides

[300]For a review of the addition of thio acids, see Janssen, M.J., in Patai, S. *The Chemistry of Carboxylic Acids and Esters*, Wiley, NY, *1969*, pp. 720–723.

[301]Haché, B.; Gareau, Y. *Tetrahedron Lett.* *1994*, *35*, 1837.

[302]See Arjona, O.; Medel, R.; Rojas, J.; Costa, A.M.; Vilarrasa, J. *Tetrahedron Lett.* *2003*, *44*, 6369.

[303]Huang, X.; Zhong, P.; Guo, W.-r. *Org. Prep. Proceed. Int.* *1999*, *31*, 201.

[304]Gabriele, B.; Salerno, G.; Fazio, A. *Org. Lett.* *2000*, *2*, 351.

[305]Usugi, S.; Yorimitsu, H.; Shinokubo, H.; Oshima, K. *Org. Lett.* *2004*, *6*, 601.

[306]Ananikov, V.P.; Beletskaya, I.P. *Org. Biomol. Chem.* *2004*, *2*, 284.

[307]Michael substrates usually give the expected orientation. For a method of reversing the orientation for RS groups (the RS group goes a to the C=O bond of a C=C–C=O system), see Gassman, P.G.; Gilbert, D.P.; Cole, S.M. *J. Org. Chem.* *1977*, *42*, 3233.

[308]Kirpichenko, S.V.; Tolstikova, L.L.; Suslova, E.N.; Voronkov, M.G. *Tetrahedron Lett.* *1993*, *34*, 3889.

[309]Kuniyasu, H.; Ogawa, A.; Sato, K.-I.; Ryu, I.; Kambe, N.; Sonoda, N. *J. Am. Chem. Soc.* *1992*, *114*, 5902.

[310]Mukaiyama, T.; Saitoh, T.; Jona, H. *Chem. Lett.* *2001*, 638.

are often produced:

As with alcohols, ketenes add thiols to give thiol esters [$R_2C=C=O + RSH \rightarrow$ $R_2CHCOSR$].[311]

Selenium compounds (RSeH) add in a similar manner to thiols.[312] Vinyl selenides can be prepared from alkynes using diphenyl diselenide and sodium borohydride.[313]

The conjugate addition of thiols to α,β-unsaturated carbonyl derivatives is discussed in **15-31**.

OS **III**, 458; **IV**, 669; **VIII**, 302. See also, OS **VIII**, 458.

D. Nitrogen or Phosphorus on the Other Side

15-8   Addition of Ammonia and Amines, Phosphines, and Related Compounds

Hydro-amino-addition
Hydro-phosphino-addition

Ammonia and primary and secondary amines add to alkenes *that are susceptible to nucleophilic attack.*[314] Ammonia and amines are much weaker acids than water, alcohols, and thiols (see **15-3, 15-5, 15-7**) and since acids turn NH_3 into the weak

---

[311]For an example, see Blake, A.J.; Friend, C.L.; Outram, R.J.; Simpkins, N.S.; Whitehead, A.J. *Tetrahedron Lett.* **2001**, *42*, 2877.

[312]Kuniyasu, H.; Ogawa, A.; Sato, K.-I.; Ryu, I.; Sonoda, N. *Tetrahedron Lett.* **1992**, *33*, 5525.

[313]Dabdoub, M.J.; Baroni, A.C.M.; Lenardão, E.J.; Gianeti, T.R.; Hurtado, G.R. *Tetrahedron* **2001**, *57*, 4271.

[314]For reviews, see Gasc, M.B.; Lattes, A.; Périé, J.J. *Tetrahedron* **1983**, *39*, 703; Pines, H.; Stalick, W.M. *Base-Catalyzed Reactions of Hydrocarbons and Related Compounds*, Academic Press, NY, *1977*, pp. 423–454; Suminov, S.I.; Kost, A.N. *Russ. Chem. Rev.* **1969**, *38*, 884; Gibson, M.S., in Patai, S. *The Chemistry of the Amino Group*, Wiley, NY, *1968*, pp. 61–65; Beller, M.; Breindl, C.; Eichberger, M.; Hartung, C.G.; Seayad, J.; Thiel, O.R.; Tillack, A.; Trauthwein, H. *Synlett* **2002**, 1579. For a discussion of Markovnikov versus anti-Markovnikov selectivity, see Tillack, A.; Khedkar, V.; Beller, M. *Tetrahedron Lett.* **2004**, *45*, 8875.

acid, the ammonium ion NH_4^+, this reaction does not occur by an electrophilic mechanism. The reaction tends to give very low yields, if any, with ordinary alkenes, unless extreme conditions are used (e.g., 178–200°C, 800–1000 atm, and the presence of metallic Na, for the reaction between NH_3 and ethylene[315]). Amine alkenes give cyclic amines as the major product, in good yield, when treated with *n*-butyllithium.[316] Ammonia gives three possible products, since the initial product is a primary amine, which may add to a second molecule of alkene, and so on. Similarly, primary amines give both secondary and tertiary products. In practice it is usually possible to control which product predominates. The mechanism is nearly always nucleophilic, and the reaction is generally performed on polyhalo alkenes[317] and alkynes.[318] Ammonia adds to alkenes photochemically.[319] Reaction of a secondary amine with butyllithium generates an amide base, which reacts with alkenes to give alkyl amines,[320] and can add intramolecularly to an alkene to form a pyrrolidine.[321] Pyrroles can be generated in this manner.[322] *N*-Chloroamines add to alkenes intramolecularly to give β-chloropyrrolidines.[323]

Conjugated carbonyl compounds react via conjugate addition with amines to give β-amino derivatives (see **15-31**)[324] As expected, on Michael-type substrates the nitrogen goes to the carbon that does not carry the Z. With substrates of the form RCH=CZZ', the same type of cleavage of the adduct can take place as in **15-3**.[325]

There are many examples of transition catalyzed addition of nitrogen compounds to alkenes, alkynes,[326] and so on. Secondary amines can be added to certain nonactivated alkenes if palladium(II) complexes are used as catalysts.[327]

[315]Howk, B.W.; Little, E.L.; Scott, S.L.; Whitman, G.M. *J. Am. Chem. Soc.* **1954**, *76*, 1899.

[316]Ates, A.; Quinet, C. *Eur. J. Org. Chem.* **2003**, 1623.

[317]For a review with respect to fluoroalkenes, see Chambers, R.D.; Mobbs, R.H. *Adv. Fluorine Chem.* **1965**, *4*, 51–112, pp. 62–68.

[318]For an intramolecular example see Cossy, J.; Belotti, D.; Bellosta, V.; Boggio, C. *Tetrahedron Lett.* **1997**, *38*, 2677. For intramolecular addition to a 1-ethoxy alkyne, see MaGee, D.I.; Ramaseshan, M. *Synlett* **1994**, 743.

[319]Yasuda, M.; Kojima, R.; Ohira, R.; Shiragami, T.; Shima, K. *Bull. Chem. Soc. Jpn.* **1998**, *71*, 1655.

[320]Hartung, C.G.; Breindl, C.; Tillack, A.; Beller, M. *Tetrahedron* **2000**, *56*, 5157.

[321]Fujita, H.; Tokuda, M.; Nitta, M.; Suginome, H. *Tetrahedron Lett.* **1992**, *33*, 6359.

[322]Dieter, R.K.; Yu, H. *Org. Lett.* **2000**, *2*, 2283.

[323]Göttlich, R. *Synthesis* **2000**, 1561; Göttlich, R.; Noack, M. *Tetrahedron Lett* **2001**, *42*, 7771 For a reaction with an *N*-bromoamine, see Outurquin, F.; Pannecoucke, X.; Berthe, B.; Paulmier, C. *Eur. J. Org. Chem.* **2002**, 1007. For a TiCl₃–AlMe₃ mediated reaction, see Sjöholm, Å.; Hemmerling, M.; Pradeille, N.; Somfai, P. *J. Chem. Soc., Perkin Trans. 1* **2001**, 891. For a variation with a sulfonamide and iodine, see Jones, A.D.; Knight, D.W.; Hibbs, D.E. *J. Chem. Soc., Perkin Trans. 1* **2001**, 1182

[324]See Cossu, S.; DeLucchi, O.; Durr, R. *Synth. Commun.* **1996**, *26*, 4597 for an example involing methyl 2-propynoate.

[325]See, for example, Bernasconi, C.F.; Murray, C.J. *J. Am. Chem. Soc.* **1986**, *108*, 5251, 5257; Bernasconi, C.F.; Bunnell, R.D. *J. Org. Chem.* **1988**, *53*, 2001.

[326]For a review, see Doye, S. *Synlett* **2004**, 1653.

[327]For a review, see Gasc, M.B.; Lattes, A.; Périé, J.J. *Tetrahedron* **1983**, *39*, 703. For a review of metal-catalyzed nucleophilic addition, see Bäckvall, J. *Adv. Met.-Org. Chem.* **1989**, *1*, 135. See Löber, O.; Kawatsura, M.; Hartwig, J.F. *J. Am. Chem. Soc.* **2001**, *123*, 4366.

The complexation lowers the electron density of the double bond, facilitating nucleophilic attack.[328] Markovnikov orientation is observed and the addition is anti.[329] Molybdenum,[330] titanium,[331] Yttrium,[332] and rhodium compounds[333] have been used in the addition of amines to alkenes. An intramolecular addition of an amine unit to an alkene to form a pyrrolidine was reported using a palladium catalyst,[334] a lanthanide reagent,[335] or an yttrium reagent.[336] Aniline reacts with dienes and a palladium catalyst to give allylic amines.[337] Diene amines react with samarium catalysts to give 2-alkenyl pyrrolidines.[338] Addition of secondary amines to dihydropyrans using a palladium catalyst gave the corresponding aminal (an α-amino ether).[339] Reduction of nitro compounds in the presence of rhodium catalysts, in the presence of alkenes, CO and H_2, leads to an amine unit adding to the alkene moiety.[340] Secondary amines react with alkenes to give the alkyl amine using a rhodium catalyst in a CO/H_2 atmosphere,[341] but modification of the chromium catalyst and conditions led to an enamine.[342] Note that the reaction of an alkene and a secondary amine with a rhodium catalyst can also give an enamine.[343]

Other nitrogen compounds, among them hydroxylamine and hydroxylamines,[344] hydrazines, and amides (**15-9**), also add to alkenes. Azodicarboxylates (Boc-N= N-Boc) react with alkenes, in the presence of $PhSiH_3$ and a cobalt catalyst, to give

[328]For a discussion of the mechanism, see Hegedus, L.S.; Åkermark, B.; Zetterberg, K.; Olsson, L.F. *J. Am. Chem. Soc.* **1984**, *106*, 7122.

[329]Åkermark, B.; Zetterberg, K. *J. Am. Chem. Soc.* **1984**, *106*, 5560; Utsunomiya, M.; Hartwig, J.F. *J. Am. Chem. Soc.* **2003**, *125*, 14286.

[330]Srivastava, R.S.; Nicholas, K.M. *Chem. Commun.* **1996**, 2335.

[331]Ackermann, L.; Kaspr, L.T.; Gschrei, C.J. *Org. Lett.* **2004**, *6*, 2515. See Castro, I.G.; Tillack, A.; Hartung, C.G.; Beller, M. *Tetrahedron Lett.* **2003**, *44*, 3217.

[332]O'Shaughnessy, P.N.; Scott, P. *Tetrahedron Asymmetry* **2003**, *14*, 1979.

[333]The anti-Markovkinov amine is produced: Utsunomiya, M.; Kuwano, R.; Kawatsura, M.; Hartwig, J.F. *J. Am. Chem. Soc.* **2003**, *125*, 5608; Utsonomiya, M.; Hartwig, J.F. *J. Am. Chem. Soc.* **2004**, *126*, 2702; Ahmed, M.; Seayad, A.M.; Jackstell, R.; Beller, M. *J. Am. Chem. Soc.* **2003**, *125*, 10311.

[334]Fix, S.R.; Brice, J.L.; Stahl, S.S. *Angew. Chem. Int. Ed.* **2002**, *41*, 164.

[335]Molander, G.A.; Dowdy, E.D. *J. Org. Chem.* **1998**, *63*, 8983; Ryu, J.-S.; Marks, T.J.; McDonald, F.E. *Org. Lett.* **2001**, *3*, 3091. The use of a chiral lanthanum catalyst led to pyrrolidines with modest asymmetric induction: Hong, S.; Tian, S.; Metz, M.V.; Marks, T.J. *J. Am. Chem. Soc.* **2003**, *125*, 14768.

[336]Kim, Y.K.; Livinghouse, T.; Bercaw, J.E. *Tetrahedron Lett.* **2001**, *42*, 2933.

[337]Minami, T.; Okamoto, H.; Ikeda, S.; Tanaka, R.; Ozawa, F.; Yoshifuji, M. *Angew. Chem. Int. Ed.* **2001**, *40*, 4501.

[338]Hong, S.; Marks, T.J. *J. Am. Chem. Soc.* **2002**, *124*, 7886.

[339]Cheng, X.; Hii, K.K. *Tetrahedron* **2001**, *57*, 5445.

[340]Rische, T.; Eilbracht, P. *Tetrahedron* **1998**, *54*, 8441; Akazome, M.; Kondo, T.; Watanabe, Y. *J. Org. Chem.* **1994**, *59*, 3375.

[341]Rische, J.; Bärfacker, L.; Eilbracht, P. *Eur. J. Org. Chem.* **1999**, 653; Lin, Y.-S.; El Ali, B.; Alper, H. *Tetrahedron Lett.* **2001**, *42*, 2423.

[342]Ahmed, M.; Seayad, A.M.; Jackstell, R.; Beller, M. *Angew. Chem. Int. Ed.* **2003**, *42*, 5615.

[343]Tillack, A.; Trauthwein, H.; Hartung, C.G.; Eichberger, M.; Pitter, S.; Jansen, A.; Beller, M. *Monat. Chem.* **2000**, *131*, 1327.

[344]Lin, X.; Stien, D.; Weinreb, S.M. *Tetrahedron Lett.* **2000**, *41*, 2333; Singh, S.; Nicholas, K.M. *Synth. Commun.* **2001**, *31*, 3087.

alkylhydrazides [RN(Boc)—NHBoc].[345] Even with amines, basic catalysts are sometimes used, so that RNH^- or R_2N^- is the actual nucleophile. Tertiary amines (except those that are too bulky) add to Michael-type substrates in a reaction that is catalyzed by acids like HCl or HNO_3 to give the corresponding quaternary ammonium salts.[346]

$$\underset{Z}{\overset{Z}{\diagdown}}C=C\diagup + R_3NH^{\oplus}\ Cl^{\ominus} \xrightarrow{HCl} Z-\overset{H}{\underset{|}{C}}-\overset{|}{\underset{|}{C}}-\overset{\oplus}{N}R_3\ Cl^{\ominus}$$

The tertiary amine can be aliphatic, cycloalkyl, or heterocyclic (including pyridine). The reaction of NaOH with an amine containing two distal alkene units, followed by addition of a neodymium catalyst leads to a bicyclic amine.[347]

Primary amines add to triple bonds[348] to give enamines that have a hydrogen on the nitrogen and (analogously to enols) tautomerize to the more stable imines, **41**.[349]

$$R-C\equiv C-R' + R^2-NH_2 \longrightarrow \underset{R}{\overset{H}{\diagdown}}C=C\underset{R'}{\overset{NHR^2}{\diagup}} \rightleftharpoons R-\overset{H}{\underset{|}{C}}-C\underset{R'}{\overset{N-R^2}{\diagdown}}$$

41

The reaction has been done with a palladium catalyst,[350] a titanium catalyst,[351] a tantalum catalyst,[352] and with a gold catalyst.[353] An intramolecular addition of amines to an alkyne unit in the presence of a palladium catalyst generated heterocyclic or cyclic amine compounds.[354] The titanium catalyzed addition of primary

[345]Waser, J.; Carreira, E.M. *J. Am. Chem. Soc.* **2004**, *126*, 5676.

[346]Le Berre, A.; Delacroix, A. *Bull. Soc. Chim. Fr.* **1973**, 640, 647. See also, Vogel, D.E.; Büchi, G. *Org. Synth.*, *66*, 29.

[347]Molander, G.A.; Pack, S.K. *J. Org. Chem.* **2003**, *68*, 9214.

[348]For a review of addition of ammonia and amines to triple bonds, see Chekulaeva, I.A.; Kondrat'eva, L.V. *Russ. Chem. Rev.* **1965**, *34*, 669. For reactions with aniline, see Haak, E.; Bytschkov, I.; Doye, S. *Angew. Chem. Int. Ed.* **1999**, *38*, 3389; Hartung, C.G.; Tillack, A.; Trauthwein, H.; Beller, M. *J. Org. Chem.* **2001**, *66*, 6339.

[349]For example, see Kruse, C.W.; Kleinschmidt, R.F. *J. Am. Chem. Soc.* **1961**, *83*, 213, 216.

[350]Kadota, I.; Shibuya, A.; Lutete, L.M.; Yamamoto, Y. *J. Org. Chem.* **1999**, *64*, 4570.

[351]Khedkar, V.; Tillack, A.; Beller, M. *Org. Lett.* **2003**, *5*, 4767; Tillack, A.; Castro, I.G.; Hartung, C.G.; Beller, M. *Angew. Chem. Int. Ed.* **2002**, *41*, 2541.

[352]Anderson, L.L.; Arnold, J.; Bergman R.G. *Org. Lett.* **2004**, *6*, 2519; Shi, Y.; Hall, C.; Ciszewski, J.T.; Cao, C.; Odom, A.L. *Chem. Commun.* **2003**, 586; Cao, C.; Li, Y.; Shi, Y.; Odom, A.L. *Chem. Commun.*, **2004**, 2002.

[353]Mizushima, E.; Hayashi, T.; Tanaka, M. *Org. Lett.* **2003**, *5*, 3349.

[354]Müller, T.E. *Tetrahedron Lett.* **1998**, *39*, 5961; Hiroya, K.; Matsumoto, S.; Sakamoto, T. *Org. Lett.* **2004**, *6*, 2953; Lutete, L.M.; Kadota, I.; Yamamoto, Y. *J. Am. Chem. Soc.* **2004**, *126*, 1622; Hiroya, K.; Itoh, S.; Ozawa, M.; Kanamori, Y.; Sakamoto, T. *Tetrahedron Lett.* **2002**, *43*, 1277. See also, Karur, S.; Kotti, S.R.S.S.; Xu, X.; Cannon, J.F.; Headley, A.; Li, G. *J. Am. Chem. Soc.* **2003**, *125*, 13340.

amines to alkynes give the enamine, which can be hydrogenated (**15-11**) to give the corresponding amine.[355] A variation treats an alkynyl imine with CuI to form pyrroles.[356] *N,N*-Diphenylhydrazine reacts with diphenyl acetylene and a titanium catalyst to give indole derivatives.[357] Treatment of an imine of 2-alkynyl benzaldehyde with iodide gave a functionalized isoquinoline.[358] When ammonia is used instead of a primary amine, the corresponding $RH_2C-\overset{\overset{\displaystyle NH}{\|}}{C}-R'$ is not stable enough for isolation, but polymerizes. Ammonia and primary amines (aliphatic and aromatic) add to conjugated diynes to give pyrroles, **42**.[359] A similar preparation of pyrroles was reported by heating non-conjugated diynes with aniline and a titanium catalyst.[360] This is not 1,4-addition, but 1,2-addition twice. Conjugated ene-ynes containing an amino group also give pyrroles with a palladium catalyst.[361]

Allenes are reaction partners,[362] and amines add to allenes in the presence of a catalytic amount of CuBr[363] or palladium compounds.[364] Intramolecular reaction of allene amines lead to dihydropyrroles, using a gold catalyst.[365]

Treatment of an allene amine with a ruthenium catalyst, 10% of TiCl$_4$ and methyl vinyl ketone to give a product of amine addition followed by Michael addition, a pyrrolidine derivative with a pendant alkenyl ketone unit.[366] Cyclic imines can be prepared from allene amines using a titanium catalyst.[367]

[355]Haak, E.; Siebeneicher, H.; Doye, S. *Org. Lett.* **2000**, *2*, 1935; Bytschkov, I.; Doye, S. *Eur. J. Org. Chem.* **2001**, 4411. For a variation using sodium cyanoborohydride and zinc chloride as the reducing agent, see Heutling, A.; Doye, S. *J. Org. Chem.* **2002**, *67*, 1961.

[356]Kel'in, A.; Sromek, A.W.; Gevorgyan, V. *J. Am. Chem. Soc.* **2001**, *123*, 2074. Another variation used a chromium carbene species to generate pyrroles from imino ene-ynes: Zhang, Y.; Herndon, J.W. *Org. Lett.* **2003**, *5*, 2043.

[357]Ackermann, L.; Born, R. *Tetrahedron Lett.* **2004**, *45*, 9541. For a different approach using hypervalent iodine, see Barluenga, J.; Trincado, M.; Rubio, E.; González, J.M. *Angew. Chem. Int. Ed.* **2003**, *42*, 2406.

[358]Huang, Q.; Hunter, J.A.; Larock, R.C *J. Org. Chem.* **2002**, *67*, 3437.

[359]Schult, K.E.; Reisch, J.; Walker, H. *Chem. Ber.* **1965**, *98*, 98.

[360]Ramanathan, B.; Keith, A.J.; Armstrong, D.; Odom, A.L. *Org. Lett.* **2004**, *6*, 2957.

[361]Gabriele, B.; Salerno, G.; Fazio, A.; Bossio, M.R. *Tetrahedron Lett.* **2001**, *42*, 1339; Gabriele, B.; Salerno, G.; Fazio, A. *J. Org. Chem.* **2003**, *68*, 7853.

[362]Meguro, M.; Yamamoto, Y. *Tetrahedron Lett.* **1998**, *39*, 5421.

[363]Geri, R.; Polizzi, C.; Lardicci, L.; Caporusso, A.M. *Gazz. Chim. Ital.*, **1994**, *124*, 241.

[364]Davies, I.W.; Scopes, D.I.C.; Gallagher, T. *Synlett* **1993**, 85.

[365]Morita, N.; Krause, N. *Org. Lett.* **2004**, *6*, 4121.

[366]Trost, B.M.; Pinkerton, A.B.; Kremzow, D. *J. Am. Chem. Soc.* **2000**, *122*, 12007.

[367]Ackermann, L.; Bergman, R.G. *Org. Lett.* **2002**, *4*, 1475; Ackerman, L.; Bergman, R.G.; Loy, R.N. *J. Am. Chem. Soc.* **2003**, *125*, 11956.

Primary and secondary amines add to ketenes to give, respectively, *N*-substituted and *N,N*-disubstituted amides:[368] and to ketenimines to give amidines, **43**.[369]

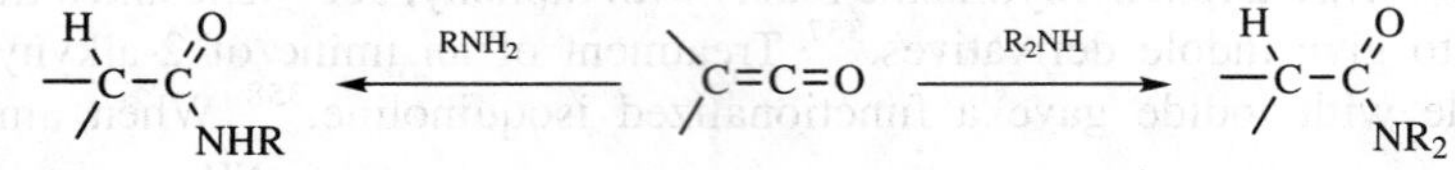

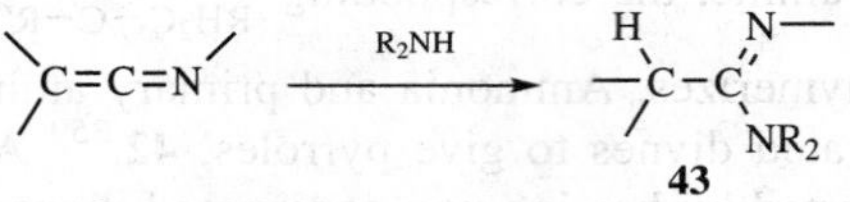

R = H or alkyl

43

NH_3 can be added to double bonds (even ordinary double bonds) in an indirect manner by the use of hydroboration (**15-16**) followed by treatment with NH_2Cl or NH_2OSO_2OH (**12-32**). This produces a primary amine with anti-Markovnikov orientation. An indirect way of adding a primary or secondary amine to a double bond consists of aminomercuration followed by reduction (see **15-3** for the analogous oxymercuration–demercuration procedure), to give amine **45**.[370]

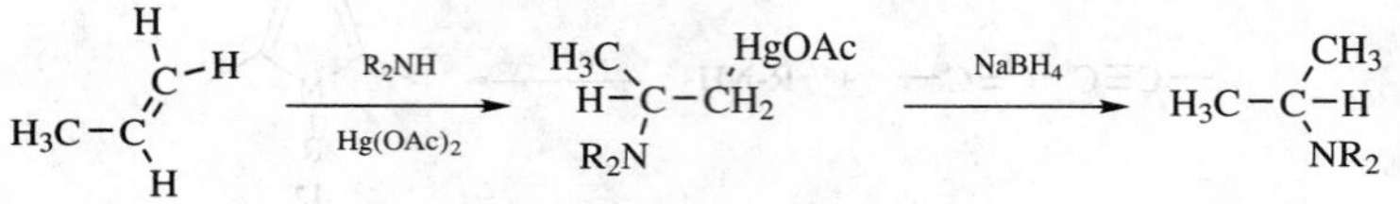

44   **45**

The addition of a secondary amine (shown above) produces a tertiary amine, while addition of a primary amine gives a secondary amine. The overall orientation follows Markovnikov's rule. For conversion of **44** to other products, see **15-53**.

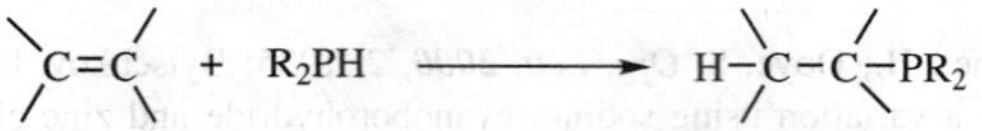

Phosphines add to alkenes to give alkyl phosphines and to alkynes to give vinyl phosphines. In the presence of an ytterbium (Yb) catalyst, diphenylphosphine added to diphenyl acetylene to give the corresponding vinyl phosphine.[371] A palladium catalyst was used for the addition *o*-diphenylphosphine to terminal alkynes, giving the anti-Markovnikov vinyl phosphine but a nickel catalyst led to the Markovnikov vinyl phosphine.[372] Alkenes also react with diarylphosphines

[368]For discussions of the mechanism of this reaction, see Briody, J.M.; Satchell, D.P.N. *Tetrahedron* **1966**, *22*, 2649; Tidwell, T.T. *Acc. Chem. Res.* **1990**, *23*, 273; Satchell, D.P.N.; Satchell, R.S. *Chem. Soc. Rev.* **1975**, *4*, 231. For an enantioselective reaction, see Hodous, B.L.; Fu, G.C. *J. Am. Chem. Soc.* **2002**, *124*, 10006.
[369]Stevens, C.L.; Freeman, R.C.; Noll, K. *J. Org. Chem.* **1965**, *30*, 3718.
[370]For a review, see Larock, R.C. *Solvation/Demercuration Reactions in Organic Synthesis*, Springer, NY, **1986**, pp. 443–504. See also, Barluenga, J.; Perez-Prieto, J.; Asensio, G. *Tetrahedron* **1990**, *46*, 2453.
[371]Takaki, K.; Koshoji, G.; Komeyama, K.; Takeda, M.; Shishido, T.; Kitani, A.; Takehira, K. *J. Org. Chem.* **2003**, *68*, 6554.
[372]Kazankova, M.A.; Efimova, I.V.; Kochetkov, A.N.; Atanas'ev, V.V.; Beletskaya, I.P.; Dixneuf, P.H. *Synlett* **2001**, 497.

and a nickel catalyst. to give the alkyl phosphine[373] Silylphosphines (R$_3$Si—PAr$_2$) react with alkenes and Bu$_4$NF to give the anti-Markovnikov ally phosphine.[374] Phosphine oxides can be prepared by the reaction of an aryl substituted alkene and diphenylphosphine oxide, Ph$_2$P(=O)H.[375] Diphenylphosphine oxide also reacted with terminal alkynes to give the anti-Markovnikov vinyl phosphine oxide using a rhodium catalyst.[376] Phosphonate esters were similar prepared from alkenes and diethyl phosphite, (EtO$_2$)P(=O)H, and a manganese catalyst in a reaction exposed to oxygen.[377] Similar addition was observed in the reaction of an alkene with NaH$_2$PO$_2$ to give the phosphinate, RCH=CH$_2$ → RCH$_2$CH$_2$PH(=O)ONa.[378] Palladium catalysts were used for the preparation of similar compounds from alkenes[379] and the reaction of terminal alkynes with dimethyl phosphite and a nickel catalyst gave the Markovnikov vinyl phosphonate ester.[380] Other phosphites were added to dienes to give an allylic phosphonate ester using a palladium catalyst.[381] Diarylphosphines react with vinyl ethers and a nickel catalyst to give α-alkoxy phosphonate esters.[382]

OS **I**, 196; **III**, 91, 93, 244, 258; **IV**, 146, 205; **V**, 39, 575, 929; **VI**, 75, 943; **VIII**,188, 190, 536; **80**, 75. See also, OS **VI**, 932.

15-9 Addition of Amides

Hydro-amido-addition

Under certain conditions, amides can add directly to alkenes to form *N*-alkylated amides. Sulfonamides react in a similar manner. 3-Pentenamide was cyclized to 5-methyl-2-pyrrolidinone by treatment with trifluorosulfonic acid.[383] Acyl hydrazine derivatives also cyclized in the presence of hypervalent iodine reagents to give lactams.[384] When a carbamate was treated with Bu$_3$SnH, and AIBN, addition to an alkene led to a bicyclic lactam.[385]

[373]Shulyupin, M.O.; Kazankova, M.A.; Beletskaya, I.P. *Org. Lett.*, **2002**, *4*, 761.
[374]Hayashi, M.; Matsuura, Y.; Watanabe, Y. *Tetrahedron Lett.* **2004**, *45*, 9167.
[375]Bunlaksananusorn, T.; Knochel, P. *J. Org. Chem.* **2004**, *69*, 4595; Rey, P.; Taillades, J.; Rossi, J.C.; Gros, G. *Tetrahedron Lett.* **2003**, *44*, 6169.
[376]Han, L.-B.; Zhao, C.-Q.; Tanaka, M. *J. Org. Chem.* **2001**, *66*, 5929.
[377]Tayama, O.; Nakano, A.; Iwahama, T.; Sakaguchi, S.; Ishii, Y. *J. Org. Chem.* **2004**, *69*, 5494.
[378]Deprèle, S.; Montchamp, J.-L. *J. Org. Chem.* **2001**, *66*, 6745.
[379]Deprèle, S.; Montchamp, J.-L. *J. Am. Chem. Soc.* **2002**, *124*, 9386.
[380]Han, L.-B.; Zhang, C.; Yazawa, H.; Shimada, S. *J. Am. Chem. Soc.* **2004**, *126*, 5080.
[381]Mirzaei, F.; Han, L.-B.; Tanaka, M. *Tetrahedron Lett.* **2001**, *42*, 297.
[382]Kazankova, M.A.; Shulyupin, M.O.; Beletskaya, I.P. *Synlett* **2003**, 2155.
[383]Marson, C.M.; Fallah, A. *Tetrahedron Lett.* **1994**, *35*, 293.
[384]Scartozzi, M.; Grondin, R.; Leblanc, Y. *Tetrahedron Lett.* **1992**, *33*, 5717.
[385]Callier, A.-C.; Quiclet-Sire, B.; Zard, S.Z. *Tetrahedron Lett.* **1994**, *35*, 6109.

The reaction can be done intramolecularly. *N*-Benzyl pent-4-ynamide reacted with tetrabutylammonium fluoride to an alkylidene lactam.[386] Similar addition of a tosylamide-alkene, with a palladium catalyst, led to a vinyl *N*-tosyl pyrrolidine.[387] Similar cyclization reactions occur with tosylamide-alkynes.[388]

Treatment of triflamide alkenes with triflic acid gives the corresponding *N*-triflyl cyclic amine.[389] Using an alkene halide and an *N*-chlorosulfonamide, an amide is generated *in situ*, and addition to the alkene gives a pyrrolidine derivative.[390] *N*-Bromocarbamates also add to alkenes, in the presence of $BF_3 \bullet OEt_2$ to give a *vic*-bromo *N*-Boc amine.[391] The titanium catalyzed reaction of alkenyl *N*-tosylamines give *N*-tosyl cyclic amines.[392]

Alkynes and allenes also react with amides. Phenylthiomethyl alkynes were converted to *N*-Boc-*N*-phenylthio allenes with Boc azide and an iron catalyst.[393] The palladium-catalyzed reaction of an allene amide, with iodobenzene, leads to *N*-sulfonyl aziridines having an allylic group at C1.[394] Other allene *N*-tosylamines similarly give *N*-tosyl tetrahydropyridines.[395]

Imides can also add to alkenes or alkynes. Ethyl 2-propynoate reacted with phthalimide, in the presence of a palladium catalyst, to give ethyl 2-phthalimido-2-propenoate.[396]

15-10   Addition of Hydrazoic Acid

Hydro-azido-addition

$$\begin{matrix} R \\ \diagdown \\ \diagup \\ \end{matrix} C = C \begin{matrix} Z \\ \diagup \\ \diagdown \\ \end{matrix} \ + \ HN_3 \ \longrightarrow \ R - \underset{\underset{H}{|}}{C} - \underset{\underset{N_3}{|}}{C} - Z$$

Hydrazoic acid (HN_3) can be added to certain Michael-type substrates (Z is as defined on p. 1007) to give β-azido compounds.[397] The reaction apparently fails if R

[386]Jacobi, P.A.; Brielmann, H.L.; Hauck, S.I. *J. Org. Chem.* **1996**, *61*, 5013.

[387]Larock, R.C.; Hightower, T.R.; Hasvold, L.A.; Peterson, K.P. *J. Org. Chem.* **1996**, *61*, 3584; Harris, Jr., G.D.; Herr, R.J.; Weinreb, S.M. *J. Org. Chem.* **1993**, *58*, 5452. See also, Pinho, P.; Minnaard, A.J.; Feringa, B.L. *Org. Lett.* **2003**, *5*, 259.

[388]Luo, F.-T.; Wang, R.-T. *Tetrahedron Lett.* **1992**, *33*, 6835.

[389]Schlummer, B.; Hartwig, J.F. *Org. Lett.* **2002**, *4*, 1471; Haskins, C.M.; Knight, D.W. *Chem. Commun.* **2002**, 2724.

[390]Minakata, S.; Kano, D.; Oderaotoshi, Y.; Komatsu, M. *Org. Lett.* **2002**, *4*, 2097.

[391]Ś liwnń ska, A.; Zwierzak, A. *Tetrahedron* **2003**, *59*, 5927.

[392]Miura, K.; Hondo, T.; Nakagawa, T.; Takahashi, T.; Hosomi, A. *Org. Lett.* **2000**, *2*, 385.

[393]Bacci, J.P.; Greenman, K.L.; van Vranken, D.L. *J. Org. Chem.* **2003**, *68*, 4955.

[394]Ohno, H.; Toda, A.; Miwa, Y.; Taga, T.; Osawa, E.; Yamaoka, Y.; Fujii, N.; Ibuka, T. *J. Org. Chem.* **1999**, *64*, 2992.

[395]Rutjes, F.P.J.T.; Tjen, K.C.M.F.; Wolf, L.B.; Karstens, W.F.J.; Schoemaker, H.E.; Hiemstra, H. *Org. Lett.* **1999**, *1*, 717; Na, S.; Yu, F.; Gao, W. *J. Org. Chem.* **2003**, *68*, 5943; Ma. S.; Gao, W. *Org. Lett.* **2002**, *4*, 2989.

[396]Trost, B.M.; Dake, G.R. *J. Am. Chem. Soc.* **1997**, *119*, 7595.

[397]Boyer, J.H. *J. Am. Chem. Soc.* **1951**, *73*, 5248; Harvey, G.R.; Ratts, K.W. *J. Org. Chem.* **1966**, *31*, 3907. For a review, see Biffin, M.E.C.; Miller, J.; Paul, D.B., in Patai, S. *The Chemistry of the Azido Group*, Wiley, NY, **1971**, pp. 120–136.

is phenyl. The HN_3 also adds to enol ethers CH_2=$CHOR$ to give CH_3–$CH(OR)N_3$, and to silyl enol ethers,[398] but it does not add to ordinary alkenes unless a Lewis acid catalyst, such as $TiCl_4$, is used, in which case good yields of azide can be obtained.[398] Hydrazoic acid can also be added indirectly to ordinary alkenes by azidomercuration, followed by demercuration,[399] analogous to the similar procedures

mentioned in **15-3**, **15-5**, **15-6**, and **15-8**. The method can be applied to terminal alkenes or strained cycloalkenes (e.g., norbornene) but fails for unstrained internal alkenes.

E. Hydrogen on Both Sides

15-11   Hydrogenation of Double and Triple Bonds[400]

Dihydro-addition

Most carbon–carbon double bonds, whether substituted by electron-donating or electron-withdrawing substituents, can be catalytically hydrogenated, usually in quantitative or near-quantitative yields.[401] Almost all known alkenes added hydrogen at temperatures between 0 and 275°C. The catalysts used can be divided into

---

[398]Hassner, A.; Fibiger, R.; Andisik, D. *J. Org. Chem.* **1984**, *49*, 4237.

[399]Heathcock, C.H. *Angew. Chem. Int. Ed.* **1969**, *8*, 134. For a review, see Larock, R.C. *Solvation/Demercuration Reactions in Organic Synthesis*, Springer, NY, **1986**, pp. 522–527.

[400]For a review, see Mitsui, S.; Kasahara, A., in Zabicky, J. *The Chemistry of Alkenes*, Vol. 2, Wiley, NY, **1970**, pp. 175–214. Also see, Smith, M.B. *Organic Synthesis*, 2nd ed., McGraw-Hill, NY, **2001**, pp. 369–382.

[401]For books on catalytic hydrogenation, see Rylander, P.N. *Hydrogenation Methods*, Academic Press, NY, **1985**; *Catalytic Hydrogenation in Organic Synthesis*, Academic Press, NY, **1979**; *Catalytic Hydrogenation over Platinum Metals*, Academic Press, NY, **1967**; Červený, L. *Catalytic Hydrogenation*, Elsevier, NY, **1986** (this book deals mostly with industrial aspects); Freifelder, M. *Catalytic Hydrogenation in Organic Synthesis*, Wiley, NY, **1978**; *Practical Catalytic Hydrogenation*, Wiley, NY, **1971**; Augustine, R.L. *Catalytic Hydrogenation*, Marcel Dekker, NY, **1965**. For reviews, see Parker, D., in Hartley, F.R. *The Chemistry of the Metal–Carbon Bond*, Vol. 4, Wiley, NY, **1987**, pp. 979–1047; Carruthers, W. *Some Modern Methods of Organic Synthesis* 3rd ed., Cambridge University Press, Cambridge, **1986**, pp. 411–431; Colquhoun, H.M.; Holton, J.; Thompson, D.J.; Twigg, M.V. *New Pathways for Organic Synthesis*, Plenum, NY, **1984**, pp. 266–300, 325–334; Kalinkin, M.I.; Kolomnikova, G.D.; Parnes, Z.N.; Kursanov, D.N. *Russ. Chem. Rev.* **1979**, *48*, 332; Candlin, J.P.; Rennie, R.A.C. in Bentley, K.W.; Kirby, G.W. *Elucidation of Organic Structures by Physical and Chemical Methods*, 2nd ed. (Vol. 4 of Weissberger, A. *Techniques of Chemistry*), pt. 2, Wiley, NY, **1973**, pp. 97–117; House, H.O. *Modern Synthetic Reactions*, 2nd ed., W.A. Benjamin, NY, **1972**, pp, 1–34.

two broad classes, both of which mainly consist of transition metals and their compounds: (*1*) catalysts insoluble in the reaction medium (*heterogeneous catalysts*). Among the most effective are Raney nickel,[402] palladium-on-charcoal (perhaps the most common),[403] NaBH$_4$-reduced nickel[404] (also called nickel boride), platinum metal or its oxide, rhodium, ruthenium, and zinc oxide.[405] (*2*) Catalysts soluble in the reaction medium (*homogeneous catalysts*).[406] An important example is chlorotris(triphenylphosphine)rhodium, RhCl(Ph$_3$P)$_3$,[407] (**100**, *Wilkinson's catalyst*),[408] which catalyzes the hydrogenation of many alkenyl compounds without disturbing such groups as COOR, NO$_2$, CN, or COR present in the same molecule.[409] Even unsaturated aldehydes can be reduced to saturated aldehydes,[410] although in this case decarbonylation (**14-32**) may be a side reaction. In general, for catalytic hydrogenation, many functional groups may be present in the molecule, for example, OH, COOH, NR$_2$ including NH$_2$, N(R)COR' including carbamates,[411] CHO, COR, COOR, or CN. Vinyl esters can be hydrogenated using homogeneous rhodium catalyst.[412] Enamides are hydrogenated, with excellent enantioselectivity, using chiral rhodium catalysts.[413] Some of these groups are also susceptible to catalytic reduction, but it is usually possible to find conditions

[402]For a review of Raney nickel, see Pizey, J.S. *Synthetic Reagents*, Vol. 2, Wiley, NY, *1974*, pp. 175–311. Double bonds have been reduced with Raney nickel alone; with no added H$_2$. The hydrogen normally present in this reagent was sufficient: Pojer, P.M. *Chem. Ind. (London) 1986*, 177.

[403]A recyclable Pd/CaCO$_3$ catalyst in polyethylene glycol (PEG) as been reported. See Chandrasekhar, S.; Narsihmulu, Ch.; Chandrashekar, G.; Shyamsunder, T. *Tetrahedron Lett. 2004*, *45*, 2421.

[404]Paul, R.; Buisson, P.; Joseph, N. *Ind. Eng. Chem. 1952*, *44*, 1006; Brown, C.A. *Chem. Commun. 1969*, 952; *J. Org. Chem. 1970*, *35*, 1900. For a review of reductions with nickel boride and related catalysts, see Ganem, B.; Osby, J.O. *Chem. Rev. 1986*, *86*, 763.

[405]For reviews of hydrogenation with metal oxides, see Minachev, Kh.M.; Khodakov, Yu.S.; Nakhshunov, V.S. *Russ. Chem. Rev. 1976*, *45*, 142; Kokes, R.J.; Dent, A.L. *Adv. Catal. 1972*, *22*, 1 (ZnO).

[406]For a monograph, see James, B.R. *Homogeneous Hydrogenation*, Wiley, NY, *1973*. For reviews, see Collman, J.P.; Hegedus, L.S.; Norton, J.R.; Finke, R.G. *Principles and Applications of Organotransition Metal Chemistry*, University Science Books, Mill Valley, CA *1987*, pp. 523–564; Birch, A.J.; Williamson, D.H. *Org. React. 1976*, *24*, 1; James, B.R. *Adv. Organomet. Chem. 1979*, *17*, 319; Harmon, R.E.; Gupta, S.K.; Brown, D.J. *Chem. Rev. 1973*, *73*, 21; Strohmeier, W. *Fortschr. Chem. Forsch. 1972*, *25*, 71; Heck, R.F. *Organotransition Metal Chemistry*, Academic Press, NY, *1974*, pp. 55–65; Rylander, P.N. *Organic Syntheses with Noble Metal Catalysts*, Academic Press, NY, *1973*, pp. 60–76; Lyons, J.E.; Rennick, L.E.; Burmeister, J.L. *Ind. Eng. Chem. Prod. Res. Dev. 1970*, *9*, 2; Vol'pin, M.E.; Kolomnikov, I.S. *Russ. Chem. Rev. 1969*, *38*, 273.

[407]Osborn, J.A.; Jardine, F.H.; Young, J.F.; Wilkinson, G. *J. Chem. Soc, A 1966*, 1711; Osborn, J.A.; Wilkinson, G. *Inorg. Synth., 1967*, *10*, 67; Biellmann, J.F. *Bull. Soc. Chim. Fr. 1968*, 3055; van Bekkum, H.; van Rantwijk, F.; van de Putte, T. *Tetrahedron Lett. 1969*, 1.

[408]For a review of Wilkinson's catalyst, see Jardine, F.H. *Prog. Inorg. Chem. 1981*, *28*, 63–202.

[409]Harmon, R.E.; Parsons, J.L.; Cooke, D.W.; Gupta, S.K.; Schoolenberg, J. *J. Org. Chem. 1969*, *34*, 3684. See also, Mohrig, J.R.; Dabora, S.L.; Foster, T.F.; Schultz, S.C. *J. Org. Chem. 1984*, *49*, 5179.

[410]Jardine, F.H.; Wilkinson, G. *J. Chem. Soc. C 1967*, 270.

[411]Hattori, K.; Sajiki, H.; Hirota, K. *Tetrahedron 2000*, *56*, 8433.

[412]Tang, W.; Liu, D.; Zhang, X *Org. Lett. 2003*, *5*, 205.

[413]Jia, X.; Guo, R.; Li, X.; Yao, X.; Chan, A.S.C. *Tetrahedron Lett. 2002*, *43*, 5541; Reetz, M.T.; Mehler, G.; Meiswinkel, A.; Sell, T. *Tetrahedron Lett. 2002*, *43*, 7941; Reetz, M.T.; Mehler, G. *Tetrahedron Lett. 2003*, *44*, 4593.

under which double bonds can be reduced selectively[414] (see Table 19.2). Controlling the solvent allows catalytic hydrogenation of an alkene in the presence of an aromatic nitro group.[415]

Among other homogeneous catalysts are chlorotris(triphenylphosphine)hydridoruthenium(II), $(Ph_3P)_3RuClH$,[416] which is specific for terminal double bonds (other double bonds are hydrogenated slowly or not at all), and pentacyanocobaltate(II), $Co(CN)_5{}^{3-}$, which is effective for double and triple bonds only when they are part of conjugated systems[417] (the conjugation may be with C=C, C=O, or an aromatic ring). Colloidal palladium has also been used as a catalyst,[418] and a polymer bound ruthenium catalyst has also been used.[419] A polymer incarcerated palladium catalyst gave the hydrogenated product in quantitative yields.[420] Rhodium on mesoporous silica can be used to hydrogenate alkenes.[421] A nanoparticulate palladium catalyst in an ionic liquid has been used for the hydrogenation of alkenes.[422]

Homogeneous catalysts often have the advantages of better catalyst reproducibility and better selectivity. They are also less susceptible to catalyst poisoning[423] (heterogeneous catalysts are usually poisoned by small amounts of sulfur, often found in rubber stoppers, or by sulfur-containing compounds, such as thiols and sulfides).[424] On the other hand, heterogeneous catalysts are usually easier to separate from the reaction mixture.

101   **102**

---

[414]For a discussion, see Rylander, P.N. *Catalytic Hydrogenation over Platinum Metals*, Academic Press, NY, *1967*, pp. 59–120. Also see, Hudlický, M. *Reductions in Organic Chemistry*, Ellis Horwood Ltd., Chichester, *1984*.

[415]Jourdant, A.; González-Zamora, E.; Zhu, J. *J. Org. Chem.* **2002**, *67*, 3163.

[416]Hallman, P.S.; McGarvey, B.R.; Wilkinson, G. *J. Chem. Soc. A* **1968**, 3143; Jardine, F.H.; McQuillin, F.J. *Tetrahedron Lett.* **1968**, 5189.

[417]Kwiatek, J.; Mador, I.L.; Seyler, J.K. *J. Am. Chem. Soc.* **1962**, *84*, 304; Jackman, L.M.; Hamilton, J.A.; Lawlor, J.M. *J. Am. Chem. Soc.* **1968**, *90*, 1914; Funabiki, T.; Matsumoto, M.; Tarama, K. *Bull. Chem. Soc. Jpn.* **1972**, *45*, 2723; Reger, D.L.; Habib, M.M.; Fauth, D.J. *Tetrahedron Lett.* **1979**, 115.

[418]Fowley, L.A.; Michos, D.; Luo, X.-L.; Crabtree, R.H. *Tetrahedron Lett.* **1993**, *34*, 3075.

[419]Taylor, R.A.; Santora, B.P.; Gagné, M.R. *Org. Lett.* **2000**, *2*, 1781.

[420]Okamoto, K.; Akiyama, R.; Kobayashi, S. *J. Org. Chem.* **2004**, *69*, 2871. See also, Bremeyer, N.; Ley, S.V.; Ramarao, C.; Shirley, I.M.; Smith, S.C. *Synlett* **2002**, 1843.

[421]Crudden, C.M.; Allen, D.; Mikoluk, M.D.; Sun, J. *Chem. Commun.* **2001**, 1154.

[422]In bmim PF_6, 1-butyl-3-methylimidazolium hexafluorophosphate: Huang, J.; Jiang, T.; Han, B.; Gao, H.; Chang, Y.; Zhao, G.; Wu, W. *Chem. Commun.* **2003**, 1654.

[423]Birch, A.J.; Walker, K.A.M. *Tetrahedron Lett.* **1967**, 1935.

[424]For a review of catalyst poisoning by sulfur, see Barbier, J.; Lamy-Pitara, E.; Marecot, P.; Boitiaux, J.P.; Cosyns, J.; Verna, F. *Adv. Catal.* **1990**, *37*, 279–318.

Unfunctionalized alkenes are hydrogenated with good diastereoselectivity and enantioselectivity using various metal catalysts and chiral ligands.[425] Soluble, chiral homogeneous catalysts are usually the best choice, especially for alkenes. The transition-metal catalyst (rhodium and ruthenium are probably the most common) is usually prepared with suitable chiral ligands prior to addition to the reaction, or an achiral catalyst, such as Wilkinson's catalyst, **100**: RhCl(Ph₃P)₃, is added along with a chiral ligand. The chiral ligand is typically a phosphine. In one case, the phosphorous may be chiral, as in **101** (called R-camp),[426] but pyramidal inversion at elevated temperatures (see p. 142) limits the utility of such ligands. The alliterative is to prepare a phosphine containing a chiral carbon, and bis(phosphines), such as **102** (called dipamp)[427] are the most common. There are many variations of chiral bis(phosphine) ligands. Mono-phosphine ligands have also been used.[428] Titanocenes[429] with chiral cyclopentadienyl ligands have given enantioselective hydrogenation of unfunctionalized alkenes, such as 2-phenyl-1-butene.[430] Chiral poisoning has been used as a strategy for asymmetric catalysis.[431]

Hydrogenations in most cases are carried out at room temperature and just above atmospheric pressure, but some double bonds are more resistant and require higher temperatures and pressures. The resistance is usually a function of increasing substitution and is presumably caused by steric factors. Trisubstituted double bonds require, say, 25°C and 100 atm, while tetrasubstituted double bonds may require 275°C and 1000 atm. Among

103

the double bonds most difficult to hydrogenate or which cannot be hydrogenated at all are those common to two rings, as in steroid **103**. Hydrogenations, even at about atmospheric pressure, are ordinarily performed in a special hydrogenator, but this is

[425]Zr: Troutman, M.V.; Appella, D.H.; Buchwald, S.L. *J. Am. Chem. Soc.* **1999**, *121*, 4916. Ir: Xu, G.; Gilbertson, S.R. *Tetrahedron Lett.* **2003**, *44*, 953; Tang, W.; Wang, W.; Zhang, X. *Angew. Chem. Int. Ed.* **2003**, *42*, 943; Cozzi, P.G.; Menges, F.; Kaiser, S. *Synlett 2003*, 833. Special ligands: Perry, M.C.; Cui, X.; Powell, M.T.; Hou, D.-R.; Reibenspies, J.H.; Burgess, K. *J. Am. Chem. Soc.* **2003**, *125*, 113.

[426]Knowles, W.S.; Sabacky, M.J.; Vineyard, B.D. *Adv. Chem. Ser 1974*, *132*, 274.

[427]Brown, J.M.; Chaloner, P.A. *J. Chem. Soc., Chem. Commun.* **1980**, 344; *1978*, 321; *Tetrahedron Lett.* *1978*, 1877; *J. Am. Chem. Soc.* **1980**, *102*, 3040.

[428]Huang, H.; Zheng, Z.; Luo, H.; Bai, C.; Hu, X.; Chen, H. *J. Org. Chem.* **2004**, *69*, 2355; Hua, Z.; Vassar, V.C.; Ojima, I. *Org. Lett.* **2003**, *5*, 3831. For a review, see Jerphagnon, T.; Renaud, J.-L.; Bruneau, C. *Tetrahedron Asymmetry 2004*, *15*, 2101.

[429]Burk, M.J.; Gross, M.F. *Tetrahedron Lett.* **1994**, *35*, 9363.

[430]Halterman, R.L.; Vollhardt, K.P.C.; Welker, M.E.; Bläser, D.; Boese, R. *J. Am. Chem. Soc.* **1987**, *109*, 8105; Lee, N.E.; Buchwald, S.L. *J. Am. Chem. Soc.* **1994**, *116*, 5985.

[431]Faller, J.W.; Parr, J. *J. Am. Chem. Soc.* **1993**, *115*, 804.

not always necessary. Both the catalyst and the hydrogen can be generated *in situ*, by treatment of H_2PtCl_6 or $RhCl_3$ with $NaBH_4$;[432] ordinary glassware can then be used. The great variety of catalysts available often allows an investigator to find one that is highly selective. For example, the catalyst Pd(salen) encapsulated in zeolites permitted the catalytic hydrogenation of 1-hexene in the presence of cyclohexene.[433] It has been shown that the pressure of the reaction can influence enantioselectivity in asymmetric catalytic hydrogenations.[434]

Triple bonds can be reduced, either by catalytic hydrogenation or by the other methods mentioned in the following two sections. The comparative reactivity of triple and double bonds depends on the catalyst. With most catalysts (e.g., Pd), triple bonds are hydrogenated more easily, and therefore it is possible to add just 1 equivalent of hydrogen and reduce a triple bond to a double bond (usually a stereoselective syn addition) or to reduce a triple bond without affecting a double bond present in the same molecule.[435] A particularly good catalyst for this purpose is the Lindlar catalyst (Pd-CaCO$_3$–PbO).[436] An alternative catalyst used for selective hydrogenation to cis-alkenes is palladium on barium sulfate (BaSO$_4$) catalyst, poisoned with quinoline[437] (sometimes called the *Rosenmund catalyst*). Palladium on calcium carbonate in polyethylene glycol (PEG) has also bee used as a recyclable catalyst system.[438] Hydrogenation using a palladium catalyst on pumice was shown to give the cis-alkene with excellent selectivity.[439] Hydrogenation of a C≡C unit occurs in the presence of other functional groups, including NR$_2$ including NH$_2$,[440] and sulfonyl.[441]

[432]Brown, C.A.; Sivasankaran, K. *J. Am. Chem. Soc. 1962, 84,* 2828; Brown, C.A.; Brown, H.C. *J. Am. Chem. Soc. 1962, 84,* 1494, 1945, 2829; *J. Org. Chem. 1966, 31,* 3989.

[433]Kowalak, S.; Weiss, R.C.; Balkus Jr., K.J. *J. Chem. Soc., Chem. Commun. 1991,* 57.

[434]Sun, Y.; Landau, R.N.; Wang, J.; LeBlond, C.; Blackmond, D.G. *J. Am. Chem. Soc. 1996, 118,* 1348.

[435]For reviews of the hydrogenation of alkynes, see Hutchins, R.O.; Hutchins, M.G.K., in Patai, S.; Rappoport, Z. *The Chemistry of Functional Groups, Supplement C* pt. 1, Wley, NY, *1983,* pp. 571–601; Marvell, E.N.; Li, T. *Synthesis 1973,* 457; Gutmann, H.; Lindlar, H., in Viehe, H.G. *Acetylenes,* Marcel Dekker, NY, *1969,* pp. 355–363.

[436]Lindlar, H.; Dubuis, R. *Org. Synth. V,* 880. See also, Rajaram, J.; Narula, A.P.S.; Chawla, H.P.S.; Dev, S. *Tetrahedron 1983, 39,* 2315; McEwen, A.B.; Guttieri, M.J.; Maier, W.F.; Laine, R.M.; Shvo, Y. *J. Org. Chem. 1983, 48,* 4436.

[437]Cram, D.J.; Allinger, N.L. *J. Am. Chem. Soc. 1956, 78,* 2518; Rosenmund, K.W. *Ber. 1918, 51,* 585; Mosettig, E.; Mozingo, R. *Org. React. 1948, 4,* 362.

[438]Chandrasekhar, S.; Narsihmulu, Ch.; Chandrashekar, G.; Shyamsunder, T. *Tetrahedron Lett. 2004, 45,* 2421.

[439]Gruttadauria, M.; Noto, R.; Deganello, G.; Liotta, L.F. *Tetrahedron Lett. 1999, 40,* 2857; Gruttadauria, M.; Liotta, L.F.; Noto, R.; Deganello, G. *Tetrahedron Lett. 2001, 42,* 2015.

[440]Campos, K.R.; Cai, D.; Journet, M.; Kowal, J.J.; Larsen, R.D.; Reider, P.J. *J. Org. Chem. 2001, 66,* 3634.

[441]Zhong, P.; Huang, X.; Ping-Guo, M. *Tetrahedron 2000, 56,* 8921.

Conjugated dienes can add hydrogen by 1,2- or 1,4-addition. Selective 1,4-addition can be achieved by hydrogenation in the presence of carbon monoxide, with bis(cyclopentadienyl)chromium as catalyst.[442] With allenes[443] catalytic hydrogenation usually reduces both double bonds.

Most catalytic reductions of double or triple bonds, whether heterogeneous or homogeneous, have been shown to be syn, with the hydrogens entering from the less-hindered side of the molecule.[444] Stereospecificity can be investigated only for tetrasubstituted alkenes (except when the reagent is D_2), which are the hardest to hydrogenate, but the results of these investigations show that the addition is usually 80–100% syn, although some of the anti addition product is normally also found and in some cases predominates. Catalytic hydrogenation of alkynes is nearly always stereoselective, giving the cis alkene (usually at least 80%), even when it is thermodynamically less stable. For example, **104** gave **105**, even although the steric hindrance is such that a planar molecule is impossible.[445] This is thus a useful method for preparing cis alkenes.[446] However, when

104   **105**

steric hindrance is too great, the trans alkene may be formed. One factor that complicates the study of the stereochemistry of heterogeneous catalytic hydrogenation is that exchange of hydrogens takes place, as can be shown by hydrogenation with deuterium.[447] Thus deuterogenation of ethylene produced all the possible deuterated ethylenes and ethanes (even C_2H_6), as well as HD.[448] With 2-butene, it was found that double-bond migration, cis–trans isomerization, and even exchange of hydrogen with groups not on the double bond could occur; for example, $C_4H_2D_8$ and C_4HD_9 were detected on treatment of cis-2-butene with deuterium and a catalyst.[449] Indeed, alkanes have been found to exchange with deuterium over a catalyst,[450] and even without deuterium, for example, $CH_4 + CD_4 \rightarrow CHD_3 + CH_3D$

[442]Miyake, A.; Kondo, H. *Angew. Chem. Int. Ed.* **1968**, *7*, 631. For other methods, with references, see Larock, R.C. *Comprehensive Organic Transformations*, 2nd ed., Wiley-VCH, NY, **1999**, pp. 403–404.

[443]For a review, see Schuster, H.F.; Coppola, G.M. *Allenes in Organoic Synthesis* Wiley, NY, **1984**, pp. 57–61.

[444]For a review of homogeneous hydrogenation directed to only one face of a substrate molecule, see Brown, J.M. *Angew. Chem. Int. Ed.* **1987**, *26*, 190.

[445]Holme, D.; Jones, E.R.H.; Whiting, M.C. *Chem. Ind. (London)* **1956**, 928.

[446]For a catalyst that leads to trans alkenes, see Burch, R.R.; Muetterties, E.L.; Teller, R.G.; Williams, J.M. *J. Am. Chem. Soc.* **1982**, *104*, 4257.

[447]For a review of the use of deuterium to study the mechanism of heterogeneous organic catalysis see Gudkov, B.S. *Russ. Chem. Rev.* **1986**, *55*, 259.

[448]Turkevich, J.; Schissler, D.O.; Irsa, P. *J. Phys. Chem.* **1951**, *55*, 1078.

[449]Wilson, J.N.; Otvos, J.W.; Stevenson, D.P.; Wagner, C.D. *Ind. Eng. Chem.* **1953**, *45*, 1480.

[450]For a review, see Gudkov, B.S.; Balandin, A.A. *Russ. Chem. Rev.* **1966**, *35*, 756. For an example of intramolecular exchange, see Lebrilla, C.B.; Maier, W.F. *Tetrahedron Lett.* **1983**, *24*, 1119. See also, Poretti, M.; Gäumann, T. *Helv. Chim. Acta* **1985**, *68*, 1160.

D in the gas phase, with a catalyst. All this makes it difficult to investigate the stereochemistry of heterogeneous catalytic hydrogenation.

The mechanism of the heterogeneous catalytic hydrogenation of double bonds is not thoroughly understood because it is a very difficult reaction to study.[451] Because the reaction is heterogeneous, kinetic data, although easy to obtain (measurement of decreasing hydrogen pressure), are difficult to interpret. Furthermore, there are the difficulties caused by the aforementioned hydrogen exchange. The currently accepted mechanism for the common two-phase reaction was originally proposed in 1934.[452] According to this, the alkene is adsorbed onto the surface of the metal, although the nature of the actual bonding is unknown,[453] despite many attempts to elucidate it.[454] In the 1934 work, the metallic site was indicated by an asterisk, but here we use $\bigcirc$. For steric reasons it is apparent that adsorption of the alkene takes place with its less-hindered side attached to the catalyst surface, probably as an η^2 complex (see p. 116). The fact that addition of hydrogen is generally also from the less-hindered side indicates that the hydrogen too is probably adsorbed on the catalyst surface before it reacts with the alkene. It is likely that as the H_2 molecule is adsorbed on (coordinated to) the metal catalyst, cleavage occurs to give η^1- coordinated hydrogen atoms (see p. $$$). Note that this model suggest a single metal particle for coordination of the alkene and the hydrogen atoms, but the hydrogen atoms and the alkene could be coordinated to different metal particles. It has been shown that platinum catalyzes homolytic cleavage of hydrogen molecules.[455] In the second step, one of the adsorbed (η^1-coordinated) hydrogen atoms becomes attached to a carbon atom, creating in effect, an alkyl radical (which is still bound to the catalyst although only by one bond, probably η^1-coordination). Transfer of a hydrogen atom to carbon opens a site on the metal catalyst for coordination to additional hydrogen atoms. Finally, another hydrogen atom (not necessarily the one originally connected to the first hydrogen) combines with the radical

[451]For reviews, see Webb, G., in Bamford, CH.; Tipper, C.F.H. *Comprehensive Chemical Kinetics*, Vol. 20, Elsevier, NY, *1978*, pp. 1–121; Clarke, J.K.A.; Rooney, J.J. *Adv. Catal.* *1976*, *25*, 125–183; Siegel, S. *Adv. Catal.* *1966*, *16*, 123–177; Burwell, Jr., R.L. *Chem. Eng. News 1966, 44(34)*, 56–67.

[452]Horiuti, I.; Polanyi, M. *Trans. Faraday Soc. 1934, 30*, 1164.

[453]See, for example, Burwell, Jr., R.L.; Schrage, K. *J. Am. Chem. Soc. 1965*, *87*, 5234.

[454]See, for example, McKee, D.W. *J. Am. Chem. Soc. 1962*, *84*, 1109; Ledoux, M.J. *Nouv. J. Chim. 1978*, *2*, 9; Bautista, F.M.; Campelo, J.M.; Garcia, A.; Guardeño, R.; Luna, D.; Marinas, J.M. *J. Chem. Soc. Perkin Trans. 2*, *1989*, 493.

[455]Krasna, A.I. *J. Am. Chem. Soc. 1961*, *83*, 289.

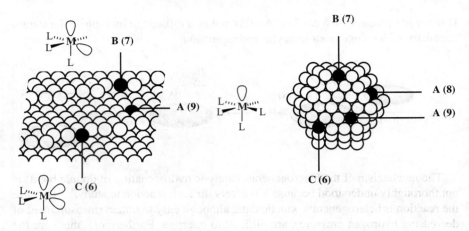

Fig. 15.1. The principal surface and particle sites for heterogeneous catalysts.

to give the reaction product, freed from the catalyst surface, and the metal catalyst that is now available for coordination of additional hydrogen atoms and/or alkenes. All the various side reactions, including hydrogen exchange and isomerism, can be explained by this type of process.[456] Although this mechanism is satisfactory as far as it goes,[457] there are still questions it does not answer, among them questions[458] involving the nature of the asterisk, the nature of the bonding, and the differences caused by the differing nature of each catalyst.[459]

Heterogeneous catalysis occurs at the surface of the metal catalyst, and there are different types of metal particles on the surface. Maier suggested the presence of **terrace-**, **step-**, and **kink**-type atoms (in Fig. 6.1)[460] on the surface of a heterogeneous catalyst. These terms refer to different atom types, characterized by the number of nearest neighbors,[460] which correspond to different transition-metal fragments, as well as to different coordination states of that metal.[461] A terrace-type atom (A in Fig. 15.1) typically has eight or nine neighbors and corresponds to a geometry shown for the ML_5 particle. The step type of atom (B) usually has seven neighbors and can be correlated with the geometry shown for the ML_4

[456]Smith, G.V.; Burwell Jr., R.L. *J. Am. Chem. Soc.* **1962**, *84*, 925.

[457]A different mechanism has been proposed by Zaera, F.; Somorjai, G.A. *J. Am. Chem. Soc.* **1984**, *106*, 2288, but there is evidence against it: Beebe, Jr., T.P.; Yates Jr., J.T. *J. Am. Chem. Soc.* **1986**, *108*, 663. See also, Thomson, S.J.; Webb, G. *J. Chem. Soc., Chem. Commun.* **1976**, 526.

[458]For discussions, see Augustine, R.L.; Yaghmaie, F.; Van Peppen, J.F. *J. Org. Chem.* **1984**, *49*, 1865; Maier, W.F. *Angew. Chem. Int. Ed.* **1989**, *28*, 135.

[459]For a study of the detailed structure of Lindlar catalysts (which were shown to consist of seven distinct chemical phases), see Schlögl, R.; Noack, K.; Zbinden, H.; Reller, A. *Helv. Chim. Acta* **1987**, *70*, 627.

[460]Maier, W.F. *Angew. Chem. Int. Ed.* **1989**, *28*, 135.

[461]Maier, W.F., in Rylander, P.N.; Greenfield, H.; Augustine, R.L. *Catalysis of Organic Reactions*, Marcel Dekker, NY, *1988*, pp. 211–231, Cf. p. 220.

particle. Finally, the kink-type atom (C) has six neighbors and corresponds to geometry shown for the ML_3 particle. In general, as the particle size increases, the relative concentration of terrace atoms will increase, whereas small particle size favors the kink type of surface atoms.

The mechanism of homogeneous hydrogenation[462] catalyzed by $RhCl(Ph_3P)_3$ (**100**, Wilkinson's catalyst)[463] involves reaction of the catalyst with hydrogen to form a metal hydride $(PPh_3)_2RhH_2Cl$ (**106**).[464] Replacement of a triphenylphosphine ligand with two toms of hydrogen constitutes an oxidative addition.

oxidative addition

Ph₃P⸴⸴Rh⸴⸴PPh₃ Cl   **100**   H₂ / −PPh₃ →   Ph₃P⸴⸴ Rh−H Ph₃P Cl   **106**   ⇌   Ph₃P⸴⸴ Rh⸴⸴ H Cl   Cl   **107**   ← H₂ / −PPh₃   Ph₃P⸴⸴Rh⸴⸴PPh₃ Cl   **100**

H₂ oxidative addition ↗

reductive elimination

insertion ↘

Ph₃P⸴⸴ Rh—Cl Ph₃P   **108**   ⇌   Ph₃P⸴⸴ Rh—C Ph₃P Cl   **109**

After coordination of the alkene to form **107**, transfer of hydrogen to carbon is an insertion process, presumably generating **109**, and a second insertion liberates the hydrogenated compound, and rhodium species **108**, which adds hydrogen by oxidative addition to give **106**. Alternatively, replacement of triphenylphosphine can lead to **107**, with two hydrogen atoms and a η^2-alkene complex. If a mixture of H_2 and D_2 is used, the product contains only dideuterated and non-deuterated compounds; no mono-deuterated products are found, indicating that (unlike the case of heterogeneous catalysis) H_2 or D_2 has been added to one alkene molecule and that no exchange takes place.[330] Although conversion of **107** to the products takes place in two steps,[465] the addition of H_2 to the double bond is syn, although bond rotation in **109** can lead to stereochemical mixtures.

The occurrence of hydrogen exchange and double-bond migration in heterogeneous catalytic hydrogenation means that the hydrogenation does not necessarily

[462]For reviews, see Crabtree, R.H. *Organometallic Chemistry of the Transition Metals*, Wiley, NY, *1988*, pp. 190–200; Jardine, F.H. in Hartley, F.R. *The Chemistry of the Metal-Carbon Bond*, Vol. 4, Wiley, NY, *1987*, pp. 1049–1071.

[463]Montelatici, S.; van der Ent, A.; Osborn, J.A.; Wilkinson, G. *J. Chem. Soc. A* **1968**, 1054; Wink, D.; Ford, P.C. *J. Am. Chem. Soc.* **1985**, *107*, 1794; Koga, N.; Daniel, C.; Han, J.; Fu, X.Y.; Morokuma, K. *J. Am. Chem. Soc.* **1987**, *109*, 3455.

[464]Tolman, C.A.; Meakin, P.Z.; Lindner, D.L.; Jesson, J.P. *J. Am. Chem. Soc.* **1976**, *96*, 2762.

[465]Biellmann, J.F.; Jung, M.J. *J. Am. Chem. Soc.* **1968**, *90*, 1673; Hussey, A.S.; Takeuchi, Y. *J. Am. Chem. Soc.* **1969**, *91*, 672; Heathcock, C.H.; Poulter, S.R. *Tetrahedron Lett.* **1969**, 2755; Smith, G.V.; Shuford, R.J. *Tetrahedron Lett.* **1970**, 525; Atkinson, J.G.; Luke, M.O. *Can. J. Chem.* **1970**, *48*, 3580.

take place by straightforward addition of two hydrogen atoms at the site of the original double bond. Consequently, this method is not synthetically useful for adding D_2 to a double or triple bond in a regioselective or stereospecific manner. However, this objective can be achieved (with syn addition) by a homogeneous catalytic hydrogenation, which usually adds D_2 without scrambling[466] or by the use of one of the diimide methods (**15-12**). Deuterium can also be regioselectively added by the hydroboration–reduction procedure previously mentioned.

Reductions of double and triple bonds are found at OS **I**, 101, 311; **II**, 191, 491; **III**, 385, 794; **IV**, 298, 304, 408; **V**, 16, 96, 277; **VI**, 68, 459; **VII**, 226, 287; **VIII**, 420. 609; **IX**, 169, 533.

Catalysts and apparatus for hydrogenation are found at OS **I**, 61, 463; **II**, 142; **III**, 176, 181, 685; **V**, 880; **VI**, 1007.

15-12 Other Reductions of Double and Triple Bonds

$$\begin{array}{c}\diagdown\diagup\\C=C\\\diagup\diagdown\end{array} \longrightarrow \begin{array}{c}H\quad H\\\diagdown|\quad|\diagup\\-C-C-\\\diagup|\quad|\diagdown\end{array}$$

Although catalytic hydrogenation is the method most often used, double or triple bonds can be reduced by other reagents, as well. Among these are sodium in ethanol, sodium and *tert*-butyl alcohol in HMPA,[467] lithium and aliphatic amines[468] (see also, **15-13**), zinc and acids, sodium hypophosphate and Pd–C,[469] $(EtO)_3$-$SiHPd(OAc)_2$,[470] triethylsilane Et_3SiH and trifluoroacetic acid[471] or palladium chloride,[472] and hydroxylamine and ethyl acetate.[473] Trialkylsilanes (R_3SiH) in conjunction with an acid will reduce double bonds.[474] Siloxanes (RO_3SiH) and a ruthenium catalyst, followed by treatment with AgF convert alkynes to *trans*-alkenes.[475] Poly(methylhydrosiloxane) was used for reduction of conjugated alkenes using a copper carbene complex.[476] Reduction of alkynes with silanes and a ruthenium catalyst, followed by treatment with CuI and Bu_4NF gave the

[466]Biellmann, J.F.; Liesenfelt, H. *Bull. Soc. Chim. Fr.* **1966**, 4029; Birch, A.J.; Walker, K.A.M. *Tetrahedron Lett.* **1966**, 4939, *J. Chem. Soc. C* **1966**, 1894; Morandi, J.R.; Jensen, H.B. *J. Org. Chem.* **1969**, *34*, 1889. See, however, Atkinson, J.G.; Luke, M.O. *Can. J. Chem.* **1970**, *48*, 3580.

[467]Angibeaud, P.; Larchevêque, M.; Normant, H.; Tchoubar, B. *Bull. Soc. Chim. Fr.* **1968**, 595; Whitesides, G.M.; Ehmann, W.J. *J. Org. Chem.* **1970**, *35*, 3565.

[468]Benkeser, R.A.; Schroll, G.; Sauve, D.M. *J. Am. Chem. Soc.* **1955**, *77*, 3378.

[469]Sala, R.; Doria, G.; Passarotti, C. *Tetrahedron Lett.* **1984**, *25*, 4565.

[470]Tour, J.M.; Pendalwar, S.L. *Tetrahedron Lett.* **1990**, *31*, 4719.

[471]For a review, see Kursanov, D.N.; Parnes, Z.N.; Loim, N.M. *Synthesis* **1974**, 633. Also see, Doyle, M.P.; McOsker, C.C. *J. Org. Chem.* **1978**, *43*, 693. For a monograph, see Kursanov, D.N.; Parnes, Z.N.; Kalinkin, M.I.; Loim, N.M. *Ionic Hydrogenation and Related Reactions*, Harwood Academic Publishers, Chur, Switzerland, **1985**.

[472]Mirza-Aghayan, M.; Boukherroub, R.; Bolourtchian, M.; Hosseini, M. *Tetrahedron Lett.* **2003**, *44*, 4579.

[473]Wade, P.A.; Amin, N.V. *Synth. Commun.* **1982**, *12*, 287.

[474]Masuno, M.N.; Molinski, T.F. *Tetrahedron Lett.* **2001**, *42*, 8263.

[475]Fürstner, A.; Radkowski, K. *Chem. Commun.* **2002**, 2182.

[476]Jurkauskas, V.; Sakighi, J.P.; Buchwald, S.L. *Org. Lett.* **2003**, *5*, 2417.

trans- alkene.[477] Samarium iodide in water and a triamine additive led to reduction of alkenes.[478] Similar reduction was reported using $Co_2(CO)_8$ and an excess of water in dimethoxyethane.[479] Reduction of an alkyne to an alkene can be done via an organometallic, by heating the alkyne with indium metal in aqueous ethanol.[480] Alkynes are reduced with palladium acetate and sodium ethoxide. In methanol the product is the alkane, whereas in THF the product is the cis-alkene.[481]

In the above-mentioned reactions with hydrazine and hydroxylamine, the actual reducing species is diimide $NH=NH$, which is formed from N_2H_4 by the oxidizing agent and from NH_2OH by the ethyl acetate.[482] The rate of this reaction has been studied.[483] Although both the syn and anti forms of diimide are produced, only the syn form reduces the double bond,[484] at least in part by a cyclic mechanism:[485]

The addition is therefore stereospecifically syn[486] and, like catalytic hydrogenation, generally takes place from the less-hindered side of a double bond, although not much discrimination in this respect is observed where the difference in bulk effects is small.[487] Diimide reductions are most successful with symmetrical multiple bonds ($C=C$, $C≡C$, $N=N$) and are not useful for those inherently polar ($C≡N$, $C=N$, $C=O$, etc.). Diimide is not stable enough for isolation at ordinary temperatures, although it has been prepared[488] as a yellow solid at $-196°C$. N-Arylsulfonylhydrazines bearing a phosphonate ester unit converted 1,1-diiodoalkenes ($C=CI_2$) to gem-diiodides, ($CH—CHI_2$).[489]

An indirect method[490] of double-bond reduction involves hydrolysis of boranes (prepared by **15-16**). Trialkylboranes can be hydrolyzed by refluxing with carboxylic

[477]Trost, B.M.; Ball, Z.T.; Jöge, T. *J. Am. Chem. Soc.* **2002**, *124*, 7922.

[478]Dahlén, A.; Hilmersson, G. *Tetrahedron Lett.* **2003**, *44*, 2661.

[479]Lee, H.-Y.; An, M. *Tetrahedron Lett.* **2003**, *44*, 2775.

[480]Ranu, B.C.; Dutta, J.; Guchhait, S.K. *J. Org. Chem.* **2001**, *66*, 5624.

[481]Wei, L.-L.; Wei, L.-M.; Pan, W.-B.; Leou, S.-P.; Wu, M.-J. *Tetrahedron Lett.* **2003**, *44*, 1979.

[482]For reviews of hydrogenations with diimide, see Pasto, D.J.; Taylor, R.T. *Org. React.* **1991**, *40*, 91; Miller, C.E. *J. Chem. Educ.* **1965**, *42*, 254; House, H.O. *Modern Synthetic Reaction*, 2nd ed., W.A. Benjamin, NY, *1972*, pp. 248–256. For reviews of diimides, see Back, R.A. *Rev. Chem. Intermed.* **1984**, *5*, 293; Hünig, S.; Müller, H.R.; Thier, W. *Angew. Chem. Int. Ed.* **1965**, *4*, 271.

[483]Nelson, D.J.; Henley, R.L.; Yao, Z.; Smith, T.D. *Tetrahedron Lett.* **1993**, *34*, 5835.

[484]Aylward, F.; Sawistowska, M.H. *J. Chem. Soc.* **1964**, 1435.

[485]van Tamelen, E.E.; Dewey, R.S.; Lease, M.F.; Pirkle, W.H. *J. Am. Chem. Soc.* **1961**, *83*, 4302; Willis, C.; Back, R.A.; Parsons, J.A.; Purdon, J.G. *J. Am. Chem. Soc.* **1977**, *99*, 4451.

[486]Corey, E.J.; Pasto, D.J.; Mock, W.L. *J. Am. Chem. Soc.* **1961**, *83*, 2957.

[487]van Tamelen, E.E.; Timmons, R.J. *J. Am. Chem. Soc.* **1962**, *84*, 1067.

[488]Wiberg, N.; Fischer, G.; Bachhuber, H. *Chem. Ber.* **1974**, *107*, 1456; *Angew. Chem. Int. Ed.* **1977**, *16*, 780. See also, Trombetti, A. *Can. J. Phys.* **1968**, *46*, 1005; Bondybey, V.E.; Nibler, J.W. *J. Chem. Phys.* **1973**, *58*, 2125; Craig, N.C.; Kliewer, M.A.; Shih, N.C. *J. Am. Chem. Soc.* **1979**, *101*, 2480.

[489]Cloarec, J.-M.; Charette, A.B. *Org. Lett.* **2004**, *6*, 4731.

[490]For a review, see Zweifel, G. *Intra-Sci. Chem. Rep.* **1973**, *7(2)*, 181–189.

acids,[491] while monoalkylboranes, RBH$_2$, can be hydrolyzed with base.[492] Triple bonds can be similarly reduced, to cis alkenes.[493] Further reduction is also possible. When an alkyne was treated with decaborane and Pd/C in methanol, two equivalents of hydrogen are transferred to give the alkane.[494] Hydrogenation with Ni$_2$B on borohydride exchange resin (BER) has also been used.[495] Reduction occurs *in situ* when an alkene is treated with NaBH$_4$, NiCl$_2$•6 H$_2$O with moist alumina.[496] Reduction of alkenes occurs with *tert*-butylamine•borane complex in methanol with 10% Pd/C.[497]

Metallic hydrides, such as lithium aluminum hydride and sodium borohydride, do not in general reduce carbon–carbon double bonds, although this can be done in special cases where the double bond is polar, as in 1,1-diarylethenes[498] and in enamines.[499] Lithium aluminum hydride reduces cyclopropenes with a pendant alcohol in the allylic position to the corresponding cyclopropane.[500]

Triple bonds can also be selectively reduced to double bonds with diisobutylaluminum hydride (Dibal-H),[501] with activated zinc (see **12-38**),[502] with hydrogen and Bi$_2$B–borohydride exchange resin,[503] or (internal triple bonds only) with alkali metals (Na, Li) in liquid ammonia or a low-molecular-weight amine.[504] Terminal alkynes are not reduced by the Na–NH$_3$ procedure because they are converted to acetylide ions under these conditions. However, terminal triple bonds can be reduced to double bonds by the addition to the Na–NH$_3$ solution of (NH$_4$)$_2$SO$_4$, which liberates the free ethynyl group.[505] The reaction of a terminal alkyne with

[491]Brown, H.C.; Murray, K.J. *Tetrahedron* **1986**, *42*, 5497; Kabalka, G.W.; Newton, Jr., R.J.; Jacobus, J. *J. Org. Chem.* **1979**, *44*, 4185.

[492]Weinheimer, A.J.; Marisco, W.E. *J. Org. Chem.* **1962**, *27*, 1926.

[493]Brown, H.C.; Zweifel, G. *J. Am. Chem. Soc.* **1959**, *81*, 1512.

[494]Lee, S.H.; Park, Y.J.; Yoon, C.M. *Tetrahedron Lett.* **2000**, *41*, 887.

[495]Choi, J.; Yoon, N.M. *Synthesis* **1996**, 597.

[496]Yakabe, S.; Hirano, M.; Morimoto, T. *Tetrahedron Lett.* **2000**, *41*, 6795.

[497]Couturier, M.; Andresen, B.M.; Tucker, J.L.; Dubé, P.; Brenek, S.J.; Negri, J.J. *Tetrahedron Lett.* **2001**, *42*, 2763.

[498]See Granoth, I.; Segall, Y.; Leader, H.; Alkabets, R. *J. Org. Chem.* **1976**, *41*, 3682.

[499]For a review of the reduction of enamines and indoles with NaBH$_4$ and a carboxylic acid, see Gribble, G.W.; Nutaitis, C.F. *Org. Prep. Proced. Int.* **1985**, *17*, 317. Enamines can also be reduced by formic acid; see Nilsson, A.; Carlson, R. *Acta Chem. Scand. Sect. B* **1985**, *39*, 187.

[500]Zohar, E.; Marek, I. *Org. Lett.* **2004**, *6*, 341.

[501]Wilke, G.; Müller, H. *Chem. Ber.* **1956**, *89*, 444, *Liebigs Ann. Chem.* **1960**, *629*, 224; Gensler, W.J.; Bruno, J.J. *J. Org. Chem.* **1963**, *28*, 1254; Eisch, J.J.; Kaska, W.C. *J. Am. Chem. Soc.* **1966**, *88*, 2213. For a catalyst with even better selectivity for triple bonds, see Ulan, J.G.; Maier, W.F.; Smith, D.A. *J. Org. Chem.* **1987**, *52*, 3132.

[502]Aerssens, M.H.P.J.; van der Heiden, R.; Heus, M.; Brandsma, L. *Synth. Commun.* **1990**, *20*, 3421; Chou, W.; Clark, D.L.; White, J.B. *Tetrahedron Lett.* **1991**, *32*, 299. See Sakai, M.; Takai, Y.; Mochizuki, H.; Sasaki, K.; Sakakibara, Y. *Bull. Chem. Soc. Jpn.* **1994**, *67*, 1984 for reduction with a NiBr$_2$–Zn reagent.

[503]Choi, J.; Yoon, N.M. *Tetrahedron Lett.* **1996**, *37*, 1057.

[504]For a list of methods of reducing triple to double bonds, with syn or anti addition, see Larock, R.C. *Comprehensive Organic Transformations*, 2nd ed., Wiley-VCH, NY, **1999**, pp. 405–410.

[505]Henne, A.L.; Greenlee, K.W. *J. Am. Chem. Soc.* **1943**, *65*, 2020.

lithium naphthalenide and $NiCl_2$ effectively reduced the alkyne unit (i.e., $PhC{\equiv}CH \rightarrow PhCH_2CH_3$).[506] This reagent is also effect for the reduction of simple alkenes.[507] A mixture of $NaBH_4$ and $BiCl_3$ also reduced certain alkenes[508] and An alkyne unit was reduced to an alkene, in the presence of a phenylthio group elsewhere in the molecule, using $Cp_2Zr(H)Cl$.[509]

Reduction of just one double bond of an allene, to give an alkene, has been accomplished by treatment with $Na–NH_3$[510] or with Dibal-H,[511] and by hydrogenation with $RhCl(PPh_3)_3$ as catalyst.[512]

When double bonds are reduced by lithium in ammonia or amines, the mechanism is similar to that of the Birch reduction (**15-13**).[513] The reduction with trifluoroacetic acid and Et_3SiH has an ionic mechanism, with H^+ coming in from the acid and H^- from the silane.[290] In accord with this mechanism, the reaction can be applied only to those alkenes, which when protonated can form a tertiary carbocation or one stabilized in some other way, for example, by a OR substitution.[514] It has been shown, by the detection of CIDNP, that reduction of α-methylstyrene by hydridopentacarbonylmanganese(I), $HMn(CO)_5$, involves free-radical addition.[515]

Catalytic hydrogenation of triple bonds and the reaction with Dibal-H usually give the cis-alkene (**15-11**). Most of the other methods of triple-bond reduction lead to the more thermodynamically stable trans alkene. However, this is not the case with the method involving hydrolysis of boranes or with the reductions with activated zinc, hydrazine, or NH_2OSO_3H, which also give the cis products.

The fact that ordinary double bonds are inert toward metallic hydrides is quite useful, since it permits reduction of, say, a carbonyl or nitro group, without disturbing a double bond in the same molecule (see Chapter 19 for a discussion of selectivity in reduction reactions). Sodium in liquid ammonia also does not reduce ordinary double bonds,[516] although it does reduce alkynes, allenes, conjugated dienes,[517] and aromatic rings (**15-13**).

[506]Alonso, F.; Yus, M. *Tetrahedron Lett.* **1997**, *38*, 149.

[507]Alonso, F.; Yus, M. *Tetrahedron Lett.* **1996**, *37*, 6925.

[508]Ren, P.-D.; Pan, S.-F.; Dong, T.-W.; Wu, S.-H. *Synth. Commun.* **1996**, *26*, 763.

[509]Lipshutz, B.H.; Lindsley, C.; Bhandari, A. *Tetrahedron Lett.* **1994**, *35*, 4669.

[510]Gardner, P.D.; Narayana, M. *J. Org. Chem.* **1961**, *26*, 3518; Vaidyanathaswamy, R.; Joshi, G.C.; Devaprabhakara, D. *Tetrahedron Lett.* **1971**, 2075.

[511]Montury, M.; Goré, J. *Tetrahedron Lett.* **1980**, *21*, 51.

[512]Bhagwat, M.M.; Devaprabhakara, D. *Tetrahedron Lett.* **1972**, 1391.

[513]For a review of the steric course of this reaction, see Toromanoff, E. *Bull. Soc. Chim. Fr.* **1987**, 893–901. For a review of this reaction as applied to α,β-unsaturated ketones, see Russell, G.A., in Patai, S.; Rappoport, Z. *The Chemistry of Enones*, pt. 2, Wiley, NY, **1989**, pp. 471–512.

[514]Parnes, Z.N.; Bolestova, G.I.; Kursanov, D.N. *Bull. Acad. Sci. USSR Div. Chem. Sci.* **1972**, *21*, 1927.

[515]Sweany, R.L.; Halpern, J. *J. Am. Chem. Soc.* **1977**, *99*, 8335. See also, Thomas, M.J.; Shackleton, T.A.; Wright, S.C.; Gillis, D.J.; Colpa, J.P.; Baird, M.C. *J. Chem. Soc., Chem. Commun.* **1986**, 312; Garst, J.F.; Bockman, T.M.; Batlaw, R. *J. Am. Chem. Soc.* **1986**, *108*, 1689; Bullock, R.M.; Samsel, E.G. *J. Am. Chem. Soc.* **1987**, *109*, 6542.

[516]There are some exceptions. See, for example, Butler, D.N. *Synth. Commun.* **1977**, *7*, 441, and references cited therein.

[517]For a review of reductions of α,β-unsaturated carbonyl compounds with metals in liquid NH_3, see Caine, D. *Org. React.* **1976**, *23*, 1–258.

Another hydrogenation method is called *transfer hydrogenation*.[518] In this meth-od the hydrogen comes from another organic molecule, which is itself oxidized. A transition-metal catalyst, heterogeneous or homogeneous, is frequently employed. Dendritic catalysts have been used for asymmetric transfer hydrogenation.[519] A common reducing agent is cyclohexene, which, when a palladium catalyst is used, is oxidized to benzene, losing 2 mol of hydrogen.

Enantioselective reduction of certain alkenes has also been achieved by reducing with baker's yeast.[520]

Reductions of double and triple bonds are found at OS **III**, 586, 742; **IV**, 136, 302, 887; **V**, 281, 993; **VII**, 524; **80**, 120.

15-13   Hydrogenation of Aromatic Rings

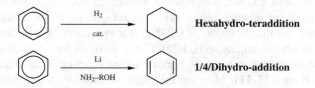

Aromatic rings can be reduced by catalytic hydrogenation,[521] but higher tem-peratures (100–200°C) are required than for ordinary double bonds.[522] although the reaction is usually carried out with heterogeneous catalysts, homogeneous catalysts have also been used; conditions are much milder with these.[523] Mild con-ditions are also successful in hydrogenations with phase transfer catalysts.[524] Hydrogenation in ionic liquids is known,[525] and also hydrogenation in supercritical ethane containing water.[526] Many functional groups, such as OH, O⁻, COOH, COOR, NH_2, do not interfere with the reaction, but some groups may be prefer-entially reduced. Among these are CH_2OH groups, which undergo hydrogenolysis to CH_3 (**19-54**). Phenols may be reduced to cyclohexanones, presumably through the enol. Heterocyclic compounds are often reduced. Thus furan gives THF. The

[518]For reviews, see Johnstone, R.A.W.; Wilby, A.H.; Entwistle, I.D. *Chem. Rev.* **1985**, *85*, 129; Brieger, G.; Nestrick, T.J. *Chem. Rev.* **1974**, *74*, 567.

[519]Chen, Y.-C.; Wu, T.-F.; Deng, J.-G.; Liu, H.; Cui, X.; Zhu, J.; Kiang, Y.-Z.; Choi, M.C.K.; Chan, A.S.C. *J. Org. Chem.* **2002**, *67*, 5301.

[520]See, for example, Ferraboschi, P.; Reza-Elahi, S.; Verza, E.; Santaniello, E. *Tetrahedron Asymmetry* **1999**, *10*, 2639. For reviews of baker's yeast, see Csuk, R.; Glänzer, B.I. *Chem. Rev.* **1991**, *91*, 49; Servi, S. *Synthesis* **1990**, 1.

[521]For reviews, see Karakhanov, E.A.; Dedov, A.G.; Loktev, A.S. *Russ. Chem. Rev.* **1985**, *54*, 171.

[522]For a highly active heterogeneous Rh catalyst, see Timmer, K.; Thewissen, D.H.M.W.; Meinema, H.A.; Bulten, E.J. *Recl. Trav. Chim. Pays-Bas* **1990**, *109*, 87.

[523]For reviews, see Bennett, M. *CHEMTECH* **1980**, *10*, 444–446; Muetterties, E.L.; Bleeke, J.R. *Acc. Chem. Res.* **1979**, *12*, 324. See also, Tsukinoki, T.; Kanda, T.; Liu, G.-B.; Tsuzuki, H.; Tashiro, M. *Tetrahedron Lett.* **2000**, *41*, 5865.

[524]Januszkiewicz, K.R.; Alper, H. *Organometallics* **1983**, *2*, 1055.

[525]In bmim BF₄, 1-butyl-3-methylimidazolium tetrafluoroborate: Dyson, P.J.; Ellis, D.J.; Parker, D.G.; Welton, T. *Chem. Commun.* **1999**, 25.

[526]Bonilla, R.J.; James, B.R.; Jessop, P.G. *Chem. Commun.* **2000**, 941.

nitrogen-containing ring of quinolines is reduced by hydrogenation using iodine and an iridium catalyst.[527] Catalytic hydrogenation of the five-membered ring in indole derivatives using a chiral rhodium catalyst gave hydroindoles with excellent enantioselectivity.[528]

With benzene rings it is usually impossible to stop the reaction after only one or two bonds have been reduced, since alkenes are more easily reduced than aromatic rings.[529] Thus, 1 equivalent of benzene, treated with 1 equivalent of hydrogen, gives no cyclohexadiene or cyclohexene, but $\frac{1}{3}$ equivalent of cyclohexane and $\frac{2}{3}$ equivalent of recovered benzene. This is not true for all aromatic systems. With anthracene, for example, it is easy to stop after only the 9,10-bond has been reduced (see p. 59). Hydrogenation of phenol derivatives can lead to conjugated cyclohexenones.[530] Hydrogenation of toluene in an ionic liquid using a ruthenium catalyst gave methyl-cyclohexane.[531]

When aromatic rings are reduced by lithium (or potassium or sodium) in liquid ammonia (such reductions are known as *dissolving metal reductions*), usually in the presence of an alcohol (often ethyl, isopropyl, or *tert*-butyl alcohol), 1,4-addition of hydrogen takes place and nonconjugated cyclohexadienes are produced.[532] This reaction is called the *Birch reduction.*[533] Heterocycles, such as pyrroles,[534] furans,[535] pyridines,[536] and indolones,[537] can be reduced using Birch reduction. Ammonia obtained commercially often has iron salts as impurities that lower the yield in the Birch reduction. Therefore it is often necessary to distill the ammonia. When substituted aromatic compounds are subjected to the Birch reduction, electron-donating groups, such as alkyl or alkoxyl decrease the rate of the reaction and are generally found on the nonreduced positions of the product. For example, anisole gives 1-methoxy-1,4-cyclohexadiene, not 3-methoxy-1,4-cyclohexadiene. On the other hand, electron-withdrawing groups, such as COOH or $CONH_2$,

---

[527]Wang, W.-B.; Lu, S.-M.; Yang, P.-Y.; Han, X.-W.; Zhou, Y.-G. *J. Am. Chem. Soc.* **2003**, *125*, 10536.

[528]Kuwano, R.; Kaneda, K.; Ito, T.; Sato, K.; Kurokawa, T.; Ito, Y. *Org. Lett.* **2004**, *6*, 2213.

[529]For an indirect method of hydrogenating benzene to cyclohexene, see Harman, W.D.; Taube, H. *J. Am. Chem. Soc.* **1988**, *110*, 7906.

[530]Higashijima, M.; Nishimura, S. *Bull. Chem. Soc. Jpn.* **1992**, *65*, 824.

[531]In bmim BF$_4$, 1-butyl-3-methylimidazolium tetrafluoroborate: Boxwell, C.J.; Dyson, P.J.; Ellis, D.J.; Welton, T. *J. Am. Chem. Soc.* **2002**, *124*, 9334.

[532]For a procedure that converts benzene to pure 1,4-cyclohexadiene, see Brandsma, L.; van Soolingen, J.; Andringa, H. *Synth. Commun.* **1990**, *20*, 2165. Also see, Weitz, I.S.; Rabinovitz, M. *J. Chem. Soc. Perkin Trans. 1*, **1993**, 117.

[533]For a monograph, see Akhrem, A.A.; Reshotova, I.G.; Titov, Yu.A. *Birch Reduction of Aromatic Compounds*, Plenum, NY, **1972**. For reviews, see Birch, A.J. *Pure Appl. Chem.* **1996**, *68*, 553; Rabideau, P.W. *Tetrahedron* **1989**, *45*, 1579; Birch, A.J.; Subba Rao, G. *Adv. Org. Chem.* **1972**, *8*, 1; Kaiser, E.M. *Synthesis* **1972**, 391; Harvey, R.G. *Synthesis* **1970**, 161; House, H.O. *Modern Synthetic Reaction*, 2nd ed., W.A. Benjamin, NY, **1972**, pp. 145–150, 173–209; Hückel, W. *Fortschr. Chem. Forsch.* **1966**, *6*, 197; Smith, M., in Augustine, R.L. *Reduction Techniques and Applications in Organic Synthesis*, Marcel Dekker, NY, **1968**, pp. 95–170.

[534]Donohoe, T.J.; House, D. *J. Org. Chem.* **2002**, *67*, 5015.

[535]Kinoshita, T.; Ichinari, D.; Sinya, J. *J. Heterocyclic Chem.* **1996**, *33*, 1313.

[536]Donohoe, T.J.; McRiner, A.J.; Helliwell, M.; Sheldrake, P. *J. Chem. Soc., Perkin Trans. 1* **2001**, 1435.

[537]Guo, Z.; Schultz, A.G. *J. Org. Chem.* **2001**, *66*, 2154.

increase the reaction rate and are found on the reduced positions of the product.[538] The regioselectivity of the reaction has been examined.[539] The mechanism involves solvated electrons,[540] which are transferred from the metal to the solvent, and hence to the ring:[541]

The sodium becomes oxidized to Na^+ and creates a radical ion (**110**).[542] There is a great deal of evidence from ESR spectra for these species.[543] The radical ion accepts a proton from the alcohol to give a radical, which is reduced to a carbanion by another sodium atom. Finally, **111** accepts another proton. Thus the function of the alcohol is to supply protons, since with most substrates ammonia is not acidic enough for this purpose. In the absence of the alcohol, products arising from dimerization of **110** are frequently obtained. There is evidence[544] at least with some substrates, for example, biphenyl, that the radical ion corresponding to **110** is converted to the carbanion corresponding to **111** by a different pathway, in which the order of the steps is reversed: first a second electron is gained to give a dianion,[542] which then acquires a proton, producing the intermediate corresponding to **111**.

Ordinary alkenes are usually unaffected by Birch-reduction conditions, and double bonds may be present in the molecule if they are not conjugated with the ring. However, phenylated alkenes, internal alkynes (**15-12**),[545] and conjugated alkenes (with C=C or C=O) are reduced under these conditions.

Note that **111** is a resonance hybrid; that is, we can write the two additional canonical forms shown. The question therefore arises: Why does the carbanion pick up a proton at the 6 position to give the 1,4-diene? Why not at the 2 position

---

[538]These regioselectivities have generally been explained by molecular-orbital considerations regarding the intermediates involved. For example, see Birch, A.J.; Hinde, A.L.; Radom, L. *J. Am. Chem. Soc.* **1980**, *102*, 3370, 4074, 6430; **1981**, *103*, 284; Zimmerman, H.E.; Wang, P.A. *J. Am. Chem. Soc.* **1990**, *112*, 1280. For methods of reversing the regioselectivities, see Epling, G.A.; Florio, E. *Tetrahedron Lett.* **1986**, *27*, 1469; Rabideau, P.W.; Karrick, G.L. *Tetrahedron Lett.* **1987**, *28*, 2481.

[539]Zimmerman, H.E.; Wang, P.A. *J. Am. Chem. Soc.* **1993**, *115*, 2205.

[540]For reviews of solvated electrons and related topics, see Dye, J.L. *Prog. Inorg. Chem.* **1984**, *32*, 327–441; Alpatova, N.M.; Krishtalik, L.I.; Pleskov, Y.V. *Top. Curr. Chem.* **1987**, *138*, 149–219.

[541]Birch, A.J.; Nasipuri, D. *Tetrahedron* **1959**, *6*, 148.

[542]For a review of radical ions and diions generated from aromatic compounds, see Holy, N.L. *Chem. Rev.* **1974**, *74*, 243.

[543]For example, see Jones, M.T., in Kaiser, E.T.; Kevan, L. *Radical Ions*, Wiley, NY, **1968**, pp. 245–274; Bowers, K.W. *Adv. Magn. Reson.*, **1965**, *1*, 317; Carrington, A. *Q. Rev. Chem. Soc.* **1963**, *17*, 67.

[544]Lindow, D.F.; Cortez, C.N.; Harvey, R.G. *J. Am. Chem. Soc.* **1972**, *94*, 5406; Rabideau, P.W.; Peters, N.K.; Huser, D.L. *J. Org. Chem.* **1981**, *46*, 1593.

[545]See Brandsma, L.; Nieuwenhuizen, W.F.; Zwikker, J.W. Mäeorg, U. *Eur. J. Org. Chem.* **1999**, 775.

to give the 1,3-diene?[546] An answer to this question has been proposed by Hine, who has suggested that this case is an illustration of the operation of the *principle of least motion*.[547] According to this principle, "those elementary reactions will be favored that involve the least change in atomic position and electronic configuration."[547] The principle can be applied to the case at hand in the following manner (simplified): The valence-bond bond orders (p. 32) for the six carbon–carbon bonds (on the assumption that each of the three forms contributes equally) are (going around the ring) $1\frac{2}{3}$, 1, 1, $1\frac{2}{3}$, $1\frac{1}{3}$, and $1\frac{1}{3}$. When the carbanion is converted to the diene, these bond orders change as follows:

It can be seen that the two bonds whose bond order is 1 are unchanged in the two products, but for the other four bonds there is a change. If the 1,4-diene is formed, the change is $\frac{1}{3}+\frac{1}{3}+\frac{1}{3}+\frac{1}{3}$, while formation of the 1,3-diene requires a change of $\frac{1}{3}+\frac{2}{3}+\frac{2}{3}+\frac{1}{3}$. Since a greater change is required to form the 1,3-diene, the principle of least motion predicts formation of the 1,4-diene. This may not be the only factor, because the ^{13}C NMR spectrum of **111** shows that the 6 position has a somewhat greater electron density than the 2 position, which presumably would make the former more attractive to a proton.[548]

Reduction of aromatic rings with lithium[549] or calcium[550] in amines (instead of ammonia: called *Benkeser reduction*) proceeds further and cyclohexenes are obtained. It is thus possible to reduce a benzene ring, by proper choice of reagent, so that one, two, or all three double bonds are reduced.[551] Lithium triethylborohydride (LiBEt$_3$H) has also been used, to reduce pyridine derivatives to piperidine derivatives.[552]

Transition metals and metal compounds can reduce aromatic rings in the proper medium. Indium metal reduces the pyridine ring in quinoline in aqueous ethanol solution[553] as well as the C=C unit in the five-membered ring of indole

[546]For a discussion of this question, see Rabideau, P.W.; Huser, D.L. *J. Org. Chem.* **1983**, *48*, 4266.

[547]Hine, J. *J. Org. Chem.* **1966**, *31*, 1236. For a review of this principle, see Hine, J. *Adv. Phys. Org. Chem.* **1977**, *15*, 1. See also, Tee, O.S. *J. Am. Chem. Soc.* **1969**, *91*, 7144; Jochum, C.; Gasteiger, J.; Ugi, I. *Angew. Chem. Int. Ed.* **1980**, *19*, 495.

[548]Bates, R.B.; Brenner, S.; Cole, C.M.; Davidson, E.W.; Forsythe, G.D.; McCombs, D.A.; Roth, A.S. *J. Am. Chem. Soc.* **1973**, *95*, 926.

[549]Reggel, L.; Friedel, R.A.; Wender, I. *J. Org. Chem.* **1957**, *22*, 891; Benkeser, R.A.; Agnihotri, R.K.; Burrous, M.L.; Kaiser, E.M.; Mallan, J.M.; Ryan, P.W. *J. Org. Chem.* **1964**, *29*, 1313; Kwart, H.; Conley, R.A. *J. Org. Chem.* **1973**, *38*, 2011.

[550]Benkeser, R.A.; Belmonte, F.G.; Kang, J. *J. Org. Chem.* **1983**, *48*, 2796. See also, Benkeser, R.A.; Laugal, J.A.; Rappa, A. *Tetrahedron Lett.* **1984**, *25*, 2089.

[551]One, two, or all three double bonds of certain aromatic nitrogen heterocycles can be reduced with metallic hydrides, such as NaBH$_4$ or LiAlH$_4$. For a review, see Keay, J.G. *Adv. Heterocycl. Chem.* **1986**, *39*, 1.

[552]Blough, B.E.; Carroll, F.I. *Tetrahedron Lett.* **1993**, *34*, 7239.

[553]Moody, C.J.; Pitts, M.R. *Synlett* **1998**, 1029.

derivatives.[554] Samarium iodide (SmI$_2$) reduces pyridine in aq. THF[555] and phenol in MeOH/KOH.[556] Ammonium formate and a Pd–C catalyst reduces pyridine *N*-oxide to piperidine in methanol.[557] The nitrogen-containing ring of quinolines is reduced with an iridium catalyst in isopropanol.[558]

OS **I**, 99, 499; **II**, 566; **III**, 278, 742; **IV**, 313, 887, 903; **V**, 398, 400, 467, 591, 670, 743, 989; **VI**, 371, 395, 461, 731, 852, 856, 996; **VII**, 249.

15-14 Reduction Of The Double or Triple Bonds Conjugated to Carbonyls, Cyano, and so on.

In certain cases,[559] metallic hydride reagents may also reduce double bonds in conjugation with C=O bonds, as well as reducing the C=O bonds, as in the conversion of cyclopentenone to cyclopentanol.[560] The reagent NaBH$_4$ has a greater tendency than LiAlH$_4$ to effect this double reduction, although even with NaBH$_4$ the product of single reduction (of the C=O bond) is usually formed in larger amount than the doubly reduced product. Lithium aluminium hydride gives significant double reduction only in cinnamyl systems, for example, with PhCH=CHCOOH.[561] Lithium aluminium hydride also reduces the double bonds of allylic alcohols[562] and NaBH$_4$ in MeOH–THF[563] or NaCNBH$_3$ on a zeolite[564] reduces α,β-unsaturated nitro compounds to nitroalkanes. The C=C unit proximal to the carbonyl in dienyl amides is selectively reduced with NaBH$_4$/I$_2$.[565] Mixed hydride reducing agents, such as NaBH$_4$–BiCl$_3$,[566] NaBH$_4$–InCl$_3$,[567] borohydride exchange resin (BER)–CuSO$_4$,[568] and Dibal–Co(acac)$_2$[569] have been

[554]Pitts, M.R.; Harrison, J.R.; Moody, C.J. *J. Chem. Soc., Perkin Trans. 1*, **2001**, 955.
[555]Kamochi, Y.; Kudo, T. *Heterocycles* **1993**, *36*, 2383.
[556]Kamochi, Y.; Kudo, T. *Tetrahedron Lett.* **1994**, *35*, 4169.
[557]Zacharie, B.; Moreau, N.; Dockendorff, C. *J. Org. Chem.* **2001**, *66*, 5264.
[558]Fujita, K.; Kitatsuji, C.; Furukawa, S.; Yamaguchi, R. *Tetrahedron Lett.* **2004**, *45*, 3215.
[559]For discussion, see Meyer, G.R. *J. Chem. Educ.* **1981**, *58*, 628.
[560]Brown, H.C.; Hess, H.M. *J. Org. Chem.* **1969**, *34*, 2206. For other methods of reducing both double bonds, see Larock, R.C. *Comprehensive Organic Transformations*, 2nd ed., Wiley-VCH, NY, **1999**, p. 1096.
[561]Nystrom, R.F.; Brown, W.G. *J. Am. Chem. Soc.* **1947**, *69*, 2548; **1948**, *70*, 3738; Gammill, R.B.; Gold, P.M.; Mizsak, S.A. *J. Am. Chem. Soc.* **1980**, *102*, 3095.
[562]For discussions of the mechanism of this reaction, see Snyder, E.I. *J. Org. Chem.* **1967**, *32*, 3531; Borden, W.T. *J. Am. Chem. Soc.* **1968**, *90*, 2197; Blunt, J.W.; Hartshorn, M.P.; Soong, L.T.; Munro, M.H.G. *Aust. J. Chem.* **1982**, *35*, 2519; Vincens, M.; Fadel, R.; Vidal, M. *Bull. Soc. Chim. Fr.* **1987**, 462.
[563]Varma, R.S.; Kabalka, G.W. *Synth. Commun.* **1985**, *15*, 151.
[564]Gupta, A.; Haque, A.; Vankar, Y.D. *Chem. Commun.* **1996**, 1653.
[565]Das, B.; Kashinatham, A.; Madhusudhan, P. *Tetrahedron Lett.* **1998**, *39*, 677.
[566]Ren, P.-D.; Pan, S.-F.; Dong, T.-W.; Wu, S.-H. *Synth. Commun.* **1995**, *25*, 3395.
[567]Ranu, B.C.; Samanta, S. *Tetrahedron Lett.* **2002**, *43*, 7405.
[568]Sim, T.B.; Yoon, N.M. *Synlett* **1995**, 726.
[569]Ikeno, T.; Kimura, T.; Ohtsuka, Y.; Yamada, T. *Synlett* **1999**, 96.

used. The $InCl_3$–$NaBH_4$ reagent was used to covert conjugated diene ketones (C=C–C=C–C=O) selectively to the nonconjugated alkenyl ketone (C=C–CH_2CH_2–C=O).[570]

Note that both $LiAlH_4$ and $NaBH_4$, as well as NaH, reduce ordinary alkenes and alkynes when complexed with transition-metal salts, such as $FeCl_2$ or $CoBr_2$.[571]

Reduction of only the C=C bond of conjugated C=C–C=O and C=C–C≡N systems[572] has been achieved by many reducing agents,[573] a few of which are H_2 and a Rh catalyst,[574] a Ru catalyst,[575] a Pd catalyst,[576] or an Ir catalyst,[577] and Raney nickel alone.[578] Reagents such as SmI_2,[579] and catecholborane[580] are effective. Conjugated ketones react with 2 equivalents of Cp_2TiCl in THF/MeOH to give the corresponding saturated ketone.[581] Indium metal in aqueous ethanol with ammonium chloride converts alkylidene dimalononitriles to the saturated dinitrile.[582] Zinc and acetic acid has been used for the conjugate reduction of dihydropyridin-4-ones.[583] Formic acid with a palladium catalysts reduced conjugated carboxylic acids.[584]

Silanes can be effective for the reduction of the C=C unit in conjugated systems in the presence of copper species.[585] $PhSiH_3$ and a nickel catalyst,[586] CuCl,[587] or a manganese catalyst.[588] In addition, PhR_2SiH with a copper catalyst,[589] and

[570]Ranu, B.C.; Samanta, S. *J. Org. Chem.* **2003**, *68*, 7130.

[571]See, for example, Ashby, E.C.; Lin, J.J. *J. Org. Chem.* **1978**, *43*, 2567; Chung, S. *J. Org. Chem.* **1979**, *44*, 1014. See also, Osby, J.O.; Heinzman, S.W.; Ganem, B. *J. Am. Chem. Soc.* **1986**, *108*, 67.

[572]For a review of the reduction of α,β-unsaturated carbonyl compounds, see Keinan, E.; Greenspoon, N., in Patai, S.; Rappoport, Z. *The Chemistry of Enones*, pt. 2, Wiley, NY, **1989**, pp. 923–1022. For a review of the stereochemistry of catalytic hydrogenation of α,β-unsaturated ketones, see Augustine, R.L. *Adv. Catal.* **1976**, *25*, 56.

[573]For a long list of these, with references, see Larock, R.C. *Comprehensive Organic Transformations*, 2nd ed., Wiley-VCH, NY, **1999**, pp. 13–27.

[574]Djerassi, C.; Gutzwiller, J. *J. Am. Chem. Soc.* **1966**, *88*, 4537; Cabello, J.A.; Campelo, J.M.; Garcia, A.; Luna, D.; Marinas, J.M. *J. Org. Chem.* **1986**, *51*, 1786; Harmon, R.E.; Parsons, J.L.; Cooke, D.W.; Gupta, S.K.; Schoolenberg, J. *J. Org. Chem.* **1969**, *34*, 3684.

[575]Chen, Y.-C.; Xue, D.; Deng, J.-G.; Cui, X.; Zhu, J.; Jiang, Y.-Z. *Tetrahedron Lett.* **2004**, *45*, 1555.

[576]Sajiki, H.; Ikawa, T.; Hirota, K. *Tetrahedron Lett.* **2003**, *44*, 8437; Nagano, H.; Yokota, M.; Iwazaki, Y. *Tetrahedron Lett.* **2004**, *45*, 3035.

[577]Yue, T.-Y.; Nugent, W.A. *J. Am. Chem. Soc.* **2002**, *124*, 13692.

[578]Barrero, A.F.; Alvarez-Manzaneda, E.J.; Chahboun, R.; Meneses, R. *Synlett* **1999**, 1663. For an ultrasound-mediated reduction with Raney nickel, see Wang, H.; Lian, H.; Chen, J.; Pan, Y.; Shi, Y. *Synth. Commun.* **1999**, *29*, 129.

[579]Cabrera, A.; Alper, H. *Tetrahedron Lett.* **1992**, *33*, 5007. See also, Guo, H.; Zhang, Y. *Synth. Commun.* **2000**, *30*, 1879.

[580]Evans, D.A.; Fu, G.C. *J. Org. Chem.* **1990**, *55*, 5678.

[581]Moisan, L.; Hardouin, C.; Rousseau, B.; Doris, E. *Tetrahedron Lett.* **2002**, *43*, 2013.

[582]Ranu, B.C.; Dutta, J.; Guchhait, S.K. *Org. Lett.* **2001**, *3*, 2603.

[583]Comins, D.L.; Brooks, C.A.; Ingalls, C.L. *J. Org. Chem.* **2001**, *66*, 2181.

[584]Arterburn, J.B.; Pannala, M.; Gonzlez, A.M.; Chamberlin, R.M. *Tetrahedron Lett.* **2000**, *41*, 7847.

[585]Mori, A.; Fujita, A.; Nishihara, Y.; Hiyama, R. *Chem. Commun.* **1997**, 2159.

[586]Boudjouk, P.; Choi, S.-B.; Hauck, B.J.; Rajkumar, A.B. *Tetrahedron Lett.* **1998**, *39*, 3951.

[587]Ito, H.; Ishizuka, T.; Arimoto, K.; Miura, K.; Hosomi, A. *Tetrahedron Lett.* **1997**, *38*, 8887.

[588]Magnus, P.; Waring, M.J.; Scott, D.A. *Tetrahedron Lett.* **2000**, *41*, 9731.

[589]Mori, A.; Fujita, A.; Kajiro, H.; Nishihara, Y.; Hiyama, T. *Tetrahedron* **1999**, *55*, 4573.

$PhSiH_3$–$Mo(CO)_6$[590] have been used. Triphenylsilane was also used for the asymmetric reduction of nitro alkenes ($C=C–NO_2$).[591] Poly(methylhydrosiloxane) with a chiral copper catalyst gave conjugate reduction of conjugated esters to give the saturated derivative with high enantioselectivity.[592]

A β-bromo conjugated lactone was reduced to the β-bromolactone with modest enantioselectivity using an excess of Ph_3SiH and a CuCl catalyst with a chiral ligand.[593] A copper complex with a chiral ligand and poly(methylhydrosiloxane) gave reduction of the C=C unit in conjugated carbonyl systems with good enantioselectivity.[594] Tributyltin hydride, in the presence of $MgBr_2$•OEt_2 gave 1,4-reduction of conjugated esters.[595]

Optically active catalysts, primarily homogeneous, have been used to achieve enantioselective hydrogenations[596] of many prochiral conjugated substrates.[597] For example,[598] hydrogenation of **112** with a suitable catalyst gives (+) or (−)

[590]Keinan, E.; Perez, D. *J. Org. Chem.* **1987**, *52*, 2576.

[591]Czekelius, C.; Carreira, E.M. *Org. Lett.* **2004**, *6*, 4575.

[592]Appella, D.H.; Moritani, Y.; Shintani, R.; Ferreira, E.M.; Buchwald, S.L. *J. Am. Chem. Soc.* **1999**, *121*, 9473.

[593]Hughes, G.; Kimura, M.; Buchwald, S.L. *J. Am. Chem. Soc.* **2003**, *125*, 11253.

[594]Jurkauskas, V.; Buchwald, S.L. *J. Am. Chem. Soc.* **2002**, *124*, 2892; Lipshutz, B.H.; Servesko, J.M.; Taft, B.R. *J. Am. Chem. Soc.* **2004**, *126*, 8352.

[595]Hirasawa, S.; Nagano, H.; Kameda, Y. *Tetrahedron Lett.* **2004**, *45*, 2207.

[596]For a discussion of the mechanism of asymmetric hydrogenation of such systems using a ruthenium catalyst, see Kitamura, M.; Tsukamoto, M.; Bessho, Y.; Yoshimura, M.; Kobs, U.; Widhalm, M.; Noyori, R. *J. Am. Chem. Soc.* **2002**, *124*, 6649. For a review of the mechanism of stereoselection in rhodium-catalyzed asymmetric hydrogenations, see Gridnev, I.D.; Imamoto, T. *Acc. Chem. Res.* **2004**, *37*, 633.

[597]For reviews, see, in Morrison, J.D. *Asymmetric Synthesis*, Vol. 5, Academic Press, NY, **1985**, the reviews by Halpern, J. pp. 41–69, Koenig, K.E. pp. 71–101, Harada, K. pp. 345–383; Ojima, I.; Clos, N.; Bastos, C. *Tetrahedron* **1989**, *45*, 6901, 6902–6916; Jardine, F.H., in Hartley, F.R. *The Chemistry of the Metal–Carbon Bond*, Vol. 4, Wiley, NY, **1987**, pp. 751–775; Nógrádi, M. *Stereoselective Synthesis*, VCH, NY, **1986**, pp. 53–87; Knowles, W.S. *Acc. Chem. Res.* **1983**, *16*, 106; Brunner, H. *Angew. Chem. Int. Ed.* **1983**, *22*, 897; Klabunovskii, E.I. *Russ. Chem. Rev.* **1982**, *51*, 630; Čaplar, V.; Comisso, G.;Šunjić, V. *Synthesis* **1981**, 85; Morrison, J.D.; Masler, W.F.; Neuberg, M.K. *Adv. Catal.* **1976**, *25*, 81; Kagan, H.B. *Pure Appl. Chem.* **1975**, *43*, 401; Bogdanović, B. *Angew. Chem. Int. Ed.* **1973**, *12*, 954. See also, Brewster, J.H. *Top. Stereochem.* **1967**, *2*, 1, *J. Am. Chem. Soc.* **1959**, *81*, 5475, 5483, 5493; Davis, D.D.; Jensen, F.R. *J. Org. Chem.* **1970**, *35*, 3410; Jullien, F.R.; Requin, F.; Stahl-Larivière, H. *Nouv. J. Chim.* **1979**, *3*, 91; Sathyanarayana, B.K.; Stevens, E.S. *J. Org. Chem.* **1987**, *52*, 3170; Wroblewski, A.E.; Applequist, J.; Takaya, A.; Honzatko, R.; Kim, S.; Jacobson, R.A.; Reitsma, B.H.; Yeung, E.S.; Verkade, J.G. *J. Am. Chem. Soc.* **1988**, *110*, 4144; Knowles, W.S. *Angew. Chem. Int. Ed.* **2002**, *41*, 1999.

[598]For some other recent examples, see Hayashi, T.; Kawamura, N.; Ito, Y. *Tetrahedron Lett.* **1988**, *29*, 5969; Muramatsu, H.; Kawano, H.; Ishii, Y.; Saburi, M.; Uchida, Y. *J. Chem. Soc., Chem. Commun.* **1989**, 769; Amrani, Y.; Lecomte, L.; Sinou, D.; Bakos, J.; Toth, I.; Heil, B. *Organometallics* **1989**, *8*, 542; Yamamoto, K.; Ikeda, K.; Lin, L.K. *J. Organomet. Chem.* **1989**, *370*, 319; Waymouth, R.; Pino, P. *J. Am. Chem. Soc.* **1990**, *112*, 4911; Ohta, T.; Takaya, H.; Noyori, R. *Tetrahedron Lett.* **1990**, *31*, 7189; Ashby, M.T.; Halpern, J. *J. Am. Chem. Soc.* **1991**, *113*, 589; Heiser, B.; Broger, E.A.; Crameri, Y. *Tetrahedron: Asymmetry* **1991**, *2*, 51; Burk, M.J. *J. Am. Chem. Soc.* **1991**, *113*, 8518.

113 (depending on which enantiomer of the catalyst is used) with an enantiomeric excess as high as 96%.[599] Prochiral substrates that give such high optical yields generally contain functional groups, such as a carbonyl group,[600] amide groups, cyano groups, or combinations of such groups as in **112**.[601] The catalyst in such cases[602] is usually a ruthenium[603] or rhodium complex with chiral phosphine ligands.[604] Iridium complexes have been used with excellent enantioselectivity.[605] Good asymmetric induction[606] has been achieved using chiral rhodium complexes with other chiral additives.[607] The role of solvent has been examined.[608] A pressure dependent enantioselective hydrogenation has been reported.[609] There are many examples for the reduction of alkylidene amino acids, amino esters or amido acids or esters that vary the catalyst and/or the chiral ligand.[610] Asymmetric catalytic hydrogenation has been reported for conjugated carboxylic acids[611] and conjugated ketones.[612] A ruthenium catalyst with a polymer supported chiral ligand has also

[599]Koenig, K.E., in Morrison, J.D. *Asymmetric Synthesis*, Vol. 5, Academic Press, NY, *1985*, p. 74.

[600]Reetz, M.T.; Mehler, G. *Angew. Chem. Int. Ed.* **2000**, *39*, 3889.

[601]For tables of substrates that have been enantioselectively hydrogenated, see Koenig, K.E., in Morrison, J.D. *Asymmetric Synthesis* Vol. 5, Academic Press, NY, *1985*, pp. 83–101.

[602]For a list of these, with references, see Larock, R.C. *Comprehensive Organic Transformations*, 2nd ed., Wiley-VCH, NY, *1999*, pp. 8–12. For reviews of optically active nickel catalysts, see Izumi, Y. *Adv. Catal.* *1983*, *32*, 215; *Angew. Chem. Int. Ed.* *1971*, *10*, 871. For a review of the synthesis of some of these phosphines, see Mortreux, A.; Petit, F.; Buono, G.; Peiffer, G. *Bull. Soc. Chim. Fr.* *1987*, 631.

[603]Wu, H.-P.; Hoge, G. *Org. Lett.* *2004*, *6*, 3645; Tang, W.; Wu, S.; Zhang, X. *J. Am. Chem. Soc.* *2003*, *125*, 9570.

[604]Lee, S.-g.; Zhang, Y.J. *Org. Lett.* *2002*, *4*, 2429; Le, J.C.D.; Pagenkopf, B.L. *J. Org. Chem.* *2004*, *69*, 4177; Fu, Y.; Guo, X.-X.; Zhu, S.-F.; Hu, A.-G.; Xie, J.-H.; Zhou, Q.-L. *J. Org. Chem.* *2004*, *69*, 4648; Yi, B.; Fan, Q.-H.; Deng, G.-J.; Li, Y.-M.; Qiu, L.-Q.; Chan, A.S.C. *Org. Lett.* *2004*, *6*, 1361.; Hoen, R.; van den Berg, M.; Bernsmann, H.; Minnaard, A.J.; de Vries, J.G.; Feringa, B.L. *Org. Lett.* *2004*, *6*, 1433; Fu, Y.; Hou, G.-H.; Xie, J.-H.; Xing, L.; Wang, L.-X.; Zhou, Q.-L. *J. Org. Chem.* *2004*, *69*, 8157; Peña, D.; Minnaard, A.J.; de Vries, J.G.; Feringa, B.L. *J. Am. Chem. Soc.* *2002*, *124*, 14552; Evans, D.A.; Michael, F.E.; Tedrow, J.S.; Campos, K.R. *J. Am. Chem. Soc.* *2003*, *125*, 3534; Hoge, G.; Wu, H.-P.; Kissel, W.S.; Pflum, D.A.; Greene, D.J.; Bao, J. *J. Am. Chem. Soc.* *2004*, *126*, 5966; Ikeda, S.-i.; Sanuki, R.; Miyachi, H.; Miyashita, H.; Taniguchi, M.; Odashima, K. *J. Am. Chem. Soc.* *2004*, *126*, 10331; Huang, H.; Liu, X.; Chen, S.; Chen, H.; Zheng, Z. *Tetrahedron Asymmetry* *2004*, *15*, 2011.

[605]Smidt, S.P.; Menges, F.; Pgaltz, A. *Org. Lett.* *2004*, *6*, 2023.

[606]Zhu, G.; Zhang, X. *J. Org. Chem.* *1998*, *63*, 9590; Burk, M.J.; Casy, G.; Johnson, N.B. *J. Org. Chem.* *1998*, *63*, 6084; Burk, M.J.; Allen, J.G.; Kiesman, W.F. *J. Am. Chem. Soc.* *1998*, *120*, 657.

[607]Noyori, R.; Hashiguchi, S. *Accts. Chem. Res.* *1997*, *30*, 97; Inoguchi, K.; Sakuraba, S.; Achiwa, K. *Synlett* *1992*, 169.

[608]Maki, S.; Harada, Y.; Matsui, R.; Okawa, M.; Hirano, T.; Niwa, H.; Koizumi, M.; Nishiki, Y.; Furuta, T.; Inoue, H.; Iwakura, C. *Tetrahedron Lett.* *2001*, *42*, 8323; Heller, D.; Drexler, H.-J.; Spannenberg, A.; Heller, B.; You, J.; Baumann, W. *Angew. Chem. Int. Ed.* *2002*, *41*, 777.

[609]Heller, D.; Holz, J.; Drexler, H.-J.; Lang, J.; Drauz, K.; Krimmer, H.-P.; Börner, A. *J. Org. Chem.* *2001*, *66*, 6816.

[610]Rhodium catalyst with a chiral bis(phosphine): Li. W.; Zhang, Z.; Xiao, D.; Zhang, X. *Tetrahedron Lett.* *1999*, *40*, 6701. New chiral phosphines: Ohashi, A.; Imamoto, T. *Org. Lett.* *2001*, *3*, 373.

[611]Uemura, T.; Zhang, X.; Matsumura, K.; Sayo, N.; Kumobayashi, H.; Ohta, T.; Nozaki, K.; Takaya, H. *J. Org. Chem.* *1996*, *61*, 5510; Suárez, A.; Pizzano, A. *Tetrahedron Asymmetry* *2001*, *12*, 2501. See, Okano, T.; Kaji, M.; Isotani, S.; Kiji, J. *Tetrahedron Lett.* *1992*, *33*, 5547 for the influence of water on the regioselectivity of this reduction.

[612]Yamaguchi, M.; Nitta, A.; Reddy, R.S.; Hirama, M. *Synlett* *1997*, 117.

been used for conjugated acids.[613] Asymmetric hydrogenation of conjugated carboxylic acids in an ionic liquid is known using a chiral ruthenium complex[614] Enamino esters have been hydrogenated with high enantioselectivity using chiral rhodium catalysts.[615]

See **19-36** for methods of reducing C=O bonds in the presence of conjugated C=C bonds.

The C=C unit of conjugated aldehydes has been reduced using $AlMe_3$ with a catalytic amount of $CuBr$[616] and with ammonium formate/Pd–C.[617] Polymer-supported formate has been used for the 1,4-reduction of conjugated ketones[618] and conjugated acids using a rhodium catalyst and microwave irradiation.[619] Selective reduction of the C=C unit in conjugated ketones was accomplished with $Na_2S_2O_4$ in aqueous dioxane, and nonconjugated alkenes were not reduced.[620] Isopropanol and an iridium catalyst gives conjugate reduction of conjugated ketones.[621] Conjugate hydrostannation by an iodotin hydride ate complex, followed by hydrolysis converts unsaturated esters to saturated esters.[622] The reaction of conjugated ketones with aluminum chlorides, followed by treatment with water generates the saturated ketone.[623]

Baker's yeast reduces conjugated nitro compounds to nitroalkanes[624] and also the C=C unit of conjugated ketones.[625] Other enzymatic reductions are possible. A reductase from *Nicotiana tabacum* reduced a conjugated ketone to the saturated ketone, with excellent enantioselectivity.[626] Enzyme YNAR-I and NADP-H reduces conjugated nitro compounds to nitroalkanes.[627]

15-15 Reductive Cleavage of Cyclopropanes

$$\triangle \xrightarrow[\text{cat.}]{H_2} CH_3CH_2CH_3$$

[613]Fan, Q.H.; Deng, G.-J.; Lin, C.-C.; Chan, A.S.C. *Tetrahedron Asymmetry* **2001**, *12*, 1241.

[614]In bmim PF_6, 1-butyl-3-methylimidazolium hexafluorophosphate: Brown, R.A.; Pollet, P.; McKoon, E.; Eckert, C.A.; Liotta, C.L.; Jessop, P.G. *J. Am. Chem. Soc.* **2001**, *123*, 1254.

[615]Hsiao, Y.; Rivera, N.R.; Rosner, T.; Krska, S.W.; Njolito, E.; Wang, F.; Sun, Y.; Armstrong III, J.D.; Grabowski, E.J.J.; Tillyer, R.D.; Spindler, F.; Malan, C. *J. Am. Chem. Soc.* **2004**, *126*, 9918.

[616]Kabbara, J.; Flemming, S.; Nickisch, K.; Neh, H.; Westermann, J. *Synlett* **1994**, 679.

[617]Ranu, B.C.; Sarkar, A. *Tetrahedron Lett.* **1994**, *35*, 8649.

[618]Basu, B.; Bhuiyan, Md.M.H.; Das, P.; Hossain, I. *Tetrahedron Lett.* **2003**, *44*, 8931.

[619]Desai, B.; Danks, T.N. *Tetrahedron Lett.* **2001**, *42*, 5963.

[620]Dhillon, R.S.; Singh, R.P.; Kaur, D. *Tetrahedron Lett.* **1995**, *36*, 1107.

[621]Sakaguchi, S.; Yamaga, T.; Ishii, Y. *J. Org. Chem.* **2001**, *66*, 4710.

[622]Shibata, I.; Suwa, T.; Ryu, K.; Baba, A. *J. Org. Chem.* **2001**, *66*, 8690.

[623]Koltunov, K.Yu.; Repinskaya, I.B.; Borodkin, G.I. *Russ. J. Org. Chem.* **2001**, *37*, 1534.

[624]Bak, R.R.; McAnda, A.F.; Smallridge, A.J.; Trewhella, M.A. *Aust. J. Chem.* **1996**, *49*, 1257; Takeshita, M.; Yoshida, S.; Kohno, Y. *Heterocycles* **1994**, *37*, 553; Kawai, Y.; Inaba,Y.; Tokitoh, N. *Tetrahedron Asymmetry* **2001**, *12*, 309.

[625]Kawai, Y.; Saitou, K.; Hida, K.; Ohno, A. *Tetrahedron Asymmetry* **1995**, *6*, 2143; Filho, E.P.S.; Rodrigues, J.A.R.; Moran, P.J.S. *Tetrahedron Asymmetry* **2001**, *12*, 847; Kawai, Y.; Hayashi, M.; Tokitoh, N. *Tetrahedron Asymmetry* **2001**, *12*, 3007.

[626]Shimoda, K.; Kubota, N.; Hamada, H. *Tetrahedron Asymmetry* **2004**, *15*, 2443; Hirata, T.; Shimoda, K.; Gondai, T. *Chem. Lett.* **2000**, 850.

[627]Kawai, Y.; Inaba, Y.; Hayashi, M.; Tokitoh, N. *Tetrahedron Lett.* **2001**, *42*, 3367.

Cyclopropanes can be cleaved by catalytic hydrogenolysis.[628] Among the catalysts used have been Ni, Pd, Rh,[629] and Pt. The reaction can often be run under mild conditions.[630] Certain cyclopropane rings, especially cyclopropyl ketones and aryl-substituted cyclopropanes,[631] can be reductively cleaved by an alkali metal (generally Na or Li) in liquid ammonia.[632] Similar reduction has been accomplished photochemically in the presence of $LiClO_4$.[633] This reaction is an excellent way to introduce a *gem*-dimethyl unit into a molecule. Hydrogenation of the cyclopropane ring in **114**, for example, gave the *gem*-dimethyl unit in **115** using PtO_2 (Adam's catalyst).[634]

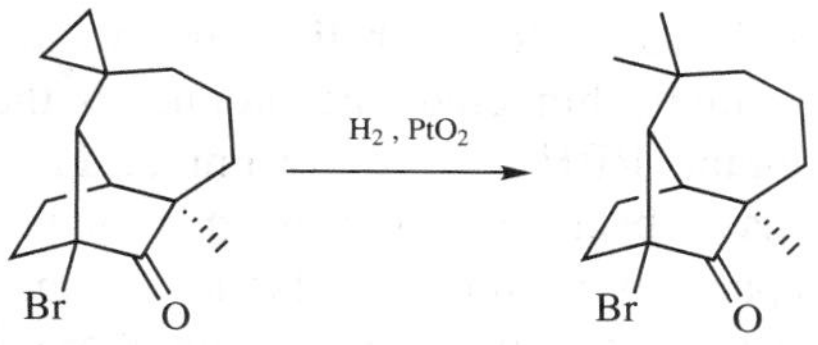

114   **115**

F. A Metal on the Other Side

15-16 Hydroboration

When alkenes are treated with borane[635] in ether solvents, BH_3 adds across the double bond.[636] Borane cannot be prepared as a stable pure compound[637] (it dimerizes to diborane, B_2H_6), but it is commercially available in the form of

---

[628]For reviews, see Charton, M., in Zabicky, J. *The Chemistry of Alkenes*, Vol. 2, Wiley, NY, *1970*, pp. 588–592; Newham, J. *Chem. Rev. 1963*, *63*, 123; Rylander, P.N. *Catalytic Hydrogenation over Platinum Metals*, Academic Press, NY, *1967*, pp. 469–474.

[629]Bart, S.C.; Chirik, P.J. *J. Am. Chem. Soc. 2003*, *125*, 886.

[630]See, for example, Woodworth, C.W.; Buss, V.; Schleyer, P.v.R. *Chem. Commun. 1968*, 569.

[631]See, for example, Walborsky, H.M.; Aronoff, M.S.; Schulman, M.F. *J. Org. Chem. 1970*, *36*, 1036.

[632]For a review, see Staley, S.W. *Sel. Org. Transform. 1972*, *2*, 309.

[633]Cossy, J.; Furet, N. *Tetrahedron Lett. 1993*, *34*, 8107.

[634]Karimi, S.; Tavares, P. *J. Nat. Prod. 2003*, *66*, 520.

[635]For a review of this reagent, see Lane, C.F., in Pizey, J.S. *Synthetic Reagents*, Vol. 3, Wiley, NY, *1977*, pp. 1–191.

[636]For books on this reaction and its many applications, see Pelter, A.; Smith, K.; Brown, H.C. *Borane Reagents*, Academic Press, NY, *1988*; Brown, H.C. *Boranes in Organic Chemistry*, Cornell University Press, Ithaca, NY, *1972*, *Organic Syntheses Via Boranes, Wiley*, NY, *1975*; Cragg, G.M.L. *Organoboranes in Organic Synthesis*, Marcel Dekker, NY, *1973*. For reviews, see Matteson, D.S., in Hartley, F.R. *The Chemistry of the Metal-Carbon Bond*, Vol. 4, Wiley, NY, *1987*, pp. 307–409, 315–337; Smith, K. *Chem. Ind. (London) 1987*, 603; Brown, H.C.; Vara Prasad, J.V.N. *Heterocycles 1987*, *25*, 641; Suzuki, A.; Dhillon, R.S. *Top. Curr. Chem. 1986*, *130*, 23.

[637]Fehlner, T.P. *J. Am. Chem. Soc. 1971*, *93*, 6366.

'ate' complexes with THF, Me$_2$S,[638] phosphines, or tertiary amines. The alkenes can be treated with a solution of one of these complexes (THF–BH$_3$ reacts at 0°C and is the most convenient to use; R$_3$N–BH$_3$ generally require temperatures of ~100°C; however, the latter can be prepared as air-stable liquids or solids, while the former can only be used as relatively dilute solutions in THF and are decomposed by moisture in air) or with a mixture of NaBH$_4$ and BF$_3$ etherate, which generates borane *in situ*.[639] With relatively unhindered alkenes, the process cannot be stopped with the addition of one molecule of BH$_3$ because the resulting RBH$_2$ adds to another molecule of alkene to give R$_2$BH, which in turn adds to a third alkene molecule, so that the isolated product is a trialkylborane R$_3$B. The reaction can be performed on alkenes with one to four substituents, including cyclic alkenes, but when the alkene is moderately hindered, the product *is* the dialkylborane R$_2$BH or even the monalkylborane RBH$_2$.[640] For example, **116** (*disiamylborane*) and **117** (*thexylborane*)[641] have been prepared in this manner. Monoalkylboranes RBH$_2$ (which can be prepared from hindered alkenes, as above) and dialkylboranes R$_2$BH also add to alkenes, to give the mixed trialkylboranes RR$'_2$B and R$_2$R$'$B, respectively. Surprisingly, when methylborane MeBH$_2$,[642] which is not a bulky molecule, adds to alkenes in the solvent THF, the reaction can be stopped with one addition to give the dialkylboranes RMeBH.[643] Reaction of this with a second alkene produces the trialkylborane RR$'$MeB.[644] Other monoalkylboranes, *i*PrBH$_2$, *n*-BuBH$_2$, *s*-BuBH$_2$, and *t*-BuBH$_2$, behave similarly with internal alkenes, but not with alkenes of the type RCH=CH$_2$.[645]

Disiamylborane, **116**   Thexylborane, **117**

[638]For a review of BH$_3$•SMe$_2$, see Hutchins, R.O.; Cistone, F. *Org. Prep. Proced. Int.* **1981**, *13*, 225. See Cadot, C.; Dalko, P.I.; Cossy, J. *Tetrahedron Lett.* **2001**, *42*, 1661.

[639]For a list of hydroborating reagents, with references, see Larock, R.C. *Comprehensive Organic Transformations*, 2nd ed., Wiley-VCH, NY, **1999**, pp. 1005–1009.

[640]Unless coordinated with a strong Lewis base such as a tertiary amine, mono and dialkylboranes actually exist as dimers, for example, R$_2$B$\overset{H}{\underset{H}{\cdots}}BR_2$ Brown, H.C.; Klender, G.J. *Inorg. Chem.* **1962**, *1*, 204.

[641]For a review of the chemistry of thexylborane, see Negishi, E.; Brown, H.C. *Synthesis* **1974**, 77.

[642]Prepared from lithium methylborohydride and HCl: Brown, H.C.; Cole, T.E.; Srebnik, M.; Kim, K. *J. Org. Chem.* **1986**, *51*, 4925.

[643]Srebnik, M.; Cole, T.E.; Brown, H.C. *J. Org. Chem.* **1990**, *55*, 5051.

[644]For a method of synthesis of RR^{1}R^{2}B, see Kulkarni, S.U.; Basavaiah, D.; Zaidlewicz, M.; Brown, H.C. *Organometallics* **1982**, *1*, 212.

[645]Srebnik, M.; Cole, T.E.; Ramachandran, P.V.; Brown, H.C. *J. Org. Chem.* **1989**, *54*, 6085.

In all cases, the boron goes to the side of the double bond that has more hydrogens, whether the substituents are aryl or alkyl.[646] This actually follows Markovnikov's rule, since boron is more positive than hydrogen. However, the regioselectivity is caused mostly by steric factors, although electronic factors also play a part. Studies of the effect of ring substituents on rates and on the direction of attack in hydroboration of substituted styrenes showed that the reaction with boron and the alkene has electrophilic character.[647] When both sides of the double bond are monosubstituted or both disubstituted, about equal amounts of each isomer are obtained. However, it is possible in such cases to make the addition regioselective by the use of a large borane molecule. For example, treatment of *i*PrCH=CHMe with borane gave 57% of product with boron on the methyl-bearing carbon and 43% of the other, while treatment with **116** gave 95% **118** and only 5% of the other isomer.[648]

116   **118**

Another reagent with high regioselectivity is 9-borabicyclo[3.3.1]nonane (9-BBN), which is prepared by hydroboration of 1,5-cyclooctadiene,[649] and has the advantage that it is stable in air. Borane is quite unselective and attacks all sorts of double bonds. Disiamylborane, 9-BBN, and similar molecules are far more selective and preferentially attack less-hindered bonds, so it is often possible to hydroborate one double bond in a molecule and leave others unaffected or to hydroborate one alkene in the presence of a less reactive alkene.[650] For example, 1-pentene can be removed from a mixture of 1- and 2-pentenes, and a cis alkene can be selectively hydroborated in a mixture of the cis and trans isomers.

9-BBN

[646]For a thorough discussion of the regioselectivity with various types of substrate and hydroborating agents, see Cragg, G.M.L.*Organoboranes in Organic Synthesis* Marcel Dekker, NY, *1973*, pp.63–84, 137–197. See also, Brown, H.C.; Vara Prasad, J.V.N.; Zee, S. *J. Org. Chem.* *1986*, *51*, 439.

[647]Brown, H.C.; Sharp, R.L. *J. Am. Chem. Soc.* *1966*, *88*, 5851; Klein, J.; Dunkelblum, E.; Wolff, M.A. *J. Organomet. Chem.* *1967*, *7*, 377. See also, Marshall, P.A.; Prager, R.H. *Aust. J. Chem.* *1979*, *32*, 1251. For a study of hyperconjugation effects in substituted methylboranes, see Mo, Y.; Jiao, H. Schleyer, P.v.R. *J. Org. Chem.* *2004*, *69*, 3493.

[648]Brown, H.C.; Zweifel, G. *J. Am. Chem. Soc.* *1961*, *83*, 1241.

[649]See Knights, E.F.; Brown, H.C. *J. Am. Chem. Soc.* *1968*, *90*, 5280, 5281; Brown, H.C.; Chen, J.C. *J. Org. Chem.* *1981*, *46*, 3978; Soderquist, J.A.; Brown, H.C. *J. Org. Chem.* *1981*, *46*, 4599.

[650]Brown, H.C.; Moerikofer, A.W. *J. Am. Chem. Soc.* *1963*, *85*, 2063; Zweifel, G.; Brown, H.C. *J. Am. Chem. Soc.* *1963*, *85*, 2066; Zweifel, G.; Ayyangar, N.R.; Brown, H.C. *J. Am. Chem. Soc.* *1963*, *85*, 2072; Brown, H.C.; Sharp, R.L. *J. Am. Chem. Soc.* *1966*, *88*, 5851; Klein, J.; Dunkelblum, E.; Wolff, M.A. *J. Organomet. Chem.* *1967*, *7*, 377.

For most substrates, the addition in hydroboration is stereospecific and syn, with attack taking place from the less-hindered side.[651] Note that organoboranes can be analyzed using ^{11}B nmr.[652] The mechanism[653] may be a cyclic four-center one:[654]

When the substrate is an allylic alcohol or amine, the addition is generally anti,[655] although the stereoselectivity can be changed to syn by the use of catecholborane and the rhodium complexes mentioned above.[656] Because the mechanism is different, use of this procedure can result in a change in regioselectivity as well, for example, styrene $PhCH=CH_2$ gave $PhCH(OH)CH_3$.[657]

Monochloroborane[658] BH_2Cl coordinated with dimethyl sulfide shows greater regioselectivity than BH_3 for terminal alkenes or those of the form $R_2C=CHR$, and the hydroboration product is a dialkylchloroborane R_2BCl.[659] For example, 1-hexene gave 94% of the anti-Markovnikov product (the boron is on the less substituted carbon) with BH_3–THF, but 99.2% with BH_2Cl–SMe_2. Treatment of alkenes with dichloroboranedimethyl sulfide $BHCl_2$–SMe_2 in the presence of BF_3[660] or with BCl_3 and Me_3SiH[661] gives alkyldichloroboranes $RBCl_2$. Extensions of this basic approach are possible with dihalo alkylboranes. The reaction of an alkene with allyl dibromoborane, incorporated an allyl group and the born on adjacent carbons.[662]

[651]Brown, H.C.; Zweifel, G. *J. Am. Chem. Soc.* **1961**, *83*, 2544; Bergbreiter, D.E.; Rainville, D.P. *J. Org. Chem.* **1976**, *41*, 3031; Kabalka, G.W.; Newton, Jr., R.J.; Jacobus, J. *J. Org. Chem.* **1978**, *43*, 1567.

[652]Medina, J.R.; Cruz, G.; Cabrera, C.R.; Soderquist, J.A. *J. Org. Chem.* **2003**, *68*, 4631.

[653]For kinetic studies, see Vishwakarma. L.C.; Fry, A. *J. Org. Chem.* **1980**, *45*, 5306; Brown, H.C.; Chandrasekharan, J.; Wang, K.K. *J. Org. Chem.* **1983**, *48*, 2901; *Pure Appl. Chem.* **1983**, *55*, 1387–1414; Nelson, D.J.; Cooper, P.J. *Tetrahedron Lett.* **1986**, *27*, 4693; Brown, H.C.; Chandrasekharan, J. *J. Org. Chem.* **1988**, *53*, 4811.

[654]Brown, H.C.; Zweifel, G. *J. Am. Chem. Soc.* **1959**, *81*, 247; Pasto, D.J.; Lepeska, B.; Balasubramaniyan, V. *J. Am. Chem. Soc.* **1972**, *94*, 6090; Pasto, D.J.; Lepeska, B.; Cheng, T. *J. Am. Chem. Soc.* **1972**, *94*, 6083; Narayana, C.; Periasamy, M. *J. Chem. Soc., Chem. Commun.* **1987**, 1857. See, however, Jones, P.R. *J. Org. Chem.* **1972**, *37*, 1886.

[655]See Still, W.C.; Barrish, J.C. *J. Am. Chem. Soc.* **1983**, *105*, 2487.

[656]See Evans, D.A.; Fu, G.C.; Hoveyda, A.H. *J. Am. Chem. Soc.* **1988**, *110*, 6917; Burgess, K.; Cassidy, J.; Ohlmeyer, M.J. *J. Org. Chem.* **1991**, *56*, 1020; Burgess, K.; Ohlmeyer, M.J. *J. Org. Chem.* **1991**, *56*, 1027.

[657]Hayashi, T.; Matsumoto, Y.; Ito, Y. *J. Am. Chem. Soc.* **1989**, *111*, 3426; Zhang, J.; Lou, B.; Guo, G.; Dai, L. *J. Org. Chem.* **1991**, *56*, 1670.

[658]For a review of haloboranes, see Brown, H.C.; Kulkarni, S.U. *J. Organomet. Chem.* **1982**, *239*, 23.

[659]Brown, H.C.; Ravindran, N.; Kulkarni, S.U. *J. Org. Chem.* **1979**, *44*, 2417.

[660]Brown, H.C.; Racherla, U.S. *J. Org. Chem.* **1986**, *51*, 895.

[661]Soundararajan, R.; Matteson, D.S. *J. Org. Chem.* **1990**, *55*, 2274.

[662]Frantz, D.E.; Singleton, D.A. *Org. Lett.* **1999**, *1*, 485.

An important use of the hydroboration reaction is oxidation of an organoborane to alcohols with hydrogen peroxide and NaOH (**12-27**). Organoboranes have been oxidized with Oxone®[663] and methanol/triethylamine/molecular oxygen.[664] The synthetic result is an indirect way of adding H_2O across a double bond in an anti-Markovnikov manner. However, boranes undergo many other reactions as well. Among other things, they react with α-halo carbonyl compounds to give alkylated products (**10-73**), with α,β-unsaturated carbonyl compounds to give Michael-type addition of R and H (**15-27**), with CO to give alcohols and ketones (**18-23–18-24**); they can be reduced with carboxylic acids, providing an indirect method for reduction of double bonds (**15-11**), or they can be oxidized with chromic acid or pyridinium chlorochromate to give ketones[665] or aldehydes (from terminal alkenes),[666] dimerized with silver nitrate and NaOH (**14-26**), isomerized (**18-11**), or converted to amines (**12-32**), halides (**12-31**), or carboxylic acids.[667] They are thus useful intermediates for the preparation of a wide variety of compounds. Intramolecular hydroboration reaction are possible.[668]

Such functional groups as OR, OH, NH_2, SMe, halogen, and COOR may be present in the molecule,[669] but not groups that are reducible by borane. Hydroboration of enamines with 9-BBN provides an indirect method for reducing an aldehyde or ketone to an alkene, e.g.[670]

Enamines can also be converted to amino alcohols via hydroboration.[671] Allene–boranes react with aldehydes to give alkyne–alcohols.[672]

Use of the reagent diisopinocampheylborane **119** (prepared by treating optically active α-pinene with BH_3) results in enantioselective hydroboration–oxidation.[673] Since both (+) and (−) α-pinene are readily available, both enantiomers

[663]Ripin, D.H.B.; Cai, W.; Brenek, S.J. *Tetrahedron Lett.* **2000**, *41*, 5817.

[664]Cadot, C.; Dalko, P.I.; Cossy, J.; Ollivier, C.; Chuard, R.; Renaud, P. *J. Org. Chem.* **2002**, *67*, 7193.

[665]Brown, H.C.; Garg, C.P. *J. Am. Chem. Soc.* **1961**, *83*, 2951; *Tetrahedron* **1986**, *42*, 5511; Rao, V.V.R.; Devaprabhakara, D.; Chandrasekaran, S. *J. Organomet. Chem.* **1978**, 162, C9; Parish, E.J.; Parish, S.; Honda, H. *Synth. Commun.* **1990**, *20*, 3265.

[666]Brown, H.C.; Kulkarni, S.U.; Rao, C.G.; Patil, V.D. *Tetrahedron* **1986**, *42*, 5515.

[667]Soderquist, J.A.; Martinez, J.; Oyola, Y.; Kock, I. *Tetrahedron Lett.* **2004**, *45*, 5541.

[668]See Shapland, P.; Vedejs, E. *J. Org. Chem.* **2004**, *69*, 4094.

[669]See, for example, Brown, H.C.; Unni, M.K. *J. Am. Chem. Soc.* **1968**, *90*, 2902; Brown, H.C.; Gallivan, Jr., R.M. *J. Am. Chem. Soc.* **1968**, *90*, 2906; Brown, H.C.; Sharp, R.L. *J. Am. Chem. Soc.* **1968**, *90*, 2915.

[670]Singaram, B.; Rangaishenvi, M.V.; Brown, H.C.; Goralski, C.T.; Hasha, D.L. *J. Org. Chem.* **1991**, *56*, 1543.

[671]Goralski, C.T.; Hasha, D.L.; Nicholson, L.W.; Singaram, B. *Tetrahedron Lett.* **1994**, *35*, 5165.

[672]Brown, H.C.; Khire, U.R.; Racherla, U.S. *Tetrahedron Lett.* **1993**, *34*, 15.

[673]Brown, H.C.; Vara Prasad, J.V.N. *J. Am. Chem. Soc.* **1986**, *108*, 2049.

can be prepared. Alcohols with moderate-to-excellent enantioselectivities have been

α-Pinene
Optically active

119

Optically active

obtained in this way.[674] However, **119** does not give good results with even moderately hindered alkenes; a better reagent for these compounds is isopinocampheylborane[675] although optical yields are lower. Limonylborane,[676] 2- and 4-dicaranylboranes,[677] a myrtanylborane,[678] and dilongifolylborane[679] have also been used. Other new asymmetric boranes have also been developed. The chiral cyclic boranes *trans*-2,15-dimethylborolanes (**51** and **52**) also add enantioselectively to alkenes (except

(R, R) **120**

121 (S, S)

alkenes of the form $RR'C=CH_2$) to give boranes of high optical purity.[680] When chiral boranes are added to trisubstituted alkenes of the form $RR'C=CHR''$, two new chiral centers are created, and, with **120** or **121**, only one of the four possible diastereomers is predominantly produced, in yields $> 90\%$.[680] This has been called *double-asymmetric synthesis*.[681] An alternative asymmetric synthesis of alcohols involves the reaction of catechol borane with an alkene in the presence

[674]For reviews of enantioselective syntheses with organoboranes, see Brown, H.C.; Singaram, B. *Acc. Chem. Res.* **1988**, *21*, 287; Srebnik, M.; Ramachandran, P.V. *Aldrichimica Acta* **1987**, *20*, 9; Brown, H.C.; Jadhav, P.K.; Singaram, B. *Mod. Synth. Methods*, **1986**, *4*, 307; Matteson, D.S. *Synthesis* **1986**, 973; Brown, H.C.; Jadhav, P.K., in Morrison, J.D. *Asymmetric Synthesis* Vol. 2, Academic Press, NY, **1983**, pp. 1–43. For a study of electronic effects, see Garner, C.M.; Chiang, S.; Nething, M.; Monestel, R. *Tetrahedron Lett.* **2002**, *43*, 8339.

[675]Brown, H.C.; Jadhav, P.K.; Mandal, A.K. *J. Org. Chem.* **1982**, *47*, 5074. See also, Brown, H.C.; Weissman, S.A.; Perumal, P.T.; Dhokte, U.P. *J. Org. Chem.* **1990**, *55*, 1217. For an improved method, see Brown, H.C.; Singaram, B. *J. Am. Chem. Soc.* **1984**, *106*, 1797; Brown, H.C.; Gupta, A.K.; Vara Prasad, J.V.N. *Bull. Chem. Soc. Jpn.* **1988**, *61*, 93. For the crystal structure of this adduct, see Soderquist, J.A.; Hwang-Lee, S.; Barnes, C.L. *Tetrahedron Lett.* **1988**, *29*, 3385.

[676]Jadhav, P.K.; Kulkarni, S.U. *Heterocycles* **1982**, *18*, 169.

[677]Brown, H.C.; Vara Prasad, J.V.N.; Zaidlewicz, M. *J. Org. Chem.* **1988**, *53*, 2911.

[678]Kiesgen de Richter, R.; Bonato, M.; Follet, M.; Kamenka, J. *J. Org. Chem.* **1990**, *55*, 2855.

[679]Jadhav, P.K.; Brown, H.C. *J. Org. Chem.* **1981**, *46*, 2988.

[680]Masamune, S.; Kim, B.M.; Petersen, J.S.; Sato, T.; Veenstra, J.S.; Imai, T. *J. Am. Chem. Soc.* **1985**, *107*, 4549.

[681]For another enantioselective hydroboration method, see p. 1082.

of a chiral rhodium catalyst, giving the alcohol enantioselectivity after the usual oxidation.[682]

The double bonds in a conjugated diene are hydroborated separately, that is, there is no 1,4-addition. However, it is not easy to hydroborate just one of a conjugated system, since conjugated double bonds are less reactive than isolated ones. Thexylborane[641] (**117**) is particularly useful for achieving the cyclic hydroboration of dienes, conjugated or nonconjugated, as in the formation of **122**.[683]

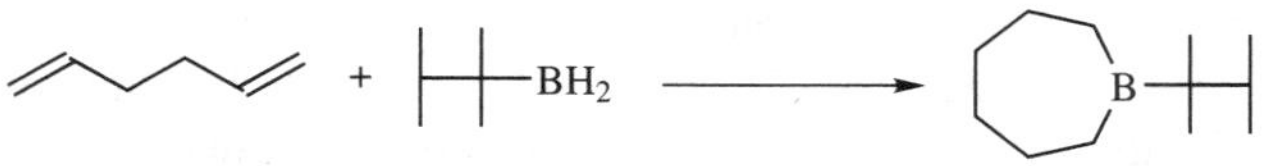

Rings of five, six, or seven members can be formed in this way. Similar cyclization can also be accomplished with other monoalkylboranes and, in some instances, with BH_3 itself.[684] One example is the formation of 9-BBN, shown above. Another is conversion of 1,5,9-cyclododecatriene to perhydro-9b-boraphenalene, **123**.[685] If a diene is treated with a diaminoborane and a samarium catalyst, oxidation leads to a carbocyclic ring with a pendant hydroxymethyl group.[686]

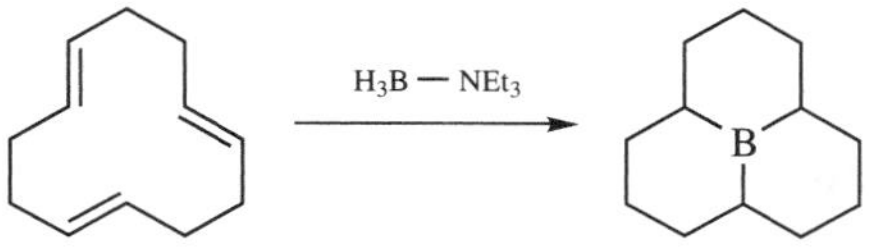

Triple bonds[687] can be monohydroborated to give vinylic boranes, which can be reduced with carboxylic acids to cis-alkenes or oxidized and hydrolyzed to aldehydes or ketones. Terminal alkynes give aldehydes by this method, in contrast to the mercuric or acid-catalyzed addition of water discussed at **15-4**. However, terminal alkynes give vinylic boranes[688] (and hence aldehydes) only when treated with a hindered borane, such as **116**, **117**, or catecholborane (p. 820),[689] or with $BHBr_2-SMe_2$.[690] The reaction between terminal alkynes and BH_3 produces

---

[682]Demay, S.; Volant, F.; Knochel, P. *Angew. Chem. Int. Ed.* **2001**, *40*, 1235.

[683]Brown, H.C.; Negishi, E. *J. Am. Chem. Soc.* **1972**, *94*, 3567.

[684]For a review of cyclic hydroboration, see Brown, H.C.; Negishi, E. *Tetrahedron* **1977**, *33*, 2331. See also, Brown, H.C.; Pai, G.G.; Naik, R.G. *J. Org. Chem.* **1984**, *49*, 1072.

[685]Rotermund, G.W.; Köster, R. *Liebigs Ann. Chem.* **1965**, *686*, 153; Brown, H.C.; Negishi, E.; Dickason, W.C. *J. Org. Chem.* **1985**, *50*, 520.

[686]Molander, G.A; Pfeiffer, D. *Org. Lett.* **2001**, *3*, 361.

[687]For a review of hydroboration of triple bonds, see Hudrlik, P.F.; Hudrlik, A.M., in Patai, S. *The Chemistry of the Carbn–Carbon Triple Bond*, pt. 1, Wiley, NY, **1978**, pp. 203–219.

[688]For a review of the preparation and reactions of vinylic boranes, see Brown, H.C.; Campbell, Jr., J.B. *Aldrichimica Acta* **1981**, *14*, 1.

[689]Brown, H.C.; Gupta, S.K. *J. Am. Chem. Soc.* **1975**, *97*, 5249. For a review of catecholborane, see Lane, C.F.; Kabalka, G.W. *Tetrahedron* **1976**, *32*, 981; Garrett, C.E.; Fu, G.C. *J. Org. Chem.* **1996**, *61*, 3224.

[690]Brown, H.C.; Campbell Jr., J.B. *J. Org. Chem.* **1980**, *45*, 389.

1,1-dibora compounds, which can be oxidized either to primary alcohols (with NaOH–H_2O_2) or to carboxylic acids (with *m*-chloroperoxybenzoic acid).[691] Double bonds can be hydroborated in the presence of triple bonds if the reagent is 9-BBN.[692] On the other hand, dimesitylborane selectively hydroborates triple bonds in the presence of double bonds.[693] Furthermore, it is often possible to hydroborate selectively one particular double bond of a nonconjugated diene.[694] A triple bond can be hydroborated in the presence of a ketone, and treatment with acetic acid reduces the C≡C unit to a cis- alkene (see **15-12**).[695] When the reagent is catecholborane, hydroboration is catalyzed by rhodium complexes,[696] such as Wilkinson's catalyst,[697] by SmI₂,[698] or lanthanide reagents.[699] Enantioselective hydroboration–oxidation has been achieved by the use of optically active rhodium complexes.[700]

A chain extension variation involved the reaction of styrene with catecholborane and then Me_3SiCHN_2.[701] Subsequent oxidation with NaOH/H_2O_2 and the reaction with Bu_4NF gave 3-phenyl-1-propanol.

An unusual extension of hydroboration involves remote C–H activation. Aryl alkenes are treated with borane and then oxidized in the usual manner. The product is a phenol and a hydroxymethyl group (Ph–C=C–CH₃ → *o*-Ph–CH–CH–CH₂OH.[702]

OS **VI**, 719, 852, 919, 943; **VII**, 164, 339, 402, 427; **VIII**, 532.

15-17 Other Hydrometalation

Hydro-metallo-addition

Metal hydrides of Groups 13 (III A) and 14 (IV B) of the periodic table (e.g., AlH₃, GaH₃) as well as many of their alkyl and aryl derivatives (e.g., R_2AlH,

[691]Zweifel, G.; Arzoumanian, H. *J. Am. Chem. Soc.* **1967**, *89*, 291.

[692]Brown, H.C.; Coleman, R.A. *J. Org. Chem.* **1979**, *44*, 2328.

[693]Pelter, A.; Singaram, S.; Brown, H.C. *Tetrahedron Lett.* **1983**, *24*, 1433.

[694]For a list of references, see Gautam, V.K.; Singh, J.; Dhillon, R.S. *J. Org. Chem.* **1988**, *53*, 187. See also, Suzuki, A.; Dhillon, R.S. *Top. Curr. Chem.* **1986**, *130*, 23.

[695]Kabalka, G.W.; Yu, S.; Li, N.-S. *Tetrahedron Lett.* **1997**, *38*, 7681.

[696]Burgess, K.; van der Donk, W.A.; Westcott, S.A.; Marder, T.B.; Baker, R.T. ; Calabrese, J.C. *J. Am. Chem. Soc.* **1992**, *114*, 9350; Wescott, S.A.; Blom, H.P.; Marder, T.B.; Baker, R.T. *J. Am. Chem. Soc.* **1992**, *114*, 8863; Evans, D.A.; Fu, G.C.; Hoveyda, A.H. *J. Am. Chem. Soc.* **1992**, *114*, 6671.

[697]Männig, D.; Nöth, H. *Angew. Chem. Int. Ed.* **1985**, *24*, 878. For a review, see Burgess, K.; Ohlmeyer, M.J. *Chem. Rev.* **1991**, *91*, 1179.

[698]Evans, D.A.; Muci, A.R.; Stürmer, R. *J. Org. Chem.* **1993**, *58*, 5307.

[699]Harrison, K.N.; Marks, T.J. *J. Am. Chem. Soc.* **1992**, *114*, 9220.

[700]Burgess, K.; Ohlmeyer, M.J. *J. Org. Chem.* **1988**, *53*, 5178; Hayashi, T.; Matsumoto, Y.; Ito, Y. *J. Am. Chem. Soc.* **1989**, *111*, 3426; Sato, M.; Miyaura, N.; Suzuki, A. *Tetrahedron Lett.* **1990**, *31*, 231; Brown, J.M.; Lloyd-Jones, G.C. *Tetrahedron: Asymmetry* **1990**, *1*, 869.

[701]Goddard, J.-P.; LeGall, T.; Mioskowski, C. *Org. Lett.* **2000**, *2*, 1455.

[702]Varela, J.A.; Peña, D.; Goldfuss, B.; Polborn, K.; Knochel, P. *Org. Lett.* **2001**, *3*, 2395.

Ar$_3$SnH) add to double bonds to give organometallic compounds.[703] The hydroboration reaction (**15-16**) is the most important example, but other important metals in this reaction are aluminum,[704] tin,[705] and zirconium[706] [a Group 4 (IV B) metal]. Some of these reactions are uncatalyzed, but in other cases various types of catalyst have been used.[707] Hydrozirconation is most commonly carried out with Cp$_2$ZrHCl (Cp = cyclopentadienyl),[708] known as *Schwartz's reagent*. The mechanism with Group 13 (III A) hydrides seems to be electrophilic (or four-centered pericyclic with some electrophilic characteristics) while with Group 14 (IV A) hydrides a mechanism involving free radicals seems more likely. Dialkylmagnesium reagents have been obtained by adding MgH$_2$ to double bonds.[709] With Grignard reagents such as RMgX, the Grignard reagent can be added to an alkene R'CH=CH$_2$ to give R'CH$_2$CH$_2$MgX, with TiCl$_4$ as a catalyst.[710] With some reagents triple bonds[711] can add 1 or 2 equivalents, to give **124** or **125**.[712]

$$R-C\equiv C-H \xrightarrow{R'_2AlH} \underset{\textbf{124}}{\overset{H \qquad\quad H}{\underset{R}{C}=\underset{AlR'_2}{C}}} \xrightarrow{R'_2AlH} \underset{\textbf{125}}{\overset{H \quad\; H}{\underset{H}{R-C}-\underset{AlR'_2}{C-AlR'_2}}}$$

[703]Negishi, E. *Adv. Met.-Org. Chem.* **1989**, *1*, 177; Eisch, J.J. *The Chemistry of Organometallic Compounds*; Macmillan, NY, **1967**, pp. 107–111. See also, Eisch, J.J.; Fichter, K.C. *J. Organomet. Chem.* **1983**, *250*, 63.

[704]For reviews of organoaluminums in organic synthesis, see Dzhemilev, U.M.; Vostrikova, O.S.; Tolstikov, G.A. *Russ. Chem. Rev.* **1990**, *59*, 1157; Maruoka, K.; Yamamoto, H. *Tetrahedron* **1988**, *44*, 5001.

[705]For a review with respect to Al, Si, and Sn, see Negishi, E. *Organometallics in Organic Synthesis*, Vol. 1, Wiley, NY, **1980**, pp. 45–48, 357–363, 406–412. For reviews of hydrosilylation, see Ojima, I. in Patai, S.; Rappoport, Z. *The Chemistry of Organic Silicon Compounds*, pt. 2, Wiley, NY, **1989**, pp. 1479–1526; Alberti, A.; Pedulli, G.F. *Rev. Chem. Intermed.* **1987**, *8*, 207; Speier, J.L. *Adv. Organomet. Chem.* **1979**, *17*, 407; Andrianov, K.A.; Souč ek, J.; Khananashvili, L.M. *Russ. Chem. Rev.* **1979**, *48*, 657.

[706]For reviews of hydrozirconation, and the uses of organozirconium compounds, see Negishi, E.; Takahashi, T. *Synthesis* **1988**, 1; Dzhemilev, U.M.; Vostrikova, O.S.; Tolstikov, G.A. *J. Organomet. Chem.* **1986**, *304*, 17; Schwartz, J.; Labinger, J.A. *Angew. Chem. Int. Ed.* **1976**, *15*, 333. Also see Hoveyda, A.H.; Morken, J.P. *J. Org. Chem.* **1993**, *58*, 4237.

[707]See, for example, Oertle, K.; Wetter, H. *Tetrahedron Lett.* **1985**, *26*, 5511; Randolph, C.L.; Wrighton, M.S. *J. Am. Chem. Soc.* **1986**, *108*, 3366; Maruoka, K.; Sano, H.; Shinoda, K.; Nakai, S.; Yamamoto, H. *J. Am. Chem. Soc.* **1986**, *108*, 6036; Miyake, H.; Yamamura, H. *Chem. Lett.* **1989**, 981; Doyle, M.P.; High, K.G.; Nesloney, C.L.; Clayton, Jr., T.W.; Lin, J. *Organometallics* **1991**, *10*, 1225.

[708]For a method of preparing this reagent (which is also available commercially), see Buchwald, S.L.; LaMaire, S.J.; Nielsen, R.B.; Watson, B.T.; King, S.M. *Tetrahedron Lett.* **1987**, *28*, 3895. It can also be generated *in situ*: Lipshutz, B.H.; Keil, R.; Ellsworth, E.L. *Tetrahedron Lett.* **1990**, *31*, 7257.

[709]For a review, see Bogdanović, B. *Angew. Chem. Int. Ed.* **1985**, *24*, 262.

[710]For a review, see Sato, F. *J. Organomet. Chem.* **1985**, *285*, 53–64. For another catalyst, see Hoveyda, A.H.; Xu, Z. *J. Am. Chem. Soc.* **1991**, *113*, 5079.

[711]For a review of the hydrometalation of triple bonds, see Hudrlik, P.F.; Hudrlik, A.M., in Patai, S. *The Chemistry of the Carbon-Carbon Triple Bond*, pt. 1, Wiley, NY, **1978**, pp. 219–232.

[712]Wilke, G.; Müller, H. *Liebigs Ann. Chem.* **1960**, *629*, 222; Eisch, J.J.; Kaska, W.C. *J. Am. Chem. Soc.* **1966**, *88*, 2213; Eisch, J.J.; Rhee, S. *Liebigs Ann. Chem.* **1975**, 565.

When 2 equivalents are added, electrophilic addition generally gives 1,1-dimetallic products **125** (as with hydroboration), while free-radical addition usually gives the 1,2-dimetallic products.

OS **VII**, 456; **VIII**, 268, 295, 507; **80**, 104. See also, OS **VIII**, 277, 381.

G. Carbon or Silicon on the Other Side

15-18   Addition of Alkanes

Hydro-alkyl-addition

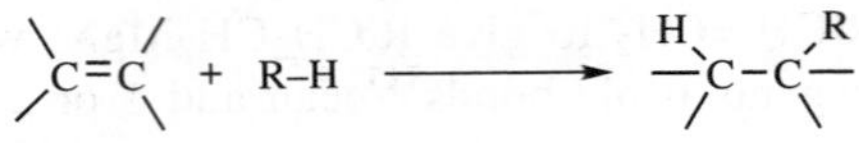

$$\diagdown C = C \diagup + \text{R–H} \longrightarrow \diagup \overset{H}{\underset{\diagdown}{C}} - \overset{R}{\underset{\diagup}{C}} \diagdown$$

There are two important ways of adding alkanes to alkenes: the thermal and the acid-catalysis method.[713] Both give chiefly mixtures, and neither is useful for the preparation of relatively pure compounds in reasonable yields. However, both are useful industrially. In the thermal method the reactants are heated to high temperatures ($\sim$500°C) at high pressures (150–300 atm) without a catalyst. As an example, propane and ethylene gave 55.5% isopentane, 7.3% hexanes, 10.1% heptanes, and 7.4% alkenes.[714] The mechanism is undoubtedly of a free-radical type and can be illustrated by one possible sequence in the reaction between propane and ethylene:

Step 1 $CH_3CH_2CH_3 + CH_2{=}CH_2 \overset{\Delta}{\longrightarrow} CH_3{-}\overset{\bullet}{CH}{-}CH_3 + CH_3CH_2\bullet$

Step 2 $CH_3{-}\overset{\bullet}{CH}{-}CH_3 + CH_2{=}CH_2 \longrightarrow (CH_3)_2CHCH_2CH_2\bullet$

Step 3 $(CH_3)_2CHCH_2CH_2\bullet + CH_3CH_2CH_3 \longrightarrow (CH_3)_2CHCH_2CH_3 + CH_3\overset{\bullet}{CH}CH_3$

There is kinetic evidence that the initiation takes place primarily by steps like 1, which are called *symproportionation* steps[715] (the opposite of disproportionation, p. 280).

In the acid-catalysis method, a proton or Lewis acid is used as the catalyst and the reaction is carried out at temperatures between -30 and 100°C. This is a Friedel–Crafts process with a carbocation mechanism[716] (illustrated for a proton

[713]For reviews, see Shuikin, N.I.; Lebedev, B.L. *Russ. Chem. Rev.* **1966**, *35*, 448; Schmerling, L., in Olah, G.A. *Friedel–Crafts and Related Reactions*, Vol. 2, Wiley, NY, **1964**, pp. 1075–1111, 1121–1122.

[714]Frey, E.J.; Hepp, H.J. *Ind. Eng. Chem.* **1936**, *28*, 1439.

[715]Metzger, J.O. *Angew. Chem. Int. Ed.* **1983**, *22*, 889; Hartmanns, J.; Klenke, K.; Metzger, J.O. *Chem. Ber.* **1986**, *119*, 488.

[716]For a review, see Mayr, H. *Angew. Chem. Int. Ed.* **1990**, *29*, 1371.

acid catalyst):

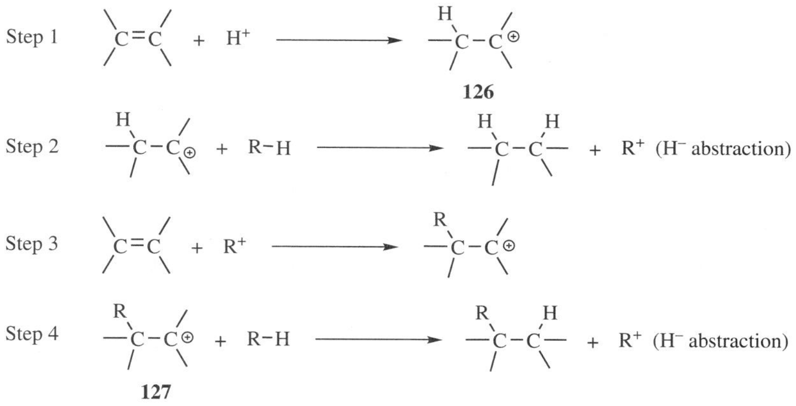

Step 1 $\begin{array}{c}\diagdown\\[-3pt]\diagup\end{array}C=C\begin{array}{c}\diagup\\[-3pt]\diagdown\end{array}$ + H$^+$ $\longrightarrow$ $\begin{array}{c}\mathrm{H}\\[-3pt]\diagdown\\[-3pt]-\overset{}{\underset{\diagup}{C}}-C\overset{\oplus}{\underset{\diagdown}{}}\end{array}$

126

Step 2 $\begin{array}{c}\mathrm{H}\\[-3pt]\diagdown\\[-3pt]-\overset{}{\underset{\diagup}{C}}-C\overset{\oplus}{\underset{}{}}\end{array}$ + R—H $\longrightarrow$ $\begin{array}{c}\mathrm{H}\ \ \ \mathrm{H}\\[-3pt]\diagdown\ \ \ \diagdown\\[-3pt]-\overset{}{\underset{\diagup}{C}}-\overset{}{\underset{\diagdown}{C}}-\end{array}$ + R$^+$ (H$^-$ abstraction)

Step 3 $\begin{array}{c}\diagdown\\[-3pt]\diagup\end{array}C=C\begin{array}{c}\diagup\\[-3pt]\diagdown\end{array}$ + R$^+$ $\longrightarrow$ $\begin{array}{c}\mathrm{R}\\[-3pt]\diagdown\\[-3pt]-\overset{}{\underset{\diagup}{C}}-C\overset{\oplus}{\underset{\diagdown}{}}\end{array}$

Step 4 $\begin{array}{c}\mathrm{R}\\[-3pt]\diagdown\\[-3pt]-\overset{}{\underset{\diagup}{C}}-C\overset{\oplus}{\underset{}{}}\end{array}$ + R—H $\longrightarrow$ $\begin{array}{c}\mathrm{R}\ \ \ \mathrm{H}\\[-3pt]\diagdown\ \ \ \diagup\\[-3pt]-\overset{}{\underset{\diagup}{C}}-\overset{}{\underset{\diagdown}{C}}-\end{array}$ + R$^+$ (H$^-$ abstraction)

127

Carbocation **127** often rearranges before it abstracts a hydride, explaining, for example, why the principal product from the reaction between isobutane and ethylene is 2,3-dimethylbutane. It is also possible for **126** (or **127**) instead of abstracting a hydride, to add to another mole of alkene, so that not only rearrangement products but also dimeric and polymeric products are frequent. If the tri- or tetrasubstituted alkenes are treated with Me_4Si, HCl, and $AlCl_3$, they become protonated to give a tertiary carbocation, which reacts with the Me_4Si to give a product that is the result of addition of H and Me to the original alkene.[717] (For a free-radical hydromethyl-addition, see **15-28**.) Addition a cation to a vinyl bromide, generated from an α-ethoxy-lactam with trifluoroacetic acid, generated a ketone.[718] An intramolecular cyclization of 1-dodecene to cyclododecane was reported using aluminum chloride in an ionic liquid.[719]

Alkanes add to alkynes under photolysis conditions to give an alkene.[720] Tetrahydrofuran adds to alkynes to give the alkene with microwave irradiation.[721]

The reaction can also be base catalyzed, in which case there is nucleophilic addition and a carbanion mechanism.[722] Carbanions most often used are those stabilized by one or more α-aryl groups. For example, toluene adds to styrene in the presence of sodium to give 1,3-diphenylpropane:[723]

$$PhCH_3 \xrightarrow{\ Na\ } PhCH_2^{\ominus} + PhCH=CH_2 \longrightarrow PhCH^{\ominus}-CH_2CH_2Ph \xrightarrow{\ solvent\ } PhCH_2CH_2CH_2Ph$$

[717]Bolestova, G.I.; Parnes, Z.N.; Kursanov, D.N. *J. Org. Chem. USSR* **1983**, *19*, 2175.
[718]Gesson, J.-P.; Jacquesy, J.-C.; Rambaud, D. *Tetrahedron* **1993**, *49*, 2239.
[719]In bmim Cl, 1-butyl-3-methylimidazolium chloride: Qiao, K.; Deng, Y. *Tetrahedron Lett.* **2003**, *44*, 2191.
[720]Geraghty, N.W.A.; Hannan, J.J. *Tetrahedron Lett.* **2001**, *42*, 3211.
[721]Zhang, Y.; Li, C.-J. *Tetrahedron Lett.* **2004**, *45*, 7581.
[722]For reviews, see Pines, H.; Stalick, W.M. *Base-Catalyzed Reactions of Hydrocarbons and Related Compounds*, Academic Press, NY, **1977**, pp. 240–422; Pines, H. *Acc. Chem. Res.* **1974**, *7*, 155; Pines, H.; Schaap, L.A. *Adv. Catal.* **1960**, *12*, 117, pp. 126.
[723]Pines, H.; Wunderlich, D. *J. Am. Chem. Soc.* **1958**, *80*, 6001.

Conjugated dienes give 1,4-addition.[724] This reaction has also been performed with salts of carboxylic acids in what amounts to a method of alkylation of carboxylic acids[725] (see also, **10-59**).

$$CH_3COOK \xrightarrow{\text{NaNH}_2} {}^{\ominus}CH_2COOK \; + \; CH_2=CH_2 \xrightarrow{\hspace{2cm}} {}^{\ominus}CH_2-CH_2CH_2COOK$$

There are transition-metal catalyzed addition reaction of alkyl units to alkenes,[726] often proceeding with metal hydride elimination to form an alkene. An intra-molecular cyclization reaction of an *N*-pyrrolidino amide alkene was reported using an iridium catalyst for addition of the carbon α to nitrogen to the alkene unit.[727]

OS **I**, 229; **IV**, 665; **VII**, 479.

15-19   Addition of Silanes

Silyl-hydro-addition

$$\underset{R'}{\overset{R}{>}}C=C\underset{\backslash}{\overset{/}{}} \; + \; (R^1)_{4-n}-SiH_n \xrightarrow{\text{catalyst}} \underset{R}{\overset{H}{>}}C-C\overset{Si-(R^1)_{4-n}}{\underset{\backslash}{}}$$

Although silanes bearing at least one Si—H unit do not generally react with alkenes or alkynes, in the presence of certain catalyst addition occurs to give the corresponding alkyl or vinyl silane. The reaction of an alkene with an yttrium,[728] ruthenium,[729] rhodium,[730] palladium,[731] lanthanum,[732] platinum[733], or samarium[734] catalyst addition occurs with high anti-Markovnikov selectivity. Silanes add to dienes with a palladium catalyst, and asymmetric induction is achieved by using a binapthyl additive.[735] Alkenes react with Li-(0) and *t*-Bu$_2$SiCl$_2$ to give a three-membered ring silane.[736] In the presence of BEt$_3$, silanes add to alkynes to give the corresponding vinyl silane[737] or to alkenes to give the alkylsilane, with

[724]Eberhardt, G.G.; Peterson, H.J. *J. Org. Chem.* **1965**, *30*, 82; Pines, H.; Stalick, W.M. *Tetrahedron Lett.* **1968**, 3723.

[725]Schmerling, L.; Toekelt, W.G. *J. Am. Chem. Soc.* **1962**, *84*, 3694.

[726]Kakiuchi, F.; Murai, S. *Acc. Chem. Res.* **2002**, *35*, 826.

[727]DeBoef, B.; Pastine, S.J.; Sames, D. *J. Am. Chem. Soc.* **2004**, *126*, 6556.

[728]Molander, G.A.; Julius, M. *J. Org. Chem.* **1992**, *57*, 6347.

[729]Glaser, P.B.; Tilley, T.D. *J. Am. Chem.Soc.* **2003**, *125*, 13640.

[730]Itami, K.; Mitsudo, K.; Nishino, A.; Yoshida, J.-i. *J. Org. Chem.* **2002**, *67*, 2645; Tsuchiya, Y.; Uchimura, H.; Kobayashi, K.; Nishiyama, H. *Synlett* **2004**, 2099.

[731]Motoda, D.; Shinokubo, H.; Oshima, K. *Synlett* **2002**, 1529.

[732]Takaki, K.; Sonoda, K.; Kousaka, T.; Koshoji, G.; Shishido, T.; Takehira, K. *Tetrahedron Lett.* **2001**, *42*, 9211.

[733]Perales, J.B.; van Vranken, D.L. *J. Org. Chem.* **2001**, *66*, 7270; Sabourault, N.; Mignani, G.; Wagner, A.; Mioskowski, C. *Org. Lett.* **2002**, *4*, 2117.

[734]Hou, Z.; Zhang, Y.; Tardif, O.; Wakatsuki, Y. *J. Am. Chem. Soc.* **2001**, *123*, 9216.

[735]Hatanaka, Y.; Goda, K.; Yamashita, F.; Hiyama, T. *Tetrahedron Lett.* **1994**, *35*, 7981.

[736]Driver, T.G.; Franz, A.K.; Woerpel, K.A. *J. Am. Chem. Soc.* **2002**, *124*, 6524.

[737]Miura, K.; Oshima, K.; Utimoto, K. *Bull. Chem. Soc. Jpn.* **1993**, *66*, 2356.

anti-Markovnikov selectivity.[738] Similar selectivity was observed when a silylated zinc reagent was added to a terminal alkyne.[739] Silanes add to alkynes to give a vinyl silane using Cp_2TiCl_2–n-butyllithium.[740] Siloxanes such as $(RO)_3SiH$ add to alkynes with a ruthenium catalyst to give the corresponding vinyl silane.[741] The reaction of Cl_2MeSiH and terminal alkynes, in ethanol–triethylamine with a ruthenium catalyst, to give primarily the Markovnikov vinyl silane.[742] However, Et_3SiH adds to terminal alkynes with a rhodium[743] or a platinum[744] catalyst to give the anti-Markovnikov vinyl silane. Using 0.5 equivalent of $HfClO_4$ with alkynes bearing a dimethylphenylsilyl unit gave a cyclic vinyl silane with transfer of the phenyl group to carbon (see **128**).[745] Dienes react with zirconium compounds and silanes to produce cyclic compounds in which the silyl group has also added to one C=C unit.[746] With an yttrium catalyst, $PhSiH_3$ reacts with nonconjugated dienes to give cyclic alkenes with a pendant CH_2SiH_2Ph group.[747] Rhodium compounds allow silanes to add to enamides to give the α-silylamide.[748] Allylsilanes add to certain allylic alcohols in the presence of Me_3SiOTf, via a S_N2'-like reaction, to give dienes.[749] Note that silanes open cyclopropane rings in the presence of 20% $AlCl_3$ to give the alkylsilane.[750] Formation of silanes via reaction with alkenes can be followed by reaction with fluoride ion and then oxidation to give an alcohol[751] (see **10-16**).

128

Silanes also add to alkenes under radical conditions (using AIBN) with high anti-Markovnikov selectivity.[752] An alternative route to alkylsilanes reacted an alkene with lithium metal in the presence of 3 equivalents of chlorotrimethylsilane, giving bis-1,2-trimethylsilyl compounds after treatment with water.[753] Silanes also

[738]Rubin, M.; Schwier, T.; Gevorgyan, V. *J. Org. Chem.* **2002**, *67*, 1936.

[739]Nakamura, S.; Uchiyama, M.; Ohwada, T. *J. Am. Chem. Soc.* **2004**, *126*, 11146.

[740]Takahashi, T.; Bao, F.; Gao, G.; Ogasawara, M. *Org. Lett.* **2003**, *5*, 3479.

[741]Trost, B.M.; Ball, Z.T. *J. Am. Chem. Soc.* **2001**, *123*, 12726.

[742]Kawanami, Y.; Sonoda, Y.; Mori, T.; Yamamoto, K. *Org. Lett.* **2002**, *4*, 2825.

[743]Sato, A.; Kinoshita, H.; Shinokubo, H.; Oshima, K. *Org. Lett.* **2004**, *6*, 2217.

[744]Wu, W.; Li, C.-J. *Chem. Commun.* **2003**, 1668.

[745]Asao, N.; Shimada, T.; Shimada, T.; Yamamoto, Y. *J. Am. Chem. Soc.* **2001**, *123*, 10899. See also, Sudo, T.; Asao, N.; Yamamoto, Y. *J. Org. Chem.* **2000**, *65*, 8919.

[746]Molander, G.A.; Corrette, C.P. *Tetrahedron Lett.* **1998**, *39*, 5011.

[747]Muci, A.R.; Bercaw, J.E. *Tetrahedron Lett.* **2000**, *41*, 7609.

[748]Murai, T.; Oda, T.; Kimura, F.; Onishi, H.; Kanda, T.; Kato, S. *J. Chem. Soc., Chem. Commun.* **1994**, 2143.

[749]Toshima, K.; Ishizuka, T.; Matsuo, G.; Nakata, M. *Tetrahedron Lett.* **1994**, *35*, 5673.

[750]Nagahara, S.; Yamakawa, T.; Yamamoto, H. *Tetrahedron Lett.* **2001**, *42*, 5057.

[751]Jensen, J.F.; Svendsen, B.H.; la Cour, T.V.; Pedersen, H.L.; Johannsen, M. *J. Am. Chem. Soc.* **2002**, *124*, 4558.

[752]Kopping, B.; Chatgilialoglu, C.; Zehnder, M.; Giese, B. *J. Org. Chem.* **1992**, *57*, 3994.

[753]Yus, M.; Martínez, P.; Guijarro, D. *Tetrahedron* **2001**, *57*, 10119.

add to alkenes to form anti-Markovnikov alkylsilane (R_3Si—C—C—R') in the presence of a hyponitrite.[754]

Vinyl silanes add to conjugated carbonyl compounds in the presence of a ruthenium catalyst,[755] or to acrylonitriles with a cobalt catalyst.[756] Silyl phosphines react with conjugated ynones directly to give an enone with an α-trimethylsilyl and a β-phosphine group.[757] Siloxanes of the type $(RO)_3SiH$ add to the α-carbon of enamines in the presence of a dirhodium catalyst.[758] The uncatalyzed reaction of trimethylsilyl cyanide and ynamines, however, gave an enamine with a β-trimethylsilyl and an α-cyano group.[759]

15-20   Addition of Alkenes and/or Alkynes to Alkenes and/or Alkynes

Hydro-alkenyl-addition

$$CH_2{=}CH_2 \ + \ CH_2{=}CH_2 \xrightarrow{\ \ H^+ \ \ } CH_2{=}CHCH_2CH_3$$

With certain substrates, alkenes can be dimerized by acid catalysts, so that the product is a dimer that contains one double bond.[760] A combination of zinc and a $CoCl_2$ catalyst accomplished the same type of coupling.[761] One alkene adds to another in the presence of a nickel catalyst.[762] Coupling conjugated alkenes with vinyl esters to give a functionalized conjugated diene is known, using a complex palladium–vanadium catalyst in an oxygen atmosphere.[763] This reaction is more often carried out internally, as in the formation of cyclohexene **129**. A palladium catalyzed cyclization is known, in which dienes are converted to cyclopentene derivatives such as **130**.[764] Ring-forming reactions with heterocyclic compounds such as indoles are known using $PtCl_2$.[765] A ruthenium catalyzed version of this reaction gave the five-membered ring with an exocyclic double bond.[766] Carbocyclization of an alkene unit to another alkene unit was reported

[754]Dang, H.-S.; Roberts, B.P. *Tetrahedron Lett.* **1995**, *36*, 2875.

[755]Kakiuchi, F.; Tanaka, Y.; Sato, T.; Chatani, N.; Murai, S. *Chem. Lett.* **1995**, 679; Trost, B.M.; Imi, K.; Davies, I.W. *J. Am. Chem. Soc.* **1995**, *117*, 5371.

[756]Tayama, O.; Iwahama, T.; Sakaguchi, S.; Ishii, Y. *Eur. J. Org. Chem.* **2003**, 2286.

[757]Reisser, M.; Maier, A.; Maas, G. *Synlett* **2002**, 1459.

[758]Hewitt, G.W.; Somers, J.J.; Sieburth, S.Mc.N. *Tetrahedron Lett.* **2000**, *41*, 10175.

[759]Lukashev, N.V.; Kazantsev, A.V.; Borisenko, A.A.; Beletskaya, I.P. *Tetrahedron* **2001**, *57*, 10309.

[760]For a review, see Onsager, O.; Johansen, J.E., in Hartley, F.R.; Patai, S. *The Chemistry of the Metal-Carbon Bond*, Vol. 3, Wiley, NY, **1985**, pp. 205–257.

[761]Wang, C.-C.; Lin, P.-S.; Cheng, C.-H. *Tetrahedron Lett.* **2004**, *45*, 6203.

[762]RajanBabu, T.V.; Nomura, N.; Jin, J.; Nandi, M.; Park, H.; Sun, X. *J. Org. Chem.* **2003**, *68*, 8431.

[763]Hatamoto, Y.; Sakaguchi, S.; Ishii, Y. *Org. Lett.* **2004**, *6*, 4623.

[764]Kisanga, P.; Goj, L.A.; Widenhoefer, R.A. *J. Org. Chem.* **2001**, *66*, 635.

[765]Liu, C.; Han, X.; Wang, X.; Widenhoefer, R.A. *J. Am. Chem. Soc.* **2004**, *126*, 3700.

[766]Yamamoto, Y.; Nakagai, Y.-i.; Ohkoshi, N.; Itoh, K. *J. Am. Chem. Soc.* **2001**, *123*, 6372; Mori, M.; Saito, N.; Tanaka, D.; Takimoto, M.; Sato, Y. *J. Am. Chem. Soc.* **2003**, *125*, 5606; Michaut, M.; Santelli, M.; Parrain, J.-L. *Tetrahedron Lett.* **2003**, *44*, 2157.

using an yttrium catalyst,[767] or a titanium catalyst.[768] In some cases, internal coupling of two alkenes can form larger rings.[769] Variations include treatment of similar dienes with $HSiMe_2OSiMe_3$ and KF-acetic acid to give a cyclopentane with a pendant trimethylsilylmethyl group trans to a methyl.[770] Exo-dig carbocyclization was reported using $HfCl_4$[771] palladium,[772] or titanium,[773] catalysts. Alkynes also add to alkenes for form rings in the presence of a palladium,[774] rhodium,[775] ruthenium,[776] iridium,[777] or a zirconium catalyst.[778] Alkene allene substrates were cyclized to form cyclic products with an exocyclic double bond using a palladium catalyst.[779] An interesting variation adds a silyl enol ether to an alkyne using $GaCl_3$ to give an unconjugated ketone ($O=C-C-C=C$).[780] Alkenes and alkynes can also add to each other to give cyclic products in other ways (see **15-63** and **15-65**).

129

130

Processes of this kind are important in the biosynthesis of steroids and tetra- and pentacyclic terpenes. For example, squalene 2,3-oxide is converted by enzymatic

[767]Molander, G.A.; Dowdy, E.D.; Schumann, H. *J. Org. Chem.* **1998**, *63*, 3386.

[768]Okamoto, S.; Livinghouse, T. *J. Am. Chem. Soc.* **2000**, *122*, 1223. See Hart, D.J.; Bennett, C.E. *Org. Lett.* **2003**, *5*, 1499.

[769]Toyota, M.; Majo, V.J.; Ihara, M. *Tetrahedron Lett.* **2001**, *42*, 1555.

[770]Pei, T.; Widenhoefer, R.A. *J. Org. Chem.* **2001**, *66*, 7639.

[771]Imamura, K.-i.; Yoshikawa, E.; Gevorgyan, V.; Yamamoto, Y. *J. Am. Chem. Soc.* **1998**, *120*, 5339.

[772]Xie, X.; Lu, X. *Synlett* **2000**, 707.

[773]Berk, S.C.; Grossman, R.B.; Buchwald, S.L. *J. Am. Chem. Soc.* **1994**, *116*, 8593; *J. Am. Chem. Soc.* **1993**, *115*, 4912.

[774]Galland, J.-C.; Savignac, M.; Genêt, J.-P. *Tetrahedron Lett.* **1997**, *38*, 8695; Widenhoefer, R.A.; Perch, N.S. *Org. Lett.* **1999**, *1*, 1103.

[775]Wender, P.A.; Dyckman, A.J. *Org. Lett.* **1999**, *1*, 2089; Cao, P.; Wang, B.; Zhang, X. *J. Am. Chem. Soc.* **2000**, *122*, 6490; Cao, P.; Zhang, X. *Angew. Chem. Int. Ed.* **2000**, *39*, 4104.

[776]Fernández-Rivas, C.; Méndez, M.; Echavarren, A.M. *J. Am. Chem. Soc.* **2000**, *122*, 1221; Fürstner, A.; Ackermann, L. *Chem. Commun.* **1999**, 95.

[777]Chatani, N.; Inoue, H.; Morimoto, T.; Muto, T.; Murai, S. *J. Org. Chem.* **2001**, *66*, 4433.

[778]Miura, K.; Funatsu, M.; Saito, H.; Ito, H.; Hosomi, A. *Tetrahedron Lett.* **1996**, *37*, 9059; Kemp, M.I.; Whitby, R.J.; Coote, S.J. *Synlett* **1994**, 451; Wischmeyer, U.; Knight, K.S.; Waymouth, R.M. *Tetrahedron Lett.* **1992**, *33*, 7735. Also see Maye, J.P.; Negishi, E. *Tetrahedron Lett.* **1993**, *34*, 3359.

[779]Iodobenzene was added and a phenyl substituent was incorporated in the product. See Ohno, H.; Takeoka, Y.; Kadoh, Y.; Miyamura, K.; Tanaka, T. *J. Org. Chem.* **2004**, *69*, 4541.

[780]Yamaguchi, M.; Tsukagoshi, T.; Arisawa, M. *J. Am. Chem. Soc.* **1999**, *121*, 4074.

catalysis to dammaradienol.

Enzyme-H   Squalene 2,3-oxide   Dammaradienol

The squalene → lanosterol biosynthesis (which is a key step in the biosynthesis of cholesterol) is similar. The idea that the biosynthesis of such compounds involves this type of multiple ring closing was proposed in 1955 and is known as the *Stork–Eschenmoser hypothesis*.[781] Such reactions can also be carried out in the laboratory, without enzymes.[782] By putting cation-stabilizing groups at positions at which positive charges develop, Johnson and co-workers have been able to close as many as four rings stereoselectively and in high yield, in one operation.[783] An example is formation of **131**,[784] also known as the *Johnson polyene cyclization*.[785]

Lewis acids can be used to initiate this cyclization,[786] including EtAlCl$_2$ used for the coupling of an alkyne and an alkene.[787] Cyclization to a tricyclic systems that included formation of a dihydropyran ring was reported using mercuric

[781]Stork, G.; Burgstahler, A.W. *J. Am. Chem. Soc.* **1955**, *77*, 5068; Eschenmoser, A.; Ruzicka, L.; Jeger, O.; Arigoni, D. *Helv. Chim. Acta* **1955**, *38*, 1890.

[782]For reviews, see Gnonlonfoun, N. *Bull. Soc. Chim. Fr.* **1988**, 862; Sutherland, J.K. *Chem. Soc. Rev.* **1980**, *9*, 265; Johnson, W.S. *Angew. Chem. Int. Ed.* **1976**, *15*, 9; *Bioorg. Chem.* **1976**, *5*, 51; *Acc. Chem. Res.* **1968**, *1*, 1; van Tamelen, E.E. *Acc. Chem. Res.* **1975**, *8*, 152. For a review of the stereochemical aspects, see Bartlett, P.A., in Morrison, J.D. *Asymmetric Synthesis* Vol. 3, Academic Press, NY, **1985**, pp. 341–409.

[783]Guay, D.; Johnson, W.S.; Schubert, U. *J. Org. Chem.* **1989**, *54*, 4731 and references cited therein.

[784]Johnson, W.S.; Gravestock, M.B.; McCarry, B.E. *J. Am. Chem. Soc.* **1971**, *93*, 4332.

[785]Johnson, W.S. *Acc. Chem. Res.* **1968**, *1*, 1; Hendrickson, J.B. *The Molecules of Nature*, W.A. Benjamin, NY, **1965**, pp. 12–57; Kametani T.; Fukumoto, K. *Synthesis* **1972**, 657.

[786]Sen, S.E.; Roach, S.L.; Smith, S.M.; Zhang, Y.Z. *Tetrahedron Lett.* **1998**, *39*, 3969. For an asymmetric version using SnCl$_4$, see Ishihara, K.; Nakamura, S.; Yamamoto, H. *J. Am. Chem. Soc.* **1999**, *121*, 4906.

[787]Asao, N.; Shimada, T.; Yamamoto, Y. *J. Am. Chem. Soc.* **1999**, *121*, 3797.

bis(trifluorosulfonate) as an initiator.[788] A radical cyclization approach (**15-30**) to polyene cyclization using a seleno-ester anchor gave a tetracyclic system.[789]

The addition of alkenes to alkenes[790] can also be accomplished by bases.[791] Coupling reactions can occur using catalyst systems[792] consisting of nickel complexes and alkylaluminum compounds (known as *Ziegler catalysts*),[793] rhodium catalysts,[794] and other transition-metal catalysts, including iron.[795] The 1,4-addition of alkenes to conjugated dienes to give nonconjugated dienes[796] occurs with various transition-metal catalysts.

and the dimerization of 1,3-butadienes to octatrienes.[797] Ethylene adds to alkenes to form a new alkene in the presence of a nickel catalyst[798] or a zirconium catalyst,[799] to alkynes in the presence of a ruthenium catalyst[800] to form a diene, and allenes add to alkynes to give a diene with a titanium catalyst.[801]

In the presence of cuprous chloride and ammonium chloride, acetylene adds to another molecule of itself to give vinylacetylene.

[788]Gopalan, A.S.; Prieto, R.; Mueller, B.; Peters, D. *Tetrahedron Lett.* **1992**, *33*, 1679.

[789]Chen, L.; Gill, G.B.; Pattenden, G. *Tetrahedron Lett.* **1994**, *35*, 2593.

[790]For a review of alkene dimerization and oligomerization with all catalysts, see Fel'dblyum, V.Sh.; Obeshchalova, N.V. *Russ. Chem. Rev.* **1968**, *37*, 789.

[791]For a review, see Pines, H. *Synthesis* **1974**, 309.

[792]For reviews, see Pillai, S.M.; Ravindranathan, M.; Sivaram, S. *Chem. Rev.* **1986**, *86*, 353; Jira, R.; Freiesleben, W. *Organomet. React.* **1972**, *3*, 1, 117; Heck, R.F. *Organotransition Metal Chemistry,* Academic Press, NY, **1974**, pp. 84–94, 150–157; Khan, M.M.T.; Martell, A.E. *Homogeneous Catalysis by Metal Complexes,* Vol. 2, Academic Press, NY, **1974**, pp. 135–15; Rylander, P.N. *Organic Syntheses with Noble Metal Catalysts,* Academic Press, NY, **1973**, pp. 175–196; Tsuji, J. *Adv. Org. Chem.* **1969**, *6*, 109, pp. 213. Also see, Kaur, G.; Manju, K.; Trehan, S. *Chem. Commun.* **1996**, 581.

[793]See, for example, Onsager, O.; Wang, H.; Blindheim, U. *Helv. Chim. Acta* **1969**, *52*, 187, 230; Fischer, K.; Jonas, K.; Misbach, P.; Stabba, R.; Wilke, G. *Angew. Chem. Int. Ed.* **1973**, *12*, 943.

[794]Cramer, R. *J. Am. Chem. Soc.* **1965**, *87*, 4717; *Acc. Chem. Res.* **1968**, *1*, 186; Kobayashi, Y.; Taira, S. *Tetrahedron* **1968**, *24*, 5763; Takahashi, N.; Okura, I.; Keii, T. *J. Am. Chem. Soc.* **1975**, *97*, 7489.

[795]Takacs, J.M.; Myoung, Y.C. *Tetrahedron Lett.* **1992**, *33*, 317.

[796]Alderson, T.; Jenner, E.L.; Lindsey, Jr., R.V. *J. Am. Chem. Soc.* **1965**, *87*, 5638; Hilt, G.; du Mesnil, F.-X.; Lüers, S. *Angew. Chem. Int. Ed.* **2001**, *40*, 387. For a review see Su, A.C.L. *Adv. Organomet. Chem.* **1979**, *17*, 269.

[797]See, for example, Denis, P.; Jean, A.; Croizy, J.F.; Mortreux, A.; Petit, F. *J. Am. Chem. Soc.* **1990**, *112*, 1292.

[798]Nomura, N.; Jin, J.; Park, H.; RajanBabu, T.V. *J. Am. Chem. Soc.* **1998**, *120*, 459; Monteiro, A.L.; Seferin, M.; Dupont, J.; de Souza, R.F. *Tetrahedron Lett.* **1996**, *37*, 1157.

[799]Takahashi, T.; Xi, Z.; Fischer, R.; Huo, S.; Xi, C.; Nakajima, K. *J. Am. Chem. Soc.* **1997**, *119*, 4561; Takahashi, T.; Xi, Z.; Rousset, C.J.; Suzuki, N. *Chem. Lett.* **1993**, 1001.

[800]Kinoshita, A.; Sakakibara, N.; Mori, M. *J. Am. Chem. Soc.* **1997**, *119*, 12388; Trost, B.M.; Indolese, A. *J. Am. Chem. Soc.* **1993**, *115*, 4361.

[801]Urabe, H.; Takeda, T.; Hideura, D.; Sato, F. *J. Am. Chem. Soc.* **1997**, *119*, 11295.

This type of alkyne dimerization is also catalyzed by nickel,[802] palladium,[803] lutetium,[804] and ruthenium catalysts.[805] Similar products are obtained by the cross-coupling an terminal alkynes with allene, using a combination of palladium and CuI catalysts.[806] The reaction has been carried out internally to convert diynes to large-ring cycloalkynes with an exocyclic double bond.[807] Diynes have also been cyclized to form cyclic enynes (an endocyclic double bond) using a diruthenium catalyst with ammonium tetrafluoroborate in methanol.[808] Enynes are similarly cyclized to cyclic alkenes with an endocyclic C=C unit, analogous to formation of **200** above, using a dicobalt catalyst.[809] A molecule containing two distal conjugated diene units was cyclized to give a bicyclic molecule with an exocyclic double bond using a palladium catalyst.[810] A nickel catalyst converted a similar system to a saturated five-membered ring containing an allylic group and a vinyl group.[811]

In another type of alkyne dimerization is the reductive coupling in which two molecules of alkyne, the same or different, give a 1,3-diene.[812]

$$R^1-C\equiv C-R^2 \ + \ R^3-C\equiv C-R^4 \longrightarrow \begin{array}{c} R^2 \quad R^3 \\ \diagdown \ / \\ C-C \\ \diagup \quad \diagdown \\ R^1-C \quad\quad C-R^4 \\ | \quad\quad | \\ H \quad\quad H \end{array}$$

In this method, one alkyne is treated with Schwartz's reagent (see **15-17**) to produce a vinylic zirconium intermediate. Addition of MeLi or MeMgBr, followed by the second alkyne, gives another intermediate, which, when treated with aqueous acid, gives the diene in moderate-to-good yields. The stereoisomer shown is the one formed in usually close to 100% purity. If the second intermediate is treated with I_2 instead of aqueous acid, the 1,4-diiodo-1,3-diene is obtained instead, in comparable yield and isomeric purity. The reaction of alkynes with two equivalents of trimethylsilyldiazomethane and a ruthenium catalyst gave a conjugated diene with trimethylsilyl groups at C-1 and C-4.[813] Alkynes can also be coupled to allylic silyl ethers with a ruthenium catalysts to give dienes.[814] Other alkyne–allylic coupling reactions are known to give dienes.[815]

[802]Ogoshi, S.; Ueta, M.; Oka, M.-A.; Kurosawa, H. *Chem. Commun.* **2004**, 2732.

[803]Rubina, M.; Gevergyan, V. *J. Am. Chem. Soc.* **2001**, *123*, 11107; Yang, C.; Nolan, S.P. *J. Org. Chem.* **2002**, *67*, 591.

[804]Nishiura, M.; Hou, Z.; Wakatsuki, Y.; Yamaki, T.; Miyamoto, T. *J. Am. Chem. Soc.* **2003**, *125*, 1184.

[805]Smulik, J.A.; Diver, S.T. *J. Org. Chem.* **2000**, *65*, 1788.

[806]Bruyere, D.; Grigg, R.; Hinsley, J.; Hussain, R.K.; Korn, S.; Del Cierva, C.O.; Sridharan, V.; Wang, J. *Tetrahedron Lett.* **2003**, *44*, 8669.

[807]Trost, B.M.; Matusbara, S.; Carninji, J.J. *J. Am. Chem. Soc.* **1989**, *111*, 8745.

[808]Nishibayashi, Y.; Yamanashi, M.; Wakiji, I.; Hidai, M. *Angew. Chem. Int. Ed.* **2000**, *39*, 2909.

[809]Ajamian, A.; Gleason, J.L. *Org. Lett.* **2003**, *5*, 2409.

[810]Takacs, J.M.; Leonov, A.P. *Org. Lett.* **2003**, *5*, 4317.

[811]Takimoto, M.; Mori, M. *J. Am. Chem. Soc.* **2002**, *124*, 10008; Takimoto, M.; Nakamura, Y.; Kimura, K.; Mori, M. *J. Am. Chem. Soc.* **2004**, *126*, 5956.

[812]Buchwald, S.L.; Nielsen, R.B. *J. Am. Chem. Soc.* **1989**, *111*, 2870.

[813]Le Paih, J.; Dérien, S.; Özdemir, I.; Dixneuf, P.H. *J. Am. Chem. Soc.* **2000**, *122*, 7400.

[814]Trost, B.M.; Surivet, J.-P.; Toste, F.D. *J. Am. Chem. Soc.* **2001**, *123*, 2897.

[815]Trost, B.M.; Pinkerton, A.B.; Toste, F.D.; Sperrle, M. *J. Am. Chem. Soc.* **2001**, *123*, 12504; Giessert, A.J.; Snyder, L.; Markham, J.; Diver, S.T. *Org. Lett.* **2003**, *5*, 1793.

This reaction can also be done intramolecularly, as in the cyclization of diyne **132** to (*E,E*)-exocyclic dienes **133** by treatment with a zirconium,[816] rhodium,[817] or platinum complex.[818] A similar reaction was reported using a titanium catalyst from a diyne amide.[819]

Rings of four, five, and six members were obtained in high yield; seven-membered rings in lower yield. When the reaction is applied to enynes, compounds similar to **133** are formed using various catalysts, but with only one double bond[820] Internal coupling of alkene–allenes and a rhodium catalyst give similar products bearing a pendant vinyl group.[821] With a $PtCl_2$ catalyst, ring closure leads to a diene in some cases.[822] Larger rings can be formed from the appropriate enyne, including forming cyclohexadiene compounds.[823] Spirocyclic compounds can be prepared from enynes in this manner using formic acid and a palladium catalyst.[824] Enynes can also be converted to bicyclo[3.1.0]hexenes[825] or nonconjugated cyclohexadienes[826] using a gold catalyst. Internal coupling of an alkyne and a vinylidene cyclopropane unit with a palladium catalyst leads to a cyclopentene derivative with an exocyclic double bond.[827] Enynes having a conjugated alkene unit also undergo this reaction

[816]Nugent, W.A.; Thorn, D.L.; Harlow, R.L. *J. Am. Chem. Soc.* **1987**, *109*, 2788. See also, Trost, B.M.; Lee, D.C. *J. Am. Chem. Soc.* **1988**, *110*, 7255; Tamao, K.; Kobayashi, K.; Ito, Y. *J. Am. Chem. Soc.* **1989**, *111*, 6478.

[817]Jang, H.-Y.; Krische, M.J. *J. Am. Chem. Soc.* **2004**, *126*, 7875.

[818]Madine, J.W.; Wang, X.; Widenhoefer, R.A. *Org. Lett.* **2001**, *3*, 385; Méndez, M.; Muñoz, M.P. ; Nevado, C.; Cárdenas, D.J.; Echavarren, A.M. *J. Am. Chem. Soc.* **2001**, *123*, 10511; Wang, X.; Chakrapani, H.; Madine, J.W.; Keyerleber, M.A.; Widenhoefer, R.A. *J. Org. Chem.* **2002**, *67*, 2778. See also, Fürstner, A.; Stelzer, F.; Szillat, H. *J. Am. Chem. Soc.* **2001**, *123*, 11863.

[819]Urabe, H.; Nakajima, R.; Sato, F. *Org. Lett.* **2000**, *2*, 3481.

[820]RajanBabu, T.V.; Nugent, W.A.; Taber, D.F.; Fagan, P.J. *J. Am. Chem. Soc.* **1988**, *110*, 7128; Méndez, M.; Muñoz, M.P.; Echavarren, A.M. *J. Am. Chem. Soc.* **2000**, *122*, 11549; Chakrapani, H.; Liu, C.; Widenhoefer, R.A. *Org. Lett.* **2003**, *5*, 157; Lei, A.; He, M.; Zhang, X. *J. Am. Chem. Soc.* **2002**, *124*, 8198; Ojima, I.; Vu, A.T.; Lee, S.-Y.; McCullagh, J.V.; Moralee, A.C.; Fujiwara, M.; Hoang, T.H. *J. Am. Chem. Soc.* **2002**, *124*, 9164; Lee, P.H.; Kim, S.; Lee, K.; Seomoon, D.; Kim, H.; Lee, S.; Kim, M.; Han, M.; Noh, K.; Livinghouse, T. *Org. Lett.* **2004**, *6*, 4825.

[821]Makino, T.; Itoh, K. *Tetrahedron Lett.* **2003**, *44*, 6335; Lei, A.; He, M.; Wu, S.; Zhang, X. *Angew. Chem. Int. Ed.* **2002**, *41*, 3457.

[822]Harrison, T.J.; Dake, G.R. *Org. Lett.* **2004**, *6*, 5023. See also, Ikeda, S.-i.; Sanuki, R.; Miyachi, H.; Miyashia, H.; Taniguchi, M.; Odashima, K. *J. Am. Chem. Soc.* **2004**, *126*, 10331.

[823]Yamamoto, Y.; Kuwabara, S.; Ando, Y.; Nagata, H.; Nishiyama, H.; Itoh, K. *J. Org. Chem.* **2004**, *69*, 6697.

[824]Hatano, M.; Mikami, K. *J. Am. Chem. Soc.* **2003**, *125*, 4704.

[825]Luzung, M.R.; Markham, J.P.; Toste, F.D. *J. Am. Chem. Soc.* **2004**, *126*, 10858.

[826]Zhang, L.; Kozmin, S.A. *J. Am. Chem. Soc.* **2004**, *126*, 11806.

[827]Delgado, A.; Rodríguez, J.R.; Castedo, L.; Mascareñas, J.L. *J. Am. Chem. Soc.* **2003**, *125*, 9282.

in the presence of $ZnBr_2$.[828] Using mercury (II) triflate in water, cyclization leads to five-membered rings having an exocyclic double bond, and a pendant alcohol group.[829] Enynes give cyclic compounds with an endocyclic double bond conjugated to another alkene unit (a conjugated diene) when treated with $GaCl_3$[830] or a platinum catalyst in an ionic liquid.[831] Allene–alkenes give a similar product with a palladium catalyst[832] or a ruthenium catalyst,[833] as do alkyne–allenes with a dirhodium catalyst.[834] Ethers having an enyne unit (propargylic and allylic) give dihydrofurans upon treatment with $Co_2(CO)_8$.[835] Amines and sulfonamides bearing two propargyl groups cyclize with a ruthenium catalyst to give the corresponding dihydropyrrole.[836]

There are many useful variations. Internal coupling of an alkyne with a vinyl halide, using triethylsilane and a palladium catalyst, gave the saturated cyclic compound with two adjacent exocyclic double bonds (a 2,3-disubstituted diene.[837] Alkynes can added to propargyl acetates using palladium catalyst to give an alkyne allene.[838] Two-Substituted malonate esters having a distal alkyne unit generated vinylidene cycloalkanes when treated with a catalytic amount of *n*-butyllithium.[839] Intramolecular coupling of alkenes and allylic sulfides using *tert*-butoxide/*n*-butyllithium, and then LiBr leads to a bicyclic compound containing a fused cyclopropane ring.[840] The reaction of an alkyne with a vinyl iodide and silver carbonate, with a palladium catalyst, gave a fulvene.[841] The reaction of a terminal alkyne and a vinyl cyclopropane, with a dirhodium catalyst, gives a cycloheptadiene.[842] Alkyne–alkenes were formed by coupling terminal alkynes and allenes in the presence of a palladium catalyst.[843] An alkyne was coupled internally to an allene using a palladium catalyst, with an exocyclic methylene

[828]Yamazaki, S.; Yamada, K.; Yamabe, S.; Yamamoto, K. *J. Org. Chem.* **2002**, *67*, 2889.

[829]Nishizawa, M.; Yadav, V.K.; Skwarczynski, M.; Takao, H.; Imagawa, H.; Sugihara, T. *Org. Lett.* **2003**, *5*, 1609.

[830]Chatani, N.; Inoue, H.; Kotsuma, T.; Morai, S. *J. Am. Chem. Soc.* **2002**, *124*, 10294.

[831]In bmim PF_6, 1-butyl-3-methylimidazolium hexafluorophosphate: Miyanohana, Y.; Inoue, H.; Chatani, N. *J. Org. Chem.* **2004**, *69*, 8541.

[832]Frazén, J.; Löfstedt, J.; Dorange, I.; Bäckvall, J.-E. *J. Am. Chem. Soc.* **2002**, *124*, 11246.

[833]Kang, S.-K.; Ko, B.-S.; Lee, D.-M. *Tetrahedron Lett.* **2002**, *43*, 6693.

[834]Brummond, K.M.; Chen, H.; Sill, P.; You, L. *J. Am. Chem. Soc.* **2002**, *124*, 15186. See also, Candran, N.; Cariou, K.; Hervé, G.; Aubert, C.; Fensterbank, L.; Malacria, M.; Marco-Contelles, J. *J. Am. Chem. Soc.* **2004**, *126*, 3408.

[835]Ajamian, A.; Gleason, J.L. *Org. Lett.* **2001**, *3*, 4161; Oh, C.H.; Han, J.W.; Kim, J.S.; Um, S.Y.; Jung, H.H.; Jang, W.H.; Won, H.S. *Tetrahedron Lett.* **2000**, *41*, 8365; Ackermann, L.; Bruneau, C.; Dixneuf, P.H. *Synlett* **2001**, 397.

[836]Trost, B.M.; Rudd, M.T. *Org. Lett.* **2003**, *5*, 1467.

[837]Oh, C.H.; Park, S.J. *Tetrahedron Lett.* **2003**, *44*, 3785.

[838]Condon-Gueugnot, S.; Linstrumelle, G. *Tetrahedron* **2000**, *56*, 1851.

[839]Kitagawa, O.; Suzuki, T.; Fujiwra, H.; Fujita, M.; Taguchi, T. *Tetrahedron Lett.* **1999**, *40*, 4585.

[840]Cheng, D.; Knox, K.R.; Cohen, T. *J. Am. Chem. Soc.* **2000**, *122*, 412.

[841]Kotora, M.; Matsumura, H.; Gao, G.; Takahashi, T. *Org. Lett.* **2001**, *3*, 3467.

[842]Wender, P.A.; Barzilay, C.M.; Dyckman, A.J. *J. Am. Chem. Soc.* **2001**, *123*, 179.

[843]Rubin, M.; Markov, J.; Chuprakov, S.; Wink, D.J.; Gevorgyan, V. *J. Org. Chem.* **2003**, *68*, 6251.

group trapped and a vinyltin derivative.[844] A similar process occurred with a rhodium catalyst, $RhCl(PPh_3)_3$, to incorporate a vinyl chloride.[845] Allene–allylic halide systems reacted with phenylboronic acid and a palladium catalyst to give cyclopentane rings with two pendant vinyl groups, one of which contained a phenyl group.[846]

In another reductive coupling, substituted alkenes ($CH_2=CH-Y$; $Y=R$, COOMe, OAc, CN, etc.) can be dimerized to substituted alkanes CH_3CHY-$CHYCH_3$ by photolysis in an H_2 atmosphere, using Hg as a photosensitizer.[847] Still another procedure involves palladium-catalyzed addition of vinylic halides to triple bonds to give 1,3-dienes.[848]

134   **135**

An important cyclization procedure involves acid-catalyzed addition of diene-ketones, such as **134**, where one conjugated alkene adds to the other conjugated alkene to form cyclopentenones (**135**). This is called the *Nazarov cyclization*.[849] Structural variations are possible that prepare a variety of cyclopentenones. When one of the C=C units is a vinyl ether, a cyclopentenone is formed with an oxygen attached to the alkenyl carbon.[850] Substituents on the C=C units, including $-CO_2Et$, lead to cyclopentenones that bear those substituents. The use of such a substrate with $AgSbF_6$, $CuBr_2$ and a chiral ligand gave the cyclopentenone with modest enantioselectivity.[851] Cyclization can also give the nonconjugated five-membered ring.[852] A reductive Nazarov cyclization was reported using $BF_3\bullet OEt_2$ and Et_3SiH, giving a cyclopentanone rather than a cyclopentenone.[853] A palladium catalyzed reaction that is related to the Nazarov cyclization converts terminal

[844]Shin, S.; RajanBabu, T.V. *J. Am. Chem. Soc.* **2001**, *123*, 8416.

[845]Tong, X.; Zhang, Z.; Zhang, X. *J. Am. Chem. Soc.* **2003**, *125*, 6370.

[846]Zhu, G.; Zhang, Z. *Org. Lett.* **2004**, *6*, 4041.

[847]Muedas, C.A.; Ferguson, R.R.; Crabtree, R.H. *Tetrahedron Lett.* **1989**, *30*, 3389.

[848]Arcadi, A.; Bernocchi, E.; Burini, A.; Cacchi, S.; Marinelli, F.; Pietroni, B. *Tetrahedron Lett.* **1989**, *30*, 3465.

[849]Nazarov, I.N.; Torgov, I.B.; Terekhova, L.N. *Izv. Akad. Nauk. SSSR otd. Khim. Nauk*, **1942**, 200; Braude, E.A.; Forbes, W.F. *J. Chem. Soc.* **1953**, 2208. Also see, Motoyoshiya, J.; Mizuno, K.; Tsuda, T.; Hayashi, S. *Synlett* **1993**, 237; Oda, M.; Yamazaki, T.; Kajioka, T.; Miyatake, R.; Kuroda, S. *Liebigs Ann. Chem.* **1997**, 2563. See Smith, D.A.; Ulmer II, C.W. *J. Org. Chem.* **1993**, *58*, 4118 for a discussion of torquoselectivity and hyperconjugation in this reaction.

[850]Using $AlCl_3$, see Liang, G.; Gradl, S.N.; Trauner, D. *Org. Lett.* **2003**, *5*, 4931. Using a $Cu(OTf)_2$ catalyst, see He, W.; Sun, X.; Frontier, A.J. *J. Am. Chem. Soc.* **2003**, *125*, 14278. Using a scandium catalyst, see Liang, G.; Trauner, D. *J. Am. Chem. Soc.* **2004**, *126*, 9544.

[851]Aggarwal, V.K.; Belfield, A.J. *Org. Lett.* **2003**, *5*, 5075.

[852]Giese, S.; West, F.G. *Tetrahedron Lett.* **1998**, *39*, 8393.

[853]Giese, S.; West, F.G. *Tetrahedron* **2000**, *56*, 10221.

alkynes to fulvenes.[854] Note that a *retro*-Nazarov is possible with α-bromocyclo-pentanones.[855] In one variation using a aluminum complex, a cyclohexenone was formed.[856]

OS **VIII**, 190, 381, 505; **IX**, 310.

15-21 Addition of Organometallics to Double and Triple Bonds Not Conjugated to Carbonyls

Hydro-alkyl-addition

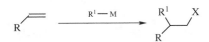

Neither Grignard reagents nor lithium dialkylcopper reagents generally add to ordinary C=C double bonds.[857] An exception is the reaction of $(PhMe_2Si)_2Cu(CN)Li$ to 8-oxabicyclo[3.2.1]oct-2-ene derivatives.[858] Grignard reagents usually add only to double bonds susceptible to nucleophilic attack, (e.g., fluoroalkenes and tetracya-noethylene).[859] However, active Grignard reagents (benzylic, allylic) also add to the double bonds of allylic amines,[860] and of allylic and homoallylic alcohols,[861] as well as to the triple bonds of propargyl alcohols and certain other alkynols.[862] Grignard reagents also add to alkynes in the presence of $MnCl_2$ at 100°C[863] and to alkenes in the presence of zirconium[864] or nickel[865] catalysts. It is likely that cyclic intermediates are involved in these cases, in which the magnesium coordinates with the heteroatom. Allylic, benzylic, and tertiary alkyl Grignard reagents also add to 1-alkenes and strained internal alkenes (e.g., norbornene), if the reaction is carried

[854]Radhakrishnan, U.; Gevorgyan, V.; Yamamoto, Y. *Tetrahedron Lett.* **2000**, *41*, 1971.

[855]Harmata, M.; Lee, D.R. *J. Am. Chem. Soc.* **2002**, *124*, 14328. For a discussion of the scope and mechanism of the retro-Nazarov reaction, see Harmata, M.; Schreiner, P.R.; Lee, D.R.; Kirchhoefer, P.L. *J. Am. Chem. Soc.* **2004**, *126*, 10954.

[856]Magomedev, N.A.; Ruggiero, P.L.; Tang, Y. *Org. Lett.* **2004**, *6*, 3373.

[857]For reviews of the addition of RM to isolated double bonds, see Wardell, J.L.; Paterson, E.S., in Hartley, F.R.; Patai, S. *The Chemistry of the Metal–Carbon Bond*, Vol. 2, Wiley, NY, **1985**, 219–338, pp. 268–296; Vara Prasad, J.V.N.; Pillai, C.N. *J. Organomet. Chem.* **1983**, *259*, 1.

[858]Lautens, M.; Belter, R.K.; Lough, A.J. *J. Org. Chem.* **1992**, *57*, 422.

[859]Gardner, H.C.; Kochi, J.K. *J. Am. Chem. Soc.* **1976**, *98*, 558.

[860]Richey, Jr., H.G.; Moses, L.M.; Domalski, M.S.; Erickson, W.F.; Heyn, A.S. *J. Org. Chem.* **1981**, *46*, 3773.

[861]Felkin, H.; Kaeseberg, C. *Tetrahedron Lett.* **1970**, 4587; Richey Jr., H.G.; Szucs, S.S. *Tetrahedron Lett.* **1971**, 3785; Eisch, J.J.; Merkley, J.H. *J. Am. Chem. Soc.* **1979**, *101*, 1148; Kang, J. *Organometallics* **1984**, *3*, 525.

[862]Eisch, J.J.; Merkley, J.H. *J. Am. Chem. Soc.* **1979**, *101*, 1148; Von Rein, F.W.; Richey Jr., H.G. *Tetrahedron Lett.* **1971**, 3777; Miller, R.B.; Reichenbach, T. *Synth. Commun.* **1976**, *6*, 319. See also, Duboudin, J.G.; Jousseaume, B. *J. Organomet. Chem.* **1979**, *168*, 1; *Synth. Commun.* **1979**, *9*, 53.

[863]Yorimitsu, H.; Tang, J.; Okada, K.; Shinokubo, H.; Oshima, K. *Chem. Lett.* **1998**, 11.

[864]Rousset, C.J.; Negishi, E.; Suzuki, N.; Takahashi, T. *Tetrahedron Lett.* **1992**, *33*, 1965; de Armas, J.; Hoveyda, A.H. *Org. Lett.* **2001**, *3*, 2097.

[865]Pellet-Rostaing, S.; Saluzzo, C.; Ter Halle, R.; Breuzard, J.; Vial, L.; LeGuyader, F.; Lemaire, M. *Tetrahedron Asymmetry* **2001**, *12*, 1983.

out in a hydrocarbon solvent, such as pentane rather than ether, or in the alkene itself as solvent, heated, under pressure if necessary, to 60–130°C.[866] Yields are variable. *Intramolecular* addition of RMgX to completely unactivated double and triple bonds has been demonstrated.[867] The reaction of tosylates bearing a remote alkene unit and a Grignard reagent leads to cyclization when a zirconium catalyst is used.[868] The intramolecular addition of a CH_2Br unit to the C=C unit of an allylic ether was accomplished using PhMgBr and a cobalt catalyst, give a functionalized tetrahydrofuran and incorporation of the phenyl group on the C=C unit as well.[869] Grignard reagents add to the C=C unit of $(MeO)_2CRCH=CH-R$ moieties to give a 3-alkyl substituted ketone with good enantioselectivity using a nickel catalyst and a chiral additive.[870] In a useful variation, vinyl epoxides react with Grignard reagents and CuBr to give an allylic alcohol via reaction at the C=C unit and concomitant opening of the epoxide.[871] Conjugated dienes react with arylmagnesium halides, Ph_3SiCl and a palladium catalyst to give a coupling product involving the reaction of two equivalents of the diene and incorporation of two $SiPh_3$ units.[872]

Organolithium reagents (primary, secondary, and tertiary alkyl and in some cases aryl) also add to the double and triple bonds of allylic and propargylic alcohols[873] (in this case tetramethylethylenediamine is a catalyst) and to certain other alkenes containing hetero groups, such as OR, NR_2, or SR. Addition of butyllithium to alkenes has been observed with good enantioselectivity when sparteine was added.[874] Mixing an organolithium reagent with transition metal compounds, such as $CeCl_3$[875] or $Fe(acac)_3$[876] leads to addition of the alkyl group. The intramolecular addition of RLi and R_2CuLi has been reported.[877] Organolithium reagents

[866]Lehmkuhl, H.; Reinehr, D. *J. Organomet. Chem.* **1970**, *25*, C47; **1973**, *57*, 29; Lehmkulhl, H.; Janssen, E. *Liebigs Ann. Chem.* **1978**, 1854. This is actually a type of ene reaction. For a review of the intramolecular version of this reaction, see Oppolzer, W. *Angew. Chem. Int. Ed.* **1989**, *28*, 38.

[867]See, for example, Richey Jr., H.G.; Rees, T.C. *Tetrahedron Lett.* **1966**, 4297; Drozd, V.N.; Ustynyuk, Yu.A.; Tsel'eva, M.A.; Dmitriev, L.B. *J. Gen. Chem. USSR* **1969**, *39*, 1951; Felkin, H.; Umpleby, J.D.; Hagaman, E.; Wenkert, E. *Tetrahedron Lett.* **1972**, 2285; Hill, E.A.; Myers, M.M. *J. Organomet. Chem.* **1979**, *173*, 1. See also, Yang, D.; Gu, S.; Yan, Y.-L.; Zhu, N.-Y.; Cheung, K.-K. *J. Am. Chem. Soc.* **2001**, *123*, 8612.

[868]Cesati III, R.R.; de Armas, J.; Hoveyda, A.H. *Org. Lett.* **2002**, *4*, 395.

[869]Wakabayashi, K.; Yorimitsu, H.; Oshima, K. *J. Am. Chem. Soc.* **2001**, *123*, 5374.

[870]Gomez-Bengoa, E.; Heron N.M.; Didiuk, M.T.; Luchaco, C.A.; Hoveyda, A.H. *J. Am. Chem. Soc.* **1998**, *120*, 7649.

[871]Taber, D.F.; Mitten, J.V. *J. Org. Chem.* **2002**, *67*, 3847.

[872]Terao, J.; Oda, A.; Kambe, N. *Org. Lett.* **2004**, *6*, 3341.

[873]For a review of the addition of organolithium compounds to double or triple bonds, see Wardell, J.L., in Zuckerman, J.J. *Inorganic Reactions and Methods*, Vol. 11; VCH, NY, **1988**, pp. 129–142. For a tandem reaction, see García, G.V.; Budelman, N.S. *Org. Prep. Proceed. Int.* **2003**, *35*, 445.

[874]Norsikian, S.; Marek, I.; Poisson, J.-F.; Normant, J.F. *J. Org. Chem.* **1997**, *62*, 4898.

[875]Bartoli, G.; Dalpozzo, R.; DeNino, A.; Procopio, A.; Sanbri, L.; Tagarelli, A. *Tetrahedron Lett.* **2001**, *42*, 8823.

[876]Hojo, M.; Murakami, Y.; Aihara, H.; Sakuragi, R.; Baba, Y.; Hosomi, A. *Angew. Chem. Int. Ed.* **2001**, *40*, 621.

[877]Wender, P.A.; White, A.W. *J. Am. Chem. Soc.* **1988**, *110*, 2218; Bailey, W.F.; Nurmi, T.T.; Patricia, J.J.; Wang, W. *J. Am. Chem. Soc.* **1987**, *109*, 2442.

containing an alkene[377,878] or alkyne[879] unit cyclize[880] at low temperatures and quenching with methanol replaces the new C—Li bond with C—H. Cyclopropane derivatives have been formed in this manner.[881] Alkyllithium and alkenyllithium derivatives containing an ester moiety can be cyclized.[882] Tandem cyclization are possible with dienes and enynes to form more than one ring,[883] including bicyclic compounds.[884] Tandem cyclization is possible with alkyne iodides[885] or alkynes with a homoallylic CH$_2$Li unit.[886] The organolithium reagents can contain heteroatoms, such as nitrogen elsewhere in the molecule, and the organolithium species can be generated from an intermediate organotin derivative.[887] Organolithium reagents add to the less substituted C=C unit of conjugated dienes.[888] The organolithium compound can be generated *in situ* by reaction of an organotin compound with butyllithium, allowing cyclization of occur upon treatment with an excess of LiCl.[889]

Ketones with an α-hydrogen add to alkenes, intramolecular, when heated in a sealed tube with CuCl$_2$ and a palladium catalyst.[890] A similar reaction was reported using Yb(OTf)$_3$ and a palladium catalyst.[891] Keto esters add to alkynes using 10% benzoic acid and a palladium catalyst,[892] or an indium catalyst.[893] 1,3-Diketones add to dienes (1,4-addition) using a palladium catalyst,[894] a AuCl$_3$/AgOTf catalyst,[895] and this addition has been done intramolecularly using 2.4 equivalents of CuCl$_2$ and a palladium catalyst.[896] A related cyclization reaction was reported for diesters having a remote terminal alkyne unit in the molecule, with a palladium

[878]Bailey, W.F.; Khanolkar, A.D. *J. Org. Chem.* **1990**, *55*, 6058 and references cited therein; Bailey, W.F.; Mealy, M.J. *J. Am. Chem. Soc* **2000**, *122*, 6787; Gil, G.S.; Groth, U.M. *J. Am. Chem. Soc.* **2000**, *122*, 6789; Bailey, W.F.; Daskapan, T.; Rampalli, S. *J. Org. Chem.* **2003**, *68*, 1334. Also see Bailey, W.F.; Carson, M.W. *J. Org. Chem.* **1998**, *63*, 361; Fretwell, P.; Grigg, R.; Sansano, J.M.; Sridharan, V.; Sukirthalingam, S.; Wilson, D.; Redpath, J. *Tetrahedron* **2000**, *56*, 7525.

[879]Bailey, W.F.; Ovaska, T.V. *J. Am. Chem. Soc.* **1993**, *115*, 3080, and references cited therein. See Funk, R.L.; Bolton, G.L.; Brummond, K.M.; Ellestad, K.E.; Stallman, J.B. *J. Am. Chem. Soc.* **1993**, *115*, 7023.

[880]A conformational radical clock has been developed to evaluate organolithium-mediated cyclization reactions. See Rychnovsky, S.D.; Hata, T.; Kim, A.I.; Buckmelter, A.J. *Org. Lett.* **2001**, *3*, 807.

[881]Gandon, V.; Laroche, C.; Szymoniak, J. *Tetrahedron Lett.* **2003**, *44*, 4827.

[882]Cooke, Jr., M.P. *J. Org. Chem.* **1992**, *57*, 1495 and references cited therein.

[883]Bailey, W.F.; Ovaska, T.V. *Chem. Lett.* **1993**, 819.

[884]Bailey, W.F.; Khanolkar, A.D.; Gavaskar, K.V. *J. Am. Chem. Soc.* **1992**, *114*, 8053.

[885]Harada, T.; Fujiwara, T.; Iwazaki, K.; Oku, A. *Org. Lett.* **2000**, *2*, 1855.

[886]Wei, X.; Taylor, R.J.K. *Angew. Chem. Int. Ed.* **2000**, *39*, 409.

[887]Coldham, I.; Hufton, R.; Rathmell, R.E. *Tetrahedron Lett.* **1997**, *38*, 7617; Coldham, I.; Lang-Anderson, M.M.S.; Rathmell, R.E.; Snowden, D.J. *Tetrahedron Lett.* **1997**, *38*, 7621.

[888]Norsikian, S.; Baudry, M.; Normant, J.F. *Tetrahedron Lett.* **2000**, *41*, 6575.

[889]Komine, N.; Tomooka, K.; Nakai, T. *Heterocycles* **2000**, *52*, 1071.

[890]Wang, X.; Pei, T.; Han, X.; Widenhoefer, R.A. *Org. Lett.* **2003**, *5*, 2699.

[891]Yang, D.; Li, J.-H.; Gao, Q.; Yan, Y.-L. *Org. Lett.* **2003**, *5*, 2869.

[892]Patil, N.T.; Yamamoto, Y. *J. Org. Chem.* **2004**, *69*, 6478.

[893]Nakamura, M.; Endo, K.; Nakamura, E. *J. Am. Chem. Soc.* **2003**, *125*, 13002.

[894]Leitner, A.; Larsen, J.; Steffens, C.; Hartwig, J.F. *J. Org. Chem.* **2004**, *69*, 7552.

[895]Yao, X.; Li, C.-J. *J. Am. Chem. Soc.* **2004**, *126*, 6884.

[896]Pei, T.; Wang, X.; Widenhoefer, R.A. *J. Am. Chem. Soc.* **2003**, *125*, 648; Pei, T.; Widenhoefer, R.A. *J. Am. Chem. Soc.* **2001**, *123*, 11290.

catalyst.[897] The intermolecular addition of diesters, such as malonates, to alkynes was accomplished in acetic acid and a palladium catalyst under microwave irradiation.[898] Enolate anions can add to allylic sulfides, forming β-lactams in some cases.[899] α-Potassio amines (from an *N*-Boc amine and KHMDS) undergoes intramolecular cyclization with an alkene unit to form a dihydropyrrole.[900] The enolate anion derived from the reaction of a nitrile with potassium *tert*-butoxide added to the less substituted carbon of the C=C unit of styrene in DMSO.[901] Similarly, the intramolecular addition of a nitrile enolate (from treatment with CsOH in *N*-methylpyrrolidinone) to an alkyne gave a cyclized product with an exocyclic methylene unit.[902] Silyl enol ethers add to alkynes using a tungsten catalyst.[903] Malonate derivatives add to alkenes in the presence of an Al(OR)$_3$ catalyst.[904]

Unactivated alkenes or alkynes[905] can react with other organometallic compounds under certain conditions. Trimethylaluminum reacts with 4-methyl-1-pentene, in the presence of Cl$_2$ZrCp$_2$, for example, and subsequent reaction with molecular oxygen leads to (2*R*),4-dimethyl-1-pentanol in good yield and 74% ee.[906] These reagents also add to alkynes.[907] Aluminum chloride mediated cyclization of α-iodo ketones to a pendant alkyne unit, in the presence of ICl, gave the spirocyclic ketone with an exocyclic C=CHI unit.[908] Isopropylchloroformate (*i*PrO$_2$CCl) reacts with an alkene, in conjunction with Et$_3$Al$_2$Cl$_3$, to add an isopropyl group.[909] Ruthenium catalysts have been used to add allylic alcohols to alkynes.[910] Samarium iodide (SmI$_2$) induces cyclization of a halide moiety to an alkyne unit[911] or an alkene unit[912] to form cyclized products. Copper complexes can catalyze similar cyclization to alkenes, even when an ester unit is present in the molecule.[913] The reaction of a dithioketal containing a remote alkene moiety, with a titanium complex, leads to cyclization and incorporation of an endocyclic C=C unit in the final product.[914] Allyl manganese compounds add to allenes to give nonconjugated dienes.[915]

[897]Liu, G.; Lu, X. *Tetrahedron Lett.* **2002**, *43*, 6791.

[898]Patil, N.T.; Khan, F.N.; Yamamoto, Y. *Tetrahedron Lett.* **2004**, *45*, 8497.

[899]Attenni, B.; Cereti, A.; D'Annibale, A.; Resta, S.; Trogolo, C. *Tetrahedron* **1998**, *54*, 12029.

[900]Green, M.P.; Prodger, J.C.; Sherlock, A.E.; Hayes, C.J. *Org. Lett.* **2001**, *3*, 3377.

[901]Rodriguez, A.L.; Bunlaksananusorn, T.; Knochel, P. *Org. Lett.* **2000**, *2*, 3285.

[902]Koradin, C.; Rodriguez, A.; Knochel, P. *Synlett* **2000**, 1452.

[903]Iwasawa, N.; Miura, T.; Kiyota, K.; Kusama, H.; Lee, K.; Lee, P.H. *Org. Lett.* **2002**, *4*, 4463.

[904]Black, P.J.; Harris, W.; Williams, J.M.J. *Angew. Chem. Int. Ed.* **2001**, *40*, 4475.

[905]See Trost, B.M.; Ball, Z.T. *Synthesis* **2005**, 853.

[906]Kondakov, D.Y.; Negishi, E. *J. Am. Chem. Soc.* **1995**, *117*, 10771. Also see, Shibata, K.; Aida, T.; Inoue, S. *Tetrahedron Lett.* **1993**, *33*, 1077 for reactions of Et$_3$Al catalyzed by a zirconium complex.

[907]Wipf, P.; Lim, S. *Angew. Chem. Int. Ed.* **1993**, *32*, 1068.

[908]Sha, C.-K.; Lee, F.-C.; Lin, H.-H. *Chem. Commun.* **2001**, 39.

[909]Biermann, U.; Metzger, J.O. *Angew. Chem. Int. Ed.* **1999**, *38*, 3675.

[910]Trost, B.M.; Indolese, A.F.; Müller, T.J.J.; Treptow, B. *J. Am. Chem. Soc.* **1995**, *117*, 615.

[911]Zhou, Z.; Larouche, D.; Bennett, S.M. *Tetrahedron* **1995**, *51*, 11623.

[912]Fukuzawa, S.; Tsuchimoto, T. *Synlett* **1993**, 803.

[913]Pirrung, F.O.H.; Hiemstra, H.; Speckamp, W.N.; Kaptein, B.; Schoemaker, H.E. *Tetrahedron* **1994**, *50*, 12415.

[914]Fujiwara, T.; Kato, Y.; Takeda, T. *Heterocycles* **2000**, *52*, 147.

[915]Nishikawa, T.; Shinokubo, H.; Oshima, K. *Org. Lett.* **2003**, *5*, 4623.

Vinyl halides add to allylic amines in the presence of Ni(cod)$_2$ where cod = 1, 5-cyclooctodine, followed by reduction with sodium borohydride.[916] Aryl iodides add to alkynes using a platinum complex in conjunction with a palladium catalyst.[917] A palladium catalyst has been used alone for the same purpose,[918] and the intramolecular addition of a arene to an alkene was accomplished with a palladium[919] or a GaCl$_3$ catalyst,[920] Alkyl iodides add intramolecularly to alkenes with a titanium catalyst,[921] or to alkynes using indium metal and additives.[922] The latter cyclization of aryl iodides to alkenes was accomplished with indium and iodine[923] or with SmI$_2$.[924]

Aromatic hydrocarbons, such as benzene add to alkenes using a ruthenium catalyst[925] a catalytic mixture of AuCl$_3$/AgSbF$_6$,[926] or a rhodium catalyst,[927] and ruthenium complexes catalyze the addition of heteroaromatic compounds, such as pyridine, to alkynes.[928] Such alkylation reactions are clearly reminiscent of the Friedel–Crafts reaction (**11-11**). Palladium catalysts can also be used to for the addition of aromatic compounds to alkynes,[929] and rhodium catalysts for addition to alkenes (with microwave irradiation).[930] Note that vinylidene cyclopropanes react with furans and a palladium catalyst to give allylically substituted furans.[931]

Arylboronic acids (p. 905) add to alkynes to give the substituted alkene using a rhodium catalyst.[932] Allenes react with phenylboronic acid and an aryl iodide, in the presence of a palladium catalyst, to give a substituted alkene.[933] 2-Bromo-1,6-dienes react with phenylboronic acid with a palladium catalyst to give a cyclopentane with an exocyclic double bond and a benzyl substituent.[934]

[916]Solé, D.; Cancho, Y.; Llebaria, A.; Moretó, J.M.; Delgado, A. *J. Am. Chem. Soc.* **1994**, *116*, 12133.

[917]Denmark, S.E.; Wang, Z. *Org. Lett.* **2001**, *3*, 1073.

[918]Wu, M.-J.; Wei, L.-M.; Lin, C.-F.; Leou, S.-P.; Wei, L.-L. *Tetrahedron* **2001**, *57*, 7839; Havránek, M.; Dvořák, D. *J. Org. Chem.* **2002**, *67*, 2125. See also, Lee, K.; Seomoon, D.; Lee, P.H. *Angew. Chem. Int. Ed.* **2002**, *41*, 3901.

[919]Huang, Q.; Fazio, A.; Dai, G.; Campo, M.A.; Larock, R.C. *J. Am. Chem. Soc.* **2004**, *126*, 7460.

[920]Inoue, H.; Chatani, N.; Murai, S. *J. Org. Chem.* **2002**, *67*, 1414.

[921]Zhou, L.; Hirao, T. *J. Org. Chem.* **2003**, *68*, 1633.

[922]Yanada, R.; Koh, Y.; Nishimori, N.; Matsumura, A.; Obika, S.; Mitsuya, H.; Fujii, N.; Takemoto, Y. *J. Org. Chem.* **2004**, *69*, 2417; Yanada, R.; Obika, S.; Oyama, M.; Takemoto, Y. *Org. Lett.* **2004**, *6*, 2825.

[923]Yanada, R.; Obika, S.; Nishimori, N.; Yamauchi, M.; Takemoto, Y. *Tetrahedron Lett.* **2004**, *45*, 2331.

[924]Dahlén, A.; Petersson, A.; Hilmersson, G. *Org. Biomol. Chem.* **2003**, *1*, 2423.

[925]Lail, M.; Arrowood, B.N.; Gunnoe, T.B. *J. Am. Chem. Soc.* **2003**, *125*, 7506.

[926]Reetz, M.T.; Sommer, K. *Eur. J. Org. Chem.* **2003**, 3485.

[927]Thalji, R.K.; Ellman, J.A.; Bergman, R.G. *J. Am. Chem. Soc.* **2004**, *126*, 7192.

[928]Murakami, M.; Hori, S. *J. Am. Chem. Soc.* **2003**, *125*, 4720. For a review, see Alonso, F.; Beletskaya, I.P.; Yus, M. *Chem. Rev.* **2004**, *104*, 3079.

[929]Tsukada, N.; Mitsuboshi, T.; Setoguchi, H.; Inoue, Y. *J. Am. Chem. Soc.* **2003**, *125*, 12102.

[930]Vo-Thanh, G.; Lahrache, H.; Loupy, A.; Kim, I.-J.; Chang, D.-H.; Jun, C.-H. *Tetrahedron* **2004**, *60*, 5539.

[931]Nakamura, I.; Siriwardana, A.I.; Saito, S.; Yamamoto, Y. *J. Org. Chem.* **2002**, *67*, 3445.

[932]Hayashi, T.; Inoue, K.; Taniguchi, N.; Ogasawara, M. *J. Am. Chem. Soc.* **2001**, *123*, 9918; Lautens, M.; Yoshida, M. *J. Org. Chem.* **2003**, *68*, 762; Genin, E.; Michelet, V.; Genêt, J.-P. *Tetrahedron Lett.* **2004**, *45*, 4157.

[933]Huang, T.-H.; Chang, H.-M.; Wu, M.-Y.; Cheng, C.-H. *J. Org. Chem.* **2002**, *67*, 99; Oh, C.H.; Ahn, T.W.; Reddy, R. *Chem. Commun.* **2003**, 2622; Yoshida, M.; Gotou, T.; Ihara, M. *Chem. Commun.* **2004**, 1124.

[934]Oh, C.H.; Sung, H.R.; Park, S.J.; Ahn, K.H. *J. Org. Chem.* **2002**, *67*, 7155.

Organomanganese reagents add to alkenes.[935] Manganese triacetate [Mn(OAc)$_3$], in the presence of cupric acetate, facilitates intramolecular cyclization of a halide unit to an alkene.[936] A combination of Mn(OAc)$_2$ and Co(OAc)$_2$ catalysts, and an oxygen atmosphere in acetic acid, leads to addition of ketones to simple alkenes, give the 2-alkyl ketone.[937] Alkynes react with indium reagents, such as (allyl)$_3$In$_2$I$_3$, to form dienes (allyl substituted alkenes from the alkyne).[938] Allylic halides add to propargyl alcohols using indium metal to form the aryl organometallic *in situ*.[939] Allyltin reagents add to alkynes in a similar manner in the presence of ZrCl$_4$.[940] Alkylzinc reagents add to alkynes to give substituted alkenes in the presence of a palladium catalyst.[941] Allylzinc reagents add to alkynes in the presence of a cobalt catalyst.[942] A variation reacts dialkylzinc compounds with a 7-oxabicyclo[2.2.1]hept-2-ene system to give incorporation of the alkyl group and opening of the ring to give a cyclohexenol derivative.[943] Vinyltellurium add to alkynes in the presence of CuI/PdCl$_2$.[944]

An indirect addition converts alkynes to an organozinc compound using a palladium catalyst, which then reacts with allylic halides.[945] Similarly, the reaction of an alkyne with Ti(OiPr)$_4$/2 iPrMgCl followed by addition of an alkyne leads to a conjugated diene.[946]

OS **81**, 121.

15-22   The Addition of Two Alkyl Groups to an Alkyne

Dialkyl-addition

$$R-C\equiv C-H \ + \ R^1CuMgBr_2 \ + \ R^2\text{-I} \xrightarrow[\text{ether–HMPA}]{(EtO)_2P} \ \underset{R^1}{\overset{R}{>}}C=C\underset{R^2}{\overset{H}{<}}$$

Two different alkyl groups can be added to a terminal alkyne[947] in one laboratory step by treatment with an alkylcopper-magnesium bromide reagent (called *Normant*

[935]Nakao, J.; Inoue, R.; Shinokubo, H.; Oshima, K. *J. Org. Chem.* **1997**, *62*, 1910.
[936]Snider, B.B.; Merritt, J.E. *Tetrahedron* **1991**, *47*, 8663.
[937]Iwahama, T.; Sakaguchi, S.; Ishii, Y. *Chem. Commun.* **2000**, 2317.
[938]Fujiwara, N.; Yamamoto, Y. *J. Org. Chem.* **1999**, *64*, 4095.
[939]Klaps, E.; Schmid, W. *J. Org. Chem.* **1999**, *64*, 7537.
[940]Asao, N.; Matsukawa, Y.; Yamamoto, Y. *Chem. Commun.* **1996**, 1513.
[941]Luo, F.-T.; Fwu, S.-L.; Huang, W.-S. *Tetrahedron Lett.* **1992**, *33*, 6839.
[942]Nishikawa, T.; Yorimitsu, H.; Oshima, K. *Synlett* **2004**, 1573.
[943]Lautens, M.; Hiebert, S. *J. Am. Chem. Soc.* **2004**, *126*, 1437.
[944]Zeni, G.; Nogueira, C.W.; Pena, J.M.; Pialssã, C.; Menezes, P.H.; Braga, A.L.; Rocha, J.B.T. *Synlett* **2003**, 579.
[945]Matsubara, S.; Ukai, K.; Toda, N.; Utimoto, K.; Oshima, K. *Synlett* **2000**, 995.
[946]Tanaka, R.; Hirano, S.; Urabe, H.; Sato, F. *Org. Lett.* **2003**, *5*, 67.
[947]For reviews of this and related reactions, see Raston, C.L.; Salem, G., in Hartley, F.R. *The Chemistry of the Metal–Carbon Bond*, Vol. 4, Wiley, NY, **1987**, pp. 159–306, pp. 233–248; Normant, J.F.; Alexakis, A. *Synthesis* **1981**, 841; Hudrlik, P.F.; Hudrlik, A.M., in Patai, S. *The Chemistry of the Carbon–Carbon Triple Bond*, pt. 1, Wiley, NY, **1978**, pp. 233–238. For a list of reagents and references for this and related reactions, see Larock, R.C. *Comprehensive Organic Transformations*, 2nd ed., Wiley-VCH, NY, **1999**, pp. 452–460.

reagents)[948] and an alkyl iodide in ether–HMPA containing triethyl phosphite.[949] The groups add stereoselectively syn. The reaction, which has been applied to primary[950] R' and to primary, allylic, benzylic, vinylic, and α-alkoxyalkyl R', involves initial addition of the alkylcopper reagent,[951] followed by a coupling reaction (**10-57**):

$$R-C\equiv C-H \xrightarrow{\text{R}^1\text{CuMgBr}_2} \underset{\textbf{136}}{\overset{\displaystyle R \quad\quad H}{\underset{\displaystyle R^1 \quad\quad Cu\text{-}MgBr_2}{C=C}}} \xrightarrow{\text{R}^2\text{-I}} \overset{\displaystyle R \quad\quad H}{\underset{\displaystyle R^1 \quad\quad R^2}{C=C}}$$

Acetylene itself (R =H) undergoes the reaction with R_2CuLi instead of the Normant reagent.[952] The use of R' containing functional groups has been reported.[953] If the alkyl iodide is omitted, the vinylic copper intermediate

$$\overset{\displaystyle R \quad\quad H}{\underset{\displaystyle R^1 \quad\quad CONHR^2}{C=C}} \xleftarrow[\text{RNCO}]{\text{P(OEt)}_2,\ \text{HMPA}} \overset{\displaystyle R \quad\quad H}{\underset{\displaystyle R^1 \quad\quad Cu\text{-}MgBr_2}{C=C}} \xrightarrow[\text{CO}_2]{\text{P(OEt)}_2,\ \text{HMPA}} \overset{\displaystyle R \quad\quad H}{\underset{\displaystyle R^1 \quad\quad COOH}{C=C}}$$

136 can be converted to a carboxylic acid by the addition of CO_2 (see **16-30**) or to an amide by the addition of an isocyanate, in either case in the presence of HMPA and a catalytic amount of triethyl phosphite.[954] The use of I_2 results in a vinylic iodide.[955]

Similar reactions, in which two alkyl groups are added to a triple bond, have been carried out with trialkylalanes (R_3Al), with zirconium complexes as catalysts.[956] Allyl ethers and iodobenzene have also been added using a

[948]For the composition of these reagents see Ashby, E.C.; Smith, R.S.; Goel, A.B. *J. Org. Chem.* **1981**, *46*, 5133; Ashby, E.C.; Goel, A.B. *J. Org. Chem.* **1983**, *48*, 2125.

[949]Gardette, M.; Alexakis, A.; Normant, J.F. *Tetrahedron* **1985**, *41*, 5887, and references cited therein. For an extensive list of references, see Marfat, A.; McGuirk, P.R.; Helquist, P. *J. Org. Chem.* **1979**, *44*, 3888.

[950]For a method of using secondary and tertiary R, see Rao, S.A.; Periasamy, M. *Tetrahedron Lett.* **1988**, *29*, 4313.

[951]The initial product, **136**, can be hydrolyzed with acid to give RR'C=CH_2. See Westmijze, H.; Kleijn, H.; Meijer, J.; Vermeer, P. *Recl. Trav. Chim. Pays-Bas* **1981**, *100*, 98, and references cited therein.

[952]Alexakis, A.; Cahiez, G.; Normant, J.F. *Synthesis* **1979**, 826; *Tetrahedron* **1980**, *36*, 1961; Furber, M.; Taylor, R.J.K.; Burford, S.C. *J. Chem. Soc. Perkin Trans. 1*, **1986**, 1809.

[953]Rao, S.A.; Knochel, P. *J. Am. Chem. Soc.* **1991**, *113*, 5735.

[954]Normant, J.F.; Cahiez, G.; Chuit, C.; Villieras, J. *J. Organomet. Chem.* **1973**, *54*, C53.

[955]Alexakis, A.; Cahiez, G.; Normant, J.F. *Org. Synth.* **VII**, 290.

[956]Negishi, E.; Van Horn, D.E.; Yoshida, T. *J. Am. Chem. Soc.* **1985**, *107*, 6639. For reviews, see Negishi, E. *Acc. Chem. Res.* **1987**, *20*, 65; *Pure Appl. Chem.* **1981**, *53*, 2333; Negishi, E.; Takahashi, T. *Aldrichimica Acta* **1985**, *18*, 31.

zirconium complex.[957] Similarly, allyl ethers and allyl chlorides have been added.[958]

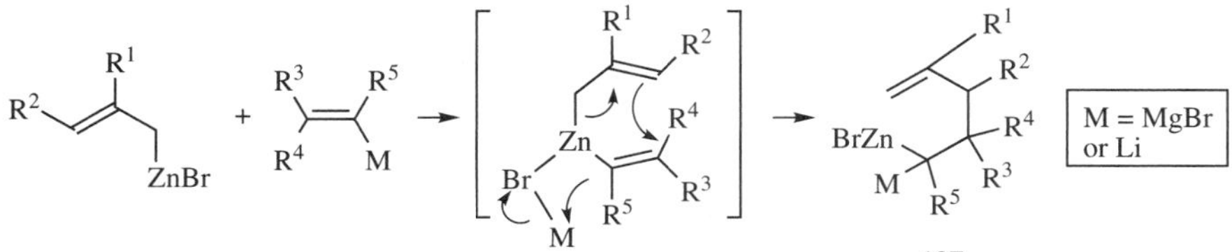

137

Allylic zinc bromides add to vinylic Grignard and lithium reagents to give the gem-dimetallo compounds **137**. The two metallo groups can be separately reacted with various nucleophiles.[959]

Arylboronic acids (p. 905) react with alkynes and 1 equivalent of an aryl iodide, with a palladium catalyst, to add two aryl groups across the triple bond.[960]

OS **VII**, 236, 245, 290.

15-23   The Ene Reaction

Hydro-allyl-addition

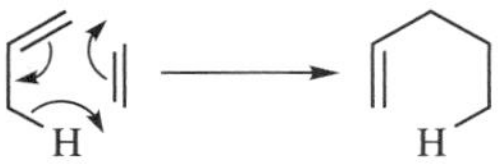

Alkenes can add to double bonds in a reaction different from those discussed in **15-20**, which, however, is still formally the addition of RH to a double bond. This is called the *ene reaction* or the *ene synthesis*.[961] For the reaction to proceed without a catalyst, one of the components must be a reactive dienophile (see **15-60** for a definition of this word), such as maleic anhydride, but the other (which supplies the hydrogen) may be a simple alkene such as propene. Rather high reaction temperatures (250–450°C) are common unless the substrates are very

[957]Hara, R.; Nishihara, Y.; Landré, P.D.; Takahashi, T. *Tetrahedron Lett.* **1997**, *38*, 447.

[958]Takahashi, T.; Kotora, M.; Kasai, K.; Suzuki, N. *Tetrahedron Lett.* **1994**, *35*, 5685.

[959]Knochel, P.; Normant, J.F. *Tetrahedron Lett.* **1986**, *27*, 1039, 1043, 4427, 4431, 5727.

[960]Zhou, C.; Emrich, D.E.; Larock, R.C. *Org. Lett.* **2003**, *5*, 1579.

[961]Alder, K.; von Brachel, H. *Liebigs Ann. Chem.* **1962**, *651*, 141. For a monograph, see Carruthers, W. *Cycloaddition Reactions in Organic Synthesis*, Pergamon, Elmsford, NY, **1990**. For reviews, see Boyd, G.V., in Patai, S. *Supplement A: The Chemistry of Double-Bonded Functional Groups*, Vol. 2, pt. 1, Wiley, NY, **1989**, pp. 477–525; Keung, E.C.; Alper, H. *J. Chem. Educ.* **1972**, *49*, 97–100; Hoffmann, H.M.R. *Angew. Chem. Int. Ed.* **1969**, *8*, 556. For reviews of intramolecular ene reactions see, Taber, D.F. *Intramolecular Diels–Alder and Alder Ene Reactions*, Springer, NY, **1984**; pp. 61–94; Oppolzer, W.; Snieckus, V. *Angew. Chem. Int. Ed.* **1978**, *17*, 476–486; Conia, J.M.; Le Perchec, P. *Synthesis 1975*, 1. See Desimoni, G.; Faita, G.; Righetti, P.P.; Sfulcini, A.; Tsyganov, D. *Tetrahedron 1994*, *50*, 1821 for solvent effects in the ene reaction.

activated. Note that steric acceleration of the uncatalyzed ene reaction is known.[962] Cyclopropene has also been used.[963] The reaction is compatible with a variety of functional groups that can be appended to the ene and dienophile.[964] N,N-Diallyl amides give an ene cyclization, for example.[965] The ene reaction is known with fullerene (see p. 94) derivatives.[966] There has been much discussion of the mechanism of this reaction, and both concerted pericyclic (as shown above) and stepwise

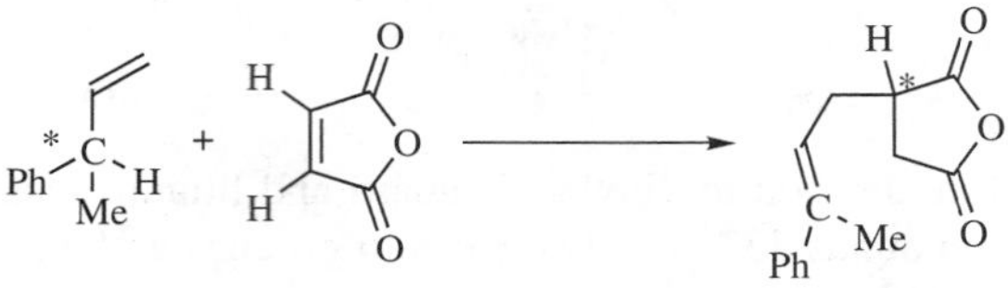

138

mechanisms have been suggested. The mechanism of the ene reaction of singlet ($^1\Delta_g$) oxygen with simple alkenes was found to involve two steps, with no intermediate.[967] A retro-ene reaction is known with allylic dithiocarbonate.[968] The reaction between maleic anhydride and optically active PhCHMeCH=CH$_2$ gave an optically active product (**138**),[969] which is strong evidence for a concerted rather than a stepwise mechanism.[970] The reaction can be highly stereoselective.[971]

The reaction can be extended to less-reactive enophiles by the use of Lewis acid catalysts, especially alkylaluminum halides.[972] Titanium catalysts,[973] Sc(OTf)$_3$,[974]

[962]Choony, N.; Kuhnert, N.; Sammes, P.G.; Smith, G.; Ward, R.W. *J. Chem. Soc., Perkin Trans. 1* **2002**, 1999.

[963]Deng, Q.; Thomas IV, B.E.; Houk, K.N.; Dowd, P. *J. Am. Chem. Soc.* **1997**, *119*, 6902.

[964]For a review of ene reactions in which one of the reactants bears a Si or Ge atom, see Dubac, J.; Laporterie, A. *Chem. Rev.* **1987**, *87*, 319.

[965]Cossy, J.; Bouzide, A. *Tetrahedron* **1997**, *53*, 5775; Oppolzer, W.; Fürstner, A. *Helv. Chim. Acta* **1993**, *76*, 2329; Oppolzer, W.; Schröder, F. *Tetrahedron Lett.* **1994**, *35*, 7939.

[966]Wu, S.; Shu, L.; Fan, K. *Tetrahedron Lett.* **1994**, *35*, 919.

[967]Singleton, D. A.; Hang, C.; Szymanski, M. J.; Meyer, M. P.; Leach, A. G.; Kuwata, K. T.; Chen, J. S.; Greer, A.; Foote, C. S.; Houk, K. N. *J. Am. Chem. Soc.* **2003**, *125*, 1319.

[968]Eto, M.; Nishimoto, M.; Kubota, S.; Matsuoka, T.; Harano, K. *Tetrahedron Lett.* **1996**, *37*, 2445.

[969]Hill, R.K.; Rabinovitz, M. *J. Am. Chem. Soc.* **1964**, *86*, 965. See also, Garsky, V.; Koster, D.F.; Arnold, R.T. *J. Am. Chem. Soc.* **1974**, *96*, 4207; Stephenson, L.M.; Mattern, D.L. *J. Org. Chem.* **1976**, *41*, 3614; Nahm, S.H.; Cheng, H.N. *J. Org. Chem.* **1986**, *51*, 5093.

[970]For other evidence for a concerted mechanism see Benn, F.R.; Dwyer, J.; Chappell, I. *J. Chem. Soc. Perkin Trans. 2*, **1977**, 533; Jenner, G.; Salem, R.B.; El'yanov, B.; Gonikberg, E.M. *J. Chem. Soc. Perkin Trans. 2*, **1989**, 1671. See Thomas IV, B.E.; Loncharich, R.J.; Houk, K.N. *J. Org. Chem.* **1992**, *57*, 1354 for transition-state structures of the intramolecular ene reaction.

[971]Cossy, J.; Bouzide, A.; Pfau, M. *Tetrahedron Lett.* **1992**, *33*, 4883; Ooi, T.; Maruoka, K.; Yamamoto, H. *Tetrahedron* **1994**, *50*, 6505; Thomas IV, B.E.; Houk, K.N. *J. Am. Chem. Soc.* **1993**, *115*, 790; Also see Masaya, K.; Tanino, K.; Kuwajima, I. *Tetrahedron Lett.* **1994**, *35*, 7965.

[972]For reviews, see Chaloner, P.A., in Hartley, F.R. *The Chemistry of the Metal–Carbon Bond*, Vol. 4, Wiley, NY, **1987**, pp. 456–460; Snider, B.B. *Acc. Chem. Res.* **1980**, *13*, 426.

[973]Waratuke, S.A.; Johnson, E.S.; Thorn, M.G.; Fanwick, P.E.; Rothwell, I.P. *Chem. Commun.* **1996**, 2617; Sturla, S.J.; Kablaoui, N.M.; Buchwald, S.L. *J. Am. Chem. Soc.* **1999**, *121*, 1976.

[974]Aggarwal, V.K.; Vennall, G.P.; Davey, P.N.; Newman, C. *Tetrahedron Lett.* **1998**, *39*, 1997.

LiClO$_4$[975] yttrium,[976] nickel catalysts,[977] as well as a combination of silver and gold catalysts[978] have also been used. A magnesium-ene cyclization stereochemically directed by an allylic oxyanionic group has been reported.[979] The Lewis acid catalyzed reaction probably has a stepwise mechanism.[980] The ene reaction has also been mediated on certain resins,[981] and using formaldehyde that was encapsulated in zeolite.[982] An iridium catalyzed ene reaction has been done in an ionic liquid.[983] The carbonyl-ene reaction is also very useful, and often gives synthetically useful yields of products when catalyzed by Lewis acids,[984] including asymmetric catalysts.[985] Among the useful Lewis acids are scandium triflate[986] and chromium complexes.[987] Carbonyl ene cyclization has been reported on silica gel at high pressure (15 kbar).[988] Ene reactions with imines,[989] nitrile oxides,[990] as well as nitroso ene reactions are known.[991]

OS **IV**, 766; **V**, 459. See also, OS **VIII**, 427.

15-24 The Michael Reaction

Hydro-bis(ethoxycarbonyl)methyl-addition, and so on

Compounds containing electron-withdrawing groups (Z is defined on p. 1007) add, in the presence of bases, to alkenes of the form C=C—Z (including quinones).

[975]Davies, A.G.; Kinart, W.J. *J. Chem. Soc. Perkin Trans. 2, 1993*, 2281.

[976]Molander, G.A.; Corrette, C.P. *J. Org. Chem. 1999, 64*, 9697.

[977]Michelet, V.; Galland, J.-C.; Charruault, L.; Savignac, M.; Genêt, J.-P. *Org. Lett. 2001, 3*, 2065.

[978]Kennedy-Smith, J.J.; Staben, S.T.; Toste, F.D. *J. Am. Chem. Soc. 2004, 126*, 4526.

[979]Cheng, D.; Zhu, S.; Yu, Z.; Cohen, T. *J. Am. Chem. Soc. 2001, 123*, 30.

[980]See Snider, B.B.; Ron E. *J. Am. Chem. Soc. 1985, 107*, 8160.

[981]Cunningham, I.D.; Brownhill, A.; Hamereton, I.; Howlin, B.J. *Tetrahedron 1997, 53*, 13473.

[982]Okachi, T.; Onaka, M. *J. Am. Chem. Soc. 2004, 126*, 2306.

[983]Shibata, T.; Yamasaki, M.; Kadowaki, S.; Takagi, K. *Synlett 2004*, 2812.

[984]See Achmatowicz, O.; Bialeck-Florjańczyk, E. *Tetrahedron 1996, 52*, 8827; Marshall, J.A.; Andersen, M.W. *J. Org. Chem. 1992, 57*, 5851 for mechanistic discussions of this reaction.

[985]Mikami, K.; Terada, M.; Narisawa, S.; Nakai, T. *Synlett 1992*, 255; Wu, X.-M.; Funakoshi, K.; Sakai, K. *Tetrahedron Lett. 1993, 34*, 5927. For a discussion of the mechanism of the chiral copper complex-catalyzed carbonyl ene reaction, see Morao, I.; McNamara, J.P.; Hillier, I.H. *J. Am. Chem. Soc. 2003, 125*, 628.

[986]See Aggarwal, V.K.; Vennall, G.P.; Davey, P.N.; Newman, C. *Tetrahedron Lett. 1998, 39*, 1997.

[987]Ruck, R.T.; Jacobsen, E.N. *J. Am. Chem. Soc. 2002, 124*, 2882.

[988]Dauben, W.G.; Hendricks, R.T. *Tetrahedron Lett. 1992, 33*, 603.

[989]Tohyama, Y.; Tanino, K.; Kuwajima, I. *J. Org. Chem. 1994, 59*, 518; Yamanaka, M.; Nishida, A.; Nakagawa, M.; *Org. Lett. 2000, 2*, 159.

[990]For a discussion of the mechanism of the intramolecular reaction, see Yu, Z.-X.; Houk, K.N. *J. Am. Chem. Soc. 2003, 125*, 13825.

[991]Lu, X. *Org. Lett. 2004, 6*, 2813. See also, Leach, A.G.; Houk, K.N. *J. Am. Chem. Soc. 2002, 124*, 14820. For a review, see Adam, W.; Krebs, O. *Chem. Rev. 2003, 103*, 4131.

This is called the *Michael reaction* and involves conjugate addition.[992] The compound RCH$_2$Z or RCHZZ′ can include aldehydes,[993] ketones,[994] esters[995] and diesters,[996] diketones,[997] keto-esters,[998] carboxylic acids and dicarboxylic acids,[999] nitriles,[1000] and nitro compounds,[1001] often with chiral catalysts or additives that give asymmetric induction. Enamines can also be used as the nucleophilic partner in Michael additions.[1002] In the most common examples, a base removes the acidic proton and then the mechanism is as outlined on p. 1008. The reaction has been carried out with conjugated substrates that include malonates, cyanoacetates, acetoacetates, other β-keto esters, and compounds of the form ZCH$_3$, ZCH$_2$R, ZCHR$_2$, and ZCHRZ′, including carboxylic esters, amides,[1003] ketones, aldehydes, nitriles, nitro compounds,[1004]

[992]For reviews, see Yanovskaya, L.A.; Kryshtal, G.V.; Kulganek, V.V. *Russ. Chem. Rev.* **1984**, *53*, 744; Bergmann, E.D.; Ginsburg, D.; Pappo, R. *Org. React.* **1959**, *10*, 179; House, H.O. *Modern Synthetic Reaction*, 2nd ed., W.A. Benjamin, NY, **1972**, pp. 595–623. The subject is also discussed at many places, in Stowell, J.C. *Carbanions in Organic Synthesis*, Wiley, NY, **1979**.

[993]Hagiwara, H.; Okabe, T.; Hakoda, K.; Hoshi, T.; Ono, H.; Kamat, V.P.; Suzuki, T.; Ando, M. *Tetrahedron Lett.* **2001**, *42*, 2705; Melchiorre, P.; Jørgensen, K.A. *J. Org. Chem.* **2003**, *68*, 4151; Shimizu, K.; Suzuki, H.; Hayashi, E.; Kodama, T.; Tsuchiya, Y.; Hagiwara, H.; Kityama, Y. *Chem. Commun.* **2002**, 1068; Willis, M.C.; McNally, S.J.; Beswick, P.J. *Angew. Chem. Int. Ed.* **2004**, *43*, 340. For an intramolecular example, see Fonseca, M.T.H.; List, B. *Angew. Chem. Int. Ed.* **2004**, *43*, 3958.

[994]Betancort, J.M.; Sakthivel, K.; Thayumanavan, R.; Barbas III, C.F. *Tetrahedron Lett.* **2001**, *42*, 4441; Enders, D.; Seki, A. *Synlett* **2002**, 26. For an example using an α-hydroxy ketone, see Andrey, O.; Alexakis, A.; Bernardinelli, G. *Org. Lett.* **2003**, *5*, 2559; Harada, S.; Kumagai, N.; Kinoshita, T.; Matsunaga, S.; Shibasaki, M. *J. Am. Chem. Soc.* **2003**, *125*, 2582.

[995]Kim, S.-G.; Ahn, K.H. *Tetrahedron Lett.* **2001**, *42*, 4175.

[996]Halland, N.; Aburel, P.S.; Jørgensen, K.A. *Angew. Chem. Int. Ed.* **2003**, *42*, 661.

[997]da silva, F.M.; Gomes, A.K.; Jones Jr., J. *Can. J. Chem.* **1999**, *77*, 624.

[998]Suzuki, T.; Torii, T. *Tetrahedron Asymmetry* **2001**, *12*, 1077; García-Gómez, G.; Moretó, J.M. *Eur. J. Org. Chem.* **2001**, 1359; Kobayashi, S.; Kakumoto, K.; Mori, Y.; Manabe, K. *Isr. J. Chem.* **2001**, *41*, 247.

[999]Méou, A.; Lamarque, L.; Brun, P. *Tetrahedron Lett.* **2002**, *43*, 5301.

[1000]Kraus, G.A.; Dneprovskaia, E. *Tetrahedron Lett.* **2000**, *41*, 21. For an example using an α-cyano amide, see Wolckenhauer, S.A.; Rychnovsky, S.D. *Org. Lett.* **2004**, *6*, 2745. For a review of conjugate addition to this relatively unreactive class of compounds, see Fleming, F.F.; Wang, Q. *Chem. Rev.* **2003**, *103*, 2035.

[1001]Sebti, S.; Boukhal, H.; Hanafi, N.; Boulaajaj, S. *Tetrahedron Lett.* **1999**, *40*, 6207; Novák, T.; Tatai, J.; Bakó, P.; Czugler, M.; Keglevich, G. Töke, L. *Synlett* **2001**, 424; Halland, N.; Hazell, R.G.; Jørgensen, K.A. *J. Org. Chem.* **2002**, *67*, 8331; Ooi, T.; Fujioka, S.; Maruoka, K. *J. Am. Chem. Soc.* **2004**, *126*, 11790. For a reaction using nitromethane and a Mg–Al hydrotalcite catalyst, see Choudary, B.M.; Kantam, M.L.; Kavita, B.; Reddy, Ch.V.; Figueras, F. *Tetrahedron* **2000**, *56*, 9357. For the importance of the aggregation state, see Strzalko, T.; Seyden-Penne, J.; Wartski, L.; Froment, F.; Corset, J. *Tetrahedron Lett.* **1994**, *35*, 3935.

[1002]Sharma, U.; Bora, U.; Boruah, R.C.; Sandhu, J.S. *Tetrahedron Lett.* **2002**, *43*, 143.

[1003]Taylor, M.S.; Jacobsen, E.N. *J. Am. Chem. Soc.* **2003**, *125*, 11204.

[1004]For reviews of Michael reactions, where Z or Z′ is nitro, see Yoshikoshi, A.; Miyashita, M. *Acc. Chem. Res.* **1985**, *18*, 284; Baer, H.H.; UrBas L., in Feuer, H. *The Chemistry of the Nitro and Nitroso Groups*, pt. 2, Wiley, NY, **1970**, pp. 130–148. See Kumar, H.M.S.; Reddy, B.V.S.; Reddy, P.T.; Yadav, J.S. *Tetrahedron Lett.* **1999**, *40*, 5387; Ji, J.; Barnes, D.M.; Zhang, J.; King, S.A.; Wittenberger, S.J.; Morton, H.E. *J. Am. Chem. Soc.* **1999**, *121*, 10215; List, B.; Pojarliev, P.; Martin, H.J. *Org. Lett.* **2001**, *3*, 2423; Alexakis, A.; Andrey, O. *Org. Lett.* **2002**, *4*, 3611; Mase, N.; Thayumanavan, R.; Tanaka, F.; Barbas III, C.F. *Org. Lett.* **2004**, *6*, 2527; Okino, T.; Hoashi, Y.; Takemoto, Y. *J. Am. Chem. Soc.* **2003**, *125*, 12672; Ishii, T.; Fujioka, S.; Sekiguchi, Y.; Kotsuki, H. *J. Am. Chem. Soc.* **2004**, *126*, 9558; Li, H.; Wang, Y.; Tang, L.; Deng, L. *J. Am. Chem. Soc.* **2004**, *126*, 9906; Watanabe, M.; Ikagawa, A.; Wang, H.; Murata, K.; Ikariya, T. *J. Am. Chem. Soc.* **2004**, *126*, 11148.

and sulfones, as well as other compounds with relatively acidic hydrogens, such as indenes and fluorenes. Vinylogous Michael reaction are well known, using a variety of nucleophilic species.[1005] Michael addition of methyl 2,2-dichloroacetate/ LiN(TMS)$_2$ where TMS = trinethysilyl, leads to formation of a cyclopropane ring.[1006] Similarly, the intramolecular Michael addition of an α-chloro ketone enolate anion, formed *in situ* using DABCO, leads to formation of a bicyclo[4.1.0] diketone.[1007] It is noted that activated aryl compounds undergo Michael addition in the presence of an imidazolidinone catalyst.[1008] Conjugate addition of nitrones using SmI$_2$ has been reported.[1009]

These reagents do not add to ordinary double bonds, except in the presence of free-radical initiators (**15-33**). 1,2 Addition (to the C=O or C≡N group) often competes and sometimes predominates (**16-38**).[1010] In particular, α,β-unsaturated *aldehydes* seldom give 1,4 addition.[1011] The Michael reaction has traditionally been performed in protic solvents, with catalytic amounts of base,[1012] but more recently better yields with fewer side reactions have been obtained in some cases by using an equimolar amount of base to convert the nucleophile to its enolate form (*preformed enolate*). In particular, preformed enolates are often used where stereoselective reactions are desired.[1013] Silyl enol ethers can also be used, and in conjunction with chiral additives.[1014] Phase-transfer catalysts have been used,[1015] and ionic liquids have been used in conjunction with phase-transfer catalysis.[1016] Michael addition has been done in ionic liquids, adding aldehydes to conjugated nitro compounds using proline as a catalyst.[1017] Transition-metal compounds, such as CeCl$_3$,[1018] Yb(OTf)$_3$,[1019] Bi(OTf)$_3$,[1020] ferric chloride hexahydrate,[1021]

[1005]Ballini, R.; Bosica, G.; Fiorini, D. *Tetrahedron Lett.* **2001**, *42*, 8471.

[1006]Escribano, A.; Pedregal, C.; González, R.; Fernádez, A.; Burton, K.; Stephenson, G.A. *Tetrahedron* **2001**, *57*, 9423.

[1007]Bremeyer, N.; Smith, S.C.; Ley, S.V.; Gaunt, M.J. *Angew. Chem. Int. Ed.* **2004**, *43*, 2681.

[1008]Paras, N.A.; MacMillan, D.W.C. *J. Am. Chem. Soc.* **2002**, *124*, 7894.

[1009]Masson, G.; Cividino, P.; Py, S.; Vallée, Y. *Angew. Chem. Int. Ed.* **2003**, *42*, 2265.

[1010]For a discussion of 1,2 versus 1,4-addition, see, Oare, D.A.; Heathcock, C.H. *Top. Stereochem.* **1989**, *19*, 227, pp. 232–236.

[1011]For reports of successful 1,4-additions to α,β-unsaturated aldehydes, see Kryshtal, G.V.; Kulganek, V.V.; Kucherov, V.F.; Yanovskaya, L.A. *Synthesis* **1979**, 107; Yamaguchi, M.; Yokota, N.; Minami, T. *J. Chem. Soc., Chem. Commun.* **1991**, 1088.

[1012]See Macquarrie, D.J. *Tetrahedron Lett.* **1998**, *39*, 4125 for the use of supported phenolates as catalysts.

[1013]For reviews of stereoselective Michael additions, see Oare, D.A.; Heathcock, C.H. *Top. Stereochem.* **1991**, *20*, 87; **1989**, *19*, 227.

[1014]Harada, T.; Adachi, S.; Wang, X. *Org. Lett.* **2004**, *6*, 4877.

[1015]Kim, D.Y.; Huh, S.C. *Tetrahedron* **2001**, *57*, 8933.

[1016]Dere, R.T.; Pal, R.R.; Patil, P.S.; Salunkhe, M.M. *Tetrahedron Lett.* **2003**, *44*, 5351.

[1017]In bmim PF$_6$, 3-butyl-1-methylimidazolium hexafluorophosphate: Kotrusz, P.; Toma, S.; Schamlz, H.-G.; Adler, A. *Eur. J. Org. Chem.* **2004**, 1577.

[1018]Boruah, A.; Baruah, M.; Prajapati, D.; Sandhu, J.S. *Synth. Commun.* **1998**, *28*, 653; Bartoli, G.; Bosco, M.; Bellucci, M.C.; Marcantoni, E.; Sambri, L.; Torregiani, E. *Eur. J. Org. Chem.* **1999**, 617.

[1019]Keller, E.; Feringa, B.L. *Tetrahedron Lett.* **1996**, *37*, 1879; Kotsuki, H.; Arimura, K. *Tetrahedron Lett.* **1997**, *38*, 7583.

[1020]Varala, R.; Alam, M.M.; Adapa, S.R. *Synlett* **2003**, 720.

[1021]For a review, see Christoffers, J. *Synlett* **2001**, 723.

copper compounds,[1022] lanthanum complexes,[1023] ruthenium complexes,[1024] or scandium complexes[1025] also induce the reaction. In many cases, such compounds lead to catalytic enantioselective Michael additions.[1026] Conjugate addition has also been promoted by Y-zeolite,[1027] and water-promoted Michael additions have also been reported.[1028] Other catalysts have also been used.[1029] Vinylzinc complexes add to conjuguated keotnes in the presence of a CuBr catlayst.[1030]

In a Michael reaction with suitably different R groups, two new stereogenic centers are created (see **139**).

$$\underset{\text{Enolate}}{\overset{\ominus}{\underset{R^1}{\overset{O}{\diagup}}}C=C\underset{R^2}{\overset{R^3}{\diagdown}}} \quad + \quad \underset{\text{Substrate}}{\overset{R^4}{\underset{R^5}{\diagup}}C=C\underset{Z}{\overset{H}{\diagdown}}} \quad \longrightarrow \quad \underset{\textbf{139}}{\overset{O}{\underset{R^1}{\overset{\|}{C}}}\underset{R^2}{\overset{R^4}{\underset{R^3}{C}}}\underset{H}{\overset{R^5}{\underset{H}{C}}}Z}$$

In a diastereoselective process, one of the two pairs is formed exclusively or predominantly, as a racemic mixture.[1031] Many such examples have been reported.[672] In many of these cases, both the enolate anion and substrate can exist as (Z) or (E) isomers. With enolates derived from ketones or carboxylic esters, The (E) enolates gave the syn pair of enantiomers (p. 166), while (Z) enolates gave the anti pair.[1032] Nitro compounds add to conjugated ketones in the presence of a dipeptide and a piperazine.[1033] Malonate derivatives also add to conjugated ketones,[1034] and keto esters add to conjugated esters.[1035] Addition of chiral additives to the reaction, such as metal–salen complexes,[1036] proline derivatives,[1037] or (−)-sparteine,[1038]

[1022]Iguchi, Y.; Itooka, R.; Miyaura, N. *Synlett* **2003**, 1040; Meyer, O.; Becht, J.-M.; Helmchen, G. *Synlett* **2003**, 1539. For a discussion of the mechanism, see Comelles, J.; Moreno-Mañas, M.; Pérez, E.; Roglans, A.; Sebastián, R.M.; Vallribera, A. *J. Org. Chem.* **2004**, *69*, 6834.

[1023]Kim, Y.S.; Matsunaga, S.; Das, J.; Sekine, A.; Ohshima, T.; Shibasaki, M. *J. Am. Chem. Soc.* **2000**, *122*, 6506.

[1024]Watanabe, M.; Murata, K.; Ikariya, T. *J. Am. Chem. Soc.* **2003**, *125*, 7508; Wadsworth, K.J.; Wood, F.K.; Chapman, C.J.; Frost, C.G. *Synlett* **2004**, 2022. For a review, see Hayashi, T.; Yamasaki, K. *Chem. Rev.* **2003**, *103*, 2829.

[1025]Mori, Y.; Kakumoto, K.; Manabe, K.; Kobayashi, S. *Tetahedron Lett.* **2000**, *41*, 3107.

[1026]For a review, see Krause, N.; Hoffmann-Röder, A. *Synthesis* **2001**, 171.

[1027]Sreekumar, R.; Rugmini, P.; Padmakumar, R. *Tetrahedron Lett.* **1997**, *38*, 6557.

[1028]Lubineau, A.; Augé, J. *Tetrahedron Lett.* **1992**, *33*, 8073.

[1029]Phosphoramidites: Grossman, R.B.; Comesse, S.; Rasne, R.M.; Hattori, K.; Delong, M.N. *J. Org. Chem.* **2003**, *68*, 871. Fluorapatite: Zahouily, M.; Abrouki, Y.; Rayadh, A.; Sebti, S.; Dhimane, H.; David, M. *Tetrahedron Lett.* **2003**, *44*, 2463.

[1030]Huang, X.; Pi, J. *Synlett* **2003**, 481.

[1031]For a more extended analysis, see Oare, D.A.; Heathcock, C.H. *Top. Stereochem.* **1989**, *19*, p. 237.

[1032]For example, see Oare, D.A.; Heathcock, C.H. *J. Org. Chem.* **1990**, *55*, 157.

[1033]Tsogoeva, S.B.; Jagtap, S.B. *Synlett* **2004**, 2624; Ballini, R.; Barboni, L.; Bosica, G.; Fiorini, D. *Synthesis* **2002**, 2725.

[1034]Zhang, Z.; Dong, Y.-W.; Wang, G.-W.; Komatsu, K. *Synlett* **2004**, 61.

[1035]Yadav, J.S.; Geetha, V.; Reddy, B.V.S. *Synth. Commun.* **2002**, *32*, 3519.

[1036]Jha, S.C.; Joshi, N.N. *Tetrahedron Asymmetry* **2001**, *12*, 2463.

[1037]Yamaguchi, M.; Shiraishi, T.; Hirama, M. *J. Org. Chem.* **1996**, *61*, 3520.

[1038]Xu, F.; Tillyer, R.D.; Tschaen, D.M.; Grabowski, E.J.J.; Reider, P.J. *Tetrahedron Assymetry* **1998**, *9*, 1651.

lead to product formation with good-to-excellent asymmetric induction. Ultrasound has also been used to promote asymmetric Michael reactions.[1039] Intramolecular versions of Michael addition are known.[1040] A double Michael process is possible, where conjugate addition to an alkynyl ketone is followed by an intramolecular Michael to form a functionalized ring.[1041]

When either or both of the reaction components has a chiral substituent, the reaction can be enantioselective (only one of the four diastereomers formed predominantly).[1042] Enantioselective addition has also been achieved by the use of a chiral catalyst[1043] and by using optically active enamines instead of enolates.[1044] Chiral imines have also been used.[1045]

Mannich bases (see **16-19**) and β-halo carbonyl compounds can also be used as substrates; these are converted to the C=C—Z compounds *in situ* by the base (**16-19**, **17-13**).[1046] Substrates of this kind are especially useful in cases where the C=C—Z compound is unstable. The reaction of C=C—Z compounds with enamines (**10-69**) can also be considered a Michael reaction. Michael reactions are reversible.

When the substrate contains *gem*-Z groups (e.g., **141**), bulky groups can be added, if the reaction is carried out under aprotic conditions. For example, addition of enolate **140** to **141** gave **142** in which two adjacent quaternary centers have been formed.[1047]

| 140 | 141 | 142 |

[1039]Mirza-Aghayan, M.; Etemad-Moghadam, G.; Zaparucha, A.; Berlan, J.; Loupy, A.; Koenig, M. *Tetrahedron Asymmetry* **1995**, *6*, 2643.

[1040]Christoffers, J. *Tetrahedron Lett.* **1998**, *39*, 7083.

[1041]Holeman, D.S.; Rasne, R.M.; Grossman, R.B. *J. Org. Chem.* **2002**, *67*, 3149.

[1042]See, for example, Töke, L.; Fenichel, L.; Albert, M. *Tetrahedron Lett.* **1995**, *36*, 5951; Corey, E.J.; Peterson, R.T. *Tetrahedron Lett.* **1985**, *26*, 5025; Calderari, G.; Seebach, D. *Helv. Chim. Acta* **1985**, *68*, 1592; Tomioka, K.; Ando, K.; Yasuda, K.; Koga, K. *Tetrahedron Lett.* **1986**, *27*, 715; Posner, G.H.; Switzer, C. *J. Am. Chem. Soc.* **1986**, *108*, 1239; Enders, D.; Demir, A.S.; Rendenbach, B.E.M. *Chem. Ber.* **1987**, *120*, 1731. Also see, Hawkins, J.M.; Lewis, T.A. *J. Org. Chem.* **1992**, *57*, 2114.

[1043]Yura, T.; Iwasawa, N.; Mukaiyama, T. *Chem. Lett.* **1988**, 1021; Yura, T.; Iwasawa, N.; Narasaka, K.; Mukaiyama, T. *Chem. Lett.* **1988**, 1025; Desimoni, G.; Quadrelli, P.; Righetti, P.P. *Tetrahedron* **1990**, *46*, 2927.

[1044]See d'Angelo, J.; Revial, G.; Volpe, T.; Pfau, M. *Tetrahedron Lett.* **1988**, *29*, 4427.

[1045]d'Angelo, J.; Desmaële, D.; Dumas, F.; Guingant, A. *Tetrahedron Asymmetry* **1992**, *3*, 459.

[1046]Mannich bases react with ketones *without* basic catalysts to give 1,5-diketones, but this process, known as the *thermal-Michael reaction*, has a different mechanism: Brown, H.L.; Buchanan, G.L.; Curran, A.C.W.; McLay, G.W. *Tetrahedron* **1968**, *24*, 4565; Gill, N.S.; James, K.B.; Lions, F.; Potts, K.T. *J. Am. Chem. Soc.* **1952**, *74*, 4923.

[1047]Holton, R.A.; Williams, A.D.; Kennedy, R.M. *J. Org. Chem.* **1986**, *51*, 5480.

In certain cases, Michael reactions can take place under acidic conditions.[1048] Michael-type addition of radicals to conjugated carbonyl compounds is also known.[1049] Radical addition can be catalyzed by Yb(OTf)$_3$,[1050] but radicals add under standard conditions as well, even intramolecularly.[1051] Electrochemical-initiated Michael additions are known.

Michael reactions are sometimes applied to substrates of the type C≡C–Z, where the coproducts are conjugated systems of the type C=C–Z.[1052] Indeed, because of the greater susceptibility of triple bonds to nucleophilic attack, it is even possible for nonactivated alkynes (e.g., acetylene), to be substrates in this reaction.[1053]

In a closely related reaction, silyl enol ethers add to α,β-unsaturated ketones and esters when catalyzed[1054] by TiCl$_4$, for example,[1055]

InCl$_3$ also catalyzes this reaction.[1056] Aluminum compounds also catalyze this reaction[1057] and the reaction has been done in neat tri-n-propylaluminum.[1058] A solid-state version of the reaction used alumina•ZnCl$_2$.[1059] This reaction, also, has been performed diastereoselectively.[1060] Tin enolates have been used.[1061]

OS **I**, 272; **II**, 200; **III**, 286; **IV**, 630, 652, 662, 776; **V**, 486, 1135; **VI**, 31, 648, 666, 940; **VII**, 50, 363, 368, 414, 443; **VIII**, 87, 210, 219, 444, 467; **IX**, 526. See also, OS **VIII**, 148.

[1048]See Hajos, Z.G.; Parrish, D.R. *J. Org. Chem.* **1974**, *39*, 1612; *Org. Synth.* **VII**, 363.

[1049]Undheim, K.; Williams, K. *J. Chem. Soc., Chem. Commun.* **1994**, 883; Bertrand, S.; Glapski, C.; Hoffmann, N.; Pete, J.-P. *Tetrahedron Lett.* **1999**, *40*, 3169.

[1050]Sibi, M.P.; Jasperse, C.P.; Ji, J. *J. Am. Chem. Soc.* **1995**, *117*, 10779. See Wu, J.H.; Radinov, R.; Porter, N.A. *J. Am. Chem. Soc.* **1995**, *117*, 11029 for a related reaction involving Zn(OTf)$_2$.

[1051]Enholm, E.J.; Kinter, K.S. *J. Org. Chem.* **1995**, *60*, 4850.

[1052]Rudorf, W.-D.; Schwarz, R. *Synlett* **1993**, 369.

[1053]See, for example, Makosza, M. *Tetrahedron Lett.* **1966**, 5489.

[1054]Other catalysts have also been used. For a list of catalysts, with references, see Larock, R.C. *Comprehensive Organic Transformations*, 2nd ed., Wiley-VCH, NY, **1999**, pp. 1576–1582. See also, Mukaiyama, T.; Kobayashi, S.; Tamura, M.; Sagawa, Y. *Chem. Lett.* **1987**, 491; Mukaiyama, T.; Kobayashi, S. *J. Organomet. Chem.* **1990**, *382*, 39.

[1055]Narasaka, K.; Soai, K.; Aikawa, Y.; Mukaiyama, T. *Bull. Chem. Soc. Jpn.* **1976**, *49*, 779; Saigo, K.; Osaki, M.; Mukaiyama, T. *Chem. Lett.* **1976**, 163; Matsuda, I. *J. Organomet. Chem.* **1987**, *321*, 307; Narasaka, K. *Org. Synth,.* **65**, 12. See also, Yoshikoshi, A.; Miyashita, M. *Acc. Chem. Res.* **1985**, *18*, 284.

[1056]Loh, T.-P.; Wei, L.-L. *Tetrahedron* **1998**, *54*, 7615.

[1057]Tucker, J.A.; Clayton, T.L.; Mordas, D.M. *J. Org. Chem.* **1997**, *62*, 4370.

[1058]Kabbara, J.; Flemming, S.; Nickisch, K.; Neh, H.; Westermann, J. *Tetrahedron* **1995**, *51*, 743.

[1059]Ranu, B.C.; Saha, M.; Bhar, S. *Tetrahedron Lett.* **1993**, *34*, 1989.

[1060]See Heathcock, C.H.; Uehling, D.E. *J. Org. Chem.* **1986**, *51*, 279; Mukaiyama, T.; Tamura, M.; Kobayashi, S. *Chem. Lett.* **1986**, 1017, 1817, 1821; **1987**, 743.

[1061]Yasuda, M.; Ohigashi, N.; Shibata, I.; Baba, A. *J. Org. Chem.* **1999**, *64*, 2180; Yasuda, M.; Chiba, K.; Ohigashi, N.; Katoh, Y.; Baba, A. *J. Am. Chem. Soc.* **2003**, *125*, 7291.

15-25   1,4 Addition of Organometallic Compounds to Activated Double Bonds

Hydro-alkyl-addition

143

Lithium dialkylcopper reagents (see **10-57**) add to α,β-unsaturated aldehydes[1062] and ketones ($R' = H$, R, Ar) to give conjugate addition products[1063] in a reaction closely related to the Michael reaction. α,β-Unsaturated esters are less reactive,[1064] and the corresponding acids do not react at all. R can be primary alkyl, vinylic,[1065] or aryl. If Me$_3$SiCl is present, the reaction takes place much faster and with higher yields; in this case the product is the silyl enol ether of **143** (see **12-17**).[1066] The use of Me$_3$SiCl also permits good yields with allylic R groups.[1067] Conjugated alkynyl-ketones also react via 1,4-addition to give substituted alkenyl-ketones.[1068]

Various functional groups, such as OH and unconjugated C=O groups, may be present in the substrate.[1069] Conjugated sulfones are also good substrates.[1070] An excess of the cuprate reagent relative to the conjugated substrate is often required. In general, only one of the R groups of R$_2$CuLi adds to the substrate; the other is wasted. This can be a limitation where the precursor (RLi or RCu, see **12-36**) is expensive or available in limited amounts, particularly if an excess of the reagent

[1062]For reviews, see Alexakis, A.; Chuit, C.; Commerçon-Bourgain, M.; Foulon, J.P.; Jabri, N.; Mangeney, P.; Normant, J.F. *Pure Appl. Chem.* **1984**, *56*, 91.

[1063]House, H.O.; Respess, W.L.; Whitesides, G.M. *J. Org. Chem.* **1966**, *31*, 3128. For reviews, see Posner, G.H. *Org. React.* **1972**, *19*, 1; House, H.O. *Acc. Chem. Res.* **1976**, *9*, 59. For a discussion of the mechanism and regioselectivity, see Yamanaka, M.; Kato, S.; Nakamura, E. *J. Am. Chem. Soc.* **2004**, *126*, 6287. For examples of the use of this reaction in the synthesis of natural products, see Posner, G.H. *An Introduction to Synthesis Using Organocopper Reagents*, Wiley, NY, **1980**, pp. 10–67. For a list of organocopper reagents that give this reaction, with references, see Larock, R.C. *Comprehensive Organic Transformations*, 2nd ed., Wiley-VCH, NY, **1999**, pp. 1599–1613, 1814–1824.

[1064]R$_2$CuLi also add to *N*-tosylated α,β-unsaturated amides: Nagashima, H.; Ozaki, N.; Washiyama, M.; Itoh, K. *Tetrahedron Lett.* **1985**, *26*, 657.

[1065]Bennabi, S.; Narkunan, K.; Rousset, L.; Bouchu, D.; Ciufolini, M.A. *Tetrahedron Lett.* **2000**, *41*, 8873.

[1066]Corey, E.J.; Boaz, N.W. *Tetrahedron Lett.* **1985**, *26*, 6019; Alexakis, A.; Berlan, J.; Besace, Y. *Tetrahedron Lett.* **1986**, *27*, 1047; Matsuza, S.; Horiguchi, Y.; Nakamura, E.; Kuwajima, I. *Tetrahedron* **1989**, *45*, 349; Horiguchi, Y.; Komatsu, M.; Kuwajima, I. *Tetrahedron Lett.* **1989**, *30*, 7087; Linderman, R.J.; McKenzie, J.R. *J. Organomet. Chem.* **1989**, *361*, 31; Bertz, S.H.; Smith, R.A.J. *Tetrahedron* **1990**, *46*, 4091. For a list of references, see Larock, R.C. *Comprehensive Organic Transformations*, 2nd ed., Wiley-VCH, NY, **1999**, pp. 1491–1492.

[1067]Lipshutz, B.H.; Ellsworth, E.L.; Dimock, S.H.; Smith, R.A.J. *J. Am. Chem. Soc.* **1990**, *112*, 4404; Lipshutz, B.H.; James, B. *Tetrahedron Lett.* **1993**, *34*, 6689.

[1068]Degl'Innocenti, A.; Stucchi, E.; Capperucci, A.; Mordini, A.; Reginato, G.; Ricci, A. *Synlett* **1992**, *329*, 332.

[1069]For the use of enol tosylates of 1,2-diketones as substrates, see Charonnat, J.A.; Mitchell, A.L.; Keogh, B.P. *Tetrahedron Lett.* **1990**, *31*, 315.

[1070]Domínguez, E.; Carretero, J.C. *Tetrahedron Lett.* **1993**, *34*, 5803.

is required. The difficulty of group transfer can be overcome by using one of the mixed reagents R(R'C≡C)CuLi,[1071] R(O—t—Bu)CuLi,[1072] R(PhS)CuLi,[1073] each of which transfers only the R group. Mixed reagents are easily prepared by the reaction of RLi with R'C≡CCu (R' =n-Pr or t-Bu), t-BuOCu, or PhSCu, respectively. A further advantage of the mixed reagents is that good yields of addition product are achieved when R is tertiary, so that use of one of them permits the introduction of a tertiary alkyl group. The mixed reagents R(CN)CuLi[1074] (prepared from RLi and CuCN) and $R_2Cu(CN)Li_2$[1075] also selectively transfer the R group.[1076] Other mixed reagents incorporate a ligand that is not easily transferred, such as $R(R'Se)Cu(CN)Li_2$, leading to selective transfer of the R group.[1077] The reaction has also been carried out[1078] with α,β-acetylenic ketones,[1079] esters, and nitriles.

Both Grignard and R_2CuLi reagents[1080] have also been added to systems of the form C≡C—C=O.[1081] Conjugate addition to α,β-unsaturated and acetylenic acids and esters, as well as ketones, can be achieved by the use of the coordinated reagents RCu•BF₃ (R = primary).[1082] Alkylcopper compounds RCu (R = primary or secondary alkyl) have also been used with tetramethylethylenediamine and Me_3SiCl to give silyl enol ethers from α,β-unsaturated ketones in high yield.[1083] Amine units have been transferred in this manner using α-lithio amides, CuCN, and additives ranging from LiCl to $Me_2NCH_2SnBu_3$, which gave conjugate addition of an amidomethyl unit, —$CH_2N(Me)Boc$.[1084] Other amino-cuprates are known to give conjugate addition reactions.[1085]

[1071]House, H.O.; Umen, M.J. *J. Org. Chem.* **1973**, *38*, 3893; Corey, E.J.; Floyd, D.; Lipshutz, B.H. *J. Org. Chem.* **1978**, *43*, 3419.

[1072]Posner, G.H.; Whitten, C.E. *Tetrahedron Lett.* **1973**, 1815.

[1073]Posner, G.H.; Whitten, C.E.; Sterling, J.J. *J. Am. Chem. Soc.* **1973**, *95*, 7788.

[1074]Gorlier, J.; Hamon, L.; Levisalles, J.; Wagnon, J. *J. Chem. Soc., Chem. Commun.* **1973**, 88. For another useful mixed reagent, see Ledlie, D.B.; Miller, G. *J. Org. Chem.* **1979**, *44*, 1006.

[1075]Lipshutz, B.H.; Wilhelm, R.S.; Kozlowski, J. *Tetrahedron Lett.* **1982**, *23*, 3755; Lipshutz, B.H. *Tetrahedron Lett.* **1983**, *24*, 127.

[1076]When the two R groups of $R_2Cu(CN)Li_2$ are different, one can be selectively transferred: Lipshutz, B.H.; Wilhelm, R.S.; Kozlowski, J.A. *J. Org. Chem.* **1984**, *49*, 3938.

[1077]Zinn, F.K.; Ramos, E.C.; Comasseto, J.V. *Tetrahedron Lett.* **2001**, *42*, 2415.

[1078]For a list of references, see Larock, R.C. *Comprehensive Organic Transformations*, 2nd ed., Wiley-VCH, NY, **1999**, pp. 456–457.

[1079]Lee, P.H.; Park, J.; Lee, K.; Kim, H.-C. *Tetrahedron Lett.* **1999**, *40*, 7109.

[1080]For example, see Corey, E.J.; Kim, C.U.; Chen, H.K.; Takeda, M. *J. Am. Chem. Soc.* **1972**, *94*, 4395; Anderson, R.J.; Corbin, V.L.; Cotterrell, G.; Cox, G.R.; Henrick, C.A.; Schaub, F.; Siddall, J.B. *J. Am. Chem. Soc.* **1975**, *97*, 1197.

[1081]For a review of the addition of organometallic reagents to conjugated enynes see Miginiac, L. *J. Organomet. Chem.* **1982**, *238*, 235.

[1082]For a review, see Yamamoto, Y. *Angew. Chem. Int. Ed.* **1986**, *25*, 947. For a discussion of the role of the BF₃, see Lipshutz, B.H.; Ellsworth, E.L.; Siahaan, T.J. *J. Am. Chem. Soc.* **1988**, *110*, 4834; **1989**, *111*, 1351.

[1083]Johnson, C.R.; Marren, T.J. *Tetrahedron Lett.* **1987**, *28*, 27.

[1084]Dieter, R.K.; Velu, S.E. *J. Org. Chem.* **1997**, *62*, 3798; Dieter, R.K.; Alexander, C.W. *Synlett* **1993**, 407; Dieter, R.K.; Alexander, C.W.; Nice, L.E. *Tetrahedron* **2000**, *56*, 2767; Dieter, R.K.;Lu, K.; Velu, S.E. *J. Org. Chem.* **2000**, *65*, 8715. See Dieter, R.K.; Topping, C.M.; Nice, L.E. *J. Org. Chem.* **2001**, *66*, 2302.

[1085]Yamamoto, Y.; Asao, N.; Uyehara, T. *J. Am. Chem. Soc.* **1992**, *114*, 5427.

There is generally little or no competition from 1,2-addition (to the C=O). However, when R is allylic, 1,4-addition is observed with some substrates and 1,2-addition with others.[1086] R_2CuLi also add to α,β-unsaturated sulfones[1087] but not to simple α,β-unsaturated nitriles.[1088] Organocopper reagents (RCu), as well as certain R_2CuLi add to α,β-unsaturated and acetylenic sulfoxides.[1089]

143

Conjugate addition of the cuprate to the α,β-unsaturated ketone leads to an eno-late ion, **143**. It is possible to have this enolate anion reacts with an electrophilic species (*tandem vicinal difunctionalization*), in some cases at the O and in other cases at the C.[1090] For example, if an alkyl halide R^2X is present (R^2 = primary alkyl or allylic), the enolate **143** can be alkylated directly.[1091] Thus, by this method, both the α and β positions of a ketone are alkylated in one synthetic operation (see also, **15-22**).

144   79% (84% ee , S)

As with the Michael reaction (**15-24**) the 1,4-addition of organometallic com-pounds has been performed diastereoselectively[1092] and enantioselectively.[1093]

[1086]House, H.O.; Fischer, Jr., W.F. *J. Org. Chem.* **1969**, *34*, 3615. See also, Daviaud, G.; Miginiac, P. *Tetrahedron Lett.* **1973**, 3345.

[1087]Posner, G.H.; Brunelle, D.J. *Tetrahedron Lett.* **1973**, 935.

[1088]House, H.O.; Umen, M.J. *J. Org. Chem.* **1973**, *38*, 3893.

[1089]Truce, W.E.; Lusch, M.J. *J. Org. Chem.* **1974**, *39*, 3174; **1978**, *43*, 2252.

[1090]For reviews of such reactions, see Chapdelaine, M.J.; Hulce, M. *Org. React.* **1990**, *38*, 225; Taylor, R.J.K. *Synthesis* **1985**, 364. For a list of references, see Larock, R.C. *Comprehensive Organic Transformations*, 2nd ed., Wiley-VCH, NY, **1999**, pp. 1609–1612, 1826.

[1091]Coates, R.M.; Sandefur, L.O. *J. Org. Chem.* **1974**, *39*, 275; Posner, G.H.; Lentz, C.M. *Tetrahedron Lett.* **1977**, 3215.

[1092]For some examples, see Isobe, M.; Funabashi, Y.; Ichikawa, Y.; Mio, S.; Goto, T. *Tetrahedron Lett.* **1984**, *25*, 2021; Kawasaki, H.; Tomioka, K.; Koga, K. *Tetrahedron Lett.* **1985**, *26*, 3031; Yamamoto, Y.; Nishii, S.; Ibuka, T. *J. Chem. Soc., Chem. Commun.* **1987**, 464, 1572; Smith III, A.B.; Dunlap, N.K.; Sulikowski, G.A. *Tetrahedron Lett.* **1988**, *29*, 439; Smith III, A.B.; Trumper, P.K. *Tetrahedron Lett.* **1988**, *29*, 443; Alexakis, A.; Sedrani, R.; Mangeney, P.; Normant, J.F. *Tetrahedron Lett.* **1988**, *29*, 4411; Larchevêque, M.; Tamagnan, G.; Petit, Y. *J. Chem. Soc., Chem. Commun.* **1989**, 31; Page, P.C.B.; Prodger, J.C.; Hursthouse, M.B.; Mazid, M. *J. Chem. Soc. Perkin Trans. 1*, **1990**, 167; Corey, E.J.; Hannon, F.J. *Tetrahedron Lett.* **1990**, *31*, 1393.

[1093]For reviews, see Posner, G.H. *Acc. Chem. Res.* **1987**, *20*, 72; in Morrison, J.D. *Assymmetric Synthesis* Vol. 2, Academic Press, NY, **1983**, the articles by Tomioka, K.; Koga, K. pp. 201–224; Posner, G. pp. 225–241.

The influence of solvent and additives on yield and selectivity has been examined.[1094] The conjugate addition of dimethyl cuprate in the presence of a chiral ligand, such as **144**, is an example.[1095] The use of chiral ligands with MgI_2/I_2 and Bu_3SnI gave conjugate addition products with α,β-unsaturated amides with good % ee.[1096] Chiral bis(oxazoline) copper catalysts have been used for the conjugate addition of indoles to α,β-unsaturated esters.[1097] Chiral templates have also been used with Grignard reagents, directly[1098] and in the presence of $AlMe_2Cl$.[1099] Many of the examples cited below involve the use of chiral additives, chiral catalysts, or chiral templates.

Grignard reagents also add to conjugated substrates such as α,β-unsaturated ketones, cyano-ketones,[1100] esters, and nitriles,[1101] but 1,2-addition may seriously compete:[1102] The product is often controlled by steric factors. Thus **145** with phenylmagnesium bromide gives 100% 1,4-addition, while **146** gives 100% 1,2-addition. In general, substitution at the carbonyl group increases 1,4-addition, while substitution at the double bond increases 1,2-addition. In most cases, both products are obtained, but α,β-unsaturated *aldehydes* nearly always give exclusive 1,2-addition when treated with Grignard reagents. However, the extent of 1,4-addition of Grignard reagents can be increased by the use of a copper ion catalyst, for example, CuCl, Cu(OAc)$_2$.[1103] A dialkyl copper–magnesium iodide complex (R$_2$Cu·MgI) has been used for conjugate addition to chiral α,β-unsaturated amides.[1104] Grignard reagents mixed with CeCl$_3$ generates a reactive species that gives primarily 1,4-addition.[1105] It is likely that alkylcopper reagents, formed from RMgX and

[1094]Christenson, B.; Ullenius, C.; Håkansson, M.; Jagner, S. *Tetrahedron* **1992**, *48*, 3623.

[1095]Kanai, M.; Koga, K.; Tomioka, K. *Tetrahedron Lett.* **1992**, *33*, 7193.

[1096]Sibi, M.P.; Ji, J.; Wu, J.H.; Gürtler, S.; Porter, N.A. *J. Am. Chem. Soc.* **1996**, *118*, 9200.

[1097]Jensen, K.B.; Thorhauge, J.; Hazell, R.G.; Jørgensen, K.A. *Angew. Chem. Int. Ed.* **2001**, *40*, 160.

[1098]Han, Y.; Hruby, V.J. *Tetrahedron Lett.* **1997**, *38*, 7317.

[1099]Bongini, A.; Cardillo, G.; Mingardi, A.; Tomasini, C. *Tetrahedron Asymmetry* **1996**, *7*, 1457.

[1100]Kung, L.-R.; Tu, C.-H.; Shia, K.-S.; Liu, H.-J. *Chem. Commun.* **2003**, 2490.

[1101]Fleming, F.F.; Wang, Q.; Zhang, Z.; Steward, O.W. *J. Org. Chem.* **2002**, *67*, 5953.

[1102]For a discussion of the factors affecting 1,2- versus 1,4-addition, see Negishi, E. *Organometallics in Organic Synthesis* Vol. 1, Wiley, NY, **1980**, pp. 127–133.

[1103]Posner, G.H. *Org. React.* **1972**, *19*, 1; López, F.; Harutyunyan, S.R.; Minnaard, A.J.; Feringa, B.L. *J. Am. Chem. Soc.* **2004**, *126*, 12784.

[1104]Schneider, C.; Reese, O. *Synthesis* **2000**, 1689.

[1105]Bartoli, G.; Bosco, M.; Sambri, L.; Marcantoni, E. *Tetrahedron Lett.* **1994**, *35*, 8651.

Cu^+ (cupric acetate is reduced to cuprous ion by excess RMgX), are the actual attacking species in these cases.[1063] Alkylidene malonic ester derivatives, $C=C(CO_2R)$, increase the facility of 1,4-addition with the two electron withdrawing groups.[1106] Alkylidene amido esters, $C=C(CO_2R)NHCOAr$, react with $EtI/Mg(ClO_4)_2$ and Bu_3SnH, in the presence of BEt_3/O_2 and a chiral ligand, to give the ethylated product $EtCHCH(CO_2R)NHCOAr$.[1107] This is probably a radical process (see **15-35**).

Organolithium reagents[1108] generally react with conjugated aldehydes, ketones and esters by 1,2-addition,[1109] but 1,4-addition was achieved with esters of the form $C=C-COOAr$, where Ar was a bulky group such as 2,6-di-*tert*-butyl-4-methoxyphenyl.[1110] Alkyllithium reagents can be made to give 1,4-addition with α,β-unsaturated ketones[1111] and aldehydes[1112] if the reactions are conducted in the presence of HMPA.[1113] Among organolithium reagents that have been found to add 1,4 in this manner are 2-lithio-1,3-dithianes (see **10-71**),[1114] vinyllithium reagents,[1115] and α-lithio allylic amides.[1116] Lithium–halogen exchange (**12-22**) generates an organolithium species that adds intramolecularly to conjugated esters to give cyclic and bicyclic products.[1117] 1,4-Addition of alkyllithium reagents to α,β-unsaturated aldehydes can also be achieved by converting the aldehyde to a benzothiazole derivative (masking the aldehyde function),[1118] from which the aldehyde group can be regenerated. When some conjugated acids are added to organolithium reagents, the conjugate addition product was isolated in good yield.[1119] α,β-Unsaturated nitro compounds undergo conjugate addition with aryllithium reagents, and subsequent treatment with acetic acid gives the α-aryl ketone.[1120]

[1106]See Kim, Y.M.; Kwon, T.W.; Chung, S.K.; Smith, M.B. *Synth. Commun.* **1999**, *29*, 343.

[1107]Sibi, M.P.; Asano, Y.; Sausker, J.B. *Angew. Chem. Int. Ed.* **2001**, *40*. 1293.

[1108]For a review of addition of organolithium compounds to double bonds, see Hunt, D.A. *Org. Prep. Proced. Int.* **1989**, *21*, 705–749.

[1109]Rozhkov, I.N.; Makin, S.M. *J. Gen. Chem. USSR* **1964**, *34*, 57. For a discussion of 1,2- versus 1,4-addition with organolithiums, see Cohen, T.; Abraham, W.D.; Myers, M. *J. Am. Chem. Soc.* **1987**, *109*, 7923.

[1110]Cooke, Jr., M.P. *J. Org. Chem.* **1986**, *51*, 1637.

[1111]Roux, M.C.; Wartski, L.; Seyden-Penne, J. *Tetrahedron* **1981**, *37*, 1927; *Synth. Commun.* **1981**, *11*, 85.

[1112]El-Bouz, M.; Wartski, L. *Tetrahedron Lett.* **1980**, *21*, 2897.

[1113]Sikorski, W.H.; Reich, H.J. *J. Am. Chem. Soc.* **2001**, *123*, 6527.

[1114]Lucchetti, J.; Dumont, W.; Krief, A. *Tetrahedron Lett.* **1979**, 2695; Brown, C.A.; Yamaichi, A. *J. Chem. Soc., Chem. Commun.* **1979**, 100; El-Bouz, M.; Wartski, L. *Tetrahedron Lett.* **1980**, *21*, 2897. See also, Bürstinghaus, R.; Seebach, D. *Chem. Ber.* **1977**, *110*, 841.

[1115]For an intramolecular example, see Maezaki, N.; Sawamoto, H.; Yuyama, S.; Yoshigami, R.; Suzuki, T.; Izumi, M.; Ohishi, H.; Tanaka, T. *J. Org. Chem.* **2004**, *69*, 6335.

[1116]For an example using sparteine as a chiral additive, see Curtis, M.D.; Beak, P. *J. Org. Chem.* **1999**, *64*, 2996.

[1117]Cooke Jr., M.P.; Gopal, D. *Tetrahedron Lett.* **1994**, *35*, 2837. For an example involving the intramolecular addition of a vinyllithium reagent, see Piers, E.; Harrison, C.L.; Zetina-Rocha, C. *Org. Lett.* **2001**, *3*, 3245.

[1118]Corey, E.J.; Boger, D.L. *Tetrahedron Lett.* **1978**, 9. For another indirect method, see Sato, T.; Okazaki, H.; Otera, J.; Nozaki, H. *Tetrahedron Lett.* **1988**, *29*, 2979.

[1119]Aurell, M.J.; Mestres, R.; Muñoz, E. *Tetrahedron Lett.* **1998**, *39*, 6351. Also see, Plunian, B.; Vaultier, M.; Mortier, J. *Chem. Commun.* **1998**, 81. For a discussion of the mechansim, see Aurell, M.J.; Bañuls, M.J.; Mestres, R.; Muñoz, E. *Tetrahedron* **2001**, *57*, 1067.

[1120]Santos, R.P.; Lopes, R.S.C.; Lopes, C.C. *Synthesis* **2001**, 845.

If the organolithium reagent is modified, 1,4-addition is more successful. The reaction of an aryllithium reagent with B(OMe)$_3$, for example, led to a rhodium-catalyzed conjugate addition with excellent enantioselectivity in when a chiral ligand was employed.[1121] Allylic tellurium reagents that are treated with lithium diisopropyl amide, and then conjugated esters give the 1,4-addition product, which cyclizes to form the corresponding cyclopropane derivative.[1122]

Boron reagents add to conjugated carbonyl compounds.[1123] Alkynyl borate esters (p. 815) give conjugate addition in the presence of boron trifluoride etherate,[1124] as do arylboronic acids (p. 815) with a rhodium,[1125] palladium,[1126] or a bismuth catalyst.[1127] Diethylzinc has also been used.[1128] Aryl boronic acids add to the double bond of vinyl sulfones in the presence of a rhodium catalyst.[1129] Similarly, LiBPh(OMe)$_3$ and a rhodium catalyst gave conjugate addition of the phenyl group to α,β-unsaturated esters.[1130] Potassium vinyltrifluoroborates (see p. 607) give 1,4-addition with a rhodium catalyst,[1131] as do aryltrifluoroborates.[1132]

Organozinc compounds add to conjugated systems. The use of chiral ligands is effective for conjugate addition of dialkylzinc compounds to α,β-unsaturated ketones, esters, and so on,[1133] including conjugated lactones.[1134] Many dialkylzinc compounds can be used, including vinylzinc compounds.[1135] Dialkylzinc

---

[1121]Takaya, Y.; Ogasawara, M.; Hayashi, T. *Tetrahedron Lett.* **1999**, *40*, 6957.

[1122]Liao, W.-W.; Li, K.; Tang, Y. *J. Am. Chem. Soc.* **2003**, *125*, 13030.

[1123]Kabalka, G.W.; Das, B.C.; Das, S. *Tetrahedron Lett.* **2002**, *43*, 2323.

[1124]Chong, J.M.; Shen, L.; Taylor, N.J. *J. Am. Chem. Soc.* **2000**, *122*, 1822.

[1125]Itooka, R.; Iguchi, Y.; Miyaura, N. *J. Org. Chem.* **2003**, *68*, 6000; Ramnauth, J.; Poulin, O.; Bratovanov, S.S.; Rakhit, S.; Maddaford, S.P. *Org. Lett.* **2001**, *3*, 2571; Reetz, M.T.; Moulin, D.; Gosberg, A. *Org. Lett.* **2001**, *3*, 4083; Kuriyama, M.; Nagai, K.; Yamada, K.-i.; Miwa, Y.; Taga, T.; Tomioka, K. *J. Am. Chem. Soc.* **2002**, *124*, 8932; Boiteau, J.-G.; Minnaard, A.J.; Feringa, B.L. *J. Org. Chem.* **2003**, *68*, 9481; Boiteau, J.-G.; Imbos, R.; Minnaard, A.J.; Feringa, B.L. *Org. Lett.* **2003**, *5*, 681; Shintani, R.; Ueyama, K.; Yamada, I.; Hayashi, T. *Org. Lett.* **2004**, *6*, 3425; Shi, Q.; Xu, L.; Li, X.; Wang, R.; Au-Yeung, T.T.-L.; Chan, A.S.C.; Hayashi, T.; Cao, R.; Hong, M. *Tetrahedron Lett.* **2003**, *44*, 6505; Amengual, R.; Michelet, V.; Genêt, J.-P. *Synlett.* **2002**, 1791. For a review, see Hayashi, T. *Synlett* **2001**, 879.

[1126]Nishikata, T.; Yamamoto, Y.; Miyaura, N. *Angew. Chem. Int. Ed.* **2003**, *42*, 2768.

[1127]Sakuma, S.; Miyaura, N. *J. Org. Chem.* **2001**, *66*, 8944.

[1128]Dong, L.; Xu, Y.-J.; Gong, L.-Z.; Mi, A.-Q.; Jiang, Y.-Z. *Synthesis* **2004**, 1057.

[1129]With a chiral ligand, see Mauleón, P.; Carretero, J.C. *Org. Lett.* **2004**, *6*, 3195.

[1130]Takaya, Y.; Senda, T.; Kurushima, H.; Ogasawara, M.; Hayashi, T. *Tetrahedron Asymmetry* **1999**, *10*, 4047.

[1131]Duursma, A.; Boiteau, J.-G.; Lefort, L.; Boogers, J.A.F.; de Vries, A.H.M.; de Vires, J.G.; Minnaard, A.J.; Feringa, B.L. *J. Org. Chem.* **2004**, *69*, 8045.

[1132]Moss, R.J.; Wadsworth, K.J.; Chapman, C.J.; Frost, C.G. *Chem. Commun.* **2004**, 1984; Pucheault, M.; Darses, S.; Genêt, J.-P. *Eur. J. Org. Chem.* **2002**, 3552.

[1133]Alexakis, A.; Burton, J.; Vastra, J.; Mangeney, P. *Tetrahedron Asymm.*, **1997**, *8*, 3987. Yan, M.; Yang, L.,-W.; Wong, K.-Y.; Chan, A.S.C. *Chem. Commun.* **1999**, 11; Tong, P.-E.; Li, P.; Chan, A.S.C. *Tetrahedron Asymmetry* **2001**, *12*, 2301; Liang, L.; Au-Yeung, T.T.-L.; Chan, A.S.C. *Org. Lett.* **2002**, *4*, 3799.

[1134]Reetz, M.T.; Gosberg, A.; Moulin, D. *Tetrahedron Lett.* **2002**, *43*, 1189.

[1135]For an example using a nickel catalyst with a chiral ligand, see Ikeda, S.-i.; Cui, D.-M.; Sato, Y. *J. Am. Chem. Soc.* **1999**, *121*, 4712.

compounds and a chiral complex leads to enantioselective conjugate addition in conjunction with $Cu(OTf)_2$[1136] or other copper compounds.[1137] Diethylzinc adds to conjugated nitro compounds in the presence of a catalytic amount of $Cu(OTf)_2$ to give the conjugate addition product.[1138] Other transition-metal compounds can be used in conjunction with dialkylzinc compounds[1139] or with arylzinc halides (ArZnCl).[1140] Reaction of alkyl iodides with Zn/CuI with ultrasound generates an organometallic that adds to conjugated esters.[1141] Diarylzinc compounds (prepared with the aid of ultrasound) in the presence of nickel acetylacetonate, undergo 1,4-addition not only to α,β-unsaturated ketones, but also to α,β-unsaturated aldehydes.[1142] Mixed alkylzinc compounds also add to conjugated systems.[1143] Functionalized allylic groups can be added to terminal alkynes with allylic halides, zinc, and ultrasound, to give 1,4-dienes.[1144] Internal alkynes undergo 1,4-addition to conjugated esters using a combination of zinc metal and a cobalt complex as catalysts.[1145]

[1136]Liang, L.; Yan, M.; Li, Y.-M.; Chan, A.S.C. *Tetrahedron Asymmetry* **2004**, *15*, 2575, and references cited therein; Pàmies, O.; Net, G.; Ruiz, A.; Claver, C.; Woodward, S. *Tetrahedron Asymmetry* **2000**, *11*, 871; Diéguez, M. Ruiz, A.; Claver, C. *Tetrahedron Asymmetry* **2001**, *12*, 2861; Arena, C.G.; Calabrò, G.P.; Franciò, G.; Faraone, F. *Tetrahedron Asymmetry* **2000**, *11*, 2387; Mandoli, A.; Arnold, L.A.; de Vries, A.H.M.; Salvadori, P.; Feringa, B.L. *Tetrahedron Asymmetry* **2001**, *12*, 1929; Martorell, A.; Naasz, R.; Feringa, B.L.; Pringle, P.G. *Tetrahedron Asymmetry* **2001**, *12*, 2497; Escher, I.H.; Pfaltz, A. *Tetrahedron* **2000**, *56*, 2879; Morimoto, T.; Yamaguchi, Y.; Suzuki, M.; Saitoh, A. *Tetrahedron Lett.* **2000**, *41*, 10025; Alexakis, A.; Polet, D.; Rosset, S.; March, S. *J. Org. Chem.* **2004**, *69*, 5660, and references cited therein; Pytkowicz, J.; Roland, S.; Mangeney, P. *Tetrahedron Asymmetry* **2001**, *12*, 2087; Zhou, H.; Wang, W.-H.; Fu, Y.; Xie, J.-H.; Shi, W.-J.; Wang, L.-X.; Zhou, Q.-L. *J. Org. Chem.* **2003**, *68*, 1582; Duncan, A.P.; Leighton, J.L. *Org. Lett.* **2004**, *6*, 4117, and references cited therein; Choi, Y.H.; Choi, J.Y.; Yang, H.-Y.; Kim, H.Y. *Tetrahedron Asymmetry* **2002**, *13*, 801; Kang, J.; Lee, J.H.; Lim, D.S. *Tetrahedron Asymmetry* **2003**, *14*, 305; Scafato, P.; Labano, S.; Cunsolo, G.; Rosini, C. *Tetrahedron Asymmetry* **2003**, *14*, 3873; Hird, A.W.; Hoveyda, A.H. *Angew. Chem. Int. Ed.* **2003**, *42*, 1276.

[1137]Delapierre, G.; Brunel, J.M.; Constantieux, T.; Buono, G. *Tetrahedron Asymmetry* **2001**, *12*, 1345; Hu, Y.; Liang, X.; Wang, J.; Zheng, Z.; Hu, X. *J. Org. Chem.* **2003**, *68*, 4542; Wan, H.; Hu, Y.; Liang, Y.; Gao, S.; Wang, J.; Zheng, Z.; Hu, X. *J. Org. Chem.* **2003**, *68*, 8277; Alexakis, A.; Polet, D.; Benhaim, C.; Rosset, S. *Tetrahedron Asymmetry* **2004**, *15*, 2199; Breit, B.; Laungani, A.Ch. *Tetrahedron Asymmetry* **2003**, *14*, 3823; Shi, M.; Wang, C.-J.; Zhang, W. *Chem. Eur. J.* **2004**, *10*, 5507.

[1138]Yan, M.; Chan, A.S.C. *Tetrahedron Lett.* **1999**, *40*, 6645; Rimkus, A.; Sewald, N. *Org. Lett.* **2002**, *4*, 3289; Choi, H.; Hua, Z.; Ojima, I. *Org. Lett.* **2004**, *6*, 2689; Mampreian, D.M.; Hoveyda, A.H. *Org. Lett.* **2004**, *6*, 2829; Duursma, A.; Minnaard, A.J.; Feringa, B.L. *J. Am. Chem. Soc.* **2003**, *125*, 3700.

[1139]With a vanadium complex: Hirao, T.; Takada, T.; Sakurai, H. *Org. Lett.* **2000**, *2*, 3659. With a nickel complex: Yin, Y.; Li, X.; Lee, D.-S.; Yang, T.-K. *Tetrahedron Asymmetry* **2000**, *11*, 3329; Shadakshari, U.; Nayak, S.K. *Tetrahedron* **2001**, *57*, 8185.

[1140]Shitani, R.; Tokunaga, N.; Doi, H.; Hayashi, T. *J. Am. Chem. Soc.* **2004**, *126*, 6240.

[1141]Sarandeses, L.A.; Mouriño, A.; Luche, J.-L. *J. Chem. Soc., Chem. Commun.* **1992**, 798. See also, Das, B.; Banerjee, J.; Mahender, G.; Majhi, A. *Org. Lett.* **2004**, *6*, 3349.

[1142]Pétrier, C.; de Souza Barboza, J.C.; Dupuy, C.; Luche, J. *J. Org. Chem.* **1985**, *50*, 5761.

[1143]Berger, S.; Langer, F.; Lutz, C.; Knochel, P.; Mobley, T.A.; Reddy, C.K. *Angew. Chem. Int. Ed.* **1997**, *36*, 1496.

[1144]Knochel, P.; Normant, J.F. *J. Organomet. Chem.* **1986**, *309*, 1.

[1145]Wang, C.-C.; Lin, P.-S.; Cheng, C.-H. *J. Am. Chem. Soc.* **2002**, *124*, 9696.

Trialkylalanes R_3Al add 1,4 to α,β-unsaturated carbonyl compounds in the presence of nickel acetylacetonate[1146] or $Cu(OTf)_2$.[1147] In the presence of aluminum chloride, benzene reacts with conjugated amides to add a phenyl group to C-4.[1148] Alkyl halides react via conjugate addition using BEt_3 or $AlEt_3$.[1149] An alkynyl group can be added to the double bond of an α,β-unsaturated ketone by use of the diethylalkynylalane reagents $Et_2AlC{\equiv}CR$.[1150] In a similar manner, the alkenyl reagents $R_2AlCH{=}CR$ transfer an alkenyl group.[1151]

Terminal alkynes add to conjugated systems when using a ruthenium,[1152] palladium,[1153] or a rhodium catalyst.[1154] Triphenylbismuth (Ph_3Bi) and a rhodium catalyst gives conjugate addition of the phenyl group upon exposure to air.[1155] Similar reactivity is observed with a palladium catalyst in aqueous media.[1156] Lithium tetraalkylgallium reagents give 1,4-addition.[1157] Trimethyl(phenyl)tin and a rhodium catalyst gives conjugate addition of a methyl group[1158] and tetraphenyltin and a palladium catalyst adds a phenyl group.[1159] Allyltin compounds add an allyl group in the presence of a scandium catalyst.[1160] Benzylic bromides add to conjugated nitriles using a 2:1 mixture of $CrCl_3$ and manganese metal.[1161] Electrochemical conjugate addition to α,β-unsaturated ketones was reported using aryl halides and a cobalt catalyst.[1162] Aryl halides add in the presence of $NiBr_2$.[1163] Vinyl zirconium complexes undergo conjugate addition when using a rhodium catalyst.[1164] Pyrrole adds to conjugated alkynyl esters in the presence

[1146]Bagnell, L.; Meisters, A.; Mole, T. *Aust. J. Chem.* **1975**, *28*, 817; Ashby, E.C.; Heinsohn, G. *J. Org. Chem.* **1974**, *39*, 3297. See also, Sato, F.; Oikawa, T.; Sato, M. *Chem. Lett.* **1979**, 167; Kunz, H.; Pees, K.J. *J. Chem. Soc. Perkin Trans. 1*, **1989**, 1168.

[1147]Su, L.; Li, X.; Chan, W.L.; Jia, X.; Chan, A.S.C. *Tetrahedron Asymmetry* **2003**, *14*, 1865.

[1148]Koltunov, K.Yu.; Walspurger, S.; Sommer, J. *Tetrahedron Lett.* **2004**, *45*, 3547.

[1149]Liu, J.-Y.,; Jang, Y.-J.; Lin, W.-W.; Liu, J.-T.; Yao, C.-F. *J. Org. Chem.* **2003**, *68*, 4030.

[1150]Hooz, J.; Layton, R.B. *J. Am. Chem. Soc.* **1971**, *93*, 7320; Schwartz, J.; Carr, D.B.; Hansen, R.T.; Dayrit, F.M. *J. Org. Chem.* **1980**, *45*, 3053.

[1151]Hooz, J.; Layton, R.B. *Can. J. Chem.* **1973**, *51*, 2098. For a similar reaction with an alkenylzirconium reagent, see Dayrit, F.M.; Schwartz, J. *J. Am. Chem. Soc.* **1981**, *103*, 4466.

[1152]With $SnCl_4$, see Trost, B.M.; Pinkerton, A.B. *J. Am. Chem. Soc.* **1999**, *121*, 1988. See Chang, S.; Na, Y.; Choi, E.; Kim, S. *Org. Lett.* **2001**, *3*, 2089.

[1153]Chen, L.; Li, C.-J. *Chem. Commun.* **2004**, 2362.

[1154]Hayashi, T.; Tokunaga, N.; Yoshida, K.; Han, J.W. *J. Am. Chem. Soc.* **2002**, *124*, 12102; Lerum, R.V.; Chisholm, J.D. *Tetrahedron Lett.* **2004**, *45*, 6591.

[1155]Venkatraman, S.; Li, C.-J. *Tetrahedron Lett.* **2001**, *42*, 781.

[1156]Nishikata, T.; Yamamoto, Y.; Miyaura, N. *Chem. Commun.* **2004**, 1822.

[1157]Han, Y.; Huang, Y.-Z.; Fang, L.; Tao, W.-T. *Synth. Commun.* **1999**, *29*, 867.

[1158]Venkatraman, S.; Meng, Y.; Li, C.-J. *Tetrahedron Lett.* **2001**, *42*, 4459; Oi, S.; Moro, M.; Ito, H.; Honma, Y.; Miyano, S.; Inoue, Y. *Tetrahedron* **2002**, *58*, 91.

[1159]Ohe, T.; Wakita, T.; Motofusa, S.-i; Cho, C.S.; Ohe, K.; Uemura, S. *Bull. Chem. Soc. Jpn.* **2000**, *73*, 2149; Ohe, T.; Uemura, S. *Tetrahedron Lett.* **2002**, *43*, 1269.

[1160]Williams, D.R.; Mullins, R.J.; Miller, N.A. *Chem. Commun.* **2003**, 2220.

[1161]Augé, J.; Gil, R.; Kalsey, S. *Tetrahedron Lett.* **1999**, *40*, 67.

[1162]Gomes, P.; Gosmini, C.; Nédélec, J.-Y.; Périchon, J. *Tetrahedron Lett.* **2000**, *41*, 3385.

[1163]Condon, S.; Dupré, D.; Falgayrac, G.; Nédélec, J.-Y. *Eur. J. Org. Chem.* **2002**, 105.

[1164]Kakuuchi, A.; Taguchi, T.; Hanzawa, Y. *Tetrahedron* **2004**, *60*, 1293.

of palladium acetate, to give the 2-alkenyl pyrrole.[1165]

147

In certain cases, Grignard reagents add 1,4 to *aromatic* systems to give **147** after tautomerization (p. $$$) of the initial formed enol.[1166] Such cyclohexadienes are easily oxidizable to benzenes (often by atmospheric oxygen), so this reaction becomes a method of alkylating and arylating suitably substituted (usually hindered) aryl ketones. A similar reaction has been reported for aromatic nitro compounds where 1,3,5-trinitrobenzene reacts with excess methylmagnesium halide to give 2,4,6-trinitro-1,3,5-trimethylcyclohexane.[1167]

The mechanisms of most of these reactions are not well known. The 1,4 uncatalyzed Grignard reaction has been postulated to proceed by the cyclic mechanism shown, but there is evidence against it.[1168] The R_2CuLi[1169] and copper-catalyzed Grignard additions may involve a number of mechanisms, since the actual attacking species and substrates are so diverse.[1170] A free-radical mechanism of some type

[1165]Lu, W.; Jia, C.; Kitamura, T.; Fujiwara, Y. *Org. Lett.* **2000**, *2*, 2927.

[1166]This example is from Schmidlin, J.; Wohl, J. *Ber.* **1910**, *43*, 1145; Mosher, W.A.; Huber, M.B. *J. Am. Chem. Soc.* **1953**, *75*, 4604. For a review of such reactions see Fuson, R.C. *Adv. Organomet. Chem.* **1964**, *1*, 221.

[1167]Severin, T.; Schmitz, R. *Chem. Ber.* **1963**, *96*, 3081. See also, Bartoli, G. *Acc. Chem. Res.* **1984**, *17*, 109; Bartoli, G.; Dalpozzo, R.; Grossi, L. *J. Chem. Soc. Perkin Trans. 2*, **1989**, 573. For a study of the mechanism, see Bartoli, G.; Bosco, M.; Cantagalli, G.; Dalpozzo, R.; Ciminale, F. *J. Chem. Soc. Perkin Trans. 2*, **1985**, 773.

[1168]House, H.O.; Thompson, H.W. *J. Org. Chem.* **1963**, *28*, 360; Klein, J. *Tetrahedron* **1964**, *20*, 465. See, however, Marets, J.; Rivière, H. *Bull. Soc. Chim. Fr.* **1970**, 4320.

[1169]See Kingsbury, C.L.; Smith, R.A.J. *J. Org. Chem.* **1997**, *62*, 4629. Also see, Bertz, S.H.; Miao, G.; Rossiter, B.E.; Snyder, J.P. *J. Am. Chem. Soc.* **1995**, *117*, 11023; Snyder, J.P. *J. Am. Chem. Soc.* **1995**, *117*, 11025; Vellekoop, A.S.; Smith, R.A.J. *J. Am. Chem. Soc.* **1994**, *116*, 2902.

[1170]For some mechanistic investigations, see Berlan, J.; Battioni, J.; Koosha, K. *J. Organomet. Chem.* **1978**, *152*, 359; *Bull. Soc. Chim. Fr.* **1979**, II-183; Four, P.; Riviere, H.; Tang, P.W. *Tetrahedron Lett.* **1977**, 3879; Casey, C.P.; Cesa, M.C. *J. Am. Chem. Soc.* **1979**, *101*, 4236; Krauss, S.R.; Smith, S.G. *J. Am. Chem. Soc.* **1981**, *103*, 141; Bartoli, G.; Bosco, M.; Dal Pozzo, R.; Ciminale, F. *J. Org. Chem.* **1982**, *47*, 5227; Corey, E.J.; Boaz, N.W. *Tetrahedron Lett.* **1985**, *26*, 6015; Yamamoto, Y.; Yamada, J.; Uyehara, T. *J. Am. Chem. Soc.* **1987**, *109*, 5820; Ullenius, C.; Christenson, B. *Pure Appl. Chem.* **1988**, *60*, 57; Christenson, B.; Olsson, T.; Ullenius, C. *Tetrahedron* **1989**, *45*, 523; Krause, N. *Tetrahedron Lett.* **1989**, *30*, 5219.

(perhaps SET) has been suggested[1171] although the fact that retention of configuration at R has been demonstrated in several cases rules out a completely free R• radical.[1172] For simple α,β-unsaturated ketones, such as 2-cyclohexenone, and Me$_2$CuLi, there is evidence[1173] for this mechanism:

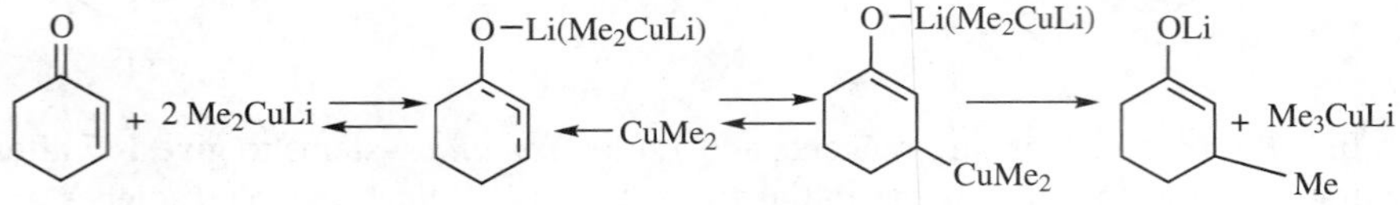

148

148 is a d,π^* complex, with bonding between copper, as a base supplying a pair of d electrons, and the enone as a Lewis acid using the π^* orbital of the allylic system.[1173] The ^{13}C NMR spectrum of an intermediate similar to **148** has been reported.[1174]

For the addition of organocopper reagents to alkynes and conjugated dienes, see **15-22**.

OS **IV**, 93; **V**, 762; **VI**, 442, 666, 762, 786; **VIII**, 112, 257, 277, 479; **IX**, 328, 350, 640.

15-26   The Sakurai Reaction

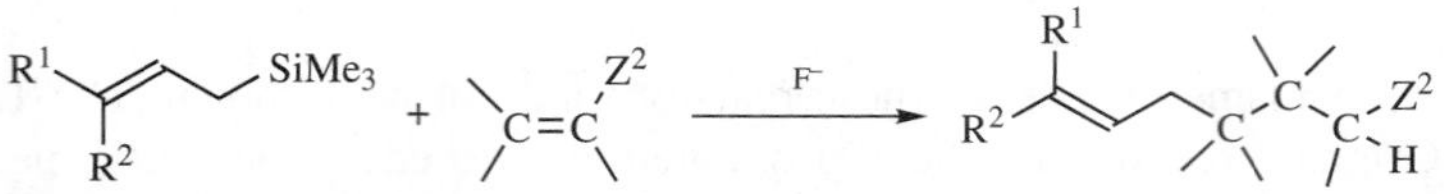

Allylic silanes R$_2$C=CHCH$_2$SiMe$_3$ can be used instead of silyl enol ethers (the *Sakurai reaction*).[1175] An allyl group can be added, to α,β-unsaturated carboxylic esters, amides and nitriles, with CH$_2$=CHCH$_2$SiMe$_3$ and F$^-$ ion (see **15-47**).[1176] This reagent gave better results than lithium diallylcuprate (**15-25**). Catalytic Sakurai reactions are known.[1177] The palladium catalyzed reaction of conjugated ketones with PhSi(OEt)$_3$ with SbCl$_3$ and Bu$_4$NF in acetic acid gave the 1,4-addition product.[1178] A similar reaction was reported using PhSi(OMe)$_3$

[1171]See, for example, Ruden, R.A.; Litterer, W.E. *Tetrahedron Lett.* **1975**, 2043; House, H.O.; Snoble, K.A.J. *J. Org. Chem.* **1976**, *41*, 3076; Wigal, C.T.; Grunwell, J.R.; Hershberger, J. *J. Org. Chem.* **1991**, *56*, 3759.

[1172]Näf, F.; Degen, P. *Helv. Chim. Acta* **1971**, *54*, 1939; Whitesides, G.M.; Kendall, P.E. *J. Org. Chem.* **1972**, *37*, 3718. See also, Ref. 1063.

[1173]Corey, E.J.; Hannon, F.J.; Boaz, N.W. *Tetrahedron* **1989**, *45*, 545.

[1174]Bertz, S.H.; Smith, R.A.J. *J. Am. Chem. Soc.* **1989**, *111*, 8276.

[1175]Hosomi, A.; Sakurai, H. *J. Am. Chem. Soc.* **1977**, *99*, 1673; Jellal, A.; Santelli, M. *Tetrahedron Lett.* **1980**, *21*, 4487; Sakurai, H.; Hosomi, A.; Hayashi, J. *Org. Synth.* **VII**, 443; Kuhnert, N.; Peverley, J.; Robertson, J. *Tetrahedron Lett.* **1998**, *39*, 3215. For a review, see Fleming, I.; Dunoguès, J.; Smithers, R. *Org. React.* **1989**, *37*, 57, see pp. 127, 335–370. For a review of intramolecular additions, see Schinzer, D. *Synthesis* **1988**, 263.

[1176]Majetich, G.; Casares, A.; Chapman, D.; Behnke, M. *J. Org. Chem.* **1986**, *51*, 1745.

[1177]InCl$_3$: Lee, P.H.; Lee, K.; Sung, S.-y.; Chang, S. *J. Org. Chem.* **2001**, *66*, 8646.

[1178]Denmark, S.E.; Amishiro, N. *J. Org. Chem.* **2003**, *68*, 6997.

with a rhodium catalyst.[1179] In a related reaction, Ph_2SiCl_2, NaF and a rhodium catalyst gives conjugate addition of a phenyl group to α,β-unsaturated ketones.[1180] An interesting rhodium-catalyzed, conjugate addition of a phenyl group was reported using a siloxane polymer bearing Si—Ph units.[1181]

Silyl ketene acetals, $RCH{=}C(OMe)OSiMe_3$, add to conjugated ketones to give δ-keto esters, in $MeNO_2$ as solvent.[1182]

15-27 Conjugate Addition of Boranes to Activated Double Bonds

Hydro-alkyl-addition (overall transformation)

$$R' = H, Me$$

Just as trialkylboranes add to simple alkenes (**15-16**), they rapidly add to the double bonds of acrolein, methyl vinyl ketone, and certain of their derivatives in THF at 25°C to give enol borinates (also see, p. 631), which can be hydrolyzed to aldehydes or ketones.[1183] If water is present in the reaction medium from the beginning, the reaction can be run in one laboratory step. Since the boranes can be prepared from alkenes (**15-16**), this reaction provides a means of lengthening a carbon chain by three or four carbons, respectively. Compounds containing a terminal alkyl group, such as crotonaldehyde ($CH_3CH{=}CHCHO$) and 3-penten-2-one, fail to react under these conditions, as does acrylonitrile, but these compounds can be induced to react by the slow and controlled addition of O_2 or by initiation with peroxides or UV light.[1184] A disadvantage is

149

[1179]Oi, S.; Honma, Y.; Inoue, Y. *Org. Lett.* **2002**, *4*, 667; Oi, S.; Taira, A.; Honma, Y.; Inoue, Y. *Org. Lett.* **2003**, *5*, 97.

[1180]Huang, T.-S.; Li, C.-J. *Chem. Commun.* **2001**, 2348.

[1181]Koike, T.; Du, X.; Mori, A.; Osakada, K. *Synlett* **2002**, 301.

[1182]RajanBabu, T.V. *J. Org. Chem.* **1984**, *49*, 2083.

[1183]Suzuki, A.; Arase, A.; Matsumoto, H.; Itoh, M.; Brown, H.C.; Rogić, M.M.; Rathke, M.W. *J. Am. Chem. Soc.* **1967**, *89*, 5708; Köster, R.; Zimmermann, H.; Fenzl, W. *Liebigs Ann. Chem.* **1976**, 1116. For reviews, see Pelter, A.; Smith, K.; Brown, H.C. *Borane Reagents*, Academic Press, NY, **1988**, pp. 301–305, 318–323; Brown, H.C.; Midland, M.M. *Angew. Chem. Int. Ed.* **1972**, *11*, 692, sse pp. 694–698; Kabalka, G.W. *Intra-Sci. Chem. Rep.* **1973**, *7(1)*, 57; Brown, H.C. *Boranes in Organic Chemistry*, Cornell University Press, Ithica, NY, **1972**, pp. 413–433.

[1184]Brown, H.C.; Kabalka, G.W. *J. Am. Chem. Soc.* **1970**, *92*, 712, 714. See also, Utimoto, K.; Tanaka, T.; Furubayashi, T.; Nozaki, H. *Tetrahedron Lett.* **1973**, 787; Miyaura, N.; Kashiwagi, M.; Itoh, M.; Suzuki, A. *Chem. Lett.* **1974**, 395.

that only one of the three R groups of R_3B adds to the substrate, so that the other two are wasted. This difficulty is overcome by the use of a β-alkyl borinate, such as **149**,[1185] which can be prepared as shown. **149** (R = *tert*-butyl) can be made by treatment of **149** (R = OMe) with *t*-BuLi. The use of this reagent permits *tert*-butyl groups to be added. β-1-Alkenyl-9-BBN compounds β-RCH = CR′-9-BBN (prepared by treatment of alkynes with 9-BBN or of RCH=CR′Li with β-methoxy-9-BBN[1186]) add to methyl vinyl ketones to give, after hydrolysis, γ,δ-unsaturated ketones,[1187] although β-R-9-BBN, where R = a saturated group, are not useful here, because the R group of these reagents does not preferentially add to the substrate.[1184] The corresponding β-1-alkynyl-9-BBN compounds also give the reaction.[1188] Like the three substrates mentioned above, 3-butyn-2-one fails to react in the absence of air, but undergoes the reaction when exposed to a slow stream of air:[1189] Since the product, **150**, is an α,β-unsaturated ketone, it can be made to react with another BR_3, the same or different, to produce a wide variety of ketones **151**.

Vinyl boranes add to conjugated ketones in the presence of a rhodium catalyst (with high asymmetric induction in the presence of BINAP).[1190] Alkynyl-boranes also add to conjugated ketones, in the presence of BF_3.[1191]

The fact that these reactions are catalyzed by free-radical initiators and inhibited by galvinoxyl[1192] (a free-radical inhibitor) indicates that free-radical mechanisms are involved.

15-28   Radical Addition to Activated Double Bonds

Hydro-alkyl-addition

[1185]Brown, H.C.; Negishi, E. *J. Am. Chem. Soc.* **1971**, *93*, 3777.

[1186]Brown, H.C.; Bhat, N.G.; Rajagopalan, S. *Organometallics* **1986**, *5*, 816.

[1187]Jacob III, P.; Brown, H.C. *J. Am. Chem. Soc.* **1976**, *98*, 7832; Satoh, Y.; Serizawa, H.; Hara, S.; Suzuki, A. *J. Am. Chem. Soc.* **1985**, *107*, 5225. See also, Molander, G.A.; Singaram, B.; Brown, H.C. *J. Org. Chem.* **1984**, *49*, 5024. Alkenyldialkoxyboranes, together with BF_3–etherate, also transfer vinylic groups: Hara, S.; Hyuga, S.; Aoyama, M.; Sato, M.; Suzuki, A. *Tetrahedron Lett.* **1990**, *31*, 247.

[1188]Sinclair, J.A.; Molander, G.A.; Brown, H.C. *J. Am. Chem. Soc.* **1977**, *99*, 954. See also, Molander, G.A.; Brown, H.C. *J. Org. Chem.* **1977**, *42*, 3106.

[1189]Suzuki, A.; Nozawa, S.; Itoh, M.; Brown, H.C.; Kabalka, G.W.; Holland, G.W. *J. Am. Chem. Soc.* **1970**, *92*, 3503.

[1190]Takaya, Y.; Ogasawara, M.; Hayashi, T. *Tetrahedron Lett.* **1998**, *39*, 8479.

[1191]Fujishima, H.; Takada, E.; Hara, S.; Suzuki, A. *Chem. Lett.* **1992**, 695.

[1192]Kabalka, G.W.; Brown, H.C.; Suzuki, A.; Honma, S.; Arase, A.; Itoh, M. *J. Am. Chem. Soc.* **1970**, *92*, 710. See also, Arase, A.; Masuda, Y.; Suzuki, A. *Bull. Chem. Soc. Jpn.* **1976**, *49*, 2275.

In a reaction similar to **15-25**, alkyl groups can be added to alkenes activated by, such groups as COR', COOR', CN, and even Ph.[1193] In the method illustrated above, the R group comes from an alkyl halide (R = primary, secondary, or tertiary alkyl; X = Br or I) and the hydrogen from the tin hydride. The reaction of *tert*-butyl bromide, Bu_3SnH and AIBN (p. 935), for example, adds a *tert*-butyl group to a conjugated ester via 1,4-addition.[1194] An alkene is converted to an alkylborane with catecholborane (p. 817) and when treated with a conjugated ketone and O_2, radical conjugate addtion leads to the β-substituted ketone.[1195] Organomercury hydrides (RHgH) generated *in situ* from RHgX and $NaBH_4$, can also be used.[1196] When the tin method is used, Bu_3SnH can also be generated in a similar way, from R_3SnX and $NaBH_4$. The tin method has a broader scope (e.g., it can be used on $CH_2=CCl_2$), but the mercury method uses milder reaction conditions. Like **15-27**, these additions have free-radical mechanisms. The reaction has been used for free-radical cyclizations of the type discussed on p. 1125.[1197] Such cyclizations normally give predominant formation of 15-membered rings, but large rings (11–20 members) have also been synthesized by this reaction.[1198]

Free-radical addition of an aryl group and a hydrogen has been achieved by treatment of activated alkenes with a diazonium salt and $TiCl_3$.[1199] The addition of R_3Al takes place by a free-radical mechanism.[1146]

OS **VII**, 105.

15-29   Radical Addition to Unactivated Double Bonds[1200]

Alkyl-hydro-addition

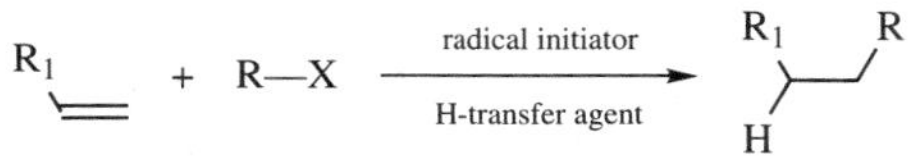

[1193]For reviews, see Giese, B. *Radicals in Organic Synthesis: Formation of Carbon–Carbon Bonds*, Pergamon, Elmsford, NY, *1986*, pp. 36–68; Giese, B. *Angew. Chem. Int. Ed. 1985*, *24*, 553; Larock, R.C. *Organomercury Compounds in Organic Synthesis*, Springer, NY, *1985*, pp. 263–273. The last review includes a table with many examples of the mercury method. For a list of reagents, with references, see Larock, R.C. *Comprehensive Organic Transformations*, 2nd ed., Wiley-VCH, NY, *1999*, pp. 1809–1813.
[1194]Hayen, A.; Koch, R.; Metzger, J.O. *Angew. Chem. Int. Ed. 2000*, *39*, 2758.
[1195]Ollivier, C.; Renaud, P. *Chem. Eur. J. 1999*, *5*, 1468.
[1196]For the use of tris(trimethylsilyl)silane, see Giese, B.; Kopping, B.; Chatgilialoglu, C. *Tetrahedron Lett. 1989*, *30*, 681.
[1197]For reviews, see Jasperse, C.P.; Curran, D.P.; Fevig, T.L. *Chem. Rev. 1991*, *91*, 1237; Curran, D.P. *Adv. Free Radical Chem. (Greenwich, Conn.) 1990*, *1*, 121; Giese, B. *Radicals in Organic Synthesis: Formation of Carbon-Carbon Bonds*, Pergamon, Elmsford, NY, *1986*, pp. 151–169. For a list of references, see Larock, R.C. *Comprehensive Organic Transformations*, 2nd ed., Wiley-VCH, NY, *1999*, pp. 413–418.
[1198]See Porter, N.A.; Chang, V.H. *J. Am. Chem. Soc. 1987*, *109*, 4976.
[1199]Citterio, A.; Vismara, E. *Synthesis 1980*, 291. For other methods of adding an alkyl or aryl group and a hydrogen to activated double bonds by free-radical processes, see Cacchi, S.; Palmieri, G. *Synthesis 1984*, 575; Lebedev, S.A.; Lopatina, V.S.; Berestova, S.S.; Petrov, E.S.; Beletskaya, I.P. *J. Org. Chem. USSR 1986*, *22*, 1238; Luche, J.L.; Allavena, C. *Tetrahedron Lett. 1988*, *29*, 5369; Varea, T.; González-Núñez, M.E.; Rodrigo-Chiner, J.; Asensio, G. *Tetrahedron Lett. 1989*, *30*, 4709; Barton, D.H.R.; Sarma, J.C. *Tetrahedron Lett. 1990*, *31*, 1965.
[1200]See Smith, M.B. *Organic Synthesis*, 2nd ed., McGraw-Hill, NY *2001*, pp. 1167–1172

Radical addition to alkenes is usually difficult, except when addition occurs to conjugated carbonyl compounds (**15-24**). An important exception involves radicals bearing a heteroatom α to the carbon bearing the radical center. These radical are much more stable and can add to alkenes, usually with anti-Markovnikov orientation, as in the radical induced addition of HBr to alkenes (**15-2**).[1201] Examples of this type of reaction include the use of alcohol-, ester-,[1202] amino-, and aldehyde-stabilized radicals.[455] Carbon tetrachloride can be cleaved homolytically to generate Cl• and Cl$_3$C•, which can add to alkenes. The alkyl group of alkyl iodides adds to alkenes with BEt$_3$/O$_2$ as the initiator and in the presence of a tetraalkylammonium hypophosphite.[1203] When chloroform was treated with a ruthenium carbene complex, Cl$_2$CH add to the less substituted carbon of an alkene, and Cl to the more substituted carbon.[1204] The radical generated from (EtO)$_2$POCH$_2$Br adds to alkenes to generate a new phosphonate ester.[1205] α-Bromo esters add to alkenes in the presence of BEt$_3$/air to give a γ-bromo ester.[1206] α-Bromo amides add the Br and the acyl carbon to an alkene using Yb(OTf)$_3$ with BEt$_3$/O$_2$ as the radical initiator.[1207] α-Iodo amides add to alkenes using a water soluble azobis initiator to give the iodo ester, which cyclizes under the reaction conditions to give a lactone.[1208] β-Keto dithiocarbonates, RC(=O)—C—SC(=S)OEt, generate the radical in the presence of a peroxide and add to alkenes.[1209] Malonate derivatives add to alkenes in the presence of a mixture of Mn/Co catalyst, in oxygenated acetic acid.[1210]

Other radicals can add to alkenes, and the rate constant for the addition of methyl radicals to alkenes has been studied,[1211] and the rate of radical additions to alkenes in general has also been studied.[1212] The kinetic and thermodynamic control of a radical addition regiochemistry has also been studied.[1213] Alkynes are generally less reactive than alkenes in radical coupling reactions.[1214] Nonradical nucleophiles usually react faster with alkynes than with alkenes, however.[1215]

[1201]See Curran, D.P. *Synthesis* **1988**, 489 (see pp. 497–498).

[1202]Deng, L.X.; Kutateladze, A.G. *Tetrahedron Lett.* **1997**, *38*, 7829.

[1203]Jang, D.O.; Cho, D.H.; Chung, C.-M. *Synlett* **2001**, 1923.

[1204]Tallarico, J.A.; Malnick, L.M.; Snapper, M.L. *J. Org. Chem.* **1999**, *64*, 344.

[1205]Baczewski, P.; Mikoajczyk, M. *Synthesis* **1995**, 392.

[1206]Yorimitsu, H.; Shinokubo, H.; Matsubara, S.; Oshima, K.; Omoto, K.; Fujimoto, H. *J. Org. Chem.* **2001**, *66*, 7776.

[1207]Mero, C.L.; Porter, N.A. *J. Am. Chem. Soc.* **1999**, *121*, 5155.

[1208]Yorimitsu, H.; Wakabayashi, K.; Shinokubo, H.; Oshima, K. *Bull. Chem. Soc. Jpn.* **2001**, *74*, 1963.

[1209]Ouvry, G.; Zard, S.Z. *Chem. Commun.* **2003**, 778.

[1210]Hirase, K.; Iwahama, T.; Sakaguchi, S.; Ishii, Y. *J. Org. Chem.* **2002**, *67*, 970.

[1211]Zytowski, T.; Fischer, H. *J. Am. Chem. Soc.* **1996** *118*, 437.

[1212]Avila, D.V.; Ingold, K.U.; Lusztyk, J.; Dolbier Jr., W.R.; Pan, H.-Q. *J. Org. Chem.* **1996**, *61*, 2027.

[1213]Leach, A.G.; Wang, R.; Wohlhieter, G.E.; Khan, S.I.; Jung, M.E.; Houk, K.N. *J. Am. Chem. Soc.* **2003**, *125*, 4271.

[1214]Giese, B.; Lachhein, S. *Angew. Chem. Int. Ed.* **1982**, *21*, 768; Giese, B.; Meixner, J. *Angew. Chem. Int. Ed.* **1979** *18* 154.

[1215]Dickstein, J.I.; Miller, G.I., in *The Chemistry of Carbon Carbon Triple Bonds*, Vol. 2, Patai, S., Ed., Wiley, NY **1978**.

15-30   Radical Cyclization[1216]

Alkyl-hydro-addition

ω-Haloalkenes generate radicals upon treatment with reagents, such as AIBN or under photolysis conditions,[1217] and the radical carbon adds to the alkene to form cyclic compounds.[1218] This intramolecular addition of a radical to an alkene is called radical cyclization. In a typical example, haloalkene **154** reacts with the radical produced by AIBN to give radical **153**. The radical can add to the more substituted carbon to give **155** via a 5-exo–trig reaction (p. 305).[1219] If the radical adds to the less substituted carbon, **156** is formed via a 6-endo–trig reaction.[1220] In both cases, the product is another radical, which must be converted to an unreactive product. This is generally accomplished by adding a hydrogen transfer agent,[1221] such as tributyltin hydride (Bu₃SnH), which reacts with **155** to form methylcyclopentane and Bu₃Sn•, or with **156** to give cyclohexane. The Bu₃Sn• formed in both cases usually dimerizes to form Bu₃SnSnBu₃. Cyclization can compete with hydrogen transfer[1222] of Bu₃SnH to **153** to give **152**, the reduction product. In general, formation of the five-membered ring dominates the cyclization, but if addition to the C=C unit is relatively slow, the reduction product is formed preferentially. Radical rearrangements can also diminish the yield of the desired product.[1223] Given a choice between a larger and a smaller ring, radical cyclization generally gives the smaller ring,[1224] but not

[1216]See Smith, M.B. *Organic Synthesis*, 2nd ed., McGraw-Hill, NY **2001**, pp. 1172–1181. For a review of radical-mediated annulation reactions, see Rheault, T.R.; Sibi, M.P. *Synthesis* **2003**, 803.

[1217]For example, see Pandey, G.; Reddy, G.D.; Chakrabarti, D. *J. Chem. Soc., Perkin Trans. 1* **1996**, 219; Abe, M.; Hayashi, T.; Kurata, T. *Chem. Lett.* **1994** 1789; Pandey, G.; Hajra, S.; Ghorai, M.K. *Tetrahedron Lett.* **1994**, *35*, 7837; Pandey, G.; Reddy, G.D. *Tetrahedron Lett.* **1992**, *33*, 6533.

[1218]Curran, D.P. *Synthesis* **1988**, 417, 489; Chang, S.-Y.; Jiang, W.-T.; Cherng, C.-D.; Tang, K.-H.; Huang, C.-H.; Tsai, Y.-M. *J. Org. Chem.* **1997**, *62*, 9089. For a review of applications to organic synthesis see McCarroll, A.J.; Walton, J.C. *J. Chem. Soc., Perkin Trans. 1* **2001**, 3215.

[1219]For a discussion of whether 5-endo–trig radical cyclizations are favored or disfavored, see Chatgilialoglu, C.; Ferreri, C.; Guerra, M.; Timokhin, V.; Froudakis, G.; Gimisis, Z.T. *J. Am. Chem. Soc.* **2002**, *124*, 10765.

[1220]For a review of 5-endo–trig radical cyclizations, see Ishibashi, H.; Sato, T.; Ikeda, M. *Synthesis* **2002**, 695.

[1221]See Ha, C.; Horner, J.H.; Newcomb, M.; Varick, T.R.; Arnold, B.R.; Lusztyk, J. *J. Org. Chem.* **1993**, *58* 1194.

[1222]For a discussion of the kinetics of radical cyclization, see Furxhi, E.; Horner, J.H.; Newcomb, M. *J. Org. Chem.* **1999**, *64*, 4064. Rate constants have been determined for selected reactions: Tauh, P.; Fallis, A.G. *J. Org. Chem.* **1999**, *64*, 6960.

[1223]Mueller, A.M.; Chen, P. *J. Org. Chem.* **1998**, *63*, 4581.

[1224]Bogen, S; Malacria, M. *J. Am. Chem. Soc.* **1996** *118*, 3992.; Beckwith, A.L.J.; Ingold, K.U., in Vol 1 of *Rearrangements in Ground States and Excited States*, de Mayo, P., Ed., Academic Press, NY **1980**, pp. 162–283. For a discussion of six- versus five-membered rings, see Gómez, A.M.; Company, M.D.; Uriel, C.; Valverde, S.; López, J.C. *Tetrahedron Lett.* **2002**, *43*, 4997.

always.[1225] The mechanism of this reaction has been discussed.[1226] Formation of other size rings is possible of course. A 4-exo–trig radical cyclization has been studied,[1227] selectivity in a 7-endo versus 6-exo cyclization,[1228] and also an 8-endo-trig reaction.[1229] In radical cyclization to form large rings, 1,5- and 1,9-hydrogen atom abstractions can pose a problem[1230]

In cases where hydrogen atom transfer gives primarily reduced products, $Bu_3Sn-SnBu_3$ under photochemical generates the radical which can cyclize (see **15-46**),[1231] but a halogen atom transfer agent, such as iodoethane, is used rather than a hydrogen-transfer agent, so the final product is an alkyl iodide.

A mixture of a Grignard reagent and $CoCl_2$ has also been used to initiate aryl radical cyclizations.[1232] Titanium(III)-mediated radical cyclizations are known,[1233] and SmI_2-mediate reactions are possible in the presence of a nickel catalyst.[1234] Organoborane-mediated radical cyclizations are known.[1235] Electrochemically generated radicals also cyclize.[1236] The influence of the halogen atom on radical cyclization has been studied.[1237] Both phenylthio[1238] and phenylseleno

[1225]Mayon, P.; Chapleur, Y. *Tetrahedron Lett.* **1994**, *35*, 3703; Marco-Contelles, J.; Sánchez, B. *J. Org. Chem.* **1993**, *58*, 4293.

[1226]Bailey, W.F.; Carson, M.W. *Tetrahedron Lett.* **1999**, *40*, 5433.

[1227]Jung, M.E.; Marquez, R.; Houk, K.N. *Tetrahedron Lett.* **1999**, *40*, 2661.

[1228]Kamimura, A.; Taguchi, Y. *Tetrahedron Lett.* **2004**, *45*, 2335.

[1229]Wang, Li.C. *J. Org. Chem.* **2002**, *67*, 1271.

[1230]Kraus, G.A.; Wu, Y. *J. Am. Chem. Soc.* **1992** *114*, 8705.

[1231]A polymer-bound tin catalyst has been used under photochemical conditions. See Hernán, A.G.; Kilburn, J.D. *Tetrahedron Lett.* **2004**, *45*, 831.

[1232]Clark, A.J.; Davies, D.I.; Jones, K.; Millbanks, C. *J. Chem. Soc., Chem. Commun.* **1994**, 41.

[1233]Barrero, A.F.; Oltra, J.E.; Cuerva, J.M.; Rosales, A. *J. Org. Chem.* **2002**, *67*, 2566.

[1234]Molander, G.A.; St. Jean, Jr., D.J. *J. Org. Chem.* **2002**, *67*, 3861.

[1235]Becattini, B.; Ollivier, C.; Renaud, P. *Synlett* **2003**, 1485.

[1236]Olivero, S.; Clinet, J.C.; Duñach, E. *Tetrahedron Lett.* **1995**, *36*, 4429; Ozaki, S.; Horiguchi, I.; Matsushita, H.; Ohmori, H. *Tetrahedron Lett.* **1994**, *35*, 725.

[1237]Tamura, O.; Matsukida, H.; Toyao, A.; Takeda, Y.; Ishibashi, H. *J. Org. Chem.* **2002**, *67*, 5537.

[1238]See, for example, Ikeda, M.; Shikaura, J.; Maekawa, N.; Daibuzono, K.; Teranishi, H.; Teraoka, Y.; Oda, N.; Ishibashi, H. *Heterocycles* **1999**, *50*, 31.

groups[1239] can be used as 'leaving groups' for radical cyclization, where sulfur or selenium atom transfer leads to formation of the radical. A seleno ester, $R_2N-CH_2C(-O)SeMe$, has also been used with $(Me_3Si)_3SiH$ (tristrimethylsilylsilane, TTMSS) and AIBN to generate R_2NCH_2•.[1240] O-Phosphonate esters have also served as the leaving group.[1241] N-(2-bromophenylbenzyl)methylamino groups have been used as leaving groups for formation of a radical.[1242] Alkenes also serve as radical precursors, adding to another alkene,[1243] including conjugated systems.[1244]

Radical cyclization reaction often proceeds with high diastereoselectivity[1245] and high asymmetric induction when chiral precursors are used. Internal alkynes are good substrates for radical cyclization,[1246] but terminal alkynes tend to give mixtures of *exo/endo–dig* products (p. 305).[1247] N-Alkenyl pyridinium salts, with ortho-halogen substituents generate the aryl radical with $Bu_3SnH/AIBN$, which cyclizes on the pendant alkene unit.[1248] Cyclization of vinyl radicals[1249] and allenyl radicals[1250] are also well known. Ring expansion during radical cyclization is possible when the terminal intermediate is a cyclobutylcarbinyl radical.[1251]

Aryl radicals participate in radical cyclization reactions when the aromatic ring has an alkene or alkyne substituent. *o*-Iodo aryl allyl ethers cyclize to benzofuran derivatives, for example, when treated with AIBN, aqueous H_3PO_2 and $NaHCO_3$ in ethanol.[1252] Cyclization of an *o*-bromo-N-acyl aniline (a methacrylic acid derivative) with $AIBN/Bu_3SnH$ gave an indolone under the typical conditions used for cyclization of alkenes.[1253]

Radical cyclization is compatible with the presence of other functional groups. Treatment of $XCH_2CON(R)-C(R^1)=CH_2$ derivatives (X = Cl, Br, I) with Ph_3SnH

[1239]See, for example, Ericsson, C.; Engman, L. *Org. Lett.* **2001**, *3*, 3459.

[1240]Quirante, J.; Vila, X.; Escolano, C.; Bonjoch, J. *J. Org. Chem.* **2002**, *67*, 2323.

[1241]Crich, D.; Ranganathan, K.; Huang, X. *Org. Lett.* **2001**, *3*, 1917.

[1242]Andrukiewicz, R.; Loska, R.; Prisyahnyuk, V.; Staliński, K. *J. Org. Chem.* **2003**, *68*, 1552.

[1243]See Jessop, C.M.; Parsons, A.F.; Routledge, A.; Irvine, D. *Tetrahedron Lett.* **2003**, *44*, 479.

[1244]Bebbington, D.; Bentley, J.; Nilsson, P.A.; Parsons, A.F. *Tetrahedron Lett.* **2000**, *41*, 8941; Menes-Arzate, M.; Martínez, R.; Cruz-Almanza, R.; Muchowski, J.M.; Osornio, Y.M.; Miranda, L.D. *J. Org. Chem.* **2004**, *69*, 4001. For a review, see Zhang, W. *Tetrahedron* **2001**, *57*, 7237.

[1245]For a discussion of stereocontrol in radical processes, see Bouvier, J.-P.; Jung, G.; Liu, Z.; Guérin, B.; Guindon, Y. *Org. Lett.* **2001**, *3*, 1391. See Bailey, W.F.; Longstaff, S.C. *Org. Lett.* **2001**, *3*, 2217; Stalinski, K.; Curran, D.P. *J. Org. Chem.* **2002**, *67*, 2982.

[1246]See Sha, C.-K.; Shen, C.-Y.; Jean, T.-S.; Chiu, R.-T.; Tseng, W.-H. *Tetrahedron Lett.* **1993**, *34*, 764.

[1247]Choi, J.-K.; Hart, D.J.; Tsai, Y.-M. *Tetrahedron Lett.* **1982**, *23*, 4765; Burnett, D.A.; Choi, J.-K.; Hart, D.-J.; Tsai, Y.-M. *J. Am. Chem. Soc.* **1984** *106*, 8201; Hart, D.J.; Tsai, Y.-M. *Ibid* **1984** *106*, 8209; Choi, J.-K.; Hart, D.J. *Tetrahedron* **1985**, *41*, 3959; Hart, D.J.; Tsai, Y.-M. *J. Am. Chem. Soc.* **1982** *104* 1430; Kano, S.; Yuasa, Y.; Asami, K.; Shibuya, S. *Chem. Lett.* **1986**, 735; Robertson, J.; Lam, H.W.; Abazi, S.; Roseblade, S.; Lush, R.K. *Tetrahedron* **2000**, *56*, 8959.

[1248]Dobbs, A.P.; Jones, K.; Veal, K.T. *Tetrahedron Lett.* **1997**, *38*, 5383.

[1249]Crich, D.; Hwang, J.-T.; Liu, H. *Tetrahedron Lett.* **1996**, *37*, 3105; Sha, C.-K.; Zhan, Z.-P.; Wang, F.-S. *Org. Lett.* **2000**, *2*, 2011.

[1250]Wartenberg, F.-H.; Junga, H.; Blechert, S. *Tetrahedron Lett.* **1993**, *34*, 5251.

[1251]Zhang, W.; Dowd, P. *Tetrahedron Lett.* **1995**, *36*, 8539.

[1252]Yorimitsu, H.; Shinokubo, H.; Oshima, K. *Chem. Lett.* **2000**, 104.

[1253]Jones, K.; Brunton, S.A.; Gosain, R. *Tetrahedron Lett.* **1999**, *40*, 8935.

and AIBN led to formation of a lactam via radical cyclization.[1254] Cyclization of
N-iodoethyl-5-vinyl-2-pyrrolidinone led to the corresponding bicyclic lactam,[1255]
and there are other examples of radical cyclization with molecules containing a
lactam unit[1256] or an amide unit.[1257] β-Lactams can be produced by radical cyclization, using Mn(OAc)$_3$.[1258] Radical cyclization occurs with enamines as well.[1259]
Photochemical irradiation of N,N-diallyl acrylamide leads to formation of a lactam
ring, and in this case thiophenol was added to generate the phenylthio derivative.[1260] Phenylseleno N-allylamines lead to cyclic amines.[1261] ω-Iodo acrylate
esters cyclize to form lactones,[1262] and allylic acetoxy compounds of the type
$C=C-C-O_2C-CH_2I$ cyclize in a similar manner to give lactones.[1263] Iodolactonization (p. 1154) occurs under standard radical cyclization conditions using allylic
acetoxy compounds[1264] and HGaCl$_2$/BEt$_3$ has been used to initiate the radical process.[1265] α-Bromo mixed acetals give α-alkoxy tetrahydrofuran derivatives[1266] and
α-iodoacetals cyclize to give similar products.[1267] The reaction of an ortho-alkynyl
aryl isonitrile with AIBN and 2.2 equivalents of Bu$_3$SnH gave an indole via
5-exo–digcyclization.[1268] Indole derivatives have also been prepared from ortho-
iodo aniline derivatives, using AIBN and tristrimethylsilylsilane (TTMSS).[1269]

Acyl radicals can be generated and they cyclize in the usual manner.[1270] A poly-
ene-cyclization reaction generated four rings, initiating the sequence by treatment

[1254]Baker, S.R.; Parsons, A.F.; Pons, J.-F.; Wilson, M. *Tetrahedron Lett.* **1998**, *39*, 7197; Sato, T.; Chono,
N.; Ishibashi, H.; Ikeda, M. *J. Chem. Soc, Perkin Trans. 1* **1995** 1115; Goodall, K.; Parsons, A.F. *J. Chem.
Soc., Perkin Trans. 1* **1994**, 3257; Sato, T.; Machigashira, N.; Ishibashi, H.; Ikeda, M. *Heterocycles* **1992**,
33 139; Bryans, J.S.; Large, J.M.; Parsons, A.F. *Tetrahedron Lett.* **1999**, *40*, 3487; Gilbert, B.C.; Kalz, W.;
Lindsay, C.I.; McGrail, P.T.; Parsons, A.F.; Whittaker, D.T.E. *J. Chem. Soc., Perkin Trans. 1* **2000**, 1187.
For a tandem cyclization to give a tricyclic compound, see Parsons, A.F.; Williams, D.A.J. *Tetrahedron*
2000, *56*, 7217; Wakabayashi, K.; Yorimitsu, H.; Shinokubo, H.; Oshima, K. *Bull. Chem. Soc. Jpn.* **2000**,
73, 2377; El Bialy, S.A.A.; Ohtani, S.; Sato, T.; Ikeda, M. *Heterocycles* **2001**, *54*, 1021; Liu, L.; Wang,
X.; Li, C. *Org. Lett.* **2003**, *5*, 361.
[1255]Keusenkothen, P.F.; Smith, M.B. *J. Chem. Soc., Perkin Trans. 1* **1994**, 2485; Keusenkothen, P.F.;
Smith, M.B. *Tetrahedron* **1992**, *48*, 2977.
[1256]Rigby, J.H.; Qabar, M.N. *J. Org. Chem.* **1993**, *58*, 4473.
[1257]Beckwith, A.L.J.; Joseph, S.P.; Mayadunne, R.T.A. *J. Org. Chem.* **1993**, *58*, 4198.
[1258]D'Annibale, A.; Nanni, D.; Trogolo, C.; Umani, F. *Org. Lett.* **2000**, *2*, 401. See Lee, E.; Kim, S.K.;
Kim, J.Y.; Lim, J. *Tetrahedron Lett.* **2000**, *41*, 5915.
[1259]Glover, S.A.; Warkentin, J. *J. Org. Chem.* **1993**, *58*, 2115.
[1260]Naito, T.; Honda, Y.; Miyata, O.; Ninomiya, I. *J. Chem. Soc., Perkin Trans. 1* **1995**, 19.
[1261]Gupta, V.; Besev, M.; Engman, L. *Tetrahedron Lett.* **1998**, *39*, 2429.
[1262]Ryu, I.; Nagahara, K.; Yamazaki, H.; Tsunoi, S.; Sonoda, N. *Synlett* **1994**, 643.
[1263]Ollivier, C.; Renaud, P. *J. Am. Chem. Soc.* **2000**, *122*, 6496.
[1264]Ollivier, C.; Bark, T.; Renaud, P. *Synthesis* **2000**, 1598.
[1265]Mikami, S.; Fujita, K.; Nakamura, T.; Yorimitsu, H.; Shinokubo, H.; Matsubara, S.; Oshima, K. *Org.
Lett.* **2001**, *3*, 1853.
[1266]Villar, F.; Equey, O.; Renaud, P. *Org. Lett.* **2000**, *2*, 1061.
[1267]Fujioka, T.; Nakamura, T.; Yorimitsu, H.; Oshima, K. *Org. Lett.* **2002**, *4*, 2257.
[1268]Rainer, J.D.; Kennedy, A.R.; Chase, E. *Tetrahedron Lett.* **1999**, *40*, 6325.
[1269]Kizil, M.; Patro, B.; Callaghan, O.; Murphy, J.A.; Hursthouse, M.B.; Hobbs, D. *J. Org. Chem.* **1999**,
64, 7856.
[1270]See Jiaang, W.-T.; Lin, H.-C.; Tang, K.-H.; Chang, L.-B.; Tsai, Y.-M. *J. Org. Chem.* **1999**, *64*, 618.

of a phenylseleno ester with $Bu_3SnH/AIBN$ to form the acyl radical, which added to the first alkene unit.[1271] The newly formed carbon radical added to the next alkene, and so on. Acyl radicals generated from $Ts(R)NCOSePh$ derivatives cyclize to form lactams.[1272]

Radical cyclization of iodo aldehydes or ketones, at the carbon of the carbonyl, is effectively an acyl addition reaction (**16-24**, **16-25**). This cyclization is often reversible, and there are many fewer examples can addition to an alkene or alkyne. In one example, a δ-iodo aldehyde was treated with BEt_3/O_2 to initiate formation of the radical, and in the presence of Bu_3SnH cyclization gave a cyclopentanol.[1273] The reaction of an aldehyde-alkene with AIBN, 0.5 $PhSiH_3$ and 0.1 Bu_3SnH generated a radical from the alkene, which cyclized at the aldehyde to give cyclopentanol derivatives.[1274] An aldehyde-O-methyloxime generated a radical adjacent to nitrogen under standard conditions, which cyclized at the carbonyl to give a cyclic α-hydroxy N-methoxyamine.[1275] Alternatively an α-bromoacetal-O-methyl oxime cyclized at the C=NOMe unit under electrolytic conditions in the presence of cobaloxime.[1276]

The attacking radical in radical cyclization reactions is not limited to a carbon, and a number of heterocycles can be prepared.[1277] Amidyl radical are known and give cyclization reactions.[1278] Aminyl radical cyclizations have been reported.[1279] N-Chloroamine-alkenes give an aminyl radical when treated with $TiCl_3•BF_3$, and cyclization give a pyrrolidine derivative with a pendant chloromethyl group.[1280] N-(S-substituted) amines give similar results using $AIBN/Bu_3SnH$.[1281] Oxime–alkenes cyclize to imines when treated with PhSSPh and TEMPO (p. 274).[1282] An oxygen radical can be generated under photochemical conditions, and they add to alkenes in a normal manner.[1283] Note that radical substitution occurs, and reaction of $Ph_3SnH/AIBN$ and an O-amidyl compound having a phosphonate ester elsewhere in the molecule gave cyclization to a tetrahydrofuran derivative.[1284]

[1271]Pattenden, G.; Roberts, L.; Blake, A.J. *J. Chem. Soc., Perkin Trans. 1* **1998**, 863; Batsanov, A.; Chen, L.; Gill, G.B.; Pattenden, G. *J. Chem. Soc., Perkin Trans. 1* **1996**, 45. Also see, Pattenden, G.; Smithies, A.J.; Tapolczay, D.; Walter, D.S. *J. Chem. Soc., Perkin Trans. 1* **1996**, 7 for a related reaction that generates a bicyclic species from an initially generated alkyl radical.

[1272]Rigby, J.H.; Danca, D.M.; Horner, J.H. *Tetrahedron Lett.* **1998**, *39*, 8413.

[1273]Devin, P.; Fensterbank, L.; Malacria, M. *Tetrahedron Lett.* **1999**, *40*, 5511.

[1274]Hays, D.S.; Fu, G.C. *Tetrahedron* **1999**, *55*, 8815.

[1275]Naito, T.; Nakagawa, K.; Nakamura, T.; Kasei, A.; Ninomiya, I.; Kiguchi, T. *J. Org. Chem.* **1999**, *64*, 2003.

[1276]Inokuchi, T.; Kawafuchi, H. *Synlett* **2001**, 421.

[1277]See Majumdar, K.C.; Basu, P.K.; Mukhopadhyay, P.P. *Tetrahedron* **2004**, *60*, 6239. For a review, see Bowman, W.R.; Cloonan, M.O.; Krintel, S.L. *J. Chem. Soc., Perkin Trans. 1* **2001**, 2885.

[1278]Clark, A.J.; Peacock, J.L. *Tetrahedron Lett.* **1998**, *39*, 6029. See Prabhakaran, E.N.; Nugent, B.M.; Williams, A.L.; Nailor, K.E.; Johnston, J.N. *Org. Lett.* **2002**, *4*, 4197.

[1279]Maxwell, B.J.; Tsanaktsidis, J. *J. Chem. Soc., Chem. Commun.* **1994**, 533.

[1280]Hemmerling, M.; Sjöholm, Å.; Somfai, P. *Tetrahedron Asymmetry* **1999**, *10*, 4091.

[1281]Guindon, Y.; Guérin, B.; Landry, S.R. *Org. Lett.* **2001**, *3*, 2293.

[1282]Lin, X.; Stien, D.; Weinreb, S.M. *Org. Lett.* **1999**, *1*, 637.

[1283]Newcomb, M.; Dhanabalasingam, B. *Tetrahedron Lett.* **1994**, *35*, 5193. For a review, see Hartung, J. *Eur. J. Org. Chem.* **2001**, 619.

[1284]Crich, D.; Huang, X.; Newcomb, M. *Org. Lett.* **1999**, *1*, 225.

15-31 Conjugate Addition With Heteroatom Nucleophiles

$$X \;+\; \underset{/}{\overset{\backslash}{C}}{=}\underset{\backslash}{\overset{Z^2}{C}} \quad\longrightarrow\quad X\text{-}\underset{/}{\overset{\backslash}{C}}\text{-}\underset{/}{\overset{\backslash}{C}}\text{-}\underset{H}{\overset{Z^2}{C}}$$

Other nucleophiles add to conjugated systems to give Michael-type products. Aniline derivatives add to conjugated aldehydes in the presence of a catalytic amount of DBU (p. 1132).[1285] Amines add to conjugated esters in the presence of InCl$_3$,[1286] Bi(NO)$_3$,[1287] Cu(OTf)$_2$,[1288] CeCl$_3$/NaI/SiO$_2$,[1289] La(OTf)$_3$,[1290] or Yb(OTf)$_3$ at 3 kbar,[1291] for example, to give β-amino esters. Palladium catalysts have been used as well.[1292] Conjugate addition of amines has also been promoted by lithium perchlorate,[1293] and by clay.[1294] This reaction can be initiated photochemically[1295] or with microwave irradiation.[1296] Lithium amides add to conjugated esters to give the β-amino ester.[1297] An intramolecular addition of an amine unit to a conjugated ketone in the presence of a palladium catalyst, or photochemically, led to cyclic amines.[1298] Amines add to conjugated thio-lactams.[1299] Chiral catalysts lead to enantioselective reactions.[1300] Chiral imines add in a highly stereoselective manner.[1301] Chiral additives, such as chiral Cinchona alkaloids[1302] or chiral naphthol derivatives,[1303] have also been used. The nitrogen of carbamates add to conjugated ketones with a platinum,[1304] palladium,[1305] copper,[1306] or with a bis-(triflamide) catalyst.[1307] The amine moiety of a carbamate adds to conjugated

[1285]Markó, I.E.; Chesney, A. *Synlett* **1992**, 275.

[1286]Loh, T.-P.; Wei, L.-L. *Synlett* **1998**, 975.

[1287]Srivastava, N.; Banik, B.K. *J. Org. Chem.* **2003**, *68*, 2109.

[1288]Xu, L.-W.; Wei, J.-W.; Xia, C.-G.; Zhou, S.-L.; Hu, X.-X. *Synlett* **2003**, 2425.

[1289]Bartoli, G.; Bosco, M.; Marcantoni, E.; Petrini, M.; Sambri, L.; Torregiani, E. *J. Org. Chem.* **2001**, *66*, 9052.

[1290]Matsubara, S.; Yoshioka, M.; Utimoto, K. *Chem. Lett.* **1994**, 827.

[1291]Jenner, G. *Tetrahedron Lett.* **1995**, *36*, 233.

[1292]Takasu, K.; Nishida, N.; Ihara, M. *Synlett* **2004**, 1844.

[1293]Azizi, N.; Saidi, M.R. *Tetrahedron* **2004**, *60*, 383.

[1294]Shaikh, N.S.; Deshpande, V.H.; Bedekar, A.V. *Tetrahedron* **2001**, *57*, 9045.

[1295]Das, S.; Kumar, J.S.D.; Shivaramayya, K.; George, M.V. *J. Chem. Soc. Perkin Trans. 1*, **1995**, 1797; Hoegy, S.E.; Mariano, P.S. *Tetrahedron Lett.* **1994**, *35*, 8319.

[1296]Moghaddam, F.M.; Mohammadi, M.; Hosseinnia, A. *Synth. Commn.* **2000**, *30*, 643.

[1297]Doi, H.; Sakai, T.; Iguchi, M.; Yamada, K.-i.; Tomioka, K. *J. Am. Chem. Soc.* **2003**, *125*, 2886.

[1298]Zhang, X.; Jung, Y.S.; Mariano, P.S.; Fox, M.A.; Martin, P.S.; Merkert, J. *Tetrahedron Lett.* **1993**, *34*, 5239.

[1299]Sośnicki, J.G.; Jagodziński, T.S.; Liebscher, J. *J. Heterocyclic Chem.* **1997**, *34*, 643.

[1300]Sugihara, H.; Daikai, K.; Jin, X.L.; Furuno, H.; Inanaga, J. *Tetrahedron Lett.* **2002**, *43*, 2735.

[1301]Ambroise, L.; Desmaële, D.; Mahuteau, J.; d'Angelo, J. *Tetrahedron Lett.* **1994**, *35*, 9705.

[1302]Jew, S.-s.; Jeong, B.S.; Yoo, M.-S.; Huh, H.; Park, H.-g. *Chem. Commun.* **2001**, 1244.

[1303]Yamagiwa, N.; Matsunaga, S.; Shibasaki, M. *J. Am. Chem. Soc.* **2003**, *125*, 16178.

[1304]Kakumoto, K.; Kobayashi, S.; Sugiura, M. *Org. Lett.* **2002**, *4*, 1319.

[1305]Gaunt, M.J.; Spencer, J.B. *Org. Lett.* **2001**, *3*, 25.

[1306]Wabnitz, T.C.; Spencer, J.B. *Tetrahedron Lett.* **2002**, *43, 3*891.

[1307]Wabnitz, T.C.; Spencer, J.B. *Org. Lett.* **2003**, *5*, 2141.

ketones with a polymer-supported acid catalyst,[1308] or with $BF_3 \cdot OEt_2$.[1309] Chiral catalysts have been used for the conjugated addition of carbamates.[1310] The reaction of ammonium formate with 1,4-diphenylbut-2-en-1,4-dione, in PEG-200 and a palladium catalyst under microwave irradiation, gave 2,5-diphenylpyrrole.[1311]

Lactams have been shown to add to conjugated esters in the presence of $Si(OEt)_4$ and CsF.[1312] Phthalimide adds to alkylidene malononitriles via 1,4-addition with a palladium catalyst, and the resulting anion can be alkylated with an added allylic halide.[1313] Alkylidene amido amides, $C=C(NHAc)CONHR$, react with secondary amines in water to give the β-amino amido amide.[1314] Amines also add in a conjugate manner to alkynyl phosphonate esters, $C\equiv C-PO(OEt)_2$, using a CuI catalyst.[1315] Hydroxylamines add to conjugated nitro compounds to give 2-nitro hydroxylamines.[1316] N,O-Trimethylsilyl hydroxylamines add to conjugated esters, via nitrogen, using a copper catalyst.[1317] Trimethylsilyl azide with acetic acid reacts with conjugated ketones to give the β-azido ketone.[1318] Sodium azide adds to conjugated ketones in aqueous acetic acid and 20% PBu_3.[1319]

Phosphines react similarly to amines under certain conditions. Conjugate addition of R_2PH and a nickel catalyst give conjugate addition to α,β-unsaturated nitriles.[1320]

Alcohols add to conjugated ketones with a PMe_3 catalyst to give the β-alkoxy ketone.[1321] The conjugate addition of peroxide anions (HOO^- and ROO^-) to α,β-unsaturated carbonyl compounds is discussed in **15-48**.

bis(Silanes) add to alkylidene malonate derivatives in the presence of a copper catalyst to give β-silyl malonates, $RCH(SiR_3)CH(CO_2Me)_2$.[1322] Alkylsilane units add using bis(trialkylsilyl)zinc reagents with a CuCN catalyst.[1323]

Thiophenol and butyllithium (lithium phenylthiolate) adds to conjugated esters.[1324] Similar addition is observed with selenium compounds RSeLi.[1325]

[1308]Wabnitz, T.C.; Yu, J.-Q.; Spencer, J.B. *Synlett* **2003**, 1070.

[1309]Xu, L.-W.; Li, L.; Xia, C.-G.; Zhou, S.-L.; Li, J.-W.; Hu, X.-X. *Synlett* **2003**, 2337.

[1310]Palomo, C.; Oiarbide, M.; Halder, R.; Kelso, M.; Gómez-Bengoa, E.; García, J.M. *J. Am. Chem. Soc.* **2004**, *126*, 9188.

[1311]Rao, H.S.P.; Jothilingam, S. *Tetrahedron Lett.* **2004**, *42*, 6595.

[1312]Ahn, K.H.; Lee, S.J. *Tetrahedron Lett.* **1994**, *35*, 1875.

[1313]Aoyagi, K.; Nakamura, H.; Yamamoto, Y. *J. Org. Chem.* **2002**, *67*, 5977.

[1314]Naidu, B.N.; Sorenson, M.E.; Connolly, J.P.; Ueda, Y. *J. Org. Chem.* **2003**, *68*, 10098.

[1315]Panarina, A.E.; Dogadina, A.V.; Zakharov, V.I.; Ionin, B.I. *Tetrahedron Lett.* **2001**, *42*, 4365.

[1316]O'Neil, I.A.; Cleator, E.; Southern, J.M.; Bickley, J.F.; Tapolczay, D.J. *Tetrahedron Lett.* **2001**, *42*, 8251.

[1317]Cardillo, G.; Gentilucci, L.; Gianotti, M.; Kim, H.; Perciaccante, R.; Tolomelli, A. *Tetrahedron Asymmetry* **2001**, *12*, 2395.

[1318]Guerin, D.J.; Horstmann, T.E.; Miller, S.J. *Org. Lett.* **1999**, *1*, 1107.

[1319]Xu, L.-W.; Xia, C.-G.; Li, J.-W.; Zhou, S.-L. *Synlett* **2003**, 2246.

[1320]Sadow, A.D.; Haller, I.; Fadini, L.; Togni, A. *J. Am. Chem. Soc.* **2004**, *126*, 14704.

[1321]Stewart, I.C.; Bergman, R.G.; Toste, F.D. *J. Am. Chem. Soc.* **2003**, *125*, 8696.

[1322]Clark, C.T.; Lake, J.F.; Scheidt, K.A. *J. Am. Chem. Soc.* **2004**, *126*, 84.

[1323]Oestreich, M.; Weiner, B. *Synlett* **2004**, 2139.

[1324]Kamimura, A.; Kawahara, F.; Omata, Y.; Murakami, N.; Morita, R.; Otake, H.; Mitsudera, H.; Shirai, M.; Kakehi, A. *Tetrahedron Lett.* **2001**, *42*, 8497.

[1325]Zeni, G.; Stracke, M.P.; Nogueira, C.W.; Braga, A.L.; Menezes, P.H.; Stefani, H.A. *Org. Lett.* **2004**, *6*, 1135.

Thiols react with conjugated amides via 1,4-addition with the addition of 10% Hf(OTf)$_4$ or other lanthanide triflates[1326] or to conjugated ketones in ionic solvents.[1327] Thiophenol adds in a similar manner in the presence of Na$_2$Ca-P$_2$O$_7$[1328] or LiAl–poly2a.[1329] Thioaryl moieties can be added in the presence of Yb[1330] or a catalytic amount of (DHQD)$_2$PYR (a dihydroquinidine, see **15-48**).[1331] Thioalkyl units, such as BuS—, add to conjugated ketones using BuS—SnBu and In—I.[1332] Addition of conjugated lactones is possible to produce β-arylthiolated lactones.[1333]

α,β-Unsaturated sulfones undergo conjugate addition of a cyano group using Et$_2$AlCN.[1334] Trimethylsilyl cyanide (Me$_3$SiCN) adds a cyano group to α,β-unsaturated amines with a specialized aluminum salen-ytterbium catalyst.[1335]

15-32 Acylation of Activated Double Bonds and of Triple Bonds

Hydro-acyl-addition

Under some conditions, acid derivatives add directly to activated double bonds. Acetic anhydride, magnesium metal, and Me$_3$SiCl reacts with conjugated esters to give a γ-keto ester.[1336] Similar reaction with vinyl phosphonate esters leads to a γ-keto phosphonate ester.[1337] Thioesters undergo conjugate addition to α,β-unsaturated ketones in the presence of SmI$_2$.[1338] Using DBU (1,8-diazabicyclo [5.4.0] undec-7-ene) (p. 1132) and a thioimidazolium salt, acyl silanes, Ar(C=O)SiMe$_3$, add in a similar manner.[1339] Under microwave irradiation, aldehydes add to conjugated ketones using DBU/Al$_2$O$_3$ and a thiazolium salt.[1340] The conjugate addition

[1326]Kobayashi, S.;Ogawa, C.; Kawamura, M.; Sugiura, M. *Synlett* **2001**, 983.

[1327]In bmim PF$_6$, 1-butyl-3-methylimidazolium hexafluorophosphate: Yadav, J.S.; Reddy, B.V.S.; Baishya, G. *J. Org. Chem.* **2003**, *68*, 7098.

[1328]Zahouily, M.; Abrouki, Y.; Rayadh, A. *Tetrahedron Lett.* **2002**, *43*, 7729.

[1329]Sundararajan, G.; Prabagaran, N. *Org. Lett.* **2001**, *3*, 389.

[1330]Taniguchi, Y.; Maruo, M.; Takaki, K.; Fujiwara, Y. *Tetrahedron Lett.* **1994**, *35*, 7789.

[1331]McDaid, P.; Chen, Y.; Deng, L. *Angew. Chem. Int. Ed.* **2002**, *41*, 338.

[1332]Ranu, B.C.; Mandal, T. *Synlett* **2004**, 1239.

[1333]Nishimura, K.; Tomioka, K. *J. Org. Chem.* **2002**, *67*, 431.

[1334]Ruano, J.L.G.; García, M.C.; Laso, N.M.; Castro, A.M.M.; Ramos, J.H.R. *Angew. Chem. Int. Ed.* **2001**, *40*, 2507.

[1335]With high enantioselectivity, see Sammis, G.M.; Danjo, H.; Jacobsen, E.N. *J. Am. Chem. Soc.* **2004**, *126*, 9928.

[1336]Ohno, T.; Sakai, M.; Ishino, Y.; Shibata, T.; Maekawa, H.; Nishiguchi, I. *Org. Lett.* **2001**, *3*, 3439.

[1337]Kyoda, M.; Yokoyama, T.; Maekawa, H.; Ohno, T.; Nishiguchi, I. *Synlett* **2001**, 1535.

[1338]Blakskjær, P.; Høj, B.; Riber, D.; Skrydstrup, T. *J. Am. Chem. Soc.* **2003**, *125*, 4030.

[1339]Mattson, A.E.; Bharadwaj, A.R.; Scheidt, K.A. *J. Am. Chem. Soc.* **2004**, *126*, 2314.

[1340]Yadav, J.S.; Anuradha, K.; Reddy, B.V.S.; Eeshwaraiah, B. *Tetrahedron Lett.* **2003**, *44*, 8959.

of acyl zirconium complexes in the presence of BF3•OEt$_2$ is catalyzed by palladium acetate.[1341]

$$\underset{/}{\overset{\backslash}{C}}=C\overset{O}{\underset{\backslash}{\overset{\|}{C}}}\ +\ \text{R-Li}\ +\ \text{Ni(CO)}_4\ \xrightarrow{\text{ether}}\ \underset{R}{\overset{O}{C}}-\underset{/}{\overset{\backslash}{C}}-\underset{H}{\overset{|}{C}}-\overset{O}{\overset{\|}{C}}$$

157

An acyl group can be introduced into the 4 position of an α,β-unsaturated ketone by treatment with an organolithium compound and nickel carbonyl.[1342] The product is a 1,4-diketone, **157**. The R group may be aryl or primary alkyl. The reaction can also be applied to alkynes (which need not be activated), in which case 2 mol add and the product is also a 1,4-diketone (e.g., R'C≡CH → RCOCHR'CH$_2$COR).[1343] In a different procedure, α,β-unsaturated ketones and aldehydes are acylated by treatment at −110°C with R$_2$(CN)CuLi$_2$ and CO. This method is successful for R = primary, secondary, and tertiary alkyl.[1344] For secondary and tertiary groups, R(CN)CuLi (which does not waste an R group) can be used instead.[1345]

Another method involves treatment with an aldehyde and cyanide ion (see **16-52**) in a polar aprotic solvent (e.g., DMF or DMSO).[1346]

$$\underset{R}{\overset{O}{\overset{\|}{C}}}{}_{H}\ +\ ^-CN\ \rightleftarrows\ \underset{H}{\overset{R}{C}}\overset{O^{\ominus}}{\underset{CN}{}}\ \rightleftarrows\ R-\overset{OH}{\underset{CN}{C^{\ominus}}}\ +\ \underset{/}{\overset{\backslash}{C}}=C\overset{O}{\overset{\|}{C}}\ \rightarrow\ \underset{NC}{\overset{HO}{R-C-C-C}}\overset{O}{\overset{\|}{C}}\ \xrightarrow{-HCN}\ \textbf{73}$$

158

This method has been applied to α,β-unsaturated ketones, esters, and nitriles to give the corresponding 1,4-diketones, γ-keto esters, and γ-keto nitriles, respectively (see also, **16-55**). The ion **158** is a synthon for the unavailable R$^{\ominus}$ C=O anion (see also, p. 634); it is a masked R$^{\ominus}$ C=O anion. Other masked carbanions that have been used in this reaction are the RC$^{\ominus}$(CN) NR ion,[1347] the EtSC$^{\ominus}$ RSOEt ion[1348] (see p. 634), the CH$_2$=C$^{\ominus}$ OEt ion,[1349] CH$_2$=C(OEt)Cu$_2$Li,[1350] CH$_2$=CMe(SiMe$_3$),[750]

[1341]Hanzawa, Y.; Tabuchi, N.; Narita, K.; Kakuuchi, A.; Yabe, M.; Taguchi, T. *Tetrahedron* **2002**, *58*, 7559.

[1342]Corey, E.J.; Hegedus, L.S. *J. Am. Chem. Soc.* **1969**, *91*, 4926.

[1343]Sawa, Y.; Hashimoto, I.; Ryang, M.; Tsutsumi, S. *J. Org. Chem.* **1968**, *33*, 2159.

[1344]Seyferth, D.; Hui, R.C. *J. Am. Chem. Soc.* **1985**, *107*, 4551. See also, Lipshutz, B.H.; Elworthy, T.R. *Tetrahedron Lett.* **1990**, *31*, 477.

[1345]Seyferth, D.; Hui, R.C. *Tetrahedron Lett.* **1986**, *27*, 1473.

[1346]For reviews, see Stetter, H.; Kuhlmann, H. *Org. React.* **1991**, *40*, 407–496; Stetter, H. *Angew. Chem. Int. Ed.* **1976**, *15*, 639. For a similar method involving thiazolium salts, see Stetter, H.; Skobel, H. *Chem. Ber.* **1987**, *120*, 643; Stetter, H.; Kuhlmann, H.; Haese, W. *Org. Synth.*, *65*, 26.

[1347]Enders, D.; Gerdes, P.; Kipphardt, H. *Angew. Chem. Int. Ed.* **1990**, *29*, 179.

[1348]Herrmann, J.L.; Richman, J.E.; Schlessinger, R.H. *Tetrahedron Lett.* **1973**, 3271, 3275.

[1349]Beockman Jr., R.K.; Bruza, K.J.; Baldwin, J.E.; Lever Jr., O.W. *J. Chem. Soc., Chem. Commun.* **1975**, 519.

[1350]Boeckman Jr., R.K ; Bruza, K.J. *J. Org. Chem.* **1979**, *44*, 4781.

and the $RC^{\ominus}(OCHMeOEt)$ CN ion[1351] (see p. 640). In the last case, best results are obtained when R is a vinylic group. Anions of 1,3-dithianes (**10-71**) do not give 1,4-addition to these substrates (except in the presence of HMPA, see **15-25**), but add 1,2 to the C=O group instead (**16-38**).

In another procedure, acyl radicals derived from phenyl selenoesters ArCOSePh (by treatment of them with Bu_3SnH) add to α,β-unsaturated esters and nitriles to give γ-keto esters and γ-keto nitriles, respectively.[1352]

OS **VI**, 866; **VIII**, 620.

15-33 Addition of Alcohols, Amines, Carboxylic Esters, Aldehydes, and so on.

Hydro-acyl-addition, and so on.

Formates, primary, and secondary alcohols, amines, ethers, alkyl halides, compounds of the type Z—CH₂— Z′, and a few other compounds add to double bonds in the presence of free-radical initiators.[1353] This is formally the addition of RH to a double bond, but the "R" is not just any carbon but one connected to an oxygen or a nitrogen, a halogen, or to two Z groups (defined as on p. 1007). Formates and formamides[1354] add similarly:

Alcohols, ethers, amines, and alkyl halides add as follows (shown for alcohols):

ZCH_2Z' compounds react at the carbon bearing the active hydrogen:[1355]

[1351]Stork, G.; Maldonado, L. *J. Am. Chem. Soc.* **1974**, *96*, 5272.

[1352]Boger, D.L.; Mathvink, R.J. *J. Org. Chem.* **1989**, *54*, 1777.

[1353]For reviews, see Giese, B. *Radicals in Organic Synthesis: Formation of Carbon-Carbon Bonds*, Pergamon, Elmsford, NY, **1986**, pp. 69–77; Vogel, H. *Synthesis* **1970**, 99; Huyser, E.S. *Free-Radical Chain Reactions*, Wiley, NY, **1970**, pp. 152–159; Elad, D. *Fortschr. Chem. Forsch.* **1967**, *7*, 528. Hyponitrites have been used to initiate this reaction; see Dang, H.-S.; Roberts, B.P. *Chem. Commun.* **1996**, 2201.

[1354]Elad, D. *Fortschr. Chem. Forsch.* **1967**, *7*, 528, see pp. 530–543.

[1355]For example, see Cadogan, J.I.G.; Hey, D.H.; Sharp, J.T. *J. Chem. Soc. C* **1966**, 1743; *J. Chem. Soc. B* **1967**, 803; Hájek, M.; Málek, J. *Coll. Czech. Chem. Commun.* **1979**, *44*, 3695.

Similar additions have been successfully carried out with carboxylic acids, anhydrides,[1356] acyl halides, carboxylic esters, nitriles, and other types of compounds.[1357]

Similar reactions have been carried out on acetylene.[1358] In an interesting variation, thiocarbonates add to alkynes in the presence of a palladium catalyst to give a β-phenylthio α,β-unsaturated ester.[1359] Aldehydes add to alkynes in the presence of a rhodium catalyst to give conjugated ketones.[1360] In a cyclic version of the addition of aldehydes, 4-pentenal was converted to cyclopentanone with a rhodium–complex catalyst.[1361] An intramolecular acyl addition to an alkyne was reported using silyl ketones, acetic aid and a rhodium catalyst.[1362] In the presence of a palladium catalyst, a tosylamide group added to an alkene unit to generate N-tosylpyrrolidine derivatives.[1363]

OS **IV**, 430; **V**, 93; **VI**, 587, 615.

15-34   Addition of Aldehydes

Alkyl-carbonyl-addition

In the presence of metal catalysts, such as rhodium compounds[1364] or Yb(OTf)$_3$,[1365] aldehydes can add directly to alkenes to form ketones. The reaction of ω-alkenyl aldehydes with rhodium catalyst leads to cyclic ketones,[1366] with high enantioselectivity if chiral ligands are employed. Aldehydes also add to vinyl esters in the presence of hyponitrites and thioglycolates.[1367] The addition of aldehydes to activated double bonds, mediated by a catalytic amount of thiazolium salt in the presence of a

[1356]de Klein, W.J. *Recl. Trav. Chim. Pays-Bas* **1975**, *94*, 48.

[1357]Allen, J.C.; Cadogan, J.I.G.; Hey, D.H. *J. Chem. Soc.* **1965**, 1918; Cadogan, J.I.G. *Pure Appl. Chem.* **1967**, *15*, 153, pp. 153–158. See also, Giese, B.; Zwick, W. *Chem. Ber.* **1982**, *115*, 2526; Giese, B.; Erfort, U. *Chem. Ber.* **1983**, *116*, 1240.

[1358]For example, see Cywinski, N.F.; Hepp, H.J. *J. Org. Chem.* **1965**, *31*, 3814; DiPietro, J.; Roberts, W.J. *Angew. Chem. Int. Ed.* **1966**, *5*, 415.

[1359]Hua, R.; Takeda, H.; Onozawa, S.-y.; Abe, Y.; Tanaka, M. *J. Am. Chem. Soc.* **2001**, *123*, 2899.

[1360]Kokubo, K.; Matsumasa, K.; Miura, M.; Nomura, M. *J. Org. Chem.* **1997**, *62*, 4564.

[1361]Fairlie, D.P.; Bosnich, B. *Organometallics* **1988**, *7*, 936, 946. Also see, Barnhart, R.W.; Wang, X.; Noheda, P.; Bergens, S.H.; Whelan, J.; Bosnich, B. *J. Am. Chem. Soc.* **1994**, *116*, 1821 for an enantioselective version of this cyclization.

[1362]Yamane, M.; Amemiya, T.; Narasaka, K. *Chem. Lett.* **2001**, 1210.

[1363]Larock, R.C.; Hightower, T.R.; Hasvold, L.A.; Peterson, K.P. *J. Org. Chem.* **1996**, *61*, 3584.

[1364]Jun, C.-H.; Lee, H.; Hong, J.-B. *J. Org. Chem.* **1997**, *62*, 1200.

[1365]Curini, M.; Epifano, F.; Maltese, F.; Rosati, O. *Synlett* **2003**, 552.

[1366]Barnhart, R.W.; McMorran, D.A.; Bosnich, B. *Chem. Commun.* **1997**, 589.

[1367]Dang, H.-S.; Roberts, B.P. *J. Chem. Soc, Perkin Trans. 1*, **1998**, 67.

weak base, is called the *Stetter reaction*,[1368] An internal addition of an alkynyl aldehyde, catalyzed by a rhodium complex, led to a cyclopentenone derivative.[1369] A similar carbonyl addition with benzaldehyde derivatives having an ortho-allylic ether led to a benzopyranone when treated with potassium hexamethyldisilazide.[1370]

These reactions are not successful when the alkene contains electron-withdrawing groups, such as halo or carbonyl groups. A free-radical initiator is required,[1371] usually peroxides or *UV* light. The mechanism is illustrated for aldehydes but is similar for the other compounds:

etc.

Polymers are often side products. Photochemical addition of aldehyde to conjugated C=C units can be efficient when a triplet sensitizer (p. 340), such as benzophenone is used.[1372]

A variation that is more of an acyl addition (**16-25**) involves the reaction of an allylic alcohol with benzaldehyde. With a ruthenium catalyst and in an ionic liquid, the C=C unit reacts with the aldehyde, with concomitant oxidation of the allylic alcohol unit, to give a β-hydroxy ketone, PhCHO +C=C—CH(OH)R → PhCH(OH)—CH(Me)COR.[1373] In another variation, formate esters add to alkenes using a ruthenium catalyst to give an alkyl ester via a formylation process.[1374]

15-35 Hydrocarboxylation

Hydro-carboxy-addition

[1368]Stetter, H.; Schreckenberg, M. *Angew. Chem., Int. Ed* **1973**, *12*, 81; Stetter, H.; Kuhlmann, H. *Angew. Chem., Int. Ed.* **1974**, *13*, 539; Stetter, H. *Angew. Chem., Int. Ed.* **1976**, *15*, 639; Stetter, H.; Haese, W. *Chem. Ber.* **1984**, *117*, 682; Stetter, H.; Kuhlmann, H. *Org. React.* **1991**, *40*, 407; Enders, D.; Breuer, K.; Runsink, J.; Teles, J.H. *Helv. Chim. Acta* **1996**, *79*, 1899; Kerr, M.S.; Rovis, T. *Synlett* **2003**, 1934; Kerr, M.S.; Rovis, T. *J. Am. Chem. Soc.* **2004**, *126*, 8876; Pesch, J.; Harms, K.; Bach, T. *Eur. J. Org. Chem.* **2004**, 2025; Mennen, S.; Blank, J.; Tran-Dube, M.B.; Imbriglio, J.E.; Miller, S.J. *Chem. Commun.* **2005**, 195. For examples of the Stetter reaction with acyl silanes, see Mattson, A.E.; Bharadwaj, A.R.; Scheidt, K.A. *J. Am. Chem. Soc.* **2004**, *126*, 2314.

[1369]Tanaka, K.; Fu, G.C. *J. Am. Chem. Soc.* **2002**, *124*, 10296.

[1370]Kerr, M.S.; de Alaniz, J.R.; Rovis, T. *J. Am. Chem. Soc.* **2002**, *124*, 10298.

[1371]See Lee, E.; Tae, J.S.; Chong, Y.H.; Park, Y.C.; Yun, M.; Kim, S. *Tetrahedron Lett.* **1994**, *35*, 129 for an example.

[1372]Kraus, G.A.; Liu, P. *Tetrahedron Lett.* **1994**, *35*, 7723.

[1373]In bmim PF$_6$, 3-butyl-1-methylimidazolium hexafluorophosphate: Yang, X.-F.; Wang, M.; Varma, R.S.; Li, C.-J. *Org. Lett.* **2003**, *5*, 657.

[1374]Na, Y.; Ko, S.; Hwang, L.K.; Chang, S. *Tetrahedron Lett.* **2003**, *44*, 4475.

The acid-catalyzed hydrocarboxylation of alkenes (the *Koch reaction*) can be performed in a number of ways.[1375] In one method, the alkene is treated with carbon monoxide and water at 100–350°C and 500–1000-atm pressure with a mineral acid catalyst. However, the reaction can also be performed under milder conditions. If the alkene is first treated with CO and catalyst and then water added, the reaction can be accomplished at 0–50°C and 1–100 atm. If formic acid is used as the source of both the CO and the water, the reaction can be carried out at room temperature and atmospheric pressure.[1376] The formic acid procedure is called the *Koch–Haaf reaction* (the Koch–Haaf reaction can also be applied to alcohols, see **10-77**). Nearly all alkenes can be hydrocarboxylated by one or more of these procedures. However, conjugated dienes are polymerized instead. Hydrocarboxylation can also be accomplished under mild conditions (160°C and 50 atm) by the use of nickel carbonyl as catalyst. Acid catalysts are used along with the nickel carbonyl, but basic catalysts can also be employed.[1377] Other metallic salts and complexes can be used, sometimes with variations in the reaction procedure, including palladium,[1378] platinum,[1379] and rhodium[1380] catalysts. The $Ni(CO)_4$-catalyzed oxidative carbonylation with CO and water as a nucleophile is often called *Reppe carbonylation*.[1381] The toxic nature of nickel

---

[1375]For reviews of hydrocarboxylation of double and triple bonds catalyzed by acids or metallic compounds, see Lapidus, A.L.; Pirozhkov, S.D. *Russ. Chem. Rev.* **1989**, *58*, 117; Anderson, G.K.; Davies, J.A., in Hartley, F.R.; Patai, S. *The Chemistry of the Metal-Carbon Bond*, Vol. 3, Wiley, NY, **1985**, pp. 335–359, 335–348; in Falbe, J. *New Syntheses with Carbon Monoxide*, Springer, NY, **1980**, the articles by Mullen, A. pp. 243–308; and Bahrmann, H. pp. 372–413; in Wender, I.; Pino, P. *Organic Syntheses via Metal Carbonyls*, Vol. 2, Wiley, NY, **1977**, the articles by Pino, P.; Piacenti, F.; Bianchi, M. pp. 233–296; and Pino, P.; Braca, G. pp. 419–516; Eidus, Ya.T.; Lapidus, A.L.; Puzitskii, K.V.; Nefedov, B.K. *Russ. Chem. Rev.* **1973**, *42*, 199; *Russ. Chem. Rev.* **1971**, *40*, 429; Falbe, J. *Carbon Monoxide in Organic Synthesis*, Springer, Berlin, **1970**, pp. 78–174.
[1376]Haaf, W. *Chem. Ber.* **1966**, *99*, 1149; Christol, H.; Solladié, G. *Bull. Soc. Chim. Fr.* **1966**, 1307.
[1377]Sternberg, H.W.; Markby, R.; Wender, P. *J. Am. Chem. Soc.* **1960**, *82*, 3638.
[1378]For reviews, see Heck, R.F. *Palladium Reagents in Organic Synthesis*, Academic Press, NY, **1985**, pp. 381–395; Bittler, K.; Kutepow, N.V.; Neubauer, D.; Reis, H. *Angew. Chem. Int. Ed.* **1968**, *7*, 329. For a review with respect to fluoroalkenes, see Ojima, I. *Chem. Rev.* **1988**, *88*, 1011, p. 1016. Seayad, A.; Jayasree, S.; Chaudhari, R.V. *Org. Lett.* **1999**, *1*, 459; Mukhopadhyay, K.; Sarkar, B.R.; Chaudhari, R.V. *J. Am. Chem. Soc.* **2002**, *124*, 9692. See also, the references cited in these latter articles.
[1379]Xu, Q.; Fujiwara, M.; Tanaka, M.; Souma, Y. *J. Org. Chem.* **2000**, *65*, 8105.
[1380]Xu, Q.; Nakatani, H.; Souma, Y. *J. Org. Chem.* **2000**, *65*, 1540.
[1381]Tsuji, J. *Palladium Reagents and Catalysts*, Wiley, NY, **1999**; Hohn, A., in *Applied Homogeneous Catalysis with Organometallic Compounds*, Vol. 1, VCH, NY, **1996**, p. 137; Beller, M.; Tafesh, A.M., in *Applied Homogeneous Catalysis with Organometallic Compounds*, Vol. 1, VCH, NY, **1996**, p. 187; Drent, E.; Jager, W.W.; Keijsper, J.J.; Niele, F.G.M., in *Applied Homogeneous Catalysis with Organometallic Compounds*, Vol. 1, VCH, NY, **1996**, p. 1119.; Parshall, G.W.; Ittel, S.D. *Homogeneous Catalysis*, 2nd ed., Wiley, NY, **1992**; Mullen, A. in *New Syntheses with Carbon Monoxide*, Springer-Verlag, NY, **1980**, p. 243; Tsuji, J. *Organic Synthesis with Palladium Compounds*, Springer-Verlag, NY, **1980**; Bertoux, F.; Monflier, E.; Castanet, Y.; Mortreux, A. *J. Mol. Catal. A: Chem.* **1999**, *143*, 11; Beller, M.; Cornils, B.; Frohning, C. D.; Kohlpaintner, C. W. *J. Mol. Catal. A: Chem.* **1995**, *104*, 17; Chiusoli, G. P. *Transition Met. Chem.* **1991**, *16*, 553; Roeper, M. *Stud. Surf. Sci. Catal.* **1991**, *64*, 381; Milstein, D. *Acc. Chem. Res.* **1988**, *21*, 428; Escaffre, P.; Thorez, A.; Kalck, P. *J. Mol. Catal.* **1985**, *33*, 87; Cassar, L.; Chiusoli, G. P.; Guerrieri, F. *Synthesis 1973*, 509; Tsuji, J. *Acc. Chem. Res.* **1969**, *2*, 144; Bird, C. W. *Chem. Rev.* **1962**, *62*, 283.

tetracarbonyl has led to development of other catalysts, including Co, Rh, Ir, Pd, and Pt, and Mo compounds.[1382] This reaction converts alkenes, alkynes and dienes and is tolerant of a wide variety of functional groups. When the additive is alcohol or acid, saturated or unsaturated acids, esters, or anhydrides are produced (see **15-36**). The transition-metal-catalyzed carbonylation has been done enantioselectively, with moderate-to-high optical yields, by the use of an optically active palladium complex catalyst.[1383] Dienes react with Cp_2TiCl_2/RMgCl and then with Me_2NCOCl to give amides.[1384] In the presence of formic acid, CO, and palladium acids can similarly be formed.[1385] Alkenes also react with $Fe(CO)_5$ and CO to give carboxylic acids.[1386] Electrochemical carboxylation procedures have been developed, including the conversion of alkenes to 1,4-butane-dicarboxylic acids.[1387]

When applied to triple bonds, hydrocarboxylation gives α,β-unsaturated acids under very mild conditions. Triple bonds give unsaturated acids and saturated dicarboxylic acids when treated with carbon dioxide and an electrically reduced nickel complex catalyst.[1388] Alkynes also react with $NaHFe(CO)_4$, followed by $CuCl_2 \bullet 2 H_2O$, to give alkenyl acid derivatives.[1389] A related reaction with CO and palladium catalysts in the presence of $SnCl_2$ also leads to conjugated acid derivatives.[1390] Terminal alkynes react with CO_2 and $Ni(cod)_2$, and subsequent treatment with DBU (p. 1132) gives the α,β-unsaturated carboxylic acid.[1391]

When acid catalysts are employed, in the absence of nickel carbonyl, the mechanism[1392] involves initial attack by a proton, followed by attack of the resulting carbocation on carbon monoxide to give an acyl cation, which, with water, gives the product, **159**. Markovnikov's rule is followed, and carbon skeleton rearrangements and double-bond isomerizations (prior to attack by CO) are frequent.

159

[1382]For a review, see Kiss, G. *Chem. Rev.* **2001**, *101*, 3435.

[1383]Alper, H.; Hamel, N. *J. Am. Chem. Soc.* **1990**, *112*, 2803.

[1384]Szymoniak, J.; Felix, D.; Moïse, C. *Tetrahedron Lett.* **1996**, *37*, 33.

[1385]Vasapollo, G.; Somasunderam, A.; El Ali, B.; Alper, H. *Tetrahedron Lett.* **1994**, *35*, 6203. See El Ali, B.; Vasapollo, G.; Alper, H. *J. Org. Chem.* **1993**, *58*, 4739 and El Ali, B.; Alper, H. *J. Org. Chem.* **1993**, *58*, 3595 for the same reaction with alkenes.

[1386]Brunet, J.-J.; Neibecker, D.; Srivastava, R.S. *Tetrahedron Lett.* **1993**, *34*, 2759.

[1387]Senboku, H.; Komatsu, H.; Fujimura, Y.; Tokuda, M. *Synlett* **2001**, 418.

[1388]Duñach, E.; Dérien, S.; Périchon, J. *J. Organomet. Chem.* **1989**, *364*, C33.

[1389]Periasamy, M.; Radhakrishnan, U.; Rameshkumar, C.; Brunet, J.-J. *Tetrahedron Lett.* **1997**, *38*, 1623.

[1390]Takeuchi, R.; Sugiura, M. *J. Chem. Soc. Perkin Trans. 1*, **1993**, 1031.

[1391]Saito, S.; Nakagawa, S.; Koizumi, T.; Hirayama, K.; Yamamoto, Y. *J. Org. Chem.* **1999**, *64*, 3975. See also, Takimoto, M.; Shimizu, K.; Mori, M. *Org. Lett.* **2001**, *3*, 3345.

[1392]For a review, see Hogeveen, H. *Adv. Phys. Org. Chem.* **1973**, *10*, 29.

For the transition metal catalyzed reactions, the nickel carbonyl reaction has been well studied and the addition is syn for both alkenes and alkynes.[1393] The following is the accepted mechanism:[785]

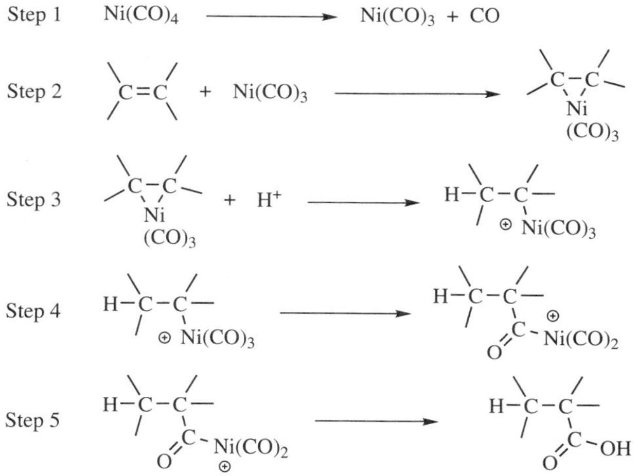
Step 1
Ni(CO)4
Ni(CO)3 + CO
Step 2
C=C
+ Ni(CO)3
C-C
Ni
(CO)3
Step 3
+ H+
H-C-C-
⊕ Ni(CO)3
Step 4
Ni(CO)2
Step 5
C-OH

Step 3 is an electrophilic substitution. The principal step of the mechanism, step 4, is a rearrangement.

An indirect method for hydrocarboxylation involves the reaction of an alkene with a borate, $(RO)_2BH$ and a rhodium catalysts. Subsequent reaction with $LiCHCl_2$ and then $NaClO_2$ gives the Markovnikov carboxylic acid (RC=C → $RC(COOH)CH_3$.[1394] When a chiral ligand is used, the reaction proceeds with good enantioselectivity.

15-36 Carbonylation, Alkoxycarbonylation and Aminocarbonylation of Double and Triple Bonds

Alkyl, Alkoxy or Amino-carbonyl-addition

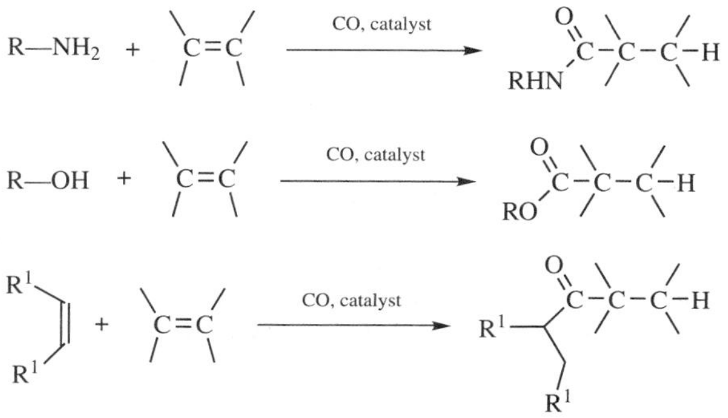
R—NH2 +
C=C
CO, catalyst
RHN
C-C-C-H
R—OH +
RO
R1
R1

[1393]Bird, C.W.; Cookson, R.C.; Hudec, J.; Williams, R.O. *J. Chem. Soc.* ***1963***, 410.
[1394]Chen, A.; Ren, L.; Crudden, C.M. *J. Org. Chem.* ***1999***, *64*, 9704.

In the presence of certain metal catalysts, alkenes and alkynes can be carbonylated or converted to amides or esters.[1395] There are several variations. The reaction of an alkyl iodide and a conjugated ester with CO, $(Me_3Si)_3SiH$ and AIBN (p. 935) in supercritical CO_2 (p. 414) gave a γ-keto ester.[1396] Terminal alkynes react with CO and methanol, and in the presence of $CuCl_2$ and $PdCl_2$ the product is a β-chloro-α-β-unsaturated methyl ester.[1397] Conjugated dienes react with thiophenol, CO and palladium(II) acetate to give the β,γ-unsaturated thioester.[1398] Allene reacts with CO, methanol and a ruthenium catalyst go give methacrylic acid.[1399] 5-Iodo-1-pentene reacted with 40 atm of CO in butanol to give a cyclopentanone with a pendant ester ($-CH_2CO_2Bu$).[1400] Alkynes react with thiophenol and CO with a palladium[1401] or platinum[1402] catalyst to give a conjugated thioester. Terminal alkynes react with CO and methanol, using a combination of a palladium (II) halide and a copper (II) halide, to give a conjugated diester, $MeO_2C-C=C-CO_2Me$.[1403] A similar reaction with alkenes using a combination of a palladium and a molybdenum catalyst led to a saturated diester, $MeO_2C-C-C-CO_2Me$.[1404] Alkenes were converted to the dimethyl ester of 1,4-butanedioic acid derivatives with CO/O_2 and a combination of $PdCl_2$ and $CuCl$ catalysts.[1405] Note that alkenes are converted to primarily the anti-Markovnikov ester upon treatment with arylmethyl formate esters ($ArCH_2OCHO$) and a ruthenium catalyst.[1406]

A bicyclic ketone was generated when 1,2-diphenylethyne was heated with CO, methanol and a dirhodium catalyst.[1407] 2-Iodostyrene reacted at 100°C with CO and a palladium catalyst to give the bicyclic ketone 1-indanone.[1408] Another variation reacted a conjugated allene–alkene with 5 atm of CO and a rhodium catalyst to give a bicyclic ketone.[1409] An intermolecular version of this reaction is known

---

[1395]For a review of carbometallation of alkenes and alkynes containing adjacent heteroatoms, see Fallis, A.G.; Forgione, P. *Tetrahedron* **2001**, *57*, 5899.

[1396]Kishimoto, Y.; Ikariya, T. *J. Org. Chem.* **2000**, *65*, 7656.

[1397]Li, J.; Jiang, H.; Feng, A.; Jia, L. *J. Org. Chem.* **1999**, *64*, 5984. See also, Clarke, M.L. *Tetrahedron Lett.* **2004**, *45*, 4043.

[1398]Xiao, W.-J.; Vasapollo, G.; Alper, H. *J. Org. Chem.* **2000**, *65*, 4138; Xiao, W.-J.; Alper, H. *J. Org. Chem.* **2001**, *66*, 6229.

[1399]Zhou, D.-Y.; Yoneda, E.; Onitsuka, K.; Takahashi, S. *Chem. Commun.* **2002**, 2868.

[1400]Ryu, I.; Kreimerman, S.; Araki, S. Nishitani, S.; Oderaotosi, Y.; Minakata, S.; Komatsu, M. *J. Am. Chem. Soc.* **2002**, *124*, 3812.

[1401]Xiao, W.-J.; Vasapollo, G.; Alper, H. *J. Org. Chem.* **1999**, *64*, 2080.

[1402]Kawakami, J.-i.; Mihara, M.; Kamiya, I.; Takeba, M.; Ogawa, A.; Sonoda, N. *Tetrahedron* **2003**, *59*, 3521.

[1403]Li, J.; Jiang, H.; Jia, L. *Synth. Commun.* **1999**, *29*, 3733; Li, J.; Jiang, H.; Chen, M. *Synth. Commun.* **2001**, *31*, 3131. For the identical reaction using only a palladium catalyst, see El Ali, B.; Tijani, J.; El-Ghanam, A.; Fettouhi, M. *Tetrahedron Lett.* **2001**, *42*, 1567.

[1404]Yokota, T.; Sakaguchi, S.; Ishii, Y. *J. Org. Chem.* **2002**, *67*, 5005.

[1405]Dai, M.; Wang, C.; Dong, G.; Xiang, J.; Luo, J.; Liang, B.; Chen, J.; Yang, Z. *Eur. J. Org. Chem.* **2003**, 4346.

[1406]Ko, S.; Na, Y.; Chang, S. *J. Am. Chem. Soc.* **2002**, *124*, 750.

[1407]Yoneda, E.; Kaneko, T.; Zhang, S.-W.; Onitsuka, K.; Takahashi, S. *Tetrahedron Lett.* **1999**, *40*, 7811.

[1408]Gagnier, S.V.; Larock, R.C. *J. Am. Chem. Soc.* **2003**, *125*, 4804.

[1409]Murakami, M.; Itami, K.; Ito, Y. *J. Am. Chem. Soc.* **1999**, *121*, 4130.

using a cobalt catalyst, giving a cyclopentenone[1410] in a reaction related to the Pauson–Khand reaction (see below). The reaction of a conjugated diene having a distal alkene unit and CO with a rhodium catalyst led to a bicyclic conjugated ketone.[1411] When a Stille coupling (**12-15**) is done in a CO atmosphere, conjugated ketones of the type C=C–CO–C=C are formed,[1412] suitable for a Nazarov cyclization (**15-20**). Alkynes were converted to cyclobutenones using $Fe_3(CO)_{12}$ to form the initial complex, followed by reaction with copper(II) chloride.[1413] An interesting variation treated cyclohexene with 5 equivalents of Oxone® and a $RuCl_3$ catalyst to give 2-hydroxycyclohexanone.[1414]

The reaction of dienes, diynes, or en-ynes with transition metals[1415] (usually cobalt)[1416] forms organometallic coordination complexes. In the presence of carbon monoxide, the metal complexes derived primarily from enynes (alkene-alkynes) form cyclopentenone derivatives in what is known as the *Pauson–Khand reaction.*[1417] The reaction involves (*1*) formation of a hexacarbonyldicobalt–alkyne complex and (*2*) decomposition of the complex in the presence of an alkene.[1418] A typical example is formation of **160**.[1419] Cyclopentenones can be prepared by an intermolecular reaction of a vinyl silane and an alkyne using CO and a ruthenium catalyst.[1420] Carbonylation of an alkene–diene using a rhodium catalyst leads to cyclization to an α-vinyl cyclopentanone.[1421] An yne–diene can also be used for the Pauson–Khand reaction.[1422]

160

[1410]Jeong, N.; Hwang, S.H. *Angew. Chem. Int. Ed.* **2000**, *39*, 636.

[1411]Lee, S.I.; Park, J.H.; Chung, Y.K.; Lee, S.-G. *J. Am. Chem. Soc.* **2004**, *126*, 2714.

[1412]Mazzola Jr., R.D.; Giese, S.; Benson, C.L.; West, F.G. *J. Org. Chem.* **2004**, *69*, 220.

[1413]Rameshkumar, C.; Periasamy, M. *Tetrahedron Lett.* **2000**, *41*, 2719.

[1414]Plietker, B. *J. Org. Chem.* **2004**, *69*, 8287.

[1415]For a discussion of catalytic precursors, see Krafft, M.E.; Hirosawa, C.; Bonaga, L.V.R. *Tetrahedron Lett.* **1999**, *40*, 9177.

[1416]For development of practical cobalt catalysts, see Krafft, M.E.; Boñaga, L.V.R.; Hirosawa, c. *J. Org. Chem.* **2001**, *66*, 3004.

[1417]Khand, I.U.; Knox, G.R.; Pauson, P.L.; Watts, W.E.; Foreman, M.I. *J. Chem. Soc. Perkin Trans. 1*, **1973**, 977; Khand, I.U.; Pauson, P.L.; Habib, M.J. *J. Chem. Res. (S)* **1978**, 348; Khand, I.U; Pauson, P.L. *J. Chem. Soc. Perkin Trans. 1*, **1976**, 30. Gibson, S.E.; Stevenazzi, A. *Angew. Chem. Int. Ed.* **2003**, *42*, 1800.

[1418]For a discussion of the reactivity of alkenes, see de Bruin, T.J.M.; Milet, A.; Greene, A.E.; Gimbert, Y. *J. Org. Chem.* **2004**, *69*, 1075. See also, Rivero, M.R.; Adrio, J.; Carretero, J.C. *Eur. J. Org. Chem.* **2002**, 2881.

[1419]Magnus, P.; Principe, L.M. *Tetrahedron Lett.* **1985**, *26*, 4851.

[1420]Itami, K.; Mitsudo, K.; Fujita, K.; Ohashi, Y.; Yoshida, J.-i. *J. Am. Chem. Soc.* **2004**, *126*, 11058.

[1421]Wender, P.A.; Croatt, M.P.; Deschamps, N.M. *J. Am. Chem. Soc.* **2004**, *126*, 5948.

[1422]Wender, P.A; Deschamps, N.M.; Gamber, G.G. *Angew. Chem. Int. Ed.* **2003**, *42*, 1853.

Rhodium,[1423] titanium,[1424] and tungsten[1425] complexes have also been used for this reaction. The reaction can be promoted photochemically[1426] and the rate is enhanced by the presence of primary amines.[1427] Coordinating ligands also accelerate the reaction,[1428] polymer-supported promoters have been developed[1429] and there are many possible variations in reaction conditions.[1430] The Pauson–Khand reaction has been done under heterogeneous reaction conditions,[1431] and with cobalt nanoparticles.[1432] A dendritic cobalt catalyst has been used.[1433] Ultrasound promoted[1434] and microwave promoted[1435] reactions have been developed. Polycyclic compounds (tricyclic and higher) are prepared in a relatively straightforward manner using this reaction.[1436] Asymmetric Pauson–Khand reactions are known.[1437]

The Pauson–Khand reaction is compatible with other groups or heteroatoms elsewhere in the molecule. These include ethers and aryl halides,[1438] esters,[1439]

[1423]Koga, Y.; Kobayashi, T.; Narasaka, K. *Chem. Lett.* **1998**, 249. An entrapped-rhodium catalyst has been used: Park, K.H.; Son, S.U.; Chung, Y.K. *Tetrahedron Lett.* **2003**, *44*, 2827.

[1424]Hicks, F.A.; Buchwald, S.L. *J. Am. Chem. Soc.* **1996**, *118*, 11688; Hicks, F.A.; Kablaoui, N.M.; Buchwald, S.L. *J. Am. Chem. Soc.* **1997**, *118*, 9450; Hicks, F.A.; Kablaoui, N.M.; Buchwald, S.L. *J. Am. Chem. Soc.* **1999**, *121*, 5881.

[1425]Hoye, T.R.; Suriano, J.A. *J. Am. Chem. Soc.* **1993**, *115*, 1154.

[1426]Pagenkopf, B.L.; Livinghouse, T. *J. Am. Chem. Soc.* **1996**, *118*, 2285.

[1427]Sugihara, T.; Yamada, M.; Ban, H.; Yamaguchi, M.; Kaneko, C. *Angew. Chem. Int. Ed.* **1997**, *36*, 2801.

[1428]Krafft, M.E.; Scott, I.L.; Romero, R.H. *Tetrahedron Lett.* **1992**, *33*, 3829.

[1429]Kerr, W.J.; Lindsay, D.M.; McLaughlin, M.; Pauson, P.L. *Chem. Commun.* **2000**, 1467; Brown, D.S.; Campbell, E. Kerr, W.J.; Lindsay, D.M.; Morrison, A.J.; Pike, K.G.; Watson, S.P. *Synlett* **2000**, 1573.

[1430]Krafft, M.E.; Boñaga, L.V.R.; Wright, J.A.; Hirosawa, C. *J. Org. Chem.* **2002**, *67*, 1233; Blanco-Urgoiti, J.; Casarrubios, L.; Domínguez, G.; Pérez-Castells, J. *Tetrahedron Lett.* **2002**, *43*, 5763. The reaction has been done in aqueous media: Krafft, M.E.; Wright, J.A.; Boñaga, L.V.R. *Tetrahedron Lett.* **2003**, *44*, 3417.

[1431]Kim, S.-W.; Son, S.U.; Lee, S.I.; Hyeon, T.; Chung, Y.K. *J. Am. Chem. Soc.* **2000**, *122*, 1550.

[1432]Kim, S.-W.; Son, S.U.; Lee, S.S.; Hyeon, T.; Chung, Y.K. *Chem. Commun.* **2001**, 2212; Son, S.U.; Lee, S.I.; Chung, Y.K.; Kim, S.-W.; Hyeon, T. *Org. Lett.* **2002**, *4*, 277.

[1433]Dahan, A.; Portnoy, M. *Chem. Commun.* **2002**, 2700.

[1434]Ford, J.G.; Kerr, W.J.; Kirk, G.G.; Lindsay, D.M.; Middlemiss, D. *Synlett* **2000**, 1415.

[1435]Iqbal, M.; Vyse, N.; Dauvergne, J.; Evans, P. *Tetrahedron Lett.* **2002**, *43*, 7859.

[1436]Ishizaki, M.; Iwahara, K.; Niimi, Y.; Satoh, H.; Hoshino, O. *Tetrahedron* **2001**, *57*, 2729; Son, S.U.; Chung, Y.K.; Lee, S.-G. *J. Org. Chem.* **2000**, *65*, 6142; Son, S.U.; Yoon, Y.A.; Choi, D.S.; Park, J.K.; Kim, B.M.; Chung, Y.K. *Org. Lett.* **2001**, *3*, 1065; Jung, J.-C.; Jung, Y.-J.; Park, O.-S. *Synth. Commun.* **2001**, *31*, 2507; Pérez-Serrano, L.; Casarrubios, L.; Domínguez, G.; Pérez-Castells, J. *Chem. Commun.*, **2001**, 2602.

[1437]Ingate, S.T.; Marco-Contelles, J. *Org. Prep. Proceed. Int.* **1998**, *30*, 121; Urabe, H.; Hideura, D.; Sato, F. *Org. Lett.* **2000**, *2*, 381; Verdaguer, X.; Moyano, A.; Pericás, M.A.; Riera, A.; Maestro, M.A.; Mahía, J. *J. Am. Chem. Soc.* **2000**, *122*, 10242l; Konya, D.; Robert, F.; Gimbert, Y.; Greene, A.E. *Tetrahedron Lett.* **2004**, *45*, 6975.

[1438]Pérez-Serrano, L.; Banco-Urgoiti, J.; Casarrubios, L.; Domínguez, G.; Pérez-Castells, J. *J. Org. Chem.* **2000**, *65*, 3513. For a review, see Suh. W.H.; Choi, M.; Lee, S.I.; Chung, Y.K. *Synthesis* **2003**, 2169.

[1439]Son, S.U.; Choi, D.S.; Chung, Y.K.; Lee, S.-G. *Org. Lett.* **2000**, *2*, 2097; Krafft, M.E.; Boñaga, L.V.R. *Angew. Chem. Int. Ed.* **2000**, *39*, 3676, and references cited therein; Jeong, N.; Sung, B.S.; Choi, Y.K. *J. Am. Chem. Soc.* **2000**, *122*, 6771; Hayashi, M.; Hashimoto, Y.; Yamamoto, Y.; Usuki, J.; Saigo, K. *Angew. Chem. Int. Ed.* **2000**, *39*, 631; Son, C.U.; Lee, S.I.; Chung, Y.K. *Angew. Chem. Int. Ed.* **2000**, *39*, 4158; Sturla, S.J.; Buchwald, S.L. *J. Org. Chem.* **2002**, *67*, 3398.

amides,[1440] alcohols,[1441] diols,[1442] and an indole unit.[1443] A silicon-tethered Pauson–Khand reaction is known.[1444] Allenes are reaction partners in the Pauson–Khand reaction.[1445] This type of reaction can be extended to form six-membered rings using a ruthenium catalyst.[1446] A double-Pauson–Khand process was reported.[1447] In some cases, an aldehyde can serve as the source of the carbonyl for carbonylation.[1448]

The accepted mechanism was proposed by Magnus,[1449] shown for the formation of **161**,[1450] and supported by Krafft's work.[1451] It has been shown that CO is lost from the Pauson–Khand complex prior to alkene coordination and insertion.[1452] Calculations concluded that the LUMO of the coordinated alkene plays a crucial role in alkene reactivity by determining the degree of back-donation in the complex.[1453]

Other carbonylation methods are available. Carbonylation occurs with conjugated ketones to give 1.4-diketones, using phenylboronic acid (**13-12**), CO and a rhodium catalyst.[1454] A non-carbonylation route treated a conjugated diene with an

[1440]Comely, A.C.; Gibson, S.E.; Stevenazzi, A.; Hales, N.J. *Tetrahedron Lett.* **2001**, *42*, 1183.

[1441]Blanco-Urgoiti, J.; Casarrubios, L.; Domínguez, G.; Pérez-Castells, J. *Tetrahedron Lett.* **2001**, *42*, 3315.

[1442]Mukai, C.; Kim, J.S.; Sonobe, H.; Hanaoka, M. *J. Org. Chem.* **1999**, *64*, 6822.

[1443]Pérez-Serrano, L.; Domínguez, G.; Pérez-Castells, J. *J. Org. Chem.* **2004**, *69*, 5413.

[1444]Brummond, K.M.; Sill, P.C.; Rickards, B.; Geib, S.J. *Tetrahedron Lett.* **2002**, *43*, 3735; Reichwein, J.F.; Iacono, S.T.; Patel, U.C.; Pagenkopf, B.L. *Tetrahedron Lett.* **2002**, *43*, 3739.

[1445]Antras, F.; Ahmar, M.; Cazes, B. *Tetrahedron Lett.* **2001**, *42*, 8153, 8157; Brummond, K.M.; Chen, H.; Fisher, K.D.; Kerekes, A.D.; Rickards, B.; Sill, P.C.; Geib, A.D. *Org. Lett.* **2002**, *4*, 1931. For an intramolecular version, see Shibata, T.; Kadowaki, S.; Hirase, M.; Takagi, K. *Synlett* **2003**, 573.

[1446]Trost, B.M.; Brown, R.E.; Toste, F.D. *J. Am. Chem. Soc.* **2000**, *122*, 5877.

[1447]Rausch, B.J.; Gleiter, R. *Tetrahedron Lett.* **2001**, *42*, 1651.

[1448]For example, see Shibata, T.; Toshida, N.; Takagi, K. *J. Org. Chem.* **2002**, *67*, 7446; Shibata, T.; Toshida, N.; Takagi, K. *Org. Lett.* **2002**, *4*, 1619; Morimoto, T.; Fuji, K.; Tsutsumi, K.; Kakiuchi, K. *J. Am. Chem. Soc.* **2002**, *124*, 3806; Fuji, K.; Morimoto, T.; Tsutsumi, K.; Kakiuchi, K. *Tetrahedron Lett.* **2004**, *45*, 9163.

[1449]Magnus, P.; Principe, L.M. *Tetrahedron Lett.* **1985**, *26*, 4851.

[1450]For a review, see Brummond, K.M.; Kent, J.L. *Tetrahedron* **2000**, *56*, 3263.

[1451]Krafft, M.E. *Tetrahedron Lett.* **1988**, *29*, 999.

[1452]Gimbert, Y.; Lesage, D.; Milet, A.; Fournier, F.; Greene, A.E.; Tabet, J.-C. *Org. Lett.* **2003**, *5*, 4073. See Robert, F.; Milet, A.; Gimbert, Y.; Konya, D.; Greene, A.E. *J. Am. Chem. Soc.* **2001**, *123*, 5396.

[1453]de Bruin, T.J.M.; Milet, A.; Greene, A.E.; Gimbert, Y. *J. Org. Chem.*, **2004** *69*, 1075.

[1454]Sauthier, M.; Castanet, Y.; Mortreux, A. *Chem. Commun.* **2004** 1520.

excess of *tert*-butyllithium and quenching with carbon dioxide led to a cyclopenta-dienone.[1455] When quenched with CO rather than CO_2, a nonconjugated cyclopen-tenone was formed.[1456] It is noted that a carbonylation reaction with CO, a diyne and an iridium catalyst[1457] or a cobalt catalyst[1458] provided similar molecules.

The reaction of a secondary amine, CO, a terminal alkyne and *t*-BuMe$_2$SiH with a rhodium catalyst led to a conjugated amide bearing the silyl group of the C=C unit.[1459] Reaction of a molecule containing an amine and an alkene unit was car-boxylated with CO in the presence of a palladium catalyst to give a lactam.[1460] A similar reaction with a molecule containing an amine and an alkyne also generated a lactam, in the presence of CO and a rhodium catalyst.[1461] An intramolecular carbonylation reaction of a conjugated imine, with CO, ethylene and a ruthenium catalyst, led to a highly substituted β,γ-unsaturated lactam.[1462]

With any method, if the alkene contains a functional group, such as OH, NH$_2$, or CONH$_2$, the corresponding lactone (**16-63**),[1463] lactam (**16-74**), or cyclic imide may be the product.[1464] Titanium,[1465] palladium,[1466] ruthenium,[1467] and rho-dium[1468] catalysts have been used to generate lactones. Allenic alcohols are con-verted to butenolides with 10 atm of CO and a ruthenium catalyst.[1469] Larger ring conjugated lactones can also be formed by this route using the appropriate alle-nic alcohol.[1470] Propargylic alcohols lead to β-lactones.[1471] Allenic tosyl-amides are converted to *N*-tosyl α,β-unsaturated pyrrolidinones using 20 atm of CO and a ruthenium catalyst.[1472] Conjugated imines are converted to similar products with CO, ethylene and a ruthenium catalyst.[1473] Propargyl alcohols are converted to

[1455]Xi, Z.; Song, Q. *J. Org. Chem.* **2000**, *65*, 9157.

[1456]Song, Q.; Chen, J.; Jin, X.; Xi, Z. *J. Am. Chem. Soc.* **2001**, *123*, 10419; Song, Q.; Li, Z.; Chen, J.; Wang, C.; Xi, Z. *Org. Lett.* **2002**, *4*, 4627.

[1457]Shibata, T.; Yamashita, K.; Ishida, H.; Takagi, K. *Org. Lett.* **2001**, *3*, 1217; Shibata, T.; Yamashita, K.; Katayama, E.; Takagi, K. *Tetrahedron* **2002**, *58*, 8661.

[1458]Sugihara, T.; Wakabayashi, A.; Takao, H.; Imagawa, H.; Nishizawa, M. *Chem. Commun.* **2001**, 2456.

[1459]Matsuda, I.; Takeuchi, K.; Itoh, K. *Tetrahedron Lett.* **1999**, *40*, 2553.

[1460]Okuro, K.; Kai, H.; Alper, H. *Tetrahedron Asymmetry* **1997**, *8*, 2307.

[1461]Hirao, K.; Morii, N.; Joh, T.; Takahashi, S. *Tetrahedron Lett.* **1995**, *36*, 6243; Shiba, T.; Zhou, D.-Y.; Onitsuka, K.; Takahashi, S. *Tetrahedron Lett.* **2004**, *45*, 3211.

[1462]Berger, D.; Imhof, W. *Tetrahedron* **2000**, *56*, 2015.

[1463]Dong, C.; Alper, H. *J. Org. Chem.* **2004**, *69*, 5011.

[1464]For reviews of these ring closures see Ohshiro, Y.; Hirao, T. *Heterocycles* **1984**, *22*, 859; Falbe, J. *New Syntheses with Carbon Monoxide*, Springer, NY, **1980**, pp. 147–174, *Angew. Chem. Int. Ed.* **1966**, *5*, 435; *Newer Methods Prep. Org. Chem.* **1971**, *6*, 193. See also, Krafft, M.E.; Wilson, L.J.; Onan, K.D. *Tetrahedron Lett.* **1989**, *30*, 539.

[1465]Kablaoui, N.M.; Hicks, F.A.; Buchwald, S.L. *J. Am. Chem. Soc.* **1997**, *119*, 4424.

[1466]El Ali, B.; Okuro, K.; Vasapollo, G.; Alper, H. *J. Am. Chem. Soc.* **1996**, *118*, 4264. Also see, Brunner, M.; Alper, H. *J. Org. Chem.* **1997**, *62*, 7565.

[1467]Kondo, T.; Kodoi, K.; Mitsudo, T.-a.; Watanabe, Y. *J. Chem. Soc., Chem. Commun.* **1994**, 755.

[1468]Yoneda, E.; Kaneko, T.; Zhang, S.-W.; Takahashi, S. *Tetrahedron Lett.* **1998**, *39*, 5061.

[1469]Yoneda, E.; Kaneko, T.; Zhang, S.-W.; Onitsuka, K.; Takahashi, S. *Org. Lett.* **2000**, *2*, 441.

[1470]Yoneda, E.; Zhang, S.-W.; Onitsuka, K.; Takahashi, S. *Tetrahedron Lett.* **2001**, *42*, 5459.

[1471]Ma, S.; Wu, B.; Zhao, S. *Org. Lett.* **2003**, *5*, 4429.

[1472]Kang, S.-K.; Kim, K.-J.; Yu, C.-M.; Hwang, J.-W.; Do, Y.-K. *Org. Lett.* **2001**, *3*, 2851.

[1473]Chatani, N.; Kamitani, A.; Murai, S. *J. Org. Chem.* **2002**, *67*, 7014.

butenolides with CO/H_2O and a rhodium catalyst.[1474] Propargyl alcohols generate lactones when treated with a chromium pentacarbonyl carbene complex.[1475] Amines add to allenes, in the presence of CO and a palladium catalyst, to form conjugated amides.[1476]

15-37 Hydroformylation

Hydro-formyl-addition

$$\ce{\underset{/}{\overset{\backslash}{C}}=\underset{\backslash}{\overset{/}{C}} + CO + H_2 ->[{[Co(CO)_4]_2}][pressure] H-\underset{/}{\overset{\backslash}{C}}-\underset{\backslash}{\overset{/}{C}}-CHO}$$

Alkenes can be hydroformylated[1477] by treatment with carbon monoxide and hydrogen over a catalyst. The most common catalysts are cobalt carbonyls (see below for a description of the mechanism) and rhodium complexes,[1478] but other transition metal compounds have also been used. Cobalt catalysts are less active than the rhodium type, and catalysts of other metals are generally less active.[1479] Commercially, this is called the *oxo process*, but it can be carried out in the laboratory in an ordinary hydrogenation apparatus. The order of reactivity is straight-chain terminal alkenes > straight-chain internal alkenes > branched-chain alkenes. With terminal alkenes, for example, the aldehyde unit is formed on both the primary and secondary carbon, but proper choice of catalyst and additive leads to selectivity for the secondary product[1480] or primary

---

[1474]Fukuta, Y.; Matsuda, I.; Itoh, K. *Tetrahedron Lett.* **2001**, *42*, 1301.

[1475]Good, G.M.; Kemp, M.I.; Kerr, W.J. *Tetrahedron Lett.* **2000**, *41*, 9323.

[1476]Grigg, R.; Monteith, M.; Sridharan, V.; Terrier, C. *Tetrahedron* **1998**, *54*, 3885.

[1477]For reviews, see Kalck, P.; Peres, Y.; Jenck, J. *Adv. Organomet. Chem.* **1991**, *32*, 121; Davies, J.A., in Hartley, F.R.; Patai, S. *The Chemistry of the Metal–Carbon Bond*, Vol. 3, Wiley, NY, **1985**, pp. 361–389; Pino, P.; Piacenti, F.; Bianchi, M., in Wender, I.; Pino, P. *Organic Syntheses via Metal Carbonyls*, Vol. 2, Wiley, NY, **1977**, pp. 43–231; Cornils, B., in Falbe, J. *New Syntheses with Carbon Monoxide*, Springer, NY, **1980**, pp. 1–225; Collman, J.P.; Hegedus, L.S.; Norton, J.R.; Finke, R.G. *Principles and Applications of Organotransition Metal Chemistry*, University Science Books, Mill Valley, CA **1987**, pp. 621–632; Pino, P. *J. Organomet. Chem.* **1980**, *200*, 223; Pruett, R.L. *Adv. Organomet. Chem.* **1979**, *17*, 1; Stille, J.K.; James, D.E., in Patai, S. *Supplement A: The Chemistry of Double-Bonded Functional Groups*, Vol. 1, pt. 2, Wiley, NY, **1977**, pp. 1099–1166; Heck, R.F. *Organotransition Metal Chemistry*, Academic Press, NY, **1974**, pp. 215–224; Khan, M.M.T.; Martell, A.E. *Homogeneous Catalysis by Metal Complexes*, Vol. 2, Academic Press, NY, **1974**, pp. 39–60; Falbe, J. *Carbon Monoxide in Organic Synthesis* Springer, NY, **1980**, pp. 3–77; Chalk, A.J.; Harrod, J.F. *Adv. Organomet. Chem.* **1968**, *6*, 119. For a review with respect to fluoroalkenes, see Ohshiro, Y.; Hirao, T. *Heterocycles* **1984**, *22*, 859.

[1478]For example, see Brown, J.M.; Kent, A.G. *J. Chem. Soc. Perkin Trans. 1*, **1987**, 1597; Hanson, B.E.; Davis, M.E. *J. Chem. Ed.*, **1987**, *64*, 928; Jackson, W.R.; Perlmutter, P.; Suh, G. *J. Chem. Soc., Chem. Commun.* **1987**, 724; Amer, I.; Alper, H. *J. Am. Chem. Soc.* **1990**, *112*, 3674. See the references cited in these papers. For a review of the rhodium-catalyzed process, see Jardine, F.H., in Hartley, F.R. *The Chemistry of the Metal–Carbon Bond*, Vol. 4, Wiley, NY, **1987**, pp. 733–818, pp. 778–784.

[1479]Collman, J.P.; Hegedus, L.S.; Norton, J.R.; Finke, R.G. *Principles and Applications of Organotransition Metal Chemistry*, University Science Books, Mill Valley, CA **1987**, p. 630.

[1480]Chan, A.S.C.; Pai, C.-C.; Yang, T.-K.; Chen, S.M. *J. Chem. Soc., Chem. Commun.* **1995**, 2031; Doyle, M.P.; Shanklin, M.S.; Zlokazov, M.V. *Synlett* **1994**, 615; Higashizima, T.; Sakai, N.; Nozaki, K.; Takaya, H. *Tetrahedron Lett.* **1994**, *35*, 2023.

product.[1481] Good yields for hydroformylation have been reported using rhodium catalysts in the presence of certain other additives.[1482] Among the side reactions are the aldol reaction (**16-34**), acetal formation, the Tishchenko reaction (**19-82**), and polymerization. In one case using a rhodium catalyst, 2-octene gave nonanal, presumably via a η^3-allyl complex (p. 116).[1483] Conjugated dienes give dialdehydes when rhodium catalysts are used[1484] but saturated mono-aldehydes (the second double bond is reduced) with cobalt carbonyls. Both 1,4- and 1,5-dienes may give cyclic ketones.[1485] Hydroformylation of triple bonds proceeds very slowly, and few examples have been reported.[1486] However, in the presence of a rhodium catalyst, the triple bond of a conjugated enyne is formylated.[1487] Many functional groups such as OH, CHO, COOR,[1488] CN, can be present in the molecule, although halogens usually interfere. Stereoselective syn addition has been reported,[1489] and also stereoselective anti addition.[1490] Asymmetric hydroformylation has been accomplished with a chiral catalyst,[1491] and in the presence of chiral additives.[1492] Cyclization to prolinal derivatives has been reported with allylic amines.[1493]

When dicobalt octacarbonyl, $[Co(CO)_4]_2$, is the catalyst, the species that actually adds to the double bond is tricarbonylhydrocobalt, $HCo(CO)_3$.[1494] Carbonylation $RCo(CO)_3 + CO \rightarrow RCo(CO)_4$ takes place, followed by a rearrangement and a

[1481]Fernández, E.; Castillón, S. *Tetrahedron Lett.* **1994**, *35*, 2361; Klein, H.; Jackstell, R.; Wiese, K.-D.; Borgmann, C.; Beller, M. *Angew. Chem. Int. Ed.* **2001**, *40*, 3408; Breit, B.; Seiche, W. *J. Am. Chem. Soc.* **2003**, *125*, 6608.

[1482]Johnson, J.R.; Cuny, G.D.; Buchwald, S.L. *Angew. Chem. Int. Ed.* **1995**, *34*, 1760.

[1483]van der Veen, L.A.; Kamer, P.C.J.; van Leeuwen, P.W.N.M. *Angew. Chem. Int. Ed.* **1999**, *38*, 336.

[1484]Fell, B.; Rupilius, W. *Tetrahedron Lett.* **1969**, 2721.

[1485]For a review of ring closure reactions with CO, see Mullen, A., in Falbe, J. *New Syntheses with Carbon Monoxide*, Springer, NY, **1980**, pp. 414–439. See also, Eilbracht, P.; Hüttmann, G.; Deussen, R. *Chem. Ber.* **1990**, *123*, 1063, and other papers in this series.

[1486]For examples with rhodium catalysts, see Fell, B.; Beutler, M. *Tetrahedron Lett.* **1972**, 3455; Botteghi, C.; Salomon, C. *Tetrahedron Lett.* **1974**, 4285. For an indirect method, see Campi, E.; Fitzmaurice, N.J.; Jackson, W.R.; Perlmutter, P.; Smallridge, A.J. *Synthesis* **1987**, 1032.

[1487]van den Hoven, B.G.; Alper, H. *J. Org. Chem.* **1999**, *64*, 3964.

[1488]For formylation at the β-carbon of methyl acrylate, see Hu, Y.; Chen, W.; Osuna, A.M.B.; Stuart, A.M.; Hope, E.G.; Xiao, J. *Chem. Commun.* **2001**, 725.

[1489]See, for example, Haelg, P.; Consiglio, G.; Pino, P. *Helv. Chim. Acta* **1981**, *64*, 1865.

[1490]Krauss, I.J.; Wang, C.C-Y.; Leighton, J.L. *J. Am. Chem. Soc.* **2001**, *123*, 11514.

[1491]For reviews, see Ojima, I.; Hirai, K., in Morrison, J.D. *Organic Synthesis*, Vol. 5, Wiley, NY, **1985**, pp. 103–145, 125–139; Consiglio, G.; Pino, P. *Top. Curr. Chem.* **1982**, *105*, 77; Breit, B.; Seiche, W. *Synthesis* **2001**, 1; Diéguez, M.; Pàmies, O.; Claver, C. *Tetrahedron Asymmetry* **2004**, *15*, 2113. See also, Hegedüs, C.; Madrász, J.; Gulyás, H.; Szöllöy, A.; Bakos, J. *Tetrahedron Asymmetry* **2001**, *12*, 2867.

[1492]Nozaki, K.; Itoi, Y.; Shibahara, F.; Shirakawa, E.; Ohta, T.; Takaya, H.; Hiyama, T. *J. Am. Chem. Soc.* **1998**, *120*, 4051; Sakai, N.; Nozaki, K.; Takaya, H. *J. Chem. Soc., Chem. Commun.* **1994**, 395; Rajan Babu, T.V.; Ayers, T.A. *Tetrahedron Lett.* **1994**, *35*, 4295. See Gladiali, S.; Bayón, J.C.; Claver, C. *Tetrahedron Asymmetry* **1995**, *6*, 1453.

[1493]Anastasiou, D.; Campi, E.M.; Chaouk, H.; Jackson, W.R.; McCubbin, Q.J. *Tetrahedron Lett.* **1992**, *33*, 2211.

[1494]Heck, R.F.; Breslow, D.S. *J. Am. Chem. Soc.* **1961**, *83*, 4023; Karapinka, G.L.; Orchin, M. *J. Org. Chem.* **1961**, *26*, 4187; Whyman, R. *J. Organomet. Chem.* **1974**, *81*, 97; Mirbach, M.F. *J. Organomet. Chem.* **1984**, *265*, 205. For discussions of the mechanism, see Orchin, M. *Acc. Chem. Res.* **1981**, *14*, 259; Versluis, L.; Ziegler, T.; Baerends, E.J.; Ravenek, W. *J. Am. Chem. Soc.* **1989**, *111*, 2018.

reduction of the C–Co bond, similar to steps 4 and 5 of the nickel carbonyl mechanism shown in **15-35**. The reducing agent in the reduction step is tetra-carbonylhydrocobalt, $HCo(CO)_4$,[1495] or, under some conditions, H_2.[1496] When $HCo(CO)_4$ was the agent used to hydroformylate styrene, the observation of CIDNP indicated that the mechanism is different, and involves free radicals.[1497] Alcohols can be obtained by allowing the reduction to continue after all the carbon monoxide is used up. It has been shown[1498] that the formation of alcohols is a second step, occurring after the formation of aldehydes, and that $HCo(CO)_3$ is the reducing agent.

OS **VI**, 338.

15-38 Addition of HCN

Hydro-cyano-addition

$$\underset{/}{\overset{\backslash}{C}}=\underset{\backslash}{\overset{/}{C}} \quad + \quad HCN \quad \longrightarrow \quad H-\underset{/}{\overset{\backslash}{C}}-\underset{\backslash}{\overset{/}{C}}-CN$$

Ordinary alkenes do not react with HCN, but polyhalo alkenes and alkenes of the form C=C–Z add HCN to give nitriles.[1499] The reaction is therefore a nucleophilic addition and is base catalyzed. When Z is COR or, more especially, CHO, 1,2-addition (**16-53**) is an important competing reaction and may be the only reaction. Triple bonds react very well when catalyzed by an aqueous solution of CuCl, NH_4Cl, and HCl or by Ni or Pd compounds.[1500] The HCN can be generated *in situ* from acetone cyanohydrin (see **16-52**), avoiding the use of the poisonous HCN.[1501] One or 2 equivalents of HCN can be added to a triple bond, since the initial product is a Michael-type substrate. Acrylonitrile is commercially prepared this way, by the addition of HCN to acetylene. Alkylaluminum cyanides, for example, Et_2AlCN, or mixtures of HCN and trialkylalanes R_3Al are especially good reagents for conjugate addition of HCN[1502] to α,β-unsaturated ketones and α,β-unsaturated acyl halides. Hydrogen cyanide can be added to ordinary alkenes in the presence of dicobalt octacarbonyl[1503] or certain other

[1495]Alemdaroğ lu, N.H.; Penninger, J.L.M.; Oltay, E. *Monatsh. Chem.* **1976**, *107*, 1153; Ungváry, F.; Markó, L. *Organometallics* **1982**, *1*, 1120.

[1496]See Kovács, I.; Ungváry, F.; Markó, L. *Organometallics* **1986**, *5*, 209.

[1497]Bockman, T.M.; Garst, J.F.; King, R.B.; Markó, L.; Ungváry, F. *J. Organomet. Chem.* **1985**, *279*, 165.

[1498]Aldridge, C.L.; Jonassen, H.B. *J. Am. Chem. Soc.* **1963**, *85*, 886.

[1499]For reviews see Friedrich, K., in Patai, S.; Rappoport, Z. *The Chemistry of Functional Groups, Supplement C*, pt. 2, Wiley, NY, **1983**, pp. 1345–1390; Nagata, W.; Yoshioka, M. *Org. React.* **1977**, *25*, 255; Brown, E.S., in Wender, I.; Pino, P. *Organic Syntheses via Metal Carbonyls*, Vol. 2, Wiley, NY, **1977**, pp. 655–672; Friedrich, K.; Wallenfels, K., in Rappoport, Z. *The Chemistry of the Cyano Group*, Wiley, NY, **1970**, pp. 68–72.

[1500]Jackson, W.R.; Lovel, C.G. *Aust. J. Chem.* **1983**, *36*, 1975.

[1501]Jackson, W.R.; Perlmutter, P. *Chem. Br.* **1986**, 338.

[1502]For a review, see Nagata, W.; Yoshioka, M. *Org. React.* **1977**, *25*, 255.

[1503]Arthur, Jr., P.; England, D.C.; Pratt, B.C.; Whitman, G.M. *J. Am. Chem. Soc.* **1954**, *76*, 5364.

transition-metal compounds.[1504] An indirect method for the addition of HCN to ordinary alkenes uses an isocyanide (RNC) and Schwartz's reagent (see **15-17**); this method gives anti-Markovnikov addition.[1505] *tert*-Butyl isocyanide and TiCl$_4$ have been used to add HCN to C=C–Z alkenes.[1506] Pretreatment with NaI/Me$_3$SiCl followed by CuCN converts alkynes to vinyl nitriles.[1507]

When an alkene is treated with Me$_3$SiCN and AgClO$_4$, followed by aq. NaHCO$_3$, the product is the isonitrile (RNC) formed with Markovnikov selectivity.[1508] An alternative reagent is the cyanohydrin of acetone, which adds to alkenes to give a nitrile in the presence of a nickel complex.[1509]

OS **I**, 451; **II**, 498; **III**, 615; **IV**, 392, 393, 804; **V**, 239, 572; **VI**, 14. For addition of ArH, see **11-12** (Friedel–Crafts alkylation).

REACTIONS IN WHICH HYDROGEN ADDS TO NEITHER SIDE

Some of these reactions are *cycloadditions* (reactions **15-50**, **15-62**, **15-54**, and **15-57–15-66**). In such cases, addition to the multiple bond closes a ring:

$$\text{C=C} \;+\; W \frown Y \longrightarrow \overset{W}{\underset{}{\text{C}}}\!-\!\overset{Y}{\underset{}{\text{C}}}$$

A. Halogen on One or Both Sides

15-39 Halogenation of Double and Triple Bonds (Addition of Halogen, Halogen)

Dihalo-Addition

$$\text{C=C} \;+\; Br_2 \longrightarrow \overset{Br}{\underset{}{\text{C}}}\!-\!\overset{Br}{\underset{}{\text{C}}}$$

[1504]For a review, see Brown, E.S., in Wender, P.; Pino, P. *Organic Syntheses via Metal Carbonyls*, Vol. 2, Wiley, NY, *1977*, pp. 658–667. For a review of the nickel-catalyzed process, see Tolman, C.A.; McKinney, R.J.; Seidel, W.C.; Druliner, J.D.; Stevens, W.R. *Adv. Catal.* *1985*, *33*, 1. For studies of the mechanism see Tolman, C.A.; Seidel, W.C.; Druliner, J.D.; Domaille, P.J. *Organometallics* *1984*, *3*, 33; Druliner, J.D. *Organometallics* *1984*, *3*, 205; Bäckvall, J.E.; Andell, O.S. *Organometallics* *1986*, *5*, 2350; McKinney, R.J.; Roe, D.C. *J. Am. Chem. Soc.* *1986*, *108*, 5167; Funabiki, T.; Tatsami, K.; Yoshida, S. *J. Organomet. Chem.* *1990*, *384*, 199. See also, Jackson, W.R.; Lovel, C.G.; Perlmutter, P.; Smallridge, A.J. *Aust. J. Chem.* *1988*, *41*, 1099.

[1505]Buchwald, S.L.; LeMaire, S.J. *Tetrahedron Lett.* *1987*, *28*, 295.

[1506]Ito, Y.; Kato, H.; Imai, H.; Saegusa, T. *J. Am. Chem. Soc.* *1982*, *104*, 6449.

[1507]Luo, F.-T.; Ko, S.-L.; Chao, D.-Y. *Tetrahedron Lett.* *1997*, *38*, 8061.

[1508]Kitano, Y.; Chiba, K.; Tada, M. *Synlett* *1999*, 288.

[1509]Yan, M.; Xu, Q.-Y.; Chan, A.S.C. *Tetrahedron Asymmetry* *2000*, *11*, 845.

Most double bonds are easily halogenated[1510] with bromine, chlorine, or inter-halogen compounds.[1511] Substitution can compete with addition in some cases.[1512] Iodination has also been accomplished, but the reaction is slower.[1513] Under free-radical conditions, iodination proceeds more easily.[1514] However, *vic*-diiodides are generally unstable and tend to revert to iodine and the alkene.

162

The mechanism is usually electrophilic (see p. 1002), involving formation of an halonium ion (**162**),[1515] followed by nucleophilic opening to give the *vic*-dihalide. Nucleophilic attack is occurs with selectivity for the less substituted carbon with unsymmetrical alkenes. When free-radical initiators (or UV light) are present, addition can occur by a free-radical mechanism.[1516] Once Br• or Cl• radicals are formed, however, substitution may compete (**14-1** and **14-3**). This is especially important when the alkene has allylic hydrogens. Under free-radical conditions (UV light) bromine or chlorine adds to the benzene ring to give, respectively, hexabromo- and hexachlorocyclohexane. These are mixtures of stereoisomers (see p. 187).[1517]

Under ordinary conditions fluorine itself is too reactive to give simple addition; it attacks other bonds and mixtures are obtained.[1518] However, F_2 has been successfully added to certain double bonds in an inert solvent at low temperatures ($-78°C$), usually by diluting the F_2 gas with Ar or N_2.[1519] Addition of fluorine has also been accomplished with other reagents (e.g., *p*-Tol—IF$_2$/Et$_3$N•5 HF),[1520] and a mixture of PbO_2 and SF_4.[1521]

[1510]For a list of reagents that have been used for di-halo-addition, with references, see Larock, R.C. *Comprehensive Organic Transformations*, 2nd ed., Wiley-VCH, NY, *1999*, pp. 629–632.

[1511]For a monograph, see de la Mare, P.B.D. *Electrophilic Halogenation*, Cambridge University Press, Cambridge, *1976*. For a review, see House, H.O. *Modern Synthetic Reaction*, 2nd ed., W.A. Benjamin, NY, *1972*, pp. 422–431.

[1512]McMillen, D.W.; Grutzner, J.B. *J. Org. Chem. 1994*, *59*, 4516.

[1513]Sumrell, G.; Wyman, B.M.; Howell, R.G.; Harvey, M.C. *Can. J. Chem. 1964*, *42*, 2710; Zanger, M.; Rabinowitz, J.L. *J. Org. Chem. 1975*, *40*, 248.

[1514]Skell, P.S.; Pavlis, R.R. *J. Am. Chem. Soc. 1964*, *86*, 2956; Ayres, R.L.; Michejda, C.J.; Rack, E.P. *J. Am. Chem. Soc. 1971*, *93*, 1389.

[1515]See Lenoir, D.; Chiappe, C. *Chem. Eur. J. 2003*, *9*, 1037.

[1516]For example, see Poutsma, M.L. *J. Am. Chem. Soc. 1965*, *87*, 2161, 2172; *J. Org. Chem. 1966*, *31*, 4167; Dessau, R.M. *J. Am. Chem. Soc. 1979*, *101*, 1344.

[1517]For a review, see Cais, M., in Patai, S. *The Chemistry of Alkenes*, Vol. 1, Wiley, NY, *1964*, p. 993.

[1518]See, for example, Fuller, G.; Stacey, F.W.; Tatlow, J.C.; Thomas, C.R. *Tetrahedron 1962*, *18*, 123.

[1519]Merritt, R.F. *J. Am. Chem. Soc. 1967*, *89*, 609; Barton, D.H.R.; Lister-James, J.; Hesse, R.H.; Pechet, M.M.; Rozen, S. *J. Chem. Soc. Perkin Trans. 1 1982*, 1105; Rozen, S.; Brand, M. *J. Org. Chem. 1986*, *51*, 3607.

[1520]Hara, S.; Nakahigashi, J.; Ishi-i, K.; Sawaguchi, M.; Sakai, H.; Fukuhara, T.; Yoneda, N. *Synlett 1998*, 495.

[1521]Bissell, E.R.; Fields, D.B. *J. Org. Chem. 1964*, *29*, 1591.

The reaction with bromine is very rapid and is easily carried out at room temperature,[1522] although the reaction is reversible under some conditions.[1523] In the case of bromine, an alkene•Br_2 complex has been detected in at least one case.[1524] Bromine is often used as a test, qualitative or quantitative, for unsaturation.[1525] The vast majority of double bonds can be successfully brominated. Even when aldehyde, ketone, amine, and so on functions are present in the molecule, they do not interfere, since the reaction with double bonds is faster. Bromination has been carried out in an ionic liquid.[1526]

Several other reagents add Cl_2 to double bonds, among them Me_3SiCl^- MnO_2,[1527] $NaClO_2$/Mn(acac)$_2$/moist Al_2O_3,[1528] $BnNEt_3MnO_4$/Me_3SiCl,[1529] and $KMnO_4$-oxalyl chloride.[1530] A convenient reagent for the addition of Br_2 to a double bond on a small scale is the commercially available pyridinium bromide perbromide $C_5H_5NH^+Br_3^-$.[1531] Potassium bromide with ceric ammonium nitrate, in water/dichloromethane, gives the dibromide.[1532] A combination of KBr and Selectfluor also give the dibromide.[1533] A combination of $CuBr_2$ in aq. THF and a chiral ligand led to the dibromide with good enantioselectivity.[1534] A mixture of (decyl)Me_3 $NMnO_4$ and Me_3SiBr is also an effective reagent.[1535] Either Br_2 or Cl_2 can also be added with $CuBr_2$ or $CuCl_2$ in the presence of a compound, such as acetonitrile, methanol, or triphenylphosphine.[1536]

[1522]See Bellucci, G.; Chiappe, C. *J. Org. Chem.* **1993**, *58*, 7120 for a study of the rate and kinetics of alkene bromination.
[1523]Zheng, C.Y.; Slebocka-Tilk, H.; Nagorski, R.W.; Alvarado, L.; Brown, R.S. *J. Org. Chem.* **1993**, *58*, 2122.
[1524]Bellucci, G.; Chiappe, C.; Bianchini, R.; Lenoir, D.; Herges, R. *J. Am. Chem. Soc.* **1995**, *117*, 12001.
[1525]For a review of this, see Kuchar, E.J., in Patai, S. *The Chemistry of Alkenes*, Vol. 1, Wiley, NY, **1964**, pp. 273–280.
[1526]In bmim Br, 1-butyl-3-methylimidazolium bromide: Chiappe, C.; Capraro, D.; Conte, V.; Picraccini, D. *Org. Lett.* **2001**, *3*, 1061.
[1527]Bellesia, F.; Ghelfi, F.; Pagnoni, U.M.; Pinetti, A. *J. Chem. Res. (S)* **1989**, *108*, 360.
[1528]Yakabe, S.; Hirano, M.; Morimoto, T. *Synth. Comun.* **1998**, *28*, 1871.
[1529]Markó, I.E.; Richardson, P.R.; Bailey, M.; Maguire, A.R.; Coughlan, N. *Tetrahedron Lett.* **1997**, *38*, 2339.
[1530]Markó, I.E.; Richardson, P.F. *Tetrahedron Lett.* **1991**, *32*, 1831.
[1531]Fieser, L.F.; Fieser, M. *Reagents for Organic Synthesis*, Vol. 1, Wiley, NY, **1967**, pp. 967–970. For a discussion of the mechanism with Br_3^-, see Bellucci, G.; Bianchini, R.; Vecchiani, S. *J. Org. Chem.* **1986**, *51*, 4224.
[1532]Nair, V.; Panicker, S.B.; Augstine, A.; George, T.G.; Thomas, S.; Vairamani, M. *Tetrahedron* **2001**, *57*, 7417.
[1533]Ye, C.; Shreeve, J.M. *J. Org. Chem.* **2004**, *69*, 8561.
[1534]El-Quisairi, A.K.; Qaseer, H.A.; Katsigras, G.; Lorenzi, P.; Tribedi, U.; Tracz, S.; Hartman, A.; Miller, J.A.; Henry, P.M. *Org. Lett.* **2003**, *5*, 439.
[1535]Hazra, B.G.; Chordia, M.D.; Bahule, B.B.; Pore, V.S.; Basu, S. *J. Chem. Soc. Perkin Trans. 1* **1994**, 1667.
[1536]Koyano, T. *Bull. Chem. Soc. Jpn.* **1970**, *43*, 1439, 3501; Uemura, S.; Tabata, A.; Kimura, Y.; Ichikawa, K. *Bull. Chem. Soc. Jpn.* **1971**, *44*, 1973; Or, A.; Levy, M.; Asscher, M.; Vofsi, D. *J. Chem. Soc. Perkin Trans. 2* **1974**, 857; Uemura, S.; Okazaki, H.; Onoe, A.; Okano, M. *J. Chem. Soc. Perkin Trans. 1* **1977**, 676; Baird, Jr., W.C.; Surridge, J.H.; Buza, M. *J. Org. Chem.* **191**, *36*, 2088, 3324.

Mixed halogenations have also been achieved, and the order of activity for some of the reagents is BrCl > ICl[1537] > Br$_2$ > IBr > I$_2$.[1538] Mixtures of Br$_2$ and Cl$_2$ have been used to give bromochlorination,[1539] as has tetrabutylammonium dichlorobromate, Bu$_4$NBrCl$_2$;[1540] iodochlorination has been achieved with KICl$_2$,[1541] CuCl$_2$, and either I$_2$, HI, or CdI$_2$; iodofluorination[1542] with mixtures of AgF and I$_2$;[1543] and mixtures of *N*-bromo amides in anhydrous HF give bromofluorination.[1544] Bromo-, iodo-, and chlorofluorination have also been achieved by treatment of the substrate with a solution of Br$_2$, I$_2$, or an *N*-halo amide in polyhydrogen fluoride-pyridine;[1545] while addition of I along with Br, Cl, or F has been accomplished with the reagent bis(pyridine)iodo(I) tetrafluoroborate I(Py)$_2$BF$_4$ and Br$^-$, Cl$^-$, or F$^-$, respectively.[1546] This reaction (which is also successful for triple bonds[1547]) can be extended to addition of I and other nucleophiles (e.g., NCO, OH, OAc, and NO$_2$).[1547] Cyclohexene is converted to *trans*-2-fluoroiodocyclohexane under electrolytic conditions using Et$_4$NI—Et$_3$N•HF in the reaction medium.[1548]

Conjugated systems give both 1,2- and 1,4-addition.[1518] Triple bonds add bromine, although generally more slowly than double bonds (see p. 1015). Molecules that contain both double and triple bonds are preferentially attacked at the double bond. Addition of 2 equivalents of bromine to triple bonds gives tetrabromo products. There is evidence that the addition of the first mole of bromine to a triple bond may take place by a nucleophilic mechanism.[1549] Molecular diiodine on Al$_2$O$_3$ adds to triple bonds to give good yields of 1,2-diiodoalkenes.[1550] Interestingly, 1,1-diiodo alkenes are prepared from an alkynyltin compound, via initial treatment with Cp$_2$Zr(H)Cl, and then 2.15 equivalents of iodine.[1551] A mixture of

[1537]For a review of ICl, see McCleland, C.W., in Pizey, J.S. *Synthetic Reagents*, Vol. 5, Wiley, NY, *1983*, pp. 85–164.

[1538]White, E.P.; Robertson, P.W. *J. Chem. Soc.* *1939*, 1509.

[1539]Buckles, R.E.; Forrester, J.L.; Burham, R.L.; McGee, T.W. *J. Org. Chem.* *1960*, 25, 24.

[1540]Negoro, T.; Ikeda, Y. *Bull. Chem. Soc. Jpn.* *1986*, 59, 3519.

[1541]Zefirov, N.S.; Sereda, G.A.; Sosounk, S.E.; Zyk, N.V.; Likhomanova, T.I. *Synthesis* *1995*, 1359.

[1542]For a review of mixed halogenations where one side is fluorine, see Sharts, C.M.; Sheppard, W.A. *Org. React.* *1974*, 21, 125, see pp. 137–157. For a review of halogen fluorides in organic synthesis, see Boguslavskaya, L.S. *Russ. Chem. Rev.* *1984*, 53, 1178.

[1543]Evans, R.D.; Schauble, J.H. *Synthesis* *1987*, 551; Kuroboshi, M.; Hiyama, T. *Synlett* *1991*, 185.

[1544]Pattison, F.L.M.; Peters, D.A.V.; Dean, F.H. *Can. J. Chem.* *1965*, 43, 1689. For other methods, see Boguslavskaya, L.S.; Chuvatkin, N.N.; Kartashov, A.V.; Ternovskoi, L.A. *J. Org. Chem. USSR* *1987*, 23, 230; Shimizu, M.; Nakahara, Y.; Yoshioka, H. *J. Chem. Soc., Chem. Commun.* *1989*, 1881.

[1545]Olah, G.A.; Nojima, M.; Kerekes, I. *Synthesis* *1973*, 780; Olah, G.A.; Welch, J.T.; Vankar, Y.D.; Nojima, M.; Kerekes, I.; Olah, J.A. *J. Org. Chem.* *1979*, 44, 3872. For other halofluorination methods, see Rozen, S.; Brand, M. *J. Org. Chem.* *1985*, 50, 3342; *1986*, 51, 222; Alvernhe, G.; Laurent, A.; Haufe, G. *Synthesis* *1987*, 562; Camps, F.; Chamorro, E.; Gasol, V.; Guerrero, A. *J. Org. Chem.* *1989*, 54, 4294; Ichihara, J.; Funabiki, K.; Hanafusa, T. *Tetrahedron Lett.* *1990*, 31, 3167.

[1546]Barluenga, J.; González, J.M.; Campos, P.J.; Asensio, G. *Angew. Chem. Int. Ed.* *1985*, 24, 319.

[1547]Barluenga, J.; Rodríguez, M.A.; González, J.M.; Campos, P.J.; Asensio, G. *Tetrahedron Lett.* *1986*, 27, 3303.

[1548]Kobayashi, S.; Sawaguchi, M.; Ayuba, S.; Fukuhara, T.; Hara, S. *Synlett* *2001*, 1938.

[1549]Sinn, H.; Hopperdietzel, S.; Sauermann, D. *Monatsh. Chem.* *1965*, 96, 1036.

[1550]Hondrogiannis, G.; Lee, L.C.; Kabalka, G.W.; Pagni, R.M. *Tetrahedron Lett.* *1989*, 30, 2069.

[1551]Dabdoub, M.J.; Dabdoub, V.B.; Baroni, A.C.M. *J. Am. Chem. Soc.* *2001*, 123, 9694.

$NaBO_3$ and NaBr adds two bromine atoms across a triple bond.[1552] With allenes it is easy to stop the reaction after only 1 equivalent has added, to give $X-C-CX=C$.[1553] Addition of halogen to ketenes gives α-halo acyl halides, but the yields are not good.

OS **I**, 205, 521; **II**, 171, 177, 270, 408; **III**, 105, 123, 127, 209, 350, 526, 531, 731, 785; **IV**, 130, 195, 748, 851, 969; **V**, 136, 370, 403, 467; **VI**, 210, 422, 675, 862, 954; **IX**, 117; **76**, 159.

15-40 Addition of Hypohalous Acids and Hypohalites (Addition of Halogen, Oxygen)

Hydroxy-chloro-addition, and so on.[1554]
Alkoxy-chloro-addition, and so on.

Hypohalous acids (HOCl, HOBr, and HOI) can be added to alkenes[1555] to produce halohydrins.[1556] Both HOBr and HOCl are often generated *in situ* by the reaction between water and Br_2 or Cl_2, respectively. HOI, generated from I_2 and H_2O, also adds to double bonds, if the reaction is carried out in tetramethylene sulfone-$CHCl_3$[1557] or if an oxidizing agent, such as HIO_3 is present.[1558] Iodine and cerium sulfate in aqueous acetonitrile generates iodohydrins,[1559] as does iodine and ammonium acetate in acetic acid,[1560] or $NaIO_4$ with sodium bisulfite.[1561] The HOBr can also be conveniently added by the use of a reagent consisting of an *N*-bromo amide

[1552]Kabalka, G.W.; Yang, K. *Synth. Commun.* **1998**, *28*, 3807; Kabalka, G.W.; Yang, K.; Reddy, N.K.; Narayana, A. *Synth. Commun.* **1998**, *28*, 925.

[1553]For a review of additions of halogens to allenes, see Jacobs, T.L., in Landor, S.R. *The Chemistry of Allenes*, Vol. 2, Academic Press, NY, *1982*, pp. 466–483.

[1554]Addends are listed in order of priority in the Cahn–Ingold–Prelog system (p. 155).

[1555]For a list of reagents used to accomplish these additions, with references, see Larock, R.C. *Comprehensive Organic Transformations*, 2nd ed., Wiley-VCH, NY, *1999*, pp. 638–642.

[1556]For a review, see Boguslavskaya, L.S. *Russ. Chem. Rev.* **1972**, *41*, 740.

[1557]Cambie, R.C.; Noall, W.I.; Potter, G.J.; Rutledge, P.S.; Woodgate, P.D. *J. Chem. Soc. Perkin Trans. 1* *1977*, 266.

[1558]See, for example, Cornforth, J.W.; Green, D.T. *J. Chem. Soc. C* *1970*, 846; Furrow, S.D. *Int. J. Chem. Kinet.* *1982*, *14*, 927; Antonioletti, R.; D'Auria, M.; De Mico, A.; Piancatelli, G.; Scettri, A. *Tetrahedron* *1983*, *39*, 1765.

[1559]Horiuchi, C.A.; Ikeda, A.; Kanamori, M.; Hosokawa, H.; Sugiyama, T.; Takahashi, T.T. *J. Chem. Res. (S)* *1997*, 60.

[1560]Myint, Y.Y.; Pasha, M.A. *Synth. Commun.* **2004**, *34*, 4477.

[1561]Masuda, H.; Takase, K.; Nishio, M.; Hasegawa, A.; Nishiyama, Y.; Ishii, Y. *J. Org. Chem.* **1994**, *59*, 5550.

(e.g., NBS or *N*-bromoacetamide) and a small amount of water in a solvent, such as DMSO or dioxane.[1562] *N*-Iodosuccinimide (NIS) in aqueous dimethoxyethane leads to the iodohydrin.[1563] An especially powerful reagent for HOCl addition is *tert*-butyl hydroperoxide (or di-*tert*-butyl peroxide) along with TiCl$_4$. This reaction is generally complete within 15 min at $-78°$C.[1564] Chlorohydrins can be conveniently prepared by treatment of the alkene with Chloramine T (TsNCl$^-$ Na$^+$)[1565] in acetone–water.[1566] The compound HOI can be added by treatment of alkenes with periodic acid and NaHSO$_3$.[1567] The reaction of an alkene with polymeric (SnO)$_n$, and then HCl with Me$_3$SiOOSiMe$_3$ leads to the chlorohydrin.[1568] Hypervalent iodine compounds react with an alkene and iodine in aqueous media to give the iodohydrin.[1569]

The compound HOF has also been added, but this reagent is difficult to prepare in a pure state and *explosions have occurred.*[1570]

The mechanism of HOX addition is electrophilic, with initial attack by the positive halogen end of the HOX dipole. Following Markovnikov's rule, the positive halogen goes to the side of the double bond that has more hydrogens (forming a more stable carbocation). This carbocation (or bromonium or iodonium ion in the absence of an aqueous solvent) reacts with $^-$OH or H$_2$O to give the product. If the substrate is treated with Br$_2$ or Cl$_2$ (or another source of positive halogen such as NBS) in an alcohol or a carboxylic acid solvent, it is possible to obtain,

directly $\mathrm{X-\overset{|}{C}-\overset{|}{C}-OR}$ or $\mathrm{X-\overset{|}{C}-\overset{|}{C}-OCOR}$, respectively (see also, **15-48**).[1571]

Even the weak nucleophile CF$_3$SO$_2$O$^-$ can participate in the second step: The addition of Cl$_2$ or Br$_2$ to alkenes in the presence of this ion resulted in the formation of some β-haloalkyl triflates.[1572] There is evidence that the mechanism with Cl$_2$ and H$_2$O is different from that with HOCl.[1573] HOCl and HOBr can be added to triple bonds to give dihalo carbonyl compounds $-$CX$_2$$-CO-$.

[1562]For examples, see Dalton, D.R.; Hendrickson, J.B.; Jones, D. *Chem. Commun.* **1966**, 591; Dalton, D.R.; Dutta, V.P. *J. Chem. Soc. B* **1971**, 85; Sisti, A.J. *J. Org. Chem.* **1970**, *35*, 2670.

[1563]Smietana, M.; Gouverneur, V.; Mioskowski, C. *Tetrahedron Lett.* **2000**, *41*, 193.

[1564]Klunder, J.M.; Caron M.; Uchiyama, M.; Sharpless, K.B. *J. Org. Chem.* **1985**, *50*, 912.

[1565]For reviews of this reagent, see Bremner, D.H., in Pizey, J.S. *Synthetic Reagents*, Vol. 6, Wiley, NY, **1985**, pp. 9–59; Campbell, M.M.; Johnson, G. *Chem. Rev.* **1978**, *78*, 65.

[1566]Damin, B.; Garapon, J.; Sillion, B. *Synthesis* **1981**, 362.

[1567]Ohta, M.; Sakata, Y.; Takeuchi, T.; Ishii, Y. *Chem. Lett.* **1990**, 733.

[1568]Sakurada, I.; Yamasaki, S.; Göttlich, R.; Iida, T.; Kanai, M.; Shibasaki, M. *J. Am. Chem. Soc.* **2000**, *122*, 1245.

[1569]DeCorso, A.R.; Panunzi, B.; Tingoli, M. *Tetrahedron Lett.* **2001**, *42*, 7245.

[1570]Migliorese, K.G.; Appelman, E.H.; Tsangaris, M.N. *J. Org. Chem.* **1979**, *44*, 1711.

[1571]For a list of reagents that accomplish alkoxy-halo-addition, with references, see Larock, R.C. *Comprehensive Organic Transformations*, 2nd ed., Wiley-VCH, NY, **1999**, pp. 642–643.

[1572]Zefirov, N.S.; Koz'min, A.S.; Sorokin, V.D.; Zhdankin, V.V. *J. Org. Chem. USSR* **1982**, *18*, 1546. For reviews of this and related reactions, see Zefirov, N.S.; Koz'min, A.S. *Acc. Chem. Res.* **1985**, *18*, 154; *Sov. Sci. Rev. Sect. B* **1985**, *7*, 297.

[1573]Buss, E.; Rockstuhl, A.; Schnurpfeil, D. *J. Prakt. Chem.* **1982**, *324*, 197.

Alcohols and halogens react with alkenes to form halo ethers. When a homoallylic alcohol is treated with bromine, cyclization occurs to give a 3-bromotetrahydrofuran derivative.[1574] *tert*-Butyl hypochlorite (Me$_3$COCl), hypobromite, and hypoiodite[1575] add to double bonds to give halogenated *tert*-butyl ethers, X–C–C–OCMe$_3$. This is a convenient method for the preparation of tertiary ethers. Iodine and ethanol convert some alkenes to iodo-ethers.[1576] Iodine, alcohol and a Ce(OTf)$_2$ catalyst also generates the iodo-ether.[1577] When Me$_3$COCl or Me$_3$COBr is added to alkenes in the presence of excess ROH, the ether produced

is $X - \overset{|}{\underset{|}{C}} - \overset{|}{\underset{|}{C}} - OR$.[1578] Vinylic ethers give β-halo acetals.[1579] A mixture of Cl$_2$ and

SO$_3$ at −78°C converts alkenes to 2-chloro chlorosulfates ClCHRCHROSO$_2$Cl, which are stable compounds.[1580] Chlorine acetate [solutions of which are prepared by treating Cl$_2$ with Hg(OAc)$_2$ in an appropriate solvent] adds to alkenes to give acetoxy chlorides.[1581] Acetoxy fluorides have been obtained by treatment of alkenes with CH$_3$COOF.[1582]

For a method of iodoacetyl addition, see **15-48**.

An oxidative variation of this reaction treats a vinyl chloride with NaOCl and acetic acid, generating an α-chloro ketone.[1583]

OS **I**, 158; **IV**, 130, 157; **VI**, 184, 361, 560; **VII**, 164; **VIII**, 5, 9.

15-41   Halolactonization and Halolactamization

Halo-alkoxylation

Halo esters can be formed by addition of halogen atoms and ester groups to an alkene. Alkene carboxylic acids give a tandem reaction of formation of a halonium ion followed by intramolecular displacement of the carboxylic group to give a halo lactone. This tandem addition of X and OCOR is called

[1574]Chirskaya, M.V.; Vasil'ev, A.A.; Sergovskaya, N.L.; Shovshinev, S.V.; Sviridov, S.I. *Tetrahedron Lett.* **2004**, *45*, 8811.

[1575]Glover, S.A.; Goosen, A. *Tetrahedron Lett.* **1980**, *21*, 2005.

[1576]Sanseverino, A.M.; de Mattos, M.C.S. *Synthesis* **1998**, 1584. See Horiuchi, C.A.; Hosokawa, H.; Kanamori, M.; Muramatsu, Y.; Ochiai, K.; Takahashi, E. *Chem. Lett.* **1995**, 13 for an example using I$_2$/MeOH/ceric ammonium nitrate.

[1577]Iranpoor, N.; Shekarriz, M. *Tetrahedron* **2000**, *56*, 5209.

[1578]Bresson, A.; Dauphin, G.; Geneste, J.; Kergomard, A.; Lacourt, A. *Bull. Soc. Chim. Fr.* **1970**, 2432; **1971**, 1080.

[1579]Weissermel, K.; Lederer, M. *Chem. Ber.* **1963**, *96*, 77.

[1580]Zefirov, N.S.; Koz'min, A.S.; Sorokin, V.D. *J. Org. Chem.* **1984**, *49*, 4086.

[1581]de la Mare, P.B.D.; O'Connor, C.J.; Wilson, M.A. *J. Chem. Soc. Perkin Trans. 2*, **1975**, 1150. For the addition of bromine acetate, see Wilson, M.A.; Woodgate, P.D. *J. Chem. Soc. Perkin Trans. 2*, **1976**, 141. For a list of reagents that accomplish acyloxy-halo-addition, with references, see Larock, R.C. *Comprehensive Organic Transformations*, 2nd ed., Wiley-VCH, NY, **1999**, pp. 643–644.

[1582]Rozen, S.; Lerman, O.; Kol, M.; Hebel, D. *J. Org. Chem.* **1985**, *50*, 4753.

[1583]Van Brunt, M.P.; Ambenge, R.O.; Weinreb, S.M. *J. Org. Chem.* **2003**, *68*, 3323.

halolactonization.[1584]

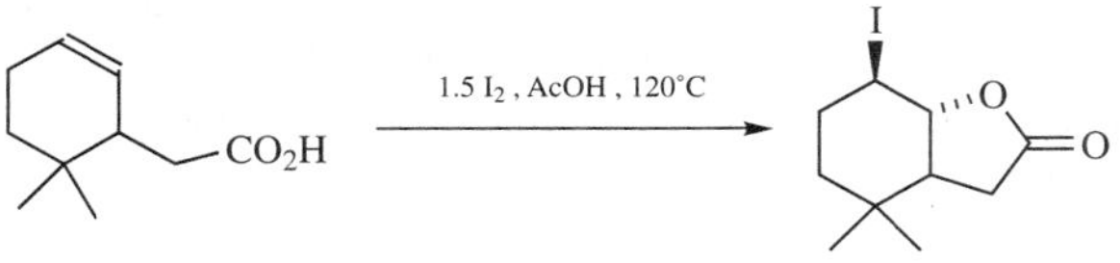

163   **164**

The most common version of this reaction is known as *iodolactonization,*[1585] and a typical example is the conversion of **163** to **164**.[1586] Bromo lactones and, to a lesser extent, chloro lactones have also been prepared. In general, addition of the halogen to an alkenyl acid, as shown, leads to the halo-lactone. Other reagents include $I^+(collidine)_2PF_6^-$,[1587] KI/sodium persulfate.[1588] Thallium reagents, along with the halogen, have also been used.[1589] When done in the presence of a chiral titanium reagent, I_2, and CuO, lactones are formed with good enantioselectivity.[1590] ICl has been used, with formation of a quaternary center at the oxygen-bearing carbon of the lactone.[1591]

In the case of γ,δ-unsaturated acids, 5-membered rings (γ-lactones) are predominantly formed (as shown above; note that Markovnikov's rule is followed), but 6-membered and even 4-membered lactones have also been made by this procedure. There is a gem-dimethyl effect that favors formation of 7–11-membered ring lactones by this procedure.[1592]

Formation of halo-lactams (**15-43**) by a similar procedure is difficult, but the problems have been overcome. Formation of a triflate followed by treatment with iodine leads to the iodo-lactam, **165**.[1593]

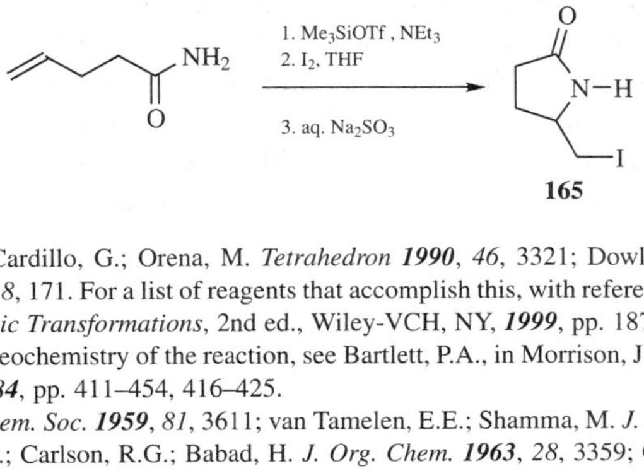

165

[1584]For reviews, see Cardillo, G.; Orena, M. *Tetrahedron* **1990**, *46*, 3321; Dowle, M.D.; Davies, D.I. *Chem. Soc. Rev.* **1979**, *8*, 171. For a list of reagents that accomplish this, with references, see Larock, R.C. *Comprehensive Organic Transformations*, 2nd ed., Wiley-VCH, NY, **1999**, pp. 1870–1876. For a review with respect to the stereochemistry of the reaction, see Bartlett, P.A., in Morrison, J.D. *Organic Synthesis*, Vol. 3, Wiley, NY, **1984**, pp. 411–454, 416–425.

[1585]Klein, J. *J. Am. Chem. Soc.* **1959**, *81*, 3611; van Tamelen, E.E.; Shamma, M. *J. Am. Chem. Soc.* **1954**, *76*, 2315; House, H.O.; Carlson, R.G.; Babad, H. *J. Org. Chem.* **1963**, *28*, 3359; Corey, E.J.; Albonico, S.M.; Koelliker, V.; Schaaf, T.K.; Varma, R.K. *J. Am. Chem. Soc.* **1971**, *93*, 1491.

[1586]Yaguchi, Y.; Akiba, M.; Harada, M.; Kato, T. *Heterocycles* **1996**, *43*, 601.

[1587]Homsi, F.; Rousseau, G. *J. Org. Chem.* **1998**, *63*, 5255; Simonet, B.; Rousseau, G. *J. Org. Chem.* **1993**, *58*, 4.

[1588]Royer, A.C.; Mebane, R.C.; Swafford, A.M. *Synlett* **1993**, 899.

[1589]See Cambie, R.C.; Rutledge, P.S.; Somerville, R.F.; Woodgate, P.D. *Synthesis* **1988**, 1009, and references cited therein.

[1590]Inoue, T.; Kitagawa, O.; Kurumizawa, S.; Ochiai, O.; Taguchi, T. *Tetrahedron Lett.* **1995**, *36*, 1479.

[1591]Haas, J.; Piguel, S.; Wirth, T. *Org. Lett.* **2002**, *4*, 297.

[1592]Simonot, B.; Rousseau, G. *Tetrahedron Lett.* **1993**, *34*, 4527.

[1593]Knapp, S.; Rodriques, K.E. *Tetrahedron Lett.* **1985**, *26*, 1803.

A related cyclization of *N*-sulfonyl-amino-alkenes and NBS gave the bromo-lactam,[1594] and a dichloro-*N,N*-bis(allylamide) was converted to a dichloro-lactam with FeCl₂.[1595] It is noted that lactone formation is possible from unsaturated amides. The reaction of 3-methyl *N,N*-dimethylpent-4-ene amide with iodine in aqueous acetonitrile, for example, gave iodolactone **166**.[1596]

166

OS IX, 516.

15-42 Addition of Sulfur Compounds (Addition of Halogen, Sulfur)

Alkylsulfonyl-chloro-addition, and so on.[1597]

Sulfonyl halides add to double bonds, to give β-halo sulfones, in the presence of free-radical initiators or UV light. A particularly good catalyst is cuprous chloride.[1598] A combination of the anion ArSO₂Na, NaI, and ceric ammonium nitrate converts alkenes to vinyl sulfones.[1599] Triple bonds behave similarly, to give β-halo-α,β-unsaturated sulfones.[1600] In a similar reaction, sulfenyl chlorides, RSCl, give β-halo thioethers.[1601] The latter may be free-radical or electrophilic additions, depending on conditions. The addition of MeS and Cl has also been accomplished by treating the alkene with Me₃SiCl and Me₂SO.[1602] The use of Me₃SiBr and Me₂SO does not give this result; dibromides (**15-39**) are formed instead. β-Iodo

[1594]Tamaru, Y.; Kawamura, S.; Tanaka, K.; Yoshida, Z. *Tetrahedron Lett.* **1984**, *25*, 1063.

[1595]Tseng, C.K.; Teach, E.G.; Simons, R.W. *Synth. Commun.* **1984**, *14*, 1027.

[1596]Ha, H.-J.; Lee, S.-Y.; Park, Y.-S. *Synth. Commun.* **2000**, *30*, 3645.

[1597]When a general group (e.g., halo) is used, its priority is that of the lowest member of its group (see Ref. 1555). Thus the general name for this transformation is halo-alkylsulfonyl-addition because "halo" has the same priority as "fluoro," its lowest member.

[1598]Asscher, M.; Vofsi, D. *J. Chem. Soc.* **1964**, 4962; Truce, W.E.; Goralski, C.T.; Christensen, L.W.; Bavry, R.H. *J. Org. Chem.* **1970**, *35*, 4217; Sinnreich, J.; Asscher, M. *J. Chem. Soc. Perkin Trans. 1*, **1972**, 1543.

[1599]Nair, V.; Augustine, A.; George, T.G.; Nair, L.G. *Tetrahedron Lett.* **2001**, *42*, 6763.

[1600]Truce, W.E.; Wolf, G.C. *J. Org. Chem.* **1971**, *36*, 1727; Amiel, Y. *J. Org. Chem.* **1974**, *39*, 3867; Zakharkin, L.I.; Zhigareva, G.G. *J. Org. Chem. USSR* **1973**, *9*, 918; Okuyama, T.; Izawa, K.; Fueno, T. *J. Org. Chem.* **1974**, *39*, 351.

[1601]For reviews, see Rasteikiene, L.; Greiciute, D.; Lin'kova, M.G.; Knunyants, I.L. *Russ. Chem. Rev.* **1977**, *46*, 548; Kühle, E. *Synthesis* **1971**, 563.

[1602]Bellesia, F.; Ghelfi, F.; Pagnoni, U.M.; Pinetti, A. *J. Chem. Res. (S)* **1987**, 238. See also, Liu, H.; Nyangulu, J.M. *Tetrahedron Lett.* **1988**, *29*, 5467.

thiocyanates can be prepared from alkenes by treatment with I_2 and isothiocyana-totributylstannane Bu_3SnNCS.[1603] Bromothiocyanation can be accomplished with Br_2 and thallium(I) thiocyanate.[1604] Lead (II) thiocyanate reacts with terminal alkynes in the presence of $PhICl_2$ to give the bis(thiocyanato) alkene, $ArC(SCN)-CHSCN$.[1605] Such compounds were also prepared from alkenes using KSCN and $FeCl_3$.[1606] β-Halo disulfides, formed by addition of arenethiosulfenyl chlorides to double-bond compounds, are easily converted to thiiranes by treatment with sodium amide or sodium sulfide.[1607]

OS **VIII**, 212. See also OS **VII**, 251.

15-43   Addition of Halogen and an Amino Group (Addition of Halogen, Nitrogen)

Dialkylamino-chloro-addition

$$\diagdown_{\diagup}C=C_{\diagdown}^{\diagup} + R_2N-Cl \xrightarrow[\text{HOAc}]{H_2SO_4} \overset{Cl}{\underset{\diagup}{\diagdown}}C-C\overset{NR_2}{\underset{\diagdown}{\diagup}}$$

The groups R_2N and Cl can be added directly to alkenes, allenes, conjugated dienes, and alkynes, by treatment with dialkyl-N-chloroamines and acids.[1608] The reaction of $TsNCl_2$ and a $ZnCl_2$ catalyst gave the chloro tosylamine.[1609] These are free-radical additions, with initial attack by the $R_2NH^{\bullet+}$ radical ion.[1610] N-Halo amides RCONHX add RCONH and X to double bonds under the influence of uv light or chromous chloride.[1611] Amines add to allenes in the presence of a palladium catalyst.[1612] A mixture of N-(2-nosyl)NCl_2 and sodium N-(2-nosyl)NH^- with a CuOTf catalyst reacted with conjugated esters to give the *vicinal* (E)-3-chloro-2-amino ester.[1613] A variation of this latter reaction was done in an ionic liquid.[1614]

[1603]Woodgate, P.D.; Janssen, S.J.; Rutledge, P.S.; Woodgate, S.D.; Cambie, R.C. *Synthesis* **1984**, 1017, and references cited therein. See also, Watanabe, N.; Uemura, S.; Okano, M. *Bull. Chem. Soc. Jpn.* **1983**, *56*, 2458.

[1604]Cambie, R.C.; Larsen, D.S.; Rutledge, P.S.; Woodgate, P.D. *J. Chem. Soc. Perkin Trans. 1*, **1981**, 58.

[1605]Prakash, O.; Sharma, V.; Batra, H.; Moriarty, R.M. *Tetrahedron Lett.* **2001**, *42*, 553.

[1606]Yadav, J.S.; Reddy, B.V.S.; Gupta, M.K. *Synthesis* **2004**, 1983.

[1607]Fujisawa, T.; Kobori, T. *Chem. Lett.* **1972**, 935. For another method of alkene-thiirane conversion, see Capozzi, F.; Capozzi, G.; Menichetti, S. *Tetrahedron Lett.* **1988**, *29*, 4177.

[1608]For reviews see Mirskova, A.N.; Drozdova, T.I.; Levkovskaya, G.G.; Voronkov, M.G. *Russ. Chem. Rev.* **1989**, *58*, 250; Neale, R.S. *Synthesis* **1971**, 1; Sosnovsky, G.; Rawlinson, D.J. *Adv. Free-Radical Chem.* **1972**, *4*, 203, see pp. 238–249.

[1609]Li, G.; Wei, H.-X.; Kim, S.H.; Neighbors, M. *Org. Lett.* **1999**, *1*, 395; Wei, H.-X.; Ki, S.H.; Li, G. *Tetrahedron* **2001**, *57*, 3869.

[1610]For a review of these species, see Chow, Y.L.; Danen, W.C.; Nelson, S.F.; Rosenblatt, D.H. *Chem. Rev.* **1978**, *78*, 243.

[1611]Tuaillon, J.; Couture, Y.; Lessard, J. *Can. J. Chem.* **1987**, *65*, 2194, and other papers in this series. For a review, see Labeish, N.N.; Petrov, A.A. *Russ. Chem. Rev.* **1989**, *58*, 1048.

[1612]Besson, L.; Goré, J.; Cazes, B. *Tetrahedron Lett.* **1995**, *36*, 3857.

[1613]Li, G.; Wei, H.-X.; Kim, S.H. *Org. Lett.* **2000**, *2*, 2249; Li, G.; Wei, H.-X.; Kim, S.H. *Tetrahedron* **2001**, *57*, 8407.

[1614]In bmim BF_4, 1-butyl-3-methylimidazolium tetrafluoroborate: Xu, X.; Kotti, S.R.S.S.; Liu, J.; Cannon, J.F.; Headley, A.D.; Li, G. *Org. Lett.* **2004**, *6*, 4881.

15-44   Addition of NOX and NO₂X (Addition of Halogen, Nitrogen)

Nitroso-chloro-addition

$$\ce{>C=C< + NO-Cl ->} \underset{\text{Cl}}{\overset{}{}}\ce{-C-C-}\underset{}{\overset{\text{N=O}}{}}$$

There are three possible products when NOCl is added to alkenes, a β-halo nitroso compound, an oxime, or a β-halo nitro compound.[1615] The initial product is always the β-halo nitroso compound,[1616] but these are stable only if the carbon bearing the nitrogen has no hydrogen. If it has, the nitroso compound tautomerizes to the oxime, H—C—N=O → C=N–OH. With some alkenes, the initial β-halo nitroso compound is oxidized by the NOCl to a β-halo nitro compound.[1617] Many functional groups can be present without interference (e.g., COOH, COOR, CN, OR). The mechanism in most cases is probably simple electrophilic addition, and the addition is usually anti, although syn addition has been reported in some cases.[1618] Markovnikov's rule is followed, the positive NO going to the carbon that has more hydrogens.

Nitryl chloride NO₂Cl also adds to alkenes, to give β-halo nitro compounds, but this is a free-radical process. The NO₂ goes to the less-substituted carbon.[1619] Nitryl chloride also adds to triple bonds to give the expected 1-nitro-2-chloro alkenes.[1620] The compound FNO₂ can be added to alkenes[1621] by treatment with HF in HNO₃[1622] or by addition of the alkene to a solution of nitronium tetrafluoroborate (NO₂⁺ BF₄⁻) (see **11-2**) in 70% polyhydrogen fluoride–pyridine solution[1623] (see also **15-37**).

OS **IV**, 711; **V**, 266, 863.

15-45   Addition of XN₃ (Addition of Halogen, Nitrogen)

Azido-iodo-addition

$$\ce{>C=C< + I-N3 ->} \underset{\text{I}}{\overset{}{}}\ce{-C-C-}\underset{}{\overset{\text{N3}}{}}$$

[1615]For a review, see Kadzyauskas, P.P.; Zefirov, N.S. *Russ. Chem. Rev.* **1968**, *37*, 543.

[1616]For a review of preparations of *C*-nitroso compounds, see Gowenlock, B.G.; Richter-Addo, G.B. *Chem. Rev.* **2004**, *104*, 3315.

[1617]For a review of the preparation of halo nitro compounds, see Shvekhgeimer, G.A.; Smirnyagin, V.A.; Sadykov, R.A.; Novikov, S.S. *Russ. Chem. Rev.* **1968**, *37*, 351.

[1618]For example, see Meinwald, J.; Meinwald, Y.C.; Baker III, T.N. *J. Am. Chem. Soc.* **1964**, *86*, 4074.

[1619]Shechter, H. *Rec. Chem. Prog.* **1964**, *25*, 55–76.

[1620]Schlubach, H.H.; Braun, A. *Liebigs Ann. Chem.* **1959**, *627*, 28.

[1621]For a review, see Sharts, C.M.; Sheppard, W.A. *Org. React.* **1974**, *21*, 125–406, see pp. 236–243.

[1622]Knunyants, I.L.; German, L.S.; Rozhkov, I.N. *Bull Acad. Sci. USSR Div. Chem. Sci.* **1963**, 1794.

[1623]Olah, G.A.; Nojima, M. *Synthesis* **1973**, 785.

The addition of iodine azide to double bonds gives β-iodo azides.[1624] The reagent can be prepared *in situ* from KI—NaN$_3$ in the presence of Oxone®-wet alumina.[1625] The addition is stereospecific and anti, suggesting that the mechanism involves a cyclic iodonium ion intermediate.[1626] The reaction has been performed on many double-bond compounds, including allenes[1627] and α,β-unsaturated ketones. Similar reactions can be performed with BrN$_3$[1628] and ClN$_3$. 1,4-Addition has been found with acyclic conjugated dienes.[1629] In the case of BrN$_3$, both electrophilic and free-radical mechanisms are important,[1630] while with ClN$_3$ the additions are chiefly free radical.[1631] The compound IN$_3$ also adds to triple bonds to give β-iodo-α,β-unsaturated azides.[1632]

β-Iodo azides can be reduced to aziridines (**167**) with LiAlH$_4$[1633] or converted to *N*-alkyl- or *N*-arylaziridines (**168**) by treatment with an alkyl- or aryldichloroborane followed by a base.[1634] In both cases, the azide is first reduced to the corresponding amine (primary or secondary, respectively) and ring closure (**10-31**) follows. With Chloramine T (TsNCl$^-$ Na$^+$) and 10% of pyridinium bromide perbromide, however, the reaction with alkenes give an *N*-tosyl aziridine directly.[1635]

OS **VI**, 893.

15-46   Addition of Alkyl Halides (Addition of Halogen, Carbon)

Alkyl-halo-addition[1062]

[1624]For reviews, see Dehnicke, K. *Angew. Chem. Int. Ed.* **1979**, *18*, 507; Hassner, A. *Acc. Chem. Res.* **1971**, *4*, 9; Biffin, M.E.C.; Miller, J.; Paul, D.B., in Patai, S. *The Chemistry of the Azido Group*, Wiley, NY, **1971**, pp. 136–147. See Nair, V.; George, T.G.; Sheeba, V.; Augustine, A.; Balagopal, L.; Nair, L.G. *Synlett* **2000**, 1597.

[1625]Curini, M.; Epifano, F.; Marcotullio, M.C.; Rosati, O. *Tetrahedron Lett.* **2002**, *43*, 1201.

[1626]See, however, Cambie, R.C.; Hayward, R.C.; Rutledge, P.S.; Smith-Palmer, T.; Swedlund, B.E.; Woodgate, P.D. *J. Chem. Soc. Perkin Trans. 1*, **1979**, 180.

[1627]Hassner, A.; Keogh, J. *J. Org. Chem.* **1986**, *51*, 2767.

[1628]Azido-bromo-addition has also been done with another reagent: Olah, G.A.; Wang, Q.; Li, X.; Prakash, G.K.S. *Synlett* **1990**, 487.

[1629]Hassner, A.; Keogh, J. *Tetrahedron Lett.* **1975**, 1575.

[1630]Hassner, A.; Teeter, J.S. *J. Org. Chem.* **1971**, *36*, 2176.

[1631]Even IN$_3$ can be induced to add by a free-radical mechanism [see, e.g., Cambie, R.C.; Jurlina, J.L.; Rutledge, P.S.; Swedlund, B.E.; Woodgate, P.D. *J. Chem. Soc. Perkin Trans. 1*, **1982**, 327]. For a review of free-radical additions of XN$_3$, see Hassner, A. *Intra-Sci. Chem. Rep.* **1970**, *4*, 109.

[1632]Hassner, A.; Isbister, R.J.; Friederang, A. *Tetrahedron Lett.* **1969**, 2939.

[1633]Hassner, A.; Matthews, G.J.; Fowler, F.W. *J. Am. Chem. Soc.* **1969**, *91*, 5046.

[1634]Levy, A.B.; Brown, H.C. *J. Am. Chem. Soc.* **1973**, *95*, 4067.

[1635]Ali, S.I.; Nikalje, M.D.; Sudalai, A. *Org. Lett.* **1999**, *1*, 705.

Alkyl halides can be added to alkenes in the presence of a Friedel–Crafts catalyst, most often AlCl$_3$.[1636] The yields are best for tertiary R. Secondary R can also be used, but primary R give rearrangement products (as with **11-11**). Methyl and ethyl halides, which cannot rearrange to a more stable secondary or tertiary carbocation, give no reaction at all. The attacking species is the carbocation formed from the alkyl halide and the catalyst (see **11-11**).[1637] The addition therefore follows Markovnikov's rule, with the cation going to the carbon with more hydrogens. Substitution is a side reaction, arising from loss of hydrogen from the carbocation formed when an additional molecule of alkene attacks the initially formed carbocation (**169**). Conjugated dienes can add 1,4.[1638] Triple bonds also undergo the reaction, to give vinylic halides.[1639]

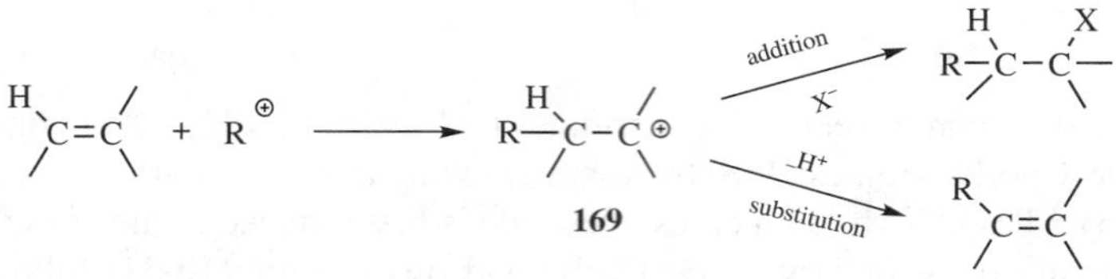

Simple polyhalo alkanes, such as CCl$_4$, BrCCl$_3$, ICF$_3$ and related molecules, add to alkenes in good yield.[1640] These are free-radical additions and require initiation, for example,[1641] by peroxides, metal halides (e.g., FeCl$_2$, CuCl),[1642] ruthenium catalysts,[1643] or UV light. The initial attack is by the carbon, and it goes to the carbon with more hydrogens, as in most free-radical attack:

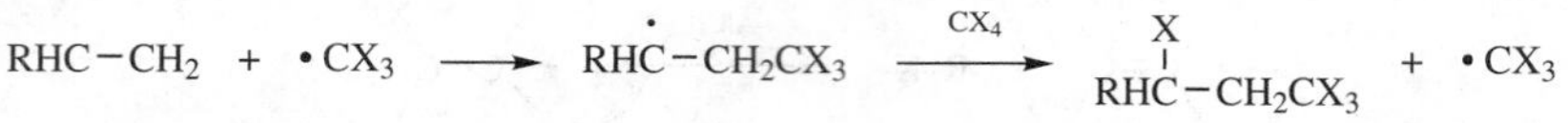

[1636]For a review, see Schmerling, L., in Olah, G.A. *Friedel–Crafts and Related Reactions*, Vol. 2, Wiley, NY, *1964*, pp. 1133–1174. See also, Mayr, H.; Schade, C.; Rubow, M.; Schneider, R. *Angew. Chem. Int. Ed.* *1987*, *26*, 1029.

[1637]For a discussion of the mechanism, see Pock, R.; Mayr, H.; Rubow, M.; Wilhelm, E. *J. Am. Chem. Soc.* *1986*, *108*, 7767.

[1638]Kolyaskina, Z.N.; Petrov, A.A. *J. Gen. Chem. USSR* *1962*, *32*, 1067.

[1639]See, for example, Maroni, R.; Melloni, G.; Modena, G. *J. Chem. Soc. Perkin Trans. 1*, *1973*, 2491; *1974*, 353.

[1640]For reviews, see Freidlina, R.Kh.; Velichko, F.K. *Synthesis* *1977*, 145; Freidlina, R.Kh.; Chukovskaya, E.C. *Synthesis* *1974*, 477.

[1641]For other initiators, see Matsumoto, H.; Nakano, T.; Takasu, K.; Nagai, Y. *J. Org. Chem.* *1978*, *43*, 1734; Tsuji, J.; Sato, K.; Nagashima, H. *Tetrahedron* *1985*, *41*, 393; Bland, W.J.; Davis, R.; Durrant, J.L.A. *J. Organomet. Chem.* *1985*, *280*, 397; Phelps, J.C.; Bergbreiter, D.E.; Lee, G.M.; Villani, R.; Weinreb, S.M. *Tetrahedron Lett.* *1989*, *30*, 3915.

[1642]For example, see Asscher, M.; Vofsi, D. *J. Chem. Soc.* *1963*, 1887, 3921; *J. Chem. Soc. B* *1968*, 947; Murai, S.; Tsutsumi, S. *J. Org. Chem.* *1966*, *31*, 3000; Martin, P.; Steiner, E.; Streith, J.; Winkler, T.; Bellus, D. *Tetrahedron* *1985*, *41*, 4057. For the addition of CH$_2$Cl$_2$ and PhBr, see Mitani, M.; Nakayama, M.; Koyama, K. *Tetrahedron Lett.* *1980*, *21*, 4457.

[1643]Simal, F.; Wlodarczak, L.; Demonceau, A.; Noels, A.F. *Eur. J. Org. Chem.* *2001*, 2689.

This type of polyhalo alkane adds to halogenated alkenes in the presence of $AlCl_3$ by an electrophilic mechanism. This is called the *Prins reaction* (not to be confused with the other Prins reaction, **16-54**).[1644]

α-Iodolactones add to alkenes in the presence of BEt_3/O_2 to give the addition product.[1645] Other α-iodoesters add under similar conditions to give the lactone.[1646] Iodoesters also add to alkenes in the presence of BEt_3 to give iodo-esters that have not cyclized.[1647]

A variant of the free-radical addition method has been used for ring closure (see **15-30**).

For another method of adding R and I to a triple bond (see **15-23**).

OS **II**, 312; **IV**, 727; **V**, 1076; **VI**, 21; **VII**, 290.

15-47   Addition of Acyl Halides (Addition of Halogen, Carbon)

Acyl-halo-addition

Acyl halides have been added to many alkenes, in the presence of Friedel–Crafts catalysts, although polymerization is a problem. The reaction has been applied to straight-chain, branched, and cyclic alkenes, but to very few containing functional groups, other than halogen.[1648] The mechanism is similar to that of **15-46**, and, as in that case, substitution competes (**12-16**). Increasing temperature favors substitution,[1649] and good yields of addition products can be achieved if the temperature is kept under 0°C. The reaction usually fails with conjugated dienes, since polymerization predominates.[1650] Iodo acetates have been formed from alkenes using iodine, $Pb(OAc)_2$ in acetic acid.[1651] The reaction can be performed on triple-bond compounds, producing compounds of the form $\overset{}{RCO-\underset{|}{C}=\underset{|}{C}-Cl}$.[1652] A *formyl* group and a halogen can be added to triple bonds by treatment with *N,N*-disubstituted formamides

[1644]For a review with respect to fluoroalkenes, see Paleta, O. *Fluorine Chem. Rev.* **1977**, *8*, 39.

[1645]Nakamura, T.; Yorimitsu, H.; Shinokubo, H.; Oshima, K. *Synlett* **1998**, 1351.

[1646]Yorimitsu, H.; Nakamura, T.; Shinokubo, H.; Oshima, K. *J. Org. Chem.* **1998**, *63*, 8604.

[1647]Baciocchi, E.; Muraglia, E. *Tetrahedron Lett.* **1994**, *35*, 2763.

[1648]For reviews, see Groves, J.K. *Chem. Soc. Rev.* **1972**, *1*, 73; House, H.O. *Modern Synthetic Reaction*, 2nd ed., W.A. Benjamin, NY, **1972**, pp. 786–797; Nenitzescu, C.D.; Balaban, A.T., in Olah, G.A. *Friedel–Crafts and Related Reactions*, Vol. 3, Wiley, NY, **1964**, pp. 1033–1152.

[1649]Jones, N.; Taylor, H.T.; Rudd, E. *J. Chem. Soc.* **1961**, 1342.

[1650]For examples of 1,4-addition at low temperatures, see Melikyan, G.G.; Babayan, E.V.; Atanesyan, K.A.; Badanyan, Sh.O. *J. Org. Chem. USSR* **1984**, *20*, 1884.

[1651]Bedekar, A.V.; Nair, K.B.; Soman, R. *Synth. Commun.* **1994**, *24*, 2299.

[1652]For example, see Nifant'ev, E.Ye.; Grachev, M.A.; Bakinovskii, L.V.; Kara-Murza, C.G.; Kochetkov, N.K. *J. Appl. Chem. USSR* **1963**, *36*, 646; Savenkov, N.F.; Khokhlov, P.S.; Nazarova, T.A.; Mochalkin, A.I. *J. Org. Chem. USSR* **1973**, *9*, 914; Martens, H.; Janssens, F.; Hoornaert, G. *Tetrahedron* **1975**, *31*, 177; Brownstein, S.; Morrison, A.; Tan, L.K. *J. Org. Chem.* **1985**, *50*, 2796.

and POCl$_3$ (Vilsmeier conditions, see **11-18**).[1653] Chloroformates add to allenes in the presence of a rhodium catalyst go give a β-chloro, β,γ-unsaturated ester.[1654]

OS **IV**, 186; **VI**, 883; **VIII**, 254.

B. Oxygen, Nitrogen, or Sulfur on One or Both Sides

15-48 Hydroxylation (Addition of Oxygen, Oxygen)

Dihydroxy-addition

$$\underset{/}{\overset{\backslash}{C}}=\underset{\backslash}{\overset{/}{C}} \quad\longrightarrow\quad \underset{/}{\overset{HO\quad OH}{\underset{\backslash}{C-C}}}$$

There are many reagents that add two OH groups to a double bond.[1655] The most common are OsO$_4$[1656] and alkaline KMnO$_4$,[1657] which give syn addition from the less-hindered side of the double bond. Less substituted double bonds are oxidized more rapidly than more substituted alkenes.[1658] Permanganate adds to alkenes to form an intermediate manganate ester (**171**), which is decomposed under alkaline conditions. Bases catalyze the decomposition of **171** by coordinating with the ester. Osmium tetroxide adds rather slowly but almost quantitatively to form a cyclic ester, such as **170**, as an intermediate, which can be isolated,[1659] but is usually decomposed solution, with sodium sulfite in ethanol or other reagents. The chief drawback to the use of OsO$_4$ is expensive and highly toxic, but the reaction is made catalytic in OsO$_4$ by using *N*-methylmorpholine-*N*-oxide (NMO),[1660] *tert*-butyl hydroperoxide in alkaline solution,[1661] H$_2$O$_2$,[1662] peroxyacid,[1663] flavin and

[1653]Yen, V.Q. *Ann. Chim. (Paris)* **1962**, *[13] 7*, 785.

[1654]Hua, R.; Tanaka, M. *Tetrahedron Lett.* **2004**, *45*, 2367.

[1655]For reviews, see Hudlický, M. *Oxidations in Organic Chemistry*, American Chemical Society, Washington, **1990**, pp. 67–73; Haines, A.H. *Methods for the Oxidation of Organic Compounds*, Academic Press, NY, **1985**, pp. 73–98, 278–294; Sheldon, R.A.; Kochi, J.K. *Metal-Catalyzed Oxidations of Organic Compounds*, Academic Press, NY, **1981**, pp. 162–171, 294–296. For a list of reagents, with references, see Larock, R.C. *Comprehensive Organic Transformations*, 2nd ed., Wiley-VCH, NY, **1999**, pp. 996–1003.

[1656]For a review, see Schröder, M. *Chem. Rev.* **1980**, *80*, 187. OsO$_4$ was first used for this purpose by Criegee, R. *Liebigs Ann. Chem.* **1936**, *522*, 75. Also see, Norrby, P.-O.; Gable, K.P. *J. Chem. Soc. Perkin Trans. 2*, **1996**, 171; Lohray, B.B.; Bhushan, V. *Tetrahedron Lett.* **1992**, *33*, 5113.

[1657]For a review, see Fatiadi, A.J. *Synthesis* **1987**, 85, 86. See Nelson, D.J.; Henley, R.L. *Tetrahedron Lett.* **1995**, *36*, 6375 for rate of oxidation of alkenes.

[1658]Crispino, G.A.; Jeong, K.-S.; Kolb, H.C.; Wang, Z.-M.; Xu, D.; Sharpless, K.B. *J. Org. Chem.* **1993**, *58*, 3785.

[1659]For a molecular-orbital study of the formation of **170**, see Jørgensen, K.A.; Hoffmann, R. *J. Am. Chem. Soc.* **1986**, *108*, 1867.

[1660]VanRheenen, V.; Kelly, R.C.; Cha, D.Y. *Tetrahedron Lett.* **1976**, 1973; Iwasawa, N.; Kato, T.; Narasaka, K. *Chem. Lett.* **1988**, 1721. See also, Ray, R.; Matteson, D.S. *Tetrahedron Lett.* **1980**, 449.

[1661]Akashi, K.; Palermo, R.E.; Sharpless, K.B. *J. Org. Chem.* **1978**, *43*, 2063.

[1662]For a review, see Rylander, P.N. *Organic Syntheses withy Noble Metal Catalysts*, Academic Press, NY, **1973**, pp. 121–133. See Venturello, C.; Gambaro, M. *Synthesis* **1989**, 295; Usui, Y.; Sato, K.; Tanaka, M. *Angew. Chem. Int. Ed.* **2003**, *42*, 5623.

[1663]Bergstad, K.; Piet, J.J.N.; Bäckvall, J.-E. *J. Org. Chem.* **1999**, *64*, 2545.

TEAA,[1664] $K_3Fe(CN)_6$[1665] and non-heme iron catalysts.[1666] Polymer-bound OsO_4,[1667] and encapsulated OsO_4 have been shown to give the diol in the presence of NMO,[1668] as well as OsO_4^{-2} on an ion exchange resin.[1669] Dihydroxylation has also been reported in ionic liquids,[1670] and with fluorous osmium tetroxide.[1671] A catalytic amount of K_2OsO_4 with a Cinchona alkaloid on a ordered inorganic support, in the presence of $K_3Fe(CN)_6$, gives the cis-diol.[1672] Oxidation of pent-4-en-1-ol to valerolactone was accomplished with Oxone® and a catalytic amount of OsO_4 in DMF.[1673]

170   171

The end-product of the reaction of either potassium permanganate or osmium tetroxide under the conditions described above is a 1,2-diol. Potassium permanganate is a strong oxidizing agent and can oxidize the glycol product[1674] (see **19-7** and **19-10**). In acid and neutral solution it always does so; hence glycols must be prepared with alkaline[1675] permanganate, but the conditions must be mild. Even so, yields are seldom >50%, although they can be improved with phase-transfer catalysis[1676] or increased stirring.[1677] The use of ultrasound with permanganate

[1664]Jonsson, S.Y.; Färnegårdh, K.; Bäckvall, J.-E. *J. Am. Chem. Soc.* **2001**, *123*, 1365.

[1665]Minato, M.; Yamamoto, K.; Tsuji, J. *J. Org. Chem.* **1990**, *55*, 766; Torii, S.; Liu, P.; Tanaka, H. *Chem. Lett.* **1995**, 319; Soderquist, J.A.; Rane, A.M.; López, C.J. *Tetrahedron Lett.* **1993**, *34*, 1893. See Corey, E.J.; Noe, M.C.; Grogan, M.J. *Tetrahedron Lett.* **1994**, *35*, 6427; Imada, Y.; Saito, T.; Kawakami, T.; Murahashi, S.-I. *Tetrahedron Lett.* **1992**, *33*, 5081 for oxidation using an asymmetric ligand.

[1666]Chen, K.; Costas, M.; Kim, J.; Tipton, A.K.; Que, Jr., L. *J. Am. Chem. Soc.* **2002**, *124*, 3026.

[1667]Cainelli, G.; Contento, M.; Manescalchi, F.; Plessi, L. *Synthesis* **1989**, 45; Ley, S.V.; Ramarao, C.; Lee, A.-L.; Østergaard, N.; Smith, S.C.; Shirley, I.M. *Org. Lett.* **2003**, *5*, 185.

[1668]Nagayama, S.; Endo, M.; Kobayashi, S. *J. Org. Chem.* **1998**, *63*, 6094.

[1669]Choudary, B.M.; Chowdari, N.S.; Jyothi, K.; Kantam, M.L. *J. Am. Chem. Soc.* **2002**, *124*, 5341.

[1670]In bmim PF_6, 1-butyl-3-methylimidazolium hexafluorophosphate: Yao, Q. *Org. Lett.* **2002**, *4*, 2197; Closson, A.; Johansson, M.; Bäckvall, J.-E. *Chem. Commun.* **2004**, 1494. In emim BF_4, 1-ethyl-3-methylimidazolium tetrafluoroborate: Yanada, R.; Takemoto, Y. *Tetrahedron Lett.* **2002**, *43*, 6849.

[1671]Huang, Y.; Meng, W.-D.; Qing, F.L. *Tetrahedron Lett.* **2004**, *45*, 1965.

[1672]Motorina, I.; Crudden, C.M. *Org. Lett.* **2001**, *3*, 2325.

[1673]Schomaker, J.M.; Travis, B.R.; Borhan, B. *Org. Lett.* **2003**, *5*, 3089.

[1674]Or give more highly oxidized products, such as α-hydroxy ketones without going through the glycols. See, for example, Wolfe, S.; Ingold, C.F.; Lemieux, R.U. *J. Am. Chem. Soc.* **1981**, *103*, 938; Wolfe, S.; Ingold, C.F. *J. Am. Chem. Soc.* **1981**, *103*, 940. Also see, Lohray, B.B.; Bhushan, V.; Kumar, R.K. *J. Org. Chem.* **1994**, *59*, 1375.

[1675]The role of the base seems merely to be to inhibit acid-promoted oxidations. The base does not appear to play any part in the mechanism: Taylor, J.E.; Green, R. *Can. J. Chem.* **1985**, *63*, 2777.

[1676]See, for example, Weber, W.P.; Shepherd, J.P. *Tetrahedron Lett.* **1972**, 4907; Ogino, T.; Mochizuki, K. *Chem. Lett.* **1979**, 443.

[1677]Taylor, J.E.; Williams, D.; Edwards, K.; Otonnaa, D.; Samanich, D. *Can. J. Chem.* **1984**, *62*, 11; Taylor, J.E. *Can. J. Chem.* **1984**, *62*, 2641.

dihydroxylation has resulted in good yields of the diol.[1678] There is evidence that cyclic esters (**171**) are intermediates for OsO_4 dihydroxylation.[1679] This reaction is the basis of the *Baeyer test* for the presence of double bonds. The oxidation is compatible with a number of functional groups, including trichloroacetamides.[1680]

Anti hydroxylation can be achieved by treatment with H_2O_2 and formic acid. In this case, epoxidation (**15-50**) occurs first, followed by an S_N2 reaction, which results in overall anti addition:

The same result can be achieved in one step with *m*-chloroperoxybenzoic acid and water.[1681] Overall anti addition can also be achieved by the method of Prévost (the *Prévost reaction*). In this method, the alkene is treated with iodine and silver benzoate in a 1:2 molar ratio. The initial addition is anti and results in a β-halo benzoate (**172**). These can be isolated, and this represents a method of addition of IOCOPh. However, under the normal reaction conditions, the iodine is replaced by a second PhCOO group. This is a nucleophilic substitution reaction, and it operates by the neighboring-group mechanism (p. 446), so the groups are still anti:

172

Hydrolysis of the ester does not change the configuration. The *Woodward modification* of the Prévost reaction is similar, but results in overall syn hydroxylation.[1682] The alkene is treated with iodine and silver acetate in a 1:1 molar ratio in acetic acid containing water. Here again, the initial product is a β-halo ester; the addition is anti and a nucleophilic replacement of the iodine occurs. However, in the presence of water, neighboring-group participation is prevented or greatly decreased by solvation of the ester function, and the mechanism is the normal S_N2 process,[1683]

[1678]Varma, R.S.; Naicker, K.P. *Tetrahedron Lett.* **1998**, *39*, 7463.

[1679]For some recent evidence, see Lee, D.G.; Chen, T. *J. Am. Chem. Soc.* **1989**, *111*, 7534; Ogino, T.; Hasegawa, K.; Hoshino, E. *J. Org. Chem.* **1990**, *55*, 2653. See, however, Freeman, F.; Chang, L.Y.; Kappos, J.C.; Sumarta, L. *J. Org. Chem.* **1987**, *52*, 1461; Freeman, F.; Kappos, J.C. *J. Org. Chem.* **1989**, *54*, 2730, and other papers in this series; Perez-Benito, J.F.; Lee, D.G. *Can. J. Chem.* **1985**, *63*, 3545.

[1680]Donohoe, T.J.; Blades, K.; Moore, P.R.; Waring, M.J.; Winter, J.J.G.; Helliwell, M.; Newcombe, N.J.; Stemp, G. *J. Org. Chem.* **2002**, *67*, 7946.

[1681]Fringuelli, F.; Germani, R.; Pizzo, F.; Savelli, G. *Synth. Commun.* **1989**, *19*, 1939.

[1682]See Brimble, M.A.; Nairn, M.R. *J. Org. Chem.* **1996**, *61*, 4801.

[1683]For another possible mechanism that accounts for the stereochemical result of the Woodward method, see Woodward, R.B.; Brutcher, Jr., F.V. *J. Am. Chem. Soc.* **1958**, *80*, 209.

so the monoacetate is syn and hydrolysis gives the glycol that is the product of overall syn addition. Although the Woodward method results in overall syn addition, the product may be different from that with OsO_4 or $KMnO_4$, since the overall syn process is from the more-hindered side of the alkene.[1684] Both the Prévost and the Woodward methods[1685] have also been carried out in high yields with thallium(I) acetate and thallium(I) benzoate instead of the silver carboxylates.[1686] Note that cyclic sulfates can be prepared from alkenes by reaction with PhIO and $SO_3 \cdot DMF$.[1687]

With suitable substrates, addition of two OH groups creates one new stereogenic center from a terminal alkene and two new stereogenic centers from internal alkenes. Addition to alkenes of the form $RCH=CH_2$ has been made enantioselective, and addition to $RCH=CHR'$ both diastereoselective[1688] and enantioselective,[1689] by using chiral additives or chiral catalysts,[1690] such as **173**, **174** (derivatives of the

173	**174**
9'-Phenanthryl ether of	Dihydroquinine
dihydroquinidine	*p*-chlorobenzoate

[1684]For another method of syn hydroxylation, which can be applied to either face, see Corey, E.J.; Das, J. *Tetrahedron Lett.* **1982**, *23*, 4217.

[1685]For some related methods, see Jasserand, D.; Girard, J.P.; Rossi, J.C.; Granger, R. *Tetrahedron Lett.* **1976**, 1581; Ogata, Y.; Aoki, K. *J. Org. Chem.* **1966**, *31*, 1625; Mangoni, L.; Adinolfi, M.; Barone, G.; Parrilli, M. *Tetrahedron Lett.* **1973**, 4485; *Gazz. Chim. Ital.* **1975**, *105*, 377; Horiuchi, C.A.; Satoh, J.Y. *Chem. Lett.* **1988**, 1209; Campi, E.M.; Deacon, G.B.; Edwards, G.L.; Fitzroy, M.D.; Giunta, N.; Jackson, W.R.; Trainor, R. *J. Chem. Soc., Chem. Commun.* **1989**, 407.

[1686]Cambie, R.C.; Hayward, R.C.; Roberts, J.L.; Rutledge, P.S. *J. Chem. Soc. Perkin Trans. 1*, **1974**, 1858, 1864; Cambie, R.C.; Rutledge, P.S. *Org. Synth.* **VI**, 348.

[1687]Robinson, R.I.; Woodward, S. *Tetrahedron Lett.* **2003**, *44*, 1655.

[1688]For diastereoselective, but not enantioselective, addition of OsO_4, see Cha, J.K.; Christ, W.J.; Kishi, Y. *Tetrahedron* **1984**, *40*, 2247; Stork, G.; Kahn, M. *Tetrahedron Lett.* **1983**, *24*, 3951; Vedejs, E.; McClure, C.K. *J. Am. Chem. Soc.* **1986**, *108*, 1094; Evans, D.A.; Kaldor, S.W. *J. Org. Chem.* **1990**, *55*, 1698.

[1689]Lohray, B.B. *Tetrahedron Asymmetry* **1992**, *3*, 1317.

[1690]For a review of enantioselective oxidation methodologies, see Bonini, C.; Righi, G. *Tetrahedron* **2002**, *58*, 4981. For a study using triazines as a new class of ligand, see McNamara, C.A.; King, F.; Bradley, M. *Tetrahedron Lett.* **2004**, *45*, 8527. See also, Kuang, Y.-Q.; Zhang, S.-Y.; Jiang, R.; Wei, L.-L. *Tetrahedron Lett.* **2002**, *43*, 3669; Jiang, R.; Kuang, Y.; Sun, X.; Zhang, S. *Tetrahedron Asymmetry* **2004**, *15*, 743.

naturally occurring quinine and quinuclidine),[1691] along with OsO$_4$, in what is called *Sharpless asymmetric dihydroxylation.*[1692] Other chiral ligands[1693] have also been used, as well as polymer[1694] and silica-bound[1695] *Cinchona* alkaloids. These amines bind to the OsO$_4$ *in situ* as chiral ligands, causing it to add asymmetrically.[1696] This has been done both with the stoichiometric and with the catalytic method.[1697] The catalytic method has been extended to conjugated dienes, which give tetrahydroxy products diastereoselectively,[1698] and to conjugated ketones.[1699] Asymmetric dihydroxylation has also been reported with chiral alkenes.[1700] Ligands **173** and **174** not only cause enantioselective addition, but also accelerate the reaction, so that they may be useful even where enantioselective addition is not required.[1701] Although **173** and **174** are not enantiomers, they give enantioselective addition to a given alkene in the opposite sense; for example, styrene predominantly gave the (*R*)-diol with **173**, and the

[1691]Wai, J.S.M.; Marko, I.; Svendsen, J.S.; Finn, M.G.; Jacobsen, E.N.; Sharpless, K.B. *J. Am. Chem. Soc.* **1989**, *111*, 1123; Kwong, H.; Sorato, C.; Ogina, Y.; Chen, H.; Sharpless, K.B. *Tetrahedron Lett.* **1990**, *31*, 2999; Shibata, T.; Gilheany, D.C.; Blackburn, B.K.; Sharpless, K.B. *Tetrahedron Lett.* **1990**, *31*, 3817; Sharpless, K.B.; Amberg, W.; Beller, M.; Chens, H.; Hartung, J.; Kawanami, Y.; Lübben, D.; Manoury, E.; Ogino, Y.; Shibata, T.; Ukita, T. *J. Org. Chem.* **1991**, *56*, 4585.

[1692]Kolb, H.C.; Van Nieuwenhze, M.S.; Sharpless, K.B. *Chem. Rev.* **1994**, *94*, 2483. Also see, Smith, M.B. *Organic Synthesis*, 2nd ed., McGraw-Hill, NY, **2001**, pp. 248–254.

[1693]Wang, L.; Sharpless, K.B. *J. Am. Chem. Soc.* **1992**, *114*, 7568; Xu, D.; Crispino, G.A.; Sharpless, K.B. *J. Am. Chem. Soc.* **1992**, *114*, 7570; Corey, E.J.; Jardine, P.D.; Virgil, S.; Yuen, P.; Connell, R.D. *J. Am. Chem. Soc.* **1989**, *111*, 9243; Corey, E.J.; Lotto, G.I. *Tetrahedron Lett.* **1990**, *31*, 2665; Tomioka, K.; Nakajima, M.; Koga, K. *J. Am. Chem. Soc.* **1987**, *109*, 6213; *Tetrahedron Lett.* **1990**, *31*, 1741; Rosini, C.; Tanturli, R.; Pertici, P.; Salvadori, P. *Tetrahedron Asymmetry* **1996**, *7*, 2971; Sharpless, K.B.; Amberg, W.; Bennani, Y.L.; Crispino, G.A.; Hartung, J.; Jeong, K.-S.; Kwong, H.-L.; Morikawa, K.; Wang, Z.-M.; Xu, D.; Zhang, X.-L. *J. Org. Chem.* **1992**, *57*, 2768.

[1694]Bolm, C.; Gerlach, A. *Eur. J. Org. Chem.* **1998**, 21; Lohray, B.B.; Nandanan, E.; Bhushan, V. *Tetrahedron Asymmetry* **1996**, *7*, 2805; Lohray, B.B.; Thomas, A.; Chittari, P.; Ahuja, J.; Dhal, P.K. *Tetrahedron Lett.* **1992**, *33*, 5453. For a review, see Karjalainen, J.K.; Hormi, O.E.O.; Sherrington, D.C. *Tetrahedron Asymmetry* **1998**, *9*, 1563.

[1695]Song, C.E.; Yang, J.W.; Ha, H.-J. *Tetrahedron Asymmetry* **1997**, *8*, 841.

[1696]For discussions of the mechanism of the enantioselectivity, see Corey, E.J.; Noe, M.C. *J. Am. Chem. Soc.* **1996**, *118*, 319; Norrby, P.-O.; Kolb, H.C.; Sharpless, K.B. *J. Am. Chem. Soc.* **1994**, *116*, 8470; Veldkamp, A.; Frenking, G. *J. Am. Chem. Soc.* **1994**, *116*, 4937; Wu, Y.-D.; Wang, Y.; Houk, K.N. *J. Org. Chem.* **1992**, *57*, 1362; Jørgensen, K.A. *Tetrahedron Lett.* **1990**, *31*, 6417. See Nelson, D.W.; Gypser, A.; Ho, P.T.; Kolb, H.C.; Kondo, T.; Kwong, H.-L.; McGrath, D.V.; Rubin, A.E.; Norrby, P.-O.; Gable, K.P.; Sharpless, K.B. *J. Am. Chem. Soc.* **1997**, *119*, 1840 for a discussion of electronic effects and Kolb, H.C.; Andersson, P.G.; Sharpless, K.B. *J. Am. Chem. Soc.* **1994**, *116*, 1278 for a kinetic study.

[1697]For other examples of asymmetric dihydroxylation, see Yamada, T.; Narasaka, K. *Chem. Lett.* **1986**, 131; Tokles, M.; Snyder, J.K. *Tetrahedron Lett.* **1986**, *27*, 3951; Annunziata, R.; Cinquini, M.; Cozzi, F.; Raimondi, L.; Stefanelli, S. *Tetrahedron Lett.* **1987**, *28*, 3139; Hirama, M.; Oishi, T.; Itô, S. *J. Chem. Soc., Chem. Commun.* **1989**, 665.

[1698]Park, C.Y.; Kim, B.M.; Sharpless, K.B. *Tetrahedron Lett.* **1991**, *32*, 1003.

[1699]Walsh, P.J.; Sharpless, K.B. *Synlett* **1993**, 605.

[1700]Oishi, T.; Iida, K.; Hirama, M. *Tetrahedron Lett.* **1993**, *34*, 3573.

[1701]See Jacobsen, E.N.; Marko, I.; France, M.B.; Svendsen, J.S.; Sharpless, K.B. *J. Am. Chem. Soc.* **1989**, *111*, 737.

(*S*)-diol with **174**.[1702] Note that ionic liquids have been used in asymmetric dihydroxylation.[1703]

175   **176**

Two phthalazine derivatives,[1704] (DHQD)$_2$PHAL (**175**) and (DHQ)$_2$PHAL (**176**) used in conjunction with an osmium reagent improve the efficiency and ease of use, and are commercial available as AD-mix-β^{TM} (using **175**) and AD-mix-α^{TM} (using **176**). Catalyst **175** is prepared from dihydroquinidine (DHQD) and 1,4-dichlorophthalazine (PHAL), and **176** is prepared from dihydroquinine (DHQ) and PHAL. The actual oxidation labeled AD-mix α- or β-uses **176** or **175**, respectively, mixed with potassium osmate [K$_2$OsO$_2$(OH)$_4$], powdered K$_3$Fe(CN)$_6$, and powdered K$_2$CO$_3$ in an aqueous solvent mixture.[1705] These additives have been used in conjunction with microencapsulated OsO$_4$,[1705] and polymer bound **175** has been used.[1706] A catalytic amount of flavin has been used.[1707] Both **175**[1708] and **176**[1709] have been used to generate diols with high enantioselectivity. Oxidation of a terminal alkene with AD-mix and then oxidation with TEMPO/NaOCl/NaOCl$_2$ leads to α-hydroxyl carboxylic acids with high enantioselectivity.[1710]

Enantioselective and diastereoselective addition have also been achieved by using preformed derivatives of OsO$_4$, already containing chiral ligands,[1711] and by the use of OsO$_4$ on alkenes that have a chiral group elsewhere in the molecule.[1712]

[1702]Jacobsen, E.N.; Marko, I.; Mungall, W.S.; Schröder, G.; Sharpless, K.B. *J. Am. Chem. Soc.* **1988**, *110*, 1968.

[1703]See Branco, L.C.; Afonso, C.A.M. *J. Org. Chem.* **2004**, *69*, 4381; Branco, L.C.; Afonso, C.A.M. *Chem. Commun.* **2002**, 3036.

[1704]Sharpless, K.B.; Amberg, W.; Bennani, Y.L.; Crispino, G.A.; Hartung, J.; Jeong, K.-S.; Kwong, H.-L.; Morikawa, K.; Wang, Z.-M.; Xu, D.; Zhang, X.-L. *J. Org. Chem.* **1992**, *57*, 2768.

[1705]Kobayashi, S.; Ishida, T.; Akiyama, R. *Org. Lett.* **2001**, *3*, 2649.

[1706]Kuang, Y.-Q.; Zhang, S.-Y.; Wei, L.-L. *Tetrahedron Lett.* **2001**, *42*, 5925.

[1707]Jonsson, S.Y.; Adolfsson, H.; Bäckvall, J.-E. *Org. Lett.* **2001**, *3*, 3463.

[1708]Krief, A.; Colaux-Castillo, C. *Tetrahedron Lett.* **1999**, *40*, 4189.

[1709]Junttila, M.H.; Hormi, O.E.O. *J. Org. Chem.* **2004**, *69*, 4816.

[1710]Aladro, F.J.; Guerra, I.M.; Moreno-Dorado, F.J.; Bustamante, J.M.; Jorge, Z.D.; Massanet, G.M. *Tetrahedron Lett.* **2000**, *41*, 3209.

[1711]Kokubo, T.; Sugimoto, T.; Uchida, T.; Tanimoto, S.; Okano, M. *J. Chem. Soc., Chem. Commun.* **1983**, 769.

[1712]Hauser, F.M.; Ellenberger, S.R.; Clardy, J.C.; Bass, L.S. *J. Am. Chem. Soc.* **1984**, *106*, 2458; Johnson, C.R.; Barbachyn, M.R. *J. Am. Chem. Soc.* **1984**, *106*, 2459.

Alkenes can also be oxidized with metallic acetates such as lead tetraacetate[1713] or thallium(III) acetate[1714] to give bis(acetates) of glycols.[1715] Oxidizing agents, such as benzoquinone, MnO_2, or O_2, along with palladium acetate, have been used to convert conjugated dienes to 1,4-diacetoxy-2-alkenes (1,4-addition).[1716] Sodium periodate and sulfuric acid in aqueous media converts conjugated esters to dihydroxy esters.[1717] Diols are also produced by the reaction of a terminal alkyne with Bu_3SnH, followed by ozonolysis, followed by reduction with $BH_3 \cdot SMe_2$.[1718] 1,2-Diols are also generated from terminal alkynes by two sequential reactions with a platinum catalyst, and then a palladium catalyst, both with $HSiCl_3$, and a final oxidation with H_2O_2–KF.[1719]

1,2-Dithiols can be prepared from alkenes.[1720]

OS **II**, 307; **III**, 217; **IV**, 317; **V**, 647; **VI**, 196, 342, 348; **IX**, 251, 383.

15-49   Dihydroxylation of Aromatic Rings

Dihydroxy-addition

One π-bond of an aromatic ring can be converted to a cyclohexadiene 1,2-diol by reaction with enzymes associated with *Pseudomonas putida*.[1721] A variety of substituted aromatic compounds can be oxidized, including bromobenzene, chlorobenzene,[1722] and toluene.[1723] In these latter cases, introduction of the hydroxyl

[1713]For a review, see Moriarty, R.M. *Sel Org. Transform.* **1972**, *2*, 183–237.

[1714]See, for example, Uemura, S.; Miyoshi, H.; Tabata, A.; Okano, M. *Tetrahedron* **1981**, *37*, 291. For a review of the reactions of thallium (III) compounds with alkenes, see Uemura, S., in Hartley, F.R. *The Chemistry of the Metal–Carbon Bond*, Vol. 4, Wiley, NY, **1987**, pp. 473–538, 497–513. For a review of thallium (III) acetate and trifluoroacetate, see Uemura, S., in Pizey, J.S. *Synthetic Reagents*, Vol. 5, Wiley, NY, **1983**, pp. 165–187.

[1715]For another method, see Fristad, W.E.; Peterson, J.R. *Tetrahedron* **1984**, *40*, 1469.

[1716]See Bäckvall, J.E.; Awasthi, A.K.; Renko, Z.D. *J. Am. Chem. Soc.* **1987**, *109*, 4750, and references cited therein. For articles on this and related reactions, see Bäckvall, J.E. *Bull. Soc. Chim. Fr.* **1987**, 665; *New. J. Chem.* **1990**, *14*, 447. For another method, see Uemura, S.; Fukuzawa, S.; Patil, S.R.; Okano, M. *J. Chem. Soc. Perkin Trans. 1*, **1985**, 499.

[1717]Plietker, B.; Niggemann, M. *Org. Lett.* **2003**, *5*, 3353.

[1718]Gómez, A.M.; Company, M.D.; Valverde, S.; López, J.C. *Org. Lett.* **2002**, *4*, 383.

[1719]Shimada, T.; Mukaide, K.; Shinohara, A.; Han, J.W.; Hayashi, T. *J. Am. Chem. Soc.* **2004**, *124*, 1584.

[1720]Elgemeie, G.H.; Sayed, S.H. *Synthesis* **2001**, 1747.

[1721]Gibson, D.T.; Koch, J.R.; Kallio, R.E. *Biochemistry* **1968**, *7*, 2653; Brown, S.M., in Hudlicky, T. *Organic Synthesis: Theory and Practice*, JAI Press, Greenwich, CT., **1993**, Vol. 2, p. 113; Carless, H.A.J. *Tetrahedron Asymmetry* **1992**, *3*, 795; Widdowson, D.A.; Ribbons, D.A.; Thomas, S.D. *Janssenchimica Acta* **1990**, *8*, 3.

[1722]Gibson, D.T.; Koch, J.R.; Schuld, C.L.; Kallio, R.E. *Biochemistry* **1968**, *7*, 3795; Hudlicky, T.; Price, J.D. *Synlett.* **1990**, 159.

[1723]Gibson, D.T.; Hensley, M.; Yoshioka, H.; Mabry, T.J. *Biochemsitry* **1970**, *9*, 1626.

groups generates a chiral molecule that can be used as a template for asymmetric syntheses.[1724]

OS **X**, 217.

15-50   Epoxidation (Addition of Oxygen, Oxygen)

epi-**Oxy-addition**

Alkenes can be epoxidized with many peroxyacids,[1725] of which *m*-chloroperoxybenzoic has been the most often used. The reaction, called the *Prilezhaev reaction*, has wide utility.[1726] Alkyl, aryl, hydroxyl, ester, and other groups may be present, although not amino groups, since these are affected by the reagent. Electron-donating groups increase the rate, and the reaction is particularly rapid with tetraalkyl alkenes. Conditions are mild and yields are high. Other peroxyacids, especially peroxyacetic and peroxybenzoic, are also used; trifluoroperoxyacetic acid[1727] and 3,5-dinitroperoxybenzoic acid[1728] are particularly reactive ones. Transition metal catalysts can facilitate epoxidation of alkenes at low temperatures or with alkenes that may otherwise react sluggishly.[1729] Magnesium monoperoxyphthalate (MMPP)[1730] is commercially available, and has been

[1724]Hudlicky, T.; Gonzalez, D.; Gibson, D.T. *Aldrichimica Acta* **1999**, *32*, 35; Hudlicky, T.; Luna, H.; Barbieri, G.; Kwart, L.D. *J. Am. Chem. Soc.* **1988**, *110*, 4735; Hudlicky, T.; Seoane, G.; Pettus, T. *J. Org. Chem.* **1989**, *54*, 4239; Ley, S.V.; Redgrave, A.J. *Synlett* **1990**, 393; Ley, S.V.; Sternfeld, F.; Taylor, S. *Tetrahedron Lett.* **1987**, *28*, 225; Hudlicky, T.; Olivo, H.F. *Tetrahedron Lett.* **1991**, *32*, 6077; Hudlicky, T.; Luna, H.; Price, J.D.; Rulin, F. *J. Org. Chem.* **1990**, *55*, 4683; Hudlicky, T.; Olivo, H.F. *J. Am. Chem. Soc.* **1992**, *114*, 9694. Also see, Smith, M.B. *Organic Synthesis*, 2nd ed., McGraw-Hill, NY, **2001**, pp. 256–258.

[1725]For a list of reagents, including peracids and others, used for epoxidation, with references, see Larock, R.C. *Comprehensive Organic Transformations*, 2nd ed., Wiley-VCH, NY, **1999**, pp. 915–927.

[1726]For reviews, see Hudlický, M. *Oxidations in Organic Chemistry*, American Chemical Society, Washington, DC, **1990**, pp. 60–64; Haines, A.H. *Methods for the Oxidation of Organic Compunds*, Academic Press, NY, **1985**, pp. 98–117, 295–303; Dryuk, V.G. *Russ. Chem. Rev.* **1985**, *54*, 986; Plesničar, B., in Trahanovsky, W.S. *Oxidation in Organic Chemistry*, pt. C, Academic Press, NY, **1978**, pp. 211–252; Swern, D., in Swern, D. *Organic Peroxides*, Vol. 2, Wiley, NY, **1971**, pp. 355–533; Metelitsa, D.I. *Russ. Chem. Rev.* **1972**, *41*, 807; Hiatt, R., in Augustine, R.L.; Trecker, D.J. *Oxidation*, Vol. 2, Marcel Dekker, NY, **1971**; pp. 113–140; House, H.O. *Modern Synthetic Reaction*, 2nd ed., W.A. Benjamin, NY, **1972**, pp. 292–321. For a review pertaining to the stereochemistry of the reaction, see Berti, G. *Top Stereochem.* **1973**, *7*, 93, p. 95.

[1727]Emmons, W.D.; Pagano, A.S. *J. Am. Chem. Soc.* **1955**, *77*, 89.

[1728]Rastetter, W.H.; Richard, T.J.; Lewis, M.D. *J. Org. Chem.* **1978**, *43*, 3163.

[1729]Cu catalysts: Andrus, M.B.; Poehlein, B.W. *Tetrahedron Lett.* **2000**, *41*, 1013. Fe catalysts: Dubois, G.; Murphy, A.; Stack, T.D.P. *Org. Lett.* **2003**, *5*, 2469. Mn catalysts: Murphy, A.; Pace, A.; Stack, T.D.P. *Org. Lett.* **2004**, *6*, 3119; Murphy, A.; Dubois, G.; Stack, T.D.P. *J. Am. Chem. Soc.* **2003**, *125*, 5250.

[1730]Brougham, P.; Cooper, M.S.; Cummerson, D.A.; Heaney, H.; Thompson, N. *Synthesis* **1987**, 1015; Querci, C.; Ricci, M. *J. Chem. Soc., Chem. Commun.* **1989**, 889. For a reaction using moist MMPP, see Foti, C.J.; Fields, J.D.; Kropp, P.J. *Org. Lett.* **1999**, *1*, 903.

shown to be a good substitute for *m*-chloroperoxybenzoic acid in a number of reactions.[1731]

177

The one-step mechanism involving a transition state, such as **177**,[1731] was proposed by Bartlett:[1732] Evidence for this concerted mechanism is as follows:[1733] (*1*) The reaction is second order. If ionization were the rate-determining step, it would be first order in peroxyacid. (*2*) The reaction readily takes place in nonpolar solvents, where formation of ions is inhibited.[1734] (*3*) Measurements of the effect on the reaction rate of changes in the substrate structure show that there is no carbocation character in the transition state.[1735] (*4*) The addition is stereospecific (i.e., a trans-alkene gives a trans-epoxide and a cis-alkene a cis-epoxide) even in cases where electron-donating substituents would stabilize a hypothetical carbocation intermediate.[1736] However, where there is an OH group in the allylic or homoallylic position, the stereospecificity diminishes or disappears, with both cis and trans isomers giving predominantly or exclusively the product where the incoming oxygen is syn to the OH group. This probably indicates a transition state in which there is hydrogen bonding between the OH group and the peroxy acid.[1737]

[1731]For discussions of the mechanism, see Dryuk, V.G. *Tetrahedron* **1976**, *32*, 2855; Finn, M.G.; Sharpless, K.B., in Morrison, J.D. *Asymmetric Synthesis*, Vol. 5, Wiley, NY, **1985**, pp. 247–308; Bach, R.D.; Canepa, C.; Winter, J.E.; Blanchette, P.E. *J. Org. Chem.* **1997**, *62*, 5191. For a review of polar mechanisms involving peroxides, see Plesnicar, B., in Patai, S. *The Chemistry of Peroxides*, Wiley, NY, **1983**, pp. 521–584. See Freccero, M.; Gandolfi, R.; Sarzi-Amadè, M.; Rastelli, A. *J. Org. Chem.* **2002**, *67*, 8519. For a discussion of arene–arene interactions as related to selectivity, see Kishikawa, K.; Naruse, M.; Kohmoto, S.; Yamamoto, M.; Yamaguchi, K. *J. Chem. Soc., Perkin Trans. 1* **2001**, 462.

[1732]Bartlett, P.D. *Rec. Chem. Prog.* **1957**, *18*, 111. For other proposed mechanisms, see Kwart, H.; Hoffman, D.M. *J. Org. Chem.* **1966**, *31*, 419; Hanzlik, R.P.; Shearer, G.O. *J. Am. Chem. Soc.* **1975**, *97*, 5231.

[1733]Ogata, Y.; Tabushi, I. *J. Am. Chem. Soc.* **1961**, *83*, 3440; Freccero, M.; Gandolfi, R.; Sarzi-Amadè, M.; Rastelli, A. *J. Org. Chem.* **2004**, *69*, 7479. See also, Woods, K.W.; Beak, P. *J. Am. Chem. Soc.* **1991**, *113*, 6281. Also see, Vedejs, E.; Dent III, W.H.; Kendall, J.T.; Oliver, P.A. *J. Am. Chem. Soc.* **1996**, *118*, 3556.

[1734]See Gisdakis, P.; Rösch, N. *Eur. J. Org. Chem.* **2001**, 719.

[1735]Khalil, M.M.; Pritzkow, W. *J. Prakt. Chem.* **1973**, *315*, 58; Schneider, H.; Becker, N.; Philippi, K. *Chem. Ber.* **1981**, *114*, 1562; Batog, A.E.; Savenko, T.V.; Batrak, T.A.; Kucher, R.V. *J. Org. Chem. USSR* **1981**, *17*, 1860.

[1736]For a computational study of facial selectivity, see Freccero, M.; Gandolfi, R.; Sarzi-Amadè, M.; Rastelli, A. *J. Org. Chem.* **2000**, *65*, 8948.

[1737]See Berti, G. *Top. Stereochem.* **1973**, *7*, 93, 130–162; Houk, K.N.; Liu, J.; DeMello, N.C.; Condroski, K.R. *J. Am. Chem. Soc.* **1997**, *119*, 10147.

In general, peroxides (HOOH and ROOH) are poor regents for epoxidation of simple alkenes since OH and OR are poor leaving groups in the concerted mechanism shown above.[1738] Aqueous hydrogen peroxide epoxidizes alkenes in the presence of fluorous compounds, such as CF_3CH_2OH[1739] or hexafluoroacetone.[1740] Transition-metal catalysts[1741] have been used with alkyl hydroperoxides.[1742] In the presence of other reagents,[1743] peroxides give good yields of the epoxide. These coreagents include dicyclohexylcarbodiimide,[1744] magnesium aluminates,[1745] metalloporphyrins,[1746] hydrotalcite[1747] with microwave irradiation,[1748] fluorous aryl selenides,[1749] and arsines in fluorous solvents.[1750] The catalyst $MeReO_3$[1751] has been used for epoxidation with sodium percarbonate and pyrazole,[1752] with hydrogen peroxide,[1753] and with urea–H_2O_2.[1754] Epoxidation occurs with $FeSO_4$/silica,[1755] and with N_2O and a zinc catalyst.[1756] Epoxidation occurs when alkenes

---

[1738]See Deubel, D.V.; Frenking, G.; Gisdakis, P.; Herrmann, W.A.; Rösch, N.; Sundermeyer, J. *Acc. Chem. Res. 2004*, *37*, 645.

[1739]Neimann, K.; Neumann, R. *Org. Lett. 2000*, *2*, 2861; van Vliet, M.C.A.; Arends, I.W.C.E.; Sheldon, R.A. *Synlett 2001*, 248.

[1740]Shu, L.; Shi, Y. *J. Org. Chem. 2000*, *65*, 8807.

[1741]**V**: Sharpless, K.B.; Verhoeven, T.R. *Aldrichimica Acta 1979*, *12*, 63; Hoshino, Y.; Yamamoto, H. *J. Am. Chem. Soc. 2000*, *122*, 10452; Lattanzi, A.; Leadbeater, N.E. *Org. Lett. 2002*, *4*, 1519; Torres, G.; Torres, W.; Prieto, J.A. *Tetrahedron 2004*, *60*, 10245. **Mn**: Lane, B.S.; Vogt, M.; De Rose, V.T.; Burgess, K. *J. Am. Chem. Soc. 2002*, *124*, 11946. **Ti**: Della Sala, G.D.; Giordano, L.; Lattanzi, A.; Proto, A.; Screttri, A. *Tetrahedron 2000*, *56*, 3567; Lattanzi, A.; Iannece, P.; Screttri, A. *Tetrahedron Lett. 2002*, *43*, 5629. **Pd**: Yu, J.-Q.; Corey, E.J. *Org. Lett. 2002*, *4*, 2727. **Ru**: Adam, W.; Alsters, P.L.; Neumann, R.; Saha-Möller, C.; Sloboda-Rozner, D.; Zhang, R. *Synlett 2002*, 2011. **La**: Nemoto, T.; Ohshima, T.; Yamaguchi, K.; Shibasaki, M. *J. Am. Chem. Soc. 2001*, *123*, 2725; Chen, R.; Qian, C.; de Vries, J.G. *Tetrahedron 2001*, *57*, 9837; Nemoto, T.; Kakei, H.; Gnanadesikan, V.; Tosaki, S.-y.; Ohshima, T.; Shibasaki, M. *J. Am. Chem. Soc. 2002*, *124*, 14544.

[1742]For a table containing several common catalysts, see Hiatt, R., in Augustine, R.L.; Trecker, D.J. *Oxidation*, Vol. 2, Marcel Dekker, NY, *1971*, p. 124.

[1743]For other methods of converting alkenes to epoxides, see Bruice, T.C. *Aldrichimica Acta 1988*, *21*, 87; Adam, W.; Curci, R.; Edwards, J.O. *Acc. Chem. Res. 1989*, *22*, 205.

[1744]Majetich, G.; Hicks, R.; Sun, G.-r.; McGill, P. *J. Org. Chem. 1998*, *63*, 2564; Murray, R.W.; Iyanar, K. *J. Org. Chem. 1998*, *63*, 1730.

[1745]Yamaguchi, K.; Ebitani, K.; Kaneda, K. *J. Org. Chem. 1999*, *64*, 2966.

[1746]Chan, W.-K.; Liu, P.; Yu, W.-Y.; Wong, M.-K.; Che, C.-M. *Org. Lett. 2004*, *6*, 1597.

[1747]For an example without microwave irradiation, see Pillai, U.R.; Sahle-Demessie, E.; Varma, R.S. *Synth. Commun. 2003*, *33*, 2017.

[1748]Pillai, U.R.; Sahle-Demessie, E.; Varma, R.S. *Tetrahedron Lett. 2002*, *43*, 2909.

[1749]Betzemeier, B.; Lhermitte, F.; Knochel, P. *Synlett 1999*, 489.

[1750]Van Vliet, M.C.A.; Arends, I.W.C.E.; Sheldon, R.A. *Tetrahedron Lett. 1999*, *40*, 5239.

[1751]For a polymer-supported $MeReO_3$ reagent, see Saladino, R.; Neri, V.; Pelliccia, A.R.; Caminiti, R.; Sadun, C. *J. Org. Chem. 2002*, *67*, 1323.

[1752]Vaino, A.R. *J. Org. Chem. 2000*, *65*, 4210.

[1753]van Vliet, M.C.A.; Arends, I.W.C.E.; Sheldon, R.A. *Chem. Commun. 1999*, 821; Adolfsson, H.; Copéret, C.; Chiang, J.P.; Yudin, A.K. *J. Org. Chem. 2000*, *65*, 8651; Iskra, J.; Bonnet-Delpon, D.; Bégué, J.-P. *Tetrahedron Lett. 2002*, *43*, 1001.

[1754]Owens, G.S.; Abu-Omar, M.M. *Chem. Commun. 2000*, 1165.

[1755]Monfared, H.H.; Ghorbani, M. *Monat. Chem. 2001*, *132*, 989.

[1756]Ben-Daniel, R.; Weiner, L.; Neumann, R. *J. Am. Chem. Soc. 2002*, *124*, 8788.

are treated with oxygen gas, *N*-hydroxyphthalimide, and a mixture of cobalt and molybdenum catalyst.[1757]

Other epoxidation methods are available. Enzymatic epoxidation[1758] and epoxidation with catalytic antibodies[1759] have been reported. Chromyl chloride (CrO_2Cl_2) reacts with alkenes, even at $-78°C$ to give an epoxide and numerous side products including chlorohydrins and dichlorides.[1760] Several mechanisms have been proposed.[1761] Epoxidation has been done in ionic liquids using 10% H_2O_2 with $MnSO_4$[1762] or an iron catalyst.[1763] Hypervalent iodine compounds, such as $PhI(OAc)_2$, in conjunction with a ruthenium catalyst in aqueous media, converts alkenes to epoxides.[1764] This reagent has been used in an ionic liquid with a manganese catalyst.[1765]

Dioxiranes,[1766] such as dimethyl dioxirane (**178**),[1767] either isolated or generated *in situ*,[1768] are important epoxidation reagents. With dimethyloxirane, C–H insertion reactions can occur preferentially.[1769] The reaction with alkenes is rapid, mild, safe, and a variety methods have been developed using an oxidant as a coreagent. The most commonly used coreagent is probably potassium peroxomonosulfate ($KHSO_5$). Oxone® (2 $KHSO_5$·$KHSO_4$·K_2SO_4) is a common source of $KHSO_5$. Oxone® reacts with ketones[1770] and sodium bicarbonate to convert an alkene

[1757]Iwahama, T.; Hatta, G.; Sakaguchi, S.; Ishii, Y. *Chem. Commun.* **2000**, 163.

[1758]**Haloperoxidases**: Hu, S.; Hager, L.P. *Tetrahedron Lett.* **1999**, *40*, 1641; Dembitsky, V.M. *Tetrahedron* **2003**, *59*, 4701. *E. coli JM109(pTAB19)*: Bernasconi, S.; Orsini, F.; Sello, G.; Colmegna, A.; Galli, E.; Bestetti, G. *Tetrahedron Lett.* **2000**, *41*, 9157. **Cyclohexanone monooxygenase**: Colonna, S.; Gaggero, N.; Carrea, G.; Ottolina, G.; Pasta, P.; Zambianchi, F. *Tetrahedron Lett.* **2002**, *43*, 1797.

[1759]Chen, Y.; Reymond, J.-L. *Synthesis* **2001**, 934.

[1760]Sharpless, K.B.; Teranishi, A.Y.; Bäckvall, J.-E. *J. Am. Chem. Soc.* **1977**, *99*, 3120.

[1761]For leading references, see Rappe, A.K.; Li, S. *J. Am. Chem. Soc.* **2003**, *125*, 11188.

[1762]In bmim BF_4, 1-butyl-3-methylimidazolium tetrafluoroborate: Tong, K.-H.; Wong, K.-Y.; Chan, T.H. *Org. Lett.* **2003**, *5*, 3423.

[1763]In bmim Br, 1-butyl-3-methylimidazolium bromide: Srinivas, K.A.; Kumar, A.; Chauhan, S.M.S. *Chem. Commun.* **2002**, 2456.

[1764]Tse, M.K.; Bhor, S.; Klawonn, M.; Döbler, C.; Beller, M. *Tetrahedron Lett.* **2003**, *44*, 7479.

[1765]In bmim PF_6, 1-butyl-3-methylimidazolium hexafluorophosphate: Li, Z.; Xia, C.-G. *Tetrahedron Lett.* **2003**, *44*, 2069.

[1766]For general leading references, see Murray, R.W. *Chem. Rev.* **1989**, *89*, 1187; Adam, W.; Curci, R.; Edwards, J.O. *Acc. Chem. Res.* **1989**, *22*, 205; Curci, R.; Dinoi, A.; Rubino, M.E. *Pure Appl. Chem.* **1995**, *67*, 811; Clennan, E.L. *Trends in Organic Chemistry*, **1995**, *5*, 231; Adam, W.; Smerz, A.K. *Bull Soc. Chim. Belg.* **1996**, *105*, 581; Denmark, S.E.; Wu, Z. *Synlett* **1999**, 847.

[1767]Frohn, M.; Wang, Z.-X.; Shi, Y. *J. Org. Chem.* **1998**, *63*, 6425. See Angelis, Y.; Zhang, X.; Organopoulos, M. *Tetrahedron Lett.* **1996**, *37*, 5991 for a discussion of the mechanism of this oxidation.

[1768]See Curci, R.; Fiorentino, M.; Troisi, L.; Edwards, J.O.; Pater, R.H. *J. Org. Chem.* **1980**, *45*, 4758; Gallopo, A.R.; Edwards, J.O. *J. Org. Chem.* **1981**, *46*, 1684; Corey, P.E; Ward, F.E. *J. Org. Chem.* **1986**, *51*, 1925; Adam, W.; Hadjiarapoglou, L.; Smerz, A. *Chem. Ber.* **1991**, *124*, 227; Yang, D.; Wong, M.K.; Yip, Y.C. *J. Org. Chem.* **1995**, *60*, 3887; Denmark, S.E.; Wu, Z. *J. Org. Chem.* **1998**, *63*, 2810, and references cited therein; Frohn, M.; Wang, Z.-X.; Shi, Y. *J. Org. Chem.* **1998**, *63*, 6425; Yang, D.; Yip, Y.-C.; Tang, M.-W.; Wong, M.-K.; Cheung, K.-K. *J. Org. Chem.* **1998**, *63*, 9888, and references cited therein.

[1769]Adam, W.; Prechtl, F.; Richter, M.J.; Smerz, A.K. *Tetrahedron Lett.* **1993**, *34*, 8427.

[1770]Ferraz, H.M.C.; Muzzi, R.M.; de O.Viera, T.; Viertler, H. *Tetrahedron Lett.* **2000**, *41*, 5021; Legros, J.; Crousse, B.; Bourdon, J.; Bonnet-Delpon, D.; Bégué, J.-P. *Tetrahedron Lett.* **2001**, *42*, 4463. For a reaction with a ketone immobilized on silica, see Sartori, G.; Armstrong, A.; Maggi, R.; Mazzacani, A.; Sartorio, R.; Bigi, F.; Dominguez-Fernandez, B. *J. Org. Chem.* **2003**, *68*, 3232.

to an epoxide. Oxone® also converts alkenes to epoxides in the presence of certain additives, such as *N,N*-dialkylalloxans.[1771] Oxone, usually with hydrogen peroxide or another similar oxidant, can be used with chiral ketones[1772] or aldehydes to convert alkenes to chiral, nonracemic epoxides.[1773] Chiral dioxiranes have reportedly given nonracemic epoxides.[1774] Hydrogen peroxide, in the presence of chiral ketones in acetonitrile (or other nitrile solvents), probably converts alkenes to epoxides with good enantioselectivity by *in situ* generation of dioxirane.[1775] Epoxidation does not occur in good yields with these reagents in most other solvents, and it is suggested that the active agent that generates dioxirane is peroxyimidic acid MeC(=NH)OOH.[1776] Note that benzaldehyde with Chloramine-M[1777] will convert alkenes to epoxides.[1778] Amines, including chiral amines can be similarly used with aldehydes with aqueous sodium bicarbonate.[1779]

178   oxirene

179

Oxone® oxidizes iminium salts to an oxaziridinium intermediate **179**, which can transfer oxygen to an alkene to form an epoxide and regenerate the iminium salt.[1780]

[1771]Carnell, A.J.; Johnstone, R.A.W.; Parsy, C.C.; Sanderson, W.R. *Tetrahedron Lett.* **1999**, *40*, 8029.

[1772]For reviews, see Shi, Y. *Acc. Chem. Res.* **2004**, *37*, 488; Yang, D. *Acc. Chem. Res.* **2004**, *37*, 497.

[1773]For leading references, see: Denmark, S.E.; Wu, Z.; Crudden, C.M.; Matsuhashi, H. *J. Org. Chem.* **1997**, *62*, 8288; Yang, D.; Yip, Y.-C.; Chen, J.; Cheung, K.-K. *J. Am. Chem. Soc.* **1998**, *120*, 7659; Daly, A.M.; Renehan, M.F.; Gilheany, D.G. *Org. Lett.* **2001**, *3*, 663; Tian, H.; She, X.; Yu, H.; Shu, L.; Shi, Y. *J. Org. Chem.* **2002**, *67*, 2435; Denmark, S.E.; Matsuhashi, H. *J. Org. Chem.* **2002**, *67*, 3479; Arsmtrong, A.; Ahmed, G.; Dominguez-Fernandez, B.; Hayter, B.R.; Wailes, J.S. *J. Org. Chem.* **2002**, *67*, 8610; Wu, X.-Y.; She, X.; Shi, Y. *J. Am. Chem. Soc.* **2002**, *124*, 8792; Matsumoto, K.; Tomioka, K. *Tetrahedron Lett.* **2002**, *43*, 631; Bez, G.; Zhao, C.-G. *Tetrahedron Lett.* **2003**, *44*, 7403. For a carbonyl derivative bound to cyclodextrin, see Chan, W.-K.; Yu, W.-y.; Che, C.-M.; Wong, M.-K. *J. Org. Chem.* **2003**, *68*, 6576.

[1774]Tian, H.; She, X.; Shu, L.; Yu, H.; Shi, Y. *J. Am. Chem. Soc.* **2000**, *122*, 11551.

[1775]Shu, L.; Shi, Y. *Tetrahedron Lett.* **1999**, *40*, 8721.

[1776]Payne, G.B.; Deming, P.H.; Williams, P.H. *J. Org. Chem.* **1961**, *26*, 659; Payne, G.B. *Tetrahedron* **1962**, *18*, 763; McIsaac, Jr., J.E.; Ball, R.E.; Behrman, E.J. *J. Org. Chem.* **1971**, *36*, 3048; Bach, R.D.; Knight, J.W. *Org. Synth.* **1981**, *60*, 63; Arias, L.A.; Adkins, S.; Nagel, C.J.; Bach, R.D. *J. Org. Chem.* **1983**, *48*, 888.

[1777]For the preparation of Chloramine-M, see Rudolph, J.; Sennhenn, P.C.; Vlaar, C.P.; Sharpless, K.B. *Angew. Chem. Int. Ed.* **1996**, *35*, 2810.

[1778]Yang, D.; Zhang, C.; Wang, X.-C. *J. Am. Chem. Soc.* **2000**, *122*, 4039.

[1779]Wong, M.-K.; Ho, L.-M.; Zheng, Y.-S.; Ho, C.-Y.; Yang, D. *Org. Lett.* **2001**, *3*, 2587.

[1780]See Lusinchi, X.; Hanquet, G. *Tetrahedron* **1997**, *53*, 13727; Hanquet, G.; Lusinchi, X.; Milliet, P. *Tetrahedron Lett.* **1988**, *29*, 3941; Bohé, L.; Kammoun, M. *Tetrahedron Lett.* **2002**, *43*, 803; Bohé, L.; Kammoun, M. *Tetrahedron Lett.* **2004**, *45*, 747.

This variation has been applied to asymmetric[1781] epoxidations using chiral iminium salt precursors.[1782]

Although cis–trans isomerization of epoxides is not formally associated with this section, it is clearly a potential problem in the conversion of an alkene to an epoxide. There are several catalysts for this process.[1783]

It would be useful if triple bonds could be similarly epoxidized to give oxirenes, but they are not stable compounds.[1784] Two of them have been trapped in solid argon matrices at very low temperatures, but they decayed on warming to 35 K.[1785] Oxirenes probably form in the reaction,[1786] but react further before they can be isolated. Note that oxirenes bear the same relationship to cyclobutadiene that furan does to benzene and may therefore be expected to be antiaromatic (see p. 38).

Conjugated dienes can be epoxidized (1,2-addition), although the reaction is slower than for corresponding alkenes, but α,β-unsaturated ketones do not generally give epoxides when treated with peroxyacids.[1787] The epoxidation of α,β-unsaturated ketones with hydrogen peroxide under basic conditions is known as the *Waits–Scheffer epoxidation*, discovered in 1921.[1788] This fundamental reaction has been extended to α,β-unsaturated ketones (including quinones), aldehydes, and sulfones.[1789] This is a nucleophilic addition by a Michael-type mechanism, involving attack by HO_2^-:[1790] This reaction is another example of 1,4-addition of a heteroatom containing species as discussed in **15-31**.

[1781] For a discussion of the origins of selectivity in these reactions, see Washington, I.; Houk, K. N. *J. Am. Chem. Soc.* **2000**, *122*, 2948.

[1782] See Jacobson, E.N., in Ojima, I. *Catalytic Asymmetric Synthesis*, VCH, NY, **1993**, pp. 159–203; Armstrong, A.; Ahmed, G.; Garnett, I.; Goacolou, K.; Wailes, J.S. *Tetrahedron* **1999**, *55*, 2341; Minakata, S.; Takemiya, A.; Nakamura, K.; Ryu, I.; Komatsu, M. *Synlett* **2000**, 1810; Page, P.C.B.; Rassias, G.A.; Barros, D.; Ardakani, A.; Buckley, B.; Bethell, D.; Smith, T.A.D.; Slawin, A.M.Z. *J. Org. Chem.* **2001**, *66*, 6926; Page, P.C.B.; Barros, D.; Buckley, B.R.; Ardakani, A.; Marples, B.A. *J. Org. Chem.* **2004**, *69*, 3595; Page, P.C.B.; Buckley, B.R.; Blacker, A.J. *Org. Lett.* **2004**, *6*, 1543; Page, P.C.B.; Rassias, G.A.; Barros, D.; Ardakani, A.; Bethell, D.; Merrifield, E. *Synlett* **2002**, 580.

[1783] Lo, C.-Y.; Pal, S.; Odedra, A.; Liu, R.-S. *Tetrahedron Lett.* **2003**, *44*, 3143.

[1784] For a review of oxirenes, see Lewars, E.G. *Chem. Rev.* **1983**, *83*, 519.

[1785] Torres, M.; Bourdelande, J.L.; Clement, A.; Strausz, O.P. *J. Am. Chem. Soc.* **1983**, *105*, 1698. See also, Laganis, E.D.; Janik, D.S.; Curphey, T.J.; Lemal, D.M. *J. Am. Chem. Soc.* **1983**, *105*, 7457.

[1786] McDonald, R.N.; Schwab, P.A. *J. Am. Chem. Soc.* **1964**, *86*, 4866; Ibne-Rasa, K.M.; Pater, R.H.; Ciabattoni, J.; Edwards, J.O. *J. Am. Chem. Soc.* **1973**, *95*, 7894; Ogata, Y.; Sawaki, Y.; Inoue, H. *J. Org. Chem.* **1973**, *38*, 1044.

[1787] A few exceptions are known. For example, see Hart, H.; Verma, M.; Wang, I. *J. Org. Chem.* **1973**, *38*, 3418.

[1788] Weitz, E.; Scheffer, A. *Ber. Dtsch. Chem. Ges.* **1921**, *54*, 2327.

[1789] For example, see Payne, G.B.; Williams, P.H. *J. Org. Chem.* **1961**, *26*, 651; Zwanenburg, B.; ter Wiel, J. *Tetrahedron Lett.* **1970**, 935.

[1790] Bunton, C.A.; Minkoff, G.J. *J. Chem. Soc.* **1949**, 665; Temple, R.D. *J. Org. Chem.* **1970**, *35*, 1275; Apeloig, Y.; Karni, M.; Rappoport, Z. *J. Am. Chem. Soc.* **1983**, *105*, 2784. For a review, see Patai, S.; Rappoport, Z., in Patai, S. *The Chemistry of Alkenes*, pt. 1, Wiley, NY, **1964**, pp. 512–517.

α,β-Unsaturated compounds can be epoxidized alkyl hydroperoxides and a base,[1791] or with H_2O_2 and a base or heteropoly acids.[1792] The reaction has been done in D_2O using sodium bicarbonate with hydrogen peroxide.[1793] The reaction has been done with LiOH and polymer-bound quaternary ammonium salts.[1794] Epoxides can also be prepared by treating alkenes with oxygen or with an alkyl peroxide[1795] catalyzed by a complex of a transition metal such as V, Mo, Ti, La,[1796] or Co.[1797] The reaction with oxygen, which can also be carried out without a catalyst, is probably a free-radical process.[1798] Conjugated ketones are oxidized to epoxy-ketones with $NaBO_3$ and tetrahexylammonium hydrogen sulfate,[1799] $KF-Al_2O_3$/*tert*-butyl hydroperoxide.[1800] α,β-Unsaturated esters react normally to give glycidic esters.[1801] When a carbonyl group is elsewhere in the molecule but not conjugated with the double bond, the Baeyer–Villiger reaction (**18-19**) may compete. Allenes[1802] are converted by peroxyacids to allene oxides[1803] or spiro dioxides, both of which species can in certain cases be isolated[1804] but more often are unstable under the reaction conditions and react further to give other products.[1805]

Asymmetric Weitz–Scheffer epoxidation is commonly used for the epoxidation of electron-poor alkenes. Cinchona-derived phase-transfer catalysts, initially used

[1791]**Organolithium reagents**: Bailey, P.L.; Clegg, W.; Jackson, R.F.W.; Meth-Cohn, O. *J. Chem. Soc. Perkin Trans. 1*, **1990**, 200. **KOH**: Adam, W.; Rao, P.B.; Degen, H.-G.; Saha-Möller, C.R. *J. Am. Chem. Soc.* **2000**, *122*, 5654. **LiOH**: Arai, S.; Tsuge, H.; Oku, M.; Miura, M.; Shioiri, T. *Tetrahedron* **2002**, *58*, 1623. **1,5,7-Triazabicyclo[4.4.0]dec-5-ene derivatives**: Genski, T.; Macdonald, G.; Wei, X.; Lewis, N.; Taylor, R.J.K. *Synlett* **1999**, 795. **DBU**: Yadav, V.K.; Kapoor, K.K. *Tetrahedron* **1995**, *51*, 8573. **NaHCO₃**: Bortolini, O.; Fogagnolo, M.; Fantin, G.; Maietti, S.; Medici, A. *Tetrahedron Asymmetry* **2001**, *12*, 1113. **Hydrotalcites**: Honma, T.; Nakajo, M.; Mizugaki, T.; Ebitani, K.; Kaneda, K. *Tetrahedron Lett.* **2002**, *43*, 6229.

[1792]Oguchi, T.; Sakata, Y.; Takeuchi, N.; Kaneda, K.; Ishii, Y.; Ogawa, M. *Chem. Lett.* **1989**, 2053.

[1793]Yao, H.; Richardson, D.E. *J. Am. Chem. Soc.* **2000**, *122*, 3220.

[1794]Anand, R.V.; Singh, V.K. *Synlett* **2000**, 807.

[1795]For example, see Gould, E.S.; Hiatt, R.R.; Irwin, K.C. *J. Am. Chem. Soc.* **1968**, *90*, 4573; Sharpless, K.B.; Michaelson, R.C. *J. Am. Chem. Soc.* **1973**, *95*, 6136; Kochi, J.K. *Organometallic Mechanisms and Catalysis*; Academic Press, NY, **1978**, pp. 69–73; Ledon, H.J.; Durbut, P.; Varescon, F. *J. Am. Chem. Soc.* **1981**, *103*, 3601; Mimoun, H.; Mignard, M.; Brechot, P.; Saussine, L. *J. Am. Chem. Soc.* **1986**, *108*, 3711; Laszlo, P.; Levart, M.; Singh, G.P. *Tetrahedron Lett.* **1991**, *32*, 3167.

[1796]Nemoto, T.; Ohshima, T.; Shibasaki, M. *J. Am. Chem. Soc.* **2001**, *123*, 9474.

[1797]For a review, see Jørgensen, K.A. *Chem. Rev.* **1989**, *89*, 431.

[1798]For reviews, see Van Santen, R.A.; Kuipers, H.P.C.E. *Adv. Catal.* **1987**, *35*, 265; Filippova, T.V.; Blyumberg, E.A. *Russ. Chem. Rev.* **1982**, *51*, 582.

[1799]Straub, T.S. *Tetrahedron Lett.* **1995**, *36*, 663.

[1800]Yadav, V.K.; Kapoor, K.K. *Tetrahedron Lett.* **1994**, *35*, 9481.

[1801]MacPeek, D.L.; Starcher, P.S.; Phillips, B. *J. Am. Chem. Soc.* **1959**, *81*, 680.

[1802]For a review of epoxidation of allenes, see Jacobs, T.L., in Landor, S.R. *The Chemistry of Allenes*, Vol. 2, Academic Press, NY, **1982**, pp. 417–510, 483–491.

[1803]For a review of allene oxides, see Chan, T.H.; Ong, B.S. *Tetrahedron* **1980**, *36*, 2269.

[1804]Camp, R.L.; Greene, F.D. *J. Am. Chem. Soc.* **1968**, *90*, 7349; Crandall, J.K.; Conover, W.W.; Komin, J.B.; Machleder, W.H. *J. Org. Chem.* **1974**, *39*, 1723; Crandall, J.K.; Batal, D.J. *J. Org. Chem.* **1988**, *53*, 1338.

[1805]For example, see Crandall, J.K.; Machleder, W.H.; Sojka, S.A. *J. Org. Chem.* **1973**, *38*, 1149; Crandall, J.K.; Rambo, E. *J. Org. Chem.* **1990**, *55*, 5929.

by Wynberg, are now common.[1806] Enantioselectivities can be significantly improved by changes of the catalyst structure as well as the type of oxidant.[1807] A Yb-BINOL complex, with *t*-BuOOH led to epoxidation of conjugated ketones with high asymmetric induction,[1808] as did a mixture of NaOCl and a *Cinchona* alkaloid.[1809] Other enantioselective methods include treatment with diethylzinc, O_2, in the presence of a chiral amino-alcohol, to give the epoxy-ketone.[1810] Similarly, treatment with aqueous NaOCl[1811] or with an alkyl hydroperoxide[1812] and a chiral phase-transfer agent leads to chiral nonracemic epoxy-ketones.

Another important asymmetric epoxidation of a conjugated systems is the reaction of alkenes with polyleucine, DBU and urea–H_2O_2, giving an epoxy-carbonyl compound with good enantioselectivity.[1813] The hydroperoxide anion epoxidation of conjugated carbonyl compounds with a polyamino acid, such as poly-L-alanine or poly-L-leucine is known as the *Juliá–Colonna epoxidation*.[1814] Epoxidation of conjugated ketones to give nonracemic epoxy-ketones was done with aq. NaOCl and a Cinchona alkaloid derivative as catalyst.[1815] A triphasic phase-transfer catalysis protocol has also been developed.[1816] β-Peptides have been used as catalysts in this reaction.[1817]

Allylic alcohols can be converted to epoxy-alcohols with *tert*-butylhydroperoxide on molecular sieves,[1818] or with peroxy acids.[1819] The addition of an appropriate chiral ligand to the metal-catalyzed hydroperoxide epoxidation of allylic alcohols leads to high enantioselectivity. This important modification is

[1806]Helder, R.; Hummelen, J.C.; Laane, R.W.P.M.; Wiering, J.S.; Wynberg, H. *Tetrahedron Lett.* *1976*, *17*, 1831; Wynberg, H.; Greijdanus, B. *J. Chem. Soc., Chem. Commun.* *1978*, 427; Wynberg, H.; Marsman, B. *J. Org. Chem.* *1980*, *45*, 158; Pluim, H.; Wynberg, H. *J. Org. Chem.* *1980*, *45*, 2498.

[1807]Arai, S.; Tsuge, H.; Shioiri, T. *Tetrahedron Lett.* *1998*, *39*, 7563; Arai, S.; Shirai, Y.; Ishida, T.; Shioiri, T. *Tetrahedron* *1999*, *55*, 6375; Corey, E.J.; Zhang, F.-Y. *Org. Lett.* *1999*, *1*, 1287; Lygo, B.; Wainwright, P.G. *Tetrahedron* *1999*, *55*, 6289. See Adam, W.; Rao, P.B.; Degen, H.-G.; Levai, A.; Patonay, T.; Saha-Moller, C.R. *J. Org. Chem.* *2002*, *67*, 259.

[1808]Watanabe, S.; Arai, T.; Sasai, H.; Bougauchi, M.; Shibasaki, M. *J. Org. Chem.* *1998*, *63*, 8090.

[1809]Lygo, B.; Wainwright, P.G. *Tetrahedron Lett.* *1998*, *39*, 1599.

[1810]Enders, D.; Zhu, J.; Kramps, L. *Liebigs Ann. Chem.* *1997*, 1101; Enders, D.; Zhu, J.; Raabe, G. *Angew. Chem. Int. Ed.* *1996*, *35*, 1725.

[1811]Lygo, B.; To, D.C.M. *Tetrahedron Lett.* *2001*, *42*, 1343.

[1812]Adam, W.; Rao, P.B.; Degen, H.-G.; Saha-Möller, C.R. *Tetrahedron Asymmetry* *2001*, *12*, 121.

[1813]Allen, J.V.; Drauz, K.-H.; Flood, R.W.; Roberts, S.M.; Skidmore, J. *Tetrahedron Lett.* *1999*, *40*, 5417; Geller, T.; Roberts, S.M. *J. Chem. Soc., Perkin Trans. 1, 1999*, 1397; Bentley, P.A.; Bickley, J.F.; Roberts, S.M.; Steiner, A. *Tetrahedron Lett.* *2001*, *42*, 3741.

[1814]Banfi, S.; Colonna, S.; Molinari, H.; Juliá, S.; Guixer, J. *Tetrahedron* *1984*, *40*, 5207. For reviews, see Lin, P. *Tetrahedron: Asymmetry* *1998*, *9*, 1457; Ebrahim, S.; Wills, M. *Tetrahedron; Asymmetry* *1997*, *8*, 3163.

[1815]Lygo, B.; Wainwright, P.G. *Tetrahedron* *1999*, *55*, 6289.

[1816]Geller, T.; Krüger, C.M.; Militzer, H.-C. *Tetrahedron Lett.* *2004*, *45*, 5069.

[1817]Coffey, P.E.; Drauz-K.-H.; Roberts, S.M.; Skidmore, J.; Smith, J.A. *Chem. Commun.* *2001*, 2330.

[1818]Antonioletti, R.; Bonadies, F.; Locati, L.; Scettri, A. *Tetrahedron Lett.* *1992*, *33*, 3205.

[1819]Fringuelli, F.; Germani, R.; Pizzo, F.; Santinelli, F.; Savelli, G. *J. Org. Chem.* *1992*, *57*, 1198.

known as the *Sharpless asymmetric epoxidation*,[1820] where allylic alcohols are converted to optically active epoxides with excellent enantioselectivity by treatment with *t*-BuOOH, titanium tetraisopropoxide and optically active diethyl tartrate.[1821] The $Ti(OCHMe_2)_4$ and diethyl tartrate can be present in catalytic amounts (15–10 mol%) if molecular sieves are present.[1822] Polymer-supported catalysts have also been reported.[1823] Both (+) and (−) diethyl tartrate are readily available, so either enantiomer of the product can be prepared. The method has been successful for a wide range of primary allylic alcohols, including substrates where the double bond is mono-, di-, tri-, and tetrasubstituted,[1824] and is highly useful in natural product synthesis. The mechanism of the Sharpless epoxidation is believed to involve attack on the substrate by a compound[1825] formed from the titanium alkoxide and the diethyl tartrate to produce a complex that also contains the substrate and the *t*-BuOOH.[1826]

Ordinary alkenes (without an allylic OH group) do not give optically active alcohols by the Sharpless protocol because binding to the catalyst is necessary for enantioselectivity. Simples alkenes can be epoxidized enantioselectively with sodium hypochlorite (NaOCl, commercial bleach) and an optically active manganese-complex catalyst.[1827] An important variation of this oxidation uses a manganese–salen complex[1828] with various oxidizing agents, in what is called

[1820]For reviews, see Pfenninger, A. *Synthesis* **1986**, 89; Rossiter, B.E., in Morrison, J.D. *Asymmetric Synthesis*, Vol. 5, Academic Press, NY, **1985**, pp. 193–246. For histories of its discovery, see Sharpless, K.B. *Chem. Br.* **1986**, 38; *CHEMTECH* **1985**, 692. Also see, Smith, M.B. *Organic Synthesis*, 2nd ed., McGraw-Hill, NY, **2001**, pp. 239–245.

[1821]Sharpless, K.B.; Woodard, S.S.; Finn, M.G. *Pure Appl. Chem.* **1983**, 55, 1823, and references cited therein.

[1822]Gao, Y.; Hanson, R.M.; Klunder, J.M.; Ko, S.Y.; Masamune, H.; Sharpless, K.B. *J. Am. Chem. Soc.* **1987**, 109, 5765. See Massa, A.; D'Ambrosi, A.; Proto, A.; Screttri, A. *Tetrahedron Lett.* **2001**, 42, 1995. For another improvement, see Wang, Z.; Zhou, W. *Tetrahedron* **1987**, 43, 2935.

[1823]Canali, L.; Karjalainen, J.K.; Sherrington, D.C.; Hormi, O. *Chem. Commun.* **1997**, 123.

[1824]See the table, in Finn, M.G.; Sharpless, K.B., in Morrison, J.D. *Asymmetric Synthesis*, Vol. 5, Academic Press, NY, **1985**, pp. 249–250. See also, Schweiter, M.J.; Sharpless, K.B. *Tetrahedron Lett.* **1985**, 26, 2543.

[1825]Very similar compounds have been prepared and isolated as solids whose structures have been determined by X-ray crystallography: Williams, I.D.; Pedersen, S.F.; Sharpless, K.B.; Lippard, S.J. *J. Am. Chem. Soc.* **1984**, 106, 6430.

[1826]For a review of the mechanism, see Finn, M.G.; Sharpless, K.B., in Morrison, J.D. *Asymmetric Synthesis*, Vol. 5, Academic Press, NY, **1985**, p. 247. For other mechanistic studies, see Jørgensen, K.A.; Wheeler, R.A.; Hoffmann, R. *J. Am. Chem. Soc.* **1987**, 109, 3240; Carlier, P.R.; Sharpless, K.B. *J. Org. Chem.* **1989**, 54, 4016; Corey, E.J. *J. Org. Chem.* **1990**, 55, 1693; Woodard, S.S.; Finn, M.G.; Sharpless, K.B. *J. Am. Chem. Soc.* **1991**, 113, 106; Finn, M.G.; Sharpless, K.B. *J. Am. Chem. Soc.* **1991**, 113, 113; Takano, S.; Iwebuchi, Y.; Ogasawara, K. *J. Am. Chem. Soc.* **1991**, 113, 2786. See Cui, M.; Adam, W.; Shen, J.H.; Luo, X.M.; Tan, X.J.; Chen, K.X.; Ji, R.Y.; Jiang, H.L. *J. Org. Chem.* **2002**, 67, 1427.

[1827]Jacobsen, E.N.; Zhang, W.; Muci, A.R.; Ecker, J.R.; Deng, L. *J. Am. Chem. Soc.* **1991**, 113, 7063. See also, Irie, R.; Noda, K.; Ito, Y.; Katsuki, T. *Tetrahedron Lett.* **1991**, 32, 1055; Halterman, R.L.; Jan, S. *J. Org. Chem.* **1991**, 56, 5253.

[1828]These complexes have been characterized. See Adam, W.; Mock-Knoblauch, C.; Saha-Moller, C.R.; Herderich, M. *J. Am. Chem. Soc.* **2000**, 122, 9685.

the *Jacobsen–Katsuki reaction.*[1829] Apart from the commonly used NaOCl, urea–H_2O_2 has been used.[1830] With this reaction, simple alkenes can be epoxidized with high enantioselectivity.[1831] The mechanism of this reaction has been examined.[1832] Radical intermediates have been suggested for this reaction,[1833] A polymer-bound $Mn^{(III)}$–salen complex, in conjunction with NaOCl, has been used for asymmetric epoxidation.[1834] Chromium–salen complexes[1835] and ruthenium–salen complexes[1836] have been used for epoxidation. Manganese porphyrin complexes have also been used.[1837] Cobalt complexes give similar results.[1838] A related epoxidation reaction used an iron complex with molecular oxygen and isopropanal.[1839] Nonracemic epoxides can be prepared from racemic epoxides with salen–cobalt(II) catalysts following a modified procedure for kinetic resolution.[1840]

In a different type of reaction, alkenes are photooxygenated (with singlet O_2, see **14-7**) in the presence of a Ti, V, or Mo complex to give epoxy alcohols, such as **180**, formally derived from allylic hydroxylation followed by epoxidation.[1841] In other cases, modification of the procedure gives simple epoxidation.[1842] Alkenes react with aldehydes and oxygen, with palladium-on-silica[1843] or a ruthenium catalyst,[1844]

[1829]Hosoya, N.; Hatayama, A.; Irie, R.; Sasaki, H.; Katsuki, T. *Tetrahedron* **1994**, *50*, 4311, and references cited therein; Brandes, B.D.; Jacobsen, E.N. *J. Org. Chem.* **1994**, *59*, 4378; Sasaki, H.; Irie, R.; Hamada, T.; Suzuki, K.; Katsuki, T. *Tetrahedron* **1994**, *50*, 11827; Brandes, B.D.; Jacobsen, E.N. *Tetrahedron Lett.* **1995**, *36*, 5123; Nishikori, H.; Ohta, C.; Katsuki, T. *Synlett* **2000**, 1557; Tangestaninejad, S.; Habibi, M.H.; Mirkhani, V.; Moghadam, M. *Synth. Commun.* **2002**, *32*, 3331.

[1830]Kureshy, R.I.; Khan, N.H.; Abdi, S.H.R.; Patel, S.T.; Jasra, R.V. *Tetrahedron Asymmetry* **2001**, *12*, 433.

[1831]For a discussion of stereocontrol factors, see Nishida, T.; Miyafuji, A.; Ito, Y.N.; Katsuki, T. *Tetrahedron Lett.* **2000**, *41*, 7053.

[1832]See Linker, T. *Angew. Chem., Int. Ed.* **1997**, *36*, 2060. See Adam, W.; Roschmann, K.J.; Saha-Möller, C.R. *Eur. J. Org. Chem.* **2000**, 3519. For the importance of electronic effects, see Cavallo, L.; Jacobsen, H. *J. Org. Chem.* **2003**, *68*, 6202.

[1833]Cavallo, L.; Jacobsen, H. *Angew. Chem. Int. Ed.* **2000**, *39*, 589.

[1834]Song, C.E.; Roh, E.J.; Yu, B.M.; Chi, D.Y.; Kim, S.C.; Lee, K.J. *Chem. Commun.* **2000**, 615; Ahn, K.-H.; Park, S.W.; Choi, S.; Kim, H.-J.; Moon, C.J. *Tetrahedron Lett.* **2001**, *42*, 2485.

[1835]Daly, A.M.; Renehan, M.F.; Gilheany, D.G. *Org. Lett.* **2001**, *3*, 663; O'Mahony, C.P.; McGarrigle, E.M.; Renehan, M.F.; Ryan, K.M.; Kerrigan, N.J.; Bousquet, C.; Gilheany, D.G. *Org. Lett.* **2001**, *3*, 3435. See the references cited therein.

[1836]Nakata, K.; Takeda, T.; Mihara, J.; Hamada, T.; Irie, R.; Katsuki, T. *Chem. Eur. J.* **2001**, *7*, 3776.

[1837]Konishi, K.; Oda, K.; Nishida, K.; Aida, T.; Inoue, S. *J. Am. Chem. Soc.* **1992**, *114*, 1313.

[1838]Takai, T.; Hata, E.; Yorozu, K.; Mukaiyama, T. *Chem. Lett.* **1992**, 2077.

[1839]Saalfrank, R.W.; Reihs, S.; Hug, M. *Tetrahedron Lett.* **1993**, *34*, 6033.

[1840]Savle, P.S.; Lamoreaux, M.J.; Berry, J.F.; Gandour, R.D. *Tetrahedron Asymmetry* **1998**, *9*, 1843.

[1841]Adam, W.; Braun, M.; Griesbeck, A.; Lucchini, V.; Staab, E.; Will, B. *J. Am. Chem. Soc.* **1989**, *111*, 203.

[1842]See Iwahama, T.; Hatta, G.; Sakaguchi, S.; Ishii, Y. *Chem. Commun.* **2000**, 163.

[1843]Gao, H.; Angelici, R.J. *Synth. Commun.* **2000**, *30*, 1239; Chen, W.; Yamada, J.; Matsumoto, K. *Synth. Commun.* **2002**, *32*, 17; Ragagnin, G.; Knochel, P. *Synlett* **2004**, 951.

[1844]Srikanth, A.; Nagendrappa, G.; Chandrasekaran, S. *Tetrahedron* **2003**, *59*, 7761; Qi, J.Y.; Qiu, L.Q.; Lam, K.H.; Yip, C.W.; Zhou, Z.Y.; Chan, A.S.C. *Chem. Commun.* **2003**, 1058.

to give the epoxide.

180

Thiiranes can be prepared directly from alkenes using specialized reagents.[1845] Thiourea with a tin catalyst gives the thiirane, for example.[1846] Interestingly, internal alkynes were converted to 1,2-dichorothiiranes by reaction with S_2Cl_2 (sulfur monochloride).[1847] It is noted that epoxides are converted to thiiranes with ammonium thiocyanate and a cerium complex.[1848] A trans-thiiration reaction occurs with a molybdenum catalyst, in which an alkene reacts with styrene thiirane to give the new thiirane.[1849]

OS **I**, 494; **IV**, 552, 860; **V**, 191, 414, 467, 1007; **VI**, 39, 320, 679, 862; **VII**, 121, 126, 461; **VIII**, 546; **IX**, 288; **X**, 29; **80**, 9.

15-51   Hydroxysulfenylation (Addition of Oxygen, Sulfur)

Hydroxy-arylthio-addition (overall transformation)

A hydroxy and an arylthio group can be added to a double bond by treatment with an aryl disulfide and lead tetraacetate in the presence of trifluoroacetic acid.[1850] Manganese and copper acetates have been used instead of $Pb(OAc)_4$.[1851] Addition of the groups OH and RSO has been achieved by treatment of alkenes with O_2 and a thiol RSH.[1852] Two RS groups were added, to give *vic*-dithiols, by treatment of the alkene with a disulfide RSSR and BF_3-etherate.[1853] This reaction

---

[1845]Capozzi, G.; Menichetti, S.; Neri, S.; Skowronska, A. *Synlett* **1994**, 267; Adam, W.; Bargon, R.M. *Eur. J. Org. Chem.* **2001**, 1959; Adam, W.; Bargon, R.M. *Chem. Commun.* **2001**, 1910.

[1846]Tangestaninejad, S.; Mirkhani, V. *Synth. Commun.* **1999**, *29*, 2079.

[1847]Nakayama, J.; Takahashi, K.; Watanabe, T.; Sugihara, Y.; Ishii, A. *Tetrahedron Lett.* **2000**, *41*, 8349.

[1848]Iranpoor, N.; Tamami, B.; Shekarriz, M. *Synth. Commun.* **1999**, *29*, 3313.

[1849]Adam, W.; Bargon, R.M.; Schenk, W.A. *J. Am. Chem. Soc.* **2003**, *125*, 3871.

[1850]Trost, B.M.; Ochiai, M.; McDougal, P.G. *J. Am. Chem. Soc.* **1978**, *100*, 7103. For a related reaction, see Zefirov, N.S.; Zyk, N.V.; Kutateladze, A.G.; Kolbasenko, S.I.; Lapin, Yu.A. *J. Org. Chem. USSR* **1986**, *22*, 190.

[1851]Bewick, A.; Mellor, J.M.; Owton, W.M. *J. Chem. Soc. Perkin Trans. 1*, **1985**, 1039; Bewick, A.; Mellor, J.M.; Milano, D.; Owton, W.M. *J. Chem. Soc. Perkin Trans. 1*, **1985**, 1045; Samii, Z.K.M.A.E.; Ashmawy, M.I.A.; Mellor, J.M. *Tetrahedron Lett.* **1986**, *27*, 5289.

[1852]Chung, M.; D'Souza, V.T.; Szmant, H.H. *J. Org. Chem.* **1987**, *52*, 1741, and other papers in this series.

[1853]Caserio, M.C.; Fisher, C.L.; Kim, J.K. *J. Org. Chem.* **1985**, *50*, 4390; Inoue, H.; Murata, S. *Heterocycles* **1997**, *45*, 847.

has been carried out internally.[1854] In a similar manner, reaction of alkenes with ceric ammonium nitrate, diphenyl diselenide in methanol leads to vicinally substituted phenylselenyl methyl ethers.[1855] Dimethyl diselenide adds to alkenes to form vicinal bis-methylselenyl compounds, in the presence of tin tetrachloride.[1856]

Halo-ethers can be formed by the reaction of alkenyl alcohols with various reagents. Hept-6-en-1-ol reacts with $(collidine)_2I^+PF_6^-$, for example, to form 2-iodomethyl-1-oxacycloheptane.[1857]

15-52 Oxyamination (Addition of Oxygen, Nitrogen)

Tosylamino-hydroxy-addition

$$\ce{\underset{/}{\overset{\backslash}{C}}=\underset{\backslash}{\overset{/}{C}} + TsNClNa\cdot3 H_2O ->[1\% OsO_4][\textit{t}-BuOH] \underset{/}{\overset{HO}{-C}}-\underset{\backslash}{\overset{NHTs}{C-}}}$$

N-Tosylated β-hydroxy alkylamines (which can be easily hydrolyzed to β-hydroxyamines[1858]) can be prepared[1859] by treatment of alkenes with the trihydrate of Chloramine-T (*N*-chloro-*p*-toluenesulfonamide sodium salt)[1566] and a catalytic amount of OsO4.[1860] In some cases, yields can be improved by the use of phase-transfer catalysis.[1861] The reaction has been carried out enantioselectively.[1862] Alkenes can be converted to amido alcohols enantioselectivity by modification of this basic scheme. The *Sharpless asymmetric aminohydroxylation* employs a catalyst consisting of *Cinchona* alkaloid derived ligands and an osmium species in combination with a stoichiometric nitrogen source that also functions as the oxidant.[1863] The reaction of a carbamate with $(DHQ)_2PHAL$ (**176**) and the osmium compound, with NaOH and *tert*-butyl hypochlorite, leads to a diastereomeric mixture of amido alcohols **181** and **182**, each formed with high enantioselectivity.[1864] In general, the nitrogen adds to the less sterically hindered carbon of the alkene to give the major product. *N*-Bromoamides, in the presence of a catalytic amount of $(DHQ)_2PHAL$

[1854]Tuladhar, S.M.; Fallis, A.G. *Tetrahedron Lett.* **1987**, *28*, 523. For a list of other examples, with references, see Larock, R.C. *Comprehensive Organic Transformations*, 2nd ed., Wiley-VCH, NY, **1999**, pp. 905–908.

[1855]Bosman, C.; D'Annibale, A.; Resta, S.; Trogolo, C. *Tetrahedron Lett.* **1994**, *35*, 6525. See Ogawa, A.; Tanaka, H.; Yokoyama, H.; Obayashi, R.; Yokoyama, K.; Sonoda, N. *J. Org. Chem.* **1992**, *57*, 111 for formation of mixed PhS–PhSe- compounds from alkenes.

[1856]Hermans, B.; Colard, N.; Hevesi, L. *Tetrahedron Lett.* **1992**, *33*, 4629.

[1857]Brunel, Y.; Rousseau, G. *Synlett* **1995**, 323.

[1858]For some reactions of the oxyamination products, see Bäckvall, J.E.; Oshima, K.; Palermo, R.E.; Sharpless, K.B. *J. Org. Chem.* **1979**, *44*, 1953.

[1859]Sharpless, K.B.; Chong, A.O.; Oshima, K. *J. Org. Chem.* **1976**, *41*, 177. See Rudolph, J.; Sennhenn, P.C.; Vlaar, C.P.; Sharpless, K.B. *Angew. Chem. Int. Ed.* **1996**, *35*, 2810 for a discussion of the influence of substituents on nitrogen in this reaction.

[1860]See Fokin, V.V.; Sharpless, K.B. *Angew. Chem. Int. Ed.* **2001**, *40*, 3455.

[1861]Herranz, E.; Sharpless, K.B. *J. Org. Chem.* **1978**, *43*, 2544.

[1862]Hassine, B.B.; Gorsane, M.; Pecher, J.; Martin, R.H. *Bull. Soc. Chim. Belg.* **1985**, *94*, 759.

[1863]For a review, see Bodkin, J.A.; McLeod, M.D. *J. Chem. Soc., Perkin Trans. 1* **2002**, 2733.

[1864]Li, G.; Chang, H.-T.; Sharpless, K.B. *Angew. Chem., Int. Ed.* **1996**, *35*, 451.

and LiOH converts conjugated esters to β-amido-α-hydroxy esters with good enantioselectivity.[1865] In another procedure, certain β-hydroxy secondary alkylamines can be prepared by treatment of alkenes with the osmium compound t-Bu—N=OsO$_3$, followed by reductive cleavage with LiAlH$_4$ of the initially formed osmic esters.[1866] It is presumed that Ts—N=OsO$_3$ is an intermediate in the Chloramine-T reaction. Another oxyamination reaction involves treatment of a palladium complex of the alkene with a secondary or primary amine, followed by lead tetraacetate or another oxidant.[1867]

The organolanthanide-catalyzed alkene hydroamination has been reported.[1868] With this approach, amino alkenes (not enamines) can be cyclized to form cyclic amines,[1869] and amino alkynes lead to cyclic imine.[1870] The use of synthesized C-1[1871] and C-2 symmetric[1872] chiral organolanthanide complexes give the amino alcohol with good enantioselectivity.

β-Amino alcohols can be prepared by treatment of an alkene with a reagent prepared from HgO and HBF$_4$ along with aniline to give an aminomercurial compound PhHN—C—C—HgBF$_4$ (aminomercuration; see **15-7**) which is hydrolyzed

---

[1865]Demko, Z.P.; Bartsch, M.; Sharpless, K.B. *Org. Lett.* **2000**, *2*, 2221.

[1866]Hentges, S.G.; Sharpless, K.B. *J. Org. Chem.* **1980**, *45*, 2257. Also see, Rubinstein, H.; Svendsen, J.S. *Acta Chem. Scand. B* **1994**, *48*, 439. For another method, in which the NH in the product is connected to an easily removable protecting group, see Herranz, E.; Sharpless, K.B. *J. Org. Chem.* **1980**, *45*, 2710.

[1867]Bäckvall, J.E.; Björkman, E.E. *Acta Chem. Scand. Ser. B* **1984**, *38*, 91; Bäckvall, J.E.; Bystrom, S.E. *J. Org. Chem.* **1982**, *47*, 1126.

[1868]Ryu, J.-S.; Li, G.Y.; Marks, T.J. *J. Am. Chem. Soc.* **2003**, *125*, 12584; Li, Y.; Marks, T.J. *Organometallics* **1996**, *15*, 3770; Gagné, M.R.; Stern, C.L.; Marks, T.J. *J. Am Chem. Soc.* **1992**, *114*, 275; Gagné, M.R.; Marks, T.J. *J. Am Chem. Soc.* **1989**, *111*, 4108. For a review, see Hong, S.; Marks, T.J. *Acc. Chem. Res.* **2004**, *37*, 673.

[1869]Gagné, M.R.; Stern, C.L.; Marks, T.J. *J. Am Chem. Soc.* **1992**, *114*, 275; Gagné, M.R.; Marks, T.J. *J. Am Chem. Soc.* **1989**, *111*, 4108.

[1870]Li, Y.; Marks, T.J. *J. Am. Chem. Soc.* **1996**, *118*, 9295; Li, Y.; Fu, P.-F.; Marks, T.J. *Organometallics* **1994**, *13*, 439; Li, Y.; Marks, T.J. *J. Am. Chem. Soc.* **1998**, *120*, 1757; Li, Y.; Marks, T.J. *J. Am. Chem. Soc.* **1996**, *118*, 707.

[1871]Douglass, M.R.; Ogasawara, M.; Hong, S.; Metz, M.V.; Marks, T.J. *Organometallics* **2002**, *21*, 283; Giardello, M.A.; Conticello, V.P.; Brard, L.; Gagné, M.R.; Marks, T.J. *J. Am. Chem. Soc.* **1994**, *116*, 10241; Giardello, M.A.; Conticello, V.P.; Brard, L.; Sabat, M.; Rheingold, A.L.; Stern, C.L.; Marks, T.J. *J. Am. Chem. Soc.* **1994**, *116*, 10212; Gagné, M.R.; Brard, L.; Conticello, V.P.; Giardello, M.A.; Stern, C.L.; Marks, T.J. *Organometallics* **1992**, *11*, 2003.

[1872]Hong, S.; Tian, S.; Metz, M.V.; Marks, T.J. *J. Am. Chem. Soc.* **2003**, *125*, 14768.

to PhHN$-$C$-$C$-$OH.[1873] The use of an alcohol instead of water gives the corresponding

amino ether. β-Azido alcohols are prepared by the reaction of an alkene with $Me_3SiOOSiMe_3$, Me_3SiN_3, and 20% $(Cl_2SnO)_n$, followed by treatment with aqueous acetic acid.[1874]

OS **VII**, 223, 375.

15-53   Diamination (Addition of Nitrogen, Nitrogen)

Di(alkylarylamino)-addition

Primary (R $=$ H) and secondary aromatic amines react with alkenes in the presence of thallium(III) acetate to give *vic*-diamines in good yields.[1875] The reaction is not successful for primary aliphatic amines. In another procedure, alkenes can be diaminated by treatment with the osmium compounds R_3NOsO (R $=$ *t*-Bu) and R_2NOsO_2,[1876] analogous to the osmium compound mentioned at **15-52**.[1877] The palladium-promoted method of **15-52** has also been extended to diamination.[1878] Alkenes can also be diaminated[1879] indirectly by treatment of the aminomercurial compound mentioned in **15-52** with a primary or secondary aromatic amine.[1880] The reaction of an alkene with *N*-arylsulfonyl dichloroamines, $ArSO_2NCl_2$, followed by reaction with aqueous Na_2SO_3, gives the *anti-vic*-diacetamde.[1881]

Two azido groups can be added to double bonds by treatment with sodium azide and iodosobenzene in acetic acid, $C=C + NaN_3 + PhIO \rightarrow N_3-C-C-N_3$.[1882]

[1873]Barluenga, J.; Alonso-Cires, L.; Asensio, G. *Synthesis* **1981**, 376.

[1874]Sakurada, I.; Yamasaki, S.; Kanai, M.; Shibasaki, M. *Tetrahedron Lett.* **2000**, *41*, 2415.

[1875]Gómez Aranda, V.; Barluenga, J.; Aznar, F. *Synthesis* **1974**, 504.

[1876]Chong, A.O.; Oshima, K.; Sharpless, K.B. *J. Am. Chem. Soc.* **1977**, *99*, 3420. See also, Sharpless, K.B.; Singer, S.P. *J. Org. Chem.* **1976**, *41*, 2504.

[1877]For a X-ray structure of the osmium intermediate, see Muñiz, K.; Iesato, A.; Nieger, M. *Chem. Eur. J.* **2003**, *9*, 5581.

[1878]Bäckvall, J. *Tetrahedron Lett.* **1978**, 163.

[1879]For other diamination methods, see Michejda, C.J.; Campbell, D.H. *J. Am. Chem. Soc.* **1979**, *101*, 7687; Becker, P.N.; White, M.A.; Bergman, R.G. *J. Am. Chem. Soc.* **1980**, *102*, 5676; Becker, P.N.; Bergman, R.G. *Organometallics* **1983**, *2*, 787; Jung, S.; Kohn, H. *Tetrahedron Lett.* **1984**, *25*, 399; *J. Am. Chem. Soc.* **1985**, *107*, 2931; Osowska-Pacewicka, K.; Zwierzak, A. *Synthesis* **1990**, 505.

[1880]Barluenga, J.; Alonso-Cires, L.; Asensio, G. *Synthesis* **1979**, 962.

[1881]Li, G.; Kim, S.H.; Wei, H.-X. *Tetrahedron Lett.* **2000**, *41*, 8699.

[1882]Moriarty, R.M.; Khosrowshahi, J.S. *Tetrahedron Lett.* **1986**, *27*, 2809. For other methods, see Minisci, F.; Galli, R. *Tetrahedron Lett.* **1962**, 533; Fristad, W.E.; Brandvold, T.A.; Peterson, J.R.; Thompson, S.R. *J. Org. Chem.* **1985**, *50*, 3647.

15-54   Formation of Aziridines (Addition of Nitrogen, Nitrogen)

epi-**Arylimino-addition**, and so on.

Triazoline

Aziridines can be prepared directly from double-bond compounds by photolysis or thermolysis of a mixture of the substrate and an azide.[1883] The reaction has been carried out with R = aryl, cyano, EtOOC, and RSO_2, as well as other groups. The reaction can take place by at least two pathways.

In one pathway a 1,3-dipolar addition (**15-58**) takes place to give a triazoline (which can be isolated), followed thermal by extrusion of nitrogen (**17-34**). Evidence for the nitrene pathway is most compelling for R = acyl groups. In the other, the azide is converted to a nitrene, which adds to the double bond in a manner analogous to that of carbene addition (**15-64**). Sulfonyloxy amines, such as $ArSO_2ONHCO_2Et$, form an aziridine when treated with CaO in the presence of a conjugated carbonyl compound.[1884] In the presence of copper,[1885] cobalt,[1886] or rhodium complexes,[1887] ethyl diazoacetate adds to imines to give aziridines. Diazirines (p. 288) with *n*-butyllithium converted conjugated amides to the α,β-aziridino amide.[1888] Calcium oxide has also been used to generate the nitrene,[1889] including nitrene precursors that have an attached chiral ester.[1890] Other specialized reagents have also been used.[1891] As discussed on p. 293, singlet nitrenes add stereospecifically while triplet nitrenes do not. Diphenyl sulfimide (Ph_2SNH) converts

[1883]For reviews, see Dermer, O.C.; Ham, G.E. *Ethylenimine and Other Aziridines*, Academic Press, NY, *1969*, pp. 68–79; Muller, L.L.; Hamer, J. *1,2-Cycloaddition Reactions*, Wiley, NY, *1967*.

[1884]Fioravanti, S.; Pellacani, L.; Tabanella, S.; Tardella, P.A. *Tetrahedron* *1998*, *54*, 14105; Fioravanti, S.; Morreale, A.; Pellacani, L.; Tardella, P.A. *Synthesis* *2001*, 1975. For an enantioselective version of this reaction using a chiral ester auxiliary, see Fioravanti, S.; Morreale, A.; Pellacani, L.; Tardella, P.A. *J. Org. Chem.* *2002*, *67*, 4972.

[1885]Li, Z.; Zheng, Z.; Chen, H. *Tetrahedron Asymmetry* *2000*, *11*, 1157; Wong, H.L.; Tian, Y.; Chan, K.S. *Tetrahedron Lett.* *2000*, *41*, 7723; Sanders, C.J.; Gillespie, K.M.; Scott, P. *Tetrahedron Asymmetry* *2001*, *12*, 1055; Ma, J.-A.; Wang, L.-X.; Zhang, W.; Zhou, W.; Zhou, Q.-L. *Tetrahedron Asymmetry* *2001*, *12*, 2801.

[1886]Ikeno, T.; Nishizuka, A.; Sato, M.; Yamada, T. *Synlett* *2001*, 406.

[1887]Mohan, J.M.; Uphade, T.S.S.; Choudhary, V.R.; Ravindranathan, T.; Sudalai, A. *Chem. Commun.* *1997*, 1429; Moran, M.; Bernardinelli, G.; Müller, P. *Helv. Chim. Acta* *1995*, *78*, 2048.

[1888]Hori, K.; Sugihara, H.; Ito, Y.N.; Katsuki, T. *Tetrahedron Lett.* *1999*, *40*, 5207; Ishihara, H.; Ito, Y.N.; Katsuki, T. *Chem. Lett.* *2001*, 984.

[1889]Carducci, M.; Fioravanti, S.; Loreta, M.A.; Pellacani, L.; Tardella, P.A. *Tetrahedron Lett.* *1996*, *37*, 3777.

[1890]Fioravanti, S.; Morreale, A.; Pellacani, L.; Tardella, P.A. *Tetrahedron Lett.* *2003*, *44*, 3031.

[1891]Aires-de-Sousa, J.; Labo, A.M.; Prabhakar, S. *Tetrahedron Lett.* *1996*, *37*, 3183.

Michael-type substrates to the corresponding aziridines.[1892] Aminonitrenes (R_2NN:) have been shown to add to alkenes[1893] to give N-substituted aziridines and to triple bonds to give 1-azirines, which arise from rearrangement of the initially formed 2-azirines.[1894] Like oxirenes (see **15-50**), 2-azirines are unstable, probably because of anti-aromaticity. 1-Azirines can be reduced to give chiral aziridines.[1895]

$$R^2-C\equiv C-R^1 \ + \ R_2N-N: \ \longrightarrow \ \left[\begin{array}{c} NR_2 \\ | \\ N \\ \diagup \backslash \\ R^2-\underset{}{C}=\underset{}{C}-R^1 \end{array} \right] \ \longrightarrow \ R^2 \diagup \overset{N}{\underset{R^1}{\overset{\parallel}{C}-C}}-NR_2$$

2-Azirine   1-Azirine

An alternative preparation of aziridines reacts an alkene with iodine and chloramine-T, generating the corresponding N-tosyl aziridine.[1896] Chloramine T and NBS also gives the N-tosyl aziridine,[1897] and bromamine-T ($TsNBr^-Na^+$) has been used in a similar manner,[1898] and also TsNIK.[1899] Diazoalkanes react with imines to give aziridines.[1900] Another useful reagent is NsN=IPh, which reacts with alkenes in the presence of rhodium compounds[1901] or $Cu(OTf)_2$[1902] to give N-Ns aziridines. Other sulfonamide reagents can be used,[1903] including PhI=NTs.[1904] Enantioselective aziridination is possible using this reaction with

[1892]Furukawa, N.; Yoshimura, T.; Ohtsu, M.; Akasaka, T.; Oae, S. *Tetrahedron* **1980**, *36*, 73. For other methods, see Groves, J.T.; Takahashi, T. *J. Am. Chem. Soc.* **1983**, *105*, 2073; Mahy, J.; Bedi, G.; Battioni, P.; Mansuy, D. *J. Chem. Soc. Perkin Trans. 2*, **1988**, 1517; Atkinson, R.S.; Kelly, B.J. *J. Chem. Soc. Perkin Trans. 1*, **1989**, 1515.

[1893]Siu, T.; Yudin, A.K. *J. Am. Chem. Soc.* **2002**, *124*, 530.

[1894]Anderson, D.J.; Gilchrist, T.L.; Rees, C.W. *Chem. Commun.* **1969**, 147.

[1895]Roth, P.; Andersson, P.G.; Somfai, P. *Chem. Commun.* **2002**, 1752.

[1896]Ando, T.; Kano, D.; Minakata, S.; Ryu, I.; Komatsu, M. *Tetrahedron* **1998**, *54*, 13485. For the use of $TsNCl_2$, see Chen, D.; Timmons, C.; Guo, L.; Xu, X.; Li, G. *Synthesis* **2004**, 2479.

[1897]Thakur, V.V.; Sudalai, A. *Tetrahedron Lett.* **2003**, *44*, 989.

[1898]Vyas, R.; Chanda, B.M.; Bedekar, A.V. *Tetrahedron Lett.* **1998**, *39*, 4715; Hayer, M.F.; Hossain, M.M. *J. Org. Chem.* **1998**, *63*, 6839. This reaction was catalyzed by $CuCl_2$ with microwave irradiation, see Chanda, B.M.; Vyas, R.; Bedekar, A.V. *J. Org. Chem.* **2001**, *66*, 30. Iron catalysts have been used, see Vyas, R.; Gao, G.-Y.; Hardin, J.D.; Zhang, X.P. *Org. Lett.* **2003**, *6*, 1907.

[1899]Jain, S.L.; Sain, B. *Tetrahedron Lett.* **2003**, *44*, 575.

[1900]Casarrubios, L.; Pérez, J.A.; Brookhart, M.; Templeton, J.L. *J. Org. Chem.* **1996**, *61*, 8358.

[1901]Müller, P.; Baud, C.; Jacquier, Y. *Tetrahedron* **1996**, *52*, 1543. Also see, Södergren, M.J.; Alonso, D.A.; Bedekar, A.V.; Andersson, P.G. *Tetrahedron Lett.* **1997**, *38*, 6897.

[1902]Knight, J.G.; Muldowney, M.P. *Synlett* **1995**, 949. See also, Dauben, P.; Sanière, L.; Tarrade, A.; Dodd, R.H. *J. Am. Chem. Soc.* **2001**, *123*, 7707; Shi, M.; Wang, C.-J.; Chan, A.S.C. *Tetrahedron Asymmetry* **2001**, *12*, 3105.

[1903]PhI=NSO$_2$CH$_2$CCl$_3$: GuthiKonda, K.; Du Bois, J. *J. Am. Chem. Soc.* **2002**, *124*, 13672. See also, Di Chenna, P.H.; Robert-Peillard, F.; Dauban, P.; Dodd, R.H. *Org. Lett.* **2004**, *6*, 4503; Kwong, H.-L.; Liu, D.; Chan, K.-Y.; Lee, C.-S.; Huang, K.-H.; Che, C.-M. *Tetrahedron Lett.* **2004**, *45*, 3965.

[1904]Vedernikov, A.N.; Caulton, K.G. *Org. Lett.* **2003**, *5*, 2591; Cui, Y.; He, C. *J. Am. Chem. Soc.* **2003**, *125*, 16202. PhI=NSO$_2$CH$_2$CH$_2$SiMe$_3$: Dauban, P.; Dodd, R.H. *J. Org. Chem.* **1999**, *64*, 5304, and see Nishimura, M.; Minakata, S.; Takahashi, T.; Oderaotoshi, Y.; Komatsu, M. *J. Org. Chem.* **2002**, *67*, 2101.

chiral ligands.[1905] This reagent has been used in ionic liquids with a copper catalyst.[1906] Such reactions are catalyzed by palladium[1907] and methyl trioxorhenium (MeReO$_3$) can be used in these reactions.[1908] Manganese–salen catalysts have also been used with this reagent.[1909] A nitrido manganese–salen complex was also used with ditosyl anhydride, converting a conjugated diene to an allylic *N*-tosylaziridine.[1910]

Nitrenes can also add to aromatic rings to give ring-expansion products analogous to those mentioned in **15-62**.[1911]

OS **VI**, 56.

15-55   Aminosulfenylation (Addition of Nitrogen, Sulfur)

Arylamino-arylthio-addition

An amino and an arylthio group can be added to a double bond by treatment with a sulfenanilide PhSNHAr in the presence of BF$_3$-etherate.[1912] The addition is anti, and the mechanism probably involves a thiiranium ion.[1913] In another aminosulfenylation procedure, the substrate is treated with dimethyl(methylthio)sulfonium fluoroborate (MeSSMe$_2$ BF$_4^-$) and ammonia or an amine,[1914] the latter acting as a nucleophile. This reaction was extended to other nucleophiles:[1915] N$_3^-$,[1916]

[1905]See Gillespie, K.M.; Sanders, C.J.; O'Shaughnessy, P.; Westmoreland, I.; Thickitt, C.P.; Cott, P. *J. Org. Chem.* **2002**, *67*, 3450.

[1906]In bmim BF$_4$, 1-butyl-3-methylimidazolium tetrafluoroborate: Kantam, M.L.; Neeraja, V.; Kavita, B.; Haritha, Y. *Synlett* **2004**, 525.

[1907]Antunes, A.M.M.; Marto, S.J.L.; Branco, P.S.; Prabhakar, S.; Lobo, A.M. *Chem. Commun.* **2001**, 405.

[1908]Jean, H.-J.; Nguyen, S.B.T. *Chem. Commun.* **2001**, 235.

[1909]O'Connor, K.J.; Wey, S.-J.; Burrows, C.J. *Tetrahedron Lett.* **1992**, *33*, 1001; Nishikori, H.; Katsuki, T. *Tetrahedron Lett.* **1996**, *37*, 9245; Noda, K.; Hosoya, N.; Irie, R.; Ito, Y.; Katsuki, T. *Synlett* **1993**, 469.

[1910]Nishimura, M.; Minakata, S.; Thonchant, S.; Ryu, I.; Komatsu, M. *Tetrahedron Lett.* **2000**, *41*, 7089.

[1911]For example, see Hafner, K.; König, C. *Angew. Chem. Int. Ed.* **1963**, *2*, 96; Lwowski, W.; Johnson, R.L. *Tetrahedron Lett.* **1967**, 891.

[1912]Benati, L.; Montavecchi, P.C.; Spagnolo, P. *Tetrahedron Lett.* **1984**, *25*, 2039. See also, Brownbridge, P. *Tetrahedron Lett.* **1984**, *25*, 3759.

[1913]See Ref. 21.

[1914]Trost, B.M.; Shibata, T. *J. Am. Chem. Soc.* **1982**, *104*, 3225; Caserio, M.C.; Kim., J.K. *J. Am. Chem. Soc.* **1982**, *104*, 3231.

[1915]Trost, B.M.; Shibata, T.; Martin, S.J. *J. Am. Chem. Soc.* **1982**, *104*, 3228; Trost, B.M.; Shibata, T. *J. Am. Chem. Soc.* **1982**, *104*, 3225. For an extension that allows A to be C≡CR, see Trost, B.M.; Martin, S.J. *J. Am. Chem. Soc.* **1984**, *106*, 4263.

[1916]Sreekumar, R.; Padmakumar, R.; Rugmini, P. *Chem. Commun.* **1997**, 1133.

NO_2^- CN^-, ^-OH, and ^-OAc to give $MeS-\overset{|}{C}-\overset{|}{C}-A$, where $A = N_3$, NO_2, CN, OH,

and OAc, respectively. An RS (R = alkyl or aryl) and an NHCOMe group have been added in an electrochemical procedure.[1917]

15-56 Acylacyloxylation and Acylamidation (Addition of Oxygen, Carbon, or Nitrogen, Carbon)

Acyl-acyloxy-addition

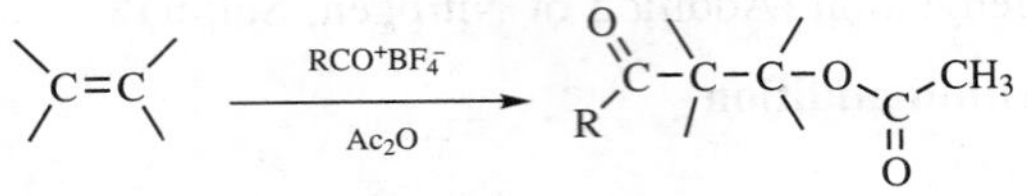

An acyl and an acyloxy group can be added to a double bond by treatment with an acyl fluoroborate and acetic anhydride.[1918] As expected, the addition follows Markovnikov's rule, with the electrophile Ac^+ going to the carbon with more hydrogens. In an analogous reaction, an acyl and an amido group can be added to give **183**, if a nitrile is used in place of the anhydride. Similarly, halo acetoxylation is known.[1919] This reaction has also been carried out on triple bonds, to give the unsaturated analogs of **183** (syn addition).[1920]

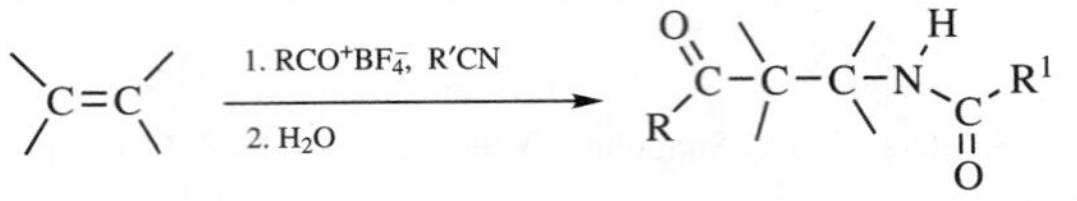

183

15-57 The Conversion of Alkenes to γ-Lactones (Addition of Oxygen, Carbon)

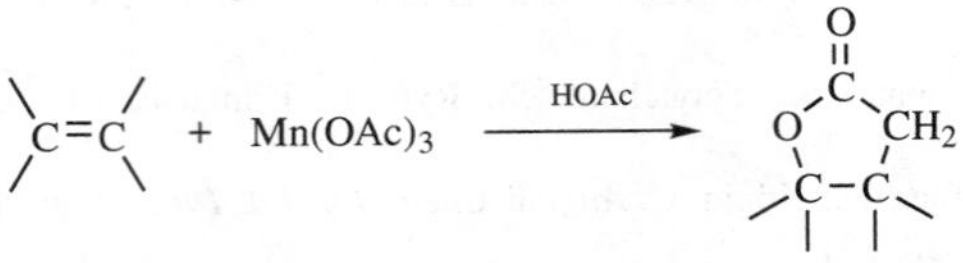

This reaction is clearly related to forming esters and lactones by reaction of carboxylic acids with alkenes (**15-6**), but the manganese reagent leads to

[1917]Bewick, A.; Coe, D.E.; Mellor, J.M.; Owton, M.W. *J. Chem. Soc. Perkin Trans. 1*, **1985**, 1033.
[1918]Shastin, A.V.; Balenkova, E.S. *J. Org. Chem. USSR* **1984**, *20*, 870.
[1919]Hashem, Md.A.; Jung, A.; Ries, M.; Kirschning, A. *Synlett* **1998**, 195.
[1920]Gridnev, I.D.; Balenkova, E.S. *J. Org. Chem. USSR* **1988**, *24*, 1447.

differences. Alkenes react with manganese(III) acetate to give γ-lactones.[1921] The mechanism is probably free radical, involving addition of •CH$_2$COOH to the double bond. Ultrasound improves the efficiency of the reaction.[1922] In a related reaction, cyclohexene reacted with MeO$_2$CCH$_2$CO$_2$K and Mn(OAc)$_3$ to give an α-carbomethoxy bicyclic lactone.[1923] The use of dimethyl malonate and ultrasound in this reaction gave the same type of product.[1924] Lactone formation has also been accomplished by treatment of alkenes with α-bromo carboxylic acids in the presence of benzoyl peroxide as catalyst,[1925] and with alkylidene chromium pentacarbonyl complexes.[1926] Alkenes can also be converted to γ- lactones by indirect routes.[1927] Chromium–carbene complexes add to alkenes to give β-lactones using ultrasound.[1928]

An intramolecular variation of this reaction is known, involving amides, which generates a lactam.[1929]

OS **VII**, 400.

For addition of aldehydes and ketones, see the Prins reaction (**16-54**), and reactions **16-95** and **16-96**.

15-58   1,3-Dipolar Addition (Addition of Oxygen, Nitrogen, Carbon)

There are a large group of reactions ([3 + 2]-cycloadditions) in which five-membered heterocyclic compounds are prepared by addition of 1,3-dipolar compounds to double bonds. This reaction is quite useful in the synthesis of alkaloids,[1930] including asymmetric syntheses.[1931] These dipolar compounds have a

[1921]Bush Jr., J.B.; Finkbeiner, H. *J. Am. Chem. Soc.* **1968**, *90*, 5903; Heiba, E.I.; Dessau, R.M.; Koehl, Jr., W.J. *J. Am. Chem. Soc.* **1968**, *90*, 5905; Heiba, E.I.; Dessau, R.M.; Rodewald, P.G. *J. Am. Chem. Soc.* **1974**, *96*, 7977; Midgley, G.; Thomas, C.B. *J. Chem. Soc. Perkin Trans. 2*, **1984**, 1537; Ernst, A.B.; Fristad, W.E. *Tetrahedron Lett.* **1985**, *26*, 3761; Shundo, R.; Nishiguchi, I.; Matsubara, Y.; Hirashima, T. *Tetrahedron* **1991**, *47*, 831. See also, Corey, E.J.; Gross, A.W. *Tetrahedron Lett.* **1985**, *26*, 4291.

[1922]D'Annibale, A.; Trogolo, C. *Tetrahdron Lett.* **1994**, *35*, 2083.

[1923]Lamarque, L.; Méou, A.; Brun, P. *Tetrahedron* **1998**, *54*, 6497.

[1924]Allegretti, M.; D'Annibale, A.; Trogolo, C. *Tetrahedron* **1993**, *49*, 10705.

[1925]Nakano, T.; Kayama, M.; Nagai, Y. *Bull. Chem. Soc. Jpn.* **1987**, *60*, 1049. See also, Kraus, G.A.; Landgrebe, K. *Tetrahedron Lett.* **1984**, *25*, 3939.

[1926]Wang, S.L.B.; Su, J.; Wulff, W.D. *J. Am. Chem. Soc.* **1992**, *114*, 10665.

[1927]See, for example, Boldt, P.; Thielecke, W.; Etzemüller, J. *Chem. Ber.* **1969**, *102*, 4157; Das Gupta, T.K.; Felix, D.; Kempe, U.M.; Eschenmoser, A. *Helv. Chim. Acta* **1972**, *55*, 2198; Bäuml, E.; Tscheschlok, K.; Pock, R.; Mayr, H. *Tetrahedron Lett.* **1988**, *29*, 6925.

[1928]Caldwell, J.J.; Harrity, J.P.A.; Heron, N.M.; Kerr, W.J.; McKendry, S.; Middlemiss, D. *Tetrahedron Lett.* **1999**, *40*, 3481; Caldwell, J.J.; Kerr, W.J.; McKendry, S. *Tetrahedron Lett.* **1999**, *40*, 3485.

[1929]Davies, D.T.; Kapur, N.; Parsons, A.F. *Tetrahedron Lett.* **1998**, *39*, 4397.

[1930]See Broggini, G.; Zecchi, G. *Synthesis* **1999**, 905.

[1931]Karlsson, S.; Högberg, H.-E. *Org. Prep. Proceed. Int.* **2001**, *33*, 103.

sequence of three atoms a–b–c, of which a has a sextet of electrons in the outer shell and c an octet with at least one unshared pair (see Table 15.3).[1932] The reaction can then be formulated as shown to generate **184**. Note that the initial reaction of potassium permanganate (**15-48**) occurs by [3 + 2]-cycloaddition to give a manganate ester (**171**).[1933] [3+2]-Cycloadditions occur with other metal oxides.[1934] Hydrazones have also been reported to give [3 + 2]-cycloadditions.[1935]

1,3-Dipoles of the type shown in Table 15.3 have an atom with six electrons in the outer shell, which is usually unstable, and such compounds will delocalize the change to alleviate this electronic arrangement (they are resonance stabilized). 1,3-Dipolar compounds can be divided into two main types:

1. Those in which the dipolar canonical form has a double bond on the sextet atom and the other canonical form has a triple bond on that atom:

$$\overset{\ominus}{a}-b=\overset{\oplus}{c}- \quad \longleftrightarrow \quad \overset{\ominus}{a}-b\equiv\overset{\oplus}{c}-$$

[1932]For a treatise, see Padwa, A. *1,3-Dipolar Cycloaddition Chemistry* 2 vols., Wiley, NY, *1984*. For general reviews, see Carruthers, W. *Cycloaddition reactins in Organic Synthesis*, Pergamon, Elmsford, NY, *1990*; Drygina, O.V.; Garnovskii, A.D. *Russ. Chem. Rev. 1986*, *55*, 851; Samuilov, Ya.D.; Konovalov, A.I. *Russ. Chem. Rev. 1984*, *53*, 332; Beltrame, P., in Bamford, C.H.; Tipper, C.F.H. *Comprhensive Chemical Kinetics*, Vol. 9, Elsevier, NY, *1973*, pp. 117–131; Huisgen, R.; Grashey, R.; Sauer, J., in Patai, S. *The Chemistry of Alkenes*, Vol. 1, Wiley, NY, *1964*, pp. 806–878; Huisgen, R. *Helv. Chim. Acta 1967*, *50*, 2421; *Bull. Soc. Chim. Fr. 1965*, 3431; *Angew. Chem. Int. Ed. 1963*, *2*, 565, 633. For specific monographs and reviews, see Torssell, K.B.G. *Nitrile Oxides, Nitrones, and Nitronates in Organic Synthesis*; VCH, NY, *1988*; Scriven, E.F.V. *Azides and Nitrenes*; Academic Press, NY, *1984*; Stanovnik, B. *Tetrahedron 1991*, *47*, 2925 (diazoalkanes); Kanemasa, S.; Tsuge, O. *Heterocycles 1990*, *30*, 719 (nitrile oxides); Paton, R.M. *Chem. Soc. Rev. 1989*, *18*, 33 (nitrile sulfides); Terao, Y.; Aono, M.; Achiwa, K. *Heterocycles 1988*, *27*, 981 (azomethine ylids); Vedejs, E. *Adv. Cycloaddit. 1988*, *1*, 33 (azomethine ylids); DeShong, P.; Lander, Jr., S.W.; Leginus, J.M.; Dicken, C.M. *Adv. Cycloaddit. 1988*, *1*, 87 (nitrones); Balasubramanian, N. *Org. Prep. Proced. Int. 1985*, *17*, 23 (nitrones); Confalone, P.N.; Huie, E.M. *Org. React. 1988*, *36*, 1 (nitrones); Padwa, A., in Horspool, W.M. *Synthetic Organic Photochemistry*, Plenum, NY, *1984*, pp. 313–374 (nitrile ylids); Bianchi, G.; Gandolfi, R.; Grünanger, P., in Patai, S.; Rappoport, Z. *The Chemistry of Functional Groups, Supplement C*, pt. 1, Wley, NY, *1983*, pp. 752–784 (nitrile oxides); Black, D.S.; Crozier, R.F.; Davis, V.C. *Synthesis 1975*, 205 (nitrones); Stuckwisch, C.G. *Synthesis 1973*, 469 (azomethine ylids, azomethine imines). For reviews of intramolecular 1,3-dipolar additions, see Padwa, A., in Padwa, A. treatise cited above, Vol. 2, pp. 277–406; Padwa, A.; Schoffstall, A.M. *Adv. Cycloaddit. 1990*, *2*, 1; Tsuge, O.; Hatta, T.; Hisano, T., in Patai, S. *Supplement A: The Chemistry of Double-bonded Functional Groups*, Vol. 2, pt. 1, Wiley, NY, *1989*, pp. 345–475; Padwa, A. *Angew. Chem. Int. Ed. 1976*, *15*, 123. For a review of azomethine ylids, see Tsuge, O.; Kanemasa, S. *Adv. Heterocycl. Chem. 1989*, *45*, 231. For reviews of 1,3-dipolar cycloreversions, see Bianchi, G.; Gandolfi, R. in Padwa, A. treatise cited above, Vol. 2, pp. 451–542; Bianchi, G.; De Micheli, C.; Gandolfi, R. *Angew. Chem. Int. Ed. 1979*, *18*, 721. For a related review, see Petrov, M.L.; Petrov, A.A. *Russ. Chem. Rev. 1987*, *56*, 152. For the use of this reaction to synthesize natural products, see papers in *Tetrahedron 1985*, *41*, 3447.

[1933]Houk, K.N.; Strassner, T. *J. Org. Chem. 1999*, *64*, 800.

[1934]See Gisdakis, P.; Rösch, N. *J. Am. Chem. Soc. 2001*, *123*, 697.

[1935]Kobayashi, S.; Hirabayashi, R.; Shimizu, H.; Ishitani, H.; Yamashita, Y. *Tetrahedron Lett. 2003*, *44*, 3351.

TABLE 15.3. Some Common 1,3-Dipolar Compounds

Type 1

Azide	$\overset{\ominus}{R-N}-N=\overset{\oplus}{N}$ ⟷ $\overset{\ominus}{R-N}-N\equiv\overset{\oplus}{N}$
Diazoalkane[1936]	$R_2\overset{\ominus}{C}-N=\overset{\oplus}{N}$ ⟷ $R_2\overset{\ominus}{C}-N\equiv\overset{\oplus}{N}$
Nitrous oxide	$\overset{\ominus}{O}-N=\overset{\oplus}{N}$ ⟷ $\overset{\ominus}{O}-N\equiv\overset{\oplus}{N}$
Nitrile imine[1937]	$R-\overset{\ominus}{N}-N=\overset{\oplus}{C}R'$ ⟷ $R-\overset{\ominus}{N}-N\equiv\overset{\oplus}{C}R'$
Nitrile ylid[1938]	$R_2\overset{\ominus}{C}-N=\overset{\oplus}{C}R'$ ⟷ $R_2\overset{\ominus}{C}-N\equiv\overset{\oplus}{C}R'$
Nitrile oxide[1939]	$\overset{\ominus}{O}-N=\overset{\oplus}{C}R$ ⟷ $\overset{\ominus}{O}-N\equiv\overset{\oplus}{C}R$

Type 2

Azomethine imine	$R_2\overset{\ominus}{C}-\underset{\underset{R^2}{\vert}}{N}-\overset{\oplus}{N}R'$ ⟷ $R_2\overset{\ominus}{C}-\underset{\underset{R^2}{\vert}}{N}=\overset{\oplus}{N}R'$
Azoxy compound	$\overset{\ominus}{O}-\underset{\underset{R}{\vert}}{N}-\overset{\oplus}{N}R'$ ⟷ $\overset{\ominus}{O}-\underset{\underset{R}{\vert}}{N}=\overset{\oplus}{N}R'$
Azomethine ylid[1940]	$R_2\overset{\ominus}{C}-\underset{\underset{R^2}{\vert}}{N}-\overset{\oplus}{C}R'_2$ ⟷ $R_2\overset{\ominus}{C}-\underset{\underset{R^2}{\vert}}{N}=\overset{\oplus}{C}R'_2$
Nitrone	$\overset{\ominus}{O}-\underset{\underset{R'}{\vert}}{N}-\overset{\oplus}{C}R_2$ ⟷ $\overset{\ominus}{O}-\underset{\underset{R'}{\vert}}{N}=\overset{\oplus}{C}R_2$
Carbonyl oxide[1941]	$\overset{\ominus}{O}-O-\overset{\oplus}{C}R_2$ ⟷ $\overset{\ominus}{O}-O=\overset{\oplus}{C}R_2$
Ozone	$\overset{\ominus}{O}-O-\overset{\oplus}{O}$ ⟷ $\overset{\ominus}{O}-O=\overset{\oplus}{O}$

If we limit ourselves to the first row of the periodic table, b can only be nitrogen, c can be carbon or nitrogen, and a can be carbon, oxygen, or

---

[1936]See Baskaran, S.; Vasu, J.; Prasad, R.; Kodukulla, K.; Trivedi, G.K. *Tetrahedron* **1996**, *52*, 4515.

[1937]Foti, F.; Grassi, G.; Risitano, F. *Tetrahedron Lett.* **1999**, *40*, 2605.

[1938]Raposo, C.; Wilcox, C.S. *Tetrahedron Lett.* **1999**, *40*, 1285.

[1939]See Nishiwaki, N.; Uehara, T.; Asaka, N.; Tohda, Y.; Ariga, M.; Kanemasa, S. *Tetrahedron Lett.* **1998**, *39*, 4851; Jung, M.E.; Vu, B.T. *Tetrahedron Lett.* **1996**, *37*, 451; Weidner-Wells, M.A.; Fraga, S.A.; Demers, J.P. *Tetrahedron Lett.* **1994**, *35*, 6473; Easton, C.J.; Hughes, C.M.; Tiekink, E.R.T.; Lubin, C.E.; Savage, G.P.; Simpson, G.W. *Tetrahedron Lett.* **1994**, *35*, 3589; Brown, F.K.; Raimondi, L.; Wu, Y.-D.; Houk, K.N. *Tetrahedron Lett.* **1992**, *33*, 4405; Raimondi, L.; Wu, Y.-D.; Brown, F.K.; Houk, K.N. *Tetrahedron Lett.* **1992**, *33*, 4409. For a synthesis of nitrile oxides, see Muri, D.; Bode, J.W.; Carreira, E.M. *Org. Lett.* **2000**, *2*, 539. Nitrolic acids are precursors, see Matt, C.; Gissot, A.; Wagner, A.; Mioskowski, C. *Tetrahedron Lett.* **2000**, *41*, 1191.

[1940]For a review, see Pearson, W.H.; Stoy, P. *Synlett* **2003**, 903. For chloroiminium salts as precursors, see Anderson, R.J.; Batsanov, A.S.; Belskaia, N.; Groundwater, P.W.; Meth-Cohn, O.; Zaytsev, A. *Tetrahedron Lett.* **2004**, *45*, 943.

[1941]See Iesce, M.R.; Cermola, F.; Giordano, F.; Scarpati, R.; Graziano, M.L. *J. Chem. Soc. Perkin Trans. 1*, **1994**, 3295; MuCullough, K.J.; Sugimoto, T.; Tanaka, S.; Kusabayashi, S.; Nojima, M. *J. Chem. Soc. Perkin Trans. 1*, **1994**, 643.

nitrogen; hence there are six types. Among these are azides ($a = b = c = N$) and diazoalkanes.

2. Those in which the dipolar canonical form has a single bond on the sextet atom and the other form has a double bond:

$$\overset{\ominus}{a}-b-\overset{\oplus}{c}- \quad \longleftrightarrow \quad \overset{\ominus}{a}-\overset{\oplus}{b}=c-$$

Here b can be nitrogen or oxygen, and a and c can be nitrogen, oxygen, or carbon, but there are only 12 types, since, for example, N–N–C is only another form of C–N–N. Examples are shown in Table 15.3.

Of the 18 systems, some of which are unstable and must be generated *in situ*,[1942] the reaction has been accomplished for at least 15, but not in all cases with a carbon–carbon double bond (the reaction also can be carried out with other double bonds[1943]). Not all alkenes undergo 1,3-dipolar addition equally well. The reaction is most successful for those that are good dienophiles in the Diels–Alder reaction (**15-60**). The addition is stereospecific and syn, and the mechanism is probably a one-step concerted process,[1944] as illustrated above,[1945] largely controlled by Frontier Molecular Orbital considerations.[1946] In-plane aromaticity has been invoked for these dipolar cycloadditions.[1947] As expected for this type of mechanism, the rates do not vary much with changes in solvent,[1948] although rate acceleration has been observed in ionic liquids.[1949] Nitrile oxide cycloadditions have also been done in supercritical carbon dioxide.[1950] There are no simple rules

---

[1942]For a review of some aspects of this, see Grigg, R. *Chem. Soc. Rev.* **1987**, *16*, 89.

[1943]For a review of 1,3-dipolar addition to other double bonds, see Bianchi, G.; De Micheli, C.; Gandolfi, R., in Patai, S. *Supplement A: The Chemistry of Double-Bonded Functional Groups*, pt. 1, Wiley, NY, **1977**, pp. 369–532. For a review of such addition to the C=S bond, see Dunn, A.D.; Rudorf, W. *Carbon Disulfide in Organic Chemistry*, Wiley, NY, **1989**, pp. 97–119.

[1944]Di Valentin, C.; Freccero, M.; Gandolfi, R.; Rastelli, A. *J. Org. Chem.* **2000**, *65*, 6112. For a theoretical study of transition states, see Lu, X.; Xu, X.; Wang, N.; Zhang, Q. *J. Org. Chem.* **2002**, *67*, 515. For a theoretical study of stepwise vs. concerted reactions, see DiValentin, C.; Freccero, M.; Gandolfi, R.; Rastelli, A. *J. Org. Chem.* **2000**, *65*, 6112. For a discussion of loss of concertedness in reactions of azomethine ylids, see Vivanco, S.; Lecea, B.; Arrieta, A.; Prieto, P.; Morao, I.; Linden, A.; Cossío, F.P. *J. Am. Chem. Soc.* **2000**, *122*, 6078.

[1945]For a review, see Huisgen, R. *Adv. Cycloaddit.* **1988**, *1*, 1. For discussions, see Huisgen, R. *J. Org. Chem.* **1976**, *41*, 403; Firestone, R.A. *Tetrahedron* **1977**, *33*, 3009; Harcourt, R.D. *Tetrahedron* **1978**, *34*, 3125; Haque, M.S. *J. Chem. Educ.* **1984**, *61*, 490; Al-Sader, B.H.; Kadri, M. *Tetrahedron Lett.* **1985**, *26*, 4661; Houk, K.N.; Firestone, R.A.; Munchausen, L.L.; Mueller, P.H.; Arison, B.H.; Garcia, L.A. *J. Am. Chem. Soc.* **1985**, *107*, 7227; Majchrzak, M.W.; Warkentin, J. *J. Phys. Org. Chem.* **1990**, *3*, 339.

[1946]Caramella, P.; Gandour, R.W.; Hall, J.A.; Deville, C.G.; Houk, K.N. *J. Am. Chem. Soc.* **1977**, *99*, 385, and references cited therein.

[1947]Morao, I.; Lecea, B.; Cossío, F.P. *J. Org. Chem.* **1997**, *62*, 7033; Cossío, F.P.; Marao, I.; Jiao, H.; Schleyer, P.v.R. *J. Am. Chem. Soc.* **1999**, *121*, 6737.

[1948]For a review of the role of solvents in this reaction, see Kadaba, P.K. *Synthesis* **1973**, 71.

[1949]Dubreuil, J.F.; Bazureau, J.P. *Tetrahedron Lett.* **2000**, *41*, 7351.

[1950]Lee, C.K.Y; Holmes, A.B.; Al-Duri, B.; Leeke, G.A.; Santos, R.C.D.; Seville, J.P.K. *Chem. Commun.* **2004**, 2622.

covering orientation in 1,3-dipolar additions. The regioselectivity has been explained by molecular-orbital treatments,[1951] where overlap of the largest orbital coefficients of the atoms forming the new bonds leads to the major regioisomer. When the 1,3-dipolar compound is a thiocarbonyl ylid ($R_2C=S^+—CH_2^-$) the addition has been shown to be nonstereospecific with certain substrates but stereospecific with others, indicating a nonsynchronous mechanism in these cases, and in fact, a diionic intermediate (see mechanism *c* on p. 1224) has been trapped in one such case.[1952] In a theoretical study of the 1,3-dipolar cycloadditions (diazomethane and ethene; fulminic acid [$H—C≡N—O$] and ethyne),[1953] calculations based on valence bond descriptions suggest that many concerted 1,3-dipolar cycloaddition reactions follow an electronic heterolytic mechanism where the movement of well-identifiable orbital pairs are retained along the entire reaction path from reactants to product.[1954]

An antibody-catalyzed [3 + 2]-cycloaddition has been reported.[1955] Metal assisted dipolar additions are also known.[1956]

Many of the cycloadducts formed from the dipoles in Table 15.3 are unstable, leading to other products. The reaction of alkyl azides with alkenes generates triazolines (**15-54**), which extrude nitrogen ($N≡N$) upon heating or photolysis to give an aziridine.

[3 + 2]-Cycloaddition reactions occur intramolecularly to generate bicyclic and polycyclic compounds.[1957] The intramolecular cycloaddition of azomethine imines give bicyclic pyrrazolidines for example.[1958] When diazoalkanes, including diazo acetates such as N_2CHCO_2Et react with an alkene and a chromium catalyst the initially formed product is a five-membered ring, a pyrazoline. Pyrazolines are generally unstable and extrusion of nitrogen leads to a cyclopropane.[1959]

There are many cases where the [3 + 2]-cycloaddition leads to cycloadducts with high enantioselectivity.[1960] Cycloaddition of diazo esters with a cobalt catalyst having a chiral ligand leads to cyclopropane derivatives with good enantioselectivity.[1961]

[1951]For a review, see Houk, K.N.; Yamaguchi, K., in Padwa, A. *1,3-Dipolar Cycloaddition Chemistry* Vol. 2, Wiley, NY, *1984*, pp. 407–450. See also, Burdisso, M.; Gandolfi, R.; Quartieri, S.; Rastelli, A. *Tetrahedron* *1987*, *43*, 159.

[1952]Huisgen, R.; Mloston, G.; Langhals, E. *J. Am. Chem. Soc.* *1986*, *108*, 6401; *J. Org. Chem.* *1986*, *51*, 4085; Mloston, G.; Langhals, E.; Huisgen, R. *Tetrahedron Lett.* *1989*, *30*, 5373; Huisgen, R.; Mloston, G. *Tetrahedron Lett.* *1989*, *30*, 7041.

[1953]Karadakov, P.B.; Cooper, D.L.; Gerratt, J. *Theor. Chem. Acc.* *1998*, *100*, 222.

[1954]Blavins, J.J.; Karadakov, P.B.; Cooper, D.L. *J. Org. Chem.* *2001*, *66*, 4285.

[1955]Toker, J.D.; Wentworth Jr., P.; Hu, Y.; Houk, K.N.; Janda, K.D. *J. Am. Chem. Soc.* *2000*, *122*, 3244.

[1956]Kanemasa, S. *Synlett* *2002*, 1371.

[1957]For reviews, see Padwa, A. *Angew. Chem. Int. Ed.* *1976*, *15*, 123; Oppolzer, W. *Angew. Chem. Int. Ed.* *1977*, *16*, 10 (see pp. 18–22).

[1958]Dolle, R.E.; Barden, M.C.; Brennan, P.E.; Ahmed, G.; Tran, V.; Ho, D.M. *Tetrahedron Lett.* *1999*, *40*, 2907.

[1959]Jan, D.; Simal, F.; Demonceau, A.; Noels, A.F.; Rufanov, K.A.; Ustynyuk, N.A.; Gourevitch, D.N. *Tetrahedron Lett.* *1999*, *40*, 5695.

[1960]Gothelf, K.V.; Jørgensen, K.A. *Chem. Rev.* *1998*, *98*, 863.

[1961]Niimi, T.; Uchida, T.; Irie, R.; Katsuki, T. *Tetrahedron Lett.* *2000*, *41*, 3647.

Cycloaddition of nitrones and pyrazolinones with a copper catalyst and a chiral ligand leads to pyrrolidine derivatives with good enantioselectivity.[1962]

Conjugated dienes generally give exclusive 1,2-addition, although 1,4 addition (a [3 + 4]-cycloaddition) has been reported.[1963] Carbon–carbon triple bonds can also undergo 1,3-dipolar addition.[1964] For example, azides react to give triazoles, **185**.

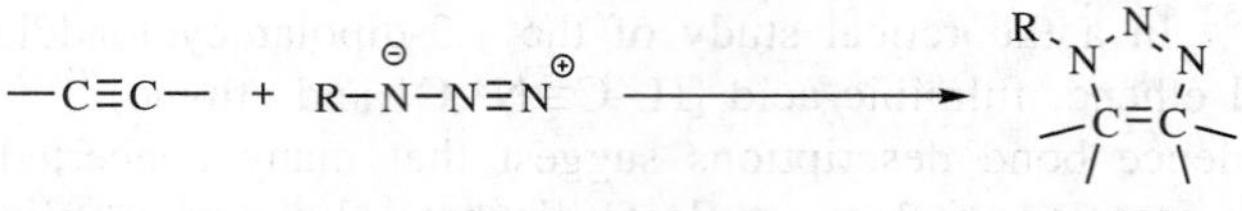

$$-C\equiv C- \ + \ R-N-N\overset{\oplus}{\equiv}N \quad \longrightarrow \quad \overset{\displaystyle R_{\diagdown}N\diagup N_{\diagdown}N}{\underset{\diagup C=C \diagdown}{|\quad\quad|}}$$

185

The 1,3-dipolar reagent can in some cases be generated by the *in situ* opening of a suitable three-membered ring system. For example, aziridines open to give a zwitterion, such as **186**, which can add to activated double bonds to give pyrrolidines.[1965]

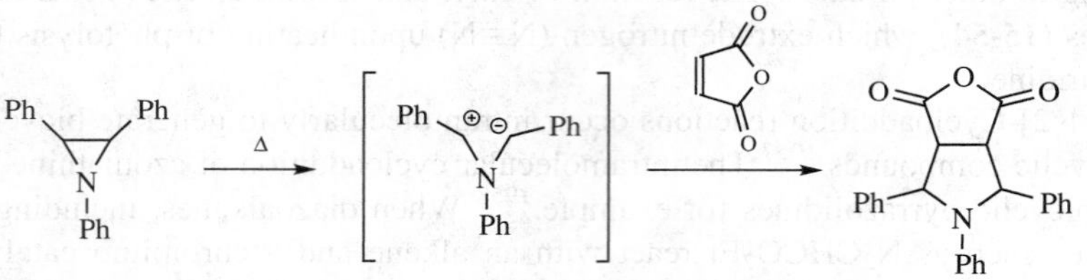

186

Aziridines also add to C≡C triple bonds as well as to other unsaturated linkages, including C=O, C=N, and C≡N.[1966] In some of these reactions it is a C–N bond of the aziridine that opens rather than the C–C bond.

For other [3 + 2]-cycloadditions, see **15-59**.

OS **V**, 957, 1124; **VI**, 592, 670; **VIII**, 231. Also see, OS **IV**, 380.

C. Carbon on Both Sides

Reactions **15-58–15-64** are cycloaddition reactions.[1967]

[1962]Sibi, M.P.; Ma, Z.; Jasperse, C.P. *J. Am. Chem. Soc.* **2004**, *126*, 718.

[1963]Baran, J.; Mayr, H. *J. Am. Chem. Soc.* **1987**, *109*, 6519.

[1964]For reviews, see Bastide, J.; Hamelin, J.; Texier, F.; Quang, Y.V. *Bull. Soc. Chim. Fr.* **1973**, 2555; 2871; Fuks, R.; Viehe, H.G., in Viehe, H.G. *Acetylenes*, Marcel Dekker, NY, **1969**, pp. 460–477.

[1965]For a review, see Lown, J.W., in Padwa, A. *1,3-Dipolar Cycloaddition Chemistry*, Vol 1. Wiley, NY, **1984**, pp. 683–732.

[1966]For reviews, see Lown, J.W. *Rec. Chem. Prog.* **1971**, *32*, 51; Gladysheva, F.N.; Sineokov, A.P.; Etlis, V.S. *Russ. Chem. Rev.* **1970**, *39*, 118.

[1967]For a system of classification of cycloaddition reactions, see Huisgen, R. *Angew. Chem. Int. Ed.* **1968**, *7*, 321. For a review of certain types of cycloadditions leading to 3- to 6-membered rings involving 2, 3, or 4 components, see Posner, G.H. *Chem. Rev.* **1986**, *86*, 831. See also, the series *Advances in Cycloaddition*.

15-59   All-Carbon [3 + 2]-Cycloadditions[1968]

Several methods have been reported for the formation of cyclopentanes by [3 + 2]-cycloadditions.[1969] Heating a conjugated ketones wtih trialkylphosneines genrates an intermdiate that adds to conjugated alkynes.[1970] One type involves reagents that produce intermediates **187** or **188**.[1971] A synthetically useful example[1972] uses 2-[(trimethylsilyl)methyl]-2-propen-1-yl acetate (**191**) (which is commercially available) and a palladium or other transition-metal catalyst to generate **187** or **188**, which adds to double bonds, to give, in

good yields, cyclopentanes with an exocyclic double bond. Note that **95** also reacts with *N*-tosyl aziridines, with 20% *n*-butyllithium and 10% of Pd(OAc)$_2$, to give a vinylidene piperidine derivative.[1973] Similar or identical intermediates generated from bicyclic azo compounds **189** (see **17-34**) or methylenecyclopropane **190**[1974] also add to activated double bonds. With suitable substrates the addition can be enantioselective.[1975]

In a different type of procedure, [3 + 2]-cycloadditions are performed with allylic anions. Such reactions are called 1,3-anionic cycloadditions.[1976] For example, α-methylstyrene adds to stilbene on treatment with the strong base LDA.[1977]

[1968]See Smith, M.B. *Organic Synthesis, 2nd* ed., McGraw-Hill, NY, *2001*, pp. 999–1010.

[1969]For a list of methods, with references, see Trost, B.M.; Seoane, P.; Mignani, S.; Acemoglu, M. *J. Am. Chem. Soc. 1989*, *111*, 7487.

[1970]Wang, J.-C.; Ng, S.-S.; Krische, M.J. *J. Am. Chem. Soc. 2003*, *125*, 3682.

[1971]For reviews, see Trost, B.M. *Pure Appl. Chem. 1988*, *60*, 1615; *Angew. Chem. Int. Ed. 1986*, *25*, 1.

[1972]See, for example, Trost, B.M.; Lynch, J.; Renaut,P.; Steinman, D.H. *J. Am. Chem. Soc. 1986*, *108*, 284.

[1973]Hedley, S.J.; Moran, W.J.; Price, D.A.; Harrity, J.P.A. *J. Org. Chem. 2003*, *68*, 4286.

[1974]See Yamago, S.; Nakamura, E. *J. Am. Chem. Soc. 1989*, *111*, 7285.

[1975]See Binger, P.; Schäfer, B. *Tetrahedron Lett. 1988*, *29*, 529; Chaigne, F.; Gotteland, J.; Malacria, M. *Tetrahedron Lett. 1989*, *30*, 1803.

[1976]For reviews, see Kauffmann, T. *Top. Curr. Chem. 1980*, *92*, 109, pp. 111–116; *Angew. Chem. Int. Ed. 1974*, *13*, 627.

[1977]Eidenschink, R.; Kauffmann, T. *Angew. Chem. Int. Ed. 1972*, *11*, 292.

The mechanism can be outlined as

192

In the case above, **192** is protonated in the last step by the acid HA, but if the acid is omitted and a suitable nucleofuge is present, it may leave, resulting in a cyclopentene.[1978] In these cases the reagent is an allylic anion, but similar [3 + 2]-cycloadditions involving allylic cations have also been reported.[1979]

OS **VIII**,173, 347.

15-60   The Diels–Alder Reaction

   (4 + 2)*cyclo*-Ethylene-1/4/addition or (4 + 2)*cyclo*-[But-2-ene-1,4-diyl]-1/2/ addition, and so on.

In the prototype *Diels–Alder reaction* the double bond of an alkene adds 1,4 to a conjugated diene (a [4 + 2]-cycloaddition),[1980] so the product is always a cyclohexene. The cycloaddition is not limited to alkenes or to dienes (see **15-61**), but the substrate that reacts with the diene is called a *dienophile*. The reaction is of

---

[1978]See, for example, Padwa, A.; Yeske, P.E. *J. Am. Chem. Soc.* **1988**, *110*, 1617; Beak, P.; Burg, D.A. *J. Org. Chem.* **1989**, *54*, 1647.

[1979]For example, see Hoffmann, H.M.R.; Vathke-Ernst, H. *Chem. Ber.* **1981**, *114*, 2208, 2898; Klein, H.; Mayr, H. *Angew. Chem. Int. Ed.* **1981**, *20*, 1027; Noyori, R.; Hayakawa, Y. *Tetrahedron* **1985**, *41*, 5879.

[1980]For a monograph, see Wasserman, A. *Diels-Alder Reactions*, Elsevier, NY, **1965**. For reviews, see Fleming, I. *Pericyclic Reactions*, Oxford University Press, Oxford, **1999**, pp. 7–30; Roush, W.R. *Adv. Cycloaddit.* **1990**, *2*, 91; Carruthers, W. *Cycloaddition Reactions in Organic Synthesis*, Pergamon, Elmsford, NY, **1990**; Brieger, G.; Bennett, J.N. *Chem. Rev.* **1980**, *80*, 63; Oppolzer, W. *Angew. Chem. Int. Ed.* **1977**, *16*, 10; Beltrame, P., in Bamford, C.H.; Tipper, C.F.H *Comprehensive Chemical Kinetics*, Vol. 9, Elsevier, NY, **1973**, pp. 94–117; Huisgen, R.; Grashey, R.; Sauer, J., in Patai, S. *The Chemistry of Alkenes*, Vol. 1, Wiley, NY, **1964**, pp. 878–929; Carruthers, W. *Some Modern Methods of Organic Synthesis*, 3rd. ed., Cambridge University Press, Cambridge, **1986**, pp. 183–244; Sauer, J. *Angew. Chem. Int. Ed.* **1966**, *5*, 211; **1967**, *6*, 16. For a monograph on intramolecular Diels–Alder reactions, see Taber, D.F. *Intramolecular Diels-Alder and Alder Ene Reactions*, Springer, NY, **1984**. For reviews, see Deslongchamps, P. *Aldrichimica Acta* **1991**, *24*, 43; Craig, D. *Chem. Soc. Rev.* **1987**, *16*, 187; Salakhov, M.S.; Ismailov, S.A. *Russ. Chem. Rev.* **1986**, *55*, 1145; Fallis, A.G. *Can. J. Chem.* **1984**, *62*, 183. For a long list of references to various aspects of the Diels–Alder reaction, see Larock, R.C. *Comprehensive Organic Transformations*, 2nd ed., Wiley-VCH, NY, **1999**, pp. 523–544.

very broad scope[1981] and reactivity of dienes and dienophiles can be predicted based on analysis of the HOMOs[1982] and LUMOs of these species (frontier molecular orbital theory).[1983] Ethylene and simple alkenes make poor dienophiles, unless high temperatures and/or pressures are used. Most dienophiles are of the form $\overset{\displaystyle -\text{C=C-Z}}{\underset{\displaystyle | \quad |}{}}$ or $\overset{\displaystyle \text{Z-C=C-Z}'}{\underset{\displaystyle | \quad |}{}}$, where Z and Z' are electron-withdrawing groups,[1984] such as CHO, COR,[1985] COOH, COOR, COCl, COAr, CN,[1986] NO_2,[1987] Ar, CH_2OH, CH_2Cl, CH_2NH_2, CH_2CN, CH_2COOH, halogen, $PO(OEt)_2$,[1988] or C=C. In the last case, the dienophile is itself a diene.[1989] Particularly common dienophiles are maleic anhydride[1990] and quinones.[1991] Triple bond compounds ($-C\equiv C-Z$ or $Z-C\equiv C-\ Z'$)

193

may be dienophiles,[1992] generating nonconjugated cyclohexadienes (**193**), and this reaction can be catalyzed by transition-metal compounds.[1993] Allenes react as dienophiles, but without activating groups are very poor dienophiles.[1994]

[1981]For a review of reactivity in the Diels–Alder reaction, see Konovalov, A.I. *Russ. Chem. Rev.* **1983**, *52*, 1064.

[1982]For a correlation of ionization potential and HOMO correlation with alkene reactions, see Nelson, D.J.; Li, R.; Brammer, C. *J. Org. Chem.* **2001**, *66*, 2422.

[1983]For a discussion of Frontier Orbital interactions, see Spino, C.; Rezaei, H.; Dory, Y.L. *J. Org. Chem.* **2004**, *69*, 757. For tables of experimentally determined HOMOs and LUMOs for dienes and dienophiles, see Smith, M.B. *Organic Synthesis*, 2nd ed., McGraw-Hill, NY, **2001**, pp. 917–940.

[1984]For a density-Functional theory analysis see Domingo, L.R. *Eur. J. Org. Chem.* **2004**, 4788.

[1985]For a review of Diels–Alder reactions with cyclic enones, see Fringuelli, F.; Taticchi, A.; Wenkert, E. *Org. Prep. Proced. Int.* **1990**, *22*, 131.

[1986]For a review of the Diels–Alder reaction with acrylonitrile, see Butskus, P.F. *Russ. Chem. Rev.* **1962**, *31*, 283. For a review of tetracyanoethylene as a dienophile, see Ciganek, E.; Linn, W.J.; Webster, O.W., in Rappoport, Z. *The Chemistry of the Cyano Group*, Wiley, NY, **1970**, pp. 449–453.

[1987]For a review of the Diels–Alder reaction with nitro compounds, see Novikov, S.S.; Shuekhgeimer, G.A.; Dudinskaya, A.A. *Russ. Chem. Rev.* **1960**, *29*, 79.

[1988]McClure, C.K.; Herzog, K.J.; Bruch, M.D. *Tetrahedron Lett.* **1996**, *37*, 2153.

[1989]Johnstone, R.A.W.; Quan, P.M. *J. Chem. Soc.* **1963**, 935.

[1990]For a review of Diels–Alder reactions with maleic anhydride see Kloetzel, M.C. *Org. React.* **1948**, *4*, 1.

[1991]For reviews of Diels–Alder reactions with quinones, see Finley, K.T., in Patai, S *The Chemistry of the Quinoid Compounds*, Vol. 1, pt. 2, Wiley, NY, **1988**, pp. 986–1018; Patai, S.; Rapaport, Z. Vol. 2, pt. 1 **1988**, 537–717, 614–645. For a review of the synthesis of quinones using Diels–Alder reactions, see Naruta, Y.; Maruyama, K. in the same treatise, Vol. 2, pt. 1, pp. 241–402, 277–303.

[1992]For reviews of triple bonds in cycloaddition reactions, see Bastide, J.; Henri-Rousseau, O., in Patai, S. *The Chemistry of the Carbon-Carbon Triple Bond*, pt. 1, Wiley, NY, **1978**, pp. 447–522, Fuks, R.; Viehe, H.G., in Viehe, H.G. *Acetylenes*, Marcel Dekker, NY, **1969**, pp. 477–508.

[1993]See Paik, S.-J.; Son, S.U.; Chung, Y.K. *Org. Lett.* **1999**, *1*, 2045.

[1994]For a review of allenes as dienes or dienophiles, see Hopf, H., in Landor, S.R. *The Chemistry of Allenes*, Vol. 2, Academic Press, NY, **1982**, pp. 563–577. See Nendel, M.; Tolbert, L.M.; Herring, L.E.; Islam, Md.N.; Houk, K.N. *J. Org. Chem.* **1999**, *64*, 976.

1196   ADDITION TO CARBON–CARBON MULTIPLE BONDS

Ketenes, however, do not undergo Diels–Alder reactions.[1995] Benzynes, although not isolable, act as dienophiles and can be trapped with dienes,[1996] for example,

194

The low reactivity of simple alkenes can be overcome by incorporating an electron-withdrawing group to facilitate the cycloaddition, but a group that can be removed after the cycloaddition. An example is phenyl vinyl sulfone $PhSO_2CH=CH_2$.[1997] The $PhSO_2$ group can be easily removed with Na–Hg after the ring-closure reaction. Similarly, phenyl vinyl sulfoxide ($PhSOCH=CH_2$) can be used as a synthon for acetylene.[1998] In this case PhSOH is lost from the sulfoxide product (**17-12**).

195   **196**   **197**

Electron-donating substituents in the diene accelerate the reaction; electron-withdrawing groups retard it.[1999] For the dienophile it is just the reverse: donating groups decrease the rate, and withdrawing groups increase it. The cisoid conformation is required for the cycloaddition,[2000] and acyclic dienes are conformationally mobile so the cisoid conformation will be available. Cyclic dienes, in which the cisoid conformation is built in, usually react faster than the corresponding open-chain compounds, which have to achieve the cisoid conformation by rotation.[2001] Dienes can be open-chain, inner-ring (e.g., **194**), outer-ring[2002] (e.g., **195**), across

[1995]Ketenes react with conjugated dienes to give 1,2-addition (see **15-49**).
[1996]For a review of benzynes as dienophiles, see Hoffmann, R.W. *Dehydrobenzene and Cycloalkynes*; Academic Press, NY, *1967*, pp. 200–239. For a review of the reactions of benzynes with heterocyclic compounds see Bryce, M.R.; Vernon, J.M. *Adv. Heterocycl. Chem.* *1981*, 28, 183–229.
[1997]Carr, R.V.C.; Williams, R.V.; Paquette, L.A. *J. Org. Chem.* *1983*, 48, 4976; Kinney, W.A.; Crouse, G.D.; Paquette, L.A. *J. Org. Chem.* *1983*, 48, 4986.
[1998]Paquette, L.A.; Moerck, R.E.; Harirchian, B.; Magnus, P.D. *J. Am. Chem. Soc.* *1978*, 100, 1597. For other acetylene synthons see De Lucchi, O.; Lucchini, V.; Pasquato, L.; Modena, G. *J. Org. Chem.* *1984*, 49, 596; Hermeling, D.; Schäfer, H.J. *Angew. Chem. Int. Ed.* *1984*, 23, 233. For a review, see De Lucchi, O.; Modena, G. *Tetrahedron* *1984*, 40, 2585. For a review of [2+2]- and [2+4]-cycloadditions of vinylic sulfides, sulfoxides, and sulfones, see De Lucchi, O.; Pasquato, L. *Tetrahedron* *1988*, 44, 6755.
[1999]For a discussion of the electrophilicity power of dienes and dienophiles, see Domingo, L.R.; Aurell, M.J.; Pérez, P.; Contreras, R. *Tetrahedron* *2002*, 58, 4417.
[2000]For a discussion of ground state conformations, see Bur, S.K.; Lynch, S.M.; Padwa, A. *Org. Lett.* *2002*, 4, 473.
[2001]Sauer, J.; Lang, D.; Mielert, A. *Angew. Chem. Int. Ed.* *1962*, 1, 268; Sauer, J.; Wiest, H. *Angew. Chem. Int. Ed.* *1962*, 1, 269. See, however, Scharf, H.; Plum, H.; Fleischhauer, J.; Schleker, W. *Chem. Ber.* *1979*, 112, 862.
[2002]For reviews of Diels-Alder reactions of some of these compounds, see Charlton, J.L.; Alauddin, M.M. *Tetrahedron* *1987*, 43, 2873; Oppolzer, W. *Synthesis* *1978*, 793.

rings (e.g., **196**), or inner-outer (e.g., **197**), except that they may not be frozen into a transoid conformation (see p. 1201). They need no special activating groups, and nearly all conjugated dienes undergo the reaction with suitable dienophiles.[2003]

In most Diels–Alder reactions, no catalyst is needed, but Lewis acids are effective catalysts in many cases,[2004] particularly those in which Z in the dienophile is a C=O or C=N group. A Lewis acid catalyst usually increases both the regioselectivity of the reaction (in the sense given above) and the extent of endo addition,[2005] and, in the case of enantioselective reactions, the extent of enantioselectivity. It has been shown that $InCl_3$ is an effective catalyst for aqueous Diels–Alder reactions,[2006] which is suitable for ionic Diels–Alder reactions,[2007] and there are other Lewis acid catalysts that are effective in water.[2008] Brønsted acids have also been used to accelerate the rate of the Diels–Alder reaction.[2009] Lanthanum triflate $[La(OTf)_3]$ has been reported as a reusable catalyst[2010] and Me_3SiNTf_2 has been used as a green Lewis acid catalyst.[2011] Cationic Diels–Alder catalysts have been developed, particularly oxazaborolidine catalysts.[2012] Some Diels–Alder reactions can also be catalyzed by the addition of a stable cation radical,[2013] for

[2003]For a monograph on dienes, with tables showing > 800 types, see Fringuelli, F.; Taticchi, A. *Dienes in the Diels–Alder Reaction*, Wiley, NY, *1990*. For a review of Diels–Alder reactions with 2-pyrones, see Shusherina, N.P. *Russ. Chem. Rev.* *1974*, *43*, 851. For reviews of dienes with hetero substituents, see Danishefsky, S. *Chemtracts: Org. Chem.* *1989*, *2*, 273; Petrzilka, M.; Grayson, J.I. *Synthesis* *1981*, 753. For dienes containing a 1-CONR$_2$ group, see Smith, M.B. *Org. Prep. Proced. Int.* *1990*, *22*, 315; Robiette, R.; Cheboub-Benchaba, K.; Peeters, D.; Marchand-Brynaert, J. *J. Org. Chem.* *2003*, *68*, 9809. For dienes containing a 1-NRCO$_2$R group, see Huang, Y.; Iwama, T.; Rawal, V.H. *J. Am. Chem. Soc.* *2002*, *122*, 5950.

[2004]Yates, P.; Eaton, P. *J. Am. Chem. Soc.* *1960*, *82*, 4436; Avalos, M.; Babiano, R.; Bravo, J.L.; Cintas, P.; Jiménez, J.L.; Palacios, J.C.; Silva, M.A. *J. Org. Chem.* *2000*, *65*, 6613. For review of the role of the catalyst in increasing reactivity, see Kiselev, V.D.; Konovalov, A.I. *Russ. Chem. Rev.* *1989*, *58*, 230. For a discussion of the transition state for the acrolein-1,3-butadiene reaction see Zheng, M.; Zhang, M.-H.; Shao, J.-G.; Zhong, Q. *Org. Prep. Proceed. Int.* *1996*, *28*, 117. For a discussion of isotope effects see Singleton, D.A.; Merrigan, S.R.; Beno, B.R.; Houk, K.N. *Tetrahedron Lett.* *1999*, *40*, 5817. For a discussion of three-center orbital interactions, see Yamabe, S.; Minato, T. *J. Org. Chem.* *2000*, *65*, 1830. Chiral silica Lewis acids are known, see Mathieu, B.; de Fays, L.; Ghosez, L. *Tetrahedron Lett.* *2000*, *41*, 9561.

[2005]For discussions see Houk, K.N.; Strozier, R.W. *J. Am. Chem. Soc.* *1973*, *95*, 4094; Alston, P.V.; Ottenbrite, R.M. *J. Org. Chem.* *1975*, *40*, 1111.

[2006]Loh, T.-P.; Pei, J.; Lin, M. *Chem. Commun.* *1996*, 2315. For a review of Lewis acid catalysis in aqueous media, see Fringuelli, F.; Piermatti, O.; Pizzo, F.; Vaccaro, L. *Eur. J. Org. Chem.* *2001*, 439.

[2007]Reddy, B.G.; Kumareswaran, R.; Vankar, Y.D. *Tetrahedron Lett.* *2000*, *41*, 10333. Iodine is a catalyst for ionic Diels–Alder reactions, see Chavan, S.P.; Sharma, P.; Krishna, G.R.; Thakkar, M. *Tetrahedron Lett.* *2003*, *44*, 3001.

[2008]Otto, S.; Engberts, J.B.F.N. *Tetrahedron Lett.* *1995*, *36*, 2645; Ward, D.E.; Gai, Y. *Tetrahedron Lett.* *1992*, *33*, 1851.

[2009]Ishihara, K.; Kurihara, H.; Yamamoto, H. *J. Am. Chem. Soc,* *1996*, *118*, 3049.

[2010]Kobayashi, S.; Hachiya, I.; Takahori, T.; Araki, M.; Ishitani, H. *Tetrahedron Lett.* *1992*, *33*, 6815.

[2011]Mathieu, B.; Ghosez, L. *Tetrahedron* *2002*, *58*, 8219.

[2012]See Sprott, K.T.; Corey, E.J. *Org. Lett.* *2003*, *5*, 2465; Corey, E.J.; Shibata, T.; Lee, T.W. *J. Am. Chem. Soc.* *2002*, *124*, 3808; Ryu, D.H.; Lee, T.W.; Corey, E.J. *J. Am. Chem. Soc.* *2002*, *124*, 9992.

[2013]Gao, D.; Bauld, N.L. *J. Org. Chem.* *2000*, *65*, 6276. See Saettel, N.J.; Oxgaard, J.; Wiest, O. *Eur. J. Org. Chem.* *2001*, 1429.

example, tris(4-bromophenyl)aminium hexachloroantimonate $Ar_3N^{\bullet+}$ $SbCl_6^-$.[2014] Carbazoles are dienophiles for cation radical Diels–Alder reactions.[2015] Zirconocene-catalyzed cationic Diels–Alder reactions are known.[2016] Certain antibodies have been developed that catalyze Diels–Alder reactions.[2017] Photochemically induced Diels–Alder reactions are also known.[2018] Cyclodextrins exhibit noncovalent catalysis of Diels–Alder reactions.[2019] There are cases of hydrogen-bonding acceleration.[2020]

A number of other methods have been reported for the acceleration of Diels–Alder reactions,[2021] including the use of microwave irradiation,[2022] ultrasound,[2023] absorption of the reactants on chromatographic absorbents,[2024] via encapsulation techniques,[2025] and the use of an ultracentrifuge[2026] (one of several ways to achieve reaction at high pressures).[2027] Solid-state Diels–Alder reactions are known.[2028] One of the most common methods is to use water as a solvent or a cosolvent (a hydrophobic effect).[2029] The influence of hydrophobicity of reactants

[2014]For a review, see Bauld, N.L. *Tetrahedron* **1989**, *45*, 5307.

[2015]Gao, D.; Bauld, N.L. *Tetrahedron Lett.* **2000**, *41*, 5997.

[2016]Wipf, P.; Xu, W. *Tetrahedron* **1995**, *51*, 4551.

[2017]Meekel, A.A.P.; Resmini, M.; Pandit, U.K. *J. Chem. Soc., Chem. Commun.* **1995**, 571; Zhang, X.; Deng, Q.; Yoo, S.H.; Houk, K.N. *J. Org. Chem.* **2002**, *67*, 9043.

[2018]Pandey, B.; Dalvi, P.V. *Angew. Chem. Int. Ed.* **1993**, *32*, 1612.

[2019]Kim, S.P.; Leach, A.G.; Houk, K.N. *J. Org. Chem.* **2002**, *67*, 4250. For a discussion of micellar catalysis see Rispens, T.; Engberts, J.B.F.N. *J. Org. Chem.* **2002**, *67*, 7369.

[2020]Pearson, R.J.; Kassianidis, E.; Philip, D. *Tetrahedron Lett.* **2004**, *45*, 4777.

[2021]See Smith, M.B. *Organic Synthesis, 2nd* ed., McGraw-Hill, NY, **2001**, pp. 944–953.

[2022]Giguere, R.J.; Bray, T.L.; Duncan, S.M.; Majetich, G. *Tetrahedron Lett.* **1986**, *27*, 4945; Berlan, J.; Giboreau, P.; Lefeuvre, S.; Marchand, C. *Tetrahedron Lett.* **1991**, *32*, 2363; DaCunha, L.; Garrigues, B. *Bull. Soc. Chim. Belg.* **1997**, *106*, 817; Jankowski, C.K.; LeClair, G.; Bélanger, J.M.R.; Paré, J.R.J.; Van Calsteren, M.-R. *Can. J. Chem.* **2001**, *79*, 1906. For a review, see de la Hoz, A.; Díaz-Ortis, A.; Moreno, A.; Langa, F. *Eur. J. Org. Chem.* **2000**, 3659. For the effect of pressure of microwave-enhanced Diels–Alder reactions, see Kaval, N.; Dehaen, W.; Kappe, C.O.; van der Eycken, E. *Org. Biomol. Chem.* **2004**, *2*, 154.

[2023]Raj. C.P.; Dhas, N.A.; Cherkinski, M.; Gedanken, A.; Braverman, S. *Tetrahedron Lett.* **1998**, *39*, 5413.

[2024]Veselovsky, V.V.; Gybin, A.S.; Lozanova, A.V.; Moiseenkov, A.M.; Smit, W.A.; Caple, R. *Tetrahedron Lett.* **1988**, *29*, 175.

[2025]Kang, J.; Hilmersson, G.; Sartamaría, J.; Rebek Jr., J. *J. Am. Chem. Soc.* **1998**, *120*, 3650. For a discussion of the Diels–Alder reaction with aqueous surfactants see Diego-Castro, M.J.; Hailes, H.C. *Tetrahedron Lett.* **1998**, *39*, 2211.

[2026]Dolata, D.P.; Bergman, R. *Tetrahedron Lett.* **1987**, *28*, 707.

[2027]For reviews, see Isaacs, N.S.; George, A.V. *Chem. Br.* **1987**, 47–54; Asano, T.; le Noble, W.J. *Chem. Rev.* **1978**, *78*, 407. See also, Firestone, R.A.; Smith, G.M. *Chem. Ber.* **1989**, *122*, 1089.

[2028]Kim, J.H.; Hubig, S.M.; Lindeman, S.V.; Kochi, J.K. *J. Am. Chem. Soc.* **2001**, *123*, 87.

[2029]Rideout, D.C.; Breslow, R. *J. Am. Chem. Soc.* **1980**, *102*, 7816. For a review, see Breslow, R. *Acc. Chem. Res.* **1991**, *24*, 159; Furlani, T.R.; Gao, J. *J. Org. Chem.* **1996**, *61*, 5492. See also, Grieco, P.A.; Garner, P.; He, Z. *Tetrahedron Lett.* **1983**, 1897; Blokzijl, W.; Blandamer, M.J.; Engberts, J.B.F.N. *J. Am. Chem. Soc.* **1991**, *113*, 4241; Breslow, R.; Rizzo, C.J. *J. Am. Chem. Soc.* **1991**, *113*, 4340; Engberts, J.B.F.N. *Pure Appl. Chem.* **1995**, *67*, 823; Pindur, U.; Lutz, G.; Otto, C. *Chem. Rev.* **1993**, *93*, 741; Otto, S.; Blokzijl, W.; Engberts, J.B.F.N. *J. Org. Chem.* **1994**, *59*, 5372; Otto, S.; Egberts, J.B.F.N. *Pure. Appl. Chem.* **2000**, *72*, 1365.

on the reaction has been examined[2030] as has micellular effects.[2031] Another alternative reaction medium is the use of 5 M $LiClO_4$ in Et_2O as solvent,[2032] An alternative to lithium perchlorate in ether is lithium triflate in acetonitrile.[2033] The addition of HPO_4^- – to an aqueous ethanol solution has also been shown to give an small rate enhancement.[2034] This appears to be the only case where an anion is responsible for a rate enhancement. The *retro*-Diels–Alder reaction has also been done in water.[2035]

It is noted that the Diels–Alder reaction has been done with supercritical CO_2[2036] and with supercritical water[2037] as solvents. Diels–Alder reactions on solid supports have also been reported,[2038] and zeolites have been used in conjunction with catalytic agents.[2039] Alumina has been used to promote Diels–Alder reactions.[2040] Diels–Alder reactions can be done in ionic liquids,[2041] including asymmetric Diels–Alder reactions.[2042]

When an unsymmetrical diene adds to an unsymmetrical dienophile, regioisomeric products (not counting stereoisomers) are possible. Rearrangements have been encountered in some cases.[2043] In simple cases, 1-substituted dienes give cyclohexenes with a 1,2- and a 1,3- substitution pattern. 2-Substituted dienes

[2030]Meijer, A.; Otto, S.; Engberts, J.B.F.N. *J. Org. Chem.* **1998**, *63*, 8989; Rizzo, C.J. *J. Org. Chem.* **1992**, *57*, 6382.

[2031]Jaeger, D.A.; Wang, J. *Tetrahedron Lett.* **1992**, *33*, 6415.

[2032]Grieco, P.A.; Nunes, J.J.; Gaul, M.D. *J. Am. Chem. Soc.* **1990**, *112*, 4595. See also, Braun, R.; Sauer, J. *Chem. Ber.* **1986**, *119*, 1269; Grieco, P.A.; Handy, S.T.; Beck, J.P. *Tetrahedron Lett.* **1994**, *35*, 2663. For the possibility of migration of terminal dienes prior to cycloaddition see Grieco, P.A.; Beck, J.P.; Handy, S.T.; Saito, N.; Daeuble, J.F. *Tetrahedron Lett.* **1994**, *35*, 6783. An alternative to this catalyst is $LiNTf_2$ in ether, see Handy, S.T.; Grieco, P.A.; Mineur, C.; Ghosez, L. *Synlett* **1995**, 565.

[2033]Augé, J.; Gil, R.; Kalsey, S.; Lubin-Germain, N. *Synlett* **2000**, 877.

[2034]Pai, C.K.; Smith, M.B. *J. Org. Chem.* **1995**, *60*, 3731; Smith, M.B.; Fay, J.N.; Son, Y.C. *Chem. Lett.* **1992**, 2451.

[2035]Wijnen, J.W.; Engberts, J.B.F.N. *J. Org. Chem.* **1997**, *62*, 2039.

[2036]Renslo, A.R.; Weinstein, R.D.; Tester, J.W.; Danheiser, R.L. *J. Org. Chem.* **1997**, *62*, 4530; Oakes, R.S.; Heppenstall, T.J.; Shezad, N.; Clifford, A.A.; Rayner, C.M. *Chem. Commun.* **1999**, 1459. For an asymmetric cycloaddition, see Fukuzawa, S.-i.; Metoki, K.; Esumi, S.-i. *Tetrahedron* **2003**, *59*, 10445.

[2037]Harano, Y.; Sato, H.; Hirata, F. *J. Am. Chem. Soc.* **2000**, *122*, 2289.

[2038]For a review, see Yli-Kauhaluoma, J. *Tetrahedron* **2001**, *57*, 7053. For silica and alumina-modified Lewis acid catalysts, see Cativiela, C.; Figueras, F.; García, J.I.; Mayoral, J.A.; Pires, E.; Royo, A.J. *Tetrahedron Asymmetry* **1993**, *4*, 621.

[2039]Eklund, L.; Axelsson, A.-K.; Nordahl, Å.; Carlson, R. *Acta Chem. Scand.* **1993**, *47*, 581.

[2040]Pagni, R.M.; Kabalka, G.W.; Hondrogiannis, G.; Bains, S.; Anosike, P.; Kurt, R. *Tetrahedron* **1993**, *49*, 6743.

[2041]In **bmim** BF_4 and ClO_4: 1-butyl-3-methylimidazolium tetrafluoroborate and perchlorate: Fischer, T.; Sethi, A.; Welton, T.; Woolf, J. *Tetrahedron Lett.* **1999**, *40*, 793. In **chloroaluminates**: Lee, C.W. *Tetrahedron Lett.* **1999**, *40*, 2461. In **phosphonium tosylates**: Ludley, P.; Karodia, N. *Tetrahedron Lett.* **2001**, *42*, 2011. In **pyridinium salts**: Xiao, Y.; Malhotra, S.V. *Tetrahedron Lett.* **2004**, *45*, 8339. In **HBuIm**, hydrogenbutylimidazolium tetrafluoroborate and **DiBuIm**, 1,3-dibutylimidazolium, tetrafluoroborate: Jaegar, D. A.; Tucker, C. E. *Tetrahedron Lett.* **1989**, *30*, 1785.

[2042]Meracz, I.; Oh, T. *Tetrahedron Lett.* **2003**, *44*, 6465.

[2043]Murali, R.; Scheeren, H.W. *Tetrahedron Lett.* **1999**, *40*, 3029.

lead to 1,4- and 1,3-disubstituted products.

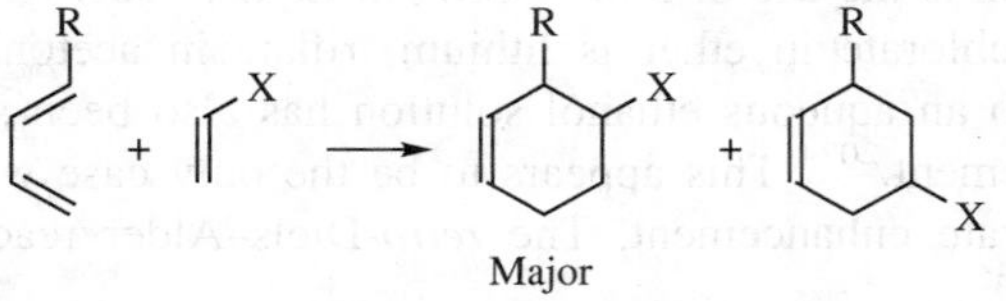

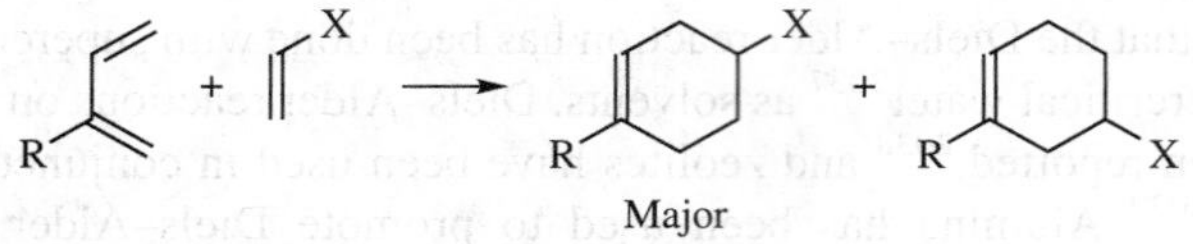

Major

Major

Although mixtures are often obtained, usually one predominates, the one indicated above, but selectivity depends on the nature of the substituents on both diene and alkene. This regioselectivity, in which the "ortho" or "para" product is favored over the "meta," has been explained by molecular-orbital considerations.[2044] When $X = NO_2$, regioselectivity to give the "ortho" or "para" product was very high at room temperature, and this method, combined with subsequent removal of the NO_2 (see **19-67**) has been used to perform regioselective Diels–Alder reactions.[2045]

The stereochemistry of the Diels–Alder reaction can be considered from several aspects:[2046]

1. With respect to the dienophile, the addition is stereospecifically syn, with very few exceptions.[2047] This means that groups that are cis in the alkene will be cis in the cyclohexene ring, (A—B and C—D) and groups that are trans in the alkene will be trans in the cyclohexene ring (A—D and C—B).

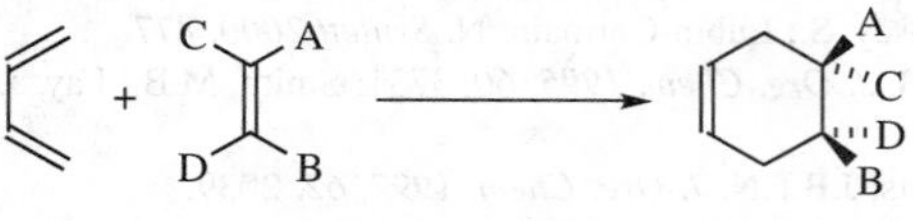

2. With respect to 1,4-disubstituted dienes, fewer cases have been investigated, but here too the reaction is stereospecific and syn. Thus, *trans, trans*-1,4-diphenylbutadiene gives *cis*-1,4-diphenylcyclohexene derivatives. This

---

[2044]Feuer, J.; Herndon, W.C.; Hall, L.H. *Tetrahedron* **1968**, *24*, 2575; Inukai, T.; Sato, H.; Kojima, T. *Bull. Chem. Soc. Jpn.* **1972**, *45*, 891; Epiotis, N.D. *J. Am. Chem. Soc.* **1973**, *95*, 5624; Sustmann, R. *Pure Appl. Chem.* **1974**, *40*, 569; Trost, B.M.; Vladuchick, W.C.; Bridges, A.J. *J. Am. Chem. Soc.* **1980**, *102*, 3554; Alston, P.V.; Gordon, M.D.; Ottenbrite, R.M.; Cohen, T. *J. Org. Chem.* **1983**, *48*, 5051; Kahn, S.D.; Pau, C.F.; Overman, L.E.; Hehre, W.J. *J. Am. Chem. Soc.* **1986**, *108*, 7381.

[2045]Danishefsky, S.; Hershenson, F.M. *J. Org. Chem.* **1979**, *44*, 1180; Ono, N.; Miyake, H.; Kamimura, A.; Kaji, A. *J. Chem. Soc. Perkin Trans. 1*, **1987**, 1929. For another method of controlling regioselectivity, see Kraus, G.A.; Liras, S. *Tetrahedron Lett.* **1989**, *30*, 1907.

[2046]See Smith, M.B. *Organic Synthesis*, 2nd ed., McGraw-Hill, NY, **2001**, pp. 933–940, 968–977; Bakalova, S.M.; Santos, A.G. *J. Org. Chem.* **2004**, *69*, 8475.

[2047]For an exception, see Meier, H.; Eckes, H.; Niedermann, H.; Kolshorn, H. *Angew. Chem. Int. Ed.* **1987**, *26*, 1046.

selectivity is predicted by disrotatory motion of the substituent in the transition state[2048] of the reaction (see **18-27**).

3. The diene must be in the cisoid conformation. If it is frozen into the transoid conformation, as in **198**, the reaction does not take place. The diene either must be frozen into the cisoid conformation or must be able to achieve it during the reaction.

198

4. When the diene is cyclic, there are two possible ways in which addition can occur if the dienophile is not symmetrical. The larger side of the dienophile may be under the ring (*endo addition*), or it may be the smaller side (*exo addition*):

Endo addition   Exo addition

Most of the time, the addition is predominantly endo; that is, the more bulky side of the alkene is under the ring, and this is probably true for open-chain dienes also.[2049] However, exceptions are known, and in many cases mixtures of exo and endo addition products are found.[2050] An imidazolidone catalyst was used to give a 1:1.3 mixture favoring the exo isomer in a reaction of conjugated aldehydes and cyclopentadiene.[2051] It has been argued that facial selectivity is not due to torsional angle decompression.[2052] Secondary orbital interactions.[2053] have been invoked, but this approach has been called into question.[2054] There has been a direct evaluation of such interactions, however.[2055] The endo/exo ratio can be influenced by the nature of the solvent.[2056]

[2048]Robiette, R.; Marchand-Brynaert, J.; Peeters, D. *J. Org. Chem.* **2002**, *67*, 6823.

[2049]See, for example, Baldwin, J.E.; Reddy, V.P. *J. Org. Chem.* **1989**, *54*, 5264. For a theoretical study for endo selectivity, see Imade, M.; Hirao, H.; Omoto, K.; Fujimoto, H. *J. Org. Chem.* **1999**, *64*, 6697.

[2050]See, for example, Alder, K.; Günzl, W. *Chem. Ber.* **1960**, *93*, 809; Stockmann, H. *J. Org. Chem.* **1961**, *26*, 2025; Jones, D.W.; Wife, R.L. *J. Chem. Soc., Chem. Commun.* **1973**, 421; Lindsay Smith, J.R.; Norman, R.O.C.; Stillings, M.R. *Tetrahedron* **1978**, *34*, 1381; Mülle, P.; Bernardinelli, G.; Rodriguez, D.; Pfyffer, J.; Schaller, J. *Chimia* **1987**, *41*, 244.

[2051]Ahrendt, K.A.; Borths, C.J.; MacMillan, D.W.C. *J. Am. Chem. Soc.* **2000**, *122*, 4243.

[2052]Hickey, E.R.; Paquette, L.A. *Tetrahedron Lett.* **1994**, *35*, 2309, 2313.

[2053]Hoffmann, R.; Woodward, R.B. *J. Am. Chem. Soc.* **1965**, *87*, 4388, 4389.

[2054]García, J.I.; Mayoral, J.A.; Salvatella, L. *Acc. Chem. Res.* **2000**, *33*, 658.

[2055]Arrieta, A.; Cossío, F.P.; Lecea, B. *J. Org. Chem.* **2001**, *66*, 6178.

[2056]Cainelli, G.; Galletti, P.; Giacomini, D.; Quintavalla, A. *Tetrahedron Lett.* **2003**, *44*, 93.

5. As we have seen, the Diels–Alder reaction can be both stereoselective and regioselective.[2057] In some cases, the Diels–Alder reaction can be made enantioselective[2058] Solvent effects are important in such reactions.[2059] The role of reactant polarity on the course of the reaction has been examined.[2060] Most enantioselective Diels–Alder reactions have used a chiral dienophile (e.g., **199**) and an achiral diene,[2061] along with a Lewis acid catalyst (see below). In such cases, addition of the diene to the two faces[2062] of **199** takes place at different rates, and **200** and **201** are formed in different amounts.[2063] An achiral compound A can be converted to a chiral compound by a chemical reaction with a compound B that is enantiopure. After the reaction, the resulting diastereomers can be separated, providing enantiopure compounds, each with a bond between molecule A and chiral compound B (a chiral auxiliary). Common chiral auxiliaries include chiral carboxylic acids, alcohols, or sultams. In the case illustrated, hydrolysis of the product removes the chiral R group, making it a chiral auxiliary in this reaction. Asymmetric Diels–Alder reactions have also been carried out with achiral dienes and dienophiles, but with an optically active catalyst.[2064] Many chiral catalysts

---

[2057]Domingo, L.R.; Picher, M.T.; Andrés, J.; Safont, V.S. *J. Org. Chem.* **1997**, *62*, 1775. Also see, Smith, M.B. *Organic Synthesis*, 2nd ed., McGraw-Hill, NY, **2001**, pp. 933–940, 968–977. See Ujaque, G.; Norton, J.E.; Houk, K.N. *J. Org. Chem.* **2002**, *67*, 7179.

[2058]See Corey, E.J.; Sarshar, S.; Lee, D.-H. *J. Am. Chem. Soc.* **1994**, *116*, 12089. For reviews, see Taschner, M.J. *Org. Synth: Theory Appl.* **1989**, *1*, 1; Helmchen, G.; Karge, R.; Weetman, J. *Mod. Synth. Methods 1986*, *4*, 261; Paquette, L.A. in Morrison, J.D. *Asymmetric Synthesis*, Vol. 3, Academic Press, NY, *1983*, pp. 455–501; Oppolzer, W. *Angew. Chem. Int. Ed. 1984*, *23*, 876. See also, the list of references in Macaulay, J.B.; Fallis, A.G. *J. Am. Chem. Soc.* **1990**, *112*, 1136.

[2059]Ruiz-López, M.F.; Assfeld, X.; García, J.I.; Mayoral, J.A.; Salvatella, L. *J. Am. Chem. Soc.* **1993**, *115*, 8780.

[2060]Sustmann, R.; Sicking, W. *J. Am. Chem. Soc.* **1996**, *118*, 12562.

[2061]For the use of chiral dienes, see Fisher, M.J.; Hehre, W.J.; Kahn, S.D.; Overman, L.E. *J. Am. Chem. Soc. 1988*, *110*, 4625; Menezes, R.F.; Zezza, C.A.; Sheu, J.; Smith, M.B. *Tetrahedron Lett.* **1989**, *30*, 3295; Charlton, J.L.; Plourde, G.L.; Penner, G.H. *Can. J. Chem.* **1989**, *67*, 1010; Tripathy, R.; Carroll, P.J.; Thornton, E.R. *J. Am. Chem. Soc.* **1990**, *112*, 6743; *1991*, *113*, 7630; Rieger, R.; Breitmaier, E. *Synthesis* *1990*, 697.

[2062]For a discussion of facial selectivity, see Xidos, J.D.; Poirier, R.A.; Pye, C.C.; Burnell, D.J. *J. Org. Chem.* **1998**, *63*, 105.

[2063]Oppolzer, W.; Kurth, M.; Reichlin, D.; Moffatt, F.*Tetrahedron Lett.* **1981**, *22*, 2545. See also, Walborsky, H.M.; Barash, L.; Davis, T.C. *Tetrahedron 1963*, *19*, 2333; Furuta, K.; Iwanaga, K.; Yamamoto, H. *Tetrahedron Lett. 1986*, *27*, 4507; Evans, D.A.; Chapman, K.T.; Bisaha, J. *J. Am. Chem. Soc. 1988*, *110*, 1238; Mattay, J.; Mertes, J.; Maas, G. *Chem. Ber. 1989*, *122*, 327; Alonso, I.; Carretero, J.C.; Garcia Ruano, J.L. *Tetrahedron Lett.* **1989**, *30*, 3853; Tomioka, K.; Hamada, N.; Suenaga, T.; Koga, K. *J. Chem. Soc. Perkin Trans. 1*, *1990*, 426; Cativiela, C.; López, P.; Mayoral, J.A. *Tetrahedron: Asymmetry 1990*, *1*, 61.

[2064]For a review, see Narasaka, K. *Synthesis 1991*, 1. For some recent examples, see Bir, G.; Kaufmann, D. *J. Organomet. Chem. 1990*, *390*, 1; Rebiere, F.; Riant, O.; Kagan, H.B. *Tetrahedron: Asymmetry 1990*, *1*, 199; Terada, M.; Mikami, K.; Nakai, T. *Tetrahedron Lett. 1991*, *32*, 935; Corey, E.J.; Imai, N.; Zhang, H. *J. Am. Chem. Soc. 1991*, *113*, 728; Narasaka, K.; Tanaka, H.; Kanai, F. *Bull. Chem. Soc. Jpn. 1991*, *64*, 387; Hawkins, J.M.; Loren, S. *J. Am. Chem. Soc. 1991*, *113*, 7794; Evans, D.A.; Barnes, D.M.; Johnson, J.S.; Lectka, T.; von Matt, P.; Miller, S.J.; Murry, J.A.; Norcross, R.D.; Shaughnessy, E.A.; Campos, K.R. *J. Am. Chem. Soc. 1999*, *121*, 7582.

have been developed.[2065] In many cases, asymmetric Lewis acids form a chiral complex with the dienophile.[2066]

200   **199**   **201**

R = (+) or (–) phenylmenthyl

Many interesting compounds can be prepared by the Diels–Alder reaction,[2067] some of which would be hard to make in any other way. Azelines react with dienes to form cyclohexene derivatives fused to a four-membered ring amine (azetidine).[2068] The C_{60}-Fullerenes undergo Diels–Alder reactions,[2069] and the reaction is reversible.[2070] Bicyclic sultams can be prepared by an intramolecular Diels–Alder reaction.[2071] Polycyclic lactones can be prepared.[2072] Aromatic compounds can behave as dienes,[2073] but benzene is very unreactive toward dienophiles,[2074] and very few dienophiles (one of them is benzyne) have been reported to give Diels–Alder adducts with it.[2075] Naphthalene and phenanthrene are also quite resistant, although naphthalene has given Diels–Alder addition at high pressures.[2076] However, anthracene and other compounds with at least three linear benzene rings give Diels–Alder reactions readily. The interesting compound triptycene can be prepared by a Diels–Alder

[2065]For a review, see Corey, E.J. *Angew. Chem. Int. Ed.* **2002**, *41*, 1651. See also, Doyle, M.P.; Phillips, I.M.; Hu, W. *J. Am. Chem. Soc.* **2001**, *123*, 5366; Owens, T.D.; Hollander, F.J.; Oliver A.G.; Ellman, J.A. *J. Am. Chem. Soc.* **2001**, *123*, 1539; Faller, J.W.; Grimmond, B.J.; D'Alliessi, D.G. *J. Am. Chem. Soc.* **2001**, *123*, 2525; Bolm, C.; Simić, O. *J. Am. Chem. Soc.* **2001**, *123*, 3830; Fukuzawa, S.; Komuro, Y.; Nakano, N.; Obara, S. *Tetrahedron Lett.* **2003**, *44*, 3671.

[2066]Hawkins, J.M.; Loren, S.; Nambu, M. *J. Am. Chem Soc.* **1994**, *116*, 1657. See Sibi, M.P.; Venkatraman, L.; Liu, M.; Jaspersé, C.P. *J. Am. Chem. Soc.* **2001**, *123*, 8444.

[2067]For a review of this reaction in synthesis, see Nicolaou, K.C.; Snyder, S.A.; Montagnon, T.; Vassilkogiannakis, G. *Angew. Chem. Int. Ed.* **2002**, *41*, 1669.

[2068]Dave, P.R.; Duddu, R.; Surapaneni, R.; Gilardi, R. *Tetrahedron Lett.* **1999**, *40*, 443.

[2069]Murata, Y.; Kato, N.; Fujiwara, K.; Komatsu, K. *J. Org. Chem.* **1999**, *64*, 3483.

[2070]Wang, G.-W.; Saunders, M.; Cross, R.J. *J. Am. Chem. Soc.* **2001**, *123*, 256.

[2071]Greig, I.R.; Tozer, M.J.; Wright, P.T. *Org. Lett.* **2001**, *3*, 369.

[2072]Vlaar, M.J.M.; Lor, M.H.; Ehlers, A.W.; Schakel, M.; Lutz, M.; Spek, A.L.; Lammertsma, K. *J. Org. Chem.* **2002**, *67*, 2485.

[2073]For a review, see Wagner-Jauregg, T. *Synthesis* **1980**, 165, 769. See also, Balaban, A.T.; Biermann, D.; Schmidt, W. *Nouv. J. Chim.* **1985**, *9*, 443.

[2074]However, see Chordia, M.D.; Smith, P.L.; Meiere, S.H.; Sabat, M.; Harman, W.D. *J. Am. Chem. Soc.* **2001**, *123*, 10756.

[2075]Miller, R.G.; Stiles, M. *J. Am. Chem. Soc.* **1963**, *85*, 1798; Meyerson, S.; Fields, E.K. *Chem. Ind. (London)* **1966**, 1230; Ciganek, E. *Tetrahedron Lett.* **1967**, 3321; Friedman, L. *J. Am. Chem. Soc.* **1967**, *89*, 3071; Liu, R.S.H.; Krespan, C.G. *J. Org. Chem.* **1969**, *34*, 1271.

[2076]Plieninger, H.; Wild, D.; Westphal, J. *Tetrahedron* **1969**, *25*, 5561.

reaction between benzyne and anthracene:[2077] For both all-carbon and hetero systems, the "diene" can be a conjugated enyne. If the geometry of the molecule is suitable, the diene can even be nonconjugated, for example,[2078]

This last reaction is known as the *homo–Diels–Alder reaction*. A similar reaction has been reported with alkynes, using a mixture of a cobalt complex, ZnI_2 and tetrabutylammonium borohydride as catalysts.[2079]

Competing reactions are polymerization of the diene or dienophile, or both, and [1,2]-cycloaddition (**15-63**). Intramolecular versions of the Diels–Alder reaction are well-known, and this is a powerful method for the synthesis of mono- and polycyclic compounds.[2080] There are many examples and variations. Internal Diels–Alder reactions can be viewed as linking the diene and alkene by a tether, usually of carbon atoms. If the tether is replaced by functional groups that allow the selectivity inherent to the intramolecular cycloaddition, but can be cleaved afterward, a powerful modification is available. Indeed, such tethered cycloaddition reactions are increasingly common. After cycloaddition, the tether can be cleaved to give a functionalized cyclohexene derivative. Such tethered reactions allow enhancement of stereoselectivity[2081] and sometimes reactivity, relative to an untethered reaction, giving an indirect method for enhancing those parameters. Tethers or linkages include $C–O–SiR_2–C$[2082] or a $C–O–SiR_2–O–C$,[2083] or hydroxamides.[2084] Transient tethers can be used, as in the reaction of a diene having an allylic alcohol unit in a reaction is allyl alcohol, with $AlMe_3$, to give the cycloadduct with good selectivity.[2085]

3-Sulfolene

[2077]Wittig, G.; Niethammer, K. *Chem. Ber.* **1960**, *93*, 944; Wittig, G.; Härle, H.; Knauss, E.; Niethammer, K. *Chem. Ber.* **1960**, *93*, 951. For a review of triptycene, see Skvarchenko, V.R.; Shalaev, V.K.; Klabunovskii, E.I. *Russ. Chem. Rev.* **1974**, *43*, 951.

[2078]See, for example, Fickes, G.N.; Metz, T.E. *J. Org. Chem.* **1978**, *43*, 4057; Paquette, L.A.; Kesselmayer, M.A.; Künzer, H. *J. Org. Chem.* **1988**, *53*, 5183.

[2079]Hilt, G.; du Mesnil, F.-X. *Tetrahedron Lett.* **2000**, *41*, 6757.

[2080]Carlson, R.G. *Ann. Rep. Med. Chem.* **1974**, *9*, 270; Oppolzer, W. *Angew. Chem. Int. Ed.* **1977**, *16*, 10 (see pp. 10–18); Brieger, G.; Bennett, J.N. *Chem. Rev.* **1980**, *80*, 63 (see p. 67); Fallis, A.G. *Can. J. Chem.* **1984**, *62*, 183; Smith, M.B. *Org. Prep. Proceed. Int.* **1990**, *22*, 315.

[2081]For a discussion of the origins of stereoselectivity in intramolecular tethered reactions, see Tantillo, D.J.; Houk, K.N.; Jung, M.E. *J. Org. Chem.* **2001**, *66*, 1938.

[2082]Stork, G.; Chan, T.Y.; Breault, G.A. *J. Am. Chem. Soc.* **1992**, *114*, 7578.

[2083]Craig, D.; Reader, J.C. *Tetrahedron Lett.* **1992**, *33*, 6165.

[2084]Ishikawa, T.; Senzaki, M.; Kadoya, R.; Morimoto, T.; Miyake, N.; Izawa, M.; Saito, S. Kobayashi, H. *J. Am. Chem. Soc.* **2001**, *123*, 4607.

[2085]Bertozzi, F.; Olsson, R.; Frejd, T. *Org. Lett.* **2000**, *2*, 1283.

The Diels–Alder reaction is usually reversible, although the retro reaction typically occurs at significantly higher temperatures than the forward reaction. However, the reversibility of the reaction and has been used to protect double bonds.[2086] A convenient substitute for butadiene in the Diels–Alder reaction is the compound 3-sulfolene since the latter is a solid which is easy to handle while the former is gas.[2087] Butadiene is generated *in situ* by a reverse Diels–Alder reaction (see **17-20**).

There are, broadly speaking, three possible mechanisms that have been considered for the uncatalyzed Diels-Alder reaction.[2088] In mechanism *a* there is a cyclic six-centered transition state and no intermediate. The reaction is concerted and occurs in one step. In mechanism *b*, one end of the diene fastens to one end of the dienophile first to give a diradical, and then, in a second step, the other ends become fastened.[2089] A diradical formed in this manner must be a singlet; that is, the two unpaired electrons must have opposite spins, by an argument similar to that outlined on p. 277. The third mechanism (*c*, not shown) is similar to mechanism *b*, but the initial bond and the subsequent bond are formed by movements of electron pairs and the intermediate is a diion. There have been many mechanistic investigations of the Diels–Alder reaction. The bulk of the evidence suggests that most Diels–Alder reactions take place by the one-step cyclic mechanism *a*,[2090] although it is possible

[2086]For reviews of the reverse Diels–Alder reaction, see Ichihara, A. *Synthesis* **1987**, 207; Lasne, M.; Ripoll, J.L. *Synthesis* **1985**, 121; Ripoll, J.L.; Rouessac, A.; Rouessac, F. *Tetrahedron* **1978**, *34*, 19; Brown, R.F.C. *Pyrolytic Methods in Organic Chemistry*, Academic Press, NY, **1980**, pp. 259–281; Kwart, H.; King, K. *Chem. Rev.* **1968**, *68*, 415.

[2087]Sample Jr., T.E.; Hatch, L.F. *Org. Synth. VI*, 454. For a review, see Chou, T.; Tso, H. *Org. Prep. Proced. Int.* **1989**, *21*, 257.

[2088]For reviews, see Sauer, J.; Sustmann, R. *Angew. Chem. Int. Ed.* **1980**, *19*, 779; Houk, K.N. *Top. Curr. Chem.* **1979**, *79*, 1; Seltzer, S. *Adv. Alicyclic Chem.* **1968**, *2*, 1; Ref. 1981. For a review of the application of quantum-chemical methods to the study of this reaction, see Babichev, S.S.; Kovtunenko, V.A.; Voitenko, Z.V.; Tyltin, A.K. *Russ. Chem. Rev.* **1988**, *57*, 397. For a discussion of synchronous versus nonsynchronous mechanisms, see Beno, B.R.; Houk, K.N.; Singleton, D.A. *J. Am. Chem. Soc.* **1996**, *118*, 9984; Singleton, D.A.; Schulmeier, B.E.; Hang, C.; Thomas, A.A.; Leung, S.-W.; Merrigan, S.R. *Tetrahedron* **2001**, *57*, 5149. Also see, Li, Y.; Houk, K.N. *J. Am. Chem. Soc.* **1993**, *115*, 7478 for the dimerization mechanism of 1,3-butadiene.

[2089]For a discussion of a diradical stepwise versus concerted mechanism for reactions with chalcogens, see Orlova, G.; Goddard, J.D. *J. Org. Chem.* **2001**, *66*, 4026.

[2090]For a contrary view, see Dewar, M.J.S.; Olivella, S.; Stewart, J.J.P. *J. Am. Chem. Soc.* **1986**, *108*, 5771. For arguments against this view, see Houk, K.N.; Lin, Y.; Brown, F.K. *J. Am. Chem. Soc.* **1986**, *108*, 554; Hancock, R.A.; Wood, Jr., B.F. *J. Chem. Soc., Chem. Commun.* **1988**, 351; Gajewski, J.J.; Peterson, K.B.; Kagel, J.R.; Huang, Y.C.J. *J. Am. Chem. Soc.* **1989**, *111*, 9078.

that a diradical[2091] or even a diion[2092] mechanism may be taking place in some cases. Radical cation Diels–Alder reactions have been considered.[2093] The main evidence in support of mechanism *a* is as follows: (*1*) The reaction is stereospecific in both the diene and dienophile. A completely free diradical or diion probably would not be able to retain its configuration. (*2*) In general, the rates of Diels–Alder reactions depend very little on the nature of the solvent. This would rule out a diion intermediate because polar solvents increase the rates of reactions that develop charges in the transition state. (*3*) It was shown that, in the decomposition of **202**, the isotope effect k_I/k_{II} was equal to 1.00 within experimental error.[2094] If bond *x* were to break before bond *y*, there

I: R = H, R′ = D

II: R = D, R′ = H

202

should surely be a secondary isotope effect. This result strongly indicates that the bond breaking of *x* and *y* is simultaneous. This is the reverse of a Diels–Alder reaction, and by the principle of microscopic reversibility, the mechanism of the forward reaction should involve simultaneous formation of bonds *x* and *y*. Subsequently, a similar experiment was carried out on the forward reaction[2095] and the result was the same. There is also other evidence for mechanism *a*.[2096] However, the fact that the mechanism is concerted does not necessarily mean that it is synchronous.[2097] In the transition state of a synchronous reaction both new σ bonds would be formed to the same extent, but a Diels–Alder reaction with non-symmetrical components might very well be non-synchronous;[2098] that is, it could have a transition state in which one

[2091]See, for example, Bartlett, P.D.; Mallet, J.J. *J. Am. Chem. Soc.* **1976**, *98*, 143; Jenner, G.; Rimmelin, J. *Tetrahedron Lett.* **1980**, *21*, 3039; Van Mele, B.; Huybrechts, G. *Int. J. Chem. Kinet.* **1987**, *19*, 363; **1989**, *21*, 967.

[2092]For a reported example, see Gassman, P.G.; Gorman, D.B. *J. Am. Chem. Soc.* **1990**, *112*, 8624.

[2093]Haberl, U.; Wiest, O.; Steckhan, E. *J. Am. Chem. Soc.* **1999**, *121*, 6730.

[2094]Seltzer, S. *J. Am. Chem. Soc.* **1963**, *85*, 1360; **1965**, *87*, 1534. For a review of isotope effect studies of Diels–Alder and other pericyclic reactions, see Gajewski, J.J. *Isot. Org. Chem.* **1987**, *7*, 115–176.

[2095]Van Sickle, D.E.; Rodin, J.O. *J. Am. Chem. Soc.* **1964**, *86*, 3091.

[2096]See, for example, Dewar, M.J.S.; Pyron R.S. *J. Am. Chem. Soc.* **1970**, *92*, 3098; Brun, C.; Jenner, G. *Tetrahedron* **1972**, *28*, 3113; Doering, W. von E.; Franck-Neumann, M.; Hasselmann, D.; Kaye, R.L. *J. Am. Chem. Soc.* **1972**, *94*, 3833; McCabe, J.R.; Eckert, C.A. *Acc. Chem. Res.* **1974**, *7*, 251; Berson, J.A.; Dervan, P.B.; Malherbe, R.; Jenkins, J.A. *J. Am. Chem. Soc.* **1976**, *98*, 5937; Rücker, C.; Lang, D.; Sauer, J.; Friege, H.; Sustmann, R. *Chem. Ber.* **1980**, *113*, 1663; Tolbert, L.M.; Ali, M.B. *J. Am. Chem. Soc.* **1981**, *103*, 2104.

[2097]For an example of a study of a reaction that is concerted but asynchronous, see Avalos, M.; Babiano, R.; Clemente, F.R.; Cintas, P.; Gordillo, R.; Jiménez, J.L.; Palacios, J.C. *J. Org. Chem.* **2000**, *65*, 8251

[2098]Woodward, R.B.; Katz, T.J. *Tetrahedron* **1959**, *5*, 70; Liu, M.T.H.; Schmidt, C. *Tetrahedron* **1971**, *27*, 5289; Dewar, M.J.S.; Pyron R.S. *J. Am. Chem. Soc.* **1970**, *92*, 3098; Papadopoulos, M.; Jenner, G. *Tetrahedron Lett.* **1982**, *23*, 1889; Houk, K.N.; Loncharich, R.J.; Blake, J.F.; Jorgensen, W.L. *J. Am. Chem. Soc.* **1989**, *111*, 9172; Lehd, M.; Jensen, F. *J. Org. Chem.* **1990**, *55*, 1034.

bond has been formed to a greater degree than the other.[2099] A biradical mechanism has been proposed for some Diels–Alder reactions.[2100]

In another aspect of the mechanism, the effects of electron-donating and electron-withdrawing substituents (p. 1196) indicate that the diene is behaving as a nucleophile and the dienophile as an electrophile. However, this can be reversed. Perchlorocyclopentadiene reacts better with cyclopentene than with maleic anhydride and not at all with tetracyanoethylene, although the latter is normally the most reactive dienophile known. It is apparent, then, that this diene is the electrophile in its Diels–Alder reactions.[2101] Reactions of this type are said to proceed with *inverse electron demand*.[2102]

We have emphasized that the Diels–Alder reaction generally takes place rapidly and conveniently. In sharp contrast, the apparently similar dimerization of alkenes to cyclobutanes (**15-63**) gives very poor results in most cases, except when photochemically induced. Woodward and Hoffmann, and Fukui have shown that these contrasting results can be explained by the *principle of conservation of orbital symmetry*,[2103] which predicts that certain reactions are allowed and others forbidden. The orbital-symmetry rules (also called the Woodward–Hoffmann rules)[2104] apply *only to concerted reactions*, for example, mechanism *a*, and are based on the principle that reactions take place in such a way as to maintain maximum bonding throughout the course of the reaction. There are several ways of applying the orbital-symmetry principle to cycloaddition reactions, three

---

[2099]For a theoretical investigation of the ionic Diels–Alder reaction, see dePascual-Teresa, B.; Houk, K.N. *Tetrahedron Lett.* **1996**, *37*, 1759. For a discussion of the origin of synchronicity in the transition state of polar Diels–Alder reactions, see Domingo, L.R.; Aurell, M.J.; Pérez, P.; Contreras, R. *J. Org. Chem.* **2003**, *68*, 3884.

[2100]de Echagüen, C.O.; Ortuño, R.M. *Tetrahedron Lett.* **1995**, *36*, 749. See Li, Y.; Padias, A.B.; Hall Jr., H.K. *J. Org. Chem.* **1993**, *58*, 7049 for a discussion of diradicals in concerted Diels–Alder reactions.

[2101]Sauer, J.; Wiest, H. *Angew. Chem. Int. Ed.* **1962**, *1*, 269.

[2102]For a review, see Boger, D.L.; Patel, M. *Prog. Heterocycl. Chem.* **1989**, *1*, 30. Also see, Pugnaud, S.; Masure, D.; Hallé, J.-C.; Chaquin, P. *J. Org. Chem,.* **1997**, *62*, 8687; Wan, Z.-K.; Snyder, J.K. *Tetrahedron Lett.* **1998**, *39*, 2487; Markó, I.E.; Evans, G.R. *Tetrahedron Lett.* **1994**, *35*, 2767, 2771.

[2103]For monographs, see Fleming, I. *Pericyclic Reactions*, Oxford University Press, Oxford, **1999**, pp. 31–56; Gilchrist, T.L.; Storr, R.C. *Organic Reactions and Orbital Symmetry*, 2nd ed., Cambridge University Press, Cambridge, **1979**; Fleming, I. *Frontier Orbitals and Organic Chemical Reactions*, Wiley, NY, **1976**; Woodward, R.B.; Hoffmann, R. *The Conservation of Orbital Symmetry*, Academic Press, NY, **1970** [the text of this book also appears in *Angew. Chem. Int. Ed.* **1969**, *8*, 781; Lehr, R.E.; Marchand, A.P. *Orbital Symmetry*, Academic Press, NY, **1972**. For reviews, see Pearson, R.G. *J. Chem. Educ.* **1981**, *58*, 753; in Klopman, G. *Chemical Reactivity and Reaction Paths*, Wiley, NY, **1974**, the articles by Fujimoto, H.; Fukui, K. pp. 23–54, Klopman, G. pp. 55–165, Herndon, W.C.; Feuer, J.; Giles, W.B.; Otteson, D.; Silber, E. pp. 275–299; Michl, J. pp. 301–338; Simonetta, M. *Top. Curr. Chem.* **1973**, *42*, 1; Houk, K.N. *Surv. Prog. Chem.* **1973**, *6*, 113; Vollmer, J.J.; Servis, K.L. *J. Chem. Educ.* **1970**, *47*, 491; Gill, G.B. *Essays Chem.* **1970**, *1*, 43; *Q. Rev. Chem. Soc.* **1968**, *22*, 338; Seebach, D. *Fortschr. Chem. Forsch.* **1969**, *11*, 177; Miller, S.I. *Adv. Phys. Org. Chem.* **1968**, *6*, 185; Miller, S.I. *Bull. Soc. Chim. Fr.* **1966**, 4031. For a review of applications to inorganic chemistry, see Pearson, R.G. *Top Curr. Chem.* **1973**, *41*, 75.

[2104]Chattaraj, P.K.; Fuentealba, P.; Gómez, B.; Contreras, R. *J. Am. Chem. Soc.* **2000**, *122*, 348.

of which are used more frequently than others.[2105] Of these three, we will discuss two: the frontier-orbital method and the Möbius–Hückel method. The third, called the correlation diagram method,[2106] is less convenient to apply than the other two.

The Frontier Orbital Method[2107]

As applied to cycloaddition reactions the rule is that *reactions are allowed only when all overlaps between the highest occupied molecular orbital (HOMO) of one reactant and the lowest unoccupied molecular orbital (LUMO) of the other are such that a positive lobe overlaps only with another positive lobe and a negative lobe only with another negative lobe.* We may recall that monoalkenes have two π molecular orbitals (p. 10) and that conjugated dienes have four (p. 38), as shown in Fig. 15.2. A concerted cyclization of two monoalkenes (a [2 + 2]-reaction) is not allowed because it would require that a positive lobe overlap with a negative lobe (Fig. 15.3). On the other hand, the Diels–Alder reaction (a [4 + 2]-reaction) is allowed, whether considered from either direction (Fig. 15.4).

These considerations are reversed when the ring closures are photochemically induced since in such cases an electron is promoted to a vacant orbital before the reaction occurs. Obviously, the [2 + 2] reaction is now allowed (Fig. 15.5) and the [4 + 2]-reaction disallowed. The reverse reactions follow the same rules, by the principle of microscopic reversibility. In fact, Diels–Alder adducts are usually cleaved quite readily, while cyclobutanes, despite the additional strain, require more strenuous conditions.

[2105]For other approaches, see Epiotis, N.D. *Theory of Organic Reactions*, Springer, NY, *1978*; Epiotis, N.D.; Shaik, S. *J. Am. Chem. Soc.* *1978*, *100*, 1, 9; Halevi, E.A. *Angew. Chem. Int. Ed.* *1976*, *15*, 593; Shen, K. *J. Chem. Educ.* *1973*, *50*, 238; Salem, L. *J. Am. Chem. Soc.* *1968*, *90*, 543, 553; Trindle, C. *J. Am. Chem. Soc.* *1970*, *92*, 3251, 3255; Mulder, J.J.C.; Oosterhoff, L.J. *Chem. Commun.* *1970*, 305, 307; Goddard III, W.A. *J. Am. Chem. Soc.* *1970*, *92*, 7520; *1972*, *94*, 793; Herndon, W.C. *Chem. Rev.* *1972*, *72*, 157; Perrin, C.L. *Chem. Br.* *1972*, *8*, 163; Langlet, J.; Malrieu, J. *J. Am. Chem. Soc.* *1972*, *94*, 7254; Pearson, R.G. *J. Am. Chem. Soc.* *1972*, *94*, 8287; Mathieu, J. *Bull. Soc. Chim. Fr.* *1973*, 807; Silver, D.M.; Karplus, M. *J. Am. Chem. Soc.* *1975*, *97*, 2645; Day, A.C. *J. Am. Chem. Soc.* *1975*, *97*, 2431; Mok, K.; Nye, M.J. *J. Chem. Soc. Perkin Trans. 2*, *1975*, 1810; Ponec, R. *Collect. Czech. Chem. Commun.* *1984*, *49*, 455; *1985*, *50*, 1121; Hua-ming, Z.; De-xiang, W. *Tetrahedron* *1986*, *42*, 515; Bernardi, F.; Olivucci, M.; Robb, M.A. *Res. Chem. Intermed.* *1989*, *12*, 217; *Acc. Chem. Res.* *1990*, *23*, 405.
[2106]For excellent discussions of this method, see Woodward, R.B.; Hoffmann, R. *The Conservation of Orbital Symmetry*, Academic Press, NY, *1970*; *Angew. Chem. Int. Ed.* *1969*, *8*, 781; Jones, R.A.Y. *Physical and Mechanistic Organic Chemistry* 2nd ed., Cambridge University Press, Cambridge, *1984*, pp. 352–366; Klumpp, G.W. *Reactivity in Organic Chemistry*, Wiley, NY, *1982*, pp. 378–389; Yates, K. *Hückel Molecular Orbital Theory*, Academic Press, NY, *1978*, pp. 263–276.
[2107]Fukui, K.; Fujimoto, H. *Bull. Chem. Soc. Jpn.* *1967*, *40*, 2018; *1969*, *42*, 3399; Fukui, K. *Fortschr. Chem. Forsch.* *1970*, *15*, 1; *Acc. Chem. Res.* *1971*, *4*, 57; Houk, K.N. *Acc. Chem. Res.* *1975*, *8*, 361. See also, Chu, S. *Tetrahedron* *1978*, *34*, 645. For a monograph on Frontier Orbitals see Fleming, I. *Pericyclic Reactions*, Oxford University Press, Oxford, *1999*. For reviews, see Fukui, K. *Angew. Chem. Int. Ed.* *1982*, *21*, 801; Houk, K.N., in Marchand, A.P.; Lehr, R.E. *Pericyclic Reactions*, Vol. 2; Academic Press, NY, *1977*, pp. 181–271.

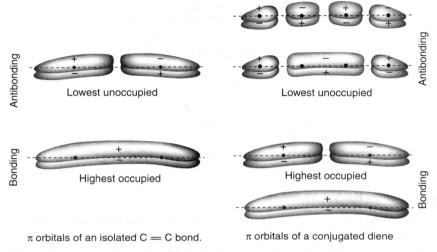

π orbitals of an isolated C = C bond.   π orbitals of a conjugated diene

Fig. 15.2. Schematic drawings of the π-orbitals of an isolated C=C bnd and a conjugated diene.

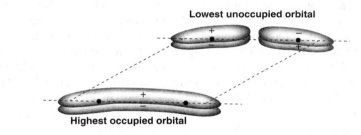

Fig. 15.3. Overlap of orbitals in a thermal [2 + 2]-cycloaddition.

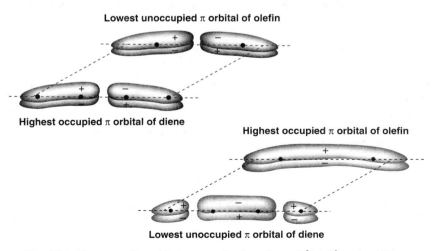

Fig. 15.4. Two ways for orbitals to overlap in a thermal [4 + 2]-cycloaddition.

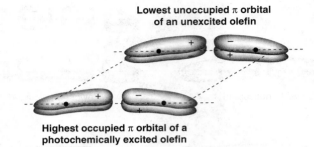

Fig. 15.5. Overlap of orbitals in a photochemical [2 + 2]-cycloaddition.

The Möbius–Hückel Method[2108]

In this method, the orbital symmetry rules are related to the Hückel aromaticity rule discussed in Chapter 2.[2109] Hückel's rule, which states that a cyclic system of electrons is aromatic (hence, stable) when it consists of $4n + 2$ electrons, applies of course to molecules in their ground states. In applying the orbital symmetry principle we are not concerned with ground states, but with transition states. In the present method, we do not examine the molecular orbitals themselves, but rather the p orbitals before they overlap to form the molecular orbitals. Such a set of p orbitals is called a *basis set* (Fig. 15.6). In investigating the possibility of a concerted reaction, we put the basis sets into the position they would occupy in the transition state. Figure 15.7 shows this for both the [2 + 2] and the [4 + 2] ring closures. What we look for are *sign inversions*. In Fig. 15.7, we can see that there are no sign inversions in either case. That is, the dashed line connects only lobes with a minus sign. Systems with *zero or an even number* of sign inversions are called *Hückel systems*. Because they have no sign inversions, both of these systems are Hückel systems. Systems with *an odd number* of sign

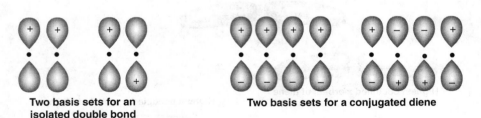

Two basis sets for an isolated double bond

Two basis sets for a conjugated diene

Fig. 15.6. Some basis sets.

[2108]Zimmerman, H.E., in Marchand, A.P.; Lehr, R.E. *Pericyclic Reactions*, Vol. 2, Academic Press, NY, *1977*, pp. 53–107; *Acc. Chem. Res.* *1971*, *4*, 272; *J. Am. Chem. Soc.* *1966*, *88*, 1564, 1566; Dewar, M.J.S. *Angew. Chem. Int. Ed.* *1971*, *10*, 761; Jefford, C.W.; Burger, U. *Chimia*, *1971*, *25*, 297; Herndon, W.C. *J. Chem. Educ.* *1981*, *58*, 371.
[2109]See Morao, I.; Cossío, F.P. *J. Org. Chem.* *1999*, *64*, 1868.

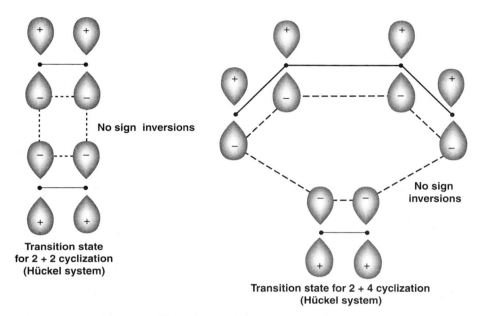

No sign inversions

Transition state
for 2 + 2 cyclization
(Hückel system)

No sign
inversions

Transition state for 2 + 4 cyclization
(Hückel system)

Fig. 15.7. Transition states illustrating Hückel–Möbius rules for cycloaddition reactions.

inversions are called *Möbius systems* (because of the similarity to the Möbius strip, which is a mathematical surface, shown in Fig. 15.8). Möbius systems do not enter into either of these reactions, but an example of such a system is shown on p. $$$.

The rule may then be stated: *A thermal pericyclic reaction involving a Hückel system is allowed only if the total number of electrons is 4n + 2. A thermal pericyclic reaction involving a Möbius system is allowed only if the total number of electrons is 4n.* For photochemical reactions these rules are reversed. Since both the [4 + 2]- and [2 + 2]-cycloadditions are Hückel systems, the Möbius–Hückel method predicts that the [4 + 2]-reaction, with 6 electrons, is thermally allowed,

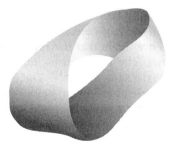

Fig. 15.8. A Möbius strip. Such a strip is easily constructed by twisting a thin strip of paper 180° and fastening the ends together.

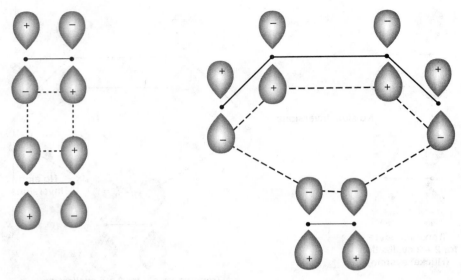

Fig. 15.9. Transition states of [2 + 2]- and [4 + 2]-cyclizations involing other basis sets.

but the [2 + 2]-reaction is not. One the other hand, the [2 + 2]-reaction is allowed photochemically, while the [4 + 2]-reaction is forbidden.

Note that both the [2 + 2] and [4 + 2] transition states are Hückel systems no matter what basis sets we chose. For example, Fig. 15.9 shows other basis sets we might have chosen. In every case there will be zero or an even number of sign inversions.

Thus, the frontier orbital and Hückel–Möbius methods (and the correlation-diagram method as well) lead to the same conclusions: thermal [4 + 2]-cycloadditions and photochemical [2 + 2]-cycloadditions (and the reverse ring openings) are allowed, while photochemical [4 + 2] and thermal [2 + 2] ring closings (and openings) are forbidden.

Application of the same procedures to other ring closures shows that [4 + 4] and [2 + 6] ring closures and openings require photochemical induction while the [4 + 6]- and [2 + 8]-reactions can take place only thermally (see **15-53**). In general, cycloaddition reactions allowed thermally are those with $4n + 2$ electrons, while those allowed photochemically have $4n$ electrons.

It must be emphasized once again that the rules apply only to cycloaddition reactions that take place by cyclic mechanisms, that is, where two σ bonds are formed (or broken) at about the same time.[2110] The rule does not apply to cases where one bond is clearly formed (or broken) before the other. It must further be emphasized that the fact that the thermal Diels–Alder reaction (mechanism *a*) is allowed by the principle of conservation of orbital symmetry does not constitute

---

[2110]For a discussion of concertedness in these reactions see Lehr, R.E.; Marchand, A.P., in Marchand, A.P.; Lehr, R.E. *Pericyclic Reactions*, Vol. 1, Academic Press, NY, *1977*, pp. 1–51.

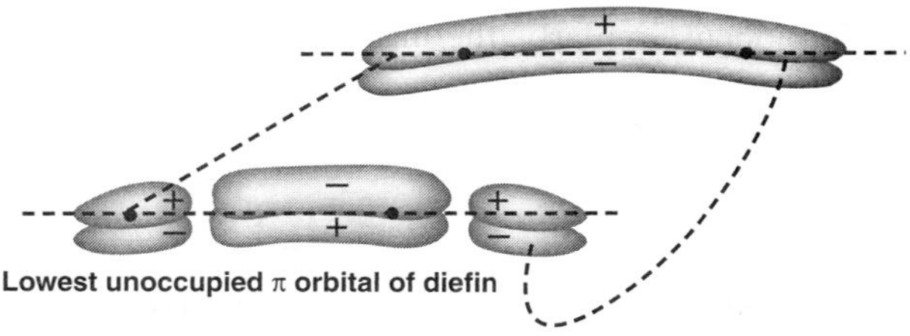

Highest occupied π orbital of olefin

Lowest unoccupied π orbital of diefin

Fig. 15.10. Overlap of orbitals in an antarafacial thermal $[4 + 2]$-cycloaddition.

proof that any given Diels–Alder reaction proceeds by this mechanism. The principle merely says the mechanism is allowed, not that it must go by this pathway. However, the principle does say that thermal $[2 + 2]$-cycloadditions in which the molecules assume a face-to-face geometry cannot[2111] take place by a cyclic mechanism because their activation energies would be too high (however, see below). As we will see (**15-62**), such reactions largely occur by two-step mechanisms. Similarly, $[4 + 2]$-photochemical cycloadditions are also known, but the fact that they are not stereospecific indicates that they also take place by the two-step diradical mechanism (mechanism *b*).[2112]

In all of the above discussion we have assumed that a given molecule forms both the new σ bonds from the same face of the π system. This manner of bond formation, called *suprafacial*, is certainly most reasonable and almost always takes place. The subscript s is used to designate this geometry, and a normal Diels–Alder reaction would be called a $[_{\pi}2_s + _{\pi}4_s]$-cycloaddition (the subscript π indicates that π electrons are involved in the cycloaddition). However, we can conceive of another approach in which the newly forming bonds of the diene lie on *opposite* faces of the π system, that is, they point in opposite directions. This type of orientation of the newly formed bonds is called *antarafacial*, and the reaction would be a $[_{\pi}2_s + _{\pi}4_a]$-cycloaddition (a stands for antarafacial). We can easily show by the frontier-orbital method that this reaction (and consequently the reverse ring-opening reactions) are thermally forbidden and photochemically allowed. Thus in order for a $[_{\pi}2_s + _{\pi}4_a]$-reaction to proceed, overlap between the highest occupied π orbital of the alkene and the lowest unoccupied π orbital of the diene would have to occur as shown in Fig. 15.10, with a + lobe

---

[2111]The possibility has been raised that some disallowed reactions may nevertheless proceed by concerted mechanisms: see Schmidt, W. *Helv. Chim. Acta* **1971**, *54*, 862; *Tetrahedron Lett.* **1972**, 581; Muszkat, K.A.; Schmidt, W. *Helv. Chim. Acta* **1971**, *54*, 1195; Baldwin, J.E.; Andrist, A.H.; Pinschmidt Jr., R.K. *Acc. Chem. Res.* **1972**, *5*, 402; Berson, J.A. *Acc. Chem. Res.* **1972**, *5*, 406; Baldwin, J.E., in Marchand, A.P.; Lehr, R.E. *Pericyclic Reactions*, Vol. 2, Academic Press, NY, **1977**, pp. 273–302.

[2112]For example, see Sieber, W.; Heimgartner, H.; Hansen, H.; Schmid, H. *Helv. Chim. Acta* **1972**, *55*, 3005. For discussions see Bartlett, P.D.; Helgeson, R.; Wersel, O.A. *Pure Appl. Chem.* **1968**, *16*, 187; Seeley, D.A. *J. Am. Chem. Soc.* **1972**, *94*, 4378; Kaupp, G. *Angew. Chem. Int. Ed.* **1972**, *11*, 313, 718.

overlapping a − lobe. Since like signs are no longer overlapping, the thermal reaction is now forbidden. Similarly, thermal $[_\pi 4_s + _\pi 2_a]$- and

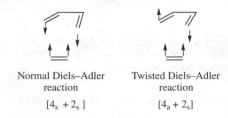

Normal Diels–Adler
reaction
$[4_s + 2_s]$

Twisted Diels–Adler
reaction
$[4_a + 2_s]$

$[_\pi 2_a + _\pi 2_a]$-cyclizations are forbidden, while thermal $[_\pi 4_a + _\pi 2_a]$- and $[_\pi 2_s + _\pi 2_a]$-cyclizations are allowed, and these considerations are reversed for the corresponding photochemical processes. Of course, an antarafacial approach is highly unlikely in a $[4 + 2]$-cyclization,[2113] but larger ring closures could take place by such a pathway, and $[2 + 2]$-thermal cyclizations, where the $[_\pi 2_s + _\pi 2_s]$ pathway is forbidden, can also do so in certain cases (see **15-63**). We therefore see that whether a given cycloaddition is allowed or forbidden depends on the geometry of approach of the two molecules involved.

Symmetry considerations have also been advanced to explain predominant endo addition.[2114] In the case of $[4 + 2]$ addition of butadiene to itself, the approach can be exo or endo. It can be seen (Fig. 15.11) that whether the HOMO of the diene overlaps with the LUMO of the alkene or vice versa, the endo orientation is stabilized by additional secondary overlap of orbitals[2115] of like sign (dashed lines between heavy dots). Addition from the exo direction has no such stabilization. Evidence for secondary orbital overlap as the cause of predominant endo orientation, at least in some cases, is that $[4 + 6]$-cycloaddition is predicted by similar considerations to proceed with predominant exo orientation, and that is what is found.[2116] However, this explanation does not account for endo orientation in cases where the dienophile does not possess additional π orbitals, and a number of alternative explanations have been offered.[2117]

[2113]A possible photochemical $[_\pi 2_a + _\pi 4_s]$-cycloaddition has been reported: Hart, H.; Miyashi, T.; Buchanan, D.N.; Sasson, S. *J. Am. Chem. Soc.* **1974**, *96*, 4857.

[2114]Hoffmann, R.; Woodward, R.B. *J. Am. Chem. Soc.* **1965**, *87*, 4388.

[2115]For reviews of secondary orbital interactions, see Ginsburg, D. *Tetrahedron* **1983**, *39*, 2095; Gleiter, R.; Paquette, L.A. *Acc. Chem. Res.* **1983**, *16*, 328. For a new secondary orbital interaction see Singleton, D.A. *J. Am. Chem. Soc.* **1992**, *114*, 6563.

[2116]See, for example, Cookson, R.C.; Drake, B.V.; Hudec, J.; Morrison, A. *Chem. Commun.* **1966**, 15; Itô, S.; Fujise, Y.; Okuda, T.; Inoue, Y. *Bull. Chem. Soc. Jpn.* **1966**, *39*, 1351; Paquette, L.A.; Barrett, J.H.; Kuhla, D.E. *J. Am. Chem. Soc.* **1969**, *91*, 3616; Houk, K.N.; Woodward, R.B. *J. Am. Chem. Soc.* **1970**, *92*, 4143, 4145; Jones, D.W.; Kneen, G. *J. Chem. Soc., Chem. Commun.* **1973**, 420. Also see Apeloig, Y.; Matzner, E. *J. Am. Chem. Soc.* **1995**, *117*, 5375.

[2117]See, for example, Houk, K.N.; Luskus, L.J. *J. Am. Chem. Soc.* **1971**, *93*, 4606; Kobuke, Y.; Sugimoto, T.; Furukawa, J.; Fueno, T. *J. Am. Chem. Soc.* **1972**, *94*, 3633; Jacobson, B.M. *J. Am. Chem. Soc.* **1973**, *95*, 2579; Mellor, J.M.; Webb, C.F. *J. Chem. Soc. Perkin Trans. 2*, **1974**, 17, 26; Fox, M.A.; Cardona, R.; Kiwiet, N.J. *J. Org. Chem.* **1987**, *52*, 1469.

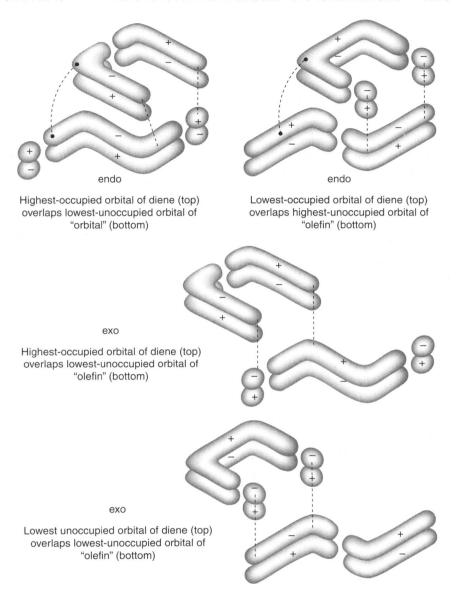

endo

Highest-occupied orbital of diene (top)
overlaps lowest-unoccupied orbital of
"orbital" (bottom)

endo

Lowest-occupied orbital of diene (top)
overlaps highest-unoccupied orbital of
"olefin" (bottom)

exo

Highest-occupied orbital of diene (top)
overlaps lowest-unoccupied orbital of
"olefin" (bottom)

exo

Lowest unoccupied orbital of diene (top)
overlaps lowest-unoccupied orbital of
"olefin" (bottom)

Fig. 15.11. Overlap of orbitals in [4 + 2]-cycloaddition of dienes.

OS **II**, 102; **III**, 310, 807; **IV**, 238, 738, 890, 964; **V**, 414, 424, 604, 985, 1037; **VI**, 82, 196, 422, 427, 445, 454; **VII**, 4, 312, 485; **VIII**, 31, 38, 298, 353, 444, 597; **IX**, 186, 722; **75**, 201; **81**, 171. For a reverse Diels–Alder reaction, see OS **VII**, 339.

15-61   Heteroatom Diels–Alder Reactions

Carbon–carbon multiple bonds are not the only units that can participate in Diels–Alder reactions. Other double- and triple-bond compounds can be dienophiles and they give rise to heterocyclic compounds.[2118] Among these are N≡C–, –N=C–,[2119] iminium salts,[2120] –N=N–, O=N–,[2121] and –C=O compounds[2122] and even molecular oxygen (**15-62**). It is noted that in the presence of a YbCl$_3$ catalyst, azirines reaction with dienes to give a 1-azabicyclo[4.1.0] heptene.[2123] Several catalysts can be used, depending on the nature of the heteroatoms incorporated into the alkene or diene.[2124] Intramolecular cycloaddition with a diene–imine substrate leads to pyrrolidines.[2125]

203

Aldehydes react with suitably functionalized dienes, such as **203**, known as *Danishefsky's diene*,[2126] and the reaction usually requires a Lewis acid catalyst

[2118]For transition structures for selected hetero Diels–Alder reactions, see McCarrick, M.A.; Wu, Y.-D.; Houk, K.N. *J. Org. Chem.* **1993**, *58*, 3330.

[2119]Nogue, D.; Paugam, R.; Wartski, L. *Tetrahedron Lett.* **1992**, *33*, 1265; Collin, J.; Jaber, N.; Lannou, M.I. *Tetrahedron Lett.* **2001**, *42*, 7405; Hedberg, C.; Pinho, P.; Roth, P.; Andersson, P.G. *J. Org. Chem.* **2000**, *65*, 2810. For a review see Buonora, P.; Olsen, J.-C.; Oh, T. *Tetrahedron* **2001**, *57*, 6099; Anniyappan, M.; Muralidharan, D.; Perumal, P.T. *Tetrahedron Lett.* **2003**, *44*, 3653.

[2120]Domingo, L.R. *J. Org. Chem.* **2001**, *66*, 3211; Chou, S.-S.P.; Hung, C.-C. *Synth. Commun.* **2001**, *31*, 1097.

[2121]Martin, S.F.; Hartmann, M.; Josey, J.A. *Tetrahedron Lett.* **1992**, *33*, 3583.

[2122]For monographs on dienes and dienophiles with heteroatoms, see Boger, D.L.; Weinreb, S.M. *Hetero Diels–Alder Methodology in Organic Synthesis*, Academic Press, NY, **1987**; Hamer, J. *1,4-Cycloaddition Reactions*, Academic Press, NY, **1967**. For reviews, see Weinreb, S.M.; Scola, P.M. *Chem. Rev.* **1989**, *89*, 1525; Boger, D.L., in Lindberg, T. *Strategies and Tactics in Organic Synthesis*, Vol. 2, Academic Press, NY, **1989**, pp. 1–56; Kametani, T.; Hibino, S. *Adv. Heterocycl. Chem.* **1987**, *42*, 245; Boger, D.L. *Tetrahedron* **1983**, *39*, 2869; Weinreb, S.M.; Staib, R.R. *Tetrahedron* **1982**, *38*, 3087; Weinreb, S.M.; Levin, J.I. *Heterocycles* **1979**, *12*, 949; Desimoni, G.; Tacconi, G. *Chem. Rev.* **1975**, *75*, 651; Kresze, G.; Firl, J. *Fortschr. Chem. Forsch.* **1969**, *11*, 245. See also, Katritzky, A.R.; Dennis, N. *Chem. Rev.* **1989**, *89*, 827; Schmidt, R.R. *Acc. Chem. Res.* **1986**, *19*, 250; Boger, D.L. *Chem. Rev.* **1986**, *86*, 781.

[2123]Ray, C.A.; Risberg, E.; Somfai, P. *Tetrahedron Lett.* **2001**, *42*, 9289.

[2124]See Molander, G.A.; Rzasa, R.M. *J. Org. Chem.* **2000**, *65*, 1215.

[2125]Amos, D.T.; Renslo, A.R.; Danhesier, R.L. *J. Am. Chem. Soc.* **2003**, *125*, 4970.

[2126]Danishefsky, S.; Kitahara, T.; Schuda, P.F.; Etheredge, S.J. *J. Am. Chem. Soc.* **1976**, *98*, 3028; Danishefsky, S.; Kitahara, T.; McKee, R.; Schuda, P.F. *J. Am. Chem. Soc.* **1976**, *98*, 6715; Danishefsky, S.; Schuda, P.F.; Kitahara, T. Etheredge, S.J. *J. Am. Chem. Soc.* **1977**, *99*, 6066.

such as lanthanide compounds. Aldehydes react using a chiral titanium[2127] or a zirconium[2128] catalyst to give the dihydropyran with good enantioselectivity. Note that the reaction of Danishefsky's diene with an imine, formed *in situ* by reaction of an aryl aldehyde and an aniline derivative, proceeds without a Lewis acid.[2129] Such reactions of aldehydes can be catalyzed with Lewis acids and transition-metal catalysts. The Diels–Alder reaction of aldehydes with dienes can be catalyzed by many transition-metal compounds, including cobalt[2130] and indium[2131] catalyst. Ketones also react with suitably functionalized dienes.[2132]

Azadienes undergo Diels–Alder reactions to form pyridine, dihydro- and tetra-hydropyridine derivatives.[2133] Aza-Diels–Alder reactions have been done in ionic liquids.[2134] Similarly, acyl iminium salts C=N(R)–C=O react with alkenes via cycloaddition.[2135] *N*-Vinyl lactim ethers undergo Diels–Alder reactions with a limited set of dienophiles.[2136] Thioketones react with dienes to give Diels–Alder cycloadducts.[2137] The carbonyl group of lactams have also been shown to be a dienophile.[2138] Certain heterocyclic aromatic rings (among them furans)[2139] can also behave as dienes in the Diels–Alder reaction. Some hetero dienes that give the reaction are –C=C–C=O, O=C–C=O, and N=C–C=N.[1999] Nitroso compounds of the type *t*-BuO$_2$C–N=O react with dienes to give the corresponding 2-azadihydropyran.[2140]

Catalysts, such as Fe(BuEtCHCO$_2$)$_3$, have been developed that are effective for the heteroatom Diels–Alder reaction.[2141] Indium trichloride (InCl$_3$) is a good catalyst for imino-Diels–Alder reactions.[2142] Hetero-Diels–Alder reactions involving carbonyls have been done in water.[2143] Ultrasound has been used to promote the Diels–Alder reactions of 1-azadienes.[2144] Polymer-supported dienes have been used.[2145]

[2127]Wang, B.; Feng, X.; Huang, Y.; Liu, H.; Cui, X.; Jiang, Y. *J. Org. Chem.* **2002**, *67*, 2175.

[2128]Yamashita, Y.; Saito, S.; Ishitani, H.; Kobayashi, S. *J. Am. Chem. Soc.* **2003**, *125*, 3793.

[2129]Yuan, Y.; Li, X.; Ding, K. *Org. Lett.* **2002**, *4*, 3309.

[2130]Kezuka, S.; Mita, T.; Ohtsuki, N.; Ikeno, T.; Yamada, T. *Bull. Chem. Soc. Jpn.* **2001**, *74*, 1333.

[2131]Ali, T.; Chauhan, K.K.; Frost, C.G. *Tetrahedron Lett.* **1999**, *40*, 5621.

[2132]Huang, Y.; Rawal, V.H.; *J. Am. Chem. Soc.* **2002**, *124*, 9662; Jørgensen, K.A. *Eur. J. Org. Chem.* **2004**, 2093.

[2133]Gilchrist, T.L.; Gonsalves, A.M. d'A.R.; Pinho e Melo, T.M.V.D. *Pure Appl. Chem.* **1996**, *68*, 859; Jayakumar, S.; Ishar, M.P.S.; Mahajan, M.P. *Tetrahedron* **2002**, *58*, 379.

[2134]Yadav, J.S.; Reddy, B.V.S.; Reddy, J.S.S.; Rao, R.S. *Tetrahedron* **2003**, *59*, 1599.

[2135]Suga, S.; Nagaki, A.; Tsutsui, Y.; Yoshida, J.-i. *Org. Lett.* **2003**, *5*, 945.

[2136]Sheu, J.; Smith, M.B.; Matsumoto, K. *Synth. Commun*, **1993**, *23*, 253.

[2137]Schatz, J.; Sauer, J. *Tetrahedron Lett.* **1994**, *35*, 4767.

[2138]Degnan, A.P.; Kim, C.S.; Stout, C.W.; Kalivretenos, A.G. *J. Org. Chem.* **1995**, *60*, 7724.

[2139]For reviews, see Katritzky, A.R.; Dennis, N. *Chem. Rev.* **1989**, *89*, 827; Schmidt, R.R. *Acc. Chem. Res.* **1986**, *19*, 250; Boger, D.L. *Chem. Rev.* **1986**, *86*, 781. See Hayashi, Y.; Nakamura, M.; Nakao, S.; Inoue, T.; Shoji, M. *Angew. Chem. Int. Ed.* **2002**, *41*, 4079.

[2140]Bach, P.; Bols, M. *Tetrahedron Lett.* **1999**, *40*, 3461.

[2141]Gorman, D.B.; Tomlinson, I.A. *Chem. Commun.* **1998**, 25.

[2142]Babu, G.; Perumal, P.T. *Tetrahedron* **1998**, *54*, 1627.

[2143]Lubineau, A.; Augé, J.; Grand, E.; Lubin, N. *Tetrahedron* **1994**, *50*, 10265.

[2144]Villacampa, M.; Pérez, J.M.; Avendaño, C.; Menéndez, J.C. *Tetrahedron* **1994**, *50*, 10047.

[2145]Pierres, C.; George, P.; van Hijfte, L.; Ducep, J.-B.; Hibert, M.; Mann, A. *Tetrahedron Lett.* **2003**, *44*, 3645.

Hetero- Diels–Alder reactions that proceed with good to excellent asymmetric induction are well known.[2146] Chiral 1-aza-dienes have been developed as substrates, for example.[2147] Chiral catalysts have been developed.[2148] Conjugated aldehydes react with vinyl ethers, with a chiral chromium catalyst, in an inverse electron demand cycloaddition that give a dihydropyran with good enantioselectivity.[2149] Vinyl sulfilimines have been used in chiral Diels–Alder reactions.[2150]

OS **IV**, 311; **V**, 60, 96; **80**, 133. See also OS **VII**, 326.

15-62   Photooxidation of Dienes (Addition of Oxygen, Oxygen)

[4 + 2] *OC,OC-cyclo-***Peroxy-1/4/addition**

204

Conjugated dienes react with oxygen under the influence of light to give cyclic peroxides **204**.[2151] The reaction has mostly[2152] been applied to cyclic dienes.[2153] Cycloaddition of furan has been reported using singlet oxygen.[2154] The scope extends to certain aromatic compounds such as phenanthrene.[2155] Besides those dienes and aromatic rings that can be photooxidized directly, there is a larger group that give the reaction in the presence of a photosensitizer such as eosin (see p. 340).

[2146]Yao, S.; Johannsen, M.; Audrain, H.; Hazell, R.G.; Jørgensen, K.A. *J. Am. Chem. Soc.* **1998**, *120*, 8599; Pouilhès, A.; Langlois, Y.; Nshimyumkiza, P.; Mbiya, K.; Ghosez, L. *Bull. Soc. Chim. Fr.* **1993**, *130*, 304.

[2147]Beaudegnies, R.; Ghosez, L. *Tetrahedron Asymmetry* **1994**, *5*, 557.

[2148]Du, H.; Long, J.; Hu, J.; Li, X.; Ding, K. *Org. Lett.* **2002**, *4*, 4349; Du, H.; Ding, K. *Org. Lett.* **2003**, *5*, 1091; Macheño, O.G.; Arrayás, R.G.; Carretero, J.C. *J. Am. Chem. Soc.* **2004**, *126*, 456.

[2149]Gademann, K.; Chavez, D.E.; Jacobsen, E.N. *Angew. Chem. Int. Ed.* **2002**, *41*, 3059.

[2150]Ruano, J.L.G.; Clemente, F.R.; Gutiérrez, L.G.; Gordillo, R.; Castro, A.M.M.; Ramos, J.H.R. *J. Org. Chem.* **2002**, *67*, 2926.

[2151]For reviews, see Clennan, E.L. *Tetrahedron* **1991**, *47*, 1343; *Adv. Oxygenated Processes* **1988**, *1*, 85; Wasserman, H.H.; Ives, J.L. *Tetrahedron* **1981**, *37*, 1825; Denny, R.W.; Nickon, A. *Org. React.* **1973**, *20*, 133; Adams, W.R. in Augustine, R.L.; Trecker, D.J. *Oxidation*, Vol. 2, Marcel Dekker, NY, **1971**, pp. 65–112; Gollnick, K. *Adv. Photochem.* **1968**, *6*, 1; Schönberg, A. *Preparative Organic Photochemistry*, Springer, NY, **1968**, pp. 382–397; Gollnick, K.; Schenck, G.O., in Hamer, T. *1,4-Cycloaddition Reactions*, Academic Press, NY, **1967**, pp. 255–344; Arbuzov, Yu.A. *Russ. Chem. Rev.* **1965**, *34*, 558.

[2152]For many examples with acyclic dienes, see Matsumoto, M.; Dobashi, S.; Kuroda, K.; Kondo, K. *Tetrahedron* **1985**, *41*, 2147.

[2153]For reviews of cyclic peroxides, see Saito, I.; Nittala, S.S., in Patai, S. *The Chemistry of Peroxides*, Wiley, NY, **1983**, pp. 311–374; Balci, M. *Chem. Rev.* **1981**, *81*, 91; Adam, W.; Bloodworth, A.J. *Top. Curr. Chem.* **1981**, *97*, 121.

[2154]Onitsuka, S.; Nishino, H.; Kurosawa, K. *Tetrahedron* **2001**, *57*, 6003.

[2155]For reviews, see in Wasserman, H.H.; Murray, R.W. *Singlet Oxygen*; Academic Press, NY, **1979**, the articles by Wasserman, H.H.; Lipshutz, B.H. pp. 429–509; Saito, I.; Matsuura, T. pp. 511–574; Rigaudy, J. *Pure Appl. Chem.* **1968**, *16*, 169.

Among these is α-terpinene, which is converted to ascaridole:

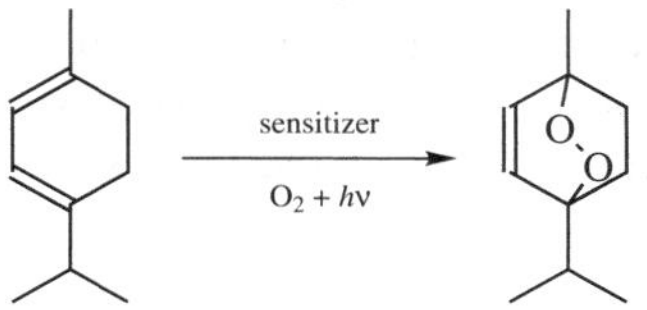

As in **14-7**, it is not the ground-state oxygen (the triplet), that reacts, but the excited singlet state,[2156,2157] so the reaction is actually a Diels–Alder reaction (see **15-60**) with singlet oxygen as dienophile:[2158]

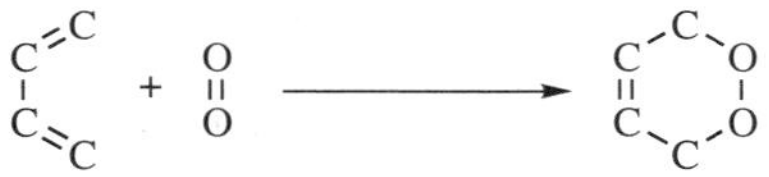

Like **15-60**, this reaction is reversible.

We have previously discussed the reaction of singlet oxygen with double-bond compounds to give

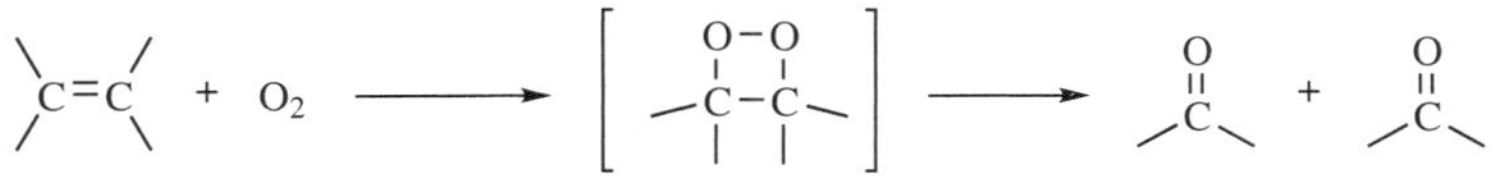

205

hydroperoxides (**14-7**), but singlet oxygen can also react with double bonds in another way to give a dioxetane intermediate[2159] (**205**), which usually cleaves to

---

[2156]For books on singlet oxygen, see Frimer, A.A. *Singlet O₂*, 4 vols., CRC Press, Boca Raton, FL, *1985*; Wasserman, H.H.; Murray, R.W. *Singlet Oxygen*, Academic Press, NY, *1979*. For reviews, see Frimer, A.A. in Patai, S. *The Chemistry of Peroxides*, Wiley, NY, *1983*, pp. 201–234; Gorman, A.A.; Rodgers, M.A.J. *Chem. Soc. Rev. 1981*, *10*, 205; Shinkarenko, N.V.; Aleskovskii, V.B. *Russ. Chem. Rev. 1981*, *50*, 220; Shlyapintokh, V.Ya.; Ivanov, V.B. *Russ. Chem. Rev. 1976*, *45*, 99; Ohloff, G. *Pure Appl. Chem. 1975*, *43*, 481; Kearns, D.R. *Chem. Rev. 1971*, *71*, 395; Wayne, R.P. *Adv. Photochem. 1969*, *7*, 311.

[2157]For reviews, see Turro, N.J.; Ramamurthy, V., in de Mayo, P. *Rearrangements in Ground and Excited States*, Vol. 3, Academic Press, NY, *1980*, pp. 1–23; Murray, R.W., in Wasserman, H.H.; Murray, R.W. *Singlet Oxygen*, Academic Press, NY, *1979*, pp. 59–114. For a general monograph, see Adam, W.; Cilento, G. *Chemical and Biological Generation of Excited States*, Academic Press, NY, *1982*.

[2158]Corey, E.J.; Taylor, W.C. *J. Am. Chem. Soc. 1964*, *86*, 3881; Foote, C.S.; Wexler, S.; Ando, W. *Tetrahedron Lett. 1965*, 4111; Monroe, B.M. *J. Am. Chem. Soc. 1981*, *103*, 7253. See also, Hathaway, S.J.; Paquette, L.A. *Tetrahedron Lett. 1985*, *41*, 2037; O'Shea, K.E.; Foote, C.S. *J. Am. Chem. Soc. 1988*, *110*, 7167.

[2159]For reviews, see Adam, W.; Cilento, G. *Angew. Chem. Int. Ed. 1983*, *22*, 529; Schaap, A.; Zaklika, K.A. in Wasserman, H.H.; Murray, R.W. *Singlet Oxygen*, Academic Press, NY, *1979*, pp. 173–242; Bartlett, P.D. *Chem. Soc. Rev. 1976*, *5*, 149. For discussions of the mechanisms see Frimer, A.A. *Chem. Rev. 1979*, *79*, 359; Clennan, E.L.; Nagraba, K. *J. Am. Chem. Soc. 1988*, *110*, 4312.

aldehydes or ketones,[2160] but has been isolated.[2161] Both the six-membered cyclic peroxides[2162] and the four-membered **205**[2163] have been formed from oxygenation reactions that do not involve singlet oxygen. If cyclic peroxides, such as **205**, are desired, better reagents[2164] are triphenyl phosphite ozonide $(PhO)_3PO_3$ and triethylsilyl hydrotrioxide $(Et_3SiOOOH)$, but yields are not high.[2165]

15-63   [2 + 2]-Cycloadditions

[2 + 2]*cyclo*-Ethylene-1/2/addition

The thermal reaction between two molecules of alkene to give cyclobutane derivatives (a [2 + 2]-cycloaddition) can be carried out where the alkenes are the same or different, but the reaction is not a general one for alkenes.[2166] The cycloaddition can be catalyzed by certain transition-metal complexes.[2167] Dimerization of like alkenes occurs with the following compounds: $F_2C=CX_2$ (X = F or Cl) and certain other

206   Biphenylene

[2160]For discussions see Kearns, D.R. *Chem. Rev.* *1971*, *71*, 395, 422–424; Foote, C.S. *Pure Appl. Chem.* *1971*, *27*, 635.

[2161]For reviews of 1,2-dioxetanes see Adam, W., in Patai, S. *The Chemistry of Peroxides*, Wiley, NY, *1983*, pp. 829–920; Bartlett, P.D.; Landis, M.E., in Wasserman, H.H.; Murray, R.W. *Singlet Oxygen*, Academic Press, NY, *1979*, pp. 243–286; Adam, W. *Adv. Heterocycl. Chem.* *1977*, *21*, 437. See also, Inoue, Y.; Hakushi, T.; Turro, N.J. *Kokagaku Toronkai Koen Yoshishu* *1979*, 150 [*C.A.* *92*, 214798q]; Adam, W.; Encarnación, L.A.A. *Chem. Ber.* *1982*, *115*, 2592; Adam, W.; Baader, W.J. *Angew. Chem. Int. Ed.* *1984*, *23*, 166.

[2162]See Nelson, S.F.; Teasley, M.F.; Kapp, D.L. *J. Am. Chem. Soc.* *1986*, *108*, 5503.

[2163]For a review, see Nelson, S.F. *Acc. Chem. Res.* *1987*, *20*, 269.

[2164]For another reagent, see Curci, R.; Lopez, L.; Troisi, L.; Rashid, S.M.K.; Schaap, A.P. *Tetrahedron Lett.* *1987*, *28*, 5319.

[2165]Posner, G.H.; Weitzberg, M.; Nelson, W.M.; Murr, B.L.; Seliger, H.H. *J. Am. Chem. Soc.* *1987*, *109*, 278.

[2166]For reviews, see Carruthers, W. *Cycloaddition Reactions in Organic Synthesis*, Pergamon, Elmsford, NY, *1990*; Reinhoudt, D.N. *Adv. Heterocycl. Chem.* *1977*, *21*, 253; Roberts, J.D.; Sharts, C.M. *Org. React.* *1962*, *12*, 1; Gilchrist, T.L.; Storr, R.C. *Organic Reactions and Orbital Symmetry* 2nd ed., Cambridge University Press, Cambridge, *1979*, pp. 173–212; Beltrame, P., in Bamford, C.H.; Tipper, C.F.H. Ref. 1, Vol. 9, pp. 131–152; Huisgen, R.; Grashey, R.; Sauer, J., in Patai, S. *The Chemistry of Alkenes*, Vol. 1, Wiley, NY, *1964*, pp. 779–802. For a review of the use of [2 + 2]-cycloadditions in polymerization reactions, see Dilling, W.L. *Chem. Rev.* *1983*, *83*, 1. For a list of references, see Larock, R.C. *Comprehensive Organic Transformations*, 2nd ed., Wiley-VCH, NY, *1999*, pp. 546–647, 1341–1344.

[2167]For an example using EtAlCl$_2$ see Takasu, K.; Ueno, M.; Inanaga, K.; Ihara, M. *J. Org. Chem.* *2004*, *69*, 517.

fluorinated alkenes (although not $F_2C=CH_2$), allenes (to give derivatives of **206**),[2168] benzynes (to give biphenylene derivatives),[2169] activated alkenes (e.g., styrene, acrylonitrile, butadiene), and certain methylenecyclopropanes.[2170] Dimerization of allenes lead to bis(alkylidene) cyclobutanes.[2171] Substituted ketenes can dimerize to give cyclobutenone derivatives, although ketene itself dimerizes in a different manner, to give an unsaturated β-lactone (**16-95**).[2172] Alkenes react with activated alkynes, with a ruthenium catalyst, to give cyclobutenes.[2173]

Intramolecular [2 + 2]-cycloadditions are common in which a diene is converted to a bicyclic compound with a four-membered ring fused to another ring. Heating *N*-vinyl imines, where the vinyl moiety is a silyl enol, gives β-lactams.[2174] Apart from photochemical initiation of such reactions, intramolecular cycloaddition of two conjugated ketone units, in the presence of $PhMeSiH_2$ and catalyzed by cobalt compounds, leads to the bicyclic compound with two ketone substituents.[2175] In a variation of this reaction, a diyne was treated with $Ti(OiPr)_4/2$ *i*-PrMgCl to generate a bicyclic cyclobutene with two vinylidene units.[2176]

Ketenes react with many alkenes to give cyclobutanone derivatives[2177] and intermolecular cycloadditions are well known.[2178] typical reaction is that of dimethylketene and ethene to give 2,2-dimethylcyclobutanone.[2179] Ketenes react with imines via [2 + 2]-cycloaddition to produce β-lactams.[2180] Cycloaddition of an imine with a conjugated ester in the presence of Et_2MeSiH and an iridium

[2168]For a review, see Fischer, H., in Patai, S. *The Chemistry of Alkenes*, Vol. 1, Wiley, NY, *1964*, pp. 1064–1067.

[2169]For cycloaddition with a pyridyne, see Mariet, N.; Ibrahim-Ouali, M.; Santelli, M. *Tetrahedron Lett.* *2002*, *43*, 5789.

[2170]Dolbier, Jr., W.R.; Lomas, D.; Garza, T.; Harmon, C.; Tarrant, P. *Tetrahedron 1972*, *28*, 3185.

[2171]Saito, S.; Hirayama, K.; Kabuto, C.; Yamamoto, Y. *J. Am. Chem. Soc. 2000*, *122*, 10776.

[2172]Farnum, D.G.; Johnson, J.R.; Hess, R.E.; Marshall, T.B.; Webster, B. *J. Am. Chem. Soc. 1965*, *87*, 5191; Dehmlow, E.V.; Pickardt, J.; Slopianka, M.; Fastabend, U.; Drechsler, K.; Soufi, J. *Liebigs Ann. Chem. 1987*, 377.

[2173]Jordan, R.W.; Tam, W. *Org. Lett. 2000*, *2*, 3031.

[2174]Bandin, E.; Favi, G.; Martelli, G.; Panunzio, M.; Piersanti, G. *Org. Lett. 2000*, *2*, 1077.

[2175]Baik, T.-G.; Luis, A.L.; Wang, L.-C.; Krische, M.J. *J. Am. Chem. Soc. 2001*, *123*, 6716.

[2176]Delas, C.; Urabe, H.; Sato, F. *Tetrahedron Lett. 2001*, *42*, 4147.

[2177]An example is de Faria, A.R.; Matos, C.R.; Correia, C.R.D. *Tetrahedron Lett. 1993*, *34*, 27.

[2178]Krepski, L.R.; Hassner, A. *J. Org. Chem. 1978*, *43*, 2879; Bak, D.A.; Brady, W.T. *J. Org. Chem. 1979*, *44*, 107; Martin, P.; Greuter, H.; Belluš, D. *Helv. Chim. Acta.*, *1984*, *64*, 64; Brady, W.T. *Synthesis 1971*, 415.

[2179]Sustmann, R.; Ansmann, A.; Vahrenholt, F. *J. Am. Chem. Soc. 1972*, *94*, 8099; Desimoni, G.; Tacconi, G.; Barco, A.; Pollini, G.P. *Natural Product Synthesis Through Pericyclic Reactions*, American Chemical Society, Washington, DC, *1983*, pp. 119–254, 39.

[2180]For reviews of the formation of β-lactams, see Brown, M.J. *Heterocycles 1989*, *29*, 2225; Isaacs, N.S. *Chem. Soc. Rev. 1976*, *5*, 181; Mukerjee, A.K.; Srivastava, R.C. *Synthesis 1973*, 327. For a review of cycloaddition reactions of imines, see Sandhu, J.S.; Sain, B. *Heterocycles 1987*, *26*, 777. For a new catalyst, see Wack, H.; France, S.; Hafez, A.M.; Drury, III, W.J.; Weatherwax, A.; Lectka, T. *J. Org. Chem. 2004*, *69*, 4531.

catalyst also gives a β-lactam.[2181]

Different alkenes combine as follows:

1. $F_2C=CX_2$ ($X = F$ or Cl), especially $F_2C=CF_2$, form cyclobutanes with many alkenes. Compounds of this type even react with conjugated dienes to give four-membered rings rather than undergoing normal Diels–Alder reactions.[2182]

2. Allenes[2183] and ketenes[2184] react with activated alkenes and alkynes. Ketenes give 1,2-addition, even with conjugated dienes.[2185] Ketenes also add to unactivated alkenes if sufficiently long reaction times are used.[2186] Allenes and ketenes also add to each other.[2187]

3. Enamines[2188] form four-membered rings with Michael-type alkenes[2189] and ketenes.[2190] In both cases, only enamines from aldehydes give stable

[2181]Townes, J.A.; Evans, M.A.; Queffelec, J.; Taylor, S.J.; Morken, J.P. *Org. Lett.* **2002**, *4*, 2537.

[2182]Bartlett, P.D.; Montgomery, L.K.; Seidel, B. *J. Am. Chem. Soc.* **1964**, *86*, 616; De Cock, C.; Piettre. S.; Lahousse, F.; Janousek, Z.; Merényi, R.; Viehe, H.G. *Tetrahedron* **1985**, *41*, 4183.

[2183]For reviews of [2 + 2]-cycloadditions of allenes, see Schuster, H.F.; Coppola, G.M. *Allenes in Organic Synthesis*, Wiley, NY, **1984**, pp. 286–317; Hopf, H., in Landor, S.R.I. *The Chemistry of Allenes*, Vol. 2, Academic Press, NY, **1982**, pp. 525–562; Ghosez, L.; O'Donnell, M.J., in Marchand, A.P.; Lehr, R.E. *Pericyclic Reactions*, Vol. 2, Academic Press, NY, **1977**, pp. 79–140; Baldwin, J.E.; Fleming, R.H. *Fortschr. Chem. Forsch.* **1970**, *15*, 281.

[2184]For reviews of cycloadditions of ketenes, see Ghosez, L.; O'Donnell, M.J. in Marchand, A.P.; Lehr, R.E. *Pericyclic Reactions*, Vol. 2, Academic Press, NY, **1977**; Brady, W.T. *Synthesis* **1971**, 415; Luknitskii, F.I.; Vovsi, B.A. *Russ. Chem. Rev.* **1969**, *38*, 487; Ulrich, H. *Cycloaddition Reactions of Heterocumulenes*, Academic Press, NY, **1967**, pp. 38–121; Holder, R.W. *J. Chem. Educ.* **1976**, *53*, 81. For a review of intramolecular cycloadditions of ketenes to alkenes, see Snider, B.B. *Chem. Rev.* **1988**, *88*, 793.

[2185]See, for example, Martin, J.C.; Gott, P.G.; Goodlett, V.W.; Hasek, R.H. *J. Org. Chem.* **1965**, *30*, 4175; Brady, W.T.; O'Neal, H.R. *J. Org. Chem.* **1967**, *32*, 2704; Huisgen, R.; Feiler, L.A.; Otto, P. *Tetrahedron Lett.* **1968**, 4491; *Chem. Ber.* **1969**, *102*, 3475. For indirect methods of the 1,4-addition of the elements of ketene to a diene, see Freeman, P.K.; Balls, D.M.; Brown, D.J. *J. Org. Chem.* **1968**, *33*, 2211; Corey, E.J.; Ravindranathan, T.; Terashima, S. *J. Am. Chem. Soc.* **1971**, *93*, 4326. For a review of ketene equivalents, see Ranganathan, S.; Ranganathan, D.; Mehrotra, A.K. *Synthesis* **1977**, 289.

[2186]Huisgen, R.; Feiler, L.A. *Chem. Ber.* **1969**, *102*, 3391; Bak, D.A.; Brady, W.T. *J. Org. Chem.* **1979**, *44*, 107.

[2187]Bampfield, H.A.; Brook, P.R.; McDonald, W.S. *J. Chem. Soc., Chem. Commun.* **1975**, 132; Gras, J.; Bertrand, M. *Nouv. J. Chim.* **1981**, *5*, 521.

[2188]For a review of cycloaddition reactions of enamines, see Cook, A.G., in Cook, A.G. *Enamines*, 2nd ed.; Marcel Dekker, NY, **1988**, pp. 347–440.

[2189]Brannock, K.C.; Bell, A.; Goodlett, V.W.; Thweatt, J.G. *J. Org. Chem.* **1964**, *29*, 813.

[2190]Berchtold, G.A.; Harvey, G.R.; Wilson, G.E. *J. Org. Chem.* **1961**, *26*, 4776; Opitz, G.; Kleeman, M. *Liebigs Ann. Chem.* **1963**, *665*, 114; Hasek, R.H.; Gott, P.G.; Martin, J.C. *J. Org. Chem.* **1966**, *31*, 1931.

four-membered rings:

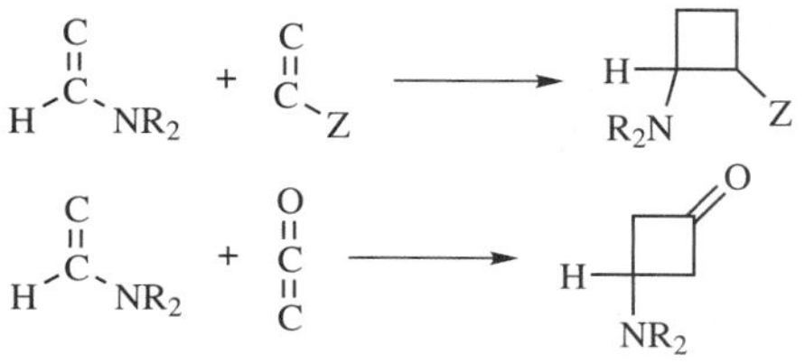

The reaction of enamines with ketenes can be conveniently carried out by generating the ketene *in situ* from an acyl halide and a tertiary amine.

4. Alkenes with electron-withdrawing groups may form cyclobutanes with alkenes containing electron-donating groups. The enamine reactions, mentioned above, are examples of this, but it has also been accomplished with tetracyanoethylene and similar molecules, which give substituted cyclobutanes when treated with alkenes of the form C=C—A, where A may be OR,[2191] SR (enol and thioenol ethers),[2192] cyclopropyl,[2193] or certain aryl groups.[2194]

Solvents are not necessary for $[2 + 2]$-cycloadditions. They are usually carried out at 100–225°C under pressure, although the reactions in Group 4 (IVB) occur under milder conditions.

It has been found that certain $[2 + 2]$-cycloadditions, which do not occur thermally can be made to take place without photochemical initiation by the use of certain catalysts, usually transition-metal compounds.[2195] Photochemical[2196] $[_{\pi}2 + _{s}2]$-cycloadditions have also been reported. Among the catalysts used are Lewis acids[2197] and phosphine–nickel complexes.[2198] Certain of the reverse cyclobutane ring openings can also be catalytically induced (**18-38**). The role of the catalyst is not certain and may be different in each case. One possibility is that the

[2191]For a review with ketene acetals R_2C=$C(OR')_2$, see Scheeren, J.W. *Recl. Trav. Chim. Pays-Bas* **1986**, *105*, 71–84.

[2192]Williams, J.K., Wiley, D.W.; McKusick, B.C. *J. Am. Chem. Soc.* **1962**, *84*, 2210.

[2193]Nishida, S.; Moritani, I.; Teraji, T. *J. Org. Chem.* **1973**, *38*, 1878.

[2194]Nagata, J.; Shirota, Y.; Nogami, T.; Mikawa, H. *Chem. Lett.* **1973**, 1087; Shirota, Y.; Yoshida, K.; Nogami, T.; Mikawa, H. *Chem. Lett.* **1973**, 1271.

[2195]For reviews, see Dzhemilev, U.M.; Khusnutdinov, R.I.; Tolstikov, G.A. *Russ. Chem. Rev.* **1987**, *56*, 36; Kricka, L.J.; Ledwith, A. *Synthesis* **1974**, 539.

[2196]Freeman, P.K.; Balls, D.M. *J. Org. Chem.* **1967**, *32*, 2354; Wiskott, E.; Schleyer, P.v.R. *Angew. Chem. Int. Ed.* **1967**, *6*, 694; Prinzbach, H.; Eberbach, W. *Chem. Ber.* **1968**, *101*, 4083; Prinzbach, H.; Sedelmeier, G.; Martin, H. *Angew. Chem. Int. Ed.* **1977**, *16*, 103.

[2197]Yamazaki, S.; Fujitsuka, H.; Yamabe, S.; Tamura, H. *J. Org. Chem.* **1992**, *57*, 5610. West, R.; Kwitowski, P.T. *J. Am. Chem. Soc.* **1968**, *90*, 4697; Lukas, J.H.; Baardman, F.; Kouwenhoven, A.P. *Angew. Chem. Int. Ed.* **1976**, *15*, 369.

[2198]See, for example, Hoover, F.W.; Lindsey Jr., R.V. *J. Org. Chem.* **1969**, *34*, 3051; Noyori, R.; Ishigami, T.; Hayashi, N.; Takaya, H. *J. Am. Chem. Soc.* **1973**, *95*, 1674; Yoshikawa, S.; Aoki, K.; Kiji, J.; Furukawa, J. *Tetrahedron* **1974**, *30*, 405.

presence of the catalyst causes a forbidden reaction to become allowed, through coordination of the catalyst to the π or s bonds of the substrate.[2199] In such a case, the reaction would of course be a concerted $[2_s + 2_s]$-process. [2200] However, the available evidence is more consistent with nonconcerted mechanisms involving metal–carbon σ-bonded intermediates, at least in most cases.[2201] For example, such an intermediate was isolated in the dimerization of norbornadiene, catalyzed by iridium complexes.[2202]

Thermal cycloadditions leading to four-membered rings can also take place between a cyclopropane ring and an alkene or alkyne[2203] bearing electron-withdrawing groups.[2204] These reactions are $[_\pi2 + _s2]$-cycloadditions. Ordinary cyclopropanes do not undergo the reaction, but it has been accomplished with strained systems such as bicyclo[1.1.0]butanes[2205] and bicyclo[2.1.0]pentanes. For example, bicyclo[2.1.0]pentane reacts with maleonitrile (or fumaronitrile) to give all three isomers of 2,3-dicyanonorbornane, as well as four other products.[2206] The lack of stereospecificity and the negligible effect of solvent on the rate indicate a diradical mechanism.

The reaction is similar to the Diels–Alder (in action, not in scope), and if dienes are involved, the latter reaction may compete, although most alkenes react with a diene either entirely by 1,2 or entirely by 1,4 addition. Three mechanisms can be proposed[2207] analogous to those proposed for the Diels–Alder reaction. Mechanism *a* is a

Mechanism *a*

Mechanism *b*

207

Mechanism *c*

208

[2199]For discussions, see Labunskaya, V.I.; Shebaldova, A.D.; Khidekel, M.L. *Russ. Chem. Rev.* **1974**, *43*, 1; Mango, F.D. *Top. Curr. Chem.* **1974**, *45*, 39; *Tetrahedron Lett.* **1973**, 1509; *Intra-Sci. Chem. Rep.* **1972**, *6 (3)*, 171; *CHEMTECH* **1971**, *1*, 758; *Adv. Catal.* **1969**, *20*, 291; Mango, F.D.; Schachtschneider, J.H. *J. Am. Chem. Soc.* **1971**, *93*, 1123; **1969**, *91*, 2484; van der Lugt, W.T.A.M. *Tetrahedron Lett.* **1970**, 2281; Wristers, J.; Brener, L.; Pettit, R. *J. Am. Chem. Soc.* **1970**, *92*, 7499.

[2200]See Bachrach, S.M.; Gilbert, J.C. *J. Org. Chem.* **2004**, *69*, 6357; Ozkan, I.; Kinal, A. *J. Org. Chem.* **2004**, *69*, 5390.

[2201]See, for example, Cassar, L.; Halpern, J. *Chem. Commun.* **1970**, 1082; Doyle, M.J.; McMeeking, J.; Binger, P. *J. Chem. Soc., Chem. Commun.* **1976**, 376; Grubbs, R.H.; Miyashita, A.; Liu, M.M.; Burk, P.L. *J. Am. Chem. Soc.* **1977**, *99*, 3863.

[2202]Fraser, A.R.; Bird, P.H.; Bezman, S.A.; Shapley, J.R.; White, R.; Osborn, J.A. *J. Am. Chem. Soc.* **1973**, *95*, 597.

[2203]Gassman, P.G.; Mansfield, K.T. *J. Am. Chem. Soc.* **1968**, *90*, 1517, 1524.

[2204]For a review, see Gassman, P.G. *Acc. Chem. Res.* **1971**, *4*, 128.

[2205]Cairncross, A.; Blanchard, E.P. *J. Am. Chem. Soc.* **1966**, *88*, 496.

[2206]Gassman, P.G.; Mansfield, K.T.; Murphy, T.J. *J. Am. Chem. Soc.* **1969**, *91*, 1684.

[2207]For a review, see Bartlett, P.D. *Q. Rev. Chem. Soc.* **1970**, *24*, 473.

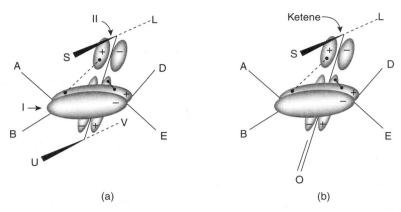

Fig. 15.12. Orbital overlap in $[_\pi 2_s + _\pi 2_s]$-cycloaddition between (*a*) two alkene molecules and (*b*) a ketene and an alkene. S and L stand for small and large

concerted pericyclic process, and mechanisms *b* and *c* are two-step reactions involving, respectively, a diradical (**207**) and a diion (**208**) intermediate. As in **15-60**, a diradical intermediate must be a singlet. In searching for ways to tell which mechanism is operating in a given case, we would expect mechanism *c* to be sensitive to changes in solvent polarity, while mechanisms *a* and *b* should be insensitive. We would also expect mechanism *a* to be stereospecific, while mechanisms *b* and *c* probably would not be stereospecific, although if the second step of these processes takes place very rapidly, before **207** or **208** has a chance to rotate about the newly formed single bond, stereospecificity might be observed. Because of entropy considerations such rapid ring closure might be more likely here than in a $[4 + 2]$-cycloaddition.

There is evidence that the reactions can take place by all three mechanisms, depending on the structure of the reactants. A thermal $[_\pi 2_s + _\pi 2_s]$ mechanism is ruled out for most of these substrates by the orbital symmetry rules, but a $[_\pi 2_s + _\pi 2_a]$ mechanism is allowed (p. 1212), and there is much evidence that ketenes and certain other linear molecules[2208] in which the steric hindrance to such an approach is minimal can and often do react by this mechanism. In a $[_\pi 2_s + _\pi 2_a]$-cycloaddition the molecules must approach each other in such a way (Fig. 15.12*a*) that the $+$ lobe of the HOMO of one molecule (I) overlaps with both $+$ lobes of the LUMO of the other (II), even although these lobes are on opposite sides of the nodal plane of II. The geometry of this approach requires that the groups S and U of molecule II project *into* the plane of molecule I. This has not been found to happen for ordinary alkenes,[2209] but if

[2208]There is evidence that a cyclopentyne (generated *in situ*) also adds to a double bond by an antarafacial process: Gilbert, J.C.; Baze, M.E. *J. Am. Chem. Soc.* **1984**, *106*, 1885.

[2209]See, for example, Padwa, A.; Koehn, W.; Masaracchia, J.; Osborn, C.L.; Trecker, D.J. *J. Am. Chem. Soc.* **1971**, *93*, 3633; Bartlett, P.D.; Cohen, G.M.; Elliott, S.P.; Hummel, K.; Minns, R.A.; Sharts, C.M.; Fukunaga, J.Y. *J. Am. Chem. Soc.* **1972**, *94*, 2899.

molecule II is a ketene (Fig. 15.12*b*), the group marked U is not present and the $[_\pi 2_s + _\pi 2_a]$-reaction can take place. Among the evidence[2210] for this mechanism[2211] is the following: (*1*) The reactions are stereospecific.[2212] (*2*) The isomer that forms is the *more-hindered one*. Thus methylketene plus cyclopentadiene gave only the endo product (**209**, A = H, R = CH_3).[2213] Even more remarkably, when

endo **209**   endo **210**

haloalkyl ketenes RXC=C=O were treated with cyclopentadiene, the endo/exo ratio of the product (**209**, **210**, A = halogen) actually *increased* substantially when R was changed from Me to *i*-Pr to *t*-Bu![2214] One would expect preferential formation of the exo products (**210**) from $[_\pi 2_s + _\pi 2_s]$-cycloadditions where the molecules approach each other face-to-face, but a $[_\pi 2_s + _\pi 2_a]$ process leads to endo products because the ketene molecule (which for steric reasons would approach with its smaller group directed toward the alkene) must twist as shown in Fig. 15.13 (L = larger; S = smaller group) in order for the + lobes to interact and this swings the larger group into the endo position.[2215] The experimental results in which the amount of endo isomer increases with the increasing size of the R group would seem to be contrary to what would be expected from

[2210]For other evidence, see Baldwin, J.E.; Kapecki, J.A. *J. Am. Chem. Soc.* **1970**, *92*, 4874; Brook, P.R.; Griffiths, J.G. *Chem. Commun.* **1970**, 1344; Egger, K.W. *Int. J. Chem. Kinet.* **1973**, *5*, 285; Moon, S.; Kolesar, T.F. *J. Org. Chem.* **1974**, *39*, 995; Isaacs, N.S.; Hatcher, B.G. *J. Chem. Soc., Chem. Commun.* **1974**, 593; Hassner, A.; Cory, R.M.; Sartoris, N. *J. Am. Chem. Soc.* **1976**, *98*, 7698; Gheorghiu, M.D.; Pârvulescu, L.; Drǎghici, C.; Elian, M. *Tetrahedron* **1981**, *37* Suppl., 143. See, however, Holder, R.W.; Graf, N.A.; Duesler, E.; Moss, J.C. *J. Am. Chem. Soc.* **1983**, *105*, 2929.

[2211]On the other hand, molecular-orbital calculations predict that the cycloaddition of ketenes to alkenes does not take place by a $[_\pi 2_s + _\pi 2_a]$ mechanism: Wang, X.; Houk, K.N. *J. Am. Chem. Soc.* **1990**, *112*, 1754; Bernardi, F.; Bottoni, A.; Robb, M.A.; Venturini, A. *J. Am. Chem. Soc.* **1990**, *112*, 2106; Valentí, E.; Pericàs, M.A.; Moyano, A. *J. Org. Chem.* **1990**, *55*, 3582.

[2212]Huisgen, R.; Feiler, L.A.; Binsch, G. *Angew. Chem. Int. Ed.* **1964**, *3*, 753; *Chem. Ber.* **1969**, *102*, 3460; Martin, J.C.; Goodlett, V.W.; Burpitt, R.D. *J. Org. Chem.* **1965**, *30*, 4309; Montaigne, R.; Ghosez, L. *Angew. Chem. Int. Ed.* **1968**, *7*, 221 Bertrand, M.; Gras, J.L.; Goré, J. *Tetrahedron* **1975**, *31*, 857; Marchand-Brynaert, J.; Ghosez, L. *J. Am. Chem. Soc.* **1972**, *94*, 2870; Huisgen, R.; Mayr, H. *Tetrahedron Lett.* **1975**, 2965, 2969.

[2213]Brady, W.T.; Hoff, E.F.; Roe, Jr., R.; Parry III, F.H. *J. Am. Chem. Soc.* **1969**, *91*, 5679; Rey, M.; Roberts, S.; Dieffenbacher, A.; Dreiding, A.S. *Helv. Chim. Acta* **1970**, *53*, 417. See also, Brady, W.T.; Parry III, F.H.; Stockton, J.D. *J. Org. Chem.* **1971**, *36*, 1486; DoMinh, T.; Strausz, O.P. *J. Am. Chem. Soc.* **1970**, *92*, 1766; Isaacs, N.S.; Stanbury, P. *Chem. Commun.* **1970**, 1061; Brook, P.R.; Harrison, J.M.; Duke, A.J. *Chem. Commun.* **1970**, 589; Dehmlow, E.V. *Tetrahedron Lett.* **1973**, 2573; Rey, M.; Roberts, S.M.; Dreiding, A.S.; Roussel, A.; Vanlierde, H.; Toppet, S.; Ghosez, L. *Helv. Chim. Acta* **1982**, *65*, 703.

[2214]Brady, W.T.; Roe Jr., R. *J. Am. Chem. Soc.* **1970**, *92*, 4618.

[2215]Brook, P.R.; Harrison, J.M.; Duke, A.J. *Chem. Commun.* **1970**, 589

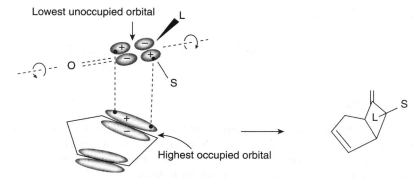

Fig. 15.13. Orbital overlap in the reaction of a ketene with cyclopentadiene. S and L stand for small and large.

considerations of steric hindrance (we may call them *masochistic steric effects*), but they are just what is predicted for a $[_\pi 2_s + _\pi 2_a]$-reaction. (*3*) There is only moderate polar solvent acceleration.[2216] (*4*) The rate of the reaction is not very sensitive to the presence of electron-withdrawing or electron-donating substituents.[2217] Because cycloadditions involving allenes are often stereospecific, it has been suggested that these also take place by the $[_\pi 2_s + _\pi 2_a]$ mechanism,[2218] but the evidence in these cases is more consistent with the diradical mechanism *b*.[2219]

The diradical mechanism *b* is most prominent in the reactions involving fluorinated alkenes.[2220] These reactions are generally not stereospecific[2221] and are insensitive to solvent effects. Further evidence that a diion is not involved is that head-to-head coupling is found when an unsymmetrical molecule is dimerized. Thus dimerization of $F_2C=CFCl$ gives **211**, not **212**. If one pair of electrons moved before the other, the positive end of one molecule would be expected to attack the

[2216]Brady, W.T.; O'Neal, H.R. *J. Org. Chem.* **1967**, *32*, 612; Huisgen, R.; Feiler, L.A.; Otto, P. *Tetrahedron Lett.* **1968**, 4485, *Chem. Ber.* **1969**, *102*, 3444; Sterk, H. *Z. NaturForsch. Teil B* **1972**, *27*, 143.

[2217]Baldwin, J.E.; Kapecki, J.A. *J. Am. Chem. Soc.* **1970**, *92*, 4868; Isaacs, N.S.; Stanbury, P. *J. Chem. Soc. Perkin Trans. 2*, **1973**, 166.

[2218]For example, see Kiefer, E.F.; Okamura, M.Y. *J. Am. Chem. Soc.* **1968**, *90*, 4187; Baldwin, J.E.; Roy, U.V. *Chem. Commun.* **1969**, 1225; Moore, W.R.; Bach, R.D.; Ozretich, T.M. *J. Am. Chem. Soc.* **1969**, *91*, 5918.

[2219]Muscio Jr., O.J.; Jacobs, T.L. *Tetrahedron Lett.* **1969**, 2867; Taylor, D.R.; Warburton, M.R.; Wright, D.B. *J. Chem. Soc. C* **1971**, 385; Dai, S.; Dolbier Jr., W.R. *J. Am. Chem. Soc.* **1972**, *94*, 3946; Duncan, W.G.; Weyler Jr., W.; Moore, H.W. *Tetrahedron Lett.* **1973**, 4391; Grimme, W.; Rother, H. *Angew. Chem. Int. Ed.* **1973**, *12*, 505; Levek, T.J.; Kiefer, E.F. *J. Am. Chem. Soc.* **1976**, *98*, 1875; Pasto, D.J.; Yang, S.H. *J. Org. Chem.* **1986**, *51*, 1676; Dolbier, D.W.; Seabury, M. *Tetrahedron Lett.* **1987**, *28*, 1491; *J. Am. Chem. Soc.* **1987**, *109*, 4393; Dolbier Jr., W.R.; Weaver, S.L. *J. Org. Chem.* **1990**, *55*, 711; Becker, D.; Denekamp, C.; Haddad, N. *Tetrahedron Lett.* **1992**, *33*, 827.

[2220]It has been argued that the mechanism here is not the diradical mechanism, but the $[_\pi 2_s + _\pi 2_a]$ mechanism: Roberts, D.W. *Tetrahedron* **1985**, *41*, 5529.

[2221]Bartlett, P.D.; Hummel, K.; Elliott, S.P.; Minns, R.A. *J. Am. Chem. Soc.* **1972**, *94*, 2898.

negative end of the other.[2222]

The diion mechanism[2223] c has been reported for at least some of the reactions[2224] in categories 3 and 4,[2225] as well as some ketene dimerizations.[2226] For example, the rate of the reaction between 1,2-bis(trifluoromethyl)-1,2-dicyanoethene and ethyl vinyl ether was strongly influenced by changes in solvent polarity.[2227] Some of these reactions are nonstereospecific, but others are stereospecific.[2228] As previously indicated, it is likely that in the latter cases the diionic intermediate closes before rotation can take place. Such rapid ring closure is more likely for a diion than for a diradical because of the attraction between the opposite charges. Other evidence for the diion mechanism in these cases is that reaction rates are greatly dependent on the presence of electron-donating and electron-withdrawing groups and that it is possible to trap the diionic intermediates.

Whether a given alkene reacts by the diradical or diion mechanism depends, among other things, on the groups attached to it. For example, phenyl and vinyl groups at the α positions of **207** or **208** help to stabilize a diradical, while donors, such as oxygen and nitrogen, favor a diion (they stabilize the positively charged end).[2229] A table on p. 451 of Ref. 2230 shows which mechanism is more likely for [2 + 2]-cycloadditions of various pairs of alkenes.

Thermal cleavage of cyclobutanes[2230] to give two alkene molecules (*cycloreversion*,[2231] the reverse of [2 + 2]-cycloaddition) operates by the diradical mechanism,

---

[2222]For additional evidence based on radical stabilities, see Silversmith, E.F.; Kitahara, Y.; Caserio, M.C.; Roberts, J.D. *J. Am. Chem. Soc.* **1958**, *80*, 5840; Bartlett, P.D.; Montgomery, L.K.; Seidel, B. *J. Am. Chem. Soc.* **1964**, *86*, 616; De Cock, C.; Piettre. S.; Lahousse, F.; Janousek, Z.; Merényi, R.; Viehe, H.G. *Tetrahedron* **1985**, *41*, 4183; Doering, W. von E.; Guyton, C.A. *J. Am. Chem. Soc.* **1978**, *100*, 3229.

[2223]For reviews of this mechanism, see Huisgen, R. *Acc. Chem. Res.* **1977**, *10*, 117, 199; Huisgen, R.; Schug, R.; Steiner, G. *Bull. Soc. Chim. Fr.* **1976**, 1813.

[2224]For a review of cycloadditions with polar intermediates, see Gompper, R. *Angew. Chem. Int. Ed.* **1969**, *8*, 312.

[2225]The reactions of ketenes with enamines are apparently not concerted, but take place by the diionic mechanism: Otto, P.; Feiler, L.A.; Huisgen, R. *Angew. Chem. Int. Ed.* **1968**, *7*, 737.

[2226]See Moore, H.W.; Wilbur, D.S. *J. Am. Chem. Soc.* **1978**, *100*, 6523.

[2227]Proskow, S.; Simmons, H.E.; Cairns, T.L. *J. Am. Chem. Soc.* **1966**, *88*, 5254. See also, Huisgen, R. *Pure Appl. Chem.* **1980**, *52*, 2283.

[2228]Proskow, S.; Simmons, H.E.; Cairns, T.L. *J. Am. Chem. Soc.* **1966**, *88*, 5254; Huisgen, R.; Steiner, G. *J. Am. Chem. Soc.* **1973**, *95*, 5054, 5055.

[2229]Hall, Jr., H.K. *Angew. Chem. Int. Ed.* **1983**, *22*, 440.

[2230]See Frey, H.M. *Adv. Phys. Org. Chem.* **1966**, *4*, 147, see pp. 170–175, 180–183.

[2231]For reviews of [2 + 2]-cycloreversions, see Schaumann, E.; Ketcham, R. *Angew. Chem. Int. Ed.* **1982**, *21*, 225; Brown, R.F.C. *Pyrolytic Methods in Organic Chemistry*, Academic Press, NY, **1980**, pp. 247–259. See also, Reddy, G.D.; Wiest, O.; Hudlický, T.; Schapiro, V.; Gonzalez, D. *J. Org. Chem.* **1999**, *64*, 2860.

and the $[_\sigma2_s + _\sigma2_a]$ pathway has not been found[2232] (the subscripts σ indicate that σ bonds are involved in this reaction).

In some cases, double bonds add to triple bonds to give cyclobutenes, apparently at about the same rate that they add to double bonds. The addition of triple bonds to triple bonds would give cyclobutadienes, and this has not been observed, except where these rearrange before they can be isolated (see **15-65**)[2233] or in the presence of a suitable coordination compound, so that the cyclobutadiene is produced in the form of a complex (p. 76).[2234]

Although thermal $[2 + 2]$-cycloaddition reactions are essentially limited to the cases described above, many (although by no means all) double-bond compounds undergo such reactions *when photochemically excited* (either directly or by a photosensitizer, see p. 340), even if they are not in the above categories.[2235] Simple alkenes absorb in the far UV (p. 332), which is difficult to reach experimentally, although this problem can sometimes be overcome by the use of suitable photosensitizers. The reaction has been applied to simple alkenes[2236] (especially to strained compounds, such as cyclopropenes and cyclobutenes), but more often the double-bond compounds involved are conjugated dienes,[2237] α,β-unsaturated ketones,[2238]

[2232]See, for example, Cocks, A.T.; Frey, H.M.; Stevens, I.D.R. *Chem. Commun.* **1969**, 458; Srinivasan, R.; Hsu, J.N.C. *J. Chem. Soc., Chem. Commun.* **1972**, 1213; Paquette, L.A.; Carmody, M.J. *J. Am. Chem. Soc.* **1976**, *98*, 8175. See however Cant, P.A.E.; Coxon, J.M.; Hartshorn, M.P. *Aust. J. Chem.* **1975**, *28*, 391; Doering, W. von E.; Roth, W.R.; Breuckmann, R.; Figge, L.; Lennartz, H.; Fessner, W.; Prinzbach, H. *Chem. Ber.* **1988**, *121*, 1.
[2233]For a review of these cases, and of cycloadditions of triple to double bonds, see Fuks, R.; Viehe, H.G., in Viehe, H.G. *Acetylenes*, Marcel Dekker, NY, **1969**, pp. 435–442.
[2234]D'Angelo, J.; Ficini, J.; Martinon, S.; Riche, C.; Sevin, A. *J. Organomet. Chem.* **1979**, *177*, 265. For a review, see Hogeveen, H.; Kok, D.M., in Patai, S.; Rappoport, Z. *The Chemistry of Functional Groups, Supplement C*, pt. 2, Wiley, NY, **1983**, pp. 981–1013.
[2235]For reviews, see Demuth, M.; Mikhail, G. *Synthesis* **1989**, 145; Ninomiya, I.; Naito, T. *Photochemical Synthesis*, Academic Press, NY, **1989**, pp. 58–109; Ramamurthy, V.; Venkatesan, K. *Chem. Rev.* **1987**, *87*, 433; Lewis, F.D. *Adv. Photochem.* **1986**, *13*, 165; Wender, P.A., in Coyle, J.D. *Photochemistry in Organic Synthesis*, Royal Society of Chemistry, London, **1986**, pp. 163–188; Schreiber, S.L. *Science*, **1985**, *227*, 857; Neckers, D.C.; Tinnemans, A.H.A., in Horspool, W.M. *Synthetic Organic Photochemistry*, Plenum, NY, **1984**, pp. 285–311; Baldwin, S.W. *Org. Photochem.* **1981**, *5*, 123; Turro, N.J. *Modern Molecular Photochemistry*, W.A. Benjamin, NY, **1978**, pp. 417–425, 458–465; Kricka, L.J.; Ledwith, A. *Synthesis* **1974**, 539; Herndon, W.C. *Top. Curr. Chem.* **1974**, *46*, 141; Sammes, P.G. *Q. Rev. Chem. Soc.* **1970**, *24*, 37, 46–55; Crowley, K.J.; Mazzocchi, P.H., in Zabicky, J. *The Chemistry of Alkenes*, Vol. 2, Wiley, NY, **1970**, 297–316; Turro, N.J.; Dalton, J.C.; Weiss, D.S. *Org. Photochem.* **1969**, *2*, 1; Trecker, D.J. *Org. Photochem.* **1969**, *2*, 63; Scharf, H. *Fortschr. Chem. Forsch.* **1969**, *11*, 216; Steinmetz, R. *Fortschr. Chem. Forsch.* **1967**, *7*, 445; Fonken, G.J. *Org. Photochem.* **1967**, *1*, 197; Chapman, O.L.; Lenz, G. *Org. Photochem.* **1967**, *1*, 283; Schönberg, A. *Preparative Organic Photochemistry*, Springer, NY, **1968**, pp. 70–96, 109–117; Warrener, R.N.; Bremner, J.B. *Rev. Pure Appl. Chem.* **1966**, *16*, 117, 122–128.
[2236]For examples of nonphotosensitized dimerization of simple alkenes, see Arnold, D.R.; Abraitys, V.Y. *Chem. Commun.* **1967**, 1053; Yamazaki, H.; Cvetanović, R.J. *J. Am. Chem. Soc.* **1969**, *91*, 520.
[2237]For a review, see Dilling, W.L. *Chem. Rev.* **1969**, *69*, 845.
[2238]For reviews of various aspects of this subject, see Cossy, J.; Carrupt, P.; Vogel, P., in Patai, S. *Supplement A: The Chemistry of Double-Bonded Functional Groups*, Vol. 2, pt. 2, Wiley, NY, **1989**, pp. 1369–1565; Kemernitskii, A.V.; Ignatov, V.N.; Levina, I.S. *Russ. Chem. Rev.* **1988**, *57*, 270; Weedon, A.C., in Horspool, W.M. *Synthetic Organic Photochemistry*, Plenum, NY, **1984**, pp. 61–143; Lenz, G.R. *Rev. Chem. Intermed.* **1981**, *4*, 369; Margaretha, P. *Chimia*, **1975**, *29*, 203; Bauslaugh, P.G. *Synthesis* **1970**, 287; Eaton, P.E. *Acc. Chem. Res.* **1968**, *1*, 50; Schuster, D.I.; Lem, G.; Kaprinidis, N.A. *Chem. Rev.* **1993**, *93*, 3; Erickson, J.A.; Kahn, S.D. *Tetrahedron* **1993**, *49*, 9699.

acids, or acid derivatives, or quinones, since these compounds, because they are conjugated, absorb at longer wavelengths (p. 332). Both dimerizations and mixed additions are common, some examples being (see also, the example on p. 347):

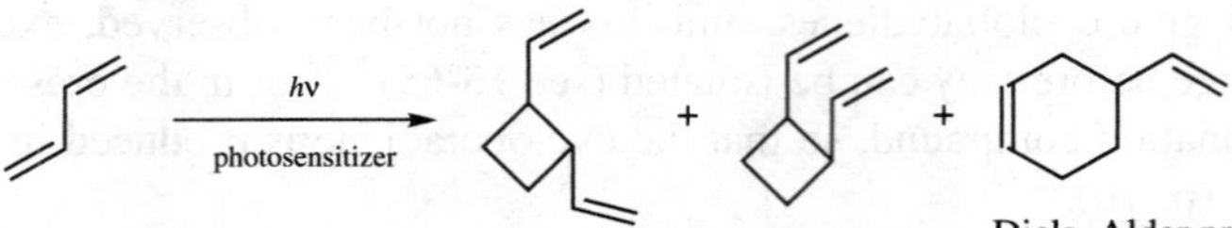

Diels–Alder product

Ref. [2239]

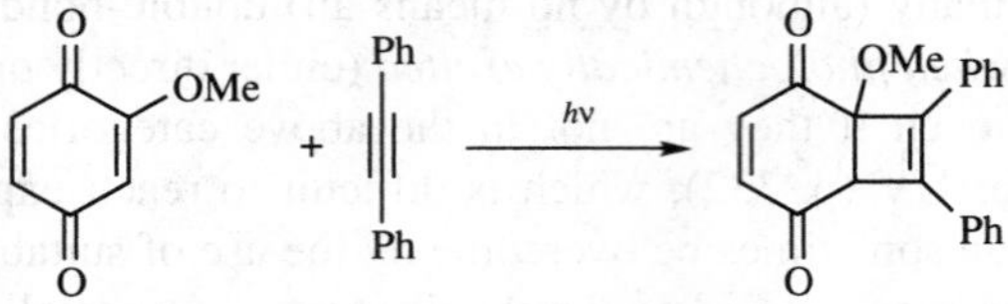

Ref. [2240]

Photochemical [2 + 2]-cycloadditions can also take place intramolecularly if a molecule has two double bonds that are properly oriented.[2241] The cyclization of the quinone dimer shown above is one example. Other examples are

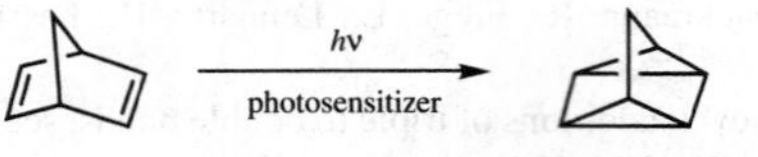

Quadricyclane

Ref. [2242]

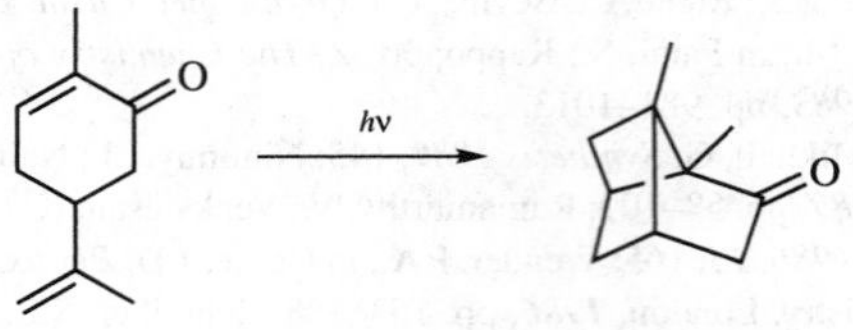

Carvone   Carvonecamphor

Ref. [2243]

It is obvious that many molecules can be constructed in this way that would be difficult to make by other procedures. However, attempted cyclizations of this kind are not always successful. In many cases, polymeric or other side products are

[2239]Liu, R.S.H.; Turro, N.J.; Hammond, G.S. *J. Am. Chem. Soc.* **1965**, *87*, 3406; Cundall, R.B.; Griffiths, P.A. *Trans. Faraday Soc.* **1965**, *61*, 1968; DeBoer, C.D.; Turro, N.J.; Hammond, G.S. *Org. Synth. V*, 528.

[2240]Pappas, S.P.; Pappas, B.C. *Tetrahedron Lett.* **1967**, 1597.

[2241]For reviews, see Becker, D.; Haddad, N. *Org. Photochem.* **1989**, *10*, 1–162; Crimmins, M.T. *Chem. Rev.* **1988**, *88*, 1453; Oppolzer, W. *Acc. Chem. Res.* **1982**, *15*, 135; Prinzbach, H. *Pure Appl. Chem.* **1968**, *16*, 17; Dilling, W.L. *Chem. Rev.* **1966**, *66*, 373.

[2242]Hammond, G.S.; Turro, N.J.; Fischer, A. *J. Am. Chem. Soc.* **1961**, *83*, 4674; Dauben, W.G.; Cargill, R.L. *Tetrahedron* **1961**, *15*, 197. See also, Cristol, S.J.; Snell, R.L. *J. Am. Chem. Soc.* **1958**, *80*, 1950.

[2243]Ciamician, G.; Silber, P. *Ber.* **1908**, *41*, 1928; Büchi, G.; Goldman, I.M. *J. Am. Chem. Soc.* **1957**, *79*, 4741.

obtained instead of the desired product.

213

The photochemical cycloaddition of a carbonyl, generally from an aldehyde or ketone, and an alkene is called the *Paternò–Büchi reaction.*[2244] This [2 + 2]-cycloaddition gives an oxetane (**213**) and the reaction is believed to proceed via a diradical intermediate. Silyl enol ethers react with aldehydes under nonphotochemical conditions using $ZnCl_2$ at 25°C or $SnCl_4$ at −78°C.[2245]

It is possible that some of these photochemical cycloadditions take place by a $[_\pi 2_s + _\pi 2_s]$ mechanism (which is of course allowed by orbital symmetry); when and if they do, one of the molecules must be in the excited singlet state (S_1) and the other in the ground state.[2246] The nonphotosensitized dimerizations of *cis-* and *trans*-2-butene are stereospecific,[2247] making it likely that the $[_\pi 2_s + _\pi 2_s]$ mechanism is operating in these reactions. However, in most cases it is a triplet excited state that reacts with the ground-state molecule; in these cases the diradical (or in certain cases, the diionic) mechanism is taking place.[2248] In one intramolecular case, the intermediate diradical has been trapped.[2249] Photosensitized [2π + 2π]-cycloadditions almost always involve the triplet state, and hence a diradical (or diionic) mechanism.

The photochemical diradical mechanism is not quite the same as the thermal diradical mechanism. In the thermal mechanism the initially formed diradical must be a singlet, but in the photochemical process a triplet excited state is adding to a ground state (which is of course a singlet). Thus, in order to conserve spin,[2250] the initially formed diradical must be a triplet; that is, the two electrons must have the same spin. Consequently, the second, or ring-closing, step of the mechanism cannot take place at once, because a new bond cannot form from a combination of two electrons with the same spin, and the diradical has a reasonably long lifetime before collisions with molecules in the environment allow a spin inversion to take

[2244]Paternò, E.; Chieffi, C. *Gazz. Chim. Ital.* **1909**, *39*, 341; Büchi, G.; Inman, C.G.; Lipinsky, E.S. *J. Am. Chem. Soc.* **1954**, *76*, 4327. See García-Expósito, E.; Bearpark, M.J.; Ortuño, R.M.; Robb, M.A.; Branchadell, V. *J. Org. Chem.* **2002**, *67*, 6070.

[2245]Wang, Y.; Zhao, C.; Romo, D. *Org. Lett.* **1999**, *1*, 1197.

[2246]We have previously seen (p. $$$) that reactions between two excited molecules are extremely rare.

[2247]Yamazaki, H.; Cvetanović, R.J. *J. Am. Chem. Soc.* **1969**, *91*, 520; Yamazaki, H.; Cvetanović, R.J.; Irwin, R.S. *J. Am. Chem. Soc.* **1976**, *98*, 2198. For other likely examples, see Lewis, F.D.; Hoyle, C.E.; Johnson, D.E. *J. Am. Chem. Soc.* **1975**, *97*, 3267; Lewis, F.D.; Kojima, M. *J. Am. Chem. Soc.* **1988**, *110*, 8660.

[2248]Maradyn, D.J.; Weedon, A.C. *Tetrahedron Lett.* **1994**, *35*, 8107.

[2249]Becker, D.; Haddad, N.; Sahali, Y. *Tetrahedron Lett.* **1989**, *30*, 2661.

[2250]This is an example of the Wigner spin conservation rule (p. 340). Note that spin conservation is something entirely different from symmetry conservation.

place and the diradical to cyclize. We would therefore predict nonstereospecificity, and that is what is found.[2251] It has been believed that at least some [2 + 2]-photo-cycloadditions take place by way of exciplex intermediates[2252] [an *exciplex*[2253] is an excited EDA complex (p. 342) that is dissociated in the ground state; in this case one double bond is the donor and the other the acceptor], but there is evidence against this.[2254]

In **15-60**, we used the principle of conservation of orbital symmetry to explain why certain reactions take place readily and others do not. The orbital-symmetry principle can also explain why certain molecules are stable although highly strained. For example, quadricyclane and hexamethylprismane[2255] are thermodynamically much less stable (because much more strained) than their corresponding isomeric dienes, norbornadiene and

| Quadricyclane | Norbornadiene | Hexamethyl-prismane | **214** (A Dewar benzene) |

hexamethylbicyclo[2.2.0]hexadiene (**214**).[2256] Yet the former two compounds can be kept indefinitely at room temperature, although in the absence of orbital-symmetry considerations it is not easy to understand why the electrons simply do not move over to give the more stable diene isomers. The reason is that both these reactions involve the conversion of a cyclobutane ring to a pair of double bonds (a $_s2 + _s2$ process) and, as we have seen, a thermal process of this sort is forbidden by the Woodward–Hoffmann rules. The process is allowed photochemically, and we are not surprised to find that both quadricyclane and hexamethylprismane are photochemically converted to the respective dienes at room temperature or below.[2257] It is also possible to conceive of simple

---

[2251]See, for example, Liu, R.S.H.; Hammond, G.S. *J. Am. Chem. Soc.* **1967**, *89*, 4936; Kramer, B.D.; Bartlett, P.D. *J. Am. Chem. Soc.* **1972**, *94*, 3934.

[2252]See, for example, Farid, S.; Doty, J.C.; Williams, J.L.R. *J. Chem. Soc., Chem. Commun.* **1972**, 711; Mizuno, K.; Pac, C.; Sakurai, H. *J. Am. Chem. Soc.* **1974**, *96*, 2993; Caldwell, R.A.; Creed, D. *Acc. Chem. Res.* **1980**, *13*, 45; Mattes, S.L.; Farid, S. *Acc. Chem. Res.* **1982**, *15*, 80; Swapna, G.V.T.; Lakshmi, A.B.; Rao, J.M.; Kunwar, A.C. *Tetrahedron* **1989**, *45*, 1777.

[2253]For a review of exciplexes, see Davidson, R.S. *Adv. Phys. Org. Chem.* **1983**, *19*, 1–130.

[2254]Schuster, D.I.; Heibel, G.E.; Brown, P.B.; Turro, N.J.; Kumar, C.V. *J. Am. Chem. Soc.* **1988**, *110*, 8261.

[2255]This compound can be prepared by photolysis of **210**, another example of an intramolecular photochemical [2 + 2]-cycloaddition: Lemal, D.M.; Lokensgard, J.P. *J. Am. Chem. Soc.* **1966**, *88*, 5934; Schäfer, W.; Criegee, R.; Askani, R.; Grüner, H. *Angew. Chem. Int. Ed.* **1967**, *6*, 78.

[2256]For a review of this compound, see Schäfer, W.; Hellmann, H. *Angew. Chem. Int. Ed.* **1967**, *6*, 518.

[2257]These conversions can also be carried out by the use of transition-metal catalysts: Hogeveen, H.; Volger, H.C. *Chem. Commun.* **1967**, 1133; *J. Am. Chem. Soc.* **1967**, *89*, 2486; Kaiser, K.L.; Childs, R.F.; Maitlis, P.M. *J. Am. Chem. Soc.* **1971**, *93*, 1270; Landis, M.E.; Gremaud, D.; Patrick, T.B. *Tetrahedron Lett.* **1982**, *23*, 375; Maruyama, K.; Tamiaki, H. *Chem. Lett.* **1987**, 683.

bond rearrangements whereby hexamethylprismane is converted to hexamethyl-benzene, which

of course is far more stable than either hexamethylprismane or **214**. It has been calculated that hexamethylbenzene is at least $90\,\text{kcal mol}^{-1}$ ($380\,\text{kJ mol}^{-1}$) more stable than hexamethylprismane. The fact that hexamethylprismane does not spontaneously undergo this reaction has prompted the observation[2258] that the prismane has "the aspect of an angry tiger unable to break out of a paper cage." However, a correlation diagram for this reaction[2259] discloses that it too is a symmetry-forbidden process. All three of these "forbidden" reactions do take place when the compounds are heated, but the diradical mechanism is likely under these conditions.[2259]

Bicyclo[2.2.0]hexadienes and prismanes are *valence isomers* of benzenes.[2260] These compounds actually have the structures that were proposed for benzenes in the nineteenth century. Prismanes have the Ladenburg formula, and bicyclo[2.2.0]hexadienes have the Dewar formula. Because of this bicyclo[2.2.0]hexadiene is often called Dewar benzene. On p. 32, it was mentioned that Dewar formulas are canonical forms (although not very important) of benzenes. Yet they also exist as separate compounds in which the positions of the nuclei are different from those of benzenes.

OS **V**, 54, 235, 277, 297, 370, 393, 424, 459, 528; **VI**, 378, 571, 962, 1002, 1024, 1037; **VII**, 177, 256, 315; **VIII**, 82, 116, 306, 377; **IX**, 28, 275; **80**, 160. For the reverse reaction, see OS **V**, 734.

15-64   The Addition of Carbenes and Carbenoids to Double and Triple Bonds

epi-**Methylene-addition**

[2258]Woodward, R.B.; Hoffmann, R. *The Conservation of Orbital Symmetry* Acaademic Press, NY, *1970*, pp. 107–112.

[2259]See, for example, Oth, J.F.M. *Recl. Trav. Chim. Pays-Bas 1968*, 87, 1185.

[2260]For reviews of valence isomers of benzene, see Kobayashi, Y.; Kumadaki, I. *Adv. Heterocycl. Chem. 1982*, *31*, 169; *Acc. Chem. Res. 1981*, *14*, 76; van Tamelen, E.E. *Acc. Chem. Res. 1972*, *5*, 186; *Angew. Chem. Int. Ed. 1965*, *4*, 738; Bolesov, I.G. *Russ. Chem. Rev. 1968*, *37*, 666; Viehe, H.G. *Angew. Chem. Int. Ed. 1965*, *4*, 746; Schäfer, W.; Hellmann, H. *Angew. Chem. Int. Ed. 1967*, *6*, 518.

Carbenes and substituted carbenes add to double bonds to give cyclopropane derivatives by what can be considered as a formal [1 + 2]-cycloaddition.[2261] Many carbene derivatives, for example, PhCH, ROCH,[2262] Me$_2$C=C, C(CN)$_2$, have been added to double bonds, but the reaction is often performed with CH$_2$ itself, with halo and dihalocarbenes,[2263] and with carbalkoxycarbenes[2264] (generated from diazoacetic esters). Alkylcarbenes HCR have been added to alkenes,[2265] but more often these rearrange to give alkenes (p. 291). The carbene can be generated in any of the ways normally used (p. 287). However, most reactions in which a cyclopropane is formed by treatment of an alkene with a carbene "precursor" do not actually involve free carbene intermediates. In some cases, it is certain that free carbenes are not involved, and in other cases there is doubt. Because of this, the term *carbene transfer* is often used to cover all reactions in which a double bond is converted to a cyclopropane, whether a carbene or a carbenoid (p. 288) is actually involved.

Carbene itself (:CH$_2$) is extremely reactive and gives many side reactions, especially insertion reactions (**12-21**), which greatly reduce yields. This competition is also true with rhodium-catalyzed diazoalkane cyclopropanations[2266] (see below). When it is desired to add :CH$_2$ for preparative purposes, free carbene is not used, but the Simmons–Smith procedure (p. 1241) or some other method that does not involve free carbenes is employed instead. Halocarbenes are less active than carbenes, and this reaction proceeds quite well, since insertion reactions do not interfere.[2267] The absolute rate constant for addition of selected alkoxychlorocarbene to butenes has been measured to range from 330 to $1 \times 10^4\, M^{-1}\, s^{-1}$.[2268] A few of the many ways[2269] in which halocarbenes or carbenoids are generated for

[2261]For reviews, see, in Rappoport, Z. *The Chemistry of the Cyclopropyl Group*, Wiley, NY, *1987*, the reviews by Tsuji, T.; Nishida, S., pt. 1, pp. 307–373; Verhé, R.; De Kimpe, N. pt. 1, pp. 445–564; Marchand, A.P., in Patai, S. *Supplement A: The Chemistry of Double-Bonded Functional Groups*, pt. 1, Wiley, NY, *1977*, pp. 534–607, 625–635; Bethell, D., in McManus, S.P. *Organic Reactive Intermediates*; Academic Press, NY, *1973*, pp. 101–113; in Patai, S. *The Chemistry of Alkenes*, Vol. 1, Wiley, NY, *1964*, the articles by Cadogan, J.I.G.; Perkins, M.J. pp. 633–671; Huisgen, R.; Grashey, R.; Sauer, J. pp. 755–776; Kirmse, W. *Carbene Chemistry* 2nd ed.; Academic Press, NY, *1971*, pp. 85–122, 267–406. For a review of certain intramolecular additions, see Burke, S.D.; Grieco, P.A. *Org. React. 1979*, 26, 361. For a list of reagents, with references, see Larock, R.C. *Comprehensive Organic Transformations*, 2nd ed., Wiley-VCH, NY, *1999*, pp. 135–153.

[2262]For a review, see Schöllkopf, U. *Angew. Chem. Int. Ed. 1968*, 7, 588.

[2263]For a review of the addition of halocarbenes, see Parham, W.E.; Schweizer, E.E. *Org. React. 1963*, 13, 55.

[2264]For a review, see Dave, V.; Warnhoff, E.W. *Org. React. 1970*, 18, 217.

[2265]For example see Frey, H.M. *J. Chem. Soc. 1962*, 2293.

[2266]Doyle, M.P.; Phillips, I.M. *Tetrahedron Lett. 2001*, 42, 3155. For a review, see Merlic, C.A.; Zechman, A.L. *Synthesis 2003*, 1137.

[2267]For reviews of carbene selectivity in this reaction, see Moss, R.A. *Acc. Chem. Res. 1989*, 22, 15; *1980*, 13, 58. For a review with respect to halocarbenes, see Kostikov, R.R.; Molchanov, A.P.; Khlebnikov, A.F. *Russ. Chem. Rev. 1989*, 58, 654.

[2268]Moss, R.A.; Ge, C.-S.; Wostowska, J.; Jang, E.G.; Jefferson, E.A.; Fan, H. *Tetrahedron Lett. 1995*, 36, 3083.

[2269]Much of the work in this field has been carried out by Seyferth, D. and co-workers; see, for example, Seyferth, D.; Haas, C.K. *J. Org. Chem. 1975*, 40, 1620; Seyferth, D.; Haas, C.K.; Dagani, D. *J. Organomet. Chem. 1976*, 104, 9.

this reaction are the following,[2270] most of which involve formal elimination (the first two steps of the S_N1cB mechanism, p. 521):

$$CH_2Cl_2 \ + \ RLi \longrightarrow CHCl$$

$$N_2CHBr \xrightarrow{h\nu} CHBr$$

$$CHCl_3 \ + \ {}^-OH \longrightarrow CCl_2$$

$$PhHgCCl_2Br \xrightarrow{\Delta} CCl_2 \quad \text{Ref. }^{2271}$$

$$Me_3SnCF_3 \ + \ NaI \longrightarrow CF_2 \quad \text{Ref. }^{2272}$$

$$Me_2CBr_2 \xrightarrow{\text{electrolysis}} Me_2C: \quad \text{Ref. }^{2273}$$

The reaction between $CHCl_3$ and $HO-$ is often carried out under phase transfer conditions.[2274] It has been shown that the reaction between $PhCHCl_2$ and t-BuOK produces a carbenoid, but when the reaction is run in the presence of a crown ether, the free $Ph(Cl)C:$ is formed instead.[2275] The reaction of iodoform and $CrCl_2$ leads to iodocyclopropanes upon reaction with alkenes.[2276] Dihalocyclopropanes are very useful compounds[2277] that can be reduced to cyclopropanes, treated with magnesium or sodium to give allenes (**18-3**), or converted to a number of other products.

Alkenes of all types can be converted to cyclopropane derivatives by this reaction, but difficulty may be encountered with sterically hindered ones.[2278]

[2270]A much longer list, with references, is given, in Kirmse, W. *Carbene Chemistry Carbene Chemistry* 2nd ed., Academic Press, NY, *1971*, pp. 313–319. See also, Larock, R.C. *Comprehensive Organic Transformations*, 2nd ed., Wiley-VCH, NY, *1999*, pp. 135–143.

[2271]For a review of the use of phenyl(trihalomethyl)mercury compounds as dihalocarbene or dihalocarbenoid precursors, see Seyferth, D. *Acc. Chem. Res. 1972*, 5, 65. For a review of the synthesis of cyclopropanes with the use of organomercury reagents, see Larock, R.C. *Organomercurcury Compounds in Organic Synthesis*, Springer, NY, *1985*, pp. 341–380.

[2272]For reviews of flourinated carbenes, see Seyferth, D., in Moss, R.A.; Jones, Jr., M. *Carbenes*, Vol. 2, Wiley, NY, *1975*, pp. 101–158; Sheppard, W.A.; Sharts, C.M. *Organic Fluorine Chemistry*, W. A. Benjamin, NY, *1969*, pp. 237–270.

[2273]Léonel, E.; Paugam, J.P.; Condon-Gueugnot, S.; Nédélec, Y.-Y. *Tetrahedron 1998*, 54, 3207.

[2274]For reviews of the use of phase-transfer catalysis in the addition of dihalocarbenes to C=C bonds, see Starks, C.M.; Liotta, C. *Phase Transfer Catalysis*, Academic Press, NY, *1978*, pp. 224–268; Weber, W.P.; Gokel, G.W. *Phase Transfer Catalysis in Organic Synthesis*, Springer, NY, *1977*, pp. 18–43, 58–62. For a discussion of the mechanism, see Gol'dberg, Yu.Sh.; Shimanskaya, M.V. *J. Org. Chem. USSR 1984*, 20, 1212.

[2275]Moss, R.A.; Pilkiewicz, F.G. *J. Am. Chem. Soc. 1974*, 96, 5632; Moss, R.A.; Lawrynowicz, W. *J. Org. Chem. 1984*, 49, 3828.

[2276]Takai, K.; Toshikawa, S.; Inoue, A.; Kokumai, R. *J. Am. Chem. Soc. 2003*, 125, 12990.

[2277]For reviews of dihalocyclopropanes, see Banwell, M.G.; Reum, M.E. *Adv. Strain Org. Chem. 1991*, 1, 19–64; Kostikov, R.R.; Molchanov, A.P.; Hopf, H. *Top. Curr. Chem. 1990*, 155, 41–80; Barlet, R.; Vo-Quang, Y. *Bull. Soc. Chim. Fr. 1969*, 3729–3760.

[2278]Dehmlow, E.V.; Eulenberger, A. *Liebigs Ann. Chem. 1979*, 1112.

Even tetracyanoethylene, which responds very poorly to electrophilic attack, gives cyclopropane derivatives with carbenes.[2279] Conjugated dienes give 1,2-addition to give a vinylcyclopropane.[2280] Addition of a second mole gives bicyclopropyl derivatives.[2281] 1,4-Addition is rare but has been reported in certain cases.[2282] Carbene adds to ketene to give cyclopropanone.[2283] Allenes react with carbenes to give cyclopropanes with exocyclic unsaturation:[2284]

$$=C= \quad \xrightarrow{:CH_2} \quad \triangleright\!=\!= \quad \xrightarrow{:CH_2} \quad \bowtie$$

A second equivalent gives spiropentanes. In fact, any size ring with an exocyclic double bond can be converted by a carbene to a spiro compound.[2285]

Free carbenes can also be avoided by using transition-metal–carbene complexes $L_nM{=}CRR'$ (L = a ligand, M = a metal),[2286] which add the group CRR' to double bonds.[2287] An example is the reaction of iron carbene **215**.[2288]

215

These complexes can be isolated in some cases; in others they are generated *in situ* from appropriate precursors, of which diazo compounds are among the

[2279]Cairns, T.L.; McKusick, B.C. *Angew. Chem.* **1961**, *73*, 520.

[2280]Woodworth, R.C.; Skell, P.S. *J. Am. Chem. Soc.* **1957**, *79*, 2542.

[2281]Orchin, M.; Herrick, E.C. *J. Org. Chem.* **1959**, *24*, 139; Nakhapetyan, L.A.; Safonova, I.L.; Kazanskii, B.A. *Bull. Acad. Sci. USSR Div. Chem. Sci.* **1962**, 840; Skattebøl, L. *J. Org. Chem.* **1964**, *29*, 2951.

[2282]Anastassiou, A.G.; Cellura, R.P.; Ciganek, E. *Tetrahedron Lett.* **1970**, 5267; Jefford, C.W.; Mareda, J.; Gehret, J-C.E.; Kabengele, T.; Graham, W.D.; Burger, U. *J. Am. Chem. Soc.* **1976**, *98*, 2585; Mayr, H.; Heigl, U.W. *Angew. Chem. Int. Ed.* **1985**, *24*, 579; Le, N.A.; Jones, Jr., M.; Bickelhaupt, F.; de Wolf, W.H. *J. Am. Chem. Soc.* **1989**, *111*, 8491; Kraakman, P.A.; de Wolf, W.H.; Bickelhaupt, F. *J. Am. Chem. Soc.* **1989**, *111*, 8534; Hudlický, T.; Seoane, G.; Price, J.D.; Gadamasetti, K.G. *Synlett* **1990**, 433; Lambert, J.B.; Ziemnicka-Merchant, B.T. *J. Org. Chem.* **1990**, *55*, 3460.

[2283]Turro, N.J.; Hammond, W.B. *Tetrahedron* **1968**, *24*, 6017; Rothgery, E.F.; Holt, R.J.; McGee, Jr., H.A. *J. Am. Chem. Soc.* **1975**, *97*, 4971. For a review of cyclopropanones, see Wasserman, H.H.; Berdahl, D.R.; Lu, T., in Rappoport, Z. *The Chemistry of the Cyclopropyl Group*, Wiley, NY, **1987**, pt. 2, pp. 1455–1532.

[2284]For reviews of the addition of carbenes and carbenoids to allenes, see Landor, S.R., in Landor, S.R. *The Chemistry of Allenes*, Vol. 2, Academic Press, NY, **1982**, pp. 351–360; Bertrand, M. *Bull. Soc. Chim. Fr.* **1968**, 3044–3054. For a review of the synthetic uses of methylenecyclopropanes and cyclopropenes, see Binger, P.; Büch, H.M. *Top. Curr. Chem.* **1987**, *135*, 77.

[2285]For a review of the preparation of spiro compounds by this reaction, see Krapcho, A.P. *Synthesis* **1978**, 77–126.

[2286]Doyle, M.P.; McKervey, M.A.; Ye, T. *Modern Catalytic Methods for Organic Synthesis with Diazo Compounds*, Wiley, NY, **1998**.

[2287]For reviews, see Helquist, P. *Adv. Met.-Org. Chem.* **1991**, *2*, 143; Brookhart, M.; Studabaker, W.B. *Chem. Rev.* **1987**, *87*, 411; Syatkovskii, A.I.; Babitskii, B.D. *Russ. Chem. Rev.* **1984**, *53*, 672.

[2288]Brookhart, M.; Tucker, J.R.; Husk, G.R. *J. Am. Chem. Soc.* **1983**, *105*, 258.

most important. Chromium complexes have been used for the cyclopropanation of alkenes.[2289]

Polymer-supported benzenesulfonyl azides have been developed as a safe diazotransfer reagent.[2290] These compounds, including CH_2N_2 and other diazoalkanes, react with metals or metal salts (copper, palladium,[2291] and rhodium are most commonly used) to give the carbene complexes that add: CRR' to double bonds.[2292] Diazoketones and diazoesters with alkenes to give the cyclopropane derivative, usually with a transition-metal catalyst, such as a copper complex.[2293] The ruthenium catalyst reaction of diazoesters with an alkyne give a cyclopropene.[2294] An X-ray structure of an osmium catalyst intermediate has been determined.[2295] Electron-rich alkenes react faster than simple alkenes.[2296]

Optically active complexes have been used for enantioselective cyclopropane synthesis.[2297] Decomposition of diazoalkanes in the presence of chiral rhodium[2298] copper,[2299] or ruthenium[2300] complexes leads to optically active cyclopropanes.

[2289]Barluenga, J.; Aznar, F.; Gutiérrez, I.; García-Granda, S. Llorca-Baragaño, M.A. *Org. Lett.* **2002**, *4*, 4233.

[2290]Green, G.M.; Peet, N.P.; Metz, W.A. *J. Org. Chem.* **2001**, *66*, 2509.

[2291]For a discussion of the mechanism of the palladium-catalyzed reaction, see Rodríguez-García, C.; Oliva, A.; Ortuño, R.M.; Branchadell, V. *J. Am. Chem. Soc.* **2001**, *123*, 6157.

[2292]For reviews, see Adams, J.; Spero, D.M. *Tetrahedron* **1991**, *47*, 1765; Collman, J.P., Hegedus, L.S.; Norton, J.R.; Finke, R.G. *Principles and Applications of Organotransition Metal Chemistry* University Science Books, Mill Valley, CA **1987**, pp. 800–806; Maas, G. *Top. Curr. Chem.* **1987**, *137*, 75; Doyle, M.P. *Chem. Rev.* **1986**, *86*, 919; *Acc. Chem. Res.* **1986**, *19*, 348; Heck, R.F. *Palladium Reagents in Organic Synthesis*, Academic Press, NY, **1985**, pp. 401–407; Wulfman, D.S.; Poling, B. *React. Intermed. (Plenum)* **1980**, *1*, 321; Müller, E.; Kessler, H.; Zeeh, B. *Fortschr. Chem. Forsch.* **1966**, *7*, 128.

[2293]Díaz-Requejo, M.M.; Belderraín, T.R.; Trofimenko, S.; Pérez, P.J. *J. Am. Chem. Soc.* **2001**, *123*, 3167. For a discussion of the mechanism and selectivity, see Bühl, M.; Terstegen, F.; Löffler, F.; Meynhardt, B.; Kierse, S.; Müller, M.; Näther, C.; Lüning, U. *Eur. J. Org. Chem.* **2001**, 2151.

[2294]Lou, Y.; Horikawa, M.; Kloster, R.A.; Hawryluk, N.A.; Corey, E.J. *J. Am. Chem. Soc.* **2004**, *126*, 8916.

[2295]Li, Y.; Huang, J.-S.; Zhou, Z.-Y.; Che, C.-M. *J. Am. Chem. Soc.* **2001**, *123*, 4843.

[2296]See Davies, H.M.L.; Xiang, B.; Kong, N.; Stafford, D.G. *J. Am. Chem. Soc.* **2001**, *123*, 7461.

[2297]Brookhart, M.; Liu, Y.; Goldman, E.W.; Timmers, D.A.; Williams, G.D. *J. Am. Chem. Soc.* **1991**, *113*, 927; Lowenthal, R.E.; Abiko, A.; Masamune, S. *Tetrahedron Lett.* **1990**, *31*, 6005; Evans, D.A.; Werpel, K.A.; Hinman, M.M.; Faul, M.M. *J. Am. Chem. Soc.* **1991**, *113*, 726; Ito, K.; Katsuki, T. *Tetrahedron Lett.* **1993**, *34*, 2661. For a review of enantioselective cyclopropanation using carbenoid chemistry see Singh, V.K.; DattaGupta, A.; Sekar, G. *Synthesis* **1997**, 137. For the effect of diazoalkane structure of stereoselectivity, see Davies, H.M.L.; Bruzinski, P.R.; Fall, M.J. *Tetrahedron Lett.* **1996**, *37*, 4133.

[2298]Davies, H.M.L.; Rusiniak, L. *Tetrahedron Lett.* **1998**, *39*, 8811; Haddad, N.; Galili, N. *Tetrahedron Asymmetry* **1997**, *8*, 3367; Ichiyanagi, T.; Shimizu, M.; Fujisawa, T. *Tetrahedron* **1997**, *53*, 9599; Fukuda, T.; Katsuki, T. *Tetrahedron* **1997**, *53*, 7201; Frauenkron M.; Berkessel, A. *Tetrahedron Lett.* **1997**, *38*, 7175; Doyle, M.P.; Zhou, Q.-L.; Charnsangavej, C.; Longoria, M.A.; McKervey, M.A.; Garcia, C.F. *Tetrahedron Lett.* **1996**, *37*, 4129.

[2299]Díaz-Requejo, M.M.; Caballero, A.; Belderraín, T.R.; Nicasio, M.C.; Trofimenko, S.; Pérez, P.J. *J. Am. Chem. Soc.* **2002**, *124*, 978.

[2300]Uchida, T.; Irie, R.; Katsuki, T. *Synlett* **1999**, 1163; Uchida, T.; Irie, R.; Katsuki, T. *Synlett* **1999**, 1793; Iwasa, S.; Takezawa, F.; Tuchiya, Y.; Nishiyama, H. *Chem. Commun.* **2001**, 59. For a discussion of the mechanism, see Oxgaard, J.; Goddard II, W.A. *J. Am. Chem. Soc.* **2004**, *126*, 442.

The use of chiral additives with a rhodium complex also leads to cyclopropanes enantioselectively.[2301] An important chiral rhodium species is $Rh_2(S\text{-}DOSP)_4$,[2302] which leads to cyclopropanes with excellent enantioselectivity in carbene cyclopropanation reactions.[2303] Asymmetric, intramolecular cyclopropanation reactions have been reported.[2304] The copper catalyzed diazoester cyclopropanation was reported in an ionic liquid.[2305] It is noted that the reaction of a diazoester with a chiral dirhodium catalyst leads to β-lactones with modest enantioselectivity.[2306] Phosphonate esters have been incorporated into the diazo compound.[2307]

Triple-bond compounds[2308] react with carbenes to give cyclopropenes, except that in the case of acetylene itself, the cyclopropenes first formed cannot be isolated because they rearrange to allenes.[2309] Cyclopropenones (p. 73) are obtained by hydrolysis of dihalocyclopropenes.[2310]

Most carbenes are electrophilic, and, in accord with this, electron-donating substituents on the alkene increase the rate of the reaction, and electron-withdrawing groups decrease it,[2311] although the range of relative rates is not very great.[2312] As discussed on p. 284, carbenes in the singlet state (which is the most common state) react stereospecifically and syn,[2313] probably by a one-step mechanism,[2314] similar

[2301]Aggarwal, V.K.; Smith, H.W.; Hynd, G. ; Jones, R.V.H.; Fieldhouse, R.; Spey, S.E. *J. Chem. Soc., Perkin Trans. 1* **2000**, 3267; Yao, X.; Qiu, M.; Lü, W.; Chen, H.; Zheng, Z. *Tetrahedron Asymmetry* **2001**, *12*, 197.

[2302]Doyle, M.P. *Pure Appl. Chem.* **1998**, *70* 1123; Doyle, M.P.; Protopopova, M.N. *Tetrahedron* **1998**, *54*, 7919; Martin, S.F.; Spaller, M.R.; Liras, L.; Hartman, B. *J. Am. Chem. Soc.* **1994**, *116*, 4493; Davies, H.M.L.; Hansen, T.; Churchill, M.R. *J. Am. Chem. Soc.* **2000**, *122*, 3063; Davies, H.M.L.; Hansen, T. *J. Am. Chem. Soc.* **1997**, *119*, 9075. See also, Davies, H.M.L. *Aldrichimica Acta* **1997**, *30*, 107. For related chiral ligands see Nagashima, T.; Davies, H.M.L. *Org. Lett.* **2002**, *4*, 1989; Davies, H.M.L.; Lee, G.H. *Org. Lett.* **2004**, *6*, 2117.

[2303]Davies, H.M.L.; Townsend, R.J. *J. Org. Chem.* **2001**, *66*, 6595; Davies, H.M.; Boebel, T.A. *Tetrahedron Lett.* **2000**, *41*, 8189.

[2304]Piqué, C.; Fähndrich, B.; Pfaltz, A. *Synlett* **1995**, 491; Barberis, M.; Pérez-Prieto, J.; Stiriba, S.-E.; Lahuerta, P. *Org. Lett.* **2001**, *3*, 3317; Saha, B.; Uchida, T.; Katsuki, T. *Synlett* **2001**, 114; Honma, M.; Sawada, T.; Fujisawa, Y.; Utsugi, M.; Watanabe, H.; Umino, A.; Matsumura, T.; Hagihara, T.; Takano, M.; Nakada, M. *J. Am. Chem. Soc.* **2003**, *125*, 2860.

[2305]In emim NTf$_2$, 1-ethyl-3-methylimidazolium triflimide: Fraile, J.M.; García, J.I.; Herrerías, C.I.; Mayoral, J.A.; Carrié, D.; Vaultier, M. *Tetrahedron Asymmetry* **2001**, *12*, 1891.

[2306]Doyle, M.P.; May, E.J. *Synlett* **2001**, 967.

[2307]Ferrand, Y.; Le Maux, P.; Simonneaux, G. *Org. Lett.* **2004**, *6*, 3211.

[2308]For reviews, see Fuks, R.; Viehe, H.G., in Viehe, H.G. *Acetylenes*, Marcel Dekker, NY, **1969**, pp. 427–434; Closs, G.L. *Adv. Alicyclic Chem.* **1966**, *1*, 53–127, see pp. 58–65.

[2309]Frey, H.M. *Chem. Ind. (London)* **1960**, 1266.

[2310]Vol'pin, M.E.; Koreshkov, Yu.D.; Kursanov, D.N. *Bull. Acad. Sci. USSR Div. Chem. Sci.* **1959**, 535.

[2311]Skell, P.S.; Garner, A.Y. *J. Am. Chem. Soc.* **1956**, *78*, 5430; Doering, W. von E.; Henderson, Jr., W.A. *J. Am. Chem. Soc.* **1958**, *80*, 5274; Mitsch, R.A.; Rodgers, A.S. *Int. J. Chem. Kinet.* **1969**, *1*, 439.

[2312]For a review of reactivity in this reaction, with many comprehensive tables of data, see Moss, R.A., in Jones, Jr. M.; Moss, R.A. *Carbenes*, Vol. 1, Wiley, NY, **1973**, pp. 153–304. See also, Cox, D.P.; Gould, I.R.; Hacker, N.P.; Moss, R.A.; Turro, N.J. *Tetrahedron Lett.* **1983**, *24*, 5313.

[2313]Woodworth, R.C.; Skell, P.S. *J. Am. Chem. Soc.* **1959**, *81*, 3383; Jones Jr., M.; Ando, W.; Hendrick, M.E.; Kulczycki Jr., A.; Howley, P.M.; Hummel, K.F.; Malament, D.S. *J. Am. Chem. Soc.* **1972**, *94*, 7469.

[2314]For evidence that at least some singlet carbenes add by a two-step mechanism, see Giese, B.; Lee, W.; Neumann, C. *Angew. Chem. Int. Ed.* **1982**, *21*, 310.

to mechanism *a* of **15-60** and **15-63**:

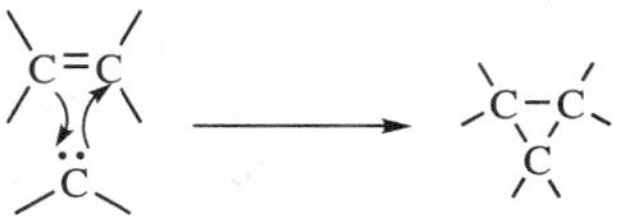

Infrared spectra of a carbene and the cyclopropane product have been observed in an argon matrix at 12–45 K.[2315] Carbenes in the triplet state react nonstereospecifically,[2316] probably by a diradical mechanism, similar to mechanism *b* of **15-49** and **15-63**:

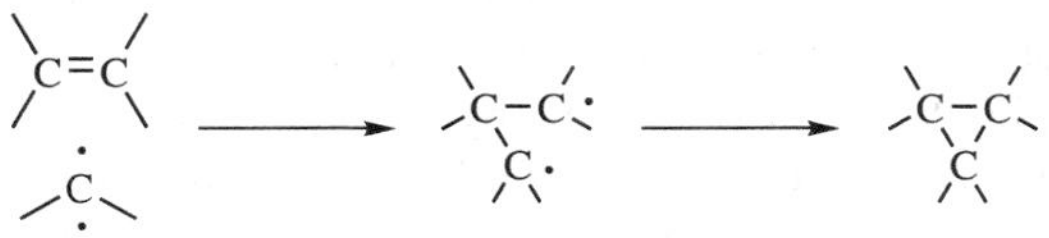

For carbenes or carbenoids of the type R—C—R′ there is another aspect of stereochemistry.[2317] When these species are added to all but symmetrical alkenes, two isomers are possible, even if the four groups originally on the double–bond carbons maintain their configurations:

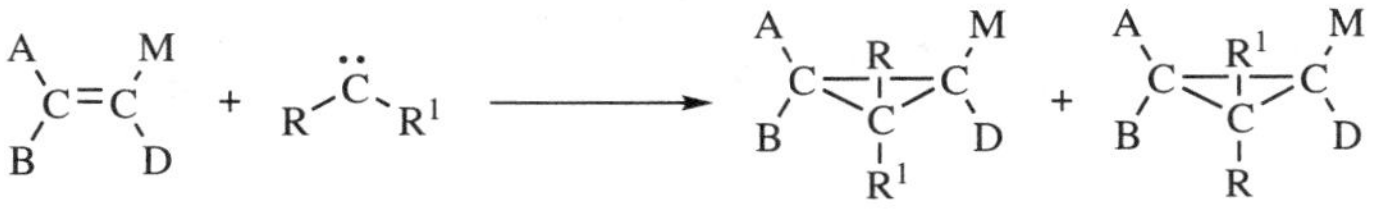

Which isomer is predominantly formed depends on R, R′, and on the method by which the carbene or carbenoid is generated. Most studies have been carried out on monosubstituted species (R′ = H), and in these studies it is found that aryl groups generally prefer the more substituted side (syn addition) while carbethoxy groups usually show anti stereoselectivity. When R = halogen, free halocarbenes show little or no stereochemical preference, while halocarbenoids exhibit a preference for syn addition. Beyond this, it is difficult to make simple generalizations.

Carbenes are so reactive that they add to the "double bonds" of aromatic rings.[2318] The products are usually unstable and rearrange to give ring expansion. Carbene reacts with benzene to give cycloheptatriene (**216**),[2319]

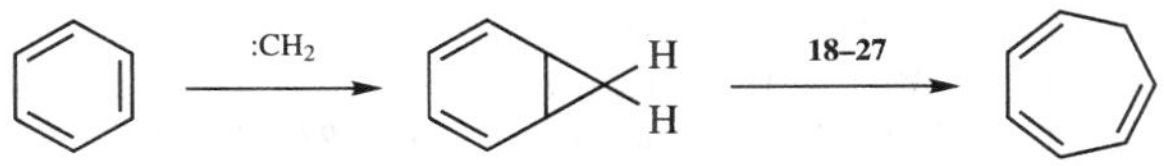

Norcaradiene   **216**

[2315]Nefedov, O.M.; Zuev, P.S.; Maltsev, A.K.; Tomilov, Y.V. *Tetrahedron Lett.* **1989**, *30*, 763.

[2316]Skell, P.S.; Klebe, J. *J. Am. Chem. Soc.* **1960**, *82*, 247. See also, Jones, Jr., M.; Tortorelli, V.J.; Gaspar, P.P.; Lambert, J.B. *Tetrahedron Lett.* **1978**, 4257.

[2317]For reviews of the stereochemistry of carbene and carbenoid addition to double bonds, see Moss, R.A. *Sel. Org. Transform.*, **1970**, *1*, 35–88; Closs, G.L. *Top Stereochem.* **1968**, *3*, 193–235. For a discussion of enantioselectivity in this reaction, see Nakamura, A. *Pure Appl. Chem.* **1978**, *50*, 37.

[2318]See Giese, C.M.; Hadad, C.M. *J. Org. Chem.* **2002**, *67*, 2532.

[2319]Doering, W. von E.; Knox, L.H. *J. Am. Chem. Soc.* **1951**, *75*, 297.

but not all carbenes are reactive enough to add to benzene. The norcaradiene intermediate cannot be isolated in this case[2320] (it undergoes an electrocyclic rearrangement, **18-27**), although certain substituted norcaradienes, for example, the product of addition of: $C(CN)_2$ to benzene,[2321] have been isolated.[2322] With: CH_2, insertion is a major side reaction, and, for example, benzene gives toluene as well as cycloheptatriene. A method of adding: CH_2 to benzene rings without the use of free carbene is the catalytic decomposition of diazomethane (CH_2N_2) in the aromatic compound as solvent with CuCl or CuBr.[2323] By this method better yields of cycloheptatrienes are obtained without insertion side products. Picosecond optical grating calorimetry has been used to investigate the photochemical decomposition of diazomethane in benzene, and it appears that a transient is formed that is consistent with a weak complex between singlet methylene and benzene.[2324] Chlorocarbene, :CHCl, is active enough to add to benzene, but dihalocarbenes do not add to benzene or toluene, only to rings with greater electron density. Pyrroles and indoles can be expanded, respectively, to pyridines and quinolines by treatment with halocarbenes[2325] via the initially formed adduct **217** in the case of the indole.

217

In such cases, a side reaction that sometimes occurs is expansion of the *six-membered* ring. Ring expansion can occur even with non–aromatic compounds, when the driving force is supplied by relief of strain (see **218**).[2326]

218

[2320]It has been detected by uv spectroscopy: Rubin, M.B. *J. Am. Chem. Soc.* **1981**, *103*, 7791.

[2321]Ciganek, E. *J. Am. Chem. Soc.* **1967**, *89*, 1454.

[2322]See, for example, Mukai, T.; Kubota, H.; Toda, T. *Tetrahedron Lett.* **1967**, 3581; Maier, G.; Heep, U. *Chem. Ber.* **1968**, *101*, 1371; Ciganek, E. *J. Am. Chem. Soc.* **1971**, *93*, 2207; Dürr, H.; Kober, H. *Tetrahedron Lett.* **1972**, 1255, 1259; Vogel, E.; Wiedemann, W.; Roth, H.D.; Eimer, J.; Günther, H. *Liebigs Ann. Chem.* **1972**, *759*, 1; Bannerman, C.G.F.; Cadogan, J.I.G.; Gosney, I.; Wilson, N.H. *J. Chem. Soc., Chem. Commun.* **1975**, 618; Takeuchi, K.; Kitagawa, T.; Senzaki, Y.; Okamoto, K. *Chem. Lett.* **1983**, 73; Kawase, T.; Iyoda, M.; Oda, M. *Angew. Chem. Int. Ed.* **1987**, *26*, 559.

[2323]Wittig, G.; Schwarzenbach, K. *Liebigs Ann. Chem.* **1961**, *650*, 1; Müller, E.; Fricke, H. *Liebigs Ann. Chem.* **1963**, *661*, 38; Müller, E.; Kessler, H.; Fricke, H.; Kiedaisch, W. *Liebigs Ann. Chem.* **1961**, *675*, 63.

[2324]Khan, M.I.; Goodman, J.L. *J. Am. Chem. Soc.* **1995**, *117*, 6635.

[2325]For a review of the reactions of heterocyclic compounds with carbenes, see Rees, C.W.; Smithen, C.E. *Adv. Heterocycl. Chem.* **1964**, *3*, 57–78.

[2326]Jefford, C.W.; Gunsher, J.; Hill, D.T.; Brun, P.; Le Gras, J.; Waegell, B. *Org. Synth.* **VI**, 142. For a review of the addition of halocarbenes to bridged bicyclic alkenes see Jefford, C.W. *Chimia*, **1970**, *24*, 357–363.

As previously mentioned, free carbene is not very useful for additions to double bonds since it gives too many side products. The *Simmons–Smith procedure* accomplishes the same result without a free carbene intermediate and without insertion side products.[2327] This procedure involves treatment of the double-bond compound with CH_2I_2 and a Zn–Cu couple and leads to cyclopropane derivatives in good yields.[2328] The Zn–Cu couple can be prepared in several ways,[2329] of which heating Zn dust with CuCl in ether under nitrogen[2330] is particularly convenient. The reaction has also been done with unactivated zinc and ultrasound.[2331] When $TiCl_4$ is used along with Zn and CuCl, CH_2I_2 can be replaced by the cheaper CH_2Br_2.[2332] The actual attacking species is an organozinc intermediate, probably $(ICH_2)_2Zn \cdot ZnI_2$, which is stable enough for isolable solutions.[2333] An X-ray crystallographic investigation of the intermediate, complexed with a diether, has been reported.[2334] The addition is stereospecifically syn, and a concerted mechanism[2335] is likely, perhaps[2336]

Asymmetric induction is possible when chiral additives are used.[2337] With the Simmons–Smith procedure, as with free carbenes, conjugated dienes give 1,2-addition,[2338] and allenes give methylenecyclopropanes or spiropentanes.[2339]

An alternative way of carrying out the Simmons–Smith reaction is by treatment of the substrate with CH_2I_2 or another dihalomethane and Et_2Zn in ether.[2340] This method can be adapted to the introduction of RCH and ArCH by the use of $RCHI_2$

[2327]For reviews, see Simmons, H.E.; Cairns, T.L.; Vladuchick, S.A.; Hoiness, C.M. *Org. React.* **1973**, *20*, 1–131; Furukawa, J.; Kawabata, N. *Adv. Organomet. Chem.* **1974**, *12*, 83–134, see pp. 84–103.

[2328]Simmons, H.E.; Smith, R.D. *J. Am. Chem. Soc.* **1959**, *81*, 4256.

[2329]Shank, R.S.; Shechter, H. *J. Org. Chem.* **1959**, *24*, 1525; LeGoff, E. *J. Org. Chem.* **1964**, *29*, 2048. For the use of a Zn—Ag couple, see Denis, J.M.; Girard, C.; Conia, J.M. *Synthesis* **1972**, 549.

[2330]Rawson, R.J.; Harrison, I.T. *J. Org. Chem.* **1970**, *35*, 2057.

[2331]Repič; O.; Lee, P.G.; Giger, U. *Org. Prep. Proced. Int.* **1984**, *16*, 25.

[2332]Friedrich, E.C.; Lunetta, S.E.; Lewis, E.J. *J. Org. Chem.* **1989**, *54*, 2388.

[2333]Blanchard, E.P.; Simmons, H.E. *J. Am. Chem. Soc.* **1964**, *86*, 1337. For an analysis of the reaction by density functional theory, see Fang, W.-H.; Phillips, D.L.; Wang, D.-q.; Li, Y.-L. *J. Org. Chem.* **2002**, *67*, 154.

[2334]Denmark, S.E.; Edwards, J.P.; Wilson, S.R. *J. Am. Chem. Soc.* **1991**, *113*, 723.

[2335]Dargel, T.K.; Koch, W. *J. Chem. Soc. Perkin Trans. 2*, **1996**, 877.

[2336]Simmons, H.E.; Blanchard, E.P.; Smith, R.D. *J. Am. Chem. Soc.* **1964**, *86*, 1347. For a discussion of the transition state and intermediate in this reaction, see Bernardi, F.; Bottoni, A.; Miscione, G.P. *J. Am. Chem. Soc.* **1997**, *119*, 12300.

[2337]Charette, A.B.; Juteau, H.; Lebel, H.; Molinaro, C. *J. Am. Chem. Soc.* **1998**, *120*, 11943; Kitajima, H.; Ito, K.; Aoki, Y.; Katsuki, T. *Bull. Chem. Soc. Jpn.* **1997**, *70*, 207; Imai, N.; Sakamoto, K.; Maeda, M.; Kouge, K.; Yoshizane, K.; Nokami, J. *Tetrahedron Lett*, **1997**, *38*, 1423; Denmark, S.E.; Edwards, J.P. *Synlett* **1992**, 229; Balsells, J.; Walsh, P.J. *J. Org. Chem.* **2000**, *65*, 5005.

[2338]Overberger, C.G.; Halek, G.W. *J. Org. Chem.* **1963**, *28*, 867.

[2339]Charette, A.B.; Jolicoeur, E.; Bydlinski, G.A.S. *Org. Lett.* **2001**, *3*, 3293.

[2340]See Charette, A.B.; Beauchemin, A.; Marcoux, J.-F. *Tetrahedron Lett.* **1999**, *40*, 33; Zhao, C.; Wang, D.; Phillips, D.L. *J. Am. Chem. Soc.* **2002**, *124*, 12903.

or $ArCHI_2$ instead of the dihalomethane.[2341] The reaction is compatible with other functionality in the carbenoid complex. The reaction of RCO_2CH_2I with diethyl zinc and an alkene under photolysis conditions give a cyclopropane.[2342] Chiral additives lead to enantioselectivity in the cyclopropanation reaction.[2343] In another method, CH_2I_2 or $MeCHI_2$ is used along with an alane R_3Al to transfer CH_2 or $CHMe$.[2344] Titanium complexes have been used similarly.[2345] Samarium and CH_2I_2 has been used for the cyclopropanation of conjugated amides.[2346] For the conversion of enolates to cyclopropanols, CH_2I_2 has been used along with SmI_2.[2347]

Other cyclopropanation techniques have been developed. Treatment of an alkene with $ArCH(SnBu_3)OCO_2Me$ and $BF_3\cdot OEt_2$ leads to the cyclopropane with high cis-selectivity.[2348] Diodomethane in the presence of isopropylmagnesium chloride has been used to cyclopropanate allyl alcohols.[2349]

The Simmons–Smith reaction is the basis of a method for the indirect α methylation of a ketone.[2350] The ketone (illustrated for cyclohexanone) is first converted to an enol ether, an enamine (**16-13**) or silyl enol ether[2351] (**12-17**) and cyclopropanation via the Simmons–Smith reaction is followed by hydrolysis to give the α methylated ketone. A related procedure using diethylzinc and diiodomethane allows ketones to be chain-extended by one carbon.[2352] In another variation, phenols can be ortho-methylated in one laboratory step, by treatment with Et_2Zn and CH_2I_2.[2353]

Diazoesters react with amines with a rhodium catalyst to give α-amino esters.[2354] Diazoesters also react with aldehydes and a rhodium catalyst, and the product is an α,β-epoxy ester.[2355] Diazoalkanes react similarly with aldehydes to give an alkene ($Me_3SiCH=N_2 + ArCHO \rightarrow ArCH=CHOSiMe_3$).[2356]

OS V, 306, 855, 859, 874; VI, 87, 142, 187, 327, 731, 913, 974; VII, 12, 200, 203; VIII, 124, 196, 321, 467; IX, 422; 76, 86.

[2341]Nishimura, J.; Kawabata, N.; Furukawa, J. *Tetrahedron* **1969**, *25*, 2647; Miyano, S.; Hashimoto, H. *Bull. Chem. Soc. Jpn.* **1973**, *46*, 892; Friedrich, E.C.; Biresaw, G. *J. Org. Chem.* **1982**, *47*, 1615.

[2342]Charette, A.B.; Beauchemin, A.; Fraancoeur, S. *J. Am. Chem. Soc.* **2001**, *123*, 8139.

[2343]Long, J.; Yuan, Y.; Shi, Y. *J. Am. Chem. Soc.* **2003**, *125*, 13632.

[2344]Maruoka, K.; Fukutani, Y.; Yamamoto, H. *J. Org. Chem.* **1985**, *50*, 4412; *Org. Synth., 67*, 176.

[2345]Charette, A.B.; Molinaro, C.; Brochu, C. *J. Am. Chem. Soc.* **2001**, *123*, 12168.

[2346]Concellón, J.M.; Rodríguez-Solla, H.; Gómez, C. *Angew. Chem. Int. Ed.* **2002**, *41*, 1917.

[2347]Imamura, T.; Takiyama, N. *Tetrahedron Lett.* **1987**, *28*, 1307. See also, Molander, G.A.; Harring, L.S. *J. Org. Chem.* **1989**, *54*, 3525.

[2348]Sugawara, M.; Yoshida, J. *J. Am. Chem. Soc.* **1997**, *119*, 11986.

[2349]Bolm, C.; Pupowicz, D. *Tetrahedron Lett.* **197**, *38*, 7349.

[2350]See Wenkert, E.; Mueller, R.A.; Reardon Jr., E.J.; Sathe, S.S.; Scharf, D.J.; Tosi, G. *J. Am. Chem. Soc.* **1970**, *92*, 7428 for the enol ether procedure; Kuehne, M.E.; King, J.C. *J. Org. Chem.* **1973**, *38*, 304 for the enamine procedure; Conia, J.M. *Pure Appl. Chem.* **1975**, *43*, 317–326 for the silyl ether procedure.

[2351]In the case of silyl enol ethers, the inner bond can be cleaved with $FeCl_3$, giving a ring-enlarged β-chloro ketone: Ito, Y.; Fujii, S.; Saegusa, T. *J. Org. Chem.* **1976**, *41*, 2073; *Org. Synth.* VI, 327.

[2352]Brogan, J.B.; Zercher, C.K. *J. Org. Chem.* **1997**, *62*, 6444.

[2353]Lehnert, E.K.; Sawyer, J.S.; Macdonald, T.L. *Tetrahedron Lett.* **1989**, *30*, 5215.

[2354]Yang, M.; Wang, X.; Li, H.; Livant, P. *J. Org. Chem.* **2001**, *66*, 6729.

[2355]Doyle, M.P.; Hu, W.; Timmons, D.J. *Org. Lett.* **2001**, *3*, 933.

[2356]Dias, E.L.; Brookhart, M.; White, P.S. *J. Am. Chem. Soc.* **2001**, *123*, 2442.

15-65   Trimerization and Tetramerization of Alkynes

Aromatic compounds can be prepared by cyclotrimerization of alkynes[2357] or triynes. Cyclotrimerization is possible by heating to 450–600°C with no catalyst.[2358] The *spontaneous* (no catalyst) trimerization of t-BuC≡CF gave 1,2,3-tri-*tert*-butyl-4,5,6-trifluorobenzene (**220**), the first time three adjacent *tert*-butyl groups had been put onto a benzene ring.[2359] The fact that this is a head-to-head joining allows formation of **220** from two alkynes. The fact that **219** (a Dewar benzene) was also isolated lends support to this scheme.[2360] Three equivalents of 3-hexyne trimerized to hexaethylbenzene at 200°C in the presence of Si_2Cl_6.[2361]

R = *tert*-butyl   **219**   **220**

When acetylene is heated with nickel cyanide, other Ni(II) or Ni(0) compounds, or similar catalysts, it gives benzene and cyclooctatetraene.[2362] It is possible to get more of either product by a proper choice of catalyst. Substituted acetylenes give substituted benzenes.[2363] This reaction has been used to prepare very crowded

[2357]For a review, see Rubin, M.; Sromek, A.W.; Gevorgyan, V. *Synlett* **2003**, 2265.

[2358]Kociolek, M.G.; Johnson, R.P. *Tetrahedron Lett.* **1999**, *40*, 4141.

[2359]Viehe, H.G.; Merényi, R.; Oth, J.F.M.; Valange, P. *Angew. Chem. Int. Ed.* **1964**, *3*, 746; Viehe, H.G.; Merényi, R.; Oth, J.F.M.; Senders, J.R.; Valange, P. *Angew. Chem. Int. Ed.* **1964**, *3*, 755.

[2360]For other reactions between cyclobutadienes and triple bonds to give Dewar benzenes, see Wingert, H.; Regitz, M. *Chem. Ber.* **1986**, *119*, 244.

[2361]Yang, J.; Verkade, J.G. *J. Am. Chem. Soc.* **1998**, *120*, 6834.

[2362]For reviews, see Winter, M.J., in Hartley, F.R.; Patai, S. *The Chemistry of the Metal–Carbon Bond*, Vol. 3, Wiley, NY, **1985**, pp. 259–294; Vollhardt, K.P.C. *Angew. Chem. Int. Ed.* **1984**, *23*, 539; *Acc. Chem. Res.* **1977**, *10*, 1; Maitlis, P.M. *J. Organomet. Chem.* **1980**, *200*, 161; *Acc. Chem. Res.* **1976**, *9*, 93; *Pure Appl. Chem.* **1972**, *30*, 427; Yur'eva, L.P. *Russ. Chem. Rev.* **1974**, *43*, 48; Khan, M.M.T.; Martell, A.E. *Homogeneous Catalysis by Metal Complexes*, Vol. 2, Academic Press, NY, **1974**, pp. 163–168; Reppe, W.; Kutepow, N.V.; Magin, A. *Angew. Chem. Int. Ed.* **1969**, *8*, 727; Fuks, R.; Viehe, H.G. in Viehe, H.G. *Acetylenes*, Marcel Dekker, NY, **1969**, pp. 450–460; Hoogzand, C.; Hübel, W., in Wender, I.; Pino, P. *Organic Syntheses Via Metal Carbonyls*, Vol. 1, Wiley, NY, **1968**, pp. 343–371; Reikhsfel'd, V.O.; Makovetskii, K.L. *Russ. Chem. Rev.* **1966**, *35*, 510. For a list of reagents, with references, see Larock, R.C. *Comprehensive Organic Transformations*, 2nd ed., Wiley-VCH, NY, **1999**, pp. 198–201. For a review of metal-catalyzed cycloadditions of alkynes to give rings of all sizes, see Schore, N.E. *Chem. Rev.* **1988**, *88*, 1081.

[2363]Sigman, M.S.; Fatland, A.W.; Eaton, B.E. *J. Am. Chem. Soc.* **1998**, *120*, 5130; Gevorgyan, V.; Takeda, A.; Yamamoto, Y. *J. Am. Chem. Soc.* **1997**, *119*, 11313; Takeda, A.; Ohno, A.; Kadota, I.; Gevorgyan, V.; Yamamoto, Y. *J. Am. Chem. Soc.* **1997**, *119*, 4547; Larock, R.C.; Tian, Q. *J. Org. Chem.* **1998**, *63*, 2002; Sato, Y.; Nishimata, T.; Mori, M. *J. Org. Chem.* **1994**, *59*, 6133; Grissom, J.W.; Calkins, T.L. *Tetrahedron Lett.* **1992**, *33*, 2315.

molecules. Diisopropylacetylene was trimerized over $CO_2(CO)_8$[2364] and over $Hg[Co(CO)_4]_2$ to hexaisopropylbenzene.[2365] The six isopropyl groups are not free to rotate but are lined up perpendicular to the plane of the benzene ring. Highly substituted benzene derivatives have also been prepared using a rhodium,[2366] nickel,[2367] titanium,[2368] molybdenum,[2369] ruthenium,[2370] cobalt,[2371] or a palladium[2372] catalyst. Alkynes react with allenes and a nickel catalyst go give highly substituted benzene derivatives.[2373] Conjugated ketones react with internal alkynes with Me_3Al and a nickel catalyst[2374] leads to an aromatic ring fused to a cyclic ketone after reaction with DBU and air.[2375] N-Aryl chloroimines react with alkynes and a rhodium catalyst to give quinolines,[2376] as do N-aryl alkynyl imines with a tungsten complex.[2377]

An intramolecular cyclotrimerization has been reported by condensation of a diyne[2378] with an alkyne in the presence of a palladium,[2379] molybdenum,[2380] nickel,[2381] rhodium,[2382] iridium,[2383] or ruthenium catalyst.[2384] Triynes have been

[2364]See Yong, L.; Butenschön, H. *Chem. Commun.* **2002**, 2852. For a modification that gives a phenol from 3,3-dimethyl-1-butyne, see Marchueta, I.; Olivella, S.; Solà, L.; Moyano, A.; Pericàs, M.A.; Riera, A. *Org. Lett.* **2001**, *3*, 3197.

[2365]Arnett, E.M.; Bollinger, J.M. *J. Am. Chem. Soc.* **1964**, *86*, 4729; Hopff, H.; Gati, A. *Helv. Chim. Acta* **1965**, *48*, 509.

[2366]Taber, D.F.; Rahimizadeh, M. *Tetrahedron Lett.* **1994**, *35*, 9139; Tanaka, K.; Shirasaka, K. *Org. Lett.* **2003**, *5*, 4697.

[2367]Mori, N.; Ikeda, S.-i.; Odashima, K. *Chem. Commun.* **2001**, 181.

[2368]Tanaka, R.; Nakano, Y.; Suzuki, D.; Urabe, H.; Sato, F. *J. Am. Chem. Soc.* **2002**, *124*, 9682.

[2369]Nishida, M.; Shiga, H.; Mori, M. *J. Org. Chem.* **1998**, *63*, 8606.

[2370]Yamamoto, Y.; Ishii, J.-i.; Nishiyama, H.; Itoh, K. *J. Am. Chem. Soc.* **2004**, *126*, 3712.

[2371]Sugihara, T.; Wakabayashi, A.; Nagai, Y.; Takao, H.; Imagawa, H.; Nishizawa, M. *Chem. Commun.* **2002**, 576.

[2372]Gevorgyan, V.; Takeda, A.; Homma, M.; Sadayori, N.; Radhakrishnan, U.; Yamamoto, Y. *J. Am. Chem. Soc.* **1999**, *121*, 6391; Gevorgyan, V.; Quan, L.G.; Yamamoto, Y. *J. Org. Chem.* **2000**, *65*, 568. For a reaction in conjunction with silver carbonate, see Kawasaki, S.; Satoh, T.; Miura, M.; Nomura, M. *J. Org. Chem.* **2003**, *68*, 6836; In conjunction with $CuCl_2$, see Li, J.-H.; Xie, Y.-X. *Synth. Commun.* **2004**, *34*, 1737.

[2373]Shanmugasundaram, M.; Wu, M.-S.; Cheng, C.-H. *Org. Lett.* **2001**, *3*, 4233.

[2374]Ikeda, S.; Kondo, H.; Arii, T.; Odashima, K. *Chem. Commun.* **2002**, 2422.

[2375]Mori, N.; Ikeda, S.-i.; Sato, Y. *J. Am. Chem. Soc.* **1999**, *121*, 2722.

[2376]Amii, H.; Kishikawa, Y.; Uneyama, K. *Org. Lett.* **2001**, *3*, 1109.

[2377]Sangu, K.; Fuchibe, K.; Akiyama, T. *Org. Lett.* **2004**, *6*, 353.

[2378]See Kawathar, S.P.; Schreiner, P.R. *Org. Lett.* **2002**, *4*, 3643.

[2379]Yamamoto, Y.; Nagata, A.; Itoh, K. *Tetrahedron Lett.* **1999**, *40*, 5035; Gevorgyan, V.; Radhakrishnan, U.; Takeda, A.; Rubina, M.; Rubin, M.; Yamamoto, Y. *J. Org. Chem.* **2001**, *66*, 2835. See also, Tsukada, N.; Sugawara, S.; Nakaoka, K.; Inoue, Y. *J. Org. Chem.* **2003**, *68*, 5961.

[2380]Hara, R.; Guo, Q.; Takahashi, T. *Chem. Lett.* **2000**, 140.

[2381]Jeevanandam, A.; Korivi, R.P.; Huang, I.-w.; Cheng, C.-H. *Org. Lett.* **2002**, *4*, 807.

[2382]Witulski, B.; Zimmermann, A. *Synlett* **2002**, 1855.

[2383]Takeuchi, R.; Tanaka, S.; Nakaya, Y. *Tetrahedron Lett.* **2001**, *42*, 2991; Shibata, T.; Fujimoto, T.; Yokota, K.; Takagi, K. *J. Am. Chem. Soc.* **2004**, *126*, 8382.

[2384]Yamamoto, Y.; Ogawa, R.; Itoh, K. *Chem. Commun.* **2000**, 549; Witulski, B.; Stengel, T.; Fernández-Hernandez, J.M. *Chem. Commun.* **2000**, 1965.

similarly condensed with a rhodium catalyst.[2385] The internal cyclotrimerization of a triyne, utilizing a siloxy tether and a cobalt catalyst has been reported.[2386] Fused ring aromatic compounds are prepared by this method. Similar results were obtained from diynes and allenes with a nickel catalyst.[2387] bis(Enynes) are cyclized to bicyclic arenes using a palladium catalyst.[2388] Diynes with nitriles and a ruthenium catalyst lead to isoquinolines.[2389] Pyridines fused to carboxylic rings can be prepared by similar methodology using a cyanoamine and a cobalt catalyst.[2390] In the presence of PhMe$_2$SiH, CO and a rhodium catalyst, a nonconjugated triyne leads to a tricyclic compound in which a benzene ring is fused to two carbocyclic rings.[2391] Internal cyclotrimerization of an aryl alkynyl ketone where the aryl group has an *ortho* trimethylsiylalkyne substituent gives a tetracyclic naphthalene derivative with a fused cyclopentanone unit.[2392] An isocyanate (Ar—N=C=O) reacts with a diyne and a ruthenium catalyst to give a bicyclic pyridone.[2393] Benzene derivatives with ortho alkyne units can be converted to naphthalene derivatives in aqueous NaOH with hydrazine, Te, NaBH$_4$ and sonication.[2394] Benzene derivatives having ortho imine and alkyne substituents give an isoquinoline when treated with iodine[2395] or with a palladium catalyst.[2396] Imino and iodo substituents with a silyl alkyne and a palladium catalyst leads to an isoquinoline.[2397] Vinyl and alkyne substituents with a ruthenium catalyst lead to naphthalene derivatives.[2398] Ortho alkynyl and epoxy substituents leads to β-naphthols using a ruthenium catalyst.[2399] Cyclotrimerization occurs with alkynyl boronic esters.[2400]

In contrast to the spontaneous reaction, the catalyzed process seldom gives the 1,2,3-trisubstituted benzene isomer from an acetylene RC≡CH. The chief product is usually the 1,2,4-isomer,[2401] with lesser amounts of the 1,3,15-isomer also generally obtained, but little if any of the 1,2,3-isomer. The mechanism of

[2385]Kinoshita, H.; Shinokubo, H.; Oshima, K. *J. Am. Chem. Soc.* **2003**, *125*, 7784.

[2386]Chouraqui, G.; Petit, M.; Aubert, C.; Malacria, M. *Org. Lett.* **2004**, *6*, 1519.

[2387]Shanmugasundaram, M.; Wu, M.-S.; Jeganmohan, M.; Huang, C.-W.; Cheng, C.-H. *J. Org. Chem.* **2002**, *67*, 7724.

[2388]Kawasaki, T.; Saito, S.; Yamamoto, Y. *J. Org. Chem.* **2002**, *67*, 2653.

[2389]Yamamoto, Y.; Okuda, S.; Itoh, K. *Chem. Commun.* **2001**, 1102; Varela, J.A.; Castedo, L.; Saá, C. *J. Org. Chem.* **2003**, *68*, 8595.

[2390]Boñaga, L.V.R.; Zhang, H.-C.; Maryanoff, B.E. *Chem. Commun.* **2004**, 2394.

[2391]Ojima, I.; Vu, A.T.; McCullagh, J.V.; Kinoshita, A. *J. Am. Chem. Soc.* **1999**, *121*, 3230.

[2392]Atienza, C.; Mateo, C.; de Frutos, Ó.; Echavarren, A.M. *Org. Lett.* **2001**, *3*, 153.

[2393]Yamamoto, Y.; Takagishi, H.; Itoh, K. *Org. Lett.* **2001**, *3*, 2117.

[2394]Landis, C.A.; Payne, M.M.; Eaton, D.L.; Anthony, J.E. *J. Am. Chem. Soc.* **2004**, *126*, 1338.

[2395]Huang, Q.; Hunter, J.A.; Larock, R.C. *Org. Lett.* **2001**, *3*, 2973.

[2396]Dai, G.; Larock, R.C. *Org. Lett.* **2001**, *3*, 4035; Dai, G.; Larock, R.C. *J. Org. Chem.* **2003**, *68*, 920; Dai, G.; Larock, R.C. *Org. Lett.* **2002**, *4*, 193.

[2397]Roesch, K.R.; Larock, R.C. *J. Org. Chem.* **2002**, *67*, 86.

[2398]Klumpp, D.A.; Beauchamp, P.S.; Sanchez, Jr., G.V.; Aguirre, S.; de Leon, S. *Tetrahedron Lett.* **2001**, *42*, 5821.

[2399]Madhusaw, R.J.; Lin, M.-Y.; Shoel, S.Md.A.; Liu, R.-S. *J. Am. Chem. Soc.* **2004**, *126*, 6895.

[2400]Gandon, V.; Leca, D.; Aechtner, T.; Vollhardt, K.P.C.; Malacria, M.; Aubert, C. *Org. Lett.* **2004**, *6*, 3405.

[2401]See Saito, S.; Kawasaki, T.; Tsuboya, N.; Yamamoto, Y. *J. Org. Chem.* **2001**, *66*, 796.

the catalyzed

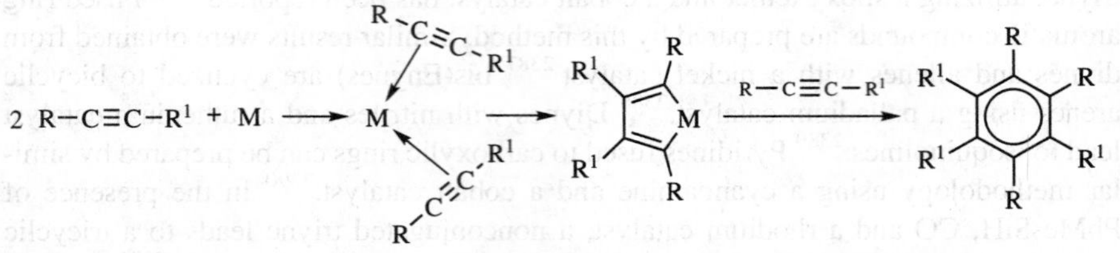

$$2 \ R-C{\equiv}C-R^1 + M \longrightarrow \mathbf{221} \longrightarrow \mathbf{222} \longrightarrow \mathbf{223}$$

reaction to form benzenes[2402] is believed to go through a species **221** in which two molecules of alkyne coordinate with the metal, and another species **222**, a five-membered heterocyclic intermediate.[2403] Such intermediates (where $M = Rh, \ Ir, \ Zr$,[2404] or Ni) have been isolated and shown to give benzenes (**223**) when treated with alkynes.[2405] Note that this pathway accounts for the predominant formation of the 1,2,4-isomer. Two possibilities for the last step are a Diels–Alder reaction, and a ring expansion, each followed by extrusion of the metal:[2406]

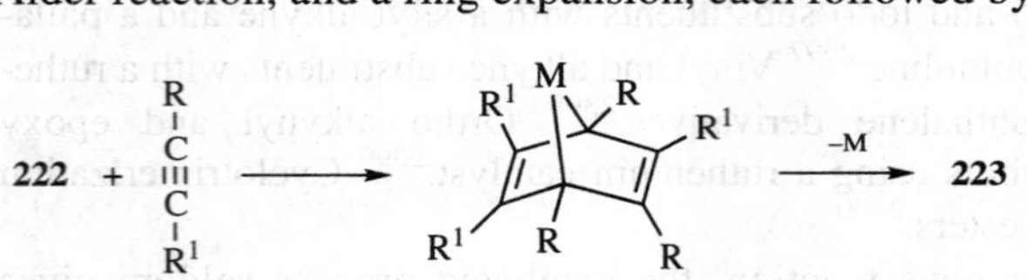

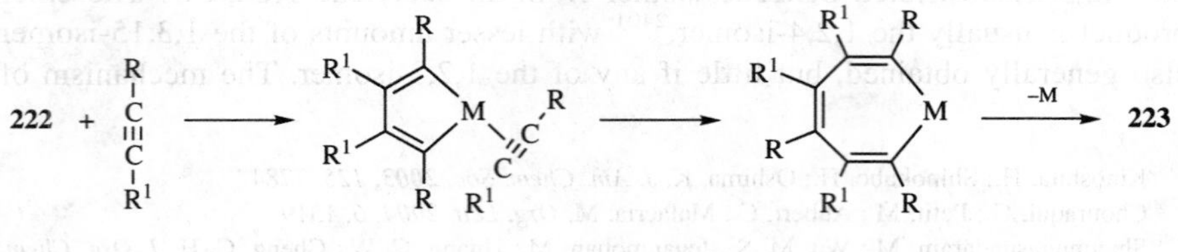

[2402]For studies of the mechanism of the reaction that produces cyclooctatetraenes, see Diercks, R.; Stamp, L.; Kopf, J.; Tom Dieck, H. *Angew. Chem. Int. Ed.* **1984**, *23*, 893; Colborn, R.E.; Vollhardt, K.P.C. *J. Am. Chem. Soc.* **1986**, *108*, 5470; Lawrie, C.J.; Gable, K.P.; Carpenter, B.K. *Organometallics* **1989**, *8*, 2274.
[2403]See, for example, Colborn, R.E.; Vollhardt, K.P.C. *J. Am. Chem. Soc.* **1981**, *103*, 6259; Kochi, J.K. *Organometallic Mechanisms and Catalysis*, Academic Press, NY, **1978**, pp. 428–432; Collman, J.P., Hegedus, L.S.; Norton, J.R.; Finke, R.G. *Principles and Applications of Organotransition Metal Chemistry*, University Science Books, Mill Valley, CA **1987**, pp. 870–877; Eisch, J.J.; Sexsmith, S.R. *Res. Chem. Intermed.* **1990**, *13*, 149–192.
[2404]Takahahsi, T.; Ishikawa, M.; Huo, S. *J. Am. Chem. Soc.* **2002**, *124*, 388.
[2405]See, for example, Collman, J.P. *Acc. Chem. Res.* **1968**, *1*, 136; Yamazaki, H.; Hagihara, N. *J. Organomet. Chem.* **1967**, *7*, P22; Wakatsuki, Y.; Kuramitsu, T.; Yamazaki, H. *Tetrahedron Lett.* **1974**, 4549; Moseley, K.; Maitlis, P.M. *J. Chem. Soc. Dalton Trans.* **1974**, 169; Müller, E. *Synthesis* **1974**, 761; Eisch, J.J.; Galle, J.E. *J. Organomet. Chem.* **1975**, *96*, C23; McAlister, D.R.; Bercaw, J.E.; Bergman, R.G. *J. Am. Chem. Soc.* **1977**, *99*, 1666.
[2406]There is evidence that the mechanism of the last step more likely resembles the Diels–Alder pathway than the ring expansion pathway: Bianchini, C.; Caulton, K.G.; Chardon, C.; Eisenstein, O.; Folting, K.; Johnson, T.J.; Meli, A.; Peruzzini, M.; Raucher, D.J.; Streib, W.E.; Vizza, F. *J. Am. Chem. Soc.* **1991**, *113*, 5127.

In at least one case the mechanism is different, going through a cyclobutadiene–nickel complex (see p. 76), which has been isolated.[2407] Similar results were obtained with a titanium complex.[2408] Using a mixture of $PdCl_2$ and $CuCl_2$, however, aliphatic alkynes are converted to the 1,3,5-trialkyl benzene derivative.[2409]

Alkoxy chromium carbenes (Fischer carbene complexes) react with phenylalkynes to give naphthalene derivatives.[2410] These chromium carbenes react with alkynyl boronates, cerium(IV) compounds, and then PhBr and a palladium catalyst to give a naphthoquinone.[2411] Diynes react to give cyclotrimerization.[2412] It is noted that vinyl chromium carbenes react directly with alkynes to give spirocyclic compounds (spiro[4.4]nona-1,3,6-trienes).[2413] Benzofurans can be prepared using methoxy carbenes.[2414] Amino-substituted chromium carbenes react with alkynes and then silica to give substituted benzene derivatives that have an aminoalkyl ($-NR_2$) substituent.[2415] Imino-substituted chromium carbenes react with alkynes to give pyrrole derivatives.[2416] Fischer carbene complexes react with alkynes to give the *Dötz benzannulation*,[2417] giving *p*-alkoxyphenol derivatives. Modification of this basic technique can lead to eight-membered ring carbocycles (see **15-66**).[2418]

When benzene, in the gas phase, was adsorbed onto a surface of 10% rhodium-on-alumina, the reverse reaction took place, and acetylene was formed.[2419]

In a related reaction, heating ketones in the presence of $TlCl_3OTf$ leads to 1,3,5-trisubstituted arenes.[2420] Heating acetophenone with $TiCl_4$ gives 1,3,5-triphenyl-benzene.[2421] Nitriles react with 2 mol of acetylene, in the presence of a cobalt catalyst, to give 2-substituted pyridines.[2422] Propargyl amines react with cyclohexanone derivatives and a gold complex give tetrahydroquinolines.[2423] Treatment of alkynes with Cp_2ZrEt_2 followed by reaction with acetonitrile and then a second alkyne with a nickel catalyst gives a highly substituted pyridine.[2424] This reaction can be done intramolecularly using a photochemically induced reaction with a

[2407]Mauret, P.; Alphonse, P. *J. Organomet. Chem.* **1984**, *276*, 249. See also, Pepermans, H.; Willem, R.; Gielen, M.; Hoogzand, C. *Bull. Soc. Chim. Belg.* **1988**, *97*, 115.

[2408]Suzuki, D.; Urabe, H.; Sato, F. *J. Am. Chem. Soc.* **2001**, *123*, 7925.

[2409]Li, J.; Jiang, H.; Chen, M. *J. Org. Chem.* **2001**, *66*, 3627.

[2410]Pulley, S.R.; Sen, S.; Vorogushin, A.; Swanson, E. *Org. Lett.* **1999**, *1*, 1721; Jackson, T.J.; Herndon, J.W. *Tetrahedron* **2001**, *57*, 3859.

[2411]Davies, M.W.; Johnson, C.N.; Harrity, J.P.A. *J. Org. Chem.* **2001**, *66*, 3525.

[2412]Jiang, M.X.-W.; Rawat, M.; Wulff, W.D. *J. Am. Chem. Soc.* **2004**, *126*, 5970.

[2413]Schirmer, H.; Flynn, B.L.; de Meijere, A. *Tetrahedron* **2000**, *56*, 4977.

[2414]Herndon, J.W.; Zhang, Y.; Wang, H.; Wang, K. *Tetrahedron Lett.* **2000**, *41*, 8687.

[2415]Barluenga, J.; López, L.A.; Martínez, S.; Tomás, M. *Tetrahedron* **2000**, *56*, 4967.

[2416]Campos, P.J.; Sampedro, D.; Rodríquez, M.A. *J. Org. Chem.* **2003**, *68*, 4674.

[2417]Dötz, K.H. *Angew. Chem. Int. Ed.* **1975**, *14*, 644.

[2418]Barluenga, J.; Aznar, F.; Palomero, M.A. *Angew. Chem. Int. Ed.* **2000**, *39*, 4346.

[2419]Parker, W.L.; Hexter, R.M.; Siedle, A.R. *J. Am. Chem. Soc.* **1985**, *107*, 4584.

[2420]Iranpoor, N.; Zeynizadeh, B. *Synlett* **1998**, 1079.

[2421]Li, Z.; Sun, W.-H.; Jin, X.; Shao, C. *Synlett* **2001**, 1947.

[2422]Heller, B.; Oehme, G. *J. Chem. Soc., Chem. Commun.* **1995**, 179.

[2423]Abbiati, G.; Arcadi, A.; Bianchi, G.; Di Giuseppe, S.; Marinelli, F.; Rossi, E. *J. Org. Chem.* **2003**, *68*, 6959.

[2424]Takahashi, T.; Tsai, F.Y.; Kotora, M. *J. Am. Chem. Soc.* **2000**, *122*, 4994.

cobalt catalyst and *p*-TolCN to give pyridines incorporated into macrocycles.[2425] Alkynyl esters react with enamino esters with a ZnBr$_2$ catalyst to give substituted pyridines.[2426] α-Halo oxime ethers react with alkynes and Grignard reagents, with a mixture of palladium and copper catalysts, to give pyrimidines.[2427] Triketones fix nitrogen gas in the presence of TiCl$_4$ and lithium metal to form bicyclic pyrrole derivatives.[2428]

OS **VII**, 256; **IX**, 1; **80**, 93.

15-66   Other Cycloaddition Reactions

cyclo-**[But-2-en-1,4-diyl]-1/4/addition**, and so on

Cycloaddition reactions other than $[4 + 2]$, $[3 + 2]$, or $[2 + 2]$ are possible, often providing synthetically useful routes to cyclic compounds. Conjugated dienes can be dimerized or trimerized at their 1,4 positions (formally, $[4 + 4]$ and $[4 + 4 + 4]$-cycloadditions) by treatment with certain complexes or other transition-metal compounds.[2429] Thus butadiene gives 1,5-cyclooctadiene and 1,5,9-cyclododecatriene.[2430] The relative amount of each product can be controlled by use of the proper catalyst. For example, Ni:P(OC$_6$H$_4$—*o*—Ph)$_3$ gives predominant dimerization, while Ni(cyclooctadiene)$_2$ gives mostly trimerization. The products arise, not by direct 1,4 to 1,4 attack, but by stepwise mechanisms involving metal–alkene complexes.[2431] The rhodium catalyzed intramolecular cycloaddition of a furan with a conjugated diazoester gives a $[3 + 4]$-cycloadduct.[2432] The suprafacial thermal addition of an allylic cation to a diene (a $[4 + 3]$-cycloaddition) is allowed by the Woodward–Hoffmann rules (this reaction would

[2425]Moretto, A.F.; Zhang, H.-C.; Maryanoff, B.E. *J. Am. Chem. Soc.* **2001**, *123*, 3157.

[2426]Bagley, M.C.; Dale, J.W.; Hughes, D.D.; Ohnesorge, M.; Philips, N.G.; Bower, J. *Synlett* **2001**, 1523.

[2427]Kikiya, H.; Yagi, K.; Shinokubo, H.; Oshima, K. *J. Am. Chem. Soc.* **2002**, *124*, 9032.

[2428]Mori, M.; Hori, M.; Sato, Y. *J. Org. Chem.* **1998**, *63*, 4832; Mori, M.; Hori, K.; Akashi, M.; Hori, M.; Sato, Y.; Nishida, M. *Angew. Chem. Int. Ed.* **1998**, *37*, 636.

[2429]For reviews, see Wilke, G. *Angew. Chem. Int. Ed.* **1988**, *27*, 186; Tolstikov, G.A.; Dzhemilev, U.M. *Sov. Sci. Rev. Sect. B* **1985**, *7*, 237, 278–290; Heimbach, P.; Schenkluhn, H. *Top. Curr. Chem.* **1980**, *92*, 45; Baker, R. *Chem. Rev.* **1973**, *73*, 487, see pp. 489–512; Semmelhack, M.F. *Org. React.* **1972**, *19*, 115, pp. 128–143; Khan, M.M.T.; Martell, A.E. *Homogeneous Catalysis by Metal Complexes*, Vol. 2, Academic Press, NY, **1974**, pp. 159–163; Heck, R.F. *Organotransition Metal Chemistry*, Academic Press, NY, **1974**, pp. 157–164.

[2430]For a review of the 1,5,9-cyclododecatrienes (there are four stereoisomers, of which the *ttt* is shown above), see Rona, P. *Intra-Sci. Chem. Chem. Rep.* **1971**, *5*, 105.

[2431]For example, see Heimbach, P.; Wilke, G. *Liebigs Ann. Chem.* **1969**, *727*, 183; Barnett, B.; Büssemeier, B.; Heimbach, P.; Jolly, P.W.; Krüger, C.; Tkatchenko, I.; Wilke, G. *Tetrahedron Lett.* **1972**, 1457; Barker, G.K.; Green, M.; Howard, J.A.K.; Spencer, J.L.; Stone, F.G.A. *J. Am. Chem. Soc.* **1976**, *98*, 3373; Graham, G.R.; Stephenson, L.M. *J. Am. Chem. Soc.* **1977**, *99*, 7098.

[2432]Davies, H.M.L.; Calvo, R.L.; Townsend, R.-J.; Ren, P.; Churchill, R.M. *J. Org. Chem.* **2000**, *65*, 4261. For reviews of [3 + 4]-cycloadditions see Mann, J. *Tetrahedron* **1986**, *42*, 4611; Hoffmann, H.M.R. *Angew. Chem. Int. Ed.* **1984**, *23*, 1; **1973**, *12*, 819; Noyori, R. *Acc. Chem. Res.* **1979**, *12*, 61.

be expected to follow the same rules as the Diels–Alder reaction[2433]). Chiral cations have been used in [4 + 3]-cycloadditions.[2434]

As we saw in **15-60**, the Woodward–Hoffmann rules allow suprafacial concerted cycloadditions to take place thermally if the total number of electrons is $4n+2$ and photochemically if the number is $4n$. Furthermore, forbidden reactions become allowed if one molecule reacts antarafacially. It would thus seem that syntheses of many large rings could easily be achieved. However, when the newly formed ring is eight-membered or greater, concerted mechanisms, although allowed by orbital symmetry for the cases stated, become difficult to achieve because of the entropy factor (the two ends of one system must simultaneously encounter the two ends of the other), unless one or both components are cyclic, in which case the molecule has many fewer possible conformations. There have been a number of reports of cycloaddition reactions leading to eight-membered and larger rings, some thermally and some photochemically induced, but (apart from the dimerization and trimerization of butadienes mentioned above, which are known not to involve direct [4 + 4]- or [4 + 4 + 4]-cycloaddition) in most cases evidence is lacking to indicate whether they are concerted or stepwise processes. Some examples are

[2433]Garst, M.E.; Roberts, V.A.; Houk, K.N.; Rondan, N.G. *J. Am. Chem. Soc.* **1984**, *106*, 3882.

[2434]Harmata, M; Jones, D.E.; Kahraman, M.; Sharma, U.; Barnes, C.L. *Tetrahedron Lett.* **1999**, *40*, 1831.

[2435]Wender, P.A.; Glorius, F.; Husfeld, C.O.; Langkopf, E.; Love, J.A. *J. Am. Chem. Soc.* **1999**, *121*, 5348. For another example, see Trost, B.M.; Toste, F.D.; Shen, H. *J. Am. Chem. Soc.* **2000**, *122*, 2379. See also, Wender, P.A.; Gamber, G.G.; Scanio, M.J.C. *Angew. Chem. Int. Ed.* **2001**, *40*, 3895; Wender, P.A.; Pedersen, T.M.; Scanio, M.J.C. *J. Am. Chem. Soc.* **2002**, *124*, 15154; Wender, P.A.; Love, J.A.; Williams, T.J. *Synlett* **2003**, 1295.

[2436]Shönberg, A. *Preparative Organic Photochemistry*, Springer, NY, **1968**, pp. 97–99. For other examples see Sieburth, S.Mc.N.; McGee, Jr., K.F.; Al-Tel, T.H. *Tetrahedron Lett.* **1999**, *40*, 4007; Sieburth, S.Mc.N.; Lin, C.H.; Rucando, D. *J. Org. Chem.* **1999**, *64*, 950, 954; Zhu, M.; Qiu, Z.; Hiel, G.P.; Sieburth, S.Mc.N. *J. Org. Chem.* **2002**, *67*, 3487.

[2437]Farrant, G.C.; Feldmann, R. *Tetrahedron Lett.* **1970**, 4979.

Benzene rings can undergo photochemical cycloaddition with alkenes.[2439] The major product is usually the 1,3-addition product **224** (in which a three-membered ring has also been formed), although some of the 1,2 product **225**

(**15-63**) is sometimes formed as well. (**225** is usually the main product where the alkene bears electron-withdrawing groups and the aromatic compound electron-donating groups, or vice versa.) The 1,4 product **226** is rarely formed. The reaction has also been run with benzenes substituted with alkyl, halo, OR, CN, and other groups, and with acyclic and cyclic alkenes bearing various groups.[2440]

A [2 + 2 + 2]-cycloaddition reaction is also known, facilitated by Ni(cod)$_2$[2441] or a cobalt catalyst.[2442] [2 + 2 + 1]-Cycloaddition is known.[2443] A cobalt catalyst is used for a [4 + 2 + 2]-cycloaddition of 1,3-butadiene and bicyclo[2.2.2]octa-2,5-diene.[2444] Eight-membered rings are products by a rhodium catalyzed [4 + 2 + 2]-cycloaddition.[2445] Chromium catalysts are available for [6 + 4]-cycloadditions.[2446]

OS **VI**, 512; **VII**, 485; **X**, 1, 336.

[2438]Rigby, J.H.; Kondratenko, M.A.; Fiedler, C. *Org. Lett.* **2000**, *2*, 3917; Rigby, J.H.; Mann, L.W.; Myers, B.J. *Tetrahedron Lett.* **2001**, *42*, 8773. See Rigby, J.H.; Ateeq, H.S.; Charles, N.R.; Henshilwood, J.A.; Short, K.M.; Sugathapala, P.M. *Tetrahedron* **1993**, *49*, 5495.

[2439]For reviews, see Wender, P.A.; Ternansky, R.; deLong, M.; Singh, S.; Olivero, A.; Rice, K. *Pure Appl. Chem.* **1990**, *62*, 1597; Gilbert, A., in Horspool, W.M. *Synthetic Organic Photochemistry*, Plenum, NY, *1984*, pp. 1–60. For a review of this and related reactions, see McCullough, J.J. *Chem. Rev.* **1987**, *87*, 811.

[2440]See the table, in Wender, P.A.; Siggel, L.; Nuss, J.M. *Org. Photochem.* **1989**, *10*, 357, pp. 384–415.

[2441]Lautens, M.; Edwards, L.G.; Tam, W.; Lough, A.J. *J. Am. Chem. Soc.* **1995**, *117*, 10276; Louie, J.; Gibby, J.E.; Farnsworth, M.V.; Tekavec, T.N. *J. Am. Chem. Soc.* **2002**, *124*, 15188.

[2442]Slowinski, F.; Aubert, C.; Malacria, M. *Tetrahedron Lett.* **1999**, *40*, 5849.

[2443]Knölker, H.-J.; Braier, A.; Bröcher, D.J.; Jones, P.G.; Piotrowski, H. *Tetrahedron Lett.* **1999**, *40*, 8075; Chatani, N.; Tobisu, M.; Asaumi, T.; Fukumoto, Y.; Murai, S. *J. Am. Chem. Soc.* **1999**, *121*, 7160.

[2444]Kiattansakul, R.; Snyder, J.K. *Tetrahedron Lett.* **1999**, *40*, 1079.

[2445]Gilbertson, S. R.; DeBoef, B. *J. Am. Chem. Soc.* **2002**, *124*, 8784.

[2446]Kündig, E.P.; Robvieux, F.; Kondratenko, M. *Synthesis* **2002**, 2053.

CHAPTER 16

Addition to Carbon–Hetero Multiple Bonds

MECHANISM AND REACTIVITY

The reactions considered in this chapter involve addition to the carbon–oxygen, carbon–nitrogen, and carbon–sulfur double bonds, and the carbon–nitrogen triple bond. The mechanistic study of these reactions is much simpler than that of the additions to carbon–carbon multiple bonds considered in Chapter 15.[1] Most of the questions that concerned us there either do not arise here or can be answered very simply. Since C=O, C=N, and C≡N bonds are strongly polar, with the carbon always the positive end (except for isocyanides, see p. 1466), there is never any doubt about the *orientation* of unsymmetrical addition to these bonds. Nucleophilic attacking species always go to the carbon and electrophilic species to the oxygen or nitrogen. Additions to C=S bonds are much less common,[2] but in these cases the addition can be in the other direction.[3] For example, thiobenzophenone ($Ph_2C=S$), when treated with phenyllithium gives, after hydrolysis, benzhydryl phenyl sulfide (Ph_2CHSPh).[4]

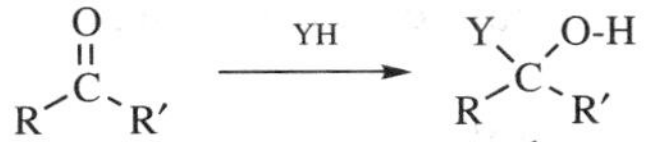

$$\underset{R \diagdown C \diagup R'}{\overset{\overset{O}{\|}}{\underset{}{C}}} \quad \overset{YH}{\longrightarrow} \quad \underset{\underset{\mathbf{1}}{R \diagdown C \diagup R'}}{\overset{Y \diagdown \diagup O\text{-}H}{C}}$$

---

[1]For a discussion, see Jencks, W.P. *Prog. Phys. Org. Chem.* **1964**, *2*, 63.

[2]For reviews of thioketones and other compounds with C=S bonds, see Schaumann, E., in Patai, S. *Supplement A: The Chemistry of Double-Bonded Functional Groups*, Vol. 2, pt. 2, Wiley, NY, **1989**, pp. 1269–1367; Ohno, A. in Oae, S. *Organic Chemistry of Sulfur*, Plenum, NY, **1977**, pp. 189–229; Mayer, R., in Janssen, M.J. *Organosulfur Chemistry*, Wiley, NY, **1967**, pp. 219–240; Campaigne, E., in Patai, S. *The Chemistry of the Carbonyl Group*, pt. 1, Wiley, NY, **1966**, pp. 917–959.

[3]For a review of additions of organometallic compounds to C=S bonds, both to the sulfur (*thiophilic addition*) and to the carbon (*carbophilic addition*), see Wardell, J.L.; Paterson, E.S., in Hartley, F.P.; Patai, S. *The Chemistry of the Metal-Carbon Bond*, Vol. 2, Wiley, NY, **1985**, pp. 219–338, 261–267.

[4]Beak, P.; Worley, J.W. *J. Am. Chem. Soc.* **1972**, *94*, 597. For some other examples, see Schaumann, E.; Walter, W. *Chem. Ber.* **1974**, *107*, 3562; Metzner, P.; Vialle, J.; Vibet, A. *Tetrahedron* **1978**, *34*, 2289.

In addition of YH to a ketone to give **1** the product has a stereogenic carbon, but unless there is chirality in R or R′ or YH is optically active, the product must be a racemic mixture because there is no facial bias about the carbonyl. The same holds true for C=N and C=S bonds, since in none of these cases can chirality be present at the heteroatom. The stereochemistry of addition of a single YH to the carbon–nitrogen triple bond could be investigated, since the product can exist in (*E*) and (*Z*) forms (p. 183), but these reactions generally give imine products that undergo further reaction. Of course, if R or R′ *is* chiral, a racemic mixture will not always arise and the stereochemistry of addition can be studied in such cases. Cram's rule (p. 168) allows us to predict the direction of attack of Y in many cases.[5] However, even in this type of study, the relative directions of attack of Y and H are not determined, but only the direction of attack of Y with respect to the rest of the substrate molecule.

On p. 1023, it was mentioned that electronic effects can play a part in determining which face of a carbon–carbon double bond is attacked. The same applies to additions to carbonyl groups. For example, in 5-substituted adamantanones (**2**) electron-withdrawing (-*I*) groups W cause the attack to come from the syn face, while electron-donating groups cause it to come from the anti face.[6] In 5,6-disubstituted norborn-2-en-7-one systems, the carbonyl appears to tilt away from the π-bond, with reduction occurring from the more hindered face.[7] An *ab initio* study of nucleophilic addition to 4-*tert*-butylcyclohexanones attempted to predict π-facial selectivity in that system.[8]

The mechanistic picture is further simplified by the fact that free-radical additions to carbon–heteroatom double bonds are not as prevalent.[9] In most cases, it

[5]For a discussion of such rules, see Eliel, E.L. *The Stereochemistry of Carbon Compounds*, McGraw-Hill, NY, *1962*, pp. 68–74. For reviews of the stereochemistry of addition to carbonyl compounds, see Bartlett, P.A. *Tetrahedron* **1980**, *36*, 2, 22; Ashby, E.C.; Laemmle, J.T. *Chem. Rev.* **1975**, *75*, 521; Goller, E.J. *J. Chem. Educ.* **1974**, *51*, 182; Toromanoff, E. *Top. Stereochem.* **1967**, *2*, 157.

[6]Cheung, C.K.; Tseng, L.T.; Lin, M.; Srivastava, S.; le Noble, W.J. *J. Am. Chem. Soc.* **1986**, *108*, 1598; Laube, T.; Stilz, H.U. *J. Am. Chem. Soc.* **1987**, *109*, 5876.

[7]Kumar, V.A.; Venkatesan, K.; Ganguly, B.; Chandrasekhar, J.; Khan, F.A.; Mehta, G. *Tetrahedron Lett.* **1992**, *33*, 3069.

[8]Yadav, V.K.; Jeyaraj, D.A. *J. Org. Chem.* **1998**, *63*, 3474. For a discussion of models, see Priyakumar, U.D.; Sastry, G.N.; Mehta, G. *Tetrahedron* **2004**, *60*, 3465.

[9]An example is found in **16-31**. For other examples, see Kaplan, L. *J. Am. Chem. Soc.* **1966**, *88*, 1833; Drew, R.M.; Kerr, J.A. *Int. J. Chem. Kinet.* **1983**, *15*, 281; Fraser-Reid, B.; Vite, G.D.; Yeung, B.A.; Tsang, R. *Tetrahedron Lett.* **1988**, *29*, 1645; Beckwith, A.L.J.; Hay, B.P. *J. Am. Chem. Soc.* **1989**, *111*, 2674; Clerici, A.; Porta, O. *J. Org. Chem.* **1989**, *54*, 3872; Cossy, J.; Pete, J.P.; Portella, C. *Tetrahedron Lett.* **1989**, *30*, 7361.

is the nucleophile that forms the first new bond to carbon, and these reactions are regarded as *nucleophilic additions*, which can be represented thus (for the C=O bond, analogously for the others):

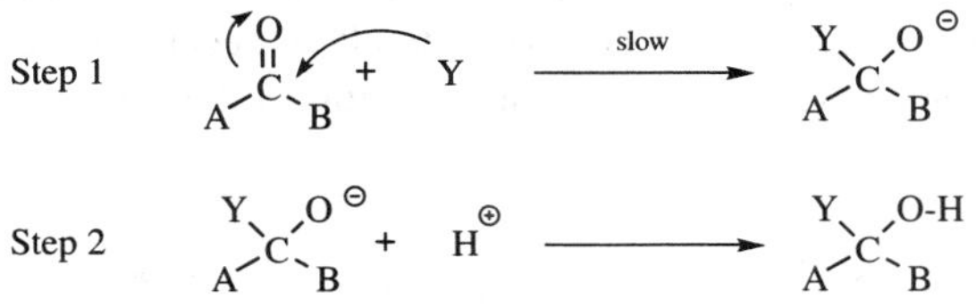

Step 1

Step 2

The electrophile shown in step 2 is the proton. In almost all the reactions considered in this chapter, the electrophilic atom is either hydrogen or carbon. Note that step 1 is exactly the same as step 1 of the tetrahedral mechanism of nucleophilic substitution at a carbonyl carbon (p. 1255), but carbon groups (A, B = H, alkyl aryl, etc.) are poor leaving groups so that substitution does not compete with addition. For carboxylic acids and their derivatives (B = OH, OR, NH_2, etc.) much better leaving groups are available and acyl substitution predominates (p. 1254). It is thus the nature of A and B that determines whether a nucleophilic attack at a carbon–heteroatom multiple bond will lead to substitution or addition.

It is also possible for the heteroatom (oxygen in a carbonyl) to react as a base, attacking the electrophilic species. This species is most often a proton and the mechanism is

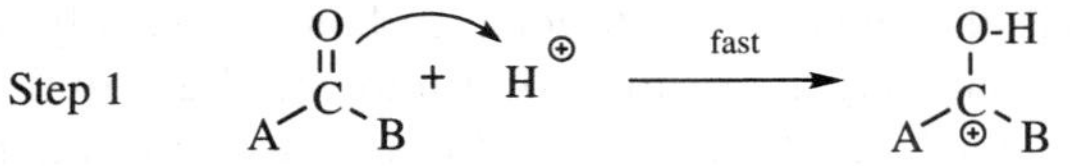

Step 1

3

Step 2

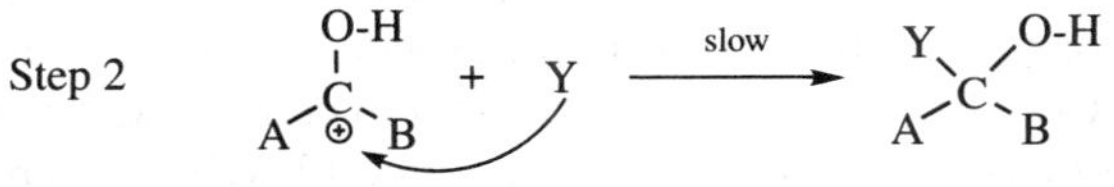

Whether the nucleophile attacks the carbon or the heteroatom attacks the electrophilic species, the rate-determining step is usually the one involving nucleophilic attack. It may be observed that many of these reactions can be catalyzed by both acids and bases.[10] Bases catalyze the reaction by converting a reagent of the form YH to the more powerful nucleophile Y^- (see p. 490). Acids catalyze it by converting the substrate to an heteroatom-stabilized cation (formation of **3**), thus making it more attractive to nucleophilic attack. Similar catalysis can also be found with metallic ions (e.g., Ag^+) which act here as Lewis acids.[11] We have mentioned before (p. 242) that ions of type **3** are comparatively stable carbocations because the positive charge is spread by resonance.

---

[10]For a discussion of acid and base catalysis in these reactions, see Jencks, W.P.; Gilbert, H.F. *Pure Appl. Chem.* **1977**, *49*, 1021.
[11]Toromanoff, E. *Bull. Soc. Chim. Fr.* **1962**, 1190.

Reactivity factors in additions to carbon–heteroatom multiple bonds are similar to those for the tetrahedral mechanism of nucleophilic substitution.[12] If A and/or B are electron-donating groups, rates are decreased. Electron-attracting substituents increase rates. This means that aldehydes are more reactive than ketones. Aryl groups are somewhat deactivating compared to alkyl, because of resonance that stabilizes the substrate molecule, but is lost on going to the intermediate:

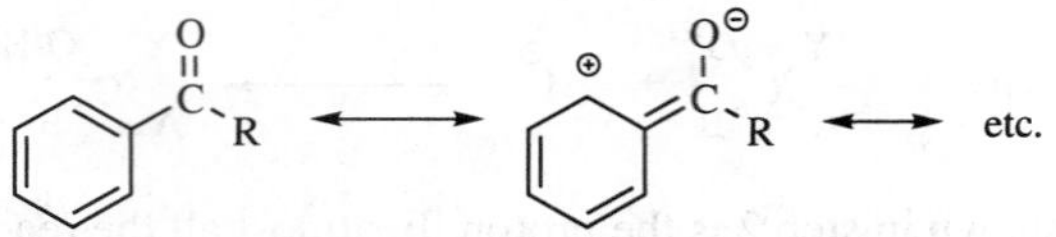

Double bonds in conjugation with the carbon–heteroatom multiple bond also lower addition rates, for similar reasons but, more important, may provide competition from 1,4-addition (p. 1008). Steric factors are also quite important and contribute to the decreased reactivity of ketones compared with aldehydes. Highly hindered ketones like hexamethylacetone and dineopentyl ketone either do not undergo many of these reactions or require extreme conditions.

Nucleophilic Substitution at an Aliphatic Trigonal Carbon: The Tetrahedral Mechanism

All the mechanisms so far discussed take place at a saturated carbon atom. Nucleophilic substitution is also important at trigonal carbons, especially when the carbon is double bonded to an oxygen, a sulfur, or a nitrogen. Substitution at a carbonyl group (or the corresponding nitrogen and sulfur analogs) most often proceeds by a second-order mechanism, which in this book is called the *tetrahedral*[13] *mechanism.*[14] The IUPAC designation is $A_N + D_N$. The S_N1 mechanisms, involving carbocations, are sometimes found with these substrates, especially with essentially ionic substrates such as $RCO^+ BF_4^-$; there is evidence that in certain cases simple S_N2 mechanisms can take place, especially with a very good leaving group such as Cl^-;[15]

[12]For a review of the reactivity of nitriles, see Schaefer, F.C., in Rappoport, Z. *The Chemistry of the Cyano Group*, Wiley, NY, *1970*, pp. 239–305.

[13]This mechanism has also been called the "additionelimination mechanism," but in this book we limit this term to the type of mechanism shown on p. $$$.

[14]For reviews of this mechanism, see Talbot, R.J.E., in Bamford, C.H.; Tipper, C.F.H. *Comprehensive Chemical Kinetics*, Vol. 10, Elsevier, NY, *1972*, pp. 209–223; Jencks, W.P. *Catalysis in Chemistry and Enzymology*, McGraw-Hill, NY, *1969*, pp. 463–554; Satchell, D.P.N.; Satchell, R.S., in Patai, S. *The Chemistry of Carboxylic Acids and Esters*, Wiley, NY, *1969*, pp. 375–452; Johnson, S.L. *Adv. Phys. Org. Chem. 1967*, *5*, 237.

[15]For a review, see Williams, A. *Acc. Chem. Res. 1989*, *22*, 387. For examples, see Kevill, D.N.; Foss, F.D. *J. Am. Chem. Soc. 1969*, *91*, 5054; Haberfield, P.; Trattner, R.B. *Chem. Commun. 1971*, 1481; De Tar, D.F. *J. Am. Chem. Soc. 1982*, *104*, 7205; Shpan'ko, I.V.; Goncharov, A.N. *J. Org. Chem. USSR 1987*, *23*, 2287; Guthrie, J.P.; Pike, D.C. *Can. J. Chem. 1987*, *65*, 1951; Kevill, D.N.; Kim, C. *Bull. Soc. Chim. Fr. 1988*, 383, *J. Chem. Soc. Perkin Trans. 2 1988*, 1353; Bentley, T.W.; Koo, I.S. *J. Chem. Soc. Perkin Trans. 2 1989*, 1385. See, however, Buncel, E.; Um, I.H.; Hoz, S. *J. Am. Chem. Soc. 1989*, *111*, 971.

and an SET mechanism has also been reported.[16] However, the tetrahedral mechanism is by far the most prevalent. Although this mechanism displays second-order kinetics, it is not the same as the S_N2 mechanism previously discussed. In the tetrahedral mechanism, first Y attacks to give an intermediate containing both X and Y (**4**), and then X leaves. This sequence, impossible at a saturated carbon, is possible at an unsaturated one because the central carbon can release a pair of electrons to the oxygen and so preserve its octet:

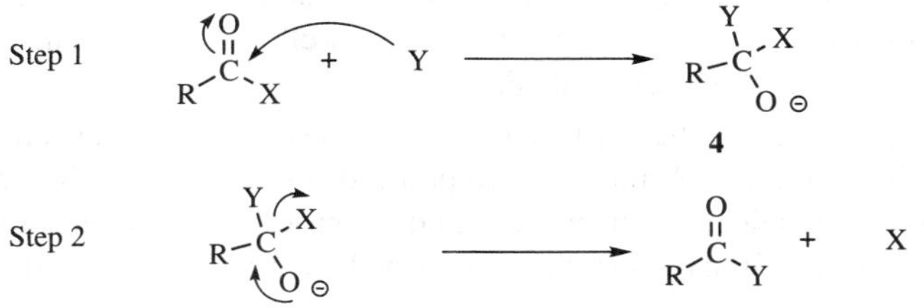

When reactions are carried out in acid solution, there may also be a preliminary and a final step:

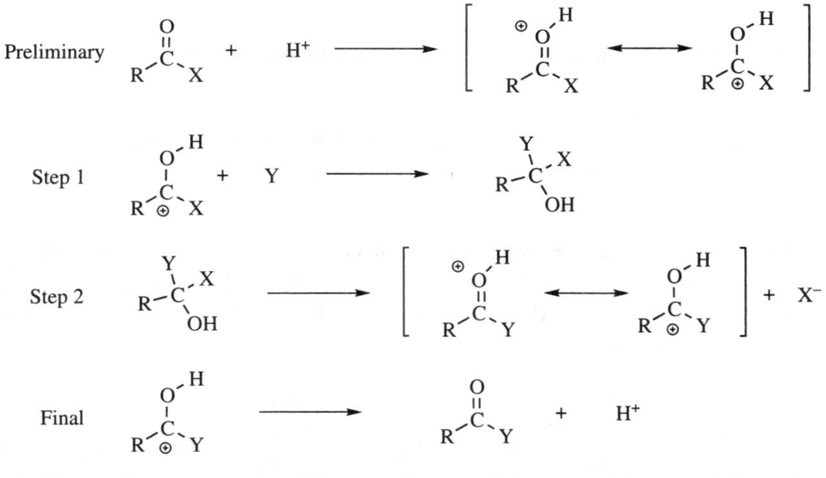

The hydrogen ion is a catalyst. The reaction rate is increased because it is easier for the nucleophile to attack the carbon when the electron density of the latter has been decreased.[17]

Evidence for the existence of the tetrahedral mechanism is as follows:[18]

1. The kinetics are first order each in the substrate and in the nucleophile, as predicted by the mechanism.

[16]Bacaloglu, R.; Blaskó, A.; Bunton, C.A.; Ortega, F. *J. Am. Chem. Soc.* **1990**, *112*, 9336.

[17]For discussions of general acid and base catalysis of reactions at a carbonyl group, see Jencks, W.P. *Acc. Chem. Res.* **1976**, *9*, 425; *Chem. Rev.* **1972**, *72*, 705.

[18]For additional evidence, see Guthrie, J.P. *J. Am. Chem. Soc.* **1978**, *100*, 5892; Kluger, R.; Chin, J. *J. Am. Chem. Soc.* **1978**, *100*, 7382; O'Leary, M.H.; Marlier, J.F. *J. Am. Chem. Soc.* **1979**, *101*, 3300.

2. There is other kinetic evidence in accord with a tetrahedral intermediate. For example, the rate "constant" for the reaction between acetamide and hydroxylamine is not constant, but decreases with increasing hydroxylamine concentration.[19] This is not a smooth decrease; there is a break in the curve. A straight line is followed at low hydroxylamine concentration and another straight line at high concentration. This means that the identity of the rate-determining step is changing. Obviously, this cannot happen if there is only one step: there must be two steps, and hence an intermediate. Similar kinetic behavior has been found in other cases as well,[20] in particular, plots of rate against pH are often bell shaped.

3. Basic hydrolysis has been carried out on carboxylic esters labeled with ^{18}O in the carbonyl group.[21] If this reaction proceeded by the normal S_N2 mechanism, all the ^{18}O would remain in the carbonyl group, even if, in an equilibrium process, some of the carboxylic acid formed went back to the starting material:

$$HO^- + \underset{R}{\overset{^{18}O}{C}}{}^{OR'} \rightleftharpoons \underset{R}{\overset{^{18}O}{C}}{}^{OH} + R'O^- \longrightarrow \underset{R}{\overset{^{18}O}{C}}{}^{O^-} + R'OH$$

On the other hand, if the tetrahedral mechanism operates

$$HO^- + \underset{R}{\overset{^{18}O}{C}}{}^{OR'} \rightleftharpoons \underset{{}^{18}O^{\ominus}}{\overset{HO}{\underset{|}{R-C-OR'}}} \overset{H_2O}{\rightleftharpoons} \underset{{}^{18}OH}{\overset{HO}{\underset{|}{R-C-OR'}}}$$
5

then the intermediate **5**, by gaining a proton, becomes converted to the symmetrical intermediate **6**. In this intermediate the OH groups are equivalent, and (except for the small $^{18}O/^{16}O$ isotope effect) either one can lose a proton with equal facility:

$$\underset{R}{\overset{^{18}O}{C}}{}^{OR'} + OH^- \rightleftharpoons \underset{{}^{18}O^{\ominus}}{\overset{HO}{\underset{|}{R-C-OR'}}} \rightleftharpoons \underset{{}^{18}OH}{\overset{HO}{\underset{|}{R-C-OR'}}} \rightleftharpoons \underset{O^{\ominus}}{\overset{HO}{\underset{|}{R-C-OR'}}} \rightleftharpoons \underset{R}{\overset{O}{C}}{}^{OR'} + {}^{18}OH^-$$
5　　　　　**6**　　　　　**7**

The intermediates **5** and **7** can now lose OR′ to give the acid (not shown in the equations given), or they can lose OH to regenerate the carboxylic ester. If **5** goes back to ester, the ester will still be labeled, but if **7** reverts to ester, the

[19]Jencks, W.P.; Gilchrist, M. *J. Am. Chem. Soc.* **1964**, *86*, 5616.
[20]Hand, E.S.; Jencks, W.P. *J. Am. Chem. Soc.* **1962**, *84*, 3505; Johnson, S.L. *J. Am. Chem. Soc.* **1964**, *86*, 3819; Fedor, L.R.; Bruice, T.C. *J. Am. Chem. Soc.* **1964**, *86*, 5697; **1965**, *87*, 4138; Kevill, D.N.; Johnson, S.L. *J. Am. Chem. Soc.* **1965**, *87*, 928; Leinhard, G.E.; Jencks, W.P. *J. Am. Chem. Soc.* **1965**, *87*, 3855; Schowen, R.L.; Jayaraman, H.; Kershner, L.D. *J. Am. Chem. Soc.* **1966**, *88*, 3373.
[21]Bender, M.L. *J. Am. Chem. Soc.* **1951**, *73*, 1626; Bender, M.L.; Thomas, R.J. *J. Am. Chem. Soc.* **1961**, *83*, 4183, 4189.

^{18}O will be lost. A test of the two possible mechanisms is to stop the reaction before completion and to analyze the recovered ester for ^{18}O. This is just what was done by Bender, who found that in alkaline hydrolysis of methyl, ethyl, and isopropyl benzoates, the esters had lost ^{18}O. A similar experiment carried out for acid-catalyzed hydrolysis of ethyl benzoate showed that here too the ester lost ^{18}O. However, alkaline hydrolysis of substituted benzyl benzoates showed *no* ^{18}O loss.[22] This result does not necessarily mean that no tetrahedral intermediate is involved in this case. If **5** and **7** do not revert to ester, but go entirely to acid, no ^{18}O loss will be found even with a tetrahedral intermediate. In the case of benzyl benzoates, this may very well be happening, because formation of the acid relieves steric strain. Another possibility is that **5** loses OR$'$ before it can become protonated to **6**.[23] Even the experiments that *do* show ^{18}O loss do not *prove* the existence of the tetrahedral intermediate, since it is possible that ^{18}O is lost by some independent process not leading to ester hydrolysis. To deal with this possibility, Bender and Heck[24] measured the rate of ^{18}O loss in the hydrolysis of ethyl trifluorothioloacetate-^{18}O:

$$\underset{F_3C}{\overset{\overset{\displaystyle ^{18}O}{\overset{\displaystyle \|}{C}}}{\diagup}}\diagdown_{SEt} \;+\; H_2O \; \underset{k_2}{\overset{k_1}{\rightleftharpoons}} \; \text{Intermediate} \; \xrightarrow{k_3} \; F_3CCOOH + EtSH$$

This reaction had previously been shown[25] to involve an intermediate by the kinetic methods mentioned on p. 1256. Bender and Heck showed that the rate of ^{18}O loss and the value of the partitioning ratio k_2/k_3 as determined by the oxygen exchange technique were exactly in accord with these values as previously determined by kinetic methods. Thus the original ^{18}O-exchange measurements showed that there is a tetrahedral species present, though not necessarily on the reaction path, while the kinetic experiments showed that there is some intermediate present, though not necessarily tetrahedral. Bender and Heck's results demonstrate that there is a tetrahedral intermediate and that it lies on the reaction pathway.

4. In some cases, tetrahedral intermediates have been isolated[26] or detected spectrally.[27]

[22]Bender, M.L.; Matsui, H.; Thomas, R.J.; Tobey, S.W. *J. Am. Chem. Soc.* **1961**, *83*, 4193. See also, Shain, S.A.; Kirsch, J.F. *J. Am. Chem. Soc.* **1968**, *90*, 5848.

[23]For evidence for this possibility, see McClelland, R.A. *J. Am. Chem. Soc.* **1984**, *106*, 7579.

[24]Bender, M.L.; Heck, H. d'A. *J. Am. Chem. Soc.* **1967**, *89*, 1211.

[25]Fedor, L.R.; Bruice, T.C. *J. Am. Chem. Soc.* **1965**, *87*, 4138.

[26]Rogers, G.A.; Bruice, T.C. *J. Am. Chem. Soc.* **1974**, *96*, 2481; Khouri, F.F.; Kaloustian, M.K. *J. Am. Chem. Soc.* **1986**, *108*, 6683.

[27]For reviews, see Capon, B.; Dosunmu, M.I.; Sanchez, M. de N de M. *Adv. Phys. Org. Chem.* **1985**, *21*, 37; McClelland, R.A.; Santry, L.J. *Acc. Chem. Res.* **1983**, *16*, 394; Capon, B.; Ghosh, A.K.; Grieve, D.M.A. *Acc. Chem. Res.* **1981**, *14*, 306. See also, Lobo, A.M.; Marques, M.M.; Prabhakar, S.; Rzepa, H.S. *J. Chem. Soc., Chem. Commun.* **1985**, 1113; van der Wel, H.; Nibbering, N.M.M. *Recl. Trav. Chim. Pays-Bas* **1988**, *107*, 479, 491.

Several studies have been made of the directionality of approach by the nucleophile.[28] Menger has proposed for reactions in general, and specifically for those that proceed by the tetrahedral mechanism, that there is no single definable preferred transition state, but rather a "cone" of trajectories. All approaches within this cone lead to reaction at comparable rates; it is only when the approach comes outside of the cone that the rate falls.

Directionality has also been studied for the second step. Once the tetrahedral intermediate (**4**) is formed, it loses Y (giving the product) or X (reverting to the starting compound). Deslongchamps has proposed that one of the factors affecting this choice is the conformation of the intermediate; more specifically, the positions of the lone pairs. In this view, a leaving group X or Y can depart only if the other two atoms on the carbon both have an orbital antiperiplanar to the C—X or C—Y bond. For example, consider an intermediate **8** formed by attack of $^-$OR on a substrate R'COX. Cleavage of the C—X bond with loss of X can take place from conformation **A**, because the two lone-pair orbitals marked * are antiperiplanar to the C—X bond, but not from **B** because only the O$^-$ has such an orbital. If the intermediate is in conformation **B**, the OR may leave (if X has a lone-pair orbital in the proper position) rather than X. This factor is called *stereoelectronic control*.[29] Of course, there is free rotation in acyclic intermediates, and many conformations are possible, but some are preferred, and cleavage reactions may take place faster than rotation, so stereoelectronic control can be a factor in some situations. Much evidence has been presented for this concept.[30] More generally, the term *stereoelectronic effects* refers to any case in which orbital

[28]For discussions, see Menger, F.M. *Tetrahedron* **1983**, *39*, 1013; Liotta, C.L.; Burgess, E.M.; Eberhardt, W.H. *J. Am. Chem. Soc.* **1984**, *106*, 4849.

[29]It has also been called the "antiperiplanar lone pair hypothesis (ALPH)." For a reinterpretation of this factor in terms of the principle of least nuclear motion (see **15-10**), see Hosie, L.; Marshall, P.J.; Sinnott, M.L. *J. Chem. Soc. Perkin Trans. 2* **1984**, 1121; Sinnott, M.L. *Adv. Phys. Org. Chem.* **1988**, *24*, 113.

[30]For monographs, see Kirby, A.J. *The Anomeric Effect and Related Stereoelectronic Effects at Oxygen*, Springer, NY, **1983**; Deslongchamps, P. *Stereoelectronic Effects in Organic Chemistry*, Pergamon, NY, **1983**. For lengthy treatments, see Sinnott, M.L. *Adv. Phys. Org. Chem.* **1988**, *24*, 113; Gorenstein, D.G. *Chem. Rev.* **1987**, *87*, 1047; Deslongchamps, P. *Heterocycles* **1977**, *7*, 1271; *Tetrahedron* **1975**, *31*, 2463. For additional evidence, see Perrin, C.L.; Arrhenius, G.M.L. *J. Am. Chem. Soc.* **1982**, *104*, 2839; Briggs, A.J.; Evans, C.M.; Glenn, R.; Kirby, A.J. *J. Chem. Soc. Perkin Trans. 2* **1983**, 1637; Ndibwami, A.; Deslongchamps, P.*Can. J. Chem.* **1986**, *64*, 1788; Hegarty, A.F.; Mullane, M. *J. Chem. Soc. Perkin Trans. 2* **1986**, 995. For evidence against the theory, see Perrin, C.L.; Nuñez, O. *J. Am. Chem. Soc.* **1986**, *108*, 5997; *1987*, *109*, 522.

position requirements affect the course of a reaction. The backside attack in the S_N2 mechanism is an example of a stereoelectronic effect.

Some nucleophilic substitutions at a carbonyl carbon are *catalyzed* by nucleophiles.[31] There occur, in effect, two tetrahedral mechanisms:

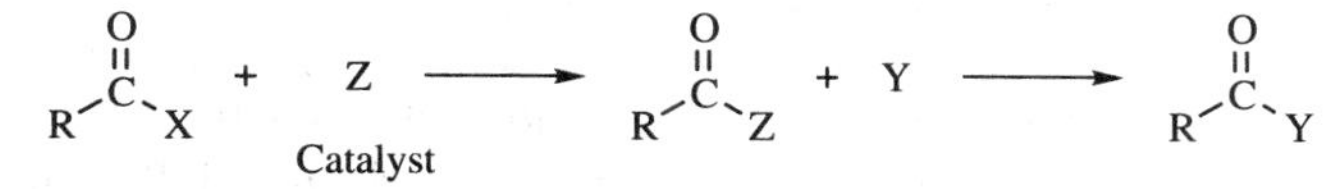

(For an example, see **16-58**). When this happens internally, we have an example of a neighboring-group mechanism at a carbonyl carbon.[32] For example, the hydrolysis of phthalamic acid (**9**) takes place as follows:

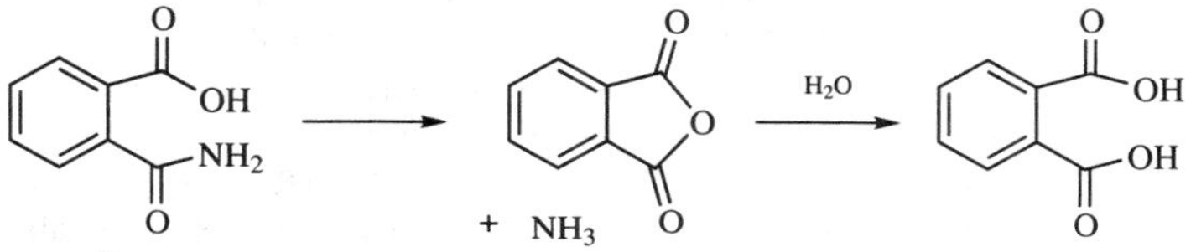

Evidence comes from comparative rate studies.[33] Thus **9** was hydrolyzed $\sim 10^5$ times faster than benzamide ($PhCONH_2$) at about the same concentration of hydrogen ions. That this enhancement of rate was not caused by the resonance or field effects of COOH (an electron-withdrawing group) was shown by the fact both *o*-nitrobenzamide and terephthalamic acid (the para isomer of **9**) were hydrolyzed more slowly than benzamide. Many other examples of neighboring-group participation at a carbonyl carbon have been reported.[34] It is likely that nucleophilic catalysis is involved in enzyme catalysis of ester hydrolysis.

The attack of a nucleophile on a carbonyl group can result in substitution or addition, though the first step of each mechanism is the same. The main factor that determines the product is the identity of the group X in RCOX. When X is alkyl or hydrogen, addition usually takes place. When X is halogen, OH, OCOR, NH_2, and so on, the usual reaction is substitution.

In both the S_N1 and S_N2 mechanisms, the leaving group departs during the rate-determining step and so directly affects the rate. In the tetrahedral mechanism at a carbonyl carbon, the bond between the substrate and leaving group is still intact during the slow step. Nevertheless, the nature of the leaving group still affects the reactivity in two ways: (*1*) By altering the

[31]For reviews of nucleophilic catalysis, see Bender, M.L. *Mechanisms of Homogeneous Catalysis from Protons to Proteins*, Wiley, NY, *1971*, pp. 147–179; Jencks, W.P. *Catalysis in Chemistry and Enzymology*, McGraw-Hill, NY, *1969*, pp. 67–77; Johnson, S.L. *Adv. Phys. Org. Chem.* *1967*, *5*, p. 271. For a review where Z = a tertiary amine (the most common case), see Cherkasova, E.M.; Bogatkov, S.V.; Golovina, Z.P. *Russ. Chem. Rev.* *1977*, *46*, 246.

[32]For reviews, see Kirby, A.J.; Fersht, A.R. *Prog. Bioorg. Chem.* *1971*, *1*, 1; Capon, B. *Essays Chem.* *1972*, *3*, 127.

[33]Bender, M.L.; Chow, Y.; Chloupek, F.J. *J. Am. Chem. Soc.* *1958*, *80*, 5380.

[34]For examples, see Bruice, T.C.; Pandit, U.K. *J. Am. Chem. Soc.* *1960*, *82*, 5858; Kluger, R.; Lam, C. *J. Am. Chem. Soc.* *1978*, *100*, 2191; Page, M.I.; Render, D.; Bernáth, G. *J. Chem. Soc. Perkin Trans. 2* *1986*, 867.

TABLE 16.1. The More Important Synthetic Reactions that Take Place by the Tetrahedral Mechanism[a]

Reaction Number	Reaction
16-57	$RCOX + H_2O \longrightarrow RCOOH$
16-58	$RCOOCOR' + H_2O \longrightarrow RCOOH + R'COOH$
16-59	$RCO_2R' + H_2O \longrightarrow RCOOH + R'OH$
16-59	$RCONR'_2 + H_2O \longrightarrow RCOOH + R_2NH$ (R' = H, alkyl, aryl)
16-61	$RCOX + R'OH \longrightarrow RCO_2R'$
16-62	$RCOOCOR + R'OH \longrightarrow RCO_2R'$
16-63	$RCOOH + R'OH \longrightarrow RCO_2R'$
16-64	$RCO_2R' + R''OH \longrightarrow RCO_2R'' + R'OH$
16-66	$RCOX + R'COO^- \longrightarrow RCOOCOR'$
10-21	$RCOX + H_2O_2 \longrightarrow RCO_3H$
16-69	$RCOX + R'SH \longrightarrow RCOSR'$
16-72	$RCOX + NHR'_2 \longrightarrow RCONR'_2$ (R' = H, alkyl, aryl)
16-73	$RCOOCOR + NHR'_2 \longrightarrow RCONR'_2$ (R' = H, alkyl, aryl)
16-74	$RCOOH + NHR'_2 \xrightarrow[\text{agent}]{\text{coupling}} RCONR'_2$ (R' = H, alkyl, aryl)
16-75	$RCO_2R' + NHR^2 \longrightarrow RCONR^2$ (R^2 = H, alkyl, aryl)
16-79	$RCOOH + SOCl_2 \longrightarrow RCOCl$
19-39	$RCOX + LiAlH(O\text{–}t\text{-Bu})_3 \longrightarrow RCHO$
19-41	$RCONR'_2 + LiAlH_4 \longrightarrow RCHO$
16-81	$RCOX + R_2CuLi \longrightarrow RCOR'$
16-85	$2RCH_2CO_2R' \longrightarrow RCH_2COCHRCO_2R'$

[a]Catalysts are not shown.

electron density at the carbonyl carbon, the rate of the reaction is affected. The greater the electron-withdrawing character of X, the greater the partial positive charge on C and the more rapid the attack by a nucleophile. (2) The nature of the leaving group affects the *position of equilibrium*. In the intermediate **4** (p. 1255), there is competition between X and Y as to which group leaves. If X is a poorer leaving group than Y, then Y will preferentially leave and **4** will revert to the starting compounds. Thus there is a partitioning factor between **4** going on to product (loss of X) or back to starting compound (loss of Y). The sum of these two factors causes the sequence of reactivity to be RCOCl > RCOOCOR' > RCOOAr > RCOOR' > RCONH$_2$ > RCONR'$_2$ > RCOO$^-$.[35] Note that this order is approximately the order of decreasing stability of the leaving-group anion. If the leaving group is bulky, it may exert a steric effect and retard the rate for this reason.

For a list of some of the more important reactions that operate by the tetrahedral mechanism, see Table 16.1, which shows the main reactions that proceed by the tetrahedral mechanism.

[35]RCOOH would belong in this sequence just after RCOOAr, but it fails to undergo many reactions for a special reason. Many nucleophiles, instead of attacking the C=O group, are basic enough to take a proton from the acid, converting it to the unreactive RCOO$^-$.

REACTIONS

Many of the reactions in this chapter are simple additions to carbon–hetero multiple bonds, with the reaction ending when the two groups have been added. But in many other cases subsequent reactions take place. We will meet a number of such reactions, but most are of two types:

Type A
$$\underset{A}{\overset{O}{\underset{}{\overset{\|}{C}}}}{}_{B} + YH_2 \longrightarrow \underset{A}{\overset{HY}{\underset{}{C}}}{\overset{O\text{-}H}{}}_{B} \xrightarrow{-H_2O} \underset{A}{\overset{Y}{\underset{}{\overset{\|}{C}}}}{}_{B}$$

Type B
$$\underset{A}{\overset{O}{\underset{}{\overset{\|}{C}}}}{}_{B} + YH_2 \longrightarrow \underset{A}{\overset{HY}{\underset{}{C}}}{\overset{O\text{-}H}{}}_{B} \xrightarrow{z} \underset{A}{\overset{HY}{\underset{}{C}}}{\overset{Z}{}}_{B}$$

In type A, the initially formed adduct loses water (or, in the case of addition to C=NH, ammonia, etc.), and the net result of the reaction is the substitution of C=Y for C=O (or C=NH, etc.). In type B, there is a rapid substitution, and the OH (or NH_2, etc.) is replaced by another group Z, which is often another YH moiety. This substitution is in most cases nucleophilic, since Y usually has an unshared pair and S_N1 reactions occur very well on this type of compound (see p. 482), even when the leaving group is as poor as OH or NH_2. In this chapter, we will classify reactions according to what is initially adding to the carbon–hetero multiple bond, even if subsequent reactions take place so rapidly that it is not possible to isolate the initial adduct.

Most of the reactions considered in this chapter can be reversed. In many cases, we will consider the reverse reactions with the forward ones, in the same section. The reverse of some of the other reactions are considered in other chapters. In still other cases, one of the reactions in this chapter is the reverse of another (e.g., **16-2** and **16-13**). For reactions that are reversible, the principle of microscopic reversibility (p. 309) applies.

First, we will discuss reactions in which hydrogen or a metallic ion (or in one case phosphorus or sulfur) adds to the heteroatom and second reactions in which carbon adds to the heteroatom. Within each group, the reactions are classified by the nature of the nucleophile. Additions to isocyanides, which are different in character, follow. Acyl substitution reactions that proceed by the tetrahedral mechanism, which mostly involve derivatives of carboxylic acids, are treated at the end.

REACTIONS IN WHICH HYDROGEN OR A METALLIC ION ADDS TO THE HETEROATOM

A. Attack by OH (Addition of H_2O)

16-1   The Addition of Water to Aldehydes and Ketones: Formation of Hydrates

O-Hydro-*C*-hydroxy-addition

The adduct formed upon addition of water to an aldehyde or ketone is called a hydrate or *gem*-diol.[36] These compounds are usually stable only in water solution and decompose on distillation; that is, the equilibrium shifts back toward the carbonyl compound. The position of the equilibrium is greatly dependent on the structure of the hydrate. Thus, formaldehyde in water at 20°C exists 99.99% in the hydrated form, while for acetaldehyde this figure is 58%, and for acetone the hydrate concentration is negligible.[37] It has been found, by exchange with ^{18}O, that the reaction with acetone is quite rapid when catalyzed by acid or base, but the equilibrium lies on the side of acetone and water.[38] Since methyl, a +*I* group, inhibits hydrate formation, it may be expected that electron-attracting groups would have the opposite effect, and this is indeed the case. The hydrate of chloral (trichloroacetaldehyde)[39] is a stable crystalline substance. In order for it to revert to chloral, ⁻OH or H_2O must leave; this is made difficult by the electron-withdrawing character of the Cl_3C group. Some other[40] polychlorinated and polyfluorinated

Chloral hydrate   Hydrate of cyclopropanone

aldehydes and ketones[41] and α-keto aldehydes also form stable hydrates, as do cyclopropanones.[42] In the last case,[43] formation of the hydrate relieves some of the *I* strain (p. 399) of the parent ketone.

[36]For reviews, see Bell, R.P. *The Proton in Chemistry*, 2nd ed., Cornell University Press, Ithaca, NY, *1973*, pp. 183–187; *Adv. Phys. Org. Chem. 1966*, *4*, 1; Le Hénaff, P. *Bull. Soc. Chim. Fr. 1968*, 4687.

[37]Bell, R.P.; Clunie, J.C. *Trans. Faraday Soc. 1952*, *48*, 439. See also, Bell, R.P.; McDougall, A.O. *Trans. Faraday Soc. 1960*, *56*, 1281.

[38]Cohn, M.; Urey, H.C. *J. Am. Chem. Soc. 1938*, *60*, 679.

[39]For a review of chloral, see Luknitskii, F.I. *Chem. Rev. 1975*, *75*, 259.

[40]For a discussion, see Schulman, E.M.; Bonner, O.D.; Schulman, D.R.; Laskovics, F.M. *J. Am. Chem. Soc. 1976*, *98*, 3793.

[41]For a review of addition to fluorinated ketones, see Gambaryan, N.P.; Rokhlin, E.M.; Zeifman, Yu.V.; Ching-Yun, C.; Knunyants, I.L. *Angew. Chem. Int. Ed. 1966*, *5*, 947.

[42]For other examples, see Krois, D.; Lehner, H. *Monatsh. Chem. 1982*, *113*, 1019.

[43]Turro, N.J.; Hammond, W.B. *J. Am. Chem. Soc. 1967*, *89*, 1028; Schaafsma, S.E.; Steinberg, H.; de Boer, T.J. *Recl. Trav. Chim. Pays-Bas 1967*, *86*, 651. For a review of cyclopropanone chemistry, see Wasserman, H.H.; Clark, G.M.; Turley, P.C. *Top. Curr. Chem. 1974*, *47*, 73.

The reaction is subject to both general-acid and general-base catalysis; the following mechanisms can be written for basic (B) and acidic (BH) catalysis, respectively:[44]

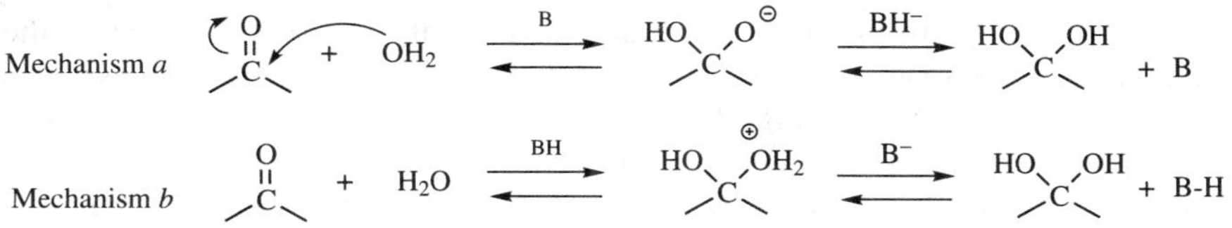

Mechanism *a*

Mechanism *b*

In mechanism *a*, as the H_2O attacks, the base pulls off a proton, and the net result is addition of ^-OH. This can happen because the base is already hydrogen bonded to the H_2O molecule before the attack. In mechanism *b*, because HB is already hydrogen bonded to the oxygen of the carbonyl group, it gives up a proton to the oxygen as the water attacks. In this way, B and HB accelerate the reaction even beyond the extent that they form ^-OH or H_3O^+ by reaction with water. Reactions in which the catalyst donates a proton to the electrophilic reagent (in this case the aldehyde or ketone) in one direction and removes it in the other are called class e reactions. Reactions in which the catalyst does the same to the nucleophilic reagent are called class n reactions.[45] Thus the acid-catalyzed process here is a class e reaction, while the base catalyzed process is a class n reaction.

For the reaction between ketones and H_2O_2, see **17-37**.

There are no OS references, but see OS **VIII**, 597, for the reverse reaction.

16-2   Hydrolysis of the Carbon-Nitrogen Double Bond[46]

Oxo-de-alkylimino-bisubstitution, and so on

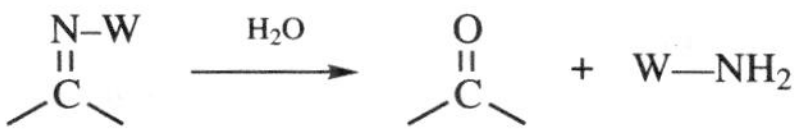

Compounds containing carbon–nitrogen double bonds can be hydrolyzed to the corresponding aldehydes or ketones.[47] For imines (W = R or H) the hydrolysis is easy and can be carried out with water. When W = H, the imine is seldom stable enough for isolation, and in aqueous media hydrolysis usually occurs *in situ*, without isolation. The hydrolysis of Schiff bases (W = Ar) is more difficult and requires

---

[44]Bell, R.P.; Rand, M.H.; Wynne-Jones, K.M.A. *Trans. Faraday Soc.* **1956**, *52*, 1093; Pocker, Y. *Proc. Chem. Soc.* **1960**, 17; Sørensen, P.E.; Jencks, W.P. *J. Am. Chem. Soc.* **1987**, *109*, 4675. For a comprehensive treatment, see Lowry, T.H.; Richardson, K.S. *Mechanism and Theory in Organic Chemistry*, 3rd ed., Harper and Row, NY, **1987**, pp. 662–680. For a theoretical treatment see Wolfe, S.; Kim, C.-K.; Yang, K.; Weinberg, N.; Shi, Z. *J. Am. Chem. Soc.* **1995**, *117*, 4240.

[45]Jencks, W.P. *Acc. Chem. Res.* **1976**, *9*, 425.

[46]For a review, see Khoee, S.; Ruoho, A.E. *Org. Prep. Proceed. Int.* **2003**, *35*, 527.

[47]The proton affinities of imines have been determined, see Hammerum, S.; Sølling, T.I. *J. Am. Chem. Soc.* **1999**, *121*, 6002.

acid or base catalysis. Oximes (W = OH), arylhydrazones (W = NHAr), and, most easily, semicarbazones (W = NHCONH$_2$) can also be hydrolyzed. Often a reactive aldehyde (e.g., formaldehyde) is added to combine with the liberated amine.

A number of other reagents[48] have been used to cleave C=N bonds, especially those not easily hydrolyzable with acidic or basic catalysts or that contain other functional groups that are attacked under these conditions. Oximes have been converted to the corresponding aldehyde or ketone[49] by treatment with, among other reagents, NBS in water,[50] glyoxylic acid (HCOCOOH),[51] Chloramine-T[52] Caro's acid on SiO$_2$,[53] HCOOH on SiO$_2$ with microwave irradiation,[54] bromosulfonamides,[55] SiBr$_4$ on wet silica,[56] and KMnO$_4$ on Al$_2$O$_3$[57] or on zeolite.[58] Chromate oxidizing agents can be quite effective, including tetraethylammonium permanganate,[59] tetraalkylammonium dichromate with microwave irradiation,[60] pyridinium fluorochromate,[61] quinolinium fluorochromate[62] or dichromate.[63] Alkaline H$_2$O$_2$,[64] iodine in acetonitrile,[65] singlet oxygen with NaOMe/MeOH[66] and with an ionic liquids on SiO$_2$[67] have also been used. Transition-metal compounds have been used, including SbCl$_5$,[68] Co$_2$(CO)$_8$[69] Hg(NO$_3$)$_2$/SiO$_2$,[70] or BiBr$_3$–Bi(OTf)$_3$ in aqueous media,[71] Bi(NO$_3$)$_3$/SiO$_2$,[72] a nickel(II) complex

[48]For a list of reagents, with references, see Ranu, B.C.; Sarkar, D.C. *J. Org. Chem.* **1988**, *53*, 878.

[49]For a review, see Corsaro, A.; Chiacchio, U.; Pistarià, V. *Synthesis* **2001**, 1903.

[50]Bandgar, B.P.; Makone, S.S. *Org. Prep. Proceed. Int.* **2000**, *32*, 391.

[51]Chavan, S.P.; Soni, P. *Tetrahedron Lett.* **2004**, *45*, 3161.

[52]Padmavathi, V.; Reddy, K.V.; Padmaja, A.; Venugopalan, P. *J. Org. Chem.* **2003**, *68*, 1567.

[53]Movassagh, B.; Lakouraj, M.M.; Ghodrati, K. *Synth. Commun.* **2000**, *30*, 4501.

[54]A solvent-free reaction. See Zhou, J.-F.; Tu, S.-J.; Feng, J.-C. *Synth.Commun.* **2002**, *32*, 959.

[55]Khazaei, A.; Vaghei, R.G.; Tajbakhsh, M. *Tetrahedron Lett.* **2001**, *42*, 5099.

[56]De, S.K. *Tetrahedron Lett.* **2003**, *44*, 9055.

[57]Chrisman, W.; Blankinship, M.J.; Taylor, B.; Harris, C.E. *Tetrahedron Lett.* **2003**, *33*, 4775; Imanzadeh, G.H.; Hajipour, A.R.; Mallakpour, S.E. *Synth. Commun.* **2003**, *33*, 735.

[58]Jadhav, V.K.; Wadgaonkar, P.P.; Joshi, P.L.; Salunkhe, M.M. *Synth. Commun.* **1999**, *29*, 1989.

[59]Bigdeli, M.A.; Nikje, M.M.A.; Heravi, M.M. *J. Chem. Res. (S)* **2001**, 496.

[60]Hajipour, A.R.; Mallakpour, S.E.; Khoee, E. *Synth. Commun.* **2002**, *32*, 9.

[61]Ganguly, N.C.; De, P.; Sukai, A.K.; De, S. *Synth. Commun.* **2002**, *32*, 1.

[62]Bose, D.S.; Narasaiah, A.V. *Synth. Commun.* **2000**, *30*, 1153. See also, Ganguly, N.C.; Sukai, A.K.; De, S.; De, P. *Synth. Commun.* **2001**, *31*, 1607.

[63]Sadeghi, M.M.; Mohammadpoor-Baltork, I.; Azarm, M.; Mazidi, M.R. *Synth. Commun.* **2001**, *31*, 435. See also, Hajipour, A.R.; Mallakpour, S.E.; Mohammadpoor-Baltork, I.; Khoee, S. *Synth. Commun.* **2001**, *31*, 1187; Tajbakhsh, M.; Heravi, M.M.; Mohanazadeh, F.; Sarabi, S.; Ghassemzadeh, M. *Monat. Chem.* **2001**, *132*, 1229; Zhang, G.-S.; Yang, D.-H.; Chen. M.-F. *Org. Prep. Proceed. Int.* **1998**, *30*, 713.

[64]Ho, T. *Synth. Commun.* **1980**, *10*, 465.

[65]Yadav, J.S.; Sasmal, P.K.; Chand, P.K. *Synth. Commun.* **1999**, *29*, 3667.

[66]Öcal, N.; Erden, I. *Tetrahedron Lett.* **2001**, *42*, 4765.

[67]BAcIm BF$_4$, 3-butyl-1-(CH$_2$COOH)imidazolium tetrafluoroborate: Li, D.; Shi, F.; Guo, S.; Deng, Y. *Tetrahedron Lett.* **2004**, *45*, 265. In Dmim BF$_4$, 1-decyl-3-methyl imidazolium tetrafluoroborate: Li, D.; Shi, F.; Deng, Y. *Tetrahedron Lett.* **2004**, *45*, 6791.

[68]Narsaiah, A.V.; Nagaiah, K. *Synthesis* **2003**, 1881.

[69]Mukai, C.; Nomura, I.; Kataoka, O.; Hanaoka, M. *Synthesis* **1999**, 1872.

[70]De, S.K. *Synth. Commun.* **2004**, *34*, 2289.

[71]Arnold, J.N.; Hayes, P.D.; Kohaus, R.L.; Mohan, R.S. *Tetrahedron Lett.* **2003**, *44*, 9173.

[72]Samajdar, S.; Basu, M.K.; Becker, F.F.; Banik, B.K. *Synth. Commun.* **2002**, *32*, 1917.

with trimethylacetaldehyde,[73] CuCl on Kieselguhr with oxygen,[74] Bi(NO$_3$)$_3$/Cu(OAc)$_2$ on Montmorillonite K10,[75] CrO$_3$-SiO$_2$,[76] and In addition, peroxomono-sulfate-SiO$_2$,[77] Cu(NO$_3$)$_2$—SiO$_2$,[78] Zn(NO$_3$)$_2$-SiO$_2$,[79] and clay (Clayan)[80] have all been used with microwave irradiation. Phenylhydrazones can be converted to a ketone using Oxone® and KHCO$_3$,[81] polymer-bound iodonium salts,[82] or KMnO$_4$ on wet SiO$_2$.[83] Dimethylhydrazones have been converted to ketones by heating with potassium carbonate in dimethyl sulfate,[84] with MeReO$_3$/H$_2$O$_2$ in acetic acid-acetonitrile,[85] with Pd(OAc)$_2$/SnCl$_2$ in aq. DMF,[86] FeSO$_4$·7 H$_2$O in chloroform,[87] Me$_3$SiCl/NaI in acetonitrile with 1% water,[88] [Ni(en)$_3$]$_2$S$_2$O$_3$, where en = ethylene-diamine, in chloroform,[89] or CeCl$_3$·7 H$_2$O—SiO$_2$ with microwave irradiation.[90]

Hydrazones, such as RAMP or SAMP (see p. 633) can be hydrolyzed with aq. CuCl$_2$.[91] Tosylhydrazones can be hydrolyzed to the corresponding ketones with aq. acetone and BF$_3$–etherate,[92] as well as with other reagents.[93] Semicarbazones have been cleaved with ammonium chlorochromates on alumina[94] (Bu$_4$N)$_2$-S$_2$O$_8$,[95] Mg(HSO$_4$)$_2$ on wet silica,[96] or by SbCl$_3$ with microwave irradiation.[97]

The hydrolysis of carbon–nitrogen double bonds involves initial addition of water and elimination of a nitrogen moiety:

[73]Blay, G.; Benach, E.; Fernández, I.; Galletero, S.; Pedro, J.R.; Ruiz, R. *Synthesis* **2000**, 403.

[74]Hashemi, M.M.; Beni, Y.A. *Synth. Commun.* **2001**, *31*, 295.

[75]Nattier, B.A.; Eash, K.J.; Mohan, R.S. *Synthesis* **2001**, 1010.

[76]Bendale, P.M.; Khadilkar, B.M. *Synth. Commun.* **2000**, *30*, 665.

[77]Bose, D.S.; Narsaiah, A.V.; Lakshminarayana, V. *Synth. Commun.* **2000**, *30*, 3121.

[78]Ghiaci, M.; Asghari, J. *Synth. Commun.* **2000**, *30*, 3865.

[79]Tamami, B.; Kiasat, A.R. *Synth. Commun.* **2000**, *30*, 4129.

[80]Meshram, H.M.; Srinivas, D.; Reddy, G.S.; Yadav, J.S. *Synth. Commun.* **1998**, *28*, 4401; 2593.

[81]Hajipour, A.R.; Mahboubghah, N. *Org. Prep Proceed. Int.* **1999**, *31*, 112.

[82]Chen, D.-J.; Cheng, D.-P.; Chen, Z.-C. *Synth. Commun.* **2001**, *31*, 3847.

[83]Hajipour, A.R.; Adibi, H.; Ruoho, A.E. *J. Org. Chem.* **2003**, *68*, 4553.

[84]Kamal, A.; Arifuddin, M.; Rao, N.V. *Synth. Commun.* **1998**, *28*, 3927.

[85]Stanković, S.; Espenson, J.H. *J. Org. Chem.* **2000**, *65*, 2218.

[86]Mino, C.; Hirota, T.; Fujita, N.; Yamashita, M. *Synthesis* **1999**, 2024.

[87]Nasreen, A.; Adapa, S.R. *Org. Prep. Proceed. Int.* **1999**, *31*, 573.

[88]Kamal, A.; Ramana, K.V.; Arifuddin, M. *Chem. Lett.* **1999**, 827.

[89]Kamal, A.; Arifuddin, M.; Rao, M.V. *Synlett* **2000**, 1482.

[90]Yadav, J.S.; Subba Reddy, B.V.; Reddy, M.S.K.; Sabitha, G. *Synlett* **2001**, 1134.

[91]Enders, D.; Hundertmark, T.; Lazny, R. *Synth. Commun.* **1999**, *29*, 27.

[92]Sacks, C.E.; Fuchs, P.L. *Synthesis* **1976**, 456.

[93]DDQ with dichloromethane/water: Chandrasekhar, S.; Reddy, Ch.R.; Reddy, M.V. *Chem. Lett.* **2000**, 430. For references, see Jiricny, J.; Orere, D.M.; Reese, C.B. *Synthesis* **1970**, 919.

[94]Zhang, G.-S.; Gong, H.; Yang, D.-H.; Chen, M.-F. *Synth. Commun.* **1999**, *29*, 1165; Gong, H.; Zhang, G.-S. *Synth. Commun.* **1999**, *29*, 2591.

[95]Chen, F.-E.; Liu, J.-P.; Fu, H.; Peng, Z.-Z.; Shao, L.-Y. *Synth. Commun.* **2000**, *30*, 2295.

[96]Shirini, F.; Zolfigol, M.A.; Mallakpour, B.; Mallakpour, S.E.; Hajipour, A.R.; Baltork, I.M. *Tetrahedron Lett.* **2002**, *43*, 1555.

[97]Mitra, A.K.; De, A.; Karchaudhuri, N. *Synth. Commun.* **2000**, *30*, 1651.

It is thus an example of reaction type A (p. 1261). The sequence shown is generalized.[98] In specific cases, there are variations in the sequence of the steps, depending on acid or basic catalysis or other conditions.[99] Which step is rate determining also depends on acidity and on the nature of W and of the groups connected to the carbonyl.[100]

$$
\left[
\begin{array}{c}
\overset{R}{\underset{\underset{C}{\|}}{\overset{\oplus}{N}}}\overset{R}{\diagdown}
\end{array}
\quad\longleftrightarrow\quad
\begin{array}{c}
\overset{R}{\underset{\underset{\overset{\oplus}{C}}{|}}{N}}\overset{R}{\diagdown}
\end{array}
\right]
$$

10

Iminium ions $(\mathbf{10})^{101}$ would be expected to undergo hydrolysis quite readily, since there is a contributing form with a positive charge on the carbon. Indeed, they react with water at room temperature.[102] Acid-catalyzed hydrolysis of enamines (the last step of the Stork reaction, **10-69** involves conversion to iminium ions:[103]

The mechanism of enamine hydrolysis is thus similar to that of vinyl ether hydrolysis (**10-6**).

OS **I**, 217, 298, 318, 381; **II**, 49, 223, 234, 284, 310, 333, 395, 519, 522; **III**, 20, 172, 626, 818; **IV**, 120; **V**, 139, 277, 736, 758; **VI**, 1, 358, 640, 751, 901, 932; **VII**, 8; **65**, 108, 183; **67**, 33; **76**, 23.

Related to this process is the hydrolysis of isocyanates or isothiocyanates[104] where addition of water to the carbon–nitrogen double bond would give an N-substituted carbamic acid (**11**). Such compounds are unstable and break down to

[98]For reviews of the mechanism, see Bruylants, A.; Feytmants-de Medicis, E., in Patai, S. *The Chemistry of the Carbon–Nitrogen Double Bond*, Wiley, NY, *1970*, pp. 465–504; Salomaa, P., in Patai, S. *The Chemistry of the Carbonyl Group* pt. 1, Wiley, NY, *1966*, pp. 199–205.

[99]For example, see Reeves, R.L. *J. Am. Chem. Soc. 1962*, 82, 3332; Sayer, J.M.; Conlon, E.H. *J. Am. Chem. Soc. 1980*, 102, 3592.

[100]Cordes, E.H.; Jencks, W.P. *J. Am. Chem. Soc. 1963*, 85, 2843.'

[101]For a review of iminium ions, see Böhme, H.; Haake, M. *Adv. Org. Chem. 1976*, 9, pt. 1, 107.

[102]Hauser, C.R.; Lednicer, D. *J. Org. Chem. 1959*, 24, 46. For a study of the mechanism, see Gopalakrishnan, G.; Hogg, J.L. *J. Org. Chem. 1989*, 54, 768.

[103]Maas, W.; Janssen, M.J.; Stamhuis, E.J.; Wynberg, H. *J. Org. Chem. 1967*, 32, 1111; Sollenberger, P.Y.; Martin, R.B. *J. Am. Chem. Soc. 1970*, 92, 4261. For a review of enamine hydrolysis, see Stamhuis, E.J.; Cook, A.G., in Cook *Enamines*, 2nd ed.; Marcel Dekker, NY, *1988*, pp. 165–180.

[104]For a study of the mechanism, see Castro, E.A.; Moodie, R.B.; Sansom, P.J. *J. Chem. Soc. Perkin Trans. 2 1985*, 737. For a review of the mechanisms of reactions of isocyanates with various nucleophiles, see Satchell, D.P.N.; Satchell, R.S. *Chem. Soc. Rev. 1975*, 4, 231.

carbon dioxide (or COS in the case of isothiocyanates) and the amine:

OS **II**, 24; **IV**, 819; **V**, 273; **VI**, 910.

16-3   Hydrolysis of Aliphatic Nitro Compounds

Oxo-de-hydro,nitro-bisubstitution

Primary or secondary aliphatic nitro compounds can be hydrolyzed, respectively, to aldehydes or ketones, by treatment of their conjugate bases with sulfuric acid. This is called the *Nef reaction*.[105] Tertiary aliphatic nitro compounds do not give the reaction because they cannot be converted to their conjugate bases. Like **16-2**, this reaction involves hydrolysis of a C=N double bond. A possible mechanism is[106]

Intermediates of type **12** have been isolated in some cases.[107]

The conversion of nitro compounds to aldehydes or ketones has been carried out with better yields and fewer side reactions by several alternative methods.[108] Among these are treatment of the nitro compound with tin complexes and $NaHSO_3$,[109] activated dry silica gel,[110] or 30% $H_2O_2–K_2CO_3$,[111] *t*-BuOOH and a catalyst,[112]

---

[105]For reviews, see Pinnick, H.W. *Org. React.* **1990**, *38*, 655; Haines, A.H. *Methods for the Oxidation of Organic Compounds*, Academic Press, NY, **1988**, pp. 220–231, 416–419.

[106]Hawthorne, M.F. *J. Am. Chem. Soc.* **1957**, *79*, 2510. A similar mechanism, but with some slight differences, was suggested earlier by van Tamelen, E.E.; Thiede, R.J. *J. Am. Chem. Soc.* **1952**, *74*, 2615. See also, Sun, S.F.; Folliard, J.T. *Tetrahedron* **1971**, *27*, 323.

[107]Feuer, H.; Spinicelli, L.F. *J. Org. Chem.* **1977**, *42*, 2091.

[108]For a review, see Ballini, R.; Petrini, M. *Tetrahedron* **2004**, *60*, 1017.

[109]Urpí, F.; Vilarrasa, J. *Tetrahedron Lett.* **1990**, *31*, 7499.

[110]Keinan, E.; Mazur, Y. *J. Am. Chem. Soc.* **1977**, *99*, 3861.

[111]Olah, G.A.; Arvanaghi, M.; Vankar, Y.D.; Prakash, G.K.S. *Synthesis* **1980**, 662.

[112]Bartlett, P.A.; Green III, F.R.; Webb, T.R. *Tetrahedron Lett.* **1977**, 331.

DBU in acetonitrile,[113] NaH and Me$_3$SiOOSiMe$_3$,[114] NaNO$_2$ in aq. DMSO,[115] or ceric ammonium nitrate (CAN).[116] The reaction of Al–NiCl$_2$·6 H$_2$O in THF converted α,β-unsaturated nitro compounds to the corresponding aldehyde, PhCH=CHNO$_2$ → PhCH$_2$CHO.[117]

When *primary* nitro compounds are treated with sulfuric acid without previous conversion to the conjugate bases, they give carboxylic acids. Hydroxamic acids are intermediates and can be isolated, so that this is also a method for preparing them.[118] Both the Nef reaction and the hydroxamic acid process involve the aci form; the difference in products arises from higher acidity, for example, a difference in sulfuric acid concentration from 2 to 15.5 *M* changes the product from the aldehyde to the hydroxamic acid.[119] The mechanism of the hydroxamic acid reaction is not known with certainty, but if higher acidity is required, it may be that the protonated aci form of the nitro compound is further protonated.

OS **VI**, 648; **VII**, 414. See also OS **IV**, 573.

16-4   Hydrolysis of Nitriles

NN-Dihydro-*C-oxo*-biaddition

$$R{-}C{\equiv}N \ + H_2O \xrightarrow{\text{H}^+ \text{ or OH}^-} R{-}\underset{\underset{NH_2}{}}{\overset{\overset{O}{\parallel}}{C}}$$

Hydroxy,oxo-de-nitrilo-tersubstitution

$$R{-}C{\equiv}N \ + \ H_2O \xrightarrow{\text{H}^+ \text{ or OH}^-} R{-}\underset{\underset{OH}{}}{\overset{\overset{O}{\parallel}}{C}} \ \text{or} \ R{-}\underset{\underset{O^{\ominus}}{}}{\overset{\overset{O}{\parallel}}{C}}$$

Nitriles can be hydrolyzed to give either amides or carboxylic acids.[120] The amide is formed initially, but since amides are also hydrolyzed with acid or basic treatment, the carboxylic acid is readily formed. When the acid is desired,[121] the reagent of choice is aq. NaOH containing ~6–12% H$_2$O$_2$, though acid-catalyzed hydrolysis is also frequently carried out. A "dry" hydrolysis of nitriles has been reported.[122] The hydrolysis of nitriles to carboxylic acids is one of the best methods for the preparation of these compounds. Nearly all nitriles give the reaction, with either acidic or basic catalysts. Hydrolysis of cyanohydrins, RCH(OH)CN, is usually carried out under acidic conditions, because basic solutions cause competing

[113]Ballini, R.; Bosica, G.; Fiorini, D.; Petrini, M. *Tetahedron Lett.* **2002**, *43*, 5233.

[114]Shahi, S.P.; Vankar, Y.D. *Synth. Commun.* **1999**, *29*, 4321.

[115]Gissot, A.; N'Gouela, S.; Matt, C.; Wagner, A.; Mioskowski, C. *J. Org. Chem.* **2004**, *69*, 8997.

[116]Olah, G.A.; Gupta, B.G.B. *Synthesis* **1980**, 44.

[117]Bezbarua, M.S.; Bez, G.; Barua, N.C. *Chem. Lett.* **1999**, 325.

[118]Hydroxamic acids can also be prepared from primary nitro compounds with SeO$_2$ and Et3N: Sosnovsky, G.; Krogh, J.A. *Synthesis* **1980**, 654.

[119]Kornblum, N.; Brown, R.A. *J. Am. Chem. Soc.* **1965**, *87*, 1742. See also, Cundall, R.B.; Locke, A.W. *J. Chem. Soc. B* **1968**, 98; Edward, J.T.; Tremaine, P.H. *Can J. Chem.* **1971**, *49*, 3483, 3489, 3493.

[120]For reviews, see Zil'berman, E.N. *Russ. Chem. Rev.* **1984**, *53*, 900; Compagnon, P.L.; Miocque, M. *Ann. Chim. (Paris)* **1970**, *[14]* 5, 11, 23.

[121]For a list of reagents, with references, see Larock, R.C. *Comprehensive Organic Transformations*, 2nd ed., Wiley-VCH, NY, *1999*, pp. 1986–1987.

[122]Chemat, F.; Poux, M.; Berlan, J. *J. Chem. Soc. Perkin Trans. 2 1996*, 1781; *1994*, 2597.

reversion of the cyanohydrin to the aldehyde and CN^-. However, cyanohydrins have been hydrolyzed under basic conditions with borax or alkaline borates.[123] Enzymatic hydrolysis with *Rhodococcus* sp AJ270 has also been reported.[124] In methanol with $BF_3 \cdot OEt_2$, benzonitrile is converted to methyl benzoate.[125]

There are a number of procedures for stopping at the amide stage,[126] among them the use of concentrated H_2SO_4; 2 equivalents of chlorotrimethylsilane followed by H_2O,[127] aq. NaOH with PEG-400 and microwave irradiation,[128] $NaBO_3$ with 4 equivalents of water and microwave irradiation,[129] heating on neutral alumina,[130] Oxone®,[131] and dry HCl followed by H_2O. The same result can also be obtained by use of water and certain metal ions or complexes;[132] a ruthenium catalyst on alumina with water,[133] MnO_2/SiO_2 with microwave irradiation,[134] $Hg(OAc)_2$ in HOAc;[135] or 2-mercaptoethanol in a phosphate buffer.[136] Nitriles can be hydrolyzed to the carboxylic acids without disturbing carboxylic ester functions also present, by the use of tetrachloro- or tetrafluorophthalic acid.[137] Nitriles are converted to thioamides $ArC(=S)NH_2$ with ammonium sulfide $(NH_4)_2S$ in methanol, with microwave irradiation.[138]

Thiocyanates are converted to thiocarbamates in a similar reaction:[139] $R-S-C\equiv N + H_2O \rightarrow R-S-C-O\ NH_2$. Hydrolysis of cyanamides gives amines, produced by the breakdown of the unstable carbamic acid intermediates: $R_2NCN \rightarrow [R_2NCOOH] \rightarrow R_2NH$.

OS **I**, 21, 131, 201, 289, 298, 321, 336, 406, 436, 451; **II**, 29, 44, 292, 376, 512, 586 (see, however, **V**, 1054), 588; **III**; 34, 66, 84, 88, 114, 221, 557, 560, 615, 851; **IV**, 58, 93, 496, 506, 664, 760, 790; **V**, 239; **VI**, 932; **76**, 169. Also see, OS **III**, 609; **IV**, 359, 502; **66**, 142.

[123]Jammot, J.; Pascal, R.; Commeyras, A. *Tetrahedron Lett.* **1989**, *30*, 563.

[124]Wang, M.-X.; Lin, S.-J. *J. Org. Chem.* **2002**, *67*, 6542.

[125]Jayachitra, G.; Yasmeen, N..; Rao, K.S.; Ralte, S.L.; Srinivasan, R.; Singh, A.K. *Synth. Commun.* **2003**, *33*, 3461.

[126]For a discussion, see Beckwith, A.L.J., in Zabicky, J. *The Chemistry of Amides*, Wiley, NY, **1970**, pp. 119–125. For a list of reagents, with references, see Larock, R.C. *Comprehensive Organic Transformations*, 2nd ed., Wiley-VCH, NY, **1999**, pp. 1988–1990.

[127]Basu, M.K.; Luo, F.-T. *Tetrahedron Lett.* **1998**, *39*, 3005.

[128]Bendale, P.M.; Khadilkar, B.M. *Synth. Commun.* **2000**, *30*, 1713.

[129]Sharifi, A.; Mohsenzadeh, F.; Mohtihedi, M.M.; Saidi, M.R.; Balalaie, S. *Synth. Commun.* **2001**, *31*, 431.

[130]Wligus, C.P.; Downing, S.; Molitor, E.; Bains, S.; Pagni, R.M.; Kabalka, G.W. *Tetrahedron Lett.* **1995**, *36*, 3469.

[131]Bose, D.S.; Baquer, S.M. *Synth. Commun.* **1997**, *27*, 3119.

[132]For example, see Bennett, M.A.; Yoshida, T. *J. Am. Chem. Soc.* **1973**, *95*, 3030; Paraskewas, S. *Synthesis* **1974**, 574; McKenzie, C.J.; Robson, R. *J. Chem. Soc., Chem. Commun.* **1988**, 112.

[133]Yamaguchi, K.; Matsushita, M.; Mizuno, N. *Angew. Chem. Int. Ed.* **2004**, *43*, 1576.

[134]A solvent-free reaction. See Khadilkar, B.M.; Madyar, V.R. *Synth. Commun.* **2002**, *32*, 1731.

[135]Plummer, B.F.; Menendez, M.; Songster, M. *J. Org. Chem.* **1989**, *54*, 718.

[136]Lee, Y.B.; Goo, Y.M.; Lee,Y.Y.; Lee, J.K. *Tetrahedron Lett.* **1989**, *30*, 7439.

[137]Rounds, W.D.; Eaton, J.T.; Urbanowicz, J.H.; Gribble, G.W. *Tetrahedron Lett.* **1988**, *29*, 6557.

[138]Bagley, M.C.; Chapaneri, K.; Glover, C.; Merritt, E.A. *Synlett* **2004** 2615.

[139]Zil'berman, E.N.; Lazaris, A.Ya. *J. Gen. Chem. USSR* **1963**, *33*, 1012.

B. Attack by OR or SR (Addition of ROH; RSH)

16-5   The Addition of Alcohols to Aldehydes and Ketones

Dialkoxy-de-oxo-bisubstitution

Dithioalkyl-de-oxo-bisubstitution

$$\underset{\underset{\textstyle C}{\|}}{\overset{\textstyle O}{\|}} + ROH \underset{}{\overset{H^+}{\rightleftharpoons}} \underset{\underset{\textstyle C}{}}{\overset{RO \quad OR}{}} + H_2O$$

Acetals and ketals are formed by treatment of aldehydes and ketones, respectively, with alcohols in the presence of acid catalysts.[140] Lewis acids such as TiCl$_4$[141] RuCl$_3$,[142] or CoCl$_2$[143] can be used in conjunction with alcohols. Dioxolanes have been prepared in ethylene glycol using microwave irradiation and *p*-toluenesulfonic acid as a catalyst.[144] This reaction is reversible, and acetals and ketals can be hydrolyzed by treatment with acid.[145] With small unbranched aldehydes the equilibrium lies to the right. If ketals or acetals of larger molecules must be prepared the equilibrium must be shifted, usually by removal of water. This can be done by azeotropic distillation, ordinary distillation, or the use of a drying agent such as Al$_2$O$_3$ or a molecular sieve.[146] The reaction is not catalyzed in either direction by bases, so most acetals and ketals are quite stable to bases, though they are easily hydrolyzed by acids. This reaction is therefore a useful method of protection of aldehyde or ketone functions from attack by bases. The reaction is of wide scope. Most aldehydes are easily converted to acetals.[147] With ketones the process is more difficult, presumably for steric reasons, and the reaction often fails, though many ketals, especially from cyclic ketones, have been made in this manner.[148] Many functional groups may be present without being affected. 1,2-Glycols and 1,3-glycols form cyclic acetals and ketals (1,3-dioxolanes[149]

---

[140]For reviews, see Meskens, F.A.J. *Synthesis* **1981**, 501; Schmitz, E.; Eichhorn, I., in Patai, S. *The Chemistry of the Ether Linkage*, Wiley, NY, **1967**, pp. 309–351.

[141]Clerici, A.; Pastori, N.; Porta, O. *Tetrahedron* **2001**, *57*, 217.

[142]De, S.K.; Gibbs, R.A. *Tetrahedron Lett.* **2004**, *45*, 8141.

[143]Velusamy, S.; Punniyamurthy, T. *Tetrahedron Lett.* **2004**, *45*, 4917.

[144]Pério, B.; Dozias, M.-J.; Jacquault, P.; Hamelin, J. *Tetrahdron Lett.* **1997**, *38*, 7867; Moghaddam, F.M.; Sharifi, A. *Synth. Commun.* **1995**, *25*, 2457.

[145]See Heravi, M.M.; Tajbakhsh, M.; Habibzadeh, S.; Ghassemzadeh, M. *Monat. Chem.* **2001**, *132*, 985.

[146]For many examples of each of these methods, see Meskens, F.A.J. *Synthesis* **1981**, 501, pp. 502–505.

[147]For other methods, see Caputo, R.; Ferreri, C.; Palumbo, G. *Synthesis* **1987**, 386; Ott, J.; Tombo, G.M.R.; Schmid, B.; Venanzi, L.M.; Wang, G.; Ward, T.R. *Tetrahedron Lett.* **1989**, *30*, 6151, Liao, Y.; Huang, Y.; Zhu, F. *J. Chem. Soc., Chem. Commun.* **1990**, 493; Chan, T.H.; Brook, M.A.; Chaly, T. *Synthesis* **1983**, 203.

[148]High pressure has been used to improve the results with ketones: Dauben, W.G.; Gerdes, J.M.; Look, G.C. *J. Org. Chem.* **1986**, *51*, 4964. For other methods, see Otera, J.; Mizutani, T.; Nozaki, H. *Organometallics*, **1989**, *8*, 2063; Thurkauf, A.; Jacobson, A.E.; Rice, K.C. *Synthesis* **1988**, 233.

[149]See Yadav, J.S.; Reddy, B.V.S.; Srinivas, R.; Ramalingam, T. *Synlett* **2000**, 701; Laskar, D.D.; Prajapati, D.; Sandhu, J.S. *Chem. Lett.* **1999**, 1283; Curini, M.; Epifano, F.; Marcotullio, M.C.; Rosati, O. *Synlett* **2001**, 1182; Kawabata, T.; Mizugaki, T.; Ebitani, K.; Kaneda, K. *Tetrahedron Lett.* **2001**, *42*, 8329; Reddy, B.M.; Reddy, V.R.; Giridhar, D. *Synth. Commun.* **2001**, *31*, 1819; Gopinath, R.; Haque, Sk.J.; Patel, B.K. *J. Org. Chem.* **2002**, *67*, 5842.

and 1,3-dioxanes,[150] respectively), and these are often used to protect aldehydes and ketones. Chiral dioxolanes have been prepared from chiral diols.[151] Dioxolanes have been prepared from ketones in ionic liquids.[152] Ketones are converted with dimethyl ketals by electrolysis with NaBr in methanol.[153]

Intramolecular reactions are possible in which a keto diol or an aldehyde diol generates a bicyclic ketal or acetal. Fused ring [2.2.0] ketals have been prepared in this manner.[154]

The mechanism, which involves initial formation of a *hemiacetal*,[155] is the reverse of that given for acetal hydrolysis:

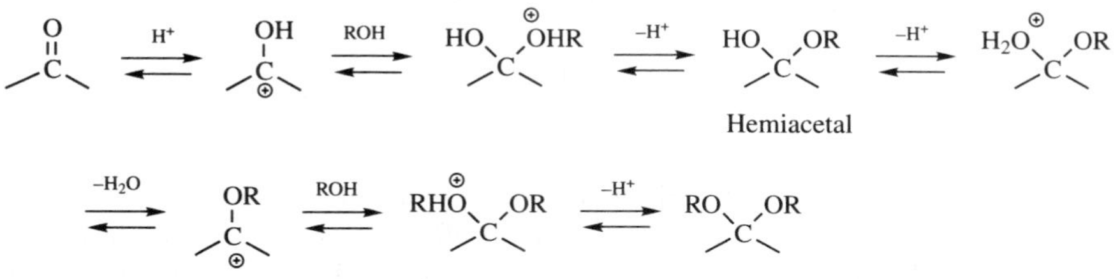

$$\begin{array}{ccccccccc}
\ce{O} & \xrightarrow{H^+} & \ce{OH} & \xleftarrow{ROH} & \ce{HO\ \overset{\oplus}{O}HR} & \xrightarrow{-H^+} & \ce{HO\ OR} & \xrightarrow{-H^+} & \ce{H2\overset{\oplus}{O}\ OR}
\end{array}$$

Hemiacetal

$$\xrightarrow{-H_2O}\quad \ce{\overset{\oplus}{C}\!-\!OR}\quad \xleftarrow{ROH}\quad \ce{RH\overset{\oplus}{O}\ OR}\quad \xrightarrow{-H^+}\quad \ce{RO\ OR}$$

In a study of the acid-catalyzed formation of the hemiacetal, Grunwald showed[156] that the data best fit a mechanism in which the three steps shown here are actually all concerted; that is, the reaction is simultaneously catalyzed by acid and base, with water acting as the base:[157]

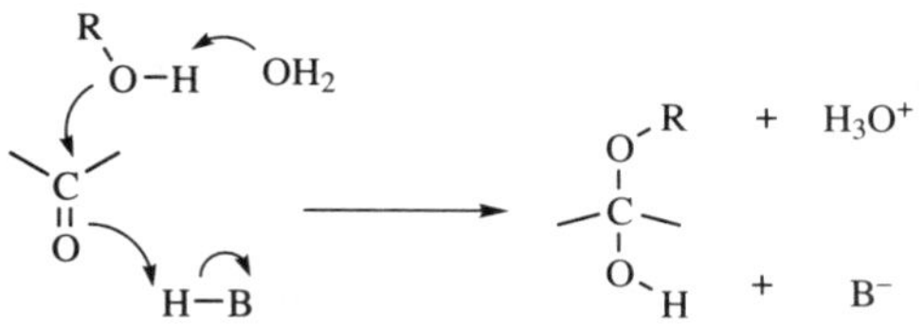

$$\begin{array}{ccc}
\ce{R-O-H \cdots OH2} & \longrightarrow & \ce{O-R} \\
\ce{C=O} & & \ce{C} \\
\ce{H-B} & & \ce{O-H}\ +\ B^-
\end{array}\quad +\ H_3O^+$$

If the original aldehyde or ketone has an α hydrogen, it is possible for water to split out in that way and enol ethers can be prepared in this manner:

$$\ce{R^2O\ \underset{R}{\overset{H}{C}}\!-\!\underset{OH}{\overset{H}{C}}\!-\!R'} \longrightarrow \ce{\underset{R}{\overset{R^2O}{C}}=\underset{H}{\overset{R'}{C}}}$$

[150]Wu, H.-H.; Yang, F.; Cui, P.; Tang, J.; He, M.-Y. *Tetrahedron Lett.* **2004**, *45*, 4963; Ishihara, K.; Hasegawa, A.; Yamamoto, H. *Synlett* **2002**, 1296.

[151]Kurihara, M.; Hakamata, W. *J. Org. Chem.* **2003**, *68*, 3413.

[152]In AmBIm Cl,: Li, D.; Shi, F.; Peng, J.; Guo, S.; Deng, Y. *J. Org. Chem.* **2004**, *69*, 3582.

[153]Elinson, M.N.; Feducovich, S.K.; Dmitriev, D.E.; Dorofeev, A.S.; Vereshchagin, A.N.; Nikishin, G.I. *Tetrahedron Lett.* **2001**, *42*, 5557.

[154]Wang, G.; Wang, Y.; Arcari, A.R.; Rheingold, A.L.; Concolino, T. *Tetrahedron Lett.* **1999**, *40*, 7051.

[155]For a review of hemiacetals, see Hurd, C.D. *J. Chem. Educ.* **1966**, *43*, 527.

[156]Grunwald, E. *J. Am. Chem. Soc.* **1985**, *107*, 4715.

[157]Grunwald also studied the mechanism of the base-catalyzed formation of the hemiacetal, and found it to be the same as that of base-catalyzed hydration (**16-1**, mechanism *a*): Grunwald, E. *J. Am. Chem. Soc.* **1985**, *107*, 4710. See also, Sørensen, P.E.; Pedersen, K.J.; Pedersen, P.R.; Kanagasabapathy, V.M.; McClelland, R.A. *J. Am. Chem. Soc.* **1988**, *110*, 5118; Leussing, D.L. *J. Org. Chem.* **1990**, *55*, 666.

Similarly, treatment with an anhydride and a catalyst can give an enol ester (see **16-6**).[158]

Hemiacetals themselves are no more stable than the corresponding hydrates (**16-1**). As with hydrates, hemiacetals of cyclopropanones[159] and of polychloro and polyfluoro aldehydes and ketones may be quite stable.

When acetals or ketals are treated with an alcohol of higher molecular weight than the one already there, it is possible to get a transacetalation (see **10-13**). In another type of transacetalation, aldehydes or ketones can be converted to acetals or ketals by treatment with another acetal or ketal or with an ortho ester,[160] in the presence of an acid catalyst (shown for an ortho ester):

$$\underset{R}{\overset{O}{\underset{\quad}{\parallel}}}\!\!-\!\!C\!\!-\!\!R^1 \;+\; \underset{R^2}{\overset{EtO}{\diagdown}}C\overset{OEt}{\diagup}OEt \;\;\rightleftharpoons\;\; \underset{R}{\overset{EtO}{\diagdown}}C\overset{OEt}{\diagup}R^1 \;+\; \underset{R^2}{\overset{O}{\underset{\quad}{\parallel}}}\!\!-\!\!C\!\!-\!\!OEt$$

This method is especially useful for the conversion of ketones to ketals, since the direct reaction of a ketone with an alcohol often gives poor results. In another method, the substrate is treated with an alkoxysilane $ROSiMe_3$ in the presence of trimethylsilyl trifluoromethanesulfonate.[161]

1,4-Diketones give furans when treated with acids. This is actually an example of an intramolecular addition of an alcohol to a ketone, since it is the enol form that adds:

$$R\!\!-\!\!\overset{O\;\;O}{\underset{H}{\diagup\diagdown}}\!\!-\!\!R \;\xrightarrow{H^+}\; R\!\!-\!\!\overset{\diagup\diagdown}{\underset{O}{\quad}}\!\!\overset{OH}{\underset{R}{\quad}} \;\xrightarrow{-H_2O}\; R\!\!-\!\!\overset{\diagup\diagdown}{\underset{O}{\quad}}\!\!-\!\!R$$

Similarly, 1,5-diketones give pyrans. Conjugated 1,4-diketones, such as 1,4-diphenylbut-2-en-1,4-dione is converted to 2,5-diphenylfuran with formic acid, 5% Pd/C, PEG-200, and a sulfuric acid catalyst with microwave irradiation.[162] Formic acid reacts with alcohols to give orthoformates. Note that alkynyl ketones are converted to furans with palladium (II) acetate.[163]

---

[158]For a list of catalysts, with references, see Larock, R.C. *Comprehensive Organic Transformations*, 2nd ed., Wiley-VCH, NY, *1999*, pp. 1484–1485.

[159]For a review, see Salaun, J. *Chem. Rev. 1983*, *83*, 619.

[160]For a review with respect to ortho esters, see DeWolfe, R.H. *Carboxylic Ortho Ester Derivatives*; Academic Press, NY, *1970*, pp. 154–164. See Karimi, B.; Ebrahimian, G.R.; Seradj, H. *Org. Lett. 1999*, *1*, 1737; Karimi, B.; Ashtiani, A.M. *Chem. Lett. 1999*, 1199; Firouzabadi, H.; Iranpoor, N.; Karimi, B. *Synlett 1999*, 321; Firouzabadi, H.; Iranpoor, N,; Karimi, B. *Synth. Commun. 1999*, 29, 2255; Leonard, N.M.; Oswald, M.C.; Freiberg, D.A.; Nattier, B.A.; Smith, R.C.; Mohan, R.S. *J. Org. Chem. 2002*, *67*, 5202.

[161]Tsunoda, T.; Suzuki, M.; Noyori, R.*Tetrahedron Lett. 1980*, *21*, 1357; Kato, J.; Iwasawa, N.; Mukaiyama, T. *Chem. Lett. 1985*, 743. See also, Torii, S.; Takagishi, S.; Inokuchi, T.; Okumoto, H. *Bull. Chem. Soc. Jpn. 1987*, *60*, 775.

[162]Rao, H.S.P.; Jothilingam, S. *J. Org. Chem. 2003*, *68*, 5392.

[163]Jeevanandam, A.; Narkunan, K.; Ling, Y.-C. *J. Org. Chem. 2001*, *66*, 6014. See Arcadi, A.; Cerichelli, G.; Chiarini, M.; Di Giuseppe, S. Marinelli, F. *Tetrahedron Lett. 2000*, *41*, 9195.

OS **I**, 1, 298, 364, 381; **II**, 137; **III**, 123, 387, 502, 536, 644, 731, 800; **IV**, 21, 479, 679; **V**, 5, 292, 303, 450, 539; **VI**, 567, 666, 954; **VII**, 59, 149, 168, 177, 241, 271, 297; **VIII**, 357. Also see OS **IV**, 558, 588; **V**, 25; **VIII**, 415.

16-6 Acylation of Aldehydes and Ketones

O-Acyl-*C*-acyloxy-addition

Aldehydes can be converted to *acylals* by treatment with an anhydride in the presence of BF_3, proton acids,[164] PCl_3,[165] NBS,[166] $LiBF_4$,[167] $FeCl_3$,[168] $InCl_3$,[169] $InBr_3$,[170] $Cu(OTf)_2$,[171] $Bi(OTf)_3$,[172] $BiCl_3$,[173] $Bi(NO_3)_3$,[174] WCl_6,[175] $ZrCl_4$,[176] ceric ammonium nitrate,[177] With Envirocat EPZ10 and microwave irradiation, acetic anhydride react with aldehydes to give the acylal.[178] Conjugated aldehydes are converted to the corresponding acylal by reaction with acetic anhydride and a $FeCl_3$ catalyst.[179] The reaction cannot normally be applied to ketones, though an exception has been reported when the reagent is trichloroacetic anhydride, which gives acylals with ketones without a catalyst.[180]

OS **IV**, 489.

16-7 Reductive Alkylation of Alcohols

C-Hydro-*O*-alkyl-addition

---

[164]For example, see Olah, G.A.; Mehrotra, A.K. *Synthesis* **1982**, 962.

[165]See Michie, J.K.; Miller, J.A. *Synthesis* **1981**, 824.

[166]Karimi, B.; Seradj, H.; Ebrahimian, G.R. *Synlett* **2000**, 623

[167]Sumida, N.; Nishioka, K.; Sato, T. *Synlett* **2001**, 1921; Yadav, J.S.; Reddy, B.V.S.; Venugapal, C.; Ramalingam, V.T. *Synlett* **2002**, 604.

[168]Li, Y.-Q. *Synth. Commun.* **2000**, *30*, 3913; Trost, B.M.; Lee, C.B. *J. Am. Chem. Soc.* **2001**, *123*, 3671; Wang, C.; Li, M. *Synth. Commun.* **2002**, *32*, 3469.

[169]Yadav, J.S.; Reddy, B.V.S.; Srinivas, Ch. *Synth. Commun.* **2002**, *32*, 1175, 2169.

[170]Yin, L.; Zhang, Z.H.; Wang, Y.-M. ; Pang, M.-L. *Synlett* **2004**, 1727.

[171]Chandra, K.L.; Saravanan, P.; Singh, V.K. *Synlett* **2000**, 359.

[172]Carrigan, M.D.; Eash, K.J.; Oswald, M.C.; Mohan, R.S. *Tetrahedron Lett.* **2001**, *42*, 8133.

[173]Mohammadpoor-Baltork, I.; Aliyan, H. *Synth. Commun.* **1999**, *29*, 2741.

[174]Aggen, D.H.; Arnold, J.N.; Hayes, P.D.; Smoter, N.J.; Mohan, R.S. *Tetrahedron* **2004**, *60*, 3675.

[175]A solvent-free reaction. See Karimi, B.; Ebrahimian, G.-R.; Seradj, H. *Synth. Commun.* **2002**, *32*, 669.

[176]Smitha, G.; Reddy, Ch.S. *Tetrahedron* **2003**, *59*, 9571.

[177]Roy, S.C.; Banerjee, B. *Synlett* **2002**, 1677.

[178]Bandgar, B.P.; Makone, S.S.; Kulkarni, S.R. *Monat. Chem.* **2000**, *131*, 417.

[179]Trost, B.M.; Lee, C.B. *J. Am. Chem. Soc.* **2001**, *123*, 3671.

[180]Libman, J.; Sprecher, M.; Mazur, Y. *Tetrahedron* **1969**, *25*, 1679.

Aldehydes and ketones can be converted to ethers by treatment with an alcohol and triethylsilane in the presence of a strong acid[181] or by hydrogenation in alcoholic acid in the presence of platinum oxide.[182] The process can formally be regarded as addition of ROH to give a hemiacetal, $RR'C(OH)OR^2$, followed by reduction of the OH. In this respect, it is similar to **16-17**. The reaction of an aldehyde with $BuOSiHMe_2$ and a Me_3SiI catalyst gives the corresponding butyl alkyl ether.[183] In a similar reaction, ketones can be converted to carboxylic esters (reductive acylation of ketones) by treatment with an acyl chloride and triphenyltin hydride.[184]

$$ \underset{R}{\overset{O}{\underset{\|}{\underset{C}{}}}}\!\!R^1 \;+\; R^2COCl \;\xrightarrow{Ph_3SnH}\; H\!-\!\underset{R}{\underset{C}{}}\!\!R^1 \; \underset{O}{\overset{R^2}{\underset{\|}{O\!-\!C}}} $$

Ethers have also been prepared by the reductive dimerization of two molecules of an aldehyde or ketone (e.g., cyclohexanone → dicyclohexyl ether). This was accomplished by treatment of the substrate with a trialkylsilane and a catalyst.[185]

16-8 The Addition of Alcohols to Isocyanates

N-Hydro-*C*-alkoxy-addition

$$ \underset{N=C=O}{\overset{R}{\diagdown}} \;+\; R'OH \;\longrightarrow\; \underset{\underset{H}{|}}{R}\diagdown_N\overset{\overset{O}{\|}}{\underset{}{C}}\diagup OR' $$

Carbamates (substituted urethanes) are prepared when isocyanates are treated with alcohols. This is an excellent reaction, of wide scope, and gives good yields. Isocyanic acid HNCO gives unsubstituted carbamates. Addition of a second equivalent of HNCO gives *allophanates*.

$$ \underset{N=C=O}{\overset{H}{\diagdown}} \;+\; R'OH \;\longrightarrow\; \underset{\underset{H}{|}}{H}\diagdown_N\overset{\overset{O}{\|}}{\underset{}{C}}\diagup OR' \;\xrightarrow{HNCO}\; H_2N\overset{\overset{O}{\|}}{\underset{}{C}}\underset{\underset{H}{|}}{N}\overset{\overset{O}{\|}}{\underset{}{C}}OR' $$

Allophanate

The isocyanate can be generated *in situ* by the reaction of an amine and oxalyl chloride, and subsequent reaction with HCl and then an alcohol gives the carbamate.[186] Polyurethanes are made by combining compounds with two NCO groups with

[181]Doyle, M.P.; DeBruyn, D.J.; Kooistra, D.A. *J. Am. Chem. Soc.* **1972**, *94*, 3659.

[182]Verzele, M.; Acke, M.; Anteunis, M. *J. Chem. Soc.* **1963**, 5598. For still another method, see Loim, L.M.; Parnes, Z.N.; Vasil'eva, S.P.; Kursanov, D.N. *J. Org. Chem. USSR* **1972**, *8*, 902.

[183]Miura, K.; Ootsuka, K.; Suda, S.; Nishikori, H.; Hosomi, A. *Synlett* **2002**, 313.

[184]Kaplan, L. *J. Am. Chem. Soc.* **1966**, *88*, 4970.

[185]Sassaman, M.B.; Kotian, K.D.; Prakash, G.K.S.; Olah, G.A. *J. Org. Chem.* **1987**, *52*, 4314. See also, Kikugawa, Y. *Chem. Lett.* **1979**, 415.

[186]Oh, L.M.; Spoors, P.G.; Goodman, R.M. *Tetrahedron Lett.* **2004**, *45*, 4769.

compounds containing two OH groups. Cyclic carbamates, such as 1,3-oxazine-2-ones, are generated by the reaction of an isocyanate with an oxetane, in the presence of a palladium catalyst.[187] Isothiocyanates similarly give thiocarbamates[188] RNHCSOR', though they react slower than the corresponding isocyanates. Isocyanates react with LiAlHSeH and then iodomethane to give the corresponding selenocarbonate (RNHCOSeMe).[189]

The details of the mechanism are poorly understood,[190] though the oxygen of the alcohol is certainly attacking the carbon of the isocyanate. Hydrogen bonding complicates the kinetic picture.[191] The addition of ROH to isocyanates can also be catalyzed by metallic compounds,[192] by light,[193] or, for tertiary ROH, by lithium alkoxides[194] or *n*-butyllithium.[195]

OS **I**, 140; **V**, 162; **VI**, 95, 226, 788, 795.

16-9   Alcoholysis of Nitriles

Alkoxy,oxo-de-nitrilo-tersubstitution

$$R-C{\equiv}N + R'OH \quad \xrightarrow{HCl} \quad \underset{R}{\overset{\overset{\displaystyle \overset{\oplus}{N}H_2}{\|}}{\underset{\displaystyle OR'}{C}}} \; Cl^{\ominus} \quad \xrightarrow[H^+]{H_2O} \quad \underset{R}{\overset{\overset{\displaystyle O}{\|}}{\underset{\displaystyle OR'}{C}}}$$

The addition of dry HCl to a mixture of a nitrile and an alcohol in the absence of water leads to the hydrochloride salt of an imino ester (imino esters are also called imidates and imino ethers). This reaction is called the *Pinner synthesis*.[196] The salt can be converted to the free imino ester by treatment with a weak base such as sodium bicarbonate, or it can be hydrolyzed with water and an acid catalyst to the corresponding carboxylic ester. If the latter is desired, water may be present from the beginning, in which case aq. HCl can be used and the need for gaseous HCl is eliminated. Imino esters can also be prepared from nitriles with basic catalysts.[197]

---

[187]Larksarp, C.; Alper, H. *J. Org. Chem.* **1999**, *64*, 4152.

[188]For a review of thiocarbamates, see Walter, W.; Bode, K. *Angew. Chem. Int. Ed.* **1967**, *6*, 281. See also, Wynne, J.H.; Jensen, S.D.; Snow, A.W. *J. Org. Chem.* **2003**, *68*, 3733.

[189]Koketsu, M.; Ishida, M.; Takakura, N.; Ishihara, H. *J. Org. Chem.* **2002**, *67*, 486.

[190]For reviews, see Satchell, D.P.N.; Satchell, R.S.*Chem. Soc. Rev.* **1975**, *4*, 231; Entelis, S.G.; Nesterov, O.V. *Russ. Chem. Rev.* **1966**, *35*, 917.

[191]See for example, Robertson, W.G.P.; Stutchbury, J.E. *J. Chem. Soc.* **1964**, 4000; Donohoe, G.; Satchell, D.P.N.; Satchell, R.S. *J. Chem. Soc. Perkin Trans. 2* **1990**, 1671 and references cited therein. See also, Sivakamasundari, S.; Ganesan, R. *J. Org. Chem.* **1984**, *49*, 720.

[192]For example, see Kim, Y.H.; Park, H.S. *Synlett* **1998**, 261; Hazzard, G.; Lammiman, S.A.; Poon, N.L.; Satchell, D.P.N.; Satchell, R.S. *J. Chem. Soc. Perkin Trans. 2* **1985**, 1029; Duggan, M.E.; Imagire, J.S. *Synthesis* **1989**, 131.

[193]McManus, S.P.; Bruner, H.S.; Coble, H.D.; Ortiz, M. *J. Org. Chem.* **1977**, *42*, 1428.

[194]Bailey, W.J.; Griffith, J.R. *J. Org. Chem.* **1978**, *43*, 2690.

[195]Nikoforov, A.; Jirovetz, L.; Buchbauer, G. *Liebigs Ann. Chem.* **1989**, 489.

[196]For a review, see Compagnon, P.L.; Miocque, M. *Ann. Chim. (Paris)* **1970**, *[14] 5*, 23, see pp. 24–26. For a review of imino esters, see Neilson, D.G., in Patai, S. *The Chemistry of Amidines and Imidates*, Wiley, NY, **1975**, pp. 385–489.

[197]Schaefer, F.C.; Peters, G.A. *J. Org. Chem.* **1961**, *26*, 412.

This reaction is of broad scope and is good for aliphatic, aromatic, and heterocyclic R and for nitriles with oxygen-containing functional groups. The application of the reaction to nitriles containing a carboxyl group constitutes a good method for the synthesis of mono esters of dicarboxylic acids with the desired group esterified and with no diester or diacid present.

Cyanogen chloride reacts with alcohols in the presence of an acid catalyst, such as dry HCl or AlCl$_3$, to give carbamates:[198]

$$\text{ClCN} + 2\,\text{ROH} \xrightarrow[\text{or AlCl}_3]{\text{HCl}} \text{ROCONH}_2 + \text{RCl}$$

ROH can also be added to nitriles in another manner (**16-91**).
OS **I**, 5, 270; **II**, 284, 310; **IV**, 645; **VI**, 507; **VIII**, 415.

16-10 The Formation of Carbonates and Xanthates

Di-C-alkoxy-addition; S-Metallo-C-alkoxy-addition

$$\underset{\text{Cl} \quad \text{Cl}}{\overset{\overset{\text{O}}{\|}}{\text{C}}} \xrightarrow{\text{ROH}} \underset{\text{RO} \quad \text{Cl}}{\overset{\overset{\text{O}}{\|}}{\text{C}}} \xrightarrow{\text{ROH}} \underset{\text{RO} \quad \text{OR}}{\overset{\overset{\text{O}}{\|}}{\text{C}}}$$

The reaction of phosgene with an alcohol generates a haloformic esters, and reaction with a second equivalent of alcohol gives a carbonate. This reaction is related to the acyl addition reactions of acyl chlorides in Reaction **16-98**. An important example is the preparation of carbobenzoxy chloride (PhCH$_2$OCOCl) from phosgene and benzyl alcohol. This compound is widely used for protection of amino groups during peptide synthesis. When an alcohol reacts with certain alkyl halides (e.g., benzyl chloride) and carbon dioxide, in the presence of Cs$_2$CO$_3$ and tetrabutylammonium iodide, a mixed carbonate is formed.[199]

$$\text{S}=\text{C}=\text{S} + \text{ROH} \xrightarrow{\text{NaOH}} \underset{\text{RO} \quad \text{S}^{\ominus} \;\text{Na}^{\oplus}}{\overset{\overset{\text{S}}{\|}}{\text{C}}}$$

The addition of alcohols to carbon disulfide in the presence of a base produces xanthates.[200] The base is often HO$^-$, but in some cases better results can be obtained by using methylsulfinyl carbanion MeSOCH$_2^-$[201] If an alkyl halide RX is present, the xanthate ester ROCSSR' can be produced directly. In a similar manner, alkoxide ions add to CO$_2$ to give carbonate ester salts (ROCOO$^-$).
OS **V**, 439; **VI**, 207, 418; **VII**, 139.

[198]Bodrikov, I.V.; Danova, B.V. *J. Org. Chem. USSR* **1968**, *4*, 1611; **1969**, *5*, 1558; Fuks, R.; Hartemink, M.A. *Bull. Soc. Chim. Belg.* **1973**, *82*, 23.
[199]Kim, S.i.; Chu, F.; Dueno, E.E.; Jung, K.W. *J. Org. Chem.* **1999**, *64*, 4578.
[200]For a review of the formation and reactions of xanthates, see Dunn, A.D.; Rudorf, W. *Carbon Disulphide in Organic Chemistry*; Ellis Horwood: Chichester, **1989**, pp. 316–367.
[201]Meurling, P.; Sjöberg, B.; Sjöberg, K. *Acta Chem. Scand.* **1972**, *26*, 279.

C. Sulfur Nucleophiles

16-11   The Addition of H_2S and Thiols to Carbonyl Compounds

***O*-Hydro-*C*-mercapto-addition**[202]

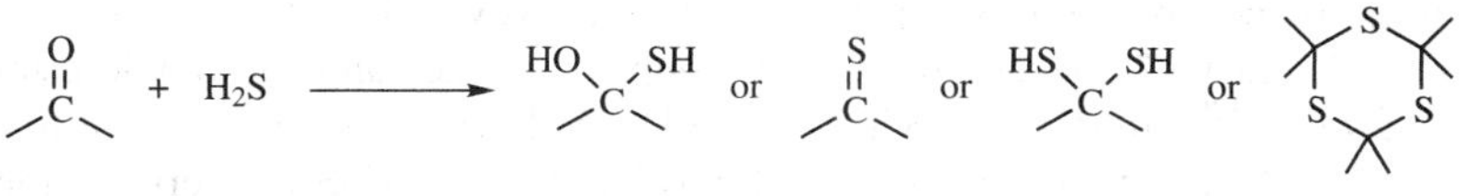

The addition of H_2S to an aldehyde or ketone can result in a variety of products. The most usual product is the trithiane **16**.[203] *gem*-Dithiols (**15**) are much more stable than the corresponding hydrates or α-hydroxy thiols.[204] They have been prepared by the treatment of ketones with H_2S under pressure[205] and under mild conditions with HCl as a catalyst.[206] Thiols add to aldehydes and ketones to give hemimercaptals, CH(OH)SR and dithioacetals, CH(SR)$_2$ (**16-5**). α-Hydroxy thiols (**13**) can be prepared from polychloro and polyfluoro aldehydes and ketones.[207] Apparently **13** are stable only when prepared from these compounds, and not even for all of them.

Thioketones[2] (**14**) can be prepared from certain ketones, such as diaryl ketones, by treatment with H_2S and an acid catalyst, usually HCl. They are often unstable and usually trimerize (to **16**) or react with air. Thioaldehydes[208] are even less stable and simple ones[209] apparently have never been isolated, though *t*-BuCHS has been prepared in

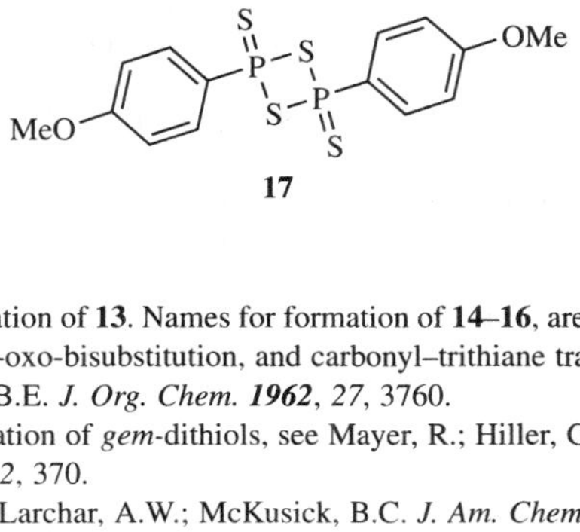

17

---

[202]This name applies to formation of **13**. Names for formation of **14–16**, are, respectively, thioxo-de-oxo-bisubstitution, dimercapto-de-oxo-bisubstitution, and carbonyl–trithiane transformation.

[203]Campaigne, E.; Edwards, B.E. *J. Org. Chem.* **1962**, *27*, 3760.

[204]For a review of the preparation of *gem*-dithiols, see Mayer, R.; Hiller, G.; Nitzschke, M.; Jentzsch, J. *Angew. Chem. Int. Ed.* **1963**, *2*, 370.

[205]Cairns, T.L.; Evans, G.L.; Larchar, A.W.; McKusick, B.C. *J. Am. Chem. Soc.* **1952**, *74*, 3982.

[206]Campaigne, E.; Edwards, B.E. *J. Org. Chem.* **1962**, *27*, 3760; Demuynck, M.; Vialle, J. *Bull. Soc. Chim. Fr.* **1967**, 1213.

[207]Harris Jr., J.F. *J. Org. Chem.* **1960**, *25*, 2259.

[208]For a review of thioaldehydes, see Usov, V.A.; Timokhina, L.V.; Voronkov, M.G. *Russ. Chem. Rev.* **1990**, *59*, 378.

[209]For the preparation and reactions of certain substituted thioaldehydes, see Hofstra, G.; Kamphuis, J.; Bos, H.J.T. *Tetrahedron Lett.* **1984**, *25*, 873; Okazaki, R.; Ishii, A.; Inamoto, N. *J. Am. Chem. Soc.* **1987**, *109*, 279; Adelaere, B.; Guemas, J.; Quiniou, H. *Bull. Soc. Chim. Fr.* **1987**, 517; Muraoka, M.; Yamamoto, T.; Enomoto, K.; Takeshima, T. *J. Chem. Soc. Perkin Trans. 1* **1989**, 1241, and references cited in these papers.

solution, where it exists for several hours at $20°C$.[210] A high-yield synthesis of thio-ketones involves treatment of acyclic[211] ketones with 2,4-bis(4-methoxyphenyl)-1,3,2,4-dithiadiphosphetane-2,4-disulfide **9**[212] (known as *Lawesson's reagent*)[213]. Thioketones can also be prepared by treatment of ketones with P_4S_{10},[214] P_4S_{10} and hexamethyldisiloxane,[215] P_4S_{10} on alumina,[216] or $CF_3SO_3SiMe_3/S(SiMe_3)_2$,[217] and from oximes or various types of hydrazone (overall conversion C=N⁻ → C=S).[218] Reagent **17** converts the C=O groups of amides and carboxylic esters[219] to C=S groups.[220] Similarly, $POCl_3$ followed by $S(TMS)_2$ converts lactams to thio-lactams[221] and treatment with triflic anhydride, and then H_2S converts amides to thioamides.[222] The reaction of an amide with triflic anhydride, and then aq. $S(NH_4)_2$ gives the corresponding thioamide.[223] The H_2S–Me_3SiCl-$(iPr)_2NLi$ com-plex converts carboxylic esters to thiono esters.[224] Lactones react with **9** in the pre-sence of hexamethyldisiloxane an microwave irradiation to give the thiolactone.[225] Carboxylic acids (RCOOH) can be converted directly to dithiocarboxylic esters (RCSSR′)[226] in moderate yield, with P_4S_{10} and a primary alcohol (R′OH).[227]

Thiols add to aldehydes and ketones to give hemimercaptals and dithioacetals. Hemimercaptals are ordinarily unstable,[228] though they are more stable than the corresponding hemiacetals and can be isolated in certain cases.[229] Dithioacetals,

[210]Vedejs, E.; Perry, D.A. *J. Am. Chem. Soc.* **1983**, *105*, 1683. See also, Baldwin, J.E.; Lopez, R.C.G. *J. Chem. Soc., Chem. Commun.* **1982**, 1029.

[211]Cyclopentanone and cyclohexanone gave different products: Scheibye, S.; Shabana, R.; Lawesson, S.; Rømming, C. *Tetrahedron* **1982**, *38*, 993.

[212]See Thomsen, I.; Clausen, K.; Scheibye, S.; Lawesson, S. *Org. Synth. VII*, 372.

[213]For reviews of this and related reagents, see Cava, M.P.; Levinson, M.I.*Tetrahedron* **1985**, *41*, 5061; Cherkasov, R.A.; Kutyrev, G.A.; Pudovik, A.N. *Tetrahedron* **1985**, *41*, 2567; Jesberger, M.; Davis, T.P.; Barner, L. *Synthesis* **2003**, 1929. For a study of the mechanism, see Rauchfuss, T.B.; Zank, G.A. *Tetrahedron Lett.* **1986**, *27*, 3445.

[214]See, for example, Scheeren, J.W.; Ooms, P.H.J.; Nivard, R.J.F. *Synthesis* **1973**, 149.

[215]Curphey, T.J. *J. Org. Chem.* **2002**, *67*, 6461.

[216]Polshettiwar, V.; Kaushik, M.P. *Tetrahedron Lett.* **2004**, *45*, 6255.

[217]Degl'Innocenti, A.; Capperucci, A.; Mordini, A.; Reginato, G.; Ricci, A.; Cerreta, F. *Tetrahedron Lett.* **1993**, *34*, 873.

[218]See for example, Kimura, K.; Niwa, H.; Motoki, S. *Bull. Chem. Soc. Jpn.* **1977**, *50*, 2751; de Mayo, P.; Petrainas, G.L.R.; Weedon, A.C. *Tetrahedron Lett.* **1978**, 4621; Okazaki, R.; Inoue, K.; Inamoto, N. *Tetrahedron Lett.* **1979**, 3673.

[219]For a review of thiono esters RC(=S)OR′, see Jones, B.A.; Bradshaw, J.S. *Chem. Rev.* **1984**, *84*, 17.

[220]Ghattas, A.A.G.; El-Khrisy, E.A.M.; Lawesson, S. *Sulfur Lett.* **1982**, *1*, 69; Yde, B.; Yousif, N.M.; Pedersen, U.S.; Thomsen, I.; Lawesson, S.-O. *Tetrahedron* **1984**, *40*, 2047; Thomsen, I.; Clausen, K.; Scheibye, S.; Lawesson, S. *Org. Synth. VII*, 372.

[221]Smith, D.C.; Lee, S.W.; Fuchs, P.L. *J. Org. Chem.* **1994**, *59*, 348.

[222]Charette, A.B.; Chua, P. *Tetrahedron Lett.* **1998**, *39*, 245.

[223]Charette, A.B.; Grenon, M. *J. Org. Chem.* **2003**, *68*, 5792.

[224]Corey, E.J.; Wright, S.W. *Tetrahedron Lett.* **1984**, *25*, 2639.

[225]Filippi, J.-J.; Fernandez, X.; Lizzani-Cuvelier, L.; Loiseau, A.-M. *Tetrahedron Lett.* **2003**, *44*, 6647.

[226]For a review of dithiocarboxylic esters, see Kato, S.; Ishida, M. *Sulfur Rep.* **1988**, *8*, 155.

[227]Davy, H.; Metzner, P. *Chem. Ind. (London)* **1985**, 824.

[228]See, for example, Fournier, L.; Lamaty, G.; Nata, A.; Roque, J.P. *Tetrahedron* **1975**, *31*, 809.

[229]For example, see Field, L.; Sweetman, B.J. *J. Org. Chem.* **1969**, *34*, 1799.

like acetals, are stable in the presence of bases, except that a strong base can remove the aldehyde proton, if there is one[230] (see **10-71**).

18

   The reaction of aldehydes or ketones with thiols, usually with a Lewis acid catalyst, leads to dithioacetals[231] or dithioketals. The most common catalyst used is probably boron trifluoride etherate ($BF_3 \cdot OEt_2$).[232] Similarly reactions that use 1,2-ethanedithiol or 1,3-propanedithiol lead to 1,3-dithiolanes, such as **18**[233] or 1,3-dithianes.[234] Dithioacetals can also be prepared from aldehydes or ketones by treatment with thiols in the presence of $TiCl_4$,[235] $SiCl_4$,[236] $LiBF_4$,[237] $Al(OTf)_3$,[238] with a disulfide RSSR (R = alkyl or aryl),[239] or with methylthiotrimethylsilane ($MeSSiMe_3$).[240]

   Dithioacetals and dithioketals are used as protecting groups for aldehydes and ketones, and after subsequent reactions involving the R or R' group, the protecting group can then be removed.[241] There are a variety of reagents that convert these compounds back to the carbonyl.[242] Simple hydrolysis is the most common method for converting thiocarbonyls to carbonyls. Stirring thioketones with 4-nitrobenzaldehyde and a catalytic amount of TMSOTf gives the ketone.[243] The reaction of a thioketone and Clayfen with microwave irradiation give the ketone.[244] Thioamides

[230]Truce, W.E.; Roberts, F.E. *J. Org. Chem.* **1963**, *28*, 961.

[231]See Samajdar, S.; Basu, M.K.; Becker, F.F.; Banik, N.K. *Tetrahedron Lett.* **2001**, *42*, 4425.

[232]Fujita, E.; Nagao, Y.; Kaneko, K. *Chem. Pharm. Bull.* 1978, *26*, 3743; Corey, E.J.; Bock, M.G. *Tetrahedron Lett.* **1975**, 2643.

[233]See Anand, R.V.; Saravanan, P.; Singh, V.K. *Synlett* **1999**, 415; Ceschi, M.A.; Felix, L.de A.; Peppe, C. *Tetrahedron Lett.* **2000**, *41*, 9695; Muthusamy, S.; Babu, S.A.; Gunanathan, C. *Tetrahedron Lett.* **2001**, *42*, 359; Yadav, J.S.; Reddy, B.V.S.; Pandey, S.K. *Synlett* **2001**, 238; Ballini, R.; Barboni, L.; Maggi, R.; Sartori, G. *Synth. Commun.* **1999**, *29*, 767; Jin, T.-S.; Sun, X.; Ma, Y.-R.; Li, T.-S. *Synth. Commun.* **2001**, *31*, 1669; Deka, N.; Sarma, J.C. *Chem. Lett.* **2001**, 794; Kamal, A.; Chouhan, G. *Synlett* **2002**, 474. For a review, see Olsen, R.K.; Currie, Jr., J.O., in Patai, S. *The Chemistry of the Thiol Group*, pt. 2, Wiley, NY, **1974**, pp. 521–532.

[234]See Firouzabadi, H.; Karimi, B.; Eslami, S. *Tetrahedron Lett.* **1999**, *40*, 4055; Firouzabadi, H.; Iranpoor, N.; Karimi, B. *Synthesis* **1999**, 58; Tietze, L.F.; Weigand, B.; Wulff, C. *Synthesis* **2000**, 69; Graham, A.E. *Synth. Commun.* **1999**, *29*, 697; Firouzabadi, H.; Eslami, S.; Karimi, B. *Bull. Chem. Soc. Jpn.* **2001**, *74*, 2401; Laskar, D.D.; Prajapati, D.; Sandhu, J.S. *J. Chem. Res. (S)* **2001**, 313; De, S.K. *Tetrahedron Lett.* **2004**, *45*, 1035, 2339.

[235]Kumar, V.; Dev, S. *Tetrahedron Lett.* **1983**, *24*, 1289.

[236]Ku, B.; Oh, D.Y. *Synth. Commun.* **1989**, 433.

[237]This reaction is done neat, see Kazaraya, K.; Tsuji, S.; Sato, T. *Synlett* **2004**, 1640.

[238]This is a solvent-free reaction. See Firouzabadi, H.; Iranpoor, N.; Kohmarch, G. *Synth. Commun.* **2003**, *33*, 167.

[239]Tazaki, M.; Takagi, M. *Chem. Lett.* **1979**, 767.

[240]Evans, D.A.; Grimm, K.G.; Truesdale, L.K. *J. Am. Chem. Soc.* **1975**, *97*, 3229.

[241]For example, see Ganguly, N.C.; Datta, M. *Synlett* **2004**, 659.

[242]Corsaro, A.; Pistarà, V. *Tetrahedron* **1998**, *54*, 15027.

[243]Ravindrananthan, T.; Chavan, S.P.; Awachat, M.M.; Kelkar, S.V. *Tetrahedron Lett.* **1995**, *36*, 2277.

[244]Varma, R.S.; Kumar, D. *Synth. Commun.* **1999**, *29*, 1333.

are converted to amides with Caro's acid on SiO_2.[245] Lewis acids, such as aluminum chloride ($AlCl_3$) and mercuric salts, are common reagents and their use is referred to as the Corey–Seebach procedure.[246] Other reagents include $BF_3\bullet OEt_2$ in aq. THF containing mercuric oxide (HgO),[247] NBS,[248] iodine in DMSO,[249] ceric ammonium nitrate, $Ce(NH_4)_2(NO_3)_6$,[250] iodomethane in aqueous media,[251] Clayfen with microwave irradiation,[252] $PhI(OAc)_2$ in aqueous acetone,[253] and NCS with silver nitrate in aqueous acetonitrile.[254] When aldehydes and ketones react with mercapto-alcohols, mixed acetals or ketals are formed. The use of 2-mercaptoethanol ($HSCH_2CH_2OH$), for example, leads to an oxathiolane[255] and 3-mercaptopropanol ($HSCH_2CH_2CH_2OH$) leads to an oxathiane. Alternatively, the dithioketal can be desulfurized with Raney nickel (**14-27**), giving the overall conversion $C{=}O \rightarrow CH_2$.

If an aldehyde or ketone possesses an α hydrogen, it can be converted to the corresponding enol thioether (**19**) by treatment with a thiol in the presence of $TiCl_4$.[256] Aldehydes and ketones have been converted to sulfides by treatment with thiols and pyridine–borane, $RCOR' + R^2SH \rightarrow RR'CHSR^2$,[257] in a reductive alkylation reaction, analogous to **16-7**.

OS **II**, 610; **IV**, 927; **VI**, 109; **VII**, 124, 372. Also see OS **III**, 332; **IV**, 967; **V**, 780; **VI**, 556; **VIII**, 302.

16-12   Formation of Bisulfite Addition Products

O-Hydro-*C*-sulfonato-addition

[245]Movassagh, B.; Lakouraj, M.M.; Ghodrati, K. *Synth. Commun.* **2000**, *30*, 2353.
[246]Seebach, D.; Corey, E.J. *J. Org. Chem.* **1975**, *40*, 231; Seebach, D. *Synthesis* **1969**, 17; Vedejs, E.; Fuchs, P.L. *J. Org. Chem.* **1971**, *36*, 366.
[247]Vedejs, E.; Fuchs, P.L. *J. Org. Chem.* **1971**, *36*, 366.
[248]Cain, E.N.; Welling, L.L. *Tetrahedron Lett.* **1975**, 1353; Corey, E.J.; Erickson, B.W. *J. Org. Chem.* **1971**, *36*, 3553.
[249]Chattopadhyaya, J.B.; Rama Rao, A.V. *Tetrahedron Lett.* **1973**, 3735.
[250]Ho, T.-L.; Ho, H.C.; Wong, C.M. *J. Chem. Soc., Chem. Commun.* **1972**, 791a.
[251]Fétizon, M.; Jurion, M. *J. Chem. Soc., Chem. Commun.* **1972**, 382; Takano, S.; Hatakeyama, S.; Ogasawara, K. *J. Chem. Soc., Chem. Commun.* **1977**, 68.
[252]Meshram, H.M.; Reddy, G.S.; Sumitra, G.; Yadav, J.S. *Synth. Commun.* **1999**, *29*, 1113.
[253]Shi, X.-X.; Wu, Q.-Q. *Synth. Commun.* **2000**, *30*, 4081.
[254]Corey, E.J.; Erickson, B.W. *J. Org. Chem.* **1971**, *36*, 3553.
[255]See Karimi, B.; Seradj, H. *Synlett* **2000**, 805; Ballini, R.; Bosica, G.; Maggi, R. ; Mazzacani, A.; Righi, P.; Sartori, G. *Synthesis* **2001**,1826; Mondal, E.; Sahu, P.R.; Khan, A.T. *Synlett* **2002**, 463.
[256]Mukaiyama, T.; Saigo, K. *Chem. Lett.* **1973**, 479.
[257]Kikugawa, Y. *Chem. Lett.* **1981**, 1157.

Bisulfite addition products are formed from aldehydes, methyl ketones, cyclic ketones (generally seven-membered and smaller rings), α-keto esters, and isocyanates, upon treatment with sodium bisulfite. Most other ketones do not undergo the reaction, probably for steric reasons. The reaction is reversible (by treatment of the addition product with either acid or base[258])[259] and is useful for the purification of the starting compounds, since the addition products are soluble in water and many of the impurities are not.[260]

OS **I**, 241, 336; **III**, 438; **IV**, 903; **V**, 437.

D. Attack by NH$_2$, NHR, or NR$_2$ (Addition of NH$_3$, RNH$_2$, R$_2$NH)

16-13   The Addition of Amines to Aldehydes and Ketones

Alkylimino-de-oxo-bisubstitution

$$R\overset{\overset{\textstyle O}{\|}}{\underset{\textstyle R^1}{C}} + R^2NH_2 \longrightarrow R\overset{\overset{\textstyle N-R^2}{\|}}{\underset{\textstyle R^1}{C}}$$

The addition of ammonia[261] to aldehydes or ketones does not generally give useful products. According to the pattern followed by analogous nucleophiles, the initial products would be expected to be *hemiaminals*,[262] but these compounds are generally unstable. Most imines with a hydrogen on the nitrogen spontaneously polymerize.[263] In the presence of an oxidizing agent, such a MnO$_2$ (see **19-3**), primary alcohols can be converted to imines.[264]

In contrast to ammonia, primary, secondary, and tertiary amines can add to aldehydes[265] and ketones to give different kinds of products. Primary amines give imines[266] and secondary amines gives enamines (**10-69**). This section will focus on imines. Reduction of ω-azido ketones leads to the amino-ketones, which cyclizes to form a 2-substituted pyrroline.[267] Reduction of nitro-ketones in the presence

[258]For cleavage with ion-exchange resins, see Khusid, A.Kh.; Chizhova, N.V. *J. Org. Chem. USSR* **1985**, *21*, 37.

[259]For a discussion of the mechanism, see Young, P.R.; Jencks, W.P. *J. Am. Chem. Soc.* **1978**, *100*, 1228.

[260]The reaction has also been used to protect an aldehyde group in the presence of a keto group: Chihara, T.; Wakabayashi, T.; Taya, K. *Chem. Lett.* **1981**, 1657.

[261]For a review of this reagent in organic synthesis, see Jeyaraman, R., in Pizey, J.S. *Synthetic Reagents*, Vol. 5, Wiley, NY, **1983**, pp. 9–83.

[262]These compounds have been detected by ^{13}C NMR: Chudek, J.A.; Foster, R.; Young, D. *J. Chem. Soc. Perkin Trans. 2* **1985**, 1285.

[263]Methanimine CH$_2$=NH is stable in solution for several hours at −95°C, but rapidly decomposes at −80°C: Braillon, B.; Lasne, M.C.; Ripoll, J.L.; Denis, J.M. *Nouv. J. Chim.*, **1982**, *6*, 121. See also, Bock, H.; Dammel, R. *Chem. Ber.* **1987**, *120*, 1961.

[264]Kanno, H.; Taylor, R.J.K. *Synlett* **2002**, 1287.

[265]For a review of the reactions between amines and formaldehyde, see Farrar, W.V. *Rec. Chem. Prog.*, **1968**, *29*, 85.

[266]For reviews of reactions of carbonyl compounds leading to the formation of C=N bonds, see Dayagi, S.; Degani, Y. in Patai, S. *The Chemistry of the Carbon–Nitrogen Double Bond*, Wiley, NY, **1970**, pp. 64–83; Reeves, R.L., in Patai, S. *The Chemistry of the Carbonyl Group*, pt. 1, Wiley, NY, **1966**, pp. 600–614.

[267]Prabhu, K.P.; Sivanand, P.S.; Chandrasekaran, S. *Synlett* **1998**, 47.

of ruthenium compounds and CO also leads to 1-substituted pyrrolines.[268] In contrast to imines in which the nitrogen is attached to a hydrogen, these imines are stable enough for isolation. However, in some cases, especially with simple R groups, they rapidly decompose or polymerize unless there is at least one aryl group on the nitrogen or the carbon. When there is an aryl group, the compounds are quite stable. They are usually called *Schiff bases*, and this reaction is the best way to prepare them.[269] The reaction is straightforward and proceeds in high yields. Even sterically hindered imines can be prepared.[270] Both imines and enamines (see below) have been prepared on clay with microwave irradiation.[271] The initial *N*-substituted hemiaminals[272] lose water to give the stable Schiff bases:

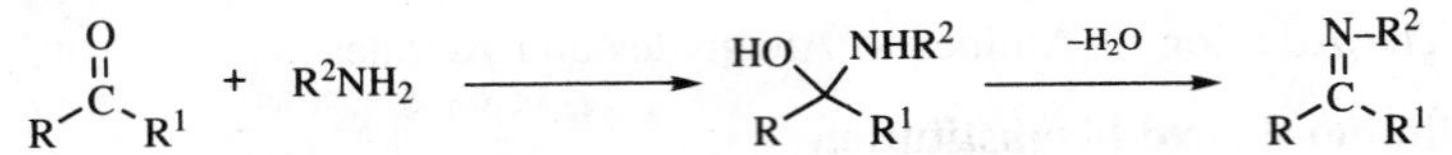

In general, ketones react more slowly than aldehydes, and higher temperatures and longer reaction times are often required.[273] In addition, the equilibrium must often be shifted, usually by removal of the water, either azeotropically by distillation, or with a drying agent, such as $TiCl_4$,[274] or with a molecular sieve.[275] Imines have been formed from aldehydes and amines in an ionic liquid.[276]

The reaction is often used to effect ring closure.[277] The *Friedländer quinoline synthesis*[278] is an example where ortho alkenyl aniline derivatives give the quinoline, **20**.[279] The alkene derivative can be prepared *in situ* from an aldehyde and a suitably functionalized ylid.[280]

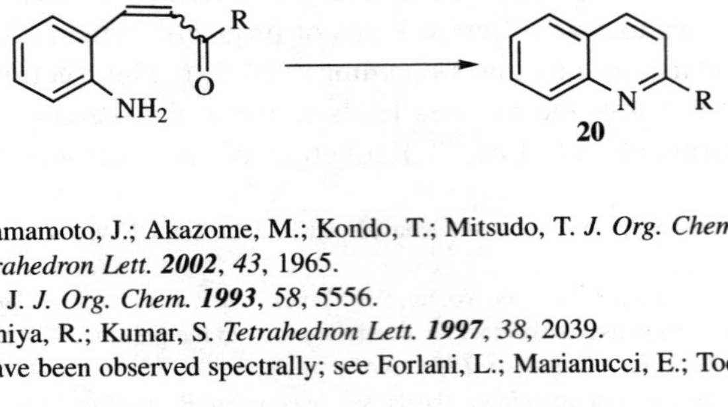

[268]Watanabe, Y.; Yamamoto, J.; Akazome, M.; Kondo, T.; Mitsudo, T. *J. Org. Chem.* **1995**, *60*, 8328.

[269]See Lai, J.T. *Tetrahedron Lett.* **2002**, *43*, 1965.

[270]Love, B.E.; Ren, J. *J. Org. Chem.* **1993**, *58*, 5556.

[271]Varma, R.S.; Dahiya, R.; Kumar, S. *Tetrahedron Lett.* **1997**, *38*, 2039.

[272]Some of these have been observed spectrally; see Forlani, L.; Marianucci, E.; Todesco, P.E. *J. Chem. Res. (S)* **1984**, 126.

[273]For improved methods, see Morimoto, T.; Sekiya, M. *Chem. Lett.* **1985**, 1371; Eisch, J.J.; Sanchez, R. *J. Org. Chem.* **1986**, *51*, 1848.

[274]Weingarten, H.; Chupp, J.P.; White, W.A. *J. Org. Chem.* **1967**, *32*, 3246.

[275]Bonnett, R.; Emerson, T.R. *J. Chem. Soc.* **1965**, 4508; Roelofsen, D.P.; van Bekkum, H. *Recl. Trav. Chim. Pays-Bas* **1972**, *91*, 605.

[276]Andrade, C.K.Z.; Takada, S.C.S.; Alves, L.M.; Rodrigues, J.P.; Suarez, P.A.Z.; Brand, R.F.; Soares, V.C.D. *Synlett* **2004**, 2135.

[277]For a review of such ring closures, see Katritzky, A.R.; Ostercamp, D.L.; Yousaf, T.I. *Tetrahedron* **1987**, *43*, 5171.

[278]For a review, see Cheng, C.; Yan, S. *Org. React.* **1982**, *28*, 37.

[279]See Arcadi, A.; Chiarini, M.; Di Giuseppe, S.; Marinelli, F. *Synlett* **2003**, 203; Yadav, J.S.; Reddy, B.V.S.; Premalatha, K. *Synlett* **2004**, 963.

[280]Hsiao, Y.; Rivera, N.R.; Yasuda, N.; Hughes, D.L.; Reider, P.J. *Org. Lett.* **2001**, *3*, 1101.

Pyrylium ions react with ammonia or primary amines to give pyridinium ions[281] (see p. 498). Primary amines react with 1,4-diketones, with microwave irradiation, to give N-substituted pyrroles.[282] Similar reactions in the presence of Montmorillonite KSF[283] or by simply heating the components with tosic acid[284] have been reported.

As mentioned, the reaction of secondary amines with ketones leads to enamines (**10-69**). When secondary amines are added to aldehydes or ketones, the initially formed N,N-disubstituted hemiaminals (**21**) cannot lose water in the same way, and in some cases it is possible to isolate them.[285] However, they are generally unstable, and under the reaction conditions usually react further. If no α hydrogen is present, **10** is converted to

| 21 | 22 (aminal) | 23 |

the more stable *aminal* (**22**).[286] However, if an α hydrogen is present, water (from **21**) or RNH_2 (from **22**) can be lost in that direction to give an enamine, **24**.[287] This is the most common method[288] for the preparation of enamines and usually takes place when an aldehyde or ketone containing an α hydrogen is treated with a secondary amine. The water is usually removed azeotropically or with a drying agent,[289] but molecular sieves can also be used.[290] Silyl carbamates, such as Me_2 NCO_2SiMe_3, have been used to convert ketones to enamines.[291] Stable primary enamines have also been prepared.[292] Enamino-ketones have been prepared from diketones and secondary amines using low molecular weight amines in water,[293] or using microwave irradiation on silica gel.[294] Secondary amine perchlorates react

---

[281]For a review, see Zvezdina, E.A.; Zhadonva, M.P.; Dorofeenko, G.N. *Russ. Chem. Rev.* **1982**, *51*, 469.

[282]Danks, T.N. *Tetrahedron Lett.* **1999**, *40*, 3957.

[283]Banik, B.K.; Samajdar, S.; Banik, I. *J. Org. Chem.* **2004**, *69*, 213.

[284]Klappa, J.J.; Rich, A.E.; McNeill, K. *Org. Lett.* **2002**, *4*, 435.

[285]For example, see Duhamel, P.; Cantacuzène, J. *Bull. Soc. Chim. Fr.* **1962**, 1843.

[286]For a review of aminals, see Duhamel, P., in Patai, S. *The Chemistry of Functional Groups, Supplement F*, pt. 2, Wiley, NY, **1982**, pp. 849–907.

[287]For reviews of the preparation of enamines, see Haynes, L.W.; Cook, A.G., in Cook, A.G. *Enamines*, 2nd. ed., Marcel Dekker, NY, **1988**, pp. 103–163; Pitacco, G.; Valentin, E., in Patai, S. *The Chemistry of Functional Groups, Supplement F*, pt. 1, Wiley, NY, **1982**, pp. 623–714.

[288]For another method, see Katritzky, A.R.; Long, Q.; Lue, P.; Jozwiak, A. *Tetrahedron* **1990**, *46*, 8153.

[289]For example, TiCl₄: White, W.A.; Weingarten, H. *J. Org. Chem.* **1967**, *32*, 213; Kuo, S.C.; Daly, W.H. *J. Org. Chem.* **1970**, *35*, 1861; Nilsson, A.; Carlson, R. *Acta Chem. Scand. Ser. B* **1984**, *38*, 523.

[290]Brannock, K.C.; Bell, A.; Burpitt, R.D.; Kelly, C.A. *J. Org. Chem.* **1964**, *29*, 801; Taguchi, K.; Westheimer, F.H. *J. Org. Chem.* **1971**, *36*, 1570; Roelofsen, D.P.; van Bekkum, H. *Recl. Trav. Chim. Pays-Bas* **1972**, *91*, 605; Carlson, R.; Nilsson, A.; Strömqvist, M. *Acta Chem. Scand. Ser. B* **1983**, *37*, 7.

[291]Kardon, F.; Mörtl, M.; Knausz, D. *Tetrahedron Lett.* **2000**, *41*, 8937.

[292]Erker, G.; Riedel, M.; Koch, S.; Jödicke, T.; Würthwein, E.-U. *J. Org. Chem.* **1995**, *60*, 5284.

[293]Stefani, H.A.; Costa, I.M.; Silva, D. de O. *Synthesis* **2000**, 1526.

[294]Rechsteiner, B.; Texier-Boullet, F.; Hamelin, J. *Tetrahedron Lett.* **1993**, *34*, 5071.

with aldehydes and ketones to give iminium salts (**10**, p. $$$).[295] Tertiary amines can only give salts (**23**). Enamines have been prepared by the reaction of an aldehyde, a secondary amine and a terminal alkyne in the presence of AgI at 100°C,[296] AgI in an ionic liquid,[297] CuI with microwave irradiation,[298] or a gold catalyst.[299]

$$ H-\overset{|}{\underset{|}{C}}-\overset{NR_2}{\underset{OH}{\overset{|}{C}}}- \longrightarrow \overset{\diagdown}{\diagup}C=C\overset{NR_2}{\diagdown} \quad \mathbf{24} $$

OS **I**, 80, 355, 381; **II**, 31, 49, 65, 202, 231, 422; **III**, 95, 328, 329, 332, 358, 374, 513, 753, 827; **IV**, 210, 605, 638, 824; **V**, 191, 277, 533, 567, 627, 703, 716, 736, 758, 808, 941, 1070; **VI**, 5, 448, 474, 496, 520, 526, 592, 601, 818, 901, 1014; **VII**, 8, 135, 144, 473; **VIII**, 31, 132, 403, 451, 456, 493, 586, 597. Also see OS **IV**, 283, 464; **VII**, 197; **VIII**, 104, 112, 241.

16-14   The Addition of Hydrazine Derivatives to Carbonyl Compounds

Hydrazono-de-oxo-bisubstitution

$$ \overset{O}{\underset{}{\overset{||}{\underset{}{C}}}} \quad \xrightarrow{RNHNH_2} \quad \overset{N\diagup^{NHR}}{\underset{}{\overset{||}{\underset{}{C}}}} $$

The product of condensation of a hydrazine and an aldehyde or ketone is called a *hydrazone*. Hydrazine itself gives hydrazones only with aryl ketones. With other aldehydes and ketones, either no useful product can be isolated, or the remaining NH_2 group condenses with a second equivalent of carbonyl compound to give an *azine*. This type of product is especially important for aromatic aldehydes:

$$ ArCH=N-NH_2 \quad + \quad ArCHO \longrightarrow ArCH=N-N=CHAr \qquad \text{An azine} $$

However, in some cases azines can be converted to hydrazones by treatment with excess hydrazine and NaOH.[300] Arylhydrazines, especially phenyl, p-nitrophenyl, and 2,4-dinitrophenyl,[301] are used much more often and give the corresponding hydrazones with most aldehydes and ketones.[302] Since these are usually solids, they make excellent derivatives and are commonly employed for this purpose. Cyclic hydrazones are also known,[303] as are conjugated hydrazones.[304] Azides react

[295]Leonard, N.J.; Paukstelis, J.V. *J. Org. Chem.* **1964**, *28*, 3021.

[296]Wei, C.; Li, Z.; Li, C.-J. *Org. Lett.* **2003**, *5*, 4473.

[297]In bmim BF$_4$, 1-butyl-3-methylimidazolium tetrafluoroborate: Li, Z.; Wei, C.; Chen, L.; Varma, R.S.; Li, C.-J. *Tetrahedron Lett.* **2004**, *45*, 2443.

[298]Shi, L.; Tu, Y.-Q.; Wang, M.; Zhang, F.-M.; Fan, C.-A. *Org. Lett.* **2004**, *6*, 1001.

[299]Wei, C.; Li, C.-J. *J. Am. Chem. Soc.* **2003**, *125*, 9584.

[300]For example, see Day, A.C.; Whiting, M.C. *Org. Synth.* **VI**, 10.

[301]For an improved procedure for the preparation of 2,4-dinitrophenylhydrazones, see Behforouz, M.; Bolan, J.L.; Flynt, M.S. *J. Org. Chem.* **1985**, *50*, 1186.

[302]For a review of arylhydrazones, see Buckingham, J. *Q. Rev. Chem. Soc.* **1969**, *23*, 37.

[303]Nakamura, E.; Sakata, G.; Kubota, K. *Tetrahedron Lett.* **1998**, *39*, 2157.

[304]Palacios, F.; Aparicio, D.; de los Santos, J.M. *Tetrahedron Lett.* **1993**, *34*, 3481.

with *N,N*-dimethylhydrazine and ferric chloride to give the *N,N*-dimethylhydra-zone.[305] Alkenes react with CO/H_2, phenylhydrazine and a diphosphine catalyst to give a regioisomeric mixture of phenylhydrazones that favored "anti-Markovni-kov" addition.[306] Oximes are converted to hydrazones with water and hydrazine in refluxing ethanol.[307]

α-Hydroxy aldehydes and ketones and α-dicarbonyl compounds give *osazones*, in which two adjacent carbons have carbon–nitrogen double bonds:

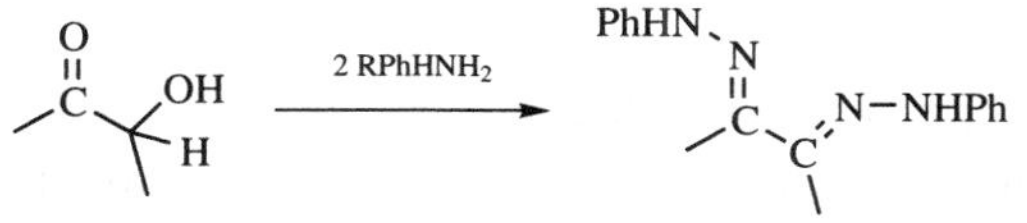

Osazones are particularly important in carbohydrate chemistry. In contrast to this behavior, β-diketones and β-keto esters give *pyrazoles* and *pyrazolones*, respectively (illustrated for β-keto esters):

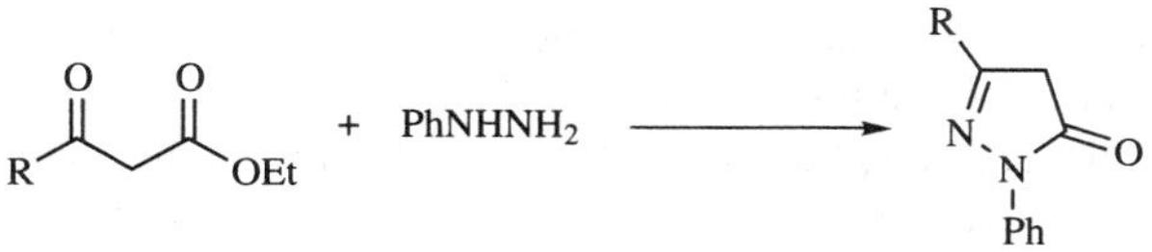

Other hydrazine derivatives frequently used to prepare the corresponding hydra-zone are semicarbazide $NH_2NHCONH_2$, in which case the hydrazone is called a semicarbazone, and *Girard's reagents T and P*, in which case the hydrazone is water soluble because of the ionic group. Girard's reagents are often used for purification of carbonyl compounds.[308]

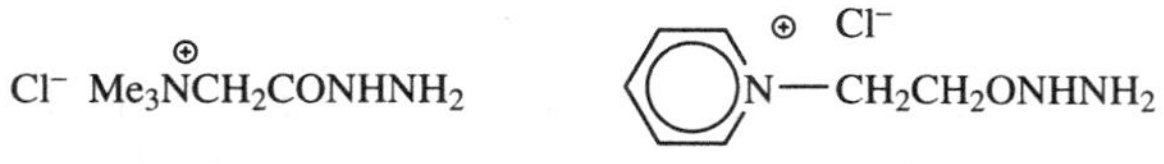

Girard's reagent T   Girard's reagent P

Simple *N*-unsubstituted hydrazones can be obtained by an exchange reaction. The *N,N*-dimethylhydrazone is prepared first, and then treated with hydrazine:[309]

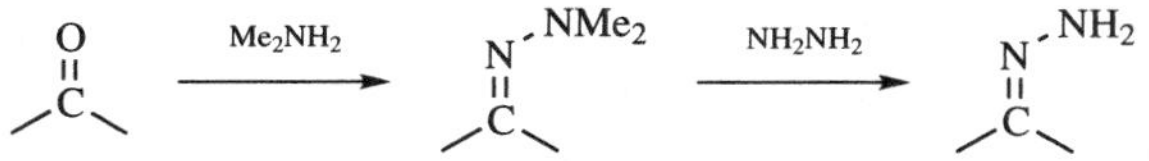

No azines are formed under these conditions.

[305]Barrett, I.C.; Langille, J.D.; Kerr, M.A. *J. Org. Chem.* **2000**, *65*, 6268.

[306]Ahmed, M.; Jackstell, R.; Seayad, A.M.; Klein, H.; Beller, M. *Tetrahedron Lett.* **2004**, *45*, 869.

[307]Pasha, M.A.; Nanjundaswamy, H.M. *Synth. Commun.* **2004**, *34*, 3827.

[308]For a study of the mechanism with Girard's reagent T, see Stachissini, A.S.; do Amaral, L. *J. Org. Chem.* **1991**, *56*, 1419.

[309]Newkome, G.R.; Fishel, D.L. *J. Org. Chem.* **1966**, *31*, 677.

OS **II**, 395; **III**, 96, 351; **IV**, 351, 377, 536, 884; **V**, 27, 258, 747, 929; **VI**, 10, 12, 62, 242, 293, 679, 791; **VII**, 77, 438. Also see OS **III**, 708; **VI**, 161; **VIII**, 597.

16-15   The Formation of Oximes

Hydroxyimino-de-oxo-bisubstitution

In a reaction very much like **16-14**, oximes can be prepared by the addition of hydroxylamine to aldehydes or ketones. Derivatives of hydroxylamine [e.g., H_2NOSO_3H and $HON(SO_3Na)_2$] have also been used. For hindered ketones, such as hexamethylacetone, high pressures (as high as 10,000 atm) may be necessary.[310] The reaction of hydroxylamine with unsymmetrical ketones or with aldehydes leads to a mixture of (*E*)- and (*Z*)-isomers. For aromatic aldehydes, heating with K_2CO_3 led to the (*E*)- isomer whereas heating with $CuSO_4$ gave the (*Z*)-hydroxylamine.[311] Hydroxylamines react with ketones in ionic liquids[312] and on silica gel.[313]

It has been shown[314] that the rate of formation of oximes is at a maximum at a pH that depends on the substrate but is usually ~4, and that the rate decreases as the pH is either raised or lowered from this point. We have previously seen (p. 1256) that bell-shaped curves like this are often caused by changes in the rate-determining step. In this case, at low pH values step 2 is rapid (because it is acid-catalyzed), and step 1

25

is slow (and rate-determining), because under these acidic conditions most of the NH_2OH molecules have been converted to the conjugate NH_3OH^+ ions, which cannot attack the substrate. As the pH is slowly increased, the fraction of free NH_2OH molecules increases and consequently so does the reaction rate, until the maximum rate is reached at ~pH 4. As the rising pH has been causing an increase in the rate of step 1, it has also been causing a *decrease* in the rate of the acid-catalyzed step 2, although this latter process has not affected the overall rate since step 2 was still faster than step 1. However, when the pH goes above ~4, step 2 becomes rate-determining, and although the rate of step 1 is still increasing (as it will until essentially all the NH_2OH is unprotonated), it is now

[310]Jones, W.H.; Tristram, E.W.; Benning, W.F. *J. Am. Chem. Soc.* **1959**, *81*, 2151.

[311]Sharghi, H.; Sarvari, M.H. *Synlett* **2001**, 99.

[312]In bmim PF$_6$, 1-butyl-3-methylimidazolium hexafluorophosphate: Ren, R.X.; Ou, W. *Tetrahedron Lett.* **2001**, *42*, 8445.

[313]Hajipour, A.R.; Mohammadpoor-Baltork, I.; Nikbaghat, K.; Imanzadeh, G. *Synth. Commun.* **1999**, *29*, 1697.

[314]Jencks, W.P. *J. Am. Chem. Soc.* **1959**, *81*, 475; *Prog. Phys. Org. Chem.* **1964**, *2*, 63.

step 2 that determines the rate, and this step is slowed by the decrease in acid concentration. Thus the overall rate decreases as the pH rises beyond ~4. It is likely that similar considerations apply to the reaction of aldehydes and ketones with amines, hydrazines, and other nitrogen nucleophiles.[315] There is evidence that when the nucleophile is 2-methylthiosemicarbazide, there is a second change in the rate-determining step: above pH ~10 *basic* catalysis of step 2 has increased the rate of this step to the point where step 1 is again rate determining.[316] Still a third change in the rate-determining step has been found at about pH 1, showing that at least in some cases step 1 actually consists of two steps: formation of a zwitterion, for example,

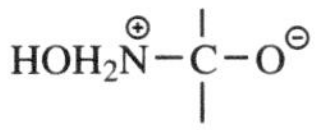

$$HOH_2\overset{\oplus}{N}-\overset{|}{\underset{|}{C}}-O^{\ominus}$$

in the case shown above, and conversion of this to **25**.[317] The intermediate **25** has been detected by nmr in the reaction between NH_2OH and acetaldehyde.[318]

In another type of process, oximes can be obtained by passing a mixture of ketone vapor, NH_3, and O_2 over a silica-gel catalyst.[319] Ketones can also be converted to oximes by treatment with other oximes, in a transoximation reaction.[320]

OS **I**, 318, 327; **II**, 70, 204, 313, 622; **III**, 690, **IV**, 229; **V**, 139, 1031; **VII**, 149. See also, OS **VI**, 670.

16-16   The Conversion of Aldehydes to Nitriles

Nitrilo-de-hydro,oxo-tersubstitution

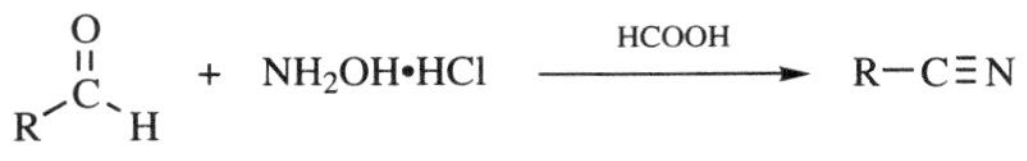

$$\underset{R}{\overset{O}{\underset{\|}{\underset{\overset{|}{C}}{}}}}\!\!\diagdown_H \;+\; NH_2OH\cdot HCl \;\xrightarrow{\;HCOOH\;}\; R{-}C{\equiv}N$$

Aldehydes can be converted to nitriles in one step by treatment with hydroxylamine hydrochloride and either formic acid,[321] $NaHSO_4 \cdot SiO_2$ with microwave irradiation,[322] or $(Bu_4N)_2S_2O_8$ with $Cu(HCO_2) \cdot Ni(COOH)_2$ and aq. KOH.[323] Heating in NMP is also effective with aryl aldehydes[324] and heating on dry alumina with

[315]For reviews of the mechanism of such reactions, see Cockerill, A.F.; Harrison, R.G., in Patai, S. *The Chemistry of Functional Groups: Supplement A*, pt. 1, Wiley, NY, *1977*, pp. 288–299; Sollenberger, P.Y.; Martin, R.B., in Patai, S. *The Chemistry of the Amino Group*, Wiley, NY, *1968*, pp. 367–392. For isotope effect studies, see Rossi, M.H.; Stachissini, A.S.; do Amaral, L. *J. Org. Chem. 1990*, 55, 1300.

[316]Sayer, J.M.; Jencks, W.P. *J. Am. Chem. Soc. 1972*, 94, 3262.

[317]Sayer, J.M.; Edman, C. *J. Am. Chem. Soc. 1979*, 101, 3010.

[318]Cocivera, M.; Fyfe, C.A.; Effio, A.; Vaish, S.P.; Chen, H.E. *J. Am. Chem. Soc. 1976*, 98, 1573; Cocivera, M.; Effio, A. *J. Am. Chem. Soc. 1976*, 98, 7371.

[319]Armor, J.N. *J. Am. Chem. Soc. 1980*, 102, 1453.

[320]For example, see Block Jr., P.; Newman, M.S. *Org. Synth. V*, 1031.

[321]Olah, G.A.; Keumi, T. *Synthesis 1979*, 112.

[322]Das, B.; Ramesh, C.; Madhusudhan, P. *Synlett 2000*, 1599.

[323]Chen, F.-E.; Fu, H.; Meng, G.; Cheng, Y.; Lü, Y.-X. *Synthesis 2000*, 1519.

[324]Kumar, H.M.S.; Reddy, B.V.S.; Reddy, P.T.; Yadav, J.S. *Synthesis 1999*, 586; Chakraborti, A.K.; Kaur, G. *Tetrahedron 1999*, 55, 13265.

aliphatic aldehyde.[325] The reaction is a combination of **16-15** and **17-29**. Direct nitrile formation has also been accomplished with certain derivatives of NH_2OH, notably, NH_2OSO_2OH.[326] Treatment with hydroxylamine and NaI[327] or certain carbonates[328] also converts aldehydes to the nitrile. Another method involves treatment with hydrazoic acid, though the Schmidt reaction (**18-16**) may compete.[329] Aromatic aldehydes have been converted to nitriles in good yield with $NH_2OH/HCOOH$ on silica gel.[330] Microwave irradiation has been used with $NH_2OH\cdot HCl$ and another reagent, which includes phthalic anhydride,[331] $Bu_2SnO\cdot Al_2O_3$,[332] or H—Y zeolite.[333] Other reagents include *N*-phenylurea with tosic acid,[334] MnO_2 and ammonia,[335] I_2 with aqueous ammonia,[336] dimethylhydrazine followed by dimethyl sulfoxide,[337] trimethylsilyl azide,[338] and with hydroxylamine hydrochloride, $MgSO_4$, and TsOH.[339] The reaction of a conjugated aldehyde with ammonia, CuCl and 50% H_2O_2 gave the conjugated nitrile.[340] Benzylic alcohols can be oxidized in the presence of ammonia to give the nitrile.[341] Trichloroisocyanuric acid with a catalytic amount of TEMPO (p. 274) converts aldehydes to nitriles at 0°C in dichloromethane.[342]

On treatment with 2 equivalents of dimethylaluminum amide Me_2AlNH_2, carboxylic esters can be converted to nitriles: $RCOOR' \rightarrow RCN$.[343] This is very likely a combination of **16-75** and **17-30**. See also, **19-5**.

OS V, 656.

16-17   Reductive Alkylation of Ammonia or Amines

Hydro,dialkylamino-de-oxo-bisubstitution

$$ \underset{R}{\overset{O}{\underset{\|}{C}}}\underset{R^1}{} + R^2_2NH + H_2 \xrightarrow{\text{catalyst}} \underset{R^1}{\overset{R}{C}}\underset{H}{\overset{NR^2_2}{}} $$

[325]Sharghi, H.; Sarvari, M.H. *Tetrahedron* **2002**, *58*, 10323. With wet alumina followed by $MeSO_2Cl$ the product is an amide.

[326]Streith, J.; Fizet, C.; Fritz, H. *Helv. Chim. Acta* **1976**, *59*, 2786.

[327]Ballini, R.; Fiorini, D.; Palmieri, A. *Synlett* **2003**, 1841.

[328]Bose, D.S.; Goud, P.R. *Synth. Commun.* **2002**, *32*, 3621.

[329]For additional methods, see Gelas-Mialhe, Y.; Vessière, R. *Synthesis* **1980**, 1005; Arques, A.; Molina, P.; Soler, A. *Synthesis* **1980**, 702; Sato, R.; Itoh, K.; Itoh, K.; Nishina, H.; Goto, T.; Saito, M. *Chem. Lett.* **1984**, 1913; Reddy, P.S.N.; Reddy, P.P. *Synth. Commun.* **1988**, *18*, 2179; Neunhoeffer, H.; Diehl, W.; Karafiat, U. *Liebigs Ann. Chem.* **1989**, 105.

[330]Kabalka, G.W.; Yang, K. *Synth. Commun.* **1998**, *28*, 3807.

[331]Veverková, E.; Toma, Š. *Synth. Commun.* **2000**, *30*, 3109.

[332]Yadav, J.S.; Reddy, B.V.S.; Madan, Ch. *J. Chem. Res. (S)* **2001**, 190.

[333]Srinivas, K.V.N.S.; Reddy, E.B.; Das, B. *Synlett* **2002**, 625.

[334]Cokun, N.; Arikan, N. *Tetrahedron* **1999**, *55*, 11943.

[335]Lai, G.; Bhamare, N.K.; Anderson, W.K. *Synlett* **2001**, 230.

[336]Talukdar, S.; Hsu, J.-L.; Chou, T.-C.; Fang, J.-M. *Tetrahedron Lett.* **2001**, *42*, 1103.

[337]Kamal, A.; Arifuddin, M.; Rao, N.V. *Synth. Commun.* **1998**, *28*, 4507.

[338]Nishiyama, K.; Oba, M.; Watanabe, A. *Tetrahedron* **1987**, *43*, 693.

[339]Ganboa, I.; Palomo, C. *Synth. Commun.* **1983**, *13*, 219.

[340]Erman, M.B.; Snow, J.W.; Williams, M.J. *Tetrahedron Lett.* **2000**, *41*, 6749.

[341]See Baxendale, I.R.; Ley, S.V.; Sneddon, H.F. *Synlett* **2002**, 775; McAllister, G.D.; Wilfred, C.D.; Taylor, R.J.K. *Synlett* **2002**, 1291.

[342]Chen, F.-E.; Kuang, Y.-Y.; Dai, H.-F.; Lu, L.; Huo, M. *Synthesis* **2003**, 2629.

[343]Wood, J.L.; Khatri, N.A.; Weinreb, S.M. *Tetrahedron Lett.* **1979**, 4907.

When an aldehyde or a ketone is treated with ammonia or a primary or secondary amine in the presence of hydrogen and a hydrogenation catalyst (heterogeneous or homogeneous),[344] *reductive alkylation* of ammonia or the amine (or *reductive amination* of the carbonyl compound) takes place.[345] The reaction can formally be regarded as occurring in the following manner (shown for a primary amine), which probably does correspond to the actual sequence of steps:[346] In this regard, the reaction of an aldehyde with an amine to give an iminium salt (**16-31**) can be followed in a second chemical step of reduction of the C=N unit (**19-42**) using NaBH$_4$ or a variety of other reagents.[347]

Primary amines have been prepared from many aldehydes with at least five carbons and from many ketones by treatment with ammonia and a reducing agent. Smaller aldehydes are usually too reactive to permit isolation of the primary amine. Secondary amines have been prepared by both possible procedures: 2 equivalents of ammonia and 1 equivalent of aldehyde or ketone, and 1 equivalent of primary amine and 1 equivalent of carbonyl compound, the latter method being better for all but aromatic aldehydes. Tertiary amines can be prepared in three ways, but the method is seldom carried out with 3 equivalents of ammonia and 1 equivalent of carbonyl compound. Much more often they are prepared from primary or secondary amines.[348] When the reagent is ammonia, it is possible for the initial product to react again and for this product to react again, so that secondary and tertiary amines are usually obtained as side products. Similarly, primary amines give tertiary as well as secondary amines. In order to minimize this, the aldehyde or ketone is treated with an excess of ammonia or primary amine (unless of course the higher amine is desired).

For ammonia and primary amines there are two possible pathways, but when secondary amines are involved, only the hydrogenolysis pathway is possible. The reaction is compatible with amino acids, giving the *N*-alkylated amino acid.[349]

[344]**Rh**: Kadyrov, R.; Riermeier, T.H.; Dingerdissen, U.; Tararov, V.; Börner, A. *J. Org. Chem.* **2003**, *68*, 4067; Gross, T.; Seayad, A.M.; Ahmad, M.; Beller, M. *Org. Lett.* **2002**, *4*, 2055. **Ir**: Chi, Y.; Zhou, Y.-G.; Zhang, X. *J. Org. Chem.* **2003**, *68*, 4120.

[345]For reviews, see Rylander, P.N. *Hydrogenation Methods*, Academic Press, NY, **1985**, pp. 82–93; Klyuev, M.V.; Khidekel, M.L. *Russ. Chem. Rev.* **1980**, *49*, 14; Rylander, P.N. *Catalytic Hydrogenation over Platinum Metals*, Academic Press, NY, **1967**, pp. 291–303.

[346]See, for example, Le Bris, A.; Lefebvre, G.; Coussemant, F. *Bull. Soc. Chim. Fr.* **1964**, 1366, 1374, 1584, 1594.

[347]For a simple example see, Bhattacharyya, S. *Synth. Commun.* **2000**, *30*, 2001.

[348]For a review of the preparation of tertiary amines by reductive alkylation, see Spialter, L.; Pappalardo, J.A. *The Acyclic Aliphatic Tertiary Amines*, Macmillan, NY, **1965**, pp. 44–52.

[349]Song, Y.; Sercel, A.D.; Johnson, D.R.; Colbry, N.L.; Sun, K.-L.; Roth, B.D. *Tetrahedron Lett.* **2000**, *41*, 8225.

Other reducing agents[350] can be used instead of hydrogen and a catalyst, among them zinc and HCl, $B_{10}H_{14}$[351] or $B_{10}H_{14}$ with Pd/C,[352] a picolinyl borane complex in acetic acid–methanol,[353] $PhSiH_3$ with 2% Bu_2SnCl_2,[354] and polymethylhydrosiloxane.[355] Several hydride reducing agents can be used, including $NaBH_4$[356] sodium borohydride with $Ti(OiPr)_4$[357] or $NiCl_2$,[358] $NaBH_4/H_3BO_4$,[359] borohydride-exchange resin,[360] sodium cyanoborohydride ($NaBH_3CN$),[361] sodium triacetoxyborohydride,[362] or a polymer-bound triethylammonium acetoxyborohydride.[363] A Hantzsch dihydropyridine in conjunction with a scandium catalyst has been used.[364] An interesting variation uses a benzylic alcohol in a reaction with a primary amine, and a mixture of MnO_2 and $NaBH_4$, giving *in situ* oxidation to the aldehyde and reductive amination to give the amine as the final product.[365]

Formic acid is commonly used for reductive amination[366] in what is called the *Wallach reaction*. Secondary amines react with formaldehyde and NaH_2PO_3 to give the *N*-methylated tertiary amine[367] and microwave irradiation has also been used.[368] Conjugated aldehydes are converted to alkenyl-amines with the amine/silica gel followed by reduction with zinc borohydride.[369] In the particular case where primary or secondary amines are reductively methylated with formaldehyde and formic acid, the method is called the *Eschweiler–Clarke procedure*. Heating with paraformaldehyde and oxalyl chloride has been used to give the same result.[370] It is

[350]For a list of many of these, with references, see Larock, R.C. *Comprehensive Organic Transformations*, 2nd ed., Wiley-VCH, NY, *1999*, pp. 835–840.

[351]Bae, J.W.; Lee, S.H.; Cho, Y.J.; Yoon, C.M. *J. Chem. Soc., Perkin Trans. 1* **2000**, 145.

[352]Jung, Y.J.; Bae, J.W.; Park, E.S.; Chang, Y.M.; Yoon, C.M. *Tetrahedron* **2003**, *59*, 10331.

[353]Sato, S.; Sakamoto, T.; Miyazawa, E.; Kitugawa, Y. *Tetrahedron* **2004**, *60*, 7899.

[354]Apodaca, R.; Xiao, W. *Org. Lett.* **2001**, *3*, 1745.

[355]Chandrasekhar, S.; Reddy, Ch.R.; Ahmed, M. *Synlett* **2000**, 1655.

[356]Sondengam, B.L.; Hentchoya Hémo, J.; Charles, G. *Tetrahedron Lett.* **1973**, 261; Schellenberg, K.A. *J. Org. Chem.* **1963**, *28*, 3259; Gribble, G.W.; Nutaitis, C.F. *Synthesis* **1987**, 709.

[357]Neidigh, K.A.; Avery, M.A.; Williamson, J.S.; Bhattacharyya, S. *J. Chem. Soc. Perkin Trans. 1* **1998**, 2527; Bhattacharyya, S. *J. Org. Chem.* **1995**, *60*, 4928.

[358]Saxena, I.; Borah, R.; Sarma, J.C. *J. Chem. Soc., Perkin Trans. 1* **2000**, 503.

[359]This is a solvent-free reaction. See Cho, B.T.; Kang, S.K. *Synlett* **2004**, 1484.

[360]Yoon, N.M.; Kim, E.G.; Son, H.S.; Choi, J. *Synth. Commun.* **1993**, *23*, 1595.

[361]Borch, R.F.; Bernstein, M.D.; Durst, H.D. *J. Am. Chem. Soc.* **1971**, *93*, 2897; Mattson, R.J.; Pham, K.M.; Leuck, D.J.; Cowen, K.A. *J. Org. Chem.* **1990**, *55*, 2552. See also, Barney, C.L.; Huber, E.W.; McCarthy, J.R. *Tetrahedron Lett.* **1990**, *31*, 5547. For reviews of $NaBH_3CN$, see Hutchins, R.O.; Natale, N.R. *Org. Prep. Proced. Int.* **1979**, *11*, 201; Lane, C.F. *Synthesis* **1975**, 135.

[362]Abdel-Magid, A.F.; Maryanoff, C.A.; Carson, K.G. *Tetrahedron Lett.* **1990**, *31*, 5595; Abdel-Magid, A.F.; Carson, K.G.; Harris, B.D.; Maryanoff, C.A.; Shah, R.D. *J. Org. Chem.* **1996**, *61*, 3849.

[363]Bhattacharyya, S.; Rana, S.; Gooding, O.W.; Labadie, J. *Tetrahedron Lett.* **2003**, *44*, 4957.

[364]Itoh, T.; Nagata, K.; Kurihara, A.; Miyazaki, M.; Ohsawa, A. *Tetrahedron lett.* **2002**, *43*, 3105; Itoh, T.; Nagata, K.; Miyazaki, M.; Ishikawa, H.; Kurihara, A.; Ohsawa, A. *Tetrahedron* **2004**, *60*, 6649.

[365]Kanno, H.; Taylor, R.J.K. *Tetrahedron Lett.* **2002**, *43*, 7337.

[366]For a microwave induced reaction see Torchy, S.; Barbry, D. *J. Chem. Res. (S)* **2001**, 292.

[367]Davis, B.A.; Durden, D.A. *Synth. Commun.* **2000**, *30*, 3353.

[368]Barbry, D.; Torchy, S. *Synth. Commun.* **1996**, *26*, 3919.

[369]Ranu, B.C.; Majee, A.; Sarkar, A. *J. Org. Chem.* **1998**, *63*, 370.

[370]Rosenau, T.; Potthast, A.; Röhrling, J.; Hofinger, A.; Sixxa, H.; Kosma, P. *Synth. Commun.* **2002**, *32*, 457.

possible to use ammonium (or amine) salts of formic acid,[371] or formamides, as a substitute for the Wallach conditions. This method is called the *Leuckart reaction*,[372] and in this case the products obtained are often the *N*-formyl derivatives of the amines instead of the free amines. A transition-metal catalyzed variation has been reported.[373] Primary and secondary amines can be *N*-ethylated (e.g., ArNHR → ArNREt) by treatment with NaBH$_4$ in acetic acid.[374] Aldehydes react with aniline in the presence of Montmorillonite K10 clay and microwaves to give the amine.[375] Tributyltin hydride is used with an ammonium salt,[376] or Bu$_2$SnClH•HMPA with an aromatic amine,[377] in the presence of a ketone to give the corresponding amine. Allylic silanes react with aldehydes and carbamates, in the presence of bismuth catalysts,[378] or BF$_3$•OEt$_2$[379] to give the corresponding allylic *N*-carbamoyl derivative, and trityl perchlorate has been used for the same purpose when *N*-trimethylsilyl carbamates are employed.[380] The reaction can be done with aromatic amines in the presence of vinyl ethers and a copper complex to give β-amino ketones.[381] Reductive amination of an aryl amine and an aryl aldehyde that contains a ortho conjugated ketone substituents gives the amine, which adds 1,4- (**15-AA**) to the α,β-unsaturated ketone unit to give a bicyclic amine.[382] Alternative methods of reductive alkylation have been developed. Alkylation of an imine formed *in situ* is also possible.[383]

Reductive alkylation has also been carried out on nitro, nitroso, azo, and other compounds that are reduced *in situ* to primary or secondary amines. Azo compounds react with aldehydes, in the presence of proline, and subsequent reduction with NaBH$_4$ gives the chiral hydrazine derivative.[384]

---

[371]For a review of ammonium formate in organic synthesis, see Ram, S.; Ehrenkaufer, R.E. *Synthesis* **1988**, 91.

[372]For a review, see Moore, M.L. *Org. React.* **1949**, 5, 301. For discussions of the mechanism, see Awachie, P.I.; Agwada, V.C. *Tetrahedron* **1990**, 46, 1899, and references cited therein. For a microwave-induced variation, see Loupy, A.; Monteux, D.; Petit, A.; Aizpurua, J.M.; Domínguez, E.; Palomo, C. *Tetrahedron Lett.* **1996**, 37, 8177. For the effects of added formamide, see Lejon, T.; Helland, I. *Acta Chem. Scand.* **1999**, 53, 76.

[373]Using a rhodium catalyst, see Kitamura, M.; Lee, D.; Hayashi, S.; Tanaka, S.; Yoshimura, M. *J. Org. Chem.* **2002**, 67, 8685. For a review of this reaction, see Riermeier, T.H.; Dingerdissen, U.; Börner, A. *Org. Prep. Proceed. Int.* **2004**, 36, 99.

[374]For a review, see Gribble, G.W.; Nutaitis, C.F. *Org. Prep. Proced. Int.* **1985**, 17, 317, pp. 336–350.

[375]Varma, R.S.; Dahiya, R. *Tetrahedron* **1998**, 54, 6293.

[376]Suwa, T.; Sugiyama, E.; Shibata, I.; Baba, A. *Synlett* **2000**, 556.

[377]Suwa, T.; Sugiyama, E.; Shibata, I.; Baba, A. *Synthesis* **2000**, 556.

[378]Ollevier, T.; Ba, T. *Tetrahedron Lett.* **2003**, 44, 9003.

[379]Billet, M.; Klotz, P.; Mann, A. *Tetrahedron lett.* **2001**, 42, 631.

[380]Niimi, L.; Serita, K.-i.; Hiraoka, S.; Yokozawa, T. *Tetrahedron Lett.* **2000**, 41, 7075.

[381]Kobayashi, S.; Ueno, M.; Suzuki, R.; Ishitani, H.; Kim, H.-S.; Wataya, Y. *J. Org. Chem.* **1999**, 64, 6833.

[382]Suwa, T.; Shibata, I.; Nishino, K.; Baba, A. *Org. Lett.* **1999**, 1, 1579.

[383]See Choudary, B.M.; Jyothi, K.; Madhi, S.; Kantam, M.L. *Synlett* **2004**, 231. For an example in the ionic liquid bmim BF$_4$, 1-butyl-3-methylimidazolium tetrafluoroborate, see Yadav, J.S.; Reddy, B.V.S.; Raju, A.K. *Synthesis* **2003**, 883.

[384]List, B. *J. Am. Chem. Soc.* **2002**, 124, 5656; Kumaragurubaran, N.; Juhl, K.; Zhuang, W.; Bøgevig, A.; Jørgensen, K.A. *J. Am. Chem. Soc.* **2002**, 124, 6254.

OS **I**, 347, 528, 531; **II**, 503; **III**, 328, 501, 717, 723; **IV**, 603; **V**, 552; **VI**, 499; **VII**, 27.

16-18   Addition of Amides to Aldehydes

Alkylamido-de-oxo-bisubstitution

Amides can add to aldehydes in the presence of bases (so the nucleophile is actually RCONH$^-$) or acids to give acylated amino alcohols, which often react further to give alkylidene or arylidene bisamides.[385] If the R$'$ group contains an α hydrogen, water may split out.

Sulfonamides add to aldehydes to give the *N*-sulfonyl imine. Benzaldehyde reacts with TsNH$_2$, for example, at 160°C in the presence of Si(OEt)$_4$,[386] with trifluoroacetic anhydride (TFAA) in refluxing dichloromethane,[387] or with TiCl$_4$ in refluxing dichloroethane,[388] to give the *N*-tosylimine, Ts—N=CHPh. In a similar manner, the reaction of TolSO$_2$Na + PhSO$_2$Na with an aldehyde in aqueous formic acid gives the *N*-phenylsulfonyl imine.[389] The reaction of an aldehyde with Ph$_3$P=NTs and a ruthenium catalyst gives the *N*-tosyl imine.[390] Reaction of aldehydes with LiAl(NHBn)$_4$ also give the corresponding *N*-benzylimine.[391]

16-19   The Mannich Reaction

Acyl,amino-de-oxo-bisubstitution, and so on

In the *Mannich reaction*, formaldehyde (or sometimes another aldehyde) is condensed with ammonia, in the form of its salt, and a compound containing an active hydrogen.[392] This can formally be considered as an addition of ammonia to give

[385]For reviews, see Challis, B.C.; Challis, J.A. in Zabicky, J. *The Chemistry of Amides*, Wiley, NY, *1970*, pp. 754–759; Zaugg, H.E.; Martin, W.B. *Org. React. 1965*, *14*, 52, 91–95, 104–112. For a discussion, see Gilbert, E.E. *Synthesis 1972*, 30.

[386]Love, B.E.; Raje, P.S.; Williams II, T.C. *Synlett 1994*, 493.

[387]Lee, K.Y.; Lee, C.G.; Kim, J.N. *Tetrahedron Lett. 2003*, *44*, 1231.

[388]Ram, R.N.; Khan, A.A. *Synth. Commun. 2001*, *31*, 841.

[389]Chemla, F.; Hebbe, V.; Normant, J.-F. *Synthesis 2000*, 75.

[390]Jain, S.L.; Sharma, V.B.; Sain, B. *Tetrahedron Lett. 2004*, *45*, 4341.

[391]Solladié-Cavallo, A.; Benchegroun, M.; Bonne, F. *Synth. Commun. 1993*, *23*, 1683.

[392]For reviews, see Tramontini, M.; Angiolini, L. *Tetrahedron 1990*, *46*, 1791; Gevorgyan, G.A.; Agababyan, A.G.; Mndzhoyan, O.L. *Russ. Chem. Rev. 1984*, *53*, 561; Tramontini, M. *Synthesis 1973*, 703; House, H.O. *Modern Synthetic Reactions*, 2nd ed., W.A. Benjamin, NY, *1972*, pp. 654–660. For reviews on the reactions of Mannich Bases, see Tramontini, M.; Angeloni, L. cited above; Gevorgyan, G.A.; Agababyan, A.G.; Mndzhoyan, O.L. *Russ. Chem. Rev. 1985*, *54*, 495.

H_2NCH_2OH, followed by a nucleophilic substitution. Instead of ammonia, the reaction can be carried out with salts of primary or secondary amines,[393] or with amides,[394] in which cases the product is substituted on the nitrogen with R, R_2, and RCO, respectively. The imine can be generated *in situ*, and the reaction of a ketone, formaldehyde, and diethylamine with microwave irradiation gave the Mannich product, a β-amino ketone.[395] Arylamines do not normally give the reaction. Hydrazines can be used.[396] The product is referred to as a *Mannich base*. Many active hydrogen compounds give the reaction, including ketones and aldehydes, esters, nitroalkanes,[397] and nitriles as well as ortho-carbons of phenols, the carbon of terminal alkynes, the oxygen of alcohols and the sulfur of thiols.[398] Vinylogous Mannich reactions are known.[399]

The Mannich base can react further in three ways. If it is a primary or secondary amine, it may condense with one or two additional molecules of aldehyde and active compound, for example,

$$H_2NCH_2CH_2COR \xrightarrow[CH_3COR]{HCHO} HN(CH_2CH_2COR)_2 \xrightarrow[CH_3COR]{HCHO} N(CH_2CH_2COR)_3$$

If the active hydrogen compound has two or three active hydrogens, the Mannich base may condense with one or two additional molecules of aldehyde and ammonia or amine, for example,

$$H_2NCH_2CH_2COR \xrightarrow[NH_3]{HCHO} (H_2NCH_2)_2CHCOR \xrightarrow[NH_3]{HCHO} (H_2NCH_2)_3CHCOR$$

Another further reaction consists of condensation of the Mannich base with excess formaldehyde:

$$H_2NCH_2CH_2COR \quad + \quad HCHO \longrightarrow H_2C=NCH_2CH_2COR$$

Sometimes it is possible to obtain these products of further condensation as the main products of the reaction. At other times they are side products.

When the Mannich base contains an amino group β to a carbonyl (and it usually does), ammonia is easily eliminated. This is a route to α,β-unsaturated aldehydes, ketones, esters, and so on.

[393]For a review where the amine component is an amino acid, see Agababyan, A.G.; Gevorgyan, G.A.; Mndzhoyan, O.L. *Russ. Chem. Rev.* **1982**, *51*, 387.

[394]Hellmann, H. *Angew. Chem.* **1957**, *69*, 463; *Newer Methods Prep. Org. Chem.* **1963**, *2*, 277.

[395]Gadhwal, S.; Baruah, M.; Prajapati, D.; Sandhu, J.S. *Synlett* **2000**, 341.

[396]El Kaim, L.; Grimaud, L.; Perroux, Y.; Tirla, C. *J. Org. Chem.* **2003**, *68*, 8733.

[397]Qian, C.; Gao, F.; Chen, R. *Tetrahedron Lett.* **2001**, *42*, 4673. See Baer, H.H.; Urbas, L., in Feuer, H. *The Chemistry of the Nitro and Nitroso Groups*, Wiley, NY, **1970**, pp. 117–130.

[398]see Massy, D.J.R. *Synthesis* **1987**, 589; Dronov, V.I.; Nikitin, Yu.E. *Russ. Chem. Rev.* **1985**, *54*, 554

[399]Bur, S.; Martin, S.F. *Tetrahedron* **2001**, *57*, 3221. For a review, see Martin, S.F. *Acc. Chem. Res.* **2002**, *35*, 895.

Studies of the reaction kinetics have led to the following proposals for the mechanism of the Mannich reaction.[400]
The base-catalyzed reaction:

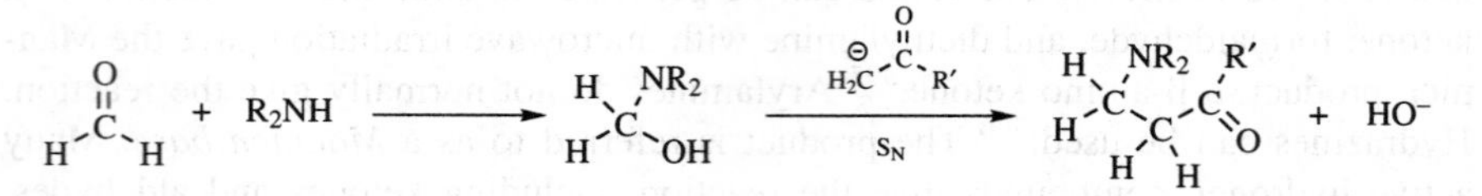

The acid-catalyzed reaction:

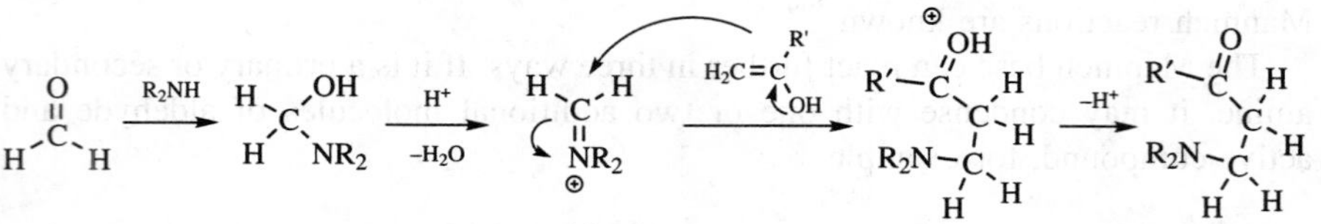

According to this mechanism, it is the free amine, not the salt that reacts, even in acid solution; and the active-hydrogen compound (in the acid-catalyzed process) reacts as the enol when that is possible. This latter step is similar to what happens in **12-4**. There is kinetic evidence for the intermediacy of the iminium ion (**26**).[401]

When an unsymmetrical ketone is used as the active-hydrogen component, two products are possible. Regioselectivity has been obtained by treatment of the ketone with pre-formed iminium ions:[402] the use of $Me_2N^+=CH_2$ $CF_3COO^\ominus$ in CF_3COOH gives substitution at the more highly substituted position, while with $(iPr)_2N=CH_2^+$ $ClO_4^\ominus$ the reaction takes place at the less highly substituted position.[403] The pre-formed iminium compound dimethyl(methylene)ammonium iodide $CH_2=N^+Me_2$ $I^\ominus$, called *Eschenmoser's salt*,[404] has also been used in Mannich reactions.[405] The analogous chloride salt has been condensed with an imine to give and a β,β'-dimethylamino ketone after acid hydrolysis.[406]

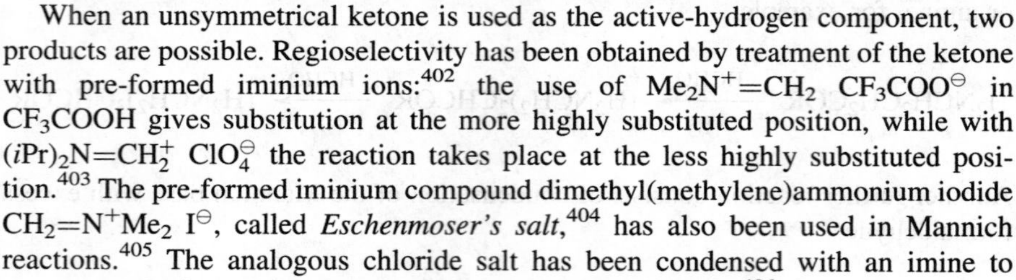

[400]Cummings, T.F.; Shelton, J.R. *J. Org. Chem.* **1960**, *25*, 419.

[401]Benkovic, S.J.; Benkovic, P.A.; Comfort, D.R. *J. Am. Chem. Soc.* **1969**, *91*, 1860.

[402]For earlier use of pre-formed iminium ions in the Mannich reaction, see Ahond, A.; Cavé, A.; Kan-Fan, C.; Potier, P. *Bull. Soc. Chim. Fr.* **1970**, 2707; Schreiber, J.; Maag, H.; Hashimoto, N.; Eschenmoser, A. *Angew. Chem. Int. Ed.* **1971**, *10*, 330.

[403]Jasor, Y.; Luche, M.; Gaudry, M.; Marquet, A. *J. Chem. Soc., Chem. Commun.* **1974**, 253; Gaudry, M.; Jasor, Y.; Khac, T.B. *Org. Synth.* **VI**, 474.

[404]Schreiber, J.; Maag, H.; Hashimoto, N.; Eschenmoser, A. *Angew. Chem. Int. Ed.* **1971**, *10*, 330.

[405]See Holy, N.; Fowler, R.; Burnett, E.; Lorenz, R. *Tetrahedron* **1979**, *35*, 613; Bryson, T.A.; Bonitz, G.H.; Reichel, C.J.; Dardis, R.E. *J. Org. Chem.* **1980**, *45*, 524, and references cited therein.

[406]Arend, M.; Risch, N. *Tetrahedron Lett.* **1999**, *40*, 6205.

Another type of pre-formed reagent (**28**) has been used to carry out diastereoselective Mannich reactions. The lithium salts **27** are treated with $TiCl_4$ to give **28**, which is then treated with the enolate of a ketone.[407] The palladium catalyzed Mannich reaction of enol ethers to imines is also known.[408] The reaction of silyl enol ethers and imines is catalyzed by HBF_4 in aqueous methanol.[409] Similarly, silyl enol ethers react with aldehydes and aniline in the presence of $InCl_3$ to give the β-amino ketone.[410] Imines react on Montmorillonite K10 clay and microwave irradiation gives β-amino esters.[411] Enol ethers react similarly in the presence of $Yb(OTf)_3$.[412]

Enantioselective Mannich reactions are known.[413] The most common method uses a chiral catalyst, including proline,[414] proline derivatives or proline analogs.[415] Chiral diamine[416] or phosphine-imine[417] ligands have been used. Chiral auxiliaries on the carbonyl fragment can be used.[418] Chiral imines, in the form of chiral hydrazones have been used with silyl enol ethers and a scandium catalyst.[419] Chiral amine react with aldehydes, with silyl enol ethers and an $InCl_3$ catalyst in ionic liquids, to give the Mannich product with good enantioselectivity.[420]

Also see, **11-22**.

OS **III**, 305; **IV**, 281, 515, 816; **VI**, 474, 981, 987; **VII**, 34. See also, OS **VIII**, 358.

16-20   The Addition of Amines to Isocyanates

N-**Hydro-*C*-alkylamino-addition**

[407]Seebach, D.; Schiess, M.; Schweizer, W.B. *Chimia* **1985**, *39*, 272. See also, Heaney, H.; Papageorgiou, G.; Wilkins, R.F. *J. Chem. Soc., Chem. Commun.* **1988**, 1161; Katritzky, A.R.; Harris, P.A. *Tetrahedron* **1990**, *46*, 987.

[408]For a discussion of the mechanism, see Fujii, A.; Hagiwara, E.; Sodeoka, M. *J. Am. Chem. Soc.* **1999**, *121*, 5450.

[409]Akiyama, T.; Takaya, J.; Kagoshima, H. *Synlett* **1999**, 1045; Akiyama, T.; Takaya, J.; Kagoshima, H. *Tetrahedron Lett.* **2001**, *42*, 4025.

[410]Loh, T.-P.; Wei, L.L. *Tetrahedron Lett.* **1998**, *39*, 323.

[411]Texier-Boullet, F.; Latouche, R.; Hamelin, J. *Tetrahedron Lett.* **1993**, *34*, 2123.

[412]Kobayashi, S.; Ishitani, H. *J. Chem. Soc., Chem. Commun.* **1995**, 1379.

[413]For a review, see Córdova, A. *Acc. Chem. Res.* **2004**, *37*, 102.

[414]List, B.; Pojarliev, P.; Biller, W.T.; Martin, H.J. *J. Am. Chem. Soc.* **2004**, *124*, 827; Ibrahem, I.; Casas, J.; Córdova, A. *Angew. Chem. Int. Ed.* **2004**, *43*, 6528.

[415]Notz, W.; Sakthivel, K.; Bui, T.; Zhong, G.; Barbas III, C.F. *Tetrahedron Lett.* **2000**, *42*, 199.

[416]Kobayashi, S.; Hamada, T.; Manabe, K. *J. Am. Chem. Soc.* **2002**, *124*, 5640; Trost, B.M.; Terrell, C.R. *J. Am. Chem. Soc.* **2003**, *125*, 338.

[417]Josephsohn, N.S.; Snapper, M.L.; Hoveyda, A.H. *J. Am. Chem. Soc.* **2004**, *126*, 3734.

[418]Hata, S.; Iguchi, M.; Iwasawa, T.; Yamada, K.-i.; Tomioka, K. *Org. Lett.* **2004**, *6*, 1721.

[419]Jacobsen, M.F.; Ionita, L.; Skrydstrup, T. *J. Org. Chem.* **2004**, *69*, 4792.

[420]In bmim BF_4, 1-butyl-3-methylimidazolium tetrafluoroborate: Sun, W.; Xia, C.-G.; Wang, H.-W. *Tetrahedron Lett.* **2003**, *44*, 2409.

Ammonia and primary and secondary amines can be added to isocyanates[421] to give substituted ureas.[422] Isothiocyanates give thioureas.[423] This is an excellent method for the preparation of ureas and thioureas, and these compounds are often used as derivatives for primary and secondary amines. Isocyanic acid (HNCO) also gives the reaction; usually its salts (e.g., NaNCO) are used. Wöhler's famous synthesis of urea involved the addition of ammonia to a salt of this acid.[424]

OS **II**, 79; **III**, 76, 617, 735; **IV**, 49, 180, 213, 515, 700; **V**, 555, 801, 802, 967; **VI**, 936, 951; **VIII**, 26.

16-21 The Addition of Ammonia or Amines to Nitriles

N-Hydro-*C*-amino-addition

$$R-C\equiv N \ + \ NH_3 \ \xrightarrow[\text{pressure}]{NH_4Cl} \ \overset{\overset{\oplus}{NH_2}}{\underset{R}{C}}\overset{Cl^{\ominus}}{\underset{NH_2}{}}$$

Unsubstituted amidines (in the form of their salts) can be prepared by addition of ammonia to nitriles.[425] Many amidines have been made in this way. Dinitriles of suitable chain length can give imidines:[426]

Primary and secondary amines can be used instead of ammonia, to give substituted amidines, but only if the nitrile contains electron-withdrawing groups; for example, Cl_3CCN gives the reaction. Ordinary nitriles do not react, and, in fact, acetonitrile is often used as a solvent in this reaction.[427] Ordinary nitriles can be converted to amidines by treatment with an alkylchloroaluminum amide, $MeAl(Cl)NR_2$ (R = H or Me).[428] The addition of ammonia to cyanamide (NH_2CN) gives guanidine, $(NH_2)_2C=NH$. Guanidines can also be formed from amines.[429]

[421]For a review of the mechanism, see Satchell, D.P.N.; Satchell, R.S. *Chem. Soc. Rev.* **1975**, *4*, 231.
[422]For a review of substituted ureas, see Vishnyakova, T.P.; Golubeva, I.A.; Glebova, E.V. *Russ. Chem. Rev.* **1985**, *54*, 249.
[423]Herr, R.J.; Kuhler, J.L.; Meckler, H.; Opalka, C.J. *Synthesis* **2000**, 1569.
[424]For a history of the investigation of the mechanism of the Wöhler synthesis, see Shorter, J. *Chem. Soc. Rev.* **1978**, *7*, 1. See also, Williams, A.; Jencks, W.P. *J. Chem. Soc. Perkin Trans. 2* **1974**, 1753, 1760; Hall, K.J.; Watts, D.W. *Aust. J. Chem.* **1977**, *30*, 781, 903.
[425]For reviews of amidines, see Granik, V.G. *Russ. Chem. Rev.* **1983**, *52*, 377; Gautier, J.; Miocque, M.; Farnoux, C.C., in Patai, S. *The Chemistry of Amidines and Imidates*, Wiley, NY, **1975**, pp. 283–348.
[426]Elvidge, J.A.; Linstead, R.P.; Salaman, A.M. *J. Chem. Soc.* **1959**, 208.
[427]Grivas, J.C.; Taurins, A. *Can. J. Chem.* **1961**, *39*, 761.
[428]Garigipati, R.S. *Tetrahedron Lett.* **1990**, *31*, 1969.
[429]Dräger, G.; Solodenko, W.; Messinger, J.; Schön, U.; Kirschning, A. *Tetrahedron Lett.* **2002**, *43*, 1401.

If water is present, in the presence of a ruthenium catalyst[430] or a platinum catalyst,[431] the addition of a primary or secondary amine to a nitrile gives an amide: $RCN + R^1NHR^2 + H_2O \rightarrow RCONR^1R^2 + NH_3$ (R^2 may be H). When benzonitrile reacts with $H_2PO_3Se^-$ in aqueous methanol, a selenoamide, $PhC{=}Se)NH_2$, is formed after treatment with aq. potassium carbonate.[432]

OS **I**, 302 [but also see OS **V**, 589]; **IV**, 245, 247, 515, 566, 769. See also, OS **V**, 39.

16-22   The Addition of Amines to Carbon Disulfide and Carbon Dioxide

S-Metallo-*C*-alkylamino-addition

$$S{=}C{=}S \;+\; RNH_2 \xrightarrow{\text{base}} \underset{RHN}{\overset{O}{\underset{}{\parallel}}} \overset{}{\underset{}{C}}{\,}_{S}{}^{\ominus}$$

Salts of dithiocarbamic acid can be prepared by the addition of primary or secondary amines to carbon disulfide.[433] This reaction is similar to **16-10**. Hydrogen sulfide can be eliminated from the product, directly or indirectly, to give isothiocyanates (RNCS). Isothiocyanates can be obtained directly by the reaction of primary amines and CS_2 in pyridine in the presence of dicyclohexylcarbodiimide.[434] Aniline derivatives react with CS_2 and NaOH, and then ethyl chloroformate to give the aryl isothiocyanate.[435] In the presence of diphenyl phosphite and pyridine, primary amines add to CO_2 and to CS_2 to give, respectively, symmetrically substituted ureas and thioureas:[436] Isoselenoureas, $R_2NC({=}NR^1)SeR^2$, can also be formed.[437]

$$RNH_2 \;+\; CO_2 \xrightarrow[\text{HPO(OPh)}_2]{\text{pyridine}} \underset{RHN}{\overset{O}{\underset{}{\parallel}}} \overset{}{\underset{}{C}}{\,}_{NHR}$$

OS **I**, 447; **III**, 360, 394, 599, 763; **V**, 223.

[430]Murahashi, S.; Naota, T.; Saito, E. *J. Am. Chem. Soc.* **1986**, *108*, 7846.

[431]Cobley, C.J.; van den Heuvel, M.; Abbadi, A.; de Vries, J.G. *Tetrahedron Lett.* **2000**, *41*, 2467.

[432]Kamiński, R.; Glass, R.S.; Skowrońska, A. *Synthesis* **2001**, 1308.

[433]For reviews, see Dunn, A.D.; Rudorf, W. *Carbon Disulphide in Organic Chemistry*, Ellis Horwood, Chichester, **1989**, pp. 226–315; Katritzky, A.R.; Faid-Allah, H.; Marson, C.M. *Heterocycles* **1987**, *26*, 1657; Yokoyama, M.; Imamoto, T. *Synthesis* **1984**, 797, see pp. 804–812. For a review of the addition of heterocyclic amines to CO_2 to give, for example, salts of pyrrole-1-carboxylic acids, see Katritzky, A.R.; Marson, C.M.; Faid-Allah, H. *Heterocycles* **1987**, *26*, 1333.

[434]Jochims, J.C. *Chem. Ber.* **1968**, *101*, 1746. For other methods, see Sakai, S.; Fujinami, T.; Aizawa, T. *Bull. Chem. Soc. Jpn.* **1975**, *48*, 2981; Gittos, M.W.; Davies, R.V.; Iddon, B.; Suschitzky, H. *J. Chem. Soc. Perkin Trans. 1* **1976**, 141; Shibanuma, T.; Shiono, M.; Mukaiyama, T. *Chem. Lett.* **1977**, 573; Molina, P.; Alajarin, M.; Arques, A. *Synthesis* **1982**, 596.

[435]Li, Z.; Qian, X.; Liu, Z.; Li, Z.; Song, G. *Org. Prep. Proceed. Int.* **2000**, *32*, 571.

[436]Yamazaki, N.; Higashi, F.; Iguchi, T. *Tetrahedron Lett.* **1974**, 1191. For other methods for the conversion of amines and CO_2 to ureas, see Ogura, H.; Takeda, K.; Tokue, R.; Kobayashi, T. *Synthesis* **1978**, 394; Fournier, J.; Bruneau, C.; Dixneuf, P.H.; Lécolier, S. *J. Org. Chem.* **1991**, *56*, 4456. See Chiarotto, I.; Feroci, M. *J. Org. Chem.* **2003**, *68*, 7137; Lemoucheux, L.; Rouden, J.; Ibazizene, M.; Sobrio, F.; Lasne, M.-C. *J. Org. Chem.* **2003**, *68*, 7289.

[437]Asanuma, Y.; Fujiwara, S.-i.; Shi-ike, T.; Kambe, N. *J. Org. Chem.* **2004**, *69*, 4845.

E. Halogen Nucleophiles

16-23 The Formation of *gem*-Dihalides from Aldehydes and Ketones

Dihalo-de-oxo-bisubstitution

$$\underset{}{\overset{O}{\underset{}{\underset{C}{\parallel}}}} + PCl_5 \longrightarrow \underset{}{\overset{Cl \quad Cl}{\underset{C}{}}}$$

Aliphatic aldehydes and ketones can be converted to *gem*-dichlorides[438] by treatment with PCl_5. The reaction fails for perhalo ketones.[439] If the aldehyde or ketone has an α hydrogen, elimination of HCl may follow and a vinylic chloride is a frequent side product:[440]

$$\underset{H}{\overset{Cl \quad Cl}{\underset{C}{\underset{|}{\overset{|}{C}}}}} \longrightarrow \underset{Cl}{\overset{}{}}C=C$$

or even the main product.[441] The PBr_5 does not give good yields of *gem*-dibromides,[442] but these can be obtained from aldehydes, by the use of Br_2 and triphenyl phosphite.[443] *gem*-Dichlorides can be prepared by reacting an aldehyde with $BiCl_3$.[444]

The mechanism of *gem*-dichloride formation involves initial attack on PCl_4^+ (which is present in solid PCl_5) at the oxygen, followed by addition of Cl^- to the carbon:[445]

$$\underset{}{\overset{O}{\underset{C}{\parallel}}} + PCl_4^{\oplus} \longrightarrow \underset{\oplus}{\overset{OPCl_4}{\underset{C}{}}} \xrightarrow{Cl^-} \underset{}{\overset{Cl \quad OPCl_4}{\underset{C}{}}} \longrightarrow \underset{\oplus}{\overset{Cl}{\underset{C}{}}} \xrightarrow{Cl^-} \underset{}{\overset{Cl \quad Cl}{\underset{C}{}}}$$

$$\mathbf{29}$$

This chloride ion may come from PCl_6^- (which is also present in solid PCl_5). There follows a two-step S_N1 process. Alternatively, **29** can be converted to the product without going through the chlorocarbocation, by an S_Ni process.

This reaction has sometimes been performed on carboxylic esters, though these compounds very seldom undergo any addition to the C=O bond. An example is the conversion of $F_3CCOOPh$ to F_3CCCl_2OPh.[446] However, formates commonly give the reaction.

[438]For a list of reagents that convert aldehydes and ketones to *gem*-dihalides or vinylic halides, with references, see Larock, R.C. *Comprehensive Organic Transformations*, 2nd ed., Wiley-VCH, NY, *1999*, pp. 719–722.

[439]Farah, B.S.; Gilbert, E.E. *J. Org. Chem.* *1965*, *30*, 1241.

[440]See, for example, Nikolenko, L.N.; Popov, S.I. *J. Gen. Chem. USSR* *1962*, *32*, 29.

[441]See, for example, Newman, M.S.; Fraenkel, G.; Kirn, W.N. *J. Org. Chem.* *1963*, *28*, 1851.

[442]For an indirect method of converting ketones to *gem*-dibromides, see Napolitano, E.; Fiaschi, R.; Mastrorilli, E. *Synthesis* *1986*, 122.

[443]Hoffmann, R.W.; Bovicelli, P. *Synthesis* *1990*, 657. See also, Lansinger, J.M.; Ronald, R.C. *Synth. Commun.* *1979*, *9*, 341.

[444]Kabalka, G.W.; Wu, Z. *Tetrahedron Lett.* *2000*, *41*, 579.

[445]Newman, M.S. *J. Org. Chem.* *1969*, *34*, 741.

[446]Kirsanov, A.V.; Molosnova, V.P. *J. Gen. Chem. USSR* *1958*, *28*, 31; Clark, R.F.; Simons, J.H. *J. Org. Chem.* *1961*, *26*, 5197.

Many aldehydes and ketones have been converted to *gem*-difluoro compounds with sulfur tetrafluoride SF$_4$,[447] including quinones, which give 1,1,4,4-tetrafluorocyclohexadiene derivatives. With ketones, yields can be raised and the reaction temperature lowered, by the addition of anhydrous HF.[448] Carboxylic acids, acyl chlorides, and amides react with SF$_4$ to give 1,1,1-trifluorides. In these cases the first product is the acyl fluoride, which then undergoes the *gem*-difluorination reaction:

$$
\underset{\substack{\\ R \quad W}}{R\text{-}\overset{\displaystyle O}{\overset{\|}{C}}\text{-}W} \;+\; SF_4 \;\longrightarrow\; R\text{-}\overset{\displaystyle O}{\overset{\|}{C}}\text{-}F \;+\; SF_4 \;\longrightarrow\; R\text{-}\overset{\displaystyle F}{\underset{\displaystyle F}{C}}\text{-}F \qquad W = OH,\ Cl,\ NH_2,\ NHR
$$

The acyl fluoride can be isolated. Carboxylic esters also give trifluorides, but more vigorous conditions are required. In this case, the carbonyl group of the ester is attacked first, and RCF$_2$OR$'$ can be isolated from RCOOR$'$[449] and then converted to the trifluoride. Anhydrides can react in either manner. Both types of intermediate are isolable under the right conditions and SF$_4$ even converts carbon dioxide to CF$_4$. A disadvantage of reactions with SF$_4$ is that they require a pressure vessel lined with stainless steel. Selenium tetrafluoride SeF$_4$ gives similar reactions, but atmospheric pressure and ordinary glassware can be used.[450] Another reagent that is often used to convert aldehydes and ketones to *gem*-difluorides is the commercially available diethylaminosulfur trifluoride (DAST, Et$_2$NSF$_3$), and CF$_2$Br$_2$ in the presence of zinc.[451] The mechanism with SF$_4$ is probably similar in general nature, if not in specific detail, to that with PCl$_5$.

Treatment with hydrazine to give the hydrazone, and then CuBr$_2$/*t*-BuOLi, generated the *gem*-dibromide.[452] Oximes gives *gem*-dichlorides upon treatment with chlorine and BF$_3$•OEt$_2$, and then HCl.[453] Some dithianes can be converted to *gem*-difluorides with a mixture of fluorine and iodine in acetonitrile.[454] Oximes give *gem*-difluorides with NO$^+$BF$_4^-$ and pyridinium polyhydrogen fluoride.[455]

In a related process, α-halo ethers can be prepared by treatment of aldehydes and ketones with an alcohol and HX. The reaction is applicable to aliphatic aldehydes and ketones and to primary and secondary alcohols. The addition of HX to an aldehyde or ketone gives α-halo alcohols, which are usually unstable, although exceptions are known, especially with perfluoro and perchloro species.[456]

[447]For reviews, see Wang, C.J. *Org. React.* **1985**, *34*, 319; Boswell, Jr., G.A.; Ripka, W.C.; Scribner, R.M.; Tullock, C.W. *Org. React.* **1974**, *21*, 1.

[448]Muratov, N.N.; Mohamed, N.M.; Kunshenko, B.V.; Burmakov, A.I.; Alekseeva, L.A.; Yagupol'skii, L.M. *J. Org. Chem. USSR* **1985**, *21*, 1292.

[449]For methods of converting RCOOR$'$ to RCF$_2$OR$'$, see Boguslavskaya, L.S.; Panteleeva, I.Yu.; Chuvatkin, N.N. *J. Org. Chem. USSR* **1982**, *18*, 198; Bunnelle, W.H.; McKinnis, B.R.; Narayanan, B.A. *J. Org. Chem.* **1990**, *55*, 768.

[450]Olah, G.A.; Nojima, M.; Kerekes, I. *J. Am. Chem. Soc.* **1974**, *96*, 925.

[451]Hu, C.-M.; Qing, F.-L.; Shen, C.-X. *J. Chem. Soc. Perkin Trans. 1* **1993**, 335.

[452]Takeda, T.; Sasaki, R.; Nakamura, A.; Yamauchi, S.; Fujiwara, T. *Synlett* **1996**, 273.

[453]Tordeux, M.; Boumizane, K.; Wakselman, C. *J. Org. Chem.* **1993**, *58*, 1939.

[454]Chambers, R.D.; Sandford, G.; Atherton, M. *J. Chem. Soc., Chem. Commun.* **1995**, 177.

[455]York, C.; Prakash, G.K.S.; Wang, Q.; Olah, G.A. *Synlett* **1994**, 425.

[456]For example, see Andreades, S.; England, D.C. *J. Am. Chem. Soc.* **1961**, *83*, 4670; Clark, D.R.; Emsley, J.; Hibbert, F. *J. Chem. Soc. Perkin Trans. 2* **1988**, 1107.

Aromatic aldehydes are converted to benzylic bromides with dibromoboranes, such as $c\text{-}C_6H_{11}BBr_2$.[457] Aldehydes are converted directly to benzylic chlorides with $HSiMe_2Cl$ and an $In(OH)_3$ catalyst.[458] The reaction of $BuBCl_2$ and oxygen gives alkylation (**16-25**) and chlorination.[459]

OS **II**, 549; **V**, 365, 396, 1082; **VI**, 505, 845; **VIII**, 247. Also see OS **I**, 506. For α-halo-ethers, see OS **I**, 377; **IV**, 101 (see, however, OS **V**, 218), 748; **VI**, 101.

F. Attack at Carbon by Organometallic Compounds[460]

16-24   The Addition of Grignard Reagents and Organolithium Reagents to Aldehydes and Ketones

O-Hydro-*C*-alkyl-addition

Organomagnesium compounds, commonly known as Grignard reagents (RMgX), are formed by the reaction of alkyl, vinyl, or aryl halides with magnesium metal, usually in ether solvents such as diethyl ether or THF (**12-38**), although the reaction can be done in water[461] under certain conditions. Halogen–magnesium exchange generates a Grignard reagent by reaction of aryl halides with reactive aliphatic Grignard reagents.[462] The addition of Grignard reagents to aldehydes and ketones [463] is known as the *Grignard reaction*.[464] The initial product is a magnesium alkoxide, requiring a hydrolysis step to generate the final alcohol product. Formaldehyde gives primary alcohols; other aldehydes give secondary alcohols; and ketones give tertiary alcohols. The reaction is of very broad scope. In many cases, the hydrolysis step is carried out with dilute HCl or H_2SO_4, but this cannot be done for tertiary alcohols in which at least one R group is alkyl because such alcohols are easily dehydrated under acidic conditions (**17-1**). In such cases (and often for other alcohols as well), an aqueous solution of ammonium chloride is used instead of a strong acid. Grignard reagents have been used in solid phase synthesis.[465]

[457]Kabalka, G.W.; Wu, Z.; Ju, Y. *Tetrahedron Lett.* **2000**, *41*, 5161.

[458]Onishi, Y.; Ogawa, D.; Yasuda, M.; Baba, A. *J. Am. Chem. Soc.* **2002**, *124*, 13690.

[459]Kabalka, G.W.; Wu, Z.; Ju, Y. *Tetrahedron Lett.* **2001**, *42*, 6239.

[460]Discussions of most of the reactions in this section are found, in Hartley, F.R.; Patai, S. *The Chemistry of the Metal-Carbon Bond*, Vols. 2–4, Wiley, NY, *1985–1987*.

[461]Li, C.-J. *Tetrahedron* **1996**, *52*, 5643.

[462]Song, J.J.; Yee, N.K.; Tan, Z.; Xu, J.; Kapadia, S.R.; Senanayake, C.H. *Org. Lett.* **2004**, *6*, 4905.

[463]For a discussion of the effect of addends on aggregation and reactivity, see Leung, S.S.-W.; Streitwieser, A. *J. Org. Chem.* **1999**, *64*, 3390.

[464]For reviews of the addition of organometallic compounds to carbonyl groups, see Eicher, T., in Patai, S. *The Chemistry of the Carbonyl Group*, pt. 1, Wiley, NY, *1966*, pp. 621–693; Kharasch, M.S.; Reinmuth, O. *Grignard Reactions of Nonmetallic Substances*, Prentice-Hall: Englewood Cliffs, NJ, *1954*, pp. 138–528. For a review of reagents that extend carbon chains by three carbons, with some functionality at the new terminus, see Stowell, J.C. *Chem. Rev.* **1984**, *84*, 409. For a computational study of this reaction, see Yamazaki, S.; Yamabe, S. *J. Org. Chem.* **2002**, *67*, 9346.

[465]Franzén, R.G. *Tetrahedron* **2000**, *56*, 685.

Alternative methods to generate the arylmagnesium compound are available, including the reaction of an aryl bromide with Bu_3MgLi in THF.[466] Subsequent addition of an aldehyde leads to addition of the aryl group to form an alcohol. An interesting method to form an alkylmagnesium halide used dibutylmagnesium (Bu_2Mg) and a chiral diamine, and subsequent reaction with an aldehyde led to the alcohol derived from acyl addition of a butyl group with good enantioselectivity.[467]

Organolithium reagents (RLi), prepared from alkyl halides and lithium metal or by exchange of an alkyl halide with a reactive organolithium (**12-38**) react with aldehydes and ketones by acyl addition to give the alcohol,[468] after hydrolysis. Organolithium reagents are more basic than the corresponding Grignard reagent, which leads to problems of deprotonation in some cases. Organolithium regents are generally more nucleophilic, and can add to hindered ketones with relative ease when compared to the analogous Grignard reagent.[469] These reagents tend to form aggregates, which influences the reactivity and selectivity of the addition reaction.[470] Alkyl, vinyl[471] and aryl organolithium reagents can be prepared and undergo this reaction. Structural variations are also possible. A lithio-epoxide was formed by treating an epoxide with *sec*-butyllithium in the presence of sparteine,[472] or with *n*-butyllithium/TMEDA,[473] and subsequent reaction with an aldehyde led to an epoxy alcohol. Treatment of an allenic silyl enol ether ($R_3SiOC=C=C$) with *tert*-butyllithium and then a ketone leads to acyl addition of a vinyllithium reagents to give a product with a conjugated ketone in which the C=C is allylic to the alcohol, $R_3SiC(=O)C(=CH_2)-C(OH)R_2$.[474] The dilithio compound $LiC\equiv CCH_2Li$ reacts with ketones via acyl addition, and an interesting workup with formaldehyde and then aqueous ammonium chloride gave the homopropargyl alcohol, $R_2C(OH)CH_2C\equiv CH$.[475] Aryl sulfonamides can be treated with 2 equivalents of *n*-butyllithium to give an ortho aryllithium which can then be added to an aldehyde to give the resulting diaryl carbinol.[476] A very interesting variation of the fundamental acyl addition reaction of organolithium reagents treated an aldehyde with an acyl-lithio amide, $LiC(=O)N(Me)CH_2Me$, to give an α-hydroxy amide derivative.[477]

The reaction of aldehydes or ketones with alkyl and aryl Grignard reagents has also been done without preliminary formation of RMgX, by mixing RX the carbonyl compound and magnesium metal in an ether solvent. This approach

[466]Inoue, A.; Kitagawa, K.; Shinokubo, H.; Oshima, K. *J. Org. Chem.* **2001**, *66*, 4333.

[467]Yong, K.H.; Taylor, N.J.; Chong, J.M. *Org. Lett.* **2002**, *4*, 3553.

[468]For a study of Hammett ρ values for this reaction, see Maclin, K.M.; Richey Jr., H.G. *J. Org. Chem.* **2002**, *67*, 4370.

[469]Lecomte, V.; Stéphan, E.; Le Bideau, F.; Jaouen, G. *Tetrahedron* **2003**, *59*, 2169.

[470]See Fressigné, C.; Maddaluno, J.; Marquez, A.; Giessner-Prettre, C. *J. Org. Chem.* **2000**, *65*, 8899; Granander, J.; Sott, R.; Hilmersson, G. *Tetrahedron* **2002**, *58*, 4717.

[471]For a discussion of selectivity, see Spino, C.; Granger, M.-C.; Tremblay, M.-C. *Org. Lett.* **2002**, *4*, 4735.

[472]Hodgson, D.M.; Reynolds, N.J.; Coote, S.J. *Org. Lett.* **2004**, *6*, 4187.

[473]Florio, S.; Aggarwal, V.; Salomone, A. *Org. Lett.* **2004**, *6*, 4191.

[474]Stergiades, I.A.; Tius, M.A. *J. Org. Chem.* **1999**, *64*, 7547.

[475]Cabezas, J.A.; Pereira, A.R.; Amey, A. *Tetrahedron Lett.* **2001**, *42*, 6819.

[476]Stanetty, P.; Emerschitz, T. *Synth. Commun.* **2001**, *31*, 961.

[477]Cunico, R.F. *Tetrahedron Lett.* **2002**, *43*, 355.

preceded Grignard's work, and is now known as the *Barbier reaction*.[478] The organolithium analog of this process is also known.[479] Yields were generally satisfactory. Carboxylic ester, nitrile, and imide groups in the R are not affected by the reaction conditions.[480] Modern versions of the Barbier reaction employ other metals and/or reaction conditions, and will be discussed in **16-25**. A retro-Barbier reaction has been reported in which a cyclic tertiary alcohol was treated to an excess of bromine and potassium carbonate to give 6-bromo-2-hexanone from 1-methylcyclopentanol.[481] This section will focus on variations of the Barbier reaction that employ Mg or Li derivatives. The reaction of allyl iodide, benzaldehyde and Mg/I_2, for example, gave the acyl addition product 1-phenylbut-3-en-1-ol.[482]

The reaction of RMgX or RLi with α,β-unsaturated aldehydes or ketones can proceed via 1,4-addition as well as normal 1,2-addition (see **15-25**).[483] In general, alkyllithium reagents give less 1,4-addition than the corresponding Grignard reagents.[484] Quinones add Grignard reagents on one or both sides or give 1,4-addition. In a compound containing both an aldehyde and a ketone it is possible to add RMgX chemoselectively to the aldehyde without significantly disturbing the carbonyl of the ketone group[485] (see also, p. 1306). In conjunction with $BeCl_2$, organolithium reagents add to conjugated ketones. In THF, 1,4- addition is observed, but in diethyl ether the 1,2-addition product is formed.[486] Organocerium reagents, generated from cerium chloride ($CeCl_3$ and a Grignard reagent or an organolithium reagent) gives an organometallic reagent that adds chemoselectively.[487] Grignard reagents with a catalytic amount of $InCl_3$ to give a mixture of 1,2- and 1,4-addition products with the 1,4-product predominating, but there was an increased 1,2-addition relative to the uncatalyzed reaction.[488]

As with the reduction of aldehydes and ketones (**19-36**), the addition of organometallic compounds to these substrates can be carried out enantioselectively and diastereoselectively.[489] Chiral secondary alcohols have been obtained with high

---

[478]Barbier, P. *Compt. Rend.*, **1899**, *128*, 110. For a review, with Mg, Li, and other metals, see Blomberg, C.; Hartog, F.A. *Synthesis* **1977**, 18. For a discussion of the mechanism, see Molle, G.; Bauer, P. *J. Am. Chem. Soc.* **1982**, *104*, 3481. For a list of Barbier-type reactions, with references, see Larock, R.C. *Comprehensive Organic Transformations*, 2nd ed., Wiley-VCH, NY, **1999**, pp. 1125–1134.

[479]Guijarro, A.; Yus, M. *Tetrahedron Lett.* **1993**, *34*, 3487; de Souza-Barboza, J.D.; Pétrier, C.; Luche, J. *J. Org. Chem.* **1988**, *53*, 1212.

[480]Yeh, M.C.P.; Knochel, P.; Santa, L.E. *Tetrahedron Lett.* **1988**, *29*, 3887.

[481]Zhang, W.-C.; Li, C.-J. *J. Org.Chem.* **2000**, *65*, 5831.

[482]Zhang, W.-C.; Li, C.-J. *J. Org. Chem.* **1999**, *64*, 3230.

[483]For a discussion of the mechanism of this reaction, see Holm, T. *Acta Chem. Scand.* **1992**, *46*, 985.

[484]An example was given on p. $$$.

[485]Vaskan, R.N.; Kovalev, B.G. *J. Org. Chem. USSR* **1973**, *9*, 501.

[486]Krief, A.; de Vos, M.J.; De Lombart, S.; Bosret, J.; Couty, F. *Tetrahedron Lett.* **1997**, *38*, 6295.

[487]Bartoli, G.; Marcantoni, E.; Petrini, M. *Angew. Chem. Int. Ed.* **1993**, *32*, 1061; Dimitrov, V.; Bratovanov, S.; Simova, S.; Kostova, K. *Tetrahedron Lett.* **1994**, *35*, 6713; Greeves, N.; Lyford, L. *Tetrahedron Lett.* **1992**, *33*, 4759.

[488]Kelly, B.G.; Gilheany; D.G. *Tetrahedron Lett.* **2002**, *43*, 887.

[489]For reviews, see Solladié, G., in Morrison, J.D. *Asymmetric Synthesis*, Vol. 2, Academic Press, NY, **1983**, pp. 157–199, 158–183; Nógrádi, M. *Stereoselective Synthesis*, VCH, NY, **1986**, pp. 160–193; Noyori, R.; Kitamura, M. *Angew. Chem. Int. Ed.* **1991**, *30*, 49.

enantioselectivity by addition of Grignard and organolithium compounds to aromatic aldehydes, in the presence of optically active amino alcohols as ligands.[490]

Diastereoselective addition[491] has been carried out with achiral reagents and chiral substrates,[492] similar to the reduction shown on p. 1802.[493] Because the attacking atom in this case is carbon, not hydrogen, it is also possible to get diastereoselective addition with an achiral substrate and an optically active reagent.[494] Use of suitable reactants creates, in the most general case, two new stereogenic centers, so the product can exist as two pairs of enantiomers:

Diastereomers

Even if the organometallic compound is racemic, it still may be possible to get a diastereoselective reaction; that is, one pair of enantiomers is formed in greater amount than the other.[495]

In some cases, the Grignard reaction can be performed intramolecularly.[496] For example, treatment of 5-bromo-2-pentanone with magnesium and a small amount

[490]Mukaiyama, T.; Soai, K.; Sato, T.; Shimizu, H.; Suzuki, K. *J. Am. Chem. Soc.* **1979**, *101*, 1455; Mazaleyrat, J.; Cram, D.J. *J. Am. Chem. Soc.* **1981**, *103*, 4585; Eleveld, M.B.; Hogeveen, H. *Tetrahedron Lett.* **1984**, *25*, 5187; Schön, M.; Naef, R. *Tetrahedron Asymmetry* **1999**, *10*, 169; Arvidsson, P.I.; Davidsson, Ö.; Hilmersson, G. *Tetrahedron Asymmetry* **1999**, *10*, 527.

[491]For a review, see Yamamoto, Y.; Maruyama, K. *Heterocycles* **1982**, *18*, 357. For a discussion of facial selectivity, see Tomoda, S.; Senju, T. *Tetrahedron* **1999**, *55*, 3871. See Schulze, V.; Nell, P.G.; Burton, A.; Hoffmann, R.W. *J. Org. Chem.* **2003**, *68*, 4546.

[492]For a review of cases in which the substrate bears a group that can influence the diastereoselectivity by chelating with the metal, see Reetz, M.T. *Angew. Chem. Int. Ed.* **1984**, *23*, 556. See also, Keck, G.E.; Castellino, S. *J. Am. Chem. Soc.* **1986**, *108*, 3847.

[493]See, for example, Eliel, E.L.; Morris-Natschke, S. *J. Am. Chem. Soc.* **1984**, *106*, 2937; Reetz, M.T.; Steinbach, R.; Westermann, J.; Peter, R.; Wenderoth, B. *Chem. Ber.* **1985**, *118*, 1441; Yamamoto, Y.; Matsuoka, K. *J. Chem. Soc., Chem. Commun.* **1987**, 923; Boireau, G.; Deberly, A.; Abenhaïm, D. *Tetrahedron Lett.* **1988**, *29*, 2175; Page, P.C.B.; Westwood, D.; Slawin, A.M.Z.; Williams, D.J. *J. Chem. Soc. Perkin Trans. 1* **1989**, 1158; Soai, K.; Niwa, S.; Hatanaka, T. *Bull. Chem. Soc. Jpn.* **1990**, *63*, 2129. For examples in which both reactants were chiral, see Roush, W.R.; Halterman, R.L. *J. Am. Chem. Soc.* **1986**, *108*, 294; Hoffmann, R.W.; Dresely, S.; Hildebrandt, B. *Chem. Ber.* **1988**, *121*, 2225; Paquette, L.A.; Learn, K.S.; Romine, J.L.; Lin, H. *J. Am. Chem. Soc.* **1988**, *110*, 879; Brown, H.C.; Bhat, K.S.; Randad, R.S. *J. Org. Chem.* **1989**, *54*, 1570.

[494]For a review of such reactions with crotylmetallic reagents, see Hoffmann, R.W. *Angew. Chem. Int. Ed.* **1982**, *21*, 555. For a discussion of the mechanism, see Denmark, S.E.; Weber, E.J. *J. Am. Chem. Soc.* **1984**, *106*, 7970. For some examples, see Greeves, N.; Pease, J.E. *Tetrahedron Lett.* **1996**, *37*, 5821; Zweifel, G.; Shoup, T.M. *J. Am. Chem. Soc.* **1988**, *110*, 5578; Gung, B.W.; Smith, D.T.; Wolf, M.A. *Tetrahedron Lett.* **1991**, *32*, 13.

[495]For examples, see Coxon, J.M.; van Eyk, S.J.; Steel, P.J. *Tetrahedron Lett.* **1985**, *26*, 6121; Mukaiyama, T.; Ohshima, M.; Miyoshi, N. *Chem. Lett.* **1987**, 1121; Masuyama, Y.; Takahara, J.P.; Kurusu, Y. *Tetrahedron Lett.* **1989**, *30*, 3437.

[496]For a list of reagents, with references, see Larock, R.C. *Comprehensive Organic Transformations*, 2nd ed., Wiley-VCH, NY, **1999**, pp. 1134–1135.

of mercuric chloride in THF produced 1-methyl-1-cyclobutanol in 60% yield.[497] Other four- and five-membered ring compounds were also prepared by this procedure. Similar closing of five- and six-membered rings was achieved by treatment of a δ- or ε-halocarbonyl compound, not with a metal, but with a dianion derived from nickel tetraphenyporphine.[498]

The *gem*-disubstituted magnesium compounds formed from CH_2Br_2 or CH_2I_2 (**12-38**) react with aldehydes or ketones to give alkenes in moderate-to-good yields.[499] Wittig type reacts also produce alkenes and are discussed in **16-44**. The reaction could not be extended to other *gem*-dihalides. Similar reactions with *gem*-dimetallic compounds prepared with metals other than magnesium have also produced alkenes.[500] An interesting variation is the reaction of methyllithium and CH_2I_2 with an aliphatic aldehyde to give an epoxide,[501] but this reagent reacted with lactones to give a cyclic hemiketal with a pendant iodomethyl unit.[502] Alkylidene oxetanes react with lithium, and then with an aldehyde to give a conjugated ketone.[503] The α,α-dimetallic derivatives of phenyl sulfones ($PhSO_2CM_2R$) (M = Li or Mg) react with aldehydes or ketones $R'COR^2$ to give good yields of the α,β-unsaturated sulfones $PhSO_2CR=CR'R^2$,[504] which can be reduced with aluminum amalgam (see **10-67**) or with $LiAlH_4\text{-}CuCl_2$ to give the alkenes $CHR=CR'R^2$.[505] On the other hand, *gem*-dihalides treated with a carbonyl compound and Li or BuLi give epoxides[506] (see also, **16-46**).

[497]Leroux, Y. *Bull. Soc. Chim. Fr.* **1968**, 359.

[498]Corey, E.J.; Kuwajima, I. *J. Am. Chem. Soc.* **1970**, *92*, 395. For another method, see Molander, G.A.; McKie, J.A. *J. Org. Chem.* **1991**, *56*, 4112, and references cited therein.

[499]Bertini, F.; Grasselli, P.; Zubiani, G.; Cainelli, G.*Tetrahedron* **1970**, *26*, 1281.

[500]For example, see Zweifel, G.; Steele, R.B. *Tetrahedron Lett.* **1966**, 6021; Cainelli, G.; Bertini, F.; Grasselli, P.; Zubiani, G. *Tetrahedron Lett.* **1967**, 1581; Takai, K.; Hotta, Y.; Oshima, K.; Nozaki, H. *Bull. Chem. Soc. Jpn.* **1980**, *53*, 1698; Knochel, P.; Normant, J.F. *Tetrahedron Lett.* **1986**, *27*, 1039; Barluenga, J.; Fernández-Simón, J.L.; Concellón, J.M.; Yus, M. *J. Chem. Soc., Chem. Commun.* **1986**, 1665; Okazoe, T.; Takai, K.; Utimoto, K. *J. Am. Chem. Soc.* **1987**, *109*, 951; Piotrowski, A.M.; Malpass, D.B.; Boleslawski, M.P.; Eisch, J.J. *J. Org. Chem.* **1988**, *53*, 2829; Tour, J.M.; Bedworth, P.V.; Wu, R. *Tetrahedron Lett.* **1989**, *30*, 3927; Lombardo, L. *Org. Synth.* 65, 81.

[501]Concellón, J.M.; Cuervo, H.; Fernández-Fano, R. *Tetrahedron* **2001**, *57*, 8983.

[502]Bessieres, B.; Morin, C. *Synlett* **2000**, 1691.

[503]Hashemsadeh, M.; Howell, A.R. *Tetrahedron Lett.* **2000**, *41*, 1855, 1859.

[504]Pascali, V.; Tangari, N.; Umani-Ronchi, A. *J. Chem. Soc. Perkin Trans. 1* **1973**, 1166.

[505]Pascali, V.; Umani-Ronchi, A. *J. Chem. Soc., Chem. Commun.* **1973**, 351.

[506]Cainelli, G.; Tangari, N.; Umani-Ronchi, A. *Tetrahedron* **1972**, *28*, 3009, and references cited therein.

In other uses of *gem*-dihalo compounds, aldehydes and ketones add the CH_2I group $[R_2CO \rightarrow R_2C(OH)CH_2I]$ when treated with CH_2I_2 in the presence of SmI_2,[507] and the CHX_2 group when treated with methylene halides and lithium dicyclohexylamide at low temperatures.[508]

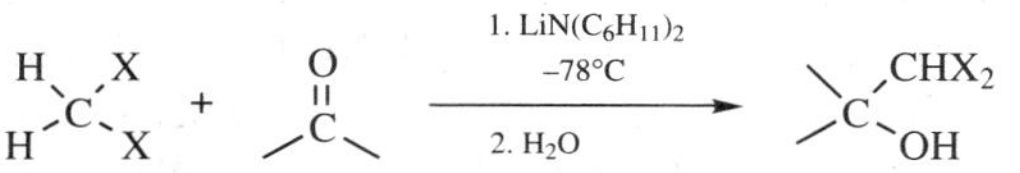

$$
\begin{array}{c}
\text{H}\diagdown\!\!\diagup\text{X} \\
\text{C} \\
\text{H}\diagup\!\!\diagdown\text{X}
\end{array}
\;+\;
\begin{array}{c}
\text{O} \\
\parallel \\
\text{C}
\end{array}
\;\xrightarrow[\text{2. H}_2\text{O}]{\substack{\text{1. LiN(C}_6\text{H}_{11})_2 \\ -78^\circ\text{C}}}\;
\begin{array}{c}
\diagdown\!\!\diagup\text{CHX}_2 \\
\text{C} \\
\diagdown\text{OH}
\end{array}
\qquad \text{X = Cl, Br, I}
$$

A hydroxymethyl group can be added to an aldehyde or ketone with the masked reagent $Me_2((iPr)O)SiCH_2MgCl$, which with R_2CO gives $R_2C(OH)CH_2$-$Si(O-(iPr))Me_2$, but with H_2O_2 give 1,2-diols $R_2C(OH)CH_2OH$.[509]

It is possible to add an acyl group to a ketone to give (after hydrolysis) an α-hydroxy ketone.[510] This can be done by adding RLi and CO to the ketone at -110°C:[511]

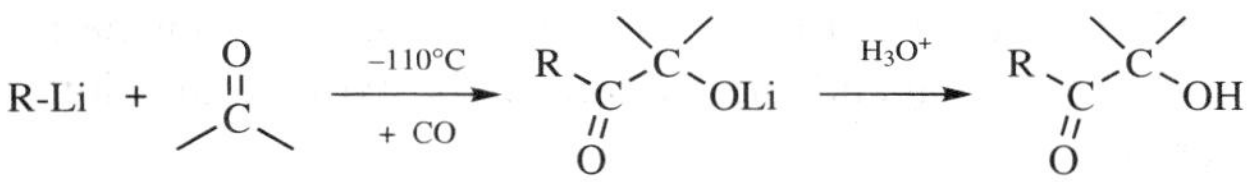

$$
\text{R-Li} \;+\;
\begin{array}{c}
\text{O} \\
\parallel \\
\text{C}
\end{array}
\;\xrightarrow[\text{+ CO}]{-110^\circ\text{C}}\;
\begin{array}{c}
\text{R}\diagdown\quad\diagup \\
\text{C}\!-\!\text{C}\!-\!\text{OLi} \\
\parallel \\
\text{O}
\end{array}
\;\xrightarrow{\text{H}_3\text{O}^+}\;
\begin{array}{c}
\text{R}\diagdown\quad\diagup \\
\text{C}\!-\!\text{C}\!-\!\text{OH} \\
\parallel \\
\text{O}
\end{array}
$$

When the same reaction is carried out with carboxylic esters ($R'COOR^2$), α-diketones ($RCOCOR'$) are obtained.[511]

Most aldehydes and ketones react with most Grignard reagents, but there are several potential side reactions[512] that occur mostly with hindered ketones and with bulky Grignard reagents. The two most important of these are *enolization* and *reduction*. The former requires that the aldehyde or ketone have an α hydrogen, and the latter requires that the Grignard reagent have a β hydrogen:

Enolization

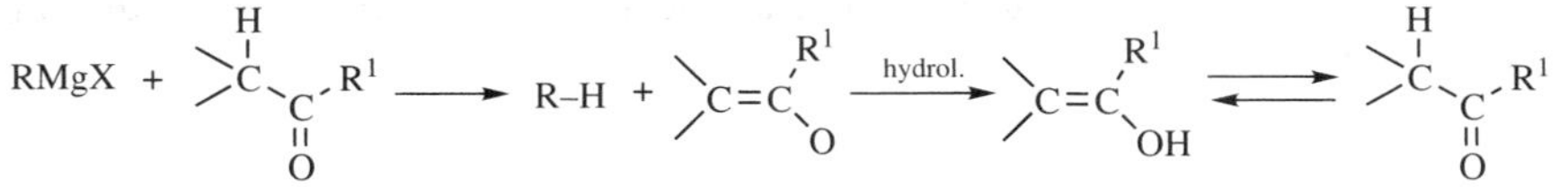

$$
\text{RMgX} +
\begin{array}{c}
\diagdown\;\;\overset{\text{H}}{\underset{|}{\text{C}}}\diagup \\
\diagup\;\text{C}\!-\!\text{R}^1 \\
\parallel \\
\text{O}
\end{array}
\longrightarrow
\text{R-H} +
\begin{array}{c}
\diagdown\quad\diagup\text{R}^1 \\
\text{C}\!=\!\text{C} \\
\diagup\quad\diagdown\text{O}
\end{array}
\xrightarrow{\text{hydrol.}}
\begin{array}{c}
\diagdown\quad\diagup\text{R}^1 \\
\text{C}\!=\!\text{C} \\
\diagup\quad\diagdown\text{OH}
\end{array}
\rightleftharpoons
\begin{array}{c}
\diagdown\;\;\overset{\text{H}}{\underset{|}{\text{C}}}\diagup \\
\diagup\;\text{C}\!-\!\text{R}^1 \\
\parallel \\
\text{O}
\end{array}
$$

Reduction

$$
\begin{array}{c}
\diagdown\;\;\overset{\text{H}}{\underset{|}{\text{C}}}\diagup \\
\diagup\;\text{C}\!-\!\text{MgX} \\
\diagup\;\diagdown
\end{array}
+
\begin{array}{c}
\text{O} \\
\parallel \\
\text{C}
\end{array}
\longrightarrow
\begin{array}{c}
\diagdown\quad\diagup \\
\text{C}\!=\!\text{C} \\
\diagup\quad\diagdown
\end{array}
+
\begin{array}{c}
\diagdown\;\;\overset{\text{H}}{\underset{|}{\text{C}}}\diagup \\
\diagup\;\text{C}\!-\!\text{OMgX}
\end{array}
\xrightarrow{\text{hydrol.}}
\begin{array}{c}
\overset{\text{H}}{\underset{|}{\text{C}}} \\
\diagup\diagdown\text{OH}
\end{array}
$$

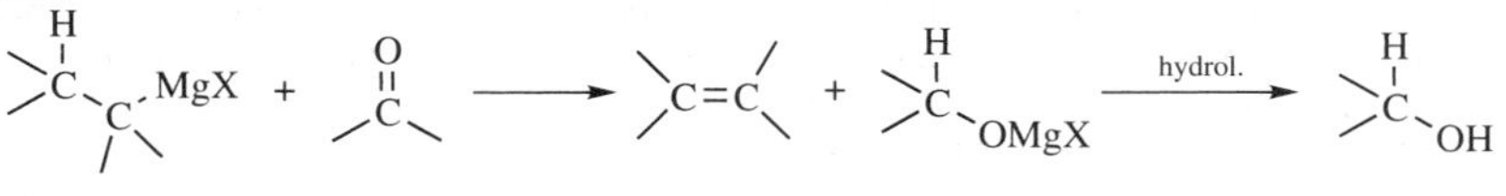

[507]Imamoto, T.; Takeyama, T.; Koto, H. *Tetrahedron Lett.* **1986**, *27*, 3243.

[508]Taguchi, H.; Yamamoto, H.; Nozaki, H. *Bull. Chem. Soc. Jpn.* **1977**, *50*, 1588.

[509]Tamao, K.; Ishida, N. *Tetrahedron Lett.* **1984**, *25*, 4245. For another method, see Imamoto, T.; Takeyama, T.; Yokoyama, M. *Tetrahedron Lett.* **1984**, *25*, 3225.

[510]For a review, see Seyferth, D.; Weinstein, R.M.; Wang, W.; Hui, R.C.; Archer, C.M. *Isr. J. Chem.* **1984**, *24*, 167.

[511]Seyferth, D.; Weinstein, R.M.; Wang, W. *J. Org. Chem.* **1983**, *48*, 1144; Seyferth, D.; Weinstein, R.M.; Wang,W.; Hui, R.C. *Tetrahedron Lett.* **1983**, *24*, 4907.

[512]Lajis, N. Hj.; Khan, M.N.; Hassan, H.A. *Tetrahedron* **1993**, *49*, 3405.

Enolization is an acid–base reaction (**12-24**) in which a proton is transferred from the α carbon to the Grignard reagent. The carbonyl compound is converted to its enolate anion form, which, on hydrolysis, gives the original ketone or aldehyde. Enolization is important not only for hindered ketones but also for those that have a relatively high percentage of enol form (e.g., β-keto esters). In reduction, the carbonyl compound is reduced to an alcohol (**16-24**) by the Grignard reagent, which itself undergoes elimination to give an alkene. Two other side reactions are condensation (between enolate ion and excess ketone) and Wurtz-type coupling (**10-64**). Such highly hindered tertiary alcohols as triisopropylcarbinol, tri-*tert*-butylcarbinol, and diisopropylneopentylcarbinol cannot be prepared (or can be prepared only in extremely low yields) by the addition of Grignard reagents to ketones, because reduction and/or enolization become prominent.[513] However, these carbinols can be prepared by the use of alkyllithium reagents at $-80°C$[514] because enolization and reduction are much less important.[515] Other methods of increasing the degree of addition at the expense of reduction include complexing the Grignard reagent with $LiClO_4$ or $Bu_4N^+ Br^-$,[516] or using benzene or toluene instead of ether as solvent.[517] Both reduction and enolization can be avoided by adding $CeCl_3$ to the Grignard reagent (see above).[518]

Another way to avoid complications is to add $(RO)_3TiCl$, $TiCl_4$,[519] $(RO)_3ZrCl$, or $(R_2N)_3TiX$ to the Grignard or lithium reagent. This produces organotitanium or organozirconium compounds that are much more selective than Grignard or organolithium reagents.[520] An important advantage of these reagents is that they do not react with NO_2 or CN functions that may be present in the substrate, as Grignard and organolithium reagents do. The reaction of a β-keto amide with $TiCl_4$, for example, gives a complex that allows selective reaction of the ketone unit with $MeMgCl–CeCl_3$ to give the corresponding alcohol.[521] Premixing an allylic Grignard reagent with $ScCl_3$ prior to reaction with the aldehyde gives direct acyl addition without allylic rearrangement as the major product, favoring the trans-alkene unit.[522]

[513]Whitmore, F.C.; George, R.S. *J. Am. Chem. Soc.* **1942**, *64*, 1239.

[514]Zook, H.D.; March, J.; Smith, D.F. *J. Am. Chem. Soc.* **1959**, *81*, 1617; Bartlett, P.D.; Tidwell, T.T. *J. Am. Chem. Soc.* **1968**, *90*, 4421. See also, Lomas, J.S. *Nouv. J. Chim.*, **1984**, *8*, 365; Molle, G.; Briand, S.; Bauer, P.; Dubois, J.E. *Tetrahedron* **1984**, *40*, 5113.

[515]Buhler, J.D. *J. Org. Chem.* **1973**, *38*, 904.

[516]Chastrette, M.; Amouroux, R. *Chem. Commun.* **1970**, 470; *Bull. Soc. Chim. Fr.* **1970**, 4348. See also, Richey Jr., H.G.; DeStephano, J.P. *J. Org. Chem.* **1990**, *55*, 3281.

[517]Canonne, P.; Foscolos, G.; Caron H.; Lemay, G. *Tetrahedron* **1982**, *38*, 3563.

[518]Imamoto, T.; Takiyama, N.; Nakamura, K.; Hatajima, T.; Kamiya, Y. *J. Am. Chem. Soc.* **1989**, *111*, 4392.

[519]See Reetz, M.T.; Kyung, S.H.; Hüllmann, M. *Tetrahedron* **1986**, *42*, 2931.

[520]For a monograph, see Reetz, M.T. *Organotitanium Reagents in Organic Synthesis*, Springer, NY, **1986**. For reviews, see Weidmann, B.; Seebach, D. *Angew. Chem. Int. Ed.* **1983**, *22*, 31; Reetz, M.T. *Top. Curr. Chem.* **1982**, *106*, 1.

[521]Bartoli, G.; Bosco, M.; Marcantoni, E.; Massaccesi, M.; Rinaldi, S.; Sambri, L. *Tetrahedron Lett.* **2001**, *42*, 6093.

[522]Matsukawa, S.; Funabashi, Y.; Imamoto, T. *Tetrahedron Lett.* **2003**, *44*, 1007.

There has been much controversy regarding the mechanism of addition of Grignard reagents to aldehydes and ketones.[523] The reaction is difficult to study because of the variable nature of the species present in the Grignard solution (p. 260) and because the presence of small amounts of impurities in the magnesium seems to have a great effect on the kinetics of the reaction, making reproducible experiments difficult.[524] There seem to be two basic mechanisms, depending on the reactants and the reaction conditions. In one of these, the R group is transferred to the carbonyl carbon with its electron pair. A detailed mechanism of this type has been proposed by Ashby and co-workers,[525] based on the discovery that this reaction proceeds by two paths: one first order in MeMgBr and the other first order in Me$_2$Mg.[526] According to this proposal, both MeMgBr and Me$_2$Mg add to the carbonyl carbon, though the exact nature of the step by which MeMgBr or Me$_2$Mg reacts with the substrate is not certain. One possibility is a four-centered cyclic transition state:[527]

$$
\begin{array}{c}
\text{Br} \\
\text{Me} - \text{Mg} \\
R \diagdown \diagup \\
\quad \text{C=O} \\
R \diagup
\end{array}
\longrightarrow
\begin{array}{c}
\text{Me} \quad \text{MgBr} \\
R - \overset{|}{\underset{|}{C}} - O \\
R
\end{array}
$$

The other type of mechanism is a single electron transfer (SET) process[528] with a ketyl intermediate:[529]

$$
R \overset{Mg}{\diagdown} X \; + \; \underset{Ar \diagup \,^{C}\diagdown Ar}{\overset{O}{\parallel}} \longrightarrow
\left[R\cdot \; + \; \underset{Ar \diagup \,^{C}\diagdown Ar}{\overset{\cdot \overset{|}{O}MgX}{}} \right]
\xrightarrow[\text{Solvent cage}]{A}
\underset{Ar \diagup \,^{C}\diagdown Ar}{\overset{R \diagdown \; \diagup OMgX}{}}
$$

Ketyl

This mechanism, which has been mostly studied with diaryl ketones, is more likely for aromatic and other conjugated aldehydes and ketones than it is for

---

[523]For reviews, see Holm, T. *Acta Chem. Scand. Ser. B* **1983**, *37*, 567; Ashby, E.C. *Pure Appl. Chem.* **1980**, *52*, 545; *Bull. Soc. Chim. Fr.* **1972**, 2133; *Q. Rev. Chem. Soc.* **1967**, *21*, 259; Ashby, E.C.; Laemmle, J.; Neumann, H.M. *Acc. Chem. Res.* **1974**, *7*, 272; Blomberg, C. *Bull. Soc. Chim. Fr.* **1972**, 2143. For a review of the stereochemistry of the reaction, see Ashby, E.C.; Laemmle, J. *Chem. Rev.* **1975**, *75*, 521. For a review of the effects of the medium and the cation, see Solv'yanov, A.A.; Beletskaya, I.P. *Russ. Chem. Rev.* **1987**, *56*, 465.

[524]See, for example, Ashby, E.C.; Neumann, H.M.; Walker, F.W.; Laemmle, J.; Chao, L. *J. Am. Chem. Soc.* **1973**, *95*, 3330.

[525]Ashby, E.C.; Laemmle, J.; Neumann, H.M. *J. Am. Chem. Soc.* **1972**, *94*, 5421.

[526]Ashby, E.C.; Laemmle, J.; Neumann, H.M. *J. Am. Chem. Soc.* **1971**, *93*, 4601; Laemmle, J.; Ashby, E.C.; Neumann, H.M. *J. Am. Chem. Soc.* **1971**, *93*, 5120.

[527]Tuulmets, A. *Org. React. (USSR)* **1967**, *4*, 5; House, H.O.; Oliver, J.E. *J. Org. Chem.* **1968**, *33*, 929; Ashby, E.C.; Yu, S.H.; Roling, P.V. *J. Org. Chem.* **1972**, *37*, 1918. See also, Billet, J.; Smith, S.G. *J. Am. Chem. Soc.* **1968**, *90*, 4108; Lasperas, M.; Perez-Rubalcaba, A.; Quiroga-Feijoo, M.L. *Tetrahedron* **1980**, *36*, 3403.

[528]For a review, see Dagonneau, M. *Bull. Soc. Chim. Fr.* **1982**, II-269.

[529]There is kinetic evidence that the solvent cage shown may not be necessary: Walling, C. *J. Am. Chem. Soc.* **1988**, *110*, 6846.

strictly aliphatic ones. Among the evidence[530] for the SET mechanism are ESR spectra[531] and the fact that

$$Ar_2C \underset{OH}{\overset{|}{-}} CAr_2 \atop OH$$

side products are obtained (from dimerization of the ketyl).[532] In the case of addition of RMgX to benzil (PhCOCOPh), esr spectra of two different ketyl radicals were observed, both reported to be quite stable at room temperature.[533] Note that a separate study failed to observe freely defusing radicals in the formation of Grignard reagents.[534] Carbon isotope effect studies with Ph^{14}COPh showed that the rate-determining step with most Grignard reagents is the carbon–carbon bond-forming step (marked *A*), though with allylmagnesium bromide it is the initial electron-transfer step.[535] In the formation of Grignard reagents from bromocyclopropane, diffusing cyclopropyl radical intermediates were found.[536] The concerted versus stepwise mechanism has been probed with chiral Grignard reagents.[537]

Mechanisms for the addition of organolithium reagents have been investigated much less.[538] Addition of a cryptand that binds Li$^+$ inhibited the normal addition reaction, showing that the lithium is necessary for the reaction to take place.[539]

There is general agreement that the mechanism leading to reduction[540] is usually as follows:

[530]For other evidence, see Savin, V.I.; Kitaev, Yu.P. *J. Org. Chem. USSR* **1975**, *11*, 2622; Ōkubo, M. *Bull. Chem. Soc. Jpn.* **1977**, *50*, 2379; Ashby, E.C.; Bowers Jr., J.R. *J. Am. Chem. Soc.* **1981**, *103*, 2242; Holm, T. *Acta Chem. Scand. Ser. B* **1988**, *42*, 685; Liotta, D.; Saindane, M.; Waykole, L. *J. Am. Chem. Soc.* **1983**, *105*, 2922; Yamataka, H.; Miyano, N.; Hanafusa, T. *J. Org. Chem.* **1991**, *56*, 2573.

[531]Fauvarque, J.; Rouget, E. *C. R. Acad. Sci., Ser C*, **1968**, *267*, 1355; Maruyama, K.; Katagiri, T. *Chem. Lett.* **1987**, 731, 735; *J. Phys. Org. Chem.* **1988**, *1*, 21.

[532]Blomberg, C.; Mosher, H.S. *J. Organomet. Chem.* **1968**, *13*, 519; Holm, T.; Crossland, I. *Acta Chem. Scand.* **1971**, *25*, 59.

[533]Maruyama, K.; Katagiri, T. *J. Am. Chem. Soc.* **1986**, *108*, 6263; *J. Phys. Org. Chem.* **1989**, *2*, 205. See also, Holm, T. *Acta Chem. Scand. Ser. B* **1987**, *41*, 278; Maruyama, K.; Katagiri, T. *J. Phys. Org. Chem.* **1991**, *4*, 158.

[534]Walter, R.I. *J. Org. Chem.* **2000**, *65*, 5014.

[535]Yamataka, H.; Matsuyama, T.; Hanafusa, T. *J. Am. Chem. Soc.* **1989**, *111*, 4912.

[536]Garst, J.F.; Ungváry, F. *Org. Lett.* **2001**, *3*, 605.

[537]Hoffmann, RW.; Hölzer, B. *Chem. Commun.* **2001**, 491.

[538]See, for example, Al-Aseer, M.A.; Smith, S.G. *J. Org. Chem.* **1984**, *49*, 2608; Yamataka, H.; Kawafuji, Y.; Nagareda, K.; Miyano, N.; Hanafusa, T. *J. Org. Chem.* **1989**, *54*, 4706.

[539]Perraud, R.; Handel, H.; Pierre, J. *Bull. Soc. Chim. Fr.* **1980**, II-283.

[540]For discussions of the mechanism of reduction, see Singer, M.S.; Salinger, R.M.; Mosher, H.S. *J. Org. Chem.* **1967**, *32*, 3821; Denise, B.; Fauvarque, J.; Ducom, J. *Tetrahedron Lett.* **1970**, 335; Cabaret, D.; Welvart, Z. *J. Organomet. Chem.* **1974**, *80*, 199; Holm, T. *Acta Chem. Scand.* **1973**, *27*, 1552; Morrison, J.D.; Tomaszewski, J.E.; Mosher, H.S.; Dale, J.; Miller, D.; Elsenbaumer, R.L. *J. Am. Chem. Soc.* **1977**, *99*, 3167; Okuhara, K. *J. Am. Chem. Soc.* **1980**, *102*, 244.

There is evidence that the mechanism leading to enolization is also cyclic, but involves prior coordination with magnesium:[541]

Aromatic aldehydes and ketones can be alkylated and reduced in one reaction vessel by treatment with an alkyl- or aryllithium, followed by lithium and ammonia and then by ammonium chloride.[542]

A similar reaction has been carried out with *N,N*-disubstituted amides: $RCONR'_2 \rightarrow RR^2CHNR'_2$.[543]

OS **I**, 188; **II**, 406, 606; **III**, 200, 696, 729, 757; **IV**, 771, 792; **V**, 46, 452, 608, 1058; **VI**, 478, 537, 542, 606, 737, 991, 1033; **VII**, 177, 271, 447; **VIII**, 179, 226, 315, 343, 386, 495, 507, 556; **IX**, 9, 103, 139, 234, 306, 391, 472; **75**, 12; **76**, 214; **X**, 200.

16-25   Addition of Other Organometallics to Aldehydes and Ketones

O-Hydro-C-alkyl-addition

A variety of organometallic reagents other than RMgX and RLi add to aldehydes and ketones. A simple example is formation of sodium, or potassium alkyne anions (e.g., RC≡CNa, **16-38**), which undergo acyl addition to ketones or aldehydes to give the propargylic alcohol. For the addition of acetylenic groups, sodium may be the metal used; while vinylic alanes (prepared as in **15-17**) are the reagents of choice for the addition of vinylic groups.[544] A variation includes the use of tetraalkylammonium hydroxide to generate the alkyne anion,[545] and terminal alkynes with CsOH react similarly.[546] A solvent-free reaction was reported that mixed a ketone, a terminal alkyne and potassium *tert*-butoxide.[547] The reagent Me₃Al/⁻C≡CH

[541]Pinkus, A.G.; Sabesan, A. *J. Chem. Soc. Perkin Trans. 2* **1981**, 273.
[542]Lipsky, S.D.; Hall, S.S. *Org. Synth. VI*, 537; McEnroe, F.J.; Sha, C.; Hall, S.S. *J. Org. Chem.* **1976**, *41*, 3465.
[543]Hwang, Y.C.; Chu, M.; Fowler, F.W. *J. Org. Chem.* **1985**, *50*, 3885.
[544]Newman, H. *Tetrahedron Lett.* **1971**, 4571. Vinylic groups can also be added with 9-vinylic-9-BBN compounds: Jacob III, P.; Brown, H.C. *J. Org. Chem.* **1977**, *42*, 579.
[545]Ishikawa, T.; Mizuta, T.; Hagiwara, K.; Aikawa, T.; Kudo, T.; Saito, S. *J. Org. Chem.* **2003**, *68*, 3702.
[546]Tzalis, D.; Knochel, P. *Angew. Chem. Int. Ed.* **1999**, *38*, 1463.
[547]Miyamoto, H.; Yasaka, S.; Tanaka, K. *Bull. Chem. Soc. Jpn.* **2001**, *74*, 185.

Na$^+$ also adds to aldehydes to give the ethynyl alcohol.[548] Dialkylzinc reagents have been used for the same purpose, and in the presence of a chiral titanium complex the propargylic alcohol was formed with good enantioselectivity.[549] Zinc(II) chloride facilitates the addition of a terminal alkyne to an aldehyde to give a propargylic alcohol.[550] Zinc(II) triflate can also be used for alkyne addition to aldehydes,[551] and in the presence of a chiral ligand leads to good enantioselectivity in the propargyl alcohol product.[552] Terminal alkynes add to aryl aldehydes in the presence of InBr$_3$ and NEt$_3$[553] or SmI$_2$.[554] 1-Iodoalkynes react with In metal and an aldehyde to give the propargylic alcohol.[555] Potassium alkynyltrifluoroborates (p. 817) react with aldehydes and a secondary amine, in an ionic liquid, to give a propargylic amine.[556]

Propargylic acetate adds to aldehydes with good anti selectivity in the presence of Et$_2$Zn and a palladium catalyst.[557] Propargylic bromide add to ketones in the presence of NaI/Dy,[558] In,[559] or Mn/Cr catalyst/TMSCl.[560] Propargylic tin compounds react with aldehydes to give the alcohol, with good antiselectivity.[561]

With other organometallic compounds, active metals, such as alkylzinc reagents,[562] are useful; and compounds such as alkylmercurys do not react. When the reagent is MeNbCl$_4$, ketones (R$_2$CO) are converted to R$_2$C(Cl)Me.[563]

[548]Joung, M.J.; Ahn, J.H.; Yoon, N.M. *J. Org. Chem.* **1996**, *61*, 4472.

[549]For a review, see Pu, L. *Tetrahedron* **2003**, *59*, 9873. For some leading references, see Gao, G.; Moore, D.; Xie, R.-G.; Pu. L. *Org. Lett.*2002, *4*, 4143; Li, Z.-B.; Pu, L. *Org. Lett.* **2004**, *6*, 1065; Dahmen, S. *Org. Lett.* **2004**, *6*, 2113; Kamble, R.M.; Singh, V.K. *Tetrahedron Lett.* **2003**, *44*, 5347; Lu, G.; Li, X.; Chen, G.; Chan, W.L.; Chan, A.S.C. *Tetrahedron Asymmetry* **2003**, *14*, 449; Kang, Y.-F.; Liu, L.; Wang, R.; Yan, W.-J.; Zhou, Y.-F. *Tetrahedron Asymmetry* **2004**, *15*, 3155; Lu, G.; Li, X.; Jia, X.; Chan, W.L.; Chan, A.S.C. *Angew. Chem. Int. Ed.* **2003**, *42*, 5057; Xu, Z.; Wang, R.; Xu, J.; Da, C.-s.; Yan, W.-j.; Chen, C. *Angew. Chem. Int. Ed.* **2003**, *42*, 5747;.

[550]Jiang, B.; Si, Y.-G. *Tetrahedron Lett.* **2002**, *43*, 8323

[551]Frantz, D.E.; Fässler, R.; Carreira, E.M. *J. Am. Chem. Soc.* **2000**, *122*, 1806.

[552]Anand, N.K.; Carreira, E.M. *J. Am. Chem. Soc.* **2001**, *123*, 9687; Sasaki, H.; Boyall, D.; Carreira, E.M. *Helv. Chim. Acta* **2001**, *84*, 964; Boyall, D.; Frantz, D.E.; Carreira, E.M. *Org. Lett.* **2002**, *4*, 2605.; Xu, Z.; Chen, C.; Xu, J.; Miao, M.; Yan, W.; Wang, R. *Org. Lett.* **2004**, *6*, 1193; Jiang, B.; Chen, Z.; Xiong, W. *Chem. Commun.* **2002**, 1524. For an example using zinc (II) diflate, see Chen, Z.; Xiong, W.; Jiang, B. *Chem. Commun.* **2002**, 2098.

[553]Sakai, N.; Hirasawa, M.; Konakahara, T. *Tetrahedron Lett.* **2003**, *44*, 4171.

[554]Kwon, D.W.; Cho, M.S.; Kim, Y.H. *Synlett* **2001**, 627.

[555]Augé, J.; Lubin-Germain, N.; Seghrouchni, L. *Tetrahedron Lett.* **2002**, *43*, 5255.

[556]In bmim BF$_4$, 1-butyl-3-methylimidazolium tetrafluoroborate: Kabalka, G.W.; Venkataiah, B.; Dong, G. *Tetrahedron Lett.* **2004**, *45*, 729.

[557]Marshall, J.A.; Adams, N.D. *J. Org. Chem.* **1999**, *64*, 5201.

[558]Li, Z.; Jia, Y.; Zhou, J. *Synth. Commun.* **2000**, *30*, 2515.

[559]In the presence of (−)-cinchonidine: Loh, T.-P.; Lin, M.-J.; Tan, K.L. *Tetrahedron Lett.* **2003**, *44*, 507.

[560]Inoue, M.; Nakada, M. *Org. Lett.* **2004**, *6*, 2977.

[561]Savall, B.M.; Powell, N.A.; Roush, W.R. *Org. Lett.* **2001**, *3*, 3057.

[562]For a review with respect to organozinc compounds, see Furukawa, J.; Kawabata, N. *Adv. Organomet. Chem.* **1974**, *12*, 103. For an example, see Sjöholm, R.; Rairama, R.; Ahonen, M. *J. Chem. Soc., Chem. Commun.* **1994**, 1217. For a review with respect to organocadmium compounds, see Jones, P.R.; Desio, P.J. *Chem. Rev.* **1978**, *78*, 491.

[563]Kauffmann, T.; Abel, T.; Neiteler, G.; Schreer, M. *Tetrahedron Lett.* **1990**, 503.

Furthermore, organotitanium reagents can be made to add chemoselectively to alde-hydes in the presence of ketones.[564] Organomanganese compounds are also chemo-selective in this way.[565] Aryl halides that have a pendant ketone unit react with a palladium catalyst to give cyclization via acyl addition.[566] Chiral amides react with aldehydes in the presence of $TiCl_4$ to give syn-selective addition products,[567] and titanium-catalyzed enantioselective additions are known.[568] An alkene-ketone, where the alkene is a vinyl bromide, reacted with $CrCl_2/NiCl_2$ to give a vinyl orga-nometallic, which cyclized to generate a cyclic allylic alcohol with the double bond within the ring.[569] Aryl halides react with a nickel complex under electrolytic con-ditions to add the aryl group to aldehydes.[570] The C-3 position of an indole adds to aldehydes in the presence of a palladium catalyst.[571] The addition of trifluoro-methyl to an aldehyde was accomplished photochemically using CF_3I and $(Me_2N)_2C{=}C(NMe_2)_2$.[572] α-Iodo phosphonate esters react with aldehydes and SmI_2 to give a β-hydroxy phosphonate ester.[573]

Dialkylzinc compounds react with aldehydes to give the secondary alcohol, and R_3ZnLi reagents also add R to a carbonyl.[574] Dimethylzinc and diethylzinc are probably the most common reagents. An intramolecular version is possible by reac-tion an allene-aldehyde with dimethylzinc. Addition to the allene in the presence of a nickel catalyst[575] or a $CeCl_3$ catalyst[576] is followed by addition of the intermedi-ate organometallic to the aldehyde to give the cyclic product. Aryl halides react with Zn–Ni complexes to give acyl addition of the aryl group to an aldehyde.[577] The reaction of an allylic halide and Zn[578] or $Zn/TMSCl$[579] leads to acyl addition of aldehydes.

[564]Reetz, M.T. *Organotitanium Reagents in Organic Synthesis*, Springer, NY, *1986* (monograph), pp. 75–86. See also, Reetz, M.T.; Maus, S. *Tetrahedron* **1987**, *43*, 101; Kim, S.-H.; Rieke, R.D. *Tetrahedron Lett.* **1999**, *40*, 4931.

[565]Cahiez, G.; Figadere, B. *Tetrahedron Lett.* **1986**, *27*, 4445. For other organometallic reagents with high selectivity towards aldehyde functions, see Kauffmann, T.; Hamsen, A.; Beirich, C. *Angew. Chem. Int. Ed.* **1982**, *21*, 144; Takai, K.; Kimura, K.; Kuroda, T.; Hiyama, T.; Nozaki, H. *Tetrahedron Lett.* **1983**, *24*, 5281; Soai, K.; Watanabe, M.; Koyano, M. *Bull. Chem. Soc. Jpn.* **1989**, *62*, 2124.

[566]Quan, L.G.; Lamrani, M.; Yamamoto, Y. *J. Am. Chem. Soc.* **2000**, *122*, 4827.

[567]Crimmins, M.T.; Chaudhary, K. *Org. Lett.* **2000** *2*, 775.

[568]Walsh, P.J. *Acc. Chem. Res.* **2003**, *36*, 739.

[569]Trost, B.M.; Pinkerton, A.B. *J. Org. Chem.* **2001**, *66*, 7714.

[570]Durandetti, M.; Nédélec, J.-Y.; Périchon, J. *Org. Lett.* **2001**, *3*, 2073.

[571]Hao, J.; Taktak, S.; Aikawa, K.; Yusa, Y.; Hatano, M.; Mikami, K. *Synlett* **2001**, 1443.

[572]Aït-Mohand, S.; Takechi, N.; Médebielle, M.; Dolbier Jr., W.R. *Org. Lett.* **2001**, *3*, 4271.

[573]Orsini, F.; Caselli, A. *Tetrahedron Lett.* **2002**, *43*, 7255.

[574]For a discussion of the electronic and steric effects, see Musser, C.A.; Richey, Jr., H.G. *J. Org. Chem.* **2000**, *65*, 7750. For a kinetic study of Et3ZnLi and di-*tert*-butyl ketone, see Maclin, K.M.; Richey, Jr., H.G. *J. Org. Chem.* **2002**, *67*, 4602.

[575]Montgomery, J.; Song, M. *Org. Lett.* **2002**, *4*, 4009.

[576]Fischer, S.; Groth, U.; Jeske, M.; Schütz, T. *Synlett* **2002**, 1922.

[577]Majumdar, K.K.; Cheng, C.-H. *Org. Lett.* **2000**, *2*, 2295.

[578]Yavari, I.; Riazi-Kermani, F. *Synth. Commun.* **1995**, *25*, 2923; Ranu, B.C.; Majee, A.; Das, A.R. *Tetrahedron Lett.* **1995**, *36*, 4885; Durant, A.; Delplancke, J.-L.; Winand, R.; Reisse, J. *Tetrahedron Lett.* **1995**, *36*, 4257; Felpin, F.-X.; Bertrand, M.-J.; Lebreton, J. *Tetrahedron* **2002**, *58*, 7381.

[579]Ito, T.; Ishino, Y.; Mizuno, T.; Ishikawa, A.; Kobyashi, J.-i. *Synlett* **2002**, 2116.

Lithium dimethylcopper (Me$_2$CuLi) reacts with aldehydes[580] and with certain ketones[581] to give the expected alcohols. The RCu(CN)ZnI reagents also react with aldehydes, in the presence of BF$_3$–etherate, to give secondary alcohols. Vinyl-tellurium compound react with BF$_3$•OEt$_2$ and cyano cuprates [R(2-thienyl)CuCN-Li$_2$] to give a reagent that adds 1,2- to the carbonyl of a conjugated ketone.[582] Vinyl tellurium compounds also react with n-butyllithium to give a reagent that adds to nonconjugated ketones.[583]

Many methods have been reported for the addition of allylic groups,[584] including enantioselective reactions.[585] One of the most common methods is the Barbier reaction, employing metals and metal compounds other than Mg or Li, although the method is not limited to allylic compounds. Allyl indium compounds[586] add to aldehydes or ketones in various solvents.[587] Indium metal is used for the acyl addition of allylic halides with a variety of aldehydes and ketones, including aliphatic aldehydes,[588] aryl aldehydes[589] and α-keto esters.[590] Indium reacts with allylic bromides and ketones in water[591] and in aqueous media. Elimination of the homoallylic alcohol to a conjugated diene can accompany the addition in some cases.[592] The reaction of a propargyl halide, In, and an aldehyde in aq. THF leads to an allenic alcohol.[593] The reaction of benzaldehyde with a pro-pargylic bromide, indium metal and water give the alcohol.[594] Allyl bromide reacts with Mn/TMSCl and an In catalyst in water to give the homoallylic

[580]Barreiro, E.; Luche, J.; Zweig, J.S.; Crabbé, P. *Tetrahedron Lett.* **1975**, 2353; Zweig, J.S.; Luche; Barreiro, E.; Crabbé, P. *Tetrahedron Lett.* **1975**, 2355; Reetz, M.T.; Rölfing, K.; Griebenow, N. *Tetrahedron Lett.* **1994**, *35*, 1969.

[581]House, H.O.; Prabhu, A.V.; Wilkins, J.M.; Lee, L.F. *J. Org. Chem.* **1976**, *41*, 3067; Matsuzawa, S.; Isaka, M.; Nakamura, E.; Kuwajima, I. *Tetrahedron Lett.* **1989**, *30*, 1975.

[582]Araújo, M.A.; Barrientos-Astigarraga, R.E.; Ellensohn, R.M.; Comasseto, J.V. *Tetrahedron Lett.* **1999**, *40*, 5115.

[583]Dabdoub, M.J.; Jacob, R.G.; Ferreira, J.T.B.; Dabdoub, V.B.; Marques, F.de.A. *Tetrahedron Lett.* **1999**, *40*, 7159.

[584]For a list of reagents and references, see Larock, R.C. *Comprehensive Organic Transformations*, 2nd ed., Wiley-VCH, NY, **1999**, pp. 1156–1170. For a discussion of a deuterium kinetic isotope effect in the addition of allylic reagents to benzaldehyde, see Gajewski, J.J.; Bocian, W.; Brichford, N.L.; Henderson, J.L. *J. Org. Chem.* **2002**, *67*, 4236.

[585]For a review, see Denmark, S.E.; Fu, J. *Chem. Rev.* **2003**, *103*, 2763.

[586]For a review, see Cintas, P. *Synlett* **1995**, 1087.

[587]Yi, X.-H.; Haberman, J.X.; Li, C.-J. *Synth. Commun.* **1998**, *28*, 2999; Lloyd-Jones, G.C.; Russell, T. *Synlett* **1998**, 903; Li, C.-J.; Lu, Y.-Q. *Tetrahedron Lett.* **1995**, *36*, 2721; Li, C.-J. *Tetrahedron Lett* **1995**, *36*, 517.

[588]Loh, T.-P.; Tan, K.-T.; Yang, J.-Y.; Xiang, C.-L. *Tetrahedron Lett.* **2001**, *42*, 8701.

[589]Khan, F.A.; Prabhudas, B. *Tetrahedron* **2000**, *56*, 7595.

[590]Loh, T.-P.; Huang, J.-M.; Xu, K.-C.; Goh, S.-H.; Vittal, J.J. *Tetrahedron Lett.* **2000**, *41*, 6511; Kumar, S.; Kaur, P.; Chimni, S.S. *Synlett* **2002**, 573.

[591]Chan, T.H.; Yang, Y. *J. Am. Chem. Soc.* **1999**, *121*, 3228; Paquette, L.A.; Bennett, G.D.; Isaac, M.B.; Chhatriwalla, A. *J. Org. Chem.* **1998**, *63*, 1836; Li, X.-R.; Loh, T.-P. *Tetrahedron Asymmetry* **1996**, *7*, 1535; Isaac, M.B.; Chan, T.-H. *Tetrahedron Lett.* **1995**, *36*, 8957.

[592]Kumar, V.; Chimni, S.; Kumar, S. *Tetrahedron Lett.* **2004**, *45*, 3409.

[593]Lin, M.-J.; Loh, T.-P. *J. Am. Chem. Soc.* **2003**, *125*, 13042.

[594]Lu, W.; Ma, J.; Yang, Y.; Chan, T.H. *Org. Lett.* **2000**, *2*, 3469.

alcohols from aldehydes.[595] When allyl iodide is mixed with In and TMSCl, reaction with a conjugated ketone proceed by 1,4-addition, but in the presence of 10% CuI, the major product is that of 1,2-addition.[596] The reaction with indium is compatible with the presence of a variety of other functional groups in the molecule, including phosphonate,[597] propargylic sulfides.[598] Ethyl α-bromoacetate reacts with aldehydes using In with ultrasound.[599] Vinyl epoxides can be added to aldehydes using InI and a palladium catalyst,[600] and vinyl acetates react with indium metal to give a reactive intermediate that adds to the carbonyl of aldehydes.[601] A tandem reaction has been reported in which a bis(indium) reagent, $Br_2InC(=CH_2)-C(=CH_2)InBr_2$, reacts with 2 equivalents of an aldehyde in the presence of ZnF_2 to give a 3-hexyne-1,6-diol derivative.[602] Alkyl halides react with Zn/Cu and an InCl catalyst, in the presence of $0.07\,M$ $Na_2Cr_2O_7$ to give an intermediate that adds to aldehydes.[603] Analogous to aldehydes, 1,1-diacetates react with In and allyl bromide in aq. THF to give a homoallylic acetate,[604] as do dimethyl ketals.[605] Allylic alcohols react with InI and a nickel catalyst to give acyl addition of an allylic group to an aldehyde, giving the homoallylic alcohol.[606]

Another important metal for Barbier-type reaction is samarium. Allyl bromide reacts with a ketone and Sm to give the homoallylic alcohol.[607] Samarium compounds, such as SmI_2,[608] can also be used with allylic halides.

Allyltin compounds readily add to aldehydes and ketones.[609] Allylic bromides react with tin to generate the organometallic *in situ*, which then adds to aldehydes.[610] Allylic chlorides react with aldehydes in the presence of ditin compounds such as $Me_3Sn-SnMe_3$ and a palladium catalyst.[611] Allyltrialkyltin compounds[612]

[595]Augé, J.; Lubin-Germain, N.; Thiaw-Woaye, A. *Tetrahedron Lett.* **1999**, *40*, 9245.

[596]Lee, P.H.; Ahn, H.; Lee. K.; Sung, S.-y.; Kim, S. *Tetrahedron Lett.* **2001**, *42*, 37.

[597]Ranu, B.C.; Samanta, S.; Hajra, A. *J. Org. Chem.* **2001**, *66*, 7519.

[598]Mitzel, T.M.; Palomo, C.; Jendza, K. *J. Org. Chem.* **2002**, *67*, 136.

[599]Lee, P.H.; Bang, K.; Lee, K.; Sung, S.-y.; Chang, S. *Synth. Commun.* **2001**, *31*, 3781.

[600]Araki, S.; Kameda, K.; Tanaka, J.; Hirashita, T.; Yamamura, H.; Kawai, M. *J. Org. Chem.* **2001**, *66*, 7919.

[601]Lombardo, M.; Girotti, R.; Morganti, S.; Trombini, C. *Org. Lett.* **2001**, *3*, 2981.

[602]Miao, W.; Lu, W.; Chan, T.H.; *J. Am. Chem. Soc.* **2003**, *125*, 2412.

[603]Keh, C.C.K.; Wei, C.; Li, C.-J. *J. Am. Chem. Soc.* **2003**, *125*, 4062.

[604]Yadav, J.S.; Reddy, B.V.S.; Reddy, G.S.K.K. *Tetrahedron Lett.* **2000**, *41*, 2695.

[605]Kwon, J.S.; Pae, A.N.; Choi, K.I.; Koh, H.Y.; Kim, Y.; Cho, Y.S. *Tetrahedron Lett.* **2001**, *42*, 1957.

[606]Hirashita, T.; Kambe, S.; Tsuji, H.; Omori, H.; Araki, S. *J. Org. Chem.* **2004**, *69*, 5054.

[607]Basu, M.K.; Banik, B.K. *Tetrahedron Lett.* **2001**, *42*, 187.

[608]Kunishima, M.; Tawaka, S.; Kono, K.; Hioki, K.; Tani, S. *Tetrahedron Lett.* **1995**, *36*, 3707; Hélion, F.; Namy, J.-L. *J. Org. Chem.* **1999**, *64*, 2944.

[609]Yasuda, M.; Kitahara, N.; Fujibayashi, T.; Baba, A. *Chem. Lett.* **1998**, 743; Marshall, J.A.; Palovich, M.R. *J. Org. Chem.* **1998**, *63*, 4381; Kobayashi, S.; Nagayama, S. *J. Org. Chem.* **1996**, *61*, 2256.

[610]Tan, K.-T.; Chng, S.-S.; Cheng, H.-S.; Loh, T.-P. *J. Am. Chem. Soc.* **2003**, *125*, 2958. For a reaction using Sn and ultrasound, see Andres, P.C.; Peatt, A.C.; Raston, C.L. *Tetraheron Lett.* **2002**, *43*, 7541.

[611]Wallner, O.A.; Szabó, K.J. *J. Org. Chem.* **2003**, *68*, 2934.

[612]For a high pressure version of this reaction, see Issacs, N.S.; Maksimovic, L.; Rintoul, G.B.; Young, D.J. *J. Chem. Soc., Chem. Commun.* **1992**, 1749; Isaacs, N.S.; Marshall, R.L.; Young, D.J. *Tetrahedron Lett.* **1992**, *33*, 3023.

and tetraallyltin react with aldehydes or ketones in the presence of BF_3–etherate,[613] $Cu(OTf)_2$,[614] $CeCl_3$ with NaI,[615] $Bi(OTf)_3$,[616] PbI_2,[617] AgOTf,[618] $Cd(ClO_4)_2$,[619] SnX_2,[620] Ti (IV),[621] $NbCl_5$,[622] $Zr(Ot-Bu)_4$,[623] or $La(OTf)_3$.[624] Tetraallyltin reacts via 1,2-addition to conjugated ketones in refluxing methanol.[625] Aluminum catalysts, such as MABR, facilitate addition of allyltributyltin to aldehydes.[626] Select-fluor has been used to induce 1,2-addition of the allyl group of allyltributyltin to a conjugated aldehyde.[627] Allyltributyltin reacts with aldehydes in the presence of aqueous trifluoromethanesulfonic acid to give the homoallylic alcohol.[628] Tetraallyltin reacts with aldehydes in ionic liquids[629] and on wet silica,[630] and allyltributyltin adds to aldehydes in ionic liquids with $InCl_3$.[631] Tetraallyltin adds to ketones or aldehydes to give homoallylic alcohols with good enantioselectivity in the presence of a chiral titanium complex.[632] Allylic alcohols and homoallylic alcohols add to aldehydes in the presence of $Sn(OTf)_2$[633] $In/InCl_3$,[634] or with a rhodium catalyst.[635] Vinyltin regents, such as (2-butadiene)tributyltin, react with aldehydes in the presence of $SnCl_4$ and 3 equivalents of DMF to

[613]Naruta, Y.; Ushida, S.; Maruyama, K. *Chem. Lett.* **1979**, 919. For a review, see Yamamoto, Y. *Aldrichimica Acta* **1987**, *20*, 45.

[614]Kamble, R.M.; Singh, V.K. *Tetrahedron Lett.* **2001**, *42*, 7525.

[615]Bartoli, G.; Bosco, M.; Giuliani, A.; Marcantoni, E.; Palmieri, A.; Petrini, M.; Sambri, L. *J. Org. Chem.* **2004**, *69*, 1290.

[616]Choudary, B.M.; Chidara, S.; Sekhar, Ch.V.R. *Synlett* **2002**, 1694.

[617]Shibata, I.; Yoshimura, N.; Yabu, M.; Baba, A. *Eurl. J. Org. Chem.* **2001**, 3207.

[618]Yanagisawa, A.; Nakashima, H.; Nakatsuka, Y.; Ishiba, A.; Yamamoto, H. *Bull. Chem. Soc. Jpn.* **2001**, *74*, 1129.

[619]Kobayashi, S.; Aoyama, N.; Manabe, K. *Synlett* **2002**, 483.

[620]$SnCl_2$: Yasuda, M.; Hirata, K.; Nishino, M.; Yamamoto, A.; Baba, A. *J. Am. Chem. Soc.* **2002**, *124*, 13442. SnI_2: Masuyama, Y.; Ito, T.; Tachi, K.; Ito, A.; Kurusu, Y. *Chem. Commun.* **1999**, 1261.

[621]Kii, S.; Maruoka, K. *Tetrahedron Lett.* **2001**, *42*, 1935.

[622]Andrade, C.K.Z.; Azevedo, N.R. *Tetrahedron Lett.* **2001**, *42*, 6473.

[623]Kurosa, M.; Lorca, M. *Tetrahedron Lett.* **2002**, *43*, 1765.

[624]Aspinall, H.C.; Bissett, J.S.; Greeves, N.; Levin, D. *Tetrahedron Lett.* **2002**, *43*, 319.

[625]Khan, A.T.; Mondal, E. *Synlett* **2003**, 694.

[626]Marx, A.; Yamamoto, H. *Synlett* **1999**, 584.

[627]Liu, J.; Wong, C.-H. *Tetrahedron Lett.* **2002**, *43*, 3915.

[628]Loh, T.-P.; Xu, J. *Tetrahedron Lett.* **1999**, *40*, 2431.

[629]In bmim BF_4, 1-butyl-3-methylimidazolium tetrafluoroborate: Gordon, C.M.; McCluskey, A. *Chem. Commun.* **1999**, 1431.

[630]Jin, Y.Z.; Yasuda, N.; Furuno, H.; Inanaga, J. *Tetrahedron Lett.* **2003**, *44*, 8765.

[631]In bmim Cl, 1-butyl-3-methylimidazolium chloride: Lu, J.; Ji, S.-J.; Qian, R.; Chen, J.-P.; Liu, Y.; Loh, T.-P. *Synlett* **2004**, 534.

[632]Waltz, K.M.; Gavenonis, J.; Walsh, P.J. *Angew. Chem. Int. Ed.* **2002**, *41*, 3697; Cunningham, A.; Woodward, S. *Synlett* **2002**, 43; Kim, J.G.; Waltz, K.M.; Garcia, I.F.; Kwiatkowski, D.; Walsh, P.J. *J. Am. Chem. Soc.* **2004**, *126*, 12580.

[633]Nokami, J.; Yoshizane, K.; Matsuura, H.; Sumida, S.-i. *J. Am. Chem. Soc.* **1998**, *120*, 6609.

[634]Jang, T.-S.; Keum, G.; Kang, S.B.; Chung, B.Y.; Kim, Y. *Synthesis* **2003**, 775.

[635]Masuyama, Y.; Kaneko, Y.; Kurusu, Y. *Tetrahedron Lett.* **2004**, *45*, 8969.

give the dienyl alcohol.[636] Allenyl tin compounds ($CH_2=C=CHSnBu_3$) also react with aldehydes in the presence of $BF_3 \cdot OEt_2$ to give a 2-dienyl alcohol.[637] The tin compound can be prepared *in situ* using an α-iodo ketone with an aldehyde and Bu_2SnI_2-LiI.[638] A similar addition occurs with $(allyl)_2SnBr_2$ in water.[639] Asymmetric induction has been reported.[640] The use of a chiral rhodium[641] or titanium[642] catalyst leads to enantioselective addition of allyltributyltin to aldehydes. Allyltributyltin reacts with aldehydes in the presence of $SiCl_4$ and a chiral phosphoramide to give the homoallylic alcohol with moderate enantioselectivity.[643] It is noted that tetraallyl germanium adds to aldehydes in a similar manner in the presence of a $Sc(OTf)_3$ catalyst.[644]

A variety of other allylic metal compounds add to aldehydes or ketones.[645] A variety of alkyl and allylic halides add to aldehydes or ketones in the presence of metals or metal compounds; the metal or compounds based on Ti,[646] Mn,[647] Fe,[648] Ga,[649]

[636]Luo, M.; Iwabuchi, Y.; Hatakeyama, S. *Synlett* **1999**, 1109.

[637]Luo, M.; Iwabuchi, Y.; Hatakeyama, S. *Chem. Commun.* **1999**, 267; Yu, C.-M.; Lee, S.-J.; Jeon, M. *J. Chem. Soc., Perkin Trans. 1* **1999**, 3557.

[638]Shibata, I.; Suwa, T.; Sakakibara, H.; Baba, A. *Org. Lett.* **2002**, *4*, 301.

[639]Chan, T.H.; Yang, Y.; Li, C.J. *J. Org. Chem.* **1999**, *64*, 4452.

[640]Yanagisawa, A.; Narusawa, H.; Ishiba, A.; Yamamoto, H. *J. Am. Chem. Soc.* **1996**, *118*, 4723; Keck, G.E.; Krishnamurthy, D.; Grier, M.C. *J. Org. Chem.* **1993**, *58*, 6543; Motoyama, Y.; Nishiyama, H. *Synlett* **2003**, 1883; Xia, G.; Shibatomi, K.; Yamamoto, H. *Synlett* **2004**, 2437.

[641]Moloyama, Y.; Narusawa, H.; Nishiyama, H. *Chem. Commun.* **1999**, 131.

[642]Doucet, H.; Santelli, M. *Tetrahedron Asymmetry* **2000**, *11*, 4163.

[643]Denmark, S.E.; Wynn, T. *J. Am. Chem. Soc.* **2001**, *123*, 6199.

[644]Akiyama, T.; Iwai, J.; Sugano, M. *Tetrahedron* **1999**, *55*, 7499.

[645]See, for example, Furuta, K.; Ikeda, Y.; Meguriya, N.; Ikeda, N.; Yamamoto H. *Bull. Chem. Soc. Jpn.* **1984**, *57*, 2781; Pétrier, C.; Luche, J.L. *J. Org. Chem.* **1985**, *50*, 910; Tanaka, H.; Yamashita, S.; Hamatani, T.; Ikemoto, Y.; Torii, S. *Chem. Lett.* **1986**, 1611; Guo, B.; Doubleday, W.; Cohen, T. *J. Am. Chem. Soc.* **1987**, *109*, 4710; Hosomi, A. *Acc. Chem. Res.* **1988**, *21*, 200; Araki, S.; Butsugan, Y. *Chem. Lett.* **1988**, 457; Minato, M.; Tsuji, J. *Chem. Lett.* **1988**, 2049; Coxon, J.M.; van Eyk, S.J.; Steel, P.J. *Tetrahedron* **1989**, *45*, 1029; Knochel, P.; Rao, S.A. *J. Am. Chem. Soc.* **1990**, *112*, 6146; Wada, M.; Ohki, H.; Akiba, K. *Bull. Chem. Soc. Jpn.* **1990**, *63*, 1738; Marton, D.; Tagliavini, G.; Zordan, M.; Wardell, J.L. *J. Organomet. Chem.* **1990**, *390*, 127; Wang, W.; Shi, L.; Xu, R.; Huang, Y. *J. Chem. Soc. Perkin Trans. 1* **1990**, 424; Shono, T.; Ishifune, M.; Kashimura, S. *Chem. Lett.* **1990**, 449.

[646]Rosales, A.; Oller-López, J.L.; Justicia, J.; Gansäuer, A.; Oltra, J.E.; Curerva, J.M. *Chem. Commun.* **2004**, 2628. For a discussion of the mechanism of titanium-mediated asymmetric addition see Balsells, J.; Davis, T.J.; Carroll, P.; Walsh, P.J. *J. Am. Chem. Soc.* **2002**, *124*, 10336; BourBouz, S.; Pradaux, F.; Cossy, J.; Ferrroud, C.; Falguières, A. *Tetrahedron Lett.* **2000**, *41*, 8877; Jana, S.; Guin, C.; Roy, S.C. *Tetrahedron Lett.* **2004**, *45*, 6575.

[647]Kakiya, H.; Nishimae, S.; Shinokubo, H.; Oshima, K. *Tetrahedron* **2001**, *57*, 8807. With a catalytic amount of $CrCl_2$ and a salen catalyst, see Berkessel, A.; Menche, D.; Sklorz, C.A.; Schröder, M.; Paterson, I. *Angew. Chem. Int. Ed.* **2003**, *42*, 1032. For a review of ligand effects for organomanganese and organocerium compounds, see Reetz, M.T.; Haning, H.; Stanchev, S. *Tetrahedron Lett.* **1992**, *33*, 6963.

[648]Chan, T.C.; Lau, C.P.; Chan, T.H. *Tetrahedron Lett.* **2004**, *45*, 4189.

[649]Han, Y.; Chi, Z.; Huang, Y.-Z. *Synth. Commun.* **1999**, *29*, 1287; Wang, Z.; Yuan, S.; Li, C.-J. *Tetrahedron Lett.* **2002**, *43*, 5097.

Ge,[650] Zr,[651] Nb,[652] Cd,[653] Sn,[654] Sb,[655] Te,[656] Ba,[657] Ce,[658] Nd,[659] Hg,[660] Bi,[661] and Pb.[662] In addition, $BiCl_3/NaBH_4$,[663] $Mg-BiCl_3$,[664] and $CrCl_2/NiCl_2$,[665] have been used. Allylic alcohols have been converted to organometallic reagents with diethyl zinc and a palladium catalyst[666] or a ruthenium catalyst[667] leading to the homoallylic alcohol upon reaction with an aldehyde. A chiral Cr/Mn complex has been used with allylic bromides in conjunction with trimethylsilyl chloride.[668] Reagents of the type R–Yb have been prepared from RMgX.[669] Vinyl bromides react with $NiBr_2/CrCl_3/TMSCl$ to give a reagent that adds to aldehydes to give the allylic alcohol.[670] Vinyl complexes generated from alkynes and SmI_2 add intramolecularly, and eight-membered rings have been formed in this way.[671] Allylic alcohols add to aldehydes in some cases, using $SnCl_2$ and a palladium catalyst.[672] Glyoxal reacted with 2 equivalents of allyl bromide and $SnCl_2$ with KI in water, to give the bis-homoallylic alcohol oct-1,7-diene-4,5-diol.[673] The alkyl group of trialkyl aluminum compounds such as $AlEt_3$ add to aldehydes, enantioselectively in the presence of chiral transition-metal complexes.[674] Certain functional groups (COOEt, CONMe$_2$, CN) can be present in the R group when

[650]Hashimoto, Y.; Kagoshima, H.; Saigo, K. *Tetrahedron Lett.* **1994**, *35*, 4805.

[651]Hanzawa, Y.; Tabuchi, N.; Saito, K.; Noguchi, S.; Taguchi, T. *Angew. Chem. Int. Ed.* **1999**, *38*, 2395.

[652]Andrade, C.K.Z.; Azevedo, N.R.; Oliveira, G.R. *Synthesis* **2002**, 928.

[653]Zheng, Y.; Bao, W.; Zhang, Y. *Synth. Commun.* **2000**, *30*, 3517.

[654]For the use of nanoparticulate tin, see Wang, Z.; Zha, Z.; Zhou, C. *Org. Lett.* **2002**, *4*, 1683.

[655]Jin, Q.-H.; Ren, P.-D.; Li, Y.-Q.; Yao, Z.-P. *Synth. Commun.* **1998**, *28*, 4151; Li, L.-H.; Chan, T.H. *Can. J. Chem.* **2001**, *79*, 1536.

[656]For an example using a cyclopropylcarbinyl tosylate and an allyl surrogate, see Avilov, D.V.; Malasare, M.G.; Arslancan, E.; Dittmer, D.L. *Org. Lett.* **2004**, *6*, 2225.

[657]Yanagisawa, A.; Habaue, S.; Yasue, K.; Yamamoto, H. *J. Am. Chem. Soc.* **1994**, *116*, 6130.

[658]Loh, T.-P.; Zhou, J.-R. *Tetrahedron Lett.* **1999**, *40*, 9115.

[659]Evans, W.J.; Workman, P.S.; Allen, N.T. *Org. Lett.* **2003**, *5*, 2041.

[660]Chan, T.H.; Yang, Y. *Tetrahedron Lett.* **1999**, *40*, 3863.

[661]Miyamoto, H.; Daikawa, N.; Tanaka, K. *Tetrahedron Lett.* **2003**, *44*, 6963; Wada, S.; Hayashi, N.; Suzuki, H. *Org. Biomol. Chem.* **2003**, *1*, 2160; Smith, K.; Lock, S.; El-Hiti, G.A.; Wada, M.; Miyoshi, N. *Org. Biomol. Chem.* **2004**, *2*, 935; Xu, X.; Zha, Z.; Miao, Q.; Wang, Z. *Synlett* **2004**, 1171.

[662]Zhou, J.-Y.; Jia, Y.; Sun, G.-F.; Wu, S.-H. *Synth. Commun.* **1997**, *27*, 1899.

[663]Ren, P.-D.; Shao, D.; Dong, T.-W. *Synth. Commun.* **1997**, *27*, 2569.

[664]Wada, M.; Fukuma, T.; Morioka, M.; Takahashi, T.; Miyoshi, N. *Tetrahedron Lett.* **1997**, *38*, 8045.

[665]Taylor, R.E.; Ciavarri, J.P. *Org. Lett.* **1999**, *1*, 467.

[666]Kimura, M.; Shimizu, M.; Shibata, K.; Tazoe, M.; Tamaru, Y. *Angew. Chem. Int. Ed.* **2003**, *42*, 3392.

[667]Wang, M.; Yang, X.-F.; Li, C.-J. *Eur. J. Org. Chem.* **2003**, 998.

[668]Inoue, M.; Suzuki, T.; Nakada, M. *J. Am. Chem. Soc.* **2003**, *125*, 1140.

[669]Matsubara, S.; Ikeda, T.; Oshima, K.; Otimoto, K. *Chem. Lett.* **2001**, 1226.

[670]Kuroboshi, M.; Tanaka, M.; Kishimoto, S.; Goto, K.; Mochizuki, M.; Tanaka, H. *Tetrahedron Lett.* **2000**, *41*, 81.

[671]Hölemann, A.; Reissig, H.-U. *Synlett* **2004**, 2732.

[672]Márquez, F.; Llebaria, A.; Delgado, A. *Tetrahedron Asymmetry* **2001**, *12*, 1625.

[673]Samoshin, V.V.; Gremyachinskiy, D.E.; Smith, L.I.; Bliznets, I.V.; Gross, P.H. *Tetrahedron Lett.* **2002**, *43*, 6329.

[674]Lu, J.-F.; You, J.-S.; Gau, H.-M. *Tetrahedron Asymmetry* **2000**, *11*, 2531.

organotin reagents $RSnEt_3$ are added to aldehydes.[675] Trimethylaluminum[676] and dimethyltitanium dichloride[677] exhaustively methylate ketones to give *gem*-dimethyl compounds[678] (see also **10-63**):

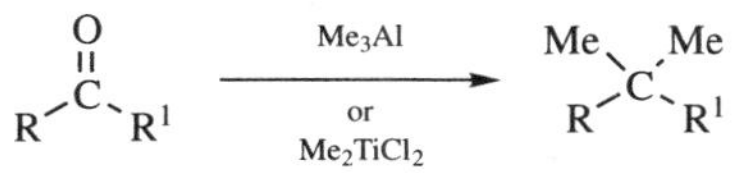

The titanium reagent also dimethylates aromatic aldehydes.[679] Triethylaluminum reacts with aldehydes, however, to give the mono-ethyl alcohol, and in the presence of a chiral additive the reaction proceeds with good asymmetric induction.[680] A complex of $Me_3Ti \cdot MeLi$ has been shown to be selective for 1,2-addition with conjugated ketones, in the presence of nonconjugated ketones.[681] In other variations, the organometallic reagent is generated *in situ*. 1,4-Dimethoxybenzene reacts with ethyl glyoxylate (EtO_2C-CHO) in the presence of 5% $Yb(OTf)_3$, to give the alcohol formed by addition of the aryl group to the aldehyde unit.[682]

High ee values have also been obtained with organometallics,[683] including organotitanium compounds (methyl, aryl, allylic) in which an optically active ligand is coordinated to the titanium,[684] allylic boron compounds, and organozinc compounds. As for the organozinc reagents, very high enantioselection was obtained from R_2Zn reagents (R = alkyl)[685] and aromatic[686] aldehydes by the use of a small

[675]Kashin, A.N.; Tulchinsky, M.L.; Beletskaya, I.P. *J. Organomet. Chem.* **1985**, *292*, 205.

[676]Meisters, A.; Mole, T. *Aust. J. Chem.* **1974**, *27*, 1655. See also, Jeffery, E.A.; Meisters, A.; Mole, T. *Aust. J. Chem.* **1974**, *27*, 2569. For discussions of the mechanism of this reaction, see Ashby, E.C.; Smith, R.S. *J. Organomet. Chem.* **1982**, *225*, 71. For a review of organoaluminum compounds in organic synthesis, see Maruoka, H.; Yamamoto, H. *Tetrahedron* **1988**, *44*, 5001.

[677]Reetz, M.T.; Westermann, J.; Kyung, S. *Chem. Ber.* **1985**, *118*, 1050.

[678]For the *gem*-diallylation of anhydrides, with an indium reagent, see Araki, S.; Katsumura, N.; Ito, H.; Butsugan, Y. *Tetrahedron Lett.* **1989**, *30*, 1581.

[679]Reetz, M.T.; Kyung, S. *Chem. Ber.* **1987**, *120*, 123.

[680]Chan, A.S.C.; Zhang, F.-Y.; Yip, C.-W. *J. Am. Chem. Soc.* **1997**, *119*, 4080.

[681]Markó, I.E.; Leung, C.W. *J. Am. Chem. Soc.* **1994**, *116*, 371.

[682]Zhang, W.; Wang, P.G. *J. Org. Chem.* **2000**, *65*, 4732.

[683]For examples involving other organometallic compounds, see Abenhaïm, D.; Boireau, G.; Deberly, A. *J. Org. Chem.* **1985**, *50*, 4045; Minowa, N.; Mukaiyama, T. *Bull. Chem. Soc. Jpn.* **1987**, *60*, 3697; Takai, Y.; Kataoka, Y.; Utimoto, K. *J. Org. Chem.* **1990**, *55*, 1707.

[684]Reetz, M.T.; Kükenhöhner, T.; Weinig, P. *Tetrahedron Lett.* **1986**, *27*, 5711; Wang, J.; Fan, X.; Feng, X.; Quian, Y. *Synthesis* **1989**, 291; Riediker, M.; Duthaler, R.O. *Angew. Chem. Int. Ed.* **1989**, *28*, 494; Riediker, M.; Hafner, A.; Piantini, U.; Rihs, G.; Togni, A. *Angew. Chem. Int. Ed.* **1989**, *30*, 499.

[685]For a discussion of transition states in the amino alcohol promoted reaction, see Rasmussen, T.; Norrby, P.-O. *J. Am. Chem. Soc.* **2001**, *123*, 2464.

[686]For catalysts that are also successful for aliphatic aldehydes, see Takahashi, H.; Kawakita, T.; Yoshioka, M.; Kobayashi, S.; Ohno, M. *Tetrahedron Lett.* **1989**, *30*, 7095; Tanaka, K.; Ushio, H.; Suzuki, H. *J. Chem. Soc., Chem. Commun.* **1989**, 1700; Soai, K.; Yokoyam, S.; Hayasaka, T. *J. Org. Chem.* **1991**, *56*, 4264.

amount of various catalysts.[687] The enantioselectivity is influenced by additives, such as LiCl.[688] Silica-immobilized chiral ligands[689] can be used in conjunction with dialkylzinc reagents, and polymer-supported ligands have been used.[690]. Chiral dendritic titanium catalysts have been used to give moderate enantioselectivity.[691]

Enantioselective reaction of a carbonyl with a dialkylzinc is possible when other functional groups are present in the molecule. Examples include keto esters.[692] An enzyme-mediated addition of dialkylzinc reagents to aldehydes has also been reported.[693] When benzaldehyde was treated with Et_2Zn in the presence of the optically active catalyst 1-piperidino-3,3-dimethyl-2-butanol, a surprising result was obtained. Although the catalyst had only 10.7% excess of one enantiomer, the

[687]For a review, see Pu, L.; Yu, H.-B. *Chem. Rev.* **2001**, *101*, 757. See also, Richmond, M.I.; Seto, C.T. *J. Org. Chem.* **2003**, *68*, 7505; Sprout, C.M.; Seto, C.T. *J. Org. Chem.* **2003**, *68*, 7788; Nugent, W.A. *Org. Lett.* **2002**, *4*, 2133; García, C.; Walsh, P.J. *Org. Lett.* **2003**, *5*, 3641; Kang, S.-W.; Ko, D.-H.; Kim, K.H.; Ha, D.-C. *Org. Lett.* **2003**, *5*, 4517; Mao, J.; Wan, B.; Wang, R.; Wu, F.; Lu, S. *J. Org. Chem.* **2004**, *69*, 9123; Superchi, S.; Giorgio, E.; Scafato, P.; Rosini, C. *Tetrahedron Asymmetry* **2002**, *13*, 1385; Bastin, S.; Ginj, M.; Brocarrd, J.; Pélinski, L.; Novogrocki, G. *Tetrahedron Asymmetry* **2003**, *14*, 1701; Joshi, S.N.; Malhotra, S.V. *Tetrahedron Asymmetry* **2003**, *14*, 1763; Prieto, O.; Ramón, D.J.; Yus, M. *Tetrahedron Asymmetry* **2003**, *14*, 1955; Lesma, G.; Danieli, B.; Passarella, D.; Sacchetti, A.; Silovani, A. *Tetrahedron Asymmetry* **2003**, *14*, 2453; Danilova, T.I.; Rozenberg, V.I.; Starikova, Z.A.; Bräse, S. *Tetrahedron Asymmetry* **2004**, *15*, 223; Scarpi, D.; Lo Galbo, F.; Occhiato, E.G.; Guarna, A. *Tetrahedron Asymmetry* **2004**, *15*, 1319; Funabashi, K.; Jachmann, M.; Kanai, M.; Shibasaki, M. *Angew. Chem. Int. Ed.* **2003**, *42*, 5489; Lake, F.; Moberg, C. *Eur. J. Org. Chem.* **2002**, 3179; Reddy, K.S.; Solà, L.; Huang, W.-S.; Pu, L. *J. Org. Chem.* **1999**, *64*, 4222; Paleo, M.R.; Cabeza, I.; Sardina, F.J. *J. Org. Chem.* **2000**, *65*, 2108; Yang, L.; Shen, J.; Da, C.; Wang, R.; Choi, M.C.K.; Yang, L.; Wong, K.-y. *Tetrahedron Asymmetry* **1999**, *10*, 133; Shi, M.; Jiang, J.-K. *Tetrahedron Asymmetry* **1999**, *10*, 1673; Kawanami, Y.; Mitsuie, T.; Miki, M.; Sakamoto, T.; Nishitani, K. *Tetrahedron* **2000**, *56*, 175; Hanyu, N.; Aoki, T.; Mino, T.; Sakamoto, M.; Fujita, T. *Tetrahedron Asymmetry* **2000**, *11*, 2971; Arroyo, N.; Haslinger, U.; Mereiter, K.; Widhalm, M. *Tetrahedron Asymmetry* **2000**, *11*, 4207; Jimeno, C.; Moyano, A.; Pericàs, M.A.; Riera, A. *Synlett* **2001**, 1155; Hermsen, P.J.; Cremers, J.G.O.; Thijs, L.; Zwanenburg, B. *Tetrahedron Lett.* **2001**, *42*, 4243; Xu, Q.; Wang, H.; Pan, X.; Chan, A.S.C.; Yang, T.-k. *Tetrahedron Lett.* **2001**, *42*, 6171; Yang, X.-w.; Shen, J.-h.; Da, C.-s.; Wang, H.-s.; Su, W.; Liu, D.-x.; Wang, R.; Choi, M.C.K.; Chan, A.S.C. *Tetrahedron Lett.* **2001**, *42*, 6573; Ohga, T.; Umeda, S.; Kawanami, Y. *Tetrahedron* **2001**, *57*, 4825; Zhao, G.; Li, X.-G.; Wang, X.-R. *Tetrahedron Asymmetry* **2001**, *12*, 399; You, J.-S.; Shao, M.-Y.; Gau, H.-M. *Tetrahedron Asymmetry* **2001**, *12*, 2971; Priego, J.; Mancheño, O.G.; Cabrera, S.; Carretero, J.C. *Chem. Commun.* **2001**, 2026; Bolm, C.; Kesselgruber, M; Hermanns, N.; Hildebrand, J.P.; Raabe, G. *Angew. Chem. Int. Ed.* **2001**, *40*, 1488; Wipf, P.; Wang, X. *Org. Lett.* **2002**, *4*, 1197; Braga, A.L.; Milani, P.; Paixão, M.W.; Zeni, G.; Rodrigues, O.E.D.; Alves, E.F. *Chem. Commun.* **2004**, 2488; Sibi, M.P.; Stanley, L.M. *Tetrahedron Asymmetry* **2004**, *15*, 3353; Tseng, S.-L.; Yang, T.-K. *Tetrahedron Asymmetry* **2004**, *15*, 3375; Harada, T.; Kanda, K.; Hiraoka, Y.; Marutani, Y.; Nakatsugawa, M. *Tetrahedron Asymmetry* **2004**, *15*, 3879; Casey, M.; Smyth, M.P. *Synlett* **2003**, 102.
[688]Sosa-Rivadeneyra, M.; Muñoz-Muñiz, O.; de Parrodi, C.A.; Quintero, L.; Juaristi, E. *J. Org. Chem.* **2003**, *68*, 2369; García, C.; La Rochelle, L.K.; Walsh, P.J. *J. Am. Chem. Soc.* **2002**, *124*, 10970; Priego, J.; Mancheño, O.G.; Cabrera, S.; Carretero, J.C. *J. Org. Chem.* **2002**, *67*, 1346.
[689]Fraile, J.M.; Mayoral, J.A.; Servano, J.; Pericàs, M.A.; Solà, L.; Castellnou, D. *Org. Lett.* **2003**, *5*, 4333.
[690]Sung, D.W.L.; Hodge, P.; Stratford, P.W. *J. Chem. Soc., Perkin Trans. 1* **1999**, 1463; Lipshutz, B.H.; Shin, Y.-J. *Tetrahedron Lett.* **2000**, *41*, 9515.
[691]Fan, O.-H.; Liu, G.-H.; Chen, X.-M.; Deng, G.-J.; Chan, A.S.C. *Tetrahedron Asymmetry* **2001**, *12*, 1559.
[692]DiMauro, E.F.; Kozlowski, M.C. *Org. Lett.* **2002**, *4*, 3781.
[693]Yang, W.K.; Cho, B.T. *Tetrahedron Asymmetry* **2000**, *11*, 2947.

product PhCH(OH)Me had an ee of 82%.[694] When the catalyst ee was increased to 20.5%, the product ee rose to 88%. The question is, how could a catalyst produce a product with an ee much higher than itself? One possible explanation[695] is that (R) and (S) molecules of the catalyst form a complex with each other, and that only the uncomplexed molecules are actually involved in the reaction. Since initially the number of (R) and (S) molecules was not the same, the (R/S) ratio of the uncomplexed molecules must be considerably higher (or lower) than that of the initial mixture.

Although organoboranes do not generally add to aldehydes and ketones,[696] allylic boranes are exceptions.[697] When they add, an allylic rearrangement always takes place. Allylic rearrangements take place with the other reagents as well. The use of a chiral catalyst leads to asymmetric induction[698] and chiral allylic boranes have been prepared.[699] It is noted that chloroboranes (R_2BCl) react with aldehydes via acyl addition of the alkyl group, giving the corresponding alcohol after treatment with water.[700] A variation is the reaction of a diketone, where one carbonyl is conjugated. Treatment with catecholborane gives addition to the conjugated ketone, and subsequent cyclization of the resulting organometallic at the nonconjugated ketone gives a cyclic alcohol with a pendant ketone unit, after treatment with methanol.[701] In the presence of ruthenium complexes, $RB(OH)_2$ and arylboronic acids $ArB(OH)_2$ (p. 815) add to aldehydes to give the corresponding alcohol.[702] Polymer-bound aryl borates add an aryl group to aldehydes in the presence of a rhodium catalyst.[703] An intramolecular version of the phenylboronic acid-induced reaction is known, where a molecule with ketone and conjugated ketone units is converted to a cyclic alcohol using a chiral rhodium catalyst.[704] Allylic boronates add to aldehydes.[705]

---

[694]Oguni, N.; Matsuda, Y.; Kaneko, T. *J. Am. Chem. Soc.* **1988**, *110*, 7877.

[695]See Wynberg, H. *Chimia* **1989**, *43*, 150.

[696]For another exception, involving a vinylic borane, see Satoh, Y.; Tayano, T.; Hara, S.; Suzuki, A. *Tetrahedron Lett.* **1989**, *30*, 5153.

[697]For reviews, see Hoffmann, R.W.; Niel, G.; Schlapbach, A. *Pure Appl. Chem.* **1990**, *62*, 1993; Pelter, A.; Smith, K.; Brown, H.C. *Borane Reagents*, Academic Press, NY, **1988**, pp. 310–318. For a review of allylic boranes, see Bubnov, Yu.N. *Pure Appl. Chem.* **1987**, *21*, 895. For an example that proceeds with asymmetric induction, see Buynak, J.D.; Geng, B.; Uang, S.; Strickland, J.B. *Tetrahedron Lett.* **1994**, *35*, 985.

[698]Ishiyama, T.; Ahiko, T.-a.; Miyaura, N. *J. Am. Chem. Soc.* **2002**, *124*, 12414; Wada, R.; Oisaki, K.; Kanai, M.; Shibasaki, M. *J. Am. Chem. Soc.* **2004**, *126*, 8910.

[699]Lachance, H.; Lu, X.; Gravel, M.; Hall, D.G. *J. Am. Chem. Soc.* **2003**, *125*, 10160; Ramachandran, P.V.; Krezeminski, M.P.; Reddy, M.V.R.; Brown, H.C. *Tetrahedron Asymmetry* **1999**, *10*, 11; Roush, W.R.; Pinchuk, A.N.; Micalizio, G.C. *Tetrahedron Lett.* **2000**, *41*, 9413; Wu, T.R.; Shen, L.; Chong, J.M. *Org. Lett.* **2004**, *6*, 2701.

[700]Kabalka, G.W.; Wu, Z.; Ju, Y. *Tetrahedron* **2001**, *57*, 1663; Kabalka, G.W.; Wu, Z.; Ju, Y. *Tetrahedron* **2002**, *58*, 3243.

[701]Huddleston, R.R.; Cauble, D.F.; Krische, M.J. *J. Org. Chem.* **2003**, *68*, 11.

[702]Ueda, M.; Miyaura, N. *J. Org. Chem.* **2000**, *65*, 4450 and references cited therein.

[703]Pourbaix, C.; Carreaux, F.; Carboni, B. *Org. Lett.* **2001**, *3*, 803; Rudolph, J.; Schmidt, F.; Bolm, C. *Synthesis* **2005**, 840.

[704]Cauble, D.F.; Gipson, J.D.; Krische, M.J. *J. Am. Chem. Soc.* **2003**, *125*, 1110.

[705]For a review of the scandium-catalyzed enantioselective addition, see Gravel, M.; Lachance, H.; Lu, X.; Hall, D.G. *Synthesis* **2004**, 1290.

A number of optically active allylic boron compounds have been used, including[706] B-allylbis(2-isocaranyl)borane (**30**),[707] (*E*)- and (*Z*)-crotyl-(*R,R*)-2,5-dimethylborolanes (**31**),[708] and the borneol

30   **31**   **32**

derivative **32**,[709] all of which add an allyl group to aldehydes, with good enantioselectivity. Where the substrate possesses an aryl group or a triple bond, enantioselectivity is enhanced by using a metal carbonyl complex of the substrate.[710]

Alkenes and alkynes add to aldehydes or ketones by conversion to a reactive organometallic. A radical-type addition is possible using alkenes with BEt_3. Benzaldehyde reacted with isoprene in the presence of BEt_3 and $Ni(acac)_2$ to give an anti-Markovnikov-type addition to the carbonyl, $C=C-C(Me)=C + PhCHO \rightarrow C=CCHMeCH_2CH(OH)Ph$.[711] Alkynes add to aldehydes elsewhere in the same molecule in the presence of BEt_3 and a nickel catalyst to give a cyclic allylic alcohol.[712] Alkene aldehydes react similarly using Me_3SiOTf.[713] In a similar manner, dienes add to aldehydes in the presence of a nickel catalyst.[714] Propargylic halides add to aldehydes to give an allenic alcohol using β-$SnO[Rd(cod)Cl]_2$.[715] Allylic acetates react with ketones to give the homoallylic alcohol under electrochemical conditions that include bipyridyl, tetrabutylammonium tetrafluoroborate and $FeBr_2$.[716] Terminal alkynes react with zirconium complexes and Me_2Zn to give an allylic tertiary alcohol.[717] Internal alkynes also give allylic alcohols in the presence of BEt_3 and a nickel catalysts.[718] Reaction of an aldehyde containing a conjugated diene unit with diethylzinc and a nickel catalyst leads to cyclic

[706]For some others, see Hoffmann, R.W. *Pure Appl. Chem.* **1988**, *60*, 123; Corey, E.J.; Yu, C.; Kim, S.S. *J. Am. Chem. Soc.* **1989**, *111*, 5495; Roush, W.R.; Ando, K.; Powers, D.B.; Palkowitz, A.D.; Halterman, R.L. *J. Am. Chem. Soc.* **1990**, *112*, 6339; Brown, H.C.; Randad, R.S. *Tetrahedron Lett.* **1990**, *31*, 455; Stürmer, R.; Hoffmann, R.W. *Synlett* **1990**, 759.

[707]Racherla, U.S.; Brown, H.C. *J. Org. Chem.* **1991**, *56*, 401, and references cited therein.

[708]Garcia, J.; Kim, B.M.; Masamune, S. *J. Org. Chem.* **1987**, *52*, 4831.

[709]Reetz, M.T.; Zierke, T. *Chem. Ind. (London)* **1988**, 663.

[710]Roush, W.R.; Park, J.C. *J. Org. Chem.* **1990**, *55*, 1143.

[711]Kimura, M.; Fujimatsu, H.; Ezoe, A.; Shibata, K.; Shimizu, M.; Matsumoto, S.; Tamaru, Y. *Angew. Chem. Int. Ed.* **1999**, *38*, 397.

[712]Miller, K.M.; Huang, W.-S.; Jamison, T.F. *J. Am. Chem. Soc.* **2003**, *125*, 3442; Miller, K.M.; Luanphaisarnnont, T.; Molinaro, C.; Jamison, T.F. *J. Am. Chem. Soc.* **2004**, *126*, 4130.

[713]Suginome, M.; Iwanami, T.; Yamamoto, A.; Ito, Y. *Synlett* **2001**, 1042.

[714]Sawaki, R.; Sato, Y.; Mori, M. *Org. Lett.* **2004**, *6*, 1131.

[715]Banerjee, M.; Roy, S. *Org. Lett.* **2004**, *6*, 2137.

[716]Durandetti, M.; Meignein, C.; Périchon, J. *J. Org. Chem.* **2003**, *68*, 3121.

[717]Li, H.; Walsh, P.J. *J. Am. Chem. Soc.* **2004**, *126*, 6538.

[718]Miller, K.M.; Jamison, T.F. *J. Am. Chem. Soc.* **2004**, *126*, 15342.

alcohols having a pendant allylic unit.[719] A similar reaction was reported using a copper catalyst.[720] Vinyl iodides also react with Cp_2ZrCl_2 to give a vinylzirconium complex that reacts with aldehydes.[721] The intramolecular addition of an alkene to an aldehyde leads to a saturated cyclic alcohol using $PhSiH_3$ and a cobalt catalyst.[722] Intramolecular addition of a conjugated ester (via the β-carbon) to an aldehyde generates a cyclic ketone.[723] This type of coupling has been called the *Stetter reaction*,[724] which actually involves the addition of aldehydes to activated double bonds (**15-34**), mediated by a catalytic amount of thiazolium salt in the presence of a weak base. The intramolecular addition of the allene moiety to an aldehyde is catalyzed by a palladium complex in the presence of Me_3SiSn-Bu_3.[725] A highly enantio- and diastereoselective intramolecular Stetter reaction has been developed.[726] Alkynyl aldehydes react with silanes such as Et_3SiH and a nickel catalyst to give a cyclic compound having a silyl ether and an exocyclic vinylidene unit.[727] Alkene-aldehydes give cyclic alcohols via intramolecular addition of the C=C unit to the carbonyl under electrolytic conditions using a phase-transfer catalyst.[728] A similar cyclization was reported using $SnCl_4$.[729] Vinylidene cycloalkanes react with aldehydes in the presence of a palladium catalyst to give a homoallylic alcohol where addition occurs at the carbon exocyclic to the ring.[730] Alkenes having an allylic hydrogen react with α-keto aldehydes, with a cobalt catalyst, to give α-hydroxy ketones where the alcohol is homoallylic relative to the C=C unit.[731] Allenes react with benzaldehyde using HCl–$SnCl_2$ with a palladium catalyst.[732] Silyl allenes react with aldehydes in the presence of a chiral scandium catalyst to give homopropargylic alcohols with good enantioselectivity.[733] Intramolecular addition of an allene to aldehyde via addition of phenyl when treated with PhI and a palladium catalyst.[734] Allenes add to ketones

[719]Shibata, K.; Kimura, M.; Shimizu, M.; Tamaru, Y. *Org. Lett.* **2001**, *3*, 2181.

[720]Agapiou, K.; Cauble, D.F.; Krische, M.J. *J. Am. Chem. Soc.* **2004**, *126*, 4528.

[721]Namba, K.; Kishi, Y. *Org. Lett.* **2004**, *6*, 5031.

[722]Baik, T.-G.; Luis, A.L.; Wang, L.-C.; Krische, M.J. *J. Am. Chem. Soc.* **2001**, *123*, 5112.

[723]Kerr, M.S.; Rovis, T. *J. Am. Chem. Soc.* **2004**, *126*, 8876.

[724]Stetter, H.; Schreckenberg, M. *Angew. Chem., Int. Ed* **1973**, *12*, 81; Stetter, H. *Angew. Chem., Int. Ed.* **1976**, *15*, 639; Stetter, H.; Kuhlmann, H. *Org. React.* **1991**, *40*,407; Enders, D.; Breuer, K.; Runsink, J.; Teles, J.H. *Helv. Chim. Acta* **1996**, *79*, 1899; Kerr, M.S.; Rovis, T. *J. Am. Chem. Soc.* **2004**, *126*, 8876; Pesch, J.; Harms, K.; Bach, T. *Eur. J. Org. Chem.* **2004**, 2025; Mennen, S.; Blank, J.; Tran-Dube, M.B.; Imbriglio, J.E.; Miller, S.J. *Chem. Commun.* **2005**, 195. For examples of the Stetter reaction with acyl silanes, see Mattson, A.E.; Bharadwaj, A.R.; Scheidt, K.A. *J. Am. Chem. Soc.* **2004**, *126*, 2314.

[725]Kang, S.-K.; Ha, Y.-H.; Ko, B.-S.; Lim, Y.; Jung, J. *Angew. Chem. Int. Ed.* **2002**, *41*, 343.

[726]Read de Alaniz, J.; Rovis, T. *J. Am. Chem. Soc.* **2005**, *127*, 6284; Kerr, M.S.; Read de Alaniz, J.; Rovis, T. *J. Am. Chem. Soc.* **2002**, *124*, 10298.

[727]Tang, X.-Q.; Montgomery, J. *J. Am. Chem. Soc.* **1999**, *121*, 6098.

[728]Locher, C.; Peerzada, N. *J. Chem. Soc., Perkin Trans. 1* **1999**, 179.

[729]Alcaide, B.; Pardo, C.; Rodríguez-Ranera, C.; Rodríguez-Vicente, A. *Org. Lett.* **2001**, *3*, 4205.

[730]Hao, J.; Hatano, M.; Mikami, K. *Org. Lett.* **2000**, *2*, 4059.

[731]Kezuka, S.; Ikeno, T.; Yamada, T. *Org. Lett.* **2001**, *3*, 1937.

[732]Chang, H.-M.; Cheng, C.-H. *Org. Lett.* **2000**, *2*, 3439.

[733]Evans, D.A.; Sweeney, Z.K.; Rovis, T.; Tedrow, J.S. *J. Am. Chem. Soc.* **2001**, *123*, 12095.

[734]Kang, S.-K.; Lee, S.-W.; Jung, J.; Lim, Y. *J. Org. Chem.* **2002**, *67*, 4376.

to give homoallylic alcohols in the presence of SmI_2 and HMPA.[735] Allenes add to carbonyl groups in the presence of 2.2 equivalents of SmI_2 and an excess of HMPA.[736] Alkenes have an allylic methyl group add to formaldehyde, in the presence of $BF_3 \cdot OEt_2$, to give a homoallylic alcohol.[737] Conjugated dienes react with aldehyde via acyl addition of a terminal carbon of the diene, in the presence of $Ni(acac)_2$ and Et_2Zn.[738] Aldehydes having an allylic acetate unit elsewhere in the molecule undergo cyclization in CO and a ruthenium catalyst to give a cyclic alcohol with a pendant vinyl group.[739]

Allylic trifluoroborates (p. 817) react with aldehydes to give the homoallylic alcohol. Pivaldehyde reacts with potassium 2-butenyltrifluoroborate and a catalytic amount of tetrabutylammonium iodide to give 2,2,4-trimethylhex-5-en-3-ol.[740] Aliphatic aldehydes react with this reagent, in the presence of $BF_3 \cdot OEt_2$, to give the homoallylic alcohol with allylic rearrangement and a preference for the syn diastereomer,[741] and aryl aldehydes react as well.[742]

16-26   Addition of Trialkylallylsilanes to Aldehydes and Ketones

O-**Hydro-***C*-**alkyl-addition**

Allylic trialkyl, trialkoxy, and trihalosilanes add to aldehydes to give the homoallylic alcohols in the presence of a Lewis acid[743] (including $TaCl_5$[744] and $YbCl_3$[745]), Me_3SiOTf,[746] fluoride ion,[747] proazaphosphatranes,[748] or a catalytic amount of iodine.[749] The mechanism of this reaction has been examined.[750] A ruthenium catalyst has also been used in conjunction with an arylsilane and an aldehyde.[751]

[735]Hölemann, A.; Reissig, H.-U. *Org. Lett.* **2003**, *5*, 1463.

[736]Hölemann, A.; Reißig, H.-U. *Chem. Eur. J.* **2004**, *10*, 5493.

[737]Okachi, T.; Fujimoto, K.; Onaka, M. *Org. Lett.* **2002**, *4*, 1667.

[738]Loh, T.-P.; Song, H.-Y.; Zhou, Y. *Org. Lett.* **2002**, *4*, 2715.

[739]Yu, C.-M.; Lee, S.; Hong, Y.-T.; Yoon, S.-K. *Tetrahedron Lett.* **2004**, *45*, 6557.

[740]Thadani, A.N.; Batey, R.A. *Org. Lett.* **2002**, *4*, 3827.

[741]Batey, R.A.; Thadani, A.N.; Smil, D.V. *Tetrahedron Lett.* **1999**, *40*, 4289.

[742]Batey, R.A.; Thadani, A.N.; Smil, D.V.; Lough, A.J. *Synthesis* **2000**, 990.

[743]For reviews, see Fleming, I.; Dunoguès, J.; Smithers, R. *Org. React.* **1989**, *37*, 57, 113–125, 290–328; Parnes, Z.N.; Bolestova, G.I. *Synthesis* **1984**, 991, see pp. 997–1000. For studies of the mechanism, see Denmark, S.E.; Weber, E.J.; Wilson, T.; Willson, T.M. *Tetrahedron* **1989**, *45*, 1053; Keck, G.E.; Andrus, M.B.; Castellino, S. *J. Am. Chem. Soc.* **1989**, *111*, 8136. For examples, see Aggarwal, V.K.; Vennall, G.P. *Tetrahedron Lett.* **1996**, *37*, 3745.

[744]Chandrasekhar, S.; Mohanty, P.K.; Raza, A. *Synth. Commun.* **1999**, *29*, 257.

[745]Fang, X.; Watkin, J.G.; Warner, B.P. *Tetrahedron Lett.* **2000**, *41*, 447.

[746]Davis, A.P.; Jaspars, M. *Angew. Chem. Int. Ed.* **1992**, *31*, 470.

[747]See Cossy, J.; Lutz, F.; Alauze, V.; Meyer, C. *Synlett* **2002**, 45.

[748]Wang, Z.; Kisanga, P.; Verkade, J.G. *J. Org. Chem.* **1999**, *64*, 6459.

[749]Yadav, J.S.; Chand, P.K.; Anjaneyulu, S. *Tetrahedron Lett.* **2002**, *43*, 3783.

[750]Bottoni, A.; Costa, A.L.; DiTommaso, D.; Rossi, I.; Tagliavini, E. *J. Am. Chem. Soc.* **1997**, *119*, 12131.

[751]Fujii, T.; Koike, T.; Mori, A.; Osakada, K. *Synlett* **2002**, 298.

The reaction of an allene aldehyde with Et_3SiH, CO and a rhodium catalyst leads to addition to the alkene followed by intramolecular addition to the aldehyde to give the cyclic alcohol.[752] Allyl(trimethoxy)silane adds an allyl group to aldehydes using a CdF_2[753] catalyst or a chiral AgF complex.[754] Allyltrichlorosilanes have also been used in addition reactions with aldehydes.[755] Hünig's base (iPr_2NEt) and a sulfoxide have also been used to facilitate the addition of an allyl group to an aldehyde from allyltrichlorosilane.[756]

Allyltrichlorosilane reacts with benzaldehyde in the presence of Bu_4NF to give 1-phenylbut-3-en-1ol,[757] and with a chiral additive the reaction proceeds with good enantioselectivity. When chiral titanium complexes are used in the reaction, allylic alcohols are produced with good asymmetric induction.[758] Other chiral additives have been used,[759] as well as chiral catalysts,[760] and chiral complexes of allyl silanes.[761] Chiral allylic silyl derivatives add to aldehydes to give the chiral homo-allylic alcohol.[762]

Allylic silanes react with *gem*-diacetates in the presence of $InCl_3$ to give a homo-allylic acetate[763] or with dimethyl acetals and TMSOTf in an ionic liquid to give the homoallylic methyl ether.[764] Allylic alcohols can be treated with TMS—Cl and NaI, and then Bi to give an organometallic reagent that adds to aldehydes.[765]

16-27   Addition of Conjugated Alkenes to Aldehydes (the Baylis–Hillman Reaction)[766]

O-**Hydro-*C*-alkenyl-addition**

33

[752]Kang, S.-K.; Hong, Y.-T.; Lee, J.-H.; Kim, W.-Y.; Lee. I.; Yu, C.-M. *Org. Lett.* **2003**, *5*, 2813.

[753]Aoyama, N.; Hamada, T.; Manabe, K.; Kobayashi, S. *Chem. Commun.* **2003**, 676.

[754]Yanagisawa, A.; Kageyama, H.; Nakatsuka, Y.; Asakawa, K.; Matsumoto, Y.; Yamamoto, H. *Angew. Chem. Int. Ed.* **1999**, *38*, 3701.

[755]Kobayashi, S.; Nishio, K. *J. Org. Chem.* **1994**, *59*, 6620.

[756]Massa, A.; Malkov, A.V.; Kočovský, P.; Scettri, A. *Tetrahedron Lett.* **2003**, *44*, 7179.

[757]Malkov, A.V.; Bell, M.; Orsini, M.; Pernazza, D.; Massa, A.; Herrmann, P.; Meghani, P.; Kočovský, P. *J. Org. Chem.* **2003**, *68*, 9659.

[758]Gauthier Jr., D.R.; Carreira, E.M. *Angew. Chem. Int. Ed.* **1996**, *35*, 2363.

[759]Angell, R.M.; Barrett, A.G.M.; Braddock, D.C.; Swallow, S.; Vickery, B.D. *Chem. Commun.* **1997**, 919.

[760]Malkov, A.V.; Dufková, L.; Farrugia, L.; Kočovský, P. *Angew. Chem Int. Ed.* **2003**, *42*, 3802; Bode, J.W.; Gauthier, Jr., D.R.; Carreira, E.M. *Chem. Commun.* **2001**, 2560.

[761]Zhang, L.C.; Sakurai, H.; Kira, M. *Chem. Commun.* **1997**, 129; Iseki, K.; Mizuno, S.; Kuroki, Y.; Kobayashi, Y. *Tetrahedron* **1999**, *55*, 977

[762]Wang, X.; Meng, Q.; Nation, A.J.; Leighton, J.L. *J. Am. Chem. Soc.* **2002**, *124*, 10672; Kubota, K.;l Leighton, J.L. *Angew. Chem. Int. Ed.* **2003**, *42*, 946; Hackman, B.M.; Lombardi, P.J.; Leighton, J.L. *Org. Lett.* **2004**, *6*, 4375.

[763]Yadav, J.S.; Reddy, B.V.S.; Madhuri, Ch.; Sabitha, G. *Chem. Lett.* **2001**, 18.

[764]In bmim PF_6, 1-butyl-3-methylimidazolium hexafluorophosphate: Zerth, H.M.; Leonard, N.M.; Mohan, R.S. *Org. Lett.* **2003**, *5*, 55.

[765]Miyoshi, N.; Nishio, M.; Murakami, S.; Fukuma, T.; Wada, M. *Bull. Chem. Soc. Jpn.* **2000**, *73*, 689.

[766]For a review, see Basavaih, D.; Rao, A.J.; Satyanarayana, T. *Chem. Rev.* **2003**, *103*, 811.

In the presence of a base,[767] such as 1,4-diazabicyclo[2.2.2]octane (DABCO) or trialkylphosphines, conjugated carbonyl compounds (ketones, esters,[768] thioesters,[769] and amides[770]) add to aldehydes via the α-carbon to give α-alkenyl-β-hydroxy esters or amides. This sequence is called the *Baylis–Hillman reaction*,[771] and a simple example is the formation of **33**.[772] It was observed that methyl vinyl ketone gave other products in the Baylis–Hillman reaction, whereas conjugated esters did not.[773] Methods that are catalytic in base have been developed for the Baylis–Hillman reaction.[774] Both microwave irradiation[775] and ultrasound[776] have been used to induce the reaction. Under certain conditions, rate enhancements have been observed.[777] Rate acceleration occurs with bis-aryl(thio)ureas in a DABCO-promoted reaction.[778] The reaction has been done in ionic liquids[779] and polyethylene glycol (PEG)[780] or sulpholane.[781] Alkynyl carbonyl compounds can be used as partners in the Baylis–Hillman reaction.[782] Transition metal compounds can facilitate the Baylis–Hillman reaction, and BF_3•OEt_2 has been used.[783] With the boron trifluoride

[767]For an example using NaOMe, see Luo, S.; Mi, X.; Xu, H.; Wang, P.G.; Cheng, J.-P. *J. Org. Chem.* **2004**, *69*, 8413.

[768]For an example with a conjugated lactone, see Karur, S.; Hardin, J.; Headley, A.; Li, G. *Tetrahedron Lett.* **2003**, *44*, 2991.

[769]Pei, W.; Wei, H.-X.; Li, G. *Chem. Commun.* **2002**, 1856.

[770]See Yu, C.; Hu, L. *J. Org. Chem.* **2002**, *67*, 219; Faltin, C.; Fleming, E.M.; Connon, S.J. *J. Org. Chem.* **2004**, *69*, 6496.

[771]Baylis, A.B.; Hillman, M.E.D. Ger. Offen. 2,155,133 *Chem. Abstr.*, **1972**, *77*, 34174q [U.S. Patent 3,743,668]; Drewes, S.E.; Roos, G.H.P. *Tetrahedron* **1988**, *44*, 4653. For a review, see Basavaiah, D.; Rao, P.D.; Hyma, R.S. *Tetrahedron* **1996**, *52*, 8001.

[772]Rafel, S.; Leahy, J.W. *J. Org. Chem.* **1997**, *62*, 1521. Also see, Drewes, S.E.; Rohwer, M.B. *Synth. Commun.* **1997**, *27*, 415.

[773]Shi, M.; Li, C.-Q.; Jiang, J.-K. *Chem. Commun.* **2001**, 833.

[774]With imidazole: Gatri, R.; El Gaïed, M.M. *Tetrahedron Lett.* **2002**, *43*, 7835. With azoles: Luo, S.; Mi, X.; Wang, P.G.; Cheng, J.-P. *Tetrahedron Lett.* **2004**, *45*, 5171. With proazaphosphatranes/TiCl₄: You, J.; Xu, J.; Verkade, J.G. *Angew. Chem. Int. Ed.* **2003**, *42*, 5054. See Leadbeater, N.E.; van der Pol, C. *J. Chem. Soc., Perkin Trans. 1* **2001**, 2831. For a discussion of pK_a and reactivity, see Aggarwal, V.K.; Emme, I.; Fulford, S.Y. *J. Org. Chem.* **2003**, *68*, 692.

[775]Kundu, M.K.; Mukherjee, S.B.; Balu, N.; Padmakumar, R.; Bhat, S.V. *Synlett* **1994**, 444.

[776]Coelho, F.; Almeida, W.P.; Veronese, D.; Mateus, C.R.; Lopes, E.C.S.; Rossi, R.C.; Silveira, G.P.C.; Pavam, C.H. *Tetrahedron* **2002**, *58*, 7437.

[777]See Rafel, S.; Leahy, J.W. *J. Org. Chem.* **1997**, *62*, 1521; Lee, W.-D.; Yang, K.-S.; Chen, K. *Chem. Commun.* **2001**, 1612. For rate acceleration in water or aqueous media, see Augé, J.; Lubin, N.; Lubineau, A. *Tetrahedron Lett.* **1994**, *35*, 7947; Luo, S.; Wang, P.G.; Cheng, J.-P. *J. Org. Chem.* **2004**, *69*, 555; Cai, J.; Zhou, Z.; Zhao, G.; Tang, C. *Org. Lett.* **2002**, *4*, 4723. For rate acceleration in polar solvents, see Aggarwal, V.K.; Dean, D.K.; Mereu, A.; Williams, R. *J. Org. Chem.* **2002**, *67*, 510. For a discussion of salt effects, see Kumar, A.; Pawar, S.S. *Tetrahedron* **2003**, *59*, 5019.

[778]Maher, D.J.; Connon, S.J. *Tetrahedron Lett.* **2004**, *45*, 1301.

[779]Rosa, J.N.; Afonso, C.A.M.; Santos, A.G. *Tetrahedron* **2001**, *57*, 4189. For an example in a chiral ionic liquid, see Pégot, B.; Vo-Thanh, G.; Gori, D.; Loupy, A. *Tetrahedron Lett.* **2004**, *45*, 6425.

[780]Chandrasekhar, S.; Narsihmulu, Ch.; Saritha, B.; Sultana, S.S. *Tetrahedron Lett.* **2004**, *45*, 5865.

[781]Krishna, P.R.; Manjuvani, A.; Kannan, V.; Sharma, G.V.M. *Tetrahedron Lett.* **2004**, *45*, 1183.

[782]Matsuya, Y.; Hayashi, K.; Nemoto, H. *J. Am. Chem. Soc.* **2003**, *125*, 646; Wei, H.-X.; Jasoni, R.L.; Hu, J.; Li, G.; Paré, P.W. *Tetrahedron* **2004**, *60*, 10233; Shi, M.; Wang, C.-J. *Helv. Chim. Acta* **2002**, *85*, 841.

[783]Walsh, L.M.; Winn, C.L.; Goodman, J.M. *Tetrahedron Lett.* **2002**, *43*, 8219.

induced reaction between an aldehyde and a conjugated ketone, a saturated β-hydroxy ketone was formed with good antiselectivity.[784] The coupling of aldehydes with conjugated ketones was accomplished with TiCl$_4$,[785] dialkylaluminum halides,[786] and with (polymethyl)hydrosiloxane and a copper catalyst.[787] Conjugated esters were coupled to aldehydes with DABCO and a lanthanum catalyst.[788] Aldehydes were coupled to conjugated nitriles with TiCl$_4$.[789] The reaction of a conjugated ester, an aldehyde and LiClO$_4$, with 15% DABCO gave the allylic alcohol product.[790] N-Tosyl imines can be used in place of aldehydes, and the reaction of the imine, a conjugated ester and DABCO gave the allylic N-tosylimine.[791] Aldehydes are coupled to conjugated esters with a chiral quinuclidine catalyst and a titanium catalyst, and in the presence of tosylamine, the final product was the allylic N-tosylamine formed with modest enantioselectivity.[792]

An intramolecular version of the Baylis–Hillman reaction generated cyclopentenone derivatives from alkyne-aldehydes and a rhodium catalyst.[793] Another intramolecular reaction gave cyclopentenols via cyclization of an aldehyde-conjugated thioester upon treatment with DBU and DMAP.[794] Cyclization of a conjugated ester using DABCO, where the "alcohol" group contained an aldehyde unit (an α-hydroxy aldehyde derivative) gave a lactone with an hydroxy unit at C3 relative to the carbonyl and an α-vinylidene.[795] A "double Baylis–Hillman" reaction has also been reported using N-tosylimines and conjugated ketones.[796]

Using a chiral auxiliary via an amide[797] or ester[798] leads to asymmetric induction.[799] Aryl aldehydes and conjugated ketones were condensed using proline, leading to modest enantioselectivity.[800] Chiral biaryl catalysts have been used with trialkylphosphines, giving good enantioselectivity.[801] Chiral quinuclidine catalysts lead to

[784]Chandrasekhar, S.; Narsihmulu, Ch.; Reddy, N.R.; Reddy, M.S. *Tetrahedron Lett.* **2003**, *44*, 2583.

[785]Li, G.; Wei, H.-X.; Gao, J.J.; Caputo, T.D. *Tetrahedron Lett.* **2000**, *41*, 1; Shi, M.; Jiang, J.-K.; Feng, Y.-S. *Org. Lett.* **2000**, *2*, 2397.

[786]Pei, W.; Wei, H.X.; Li, G. *Chem. Commun.* **2002**, 2412.

[787]Arnold, L.A.; Imbos, R.; Mandoli, A.; de Vries, A.H.M.; Naasz, R.; Feringa, B.L. *Tetrahedron* **2000**, *56*, 2865.

[788]Balan, D.; Adolfsson, H. *J. Org. Chem.* **2001**, *66*, 6498; Yang, K.-S.; Lee, W.-D.; Pan, J.-F.; Chen, K. *J. Org. Chem.* **2003**, *68*, 915.

[789]Shi, M.; Feng, Y.-S. *J. Org. Chem.* **2001**, *66*, 406.

[790]Kawamura, M.; Kobayashi, S. *Tetrahedron Lett.* **1999**, *40*, 1539.

[791]Xu, Y.-M.; Shi, M. *J. Org. Chem.* **2004**, *69*, 417.

[792]Balan, D.; Adolfsson, H. *Tetrahedron Lett.* **2003**, *44*, 2521.

[793]Tanaka, K.; Fu, G.C. *J. Am. Chem. Soc.* **2001**, *123*, 11492.

[794]Keck, G.E.; Welch, D.S. *Org. Lett.* **2000**, *4*, 3687.

[795]Krishna, P.R.; Kannan, V.; Sharma, G.V.M. *J. Org. Chem.* **2004**, *69*, 6467.

[796]Shi, M.; Xu, Y.-M. *J. Org. Chem.* **2003**, *68*, 4784.

[797]Brzezinski, L.J.; Rafel, S.; Leahy, J.W. *J. Am. Chem. Soc.* **1997**, *119*, 4317.

[798]Perlmutter, P.; Puniani, E.; Westman, G. *Tetrahedron Lett.* **1996**, *37*, 1715; Wei, H.-X.; Chen, D.; Xu, X.; Li, G.; Paré, P.W. *Tetrahedron Asymmetry* **2003**, *14*, 971.

[799]Also see, Markó, I.E.; Giles, P.R.; Hindley, N.J. *Tetrahedron* **1997**, *53*, 1015.

[800]Imbriglio, J.E.; Vasbinder, M.M.; Miller, S.J. *Org. Lett.* **2003**, *5*, 3741. See also, Shi, M.; Jiang, J.-K.; L:i, C.-Q. *Tetrahedron Lett.* **2002**, *43*, 127.

[801]McDougal, N.T.; Schaus, S.E. *J. Am. Chem. Soc.* **2003**, *125*, 12094.

asymmetric induction.[802] A combination of a chiral sulfinamide, an In catalyst and 3-hydroxyquinuclidine led to the allylic amine derivative with modest enantios-electivity.[803] Sugars have been used as ester auxiliaries, and in reaction with aryl alde-hydes and 20% DABCO gave the allylic alcohol with modest enantioselectivity.[804]

The reaction can be modified to give additional products, as with the reaction of *o*-hydroxybenzaldehyde and methyl vinyl ketone with DABCO, where the initial Baylis–Hillman product cyclized via conjugate addition of the phenolic oxygen to the conjugated ketone (**15-31**).[805] Aldehydes and conjugated esters can be coupled with a sulfonamide to give an allylic amine.[806]

A variant of this reaction couples halides with alkenes. α-Bromomethyl esters react with conjugated ketones and DABCO to give a coupling product, **34**.[807] A similar DBU-induced reaction was reported using α-bromomethyl esters and con-jugated nitro compounds.[808]

16-28 The Reformatsky Reaction

***O*-Hydro-*C*-α-ethoxycarbonylalkyl-addition**

The *Reformatsky reaction*[809] is very similar to **16-24**. An aldehyde or ketone is treated with zinc and a halide; the halide is usually an α-halo ester or a vinylog of an α-halo ester (e.g., RCHBrCH=CHCOOEt), though α-halo nitriles,[810] α-halo ketones,[811] and α-halo *N,N*-disubstituted amides have also been used. Especially

[802]Shi, M.; Jiang, J.-K. *Tetrahedron Asymmetry* **2002**, *13*, 1941. See Shi, M.; Xu, Y.-M. *Angew. Chem. Int. Ed.* **2002**, *41*, 4507.

[803]Aggarwal, V.K.; Castro, A.M.M.; Mereu, A.; Adams, H. *Tetrahedron Lett.* **2002**, *43*, 1577.

[804]Filho, E.P.S.; Rodrigues, J.A.R.; Moran, P.J.S. *Tetrahedron Asymmetry* **2001**, *12*, 847.

[805]Kaye, P.T.; Nocanda, X.W. *J. Chem. Soc., Perkin Trans. 1* **2000**, 1331.

[806]Balan, D.; Adolfsson, H. *J. Org. Chem.* **2002**, *67*, 2329.

[807]Basavaiah, D.; Sharada, D.S.; Kumaragurubaran, N.; Reddy, R.M. *J. Org. Chem.* **2002**, *67*, 7135.

[808]Ballini, R.; Barboni, L.; Bosica, G.; Fiorini, D.; Mignini, E.; Palmieri, A. *Tetrahedron* **2004**, *60*, 4995.

[809]For reviews, see Fürstner, A. *Synthesis* **1989**, 571; Rathke, M.W. *Org. React.* **1975**, *22*, 423; Gaudemar, M. *Organomet. Chem. Rev. Sect. A* **1972**, *8*, 183. For a review of the Reformatsky reaction in synthesis, see Ocampo, R.; Dolbier, Jr., W.R. *Tetrahedron* **2004**, *60*, 9325.

[810]Vinograd, L.Kh.; Vul'fson, N.S. *J. Gen. Chem. USSR* **1959**, *29*, 248, 1118, 2656, 2659; Palomo, C.; Aizpurua, J.M.; López, M.C.; Aurrekoetxea, N. *Tetrahedron Lett.* **1990**, *31*, 2205; Zheng, J.; Yu, Y.; Shen, Y. *Synth. Commun.* **1990**, *20*, 3277.

[811]For examples (with R₃Sb and CrCl₂, respectively, instead of Zn), see Huang, Y.; Chen, C.; Shen, Y. *J. Chem. Soc. Perkin Trans. 1* **1988**, 2855; Dubois, J.E.; Axiotis, G.; Bertounesque, E. *Tetrahedron Lett.* **1985**, *26*, 4371.

high reactivity can be achieved with activated zinc,[812] with zinc/silver-graphite,[813] and with zinc and ultrasound.[814] The reaction is catalytic in zinc in the presence of iodine and ultrasound.[815] Metals other than zinc can be used, including In,[816] Mn,[817] low valent Ti,[818] and metal compounds, such as TiI$_4$,[819] TiCl$_2$,[820] Cp$_2$TiCl$_2$,[821] (Bu$_3$Sn)$_2$/Bu$_2$SnCl$_2$,[822] SmI$_2$,[823] and Sc(OTf)$_3$/PPh$_3$.[824] A combination of Zn and an α-bromo ester can be used in conjunction with BF$_3$•OEt$_2$, followed by reaction with dibenzoyl peroxide.[825] The aldehyde or ketone can be aliphatic, aromatic, or heterocyclic or contain various functional groups. Solvents used are generally ethers, including Et$_2$O, THF, and 1,4-dioxane, although the reaction can be done in water[826] using dibenzoyl peroxide and MgClO$_4$.

Dialkylzinc compounds are an alternative source of zinc in the Reformatsky reaction. When an α-bromo ester, an aldehyde, and diethylzinc were reacted in THF with a rhodium catalyst, the β-hydroxy ester was formed.[827]

The use of additives, such as germanium, can lead to highly diastereoselective reactions.[828] Using chiral auxiliaries[829] or chiral additives,[830] good enantioselectivity[831] can be achieved.

Formally, the reaction can be regarded as if it were analogous to the Grignard reaction (**16-24**), with

$$\overset{\displaystyle |}{\underset{\displaystyle |}{EtOOC-C-ZnBr}}$$

[812]Rieke, R.D.; Uhm, S.J. *Synthesis* **1975**, 452; Bouhlel, E.; Rathke, M.W. *Synth. Commun.* **1991**, *21*, 133.

[813]Csuk, R.; Fürstner, A.; Weidmann, H. *J. Chem. Soc., Chem. Commun.* **1986**, 775.

[814]Han, B.; Boudjouk, P. *J. Org. Chem.* **1982**, *47*, 5030.

[815]Ross, N.A.; Bartsch, R.A. *J. Org. Chem.* **2003**, *68*, 360.

[816]Araki, S.; Yamada, M.; Butsugan, Y. *Bull. Chem. Soc. Jpn.* **1994**, *67*, 1126.

[817]Cahiez, G.; Chavant, P. *Tetrahedron Lett.* **1989**, *30*, 7373; Suh, Y.S.; Rieke, R.D. *Tetrahedron Lett.* **2004**, *45*, 1807.

[818]Aoyagi, Y.; Tanaka, W.; Ohta, A. *J. Chem. Soc., Chem. Commun.* **1994**, 1225.

[819]Shimizu, M.; Kobayashi, F.; Hayakawa, R. *Tetrahedron* **2001**, *57*, 9591.

[820]Kagayama, A.; Igarashi, K.; Shiina, I.; Mukaiyama, T. *Bull. Chem. Soc. Jpn.* **2000**, *73*, 2579.

[821]Parrish, J.D.; SheHon, D.R.; Little, R.D. *Org. Lett.* **2003**, *5*, 3615.

[822]Shibata, I.; Kawasaki, M.; Yasuda, M.; Baba, A. *Chem. Lett.* **1999**, 689.

[823]Utimoto, K.; Matsui, T.; Takai, T.; Matsubara, S. *Chem. Lett.* **1995**, 197; Arime, T.; Takahashi, H.; Kobayashi, S.; Yamaguchi, S.; Mori, N. *Synth. Commun.* **1995**, *25*, 389; Park, H.S.; Lee, I.S.; Kim, Y.H. *Tetrahedron Lett.* **1995**, *36*, 1673; Molander, G.A.; Etter, J.B. *J. Am. Chem. Soc.* **1987**, *109*, 6556.

[824]Kagoshima, H.; Hashimoto, Y.; Saigo, K. *Tetrahedron Lett.* **1998**, *39*, 8465.

[825]Chattopadhyay, A.; Salaskar, A. *Synthesis* **2000**, 561.

[826]Bieber, L.W.; Malvestiti, I.; Storch, E.C. *J. Org. Chem.* **1997**, *62*, 9061.

[827]Kanai, K.; Wakabayashi, H.; Honda, T. *Org. Lett.* **2000**, *2*, 2549.

[828]Kagoshima, H.; Hashimoto, Y.; Oguro, D.; Saigo, K. *J. Org. Chem.* **1998**, *63*, 691.

[829]Fukuzawa, S.-i.; Tatsuzawa, M.; Hirano, K. *Tetrahedron Lett.* **1998**, *39*, 6899.

[830]Soai, K.; Hirose, Y.; Sakata, S. *Tetrahedron Asymmetry* **1992**, *3*, 677.

[831]Pini, D.; Uccello-Barretta, G.; Mastantuono, A.; Salvadori, P. *Tetrahedron* **1997**, *53*, 6065; Andrés, J.M.; Martín, Y.; Pedrosa, R.; Pérez-Encabo, A. *Tetrahedron* **1997**, *53*, 3787; Mi, A.; Wang, Z.; Zhang, J.; Jiang, Y. *Synth. Commun.* **1997**, *27*, 1469.; Ribeiro, C.M.R.; de S. Santos, E.; de O. Jardim, A.H.; Maia, M.P.; da Silva, F.C.; Moreira, A.P.D.; Ferreira, V.F. *Tetrahedron Asymmetry* **2002**, *13*, 1703.

(**35**) as an intermediate analogous to RMgX.[832] There *is* an intermediate derived from zinc and the ester, the structure of which has been shown to be **36**, by X-ray crystallography of the solid intermediate prepared from *t*-BuOCOCH$_2$Br and Zn.[833] As can be seen, it has some of the characteristics of **35**.

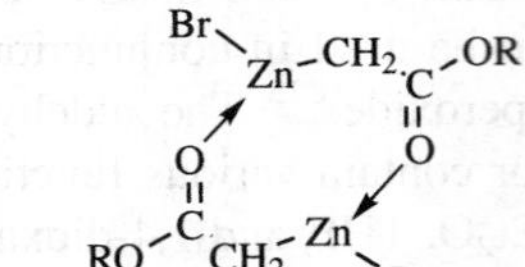

36

After hydrolysis, the alcohol is the usual product, but sometimes (especially with aryl aldehydes) elimination follows directly and the product is an alkene. By the use of Bu$_3$P along with Zn, the alkene can be made the main product,[834] making this an alternative to the Wittig reaction (**16-44**). The alkene is also the product when K$_2$CO$_3$/NaHCO$_3$ is used with 2% PEG–telluride.[835] Since Grignard reagents cannot be formed from α-halo esters, the method is quite useful, though there are competing reactions and yields are sometimes low. A similar reaction (called the *Blaise reaction*) has been carried out on nitriles:[836]

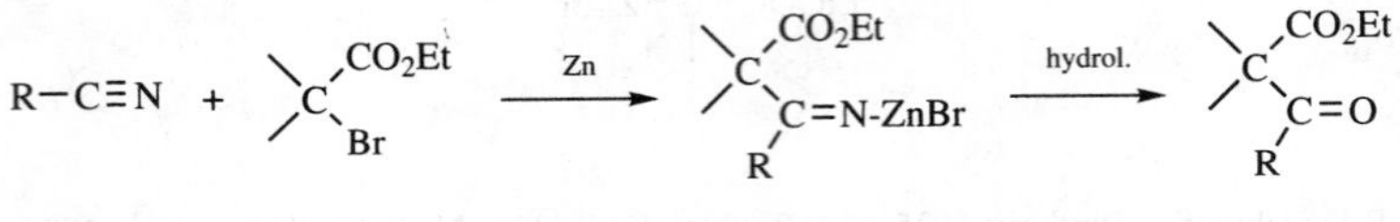

Carboxylic esters have also been used as substrates, but then, as might be expected (p. 1252), the result is substitution and not addition:

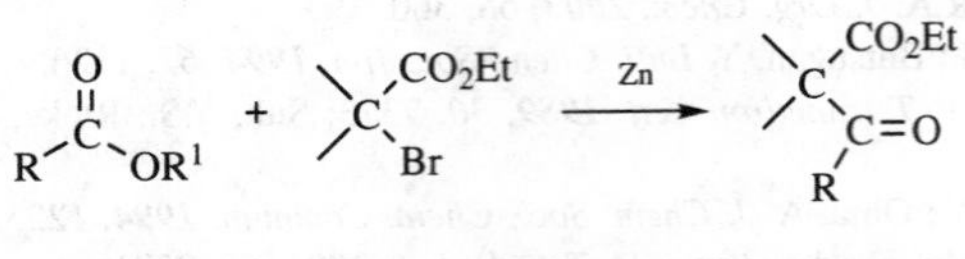

The product in this case is the same as with the corresponding nitrile, though the pathways are different. The reaction is compatible with amine substituents, and α-(*N,N*-dibenzyl)amino aldehydes have been used to prepare β-hydroxy-γ-(*N,N*-dibenzylamino) esters with good anti-selectivity.[837]

[832]For a study of transition structures, see Maiz, J.; Arrieta, A.; Lopez, X.; Ugalde, J.M.; Cossio, F.P.; Fakultatea, K.; Unibertsitatea, E.H.; Lecea, B. *Tetrahedron Lett.* **1993**, *34*, 6111.

[833]Dekker, J.; Budzelaar, P.H.M.; Boersma, J.; van der Kerk, G.J.M.; Spek, A.L. *Organometallics*, **1984**, *3*, 1403.

[834]Shen, Y.; Xin, Y.; Zhao, J. *Tetrahedron Lett.* **1988**, *29*, 6119. For another method, see Huang, Y.; Shi, L.; Li, S.; Wen, X. *J. Chem. Soc. Perkin Trans. 1* **1989**, 2397.

[835]Huang, Z.-Z.; Ye, S.; Xia, W.; Yu, Y.-H.; Tang, Y. *J. Org. Chem.* **2002**, *67*, 3096.

[836]See Cason, J.; Rinehart Jr., K.L.; Thornton, S.D. *J. Org. Chem.* **1953**, *18*, 1594; Bellassoued, M.; Gaudemar, M. *J. Organomet. Chem.* **1974**, *81*, 139; Hannick, S.M.; Kishi, Y. *J. Org. Chem.* **1983**, *48*, 3833.

[837]Andrés, J.M.; Pedrosa, R.; Pérez, A.; Pérez-Encabo, A. *Tetrahedron* **2001**, *57*, 8521.

For an alternative approach involving ester enolates (see **16-36**).
OS **III**, 408; **IV**, 120, 444; **IX**, 275.

16-29   The Conversion of Carboxylic Acid Salts to Ketones with Organometallic Compounds

Alkyl-de-oxido-substitution

Good yields of ketones can often be obtained by treatment of the lithium salt of a carboxylic acid with an alkyllithium reagent, followed by hydrolysis.[838] The R' group may be aryl or primary, secondary, or tertiary alkyl and R may be alkyl or aryl. The compounds MeLi and PhLi have been employed most often. Tertiary alcohols are side products. Lithium acetate can be used, but generally gives low yields.

A variation of this transformation reacts the acid with lithium naphthalenide in the presence of 1-chlorobutane. The product is the ketone.[839] A related reaction treats the lithium carboxylate with lithium metal and the alkyl halide, with sonication, to give the ketone.[840] Phenylboronic acid (p. 815) reacts with aryl carboxylic acids in the presence of a palladium catalyst and disuccinoyl carbonate to give a diaryl ketone.[841]

OS **V**, 775.

16-30   The Addition of Organometallic Compounds to CO_2 and CS_2

C-Alkyl-O-halomagnesio-addition

Grignard reagents add to one C=O bond of CO_2 exactly as they do to an aldehyde or a ketone.[842] Here, of course, the product is the salt of a carboxylic acid. The reaction is usually performed by adding the Grignard reagent to dry ice. Many carboxylic acids have been prepared in this manner, and this constitutes an important way of increasing a carbon chain by one unit. Since labeled CO_2 is commercially

---

[838]For a review, see Jorgenson, M.J. *Org. React.* **1970**, *18*, 1. For an improved procedure, see Rubottom, G.M.; Kim, C. *J. Org. Chem.* **1983**, *48*, 1550.

[839]Alonso, F.; Lorenzo, E.; Yus, M. *J. Org. Chem.*, **1996**, *61*, 6058.

[840]Aurell, M.J.; Danhui, Y.; Einhorn, J.; Einhorn, C.; Luche, J.L. *Synlett* **1995**, 459. Also see, Aurell, M.J.; Einhorn, C.; Einhorn, J.; Luche, J.L. *J. Org. Chem.* **1995**, *60*, 8.

[841]Gooßen, L.J.; Ghosh, K. *Chem. Commun.* **2001**, 2084.

[842]For reviews of the reaction between organometallic compounds and CO_2, see Volpin, M.E.; Kolomnikov, I.S. *Organomet. React.* **1975**, *5*, 313; Sneeden, R.P.A., in Patai, S. *The Chemistry of Carboxylic Acids and Esters*, Wiley, NY, **1969**, pp. 137–173; Kharasch, M.S.; Reinmuth, O. *Grignard Reactions of Nonmetallic Substances*, Prentice-Hall, Englewood Cliffs, NJ, **1954**, pp. 913–948. For a more general review, see Lapidus, A.L.; Ping, Y.Y. *Russ. Chem. Rev.* **1981**, *50*, 63.

available, this is a good method for the preparation of carboxylic acids labeled in the carboxyl group. Other organometallic compounds have also been used (RLi,[843] RNa, RCaX, RBa,[844] etc.), but much less often. The formation of the salt of a carboxylic acid after the addition of CO_2 to a reaction mixture is regarded as a positive test for the presence of a carbanion or of a reactive organometallic intermediate in that reaction mixture (see also, **16-42**).

When chiral additives, such as (−)-sparteine, have added to the initial reaction with the organolithium reagent, quenching with CO_2 produces carboxylic acids with good asymmetric induction.[845]

In a closely related reaction, Grignard reagents add to CS_2 to give salts of dithiocarboxylic acids.[846] These salts can be trapped with amines to form thioamides.[847] Two other reactions are worthy of note. (*1*) Lithium dialkylcopper reagents react with dithiocarboxylic esters to give tertiary thiols[848] (*2*) Thiono lactones can be converted to cyclic ethers,[849] for example:

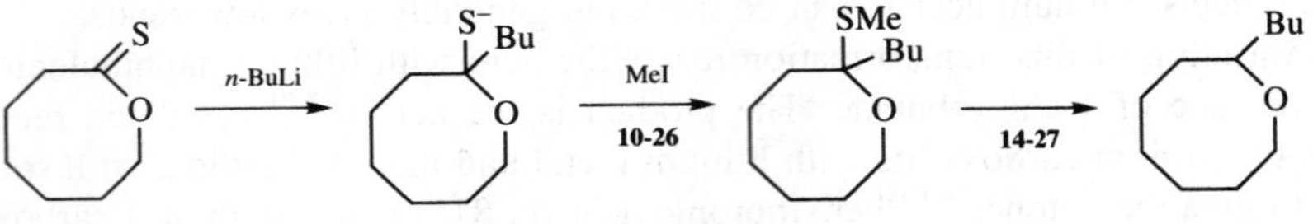

This is a valuable procedure because medium and large ring ethers are not easily made, while the corresponding thiono lactones can be prepared from the readily available lactones (see, e.g., **16-63**) by reaction **16-11**.

A terminal alkyne can be converted to the anion under electrolytic conditions, in the presence of CO_2, to give propargylic acids, R–C≡C–COOH.[850]

OS **I**, 361, 524; **II**, 425; **III**, 413, 553, 555; **V**, 890, 1043; **VI**, 845; **IX**, 317.

16-31   The Addition of Organometallic Compounds to C=N Compounds

N-Hydro-*C*-alkyl-addition

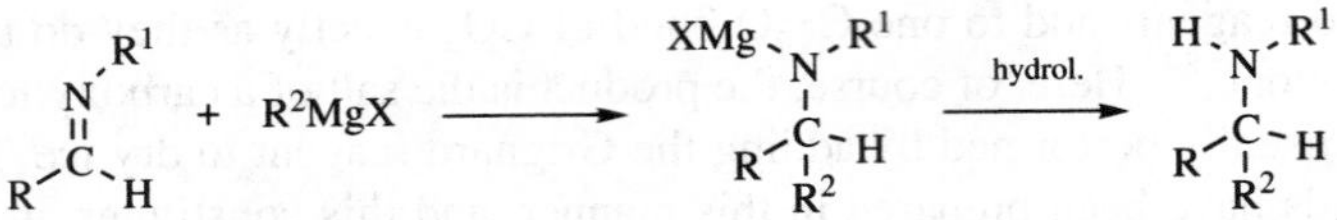

[843]The kinetics of this reaction have been studied, see Nudelman, N.S.; Doctorovich, F. *J. Chem. Soc. Perkin Trans. 2* **1994**, 1233.

[844]Yanagisawa, A.; Yasue, K.; Yamamoto, H. *Synlett* **1992**, 593.

[845]Park, Y.S.; Beak, P. *J. Org. Chem.* **1997**, *62*, 1574.

[846]For a review of the addition of Grignard reagents to C=S bonds, see Paquer, D. *Bull. Soc. Chim. Fr.* **1975**, 1439. For a review of the synthesis of dithiocarboxylic acids and esters, see Ramadas, S.R.; Srinivasan, P.S.; Ramachandran, J.; Sastry, V.V.S.K. *Synthesis* **1983**, 605.

[847]Katritzky, A.R.; Moutou, J.-L.; Yang, Z. *Synlett* **1995**, 99.

[848]Bertz, S.H.; Dabbagh, G.; Williams, L.M. *J. Org. Chem.* **1985**, *50*, 4414.

[849]Nicolaou, K.C.; McGarry, D.G.; Somers, P.K.; Veale, C.A.; Furst, G.T. *J. Am. Chem. Soc.* **1987**, *109*, 2504.

[850]Köster, F.; Dinjus, E.; Duñach, E. *Eur. J. Org. Chem.* **2001**, 2507.

Aldimines can be converted to secondary amines by treatment with Grignard reagents.[851] Ketimines generally give reduction instead of addition. However, organolithium compounds give the normal addition product with both aldimines and ketimines.[852] For the addition of an organometallic compound to an imine to give a primary amine, R′ in RCH=NR′ would have to be H, and such compounds are seldom stable. However, the conversion has been done, for R = aryl, by the use of the masked reagents $(ArCH=N)_2SO_2$ [prepared from an aldehyde RCHO and sulfamide $(NH_2)_2SO_2$]. Addition of R^2MgX or R^2Li to these compounds gives $ArCHR^2NH_2$ after hydrolysis.[853] An intramolecular version of the addition or organolithium reagents is known, and treatment of the N-(3-chloropropyl)aldimine of benzaldehyde with lithium and DTBB, followed by hydrolysis with water, gave 2-phenylpyrrolidine.[854] Grignard regents add to imines in the presence of various transition metal catalysts, including $Sc(OTf)_3$[855] or Cp_2ZrCl_2.[856] When chiral additives are used in conjunction with the organolithium reagent, chiral amines are produced[857] with good asymmetric induction.[858] Chiral auxiliaries have been used in addition reactions to imines,[859] and to oxime derivatives.[860] Chiral catalysts lead to enantioselective addition of alkynes to imines to give a homopropargylic amine.[861]

Zinc metal reacts with allylic bromides to form an allylic zinc complex, which reacts with imines to give the homoallylic amine.[862] This reaction is catalyzed by TMSCl.[863] Allylzinc bromide adds to imines.[864] Dialkylzinc reagents add to imines to give the amine, and in the presence of a chiral ligand the reaction proceeds with

[851]For reviews of the addition of organometallic reagents to C=N bonds, see Harada, K., in Patai, S. *The Chemistry of the Carbon–Nitrogen Double Bond*, Wiley, NY, *1970*, pp. 266–272; Kharasch, M.S.; Reinmuth, O. *Grignard Reactions of Nonmetallic Substances*, Prentice-Hall, Englewood Cliffs, NJ, *1954*, pp. 1204–1227. For recent examples, see Wang, D.-K.; Dai, L.-X.; Hou, X.-L.; Zhang, Y. *Tetrahedron Lett. 1996*, *37*, 4187; Bambridge, K.; Begley, M.J.; Simpkins, N.S. *Tetrahedron Lett. 1994*, *35*, 3391.

[852]Huet, J. *Bull. Soc. Chim. Fr. 1964*, 952, 960, 967, 973.

[853]Davis, F.A.; Giangiordano, M.A.; Starner, W.E. *Tetrahedron Lett. 1986*, *27*, 3957.

[854]Yus, M.; Soler, T.; Foubelo, F. *J. Org. Chem. 2001*, *66*, 6207.

[855]Saito, S.; Hatanaka, K.; Yamamoto, H. *Synlett 2001*, 1859.

[856]Gandon, V.; Bertus, P.; Szymoniak, J. *Eur. J. Org. Chem. 2001*, 3677.

[857]For a review see Enders, D.; Reinhold, U. *Tetrahedron Asymmetry, 1997*, *8*, 1895.

[858]Andersson, P.G.; Johansson, F.; Tanner, D. *Tetrahedron 1998*, *54*, 11549; Tomioka, K.; Inoue, I.; Shindo, M.; Koga, K. *Tetrahedron Lett. 1991*, *32*, 3095; Denmark, S.E.; Stiff, C.M. *J. Org. Chem. 2000*, *65*, 5875; Chrzanowska, M.; Sokołowska, J. *Tetrahedron Asymmetry 2001*, *12*, 1435.

[859]Hashimoto, Y.; Kobayashi, N.; Kai, A.; Saigo, K. *Synlett 1995*, 961.

[860]Dieter, R.K.; Datar, R. *Can. J. Chem. 1993*, *71*, 814.

[861]Benaglia, M.; Negri, D.; Dell'Anna, G. *Tetrahedron Lett. 2004*, *45*, 8705.

[862]Lee, C.-L.K.; Ling, H.-Y.; Loh, T.-P. *J. Org. Chem. 2004*, *69*, 7787.

[863]Legros, J.; Meyer, F.; Coliboeuf, M.; Crousse, B.; Bonnet-Delpon, D.; Bégué, J.-P. *J. Org. Chem. 2003*, *68*, 6444.

[864]van der Sluis, M.; Dalmolen, J.; de Lange, B.; Kaptein, B.; Kellogg, R.M.; Broxterman, Q.B. *Org. Lett. 2001*, *3*, 3943.

good enantioselectivity.[865] Dialkylzinc reagents add to N-tosyl imines using a copper catalyst, and with a chiral ligand leads to good enantioselectivity.[866] α-Bromo esters are converted to an organometallic reagent with Zn/Cu, and addition to N-arylimines gives N-aryl β-amino esters.[867] The reaction of imines, such as ArN=CHCO$_2$Et, where R = a chiral benzylic substituent, and ZnBr$_2$, followed by R′ZnBr leads to a chiral α-amino ester.[868] Terminal alkynes add to imines using ZnCl$_2$ and TMSCl, and with a chiral ligand attached to nitrogen the reaction proceeds with some enantioselectivity.[869]

Other organometallic compounds,[870] including allylic stannanes,[871] allylic samarium,[872] allylic germanium,[873] and allylic indium compounds[874] add to aldimines in the same manner. Aryltrialkylstannanes also add the aryl group to N-tosyl imines using a rhodium catalyst and sonication.[875] Catalytic enantioselective addition reactions are well known,[876] including reactions in an ionic liquid.[877] Allylic

[865]For an example using a Zr catalyst with a chiral ligand, see Porter J.R.; Traverse, J.F.; Hoveyda, A.H.; Snapper, M.L. *J. Am. Chem. Soc.* **2001**, *123*, 984. With a Pd catalyst, see Inoue, A.; Shinokubo, H.; Oshima, K. *J. Am. Chem. Soc.* **2003**, *125*, 1484. With a Zr catalyst, see Porter, J.R.; Traverse, J.F.; Hoveyda, A.H.; Snapper, M.L. *J. Am. Chem. Soc.* **2001**, *123*, 10409. See also, Zhang, X.-M.; Zhang, H.-L.; Lin, W.-Q.; Gong, L.-Z.; Mi, A.-Q.; Cui, X.; Jiang, Y.-Z.; Yu, K.B. *J. Org. Chem.* **2003**, *68*, 4322; Jensen, D.R.; Schultz, M.J.; Mueller, J.A.; Sigman, M.S. *Angew. Chem. Int. Ed.* **2003**, *42*, 3810.

[866]Fujihara, H.; Nagai, K.; Yomioka, K. *J. Am. Chem. Soc.* **2000**, *122*, 12055. See Wang, C.-J.; Shi, M. *J. Org. Chem.* **2003**, *68*, 6229.

[867]Adrian Jr., J.C.; Barkin, J.L.; Hassib, L. *Tetrahedron Lett.* **1999**, *40*, 2457.

[868]Chiev, K.P.; Roland, S.; Mangeney, P. *Tetrahedron Asymmetry* **2001**, *13*, 2205.

[869]Jiang, B.; Si, Y.-G. *Tetrahedron Lett.* **2003**, *44*, 6767.

[870]For a list of reagents, with references, see Larock, R.C. *Comprehensive Organic Transformations*, 2nd ed., Wiley-VCH, NY, **1999**, pp. 847–863.

[871]Keck, G.E.; Enholm, E.J. *J. Org. Chem.* **1985**, *50*, 146; Nakamura, H.; Nakamura, K.; Yamamoto, Y. *J. Am. Chem. Soc.* **1998**, *120*, 4242; Kobayashi, S.; Iwamoto, S.; Nagayama, S. *Synlett* **1997**, 1099; Wang, D.-K.; Dai, L.-X.; Hou, X.-L. *Tetrahedron Lett.* **1995**, *36*, 8649. Catalytic amounts of metal compounds can be used with the allyl stannane, including: **Pd**: Nakamura, H.; Iwama, H.; Yamamoto, Y. *Chem. Commun.* **1996**, 1459 and Fernandes, R.A.; Yamamoto, Y. *J. Org. Chem.* **2004**, *69*, 3562; **Zr**: Gastner, T.; Ishitani, H.; Akiyama, R.; Kobayashi, S. *Angew. Chem. Int. Ed.* **2001**, *40*, 1896; **Ta**: Shibata, I.; Nose, K.; Sakamoto, K.; Yasuda, M.; Baba, A. *J. Org. Chem.* **2004**, *69*, 2185; **La**: Aspinall, H.C.; Bissett, J.S.; Greeves, N.; Levin, D. *Tetrahedron Lett.* **2002**, *43*, 323; **Al**: Niwa, Y.; Shimizu, M. *J. Am. Chem. Soc.* **2003**, *125*, 3720; **Nb**: Andrade, C.K.Z.; Oliveira, G.R. *Tetrahedron Lett.* **2002**, *43*, 1935; Akiyama, T.; Onuma, Y. *J. Chem. Soc., Perkin Trans. 1* **2002**, 1157. With **LiClO$_4$**: Yadav, J.S.; Reddy, B.V.S.; Reddy, P.S.R.; Rao, M.S. *Tetrahedron Lett.* **2002**, *43*, 6245.

[872]Wang, J.; Zhang, Y.; Bao, W. *Synth. Commun.* **1996**, *26*, 2473. For an example using an allylic bromide with SmI$_2$, see Kim, B.; Han, R.; Park, R.; Bai, K.; Jun, Y.; Baik, W. *Synth. Commun.* **2001**, *31*, 2297.

[873]Akiyama, T.; Iwai, J.; Onuma, Y.; Kagoshima, H. *Chem. Commun.* **1999**, 2191.

[874]Chan, T.H.; Lu, W. *Tetrahedron Lett.* **1998**, *39*, 8605; Jin, S.-J.; Araki, S.; Butsugan, Y. *Bull Chem. Soc. Jpn.* **1993**, *66*, 1528; Beuchet, P.; Le Marrec, N.; Mosset, P. *Tetrahedron Lett.* **1992**, *33*, 5959; Vilaivan, T.; Winotapan, C.; Shinada, T.; Ohfune, Y. *Tetrahedron Lett.* **2001**, *42*, 9073.

[875]Ding, R.; Zhao, C.H.; Chen, Y.J.; Lu, L.; Wang, D.; Li, C.J. *Tetrahedron Lett.* **2004**, *45*, 2995.

[876]For a review, see Kobayashi, Sh.; Ishitani, H. *Chem. Rev.* **1999**, *99*, 1069.

[877]In bmim BF$_4$, 1-butyl-3-methylimidazolium tetrafluoroborate: Chowdari, N.S.; Ramachary, D.B.; Barbas III, C.F. *Synlett* **2003**, 1906.

halides react with imines in the presence of indium metal[878] or $InCl_3$[879] to give the homoallylic amine, and with N-sulfonyl imines to give the homoallylic sulfonamide.[880] In this latter reaction, antiselectivity was observed when the reaction was done in water, and syn selectivity when done in aqueous THF.[881] Propargylic halides add to imines in the presence of indium metal, in aq. THF.[882] Imines react with allylic halides and gallium metal, with ultrasound.[883] Imines also react with allylic halides and Yb, in the presence of Me_3SiCl.[884] Aryl iodides add to N-aryl imines in the presence of a rhodium catalyst.[885] Titanium enolates add to imines to give β-amino esters.[886] Terminal alkynes react with aryl aldehydes and aryl amines to give propargylic amine without a catalyst,[887] and an iridium[888] or a copper catalyst[889] also leads to a propargylic amine.[890] Terminal alkynes add to imines to give a propargylic amine with high enantioselectivity using a chiral copper complex.[891] Triethylaluminum adds an ethyl group to an imine in the presence of a europium catalyst. Reaction with $PhSnMe_3$ and N-tosylimines with a rhodium catalyst, for example, leads to addition of a phenyl group to the carbon of the C=N bond.[892] Other N-sulfonyl imines react similarly to give the corresponding sulfonamide, and in the presence of a chiral ligand the reaction proceeds to good enantioselectivity.[893] N-Tosyl imines also react with dialkylzinc reagents, giving the sulfonamide with modest enantioselectivity.[894] N-Sulfinyl imines, $R_2CH=NS(=O)R'$,[895] react with Grignard reagents (R^2MgX) to give the corresponding N-sulfinylamine, $R_2CH(R^2)NHS(=O)R'$.[896] Enolate anions, generated by reaction of dimethylaminopyridine and a conjugated ketone, add to N-tosylimines.[897] N-Carbamoyl imines add acetonitrile (via carbon) using DBU and a ruthenium catalyst.[898]

[878]Choucair, B.; Léon, H.; Miré, M.-A.; Lebreton, C.; Mosset, P. *Org. Lett.* **2000**, *2*, 1851. See Hirashita, T.; Hayashi, Y.; Mitsui, K.; Araki, S. *J. Org. Chem.* **2003**, *68*, 1309.
[879]Under electrolysis conditions, see Hilt, G.; Smolko, K.I.; Waloch, C. *Tetrahedron Lett.* **2002**, *43*, 1437.
[880]Lu, W.; Chan, T.H. *J. Org. Chem.* **2000**, *65*, 8589.
[881]Lu, W.; Chan, T.H. *J. Org. Chem.* **2001**, *66*, 3467.
[882]Prajapati, D.; Laskar, D.D.; Gogoi, B.J.; Devi, G. *Tetrahedron Lett.* **2003**, *44*, 6755.
[883]Andrews, P.C.; Peatt, A.C.; Raston, C.L. *Tetrahedron Lett.* **2004**, *45*, 243.
[884]Su, W.; Li, J.; Zhang, Y. *Synth. Commun.* **2001**, *31*, 273.
[885]Ishiyama, T.; Hartwig, J. *J. Am. Chem. Soc.* **2000**, *122*, 12043.
[886]Adrian, Jr., J.C.; Barkin, J.L.; Fox, R.J.; Chick, J.E.; Hunter, A.D.; Nicklow, R.A. *J. Org. Chem.* **2000**, *65*, 6264.
[887]Li, C.-J.; Wei, C. *Chem. Commun.* **2002**, 268.
[888]Fischer, C.; Carreira, E.M. *Org. Lett.* **2001**, *3*, 4319; Fischer, C.; Carreira, E.M. *Synthesis* **2004**, 1497.
[889]Koradin, C.; Gommermann, N.; Polborn, K.; Knochel, P. *Chem. Eur. J.* **2003**, *9*, 2797.
[890]Wei, C.; Li, C.-J. *J. Am. Chem. Soc.* **2002**, *124*, 5638.
[891]Tsvelikhovsky, D.; Gelman, D.; Molander, G.A.; Blum, J. *Org. Lett.* **2004**, *6*, 1995.
[892]Oi, S.; Moro, M.; Fukuhara, H.; Kawanishi, T.; Inoue, Y. *Tetrahedron Lett.* **1999**, *40*, 9259.
[893]Hayashi, T.; Ishigedani, M. *J. Am. Chem. Soc.* **2000**, *122*, 976.
[894]Soeta, T.; Nagai, K.; Fujihara, H.; Kuriyama, M.; Tomioka, K. *J. Org. Chem.* **2003**, *68*, 9723.
[895]For a review of these reagents, see Ellman, J.A.; Owens, T.D.; Tang, T.P. *Acc. Chem. Res.* **2002**, *35*, 984.
[896]Tang, T.P.; Volkman, S.K.; Ellman, J.A. *J. Org. Chem.* **2001**, *66*, 8772.
[897]Shi, M.; Xu, Y.-M. *Chem. Commun.* **2001**, 1876.
[898]Kumagai, N.; Matsunaga, S.; Shibasaki, M. *J. Am. Chem. Soc.* **2004**, *126*, 13632.

Activated aromatic compounds add to *N*-carbamoyl imines in the presence of copper catalysts, and with good enantioselectivity when a chiral catalyst is used.[899] A combination of AuCl₃/AgOTf facilitates the addition of arenes to *N*-tosyl imines.[900] Furan derivatives add via C-2 with good enantioselectivity using a chiral phosphoric acid catalyst.[901] Alkenes add to *N*-tosyl imines with a Yb catalyst[902] an allenes add to *N*-carbamoyl imines in the presence of vanadium catalyst.[903] *N*-Carbamoyl imines, formed *in situ*, react with allylic silanes in the presence of an iodine catalyst.[904] The intramolecular addition of an alkene to an imine, facilitated by Cp₂ZrBu₂, gave a cycloalkyl amine.[905]

Arylboronates (p. 815) add to *N*-sulfonyl imines in the presence of a rhodium catalyst to give the corresponding sulfonamide.[906] Vinyl boronates also add to nitrones in the presence of Me₂Zn, transferring the vinyl group to the C=N unit.[907] Aryl boronic acids (p. 815) add the aryl group to *N*-tosyl imines using a rhodium catalyst.[908] Allylic boronates also add to aldehydes, and subsequent treatment with ammonia give the homoallylic amine.[909]

Allylic silanes, such as allyltrimethylsilane, add to *N*-substituted imines in the presence of a palladium catalyst to give the homoallylic amine.[910] Similar results are obtained when the allylic silane and imine are treated with a catalytic amount of tetrabutylammonium fluoride.[911] *N*-Tosyl imines also react with allylic silanes, and the reaction of EtO₂C—CH=NTs and allyltrimethylsilane with a chiral copper catalyst gave EtO₂C—CH(NHTs)CH₂CH=CH₂, albeit in poor yield with modest enantioselectivity.[912] Another addition reaction converts aryl aldehydes to the imine using Me₂N—SiMe₃ and LiClO₄, and subsequent reaction with Me₂PhSiCl gave the corresponding amine, ArCH(SiMe₂Ph)NMe₂.[913] Allylic trichlorosilanes add to hydrazones to give homoallylic hydrazine derivatives with excellent anti-selectivity[914]

[899]Saaby, S.; Fang, X.; Gathergood, N.; Jørgensen, K.A. *Angew. Chem. Int. Ed.* **2000**, *39*, 4114. See also, Saaby, S.; Bayón, P.; Aburel, P.S.; Jørgensen, K.A. *J. Org. Chem.* **2002**, *67*, 4352.

[900]Luo, Y.; Li, C.-J. *Chem. Commun.* **2004**, 1930.

[901]Uraguchi, D.; Sorimachi, K.; Terada, M. *J. Am. Chem. Soc.* **2004**, *126*, 11804. See also, Spanedda, M.V.; Ourévitch, M.; Crouse, B.; Bégué, J.-P.; Bonnet-Delpon, D. *Tetrahedron Lett.* **2004**, *45*, 5023.

[902]Yamanaka, M.; Nishida, A.; Nakagawa, M. *J. Org. Chem.* **2003**, *68*, 3112.

[903]Trost, B.M.; Jonasson, C. *Angew. Chem. Int. Ed.* **2003**, *42*, 2063.

[904]Phukan, P. *J. Org. Chem.* **2004**, *69*, 4005.

[905]Makabe, M.; Sato, Y.; Mori, M. *J. Org. Chem.* **2004**, *69*, 6238.

[906]Ueda, M.; Saito, A.; Miyaura, N. *Synlett* **2000**, 1637.

[907]Pandya, A.; S.U.; Pinet, S.; Chavant, P.Y.; Vallée, Y. *Eur. J. Org. Chem.* **2003**, 3621.

[908]Kuriyama, M.; Soeta, T.; Hao, X.; Chen, Q.; Tomioka, K. *J. Am. Chem. Soc.* **2004**, *126*, 8128.

[909]Sugiura, M.; Hirano, K.; Kobayashi, S. *J. Am. Chem. Soc.* **2004**, *126*, 7182.

[910]Nakamura, K.; Nakamura, H.; Yamamoto, Y. *J. Org. Chem.* **1999**, *64*, 2614.

[911]Wang, D.-K.; Zhou, Y.-G.; Tang, Y.; Hou, X.-L.; Dai, L.-X. *J. Org. Chem.* **1999**, *64*, 4233. Tetraallylsilane and TBAF, with a chiral palladium catalyst, gives chiral homoallylic amines, see Fernandes, R.A.; Yamamoto, Y. *J. Org. Chem.* **2004**, *69*, 735.

[912]Fang, X.; Johannsen, M.; Yao, S.; Gathergood, N.; Hazell, R.G.; Jørgensen, K.A. *J. Org. Chem.* **1999**, *64*, 4844.

[913]Naimi-Jamal, M.R.; Mojtahedi, M.M.; Ipaktschi, J.; Saidi, M.R. *J. Chem. Soc., Perkin Trans. 1* **1999**, 3709.

[914]Hirabayashi, R.; Ogawa, C.; Sugiura, M.; Kobayashi, S. *J. Am. Chem. Soc.* **2001**, *123*, 9493.

and with good enantioselectivity using a chiral ligand.[915] Chiral allyl silane derivatives have been developed, and add to hydrazones with good enantioselectivity.[916]

Aldehydes add via the α-carbon using proline, to give β-amino aldehydes with good selectivity to give chiral β-amino aldehydes.[917] Silyl enol ethers add to hydrazones in the presence of ZnF_2 and a chiral ligand to give chiral β-hydrazino ketones.[918] Nitro compounds add to N-carbamoyl imines with a chiral diamine catalyst with some enantioselectivity.[919] Nitro compounds add via carbon using a copper catalyst, and with good enantioselectivity when a chiral ligand is used.[920] Similar addition to imine derivatives was accomplished using ketene silyl acetals and Amberlyst-15.[921] Alternatively, an imine is reacted first with $Zn(OTf)_2$ and then with a ketene silyl acetal.[922] The conjugate bases of nitro compounds (formed by treatment of the nitro compound with BuLi) react with Grignard reagents in the presence of $ClCH=NMe_2^+$ Cl^- to give oximes: $RCH=N(O)OLi + R'MgX \rightarrow RR'C=NOH$.[923]

Many other C=N systems (phenylhydrazones, oxime ethers, etc.) give normal addition when treated with Grignard reagents; others give reductions; others give miscellaneous reactions. Organocerium reagents add to hydrazones.[924] Oximes can be converted to hydroxylamines (**37**) by treatment with 2 equivalents of an alkyllithium reagent, followed by methanol.[925] Oxime ethers add an allyl group upon reaction with allyl bromide and indium metal in water.[926] Nitrones, $R_2C=N^+(R')-O^-$, react with allylic bromides and Sm to give homoallylic oximes,[927] and with terminal alkynes and a zinc catalyst to give propargylic oximes.[928]

[915]Kobayashi, S.; Ogawa, C.; Konishi, H.; Sugiura, M. *J. Am. Chem. Soc.* **2003**, *125*, 6610.
[916]Berger, R.; Duff, K.; Leighton, J.L. *J. Am. Chem. Soc.* **2004**, *126*, 5686.
[917]Córdova, A.; Barbas III, C.F. *Tetrahedron Lett.* **2003**, *44*, 1923; Notz, W.; Tanaka, F.; Watanabe, S.; Chowdari, N.S.; Turner, J.M.; Thayumanavan, R.; Barbas III, C.F. *J. Org. Chem.* **2003**, *68*, 9624; Chowdari, N.S.; Suri, J.T.; Barbas III, C.F. *Org. Lett.* **2004**, *6*, 2507.
[918]Hamada, T.; Manabe, K.; Kobayashi, S. *J. Am. Chem. Soc.* **2004**, *126*, 7768. For a similar reaction using a bismuth catalyst, see Ollevier, T.; Nadeau, E. *J. Org. Chem.* **2004**, *69*, 9292.
[919]Nugent, B.M.; Yoder, R.A.; Johnston, J.N. *J. Am. Chem. Soc.* **2004**, *126*, 3418.
[920]Nishiwaki, N.; Knudson, K.R.; Gothelf, K.V.; Jørgensen, K.A. *Angew. Chem. Int. Ed.* **2001**, *40*, 2992.
[921]Shimizu, M.; Itohara, S.; Hase, E. *Chem. Commun.* **2001**, 2318.
[922]Ishimaru, K.; Kojima, T. *J. Org. Chem.* **2003**, *68*, 4959.
[923]Fujisawa, T.; Kurita, Y.; Sato, T. *Chem. Lett.* **1983**, 1537.
[924]Denmark, S.E.; Edwards, J.P.; Nicaise, O. *J. Org. Chem.* **1993**, *58*, 569.
[925]Richey Jr., H.G.; McLane, R.C.; Phillips, C.J. *Tetrahedron Lett.* **1976**, 233.
[926]Bernardi, L.; Cerè, V.; Femoni, C.; Pollicino, S.; Ricci, A. *J. Org. Chem.* **2003**, *68*, 3348.
[927]Laskar, D.D.; Prajapati, D.; Sandu, J.S. *Tetrahedron Lett.* **2001**, *42*, 7883.
[928]Frantz, D.E.; Fässler, R.; Carreira, E.M. *J. Am. Chem. Soc.* **1999**, *121*, 11245. See Pinet, S.; Pandya, S.U.; Chavant, P.Y.; Ayling, A.; Vallee, Y. *Org. Lett.* **2002**, *4*, 1463.

Grignard reagents also add to nitrones.[929] Nitrones react with $CH_2=CHCH_2InBr$ in aq. DMF to give the homoallylic oxime[930] and silyl ketene acetals add in the presence of a chiral titanium catalyst to good enantioselectivity.[931] Hydrazone derivatives react with iodoalkenes in the presence of $InCl_3$ and $Mn_2(CO)_{10}$ under photochemical conditions to give the hydrazine derivative.[932] Indium metal promotes the addition of alkyl iodides to hydrazones.[933] A hydrazone can be formed *in situ* by reacting an aldehyde with a hydrazine derivative, and in the presence of tetrallyltin and a scandium catalysts, homoallylic hydrazine derivatives are formed.[934] Ketene dithioacetals add to hydrazones using a chiral zirconium catalyst to give a pyrazolidine.[935]

Radical addition to imines is known. Carbon-centered radicals add to imines.[936] The reaction of an alkyl halide with BEt_3 in aqueous methanol, for example, gives the imine addition product, an alkylated amine.[937] Secondary alkyl iodides add to *O*-alkyl oximes in the presence of BEt_3 and AIBN, and this methodology was used to convert $MeO_2C-CH=NOBn$ to $MeO_2C-CH(R)NOBn$.[938] Benzylic halides adds to imines under photochemical conditions, and in the presence of 1-benzyl-1,4-dihydronicotinamide[939] or with BEt_3 in aqueous methanol.[940] Tertiary alkyl iodides add to oxime ethers using $BF_3 \cdot OEt_2$ in the presence of BEt_3/O_2.[941]

Iminium salts[942] give tertiary amines directly, with just R adding:

$$ \underset{\underset{\displaystyle C}{\overset{\displaystyle \|}{N}}}{R^1 \diagdown \overset{\oplus}{} \diagup R^1} \quad + \quad R\text{-}MgX \quad \longrightarrow \quad \underset{\underset{\displaystyle C}{\overset{\displaystyle |}{N}}}{R \diagdown \diagup R^1} R^1 $$

Chloroiminium salts $ClCH=NR'_2\,Cl^-$ (generated *in situ* from an amide $HCONR'_2$ and phosgene $COCl_2$) react with 2 equivalents of a Grignard reagent RMgX, one adding to the C=N and the other replacing the Cl, to give tertiary amines $R_2CHNR'_2$.[943]

OS **IV**, 605; **VI**, 64. Also see OS **III**, 329.

[929]See Merino, P.; Tejero, T. *Tetrahedron* **2001**, *57*, 8125.

[930]Kumar, H.M.S.; Anjaneyulu, S.; Reddy, E.J.; Yadav, J.S. *Tetrahedron Lett.* **2000**, *41*, 9311.

[931]Murahashi, S.-I.; Imada, Y.; Kawakami, T.; Harada, K.; Yonemushi, Y.; Tomita, N. *J. Am. Chem. Soc.* **2002**, *124*, 2888.

[932]Friedstad, G.K.; Qin, J. *J. Am. Chem. Soc.* **2001**, *123*, 9922.

[933]Miyabe, H.; Ueda, M.; Nishimura, A.; Naito, T. *Tetrahedron* **2004**, *60*, 4227.

[934]Kobayashi, S.; Hamada, T.; Manabe, K. *Synlett* **2001**, 1140.

[935]Yamshita, Y.; Kobayashi, S. *J. Am. Chem. Soc.* **2004**, *126*, 11279.

[936]For a review, see Friestad, G.K. *Tetrahedron* **2001**, *57*, 5461.

[937]Miyabe, H.; Ueda, M.; Naito, T. *J. Org. Chem.* **2000**, *65*, 5043.

[938]Miyabe, H.; Ueda, M.; Yoshioka, N.; Yamakawa, K.; Naito, T. *Tetrahedron* **2000**, *56*, 2413.

[939]Jin, M.; Zhang, D.; Yang, L.; Liu, Y.; Liu, Z. *Tetrahedron Lett.* **2000**, *41*, 7357.

[940]McNabb, S.B.; Ueda, M.; Naito, T. *Org. Lett.* **2004**, *6*, 1911.

[941]Halland, N.; Jørgensen, K.A. *J. Chem. Soc., Perkin Trans. 1* **2001**, 1290.

[942]For a review of nucleophilic addition to iminium salts, see Paukstelis, J.V.; Cook, A.G., in Cook, A.G. *Enamines*, 2nd ed., Marcel Dekker, NY, *1988*, pp. 275–356.

[943]Wieland, G.; Simchen, G. *Liebigs Ann. Chem.* **1985**, 2178.

16-32   Addition of Carbenes and Diazoalkanes to C=N Compounds

In the presence of metal catalysts such as Yb(OTf)$_3$, diazoalkanes add to imines to generate aziridines. An example is:[944]

The reaction is somewhat selective for the cis-diastereomer. The use of chiral additives in this reaction leads to aziridines enantioselectively.[945] Imines can be formed by the reaction of an aldehyde and an amine, and subsequent treatment with Me$_3$SiI and butyllithium gives an aziridine.[946] N-Tosyl imines react with diazoalkenes to form N-tosyl aziridines, with good cis-selectivity[947] and modest enantioselectivity in the presence of a chiral copper catalyst,[948] but excellent enantioselectivity with a chiral rhodium catalyst.[949]. It is noted that N-tosyl aziridines are formed by the reaction of an alkene with PhI=NTs and a copper catalyst.[950] The reaction of alkenes with diazo compounds is discussed in **15-53**.

16-33   The Addition of Grignard Reagents to Nitriles and Isocyanates

Alkyl,oxo-de-nitrilo-tersubstitution (Overall transformation)

N-**Hydro**-*C*-**alkyl-addition**

Ketones can be prepared by addition of Grignard reagents to nitriles, followed by hydrolysis of the initially formed imine anion. Many ketones have been made in this manner, though when both R groups are alkyl, yields are not high.[951] Yields

---

[944]Nagayama, S.; Kobayashi, S. *Chem Lett.* **1998**, 685. Also see, Rasmussen, K.G.; Jørgensen, K.A. *J. Chem. Soc., Chem. Commun.* **1995**, 1401.

[945]Hansen, K.B.; Finney, N.S.; Jacobsen, E.N. *Angew. Chem. Int. Ed.* **1995**, *34*, 676.

[946]Reetz, M.T.; Lee, W.K. *Org. Lett.* **2001**, *3*, 3119.

[947]Aggarwal, V.K.; Ferrara, M. *Org. Lett.* **2000**, *2*, 4107; Hori, R.; Aoyama, T.; Shioiri, T. *Tetrahedron Lett.* **2000**, *41*, 9455; Krumper, J.R.; Gerisch, M.; Suh, J.M.; Bergman, R.G.; Tilley, T.D. *J. Org. Chem.* **2003**, *68*, 9705; Williams, A.L.; Johnston, J.N. *J. Am. Chem. Soc.* **2004**, *126*, 1612; Sun, W.; Xia, C.-G.; Wang, H.-W. *Tetrahedron Lett.* **2003**, *44*, 2409.

[948]Juhl, K.; Hazell, R.G.; Jørgensen, K.A. *J. Chem. Soc., Perkin Trans. 1* **1999**, 2293.

[949]Aggarwal, V.K.; Alonso, E.; Fang, G.; Ferrara, M.; Hynd, G.; Porcelloni, M. *Angew. Chem. Int. Ed.* **2001**, *40*, 1433.

[950]Handy, S.T.; Czopp, M. *Org. Lett.* **2001**, *3*, 1423.

[951]For a review, see Kharasch, M.S.; Reinmuth, O. *Grignard Reactions of Nonmetallic Substances*, Prentice-Hall, Englewood Cliffs, NJ, **1954**, pp. 767–845.

can be improved by the use of Cu(I) salts[952] or by using benzene containing one equivalent of ether as the solvent, rather than ether alone.[953] The ketimine salt does not in general react with Grignard reagents: Hence tertiary alcohols or tertiary alkyl amines are not often side products.[954] By careful hydrolysis of the salt it is sometimes possible to isolate ketimines,[955]

$$R-\underset{\underset{NH}{\|}}{C}-R'$$

especially when R and R′ = aryl. The addition of Grignard reagents to the C≡N group is normally slower than to the C=O group, and cyano group containing aldehydes add the Grignard reagent without disturbing the CN group.[956] Other metal compounds have been used, including Sm with allylic halides[957] and organocerium compounds such as $MeCeCl_2$.[958] Allylic halides react with an excess of zinc metal in the presence of 40% $AlCl_3$, and in the presence of a nitrile homoallylic ketones are produced after hydrolysis.[959] Benzonitrile reacts as with iodopropane and a mixture of SmI_2 and NiI_2 catalysts to give 1-phenyl-1-butanone.[960]

Addition of Grignard reagents[961] or organolithium reagents[962] to ω-halo nitriles leads to 2-substituted cyclic imines.

The following mechanism has been proposed for the reaction of the methyl Grignard reagent with benzonitrile:[963]

Arenes add to nitriles in the presence of a palladium catalyst in DMSO/trifluoroacetic acid to give a diaryl ketone.[964]

The addition of Grignard reagents to isocyanates gives, after hydrolysis, N-substituted amides.[965] This is a very good reaction and can be used to prepare

[952]Weiberth, F.J.; Hall, S.S. *J. Org. Chem.* **1987**, *52*, 3901.

[953]Canonne, P.; Foscolos, G.B.; Lemay, G. *Tetrahedron Lett.* **1980**, 155.

[954]For examples where tertiary amines have been made the main products, see Alvernhe, G.; Laurent, A. *Tetrahedron Lett.* **1973**, 1057; Gauthier, R.; Axiotis, G.P.; Chastrette, M. *J. Organomet. Chem.* **1977**, *140*, 245.

[955]Pickard, P.L.; Toblert, T.L. *J. Org. Chem.* **1961**, *26*,4886.

[956]Cason, J.; Kraus, K.W.; McLeod Jr., W.D. *J. Org. Chem.* **1959**, *24*, 392.

[957]Yu, M.; Zhang, Y.; Guo, H. *Synth. Commun.* **1997**, *27*, 1495.

[958]Ciganek, E. *J. Org. Chem.* **1992**, *57*, 4521.

[959]Lee, A.S.-Y.; Lin, L.-S. *Tetrahedron Lett.* **2000**, *41*, 8803.

[960]Kang, H.-Y.; Song, S.-E. *Tetrahedron Lett.* **2000**, *41*, 937.

[961]Fry, D.F.; Fowler, C.B.; Dieter, R.K. *Synlett* **1994**, 836.

[962]Gallulo, V.; Dimas, L.; Zezza, C.A.; Smith, M.B. *Org. Prep. Proceed. Int.* **1989**, *21*, 297.

[963]Ashby, E.C.; Chao, L.; Neumann, H.M. *J. Am. Chem. Soc.* **1973**, *95*, 4896, 5186.

[964]Zhou, C.; Larock, R.C. *J. Am. Chem. Soc.* **2004**, *126*, 2302.

[965]For a review of this and related reactions, see Screttas, C.G.; Steele, B.R. *Org. Prep. Proceed. Int.* **1990**, *22*, 271.

derivatives of alkyl and aryl halides. The reaction has also been performed with alkyllithium compounds.[966] Isothiocyanates give *N*-substituted thioamides. Other organometallic compounds add to isocyanates. Vinyltin reagents lead to conjugated amides.[967]

It is noted that terminal alkynes add to the carbon of an isonitrile in the presence of a uranium complex, giving a propargylic imine.[968]

OS **III**, 26, 562; **V**, 520.

G. Carbon Attack by Active Hydrogen Compounds

Reactions **16-34**–**16-50** are base-catalyzed condensations (although some of them are also catalyzed to acids).[969] In **16-34**–**16-44**, a base removes a C—H proton to give a carbanion, which then adds to a C=O. The oxygen acquires a proton, and the resulting alcohol may or may not be dehydrated, depending on whether an α hydrogen is present and on whether the new double bond would be in conjugation with double bonds already present:

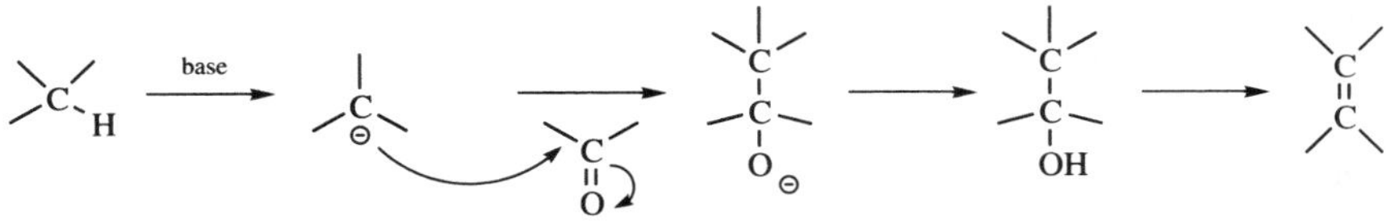

The reactions differ in the nature of the active hydrogen component and the carbonyl component. Table 16.2 illustrates the differences. Reaction **16-50** is an analogous reaction involving addition to C≡N.

16-34   The Aldol Reaction[970]

O-**Hydro-*C*-(α-acylalkyl)-addition; α-Acylalkylidine-de-oxo-bisubstitution**

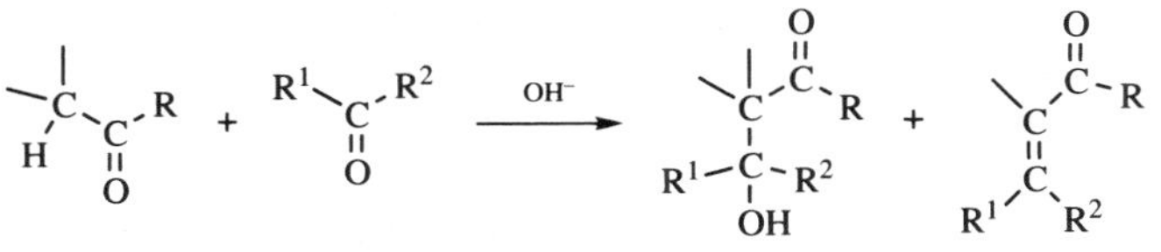

(If α H was present)

[966]LeBel, N.A.; Cherluck, R.M.; Curtis, E.A. *Synthesis* **1973**, 678; Cooke, Jr., M.P.; Pollock, C.M. *J. Org. Chem.* **1993**, *58*, 7474. For another method, see Einhorn, J.; Luche, J.L. *Tetrahedron Lett.* **1986**, *27*, 501.

[967]Niestroj, M.; Neumann, W.P.; Thies, O. *Chem. Ber.* **1994**, *127*, 1131.

[968]Barnea, E.; Andrea, T.; Kapon, M.; Berthet, J.-C.; Ephritikhine, M.; Eisen, M.S. *J. Am. Chem. Soc.* **2004**, *126*, 10860.

[969]For reviews, see House, H.O. *Modern Synthetic Reactions*, 2nd ed., W.A. Benjamin, NY, **1972**, pp. 629–682; Reeves, R.L., in Patai, S. *The Chemistry of the Carbonyl Group*, pt. 1, Wiley, NY, **1966**, pp. 567–619. See also, Stowell, J.C. *Carbanions in Organic Synthesis*, Wiley, NY, **1979**.

[970]See Smith, M.B. *Organic Synthesis*, 2nd ed., McGraw-Hill, NY, **2001**, pp. 740–745.

TABLE 16.2. Base-Catalyzed Condensations Showing the Active-Hydrogen Components and the Carbonyl Compounds

Reaction	Active-Hydrogen Component		Carbonyl Component	Subsequent Reaction
16-34	Aldehyde	$\overset{\displaystyle}{\underset{\displaystyle}{\overset{H}{\underset{CHO}{C}}}}$	Aldehyde, ketone	Dehydration may follow
	Ketone	$\overset{H}{\underset{\overset{\|}{\underset{O}{C}}-R}{C}}$		
Aldol reaction				
16-36	Ester	$\overset{H}{\underset{\overset{\|}{\underset{O}{C}}-OR}{C}}$	Aldehyde, ketone (usually without α-hydrogens)	Dehydration may follow
16-38	$\overset{H}{\underset{H}{C}}\overset{Z}{\underset{Z^1}{}}$, $\overset{R}{\underset{H}{C}}\overset{Z}{\underset{Z^1}{}}$		Aldehyde, ketone (usually without α-hydrogens)	Dehydration (usually follows)
	and similar molecules			
Knoevenagel reaction				
16-41	$\underset{\ominus}{\overset{H}{\underset{\|}{C}}}-SiMe_3$		Aldehyde, ketone	Dehydration may follow
Peterson reaction				
16-42	$\overset{H}{\underset{Z}{C}}$	Z = COR, COOR, NO$_2$	CO$_2$, CS$_2$	
16-39	Anhydride	$\overset{H}{\underset{\overset{\|}{\underset{O}{C}}-O-\overset{\|}{\underset{O}{C}}-OR}{C}}$	Aromatic aldehyde usually follows	Dehydration
Perkin reaction				
16-40	α-Halo Ester	$\overset{X}{\underset{\overset{\|}{\underset{O}{C}}-OR}{\overset{H}{C}}}$	Aldehyde, Ketone (S$_N$ reaction) follows	Epoxidation
Darzen's reaction				
16-43	Aldehyde	$\overset{H}{\underset{CHO}{C}}$	Formaldehyde reaction follows	Crossed-Cannizzaro
	Ketone	$\overset{H}{\underset{\overset{\|}{\underset{O}{C}}-R}{C}}$		

TABLE 16.2. (*Continued*)

Reaction	Active-Hydrogen Component	Carbonyl Component	Subsequent Reaction
Tollens' reaction			
16-44	Phosphorous ylid 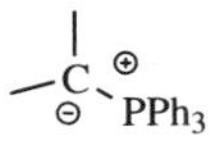	Aldehyde, ketone	"Dehydration" always follows
Wittig reaction			
16-50	Nitrile 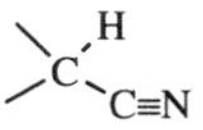		Nitrile
Thorpe reaction			

In the *aldol reaction*,[971] the α carbon of one aldehyde or ketone molecule adds to the carbonyl carbon of another.[972] Although acid-catalyzed aldol reactions are known,[973] the most common form of the reaction uses a base. There is evidence that an SET mechanism can intervene when the substrate is an aromatic ketone.[974] Although hydroxide was commonly used in early versions of this reaction, stronger bases, such as alkoxides (RO$^-$) or amides (R$_2$N$^-$), are also common. Amine bases have been used.[975] Hydroxide ion is not a strong enough base to convert substantially all of an aldehyde or ketone molecule to the corresponding enolate ion, that is., the equilibrium lies well to the left, for both aldehydes and

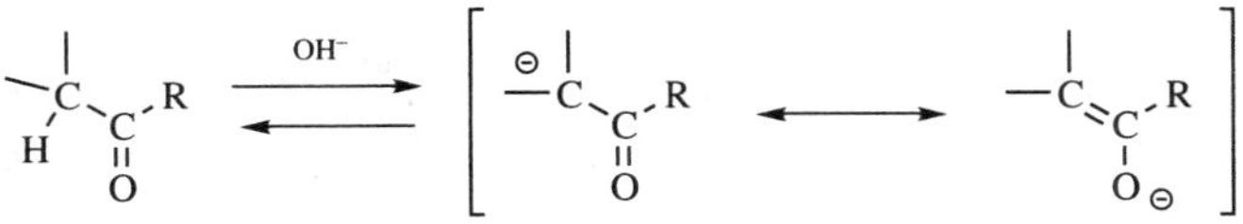

ketones. Nevertheless, enough enolate ion is present for the reaction to proceed:

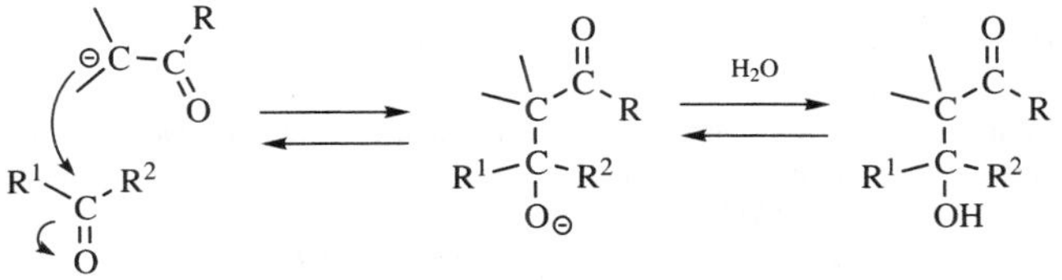

[971]This reaction is also called the *aldol condensation*, though, strictly speaking, this term applies to the formation only of the α,β-unsaturated product, and not the aldol.
[972]For reviews, see Thebtaranonth, C.; Thebtaranonth, Y., in Patai, S.; Rappoport, Z. *The Chemistry of Enones*, pt. 1, Wiley, NY, *1989*, pp. 199–280, 199–212; Hajos, Z.G., in Augustine, R.L. *Carbon–Carbon Bond Formation*, Vol. 1; Marcel Dekker, NY, *1979*; pp. 1–84; Nielsen, A.T.; Houlihan, W.J. *Org. React.* *1968*, *16*, 1.
[973]For example, see Mahrwald, R.; Gündogan, B. *J. Am. Chem. Soc.* *1998*, *120*, 413.
[974]Ashby, E.C.; Argyropoulos, J.N. *J. Org. Chem.* *1986*, *51*, 472.
[975]Trost, B.M.; Silcoff, E.R.; Ito, H. *Org. Lett.* *2001*, *3*, 2497.

This equilibrium lies further to the right with alkoxide and especially with amide bases, depending on the solvent. Protic solvents, such as water or alcohol, are acidic enough to react with the enolate anion and shift the equilibrium to the left. In an aprotic solvent, such as ether or THF, with a strong amide base, such as lithium diisopropylamide (LDA, p. 389), the equilibrium lies more to the right.[976] A variety of amide bases can be used to deprotonate the ketone or aldehyde, and in the case of an unsymmetrical ketone removal of the more acidic proton leads to the kinetic enolate anion.[977] Note that a polymer-bound amide base has been used[978] and solid-phase chiral lithium amides are known.[979] A polymer-supported phosphoramide has been used as a catalyst for the aldol condensation.[980] The product is a β-hydroxy aldehyde (called an *aldol*) or ketone, which in some cases is dehydrated during the course of the reaction. In aprotic solvents with a mild workup procedure, however, the aldol is readily isolated unless the substrate is an aromatic aldehyde or ketone. The aldol reaction has been done in ionic liquids.[981] Even if the dehydration is not spontaneous, it can usually be done easily, since the new double bond is in conjugation with the C=O bond; so that this is a method of preparing α,β-unsaturated aldehydes and ketones, as well as β-hydroxy aldehydes and ketones. One-pot procedures have been reported to give the conjugated product.[982] The entire reaction is an equilibrium (including the dehydration step), and α,β-unsaturated and β-hydroxy aldehydes and ketones can be cleaved by treatment with ⁻OH (the *retrograde aldol reaction*). The *retro*-aldol condensation has been exploited for crossed-aldol reactions.[983] A vinylogous aldol reaction is known[984] as is a 1"double" aldol.[985] Enzyme-mediated aldol reactions have been reported using two aldehydes, including formaldehyde.[986]

Under the principle of vinylogy, the active hydrogen can be one in the γ position of an α,β-unsaturated carbonyl compound:

After hydrolysis

[976]For a discussion of solvent and temperature effects, see Cainelli, G.; Galletti, P.; Giacomini, D.; Orioli, P. *Tetrahedron Lett.* **2001**, *42*, 7383.

[977]See Xie, L.; Vanlandeghem, K.; Isenberger, K.M.; Bernier, C. *J. Org. Chem.* **2003**, *68*, 641; Zhao, P.; Lucht, B.L.; Kenkre, S.L.; Collum, D.B. *J. Org. Chem.* **2004**, *69*, 242; Zhao, P.; Condo, A.; Keresztes, I.; Collum, D.B. *J. Am. Chem. Soc.* **2004**, *126*, 3113.

[978]Seki, A.; Ishiwata, F.; Takizawa, Y.; Asami, M. *Tetrahedron* **2004**, *60*, 5001.

[979]Johansson, A.; Abrahamsson, P.; Davidsson, Ö. *Tetrahedron Asymmetry* **2003**, *14*, 1261.

[980]Flowers II, R.A.; Xu, X.; Timmons, C.; Li, G. *Eur. J. Org. Chem.* **2004**, 2988.

[981]In bmim BF₄, 1-butyl-3-methylimidazolium tetrafluoroborate: Zheng, X.; Zhang, Y. *Synth. Commun.* **2003**, 161.

[982]Kourouli, T.; Kefalas, P.; Ragoussis, N.; Ragoussis, V. *J. Org. Chem.* **2002**, *67*, 4615.

[983]For an example, see Simpura, I.; Nevalainen, V. *Angew. Chem. Int. Ed.* **2000**, *39*, 3422.

[984]For reviews, see Casiraghi, G.; Zanardi, F.; Appendino, G.; Rassu, G.; *Chem. Rev.* **2000**, *100*, 1929; Casiraghi, G.; Zanardi, E.; Rassu, G. *Pure Appl. Chem.* **2000**, *72*, 1645.

[985]For a discussion of the mechanism of this reaction see Abiko, A.; Inoue, T.; Masamune, S. *J. Am. Chem. Soc.* **2002**, *124*, 10759.

[986]Demir, A.S.; Ayhan, P.; Igdir, A.C.; Duygu, A.N. *Tetrahedron* **2004**, *60*, 6509.

The scope of the aldol reaction may be discussed under five headings:

1. *Reaction between Two Molecules of the Same Aldehyde.* Hydroxide or alkoxide bases are used in protic solvents,[987] and the reaction is quite feasible. Many aldehydes have been converted to aldols and/or their dehydration products in this manner. The most effective catalysts are basic ion-exchange resins. Of course, the aldehyde must possess an α hydrogen.

2. *Reaction between Two Molecules of the Same Ketone.* With hydroxide or alkoxide bases in protic solvents the equilibrium lies well to the left,[988] and the reaction is feasible only if the equilibrium can be shifted. This can often be done by allowing the reaction to proceed in a Soxhlet extractor (e.g., see OS **I**, 199). Two molecules of the same ketone can also be condensed without a Soxhlet extractor,[989] by treatment with basic Al_2O_3.[990] Unsymmetrical ketones condense on the side that has more hydrogens. An exception is butanone, which reacts at the CH_2 group with acid catalysts, though with basic catalysts, it too reacts at the CH_3 group.

   Alternatively, the use of an amide base, such as LDA or lithium hexamethyldisilazide (p. 389), in aprotic solvents, such as ether or THF, at low temperatures, generates an enolate anion under conditions where the equilibrium lies more to the right. A second equivalent of the ketone can then be added. Clearly, this technique is effective in reactions of aldehydes.

3. *Reaction between Two Different Aldehydes.* In the most general case, this will produce a mixture of four products (eight, if the alkenes are counted). However, if one aldehyde does not have an α hydrogen, only two aldols are possible, and in many cases the crossed product is the main one. The crossed-aldol reaction is often called the *Claisen–Schmidt reaction.*[991] The crossed aldol is readily accomplished using amide bases in aprotic solvent. The first aldehyde is treated with LDA in THF at −78°C, for example, to form the enolate anion. Subsequent treatment with a second aldehyde leads to the mixed aldol product. The crossed aldol of two aldehydes has been done using potassium *tert*-butoxide and $Ti(OBu)_4$.[992]

4. *Reaction between Two Different Ketones.* This is seldom attempted with hydroxide or alkoxide bases in protic solvents since similar considerations apply to those discussed for aldehydes. This reaction is commonly done with amide bases in aprotic solvents, but with somewhat more difficulty than with aldehydes.

[987]For discussions of equilibrium constants in aldol reactions, see Guthrie, J.P.; Wang, X. *Can. J. Chem.* **1991**, *69*, 339; Guthrie, J.P. *J. Am. Chem. Soc.* **1991**, *113*, 7249, and references cited therein.

[988]The equilibrium concentration of the product from acetone in pure acetone was determined to be 0.01%: Maple, S.R.; Allerhand, A. *J. Am. Chem. Soc.* **1987**, *109*, 6609.

[989]For another method, see Barot, B.C.; Sullins, D.W.; Eisenbraun, E.J. *Synth. Commun.* **1984**, *14*, 397.

[990]Muzart, J. *Synthesis* **1982**, 60; *Synth. Commun.* **1985**, *15*, 285.

[991]For an aqueous version, see Buonora, P.T.; Rosauer, K.G.; Dai, L. *Tetrahedron Lett.* **1995**, *36*, 4009.

[992]Han, Z.; Yorimitsu, H.; Shinokubo, H.; Oshima, K. *Tetrahedron Lett.* **2000**, *41*, 4415.

5. *Reaction between an Aldehyde and a Ketone.* This is usually feasible with hydroxide or alkoxides bases in protic solvents, particularly when the aldehyde has no α hydrogen, since there is no competition from ketone condensing with itself.[993] This is also called the *Claisen–Schmidt reaction.* Even when the aldehyde has an α hydrogen, it is generally the α carbon of the ketone that adds to the carbonyl of the aldehyde, not the other way around. Mixtures are usually produced, however. If the ketone or the aldehyde is treated with an amide base in aprotic solvents, a second aldehyde or ketone can be added to give the aldolate with high regioselectivity. The reaction can be also made regioselective by preparing an enol derivative of the ketone separately[994] and then adding this to the aldehyde (or ketone). Other types of preformed derivatives that react with aldehydes and ketones are enamines (with a Lewis acid catalyst),[995] and enol borinates $R'CH{=}CR^2{-}OBR_2$[996] (which can be synthesized by **15-27**) or directly from an aldehyde or ketone[997]). Preformed metallic enolates are also used. For example, lithium enolates[998] (prepared by **12-23**) react with the substrate in the presence of $ZnCl_2$;[999] in this case the aldol product is stabilized by chelation of its two oxygen atoms with the zinc ion.[1000] Other metallic enolates can be used for aldol reactions, either preformed or generated *in situ* with a catalytic amount of a metal compound. Metals used for this purpose include Mg,[1001] Ti,[1002] Zr,[1003] Pd,[1004]

---

[993]For a study of the rate and equilibrium constants in the reaction between acetone and benzaldehyde, see Guthrie, J.P.; Cossar, J.; Taylor, K.F. *Can. J. Chem.* **1984**, *62*, 1958. For a microwave induced reaction using aqueous NaOH, see Kad, G.L.; Kaur, K.P.; Singh, V.; Singh, J. *Synth. Commun.* **1999**, *29*, 2583.

[994]For some other aldol reactions with preformed enol derivatives, see Mukaiyama, T. *Isr. J. Chem.* **1984**, *24*, 162; Caine, D., in Augustine, R.L., *Carbon–Carbon Bond Formation*, Vol. 1, Marcel Dekker, NY, **1979**, pp. 264–276.

[995]Takazawa, O.; Kogami, K.; Hayashi, K. *Bull. Chem. Soc. Jpn.* **1985**, *58*, 2427.

[996]Inoue, T.; Mukaiyama, T. *Bull. Chem. Soc. Jpn.* **1980**, *53*, 174; Hooz, J.; Oudenes, J.; Roberts, J.L.; Benderly, A. *J. Org. Chem.* **1987**, *52*, 1347; Nozaki, K.; Oshima, K.; Utimoto, K. *Tetrahedron Lett.* **1988**, *29*, 1041. For a review, see Pelter, A.; Smith, K.; Brown, H.C. *Borane Reagents*, Academic Press, NY, **1988**, pp. 324–333. For an *ab initio* study see Murga, J.; Falomir, E.; Carda, M.; Marco, J.A. *Tetrahedron* **2001**, *57*, 6239.

[997]For conversion of ketones to either (*Z*) or (*E*) enol borinates, see, for example, Evans, D.A.; Nelson, J.V.; Vogel, E.; Taber, T.R. *J. Am. Chem. Soc.* **1981**, *103*, 3099; Brown, H.C.; Dhar, R.K.; Bakshi, R.K.; Pandiarajan, P.K.; Singaram, B. *J. Am. Chem. Soc.* **1989**, *111*, 3441; Brown, H.C.; Ganesan, K. *Tetrahedron Lett.* **1992**, *33*, 3421.

[998]For a complete structure–energy analysis of one such reaction, see Arnett, E.M.; Fisher, F.J.; Nichols, M.A.; Ribeiro, A.A. *J. Am. Chem. Soc.* **1990**, *112*, 801.

[999]House, H.O.; Crumrine, D.S.; Teranishi, A.Y.; Olmstead, H.D. *J. Am. Chem. Soc.* **1973**, *95*, 3310.

[1000]It has been contended that such stabilization is not required: Mulzer, J.; Brüntrup, G.; Finke, J.; Zippel, M. *J. Am. Chem. Soc.* **1979**, *101*, 7723.

[1001]Wei, H.-X.; Jasoni, R.L.; Shao, H.; Hu, J.; Paré, P.W. *Tetrahedron* **2004**, *60*, 11829.

[1002]Stille, J.R.; Grubbs, R.H. *J. Am. Chem. Soc.* **1983**, *105*, 1664; Mahrwald, R.; Costisella, B.; Gündogan, B. *Tetrahedron Lett.* **1997**, *38*, 4543. For the use of Ti(O*i*Pr)₄ to modify syn/anti ratios of aldol products, see Mahrwald, R.; Costisella, B.; Gündogan, B. *Synthesis* **1998**, 262.

[1003]Evans, D.A.; McGee, L.R. *Tetrahedron Lett.* **1980**, *21*, 3975; *J. Am. Chem. Soc.* **1981**, *103*, 2876.

[1004]Nokami, J.; Mandai, T.; Watanabe, H.; Ohyama, H.; Tsuji, J. *J. Am. Chem. Soc.* **1989**, *111*, 4126.

In,[1005] Sn,[1006] La,[1007] and Sm,[1008] all of which give products with moderate to excellent diastereoselectivity[1009] and regioselectivity. α-Alkoxy ketones react with lithium enolates particularly rapidly.[1010] A bis(aldol) condensation has been reported with epoxy ketones and aldehydes using SmI$_2$.[1011] Vinyl silanes react with aldehydes in the presence of a copper catalyst to vie the aldol product.[1012]

The reactions with preformed enol derivatives provide a way to control the stereoselectivity of the aldol reaction.[1013] As with the Michael reaction (**15-24**), the aldol reaction creates two new stereogenic centers, and, in the most general case, there are four stereoisomers of the aldol product (two racemic diastereomers), which can be represented as

| syn (or erythro) ($\pm$) pair | anti (or threo) ($\pm$) pair |

Among the preformed enol derivatives used for diastereoselective aldol condensations have been enolates of Li,[1014] Mg, Ti,[1015] Zr,[343] and Sn,[1016] silyl enol

---

[1005]Loh, T.-P.; Wei, L.-L.; Feng, L.-C. *Synlett* **1999**, 1059. For an example using ultrasound and InCl$_3$, see Loh, T.-P.; Feng, L.-C.; Wei, L.-L. *Tetrahedron* **2001**, *57*, 4231.

[1006]Yanagisawa, A.; Kimura, K.; Nakatsuka, Y.; Yamamoto, H. *Synlett* **1998**, 958.

[1007]Kobayashi, S.; Hachiya, I.; Takahori, T. *Synthesis* **1993**, 371.

[1008]Yokoyama, Y.; Mochida, K. *Synlett* **1996**, 445; Sasai, H.; Arai, S.; Shibasaki, M. *J. Org.Chem.* **1994**, *59*, 2661. Also see, Bao, W.; Zhang, Y.; Wang, J. *Synth. Commun.* **1996**, *26*, 3025.

[1009]For a review, see Mahrwald, R. *Chem. Rev.* **1999**, *99*, 1095.

[1010]Das, G.; Thornton, E.R. *J. Am. Chem. Soc.* **1990**, *112*, 5360.

[1011]Mukaiyama, T.; Arai, H.; Shiina, I. *Chem. Lett.* **2000**, 580.

[1012]Yang, B.-Y.; Chen, X.-M.; Deng, G.-J.; Zhang, Y.-L.; Fan, Q.-H. *Tetrahedron Lett.* **2003**, *44*, 3535.

[1013]For reviews, see Heathcock, C.H. *Aldrichimica Acta* **1990**, *23*, 99; *Science* **1981**, *214*, 395; Nógrádi, M. *Stereoselective Synthesis*, VCH, NY, **1986**, pp. 193–220; Heathcock, C.H., in Morrison, J.D. *Asymmetric Synthesis*, Vol. 3, Academic Press, NY, **1984**, pp. 111–212; Heathcock, C.H., in Buncel, E.; Durst, T. *Comprehensive Carbanion Chemistry*, pt. B, Elsevier, NY, **1984**, pp. 177–237; Evans, D.A.; Nelson, J.V.; Taber, T.R. *Top. Stereochem.* **1982**, *13*, 1; Evans, D.A. *Aldrichimica Acta* **1982**, *15*, 23; Braun, M.; Sacha, H.; Galle, D.; Baskaran, S. *Pure Appl. Chem.* **1996**, *68*, 561. For a discussion of how configuration and conformation influence the stereochemistry of aldols, see Kitamura, M.; Nakano, K.; Miki, T.; Okada, M.; Noyori, R. *J. Am. Chem. Soc.* **2001**, *123*, 8939.

[1014]Fellmann, P.; Dubois, J.E. *Tetrahedron* **1978**, *34*, 1349; Heathcock, C.H.; Pirrung, M.C.; Montgomery, S.H.; Lampe, J. *Tetrahedron* **1981**, *37*, 4087; Masamune, S.; Ellingboe, J.W.; Choy, W. *J. Am. Chem. Soc.* **1982**, *104*, 5526; Ertas, M.; Seebach, D. *Helv. Chim. Acta* **1985**, *68*, 961.

[1015]Nerz-Stormes, M.; Thornton, E.R. *Tetrahedron Lett.* **1986**, 897; Evans, D.A.; Rieger, D.L.; Bilodeau, M.T.; Urpí, F. *J. Am. Chem. Soc.* **1991**, *113*, 1047; Cosp. A.; Larrosa, I.; Vilasís, I.; Romea, P.; Urpí, F.; Vilarrasa, J. *Synlett* **2003**, 1109.

[1016]Mukaiyama, T.; Iwasawa, N.; Stevens, R.W.; Haga, T. *Tetrahedron* **1984**, *40*, 1381; Labadie, S.S.; Stille, J.K. *Tetrahedron* **1984**, *40*, 2329; Yura, T.; Iwasawa, N.; Mukaiyama, T. *Chem. Lett.* **1986**, 187. See also, Nakamura, E.; Kuwajima, I. *Tetrahedron Lett.* **1983**, *24*, 3347.

ethers,[1017] enol borinates,[1018] and enol borates $R'CH=CR^2-OB(OR)_2$.[1019] The nucleophilicity of silyl enol ethers has been examined.[1020] Base-induced formation of the enolate anion generally leads to a mixture of (*E*)- and (*Z*)-isomers, and dialkyl amide bases are used in most cases. The (*E*/*Z*) stereoselectivity depends on the structure of the lithium dialkylamide base, with the highest (*E*/*Z*) ratios obtained with LiTMP-butyllithium mixed aggregates in THF.[1021] The use of LiHMDS resulted in a reversal of the (*E*/*Z*) selectivity. In general, metallic (*Z*) enolates give the syn (or erythro) pair, and this reaction is highly useful for the diastereoselective synthesis of these products.[1022] The (*E*) isomers generally react nonstereoselectively. However, anti (or threo) stereoselectivity has been achieved in a number of cases, with titanium enolates,[1023] with magnesium enolates,[1024] with certain enol borinates,[1025] and with lithium enolates at −78°C.[1026] Enolization accounts for syn–anti isomerization of aldols.[1027] In another variation, a β-keto Weinreb amide (see **16-82**) reacted with TiCl$_4$ and Hünig's base (*i*Pr$_2$NEt) and then an aldehyde to give the β-hydroxy ketone.[1028]

[1017]Matsuda, I.; Izumi, Y. *Tetrahedron Lett.* **1981**, *22*, 1805; Yamamoto, Y.; Maruyama, K.; Matsumoto, K. *J. Am. Chem. Soc.* **1983**, *105*, 6963; Sakurai, H.; Sasaki, K.; Hosomi, A. *Bull. Chem. Soc. Jpn.* **1983**, *56*, 3195; Hagiwara, H.; Kimura, K.; Uda, H. *J. Chem. Soc., Chem. Commun.* **1986**, 860.

[1018]Evans, D.A.; Nelson, J.V.; Vogel, E.; Taber, T.R. *J. Am. Chem. Soc.* **1981**, *103*, 3099; Evans, D.A.; Bartroli, J.; Shih, T.L. *J. Am. Chem. Soc.* **1981**, *103*, 2127; Masamune, S.; Choy, W.; Kerdesky, F.A.J.; Imperiali, B. *J. Am. Chem. Soc.* **1981**, *103*, 1566; Paterson, I.; Goodman, J.M.; Lister, M.A.; Schumann, R.C.; McClure, C.K.; Norcross, R.D. *Tetrahedron* **1990**, *46*, 4663; Walker, M.A.; Heathcock, C.H. *J. Org. Chem.* **1991**, *56*, 5747. For reviews, see Paterson, I. *Chem. Ind. (London)* **1988**, 390; Pelter, A.; Smith, K.; Brown, H.C. *Borane Reagents*, Academic Press, NY, **1988**, p. 324.

[1019]Hoffmann, R.W.; Ditrich, K.; Fröch, S. *Liebigs Ann. Chem.* **1987**, 977.

[1020]Patz, M.; Mayr, H. *Tetrahedron Lett.* **1993**, *34*, 3393.

[1021]Pratt, L. M.; Newman, A.; Cyr, J. S.; Johnson, H.; Miles, B.; Lattier, A.; Austin, E.; Henderson, S.; Hershey, B.; Lin, M.; Balamraju, Y.; Sammonds, L.; Cheramie, J.; Karnes, J.; Hymel, E.; Woodford, B.; Carter, C. *J. Org. Chem.* **2003**, *68*, 6387.

[1022]For discussion of transition-state geometries in this reaction, see Hoffmann, R.W.; Ditrich, K.; Froech, S.; Cremer, D. *Tetrahedron* **1985**, *41*, 5517; Anh, N.T.; Thanh, B.T. *Nouv. J. Chim.*, **1986**, *10*, 681; Li, Y.; Paddon-Row, M.N.; Houk, K.N. *J. Org. Chem.* **1990**, *55*, 481; Denmark, S.E.; Henke, B.R. *J. Am. Chem. Soc.* **1991**, *113*, 2177.

[1023]See Murphy, P.J.; Procter, G.; Russell, A.T. *Tetrahedron Lett.* **1987**, *28*, 2037; Nerz-Stormes, M.; Thornton, E.R. *J. Org. Chem.* **1991**, *56*, 2489.

[1024]Swiss, K.A.; Choi, W.; Liotta, D.; Abdel-Magid, A.F.; Maryanoff, C.A. *J. Org. Chem.* **1991**, *56*, 5978.

[1025]Masamune, S.; Sato, T.; Kim, B.M.; Wollmann, T.A. *J. Am. Chem. Soc.* **1986**, *108*, 8279; Danda, H.; Hansen, M.M.; Heathcock, C.H. *J. Org. Chem.* **1990**, *55*, 173. See also, Corey, E.J.; Kim, S.S. *Tetrahedron Lett.* **1990**, *31*, 3715.

[1026]Hirama, M.; Noda, T.; Takeishi, S.; Itô, S. *Bull. Chem. Soc. Jpn.* **1988**, *61*, 2645; Majewski, M.; Gleave, D.M. *Tetrahedron Lett.* **1989**, *30*, 5681.

[1027]Ward, D.E.; Sales, M.; Sasmal, P.K. *Org. Lett.* **2001**, *3*, 3671; Ward, D.E.; Sales, M.; Sasmal, P.K. *J. Org. Chem.* **2004**, *69*, 4808.

[1028]Calter, M.A.; Guo, X.; Liao, W. *Org. Lett.* **2001**, *3*, 1499.

These reactions can also be made enantioselective[1029] (in which case only one of the four isomers predominates)[1030] by using[1031] chiral enol derivatives,[1032] chiral aldehydes or ketones,[1033] or both.[1034] Chiral bases[1035] can be used, such as proline,[1036] proline derivatives,[1037] or chiral additives, used in conjunction with the base.[1038] A chiral binaphthol dianion has been used to catalyze the reaction.[1039] Chiral auxiliaries[1040] have been developed that can be used in conjunction with the aldol condensation, as well as chiral catalysts[1041] and chiral ligands[1042]

[1029]For a review, see Allemann, C.; Gordillo, R.; Clemente, F.R.; Cheong, P.H.-Y.; Houk, K.N. *Acc. Chem. Res.* **2004**, *37*, 558; Saito, S.; Yamamoto, H. *Acc. Chem. res.* **2004**, *37*, 570. For a discussion of chelation versus nonchelation control, see Yan, T.-H.; Tan, C.-W.; Lee, H.-C.; Lo, H.-C.; Huang, T.-Y. *J. Am. Chem. Soc.* **1993**, *115*, 2613. For the effects of lithium salts on enantioselective deprotonation, see Majewski, M.; Lazny, R.; Nowak, P. *Tetrahedron Lett.* **1995**, *36*, 5465. Also see, Smith, M.B. *Organic Synthesis*, 2nd ed., McGraw-Hill, NY, **2001**, pp. 779–790.

[1030]For anti-selective aldol reactions, see Oppolzer, W.; Lienard, P. *Tetrahedron Lett.* **1993**, *34*, 4321. For a "non-Evans" *syn*-aldol, see Yan, T.-H.; Lee, H.-C.; Tan, C.-W. *Tetrahedron Lett.* **1993**, *34*, 3559.

[1031]For reviews, see Klein, J., in Patai, S. *Supplement A: The Chemistry of Double-Bonded Functional Groups*, Vol. 2, pt. 1, Wiley, NY, **1989**, pp. 567–677; Braun, M. *Angew. Chem. Int. Ed.* **1987**, *26*, 24.

[1032]For examples, see Eichenauer, H.; Friedrich, E.; Lutz, W.; Enders, D. *Angew. Chem. Int. Ed.* **1978**, *17*, 206; Meyers, A.I.; Yamamoto, Y. *Tetrahedron* **1984**, *40*, 2309; Ando, A.; Shioiri, T. *J. Chem. Soc., Chem. Commun.* **1987**, 1620; Muraoka, M.; Kawasaki, H.; Koga, K. *Tetrahedron Lett.* **1988**, *29*, 337; Paterson, I.; Goodman, J.M. *Tetrahedron Lett.* **1989**, *30*, 997; Siegel, C.; Thornton, E.R. *J. Am. Chem. Soc.* **1989**, *111*, 5722; Gennari, C.; Molinari, F.; Cozzi, P.; Oliva, A. *Tetrahedron Lett.* **1989**, *30*, 5163; Faunce, J.A.; Grisso, B.A.; Mackenzie, P.B. *J. Am. Chem. Soc.* **1991**, *113*, 3418.

[1033]For example, see Ojima, I.; Yoshida, K.; Inaba, S. *Chem. Lett.* **1977**, 429; Heathcock, C.H.; Flippin, L.A. *J. Am. Chem. Soc.* **1983**, *105*, 1667; Reetz, M.T.; Kesseler, K.; Jung, A. *Tetrahedron* **1984**, *40*, 4327.

[1034]For example, see Heathcock, C.H.; White, C.T.; Morrison, J.J.; VanDerveer, D. *J. Org. Chem.* **1981**, *46*, 1296; Short, R.P.; Masamune, S. *Tetrahedron Lett.* **1987**, *28*, 2841.

[1035]For a review, see Notz, W.; Tanaka, F.; Barbas III, C.F. *Acc. Chem. Res.* **2004**, *37*, 580.

[1036]Notz, W.; List, B. *J. Am. Chem. Soc.* **2000**, *122*, 7386; List, B.; Pojarliev, P.; Castello, C. *Org. Lett.* **2001**, *3*, 573; Sakthivel, K.; Notz, W.; Bui, T.; Barbas III, C.F. *J. Am. Chem. Soc.* **2001**, *123*, 5260; Northrup, A.B.; MacMillan, D.W.C. *J. Am. Chem. Soc.* **2002**, *124*, 6798. See Peng, Y.-Y.; Ding, Q.-P.; Li, Z.; Wang, P.G.; Cheng, J.-P. *Tetrahedron Lett.* **2003**, *44*, 3871; Darbre, T.; Machuqueiro, M. *Chem. Commun.* **2003**, 1090; Nyberg, A.I.; Usano, A.; Pihko, P.M. *Synlett* **2004**, 1891. For an example with formaldehyde, see Casas, J.; Sundén, H.; Córdova, A. *Tetrahedron Lett.* **2004**, *45*, 6117. For a proline-catalyzed high pressure reaction, see Sekiguchi, Y.; Sasaoka, A.; Shimomoto, A.; Fujioka, S.; Kotsuki, H. *Synlett* **2003**, 1655.

[1037]Tang, Z.; Jiang, F.; Yu, L.-T.; Cui, X.; Gong, L.-Z.; Mi, A.-Q.; Jiang, Y.-Z.; Wu, Y.-D. *J. Am. Chem. Soc.* **2003**, *125*, 5262; Zhong, G.; Fan, J.; Barbas III, C.F. *Tetrahedron Lett.* **2004**, *45*, 5681.

[1038]See Mahrwald, R. *Org. Lett.* **2000**, *2*, 4011.

[1039]Nakajima, M.; Orito, Y.; Ishizuka, T.; Hashimoto, S. *Org. Lett.* **2004**, *6*, 3763.

[1040]Hein, J.E.; Hultin, P.G. *Synlett* **2003**, 635.

[1041]Suzuki, T.; Yamagiwa, N.; Matsuo, Y.; Sakamoto, S.; Yamaguchi, K.; Shibasaki, M.; Noyori, R. *Tetrahedron Lett.* **2001**, *42*, 4669. For a review, see Alcaidi, B.; Almendros, P. *Eur. J. Org. Chem.* **2002**, 1595.

[1042]Trost, B.M.; Ito, H. *J. Am. Chem. Soc.* **2000**, *122*, 12003.

in catalytic reactions. Aldehydes are condensed with ketones with potassium hexamethyldisilazide (KHMDS) and 8% of a chiral lithium catalyst, giving the aldol product with moderate enantioselectivity.[1043] Structural variations in the aldehyde or ketone are compatible with many enantioselective condensation reactions. An α-hydroxy ketone was condensed with an aldehyde using a chiral zinc catalyst to give the aldol (an α,β-dihydroxy ketone) with good syn selectivity and good enantioselectivity.[1044] A catalytic amount of a nicotine metabolite allows an enantioselective reaction in aqueous media.[1045] Chiral vinylogous aldol reactions have been reported.[1046]

Silyl enol ethers react with aldehydes in the presence of chiral boranes[1047] or other additives[1048] to give aldols with good asymmetric induction (see the Mukaiyama aldol reaction in **16-35**). Chiral boron enolates have been used.[1049] Since both new stereogenic centers are formed enantioselectively, this kind of process is called *double asymmetric synthesis*.[1050] Where both the enolate derivative and substrate were achiral, carrying out the reaction in the presence of an optically active boron compound[1051] or a diamine coordinated with a tin compound[1052] gives the aldol product with excellent enantioselectivity for one stereoisomer. Formation of the magnesium enolate anion of a chiral amide, adds to aldehydes to give the alcohol enantioselectively.[1053]

Diamine protonic acids have been used for catalytic asymmetric aldol reaction.[1054] Boron triflate derivatives, R_2BOTf, have been used for the condensation of ketals and ketone to give β-alkoxy ketones.[1055]

[1043]Yoshikawa, N.; Yamada, Y.M.A.; Das, J.; Sasai, H.; Shibasaki, M. *J. Am. Chem. Soc.* **1999**, *121*, 4168.

[1044]Kumagai, N.; Matsunaga, S.; Yoshikawa, N.; Ohshima, T.; Shibasaki, M. *Org. Lett.* **2001**, *3*, 1539; Yoshikawa, N.; Kumagai, N.; Matsunaga, S.; Moll, G.; Ohshma, T.; Suzuki, T.; Shibasaki, M. *J. Am. Chem. Soc.* **2001**, *123*, 2466; Trost, B.M.; Ito, H.; Silcoff, E.R. *J. Am. Chem. Soc.* **2001**, *123*, 3367.

[1045]Dickerson, T.J.; Janda, K.D. *J. Am. Chem. Soc.* **2002**, *124*, 3220.

[1046]Takikawa, H.; Ishihara, K.; Saito, S.; Yamamoto, H. *Synlett* **2004**, 732; Denmark, S.E.; Heemstra, Jr., J.R. *Synlett* **2004**, 2411.

[1047]Ishihara, K.; Maruyama, T.; Mouri, M.; Gao, Q.; Furuta, K.; Yamamoto, H. *Bull. Chem. Soc. Jpn.* **1993**, *66*, 3483.

[1048]Corey, E.J.; Cywin, C.L.; Roper, T.D. *Tetrahedron Lett.* **1992**, *33*, 6907.

[1049]See Yoshida, K.; Ogasawara, M.; Hayashi, T. *J. Org. Chem.* **2003**, *68*, 1901.

[1050]For a review, see Masamune, S.; Choy, W.; Petersen, J.S.; Sita, L.R. *Angew. Chem. Int. Ed.* **1985**, *24*, 1.

[1051]Corey, E.J.; Kim, S.S. *J. Am. Chem. Soc.* **1990**, *112*, 4976; Furuta, K.; Maruyama, T.; Yamamoto, H. *J. Am. Chem. Soc.* **1991**, *113*, 1041; Kiyooka, S.; Kaneko, Y.; Komura, M.; Matsuo, H.; Nakano, M. *J. Org. Chem.* **1991**, *56*, 2276. For a review, see Bernardi, A.; Gennari, C.; Goodman, J.M.; Paterson, I. *Tetrahedron Asymmetry* **1995**, *6*, 2613.

[1052]Mukaiyama, T.; Uchiro, H.; Kobayashi, S. *Chem. Lett.* **1990**, 1147.

[1053]Evans, D.A.; Tedrow, J.S.; Shaw, J.T.; Downey, C.W. *J. Am. Chem. Soc.* **2002**, *124*, 392.

[1054]Saito, S.; Nakadai, M.; Yamamoto, H. *Synlett* **2001**, 1245; Trost, B.M.; Fettes, A.; Shireman, B.T. *J. Am. Chem. Soc.* **2004**, *126*, 2660.

[1055]Li, L.-S.; Das, S.; Sinha, S.C. *Org. Lett.* **2004**, *6*, 127.

It is possible to make the α carbon of the aldehyde add to the carbonyl carbon of the ketone, by using an imine instead of an aldehyde, and LiN(*i*Pr)$_2$ as the base:[1056]

This is known as a *directed aldol reaction*. Similar reactions have been performed with α-lithiated dimethylhydrazones of aldehydes or ketones[1057] and with α-lithiated aldoximes.[1058]

The aldol reaction can also be performed with acid catalysts, as mentioned above, in which case dehydration usually follows. Here, there is initial protonation of the carbonyl group, which attacks the α carbon of the *enol* form of the other molecule:[1059]

With respect to the enol, this mechanism is similar to that of halogenation (**12-4**). A side reaction that is sometimes troublesome is further condensation, since the product of an aldol reaction is still an aldehyde or ketone. The aldol condensation of aldehydes has also been done using a mixture of pyrrolidine and benzoic acid.[1060]

The intramolecular aldol condensation is well known, and aldol reactions are often used to close five- and six-membered rings. Because of the favorable entropy (p. 303), such ring closures generally take place with ease[1061] when using hydroxide or alkoxide bases in protic solvents. In aprotic solvents with amide bases,

[1056]Wittig, G.; Frommeld, H.D.; Suchanek, P. *Angew. Chem. Int. Ed.* **1963**, *2*, 683. For reviews, see Mukaiyama, T. *Org. React.* **1982**, *28*, 203; Wittig, G. *Top. Curr. Chem.* **1976**, *67*, 1; *Rec. Chem. Prog.* **1967**, *28*, 45; Wittig, G.; Reiff, H. *Angew. Chem. Int. Ed.* **1968**, *7*, 7; Reiff, H. *Newer Methods Prep. Org. Chem.* **1971**, *6*, 48.

[1057]Corey, E.J.; Enders, D. *Tetrahedron Lett.* **1976**, 11. See also, Beam, C.F.; Thomas, C.W.; Sandifer, R.M.; Foote, R.S.; Hauser, C.R. *Chem. Ind. (London)* **1976**, 487; Sugasawa, T.; Toyoda, T.; Sasakura, K. *Synth. Commun.* **1979**, *9*, 515; Depezay, J.; Le Merrer, Y. *Bull. Soc. Chim. Fr.* **1981**, II-306.

[1058]Hassner, A.; Näumann, F. *Chem. Ber.* **1988**, *121*, 1823.

[1059]There is evidence (in the self-condensation of acetaldehyde) that a water molecule acts as a base (even in concentrated H$_2$SO$_4$) in assisting the addition of the enol to the protonated aldehyde: Baigrie, L.M.; Cox, R.A.; Slebocka-Tilk, H.; Tencer, M.; Tidwell, T.T. *J. Am. Chem. Soc.* **1985**, *107*, 3640.

[1060]Ishikawa, T.; Uedo, E.; Okada, S.; Saito, S. *Synlett* **1999**, 450.

[1061]For rate and equilibrium constants, see Guthrie, J.P.; Guo, J. *J. Am. Chem. Soc.* **1996**, *118*, 11472. For neighboring-group effects, see Eberle, M.K. *J. Org. Chem.* **1996**, *61*, 3844.

formation of the enolate anion occurs by deprotonation of the more acidic site, followed by cyclization to the second carbonyl. The acid-catalyzed intramolecular aldol condensation is known, and the mechanism has been studied.[1062] Stereoselective proline-catalyzed intramolecular aldol reactions give the cyclize product with good enantioselectivity.[1063]

An important extension of the intramolecular aldol condensation is the *Robinson annulation* reaction,[1064] which has often been used in the synthesis of steroids and terpenes. In original versions of this reaction, a cyclic ketone is converted to another cyclic ketone under equilibrium conditions using hydroxide or alkoxide bases in a protic solvent, forming one additional six-membered ring containing a double bond. The reaction can be done in a stepwise manner using amide bases in aprotic solvents. In the reaction with hydroxide or alkoxide bases in alcohol or water solvents, the substrate is treated with methyl vinyl ketone (or a simple derivative of methyl vinyl ketone) and a base.[1065] The enolate ion of the substrate adds to the methyl vinyl ketone in a Michael reaction (**15-24**) to give a diketone that undergoes or is made to undergo an internal aldol

reaction and subsequent dehydration to give the product.[1066] The Robinson annulation can be combined with alkylation.[1067] Enantioselective Robinson annulation techniques have been developed, including a proline-catalyzed reaction.[1068] The Robinson annulation has been done in ionic liquids[1069] and a solvent-free version of the reaction is known.[1070]

Because methyl vinyl ketone has a tendency to polymerize, precursors are often used instead, that is., compounds that will give methyl vinyl ketone when treated with a base. One common example, $MeCOCH_2CH_2NEt_2Me^+ \ I^-$ (see **17-9**), is easily prepared by quaternization of $MeCOCH_2CH_2NEt_2$, which itself is prepared

[1062]Bouillon, J.-P.; Portella, C.; Bouquant, J.; Humbel, S. *J. Org. Chem.* **2000**, *65*, 5823.

[1063]Bahmanyar, S.; Houk, K.N. *J. Am. Chem. Soc.* **2001**, *123*, 12911; Pidathala, C.; Hoang, L.; Vignola, N.; List, B. *Angew. Chem. Int. Ed.* **2003**, *42*, 2785.

[1064]For reviews of this and related reactions, see Gawley, R.E. *Synthesis* **1976**, 777; Jung, M.E. *Tetrahedron* **1976**, *32*, 1; Mundy, B.P. *J. Chem. Educ.* **1973**, *50*, 110. For a list of references, see Larock, R.C. *Comprehensive Organic Transformations*, 2nd ed., Wiley-VCH, NY, **1999**, pp. 1356–1358.

[1065]Acid catalysis has also been used: see Heathcock, C.H.; Ellis, J.E.; McMurry, J.E.; Coppolino, A. *Tetrahedron Lett.* **1971**, 4995.

[1066]For improved procedures, see Sato, T.; Wakahara, Y.; Otera, J.; Nozaki, H. *Tetrahedron Lett.* **1990**, *31*, 1581, and references cited therein.

[1067]Tai, C.-L.; Ly, T.W.; Wu, J.-D.; Shia, K.-S.; Liu, H.-J. *Synlett* **2001**, 214.

[1068]Bui, T.; Barbas III, C.F. *Tetrahedron Lett.* **2000**, *41*, 6951; Rajagopal, D.; Narayanan, R.; Swaminathan, S. *Tetrahedron Lett.* **2001**, *42*, 4887.

[1069]Morrison, D.W.; Forbes, D.C.; Davis Jr., J.H. *Tetrahedron Lett.* **2001**, *42*, 6053.

[1070]Miyamoto, H.; Kaneteka, S.; Tanaka, K.; Yoshizawa, K.; Toyota, S.; Toda, F. *Chem. Lett.* **2000**, 888.

by a Mannich reaction (**16-19**) involving acetone, formaldehyde, and diethylamine. α-Silylated vinyl ketones $RCOC(SiMe_3)=CH_2$ have also been used successfully in annulation reactions.[1071] The $SiMe_3$ group is easily removed. 1,5-Diketones prepared in other ways are also frequently cyclized by internal aldol reactions. When the ring closure of a 1,5-diketone is catalyzed by the amino acid (*S*)-proline, the product is optically active with high enantiomeric excess.[1072] *Stryker's reagent* [1073]$[(Ph_3P)CuH]_6$ has been used for an intramolecular addition where ketone enolate anion to a conjugated ketone, giving cyclic alcohol with a pendant ketone unit.[1074]

OS **I**, 77, 78, 81, 199, 283, 341; **II**, 167, 214; **III**, 317, 353, 367, 747, 806, 829; **V**, 486, 869; **VI**, 496, 666, 692, 781, 901; **VII**, 185, 190, 332, 363, 368, 473; **VIII**, 87, 208, 241, 323, 339, 620; **IX**, 432, 610; **X**, 339.

16-35   Mukaiyama Aldol and Related Reactions[1075]

O-Hydro-*C*-(α-acylalkyl)-addition

An important variation of the aldol condensation involves treatment of an aldehyde or ketone with a silyl ketene acetal $R_2C=C(OSiMe_3)OR'$[1076] in the presence of $TiCl_4$,[1077] to give **38**. The silyl ketene acetal can be considered a preformed enolate that gives aldol product

with $TiCl_4$ in aqueous solution, or with no catalyst at all.[1078] A combination of $TiCl_4$ and a *N*-tosyl imine has also been used to facilitate the Mukaiyama aldol

---

[1071]Stork, G.; Singh, J. *J. Am. Chem. Soc.* **1974**, *96*, 6181; Boeckman, Jr., R.K. *J. Am. Chem. Soc.* **1974**, *96*, 6179.

[1072]Eder, U.; Sauer, G.; Wiechert, R. *Angew. Chem. Int. Ed.* **1971**, *10*, 496; Hajos, Z.G.; Parrish, D.R. *J. Org. Chem.* **1974**, *39*, 1615. For a review of the mechanism, see Agami, C. *Bull. Soc. Chim. Fr.* **1988**, 499.

[1073]Mahoney, W.S.; Brestensky, D.M.; Stryker, J.M. *J. Am. Chem. Soc.* **1988**, *110*, 291; Brestensky, D.M.; Stryker, J.M. *Tetrahedron Lett.* **1989**, *30*, 5677.

[1074]Chiu, P.; Szeto, C.-P.; Geng, Z.; Cheng, K.-F. *Org. Lett.* **2001**, *3*, 1901.

[1075]See Smith, M.B. *Organic Synthesis*, 2nd ed., McGraw-Hill, NY, **2001**, pp.755–759.

[1076]For a list of references, see Larock, R.C. *Comprehensive Organic Transformations*, 2nd ed., Wiley-VCH, NY, **1999**, pp. 1745–1752. For methods of preparing silyl ketene acetals, see Revis, A.; Hilty, T.K. *Tetrahedron Lett.* **1987**, *28*, 4809, and references cited therein.

[1077]Mukaiyama, T. *Pure Appl. Chem.* **1983**, *55*, 1749; Kohler, B.A.B. *Synth. Commun.* **1985**, *15*, 39; Mukaiyama, T.; Narasaka, K. *Org. Synth.*, *65*, 6. For a discussion of the mechanism, see Gennari, C.; Colombo, L.; Bertolini, G.; Schimperna, G. *J. Org. Chem.* **1987**, *52*, 2754. For a review of this and other applications of $TiCl_4$ in organic synthesis, see Mukaiyama, T. *Angew. Chem. Int. Ed.* **1977**, *16*, 817. See also, Reetz, M.T. *Organotitanium Reagents in Organic Synthesis*, Spinger, NY, **1986**.

[1078]Lubineau, A.; Meyer, E. *Tetrahedron* **1988**, *44*, 6065; Miura, K.; Sato, H.; Tamaki, K.; Ito, H.; Hosomi, A. *Tetrahedron Lett.* **1998**, *39*, 2585. For an uncatalyzed reaction under high pressure, see Bellassoued, M.; Reboul, E.; Dumas, F. *Tetrahedron Lett.* **1997**, *38*, 5631.

reaction.[1079] The mechanism of this reaction has been explored.[1080] Other catalysts have been used for this reaction as well, including $InCl_3$,[1081] SmI_2,[1082] $Sc(OTf)_3$,[1083] HgI_2,[1084] $Yb(OTf)_3$,[1085] $Cu(OTf)_2$,[1086] $[Cp_2Zr(O\text{-}t\text{-}Bu)THF]^+[BPh_4]^-$,[1087] $LiClO_4$,[1088] $VOCl_3$,[1089] an iron catalyst,[1090] and $Bi(OTf)_3$.[1091] The reaction can be done in water using a scandium catalyst[1092] or a Montmorillonite K10 clay.[1093] Silyl enol ethers react with aqueous formaldehyde in the presence of TBAF to give the aldol product.[1094] A catalytic amount of Me_3SiCl facilitates the titanium mediated reaction.[1095] Sulfonamides, such as $HNTf_2$, have been used as a catalyst[1096] as has pyridine N-oxide.[1097] A combination of Ph_2BOH and benzoic acid in water catalyzes the reaction.[1098] Lithium perchlorate in acetonitrile (5 M) can be used for the reaction of an aldehyde and a silyl enol ether.[1099] When the catalyst is dibutyltin bis (triflate) $Bu_2Sn(OTf)_2$, aldehydes react, but not their acetals, while acetals of ketones react, but not the ketones themselves.[1100] Reaction at the carbonyl of saturated carbonyl compounds is significantly faster than 1,2-addition to unsaturated carbonyl compounds.[1101] Propargylic acetals react with silyl enol ethers and a scandium catalyst to give β-alkoxy ketones.[1102] Imines react with silyl enol ethers n the presence of $BF_3 \cdot OEt_2$ to give β-amino ketones.[1103] α-Silyl silyl enol ethers

[1079]Miura, K.; Nakagawa, T.; Hosomi, A. *J. Am. Chem. Soc.* **2002**, *124*, 536.

[1080]Hollis, T.K.; Bosnich, B. *J. Am. Chem. Soc.* **1995**, *117*, 4570. For the transition-state geometry, see Denmark, S.E.; Lee, W. *J. Org. Chem.* **1994**, *59*, 707.

[1081]Loh, T.-P.; Pei, J.; Cao, G.-Q. *Chem. Commun.* **1996**, 1819; Kobayashi, S.; Busujima, T.; Nagayama, S. *Tetrahedron Lett.* **1998**, *39*, 1579. Both $InCl_3$ and $CeCl_3$ have been used in aqueous media, see Muñoz-Muñiz, O.; Quintanar-Audelo, M.; Juaristi, E. *J. Org. Chem.* **2003**, *68*, 1622.

[1082]Van de Weghe, P.; Collin, J. *Tetrahedron Lett.* **1993**, *34*, 3881.

[1083]Kobayashi, S.; Wakabayashi, T.; Nagayama, S.; Oyamada, H. *Tetrahedron Lett.* **1997**, *38*, 4559; Komoto, I.; Kobayashi, S. *Chem. Commun.* **2001**, 1842; Komoto, I.; Kobayashi, S. *J. Org. Chem.* **2004**, *69*, 680.

[1084]Dicker, I.B. *J. Org. Chem.* **1993**, *58*, 2324.

[1085]This catalyst is tolerated in water. See Kobayashi, S.; Hachiya, I. *J. Org. Chem.* **1994**, *59*, 3590.

[1086]Kobayashi, S.; Nagayama, S.; Busujima, T. *Chem. Lett.* **1997**, 959.

[1087]Hong, Y.; Norris, D.J.; Collins, S. *J. Org. Chem.* **1997**, *58*, 3591.

[1088]Reetz, M.T.; Fox, D.N.A. *Tetrahedron Lett.* **1993**, *34*, 1119.

[1089]Kurihara, M.; Hayshi, T.; Miyata, N. *Chem. Lett.* **2001**, 1324.

[1090]Bach, T.; Fox, D.N.A.; Reetz, M.T. *J. Chem. Soc., Chem. Commun.* **1992**, 1634.

[1091]LeRoux, C.; Ciliberti, L.; Laurent-Robert, H.; Laporterie, A.; Dubac, J. *Synlett* **1998**, 1249.

[1092]Manabe, K.; Kobayashi, S. *Tetrahedron Lett.* **1999**, *40*, 3773. For a discussion of the effect of surfactants on this reaction, see Tian, H.-Y.; Chen, Y.-J.; Wang, D.; Bu, Y.-P.; Li, C.-J. *Tetrahedron Lett.* **2001**, *42*, 1803.

[1093]Loh, T.-P.; Li, X.-R. *Tetrahedron* **1999**, *55*, 10789.

[1094]Ozasa, N.; Wadamoto, M.; Ishihara, K.; Yamamoto, H. *Synlett* **2003**, 2219.

[1095]Yoshida, Y.; Matsumoto, N.; Hamasaki, R.; Tanabe, Y. *Tetrahedron Lett* **1999**, *40*, 4227.

[1096]Ishihara, K.; Hiraiwa, Y.; Yamamoto, H. *Synlett* **2001**, 1851.

[1097]Denmark, S.E.; Fan, Y. *J. Am. Chem. Soc.* **2002**, *124*, 4233.

[1098]Mori, Y.; Kobayashi, J.; Manabe, K.; Kobayashi, S. *Tetrahedron* **2002**, *58*, 8263.

[1099]Sudha, R.; Sankararaman, S. *J. Chem. Soc., Perkin Trans. 1* **1999**, 383.

[1100]Sato, T.; Otera, J.; Nozaki, H. *J. Am. Chem. Soc.* **1990**, *112*, 901.

[1101]Shirakawa, S.; Maruoka, K. *Tetrahedron Lett.* **2002**, *43*, 1469.

[1102]Yoshimatsu, M.; Kuribayashi, M.; Koike, T. *Synlett* **2001**, 1799.

[1103]Akiyama, T.; Takaya, J.; Kagoshima, H. *Chem. Lett.* **1999**, 947.

$RCH=CH(OTMS)SiMe_3$ react with acetals in the presence of $SnCl_4$ to give β-alkoxy silyl ketones.[1104]

An interesting variation in this reaction combined an intermolecular Mukaiyama aldol followed by an intramolecular reaction (a "domino" Mukaiyama aldol) that gave cyclic conjugated ketone products.[1105] Borane derivatives such as $C=C-OB(NMe_2)_2$ react with aldehydes to give β-amino ketones.[1106]

Silyl enol ethers[1107] derived from esters (silyl ketene acetals) react with aldehydes in the presence of various catalysts to give β-hydroxy esters. Water accelerates the reaction of an aldehyde and a ketene silyl acetal with no other additives.[1108] The reaction is catalyzed by triphenylphosphine[1109] and also by $SiCl_4$ with a chiral bis(phosphoramide) catalyst.[1110] The reaction was done without a catalyst in an ionic liquid.[1111] A vinylogous reaction is known that gives δ-hydroxy-α,β-unsaturated esters.[1112] Under different conditions, silyl ketene acetals of conjugated esters react with aldehydes to give conjugated lactones.[1113] Imines react with silyl ketene acetals in the presence of SmI_3 to give β-amino esters.[1114] Another variation converted a *N*-(1-trimethylsilyloxyvinyl) imine to a conjugated amide by initial reaction with 2 equivalents of *n*-butyllithium and a zirconium complex followed by reaction with an aldehyde.[1115] Silyl ketene acetals also undergo conjugate addition in reactions with conjugated ketones.[1116] Silyl ketene acetals of thio esters also react with aldehydes to give β-hydroxy thioesters.[1117]

Asymmetric Mukaiyama aldol reactions and reactions of silyl ketene acetals have been reported, [1118] usually using chiral additives[1119] although chiral auxiliaries

[1104]Honda, M.; Oguchi, W.; Segi, M.; Nakajima, T. *Tetrahedron* **2002**, *58*, 6815.

[1105]Langer, P.; Köhler, V. *Org. Lett.* **2000**, *2*, 1597.

[1106]Suginome, M.; Uehlin, L.; Yamamoto, A.; Murakami, M. *Org. Lett.* **2004**, *6*, 1167.

[1107]For a discussion of enantioselective deprotonation to form chiral silyl enol ethers, see Carswell, E.L.; Hayes, D.; Henderson, K.W.; Kerr, W.J.; Russell, C.J. *Synlett* **2003**, 1017.

[1108]Loh, T.-P.; Feng, L.-C.; Wei, L.-L. *Tetrahedron* **2000**, *56*, 7309.

[1109]Matsukawa, S.; Okano, N.; Imamoto, T. *Tetrahedron Lett.* **2000**, *41*, 103.

[1110]Denmark, S.E.; Heemstra, Jr., J.R. *Org. Lett.* **2003**, *5*, 2303; Denmark, S.E.; Wynn, T.; Beutner, G.L. *J. Am. Chem. Soc.* **2002**, *124*, 13405.

[1111]In omim Cl, 1-octyl-3-methylimidazolium chloride or in bmim PF₆, 1-butyl-3-methylimidazolium hexafluorophosphate: Chen, S.-L.; Ji, S.-J.; Loh, T.-P. *Tetrahedron Lett.* **2004**, *45*, 375.

[1112]Bluet, G.; Campagne, J.-M. *J. Org. Chem.* **2001**, *66*, 4293; Christmann, M.; Kalesse, M. *Tetrahedron Lett.* **2001**, *42*, 1269.

[1113]Bluet, G.; Bazán-Tejeda, B.; Campagne, J.-M. *Org. Lett.* **2001**, *3*, 3807.

[1114]Hayakawa, R.; Shimizu, M. *Chem. Lett.* **1999**, 591.

[1115]Gandon, V.; Bertus, P.; Szymoniak, J. *Tetrahedron* **2000**, *56*, 4467.

[1116]Harada, T.; Iwai, H.; Takatsuki, H.; Fujita, K.; Kubu, M.; Oku, A. *Org. Lett.* **2001**, *3*, 2101.

[1117]Hamada, T.; Manabe, K.; Ishikawa, S.; Nagayama, S.; Shiro, M.; Kobayashi, S. *J. Am. Chem. Soc.* **2003**, *125*, 2989.

[1118]Bach, T. *Angew. Chem. Int. Ed.* **1994**, *33*, 417. For a discussion of stereocontrol, see Annunziata, R.; Cinquini, M.; Cozzi, F.; Cozzi, P.G.; Consolandi, E. *J. Org. Chem.* **1992**, *57*, 456.

[1119]For examples, see Kobayashi, S.; Kawasuji, T.; Mori, N. *Chem. Lett.* **1994**, 217; Kobayashi, S.; Uchiro, H.; Shiina, I.; Mukaiyama, T. *Tetrahedron* **1993**, *49*, 1761; Mikami, K.; Matsukawa, S. *J. Am. Chem. Soc.* **1994**, *116*, 4077; Kaneko, Y.; Matsuo, T.; Kiyooka, S. *Tetrahedron Lett.* **1994**, *35*, 4107; Kiyooka, S.; Kido, Y.; Kaneko, Y. *Tetrahedron Lett.* **1994**, *35*, 5243.

have also been used.[1120] Chiral catalysts, usually transition-metal complexes using chiral ligands, are quite effective,[1121] but chiral bis(oxazolones)[1122] and chiral quaternary ammonium salts[1123] have also been used. A zirconium BINOL complex gave good enantioselectivity in reactions of silyl ketene acetals, and also good anti selectivity in the product.[1124] This reaction can also be run with the aldehyde or ketone in the form of its acetal $R^3R^4C(OR')_2$, in which case the product is the ether $R^1COCHR_2CR^3R^4OR'$ instead of **38**.[1125] Trichlorosilyl enol ethers react with aldehydes directly in the presence of a chiral phosphoramide to give the aldol with good syn selectivity and good enantioselectivity.[1126] Vinylogous silyl ketene acetals with a chiral oxazolidinone auxiliary attached to the α-vinylic carbon react with aldehydes and $TiCl_4$ to give a δ-hydroxy-α,β-unsaturated amide (an acyl oxazolidinone).[1127]

Enol acetates and enol ethers also give this product when treated with acetals and $TiCl_4$ or a similar catalyst.[1128] A variation of this condensation uses an enol acetate with an aldehyde in the presence of Et_2AlOEt to give the aldol product.[1129]

16-36 Aldol-Type Reactions between Carboxylic Acid Derivatives and Aldehydes or Ketones

O-Hydro-C-(α-alkoxycarbonylalkyl)-addition; α-Alkoxycarbonylalkylidene-de-oxo-bisubstitution

(if α-H present)

[1120]For an example, see Vasconcellos, M.L.; Desmaële, D.; Costa, P.R.R.; d'Angelo, J. *Tetrahedron Lett.* **1992**, *33*, 4921.

[1121]Titanium complexes: Imashiro, R.; Kuroda, T. *J. Org. Chem.* **2003**, *68*, 974. Copper complexes: Kobayashi, S.; Nagayama, S.; Busujima, T. *Tetrahedron* **1999**, *55*, 8739. Lead complexes: Nagayama, S.; Kobayashi, S. *J. Am. Chem. Soc* **2000**, *122*, 11531. Cerium complexes: Kobayashi, S.; Hamada, T.; Nagayama, S.; Manabe, K. *Org. Lett.* **2001**, *3*, 165. Silver complexes: Yanagisawa, A.; Nakatsuka, Y.; Asakawa, K.; Kageyama, H.; Yamamoto, H. *Synlett* **2001**, 69; Yanigisawa, A.; Nakatsuka, Y.; Asakawa, K.; Wadamoto, M.; Kageyama, H.; Yamamoto, H. *Bull. Chem. Soc. Jpn.* **2001**, *74*, 1477; Wadamoto, M.; Ozasa, N.; Yanigisawa, A.; Yamamoto, H. *J. Org. Chem.* **2003**, *68*, 5593. Zirconium complexes: Kobayashi, S.; Ishitani, H.; Yamashita, Y.; Ueno, M.; Shimizu, H. *Tetrahedron* **2001**, *57*, 861. Scandium complexes: Ishikawa, S.; Hamada, T.; Manabe, K.; Kobayashi, S. *J. Am. Chem. Soc.* **2004**, *126*, 12236.

[1122]Kobayashi, S.; Nagayama, S.; Busujima, T. *Chem. Lett.* **1999**, 71.

[1123]Zhang, F.-Y.; Corey, E.J. *Org. Lett.* **2001**, *3*, 639.

[1124]Ishitani, H.; Yamashita, Y.; Shimizu, H.; Kobayashi, S. *J. Am. Chem. Soc.* **2000**, *122*, 5403.

[1125]Mukaiyama, T.; Kobayashi, S.; Murakami, M. *Chem. Lett.* **1984**, 1759; Murata, S.; Suzuki, M.; Noyori, R. *Tetrahedron* **1988**, *44*, 4259. For a review of cross-coupling reactions of acetals, see Mukaiyama, T.; Murakami, M. *Synthesis* **1987**, 1043.

[1126]Denmark, S.E.; Pham, S.M. *J. Org. Chem.* **2003**, *68*, 5045; Denmark, S.E.; Stavenger, R.A. *J. Am. Chem. Soc.* **2000**, *122*, 8837; Denmark, S.E.; Ghosh, S.K. *Angew. Chem. Int. Ed.* **2001**, *40*, 4759.

[1127]Shirokawa, S.-i.; Kamiyama, M.; Nakamura, T.; Okada, M.; Nakazaki, A.; Hosokawa, S.; Kobayashi, S. *J. Am. Chem. Soc.* **2004**, *126*, 13604.

[1128]Kitazawa, E.; Imamura, T.; Saigo, K.; Mukaiyama, T. *Chem. Lett.* **1975**, 569.

[1129]Mukaiyama, T.; Shibata, J.; Shimamura, T.; Shiina, I. *Chem. Lett.* **1999**, 951.

In the presence of a strong base, the α carbon of a carboxylic ester or other acid derivative can condense with the carbonyl carbon of an aldehyde or ketone to give a β-hydroxy ester,[1130] amide, and so on., which may or may not be dehydrated to the α,β-unsaturated derivative. This reaction is sometimes called the *Claisen reaction*,[1131] an unfortunate usage since that name is more firmly connected to **16-85**. Early reactions used hydroxide or an alkoxide base in water or alcohol solvents, where self-condensation was the major process. Under such conditions, the aldehyde or ketone was usually chosen for its lack of an α-proton. Much better control of the reaction was achieved when amide bases in aprotic solvents, such as ether or THF, were used. The reaction of *tert*-butyl acetate and LDA[1132] in hexane or more commonly THF at $-78°C$ gives the enolate anion of *tert*-butyl acetate,[1133] (**12-23**, e.g., although self-condensation is occasionally a problem even here. Subsequent reaction a ketone provides a simple rapid alternative to the Reformatsky reaction (**16-28**) as a means of preparing β-hydroxy *tert*-butyl esters. It is also possible for the α carbon of an aldehyde or ketone to add to the carbonyl carbon of a carboxylic ester, but this is a different reaction (**16-86**) involving nucleophilic substitution and not addition to a C=O bond. It can, however, be a side reaction if the aldehyde or ketone has an α hydrogen.

Transition-metal mediated condensation of esters and aldehydes is known. The reaction of a thioester and an aryl aldehyde with $TiCl_4$—NBu_3, for example, gave a β-hydroxy thioester with good syn selectivity.[1134] Selenoamides $[RCH_2C(=Se)NR_2']$ react with LDA and then an aldehyde to give β-hydroxy selenoamides.[1135]

Besides ordinary esters (containing an α hydrogen), the reaction can also be carried out with lactones and, as in **16-34**, with the γ position of α,β-unsaturated esters (vinylogy). The enolate anion of an amide can be condensed with an aldehyde.[1136]

There are a number of variations of the condensation reaction of acid derivatives. The reaction between a cyclic ketone having a pendant alkynyl ester unit and tetrabutylammonium fluoride leads to cyclization to a bicyclic alcohol with an exocyclic allene moiety.[1137] A chain-extension reaction culminates in acyl addition of an ester enolate. The reaction of a β-keto ester, such as methyl 3-oxobutanote and $EtZnCH_2I$, leads to chain extension via a carbenoid-like insertion reaction (p. 803), which reacts with an aldehyde in a second step to give a methyl 3-oxopentanoate derivative with a $-CH(OH)R$ group at C-2 relative to the ester carbonyl.[1138]

[1130]If the reagent is optically active because of the presence of a chiral sulfoxide group, the reaction can be enantioselective. For a review of such cases, see Solladié, G. *Chimia* **1984**, *38*, 233.

[1131]Because it was discovered by Claisen, L. *Ber.* **1890**, *23*, 977.

[1132]Huerta, F.F.; Bäckvall, J.-E. *Org. Lett.* **2001**, *3*, 1209.

[1133]Rathke, M.W.; Sullivan, D.F. *J. Am. Chem. Soc.* **1973**, *95*, 3050.

[1134]Tanabe, Y.; Matsumoto, N.; Funakoshi, S.; Manta, N. *Synlett* **2001**, 1959.

[1135]Murai, T.; Suzuki, A.; Kato, S. *J. Chem. Soc., Perkin Trans. 1* **2001**, 2711.

[1136]For a case using $CeCl_3$ to promote the reaction, see Shang, X.; Liu, H.-J. *Synth. Commun.* **1994**, *24*, 2485.

[1137]Wendling, F.; Miesch, M. *Org. Lett.* **2001**, *3*, 2689.

[1138]Lai, S.; Zercher, C.K.; Jasinski, J.P.; Reid, S.N.; Staples, R.J. *Org. Lett.* **2001**, *3*, 4169.

For most esters, a much stronger base is needed than for aldol reactions; $(i\text{Pr})_2$-NLi (LDA, p. 389), Ph_3CNa and LiNH_2 are among those employed. However, one type of ester reacts more easily, and such strong bases are not needed: diethyl succinate and its derivatives condense with aldehydes and ketones in the presence of bases such as NaOEt, NaH, or KOCMe_3. This reaction is called the *Stobbe condensation*.[1139] One of the ester groups (sometimes both) is hydrolyzed in the course of the reaction. The following mechanism accounts for (*1*) the fact the succinic esters react so much better than others; (*2*) one ester group is always cleaved; and (*3*) the alcohol is not the product but the alkene. In addition, intermediate lactones **39** have been isolated from the mixture.[1140] The Stobbe condensation has been extended to di-*tert*-butyl esters of glutaric acid.[1141] The boron-mediated reaction is known.[1142]

Chiral additives, such as diazaborolidines can be added to an ester, and subsequent treatment with a base and then an aldehyde leads to a chiral β-hydroxy ester.[1143] A variety of chiral amide or oxazolidinone derivatives have been used to form amide linkages to carboxylic acid derivatives. These chiral auxiliaries lead to chirality transfer from the enolate anion of such derivatives, in both alkylation reactions and acyl substitution reactions with aldehydes and ketones. The so-called Evans auxiliaries (**40-42**) are commonly used and give good enantioselectivity.[1144] A variation is the magnesium halide-catalyzed anti-aldol reaction of chiral *N*-acylthiazolidinethiones (see **43**).[1145] The use of chiral *N*-acyloxazolidinthiones with TiCl_4 and sparteine also gave good selectivity in the acyl addition.[1146] Chiral diazaboron derivatives have also been used to facilitate the condensation of a α-phenylthio ester with an aldehyde.[1147]

40   **41**   **42**   **43**

[1139]For a review, see Johnson, W.S.; Daub, G.H. *Org. React.* **1951**, *6*, 1.

[1140]Robinson, R.; Seijo, E. *J. Chem. Soc.* **1941**, 582.

[1141]Puterbaugh, W.H. *J. Org. Chem.* **1962**, *27*, 4010. See also, El-Newaihy, M.F.; Salem, M.R.; Enayat, E.I.; El-Bassiouny, F.A. *J. Prakt. Chem.* **1982**, *324*, 379.

[1142]For a review, see Abiko, A. *Acc. Chem. Res.* **2004**, *37*, 387.

[1143]Corey, E.J.; Choi, S. *Tetrahedron Lett.* **2000**, *41*, 2769.

[1144]Evans, D.A.; Takacs, J.M. *Tetrahedron Lett.* **1980**, *21*, 4233; Sonnet, P.E.; Heath, R.R. *J. Org. Chem.* **1980**, *45*, 3137; Evans, D.A.; Chapman, K.T.; Bisaha, J. *Tetrahedron Lett.* **1984**, *25*, 4071.

[1145]Evans, D.A.; Downey, C.W.; Shaw, J.T.; Tedrow, J.S. *Org. Lett.* **2002**, *4*, 1127.

[1146]Crimmins, M.T.; McDougall, P.J. *Org. Lett.* **2003**, *5*, 591.

[1147]Corey, E.J.; Choi, S. *Tetrahedron Lett.* **2000**, *41*, 2769.

The condensation of an ester enolate and a ketone[1148] can be used as part of a Robinson annulation-like sequence (see **16-34**).

OS **I**, 252; **III**, 132; **V**, 80, 564; **70**, 256; **X**, 437; **81**, 157. Also see OS **IV**, 278, 478; **V**, 251.

16-37   The Henry Reaction[1149]

$$\text{CH}_3\text{NO}_2 \quad + \quad \text{HCHO} \quad \xrightarrow{\ ^-\text{OH}\ } \quad \text{HOCH}_2\text{CH}_2\text{NO}_2$$

When aliphatic nitro compounds are used instead of aldehydes or ketones, no reduction occurs, and the reaction has been referred to as a Tollens' reaction (see **16-43**). However, the classical condensation of an aliphatic nitro compound with an aldehyde or ketone is usually called the *Henry reaction*[1150] or the *Kamlet reaction*, and is essentially a nitro aldol reaction. A variety of conditions have been reported, including the use of a silica catalyst,[1151] Mg—Al hydrotalcite,[1152] a tetraalkylammonium hydroxide,[1153] proazaphosphatranes,[1154] or an ionic liquid.[1155] A solvent free Henry reaction was reported in which a nitroalkane and an aldehyde were reacted on KOH powder.[1156] Potassium phosphate has been used with nitromethane and aryl aldehydes.[1157] The Henry reaction has been done using $ZnEt_2$ and 20% ethanolamine.[1158] A gel-entrapped base has been used to catalyze this reaction.[1159]

[1148]Posner, G.H.; Lu, S.; Asirvatham, E.; Silversmith, E.F.; Shulman, E.M. *J. Am. Chem. Soc.* **1986**, *108*, 511. For an extension of this work to the coupling of four components, see Posner, G.H.; Webb, K.S.; Asirvatham, E.; Jew, S.; Degl'Innocenti, A. *J. Am. Chem. Soc.* **1988**, *110*, 4754.

[1149]For a review of this reaction with respect to nitroalkanes (the *Henry reaction*, **16-37**), see Baer, H.H.; Urbas, L., in Feuer, H. *The Chemistry of the Nitro and Nitroso Groups*, Wiley, NY, **1970**, pp. 76–117. See also, Rosini, G.; Ballini, R.; Sorrenti, P. *Synthesis* **1983**, 1014; Matsumoto, K. *Angew. Chem. Int. Ed.* **1984**, *23*, 617; Eyer, M.; Seebach, D. *J. Am. Chem. Soc.* **1985**, *107*, 3601. For reviews of the nitroalkenes that are the products of this reaction, see Barrett, A.G.M.; Graboski, G.G. *Chem. Rev.* **1986**, *86*, 751; Kabalka, G.W.; Varma, R.S. *Org. Prep. Proced. Int.* **1987**, *19*, 283.

[1150]Henry, L. *Compt. Rend.* **1895**, *120*, 1265; Kamlet, J. *U.S. Patent 2,151,171* **1939** [*Chem. Abstr., 33*: 5003' **1939**]; Hass, H.B.; Riley, E.F. *Chem. Rev.* **1943**, *32*, 373 (see p. 406); Lichtenthaler, F.W. *Angew. Chem. Int. Ed.* **1964**, *3*, 211. For a review, see Luzzio, F.A. *Tetrahedron* **2001**, *57*, 915.

[1151]Demicheli, G.; Maggi, R.; Mazzacani, A.; Righi, P.; Sartori, G.; Bigi, F. *Tetrahedron Lett.* **2001**, *42*, 2401.

[1152]Bulbule, V.J.; Deshpande, V.H.; Velu, S.; Sudalai, A.; Sivasankar, S.; Sathe, V.T. *Tetrahedron* **1999**, *55*, 9325.

[1153]Bulbule, V.J.; Jnaneshwara, G.K.; Deshmukh, R.R.; Borate, H.B.; Deshpande, V.H. *Synth. Commun.* **2001**, *31*, 3623.

[1154]Kisanga, P.B.; Verkade, J.G. *J. Org. Chem.* **1999**, *64*, 4298.

[1155]In TMG Lac, tetramethylguanidinium lactate: Jiang, T.; Gao, H.; Han, B.; Zhao, G.; Chang, Y.; Wu, W.; Gao, L.; Yang, G. *Tetrahedron Lett.* **2004**, *45*, 2699.

[1156]Ballini, R.; Bosica, G.; Parrini, M. *Chem. Lett.* **1999**, 1105.

[1157]Desai, U.V.; Pore, D.M.; Mane, R.B.; Solabannavar, S.B.; Wadgaonkar, P.P. *Synth. Commun.* **2004**, *34*, 19.

[1158]Klein, G.; Pandiaraju, S.; Reiser, O. *Tetrahedron Lett.* **2002**, *43*, 7503.

[1159]Bandgar, B.P.; Uppalla, L.S. *Synth. Commun.* **2000**, *30*, 2071.

Catalytic enantioselective Henry reactions are known,[1160] such as the use of a chiral copper catalyst[1161] or a zinc catalyst.[1162] The Henry reaction of nitromethane an a chiral aldehyde under high pressure gives the β-nitro alcohol with excellent enantioselectivity.[1163]

A variation of this reaction converts nitro compounds to nitronates $RCH=N^+(OTMS)—O^-$, which react with aldehydes in the presence of a copper catalyst to give the β-nitro alcohol.[1164]

16-38 The Knoevenagel Reaction

Bis(ethoxycarbonyl)methylene-de-oxo-bisubstitution, and so on

The condensation of aldehydes or ketones, usually not containing an α hydrogen, with compounds of the form $Z—CH_2—Z'$ or $Z—CHR—Z'$ is called the *Knoevenagel reaction.*[1165] Both Z and Z' may be CHO, COR, COOH, COOR, CN, NO_2, SOR, SO_2R, SO_2OR, or similar groups. Such compounds have a significantly higher enol content[1166] and the α-proton is much more acidic (Table 8.1 on p. 360). When Z = COOH, decarboxylation of the product often takes place *in situ.*[1167] If a strong enough base is used, the reaction can be performed on compounds possessing only a single Z (e.g., CH_3Z or RCH_2Z). Other active hydrogen compounds can also be employed, among them CHCl_3, 2-methylpyridines, terminal acetylenes, cyclopentadienes, and so on.; in fact any compound that contains a C—H bond the hydrogen of which can be removed by a base. As shown in the example, the reaction of β-keto esters and aldehydes to give **44** is promoted by diethylamine at 0°C. Nitroalkanes[1149] as well as β-keto sulfoxides[1168] undergo the reaction.

44

[1160]Christensen, C.; Juhl, K.; Jørgensen, K.A. *Chem. Commun.* **2001**, 2222; Christensen, C.; Juhl, K.; Hazell, R.G.; Jørgensen, K.A. *J. Org. Chem.* **2002**, *67*, 4875.

[1161]Evans, D.A.; Seidel, D.; Rueping, M.; Lam, H.W.; Shaw, J.T.; Downey, C.W. *J. Am. Chem. Soc.* **2003**, *125*, 12692.

[1162]Trost, B.M.; Yeh, V.S.C. *Angew. Chem. Int. Ed.* **2002**, *41*, 861.

[1163]Misumi, Y.; Matsumoto, K. *Angew. Chem. Int. Ed.* **2002**, *41*, 1031.

[1164]Risgaard, T.; Gothelf, K.V.; Jørgensen, K.A. *Org. Biomol. Chem.* **2003**, *1*, 153.

[1165]For reviews, see Jones, G. *Org. React.* **1967**, *15*, 204; Wilk, B.K. *Tetrahedron* **1997**, *53*, 7097.

[1166]Rochlin, E.; Rappoport, Z. *J. Org. Chem.* **2003**, *68*, 1715.

[1167]For a discussion of the mechanism when the reaction is accompanied by decarboxylation, see Tanaka, M.; Oota, O.; Hiramatsu, H.; Fujiwara, K. *Bull. Chem. Soc. Jpn.* **1988**, *61*, 2473.

[1168]Kuwajima, I.; Iwasawa, H. *Tetrahedron Lett.* **1974**, 107. See also, Huckin, S.N.; Weiler, L. *Can. J. Chem.* **1974**, *52*, 2157.

As with **16-34**, these reactions have sometimes been performed with acid cata-lysts.[1169] Ionic liquid solvents have been used,[1170] and heating on quaternary ammonium salts without solvent leads to a Knoevenagel reaction.[1171] Other sol-vent-free reactions are known.[1172] Ultrasound has been used to promote the reac-tion,[1173] and it has also been done using microwave irradiation[1174] or on silica,[1175] with microwave irradiation. Another solid-state variation is done on moist LiBr,[1176] heating with sodium carbonate and molecular sieves 4 Å promotes the reaction,[1177] as do zeolites.[1178] High-pressure conditions have been used.[1179] Transition-metal compounds such as palladium complexes,[1180] SmI$_2$[1181] or BiCl$_3$[1182] have been used to promote the Knoevenagel reaction.

In the reaction with terminal acetylenes,[1183] sodium acetylides are the most common reagents (when they are used, the reaction is often called the *Nef reaction*), but lithium,[1184] magnesium, and other metallic acetylides have also been used. A particularly convenient reagent is lithium acetylide–ethylenediamine complex,[1185] a stable, free-flowing powder that is commercially available. Alter-natively, the substrate may be treated with the alkyne itself in the presence of

[1169]For example, see Rappoport, Z.; Patai, S. *J. Chem. Soc.* **1962**, 731.

[1170]In bmim Cl, 1-butyl-3-methylimidazolium chloride, with AlCl$_3$: Harjani, J.R.; Nara, S.J.; Salunkhe, M.M. *Tetrahedron Lett.* **2002**, *43*, 1127. See Morrison, D.W.; Forbes, D.C.; Davis Jr., J.H. *Tetrahedron Lett.* **2001**, *42*, 6053. In bmim BF$_4$, 1-butyl-3-methylimidazolium tetrafluoroborate: Su, C.; Chen, Z.-C.; Zheng, Q.G. *Synthesis* **2003**, 555.

[1171]Bose, D.S.; Narsaiah, A.V. *J. Chem. Res. (S)* **2001**, 36.

[1172]McCluskey, A.; Robinson, P.J.; Hill, T.; Scott, J.L.; Edwards, J.K. *Tetrahedron Lett.* **2002**, *43*, 3117; Ren, Z.; Cao, W.; Tong, W. *Synth. Commun.* **2002**, *32*, 3475; Mogilaiah, K.; Prashanthi, M.; Reddy, G.R.; Reddy, Ch.S.; Reddy, N.V. *Synth. Commun.* **2003**, *33*, 2309; Zuo, W.-X.; Hua, R.; Qiu, X. *Synth. Commun.* **2004**, *34*, 3219.

[1173]McNulty, J.; Steeve, J.A.; Wolf, S. *Tetrahedron Lett.* **1998**, *39*, 8013; Li, J.-T.; Zang, H.-J.; Feng, Y.-Y.; Li, L.-J.; Li, T.-S. *Synth. Commun.* **2001**, *31*, 653.

[1174]de la Cruz, P.; Díez-Barra, E.; Loupy, A.; Langa, F. *Tetrahedron Lett.* **1996**, *37*, 1113; Mitra, A.K.; De, A.; Karchaudhuri, N. *Synth. Commun.* **1999**, *29*, 2731; Balalaie, S.; Nemati, N. *Synth. Commun.* **2000**, *30*, 869; Loupy, A.; Song, S.-J.; Sohn, S.-M.; Lee, Y.-M.; Kwon, T.W.; *J. Chem. Soc., Perkin Trans. 1* **2001**, 1220; Yadav, J.S.; Reddy, B.V.S.; Basak, A.K.; Visali, B.; Narsaiah, A.V.; Nagaiah, K. *Eur. J. Org. Chem.* **2004**, 546.

[1175]Kumar, H.M.S.; Reddy, B.V.S.; Reddy, P.T.; Srinivas, D.; Yadav, J.S. *Org. Prep. Proceed. Int.* **2000**, *32*, 81; Peng, Y.; Song, G.; Qian, X. *J. Chem. Res. (S)* **2001**, 188.

[1176]Prajapati, D.; Lekhok, K.C.; Sandhu, J.S.; Ghosh, A.C. *J. Chem. Soc. Perkin Trans. 1* **1996**, 959.

[1177]Siebenhaar, B.; Casagrande, B.; Studer, M.; Blaser, H.-U. *Can. J. Chem.* **2001**, *79*, 566.

[1178]Reddy, T.I.; Varma, R.S. *Tetrahedron Lett.* **1997**, *38*, 1721.

[1179]Jenner, G. *Tetrahedron Lett.* **2001**, *42*, 243.

[1180]You, J.; Verkade, J.G. *J. Org. Chem.* **2003**, *68*, 8003.

[1181]Chandrasekhar, S.; Yu, J.; Falck, J.R.; Mioskowski, C. *Tetrahedron Lett.* **1994**, *35*, 5441.

[1182]This catalyst was used in the reaction without solvent. See Prajapati, D.; Sandhu, J.S. *Chem. Lett.* **1992**, 1945.

[1183]For reviews, see Ziegenbein, W., in Viehe, H.G. *Acetylenes*, Marcel Dekker, NY, **1969**, pp. 207–241; Ried, W. *Newer Methods Prep. Org. Chem.* **1968**, *4*, 95.

[1184]See Midland, M.M. *J. Org. Chem.* **1975**, *40*, 2250, for the use of amine-free monolithium acetylide.

[1185]Beumel Jr., O.F.; Harris, R.F. *J. Org. Chem.* **1963**, *28*, 2775.

a base, so that the acetylide is generated *in situ*. This procedure is called the *Favorskii reaction*, not to be confused with the Favorskii rearrangement (**18-7**).[1186]

With most of these reagents the alcohol is not isolated (only the alkene) *if* the alcohol has a hydrogen in the proper position.[1187] However, in some cases the alcohol is the major product. A β-keto allylic ester was shown to react with an aldehyde to give a β-hydroxy ketone, with loss of the allyl ester moiety, upon treatment with YbCl₃ and a palladium catalyst.[1188] With suitable reactants, the Knoevenagel reaction, like the aldol (**16-2**), has been carried out diastereoselectively[1189] and enantioselectively.[1190] When the reactant is of the form ZCH₂Z′, aldehydes react much better than ketones and few successful reactions with ketones have been reported. However, it is possible to get good yields of alkene from the condensation of diethyl malonate, CH₂(COOEt)₂, with ketones, as well as with aldehydes, if the reaction is run with TiCl₄ and pyridine in THF.[1191] In reactions with ZCH₂Z′, the catalyst is most often a secondary amine (piperidine is the most common, but see formation of **44**), though many other catalysts have been used. When the catalyst is pyridine (to which piperidine may or may not be added) the reaction is known as the *Doebner modification* of the Knoevenagel reaction and the product is usually the conjugated acid **45**. Alkoxides are also common catalysts. Microwave-induced Doebner condensation reactions are known.[1192]

$$\text{PrCHO} + \underset{\text{COOH}}{\overset{\text{COOH}}{\diagup\hspace{-6pt}\diagdown}} \xrightarrow[\text{piperidine}]{\text{pyridine}} \text{PrHC}\!\!=\!\!\underset{\text{COOH}}{\overset{\text{H}}{\diagdown\hspace{-6pt}\diagup}}$$
45

A number of special applications of the Knoevenagel reaction follow:

1. The dilithio derivative of *N*-methanesulfinyl-*p*-toluidine[1193] (**46**) adds to aldehydes and ketones to give, after hydrolysis, the hydroxysulfinamides

[1186]For a discussion of the mechanism of the Favorskii addition reaction, see Kondrat'eva, L.A.; Potapova, I.M.; Grigina, I.N.; Glazunova, E.M.; Nikitin, V.I. *J. Org. Chem. USSR* **1976**, *12*, 948.

[1187]For lists of reagents (with references) that condense with aldehydes and ketones to give alkene products, see Larock, R.C. *Comprehensive Organic Transformations*, 2nd ed., Wiley-VCH, NY, **1999**, pp. 317–325, 341–350. For those that give the alcohol product, see Larock, R.C. *Comprehensive Organic Transformations*, 2nd ed., Wiley-VCH, NY, **1999**, pp. 1178–1179, 1540–1541, 1717–1724, 1727, 1732–1736, 1778–1780, 1801–1805.

[1188]Lou, S.; Westbrook J.A.; Schaus, S.E. *J. Am. Chem. Soc.* **2004**, *126*, 11440.

[1189]See, for example, Trost, B.M.; Florez, J.; Jebaratnam, D.J. *J. Am. Chem. Soc.* **1987**, *109*, 613; Mahler, U.; Devant, R.M.; Braun, M. *Chem. Ber.* **1988**, *121*, 2035; Ronan, B.; Marchalin, S.; Samuel, O.; Kagan, H.B. *Tetrahedron Lett.* **1988**, *29*, 6101; Barrett, A.G.M.; Robyr, C.; Spilling, C.D. *J. Org. Chem.* **1989**, *54*, 1233; Pyne, S.G.; Boche, G. *J. Org. Chem.* **1989**, *54*, 2663.

[1190]See, for example, Enders, D.; Lotter, H.; Maigrot, N.; Mazaleyrat, J.; Welvart, Z. *Nouv. J. Chim.*, **1984**, *8*, 747; Ito, Y.; Sawamura, M.; Hayashi, T. *J. Am. Chem. Soc.* **1986**, *108*, 6405; Togni, A.; Pastor, S.D. *J. Org. Chem.* **1990**, *55*, 1649; Sakuraba, H.; Ushiki, S. *Tetrahedron Lett.* **1990**, *31*, 5349; Niwa, S.; Soai, K. *J. Chem. Soc. Perkin Trans. 1* **1990**, 937.

[1191]Lehnert, W. *Tetrahedron* **1973**, *29*, 635; *Synthesis* **1974**, 667, and references cited therein.

[1192]Mitra, A.K.; De, A.; Karchaudhuri, N. *Synth. Commun.* **1999**, *29*, 573; Pellón, R.F.; Mamposo, T.; González, E.; Calderón, O. *Synth. Commun.* **2000**, *30*, 3769.

[1193]For a method of preparing **46**, see Bowlus, S.B.; Katzenellenbogen, J.A. *Synth. Commun.* **1974**, *4*, 137.

47, which, upon heating, undergo stereospecifically syn eliminations to give alkenes.[1194] The reaction is thus a method for achieving the conversion $RR'CO \rightarrow RR'C=CH_2$ and represents an alternative to the Wittig reaction.[1195] Note that sulfones with an amide group at the α-position, $ArSO_2CH(R)N(R)C=O$, react with ketones via acyl addition in the presence of SmI_2.[1196]

2. The reaction of ketones with tosylmethylisocyanide (**48**) gives different products,[1197] depending on the reaction conditions.

[1194]Corey, E.J.; Durst, T. *J. Am. Chem. Soc.* **1968**, *90*, 5548, 5553.

[1195]For similar reactions, see Jung, F.; Sharma, N.K.; Durst, T. *J. Am. Chem. Soc.* **1973**, *95*, 3420; Kuwajima, I.; Uchida, M. *Tetrahedron Lett.* **1972**, 649; Johnson, C.R.; Shanklin, J.R.; Kirchhoff, R.A. *J. Am. Chem. Soc.* **1973**, *95*, 6462; Lau, P.W.K.; Chan, T.H. *Tetrahedron Lett.* **1978**, 2383; Yamamoto, K.; Tomo, Y.; Suzuki, S. *Tetrahedron Lett.* **1980**, *21*, 2861; Martin, S.F.; Phillips, G.W.; Puckette, T.A.; Colapret, J.A. *J. Am. Chem. Soc.* **1980**, *102*, 5866; Arenz, T.; Vostell, M.; Frauenrath, H. *Synlett* **1991**, 23.

[1196]Yoda, H.; Ujihara, Y.; Takabe, K. *Tetrahedron Lett.* **2001**, *42*, 9225.

[1197]For reviews of α-metalated isocyanides, see Schöllkopf, U. *Pure Appl. Chem.* **1979**, *51*, 1347; *Angew. Chem. Int. Ed.* **1977**, *16*, 339; Hoppe, D. *Angew. Chem. Int. Ed.* **1974**, *13*, 789.

When the reaction is run with potassium *tert*-butoxide in THF at $-5°C$, one obtains (after hydrolysis) the normal Knoevenagel product **49**, except that the isocyano group has been hydrated (**16-97**).[1198] With the same base but with 1,2-dimethoxyethane (DME) as solvent the product is the nitrile **50**.[1199] When the ketone is treated with **48** and thallium(I) ethoxide in a 4:1 mixture of absolute ethanol and DME at room temperature, the product is a 4-ethoxy-2-oxazoline **51**.[1200] Since **50** can be hydrolyzed[1201] to a carboxylic acid[1198] and **51** to an α-hydroxy aldehyde,[1200] this versatile reaction provides a means for achieving the conversion of RCOR′ to RCHR′COOH, RCHR′CN, or RCR′(OH)CHO. The conversions to RCHR′COOH and to RCHR′CN[1202] have also been carried out with certain aldehydes (R′ = H).

3. Aldehydes and ketones RCOR′ react with α-methoxyvinyllithium, $CH_2=C(Li)OMe$, to give hydroxy enol ethers, $RR′C(OH)C(OMe)=CH_2$, which are easily hydrolyzed to acyloins, $RR′C(OH)COMe$.[1203] In this reaction, the $CH_2=C(Li)OMe$ is a synthon for the unavailable $H_3C-\overset{\ominus}{C}=O$,[1204] and is termed an *acyl anion equivalent*. The reagent also reacts with esters RCOOR′ to give $RC(OH)(COMe=CH_2)_2$. A synthon for the $Ph-C=O$ ion is $PhC(CN)OSiMe_3$, which adds to aldehydes and ketones RCOR′ to give, after hydrolysis, the α-hydroxy ketones, $RR′C(OH)$ *C(OH)COPh*.[1205]

4. Lithiated allylic carbamates (**52**) (prepared as shown) react with aldehydes or ketones (R^6COR^7), in a reaction accompanied by an allylic rearrangement, to give (after hydrolysis) γ-hydroxy aldehydes or ketones.[1206] The reaction is called *the homoaldol reaction*, since the product is a homolog of the product of **16-34**. The reaction has been performed enantioselectively.[1207]

[1198]Schöllkopf, U.; Schröder, U.; Blume, E. *Liebigs Ann. Chem.* **1972**, *766*, 130; Schöllkopf, U.; Schröder, U. *Angew. Chem. Int. Ed.* **1972**, *11*, 311.

[1199]Oldenziel, O.H.; van Leusen, D.; van Leusen, A.M. *J. Org. Chem.* **1977**, *42*, 3114.

[1200]Oldenziel, O.H.; van Leusen, A.M. *Tetrahedron Lett.* **1974**, 163, 167. For conversions to α,β-unsaturated ketones and diketones, see, respectively, Moskal, J.; van Leusen, A.M. *Tetrahedron Lett.* **1984**, *25*, 2585; van Leusen, A.M.; Oosterwijk, R.; van Echten, E.; van Leusen, D. *Recl. Trav. Chim. Pays-Bas* **1985**, *104*, 50.

[1201]Compound **49** can also be converted to a nitrile; see **17-30**.

[1202]van Leusen, A.M.; Oomkes, P.G. *Synth. Commun.* **1980**, *10*, 399.

[1203]Baldwin, J.E.; Höfle, G.A.; Lever Jr., O.W. *J. Am. Chem. Soc.* **1974**, *96*, 7125. For a similar reaction, see Tanaka, K.; Nakai, T.; Ishikawa, N. *Tetrahedron Lett.* **1978**, 4809.

[1204]For a synthon for the $\ominus$COCOOEt ion, see Reetz, M.T.; Heimbach, H.; Schwellnus, K. *Tetrahedron Lett.* **1984**, *25*, 511.

[1205]Hünig, S.; Wehner, G. *Synthesis* **1975**, 391.

[1206]For a review, see Hoppe, D. *Angew. Chem. Int. Ed.* **1984**, *23*, 932.

[1207]Krämer, T.; Hoppe, D. *Tetrahedron Lett.* **1987**, *28*, 5149.

5. The lithium salt of an active hydrogen compound adds to the lithium salt of the tosylhydrazone of an aldehyde to give product **53**. If X = CN, SPh, or SO$_2$R, **53** spontaneously loses N$_2$ and LiX to give the alkene **54**. The entire process is done in one reaction vessel: The active hydrogen compound is mixed with the tosylhydrazone and the mixture is treated with (iPr)$_2$NLi to form both salts at once.[1208] This process is another alternative to the Wittig reaction for forming double bonds.

OS **I**, 181, 290, 413; **II**, 202; **III**, 39, 165, 317, 320, 377, 385, 399, 416, 425, 456, 479, 513, 586, 591, 597, 715, 783; **IV**, 93, 210, 221, 234, 293, 327, 387, 392, 408, 441, 463, 471, 549, 573, 730, 731, 777; **V**, 130, 381, 572, 585, 627, 833, 1088, 1128; **VI**, 41, 95, 442, 598, 683; **VII**, 50, 108, 142, 276, 381, 386, 456; **VIII**, 258, 265, 309, 353, 391, 420; **X**, 271. Also see, OS **III**, 395; **V**, 450.

16-39   The Perkin Reaction

α-Carboxyalkylidene-de-oxo-bisubstitution

The condensation of aromatic aldehydes with anhydrides is called the *Perkin reaction*.[1209] When the anhydride has two α hydrogens (as shown), dehydration

---

[1208]Vedejs, E.; Dolphin, J.M.; Stolle, W.T. *J. Am. Chem. Soc.* **1979**, *101*, 249.

[1209]For a review, see Johnson, J.R. *Org. React.* **1942**, *1*, 210.

always occurs; the β-hydroxy acid salt is never isolated. In some cases, anhydrides of the form $(R_2CHCO)_2O$ have been used, and then the hydroxy compound is the product since dehydration cannot take place. The base in the Perkin reaction is nearly always the salt of the acid corresponding to the anhydride. Although the Na and K salts have been most frequently used, higher yields and shorter reaction times have been reported for the Cs salt.[1210] Besides aromatic aldehydes, their vinylogs ArCH=CHCHO also give the reaction. Otherwise, the reaction is not suitable for aliphatic aldehydes.[1211]

OS **I**, 398; **II**, 61, 229; **III**, 426.

16-40 Darzens Glycidic Ester Condensation

(2 + 1)*OC,CC-cyclo-α*-Alkoxycarbonylmethylene-addition

Aldehydes and ketones condense with α-halo esters in the presence of bases to give α,β-epoxy esters, called *glycidic esters*. This is called *the Darzens condensation*.[1212] The reaction consists of an initial Knoevenagel-type reaction (**16-38**), followed by an internal S_N^2 reaction (**10-9**):[1213]

Although the intermediate halo alkoxide is generally not isolated,[1214] it has been done, not only with α-fluoro esters (since fluorine is such a poor leaving group in nucleophilic substitutions), but also with α-chloro esters.[1215] This is only one of several types of evidence that rule out a carbene intermediate.[1216] Sodium ethoxide is often used as the base, though other bases, including sodium amide, are sometimes used. Aromatic aldehydes and ketones give good yields, but aliphatic aldehydes react poorly. However, the reaction can be made to give good yields

[1210]Koepp, E.; Vögtle, F. *Synthesis* **1987**, 177.

[1211]Crawford, M.; Little, W.T. *J. Chem. Soc.* **1959**, 722.

[1212]For a review, see Berti, G. *Top. Stereochem.* **1973**, 7, 93, pp. 210–218. Also see Bakó, P.; Szöllősy, Á; Bombicz, P.; Töke, L. *Synlett* **1997**, 291.

[1213]For discussions of the mechanism of the reaction, and especially of the stereochemistry, see Roux-Schmitt, M.; Seyden-Penne, J.; Wolfe, S. *Tetrahedron* **1972**, 28, 4965; Bansal, R.K.; Sethi, K. *Bull. Chem. Soc. Jpn.* **1980**, 53, 1197.

[1214]The transition state for this reaction has been examined. See Yliniemelä, A.; Brunow, G.; Flügge, J.; Teleman, O. *J. Org. Chem.* **1996**, 61, 6723.

[1215]Ballester, M.; Pérez-Blanco, D. *J. Org. Chem.* **1958**, 23, 652; Martynov, V.F.; Titov, M.I. *J. Gen. Chem. USSR* **1963**, 33, 1350; **1964**, 34, 2139; Elkik, E.; Francesch, C. *Bull. Soc. Chim. Fr.* **1973**, 1277, 1281.

[1216]Another, based on the stereochemistry of the products, is described by Zimmerman, H.E.; Ahramjian, L. *J. Am. Chem. Soc.* **1960**, 82, 5459.

($\sim$80%) with simple aliphatic aldehydes, as well as with aromatic aldehydes and ketones by treatment of the α-halo ester with the base lithium bis(trimethylsilyl) amide, $LiN(SiMe_3)_2$, in THF at $-78°C$ (to form the conjugate base of the ester) and addition of the aldehyde or ketone to this solution.[1217] If a preformed dianion of an α-halo carboxylic acid $Cl-\overset{\ominus}{C}R-COO^{\ominus}$ is used instead, α,β-epoxy acids are produced directly.[1218] The Darzens reaction has also been carried out on α-halo ketones, α-halo nitriles,[1219] α-halo sulfoxides[1220] and sulfones,[1221] α-halo N,N-disubstituted amides,[1222] α-halo ketimines,[1223] and even on allylic[1224] and benzylic halides. Phase-transfer catalysis has been used.[1225] Note that the reaction of a β-bromo-α-oxo ester and a Grignard reagent leads to the glycidic ester.[1226] Acid-catalyzed Darzens reactions have also been reported.[1227] (see also, **16-46**).

The Darzens reaction has been performed enantioselectively, by coupling optically active α-bromo-β-hydroxy esters with aldehydes.[1228] Chiral phase-transfer agents have been used to give epoxy ketones with modest enantioselectivity.[1229] Chiral additives have proven to be effective.[1230]

Glycidic esters can easily be converted to aldehydes (**12-40**). The reaction has been extended to the formation of analogous aziridines by treatment of an imine with an α-halo ester or an α-halo N,N-disubstituted amide and t-BuOK in the solvent 1,2-dimethoxyethane.[1231] However, yields were not high.

OS **III**, 727; **IV**, 459, 649.

16-41   The Peterson Alkenylation Reaction

Alkylidene-de-oxo-bisubstitution

[1217]Borch, R.F. *Tetrahedron Lett.* **1972**, 3761.

[1218]Johnson, C.R.; Bade, T.R. *J. Org. Chem.* **1982**, *47*, 1205.

[1219]See White, D.R.; Wu, D.K. *J. Chem. Soc., Chem. Commun.* **1974**, 988.

[1220]Satoh, T.; Sugimoto, A.; Itoh, M.; Yamakawa, K. *Tetrahedron Lett.* **1989**, *30*, 1083.

[1221]Arai, S.; Ishida, T.; Shioiri, T. *Tetrahedron Lett.* **1998**, *39*, 8299.

[1222]Tung, C.C.; Speziale, A.J.; Frazier, H.W. *J. Org. Chem.* **1963**, *28*, 1514.

[1223]Mauzé, B. *J. Organomet. Chem.* **1979**, *170*, 265.

[1224]Sulmon, P.; De Kimpe, N.; Schamp, N.; Declercq, J.; Tinant, B. *J. Org. Chem.* **1988**, *53*, 4457.

[1225]See Jończyk, A.; Kwast, A.; Makosza, M. *J. Chem. Soc., Chem. Commun.* **1977**, 902; Gladiali, S.; Soccolini, F. *Synth. Commun.* **1982**, *12*, 355; Arai, S.; Suzuki, Y.; Tokumaru, K.; Shioiri, T. *Tetrahedron Lett.* **2002**, *43*, 833. See Starks, C.M.; Liotta, C. *Phase Transfer Catalysis*, Academic Press, NY, **1978**, pp. 197–198.

[1226]Jung, M.E.; Mengel, W.; Newton, T.W. *Synth. Commun.* **1999**, *29*, 3659.

[1227]Sipos, G.; Schöbel, G.; Sirokmán, F. *J. Chem. Soc. Perkin Trans. 2* **1975**, 805.

[1228]Corey, E.J.; Choi, S. *Tetrahedron Lett.* **1991**, *32*, 2857. For a review, see Ohkata, K.; Kimura, J.; Shinohara, Y.; Takagi, R.; Hiraga, Y. *Chem. Commun.* **1996**, 2411.

[1229]Arai, S.; Shirai, Y.; Ishida, T.; Shioiri, T. *Tetrahedron* **1999**, *55*, 6375.

[1230]Aggarwal, V.K.; Hynd, G.; Picoul, W.; Vasse, J.-L. *J. Am. Chem. Soc.* **2002**, *124*, 9964.

[1231]Deyrup, J.A. *J. Org. Chem.* **1969**, *34*, 2724.

In the *Peterson alkenylation reaction*[1232, 1233] the lithio (or sometimes magnesio) derivative of a trialkylsilane adds to an aldehyde or ketone to give a β-hydroxysilane, which spontaneously eliminates water, or can be made to do so by treatment with acid or base, to produce an alkene. This reaction is still another alternative to the Wittig reaction (**16-44**), and is sometimes called the *silyl-Wittig reaction*.[1234] The R group can also be a COOR group, in which case the product is an α,β-unsaturated ester,[1235] or an SO$_2$Ph group, in which case the product is a vinylic sulfone.[1236] The stereochemistry of the product can often be controlled by whether an acid or a base is used to achieve elimination. The role of Si—O interactions has also been examined.[1237] Use of a base generally gives syn elimination (E^i mechanism, see p. 1507), while an acid usually results in anti elimination (E2 mechanism, see p. 1478).[1238] Samarium(II) iodide in HMPA has also been used for elimination of the hydroxy sulfone.[1239] α-Alkoxy benzotriazoyl sulfones (ROCH$_2$SO$_2$Bt, where Bt = benzothiazole, reacts with lithium hexamethyldisilazide and an aldehyde to give a vinyl ether.[1240]

[1232]Peterson, D.J. *J. Org. Chem.* **1968**, *33*, 780. For reviews, see Ager, D.J. *Org. React.* **1990**, *38*, 1; *Synthesis* **1984**, 384; Colvin, E.W. *Silicon Reagents in Organic Synthesis*, Academic Press, NY, **1988**, pp. 63–75; Weber, W.P. *Silicon Reagents for Organic Synthesis*, Springer, NY, **1983**, pp. 58–78; Magnus, P. *Aldrichimica Acta* **1980**, *13*, 43; Chan, T. *Acc. Chem. Res.* **1977**, *10*, 442. For a list of references, see Larock, R.C. *Comprehensive Organic Transformations*, 2nd ed., Wiley-VCH, NY, **1999**, pp. 337–341.

[1233]For reviews of these compounds, see Poirier, J. *Org. Prep. Proced. Int.* **1988**, *20*, 319; Brownbridge, P. *Synthesis* **1983**, 1–28, 85; Rasmussen, J.K. *Synthesis* **1977**, 91. For monographs on silicon reagents in organic synthesis see Colvin, E.W. *Silicon Reagents in Organic Synthesis*, Academic Press, NY, **1988**. For reviews, see Colvin, E.W., in Hartley, F.R.; Patai, S. *The Chemistry of the Metal–Carbon Bond*, Vol. 4, Wiley, NY, pp. 539–621; Ager, D.J. *Chem. Soc. Rev.* **1982**, *11*, 493; Colvin, E.W. *Chem. Soc. Rev.* **1978**, 7, 15, pp. 43–50.

[1234]For discussions of the mechanism, see Bassindale, A.R.; Ellis, R.J.; Lau, J.C.; Taylor, P.G. *J. Chem. Soc. Perkin Trans. 2* **1986**, 593; Hudrlik, P.F.; Agwaramgbo, E.L.O.; Hudrlik, A.M. *J. Org. Chem.* **1989**, *54*, 5613.

[1235]Hartzell, S.L.; Sullivan, D.F.; Rathke, M.W. *Tetrahedron Lett.* **1974**, 1403; Shimoji, K.; Taguchi, H.; Oshima, K.; Yamamoto, H.; Nozaki, H. *J. Am. Chem. Soc.* **1974**, *96*, 1620; Chan, T.H.; Moreland, M. *Tetrahedron Lett.* **1978**, 515; Strekowski, L.; Visnick, M.; Battiste, M.A. *Tetrahedron Lett.* **1984**, *25*, 5603.

[1236]Craig, D.; Ley, S.V.; Simpkins, N.S.; Whitham, G.H.; Prior, M.J. *J. Chem. Soc. Perkin Trans. 1* **1985**, 1949.

[1237]Bassindale, A.R.; Ellis, R.J.; Taylor, P.G. *J. Chem. Res. (S)* **1996**, 34.

[1238]See Colvin, E.W. *Silicon Reagents in Organic Synthesis*, Academic Press, NY, **1988**, pp. 65–69.

[1239]Markò, I.E.; Murphy, F.; Kumps, L.; Ates, A.; Touillaux, R.; Craig, D.; Carballares, S.; Dolan, S. *Tetrahedron* **2001**, *57*, 2609.

[1240]Surprenant, S.; Chan, W.Y.; Berthelette, C. *Org. Lett.* **2003**, *5*, 4851.

When aldehydes or ketones are treated with reagents of the form **55**, the product is an epoxy silane (**16-46**), which can be hydrolyzed to a methyl ketone.[1241] For aldehydes, this is a method for converting RCHO to a methyl ketone RCH_2COMe.

The reagents $Me_3SiCHRM$ (M = Li or Mg) are often prepared from $Me_3SiCHRCl$[1242] (by **12-38** or **12-39**), but they have also been made by **12-22** and by other procedures.[1243]

A new version of the reaction has been developed, reacting $Me_3SiCH_2CO_2Et$ with an aldehyde and a catalytic amount of CsF in DMSO.[1244] A seleno-amide derivative has been used in a similar manner.[1245]

There are no references in *Organic Syntheses*, but see OS **VIII**, 602, for a related reaction.

16-42   The Addition of Active Hydrogen Compounds to CO_2 and CS_2

α-Acylalkyl-de-methoxy-substitution (Overall reaction)

Ketones of the form $RCOCH_3$ and $RCOCH_2R'$ can be carboxylated indirectly by treatment with magnesium methyl carbonate **56**.[1246] Because formation of the chelate **57** provides the driving force of the reaction, carboxylation cannot be achieved at a disubstituted α position. The reaction has also been performed on CH_3NO_2 and compounds of the form RCH_2NO_2[1247] and on certain lactones.[1248] Direct carboxylation has been reported in a number of instances. Ketones have

---

[1241]Cooke, F.; Roy, G.; Magnus, P. *Organometallics* **1982**, *1*, 893.

[1242]For a review of these reagents, see Anderson, R. *Synthesis* **1985**, 717.

[1243]See, for example, Ager, D.J. *J. Chem. Soc. Perkin Trans. 1* **1986**, 183; Barrett, A.G.M.; Flygare, J.A. *J. Org. Chem.* **1991**, *56*, 638.

[1244]Bellassoued, M.; Ozanne, N. *J. Org. Chem.* **1995**, *60*, 6582.

[1245]Murai, T.; Fujishima, A.; Iwamoto, C.; Kato, S. *J. Org. Chem.* **2003**, *68*, 7979.

[1246]Stiles, M. *J. Am. Chem. Soc.* **1959**, *81*, 2598; *Ann. N.Y. Acad. Sci.* **1960**, *88*, 332; Crombie, L.; Hemesley, P.; Pattenden, G. *Tetrahedron Lett.* **1968**, 3021.

[1247]Finkbeiner, H.L.; Stiles, M. *J. Am. Chem. Soc.* **1963**, *85*, 616; Finkbeiner, H.L.; Wagner, G.W. *J. Org. Chem.* **1963**, *28*, 215.

[1248]Martin, J.; Watts, P.C.; Johnson, F. *Chem. Commun.* **1970**, 27.

been carboxylated in the α position to give β-keto acids.[1249] The base here was lithium 4-methyl-2,16-di-*tert*-butylphenoxide.

Ketones $RCOCH_2R'$ (as well as other active hydrogen compounds) undergo base-catalyzed addition to CS_2[1250] to give a dianion intermediate $RCOC^-R'CSS^{2-}$, which can be dialkylated with a halide R^2X to produce α-dithiomethylene ketones, $RCOCR'=C(SR^2)_2$.[1251] Compounds of the form ZCH_2Z' also react with bases and CS_2 to give analogous dianions.[1252]

Although reactions with $N=O$ derivatives do not formally fall into this category of reactions, it is somewhat related. Nitroso compounds react with activated nitriles in the presence of LiBr and microwave irradiation to give a cyano imine, $ArN=C(CN)Ar$.[1253] This transformation has been called the *Ehrlich–Sachs reaction*.[1254]

OS **VII**, 476. See also, OS **VIII**, 578.

16-43 Tollens' Reaction

O-Hydro-*C*(β-hydroxyalkyl)-addition

In the *Tollens' reaction* an aldehyde or ketone containing an α hydrogen is treated with formaldehyde in the presence of $Ca(OH)_2$ or a similar base. The first step is a mixed aldol reaction (**16-34**).

The reaction can be stopped at this point, but more often a second equivalent of formaldehyde is permitted to reduce the newly formed aldol to a 1,3-diol, in a crossed Cannizzaro reaction (**19-81**). If the aldehyde or ketone has several α hydrogens, they can all be replaced. An important use of the reaction is to prepare pentaerythritol from acetaldehyde:

$$CH_3CHO + 4\ HCHO \longrightarrow C(CH_2OH)_4 + HCOOH$$

[1249]Tirpak, R.E.; Olsen, R.S.; Rathke, M.W. *J. Org. Chem.* **1985**, *50*, 4877. For an enantioselective version, see Hogeveen, H.; Menge, W.M.P.B. *Tetrahedron Lett.* **1986**, *27*, 2767.

[1250]For reviews of the reactions of CS_2 with carbon nucleophiles, see Dunn, A.D.; Rudorf, W. *Carbon Disulphide in Organic Chemistry*, Ellis Horwood, Chichester, **1989**, pp. 120–225; Yokoyama, M.; Imamoto, T. *Synthesis* **1984**, 797, pp. 797–804.

[1251]See, for example, Corey, E.J.; Chen, R.H.K. *Tetrahedron Lett.* **1973**, 3817.

[1252]Jensen, L.; Dalgaard, L.; Lawesson, S. *Tetrahedron* **1974**, *30*, 2413; Konen, D.A.; Pfeffer, P.E.; Silbert, L.S. *Tetrahedron* **1976**, *32*, 2507, and references cited therein.

[1253]Laskar, D.D.; Prajapati, D.; Sandhu, J.S. *Synth. Commun.* **2001**, *31*, 1427.

[1254]Ehrlich, P.; Sachs, F. *Chem. Ber.* **1899**, *32*, 2341

OS **I**, 425; **IV**, 907; **V**, 833.

16-44   The Wittig Reaction

Alkylidene-de-oxo-bisubstitution

$$\underset{\text{}}{\overset{\text{O}}{\underset{}{\parallel}}}\underset{}{\text{C}} \;+\; \overset{\oplus}{\underset{Ph_3P}{}}\overset{R'}{\underset{R}{C}}{}^{\ominus} \;\longrightarrow\; \overset{R}{\underset{R'}{C}}=C \;+\; Ph_3P{=}O$$

In the *Wittig reaction* an aldehyde or ketone is treated with a *phosphorus ylid* (also spelled ylide and called a *phosphorane*) to give an alkene.[1255] The conversion of a carbonyl compound to an alkene with a phosphorus ylid is called the *Wittig reaction*. Phosphorus ylids are usually prepared by treatment of a phosphonium salt with a base,[1256] and phosphonium salts are usually prepared from a triaryl phosphine and an alkyl halide (**10-31**):

$$Ph_3P \;+\; \overset{X}{\underset{X}{\overset{R}{C}{\underset{R'}{}}}} \;\xrightarrow{\text{10-31}}\; \overset{\oplus}{Ph_3P}{-}\underset{\underset{\mathbf{58}}{X^-}}{\overset{R}{\underset{R'}{C}}}{-}H \;\xrightarrow{\text{BuLi}}\; \left[\overset{\oplus}{Ph_3P}{-}\overset{R}{\underset{R'}{C}}{}^{\ominus} \;\longleftrightarrow\; Ph_3P{=}\overset{R}{\underset{R'}{C}} \right]$$

Phosphonium salt   Ylid

The reaction of triphenylphosphine and an alkyl halides is facilitated by the use of microwave irradiation.[1257] Indeed, the Wittig reaction itself is assisted by microwave irradiation.[1258] Phosphonium salts are also prepared by addition of phosphines to Michael alkenes (like **15-8**) and in other ways. The phosphonium salts are most often converted to the ylids by treatment with a strong base such as

---

[1255]For a general treatise, see Cadogan, J.I.G. *Organophosphorus Reagents in Organic Synthesis*, Academic Press, NY, *1979*. For a monograph on the Wittig reaction, see Johnson, A.W. *Ylid Chemistry*, Academic Press, NY, *1966*. For reviews, see Maryanoff, B.E.; Reitz, A.B. *Chem. Rev. 1989*, *89*, 863; Bestmann, H.J.; Vostrowsky, O. *Top. Curr. Chem. 1983*, *109*, 85; Pommer, H.; Thieme, P.C. *Top. Curr. Chem. 1983*, *109*, 165; Pommer, H. *Angew. Chem. Int. Ed. 1977*, *16*, 423; Maercker, A. *Org. React. 1965*, *14*, 270; House, H.O. *Modern Synthetic Reactions*, 2nd ed., W.A. Benjamin, NY, *1972*, pp. 682–709; Lowe, P.A. *Chem. Ind. (London) 1970*, 1070; Bergelson, L.D.; Shemyakin, M.M., in Patai, S. *The Chemistry of Carboxylic Acids and Esters*, Wiley, NY, *1969*, pp. 295–340; *Newer Methods Prep. Org. Chem. 1968*, *5*, 154. For related reviews, see Tyuleneva, V.V.; Rokhlin, E.M.; Knunyants, I.L. *Russ. Chem. Rev. 1981*, *50*, 280; Starks, C.M.; Liotta, C. *Phase Transfer Catalysis*, Academic Press, NY, *1978*, pp. 288–297; Weber, W.P.; Gokel, G.W. *Phase Transfer Catalysis in Organic Synthesis*, Springer, NY, *1977*; pp. 234–241; Zbiral, E. *Synthesis 1974*, 775; Bestmann, H.J. *Bull. Soc. Chim. Fr. 1971*, 1619; *Angew. Chem. Int. Ed. 1965*, *4*, 583, 645–660, 830–838; *Newer Methods Prep. Org. Chem. 1968*, *5*, 1; Horner, L. *Fortschr. Chem. Forsch., 1966*, *7*, 1. For a historical background, see Wittig, G. *Pure Appl. Chem. 1964*, *9*, 245. For a list of reagents and references for the Wittig and related reactions, see Larock, R.C. *Comprehensive Organic Transformations*, 2nd ed., Wiley-VCH, NY, *1999*, pp. 327–337.

[1256]When phosphonium *fluorides* are used, no base is necessary, as these react directly with the substrate to give the alkene: Schiemenz, G.P.; Becker, J.; Stöckigt, J. *Chem. Ber. 1970*, *103*, 2077.

[1257]Kiddle, J.J. *Tetrahedron Lett. 2000*, *41*, 1339.

[1258]Frattini, S.; Quai, M.; Cereda, E. *Tetrahedron Lett. 2001*, *42*, 6827.; Wu, J.; Wu, H.; Wei, S.; Dai, W.-M. *Tetrahedron Lett. 2004*, *45*, 4401.

butyllithium, sodium amide,[1259] sodium hydride, or a sodium alkoxide, though weaker bases can be used if the salt is acidic enough. Unusual bases such as 1,5,7-triazabicyclo [4.4.0]dec-5-ene have been used to promote the Wittig reaction.[1260] In some cases, and excess of fluoride ion is sufficient.[1261] For $(Ph_3P^+)_2CH_2$, sodium carbonate is a strong enough base.[1262] When the base used does not contain lithium, the ylid is said to be prepared under "salt-free" conditions[1263] because the lithium halide (where the halide counterion comes from the phosphonium salt) is absent.

When the phosphorus ylid reacts with the aldehyde or ketone to form an alkene, a phosphine oxide is also formed. When triphenylphosphine is used to give $Ph_3P=CRR'$, for example, the by-product is triphenylphosphine oxide, Ph_3PO, which is sometimes difficult to separate from the other reaction products. Ylids are usually prepared from triphenylphosphine, but other triarylphosphines,[1264] trialkylphosphines,[1265] and triphenylarsine[1266] have also been used. Tellurium ylids have been prepared *in situ* from α-halo esters and $BrTeBu_2OTeBu_2Br$ and react with aldehydes to give conjugated esters.[1267] Polymer-bound aryldiphenylphosphino compounds[1268] have been used in reactions with alkyl halides to complete a Wittig reaction. Phosphines that have an α-hydrogen should be avoided, so that reaction with the chosen alkyl halide will lead to a phosphonium salt (**58**) with the α-proton at the desired position. This limitation is essential if a specific ylid is to be formed from the alkyl halide precursor. The Wittig reaction has been carried out with polymer-supported ylids.[1269] It has also been done on silica gel.[1270]

If we view the Wittig reaction from an alkyl halide starting material (alkyl halide phosphonium salt → phosphorus ylid → alkene), the halogen-bearing carbon of an alkyl halide must contain at least one hydrogen as in **59** (for deprotonation at the phosphonium salt stage).

59

[1259]For a convenient method of doing this that results in high yields, see Schlosser, M.; Schaub, B. *Chimia* **1982**, *36*, 396.

[1260]Simoni, D.; Rossi, M.; Rondanin, R.; Mazzali, A.; Baruchello, R.; Malagutti, C.; Roberti, M.; Invidiata, F.P. *Org. Lett.* **2000**, *2*, 3765.

[1261]Kobayashi, T.; Eda, T.; Tamura, O.; Ishibashi, H. *J. Org. Chem.* **2002**, *67*, 3156.

[1262]Ramirez, F.; Pilot, J.F.; Desai, N.B.; Smith, C.P.; Hansen, B.; McKelvie, N. *J. Am. Chem. Soc.* **1967**, *89*, 6273.

[1263]Bestmann, H.J. *Angew. Chem. Int. Ed.* **1965**, *4*, 586.

[1264]Schiemenz, G.P.; Thobe, J. *Chem. Ber.* **1966**, *99*, 2663.

[1265]For example, see Johnson, A.W.; LaCount, R.B. *Tetrahedron* **1960**, *9*, 130; Bestmann, H.J.; Kratzer, O. *Chem. Ber.* **1962**, *95*, 1894.

[1266]An arsenic ylid has been used in a catalytic version of the Wittig reaction; that is, the R_3AsO product is constantly regenerated to produce more arsenic ylid: Shi, L.; Wang, W.; Wang, Y.; Huang, Y. *J. Org. Chem.* **1989**, *54*, 2027; Huang, Z.-Z.; Huang, X.; Huang, Y.-Z. *Tetrahedron Lett.* **1995**, *36*, 425.

[1267]Huang, Z.-Z.; Tang, Y. *J. Org. Chem.* **2002**, *67*, 5320.

[1268]Betancort, J.M.; Barbas III, C.F. *Org. Lett.* **2001**, *3*, 3737.

[1269]Bernard, M.; Ford, W.T.; Nelson, E.C. *J. Org. Chem.* **1983**, *48*, 3164.

[1270]Patil, V.J.; Mävers, U. *Tetrahedron Lett.* **1996**, *37*, 1281.

The reaction is very general.[1271] The aldehyde or ketone may be aliphatic, alicyclic, or aromatic (including diaryl ketones). Wittig reactions in which the ylid and/or the carbonyl substrate contain double or triple bonds; it may contain various functional groups, such as OH, OR, NR_2, aromatic nitro or halo, acetal, amide,[1272] or even ester groups.[1273] Note, however, that a Wittig reaction has been reported in which the carbonyl group of an ester was converted to a vinyl ether.[1274] An important advantage of the Wittig reaction is that the *position* of the new double bond is always certain, in contrast to the result in most of the base-catalyzed condensations (**16-34–16-43**). Ylids have been shown to react with lactones, however, to form ω-alkenyl alcohols.[1275] β-Lactams have also been converted to alkenyl-azetidine derivatives using phosphorus ylids.[1276] Double or triple bonds *conjugated* with the carbonyl also do not interfere, the attack being at the C=O carbon. The carbonyl partner can be generated *in situ*, in the presence of an ylid; the reaction of an alcohol with a mixture of an oxidizing agent and an ylid generates an alkene. Oxidizing agents used in this manner include $BaMnO_4$,[1277] MnO_2,[1278] and $PhI(OAc)_2$.[1279] Polyhalomethanes, such as CBr_3F, react with triphenylphosphine in the presence of diethylzinc and an aldehyde or ketone to give the *gem*-dihaloalkene, $RCH(R')=CF(Br)$.[1280]

The phosphorus ylid may also contain double or triple bonds and certain functional groups. Simple ylids (R, R′ = hydrogen or alkyl) are highly reactive, reacting with oxygen, water, hydrohalic acids, and alcohols, as well as carbonyl compounds and carboxylic esters, so the reaction must be run under conditions where these materials are absent. When an electron-withdrawing group, for example, COR, CN, COOR, CHO, is present in the α position, the ylids are much more stable, because the charge on the carbon is delocalized by resonance as in **60**.

[1271]For a discussion of a cooperative ortho effect, see Dunne, E.C.; Coyne, É.J.; Crowley, P.B.; Gilheany, D.G. *Tetrahedron Lett.* **2002**, *43*, 2449.

[1272]Smith, M.B.; Kwon, T.W. *Synth. Commun.* **1992**, *22*, 2865. For the reaction of an acyl imidazole ylid, see Matsunaga, S.; Kinoshita, T.; Okada, S.; Harada, S.; Shibasaki, M. *J. Am. Chem. Soc.* **2004**, *126*, 7559.

[1273]For an example, see Harcken, C.; Martin, S.F. *Org. Lett.* **2001**, *3*, 3591; Yu, X.; Huang, X. *Synlett 2002*, 1895. Although phosphorus ylids also react with esters, that reaction is too slow to interfere: Greenwald, R.; Chaykovsky, M.; Corey, E.J. *J. Org. Chem.* **1963**, *28*, 1128.

[1274]Tsunoda, T.; Takagi, H.; Takaba, D.; Kaku, H.; Itô, S. *Tetrahedron Lett.* **2000**, *41*, 235.

[1275]Brunel, Y.; Rousseau, G. *Tetrahedron Lett.* **1996**, *37*, 3853.

[1276]Baldwin, J.E.; Edwards, A.J.; Farthing, C.N.; Russell, A.T. *Synlett 1993*, 49.

[1277]Shuto, S.; Niizuma, S.; Matsuda, A. *J. Org. Chem.* **1998**, *63*, 4489.

[1278]Reid, M.; Rowe, D.J.; Taylor, R.J.K. *Chem. Commun.* **2003**, 2284; Blackburn, L.; Pei, C.; Taylor, R.J.K. *Synlett 2002*, 215; Raw, S.A.; Reid, M.; Roman, E.; Taylor, R.J.K. *Synlett 2004*, 819.

[1279]Zhang, P.-F.; Chen, Z.-C. *Synth. Commun.* **2001**, *31*, 1619.

[1280]Lei, X.; Dutheuil, G.; Pannecoucke, X.; Quirion, J.-C. *Org. Lett.* **2004**, *6*, 2101.

Such ylids react readily with aldehydes, but slowly or not at all with ketones.[1281] In extreme cases (e.g., **61**), the

61

ylid does not react with ketones *or* aldehydes. Besides these groups, the ylid may contain one or two α halogens[1282] or an α OR or OAr group. In the latter case, the product is an enol ether, which can be hydrolyzed

(**10-6**) to an aldehyde,[1283] so that this reaction is a means of achieving the conversion RCOR′ → RR′CHCHO.[1284] However, the ylid may not contain an α nitro group. If the phosphonium salt contains a potential leaving group, such as Br or OMe, in the β position, treatment with a base gives elimination, instead of the ylid:

$$\overset{\oplus}{Ph_3}PCH_2CH_2Br \xrightarrow{\text{base}} \overset{\oplus}{Ph_3}PCH=CH_2$$

However, a β COO⁻ group may be present, and the product is a β,γ-unsaturated acid:[1285] This is the only convenient way to make these compounds, since elimination by any other route gives the thermodynamically more stable α,β-unsaturated isomers. This is an illustration of the utility of the Wittig method for the specific location of a double bond. Another illustration is the conversion of cyclohexanones to alkenes containing exocyclic double bonds, for example,[1286]

[1281]For successful reactions of stabilized ylids with ketones, under high pressure, see Isaacs, N.S.; El-Din, G.N. *Tetrahedron Lett.* **1987**, *28*, 2191. See also, Dauben, W.G.; Takasugi, J.J. *Tetrahedron Lett.* **1987**, *28*, 4377.

[1282]Seyferth, D.; Heeren, J.K.; Singh, G.; Grim, S.O.; Hughes, W.B. *J. Organomet. Chem.* **1966**, *5*, 267; Schlosser, M.; Zimmermann, M. *Synthesis* **1969**, 75; Burton, D.J.; Greenlimb, P.E. *J. Fluorine Chem.* **1974**, *3*, 447; Smithers, R.H. *J. Org. Chem.* **1978**, *43*, 2833; Miyano, S.; Izumi, Y.; Fujii, K.; Ohno, Y.; Hashimoto, H. *Bull. Chem. Soc. Jpn.* **1979**, *52*, 1197; Stork, G.; Zhao, K. *Tetrahedron Lett.* **1989**, *30*, 2173.

[1283]For references to the use of the Wittig reaction to give enol ethers or enol thioethers, which are then hydrolyzed, see Larock, R.C. *Comprehensive Organic Transformations*, 2nd ed., Wiley-VCH, NY, **1999**, pp. 1441–1444, 1457–1458.

[1284]For other methods of achieving this conversion via Wittig-type reactions, see Ceruti, M.; Degani, I.; Fochi, R. *Synthesis* **1987**, 79; Moskal, J.; van Leusen, A.M. *Recl. Trav. Chim. Pays-Bas* **1987**, *106*, 137; Doad, G.J.S. *J. Chem. Res. (S)* **1987**, 370.

[1285]Corey, E.J.; McCormick, J.R.D.; Swensen, W.E. *J. Am. Chem. Soc.* **1964**, *86*, 1884.

[1286]Wittig, G.; Schöllkopf, U. *Chem. Ber.* **1954**, *87*, 1318.

Still another example is the easy formation of anti-Bredt bicycloalkenones[1287] (see p. 229). As indicated above, α,α'-dihalophosphoranes can be used to prepare 1,1-dihaloalkenes. Another way to prepare such compounds[1288] is to treat the carbonyl compound with a mixture of CX_4 (X = Cl, Br, or I) and triphenylphosphine, either with or without the addition of zinc dust (which allows less Ph_3P to be used).[1289] Aryl aldehydes react with these dihalophosphoranes to give aryl alkynes after treatment of the initially formed vinyl halide with potassium *tert*-butoxide.[1290] Formamides have been converted to ynamines by reaction with a mixture of PPh_3/CCl_4 followed by *n*-butyllithium.[1291] The carbonyl compound can be generated *in situ*, in the presence of the phosphorane. A cyclopropylcarbonyl alcohol was converted to a β-cyclopropyl-α,β-unsaturated ester by reaction with MnO_2 in the presence of $Ph_3P{=}CHCO_2Me$.[1292]

The mechanism[1293] of the key step of the Wittig reaction is as follows:[1294]

Oxaphosphetane

The energetics of ylid formation and their reaction is solution has been studied.[1295] For many years it was assumed that a diionic compound, called a *betaine*, is an intermediate on the pathway from the starting compounds

---

[1287]Bestmann, H.J.; Schade, G. *Tetrahedron Lett.* **1982**, *23*, 3543.

[1288]For a list of references to the preparation of haloalkenes by Wittig reactions, with references, see Larock, R.C. *Comprehensive Organic Transformations*, 2nd ed., Wiley-VCH, NY, **1999**, pp. 725–727.

[1289]See, for example, Rabinowitz, R.; Marcus, R. *J. Am. Chem. Soc.* **1962**, *84*, 1312; Ramirez, F.; Desai, N.B.; McKelvie, N. *J. Am. Chem. Soc.* **1962**, *84*, 1745; Corey, E.J.; Fuchs, P.L. *Tetrahedron Lett.* **1972**, 3769; Posner, G.H.; Loomis, G.L.; Sawaya, H.S. *Tetrahedron Lett.* **1975**, 1373; Suda, M.; Fukushima, A. *Tetrahedron Lett.* **1981**, *22*, 759; Gaviña, F.; Luis, S.V.; Ferrer, P.; Costero, A.M.; Marco, J.A. *J. Chem. Soc., Chem. Commun.* **1985**, 296; Li, P.; Alper, H. *J. Org. Chem.* **1986**, *51*, 4354.

[1290]Michel, P.; Gennet, D.; Rassat, A. *Tetrahedron Lett.* **1999**, *40*, 8575. See Michael, P.; Rassat, A. *Tetrahedron Lett.* **1999**, *40*, 8579.

[1291]Brückner, D. *Synlett* **2000**, 1402.

[1292]Blackburn, L.; Wei, X.; Taylor, R.J.K. *Chem. Commun.* **1999**, 1337.

[1293]For a review of the mechanism, see Cockerill, A.F.; Harrison, R.G., in Patai, S. *The Chemistry of Functional Groups: Supplement A*, pt. 1, Wiley, NY, **1977**, pp. 232–240. For a thorough discussion, see Vedejs, E.; Marth, C.F. *J. Am. Chem. Soc.* **1988**, *110*, 3948.

[1294]It has been contended that another mechanism, involving single electron transfer, may be taking place in some cases: Olah, G.A.; Krishnamurthy, V.V. *J. Am. Chem. Soc.* **1982**, *104*, 3987; Yamataka, H.; Nagareda, K.; Hanafusa, T.; Nagase, S. *Tetrahedron Lett.* **1989**, *30*, 7187. A diradical mechanism has also been proposed for certain cases: Ward, Jr., W.J.; McEwen, W.E. *J. Org. Chem.* **1990**, *55*, 493.

[1295]Arnett, E.M.; Wernett, P.C. *J. Org. Chem.* **1993**, *58*, 301.

to the oxaphosphetane, and in fact it may be so, but there is little evidence for it.[1296]

$$
\begin{array}{c}
\overset{\ominus}{O} - \overset{|}{\underset{|}{C}} \diagup \\
Ph_3 \overset{\oplus}{P} - \overset{|}{\underset{R'}{C}} - R
\end{array}
$$

Betaine

"Betaine" precipitates have been isolated in certain Wittig reactions,[1297] but these are betaine–lithium halide adducts, and might just as well have been formed from the oxaphosphetane as from a true betaine.[1298] However, there is one report of an observed betaine lithium salt during the course of a Wittig reaction.[1299] An X-ray structure was determined for a gauche betaine from a thio-Wittig reaction.[1300] In contrast, there is much evidence for the presence of the oxaphosphetane intermediates, at least with unstable ylids. For example, ^{31}P NMR spectra taken of the reaction mixtures at low temperatures[1301] are compatible with an oxaphosphetane structure that persists for some time but not with a tetra-coordinated phosphorus species. Since a betaine, an ylid, and a phosphine oxide all have tetracoordinated phosphorus, these species could not be causing the spectra, leading to the conclusion that an oxaphosphetane intermediate is present in the solution. In certain cases oxaphosphetanes have been isolated.[1302] It has even been possible to detect cis and trans isomers of the intermediate oxaphosphetanes by NMR spectroscopy.[1303] According to this mechanism, an optically active phosphonium salt $RR'R^2P^+CHR^2$ should retain its configuration all the way through the reaction, and it should be preserved in the phosphine oxide $RR'R^2PO$. This has been shown to be the case.[1304]

The proposed betaine intermediates can be formed, in a completely different manner, by nucleophilic substitution by a phosphine on an epoxide (**10-35**):

[1296]See Vedejs, E.; Marth, C.F. *J. Am. Chem. Soc.* **1990**, *112*, 3905.

[1297]Wittig, G.; Weigmann, H.; Schlosser, M. *Chem. Ber.* **1961**, *94*, 676; Schlosser, M.; Christmann, K.F. *Liebigs Ann. Chem.* **1967**, *708*, 1.

[1298]Maryanoff, B.E.; Reitz, A.B. *Chem. Rev.* **1989**, *89*, 863, see p. 865.

[1299]Neumann, R.A.; Berger, S. *Eur. J. Org. Chem.* **1998**, 1085.

[1300]Puke, C.; Erker, G.; Wibbeling, B.; Fröhlich, R. *Eur. J. Org. Chem.* **1999**, 1831.

[1301]Vedejs, E.; Meier, G.P.; Snoble, K.A.J. *J. Am. Chem. Soc.* **1981**, *103*, 2823. See also, Nesmayanov, N.A.; Binshtok, E.V.; Reutov, O.A. *Doklad. Chem.* **1973**, *210*, 499.

[1302]Birum, G.H.; Matthews, C.N. *Chem. Commun.* **1967**, 137; Mazhar-Ul-Haque; Caughlan, C.N.; Ramirez, F.; Pilot, J.F.; Smith, C.P. *J. Am. Chem. Soc.* **1971**, *93*, 5229.

[1303]Maryanoff, B.E.; Reitz, A.B.; Mutter, M.S.; Inners, R.R.; Almond Jr., H.R.; Whittle, R.R.; Olofson, R.A. *J. Am. Chem. Soc.* **1986**, *108*, 7664. See also, Pískala, A.; Rehan, A.H.; Schlosser, M. *Coll. Czech. Chem. Commun.* **1983**, *48*, 3539.

[1304]McEwen, W.E.; Kumli, K.F.; Bladé-Font, A.; Zanger, M.; VanderWerf, C.A. *J. Am. Chem. Soc.* **1964**, *86*, 2378.

Betaines formed in this way can then be converted to the alkene, and this is one reason why betaine intermediates were long accepted in the Wittig reaction.

The Wittig reaction has also been carried out with phosphorus ylids other than phosphoranes, the most important being prepared from phosphonates, such as **62**.[1305]

62

This method, sometimes called the *Horner–Emmons, Wadsworth–Emmons*, or *Wittig–Horner reaction*,[1306] has several advantages over the use of phosphoranes, including selectivity.[1307] These ylids are more reactive than the corresponding phosphoranes, and when R^1 or R^2 is an electron-withdrawing group, these compounds often react with ketones that are inert to phosphoranes. High pressure has been used to facilitate this reaction.[1308] In addition, the phosphorus product is a phosphate ester and hence soluble in water, unlike Ph_3PO, which makes it easy to separate it from the alkene product. Phosphonates are also cheaper than phosphonium salts and can easily be prepared by the *Arbuzov reaction*:[1309]

Phosphonates have also been prepared from alcohols and $(ArO)_2P(=O)Cl$, NEt_3 and a $TiCl_4$ catalyst.[1310] The reaction of $(RO)_2P(=O)H$ and aryl iodides with a CuI catalyst leads to aryl phosphonates.[1311] Polymer-bound phosphonate esters have been used for olefination.[1312] Dienes are produced when allylic phosphonate esters react with aldehydes.[1313]

[1305]Horner, L.; Hoffmann, H.; Wippel, H.G.; Klahre, G. *Chem. Ber.* **1959**, *92*, 2499; Wadsworth, Jr., W.S.; Emmons, W.D. *J. Am. Chem. Soc.* **1961**, *83*, 1733.

[1306]For reviews, see Wadsworth, Jr., W.S. *Org. React.* **1977**, *25*, 73; Stec, W.J. *Acc. Chem. Res.* **1983**, *16*, 411; Walker, B.J., in Cadogan, J.I.G. *Organophosphorous Reagents in Organic Synthesis*, Academic Press, NY, **1979**, pp. 156–205; Dombrovskii, A.V.; Dombrovskii, V.A. *Russ. Chem. Rev.* **1966**, *35*, 733; Boutagy, J.; Thomas, R. *Chem. Rev.* **1974**, *74*, 87. For a convenient method of carrying out this reaction, see Seguineau, P.; Villieras, J. *Tetrahedron Lett.* **1988**, *29*, 477, and other papers in this series.

[1307]Motoyoshiya, J.; Kasaura, T.; Kokin, K.; Yokoya, S.-i.; Takaguchi, Y.; Narita, S.; Aoyama, H. *Tetrahedron* **2001**, *57*, 1715.

[1308]Has-Becker, S.; Bodmann, K.; Kreuder, R.; Santoni, G.; Rein, T.; Reiser, O. *Synlett* **2001**, 1395.

[1309]Also known as the *Michaelis-Arbuzov rearrangement*. For reviews, see Petrov, A.A.; Dogadina, A.V.; Ionin, B.I.; Garibina, V.A.; Leonov, A.A. *Russ. Chem. Rev.* **1983**, *52*, 1030; Bhattacharya, A.K.; Thyagarajan, G. *Chem. Rev.* **1981**, *81*, 415. For related reviews, see Shokol, V.A.; Kozhushko, B.N. *Russ. Chem. Rev.* **1985**, *53*, 98; Brill, T.B.; Landon, S.J. *Chem. Rev.* **1984**, *84*, 577. See also, Kaboudin, B.; Balakrishna, M.S. *Synth. Commun.* **2001**, *31*, 2773.

[1310]Jones, S.; Selitsianos, D. *Org. Lett.* **2002**, *4*, 3671.

[1311]Gelman, D.; Jiang, L.; Buchwald, S.L. *Org. Lett.* **2003**, *5*, 2315.

[1312]Barrett, A.G.M.; Cramp, S.M.; Roberts, R.S.; Zecri, F.J. *Org. Lett.* **1999**, *1*, 579.

[1313]Wang, Y.; West, F.G. *Synthesis* **2002**, 99.

Stereoselective alkenylation reactions have been achieved using chiral additives[1314] or auxiliaries.[1315] Ylids formed from phosphine oxides,

$$Ar_2P-CHRR'$$
$$\overset{\|}{O}$$

phosphonic acid bisamides, $(R_2^2N)_2POCHRR'$,[1316] and alkyl phosphonothionates, $(MeO)_2PSCHRR'$,[1317] share some of these advantages. Reagents, such as $Ph_2POCH_2NR_2'$, react with aldehydes or ketones (R_2COR^3) to give good yields of enamines $(R^2R^3C=CHNR)$.[1318] (Z)-Selective reagents are also known,[1319] including the use of a di(2,2,2-trifluoroethoxy)phosphonate with KHMDS and 18-crown-6.[1320] An interesting intramolecular version of the Horner–Emmons reaction leads to alkynes.[1321] The reaction of a functionalized aldehyde (R—CHO) with $(MeO)_2POCHN_2$, leads to the alkyne (R—C≡CH).[1322]

Some Wittig reactions give the (Z)-alkene; some the (E), and others give mixtures, and the question of which factors determine the stereoselectivity has been much studied.[1323] It is generally found that ylids containing stabilizing groups or formed from trialkylphosphines give (E)-alkenes. However, ylids formed from triarylphosphines and not containing stabilizing groups often give (Z) or a mixture of (Z) and (E)-alkenes.[1324] One explanation for this[1193] is that the reaction of the ylid with the carbonyl compound is a [2 + 2]-cycloaddition, which in order to be concerted must adopt the $[_{\pi}2_s + _{\pi}2_a]$ pathway. As we have seen earlier (p. 1225), this pathway leads to the formation of the more sterically crowded product, in this case the Z alkene. If this explanation is correct, it is not easy to explain the predominant formation of (E)

[1314]Mizuno, M.; Fujii, K.; Tomioka, K. *Angew. Chem. Int. Ed.* **1998**, *37*, 515. Also see, Arai, S.; Hamaguchi, S.; Shioiri, T. *Tetrahedron Lett.* **1998**, *39*, 2997. For a review of asymmetric Wittig-type reactions see Rein, T.; Pedersen, T.M. *Synthesis* **2002**, 579.

[1315]Abiko, A.; Masamune, S. *Tetrahedron Lett.* **1996**, *37*, 1077.

[1316]Corey, E.J.; Kwiatkowski, G.T. *J. Am. Chem. Soc.* **1968**, *90*, 6816; Corey, E.J.; Cane, D.E. *J. Org. Chem.* **1969**, *34*, 3053. For a chiral derivative, see Hanessian, S.; Beaudoin, S. *Tetrahedron Lett.* **1992**, *33*, 7655, 7659.

[1317]Corey, E.J.; Kwiatkowski, G.T. *J. Am. Chem. Soc.* **1966**, *88*, 5654.

[1318]Broekhof, N.L.J.M.; van der Gen, A. *Recl. Trav. Chim. Pays-Bas* **1984**, *103*, 305; Broekhof, N.L.J.M.; van Elburg, P.; Hoff, D.J.; van der Gen, A. *Recl. Trav. Chim. Pays-Bas* **1984**, *103*, 317.

[1319]Ando, K. *Tetrahedron Lett.* **1995**, *36*, 4105.

[1320]Yu, W.; Su, M.; Jin, Z. *Tetrahedron Lett.* **1999**, *40*, 6725.

[1321]Nangia, A.; Prasuna, G.; Rao, P.B. *Tetrahedron Lett.* **1994**, *35*, 3755; Couture, A.; Deniau, E.; Gimbert, Y.; Grandclaudon, P. *J. Chem. Soc. Perkin Trans. 1* **1993**, 2463.

[1322]Hauske, J.R.; Dorff, P.; Julin, S.; Martinelli, G.; Bussolari, J. *Tetrahedron Lett.* **1992**, *33*, 3715.

[1323]For reviews of the stereochemistry of the Wittig reactions, see Maryanoff, B.E.; Reitz, A.B. Chem. Rev. **1989**, *89*, 863; Gosney, I.; Rowley, A.G., in Cadogan, J.I.G. *Organophosphorous Reagents in Organic Synthesis*, Academic Press, NY, **1979**, pp. 17–153; Reucroft, J.; Sammes, P.G. *Q. Rev. Chem. Soc.* **1971**, *25*, 135, see pp. 137–148, 169; Schlosser, M. *Top. Stereochem.* **1970**, *5*, 1. Also see Takeuchi, K.; Paschal, J.W.; Loncharich, R.J. *J. Org. Chem.* **1995**, *60*, 156.

[1324]For cases where such an ylid gave (E)-alkenes, see Maryanoff, B.E.; Reitz, A.B.; Duhl-Emswiler, B.A. *J. Am. Chem. Soc.* **1985**, *107*, 217; Le Bigot, Y.; El Gharbi, R.; Delmas, M.; Gaset, A. *Tetrahedron* **1986**, *42*, 3813. For guidance in how to obtain the maximum yields of the Z product, see Schlosser, M.; Schaub, B.; de Oliveira-Neto, J.; Jeganathan, S. *Chimia* **1986**, *40*, 244.

products from stable ylids, but (*E*) compounds are of course generally thermodynamically more stable than the (*Z*) isomers, and the stereochemistry seems to depend on many factors.

The (*E/Z*) ratio of the product can often be changed by a change in solvent or by the addition of salts.[1325] Another way of controlling the stereochemistry of the product is by use of the aforementioned phosphonic acid bisamides. In this case, the betaine (**63**) does form and when treated with water gives the β-hydroxyphosphonic acid bisamides **64**, which can be crystallized and then cleaved to $R^1R^2C=CR^3R^4$ by refluxing in benzene or toluene in the presence of silica gel.[1316] **64** are generally formed as mixtures of diastereomers, and these mixtures can be separated by recrystallization. Cleavage of the two diastereomers gives the two isomeric alkenes. Optically active phosphonic acid bisamides have been used to give optically active alkenes.[1326] Another method of controlling the stereochemistry of the alkene [to obtain either the (*Z*) or (*E*) isomer] starting with a phosphine oxide (Ph_2POCH_2R), has been reported.[1327]

In reactions where the betaine–lithium halide intermediate is present, it is possible to extend the chain further if a hydrogen is present α to the phosphorus. For example, reaction of ethylidnetriphenylphosphorane with heptanal at −78°C gave **65**, which with butyllithium gave the ylid **66**. Treatment of this with an aldehyde R′CHO gave the intermediate **67**, which after workup gave **68**.[1328] This reaction gives the unsaturated alcohols **68** stereoselectively. **66** also reacts with other electrophiles. For example, treatment of **66** with *n*-chlorosuccinimide or $PhICl_2$ gives the vinylic chloride RCH=CMeCl stereoselectively: NCS giving the cis and $PhICl_2$

[1325]See, for example, Reitz, A.B.; Nortey, S.O.; Jordan, Jr., A.D.; Mutter, M.S.; Maryanoff, B.E. *J. Org. Chem.* **1986**, *51*, 3302.

[1326]Hanessian, S.; Delorme, D.; Beaudoin, S.; Leblanc, Y. *J. Am. Chem. Soc.* **1984**, *106*, 5754; Rein, T.; Reiser, O. *Acta Chem. Scand. B*, **1996**, *50*, 369. For a review of asymmetric ylid reactions, see Li, A.-H.; Dai, L.-X.; Aggarwal, V.K. *Chem. Rev.* **1997**, *97*, 2341.

[1327]Ayrey, P.M.; Warren, S. *Tetrahedron Lett.* **1989**, *30*, 4581.

[1328]Corey, E.J.; Yamamoto, H. *J. Am. Chem. Soc.* **1970**, *92*, 226; Schlosser, M.; Coffinet, D. *Synthesis* **1972**, 575; Corey, E.J.; Ulrich, P.; Venkateswarlu, A. *Tetrahedron Lett.* **1977**, 3231; Schlosser, M.; Tuong, H.B.; Respondek, J.; Schaub, B. *Chimia* **1983**, *37*, 10.

the trans isomer.[1329] The use of Br_2 and $FClO_3$ (see **12-4** for the explosive nature of this reagent) gives the corresponding bromides and fluorides, respectively.[1330] Reactions of **66** with electrophiles have been called *scoopy* reactions (α *s*ubstitution plus *c*arbonyl alkeneylation via β-*o*xido *p*hosphorus *y*lids).[1331]

The reaction of a phosphonate ester, DBU, NaI, and HMPA with an aldehyde leads to a conjugated ester with excellent (*Z*)-selectivity.[1332] A (*Z*)-selective reaction was reported using a trifluoroethyl phosphonate in a reaction with an aldehyde and potassium *tert*-butoxide.[1333]

The Wittig reaction has been carried out intramolecularly, to prepare rings containing from 5 to 16 carbons,[1334] both by single ring closure

and double ring closure.[1335]

The Wittig reaction has proved very useful in the synthesis of natural products, some of which are quite difficult to prepare in other ways.[1336] One example out of many is the synthesis of β-carotene:[1337]

β-Carotene

[1329]Schlosser, M.; Christmann, K. *Synthesis* **1969**, 38; Corey, E.J.; Shulman, J.I.; Yamamoto, H. *Tetrahedron Lett.* **1970**, 447.

[1330]Schlosser, M.; Christmann, K.-F. *Synthesis* **1969**, 38.

[1331]Schlosser, M. *Top. Stereochem.* **1970**, 5, 1, p. 22.

[1332]Ando, K.; Oishi, T.; Hirama, M.; Ohno, H.; Ibuka, T. *J. Org. Chem.* **2000**, 65, 4745.

[1333]Touchard, F.P. *Tetrahedron Lett.* **2004**, 45, 5519.

[1334]For a review, see Becker, K.B. *Tetrahedron* **1980**, 36, 1717.

[1335]For a review of these double-ring closures, see Vollhardt, K.P.C. *Synthesis* **1975**, 765.

[1336]For a review of applications of the Wittig reaction to the synthesis of natural products, see Bestmann, H.J.; Vostrowsky, O. *Top. Curr. Chem.* **1983**, 109, 85.

[1337]Wittig, G.; Pommer, H. German patent **1956**, 954,247, [*Chem. Abstr.* **1959**, 53, 2279].

Phosphorus ylids also react in a similar manner with the C=O bonds of ketenes,[1338] isocyanates,[1339] certain anhydrides[1340] lactones,[1341] and imides,[1342] the N=O of nitroso groups, and the C=N of imines,[1343] for example,

Phosphorus ylids react with carbon dioxide to give the isolable salts **69**,[1344] which can be hydrolyzed to the carboxylic acids **70** (thus achieving the conversion

---

[1338]For example, see Aksnes, G.; Frøyen, P. *Acta Chem. Scand.* **1968**, *22*, 2347.

[1339]For example, see Frøyen, P. *Acta Chem. Scand. Ser. B* **1974**, *28*, 586.

[1340]See, for example, Abell, A.D.; Massy-Westropp, R.A. *Aust. J. Chem.* **1982**, *35*, 2077; Kayser, M.M.; Breau, L. *Can. J. Chem.* **1989**, *67*, 1401. For a study of the mechanism, see Abell, A.D.; Clark, B.M.; Robinson, W.T. *Aust. J. Chem.* **1988**, *41*, 1243.

[1341]With microwave irradiation, see Sabitha, G.; Reddy, M.M.; Srinivas, D.; Yadov, J.S. *Tetrahedron Lett.* **1999**, *40*, 165.

[1342]For a review of the reactions with anhydrides and imides (and carboxylic esters, thiol esters, and amides), see Murphy, P.J.; Brennan, J. *Chem. Soc. Rev.* **1988**, *17*, 1. For a review with respect to imides, see Flitsch, W.; Schindler, S.R. *Synthesis* **1975**, 685.

[1343]Bestmann, H.J.; Seng, F. *Tetrahedron* **1965**, *21*, 1373.

[1344]Bestmann, H.J.; Denzel, T.; Salbaum, H. *Tetrahedron Lett.* **1974**, 1275.

RR'CHX → RR'CHCOOH) or (if neither R nor R' is hydrogen) dimerized to allenes.

A useful alternative to phosphorus ylids are the titanium reagents, such as, **71**, prepared from dicyclopentadienyltitanium dichloride and trimethylaluminum.[1347] Treatment of a carbonyl compound with the titanium cyclopentadienide complex **71** (*Tebbe's reagent*) in toluene–THF containing a small amount of pyridine[1348] leads to the alkene. Dimethyltitanocene (Me$_2$TiCp$_2$), called the *Petasis reagent*, is a convenient and highly useful alternative to **71**.[1349] The mechanism of Petasis olefination has been examined.[1350] Tebbe's reagent and the Petasis reagent give good results with ketones.[1351] An important feature of these new reagents is that

Although phosphorus ylids are most commonly used to alkenylation reactions, nitrogen ylids can occasionally be used. As an example, the reaction of N-benzyl-N-phenylpiperidinium bromide with base generated a N-ylid, which reacted with benzaldehyde to form styrene.[1345] The structure has been determined for an intermediate in an aza-Wittig reaction.[1346]

OS **V**, 361, 390, 499, 509, 547, 751, 949, 985; **VI**, 358; **VII**, 164, 232; **VIII**, 265, 451; **75**, 139, **OS IX**, 39, 230.

16-45   Tebbe, Petasis and Alternative Alkenylations

Methylene-de-oxo-bisubstitution

[1345]Lawrence, N.J.; Beynek, H. Synlett **1998**, 497.
[1346]Kano, N.; Hua, X.J.; Kawa, S.; Kawashima, T. Tetrahedron Lett. **2000**, 41, 5237.
[1347]For a method of generating this reagent in situ, see Cannizzo, L.F.; Grubbs, R.H. J. Org. Chem. **1985**, 50, 2386.
[1348]Tebbe, F.N.; Parshall, G.W.; Reddy, G.S. J. Am. Chem. Soc. **1978**, 100, 3611; Pine, S.H.; Pettit, R.J.; Geib, G.D.; Cruz, S.G.; Gallego, C.H.; Tijerina, T.; Pine, R.D. J. Org. Chem. **1985**, 50, 1212. See also, Clawson, L.; Buchwald, S.L.; Grubbs, R.H. Tetrahedron Lett. **1984**, 25, 5733; Clift, S.M.; Schwartz, J. J. Am. Chem. Soc. **1984**, 106, 8300.
[1349]Petasis N.A.; Bzowej, E.I. J. Am. Chem. Soc. **1990**, 112, 6392.
[1350]Meurer, E.C.; Santos, L.S.; Pilli, R.A.; Eberlin, M.N. Org. Lett. **2003**, 5, 1391.
[1351]Pine, S.H.; Shen, G.S.; Hoang, H. Synthesis **1991**, 165.

carboxylic esters and lactones[1352] can be converted in good yields to the corresponding enol ethers. The enol ether can be hydrolyzed to a ketone (**10-6**), so this is also an indirect method for making the conversion $RCOOR' \rightarrow RCOCH_3$ (see also, **16-82**). Conjugated esters are converted to alkoxy-dienes with this reagent.[1353] Lactams, including β-lactams, are converted with alkylidene cycloamines (alkylidene azetidines from β-lactams, which are easily hydrolyzed to β-amino ketones).[1354]

Besides stability and ease of preparation, another advantage of the Petasis reagent is that structural analogs can be prepared, including $Cp_2Ti(C_3H_5)_2$[1355] (C_3H_5 = cyclopropyl), $CpTi(CH_2SiMe_3)_3$,[1356] and $Cp_2TiMe(CH=CH_2)$.[1357] In another variation, 2 equivalents of $Cp_2Ti[P(OEt)_3]_2$ reacted with a ketone in the presence of 1,1-diphenylthiocyclobutane to give the alkenylcyclobutane derivative.[1358] An alternative titanium reagent was prepared using $TiCl_4$, magnesium metal and dichloromethane, reacting with both ketones[1359] and esters[1360] to give alkenes or vinyl ethers, respectively. Alkenes are generated form ketones and alkyl iodides in the presence of a catalytic amount of $Cp_2Ti[POEt)_3]_2$.[1361]

α,α-Dibromosulfones ($ArSO_2SHBr_2$) react with ketones in the presence of Sm/SmI_2 and a $CrCl_3$ catalyst gives to corresponding vinyl sulfone.[1362] Imides are converted to alkylidene lactams when treated with an alkyl halide, 2.5 equivalents of SmI_2 and a NiI_2 catalyst.[1363]

Carboxylic esters undergo the conversion $C=O \rightarrow C=CHR$ (R = primary or secondary alkyl) when treated with $RCHBr_2$, Zn,[1364] and $TiCl_4$ in the presence of N,N,N',N'-tetramethylethylenediamine.[1365] Metal carbene complexes[1366] $R_2C = ML_n$ (L = ligand), where M is a transition metal, such as Zr, W, or Ta, have also been

[1352]See Martínez, I.; Andrews, A.E.; Emch, J.D.; Ndakala, A.J.; Wang, J.; Howell, A.R.; Rheingold, A.L.; Figuero, J.S. *Org. Lett.* **2003**, *5*, 399; Dollinger, L.M.; Ndakala, A.J.; Hashemzadeh, M.; Wang, G.; Wang, Y.; Martínez, K.; Arcari, J.T.; Galluzzo, D.J.; Howell, A.R.; Rheingold, A.L. Figuero. J.S.; *J. Org. Chem.* **1999**, *64*, 7074.

[1353]Petasis N.A.; Lu, S.-P. *Tetrahedron Lett.* **1995**, *36*, 2393.

[1354]Tehrani, K.A.; De Kimpe, N. *Tetrahedron Lett.* **2000**, *41*, 1975. See Martínez, I.; Howell, A.R. *Tetrahedron Lett.* **2000**, *41*, 5607.

[1355]Petasis N.A.; Browej, E.I. *Tetrahedron Lett.* **1993**, *34*, 943.

[1356]Petasis N.A.; Akritopoulou, I. *Synlett* **1992**, 665.

[1357]Petasis N.A.; Hu, Y.-H. *J. Org. Chem.* **1997**, *62*, 782. Also see, Petasis N.A.; Straszewski, J.P.; Fu, D.-K. *Tetrahedron Lett.* **1995**, *36*, 3619; Rahim, Md.A.; Taguchi, H.; Watanabe, M.; Fujiwara, T.; Takeda, T. *Tetrahedron Lett.* **1998**, *39*, 2153; Petasis N.A.; Browej, E.I. *J. Org. Chem.* **1992**, *57*, 1327.

[1358]Fujiwara, T.; Iwasaki, N.; Takeda, T. *Chem. Lett.* **1998**, 741. For an example using a *gem*-dichloride, see Takeda, T.; Sasaki, R.; Fujiwara, T. *J. Org. Chem.* **1998**, *63*, 7286.

[1359]Yan, T.H.; Tsai, C.-C.; Chien, C.-T.; Cho, C.-C. Huang, P.-C. *Org. Lett.* **2004**, *6*, 4961.

[1360]Yan, T.-H.; Chien, C.-T.; Tsai, C.-C.; Lin, K.-W.; Wu, Y.-H. *Org. Lett.* **2004**, *6*, 4965.

[1361]Takeda, T.; Shimane, K.; Ito, K.; Saeki, N.; Tsubouchi, A. *Chem. Commun.* **2002**, 1974.

[1362]Liu, Y.; Wu, H.; Zhang, Y. *Synth. Commun.* **2001**, *31*, 47.

[1363]Farcas, S.; Namy, J.-L. *Tetrahedron Lett.* **2001**, *42*, 879.

[1364]Ishino, Y.; Mihara, M.; Nishihama, S.; Nishiguchi, I. *Bull. Chem. Soc. Jpn.* **1998**, *71*, 2669.

[1365]Okazoe, T.; Takai, K.; Oshima, K.; Utimoto, K. *J. Org. Chem.* **1987**, *52*, 4410. For the reaction with $CH_2(ZnI)_2$ with $TiCl_2$, see Matsubra, S.; Ukai, K.; Mizuno, T.; Utimoto, K. *Chem. Lett.* **1999**, 825. This procedure is also successful for silyl esters, to give silyl enol ethers: Takai, K.; Kataoka, Y.; Okazoe, T.; Utimoto, K. *Tetrahedron Lett.* **1988**, *29*, 1065.

[1366]For a review of the synthesis of such complexes, see Aguero, A.; Osborn, J.A. *New J. Chem.* **1988**, *12*, 111.

used to convert the C=O of carboxylic esters and lactones to CR$_2$.[1367] It is likely that the complex Cp$_2$Ti = CH$_2$ is an intermediate in the reaction with Tebbe's reagent.

There are a few other methods for converting ketones or aldehydes to alkenes. When a ketone is treated with CH$_3$CHBr$_2$/Sm/SmI$_2$, with a catalytic amount of CrCl$_3$, for example, the alkene is formed.[1368] α-Halo esters also react with CrCl$_2$ in the presence of a ketone to give vinyl halides.[1369] In another reaction, an aldehydes reacted with EtCHBr(OAc) in the presence of Zn/CrCl$_3$ to give the alkene.[1370] α-Diazo esters react with ketones in the presence of an iron catalyst to give the corresponding alkene.[1371] α-Diazo silylalkanes react similarly in the presence of a rhodium catalyst.[1372] Benzylic alcohols also react with α-diazo silylalknes in the presence of a rhodium catalyst, to give alkenes after pretreatment with oxygen and a palladium catalyst.[1373] The react of aryl aldehydes and MeC(CO$_2$Et)$_3$ with a catalytic amount of phenol leads to the corresponding conjugated ethyl ester (ArCH=CHCO$_2$Et).[1374]

OS **VIII**, 512, **IX**, 404; **X**, 355.

16-46   The Formation of Epoxides from Aldehydes and Ketones

(1 + 2)*OC,CC-cyclo*-Methylene-addition

72

Aldehydes and ketones can be converted to epoxides[1375] in good yields with the sulfur ylids dimethyloxosulfonium methylid (**72**) and dimethylsulfonium

[1367]See, for example, Schrock, R.R. *J. Am. Chem. Soc.* **1976**, *98*, 5399; Aguero, A.; Kress, J.; Osborn, J.A. *J. Chem. Soc., Chem. Commun.* **1986**, 531; Hartner, Jr., F.W.; Schwartz, J.; Clift, S.M. *J. Am. Chem. Soc.* **1990**, *105*, 640.

[1368]Matsubara, S.; Horiuchi, M.; Takai, K.; Utimoto, K. *Chem. Lett.* **1995**, 259.

[1369]Barma, D.K.; Kundu, A.; Zhang, H.; Mioskowski, C.; Falck, J.R. *J. Am. Chem. Soc.* **2003**, *125*, 3218.

[1370]Knecht, M.; Boland, W. *Synlett* **1993**, 837.

[1371]Chen, Y.; Huang, L.; Zhang, X.P. *Org. Lett.* **2003**, *5*, 2493; Mirafzal, G.A.; Cheng, G.; Woo, L.K. *J. Am. Chem. Soc.* **2002**, *124*, 176; Aggarwal, V.K.; Fulton, J.R.; Sheldon, C.G.; de Vincente, J. *J. Am. Chem. Soc.* **2003**, *125*, 6034.

[1372]Lebel, H.; Guay, D.; Paquet, V.; Huard, K. *Org. Lett.* **2004**, *6*, 3047. For a synthesis of dienes from conjugated aldehydes, see Lebel, H.; Paquet, V. *J. Am. Chem. Soc.* **2004**, *126*, 320.

[1373]Lebel, H.; Paquet, V. *J. Am. Chem. Soc.* **2004**, *126*, 11152.

[1374]Kumar, H.M.S.; Rao, M.S.; Joyasawal, S.; Yadav, J.S. *Tetrahedron Lett.* **2003**, *44*, 4287.

[1375]For reviews, see Block, E. *Reactions of Organosulfur Compounds*, Academic Press, NY, **1978**, pp. 101–105; Berti, G. *Top. Stereochem.* **1973**, *7*, 93, 218–232. For a list of reagents, with references, see Larock, R.C. *Comprehensive Organic Transformations*, 2nd ed., Wiley-VCH, NY, **1999**, pp. 944–951.

methylid (**73**).[1376] For most purposes, **72** is the

$$
\left[\begin{array}{c} \overset{O}{\overset{\|}{Me-S=CH_2}} \\ | \\ Me \end{array} \quad \longleftrightarrow \quad \begin{array}{c} \overset{O}{\overset{\|}{Me-S}{\oplus}{}^{\ominus}CH_2} \\ | \\ Me \end{array} \right] \qquad \left[\begin{array}{c} \overset{Me}{\underset{Me}{>}}S=CH_2 \\ \end{array} \quad \longleftrightarrow \quad \begin{array}{c} \overset{Me}{\underset{Me}{>}}\overset{\oplus}{S}{-}\overset{\ominus}{CH_2} \\ \end{array} \right]
$$

72   73

reagent of choice, because **73** is much less stable and ordinarily must be used as soon as it is formed, while **72** can be stored several days at room temperature. When diastereomeric epoxides can be formed, **73** usually attacks from the more hindered and **72** from the less-hindered side. Thus, 4-*tert*-butylcyclohexanone, treated with **72** gave exclusively **75** while **73** gave mostly **74**.[1377] Another difference in behavior between the

New bond is axial   New bond is equatorial

74   75

two reagents is that with α,β-unsaturated ketones, **72** gives only cyclopropanes (reaction **15-64**), while **73** gives oxirane formation. Other sulfur ylids have been used in an analogous manner, to transfer CHR or CR$_2$.[1378] High yields have been achieved by the use of sulfonium ylids anchored to insoluble polymers under phase-transfer conditions.[1379] A solvent-free version of this reaction has been developed using powdered K *tert*-butoxide and Me$_3$S$^+$I$^-$.[1380] Note that treatment of epoxides with 2 equivalents of Me$_2$S=CH$_2$ leads to allylic alcohols.[1381] Other sulfur ylids convert aldehydes to epoxides, including the one generated *in situ* from RR$'$S$^+$CH$_2$COO$^-$.[1382] Chiral sulfur ylids[1383] have been prepared, giving

[1376]For reviews, see House, H.O. *Modern Synthetic Reactions*, 2nd ed., W.A. Benjamin, NY, *1972*, pp. 709–733; Durst, T. *Adv. Org. Chem. 1969*, 6, 285, see pp. 321–330. For a monograph on sulfur ylids, see Trost, B.M.; Melvin, Jr., L.S. *Sulfur Ylids*; Academic Press, NY, *1975*.

[1377]Corey, E.J.; Chaykovsky, M. *J. Am. Chem. Soc. 1965*, 87, 1353.

[1378]Adams, J.; Hoffman, Jr., L.; Trost, B.M. *J. Org. Chem. 1970*, 35, 1600; Yoshimine, M.; Hatch, M.J. *J. Am. Chem. Soc. 1967*, 89, 5831; Braun, H.; Huber, G.; Kresze, G. *Tetrahedron Lett. 1973*, 4033; Corey, E.J.; Jautelat, M.; Oppolzer, W. *Tetrahedron Lett. 1967*, 2325.

[1379]Farrall, M.J.; Durst, T.; Fréchet, J.M.J. *Tetrahedron Lett. 1979*, 203.

[1380]Toda, F.; Kanemoto, K. *Heterocycles 1997*, 46, 185.

[1381]Harnett, J.J.; Alcaraz, L.; Mioskowski, C.; Martel, J.P.; Le Gall, T.; Shin, D.-S.; Falck, J.R. *Tetrahedron Lett. 1994*, 35, 2009; Alcaraz, L.; Harnett, J.J.; Mioskowski, C.; Martel, J.P.; Le Gall, T.; Shin, D.-S.; Falck, J.R. *Tetrahedron Lett. 1994*, 35, 5449. Also see, Alcaraz, L.; Harnett, J.J.; Mioskowski, C.; Martel, J.P. Le Gall, T.; Shin, D.-S.; Falck, J.R. *Tetrahedron Lett. 1994*, 35, 5453 for generation of alkenes from Me$_2$S=CH$_2$ and alkyl halides or mesylates.

[1382]Forbes, D.C.; Standen, M.C.; Lewis, D.L. *Org. Lett. 2003*, 5, 2283.

[1383]See Aggarwal, V.K.; Angelaud, R.; Bihan, D.; Blackburn, P.; Fieldhouse, R.; Fonguerna, S.J.; Ford, G.D.; Hynd, G.; Jones, E.; Jones, R.V.H.; Jubault, P.; Palmer, M.J.; Ratcliffe, P.D.; Adams, H. *J. Chem. Soc., Perkin Trans. 1 2001*, 2604.

the epoxide with good asymmetric induction.[1384] Chiral selenium ylids have been used in a similar manner.[1385]

76

The generally accepted mechanism for the reaction between sulfur ylids and aldehydes or ketone is formation of **76**, with displacement of the Me_2S leaving group by the alkoxide.[1386] This mechanism is similar to that of the reaction of sulfur ylids with C=C double bonds (**15-64**).[1387] The stereochemical difference in the behavior of **72** and **73** has been attributed to formation of the betaine **76** being reversible for **72**, but not for the less stable **73**, so that the more-hindered product is the result of kinetic control and the less-hindered of thermodynamic control.[1388]

Phosphorus ylids do not give this reaction, but give **16-44** instead.

Aldehydes and ketones can also be converted to epoxides by treatment with a diazoalkane,[1389] most commonly diazomethane, but an important side reaction is the formation of an aldehyde or ketone with one more carbon than the starting compound (reaction **18-9**). The reaction can be carried out with many aldehydes, ketones, and quinones, usually with a rhodium catalyst.[1390] A mechanism that accounts for both products is

rearrangement (**18-9**)

77

Compound **77** or nitrogen-containing derivatives of it have sometimes been isolated.

An alternative route to epoxides from ketones uses α-chloro sulfones and potassium *tert*-butoxide to give α,β-epoxy sulfones.[1391] A similar reaction was reported

[1384]Baird, C.P.; Taylor, P.C. *J. Chem. Soc. Perkin Trans. 1* **1998**, 3399; Domingo, V.M.; Castañer, J. *J. Chem. Soc., Chem. Commun.* **1995**, 893; Hayakawa, R.; Shimizu, M. *Synlett* **1999**, 1328; Zanardi, J.; Leriverend, C.; Aubert, D.; Julienne, K.; Metzner, P. *J. Org. Chem.* **2001**, *66*, 5620; Saito, T.; Akiba, D.; Sakairi, M.; Kanazawa, S. *Tetrahedron Lett.* **2001**, *42*, 57; Winn, C.L.; Bellenie, B.R.; Goodman, J.M. *Tetrahedron Lett.* **2002**, *43*, 5427.

[1385]See Takada, H.; Metzner, P.; Philouze, C. *Chem. Commun.* **2001**, 2350.

[1386]See Aggarwal, V.K.; Harvery, J.N.; Richardson, J. *J. Am. Chem. Soc.* **2002**, *124*, 5747.

[1387]See, for example, Townsend, J.M.; Sharpless, K.B. *Tetrahedron Lett.* **1972**, 3313; Johnson, C.R.; Schroeck, C.W.; Shanklin, J.R. *J. Am. Chem. Soc.* **1973**, *95*, 7424.

[1388]Johnson, C.R.; Schroeck, C.W.; Shanklin, J.R. *J. Am. Chem. Soc.* **1973**, *95*, 7424.

[1389]For a review, see Gutsche, C.D. *Org. React.* **1954**, *8*, 364.

[1390]See Davies, H.M.L.; De Meese, J. *Tetrahedron Lett.* **2001**, *42*, 6803.

[1391]Mąkosza, M.; Urbańska, N.; Chesnokov, A.A. *Tetrahedron Lett.* **2003**, *44*, 1473.

using KOH and 10% of a chiral phase-transfer agent, giving moderate enantioselectivity in the epoxy sulfone product.[1392]

Dihalocarbenes and carbenoids, which readily add to C=C bonds (**15-64**), do not generally add to the C=O bonds of ordinary aldehydes and ketones.[1393] See also, **16-91**.

OS **V**, 358, 755.

16-47   The Formation of Aziridines from Imines

(1 + 2)*NC,CC-cyclo***-Methylene-addition**

72

Just as sulfur ylids react with the carbonyl of an aldehyde or ketone to give an epoxide, tellurium ylids react with imines to give an aziridine. The reaction of an allylic tellurium salt, $RCH=CHCH_2Te^+Bu_2\ Br^-$, with lithium hexamethyldisilazide in HMPA/toluene leads to the tellurium ylid via deprotonation. In the presence of an imine, the ylid add to the imine and subsequent displacement of Bu_2Te generates an aziridine with a pendant vinyl group.[1394]

16-48   The Formation of Episulfides and Episulfones[1395]

Epoxides can be converted directly to episulfides by treatment with NH_4SCN and ceric ammonium nitrate.[1396] Diazoalkanes, treated with sulfur, give episulfides.[1397] It is likely that $R_2C=S$ is an intermediate, which is attacked by another molecule of diazoalkane, in a process similar to that shown in **16-46**. Thioketones *do* react with diazoalkanes to give episulfides.[1398] Thioketones have also been converted to episulfides with sulfur ylids.[1377] Carbenes, such as the dichlorocarbene from $CHCl_3$ and base, react with thioketones to give an

[1392]Arai, S.; Shioiri, T. *Tetrahedron* **2002**, *58*, 1407.

[1393]For exceptions, see Greuter, H.; Winkler, T.; Bellus, D. *Helv. Chim. Acta* **1979**, *62*, 1275; Sadhu, K.M.; Matteson, D.S. *Tetrahedron Lett.* **1986**, *27*, 795; Araki, S.; Butsugan, Y. *J. Chem. Soc., Chem. Commun.* **1989**, 1286.

[1394]Liao, W.-W.; Deng, X.-M.; Tang, Y. *Chem. Commun.* **2004**, 1516.

[1395]For a review, see Muller, L.L.; Hamer, J. *1,2-Cycloaddition Reactions*, Wiley, NY, **1967**, pp. 57–86.

[1396]Iranpoor, N.; Kazemi, F. *Synthesis* **1996**, 821.

[1397]Schönberg, A.; Frese, E. *Chem. Ber.* **1962**, *95*, 2810.

[1398]For example, see Beiner, J.M.; Lecadet, D.; Paquer, D.; Thuillier, A. *Bull. Soc. Chim. Fr.* **1973**, 1983.

α,α-dichloro episufide.[1399]

$$\text{RCH}_2\text{SO}_2\text{Cl} \xrightarrow{\text{R'}_3\text{N}} \text{RCH=SO}_2 \xrightarrow{\text{CH}_2\text{N}_2} \underset{\substack{\mathbf{79}}}{\overset{\displaystyle R}{\underset{H}{C}}\!\!\backslash\!\!\underset{\substack{S\\ O_2}}{}\!\!\diagup\!\!\overset{CH_2}{}} \xrightarrow{\Delta} \text{RCH=CH}_2$$

$$\mathbf{78}$$

Alkanesulfonyl chlorides, when treated with diazomethane in the presence of a base (usually a tertiary amine), give episulfones (**79**).[1400] The base removes HCl from the sulfonyl halide to produce the highly reactive sulfene (**78**) (17-14), which then adds CH_2. The episulfone can then be heated to give off SO_2 (17-20), making the entire process a method for achieving the conversion $RCH_2SO_2Cl \rightarrow$ $RCH=CH_2$.[1401]

OS **V**, 231, 877.

16-49 Cyclopropanation of Conjugated Carbonyl Compounds

Double-bond compounds that undergo the Michael reaction (**15-24**) can be converted to cyclopropane derivatives with sulfur ylids.[1402] Among the most common of these is dimethyloxosulfonium methylid

$$\underset{\substack{\text{Me}}}{\overset{\displaystyle O}{\overset{\|}{S}}}\!\!-\!\overset{\ominus}{CH_2} \; + \; \overset{\diagdown}{\diagup}C\!=\!C\overset{Z}{\underset{\diagdown}{\diagup}} \longrightarrow \underset{\substack{\text{Me}}}{\overset{\displaystyle O}{\overset{\|}{S}}}\!\!-\!\overset{H}{\underset{\diagdown}{C}}\!\!-\!\overset{Z}{\underset{\diagup}{C}}\overset{\ominus}{} \longrightarrow \overset{H \; H}{\underset{\diagup \; \diagdown}{C}} \; + \; \text{DMSO}$$
$$Z\!\!-\!\overset{}{\underset{\diagup}{C}}\!\!-\!\overset{}{\underset{\diagdown}{C}}$$

72

72,[1403] which is widely used to transfer CH_2 to activated double bonds, but other sulfur ylids

$$\text{Ph}\!-\!\underset{\substack{\text{NMe}_2}}{\overset{\displaystyle O}{\overset{\|}{\overset{\oplus}{S}}}}\!\!-\!\overset{\ominus}{CR_2}$$

80

[1399]Mlosteń, G.; Romański, J.; Swiątek, A.; Hemgartner, H. *Helv. Chim. Acta* **1999**, *82*, 946.
[1400]Opitz, G.; Fischer, K. *Angew. Chem. Int. Ed.* **1965**, *4*, 70.
[1401]For a review of this process, see Fischer, N.S. *Synthesis* **1970**, 393.
[1402]For a monograph on sulfur ylids, see Trost, B.M.; Melvin Jr., L.S. *Sulfur Ylids*, Academic Press, NY, **1975**. For reviews, see Fava, A., in Bernardi, F.; Csizmadia, I.G.; Mangini, A. *Organic Sulfur Chemistry*, Elsevier, NY, **1985**, pp. 299–354; Belkin, Yu.V.; Polezhaeva, N.A. *Russ. Chem. Rev.* **1981**, *50*, 481; Block, E., in Stirling, C.J.M. *The Chemistry of the Sulphonium Group*, pt. 2, Wiley, NY, **1981**, pp. 680–702; Block, E. *Reactions of Organosulfur Compounds*, Academic Press, NY, **1978**, pp. 91–127. See also, Mamai, A.; Madalengoitia, J.S. *Tetrahedron Lett.* **2000**, *41*, 9009.
[1403]Truce, W.E.; Badiger, V.V. *J. Org. Chem.* **1964**, *29*, 3277; Corey, E.J.; Chaykovsky, M. *J. Am. Chem. Soc.* **1965**, *87*, 1353; Agami, C.; Prevost, C. *Bull. Soc. Chim. Fr.* **1967**, 2299. For a review of this reagent, see Gololobov, Yu.G.; Nesmeyanov, A.N.; Lysenko, V.P.; Boldeskul, I.E. *Tetrahedron* **1987**, *43*, 2609.

have also been used. A combination of DMSO and KOH in an ionic liquid converts conjugated ketones to α,β-cyclopropyl ketones.[1404] Both CHR and CR$_2$ can be added in a similar manner with certain nitrogen-containing compounds. For example, ylids,[1405] such as **80**, add various groups to activated double bonds.[1406] Sulfur ylids react with allylic alcohols in the presence of MnO$_2$ and molecular sieve 4 Å to give the cyclopropyl aldehyde.[1407] Similar reactions have been performed with phosphorus ylids,[1408] with pyridinium ylids,[1409] and with the compounds (PhS)$_3$CLi and Me$_3$Si(PhS)$_2$CLi.[1410] The reactions with ylids such as these involve of course nucleophilic acyl addition.

Other reagents can be used to convert an aldehyde or ketone to a cyclopropane derivative. Conjugated ketones react with Cp$_2$Zr(CH$_2$—CH$_2$) and PMe$_3$ to give a vinyl cyclopropane derivative after treatment with aqueous sulfuric acid.[1411]

16-50   The Thorpe Reaction

N-**Hydro-*C*-(α-cyanoalkyl)-addition**

$$H-\overset{\displaystyle |}{\underset{\displaystyle |}{C}}-C\equiv N \;+\; H-\overset{\displaystyle |}{\underset{\displaystyle |}{C}}-C\equiv N \xrightarrow{\text{EtO}^-} H-\overset{\displaystyle |}{\underset{\displaystyle |}{C}}-\overset{\displaystyle \overset{\textstyle C-C\equiv N}{|}}{\underset{\displaystyle N_\ominus}{C}}$$

In the *Thorpe reaction*, the α carbon of one nitrile molecule is added to the CN carbon of another, so this reaction is analogous to the aldol reaction (**16-34**). The C=NH bond is, of course, hydrolyzable (**16-2**), so β-keto nitriles can be prepared in this manner. The Thorpe reaction can be done intramolecularly, in which case it is

81

[1404] In bmim PF$_6$, 1-butyl-3-methylimidazolium hexafluorophosphate: Chandrasekhar, S.; Jagadeshwar, N.V.; Reddy, K.V. *Tetrahedron Lett.* **2003**, *44*, 3629.

[1405] For a review of sulfoximides (R$_2$S(O)NR$_2$) and ylids derived from them, see Kennewell, P.D.; Taylor, J.B. *Chem. Soc. Rev.* **1980**, *9*, 477.

[1406] For reviews, see Johnson, C.R. *Aldrichimica Acta* **1985**, *18*, 1; *Acc. Chem. Res.* **1973**, *6*, 341; Kennewell, P.D.; Taylor, J.B. *Chem. Soc. Rev.* **1975**, *4*, 189; Trost, B.M. *Acc. Chem. Res.* **1974**, *7*, 85.

[1407] Oswald, M.F.; Raw, S.A.; Taylor, R.J.K. *Org. Lett.* **2004**, *6*, 3997.

[1408] Bestmann, H.J.; Seng, F. *Angew. Chem. Int. Ed.* **1962**, *1*, 116; Grieco, P.A.; Finkelhor, R.S. *Tetrahedron Lett.* **1972**, 3781.

[1409] Shestopalov, A.M.; Sharanin, Yu.A.; Litvinov, V.P.; Nefedov, O.M. *J. Org. Chem. USSR* **1989**, *25*, 1000.

[1410] Cohen, T.; Myers, M. *J. Org. Chem.* **1988**, *53*, 457.

[1411] Bertus, P.; Gandon, V.; Szymoniak, J. *Chem. Commun.* **2000**, 171.

called the *Thorpe-Ziegler reaction.*[1412] This is a useful method for closing large rings. Yields are high for five- to eight-membered rings, fall off to about zero for rings of nine to thirteen members, but are high again for fourteen-membered and larger rings, if high-dilution techniques are employed. The product in the Thorpe–Ziegler reaction is not the imine, but the tautomeric enamine, for example, **81**; if desired this can be hydrolyzed to an α-cyano ketone (**16-2**), which can in turn be hydrolyzed and decarboxylated (**16-4, 12-40**). Other active-hydrogen compounds can also be added to nitriles.[1413]

OS **VI**, 932.

H. Other Carbon or Silicon Nucleophiles

16-51   Addition of Silanes

O-Hydro-*C*-alkyl-addition

$$R^1\diagdown\diagup\diagdown SiR_3 \ + \ R^2{-}CHO \ \xrightarrow{\text{Lewis acid}} \ R^1\diagdown\diagup\diagdown\underset{\underset{R^2}{|}}{\overset{OH}{C}}$$

Allylic silanes react with aldehydes, in the presence of Lewis acids, to give a homoallylic alcohol.[1414] In the case of benzylic silanes, this addition reaction has been induced with $Mg(ClO_4)_2$ under photochemical conditions.[1415] Cyclopropyl-carbinyl silanes add to acetals in the presence of TMSOTf to give a homoallylic alcohol.[1416] Allyltrichlorosilane adds an allyl group to an aldehyde in the presence of a cyclic urea and AgOTf.[1417] The addition of chiral additives leads to the alcohol with good asymmetric induction.[1418] In a related reaction, allylic silanes react with acyl halides to produce the corresponding carbonyl derivative. The reaction of phenyl chloroformate, allyltrimethylsilane and $AlCl_3$, for example, gave phenyl but-3-enoate.[1419]

Allylic silanes also add to imines, in the presence of $TiCl_4$, to give amines.[1420]

[1412]For a monograph, see Taylor, E.C.; McKillop, A. *The Chemistry of Cyclic Enaminonitriles and ortho-Amino Nitriles*, Wiley, NY, *1970*. For a review, see Schaefer, J.P.; Bloomfield, J.J. *Org. React.* *1967*, *15*, 1.

[1413]See, for example, Josey, A.D. *J. Org. Chem.* *1964*, *29*, 707; Barluenga, J.; Fustero, S.; Rubio, V.; Gotor, V. *Synthesis 1977*, 780; Hiyama, T.; Kobayashi, K. *Tetrahedron Lett.* *1982*, *23*, 1597; Gewald, K.; Bellmann, P.; Jänsch, H. *Liebigs Ann. Chem.* *1984*, 1702; Page, P.C.B.; van Niel, M.B.; Westwood, D. *J. Chem. Soc. Perkin Trans. 1 1988*, 269.

[1414]Panek, J.S.; Liu, P. *Tetrahedron Lett.* *1997*, *38*, 5127.

[1415]Fukuzumi, S.; Okamoto, T.; Otera, J. *J. Am. Chem. Soc.* *1994*, *116*, 5503.

[1416]Braddock, D.C.; Badine, D.M.; Gottschalk, T. *Synlett 2001*, 1909.

[1417]Chataigner, I.; Piarulli, U.; Gennari, C. *Tetrahedron Lett.* *1999*, *40*, 3633.

[1418]Ishihara, K.; Mouri, M.; Gao, Q.; Maruyama, T.; Furuta, K.; Yamamoto, H. *J. Am. Chem. Soc. 1993*, *115*, 11490.

[1419]Mayr, H.; Gabriel, A.O.; Schumacher, R. *Liebigs Ann. Chem.* *1995*, 1583.

[1420]Kercher, T.; Livinghouse, T. *J. Am. Chem. Soc. 1996*, *118*, 4200.

16-52   The Formation of Cyanohydrins

O-Hydro-*C*-cyano-addition

$$\underset{/}{\overset{\displaystyle O}{\underset{C}{\parallel}}}\diagdown \;\; + \;\; HCN \;\; \longrightarrow \;\; \underset{/}{\overset{\displaystyle \diagdown}{C}}\underset{\diagdown OH}{\overset{\diagup CN}{}}$$

The addition of HCN to aldehydes or ketones produces cyanohydrins.[1421] This is an equilibrium reaction, and for aldehydes and aliphatic ketones the equilibrium lies to the right; therefore the reaction is quite feasible, except with sterically hindered ketones such as diisopropyl ketone. However, ketones ArCOR give poor yields, and the reaction cannot be carried out with ArCOAr since the equilibrium lies too far to the left. With aromatic aldehydes the benzoin condensation (**16-55**) competes. With α,β-unsaturated aldehydes and ketones, 1,4-addition competes (**15-38**). The reaction has been carried out enantioselectively: optically active cyanohydrins were prepared with the aid of optically active catalysts.[1422] Hydrogen cyanide adds to aldehydes in the presence of a lyase to give the cyanohydrin with good enantioselectivity.[1423] Cyanohydrins have been formed using a lyase in an ionic liquid.[1424]

$$\underset{R\diagup\overset{\displaystyle C}{\underset{\diagdown}{}}\diagdown R'}{\overset{\displaystyle O}{}}\; + \; Me_3Si\text{-}CN \;\; \xrightarrow[\text{acid}]{\text{Lewis}} \;\; \underset{R'\diagup \diagdown OSiMe_3}{\overset{R\diagdown \diagup CN}{C}} \;\; \longrightarrow \;\; \underset{R'\diagup \diagdown OH}{\overset{R\diagdown \diagup CN}{C}}$$

82

Ketones of low reactivity, such as ArCOR, can be converted to cyanohydrins by treatment with diethylaluminum cyanide (Et$_2$AlCN) (see OS **VI**, 307) or, indirectly, with cyanotrimethylsilane (Me$_3$SiCN)[1425] in the presence of a Lewis acid or base,[1426]

[1421]For reviews, see Friedrich, K., in Patai, S.; Rappoport, Z. *The Chemistry of Functional Groups, Supplement C*, pt. 2, Wiley, NY, **1983**, pp. 1345–1390; Friedrich, K.; Wallenfels, K., in Rappoport, Z. *The Chemistry of the Cyano Group*, Wiley, NY, **1970**, pp. 72–77.

[1422]See Minamikawa, H.; Hayakawa, S.; Yamada, T.; Iwasawa, N.; Narasaka, K. *Bull. Chem. Soc. Jpn.* **1988**, *61*, 4379; Jackson, W.R.; Jayatilake, G.S.; Matthews, B.R.; Wilshire, C. *Aust. J. Chem.* **1988**, *41*, 203; Garner, C.M.; Fernández, J.M.; Gladysz, J.A. *Tetrahedron Lett.* **1989**, *30*, 3931; Mori, A.; Ikeda, Y.; Kinoshita, K.; Inoue, S. *Chem. Lett.* **1989**, 2119; Kobayashi, S.; Tsuchiya, Y.; Mukaiyama, T. *Chem. Lett.* **1991**, 541; Gröger, H.; Capan, E.; Barthuber, A.; Vorlop, K.-D. *Org. Lett.* **2001**, *3*, 1969, and references cited therein. For a review, see Brune, J.-M.; Holmes, I.P. *Angew. Chem. Int. Ed.* **2004**, *43*, 2752.

[1423]Gerrits, P.J.; Marcus, J.; Birikaki, L.; van der Gen, A. *Tetrahedron Asymmetry* **2001**, *12*, 971.

[1424]In bmim BF$_4$, 1-butyl-3-methylimidazolium tetrafluoroborate: Gaisberger, R.P.; Fechter, M.H.; Griengl, H. *Tetrahedron Asymmetry* **2004**, *15*, 2959.

[1425]For reviews of Me$_3$SiCN and related compounds, see Rasmussen, J.K.; Heilmann, S.M.; Krepski, L. *Adv. Silicon Chem.* **1991**, *1*, 65; Groutas, W.C.; Felker, D. *Synthesis* **1980**, 861. For procedures using Me$_3$SiCl and $^-$CN instead of Me$_3$SiCN, see Yoneda, R.; Santo, K.; Harusawa, S.; Kurihara, T. *Synthesis* **1986**, 1054; Sukata, K. *Bull. Chem. Soc. Jpn.* **1987**, *60*, 3820.

[1426]Kobayashi, S.; Tsuchiya, Y.; Mukaiyama, T. *Chem. Lett.* **1991**, 537; Belokon', Y.; Flego, M.; Ikonnikov, N.; Moscalenko, M.; North, M.; Orizu, C.; Tararov, V.; Tasinazzo, M. *J. Chem. Soc. Perkin Trans. 1* **1997**, 1293; Wada, M.; Takahashi, T.; Domae, T.; Fukuma, T.; Miyoshi, N.; Smith, K. *Tetrahedron Asymmetry*, **1997**, *8*, 3939; Kanai, M.; Hamashima, Y.; Shibasaki, M. *Tetrahedron Lett.* **2000**, *41*, 2405. The reaction works in some cases without a Lewis acid, see Manju, K.; Trehan, S. *J. Chem. Soc. Perkin Trans. 1* **1995**, 2383.

followed by hydrolysis of the resulting *O*-trimethylsilyl cyanohydrin **82**. Solvent-free conditions have been reported using TMSCN, an aldehydes and potassium carbonate.[1427] Amine *N*-oxides catalyze the reaction[1428] as does tetra-butylammonium cyanide.[1429] Lithium perchlorate in ether facilitates this reaction.[1430] With MgBr$_2$ as a catalyst, the reaction proceeds to good syn selectivity. [1431] Other useful catalysts include platinum complexes,[1432] Ti(O*i*Pr)$_4$,[1433] and InBr$_3$.[1434] When TiCl$_4$ is used, the reaction between Me$_3$SiCN and aromatic aldehydes or ketones gives α-chloro nitriles (ClCRR'—CN).[1435] The use of chiral additives in this latter reaction leads to cyanohydrins with good asymmetric induction.[1436] Sulfoximine–titanium reagents have been used in enantioselective trimethylsilyl cyanations of aldehydes. [1437] Chiral transition-metal catalysts have been used to give *O*-trialkylsilyl cyanohydrins with good enantioselectivity, including titanium complexes[1438] as well as complexes of other metals.[1439] A vanadium catalysts has been used in an ionic liquid.[1440] Note that the reaction of an aldehyde and TMSCN in the presence of aniline and a BiCl$_3$ catalyst leads to an α-cyano amine.[1441] α-Cyano amines are also formed by the reaction of an aldehyde with (Et$_2$N)$_2$BCN.[1442]

[1427]He, B.; Li, Y.; Feng, X.; Zhang, G. *Synlett* **2004**, 1776.

[1428]Shen, Y.; Feng, X.; Li, Y.; Zhang, G.; Jiang, Y. *Tetrahedron* **2003**, *59*, 5667. See Bakendale, I.R.; Ley, S.V.; Sneddon, H.F. *Synlett* **2002**, 775; Shen, Y.; Feng, X.; Li, Y.; Zhang, G.; Jiang, Y. *Eur. J. Org. Chem.* **2004**, 129.

[1429]Amurrio, I.; Córdoba, R.; Csákÿ, A.G.; Plumet, J. *Tetrahedron* **2004**, *60*, 10521.

[1430]Jenner, G. *Tetrahedron Lett.* **1999**, *40*, 491.

[1431]Ward, D.E.; Hrapchak, M.J.; Sales, M. *Org. Lett.* **2000**, *2*, 57.

[1432]Fossey, J.S.; Richards, C.J. *Tetrahedron Lett.* **2003**, *44*, 8773.

[1433]Huang, W.; Song, Y.; Bai, C.; Cao, G.; Zheng, Z. *Tetrahedron Lett.* **2004**, *45*, 4763; He, B.; Chen, F.-X.; Li, Y.; Feng, X.; Zhang, G. *Tetrahedron Lett.* **2004**, *45*, 5465.

[1434]Bandini, M.; Cozzi, P.G.; Melchiorre, P.; Umani-Ronchi, A. *Tetrahedron Lett* **2001**, *42*, 3041.

[1435]Kiyooka, S.; Fujiyama, R.; Kawaguchi, K. *Chem. Lett.* **1984**, 1979.

[1436]Tararov, V.I.; Hibbs, D.E.; Hursthouse, M.B.; Ikonnikov, N.S.; Malik, K.M.A.; North, M.; Orizu, C.; Belokon, Y.N. *Chem. Commun.* **1998**, 387; Bolm, C.; Müller, P. *Tetrahedron Lett.* **1995**, *36*, 1625; Ryu, D.H.; Corey, E.J. *J. Am. Chem. Soc.* **2004**, *126*, 8106.

[1437]Bolm, C.; Müller, P.; Harms, K. *Acta Chem. Scand. B*, **1996**, *50*, 305.

[1438]Hamashima, Y.; Kanai, M.; Shibasaki, M. *J. Am. Chem.Soc.* **2000**, *122*, 7412; Belokon, Y.N.; Green, B.; Ikonnikov, N.S.; North, M.; Parsons, T.; Tararov, V.I. *Tetrahedron* **2001**, *57*, 771 and references cited therein; Liang, S.; Bu, X.R. *J. Org. Chem.* **2002**, *67*, 2702; Li, Y.; He, B.; Qin, B.; Feng, X.; Zhang, G. *J. Org. Chem.* **2004**, *69*, 7910; Chen, F.-X.; Qin, B.; Feng, X.; Zhang, G.; Jiang, Y. *Tetrahedron* **2004**, *60*, 10449; Uang, B.-J.; Fu, I.-P.; Hwang, C.-D.; Chang, C.-W.; Yang, C.-T.; Hwang, D.-R. *Tetrahedron* **2004**, *60*, 10479; Gama, A.; Flores-López, L.-Z.; Aguirre, G.; Parra-Hake, M.; Somanathan, R.; Walsh, P.J. *Tetrahedron Asymmetry* **2002**, *13*, 149.

[1439]Transition metals used include **Yb**: Aspinall, H.C.; Greeves, N.; Smith, P.M. *Tetrahedron Lett.* **1999**, *40*, 1763. **Al**: Hamashima, Y.; Sawada, D.; Nogami, H.; Kanai, M.; Shibasaki, M. *Tetrahedron* **2001**, *57*, 805; Deng, H.; Isler, M.P.; Snapper, M.L.; Hoveyda, A.H. *Angew. Chem. Int. Ed.* **2002**, *41*, 1009. **Sc**: Karimi, B.; Ma'Mani, L. *Org. Lett.* **2004**, *6*, 4813.

[1440]In bmim PF$_6$, 1-butyl-3-methylimidazolium hexafluorophosphate: Baleizão, C.; Gigante, B.; Garcia, H.; Corma, A. *Tetrahedron Lett.* **2003**, *44*, 6813.

[1441]De, S.K.; Gibbs, R.A. *Tetrahedron Lett.* **2004**, *45*, 7407.

[1442]Suginome, M.; Yamamoto, A.; Ito, Y. *Chem. Commun.* **2002**, 1392.

Ketones can be converted to cyanohydrin O-carbonates, $R_2C(CN)OCO_2R'$, by reaction with EtO_2C-CN. In the presence of a Cinchona alkaloid, the product is formed with good enantioselectivity.[1443] Potassium cyanide and acetic anhydride reacts with an aldehyde in the presence of a chiral titanium catalyst to give an α-acetoxy nitrile.[1444]

Rather than direct reaction with an aldehyde or ketone, the bisulfite addition product is often treated with cyanide. The addition is nucleophilic and the actual nucleophile is ^-CN, so the reaction rate is increased by the addition of base.[1445] This was demonstrated by Lapworth in 1903, and consequently this was one of the first organic mechanisms to be known.[1446] This method is especially useful for aromatic aldehydes, since it avoids competition from the benzoin condensation. If desired, it is possible to hydrolyze the cyanohydrin *in situ* to the corresponding α-hydroxy acid. This reaction is important in the *Kiliani–Fischer* method of extending the carbon chain of a sugar.

A particularly useful variation of this reaction uses cyanide rather than HCN. α-Amino nitriles[1447] can be prepared in one step by the treatment of an aldehyde or ketone with NaCN and NH_4Cl. This is called the *Strecker synthesis*;[1448] and it is a special case of the Mannich reaction (**16-19**). Since the CN is easily hydrolyzed to the acid, this is a convenient method for the preparation of α-amino acids. The reaction has also been carried out with $NH_3 + HCN$ and with NH_4CN. Salts of primary and secondary amines can be used instead of NH_4^+ to obtain *N*-substituted and *N,N*-disubstituted α-amino nitriles. Unlike **16-52**, the Strecker synthesis is useful for aromatic as well as aliphatic ketones. As in **16-52**, the Me_3SiCN method has been used; **76** is converted to the product with ammonia or an amine.[1449] The effect of pressure on the Strecker synthesis has been studied.[1450]

OS **I**, 336; **II**, 7, 29, 387; **III**, 436; **IV**, 58, 506; **VI**, 307; **VII**, 20, 381, 517, 521. For the reverse reaction, see OS **III**, 101. For the Strecker synthesis, see OS **I**, 21, 355; **III**, 66, 84, 88, 275; **IV**, 274; **V**, 437; **VI**, 334.

[1443]Tian, S.-K.; Deng, L. *J. Am. Chem. Soc.* **2001**, *123*, 6195.

[1444]Belokon, Y.N.; Gutnov, A.V.; Moskalenko, M.A.; Yashkina, L.V.; Lesovoy, D.E.; Ikonnikov, N.S.; Larichev, V.S.; North, M. *Chem. Commun.* **2002**, 244; Kawasaki, Y.; Fujii, A.; Nakano, Y.; Sakaguchi, S.; Ishii, Y. *J. Org. Chem.* **1999**, *64*, 4214.

[1445]For a review, see Ogata, Y.; Kawasaki, A., in Zabicky, J. *The Chemistry of the Carbonyl Group*, Vol. 2, Wiley, NY, *1970*, pp. 21–32. See also, Okano, V.; do Amaral, L.; Cordes, E.H. *J. Am. Chem. Soc.* **1976**, *98*, 4201; Ching, W.; Kallen, R.G. *J. Am. Chem. Soc.* **1978**, *100*, 6119.

[1446]Lapworth, A. *J. Chem. Soc.* **1903**, *83*, 998.

[1447]For a review of α-amino nitriles, see Shafran, Yu.M.; Bakulev, V.A.; Mokrushin, V.S. *Russ. Chem. Rev.* **1989**, *58*, 148.

[1448]For a review of asymmetric Strecker syntheses, see Williams, R.M. *Synthesis of Optically Active* α-*Amino Acids*, Pergamon, Elmsford, NY, *1989*, pp. 208–229; Yet, L. *Angew. Chem. Int. Ed.* **2001**, *40*, 875; Gröger, H. *Chem. Rev.* **2003**, *103*, 2795.

[1449]See Mai, K.; Patil, G. *Tetrahedron Lett.* **1984**, *25*, 4583; *Synth. Commun.* **1985**, *15*, 157.

[1450]Jenner, G.; Salem, R.B.; Kim, J.C.; Matsumoto, K. *Tetrahedron Lett.* **2003**, *44*, 447.

16-53 The Addition of HCN to C=N and C≡N Bonds

N-Hydro-C-cyano-addition

$$ \underset{C}{\overset{W}{N}} + HCN \longrightarrow \underset{C}{\overset{CN}{\underset{H}{N-W}}} \quad W=H, R, Ar, OH, NHAr, etc. $$

HCN adds to imines, Schiff bases, hydrazones, oximes, and similar compounds. Cyanide can be added to iminium ions to give α-cyano amines (**83**).

$$ \underset{C}{\overset{\oplus}{N}}\overset{R'}{} + {}^{-}CN \longrightarrow \underset{C}{\overset{CN}{\underset{R}{N-R'}}} $$
83

As in **16-50**, the addition to imines has been carried out enantioselectively.[1451] Chiral ammonium salts have been used with HCN.[1452] TMSCN reacts with N-tosyl imines in the presence of $BF_3 \cdot OEt_2$ to give the α-cyano N-tosyl amine.[1453] In the presence of a chiral zirconium[1454] or aluminum[1455] catalyst, Bu_3SnCN react with imines to give α-cyanoamines enantioselectively. The reaction of an imine and TMSCN gives the cyano amine with good enantioselectivity using a chiral scandium catalyst.[1456] Titanium catalysts have been used in the presence of a chiral Schiff base.[1457] Treatment of an imine with a chiral 1,4,6- triazabicyclo[3.3.0]oct-4-ene and then HCN give the α-cyano amine with good enantioselectivity.[1458]

The addition of KCN to triisopropylbenzenesulfonyl hydrazones **84** provides an indirect method for achieving the conversion $RR'CO \rightarrow RR'CHCN$.[1459] The reaction is successful for hydrazones of aliphatic aldehydes and ketones.

$$ RR'C=NNHSO_2Ar + KCN \xrightarrow{MeOH} RR'CHCN \quad Ar = 2,4,6\text{-}(i\text{-}Pr)_3C_6H_2 $$
84

[1451]Saito, K.; Harada, K. *Tetrahedron Lett.* **1989**, *30*, 4535.

[1452]Huang, J.; Corey, E.J. *Org. Lett.* **2004**, *6*, 5027.

[1453]Prasad, B.A.B.; Bisai, A.; Singh, V.K. *Tetrahedron Lett.* **2004**, *45*, 9565.

[1454]Ishitani, H.; Komiyama, S.; Hasegawa, Y.; Kobayashi, S. *J. Am. Chem. Soc.* **2000**, *122*, 762.

[1455]Nakamura, S.; Sato, N.; Sugimoto, M.; Toru, T. *Tetrahedron Asymmetry* **2004**, *15*, 1513.

[1456]Chavarot, M.; Byrne, J.J.; Chavant, P.Y.; Vallée, Y. *Tetrahedron Asymmetry* **2001**, *12*, 1147.

[1457]Krueger, C.A.; Kuntz, K.W.; Dzierba, C.D.; Wirschun, W.G.; Gleason, J.D.; Snapper, M.L.; Hoveyda, A.H. *J. Am. Chem. Soc.* **1999**, *121*, 4284.

[1458]Corey, E.J.; Grogan, M.J. *Org. Lett.* **1999**, *1*, 157.

[1459]Jiricny, J.; Orere, D.M.; Reese, C.B. *J. Chem. Soc. Perkin Trans. 1 1980*, 1487. For other methods of achieving this conversion, see Ziegler, F.E.; Wender, P.A. *J. Org. Chem.* **1977**, *42*, 2001; Cacchi, S.; Caglioti, L.; Paolucci, G. *Synthesis 1975*, 120; Yoneda, R.; Harusawa, S.; Kurihara, T. *Tetrahedron Lett.* **1989**, *30*, 3681; Okimoto, M.; Chiba, T. *J. Org. Chem.* **1990**, *55*, 1070.

HCN can also be added to the C≡N bond to give iminonitriles or α-aminomalononitriles.[1460]

$$R\text{-}CN \xrightarrow[^-CN]{HCN} \underset{R}{\overset{N-H}{\underset{\|}{\overset{\|}{C}}}}\underset{CN} \xrightarrow[^-CN]{HCN} \underset{R}{\overset{H_2N}{\overset{}{C}}}\underset{CN}$$

OS **V**, 344. See also, OS **V**, 269.

16-54 The Prins Reaction

The addition of an alkene to formaldehyde in the presence of an acid[1461] catalyst is called the *Prins reaction*.[1462] Three main products are possible; which one predominates depends on the alkene and the conditions. When the product is the 1,3-diol or the dioxane,[1463] the reaction involves addition to the C=C as well as to the C=O. The mechanism is one of electrophilic attack on both double bonds. The acid first protonates the C=O, and the resulting carbocation is attacked by the C=C to give **85**.

The cation product **85** can undergo loss of H$^+$ to give the alkene or add water to give the diol.[1464] It has been proposed that **85** is stabilized by neighboring-group

---

[1460]For an example, see Ferris, J.P.; Sanchez, R.A. *Org. Synth.* **V**, 344.

[1461]The Prins reaction has also been carried out with basic catalysts: Griengl, H.; Sieber, W. *Monatsh. Chem.* **1973**, *104*, 1008, 1027.

[1462]For reviews, see Adams, D.R.; Bhatnagar, S.P. *Synthesis* **1977**, 661; Isagulyants, V.I.; Khaimova, T.G.; Melikyan, V.R.; Pokrovskaya, S.V. *Russ. Chem. Rev.* **1968**, *37*, 17. For a list of references, see Larock, R.C. *Comprehensive Organic Transformations*, 2nd ed., Wiley-VCH, NY, **1999**, p. 248.

[1463]The reaction to produce dioxanes has also been carried out with equimolar mixtures of formaldehyde and another aldehyde RCHO. The R appears in the dioxane on the carbon between the two oxygens: Safarov, M.G.; Nigmatullin, N.G.; Ibatullin, U.G.; Rafikov, S.R. *Doklad. Chem.* **1977**, *236*, 507.

[1464]Hellin, M.; Davidson, M.; Coussemant, F. *Bull. Soc. Chim. Fr.* **1966**, 1890, 3217.

attraction, with either the oxygen[1465] or a carbon[1466] stabilizing the charge (**86** and

87, respectively). This stabilization is postulated to explain the fact that with 2-butenes[1467] and with cyclohexenes the addition is anti. A backside attack of H_2O on the three- or four-membered ring would account for it. Other products are obtained too, which can be explained on the basis of **86** or **87**.[1465,1466] Additional evidence for the intermediacy of **86** is the finding that oxetanes (**88**) subjected to the reaction conditions (which would protonate **88** to give **86**) give essentially the same product ratios as the corresponding alkenes.[1468] An argument against the intermediacy of **86** and **87** is that not all alkenes show the anti-stereoselectivity mentioned above. Indeed, the stereochemical results are often quite complex, with syn, anti, and nonstereoselective addition reported, depending on the nature of the reactants and the reaction conditions.[1469] Since addition to the C=C bond is electrophilic, the reactivity of the alkene increases with alkyl substitution and Markovnikov's rule is followed. The dioxane product may arise from a reaction between the 1,3-diol and formaldehyde[1470] (**16-5**) or between **86** and formaldehyde.

Lewis acids, such as $SnCl_4$, also catalyze the reaction, in which case the species that adds to the alkenes is $H_2C^+-O^--SnCl_4$.[1471] Montmorillonite K10 clay containing zinc (IV) has been used to promote the reaction.[1472] The reaction can also be catalyzed by peroxides, in which case the mechanism is probably a free-radical one. Other transition metal complexes can be used to form homoallylic alcohols. A typical example is the reaction of methylenecyclohexane with an aryl aldehyde to give **89**.[1473]

[1465]Blomquist, A.T.; Wolinsky, J. *J. Am. Chem. Soc.* **1957**, *79*, 6025; Schowen, K.B.; Smissman, E.E.; Schowen, R.L. *J. Org. Chem.* **1968**, *33*, 1873.

[1466]Dolby, L.J.; Lieske, C.N.; Rosencrantz, D.R.; Schwarz, M.J. *J. Am. Chem. Soc.* **1963**, *85*, 47; Dolby, L.J.; Schwarz, M.J. *J. Org. Chem.* **1963**, *28*, 1456; Safarov, M.G.; Isagulyants, V.I.; Nigmatullin, N.G. *J. Org. Chem. USSR* **1974**, *10*, 1378.

[1467]Fremaux, B.; Davidson, M.; Hellin, M.; Coussemant, F. *Bull. Soc. Chim. Fr.* **1967**, 4250.

[1468]Meresz, O.; Leung, K.P.; Denes, A.S. *Tetrahedron Lett.* **1972**, 2797.

[1469]For example, see LeBel, N.A.; Liesemer, R.N.; Mehmedbasich, E. *J. Org. Chem.* **1963**, *28*, 615; Portoghese, P.S.; Smissman, E.E. *J. Org. Chem.* **1962**, *27*, 719; Wilkins, C.L.; Marianelli, R.S. *Tetrahedron* **1970**, *26*, 4131; Karpaty, M.; Hellin, M.; Davidson, M.; Coussemant, F. *Bull. Soc. Chim. Fr.* **1971**, 1736; Coryn, M.; Anteunis, M. *Bull. Soc. Chim. Belg.* **1974**, *83*, 83.

[1470]Hellin, M.; Davidson, M; Coussemant, F *Bull. Soc. Chim. Fr.* **1966**, 1890, 3217; Isagulyants, V.I.; Isagulyants, G.V.; Khairudinov, I.R.; Rakhmankulov, D.L. *Bull. Acad. Sci. USSR. Div. Chem. Sci.*, **1973**, *22*, 1810; Sharf, V.Z.; Kheifets, V.I.; Freidlin, V.I. *Bull. Acad. Sci. USSR Div. Chem. Sci.*, **1974**, *23*, 1681.

[1471]Yang, D.H.; Yang, N.C.; Ross, C.B. *J. Am. Chem. Soc.* **1959**, *81*, 133.

[1472]Tateiwa, J.-i.; Kimura, A.; Takasuka, M.; Uemura, S. *J. Chem. Soc. Perkin Trans. 1* **1997**, 2169.

[1473]Ellis, W.W.; Odenkirk, W.; Bosnich, B. *Chem. Commun.* **1998**, 1311.

Samarium iodide promotes this addition reaction.[1474] In a related reaction, simple alkene units add to esters in the presence of sodium and liquid ammonia to give an alcohol.[1475] Structural variations in the alkene lead to different products. Homoallylic alcohols react with aldehydes in the presence of Montmorillonite KSF clay to give 4-hydroxytetrahydropyrans.[1476] A variation of this reaction converts an aryl aldehyde and a homoallylic alcohol to a 4-chlorotetrahydropyran in the presence of $InCl_3$.[1477] Homoallylic alcohols, protected as —O(CHMeOAc) react with $BF_3 \cdot OEt_2$ and acetic acid to give 4-acetoxytetrahydropyrans or with $SnBr_4$ to give 4-bromotetrahydropyrans.[1478] Homoallylic alcohols with a vinyl silane moiety react with $InCl_3$ and an aldehyde to give a dihydropyran.[1479]

A closely related reaction has been performed with activated aldehydes or ketones; without a catalyst such as chloral and acetoacetic ester, but with heat.[1480] The product in these cases is a β-hydroxy alkene, and the mechanism is pericyclic:[1481]

This reaction is reversible and suitable β-hydroxy alkenes can be cleaved by heat (**17-32**). There is evidence that the cleavage reaction occurs by a cyclic mechanism (p. 1551), and, by the principle of microscopic reversibility, the addition mechanism should be cyclic too.[1482] Note that this reaction is an oxygen analog of the ene synthesis (**15-23**). This reaction can also be done with unactivated aldehydes[1483] and ketones[1484] if Lewis-acid catalysts such as dimethylaluminum chloride (Me_2AlCl) or ethylaluminum dichloride ($EtAlCl_2$) are used.[1485] Lewis acid catalysts

[1474]Sarkar, T.K.; Nandy, S.K. *Tetrahedron Lett.* **1996**, *37*, 5195.

[1475]Cossy, J.; Gille, B.; Bellosta, V. *J. Org. Chem.* **1998**, *63*, 3141.

[1476]Yadav, J.S.; Reddy, B.V.S.; Kumar, G.M.; Murthy, Ch.V.S.R. *Tetrahedron Lett.* **2001**, *42*, 89.

[1477]Yang, J.; Viswanathan, G.S.; Li, C.-J. *Tetrahedron Lett.* **1999**, *40*, 1627.

[1478]Jaber, J.J.; Mitsui, K.; Rychnovsky, S.D. *J. Org. Chem.* **2001**, *66*, 4679.

[1479]Dobbs, A.P.; Martinović, S. *Tetrahedron Lett.* **2002**, *43*, 7055.

[1480]Arnold, R.T.; Veeravagu, P. *J. Am. Chem. Soc.* **1960**, *82*, 5411; Klimova, E.I.; Abramov, A.I.; Antonova, N.D.; Arbuzov, Yu.A. *J. Org. Chem. USSR* **1969**, *5*, 1308; Klimova, E.I.; Antonova, N.D.; Arbuzov, Yu.A. *J. Org. Chem. USSR* **1969**, *5*, 1312, 1315.

[1481]See, for example, Achmatowicz, Jr., O.; Szymoniak, J. *J. Org. Chem.* **1980**, *45*, 1228; Ben Salem, R.; Jenner, G. *Tetrahedron Lett.* **1986**, *27*, 1575. There is evidence that the mechanism is somewhat more complicated than shown here: Kwart, H.; Brechbiel, M. *J. Org. Chem.* **1982**, *47*, 3353.

[1482]For other evidence, see Achmatowicz Jr., O.; Szymoniak, J. *J. Org. Chem.* **1980**, *45*, 1228; Ben Salem, R.; Jenner, G. *Tetrahedron Lett.* **1986**, *27*, 1575; Papadopoulos, M.; Jenner, G. *Tetrahedron Lett.* **1981**, *22*, 2773.

[1483]Snider, B.B. *Acc. Chem. Res.* **1980**, *13*, 426; Cartaya-Marin, C.P.; Jackson, A.C.; Snider, B.B. *J. Org. Chem.* **1984**, *49*, 2443.

[1484]Jackson, A.C.; Goldman, B.E.; Snider, B.B. *J. Org. Chem.* **1984**, *49*, 3988.

[1485]For discussions of the mechanism with Lewis acid catalysts, see Stephenson, L.M.; Orfanopoulos, M. *J. Org. Chem.* **1981**, *46*, 2200; Kwart, H.; Brechbiel, M. *J. Org. Chem.* **1982**, *47*, 5409; Song, Z.; Beak, P. *J. Org. Chem.* **1990**, *112*, 8126.

also increase rates with activated aldehydes.[1486] The use of optically active catalysts has given optically active products with high ee.[1487]

90

In a related reaction, alkenes can be added to aldehydes and ketones to give reduced alcohols **90**. This has been accomplished by several methods,[1488] including treatment with SmI_2[1489] or Zn and Me_3SiCl,[1490] and by electrochemical[1491] and photochemical[1492] methods. Most of these methods have been used for intramolecular addition and most or all involve free-radical intermediates.

OS **IV**, 786. See also, OS **VII**, 102.

16-55   The Benzoin Condensation

Benzoin aldehyde condensation

$$2\ ArCHO\ +\ KCN\ \longrightarrow$$

91

When certain aldehydes are treated with cyanide ion, *benzoins* (**91**) are produced in a reaction called the *benzoin condensation*. The condensation can be regarded as involving the addition of one molecule of aldehyde to the C=O group of another. The reaction only occurs with aromatic aldehydes, but not all of them,[1493] and for glyoxals RCOCHO. The two molecules of aldehyde obviously perform different functions. The one that no longer has a C—H bond in the product is called the *donor*, because it has "donated" its hydrogen to the oxygen of the other molecule, the *acceptor*. Some aldehydes can perform only one of these functions, and hence cannot be self-condensed, though they can often be condensed with a different aldehyde. For example, *p*-dimethylaminobenzaldehyde is not an acceptor but only a donor. Thus it cannot condense with itself, but it can condense with benzaldehyde, which can perform both functions, but is a better acceptor than it is a donor.

[1486]Benner, J.P.; Gill, G.B.; Parrott, S.J.; Wallace, B. *J. Chem. Soc. Perkin Trans. 1* **1984**, 291, 315, 331.
[1487]Maruoka, K.; Hoshino, Y.; Shirasaka, T.; Yamamoto, H. *Tetrahedron Lett.* **1988**, *29*, 3967; Mikami, K.; Terada, M.; Nakai, T. *J. Am. Chem. Soc.* **1990**, *112*, 3949.
[1488]For references, see Ujikawa, O.; Inanaga, J.; Yamaguchi, M. *Tetrahedron Lett.* **1989**, *30*, 2837; Larock, R.C. *Comprehensive Organic Transformations*, 2nd ed., Wiley-VCH, NY, **1999**, pp. 1178–1179.
[1489]Ujikawa, O.; Inanaga, J.; Yamaguchi, M. *Tetrahedron Lett.* **1989**, *30*, 2837.
[1490]Corey, E.J.; Pyne, S.G. *Tetrahedron Lett.* **1983**, *24*, 2821.
[1491]See Shono, T.; Kashimura, S.; Mori, Y.; Hayashi, T.; Soejima, T.; Yamaguchi, Y. *J. Org. Chem.* **1989**, *54*, 6001.
[1492]See Belotti, D.; Cossy, J.; Pete, J.P.; Portella, C. *J. Org. Chem.* **1986**, *51*, 4196.
[1493]For a review, see Ide, W.S.; Buck, J.S. *Org. React.* **1948**, *4*, 269.

The following is the accepted mechanism[1494] for this reversible reaction, which was originally proposed by Lapworth in 1903:[1495]

The key step, the loss of the aldehydic proton, can take place because the acidity of this C—H bond is increased by the electron-withdrawing power of the CN group. Thus, cyanide is a highly specific catalyst for this reaction, because, almost uniquely, it can perform three functions: (*1*) It acts as a nucleophile; (*2*) its electron-withdrawing ability permits loss of the aldehydic proton; and (*3*) having done this, it then acts as a leaving group. Certain thiazolium salts can also catalyze the reaction.[1496] In this case, aliphatic aldehydes can also be used[1497] (the products are called *acyloins*), and mixtures of aliphatic and aromatic aldehydes give mixed α-hydroxy ketones.[1498] The reaction has also been carried out without cyanide, by using the benzoylated cyanohydrin as one of the components in a phase-transfer catalyzed process. By this means, products can be obtained from aldehydes that normally fail to self-condense.[1499] The condensation has also been done with excellent enantioselectivity using benzoylformate decarboxylase.[1500] Using aryl silyl ketones, $ArC(=O)SiMe_2Ph$, and aldehydes with a lanthanum catalyst, a 'mixed' benzoin condensation has been accomplished.[1501]

OS **I**, 94; **VII**, 95.

[1494]For a discussion, See Kuebrich, J.P.; Schowen, R.L.; Wang, M.; Lupes, M.E. *J. Am. Chem. Soc.* ***1971***, *93*, 1214.

[1495]Lapworth, A. *J. Chem. Soc.* ***1903***, *83*, 995; ***1904***, *85*, 1206.

[1496]See Ugai, T.; Tanaka, S.; Dokawa, S. *J. Pharm. Soc. Jpn.* ***1943***, *63*, 296 [*Chem. Abstr. 45*, 5148]; Breslow, R. *J. Am. Chem. Soc.* ***1958***, *80*, 3719; Breslow, R.; Kool, E. *Tetrahedron Lett.* ***1988***, *29*, 1635; Castells, J.; López-Calahorra, F.; Domingo, L. *J. Org. Chem.* ***1988***, *53*, 4433; Diederich, F.; Lutter, H. *J. Am. Chem. Soc.* ***1989***, *111*, 8438. For another catalyst, see Lappert, M.F.; Maskell, R.K. *J. Chem. Soc., Chem. Commun.* ***1982***, 580.

[1497]Stetter, H.; Rämsch, R.Y.; Kuhlmann, H. *Synthesis* ***1976***, 733; Stetter, H.; Kuhlmann, H. *Org. Synth.* **VII**, 95; Matsumoto, T.; Ohishi, M.; Inoue, S. *J. Org. Chem.* ***1985***, *50*, 603.

[1498]Stetter, H.; Dämbkes, G. *Synthesis* ***1977***, 403.

[1499]Rozwadowska, M.D. *Tetrahedron* ***1985***, *41*, 3135.

[1500]Demir, A.S.; Dünnwald, T.; Iding, H.; Pohl, M.; Müller, M. *Tetrahedron Asymmetry* ***1999***, *10*, 4769.

[1501]Bausch, C.C.; Johnson, J.S. *J. Org. Chem.* ***2004***, *69*, 4283.

16-56 Addition of Radicals to C=O, C=S, C=N Compounds

Radical cyclization is not limited to reaction with a C=C unit (see **15-29** and **15-30**), and reactions with both C=N and C=O moieties are known. Reaction of MeON=CH(CH$_2$)$_3$CHO with Bu$_3$SnH and AIBN, for example, led to *trans*-2-(methoxyamino)cyclopentanol in good yield.[1502] Conjugated ketones add to aldehyde via the β-carbon under radical conditions (2 equivalents of Bu$_3$SnH and 0.1 equivalent of CuCl) to give a β-hydroxy ketone.[1503] Addition of radical to the C=N unit of R–C=N–SPh[1504] or R–C=N–OBz[1505] led to cyclic imines. Radical addition to simple imines leads to aminocycloalkenes.[1506] Radical also add to the carbonyl unit of phenylthio esters to give cyclic ketones.[1507]

N,N-Dimethylaniline reacts with aldehydes under photochemical conditions to give acyl addition via the carbon atom of one of the methyl groups.[1508] The reaction of PhNMe$_2$ and benzaldehyde, for example, gave PhN(Me)CH$_2$CH(OH)Ph upon photolysis.

ACYL SUBSTITUTION REACTIONS

A. O, N, and S Nucleophiles

16-57 Hydrolysis of Acyl Halides

Hydroxy-de-halogenation

$$RCOCl + H_2O \longrightarrow RCOOH$$

Acyl halides are so reactive that hydrolysis is easily carried out.[1509] In fact, most simple acyl halides must be stored under anhydrous conditions lest they react with water in the air. Consequently, water is usually a strong enough nucleophile for the reaction, though in difficult cases hydroxide ion may be required. The reaction is seldom synthetically useful, because acyl halides are normally prepared from acids. The reactivity order is F < Cl < Br < I.[1510] If a carboxylic acid is used as the nucleophile, an exchange may take place (see **16-79**). The mechanism[1510] of hydrolysis can be either S$_N$1 or tetrahedral, the former occurring in highly polar solvents

[1502]Tormo, J.; Hays, D.S.; Fu, G.C. *J. Org. Chem.* **1998**, *63*, 201.

[1503]Ooi, T.; Doda, K.; Sakai, D.; Maruoka, K. *Tetrahedron Lett.* **1999**, *40*, 2133.

[1504]Boivin, J.; Fouquet, E.; Zard, S.Z. *Tetrahedron* **1994**, *50*, 1745.

[1505]Boivin, J.; Schiano, A.-M.; Zard, S.Z. *Tetrahedron Lett.* **1994**, *35*, 249.

[1506]Bowman, W.R.; Stephenson, P.T.; Terrett, N.K.; Young, A.R. *Tetrahedron Lett.* **1994**, *35*, 6369.

[1507]Kim, S.; Jon, S.Y. *Chem. Commun.* **1996**, 1335.

[1508]Kim, S.S.; Mah, Y.J.; Kim, A.R. *Tetrahedron Lett.* **2001**, *42*, 8315.

[1509]See Bentley, T.W.; Shim, C.S. *J. Chem. Soc. Perkin Trans. 2* **1993**, 1659 for a discussion on the solvolysis of acyl chlorides.

[1510]For a review, see Talbot, R.J.E., in Bamford, C.H.; Tipper, C.F.H. *Comprehensive Chemical Kinetics*, Vol. 10; Elsevier, NY, *1972*, pp. 226–257. For a review of the mechanisms of reactions of acyl halides with water, alcohols, and amines, see Kivinen, A., in Patai, S. *The Chemistry of Acyl Halides*, Wiley, NY, *1972*, pp. 177–230.

and in the absence of strong nucleophiles.[1511] There is also evidence for the S_N2 mechanism in some cases.[1512]

Hydrolysis of acyl halides is not usually catalyzed by acids, except for acyl fluorides, where hydrogen bonding can assist in the removal of F.[1513] There are several methods available for the hydrolysis of acyl fluorides.[1514]

OS **II**, 74.

16-58 Hydrolysis of Anhydrides

Hydroxy-de-acyloxy-substitution

$$\underset{\text{O}}{\overset{\text{R}}{\|}}\text{O}\underset{\text{O}}{\overset{\text{R}'}{\|}} + H_2O \longrightarrow \underset{\text{O}}{\overset{\text{R}}{\|}}\text{OH} + HO\underset{\text{O}}{\overset{\text{R}'}{\|}}$$

Anhydrides are somewhat more difficult to hydrolyze than acyl halides, but here too water is usually a strong enough nucleophile. The mechanism is usually tetrahedral.[1515] Only under acid catalysis does the S_N1 mechanism occur and seldom even then.[1516] Anhydride hydrolysis can also be catalyzed by bases. Of course, hydroxide ion attacks more readily than water, but other bases can also catalyze the reaction. This phenomenon, called *nucleophilic catalysis* (p. 1258), is actually the result of two successive tetrahedral mechanisms. For example, pyridine catalyzes the hydrolysis of acetic anhydride in this manner.[1517]

Many other nucleophiles similarly catalyze the reaction.

OS **I**, 408; **II**, 140, 368, 382; **IV**, 766; **V**, 8, 813.

[1511]Bender, M.L.; Chen, M.C. *J. Am. Chem. Soc.* **1963**, *85*, 30. See also, Song, B.D.; Jencks, W.P. *J. Am. Chem. Soc.* **1989**, *111*, 8470; Bentley, T.W.; Koo, I.S.; Norman, S.J. *J. Org. Chem.* **1991**, *56*, 1604.

[1512]Bentley, T.W.; Carter, G.E.; Harris, H.C.*J. Chem. Soc. Perkin Trans. 2* **1985**, 983; Guthrie, J.P.; Pike, D.C. *Can. J. Chem.* **1987**, *65*, 1951. See also, Lee, I.; Sung, D.D.; Uhm, T.S.; Ryu, Z.H. *J. Chem. Soc. Perkin Trans. 2* **1989**, 1697.

[1513]Bevan, C.W.L.; Hudson, R.F. *J. Chem. Soc.* **1953**, 2187; Satchell, D.P.N. *J. Chem. Soc.* **1963**, 555.

[1514]Motie, R.E.; Satchell, D.P.N.; Wassef, W.N. *J. Chem. Soc. Perkin Trans. 2* **1992**, 859; **1993**, 1087.

[1515]The kinetics of the acid hydrolysis has been determined. See Satchell, D.P.N.; Wassef, W.N.; Bhatti, Z.A. *J. Chem. Soc. Perkin Trans. 2* **1993**, 2373.

[1516]Satchell, D.P.N. *Q. Rev. Chem. Soc.* **1963**, *17*, 160, 172–173. For a review of the mechanism, see Talbot, R.J.E., in Bamford, C.H.; Tipper, C.F.H. *Comprehensive Chemical Kinetics*, Vol. 10, Elsevier, NY, **1972**, pp. 280–287.

[1517]Butler, A.R.; Gold, V. *J. Chem. Soc.* **1961**, 4362; Fersht, A.R.; Jencks, W.P. *J. Am. Chem. Soc.* **1970**, *92*, 5432, 5442; Deady, L.W.; Finlayson, W.L. *Aust. J. Chem.* **1983**, *36*, 1951.

16-59   Hydrolysis of Carboxylic Esters

Hydroxy-de-alkoxylation

$$RCOO^- + R'OH \xrightleftharpoons[^-OH,\,H_2O]{} \underset{O}{\overset{R \quad OR'}{\text{C}}} \xrightleftharpoons[]{H^+,\,H_2O} RCOOH + R'OH$$

Ester hydrolysis is usually catalyzed by acids or bases. Since OR is a much poorer leaving group than halide or OCOR, water alone does not hydrolyze most esters. When bases catalyze the reaction, the attacking species is the more powerful nucleophile $^-$OH. This reaction is called *saponification* and gives the salt of the acid. Acids catalyze the reaction by making the carbonyl carbon more positive and therefore more susceptible to attack by the nucleophile. Both reactions are equilibrium reactions, so they are practicable only when there is a way of shifting the equilibrium to the right. Since formation of the salt does just this, ester hydrolysis is almost always done for preparative purposes in basic solution, unless the compound is base sensitive. Even in the case of **92**, however, selective base hydrolysis of the ethyl ester gave an 80%

$$\underset{\textbf{92}}{\overset{MeO_2C}{\underset{MeO_2C}{>}}CH\text{-}(CH_2)_3\overset{CO_2Et}{}} \xrightarrow[80\%]{t\text{-BuOK},\,H_2O,\,THF} \underset{\textbf{93}}{\overset{MeO_2C}{\underset{MeO_2C}{>}}CH\text{-}(CH_2)_3\overset{CO_2H}{}}$$

yield of the acid-dimethyl ester (**93**).[1518] Ester hydrolysis can also be catalyzed[1519] by metal ions, by cyclodextrins,[1520] by enzymes,[1521] and by nucleophiles.[14] Other reagents used to cleave carboxylic esters include Dowex-50,[1522] Me$_3$SiI,[1523] and InCl$_3$ on moist silica gel using microwave irradiation.[1524] Cleavage of phenolic esters is usually faster than carboxylic esters derived from aliphatic acids. The reagent Sm/I$_2$ at $-78°$C has been used,[1525] ammonium acetate in aqueous methanol,[1526]

[1518]Wilk, B.K. *Synth. Commun.* **1996**, *26*, 3859.

[1519]For a list of catalysts and reagents that have been used to convert carboxylic esters to acids, with references, see Larock, R.C. *Comprehensive Organic Transformations*, 2nd ed., Wiley-VCH, NY, **1999**, pp. 1959-1968.

[1520]See Bender, M.L.; Komiyama, M. *Cyclodextrin Chemistry*, Springer, NY, **1978**, pp. 34–41. The mechanism is shown in Saenger, W. *Angew. Chem. Int. Ed.* **1980**, *19*, 344.

[1521]For reviews of ester hydrolysis catalyzed by pig liver esterase, see Zhu, L.; Tedford, M.C. *Tetrahedron* **1990**, *46*, 6587; Ohno, M.; Otsuka, M. *Org. React.* **1989**, *37*, 1. For reviews of enzymes as catalysts in synthetic organic chemistry, see Wong, C. *Chemtracts: Org. Chem.* **1990**, *3*, 91; *Science* **1989**, *244*, 1145; Whitesides, G.M.; Wong, C. *Angew. Chem. Int. Ed.* **1985**, *24*, 617. Addition of crown ethers can enhance the rate of hydrolysis, see Itoh, T.; Hiyama, Y.; Betchaku, A.; Tsukube, H. *Tetrahedron Lett.* **1993**, *34*, 2617.

[1522]Basu, M.K.; Sarkar, D.C.; Ranu, B.C. *Synth. Commun.* **1989**, *19*, 627.

[1523]See Olah, G.A.; Narang, S.C. *Tetrahedron* **1982**, *38*, 2225; Olah, G.A.; Husain, A.; Singh, B.P.; Mehrotra, A.K. *J. Org. Chem.* **1983**, *48*, 3667.

[1524]Ranu, B.C.; Dutta, P.; Sarkar, A. *Synth. Commun.* **2000**, *30*, 4167.

[1525]Yanada, R.; Negoro, N.; Bessho, K.; Yanada, K. *Synlett* **1995**, 1261.

[1526]Ramesh, C.; Mahender, G.; Ravindranath, N.; Das, B. *Tetrahedron* **2003**, *59*, 1049.

Amberlyst 15 in methanol,[1527] and phenolic esters have been selectively hydrolyzed in the presence of alkyl esters on alumina with microwave irradiation.[1528] Thiophenol with K_2CO_3 in NMP quantitatively converted methyl benzoate to benzoic acid.[1529] Allylic esters were cleaved with 2% Me_3SiOTf in dichloromethane,[1530] with $CeCl_3 \cdot 7 H_2O$—NaI,[1531] and with $NaHSO_4 \cdot$ silica gel.[1532] Lactones also undergo the reaction[1533] (though if the lactone is five- or six-membered, the hydroxy acid often spontaneously reforms the lactone) and thiol esters (RCOSR′) give thiols R′SH. Typical reagents for this latter transformation include NaSMe in methanol,[1534] borohydride exchange resin-Pd(OAc)$_2$ for reductive cleavage of thiol esters to thiols,[1535] and $TiCl_4$/Zn for the conversion of phenylthioacetates to thiophenols.[1536] Sterically hindered esters are hydrolyzed with difficulty (p. 479), but reaction of 2 equivalents of t-BuOK with 1 equivalent of water is effective.[1537] Hindered esters can also be cleaved by sequential treatment with zinc bromide and then water,[1538] with silica gel in refluxing toluene,[1539] and on alumina when irradiated with microwaves.[1540] For esters insoluble in water the rate of two-phase ester saponification can be greatly increased by the application of ultrasound,[1541] and phase-transfer techniques have been applied.[1542] Enzymatic hydrolysis of diesters with esterase has been shown to give the hydroxy ester,[1543] and selective hydrolysis of dimethyl succinate to monomethyl succinic acid was accomplished with aq. NaOH in THF.[1544] Hydrolysis of vinyl esters leads to ketones, and the reaction of C-substituted vinyl acetates with an esterase derived from *Marchantia polymorpha* gave substituted ketones with high enantioselectivity.[1545] Scandium triflate was shown to hydrolyze α-acetoxy ketones to α-hydroxy ketones.[1546]

[1527]Das, B.; Banerjee, J.; Ramu, R.; Pal, R.; Ravindranath, N.; Ramesh, C. *Tetrahedron Lett.* **2003**, *44*, 5465.

[1528]Varma, R.S.; Varma, M.; Chatterjee, A.K. *J. Chem. Soc. Perkin Trans. 1* **1993**, 999; Blay, G.; Cardona, L.; Garcia, B.; Pedro, J.R. *Synthesis* **1989**, 438.

[1529]Sharma, L.; Nayak, M.K.; Chakraborti, A.K. *Tetrahedron* **1999**, *55*, 9595.

[1530]Nishizawa, M.; Yamamoto, H.; Seo, K.; Imagawa, H.; Sugihara, T. *Org. Lett.* **2002**, *4*, 1947.

[1531]Yadav, J.S.; Reddy, B.V.S.; Rao, C.V.; Chand, P.K.; Prasad, A.R. *Synlett* **2002**, 137.

[1532]Ramesh, C.; Mahender, G.; Ravindranath, N.; Das, B. *Tetrahedron Lett.* **2003**, *44*, 1465.

[1533]For a review of the mechanisms of lactone hydrolysis, see Kaiser, E.T.; Kézdy, F.J. *Prog. Bioorg. Chem.* **1976**, *4*, 239, pp. 254–265.

[1534]Wallace, O.B.; Springer, D.M. *Tetrahedron Lett.* **1998**, *39*, 2693.

[1535]Choi, J.; Yoon, N.M. *Synth. Commun.* **1995**, *25*, 2655.

[1536]Jin, C.K.; Jeong, H.J.; Kim, M.K.; Kim, J.Y.; Yoon, Y.-J.; Lee, S.-G. *Synlett* **2001**, 1956.

[1537]Gassman, P.G.; Schenk, W.N. *J. Org. Chem.* **1977**, *42*, 918.

[1538]Wu, Y.-g.; Limburg, D.C.; Wilkinson, D.E.; Vaal, M.J.; Hamilton, G.S. *Tetrahedron Lett.* **2000**, *41*, 2847.

[1539]Jackson, R.W. *Tetrahedron Lett.* **2001**, *42*, 5163.

[1540]Ley, S.V.; Mynett, D.M. *Synlett* **1993**, 793.

[1541]Moon, S.; Duchin, L.; Cooney, J.V. *Tetrahedron Lett.* **1979**, 3917.

[1542]Loupy, A.; Pedoussaut, M.; Sansoulet, J. *J. Org. Chem.* **1986**, *51*, 740.

[1543]Houille, O.; Schmittberger, T.; Uguen, D. *Tetrahedron Lett.* **1996**, *37*, 625; Nair, R.V.; Shukla, M.R.; Patil, P.N.; Salunkhe, M.M. *Synth. Commun.* **1999**, *29*, 1671.

[1544]Niwayama, S. *J. Org. Chem.* **2000**, *65*, 5834.

[1545]Hirata, T.; Shimoda, K.; Kawano, T. *Tetrahedron Asymmetry* **2000**, *11*, 1063.

[1546]Kajiro, H.; Mitamura, S.; Mori, A.; Hiyama, T. *Bull. Chem. Soc. Jpn.* **1999**, *72*, 1553.

Ingold[1547] has classified the acid- and base-catalyzed hydrolyses of esters (and the formation of esters, since these are reversible reactions and thus have the same mechanisms) into eight possible mechanisms (Table 16.3), depending on the

TABLE 16.3. Classification of the Eight Mechanisms for Ester Hydrolysis and Formation[1547]

	Name		Type
Ingold	IUPAC[1548]		
$A_{AC}1$	$A_h + D_N + A_N + D_h$		S_N1

A

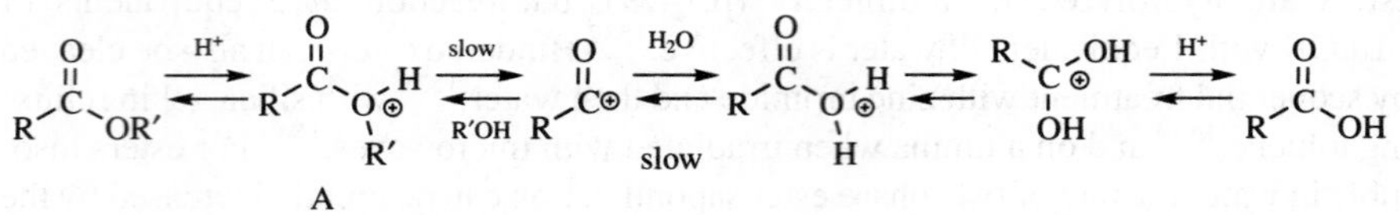

| $A_{AC}2$ | $A_h + A_N + A_hD_h + D_h$ | | Tetrahedral |

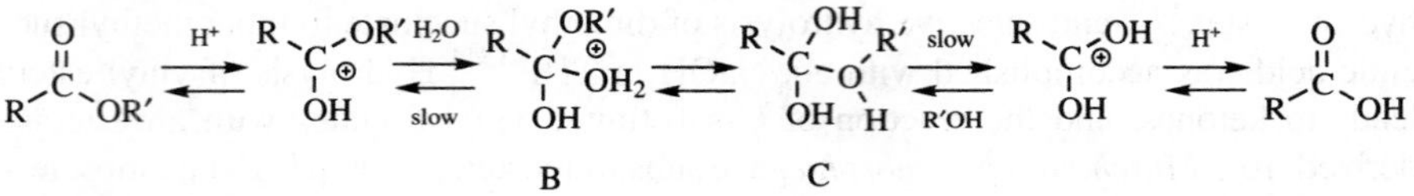

B   C

| $A_{AL}1$ | $A_h + D_N + A_N + D_h$ | | S_N1 |

| $A_{AL}2$ | $A_h + A_ND_N + D_h$ | | S_N2 |

| $B_{AC}1$ | $D_N + A_N + A_{xh}D_h$ | | S_N1 |

| $B_{AC}2$ | $A_N + D_N + A_{xh}D_h$ | | Tetrahedral |

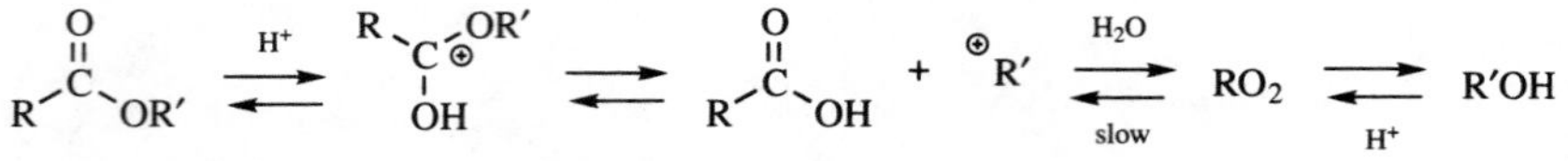

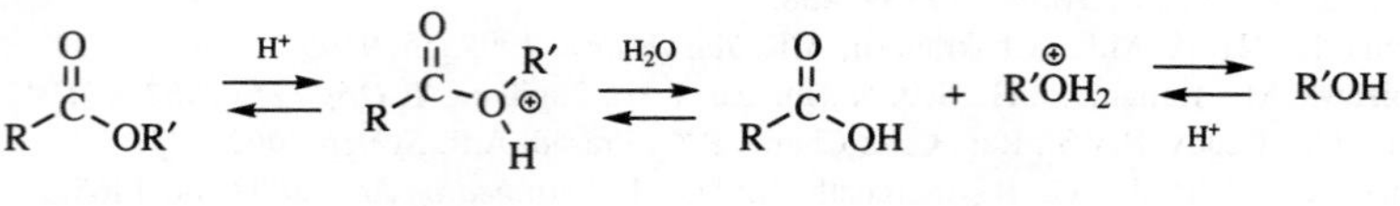

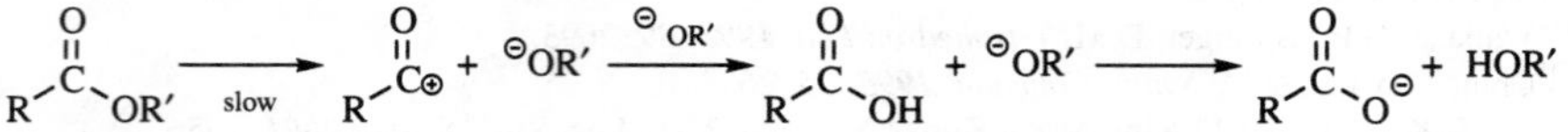

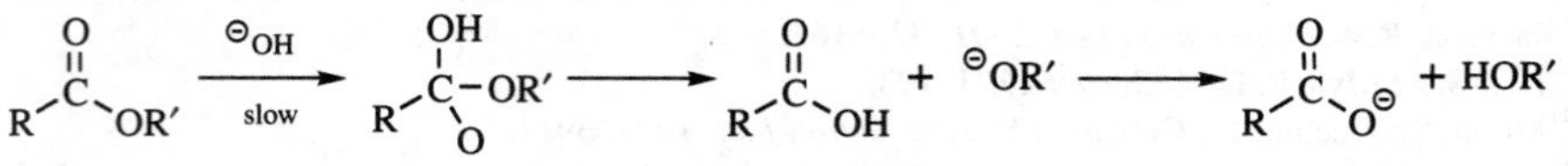

[1547]Ingold, C.K. *Structure and Mechanism in Organic Chemistry*, 2nd ed., Cornell University Press, Ithaca, NY, *1969*, pp. 1129–1131.

[1548]As given here, the IUPAC designations for $B_{AC}1$ and $B_{AL}1$ are the same, but Rule A.2 adds further symbols so that they can be distinguished: Su-AL for $B_{AL}1$ and Su-AC for $B_{AC}1$. See the IUPAC rules: Guthrie, R.D. *Pure Appl. Chem.* **1989**, *61*, 23, see p. 49.

TABLE 16.3. (*Continued*)

	Name	
Ingold	IUPAC	Type
$B_{AL}1$	$D_N + A_N + A_{xh}D_h$	S_N1

| $B_{AL}2$ | A_ND_N | S_N2 |

following criteria: (*1*) acid- or base-catalyzed, (*2*) unimolecular or bimolecular, and (*3*) acyl cleavage or alkyl cleavage.[1549] All eight of these are S_N1, S_N2, or tetrahedral mechanisms. The acid-catalyzed mechanisms are shown with reversible arrows. They are not only reversible, but symmetrical; that is, the mechanisms for ester formation are exactly the same as for hydrolysis, except that H replaces R. Internal proton transfers, such as shown for **B** and **C**, may not actually be direct but may take place through the solvent. There is much physical evidence to show that esters are initially protonated on the carbonyl and not on the alkyl oxygen (Chapter 8, Ref. 17). We have nevertheless shown the $A_{AC}1$ mechanism as proceeding through the ether-protonated intermediate **A**, since it is difficult to envision OR' as a leaving group here. It is of course possible for a reaction to proceed through an intermediate even if only a tiny concentration is present. The designations $A_{AC}1$, and so on., are those of Ingold. The $A_{AC}2$ and $A_{AC}1$ mechanisms are also called A2 and A1, respectively. Note that the $A_{AC}1$ mechanism is actually the same as the S_N1cA mechanism for this type of substrate and that $A_{AL}2$ is analogous to S_N2cA. Some authors use A1 and A2 to refer to all types of nucleophilic substitution in which the leaving group first acquires a proton. The base-catalyzed reactions are not shown with reversible arrows, since they are reversible only in theory and not in practice. Hydrolyses taking place under neutral conditions are classified as B mechanisms.

Of the eight mechanisms, seven have actually been observed in hydrolysis of carboxylic esters. The one that has not been observed is the $B_{AC}1$ mechanism.[1550] The most common mechanisms are the $B_{AC}2$ for basic catalysis and the $A_{AC}2$[1551]

[1549]For reviews of the mechanisms of ester hydrolysis and formation, see Kirby, A.J., in Bamford, C.H.; Tipper, C.F.H., *Comprehensive Chemical Kinetics*, Vol. 10, *1972*, pp. 57–207; Euranto, E.K., in Patai, S. *The Chemistry of Carboxylic Acids and Esters*, Wiley, NY, *1969*, pp. 505–588.
[1550]This is an S_N1 mechanism with OR' as leaving group, which does not happen.
[1551]For a discussion of this mechanism with specific attention to the proton transfers involved, see Zimmermann, H.; Rudolph, J. *Angew. Chem. Int. Ed. 1965*, *4*, 40.

for acid catalysis, that is, the two tetrahedral mechanisms. Both involve acyl-oxygen cleavage. The evidence is (*1*) hydrolysis with $H_2^{18}O$ results in the ^{18}O appearing in the acid and not in the alcohol;[1552] (*2*) esters with chiral R′ groups give alcohols with *retention* of configuration;[1553] (*3*) allylic R′ gives no allylic rear-rangement;[1554] (*4*) neopentyl R′ gives no rearrangement;[1555] all these facts indicate that the O—R′ bond is not broken. It has been concluded that two molecules of water are required in the $A_{AC}2$ mechanism.

If this is so, the protonated derivatives **B** and **C** would not appear at all. This con-clusion stems from a value of *w* (see p. 371) of ~5, indicating that water acts as a proton donor here, as well as a nucleophile.[1556] Termolecular processes are rare, but in this case the two water molecules are already connected by a hydrogen bond. (A similar mechanism, called $B_{AC}3$, also involving two molecules of water, has been found for esters that hydrolyze without a catalyst.[1557] Such esters are mostly those containing halogen atoms in the R group.)

The other mechanism involving acyl cleavage is the $A_{AC}1$ mechanism. This is rare, being found only where R is very bulky, so that bimolecular attack is sterically hindered, and only in ionizing solvents. The mechanism has been demonstrated for esters of 2,4,6-trimethylbenzoic acid (mesitoic acid). This acid depresses the freez-ing point of sulfuric acid four times as much as would be predicted from its mole-cular weight, which is evidence for the equilibrium

$$ArCOOH + 2 H_2SO_4 \rightleftharpoons ArCO^{\oplus} + H_3O^+ + 2 HSO_4^-$$

In a comparable solution of benzoic acid the freezing point is depressed only twice the predicted amount, indicating only a normal acid-base reaction. Further, a sulfu-ric acid solution of methyl mesitoate when poured into water gave mesitoic acid, while a similar solution of methyl benzoate similarly treated did not.[1558] The $A_{AC}1$ mechanism is also found when acetates of phenols or of primary alcohols are

[1552]For one of several examples, see Polanyi, M.; Szabo, A.L. *Trans. Faraday Soc.* **1934**, *30*, 508.

[1553]Holmberg, B. *Ber.* **1912**, *45*, 2997.

[1554]Ingold, C.K.; Ingold, E.H. *J. Chem. Soc.* **1932**, 758.

[1555]Norton, H.M.; Quayle, O.R. *J. Am. Chem. Soc.* **1940**, *62*, 1170.

[1556]Martin, R.B. *J. Am. Chem. Soc.* **1962**, *84*, 4130. See also, Lane, C.A.; Cheung, M.F.; Dorsey, G.F. *J. Am. Chem. Soc.* **1968**, *90*, 6492; Yates, K. *Acc. Chem. Res.* **1971**, *6*, 136; Huskey, W.P.; Warren, C.T.; Hogg, J.L. *J. Org. Chem.* **1981**, *46*, 59.

[1557]Euranto, E.K.; Kanerva, L.T.; Cleve, N.J. *J. Chem. Soc. Perkin Trans. 2* **1984**, 2085; Neuvonen, H. *J. Chem. Soc. Perkin Trans. 2* **1986**, 1141; Euranto, E.K.; Kanerva, L.T. *Acta Chem. Scand. Ser. B* **1988**, 42 717.

[1558]Treffers, H.P.; Hammett, L.P. *J. Am. Chem. Soc.* **1937**, *59*, 1708. For other evidence for this mechanism, see Bender, M.L.; Chen, M.C. *J. Am. Chem. Soc.* **1963**, *85*, 37.

hydrolyzed in concentrated (>90%) H_2SO_4 (the mechanism under the more usual dilute acid conditions is the normal $A_{AC}2$).[1559]

The mechanisms involving alkyl-oxygen cleavage are ordinary S_N1 and S_N2 mechanisms in which OCOR (an acyloxy group) or its conjugate acid is the leaving group. Two of the three mechanisms, the $B_{AL}1$ and $A_{AL}1$ mechanisms, occur most readily when R′ comes off as a stable carbocation, that is, when R′ is tertiary alkyl, allylic, benzylic, and so on. For acid catalysis, most esters with this type of alkyl group (especially tertiary alkyl) cleave by this mechanism, but even for these substrates, the $B_{AL}1$ mechanism occurs only in neutral or weakly basic solution, where the rate of attack by hydroxide is so slowed that the normally slow (by comparison) unimolecular cleavage takes over. These two mechanisms have been established by kinetic studies, ^{18}O labeling, and iso-merization of R′.[1560] Secondary and benzylic acetates hydrolyze by the $A_{AC}2$ mechanism in dilute H_2SO_4, but in concentrated acid the mechanism changes to $A_{AL}1$.[1559] Despite its designation, the $B_{AL}1$ mechanism is actually uncatalyzed (as is the unknown $B_{AC}1$ mechanism).

The two remaining mechanisms, $B_{AL}2$ and $A_{AL}2$, are very rare, the $B_{AL}2$ because it requires hydroxide ion to attack an alkyl carbon when an acyl carbon is also available,[1561] and the $A_{AL}2$ because it requires water to be a nucleophile in an S_N2 process. Both have been observed, however. The $B_{AL}2$ has been seen in the hydrolysis of β-lactones under neutral conditions[1562] (because cleavage of the C—O bond in the transition state opens the four-membered ring and relieves strain), the alkaline hydrolysis of methyl 2,4,6-tri-*tert*-butyl benzoate,[1563] and in the unusual reaction[1564]

$$ArCOOMe + RO^- \longrightarrow ArCOO^- + ROMe$$

When it does occur, the $B_{AL}2$ mechanism is easy to detect, since it is the only one of the base-catalyzed mechanisms that requires inversion at R′. However, in the last example given, the mechanism is evident from the nature of the product, since the ether could have been formed in no other way. The $A_{AL}2$ mechanism has been reported in the acid cleavage of γ-lactones.[1565]

To sum up the acid-catalysis mechanisms, $A_{AC}2$ and $A_{AL}1$ are the common mechanisms, the latter for R′ that give stable carbocations, the former for practically

[1559]Yates, K. *Acc. Chem. Res.* **1971**, *6*, 136; Al-Shalchi, W.; Selwood, T.; Tillett J.G. *J. Chem. Res. (S)* **1985**, 10.

[1560]For discussions, see Kirby, A.J., in Bamford, C.H.; Tipper, C.F.H. *Comprehensive Chemical Kinetics*, Vol. 9, Elsevier, NY, *1973*, pp. 86–101; Ingold, C.K. *Structure and Mechanism in Organic Chemistry*, 2nd ed., Cornell University Press, Ithaca, NY, *1969*, pp. 1137–1142, 1157–1163.

[1561]Douglas, J.E.; Campbell, G.; Wigfield, D.C. *Can. J. Chem.* **1993**, *71*, 1841.

[1562]Cowdrey, W.A.; Hughes, E.D.; Ingold, C.K.; Masterman, S.; Scott, A.D. *J. Chem. Soc.* **1937**, 1264; Long, F.A.; Purchase, M. *J. Am. Chem. Soc.* **1950**, *73*, 3267.

[1563]Barclay, L.R.C.; Hall, N.D.; Cooke, G.A. *Can. J. Chem.* **1962**, *40*, 1981.

[1564]Sneen, R.A.; Rosenberg, A.M. *J. Org. Chem.* **1961**, *26*, 2099. See also, Müller, P.; Siegfried, B. *Helv. Chim. Acta* **1974**, *57*, 987.

[1565]Moore, J.A.; Schwab, J.W. *Tetrahedron Lett.* **1991**, *32*, 2331.

all the rest. The $A_{AC}1$ mechanism is rare, being found mostly with strong acids and sterically hindered R. The $A_{AL}2$ mechanism is even rarer. For basic catalysis, $B_{AC}2$ is almost universal; $B_{AL}1$ occurs only with R' that give stable carbocations and then only in weakly basic or neutral solutions; $B_{AL}2$ is very rare; and $B_{AC}1$ has never been observed.

The above results pertain to reactions in solution. In the gas-phase[1566] reactions can take a different course, as illustrated by the reaction of carboxylic esters with MeO^-, which in the gas phase was shown to take place only by the $B_{AL}2$ mechanism,[1567] even with aryl esters,[1568] where this means that an S_N2 mechanism takes place at an aryl substrate. However, when the gas-phase reaction of aryl esters was carried out with MeO^- ions, each of which was solvated with a single molecule of MeOH or H_2O, the $B_{AC}2$ mechanism was observed.[1567]

In the special case of alkaline hydrolysis of N-substituted aryl carbamates, there is another mechanism[1569] involving elimination–addition:[1570]

This mechanism does not apply to unsubstituted or N,N-disubstituted aryl carbamates, which hydrolyze by the normal mechanisms. Carboxylic esters substituted in the a position by an electron-withdrawing group (e.g., CN or COOEt) can also hydrolyze by a similar mechanism involving a ketene intermediate.[1571]

[1566]Takashima, K.; José, S.M.; do Amaral, A.T.; Riveros, J.M. *J. Chem. Soc., Chem. Commun.* **1983**, 1255.

[1567]Comisarow, M. *Can. J. Chem.* **1977**, *55*, 171.

[1568]Fukuda, E.K.; McIver Jr., R.T. *J. Am. Chem. Soc.* **1979**, *101*, 2498.

[1569]For a review of elimination–addition mechanisms at a carbonyl carbon, see Williams, A.; Douglas, K.T. *Chem. Rev.* **1975**, *75*, 627649.

[1570]Bender, M.L.; Homer, R.B. *J. Org. Chem.* **1965**, *30*, 3975; Williams, A. *J. Chem. Soc. Perkin Trans. 2* **1972**, 808; **1973**, 1244; Hegarty, A.F.; Frost, L.N. *J. Chem. Soc. Perkin Trans. 2* **1973**, 1719; Menger, F.M.; Glass, L.E. *J. Org. Chem.* **1974**, *39*, 2469; Sartoré, G.; Bergon, M.; Calmon, J.P. *J. Chem. Soc. Perkin Trans. 2* **1977**, 650; Moravcová, J.; Večeřa, M. *Collect. Czech. Chem. Commun.* **1977**, *42*, 3048; Broxton, T.J.; Chung, R.P. *J. Org. Chem.* **1986**, *51*, 3112.

[1571]Casanova, J.; Werner, N.D.; Kiefer, H.R. *J. Am. Chem. Soc.* **1967**, *89*, 2411; Holmquist, B.; Bruice, T.C. *J. Am. Chem. Soc.* **1969**, *91*, 2993, 3003; Campbell, D.S.; Lawrie, C.W. *Chem. Commun.* **1971**, 355; Kirby, A.J.; Lloyd, G.J. *J. Chem. Soc. Perkin Trans. 2* **1976**, 1762; Broxton, T.J.; Duddy, N.W. *J. Org. Chem.* **1981**, *46*, 1186; Inoue, T.C.; Bruice, T.C. *J. Am. Chem. Soc.* **1982**, *104*, 1644; *J. Org. Chem.* **1983**, *48*, 3559; **1986**, *51*, 959; Alborz, M.; Douglas, K.T. *J. Chem. Soc. Perkin Trans. 2* **1982**, 331; Thea, S.; Cevasco, G.; Guanti, G.; Kashefi-Naini, N.; Williams, A. *J. Org. Chem.* **1985**, *50*, 1867; Isaacs, N.S.; Najem, T.S. *Can. J. Chem.* **1986**, *64*, 1140; *J. Chem. Soc. Perkin Trans. 2* **1988**, 557.

These elimination–addition mechanisms usually are referred to as E1cB mechanisms, because that is the name given to the elimination portion of the mechanism (p. 1488).

The acid-catalyzed hydrolysis of enol esters RCOOCR'=CR can take place either by the normal $A_{AC}2$ mechanism or by a mechanism involving initial protonation on the double-bond carbon, similar to the mechanism for the hydrolysis of enol ethers given in **10-6**,[1572] depending on reaction conditions.[1573] In either case, the products are the carboxylic acid RCOOH and the aldehyde or ketone R—CHCOR'.

OS **I**, 351, 360, 366, 379, 391, 418, 523; **II**, 1, 5, 53, 93, 194, 214, 258, 299, 416, 422, 474, 531, 549; **III**, 3, 33, 101, 209, 213, 234, 267, 272, 281, 300, 495, 510, 526, 531, 615, 637, 652, 705, 737, 774, 785, 809 (but see OS **V**, 1050), 833, 835; **IV**, 15, 55, 169, 317, 417, 444, 532, 549, 555, 582, 590, 608, 616, 628, 630, 633, 635, 804; **V**, 8, 445, 509, 687, 762, 887, 985, 1031; **VI**, 75, 121, 560, 690, 824, 913, 1024; **VII**, 4, 190, 210, 297, 319, 323, 356, 411; **VIII**, 43, 141, 219, 247, 258, 263, 298, 486, 516, 527. Ester hydrolyses with concomitant decarboxylation are listed at reaction **12-40**.

16-60   Hydrolysis of Amides

Hydroxy-de-amination

$$NH_3 + \underset{R}{\overset{O}{\underset{\|}{C}}}{\overset{}{\diagdown}}O^{\ominus} \quad \underset{H_2O}{\overset{OH^-}{\longleftarrow}} \quad \underset{R}{\overset{O}{\underset{\|}{C}}}{\overset{}{\diagdown}}NH_2 \quad \underset{H_2O}{\overset{H^+}{\longrightarrow}} \quad \underset{R}{\overset{O}{\underset{\|}{C}}}{\overset{}{\diagdown}}OH + NH_4^{\oplus}$$

Unsubstituted amides ($RCONH_2$) can be hydrolyzed with either acidic or basic catalysis, the products being, respectively, the free acid and the ammonium ion or the salt of the acid and ammonia. N-Substituted (RCONHR') and N,N-disubstituted ($RCONR_2''$) amides can be hydrolyzed analogously, with the primary or secondary amine, respectively (or their salts), being obtained instead of ammonia. Lactams, imides, cyclic imides, hydrazides, and so on., also undergo the reaction. Water alone is not sufficient to hydrolyze most amides,[1574] since NH_2 is even a poorer leaving group than OR.[1575] Prolonged heating is often required, even with acidic or basic catalysts.[1576] Treatment of primary

---

[1572]Alkynyl esters also hydrolyze by this mechanism; see Allen, A.D.; Kitamura, T.; Roberts, K.A.; Stang, P.J.; Tidwell, T.T. *J. Am. Chem. Soc.* **1988**, *110*, 622.

[1573]See, for example, Noyce, D.S.; Pollack, R.M. *J. Am. Chem. Soc.* **1969**, *91*, 119, 7158; Monthéard, J.; Camps, M.; Chatzopoulos, M.; Benzaïd, A. *Bull. Soc. Chim. Fr.* **1984**, II-109. For a discussion, see Euranto, E.K. *Pure Appl. Chem.* **1977**, *49*, 1009.

[1574]See Zahn, D. *Eur. J. Org. Chem.* **2004**, 4020.

[1575]The very low rate of amide hydrolysis by water alone has been measured: Kahne, D.; Still, W.C. *J. Am. Chem. Soc.* **1988**, *110*, 7529.

[1576]For a list of catalysts and reagents that have been used to hydrolyze amides, with references, see Larock, R.C. *Comprehensive Organic Transformatinos*, 2nd ed., Wiley-VCH, NY, **1999**, pp. 1976–1977. Also see, Bagno, A.; Lovato, G.; Scorrano, G. *J. Chem. Soc. Perkin Trans. 2* **1993**, 1091.

amides with phthalic anhydride at 250°C and 4 atm gives the carboxylic acid and phthalimide.[1577] The conversion of acylhydrazine derivatives to the corresponding carboxylic acid with PhI(OH)(OTs) and water is a variation of amide hydrolysis.[1578] Hydrolysis of carbamates (RNHCO$_2$R) to the corresponding amine can be categorized in this section. Although the product is an amine and the carboxyl unit fragments, this reaction is simply a variation of amide hydrolysis. Strong acids, such as trifluoroacetic acid (in dichloromethane), are usually employed.[1579] Treatment of *N*-Boc derivatives (RNHCO$_2$*t*-Bu) with AlCl$_3$[1580] or with aqueous sodium *tert*-butoxide[1581] gave the amine. The by-products of this reaction are typically carbon dioxide and isobutylene.

$$\underset{R}{\overset{O}{\underset{}{\parallel}}}\underset{NH_2}{C} + HONO \longrightarrow \underset{R}{\overset{O}{\underset{}{\parallel}}}\underset{OH}{C} + N_2$$

In difficult cases, nitrous acid, NOCl, N$_2$O$_4$,[1582] or a similar compound can be used (unsubstituted amides only[1583]).These reactions involve a diazonium ion (see **13-19**) and are much faster than ordinary hydrolysis; for benzamide the nitrous acid reaction took place 2.5×10^7 times faster than ordinary hydrolysis.[1584] Another procedure for difficult cases involves treatment with aqueous sodium peroxide.[1585] In still another method, the amide is treated with water and *t*-BuOK at room temperature.[1586] The strong base removes the proton from **107**, thus preventing the reaction marked k_{-1}. A kinetic study has been done on the alkaline hydrolyses of *N*-trifluoroacetyl aniline derivatives.[1587] Amide hydrolysis can also be catalyzed by nucleophiles (see p. 1259).

The same framework of eight possible mechanisms that was discussed for ester hydrolysis can also be applied to amide hydrolysis.[1588] Both the acid- and base-catalyzed hydrolyses are essentially irreversible, since salts are formed in both

[1577]Chemat, F. *Tetrahedron Lett.* **2000**, *41*, 3855.

[1578]Wuts, P.G.M.; Goble, M.P. *Org. Lett.* **2000**, *2*, 2139.

[1579]Schwyzer, R.; Costopanagiotis, A.; Sieber, P. *Helv. Chim. Acta* **1963**, *46*, 870.

[1580]Bose, D.S.; Lakshminarayana, V. *Synthesis* **1999**, 66.

[1581]Tom, N.J.; Simon, W.M.; Frost, H.N.; Ewing, M. *Tetrahedron Lett.* **2004**, *45*, 905.

[1582]Kim, Y.H.; Kim, K.; Park, Y.J. *Tetrahedron Lett.* **1990**, *31*, 3893.

[1583]*N*-Substituted amides can be converted to *N*-nitrosoamides, which are more easily hydrolyzable than the orginal amide. For example, see Rull, M.; Serratosa, F.; Vilarrasa, J. *Tetrahedron Lett.* **1977**, 4549. For another method of hydrolyzing *N*-substituted amides, see Flynn, D.L.; Zelle, R.E.; Grieco, P.A. *J. Org. Chem.* **1983**, *48*, 2424.

[1584]Ladenheim, H.; Bender, M.L. *J. Am. Chem. Soc.* **1960**, *82*, 1895.

[1585]Vaughan, H.L.; Robbins, M.D. *J. Org. Chem.* **1975**, *40*, 1187.

[1586]Gassman, P.G.; Hodgson, P.K.G.; Balchunis, R.J. *J. Am. Chem. Soc.* **1976**, *98*, 1275.

[1587]Hibbert, F.; Malana, M.A. *J. Chem. Soc. Perkin Trans. 2* **1992**, 755.

[1588]For reviews, see O'Connor, C. *Q. Rev. Chem. Soc.* **1970**, *24*, 553; Talbot, R.J.E., in Bamford, C.H.; Tipper, C.F.H. *Comprehensive Chemical Kinetics*, Vol. 9, Elsevier, NY, **1973**, pp. 257–280; Challis, B.C.; Challis, J.C., in Zabicky, J. *The Chemistry of Amides*, Wiley, NY, **1970**, pp. 731–857.

cases. For basic catalysis,[1589] the mechanism is $B_{AC}2$.

$$
\underset{R}{\overset{O}{\underset{\quad}{\parallel}}}\!\!\!C\!\!-\!NR'_2 \; + \; ^\ominus OH \; \underset{k_{-1}}{\overset{k_1}{\rightleftharpoons}} \; R\!-\!\overset{OH}{\underset{O_\ominus}{\underset{|}{\overset{|}{C}}}}\!\!-\!NR'_2 \; \overset{k_2}{\longrightarrow} \; \underset{R}{\overset{O}{\underset{\quad}{\parallel}}}\!\!\!C\!\!-\!OH \; + \; R'_2N^\ominus \; \longrightarrow \; \underset{R}{\overset{O}{\underset{\quad}{\parallel}}}\!\!\!C\!\!-\!O^\ominus \; + \; R'_2NH
$$

94

(slow over the first equilibrium arrow)

There is much evidence for this mechanism, similar to that discussed for ester hydrolysis. A molecular-orbital study on the mechanism of amide hydrolysis is available.[1590] In certain cases, kinetic studies have shown that the reaction is second order in OH$^-$, indicating that **94** can lose a proton to give **95**.[1591] Depending on the nature

$$
R\!-\!\overset{OH}{\underset{O_\ominus}{\underset{|}{\overset{|}{C}}}}\!\!-\!NR'_2 \; \longrightarrow \; R\!-\!\overset{O^\ominus}{\underset{O_\ominus}{\underset{|}{\overset{|}{C}}}}\!\!-\!NR'_2
$$

94   **95**

path *a*: $\underset{R}{\overset{O}{\underset{\quad}{\parallel}}}\!\!\!C\!\!-\!O^\ominus \; + \; R'_2N^\ominus$

path *b* (BH$^+$): $\underset{R}{\overset{O}{\underset{\quad}{\parallel}}}\!\!\!C\!\!-\!O^\ominus \; + \; R'_2NH$

of R', **95** can cleave directly to give the two negative ions (path *a*) or become *N*-protonated prior to or during the act of cleavage (path *b*), in which case the products are obtained directly and a final proton transfer is not necessary.[1592] Studies of the effect, on the rate of hydrolysis and on the ratio k_{-1}/k_2, of substituents on the aromatic rings in a series of amides $CH_3CONHAr$ led to the conclusion that path *a* is taken when Ar contains electron-withdrawing substituents and path *b* when electron-donating groups are present.[1593] The presence of electron-withdrawing groups helps stabilize the negative charge on the nitrogen, so that NR'^-_2 can be a leaving group (path *a*). Otherwise, the C—N bond does not cleave until the nitrogen is protonated (either prior to or in the act of cleavage), so that the leaving group, *even in the base-catalyzed reaction*, is not NR'^-_2 but the conjugate NHR'_2(path *b*). Though we have shown formation of **94** as the rate-determining step in the $B_{AC}2$ mechanism, this is true only at high base concentrations. At lower concentrations of base, the cleavage of **107** or **95** becomes rate determining.[1594]

[1589]For a comprehensive list of references, see DeWolfe, R.H.; Newcomb, R.C. *J. Org. Chem.* **1971**, *36*, 3870.

[1590]Hori, K.; Kamimura, A.; Ando, K.; Mizumura, M.; Ihara, Y. *Tetrahedron* **1997**, *53*, 4317.

[1591]Biechler, S.S.; Taft, R.W. *J. Am. Chem. Soc.* **1957**, *79*, 4927. For evidence that a similar intermediate can arise in base-catalyzed ester hydrolysis see Khan, M.N.; Olagbemiro, T.O. *J. Org. Chem.* **1982**, *47*, 3695.

[1592]Eriksson, S.O. *Acta Chem. Scand.* **1968**, *22*, 892; *Acta Pharm. Suec.*, **1969**, *6*, 139.

[1593]Schowen, R.L.; Hopper, C.R.; Bazikian, C.M. *J. Am. Chem. Soc.* **1972**, *94*, 3095. Gani, V.; Viout, P. *Tetrahedron Lett.* **1972**, 5241; Menger, F.M.; Donohue, J.A. *J. Am. Chem. Soc.* **1973**, *95*, 432; Pollack, R.M.; Dumsha, T.C. *J. Am. Chem. Soc.* **1973**, *95*, 4463; Kijima, A.; Sekiguchi, S. *J. Chem. Soc. Perkin Trans. 2* **1987**, 1203.

[1594]Schowen, R.L.; Jayaraman, H.; Kershner, L. *J. Am. Chem. Soc.* **1966**, *88*, 3373. See also, Gani, V.; Viout, P. *Tetrahedron* **1976**, *32*, 1669, 2883; Bowden, K.; Bromley, K. *J. Chem. Soc. Perkin Trans. 2* **1990**, 2103.

For acid catalysis, matters are less clear. The reaction is generally second order, and it is known that amides are primarily protonated on the oxygen (Chapter 8, Ref. 24). Because of these facts it has been generally agreed that most acid-catalyzed amide hydrolysis takes place by the $A_{AC}2$ mechanism.

Further evidence for this mechanism is that a small but detectable amount of ^{18}O exchange (see p. 1256) has been found in the acid-catalyzed hydrolysis of benzamide.[1595] (^{18}O exchange has also been detected for the base-catalyzed process,[1596] in accord with the $B_{AC}2$ mechanism). Kinetic data have shown that three molecules of water are involved in the rate-determining step,[1597] suggesting that, as in the $A_{AC}2$ mechanism for ester hydrolysis (**16-59**), additional water molecules take part in a process, such as

The four mechanisms involving alkyl–N cleavage (the AL mechanisms) do not apply to this reaction. They are not possible for unsubstituted amides, since the only N–C bond is the acyl bond. They are possible for N-substituted and N,N-disubstituted amides, but in these cases they give entirely different products and are not amide hydrolyses at all.

[1595]McClelland, R.A. *J. Am. Chem. Soc.* **1975**, *97*, 5281; Bennet, A.J.; Ślebocka-Tilk, H.; Brown, R.S.; Guthrie, J.P.; Jodhan, A. *J. Am. Chem. Soc.* **1990**, *112*, 8497.

[1596]Bender, M.L.; Thomas, R.J. *J. Am. Chem. Soc.* **1961**, *83*, 4183, Bunton, C.A.; Nayak, B.; O'Connor, C. *J. Org. Chem.* **1968**, *33*, 572; lebocka-Tilk, H.; Bennet, A.J.; Hogg, H.J.; Brown, R.S. *J. Am. Chem. Soc.* **1991**, *113*, 1288; McClelland, R.A. *J. Am. Chem. Soc.* **1975**, *97*, 5281; Bennet, A.J.; Ślebocka-Tilk, H.; Brown, R.S.; Guthrie, J.P.; Jodhan, A. *J. Am. Chem. Soc.* **1990**, *112*, 8497.

[1597]Moodie, R.B.; Wale, P.D.; Whaite, K. *J. Chem. Soc.* **1963**, 4273; Yates, K.; Stevens, J.B. *Can. J. Chem.* **1965**, *43*, 529; Yates, K.; Riordan, J.C. *Can. J. Chem.* **1965**, *43*, 2328.

This reaction, while rare, has been observed for various *N-tert*-butylamides in 98% sulfuric acid, where the mechanism was the $A_{AL}1$ mechanism,[1598] and for certain amides containing an azo group, where a $B_{AL}1$ mechanism was postulated.[1599] Of the two first-order acyl cleavage mechanisms, only the $A_{AC}1$ has been observed, in concentrated sulfuric acid solutions.[1600] Of course, the diazotization of unsubstituted amides might be expected to follow this mechanism, and there is evidence that this is true.[1584]

OS **I**, 14, 111, 194, 201, 286; **II**, 19, 25, 28, 49, 76, 208, 330, 374, 384, 457, 462, 491, 503, 519, 612; **III**, 66, 88, 154, 256, 410, 456, 586, 591, 661, 735, 768, 813; **IV**, 39, 42, 55, 58, 420, 441, 496, 664; **V**, 27, 96, 341, 471, 612, 627; **VI**, 56, 252, 507, 951, 967; **VII**, 4, 287; **VIII**, 26, 204, 241, 339, 451.

The oxidation of aldehydes to carboxylic acids can proceed by a nucleophilic mechanism, but more often it does not. The reaction is considered in Chapter 19 (**19-23**). Basic cleavage of β-keto esters and the haloform reaction could be considered at this point, but they are also electrophilic substitutions and are treated in Chapter 12 (**12-43** and **12-44**).

B. Attack by OR at an Acyl Carbon

16-61   Alcoholysis of Acyl Halides

Alkoxy-de-halogenation

$$\underset{R}{\overset{\overset{\displaystyle O}{\parallel}}{\underset{}{C}}}\diagdown X \;+\; ROH \;\longrightarrow\; \underset{R}{\overset{\overset{\displaystyle O}{\parallel}}{\underset{}{C}}}\diagdown OR'$$

The reaction between acyl halides and alcohols or phenols is the best general method for the preparation of carboxylic esters. It is believed to proceed by a S_N2 mechanism.[1601] As with **16-57**, the mechanism can be S_N1 or tetrahedral.[1510] Pyridine catalyzes the reaction by the nucleophilic catalysis route (see **16-58**). Lewis acids such as lithium perchlorate can be used.[1602] The reaction is of wide scope, and many functional groups do not interfere. A base is frequently added to combine with the HX formed. When aqueous alkali is used, this is called the *Schotten–Baumann procedure*, but pyridine is also frequently used. Both R and R' may be primary, secondary, or tertiary alkyl or aryl. Enol esters can also be prepared by this method, though *C*-acylation competes in these cases. In difficult cases, especially with hindered acids or tertiary R', the alkoxide can be used instead of the alcohol.[1603] Activated alumina has also been used as a catalyst, for

[1598]Lacey, R.N. *J. Chem. Soc.* **1960**, 1633; Druet, L.M.; Yates, K. *Can. J. Chem.* **1984**, *62*, 2401.

[1599]Stodola, F.H. *J. Org. Chem.* **1972**, *37*, 178.

[1600]Duffy, J.A.; Leisten, J.A. *J. Chem. Soc.* **1960**, 545, 853; Barnett, J.W.; O'Connor, C.J. *J. Chem. Soc., Chem. Commun.* **1972**, 525; *J. Chem. Soc. Perkin Trans. 2* **1972**, 2378.

[1601]Bentley, T.W.; Llewellyn, G.; McAlister, J.A. *J. Org. Chem.* **1996**, *61*, 7927; Kevill, D.N.; Knauss, D.C. *J. Chem. Soc. Perkin Trans. 2* **1993**, 307; Fleming, I.; Winter, S.B.D. *Tetrahedron Lett.* **1993**, *34*, 7287.

[1602]Bandgar, B.P.; Kamble, V.T.; Sadavarte, V.S.; Uppalla, L.S. *Synlett* **2002**, 735.

[1603]For an example, see Kaiser, E.M.; Woodruff, R.A. *J. Org. Chem.* **1970**, *35*, 1198.

tertiary R'.[1604] Thallium salts of phenols give very high yields of phenolic esters,[1605] and BiOCl is very effective for the preparation of phenolic acetates.[1606] Phase-transfer catalysis has been used for hindered phenols.[1607] Zinc has been used to couple alcohols and acyl chlorides,[1608] and catalytic Cu(acac)$_2$ and benzoyl chloride was used to prepare the mono-benzoate of ethylene glycol.[1609] Selective acylation is possible in some cases.[1610]

Acyl halides react with thiols, in the presence of zinc, to give the corresponding thio-ester.[1611] The reaction of acid chlorides or anhydrides (see **16-62**) with diphenyldiselenide, in the presence of Sm/CoCl$_2$[1612] or Sm/CrCl$_3$[1613] gave the corresponding seleno ester (PhSeCOMe).

Acyl halides can also be converted to carboxylic acids by using ethers instead of alcohols, in MeCN in the presence of certain catalysts such as cobalt(II) chloride.[1614] A variation of this reaction has been reported that uses acetic anhydride.[1615]

$$\underset{R}{\overset{O}{\|}}\!\!-\!\!\overset{O}{C}\!\!-\!\!X \;\; + \;\; R'\!-\!O\!-\!R^2 \xrightarrow[\text{MeCN}]{\text{CoCl}_2} \underset{R}{\overset{O}{\|}}\!\!-\!\!\overset{O}{C}\!\!-\!\!OR' \;\; + \;\; R^2Cl$$

This is a method for the cleavage of ethers (see also, **10-49**).

OS **I**, 12; **III**, 142, 144, 167, 187, 623, 714; **IV**, 84, 263, 478, 479, 608, 616, 788; **V**, 1, 166, 168, 171; **VI**, 199, 259, 312, 824; **VII**, 190; **VIII**, 257, 516.

16-62 Alcoholysis of Anhydrides

Alkoxy-de-acyloxy-substitution

$$R\!-\!\overset{O}{\overset{\|}{C}}\!-\!O\!-\!\overset{O}{\overset{\|}{C}}\!-\!R^2 \;\; + \;\; R'OH \;\;\rightleftharpoons\;\; R\!-\!\overset{O}{\overset{\|}{C}}\!-\!OR' \;\; + \;\; HO\!-\!\overset{O}{\overset{\|}{C}}\!-\!R^2$$

The scope of this reaction is similar to that of **16-61**. Anhydrides are somewhat less reactive than acyl halides, and they are often used to prepare carboxylic esters. Benzyl acetates have been prepared via microwave irradiation of benzylic alcohols

[1604]Nagasawa, K.; Yoshitake, S.; Amiya, T.; Ito, K. *Synth. Commun.* **1990**, *20*, 2033.

[1605]Taylor, E.C.; McLay, G.W.; McKillop, A. *J. Am. Chem. Soc.* **1968**, *90*, 2422.

[1606]Ghosh, R.; Maiti, S.; Chakraborty, A. *Tetrahedron Lett.* **2004**, *45*, 6775.

[1607]Illi, V.O. *Tetrahedron Lett.* **1979**, 2431. For another method, see Nekhoroshev, M.V.; Ivakhnenko, E.P.; Okhlobystin, O.Yu. *J. Org. Chem. USSR* **1977**, *13*, 608.

[1608]Yadav, J.S.; Reddy, G.S.; Svinivas, D.; Himabindu, K. *Synth. Commun.* **1998**, *28*, 2337.

[1609]Sirkecioglu, O.; Karliga, B.; Talinli, N. *Tetrahedron Lett.* **2003**, *44*, 8483.

[1610]Srivastava, V.; Tandon, A.; Ray, S. *Synth. Commun.* **1992**, *22*, 2703.

[1611]Meshram, H.M.; Reddy, G.S.; Bindu, K.H.; Yadav, J.S. *Synlett* **1998**, 877.

[1612]Chen, R.; Zhang, Y. *Synth. Commun.* **2000**, *30* , 1331.

[1613]Liu, Y.; Zhang, Y. *Synth. Commun.* **1999**, *29* , 4043.

[1614]See Ahmad, S.; Iqbal, J. *Chem. Lett.* **1987**, 953, and references cited therein.

[1615]Lakouraj, M.; Movassaghi, B.; Fasihi, J. *J. Chem. Res. (S)* **2001**, 378.

and acetic anhydride.[1616] Acids,[1617] Lewis acids,[1618] and bases, such as pyridine are often used as catalysts.[1619] Acetic anhydride and $NiCl_2$ with microwave irradiation converts benzylic alcohols to the corresponding acetate.[1620] The monoacetate of 1,2-diols have been prepared using $CeCl_3$ as a catalyst.[1621] Pyridine is a nucleophilic-type catalyst (see **16-58**). 4-(N,N-Dimethylamino)pyridine is a better catalyst and can be used in cases where pyridine fails.[1622] N-Bromosuccinimide has been shown to catalyzed esterification of alcohols with acetic anhydride.[1623] Formic anhydride is not a stable compound but esters of formic acid can be prepared by treating alcohols[1624] or phenols[1625] with acetic-formic anhydride. Cyclic anhydrides give mono-esterified dicarboxylic acids, such as **96**.[1626] The asymmetric alcoholysis of cyclic anhydrides has been reviewed.[1627]

Alcohols can also be acylated by mixed organic–inorganic anhydrides, such as acetic-phosphoric anhydride, $MeCOOPO(OH)_2$[1628] (see **16-68**). Thioesters of the type $ArS(C{=}O)Me$ have been prepared from diphenyl disulfide and PBu_3, followed by treatment with acetic anhydride.[1629]

[1616]Bandgar, B.P.; Kasture, S.P.; Kamble, V.T. *Synth. Commun.* **2001**, *31*, 2255.

[1617]Nafion-H has been used: Kumareswaran, R.; Pachamuthu, K.; Vankar, Y.D. *Synlett* **2000**, 1652.

[1618]Some of the catalysts used are **Cu(OTf)₂**: Saravanan, P.; Singh, V.K. *Tetrahedron Lett.* **1999**, *40*, 2611. In(OTf)₃: Barrero, A.F.; Alvarez-Manzaneda, E.J.; Chahboun, R.; Meneses, R. *Synlett* **1999**, 1663. InCl₃: Chakraborti, A.K.; Gulhane, R. *Tetrahedron Lett.* **2003**, *44*, 6749. TiCl₄: Chandrasekhar, S.; Ramachandar, T.; Reddy, M.V.; Takhi, M. *J. Org. Chem.* **2000**, *65*, 4729. LiClO₄: Nakae, Y.; Kusaki, I.; Sato, T. *Synlett* **2001**, 1584; Ce(OTf)₃: Dalpozzo, R.; DeNino, A.; Maiuolo, L.; Procopio, A.; Nardi, M.; Bartoli, G.; Romeo, R. *Tetrahedron Lett.* **2003**, *44*, 5621. Yb(OTf)₃: Dumeunier, R.; Markó, I.E. *Tetrahedron Lett.* **2004**, *45*, 825. RuCl₃: De, S.K. *Tetrahedron Lett.* **2004**, *45*, 2919. Mg(ClO₄)₂: Bartoli, G.; Bosco, M.; Dalpozzo, R.; Marcantoni, E.; Massaccesi, M.; Sambri, L. *Eur. J. Org. Chem.* **2003**, 4611.

[1619]For a list of catalysts, with references, see Larock, R.C. *Comprehensive Organic Transformations*, 2nd ed., Wiley-VCH, NY, **1999**, pp. 1955–1957.

[1620]Constantinou-Kokotou, V.; Peristeraki, A. *Synth. Commun.* **2004**, *34*, 4227.

[1621]Clarke, P.A.; Kayaleh, N.E.; Smith, M.A.; Baker, J.R.; Bird, S.J.; Chan, C. *J. Org. Chem.* **2002**, *67*, 5226; Clarke, P.A. *Tetrahedron Lett.* **2002**, *43*, 4761.

[1622]For reviews, see Scriven, E.F.V. *Chem. Soc. Rev.* **1983**, *12*, 129; Höfle, G.; Steglich, W.; Vorbrüggen, H. *Angew. Chem. Int. Ed.* **1978**, *17*, 569.

[1623]Karimi, B.; Seradj, H. *Synlett* **2001**, 519.

[1624]For example, see Stevens, W.; van Es, A. *Recl. Trav. Chim. Pays-Bas* **1964**, *83*, 1287; van Es, A.; Stevens, W. *Recl. Trav. Chim. Pays-Bas* **1965**, *84*, 704.

[1625]For example, see Stevens, W.; van Es, A. *Recl. Trav. Chim. Pays-Bas* **1964**, *83*, 1294; Sōfuku, S.; Muramatsu, I.; Hagitani, A. *Bull. Chem. Soc. Jpn.* **1967**, *40*, 2942.

[1626]For conversion of an anhydride to a mono-ester with high enantioselectivity, see Chen, Y.; Tian, S.-K.; Deng, L. *J. Am. Chem. Soc.* **2000**, *122*, 9542.

[1627]Chen, Y.; McDaid, P.; Deng, L. *Chem. Rev.* **2003**, *103*, 2965.

[1628]Fatiadi, A.J. *Carbohydr. Res.* **1968**, *6*, 237.

[1629]Ayers, J.T.; Anderson, S.R. *Synth. Commun* **1999**, *29*, 351. This transformation has also been accomplished with Zn/AlCl₃: see Movassagh, B. Lakouraj, M.M.; Fadaei, Z. *J. Chem. Res. (S)* **2001**, 22.

OS **I**, 285, 418; **II**, 69, 124; **III**, 11, 127, 141, 169, 237, 281, 428, 432, 690, 833; **IV**, 15, 242, 304; **V**, 8, 459, 591, 887; **VI**, 121, 245, 560, 692; 486; **VIII**, 141, 258.

16-63   Esterification of Carboxylic Acids

Alkoxy-de-hydroxylation

$$RCOOH + R'OH \underset{}{\overset{H+}{\rightleftharpoons}} RCOOR' + H_2O$$

The esterification of carboxylic acids with alcohols[1630] is the reverse of **16-60** and can be accomplished only if a means is available to drive the equilibrium to the right.[1631] There are many ways of doing this, among which are (*1*) addition of an excess of one of the reactants, usually the alcohol; (*2*) removal of the ester or the water by distillation; (*3*) removal of water by azeotropic distillation; and (*4*) removal of water by use of a dehydrating agent, silica gel,[1632] or a molecular sieve. When R′ is methyl, the most common way of driving the equilibrium is by adding excess MeOH; when R′ is ethyl, it is preferable to remove water by azeotropic distillation.[1633] The most common catalysts are H_2SO_4 and TsOH, although some reactive acids (e.g., formic,[1634] trifluoroacetic[1635]) do not require a catalyst. Ammonium salts have been used to initiate esterification,[1636] and boric acid has been used to esterify α-hydroxy acids.[1637] The R′ group may be primary or secondary alkyl groups other than methyl or ethyl, but tertiary alcohols usually give carbocations and elimination. Phenols can sometimes be used to prepare phenolic esters, but yields are generally very low. Selective esterification of an aliphatic carboxylic acid in the presence of an aromatic acid was accomplished with $NaHSO_4 \cdot SiO_2$ and methanol.[1638]

Diphenylammonium triflate was useful for direct esterification of carboxylic acids with longer chain aliphatic alcohols.[1639] Photoirradiation of carboxylic acid with CBr_4[1640] or CCl_4[1641] in methanol was shown to give the methyl ester, with high selectivity for nonconjugated acids in the case of CBr_4. O-Alkylisoureas react

[1630]For a review of some methods, see Haslam, E. *Tetrahedron* **1980**, *36*, 2409.

[1631]For a list of reagents, with references, see Larock, R.C. *Comprehensive Organic Transformations*, 2nd ed., Wiley-VCH, NY, *1999*, pp. 1932–1941.

[1632]Nascimento, M.de G.; Zanotto, S.P.; Scremin, M.; Rezende, M.C. *Synth. Commun.* **1996**, *26*, 2715.

[1633]Newman, M.S. *An Advanced Organic Laboratory Course*; Macmillan, NY, *1972*, pp. 8–10.

[1634]Formates can be prepared if diisopropyl ether is used to remove water by azeotropic distillation: Werner, W. *J. Chem. Res. (S) 1980*, 196. For an alternative synthesis of formate esters using trifluoroethyl formate, see Hill, D.R.; Hsiao, C.-N.; Kurukulasuriya, R.; Wittenberger, S.J. *Org. Lett.* **2002**, *4*, 111.

[1635]Johnston, B.H.; Knipe, A.C.; Watts, W.E. *Tetrahedron Lett.* **1979**, 4225.

[1636]Wakasugi, K.; Nakamura, A.; Tanabe, Y. *Tetrahedron Lett.* **2001**, *42*, 7427; Gacem, B.; Jenner, G. *Tetrahedron Lett.* **2003**, *44*, 1391.

[1637]Houston, T.A.; Wilkinson, B.L.; Blanchfield, J.T. *Org. Lett.* **2004**, *6*, 679.

[1638]Das, B.; Venkataiah, B.; Madhsudhan, P. *Synlett 2000*, 59.

[1639]Wakasugi, K.; Misaki, T.; Yamada, K.; Tanabe, Y. *Tetrahedron Lett.* **2000**, *41*, 5249.

[1640]Lee, A.S.-Y.; Yang, H.-C.; Su, F.-Y. *Tetrahedron Lett.* **2001** *42*, 301.

[1641]Hwu, J.R.; Hsu, C.-Y.; Jain, M.L. *Tetrahedron Lett.* **2004** *45*, 5151.

with conjugated carboxylic acids to give the corresponding ester with microwave irradiation,[1642] and a polymer-bound O-alkylurea has been used as well.[1643]

Mixing the carboxylic acid and alcohol with p-toluenesulfonic acid (neat), gave the ester in 3 min with microwave irradiation.[1644] Esterification has also been accomplished using ionic liquids as the reaction medium,[1645] and a solid-state esterification was reported on P_2O_5/SiO_2.[1646] Diols are converted to the mono-acetate by heating with acetic acid on a zeolite.[1647]

Both γ- and δ-hydroxy acids such as **97** are easily converted to a lactone by treatment with acids, or often simply on standing, but larger and smaller lactone rings cannot be made in this manner, because polyester formation occurs more readily.[1648] Often the conversion of a group, such as keto or halogen, γ or δ to a carboxyl group, to a hydroxyl group gives the lactone directly, since the hydroxy acid cyclizes too rapidly for isolation. β-Substituted β-hydroxy acids can be converted to β-lactones by treatment with benzenesulfonyl chloride in pyridine at 0–5°C.[1649] ε-Lactones (seven-membered rings) have been made by cyclization of ε-hydroxy acids at high dilution.[1650] Macrocyclic lactones[1651] can be prepared indirectly in very good yields by conversion of the hydroxy acids to 2-pyridinethiol esters and adding these to refluxing xylene.[1652] Palladium-catalyzed aromatic carboxylation

[1642]Crosignani, S.; White, P.D.; Linclau, B. *Org. Lett.* **2002**, *4*, 2961.

[1643]Crosignani, S.; White, P.D.; Steinauer, R.; Linclau, B. *Org. Lett.* **2003**, *5*, 853; Crosignani, S.; White, P.D.; Linclau, B. *J. Org. Chem.* **2004**, *69*, 5897.

[1644]Loupy, A.; Petit, A.; Ramdan, M.; Yvanaeff, C.; Majdoub, M.; Labiad, B.; Villemin, D. *Can. J. Chem.* **1993**, *71*, 90. See also, Zhang, Z.; Zhou, L.; Zhang, M.; Wu, H.; Chen, Z. *Synth. Commun.* **2001**, *31*, 2435; Fan, X.; Yuan, K.; Hao, C. Li, N.; Tan, G.; Yu, X. *Org. Prep. Proceed. Int.* **2000**, *32*, 287.

[1645]Isobe, T.; Ishikawa, T. *J. Org. Chem.* **1999**, *64*, 6984.

[1646]Eshghi, H.; Rafei, M.; Karimi, M.H. *Synth. Commun.* **2001**, *31*, 771.

[1647]Srinivas, K.V.N.S.; Mahender, I.; Das, B. *Synlett 2003*, 2419.

[1648]For a review of the synthesis of lactones and lactams, see Wolfe, J.F.; Ogliaruso, M.A., in Patai, S. *The Chemistry of Acid Derivatives*, pt. 2, Wiley, NY, *1979*, pp. 1062–1330. For a list of methods for converting hydroxy acids to lactones, with references, see Larock, R.C. *Comprehensive Organic Transformations*, 2nd ed., Wiley-VCH, NY, *1989*, pp. 1861–1867.

[1649]Adam, W.; Baeza, J.; Liu, J. *J. Am. Chem. Soc.* **1972**, *94*, 2000. For other methods of converting β-hydroxy acids to β-lactones, see Merger, F. *Chem. Ber.* **1968**, *101*, 2413; Blume, R.C. *Tetrahedron Lett.* **1969**, 1047.

[1650]Lardelli, G.; Lamberti,V.; Weller, W.T.; de Jonge, A.P. *Recl. Trav. Chim. Pays-Bas 1967*, *86*, 481.

[1651]For reviews on the synthesis of macrocyclic lactones, see Nicolaou, K.C. *Tetrahedron 1977*, *33*, 683; Back, T.G. *Tetrahedron 1977*, *33*, 3041; Masamune, S.; Bates, G.S.; Corcoran, J.W. *Angew. Chem. Int. Ed. 1977*, *16*, 585.

[1652]Corey, E.J.; Brunelle, D.J.; *Tetrahedron Lett. 1976*, 3409; Wollenberg, R.H.; Nimitz, J.S.; Gokcek, D.Y. *Tetrahedron Lett. 1980*, *21*, 2791; Thalmann, A.; Oertle, K.; Gerlach, H. *Org. Synth. VII*, 470. See also, Schmidt, U.; Heermann, D. *Angew. Chem. Int. Ed. 1979*, *18*, 308. For a ruthenium-catalyzed macrocyclization see Trost, B.M.; Chisholm, J.D. *Org. Lett. 2002*, *4*, 3743.

reactions generated carboxylic acids *in situ*, and when an alcohol unit is present elsewhere in the molecule cyclization gives the corresponding lactone.[1653]

A closely related method, which often gives higher yields of a macrocyclic lactone, involves treatment of the hydroxy acids with 1-methyl- or 1-phenyl-2-halopyridinium salts, especially 1-methyl-2-chloropyridinium iodide (*Mukaiyama's reagent*).[1654] Another method uses organotin oxides[1655] and both $TiCl_4/AgClO_4$[1656] and $TiCl_2(OSO_2CF_3)_2$[1657] have been used.

Esterification is catalyzed by acids (not bases) in ways that were discussed on p. 1402.[1549] The mechanisms are usually $A_{AC}2$, but $A_{AC}1$ and $A_{AL}1$ have also been observed.[1658] Certain acids, such as 2,6-di-ortho-substituted benzoic acids, cannot be esterified by the $A_{AC}2$ mechanism because of steric hindrance (p. 481). In such cases, esterification can be accomplished by dissolving the acid in 100% H_2SO_4 (forming the ion RCO^+) and pouring the solution into the alcohol ($A_{AC}1$ mechanism). The reluctance of hindered acids to undergo the normal $A_{AC}2$ mechanism can sometimes be put to advantage when, in a molecule containing two COOH groups,

[1653]Kayaki, Y.; Noguchi, Y.; Iwasa, S.; Ikariya, T.; Noyori, R. *Chem. Commun.* **1999**, 1235.

[1654]For a review of reactions with this and related methods, see Mukaiyama, T. *Angew. Chem. Int. Ed.* **1979**, *18*, 707. For a polymer-supported Mukaiyama reagent, see Convers, E.; Tye, H.; Whittaker, M. *Tetrahedron Lett.* **2004**, *45*, 3401.

[1655]Steliou, K.; Szczygielska-Nowosielska, A.; Favre, A.; Poupart, M.A.; Hanessian, S. *J. Am. Chem. Soc.* **1980**, *102*, 7578; Steliou, K.; Poupart, M.A. *J. Am. Chem. Soc.* **1983**, *105*, 7130. For some other methods, see Masamune, S.; Kamata, S.; Schilling, W. *J. Am. Chem. Soc.* **1975**, *97*, 3515; Scott, L.T.; Naples, J.O. *Synthesis* **1976**, 738; Kurihara, T.; Nakajima, Y.; Mitsunobu, O. *Tetrahedron Lett.* **1976**, 2455; Corey, E.J.; Brunelle, D.J.; Nicolaou, K.C. *J. Am. Chem. Soc.* **1977**, *99*, 7359; Vorbrüggen, H.; Krolikiewicz, K. *Angew. Chem. Int. Ed.* **1977**, *16*, 876; Nimitz, J.S.; Wollenberg, R.H. *Tetrahedron Lett.* **1978**, 3523; Inanaga, J.; Hirata, K.; Saeki, H.; Katsuki, T.; Yamaguchi, M. *Bull. Chem. Soc. Jpn.* **1979**, *52*, 1989; Venkataraman, K.; Wagle, D.R. *Tetrahedron Lett.* **1980**, *21*, 1893; Schmidt, U.; Dietsche, M. *Angew. Chem. Int. Ed.* **1981**, *20*, 771; Taniguchi, N.; Kinoshita, H.; Inomata, K.; Kotake, H. *Chem. Lett.* **1984**, 1347; Cossy, J.; Pete, J. *Bull. Soc. Chim. Fr.* **1988**, 989.

[1656]Shiina, I.; Miyoshi, S.; Miyashita, M.; Mukaiyama, T. *Chem. Lett.* **1994**, 515; Mukaiyama, T.; Izumi, J.; Miyashita, M.; Shiina, I. *Chem. Lett.* **1993**, 907.

[1657]Hojo, M.; Nagayoshi, M.; Fujii, A.; Yanagi, T.; Ishibashi, N.; Miura, K.; Hosomi, A. *Chem. Lett.* **1994**, 719.

[1658]For a review of aspects of the mechanism, see Salomaa, P.; Kankaanperä, A.; Pihlaja, K., in Patai, S. *The Chemistry of the Hydroxyl Group, pt. 1*, Wiley, NY, **1971**, pp. 466–481.

only the less hindered one is esterified. The $A_{AC}1$ pathway cannot be applied to unhindered carboxylic acids.

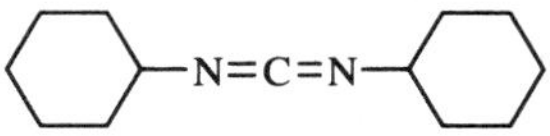

DCC 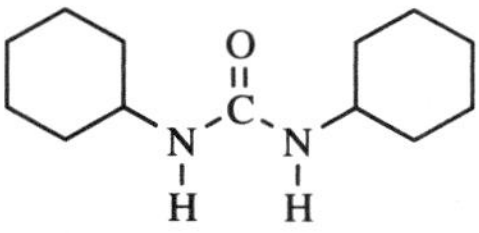   DHU

   Another way to esterify a carboxylic acid is to treat it with an alcohol in the presence of a dehydrating agent.[1631] One of these is dicyclohexylcarbodiimide (DCC), which is converted in the process to dicyclohexylurea (DHU). The mechanism[1659] has much in common with the nucleophilic catalysis mechanism; the acid is converted to a compound with a better leaving group. However, the conversion is not by a tetrahedral mechanism (as it is in nucleophilic catalysis), since the C—O bond remains intact during this step:

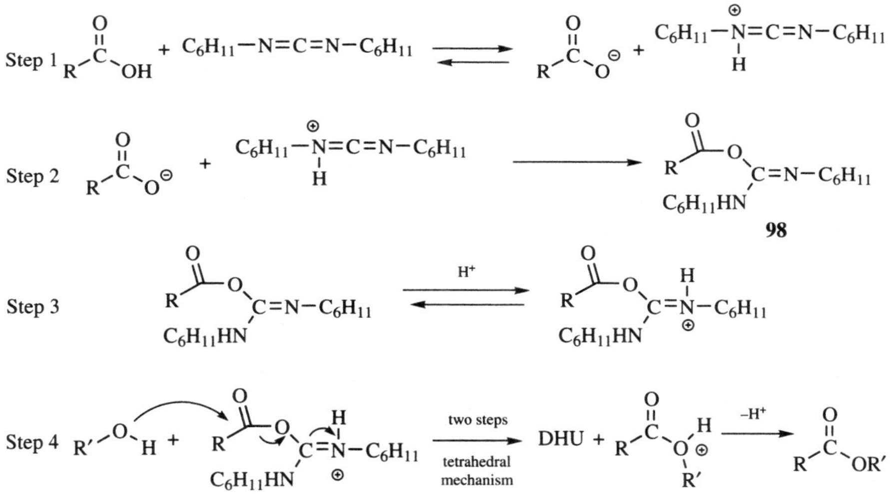

Evidence for this mechanism was the preparation of O-acylureas similar to **98** and the finding that when catalyzed by acids they react with alcohols to give esters.[1660] Hindered tertiary alcohols can be coupled via DCC to give the hindered ester.[1661] A polymer-bound carbodiimide has been used to prepare macrocyclic lactones.[1662] In at least one case, the reaction of $HOOCCH_2CN$ with DCC and *tert*-butanol gave the *tert*-butyl ester via a ketene intermediate.[1663]

[1659]Smith, M.; Moffatt, J.G.; Khorana, H.G. *J. Am. Chem. Soc.* **1958**, *80*, 6204; Balcom, B.J.; Petersen, N.O. *J. Org. Chem.* **1989**, *54*, 1922.
[1660]Doleschall, G.; Lempert, K. *Tetrahedron Lett.* **1963**, 1195.
[1661]Shimizu, T.; Hiramoto, K.; Nakata, T. *Synthesis* **2001**, 1027.
[1662]Keck, G.E.; Sanchez, C.; Wager, C.A. *Tetrahedron Lett.* **2000**, *41*, 8673.
[1663]Nahmany, M.; Melman, A. *Org. Lett.* **2001**, *3*, 3733.

There are limitations to the use of DCC; yields are variable and N-acylureas are side products. Many other dehydrating agents[1664] have been used, including DCC and an aminopyridine,[1665] Amberlyst-15,[1666] chlorosilanes,[1667] $MeSO_2Cl$-Et_3N,[1668] and N,N'-carbonyldiimidazole(**99**).[1669] In the latter case, imidazolides (**100**) are intermediates that readily react with alcohols.

99   **100**

It is known that the Lewis acid BF_3 promotes the esterification by converting the acid to RCO^+—BF_3 ^-OH, so the reaction proceeds by an $A_{AC}1$ type of mechanism. The use of BF_3-etherate is simple and gives high yields.[1670] Other Lewis acids can be used.[1671] Esterification has been done using a LaY zeolite.[1672]

Carboxylic esters can also be prepared by treating carboxylic acids with *tert*-butyl ethers and acid catalysts.[1673]

$$RCOOH \ + \ t\text{-Bu-OR}' \longrightarrow RCOOR' \ + \ H_2C=CMe_2 \ + \ H_2O$$

Carboxylic acids can be converted to *tert*-butyl esters by treatment with *tert*-butyl 2,2,2-trichloroacetimidate (see **10-10**) and $BF_3 \cdot OEt_2$.[1173] Carboxylic esters can be formed from the carboxylate anion and a suitable alkylating agent (**10-26**).

Thioesters of the type $RSC(=S)R'$ (a dithiocarboxylic ester) and $RSC(C=O)R'$ (a thiocarboxylic ester) can be generated by reaction of carboxylic acids with thiols.

[1664]For a list of many of these with references, see Arrieta, A.; García, T.; Lago, J.M.; Palomo, C. *Synth. Commun.* **1983**, *13*, 471.

[1665]Hassner, A.; Alexanian, V. *Tetrahedron Lett.* **1978**, 4475; Neises, B.; Steglich, W. *Angew. Chem. Int. Ed.* **1978**, *17*, 522; Boden, E.P.; Keck, G.E. *J. Org. Chem.* **1985**, *50*, 2394.

[1666]Petrini, M.; Ballini, R.; Marcantoni, E.; Rosini, G. *Synth. Commun.* **1988**, *18*, 847.

[1667]Nakao, R.; Oka, K.; Fukumoto, T. *Bull. Chem. Soc. Jpn.* **1981**, *54*, 1267; Brook, M.A.; Chan, T.H. *Synthesis* **1983**, 201.

[1668]Chandrasekaran, S.; Turner, J.V. *Synth. Commun.* **1982**, *12*, 727.

[1669]For a review, see Staab, H.A.; Rohr, W. *Newer Methods Prep. Org. Chem.* **1968**, *5*, 61. See also, Morton, R.C.; Mangroo, D.; Gerber, G.E. *Can. J. Chem.* **1988**, *66*, 1701.

[1670]For examples, see Marshall, J.L.; Erickson, K.C.; Folsom, T.K. *Tetrahedron Lett.* **1970**, 4011; Kadaba, P.K. *Synthesis* **1972**, 628; *Synth. Commun.* **1974**, *4*, 167.

[1671]Lewis acids used for esterification include FeCl₃: Sharma, G.V.M.; Mahalingam, A.K.; Nagarajan, M.; Ilangovan, P.; Radhakrishna, P. *Synlett* **1999**, 1200. **Hf(Cl₄(thf)₂**: Ishihara, K.; Nakayama, M.; Ohara, S.;Yamamoto, H. *Synlett* **2001**, 1117 and Ishihara, K.; Nakayama, M.; Ohara, S.; Yamamoto, H. *Tetrahedron* **2002**, *58*, 8179; **Bi(OTf)₃•x H₂O**: Carrigan,. D.; Freiberg, D.A.; Smith, R.C.; Zerth, H.M.; Mohan, R.S. *Synthesis* **2001**, 2091; **BiCl₃**: Mohammadpoor-Baltork, I.; Khosropour, A.R.; Aliyan, H. *J. Chem. Res* **2001**, 280; **Fe₂(SO₄)₃•x H₂O**: Zhang, G.-S. *Synth. Commun.* **1999**, *29*, 607. **Ceric ammonium nitrate**: Pan, W.-B.; Chang, F.-R.; Wei, L.-M.; Wu, M.J.; Wu, Y.-C. *Tetrahedron Lett.* **2003**, *44*, 331.

[1672]Narender, N.; Srinivasu, P.; Kulkarni, S.J.; Raghavan, K.V. *Synth. Commun.* **2000**, *30* , 1887.

[1673]Derevitskaya, V.A.; Klimov, E.M.; Kochetkov, N.K. *Tetrahedron Lett.* **1970**, 4269. See also, Mohacsi, E. *Synth. Commun.* **1982**, *12*, 453.

In one example, phosphorous pentasulfide was used in conjunction with a thiol to make dithiocarboxylic esters[1674] or thiocarboxylic esters.[1675] Thiocarboxylic esters were prepared from thiols and triflic acid.[1676]

OS **I**, 42, 138, 237, 241, 246, 254, 261, 451; **II**, 260, 264, 276, 292, 365, 414, 526; **III**, 46, 203, 237, 381, 413, 526, 531, 610; **IV**, 169, 178, 302, 329, 390, 398, 427, 506, 532, 635, 677; **V**, 80, 762, 946; **VI**, 471, 797; **VII**, 93, 99, 210, 319, 356, 386, 470; **VIII**, 141, 251, 597; **IX**, 24, 58; **75**, 116; **75**, 129. Also see, OS **III**, 536, 742.

16-64   Transesterification

Alkoxy-de-alkoxylation

$$\underset{R}{\overset{O}{\underset{\|}{C}}}\!\!-\!OR^1 \;+\; R^2OH \;\underset{\longleftarrow}{\overset{H^+ \text{ or } {}^-OH}{\longrightarrow}}\; \underset{R}{\overset{O}{\underset{\|}{C}}}\!\!-\!OR^2 \;+\; R^1OH$$

Transesterification[1677] is catalyzed[1678] by acids[1679] or bases,[1680] or performed under neutral conditions.[1681] It is an equilibrium reaction and must be shifted in the desired direction. In many cases low-boiling esters can be converted to higher boiling ones by the distillation of the lower boiling alcohol as fast as it is formed. Reagents used to catalyze[1682] transesterification include Montmorillonite K10[1683] and various Lewis acids.[1684] A polymer-bound siloxane has been used to induce transesterification.[1685] This reaction has been used as a method for the acylation of a primary OH in the presence of a secondary OH.[1686] Regioselectivity has

[1674]Sudalai, A.; Kanagasabapathy, S.; Benicewicz, B.C. *Org. Lett.* **2000**, *2*, 3213.

[1675]Curphey, T.J. *Tetrahedron Lett.* **2002**, *43*, 371.

[1676]Iimura, S.; Manabe, K.; Kobayashi, S. *Chem. Commun.* **2002**, 94.

[1677]Otera, J. *Chem. Rev.* **1993**, *93*, 1449.

[1678]For a list of catalysts, with references, see Larock, R.C. *Comprehensive Organic Transformations*, 2nd ed., Wiley-VCH, NY, **1999**, pp. 1969–1973.

[1679]Catalysts other than mineral acids can be used, for example, Amberlyst-15 resin. See Chavan, S.P.; Subbarao, Y.T.; Dantale, S.W.; Sivappa, R. *Synth. Commun.* **2001**, *31*, 289.

[1680]Stanton, M.G.; Gagné, M.R. *J. Org. Chem.* **1997**, *62*, 8240; Vasin, V.A.; Razin, V.V. *Synlett* **2001**, 658.

[1681]For some methods of transesterification under neutral conditions, see Otera, J.; Yano, T.; Kawabata, A.; Nozaki, H. *Tetrahedron Lett.* **1986**, *27*, 2383; Imwinkelried, R.; Schiess, M.; Seebach, D. *Org. Synth.*, *65*, 230; Bandgar, B.P.; Uppalla, L.S.; Sadavarte, V.S. *Synlett* **2001**, 1715.

[1682]For a review see Grasa, G.A.; Singh, R.; Nolan, S.P. *Synthesis* **2004**, 971.

[1683]Ponde, D.E.; Deshpande, V.H.; Bulbule, V.J.; Sudalai, A.; Gajare, A.S. *J. Org. Chem.* **1998**, *63*, 1058.

[1684]Lewis acids used for this reaction include **Ti(OEt)₄**: Krasik, P. *Tetrahedron Lett.* **1998**, *39*, 4223. **TlCl₄**: Mahrwald, R.; Quint, S. *Tetrahedron* **2000**, *56*, 7463. **Cu(NO₃)₂**: Iranpoor, N.; Firouzabadi, H.; Zolfigol, M.A. *Synth. Commun.* **1998**, *28*, 1923. **Sn(OTf)₂**: Mukaiyama, T.; Shiina, I.; Miyashita, M. *Chem. Lett.* **1992**, 625. **Yb(OTf)₃**: Sharma, G.V.M.; Ilangovan, A. *Synlett* **1999**, 1963. **LiClO₄**: Bandgar, B.P.; Sadavarte, V.S.; Uppalla, L.S. *Synlett* **2001**, 1338. **FeSO₄**: Bandgar, B.P.; Sadavarte, V.S.; Uppalla, L.S. *Synth. Commun.* **2001**, *31*, 2063. **Ceric ammonium nitrate**: Štefane, B.; Kočevar, M.; Polanc, S. *Synth. Commun.* **2002**, *32*, 1703.

[1685]Hagiwara, H.; Koseki, A.; Isobe, K.; Shimizu, K.-i.; Hoshi, T.; Suzuki, T. *Synlett* **2004**, 2188.

[1686]Yamada, S. *Tetrahedron Lett.* **1992**, *33*, 2171. See also, Costa, A.; Riego, J.M. *Can. J. Chem.* **1987**, *65*, 2327.

also been accomplished by using enzymes (lipases) as catalysts.[1687] Lactones, such as **101**, are easily opened by treatment with alcohols[1688] to give open-chain hydroxy esters.

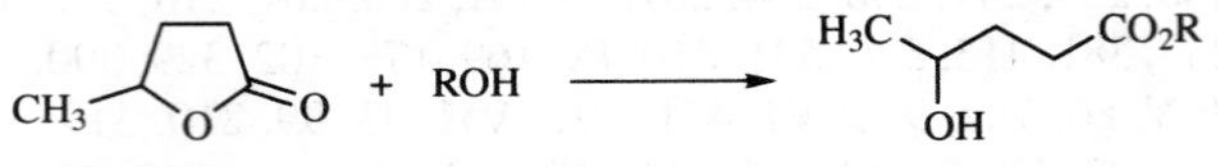

Transesterification has been carried out with phase-transfer catalysts, without an added solvent.[1689] Nonionic superbases (see p. 365) of the type $P(RNCH_2CH_2)_3N$ catalyze the transesterification of carboxylic acid esters at 25°C.[1690] Silyl esters ($R'CO_2SiR_3$) have been converted to alkyl esters ($R'CO_2R$) via reaction with alkyl halides and tetrabutylammonium fluoride.[1691] Thioesters are converted to phenolic esters by treatment with triphosgene–pyridine and then phenol.[1692]

Transesterification occurs by mechanisms[1693] that are identical with those of ester hydrolysis, except that ROH replaces HOH (by the acyl-oxygen fission mechanisms). When alkyl fission takes place, the products are the *acid* and the *ether*:

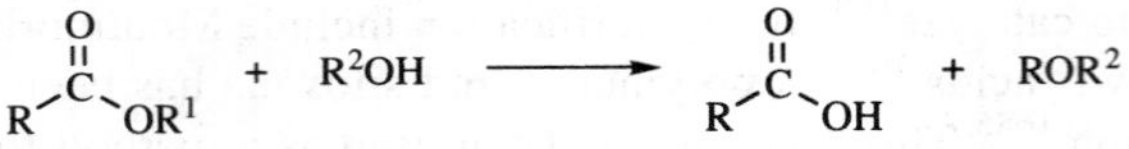

Therefore, transesterification reactions frequently fail when R' is tertiary, since this type of substrate most often reacts by alkyl–oxygen cleavage. In such cases, the reaction is of the Williamson type with OCOR as the leaving group (see **10-10**).

With enol esters such as **102**, reaction with an alcohol gives an ester and the enol of a ketone, which readily tautomerizes to the ketone as shown. Hence, enol esters are good acylating agents for alcohols.[1694] This transformation has been

[1687]Wong, C.H.; Whitesides, G. M. in Baldwin, J.E. *Enzymes in Synthetic Organic Chemistry, Tetrahedron Organic Chemistry Series* Vol. 12, Pergamon Press, NY, *1994*; Faber, K. *Biotransformations in Organic Chemistry. A Textbook*, 2nd ed; Springer-Verlag, NY, *1995*; Córdova, A.; Janda, K.D. *J. Org. Chem.* *2001*, 66, 1906; Ciuffreda, P.; Casati, S.; Santaniello, E. *Tetrahedron Lett.* *2003*, 44, 3663.

[1688]Anand, R.C.; Sevlapalam, N. *Synth. Commun.* *1994*, 24, 2743.

[1689]Barry, J.; Bram, G.; Petit, A. *Tetrahedron Lett.* *1988*, 29, 4567. See also, Nishiguchi, T.; Taya, H. *J. Chem. Soc. Perkin Trans. 1* *1990*, 172.

[1690]Ilankumaran, P.; Verkade, J.G. *J. Org. Chem.* *1999*, 64, 3086.

[1691]Ooi, T.; Sugimoto, H.; Maruoka, K. *Heterocycles* *2001*, 54, 593.

[1692]Joshi, U.M.; Patkar, L.N.; Rajappa, S. *Synth. Commun.* *2004*, 34, 33.

[1693]For a review, see Koskikallio, E.A., in Patai, S. *The Chemistry of Carboxylic Acids and Esters*, Wiley, NY, *1969*, pp. 103–136.

[1694]Rothman, E.S.; Hecht, S.S.; Pfeffer, P.E.; Silbert, L.S. *J. Org. Chem.* *1972*, 37, 3551; Ilankumaran, P.; Verkade, J.G. *J. Org. Chem.* *1999*, 64, 9063.

accomplished in ionic liquid media,[1695] and there is a $PdCl_2/CuCl_2$ mediated version.[1696] Isopropenyl acetate can also be used to convert other ketones to the corresponding enol acetates in an exchange reaction:[1697]

Enol esters can also be prepared in the opposite type of exchange reaction, catalyzed by mercuric acetate[1698] or Pd(II) chloride,[1699] for example,

$$RCOOH \ + \ R'COOCH=CH_2 \ \underset{}{\overset{Hg(OAc)_2 \atop H_2SO_4}{\rightleftharpoons}} \ RCOOCH=CH_2 \ + \ R'COOH$$

A closely related reaction is equilibration of a dicarboxylic acid and its diester to produce monoesters: The reaction of a carboxylic acid with ethyl acetate, in the presence of $NaHSO_4 \cdot SiO_2$, was shown to give the corresponding ethyl ester.[1700] Iodine catalyzes the transesterification of β-keto esters.[1701]

OS **II**, 5, 122, 360; **III**, 123, 146, 165, 231, 281, 581, 605; **IV**, 10, 549, 630, 977; **V**, 155, 545, 863; **VI**, 278; **VII**, 4, 164, 411; **VIII**, 155, 201, 235, 263, 350, 444, 528. See also, OS **VII**, 87; **VIII**, 71.

16-65   Alcoholysis of Amides

Alkoxy-de-amidation

Alcoholysis of amides is possible,[1702] although it is usually difficult. It has been most common with the imidazolide type of amides (e.g., **100**). For other amides, an activating agent is usually necessary before the alcohol will replace the NR_2 unit. Dimethylformamide, however, reacted with primary alcohols in the presence of 2,4,6-trichloro-1,3,5-pyrazine (cyanuric acid) to give the corresponding formate ester.[1703] Treatment of an amide with triflic anhydride ($CF_3SO_2OSO_2CF_3$) in the

[1695]Grasa, G.A.; Kissling, R.M.; Nolan, S.P. *Org. Lett.* **2002**, *4*, 3583.

[1696]Bosco, J.W.J.; Saikia, A.K. *Chem. Commun.* **2004**, 1116.

[1697]For examples, see Deghenghi, R.; Engel, C.R. *J. Am. Chem. Soc.* **1960**, *82*, 3201; House, H.O.; Trost, B.M. *J. Org. Chem.* **1965**, *30*, 2502.

[1698]For example, see Hopff, H.; Osman, M.A. *Tetrahedron* **1968**, *24*, 2205, 3887; Mondal, M.A.S.; van der Meer, R.; German, A.L.; Heikens, D. *Tetrahedron* **1974**, *30*, 4205.

[1699]Henry, P.M. *J. Am. Chem. Soc.* **1971**, *93*, 3853; *Acc. Chem. Res.* **1973**, *6*, 16.

[1700]Das, B.; Venkataiah, B. *Synthesis* **2000**, 1671.

[1701]Chavan, S.P.; Kale, R.R.; Shivasankar, K.; Chandake, S.I.; Benjamin, S.B. *Synthesis* **2003**, 2695.

[1702]For example, see Czarnik, A.W. *Tetrahedron Lett.* **1984**, *25*, 4875. For a list of references, see Larock, R.C. *Comprehensive Organic Transformations*, 2nd ed., Wiley-VCH, NY, **1999**, pp. 197–1978.

[1703]DeLuca, L.; Giacomelli, G.; Porcheddu, A. *J. Org. Chem.* **2002**, *67*, 5152.

presence of pyridine and then with an excess of alcohol leads to the ester,[1704] as does treatment with $Me_2NCH(OMe)_2$ followed by the alcohol.[1705] Trimethyloxonium tetrafluoroborate converted primary amides to methyl esters.[1706] The reaction of acetanilide derivatives with sodium nitrite in the presence of acetic anhydride–acetic acid leads to phenolic acetates.[1707] Acyl hydrazides $(RCONHNH_2)$ were converted to esters by reaction with alcohols and various reagents,[1708] and methoxyamides (RCONHOMe) were converted to esters with $TiCl_4/ROH$.[1709] The reaction of an oxazolidinone amide **103** with methanol and 10% $MgBr_2$ gave the corresponding methyl ester.[1710]

103

C. Attack by OCOR at an Acyl Carbon

16-66 Acylation of Carboxylic Acids With Acyl Halides

Acyloxy-de-halogenation

$$RCOCl \quad + \quad R'COO^- \quad \longrightarrow \quad RCOOCOR'$$

Unsymmetrical, as well as symmetrical, anhydrides are often prepared by the treatment of an acyl halide with a carboxylic acid salt. Cobalt(II) chloride $(CoCl_2)$ has been used as a catalyst.[1711] If a metallic salt is used, Na^+, K^+, or Ag^+ are the most common cations, but more often pyridine or another tertiary amine is added to the free acid and the resulting salt is subsequently treated with the acyl halide. Mixed formic anhydrides are prepared from sodium formate and an aryl halide, by use of a solid-phase copolymer of pyridine-1-oxide.[1712] Symmetrical anhydrides can be prepared by reaction of the acyl halide with aq. NaOH or

[1704]Charette, A.B.; Chua, P. *Synlett* **1998**, 163.

[1705]Anelli, P.L.; Brocchetta, M.; Palano, D.; Visigalli, M. *Tetrahedron Lett.* **1997**, *38*, 2367.

[1706]Kiessling, A.J.; McClure, C.K. *Synth. Commun.* **1997**, *27*, 923.

[1707]Glatzhofer, D.T.; Roy, R.R.; Cossey, K.N. *Org. Lett.* **2002**, *4*, 2349.

[1708]Prakash, O.; Sharma, V.; Sadana, A. *J. Chem. Res. (S)* **1996**, 100; Štefane, B.; Koevar, M.; Polanc, S. *Tetrahedron Lett.* **1999**, *40*, 4429; Yamaguchi, J.-i.; Aoyagi, T.; Fujikura, R.; Suyama, T. *Chem. Lett.* **2001**, 466.

[1709]Fisher, L.E.; Caroon, J.M.; Stabler, S.R.; Lundberg, S.; Zaidi, S.; Sorensen, C.M.; Sparacino, M.L.; Muchowski, J.M. *Can. J. Chem.* **1994**, *72*, 142.

[1710]Orita, A.; Nagano, Y.; Hirano, J.; Otera, J. *Synlett* **2001**, 637.

[1711]Srivastava, R.R.; Kabalka, G.W. *Tetrahedron Lett.* **1992**, *33*, 593.

[1712]Fife, W.K.; Zhang, Z. *J. Org. Chem.* **1986**, *51*, 3744. See also, Fife, W.K.; Zhang, Z., *Tetrahedron Lett.* **1986**, *27*, 4933, 4937. For a review of acetic formic anhydride see Strazzolini, P.; Giumanini, A.G.; Cauci, S. *Tetrahedron* **1990**, *46* 1081.

$NaHCO_3$ under phase-transfer conditions,[1713] or with sodium bicarbonate with ultrasound.[1714]

OS **III**, 28, 422, 488; **IV**, 285; **VI**, 8, 910; **VIII**, 132. See also, OS **VI**, 418.

16-67   Acylation of Carboxylic Acids With Carboxylic Acids

Acyloxy-de-hydroxylation

$$2\ RCOOH \xrightleftharpoons{P_2O_5} (RCO)_2O\ +\ H_2O$$

Anhydrides can be formed from two molecules of an ordinary carboxylic acid only if a dehydrating agent is present so that the equilibrium can be driven to the right. Common dehydrating agents[1715] are acetic anhydride, trifluoroacetic anhydride, dicyclohexylcarbodiimide,[1716] and P_2O_5. Triphenylphosphine/CCl_3CN with triethylamine has also been used with benzoic acid derivatives.[1717] The method is very poor for the formation of mixed anhydrides, which in any case generally undergo disproportionation to the two simple anhydrides when they are heated. However, simple heating of dicarboxylic acids does give cyclic anhydrides, provided that the ring formed contains five, six, or seven members, for example,

Malonic acid and its derivatives, which would give four-membered cyclic anhydrides, do not give this reaction when heated but undergo decarboxylation (**12-40**) instead.

Carboxylic acids exchange with amides and esters; these methods are sometimes used to prepare anhydrides if the equilibrium can be shifted, for example,

[1713]Plusquellec, D.; Roulleau, F.; Lefeuvre, M.; Brown, E. *Tetrahedron* **1988**, *44*, 2471; Wang, J.; Hu, Y.; Cui, W. *J. Chem. Res. (S)* **1990**, 84.

[1714]Hu, Y.; Wang, J.-X.; Li, S. *Synth. Commun.* **1997**, *27*, 243.

[1715]For lists of other dehydrating agents with references, see Larock, R.C. *Comprehensive Organic Transformations*, 2nd ed., Wiley-VCH, NY, **1999**, pp. 1930–1932; Ogliaruso, M.A.; Wolfe, J.F., in Patai, S. *The Chemistry of Acid Derivatives*, pt.1, Wiley, NY, **1979**, pp. 437–438.

[1716]For example, see Schüssler, H.; Zahn, H. *Chem. Ber.* **1962**, *95*, 1076; Rammler, D.H.; Khorana, H.G. *J. Am. Chem. Soc.* **1963**, *85*, 1997. See also, Hata, T.; Tajima, K.; Mukaiyama, T. *Bull. Chem. Soc. Jpn.* **1968**, *41*, 2746.

[1717]Kim, J.; Jang, D.O. *Synth. Commun.* **2001**, *31*, 395.

Enolic esters are especially good for this purpose, because the equilibrium is shifted by formation of the ketone.

$$
\underset{R}{\overset{O}{\underset{||}{C}}}\!\!-\!OH \;+\; R^1\!\!-\!\!\underset{O}{\overset{O}{\underset{||}{C}}}\!\!-\!O\!\!-\!\!\underset{CH_2}{\overset{CH_2}{\underset{||}{C}}}\!\!-\!CH_3 \longrightarrow \underset{R}{\overset{O}{\underset{||}{C}}}\!\!-\!O\!\!-\!\!\underset{R^1}{\overset{O}{\underset{||}{C}}} \;+\; \underset{H_3C}{\overset{O}{\underset{||}{C}}}\!\!-\!CH_3
$$

The combination of KF with 2-acetoxypropene under microwave conditions was effective.[1718] Carboxylic acids also exchange with anhydrides; indeed, this is how acetic anhydride acts as a dehydrating agent in this reaction.

Anhydrides can be formed from certain carboxylic acid salts; for example, by treatment of trimethylammonium carboxylates with phosgene:[1719]

$$
2\,RCOO^{\ominus}\;\overset{\oplus}{N}HEt_3 \;\xrightarrow{\;COCl_2\;}\; RCOOCOR \;+\; 2\,\overset{\oplus}{N}HEt_3\;Cl^{\ominus} \;+\; CO_2
$$

or of thallium(I) carboxylates with thionyl chloride,[1605] or of sodium carboxylates with CCl_4 and a catalyst such as CuCl or $FeCl_2$.[1720]

OS **I**, 91, 410; **II**, 194, 368, 560; **III**, 164, 449; **IV**, 242, 630, 790; **V**, 8, 822; **IX**, 151. Also see, OS **VI**, 757; **VII**, 506.

16-68   Preparation of Mixed Organic–Inorganic Anhydrides

Nitrooxy-de-acyloxy-substitution

$$
(RCO)_2O \;+\; HONO_2 \;\longrightarrow\; RCOONO_2
$$

Mixed organic–inorganic anhydrides are seldom isolated, though they are often intermediates when acylation is carried out with acid derivatives catalyzed by inorganic acids. Sulfuric, perchloric, phosphoric, and other acids form similar anhydrides, most of which are unstable or not easily obtained because the equilibrium lies in the wrong direction. These intermediates are formed from amides, carboxylic acids, and esters, as well as anhydrides. Organic anhydrides of phosphoric acid are more stable than most others and, for example, $RCOOPO(OH)_2$ can be prepared in the form of its salts.[1721] Mixed anhydrides of carboxylic and sulfonic acids $(RCOOSO_2R')$ are obtained in high yields by treatment of sulfonic acids with acyl halides or (less preferred) anhydrides.[1722]

OS **I**, 495; **VI**, 207; **VII**, 81.

[1718]Villemin, D.; Labiad, B.; Loupy, A. *Synth. Commun.* **1993**, *23*, 419.
[1719]Rinderknecht, H.; Ma, V. *Helv. Chim. Acta* **1964**, *47*, 152. See also, Nangia, A.; Chandrasekaran, S. *J. Chem. Res. (S)* **1984**, 100.
[1720]Weiss, J.; Havelka, F.; Nefedov, B.K. *Bull. Acad. Sci. USSR Div. Chem. Sci.* **1978**, *27*, 193.
[1721]Avison, A.W.D. *J. Chem. Soc.* **1955**, 732.
[1722]Karger, M.H.; Mazur, Y. *J. Org. Chem.* **1971**, *36*, 528.

16-69 Attack by SH or SR at an Acyl Carbon[1723]

Thiol acids and thiol esters[1724] can be prepared in this manner, which is analogous to **16-57** and **16-64**. Anhydrides[1725] and aryl esters (RCOOAr)[1726] are also used as substrates, but the reagents in these cases are usually HS^- and RS^-. Thiol esters can also be prepared by treatment of carboxylic acids with P_4S_{10}–Ph_3SbO,[1727] or with a thiol RSH and either polyphosphate ester or phenyl dichlorophosphate $PhOPOCl_2$.[1728] Esters RCOOR' can be converted to thiol esters $RCOSR^2$ by treatment with trimethylsilyl sulfides Me_3SiSR^2 and $AlCl_3$.[1729]

Alcohols, when treated with a thiol acid and zinc iodide, give thiol esters (R'COSR)[1730]

OS **III**, 116, 599; **IV**, 924, 928; **VII**, 81; **VIII**, 71.

16-70 Transamidation

Alkylamino-de-amidation

It is sometimes necessary to replace one amide group with another, particularly when the group attached to nitrogen functions as a protecting group[1731] N-Benzyl amides can be converted to the corresponding N-allyl amide with allylamine and titanium catalysts.[1732] Reaction of N-Boc 2-phenylethylamine (Boc = tert-butoxy carbonyl) with $Ti(OiPr)_4$ and benzyl alcohol, for example, gives the N-Cbz derivative (Cbz = carbobenzoylcarbonyl).[1733] N-Carbamoyl amines were converted to

[1723]For a review, see Satchell, D.P.N. *Q. Rev. Chem. Soc.* **1963**, *17*, 160, pp. 182–184.

[1724]For a review of these compounds, see Scheithauer, S.; Mayer, R. *Top. Sulfur Chem.* **1979**, *4*, 1.

[1725]Ahmad, S.; Iqbal, J. *Tetrahedron Lett.* **1986**, *27*, 3791.

[1726]Hirabayashi, Y.; Mizuta, M.; Mazume, T. *Bull. Chem. Soc. Jpn.* **1965**, *38*, 320.

[1727]Nomura, R.; Miyazaki, S.; Nakano, T.; Matsuda, H. *Chem. Ber.* **1990**, *123*, 2081.

[1728]Imamoto, T.; Kodera, M.; Yokoyama, M. *Synthesis* **1982**, 134; Liu, H.; Sabesan, S.I. *Can. J. Chem.* **1980**, *58*, 2645. For other methods of converting carboxylic acids to thiol esters, see the references given in these papers. See also, Dellaria, Jr., F.F.; Nordeen, C.; Swett, L.R. *Synth. Commun.* **1986**, *16*, 1043.

[1729]Mukaiyama, T.; Takeda, T.; Atsumi, K. *Chem. Lett.* **1974**, 187. See also, Hatch, R.P.; Weinreb, S.M. *J. Org. Chem.* **1977**, *42*, 3960; Cohen, T.; Gapinski, R.E. *Tetrahedron Lett.* **1978**, 4319.

[1730]Gauthier, J.Y.; Bourdon, F.; Young, R.N. *Tetrahedron Lett.* **1986**, *27*, 15.

[1731]See, for example, Swain, C.G.; Ketley, A.D.; Bader, R.F.W. *J. Am. Chem. Soc.* **1959**, *81*, 2353; Knipe, A.C. *J. Chem. Soc. Perkin Trans. 2* **1973**, 589.

[1732]Eldred, S.E.; Stone, D.A.; Gellman, S.H.; Stahl, S.S. *J. Am. Chem. Soc.* **2003**, *125*, 3422.

[1733]Shapiro, G.; Marzi, M. *J. Org. Chem.* **1997**, *62*, 7096.

N-acetyl amines with acetic anhydride, Bu$_3$SnH, and Pd(PPh$_3$)$_4$.[1734] Triethylaluminum converts methyl carbamates (ArNHCO$_2$Me) to the corresponding propanamide.[1735]

A related process reacts acetamide with amines and aluminum chloride to give the *N*-acetyl amine.[1736] Another related process converted imides to *O*-benzyloxy amides by the samarium-catalyzed reaction with *O*-benzylhydroxylamine.[1737]

Thioamides can be prepared from amide by reaction with an appropriate sulfur reagent. The reaction of *N,N*-dimethylacetamide under microwave irradiation, with the polymer-bound reagent **104** gave **105**.[1738] Reaction of the thioamide with Bi(NO$_3$)$_3$·5 H$_2$O converts regenerates the amide.[1739] Oxone® and a thioamide, on the solid-phase, regenerates the amide.[1740] Selenoamides (RC(=Se)NR$_2'$ have also been prepared from amides.[1741]

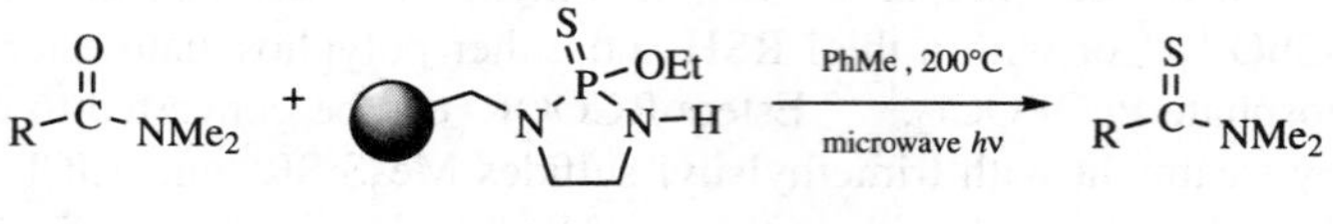

D. Attack by Halogen

16-71   The Conversion of Carboxylic Acids to Halides

Halo-de-oxido,oxo-tersubstitution

$$\text{RCOOH} \longrightarrow \text{R–X}$$

In certain cases, carboxyl groups can be replaced by halide. Acrylic acid derivatives ArCH=CHCOOH, for example, react with 3 equivalents of Oxone in the presence of NaBr to give a vinyl bromide ArCH=CHBr.[1742] In other cases, conjugated acids, such as, **106**, have been converted to the bromide by reaction with *N*-bromosuccinimide (NBS, p. 962) and LiOAc.[1743]

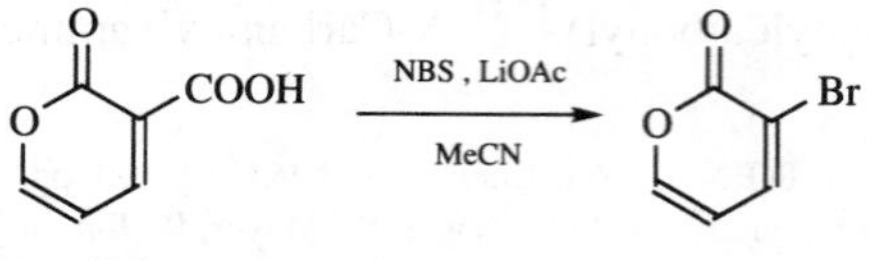

106

[1734]Roos, E.C.; Bernabé, P.; Hiemstra, H.; Speckamp, W.N.; Kaptein, B.; Boesten, W.H.J. *J. Org. Chem.* **1995**, *60*, 1733.
[1735]El Kaim, L.; Grimaud, L.; Lee, A.; Perroux, Y.; Tiria, C. *Org. Lett.* **2004**, *6*, 381.
[1736]Bon, E.; Bigg, D.C.H.; Bertrand, G. *J. Org. Chem.* **1994**, *59*, 4035.
[1737]Sibi, M.P.; Hasegawa, H.; Ghorpade, S.R. *Org. Lett.* **2002**, *4*, 3343.
[1738]Ley, S.V.; Leach, A.G.; Storer, R.I. *J. Chem. Soc. Perkin Trans. 1* **2001**, 358.
[1739]Mohammadpoor-Baltork, I.; Khodaei, M.M.; Nikoofar, K. *Tetrahedron Lett.* **2003**, *44*, 591.
[1740]Mohammadpoor-Baltork, I.; Sadeghi, M.M.; Esmayilpour, K. *Synth. Commun.* **2003**, *33*, 953.
[1741]Saravanan, V.; Mukherjee, C.; Das, S.; Chandrasekaran, S. *Tetrahedron Lett.* **2004**, *45*, 681.
[1742]You, H.-W.; Lee, K.-J. *Synlett* **2001**, 105.
[1743]Cho, C.-G.; Park, J.-S.; Jung, I.-H.; Lee, H. *Tetrahedron Lett.* **2001**, *42*, 1065.

E. Attack by Nitrogen at an Acyl Carbon[1744]

16-72 Acylation of Amines by Acyl Halides

Amino-de-halogenation

$$RCOX \quad + \quad NH_3 \quad \longrightarrow \quad RCONH_2 \quad + \quad HX$$

The treatment of acyl halides with ammonia or amines is a very general reaction for the preparation of amides.[1745] The reaction is highly exothermic and must be carefully controlled, usually by cooling or dilution. Ammonia gives unsubstituted amides, primary amines give *N*-substituted amides,[1746] and secondary amines give *N,N*-disubstituted amides. Arylamines can be similarly acylated. Hydroxamic acids have been prepared by this route.[1747] In some cases, aqueous alkali is added to combine with the liberated HCl. This is called the *Schotten–Baumann procedure*, as in **16-61**. Activated zinc can be used to increase the rate of amide formation when hindered amines and/or acid chlorides are used.[1748] An indium-mediated amidation reaction[1749] and a BiOCl-mediated reaction[1750] have been reported. A variation of this basic reaction uses DMF with acyl halides to give *N,N*-dimethylamides.[1751] A solvent-free reaction was reported using DABCO and methanol.[1752]

Hydrazine and hydroxylamine also react with acyl halides to give, respectively, hydrazides $(RCONHNH_2)$[1753] and hydroxamic acids $(RCONHOH)$.[1754] When phosgene is the acyl halide, both aliphatic and aromatic primary amines give chloroformamides (ClCONHR) that lose HCl to give isocyanates (RNCO).[1755] This is one of the most common methods for

[1744]For a review, see Challis, M.S.; Butler, A.R., in Patai, S. *The Chemistry of the Amino Group*, Wiley, NY, *1968*, pp. 279–290.

[1745]For a review, see Beckwith, A.L.J., in Zabicky, J. *The Chemistry of Amides*, Wiley, NY, *1970*, pp. 73–185. See Jedrzejczak, M.; Motie, R.E.; Satchell, D.P.N. *J. Chem. Soc. Perkin Trans. 2 1993*, 599 for a discussion of the kinetics of this reaction.

[1746]See Bhattacharyya, S.; Gooding, O.W.; Labadie, J. *Tetrahedron Lett. 2003*, *44*, 6099.

[1747]Reddy, A.S.; Kumar, M.S.; Reddy, G.R. *Tetrahedron Lett. 2000*, *41*, 6285.

[1748]Meshram, H.M.; Reddy, G.S.; Reddy, M.M.; Yadav, J.S. *Tetrahedron Lett. 1998*, *39*, 4103.

[1749]Cho, D.H.; Jang, D.O. *Tetrahedron Lett. 2004*, *45*, 2285.

[1750]Ghosh, R.; Maiti, S.; Chakraborty, A. *Tetrahedron Lett. 2004*, *45*, 6775.

[1751]Lee, W.S.; Park, K.H.; Yoon, Y-J. *Synth. Commun. 2000*, *30*, 4241.

[1752]Hajipour, A.R.; Mazloumi, Gh. *Synth. Commun. 2002*, *32*, 23.

[1753]For a review of hydrazides, see Paulsen, H.; Stoye, D., in Zabicky, J. *The Chemistry of Amides*, Wiley, NY, *1970*, pp. 515–600.

[1754]For an improved method, see Ando, W.; Tsumaki, H. *Synth. Commun. 1983*, *13*, 1053.

[1755]For reviews of the preparation and reactions of isocyanates and isothiocyanates, see, respectively, the articles by Richter, R.; Ulrich, H. pp. 619–818, and Drobnica, L.; Kristián, P.; Augustín, J. pp. 1003–1221, in Patai S. *The Chemistry of Cyanates and Their Thio Derivatives*, pt. 2, Wiley, NY, *1977*.

the preparation of isocyanates.[1756] Thiophosgene,[1757] similarly treated, gives iso-thiocyanates. A safer substitute for phosgene in this reaction is trichloromethyl chloroformate CCl_3OCOCl.[1758] When chloroformates ROCOCl are treated with primary amines, carbamates ROCONHR' are obtained.[1759] An example of this reaction is the use of benzyl chloroformate to protect the amino group of amino acids and peptides.

107

The $PhCH_2OCO$ group in **107** is called the carbobenzoxy group,[1760] and is often abbreviated Cbz or Z. Another important group similarly used is the *tert*-butoxycarbonyl group Me_3COCO, abbreviated as Boc. In this case, the chloride ($Me_3COCOCl$) is unstable, so the anhydride, $(Me_3COCO)_2O$, is used instead, in an example of **16-73**. Amino groups in general are often protected by conversion to amides.[1761] The treatment of acyl halides with lithium nitride gives *N,N*-diacyl amides (triacylamines), **108**.[1762] The reactions proceed by the tetrahedral mechanism.[1763]

$$3\ RCOCl\ +\ Li_3N\ \longrightarrow\ (RCO)_3N$$
108

A novel variation of this reaction uses nitrogen gas as the nitrogen source in the amide. The reaction of benzoyl chloride with $TiCl_4/Li/Me_3SiCl/CsF$ and N_2, gave a 77% yield of benzamide.[1764]

An interesting variation of this transformation reacts carbamoyl chlorides with organocuprates to give the corresponding amide.[1765]

[1756]For examples, see Ozaki, S. *Chem. Rev.* **1972**, *72*, 457, see pp. 457–460. For a review of the industrial preparation of isocyanates by this reaction, see Twitchett, H.J. *Chem. Soc. Rev.* **1974**, *3*, 209.

[1757]For a review of thiophosgene, see Sharma, S. *Sulfur Rep.* **1986**, *5*, 1.

[1758]Kurita, K.; Iwakura, Y. *Org. Synth.* **VI**, 715.

[1759]For example see Ariza, X.; Urpí, F.; Vilarrasa, J. *Tetrahedron Lett.* **1999**, *40*, 7515. See also, Mormeneo, D.; Llebaria, A.; Delgado, A. *Tetrahedron Lett.* **2004**, *45*, 6831. For a variation involving azide and a palladium catalyst, see Okumoto, H.; Nishihara, S.; Yamamoto, S.; Hino, H.; Nozawa, A.; Suzuki, A. *Synlett* **2000**, 991.

[1760]For an alternative reagent to prepare *N*- Cbz derivatives, see Yasuhara, T.; Nagaoka, Y.; Tomioka, K. *J. Chem. Soc. Perkin Trans. 1* **1999**, 2233.

[1761]Greene, T.W. *Protective Groups in Organic Synthesis*, Wiley, NY, **1980**, pp. 222–248, 324–326; Wuts, P.G.M.; Greene, T.W. *Protective Groups in Organic Synthesis*, 2nd ed., Wiley, NY, **1991**, pp. 327–330; Wuts, P.G.M.; Greene, T.W. *Protective Groups in Organic Synthesis*, 3rd ed., Wiley, NY, **1999**, pp. 518–525; 737–739.

[1762]Baldwin, F.P.; Blanchard, E.J.; Koening, P.E. *J. Org. Chem.* **1965**, *30*, 671.

[1763]Kivinen, A., in Patai, S. *The Chemistry of Acyl Halides*, Wiley, NY, **1972**; Bender, M.L.; Jones, J.M. *J. Org. Chem.* **1962**, *27*, 3771. See also, Song, B.D.; Jencks, W.P. *J. Am. Chem. Soc.* **1989**, *111*, 8479.

[1764]Kawaguchi, M.; Hamaoka, S.; Mori, M. *Tetrahedron Lett.* **1993**, *34*, 6907.

[1765]Lemoucheux, L.; Seitz, T.; Rouden, J.; Lasne, M.-C. *Org. Lett.* **2004**, *6*, 3703.

OS **I**, 99, 165; **II**, 76, 208, 278, 328, 453; **III**, 167, 375, 415, 488, 490, 613; **IV**, 339, 411, 521, 620, 780; **V**, 201, 336; **VI**, 382, 715; **VII**, 56, 287, 307; **VIII**,16, 339; **IX**, 559; **81**, 254. See also, OS **VII**, 302.

16-73 Acylation of Amines by Anhydrides

Amino-de-acyloxy-substitution

$$\underset{R}{\overset{O}{\underset{}{\parallel}}}\text{C}-\text{O}-\underset{R'}{\overset{O}{\underset{}{\parallel}}}\text{C} \;+\; NH_3 \quad\longrightarrow\quad \underset{R}{\overset{O}{\underset{}{\parallel}}}\text{C}-NH_2 \;+\; R'COOH$$

This reaction, similar in scope and mechanism[1766] to **16-72**, can be carried out with ammonia or primary or secondary amines.[1767] Note that there is a report where a tertiary amine (an *N*-alkylpyrrolidine) reacted with acetic anhydride at 120°C, in the presence of a BF$_3$•etherate catalyst, to give *N*-acetylpyrrolidine (an acylative dealkylation).[1768] Amino acids can be *N*-acylated using acetic anhydride and ultrasound.[1769] However, ammonia and primary amines can also give imides, in which two acyl groups are attached to the nitrogen. The conversion of cyclic anhydrides to cyclic imides is generally facile,[1770] although elevated temperatures are occasionally required to generate the imide.[1771] Microwave irradiation of formamide and a cyclic anhydride generates the cyclic imide.[1772] Cyclic imides have also been formed in ionic liquids.[1773] Cyclic imides were also formed by microwave irradiation of a polymer-bound phthalate after initial reaction with an amine.[1774]

The second step for imide formation, which is much slower than the first, is the attack of the amide nitrogen on the carboxylic carbon. Unsubstituted and *N*-substituted amides have been used instead of ammonia. Since the other product

[1766]For a discussion of the mechanism, see Kluger, R.; Hunt, J.C. *J. Am. Chem. Soc.* **1989**, *111*, 3325.

[1767]For a review, see Beckwith, A.L.J., in Zabicky, J. *The Chemistry of Amides*, Wiley, NY, **1970**, pp. 86–96. See also, Naik, S.; Bhattacharjya, G.; Talukdar, B.; Patel, B.K. *Eur. J. Org. Chem.* **2004**, 1254.

[1768]Dave, P. R.; Kumar, K. A.; Duddu, R.; Axenrod, T.; Dai, R.; Das, K. K.; Guan, X.-P.; Sun, J.; Trivedi, N. J.; Gilardi, R. D. *J. Org. Chem.* **2000**, *65*, 1207.

[1769]Anuradha, M.V.; Ravindranath, B. *Tetrahedron* **1997**, *53*, 1123.

[1770]For reviews of imides, see Wheeler, O.H.; Rosado, O., in Zabicky, J. *The Chemistry of Amides*, Wiley, NY, **1970**, pp. 335–381; Hargreaves, M.K.; Pritchard, J.G.; Dave, H.R. *Chem. Rev.* **1970**, *70*, 439 (cyclic imides).

[1771]Tsubouchi, H.; Tsuji, K.; Ishikawa, H. *Synlett* **1994**, 63.

[1772]Peng, Y.; Song, G.; Qian, X. *Synth. Commun.* **2001**, *31*, 1927; Kacprzak, K. *Synth. Commun.* **2003**, *33*, 1499.

[1773]In bmim PF$_6$, 1-butyl-3-methylimidazolium hexafluorophosphate: Le, Z.-G.; Chen, Z.-C.; Hu, Y.; Zheng, Q.-G. *Synthesis* **2004**, 995.

[1774]Martin, B.; Sekljic, H.; Chassaing, C. *Org. Lett.* **2003**, *5*, 1851.

of this reaction is RCOOH, this is a way of "hydrolyzing" such amides in the absence of water.[1775]

Even though formic anhydride is not a stable compound (see p. 723), amines can be formylated with the mixed anhydride of acetic and formic acids (HCOO-COMe)[1776] or with a mixture of formic acid and acetic anhydride. Acetamides are not formed with these reagents. Secondary amines can be acylated in the presence of a primary amine by conversion to their salts and addition of 18-crown-6.[1777] The crown ether complexes the primary ammonium salt, preventing its acylation, while the secondary ammonium salts, which do not fit easily into the cavity, are free to be acylated. Dimethyl carbonate can be used to prepare methyl carbamates in a related procedure.[1778] N-Acetylsulfonamides were prepared from acetic anhydride and a primary sulfonamide, catalyzed by Montmorillonite K10–FeO[1779] or sulfuric acid.[1780]

The reaction of anhydrides with aryl azides, in the presence of Me$_3$SiCl and NaI, gives N-aryl imides.[1781]

OS **I**, 457; **II**, 11; **III**, 151, 456, 661, 813; **IV**, 5, 42, 106, 657; **V**, 27, 373, 650, 944, 973; **VI**, 1; **VII**, 4, 70; **VIII**, 132; **76**, 123.

16-74 Acylation of Amines by Carboxylic Acids

Amino-de-hydroxylation

$$RCOOH + NH_3 \longrightarrow RCOO^-NH_4^+ \xrightarrow{pyrolysis} RCONH_2$$

When carboxylic acids are treated with ammonia or amines, salts are obtained. The salts of ammonia or primary or secondary amines can be pyrolyzed to give amides,[1782] but the method is less convenient than **16-72**, **16-73**, and **16-75** and is seldom of preparative value.[1783] Heating in the presence of a base such as hexamethyldisilazide makes the amide-forming process more efficient.[1784] Boronic acids catalyze the direct conversion of carboxylic acid and amine to amides.[1785]

[1775]Eaton, J.T.; Rounds, W.D.; Urbanowicz, J.H.; Gribble, G.W. *Tetrahedron Lett.* **1988**, *29*, 6553.
[1776]For the formylation of amines with the mixed anhydride of formic and trimethylacetic acid, see Vlietstra, E.J.; Zwikker, J.W.; Nolte, R.J.M.; Drenth, W. *Recl. Trav. Chim. Pays-Bas* **1982**, *101*, 460.
[1777]Barrett, A.G.M.; Lana, J.C.A. *J. Chem. Soc., Chem. Commun.* **1978**, 471.
[1778]Vauthey, I.; Valot, F.; Gozzi, C.; Fache, F.; Lemaire, M. *Tetrahedron Lett.* **2000**, *41*, 6347.
[1779]Singh, D.U.; Singh, P.R.; Samant, S.D. *Tetrahedron Lett.* **2004**, *45*, 4805.
[1780]Martin, M.T.; Roschangar, F.; Eaddy, J.F. *Tetrahedron Lett.* **2003**, *44*, 5461.
[1781]Kamal, A.; Laxman, E.; Laxman, N.; Rao, N.V. *Tetrahedron Lett.* **1998**, *39*, 8733.
[1782]For example, see Mitchell, J.A.; Reid, E.E. *J. Am. Chem. Soc.* **1931**, *53*, 1879. Also see, Jursic, B.S.; Zdravkovski, Z. *Synth. Commun.* **1993**, *23*, 2761.
[1783]For a review of amide formation from carboxylic acids, see Beckwith, A.L.J., in Zabicky, J. *The Chemistry of Amides*, Wiley, NY, **1970**, pp. 105–109.
[1784]Chou, W.-C.; Chou, M.-C.; Lu, Y.-Y.; Chen, S.-F. *Tetrahedron Lett.* **1999**, *40*, 3419. For alternative approaches using specialized reagents, see Jang, D.O.; Park, D.J.; Kim, J. *Tetrahedron Lett.* **1999**, *40*, 5323; Bailén, M.A.; Chinchilla, R.; Dodsworth, D.J.; Nájera, C. *Tetrahedron Lett.* **2000**, *41*, 9809 and **2001**, *42*, 5013; White, J.M.; Tunoori, A.R.; Turunen, B.J.; Georg, G.I *J. Org. Chem.* **2004**, *69*, 2573.
[1785]Ishihara, K.; Kondo, S.; Yamamoto, H. *Synlett* **2001**, 1371.

Polymer-bound reagents have also been used.[1786] The synthetically important Weinreb amides [RCON(Me)OMe, see **16-82**] can be prepared from the carboxylic acid and MeO(Me)NH•HCl in the presence of tributylphosphine and 2-pyridine-*N*-oxide disulfide.[1787] Di(2-pyridyl)carbonate has been used in a related reaction that generates amides directly.[1788] The reaction of a carboxylic acid and imidazole under microwave irradiation gives the amide.[1789] Microwave irradiation of a secondary amine, formic acid, 2-chloro-4,6-dimethoxy[1,3,5]triazine, and a catalytic amount of DMAP (4-dimethylaminopyridine) leads to the formamide.[1790] Ammonium bicarbonate and formamide converts acids to amides with microwave irradiation.[1791] Lactams are readily produced from γ- or δ-amino acids,[1792] for example,

This lactamization process can be promoted by enzymes, such as pancreatic porcine lipase.[1793] Reduction of ω-azido carboxylic acids leads to macrocyclic lactams.[1794]

Although treatment of carboxylic acids with amines does not directly give amides, the reaction can be made to proceed in good yield at room temperature or slightly above by the use of coupling agents,[1795] the most important of which is dicyclohexylcarbodiimide. This reagent is very convenient and is used[1796] a great deal in peptide synthesis.[1797] A polymer-supported carbodiimide has been used.[1798] The mechanism is probably the same as in **16-63** up to the formation

[1786]Buchstaller, H.P.; Ebert, H.M.; Anlauf, U. *Synth. Commun.* **2001**, *31*, 1001; Crosignani, S.; Gonzalez, J.; Swinnen, D. *Org. Lett.* **2004**, *6*, 4579; Chichilla, R.; Dodsworth, D.J.; Nájera, C.; Soriano, J.M. *Tetrahedron Lett.* **2003**, *44*, 463.

[1787]Banwell, M.; Smith, J. *Synth. Commun.* **2001**, *31*, 2011. For another procedure, see Kim, M.; Lee, H.; Han, K.-J.; Kay,K.-Y. *Synth. Commun.* **2003**, *33*, 4013.

[1788]Shiina, I.; Suenaga, Y.; Nakano, M.; Mukaiyama, T. *Bull. Chem. Soc. Jpn.* **2000**, *73*, 2811.

[1789]Khalafi-Nezhad, A.; Mokhtari, B.; Rad, M.N.S. *Tetrahedron Lett.* **2003**, *44*, 7325; Perreux, L.; Loupy, A.; Volatron, F. *Tetrahedron* **2002**, *58*, 2155.

[1790]De Lucca, L.; Giacomelli, G.; Porcheddu, A.; Salaris, M. *Synlett* **2004**, 2570.

[1791]Peng, Y.; Song, G. *Org. Prep. Proceed. Int.* **2002**, *34*, 95.

[1792]See, for example, Bladé-Font, A. *Tetrahedron Lett.* **1980**, *21*, 2443. See Wei, Z.-Y.; Knaus, E.E. *Tetrahedron Lett.* **1993**, *34*, 4439 for a variation of this reaction.

[1793]Gutman, A.L.; Meyer, E.; Yue, X.; Abell, C. *Tetrahedron Lett.* **1992**, *33*, 3943.

[1794]Bosch, I.; Romea, P.; Urpí, F.; Vilarrasa, J. *Tetrahedron Lett.* **1993**, *34*, 4671. See Bai, D.; Shi, Y. *Tetrahedron Lett.* **1992**, *33*, 943 for the preparation of lactam units in para-cyclophanes.

[1795]For a review of peptide synthesis with dicyclohexylcarbodiimide and other coupling agents, see Klausner, Y.S.; Bodansky, M. *Synthesis* **1972**, 453.

[1796]It was first used this way by Sheehan, J.C.; Hess, G.P. *J. Am. Chem. Soc.* **1955**, *77*, 1067.

[1797]For a treatise on peptide synthesis, see Gross, E.; Meienhofer, J. *The Peptides*, 3 vols., Academic Press, NY, **1979–1981**. For a monograph, see Bodanszky, M.; Bodanszky, A. *The Practice of Peptide Synthesis*, Springer, NY, **1984**.

[1798]Feuerstein, M.; Doucet, H.; Santelli, M. *Tetrahedron Lett.* **2001**, *42*, 6667.

of **109**. This intermediate is then attacked by another molecule of RCOO⁻ to give the anhydride $(RCO)_2O$, which is the actual species that reacts with the amine:

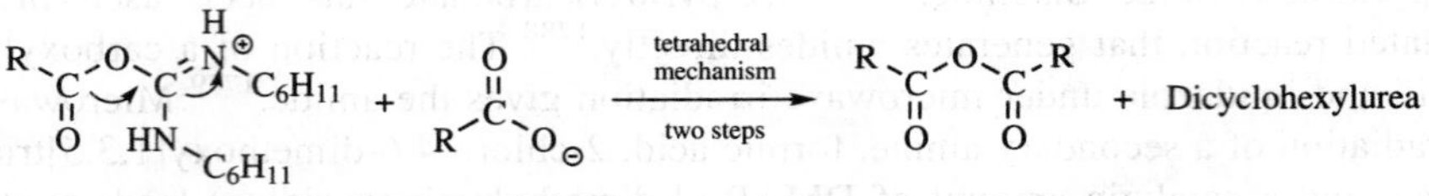

109

The anhydride has been isolated from the reaction mixture and then used to acylate an amine.[1799] Other promoting agents[1800] are $ArB(OH)_2$ reagents,[1801] $Sn[N(TMS)_2]_2$,[1802] N,N'-carbonyldiimidazole (**110**, p. 1418),[1803] which behaves as in reaction **16-63**,[1804]$TiCl_4$,[1805] molecular sieves,[1806] Lawesson's reagent (p. 1278),[1807] and $(MeO)_2POCl$.[1808] Certain dicarboxylic acids form amides simply on treatment with primary aromatic amines. In these cases, the cyclic anhydride is an intermediate and is the species actually attacked by the amine.[1809] Carboxylic acids can also be converted to amides by heating with amides of carboxylic acids (exchange),[1810] sulfonic acids, or phosphoric acids, for example,[1811]

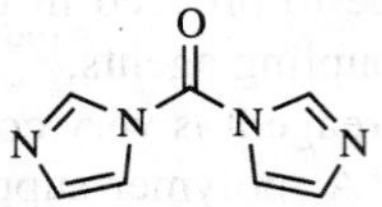

110

$$RCOOH + Ph_2PONH_2 \longrightarrow RCONH_2 + Ph_2POOH$$

[1799]Schüssler, H.; Zahn, H. *Chem. Ber.* **1962**, *95*, 1076; Rebek, J.; Feitler, D. *J. Am. Chem. Soc.* **1974**, *96*, 1606. There is evidence that some of the **98** is converted to products by another mechanism. See Rebek, J.; Feitler, D. *J. Am. Chem. Soc.* **1973**, *95*, 4052.

[1800]For a list of reagents, with references, see Larock, R.C. *Comprehensive Organic Transformations*, 2nd ed., Wiley-VCH, NY, **1999**, pp. 1941–1949.

[1801]Ishihara, K.; Ohara, S.; Yamamoto, H. *J. Org. Chem.* **1996**, *61*, 4196.

[1802]Burnell-Curty, C.; Roskamp, E.J. *Tetrahedron Lett.* **1993**, *34*, 5193.

[1803]See Vaidyanathan, R.; Kalthod, V.G.; Ngo, D.; Manley, J.M.; Lapekas, S.P. *J. Org. Chem.* **2004**, *69*, 2565. A modified but related reagent has also been used. See Grzyb, J.A.; Batey, R.A. *Tetrahedron Lett.* **2003**, *44*, 7485.

[1804]Klosa, J. *J. Prakt. Chem.* **1963**, *[4] 19*, 45.

[1805]Wilson, J.D.; Weingarten, H. *Can. J. Chem.* **1970**, *48*, 983.

[1806]Cossy, J.; Pale-Grosdemange, C. *Tetrahedron Lett.* **1989**, *30*, 2771.

[1807]Thorsen, M.; Andersen, T.P.; Pedersen, U.; Yde, B.; Lawesson, S. *Tetrahedron* **1985**, *41*, 5633.

[1808]Jászay, Z.M.; Petneházy, I.; Tőke, L. *Synth. Commun.* **1998**, *28*, 2761.

[1809]Higuchi, T.; Miki, T.; Shah, A.C.; Herd, A.K. *J. Am. Chem. Soc.* **1963**, *85*, 3655.

[1810]For example, see Schindbauer, H. *Monatsh. Chem.* **1968**, *99*, 1799.

[1811]Zhmurova, I.N.; Voitsekhovskaya, I.Yu.; Kirsanov, A.V. *J. Gen. Chem. USSR* **1959**, *29*, 2052. See also, Kopecky, J.; Smejkal, J. *Chem. Ind. (London)* **1966**, 1529; Liu, H.; Chan, W.H.; Lee, S.P. *Synth. Commun.* **1979**, *9*, 31.

or by treatment with trisalkylaminoboranes, $B(NHR')_3$, with trisdialkylaminobor-anes, $B(NR_2')_3$,[1812]

$$RCOOH \quad + \quad B(NR_2')_3 \quad \longrightarrow \quad RCONR_2'$$

or with bis(diorganoamino)magnesium reagents $(R_2N)_2Mg$.[1813] The reaction of thiocarboxylic acids and azides, in the presence of triphenylphosphine, gives the corresponding amide.[1814]

An important technique, discovered by R.B. Merrifield in 1963[1815] and since used for the synthesis of many peptides,[1816] is called *solid phase synthesis* or *poly-mer-supported synthesis*.[1817] The reactions used are the same as in ordinary synth-esis, but one of the reactants is anchored onto a solid polymer. For example, if it is desired to couple two amino acids (to form a dipeptide), the polymer selected might be polystyrene with CH_2Cl side chains. One of the amino acids, protected by a *tert*-butoxycarbonyl group (Boc), would then be coupled to the side chains. It is not necessary that all the side chains be converted, but a random selection will be. The Boc group is then removed by hydrolysis with trifluoroacetic acid in CH_2Cl_2 and the second amino acid is coupled to the first, using DCC or some other coupling agent. The second Boc group is removed, resulting in a dipeptide that is still anchored to the polymer. If this dipeptide is the desired product, it can be cleaved from the polymer by various methods,[1818] one of which is treatment with HF. If a longer peptide is wanted, additional amino acids can be added by repeating the requisite steps.

[1812]Pelter, A.; Levitt, T.E.; Nelson, P. *Tetrahedron* **1970**, *26*, 1539; Pelter, A.; Levitt, T.E. *Tetrahedron* **1970**, *26*, 1545, 1899.

[1813]Sanchez, R.; Vest, G.; Despres, L. *Synth. Commun.* **1989**, *19*, 2909.

[1814]Park, S.-D.; Oh, J.-H.; Lim, D. *Tetrahedron Lett.* **2002**, *43*, 6309.

[1815]Merrifield, R.B. *J. Am. Chem. Soc.* **1963**, *85*, 2149.

[1816]For a monograph on solid-state peptide synthesis, see Birr, C. *Aspects of the Merrifield Peptide Synthesis*, Springer, NY, **1978**. For reviews, see Bayer, E. *Angew. Chem. Int. Ed.* **1991**, *30*, 113; Kaiser, E.T. *Acc. Chem. Res.* **1989**, *22*, 47; Jacquier, R. *Bull. Soc. Chim. Fr.* **1989**, 220; Barany, G.; Kneib-Cordonier, N.; Mullen, D.G. *Int. J. Pept. Protein Res.* **1987**, *30*, 705; Andreev, S.M.; Samoilova, N.A.; Davidovich, Yu.A.; Rogozhin, S.V. *Russ. Chem. Rev.* **1987**, *56*, 366; Gross, E.; Meienhofer, J. *The Peptides*, Vol. 2, Academic Press, NY, **1980**, the articles by Barany, G.; Merrifield, R.B. pp. 1–184, Fridkin, M. pp. 333–363; Erickson, B.W.; Merrifield, R.B. in Neurath, H.; Hill, R.L.; Boeder, C.-L. *The Proteins*, 3rd ed., Vol. 2, Academic Press, NY, **1976**, pp. 255–527. For R. B. Merrifield's Nobel Prize lecture, see Merrifield, R.B. *Angew. Chem. Int. Ed.* **1985**, *24*, 799; *Chem. Scr.* **1985**, *25*, 121.

[1817]For monographs on solid-phase synthesis in general, see Laszlo, P. *Preparative Organic Chemistry Using Supported Reagents*, Academic Press, NY, **1987**; Mathur, N.K.; Narang, C.K.; Williams, R.E. *Polymers as Aids in Organic Chemistry*, Academic Press, NY **1980**; Hodge, P.; Sherrington, D.C. *Polymer-Supported Reactions in Organic Synthesis*, Wiley, NY, **1980**. For reviews, see Pillai, V.N.R.; Mutter, M. *Top. Curr. Chem.* **1982**, *106*, 119; Akelah, A.; Sherrington, D.C. *Chem. Rev.* **1981**, *81*, 557; Akelah, A. *Synthesis* **1981**, 413; Rebek, J. *Tetrahedron* **1979**, *35*, 723; McKillop, A.; Young, D.W. *Synthesis* **1979**, 401, 481; Crowley, J.I.; Rapoport, H. *Acc. Chem. Res.* **1976**, *9*, 135; Patchornik, A.; Kraus, M.A. *Pure Appl. Chem.* **1975**, *43*, 503.

[1818]For some of these methods, see Whitney, D.B.; Tam, J.P.; Merrifield, R.B. *Tetrahedron* **1984**, *40*, 4237.

The basic advantage of the polymer-support techniques is that the polymer (including all chains attached to it) is easily separated from all other reagents, because it is insoluble in the solvents used. Excess reagents, other reaction products (e.g., dicyclohexylurea), side products, and the solvents themselves are quickly washed away. Purification of the polymeric species is rapid and complete. The process can even be automated,[1819] to the extent that six or more amino acids can be added to a peptide chain in one day. Commercial automated peptide synthesizers are now available.[1820]

Although the solid-phase technique was first developed for the synthesis of peptide chains and has seen considerable use for this purpose, it has also been used to synthesize chains of polysaccharides and polynucleotides; in the latter case, solid-phase synthesis has almost completely replaced synthesis in solution.[1821] The technique has been applied less often to reactions in which only two molecules are brought together (nonrepetitive syntheses), but many examples have been reported.[1822] Combinatorial chemistry had its beginning with the Merrifield synthesis, particularly when applied to peptide synthesis, and continues as an important part of modern organic chemistry.[1823]

OS **I**, 3, 82, 111, 172, 327; **II**, 65, 562; **III**, 95, 328, 475, 590, 646, 656, 768; **IV**, 6, 62, 513; **V**, 670, 1070; **VIII**, 241; **81**, 262. Also see OS **III**, 360; **VI**, 263; **VIII**, 68.

16-75 Acylation of Amines by Carboxylic Esters

Amino-de-alkoxylation

$$RCOOR' + NH_3 \longrightarrow RCONH_2 + R'OH$$

The conversion of carboxylic esters to amides is a useful reaction, and unsubstituted, *N*-substituted, and *N,N*-disubstituted amides can be prepared this way from the appropriate amine.[1824] Both R and R' can be alkyl or aryl, but an especially

[1819]This was first reported by Merrifield, R.B.; Stewart, J.M.; Jernberg, N. *Anal. Chem.* **1966**, *38*, 1905.

[1820]For a discussion of automated organic synthesis, see Frisbee, A.R.; Nantz, M.H.; Kramer, G.W.; Fuchs, P.L. *J. Am. Chem. Soc.* **1984**, *106*, 7143. For an improved method, see Schnorrenberg, G.; Gerhardt, H. *Tetrahedron* **1989**, *45*, 7759.

[1821]For a review, see Bannwarth, W. *Chimia* **1987**, *41*, 302.

[1822]For reviews, see Fréchet, J.M.J. *Tetrahedron* **1981**, *37*, 663; Fréchet, J.M.J. in Hodge, P.; Sherrington, D.C. *Polymer-Supported Reactions in Organic Synthesis*, Wiley, NY, **1980**, pp. 293–342, Leznoff, C.C. *Acc. Chem. Res.* **1978**, *11*, 327; *Chem. Soc. Rev.* **1974**, *3*, 64.

[1823]Czarnik, A.W.; DeWitt, S.H. *A Practical Guide to Combinatorial Chemistry*, American Chemical Society, Washington, DC, **1997**; Chaiken, I.N.; Janda, K.D. *Molecular Diversity and Combinatorial Chemistry: Libraries and Drug Discovery*, American Chemical Society, Washington, DC **1996**; Balkenhol, F.; von dem Bussche-Hünnefeld, C.; Lansky, A.; Zechel, C. *Angew. Chem. Int. Ed.* **1996**, *35*, 2289; Thompson, L.A.; Ellman, J.A. *Chem. Rev.* **1996**, 96, 555; Pavia, M.R.; Sawyer, T.K.; Moos, W.H. *Bioorg. Med. Chem. Lett. Symposia–in–print no. 4* **1993**, *3*, 387; Crowley, J.I.; Rapoport, H. *Acc. Chem. Res.* **1976**, *9*, 135; Leznoff, C.C. *Acc. Chem. Res.* **1978**, *11*, 327.

[1824]For a review, see Beckwith, A.L.J., in Zabicky, J. *The Chemistry of Amides*, Wiley, NY, **1970**, pp. 96–105. For a list of reagents, with references, see Larock, R.C. *Comprehensive Organic Transformations*, 2nd ed., Wiley-VCH, NY, **1999**, pp. 1973–1976.

good leaving group is p-nitrophenyl. Ethyl trifluoroacetate was found to react selectively with primary amines to form the corresponding trifluoroacetyl amide.[1825] Many simple esters (R = Me, Et, etc.) are not very reactive, and strongly basic catalysis has been used in such cases,[1826] but catalysis by cyanide ion[1827] MgBr$_2$,[1828] InI$_3$,[1829] and acceleration by high pressure[1830] have been reported. Methyl esters have been converted to the corresponding amide under microwave irradiation,[1831] and also ethyl esters.[1832] Lithium amides have been used to convert esters to amides as well.[1833] β-Keto esters undergo the reaction especially easily.[1834] In another procedure, esters are treated with dimethylaluminum amides (Me$_2$AlNRR′) to give good yields of amides under mild conditions.[1835] The reagents are easily prepared from Me$_3$Al and NH$_3$ or a primary or secondary amine or their salts. This is particularly effective when a reactive substituent, such as a primary halide, is present elsewhere in the molecule.[1836] Tin reagents, such as Sn[N(TMS)$_2$]$_2$, in the presence of an amine can also be use to convert an ester to an amide.[1837] This reagent can also be used to convert β-amino esters to β-lactams.[1838] Aniline was treated with n-butyllithium to form the lithium amide, which reacted with an ester to give the amide.[1839] The ester-to-amide conversion has also been accomplished electrochemically, by passing electric current in the cathodic compartment.[1840] An enzyme-mediated amidation is known using amino cyclase I.[1841] The reaction of dimethyl carbonate and an amine is an effective way to prepare methyl carbamates.[1842]

[1825]Xu, D.; Prasad, K.; Repic, O.; Blacklock, T.J. *Tetrahedron Lett.* **1995**, *36*, 7357.

[1826]For references, see Matsumoto, K.; Hashimoto, S.; Uchida, T.; Okamoto, T.; Otani, S. *Chem. Ber.* **1989**, *122*, 1357.

[1827]Högberg, T.; Ström, P.; Ebner, M.; Rämsby, S. *J. Org. Chem.* **1987**, *52*, 2033.

[1828]Guo, Z.; Dowdy, E.D.; Li, W.-S.; Polniaszek, R.; Delaney, E. *Tetrahedron Lett.* **2001**, *42*, 1843.

[1829]Ranu, B.C.; Dutta, P. *Synth. Commun.* **2003**, *33*, 297.

[1830]Matsumoto, K.; Hashimoto, S.; Uchida, T.; Okamoto, T.; Otani, S. *Chem. Ber.* **1989**, *122*, 1357.

[1831]Varma, R.S.; Naicker, K.P. *Tetrahedron Lett.* **1999**, *40*, 6177.

[1832]Suri, O.P.; Satti, N.K.; Suri, K.A. *Synth. Commun.* **2000**, *30*, 3709; Zradni, F.-Z.; Hamelin, J.; Derdour, A. *Synth. Commun.* **2002**, *32*, 3525.

[1833]See Wang, J.; Rosingana, M.; Discordia, R.P.; Soundararajan, N.; Polniaszek, R. *Synlett* **2001**, 1485.

[1834]Labelle, M.; Gravel, D. *J. Chem. Soc., Chem. Commun.* **1985**, 105.

[1835]Basha, A.; Lipton, M.; Weinreb, S.M. *Org. Synth. VI*, 492; Levin, J.I.; Turos, E.; Weinreb, S.M. *Synth. Commun.* **1982**, *12*, 989; Barrett, A.G.M.; Dhanak, D. *Tetrahedron Lett.* **1987**, *28*, 3327. For the extension of this method to the formation of hydrazides, see Benderly, A.; Stavchansky, S. *Tetrahedron Lett.* **1988**, *29*, 739.

[1836]Shimizu, T.; Osako, K.; Nakata, T. *Tetrahedron Lett.* **1997**, *38*, 2685.

[1837]Smith, L.A.; Wang, W.-B.; Burnell-Curty, C.; Roskamp, E.J. *Synlett* **1993**, 850; Wang, W.-B.; Roskamp, E.J. *J. Org. Chem.* **1992**, *57*, 6101.

[1838]Wang, W.-B.; Roskamp, E.J. *J. Am. Chem. Soc.* **1993**, *115*, 9417.

[1839]Ooi, T.; Tayama, E.; Yamada, M.; Maruoka, K. *Synlett.* **1999**, 729.

[1840]Arai, K.; Shaw, C.; Nozawa, K.; Kawai, K.; Nakajima, S. *Tetrahedron Lett.* **1987**, *28*, 441.

[1841]Youshko, M.I.; van Rantwijk, F.; Sheldon, R.A. *Tetrahedron Asymmetry* **2001**, *12*, 3267.

[1842]Distaso, M.; Quaranta, E. *Tetrahedron* **2004**, *60*, 1531; Curini, M.; Epifano, F.; Maltese, F.; Rosati, O. *Tetrahedron Lett.* **2002**, *43*, 4895.

As in **16-72**, hydrazides and hydroxamic acids can be prepared from carboxylic esters, with hydrazine and hydroxylamine, respectively. Both hydrazine and hydroxylamine react more rapidly than ammonia or primary amines (the alpha effect, p. 495). Imidates $RC(=NH)OR'$ give amidines $RC(=NH)NH_2$. Lactones, when treated with ammonia or primary amines, give lactams. Lactams are also produced from γ- and δ-amino esters in an internal example of this reaction. Isopropenyl formate is a useful compound for the formylation of primary and secondary amines.[1843]

$$R_2NH + HCOOCMe = CH_2 \longrightarrow R_2NCHO + CH_2$$
$$= CMeOH \longrightarrow MeCOMe$$

Although more studies have been devoted to the mechanism of the acylation of amines with carboxylic esters than with other reagents, the mechanistic details are not yet entirely clear.[1844] In its broad outlines, the mechanism appears to be essentially $B_{AC}2$.[1845] Under the normal basic conditions, the reaction is general base-catalyzed,[1846] indicating that a proton is being transferred in the rate-determining step and that two molecules of amine are involved.[1847]

Alternatively, another base, such as H_2O or OH^-, can substitute for the second molecule of amine. With some substrates and under some conditions, especially at low pH, the breakdown of **111** can become rate determining.[1848] The reaction also takes place under acidic conditions and is general acid catalyzed, so that

[1843]van Melick, J.E.W.; Wolters, E.T.M. *Synth. Commun.* **1972**, *2*, 83.

[1844]For a discussion of the mechanism, see Satchell, D.P.N.; Satchell, R.S., in Patai, S. *The Chemistry of Carboxylic Acids and Esters*, Wiley, NY, **1969**, pp. 410–431. For a computational study see Ilieva, S.; Galabov, B.; Musaev, D.G.; Morokuma, K.; Schaefer III, H.F. *J. Org. Chem.* **2003**, *68*, 1496.

[1845]Bunnett, J.F.; Davis, G.T. *J. Am. Chem. Soc.* **1960**, *82*, 665; Bruice, T.C.; Donzel, A.; Huffman, R.W.; Butler, A.R. *J. Am. Chem. Soc.* **1967**, *89*, 2106.

[1846]Bunnett, J.F.; Davis, G.T. *J. Am. Chem. Soc.* **1960**, *82*, 665, Jencks, W.P.; Carriuolo, J. *J. Am. Chem. Soc.* **1960**, *82*, 675; Bruice, T.C.; Mayahi, M.F. *J. Am. Chem. Soc.* **1960**, *82*, 3067.

[1847]Blackburn, G.M.; Jencks, W.P. *J. Am. Chem. Soc.* **1968**, *90*, 2638; Bruice, T.C.; Felton, S.M. *J. Am. Chem. Soc.* **1969**, *91*, 2799; Felton, S.M.; Bruice, T.C. *J. Am. Chem. Soc.* **1969**, *91*, 6721; Nagy, O.B.; Reuliaux,V.; Bertrand, N.; Van Der Mensbrugghe, A.; Leseul, J.; Nagy, J.B. *Bull. Soc. Chim. Belg.* **1985**, *94*, 1055.

[1848]Hansen, B. *Acta Chem. Scand.* **1963**, *17*, 1307; Gresser, M.J.; Jencks, W.P. *J. Am. Chem. Soc.* **1977**, *99*, 6963, 6970. See also, Yang, C.C.; Jencks, W.P. *J. Am. Chem. Soc.* **1988**, *110*, 2972.

breakdown of **111** is rate determining and proceeds as follows:[1849]

$$
\underset{\textbf{111}}{R-\overset{\displaystyle OR^1}{\underset{\displaystyle O^{\ominus}}{C}}{-}{-}NHR^2} \quad\xrightarrow[\text{slow}]{\quad\quad}\quad \underset{O}{\overset{\displaystyle R\diagdown}{\underset{\displaystyle \|}{C}}}{-}NHR^2 \;+\; R^1OH \;+\; A^{\ominus}
$$

HA may be $R^2NH_3^+$ or another acid. Intermediate **111** may or may not be further protonated on the nitrogen. Even under basic conditions, a proton donor may be necessary to assist leaving-group removal. Evidence for this is that the rate is lower with NR_2^- in liquid ammonia than with NHR_2 in water, apparently owing to the lack of acids to protonate the leaving oxygen.[1850]

In the special case of β-lactones, where small-angle strain is an important factor, alkyl–oxygen cleavage is observed ($B_{AL}2$ mechanism, as in the similar case of hydrolysis of β-lactones, **16-59**), and the product is not an amide but a β-amino acid (β-alanine).

$$
H_3N \;+\; \text{(β-lactone)} \quad\xrightarrow{\quad\quad}\quad \overset{\oplus}{H_3N}\diagdown\diagup COO^{\ominus}
$$

β-Alanine

A similar result has been found for certain sterically hindered esters.[1851] This reaction is similar to **10-31**, with OCOR as the leaving group. Other lactones have been opened to ω-hydroxy amides with Dibal:$BnNH_2$.[1852]

OS **I**, 153, 179; **II**, 67, 85; **III**, 10, 96, 108, 404, 440, 516, 536, 751, 765; **IV**, 80, 357, 441, 486, 532, 566, 819; **V**, 168, 301, 645; **VI**, 203, 492, 620, 936; **VII**, 4, 30, 41, 411; **VIII**, 26, 204, 528. Also see, OS **I**, 5; **V**, 582; **VII**, 75.

16-76   Acylation of Amines by Amides

Alkylamino-de-amination

$$
RCONH_2 \;+\; R'\overset{\oplus}{N}H_3 \quad\xrightarrow{\quad\quad}\quad RCONHR' \;+\; NH_4^+
$$

This is an exchange reaction and is usually carried out with the salt of the amine.[1853] The leaving group is usually NH_2 rather than NHR or NR_2 and primary

[1849]Blackburn, G.M.; Jencks, W.P. *J. Am. Chem. Soc.* **1968**, *90*, 2638.

[1850]Bunnett, J.F.; Davis, G.T. *J. Am. Chem. Soc.* **1960**, *82*, 665.

[1851]Zaugg, H.E.; Helgren, P.F.; Schaefer, A.D. *J. Org. Chem.* **1963**, *28*, 2617. See also, Weintraub, L.; Terrell, R. *J. Org. Chem.* **1965**, *30*, 2470; Harada, R.; Kinoshita, Y. *Bull. Chem. Soc. Jpn.* **1967**, *40*, 2706.

[1852]Huang, P.-Q.; Zheng, X.; Deng, X.-M. *Tetrahedron Lett.* **2001**, *42*, 9039. See also, Taylor, S.K.; Ide, N.D.; Silver, M.E.; Stephan, M. *Synth. Commun.* **2001**, *31*, 2391.

[1853]For a list of procedures, with references, see Larock, R.C. *Comprehensive Organic Transformations*, 2nd ed., Wiley-VCH, NY, **1999**, pp. 1978–1982.

amines (in the form of their salts) are the most common reagents. Boron trifluoride can be added to complex with the leaving ammonia. Neutral amines also react in some cases to give the new amide.[1854] The reaction is often used to convert urea to substituted ureas: $NH_2CONH_2 + RNH_3^+ \rightarrow NH_2CONHR + NH_4^+$.[1855] An *N*-aryl group of a urea can be converted to a *N,N*-dialkyl group by heating the urea with the amine in an autoclave.[1856] *N*-R-Substituted amides are converted to *N*-R'-substituted amides by treatment with N_2O_4 to give an *N*-nitroso compound, followed by treatment of this with a primary amine $R'NH_2$.[1857] Lactams can be converted to ring-expanded lactams if

a side chain containing an amino group is present on the nitrogen. A strong base is used to convert the NH_2 to NH^-, which then acts as a nucleophile, expanding the ring by means of a transamidation.[1858] The discoverers call it the *Zip reaction*, by analogy with the action of zippers.[1859]

Lactams can be opened to ω-amino amides by reaction with amines at 10 kbar.[1860]

OS **I**, 302 (but see **V**, 589), 450, 453; **II**, 461; **III**, 151, 404; **IV**, 52, 361. See also, OS **VIII**, 573.

16-77 Acylation of Amines by Other Acid Derivatives

Acylamino-de-halogenation or dealkoxlaton

$$RCOCl + H_2NCOR' \longrightarrow RCONHCOR'$$

Acid derivatives that can be converted to amides include thiol acids RCOSH, thiol esters RCOSR,[1861] acyloxyboranes $RCOB(OR')_2$,[1862] silicic esters $(RCOO)_4Si$,

[1854]Murakami, Y.; Kondo, K.; Miki, K.; Akiyama, Y.; Watanabe, T.; Yokoyama, Y. *Tetrahedron Lett.* **1997**, *38*, 3751.

[1855]For a discussion of the mechanism, see Chimishkyan, A.L.; Snagovskii, Yu.S.; Gulyaev, N.D.; Leonova, T.V.; Kusakin, M.S. *J. Org. Chem. USSR* **1985**, *21*, 1955.

[1856]Yang, Y.; Lu, S. *Org. Prep. Proceed. Int.* **1999**, *31*, 559.

[1857]Garcia, J.; Vilarrasa, J. *Tetrahedron Lett.* **1982**, *23*, 1127.

[1858]Askitoğlu, E.; Guggisberg, A.; Hesse, M. *Helv. Chim. Acta* **1985**, *68*, 750, and references cited therein. For a carbon analog, see Süsse, M.; Hájiček, J.; Hesse, M. *Helv. Chim. Acta* **1985**, *68*, 1986.

[1859]For a review of this reaction, and of other ring expansions to form macrocyclic rings, see Stach, H.; Hesse, M. *Tetrahedron* **1988**, *44*, 1573.

[1860]Kotsuki, H.; Iwasaki, M.; Nishizawa, H. *Tetrahedron Lett.* **1992**, *33*, 4945.

[1861]For a discussion of the mechanism, see Douglas, K.T. *Acc. Chem. Res.* **1986**, *19*, 186.

[1862]The best results are obtained when the acyloxyboranes are made from a carboxylic acid and catecholborane (p. 1123): Collum, D.B.; Chen, S.; Ganem, B. *J. Org. Chem.* **1978**, *43*, 4393.

1,1,1-trihalo ketones $RCOCX_3$,[1863] α-keto nitriles, acyl azides, and non-enolizable ketones (see the Haller–Bauer reaction **12-34**). A polymer-bound acyl derivative was converted to an amide using tributylvinyl tin, trifluoroacetic acid, $AsPh_3$ and a palladium catalyst.[1864] The source of amine in this reaction was the polymer itself, which was an amide resin. N-Acylsulfonamides react with primary amines to the amide (AcNHR).[1865] Aniline derivatives are converted to acetamides with N-acyl oxymethylpyradazin-3-ones in dichloromethane.[1866] Carbonylation reactions can be used to prepare amides and related compounds. The reaction of a primary amine, an alkyl halide with CO_2, in the presence of Cs_2CO_3/Bu_4NI, gave the corresponding carbamate.[1867]

OS **III**, 394; **IV**, 6, 569; **V**, 160, 166; **VI**, 1004.

Imides can be prepared by the attack of amides or their salts on acyl halides, anhydrides, and carboxylic acids or esters.[1868] The best synthetic method for the preparation of acyclic imides is the reaction between an amide and an anhydride at 100°C catalyzed by H_2SO_4.[1869] When acyl chlorides are treated with amides in a 2:1 molar ratio at low temperatures in the presence of pyridine, the products are N,N-diacylamides, $(RCO)_3N$.[1870]

This reaction is often used to prepare urea derivatives, an important example being the preparation of barbituric acid, **112**.[1871]

$$
\begin{array}{ccc}
\ce{CO2Et} \\
\ce{CO2Et}
\end{array}
+
\begin{array}{c}
\ce{H2N} \\
\ce{H2N}
\end{array}
\!\!=\!\!O
\quad\xrightarrow{\ \ce{OEt^-}\ }\quad
\underset{\textbf{112}}{\text{barbituric acid ring}}
$$

When the substrate is oxalyl chloride (ClCOCOCl) and the reagent an unsubstituted amide, an acyl isocyanate (RCONCO) is formed. The "normal" product (RCONH-COCOCl) does not form, or if it does, it rapidly loses CO and HCl.[1872]

OS **II**, 60, 79, 422; **III**, 763; **IV**, 245, 247, 496, 566, 638, 662, 744; **V**, 204, 944.

[1863]See, for example, Salim, J.R.; Nome, F.; Rezende, M.C. *Synth. Commun.* **1989**, *19*, 1181; Druzian, J.; Zucco, C.; Rezende, M.C.; Nome, F. *J. Org. Chem.* **1989**, *54*, 4767.

[1864]Deshpande, M.S. *Tetrahedron Lett.* **1994**, *35*, 5613.

[1865]Coniglio, S.; Aramini, A.; Cesta, M.C.; Colagioia, S.; Curti, R.; D'Alessandro, F.; D'anniballe, G.; D'Elia, V.; Nano, G.; Orlando, V.; Allegretti, M. *Tetrahedron Lett.* **2004**, *45*, 5375.

[1866]Kang, Y.-J.; Chung, H.-A.; Kim, J.-J.; Yoon, Y.-J. *Synthesis* **2002**, 733.

[1867]Salvatore, R.N.; Shin, S.I.; Nagle, A.S.; Jung, K.W. *J. Org. Chem.* **2001**, *66*, 1035.

[1868]For a review, see Challis, B.C.; Challis, J.A., in Zabicky, J. *The Chemistry of Amides*, Wiley, NY, **1970**, pp. 759–773.

[1869]Baburao, K.; Costello, A.M.; Petterson, R.C.; Sander, G.E. *J. Chem. Soc. C* **1968**, 2779; Davidson, D.; Skovronek, H. *J. Am. Chem. Soc.* **1958**, *80*, 376.

[1870]For example, see LaLonde, R.T.; Davis, C.B. *J. Org. Chem.* **1970**, *35*, 771.

[1871]For a review of barbituric acid, see Bojarski, J.T.; Mokrosz, J.L.; Barto, H.J.; Paluchowska, M.H. *Adv. Heterocycl. Chem.* **1985**, *38*, 229.

[1872]Speziale, A.J.; Smith, L.R.; Fedder, J.E. *J. Org. Chem.* **1965**, *30*, 4306.

16-78 Acylation of Azides

$$\underset{R}{\overset{O}{\underset{\|}{C}}}\!\!-\!H \quad \xrightarrow[\text{NaN}_3]{\text{PhI(OAc)}_2} \quad \underset{R}{\overset{O}{\underset{\|}{C}}}\!\!-\!N_3$$

The reaction of an aldehyde with sodium azide and Et_4 $I(OAc)_2$ or polymer-bound $PhI(OAc)_2$ leads to an acyl azide.[1873]

F. Attack by Halogen at an Acyl Carbon

16-79 Formation of Acyl Halides from Carboxylic Acids

Halo-de-hydroxylation

$$RCOOH \quad + \quad \underset{\text{agent}}{\text{Halogenating}} \quad \longrightarrow \quad RCOX$$

Halogenating agent $= SOCl_2$, $SOBr_2$, PCl_3, $POCl_3$, PBr_3, and so on.

The same inorganic acid halides that convert alcohols to alkyl halides (**10-48**) also convert carboxylic acids to acyl halides.[1874] The reaction is the best and the most common method for the preparation of acyl chlorides. Bromides and iodides[1875] are also made in this manner, but much less often. Acyl bromides can be prepared with BBr_3 on alumina.[1876] Thionyl chloride[1877] is a good reagent, since the by-products are gases and the acyl halide is easily isolated, but PX_3 and PX_5 ($X = Cl$ or Br) are also commonly used.[1878] Hydrogen halides do not give the reaction. A particularly mild procedure, similar to one mentioned in **10-48**, involves reaction of the acid with Ph_3P in CCl_4, whereupon acyl chlorides are produced without obtaining any acidic compound as a by-product.[1879] Acyl fluorides can be prepared by treatment of carboxylic acids with cyanuric fluoride.[1880] Acid salts

---

[1873]Marinescu, L.G.; Pedersen, C.M.; Bols, M. *Tetrahedron* **2005**, *61*, 123. Aldehydes are converted to acyl azides by reaction with IN_3, see Marinescu, L.; Thinggaard, J.; Thomsen, I. B.; Bols, M. *J. Org. Chem.* **2003**, *68*, 9453. See Hünig, S.; Schaller, R. *Angew. Chem. Int. Ed.* **1982**, *21*, 36.

[1874]For a review, see Ansell, M.F., in Patai, S. *The Chemistry of Acyl Halides*, Wiley, NY, *1972*, pp. 35–68.

[1875]Carboxylic acids and some of their derivatives react with diiodosilane SiH_2I_2 to give good yields of acyl iodides: Keinan, E.; Sahai, M. *J. Org. Chem.* **1990**, *55*, 3922.

[1876]Bains, S.; Green, J.; Tan, L.C.; Pagni, R.M.; Kabalka, G.W. *Tetrahedron Lett.* **1992**, *33*, 7475.

[1877]For a review of thionyl chloride ($SOCl_2$), see Pizey, J.S. *Synthetic Reagents*, Vol. 1, Wiley, NY, *1974*, pp. 321–357. See Mohanazadeh, F.; Momeni, A.R. *Org. Prep. Proceed. Int.* **1996**, *28*, 492 for the use of $SOCl_2$ on silica gel.

[1878]For a list of reagents, with references, see Larock, R.C. *Comprehensive Organic Transformations*, 2nd ed., Wiley-VCH, NY, *1999*, pp. 1929–1930.

[1879]Lee, J.B. *J. Am. Chem. Soc.* **1966**, *88*, 3440. For other methods of preparing acyl chlorides, see Venkataraman, K.; Wagle, D.R. *Tetrahedron Lett.* **1979**, 3037; Devos, A.; Remion, J.; Frisque-Hesbain, A.; Colens, A.; Ghosez, L. *J. Chem. Soc., Chem. Commun.* **1979**, 1180.

[1880]Olah, G.A.; Nojima, M.; Kerekes, I. *Synthesis* **1973**, 487. For other methods of preparing acyl fluorides, see Mukaiyama, T.; Tanaka, T. *Chem. Lett.* **1976**, 303; Ishikawa, N.; Sasaki, S. *Chem. Lett.* **1976**, 1407.

are also sometimes used as substrates. Acyl halides are also used as reagents in an exchange reaction:

$$RCOOH \; + \; R'COCl \; \rightleftharpoons \; RCOCl \; + \; R'COOH$$

which probably involves an anhydride intermediate. This is an equilibrium reaction that must be driven to the desired side.

A mild, and often superior reagent is oxalyl chloride (**113**) and oxalyl bromide, since oxalic acid decomposes to CO and CO_2, and the equilibrium is thus driven to the side of the other acyl halide.[1881] These reagents are commonly the reagent of choice, particularly when sensitive functionality is present elsewhere in the molecule.

113

OS **I**, 12, 147, 394; **II**, 74, 156, 169, 569; **III**, 169, 490, 547, 555, 613, 623, 712, 714; **IV**, 34, 88, 154, 263, 339, 348, 554, 608, 616, 620, 715, 739, 900; **V**, 171, 258, 887; **VI**, 95, 190, 549, 715; **VII**, 467; **VIII**, 441, 486, 498.

16-80   Formation of Acyl Halides from Acid Derivatives

Halo-de-acyloxy-substitution

Halo-de-halogenation

$$(RCO)_2O \; + \; HF \; \longrightarrow \; RCOF$$
$$RCOCl \; + \; HF \; \longrightarrow \; RCOF$$

These reactions are most important for the preparation of acyl fluorides.[1882] Acyl chlorides and anhydrides can be converted to acyl fluorides by treatment with poly-hydrogen fluoride–pyridine solution[1883] or with liquid HF at $-10°C$.[1884] Formyl fluoride, which is a stable compound, was prepared by the latter procedure from the mixed anhydride of formic and acetic acids.[1885] Acyl fluorides can also be obtained by reaction of acyl chlorides with KF in acetic acid[1886] or with diethyl-aminosulfur trifluoride (DAST).[1887] Carboxylic esters and anhydrides can be

---

[1881]Adams, R.; Ulich, L.H., *J. Am. Chem. Soc.* **1920**, *42*, 599; Wood, T.R.; Jackson, F.L.; Baldwin, A.R.; Longenecker, H.E. *J. Am. Chem. Soc.* **1944**, *66*, 287. For a typical example see Zhang, A.; Nie, J. *J. Agric. Food Chem.* **2005**, *53*, 2451.

[1882]For lists of reagents converting acid derivatives to acyl halides, see Larock, R.C. *Comprehensive Organic Transformations*, 2nd ed., Wiley-VCH, NY, **1999**, pp. 1950–1951, 1955, 1968.

[1883]Olah, G.A.; Welch, J.; Vankar, Y.D.; Nojima, M.; Kerekes, I.; Olah, J.A. *J. Org. Chem.* **1979**, *44*, 3872. See also, Yin, J.; Zarkowsky, D.S.; Thomas, D.W.; Zhao, M.W.; Huffman, M.A. *Org. Lett.* **2004**, *6*, 1465.

[1884]Olah, G.A.; Kuhn, S.J. *J. Org. Chem.* **1961**, *26*, 237.

[1885]Olah, G.A.; Kuhn, S.J. *J. Am. Chem. Soc.* **1960**, *82*, 2380.

[1886]Emsley, J.; Gold, V.; Hibbert, F.; Szeto, W.T.A. *J. Chem. Soc. Perkin Trans. 2* **1988**, 923.

[1887]Markovski, L.N.; Pashinnik, V.E. *Synthesis* **1975**, 801.

converted to acyl halides other than fluorides by the inorganic acid halides mentioned in **16-79**, as well as with Ph_3PX_2 (X = Cl or Br),[1888] but this is seldom done. Halide exchange can be carried out in a similar manner. When halide exchange is done, it is always acyl bromides and iodides that are made from chlorides, since chlorides are by far the most readily available.[1889]

OS **II**, 528; **III**, 422; **V**, 66, 1103; **IX**, 13. See also, OS **IV**, 307.

G. Attack by Carbon at an Acyl Carbon[1890]

16-81   The Conversion of Acyl Halides to Ketones With Organometallic Compounds[1891]

Alkyl-de-halogenation

Acyl halides react cleanly and under mild conditions with lithium dialkylcopper reagents (see **10-58**)[1892] to give high yields of ketones.[1893] The R′ group may be primary, secondary, or tertiary alkyl or aryl and may contain iodo, keto, ester, nitro, or cyano groups. The R groups that have been used successfully are methyl, primary alkyl, and vinylic. Secondary and tertiary alkyl groups can be introduced by the use of PhS(R)CuLi (p. 602) instead of R_2CuLi,[1894] or by the use of either the mixed homocuprate $(R'SO_2CH_2CuR)^-$ Li^+,[1895] or a magnesium dialkylcopper reagent "RMeCuMgX."[1896] Secondary alkyl groups can also be introduced with the copper–zinc reagents RCu(CN)ZnI.[1897] The R group may be alkynyl if a

[1888]Burton, D.J.; Koppes, W.M. *J. Chem. Soc., Chem. Commun.* **1973**, 425; *J. Org. Chem.* **1975**, *40*, 3026; Anderson Jr., A.G.; Kono, D.H. *Tetrahedron Lett.* **1973**, 5121.

[1889]For methods of converting acyl chlorides to bromides or iodides, see Schmidt, A.H.; Russ, M.; Grosse, D. *Synthesis* **1981**, 216; Hoffmann, H.M.R.; Haase, K. *Synthesis* **1981**, 715.

[1890]For a discussion of many of the reactions in this section, see House, H.O. *Modern Synthetic Reactions*, 2nd ed., W.A. Benjamin, NY, **1972**, pp. 691–694, 734–765.

[1891]For a review, see Cais, M.; Mandelbaum, A., in Patai, S. *The Chemistry of the Carbonyl Group*, Vol. 1, Wiley, NY, **1966**, Vol. 1, pp. 303–330.

[1892]See Posner, G.H. *An Introduction to Synthesis Using Organocopper Reagents*,Wiley, NY, **1980**, pp. 81–85. Ryu, I.; Ikebe, M.; Sonoda, N.; Yamamoto, S.-y.; Yamamura, G.-h.; Komatsu, M. *Tetrahedron Lett.* **2002**, *43*, 1257.

[1893]Vig, O.P.; Sharma, S.D.; Kapur, J.C. *J. Indian Chem. Soc.* **1969**, *46*, 167; Jukes, A.E.; Dua, S.S.; Gilman, H. *J. Organomet. Chem.* **1970**, *21*, 241; Posner, G.H.; Whitten, C.E.; McFarland, P.E. *J. Am. Chem. Soc.* **1972**, *94*, 5106; Luong-Thi, N.; Rivière, H. *J. Organomet. Chem.* **1974**, *77*, C52.

[1894]Posner, G.H.; Whitten, C.E.; Sterling, H.J. *J. Am. Chem. Soc.* **1973**, *95*, 7788; Posner, G.H.; Whitten, C.E. *Tetrahedron Lett.* **1973**, 1815; Bennett, G.B.; Nadelson, J.; Alden, L.; Jani, A. *Org. Prep. Proced. Int.* **1976**, *8*, 13.

[1895]Johnson, C.R.; Dhanoa, D.S. *J. Org. Chem.* **1987**, *52*, 1885.

[1896]Bergbreiter, D.E.; Killough, J.M. *J. Org. Chem.* **1976**, *41*, 2750.

[1897]Knochel, P.; Yeh, M.C.P.; Berk, S.C.; Talbert, J. *J. Org. Chem.* **1988**, *53*, 2390.

cuprous acetylide $R^2C\equiv CCu$ is the reagent.[1898] Organocopper reagents generated *in situ* from highly reactive copper, and containing such functional groups as cyano, chloro, and ester, react with acyl halides to give ketones.[1899]

Many other organometallic reagents[1900] give good yields of ketones when treated with acyl halides because, as with R_2CuLi, R_2Cd, these compounds do not generally react with the ketone product. A particularly useful class of organometallic reagent are organocadmium reagents R_2Cd, prepared from Grignard reagents (**12-22**). In this case, R may be aryl or primary alkyl. In general, secondary and tertiary alkylcadmium reagents are not stable enough to be useful in this reaction.[1901] An ester group may be present in either $R'COX$ or R_2Cd. Direct treatment of the acid chloride with an alkyl halide and cadmium metal leads to the ketone in some cases.[1902] Organozinc compounds behave similarly to dialkylcadmium reagents, but are used less often.[1903] Organotin reagents R_4Sn react with acyl halides to give high yields of ketones, if a Pd complex is present.[1904] Organolead reagents R_4Pb behave similarly.[1905] Allylic halides and indium metal react with acyl chlorides to give the ketone.[1906] Various other groups, for example, nitrile, ester, and aldehyde can be present in the acyl halide without interference. Other reagents include organomanganese compounds[1907] (R can be primary, secondary, or tertiary alkyl, vinylic, alkynyl, or aryl), organozinc,[1908] and organothallium compounds (R can be primary alkyl or aryl).[1909] The reaction of an α-halo-ketone and an acyl chloride with SmI_2 leads to a β-diketone.[1910] Initial reaction of an acyl chloride with palladium(0), followed by reaction with potassium acetate and then a trialkylborane gave a ketone.[1911] Arylboronic acids and acid chloride give the ketone in

[1898]Castro, C.E.; Havlin, R.; Honwad, V.K.; Malte, A.; Mojé, S. *J. Am. Chem. Soc.* **1969**, *91*, 6464. For methods of preparing acetylenic ketones, see Verkruijsse, H.D.; Heus-Kloos, Y.A.; Brandsma, L. *J. Organomet. Chem.* **1988**, *338*, 289.

[1899]Wehmeyer, R.M.; Rieke, R.D. *Tetrahedron Lett.* **1988**, *29*, 4513; Stack, D.E.; Dawson, B.T.; Rieke, R.D. *J. Am. Chem. Soc.* **1992**, *114*, 5110.

[1900]For a list of reagents, with references, see Larock, R.C. *Comprehensive Organic Transformations*, 2nd ed., Wiley-VCH, NY, **1999**, pp. 1389–1400.

[1901]Cason, J.; Fessenden, R. *J. Org. Chem.* **1960**, *25*, 477.

[1902]Baruah, B.; Boruah. A.; Prajapati, D.; Sandhu, J.S. *Tetrahedron Lett.* **1996**, *37*, 9087.

[1903]For examples, see Grey, R.A. *J. Org. Chem.* **1984**, *49*, 2288; Tamaru, Y.; Ochiai, H.; Nakamura, T.; Yoshida, Z. *Org. Synth. 67*, 98.

[1904]Labadie, J.W.; Stille, J.K. *J. Am. Chem. Soc.* **1983**, *105*, 669, 6129; Labadie, J.W.; Tueting, D.; Stille, J.K. *J. Org. Chem.* **1983**, *48*, 4634. For a Me_3SiSnR_3 reagent, see Geng, F.; Maleczka, Jr., R.E. *Tetrahedron Lett.* **1999**, *40*, 3113. For an allylic SnR_3 reagent, see Inoue, K.; Shimizu, Y.; Shibata, I.; Baba, A. *Synlett* **2001**, 1659.

[1905]Yamada, J.; Yamamoto, Y. *J. Chem. Soc., Chem. Commun.* **1987**, 1302.

[1906]Yadav, J.S.; Srinivas, D.; Reddy, G.S.; Bindu, K.H. *Tetrahedron Lett.* **1997**, *38*, 8745. Also see, Bryan, V.J.; Chan, T.-H. *Tetrahedron Lett.* **1997**, *38*, 6493 for a similar reaction with an acyl imidazole.

[1907]Kim, S.-H.; Rieke, R.D. *J. Org. Chem.* **1998**, *63*, 6766; Cahiez, G.; Martin, A.; Delacroix, T. *Tetrahedron Lett.* **1999**, *40*, 6407.

[1908]Hanson, M.V.; Brown, J.D.; Rieke, R.D.; Niu, Q.J. *Tetrahedron Lett.* **1994**, *35*, 7205; Filon, H.; Gosmini, C.; Périchon, J. *Tetrahedron* **2003**, *59*, 8199.

[1909]Markó, I.E.; Southern, J.M. *J. Org. Chem.* **1990**, *55*, 3368.

[1910]Ying, T.; Bao, W.; Zhang, Y.; Xu, W. *Tetrahedron Lett.* **1996**, *37*, 3885.

[1911]Kabalka, G.W.; Malladi, R.R.; Tejedor, D.; Kelley, S. *Tetrahedron Lett.* **2000**, *41*, 999.

the presence of a palladium catalyst.[1912] Similar reaction of acid chlorides, NaBPh$_4$, KF, and a palladium catalyst gave the aryl ketone.[1913] Antimony alkynes such as Ph$_2$Sb–C≡C–Ph react with acid chloride in the presence of a palladium catalyst to give the conjugated alkynyl ketone.[1914] Such conjugated ketones can also be prepared from an acyl halide, a terminal alkyne and a CuI catalyst[1915] a palladium catalyst,[1916] or with indium metal.[1917] Terminal alkynes react with chloroformates and a palladium catalyst to give the corresponding propargyl ester.[1918] Similar reaction of an alkyne with an acid chloride and a palladium–copper[1919] or CuI catalyst,[1920] both with microwave irradiation, gave the alkynyl ketones.

When the organometallic compound is a Grignard reagent,[1921] ketones are generally not obtained because the initially formed ketone reacts with a second molecule of RMgX to give the salt of a tertiary alcohol (**16-82**). Ketones *have* been prepared in this manner by the use of low temperatures, inverse addition (i.e., addition of the Grignard reagent to the acyl halide rather than the other way), excess acyl halide, and so on., but the yields are usually low, though high yields have been reported in THF at −78°C.[1922] Pretreatment with a trialkylphosphine and then the Grignard reagent can lead to the ketone.[1923] Using CuBr[1924] or a nickel catalyst[1925] with the Grignard reagent can lead to the ketone. Some ketones are unreactive toward Grignard reagents for steric or other reasons; these can be prepared in this way.[1926] Other methods involve running the reaction in the presence of Me$_3$SiCl[1927] (which reacts with the initial adduct in the tetrahedral mechanism, p. 1254), and the use of a combined Grignard-lithium diethylamide reagent.[1928] Also, certain metallic halides, notably ferric and cuprous halides, are catalysts

[1912]Urawa, Y.; Ogura, K. *Tetrahedron Lett.* **2003**, *44*, 271.

[1913]Wang, J.-X.; Wei, B.; Hu, Y.; Liu, Z.; Yang, Y. *Synth. Commun.* **2001**, *31*, 3885.

[1914]Kakusawa, N.; Yamaguchi, K.; Kurita, J.; Tsuchiya, T. *Tetrahedron Lett.* **2000**, *41*, 4143.

[1915]Chowdhury, C.; Kundu, N.G. *Tetrahedron* **1999**, *55*, 7011; Wang, J.-X.; Wei, B.; Hu, Y.; Liua, Z.; Kang, L. *J. Chem. Res. (S)* **2001**, 146.

[1916]Karpov, A.S.; Müller, T.J.J. *Org. Lett.* **2003**, *5*, 3451.

[1917]Augé, J.; Lubin-Germain, N.; Seghrouchni, L. *Tetrahedron Lett.* **2003**, *44*, 819.

[1918]Böttcher, A.; Becker, H.; Brunner, M.; Preiss, T.; Henkelmann, J..; De Bakker, C.; Gleiter, R. *J. Chem. Soc., Perkin Trans.1* **1999**, 3555.

[1919]Wang, J.-x.; Wei, B.; Huang, D.; Hu, Y.; Bai, L. *Synth. Commun.* **2001**, *31*, 3337.

[1920]Wang, J.-X.; Wei, B.; Hu, Y.; Liu, Z.; Fu, Y. *Synth. Commun.* **2001**, *31*, 3527.

[1921]For a review, see Kharasch, M.S.; Reinmuth, O. *Grignard Reactions of Nonmetallic Substances*, Prentice-Hall, Englewood, NJ, *1954*, pp. 712–724.

[1922]Sato, M.; Inoue, M.; Oguro, K.; Sato, M. *Tetrahedron Lett.* **1979**, 4303; Eberle, M.K.; Kahle, G.G. *Tetrahedron Lett.* **1980**, *21*, 2303; Föhlisch, B.; Flogaus, R. *Synthesis* **1984**, 734.

[1923]Maeda, H.; Okamoto, J.; Ohmori, H. *Tetrahedron Lett.* **1996**, *37*, 5381.

[1924]Babudri, F.; Fiandanese, V.; Marchese, G.; Punzi, A. *Tetrahedron* **1996**, *52*, 13513; *Tetrahedron Lett.* **1995**, *36*, 7305.

[1925]Malanga, C.; Aronica, L.A.; Lardicci, L. *Tetrahedron Lett.* **1995**, *36*, 9185. For the preparation of an amide from a *N,N*-dialkylcarbamyl chloride (R$_2$NCOCl) and a Grignard reagent, with a nickel catalyst, see Lemoucheux, L.; Rouden, J.; Lasne, M.-C. *Tetrahedron Lett.* **2000**, *41*, 9997.

[1926]For example, see Lion, C.; Dubois, J.E.; Bonzougou, Y. *J. Chem. Res. (S)* **1978**, 46; Dubois, J.E.; Lion, C.; Arouisse, A. *Bull. Soc. Chim. Belg.* **1984**, *93*, 1083.

[1927]Cooke, Jr., M.P. *J. Org. Chem.* **1986**, *51*, 951.

[1928]Fehr, C.; Galindo, J.; Perret, R. *Helv. Chim. Acta* **1987**, *70*, 1745.

that improve the yields of ketone at the expense of tertiary alcohol.[1929] For these catalysis, both free radical and ionic mechanisms have been proposed.[1930]

Grignard reagents react with ethyl chloroformate to give carboxylic esters $EtOCOCl + RMgX \rightarrow EtOCOR$. Acyl halides can also be converted to ketones by treatment with $Na_2Fe(CO)_4$ followed by $R'X$ (**10-76**).

OS **II**, 198; **III**, 601; **IV**, 708; **VI**, 248, 991; **VII**, 226, 334; **VIII**, 268, 274, 371, 441, 486.

16-82   The Conversion of Anhydrides, Carboxylic Esters, or Amides to Ketones With Organometallic Compounds[1931]

Dialkyl,hydroxy-de-alkoxy,oxo-tersubstitution; Alkyl-de-acyloxy- or de-amido substitution

$$
\underset{R \quad OR^1}{\overset{O}{\underset{\|}{C}}} + 2\ R^2\text{-MgX} \longrightarrow \underset{R \quad O\text{-MgX}}{\overset{R^2 \quad R^2}{C}} \xrightarrow{\text{hydrol.}} \underset{R \quad OH}{\overset{R^2 \quad R^2}{C}}
$$

When carboxylic esters are treated with Grignard reagents, addition to the carbonyl (**16-24**) generates a ketone. Under the reaction conditions, the initially formed ketones usually undergoes acyl substitution of R^2 for OR' (**16-81**), so that tertiary alcohols are formed in which two R groups are the same. Isolation of the ketone as the major product is possible in some cases, particularly when the reaction is done at low temperature[1932] and when there is steric hindrance to the carbonyl in the first-formed ketone. Esters, RCO_2Me, react with $Zn(BH_4)_2/$ EtMgBr to give an alcohol, $RCH(OH)Et$.[1933] Formates give secondary alcohols and carbonates give tertiary alcohols in which all three R groups are the same: $(EtO)_2C{=}O + RMgX \rightarrow R_3COMgX$. Acyl halides and anhydrides behave similarly, though these substrates are employed less often.[1934] Many side reactions are possible, especially when the acid derivative or the Grignard reagent is branched: enolizations, reductions (not for esters, but for halides), condensations, and cleavages, but the most important is simple substitution (**16-81**), which in some cases can be made to predominate. When 1,4-dimagnesium compounds are used, carboxylic esters are converted to cyclopentanols.[1935] 1,5-Dimagnesium

---

[1929]For examples, see Cason, J.; Kraus, K.W. *J. Org. Chem.* **1961**, *26*, 1768, 1772; MacPhee, J.A.; Dubois, J.E. *Tetrahedron Lett.* **1972**, 467; Cardellicchio, C.; Fiandanese, V.; Marchese, G.; Ronzini, L. *Tetrahedron Lett.* **1987**, *28*, 2053; Fujisawa, T.; Sato, T. *Org. Synth. 66*, 116; Babudri, F.; D'Ettole, A.; Fiandanese, V.; Marchese, G.; Naso, F. *J. Organomet. Chem.* **1991**, *405*, 53.

[1930]For example, see Dubois, J.E.; Boussu, M. *Tetrahedron Lett.* **1970**, 2523; *Tetrahedron* **1973**, *29*, 3943; MacPhee, J.A.; Boussu, M.; Dubois, J.E. *J. Chem. Soc. Perkin Trans. 2* **1974**, 1525.

[1931]For a review, see Kharasch, M.S.; Reinmuth, O. *Grignard Reactions of Nonmetallic Substances*, Prentice-Hall, Englewood, NJ, **1954**, pp. 561–562, 846–908.

[1932]Deskus, J.; Fan, D.; Smith, M.B. *Synth. Commun.* **1998**, *28*, 1649.

[1933]Hallouis, S.; Saluzzo, C.; Amouroux, R. *Synth. Commun.* **2000**, *30*, 313.

[1934]For a review of these reactions, see Kharasch, M.S.; Reinmuth, O. *Grignard Reactions of Nonmetallic Substances*, Prentice-Hall, Englewood Cliffs, NJ, **1954**, pp. 549–766, 846–869.

[1935]Canonne, P.; Bernatchez, M. *J. Org. Chem.* **1986**, *51*, 2147; **1987**, *52*, 4025.

compounds give cyclohexanols, but in lower yields.[1936]

As is the case with acyl halides (**16-81**), anhydrides and carboxylic esters give tertiary alcohols (**16-82**) when treated with Grignard reagents. Low temperatures,[1937] the solvent HMPA,[1938] and inverse addition have been used to increase the yields of ketone.[1939] Amides give better yields of ketone at room temperature, but still not very high.[1940] Anhydrides can react with arylmagnesium halides at low temperature, and in the presence of (−)-sparteine, to give a keto acid with good enantioselectivity.[1941] Organocadmium reagents are less successful with these substrates than with acyl halides (**16-81**). Esters of formic acid, dialkylformamides, and lithium or sodium formate[1942] give good yields of aldehydes, when treated with Grignard reagents.

Alkyllithium compounds have been used to give ketones from carboxylic esters. The reaction must be carried out in a high-boiling solvent, such as toluene, since reaction at lower temperatures gives tertiary alcohols.[1943] Alkyllithium reagents also give good yields of carbonyl compounds with *N,N*-disubstituted amides.[1944] Dialkylformamides give aldehydes and other disubstituted amides give ketones and other acid derivatives have been used.[1945]

[1936]Kresge, A.J.; Weeks, D.P. *J. Am. Chem. Soc.* **1984**, *106*, 7140. See also, Fife, T.H. *J. Am. Chem. Soc.* **1967**, *89*, 3228; Craze, G.; Kirby, A.J.; Osborne, R. *J. Chem. Soc. Perkin Trans. 2* **1978**, 357; Amyes, T.L.; Jencks, W.P. *J. Am. Chem. Soc.* **1989**, *111*, 7888, 7900.

[1937]See, for example, Newman, M.S.; Smith, A.S. *J. Org. Chem.* **1948**, *13*, 592; Edwards, Jr., W.R.; Kammann Jr., K.P. *J. Org. Chem.* **1964**, *29*, 913; Araki, M.; Sakat, S.; Takei, H.; Mukaiyama, T. *Chem. Lett.* **1974**, 687.

[1938]Huet, F.; Pellet, M.; Conia, J.M. *Tetrahedron Lett.* **1976**, 3579.

[1939]For a list of preparations of ketones by the reaction of organometallic compounds with carboxylic esters, salts, anhydyrides, or amides, with references, see Larock, R.C. *Comprehensive Organic Transformations*, 2nd ed., Wiley-VCH, NY, **1999**, pp. 1386–1389, 1400–1419.

[1940]For an improved procedure with amides, see Olah, G.S.; Prakash, G.K.S.; Arvanaghi, M. *Synthesis* **1984**, 228. See Martín, R.; Romea, P.; Tey, C.; Urpí, F.; Vilarrasa, J. *Synlett* **1997**, 1414 for reaction with an amide derived from morpholine and Grignard reagents, which gives the ketone in good yield. See Kashima, C.; Kita, I.; Takahashi, K.; Hosomi, A. *J. Heterocyclic Chem.* **1995**, *32*, 25 for a related reaction.

[1941]Shintani, R.; Fu, G.C. *Angew. Chem. Int. Ed.* **2002**, *41*, 1057.

[1942]Bogavac, M.; Arsenijević, L.; Pavlov, S.; Arsenijević, V. *Tetrahedron Lett.* **1984**, *25*, 1843.

[1943]Petrov, A.D.; Kaplan, E.P.; Tsir, Ya. *J. Gen. Chem. USSR* **1962**, *32*, 691.

[1944]Evans, E.A. *J. Chem. Soc.* **1956**, 4691. For the synthesis of a silyl ketone from a silyllithium reagent, see Clark, C.T.; Milgram, B.C.; Scheidt, K.A. *Org. Lett.* **2004**, *6*, 3977. For a review, see Wakefield, B.J. *Organolithium Methods*, Academic Press, NY, **1988**, pp. 82–88.

[1945]Mueller-Westerhoff, U.T.; Zhou, M. *Synlett* **1994**, 975.

Ketones can also be obtained by treatment of the lithium salt of a carboxylic acid with an alkyllithium reagent (**16-28**). For an indirect way to convert carboxylic esters to ketones, see **16-82**. A similar reaction with hindered aryl carboxylic acids has been reported.[1946] Treatment of a β-amido acid with two equivalents of *n*-butyllithium, followed by reaction with an acid chloride leads to a β-keto amide.[1947] Carboxylic acids can be treated with 2-chloro-4,6-dimethoxy[1,3,5]-triazine and the RMgX/CuI to give ketones.[1948]

Disubstituted formamides can give addition of 2 equivalents of Grignard reagent. The products of this reaction (called *Bouveault reaction*) are an aldehyde and a tertiary amine.[1949] The use of an amide other than a formamide

$$
\begin{array}{c}
\overset{\displaystyle O}{\underset{R}{\overset{\|}{C}}}\!-\!NR_2 \;+\; 3\ R'\text{-MgX} \;\longrightarrow\; \underset{R}{\overset{R'\diagdown\ \diagup R'}{\underset{NR_2}{C}}} \;+\; R'CHO
\end{array}
$$

can give a ketone instead of an aldehyde, but yields are generally low. The addition of 2 equivalents of phenyllithium to a carbamate gave good yields of the ketone, however.[1950] When butyllithium reacted with an α-carbamoyl secondary amide [RCH(NHCO$_2$R′)C(=O)NR$_2^2$] at −78°C, the amide reacted preferentially to give the α-carbamoyl ketone.[1951] The reaction of N-(3-bromopropyl) lactams with *tert*-butyllithium gave cyclization to the bicyclic amino alcohol, and subsequent reduction with LiAlH$_4$ (**19-64**) gave the bicyclic amine.[1952] Ketones can also be prepared by treatment of thioamides with organolithium compounds (alkyl or aryl).[1953] Cerium reagents, such as MeCeCl$_2$, also add two R groups to an amide.[1954] More commonly, an organolithium reagent is treated with CeCl$_3$ to generate the organocerium reagent *in situ*.[1955] It has proven possible to add two different R groups by sequential addition of two Grignard reagents.[1956] Diketones have also been produced by using the bis(imidazole) derivative of oxalic acid.[1957] Alternatively, if R′ contains an α hydrogen, the product may be an enamine, and enamines have been synthesized in goods yields by this method.[1958] When an amide having a *gem*-dibromocyclopropyl unit elsewhere in the molecule was treated

[1946]Zhang, P.; Terefenko, E.A.; Slavin, J. *Tetrahedron Lett.* **2001**, *42*, 2097.

[1947]Chen, Y.; Sieburth, S.Mc.N. *Synthesis* **2002**, 2191.

[1948]DeLuca, L.; Giacomelli, G.; Porcheddu, A. *Org. Lett.* **2001**, *3*, 1519.

[1949]For a review, see Spialtr, L.; Pappalardo, J.A. *The Acyclic Aliphatic Tertiary Amines*, Macmillan, NY, **1965**, pp. 59–63.

[1950]Prakash, G.K.S.; York, C.; Liao, Q.; Kotian, K.; Olah, G.A. *Heterocycles* **1995**, *40*, 79.

[1951]Sengupta, S.; Mondal, S.; Das, D. *Tetrahedron Lett.* **1999**, *40*, 4107.

[1952]Jones, K.; Storey, J.M.D. *J. Chem. Soc., Perkin Trans. 1* **2000**, 769.

[1953]Tominaga, Y.; Kohra, S.; Hosomi, A. *Tetrahedron Lett.* **1987**, *28*, 1529.

[1954]Calderwood, D.J.; Davies, R.V.; Rafferty, P.; Twigger, H.L.; Whelan, H.M. *Tetrahedron Lett.* **1997**, *38*, 1241.

[1955]Ahn, Y.; Cohen, T. *Tetrahedron Lett.* **1994**, *35*, 203.

[1956]Comins, D.L.; Dernell, W. *Tetrahedron Lett.* **1981**, *22*, 1085.

[1957]Mitchell, R.H.; Iyer, V.S. *Tetrahedron Lett.* **1993**, *34*, 3683. Also see, Sibi, M.P.; Sharma, R.; Paulson, K.L. *Tetrahedron Lett.* **1992**, *33*, 1941.

[1958]Hansson, C.; Wickberg, B. *J. Org. Chem.* **1973**, *38*, 3074.

with methyllithium, Li—Br exchange was accompanied by intramolecular acyl addition to the amide carbonyl, giving a bicyclic amino alcohol.[1959]

N-Methoxy-*N*-methyl amides, such as, **114**, are referred to as a **Weinreb amide**.[1960] When a Weinreb amide reacts with a Grignard reagent or an organolithium reagent,[1961] the product is the ketone. The reaction of **56** with 3-butenylmagnesium bromide to give ketone **115** is a typical example.[1962] Aryloxy carbamates with a Weinreb amide unit, ArO$_2$C—NMe(OMe), react with RMgBr and then R'Li to give an unsymmetrical ketone, RC(=O)R'.[1963] Intramolecular displacement of a Weinreb amide by an organolithium reagent generated *in situ* from an iodide precursor leads to cyclic ketones.[1964] Reaction with vinylmagnesium bromide led to a β-*N*-methoxy-*N*-methylamino ketone, presumably by initial formation of the conjugated ketone followed by Michael addition (**15-24**) of the liberated amine.[1965] An interesting extension of this acyl substitution reaction coupled vinylmagnesium bromide with a Weinreb amide to give a conjugated ketone, which reacted with a secondary amine in a second step (see **15-31**) to give a β-amino ketone.[1966]

By the use of the compound *N*-methoxy-*N*,*N'*,*N'*-trimethylurea, it is possible to add two R groups as RLi, the same or different, to a CO group.[1967]

N,*N*-Disubstituted amides can be converted to alkynyl ketones by treatment with alkynylboranes: RCONR$_2^2$ + (R'C≡C)$_3$B → RCOC≡CR'.[1968] Lactams react with triallylborane to give cyclic 2,2-diallyl amines after treatment with methanol, and then aqueous hydroxide.[1969] Triallylborane reacts with the carbonyl group of lactams, and after treatment with methanol and then aqueous NaOH gives the

[1959]Baird, M.S.; Huber, F.A.M.; Tverezovsky, V.V.; Bolesov, I.G. *Tetrahedron* **2001**, *57*, 1593.

[1960]Nahm, S.; Weinreb, S.M. *Tetrahedron Lett.* **1981**, *22*, 3815.

[1961]See Tallier, C.; Bellosta, V.; Meyer, C.; Cossy, J. *Org. Lett.* **2004**, *6*, 2145.

[1962]Xie, W.; Zou, B.; Pei, D.; Ma, D. *Org. Lett.* **2005**, *7*, 2775. For other exmaples see Andrés, J.M.; Pedrosa, R. Pérez-Encabo, A. *Tetrahedron* **2000**, *56*, 1217.

[1963]Lee, N.R.; Lee, J.I. *Synth. Commun.* **1999**, *29*, 1249.

[1964]Ruiz, J.; Sotomayor, N.; Lete, E. *Org. Lett.* **2003**, *5*, 1115.

[1965]Gomtsyan, A. *Org. Lett.* **2000**, *2*, 11. For a reaction with methyl esters with an excess of vinylmagnesium halide and a copper catalyst to give a 3-butenyl ketone by a similar acyl substitution–Michael addition route, see Hansford, K.A.; Dettwiler, J.E.; Lubell, W.D. *Org. Lett.* **2003**, *5*, 4887.

[1966]Gomtsyan, A.; Koenig, R.J.; Lee, C.-H. *J. Org. Chem.* **2001**, *66*, 3613.

[1967]Hlasta, D.J.; Court, J.J. *Tetrahedron Lett.* **1989**, *30*, 1773. See also, Nahm, S.; Weinreb, S.M. *Tetrahedron Lett.* **1981**, *22*, 3815.

[1968]Yamaguchi, M.; Waseda, T.; Hirao, I. *Chem. Lett.* **1983**, 35.

[1969]Bubnov, Y.N.; Pastukhov, F.V.; Yampolsky, I.V.; Ignatenko, A.V. *Eur. J. Org. Chem.* **2000**, 1503; Li, Z.; Zhang, Y. *Tetrahedron Lett.* **2001**, *42*, 8507.

gem-diallyl amine: 2-pyrrolidinone → 2,2-diallylpyrrolidine.[1970] *N,N*-Disubstituted carbamates (X = OR^2) and carbamoyl chlorides (X = Cl) react with 2 equivalents of an alkyl- or aryllithium or Grignard reagent to give symmetrical ketones, in which both R groups are derived from the organometallic compound: $R_2'NCOX + 2\,RMgX \rightarrow R_2CO$.[1971] *N,N*-Disubstituted amides give ketones in high yields when treated with alkyllanthanum triflates $RLa(OTf)_2$.[1972]

Other organometallic reagents give acyl substitution. Sodium naphthalenide reacts with esters to give naphthyl ketones.[1973] Trimethylaluminum, which exhaustively methylates ketones (**16-24**), also exhaustively methylates carboxylic acids to give *tert*-butyl compounds[1974] (see also, **10-63**). Trimethylaluminum reacts with esters to form ketones, in the presence of *N,N'*-dimethylethyelnediamine.[1975] Thioesters (RCOSR') react with arylboronic acids, in the presence of a palladium catalyst, to give the corresponding ketone,[1976] and esters react similarly with arylboronic acids (a palladium catalyst)[1977] or arylboronates (a ruthenium catalyst).[1978] Trialkylboranes have been similarly used to convert thioesters to ketones.[1979] Thioesters give good yields of ketones when treated with lithium dialkylcopper reagents R_2^2CuLi (R'' = primary or secondary alkyl or aryl).[1980] Organozinc reagents also convert thioesters to ketones.[1981] Arylboronic acids also react with dialkyl anhydrides, with a rhodium catalyst[1982] or a palladium catalyst,[1983] to give the ketone. Aryl iodides react with acetic anhydride, with a palladium catalyst, to give the aryl methyl ketone.[1984] Diaryl- or dialkylzinc reagents react with anhydrides and a palladium catalyst[1985] or a nickel catalyst[1986] to give the ketone. Note that in the presence of a SmI_2 catalyst and 2 equivalents of allyl bromide, lactones were converted to the diallyl diol.[1987] *N*-(3-Iodopropyl)succinimide derivatives react with SmI_2 and an iron catalyst to give bicyclic pyrrolizidinone derivatives

[1970]Bubnov, Yu.N.; Klimkina, E.V.; Zhun', I.V.; Pastukhov, F.V.; Yampolsky, I.V. *Pure Appl. Chem.* **2000**, *72*, 1641.

[1971]Michael, U.; Hörnfeldt, A. *Tetrahedron Lett.* **1970**, 5219; Scilly, N.F. *Synthesis* **1973**, 160.

[1972]Collins, S.; Hong, Y. *Tetrahedron Lett.* **1987**, *28*, 4391.

[1973]Periasamy, M.; Reddy, M.R.; Bharathi, P. *Synth. Commun.* **1999**, *29*, 677.

[1974]Meisters, A.; Mole, T. *Aust. J. Chem.* **1974**, *27*, 1665.

[1975]Chung, E.-A.; Cho, C.-W.; Ahn, K.H. *J. Org. Chem.* **1998**, *63*, 7590.

[1976]Liebeskind, L.S.; Srogl, J. *J. Am. Chem. Soc.* **2000**, *122*, 11260; Wittenberg, R.; Srogi, J.; Egi, M.; Liebeskind, L.S. *Org. Lett.* **2003**, *5*, 3033.

[1977]Tatanidani, H.; Kakiuchi, F.; Chatani, N. *Org. Lett.* **2004**, *6*, 3597.

[1978]Tatanidani, H.; Yokota, K.; Kakiuchi, F.; Chatani, N. *J. Org. Chem.* **2004**, *69*, 5615.

[1979]Yu, Y.; Liebeskind, L.S. *J. Org. Chem.* **2004**, *69*, 3554.

[1980]Anderson, R.J.; Henrick, C.A.; Rosenblum, L.D. *J. Am. Chem. Soc.* **1974**, *96*, 3654. See also, Kim, S.; Lee, J.I. *J. Org. Chem.* **1983**, *48*, 2608.

[1981]Shimizu, T.; Seki, M. *Tetrahedron Lett.* **2002**, *43*, 1039.

[1982]Frost, C.G.; Wadsworth, K.J. *Chem. Commun.* **2001**, 2316.

[1983]Gooßen, L.J.; Ghosh, K. *Eur. J. Org. Chem.* **2002**, 3254.

[1984]Cacchi, S.; Fabrizi, G.; Gavazza, F.; Goggiamani, A. *Org. Lett.* **2003**, *5*, 289.

[1985]Wang, D.; Zhang, Z. *Org. Lett.* **2003**, *5*, 4645; Bercot, E.A.; Rovis, T. *J. Am. Chem. Soc.* **2004**, *126*, 10248.

[1986]Bercot, E.A.; Rovis, T. *J. Am. Chem. Soc.* **2002**, *124*, 174; O'Brien, E.M.; Bercot, E.A.; Rovis, T. *J. Am. Chem. Soc.* **2003**, *125*, 10498.

[1987]Lannou, M.-I.; Hélion, F.; Namy, J.-L. *Tetrahedron Lett.* **2002**, *43*, 8007.

via intramolecular addition of the organometallic intermediate to one carbonyl.[1988] The reaction of alkylzinc halides and thioesters leads to ketones in the presence of 1.5% Pd/C,[1989] in what has been called *Fukuyama coupling*.[1990]

Carboxylic esters can be converted to their homologs (RCOOEt → RCH$_2$-COOEt) by treatment with Br$_2$CHLi followed by BuLi at −90°C. The ynolate RC≡COLi is an intermediate.[1991] If the ynolate is treated with 1,3-cyclohexadiene, followed by NaBH$_4$, the product is the alcohol RCH$_2$CH$_2$OH.[1992]

Note that acyl benzotriazoles react with β-keto esters to give diketones via acyl substitution.[1993] Acyl cyanides, RC(=O)CN, react with allylic bromides and indium metal to give the corresponding ketone.[1994] Coupling an acid chloride and a silyl amide, R$_3$SiC(=O)NR$'_2$, leads to an α-keto amide.[1995] Acyl benzotriazoles have been coupled with SmI$_2$ to give the 1,2-diketone.[1996] α-Cyanoketone (acyl nitriles) were coupled with YbI$_2$ in a similar manner.[1997] α-Keto phosphonate esters undergo radical acyl substitution to give cyclic ketones under photochemical conditions.[1998]

Silyl esters are converted to the enolate anion upon treatment with *n*-butyllithium, and subsequent addition of an aldehyde followed by saponification leads to the β-hydroxy acid.[1999]

Vinyl organometallic reagents can be added to acyl derivatives. Reaction of an alkyne with Cp$_2$ZrEt$_2$ generates the vinyl zirconium reagent, which react with ethyl chloroformate to give an α,β-unsaturated ester.[2000]

OS **I**, 226; **II**, 179, 602; **III**, 237, 831, 839; **IV**, 601; **VI**, 240, 278; **VIII**, 474, 505.
OS **II**, 282; **72**, 32; **III**, 353; **IV**, 285; **VI**, 611; **VII**, 323, 451; **81**, 14.

16-83 The Coupling of Acyl Halides

De-halogen-coupling

$$2\,RCOCl \xrightarrow{\text{pyrophoric \textbf{Pb}}} RCOCOR$$

[1988]Ha, D.-C.; Yun, C.-S.; Lee, Y. *J. Org. Chem.* **2000**, *65*, 621.

[1989]Shimizu, T.; Seki, M. *Tetrahedron Lett.* **2001**, *42*, 429.

[1990]See Tokuyama, H.; Yokoshima, S.; Yamashita, T.; Fukuyama, T. *Tetrahedron Lett.* **1998**, *39*, 3189; Mori, Y.; Seki, M. *Tetrahedron Lett.* **2004**, *45*, 7343. For a different but related cross-coupling, see Zhang, Y.; Rovis, T. *J. Am. Chem. Soc.* **2004**, *126*, 15964.

[1991]Kowalski, C.J.; Haque, M.S.; Fields, K.W. *J. Am. Chem. Soc.* **1985**, *107*, 1429; Kowalski, C.J.; Haque, M.S. *J. Org. Chem.* **1985**, *50*, 5140.

[1992]Kowalski, C.J.; Haque, M.S. *J. Am. Chem. Soc.* **1986**, *108*, 1325.

[1993]Katritzky, A.R.; Wang, Z.; Wang, M.; Wilkerson, C.R.; Hall, C.D.; Akhmedov, N.G. *J. Org. Chem.* **2004**, *69*, 6617.

[1994]Yoo, B.W.; Choi, K.H.; Lee, S.J.; Nam, G.S.; Chang, K.Y.; Kim, S.H.; Kim, J.H. *Synth. Commun.* **2002**, *32*, 839.

[1995]Chen, J.; Cunico, R.F. *J. Org. Chem.* **2004**, *69*, 5509.

[1996]Wang, X.; Zhang, Y. *Tetrahedron Lett.* **2002**, *43*, 5431.

[1997]Saikia, P.; Laskar, D.D.; Prajapati, D.; Sandhu, J.S. *Tetrahedron Lett.* **2002**, *43*, 7525.

[1998]Kim, S.; Cho, C.H.; Lim, C.J. *J. Am. Chem. Soc.* **2003**, *125*, 9574.

[1999]Bellassoued, M.; Grugier, J.; Lensen, N.; Catheline, A. *J. Org. Chem.* **2002**, *67*, 5611.

[2000]Takahashi, T.; Xi, C.; Ura, Y.; Nakajima, K. *J. Am. Chem. Soc.* **2000**, *122*, 3228.

Acyl halides can be coupled with pyrophoric lead to give symmetrical α-diketones in a Wurtz-type reaction.[2001] The reaction has been performed with R = Me and Ph. Samarium iodide SmI_2[2002] gives the same reaction. Benzoyl chloride was coupled to give benzil by subjecting it to ultrasound in the presence of Li wire:

$$2\,PhCOCl + Li \longrightarrow PhCOCOPh.^{2003}$$

Unsymmetrical α-diketones, RCOCOR′, have been prepared by treatment of an acyl halide RCOCl with an acyltin reagent (R′COSnBu₃), with a palladium complex catalyst.[2004]

16-84   Acylation at a Carbon Bearing an Active Hydrogen

Bis(ethoxycarbonyl)methyl-de-halogenation, and so on

This reaction is similar to **10-67**, but there are fewer examples.[2005] Either Z or Z′ may be any of the groups listed in **10-67**.[2006] Anhydrides react similarly but are used less often. The product contains three Z groups, since RCO is a Z group. One or two of these can be cleaved (**12-40**, **12-43**). In this way, a compound ZCH₂Z′ can be converted to ZCH₂Z² or an acyl halide (RCOCl) to a methyl ketone (RCOCH₃). O-Acylation is sometimes a side reaction.[2007] When thallium(I) salts of ZCH₂Z′ are used, it is possible to achieve regioselective acylation at either the C or the O position. For example, treatment of the thallium(I) salt of MeCOCH₂COMe with acetyl chloride at −78°C gave >90% O-acylation, while acetyl fluoride at room temperature gave >95% C-acylation.[2008] The use of an alkyl chloroformate gives triesters.[2009]

The application of this reaction to simple ketones[2010] (in parallel with **10-68**) requires a strong base, such as $NaNH_2$ or Ph_3CNa, and is often complicated by

[2001]Mészáros, L. *Tetrahedron Lett.* **1967**, 4951.

[2002]Souppe, J.; Namy, J.; Kagan, H.B. *Tetrahedron Lett.* **1984**, 25, 2869. See also, Collin, J.; Namy, J.; Dallemer, F.; Kagan, H.B. *J. Org. Chem.* **1991**, 56, 3118.

[2003]Han, B.H.; Boudjouk, P. *Tetrahedron Lett.* **1981**, 22, 2757.

[2004]Verlhac, J.; Chanson, E.; Jousseaume, B.; Quintard, J. *Tetrahedron Lett.* **1985**, 26, 6075. For another procedure, see Olah, G.A.; Wu, A. *J. Org. Chem.* **1991**, 56, 902.

[2005]For examples of reactions in this section, with references, see Larock, R.C. *Comprehensive Organic Transformations*, 2nd ed., Wiley-VCH, NY, **1999**, pp. 1484–1485, 1522–1527.

[2006]For an improved procedure, see Rathke, M.W.; Cowan, P.J. *J. Org. Chem.* **1985**, 50, 2622.

[2007]When phase-transfer catalysts are used, O-acylation becomes the main reaction: Jones, R.A.; Nokkeo, S.; Singh, S. *Synth. Commun.* **1977**, 7, 195.

[2008]Taylor, E.C.; Hawks III, G.H.; McKillop, A. *J. Am. Chem. Soc.* **1968**, 90, 2421.

[2009]See, for example, Skarżewski, J. *Tetrahedron* **1989**, 45, 4593. For a review of triesters, see Newkome, G.R.; Baker, G.R. *Org. Prep. Proced. Int.* **1986**, 19, 117.

[2010]Hegedus, L.S.; Williams, R.E.; McGuire, M.A.; Hayashi, T. *J. Am. Chem. Soc.* **1980**, 102, 4973.

O-acylation, which in many cases becomes the principal pathway because acylation at the oxygen is usually much faster. It is possible to increase the proportion of *C*-acylated product by employing an excess (2–3 equivalents) of enolate anion (and adding the substrate to this, rather than vice versa), by the use of a relatively non-polar solvent and a metal ion (e.g., Mg^{2+}), which is tightly associated with the enolate oxygen atom, by the use of an acyl halide rather than an anhydride,[2011] and by working at low temperatures.[2012] In cases where the use of an excess of enolate anion results in *C*-acylation, it is because *O*-acylation takes place first, and the *O*-acylated product (an enol ester) is then *C*-acylated. Simple ketones can also be acylated by treatment of their silyl enol ethers with an acyl chloride in the presence of $ZnCl_2$ or $SbCl_3$.[2013] Ketones can be acylated by anhydrides to give β-diketones, with BF_3 as catalyst.[2014] Simple esters RCH_2COOEt can be acylated at the a carbon (at $-78°C$) if a strong base such as lithium *N*-isopropylcyclohexylamide is used to remove the proton.[2015]

OS **II**, 266, 268, 594, 596; **III**, 16, 390, 637; **IV**, 285, 415, 708; **V**, 384, 937; **VI**, 245; **VII**, 213, 359; **VIII**, 71, 326, 467. See also, OS **VI**, 620.

16-85 Acylation of Carboxylic Esters by Carboxylic Esters: The Claisen and Dieckmann Condensations

Alkoxycarbonylalkyl-de-alkoxy-substitution

When carboxylic esters containing an α hydrogen are treated with a strong base, such as sodium ethoxide, a condensation occurs to give a β-keto ester via an ester enolate anion.[2016] This reaction is called the *Claisen condensation*. When it is carried out with a mixture of two different esters, each of which possesses an α hydrogen (this reaction is called a mixed Claisen or a crossed Claisen condensation), a mixture of all four products is generally obtained and the reaction is seldom useful synthetically.[2017] However, if only one of the esters has an

[2011]See House, H.O. *Modern Synthetic Reactions*, 2nd ed., W.A. Benjamin, NY, *1972*, pp. 762–765; House, H.O.; Auerbach, R.A.; Gall, M.; Peet, N.P. *J. Org. Chem.* *1973*, *38*, 514.

[2012]Seebach, D.; Weller, T.; Protschuk, G.; Beck, A.K.; Hoekstra, M.S. *Helv. Chim. Acta* *1981*, *64*, 716.

[2013]Tirpak, R.E.; Rathke, M.W. *J. Org. Chem.* *1982*, *47*, 5099.

[2014]For a review, see Hauser, C.R.; Swamer, F.W.; Adams, J.T. *Org. React.* *1954*, *8*, 59, 98–106.

[2015]For example, see Rathke, M.W.; Deitch, J. *Tetrahedron Lett.* *1971*, 2953; Logue, M.W. *J. Org. Chem.* *1974*, *39*, 3455; Couffignal, R.; Moreau, J. *J. Organomet. Chem.* *1977*, *127*, C65; Ohta, S.; Shimabayashi, A.; Hayakawa, S.; Sumino, M.; Okamoto, M. *Synthesis* *1985*, 45; Hayden, W.; Pucher, R.; Griengl, H. *Monatsh. Chem.* *1987*, *118*, 415.

[2016]For a study of ester and amide enolate stabilization, see Rablen, P.R.; Bentrup, KL.H. *J. Am. Chem. Soc.* *2003*, *125*, 2142.

[2017]For a method of allowing certain crossed-Claisen reactions to proceed with good yields, see Tanabe, Y. *Bull. Chem. Soc. Jpn.* *1989*, *62*, 1917.

α hydrogen, the mixed reaction is frequently satisfactory. Among esters lacking a hydrogens (hence acting as the substrate ester) that are commonly used in this way are esters of aromatic acids, and ethyl carbonate and ethyl oxalate. When the ester enolate reacts with ethyl carbonate, the product is a malonic ester, and reaction with ethyl formate introduces a formyl group. Claisen condensation of phenyl esters with $ZrCl_4$ and diisopropylethylamine (Hünigs base) give the corresponding keto ester.[2018]

As with ketone enolate anions (see **16-34**), the use of amide bases under kinetic control conditions (strong base with a weak conjugate acid, aprotic solvents, low temperatures), allows the mixed Claisen condensation to proceed. Self-condensation of the lithium enolate with the parent ester is a problem when LDA is used as a base,[2019] but this is minimized with LICA (lithium isopropylcyclohexyl amide).[2020] Note that solvent-free Claisen condensation reactions have been reported.[2021]

When the two ester groups involved in the condensation are in the same molecule, the product is a cyclic β-keto ester and the reaction is called the *Dieckmann condensation.*[2022]

The Dieckmann condensation is most successful for the formation of five-, six-, and seven-membered rings. Yields for rings of 9–12 members are very low or nonexistent; larger rings can be closed with high-dilution techniques. Reactions in which large rings are to be closed are generally assisted by high dilution, since one end of the molecule has a better chance of finding the other end than of finding another molecule. A solvent-free Dieckmann condensation has been reported on solid potassium *tert*-butoxide.[2023] Dieckmann condensation of unsymmetrical substrates can be made regioselective (unidirectional) by the use of solid-phase supports.[2024]

The mechanism of the Claisen and Dieckmann reactions is the ordinary tetrahedral mechanism,[2025] with one molecule of ester being converted to a nucleophile by

[2018]Tanabe, Y.; Hamasaki, R.; Funakoshi, S. *Chem. Commun.* **2001**, 1674.

[2019]Rathke, M.W.; Sullivan, D.F. *J. Am. Chem. Soc. **1973**, 95*, 3050; Lochmann, L.; Lím, D. *J. Organomet. Chem. **1973**, 50*, 9; Sullivan, D.F.; Woodbury, R.P.; Rathke, M.W. *J. Org. Chem. **1977**, 42*, 2038.

[2020]Rathke, M.W.; Lindert, A. *J. Am. Chem. Soc. **1971**, 93*, 2318.

[2021]Yoshizawa, K.; Toyota, S.; Toda, F. *Tetrahedron Lett. **2001**, 42*, 7983.

[2022]For a review, see Schaefer, J.P.; Bloomfield, J.J. *Org. React. **1967**, 15*, 1.

[2023]Toda, F.; Suzuki, T.; Higa, S. *J. Chem. Soc. Perkin Trans. 1 **1998**, 3521.

[2024]Crowley, J.I.; Rapoport, H. *J. Org. Chem. **1980**, 45*, 3215. For another method, see Yamada, Y.; Ishii, T.; Kimura, M.; Hosaka, K. *Tetrahedron Lett. **1981**, 22*, 1353.

[2025]There is evidence that, at least in some cases, an SET mechanism is involved: Ashby, E.C.; Park, W. *Tetrahedron Lett. **1983**, 1667. The transition structures have also been examined by Nishimura, T.; Sunagawa, M.; Okajima, T.; Fukazawa, Y. *Tetrahedron Lett. **1997**, 38*, 7063.

the base and the other serving as the substrate.

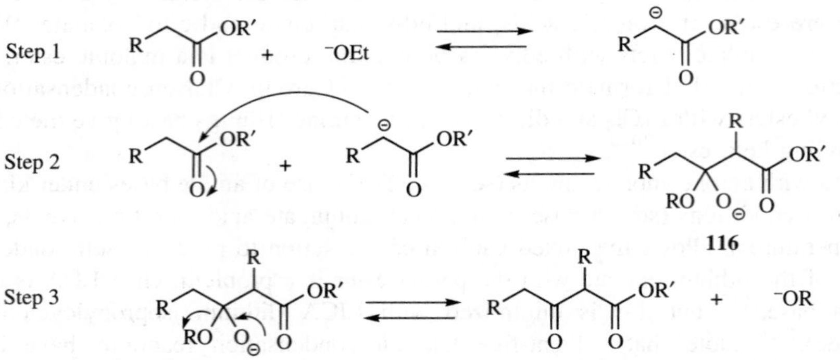

This reaction illustrates the striking difference in behavior between carboxylic esters on the one hand and aldehydes and ketones on the other. When a carbanion, such as an enolate anion, is added to the carbonyl group of an aldehyde or ketone (**16-38**), the H or R is not lost, since these groups are much poorer leaving groups than OR. Instead the intermediate similar to **116** adds a proton at the oxygen to give a hydroxy compound.

In contrast to **10-67** ordinary esters react quite well, that is, two Z groups are not needed. A lower degree of acidity is satisfactory because it is not necessary to convert the attacking ester entirely to its ion. Step 1 is an equilibrium that lies well to the left. Nevertheless, the small amount of enolate anion formed is sufficient to attack the readily approachable ester substrate. All the steps are equilibria. The reaction proceeds because the product is converted to its conjugate base by the base present (i.e., a β-keto ester is a stronger acid than an alcohol):

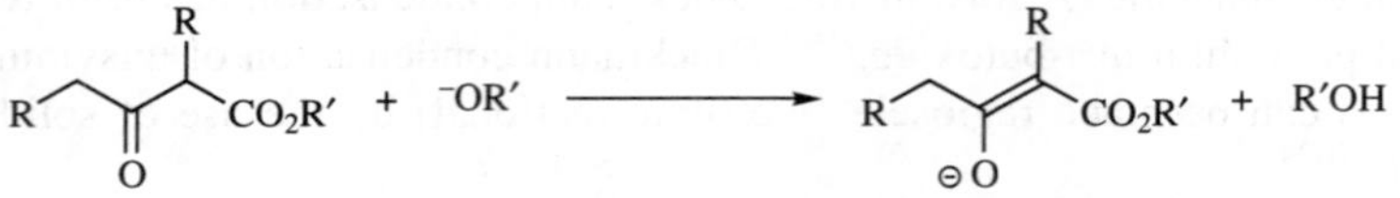

The use of a stronger base, such as $NaNH_2$, NaH, or KH,[2026] often increases the yield. For some esters stronger bases *must* be used, since sodium ethoxide is ineffective. Among these are esters of the type $R_2CHCOOEt$, the products of which ($R_2CHCOCR_2COOEt$) lack an acidic hydrogen, so that they cannot be converted to enolate anions by sodium ethoxide.[2027] The Dieckmann condensation has also been done using $TiCl_3/NBu_3$ with a TMSOTf catalyst.[2028] A Dieckmann-like condensation was reported where an α,ω-dicarboxylic acid was hated to 450°C on graphite, with microwave irradiation, to give the cyclic ketone.[2029]

[2026]Brown, C.A. *Synthesis* **1975**, 326.
[2027]For a discussion, see Garst, J.F. *J. Chem. Educ.* **1979**, *56*, 721.
[2028]Yoshida, Y.; Hayashi, R.; Sumihara, H.; Tanabe, Y. *Tetrahedron Lett.* **1997**, *38*, 8727.
[2029]Marquié, J.; Laporterie, A.; Dubac, J.; Roques, N. *Synlett* **2001**, 493.

OS **I**, 235; **II**, 116, 194, 272, 288; **III**, 231, 300, 379, 510; **IV**, 141; **V**, 288, 687, 989; **VIII**, 112.

16-86 Acylation of Ketones and Nitriles by Carboxylic Esters

α-Acylalkyl-de-alkoxy-substitution

Carboxylic esters can be treated with ketones to give β-diketones. The reaction is so similar that it is sometimes also called the Claisen reaction, though this usage may be confusing. A strong base, such as sodium amide or sodium hydride, is required. Yields can be increased by the catalytic addition of crown ethers.[2030] Esters of formic acid (R = H) give β-keto aldehydes and ethyl carbonate gives β-keto esters. β-Keto esters can also be obtained by treating the lithium enolates of ketones with methyl cyanoformate MeOCOCN[2031] (in this case CN is the leaving group) and by treating ketones with KH and diethyl dicarbonate, $(EtOCO)_2O$.[2032]

In the case of unsymmetrical ketones, the attack usually comes from the less highly substituted side, so that CH_3 is more reactive than RCH_2, and the R_2CH group rarely attacks. This reaction has been used to effect cyclization, especially to prepare five- and six-membered rings. Nitriles are frequently used instead of ketones, the products being β-keto nitriles.

Other nucleophilic carbon reagents, such as acetylide ions, and ions derived from α-methylpyridines have also been used. A particularly useful nucleophile is the methylsulfinyl carbanion, CH_3SOCH_2-,[2033] the conjugate base of DMSO, since the β-keto sulfoxide produced can easily be reduced to a methyl ketone (p. 624). The methylsulfonyl carbanion $(CH_3SO_2CH_2^-)$, the conjugate base of dimethyl sulfone, behaves similarly,[2034] and the product can be similarly reduced. Certain carboxylic esters, acyl halides, and DMF will acylate 1,3-dithianes[2035] (see **10-71**) to give, after oxidative hydrolysis with NBS or NCS, α-keto aldehydes or

[2030]Popik, V.V.; Nikolaev, V.A. *J. Org. Chem. USSR* **1989**, *25*, 1636.

[2031]Mander, L.N.; Sethi, P. *Tetrahedron Lett.* **1983**, *24*, 5425.

[2032]Hellou, J.; Kingston, J.F.; Fallis, A.G. *Synthesis* **1984**, 1014.

[2033]See Durst, T. *Adv. Org. Chem.* **1969**, *6*, 285, pp. 296–301.

[2034]Schank, K.; Hasenfratz, H.; Weber, A. *Chem. Ber.* **1973**, *106*, 1107, House, H.O.; Larson, J.K. *J. Org. Chem.* **1968**, *33*, 61.

[2035]Corey, E.J.; Seebach, D. *J. Org. Chem.* **1975**, *40*, 231

α-diketones,[254] for example,

As in **10-67**, a ketone attacks with its second most acidic position if 2 equivalents of base are used. Thus, β-diketones have been converted to 1,3,5-triketones.[2036]

Side reactions are condensation of the ketone with itself (**16-34**), of the ester with itself, and of the ketone with the ester, but with the ester supplying the a position (**16-36**). The mechanism is the same as in **16-85**.[2037]

OS **I**, 238; **II**, 126, 200, 287, 487, 531; **III**, 17, 251, 291, 387, 829; **IV**, 174, 210, 461, 536; **V**, 187, 198, 439, 567, 718, 747; **VI**, 774; **VII**, 351.

16-87   Acylation of Carboxylic Acid Salts

α-Carboxyalkyl-de-alkoxy-substitution

$$RCH_2COO^- \xrightarrow{(iPr)_2NLi} R\overset{\ominus}{C}HCOO^{\ominus} \xrightarrow{R'COOMe} R' \underset{\underset{O}{\overset{\parallel}{C}}}{\,} \overset{R}{\underset{\,}{\overset{|}{C}}} \overset{H}{\underset{COO^{\ominus}}{\,}}$$

We have previously seen (**10-70**) that dianions of carboxylic acids can be alkylated in the α position. These ions can also be acylated on treatment with a carboxylic ester[2038] to give salts of β-keto acids. As in **10-70**, the carboxylic acid can be of the form RCH_2COOH or $RR^2CHCOOH$. Since β-keto acids are so easily converted to ketones (**12-40**), this is also a method for the preparation of ketones $R'COCH_2R$ and $R'COCHRR^2$, where R' can be primary, secondary, or tertiary alkyl, or aryl. If the ester is ethyl formate, an α-formyl carboxylate salt ($R' = H$) is formed, which on acidification spontaneously decarboxylates into an aldehyde.[2039] This method accomplishes the conversion $RCH_2COOH \rightarrow RCH_2CHO$, and is an alternative to the reduction methods discussed in **19-39**. When the carboxylic acid is of the form $RR''CHCOOH$, better yields are obtained by acylating with acyl halides rather than esters.[2040]

[2036]Miles, M.L.; Harris, T.M.; Hauser, C.R. *J. Org. Chem.* **1965**, *30*, 1007.

[2037]Hill, D.G.; Burkus, T.; Hauser, C.R. *J. Am. Chem. Soc.* **1959**, *81*, 602.

[2038]Kuo, Y.; Yahner, J.A.; Ainsworth, C. *J. Am. Chem. Soc.* **1971**, *93*, 6321; Angelo, B. *C.R. Seances Acad. Sci. Ser. C* **1973**, *276*, 293.

[2039]Pfeffer, P.E.; Silbert, L.S. *Tetrahedron Lett.* **1970**, 699; Koch, G.K.; Kop, J.M.M. *Tetrahedron Lett.* **1974**, 603.

[2040]Krapcho, A.P.; Kashdan, D.S.; Jahngen, Jr., E.G.E.; Lovey, A.J. *J. Org. Chem.* **1977**, *42*, 1189; Lion, C.; Dubois, J.E. *J. Chem. Res. (S)* **1980**, 44.

16-88   Preparation of Acyl Cyanides

Cyano-de-halogenation

$$RCOX \quad + \quad CuCN \quad \longrightarrow \quad RCOCN$$

Acyl cyanides[2041] can be prepared by treatment of acyl halides with copper cyanide. The mechanism could be free-radical or nucleophilic substitution. The reaction has also been accomplished with thallium(I) cyanide,[2042] with Me_3SiCN and an $SnCl_4$ catalyst,[2043] and with Bu_3SnCN,[2044] but these reagents are successful only when R = aryl or tertiary alkyl. KCN has also been used, along with ultrasound,[2045] as has NaCN with phase-transfer catalysts.[2046]
OS **III**, 119.

16-89   Preparation of Diazo Ketones

Diazomethyl-de-halogenation

$$RCOX \quad + \quad CH_2N_2 \quad \longrightarrow \quad RCOCHN_2$$

The reaction between acyl halides and diazomethane is of wide scope and is the best way to prepare diazo ketones.[2047] Diazomethane must be present in excess or the HX produced will react with the diazo ketone (**10-52**). This reaction is the first step of the *Arndt–Eistert synthesis* (**18-8**). Diazo ketones can also be prepared directly from a carboxylic acid and diazomethane or diazoethane in the presence of DCC.[2048]
OS **III**, 119; **VI**, 386, 613; **VIII**, 196.

16-90   Ketonic Decarboxylation[2049]

Alkyl-de-hydroxylation

$$2\ RCOOH \quad \xrightarrow[\text{ThO}_2]{400\text{–}500°C} \quad RCOR \quad + \quad CO_2$$

[2041]For a review of acyl cyanides, see Hünig, S.; Schaller, R. *Angew. Chem. Int. Ed.* **1982**, *21*, 36.

[2042]Taylor, E.C.; Andrade, J.G.; John, K.C.; McKillop, A. *J. Org. Chem.* **1978**, *43*, 2280.

[2043]Olah, G.A.; Arvanaghi, M.; Prakash, G.K.S. *Synthesis* **1983**, 636.

[2044]Tanaka, M. *Tetrahedron Lett.* **1980**, *21*, 2959. See also Tanaka, M.; Koyanagi, M. *Synthesis* **1981**, 973.

[2045]Ando, T.; Kawate, T.; Yamawaki, J.; Hanafusa, T. *Synthesis* **1983**, 637.

[2046]Koenig, K.E.; Weber, W.P. *Tetrahedron Lett.* **1974**, 2275. See also, Sukata, K. *Bull. Chem. Soc. Jpn.* **1987**, *60*, 1085.

[2047]For reviews, see Fridman, A.L.; Ismagilova, G.S.; Zalesov, V.S.; Novikov, S.S. *Russ. Chem. Rev.* **1972**, *41*, 371; Ried, W.; Mengler, H. *Fortshr. Chem. Forsch.*, **1965**, *5*, 1.

[2048]Hodson, D.; Holt, G.; Wall, D.K. *J. Chem. Soc. C* **1970**, 971.

[2049]For a review, see Kwart, H.; King, K., in Patai, S. *The Chemistry of Carboxylic Acids and Esters*, Wiley, NY, **1969**, pp. 362–370.

Carboxylic acids can be converted to symmetrical ketones by pyrolysis in the presence of thorium oxide. In a mixed reaction, formic acid and another acid heated over thorium oxide give aldehydes. Mixed alkyl aryl ketones have been prepared by heating mixtures of ferrous salts.[2050] When the R group is large, the methyl ester rather than the acid can be decarbmethoxylated over thorium oxide to give the symmetrical ketone.

The reaction has been performed on dicarboxylic acids, whereupon cyclic ketones are obtained:

$$\underset{\text{COOH}}{\overset{\text{COOH}}{(CH_2)_n}} \quad \xrightarrow[\Delta]{\text{ThO}_2} \quad (CH_2)_n \quad C{=}O$$

This process, called *Ruzicka cyclization*, is good for the preparation of rings of six and seven members and, with lower yields, of C_8 and C_{10}–C_{30} cyclic ketones.[2051]

Not much work has been done on the mechanism of this reaction. However, a free-radical mechanism has been suggested on the basis of a thorough study of all the side products.[2052]

OS **I**, 192; **II**, 389; **IV**, 854; **V**, 589. Also see, OS **IV**, 55, 560.

REACTIONS IN WHICH CARBON ADDS TO THE HETEROATOM

A. Oxygen Adding to the Carbon

16-91   The Ritter Reaction
N-Hydro,N-alkyl-C-oxo-biaddition

$$R{-}C{\equiv}N \;+\; R'OH \;\xrightarrow{H^+}\; \underset{\underset{H}{|}}{R}\overset{\overset{O}{\|}}{{-}C{-}N}{-}R'$$

Alcohols can be added to nitriles in an entirely different manner from that of reaction **16-9**. In this reaction, the alcohol is converted by a strong acid to a carbocation, which is attacked by the nucleophilic nitrogen atom to give **117**. Subsequent addition of water to the electrophilic carbon atom leads to the enol form of the amide (see **118**), which tautomerizes (p. 98) to the N-alkyl amide.

$$R^1{-}OH \;\xrightarrow{H^+}\; {\oplus}R^1 + R{-}C{\equiv}N \;\longrightarrow\; \underset{\textbf{117}}{R{-}C{=}\overset{\oplus}{N}{}^{R^1}} \;\xrightarrow{H_2O}\; \underset{\textbf{118}}{R{-}\overset{\overset{N{-}R^1}{\|}}{C}_{OH}}$$

[2050]Granito, C.; Schultz, H.P. *J. Org. Chem.* **1963**, *28*, 879.
[2051]See, for example, Ruzicka, L.; Stoll, M.; Schinz, H. *Helv. Chim. Acta* **1926**, *9*, 249; **1928**, *11*, 1174; Ruzicka, L.; Brugger, W.; Seidel, C.F.; Schinz, H. *Helv. Chim. Acta* **1928**, *11*, 496.
[2052]Hites, R.A.; Biemann, K. *J. Am. Chem. Soc.* **1972**, *94*, 5772. See also, Bouchoule, C.; Blanchard, M.; Thomassin, R. *Bull. Soc. Chim. Fr.* **1973**, 1773.

Only alcohols that give rise to fairly stable carbocations react (secondary, tertiary, benzylic, etc.); non-benzylic primary alcohols do not give the reaction. The carbocation need not be generated from an alcohol, but may come from protonation of an alkene or from other sources. In any case, the reaction is called the *Ritter reaction*.[2053] Lewis acids, such as $Mg(HSO_4)_2$, have been used to promote the reaction.[2054] Highly sterically hindered nitriles have been converted to *N*-methyl amides by heating with methanol and sulfuric acid.[2055] HCN also gives the reaction, the product being a formamide. Trimethylsilyl cyanide has also been used.[2056]

Since the amides (especially the formamides) are easily cleaved under hydrolysis conditions to amines, the Ritter reaction provides a method for achieving the conversions $R'OH \rightarrow R'NH_2$ (see **10-32**) and alkene $\rightarrow R'NH_2$ (see **15-8**) in those cases where R' can form a relatively stable carbocation. The reaction is especially useful for the preparation of tertiary alkyl amines because there are few alternate ways of preparing these compounds. The reaction can be extended to primary alcohols by treatment with triflic anhydride[2057] or Ph_2CCl^+ $SbCl_6^-$ or a similar salt[2058] in the presence of the nitrile.

Alkenes of the form $RCH=CHR'$ and $RR'C=CH_2$ add to nitriles in the presence of mercuric nitrate to give, after treatment with $NaBH_4$, the same amides that would be obtained by the Ritter reaction.[2059] This method has the advantage of avoiding strong acids.

Benzylic compounds, such as ethylbenzene, react with alkyl nitriles, ceric ammonium nitrate, and a catalytic amount of *N*-hydroxysuccinimide to give the Ritter product, the amide.[2060]

The Ritter reaction can be applied to cyanamides RNHCN to give ureas RNHCONHR'.[2061]

OS **V**, 73, 471.

[2053]Ritter, J.J.; Minieri, P.P. *J. Am. Chem. Soc.* *1948*, *70*, 4045. For reviews, see Krimen, L.I.; Cota, D.J. *Org. React.* *1969*, *17*, 213; Beckwith, A.L.J., in Zabicky, J. *The Chemistry of Amides*, Wiley, NY, *1970*, pp. 125–130; Johnson, F.; Madroñero, R. *Adv. Heterocycl. Chem.* *1966*, *6*, 95; Tongco, E.C.; Prakash, G.K.S.; Olah, G.A. *Synlett* *1997*, 1193.

[2054]Salehi, P.; Khodaei, M.M.; Zolfigol, M.A.; Keyvan, A. *Synth. Commun.* *2001*, *31*, 1947.

[2055]Lebedev, M.Y.; Erman, M.B. *Tetrahedron Lett.* *2002*, *43*, 1397.

[2056]Chen, H.G.; Goel, O.P.; Kesten, S.; Knobelsdorf, J. *Tetrahedron Lett.* *1996*, *37*, 8129.

[2057]Martinez, A.G.; Alvarez, R.M.; Vilar, E.T.; Fraile, A.G.; Hanack, M.; Subramanian, L.R. *Tetrahedron Lett.* *1989*, *30*, 581.

[2058]Barton, D.H.R.; Magnus, P.D.; Garbarino, J.A.; Young, R.N. *J. Chem. Soc. Perkin Trans. 1 1974*, 2101. See also, Top, S.; Jaouen, G. *J. Org. Chem.* *1981*, *46*, 78.

[2059]Sokolov, V.I.; Reutov, O.A. *Bull. Acad. Sci. USSR Div. Chem. Sci.* *1968*, 225; Brown, H.C.; Kurek, J.T. *J. Am. Chem. Soc.* *1969*, *91*, 5647; Chow, D.; Robson, J.H.; Wright, G.F. *Can. J. Chem.* *1965*, *43*, 312; Fry, A.J.; Simon, J.A. *J. Org. Chem.* *1982*, *47*, 5032.

[2060]Sakaguchi, S.; Hirabayashi, T.; Ishii, Y. *Chem. Commun.* *2002*, 516.

[2061]Anatol, J.; Berecoechea, J. *Bull. Soc. Chim. Fr.* *1975*, 395; *Synthesis* *1975*, 111.

16-92   The Addition of Aldehydes to Aldehydes

$$3 \ RCHO \xrightarrow{\ H^+\ } \quad$$

When catalyzed by acids, low-molecular-weight aldehydes add to each other to give cyclic acetals, the most common product being the trimer.[2062] The cyclic trimer of formaldehyde is called *trioxane*,[2063] and that of acetaldehyde is known as *paraldehyde*. Under certain conditions, it is possible to get tetramers[2064] or dimers. Aldehydes can also polymerize to linear polymers, but here a small amount of water is required to form hemiacetal groups at the ends of the chains. The linear polymer formed from formaldehyde is called *paraformaldehyde*. Since trimers and polymers of aldehydes are acetals, they are stable to bases, but can be hydrolyzed by acids. Because formaldehyde and acetaldehyde have low boiling points, it is often convenient to use them in the form of their trimers or polymers.

Aryl aldehydes condense with aliphatic aldehydes in the presence of benzoylformate decarboxylase and thiamin diphosphate to give an α-hydroxy ketone with god enantioselectivity.[2065]

A slightly related reaction involves nitriles, which can be trimerized with various acids, bases, or other catalysts to give triazines (see OS **III**, 71).[2066] Here HCl is most often used. Most nitriles with an α hydrogen do not give the reaction.

B. Nitrogen Adding to the Carbon

16-93   The Addition of Isocyanates to Isocyanates (Formation of Carbodiimides)

Alkylimino-de-oxo-bisubstitution

$$2 \ \ R{\diagdown}N{=}C{=}O \ + \quad$$

119

The treatment of isocyanates with 3-methyl-1-ethyl-3-phospholene-1-oxide (**119**) is a useful method for the synthesis of carbodiimides[2067] in good

[2062]For a review, see Bevington, J.C. *Q. Rev. Chem. Soc.* **1952**, *6*, 141.

[2063]For a synthesis of trioxanes using bentonitic earth catalysts, see Camarena, R.; Cano, A.C.; Delgado, F.; Zúñiga, N.; Alvarez, C. *Tetrahedron Lett.* **1993**, *34*, 6857.

[2064]Barón, M.; de Manderola, O.B.; Westerkamp, J.F. *Can. J. Chem.* **1963**, *41*, 1893.

[2065]Dünnwald, T.; Demir, A.S.; Siegert, P.; Pohl, M.; Müller, M. *Eur. J. Org. Chem.* **2000**, 2161.

[2066]For a review, see Martin, D.; Bauer, M.; Pankratov, V.A. *Russ. Chem. Rev.* **1978**, *47*, 975. For a review with respect to cyanamides RNH–CN, see Pankratov, V.A.; Chesnokova, A.E. *Russ. Chem. Rev.* **1989**, *58*, 879. For reviews of the chemistry of carb.

[2067]For reviews of the chemistry of carbodiimides, see Williams, A.; Ibrahim, I.T. *Chem. Rev.* **1981**, *81*, 589; Mikołajczyk, M.; Kiełbasiński, P. *Tetrahedron* **1981**, *37*, 233; Kurzer, F.; Douraghi-Zadeh, K. *Chem. Rev.* **1967**, *67*, 107.

yields.[2068] The mechanism does not simply involve the addition of one molecule of isocyanate to another, since the kinetics are first order in isocyanate and first order in catalyst. The following mechanism has been proposed (the catalyst is here represented as R_3P^+—O^-:[2069]

According to this mechanism, one molecule of isocyanate undergoes addition to C=O, and the other addition to C=N. Evidence is that ^{18}O labeling experiments have shown that each molecule of CO_2 produced contains one oxygen atom derived from the isocyanate and one from **119**,[2070] precisely what is predicted by this mechanism. Certain other catalysts are also effective.[2071] High-load, soluble oligomeric carbodiimides have been prepared.[2072]

OS **V**, 501.

16-94   The Conversion of Carboxylic Acid Salts to Nitriles

Nitrilo-de-oxido,oxo-tersubstitution

$$RCOO^- \ + \ BrCN \ \xrightarrow{250-300°C} \ RCN \ + \ CO_2$$

Salts of aliphatic or aromatic carboxylic acids can be converted to the corresponding nitriles by heating with BrCN or ClCN. Despite appearances, this is not a substitution reaction. When $R^{14}COO^-$ was used, the label appeared in the nitrile, not in the CO_2,[2073] and optical activity in R was retained.[2074] The acyl isocyanate RCON=C=O could be isolated from the reaction mixture; hence the

[2068]Campbell, T.W.; Monagle, J.J.; Foldi, V.S. *J. Am. Chem. Soc.* **1962**, *84*, 3673.

[2069]Monagle, J.J.; Campbell, T.W.; McShane Jr., H.F. *J. Am. Chem. Soc.* **1962**, *84*, 4288.

[2070]Monagle, J.J.; Mengenhauser, J.V. *J. Org. Chem.* **1966**, *31*, 2321.

[2071]Monagle, J.J. *J. Org. Chem.* **1962**, *27*, 3851; Appleman, J.O.; DeCarlo, V.J. *J. Org. Chem.* **1967**, *32*, 1505; Ulrich, H.; Tucker, B.; Sayigh, A.A.R. *J. Org. Chem.* **1967**, *32*, 1360; *Tetrahedron Lett.* **1967**, 1731; Ostrogovich, G.; Kerek, F.; Buzás, A.; Doca, N. *Tetrahedron* **1969**, *25*, 1875.

[2072]Zhang, M.; Vedantham, P.; Flynn, D.L.; Hanson, P.R. *J. Org. Chem.* **2004**, *69*, 8340.

[2073]Douglas, D.E.; Burditt, A.M. *Can. J. Chem.* **1958**, *36*, 1256.

[2074]Barltrop, J.A.; Day, A.C.; Bigley, D.B. *J. Chem. Soc.* **1961**, 3185.

following mechanism was proposed:[2073]

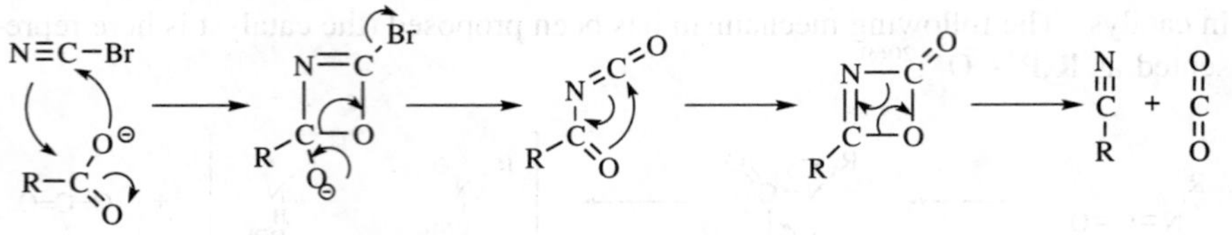

C. Carbon Adding to the Carbon.

The reactions in this group are cycloadditions.

16-95   The Formation of β-Lactones and Oxetanes

(2+2)*OC*,*CC*-*cyclo*-[oxoethylene]-1/2/addition

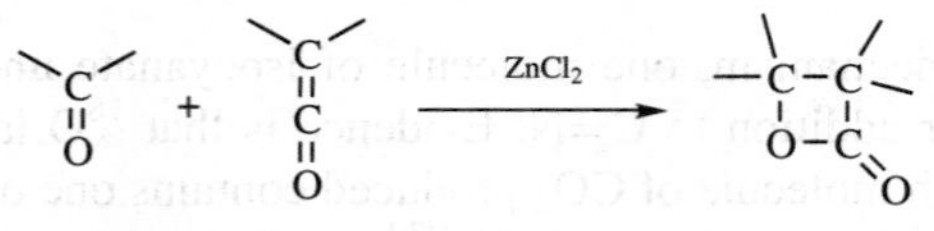

Aldehydes, ketones, and quinones react with ketenes to give β-lactones,[2075] diphenylketene being used most often.[2076] The reaction is catalyzed by Lewis acids, and without them most ketenes do not give adducts because the adducts decompose at the high temperatures necessary when no catalyst is used. When ketene was added to chloral (Cl_3CCHO) in the presence of the chiral catalyst (+)-quinidine, one enantiomer of the β-lactone was produced with excellent enantioselectivity.[2077] The use of a chiral aluminum catalyst also led to β-lactones with good syn selectivity and good enantioselectivity.[2078] Other di- and trihalo aldehydes and ketones also give the reaction enantioselectively, with somewhat lower enantioselectivity.[2079] Ketene adds to another molecule of itself:

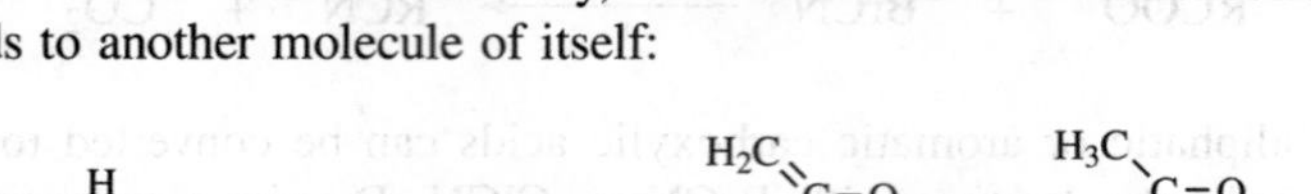

This dimerization is so rapid that ketene does not form β-lactones with aldehydes or ketones, except at low temperatures. Other ketenes dimerize more slowly. In

[2075]See Nelson, S.G.; Wan, Z.; Peclen, Y.J.; Spencer, K.L. *Tetrahedron Lett.* **1999**, *40*, 6535; Cortez, G.S.; Tennyson, R.L.; Romo, D. *J. Am. Chem. Soc.* **2001**, *123*, 7945.

[2076]For reviews, see Muller, L.L.; Hamer, J. *1,2-Cycloaddition Reactions*, Wiley, NY, **1967**, pp. 139–168; Ulrich, H. *Cycloaddition Reactions of Heterocumulenes*; Academic Press, NY, **1967**, pp. 39–45, 64–74.

[2077]Wynberg, H.; Staring, E.G.J. *J. Am. Chem. Soc.* **1982**, *104*, 166; *J. Chem. Soc., Chem. Commun.* **1984**, 1181.

[2078]Nelson, S.G.; Zhu, C.; Shen, X. *J. Am. Chem. Soc.* **2004**, *126*, 14.

[2079]Wynberg, H.; Staring, E.G.J. *J. Org. Chem.* **1985**, *50*, 1977.

these cases, the major dimerization product is not the β-lactone, but a cyclobutane-dione (see **15-63**). However, the proportion of ketene that dimerizes to β-lactone can be increased by the addition of catalysts, such as triethylamine or triethyl phosphite.[2080] Ketene acetals $R_2C{=}C(OR')_2$ add to aldehydes and ketones in the presence of $ZnCl_2$ to give the corresponding oxetanes.[2081]

Ordinary aldehydes and ketones can add to alkenes, under the influence of *UV* light, to give oxetanes. Quinones also react to give spirocyclic oxetanes.[2082] This reaction, called the *Paterno–Büchi reaction*,[2083] is similar to the photochemical dimerization of alkenes discussed at **15-63**. In general, the mechanism consists of the addition of an excited state of the carbonyl compound to the ground state of the alkene. Both singlet (S_1)[2084] and n,π^* triplet[2085] states have been shown to add to alkenes to give oxetanes. A diradical intermediate[2086]

has been detected by spectroscopic methods.[2087] Yields in the Paterno–Büchi reaction are variable, ranging from very low to fairly high (90%). The reaction can be

---

[2080]Farnum, D.G.; Johnson, J.R.; Hess, R.E.; Marshall, T.B.; Webster, B. *J. Am. Chem. Soc.* **1965**, *87*, 5191; Elam, E.U. *J. Org. Chem.* **1967**, *32*, 215.

[2081]Aben, R.W.; Hofstraat, R.; Scheeren, J.W. *Recl. Trav. Chim. Pays-Bas* **1981**, *100*, 355. For a discussion of oxetane cycloreversion, see Miranda, M.A.; Izquierdo, M.A.; Galindo, F. *Org. Lett.* **2001**, *3*, 1965.

[2082]Ciufolini, M.A.; Rivera-Fortin, M.A.; Byrne, N.E. *Tetrahedron Lett.* **1993**, *34*, 3505.

[2083]For reviews, see Ninomiya, I.; Naito, T. *Photochemical Synthesis*, Academic Press, NY, **1989**, pp. 138–152; Carless, H.A.J., in Coyle, J.D. *Photochemistry in Organic Synthesis*, Royal Society of Chemistry, London, **1986**, pp. 95–117; Carless, H.A.J., in Horspool, W.M. *Synthetic Organic Photochemistry*, Plenum, NY, **1984**, pp. 425–487; Jones II, M. *Org. Photochem.* **1981**, *5*, 1; Arnold, D.R. *Adv. PhotoChem.* **1968**, *6*, 301–423; Chapman, O.L.; Lenz, G. *Org. Photochem.* **1967**, *1*, 283, pp. 283–294; Muller, L.L.; Hamer, J. *1,2-Cycloaddition Reactions*, Wiley, NY, **1967**, pp. 111–139. Also see, Bosch, E.; Hubig, S.M.; Kochi, J.K. *J. Am. Chem. Soc.* **1998**, *120*, 386; Bach, T.; Jödicke, K.; Kather, K.; Frölich, R. *J. Am. Chem. Soc.* **1997**, *119*, 2437; Hu, S.; Neckers, D.C. *J. Org. Chem.* **1997**, *62*, 564.

[2084]See, for example, Turro, N.J. *Pure Appl. Chem.* **1971**, *27*, 679; Yang, N.C.; Kimura, M.; Eisenhardt, W. *J. Am. Chem. Soc.* **1973**, *95*, 5058; Singer, L.A.; Davis, G.A.; Muralidharan, V.P. *J. Am. Chem. Soc.* **1969**, *91*, 897; Barltrop, J.A.; Carless, H.A.J. *J. Am. Chem. Soc.* **1972**, *94*, 1951, 8761.

[2085]Arnold, D.R.; Hinman, R.L.; Glick, A.H. *Tetrahedron Lett.* **1964**, 1425; Yang, N.C.; Nussim, M.; Jorgenson, M.J.; Murov, S. *Tetrahedron Lett.* **1964**, 3657.

[2086]For other evidence for these diradical intermediates, see references cited in Griesbeck, A.G.; Stadmüller, S. *J. Am. Chem. Soc.* **1990**, *112*, 1281. See also, Kutateladze, A.G. *J. Am. Chem. Soc.* **2001**, *123*, 9279.

[2087]Freilich, S.C.; Peters, K.S. *J. Am. Chem. Soc.* **1981**, *103*, 6255; **1985**, *107*, 3819. For a review, see Griesbeck, A.G.; Mauder, H.; Stadmüller, S. *Accts. Chem. Res.* **1994**, *27*, 70.

highly diastereoselective,[2088] and allylic alcohols were shown to react with aldehydes to give an oxetane with syn selectivity.[2089] There are several side reactions. When the reaction proceeds through a triplet state, it can in general be successful only when the alkene possesses a triplet energy comparable to, or higher than, the carbonyl compound; otherwise energy transfer from the excited carbonyl group to the ground-state alkene can take place (triplet-triplet photosensitization, see p. 340).[2090] In most cases, quinones react normally with alkenes, giving oxetane products, but other α,β-unsaturated ketones usually give preferential cyclobutane formation (**15-63**). Aldehydes and ketones also add photochemically to allenes to give the corresponding alkylideneoxetanes and dioxaspiro compounds:[2091] Aldehydes add to silyl enol ethers.[2092] An intramolecular reaction of ketones was reported to give a bicyclic oxetane via photolysis on the solid state.[2093]

OS **III**, 508; **V**, 456. For the reverse reaction, see OS **V**, 679.

16-96   The Formation of β-Lactams

(2+2)*NC,CC-cyclo*-[oxoethylene]-1/2/addition

Ketenes add to imines to give β-lactams.[2094] The reaction is generally carried out with ketenes of the form $R_2C=C=O$. It has not been successfully applied to

[2088]Bach, T.; Jödicke, K.; Wibbeling, B. *Tetrahedron* **1996**, *52*, 10861; Fleming, S.A.; Gao, J.J. *Tetrahedron Lett.* **1997**, *38*, 5407; Vasudevan, S.; Brock, C.P.; Watt, D.S.; Morita, H. *J. Org. Chem.* **1994**, *59*, 4677; Adam, W.; Stegmann, V.R.; Weinkötz, S. *J. Am. Chem. Soc.* **2001**, *123*, 2452; Adam, W.; Stegmann, V.R. *J. Am. Chem. Soc.* **2002**, *124*, 3600. For a discussion of the origins of regioselectivity, see Ciufolini, M.A.; Rivera-Fortin, M.A.; Zuzukin, V.; Whitmire, K.H. *J. Am. Chem. Soc.* **1994**, *116*, 1272.
[2089]Greisbeck, A.G.; Bondock, S. *J. Am. Chem. Soc.* **2001**, *123*, 6191. See also, Adam, W.; Stegmann, V.R. *Synthesis* **2001**, 1203.
[2090]For a spin-directed reaction, see Griesbeck, A.G.; Fiege, M.; Bondock, S.; Gudipati, M.S. *Org. Lett.* **2000**, *2*, 3623.
[2091]Howell, A.R.; Fan, R.; Truong, A. *Tetrahedron Lett.* **1996**, *37*, 8651. For a review of the formation of heterocycles by cycloadditions of allenes, see Schuster, H.F.; Coppola, G.M. *Allenes in Organic Synthesis*, Wiley, NY, **1984**, pp. 317–326.
[2092]Abe, M.; Tachibana, K.; Fujimoto, K.; Nojima, M. *Synthesis* **2001**, 1243.
[2093]Kang, T.; Scheffer, J.R. *Org. Lett.* **2001**, *3*, 3361.
[2094]For a list of references, see Larock, R.C. *Comprehensive Organic Transformations*, 2nd ed., Wiley-VCH, NY, **1999**, pp. 1919–1921. For reviews of the formation of β-lactams, see Brown, M.J. *Heterocycles* **1989**, *29*, 2225; Isaacs, N.S. *Chem. Soc. Rev.* **1976**, *5*, 181; Mukerjee, A.K.; Srivastava, R.C. *Synthesis* **1973**, 327; Muller, L.L.; Hamer, J. *1,2-Cycloaddition Reactions*, Wiley, NY, **1967**, pp. 173–206; Ulrich, H. *Cycloaddition Reactions of Heterocumulenes*, Academic Press, NY, **1967**, pp. 75–83, 135–152; Anselme, J., in Patai, S. *The Chemistry of the Carbon–Nitrogen Double Bond*, Wiley, NY, **1970**, pp. 305–309. For a review of cycloaddition reactions of imines, see Sandhu, J.S.; Sain, B. *Heterocycles* **1987**, *26*, 777.

RCH=C=O, except when these are generated *in situ* by decomposition of a diazo ketone (the Wolff rearrangement, **18-8**) in the presence of the imine. It has been done with ketene, but the more usual course with this reagent is an addition to the enamine tautomer of the substrate. Thioketenes[2095] ($R_2C=C=S$) give β-thio-lactams.[2096] Imines also form β-lactams when treated with (*1*) zinc (or another metal[2097]) and an α-bromo ester (Reformatsky conditions, **16-28**),[2098] or (*2*) the chromium carbene complexes $(CO)_5Cr=C(Me)OMe$.[2099] The latter method has been used to prepare optically active β-lactams.[2100] Ketenes have also been added to certain hydrazones (e.g., $PhCH=NNMe_2$) to give *N*-amino β-lactams.[2101] A polymer-bound pyridinium salt facilitates β-lactam formation from carboxylic acids and imines.[2102]

N-Tosyl imines react with ketenes, Proton Sponge (p. 365) and a chiral amine to give the *N*-tosyl β-lactam with good enantioselectivity.[2103] A chiral ferrocenyl catalyst also gives good enantioselectivity,[2104] and chiral ammonium salts have been used as catalysts.[2105] A catalytic amount of benzoyl quinine gives β-lactams with good enantioselectivity.[2106] An intramolecular version of this ketene-imine reaction is known.[2107]

Like the similar cycloaddition of ketenes to alkenes (**15-63**), most of these reactions probably take place by the diionic mechanism *c* (p. 1224).[2108] β-Lactams have also been prepared in the opposite manner: by the addition of enamines to isocyanates:[2109]

[2095]For a review of thioketenes, see Schaumann, E. *Tetrahedron* **1988**, *44*, 1827.

[2096]Schaumann, E. *Chem. Ber.* **1976**, *109*, 906.

[2097]With In: Banik, B.K.; Ghatak, A.; Becker, F.F. *J. Chem. Soc., Perkin Trans. 1* **2000**, 2179.

[2098]For a review, see Hart, D.J.; Ha, D. *Chem. Rev.* **1989**, *89*, 1447.

[2099]Hegedus, L.S.; McGuire, M.A.; Schultze, L.M.; Yijun, C.; Anderson, O.P. *J. Am. Chem. Soc.* **1984**, *106*, 2680; Hegedus, L.S.; McGuire, M.A.; Schultze, L.M. *Org. Synth.* 65, 140.

[2100]Hegedus, L.S.; Imwinkelried, R.; Alarid-Sargent, M.; Dvorak, D.; Satoh, Y. *J. Am. Chem. Soc.* **1990**, *112*, 1109.

[2101]Sharma, S.D.; Pandhi, S.B. *J. Org. Chem.* **1990**, *55*, 2196.

[2102]Donati, D.; Morelli, C.; Porcheddu, A.; Taddei, M. *J. Org. Chem.* **2004**, *69*, 9316.

[2103]Taggi, A.E.; Hafez, A.M.; Wack, H.; Young, B.; Drury III, W.J.; Lectka, T. *J. Am. Chem. Soc.* **2000**, *122*, 7831.

[2104]Hodous, B.L.; Fu, G.C. *J. Am. Chem. Soc.* **2002**, *124*, 1578.

[2105]Taggi, A.E.; Hafez, A.M.; Wack, H.; Young, B.; Ferraris, D.; Lectka, T. *J. Am. Chem. Soc.* **2002**, *124*, 6626.

[2106]Shah, M.H.; France, S.; Lectka, T. *Synlett* **2003**, 1937.

[2107]Clark, A.J.; Battle, G.M.; Bridge, A. *Tetrahedron Lett.* **2001**, *42*, 4409.

[2108]See Moore, H.W.; Hernandez Jr., L.; Chambers, R. *J. Am. Chem. Soc.* **1978**, *100*, 2245; Pacansky, J.; Chang, J.S.; Brown, D.W.; Schwarz, W. *J. Org. Chem.* **1982**, *47*, 2233; Brady, W.T.; Shieh, C.H. *J. Org. Chem.* **1983**, *48*, 2499.

[2109]For example, see Perelman, M.; Mizsak, S.A. *J. Am. Chem. Soc.* **1962**, *84*, 4988; Opitz, G.; Koch, J. *Angew. Chem. Int. Ed.* **1963**, *2*, 152.

The reactive compound chlorosulfonyl isocyanate $(ClSO_2NCO)$[2110] forms β-lactams even with unactivated alkenes,[2111] as well as with imines,[2112] allenes,[2113] conjugated dienes,[2114] and cyclopropenes.[2115] With microwave irradiation, alkyl isocyanates also react.[2116]

α-Diazo ketones react with imines and microwave irradiation to give β-lactams.[2117] Allylic phosphonate esters react with imines, in the presence of a palladium catalyst, to give β-lactams.[2118] Alkynyl reagents, such as $BuC≡CO—Li^+$, react with imines to form β-lactams.[2119] Imines and benzylic halides react to give β-lactams in the presence of CO and a palladium catalyst.[2120] Conjugated amides react with NBS and 20% sodium acetate to give an α-bromo β-lactam.[2121] A different approach to β-lactams heated aziridines with CO and a cobalt catalyst.[2122] Aziridines also react with CO and a dendrimer catalyst to go a β-lactam.[2123]

OS **V**, 673; **VIII**, 3, 216.

ADDITION TO ISOCYANIDES[2124]

Addition to $R—^+N≡C^-$ is not a matter of a species with an electron pair adding to one atom and a species without a pair adding to the other, as is addition to the other types of double and triple bonds in this chapter and Chapter 15. In these additions, the electrophile and the nucleophile *both add to the carbon*. No species

[2110]For reviews of this compound, see Kamal, A.; Sattur, P.B. *Heterocycles* **1987**, *26*, 1051; Szabo, W.A. *Aldrichimica Acta* **1977**, *10*, 23; Rasmussen, J.K.; Hassner, A. *Chem. Rev.* **1976**, *76*, 389; Graf, R. *Angew. Chem. Int. Ed.* **1968**, *7*, 172.

[2111]Graf, R. *Liebigs Ann. Chem.* **1963**, *661*, 111; Bestian, H. *Pure Appl. Chem.* **1971**, *27*, 611. See also, Barrett, A.G.M.; Betts, M.J.; Fenwick, A. *J. Org. Chem.* **1985**, *50*, 169.

[2112]McAllister, M.A.; Tidwell, T.T. *J. Chem. Soc. Perkin Trans. 2* **1994**, 2239; Sordo, J.A.; González, J.; Sordo, T.L. *J. Am. Chem. Soc.* **1992**, *114*, 6249.

[2113]Moriconi, E.J.; Kelly, J.F. *J. Org. Chem.* **1968**, *33*, 3036. See also, Martin, J.C.; Carter, P.L.; Chitwood, J.L. *J. Org. Chem.* **1971**, *36*, 2225.

[2114]Moriconi, E.J.; Meyer, W.C. *J. Org. Chem.* **1971**, *36*, 2841; Malpass, J.R.; Tweddle, N.J. *J. Chem. Soc. Perkin Trans. 1* **1977**, 874.

[2115]Moriconi, E.J.; Kelly, J.F.; Salomone, R.A. *J. Org. Chem.* **1968**, *33*, 3448.

[2116]Taguchi, Y.; Tsuchiya, T.; Oishi, A.; Shibuya, I. *Bull. Chem. Soc. Jpn.* **1996**, *69*, 1667.

[2117]Linder, M.R.; Podlech, J. *Org. Lett.* **2001**, *3*, 1849.

[2118]Torii, S.; Okumoto, H.; Sadakane, M.; Hai, A.K.M.A.; Tanaka, H. *Tetrahedron Lett.* **1993**, *34*, 6553.

[2119]Shindo, M.; Oya, S.; Sato, Y.; Shishido, K. *Heterocycles* **1998**, *49*, 113.

[2120]Cho, C.S.; Jiang, L.H.; Shim, S.C. *Synth. Commun.* **1999**, *29*, 2695.

[2121]Naskar, D.; Roy, S. *J. Chem. Soc., Perkin Trans. 1* **1999**, 2435.

[2122]Davoli, P.; Forni, A.; Moretti, I.; Prati, F.; Torre, G. *Tetrahedron* **2001**, *57*, 1801; Davoli, P.; Prati, F. *Heterocycles* **2000**, *53*, 2379.

[2123]Lu, S.-M.; Alper, H. *J. Org. Chem.* **2004**, *69*, 3558.

[2124]For a monograph, see Ugi, I. *Isonitrile Chemistry*, Academic Press, NY, **1971**. For reviews, see Walborsky, H.M.; Periasamy, M.P., in Patai, S.; Rappoport, Z. *The Chemistry of Functional Groups, Supplement C*, pt. 2, Wiley, NY, **1983**, pp. 835–887; Hoffmann, P.; Marquarding, D.; Kliimann, H.; Ugi, I., in Rappoport, Z. *The Chemistry of the Cyano Group*, Wiley, NY, **1970**, pp. 853–883.

add to the

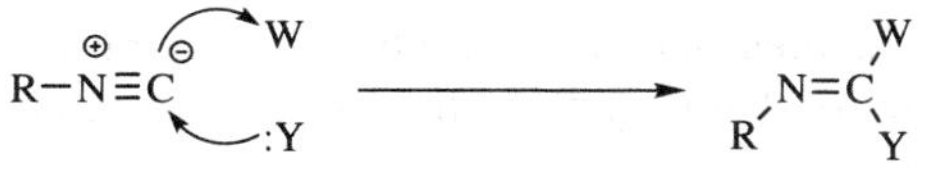

120

nitrogen, which, however, loses its positive charge by obtaining as an unshared pair one of the triple-bond pairs of electrons to give **120**. In most of the reactions considered below, **120** undergoes a further reaction, so the product is of the form.

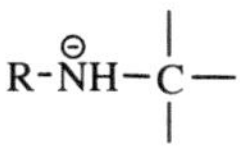

16-97   The Addition of Water to Isocyanides

1/*N*,2/*C*-Dihydro-2/*C*-oxo-biaddition

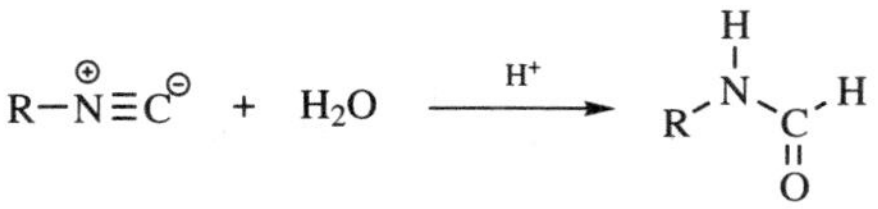

Formamides can be prepared by the acid-catalyzed addition of water to isocyanides. The mechanism is probably[2125]

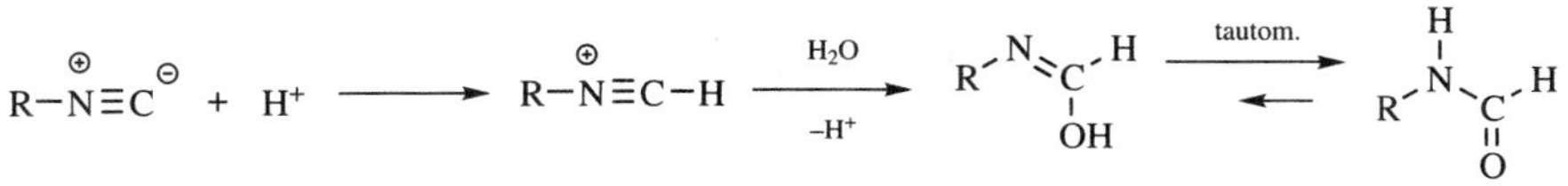

The reaction has also been carried out under alkaline conditions, with hydroxide in aqueous dioxane.[2126] The mechanism here involves nucleophilic attack by hydroxide at the carbon atom. An intramolecular addition of an alkyne (in an ortho alkynyl phenyl isonitrile) to the carbon of an isonitrile occurred with heating in methanol to give quinoline derivatives.[2127]

16-98   The Passerini and Ugi Reactions[2128]

1/*N*-Hydro-2/*C*-(α-acyloxyalkyl),2/*C*-oxo-biaddition

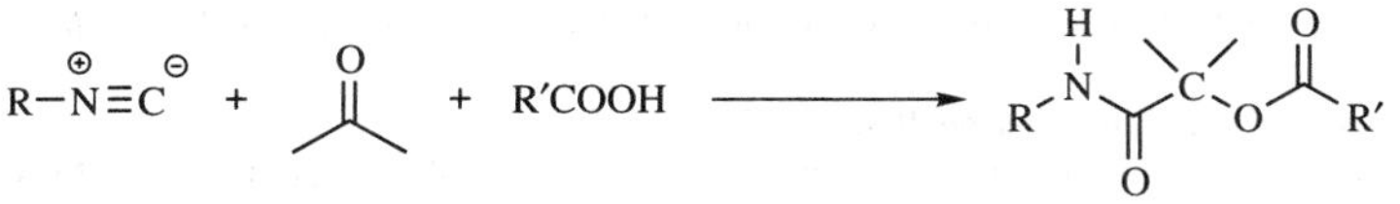

[2125]Drenth, W. *Recl. Trav. Chim. Pays-Bas* **1962**, *81*, 319; Lim, Y.Y.; Stein, A.R. *Can. J. Chem.* **1971**, *49*, 2455.

[2126]Cunningham, I.D.; Buist, G.J.; Arkle, S.R. *J. Chem. Soc. Perkin Trans. 2* **1991**, 589.

[2127]Suginome, M.; Fukuda, T.; Ito, Y. *Org. Lett.* **1999**, *1*, 1977.

[2128]For reviews, see Ugi, I. *Angew. Chem. Int. Ed.* **1982**, *21*, 810; Marquarding, D.; Gokel, G.W.; Hoffmann, P.; Ugi, I. in Ugi, I. *Isonitrile Chemistry*, Academic Press, NY, **1971**, pp. 133–143, Gokel, G.W.; Lüdke, G.; Ugi, I. in Ugi, I. Ref. 936, pp. 145–199, 252–254.

When an isocyanide is treated with a carboxylic acid and an aldehyde or ketone, an α-acyloxy amide is prepared. This is called the *Passerini reaction*. A SiCl$_4$-mediated reaction in the presence of a chiral bis-phosphoramide gives an α-hydroxy amide with good enantioselectivity.[2129] The following mechanism has been postulated for the basic reaction:[2130]

If ammonia or an amine is also added to the mixture (in which case the reaction is known as the *Ugi reaction*, or the *Ugi four-component condensation*, abbreviated 4 CC), the product is the corresponding bis(amide)

(from NH$_3$) or

(from a primary amine R^{2}NH$_2$).

Repetitive Ugi reactions are known.[2131] This product probably arises from a reaction between the carboxylic acid, the isocyanide, and the *imine* formed from the aldehyde or ketone and ammonia or the primary amine. The use of an *N*-protected amino acid[2132] or peptide as the carboxylic acid component and/or the use of an isocyanide containing a C-protected carboxyl group allows the reaction to be used for peptide synthesis.[2133]

[2129]Denmark, S.E.; Fan, Y. *J. Am. Chem. Soc.* **2003**, *125*, 7825.

[2130]For the effect of high pressure of sterically hindered reactions, see Jenner, G. *Tetrahedron Lett.* **2000**, *43*, 1235.

[2131]Constabel, F.; Ugi, I. *Tetrahedron* **2001**, *57*, 5785.

[2132]See, for example, Godet, T.; Bovin, Y.; Vincent, G.; Merle, D.; Thozet, A.; Ciufolini, M.A. *Org. Lett.* **2004**, *6*, 3281.

[2133]For reviews, see Ugi, I., in Gross, E.; Meienhofer, J. *The Peptides*, Vol. 2, Academic Press, NY, **1980**, pp. 365–381, *Intra-Sci. Chem. Rep.* **1971**, *5*, 229; *Rec. Chem. Prog.* **1969**, *30*, 289; Gokel, G.W.; Hoffmann, P.; Kleimann, H.; Klusacek, H.; Lüdke, G.; Marquarding, D.; Ugi, I., in Ugi, I. *Isonitrile Chemistry*, Academic Press, NY, **1971**, pp. 201–215. See also, Kunz, H.; Pfrengle, W. *J. Am. Chem. Soc.* **1988**, *110*, 651.

16-99   The Formation of Metalated Aldimines

1/1/Lithio-alkyl-addition

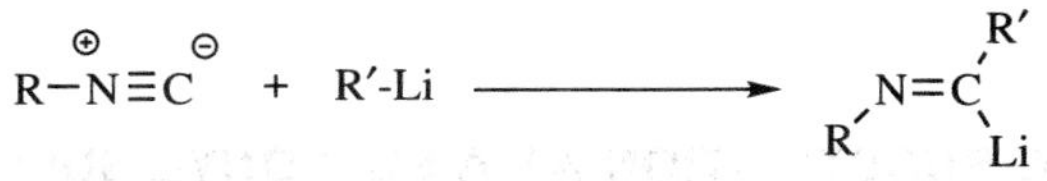

$$R-\overset{\oplus}{N}\equiv\overset{\ominus}{C} \;+\; R'\text{-Li} \;\longrightarrow\; \underset{R}{\overset{R'}{\diagdown}}N=\overset{R'}{\underset{Li}{\overset{|}{C}}}$$

Isocyanides that do not contain an α hydrogen react with alkyllithium compounds,[2134] as well as with Grignard reagents, to give lithium (or magnesium) aldimines.[2135] These metalated aldimines are versatile nucleophiles and react with various substrates as follows:

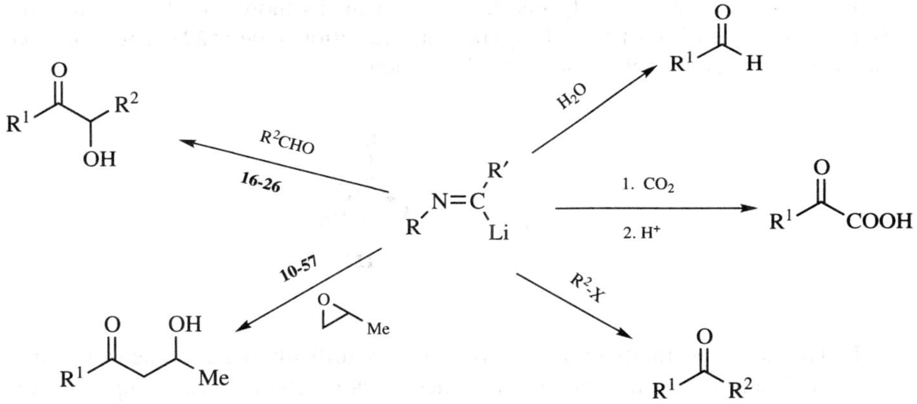

The reaction therefore constitutes a method for converting an organometallic compound R'M to an aldehyde R'CHO (see also, **12-33**), an α-keto acid,[2136] a ketone R'COR (see also, **12-33**), an α-hydroxy ketone, or a β-hydroxy ketone. In each case, the C=N bond is hydrolyzed to a C=O bond (**16-2**).

[2134]For a review of other metallation reactions of isocyanides, see Ito, Y.; Murakami, M. *Synlett* **1990**, 245.

[2135]Niznik, G.E.; Morrison III, W.H.; Walborsky, H.M. *J. Org. Chem.* **1974**, *39*, 600; Marks, M.J.; Walborsky, H.M. *J. Org. Chem.* **1981**, *46*, 5405; **1982**, *47*, 52. See also, Walborsky, H.M.; Ronman, P. *J. Org. Chem.* **1978**, *43*, 731. For the formation of zinc aldimines, see Murakami, H.; Ito, H.; Ito, Y. *J. Org. Chem.* **1988**, *53*, 4158.

[2136]For a review of the synthesis and properties of α-keto acids, see Cooper, A.J.L.; Ginos, J.Z.; Meister, A. *Chem. Rev.* **1983**, *83*, 321.

In a related reaction, isocyanides can be converted to aromatic aldimines by treatment with an iron complex followed by irradiation in benzene solution: $RNC + C_6H_6 \rightarrow PhCH{=}NR$.[2137]

OS **VI**, 751.

NUCLEOPHILIC SUBSTITUTION AT A SULFONYL SULFUR ATOM[2138]

Nucleophilic substitution at RSO_2X is similar to attack at $RCOX$. Many of the reactions are essentially the same, though sulfonyl halides are less reactive than halides of carboxylic acids.[2139] The mechanisms[2140] are not identical, because a "tetrahedral" intermediate in this case (**121**) would have five groups on the central atom. This is possible since sulfur can accommodate up to 12 electrons in its valence shell, but it seems more likely that these mechanisms more closely resemble the S_N2 mechanism, with a trigonal-bipyramidal transition state (**122**). There are two major experimental results leading to this conclusion.

1. The stereospecificity of this reaction is more difficult to determine than that of nucleophilic substitution at a saturated carbon, where chiral compounds are relatively easy to prepare, but it may be recalled (p. 142) that optical activity is possible in a compound of the form RSO_2X if one oxygen is ^{16}O and the other ^{18}O. When a sulfonate ester possessing this type of chirality was converted to a sulfone with a Grignard reagent (**16-105**), inversion of configuration was found.[2141] This is not incompatible with an intermediate such as **121** but it is also in good accord with an S_N2-like mechanism with backside attack.

[2137]Jones, W.D.; Foster, G.P.; Putinas, J.M. *J. Am. Chem. Soc.* **1987**, *109*, 5047.

[2138]For a review of mechanisms of nucleophilic substitutions at di-, tri-, and tetracoordinated sulfur atoms, see Ciuffarin, E.; Fava, A. *Prog. Phys. Org. Chem.* **1968**, *6*, 81.

[2139]For a comparative reactivity study, see Hirata, R.; Kiyan, N.Z.; Miller, J. *Bull. Soc. Chim. Fr.* **1988**, 694.

[2140]For a review of mechanisms of nucleophilic substitution at a sulfonyl sulfur, see Gordon, I.M.; Maskill, H.; Ruasse, M. *Chem. Soc. Rev.* **1989**, *18*, 123.

[2141]Sabol, M.A.; Andersen, K.K. *J. Am. Chem. Soc.* **1969**, *91*, 3603. See also, Jones, M.R.; Cram, D.J. *J. Am. Chem. Soc.* **1974**, *96*, 2183.

2. More direct evidence against **121** (though still not conclusive) was found in an experiment involving acidic and basic hydrolysis of aryl arenesulfonates, where it has been shown by the use of ^{18}O that an intermediate like **121** is not reversibly formed, since ester recovered when the reaction was stopped before completion contained no ^{18}O when the hydrolysis was carried out in the presence of labeled water.[2142]

Other evidence favoring the S_N2-like mechanism comes from kinetics and substituent effects.[2143] However, evidence for the mechanism involving **121** is that the rates did not change much with changes in the leaving group[2144] and the ρ values were large, indicating that a negative charge builds up in the transition state.[2145]

In certain cases in which the substrate carries an α hydrogen, there is strong evidence[2146] that at least some of the reaction takes place by an elimination-addition mechanism (E1cB, similar to the one shown on p. 1406), going through a *sulfene* intermediate,[2147] for example, the reaction between methanesulfonyl chloride and aniline.

$$CH_3{-}SO_2Cl \xrightarrow{\text{base}} CH_2{=}SO_2 \xrightarrow{PhNH_2} CH_3{-}SO_2{-}NHPh$$
A sulfene

[2142]Christman, D.R.; Oae, S. *Chem. Ind. (London) 1959*, 1251; Oae, S.; Fukumoto, T.; Kiritani, R. *Bull. Chem. Soc. Jpn. 1963, 36*, 346; Kaiser, E.T.; Zaborsky, O.R. *J. Am. Chem. Soc. 1968, 90*, 4626.

[2143]See, for example, Robertson, R.E.; Rossall, B. *Can. J. Chem. 1971, 49*, 1441; Rogne, O. *J. Chem. Soc. B 1971*, 1855; *J. Chem. Soc. Perkin Trans. 2 1972*, 489; Gnedin, B.G.; Ivanov, S.N.; Spryskov, A.A. *J. Org. Chem. USSR 1976, 12*, 1894; Banjoko, O.; Okwuiwe, R. *J. Org. Chem. 1980, 45*, 4966; Ballistreri, F.P.; Cantone, A.; Maccarone, E.; Tomaselli, G.A.; Tripolone, M. *J. Chem. Soc. Perkin Trans. 2 1981*, 438; Suttle, N.A.; Williams, A. *J. Chem. Soc. Perkin Trans. 2 1983*, 1563; D'Rozario, P.; Smyth, R.L.; Williams, A. *J. Am. Chem. Soc. 1984, 106*, 5027; Lee, I.; Kang, H.K.; Lee, H.W. *J. Am. Chem. Soc. 1987, 109*, 7472; Arcoria, A.; Ballistreri, F.P.; Spina, E.; Tomaselli, G.A.; Maccarone, E. *J. Chem. Soc. Perkin Trans. 2 1988*, 1793; Gnedin, B.G.; Ivanov, S.N.; Shchukina, M.V. *J. Org. Chem. USSR 1988, 24*, 731.

[2144]Ciuffarin, E.; Senatore, L.; Isola, M. *J. Chem. Soc. Perkin Trans. 2 1972*, 468.

[2145]Ciuffarin, E.; Senatore, L. *Tetrahedron Lett. 1974*, 1635.

[2146]For a review, see Opitz, G. *Angew. Chem. Int. Ed. 1967, 6*, 107. See also, King, J.F.; Lee, T.W.S. *J. Am. Chem. Soc. 1969, 91*, 6524; Skrypnik, Yu.G.; Bezrodnyi, V.P. *Doklad. Chem. 1982, 266*, 341; Farng, L.O.; Kice, J.L. *J. Am. Chem. Soc. 1981, 103*, 1137; Thea, S.; Guanti, G.; Hopkins, A.; Williams, A. *J. Am. Chem. Soc. 1982, 104*, 1128, *J. Org. Chem. 1985, 50*, 5592; Bezrodnyi, V.P.; Skrypnik, Yu.G. *J. Org. Chem. USSR 1984, 20*, 1660, 2349; King, J.F.; Skonieczny, S. *Tetrahedron Lett. 1987, 28*, 5001; Pregel, M.J.; Buncel, E. *J. Chem. Soc. Perkin Trans. 2 1991*, 307.

[2147]For reviews of sulfenes, see King, J.F. *Acc. Chem. Res. 1975, 8*, 10; Nagai, T.; Tokura, N. *Int. J. Sulfur Chem. Part B 1972*, 207; Truce, W.E.; Liu, L.K. *Mech. React. Sulfur Compd. 1969, 4*, 145; Opitz, G. *Angew. Chem. Int. Ed. 1967, 6*, 107; Wallace, T.J. *Q. Rev. Chem. Soc. 1966, 20*, 67.

In the special case of nucleophilic substitution at a sulfonic ester RSO_2OR', where R' is alkyl, R'—O cleavage is much more likely than S—O cleavage because the OSO_2R group is such a good leaving group (p. 497).[2148] Many of these reactions have been considered previously (e.g., **10-4**, **10-10**), because they are nucleophilic substitutions at an alkyl carbon atom and not at a sulfur atom. However, when R' is aryl, then the S—O bond is much more likely to cleave because of the very low tendency aryl substrates have for nucleophilic substitution.[2149]

The order of nucleophilicity toward a sulfonyl sulfur has been reported as $OH^- > RNH_2 > N_3^- > F^- > AcO^- > Cl^- > H_2O > I^-$.[2150] This order is similar to that at a carbonyl carbon (p. $$$). Both of these substrates can be regarded as relatively hard acids, compared to a saturated carbon which is considerably softer and which has a different order of nucleophilicity (p. 494).

16-100 Attack by OH: Hydrolysis of Sulfonic Acid Derivatives

S-Hydroxy-de-chlorination, etc.

$$RSO_2X \xrightarrow[\text{H}_2\text{O} + \text{H}^+]{\text{H}_2\text{O} \quad \text{or}} RSO_2OH \quad (X = Cl, OR', NR'_2)$$

Sulfonyl chlorides as well as esters and amides of sulfonic acids can be hydrolyzed to the corresponding acids. Sulfonyl chlorides can by hydrolyzed with water or with an alcohol in the absence of acid or base. Basic catalysis is also used, though of course the salt is the product obtained. Esters are readily hydrolyzed, many with water or dilute alkali. This is the same reaction as **10-4**, and usually involves R'—O cleavage, except when R' is aryl. However, in some cases retention of configuration has been shown at alkyl R', indicating S—O cleavage in these cases.[2151] Sulfonamides are generally *not* hydrolyzed by alkaline treatment, not even with hot concentrated alkali. Acids, however, do hydrolyze sulfonamides, but less readily than they do sulfonyl halides or sulfonic esters. Of course, ammonia or the amine appears as the salt. However, sulfonamides can be hydrolyzed with base if the solvent is HMPA.[2152]

Magnesium in methanol has been used to convert sulfonate esters to the parent alcohol.[2153] Likewise, $CeCl_3 \cdot 7\,H_2O$—NaI in acetonitrile converted aryl tosylates to the parent phenol derivative.[2154]

[2148]A number of sulfonates in which R contains a branching, for example, $Ph_2C(CF_3)SO_2OR'$, can be used to ensure that there will be no S—O cleavage: Netscher, T.; Prinzbach, H. *Synthesis* **1987**, 683.

[2149]See Tagaki, W.; Kurusu, T.; Oae, S. *Bull. Chem. Soc. Jpn.* **1969**, *42*, 2894.

[2150]Kice, J.L.; Kasperek, G.J.; Patterson, D. *J. Am. Chem. Soc.* **1969**, *91*, 5516; Rogne, O. *J. Chem. Soc. B* **1970**, 1056; Kice, J.L.; Legan, E. *J. Am. Chem. Soc.* **1973**, *95*, 3912.

[2151]Chang, F.C. *Tetrahedron Lett.* **1964**, 305.

[2152]Cuvigny, T.; Larchevêque, M. *J. Organomet. Chem.* **1974**, *64*, 315.

[2153]Sridhar, M.; Kumar, B.A.; Narender, R. *Tetrahedron Lett.* **1998**, *39*, 2847.

[2154]Reddy, G.S.; Mohan, G.H.; Iyengar, D.S. *Synth. Commun.* **2000**, *30*, 3829.

OS **I**, 14; **II**, 471; **III**, 262; **IV**, 34; **V**, 406; **VI**, 652, 727. Also see, OS **V**, 673; **VI**, 1016.

16-101   Attack by OR. Formation of Sulfonic Esters

*S***-Alkoxy-de-chlorination**, and so on

$$RSO_2Cl \; + \; R'OH \; \xrightarrow{\text{base}} \; RSO_2OR'$$

$$RSO_2NR_2'' \; + \; R'OH \; \xrightarrow{\text{base}} \; RSO_2OR' \; + \; NHR_2''$$

Sulfonic esters are most frequently prepared by treatment of the corresponding sulfonyl halides with alcohols in the presence of a base. This procedure is the most common method for the conversion of alcohols to tosylates, brosylates, and similar sulfonic esters. Both R and R' may be alkyl or aryl. The base is often pyridine, which functions as a nucleophilic catalyst,[2155] as in the similar alcoholysis of carboxylic acyl halides (**16-61**). Propylenediamines have also been used to facilitate tosylation of an alcohol.[2156] Silver oxide has been used, in conjunction with KI.[2157] Primary alcohols react the most rapidly, and it is often possible to sulfonate selectively a primary OH group in a molecule that also contains secondary or tertiary OH groups. The reaction with sulfonamides has been much less frequently used and is limited to *N,N*-disubstituted sulfonamides; that is, R— may not be hydrogen. However, within these limits it is a useful reaction. The nucleophile in this case is actually RO⁻. However, R' may be hydrogen (as well as alkyl) if the nucleophile is a phenol, so that the product is RSO₂OAr. Acidic catalysts are used in this case.[2158] Sulfonic acids have been converted directly to sulfonates by treatment with triethyl or trimethyl orthoformate, HC(OR)₃, without catalyst or solvent;[2159] and with a trialkyl phosphite, P(OR)₃.[2160]

Mono-tosylation of a 1,2-diol was achieved using tosyl chloride and triethylamine, with a tin oxide catalyst.[2161]

OS **I**, 145; **III**, 366; **IV**, 753; **VI**, 56, 482, 587, 652; **VII**, 117; **66**, 1; **68**, 188. Also see, OS **IV**, 529; **VI**, 324, 757; **VII**, 495; **VIII**, 568.

[2155]Rogne, O. *J. Chem. Soc. B 1971*, 1334. See also, Litvinenko, M.; Shatskaya, V.A.; Savelova, V.A. *Doklad. Chem. 1982, 265*, 199.

[2156]Yoshida, Y; Shimonishi, K.; Sakakura, Y.; Okada, S.; Aso, N.; Tanabe, Y. *Synthesis 1999*, 1633.

[2157]Bouzide, A.; LeBerre, N.; Sauvé, G. *Tetrahedron Lett. 2001, 42*, 8781.

[2158]Klamann, D.; Fabienke, E. *Chem. Ber. 1960, 93*, 252.

[2159]Padmapriya, A.A.; Just, G.; Lewis, N.G. *Synth. Commun. 1985, 15*, 1057.

[2160]Karaman, R.; Leader, H.; Goldblum, A.; Breuer, E. *Chem. Ind. (London) 1987*, 857.

[2161]Martinelli, M.J.; Vaidyanathan, R.; Khau, V.V. *Tetrahedron Lett. 2000, 41*, 3773; Bucher, B.; Curran, D.P. *Tetrahedron Lett. 2000, 41*, 9617.

16-102 Attack by Nitrogen: Formation of Sulfonamides

S-Amino-de-chlorination

$$RSO_2Cl \quad + \quad NH_3 \quad \longrightarrow \quad RSO_2NH_2$$

The treatment of sulfonyl chlorides with ammonia or amines is the usual way of preparing sulfonamides. Primary amines give *N*-alkyl sulfonamides, and secondary amines give *N,N*-dialkyl sulfonamides. The reaction is the basis of the *Hinsberg test* for distinguishing between primary, secondary, and tertiary amines. *N*-Alkyl sulfonamides, having an acidic hydrogen, are soluble in alkali, while *N,N*-dialkyl sulfonamides are not. Since tertiary amines are usually recovered unchanged, primary, secondary, and tertiary amines can be told apart. However, the test is limited for at least two reasons.[2162] (*1*) Many *N*-alkyl sulfonamides in which the alkyl group has six or more carbons are insoluble in alkali, despite their acidic hydrogen,[2163] so that a primary amine may appear to be a secondary amine. (*2*) If the reaction conditions are not carefully controlled, tertiary amines may not be recovered unchanged.[2160]

A primary or a secondary amine can be protected by reaction with phenacylsulfonyl chloride, (PhCOCH$_2$SO$_2$Cl), to give a sulfonamide, RNHSO$_2$CH$_2$COPh or R$_2$NSO$_2$CH$_2$COPh.[2164] The protecting group can be removed when desired with zinc and acetic acid. Sulfonyl chlorides react with azide ion to give sulfonyl azides (RSO$_2$N$_3$).[2165] Chlorothioformates, ROC(=S)Cl, react with triethylamine to give the *N,N*-diethylthioamide.[2166]

A quite different synthesis of sulfonamides treated allyltributyltin with PhI=NTs, in the presence of copper (II) triflate.[2167] another alternative method treats silyl enol ethers with sulfur dioxide, and subsequent and reaction with a secondary amine gave the β-sulfonamido ester.[2168]

OS **IV**, 34, 943; **V**, 39, 179, 1055; **VI**, 78, 652; **VII**, 501; **VIII**, 104. See also, OS **VI**, 788.

16-103 Attack by Halogen: Formation of Sulfonyl Halides

S-Halo-de-hydroxylation

$$RSO_2OH \quad + \quad PCl_5 \quad \longrightarrow \quad RSO_2Cl$$

[2162]For directions for performing and interpreting the Hinsberg test, see Gambill, C.R.; Roberts, T.D.; Shechter, H. *J. Chem. Educ.* **1972**, *49*, 287.

[2163]Fanta, P.E.; Wang, C.S. *J. Chem. Educ.* **1964**, *41*, 280.

[2164]Hendrickson, J.B.; Bergeron R. *Tetrahedron Lett.* **1970**, 345.

[2165]For an example, see Regitz, M.; Hocker, J.; Liedhegener, A. *Org. Synth. V*, 179.

[2166]Milan, D.S.; Prager, R.H. *Aust. J. Chem.* **1999**, *52*, 841.

[2167]Kim, D.Y.; Kim. H.S.; Choi, Y.J.; Mang, J.Y.; Lee, K. *Synth. Commun.* **2001**, *31*, 2463.

[2168]Bouchez, L.C.; Dubbaka, S.R.; Urks, M.; Vogel, P. *J. Org. Chem.* **2004**, *69*, 6413.

This reaction, parallel with **16-79**, is the standard method for the preparation of sulfonyl halides. Also used are PCl_3 and $SOCl_2$, and sulfonic acid salts can also serve as substrates. Cyanuric acid (2,4,6-trichloro[1,3,5]triazene) also serves as a chlorinating agent.[2169] Sulfonyl bromides and iodides have been prepared from sulfonyl hydrazides ($ArSO_2NHNH_2$, themselves prepared by **16-102**) by treatment with bromine or iodine.[2170] Sulfonyl fluorides are generally prepared from the chlorides, by halogen exchange.[2171]

OS **I**, 84; **IV**, 571, 693, 846, 937; **V**, 196. See also, OS **VII**, 495.

16-104   Attack by Hydrogen: Reduction of Sulfonyl Chlorides

S-Hydro-de-chlorination or **S-Dechlorination**

$$2\ RSO_2Cl\ +\ Zn\ \longrightarrow\ (RSO_2)_2Zn\ \xrightarrow{\text{H+}}\ 2\ RSO_2H$$

Sulfinic acids can be prepared by reduction of sulfonyl chlorides. Though mostly done on aromatic sulfonyl chlorides, the reaction has also been applied to alkyl compounds. Besides zinc, sodium sulfite, hydrazine, sodium sulfide, and other reducing agents have been used. For reduction of sulfonyl chlorides to thiols, see **19-78**.

OS **I**, 7, 492; **IV**, 674.

16-105   Attack by Carbon: Preparation of Sulfones

S-Aryl-de-chlorination

$$ArSO_2Cl\ +\ Ar'MgX\ \longrightarrow\ ArSO_2Ar'$$

Grignard reagents convert aromatic sulfonyl chlorides or aromatic sulfonates to sulfones. Organolithium reagents react with sulfonyl fluorides at $-78°C$ to give the corresponding sulfone.[2172] Aromatic sulfonates have also been converted to sulfones with organolithium compounds,[2173] with aryltin compounds,[2174] and with alkyl halides and Zn metal.[2175] Vinylic and allylic sulfones have been prepared by treatment of sulfonyl chlorides with a vinylic or allylic stannane and a palladium complex catalyst.[2176] Alkynyl sulfones can be prepared by treatment of sulfonyl chlorides with trimethylsilylalkynes, with an $AlCl_3$ catalyst.[2177] Note that

[2169]Blotny, G. *Tetrahedron Lett.* **2003**, *44*, 1499.

[2170]Poshkus, A.C.; Herweh, J.E.; Magnotta, F.A. *J. Org. Chem.* **1963**, *28*, 2766; Litvinenko, L.M.; Dadali, V.A.; Savelova, V.A.; Krichevtsova, T.I. *J. Gen. Chem. USSR* **1964**, *34*, 3780.

[2171]See Bianchi, T.A.; Cate, L.A. *J. Org. Chem.* **1977**, *42*, 2031, and references cited therein.

[2172]Frye, L.L.; Sullivan, E.L.; Cusack, K.P.; Funaro, J.M. *J. Org. Chem,* **1992**, *57*, 697.

[2173]Baarschers, W.H. *Can. J. Chem.* **1976**, *54*, 3056.

[2174]Neumann, W.P.; Wicenec, C. *Chem. Ber.* **1993**, *126*, 763.

[2175]Sun, X.; Wang, L.; Zhang, Y. *Synth. Commun.* **1998**, *28*, 1785.

[2176]Labadie, S.S. *J. Org. Chem.* **1989**, *54*, 2496.

[2177]See Waykole, L.; Paquette, L.A. *Org. Synth. 67*, 149.

trifluoromethylsulfones were converted to methyl sulfones by reaction with methyl-magneisum bromide.[2178]

Arylboronic acids (p. 815) react with sulfonyl chlorides in the presence of $PdCl_2$ to give the corresponding sulfone.[2179] arylboronic acids also react with sulfinate anions (RSO_2Na) in the presence of $Cu(OAc)_2$ to give the sulfone.[2180]

OS **VIII**, 281.

[2178]Steensma, R.W.; Galabi, S.; Tagat, J.R.; McCombie, S.W. *Tetrahedron Lett.* **2001**, *42*, 2281.
[2179]Bandgar, B.P.; Bettigeri, S.V.; Phopase, J. *Org. Lett.* **2004**, *6*, 2105.
[2180]Beaulieu, C.; Guay, D.; Wang, Z.; Evans, D.A. *Tetrahedron Lett.* **2004**, *45*, 3233.

CHAPTER 17

Eliminations

When two groups are lost from adjacent atoms so that a new double[1]

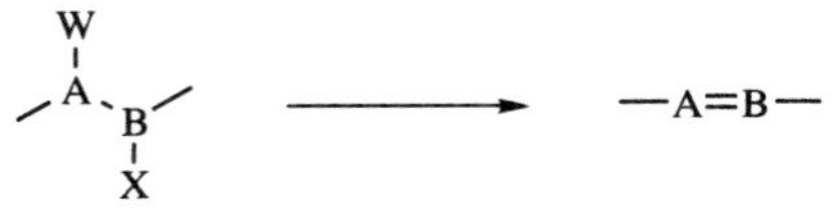

(or triple) bond is formed the reaction is called β-*elimination*; one atom is the α, the other the β atom. In an α elimination, both groups are lost from the same atom to give a carbene (or a nitrene):

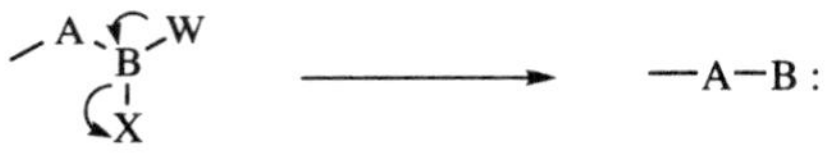

In a γ elimination, a three-membered ring is formed:

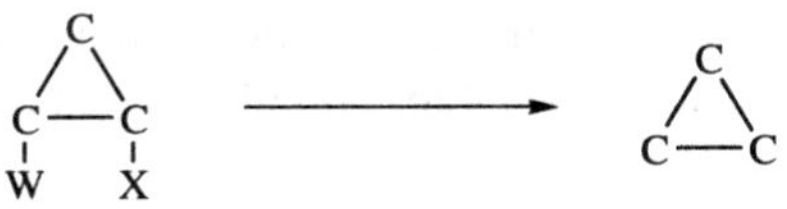

Some of these processes were discussed in Chapter 10. Another type of elimination involves the expulsion of a fragment from within a chain or ring (X–Y–Z → X–Z + Y). Such reactions are called *extrusion reactions*. This chapter discusses β-elimination and (beginning on p. 1553) extrusion reactions; however, β-elimination in which both X and W are hydrogens are oxidation reactions and are treated in Chapter 19.

[1]See Williams, J.M.J. *Preparation of Alkenes, A Practical Approach*, Oxford University Press, Oxford, *1996*.

March's Advanced Organic Chemistry: Reactions, Mechanisms, and Structure, Sixth Edition, by
Michael B. Smith and Jerry March
Copyright © 2007 John Wiley & Sons, Inc.

MECHANISMS AND ORIENTATION

β-Elimination reactions may be divided into two types; one type taking place largely in solution, the other (pyrolytic eliminations) mostly in the gas phase. In the reactions in solution, one group leaves with its electrons and the other without, the latter most often being hydrogen. In these cases, we refer to the former as the leaving group or nucleofuge. For pyrolytic eliminations, there are two principal mechanisms, one pericyclic and the other a free-radical pathway. A few photochemical eliminations are also known (the most important is Norrish type II cleavage of ketones, p. 344), but these are not generally of synthetic importance[2] and will not be discussed further. In most β-eliminations the new bonds are C=C or C≡C; our discussion of mechanisms is largely confined to these cases.[3] Mechanisms in solution (E2, E1)[4] and E1cB are discussed first.

The E2 Mechanism

In the E2 mechanism (elimination, bimolecular), the two groups depart simultaneously, with the proton being pulled off by a base:

The mechanism thus takes place in one step and kinetically is second order: first order in substrate and first order in base. An *ab initio* study has produced a model for the E2 transition state geometry.[5] The IUPAC designation is $A_{xH}D_HD_N$, or more generally (to include cases where the electrofuge is not hydrogen), $A_nD_ED_N$. It is analogous to the S_N2 mechanism (p. 426) and often competes with it. With respect

[2]For synthetically useful examples of Norrish type II cleavage, see Neckers, D.C.; Kellogg, R.M.; Prins, W.L.; Schoustra, B. *J. Org. Chem.* **1971**, *36*, 1838.
[3]For a monograph on elimination mechanisms, see Saunders, Jr., W.H.; Cockerill, A.F. *Mechanisms of Elimination Reactions*, Wiley, NY, **1973**. For reviews, see Gandler, J.R., in Patai, S. *Supplement A: The Chemistry of Double-Bonded Functional Groups*, Vol. 2, pt. 1, Wiley, NY, **1989**, pp. 733–797; Aleskerov, M.A.; Yufit, S.S.; Kucherov, V.F. *Russ. Chem. Rev.* **1978**, *47*, 134; Cockerill, A.F.; Harrison, R.G., in Patai, S. *The Chemistry of Functional Groups, Supplement A* pt. 1, Wiley, NY, **1977**, pp. 153–221; Willi, A.V. *Chimia*, **1977**, *31*, 93; More O'Ferrall, R.A., in Patai, S. *The Chemistry of the Carbon-Halogen Bond*, pt. 2, Wiley, NY, **1973**, pp. 609–675; Cockerill, A.F., in Bamford, C.H.; Tipper, C.F.H. *Comprehensive Chemical Kinetics*, Vol. 9, Elsevier, NY, **1973**, pp. 163–372; Saunders, Jr., W.H. *Acc. Chem. Res.* **1976**, *9*, 19; Stirling, C.J.M. *Essays Chem.* **1973**, *5*, 123; Bordwell, F.G. *Acc. Chem. Res.* **1972**, *5*, 374; Fry, A. *Chem. Soc. Rev.* **1972**, *1*, 163; LeBel, N.A. *Adv. Alicyclic Chem.* **1971**, *3*, 195; Bunnett, J.F. *Surv. Prog. Chem.* **1969**, *5*, 53; in Patai, S. *The Chemistry of Alkenes*, Vol. 1, Wiley, NY, **1964**, the articles by Saunders, Jr., W.H. pp. 149–201 (eliminations in solution); and by Maccoll, A. pp. 203–240 (pyrolytic eliminations); Köbrich, G. *Angew. Chem. Int. Ed.* **1965**, *4*, 49, pp. 59–63 (for the formation of triple bonds).
[4]Thibblin, A. *Chem. Soc. Rev.* **1993**, *22*, 427.
[5]Schrøder, S.; Jensen, F. *J. Org. Chem.* **1997**, *62*, 253.

to the substrate, the difference between the two pathways is whether the species with the unshared pair attacks the carbon (and thus acts as a nucleophile) or the hydrogen (and thus acts as a base). As in the case of the S_N2 mechanism, the leaving group may be positive or neutral and the base may be negatively charged or neutral.

Among the evidence for the existence of the E2 mechanism are (1) the reaction displays the proper second-order kinetics; (2) when the hydrogen is replaced by deuterium in second-order eliminations, there is an isotope effect of from 3 to 8, consistent with breaking of this bond in the rate-determining step.[6] However, neither of these results alone could prove an E2 mechanism, since both are compatible with other mechanisms also (e.g., see E1cB p. 1488). The most compelling evidence for the E2 mechanism is found in stereochemical studies.[7] As will be illustrated in the examples below, the E2 mechanism is stereospecific: The five atoms involved (including the base) in the transition state must be in one plane. There are two ways for this to happen. The H and X may be

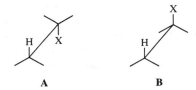

A   **B**

trans to one another (**A**) with a dihedral angle of 180°, or they may be cis (**B**) with a dihedral angle of 0°.[8] Conformation **A** is called *anti-periplanar*, and this type of elimination, in which H and X depart in opposite directions, is called *anti-elimination*. Conformation **B** is *syn-periplanar*, and this type of elimination, with H and X leaving in the same direction, is called *syn-elimination*. Many examples of both kinds have been discovered. In the absence of special effects (discussed below) anti-elimination is usually greatly favored over syn-elimination, probably because **A** is a staggered conformation (p. 199) and the molecule requires less energy to reach this transition state than it does to reach the eclipsed transition state **B**. A few of the many known examples of predominant or exclusive anti-elimination follow.

[6]See, for example, Saunders, Jr., W.H.; Edison, D.H. *J. Am. Chem. Soc.* *1960*, *82*, 138; Shiner, Jr., V.J.; Smith, M.L. *J. Am. Chem. Soc.* *1958*, *80*, 4095; *1961*, *83*, 593. For a review of isotope effects in elimination reactions, see Fry, A. *Chem. Soc. Rev.* *1972*, *1*, 163.

[7]For reviews, see Bartsch, R.A.; Závada, J. *Chem. Rev.* *1980*, *80*, 453; Coke, J.L. *Sel. Org. Transform.* *1972*, *2*, 269; Sicher, J. *Angew. Chem. Int. Ed.* *1972*, *11*, 200; *Pure Appl. Chem.* *1971*, *25*, 655; Saunders, Jr., W.H.; Cockerill, A.F. *Mechanisms of Elimination Reactions*, Wiley, NY, *1973*, pp. 105–163; Cockerill, A.F., in Bamford, C.H.; Tipper, C.F.H. *Comprehensive Chemical Kinetics*, Vol. 9, Elsevier, NY, *1973*, pp. 217–235; More O'Ferrall, R.A., in Patai, S. *The Chemistry of the Carbon–Halogen Bond*, pt. 2, Wiley, NY, *1973*, pp. 630–640.

[8]DePuy, C.H.; Morris, G.F.; Smith, J.S.; Smat, R.J. *J. Am. Chem. Soc.* *1965*, *87*, 2421.

1. Elimination of HBr from *meso*-1,2-dibromo-1,2-diphenylethane gave *cis*-2-bromostilbene, while the (+) or (−) isomer gave the trans alkene. This stereospecific result, which

was obtained in 1904,[9] demonstrates that in this case elimination is anti. Many similar examples have been discovered since. Obviously, this type of experiment need not be restricted to compounds that have a meso form. Anti-elimination requires that an erythro *dl* pair (or either isomer) give the cis alkene, and the threo *dl* pair (or either isomer) give the trans isomer, and this has been found many times. Anti-elimination has also been demonstrated in cases where the electrofuge is not hydrogen. In the reaction of 2,3-dibromobutane with iodide ion, the two bromines are removed (**17-22**). In this case, the meso compound gave the trans alkene and the *dl* pair the cis:[10]

2. In open-chain compounds, the molecule can usually adopt that conformation in which H and X are anti-periplanar. However, in cyclic systems this is not always the case. There are nine stereoisomers of 1,2,3,4,5,6-hexachlorocyclohexane: seven meso forms and a *dl* pair (see p. 165). Four of the meso compounds and the *dl* pair (all that were then known) were subjected to

[9]Pfeiffer, P. *Z. Phys. Chem.* **1904**, *48*, 40.
[10]Winstein, S.; Pressman, D.; Young, W.G. *J. Am. Chem. Soc.* **1939**, *61*, 1645.

elimination of HCl. Only one of these (**1**) has no Cl trans to an H. Of the other isomers, the fastest elimination rate was about three times as fast as the slowest, but the rate for **1** was 7000 times slower than that of the slowest of the other isomers.[11] This result demonstrates that with these compounds anti elimination is greatly favored over syn elimination, although the latter must be taking place on **1**, very slowly, to be sure.

1

3. The preceding result shows that elimination of HCl in a six-membered ring proceeds best when the H and X are trans to each other. However, there is an additional restriction. Adjacent trans groups on a six-membered ring can be diaxial or diequatorial (p. 204) and the molecule is generally free to adopt either conformation, although one may have a higher energy than the other. Anti-periplanarity of the leaving groups requires that they be diaxial, even if this is the conformation of higher energy. The results with menthyl and neomenthyl chlorides are easily

[11]Cristol, S.J.; Hause, N.L.; Meek, J.S. *J. Am. Chem. Soc.* **1951**, *73*, 674.

interpretable on this basis. Menthyl chloride has two chair conformations, **2** and **3**. Compound **3**, in which the three substituents are all equatorial, is the more stable. The more stable chair conformation of neomenthyl chloride is **4**, in which the chlorine is axial; there are axial hydrogens on both C-2 and C-4. The results are: neomenthyl chloride gives rapid E2 elimination and the alkene produced is predominantly **6** (**6/5** ratio is 3:1) in accord with Zaitsev's rule (p. 767). Since an axial hydrogen is available on both sides, this factor does not control the direction of elimination and Zaitsev's rule is free to operate. However, for menthyl chloride, elimination is much slower and the product is entirely the anti-Zaitsev, **5**. It is slow because the unfavorable conformation **2** has to be achieved before elimination can take place, and the product is **5** because only on this side is there an axial hydrogen.[12]

4. That anti-elimination also occurs in the formation of triple bonds is shown by elimination from *cis-* and *trans*-HOOC–CH=C(Cl)COOH. In this case, the product in both cases is HOOCC≡CCOOH, but the trans isomer reacts ~50 times faster than the cis compound.[13]

Some examples of syn-elimination have been found in molecules where H and X could not achieve an anti-periplanar conformation.

1. The deuterated norbornyl bromide (**7**, X = Br) gave 94% of the product containing no deuterium.[14] Similar results were obtained with other leaving groups and with bicyclo[2.2.2] compounds.[15] In these cases the exo X group cannot achieve a dihedral angle of 180° with the endo β hydrogen because of the rigid structure of the molecule. The dihedral angle here is ~120°. These leaving groups prefer syn-elimination with a dihedral angle of ~0° to anti-elimination with an angle of ~120°.

7   8

[12]Hughes, E.D.; Ingold, C.K.; Rose, J.B. *J. Chem. Soc.* **1953**, 3839.

[13]Michael, A. *J. Prakt. Chem.* **1895**, *52*, 308. See also, Marchese, G.; Naso, F.; Modena, G. *J. Chem. Soc. B* **1968**, 958.

[14]Kwart, H.; Takeshita, T.; Nyce, J.L. *J. Am. Chem. Soc.* **1964**, *86*, 2606.

[15]For example, see Bird, C.W.; Cookson, R.C.; Hudec, J.; Williams, R.O. *J. Chem. Soc.* **1963**, 410; Stille, J.K.; Sonnenberg, F.M.; Kinstle, T.H. *J. Am. Chem. Soc.* **1966**, *88*, 4922; Coke, J.L.; Cooke, Jr., M.P. *J. Am. Chem. Soc.* **1967**, *89*, 6701; DePuy, C.H.; Naylor, C.G.; Beckman, J.A. *J. Org. Chem.* **1970**, *35*, 2750; Brown, H.C.; Liu, K. *J. Am. Chem. Soc.* **1970**, *92*, 200; Sicher, J.; Pánkova, M.; Závada, J.; Kniežo, L.; Orahovats, A. *Collect. Czech. Chem. Commun.* **1971**, *36*, 3128; Bartsch, R.A.; Lee, J.G. *J. Org. Chem.* **1991**, *56*, 212, 2579.

2. Molecule **8** is a particularly graphic example of the need for a planar transition state. In **8**, each Cl has an adjacent hydrogen trans to it, and if planarity of leaving groups were not required, anti-elimination could easily take place. However, the crowding of the rest of the molecule forces the dihedral angle to be ∼120°, and elimination of HCl from **8** is much slower than from corresponding nonbridged compounds.[16] (Note that syn elimination from **8** is even less likely than anti-elimination.) Syn-elimination can take place from the trans isomer of **8** (dihedral angle ∼0°); this isomer reacted about eight times faster than **8**.[16]

The examples so far given illustrate two points. (*1*) Anti-elimination *requires* a dihedral angle of 180°. When this angle cannot be achieved, anti-elimination is greatly slowed or prevented entirely. (*2*) For the simple systems so far discussed syn-elimination is not found to any significant extent unless anti elimination is greatly diminished by failure to achieve the 180° angle.

As noted in Chapter 4 (p. 223), six-membered rings are the only ones among rings of 4–13 members in which strain-free anti-periplanar conformations can be achieved. It is not surprising, therefore, that syn elimination is least common in six-membered rings. Cooke and Coke subjected cycloalkyltrimethylammonium hydroxides to elimination (**17-7**) and found the following percentages of syn-elimination with ring size: four-membered, 90%; five-membered, 46%; six-membered, 4% seven-membered, 31 to 37%.[17] Note that the NMe_3^+ group has a greater tendency to syn-elimination than do other common leaving groups, such as OTs, Cl, and Br.

Other examples of syn-elimination have been found in medium-ring compounds, where both cis and trans alkenes are possible (p. 184). As an illustration, we can look at experiments performed by, Svoboda, and Sicher.[18] These workers subjected 1,1,4,4-tetramethyl-7-cyclodecyltrimethylammonium chloride (**9**) to

                                          trans and cis Alkenes

9

elimination and obtained mostly *trans*-, but also some *cis*-tetramethylcyclodecenes as products. (Note that *trans*-cyclodecenes, although stable, are less stable than the cis isomers). In order to determine the stereochemistry of the reaction, they repeated the elimination, this time using deuterated substrates. They found that

[16]Cristol, S.J.; Hause, N.L. *J. Am. Chem. Soc.* **1952**, *74*, 2193.

[17]Cooke, Jr., M.P.; Coke, J.L. *J. Am. Chem. Soc.* **1968**, *90*, 5556. See also, Coke, J.L.; Smith, G.D.; Britton, Jr., G.H. *J. Am. Chem. Soc.* **1975**, *97*, 4323.

[18]Závada, J.; Svoboda, M.; Sicher, J. *Tetrahedron Lett.* **1966**, 1627; *Collect. Czech. Chem. Commun.* **1968**, *33*, 4027.

when **9** was deuterated in the trans position ($H_t = D$), there was a substantial isotope effect in the formation of *both* cis and trans alkenes, but when **9** was deuterated in the cis position ($H_c = D$), there was *no* isotope effect in the formation of either alkene. Since an isotope effect is expected for an E2 mechanism,[19] these results indicated that *only* the trans hydrogen (H_t) was lost, whether the product was the cis or the trans isomer.[20] This in turn means that *the cis isomer must have been formed by anti-elimination and the trans isomer by syn-elimination.* (Anti-elimination could take place from approximately the conformation shown, but for syn elimination the molecule must twist into a conformation in which the C–H_t and C–NMe_3^+ bonds are syn-periplanar.) This remarkable result, called the *syn–anti dichotomy*, has also been demonstrated by other types of evidence.[21] The fact that syn-elimination in this case predominates over anti (as indicated by the formation of trans isomer in greater amounts than cis) has been explained by conformational factors.[22] The syn–anti dichotomy has also been found in other medium-ring systems (8–12 membered),[23] although the effect is greatest for 10-membered rings. With leaving groups,[24] the extent of this behavior decreases in the order $^+NMe_3 > OTs > Br > Cl$, which parallels steric requirements. When the leaving group is uncharged, syn-elimination is favored by strong bases and by weakly ionizing solvents.[25]

Syn-elimination and the syn—anti dichotomy have also been found in open-chain systems, although to a lesser extent than in medium-ring compounds. For example, in the conversion of 3-hexyl-4-*d*-trimethylammonium ion to 3-hexene with potassium *sec*-butoxide, ~67% of the reaction followed the syn–anti dichotomy.[26] In general syn-elimination in open-chain systems is only important in cases where certain types of steric effect are present. One such type is compounds in which substituents are found on both the β' and the γ carbons (the unprimed letter refers to the branch in which the elimination takes place). The factors that cause these results are not

---

[19]Other possible mechanisms, such as E1cB (p. 1488) or α',β elimination (p. 1524), were ruled out in all these cases by other evidence.

[20]This conclusion has been challenged by Coke, J.L. *Sel. Org. Transform* **1972**, *2*, 269.

[21]Sicher, J.; Závada, J. *Collect. Czech. Chem. Commun.* **1967**, *32*, 2122; Závada, J.; Sicher, J. *Collect. Czech. Chem. Commun.* **1967**, *32*, 3701. For a review, see Bartsch, R.A.; Závada, J. *Chem. Rev.* **1980**, *80*, 453.

[22]For discussions, see Bartsch, R.A.; Závada, J. *Chem. Rev.* **1980**, *80*, 453; Coke, J.L. *Sel. Org. Transform.* **1972**, *2*, 269; Sicher, J. *Angew. Chem. Int. Ed.* **1972**, *11*, 200; *Pure Appl. Chem.* **1971**, *25*, 655.

[23]For example, see Coke, J.L.; Mourning, M.C. *J. Am. Chem. Soc.* **1968**, *90*, 5561, where the experiment was performed on cyclooctyltrimethylammonium hydroxide, and *trans*-cyclooctene was formed by a 100% syn mechanism, and *cis*-cyclooctene by a 51% syn and 49% anti mechanism.

[24]For examples with other leaving groups, see Sicher, J.; Jan, G.; Schlosser, M. *Angew. Chem. Int. Ed.* **1971**, *10*, 926; Závada, J.; Pánková, M. *Collect. Czech. Chem. Commun.* **1980**, *45*, 2171, and references cited therein.

[25]See, for example, Sicher, J.; Závada, J. *Collect. Czech. Chem. Commun.* **1968**, *33*, 1278.

[26]Bailey, D.S.; Saunders Jr., W.H. *J. Am. Chem. Soc.* **1970**, *92*, 6904. For other examples of syn-elimination and the syn-anti dichotomy in open-chain systems, see Pánková, M.; Vítek, A.; Vasíšková, S.; Řeřicha, R.; Závada, J. *Collect. Czech. Chem. Commun.* **1972**, *37*, 3456; Schlosser, M.; An, T.D. *Helv. Chim. Acta* **1979**, *62*, 1194; Sugita, T.; Nakagawa, J.; Nishimoto, K.; Kasai, Y.; Ichikawa, K. *Bull. Chem. Soc. Jpn.* **1979**, *52*, 871; Pánková, M.; Kocián, O.; Krupička, J.; Závada, J. *Collect. Czech. Chem. Commun.* **1983**, *48*, 2944.

completely understood, but the following conformational effects have been proposed as a partial explanation.[27] The two anti- and two syn-periplanar conformations are, for a quaternary ammonium salt:

C	D	E	F
anti ⟶ trans	anti ⟶ cis	syn ⟶ trans	syn ⟶ cis

In order for an E2 mechanism to take place, a base must approach the proton marked *. In **C**, this proton is shielded on both sides by R and R'. In **D**, the shielding is on only one side. Therefore, when anti-elimination does take place in such systems, it should give more cis product than trans. Also, when the normal anti elimination pathway is hindered sufficiently to allow the syn pathway to compete, the anti → trans route should be diminished more than the anti → cis route. When syn-elimination begins to appear, it seems clear that **E**, which is less eclipsed than **F**, should be the favored pathway and syn-elimination should generally give the trans isomer. In general, deviations from the syn–anti dichotomy are greater on the trans side than on the cis. Thus, trans alkenes are formed partly or mainly by syn-elimination, but cis alkenes are formed entirely by anti-elimination. Predominant syn-elimination has also been found in compounds of the form $R^1R^2CHCHDNMe_3^+$, where R^1 and R^2 are both bulky.[28] In this case, the conformation leading to syn-elimination (**H**) is also less strained than **G**, which gives anti-elimination. The **G** compound has three bulky groups (including NMe_3^+) in the *gauche* position to each other.

G	H

It was mentioned above that weakly ionizing solvents promote syn-elimination when the leaving group is uncharged. This is probably caused by ion pairing, which

[27]Bailey, D.S.; Saunders, Jr., W.H. *J. Am. Chem. Soc.* **1970**, *92*, 6904; Chiao, W.; Saunders, Jr., W.H. *J. Am. Chem. Soc.* **1977**, *99*, 6699.
[28]Dohner, B.R.; Saunders Jr., W.H. *J. Am. Chem. Soc.* **1986**, *108*, 245.

is greatest in nonpolar solvents.[29] Ion pairing can

$$
\begin{array}{c}
\overset{C-C}{H \diagdown\diagup X} \\
\diagup \\
R-O^{\ominus}\ K^{\oplus}
\end{array}
$$

10

cause syn-elimination with an uncharged leaving group by means of the transition state shown in **10**. This effect was graphically illustrated by elimination from 1,1,4,4-tetramethyl-7-cyclodecyl bromide.[30] The ratio of syn-to-anti-elimination when this compound was treated with *t*-BuOK in the nonpolar benzene was 55.0. But when the crown ether dicyclohexano-18-crown-6 was added (this compound selectively removes K^+ from the *t*-BuO$^-$ K^+ ion pair and thus leaves *t*-BuO$^-$ as a free ion), the syn/anti ratio decreased to 0.12. Large decreases in the syn/anti ratio on addition of the crown ether were also found with the corresponding tosylate and with other nonpolar solvents.[31] However, with positively charged leaving groups the effect is reversed. Here, ion pairing *increases* the amount of anti-elimination.[32] In this case, a relatively free base (e.g., PhO$^-$) can be attracted to the leaving group, putting it in a favorable position for attack on the syn β hydrogen, while ion pairing would reduce this attraction.

$$
\begin{array}{c}
\diagdown \quad \diagup \\
-C-C- \\
H \diagup \diagdown NMe_3 \\
\quad O^{\oplus} \\
\quad \overset{|}{R}
\end{array}
$$

We can conclude that anti-elimination is generally favored in the E2 mechanism, but that steric (inability to form the anti-periplanar transition state), conformational, ion pairing, and other factors cause syn-elimination to intervene (and even predominate) in some cases.

---

[29]For reviews of ion pairing in this reaction, see Bartsch, R.A.; Závada, J. *Chem. Rev.* **1980**, *80*, 453; Bartsch, R.A. *Acc. Chem. Res.* **1975**, *8*, 239.

[30]Svoboda, M.; Hapala, J.; Závada, J. *Tetrahedron Lett.* **1972**, 265.

[31]For other examples of the effect of ion pairing, see Bayne, W.F.; Snyder, E.I. *Tetrahedron Lett.* **1971**, 571; Bartsch, R.A.; Wiegers, K.E. *Tetrahedron Lett.* **1972**, 3819; Fiandanese, V.; Marchese, G.; Naso, F.; Sciacovelli, O. *J. Chem. Soc. Perkin Trans. 2* **1973**, 1336; Borchardt, J.K.; Swanson, J.C.; Saunders, Jr., W.H. *J. Am. Chem. Soc.* **1974**, *96*, 3918; Mano, H.; Sera, A.; Maruyama, K. *Bull. Chem. Soc. Jpn.* **1974**, *47*, 1758; Závada, J.; Pánková, M.; Svoboda, M. *Collect. Czech. Chem. Commun.* **1976**, *41*, 3778; Baciocchi, E.; Ruzziconi, R.; Sebastiani, G.V. *J. Org. Chem.* **1979**, *44*, 3718; Croft, A.P.; Bartsch, R.A. *Tetrahedron Lett.* **1983**, *24*, 2737; Kwart, H.; Gaffney, A.H.; Wilk, K.A. *J. Chem. Soc. Perkin Trans. 2* **1984**, 565.

[32]Borchardt, J.K.; Saunders Jr., W.H. *J. Am. Chem. Soc.* **1974**, *96*, 3912.

The E1 Mechanism

The E1 mechanism is a two-step process in which the rate-determining step is ionization of the substrate to give a carbocation that rapidly loses a β proton to a base, usually the solvent:

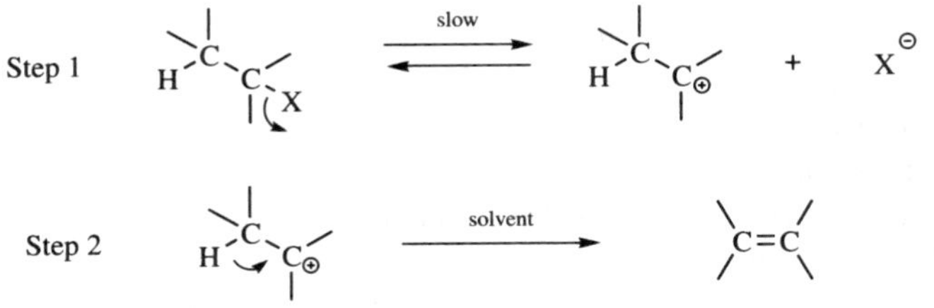

The IUPAC designation is $D_N + D_E$ (or $D_N + D_H$). This mechanism normally operates without an *added* base. Just as the E2 mechanism is analogous to and competes with the S_N2, so is the E1 mechanism related to the S_N1. In fact, the first step of the E1 is exactly the same as that of the S_N1 mechanism. The second step differs in that the solvent pulls a proton from the β carbon of the carbocation rather than attacking it at the positively charged carbon, as in the S_N1 process. In a pure E1 reaction (without ion pairs, etc.), the product should be completely nonstereospecific, since the carbocation is free to adopt its most stable conformation before giving up the proton.

Some of the evidence for the E1 mechanism is as follows:

1. The reaction exhibits first-order kinetics (in substrate) as expected. Of course, the solvent is not expected to appear in the rate equation, even if it were involved in the rate-determining step (p. 316), but this point can be easily checked by adding a small amount of the conjugate base of the solvent. It is generally found that such an addition does not increase the rate of the reaction. If this more powerful base does not enter into the rate-determining step, it is unlikely that the solvent does. An example of an E1 mechanism with a rate-determining second step (proton transfer) has been reported.[33]

2. If the reaction is performed on two molecules that differ only in the leaving group (e.g., t-BuCl and t-BuSMe₂⁺), the rates should obviously be different, since they depend on the ionizing ability of the molecule. However, once the carbocation is formed, if the solvent and the temperature are the same, it should suffer the same fate in both cases, since the nature of the leaving group does not affect the second step. This means that *the ratio of elimination to substitution should be the same*. The compounds mentioned in the example were solvolyzed at 65.3°C in 80% aqueous ethanol with the following results:[34]

[33]Baciocchi, E.; Clementi, S.; Sebastiani, G.V.; Ruzziconi, R. *J. Org. Chem.* **1979**, *44*, 32.
[34]Cooper, K.A.; Hughes, E.D.; Ingold, C.K.; MacNulty, B.J. *J. Chem. Soc.* **1948**, 2038.

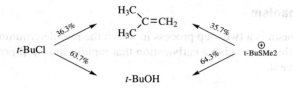

Although the rates were greatly different (as expected with such different leaving groups), the product ratios were the same, within 1%. If this had taken place by a second-order mechanism, the nucleophile would not be expected to have the same ratio of preference for attack at the β hydrogen compared to attack at a *neutral* chloride as for attack at the β hydrogen compared to attack at a *positive* SMe₂ group.

3. Many reactions carried out under first-order conditions on systems where E2 elimination is anti proceed quite readily to give alkenes where a cis hydrogen must be removed, often in preference to the removal of a trans hydrogen. For example, menthyl chloride (**2**, p. 1482), which by the E2 mechanism gave only **5**, under E1 conditions gave 68% **6** and 32% **5**, since the steric nature of the hydrogen is no longer a factor here, and the more stable alkene (Zaitsev's rule, p. 1482) is predominantly formed.

4. If carbocations are intermediates, we should expect rearrangements with suitable substrates. These have often been found in elimination reactions performed under E1 conditions.

E1 reactions can involve ion pairs, just as is true for S_N1 reactions (p. 437).[35] This effect is naturally greatest for nondissociating solvents: It is least in water, greater in ethanol, and greater still in acetic acid. It has been proposed that the ion-pair mechanism (p. 439) extends to elimination reactions too, and that the S_N1, S_N2, E1, and E2 mechanisms possess in common an ion-pair intermediate, at least occasionally.[36]

The E1cB Mechanism[37]

In the E1 mechanism, X leaves first and then H. In the E2 mechanism, the two groups leave at the same time. There is a third possibility: The H leaves first,

[35]Cocivera, M.; Winstein, S. *J. Am. Chem. Soc.* **1963**, *85*, 1702; Smith, S.G.; Goon, D.J.W. *J. Org. Chem.* **1969**, *34*, 3127; Bunnett, J.F.; Eck, D.L. *J. Org. Chem.* **1971**, *36*, 897; Sridharan, S.; Vitullo, V.P. *J. Am. Chem. Soc.* **1977**, *99*, 8093; Seib. R.C.; Shiner Jr., V.J.; Sendijarević, V.; Humski, K. *J. Am. Chem. Soc.* **1978**, *100*, 8133; Jansen, M.P.; Koshy, K.M.; Mangru, N.N.; Tidwell, T.T. *J. Am. Chem. Soc.* **1981**, *103*, 3863; Coxon, J.M.; Simpson, G.W.; Steel, P.J.; Whiteling, S.C. *Tetrahedron* **1984**, *40*, 3503; Thibblin, A. *J. Am. Chem. Soc.* **1987**, *109*, 2071; *J. Phys. Org. Chem.* **1989**, *2*, 15.
[36]Sneen, R.A. *Acc. Chem. Res.* **1973**, *6*, 46; Thibblin, A.; Sidhu, H. *J. Chem. Soc. Perkin Trans. 2* **1994**, 1423. See, however, McLennan, D.J. *J. Chem. Soc. Perkin Trans. 2* **1972**, 1577.
[37]For reviews, see Cockerill, A.F.; Harrison, R.G., in Bamford, C.H.; Tipper, C.F.H. *Comprehensive Chemical Kinetics*, Vol. 9, Elsevier, NY, **1973**, pp. 158–178; Hunter, D.H. *Intra-Sci. Chem. Rep.* **1973**, *7(3)*, 19; McLennan, D.J. *Q. Rev. Chem. Soc.* **1967**, *21*, 490. For a general discussion, see Koch, H.F. *Acc. Chem. Res.* **1984**, *17*, 137.

and then the X. This is a two-step process, called the E1cB *mechanism*,[38] or the *carbanion mechanism*, since the intermediate is a carbanion:

Step 1

11

Step 2

The name E1cB comes from the fact that it is the conjugate base of the substrate that is giving up the leaving group (see the S_N1cB mechanism, p. 521). The IUPAC designation is $A_nD_E + D_N$ or $A_{xh}D_H + D_N$ (see p. 420). We can distinguish three limiting cases: (*1*) The carbanion returns to starting material faster than it forms product: step 1 is reversible; step 2 is slow. (*2*) Step 1 is the slow step, and formation of product is faster than return of the carbanion to starting material. In this case, step 1 is essentially irreversible. (*3*) Step 1 is rapid, and the carbanion goes slowly to product. This case occurs only with the most stable carbanions. Here, too, step 1 is essentially irreversible. These cases have been given the designations: (*1*) $(E1cB)_R$, (*2*) $(E1cB)_I$ (or $E1cB_{irr}$), and (*3*) $(E1)_{anion}$. Their characteristics are listed in Table 17.1.[39] Investigations of the reaction order are generally not very useful (except for case 3, which is first order), because cases 1 and 2 are second order and thus difficult or impossible to distinguish from the E2 mechanism by this procedure.[40] We would expect the greatest likelihood of finding the E1cB mechanism in substrates that have (a) a poor nucleofuge and (b) an acidic hydrogen, and most investigations have concerned such substrates. The following is some of the evidence in support of the E1cB mechanism:

1. The first step of the $(E1cB)_R$ mechanism involves a reversible exchange of protons between the substrate and the base. In that case, if deuterium is present in the base, recovered starting material should contain deuterium. This was found to be the case in the treatment of $Cl_2C=CHCl$ with NaOD to give $ClC\equiv CCl$. When the reaction was stopped before completion, there *was*

[38]For a discussion, see Ryberg, P.; Matsson, O. *J. Org. Chem.* **2002**, *67*, 811.
[39]This table, which appears in Cockerill, A.F.; Harrison, R.G. in Bamford, C.H.; Tipper, C.F.H. *Comprehensive Chemical Kinetics*, Vol. 9, Elsevier, NY, **1973**, p. 161, was adapted from a longer one, in Bordwell, F.G. *Acc. Chem. Res.* **1972**, *5*, 374, see p. 375.
[40]$(E1cB)_I$ cannot be distinguished from E2 by this means, because it has the identical rate law: Rate = k[substrate][B$^-$]. The rate law for $(E1cB)_R$ is different: Rate = k[substrate][B$^-$]/[BH], but this is often not useful because the only difference is that the rate is also dependent (inversely) on the concentration of the conjugate acid of the base, and this is usually the solvent, so that changes in its concentration cannot be measured.

TABLE 17.1. Kinetic Predictions for Base-Induced β-Eliminations[39]

$$\text{B:} \quad + \quad \text{(D) } H-\overset{\overset{\displaystyle \beta}{|}}{\underset{\underset{\displaystyle \alpha}{|}}{C}}-\overset{|}{\underset{|}{C}}-X \quad \longrightarrow \quad B-H \quad + \quad \overset{\backslash}{\underset{/}{C}}=\overset{/}{\underset{\backslash}{C}} \quad + \quad X^{\ominus}$$

Mechanism	Kinetic[a] Order	β-Hydrogen Exchange Faster Than Elimination	General or Specific Base Catalysis	k_H/k_D	Electron Withdrawal at Cβ[d]	Electron release at Cα[d]	Leaving-Group Isotope Effect or Element Effect
(E1)$_{anion}$	1	Yes	General[c]	1.0	Rate decrease	Rate increase	Substantial
(E1cB)$_R$	2	Yes	Specific	1.0	Small rate increase	Small rate increase	Substantial
(E1cB)$_{ip}$	2	No	General[e]	$1.0 \rightarrow 1.2$	Small rate increase	Small rate increase	Substantial
(E1cB)$_I$	2	No	General	$2 \rightarrow 8$	Rate increase	Little effect	Small to negligible
E2[b]	2	No	General	$2 \rightarrow 8$	Rate increase	Small rate increase	Small

[a]All mechanism exhibit first-order kinetics in substrate.
[b]Only transition states with considerable carbanion character considered in this table.
[c]Specific base catalysis predicted if extent of substrate ionization reduced from almost complete.
[d]Effect on rate assuming no change in mechanism is caused; steric factors upon substitution at Cα and rise to Cβ have not been considered. The rate reductions are geared to substituent effects such as those giving rise to Hammett reaction constants on β- and α-aryl substitution.
[e]Depends on whether an ion pair assists in removal of leaving group.

deuterium in the recovered alkene.[41] A similar result was found for pentahaloethanes.[42] These substrates are relatively acidic. In both cases the electron-withdrawing halogens increase the acidity of the hydrogen, and in the case of trichloroethylene there is the additional factor that a hydrogen on an sp^2 carbon is more acidic than one on an sp^3 carbon (p. 388). Thus, the E1cB mechanism is more likely to be found in eliminations yielding triple bonds than in those giving double bonds. Another likely place for the E1cB mechanism should be in reaction of a substrate like PhCH$_2$CH$_2$Br, since the carbanion is stabilized by resonance with the phenyl group. Nevertheless, no deuterium exchange was found here.[43] If this type of evidence is a guide, then it may be inferred that the (E1cB)$_R$ mechanism is quite rare, at least for eliminations with common leaving groups such as Br, Cl, or OTs, which yield C=C double bonds.

[41]Houser, J.J.; Bernstein, R.B.; Miekka, R.G.; Angus, J.C. *J. Am. Chem. Soc.* **1955**, *77*, 6201.
[42]Hine, J.; Wiesboeck, R.; Ghirardelli, R.G. *J. Am. Chem. Soc.* **1961**, *83*, 1219; Hine, J.; Wiesboeck, R.; Ramsay, O.B. *J. Am. Chem. Soc.* **1961**, *83*, 1222.
[43]Skell, P.S.; Hauser, C.R. *J. Am. Chem. Soc.* **1945**, *67*, 1661.

2. When the reaction

$$p\text{-}NO_2C_6H_4\text{---}CH_2\text{---}\overset{\oplus}{CH_2}\text{---}NR_4 \ + \ \overset{\ominus}{B} \longrightarrow \ p\text{-}NO_2C_6H_4\text{---}CH_2{=}CH_2 \ + \ BH \ + \ NR_3$$

was carried out in water containing acetohydroxamate buffers, a plot of the rate against the buffer concentration was curved and the rate leveled off at high buffer concentrations, indicating a change in rate-determining step.[44] This rules out an E2 mechanism, which has only one step.[45] When D_2O was used instead of H_2O as solvent, there was an initial inverse solvent isotope effect of 7.7 (the highest inverse solvent isotope effect yet reported).

That is, the reaction took place faster in D_2O than in H_2O. This is compatible only with an E1cB mechanism in which the proton-transfer step is not entirely rate determining. The isotope effect arises from a partitioning of the carbanion intermediate **11**. This intermediate either can go to product or it can revert to starting compound, which requires taking a proton from the solvent. In D_2O, the latter process is slower (because the O–D bond of D_2O cleaves less easily than the O–H bond of H_2O), reducing the rate at which **11** returns to starting compound. With the return reaction competing less effectively, the rate of conversion of **11** to product is increased.

3. We have predicted that the E1cB mechanism would most likely be found with substrates containing acidic hydrogens and poor leaving groups. Compounds of the type ZCH_2CH_2OPh, where Z is an electron-withdrawing group (e.g., NO_2, $SMe_2{}^+$, $ArSO_2$, CN, $COOR$), belong to this category, because OPh is a very poor leaving group (p. 438). There is much evidence to show that the mechanism here is indeed E1cB.[46] Isotope effects, measured for $MeSOCD_2CH_2OPh$ and $Me_2S^+CD_2CH_2OPh$ with NaOD in D_2O, are ~0.7. This is compatible with an $(E1cB)_R$ mechanism, but not with an E2 mechanism for which an isotope effect of perhaps 5 might be expected (of course, an E1 mechanism is precluded by the extremely poor nucleofugal ability of OPh). The fact that k_H/k_D is less than the expected value of 1 is attributable to solvent and secondary isotope effects. Among other evidence for an E1cB mechanism in these systems is that changes in the identity of Z had a dramatic effect on the relative rates: a span of 10^{11} between NO_2 and COO^-. Note that elimination from substrates of the type $RCOCH_2CH_2Y$ is the reverse of Michael-type addition to C=C bonds. We have seen (p. $\$\$\$$) that such addition involves initial attack by a nucleophile Y and subsequent attack by a proton. Thus the initial loss of a proton from substrates of this type (i.e., an E1cB mechanism) is in accord with the principle of microscopic

[44]Keeffe, J.R.; Jencks, W.P. *J. Am. Chem. Soc.* **1983**, *105*, 265.

[45]For a borderline E1cB–E2 mechanism, see Jia, Z.S.; Rudziń sci, J.; Panethy, P.; Thibblin, A. *J. Org. Chem.* **2002**, *67*, 177.

[46]Cann, P.F.; Stirling, C.J.M. *J. Chem. Soc. Perkin Trans. 2* **1974**, 820. For other examples; see Fedor, L.R. *J. Am. Chem. Soc.* **1969**, *91*, 908; More O'Ferrall, R.A.; Slae, S. *J. Chem. Soc. B* **1970**, 260; Kurzawa, J.; Leffek, K.T. *Can. J. Chem.* **1977**, *55*, 1696.

reversibility.[47] It may also be recalled that benzyne formation (p. 859) can occur by such a process. It has been suggested that all base-initiated eliminations wherein the proton is activated by a strong electron-with-drawing group are E1cB reactions,[48] but there is evidence that this is not the case that when there is a good nucleofuge, the mechanism is E2 even when strong electron-withdrawing groups are present.[49] On the other hand, Cl⁻ has been found to be a leaving group in an E1cB reaction.[50]

Of the three cases of the E1cB mechanism, the one most difficult to distinguish from E2 is (E1cB)$_I$. One way to make this distinction is to study the effect of a change in leaving group. This was done in the case of the three acenaphthylenes **12**, where it was found that (*1*) the three rates were fairly similar, the largest being only about

H X Y H   X   H   X Y

t-BuOK
in
t-BuOH

a Br Cl
b Cl Cl
c Cl F

12

four times that of the smallest, and (*2*) in compound *c* (X = Cl, Y = F), the only product contained Cl and no F, that is, only the poorer nucleofuge F departed while Cl remained.[51] Result (*1*) rules out all the E1cB mechanisms except (E1cB)$_I$, because the others should all have considerable leaving group effects (Table 17.1). An ordinary E2 mechanism should also have a large leaving group effect, but an E2 mechanism with substantial carbanionic character (see the next section) might not. However, no E2 mechanism can explain result (*2*), which can be explained by the fact that an a Cl is more effective than an a F in stabilizing the planar carbanion that remains when the proton is lost. Thus (as in the somewhat similar case of aromatic nucleophilic substitution, see p. 868), when X⁻ leaves in the second step, the one that leaves is not determined by which is the better nucleofuge, but by which has had its β hydrogen removed.[52] Additional evidence for the existence of the

[47]Patai, S.; Weinstein, S.; Rappoport, Z. *J. Chem. Soc.* *1962*, 1741. See also, Hilbert, J.M.; Fedor, L.R. *J. Org. Chem.* *1978*, *43*, 452.

[48]Bordwell, F.G.; Vestling, M.M.; Yee, K.C. *J. Am. Chem. Soc.* *1970*, *92*, 5950; Bordwell, F.G. *Acc. Chem. Res.* *1972*, *5*, 374.

[49]Marshall, D.R.; Thomas, P.J.; Stirling, C.J.M. *J. Chem. Soc. Perkin Trans. 2* *1977*, 1898, 1914; Banait, N.S.; Jencks, W.P. *J. Am. Chem. Soc.* *1990*, *112*, 6950.

[50]Ölwegård, M.; McEwen, I.; Thibblin, A.; Ahlberg, P. *J. Am. Chem. Soc.* *1985*, *107*, 7494.

[51]Baciocchi, E.; Ruzziconi, R.; Sebastiani, G.V. *J. Org. Chem.* *1982*, *47*, 3237.

[52]For other evidence for the existence of the (E1cB)$_I$ mechanism, see Bordwell, F.G.; Vestling, M.M.; Yee, K.C. *J. Am. Chem. Soc.* *1970*, *92*, 5950; Fedor, L.R.; Glave, W.R. *J. Am. Chem. Soc.* *1971*. *93*, 985; Redman, R.P.; Thomas, P.J.; Stirling, C.J.M. *J. Chem. Soc. Perkin Trans. 2* *1978*. 1135; Thibblin, A. *Chem. Scr.* *1980*, *15*, 121; Carey, E.; More O'Ferrall, R.A.; Vernon, N.M. *J. Chem. Soc. Perkin Trans. 2* *1982* 1581; Baciocchi, E.; Ruzziconi, R. *J. Org. Chem.* *1984*, *49*, 3395; Jarczewski, A.; Waligorska, M.; Leffek, K.T. *Can. J Chem.* *1985*, *63*, 1194; Gula, M.J.; Vitale, D.E.; Dostal, J.M.; Trometer, J.D.; Spencer, T.A. *J. Am. Chem. Soc.* *1988* *110*, 4400; Garay, R.O.; Cabaleiro, M.C. *J. Chem. Res. (S)*, *1988*, 388; Gandler, J.R.; Storer, J.W.; Ohlberg, D.A.A. *J. Am. Chem. Soc.* *1990*, *112*, 7756.

(E1cB)$_I$ mechanism was the observation of a change in the rate-determining step in the elimination reaction of *N*-(2-cyanoethyl)pyridinium

13

ions **13**, treated with base, when X was changed.[53] Once again, the demonstration that two steps are involved precludes the one-step E2 mechanism.

4. An example of an (E1)$_{anion}$ mechanism has been found with the substrate **14**, which when treated with methoxide ion undergoes elimination to **16**, which is unstable under the reaction conditions and rearranges as

14   **15**   **16**

shown.[54] Among the evidence for the proposed mechanism in this case were kinetic and isotope-effect results, as well as the spectral detection of **15**.[55]

5. In many eliminations to form C=O and C≡N bonds the initial step is loss of a positive group (normally a proton) from the oxygen or nitrogen. These may also be regarded as E1cB processes.

There is evidence that some E1cB mechanisms can involve carbanion ion pairs, for example,[56]

This case is designated (E1cB)$_{ip}$; its characteristics are shown in Table 17.1.

[53]Bunting, J.W.; Toth, A.; Heo, C.K.M.; Moors, R.G. *J. Am. Chem. Soc.* ***1990***, *112*, 8878. See also, Bunting, J.W.; Kanter, J.P. *J. Am. Chem. Soc.* ***1991***, *113*, 6950.

[54]Bordwell, F.G.; Yee, K.C.; Knipe, A.C. *J. Am. Chem. Soc.* ***1970***, *92*, 5945.

[55]For other examples of this mechanism, see Berndt, A. *Angew. Chem. Int. Ed.* ***1969***, *8*, 613; Albeck, M.; Hoz, S.; Rappoport, Z. *J. Chem. Soc. Perkin Trans. 2* ***1972***, 1248; ***1975***, 628.

[56]Kwok, W.K.; Lee, W.G.; Miller, S.I. *J. Am. Chem. Soc.* ***1969***, *91*, 468. See also Lord, E.; Naan, M.P.; Hall, C.D. *J. Chem. Soc. B* ***1971***, 220; Rappoport, Z.; Shohamy, E. *J. Chem. Soc. B* ***1971***, 2060; Fiandanese, V.; Marchese, G.; Naso, F. *J. Chem. Soc., Chem. Commun.* ***1972***, 250; Koch, H.F.; Dahlberg, D.B.; Toczko, A.G.; Solsky, R.L. *J. Am. Chem. Soc.* ***1973***, *95*, 2029; Hunter, D.H.; Shearing, D.J. *J. Am. Chem. Soc.* ***1973***, *95*, 8333; Thibblin, A.; Ahlberg, P. *J. Am. Chem. Soc.* ***1979***, *101*, 7311; Petrillo, G.; Novi, M.; Garbarino, G.; Dell'Erba, C.; Mugnoli, A. *J. Chem. Soc. Perkin Trans. 2* ***1985***, 1291.

The E1-E2-E1cB Spectrum

In the three mechanisms so far considered, the similarities are greater than the differences. In each case, there is a leaving group that comes off with its pair of electrons and another group (usually hydrogen) that comes off without them. The only difference is in the order of the steps. It is now generally accepted that there is a spectrum of mechanisms ranging from one extreme, in which the leaving group departs well before the proton (pure E1), to the other extreme, in which the proton comes off first and then, after some time, the leaving group follows (pure E1cB). The *pure* E2 case would be somewhere in the middle, with both groups leaving simultaneously. However, most E2 reactions are not exactly in the middle, but somewhere to one side or the other. For example, the nucleofuge might depart just before the proton. This case may be described as an E2 reaction with a small amount of E1 character. The concept can be expressed by the question: In the transition state, which bond (C–H or C–X) has undergone more cleavage?[57]

One way to determine just where a given reaction stands on the E1-E2-E1cB spectrum is to study isotope effects, which ought to tell something about the behavior of bonds in the transition state.[58] For example, $CH_3CH_2NMe_3^+$ showed a nitrogen isotope effect (k^{14}/k^{15}) of 1.017, while $PhCH_2CH_2NMe_3^+$ gave a corresponding value of 1.009.[59] It would be expected that the phenyl group would move the reaction toward the E1cB side of the line, which means that for this compound the C–N bond is not as greatly broken in the transition state as it is for the unsubstituted one. The isotope effect bears this out, for it shows that in the phenyl compound, the mass of the nitrogen has less effect on the reaction rate than it does in the unsubstituted compound. Similar results have been obtained with SR_2^+ leaving groups by the use of $^{32}S/^{34}S$ isotope effects[60] and with Cl ($^{35}Cl/^{37}Cl$).[61] The position of reactions along the spectrum has also been studied from the other side of the newly forming double bond by the use of H/D and H/T isotope effects,[62] although interpretation of these results is clouded by the fact that β hydrogen isotope effects are expected to change smoothly from small to large to small again as the degree of transfer of the

[57]For discussions, see Cockerill, A.F.; Harrison, R.G., in Bamford, C.H.; Tipper, C.F.H. *Comprehensive Chemical Kinetics*, Vol. 9, Elsevier, NY, *1973*, pp. 178–189; Saunders, Jr., W.H. *Acc. Chem. Res.* **1976**, *9*, 19; Bunnett, J.F. *Surv. Prog. Chem.* **1969**, *5*, 53; Saunders, Jr., W.H.; Cockerill, A.F. *Mechanisms of Elimination Reactions*, Wiley, NY, *1973*, pp. 47–104; Bordwell, F.G. *Acc. Chem. Res.* **1972**, *5*, 374.

[58]For a review, see Fry, A. *Chem. Soc. Rev.* **1972**, *1*, 163. See also Hasan, T.; Sims, L.B.; Fry, A. *J. Am. Chem. Soc.* **1983**, *105*, 3967; Pulay, A.; Fry, A. *Tetrahedron Lett.* **1986**, *27*, 5055.

[59]Ayrey, G.; Bourns, A.N.; Vyas, V.A. *Can. J. Chem.* **1963**, *41*, 1759. Also see, Simon, H.; Müllhofer, G. *Chem. Ber.* **1963**, *96*, 3167; **1964**, *97*, 2202; *Pure Appl. Chem.* **1964**, *8*, 379, 536; Smith, P.J.; Bourns, A.N. *Can. J. Chem.* **1970**, *48*, 125.

[60]Wu, S.; Hargreaves, R.T.; Saunders Jr., W.H. *J. Org. Chem.* **1985**, *50*, 2392, and references cited therein.

[61]Grout, A.; McLennan, D.J.; Spackman, I.H. *J. Chem. Soc. Perkin Trans. 2* **1977**, 1758.

[62]For example, see Hodnett, E.M.; Sparapany, J.J. *Pure Appl. Chem.* **1964**, *8*, 385, 537; Finley, K.T.; Saunders, Jr., W.H. *J. Am. Chem. Soc.* **1967**, *89*, 898; Ghanbarpour, A.; Willi, A.V. *Liebigs Ann. Chem.* **1975**, 1295; Simon, H.; Müllhofer, G. *Chem. Ber.* **1964**, *97*, 2202; Thibblin, A. *J. Am. Chem. Soc.* **1988**, *110*, 4582; Smith, P.J.; Amin, M. *Can. J. Chem.* **1989**, *67*, 1457.

β hydrogen from the β carbon to the base increases[63] (recall, p. $$$, that isotope effects are greatest when the proton is half-transferred in the transition state), by the possibility of secondary isotope effects (e.g., the presence of a β deuterium or tritium may cause the leaving group to depart more slowly), and by the possibility of tunneling.[64] Other isotope-effect studies have involved labeled a or β carbon, labeled a hydrogen, or labeled base.[58]

Another way to study the position of a given reaction on the spectrum involves the use of β-aryl substitution. Since a positive Hammet ρ value is an indication of a negatively charged transition state, the ρ value for substituted β-aryl groups should increase as a reaction moves from E1- to E1cB-like along the spectrum. This has been shown to be the case in a number of studies;[65] for example, ρ values of $ArCH_2CH_2X$ increase as the leaving-group ability of X decreases. A typical set of ρ values was X = I, 2.07; Br, 2.14; Cl, 2.61; SMe_2^+, 2.75; F, 3.12.[66] As we have seen, decreasing leaving-group ability correlates with increasing E1cB character.

Still another method measures volumes of activation.[67] These are negative for E2 and positive for E1cB mechanisms. Measurement of the activation volume therefore provides a continuous scale for deciding just where a reaction lies on the spectrum.

The E2C Mechanism[68]

Certain alkyl halides and tosylates undergo E2 eliminations faster when treated with such weak bases as Cl^- in polar aprotic solvents or PhS^- than with the usual E2 strong bases, such as RO^- in ROH.[69] In order to explain these results, Parker

[63]There is controversy as to whether such an effect has been established in this reaction: See Cockerill, A.F. *J. Chem. Soc. B 1967*, 964; Blackwell, L.F. *J. Chem. Soc. Perkin Trans. 2 1976*, 488.

[64]For examples of tunneling in elimination reactions, see Miller, D.J.; Saunders, Jr., W.H. *J. Org. Chem. 1981, 46*, 4247 and previous papers in this series. See also, Shiner, Jr., V.J.; Smith, M.L. *J. Am. Chem. Soc. 1961, 83*, 593; McLennan, D.J. *J. Chem. Soc. Perkin Trans. 2 1977*, 1753; Fouad, F.M.; Farrell, P.G. *Tetrahedron Lett. 1978*, 4735; Koth, H.F.; McLennan, D.J.; Koch, J.G.; Tumas, W.; Dobson, B.; Koch, J.G. *J. Am. Chem. Soc. 1983, 105*, 1930; Kwart, H.; Wilk, K.A. *J. Org. Chem. 1985, 50*, 817; Amin, M.; Price, R.C.; Saunders, Jr., W.H. *J. Am. Chem. Soc. 1990, 112*, 4467.

[65]Saunders Jr., W.H.; Bushman, D.G.; Cockerill, A.F. *J. Am. Chem. Soc. 1968, 90*, 1775; Yano, Y.; Oae, S. *Tetrahedron 1970, 26*, 27, 67; Blackwell, L.F.; Buckley, P.D.; Jolley, K.W.; MacGibbon, A.K.H. *J. Chem. Soc. Perkin Trans. 2 1973*, 169; Smith, P.J.; Tsui, S.K. *J. Am. Chem. Soc. 1973, 95*, 4760; *Can. J. Chem. 1974, 52*, 749.

[66]DePuy, C.H.; Froemsdorf, D.H. *J. Am. Chem. Soc. 1957, 79*, 3710; DePuy, C.H.; Bishop, C.A. *J. Am. Chem. Soc. 1960, 82*, 2532, 2535.

[67]Brower, K.R.; Muhsin, M.; Brower, H.E. *J. Am. Chem. Soc. 1976, 98*, 779. For a review, see van Eldik, R.; Asano, T.; le Noble, W.J. *Chem. Rev. 1989, 89*, 549.

[68]For reviews, see McLennan, D.J. *Tetrahedron 1975, 31*, 2999; Ford, W.T. *Acc. Chem. Res. 1973, 6*, 410; Parker, A.J. *CHEMTECH 1971*, 297.

[69]For example; see Winstein, S.; Darwish, D.; Holness, N.J. *J. Am. Chem. Soc. 1956, 78*, 2915; de la Mare, P.B.D.; Vernon, C.A. *J. Chem. Soc. 1956*, 41; Eliel, E.L.; Ro, R.S. *Tetrahedron 1958, 2*, 353; Bunnett, J.F.; Davis, G.T.; Tanida, H. *J. Am. Chem. Soc. 1962, 84*, 1606; McLennan, D.J. *J. Chem. Soc. B 1966*, 705, 709; Hayami, J.; Ono, N.; Kaji, A. *Bull. Chem. Soc. Jpn. 1971, 44*, 1628.

and co-workers proposed[70] that there is a spectrum[71] of E2 transition states in which the base can interact in the transition state with the α carbon, as well as with the β hydrogen. At one end of this spectrum is

$$
\begin{array}{cccc}
\text{E2C} & \mathbf{17} & \text{E2H} & \mathbf{18}
\end{array}
$$

a mechanism (called E2C) in which, in the transition state, the base interacts mainly with the carbon. The E2C mechanism is characterized by strong nucleophiles that are weak bases. At the other extreme is the normal E2 mechanism, here called E2H to distinguish it from E2C, characterized by strong bases. Transition state **17** represents a transition state between these extremes. Additional evidence[72] for the E2C mechanism is derived from Brønsted equation considerations (p. 373), from substrate effects, from isotope effects, and from the effects of solvents on rates.

However, the E2C mechanism has been criticized, and it has been contended that all the experimental results can be explained by the normal E2 mechanism.[73] McLennan has suggested that the transition state is that shown as **18**.[74] An ion-pair mechanism has also been proposed.[75] Although the actual mechanisms involved may be a matter of controversy, there is no doubt that a class of elimination reactions exists that is characterized by second-order attack by weak bases.[76] These reactions also have the following general characteristics:[77] (*1*) they are favored by good leaving groups; (*2*) they are favored by polar aprotic solvents;

[70]Parker, A.J.; Ruane, M.; Biale, G.; Winstein, S. *Tetrahedron Lett.* **1968**, 2113.

[71]This is apart from the E1-E2-E1cB spectrum.

[72]Lloyd, D.J.; Parker, A.J. *Tetrahedron Lett.* **1968**, 5183; **1970**, 5029; Alexander, R.; Ko, E.C.F.; Parker, A.J.; Broxton, T.J. *J. Am. Chem. Soc.* **1968**, *90*, 5049; Ko, E.C.F.; Parker, A.J. *J. Am. Chem. Soc.* **1968**, *90*, 6447; Parker, A.J.; Ruane, M.; Palmer, D.A.; Winstein, S. *J. Am. Chem. Soc.* **1972**, *94*, 2228; Biale, G.; Parker, A.J.; Stevens, I.D.R.; Takahashi, J.; Winstein, S. *J. Am. Chem. Soc.* **1972**, *94*, 2235; Cook, D. *J. Org. Chem.* **1976**, *41*, 2173, and references cited therein; Muir, D.M.; Parker, A.J. *Aust. J. Chem.* **1983**, *36*, 1667; Kwart, H.; Wilk, K.A. *J. Org. Chem.* **1985**, *50*, 3038.

[73]McLennan, D.J.; Wong, R.J. *J. Chem. Soc. Perkin Trans. 2* **1974**, 1818, and references cited therein; Ford, W.T.; Pietsek, D.J.J. *J. Am. Chem. Soc.* **1975**, *97*, 2194; Loupy, A. *Bull. Soc. Chim. Fr.* **1975**, 2662; Miller, D.J.; Saunders Jr., W.H. *J. Am. Chem. Soc.* **1979**, *101*, 6749; Bordwell, F.G.; Mrozack, S.R. *J. Org. Chem.* **1982**, *47*, 4813; Bunnett, J.F.; Migdal, C.A. *J. Org. Chem.* **1989**, *54*, 3037, 3041, and references cited therein.

[74]McLennan, D.J.; Lim, G. *Aust. J. Chem.* **1983**, *36*, 1821. For an opposing view, see Kwart, H.; Gaffney, A. *J. Org. Chem.* **1983**, *48*, 4502.

[75]Ford, W.T. *Acc. Chem. Res.* **1973**, *6*, 410.

[76]For convenience, we will refer to this class of reactions as E2C reactions, though the actual mechanism is in dispute.

[77]Beltrame, P.; Biale, G.; Lloyd, D.J.; Parker, A.J.; Ruane, M.; Winstein, S. *J. Am. Chem. Soc.* **1972**, *94*, 2240; Beltrame, P.; Ceccon, A.; Winstein, S. *J. Am. Chem. Soc.* **1972**, *94*, 2315.

(*3*) the reactivity order is tertiary > secondary > primary, the opposite of the normal E2 order (p. 1503); (*4*) the elimination is always anti- (syn-elimination is not found), but in cyclohexyl systems, a diequatorial anti-elimination is about as favorable as a diaxial anti-elimination (unlike the normal E2 reaction, p. 1481); (*5*) they follow Zaitsev's rule (see below), where this does not conflict with the requirement for anti-elimination.

Regiochemistry of the Double Bond

With some substrates, a β hydrogen is present on only one carbon and (barring rearrangements) there is no doubt as to the identity of the product. For example, $PhCH_2CH_2Br$ can give only $PhCH=CH_2$. However, in many other cases two or three alkenyl products are possible. In the simplest such case, a *sec*-butyl compound can give either 1- or 2-butene. There are a number of rules that enable us to predict, in many instances, which product will predominantly form.[78]

1. No matter what the mechanism, a double bond does not go to a bridgehead carbon unless the ring sizes are large enough (Bredt's rule, see p. 229). This means, for example, not only that **19** gives only **20** and not **21** (indeed **21** is not a known compound), but also that **22** does not undergo elimination.

21	**19**	**20**	**22**

2. No matter what the mechanism, if there is a double bond ($C=C$ or $C=O$) or an aromatic ring already in the molecule that can be in conjugation with the new double bond, the conjugated product usually predominates, sometimes even when the stereochemistry is unfavorable (for an exception, see p. 1501).

3. In the E1 mechanism the leaving group is gone before the choice is made as to which direction the new double bond takes. Therefore the direction is determined almost entirely by the relative stabilities of the two (or three) possible alkenes. In such cases, *Zaitsev's rule*[79] operates. This rule states that *the double bond goes mainly toward the most highly substituted carbon*. That is, a *sec*-butyl compound gives more 2-butene than 1-butene, and 3-bromo-2,3-dimethylpentane gives more 2,3-dimethyl-2-pentene than either 3,4-dimethyl-2-pentene or 2-ethyl-3-methyl-1-butene. Thus Zaitsev's rule predicts that the alkene predominantly formed will be the one with the largest possible number of alkyl groups on the $C=C$ carbons, and in most cases this is what is found. From heat of combustion data (see p. 29) it is known that

[78]For a review of orientation in cycloalkyl systems, see Hückel, W.; Hanack, M. *Angew. Chem. Int. Ed.* **1967**, *6*, 534.
[79]Often given the German spelling: Saytzeff.

alkene stability increases with alkyl substitution, although just why this should be is a matter of conjecture. The most common explanation is hyperconjugation. For E1 eliminations, Zaitsev's rule governs the orientation whether the leaving group is neutral or positive, since, as already mentioned, the leaving group is not present when the choice of direction is made. This statement does not hold for E2 eliminations, and it may be mentioned here, for contrast with later results, that E1 elimination of $Me_2CHCHMeSMe_2^+$ gave 91% of the Zaitsev product and 9% of the other.[80] However, there *are* cases in which the leaving group affects the direction of the double bond in E1-eliminations.[81] This may be attributed to ion pairs; that is, the leaving group is not completely gone when the hydrogen departs. Zaitsev's rule breaks down in cases where the non-Zaitsev product is more stable for steric reasons. For example, E1 or E1-like eliminations of 1,2-diphenyl-2-X-propanes $PhMeCXCH_2Ph$ were reported to give $\sim$50% $CH_2{=}CPhCH_2Ph$, despite the fact that the double bond of the Zaitsev product ($PhMeC{=}CHPh$) is conjugated with two benzene rings.[82]

4. For the anti E2 mechanism a trans β proton is necessary; if this is available in only one direction, that is the way the double bond will form. Because of the free rotation in acyclic systems (except where steric hindrance is great), this is a factor only in cyclic systems. Where trans β hydrogens are available on two or three carbons, two types of behavior are found, depending on substrate structure and the nature of the leaving group. Some compounds follow Zaitsev's rule and give predominant formation of the most highly substituted alkene, but others follow *Hofmann's rule: The double bond goes mainly toward the least highly substituted carbon.* although many exceptions are known, the following general statements can be made: In most cases, compounds containing uncharged nucleofuges (those that come off as negative ions) follow Zaitsev's rule, just as they do in E1 elimination, no matter what the structure of the substrate. However, elimination from compounds with charged nucleofuges, for example, NR_3^+, SR_2^+ (those that come off as neutral molecules), follow Hofmann's rule if the substrate is acyclic,[83] but Zaitsev's rule if the leaving group is attached to a six-membered ring.[84]

Much work has been devoted to searching for the reasons for the differences in orientation. Since Zaitsev orientation almost always gives the

[80]de la Mare, P.B.D. *Prog. Stereochem.* **1954**, *1*, 112.

[81]Cram, D.J.; Sahyun, M.R.V. *J. Am. Chem. Soc.* **1963**, *85*, 1257; Silver, M.S. *J. Am. Chem. Soc.* **1961**, *83*, 3482.

[82]Ho, I.; Smith, J.G. *Tetrahedron* **1970**, *26*, 4277.

[83]An example of an acyclic quaternary ammonium salt that follows Zaitsev's rule is found, in Feit, I.N.; Saunders, Jr., W.H. *J. Am. Chem. Soc.* **1970**, *92*, 5615.

[84]For examples where Zaitsev's rule is followed with charged leaving groups in cyclohexyl systems, see Gent, B.B.; McKenna, J. *J. Chem. Soc.* **1959**, 137; Hughes, E.D.; Wilby, J. *J. Chem. Soc.* **1960**, 4094; Brownlee, T.H.; Saunders Jr., W.H. *Proc. Chem. Soc.* **1961**, 314; Booth, H.; Franklin, N.C.; Gidley, G.C. *J. Chem. Soc. C* **1968**, 1891. For a discussion of the possible reasons for this, see Saunders, Jr., W.H.; Cockerill, A.F. *Mechanisms of Elimination Reactions*, Wiley, NY, **1973**, pp. 192–193.

thermodynamically more stable isomer, what needs to be explained is why in some cases the less stable Hofmann product predominates. Three explanations have been offered for the change in orientation in acyclic systems with a change from uncharged to charged nucleofuges. The first of these, by Hughes and Ingold,[85] is that Hofmann orientation is caused by the fact that the acidity of the β hydrogen is decreased by the presence of the electron-donating alkyl groups. For example, under E2 conditions $Me_2CHCHMeSMe_2^+$ gives more of the Hofmann product; it is the more acidic hydrogen that is removed by the base.

Zaitsev product   Less acidic   Hofmann product

Of course, the CH_3 hydrogens would still be more acidic than the Me_2CH hydrogen even if a neutral leaving group were present, but the explanation of Hughes and Ingold is that acidity matters with charged and not with neutral leaving groups, because the charged groups exert a strong electron-withdrawing effect, making differences in acidity greater than they are with the less electron-withdrawing neutral groups.[85] The explanation of Bunnett[86] is similar. According to this, the change to a positive leaving group causes the mechanism to shift toward the E1cB end of the spectrum, where there is more C–H bond breaking in the rate-determining step and where, consequently, acidity is more important. In this view, when there is a neutral leaving group, the mechanism is more E1-like, C–X bond breaking is more important, and alkene stability determines the direction of the new double bond. The third explanation, by H.C. Brown, is completely different. In this picture, field effects are unimportant, and the difference in orientation is largely a steric effect caused by the fact that charged groups are usually larger than neutral ones. A CH_3 group is more open to attack than a CH_2R group and a CHR_2 group is still less easily attacked. Of course, these considerations also apply when the leaving group is neutral, but, according to Brown, they are much less important here because the neutral groups are smaller and do not block access to the hydrogens as much. Brown showed that Hofmann elimination increases with the size of the leaving group. Thus the percentage of 1-ene obtained from $CH_3CH_2CH_2CHXCH_3$ was as follows (X listed in order of increasing size): Br, 31%; I, 30%; OTs, 48%; SMe_2^+, 87%; SO_2Me, 89%;

[85]For summaries of this position, see Ingold, C.K. *Proc. Chem. Soc.* **1962**, 265; Banthorpe, D.V.; Hughes, E.D.; Ingold, C.K. *J. Chem. Soc.* **1960**, 4054.
[86]Bunnett, J.F. *Surv. Prog. Chem.* **1969**, 5, 53.

NMe_3^+, 98%.[87] Hofmann elimination was also shown to increase with increase in bulk of the substrate.[88] With large enough compounds, Hofmann orientation can be obtained even with halides, for example, *tert*-amyl bromide ogave 89% of the Hofmann product. Even those who believe in the acidity explanations concede that these steric factors operate in extreme cases.[89]

There is one series of results incompatible with the steric explanation E2 elimination from the four 2-halopentanes gave the following percentages of 1-pentene: F, 83%; Cl, 37%; Br, 25%; I, 20%.[90] The same order was found for the four-2-halohexanes.[91] Although there is some doubt about the relative steric requirements of Br, Cl, and I, there is no doubt that F is the smallest of the halogens, and if the steric explanation were the only valid one, the fluoroalkanes could not give predominant Hofmann orientation. Another result that argues against the steric explanation is the effect of changing the nature of the base. An experiment in which the effective size of the base was kept constant while its basicity was increased (by using as bases a series of $XC_6H_4O^-$ ions) showed that the percentage of Hofmann elimination increased with increasing base strength, although the size of the base did not change.[92] These results are in accord with the explanation of Bunnett, since an increase in base strength moves an E2 reaction closer to the E1cB end of the spectrum. In further experiments, a large series of bases of different kinds was shown to obey linear free-energy relationships between basicity and percentage of Hofmann elimination,[93] although certain very large bases (e.g., 2,6-di-*tert*-butyl-phenoxide) did not obey the relationships, steric effects becoming important in these cases. How large the base must be before steric effects are observed depends on the pattern of alkyl substitution in the substrate, but not on the nucleofuge.[94] One further result may be noted. In the gas phase, elimination of H and BrH^+ or H and ClH^+ using Me_3N as the base predominantly followed Hofmann's rule,[95] although BrH^+ and ClH^+ are not very large.

[87]Brown, H.C.; Wheeler, O.H. *J. Am. Chem. Soc.* **1956**, *78*, 2199.

[88]Brown, H.C.; Moritani, I.; Nakagawa, M. *J. Am. Chem. Soc.* **1956**, *78*, 2190; Brown, H.C.; Moritani, I. *J. Am. Chem. Soc.* **1956**, *78*, 2203; Bartsch, R.A. *J. Org. Chem.* **1970**, *35*, 1334. See also, Charton, M. *J. Am. Chem. Soc.* **1975**, *97*, 6159.

[89]For example, see Banthorpe, D.V.; Hughes, E.D.; Ingold, C.K. *J. Chem. Soc.* **1960**, 4054.

[90]Saunders, Jr., W.H.; Fahrenholtz, S.R.; Caress, E.A.; Lowe, J.P.; Schreiber, M.R. *J. Am. Chem. Soc.* **1965**, *87*, 3401. Similar results were obtained by Brown, H.C.; Klimisch, R.L. *J. Am. Chem. Soc.* **1966**, *88*, 1425.

[91]Bartsch, R.A.; Bunnett, J.F. *J. Am. Chem. Soc.* **1968**, *90*, 408.

[92]Froemsdorf, D.H.; Robbins, M.D. *J. Am. Chem. Soc.* **1967**, *89*, 1737. See also, Froemsdorf, D.H.; Dowd, W.; Leimer, K.E. *J. Am. Chem. Soc.* **1966**, *88*, 2345; Bartsch, R.A.; Kelly, C.F.; Pruss, G.M. *Tetrahedron Lett.* **1970**, 3795; Feit, I.N.; Breger, I.K.; Capobianco, A.M.; Cooke, T.W.; Gitlin, L.F. *J. Am. Chem. Soc.* **1975**, *97*, 2477; Feit, I.N.; Suanders, Jr., W.H. *J. Am. Chem. Soc.* **1970**, *92*, 5615.

[93]Bartsch, R.A.; Roberts, D.K.; Cho, B.R. *J. Org. Chem.* **1979**, *44*, 4105.

[94]Bartsch, R.A.; Read, R.A.; Larsen, D.T.; Roberts, D.K.; Scott, K.J.; Cho, B.R. *J. Am. Chem. Soc.* **1979**, *101*, 1176.

[95]Angelini, G.; Lilla, G.; Speranza, M. *J. Am. Chem. Soc.* **1989**, *111*, 7393.

5. Only a few investigations on the orientation of syn E2 eliminations have been carried out, but these show that Hofmann orientation is greatly favored over Zaitsev.[96]

6. In the E1cB mechanism the question of orientation seldom arises because the mechanism is generally found only where there is an electron-withdrawing group in the β position, and that is where the double bond goes.

7. As already mentioned, E2C reactions show a strong preference for Zaitsev orientation.[97] In some cases, this can be put to preparative use. For example, the compound $PhCH_2CHOTsCHMe_2$ gave ~98% $PhCH=CHCHMe_2$ under the usual E2 reaction conditions (t-BuOK in t-BuOH). In this case, the double bond goes to the side with more hydrogens because on that side it will be able to conjugate with the benzene ring. However, with the weak base Bu_4N^+ Br^- in acetone the Zaitsev product $PhCH_2CH=CMe_2$ was formed in 90% yield.[98]

Stereochemistry of the Double Bond

When elimination takes place on a compound of the form CH_3–CABX or CHAB–CGGX, the new alkene does not have cis–trans isomerism, but for compounds of the form CHEG–CABX (E and G *not* H) (**23**) and CH_2E–CABX (**24**), cis and trans isomers are possible. When the anti E2 mechanism is in operation, **23** gives the isomer

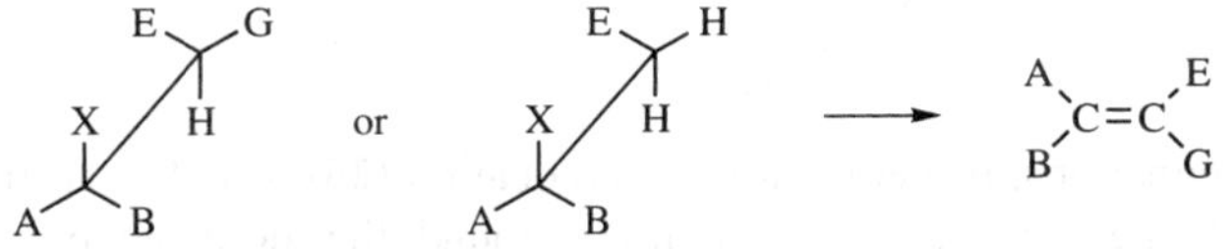

arising from trans orientation of X and H and, as we have seen before (p. 1478), an erythro compound gives the cis alkene and a threo compound the trans. For **24**, two conformations are possible for the transition state; these lead to different isomers and often both are obtained. However, the one that predominates is often determined by an eclipsing effect.[99] For example, Zaitsev elimination from 2-bromopentane can occur as follows:

[96]Sicher, J.; Svoboda, M.; Pánková, M.; Závada, J. *Collect. Czech. Chem. Commun.* **1971**, *36*, 3633; Bailey, D.S.; Saunders, Jr., W.H. *J. Am. Chem. Soc.* **1970**, *92*, 6904.

[97]For example; see Ono, N. *Bull. Chem. Soc. Jpn.* **1971**, *44*, 1369; Bailey, D.S.; Saunders, Jr., W.H. *J. Org. Chem.* **1973**, *38*, 3363; Muir, D.M.; Parker, A.J. *J. Org. Chem.* **1976**, *41*, 3201.

[98]Lloyd, D.J.; Muir, D.M.; Parker, A.J. *Tetrahedron Lett.* **1971**, 3015

[99]See Cram, D.J.; Greene, F.D.; DePuy, C.H. *J. Am. Chem. Soc.* **1956**, *78*, 790; Cram, D.G., in Newman, M.S. *Steric Effects in Organic Chemistry*, Wiley, NY, **1956**, pp. 338–345.

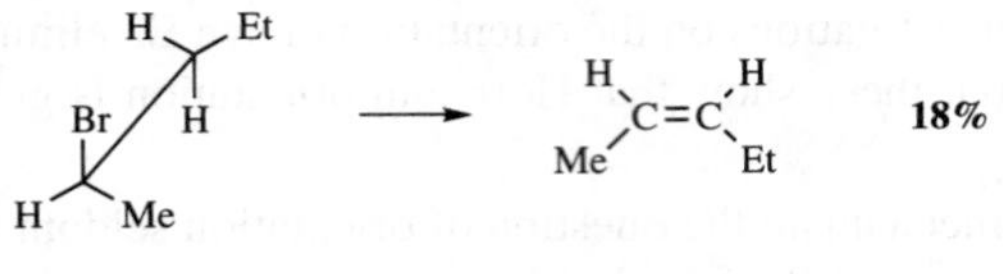

In conformation **I**, the ethyl group is between Br and Me, while in **J** it is between Br and H. This means that **J** is more stable, and most of the elimination should occur from this conformation. This is indeed what happens, and 51% of the trans isomer is formed (with KOEt) compared to 18% of the cis (the rest is the Hofmann product).[100] These effects become larger with increasing size of A, B, and E.

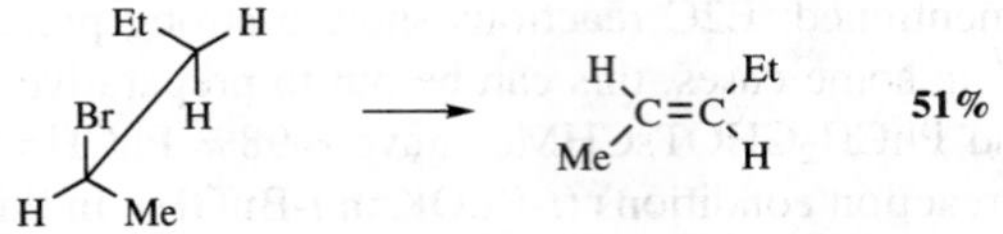

However, eclipsing effects are not the only factors that affect the cis/trans ratio in anti E2 eliminations. Other factors are the nature of the leaving group, the base, the solvent, and the substrate. Not all these effects are completely understood.[101]

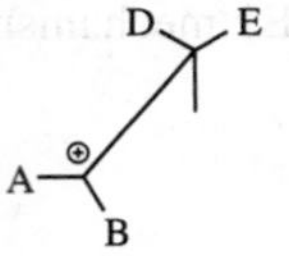

(D may be H)

25

For E1 eliminations, if there is a free carbocation (**25**), it is free to rotate, and no matter what the geometry of the original compound, the more stable situation is the one where the larger of the D–E pair is opposite the smaller of the A–B pair and the corresponding alkene should form. If the carbocation is not completely free, then to that extent, E2-type products are formed. Similar considerations apply in E1cB eliminations.[102]

REACTIVITY

In this section, we examine the effects of changes in the substrate, base, leaving group, and medium on (1) overall reactivity, (2) E1 versus E2 versus E1cB,[103] and (3) elimination versus substitution.

[100]Brown, H.C.; Wheeler, O.H. *J. Am. Chem. Soc.* **1956**, 78 2199.

[101]For discussions, see Bartsch, R.A.; Bunnett, J.F. *J. Am. Chem. Soc.* **1969**, 91, 1376, 1382; Feit, I.N.; Saunders, Jr., W.H. *J. Am. Chem. Soc.* **1970**, 92, 1630, 5615; Alunni, S.; Baciocchi, E. *J. Chem. Soc. Perkin Trans. 2* **1976**, 877; Saunders, Jr., W.H.; Cockerill, A.F. *Mechanisms of Elimination Reactions*, Wiley, NY, **1973**, pp. 165–193.

[102]See, for example, Redman, R.P.; Thomas, P.J.; Stirling, C.J.M. *J. Chem. Soc., Chem. Commun.* **1978**, 43.

[103]For discussions, see Cockerill, A.F.; Harrison, R.G., in Patai, S. *The Chemistry of Functional Groups*, Supplement A, pt. 1, Wiley, NY, **1977**, pp. 178–189.

Effect of Substrate Structure

1. *Effect on Reactivity.* We refer to the carbon containing the nucleofuge (X) as the a carbon and to the carbon that loses the positive species as the β carbon. Groups attached to the α or β carbons can exert at least four kinds of influence:

   a. They can stabilize or destabilize the incipient double bond (both α and β groups).
   b. They can stabilize or destabilize an incipient negative charge, affecting the acidity of the proton (β groups only).
   c. They can stabilize or destabilize an incipient positive charge (α groups only).
   d. They can exert steric effects (e.g., eclipsing effects) (both α and β groups).

   Effects a and d can apply in all three mechanisms, although steric effects are greatest for the E2 mechanism. Effect b does not apply in the E1 mechanism, and effect c does not apply in the E1cB mechanism. Groups, such as Ar and C=C, increase the rate by any mechanism, except perhaps when formation of the C=C bond is not the rate-determining step, whether they are α or β (effect a). Electron-withdrawing groups increase the acidity when in the β position, but have little effect in the a position unless they also conjugate with the double bond. Thus Br, Cl, CN, Ts, NO_2, CN, and SR in the β position all increase the rate of E2 eliminations.

2. *Effect on E1 versus E2 versus E1cB.* The α alkyl and α aryl groups stabilize the carbocation character of the transition state, shifting the spectrum toward the E1 end. β alkyl groups also shift the mechanism toward E1, since they *decrease* the acidity of the hydrogen. However, β aryl groups shift the mechanism the other way (toward E1cB) by stabilizing the carbanion. Indeed, as we have seen (p. $$$), all electron-withdrawing groups in the β position shift the mechanism toward E1cB.[104] α alkyl groups also increase the extent of elimination with weak bases (E2C reactions).

3. *Effect on Elimination versus Substitution.* Under second-order conditions, a branching increases elimination, to the point where tertiary substrates undergo few S_N2 reactions, as we saw in Chapter 10. For example, Table 17.2 shows results on some simple alkyl bromides. Similar results were obtained with SMe_2^+ as the leaving group.[105] Two reasons can be presented for this trend. One is statistical: As a branching increases, there are usually more hydrogens for the base to attack. The other is that a branching presents steric hindrance to attack of the base at the carbon. Under first-order conditions, increased a branching also increases the amount of elimination (E1 vs. S_N1), although not

[104]For a review of eliminations with COOH, COOR, $CONH_2$, and CN groups in the β position, see Butskus, P.F.; Denis, G.I. *Russ. Chem. Rev.* **1966**, *35*, 839.
[105]Dhar, M.L.; Hughes, E.D.; Ingold, C.K.; Masterman, S. *J. Chem. Soc.* **1948**, 2055.

TABLE 17.2. The Effect of α and β Branching on the Rate of E2 Elimination and the Amount of Alkene Formed[a]

Substrate	Temperature, °C	Alkene, %	Rate × 10^5 of E2 Reaction	Reference
CH_3CH_2Br	55	0.9	1.6	108
$(CH_3)_2CHBr$	24	80.3	0.237	109
$(CH_3)_3CBr$	25	97	4.17	107
$CH_3CH_2CH_2Br$	55	8.9	5.3	105
$(CH_3)_2CHCH_2Br$	55	59.5	8.5	105

[a]The reactions were between the alkyl bromide and ⁻OEt The rate for isopropyl bromide was actually greater than that for ethyl bromide, if the temperature difference is considered. Neopentyl bromide, the next compound in the β-branching series, cannot be compared because it has no β-hydrogen and cannot give an elimination product without rearrangement.

so much, and usually the substitution product predominates. For example, solvolysis of *tert*-butyl bromide gave only 19% elimination[106] (cf. with Table 17.2). β Branching also increases the amount of E2 elimination with respect to S_N2 substitution (Table 17.2), not because elimination is faster, but because the S_N2 mechanism is so greatly slowed (p. 478). Under first-order conditions too, β branching favors elimination over substitution, probably for steric reasons.[107] However, E2 eliminations from compounds with charged leaving groups are slowed by β branching. This is related to Hofmann's rule (p. 1498). Electron-withdrawing groups in the β position not only increase the rate of E2 eliminations and shift the mechanisms toward the E1cB end of the spectrum, but also increase the extent of elimination as opposed to substitution.

Effect of the Attacking Base

1. *Effect on* E1 *versus* E2 *versus* E1cB. In the E1 mechanism, an external base is generally not required: The solvent acts as the base. Hence, when external bases are added, the mechanism is shifted toward E2. Stronger bases and higher base concentrations cause the mechanism to move toward the E1cB end of the E1-E2-E1cB spectrum.[110] However, weak bases in polar aprotic solvents can also be effective in elimination reactions with certain substrates (the E2C reaction). Normal E2 elimination has been accomplished with the following bases:[111] H_2O, NR_3, ⁻OH, ⁻OAc, ⁻OR, ⁻OAr, ⁻NH₂, CO_3^{2-},

[106]Dhar, M.L.; Hughes, E.D.; Ingold, C.K. *J. Chem. Soc.* **1948**, 2058.

[107]Hughes, M.L.; Ingold, C.K.; Maw, G.A. *J. Chem. Soc.* **1948**, 2065.

[108]Hughes, E.D.; Ingold, C.K.; Maw, G.A. *J. Chem. Soc.* **1948**, 2072; Hughes, E.D.; Ingold, C.K.; Woolf, L.I. *J. Chem. Soc.* **1948**, 2084.

[109]Brown, H.C.; Berneis, H.L. *J. Am. Chem. Soc.* **1953**, 75, 10.

[110]For a review, see Baciocchi, E. *Acc. Chem. Res.* **1979**, 12, 430. See also, Baciocchi, E.; Ruzziconi, R.; Sebastiani, G.V. *J. Org. Chem.* **1980**, 45, 827.

[111]This list is from Banthorpe, D.V. *Elimination Reactions*; Elsevier, NY, **1963**, p. 4.

LiAlH$_4$, I$^-$, CN$^-$, and organic bases. However, the only bases of preparative importance in the normal E2 reaction are OH$^-$, OR$^-$, and NH$_2^-$, usually in the conjugate acid as solvent, and certain amines. Weak bases effective in the E2C reaction are Cl$^-$, Br$^-$ F$^-$, $^-$OAc, and RS$^-$. These bases are often used in the form of their R$_4$N$^+$ salts.

2. *Effect on Elimination versus Substitution.* Strong bases not only benefit E2 as against E1, but also benefit elimination as against substitution. With a high concentration of strong base in a non-ionizing solvent, bimolecular mechanisms are favored and E2 predominates over S$_N$2. At low base concentrations, or in the absence of base altogether, in ionizing solvents, unimolecular mechanisms are favored, and the S$_N$1 mechanism predominates over the E1. In Chapter 10, it was pointed out that some species are strong nucleophiles, but weak bases (p. 490). The use of these obviously favors substitution, except that, as we have seen, elimination can predominate if polar aprotic solvents are used. It has been shown for the base cyanide that in polar aprotic solvents, the less the base is encumbered by its counterion in an ion pair (i.e., the freer the base), the more substitution is favored at the expense of elimination.[112]

Effect of the Leaving Group

1. *Effect on Reactivity.* The leaving groups in elimination reactions are similar to those in nucleophilic substitution. The E2 eliminations have been performed with the following groups: NR$_3^+$, PR$_3^+$, SR$_2^+$, OHR$^+$, SO$_2$R, OSO$_2$R, OCOR, OOH, OOR, NO$_2$,[113] F, Cl, Br, I, and CN (*not* OH$_2^+$). The E1 eliminations have been carried out with: NR$_3^+$, SR$_2^+$, OH$_2^+$, OHR$^+$, OSO$_2$R, OCOR, Cl, Br, I, and N$_2^+$.[114] However, the major leaving groups for preparative purposes are OH$_2^+$ (always by E1) and Cl, Br, I, and NR$_3^+$ (usually by E2).

2. *Effect on* E1 *versus* E2 *versus* E1cB. Better leaving groups shift the mechanism toward the E1 end of the spectrum, since they make ionization easier. This effect has been studied in various ways. One way already mentioned was a study of ρ values (p. 1495). Poor leaving groups and positively charged leaving groups shift the mechanism toward the E1cB end of the spectrum because the strong electron-withdrawing field effects increase the acidity of the β hydrogen.[115] The E2C reaction is favored by good leaving groups.

3. *Effect on Elimination versus Substitution.* As we have already seen (p. 1487), for first-order reactions the leaving group has nothing to do with the

[112]Loupy, A.; Seyden-Penne, J. *Bull. Soc. Chim. Fr. 1971*, 2306.
[113]For a review of eliminations in which NO$_2$ is a leaving group, see Ono, N., in Feuer, H.; Nielsen, A.T. *Nitro Compounds; Recent Advances in Synthesis and Chemistry*, VCH, NY, *1990*, pp. 1–135, 86–126.
[114]These lists are from Banthorpe, D.V. *Elimination Reactions*, Elsevier, NY, *1963*, pp. 4, 7.
[115]For a discussion of leaving-group ability, see Stirling, C.J.M. *Acc. Chem. Res. 1979*, *12*, 198. See also, Varma, M.; Stirling, C.J.M. *J. Chem. Soc., Chem. Commun. 1981*, 553.

competition between elimination and substitution, since it is gone before the decision is made as to which path to take. However, where ion pairs are involved, this is not true, and results have been found where the nature of the leaving group does affect the product.[116] In second-order reactions, the elimination/substitution ratio is not greatly dependent on a halide leaving group, although there is a slight increase in elimination in the order $I > Br > Cl$. When OTs is the leaving group, there is usually much more substitution. For example, $n\text{-}C_{18}H_{37}Br$ treated with $t\text{-}BuOK$ gave 85% elimination, while $n\text{-}C_{18}H_{37}OTs$ gave, under the same conditions, 99% substitution.[117] On the other hand, positively charged leaving groups increase the amount of elimination.

Effect of the Medium

1. *Effect of Solvent on* E1 *versus* E2 *versus* E1cB. With any reaction a more polar environment enhances the rate of mechanisms that involve ionic intermediates. For neutral leaving groups, it is expected that E1 and E1cB mechanisms will be aided by increasing polarity of solvent and by increasing ionic strength. With certain substrates, polar aprotic solvents promote elimination with weak bases (the E2C reaction).

2. *Effect of Solvent on Elimination versus Substitution.* Increasing polarity of solvent favors S_N2 reactions at the expense of E2. In the classical example, alcoholic KOH is used to effect elimination, while the more polar aqueous KOH is used for substitution. Charge-dispersal discussions, similar to those on p. 503,[118] only partially explain this. In most solvents, S_N1 reactions are favored over E1. The E1 reactions compete best in polar solvents that are poor nucleophiles, especially dipolar aprotic solvents.[119] A study made in the gas phase, where there is no solvent, has shown that when 1-bromopropane reacts with MeO^- only elimination takes place (no substitution) even with this primary substrate.[120]

3. *Effect of Temperature.* Elimination is favored over substitution by increasing temperature, whether the mechanism is first or second order.[121] The reason is that the activation energies of eliminations are higher than those of substitutions (because eliminations have greater changes in bonding).

[116]For example, see Skell, P.S.; Hall, W.L. *J. Am. Chem. Soc.* *1963*, 85 2851; Cocivera, M.; Winstein, S. *J. Am. Chem. Soc. 1963*, 85, 1702; Feit, I.N.; Wright, D.G. *J. Chem. Soc., Chem. Commun. 1975*, 776. See, however, Cavazza, M. *Tetrahedron Lett. 1975*, 1031.

[117]Veeravagu, P.; Arnold, R.T.; Eigenmann, E.W. *J. Am. Chem. Soc. 1964*, 86, 3072.

[118]Cooper, K.A.; Dhar, M.L.; Hughes, E.D.; Ingold, C.K.; MacNulty, B.J.; Woolf, L.I. *J. Chem. Soc. 1948*, 2043.

[119]Aksnes, G.; Stensland, P. *Acta Chem. Scand. 1989*, 43, 893, and references cited therein.

[120]Jones, M.E.; Ellison, G.B. *J. Am. Chem. Soc. 1989*, 111, 1645. For a different result with other reactants, see Lum, R.C.; Grabowski, J.J. *J. Am. Chem. Soc. 1988*, 110, 8568.

[121]Cooper, K.A.; Hughes, E.D.; Ingold, C.K.; Maw, G.A.; MacNulty, B.J. *J. Chem. Soc. 1948*, 2049.

MECHANISMS AND ORIENTATION IN PYROLYTIC ELIMINATIONS

Mechanisms[122]

Several types of compound undergo elimination on heating, with no other reagent present. Reactions of this type are often run in the gas phase. The mechanisms are obviously different from those already discussed, since all those require a base (which may be the solvent) in one of the steps, and there is no base or solvent present in pyrolytic elimination. Two mechanisms have been found to operate. One involves a cyclic transition state, which may be four, five, or six membered. Examples of each size are

In this mechanism, the two groups leave at about the same time and bond to each other as they are doing so. The designation is E^i in the Ingold terminology and $cyclo\text{-}D_E D_N A_n$ in the IUPAC system. The elimination must be syn and, for the four- and five-membered transition states, the four or five atoms making up the ring must be coplanar. Coplanarity is not required for the six-membered transition state, since there is room for the outside atoms when the leaving atoms are staggered.

[122]For reviews, see Taylor, R., in Patai, S. *The Chemistry of Functional Groups, Supplement B*, pt. 2, Wiley, NY, *1979*, pp. 860–914; Smith, G.G.; Kelly, F.W. *Prog. Phys. Org. Chem.* *1971*, 8, 75, pp. 76–143, 207–234; in Bamford, C.H.; Tipper, C.F.H. *Comprehensive Chemical Kinetics*, Vol. 5, Elsevier, NY, *1972*, the articles by Swinbourne, E.S. pp. 149–233 (pp. 158–188), and by Richardson, W.H.; O'Neal, H.E. pp. 381–565 (pp. 381–446); Maccoll, A. *Adv. Phys. Org. Chem.* *1965*, 3, 91. For reviews of mechanisms in pyrolytic eliminations of halides, see Egger, K.W.; Cocks, A.T., in Patai's. *The Chemistry of the Carbon-Halogen Bond*, pt. 2, Wiley, NY, *1973*, pp. 677–745; Maccoll, A. *Chem. Rev.* *1969*, 69, 33.

As in the E2 mechanism, it is not necessary that the C–H and C–X bond be equally broken in the transition state. In fact, there is also a spectrum of mechanisms here, ranging from a mechanism in which C–X bond breaking is a good deal more advanced than C–H bond breaking to one in which the extent of bond breaking is virtually identical for the two bonds. Evidence for the existence of the E^i mechanism is

1. The kinetics are first order, so only one molecule of the substrate is involved in the reaction (i.e., if one molecule attacked another, the kinetics would be second order in substrate).[123]

2. Free-radical inhibitors do not slow the reactions, so no free-radical mechanism is involved.[124]

3. The mechanism predicts exclusive syn elimination, and this behavior has been found in many cases.[125] The evidence is inverse to that for the anti E2 mechanism and generally involves the following facts: (*1*) an erythro isomer gives a trans alkene and a threo isomer gives a cis alkene; (*2*) the reaction takes place only when a cis β hydrogen is available; (*3*) if, in a cyclic compound, a cis hydrogen is available on only one side, the elimination goes in that direction. Another piece of evidence involves a pair of steroid molecules. In 3β-acetoxy-(*R*)-5α-methylsulfinylcholestane (**26** shows rings A and B of this compound) and in 3β-acetoxy-(*S*)-5α-methylsulfinylcholestane (**27**: rings A and B), the *only* difference is the configuration of

oxygen and methyl about the sulfur. Yet pyrolysis of **26** gave only elimination to the 4-side (86% 4-ene), while **27** gave predominant elimination to the 6-side (65% 5-ene and 20% 4-ene).[126] Models show that interference from the 1- and 9-hydrogens causes the two groups on the sulfur to lie *in front of it* with respect to the rings, rather than behind it. Since the sulfur is chiral, this

[123]O'Connor, G.L.; Nace, H.R. *J. Am. Chem. Soc.* **1953**, *75*, 2118.

[124]Barton, D.H.R.; Head, A.J.; Williams, R.J. *J. Chem. Soc.* **1953**, 1715.

[125]In a few instances anti or nonstereoselective elimination has been found; this behavior is generally ascribed to the intervention of other mechanisms. For example, see Bordwell, F.G.; Landis, P.S. *J. Am. Chem. Soc.* **1958**, *80*, 2450, 6383; Briggs, W.S.; Djerassi, C. *J. Org. Chem.* **1968**, *33*, 1625; Smissman, E.E.; Li, J.P.; Creese, M.W. *J. Org. Chem.* **1970**, *35*, 1352.

[126]Jones, D.N.; Saeed, M.A. *Proc. Chem. Soc.* **1964**, 81. See also, Goldberg, S.I.; Sahli, M.S. *J. Org. Chem.* **1967**, *32*, 2059.

means that in **26** the oxygen is near the 4-hydrogen, while in **27** it is near the 6-hydrogen. This experiment is compatible only with syn-elimination.[127]

4. The ^{14}C isotope effects for the Cope elimination (**17-9**) show that both the C–H and C–N bonds have been extensively broken in the transition state.[128]

5. Some of these reactions have been shown to exhibit negative entropies of activation, indicating that the molecules are more restricted in geometry in the transition state than they are in the starting compound.

Where a pyrolytic elimination lies on the mechanistic spectrum seems to depend mostly on the leaving group. When this is halogen, all available evidence suggests that in the transition state the C–X bond is cleaved to a much greater extent than the C–H bond, that is, there is a considerable amount of carbocation character in the transition state. This is in accord with the fact that a completely nonpolar four-membered cyclic transition state violates the Woodward–Hoffmann rules (see the similar case of **15-63**). Evidence for the carbocation-like character of the transition state when halide is the leaving group is that relative rates are in the order $I > Br > Cl$[129] (see p. 496), and that the effects of substituents on reaction rates are in accord with such a transition state.[130] Rate ratios for pyrolysis of some alkyl bromides at 320°C were ethyl bromide, 1; isopropyl bromide, 280; *tert*-butyl bromide, 78,000. Also, α-phenylethyl bromide had about the same rate as *tert*-butyl bromide. On the other hand, β-phenylethyl bromide was only slightly faster than ethyl bromide.[131] This indicates that C–Br cleavage was much more important in the transition state than C–H cleavage, since the incipient carbocation was stabilized by a alkyl and α-aryl substitution, while there was no incipient carbanion to be stabilized by β-aryl substitution. These substituent effects, as well as those for other groups, are very similar to the effects found for the S_N1 mechanism and thus in very good accord with a carbocation-like transition state.

For carboxylic esters, the rate ratios were much smaller,[132] although still in the same order, so that this reaction is closer to a pure E^i mechanism, although the transition state still has some carbocationic character. Other evidence for a greater initial C–O cleavage with carboxylic esters is that a series of 1-arylethyl acetates followed σ^+ rather than σ, showing carbocationic character at the 1 position.[133]

[127]For other evidence for syn-elimination, see Curtin, D.Y.; Kellom, D.B. *J. Am. Chem. Soc.* **1953**, *75*, 6011; Skell, P.S.; Hall, W.L. *J. Am. Chem. Soc.* **1964**, *86*, 1557; Bailey, W.J.; Bird, C.N. *J. Org. Chem.* **1977**, *42*, 3895.

[128]Wright, D.R.; Sims, L.B.; Fry, A. *J. Am. Chem. Soc.* **1983**, *105*, 3714.

[129]Maccoll, A., in Patai, S. *The Chemistry of Alkenes*, Vol. 1, Wiley, NY, **1964**, pp. 215–216.

[130]For reviews of such studies, see Maccoll, A. *Chem. Rev.* **1969**, *69*, 33.

[131]For rate studies of pyrolysis of some β-alkyl substituted ethyl bromides, see Chuchani, G.; Rotinov, A.; Dominguez, R.M.; Martin, I. *Int. J. Chem. Kinet.* **1987**, *19*, 781.

[132]For example, see Scheer, J.C.; Kooyman, E.C.; Sixma, F.L.J. *Recl. Trav. Chim. Pays-Bas* **1963**, *82*, 1123. See also, Louw, R.; Vermeeren, H.P.W.; Vogelzang, M.W. *J. Chem. Soc. Perkin Trans. 2* **1983**, 1875.

[133]Taylor, R.; Smith, G.G.; Wetzel, W.H. *J. Am. Chem. Soc.* **1962**, *84*, 4817; Smith, G.G.; Jones, D.A.K.; Brown, D.F. *J. Org. Chem.* **1963**, *28*, 403; Taylor, R. *J. Chem. Soc. Perkin Trans. 2* **1978**, 1255. See also, Ottenbrite, R.M.; Brockington, J.W. *J. Org. Chem.* **1974**, *39*, 2463; Jordan, E.A.; Thorne, M.P. *J. Chem. Soc. Perkin Trans. 2* **1984**, 647; August, R.; McEwen, I.; Taylor, R. *J. Chem. Soc. Perkin Trans. 2* **1987**, 1683, and other papers in this series; Al-Awadi, N.A. *J. Chem. Soc. Perkin Trans. 2* **1990**, 2187.

The extent of E1 character in the transition state increases in the following order of ester types: acetate < phenylacetate < benzoate < carbamate < carbonate.[134] Cleavage of xanthates (**17-5**), cleavage of sulfoxides (**17-12**), the Cope reaction (**17-9**), and reaction **17-8** are probably very close to straight E^i mechanisms.[135]

The second type of pyrolysis mechanism is completely different and involves free radicals. Initiation occurs by pyrolytic homolytic cleavage. The remaining steps may vary, and a few are shown

$$\text{Initiation} \qquad R_2CHCH_2X \longrightarrow R_2CHCH_2\bullet + X\bullet$$

$$\text{Propagation} \qquad R_2CHCH_2X + X\bullet \longrightarrow R_2\overset{\bullet}{C}CH_2X + HX$$

$$R_2\overset{\bullet}{C}CH_2X \longrightarrow CH_2 = CH_2X\bullet$$

$$\text{Termination (disproportionation)} \qquad 2R_2\overset{\bullet}{C}CH_2X \longrightarrow R_2C = CH_2 + R_2CXCH_2X$$

Free-radical mechanisms are mostly found in pyrolyses of polyhalides and of primary monohalides,[136] although they also have been postulated in pyrolysis of certain carboxylic esters.[137] β-Elimination of tosyl radicals is known.[138] Much less is known about these mechanisms and we will not consider them further. Free-radical eliminations in solution are also known, but are rare.[139]

Orientation in Pyrolytic Eliminations

As in the E1-E2-E1cB mechanistic spectrum, Bredt's rule applies; and if a double bond is present, a conjugated system will be preferred, if sterically possible. Apart from these considerations, the following statements can be made for E^i eliminations:

1. In the absence of considerations mentioned below, orientation is statistical and is determined by the number of β hydrogens available (therefore *Hofmann's rule* is followed). For example, *sec*-butyl acetate gives 55–62%

[134]Taylor, R. *J. Chem. Soc. Perkin Trans. 2* **1975**, 1025.

[135]For a review of the mechanisms of **17-12**, **17-9**, and the pyrolysis of sulfilimines, see Oae, S.; Furukawa, N. *Tetrahedron* **1977**, *33*, 2359.

[136]For example, see Barton, D.H.R.; Howlett, K.E. *J. Chem. Soc.* **1949**, 155, 165.

[137]For example, see Rummens, F.H.A. *Recl. Trav. Chim. Pays-Bas* **1964**, *83*, 901; Louw, R.; Kooyman, E.C. *Recl. Trav. Chim. Pays-Bas* **1965**, *84*, 1511.

[138]Timokhin, V.I.; Gastaldi, S.; Bertrand, M.P.; Chatgilialoglu, C. *J. Org. Chem.* **2003**, *68*, 3532.

[139]For examples; see Kampmeier, J.A.; Geer, R.P.; Meskin, A.J.; D'Silva, R.M. *J. Am. Chem. Soc.* **1966**, *88*, 1257; Kochi, J.K.; Singleton, D.M.; Andrews, L.J. *Tetrahedron* **1968**, *24*, 3503; Boothe, T.E.; Greene Jr., J.L.; Shevlin, P.B. *J. Org. Chem.* **1980**, *45*, 794; Stark, T.J.; Nelson, N.T.; Jensen, F.R. *J. Org. Chem.* **1980**, *45*, 420; Kochi, J.K. *Organic Mechanisms and Catalysis*, Academic Press, NY, **1978**, pp. 346–349; Kamimura, A.; Ono, N. *J. Chem. Soc., Chem. Commun.* **1988**, 1278.

1-butene and 38–45% 2-butene,[140] which is close to the 3:2 distribution predicted by the number of hydrogens available.[141]

2. A cis β hydrogen is required. Therefore in cyclic systems, if there is a cis hydrogen on only one side, the double bond will go that way. However, when there is a six-membered transition state, this does not necessarily mean that the leaving groups must be cis to each other, since such transition states need not be completely coplanar. If the leaving group is axial, then the hydrogen obviously must be equatorial (and consequently cis to the leaving group), since the transition state cannot be realized when the groups are both axial. But if the leaving group is equatorial, it can form a transition state with a β hydrogen that is either axial (hence, cis) or equatorial (hence, trans). Thus **28**, in which the leaving group is most likely axial, does not form a double bond in the

$$\text{28} \xrightarrow{\hspace{2cm}} \text{29} \hspace{2cm} (17\text{-}4)$$

$$\text{30} \xrightarrow{\hspace{2cm}} \qquad \text{Me} \quad + \qquad \text{Me} \hspace{1cm} (17\text{-}5)$$

direction of the carbethoxyl group, even although that would be conjugated, because there is no equatorial hydrogen on that side. Instead it gives 100% **29**.[142] On the other hand, **30**, with an equatorial leaving group, gives ~50% of each alkene, even although, for elimination to the 1-ene, the leaving group must go off with a trans hydrogen.[143]

3. In some cases, especially with cyclic compounds, the more stable alkene forms and Zaitsev's rule applies. For example, menthyl acetate gives 35% of the Hofmann product and 65% of the Zaitsev, even although a cis β hydrogen is present on both sides and the statistical distribution is the other way. A similar result was found for the pyrolysis of menthyl chloride.[144]

4. There are also steric effects. In some cases, the direction of elimination is determined by the need to minimize steric interactions in the transition state or to relieve steric interactions in the ground state.

[140]Froemsdorf, D.H.; Collins, C.H.; Hammond, G.S.; DePuy, C.H. *J. Am. Chem. Soc.* *1959*, *81*, 643; Haag, W.O.; Pines, H. *J. Org. Chem.* *1959*, 24, 877.

[141]DePuy, C.H.; King, R.W. *Chem. Rev.* *1960*, 60, 431, have tables showing the product distribution for many cases.

[142]Bailey, W.J.; Baylouny, R.A. *J. Am. Chem. Soc.* *1959*, *81*, 2126.

[143]Botteron D.G.; Shulman, G.P. *J. Org. Chem.* *1962*, 27, 2007.

[144]Barton, D.H.R.; Head, A.J.; Williams, R.J. *J. Chem. Soc.* *1952*, 453; Bamkole, T.; Maccoll, A. *J. Chem. Soc. B* *1970*, 1159.

1,4 Conjugate Eliminations[145]

1,4-Eliminations of the type

$$H-C-C=C-C-X \longrightarrow C=C-C=C$$

are much rarer than conjugate additions (Chapter 15), but some examples are known.[146] One such is[147]

REACTIONS

First, we consider reactions in which a C=C or a C≡C bond is formed. From a synthetic point of view, the most important reactions for the formation of double bonds are **17-1** (usually by an E1 mechanism), **17-7**, **17-13**, and **17-22** (usually by an E2 mechanism), and **17-4**, **17-5**, and **17-9** (usually by an E^i mechanism). The only synthetically important method for the formation of triple bonds is **17-13**.[148] In the second section, we treat reactions in which C≡N bonds and C=N bonds are formed, and then eliminations that give C=O bonds and diazoalkanes. Finally, we discuss extrusion reactions.

REACTIONS IN WHICH C=C AND C≡C BONDS ARE FORMED

A. Reactions in which Hydrogen Is Removed from One Side

In **17-1–17-6**, the other leaving atom is oxygen. In **17-7–17-11**, it is nitrogen. For reactions in which hydrogen is removed from both sides, see **19-1–19-6**.

[145]Taylor, R., in Patai *The Chemistry of Functional Groups, Supplement B*, pt. 2, Wiley, NY, *1979*, pp. 885–890; Smith, G.G.; Mutter, L.; Todd, G.P. *J. Org. Chem.* **1977**, *42*, 44; Chuchani, G.; Dominguez, R.M. *Int. J. Chem. Kinet.* **1981**, *13*, 577; Hernández, A.; Chuchani, G. *Int. J. Chem. Kinet.* **1983**, *15*, 205.
[146]For a review of certain types of 1,4- and 1,6-eliminations, see Wakselman, M. *Nouv. J. Chem.* **1983**, *7*, 439.
[147]Thibblin, A. *J. Chem. Soc. Perkin Trans. 2 1986*, 321; Ölwegård, M.; Ahlberg, P. *Acta Chem. Scand.* **1990**, *44*, 642. For studies of the stereochemistry of 1,4-eliminations, see Hill, R.K.; Bock, M.G. *J. Am. Chem. Soc.* **1978**, *100*, 637; Moss, R.J.; Rickborn, B. *J. Org. Chem.* **1986**, *51*, 1992; Ölwegård, M.; Ahlberg, P. *J. Chem. Soc., Chem. Commun.* **1989**, 1279.
[148]For reviews of methods for preparing alkynes, see Friedrich, K., in Patai, S.; Rappoport, Z. *The Chemistry of Functional Groups, Supplement C*, pt. 2, Wiley, NY, *1983*; pp. 1376–1384; Ben-Efraim, D.A., in Patai, S. *The Chemistry of the Carbon–Carbon Triple Bond*, pt. 2, Wiley, NY, *1978*, pp. 755–790. For a comparative study of various methods, see Mesnard, D.; Bernadou, F.; Miginiac, L. *J. Chem. Res. (S) 1981*, 270, and references cited therein.

17-1   Dehydration of Alcohols

Hydro-hydroxy-elimination

$$\underset{\underset{H}{\big|}}{-}\overset{}{\underset{}{C}}-\overset{}{\underset{\underset{OH}{\big|}}{C}}- \quad\xrightarrow[\text{Al}_2\text{O}_3,\ 300^{\circ}\text{C}]{\text{H}_2\text{SO}_4\ \text{or}}\quad \underset{}{\overset{}{C}}=\underset{}{\overset{}{C}}$$

Dehydration of alcohols can be accomplished in several ways. Both H_2SO_4 and H_3PO_4 are common reagents, but in many cases these lead to rearrangement products and to ether formation (**10-12**). If the alcohol can be evaporated, vapor-phase elimination over Al_2O_3 is an excellent method since side reactions are greatly reduced. This method has even been applied to such high-molecular-weight alcohols as 1-dodecanol.[149] Other metallic oxides (e.g., Cr_2O_3, TiO_2, WO_3) have also been used, as have been sulfides, other metallic salts, and zeolites. The presence of an electron-withdrawing group usually facilitates elimination of water, as in the aldol condensation (**16-35**). 2-Nitroalcohols, for example, give conjugated nitro compounds when heated with zeolite Y–Y.[150] Treating a 4-hydroxy lactam with DMAP (*N,N*-dimethylaminopyridine) and Boc anhydride leads to the conjugated lactam.[151] Elimination of serine derivatives to α-alkylidene amino acid derivatives was accomplished with $(EtO)_2POCl$.[152] Another method of avoiding side reactions is the conversion of alcohols to esters, and the pyrolysis of these (**17-4–17-6**). The ease of dehydration increases with α branching, and tertiary alcohols are dehydrated so easily with only a trace of acid that it sometimes happens even when the investigator desires otherwise. It may also be recalled that the initial alcohol products of many base-catalyzed condensations dehydrate spontaneously (Chapter 16) because the new double bond can be in conjugation with one already there. Many other dehydrating agents[153] have been used on occasion: P_2O_5, I_2, $ZnCl_2$, Ph_3BiBr_2/I_2,[154] $PPh_3–I_2$,[155] BF_3–etherate, DMSO, $SiO_2–Cl/Me_3SiCl$,[156] $KHSO_4$, anhydrous $CuSO_4$, and phthalic anhydride, among others. Secondary and tertiary alcohols can also be dehydrated, without rearrangements, simply on refluxing in HMPA.[157] With nearly all reagents, dehydration follows Zaitsev's rule.

[149]For example, see Spitzin, V.I.; Michailenko, I.E.; Pirogowa, G.N. *J. Prakt. Chem.* **1964**, *[4] 25*, 160; Bertsch, H.; Greiner, A.; Kretzschmar, G.; Falk, F. *J. Prakt. Chem.* **1964**, *[4] 25*, 184.
[150]Anbazhagan, M.; Kumaran, G.; Sasidharan, M. *J. Chem. Res. (S)* **1997**, 336.
[151]Mattern, R.-H. *Tetrahedron Lett.* **1996**, *37*, 291.
[152]Berti, F.; Ebert, C.; Gardossi, L. *Tetrahedron Lett.* **1992**, *33*, 8145.
[153]For a list of reagents, with references, see Larock, R.C. *Comprehensive Organic Transformations*, 2nd ed., Wiley-VCH, NY, **1999**, pp. 291–294.
[154]Dorta, R.L.; Suárez, E.; Betancor, C. *Tetrahedron Lett.* **1994**, *35*, 5035.
[155]Alvarez-Manzaneda, E.J.; Chahboun, R.; Torres, E.C.; Alvarez, E.; Alvarez-Manzaneda, R.; Haidour, A.; Ramos, J. *Tetrahedron Lett.* **2004**, *45*, 4453.
[156]Firouzabadi, H.; Iranpoor, N.; Hazarkhani, H.; Karimi, B. *Synth. Commun.* **2003**, *33*, 3653.
[157]Monson, R.S. *Tetrahedron Lett.* **1971**, 567; Monson, R.S.; Priest, D.N. *J. Org. Chem.* **1971**, *36*, 3826; Lomas, J.S.; Sagatys, D.S.; Dubois, J.E. *Tetrahedron Lett.* **1972**, 165.

An exception involves the passage of hot alcohol vapors over thorium oxide at 350–450°C, under which conditions Hofmann's rule is followed,[158] and the mechanism is probably different. Cyclobutanol derivatives can be opened in the presence of a palladium catalyst. 2-Phenylbicyclo[3.2.0]octan-2-ol, for example, reacted with a catalytic amount of palladium acetate in the presence of pyridine and oxygen to give phenyl methylenecyclohexane ketone.[159]

Transition metals can induce the dehydration of certain alcohols. β-Hydroxy ketones are converted to conjugated ketones by treatment with $CeCl_3$ and NaI.[160] In the presence of a palladium complex, alkyl cyclopropanols undergo a dehydration reaction to give a conjugated ketone.[161] A δ-hydroxy-α,β-unsaturated aldehyde was converted to a dienyl aldehyde with a hafnium catalyst.[162] β-Hydroxy esters are converted to conjugated esters when treated with 2 equivalents of SmI_2.[163] The reaction of a β-hydroxy nitrile with methylmagneisum chloride[164] or with MgO[165] leads to a conjugated nitrile. In another variation of the dehydration reaction, vicinal bromohydrins are converted to alkenes upon treatment with In, $InCl_3$, and a palladium catalyst.[166] Chlorohydrins react similarly when treated with samarium, and then diiodomethane.[167]

Carboxylic acids can be dehydrated by pyrolysis, the product being a ketene:

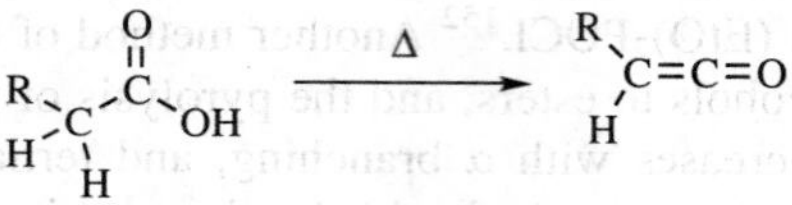

Ketene itself is commercially prepared in this manner. Carboxylic acids have also been converted to ketenes by treatment with certain reagents, among them TsCl,[168] dicyclohexylcarbodiimide,[169] and 1-methyl-2-chloropyridinium iodide (*Mukaiyama's reagent*).[170] Analogously, amides can be dehydrated with P_2O_5, pyridine,

[158]Lundeen, A.J.; Van Hoozer, R. *J. Am. Chem. Soc.* **1963**, *85*, 2180; *J. Org. Chem.* **1967**, *32*, 3386. See also, Davis, B.H. *J. Org. Chem.* **1982**, *47*, 900; Iimori, T.; Ohtsuka, Y.; Oishi, T. *Tetrahedron Lett.* **1991**, *32*, 1209.

[159]Nishimura, T.; Ohe, K.; Uemura, S. *J. Am. Chem. Soc.* **1999**, *121*, 2645.

[160]Bartoli, G.; Bellucci, M.C.; Petrini, M.; Marcantoni, E.; Sambri, L.; Torregiani, E. *Org. Lett.* **2000**, *2*, 1791.

[161]Okumoto, H.; Jinnai, T.; Shimizu, H.; Harada, Y.; Mishima, H.; Suzuki, A. *Synlett* **2000**, 629.

[162]Saito, S.; Nagahara, T.; Yamamoto, H. *Synlett* **2001**, 1690.

[163]Concellón, J.M.; Pérez-Andrés, J.A.; Rodríguez-Solla, H. *Angew. Chem. Int. Ed.* **2000**, *39*, 2773.

[164]Fleming, F.F.; Shook, B.C. *Tetrahedron Lett.* **2000**, *41*, 8847.

[165]Fleming, F.F.; Shook, B.C. *J. Org. Chem.* **2002**, *67*, 3668.

[166]Cho, S.; Kang, S.; Keum, G.; Kang, S.B.; Han, S.-Y.; Kim, Y. *J. Org. Chem.* **2003**, *68*, 180.

[167]Concellón, J.M.; Rodríguez-Solla, H.; Huerta, M.; Pérez-Andrés, J.A. *Eur. J. Org. Chem.* **2002**, 1839.

[168]Brady, W.T.; Marchand, A.P.; Giang, Y.F.; Wu, A. *Synthesis* **1987**, 395; *J. Org. Chem.* **1987**, *52*, 3457.

[169]Olah, G.A.; Wu, A.; Farooq, O. *Synthesis* **1989**, 568.

[170]Brady, W.T.; Marchand, A.P.; Giang, Y.F.; Wu, A. *J. Org. Chem.* **1987**, *52*, 3457; Funk, R.L.; Abelman, M.M.; Jellison, K.M. *Synlett* **1989**, 36.

and Al_2O_3 to give ketenimines:[171]

$$
\underset{\underset{H}{\overset{R}{\underset{|}{\overset{|}{C}}}}{\overset{R}{\underset{}{C}}}\overset{\overset{O}{\|}}{C}\text{—NHR'} \xrightarrow[\text{pyridine}]{P_2O_5,\ Al_2O_3} \underset{R}{\overset{R}{C}}{=}C{=}NR'
$$

There is no way in which dehydration of alcohols can be used to prepare triple bonds: *gem*-diols and vinylic alcohols are not normally stable compounds and *vic*-diols[172] give either conjugated dienes or lose only 1 equivalent of water to give an aldehyde or ketone. Dienes can be prepared, however, by heating alkynyl alcohols with triphenyl phosphine.[173]

When proton acids catalyze alcohol dehydration, the mechanism is E1.[174] The principal process involves conversion of ROH to ROH_2^+ and cleavage of the latter to R^+ and H_2O, although with some acids a secondary process probably involves conversion of the alcohol to an inorganic ester and ionization of *this* (illustrated for H_2SO_4):

$$
\text{ROH} \xrightarrow{H_2SO_4} \text{ROSO}_2\text{OH} \longrightarrow R^+ \ + \ HSO_4^-
$$

Note that these mechanisms are the reverse of those involved in the acid-catalyzed hydration of double bonds (**15-3**), in accord with the principle of microscopic reversibility. With anhydrides (e.g., P_2O_5, phthalic anhydride), as well as with some other reagents, such as HMPA,[175] it is likely that an ester is formed, and the leaving group is the conjugate base of the corresponding acid. In these cases, the mechanism can be E1 or E2. The mechanism with Al_2O_3 and other solid catalysts has been studied extensively, but is poorly understood.[176]

Magnesium alkoxides (formed by $ROH + Me_2Mg \rightarrow ROMgMe$) have been decomposed thermally, by heating at 195–340°C to give the alkene, CH_4, and MgO.[177] Syn-elimination is found and an E^i mechanism is likely. Similar decomposition of aluminum and zinc alkoxides has also been accomplished.[178,189]

[171]Stevens, C.L.; Singhal, G.H. *J. Org. Chem.* **1964**, *29*, 34.
[172]For a review on the dehydration of 1,2- and 1,3-diols, see Bartók, M.; Molnár, A., in Patai, S. *The Chemistry of Functional Groups, Supplement E*, pt. 2, Wiley, NY, **1980**, pp. 721–760.
[173]Guo, C.; Lu, X. *J. Chem. Soc., Chem. Commun.* **1993**, 394.
[174]For reviews of dehydration mechanisms, see Vinnik, M.I.; Obraztsov, P.A. *Russ. Chem. Rev.* **1990**, *59*, 63; Saunders, Jr., W.H.; Cockerill, A.F. *Mechanisms of Elimination Reactions*, Wiley, NY, **1973**, pp. 221–274, 317–331; Knözinger, H., in Patai, S. *The Chemistry of the Hydroxyl Group*, pt. 2, Wiley, NY, **1971**, pp. 641–718.
[175]See, for example, Kawanisi, M.; Arimatsu, S.; Yamaguchi, R.; Kimoto, K. *Chem. Lett.* **1972**, 881.
[176]For reviews, see Beránek, L.; Kraus, M., in Bamford, C,H; Tipper, C.F.H. *Comprehensive Chemical Kinetics*, Vol. 20, Elsevier, NY, **1978**, pp. 274–295; Pines, H. *Intra-Sci. Chem. Rep.* **1972**, *6*(2), 1, pp. 17–21; Noller, H.; Andréu, P.; Hunger, M. *Angew. Chem. Int. Ed.* **1971**, *10*, 172; Knözinger, H. *Angew. Chem. Int. Ed.* **1968**, *7*, 791. See also, Berteau, P.; Ruwet, M.; Delmon, B. *Bull. Soc. Chim. Belg.* **1985**, *94*, 859.
[177]Ashby, E.C.; Willard, G.F.; Goel, A.B. *J. Org. Chem.* **1979**, *44*, 1221.
[178]Brieger, G.; Watson, S.W.; Barar, D.G.; Shene, A.L. *J. Org. Chem.* **1979**, *44*, 1340.

OS **I**, 15, 183, 226, 280, 345, 430, 473, 475; **II**, 12, 368, 408, 606; **III**, 22, 204, 237, 312, 313, 353, 560, 729, 786; **IV**, 130, 444, 771; **V**, 294; **VI**, 307, 901; **VII**, 210, 241, 363, 368, 396; **VIII**, 210, 444. See also, OS **VII**, 63; **VIII**, 306, 474. No attempt has been made to list alkene-forming dehydration reactions accompanying condensations or rearrangements.

17-2   Cleavage of Ethers to Alkenes

Hydro-alkoxy-elimination

Alkenes can be formed by the treatment of ethers with very strong bases, such as alkylsodium or alkyllithium[179] compounds, sodium amide,[180] or LDA,[181] although there are side reactions with many of these reagents. The reaction is aided by electron-withdrawing groups in the β position, and, for example, $EtOCH_2CH(COOEt)_2$ can be converted to $CH_2=C(COOEt)_2$ without any base at all, but simply on heating.[182] *tert*-Butyl ethers are cleaved more easily than others. Several mechanisms are possible. In many cases, the mechanism is probably E1cB or on the E1cB side of the mechanistic spectrum,[183] since the base required is so strong, but it has been shown (by the use of $PhCD_2OEt$) that $PhCH_2OEt$ reacts by the five-membered E^i mechanism:[184] Propargylic benzyl ethers are converted to conjugated dienes by heating with a ruthenium catalyst.[185]

Ethers have also been converted to alkenes and alcohols by passing vapors over hot P_2O_5 or Al_2O_3 (this method is similar to **17-1**), but this is not a general reaction. Cyclic ethers, such as THF, react slowly with organolithium reagents with cleavage that produces a C=C unit.[186] Fragmentation of 2,5-dihydrofuran with ethylmagnesium chloride and a chiral zirconium catalyst leads to a chiral, homoallylic alcohol.[187] However, acetals can be converted to enol ethers (**31**) in this manner.

[179] Hodgson, D.M.; Stent, M.A.H.; Wilson, F.X. *Org. Lett.* **2001**, *3*, 3401.

[180] For a review, see Maercker, A. *Angew. Chem. Int. Ed.* **1987**, *26*, 972.

[181] Fleming, F.F.; Wang, Q.; Steward, O.W. *J. Org. Chem.* **2001**, *66*, 2171.

[182] Feely, W.; Boekelheide, V. *Org Synth.* **IV**, 298.

[183] For an investigation in the gas phase, see DePuy, C.H.; Bierbaum, V.M. *J. Am. Chem. Soc.* **1981**, *103*, 5034.

[184] Letsinger, R.L.; Pollart, D.F. *J. Am. Chem. Soc.* **1956**, *78*, 6079.

[185] Yeh, K.-L.; Liu, B.; Lo, C.-Y.; Huang, H.-L.; Liu, R.-S. *J. Am. Chem. Soc.* **2002**, *124*, 6510.

[186] For the mechanism of *n*-butyllithium cleavage of 2-methyltetrahydrofuran, see Cohen, T.; Stokes, S. *Tetrahedron Lett.* **1993**, *34*, 8023.

[187] Morken, J.P.; Didiuk, M.T.; Hoveyda, A.H. *J. Am. Chem. Soc.* **1993**, *115*, 6997.

When ketals react with 2 equivalents of triisobutylaluminum, the product is a vinyl ether.[188]

This can also be done at room temperature by treatment with trimethylsilyl triflate and a tertiary amine[189] or with Me_3SiI in the presence of hexamethyldisilazane.[190]

Enol ethers can be pyrolyzed to alkenes and aldehydes in a manner similar to that of **17-4**

The rate of this reaction for $R-O-CH=CH_2$ increased in the order $Et < iPr < t\text{-Bu}$.[191] The mechanism is similar to that of **17-4**.

OS **IV**, 298, 404; **V**, 25, 642, 859, 1145; **VI**, 491, 564, 584, 606, 683, 948; **VIII**, 444.

17-3   The Conversion of Epoxides and Episulfides to Alkenes

epi-**Oxy-elimination**

Epoxides can be converted to alkenes[192] by treatment with triphenylphosphine[193] or triethyl phosphite $P(OEt)_3$.[194] The first step of the mechanism is nucleophilic substitution (**10-35**), followed by a four-center elimination. Since inversion accompanies the substitution, the overall elimination is anti, that is, if two groups A and C are cis in the epoxide, they will be trans in the alkene:

Betaine

[188]Cabrera, G.; Fiaschi, R.; Napolitano, E. *Tetrahedron Lett.* **2001**, *42*, 5867.

[189]Gassman, P.G.; Burns, S.J. *J. Org. Chem.* **1988**, *53*, 5574.

[190]Miller, R.D.; McKean, D.R. *Tetrahedron Lett.* **1982**, *23*, 323. For another method, see Marsi, M.; Gladysz, J.A. *Organometallics* **1982**, *1*, 1467.

[191]McEwen, I.; Taylor, R. *J. Chem. Soc. Perkin Trans. 2* **1982**, 1179. See also Taylor, R. *J. Chem. Soc. Perkin Trans. 2* **1988**, 737.

[192]For reviews, see Wong, H.N.C.; Fok, C.C.M.; Wong, T. *Heterocycles* **1987**, *26*, 1345; Sonnet, P.E. *Tetrahedron* **1980**, *36*, 557, pp. 576.

[193]Wittig, G.; Haag, W. *Chem. Ber.* **1955**, *88*, 1654.

[194]Scott, C.B. *J. Org. Chem.* **1957**, *22*, 1118.

Alternatively, the epoxide can be treated with lithium diphenylphosphide, Ph_2PLi, and the product quaternized with methyl iodide.[195] Alkenes have also been obtained from epoxides by reaction with a large number of reagents,[196] among them Li in THF,[197] TsOH and NaI,[198] trimethylsilyl iodide,[199] PI_3,[200] $F_3COOH-NaI$,[201] SmI_2,[202] $Mo(CO)_6$, $TpReO_3$, where Tp is a pyrazolyl borate,[203] and the tungsten reagents mentioned in **17-18**. Some of these methods give syn elimination. Treatment of cyclooctane oxide with $Ph_3P-OPPh_3$ and NEt_3 gave cyclooctadiene.[204] Sodium amalgam with a cobalt–salen complex converted epoxides to alkenes.[205]

Epoxides can be converted to allylic alcohols[206] by treatment with several reagents, including sec-butyllithium,[207] $tert$-butyldimethylsilyl iodide,[208] and iPr_2N-Li-t-BuOK (the *LIDAKOR reagent*).[209] Phenyllithium reacts with epoxides in the presence of lithium tetramethylpiperidide (LTMP) to give a trans alkene.[210] Sulfur ylids, such as $Me_2S=CH_2$, also convert epoxides to allylic alcohols.[211] Bromomethyl epoxides react with $InCl_3/NaBH_4$ to give an allylic alcohol.[212] α,β-Epoxy ketones are converted to conjugated ketones by treatment with NaI in acetone in the presence of Amberlyst 15,[213] or with 2.5 equivalents of SmI_2.[214] Cyclic epoxides are converted to conjugated dienes by heating with $(NMe_2)_2P(=O)Cl$ and H_2O.[215]

[195]Vedejs, E.; Fuchs, P.L. *J. Am. Chem. Soc.* *1971*, *93*, 4070; *1973*, *95*, 822.

[196]For a list of reagents, with references, see Larock, R.C. *Comprehensive Organic Transformations*, 2nd ed., Wiley-VCH, NY, *1999*, pp. 272–277.

[197]Gurudutt, K.N.; Ravindranath, B. *Tetrahedron Lett.* *1980*, *21*, 1173.

[198]Baruah, R.N.; Sharma, R.P.; Baruah, J.N. *Chem. Ind. (London)* *1983*, 524.

[199]Denis, J.N.; Magnane, R.; Van Eenoo, M.; Krief, A. *Nouv. J. Chim.* *1979*, *3*, 705. For other silyl reagents, see Reetz, M.T.; Plachky, M. *Synthesis 1976*, 199; Dervan, P.B.; Shippey, M.A. *J. Am. Chem. Soc.* *1976*, *98*, 1265; Caputo, R.; Mangoni, L.; Neri, O.; Palumbo, G. *Tetrahedron Lett.* *1981*, *22*, 3551.

[200]Denis, J.N.; Magnaane, R.; Van Eenoo, M.; Krief, A. *Nouv. J. Chim.* *1979*, *3*, 705.

[201]Sarma, D.N.; Sharma, R.P. *Chem. Ind. (London)* *1984*, 712.

[202]Girard, P.; Namy, J.L.; Kagan, H.B. *J. Am. Chem. Soc.* *1980*, *102*, 2693; Matsukawa, M.; Tabuchi, T.; Inanaga, J.; Yamaguchi, M. *Chem. Lett.* *1987*, 2101.

[203]Gable, K.P.; Brown, E.C. *Synlett 2003*, 2243.

[204]Hendrickson, J.B.; Walker, M.A.; Varvak, A.; Hussoin, Md.S. *Synlett 1996*, 661.

[205]Isobe, H.; Branchaud, B.P. *Tetrahedron Lett.* *1999*, *40*, 8747.

[206]For reviews, see Smith, J.G. *Synthesis 1984*, 629, pp. 637–642; Crandall, J.K.; Apparu, M. *Org. React.* *1983*, *29*, 345. For a list of reagents, with references, see Larock, R.C. *Comprehensive Organic Transformations*, 2nd ed., Wiley-VCH, NY, *1999*, pp. 231–233. See also, Okovytyy, S.; Gorb, L.; Leszczynski, J. *Tetrahedron 2001*, *57*, 1509.

[207]Doris, E.; Dechoux, L.; Mioskowski, C. *Tetrahedron Lett.* *1994*, *35*, 7943.

[208]Detty, M.R. *J. Org. Chem.* *1980*, *45*, 924. For another silyl reagent, see Murata, S.; Suzuki, M.; Noyori, R. *J. Am. Chem. Soc.* *1979*, *101*, 2738.

[209]Mordini, A.; Ben Rayana, E.; Margot, C.; Schlosser, M. *Tetrahedron 1990*, *46*, 2401; Degl'Innocenti, A.; Mordini, A.; Pecchi, S.; Pinzani, D.; Reginato, G.; Ricci, A. *Synlett 1992*, 753, 803; Thurner, A.; Faigl, F.; Töke, L.; Mordini, A.; Valacchi, M.; Reginato, G.; Czira, G. *Tetrahedron 2001*, *57*, 8173.

[210]Hodgson, D.M.; Fleming, M.J.; Stanway, S.J. *J. Am. Chem. Soc.* *2004*, *126*, 12250.

[211]Alcaraz, L.; Cridland, A.; Kinchin, E. *Org. Lett.* *2001*, *3*, 4051.

[212]Ranu, B.C.; Banerjee, S.; Das, A. *Tetrahedron Lett.* *2004*, *45*, 8579.

[213]Righi, G.; Bovicelli, P.; Sperandio, A. *Tetrahedron 2000*, *56*, 1733.

[214]Concellón, J.M.; Bardales, E. *J. Org. Chem.* *2003*, *68*, 9492; Concellón, J.M.; Bardales, E. *Org. Lett.* *2002*, *4*, 189. In a similar manner, epoxy amides are converted to conjugated amides, see Concellón, J.M.; Bardales, E. *Eur. J. Org. Chem.* *2004*, 1523.

[215]Demir, A.S. *Tetrahedron 2001*, *57*, 227.

When an optically active reagent is used, optically active allylic alcohols can be produced from achiral epoxides.[216] Sparteine and *sec*-butyllithium generate a chiral base that leads to formation of chiral allylic alcohols.[217] Chiral diamines react with organolithium reagents to produce chiral bases that convert epoxides to allylic alcohols with good enantioselectivity.[218] Chiral diamines with a mixture of LDA and DBU (p. 1132) give similar results.[219]

Episulfides[220] can be converted to alkenes.[221] However, in this case the elimination is syn, so the mechanism cannot be the same as that for conversion of epoxides. The phosphite attacks sulfur rather than carbon. Among other reagents that convert episulfides to alkenes are Bu_3SnH,[222] certain rhodium complexes,[223] $LiAlH_4$[224] (this compound behaves quite differently with epoxides, see **19-35**), and meI.[225] The reaction of H_2S/PPh_3 and $MeReO_3$ converts episulfides to alkenes.[226] Episulfoxides can be converted to alkenes and sulfur monoxide simply by heating.[227]

17-4 Pyrolysis of Carboxylic Acids and Esters of Carboxylic Acids

Hydro-acyloxy-elimination

[216]Su, H.; Walder, L.; Zhang, Z.; Scheffold, R. *Helv. Chim. Acta* **1988**, *71*, 1073, and references cited therein. Also see, Asami, M.; Suga, T.; Honda, K.; Inoue, S. *Tetrahedron Lett.* **1997**, *38*, 6425.; Lill, S.O.N.; Pettersen, D.; Amedjkouh, M.; Ahlberg, P. *J. Chem. Soc., Perkin Trans. 1* **2001**, 3054; Brookes, P.C.; Milne, D.J.; Murphy, P.J.; Spolaore, B. *Tetrahedron* **2002**, *58*, 4675.

[217]Alexakis, A.; Vrancken, E.; Mangeney, P. *J. Chem. Soc. Perkin Trans. 1* **2000**, 3354.

[218]de Sousa, S.E.; O'Brien, P.; Steffens, H.C. *Tetrahedron Lett.* **1999**, *40*, 8423; Equey, O.; Alexakis, A. *Tetrahedron Asymmetry* **2004**, *15*, 1069.

[219]Bertilsson, S.K.; Södergren, M.J.; Andersson, P.G. *J. Org. Chem.* **2002**, *67*, 1567; Bertilsson, S.K.; Andersson, P.G. *Tetrahedron* **2002**, *58*, 4665.

[220]For a review of this reaction, see Sonnet, P.E. *Tetrahedron* **1980**, *36*, 557, see p. 587. For a review of episulfides, see Goodman, L.; Reist, E.J., in Kharasch; Meyers *The Chemistry of Organic Sulfur Compounds*, Vol. 2; Pergamon: Elmsford, NY, **1966**, pp. 93–113.

[221]Neureiter, N.P.; Bordwell, F.G. *J. Am. Chem. Soc.* **1959**, *81*, 578; Davis, R.E. *J. Org. Chem.* **1957**, *23*, 1767.

[222]Schauder, J.R.; Denis, J.N.; Krief, A. *Tetrahedron Lett.* **1983**, *24*, 1657.

[223]Calet, S.; Alper, H. *Tetrahedron Lett.* **1986**, *27*, 3573.

[224]Lightner, D.A.; Djerassi, C. *Chem. Ind. (London)* **1962**, 1236; Latif, N.; Mishriky, N.; Zeid, I. *J. Prakt. Chem.* **1970**, *312*, 421.

[225]Culvenor, C.J.; Davies, W.; Heath, N.S. *J. Chem. Soc.* **1949**, 282; Helmkamp, G.K.; Pettitt, D.J. *J. Org. Chem.* **1964**, *29*, 3258.

[226]Jacob, J.; Espenson, J.H. *Chem. Commun.* **1999**, 1003.

[227]Hartzell, G.E.; Paige, J.N. *J. Am. Chem. Soc.* **1966**, *88*, 2616, *J. Org. Chem.* **1967**, *32*, 459; Aalbersberg, W.G.L.; Vollhardt, K.P.C. *J. Am. Chem. Soc.* **1977**, *99*, 2792.

Direct elimination of a carboxylic acid to an alkene has been accomplished by heating in the presence of palladium catalysts.[228] Carboxylic esters in which the alkyl group has a β hydrogen can be pyrolyzed, most often in the gas phase, to give the corresponding acid and an alkene.[229] No solvent is required. Since rearrangement and other side reactions are few, the reaction is synthetically very useful and is often carried out as an indirect method of accomplishing **17-1**. The yields are excellent and the workup is easy. Many alkenes have been prepared in this manner. For higher alkenes (above ~C_{10}) a better method is to pyrolyze the alcohol in the presence of acetic anhydride.[230]

The mechanism is E^i (see p. 1507). Lactones can be pyrolyzed to give unsaturated acids, provided that the six-membered transition state required for E^i reactions is available (it is not available for five- and six-membered lactones, but it is for larger rings[231]). Amides give a similar reaction, but require higher temperatures.

Allylic acetates give dienes when heated with certain palladium[232] or molybdenum[233] compounds.

OS **III**, 30; **IV**, 746; **V**, 235; **IX**, 293.

17-5 The Chugaev Reaction

Methyl xanthates are prepared by treatment of alcohols with NaOH and CS_2 to give RO–C(=S)–SNa, followed by treatment of this with methyl iodide.[234] Pyrolysis of the xanthate to give the alkene, COS, and the thiol is called the *Chugaev reaction*.[235] The reaction is thus, like **17-4**, an indirect method of accomplishing **17-2**. The temperatures required with xanthates are lower than with ordinary esters, which is advantageous because possible isomerization of the resulting alkene is minimized. The mechanism is E^i, similar to that of **17-4**. For a time there was doubt as to which sulfur atom closed the ring, but now there is much evidence, including

[228]Miller, J.A.; Nelson, J.A.; Byrne, M.P. *J. Org. Chem.* **1993**, *58*, 18; Gooßen, L.J.; Rodríguez, N. *Chem. Commun.* **2004**, 724.

[229]For a review, see DePuy, C.H.; King, R.W. *Chem. Rev.* **1960**, *60*, 431, 432. For some procedures, see Jenneskens, L.W.; Hoefs, C.A.M.; Wiersum, U.E. *J. Org. Chem.* **1989**, *54*, 5811, and references cited therein.

[230]Aubrey, D.W.; Barnatt, A.; Gerrard, W. *Chem. Ind. (London)* **1965**, 681.

[231]See, for example, Bailey, W.J.; Bird, C.N. *J. Org. Chem.* **1977**, *42*, 3895.

[232]For a review, see Heck, R.F. *Palladium Reagents in Organic Synthesis*; Academic Press, NY, **1985**, pp. 172–178.

[233]Trost, B.M.; Lautens, M.; Peterson, B. *Tetrahedron Lett.* **1983**, *24*, 4525.

[234]For a method of preparing xanthates from alcohols in one laboratory step, see Lee, A.W.M.; Chan, W.H.; Wong, H.C.; Wong, M.S. *Synth. Commun.* **1989**, *19*, 547; Nagle, A.S.; Salvataore, R.N.; Cross, R.M.; Kapxhiu, E.A.; Sahab, S.; Yoon, C.H.; Jung, K.W. *Tetrahedron Lett.* **2003**, *44*, 5695.

[235]For reviews, see DePuy, C.H.; King, R.W. *Chem. Rev.* **1960**, *60*, 431, see p. 444; Nace, H.R. *Org. React.* **1962**, *12*, 57.

the study of ^{34}S and ^{13}C isotope effects, to show that it is the C=S sulfur:[236] In a structural variation of this reaction, heating a propargylic xanthate with 2,4,6-trimethylpyridinium trifluoromethyl sulfonate leads to formation of an alkene.[237]

The mechanism is thus exactly analogous to that of **17-5**.
   OS **VII**, 139.

17-6   Decomposition of Other Esters

Hydro-tosyloxy-elimination

Several types of inorganic ester can be cleaved to alkenes by treatment with bases. Esters of sulfuric, sulfurous, and other acids undergo elimination in solution by E1 or E2 mechanisms, as do tosylates and other esters of sulfonic acids.[238] It has been shown that bis(tetra-*n*-butylammonium) oxalate, $(Bu_4N^+)_2 (COO^-)_2$, is an excellent reagent for inducing tosylates to undergo elimination rather than substitution.[239] Aryl sulfonates have also been cleaved without a base. Esters of 2-pyridinesulfonic acid and 8-quinolinesulfonic acid gave alkenes in high yields simply on heating, without a solvent.[240] Phosphonate esters have been cleaved to alkenes by treatment with Lawesson's reagent.[241] Esters of $PhSO_2OH$ and TsOH behaved similarly when heated in a dipolar aprotic solvent, such as Me_2SO or HMPA.[242]
   OS, **VI**, 837; **VII**, 117.

17-7   Cleavage of Quaternary Ammonium Hydroxides

Hydro-trialkylammonio-elimination

[236]Bader, R.F.W.; Bourns, A.N. *Can. J. Chem.* **1961**, *39*, 348.

[237]Fauré-Tromeur, M.; Zard, S.Z. *Tetrahedron Lett.* **1999**, *40*, 1305.

[238]For a list of reagents used for sulfonate cleavages, with references, see Larock, R.C. *Comprehensive Organic Transformations*, 2nd ed., Wiley-VCH, NY, **1999**, pp. 294–295.

[239]Corey, E.J.; Terashima, S. *Tetrahedron Lett.* **1972**, 111.

[240]Corey, E.J.; Posner, G.G.; Atkinson, R.F.; Wingard, A.K.; Halloran, D.J.; Radzik, D.M.; Nash, J.J. *J. Org. Chem.* **1989**, *54*, 389.

[241]Shimagaki, M.; Fujieda, Y.; Kimura, T.; Nakata, T. *Tetrahedron Lett.* **1995**, *36*, 719.

[242]Nace, H.R. *J. Am. Chem. Soc.* **1959**, *81*, 5428.

Cleavage of quaternary ammonium hydroxides is the final step of the process known as *Hofmann exhaustive methylation or Hofmann degradation*.[243] In the first step, a primary, secondary, or tertiary amine is treated with enough methyl iodide to convert it to the quaternary ammonium iodide (**10-31**). In the second step, the iodide is converted to the hydroxide by treatment with silver oxide. In the cleavage step, an aqueous or alcoholic solution of the hydroxide is distilled, often under reduced pressure. The decomposition generally takes place at a temperature between 100 and 200°C. Alternatively, the solution can be concentrated to a syrup by distillation or freeze-drying.[244] When the syrup is heated at low pressures, the cleavage reaction takes place at lower temperatures than are required for the reaction in the ordinary solution, probably because the base (HO^- or RO^-) is less solvated.[245] The reaction has never been an important synthetic tool, but in the nineteenth century and the first part of the twentieth century, it saw much use in the determination of the structure of unknown amines, especially alkaloids. In many of these compounds, the nitrogen is in a ring, or even at a ring junction, and in such cases the alkene still contains nitrogen. Repetitions of the process are required to remove the nitrogen completely, as in the conversion of 2-methylpiperidine to 1,5-hexadiene by two rounds of exhaustive methylation followed by pyrolysis.

A side reaction involving nucleophilic substitution to give an alcohol (R_4N^+ $^-OH \rightarrow ROH + R_3N$) generally accompanies the normal elimination reaction,[246] but seldom causes trouble. However, when none of the four groups on the nitrogen has a β hydrogen, substitution is the only reaction possible. On heating Me_4N^+ ^-OH in water, methanol is obtained, although without a solvent the product is not methanol, but dimethyl ether.[247]

The mechanism is usually E2. Hofmann's rule is generally obeyed by acyclic and Zaitsev's rule by cyclohexyl substrates (p. 1498). In certain cases, where the molecule is highly hindered, a five-membered E^i mechanism, similar to that in **17-8**, has been shown to operate. That is, the hydroxide in these cases does not attract the β hydrogen, but instead removes one of the methyl hydrogens:

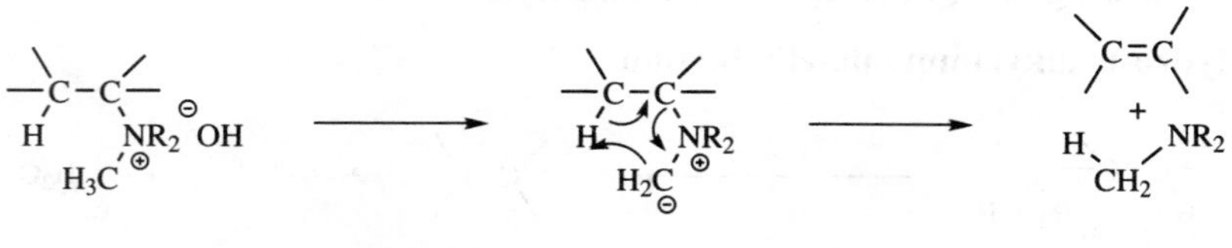

[243]For reviews, see Bentley, K.W., in Bentley, K.W.; Kirby, G.W *Elucidation of Organic Structures by Physical and Chemical Methods*, 2nd ed. (Vol. 4 of Weissberger *Techniques of Chemistry*), pt. 2, Wiley, NY, *1973*, pp. 255–289; White, E.H.; Woodcock, D.J., in Patai, S. *The Chemistry of the Amino Group*, Wiley, NY, *1968*, pp. 409–416; Cope, A.C.; Trumbull, E.R. *Org. React.* *1960*, *11*, 317.

[244]Archer, D.A. *J. Chem. Soc. C* *1971*, 1327.

[245]Saunders, Jr., W.H.; Cockerill, A.F. *Mechanisms of Elimination Reactions*, Wiley, NY, *1973*, pp. 4–5.

[246]Baumgarten, R.J. *J. Chem. Educ.* *1968*, *45*, 122.

[247]Musker, W.K. *J. Chem. Educ.* *1968*, *45*, 200; Musker, W.K.; Stevens, R.R. *J. Am. Chem. Soc.* *1968*, *90*, 3515; Tanaka, J.; Dunning, J.E.; Carter, J.C. *J. Org. Chem.* *1966*, *31*, 3431.

The obvious way to distinguish between this mechanism and the ordinary E2 mechanism is by the use of deuterium labeling. For example, if the reaction is carried out on a quaternary hydroxide deuterated on the β carbon ($R_2CDCH_2NMe_3^+$ ^-OH), the fate of the deuterium indicates the mechanism. If the E2 mechanism is in operation, the trimethylamine produced would contain no deuterium (which would be found only in the water). But if the mechanism is E^i, the amine would contain deuterium. In the case of the highly hindered compound $(Me_3C)_2CDCH_2NMe_3^+$ ^-OH, the deuterium did appear in the amine, demonstrating an Ei mechanism for this case.[248] With simpler compounds, the mechanism is E2, since here the amine was deuterium-free.[249]

When the nitrogen bears more than one group possessing a β hydrogen, which group cleaves? The Hofmann rule says that *within* a group the hydrogen on the least alkylated carbon cleaves. This tendency is also carried over to the choice of which group cleaves: thus ethyl with three β hydrogens cleaves more readily than any longer *n*-alkyl group, all of which have two β hydrogens. "The β hydrogen is removed most readily if it is located on a methyl group, next from RCH_2, and least readily from R_2CH."[250] In fact, the Hofmann rule as first stated[251] in 1851 applied only to which group cleaved, not to the orientation within a group; the latter could not have been specified in 1851, since the structural theory of organic compounds was not formulated until 1857–1860. Of course, the Hofmann rule (applied to which group cleaves *or* to orientation within a group) is superseded by conjugation possibilities. Thus $PhCH_2CH_2NMe_2Et^+$ ^-OH gives mostly styrene instead of ethylene.

Triple bonds have been prepared by pyrolysis of 1,2-bis(ammonium) salts.[252]

$$HO^\ominus \ \overset{\oplus}{R_3N}-\overset{\underset{|}{H}}{\underset{|}{C}}-\overset{\underset{|}{H}}{\underset{|}{C}}-\overset{\oplus}{N}R_3 \ \ ^\ominus OH \longrightarrow -C\equiv C-$$

OS **IV**, 980; **V**, 315, 608; **VI**, 552. Also see, OS **V**, 621, 883; **VI**, 75.

17-8   Cleavage of Quaternary Ammonium Salts With Strong Bases

Hydro-trialkylammonio-elimination

$$-\overset{\underset{|}{H}}{\underset{|}{C}}-\overset{\underset{|}{\underset{|}{\overset{\oplus}{N}R_2-CH_2R'}}{\underset{Cl^\ominus}{}}}{C}- \xrightarrow{PhLi} \ \ \overset{}{\underset{}{C}}=\overset{}{\underset{}{C}} \ + \ Ph\text{-}H \ + \ R_2N\text{-}CH_2R' \ + \ LiCl$$

[248]Cope, A.C.; Mehta, A.S. *J. Am. Chem. Soc.* **1963**, *85*, 1949. See also, Baldwin, M.A.; Banthorpe, D.V.; Loudon, A.G.; Waller, F.D. *J. Chem. Soc. B* **1967**, 509.

[249]Cope, A.C.; LeBel, N.A.; Moore, P.T.; Moore, W.R. *J. Am. Chem. Soc.* **1961**, *83*, 3861.

[250]Cope, A.C.; Trumbull, E.R. *Org. React.* **1960**, *11*, 317, see p. 348.

[251]Hofmann, A.W. *Liebigs Ann. Chem.* **1851**, *78*, 253.

[252]For a review, see Franke, W.; Ziegenbein, W.; Meister, H. *Angew. Chem.* **1960**, *72*, 391, see p. 397–398.

When quaternary ammonium halides are treated with strong bases (e.g., PhLi, KNH_2 in liquid NH_3[253]), an elimination can occur that is similar in products, although not in mechanism, to **17-7**. This is an alternative to **17-7** and is done on the quaternary ammonium halide, so that it is not necessary to convert this to the hydroxide. The mechanism is E^i:

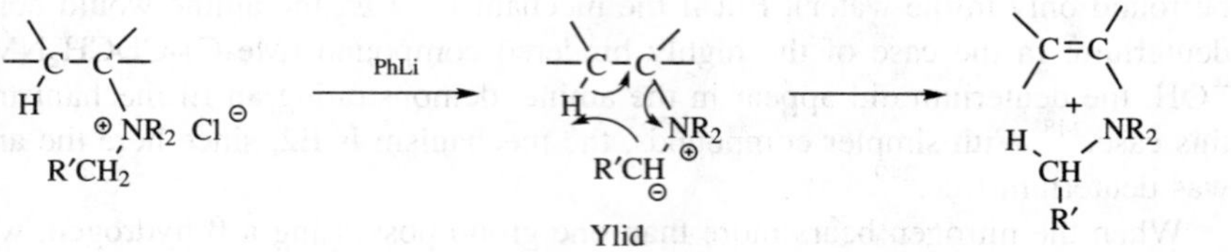

An α' hydrogen is obviously necessary in order for the ylid to be formed. This type of mechanism is called α',β elimination, since a β hydrogen is removed by the α' carbon. The mechanism has been confirmed by labeling experiments similar to those described at **17-7**,[254] and by isolation of the intermediate ylids.[255] An important synthetic difference between this and most instances of **17-7** is that syn-elimination is observed here and anti-elimination in **17-7**, so products of opposite configuration are formed when the alkene exhibits cis–trans isomerism.

An alternative procedure that avoids the use of a very strong base is heating the salt with KOH in polyethylene glycol monomethyl ether.[256]

Benzotriazole has been shown to be a good leaving group for elimination reactions. The reaction of an allylic benzotriazole (3-benzotriazoyl-4-trimethylsilyl-1-butene) with *n*-butyllithium, and then an alkyl halide leads to an alkylated 1,3-diene upon heating.[257]

17-9 Cleavage of Amine Oxides

Hydro-(Dialkyloxidoammonio)-elimination

Cleavage of amine oxides to produce an alkene and a hydroxylamine is called the *Cope reaction* or *Cope elimination* (not to be confused with the Cope *rearrangement*, **18-32**). It is an alternative to **17-7** and **17-8**.[258] The reaction is usually

[253]Bach, R.D.; Bair, K.W.; Andrzejewski, D. *J. Am. Chem. Soc.* **1972**, *94*, 8608; *J. Chem. Soc., Chem. Commun.* **1974**, 819.
[254]Weygand, F.; Daniel, H.; Simon, H. *Chem. Ber.* **1958**, *91*, 1691; Bach, R.D.; Knight, J.W. *Tetrahedron Lett.* **1979**, 3815.
[255]Wittig, G.; Burger, T.F. *Liebigs Ann. Chem.* **1960**, *632*, 85.
[256]Hünig, S.; Öller, M.; Wehner, G. *Liebigs Ann. Chem.* **1979**, 1925.
[257]Katritzky, A.R.; Serdyuk, L.; Toader, D.; Wang, X. *J. Org. Chem.* **1999**, *64*, 1888.
[258]For reviews, see Cope, A.C.; Trumbull, E.R. *Org. React.* **1960**, *11*, 317, see p. 361; DePuy, C.H.; King, R.W. *Chem. Rev.* **1960**, *60*, 431, see pp. 448–451.

performed with a mixture of amine and oxidizing agent (see **19-29**) without isolation of the amine oxide. Because of the mild conditions side reactions are few, and the alkenes do not usually rearrange. The reaction is thus very useful for the preparation of many alkenes. A limitation is that it does not open six-membered rings containing nitrogen, although it does open rings of 5 and 7–10 members.[259] Rates of the reaction increase with increasing size of α and β-substituents.[260] The reaction can be carried out at room temperature in dry Me_2SO or THF.[261] The elimination is a stereoselective syn process,[262] and the five-membered E^i mechanism operates:

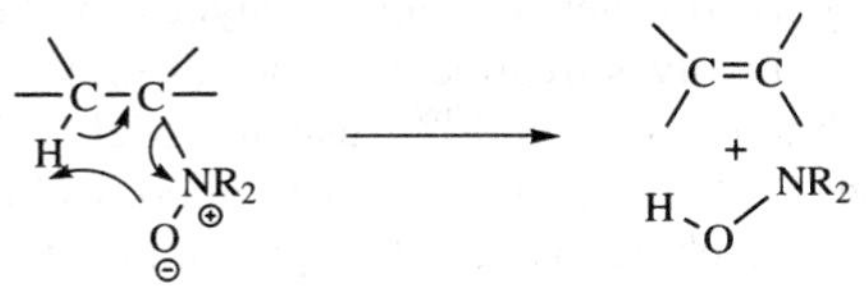

Almost all evidence indicates that the transition state must be planar. Deviations from planarity as in **17-4** (see p. 1507) are not found here, and indeed this is why six-membered heterocyclic nitrogen compounds do not react. Because of the stereoselectivity of this reaction and the lack of rearrangement of the products, it is useful for the formation of trans-cycloalkenes (eight-membered and higher). A polymer-bound Cope elimination reaction has been reported.[263]

OS **IV**, 612.

17-10   Pyrolysis of Keto-ylids

Hydro-(oxophosphoryl)-elimination

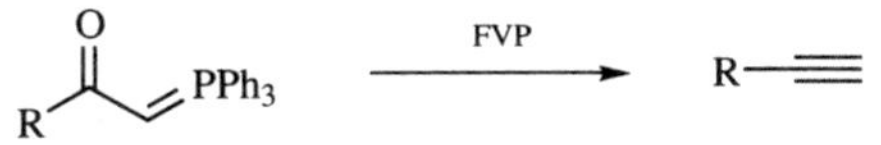

Phosphorus ylids are quite common (see **16-44**) and keto-phosphorus ylids [RCOCH=PPh₃] are also known. When these compounds are heating (flash vacuum pyrolysis, FVP) to > 500°C, alkynes are formed. Simple alkynes[264] can be formed as well as keto-alkynes[265] and en-ynes.[266] Rearrangement from ylids derived from tertiary amines an α-diazo ketones is also known.[267]

[259]Cope, A.C.; LeBel, N.A. *J. Am. Chem. Soc.* **1960**, *82*, 4656; Cope, A.C.; Ciganek, E.; Howell, C.F.; Schweizer, E.E. *J. Am. Chem. Soc.* **1960**, *82*, 4663.
[260]Závada, J.; Pánková, M.; Svoboda, M. *Collect. Czech. Chem. Commun.* **1973**, *38*, 2102.
[261]Cram, D.J.; Sahyun, M.R.V.; Knox, G.R. *J. Am. Chem. Soc.* **1962**, *84*, 1734.
[262]See, for example, Bach, R.D.; Andrzejewski, D.; Dusold, L.R. *J. Org. Chem.* **1973**, *38*, 1742.
[263]Sammelson, R.E.; Kurth, M.J. *Tetrahedron Lett.* **2001**, *42*, 3419.
[264]Aitken, R.A.; Atherton, J.I. *J. Chem. Soc. Perkin Trans. 1* **1994**, 1281.
[265]Aitken, R.A.; Hérion, H.; Janosi, A.; Karodia, N.; Raut, S.V.; Seth, S.; Shannon, I.J.; Smith, F.C. *J. Chem. Soc. Perkin Trans. 1* **1994**, 2467.
[266]Aitken, R.A.; Boeters, C; Morrison, J.J. *J. Chem. Soc. Perkin Trans. 1* **1994**, 2473.
[267]DelZotto, A.; Baratta, W.; Miani, F.; Verardo, G.; Rigo, P. *Eur. J. Org. Chem.* **2000**, 3731.

17-11 Decomposition of Toluene-*p*-sulfonylhydrazones

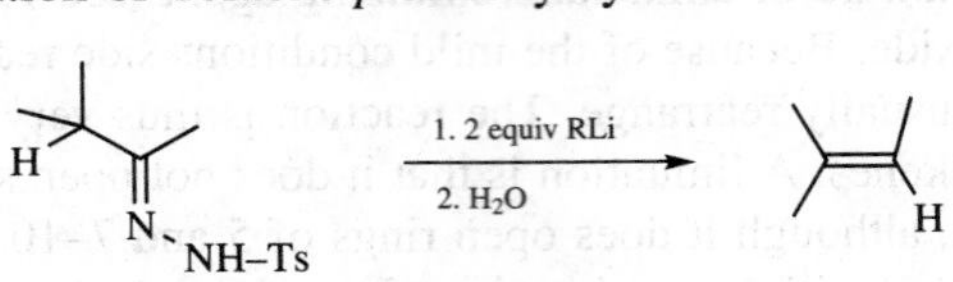

Treatment of the tosylhydrazone of an aldehyde or a ketone with a strong base leads to the formation of an alkene, the reaction being formally an elimination accompanied by a hydrogen shift.[268] The reaction (called the *Shapiro reaction*) has been applied to tosylhydrazones of many aldehydes and ketones. The most useful method synthetically involves treatment of the substrate with at least 2 equivalents of an organolithium compound[269] (usually MeLi) in ether, hexane, or tetramethylenediamine.[270] This procedure gives good yields of alkenes without side reactions and, where a choice is possible, predominantly gives the less highly substituted alkene. Tosylhydrazones of α,β-unsaturated ketones give conjugated dienes.[271] The mechanism[272] has been formulated as:

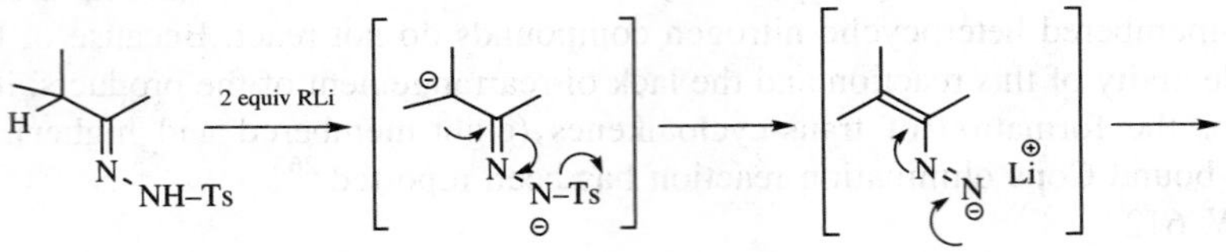

Evidence for this mechanism is (*1*) 2 equivalents of RLi are required; (*2*) the hydrogen in the product comes from the water and not from the adjacent carbon, as shown by deuterium labeling;[273] and (*3*) the intermediates **32–34** have been trapped.[274] This reaction, when performed in tetramethylenediamine, can be a synthetically useful method[275] of generating vinylic lithium compounds (**34**), which

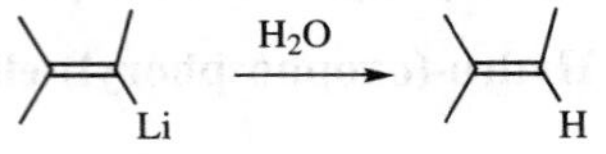

[268]For reviews, see Adlington, R.M.; Barrett, A.G.M. *Acc. Chem. Res.* **1983**, *16*, 55; Shapiro, R.H. *Org. React.* **1976**, *23*, 405.

[269]Shapiro, R.H.; Heath, M.J. *J. Am. Chem. Soc.* **1967**, *89*, 5734; Kaufman, G.; Cook, F.; Shechter, H.; Bayless, J.; Friedman, L. *J. Am. Chem. Soc.* **1967**, *89*, 5736; Shapiro, R.H. *Tetrahedron Lett.* **1968**, 345; Meinwald, J.; Uno, F. *J. Am. Chem. Soc.* **1968**, *90*, 800.

[270]Stemke, J.E.; Bond, F.T. *Tetrahedron Lett.* **1975**, 1815.

[271]See Dauben, W.G.; Rivers, G.T.; Zimmerman, W.T. *J. Am. Chem. Soc.* **1977**, *99*, 3414.

[272]For a review of the mechanism, see Casanova, J.; Waegell, B. *Bull. Soc. Chim. Fr.* **1975**, 922.

[273]Ref. 269; Shapiro, R.H.; Hornaman, E.C. *J. Org. Chem.* **1974**, *39*, 2302.

[274]Lipton, M.F.; Shapiro, R.H. *J. Org. Chem.* **1978**, *43*, 1409.

[275]See Traas, P.C.; Boelens, H.; Takken, H.J. *Tetrahedron Lett.* **1976**, 2287; Stemke, J.E.; Chamberlin, A.R.; Bond, F.T. *Tetrahedron Lett.* **1976**, 2947.

can be trapped by various electrophiles[276] such as D_2O (to give deuterated alkenes), CO_2 (to give α,β-unsaturated carboxylic acids, **16-30**), or DMF (to give α,β-unsaturated aldehydes, **16-82**). Treatment of *N*-aziridino hydrazones with LDA leads to alkenes with high cis selectivity.[277]

The reaction also takes place with other bases (e.g., LiH,[278] Na in ethylene glycol, NaH, NaNH$_2$) or with smaller amounts of RLi, but in these cases side reactions are common and the orientation of the double bond is in the other direction (to give the more highly substituted alkene). The reaction with Na in ethylene glycol is called the *Bamford–Stevens reaction*.[279] For these reactions two mechanisms are possible: a carbenoid and a carbocation mechanism.[280] The side reactions found are those expected of carbenes and carbocations. In general, the carbocation mechanism is chiefly found in protic solvents and the carbenoid mechanism in aprotic solvents. Both routes involve formation of a diazo compound (**35**) which in some cases can be isolated.

In fact, this reaction has been used as a synthetic method for the preparation of diazo compounds.[281] In the absence of protic solvents, **36** loses N$_2$, and hydrogen migrates, to give the alkene product. The migration of

[276]For a review, see Chamberlin, A.R.; Bloom, S.H. *Org. React.* **1990**, *39*, 1.
[277]Maruoka, K.; Oishi, M.; Yamamoto, H. *J. Am. Chem. Soc.* **1996**, *118*, 2289.
[278]Biellmann, J.F.; Pète, J. *Bull. Soc. Chim. Fr.* **1967**, 675.
[279]Bamford, W.R.; Stevens, R.R. *J. Chem. Soc.* **1952**, 4735. For a tandem Bamford-Stevens–Claisen rearrangement, see May, J.A.; Stoltz, B.M. *J. Am. Chem. Soc.* **2002**, *124*, 12426.
[280]Powell, J.W.; Whiting, M.C. *Tetrahedron* **1959**, *7*, 305; **1961**, *12* 168; DePuy, C.H.; Froemsdorf, D.H. *J. Am. Chem. Soc.* **1960**, *82*, 634; Bayless, J.H.; Friedman, L.; Cook, F.B.; Shechter, H. *J. Am. Chem. Soc.* **1968**, *90*, 531; Nickon, A.; Werstiuk, N.H. *J. Am. Chem. Soc.* **1972**, *94*, 7081.
[281]For a review, see Regitz, M.; Maas, G. *Diazo Compounds*; Academic Press, NY, **1986**, pp. 257–295. For an improved procedure, see Wulfman, D.S.; Yousefian, S.; White, J.M. *Synth. Commun.* **1988**, *18*, 2349.

hydrogen may immediately follow, or be simultaneous with, the loss of N_2. In a protic solvent, **35** becomes protonated to give the diazonium ion **36**, which loses N_2 to give the corresponding carbocation, that may then undergo elimination or give other reactions characteristic of carbocations. A diazo compound is an intermediate in the formation of alkenes by treatment of *N*-nitrosoamides with a rhodium(II) catalyst.[282]

OS **VI**, 172; **VII**, 77; **IX**, 147. For the preparation of a diazo compound, see OS **VII**, 438.

17-12   Cleavage of Sulfoxides, Selenoxides, and Sulfones

Sulfonium compounds ($-C-^+SR_2$) undergo elimination similar to that of their ammonium counterparts (**17-7** and **17-8**) in scope and mechanism but this reaction is not of great synthetic importance. These syn-elimination reactions are related to the Cope elimination (**17-9**) and the Hofmann elimination (**17-7**).[283]

Sulfones and sulfoxides[284] with a β hydrogen, on the other hand, undergo elimination on treatment with an alkoxide or, for sulfones,[285] even with hydroxide.[286] Sulfones also eliminate in the presence of an organolithium reagent and a palladium catalyst.[287] Mechanistically, these reactions belong on the E1-E2-E1cB spectrum.[288] Although the leaving groups are uncharged, the orientation follows Hofmann's rule, not Zaitsev's. Sulfoxides (but not sulfones) also undergo elimination

[282]Godfrey, A.G.; Ganem, B. *J. Am. Chem. Soc.* **1990**, *112*, 3717.

[283]For a discussion and leading references of this class of eliminations, see Smith, M.B. *Organic Synthesis*, 2nd ed., McGraw-Hill, NY, **2001**, pp. 135–141.

[284]See Cubbage, J.W.; Guo, Y.; McCulla, R.D.; Jenks, W.S. *J. Org. Chem.* **2001**, *66*, 8722.

[285]Certain sulfones undergo elimination with 5% HCl in THF: Yoshida, T.; Saito, S. *Chem. Lett.* **1982**, 165.

[286]Hofmann, J.E.; Wallace, T.J.; Argabright, P.A.; Schriesheim, A. *Chem. Ind. (London)* **1963**, 1234.

[287]Gai, Y.; Jin, L.; Julia, M.; Verpeaux, J.-N. *J. Chem. Soc., Chem. Commun.* **1993**, 1625.

[288]Hofmann, J.E.; Wallace, T.L.; Schriesheim, A. *J. Am. Chem. Soc.* **1964**, *86*, 1561.

on pyrolysis at $\sim 80°C$ in a manner analogous to **17-9**. The mechanism is also analogous, being the five-membered E^i mechanism with syn elimination.[289]

Selenoxides[290] and sulfinate esters $R_2CH–CHR–SO–OMe$[291] also undergo elimination by the E^i mechanism, the selenoxide reaction taking place at room temperature. The reaction with selenoxides has been extended to the formation of triple bonds.[292]

Both the selenoxide[293] and sulfoxide[294] reactions have been used in a method for the conversion of ketones, aldehydes, and carboxylic esters to their α,β-unsaturated derivatives (illustrated for the selenoxide).

Because of the mildness of the procedure, this is probably the best means of accomplishing this conversion. Treatment of ketones with LDA and then $PhClS=Nt\text{-}Bu$ leads to the conjugated ketone.[295] Allylic sulfoxides undergo 1,4-elimination to give dienes.[296] Ketones also react with hypervalent iodine cmpound in DMSO to give conjugated ketone.[297] In a simialr manner, keotnes are converted to conjugated keotnes by heating with HIO_5/I_2O_5 in DMSO.[298]

[289]Schmitz, C.; Harvey, J.N.; Viehe, H.G. *Bull. Soc. Chim. Belg.* **1994**, *103*, 105; Yoshimura, T.; Tsukurimichi, E.; Iizuka, Y.; Mizuno, H.; Isaji, H.; Shimasaki, C. *Bull. Chem. Soc. Jpn.* **1989**, *62*, 1891.

[290]For reviews, see Back, T.G., in Patai, S. *The Chemistry of Organic Selenium and Telurium Compounds,* Vol. 2, Wiley, NY, **1987**, pp. 91–213, 95–109; Paulmier, C. *Selenium Reagents and Intermediates in Organic Synthesis,* Pergamon, Elmsford, NY, **1986**, pp. 132–143; Reich, H.J. *Acc. Chem. Res. 1979, 12,* 22, in Trahanovsky, W.S. *Oxidation in Organic Chemistry,* pt. C, Academic Press, NY, **1978**, pp. 15–101; Sharpless, K.B.; Gordon, K.M.; Lauer, R.F.; Patrick, D.W.; Singer, S.P.; Young, M.W. *Chem. Scr.* **1975**, *8A*; 9. See also, Liotta, D. *Organoselenium Chemistry,* Wiley, NY, **1987**.

[291]Jones, D.N.; Higgins, W. *J. Chem. Soc. C 1970*, 81.

[292]Reich, H.J.; Willis, Jr., W.W. *J. Am. Chem. Soc.* **1980**, *102*, 5967.

[293]Clive, D.L.J. *J. Chem. Soc., Chem. Commun. 1973*, 695; Reich, H.J.; Renga, J.M.; Reich, I.L. *J. Am. Chem. Soc. 1975*, *97*, 5434, and references cited therein; Sharpless, K.B.; Lauer, R.F.; Teranishi, A.Y. *J. Am. Chem. Soc. 1973*, *95*, 6137; Grieco, P.A.; Miyashita, M. *J. Org. Chem. 1974*, *39*, 120; Crich, D.; Barba, G.R. *Org. Lett.* **2000**, *2*, 989. For lists of reagents, with references, see Larock, R.C. *Comprehensive Organic Transformations,* 2nd ed., Wiley-VCH, NY, **1999**, pp. 287–290. For a discussion of the effect of ortho substituents, see Sayama, S.; Onami, T. *Tetrahedron Lett. 2000*, *41*, 5557.

[294]Trost, B.M.; Salzmann, T.N.; Hiroi, K. *J. Am. Chem. Soc. 1976*, *98*, 4887. For a review of this and related methods, see Trost, B.M. *Acc. Chem. Res. 1978*, *11*, 453.

[295]Mukaiyama, T.; Matsuo, J.-i.; Kitgawa, H. *Chem. Lett.* **2000**, 1250.

[296]de Groot, A.; Jansen, B.J.M.; Reuvers, J.T.A.; Tedjo, E.M. *Tetrahedron Lett. 1981*, *22*, 4137.

[297]Nicolaou, K.C.; Montagnon, T.; Baran, P.S.; Zhong, Y.-L. *J. Am. Chem. Soc.* **2002**, *124*, 2245; Nicolaou, K.C.; Gray, D.L.F.; Montagnon, T.; Harrison, S.T. *Angew. Chem. Int. Ed.* **2002**, *41*, 996.

[298]Nicolaou, K.C.; Montagnon, T.; Baran, P.S. *Angew. Chem. Int. Ed.* **2002**, *41*, 1386.

A radical elimination reaction generates alkenes from sulfoxides. The reaction of a 2-bromophenyl alkylsulfoxide with Bu_3SnH and AIBN (see p. 935 for a discussion of these standard radical conditions) leads to an alkene.[299]

OS **VI**, 23, 737; **VIII**, 543; **IX**, 63.

17-13   Dehydrohalogenation of Alkyl Halides

Hydro-halo-elimination

$$\overset{\displaystyle\ \ |}{\underset{\displaystyle H}{-}}\overset{\displaystyle |\ \ }{\underset{\displaystyle X}{C}}-\overset{\displaystyle |\ \ }{\underset{\displaystyle\ \ |}{C}}- \quad \xrightarrow[\text{alcohol}]{^-OH} \quad \overset{\displaystyle\ \ /}{\underset{\displaystyle /\ \ }{}}C=C\overset{\displaystyle\ \ /}{\underset{\displaystyle\ \ \backslash}{}}$$

The elimination of HX from an alkyl halide is a very general reaction and can be accomplished with chlorides, fluorides, bromides, and iodides.[300] Hot alcoholic KOH is the most frequently used base, although stronger bases[301] (^-OR, $^-NH_2$, etc.) or weaker ones (e.g., amines) are used where warranted.[302] The bicyclic amidines 1,5-diazabicyclo[3.4.0]nonene-5 (DBN)[303] and 1,8-diazabicyclo[5.4.0]undecene-7 (DBU)[304] are good reagents for difficult cases.[305] Dehydrohalogenation with the non-ionic base $(Me_2N)_3P=N-P(NMe_2)_2=NMe$ is even faster.[306] Phase-transfer catalysis has been used with hydroxide as base.[307] As previously mentioned (p. 1495), certain weak bases in dipolar aprotic solvents are effective reagents for dehydrohalogenation. Among those most often used for synthetic purposes are LiCl or $LiBr-LiCO_3$ in DMF.[308] Dehydrohalogenation has also been effected by heating of the alkyl halide in HMPA with no other reagent present.[309] As in

[299]Imboden, C.; Villar, F.; Renaud, P. *Org. Lett.* **1999**, *1*, 873.

[300]For a review of eliminations involving the carbon–halogen bond, see Baciocchi, E., in Patai, S.; Rappoport, Z. *The Chemistry of Functional Groups, Supplement D*, pt. 2, Wiley, NY, **1983**, pp. 1173–1227.

[301]Triphenylmethylpotassium rapidly dehydrohalogenates secondary alkyl bromides and iodides, in >90% yields, at 0°C: Anton, D.R.; Crabtree, R.H. *Tetrahedron Lett.* **1983**, *24*, 2449.

[302]For a list of reagents, with references, see Larock, R.C. *Comprehensive Organic Transformations*, 2nd ed., Wiley-VCH, NY, **1999**, pp. 256–258.

[303]Truscheit, E.; Eiter, K. *Liebigs Ann. Chem.* **1962**, *658*, 65; Oediger, H.; Kabbe, H.; Möller, F.; Eiter, K. *Chem. Ber.* **1966**, *99*, 2012; Vogel, E.; Klärner, F. *Angew. Chem. Int. Ed.* **1968**, *7*, 374.

[304]Oediger, H.; Möller, F. *Angew. Chem. Int. Ed.* **1967**, *6*, 76; Wolkoff, P. *J. Org. Chem.* **1982**, *47*, 1944.

[305]For a review of these reagents, see Oediger, H.; Möller, F.; Eiter, K. *Synthesis* **1972**, 591.

[306]Schwesinger, R.; Schlemper, H. *Angew. Chem. Int. Ed.* **1987**, *26*, 1167.

[307]Kimura, Y.; Regen, S.L. *J. Org. Chem.* **1983**, *48*, 195; Halpern, M.; Zahalka, H.A.; Sasson, Y.; Rabinovitz, M. *J. Org. Chem.* **1985**, *50*, 5088. See also, Barry, J.; Bram, G.; Decodts, G.; Loupy, A.; Pigeon, P.; Sansoulet, J. *J. Org. Chem.* **1984**, *49*, 1138.

[308]For a discussion, see Fieser, L.F.; Fieser, M. *Reagents for Organic Syntheses*, Vol. 1, Wiley, NY, **1967**, pp. 606–609. For a review of alkali-metal fluorides in this reaction, see Yakobson, G.G.; Akhmetova, N.E. *Synthesis* **1983**, 169, see pp. 170–173.

[309]Hanna, R. *Tetrahedron Lett.* **1968**, 2105; Monson, R.S. *Chem. Commun.* **1971**, 113; Hutchins, R.O.; Hutchins, M.G.; Milewski, C.A. *J. Org. Chem.* **1972**, *37*, 4190; Hoye, T.R.; van Deidhuizen, J.J.; Vos, T.J.; Zhao, P. *Synth. Commun.* **2001**, *31*, 1367.

nucleophilic substitution (p. 496), the order of leaving group reactivity is I > Br > Cl > F.[310]

DBN   DBU

Tertiary halides undergo elimination most easily. Eliminations of chlorides, bromides, and iodides follow Zaitsev's rule, except for a few cases where steric effects are important (for an example, see p. 1499). Eliminations of fluorides follow Hofmann's rule (p. 1500).

This reaction is by far the most important way of introducing a triple bond into a molecule.[311] Alkyne formation can be accomplished with substrates of the types:[312]

When the base is $NaNH_2$ 1-alkynes predominate (where possible), because this base is strong enough to form the salt of the alkyne, shifting any equilibrium between 1- and 2-alkynes. When the base is ^-OH or ^-OR, the equilibrium tends to be shifted to the internal alkyne, which is thermodynamically more stable. If another hydrogen is suitably located (e.g., $-CRH-CX_2-CH_2-$), allene formation can compete, although alkynes are usually more stable. 1,1,2-Trihalocyclopropanes are converted to alkynes by ring opening reactions.[313]

Dehydrohalogenation is generally carried out in solution, with a base, and the mechanism is usually E2, although the E1 mechanism has been demonstrated in some cases. However, elimination of HX can be accomplished by pyrolysis of the halide, in which case the mechanism is E^i (p. 1507) or, in some instances, the free-radical mechanism (p. 1510). Pyrolysis is normally performed without a catalyst at ~400°C. The pyrolysis reaction is not generally useful synthetically, because of its reversibility. Less work has been done on pyrolysis with a catalyst[314] (usually a metallic oxide or salt), but the mechanisms here are probably E1 or E2.

[310]Matsubara, S.; Matsuda, H.; Hamatani, T.; Schlosser, M. *Tetrahedron* **1988**, *44*, 2855.

[311]For reviews, see Ben-Efraim, D.A. in Patai, S. *The Chemistry of the Carbon-Carbon Triple Bond*, pt. 2, Wiley, NY, **1978**, p. 755; Köbrich, G.; Buck, P., in Viehe, H. G. *Acetylenes*, Marcel Dekker, NY, **1969**, pp. 100–134; Franke, W.; Ziegenbein, W.; Meister, H. *Angew. Chem.* **1960**, *72*, 391, see p. 391; Köbrich, G. *Angew. Chem. Int. Ed.* **1965**, *4*, 49, see pp. 50–53.

[312]For a list of reagents, with references, see Larock, R.C. *Comprehensive Organic Transformations*, 2nd ed., Wiley-VCH, NY, **1999**, pp. 569–571.

[313]For a review, see Sydnes, L.K. *Eur. J. Org. Chem.* **2000**, 3511.

[314]For a review, see Noller, H.; Andréu, P.; Hunger, M. *Angew. Chem. Int. Ed.* **1971**, *10*, 172.

37

In the special case of the prochiral carboxylic acids **37**, dehydrohalogenation with an optically active lithium amide gave an optically active product with ee as high as 82%.[315]

Other regents lead to dehydrohalogenation. 1,1,1-Trichloro compounds are converted to vinyl chlorides with $CrCl_2$.[316]

OS **I**, 191, 205, 209, 438; **II**, 10, 17, 515; **III**, 125, 209, 270, 350, 506, 623, 731, 785; **IV**, 128, 162, 398, 404, 555, 608, 616, 683, 711, 727, 748, 755, 763, 851, 969; **V**, 285, 467, 514; **VI**, 87, 210, 327, 361, 368, 427, 462, 505, 564, 862, 883, 893, 954, 991, 1037; **VII**, 126, 319, 453, 491; **VIII**, 161, 173, 212, 254; **IX**, 191, 656, 662. Also see, OS **VI**, 968.

17-14 Dehydrohalogenation of Acyl Halides and Sulfonyl Halides

Hydro-halo-elimination

Ketenes can be prepared by treatment of acyl halides with tertiary amines[317] or with NaH and a crown ether.[318] The scope is broad, and most acyl halides possessing an α hydrogen give the reaction, but if at least one R is hydrogen, only the ketene dimer, not the ketene, is isolated. However, if it is desired to use a reactive ketene in a reaction with a given compound, the ketene can be generated *in situ* in the presence of the given compound.[319]

$$RCH_2SO_2Cl \xrightarrow{R_3N} [\ RCH{=}SO_2\ ] \xrightarrow{} RCH{=}CHR \ + \ \text{Other products}$$
$$\text{Sulfene}$$

Closely related is the reaction of tertiary amines with sulfonyl halides that contain an α hydrogen. In this case, the initial product is the highly reactive sulfene, which cannot be isolated but reacts further to give products, one of which may be the alkene that is the dimer of RCH.[320] Reactions of sulfenes *in situ* are also common (e.g., see **16-48**).

OS **IV**, 560; **V**, 294, 877; **VI**, 549, 1037; **VII**, 232; **VIII**, 82.

[315]Duhamel, L.; Ravard, A.; Plaquevent, J.C.; Plé, G.; Davoust, D. *Bull. Soc. Chim. Fr.* **1990**, 787.

[316]Baati, R.; Barma, D.K.; Krishna, U.M.; Mioskowski, C.; Falck, J.R. *Tetrahedron Lett.* **2002**, *43*, 959.

[317]For a monograph on the chemistry of ketenes, see Tidwell, T.T. *Ketenes*, Wiley, NY, **1995**.

[318]Taggi, A.E.; Wack, H.; Hafez, A.M.; France, S.; Lectka, T. *Org. Lett.* **2002**, *4*, 627.

[319]For a review of this procedure, see Luknitskii, F.I.; Vovsi, B.A. *Russ. Chem. Rev.* **1969**, *38*, 487.

[320]For reviews of sulfenes, see King, J.F. *Acc. Chem. Res.* **1975**, *8*, 10; Nagai, T.; Tokura, N. *Int. J. Sulfur Chem. Part B* **1972**, 207; Truce, W.E.; Liu, L.K. *Mech. React. Sulfur Compd.* **1969**, *4*, 145; Opitz, G. *Angew. Chem. Int. Ed.* **1967**, *6*, 107; Wallace, T.J. *Q. Rev. Chem. Soc.* **1966**, *20*, 67.

17-15   Elimination of Boranes

Hydro-boranetriyl-elimination

$$(R_2CH-CH_2)_3B \;+\; 3\;1\text{-Decene} \;\rightleftharpoons\; 3\;R_2C{=}CH_2 \;+\; [CH_3(CH_2)_8CH_2]_3B$$

Trialkylboranes are formed from an alkene and BH_3 (**15-16**). When the resulting borane is treated with another alkene, an exchange reaction occurs.[321] This is an equilibrium process that can be shifted by using a large excess of alkene, by using an unusually reactive alkene, or by using an alkene with a higher boiling point than the displaced alkene and removing the latter by distillation. The reaction is useful for shifting a double bond in the direction opposite to that resulting from normal isomerization methods (**12-2**). This cannot be accomplished simply by treatment of a borane, such as **39**, with an alkene, because elimination in this reaction follows Zaitsev's rule: It is in the direction of the most stable alkene, and the product would be **38**, not **41**. However, if it is desired to convert **38** to **41**, this can be accomplished by converting **38** to **39**, isomerizing **39** to **40** (**18-11**) and then subjecting **40** to the exchange reaction with a higher boiling alkene (e.g., 1-decene), whereupon **41** is produced. In the usual isomerizations (**12-2**), **41** could be isomerized to **38**, but not the other way around. The reactions **39** → **40** and **40** → **41** proceed essentially without rearrangement. The mechanism is probably the reverse of borane addition (**15-16**).

38   **39**   **40**   **41**

A similar reaction, but irreversible, has been demonstrated for alkynes.[322]

$$(R_2CH-CH_2)_3B \;+\; R'C{\equiv}CR' \;\longrightarrow\; 3\;R_2C{=}CH_2 \;+\; (R'CH{=}CR')_3B$$

17-16   Conversion of Alkenes to Alkynes

Hydro-methyl-elimination

Alkenes of the form shown lose the elements of methane when treated with sodium nitrite in acetic acid and water, to form alkynes in moderate-to-high yields.[323] The R may contain additional unsaturation, as well as OH, OR, OAc,

[321]Brown, H.C.; Bhatt, M.V.; Munekata, T.; Zweifel, G. *J. Am. Chem. Soc.* **1967**, *89*, 567; Taniguchi, H. *Bull. Chem. Soc. Jpn.* **1979**, *52*, 2942.
[322]Hubert, A.J. *J. Chem. Soc.* **1965**, 6669.
[323]Abidi, S.L. *Tetrahedron Lett.* **1986**, *27*, 267; *J. Org. Chem.* **1986**, *51*, 2687.

C=O, and other groups, but the $Me_2C=CHCH_2$ portion of the substrate is necessary for the reaction to take place. The mechanism is complex, beginning with a nitration that takes place with allylic rearrangement $[Me_2C=CHCH_2R \rightarrow H_2 C=CMeCH(NO_2)CH_2R]$, and involving several additional intermediates.[324] The CH_3 lost from the substrate appears as CO_2, as demonstrated by the trapping of this gas.[324]

1,1-Dibromoalkenes are converted to alkynes when treated with *n*-butyllithium.[325] This transformation is a modification of the *Fritsch–Buttenberg–Wiechell rearrangement*.[326] Vinyl sulfoxides that contain a leaving group, such as chloride on the double bond, react with *tert*-butyllithium to give a lithio alkyne, and hydrolysis leads to the final product, an alkyne.

17-17 Decarbonylation of Acyl Halides

Hydro-chloroformyl-elimination

$$R\text{-}CH_2\text{-}CH_2\text{-}\overset{\displaystyle O}{\underset{}{C}}\text{-}Cl \xrightarrow[\Delta]{RhCl(PPh_3)_3} \overset{R}{\underset{H}{>}}C=C\overset{H}{\underset{H}{<}} + HCl + RhCl(CO)(PPh_3)_2$$

Acyl chlorides containing an a hydrogen are smoothly converted to alkenes, with loss of HCl and CO, on heating with chlorotris(triphenylphosphine)rhodium, with metallic platinum, or with certain other catalysts.[327] The mechanism probably involves conversion of RCH_2CH_2COCl to RCH_2CH_2–$RhCO(Ph_3P)_2Cl_2$ followed by a concerted syn elimination of Rh and H[328] (see also, **14-32** and **19-12**).

B. Reactions in Which Neither Leaving Atom Is Hydrogen

17-18 Deoxygenation of Vicinal Diols

Dihydroxy-elimination

$$\overset{|}{\underset{HO}{C}}\text{-}\overset{|}{\underset{OH}{C}} \xrightarrow[THF]{2\ MeLi} \overset{|}{\underset{\ominus O}{C}}\text{-}\overset{|}{\underset{O\ominus}{C}} \xrightarrow[THF\ reflux]{K_2WCl_4} \overset{|}{\underset{}{C}}=\overset{|}{\underset{}{C}}$$

vic-Diols can be deoxygenated by treatment of the dilithium dialkoxide with the tungsten halide (K_2WCl_6), or with certain other tungsten reagents, in refluxing THF.[329] Tetrasubstituted diols react most rapidly. The elimination is largely, but

---

[324]Corey, E.J.; Seibel, W.L.; Kappos, J.C. *Tetrahedron Lett.* **1987**, *28*, 4921.

[325]Chernick, E.T.; Eisler, S.; Tykwinski, R.R. *Tetrahedron Lett.* **2001**, *42*, 8575.

[326]Fritsch, P. *Ann.* **1894**, *279*, 319; Buttenberg, W.P. *Ann.*, **1894**, *279*, 324; Wiechell, H. *Ann.* **1894**, *279*, 337; Stang, P.J.; Fox, D.P.; Collins, C.J.; Watson, Jr., C.R. *J. Org. Chem.* **1978**, *43*, 364. For a review, see Stang, P.J. *Chem. Rev.* **1978**, *78*, 383.

[327]For a review, see Tsuji, J.; Ohno, K. *Synthesis* **1969**, 157. For extensions to certain other acid derivatives, see Minami, I.; Nisar, M.; Yuhara, M.; Shimizu, I.; Tsuji, J. *Synthesis* **1987**, 992.

[328]Lau, K.S.Y.; Becker, Y.; Huang, F.; Baenziger, N.; Stille, J.K. *J. Am. Chem. Soc.* **1977**, *99*, 5664.

[329]Sharpless, K.B.; Flood, T.C. *J. Chem. Soc., Chem. Commun.* **1972**, 370; Sharpless, K.B.; Umbreit, M.A.; Nieh, T.; Flood, T.C. *J. Am. Chem. Soc.* **1972**, *94*, 6538.

not entirely, syn. Several other methods have been reported,[330] in which the diol is deoxygenated directly, without conversion to the dialkoxide. These include treatment with titanium metal,[331] with TsOH–NaI,[332] and by heating with $CpReO_3$, where Cp is cyclopentadienyl.[333]

vic-Diols can also be deoxygenated indirectly, through sulfonate ester derivatives. For example, *vic*-dimesylates and *vic*-ditosylates have been converted to alkenes by treatment, respectively, with naphthalene-sodium[334] and with NaI in DMF.[335] In another procedure, the diols are converted to bisdithiocarbonates (bis xanthates), which undergo elimination (probably by a free-radical mechanism) when

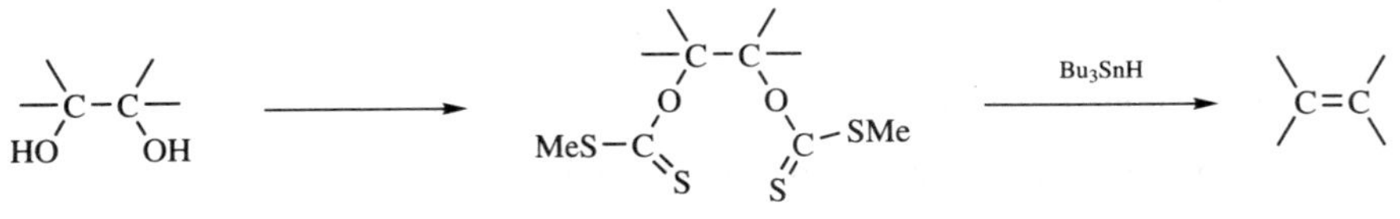

treated with tri-*n*-butylstannane in toluene or benzene.[336] *vic*-Diols can also be deoxygenated through cyclic derivatives (**17-19**).

17-19   Cleavage of Cyclic Thionocarbonates

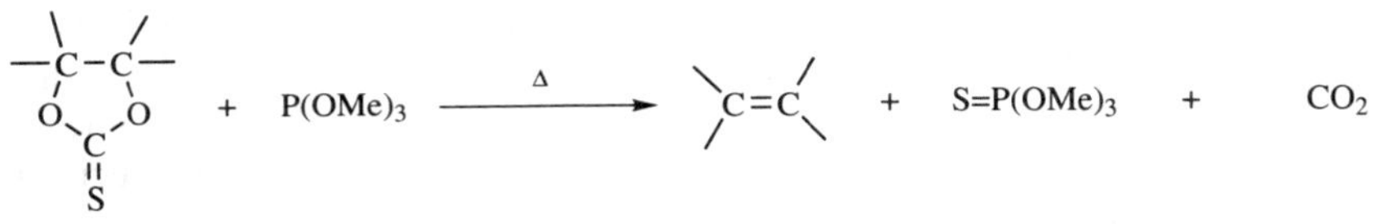

42

Cyclic thionocarbonates (**42**) can be cleaved to alkenes (the *Corey–Winter reaction*)[337] by heating with trimethyl phosphite[338] or other trivalent phosphorus compounds[339] or by treatment with bis(1,5-cyclooctadiene)nickel.[340] The

---

[330]For a list of reagents, with references, see Larock, R.C. *Comprehensive Organic Transformations*, 2nd ed., Wiley-VCH, NY, *1999*, pp. 297–299.

[331]McMurry, J.E. *Acc. Chem. Res. 1983*, *16*, 405, and references cited therein.

[332]Sarma, J.C.; Sharma, R.P. *Chem. Ind. (London) 1987*, 96.

[333]Cook, G.K.; Andrews, M.A. *J. Am. Chem. Soc. 1996*, *118*, 9448.

[334]Carnahan Jr., J.C.; Closson, W.D. *Tetrahedron Lett. 1972*, 3447.

[335]Dafaye, J. *Bull. Soc. Chim. Fr. 1968*, 2099.

[336]Barrett, A.G.M.; Barton, D.H.R.; Bielski, R. *J. Chem. Soc. Perkin Trans. 1 1979*, 2378.

[337]For reviews, see Block, E. *Org. React. 1984*, *30*, 457; Sonnet, P.E. *Tetrahedron 1980*, *36*, 557, 593–598; Mackie, R.K., in Cadogan, J.I.G. *Organophosphorus Reagents in Organic Synthesis*, Academic Press, NY, *1979*, pp. 354–359.

[338]Corey, E.J.; Winter, R.A.E. *J. Am. Chem. Soc. 1963*, *85*, 2677.

[339]Corey, E.J. *Pure Appl. Chem. 1967*, *14*, 19, see pp. 32–33.

[340]Semmelhack, M.F.; Stauffer, R.D. *Tetrahedron Lett. 1973*, 2667. For another method, see Vedejs, E.; Wu, E.S.C. *J. Org. Chem. 1974*, *39*, 3641.

thionocarbonates can be prepared by treatment of 1,2-diols with thiophosgene and 4-dimethylaminopyridine (DMAP):[341]

$$ \text{HO}-\overset{|}{\underset{}{C}}-\overset{|}{\underset{OH}{C}}- \quad + \quad \overset{S}{\underset{Cl}{\overset{||}{C}}}\text{Cl} \quad \xrightarrow{\text{DMAP}} \quad \textbf{42} $$

The elimination is of course syn, so the product is sterically controlled. Alkenes that are not sterically favored can be made this way in high yield, (e.g., *cis*-PhCH$_2$CH=CHCH$_2$Ph).[342] Certain other five-membered cyclic derivatives of 1,2-diols can also be converted to alkenes.[343]

17-20 The Ramberg–Bäcklund Reaction

Ramberg–Bäcklund halosulfone transformation

$$ \underset{CH_2}{R}\overset{O\;\;O}{\underset{H}{\overset{\backslash\!/}{S}}}\overset{R}{\underset{Cl}{C}} \quad \xrightarrow{^-OH} \quad \overset{H}{\underset{R}{\underset{}{C}}}=\overset{H}{\underset{R}{C}} $$

The reaction of an α-halo sulfone with a base to give an alkene is called the *Ramberg–Bäcklund reaction*.[344] The reaction is quite general for α-halo sulfones with an α' hydrogen, despite the unreactive nature of α-halo sulfones in normal S$_N$2 reactions (p. 486). Halogen reactivity is in the order I > Br ≫ Cl. Phase-transfer catalysis has been used.[345] In general, mixtures of cis and trans isomers are obtained, but usually the less stable cis isomer predominates. The mechanism involves formation of an episulfone, and then elimination of SO$_2$. There is much

$$ \underset{H}{\overset{H\quad R}{\underset{\overset{||}{O}\;\;\overset{||}{O}}{\overset{|\;\;\;|}{C-C}}}}\overset{H}{\underset{Cl}{}} \xrightarrow{^-OH} \underset{H}{\overset{R}{\underset{\overset{||}{O}\;\;\overset{||}{O}}{C-C}}}\overset{R}{\underset{Cl}{}}H \longrightarrow \underset{H}{\overset{R}{\underset{\overset{||}{O}\;\;\overset{||}{O}}{C-C}}}\overset{R}{\underset{}{}}H \longrightarrow \underset{H}{\overset{R}{C}}=\underset{H}{\overset{R}{C}} + O{=}S{=}O $$

[341]Corey, E.J.; Hopkins, P.B. *Tetrahedron Lett.* **1982**, *23*, 1979.

[342]Corey, E.J.; Carey, F.A.; Winter, R.A.E. *J. Am. Chem. Soc.* **1965**, *87*, 934.

[343]See Hines, J.N.; Peagram, M.J.; Whitham, G.H.; Wright, M. *Chem. Commun.* **1968**, 1593; Josan, J.S.; Eastwood, F.W. *Aust. J. Chem.* **1968**, *21*, 2013; Hiyama, T.; Nozaki, H. *Bull. Chem. Soc. Jpn.* **1973**, *46*, 2248; Marshall, J.A.; Lewellyn, M.E. *J. Org. Chem.* **1977**, *42*, 1311; Breuer, E.; Bannet, D.M. *Tetrahedron* **1978**, *34*, 997; Hanessian, S.; Bargiotti, A.; LaRue, M. *Tetrahedron Lett.* **1978**, 737; Hatanaka, K.; Tanimoto, S.; Oida, T.; Okano, M. *Tetrahedron Lett.* **1981**, *22*, 5195; Ando, M.; Ohhara, H.; Takase, K. *Chem. Lett.* **1986** 879; King, J.L.; Posner, B.A.; Mak, K.T.; Yang, N.C. *Tetrahedron Lett.* **1987**, *28*, 3919; Beels, C.M.D.; Coleman, M.J.; Taylor, R.J.K. *Synlett* **1990**, 479.

[344]For reviews, see Paquette, L.A. *Org. React.* **1977**, *25*, 1; *Mech. Mol. Migr.* **1968**, *1*, 121; *Acc. Chem. Res.* **1968**, *1*, 209; Meyers, C.Y.; Matthews, W.S.; Ho, L.L.; Kolb, V.M.; Parady, T.E., in Smith, G.V. *Catalysis in Organic Synthesis*, Academic Press, NY, **1977**, pp. 197–278; Rappe, C., in Patai, S. *The Chemistry of the Carbon-Halogen Bond*, pt. 2, Wiley, NY, **1973**, pp. 1105–1110; Bordwell, F.G. *Acc. Chem. Res.* **1970**, *3*, 281, pp. 285–286; in Janssen, M.J. *Organosulfur Chemistry*, Wiley, NY, **1967**, pp. 271–284.

[345]Hartman, G.D.; Hartman, R.D. *Synthesis* **1982**, 504.

evidence for this mechanism,[346] including the isolation of the episulfone intermediate,[347] and the preparation of episulfones in other ways and the demonstration that they give alkenes under the reaction conditions faster than the corresponding α-halo sulfones.[348] Episulfones synthesized in other ways (e.g., **16-48**) are reasonably stable compounds, but eliminate SO_2 to give alkenes when heated or treated with base.

If the reaction is run on the unsaturated bromo sulfones $RCH_2CH=CHSO_2$ CH_2Br (prepared by reaction of $BrCH_2SO_2Br$ with $RCH_2CH=CH_2$ followed by treatment with Et_3N), the dienes $RCH=CHCH=CH_2$ are produced in moderate-to-good yields.[349] The compound mesyltriflone $CF_3SO_2CH_2SO_2CH_3$ can be used as a synthon for the tetraion $^{2-}C=C^{2-}$. Successive alkylation (**10-67**) converts it to $CF_3SO_2CR^1R^2SO_2CHR^3R^4$ (anywhere from one to four alkyl groups can be put in), which, when treated with base, gives $R^1R^2C=CR^3R^4$.[350] The nucleofuge here is the $CF_3SO_2^-$ ion.

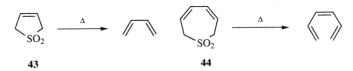

2,5-Dihydrothiophene-1,1-dioxides (**43**) and 2,17-dihydrothiepin-1,1-dioxides (**44**) undergo analogous 1,4- and 1,6-eliminations, respectively (see also, **17-36**). These are concerted reactions and, as predicted by the orbital-symmetry rules (p. 1207), the former[351] is a suprafacial process and the latter[352] an antarafacial process. The rules also predict that elimination of SO_2 from episulfones cannot take place by a concerted mechanism (except antarafacially, which is unlikely for such a small ring), and the evidence shows that this reaction occurs by a nonconcerted pathway.[353] The eliminations of SO_2 from **43** and **44** are examples of *cheletropic reactions*,[354] which are defined as reactions in which two σ bonds that terminate

[346]See, for example, Paquette, L.A. *J. Am. Chem. Soc.* **1964**, *86*, 4089; Neureiter, N.P. *J. Am. Chem. Soc.* **1966**, *88*, 558; Bordwell, F.G.; Wolfinger, M.D. *J. Org. Chem.* **1974**, *39*, 2521; Bordwell, F.G.; Doomes, E. *J. Org. Chem.* **1974**, *39*, 2526, 2531.

[347]Sutherland, A.G.; Taylor, R.J.K. *Tetrahedron Lett.* **1989**, *30*, 3267.

[348]Bordwell, F.G.; Williams Jr., J.M.; Hoyt, Jr., E.B.; Jarvis, B.B. *J. Am. Chem. Soc.* **1968**, *90*, 429; Bordwell, F.G.; Williams, Jr., J.M. *J. Am. Chem. Soc.* **1968**, *90*, 435.

[349]Block, E.; Aslam, M.; Eswarakrishnan, V.; Gebreyes, K.; Hutchinson, J.; Iyer, R.; Laffitte, J.; Wall, A. *J. Am. Chem. Soc.* **1986**, *108*, 4568.

[350]Hendrickson, J.B.; Boudreaux, G.J.; Palumbo, P.S. *J. Am. Chem. Soc.* **1986**, *108*, 2358.

[351]Mock, W.L. *J. Am. Chem. Soc.* **1966**, *88*, 2857; McGregor, S.D.; Lemal, D.M. *J. Am. Chem. Soc.* **1966**, *88*, 2858.

[352]Mock, W.L. *J. Am. Chem. Soc.* **1969**, *91*, 5682.

[353]Bordwell, F.G.; Williams, Jr., J.M.; Hoyt, Jr., E.B.; Jarvis, B.B. *J. Am. Chem. Soc.* **1968**, *90*, 429; Bordwell, F.G.; Williams Jr., J.M. *J. Am. Chem. Soc.* **1968**, *90*, 435. See also, Vilsmaier, E.; Tropitzsch, R.; Vostrowsky, O. *Tetrahedron Lett.* **1974**, 3987.

[354]For a review, see Mock, W.L., in Marchand, A.P.; Lehr, R.E. *Pericyclic Reactions*, Vol. 2, Academic Press, NY, **1977**, pp. 141–179.

at a single atom (in this case the sulfur atom) are made or broken in concert.[355]

α,α-Dichlorobenzyl sulfones (**45**) react with an excess of the base triethylenediamine (TED) in DMSO at room temperature to give 2,3-diarylthiiren-1,1-dioxides (**46**), which can be isolated.[356] Thermal decomposition of **46** gives the alkynes **47**.[357]

A Ramberg–Bäcklund-type reaction has been carried out on the α-halo *sulfides* (ArCHClSCH$_2$Ar), which react with *t*-BuOK and PPh$_3$ in refluxing THF to give the alkenes (ArCH=CHAr).[358] Cyclic sulfides lead to ring-contracted cyclic alkenes upon treatment with NCS in CCl$_4$ followed by oxidation with *m*-chloroperoxybenzoic acid.[359]

The Ramberg–Bäcklund reaction can be regarded as a type of extrusion reaction (see p. 1553).

OS **V**, 877; **VI**, 454, 555; **VIII**, 212.

17-21   The Conversion of Aziridines to Alkenes

epi-Imino-elimination

Aziridines not substituted on the nitrogen atom react with nitrous acid to produce alkenes.[360] An *N*-nitroso compound is an intermediate (**12-50**); other reagents that produce such intermediates also give alkenes. The reaction is stereospecific: cis aziridines give cis alkenes and trans aziridines give trans alkenes.[361] Aziridines carrying *N*-alkyl substituents can be converted to alkenes by treatment with ferrous iodide[362] or with *m*-chloroperoxybenzoic acid.[363] An *N*-oxide intermediate

[355]Woodward, R.B.; Hoffmann, R. *The Conservation of Orbital Symmetry*; Academic Press, NY, *1970*, pp. 152–163.

[356]Philips, J.C.; Swisher, J.V.; Haidukewych, D.; Morales, O. *Chem. Commun.* *1971*, 22.

[357]Carpino, L.A.; McAdams, L.V.; Rynbrandt, R.H.; Spiewak, J.W. *J. Am. Chem. Soc.* *1971*, *93*, 476; Philips, J.C.; Morales, O. *J. Chem. Soc., Chem. Commun.* *1977*, 713.

[358]Mitchell, R.H. *Tetrahedron Lett.* *1973*, 4395. For a similar reaction without base treatment, see Pommelet, J.; Nyns, C.; Lahousse, F.; Merényi, R.; Viehe, H.G. *Angew. Chem. Int. Ed.* *1981*, *20*, 585.

[359]MacGee, D.I.; Beck, E.J. *J. Org. Chem.* *2000*, *65*, 8367.

[360]For reviews, see Sonnet, P.E. *Tetrahedron* *1980*, *36*, 557, see p. 591; Dermer, O.C.; Ham, G.E. *Ethylenimine and other Aziridines*, Academic Press, NY, *1969*, pp. 293–295.

[361]Clark, R.D.; Helmkamp, G.K. *J. Org. Chem.* *1964*, *29*, 1316; Carlson, R.M.; Lee, S.Y. *Tetrahedron Lett.* *1969*, 4001.

[362]Imamoto, T.; Yukawa, Y. *Chem. Lett.* *1974*, 165.

[363]Heine, H.W.; Myers, J.D.; Peltzer III, E.T. *Angew. Chem. Int. Ed.* *1970*, *9*, 374.

(**19-29**) is presumably involved in the latter case. *N*-Tosyl aziridines are converted to *N*-tosyl imines when treated with boron trifluoride.[364] 2-Tosylmethyl *N*-tosylazir-idines react with Te^{2-} in the presence of Adogen 464 to give allylic *N*-tosyl amines.[365] 2-Halomethyl *N*-tosyl aziridines also react with indium metal in methanol to give *N*-tosyl allylic amines.[366]

17-22   Elimination of Vicinal Dihalides

Dihalo-elimination

$$\underset{X\quad X}{-\overset{|}{\underset{|}{C}}-\overset{|}{\underset{|}{C}}-} \quad \xrightarrow{\text{Zn}} \quad \overset{}{\underset{}{C}}=\overset{}{\underset{}{C}}$$

Dehalogenation has been accomplished with many reagents, the most common being zinc, magnesium, and iodide ion.[367] Heating in HMPA is often enough to convert a *vic*-dibromide to an alkene.[368] Among reagents used less frequently have been phenyllithium, phenylhydrazine, $CrCl_2$, Na_2S in DMF,[369] and $LiAlH_4$.[370] Electrochemical reduction has also been used.[371] Treatment with In[372] or Sm[373] metal in CH_3OH, $InCl_3/NaBH_4$,[374] a Grignard reagent and Ni(dppe)Cl_2, (dppe = 1, 2-diphenylphosphinoethane),[375] nickel compounds with Bu_3SnH,[376] or SmI_2[377] leads to the alkene. Although the reaction usually gives good yields, it is not very useful because the best way to prepare *vic*-dihalides is by the addition of halogen to a double bond (**15-39**). One useful feature of this reaction is that there is no doubt about the *position* of the new double bond, so that it can be used to give double bonds exactly where they are wanted. For example, allenes, which are not easily prepared by other methods, can be prepared from $X–C–CX_2–C–X$ or $X–C–CX=C–$ systems.[378] Cumulenes

[364]Sugihara, Y.; Iimura, S.; Nakayama, J. *Chem. Commun.* **2002**, 134.

[365]Chao, B.; Dittmer, D.C. *Tetrahedron Lett.* **2001**, *42*, 5789.

[366]Yadav, J.S.; Bandyapadhyay, A.; Reddy, B.V.S. *Synlett* **2001**, 1608.

[367]For a review of this reaction, see Baciocchi, E., in Patai, S.; Rappoport, Z. *The Chemistry of Functional Groups, Supplement D*, pt. 1, Wiley, NY, **1983**, pp. 161–201. Also see, Bosser, G.; Paris, J. *J. Chem. Soc. Perkin Trans. 2* **1992**, 2057.

[368]Khurana, J.M.; Bansal, G.; Chauhan, S. *Bull. Chem. Soc. Jpn.* **2001**, *74*, 1089.

[369]Fukunaga, K.; Yamaguchi, H. *Synthesis* **1981**, 879. See also, Nakayama, J.; Machida, H.; Hoshino, M. *Tetrahedron Lett.* **1983**, *24* 3001; Landini, D.; Milesi, L.; Quadri, M.L.; Rolla, F. *J. Org. Chem.* **1984**, *49*, 152.

[370]For a lists of reagents, with references, see Larock, R.C. *Comprehensive Organic Transformations*, 2nd ed., Wiley-VCH, NY, **1999**, pp. 259–263.

[371]See Shono, T. *Electroorganic Chemistry as a New Tool in Organic Synthesis*, Springer, NY, **1984**, pp. 145–147; Fry, A.J. *Synthetic Organic Electrochemistry*, 2nd ed., Wiley, NY, **1989**, pp. 151–154.

[372]Ranu, B.C.; Guchhait, S.K.; Sarkar, A. *Chem. Commun.* **1998**, 2113.

[373]Yanada, R.; Negoro, N.; Yanada, K.; Fujita, T. *Tetrahedron Lett.* **1996**, *37*, 9313.

[374]Ranu, B.C.; Das, A.; Hajra, A. *Synthesis* **2003**, 1012.

[375]Malanga, C.; Aronica, L.A.; Lardicci, L. *Tetrahedron Lett.* **1995**, *36*, 9189.

[376]Malanga, C.; Mannucci, S.; Lardicci, L. *Tetrahedron* **1998**, *54*, 1021.

[377]Yanada, R.; Bessho, K.; Yanada, K. *Chem. Lett,* **1994**, 1279.

[378]For reviews of allene formation, see Schuster, H.F.; Coppola, G.M. *Allenes in Organic Synthesis*, Wiley, NY, **1984**, pp. 9–56; Landor, P.D., in Landor, S.R. *The Chemistry of the Allenes*, Vol. 1, Academic Press, NY, **1982**; pp. 19–233; Taylor, D.R. *Chem. Rev.* **1967**, *67*, 317.

have been obtained from 1,4-elimination:

$$BrCH_2-C\equiv C-CH_2Br \quad + \quad Zn \quad \longrightarrow \quad CH_2=C=C=CH_2$$

Cumulenes have also been prepared by treating alkynyl epoxides with boron trifluoride.[379] 1,4-Elimination of BrC–C=C–CBr has been used to prepare conjugated dienes C=C–C=C.[380] Allenes are formed by heating propargylic alcohols with arylboronic acids (p. 815) and a palladium catalyst.[381] Allenes are also formed from propargylic amines using a CuI and a palladium catalyst.[382]

The reaction of a vicinal dibromide with triethylamine and DMF with microwave irradiation leads to vinyl bromide.[383] Alkenes are formed from vicinal bromides by heating with iron in methanol[384] or samarium in the presence of TMSCl and a trace of water.[385] α,β-Dibromo amides are converted to conjugated amides upon photolysis in methanol.[386]

The reaction can be carried out for any combination of halogens, except where one is fluorine. Mechanisms are often complex and depend on the reagent and reaction conditions.[387] For different reagents, mechanisms involving carbocations, carbanions, and free-radical intermediates, as well as concerted mechanisms, have been proposed. When the reagent is zinc, anti stereospecificity has been observed in some cases,[388] but not in others.[389]

Note that geminal dibromo cyclopropanes (1,1-dibromocyclopropanes) are opened to conjugated dienes by heating to 500°C.[390]

OS **III**, 526, 531; **IV**, 195, 268; **V**, 22, 255, 393, 901; **VI**, 310, **VII**, 241. Also see, OS **IV**, 877, 914, 964.

17-23   Dehalogenation of α-Halo Acyl Halides

Dihalo-elimination

[379]Wang, X.; Ramos, B.; Rodriguez, A. *Tetrahedron Lett.* **1994**, *35*, 6977.

[380]Engman, L.; Byström, S.E. *J. Org. Chem.* **1985**, *50*, 3170.

[381]Yoshida, M.; Gotou, T.; Ihara, M. *Tetrahedron Lett.* **2004**, *45*, 5573.

[382]Nakmura, H.; Kamakura, T.; Ishikura, M.; Biellmann, J.-F. *J. Am. Chem. Soc.* **2004**, *126*, 5958.

[383]Kuang, C.; Senboku, H.; Tokuda, M. *Tetrahedron Lett.* **2001**, *42*, 3893.

[384]Thakur, A.J.; Boruah, A.; Baruah, B.; Sandhu, J.S. *Synth. Commun.* **2000**, *30*, 157.

[385]Xu, X.; Lu, P.; Zhang, Y. *Synth. Commun.* **2000**, *30*, 1917.

[386]Aruna, S.; Kalyanakumar, R.; Ramakrishnan, V.T. *Synth. Commun.* **2001**, *31*, 3125.

[387]For discussion, see Saunders, Jr., W.H.; Cockerill, A.F. *Mechanisms of Elimination Reactions*, Wiley, NY, **1973**, pp. 332–368; Baciocchi, W., in Patai, S.; Rappoport, Z. *The Chemistry of Functional Grups, Supplement D*, pt. 2, Wiley, NY, **1983**, p. 161.

[388]For example, see House, H.O.; Ro, R.S. *J. Am. Chem. Soc.* **1958**, *80*, 182; Gordon, M.; Hay, J.V. *J. Org. Chem.* **1968**, *33*, 427.

[389]For example, see Stevens, C.L.; Valicenti, J.A. *J. Am. Chem. Soc.* **1965**, *87*, 838; Sicher, J.; Havel, M.; Svoboda, M. *Tetrahedron Lett.* **1968**, 4269.

[390]Werstiuk, N.H.; Roy, C.D. *Tetrahedron Lett.* **2001**, *42*, 3255.

Ketenes can be prepared by dehalogenation of α-halo acyl halides with zinc or with triphenylphosphine.[391] The reaction generally gives good results when the two R groups are aryl or alkyl, but not when either one is hydrogen.[392]
    OS **IV**, 348; **VIII**, 377.

17-24   Elimination of a Halogen and a Hetero Group

Alkoxy-halo-elimination

$$\overset{|}{\underset{X}{C}}-\overset{|}{\underset{OR}{C}} \quad \xrightarrow{Zn} \quad \overset{\diagdown}{\diagup}C=C\overset{\diagup}{\diagdown}$$

The elimination of OR and halogen from β-halo ethers is called the *Boord reaction*. It can be carried out with zinc, magnesium, sodium, or certain other reagents.[393] The yields are high and the reaction is of broad scope. β-Halo acetals readily yield vinylic ethers

$$X-\overset{|}{\underset{|}{C}}-\overset{|}{C}(OR)_2 \longrightarrow -\overset{|}{C}=\overset{|}{C}-OR$$

and 2 equivalents of SmI_2 in HMPA is effective.[394] Besides β-halo ethers, the reaction can also be carried out on compounds of the formula

$$Z-\overset{|}{\underset{|}{C}}-\overset{|}{\underset{|}{C}}-Z$$

where X is halogen and Z is OCOR, OTs,[395] NR_2,[396] or SR.[397] When X = Cl and Z = OAc, heating in THF with an excess of SmI_2 followed by treatment with dilute aq. HCl gives an alkene.[398] When Z = I and the other Z is an oxygen of an oxazolone (a carbamate unit), heating with indium metal in methanol leads to an allylic amine.[399] The Z group may also be OH, but then X is limited to Br and I.[400] Like **17-22**, this method ensures that the new double bond will be in a specific position.

[391]Darling, S.D.; Kidwell, R.L. *J. Org. Chem.* **1968**, *33*, 3974.

[392]For a procedure that gives 60–65% yields when one R = H, see McCarney, C.C.; Ward, R.S. *J. Chem. Soc. Perkin Trans. 1* **1975**, 1600. See also, Masters, A.P.; Sorensen, T.S.; Ziegler, T. *J. Org. Chem.* **1986**, *51*, 3558.

[393]See Larock, R.C. *Comprehensive Organic Transformations*, 2nd ed., Wiley-VCH, NY, **1999**, pp. 263–267, for reagents that produce olefins from β-halo ethers and esters, and from halohydrins.

[394]Park, H.S.; Kim, S.H.; Park, M.Y.; Kim, Y.H. *Tetrahedron Lett.* **2001**, *42*, 3729.

[395]Cristol, S.J.; Rademacher, L.E. *J. Am. Chem. Soc.* **1959**, *81*, 1600; Reeve, W.; Brown, R.; Steckel, T.F. *J. Am. Chem. Soc.* **1971**, *93*, 4607.

[396]Gurien, H. *J. Org. Chem.* **1963**, *28*, 878.

[397]Amstutz, E.D. *J. Org. Chem.* **1944**, *9*, 310.

[398]Concellón, J.M.; Bernad, P.L.; Bardales, E. *Org. Lett.* **2001**, *3*, 937.

[399]Yadav, J.S.; Bandyopadhyay, A.; Reddy, B.V.S. *Tetrahedron Lett.* **2001**, *42*, 6385.

[400]Concellón, J.M.; Pérez-Andrés, J.A.; Rodríguez-Solla, H. *Chem. Eur. J.* **2001**, *7*, 3062.

The fact that magnesium causes elimination in these cases limits the preparation of Grignard reagents from these compounds. It has been shown that treatment of β-halo ethers and esters with zinc gives nonstereospecific elimination,[401] so the mechanism was not E2. An E1cB mechanism was postulated because of the poor leaving-group ability of OR and OCOR. Bromohydrins can be converted to alkenes (elimination of Br, OH) in high yields by treatment with $LiAlH_4$–$TiCl_3$.[402]

OS **III**, 698, **IV**, 748; **VI**, 675.

FRAGMENTATIONS

When carbon is the positive leaving group (the electrofuge) in an elimination, the reaction is called *fragmentation*.[403] These processes occur on substrates of the form W–C–C–X, where X is a normal nucleofuge (e.g., halogen, OH_2^+, OTs, NR_3^+) and W is a positive-carbon electrofuge. In most of the cases, W is HO–C– or R_2N–C–, so that the positive charge on the carbon atom is stabilized by the unshared pair of the oxygen or nitrogen, for example,

The mechanisms are mostly E1 or E2. We will discuss only a few fragmentations, since many are possible and not much work has been done on most of them. Reactions **17-25**–**17-28** and **17-30** may be considered fragmentations (see also **19-12** and **19-13**).

17-25 1,3-Fragmentation of γ-Amino, γ-Hydroxy Halides, and 1,3-Diols

Dialkylaminoalkyl-halo-elimination, and so on

Hydroxyalkyl-hydroxy-elimination

---

[401] House, H.O.; Ro, R.S. *J. Am. Chem. Soc.* **1965**, *87*, 838.

[402] McMurry, J.E.; Hoz, T. *J. Org. Chem.* **1975**, *40*, 3797.

[403] For reviews, see Becker, K.B.; Grob, C.A., in Patai, S. *The Chemistry of Functional Groups, Supplement A*, pt. 2, Wiley, NY, **1977**, pp. 653–723; Grob, C.A. *Angew. Chem. Int. Ed.* **1969**, *8*, 535; Grob, C.A.; Schiess, P.W. *Angew. Chem. Int. Ed.* **1967**, *6*, 1.

γ-Dialkylamino halides undergo fragmentation when heated with water to give an alkene and an iminium salt, which under the reaction conditions is hydrolyzed to an aldehyde or ketone (**16-2**).[404] γ-Hydroxy halides and tosylates are fragmented with base. In this instance, the base does not play its usual role in elimination reactions, but instead serves to remove a proton from the OH group, which enables the carbon leaving group to come off more easily:

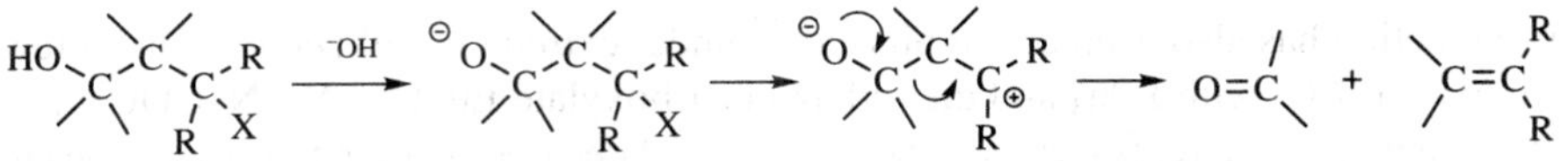

The mechanism of these reactions is often E1. However, in at least some cases, an E2 mechanism operates.[405] It has been shown that stereoisomers of cyclic γ-amino halides and tosylates in which the two leaving groups can assume an antiperiplanar conformation react by the E2 mechanism, while those isomers in which the groups cannot assume such a conformation either fragment by the E1 mechanism or do not undergo fragmentation at all, but in either case give rise to side products characteristic of carbocations.[406]

γ-Dialkylamino alcohols do not give fragmentation, since for ionization the OH group must be converted to OH_2^+ and this would convert NR_2 to NR_2H^+, which does not have the unshared pair necessary to form the double bond with the carbon.[407]

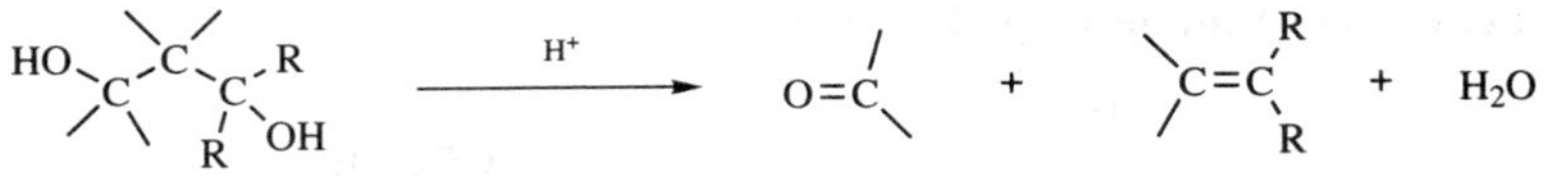

1,3-Diols in which at least one OH group is tertiary or is located on a carbon with aryl substituents can be cleaved by acid treatment.[408] The reaction is most useful synthetically when at least one of the OH groups is on a ring.[409]

17-26   Decarboxylation of β-Hydroxy Carboxylic Acids and of β-Lactones

Carboxy-hydroxy-elimination

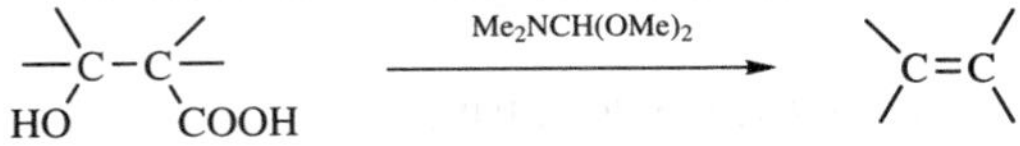

An OH and a COOH group can be eliminated from β-hydroxy carboxylic acids by refluxing with excess dimethylformamide dimethyl acetal.[410] Mono-, di-, tri-, and tetrasubstituted alkenes have been prepared by this method in good yields.[411]

[404]Grob, C.A.; Ostermayer, F.; Raudenbusch, W. *Helv. Chim. Acta* **1962**, *45*, 1672.

[405]Fischer, W.; Grob, C.A. *Helv. Chim. Acta* **1978**, *61*, 2336, and references cited therein.

[406]Geisel, M.; Grob, C.A.; Wohl, R.A. *Helv. Chim. Acta* **1969**, *52*, 2206, and references cited therein.

[407]Grob, C.A.; Hoegerle, R.M.; Ohta, M. *Helv. Chim. Acta* **1962**, *45*, 1823.

[408]Zimmerman, H.E.; English, Jr., J. *J. Am. Chem. Soc.* **1954**, *76*, 2285, 2291, 2294.

[409]For a review of such cases, see Caine, D. *Org. Prep. Proced. Int.* **1988**, *20*, 1.

[410]Hara, S.; Taguchi, H.; Yamamoto, H.; Nozaki, H. *Tetrahedron Lett.* **1975**, 1545.

[411]For a 1,4 example of this reaction, see Rüttimann, A.; Wick, A.; Eschenmoser, A. *Helv. Chim. Acta* **1975**, *58*, 1450.

There is evidence that the mechanism involves E1 or E2 elimination from the zwit-terionic intermediate[412]

$$\ominus O_2C-\overset{|}{\underset{|}{C}}-\overset{|}{\underset{|}{C}}-OC\overset{\oplus}{=}NMe_2$$

The reaction has also been accomplished[413] under extremely mild conditions (a few seconds at 0°C) with PPh$_3$ and diethyl azodicarboxylate EtOOC–N=N–COOEt.[414] In a related procedure, β-lactones undergo thermal decarboxylation to give alkenes in high yields. The reaction has been shown to be a stereospecific syn-elimination.[415] There is evidence that this reaction also involves a zwitterionic intermediate.[416]

There are no OS references, but see OS **VII**, 172, for a related reaction.

17-27 Fragmentation of α,β-Epoxy Hydrazones

Eschenmoser–Tanabe ring cleavage

Cyclic α,β-unsaturated ketones[417] can be cleaved by treatment with base of their epoxy tosylhydrazone derivatives to give acetylenic ketones.[418] The reaction can be applied to the formation of acetylenic aldehydes (R = H) by using the

[412]Mulzer, J.; Brüntrup, G. *Tetrahedron Lett.* **1979**, 1909.

[413]For another method, see Tanzawa, T.; Schwartz, J. *Organometallics* **1990**, 9, 3026.

[414]Mulzer, J.; Brüntrup, G. *Angew. Chem. Int. Ed.* **1977**, 16, 255; Mulzer, J.; Lammer, O. *Angew. Chem. Int. Ed.* **1983**, 22, 628.

[415]Noyce, D.S.; Banitt, E.H. *J. Org. Chem.* **1966**, 31, 4043; Adam, W.; Baeza, J.; Liu, J. *J. Am. Chem. Soc.* **1972**, 94, 2000; Krapcho, A.P.; Jahngen, Jr., E.G.E. *J. Org. Chem.* **1974**, 39, 1322, 1650; Mageswaran, S.; Sultanbawa, M.U.S. *J. Chem. Soc. Perkin Trans. 1* **1976**, 884; Adam, W.; Martinez, G.; Thompson, J.; Yany, F. *J. Org. Chem.* **1981**, 46, 3359.

[416]Mulzer, J.; Zippel, M.; Brüntrup, G. *Angew. Chem. Int. Ed.* **1980**, 19, 465; Mulzer, J.; Zippel, M. *Tetrahedron Lett.* **1980**, 21, 751. See also, Moyano, A.; Pericàs, M.A.; Valentí, E. *J. Org. Chem.* **1989**, 573.

[417]For other methods of fragmentation of an α,β-epoxy ketone derivatives, see MacAlpine, G.A.; Warkentin, J. *Can. J. Chem.* **1978**, 56, 308, and references cited therein.

[418]Eschenmoser, A.; Felix, D.; Ohloff, G. *Helv. Chim. Acta* **1967**, 50, 708; Tanabe, M.; Crowe, D.F.; Dehn, R.L.; Detre, G. *Tetrahedron Lett.* **1967**, 3739; Tanabe, M.; Crowe, D.F.; Dehn, R.L. *Tetrahedron Lett.* **1967**, 3943.

corresponding, 2,4-dinitro-tosylhydrazone derivatives.[419] Hydrazones (e.g., **48**) prepared from epoxy ketones and ring-substituted *N*-aminoaziridines undergo similar fragmentation when heated.[420]

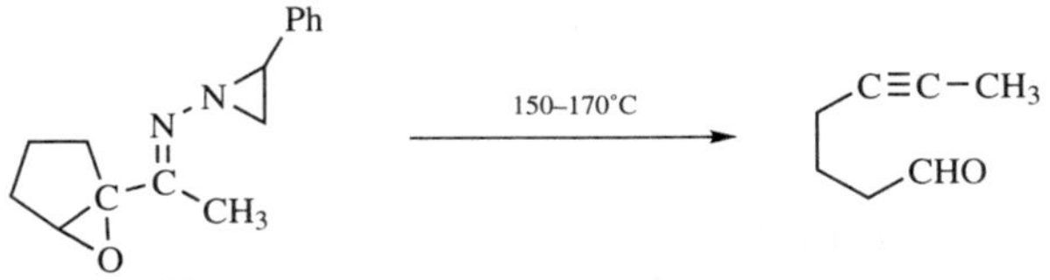

48

OS **VI**, 679.

17-28   Elimination of CO and CO_2 from Bridged Bicyclic Compounds

seco-**Carbonyl-1/4/elimination**

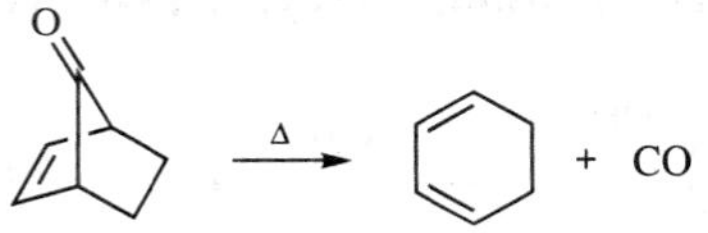

49

On heating, bicyclo[2.2.1]hept-2,3-en-17-ones (**49**) usually lose CO to give cyclohexadienes,[421] in a type of reverse Diels–Alder reaction. Bicyclo[2.2.1]hepta-dienones (**50**) undergo the reaction so readily (because of the

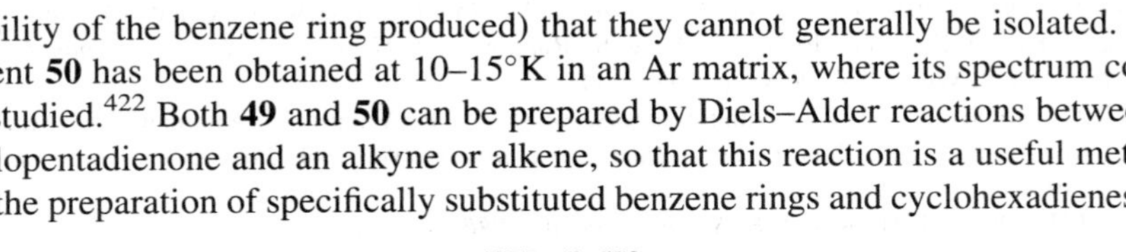

50

stability of the benzene ring produced) that they cannot generally be isolated. The parent **50** has been obtained at 10–15°K in an Ar matrix, where its spectrum could be studied.[422] Both **49** and **50** can be prepared by Diels–Alder reactions between a cyclopentadienone and an alkyne or alkene, so that this reaction is a useful method for the preparation of specifically substituted benzene rings and cyclohexadienes.[423]

[419]Corey, E.J.; Sachdev, H.S. *J. Org. Chem.* **1975**, *40*, 579.

[420]Felix, D.; Müller, R.K.; Horn, U.; Joos, R.; Schreiber, J.; Eschenmoser, A. *Helv. Chim. Acta* **1972**, *55*, 1276.

[421]For a review, see Stark, B.P.; Duke, A.J. *Extrusion Reactions*, Pergamon, Elmsford, NY, **1967**, pp. 16–46.

[422]LeBlanc, B.F.; Sheridan, R.S. *J. Am. Chem. Soc.* **1985**, *107*, 4554; Birney, D.M.; Wiberg, K.B.; Berson, J.A. *J. Am. Chem. Soc.* **1988**, *110*, 6631.

[423]For a review with many examples; see Ogliaruso, M.A.; Romanelli, M.G.; Becker, E.I. *Chem. Rev.* **1965**, *65*, 261, 300–348. For references to this and related reactions, see Larock, R.C. *Comprehensive Organic Transformations*, 2nd ed., Wiley-VCH, NY, **1999**, pp. 207–213.

Unsaturated bicyclic lactones of the type **51** can also undergo the reaction, losing CO_2 (see also **17-35**).

51

OS **III**, 807; **V**, 604, 1037.

Reversal of the Diels–Alder reaction may be considered a fragmentation (see **15-50**).

REACTIONS IN WHICH C≡N OR C=N BONDS ARE FORMED

17-29   Dehydration of Oximes and Similar Compounds

C-Hydro-*N*-hydroxy-elimination; *C*-Acyl-*N*-hydroxy-elimination

Aldoximes can be dehydrated to nitriles[424] by many dehydrating agents, of which acetic anhydride is the most common. Among reagents that are effective under mild conditions[425] (room temperature) are $Ph_3P–CCl_4$,[426] SeO_2,[427] Me_2t-BuSiCl/imidazole,[428] ferric sulfate,[429] $SOCl_2$/benzotriazole,[430] $TiCl_3(OTf)$,[431] CS_2, and Amberlyst A26 (^-OH),[432] Montmorillonite KSF clay,[433] (*S,S*)-dimethyl dithiocarbonates,[434] and chloromethylene dimethylammonium chloride $Me_2N=CHCl^+$ Cl^-.[435] Heating an oxime with a ruthenium catalyst gives the nitrile.[436] Heating with the *Burgess reagent* [Et_3N^+ $^-SO_2N–CO_2Me$] in polyethylene glycol

---

[424]For reviews, see Friedrich, K., in Patai, S.; Rappoport, Z. *The Chemistry of the Carbon–Carbon Triple Bond*, pt. 2, Wiley, NY, *1978*, pp. 1345–1390; Friedrich, K.; Wallenfels, K., in Rappoport, Z. *The Chemistry of the Cyano Group*, Wiley, NY, *1970*, pp. 92–96. For a review of methods of synthesizing nitriles, see Fatiadi, K., in Friedrich, K. in Patai, S.; Rappoport, Z. *The Chemistry of the Carbon–Carbon Triple Bond*, pt. 2, Wiley, NY, *1978*, pp. 1057–1303.

[425]For lists of some other reagents, with references, see Molina, P.; Alajarin, M.; Vilaplana, M.J. *Synthesis* *1982*, 1016; Attanasi, O.; Palma, P.; Serra-Zanetti, F. *Synthesis 1983*, 741; Jurš ić, B. *Synth. Commun.* *1989*, *19*, 689.

[426]Kim, J.N.; Chung, K.H.; Ryu, E.K. *Synth. Commun. 1990*, *20*, 2785.

[427]Shinozaki, H.; Imaizumi, M.; Tajima, M. *Chem. Lett. 1983*, 929.

[428]Ortiz-Marciales, M.; Piñero, L.; Ufret, L.; Algarín, W.; Morales, J. *Synth. Commun. 1998*, *28*, 2807.

[429]Desai, D.G.; Swami, S.S.; Mahale, G.D. *Synth. Commun. 2000*, *30*, 1623.

[430]Chaudhari, S.S.; Akamanchi, K.G. *Synth. Commun. 1999*, *29*, 1741.

[431]Iranpoor, N.; Zeynizadeh, B. *Synth. Commun. 1999*, *29*, 2747.

[432]Tamami, B.; Kiasat, A.R. *Synth. Commun. 2000*, *30*, 235.

[433]Meshram, H.M. *Synthesis 1992*, 943.

[434]Khan, T.A.; Peruncheralathan, S.; Ila, H.; Junjappa, H. *Synlett 2004*, 2019.

[435]See Shono, T.; Matsumura, Y.; Tsubata, K.; Kamada, T.; Kishi, K. *J. Org. Chem. 1989*, *54*, 2249.

[436]Yang, S.H.; Chang, S. *Org. Lett. 2001*, *3*, 4209.

is effective for this transformation.[437] Microwave irradiation on EPZ-10[438] or sulfuric acid impregnated silica gel[439] gives the nitrile, as does microwave irradiation of an oxime with tetrachloropyridine on alumina.[440] Aldehydes can be converted to oximes *in situ* and microwave irradiation on alumina[441] or with ammonium acetate[442] gives the nitrile. Solvent-free reactions are known.[443] Electrochemical synthesis has also been used.[435] The reaction is most successful when the H and OH are anti. Various alkyl and acyl derivatives of aldoximes, for example, $RCH=NOR$, $RCH=NOCOR$, $RCH=NOSO_2Ar$, and so on, also give nitriles, as do chlorimines $RCH=NCl$ (the latter with base treatment).[444] *N,N*-Dichloro derivatives of primary amines give nitriles on pyrolysis: $RCH_2NCl_2 \rightarrow RCN$.[445]

$$\underset{R}{\overset{\overset{\displaystyle \overset{\oplus}{N}-NR_3}{\parallel}}{\underset{\displaystyle C}{}}}\overset{}{\underset{H}{}} \quad \xrightarrow[\text{or DBU}]{^-OEt} \quad R-C≡N \ + \ NR_3 \ + \ EtOH$$

Quaternary hydrazonium salts (derived from aldehydes) give nitriles when treated with ^-OEt[446] or DBU (p. 1132):[447] as do dimethylhydrazones, $RCH=NNMe_2$, when treated with Et_2NLi and HMPA.[448] All these are methods of converting aldehyde derivatives to nitriles. For the conversion of aldehydes directly to nitriles, without isolation of intermediates (see **16-16**).

Hydroxylamines that have an α-proton are converted to nitrones when treated with a manganese salen complex.[449]

$$\underset{R}{\overset{\overset{\displaystyle N-OH}{\parallel}}{\underset{\displaystyle C}{}}}\overset{}{\underset{\overset{\displaystyle C}{\underset{\displaystyle \overset{\parallel}{O}}{}}}{}}R' \quad \xrightarrow{SOCl_2} \quad R-C≡N \ + \ R'COO^{\ominus}$$

Certain ketoximes can be converted to nitriles by the action of proton or Lewis acids.[450] Among these are oximes of α-diketones (illustrated above), α-keto acids,

[437]Miller, C.P.; Kaufman, D.H. *Synlett* **2000**, 1169.

[438]Bandgar, B.P.; Sadavarte, V.S.; Sabu, K.R. *Synth. Commun.* **1999**, *29*, 3409.

[439]Kumar, H.M.S.; Mohanty, P.K.; Kumar, M.S.; Yadav, J.S. *Synth. Commun.* **1997**, *27*, 1327; Sarvari, M.H. *Synthesis* **2005**, 787.

[440]Lingaiah, N.; Narender, R. *Synth. Commun.* **2002**, *32*, 2391.

[441]Bose, D.S.; Narsaiah, A.V. *Tetrahedron Lett.* **1998**, *39*, 6533.

[442]Das, B.; Ramesh, C.; Madhusudhan, P. *Synlett* **2000**, 1599.

[443]See Sharghi, H.; Sarvari, M.H. *Synthesis* **2003**, 243.

[444]Hauser, C.R.; Le Maistre, J.W.; Rainsford, A.E. *J. Am. Chem. Soc.* **1935**, *57*, 1056.

[445]Roberts, J.T.; Rittberg, B.R.; Kovacic, P. *J. Org. Chem.* **1981**, *46*, 4111.

[446]Smith, R.F.; Walker, L.E. *J. Org. Chem.* **1962**, *27*, 4372; Grandberg, I.I. *J. Gen. Chem. USSR*, **1964**, *34*, 570; Grundon, M.F.; Scott, M.D. *J. Chem. Soc.* **1964**, 5674; Ioffe, B.V.; Zelenina, N.L. *J. Org. Chem. USSR*, **1968**, *4*, 1496.

[447]Moore, J.S.; Stupp, S.I. *J. Org. Chem.* **1990**, *55*, 3374.

[448]Cuvigny, T.; Le Borgne, J.F.; Larchevêque, M.; Normant, H. *Synthesis* **1976**, 237.

[449]Cicchi, S.; Cardona, F.; Brandi, A.; Corsi, M.; Goti, A. *Tetrahedron Lett.* **1999**, *40*, 1989.

[450]For reviews, see Gawley, R.E. *Org. React.* **1988**, *35*, 1; Conley, R.T.; Ghosh, S. *Mech. Mol. Migr.* **1971**, *4*, 197, 197–251; McCarty, C.G., in Patai, S. *The Chemistry of the Carbon–Nitrogen Double Bond*, Wiley, NY, **1970**, pp. 416–439; Casanova, J., in Rappoport, Z. *The Chemistry of the Cyano Group*, Wiley, NY, **1970**, pp. 915–932.

α-dialkylamino ketones, α-hydroxy ketones, β-keto ethers, and similar compounds.[451] These are fragmentation reactions, analogous to **17-25**. For example, α-dialkylamino ketoximes also give amines and aldehydes or ketones besides nitriles:[452]

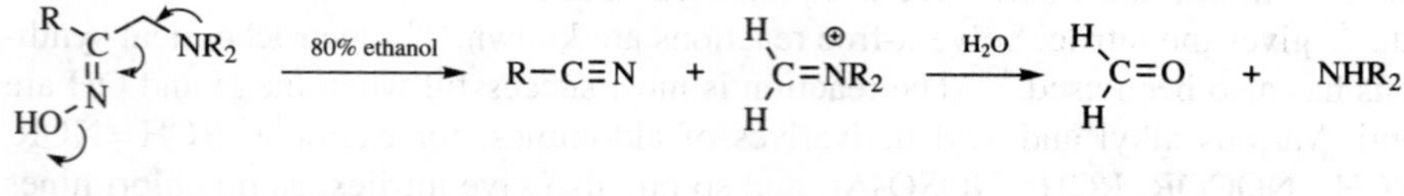

The reaction that normally occurs on treatment of a ketoxime with a Lewis or proton acid is the Beckmann rearrangement (**18-17**); fragmentations are considered side reactions, often called "abnormal" or "second-order" Beckmann rearrangements.[453] Obviously, the substrates mentioned are much more susceptible to fragmentation than are ordinary ketoximes, since in each case an unshared pair is available to assist in removal of the group cleaving from the carbon. However, fragmentation is a side reaction even with ordinary ketoximes[454] and, in cases where a particularly stable carbocation can be cleaved, may be the main reaction:[455]

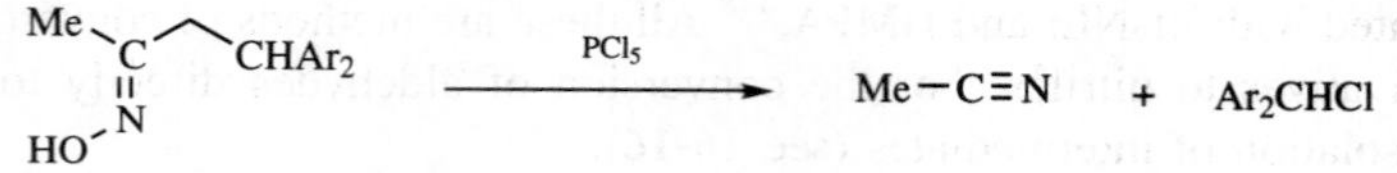

There are indications that the mechanism at least in some cases first involves a rearrangement and then cleavage. The ratio of fragmentation to Beckmann rearrangement of a series of oxime tosylates, RC(=NOTs)Me, was not related to the solvolysis rate but *was* related to the stability of R^+ (as determined by the solvolysis rate of the corresponding RCl), which showed that fragmentation did not take place in the rate-determining step.[456] It may be postulated then that the first step in the fragmentation and in the rearrangement is the same and that this is the rate-determining step. The product is determined in the second step:

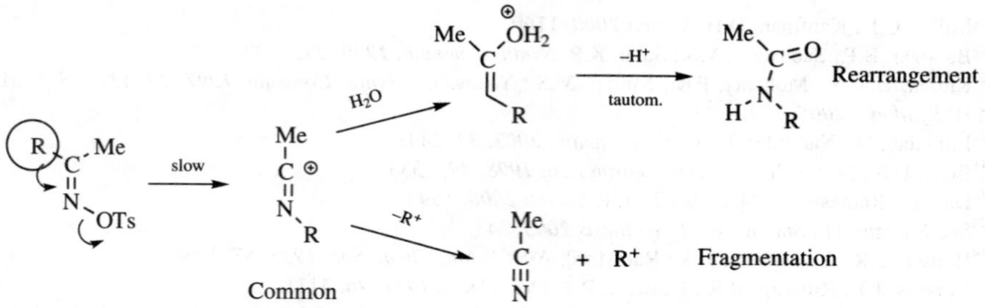

[451]For more complete lists with references, see Olah, G.A.; Vankar, Y.D.; Berrier, A.L. *Synthesis* **1980**, 45; Conley, R.T.; Ghosh, S. *Mech. Mol. Migr.* **1971**, 4, 197.

[452]Fischer, H.P.; Grob, C.A.; Renk, E. *Helv. Chim. Acta* **1962**, 45, 2539; Fischer, H.P.; Grob, C.A. *Helv. Chim. Acta* **1963**, 46, 936.

[453]See the discussion in Ferris, A.F. *J. Org. Chem.* **1960**, 25, 12.

[454]See, for example, Hill, R.K.; Conley, R.T. *J. Am. Chem. Soc.* **1960**, 82, 645.

[455]Hassner, A.; Nash, E.G. *Tetrahedron Lett.* **1965**, 525.

[456]Grob, C.A.; Fischer, H.P.; Raudenbusch, W.; Zergenyi, J. *Helv. Chim. Acta* **1964**, 47, 1003.

However, in other cases the simple E1 or E2 mechanisms operate.[457]
OS **V**, 266; **IX**, 281; OS **II**, 622; **III**, 690.

17-30   Dehydration of Unsubstituted Amides

N,N-Dihydro-*C*-oxo-bielimination

$$\underset{\underset{R}{\overset{\displaystyle\parallel}{\overset{\textstyle O}{}}}{C}{-}NH_2} \xrightarrow{P_2O_5} R{-}C{\equiv}N$$

Unsubstituted amides can be dehydrated to nitriles.[458] Phosphorous pentoxide is the most common dehydrating agent for this reaction, but many others, including POCl$_3$, PCl$_5$, CCl$_4$-Ph$_3$P,[459] HMPA,[460] LiCl with a zirconium catalyst,[461] MeOOCNSO$_2$NEt$_3$ (the Burgess reagent),[462] Me$_2$N=CHCl$^+$ Cl$^-$,[463] AlCl$_3$/KI/ H$_2$O,[464] Bu$_2$SnO with microwave irradiation,[465] PPh$_3$/NCS,[466] triflic anhydride,[467] oxalyl chloride/DMSO/$-78°C$[468] (Swern conditions, see **19-3**), and SOCl$_2$ have also been used.[469] Heating an amide with paraformaldehyde and formic acid gives the nitrile.[470] Treatment with benzotriazol-1-yloxytris(pyrrolidino)phosphonium hexafluorophosphate converts amides to nitriles.[471] It is possible to convert an acid to the nitrile, without isolation of the amide, by heating its ammonium salt with the dehydrating agent,[472] or by other methods.[473] Acyl halides can also be directly converted to nitriles by heating with sulfamide (NH$_2$)$_2$SO$_2$.[474] The reaction may be formally looked on as a β-elimination from the enol form of the amide RC(OH)=NH, in which case it is like **17-29**, except that H and OH have changed

[457]Ahmad, A.; Spenser, I.D. *Can. J. Chem.* **1961**, *39*, 1340; Ferris, A.F.; Johnson, G.S.; Gould, F.E. *J. Org. Chem.* **1960**, *25*, 1813; Grob, C.A.; Sieber, A. *Helv. Chim. Acta* **1967**, *50*, 2520; Green, M.; Pearson, S.C. *J. Chem. Soc. B* **1969**, 593.

[458]For reviews, see Bieron J.F.; Dinan, F.J., in Zabicky, J. *The Chemistry of Amides*, Wiley, NY, **1970**, pp. 274–283; Friedrich, K.; Wallenfels, K., in Rappoport, Z. *The Chemistry of the Cyano Group*, Wiley, NY, **1970**, pp. 96–103; Friedrich, K., in Patai, S.; Rapoport, Z. *The Chemistry of Functional Groups, Supplement C*, pt. 2, Wiley, NY, **1978**, p. 1345.

[459]Yamato, E.; Sugasawa, S. *Tetrahedron Lett.* **1970**, 4383; Appel, R.; Kleinstück, R.; Ziehn, K. *Chem. Ber.* **1971**, *104*, 1030; Harrison, C.R.; Hodge, P.; Rogers, W.J. *Synthesis* **1977**, 41.

[460]Monson, R.S.; Priest, D.N. *Can. J. Chem.* **1971**, *49*, 2897.

[461]Ruck, R.T.; Bergman, R.G. *Angew. Chem. Int. Ed.* **2004**, *43*, 5375.

[462]Claremon, D.A.; Phillips, B.T. *Tetrahedron Lett.* **1988**, *29*, 2155.

[463]Barger, T.M.; Riley, C.M. *Synth. Commun.* **1980**, *10*, 479.

[464]Boruah, M.; Konwar, D. *J. Org. Chem.* **2002**, *67*, 7138.

[465]Bose, D.S.; Jayalakshmi, B. *J. Org. Chem.* **1999**, *64*, 1713.

[466]Iranpoor, N.; Firouzabadi, H.; Aghapoor, G. *Synth. Commun.* **2002**, *32*, 2535.

[467]Bose, D.S.; Jayalakshmi, B. *Synthesis* **1999**, 64.

[468]Nakajima, N.; Ubukata, M. *Tetrahedron Lett.* **1997**, *38*, 2099.

[469]For a list of reagents, with references, see Larock, R.C. *Comprehensive Organic Transformations*, 2nd ed., Wiley-VCH, NY, **1999**, pp. 1983–1985.

[470]Heck, M.-P.; Wagner, A.; Mioskowski, C. *J. Org. Chem.* **1996**, *61*, 6486.

[471]Bose, D.S.; Narsaiah, A.V. *Synthesis* **2001**, 373.

[472]See, for example, Imamoto, T.; Takaoka, T.; Yokoyama, M. *Synthesis* **1983**, 142.

[473]For a list of methods, with references, see Larock, R.C. *Comprehensive Organic Transformations*, 2nd ed., Wiley-VCH, NY, **1999**, pp. 1949–1950.

[474]Hulkenberg, A.; Troost, J.J. *Tetrahedron Lett.* **1982**, *23*, 1505.

places. In some cases, for example, with $SOCl_2$, the mechanism probably *is* through the enol form, with the dehydrating agent forming an ester with the OH group, for example, $RC(OSOCl)=NH$, which undergoes elimination by the E1 or E2 mechanism.[475] *N,N*-Disubstituted ureas give cyanamides ($R_2N–CO–NH_2 \rightarrow R_2N–CN$) when dehydrated with $CHCl_3–NaOH$ under phase-transfer conditions.[476] Treatment of an amide with aqueous NaOH and ultrasound leads to the nitrile.[477]

N-Alkyl-substituted amides can be converted to nitriles and alkyl chlorides by treatment with PCl_5. This is called the *von Braun reaction* (not to be confused with the other von Braun reaction, **10-54**).

$$R'CONHR + PCl_5 \longrightarrow R'CN + RCl$$

OS **I**, 428; **II**, 379; **III**, 493, 535, 584, 646, 768; **IV**, 62, 144, 166, 172, 436, 486, 706; **VI**, 304, 465.

17-31   Conversion of *N*-Alkylformamides to Isonitriles (Isocyanides)

CN-Dihydro-*C*-oxo-bielimination

Isocyanides (isonitriles) can be prepared by elimination of water from *N*-alkylformamides[478] with phosgene and a tertiary amine.[479] Other reagents, among them TsCl in quinoline, $POCl_3$ and a tertiary amine,[480] $Me_2N=CHCl^+ \, Cl^-$,[481] triflic anhydride-$(iPr)_2NEt$,[482] $PhOC(=S)Cl$,[483] and $Ph_3P–CCl_4–Et_3N$[484] have also been employed. Formamides react with thionyl chloride (two sequential treatments) to give an intermediate that gives an isonitrile upon electrolysis in DMF with $LiClO_4$.[485]

A variation of this process uses carbodiimides,[486] which can be prepared by the dehydration of *N,N'*-disubstituted ureas with various dehydrating agents,[487] among

[475]Rickborn, B.; Jensen, F.R. *J. Org. Chem.* **1962**, *27*, 4608.

[476]Schroth, W.; Kluge, H.; Frach, R.; Hodek, W.; Schädler, H.D. *J. Prakt. Chem.* **1983**, *325*, 787.

[477]Sivakumar, M.; Senthilkumar, P.; Pandit, A.B. *Synth. Commun.* **2001**, *31*, 2583.

[478]For a new synthesis see Creedon, S.M.; Crowley, H.K.; McCarthy, D.G. *J. Chem. Soc. Perkin Trans. 1* **1998**, 1015.

[479]For reviews, see Hoffmann, P.; Gokel, G.W.; Marquarding, D.; Ugi, I., in Ugi, I. *Isonitrile Chemistry*, Academic Press, NY, **1971**, pp. 10–17; Ugi, I.; Fetzer, U.; Eholzer, U.; Knupfer, H.; Offermann, K. *Angew. Chem. Int. Ed.* **1965**, *4*, 472; *Newer Methods Prep. Org. Chem.* **1968**, *4*, 37.

[480]See Obrecht, R.; Herrmann, R.; Ugi, I. *Synthesis* **1985**, 400.

[481]Walborsky, H.M.; Niznik, G.E. *J. Org. Chem.* **1972**, *37*, 187.

[482]Baldwin, J.E.; O'Neil, I.A. *Synlett* **1991**, 603.

[483]Bose, D.S.; Goud, P.R. *Tetrahedron Lett.* **1999**, *40*, 747.

[484]Appel, R.; Kleinstück, R.; Ziehn, K. *Angew. Chem. Int. Ed.* **1971**, *10*, 132.

[485]Guirado, A.; Zapata, A.; Gómez, J.L.; Trebalón, L.; Gálvez, J. *Tetrahedron* **1999**, *55*, 9631.

[486]For a review of the reactions in this section, see Bocharov, B.V. *Russ. Chem. Rev.* **1965**, *34*, 212. For a review of carbodiimide chemistry; see Williams, A.; Ibrahim, I.T. *Chem. Rev.* **1981**, *81*, 589.

[487]For some others not mentioned here, see Sakai, S.; Fujinami, T.; Otani, N.; Aizawa, T. *Chem. Lett.* **1976**, 811; Shibanuma, T.; Shiono, M.; Mukaiyama, T. *Chem. Lett.* **1977**, 575; Kim, S.; Yi, K.Y. *J. Org. Chem.* **1986**, *51*, 2613, *Tetrahedron Lett.* **1986**, *27*, 1925.

which are TsCl in pyridine, POCl$_3$, PCl$_5$, P$_2$O$_5$–pyridine, TsCl (with phase-transfer catalysis),[488] and Ph$_3$PBr$_2$–Et$_3$N.[489] Hydrogen sulfide can be removed from the corresponding thioureas by treatment with HgO, NaOCl, or diethyl azodicarboxylate–triphenylphosphine.[490]

OS **V**, 300, 772; **VI**, 620, 751, 987. See also OS **VII**, 27. For the carbodiimide/thiourea dehydration, see OS **V**, 555; **VI**, 951.

REACTIONS IN WHICH C=O BONDS ARE FORMED

Many elimination reactions in which C=O bonds are formed were considered in Chapter 16, along with their more important reverse reactions (also see, **12-40** and **12-41**).

17-32   Pyrolysis of β-Hydroxy Alkenes

O-**Hydro-*C*-allyl-elimination**

When pyrolyzed, β-hydroxy alkenes cleave to give alkenes and aldehydes or ketones.[491] Alkenes produced this way are quite pure, since there are no side reactions. The mechanism has been shown to be pericyclic, primarily by observations that the kinetics are first order[492] and that, for ROD, the deuterium appeared in the allylic position of the new alkene.[493] This mechanism is the reverse of that for the oxygen analog of the ene synthesis (**16-54**). β-Hydroxyacetylenes react similarly to give the corresponding allenes and carbonyl compounds.[494] The mechanism is the same despite the linear geometry of the triple bonds.

[488]Jászay, Z.M.; Petneházy, I.; Töke, L.; Szajáni, B. *Synthesis* **1987**, 520.

[489]Bestmann, H.J.; Lienert, J.; Mott, L. *J.L. Liebigs Ann. Chem.* **1968**, *718*, 24.

[490]Mitsunobu, O.; Kato, K.; Tomari, M. *Tetrahedron* **1970**, *26*, 5731.

[491]Arnold, R.T.; Smolinsky, G. *J. Am. Chem. Soc.* **1959**, *81*, 6643. For a review, see Marvell, E.N.; Whalley, W., in Patai, S. *The Chemistry of the Hydroxyl Group*, pt. 2, Wiley, NY, **1971**, pp. 729–734.

[492]Voorhees, K.J.; Smith, G.G. *J. Org. Chem.* **1971**, *36*, 1755.

[493]Arnold, R.T.; Smolinsky, G. *J. Org. Chem.* **1960**, *25*, 128; Smith, G.G.; Taylor, R. *Chem. Ind. (London)* **1961**, 949.

[494]Viola, A.; Proverb, R.J.; Yates, B.L.; Larrahondo, J. *J. Am. Chem. Soc.* **1973**, *95*, 3609.

In a related reaction, pyrolysis of allylic ethers that contain at least one α hydrogen gives alkenes and aldehydes or ketones. The mechanism is also pericyclic[495]

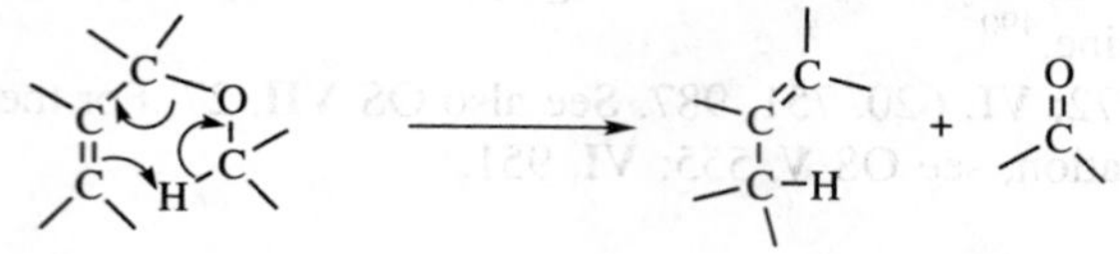

REACTIONS IN WHICH N=N BONDS ARE FORMED

17-33   Eliminations to Give Diazoalkanes

N-Nitrosoamine-diazoalkane transformation

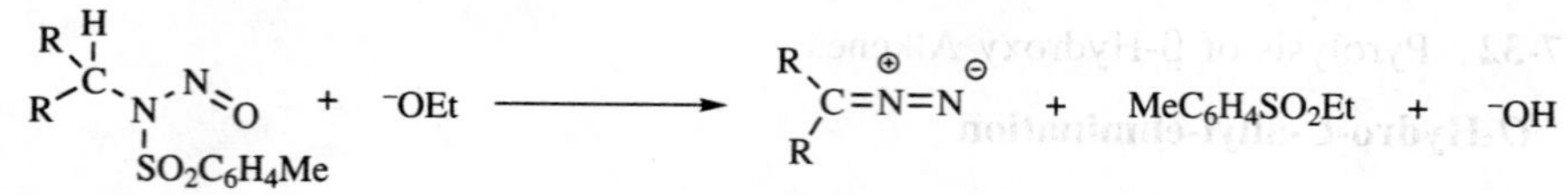

Various *N*-nitroso-*N*-alkyl compounds undergo elimination to give diazoalkanes.[496] One of the most convenient methods for the preparation of diazomethane involves base treatment of *N*-nitroso-*N*-methyl-*p*-toluenesulfonamide (illustrated above, with R = H).[497] However, other compounds commonly used are (base treatment is required in all cases):

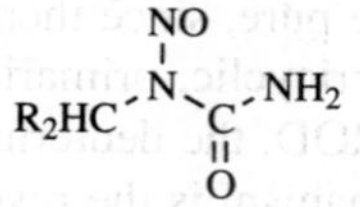

N-Nitroso-*N*-alkylureas 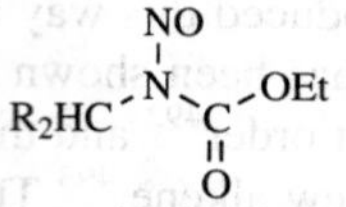   *N*-Nitroso-*N*-alkylcarbamates

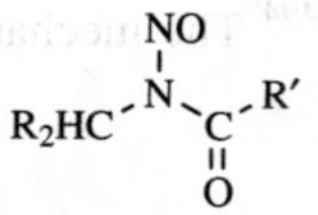

N-Nitroso-*N*-alkyl amides

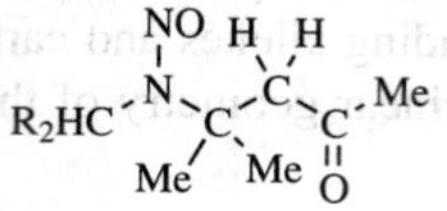

N-Nitroso-*N*-alkyl-4-amino-4-methyl-2-pentanone

All these compounds can be used to prepare diazomethane, although the sulfonamide, which is commercially available, is most satisfactory. *N*-Nitroso-*N*-methylcarbamate and *N*-nitroso-*N*-methylurea give good yields, but are highly irritating and carcinogenic.[498] For higher diazoalkanes the preferred substrates are nitrosoalkylcarbamates.

[495]Cookson, R.C.; Wallis, S.R. *J. Chem. Soc. B* **1966**, 1245; Kwart, H.; Slutsky, J.; Sarner, S.F. *J. Am. Chem. Soc.* **1973**, *95*, 5242; Egger, K.W.; Vitins, P. *Int. J. Chem. Kinet.* **1974**, *6*, 429.

[496]For a review, see Regitz, M.; Maas, G. *Diazo Compounds*; Academic Press, NY, **1986**, pp. 296–325. For a review of the preparation and reactions of diazomethane, see Black, T.H. *Aldrichimica Acta* **1983**, *16*, 3. For discussions, see Cowell, G.W.; Ledwith, A. *Q. Rev. Chem. Soc.* **1970**, *24*, 119, pp. 126–131; Smith, P.A.S. *Open-chain Nitrogen Compounds*; W. A. Benjamin, NY, **1966**, especially pp. 257–258, 474–475, in Vol. 2.

[497]de Boer, T.J.; Backer, H.J. *Org. Synth.* **IV**, 225, 250; Hudlicky, M. *J. Org. Chem.* **1980**, *45*, 5377.

[498]Searle, C.E. *Chem. Br.* **1970**, *6*, 5.

Most of these reactions probably begin with a 1,3 nitrogen-to-oxygen rearrangement, followed by the actual elimination (illustrated for the carbamate):

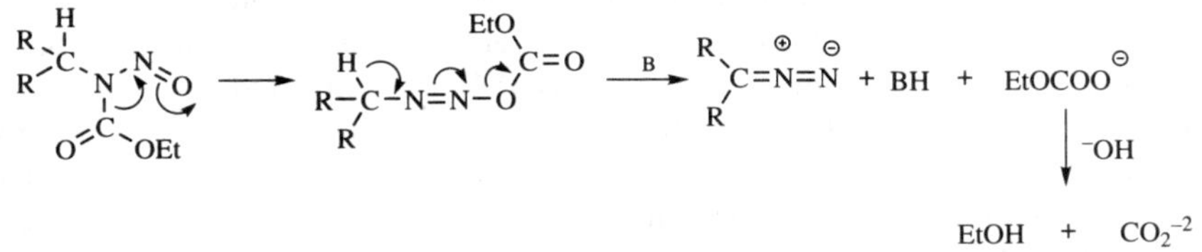

$$EtOH \quad + \quad CO_2^{-2}$$

OS **II**, 165; **III**, 119, 244; **IV**, 225, 250; **V**, 351; **VI**, 981.

EXTRUSION REACTIONS

We consider an *extrusion reaction*[499] to be one in which an atom or group Y connected to two other atoms X and Z is lost from a molecule, leading to a product in which X is bonded directly to Z.

$$X-Y-Z \longrightarrow X-Z + Y$$

Reactions **14-32** and **17-20** also fit this definition. Reaction **17-28** does not fit the definition, but is often also classified as an extrusion reaction. An extrusibility scale has been developed, showing that the ease of extrusion of the common Y groups is in the order: $-N=N- > -COO- > -SO_2- > -CO-$.[500]

17-34   Extrusion of N_2 from Pyrazolines, Pyrazoles, and Triazolines

Azo-extrusion

52   53

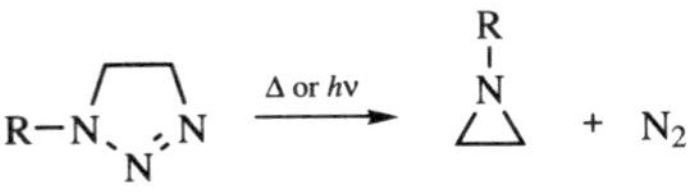

54

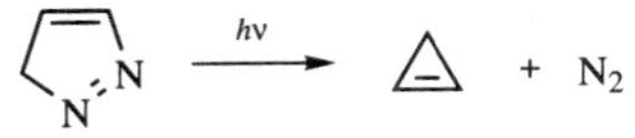

55

---

[499]For a monograph, see Stark, B.P.; Duke, A.J. *Extrusion Reactions*, Pergamon, Elmsford, NY, *1967*. For a review of extrusions that are photochemically induced, see Givens, R.S. *Org. Photochem.* **1981**, *5*, 227.
[500]Paine, A.J.; Warkentin, J. *Can. J. Chem.* **1981**, *59*, 491.

1-Pyrazolines (**52**) can be converted to cyclopropane and N_2 on photolysis[501] or pyrolysis.[502] The tautomeric 2-pyrazolines (**53**), which are more stable than **52** also give the reaction, but in this case an acidic or basic catalyst is required, the function of which is to convert **53** to **52**.[503] In the absence of such catalysts, **53** do not react.[504] In a similar manner, triazolines (**54**) are converted to aziridines.[505] Side reactions are frequent with both **52** and **54**, and some substrates do not give the reaction at all. However, the reaction has proved synthetically useful in many cases. In general, photolysis gives better yields and fewer side reactions than pyrolysis with both **52** and **54**. 3H-Pyrazoles[506] (**55**) are stable to heat, but in some cases can be converted to cyclopropenes on photolysis,[507] although in other cases other types of products are obtained.

There is much evidence that the mechanism[508] of the 1-pyrazoline reactions generally involves diradicals, although the mode of formation and detailed structure (e.g., singlet vs. triplet) of these radicals may vary with the substrate and reaction conditions. The reactions of the 3H-pyrazoles have been postulated to proceed through a diazo compound that loses N_2 to give a vinylic carbene.[509]

OS **V**, 96, 929. See also, OS **VIII**, 597.

[501]Van Auken, T.V.; Rinehart Jr., K.L. *J. Am. Chem. Soc.* **1962**, *84*, 3736.

[502]For reviews of the reactions in this section, see Adam, W.; De Lucchi, O. *Angew. Chem. Int. Ed.* **1980**, *19*, 762; Meier, H.; Zeller, K. *Angew. Chem. Int. Ed.* **1977**, *16*, 835; Stark, B.P.; Duke, A.J. *Extrusion Reactions*, Pergamon, Elmsford, NY, **1967**, pp. 116–151. For a review of the formation and fragmentation of cyclic azo compounds, see Mackenzie, K., in Patai, S. *The Chemistry of the Hydrazo, Azo, and Azoxy Groups*, pt. 1, Wiley, NY, **1975**, pp. 329–442.

[503]For example, see Jones, W.M.; Sanderfer, P.O.; Baarda, D.G. *J. Org. Chem.* **1967**, *32*, 1367.

[504]McGreer, D.E.; Wai, W.; Carmichael, G. *Can. J. Chem.* **1960**, *38*, 2410; Kocsis K.; Ferrini, P.G.; Arigoni, D.; Jeger, O. *Helv. Chim. Acta* **1960**, *43*, 2178.

[505]For a review, see Scheiner, P. *Sel. Org. Transform.* **1970**, *1*, 327.

[506]For a review of 3H-pyrazoles, see Sammes, M.P.; Katritzky, A.R. *Adv. Heterocycl. Chem.* **1983**, *34*, 2.

[507]Ege, G.*Tetrahedron Lett.* **1963**, 1667; Closs, G.L.; Böll, W.A.; Heyn, H.; Dev, V. *J. Am. Chem. Soc.* **1968**, *90*, 173; Franck-Neumann, M.; Buchecker, C. *Tetrahedron Lett.* **1969**, 15; Pincock, J.A.; Morchat, R.; Arnold, D.R. *J. Am. Chem. Soc.* **1973**, *95*, 7536.

[508]For a review of the mechanism; see Engel, P.S. *Chem. Rev.* **1980**, *80*, 99. See also, Engel, P.S.; Nalepa, C.J. *Pure Appl. Chem.* **1980**, *52*, 2621; Engel, P.S.; Gerth, D.B. *J. Am. Chem. Soc.* **1983**, *105*, 6849; Reedich, D.E.; Sheridan, R.S. *J. Am. Chem. Soc.* **1988**, *110*, 3697.

[509]Closs, G.L.; Böll, W.A.; Heyn, H.; Dev, V. *J. Am. Chem. Soc.* **1968**, *90*, 173; Pincock, J.A.; Morchat, R.; Arnold, D.R. *J. Am. Chem. Soc.* **1973**, *95*, 7536.

17-35 Extrusion of CO or CO$_2$

Carbonyl-extrusion

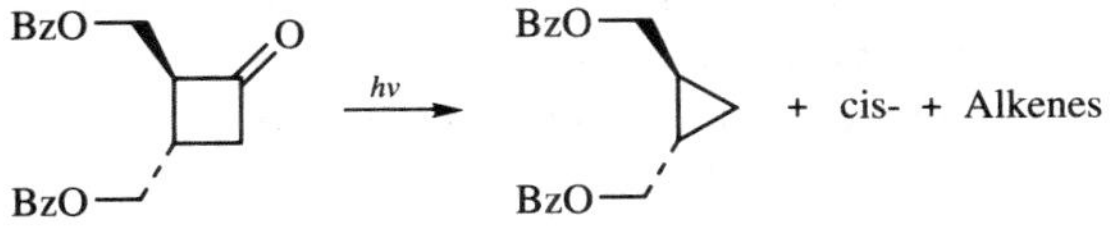

56   **57**

Although the reaction is not general, certain cyclic ketones can be photolyzed to give ring-contracted products.[510] In the example above, the cyclobutanone **56** was photolyzed to give **57**.[511] This reaction was used to synthesize tetra-*tert*-butyltetrahedrane, **58**.[512]

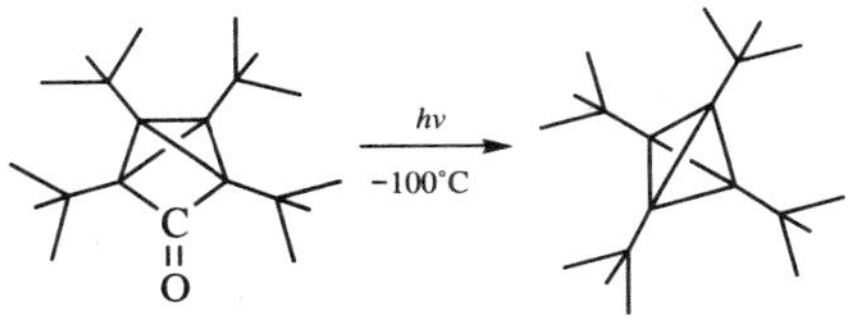

58

The mechanism probably involves a Norrish type I cleavage (p. 343), loss of CO from the resulting radical, and recombination of the radical fragments.

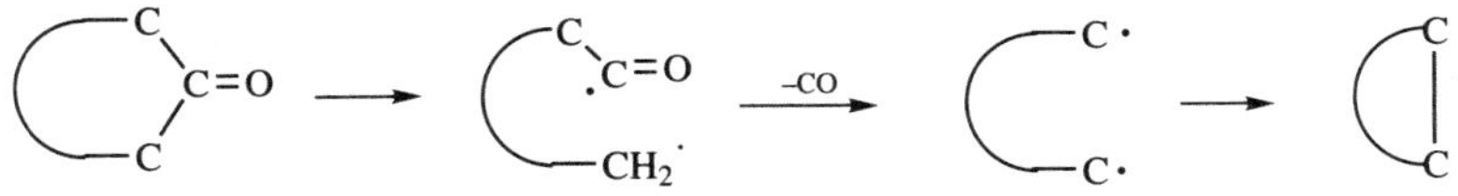

Certain lactones extrude CO$_2$ on heating or on irradiation, such as the pyrolysis of **59**.[513]

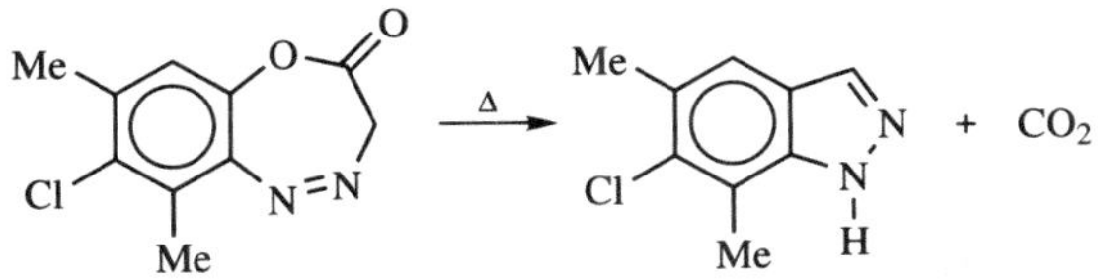

59

[510]For reviews of the reactions in this section, see Redmore, D.; Gutsche, C.D. *Adv. Alicyclic Chem.* **1971**, *3*, 1, see pp. 91–107; Stark, B.P.; Duke, A.J. *Extrusion Reactions*, Pergamon, Elmsford, NY, **1967**, pp. 47–71.
[511]Ramnauth, J.; Lee-Ruff, E. *Can. J. Chem.* **1997**, *75*, 518. See also, Ramnauth, J.; Lee-Ruff, E. *Can. J. Chem.* **2001**, *79*, 114.
[512]Maier, G.; Pfriem, S.; Schäfer, U.; Matusch, R. *Angew. Chem. Int. Ed.* **1978**, *17*, 520.
[513]Ried, W.; Wagner, K. *Liebigs Ann. Chem.* **1965**, *681*, 45.

Decarboxylation of β-lactones (see **17-26**) may be regarded as a degenerate example of this reaction. Unsymmetrical diacyl peroxides RCO–OO–COR' lose two molecules of CO_2 when photolyzed in the solid state to give the product RR'.[514] Electrolysis was also used, but yields were lower. This is an alternative to the Kolbe reaction (**11-34**) (see also **17-28** and **17-38**).

There are no OS references, but see OS **VI**, 418, for a related reaction.

17-36 Extrusion of SO₂

Sulfonyl-extrusion

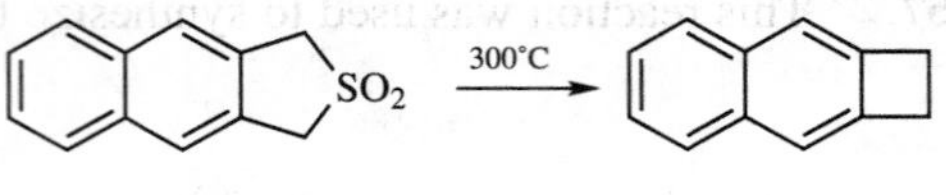

In a reaction similar to **17-35**, certain sulfones, both cyclic and acyclic,[515] extrude SO_2 on heating or photolysis to give ring-contracted products.[516] An example is the preparation of naphtho(b)cyclobutene shown above.[517] In a different kind of reaction, five-membered cyclic sulfones can be converted to cyclobutenes by treatment with butyllithium followed by $LiAlH_4$,[518] for example,

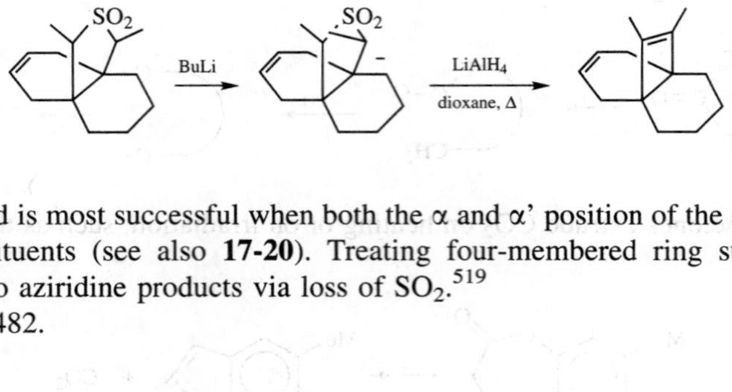

This method is most successful when both the α and α' position of the sulfone bear alkyl substituents (see also **17-20**). Treating four-membered ring sultams with $SnCl_2$ led to aziridine products via loss of SO_2.[519]

OS **VI**, 482.

[514]Lomölder, R.; Schäfer, H.J. *Angew. Chem. Int. Ed.* **1987**, *26*, 1253.

[515]See, for example, Gould, I.R.; Tung, C.; Turro, N.J.; Givens, R.S.; Matuszewski, B. *J. Am. Chem. Soc.* **1984**, *106*, 1789.

[516]For reviews of extrusions of SO₂, see Vögtle, F.; Rossa, L. *Angew. Chem. Int. Ed.* **1979**, *18*, 515; Stark, B.P.; Duke, A.J. *Extrusion Reactions*, Pergamon, Elmsford, NY, **1967**, pp. 72–90; Kice, J.L., in Kharasch, N.; Meyers, C.Y. *The Chemisry of Organic Sulfur Compounds*, Vol. 2, Pergamon, Elmsford, NY, **1966**, pp. 115–136. For a review of extrusion reactions of S, Se, and Te compounds, see Guziec, Jr., F.S.; SanFilippo, L.J. *Tetrahedron* **1988**, *44*, 6241.

[517]Cava, M.P.; Shirley, R.L. *J. Am. Chem. Soc.* **1960**, *82*, 654.

[518]Photis, J.M.; Paquette, L.A. *J. Am. Chem. Soc.* **1974**, *96*, 4715.

[519]Kataoka, T.; Iwama, T. *Tetrahedron Lett.* **1995**, *36*, 5559.

17-37   The Story Synthesis

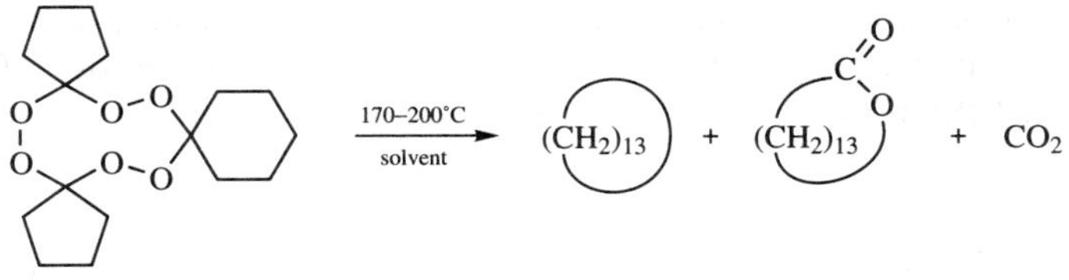

60

When cycloalkylidine peroxides (e.g., **60**) are heated in an inert solvent (e.g., decane), extrusion of CO_2 takes place; the products are the cycloalkane containing three carbon atoms less than the starting peroxide and the lactone containing two carbon atoms less[520] (the *Story synthesis*).[521] The two products are formed in comparable yields, usually ~15–25% each. Although the yields are low, the reaction is useful because there are not many other ways to prepare large rings. The reaction is versatile, having been used to prepare rings of every size from 8 to 33 members.

Both dimeric and trimeric cycloalkylidine peroxides can be synthesized[522] by treatment of the corresponding cyclic ketones with H_2O_2 in acid solution.[523] The trimeric peroxide is formed first and is subsequently converted to the dimeric compound.[524]

17-38   Alkene Synthesis by Twofold Extrusion

Carbon dioxide, thio-extrusion

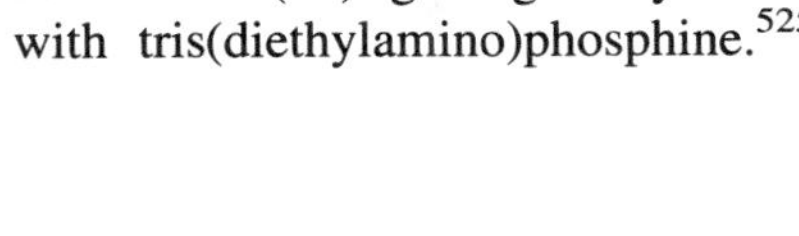

61

4,4-Diphenyloxathiolan-5-ones (**61**) give good yields of the corresponding alkenes when heated with tris(diethylamino)phosphine.[525] This reaction is an

---

[520]Sanderson, J.R.; Story, P.R.; Paul, K. *J. Org. Chem.* **1975**, *40*, 691; Sanderson, J.R.; Paul, K.; Story, P.R. *Synthesis* **1975**, 275.

[521]For a review, see Story, P.R.; Busch, P. *Adv. Org. Chem.* **1972**, *8*, 67, see pp. 79–94.

[522]For synthesis of mixed trimeric peroxides (e.g., **60**), see Sanderson, J.R.; Zeiler, A.G. *Synthesis* **1975**, 388; Paul, K.; Story, P.R.; Busch, P.; Sanderson, J.R. *J. Org. Chem.* **1976**, *41*, 1283.

[523]Kharasch, M.S.; Sosnovsky, G. *J. Org. Chem.* **1958**, *23*, 1322; Ledaal, T. *Acta Chem. Scand.*, **1967**, *21*, 1656. For another method, see Sanderson, J.R.; Zeiler, A.G. *Synthesis* **1975**, 125.

[524]Story, P.R.; Lee, B.; Bishop, C.E.; Denson, D.D.; Busch, P. *J. Org. Chem.* **1970**, *35*, 3059. See also, Sanderson, J.R.; Wilterdink, R.J.; Zeiler, A.G. *Synthesis* **1976**, 479.

[525]Barton, D.H.R.; Willis, B.J. *J. Chem. Soc. Perkin Trans. 1* **1972**, 305.

example of a general type: alkene synthesis by twofold extrusion of X and Y from a molecule of the type **62**.[526] Other examples are photolysis of 1,4-diones[527] (e.g., **63**) and treatment of acetoxy sulfones [$RCH(OAc)CH_2SO_2Ph$] with Mg/EtOH and a catalytic amount of $HgCl_2$.[528] **61** can be prepared by the condensation of thiobenzilic acid $Ph_2C(SH)COOH$ with aldehydes or ketones.

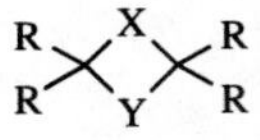

62 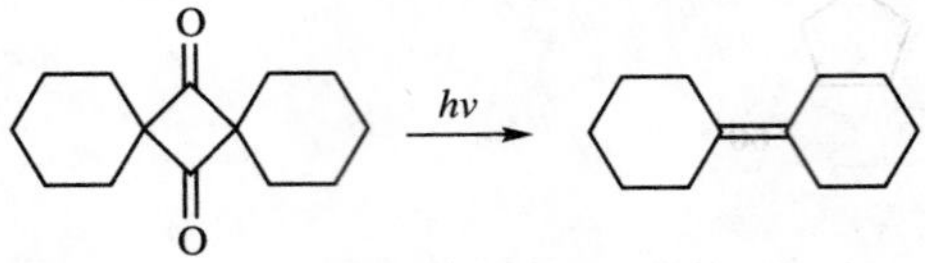   **63**

OS **V**, 297.

---

[526]For a review of those in which X or Y contains S, Se, or Te, see Guziec, Jr., F.S.; SanFilippo, L.J. *Tetrahedron* **1988**, *44*, 6241.

[527]Turro, N.J.; Leermakers, P.A.; Wilson, H.R.; Neckers, D.C.; Byers, G.W.; Vesley, G.F. *J. Am. Chem. Soc.* **1965**, *87*, 2613.

[528]Lee, G.H.; Lee, H.K.; Choi, E.B.; Kim, B.T.; Pak, C.S. *Tetrahedron Lett.* **1995**, *36*, 5607.

CHAPTER 18

Rearrangements

In a rearrangement reaction a group moves from one atom to another in the same molecule.[1] Most are migrations from an atom to an adjacent one (called 1,2-shifts), but some are over longer distances. The migrating group (W)

$$\begin{array}{ccc} \overset{W}{\underset{\diagdown}{A}}\;_B & \longrightarrow & \overset{W}{\underset{\diagup}{B}}_{\,A} \end{array}$$

may move with its electron pair (these can be called *nucleophilic* or *anionotropic* rearrangements; the migrating group can be regarded as a nucleophile), without its electron pair (*electrophilic* or *cationotropic* rearrangements; in the case of migrating hydrogen, *prototropic* rearrangements), or with just one electron (free-radical rearrangements). The atom A is called the *migration origin* and B is the *migration terminus*. However, there are some rearrangements that do not lend themselves to neat categorization in this manner. Among these are those with cyclic transition states (**18-27–18-36**).

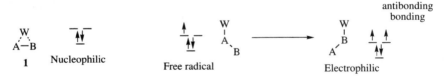

As we will see, nucleophilic 1,2-shifts are much more common than electrophilic or free-radical 1,2-shifts. The reason for this can be seen by a consideration of the transition states (or in some cases intermediates) involved. We represent the transition state or intermediate for all three cases by **1**, in which the two-electron

---

[1]For books, see de Mayo, P. *Rearrangements in Ground and Excited States*, 3 vols., Academic Press, NY, *1980*; Stevens, T.S.; Watts, W.E. *Selected Molecular Rearrangements*, Van Nostrand-Reinhold, Princeton, NJ, *1973*. For a review of many of these rearrangements, see Collins, C.J.; Eastham, J.F., in Patai, S. *The Chemistry of the Carbonyl Group*, Vol. 1, Wiley, NY, *1966*, pp. 761–821. See also, the series *Mechanisms of Molecular Migrations*.

March's Advanced Organic Chemistry: Reactions, Mechanisms, and Structure, Sixth Edition, by Michael B. Smith and Jerry March
Copyright © 2007 John Wiley & Sons, Inc.

A–W bond overlaps with the orbital on atom B, which contains zero, one, and two electrons, in the case of nucleophilic, free-radical, and electrophilic migration, respectively. The overlap of these orbitals gives rise to three new orbitals, which have an energy relationship similar to those on p. 72 (one bonding and two degenerate antibonding orbitals). In a nucleophilic migration, where only two electrons are involved, both can go into the bonding orbital and **1** is a low-energy transition state; but in a free-radical or electrophilic migration, there are, respectively, three or four electrons that must be accommodated, and antibonding orbitals must be occupied. It is not surprising therefore that, when 1,2-electrophilic or free-radical shifts are found, the migrating group W is usually aryl or some other group that can accommodate the extra one or two electrons and thus effectively remove them from the three-membered transition state or intermediate (see **41** on p. 1577).

In any rearrangement, we can in principle distinguish between two possible modes of reaction: In one of these, the group W becomes completely detached from A and may end up on the B atom of a different molecule (*intermolecular* rearrangement); in the other W goes from A to B in the *same* molecule (*intramolecular* rearrangement), in which case there must be some continuing tie holding W to the A–B system, preventing it from coming completely free. Strictly speaking, only the intramolecular type fits our definition of a rearrangement, but the general practice, which is followed here, is to include under the title "rearrangement" all net rearrangements whether they are inter- or intramolecular. It is usually not difficult to tell whether a given rearrangement is inter- or intramolecular. The most common method involves the use of *crossover* experiments. In this type of experiment, rearrangement is carried out on a mixture of W–A–B and V–A–C, where V is closely related to W (say, methyl vs. ethyl) and B to C. In an intramolecular process only A–B–W and A–C–V are recovered, but if the reaction is intermolecular, then not only will these two be found, but also A–B–V and A–C–W.

MECHANISMS

Nucleophilic Rearrangements[2]

Broadly speaking, such rearrangements consist of three steps, of which the actual migration is the second:

$$\underset{\diagup}{\overset{\displaystyle \curvearrowright}{\underset{A-B}{\overset{W}{}}}}\diagdown \quad\longrightarrow\quad \underset{\diagup}{\overset{W}{\underset{A-B}{}}}\diagdown$$

[2]For reviews, see Vogel, P. *Carbocation Chemistry*; Elsevier, NY, *1985*, pp. 323–372; Shubin, V.G. *Top. Curr. Chem.* *1984*, *116/117*, 267; Saunders, M.; Chandrasekhar, J.; Schleyer, P.v.R., in de Mayo, P. *Rearrangements in Ground and Excited States*, Vol. 1, Academic Press, NY, *1980*, pp. 1–53; Kirmse, W. *Top. Curr. Chem.* *1979*, *80*, 89. For reviews of rearrangements in vinylic cations, see Shchegolev, A.A.; Kanishchev, M.I. *Russ. Chem. Rev.* *1981*, *50*, 553; Lee, C.C. *Isot. Org. Chem.* *1980*, *5*, 1.

This process has been called the *Whitmore 1,2-shift*.[3] Since the migrating group carries the electron pair with it, the migration terminus B must be an atom with only six electrons in its outer shell (an open sextet). The first step therefore is creation of a system with an open sextet. Such a system can arise in various ways, but two of these are the most important:

1. *Formation of a Carbocation.* These can be formed in a number of ways (see p. 247), but one of the most common methods when a rearrangement is desired is the acid treatment of an alcohol to give **2** from an intermediate oxonium ion. These two steps are of course the same as the first two steps of the S_N1cA or the E1 reactions of alcohols.

2. *Formation of a Nitrene.* The decomposition of acyl azides is one of several ways in which acyl nitrenes **3** are formed (see p. 293). After the migration has taken place, the atom at the migration origin (A) must necessarily have an open sextet. In the third step, this atom acquires an octet. In the case of carbocations, the most common third steps are combinations with a nucleophile (rearrangement with substitution) and loss of H^+ (rearrangement with elimination).

Although we have presented this mechanism as taking place in three steps, and some reactions do take place in this way, in many cases two or all three steps are simultaneous. For example, in the nitrene example above, as the R migrates, an electron pair from the nitrogen moves into the C–N bond to give a stable isocyanate, **4**.

In this example, the second and third steps are simultaneous. It is also possible for the second and third steps to be simultaneous even when the "third" step involves more than just a simple motion of a pair of electrons. Similarly, there are many reactions in which the first two steps are simultaneous; that is, there is no actual formation of a species, such as **2** or **3**. In these instances, it may be said that

[3]It was first postulated by Whitmore, F.C. *J. Am. Chem. Soc.* **1932**, *54*, 3274.

R assists in the removal of the leaving group, with migration of R and the removal of the leaving group taking place simultaneously. Many investigations have been carried out in attempts to determine, in various reactions, whether such intermediates as **2** or **3** actually form, or whether the steps are simultaneous (see, e.g., the discussions on pp. 1381, 1563), but the difference between the two possibilities is often subtle, and the question is not always easily answered.[4]

Evidence for this mechanism is that rearrangements of this sort occur under conditions where we have previously encountered carbocations: S_N1 conditions, Friedel–Crafts alkylation, and so on. Solvolysis of neopentyl bromide leads to rearrangement products, and the rate increases with increasing ionizing power of the solvent but is unaffected by concentration of base,[5] so that the first step is carbocation formation. The same compound under S_N2 conditions gave no rearrangement, but only ordinary substitution, though slowly. Thus with neopentyl bromide, formation of a carbocation leads only to rearrangement. Carbocations usually rearrange to more stable carbocations. Thus the direction of rearrangement is usually primary → secondary → tertiary. Neopentyl (Me_3CCH_2), neophyl ($PhCMe_2CH_2$), and norbornyl (e.g., **5**) type systems are especially prone to carbocation rearrangement reactions. It has been shown that the rate of migration increases with the degree of electron deficiency at the migration terminus.[6]

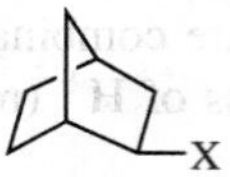

5

We have previously mentioned (p. 236) that stable tertiary carbocations can be obtained, in solution, at very low temperatures. The NMR studies have shown that when these solutions are warmed, rapid migrations of hydride and of alkyl groups take place, resulting in an equilibrium mixture of structures.[7] For example, the *tert*-pentyl cation (**5**)[8] equilibrates as follows:

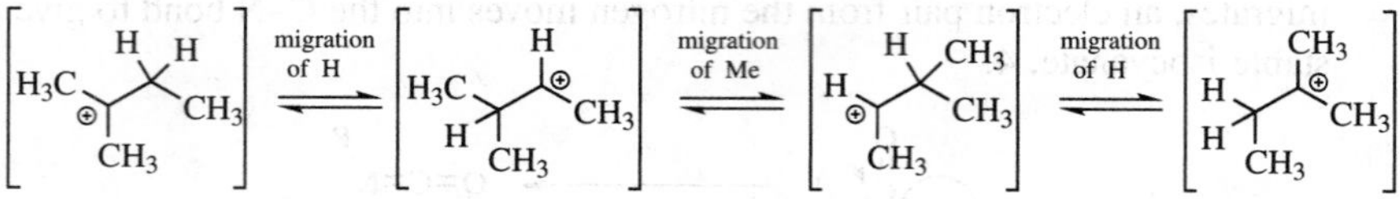

[4]The IUPAC designations depend on the nature of the steps. For the rules, see Guthrie, R.D. *Pure Appl. Chem.* **1989**, *61*, 23, 44–45.

[5]Dostrovsky, I.; Hughes, E.D. *J. Chem. Soc.* **1946**, 166.

[6]Borodkin, G.I.; Shakirov, M.M.; Shubin, V.G.; Koptyug, V.A. *J. Org. Chem. USSR* **1978**, *14*, 290, 924.

[7]For reviews, see Brouwer, D.M.; Hogeveen, H. *Prog. Phys. Org. Chem.* **1972**, *9*, 179, see pp. 203–237; Olah, G.A.; Olah, J.A., in Olah, G.A.; Schleyer, P.V.R. *Carbonium Ions*, Vol. 2, Wiley, NY, **1970**, pp. 751–760, 766–778. For a discussion of the rates of these reactions, see Sorensen, T.S. *Acc. Chem. Res.* **1976**, *9*, 257.

[8]Brouwer, D.M. *Recl. Trav. Chim. Pays-Bas* **1968**, *87*, 210; Saunders, M.; Hagen, E.L. *J. Am. Chem. Soc.* **1968**, *90*, 2436.

Carbocations that rearrange to give products of identical structure (e.g., $6 \rightleftharpoons 6',7 \rightleftharpoons 7'$) are called *degenerate carbocations* and such rearrangements are *degenerate rearrangements*. Many examples are known.[9]

The Actual Nature of the Migration

Most nucleophilic 1,2-shifts are intramolecular. The W group does not become free, but always remains connected in some way to the substrate. Apart from the evidence from crossover experiments, the strongest evidence is that when the W group is chiral, the configuration is *retained* in the product. For example, (+)-PhCHMe-COOH was converted to (−)-PhCHMeNH$_2$ by the Curtius (**18-14**), Hofmann (**18-13**), Lossen (**18-15**), and Schmidt (**18-16**) reactions.[10] In these reactions, the extent of retention varied from 95.8 to 99.6%. Retention of configuration in the migrating group has been shown many times since.[11] Another experiment demonstrating retention was the

easy conversion of **8** to **9**.[11] Neither inversion nor racemization could take place at a bridgehead. There is much other evidence that retention of configuration usually occurs in W, and inversion never.[12] However, this is not the state of affairs at A and B. In many reactions, of course, the structure of W–A–B is such that the product has only one steric possibility at A or B or both, and in most of these cases nothing can be learned. But in cases where the steric nature of A or B can be investigated, the results are mixed. It has been shown that either inversion or racemization can occur at A or B. Thus the following conversion proceeded with inversion at B:[13]

[9]For reviews, see Ahlberg, P.; Jonsäll, G.; Engdahl, C. *Adv. Phys. Org. Chem.* **1983**, *19*, 223; Leone, R.E.; Barborak, J.C.; Schleyer, P.v.R., in Olah, G.A.; Schleyer, P.v.R. *Carbonium Ions*, Vol. 4, Wiley, NY, **1970**, pp. 1837–1939; Leone, R.E.; Schleyer, P.v.R. *Angew. Chem. Int. Ed.* **1970**, *9*, 860.

[10]Campbell, A.; Kenyon, J. *J. Chem. Soc.* **1946**, 25, and references cited therein.

[11]For retention of migrating group configuration in the Wagner–Meerwein and pinacol rearrangements, see Beggs, J.J.; Meyers, M.B. *J. Chem. Soc. B* **1970**, 930; Kirmse, W.; Gruber, W.; Knist, J. *Chem. Ber.* **1973**, *106*, 1376; Shono, T.; Fujita, K.; Kumai, S. *Tetrahedron Lett.* **1973**, 3123; Borodkin, G.I.; Panova, Y.B.; Shakirov, M.M.; Shubin, V.G. *J. Org. Chem. USSR* **1983**, *19*, 103.

[12]See Cram, D.J., in Newman *Steric Effects in Organic Chemistry*, Wiley, NY, **1956**; pp. 251–254; Wheland, G.W. *Advanced Organic Chemistry*, 3rd ed., Wiley, NY, **1960**, pp. 597–604.

[13]Bernstein, H.I.; Whitmore, F.C. *J. Am. Chem. Soc.* **1939**, *61*, 1324. For other examples, see Tsuchihashi, G.; Tomooka, K.; Suzuki, K. *Tetrahedron Lett.* **1984**, *25*, 4253.

and inversion at A has been shown in other cases.[14] However, in many other cases, racemization occurs at A or B or both.[15] It is not always necessary for the product to have two steric possibilities in order to investigate the stereochemistry at A or B. Thus, in most Beckmann rearrangements (**18-17**), only the group trans (usually called *anti*) to the hydroxyl group migrates:

$$
\begin{array}{c}
R' \\
\quad\;\; C=N \\
R \;\;\;\;\;\; OH
\end{array}
\longrightarrow
\begin{array}{c}
R' \\
\quad\; C-NHR \\
\quad\;\; \| \\
\quad\;\; O
\end{array}
$$

showing inversion at B.

This information tells us about the degree of concertedness of the three steps of the rearrangement. First consider the migration terminus B. If racemization is found at B, it is probable that the first step takes place before the second and that a positively charged carbon (or other sextet atom) is present at B:

$$
\begin{array}{c}
R \\
\backslash \\
A-B-X
\end{array}
\longrightarrow
\begin{array}{c}
R \\
\backslash \\
A-B^+
\end{array}
\longrightarrow
\begin{array}{c}
\quad\;\;\; R \\
\quad\; \nearrow \\
^+A-B
\end{array}
\longrightarrow \text{Third step}
$$

With respect to B this is an S_N1-type process. If inversion occurs at B, it is likely that the first two steps are concerted, that a carbocation is *not* an intermediate, and that the process is S_N2-like:

$$
\begin{array}{c}
\overset{\frown}{R} \\
A-B-X
\end{array}
\longrightarrow
\begin{array}{c}
R \\
/\backslash \\
A\overset{\oplus}{-}B
\end{array}
\longrightarrow
\begin{array}{c}
\quad\;\;\; R \\
\quad\; \nearrow \\
^+A-B
\end{array}
\longrightarrow \text{Third step}
$$

10

In this case, participation by R assists in removal of X in the same way that neighboring groups do (p. 446). Indeed, R *is* a neighboring group here. The only difference is that, in the case of the neighboring-group mechanism of nucleophilic substitution, R never becomes detached from A, while in a rearrangement the bond between R and A is broken. In either case, the anchimeric assistance results in an increased rate of reaction. Of course, for such a process to take place, R must be in a favorable geometrical position (R and X antiperiplanar). Intermediate **10** may be a true intermediate or only a transition state, depending on what migrates. In certain cases of the S_N1-type process, it is possible for migration to take place with net retention of configuration at the migrating terminus because of conformational effects in the carbocation.[16]

We may summarize a few conclusions:

1. The S_N1-type process occurs mostly when B is a tertiary atom or has one aryl group and at least one other alkyl or aryl group. In other cases, the S_N2-type

---

[14]See Meerwein, H.; van Emster, K. *Ber.* **1920**, *53*, 1815; **1922**, *55*, 2500; Meerwein, H.; Gérard, L. *Liebigs Ann. Chem.* **1923**, *435*, 174.

[15]For example, see Winstein, S.; Morse, B.K. *J. Am. Chem. Soc.* **1952**, *74*, 1133.

[16]Collins, C.J.; Benjamin, B.M. *J. Org. Chem.* **1972**, *37*, 4358, and references cited therein.

process is more likely. Inversion of configuration (indicating an S_N2-type process) has been shown for a neopentyl substrate by the use of the chiral neopentyl-1-*d* alcohol.[17] On the other hand, there is other evidence that neopentyl systems undergo rearrangement by a carbocation (S_N1-type) mechanism.[18]

2. The question as to whether **10** is an intermediate or a transition state has been much debated. When R is aryl or vinyl, then **10** is probably an intermediate and the migrating group lends anchimeric assistance[19] (see p. 459 for resonance stabilization of this intermediate, when R is aryl). When R is alkyl, **10** is a protonated cyclopropane (edge- or corner-protonated; see p. 1026). There is much evidence that in simple migrations of a methyl group, the bulk of the products formed do not arise from protonated cyclopropane *intermediates*. Evidence for this statement has already been given (p. 467). Further evidence was obtained from experiments involving labeling.

11   12   13

(hypothetical)   14

Rearrangement of the neopentyl cation labeled with deuterium in the 1 position (**11**) gave only *tert*-pentyl products with the label in the 3 position (derived from **13**), though if **12** were an intermediate, the cyclopropane ring could just as well cleave the other way to give *tert*-pentyl derivatives labeled in the 4 position (derived from **14**).[20] Another experiment that led to the same conclusion was the generation, in several ways, of $Me_3C^{13}CH_2^+$. In this case, the only *tert*-pentyl products isolated were labeled in C-3, that is, $Me_2C^+-^{13}CH_2CH_3$ derivatives; no derivatives of $Me_2C^+-CH_2^{13}CH_3$ were found.[21]

Although the bulk of the products are not formed from protonated cyclopropane intermediates, there is considerable evidence that at least in 1-propyl

[17]Sanderson, W.A.; Mosher, H.S. *J. Am. Chem. Soc.* **1966**, *88*, 4185; Mosher, H.S. *Tetrahedron* **1974**, *30*, 1733. See also, Guthrie, R.D. *J. Am. Chem. Soc.* **1967**, *89*, 6718.

[18]Nordlander, J.E.; Jindal, S.P.; Schleyer, P.v.R.; Fort Jr., R.C.; Harper, J.J.; Nicholas, R.D. *J. Am. Chem. Soc.* **1966**, *88*, 4475; Shiner, Jr., V.J.; Imhoff, M.A. *J. Am. Chem. Soc.* **1985**, *107*, 2121.

[19]For example, see Rachon, J.; Goedkin, V.; Walborsky, H.M. *J. Org. Chem.* **1989**, *54*, 1006. For an opposing view, see Kirmse, W.; Feyen, P. *Chem. Ber.* **1975**, *108*, 71; Kirmse, W.; Plath, P.; Schaffrodt, H. *Chem. Ber.* **1975**, *108*, 79.

[20]Skell, P.S.; Starer, I.; Krapcho, A.P. *J. Am. Chem. Soc.* **1960**, *82*, 5257.

[21]Karabatsos, G.J.; Orzech Jr., C.E.; Meyerson, S. *J. Am. Chem. Soc.* **1964**, *86*, 1994.

systems, a small part of the product can in fact arise from such intermediates.[22] Among this evidence is the isolation of 10–15% cyclopropanes (mentioned on p. 467). Additional evidence comes from propyl cations generated by diazotization of labeled amines ($CH_3CH_2CD_2^+$, $CH_3CD_2CH_2^+$, $CH_3CH_2{}^{14}CH_2^+$), where isotopic distribution in the products indicated that a small amount (~5%) of the product had to be formed from protonated cyclopropane intermediates, for example,[23]

$$CH_3CH_2CD_2NH_2 \xrightarrow{\text{HONO}} -1\% \quad C_2H_4D\!\!-\!\!CHD\!\!-\!\!OH$$

$$CH_3CD_2CH_2NH_2 \xrightarrow{\text{HONO}} -1\% \quad C_2H_4D\!\!-\!\!CHD\!\!-\!\!OH$$

$$CH_3CH_2{}^{14}CH_2NH_2 \xrightarrow{\text{HONO}} -2\% \ {}^{14}CH_3CH_2CH_2OH \ + \ -2\% \ CH_3{}^{14}CH_2CH_2OH$$

Even more scrambling was found in trifluoroacetolysis of 1-propyl-1-^{14}C-mercuric perchlorate.[24] However, protonated cyclopropane intermediates accounted for <1% of the products from diazotization of labeled isobutylamine[25] and from formolysis of labeled 1-propyl tosylate.[26]

It is likely that protonated cyclopropane transition states or intermediates are also responsible for certain non-1,2 rearrangements. For example, in super acid solution, the ions **15** and **17** are in equilibrium. It is not possible for these to interconvert solely by 1,2-alkyl or hydride shifts unless primary carbocations (which are highly unlikely) are intermediates. However, the reaction can be explained[27] by postulating that (in the forward reaction) it is the 1,2 bond

[22]For reviews, see Saunders, M.; Vogel, P.; Hagen, E.L.; Rosenfeld, J. *Acc. Chem. Res.* **1973**, *6*, 53; Lee, C.C. *Prog. Phys. Org. Chem.* **1970**, *7*, 129; Collins, C.J. *Chem. Rev.* **1969**, *69*, 543. See also, Cooper, C.N.; Jenner, P.J.; Perry, N.B.; Russell-King, J.; Storesund, H.J.; Whiting, M.C. *J. Chem. Soc. Perkin Trans. 2* **1982**, 605.

[23]Lee, C.C.; Kruger, J.E. *Tetrahedron* **1967**, *23*, 2539; Lee, C.C.; Wan, K. *J. Am. Chem. Soc.* **1969**, *91*, 6416; Karabatsos, G.J.; Orzech, Jr., C.E.; Fry, J.L.; Meyerson, S. *J. Am. Chem. Soc.* **1970**, *92*, 606.

[24]Lee, C.C.; Cessna, A.J.; Ko, E.C.F.; Vassie, S. *J. Am. Chem. Soc.* **1973**, *95*, 5688. See also, Lee, C.C.; Reichle, R. *J. Org. Chem.* **1977**, *42*, 2058, and references cited therein.

[25]Karabatsos, G.J.; Hsi, N.; Meyerson, S. *J. Am. Chem. Soc.* **1970**, *92*, 621. See also, Karabatsos, G.J.; Anand, M.; Rickter, D.O.; Meyerson, S. *J. Am. Chem. Soc.* **1970**, *92*, 1254.

[26]Lee, C.C.; Kruger, J.E. *Can. J. Chem.* **1966**, *44*, 2343; Shatkina, T.N.; Lovtsova, A.N.; Reutov, O.A. *Bull. Acad. Sci. USSR Div. Chem. Sci.* **1967**, 2616; Karabatsos, G.J.; Fry, J.L.; Meyerson, S. *J. Am. Chem. Soc.* **1970**, *92*, 614. See also, Lee, C.C.; Zohdi, H.F. *Can. J. Chem.* **1983**, *61*, 2092.

[27]Brouwer, D.M.; Oelderik, J.M. *Recl. Trav. Chim. Pays-Bas* **1968**, *87*, 721; Saunders, M.; Jaffe, M.H.; Vogel, P. *J. Am. Chem. Soc.* **1971**, *93*, 2558; Saunders, M.; Vogel, P. *J. Am. Chem. Soc.* **1971**, *93*, 2559, 2561; Kirmse, W.; Loosen, K.; Prolingheuer, E. *Chem. Ber.* **1980**, *113*, 129.

of the intermediate or transition state **16** that opens up rather than the 2,3 bond, which is the one that would open if the reaction were a normal 1,2-shift of a methyl group. In this case, opening of the 1,2 bond produces a tertiary cation, while opening of the 2,3 bond would give a secondary cation. (In the reaction **17** → **15**, it is of course the 1,3 bond that opens).

3. There has been much discussion of H as migrating group. There is no conclusive evidence that **10** in this case is or is not a true intermediate, although both positions have been argued (see p. 467).

The stereochemistry at the migration origin A is less often involved, since in most cases it does not end up as a tetrahedral atom; but when there is inversion here, there is an S_N2-type process at the beginning of the migration. This may or may not be accompanied by an S_N2 process at the migration terminus B:

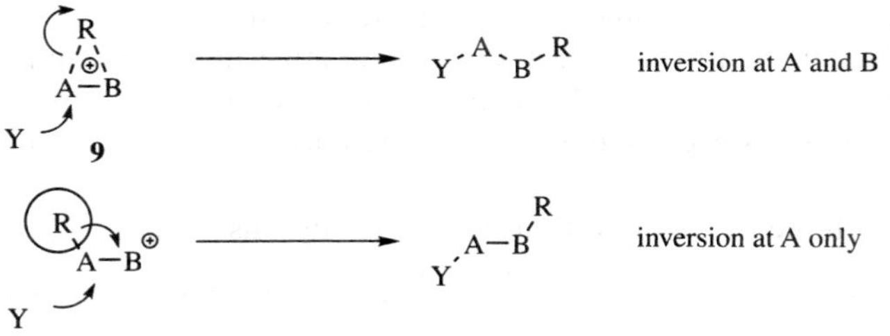

inversion at A and B

inversion at A only

In some cases, it has been found that, when H is the migrating species, the configuration at A may be *retained*.[28]

There is evidence that the configuration of the molecule may be important even where the leaving group is gone long before migration takes place. For example, the 1-adamantyl cation (**18**) does not equilibrate intramolecularly, even at temperatures up to 130°C,[29] though open-chain (e.g., **6** ⇌ **6'**) and cyclic tertiary

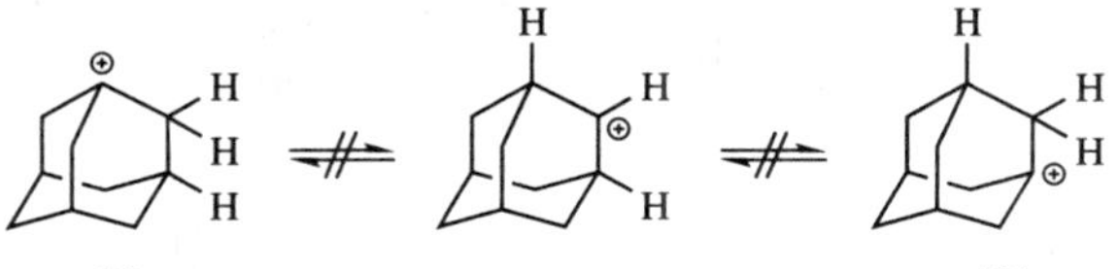

| **18** | **18'** |

carbocations undergo such equilibration at 0°C or below. On the basis of this and other evidence it has been concluded that for a 1,2-shift of hydrogen or methyl to proceed as smoothly as possible, the vacant *p* orbital of the carbon bearing the positive charge and the sp^3 orbital carrying the migrating group must be coplanar,[29] which is not possible for **18**.

[28]Winstein, S.; Holness, N.J. *J. Am. Chem. Soc.* **1955**, *77*, 5562; Cram, D.J.; Tadanier, J. *J. Am. Chem. Soc.* **1959**, *81*, 2737; Bundel', Yu.G.; Pankratova, K.G.; Gordin, M.B.; Reutov, O.A. *Doklad. Chem.* **1971**, *199*, 700; Kirmse, W.; Ratajczak, H.; Rauleder, G. *Chem. Ber.* **1977**, *110*, 2290.
[29]Brouwer, D.M.; Hogeveen, H. *Recl. Trav. Chim. Pays-Bas* **1970**, *89*, 211; Majerski, Z.; Schleyer, P.v.R.; Wolf, A.P. *J. Am. Chem. Soc.* **1970**, *92*, 5731.

Migratory Aptitudes[30]

In many reactions, there is no question about which group migrates. For example, in the Hofmann, Curtius, and similar reactions there is only one possible migrating group in each molecule, and one can measure migratory aptitudes only by comparing the relative rearrangement rates of different compounds. In other instances, there are two or more potential migrating groups, but which migrates is settled by the geometry of the molecule. The Beckmann rearrangement (**18-17**) provides an example. As we have seen, only the group trans to the OH migrates. In compounds whose geometry is not restricted in this manner, there still may be eclipsing effects (see p. 1502), so that the choice of migrating group is largely determined by which group is in the right place in the most stable conformation of the molecule.[31] However, in some reactions, especially the Wagner–Meerwein (**18-1**) and the pinacol (**18-2**) rearrangements, the molecule may contain several groups that, geometrically at least, have approximately equal chances of migrating, and these reactions have often been used for the direct study of relative migratory aptitudes. In the pinacol rearrangement, there is the additional question of which OH group leaves and which does not, since a group can migrate only if the OH group on *the other* carbon is lost.

We deal with the second question first. To study this question, the best type of substrate to use is one of the form $\begin{smallmatrix} R_2C - CR'_2 \\ | \quad | \\ OH \ OH \end{smallmatrix}$, since the only thing that determines migratory aptitude is which OH group comes off. Once the OH group is gone, the migrating group is determined. As might be expected, the OH that leaves is the one whose loss gives rise to the more stable carbocation. Thus 1,1-diphenylethanediol (**19**) gives diphenylacetaldehyde (**20**), not phenylacetophenone (**21**). Obviously, it does not matter in this case whether phenyl has a greater

inherent migratory aptitude than hydrogen or not. Only the hydrogen can migrate because **22** is not formed. As we know, carbocation stability is enhanced by

[30]For discussions, see Koptyug, V.A.; Shubin, V.G. *J. Org. Chem. USSR* **1980**, *16*, 1685; Wheland, G.W. *Advanced Organic Chemistry*, 3rd ed., Wiley, NY, **1960**, pp. 573–597.

[31]For a discussion, see Cram, D.J., in Newman, M.S. *Steric Effects in Organic Chemistry*, Wiley, NY, **1956**, pp. 270–276. For an interesting example, see Nickon, A.; Weglein, R.C. *J. Am. Chem. Soc.* **1975**, *97*, 1271.

groups in the order aryl > alkyl > hydrogen, and this normally determines which side loses the OH group. However, exceptions are known, and which group is lost may depend on the reaction conditions (e.g., see the reaction of **53**, p. 1586).

In order to answer the question about inherent migratory aptitudes, the obvious type of substrate to use (in the pinacol rearrangement) is $\begin{smallmatrix} R'RC-CRR' \\ | \quad\;\; | \\ OH \;\; OH \end{smallmatrix}$, since the same carbocation is formed no matter which OH leaves, and it would seem that a direct comparison of the migratory tendencies of R and R' is possible. On closer inspection, however, we can see that several factors are operating. Apart from the question of possible conformational effects, already mentioned, there is also the fact that whether the group R or R' migrates is determined not only by the relative inherent migrating abilities of R and R', but also by whether the group that does *not* migrate is better at stabilizing the positive charge that will now be found at the migration origin.[32] Thus, migration of R gives rise to the cation $R'C^+(OH)CR_2R'_2$, while migration of R' gives the cation $R^+C(OH)CRR'_2$, and these cations have different stabilities. It is possible that in a given case R might be found to migrate less than R', not because it actually has a lower inherent migrating tendency, but because it is much better at stabilizing the positive charge. In addition to this factor,

migrating ability of a group is also related to its capacity to render anchimeric assistance to the departure of the nucleofuge. An example of this effect is the finding that in the decomposition of tosylate **23** only the phenyl group migrates, while in acid treatment of the corresponding alkene **24**, there is competitive migration of both methyl and phenyl (in these reactions ^{14}C labeling is necessary to determine which group has migrated).[33] Both **23** and **24** give the same carbocation; the differing results must be caused by the fact that in **23** the phenyl group can assist the leaving group, while no such process is possible for **24**. This example clearly illustrates the difference between migration to a relatively

[32]For example, see McCall, M.J.; Townsend, J.M.; Bonner, W.A. *J. Am. Chem. Soc.* **1975**, *97*, 2743; Brownbridge, P.; Hodgson, P.K.G.; Shepherd, R.; Warren, S. *J. Chem. Soc. Perkin Trans. 1* **1976**, 2024.
[33]Grimaud, J.; Laurent, A. *Bull. Soc. Chim. Fr.* **1967**, 3599.

free terminus and one that proceeds with the migrating group lending anchimeric assistance.[34]

It is not surprising therefore that clear-cut answers as to relative migrating tendencies are not available. More often than not migratory aptitudes are in the order aryl > alkyl, but exceptions are known, and the position of hydrogen in this series is often unpredictable. In some cases, migration of hydrogen is preferred to aryl migration; in other cases, migration of alkyl is preferred to that of hydrogen. Mixtures are often found and the isomer that predominates often depends on conditions. For example, the comparison between methyl and ethyl has been made many times in various systems, and in some cases methyl migration and in others ethyl migration has been found to predominate.[35] However, it can be said that among aryl migrating groups, electron-donating substituents in the para and meta positions increase the migratory aptitudes, while the same substituents in the ortho positions decrease them. Electron-withdrawing groups decrease migrating ability in all positions. The following are a few of the relative migratory aptitudes determined for aryl groups by Bachmann and Ferguson:[36] *p*-anisyl, 500; *p*-tolyl, 15.7; *m*-tolyl, 1.95; phenyl, 1.00; *p*-chlorophenyl, 0.7; *o*-anisyl, 0.3. For the *o*-anisyl group, the poor migrating ability probably has a steric cause, while for the others there is a fair correlation with activation or deactivation of electrophilic aromatic substitution, which is what the process is with respect to the benzene ring. It has been reported that at least in certain systems acyl groups have a greater migratory aptitude than alkyl groups.[37]

Memory Effects[38]

Solvolysis of the endo bicyclic compound **25** (X = ONs, p. 497, or Br) gave mostly the bicyclic allylic alcohol, **28**, along with a smaller amount of the tricyclic alcohol **32**, while solvolysis of the exo isomers, **29**, gave mostly **32**, with smaller amounts of **28**.[39] Thus the two isomers gave entirely different ratios of products, although

[34]A number of studies of migratory aptitudes in the dienone-phenol rearrangement (**18-5**) are in accord with the above. For a discussion, see Fischer, A.; Henderson, G.N. *J. Chem. Soc., Chem. Commun.* **1979**, 279, and references cited therein. See also, Palmer, J.D.; Waring, A.J. *J. Chem. Soc. Perkin Trans. 2* **1979**, 1089; Marx, J.N.; Hahn, Y.P. *J. Org. Chem.* **1988**, *53*, 2866.

[35]For examples, see Cram, D.J.; Knight, J.D. *J. Am. Chem. Soc.* **1952**, *74*, 5839; Stiles, M.; Mayer, R.P. *J. Am. Chem. Soc.* **1959**, *81*, 1497; Heidke, R.L.; Saunders, Jr., W.H. *J. Am. Chem. Soc.* **1966**, *88*, 5816; Dubois, J.E.; Bauer, P. *J. Am. Chem. Soc.* **1968**, *90*, 4510, 4511; Bundel', Yu. G.; Levina, I.Yu.; Reutov, O.A. *J. Org. Chem. USSR* **1970**, *6*, 1; Pilkington, J.W.; Waring, A.J. *J. Chem. Soc. Perkin Trans. 2* **1976**, 1349; Korchagina, D.V.; Derendyaev, B.G.; Shubin, V.G.; Koptyug, V.A. *J. Org. Chem. USSR* **1976**, *12*, 378; Wistuba, E.; Rüchardt, C. *Tetrahedron Lett.* **1981**, *22*, 4069; Jost, R.; Laali, K.; Sommer, J. *Nouv. J. Chim.* **1983**, *7*, 79

[36]Bachmann, W.E.; Ferguson, J.W. *J. Am. Chem. Soc.* **1934**, *56*, 2081.

[37]Le Drian, C.; Vogel, P. *Helv. Chim. Acta* **1987**, *70*, 1703; *Tetrahedron Lett.* **1987**, *28*, 1523.

[38]For a review, see Berson, J.A. *Angew. Chem. Int. Ed.* **1968**, *7*, 779.

[39]Berson, J.A.; Poonian, M.S.; Libbey, W.J. *J. Am. Chem. Soc.* **1969**, *91*, 5567; Berson, J.A.; Donald, D.S.; Libbey, W.J. *J. Am. Chem. Soc.* **1969**, *91*, 5580; Berson, J.A.; Wege, D.; Clarke, G.M.; Bergman, R.G. *J. Am. Chem. Soc.* **1969**, *91*, 5594, 5601.

the carbocation initially formed (**26** or **30**) seems to be the same for each. In the case of **26**, a second rearrangement (a shift of the 1,7 bond) follows, while with **30** what follows is an intramolecular addition of the positive carbon to the double bond.

It seems as if **26** and **30** "remember" how they were formed before they go on to give the second step. Such effects are called *memory effects* and other such cases are known.[40] The causes of these effects are not well understood, though there has been much discussion. One possible cause is differential solvation of the apparently identical ions **26** and **30**. Other possibilities are (*1*) that the ions have geometrical structures that are twisted in opposite senses (e.g., a twisted **30** might have its positive carbon closer to the double

Twisted **30**　　　Twisted **26**

bond than a twisted **26**); (*2*) that ion pairing is responsible;[41] and (*3*) that nonclassical carbocations are involved.[42] One possibility that has been ruled out is that the steps **25** → **26** → **27** and **29** → **30** → **31** are concerted, so that **26** and **30** never exist at all. This possibility has been excluded by several kinds of evidence, including the fact that **25** gives not only **28**, but also some **32**; and **29** gives some **28**

---

[40]For examples of memory effects in other systems, see Berson, J.A.; Luibrand, R.T.; Kundu, N.G.; Morris, D.G. *J. Am. Chem. Soc.* **1971**, *93*, 3075; Collins, C.J. *Acc. Chem. Res.* **1971**, *4*, 315; Collins, J.A.; Glover, I.T.; Eckart, M.D.; Raaen, V.F.; Benjamin, B.M.; Benjaminov, B.S. *J. Am. Chem. Soc.* **1972**, *94*, 899; Svensson, T. *Chem. Scr.,* **1974**, *6*, 22.

[41]See Collins, C.J. *Chem. Soc. Rev.* **1975**, *4*, 251.

[42]See, for example, Seybold, G.; Vogel, P.; Saunders, M.; Wiberg, K.B. *J. Am. Chem. Soc.* **1973**, *95*, 2045; Kirmse, W.; Günther, B. *J. Am. Chem. Soc.* **1978**, *100*, 3619.

along with **32**. This means that some of the **26** and **30** ions interconvert, a phenomenon known as *leakage*.

Longer Nucleophilic Rearrangements

The question as to whether a group can migrate with its electron pair from A to C in W–A–B–C or over longer distances has been much debated. Although claims have been made that alkyl groups can migrate in this way, the evidence is that such migration is extremely rare, if it occurs at all. One experiment that demonstrated this was the generation of the 3,3-dimethyl-1-butyl cation $Me_3CCH_2CH_2{}^+$. If 1,3-methyl migrations are possible, this cation would appear to be a favorable substrate, since such a migration would convert a primary cation into the tertiary 2-methyl-2-pentyl cation $Me_2CCH_2CH_2CH_3$, while the only possible 1,2 migration (of hydride) would give only a secondary cation. However, no products arising from the 2-methyl-2-pentyl cation were found, the only rearranged products being those formed by the 1,2 hydride migration.[43] 1,3 Migration of bromine has been reported.[44]

However, most of the debate over the possibility of 1,3 migrations has concerned not methyl or bromine, but 1,3 hydride shifts.[45] There is no doubt that *apparent* 1,3 hydride shifts take place (many instances have been found), but the question is whether they are truly direct hydride shifts or whether they occur by another

A   **B**

mechanism. There are at least two ways in which indirect 1,3-hydride shifts can take place: (*1*) by successive 1,2-shifts or (*2*) through the intervention of protonated cyclopropanes (see p. 1565). A direct 1,3-shift would have the transition state **A**, while the transition state for a 1,3-shift involving a protonated cyclopropane intermediate would resemble **B**. The evidence is that most reported 1,3 hydride shifts are actually the result of successive 1,2 migrations,[46] but that in some cases small amounts of products cannot be accounted for in this way. For example, the reaction of 2-methyl-1-butanol with KOH and bromoform gave a mixture of alkenes, nearly all of which could have arisen from simple

[43]Skell, P.S.; Reichenbacher, P.H. *J. Am. Chem. Soc.* **1968**, *90*, 2309.

[44]Reineke, C.E.; McCarthy, Jr., J.R. *J. Am. Chem. Soc.* **1970**, *92*, 6376; Smolina, T.A.; Gopius, E.D.; Gruzdneva, V.N.; Reutov, O.A. *Doklad. Chem.* **1973**, *209*, 280.

[45]For a review, see Fry, J.L.; Karabatsos, G.J., in Olah, G.A.; Schleyer, P.v.R. *Carbonium Ions*, Vol. 2, Wiley, NY, **1970**, p. 527.

[46]For example, see Bundel', Yu.G.; Levina, I.Yu.; Krzhizhevskii, A.M.; Reutov, O.A. *Doklad. Chem.* **1968**, *181*, 583; Fărcaşiu, D.; Kascheres, C.; Schwartz, L.H. *J. Am. Chem. Soc.* **1972**, *94*, 180; Kirmse, W.; Knist, J.; Ratajczak, H. *Chem. Ber.* **1976**, *109*, 2296.

elimination or 1,2-shifts of hydride or alkyl. However, 1.2% of the product was **33**:[47]

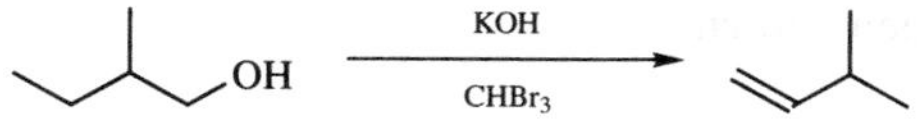

33

Hypothetically, **33** could have arisen from a 1,3-shift (direct or through a protonated cyclopropane) or from two successive 1,2-shifts:

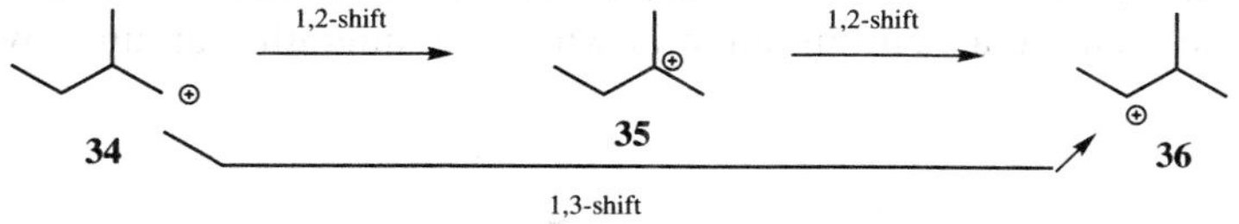

However, the same reaction applied to 2-methyl-2-butanol gave no **33**, which demonstrated that **36** was not formed from **35**. The conclusion made was that **36** was formed directly from **34**. This experiment does not answer the question as to whether **36** was formed by a direct shift or through a protonated cyclopropane, but from other evidence[48] it appears that 1,3 hydride shifts that do not result from successive 1,2 migrations usually take place through protonated cyclopropane intermediates (which, as we saw on p. 1565, account for only a small percentage of the product in any case). However, there is evidence that direct 1,3 hydride shifts by way of **A** may take place in super acid solutions.[49] Although direct nucleophilic rearrangements over distances >1,2 are rare (or perhaps nonexistent) when the migrating atom or group must move along a chain, this is not so for a shift across a ring of 8–11 members. Many such transannular rearrangements are known.[50] Several examples are given on p. 223. This is the mechanism of one of these:[51]

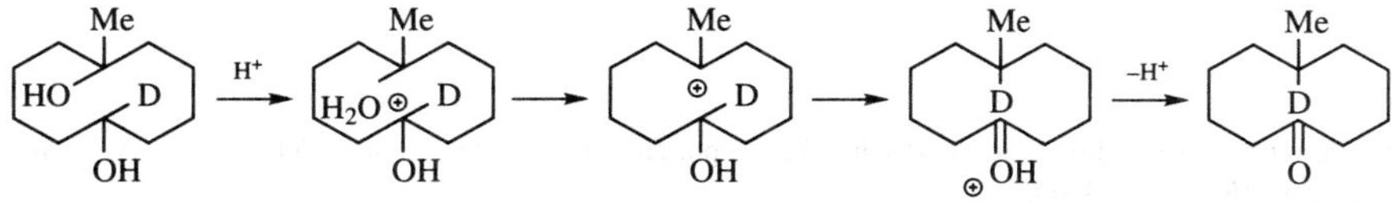

[47]Skell, P.S.; Maxwell, R.J. *J. Am. Chem. Soc.* **1962**, *84*, 3963. See also, Skell, P.S.; Starer, I. *J. Am. Chem. Soc.* **1962**, *84*, 3962.

[48]For example, see Brouwer, D.M.; van Doorn, J.A. *Recl. Trav. Chim. Pays-Bas* **1969**, *8*, 573; Dupuy, W.E.; Goldsmith, E.A.; Hudson, H.R. *J. Chem. Soc. Perkin Trans. 2* **1973**, 74; Hudson, H.R.; Koplick, A.J.; Poulton, D.J. *Tetrahedron Lett.* **1975**, 1449; Fry, J.L.; Karabatsos, G.J., in Olah, G.A.; Schleyer, P.v.R. *Carbonium Ions*, Vol. 2, Wiley, NY, **1970**, p. 527.

[49]Saunders, M.; Stofko Jr., J.J. *J. Am. Chem. Soc.* **1973**, *95*, 252.

[50]For reviews, see Cope, A.C.; Martin, M.M.; McKervey, M.A. *Q. Rev. Chem. Soc.* **1966**, *20*, 119. For many references, see Blomquist, A.T.; Buck, C.J. *J. Am. Chem. Soc.* **1951**, *81*, 672.

[51]Prelog, V.; Küng, W. *Helv. Chim. Acta* **1956**, *39*, 1394.

It is noteworthy that the *methyl* group does not migrate in this system. It is generally true that alkyl groups do not undergo transannular migration.[52] In most cases, it is hydride that undergoes this type of migration, though a small amount of phenyl migration has also been shown.[53]

Free-Radical Rearrangements[54]

1,2-Free-radical rearrangements are much less common than the nucleophilic type previously considered, for the reasons mentioned on p. 1559. Where they do occur, the general pattern is similar. There must first be generation of a free radical, and then the actual migration in which the migrating group moves with one electron:

$$\underset{A-B^\bullet}{\overset{R}{\diagdown}} \longrightarrow ^\bullet A - B \overset{R}{\diagup}$$

Finally, the new free radical must stabilize itself by a further reaction. The order of radical stability leads us to predict that here too, as with carbocation rearrangements, any migrations should be in the order primary → secondary → tertiary, and that the logical place to look for them should be in neopentyl and neophyl systems. The most common way of generating free radicals for the purpose of detection of rearrangements is by decarbonylation of aldehydes (**14-32**). In this manner, it was found that neophyl radicals *do* undergo rearrangement. Thus, $PhCMe_2CH_2CHO$ treated with di-*tert*-butyl peroxide gave about equal amounts of the normal product $PhCMe_2CH_3$ and the product arising from migration of phenyl:[55]

[52]For an apparent exception, see Fărcaşiu, D.; Seppo, E.; Kizirian, M.; Ledlie, D.B.; Sevin, A. *J. Am. Chem. Soc.* **1989**, *111*, 8466.

[53]Cope, A.C.; Burton, P.E.; Caspar, M.L. *J. Am. Chem. Soc.* **1962**, *84*, 4855.

[54]For reviews, see Beckwith, A.L.J.; Ingold, K.U. in de Mayo, P. *Rearrangements in Ground and Excited States*, Vol. 1, Academic Press, NY, **1980**, pp. 161–310; Wilt, J.W., in Kochi, J.K. *Free Radicals*, Vol. 1, Wiley, NY, **1973**, pp. 333–501; Stepukhovich, A.D.; Babayan, V.I. *Russ. Chem. Rev.* **1972**, *41*, 750; Nonhebel, D.C.; Walton, J.C. *Free-Radical Chemistry*, Cambridge University Press, London, **1974**, pp. 498–552; Huyser, E.S. *Free-Radical Chain Reactions*, Wiley, NY, **1970**, pp. 235–255; Freidlina, R.Kh. *Adv. Free-Radical Chem.* **1965**, *1*, 211–278; Pryor, W.A. *Free Radicals*, McGraw-Hill, NY, **1966**, pp. 266–284.

[55]Winstein, S.; Seubold, Jr., F.H. *J. Am. Chem. Soc.* **1947**, *69*, 2916; Seubold, Jr., F.H. *J. Am. Chem. Soc.* **1953**, *75*, 2532. For the observation of this rearrangement by esr, see Hamilton, Jr., E.J.; Fischer, H. *Helv. Chim. Acta* **1973**, *56*, 795.

Many other cases of free-radical migration of aryl groups have been found.[56] Intramolecular radical rearrangements are known.[57] The C-4 radicals of α- and β-thujone undergo two distinct rearrangement reactions, and it has been proposed that these could serve as simultaneous, but independent radical clocks.[58]

A 1,2-shift has been observed in radicals bearing an OCOR group at the β-carbon where the oxygen group migrates as shown in the interconversion of **37** and **38**. This has been proven by ^{18}O isotopic labeling experiments[59] and other mechanistic explorations.[60] A similar rearrangement was observed with phosphatoxy alkyl radicals, such as **39**.[61] A 1,2-shift of hydrogen atoms has been observed in aryl radicals.[62]

A C → N 1,2-aryl rearrangement was observed when alkyl azides were treated with n-Bu$_3$SnH, proceeding via an C–N$^{•}$–SnBu$_3$ species to give an imine.[63]

It is noteworthy that the extent of migration is much less than with corresponding carbocations: Thus in the example given, there was only ~50% migration, whereas the carbocation would have given much more. Also noteworthy is that there was no migration of the methyl group. In general, it may be said that free-radical migration of alkyl groups does not occur at ordinary temperatures. Many attempts have been made to detect such migration on the traditional neopentyl and bornyl types of substrates. However, alkyl migration is not observed, even in substrates where the corresponding carbocations undergo facile rearrangement.[64] Another type of migration that is very common for carbocations, but not observed

[56]For example, see Curtin, D.Y.; Hurwitz, M.J. *J. Am. Chem. Soc.* **1952**, *74*, 5381; Wilt, J.K.; Philip, H. *J. Org. Chem.* **1959**, *24*, 441; **1960**, *25*, 891; Pines, H.; Goetschel, C.T. *J. Am. Chem. Soc.* **1964**, *87*, 4207; Goerner Jr., R.N.; Cote, P.N.; Vittimberga, B.M. *J. Org. Chem.* **1977**, *42*, 19; Collins, C.J.; Roark, W.H.; Raaen, V.F.; Benjamin, B.M. *J. Am. Chem. Soc.* **1979**, *101*, 1877; Walter, D.W.; McBride, J.M. *J. Am. Chem. Soc.* **1981**, *103*, 7069, 7074. For a review, see Studer, A.; Bossart, M. *Tetrahedron* **2001**, *57*, 9649.
[57]Prévost, N.; Shipman, M. *Org. Lett.* **2001**, *3*, 2383.
[58]He, X.; Ortiz de Montellano, P.R. *J. Org. Chem.* **2004**, *69*, 5684.
[59]Crich, D.; Filzen, G.F. *J. Org. Chem.* **1995**, *60*, 4834.
[60]Beckwith, A.L.J.; Duggan, P.J. *J. Chem. Soc. Perkin Trans. 2* **1992**, 1777; **1993**, 1673.
[61]Crich, D.; Yao, Q. *Tetrahedron Lett.* **1993**, *34*, 5677. See Ganapathy, S.; Cambron R.T.; Dockery, K.P.; Wu, Y.-W.; Harris, J.M.; Bentrude, W.G. *Tetrahedron Lett.* **1993**, *34*, 5987 for a related triplet sensitized rearrangement of allylic phosphites and phosphonates.
[62]Brooks, M.A.; Scott, L.T. *J. Am. Chem. Soc.* **1999**, *121*, 5444.
[63]Kim, S.; Do, J.Y. *J. Chem. Soc., Chem. Commun.* **1995**, 1607.
[64]For a summary of unsuccessful attempts, see Slaugh, L.H.; Magoon, E.F.; Guinn, V.P. *J. Org. Chem.* **1963**, *28*, 2643.

for free radicals, is 1,2 migration of hydrogen. We confine ourselves to a few examples of the lack of migration of alkyl groups and hydrogen:

1. 3,3-Dimethylpentanal ($EtCMe_2CH_2CHO$) gave no rearranged products on decarbonylation.[65]

2. Addition of RSH to norbornene gave only *exo*-norbornyl sulfides, though **40** is an intermediate, and the corresponding carbocation cannot be formed without rearrangement.[66]

40

3. The cubylcarbinyl radical did not rearrange to the 1-homocubyl radical, though doing so would result in a considerable decrease in strain.[67]

Cubylcarbinyl   1-Homocubyl
radical   radical

4. It was shown[68] that no rearrangement of isobutyl radical to *tert*-butyl radical (which would involve the formation of a more stable radical by a hydrogen shift) took place during the chlorination of isobutane.

However, 1,2 migration of alkyl groups has been shown to occur in certain *diradicals*.[69] For example, the following rearrangement has been established by tritium labeling.[70]

In this case, the fact that migration of the methyl group leads directly to a compound in which all electrons are paired undoubtedly contributes to the driving force of the reaction.

[65]Seubold, Jr., F.H. *J. Am. Chem. Soc.* **1954**, *76*, 3732.
[66]Cristol, S.J.; Brindell, G.D. *J. Am. Chem. Soc.* **1954**, *76*, 5699.
[67]Eaton, P.E.; Yip, Y. *J. Am. Chem. Soc.* **1991**, *113*, 7692.
[68]Brown, H.C.; Russel, G.A. *J. Am. Chem. Soc.* **1952**, *74*, 3995. See also, Desai, V.R.; Nechvatal, A.; Tedder, J.M. *J. Chem. Soc. B* **1970**, 386.
[69]For a review, see Freidlina, R.Kh.; Terent'ev, A.B. *Russ. Chem. Rev.* **1974**, *43*, 129.
[70]McKnight, C.; Rowland, F.S. *J. Am. Chem. Soc.* **1966**, *88*, 3179. For other examples, see Greene, F.D.; Adam, W.; Knudsen Jr., G.A. *J. Org. Chem.* **1966**, *31*, 2087; Gajewski, J.J.; Burka, L.T. *J. Am. Chem. Soc.* **1972**, *94*, 8857, 8860, 8865; Adam, W.; Aponte, G.S. *J. Am. Chem. Soc.* **1971**, *93*, 4300.

The fact that aryl groups migrate, but alkyl groups and hydrogen generally do not, leads to the proposition that **41**, in which the odd electron is not found in the three-membered ring, may be an intermediate. There has been much controversy on this point, but the bulk of the evidence indicates that **41** is a transition state, not an intermediate.[71] Among the evidence is the failure to observe **41** either by ESR[72] or CIDNP.[73] Both of these techniques can detect free radicals with extremely short lifetimes (pp. 266–268).[74]

41

Besides aryl, vinylic[75] and acetoxy groups[76] also migrate. Vinylic groups migrate by way of a cyclopropylcarbinyl radical intermediate (**42**),[77] while the migration of acetoxy groups may involve the charge-separated structure shown.[78] Thermal isomerization of 1-(3-butenyl)cyclopropane at 415°C leads to bicyclo[2.2.1]heptane.[79] Migration has been observed for chloro (and to a much lesser extent

bromo) groups. For example, in the reaction of $Cl_3CCH=CH_2$ with bromine under the influence of peroxides, the products were 47% $Cl_3CCHBrCH_2Br$

---

[71]For molecular-orbital calcualtions indicating that **41** is an intermediate, see Yamabe, S. *Chem. Lett.* **1989**, 1523.

[72]Edge, D.J.; Kochi, J.K. *J. Am. Chem. Soc.* **1972**, *94*, 7695.

[73]Shevlin, P.B.; Hansen, H.J. *J. Org. Chem.* **1977**, *42*, 3011; Olah, G.A.; Krishnamurthy, V.V.; Singh, B.P.; Iyer, P.S. *J. Org. Chem.* **1983**, *48*, 955. **37** has been detected as an intemediate in a different reaction: Effio, A.; Griller, D.; Ingold, K.U.; Scaiano, J.C.; Sheng, S.J. *J. Am. Chem. Soc.* **1980**, *102*, 6063; Leardini, R.; Nanni, D.; Pedulli, G.F.; Tundo, A.; Zanardi, G.; Foresti, E.; Palmieri, P. *J. Am. Chem. Soc.* **1989**, *111*, 7723.

[74]For other evidence, see Martin, M.M. *J. Am. Chem. Soc.* **1962**, *84*, 1986; Rüchardt, C.; Hecht, R. *Chem. Ber.* **1965**, *98*, 2460, 2471; Rüchardt, C.; Trautwein, H. *Chem. Ber.* **1965**, *98*, 2478.

[75]For example, see Slaugh, L.H. *J. Am. Chem. Soc.* **1965**, *87*, 1522; Newcomb, M.; Glenn, A.G.; Williams, W.G. *J. Org. Chem.* **1989**, *54*, 2675.

[76]Surzur, J.; Teissier, P. *Bull. Soc. Chim. Fr.* **1970**, 3060; Tanner, D.D.; Law, F.C.P. *J. Am. Chem. Soc.* **1969**, *91*, 7535; Julia, S.; Lorne, R. *C. R. Acad. Sci. Ser. C* **1971**, *273*, 174; Lewis, S.N.; Miller, J.J.; Winstein, S. *J. Org. Chem.* **1972**, *37*, 1478.

[77]For evidence for this species, see Montgomery, L.K.; Matt, J.W.; Webster, J.R. *J. Am. Chem. Soc.* **1967**, *89*, 923; Montgomery, L.K.; Matt, J.W. *J. Am. Chem. Soc.* **1967**, *89*, 934, 6556; Giese, B.; Heinrich, N.; Horler, H.; Koch, W.; Schwarz, H. *Chem. Ber.* **1986**, *119*, 3528.

[78]Beckwith, A.L.J.; Thomas, C.B. *J. Chem. Soc. Perkin Trans. 2* **1973**, 861; Barclay, L.R.C.; Lusztyk, J.; Ingold, K.U. *J. Am. Chem. Soc.* **1984**, *106*, 1793.

[79]Baldwin, J.E.; Burrell, R.C.; Shukla, R. *Org. Lett.* **2002**, *4*, 3305.

(the normal addition product) and 53% $BrCCl_2CHClCH_2Br$, which arose by rearrangement:

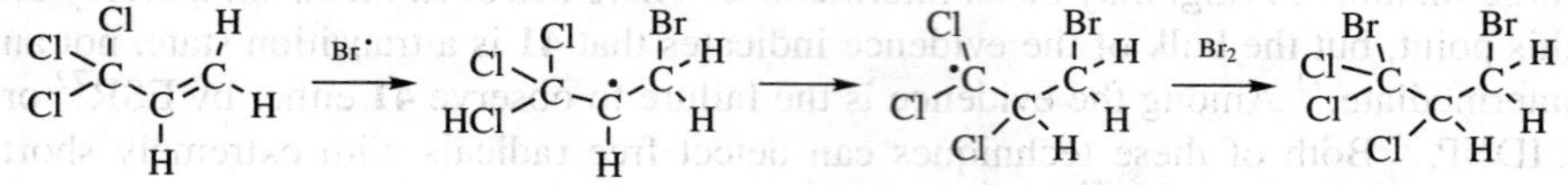

In this particular case, the driving force for the rearrangement is the particular stability of dichloroalkyl free radicals. Nesmeyanov, Freidlina, and co-workers have extensively studied reactions of this sort.[80] It has been shown that the 1,2 migration of Cl readily occurs if the migration origin is tertiary and the migration terminus primary.[81] Migration of Cl and Br could take place by a transition state in which the odd electron is accommodated in a vacant d orbital of the halogen.

Migratory aptitudes have been measured for the phenyl and vinyl groups, and for three other groups, using the system $RCMe_2CH_2\bullet \rightarrow Me_2\dot{C}CH_2R$. These were found to be in the order $R = H_2C=CH_2 > Me_3CC=O > Ph > Me_3C\equiv C > CN$.[82]

In summary then, 1,2 free-radical migrations are much less prevalent than the analogous carbocation processes, and are important only for aryl, vinylic, acetoxy, and halogen migrating groups. The direction of migration is normally toward the more stable radical, but "wrong-way" rearrangements are also known.[83]

Despite the fact that hydrogen atoms do not migrate 1,2, longer free-radical migrations of hydrogen are known.[84] The most common are 1,5-shifts, but 1,6 and longer shifts have also been found. The possibility of 1,3 hydrogen shifts has been much investigated, but it is not certain if any actually occur. If they do they are rare, presumably because the most favorable geometry for $C\cdots H\cdots C$ in the transition state is linear and this geometry cannot be achieved in a 1,3-shift. 1,4-Shifts are definitely known, but are still not very common. These long shifts are best regarded as internal abstractions of hydrogen (for reactions involving them, see **14-6** and **18-40**):

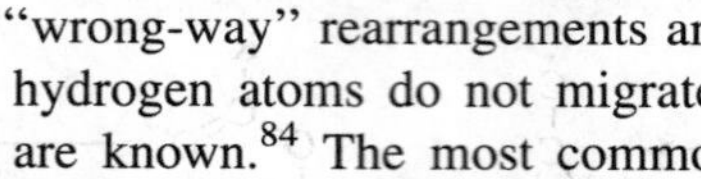

Transannular shifts of hydrogen atoms have also been observed.[85]

[80]For reviews, see Freidlina, R.Kh.; Terent'ev, A.B. *Russ. Chem. Rev.* **1979**, *48*, 828; Freidlina, R.Kh. *Adv. Free-Radical Chem.* **1965**, *1*, 211, 231–249.

[81]See, for example, Skell, P.S.; Pavlis, R.R.; Lewis, D.C.; Shea, K.J. *J. Am. Chem. Soc.* **1973**, *95*, 6735; Chen, K.S.; Tang, D.Y.H.; Montgomery, L.K.; Kochi, J.K. *J. Am. Chem. Soc.* **1974**, *96*, 2201.

[82]Lindsay, D.A.; Lusztyk, J.L.; Ingold, K.U. *J. Am. Chem. Soc.* **1984**, *106*, 7087.

[83]Slaugh, L.H.; Raley, J.H. *J. Am. Chem. Soc.* **1960**, *82*, 1259; Bonner, W.A.; Mango, F.D. *J. Org. Chem.* **1964**, *29*, 29; Dannenberg, J.J.; Dill, K. *Tetrahedron Lett.* **1972**, 1571.

[84]For a discussion, see Freidlina, R.Kh.; Terent'ev, A.B. *Acc. Chem. Res.* **1977**, *10*, 9.

[85]Heusler, K.; Kalvoda, J. *Tetrahedron Lett.* **1963**, 1001; Cope, A.C.; Bly, R.S.; Martin, M.M.; Petterson, R.C. *J. Am. Chem. Soc.* **1965**, *87*, 3111; Fisch, M.; Ourisson, G. *Chem. Commun.* **1965**, 407; Traynham, J.G.; Couvillon, T.M. *J. Am. Chem. Soc.* **1967**, *89*, 3205.

Carbene Rearrangements[86]

Carbenes can rearrange to alkenes in many cases.[87] A 1,2-hydrogen shift leads to an alkene, and this is often competitive with insertion reactions.[88] Benzylchloro-carbene (**43**) rearranges via a 1,2 hydrogen shift to give the alkene.[89] Similarly, carbene **44** rearranges to alkene **45**, and replacement of H on the α-carbon with D showed a deuterium isotope effect of ~5.[90] Vinylidene carbene ($H_2C=C:$) rearranges to acetylene.[91] Rearrangement of alkylidene carbene **46** has been calculated to give the highly unstable cyclopentyne (**47**), which cannot be isolated, but can give a [2 + 2]-cycloaddition product when generated in the presence of a simple alkene.[92] The spiro carbenes undergo rearrangement reactions.[93]

| **43** | **44** | **45** | **46** | **47** |

Electrophilic Rearrangements[94]

Rearrangements in which a group migrates without its electrons are much rarer than the two kinds previously considered, but the general principles are the same. A carbanion (or other negative ion) is created first, and the actual rearrangement step involves migration of a group without its electrons:

The product of the rearrangement may be stable or may react further, depending on its nature (see also, pp. 1585). An *ab initio* study predicts that a [1,2]-alkyl shift in alkyne anions should be facile.[95]

---

[86]For a review of thermally induced cyclopropane–carbene rearrangements, see Baird, M.S. *Chem. Rev.* **2003**, *103*, 1271.

[87]de Meijere, A.; Kozhushkov, S.I.; Faber, D.; Bagutskii, V.; Boese, R.; Haumann, T.; Walsh, R. *Eur. J. Org. Chem.* **2001**, 3607.

[88]Nickon, A.; Stern, A.G.; Ilao, M.C. *Tetrahedron Lett.* **1993**, *34*, 1391.

[89]Merrer, D.C.; Moss, R.A.; Liu, M.T.H.; Banks, J.-T.; Ingold, K.U. *J. Org. Chem.* **1998**, *63*, 3010.

[90]Moss, R.A.; Ho, C.-J.; Liu, W.; Sierakowski, C. *Tetrahedron Lett.* **1992**, *33*, 4287.

[91]Hayes, R.L.; Fattal, E.; Govind, N.; Carter, E.A. *J. Am. Chem. Soc.* **2001**, *123*, 641.

[92]Gilbert, J.C.; Kirschner, S. *Tetrahedron Lett.* **1993**, *34*, 599, 603.

[93]Moss, R.A.; Zheng, F.; Krough-Jespersen, K. *Org. Lett.* **2001**, *3*, 1439.

[94]For reviews, see Hunter, D.H.; Stothers, J.B.; Warnhoff, E.W. in de Mayo, P. *Rearrangments in Ground and Excited States*, Vol. 1, Academic Press, NY, *1980*, pp. 391–470; Grovenstein, Jr., E. *Angew. Chem. Int. Ed.* **1978**, *17*, 313; *Adv. Organomet. Chem.* **1977**, *16*, 167; Jensen, F.R.; Rickborn, B. *Electrophilic Substitution of Organomercurials*, McGraw-Hill, NY, *1968*, pp. 21–30; Cram, D.J. *Fundamentals of Carbanion Chemistry*, Academic Press, NY, *1965*, pp. 223–243.

[95]Borosky, G.L. *J. Org. Chem.* **1998**, *63*, 3337.

REACTIONS

The reactions in this chapter are classified into three main groups and 1,2-shifts are considered first. Within this group, reactions are classified according to (*1*) the identity of the substrate atoms A and B and (*2*) the nature of the migrating group W. In the second group are the cyclic rearrangements. The third group consists of rearrangements that cannot be fitted into either of the first two categories.

Reactions in which the migration terminus is on an aromatic ring have been treated under aromatic substitution. These are **11-27–11-32**, **11-36**, **13-30–13-32**, and, partially, **11-33**, **11-38**, and **11-39**. Double-bond shifts have also been treated in other chapters, though they may be considered rearrangements (p. $$$, p. $$$, and **12-2**). Other reactions that may be regarded as rearrangements are the Pummerer (**19-83**) and Willgerodt (**19-84**) reactions.

1,2-REARRANGEMENTS

A. Carbon-to-Carbon Migrations of R, H, and Ar

18-1   Wagner–Meerwein and Related Reactions

1/Hydro,1/hydroxy-(2/ → 1/alkyl)-*migro*-elimination, and so on

isoborneol   **48**   **49**   camphene

Wagner–Meerwein rearrangements were first discovered in the bicyclic terpenes, and most of the early development of this reaction was with these compounds.[96] An example is the conversion of isoborneol to camphene. It fundamentally involves a 1,2 alkyl shift of an intermediate carbocation, such as **48** → **49**. When alcohols are treated with acids, simple substitution (e.g., **10-48**) or elimination (**17-1**) usually accounts for most or all of the products. But in many cases, especially where two or three alkyl or aryl groups are on the β carbon, some or all of the product is rearranged. These rearrangements have been called *Wagner–Meerwein rearrangements*, although this term is nowadays reserved for relatively specific transformations, such as isoborneol to camphene and related reactions. As pointed out previously, the carbocation that is a direct product of the rearrangement must stabilize itself, and most often it does this by the loss

[96]For a review of rearrangements in bicyclic systems, see Hogeveen, H.; van Kruchten, E.M.G.A. *Top. Curr. Chem.* **1979**, *80*, 89. For reviews concerning caranes and pinanes see, respectively, Arbuzov, B.A.; Isaeva, Z.G. *Russ. Chem. Rev.* **1976**, *45*, 673; Banthorpe, D.V.; Whittaker, D. *Q. Rev. Chem. Soc.* **1966**, *20*, 373.

of a hydrogen β to it, so the rearrangement product is usually an alkene.[97] If there is a choice of protons, Zaitsev's rule (p. 1482) governs the direction, as we might expect. Sometimes a different positive group is lost instead of a proton. Less often, the new carbocation stabilizes itself by combining with a nucleophile instead of losing a proton. The nucleophile may be the water that is the original leaving group, so that the product is a rearranged alcohol, or it may be some other species present (solvent, added nucleophile, etc.). Rearrangement is usually predominant in neopentyl and neophyl types of substrates, and with these types normal nucleophilic substitution is difficult (normal elimination is of course impossible). Under S_N2 conditions, substitution is extremely slow;[98] and under S_N1 conditions, carbocations are formed that rapidly rearrange. However, free-radical substitution, unaccompanied by rearrangement, can be carried out on neopentyl systems, though, as we have seen (p. 1574), neophyl systems undergo rearrangement as well as substitution.

Examples of Wagner–Meerwein-type rearrangements are found in simpler systems, such as neopentyl chloride (example a) and even 1-bromopropane (example b). These two examples illustrate the following points:

1. Hydride ion can migrate. In example b, it was hydride that shifted, not bromine:

2. The leaving group does not have to be H_2O, but can be any departing species whose loss creates a carbocation, including N_2 from aliphatic diazonium ions[99] (see the section on leaving groups in nucleophilic substitution, p. 438). Also, rearrangement may follow when the carbocation is created by addition of a proton or other positive species to a double bond. Even alkanes give

---

[97]For a review of such rearrangements, see Kaupp, G. *Top. Curr. Chem.* **1988**, *146*, 57.

[98]See, however, Lewis, R.G.; Gustafson, D.H.; Erman, W.F. *Tetrahedron Lett.* **1967**, 401; Paquette, L.A.; Philips, J.C. *Tetrahedron Lett.* **1967**, 4645; Anderson, P.H.; Stephenson, B.; Mosher, H.S. *J. Am. Chem. Soc.* **1974**, *96*, 3171.

[99]For reviews of rearrangements arising from diazotization of aliphatic amines, see, in Patai, S. *The Chemistry of the Amino Group*, Wiley, NY, **1968**, the articles by White, E.H.; Woodcock, D.J. pp. 407–497 (473–483) and by Banthorpe, D.V. pp. 585–667 (586–612).

rearrangements when heated with Lewis acids, provided some species is initially present to form a carbocation from the alkane.

3. Example *b* illustrates that the last step can be substitution instead of elimination.

4. Example *a* illustrates that the new double bond is formed in accord with Zaitsev's rule.

2-Norbornyl cations (see **48**), besides displaying the 1,2-shifts of a CH_2 group previously illustrated for the isoborneol → camphene conversion, are also prone to rapid hydride shifts from the 3 to the 2 position (known as 3,2-shifts). These 3,2-shifts usually take place from the exo side;[100] that is, the 3-exo hydrogen migrates to the 2-exo position.[101] This stereoselectivity is analogous to the behavior we have previously seen for norbornyl

systems, namely, that nucleophiles attack norbornyl cations from the exo side (p. 461) and that addition to norbornenes is also usually from the exo direction (p. 1023).

For rearrangements of alkyl carbocations, the direction of rearrangement is usually toward the most stable carbocation (or radical), which is tertiary > secondary > primary, but rearrangements in the other direction have also been found,[102] and often the product is a mixture corresponding to an equilibrium mixture of the possible carbocations. In the Wagner–Meerwein rearrangement, the rearrangement has been observed for a secondary to a secondary carbocation rearrangement, leading to some controversy. Winstein[103] described norbornyl cations in terms of the resonance structures represented by the nonclassical ion **50**.[104] This view was questioned, primarily by Brown,[105] who suggested that the facile rearrangements could be explained by a series of fast 1,3-Wagner–Meerwein shifts.[106]

[100]For example, see Kleinfelter, D.C.; Schleyer, P.v.R. *J. Am. Chem. Soc.* **1961**, *83*, 2329; Collins, C.J.; Cheema, Z.K.; Werth, R.G.; Benjamin, B.M. *J. Am. Chem. Soc.* **1964**, *86*, 4913; Berson, J.A.; Hammons, J.H.; McRowe, A.W.; Bergman, R.G.; Remanick, A.; Houston, D. *J. Am. Chem. Soc.* **1967**, *89*, 2590.
[101]For examples of 3,2-endo shifts, see Bushell, A.W.; Wilder, Jr., P. *J. Am. Chem. Soc.* **1967**, *89*, 5721; Wilder, Jr., P.; Hsieh, W. *J. Org. Chem.* **1971**, *36*, 2552.
[102]See, for example, Cooper, C.N.; Jenner, P.J.; Perry, N.B.; Russell-King, J.; Storesund, H.J.; Whiting, M.C. *J. Chem. Soc. Perkin Trans. 2* **1982**, 605.
[103]Winstein, S. *Quart. Rev. Chem. Soc.* **1969**, *23*, 141; Winstein, S.; Trifan, D.S. *J. Am. Chem. Soc.* **1949**, *71*, 2953; Winstein, S.; Trifan, D.S. *J. Am. Chem. Soc.* **1952**, *74*, 1154.
[104]Berson, J.A., in de Mayo, P. *Molecular Rearrangements*, Vol. 1, Academic Press, NY, **1980**, p. 111; Sargent, G.D. *Quart. Rev. Chem. Soc.* **1966**, *20*, 301; Olah, G.A. *Acc. Chem. Res.* **1976**, *9*, 41; Scheppele, S.E. *Chem. Rev.* **1972**, *72*, 511.
[105]Brown, H.C. *The Non–Classical Ion Problem*, Plenum, New York, **1977**; Brown, H.C. *Tetrahedron* **1976**, *32*, 179; Brown, H.C.; Kawakami, J.H. *J. Am. Chem. Soc.* **1970**, *92*, 1990. See also, Story, R.R.; Clark, B.C., in Olah, G.A.; Schleyer, P.v.R. *Carbonium Ions*, Vol. 3, Wiley, New York, **1972**, p. 1007.
[106]Brown, H.C.; Ravindranathan, M. *J. Am. Chem. Soc.* **1978**, *100*, 1865.

There is considerable evidence, however, that the norbornyl cation rearranges with σ-participation,[107] and there is strong NMR evidence for the nonclassical ion in super acids at low temperatures.[108]

50

As alluded to above, the term "Wagner–Meerwein rearrangement" is not precise. Some use it to refer to all the rearrangements in this section and in **18-2**. Others use it only when an alcohol is converted to a rearranged alkene. Terpene chemists call the migration of a methyl group the *Nametkin rearrangement*. The term *retropinacol rearrangement* is often applied to some or all of these. Fortunately, this disparity in nomenclature does not seem to cause much confusion.

Sometimes several of these rearrangements occur in one molecule, either simultaneously or in rapid succession. A spectacular example is found in the triterpene series. Friedelin is a triterpenoid ketone found in cork. Reduction gives 3β-friedelanol (**51**). When this compound is treated with acid, 13(18)-oleanene (**52**) is formed.[109] In this case, *seven* 1,2-shifts take place. On removal of H_2O from position 3 to leave a positive

51   **52**

charge, the following shifts occur: hydride from 4 to 3; methyl from 5 to 4; hydride from 10 to 5; methyl from 9 to 10; hydride from 8 to 9; methyl from 14 to 8; and methyl from 13 to 14. This leaves a positive charge at position 13, which is stabilized by loss of the proton at the 18 position to give **52**. All these shifts are stereospecific, the group always migrating on the side of the ring system on which it is located; that is, a group above the "plane" of the ring system (indicated by a solid line in **51**) moves above the plane, and a group below the plane (dashed line) moves

[107]Coates, R.M.; Fretz, E.R. *J. Am. Chem. Soc.* **1977**, *99*, 297; Brown, H.C.; Ravindranathan, M. *J. Am. Chem. Soc.* **1977**, *99*, 299.
[108]Olah, G.A. *Carbocations and Electrophilic Reactions*, Verlag Chemie/Wiley, New York, *1974*, pp. 80–89; Olah, G.A.; White, A.M.; DeMember, J.R.; Commeyras, A.; Lui, C.Y. *J. Am. Chem. Soc.* **1970**, *92*, 4627.
[109]Corey, E.J.; Ursprung, J.J. *J. Am. Chem. Soc.* **1956**, *78*, 5041.

below it. It is probable that the seven shifts are not all concerted, although some of them may be, for intermediate products can be isolated.[110] As an illustration of point 2 (p. 1581), it may be mentioned that friedelene, derived from dehydration of **51**, also gives **52** on treatment with acid.[111]

It was mentioned above that even alkanes undergo Wagner–Meerwein rearrangements if treated with Lewis acids and a small amount of initiator. Catalytic asymmetric Wagner–Meerwein shifts have been observed.[112] An interesting application of this reaction is the conversion of tricyclic molecules to adamantane and its derivatives.[113] It has been found that *all* tricyclic alkanes containing 10 carbons are converted to adamantane by treatment with a Lewis acid, such as $AlCl_3$. If the substrate contains >10 carbons, alkyl-substituted adamantanes are produced. The IUPAC name for these reactions is *Schleyer adamantization*. Two examples are

If 14 or more carbons are present, the product may be diamantane or a substituted diamantane.[114] These reactions are successful because of the high thermodynamic stability of adamantane, diamantane, and similar diamond-like molecules. The most stable of a set of C_nH_m isomers (called the *stabilomer*) will be the end product if the reaction reaches equilibrium.[115] Best yields are obtained by the use of "sludge" catalysts[116] (i.e., a mixture of AlX_3 and *tert*-butyl bromide or *sec*-butyl bromide).[117] Though it is certain that these adamantane-forming reactions take place by nucleophilic 1,2-shifts, the exact pathways are not easy to unravel

[110]For a discussion, see Whitlock Jr., H.W.; Olson, A.H. *J. Am. Chem. Soc.* **1970**, *92*, 5383.

[111]Dutler, H.; Jeger, O.; Ruzicka, L. *Helv. Chim. Acta* **1955**, *38*, 1268; Brownlie, G.; Spring, F.S.; Stevenson, R.; Strachan, W.S. *J. Chem. Soc.* **1956**, 2419; Coates, R.M. *Tetrahedron Lett.* **1967**, 4143.

[112]Trost, B.M.; Yasukata, T. *J. Am. Chem. Soc.* **2001**, *123*, 7162.

[113]For reviews, see McKervey, M.A.; Rooney, J.J., in Olah, G.A. *Cage Hydrocarbons*, Wiley, NY, **1990**, pp. 39–64; McKervey, M.A. *Tetrahedron* **1980**, *36*, 971; *Chem. Soc. Rev.* **1974**, *3*, 479; Greenberg, A.; Liebman, J.F. *Strained Organic Molecules*, Academic Press, NY, **1978**, pp. 178–202; Bingham, R.C.; Schleyer, P.v.R. *Fortschr. Chem. Forsch.* **1971**, *18*, 1, 3–23.

[114]See Gund, T.M.; Osawa, E.; Williams, Jr., V.Z.; Schleyer, P.v.R. *J. Org. Chem.* **1974**, *39*, 2979.

[115]For a method for the prediction of stabilomers, see Godleski, S.A.; Schleyer, P.v.R.; Ōsawa, E.; Wipke, W.T. *Prog. Phys. Org. Chem.* **1981**, *13*, 63.

[116]Schneider, A.; Warren, R.W.; Janoski, E.J. *J. Org. Chem.* **1966**, *31*, 1617; Williams, Jr., V.Z.; Schleyer, P.v.R.; Gleicher, G.J.; Rodewald, L.B. *J. Am. Chem. Soc.* **1966**, *88*, 3862; Robinson, M.J.T.; Tarratt, H.J.F. *Tetrahedron Lett.* **1968**, 5.

[117]For other methods, see Johnston, D.E.; McKervey, M.A.; Rooney, J.J. *J. Am. Chem. Soc.* **1971**, *93*, 2798; Olah, G.A.; Wu, A.; Farooq, O.; Prakash, G.K.S. *J. Org. Chem.* **1989**, *54*, 1450.

because of their complexity.[118] Treatment of adamantane-2-^{14}C with $AlCl_3$ results in total carbon scrambling on a statistical basis.[119]

As already indicated, the mechanism of the Wagner–Meerwein rearrangement is usually nucleophilic. Free-radical rearrangements are also known (see the mechanism section of this chapter), though virtually only with aryl migration. However, carbanion mechanisms (electrophilic) have also been found.[94] Thus Ph_3CCH_2Cl treated with sodium gave Ph_2CHCH_2Ph along with unrearranged products.[120] This is called the *Grovenstein–Zimmerman rearrangement*. The intermediate is $Ph_3CCH_2^-$, and the phenyl moves without its electron pair. Only aryl and vinylic,[121] and not alkyl, groups migrate by the electrophilic mechanism (p. $$$) and transition states or intermediates analogous to **41** and **42** are likely.[122]

OS **V**, 16, 194; **VI**, 378, 845.

18-2   The Pinacol Rearrangement

1/O-Hydro,3/hydroxy-(2/ → 3/alkyl)-*migro*-elimination

R = alkyl, aryl, or hydrogen

When *vic*-diols (glycols) are treated with acids,[123] they can be rearranged to give aldehydes or ketones, although elimination without rearrangement can also be accomplished. This reaction is called the *pinacol rearrangement*; the reaction gets its name from a prototype compound pinacol ($Me_2COHCOHMe_2$), which is rearranged to pinacolone (Me_3CCOCH_3).[124] In this type of reaction, reduction can compete with rearrangement.[125] The reaction has been accomplished many times, with alkyl, aryl, hydrogen, and even ethoxycarbonyl (COOEt)[126] as migrating

---

[118]See, for example, Engler, E.M.; Fǎrcaşiu, M.; Sevin, A.; Cense, J.M.; Schleyer, P.v.R. *J. Am. Chem. Soc.* **1973**, *95*, 5769; Klester, A.M.; Ganter, C. *Helv. Chim. Acta* **1983**, *66*, 1200; **1985**, *68*, 734.

[119]Majerski, Z.; Liggero, S.H.; Schleyer, P.v.R.; Wolf, A.P. *Chem. Commun.* **1970**, 1596.

[120]Grovenstein, Jr., E. *J. Am. Chem. Soc.* **1957**, *79*, 4985; Grovenstein, Jr., E.; Williams Jr., L.P. *J. Am. Chem. Soc.* **1961**, *83*, 412; Zimmerman, H.E.; Zweig, A. *J. Am. Chem. Soc.* **1961**, *83*, 1196. See also, Crimmins, T.F.; Murphy, W.S.; Hauser, C.R. *J. Org. Chem.* **1966**, *31*, 4273; Grovenstein, Jr., E.; Cheng, Y. *J. Am. Chem. Soc.* **1972**, *94*, 4971.

[121]See Grovenstein, Jr., E.; Black, K.W.; Goel, S.C.; Hughes, R.L.; Northrop, J.H.; Streeter, D.L.; VanDerveer, D. *J. Org. Chem.* **1989**, *54*, 1671, and references cited therein.

[122]Bertrand, J.A.; Grovenstein, Jr., E.; Lu, P.; VanDerveer, D. *J. Am. Chem. Soc.* **1976**, *98*, 7835.

[123]For a reaction initiated by iminium salts, see Lopez, L.; Mele, G.; Mazzeo, C. *J. Chem. Soc. Perkin Trans. 1* **1994**, 779. For reactions initiated by radical cations, see de Sanabia, J.A.; Carrión, A.E. *Tetrahedron Lett.* **1993**, *34*, 7837. $SbCl_5$ has been used: see Harada, T.; Mukaiyama, T. *Chem. Lett.* **1992**, 81.

[124]For reviews, see Bartók, M.; Molnár, A., in Patai, S. *The Chemistry of Functional Groups, Supplement E*, Wiley, NY, **1980**, pp. 722–732; Collins, C.J.; Eastham, J.F., in Patai, S. *The Chemistry of the Carbonyl Group*, Vol. 1, Wiley, NY, **1966**, pp. 762–771.

[125]Grant, A.A.; Allukian, M.; Fry, A.J. *Tetrahedron Lett.* **2002**, *43*, 4391.

[126]Kagan, J.; Agdeppa Jr., D.A.; Mayers, D.A.; Singh, S.P.; Walters, M.J.; Wintermute, R.D. *J. Org. Chem.* **1976**, *41*, 2355. COOH has been found to migrate in a Wagner–Meerwein reaction: Berner, D.; Cox, D.P.; Dahn, H. *J. Am. Chem. Soc.* **1982**, *104*, 2631.

groups. In most cases, each carbon has at least one alkyl or aryl group, and the reaction is most often carried out with tri- and tetrasubstituted glycols. As mentioned earlier, glycols in which the four R groups are not identical can give rise to more than one product, depending on which group migrates (see p. 1568 for a discussion of migratory aptitudes). A noncatalytic reaction is possible in supercritical water.[127]

Stereodifferentiation is possible in this reaction.[128] When TMSOTf was used to initiate the reaction, it was shown to be highly regioselective.[129] Mixtures are often produced, and which group preferentially migrates may depend on the reaction conditions, as well as on the nature of the substrate. Thus the

action of cold, concentrated sulfuric acid on **53** produces mainly the ketone **54** (methyl migration), while treatment of **53** with acetic acid containing a trace of sulfuric acid gives mostly **55** (phenyl migration).[130] If at least one R is hydrogen, aldehydes can be produced as well as ketones. Generally, aldehyde formation is favored by the use of mild conditions (lower temperatures, weaker acids), because under more drastic conditions the aldehydes may be converted to ketones (**18-4**). The reaction has been carried out in the solid state, by treating solid substrates with HCl gas or with an organic solid acid.[131]

The mechanism involves a simple 1,2-shift. The ion **56** (where all four R groups are Me) has been trapped by the addition of tetrahydrothiophene.[132] It may seem odd that a migration takes place when the positive charge is already at a tertiary position, but carbocations stabilized by an oxygen atom are even more stable than tertiary alkyl cations (p. 242). There is also the driving force supplied by the fact that the new carbocation can immediately stabilize itself by losing a proton.

It is obvious that other compounds in which a positive charge can be placed on a carbon α to one bearing an OH group can also give this rearrangement. This is true for β-amino alcohols, which rearrange on treatment with nitrous acid (this is called

[127]Ikushima, Y.; Hatakeda, K.; Sato, O.; Yokoyama, T.; Arai, M. *J. Am. Chem. Soc.* **2000**, *122*, 1908.
[128]Paquette, L.A.; Lanter, J.C.; Johnston, J.N. *J. Org. Chem.* **1997**, *62*, 1702.
[129]Kudo, K.; Saigo, K.; Hashimoto, Y.; Saito, K.; Hasegawa, M. *Chem. Lett.* **1992**, 1449.
[130]Ramart-Lucas, P.; Salmon-Legagneur, F. *C. R. Acad. Sci.* **1928**, *188*, 1301.
[131]Toda, F.; Shigemasa, T. *J. Chem. Soc. Perkin Trans. 1* **1989**, 209.
[132]Bosshard, H.; Baumann, M.E.; Schetty, G. *Helv. Chim. Acta* **1970**, *53*, 1271.

the *semipinacol* rearrangement), iodohydrins, for which the reagent is mercuric oxide or silver nitrate, β-hydroxyalkyl selenides, $R^1R^2C(OH)C(SeR^5)R^3R^4$,[133] and allylic alcohols,[134] which can rearrange on treatment with a strong acid that protonates the double bond.

A similar rearrangement is given by epoxides,[135]

$$
\underset{R^2}{\overset{R^1}{}} \underset{O}{\overset{R^3}{C\!-\!C}} \underset{R^4}{} \quad \xrightarrow[\underset{MgBr_2-Et_2O}{or}]{BF_3-Et_2O} \quad \underset{R^3}{\overset{R^2}{}} \overset{R^1}{\underset{\underset{O}{\overset{\|}{C}}}{C}} R^4 \qquad R = \text{alkyl, aryl, or hydrogen}
$$

when treated with acidic[136] reagents, such as BF_3–etherate or $MgBr_2$–etherate, 5 M $LiClO_4$ in ether,[137] $InCl_3$,[138] $Al(OC_6F_3)_3$,[139] $Bi(OTf)_3$,[140] $VO(OEt)Cl_2$,[141] or sometimes by heat alone.[142] Epoxides are converted to aldehydes or ketones on treatment with certain metallic catalysts[143] including treatment with iron complexes in refluxing dioxane,[144] $IrCl_3$,[145] or with $BiOClO_4$ in dichloromethane.[146] A related rearrangement called the *Meinwald rearrangement* was induced by the enzyme pig liver esterase.[147] It has been shown that epoxides are intermediates in the pinacol rearrangements of certain glycols.[148] Among the evidence for the mechanism given is that $Me_2COHCOHMe_2$, $Me_2COHCNH_2Me_2$, and $Me_2COHCClMe_2$ gave the reaction at different rates (as expected), but yielded the *same mixture* of two products pinacol and pinacolone indicating a common intermediate.[149]

[133]For a review, see Krief, A.; Laboureur, J.L.; Dumont, W.; Labar, D. *Bull. Soc. Chim. Fr.* **1990**, 681.

[134]See Wang, B.M.; Song, Z.L.; Fan, C.A.; Tu, Y.Q.; Chen, W.M. *Synlett* **2003**, 1497; Hurley, P.B.; Dake, G.R. *Synlett* **2003**, 2131.

[135]For a discussion of the mechanism, see Hodgson, D.M.; Robinson, L.A.; Jones, M.L. *Tetrahedron Lett.* **1999**, *40*, 8637.

[136]Epoxides can also be rearranged with basic catalysts, though the products are usually different. For a review, see Yandovskii, V.N.; Ershov, B.A. *Russ. Chem. Rev.* **1972**, *41*, 403, 410.

[137]Sudha, R.; Narashimhan, K.M.; Saraswathy, V.G.; Sankararaman, S. *J. Org. Chem.* **1996**, *61*, 1877; Sankararaman, S.; Nesakumar, J.E. *J. Chem. Soc., Perkin Trans. 1* **1999**, 3173.

[138]Ranu, B.C.; Jana, U. *J. Org. Chem.* **1998**, *63*, 8212.

[139]Kita, Y.; Furukawa, A.; Futamura, J.; Ueda, K.; Sawama, Y.; Hamamoto, H.; Fujioka, H. *J. Org. Chem.* **2001**, *66*, 8779.

[140]Bhatia, K.A.; Eash, K.J.; Leonard, N.M.; Oswald, M.C.; Mohan, R.S. *Tetrahedron Lett.* **2001**, *42*, 8129.

[141]Martínez, F.; del Campo, C.; Llama, E.F. *J. Chem. Soc., Perkin Trans. 1* **2000**, 1749.

[142]For a list of reagents that accomplish this transformation, with references, see Larock, R.C. *Comprehensive Organic Transformations*; 2nd ed., Wiley-VCH, NY, **1999**, pp. 1277–1280.

[143]For example, see Alper, H.; Des Roches, D.; Durst, T.; Legault, R. *J. Org. Chem.* **1976**, *41*, 3611; Milstein, D.; Buchman, O.; Blum, J. *J. Org. Chem.* **1977**, *42*, 2299; Prandi, J.; Namy, J.L.; Menoret, G.; Kagan, H.B. *J. Organomet. Chem.* **1985**, *285*, 449; Miyashita, A.; Shimada, T.; Sugawara, A.; Nohira, H. *Chem. Lett.* **1986**, 1323; Maruoka, K.; Nagahara, S.; Ooi, T.; Yamamoto, H. *Tetrahedron Lett.* **1989**, *30*, 5607.

[144]Suda, K.; Baba, K.; Nakajima, S.-I.; Takanami, T. *Tetrahedron Lett.* **1999**, *40*, 7243.

[145]Karamé, I.; Tommasino, M.L.; LeMaire, M. *Tetrahedron Lett.* **2003**, *44*, 7687.

[146]Anderson, A.M.; Blazek, J.M.; Garg, P.; Payne, B.J.; Mohan, R.S. *Tetrahedron Lett.* **2000**, *41*, 1527.

[147]Niwayama, S.; Noguchi, H.; Ohno, M.; Kobayashi, S. *Tetrahedron Lett.* **1993**, *34*, 665.

[148]See, for example, Matsumoto, K. *Tetrahedron* **1968**, *24*, 6851; Pocker, Y.; Ronald, B.P. *J. Am. Chem. Soc.* **1970**, *92*, 3385; *J. Org. Chem.* **1970**, *35*, 3362; Tamura, K.; Moriyoshi, T. *Bull. Chem. Soc. Jpn.* **1974**, *47*, 2942.

[149]Pocker, Y. *Chem. Ind.* (*London*), **1959**, 332. See also, Herlihy, K.P. *Aust. J. Chem.* **1981**, *34*, 107.

A good way to prepare β-diketones consists of heating α,β-epoxy ketones at 80–140°C in toluene with small amounts of $(Ph_3P)_4Pd$ and 1,2-bis(diphenylphosphino)ethane.[150] Epoxides are converted to 1,2-diketones with Bi, DMSO, O_2, and a catalytic amounts of $Cu(OTf)_2$ at 100°C.[151] α,β–Epoxy ketones are also converted to 1,2-diketones with a ruthenium catalyst[152] or an iron catalyst.[153] Epoxides with an α-hydroxyalkyl substituent give a pinacol rearrangement product in the presence of a $ZnBr_2$[154] or $Tb(OTf)_3$[155] catalyst to give a γ-hydroxy ketone.

Oxaziridines are converted to ring-expanded lactams under photochemical conditions.[156] N-Tosyl aziridines with an α-hydroxyalkyl substituent give a pinacol rearrangement product in the presence of Lewis acids, such as SmI_2, in this case a keto-N-tosyl amide.[157]

β-Hydroxy ketones can be prepared by treating the silyl ethers (**57**) of α,β-epoxy alcohols with $TiCl_4$.[158]

57

OS **I**, 462; **II**, 73, 408; **III**, 312; **IV**, 375, 957; **V**, 326, 647; **VI**, 39, 320; **VII**, 129. See also, OS **VII**, 456.

18-3 Expansion and Contraction of Rings

Demyanov ring contraction; Demyanov ring expansion

When a positive charge is formed on an alicyclic carbon, migration of an alkyl group can take place to give ring contraction, producing a ring that is one carbon smaller than the original, as in the interconversion of the cyclobutyl cation and the

[150]Suzuki, M.; Watanabe, A.; Noyori, R. *J. Am. Chem. Soc.* **1980**, *102*, 2095.

[151]Antoniotti, S.; Duñach, E. *Chem. Commun.* **2001**, 2566.

[152]Chang, C.-L.; Kumar, M.P.; Liu, R.-S. *J. Org. Chem.* **2004**, *69*, 2793.

[153]Suda, K.; Baba, K.; Nakajima, S.; Takanami, T. *Chem. Commun.* **2002**, 2570.

[154]Tu, Y.Q.; Fan, C.A.; Ren, S.K.; Chan, A.S.C. *J. Chem. Soc., Perkin Trans. 1* **2000**, 3791.

[155]Bickley, J.F.; Hauer, B.; Pena, P.C.A.; Roberts, S.M.; Skidmore, J. *J. Chem. Soc., Perkin Trans. 1* **2001**, 1253.

[156]Bourguet, E.; Baneres, J.-L.; Girard, J.-P.; Parello, J.; Vidal, J.-P.; Lusinchi, X.; Declerzq, J.-P. *Org. Lett.* **2001**, *3*, 3067.

[157]Wang, B.M.; Song, Z.L.; Fan, C.A.; Tu, Y.Q.; Shi, Y. *Org. Lett.* **2002**, *4*, 363.

[158]Maruoka, K.; Hasegawa, M.; Yamamoto, H.; Suzuki, K.; Shimazaki, M.; Tsuchihashi, G. *J. Am. Chem. Soc.* **1986**, *108*, 3827. For a different rearrangement of **53**, see Maruoka, K.; Ooi, T.; Yamamoto, H. *J. Am. Chem. Soc.* **1989**, *111*, 6431.

cyclopropylcarbinyl cation.

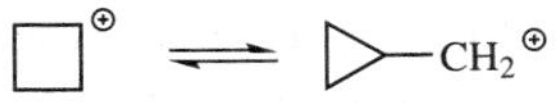

$$\square^{\oplus} \;\rightleftharpoons\; \triangleright\!-\!CH_2^{\oplus}$$

Note that this change involves conversion of a secondary to a primary carbo-cation. In a similar manner, when a positive charge is placed on a carbon a to an alicyclic ring, ring expansion can take place.[159] The new carbocation, and the old one, can then give products by combination with a nucleophile (e.g., the alcohols shown above), or by elimination, so that this reaction is a special case of **18-1**. Often, both rearranged and unrearranged products are formed, so that, for example, cyclobutylamine and cyclopropylmethylamine give similar mixtures of the two alcohols shown above on treatment with nitrous acid (a small amount of 3-buten-1-ol is also produced). When the carbocation is formed by diazotization of an amine, the reaction is called the *Demyanov rearrangement,*[160] but of course similar products are formed when the carbocation is generated in other ways. The expansion reaction has been performed on rings of C_3–C_8,[161] but yields are best with the smaller rings, where relief of small-angle strain provides a driving force for the reaction. The contraction reaction has been applied to four-membered rings and to rings of C_6–C_8, but contraction of a cyclo-pentyl cation to a cyclobutylmethyl system is generally not feasible because of the additional strain involved. Strain is apparently much less of a factor in the cyclobutyl–cyclopropylmethyl interconversion (for a discussion of this inter-conversion, see p. 450). The influence of substituents on this rearrangement has been examined.[162]

Ring expansions of certain hydroxyamines, such as **58**

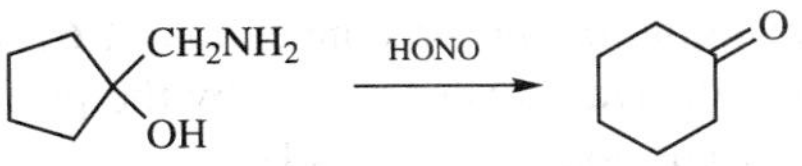

58

---

[159]For monographs on ring expansions, see Hesse, M. *Ring Enlargement in Organic Chemistry*, VCH, NY, *1991*; Gutsche, C.D.; Redmore, D. *Carbocyclic Ring Expansion Reactions*, Academic Press, NY, *1968*. For a review of ring contractions, see Redmore, D.; Gutsche, C.D. *Adv. Alicyclic Chem. 1971*, *3*, 1. For reviews of ring expansions in certain systems, see Baldwin, J.E.; Adlington, R.M.; Robertson, J. *Tetrahedron 1989*, *45*, 909; Stach, H.; Hesse, M. *Tetrahedron 1988*, *44*, 1573; Dolbier Jr., W.R. *Mech. Mol. Migr. 1971*, *3*, 1. For reviews of expansions and contractions of three- and four membered rings, see Salaün, J., in Rappoport, Z. *The Chemistry of the Cyclopropyl Group*, pt. 2, Wiley, NY, *1987*, pp. 809–878; Conia, J.M.; Robson, M.J. *Angew. Chem. Int. Ed. 1975*, *14*, 473. For a list of ring expansions and contractions, with references, see Larock, R.C. *Comprehensive Organic Transformation*, 2nd ed., Wiley-VCH, NY, *1999*, pp. 1283–1302.

[160]For a review, see Smith, P.A.S.; Baer, D.R. *Org. React. 1960*, *11*, 157. See also, Chow, L.; McClure, M.; White, J. *Org. Biomol. Chem. 2004*, *2*, 648.

[161]For a review concerning three-membered rings, see Wong, H.N.C.; Hon, M.; Tse, C.; Yip, Y.; Tanko, J.; Hudlicky, T. *Chem. Rev. 1989*, *89*, 165, see pp. 182–186. For a review concerning three- and four-membered rings, see Breslow, R., in Mayo, P. *Molecular Rearrangements*, Vol. 1, Wiley, NY, *1963*, pp. 233–294.

[162]Wiberg, K.B.; Shobe, D.; Nelson, G.C. *J. Am. Chem. Soc. 1993*, *115*, 10645.

are analogous to the semipinacol rearrangement (**18-2**). This reaction is called the *Tiffeneau–Demyanov ring expansion*. These have been performed on rings of C_4–C_8 and the yields are better than for the simple Demyanov ring expansion. A similar reaction has been used to expand rings of from five to eight members.[163] In this case, a cyclic bromohydrin of the form **59** is treated with a Grignard reagent which, acting as a base, removes the OH proton to give the alkoxide **60**. Refluxing of **60** brings about the ring enlargement. The reaction has been accomplished for **59** in which at least one R group is phenyl or methyl,[164] but fails when both R groups are hydrogen.[165]

A positive charge generated on a three-membered ring gives "contraction" to an allylic cation.[166]

We have previously seen (p. 487) that this is the reason nucleophilic substitutions are not feasible at a cyclopropyl substrate. The reaction is often used to convert cyclopropyl halides and tosylates to allylic products, especially for the purpose of ring expansion, an example being the conversion of **61–62**.[167] The stereochemistry of these cyclopropyl cleavages is governed by the principle of orbital symmetry conservation (for a discussion, see p. 1644).

Three-membered rings can also be cleaved to unsaturated products in at least two other ways. (*1*) On pyrolysis, cyclopropanes can undergo "contraction" to

[163]Sisti, A.J. *Tetrahedron Lett.* **1967**, 5327; *J. Org. Chem.* **1968**, *33*, 453. See also, Sisti, A.J.; Vitale, A.C. *J. Org. Chem.* **1972**, *37*, 4090.

[164]Sisti, A.J.; Meyers, M. *J. Org. Chem.* **1973**, *38*, 4431; Sisti, A.J.; Rusch, G.M. *J. Org. Chem.* **1974**, *39*, 1182.

[165]Sisti, A.J. *J. Org. Chem.* **1968**, *33*, 3953.

[166]For reviews, see Marvell, E.N. *Thermal Electrocyclic Reactions*, Academic Press, NY, **1980**, pp. 23–53; Sorensen, T.S.; Rauk, A., in Marchand, A.P.; Lehr, R.E. *Pericyclic Reactions*, Vol. 2, Academic Press, NY, **1977**, pp. 1–78.

[167]Skell, P.S.; Sandler, S.R. *J. Am. Chem. Soc.* **1958**, *80*, 2024.

propenes.[168] In the simplest case, cyclopropane gives propene when heated to 400–500°C. The mechanism is generally regarded[169] as involving a diradical

intermediate[170] (recall that free-radical 1,2 migration is possible for diradicals, p. 1574). (2) The generation of a carbene or carbenoid carbon in a three-membered ring can lead to allenes, and allenes are often prepared in this

way.[171] Flash vacuum pyrolysis of 1-chlorocyclopropene thermally rearranges to chloroallene.[172] One way to generate, such a species is treatment of a 1,1-dihalo-cyclopropane with an alkyllithium compound (**12-39**).[173] In contrast, the generation of a carbene or carbenoid at a cyclopropylmethyl carbon gives ring expansion.[174]

Some free-radical ring enlargements are also known, an example being:[175]

[168]For reviews, see Berson, J.A., in de Mayo, P. *Rearrangaements in Ground and Excited States*, Vol. 1, Academic Press, NY, *1980*, pp. 324–352; *Ann. Rev. Phys. Chem. 1977*, *28*, 111; Bergman, R.G., in Kochi, J.K. *Free Radicals*, Vol. 1, Wiley, NY, *1973*, pp. 191–237; Frey, H.M. *Adv. Phys. Org. Chem. 1966*, *4*, 147, see pp. 148–170.

[169]For evidence that diradical intermediates may not be involved, at least in some cases, see Fields, R.; Haszeldine, R.N.; Peter, D. *Chem. Commun. 1967*, 1081; Parry, K.A.W.; Robinson, P.J. *Chem. Commun. 1967*, 1083; Clifford, R.P.; Holbrook, K.A. *J. Chem. Soc. Perkin Trans. 2 1972*, 1972; Baldwin, J.E.; Grayston, M.W. *J. Am. Chem. Soc. 1974*, *96*, 1629, 1630.

[170]We have seen before that such diradicals can close up to give cyclopropanes (**17-34**). Therefore, pyrolysis of cyclopropanes can produce not only propenes, but also isomerized (cis → trans or optically active → inactive) cyclopropanes. See, for example, Berson, J.A.; Balquist, J.M. *J. Am. Chem. Soc. 1968*, *90*, 7343; Bergman, R.G.; Carter, W.L. *J. Am. Chem. Soc. 1969*, *91*, 7411.

[171]For reviews, see Schuster, H.F.; Coppola, G.M. *Allenes in Organic Synthesis*, Wiley, NY, *1984*, pp. 20–23; Kirmse, W. *Carbene Chemistry*, 2nd ed., Academic Press, NY, *1971*, pp. 462–467.

[172]Billups, W.E.; Bachman, R.E. *Tetrahedron Lett. 1992*, *33*, 1825.

[173]See Baird, M.S.; Baxter, A.G.W. *J. Chem. Soc. Perkin Trans. 1 1979*, 2317, and references cited therein.

[174]For a review, see Gutsche, C.D.; Redmore, D. *Carbocyclic Ring Expansion Reactions*, Academic Press, NY, *1968*, pp. 111–117.

[175]Dowd, P.; Choi, S. *Tetrahedron Lett. 1991*, *32*, 565; *Tetrahedron 1991*, *47*, 4847. For a related ring expansion, see Baldwin, J.E.; Adlington, R.M.; Robertson, J. *J. Chem. Soc., Chem. Commun. 1988*, 1404.

This reaction has been used to make rings of 6, 7, 8, and 13 members. A possible mechanism is

This reaction has been extended to the expansion of rings by three or four carbons, by the use of a substrate containing $(CH_2)_nX$ ($n = 3$ or 4) instead of CH_2Br.[176] By this means, 5-, 6-, and 7-membered rings were enlarged to 18–11-membered rings.

OS **III**, 276; **IV**, 221, 957; **V**, 306, 320; **VI**, 142, 187; **VII**, 12, 114, 117, 129, 135; **VIII**, 179, 467, 556, 578.

18-4 Acid-Catalyzed Rearrangements of Aldehydes and Ketones

1/Alkyl,2/alkyl-interchange, and so on

Rearrangements of this type, where a group α to a carbonyl "changes places" with a group attached to the carbonyl carbon, occur when migratory aptitudes are favorable.[177] The R^2, R^3, and R^4 groups may be alkyl or hydrogen. Certain aldehydes have been converted to ketones, and ketones to other ketones (though more drastic conditions are required for the latter), but no rearrangement of a ketone to an aldehyde ($R^1 = H$) has so far been reported. There are two mechanisms,[178] each beginning with protonation of the oxygen and each involving two migrations. In one pathway, the migrations are in opposite directions:[179]

[176]Dowd, P.; Choi, S. *J. Am. Chem. Soc.* **1987**, *109*, 6548; *Tetrahedron Lett.* **1991**, *32*, 565.
[177]For reviews, see Fry, A. *Mech. Mol. Migr.* **1971**, *4*, 113; Collins, C.J.; Eastham, J.F., in Patai, S. *The Chemistry of the Carbonyl Group*, Vol. 1, Wiley, NY, **1966**, pp. 771–790.
[178]Favorskii, A.; Chilingaren, A. *C. R. Acad. Sci.* **1926**, *182*, 221.
[179]Kendrick Jr., L.W.; Benjamin, B.M.; Collins, C.J. *J. Am. Chem. Soc.* **1958**, *80*, 4057; Rothrock, T.S.; Fry, A. *J. Am. Chem. Soc.* **1958**, *80*, 4349; Collins, C.J.; Bowman, N.S. *J. Am. Chem. Soc.* **1959**, *81*, 3614.

In the other pathway, the migrations are in the same direction. The actual mechanism of this pathway is not certain, but an epoxide (protonated) intermediate[180] is one possibility:[181]

If the reaction is carried out with ketone labeled in the C=O group with ^{14}C, the first pathway predicts that the product will contain all the ^{14}C in the C=O carbon, while in the second pathway the label will be in the α carbon (demonstrating migration of oxygen). The results of such experiments[182] have shown that in some cases only the C=O carbon was labeled, in other cases only the a carbon, while in still others both carbons bore the label, indicating that in these cases both pathways were in operation. With α-hydroxy aldehydes and ketones, the process may stop after only one migration (this is called the α-*ketol* rearrangement).

The α-ketol rearrangement can also be brought about by base catalysis, but only if the alcohol is tertiary, since if R^1 or R^2 = hydrogen, enolization of the substrate is more favored than rearrangement.

18-5   The Dienone–Phenol Rearrangement

$2/C \rightarrow 5/O$-Hydro,$1/C \rightarrow 2/C$-alkyl-bis-migration

[180]Zook, H.D.; Smith, W.E.; Greene, J.L. *J. Am. Chem. Soc.* **1957**, *79*, 4436.

[181]Some such pathway is necessary to account for the migration of oxygen that is found. It may involve a protonated epoxide, a 1,2-diol, or simply a [1,2]-shift of an OH group.

[182]See, for example, Barton, S.; Porter, C.R. *J. Chem. Soc.* **1956**, 2483; Zalesskaya, T.E.; Remizova, T.B. *J. Gen. Chem. USSR* **1965**, *35*, 29; Fry, A.; Oka, M. *J. Am. Chem. Soc.* **1979**, *101*, 6353.

Cyclohexadienone derivatives that have two alkyl groups in the 4 position undergo, on acid treatment,[183] 1,2 migration of one of these groups from **64** to give the phenol. Note that a photochemical version of this reaction has been observed.[184]

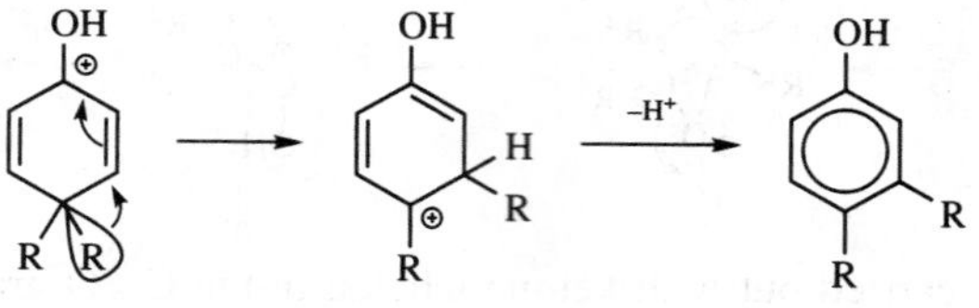

63   **64**

The driving force in the overall reaction (the *dienone–phenol rearrangement*) is of course creation of an aromatic system.[185] Note that **63** and **64** are arenium ions (p. 240), the same as those generated by attack of a phenol on an electrophile.[186] Sometimes, in the reaction of a phenol with an electrophile, a kind of reverse rearrangement (called the *phenol–dienone rearrangement*) takes place, though without an actual migration.[187] An example is

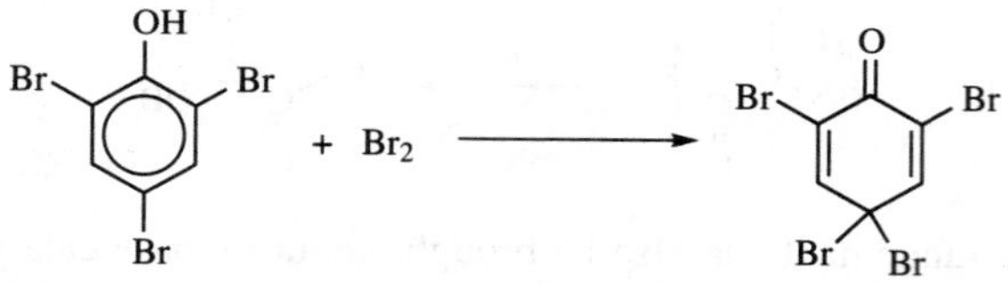

18-6   The Benzil–Benzilic Acid Rearrangement

1/*O*-Hydro,3/oxido-(1/ → 2/aryl)-*migro*-addition

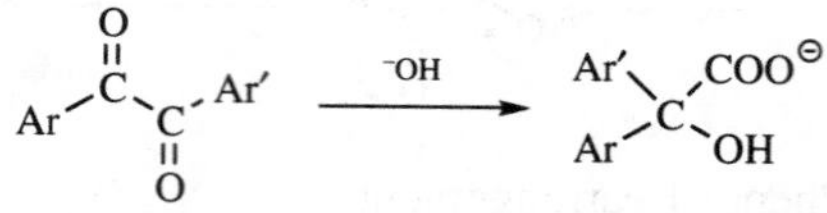

When treated with base, α-diketones rearrange to give the salts of α-hydroxy acids, a reaction known as the *benzil-benzilic acid rearrangement* (benzil is

---

[183]For a reagent that greatly accelerates this reaction, see Chalais, S.; Laszlo, P.; Mathy, A. *Tetrahedron Lett.* **1986**, *27*, 2627.

[184]Guo, Z.; Schultz, A.G. *Org. Lett.* **2001**, *3*, 1177.

[185]For reviews, see Perkins, M.J.; Ward, P. *Mech. Mol. Migr.* **1971**, *4*, 55, 90–103; Miller, B. *Mech. Mol. Migr.* **1968**, *1*, 247; Shine, H.J. *Aromatic Rearrangements*, Elsevier, NY, **1967**, pp. 55–68; Waring, A.J. *Adv. Alicyclic Chem.* **1966**, *1*, 129, 207–223. For a review of other rearrangements of cyclohexadienones, see Miller, B. *Acc. Chem. Res.* **1975**, *8*, 245.

[186]For evidence that these ions are indeed intermediates in this rearrangement, see Vitullo, V.P.; Grossman, N. *J. Am. Chem. Soc.* **1972**, *94*, 3844; Planas, A.; Tomás, J.; Bonet, J. *Tetrahedron Lett.* **1987**, *28*, 471.

[187]For a review, see Ershov, V.V.; Volod'kin, A.A.; Bogdanov, G.N. *Russ. Chem. Rev.* **1963**, *32*, 75.

PhCOCOPh; benzilic acid is Ph$_2$COHCOOH).[188] A rhodium catalyzed version of this reaction has also been reported.[189] Though the reaction is usually illustrated with aryl groups, it can also be applied to aliphatic diketones[190] and to α-keto aldehydes. The use of an alkoxide instead of hydroxide gives the corresponding ester directly,[191] though alkoxide ions that are readily oxidized (e.g., OEt$^-$ or OCHMe$_2^-$) are not useful here, since they reduce the benzil to a benzoin. The mechanism is similar to the rearrangements in **18-1–18-4**, but there is a difference: The migrating group does not move to a carbon with an open sextet. The carbon makes room for the migrating group by releasing a pair of π electrons from the C=O bond to the oxygen. The first step is attack of the base at the carbonyl group, the same as the first step of the tetrahedral mechanism of nucleophilic substitution (p. 1254) and of many additions to the C=O bond (Chapter 16):

The mechanism has been intensely studied,[188] and there is much evidence for it.[192] The reaction is irreversible.

OS **I**, 89.

18-7 The Favorskii Rearrangement

2/Alkoxy-de-chloro(2/ → 1/alkyl)-*migro*-substitution

The reaction of α-halo ketones (chloro, bromo, or iodo) with alkoxide ions[193] to give rearranged esters is called the *Favorskii rearrangement*.[194] The use of

---

[188]For a review, see Selman, S.; Eastham, J.F. *Q. Rev. Chem. Soc.* **1960**, *14*, 221.

[189]Shimizu, I.; Tekawa, M.; Maruyama, Y.; Yamamoto, A. *Chem. Lett.* **1992**, 1365.

[190]For an example, see Schaltegger, A.; Bigler, P. *Helv. Chim. Acta* **1986**, *69*, 1666.

[191]Doering, W. von E.; Urban, R.S. *J. Am. Chem. Soc.* **1956**, *78*, 5938.

[192]However, some evidence for an SET pathway has been reported: Screttas, C.G.; Micha-Screttas, M.; Cazianis, C.T. *Tetrahedron Lett.* **1983**, *24*, 3287.

[193]The reaction has also been reported to take place with BF$_3$–MeOH and Ag$^+$: Giordano, C.; Castaldi, G.; Casagrande, F.; Abis, L. *Tetrahedron Lett.* **1982**, *23*, 1385.

[194]For reviews, see Boyer, L.E.; Brazzillo, J.; Forman, M.A.; Zanoni, B. *J. Org. Chem.* **1996**, *61*, 7611; Hunter, D.H.; Stothers, J.B.; Warnhoff, E.W., in de Mayo, P. *Rearrangements in Ground and Excited States*, Vol. 1, Academic Press, NY, **1980**, pp. 437–461; Chenier, P.J. *J. Chem. Educ.* **1978**, *55*, 286; Rappe, C., in Patai, S. *The Chemistry of the Carbon–Halogen Bond*, pt. 2, Wiley, NY, **1973**, pp. 1084–1101; Redmore, D.; Gutsche, C.D. *Carbocyclic Ring Expansion Reactions*, Academic Press, NY, **1968**, pp.46–69; Akhrem, A.A.; Ustynyuk, T.K.; Titov, Yu.A. *Russ. Chem. Rev.* **1970**, *39*, 732. For an asymmetric version, see Satoh, T.; Motohashi, S.; Kimura, S.; Tokutake, N.; Yamakawa, K. *Tetrahedron Lett.* **1993**, *34*, 4823.

hydroxide ions or amines as bases leads to the free carboxylic acid (salt) or amide, respectively, instead of the ester. Cyclic α-halo ketones give ring contraction, as in the conversion of **65–66**.

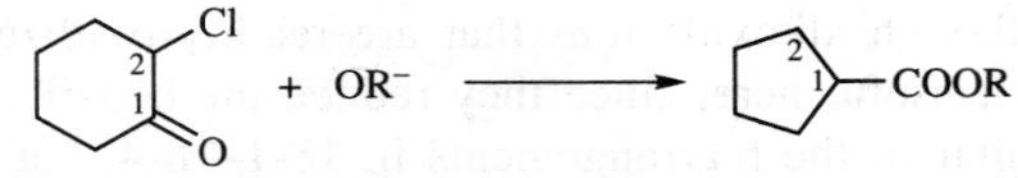

The reaction has also been carried out on α-hydroxy ketones[195] and on α,β-epoxy ketones, which give β-hydroxy acids.[196] The fact that an epoxide gives a reaction analogous to a halide indicates that the oxygen and halogen are leaving groups in a nucleophilic substitution step.

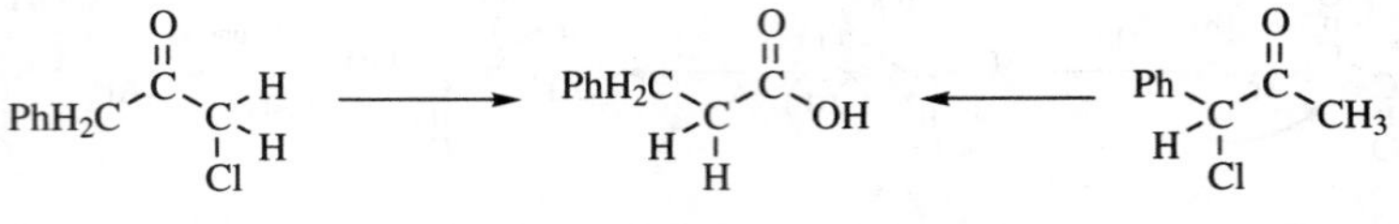

Through the years, the mechanism[197] of the Favorskii rearrangement has been the subject of much investigation; at least five different mechanisms have been proposed. However, the finding[198] that **67** and **68** *both* give **69** (this behavior is typical) shows that any mechanism where the halogen leaves and R[1] takes its place is invalid, since in such a case **67** would be expected to give **69** (with PhCH$_2$ migrating), but **68** should give PhCHMeCOOH (with CH$_3$ migrating). That is, in the case of **68**, it was PhCH that migrated and not methyl. Another important result was determined by radioactive labeling. **65**, in which C-1 and C-2 were equally labeled with ^{14}C, was converted to **66**. The product was found to contain 50% of the label on the carbonyl carbon, 25% on C-1, and 25% on C-2.[199] Now the carbonyl carbon, which originally carried half of the radio-activity, still had this much, so the rearrangement did not directly affect *it*. How-ever, if the C-6 carbon had migrated to C-2, the other half of the radioactivity would be only on C-1 of the product:

[195]Craig, J.C.; Dinner, A.; Mulligan, P.J. *J. Org. Chem.* **1972**, *37*, 3539.
[196]See, for example, House, H.O.; Gilmor, W.F. *J. Am. Chem. Soc.* **1961**, *83*, 3972; Mouk, R.W.; Patel, K.M.; Reusch, W. *Tetrahedron* **1975**, *31*, 13.
[197]For a review of the mechanism, see Baretta, A.; Waegell, B. *React. Intermed. (Plenum)* **1982**, *2*, 527.
[198]McPhee, W.D.; Klingsberg, E. *J. Am. Chem. Soc.* **1944**, *66*, 1132; Bordwell, F.G.; Scamehorn, R.G.; Springer, W.R. *J. Am. Chem. Soc.* **1969**, *91*, 2087.
[199]Loftfield, R.B. *J. Am. Chem. Soc.* **1951**, *73*, 4707.

On the other hand, if the migration had gone the other way: If the C-2 carbon had migrated to C-6–then this half of the radioactivity would be found solely on C-2 of the product:

65

The fact that C-1 and C-2 were found to be equally labeled showed that *both migrations occurred*, with equal probability. Since C-2 and C-6 of **65** are not equivalent, this means that there must be a symmetrical intermediate.[200] The type of intermediate that best fits the circumstances is a cyclopropanone,[201] and the mechanism (for the general case) is formulated (replacing R^1 of our former symbolism with CHR^5R^6, since it is obvious that for this mechanism an α hydrogen is required on the non-halogenated side of the carbonyl):

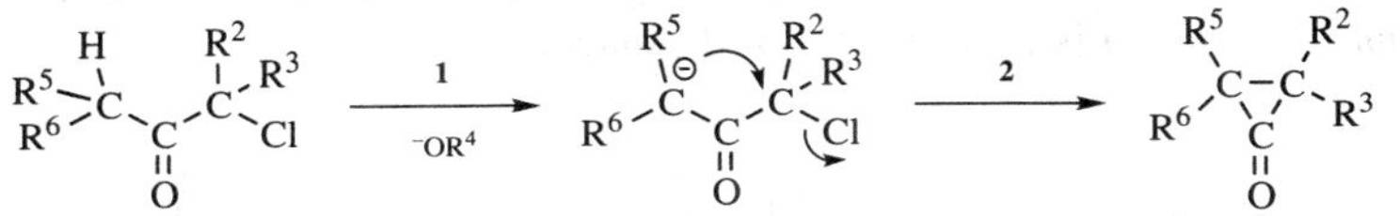

70   **71**

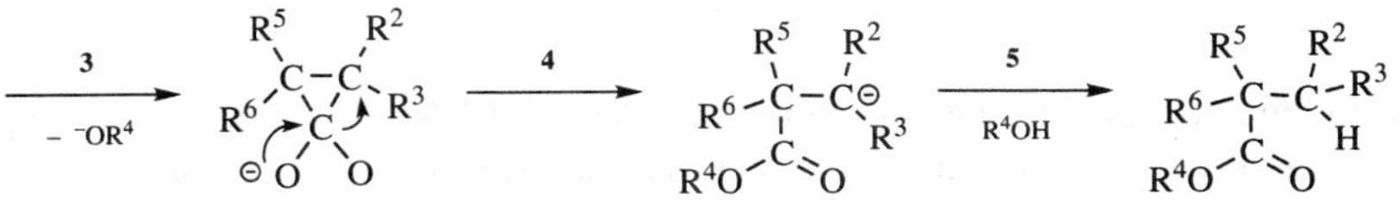

The intermediate corresponding to **71** in the case of **65** is a symmetrical compound, and the three-membered ring can be opened with equal probability on either side of the carbonyl, accounting for the results with ^{14}C. In the general case, **71** is not symmetrical and should open on the side that gives the more stable carbanion.[202] This accounts for the fact that **67** and **68** give the same product. The intermediate in both cases is **70**, which always opens to give the carbanion stabilized by resonance. The cyclopropanone intermediate (**71**) has been isolated in the case where $R^2 = R^5 = t$-Bu and $R^3 = R^6 = H$,[203] and it

---

[200]A preliminary migration of the chlorine from C-2 to C-6 was ruled out by the fact that recovered **65** had the same isotopic distribution as the starting **65**.

[201]Although cyclopropanones are very reactive compounds, several of them have been isolated. For reviews of cyclopropanone chemistry, see Wasserman, H.H.; Clark, G.M.; Turley, P.C. *Top. Curr. Chem.* **1974**, *47*, 73; Turro, N.J. *Acc. Chem. Res.* **1969**, *2*, 25.

[202]Factors other than carbanion stability (including steric factors) may also be important in determining which side of an unsymmetrical **71** is preferentially opened. See, for example, Rappe, C.; Knutsson, L. *Acta Chem. Scand.*, **1967**, *21*, 2205; Rappe, C.; Knutsson, L.; Turro, N.J.; Gagosian, R.B. *J. Am. Chem. Soc.* **1970**, *92*, 2032.

[203]Pazos, J.F.; Pacifici, J.G.; Pierson, G.O.; Sclove, D.B.; Greene, F.D. *J. Org. Chem.* **1974**, *39*, 1990.

has also been trapped.[204] Also, cyclopropanones synthesized by other methods have been shown to give Favorskii products on treatment with NaOMe or other bases.[205]

The mechanism discussed is in accord with all the facts when the halo ketone contains an α hydrogen on the other side of the carbonyl group. However, ketones that do not have a hydrogen there also rearrange to give the same type of product. This is usually called the *quasi-Favorskii rearrangement*. An example is found in the preparation of Demerol:[206]

Demerol

The quasi-Favorskii rearrangement obviously cannot take place by the cyclopropanone mechanism. The mechanism that is generally accepted (called the *semibenzilic mechanism*[207]) is a base-catalyzed pinacol

rearrangement-type mechanism similar to that of **18-6**. This mechanism requires inversion at the migration terminus and this has been found.[208] It has been shown that even where there *is* an appropriately situated a hydrogen, the semibenzilic mechanism may still operate.[209]

An interesting analog of the Favorskii rearrangement treats a ketone, such as 4-*tert*-butylcyclohexanone, without an α-halogen with $Tl(NO_3)_3$ to give 3-*tert*-butylcyclopentane-1-carboxylic acid.[210]

OS **IV**, 594; **VI**, 368, 711.

[204]Fort, A.W. *J. Am. Chem. Soc.* **1962**, *84*, 4979; Cookson, R.C.; Nye, M.J. *Proc. Chem. Soc.* **1963**, 129; Breslow, R.; Posner, J.; Krebs, A. *J. Am. Chem. Soc.* **1963**, *85*, 234; Baldwin, J.E.; Cardellina, J.H.I. *Chem. Commun.* **1968**, 558.

[205]Crandall, J.K.; Machleder, W.H. *J. Org. Chem.* **1968**, *90*, 7347; Turro, N.J.; Gagosian, R.B.; Rappe, C.; Knutsson, L. *Chem. Commun.* **1969**, 270; Wharton, P.S.; Fritzberg, A.R. *J. Org. Chem.* **1972**, *37*, 1899.

[206]Smisman, E.E.; Hite, G. *J. Am. Chem. Soc.* **1959**, *81*, 1201.

[207]Tchoubar, B.; Sackur, O. *C. R. Acad. Sci.* **1939**, *208*, 1020.

[208]Baudry, D.; Bégué, J.; Charpentier-Morize, M. *Bull. Soc. Chim. Fr.* **1971**, 1416; *Tetrahedron Lett.* **1970**, 2147.

[209]For example, see Salaun, J.R.; Garnier, B.; Conia, J.M. *Tetrahedron* **1973**, *29*, 2895; Rappe, C.; Knutsson, L. *Acta Chem. Scand.*, **1967**, *21*, 163; Warnhoff, E.W.; Wong, C.M.; Tai, W.T. *J. Am. Chem. Soc.* **1968**, *90*, 514.

[210]Ferraz, H.M.; Silva, Jr., J.F. *Tetrahedron Lett.* **1997**, *38*, 1899.

18-8   The Arndt–Eistert Synthesis

In the *Arndt–Eistert synthesis*, an acyl halide is converted to a carboxylic acid with one additional carbon.[211] The first step of this process is reaction **16-89**. The actual rearrangement occurs in the second step on treatment of the diazo ketone with water and silver oxide or with silver benzoate and triethylamine. This rearrangement is called the *Wolff rearrangement*.[212] It is the best method of increasing a carbon chain by one if a *carboxylic acid* is available (**10-75** and **16-30** begin with alkyl halides). If an alcohol R'OH is used instead of water, the ester RCH$_2$COOR' is isolated directly.[213] Similarly, ammonia gives the amide. Other catalysts are sometimes used (e.g., colloidal platinum, copper, etc.), but occasionally the diazo ketone is simply heated or photolyzed in the presence of water, an alcohol, or ammonia, with no catalyst at all.[214] The photolysis method[215] often gives better results than the silver catalysis method. Of course, diazo ketones prepared in any other way also give the rearrangement.[216] The reaction is of wide scope. The R group may be alkyl or aryl and may contain many functional groups including unsaturation, but not including groups acidic enough to react with CH$_2$N$_2$ or diazo ketones (e.g., **10-5** and **10-19**). Sometimes the reaction is performed with other diazoalkanes (i.e., R'CHN$_2$) to give RCHR'COOH. The reaction has often been used for ring contraction of cyclic diazo ketones,[217] such as **72**.[218]

72

[211]For reviews, see Meier, H.; Zeller, K. *Angew. Chem. Int. Ed.* **1975**, *14*, 32; Kirmse, W. *Carbene Chemistry*, 2nd ed., Academic Press, NY, **1971**, pp. 475–493; Rodina, L.L.; Korobitsyna, I.K. *Russ. Chem. Rev.* **1967**, *36*, 260; For a review of rearrangements of diazo and diazonium compounds, see Whittaker, D., in Patai, S. *The Chemistry of Diazonium and Diazo Compounds*, pt. 2, Wiley, NY, **1978**, pp. 593–644.

[212]For a review, see Kirmse, W. *Eur. J. Org. Chem.* **2002**, 2193. For a microwave-induced Wolff rearrangement, see Sudrik, S.G.; Chavan, S.P.; Chandrakumar, K.R.S.; Pal, S.; Date, S.K.; Chavan, S.P.; Sonawane, H.R. *J. Org. Chem.* **2002**, *67*, 1574.

[213]For an ultrasound-induced version of this variation, see Winum, J.-Y.; Kamal, M.; Leydet, A.; Roque, J.-P.; Montero, J.-L. *Tetrahedron Lett.* **1996**, *37*, 1781.

[214]For a list of methods, with references, see Larock, R.C. *Comprehensive Organic Transformations*, 2nd ed., Wiley-VCH, NY, **1999**, pp. 1850–1851.

[215]For reviews of the photolysis method, see Regitz, M.; Maas, G. *Diazo Compounds*, Academic Press, NY, **1986**, pp. 185–195; Ando, W., in Patai, S. *The Chemistry of the Carbonyl Group*, Vol. 1, Wiley, NY, **1966**,78, pp. 458–475.

[216]For a method of conducting the reaction with trimethylsilyldiazomethane instead of CH$_2$N$_2$, see Aoyama, T.; Shioiri, T. *Tetrahedron Lett.* **1980**, *21*, 4461.

[217]For a review, see Redmore, D.; Gutsche, C.D. *Carbocyclic Ring Expansion Reactions*, Academic Press, NY, **1968**, pp. 125–136.

[218]Korobitsyna, I.K.; Rodina, L.L.; Sushko, T.P. *J. Org. Chem. USSR* **1968**, *4*, 165; Jones, Jr., M.; Ando, W. *J. Am. Chem. Soc.* **1968**, *90*, 2200. See Lee, Y.R.; Suk, J.Y.; Kim, B.S. *Tetrahedron Lett.* **1999**, *40*, 8219.

The mechanism is generally regarded as involving formation of a carbene.[219] It is the divalent carbon that has the open sextet and to which the migrating group brings its electron pair:

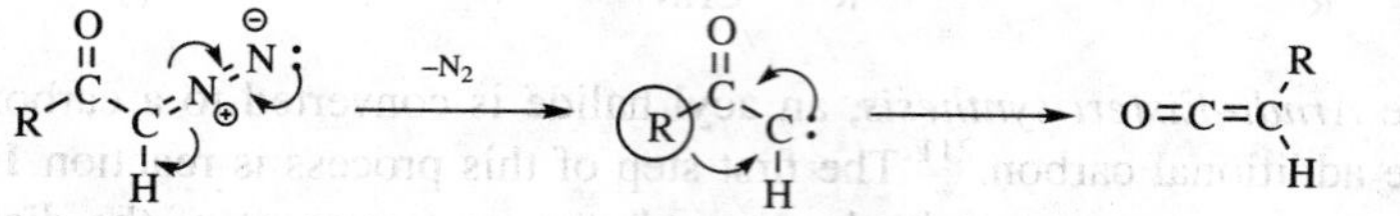

The actual product of the reaction is thus the ketene, which then reacts with water (**15-3**), an alcohol (**15-5**), or ammonia or an amine (**15-8**). Particularly stable ketenes[220] (e.g., $Ph_2C=C=O$) have been isolated and others have been trapped in other ways (e.g., as β-lactams,[221] **16-96**). The purpose of the catalyst is not well understood, though many suggestions have been made. This mechanism is strictly analogous to that of the Curtius rearrangement (**18-14**). Although the mechanism as shown above involves a free carbene and there is much evidence to support this,[222] it is also possible that at least in some cases the two steps are concerted and a free carbene is absent.

When the Wolff rearrangement is carried out photochemically, the mechanism is basically the same,[215] but another pathway can intervene. Some of the ketocarbene originally formed can undergo a carbene–carbene rearrangement, through an oxirene intermediate.[223] This was shown by ^{14}C labeling experiments, where

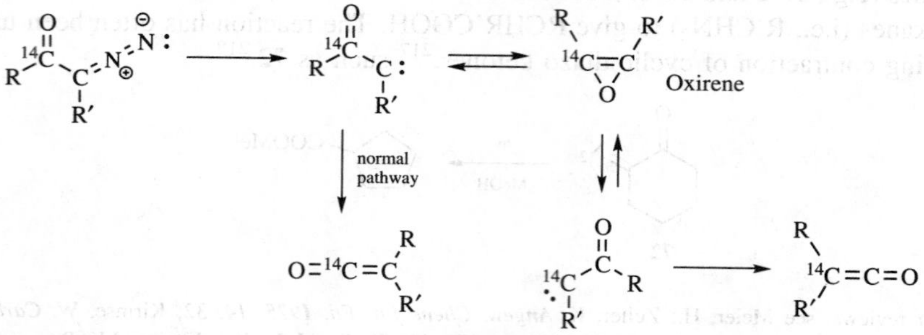

diazo ketones labeled in the carbonyl group gave rise to ketenes that bore the label at both C=C carbons.[224] In general, the smallest degree of scrambling (and thus of

[219]See Scott, A.P.; Platz, M.S.; Radom, L. *J. Am. Chem. Soc.* **2001**, *123*, 6069.

[220]In some cases, ketenes are subject to rearrangement, see Farlow, R.A.; Thamatloor, D.A.; Sunoj, R.B.; Hadad, C.M. *J. Org. Chem.* **2002**, *67*, 3257.

[221]Kirmse, W.; Horner, L. *Chem. Ber.* **1956**, *89*, 2759; also see, Horner, L.; Spietschka, E. *Chem. Ber.* **1956**, *89*, 2765.

[222]For a summary of evidence on both sides of the question, see Kirmse, W. *Carbene Chemistry*, 2nd ed., Academic Press, NY, **1971**, pp. 476–480. See also, Torres, M.; Ribo, J.; Clement, A.; Strausz, O.P. *Can. J. Chem.* **1983**, *61*, 996; Tomoika, H.; Hayashi, N.; Asano, T.; Izawa, Y. *Bull. Chem. Soc. Jpn.* **1983**, *56*, 758.

[223]For a review of oxirenes, see Lewars, Y. *Chem. Rev.* **1983**, *83*, 519.

[224]Fenwick, J.; Frater, G.; Ogi, K.; Strausz, O.P. *J. Am. Chem. Soc.* **1973**, *95*, 124; Zeller, K. *Chem. Ber.* **1978**, *112*, 678. See also, Thornton, D.E.; Gosavi, R.K.; Strausz, O.P. *J. Am. Chem. Soc.* **1970**, *92*, 1768; Russell, R.L.; Rowland, F.S. *J. Am. Chem. Soc.* **1970**, *92*, 7508; Majerski, Z.; Redvanly, C.S. *J. Chem. Soc., Chem. Commun.* **1972**, 694.

the oxirene pathway) was found when $R' = H$. An intermediate believed to be an oxirene has been detected by laser spectroscopy.[225] The oxirene pathway is not found in the thermal Wolff rearrangement. It is likely that an excited singlet state of the carbene is necessary for the oxirene pathway to intervene.[226] In the photochemical process, ketocarbene intermediates, in the triplet state, have been isolated in an Ar matrix at 10–15 K, where they have been identified by UV–visible, IR, and esr spectra.[227] These intermediates went on to give the rearrangement via the normal pathway, with no evidence for oxirene intermediates.

$$s\text{-}(Z) \qquad\qquad s\text{-}(E)$$

The diazo ketone can exist in two conformations, called s-(E) and s-(Z). Studies have shown that Wolff rearrangement takes place preferentially from the s-(Z) conformation.[228]

OS **III**, 356; **VI**, 613, 840.

18-9   Homologation of Aldehydes and Ketones

Methylene-insertion

Aldehydes and ketones[229] can be converted to their homologs with diazomethane.[230] Several other reagents[231] are also effective, including Me_3SiI, and then silica gel,[232] or $LiCH(B\text{–}OCH_2CH_2O\text{–})_2$.[233] With the diazomethane reaction,

[225]Tanigaki, K.; Ebbesen, T.W. *J. Am. Chem. Soc.* **1987**, *109*, 5883. See also, Bachmann, C.; N'Guessan, T.Y.; Debû, F.; Monnier, M.; Pourcin, J.; Aycard, J.; Bodot, H. *J. Am. Chem. Soc.* **1990**, *112*, 7488.

[226]Csizmadia, I.G.; Gunning, H.E.; Gosavi, R.K.; Strausz, O.P. *J. Am. Chem. Soc.* **1973**, *95*, 133.

[227]McMahon, R.J.; Chapman, O.L.; Hayes, R.A.; Hess, T.C.; Krimmer, H. *J. Am. Chem. Soc.* **1985**, *107*, 7597.

[228]Kaplan, F.; Mitchell, M.L. *Tetrahedron Lett.* **1979**, 759; Tomioka, H.; Okuno, H.; Izawa, Y. *J. Org. Chem.* **1980**, *45*, 5278.

[229]For a homologation of carboxylic esters RCOOEt → RCH₂COOEt, which goes by an entirely different pathway, see Kowalski, C.J.; Haque, M.S.; Fields, K.W. *J. Am. Chem. Soc.* **1985**, *107*, 1429. Also see, Yamamoto, M.; Nakazawa, M.; Kishikawa, K.; Kohmoto, S. *Chem. Commun.* **1996**, 2353.

[230]For a review, see Gutsche, C.D. *Org. React.* **1954**, *8*, 364.

[231]See Taylor, E.C.; Chiang, C.; McKillop, A. *Tetrahedron Lett.* **1977**, 1827; Villieras, J.; Perriot, P.; Normant, J.F. *Synthesis* **1979**, 968; Aoyama, T.; Shioiri, T. *Synthesis* **1988**, 228.

[232]Lemini, C.; Ordoñez, M.; Pérez-Flores, J.; Cruz-Almanza, R. *Synth. Commun.* **1995**, *25*, 2695.

[233]Schummer, D.; Höfle, G. *Tetrahedron* **1995**, *51*, 11219.

formation of an epoxide (**16-46**) is a side reaction. Although this reaction appears superficially to be similar to the insertion of carbenes into C–H bonds, **12-21** (and IUPAC names it as an insertion), the mechanism is quite different. This is a true rearrangement and no free carbene is involved. The first step is an addition to the C=O bond:

$$
\underset{73}{\text{(reaction scheme)}}
$$

The betaine **73** can sometimes be isolated. As shown in **16-46**, intermediate **73** can also go to the epoxide. The evidence for this mechanism is summarized in the review by Gutsche.[230] Note that this mechanism is essentially the same as in the apparent "insertions" of oxygen (**18-19**) and nitrogen (**18-16**) into ketones.

Aldehydes give fairly good yields of methyl ketones; that is, hydrogen migrates in preference to alkyl. The most abundant side product is not the homologous aldehyde, but the epoxide. However, the yield of aldehyde at the expense of methyl ketone can be increased by the addition of methanol. If the aldehyde contains electron-withdrawing groups, the yield of epoxides is increased and the ketone is formed in smaller amounts, if at all. Ketones give poorer yields of homologous ketones. Epoxides are usually the predominant product here, especially when one or both R groups contain an electron-withdrawing group. The yield of ketones also decreases with increasing length of the chain. The use of a Lewis acid increases the yield of ketone.[234] Cyclic ketones,[235] three-membered[236] and larger, behave particularly well and give good yields of ketones with the ring expanded by one.[237] Aliphatic diazo compounds ($RCHN_2$ and R_2CN_2) are sometimes used instead of diazomethane, with the expected results.[238] Ethyl diazoacetate can be used analogously, in the presence of a Lewis acid or of triethyloxonium fluoroborate,[239] to

[234]For a review of homologations catalyzed by Lewis acids, see Müller, E.; Kessler, H.; Zeeh, B. *Fortschr. Chem. Forsch.* **1966**, *7*, 128, see pp. 137–150.

[235]For other methods for the ring enlargement of cyclic ketones, see Krief, A.; Laboureur, J.L. *Tetrahedron Lett.* **1987**, *28*, 1545; Krief, A.; Laboureur, J.L.; Dumont, W. *Tetrahedron Lett.* **1987**, *28*, 1549; Abraham, W.D.; Bhupathy, M.; Cohen, T. *Tetrahedron Lett.* **1987**, *28*, 2203; Trost, B.M.; Mikhail, G.K. *J. Am. Chem. Soc.* **1987**, *109*, 4124.

[236]For example, see Turro, N.J.; Gagosian, R.B. *J. Am. Chem. Soc.* **1970**, *92*, 2036.

[237]For a review, see Gutsche, C.D.; Redmore, D. *Carbocyclic Ring Expansion Reactions*, Academic Press, NY, **1968**, pp. 81–98. For a review pertaining to bridged bicyclic ketones, see Krow, G.R. *Tetrahedron* **1987**, *43*, 3.

[238]For example, see Smith, R.F. *J. Org. Chem.* **1960**, *25*, 453; Warner, C.R.; Walsh, Jr., E.J.; Smith, R.F. *J. Chem. Soc.* **1962**, 1232; Loeschorn, C.A.; Nakajima, M.; Anselme, J. *Bull. Soc. Chim. Belg.* **1981**, *90*, 985.

[239]Mock, W.L.; Hartman, M.E. *J. Org. Chem.* **1977**, *42*, 459, 466; Baldwin, S.W.; Landmesser, N.G. *Synth. Commun.* **1978**, *8*, 413.

give a β-keto ester, such as **74**.

74

When unsymmetrical ketones were used in this reaction (with BF_3 as catalyst), the less highly substituted carbon preferentially migrated.[240] The reaction can be made regioselective by applying this method to the α-halo ketone, in which case only the other carbon migrates.[241] The ethyl diazoacetate procedure has also been applied to the acetals or ketals of α,β-unsaturated aldehydes and ketones.[242]

Bicyclic ketones can be expanded to form monocyclic ketones in the presence of certain reagents. Treatment of a bicyclo[4.1.0]hexan-4-one derivative with SmI_2 led to a cyclohexanone.[243] The SmI_2 also converts α-halomethyl cyclic ketones to the next larger ring ketone[244] and cyclic ketones to the next larger ring ketone in the presence of CH_2I_2.[245] α-Chloro-α-3-iodopropylcyclobutanones were converted to cycloheptanones using radical conditions ($Bu_3SnH/AIBN$).[246]

Another homologation reaction converts an aldehyde to its tosyl hydrazone, and subsequent reaction with an aldehyde and NaOEt/EtOH give a ketone.[247] The reaction of an aldehyde with vinyl acetate and barium hydroxide gives the homologated conjugated aldehyde.[248]

OS **IV**, 225, 780. For homologation of carboxyl acid derivatives, see OS **IX**, 426

B. Carbon-to-Carbon Migrations of Other Groups

18-10   Migrations of Halogen, Hydroxyl, Amino, and so on

Hydroxy-de-bromo-*cine*-substitution, and so on

When a nucleophilic substitution is carried out on a substrate that has a neighboring group (p. 446) on the adjacent carbon, a cyclic intermediate can be generated

[240]Liu, H.J.; Majumdar, S.P. *Synth. Commun.* **1975**, *5*, 125.

[241]Dave, V.; Warnhoff, E.W. *J. Org. Chem.* **1983**, *48*, 2590.

[242]Doyle, M.P.; Trudell, M.L.; Terpstra, J.W. *J. Org. Chem.* **1983**, *48*, 5146.

[243]Lee, P.H.; Lee, J. *Tetrahedron Lett.* **1998**, *39*, 7889.

[244]Hasegawa, E.; Kitazume, T.; Suzuki, K.; Tosaka, E. *Tetrahedron Lett.* **1998**, *39*, 4059.

[245]Fukuzawa, S.; Tsuchimoto, T. *Tetrahedron Lett.* **1995**, *36*, 5937.

[246]Zhang, W.; Dowd, P. *Tetrahedron Lett.* **1992**, *33*, 3285. For an example generating larger rings, see Dowd, P.; Choi, S.-C. *Tetrahedron* **1992**, *48*, 4773.

[247]Angle, S.R.; Neitzel, M.L. *J. Org. Chem.* **2000**, *65*, 6458.

[248]Mahata, P.K.; Barun, O.; Ila, H.; Junjappa, H. *Synlett* **2000**, 1345.

that is opened on the opposite side, resulting in migration of the neighboring group. In the example shown above (NR_2 = morpholino),[249] the reaction took place via an aziridinium salt **75** to give an α-amino-β-hydroxy ketone.

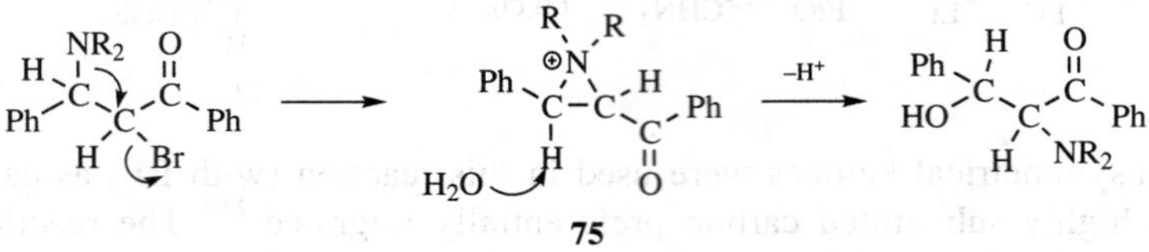

75

Sulfonate esters and halides can also migrate in this reaction.[250] α-Halo and α-acyloxy epoxides undergo ready rearrangement to α-halo and α-acyloxy ketones, respectively.[251] These substrates are very prone to rearrange, and often do so on standing without a catalyst, though in some cases an acid catalyst is necessary. The reaction is essentially the same as the rearrangement of epoxides shown in **18-2**, except that in this case halogen or acyloxy is the migrating group (as shown above; however, it is also possible for one of the R groups (alkyl, aryl, or hydrogen) to migrate instead, and mixtures are sometimes obtained).

18-11 Migration of Boron

Hydro,dialkylboro-interchange, and so on

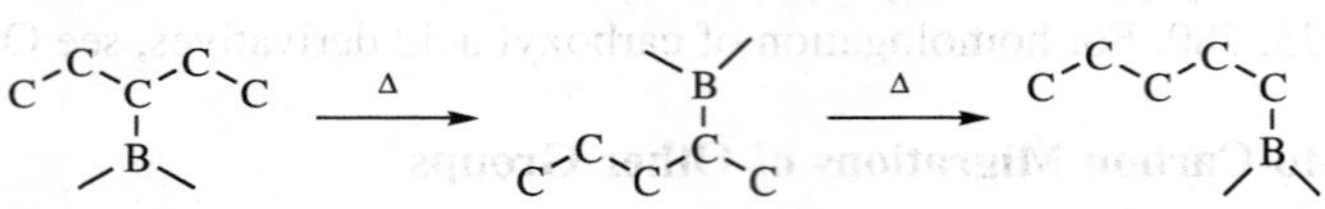

Boranes are prepared by the reaction of $BH_3(B_2H_6)$ or an alkylborane with an alkene (**15-16**). When a nonterminal borane is heated at temperatures ranging from 100 to 200°C, the boron moves toward the end of the chain.[252] The reaction is catalyzed by small amounts of borane or other species containing B–H bonds.

[249]Southwick, P.L.; Walsh, W.L. *J. Am. Chem. Soc.* **1955**, *77*, 405. See also, Suzuki, K.; Okano, K.; Nakai, K.; Terao, Y.; Sekiya, M. *Synthesis* **1983**, 723.

[250]For a review of Cl migrations, see Peterson, P.E. *Acc. Chem. Res.* **1971**, *4*, 407. See also, Loktev, V.F.; Korchagina, D.V.; Shubin, V.G.; Koptyug, V.A. *J. Org. Chem. USSR* **1977**, *13*, 201; Dobronravov, P.N.; Shteingarts, V.D. *J. Org. Chem. USSR* **1977**, *13*, 420. For examples of Br migration, see Gudkova, A.S.; Uteniyazov, K.; Reutov, O.A. *Doklad. Chem.* **1974**, *214*, 70; Brusova, G.P.; Gopius, E.D.; Smolina, T.A.; Reutov, O.A. *Doklad. Chem.* **1980**, *253*, 334. For a review of F migration (by several mechanisms) see Kobrina, L.S.; Kovtonyuk, V.N. *Russ. Chem. Rev.* **1988**, *57*, 62. For an example OH migration, see Cathcart, R.C.; Bovenkamp, J.W.; Moir, R.Y.; Bannard, R.A.B.; Casselman, A.A. *Can. J. Chem.* **1977**, *55*, 3774. For a review of migrations of ArS and $Ar_2P(O)$, see Warren, S. *Acc. Chem. Res.* **1978**, *11*, 403. See also, Aggarwal, V.K.; Warren, S. *J. Chem. Soc. Perkin Trans. 1* **1987**, 2579.

[251]For a review, see McDonald, R.N. *Mech. Mol. Migr.* **1971**, *3*, 67.

[252]Brown, H.C. *Hydroboration*, W. A. Benjamin, NY, **1962**, pp. 136–149, Brown, H.C.; Zweifel, G. *J. Am. Chem. Soc.* **1966**, *88*, 1433. See also, Brown, H.C.; Racherla, U.S. *J. Organomet. Chem.* **1982**, *241*, C37.

The boron can move past a branch, for example,

but not past a double branch, for example,

The reaction is an equilibrium: **76**, **77**, and **78** each gave a mixture containing ~40% **76**, 1% **77**, and 59% **78**. The migration can go quite a long distance. Thus $(C_{11}H_{23}CHC_{11}H_{23})_3B$ was completely converted to $(C_{23}H_{47})_3B$, involving a migration of 11 positions.[253] If the boron is on a cycloalkyl ring, it can move around

76   **77**   **78**

the ring; if any alkyl chain is also on the ring, the boron may move from the ring to the chain, ending up at the end of the chain.[254] The reaction is useful for the migration of double bonds in a controlled way (see **12-2**). The mechanism may involve a π complex, at least partially.[255]

18-12   The Neber Rearrangement

Neber oxime tosylate-amino ketone rearrangement

α-Amino ketones can be prepared by treatment of ketoxime tosylates with a base, such as ethoxide ion or pyridine.[256] This reaction is called the *Neber rearrangement*. The R group is usually aryl, though the reaction has been carried out with

[253]Logan, T.J. *J. Org. Chem.* **1961**, *26*, 3657.
[254]Brown, H.C.; Zweifel, G. *J. Am. Chem. Soc.* **1967**, *89*, 561.
[255]See Wood, S.E.; Rickborn, B. *J. Org. Chem.* **1983**, *48*, 555; Field, L.D.; Gallagher, S.P. *Tetrahedron Lett.* **1985**, *26*, 6125.
[256]For a review, see Conley, R.T.; Ghosh, S. *Mech. Mol. Migr.* **1971**, *4*, 197, pp. 289–304.

R = alkyl or hydrogen. The R' group may be alkyl, or aryl but not hydrogen. The Beckmann rearrangement (**18-17**) and the abnormal Beckmann reaction (elimination to the nitrile, **17-30**) may be side reactions, although these generally occur in acid media. A similar rearrangement is given by *N,N*-dichloroamines of the type RCH$_2$CH(NCl$_2$)R', where the product is also RCH(NH$_2$)COR'.[257] The mechanism of the Neber rearrangement is via an azirine intermediate **79**.[258]

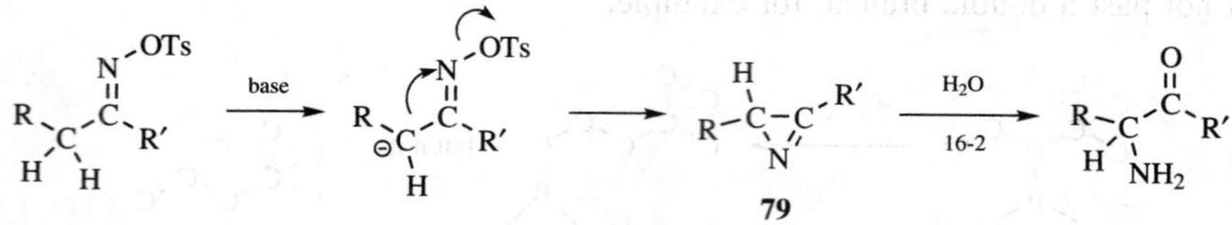

The best evidence for this mechanism is that the azirine intermediate has been isolated.[258,259] In contrast to the Beckmann rearrangement, this one is sterically indiscriminate:[260] Both a syn and an anti ketoxime give the same product. The mechanism as shown above consists of three steps. However, it is possible that the first two steps are concerted, and it is also possible that what is shown as the second step is actually two steps: loss of OTs to give a nitrene, and formation of the azirine. In the case of the dichloroamines, HCl is first lost to give RCH$_2$C(=NCl)R', which then behaves analogously.[261] *N*-Chloroimines prepared in other ways also give the reaction.[262]

OS **V**, 909; **VII**, 149.

C. Carbon-to-Nitrogen Migrations of R and AR

The reactions in this group are nucleophilic migrations from a carbon to a nitrogen atom. In each case the nitrogen atom either has six electrons in its outer shell (and thus invites the migration of a group carrying an electron pair) or else loses a nucleofuge concurrently with the migration (p. 1560). Reactions **18-13–18-16** are used to prepare amines from acid derivatives. Reactions **18-16** and **18-17** are used to prepare amines from ketones. The mechanisms of **18-13–18-16** (with carboxylic acids) are very similar and follow one of two patterns:

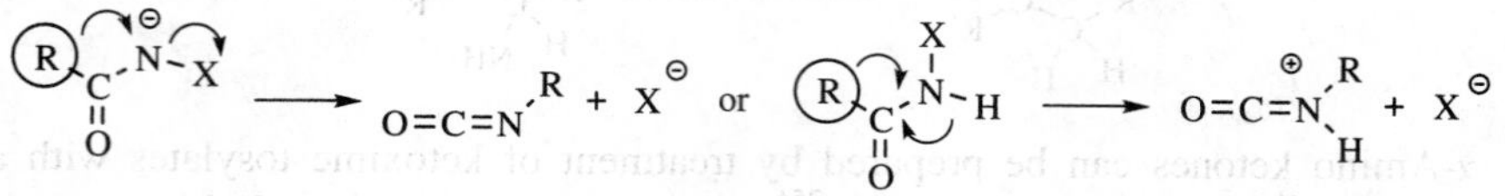

[257]Baumgarten, H.E.; Petersen, H.E. *J. Am. Chem. Soc.* **1960**, *82*, 459, and references cited therein.
[258]Cram, D.J.; Hatch, M.J. *J. Am. Chem. Soc.* **1953**, *75*, 33; Hatch, M.J.; Cram, D.J. *J. Am. Chem. Soc.* **1953**, *75*, 38.
[259]Neber, P.W.; Burgard, A. *Liebigs Ann. Chem.* **1932**, *493*, 281; Parcell, R.F. *Chem. Ind. (London)* **1963**, 1396.
[260]House, H.O.; Berkowitz, W.F. *J. Org. Chem.* **1963**, *28*, 2271.
[261]For example, see Nakai, M.; Furukawa, N.; Oae, S. *Bull. Chem. Soc. Jpn.* **1969**, *42*, 2917.
[262]Baumgarten, H.E.; Petersen, J.M.; Wolf, D.C. *J. Org. Chem.* **1963**, *28*, 2369.

Some of the evidence[263] is (*1*) configuration is retained in R (p. 1563); (*2*) the kinetics are first order; (*3*) intramolecular rearrangement is shown by labeling; and (*4*) no rearrangement occurs *within* the migrating group, for example, a neopentyl group on the carbon of the starting material is still a neopentyl group on the nitrogen of the product.

In many cases, it is not certain whether the nucleofuge X is lost first, creating an intermediate nitrene[264] or nitrenium ion, or whether migration and loss of the nucleofuge are simultaneous, as shown above.[265] It is likely that both possibilities can exist, depending on the substrate and reaction conditions.

18-13 The Hofmann Rearrangement

Bis(hydrogen)-(2/ 1/*N*-alkyl)-*migro*-detachment (formation of isocyanate)

$$RCONH_2 + NaOBr \xrightarrow{\hspace{2cm}} R{-}N{=}C{=}O \xrightarrow[16\text{-}2]{hydrolysis} RNH_2$$

In the *Hofmann rearrangement*, an unsubstituted amide is treated with sodium hypobromite (or sodium hydroxide and bromine, which is essentially the same thing) to give a primary amine that has one carbon fewer than the starting amide.[266] The actual product is the isocyanate, but this compound is seldom isolated[267] since it is usually hydrolyzed under the reaction conditions. The R group may be alkyl or aryl, but if it is an alkyl group of more than about six or seven carbons, low yields are obtained unless Br_2 and NaOMe are used instead of Br_2 and NaOH.[268] Another modification uses NBS/NaOMe.[269] Under these conditions the product of addition to the isocyanate is the carbamate RNHCOOMe (**16-8**), which is easily isolated or can be hydrolyzed to the amine. Side reactions when NaOH is the base are formation of ureas RNHCONHR and acylureas RCONH-CONHR by addition, respectively, of RNH_2 and $RCONH_2$ to RNCO (**16-20**). If acylureas are desired, they can be made the main products by using only one-half of the usual quantities of Br_2 and NaOH. Another side product, but only from primary R, is the nitrile derived from oxidation of RNH_2 (**19-5**). Imides react to give amino acids, for example, phthalimide gives *o*-aminobenzoic acid. α-Hydroxy and α-halo amides give aldehydes and ketones by way of the unstable α-hydroxy- or α-haloamines. However, a side product with an α-halo amide is a *gem*-dihalide. Ureas analogously give hydrazines.

---

[263]For a discussion of this mechanism and the evidence for it, see Smith, P.A.S., in de Mayo, P. *Molecular Rearrangements*, Vol. 1, Wiley, NY, *1963*, Vol. 1, pp. 258–550.

[264]For a review of rearrangements involving nitrene intermediates, see Boyer, J.H. *Mech. Mol. Migr. 1969*, 2, 267. See also, Ref. 282.

[265]The question is discussed by Lwowski, W., in Lwowski *Nitrenes*, Wiley, NY, *1970*, pp. 217–221.

[266]For a review, see Wallis, E.S.; Lane, J.F. *Org. React. 1946*, 3, 267.

[267]If desired, the isocyanate can be isolated by the use of phase-transfer conditions: see Sy, A.O.; Raksis, J.W. *Tetrahedron Lett. 1980*, 21, 2223.

[268]For an example of the use of this method at low temperatures, see Radlick, P.; Brown, L.R. *Synthesis 1974*, 290.

[269]Huang, X.; Keillor, J.W. *Tetrahedron Lett. 1997*, 38, 313.

The mechanism follows the pattern outlined on p. 1606.

80

The first step is an example of **12-52** and intermediate *N*-halo amides (**80**) have been isolated. In the second step, **80** lose a proton to the base. Compound **80** is acidic because of the presence of two electron-withdrawing groups (acyl and halo) on the nitrogen. It is possible that the third step is actually two steps: loss of bromide to form a nitrene, followed by the actual migration, but most of the available evidence favors the concerted reaction.[270] A similar reaction can be effected by the treatment of amides with lead tetraacetate.[271] Among other reagents that convert RCONH$_2$ to RNH$_2$ (R = alkyl, but not aryl) are phenyl-iodosyl bis(trifluoroacetate) PhI(OCOCF$_3$)$_2$[272] and hydroxy(tosyloxy)iodobenzene PhI(OH)OTs.[273] A mixture of NBS, Hg(OAc)$_2$, and R'OH is one of several reagent mixtures that convert an amide RCONH$_2$ to the carbamate RNHCOOR' (R = primary, secondary, or tertiary alkyl or aryl) in high yield.[274] A mixture of NBS and DBU (p. 1132) in methanol gives the carbamate,[275] as does electrolysis in methanol.[276]

A variation of the Hofmann rearrangement treated a β-hydroxy primary amide with PhI(O$_2$CCF$_3$)$_2$ in aqueous acetonitrile, giving an isocyanate via –CON–I, which reacts with the hydroxyl group intramolecularly to give a cyclic carbamate.[277] Note that carbamates are converted to isocyanates by heating with Montmorillonite K10.[278]

OS **II**, 19, 44, 462; **IV**, 45; **VIII**, 26, 132.

18-14   The Curtius Rearrangement

Dinitrogen-(2/ → 1/*N*-alkyl)-*migro*-detachment

$$RCON_3 \xrightarrow{\Delta} R\text{—}N=C=O$$

[270]See, for example, Imamoto, T.; Tsuno, Y.; Yukawa, Y. *Bull. Chem. Soc. Jpn.* **1971**, *44*, 1632, 1639, 1644; Imamoto, T.; Kim, S.; Tsuno, Y.; Yukawa, Y. *Bull. Chem. Soc. Jpn.* **1971**, *44*, 2776.

[271]Acott, B.; Beckwith, A.L.J.; Hassanali, A. *Aust. J. Chem.* **1968**, *21*, 185, 197; Baumgarten, H.E.; Smith, H.L.; Staklis, A. *J. Org. Chem.* **1975**, *40*, 3554.

[272]Loudon, G.M.; Radhakrishna, A.S.; Almond, M.R.; Blodgett, J.K.; Boutin, R.H. *J. Org. Chem.* **1984**, *49*, 4272; Boutin, R.H.; Loudon, G.M. *J. Org. Chem.* **1984**, *49*, 4277; Pavlides, V.H.; Chan, E.D.; Pennington, L.; McParland, M.; Whitehead, M.; Coutts, I.G.C. *Synth. Commun.* **1988**, *18*, 1615.

[273]Vasudevan, A.; Koser, G.F. *J. Org. Chem.* **1988**, *53*, 5158.

[274]Jew, S.; Park, H.G.; Park, H.; Park, M.; Cho, Y. *Tetrahedron Lett.* **1990**, *31*, 1559.

[275]Huang, X.; Seid, M.; Keillor, J.W. *J. Org. Chem.* **1997**, *62*, 7495.

[276]Matsumura, Y.; Maki, T. ; Satoh, Y. *Tetrahedron Lett.* **1997**, *38*, 8879.

[277]Yu, C.; Jiang, Y.; Liu, B.; Hu, L. *Tetrahedron Lett.* **2001**, *42*, 1449.

[278]Uriz, P.; Serra, M.; Salagre, P.; Castillon, S.; Claver, C.; Fernandez, E. *Tetrahedron Lett.* **2002**, *43*, 1673.

The *Curtius rearrangement* involves the pyrolysis of acyl azides to yield isocyanates.[279] The reaction gives good yields of isocyanates, since no water is present to hydrolyze them to the amine. Of course, they can be subsequently hydrolyzed, and indeed the reaction *can* be carried out in water or alcohol, in which case the products are amines, carbamates, or acylureas, as in **18-13**.[280] This is a very general reaction and can be applied to almost any carboxylic acid: aliphatic, aromatic, alicyclic, heterocyclic, unsaturated, and containing many functional groups. Acyl azides can be prepared as in **10-43** or by treatment of acylhydrazines (hydrazides) with nitrous acid (analogous to **12-49**). The Curtius rearrangement is catalyzed by Lewis or protic acids, but these are usually not necessary for good results.

The mechanism is similar to that in **18-13** to give an isocyanate. Also note the exact analogy between this reaction and **18-8**. However, in this case, there is no evidence for a free nitrene and it is probable that the steps are concerted.[281]

Alkyl azides can be similarly pyrolyzed to give imines, in an analogous reaction:[282]

$$R_3CN_3 \xrightarrow{\Delta} R_2C=NR$$

The R groups may be alkyl, aryl, or hydrogen, though if hydrogen migrates, the product is the unstable $R_2C=NH$. The mechanism is essentially the same as that of the Curtius rearrangement. However, in pyrolysis of tertiary alkyl azides, there is evidence that free alkyl nitrenes are intermediates.[283] The reaction can also be carried out with acid catalysis, in which case lower temperatures can be used, though the acid may hydrolyze the imine (**16-2**). Cycloalkyl azides give

[279]For a review, see Banthorpe, D.V., in Patai, S. *The Chemistry of the Azido Group*, Wiley, NY, *1971*, pp. 397–405.

[280]For a variation that conveniently produces the amine directly, see Pfister, J.R.; Wyman, W.E. *Synthesis* *1983*, 38. See also, Capson, T.L.; Poulter, C.D. *Tetrahedron Lett.* *1984*, 25, 3515.

[281]See, for example, Lwowski, W. *Angew. Chem. Int. Ed.* *1967*, 6, 897; Linke, S.; Tissue, G.T.; Lwowski, W. *J. Am. Chem. Soc.* *1967*, 89, 6308; Smalley, R.K.; Bingham, T.E. *J. Chem. Soc. C* *1969*, 2481.

[282]For a treatise on azides, which includes discussion of rearrangement reactions, see Scriven, E.F.V. *Azides and Nitrenes*, Academic Press, NY, *1984*. For a review of rearrangements of alkyl and aryl azides, see Stevens, T.S.; Watts, W.E. *Selected Molecular Rearrangements*, Van Nostrand-Reinhold, Princeton, NJ, *1973*, pp. 45–52. For reviews of the formation of nitrenes from alkyl and aryl azides, see, in Lwowski, W. *Nitrenes*, Wiley, NY, *1970*, the chapters by Lewis, F.D.; Saunders, Jr., W.H. pp. 47–97, 47–78 and by Smith, P.A.S. pp. 99–162.

[283]Abramovitch, R.A.; Kyba, E.P. *J. Am. Chem. Soc.* *1974*, 96, 480; Montgomery, F.C.; Saunders, Jr., W.H. *J. Org. Chem.* *1976*, 41, 2368.

ring expansion.[284]

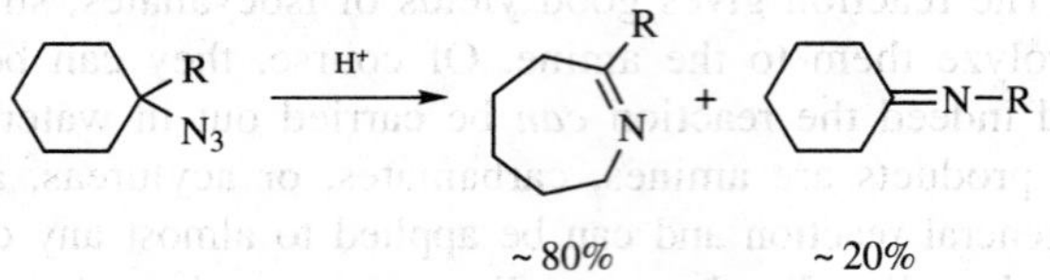

~80%   ~20%

Aryl azides also give ring expansion on heating, for example,[285]

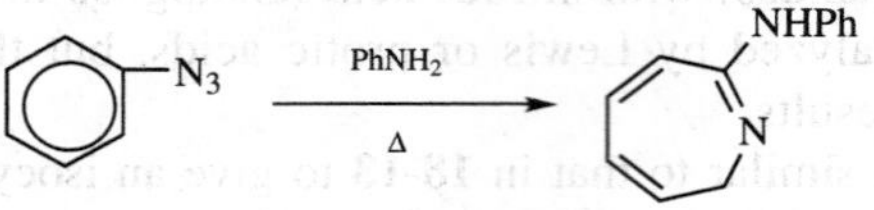

OS **III**, 846; **IV**, 819; **V**, 273; **VI**, 95, 910. Also see, OS **VI**, 210.

18-15   The Lossen Rearrangement

Hydro,acetoxy-(2/ → 1N-alkyl)-*migro*-detachment

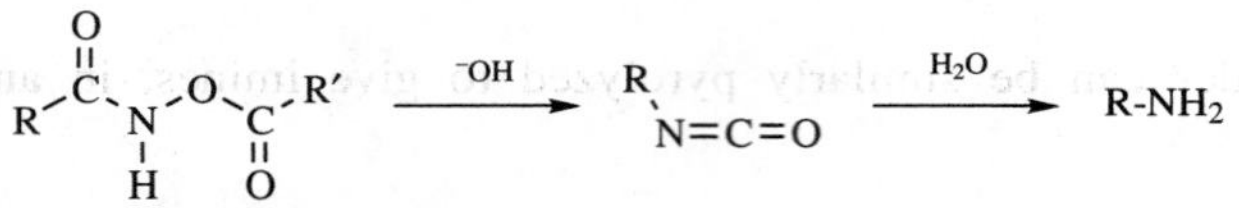

The *O*-acyl derivatives of hydroxamic acids[286] give isocyanates when treated with bases or sometimes even just on heating, in a reaction known as the *Lossen rearrangement*.[287] The mechanism is similar to that of **18-13** and **18-14**:

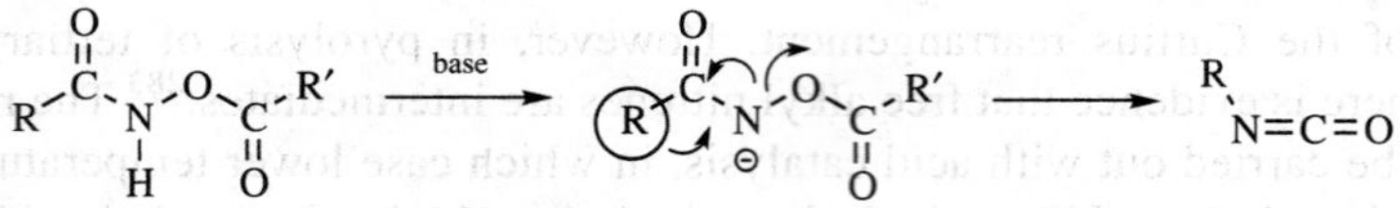

In a similar reaction, aromatic acyl halides are converted to amines in one laboratory step by treatment with hydroxylamine-*O*-sulfonic acid.[288]

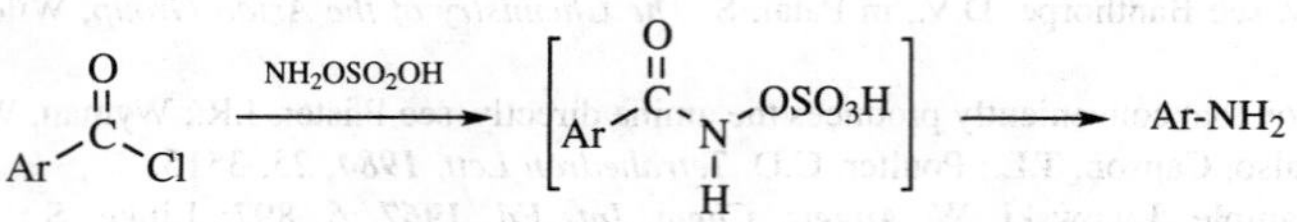

A chiral Lossen rearrangement is known.[289]

[284]Smith, P.A.S.; Lakritz, J. cited in Smith, P.A.S., in de Mayo, P. *Molecular Rearrangments*, Vol. 1, Wiley, NY, *1963*, p. 474.

[285]Huisgen, R.; Vossius, D.; Appl, M. *Chem. Ber. 1958*, *91*, 1,12.

[286]For a review of hydroxamic acids, see Bauer, L.; Exner, O. *Angew. Chem. Int. Ed. 1974*, *13*, 376.

[287]For an example, see Salomon, C.J.; Breuer, E. *J. Org. Chem*, *1997*, *62*, 3858.

[288]Wallace, R.G.; Barker, J.M.; Wood, M.L. *Synthesis 1990*, 1143.

[289]Chandrasekhar, S.; Sridhar, M. *Tetrahedron Asymmetry 2000*, *11*, 3467.

18-16   The Schmidt Reaction

$$RCOOH \ + \ HN_3 \ \xrightarrow{H^+} \ R—N=C=O \ \xrightarrow{H_2O} \ RNH_2$$

There are actually three reactions called by the name *Schmidt reaction*, involving the addition of hydrazoic acid to carboxylic acids, aldehydes and ketones, and alcohols and alkenes.[290] The most common is the reaction with carboxylic acids, illustrated above.[291] Sulfuric acid is the most common catalyst, but Lewis acids have also been used. Good results are obtained for aliphatic R, especially for long chains. When R is aryl, the yields are variable, being best for sterically hindered compounds like mesitoic acid. This method has the advantage over **18-13** and **18-14** in that there is just one laboratory step from the acid to the amine, but conditions are more drastic.[292] Under the acid conditions employed, the isocyanate is virtually never isolated.

The reaction between a ketone and hydrazoic acid is a method for "insertion" of NH between the carbonyl group and one R group, converting a ketone into an amide.[293]

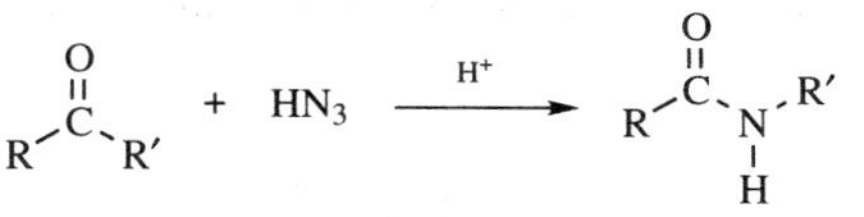

Either or both of the R groups may be aryl. In general, dialkyl ketones and cyclic ketones react more rapidly than alkyl aryl ketones, and these more rapidly than diaryl ketones. The latter require sulfuric acid and do not react in concentrated HCl, which is strong enough for dialkyl ketones. Dialkyl and cyclic ketones react sufficiently faster than diaryl or aryl alkyl ketones or carboxylic acids or alcohols so that these functions may be present in the same molecule without interference. Cyclic ketones give lactams:[294]

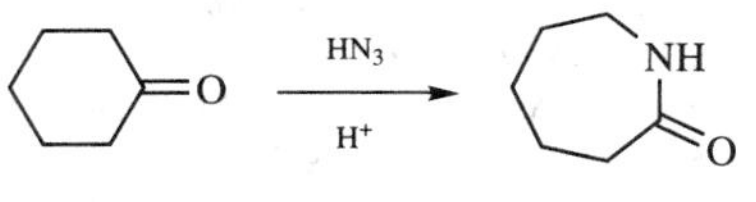

[290]For a review, see Banthorpe, D.V., in Patai, S. *The Chemistry of the Azido Group*, Wiley, NY, *1971*, pp. 405–434.
[291]For a review, see Koldobskii, G.I.; Ostrovskii, V.A.; Gidaspov, B.V. *Russ. Chem. Rev. 1978*, 47, 1084.
[292]For a comparision of reactions **18-13–18-16** as methods for converting an acid to an amine, see Smith, P.A.S. *Org. React. 1946*, 3, 337, 363–366.
[293]For reviews, see Koldobskii, G.I.; Tereschenko, G.F.; Gerasimova, E.S.; Bagal, L.I. *Russ. Chem. Rev. 1971*, 40, 835; Beckwith, A.L.J., in Zabicky, J. *The Chemistry of Amides*, Wiley, NY, *1970*, pp. 137–145.
[294]For a review with respect to bicyclic ketones, see Krow, G.R. *Tetrahedron 1981*, 37, 1283.

With alkyl aryl ketones, it is the aryl group that generally migrates to the nitrogen, except when the alkyl group is bulky.[295] The reaction has been applied to a few aldehydes, but rarely. With aldehydes the product is usually the nitrile (**16-16**). Even with ketones, conversion to the nitrile is often a side reaction, especially with the type of ketone that gives **17-30**. A useful variation of the Schmidt reaction treats a cyclic ketone with an alkyl azide (RN_3)[296] in the presence of $TiCl_4$, generating a lactam.[297] An intramolecular Schmidt reaction gives bicyclic amines by treatment of a cyclic alkene having a pendant azidoalkyl group with $Hg(ClO_4)_2$, and then $NaBH_4$.[298] Another variation treats a silyl enol ether of a cyclic ketone with $TMSN_3$ and photolyzes the product with UV light to give a lactam.[299] αAzido cyclic ketones rearrangement to lactams under radical conditions ($Bu_3SnH/AIBN$).[300]

Alcohols and alkenes react with HN_3 to give alkyl azides,[301] which in the course of reaction rearrange in the same way as discussed in reaction **18-14**.[282] The Mitsunobu reaction (**10-17**) can be used to convert alcohols to alkyl azides, and an alternative reagent for azides, $(PhO)_2PON_3$, for use in the Mitsunobu is now available.[302]

There is evidence that the mechanism with carboxylic acids[293] is similar to that of **18-14**, except that it is the protonated azide that undergoes the rearrangement:[303]

The first step is the same as that of the $A_{AC}1$ mechanism (**16-59** which explains why good results are obtained with hindered substrates. The mechanism with ketones

[295]Exceptions to this statement have been noted in the case of cyclic aromatic ketones bearing electron-donating groups in ortho and para positions: Bhalerao, U.T.; Thyagarajan, G. *Can. J. Chem.* **1968**, *46*, 3367; Tomita, M.; Minami, S.; Uyeo, S. *J. Chem. Soc. C* **1969**, 183.

[296]See Furness, K.; Aubé, J. *Org. Lett.* **1999**, *1*, 495.

[297]Desai, P.; Schildknegt, K.; Agrios, K.A.; Mossman, C.; Milligan, G.L.; Aubé, J. *J. Am. Chem. Soc.* **2000**, *122*, 7226; Sahasrabudhe, K.; Gracias, V.; Furness, K.; Smith, B.T.; Katz, C.E.; Reddy, D.S.; Aubé, J. *J. Am. Chem. Soc.* **2003**, *125*, 7914. For a variation using a ketal with TMSOTf see Mossman, C.J.; Aubé, J. *Tetrahedron*, **1996**, *52*, 3403.

[298]Pearson, W.H.; Hutta, D.A.; Fang, W.-k. *J. Org. Chem.* **2000**, *65*, 8326. See also, Wrobleski, A.; Aubé, J. *J. Org. Chem.* **2001**, *66*, 886.

[299]Evans, P.A.; Modi, D.P. *J. Org. Chem.* **1995**, *60*, 6662.

[300]Benati, L.; Nanni, D.; Sangiorgi, C.; Spagnolo, P. *J. Org. Chem.* **1999**, *64*, 7836.

[301]For an example, see Kumar, H.M.S.; Reddy, B.V.S.; Anjaneyulu, S.; Yadav, J.S. *Tetrahedron Lett.* **1998**, *39*, 7385. Also see, Saito, A.; Saito, K.; Tanaka, A.; Oritani, T. *Tetrahedron Lett.* **1997**, *38*, 3955.

[302]Thompson, A.S.; Humphrey, G.R.; DeMarco, A.M.; Mathre, D.J.; Grabowski, E.J.J. *J. Org. Chem.* **1993**, *58*, 5886.

[303]There has been some controversy about this mechanism. For a discussion, see Vogler, E.A.; Hayes, J.M. *J. Org. Chem.* **1979**, *44*, 3682.

involves formation of a nitrilium ion **82**, which reacts with water.

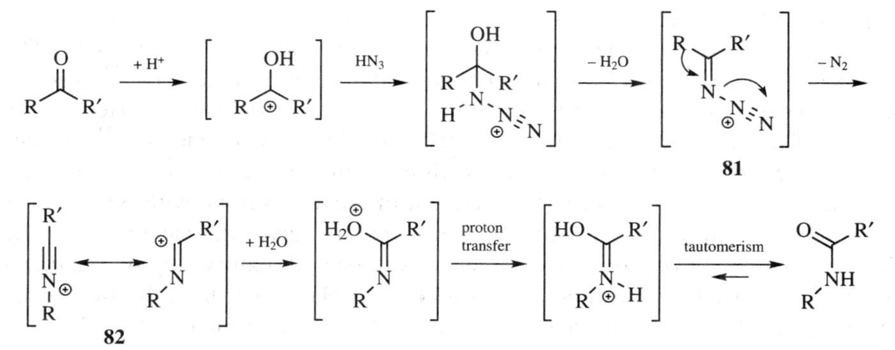

81

82

The intermediates **81** have been independently generated in aqueous solution.[304] Note the similarity of this mechanism to those of "insertion" of CH_2 (**18-9**) and of O (**18-19**). The three reactions are essentially analogous, both in products and in mechanism.[293,305] Also note the similarity of the latter part of this mechanism to that of the Beckmann rearrangement (**18-17**).

OS **V**, 408; **VI**, 368; **VII**, 254; **X**, 207. See also, OS **V**, 623.

18-17   The Beckmann Rearrangement

Beckmann oxime-amide rearrangement

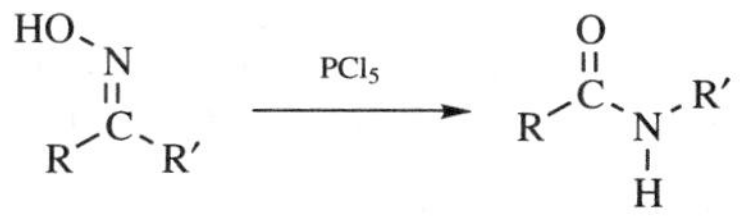

When oximes are treated with PCl_5 or a number of other reagents, they rearrange to substituted amides in a reaction called the *Beckmann rearrangement*.[306] Among other reagents used have been concentrated H_2SO_4, formic acid, liquid SO_2, $SOCl_2$,[307] silica gel,[308] MoO_3 on silica gel,[309] $RuCl_3$,[310] $Y(OTf)_3$,[311]

---

[304]Amyes, T.L.; Richard, J.P. *J. Am. Chem. Soc.* **1991**, *113*, 1867.

[305]For evidence for this mechanism, see Ostrovskii, V.A.; Koshtaleva, T.M.; Shirokova, N.P.; Koldobskii, G.I.; Gidaspov, B.V. *J. Org. Chem. USSR* **1974**, *10*, 2365, and references cited therein.

[306]For reviews, see Gawley, R.E. *Org. React.* **1988**, *35*, 1; McCarty, C.G., in Patai, S. *The Chemistry of the Carbon-Nitrogen Double Bond*, Wiley, NY, **1970**, pp. 408–439. Also see, Nguyen, M.T.; Raspoet, G.; Vanquickenborne, L.G. *J. Am. Chem. Soc.* **1997**, *119*, 2552.

[307]Butler, R.N.; O'Donoghue, D.A. *J. Chem. Res. (S)*, **1983**, 18.

[308]Costa, A.; Mestres, R.; Riego, J.M. *Synth. Commun.* **1982**, *12*, 1003. On silica with microwave irradiation, see Loupy, A.; Régnier, S. *Tetrahedron Lett.* **1999**, *40*, 6221.

[309]Dongare, M.K.; Bhagwat, V.V.; Ramana, C.V.; Gurjar, M.K. *Tetrahedron Lett.* **2004**, *45*, 4759.

[310]De, S.K. *Synth. Commun.* **2004**, *34*, 3431.

[311]De, S.K. *Org. Prep. Proceed. Int.* **2004**, *36*, 383.

HCl–HOAc-Ac$_2$O, POCl$_3$,[312] BiCl$_3$,[313] neat with FeCl$_3$,[314] and polyphosphoric acid.[315] The reaction has been done in supercritical water[316] and in ionic liquids.[317] A polymer-bound Beckman rearrangement has been reported.[318] Simply heating the oxime of benzophenone neat leads to *N*-phenyl benzamide.[319] The oximes of cyclic ketones give ring enlargement and form the lactam,[320] as in the formation of caprolactam (**83**) from the oxime of cyclohexanone. Heating an oxime of a cyclic ketone, *neat*, with AlCl$_3$ also leads to the lactam,[321] as does microwave irradiation of an oxime on Montmorillonite K10 clay.[322] Other solvent-free reactions are known.[323] Treatment of a cyclic ketone with NH$_2$OSO$_3$H on silica gel followed by microwave irradiation also gives the lactam.[324] Cyclic ketones can be converted directly to lactams in one laboratory step by treatment with NH$_2$OSO$_2$OH and formic acid (**16-14** takes place first, then the Beckmann rearrangement).[325] Heating a ketone with hydroxylamine HCl and oxalic acid also gives the amide.[326] Note that the reaction of an imine with BF$_3$•OEt$_2$ and *m*-chloroperoxybenzoic acid leads to a formamide.[327]

83

Of the groups attached to the carbon of the C=N unit, the one that migrates in the Beckman rearrangement is generally the one anti to the hydroxyl, and this is

[312]Majo, V.J.; Venugopal, M.; Prince, A.A.M.; Perumal, P.T. *Synth. Commun.* **1995**, *25*, 3863.

[313]With microwave irradiation, see Thakur, A.J.; Boruah, A.; Prajapati, D.; Sandhu, J.S. *Synth. Commun.* **2000**, *30*, 2105.

[314]Khodaei, M.M.; Meybodi, F.A.; Rezai, N.; Salehi, P. *Synth. Commun.* **2001**, *31*, 2047.

[315]For a review of Beckmann rearrangements with polyphosphoric acid, see Beckwith, A.L.J., in Zabicky, J. *The Chemistry of Amides*, Wiley, NY, **1970**, pp. 131–137.

[316]Ikushima, Y.; Hatakeda, K.; Sato, O.; Yokoyama, T.; Arai, M. *J. Am. Chem. Soc.* **2000**, *122*, 1908; Boero, M.; Ikeshoji, T.; Liew, C.C.; Terakura, K.; Parrinello, M. *J. Am. Chem. Soc.* **2004**, *126*, 6280.

[317]In BPy BF$_4$, butylpyridinium tetrafluoroborate: Peng, J.; Deng, Y. *Tetrahedron Lett.* **2001**, *42*, 403. In bmim PF$_6$, 1-butyl-3-methylimidazolium hexafluorophosphate: Ren, R.X.; Zueva, L.D.; Ou, X. *Tetrahedron Lett.* **2001**, *42*, 8441.

[318]His, S.; Meyer, C.; Cossy, J.; Emeric, G.; Greiner, A. *Tetrahedron Lett.* **2003**, *44*, 8581.

[319]Chandrasekhar, S.; Gopalaiah, K. *Tetrahedron Lett.* **2001**, *42*, 8123.

[320]For a review of such ring enlargements, see Vinnik, M.I.; Zarakhani, N.G. *Russ. Chem. Rev.* **1967**, *36*, 51. For a review with respect to bicyclic oximes, see Krow, G.R. *Tetrahedron* **1981**, *37*, 1283.

[321]Ghiaci, M.; Imanzadeh, G.H. *Synth. Commun.* **1998**, *28*, 2275. See Moghaddam, F.M.; Rad, A.A.R.; Zali-Boinee, H. *Synth. Commun.* **2004**, *34*, 2071.

[322]Bosch, A.I.; de la Cruz, P.; Diez-Barra, E.; Loupy, A.; Langa, F. *Synlett* **1995**, 1259.

[323]Sharghi, H.; Hosseini, M. *Synthesis* **2002**, 1057; Eshghi, H.; Gordi, Z. *Synth. Commun.* **2003**, *33*, 2971.

[324]Laurent, A.; Jacquault, P.; DiMarino, J.-L.; Hamelin, J. *J. Chem. Soc., Chem. Commun.* **1995**, 1101.

[325]Olah, G.A.; Fung, A.P. *Synthesis* **1979**, 537. See also, Novoselov, E.F.; Isaev, S.D.; Yurchenko, A.G.; Vodichka, L.; Trshiska, Ya. *J. Org. Chem. USSR* **1981**, *17*, 2284.

[326]Chandrassekhar, S.; Gopalaiah, K. *Tetrahedron Lett.* **2003**, *44*, 7437.

[327]An, G.-i.; Kim, M.; Kim. J.Y.; Rhee, H. *Tetrahedron Lett.* **2003**, *44*, 2183.

often used as a method of determining the configuration of the oxime. However, it is not unequivocal. It is known that with some oximes the syn group migrates and that with others, especially where R and R' are both alkyl, mixtures of the two possible amides are obtained. However, this behavior does not necessarily mean that the syn group actually undergoes migration. In most cases, the oxime undergoes isomerization under the reaction conditions *before* migration takes place.[328] The scope of the reaction is quite broad and R and R' may be alkyl, aryl, or hydrogen. However, hydrogen very seldom *migrates*, so the reaction is not generally a means of converting aldoximes to unsubstituted amides ($RCONH_2$). This latter conversion can be accomplished, however, by treatment of the aldoxime with nickel acetate under neutral conditions[329] or by heating the aldoxime for 60 h at $100°C$ after it has been adsorbed onto silica gel.[330] As in the case of the Schmidt rearrangement, when the oxime is derived from an alkyl aryl ketone, it is generally the aryl group that preferentially migrates.[331]

Not only do oximes undergo the Beckmann rearrangement, but so also do esters of oximes with many acids, organic and inorganic. A side reaction with many substrates is the formation of nitriles (the "abnormal" Beckmann rearrangement, **17-30**). The other reagents convert OH to an ester leaving group (e.g., $OPCl_4$ from PCl_5 and OSO_2OH from concentrated H_2SO_4[332]). The *O*-carbonates of imines, such as $Ph_2C=N–OCO_2Et$, react with $BF_3•OEt_2$ to give the corresponding amide, in this case *N*-phenyl benzamide.[333]

In the first step of the mechanism, the OH group is converted by the reagent to a better leaving group, for example, proton acids convert it to OH_2^+. After that, the mechanism[334] follows a course analogous to that for the Schmidt reaction of ketones (**18-16**) from the formation of nitrilium ion **82** on:[335] Alternatively, the attack on **82** can be by the leaving group, if different from H_2O. For example, when PCl_5 is used to induce the reaction, a N–O–PCl_4 species is formed, which generates **82**. Intermediates of the form **82** have been detected by nmr and uv spectroscopy.[336] The rearrangement has also been found to take place by a different mechanism, involving formation of a nitrile by fragmentation, and then addition by

[328]Lansbury, P.T.; Mancuso, N.R. *Tetrahedron Lett.* **1965**, 2445 have shown that some Beckmann rearrangements are *authentically* nonstereospecific.
[329]Field, L.; Hughmark, P.B.; Shumaker, S.H.; Marshall, W.S. *J. Am. Chem. Soc.* **1961**, *83*, 1983. See also, Leusink, A.J.; Meerbeek, T.G.; Noltes, J.G. *Recl. Trav. Chim. Pays-Bas* **1976**, *95*, 123; **1977**, *96*, 142.
[330]Chattopadhyaya, J.B.; Rama Rao, A.V. *Tetrahedron* **1974**, *30*, 2899.
[331]See Arisawa, M.; Yamaguchi, M. *Org. Lett.* **2001**, *3*, 311.
[332]Gregory, B.J.; Moodie, R.B.; Schofield, K. *J. Chem. Soc. B* **1970**, 338; Kim, S.; Kawakami, T.; Ando, T.; Yukawa, Y. *Bull. Chem. Soc. Jpn.* **1979**, *52*, 1115.
[333]Anilkumar, R.; Chandrasekhar, S. *Tetrahedron Lett.* **2000**, *41*, 5427.
[334]For a discussion of the gas-phase reaction mechanism, see Nguyen, M.T.; Vanquickenborne, L.G. *J. Chem. Soc. Perkin Trans. 2* **1993**, 1969.
[335]For summaries of the considerable evidence for this mechanism, see Donaruma, L.G.; Heldt, W.Z. *Org. React.* **1960**, *11*, 1, 5–14; Smith, P.A.S., in de Mayo, P. *Molecular Rearrangements*, Vol. 1, Wiley, NY, **1963**, 483–507, p. 488–493.
[336]Gregory, B.J.; Moodie, R.B.; Schofield, K. *J. Chem. Soc. B* **1970**, 338.

a Ritter reaction (**16-91**).[337] Beckmann rearrangements have also been carried out photochemically.[338]

If the rearrangement of oxime sulfonates is induced by organoaluminum reagents,[339] the nitrilium ion intermediate **82** is captured by the nucleophile originally attached to the Al. By this means an oxime can be converted to an imine, an imino thioether (R–N=C–SR), or an imino nitrile (R–N=C–CN).[340] In the last case, the nucleophile comes from added trimethylsilyl cyanide. The imine-producing reaction can also be accomplished with a Grignard reagent in benzene or toluene.[341]

In a related reaction, treatment of spirocyclic oxaziridines with MnCl(TPP)[342] or photolysis[343] leads to a lactam.

OS **II**, 76, 371; **VIII**, 568.

18-18   Stieglitz and Related Rearrangements

Methoxy-de-*N*-chloro-(2/ → 1/*N*-alkyl)-*migro*-substitution, and so on

Besides the reactions discussed at **18-13–18-17**, a number of other rearrangements are known in which an alkyl group migrates from C to N. Certain bicyclic *N*-haloamines, for example *N*-chloro-2-azabicyclo[2.2.2]octane (above), undergo

[337]Hill, R.K.; Conley, R.T.; Chortyk, O.T. *J. Am. Chem. Soc.* **1965**, *87*, 5646; Palmere, R.M.; Conley, R.T.; Rabinowitz, J.L. *J. Org. Chem.* **1972**, *37*, 4095.

[338]See, for example, Izawa, H.; de Mayo, P.; Tabata, T. *Can. J. Chem.* **1969**, *47*, 51; Cunningham, M.; Ng Lim, L.S.; Just, T. *Can. J. Chem.* **1971**, *49*, 2891; Suginome, H.; Yagihashi, F. *J. Chem. Soc. Perkin Trans. 1* **1977**, 2488.

[339]For a review, see Maruoka, K.; Yamamoto, H. *Angew. Chem. Int. Ed.* **1985**, *24*, 668.

[340]Maruoka, K.; Miyazaki, T.; Ando, M.; Matsumura, Y.; Sakane, S.; Hattori, K.; Yamamoto, H. *J. Am. Chem. Soc.* **1983**, *105*, 2831; Maruoka, K.; Nakai, S.; Yamamoto, H. *Org. Synth. 66*, 185.

[341]Hattori, K.; Maruoka, K.; Yamamoto, H. *Tetrahedron Lett.* **1982**, *23*, 3395.

[342]Suda, K.; Sashima, M.; Izutsu, M.; Hino, F. *J. Chem. Soc., Chem. Commun.* **1994**, 949.

[343]Post, A.J.; Nwaukwa, S.; Morrison, H. *J. Am. Chem. Soc.* **1994**, *116*, 6439.

rearrangement when solvolyzed in the presence of silver nitrate.[344] This reaction is similar to the Wagner–Meerwein rearrangement (**18-1**) and is initiated by the silver-catalyzed departure of the chloride ion.[345] Similar reactions have been used for ring expansions and contractions, analogous to those discussed for reaction **18-3**.[346] An example is the conversion of 1-(*N*-chloroamino)cyclopropanols to β-lactams.[347] Methyl prolinate was converted to the lactam 2-piperidone upon treatment with SmI$_2$ and pivalic acid–THF.[348]

The name *Stieglitz rearrangement* is generally applied to the rearrangements of trityl *N*-haloamines and

$$Ar_3CNHX \xrightarrow{\text{base}} Ar_2C=NAr$$

$$Ar_3CNHOH \xrightarrow{PCl_5} Ar_2C=NAr$$

hydroxylamines. These reactions are similar to the rearrangements of alkyl azides (**18-14**), and the name Stieglitz rearrangement is also given to the rearrangement of trityl azides. Another similar reaction is the rearrangement undergone by trityl-amines when treated with lead tetraacetate:[349]

$$Ar_3CNH_2 \xrightarrow{Pb(OAc)_4} Ar_2C=NAr$$

D. Carbon-to-Oxygen Migrations of R and AR

18-19 The Baeyer–Villiger Rearrangement[350]

Oxy-insertion

[344]Gassman, P.G.; Fox, B.L. *J. Am. Chem. Soc.* **1967**, *89*, 338. See also, Schell, F.M.; Ganguly, R.N. *J. Org. Chem.* **1980**, *45*, 4069; Davies, J.W.; Malpass, J.R.; Walker, M.P. *J. Chem. Soc., Chem. Commun.* **1985**, 686; Hoffman, R.V.; Kumar, A.; Buntain, G.A. *J. Am. Chem. Soc.* **1985**, *107*, 4731.

[345]For C → N rearrangements induced by AlCl$_3$, see Kovacic, P.; Lowery, M.K.; Roskos, P.D. *Tetrahedron* **1970**, *26*, 529.

[346]Gassman, P.G.; Carrasquillo, A. *Tetrahedron Lett.* **1971**, 109; Hoffman, R.V.; Buntain, G.A. *J. Org. Chem.* **1988**, *53*, 3316.

[347]Wasserman, H.H.; Adickes, H.W.; Espejo de Ochoa, O. *J. Am. Chem. Soc.* **1971**, *93*, 5586; Wasserman, H.H.; Glazer, E.A.; Hearn, M.J. *Tetrahedron Lett.* **1973**, 4855.

[348]Honda, T.; Ishikawa, F. *Chem. Commun.* **1999**, 1065.

[349]Sisti, A.J.; Milstein, S.R. *J. Org. Chem.* **1974**, *39*, 3932.

[350]For a review, see Renz, M.; Meunier, B. *Eur. J. Org. Chem.* **1999**, 737. For a review of green procedures, see Ten Brink, G.-J.; Arends, W.C.E.; Sheldon, R.A. *Chem. Rev.* **2004**, *104*, 4105.

The treatment of ketones with peroxyacids, such as peroxybenzoic or peroxyacetic acid, or with other peroxy compounds in the presence of acid catalysts, gives carboxylic esters by "insertion" of oxygen[351] and the carboxylic acid parent of the peroxyacid as a by-product. The reaction is called the *Baeyer–Villiger rearrangement*.[352] A particularly good reagent is peroxytrifluoroacetic acid. Reactions with this reagent are rapid and clean, giving high yields of product, though it is often necessary to add a buffer, such as Na_2HPO_4, to prevent transesterification of the product with trifluoroacetic acid that is also formed during the reaction. The reaction is often applied to cyclic ketones to give lactones.[353] Hydrogen peroxide has been used to convert cyclic ketones to lactones using a catalytic amount of $MeReO_3$[354] or a diselenide catalyst.[355] Hydrogen peroxide and a $MeReO_3$ catalyst has been used in an ionic liquid.[356] Transition-metal catalysts have been used with peroxyacids to facilitate the oxidation.[357] Hydrogen peroxide and $PhAsO_3H_2$ in hexafluoro-1-propanol can be used.[358] Polymer-supported peroxy acids have been used,[359] and solvent-free Bayer–Villiger reactions are known.[360] Enantioselective synthesis[361] of chiral lactones from achiral ketones has been achieved by the use of enzymes[362]

[351]For a list of reagents, with references, see Larock, R.C. *Comprehensive Organic Transformations*; 2nd ed., Wiley-VCH, NY, *1999*, pp. 1665–1667.

[352]For reviews, see Hudlický, M. *Oxidations in Organic Chemistry*, American Chemical Society, Washington, DC, *1990*, pp. 186–195; Plesničar, B., in Trahanovsky, W.S. *Oxidation in Organic Chemistry*, pt. C, Academic Press, NY, *1978*, pp. 254–267; House, H.O. *Modern Synthetic Reactions*, 2nd ed., W.A. Benjamin, NY, *1972*, pp. 321–329; Lewis, S.N., in Augustine, R.L. *Oxidation*, Vol. 1, Marcel Dekker, NY, *1969*, pp. 237–244; Lee, J.B.; Uff, B.C. *Q. Rev. Chem. Soc. 1967*, *21*, 429, see pp. 449–453. Also see, Mino, T.; Masuda, S.; Nishio, M.; Yamashita, M. *J. Org. Chem. 1997*, *62*, 2633. For a discussion of uncatalyzed versus BF_3-assisted reactions, see Carlqvist, P.; Eklund, R.; Brinck, T. *J. Org. Chem. 2001*, *66*, 1193.

[353]For a review of the reaction as applied to bicyclic ketones, see Krow, G.R. *Tetrahedron 1981*, *37*, 2697.

[354]Phillips, A.M.F.; Romão, C. *Eur. J. Org. Chem. 1999*, 1767.

[355]ten Brink, G.-J.; Vis, J.-M.; Arends, I.W.C.E.; Sheldon, R.A. *J. Org. Chem. 2001*, *66*, 2429.

[356]In bmim BF_4, 1-butyl-3-methylimidazoliuum tetrafluoroborate: Bernini, R.; Coratti, A.; Fabrizi, G.; Goggiamani, A. *Tetrahedron Lett. 2003*, *44*, 8991.

[357]Kotsuki, H.; Arimura, K.; Araki, T.; Shinohara, T. *Synlett 1999*, 462; Alam, M.M.; Varala, R.; Adapa, S.R. *Synth. Commun. 2003*, *33*, 3035.

[358]Berkessel, A.; Andreae, M.R.M. *Tetrahedron Lett. 2001*, *42*, 2293.

[359]Lambert, A.; Elings, J.A.; Macquarrie, D.J.; Carr, G.; Clark, J.H. *Synlett 2000*, 1052. For a discussion of selectivity in solid-state Bayer-Villiger reactions, see Hagiwara, H.; Nagatomo, H.; Yoshii, F.; Hoshi, T.; Suzuki, T.; Ando, M. *J. Chem. Soc., Perkin Trans. 1 2000*, 2645.

[360]Yakura, T.; Kitano, T.; Ikeda, M.; Uenishi, J. *Tetrahedron Lett. 2002*, *43*, 6925.

[361]See Bolm, C.; Beckmann, O.; Kühn, T.; Palazzi, C.; Adam, W.; Rao, P.B.; Saha-Möller, C.R. *Tetrahedron Asymmetry 2001*, *12*, 2441; Bolm, C.; Frison J.-C.; Zhang, Y.; Wulff, W.D. *Synlett 2004*, 1619.

[362]See Taschner, M.J.; Black, D.J. *J. Am. Chem. Soc. 1988*, *110*, 6892; Taschner, M.J.; Peddada, L. *J. Chem. Soc., Chem. Commun. 1992*, 1384; Pchelka, B.K.; Gelo Pujic, M.; Guibé-Jampel, E. *J. Chem. Soc. Perkin Trans. 1 1998*, 2625; Stewart, J.D.; Reed, K.W.; Martinez, C.A.; Zhu, J.; Chen, G.; Kayser, M.M. *J. Am. Chem. Soc. 1998*, *120*, 3541; Lemoult, S.C.; Richardson, P.F.; Roberts, S.M. *J. Chem. Soc. Perkin Trans. 1 1995*, 89; Mihovilovic, M.D.; Müller, B.; Kayser, M.M.; Stewart, J.D. *Synlett 2002*, 703. For a review of enzyme-catalyzed Baeyer–Villiger rearrangements, see Walsh, C.T.; Chen, Y.J. *Angew. Chem. Int. Ed. 1988*, *27*, 333. For a review of monooxygenase-mediated Baeyer–Villiger rearrangements, see Mihovilovic, M.D.; Müller, B.; Stanetty, P. *Eur. J. Org. Chem. 2002*, 3711. For a reaction using engineered *E. coli* cells, see Mihovilovic, M.D.; Chen, G.; Wang, S.; Kyte, B.; Rochon, F.; Kayser, M.M.; Stewart, J.D. *J. Org. Chem. 2001*, *66*, 733.

and other asymmetric reactions are known.[363] Oxidation of chiral substrates with *m*-chloroperoxybenzoic acid (mcpba) also leads to chiral lactones.[364]

For acyclic compounds, R' must usually be secondary, tertiary, or vinylic, although primary R' has been rearranged with peroxytrifluoroacetic acid,[365] with BF_3–H_2O_2,[366] and with $K_2S_2O_8$–H_2SO_4.[367] For unsymmetrical ketones the approximate order of migration is tertiary alkyl > secondary alkyl, aryl > primary alkyl > methyl. Since the methyl group has a low migrating ability, the reaction provides a means of cleaving a methyl ketone R'COMe to produce an alcohol or phenol (R'OH) (by hydrolysis of the ester R'OCOMe). The migrating ability of aryl groups is increased by electron-donating and decreased by electron-withdrawing substituents.[368] There is a preference of anti over gauche migration.[369] Enolizable β-diketones do not react. α-Diketones can be converted to anhydrides.[370] With aldehydes, migration of hydrogen gives the carboxylic acid, and this is a way of accomplishing **19-23**. Migration of the other group would give formates, but this seldom happens, though aryl aldehydes have been converted to formates with H_2O_2 and a selenium compound[371] (see also the Dakin reaction in **19-11**).

The mechanism[372] is similar to those of the analogous reactions with hydrazoic acid (**18-16** with ketones) and diazomethane (**18-8**):

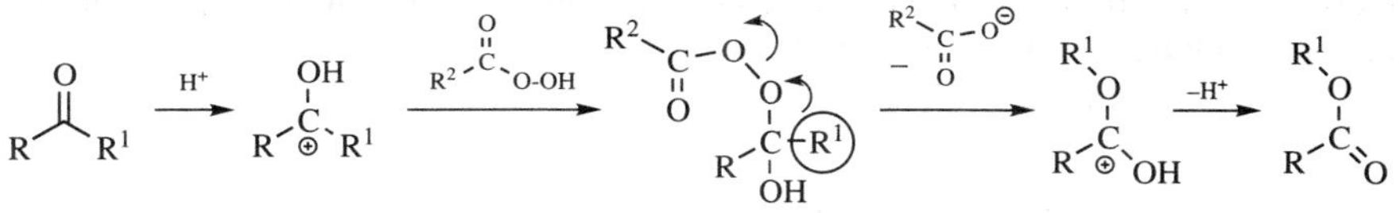

One important piece of evidence for this mechanism was that benzophenone–[18]O gave ester entirely labeled in the carbonyl oxygen, with none in the alkoxyl oxygen.[373] Carbon-14 isotope-effect studies on acetophenones have shown that

[363]For example, see Sugimura, T.; Fujiwara, Y.; Tai, A. *Tetrahedron Lett.* **1997**, *38*, 6019; Bolm, C.; Schlingloff, G.; Weickhardt, K. *Angew. Chem. Int. Ed.* **1994**, *33*, 1848; Bolm, C.; Schlingloff, G. *J. Chem. Soc., Chem. Commun.* **1995**, 1247; Bolm, C.; Beckmann, O.; Cosp, A.; Palazzi, C. *Synlett* **2001**, 1461; Bolm, C.; Beckmann, O.; Palazzi, C. *Can. J. Chem.* **2001**, *79*, 1593; Shinohara, T.; Fujioka, S.; Kotsuki, H. *Heterocycles* **2001**, *55*, 237; Watanabe, A.; Uchida, T.; Ito, K.; Katsuki, T. *Tetrahedron Lett.* **2002**, *43*, 4481; Murhashi, S.-I.; Ono, S.; Imada, Y. *Angew. Chem. Int. Ed.* **2002**, *41*, 2366.

[364]Hunt, KW.; Grieco, P.A. *Org. Lett.* **2000**, *2*, 1717.

[365]Emmons, W.D.; Lucas, G.B. *J. Am. Chem. Soc.* **1955**, *77*, 2287.

[366]McClure, J.D.; Williams, P.H. *J. Org. Chem.* **1962**, *27*, 24.

[367]Deno, N.C.; Billups, W.E.; Kramer, K.E.; Lastomirsky, R.R. *J. Org. Chem.* **1970**, *35*, 3080.

[368]For as report of substituent effects in the α, β, and γ position of alkyl groups, see Noyori, R.; Sato, T.; Kobayashi, H. *Bull. Chem. Soc. Jpn.* **1983**, *56*, 2661.

[369]Snowden, M.; Bermudez, A.; Kelly, D.R.; Radkiewicz-Poutsma, J.L. *J. Org. Chem.* **2004**, *69*, 7148.

[370]For a study of the mechanism of this conversion, see Cullis, P.M.; Arnold, J.R.P.; Clarke, M.; Howell, R.; DeMira, M.; Naylor, M.; Nicholls, D. *J. Chem. Soc., Chem. Commun.* **1987**, 1088.

[371]Syper, L. *Synthesis* **1989**, 167. See also, Godfrey, I.M.; Sargent, M.V.; Elix, J.A. *J. Chem. Soc. Perkin Trans. 1* **1974**, 1353.

[372]Proposed by Criegee, R. *Liebigs Ann. Chem.* **1948**, *560*, 127.

[373]Doering, W. von E.; Dorfman, E. *J. Am. Chem. Soc.* **1953**, *75*, 5595. For summaries of the other evidence, see Smith, P.A.S., in de Mayo, P. *Molecular Rearrangements*, Vol. 1, Wiley, NY, **1963**, pp. 578–584.

migration of aryl groups takes place in the rate-determining step,[374] demonstrating that migration of Ar is concerted with departure of OCOR^2.[375] It is hardly likely that migration would be the slow step if the leaving group departed first to give an ion with a positive charge on an oxygen atom, which would be a highly unstable species.

18-20   Rearrangement of Hydroperoxides

C-Alkyl-O-hydroxy-elimination

$$R_3C\text{-O-O-H} \xrightarrow{H^+} R_2C{=}O + ROH$$

Hydroperoxides (R = alkyl, aryl, or hydrogen) can be cleaved by proton or Lewis acids in a reaction whose principal step is a rearrangement.[376] The reaction has also been applied to peroxy esters (R$_3$COOCOR'), but less often. When aryl and alkyl groups are both present, migration of aryl dominates. It is not necessary actually to prepare and isolate hydroperoxides. The reaction takes place when the alcohols are treated with H$_2$O$_2$ and acids. Migration of an alkyl group of a primary hydroperoxide provides a means for converting an alcohol to its next lower homolog (RCH$_2$OOH → CH$_2$=O + ROH).

The mechanism is as follows:[377]

72   **84**

The last step is hydrolysis of the unstable hemiacetal. Alkoxycarbocation intermediates (**84**, R = alkyl) have been isolated in super acid solution[378] at

[374]Palmer, B.W.; Fry, A. *J. Am. Chem. Soc.* **1970**, *92*, 2580. See also, Mitsuhashi, T.; Miyadera, H.; Simamura, O. *Chem. Commun.* **1970**, 1301. For secondary isotope-effect studies, see Winnik, M.A.; Stoute, V.; Fitzgerald, P. *J. Am. Chem. Soc.* **1974**, *96*, 1977.

[375]In some cases, the rate-determining step has been shown to be the addition of peracid to the substrate (see, e.g., Ogata, Y.; Sawaki, Y. *J. Org. Chem.* **1972**, *37*, 2953). Even in these cases it is still highly probable that migration is concerted with departure of the nucleofuge.

[376]For reviews, see Yablokov, V.A. *Russ. Chem. Rev.* **1980**, *49*, 833; Lee, J.B.; Uff, B.C. *Q. Rev. Chem. Soc.* **1967**, *21*, 429, 445–449.

[377]For a discussion of the transition state involved in the migration step, see Wistuba, E.; Rüchardt, C. *Tetrahedron Lett.* **1981**, *22*, 3389.

[378]For a review of peroxy compounds in super acids, see Olah, G.A.; Parker, D.G.; Yoneda, N. *Angew. Chem. Int. Ed.* **1978**, *17*, 909.

low temperatures, and their structures proved by nmr.[379] The protonated hydroperoxides could not be observed in these solutions, evidently reacting immediately on formation.

OS **V**, 818.

E. Nitrogen-to-Carbon, Oxygen-to-Carbon, and Sulfur-to-Carbon Migration

18-21   The Stevens Rearrangement

Hydron-(2/N → 1/alkyl)-*migro*-detachment

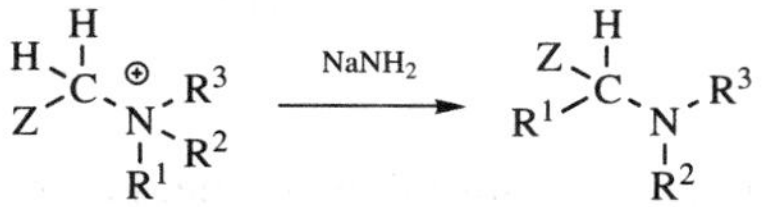

In the *Stevens rearrangement*, a quaternary ammonium salt containing an electron-withdrawing group Z on one of the carbons attached to the nitrogen is treated with a strong base (e.g., NaOR or NaNH$_2$) to give a rearranged tertiary amine. The Z group may be RCO, ROOC, or phenyl.[380] The most common migrating groups are allylic, benzylic, benzhydryl, 3-phenylpropargyl, and phenacyl, though even methyl migrates to a sufficiently negative center.[381] When an allylic group migrates, it may or may not involve an allylic rearrangement within the migrating group (see **18-35**), depending on the substrate and reaction conditions. The reaction has been used for ring enlargement,[382] for example:

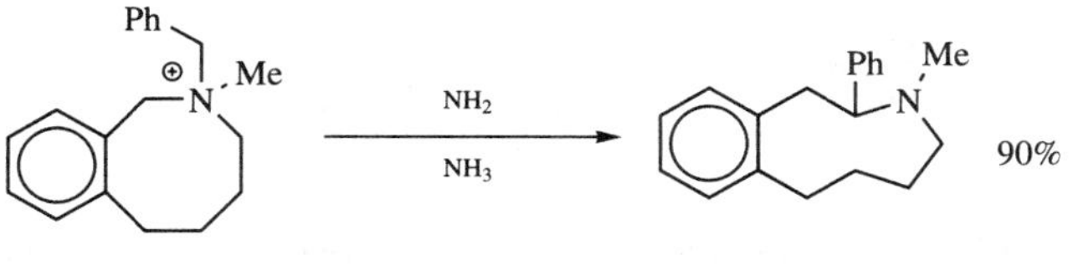

The mechanism has been the subject of much study.[383] That the rearrangement is intramolecular was shown by crossover experiments, by ^{14}C labeling,[384] and by the

---

[379]Sheldon, R.A.; van Doorn, J.A. *Tetrahedron Lett.* **1973**, 1021.

[380]For reviews of the Stevens rearrangement, see Lepley, A.R.; Giumanini, A.G. *Mech. Mol. Migr.* **1971**, *3*, 297; Pine, S.H. *Org. React.* **1970**, *18*, 403. For reviews of the Stevens and the closely related Wittig rearrangement (**18-22**), see Stevens, T.S.; Watts, W.E. *Selected Molecular Rearrangements*, Van Nostrand-Reinhold, Princeton, NJ, **1973**, pp. 81–116; Wilt, J.W., in Kochi, J.K. *Free Radicals*, Vol. 1, Wiley, NY, **1973**, pp. 448–458; Iwai, I. *Mech. Mol. Migr.* **1969**, *2*, 73, see pp. 105–113; Stevens, T.S. *Prog. Org. Chem.* **1968**, *7*, 48.

[381]Migration of aryl is rare, but has been reported: Heaney, H.; Ward, T.J. *Chem. Commun.* **1969**, 810; Truce, W.E.; Heuring, D.L. *Chem. Commun.* **1969**, 1499.

[382]Elmasmodi, A.; Cotelle, P.; Barbry, D.; Hasiak, B.; Couturier, D. *Synthesis* **1989**, 327.

[383]For example, see Pine, S.H. *J. Chem. Educ.* **1971**, *48*, 99; Heard, G.L.; Yates, B.F. *Aust. J. Chem.* **1994**, *47*, 1685.

[384]Stevens, T.S. *J. Chem. Soc.* **1930**, 2107; Johnstone, R.A.W.; Stevens, T.S. *J. Chem. Soc.* **1955**, 4487.

fact that retention of configuration is found at R^1.[385] The first step is loss of the acidic proton to give the ylid **85**, which has been isolated.[386] The finding[387] that CIDNP spectra[388] could be obtained in many instances shows that in these cases the product is formed directly from a free-radical precursor. The following radical pair mechanism was proposed:[389]

Mechanism *a*

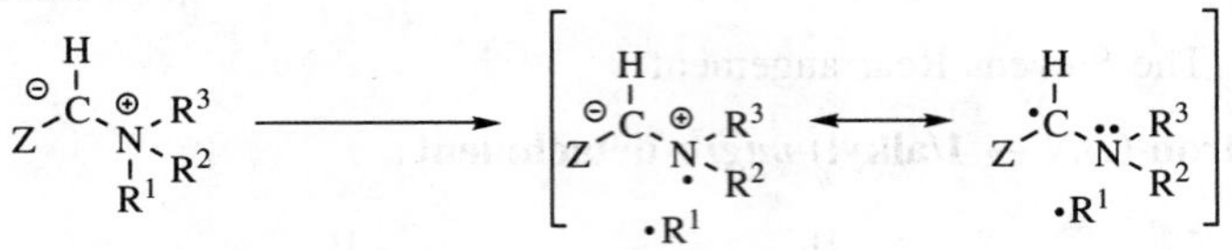

85   Solvent cage

The radicals do not drift apart because they are held together by the solvent cage. According to this mechanism, the radicals must recombine rapidly in order to account for the fact that R^1 does not racemize. Other evidence in favor of mechanism *a* is that in some cases small amounts of coupling products (R^1–R^1) have been isolated,[390] which would be expected if some •R^1 leaked from the solvent cage. However, not all the evidence is easily compatible with mechanism *a*.[391] It is possible that another mechanism (*b*) similar to mechanism *a*, but involving

Mechanism *b*

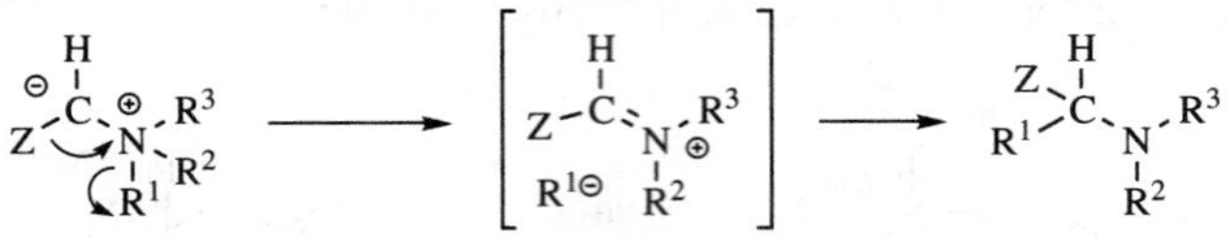

85   Solvent cage

ion pairs in a solvent cage instead of radical pairs, operates in some cases. A third possible mechanism would be a concerted 1,2-shift,[392] but the orbital symmetry

[385]Brewster, J.H.; Kline, M.W. *J. Am. Chem. Soc.* **1952**, *74*, 5179; Schöllkopf, U.; Ludwig, U.; Ostermann, G.; Patsch, M. *Tetrahedron Lett.* **1969**, 3415.

[386]Jemison, R.W.; Mageswaran, S.; Ollis, W.D.; Potter, S.E.; Pretty, A.J.; Sutherland, I.O.; Thebtaranonth, Y. *Chem. Commun.* **1970**, 1201.

[387]Lepley, A.R.; Becker, R.H.; Giumanini, A.G. *J. Org. Chem.* **1971**, *36*, 1222; Baldwin, J.E.; Brown, J.E. *J. Am. Chem. Soc.* **1969**, *91*, 3646; Jemison, R.W.; Morris, D.G. *Chem. Commun.* **1969**, 1226; Schöllkopf, U.; Ludwig, U.; Ostermann, G.; Patsch, M. *Tetrahedron Lett.* **1969**, 3415.

[388]For a review of the application of CIDNP to rearrangement reactions, see Lepley, A.R., in Lepley, A.R.; Closs, G.L. *Chemically Induced Magnetic Polarization*, Wiley, NY, **1973**, pp. 323–384.

[389]Schöllkopf, U.; Ludwig, U. *Chem. Ber.* **1968**, *101*, 2224; Ollis, W.D.; Rey, M.; Sutherland, I.O. *J. Chem. Soc. Perkin Trans. 1* **1983**, 1009, 1049.

[390]Schöllkopf, U.; Ludwig, U.; Ostermann, G.; Patsch, M. *Tetrahedron Lett.* **1969**, 3415; Hennion, G.F.; Shoemaker, M.J. *J. Am. Chem. Soc.* **1970**, *92*, 1769.

[391]See, for example, Pine, S.H.; Catto, B.A.; Yamagishi, F.G. *J. Org. Chem.* **1970**, *35*, 3663.

[392]For evidence against this mechanism, see Jenny, E.F.; Druey, J. *Angew. Chem. Int. Ed.* **1962**, *1*, 155.

principle requires that this take place with inversion at R^1.[393] (see p. 1654.) Since the actual migration takes place with retention, it cannot, according to this argument, proceed by a concerted mechanism. However, in the case where the migrating group is allylic, a concerted mechanism can also operate (**18-35**). An interesting finding compatible with all three mechanisms is that optically active allylbenzylmethylphenylammonium iodide (asymmetric nitrogen, see p. 142) gave an optically active product:[394]

The *Sommelet–Hauser rearrangement* competes when Z is an aryl group (see **13-31**). *Hofmann elimination* competes when one of the R groups contains a β hydrogen atom (**17-7** and **17-8**).

Sulfur ylids containing a Z group give an analogous rearrangement, often also referred to as a Stevens

rearrangement.[395] In this case too, there is much evidence (including CIDNP) that a radical-pair cage mechanism is operating,[396] except that when the migrating group is allylic, the mechanism may be different (see **18-35**). Another reaction with a similar mechanism[397] is the *Meisenheimer rearrangement*,[398] in which certain tertiary

[393]Woodward, R.B.; Hoffmann, R. *The Conservation of Orbital Symmetry*, Academic Press, NY, *1970*, p. 131.

[394]Hill, R.K.; Chan, T. *J. Am. Chem. Soc. 1966, 88,* 866.

[395]For a review, see Olsen, R.K.; Currie, Jr., J.O., in Patai, S. *The Chemistry of The Thiol Group*, pt. 2, Wiley, NY, *1974*, pp. 561–566.

[396]See, for example, Baldwin, J.E.; Erickson, W.F.; Hackler, R.E.; Scott, R.M. *Chem. Commun. 1970,* 576; Schöllkopf, U.; Schossig, J.; Ostermann, G. *Liebigs Ann. Chem. 1970, 737,* 158; Iwamura, H.I.; Iwamura, M.; Nishida, T.; Yoshida, M.; Nakayama, T. *Tetrahedron Lett. 1971,* 63.

[397]For some of the evidence, see Ostermann, G.; Schöllkopf, U. *Liebigs Ann. Chem. 1970, 737,* 170; Lorand, J.P.; Grant, R.W.; Samuel, P.A.; O'Connell, E.; Zaro, J. *Tetrahedron Lett. 1969,* 4087.

[398]For a review, see Johnstone, R.A.W. *Mech. Mol. Migr. 1969, 2,* 249. See Buston, J.E.H.; Coldham, I.; Mulholland, K.R. *J. Chem. Soc., Perkin Trans. 1 1999,* 2327.

amine oxides rearrange on heating to give substituted hydroxylamines.[399] The migrating group R^1 is almost always allylic or benzilic.[400] R^2 and R^3 may be alkyl or aryl, but if one of the R groups contains a β hydrogen, Cope elimination (**17-9**) often competes. In a related reaction, when 2-methylpyridine N-oxides are treated with trifluoroacetic anhydride, the *Boekelheide reaction* occurs to give 2-hydroxymethylpyridines.[401]

Certain tertiary benzylic amines, when treated with BuLi, undergo a rearrangement analogous to the Wittig rearrangement (**18-22**), for example, $PhCH_2NPh_2 \rightarrow Ph_2CHNHPh$.[402] Only aryl groups migrate in this reaction.

Isocyanides, when heated in the gas phase or in nonpolar solvents, undergo a 1,2-intramolecular rearrangement to nitriles: $RNC \rightarrow RCN$.[403] In polar solvents the mechanism is different.[404]

18-22 The Wittig Rearrangement[405]

Hydron-(2/*O* → 1/alkyl)-*migro*-detachment

The rearrangement of ethers with alkyllithium reagents is called the *Wittig rearrangement* (not to be confused with the Wittig reaction, **16-44**) and is similar to **18-21**.[380] However, a stronger base is required (e.g., phenyllithium or sodium amide). The R and R' groups, may be alkyl,[406] aryl, or vinylic.[407] Also, one of the hydrogens may be replaced by an alkyl or aryl group, in which case the product is the salt of a tertiary alcohol. Migratory aptitudes

Solvent cage

[399]For example, see Buston, J.E.H.; Coldham, I.; Mulholland, K.R. *Tetrahedron Asymmetry,* **1998**, *9*, 1995.

[400]Migration of aryl and of certain alkyl groups has also been reported. See Khuthier, A.; Al-Mallah, K.Y.; Hanna, S.Y.; Abdulla, N.I. *J. Org. Chem.* **1987**, *52*, 1710, and references cited therein.

[401]Fontenas, C.; Bejan, E.; Haddon, H.A.; Balavoine, G.G.A. *Synth. Commun.* **1995**, *25*, 629.

[402]Eisch, J.J.; Kovacs, C.A.; Chobe, P. *J. Org. Chem.* **1989**, *54*, 1275.

[403]See Pakusch, J.; Rüchardt, C. *Chem. Ber.* **1991**, *124*, 971, and references cited therein.

[404]Meier, M.; Rüchardt, C. *Chimia* **1986**, *40*, 238.

[405]See Hiersemann, M.; Abraham, L.; Pollex, A. *Synlett* **2003**, 1088.

[406]See Bailey, W.F.; England, M.D.; Mealy, M.J.; Thongsornkleeb, C.; Teng, L. *Org. Lett.* **2000**, *2*, 489.

[407]For migration of vinyl, see Rautenstrauch, V.; Büchi, G.; Wüest, H. *J. Am. Chem. Soc.* **1974**, *96*, 2576. For rearrangment of an α-trimethylsilyl allyl ether, see Maleczka, Jr., R.E.; Geng, F. *Org. Lett.* **1999**, *1*, 1115.

here are allylic, benzylic > ethyl > methyl > phenyl.[408] The stereospecificity of the 1,2-Wittig rearrangement has been discussed.[409] The following radical-pair mechanism[410] (similar to mechanism *a* of **18-21**) is likely, after removal of the proton by the base. One of the radicals in the radical pair is a ketyl. Among the evidence for this mechanism is (*1*) the rearrangement is largely intramolecular; (*2*) migratory aptitudes are in the order of free-radical stabilities, not of carbanion stabilities[411] (which rules out an ion-pair mechanism similar to mechanism *b* of **18-21**); (*3*) aldehydes are obtained as side products;[412] (*4*) partial racemization of R' has been observed[413] (the remainder of the product retained its configuration); (*5*) crossover products have been detected;[414] and (*6*) when ketyl radicals and R radicals from different precursors were brought together, similar products resulted.[415] However, there is evidence that at least in some cases the radical-pair mechanism accounts for only a portion of the product, and some kind of concerted mechanism can also take place.[416] Most of the above investigations were carried out with systems where R' is alkyl, but a radical-pair mechanism has also been suggested for the case where R' is aryl.[417] When R' is allylic a concerted mechanism can operate (**18-35**).

When R is vinylic it is possible, by using a combination of an alkyllithium and *t*-BuOK, to get migration to the γ carbon (as well as to the α carbon), producing an enolate that, on hydrolysis, gives an aldehyde:[418]

$$CH_2=CH-CH_2-OR' \longrightarrow R'CH_2-CH=CH-OLi \longrightarrow R'CH_2CH_2CHO$$

An aza-Wittig rearrangement is also known.[419]
There are no OS references, but see OS **VIII**, 501, for a related reaction.

[408]Wittig, G. *Angew. Chem.* **1954**, *66*, 10; Solov'yanov, A.A.; Ahmed, E.A.A.; Beletskaya, I.P.; Reutov, O.A. *J. Chem. Soc., Chem. Commun.* **1987**, *23*, 1232.

[409]Maleczka Jr., R.E.; Geng, F. *J. Am. Chem. Soc.* **1998**, *120*, 8551.

[410]For a review of the mechanism, see Schöllkopf, U. *Angew. Chem. Int. Ed.* **1970**, *9*, 763.

[411]Lansbury, P.T.; Pattison, V.A.; Sidler, J.D.; Bieber, J.B. *J. Am. Chem. Soc.* **1966**, *88*, 78; Schäfer, H.; Schöllkopf, U.; Walter, D. *Tetrahedron Lett.* **1968**, 2809.

[412]For example, see Hauser, C.R.; Kantor, S.W. *J. Am. Chem. Soc.* **1951**, *73*, 1437; Cast, J.; Stevens, T.S.; Holmes, J. *J. Chem. Soc.* **1960**, 3521.

[413]Schöllkopf, U.; Schäfer, H. *Liebigs Ann. Chem.* **1963**, *663*, 22; Felkin, H.; Frajerman, C. *Tetrahedron Lett.* **1977**, 3485; Hebert, E.; Welvart, Z. *J. Chem. Soc., Chem. Commun.* **1980**, 1035; *Nouv. J. Chim.* **1981**, *5*, 327.

[414]Lansbury, P.T.; Pattison, V.A. *J. Org. Chem.* **1962**, *27*, 1933; *J. Am. Chem. Soc.* **1962**, *84*, 4295.

[415]Garst, J.F.; Smith, C.D. *J. Am. Chem. Soc.* **1973**, *95*, 6870.

[416]Garst, J.F.; Smith, C.D. *J. Am. Chem. Soc.* **1976**, *98*, 1526. For evidence against this, see Hebert, E.; Welvart, Z.; Ghelfenstein, M.; Szwarc, H. *Tetrahedron Lett.* **1983**, *24*, 1381.

[417]Eisch, J.J.; Kovacs, C.A.; Rhee, S. *J. Organomet. Chem.* **1974**, *65*, 289.

[418]Schlosser, M.; Strunk, S. *Tetrahedron* **1989**, *45*, 2649.

[419]Coldham, I. *J. Chem. Soc. Perkin Trans. 1* **1993**, 1275; Anderson, J.C.; Siddons, D.C.; Smith, S.C.; Swarbrick, M.E. *J. Chem. Soc., Chem. Commun.* **1995**, 1835; Ahman, J.; Somfai, P. *J. Am. Chem. Soc.* **1994**, *116*, 9781.

F. Boron-to-Carbon Migrations[420]

For another reaction involving boron-to-carbon migration, see **10-73**.

18-23 Conversion of Boranes to Alcohols

$$R_3B \ + \ CO \ \xrightarrow[\text{100–125°C}]{\text{HOCH}_2\text{CH}_2\text{OH}} \ R_3C-B\underset{O}{\overset{O}{\big\langle}} \ \xrightarrow[\text{NaOH}]{\text{H}_2\text{O}_2} \ R_3C\text{-OH}$$

86

Trialkylboranes (which can be prepared from alkenes by **15-16**) react with carbon monoxide[421] at 100–125°C in the presence of ethylene glycol to give the 2-bora-1,3-dioxolanes (**86**), which are easily oxidized (**12-27**) to tertiary alcohols.[422] The R groups may be primary, secondary, or tertiary, and may be the same or different.[423] Yields are high and the reaction is quite useful, especially for the preparation of sterically hindered alcohols, such as tricyclohexyl-carbinol (**87**) and tri-2-norbornylcarbinol (**88**), which are difficult to prepare by **16-24**. Heterocycles in which boron is a ring atom react similarly (except that high CO pressures are required), and cyclic alcohols can be obtained from these substrates.[424] The preparation of such heterocyclic boranes was discussed at **15-16**. The overall conversion of a diene or triene to a cyclic alcohol has been described by H.C. Brown as "stitching" with boron and "riveting" with carbon.

87

88

[420]For reviews, see Matteson, D.S., in Hartley, F.R. *The Chemistry of the Metal-Carbon Bond*, Vol. 4, Wiley, NY, *1984*, pp. 307–409, 346–387; Pelter, A.; Smith, K.; Brown, H.C. *Borane Reagents*, Academic Press, NY, *1988*, pp. 256–301; Negishi, E.; Idacavage, M.J. *Org. React.* *1985*, *33*, 1; Suzuki, A *Top. Curr. Chem.* *1983*, *112*, 67; Pelter, A., in de Mayo, P. *Rearrangements in Ground and Excited States*, Vol. 2, Academic Press, NY, *1980*, pp. 95–147; *Chem. Soc. Rev.* *1982*, *11*, 191; Cragg, G.M.L.; Koch, K.R. *Chem. Soc. Rev.* *1977*, *6*, 393; Weill-Raynal, J. *Synthesis* *1976*, 633; Cragg, G.M.L. *Organoboranes in Organic Synthesis*; Marcel Dekker, NY, *1973*, pp. 249–300; Paetzold, P.I.; Grundke, H. *Synthesis* *1973*, 635.

[421]For discussions of the reaction of boranes with CO, see Negishi, E. *Intra-Sci. Chem. Rep.* *1973*, *7(1)*, 81; Brown, H.C. *Boranes in Organic Chemistry*, Cornell University Press, Ithaca, NY, *1972*, pp. 343–371; *Acc. Chem. Res.* *1969*, *2*, 65.

[422]Hillman, M.E.D. *J. Am. Chem. Soc.* *1962*, *84*, 4715; *1963*, *85*, 982; Brown, H.C.; Rathke, M.W. *J. Am. Chem. Soc.* *1967*, *89*, 2737; Puzitskii, K.V.; Pirozhkov, S.D.; Ryabova, K.G.; Pastukhova, I.V.; Eidus, Ya.T. *Bull. Acad. Sci. USSR Div. Chem. Sci.* *1972*, *21*, 1939; *1973*, *22*, 1760; Brown, H.C.; Cole, T.E.; Srebnik, M.; Kim, K. *J. Org. Chem.* *1986*, *51*, 4925.

[423]Brown, H.C.; Gupta, S.K. *J. Am. Chem. Soc.* *1971*, *93*, 1818; Negishi, E.; Brown, H.C. *Synthesis* *1972*, 197.

[424]Brown, H.C.; Negishi, E.; Dickason, W.C. *J. Org. Chem.* *1985*, *50*, 520, and references cited therein.

Though the mechanism has not been investigated thoroughly, it has been shown to be intramolecular by the failure to find crossover products when mixtures of boranes are used.[425] The following scheme, involving three boron-to-carbon migrations, has been suggested.

$$R_3B \longrightarrow R-\overset{R}{\underset{R'}{B}}-\overset{\ominus}{C}=O \longrightarrow \underset{\mathbf{89}}{\overset{R}{\underset{R'}{B}}-\overset{R}{\underset{O}{C}}} \longrightarrow \overset{R}{\underset{O}{B}}-\overset{R}{\underset{R}{C}} \longrightarrow \underset{\mathbf{90}}{R-\overset{R}{\underset{R}{C}}-B=O} \xrightarrow{HOCH_2CH_2OH} \mathbf{86}$$

The purpose of the ethylene glycol is to intercept the boronic anhydride **90**, which otherwise forms polymers that are difficult to oxidize. As we will see in **18-23** and **18-24**, it is possible to stop the reaction after only one or two migrations have taken place.

Method 1 $R_3B + CHCl_2OMe \xrightarrow[\text{2. } H_2O_2-NaOH]{\text{1. LiOCEt}_3-THF} R_3COH$

Method 2 $R_3B + {}^{\ominus}CN \xrightarrow{THF} \underset{\mathbf{91}}{R_3\overset{\ominus}{B}-CN} \xrightarrow[\text{2. NaOH}-H_2O_2]{\text{1. excess } (CF_3CO)_2O} R_3COH$

There are two other methods for achieving the conversion $R_3B \rightarrow R_3COH$, which often give better results: (1) treatment with α,α-dichloromethyl methyl ether and the base lithium triethylcarboxide[426] (2) treatment with a suspension of sodium cyanide in THF followed by reaction of the resulting trialkylcyanoborate **91** with an excess (>2 equivalents) of trifluoroacetic anhydride.[427] All the above migrations take place with retention of configuration at the migrating carbon.[428]

Several other methods for the conversion of boranes to tertiary alcohols are also known.[429]

If the reaction between trialkylboranes and carbon monoxide (**18-23**) is carried out in the presence of water followed by addition of NaOH, the product is a secondary alcohol. If H_2O_2 is added along with the NaOH, the corresponding ketone is obtained instead.[430] Various functional groups (e.g., OAc, COOR, CN) may be

[425]Brown, H.C.; Rathke, M.W. *J. Am. Chem. Soc.* **1967**, *89*, 4528.

[426]Brown, H.C.; Carlson, B.A. *J. Org. Chem.* **1973**, *38*, 2422; Brown, H.C.; Katz, J.; Carlson, B.A. *J. Org. Chem.* **1973**, *38*, 3968.

[427]Pelter, A.; Hutchings, M.G.; Smith, K.; Williams, D.J. *J. Chem. Soc. Perkin Trans. 1* **1975**, 145, and references cited therein.

[428]See however Pelter, A.; Maddocks, P.J.; Smith, K. *J. Chem. Soc., Chem. Commun.* **1978**, 805.

[429]See, for example, Brown, H.C.; Lane, C.F. *Synthesis* **1972**, 303; Yamamoto, Y.; Brown, H.C. *J. Org. Chem.* **1974**, *39*, 861; Zweifel, G.; Fisher, R.P. *Synthesis* **1974**, 339; Midland, M.M.; Brown, H.C. *J. Org. Chem.* **1975**, *40*, 2845; Levy, A.B.; Schwartz, S.J. *Tetrahedron Lett.* **1976**, 2201; Baba, T.; Avasthi, K.; Suzuki, A. *Bull. Chem. Soc. Jpn.* **1983**, *56*, 1571; Pelter, A.; Rao, J.M. *J. Organomet. Chem.* **1985**, *285*, 65; Junchai, B.; Hongxun, D. *J. Chem. Soc., Chem. Commun.* **1990**, 323.

[430]Brown, H.C.; Rathke, M.W. *J. Am. Chem. Soc.* **1967**, *89*, 2738.

present in R without being affected,[431] though if they are in the α or β position relative to the boron atom, difficulties may

$$\underset{\mathbf{91}}{R_3B-\overset{\ominus}{C}N} \quad \xrightarrow[\text{2. } H_2O_2-OH^-]{\text{1. } (CF_3CO)_2O} \quad RCOR$$

be encountered. The use of an equimolar amount of trifluoroacetic anhydride leads to the ketone rather than the tertiary alcohol.[427,432] By this procedure, thexylboranes (RR'R²B, where R² = thexyl) can be converted to unsymmetrical ketones (RCOR').[433] Variations of this methodology have been used to prepare optically active alcohols.[434]

For another conversion of trialkylboranes to ketones (see **18-26**).[435] Other conversions of boranes to secondary alcohols are also known.[436]

OS **VII**, 427. Also see, OS **VI**, 137.

18-24   Conversion of Boranes to Primary Alcohols, Aldehydes, or Carboxylic Acids

$$R_3B + CO \quad \xrightarrow[\text{2. } H_2O_2\text{-}OH^-]{\text{1. } LiBH_4} \quad RCH_2OH$$

$$R_3B + CO \quad \xrightarrow[\text{2. } H_2O_2\text{-}NaH_2PO_4\text{-}Na_2HPO_4]{\text{1. } LiAlH(OMe)_3} \quad RCHO$$

When the reaction between a trialkylborane and carbon monoxide (**18-23**) is carried out in the presence of a reducing agent such as lithium borohydride or potassium triisopropoxyborohydride, the reduction agent intercepts the intermediate **89**, so that only one boron-to-carbon migration takes place, and the product is hydrolyzed to a primary alcohol or oxidized to an aldehyde.[437] This procedure wastes two of the three R groups, but this problem can be avoided by the use of

[431]Brown, H.C.; Kabalka, G.W.; Rathke, M.W. *J. Am. Chem. Soc.* **1967**, *89*, 4530.

[432]Pelter, A.; Smith, K.; Hutchings, M.G.; Rowe, K. *J. Chem. Soc. Perkin Trans. 1* **1975**, 129; See also, Mallison, P.R.; White, D.N.J.; Pelter, A.; Rowe, K.; Smith, K. *J. Chem. Res. (S)*, **1978**, 234.

[433]This has been done enantioselectively: Brown, H.C.; Bakshi, R.K.; Singaram, B. *J. Am. Chem. Soc.* **1988**, *110*, 1529.

[434]For reviews, see Matteson, D.S. *Mol. Struct. Energ.* **1988**, *5*, 343; *Acc. Chem. Res.* **1988**, *21*, 294; *Synthesis* **1986**, 973, 980–983.

[435]For still other methods, see Brown, H.C.; Levy, A.B.; Midland, M.M. *J. Am. Chem. Soc.* **1975**, *97*, 5017; Ncube, S.; Pelter, A.; Smith, K. *Tetrahedron Lett.* **1979**, 1893; Pelter, A.; Rao, J.M. *J. Organomet. Chem.* **1985**, *285*, 65; Yogo, T.; Koshino, J.; Suzuki, A. *Chem. Lett.* **1981**, 1059; Brown. H.C.; Bhat, N.G.; Basavaiah, D. *Synthesis* **1983**, 885; Narayana, C.; Periasamy, M. *Tetrahedron Lett.* **1985**, *26*, 6361.

[436]See, for example, Zweifel, G.; Fisher, R.P. *Synthesis* **1974**, 339; Brown, H.C.; DeLue, N.R. *J. Am. Chem. Soc.* **1974**, *96*, 311; Hubbard, J.L.; Brown, H.C. *Synthesis* **1978**, 676; Uguen, D. *Bull. Soc. Chim. Fr.* **1981**, II-99.

[437]Brown, H.C.; Hubbard, J.L.; Smith, K. *Synthesis* **1979**, 701, and references cited therein. For discussions of the mechanism, see Brown, H.C.; Hubbard, J.L. *J. Org. Chem.* **1979**, *44*, 467; Hubbard, J.L.; Smith, K. *J. Organomet. Chem.* **1984**, *276*, C41.

B-alkyl-9-BBN derivatives (p. 1077). Since only the 9-alkyl group migrates, this method permits the conversion in high yield of an alkene to a primary alcohol or aldehyde containing one more carbon.[438] When B-alkyl-9-BBN derivatives are treated with CO and lithium tri-*tert*-butoxyaluminum hydride,[439] other functional groups (e.g., CN and ester) can be present in the alkyl group without being reduced.[440] Boranes can be directly converted to carboxylic acids by reaction with the dianion of phenoxyacetic acid.[441]

$$R_3B \;+\; PhO\overset{\ominus}{\frown}COO^{\ominus} \longrightarrow \quad \underset{PhO\frown COO^{\ominus}}{R-\overset{R}{\underset{\ominus}{\overset{|}{B}}}-R} \xrightarrow{-{}^-OPh} \quad \underset{R}{\overset{R}{\underset{COO^{\ominus}}{R-B}}}\!\!{}^{R} \xrightarrow{H^+} R\overset{}{\frown}COOH$$

Boronic esters RB(OR')$_2$ react with methoxy(phenylthio)methyllithium LiCH(OMe)SPh to give salts, which, after treatment with HgCl$_2$, and then H$_2$O$_2$, yield aldehydes.[442] This synthesis has been made enantioselective, with high ee values (>99%), by the use of an optically pure boronic ester,[443] for example:

$$\text{(R) or (S)} \xrightarrow{\text{LiCH(OMe)SPh}} \left[\begin{array}{c} \text{O} \\ B-O \\ \underset{PhS\ H}{C-OMe} \end{array} \right]^{-} Li^+ \xrightarrow{HgCl_2} \underset{H\ \ O}{\overset{OMe\ \ O}{C-B}} \xrightarrow[HO^-]{H_2O_2} \text{CHO}$$

18-25 Conversion of Vinylic Boranes to Alkenes

$$\underset{R}{\overset{H}{\diagdown}}C=C\underset{R^2}{\overset{BR'_2}{\diagup}} \xrightarrow[I_2]{NaOH} \underset{R}{\overset{H}{\diagdown}}C=C\underset{R'}{\overset{R^2}{\diagup}}$$

The reaction between trialkylboranes and iodine to give alkyl iodides was mentioned at **12-31**. When the substrate contains a vinylic group, the reaction takes a different course,[444] with one of the R' groups migrating to the carbon, to give alkenes.[445] The reaction is stereospecific in two senses: (*1*) if the groups

[438]Brown, H.C.; Knights, E.F.; Coleman, R.A. *J. Am. Chem. Soc.* **1969**, *91*, 2144.

[439]Brown, H.C.; Coleman, R.A. *J. Am. Chem. Soc.* **1969**, *91*, 4606.

[440]For other methods of converting boranes to aldehydes, see Yamamoto, S.; Shiono, M.; Mukaiyama, T. *Chem. Lett.* **1973**, 961; Negishi, E.; Yoshida, T.; Silveira, Jr., A.; Chiou, B.L. *J. Org. Chem.* **1975**, *40*, 814.

[441]Hara, S.; Kishimura, K.; Suzuki, A.; Dhillon, R.S. *J. Org. Chem.* **1990**, *55*, 6356. See also, Brown, H.C.; Imai, T. *J. Org. Chem.* **1984**, *49*, 892.

[442]Brown, H.C.; Imai, T. *J. Am. Chem. Soc.* **1983**, *105*, 6285. For a related method that produces primary alcohols, see Brown, H.C.; Imai, T.; Perumal, P.T.; Singaram, B. *J. Org. Chem.* **1985**, *50*, 4032.

[443]Brown, H.C.; Imai, T.; Desai, M.C.; Singaram, B. *J. Am. Chem. Soc.* **1985**, *107*, 4980.

[444]Zweifel, G.; Fisher, R.P. *Synthesis* **1975**, 376; Brown, H.C.; Basavaiah, D.; Kulkarni, S.U.; Bhat, N.G.; Vara Prasad, J.V.N. *J. Org. Chem.* **1988**, *53*, 239.

[445]For a list of methods of preparing alkenes using boron reagents, with references, see Larock, R.C. *Comprehensive Organic Transformations*, 2nd ed., Wiley-VCH, NY, **1999**, pp. 421–427.

R and R″ are cis in the starting compound, they will be trans in the product; (2) there is retention of configuration within the migrating group R′.[446] Since vinylic boranes can be prepared from alkynes (15-16), this is a method for the addition of R′ and H to a triple bond. If $R^2 = H$, the product is a (Z)-alkene. The mechanism is believed to involve an iodonium intermediate, such as **92**, and attack by iodide on boron. When R′ is vinylic, the product is a conjugated diene.[447]

In another procedure, the addition of a dialkylborane to a 1-haloalkyne produces an α-halo vinylic borane (**93**).[448] Treatment of this with NaOMe gives the rearrangement shown, and protonolysis of the product

produces the (E)-alkene.[446] If R is a vinylic group the product is a 1,3-diene.[449] If one of the groups is thexyl, the other migrates.[450] This extends the scope of the synthesis, since dialkylboranes where one R group is thexyl are easily prepared. A combination of both of the procedures described above results in the preparation of trisubstituted alkenes.[451] The entire conversion of haloalkyne to alkene can be carried out in one reaction vessel, without isolation of intermediates. An aluminum counterpart of the α-halo vinylic borane procedure has been reported.[452]

---

[446]Zweifel, G.; Fisher, R.P.; Snow, J.T.; Whitney, C.C. *J. Am. Chem. Soc.* **1971**, *93*, 6309.

[447]Zweifel, G.; Polston, N.L.; Whitney, C.C. *J. Am. Chem. Soc.* **1968**, *90*, 6243; Brown, H.C.; Ravindran, N. *J. Org. Chem.* **1973**, *38*, 1617; Hyuga, S.; Takinami, S.; Hara, S.; Suzuki, A. *Tetrahedron Lett.* **1986**, *27*, 977.

[448]For improvements in this method, see Brown, H.C.; Basavaiah, D.; Kulkarni, S.U.; Lee, H.D.; Negishi, E.; Katz, J. *J. Org. Chem.* **1986**, *51*, 5270.

[449]Negishi, E.; Yoshida, T. *J. Chem. Soc. Chem. Commun.* **1973**, 606; See also, Negishi, E.; Yoshida, T.; Abramovitch, A.; Lew, G.; Williams, R.H. *Tetrahedron* **1991**, *47*, 343.

[450]Corey, E.J.; Ravindranathan, T. *J. Am. Chem. Soc.* **1972**, *94*, 4013; Negishi, E.; Katz, J.; Brown, H.C. *Synthesis* **1972**, 555.

[451]Zweifel, G.; Fisher, R.P. *Synthesis* **1972**, 557.

[452]Miller, J.A. *J. Org. Chem.* **1989**, *54*, 998.

18-26   Formation of Alkynes, Alkenes, and Ketones from Boranes and Acetylides

$$R_3B \ + \ RC{\equiv}CLi \ \longrightarrow \ RC{\equiv}C{-}\overset{\ominus}{B}R_3'\ Li^{\oplus} \ \xrightarrow{\ I_2\ } \ RC{\equiv}CR'$$

94

A hydrogen directly attached to a triple-bond carbon can be replaced in high yield by an alkyl or an aryl group, by treatment of the lithium acetylide with a trialkyl- or triarylborane, followed by reaction of the lithium alkynyltrialkylborate **94** with iodine.[453] The R' group may be primary or secondary alkyl as well as aryl, so the reaction has a broader scope than the older reaction **10-74**.[454] The R group may be alkyl, aryl, or hydrogen, though in the last-mentioned case satisfactory yields are obtained only if lithium acetylide–ethylenediamine is used as the starting

94 $\xrightarrow{\ R^2COOH\ }$ [vinyl borate intermediate] $\longrightarrow$ [vinylborane] $\xrightarrow{\ R^2COOH\ }$ [alkene]

compound.[455] Optically active alkynes can be prepared by using optically active thexylborinates (RR^{2}BOR', R^2 = thexyl), where R is chiral, and LiC{\equiv}CSiMe$_3$.[456] The reaction can be adapted to the preparation of alkenes[414] by treatment of **94** with an electrophile such as propanoic acid[457] or tributyltin chloride.[458] The reaction with Bu$_3$SnCl produces the (Z)-alkene stereoselectively.

Treatment of **94** with an electrophile, such as methyl sulfate, allyl bromide, or triethyloxonium borofluoride, followed by oxidation of the resulting vinylic borane gives a ketone (illustrated for methyl sulfate):[459]

[reaction scheme: $R{-}C{\equiv}C{-}\overset{\ominus}{B}{-}R'$ Li$^+$ with Me$-$O$-$SO$_2$$-$OMe $\longrightarrow$ vinylborane $\xrightarrow[\text{NaOH}]{H_2O_2}$ ketone]

[453]Suzuki, A.; Miyaura, N.; Abiko, S.; Itoh, M.; Brown, H.C.; Sinclair, J.A.; Midland, M.M. *J. Org. Chem.* **1986**, *51*, 4507; Sikorski, J.A.; Bhat, N.G.; Cole, T.E.; Wang, K.K.; Brown, H.C. *J. Org. Chem.* **1986**, *51*, 4521. For a review of reactions of organoborates, see Suzuki, A. *Acc. Chem. Res.* **1982**, *15*, 178.
[454]For a study of the relative migratory aptitudes of R', see Slayden, S.W. *J. Org. Chem.* **1981**, *46*, 2311.
[455]Midland, M.M.; Sinclair, J.A.; Brown, H.C. *J. Org. Chem.* **1974**, *39*, 731.
[456]Brown, H.C.; Mahindroo, V.K.; Bhat, N.G.; Singaram, B. *J. Org. Chem.* **1991**, *56*, 1500.
[457]Miyaura, N.; Yoshinari, T.; Itoh, M.; Suzuki, A. *Tetrahedron Lett.* **1974**, 2961; Pelter, A.; Gould, K.J.; Harrison, C.R. *Tetrahedron Lett.* **1975**, 3327.
[458]Hooz, J.; Mortimer, R. *Tetrahedron Lett.* **1976**, 805; Wang, K.K.; Chu, K. *J. Org. Chem.* **1984**, *49*, 5175.
[459]Pelter, A.; Drake, R.A. *Tetrahedron Lett.* **1988**, *29*, 4181.

Note that there are reactions that involve N → O rearrangements, including those mediated by silicon.[460]

NON-1,2 REARRANGEMENTS

A. Electrocyclic Rearrangements

18-27 Electrocyclic Rearrangements of Cyclobutenes and 1,3-Cyclohexadienes

(4)*seco***-1/4/Detachment; (4)***cyclo***-1/4/Attachment**
(6)*seco***-1,6/Detachment; (6)***cyclo***-1/6/Attachment**

Cyclobutenes and 1,3-dienes can be interconverted by treatment with uv light or with heat.[461] The thermal reaction is generally not reversible (although exceptions[462] are known), and many cyclobutenes have been converted to 1,3-dienes by heating at temperatures between 100 and 200°C. The photochemical conversion can in principle be carried out in either direction, but most often 1,3-dienes are converted to cyclobutenes rather than the reverse, because the dienes are stronger absorbers of light at the wavelengths used.[463] In a similar reaction, 1,3-cyclohexadienes interconvert with 1,3,5-trienes, but in this case the ring-closing process is generally favored thermally and the ring-opening process photochemically, though exceptions are known in both directions.[464] Substituent effects can lead to acceleration of the electrocyclization process.[465] Torquoselectivity in cyclobutene ring opening reaction has been examined.[466]

[460]Talami, S.; Stirling, C.J.M. *Can. J. Chem.* **1999**, *77*, 1105.

[461]See Dolbier Jr., W.R.; Koroniak, H.; Houk, K.N.; Sheu, C. *Acc. Chem. Res.* **1996**, *29*, 471; Niwayama, S.; Kallel, E.A.; Spellmeyer, D.C.; Sheu, C.; Houk, K.N. *J. Org. Chem.* **1996**, *61*, 2813. The effect of pressure on this reaction has been discussed, see Jenner, G. *Tetrahedron* **1998**, *54*, 2771.

[462]For example; see Shumate, K.M.; Neuman, P.N.; Fonken, G.J. *J. Am. Chem. Soc.* **1965**, *87*, 3996; Gil-Av, E.; Herling, J. *Tetrahedron Lett.* **1967**, 1; Doorakian, G.A.; Freedman, H.H. *J. Am. Chem. Soc.* **1968**, *90*, 3582; Brune, H.A.; Schwab, W. *Tetrahedron* **1969**, *25*, 4375; Steiner, R.P.; Michl, J. *J. Am. Chem. Soc.* **1978**, *100*, 6413.

[463]For examples of photochemical conversion of a cyclobutene to a 1,3-diene, see Scerer, Jr., K.V. *J. Am. Chem. Soc.* **1968**, *90*, 7352; Saltiel, J.; Lim, L.N. *J. Am. Chem. Soc.* **1969**, *91*, 5404; Adam, W.; Oppenländer, T.; Zang, G. *J. Am. Chem. Soc.* **1985**, *107*, 3921; Dauben, W.G.; Haubrich, J.E. *J. Org. Chem.* **1988**, *53*, 600.

[464]For a review of photochemical rearrangements in trienes, see Dauben, W.G.; McInnis, E.L.; Michno, D.M., in de Mayo, P. *Rearrangements in Ground and Excited States*, Vol. 3, Academic Press, NY, *1980*, pp. 91–129. For an *ab initio* study see Rodríguez-Otero, J. *J. Org. Chem.* **1999**, *64*, 6842.

[465]Tanaka, K.; Mori, H.; Yamamoto, M.; Katsumura, S. *J. Org. Chem.* **2001**, *66*, 3099.

[466]Yasui, M.; Naruse, Y.; Inagaki, S. *J. Org. Chem.* **2004**, *69*, 7246.

Some examples are

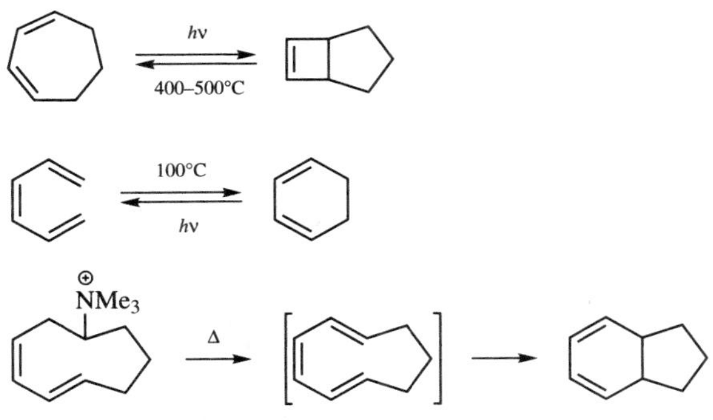

Ref.[467]

Not isolated

An interesting example of 1,3-cyclohexadiene–1,3,5-triene interconversion is the reaction of norcaradienes to give cycloheptatrienes.[468] Norcaradienes give this reaction so readily (because they are *cis*-1,2-divinylcyclopropanes, see p. 1661) that they cannot generally be isolated, though some exceptions are known[469,470] (see also, p. 1239).

Norcaradiene

[467]Dauben, W.G.; Cargill, R.L. *Tetrahedron* **1961**, *12*, 186; Chapman, O.L.; Pasto, D.J.; Borden, G.W.; Griswold, A.A. *J. Am. Chem. Soc.* **1962**, *84*, 1220.
[468]For reviews of the norcaradiene-cycloheptatriene interconversion and the analogous benzene oxide–oxepin interconversion, see Maier, G. *Angew. Chem. Int. Ed.* **1967**, *6*, 402; Vogel, E.; Günther, H. *Angew. Chem. Int. Ed.* **1967**, *6*, 385; Vogel, E. *Pure Appl. Chem.* **1969**, *20*, 237.
[469]Ciganek, E. *J. Am. Chem. Soc.* **1967**, *89*, 1454; Mukai, T.; Kubota, H.; Toda, T. *Tetrahedron Lett.* **1967**, 3581; Maier, G.; Heep, U. *Chem. Ber.* **1968**, *101*, 1371; Ciganek, E. *J. Am. Chem. Soc.* **1971**, *93*, 2207; Dürr, H.; Kober, H. *Tetrahedron Lett.* **1972**, 1255, 1259; Vogel, E.; Wiedemann, W.; Roth, H.D.; Eimer, J.; Günther, H. *Liebigs Ann. Chem.* **1972**, *759*, 1; Bannerman, C.G.F.; Cadogan, J.I.G.; Gosney, I.; Wilson, N.H. *J. Chem. Soc., Chem. Commun.* **1975**, 618; Takeuchi, K.; Kitagawa, T.; Senzaki, Y.; Okamoto, K. *Chem. Lett.* **1983**, 73; Kawase, T.; Iyoda, M.; Oda, M. *Angew. Chem. Int. Ed.* **1987**, *26*, 559.
[470]See, for example, Ciganek, E. *J. Am. Chem. Soc.* **1967**, *89*, 1454; Mukai, T.; Kubota, H.; Toda, T. *Tetrahedron Lett.* **1967**, 3581; Maier, G.; Heep, U. *Chem. Ber.* **1968**, *101*, 1371; Ciganek, E. *J. Am. Chem. Soc.* **1971**, *93*, 2207; Dürr, H.; Kober, H. *Tetrahedron Lett.* **1972**, 1255, 1259; Vogel, E.; Wiedemann, W.; Roth, H.D.; Eimer, J.; Günther, H. *Liebigs Ann. Chem.* **1972**, *759*, 1; Bannerman, C.G.F.; Cadogan, J.I.G.; Gosney, I.; Wilson, N.H. *J. Chem. Soc., Chem. Commun.* **1975**, 618; Takeuchi, K.; Kitagawa, T.; Senzaki, Y.; Okamoto, K. *Chem. Lett.* **1983**, 73; Kawase, T.; Iyoda, M.; Oda, M. *Angew. Chem. Int. Ed.* **1987**, *26*, 559.

These reactions, called *electrocyclic rearrangements*,[471] take place by pericyclic mechanisms. The evidence comes from stereochemical studies, which show a remarkable stereospecificity whose direction depends on whether the reaction is induced by heat or light. For example, it was found for the thermal reaction that *cis*-3,4-dimethylcyclobutene gave only *cis,trans*-2,4-hexadiene, while the trans isomer gave only the trans–trans diene:[472]

This is evidence for a four-membered cyclic transition state and arises from conrotatory motion about the C-3–C-4 bond.[473] It is called conrotatory because both movements are clockwise (or both counterclockwise). Because both rotate in the same direction, the cis isomer gives the cis–trans diene:[474]

The other possibility (*disrotatory* motion) would have one moving clockwise while the other moves counterclockwise; the cis isomer would have given the cis–cis

[471]For a monograph on thermal isomerizations, which includes electrocyclic and sigmatropic rearrangements, as well as other types, see Gajewski, J.J. *Hydrocarbon Thermal Isomerizations*, Academic Press, NY, *1981*. For a monograph on electrocyclic reactions, see Marvell, E.N. *Thermal Electrocyclic Reactions*, Academic Press, NY, *1980*. For reviews, see Dolbier, W.R.; Koroniak, H. *Mol. Struct. Energ.*, *1988*, *8*, 65; Laarhoven, W.H. *Org. Photochem.* *1987*, *9*, 129; George, M.V.; Mitra, A.; Sukumaran, K.B. *Angew. Chem. Int. Ed.* *1980*, *19*, 973; Jutz, J.C. *Top. Curr. Chem.* *1978*, *73*, 125; Gilchrist, T.L.; Storr, R.C. *Organic Reactions and Orbital Symmetry*, Cambridge University Press, Cambridge, *1972*, pp. 48–72; DeWolfe, R.H. in Bamford, C.H.; Tipper, C.F.H. *Comprehensive Chemical Kinetics*, Vol. 9, Elsevier, NY, *1973*; pp. 461–470; Crowley, K.J.; Mazzocchi, P.H., in Zabicky, J. *The Chemistry of Alkenes*, Vol. 2, Wiley, NY, *1970*, pp. 284–297; Criegee, R. *Angew. Chem. Int. Ed.* *1968*, *7*, 559; Vollmer, J.J.; Servis, K.L. *J. Chem. Educ.* *1968*, *45*, 214. For a review of isotope effects in these reactions, see Gajewski, J.J. *Isot. Org. Chem.* *1987*, *7*, 115. For a related review, see Schultz, A.G.; Motyka, L. *Org. Photochem.* *1983*, *6*, 1.
[472]Winter, R.E.K. *Tetrahedron Lett.* *1965*, 1207. Also see, Vogel, E. *Liebigs Ann. Chem.* *1958*, *615*, 14; Criegee, R.; Noll, K. *Liebigs Ann. Chem.* *1959*, *627*, 1.
[473]The mechanism of cyclobutene thermal isomerization has been examined. See Baldwin, J.E.; Gallagher, S.S.; Leber, P.A.; Raghavan, A.S.; Shukla, R. *J. Org. Chem.* *2004*, *69*, 7212.
[474]This picture is from Woodward, R.B.; Hoffmann, R. *J. Am. Chem. Soc.* *1965*, *87*, 395, who coined the terms, *conrotatory* and *disrotatory*.

diene (shown) or the trans–trans diene:

If the motion had been disrotatory, this would still have been evidence for a cyclic mechanism. If the mechanism were a diradical or some other kind of noncyclic process, it is likely that no stereospecificity of either kind would have been observed. The reverse reaction is also conrotatory. In contrast, the photochemical cyclobutene: 1,3-Diene interconversion is *disrotatory* in either direction.[475] On the other hand, the cyclohexadiene: 1,3,5-Triene interconversion shows precisely the opposite behavior. The thermal process is *disrotatory*, while the photochemical process is *conrotatory* (in either direction). These startling results are a consequence of the symmetry rules mentioned in Chapter 15 (p. 1208).[476] As in the case of cycloaddition reactions, we will use the frontier orbital and Möbius–Hückel approaches.[477]

The Frontier Orbital Method[478]

As applied to these reactions, the frontier orbital method may be expressed: *A σ bond will open in such a way that the resulting p orbitals will have the symmetry of the highest occupied π orbital of the product*. In the case of cyclobutenes, the HOMO of the product in the thermal reaction is the χ_2 orbital (Fig. 18.1).

[475]Photochemical ring opening of cyclobutenes can also be nonstereospecific. See Leigh, W.J.; Zheng, K. *J. Am. Chem. Soc.* **1991**, *113*, 4019; Leigh, W.J.; Zheng, K.; Nguyen, N.; Werstiuk, N.H.; Ma, J. *J. Am. Chem. Soc.* **1991**, *113*, 4993, and references cited therein.

[476]Woodward, R.B.; Hoffmann, R. *J. Am. Chem. Soc.* **1965**, *87*, 395. Also see, Longuet-Higgins, H.C.; Abrahamson, E.W. *J. Am. Chem. Soc.* **1965**, *87*, 2045; Fukui, K. *Tetrahedron Lett.* **1965**, 2009.

[477]For the correlation diagram method, see Jones, R.A.Y. *Physical and Mechanistic Organic Chemistry*, 2nd ed., Cambridge University Press, Cambridge, **1984**, pp. 352–359; Yates, K. *Hückel Molecular Orbital Theory*, Academic Press, NY, **1978**, pp. 250–263. Also see, Zimmerman, H.E., in Marchand, A.P.; Lehr, R.E. *Pericyclic Reactions*, Vol. 2, Academic Press, NY, **1977**, pp. 53–107; *Acc. Chem. Res.* **1971**, *4*, 272; *J. Am. Chem. Soc.* **1966**, *88*, 1564, 1566; Dewar, M.J.S. *Angew. Chem. Int. Ed.* **1971**, *10*, 761; Jefford, C.W.; Burger, U. *Chimia* **1971**, *25*, 297; Herndon, W.C. *J. Chem. Educ.* **1981**, *58*, 371.

[478]Fukui, K.; Fujimoto, H. *Bull. Chem. Soc. Jpn.* **1967**, *40*, 2018; **1969**, *42*, 3399; Fukui, K. *Fortschr. Chem. Forsch.* **1970**, *15*, 1; *Acc. Chem. Res.* **1971**, *4*, 57; Houk, K.N. *Acc. Chem. Res.* **1975**, *8*, 361. See also, Chu, S. *Tetrahedron* **1978**, *34*, 645. For a monograph on frontier orbitals see Fleming, I. *Pericyclic Reactions*, Oxford University Press, Oxford, **1999**. For reviews, see Fukui, K. *Angew. Chem. Int. Ed.* **1982**, *21*, 801; Houk, K.N., in Marchand, A.P.; Lehr, R.F. *Pericyclic Reactions*, Vol. 2, Academic Press, NY, **1977**, pp. 181–271.

X_2 $X_3{}^*$

Fig. 18.1. Symmetries of the X_2 and X_3* orbitals of a conjugated diene.

Therefore, in a thermal process, the cyclobutene must open so that on one side the positive lobe lies above the plane, and on the other side below it. Thus the substituents are forced into conrotatory motion (Fig. 18.2). On the other hand, in the photochemical process, the HOMO of the product is now the χ_3 orbital (Fig. 18.1), and in order for the *p* orbitals to achieve this symmetry (the two plus lobes on the same side of the plane), the substituents are forced into disrotatory motion.

We may also look at this reaction from the opposite direction (ring closing). For this direction, the rule is that *those lobes of orbitals that overlap (in the HOMO) must be of the same sign.* For thermal cyclization of butadienes, this requires conrotatory motion (Fig. 18.3). In the photochemical process the HOMO is the χ_3 orbital, so that disrotatory motion is required for lobes of the same sign to overlap.

The Möbius–Hückel Method[481]

As we saw on p. 1210, in this method we choose a basis set of *p* orbitals and look for sign inversions in the transition state. Figure 18.4 shows a basis set for a 1,3-diene. It is seen that disrotatory ring closing (Fig. 18.4*a*) results in overlap of plus lobes only, while in conrotatory closing (Fig. 18.4*b*) there is one overlap of a plus

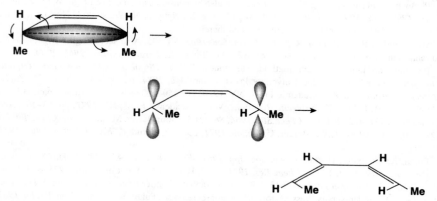

Fig. 18.2. Thermal opening of 1,2-dimethylcyclobutene. The two hydrogens and two methyls are forced into conrotatory motion so that the resulting *p* orbitals have the symmetry of the HOMO of the diene.

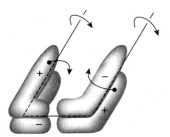

Fig. 18.3. Thermal ring closing of a 1,3-diene. Conrotatory motion is required for two + lobes to overlap.

with a minus lobe. In the first case, we have zero sign inversions, while in the second there is one sign inversion. With zero (or an even number of) sign inversions, the disrotatory transition state is a Hückel system, and so is allowed thermally only if the total number of electrons is $4n + 2$ (p. 1211). Since the total here is 4, the

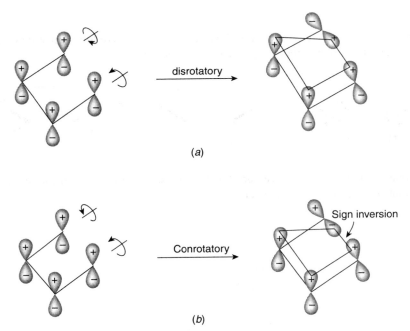

Fig. 18.4. The 1,3-diene–cyclobutene interconversion. The orbitals shown are *not* molecular orbitals, but a basis set of *p*-atomic orbitals. (*a*) Disrotatory ring closure gives zero sign inversion. (*b*) Conrotatory ring closure gives one sign inversion. We could have chosen to show any other basis set (e.g., another basis set would have two plus lobes above the plane and two below, etc.). This would change the number of sign inversion, but the disrotatory mode would still have an even number of sign inversions, and the conrotatory mode an odd number, whichever basis set was chosen.

disrotatory process is not allowed. On the other hand, the conrotatory process, with one sign inversion, is a Möbius system, which is thermally allowed if the total number is $4n$. The conrotatory process is therefore allowed thermally. For the photochemical reactions, the rules are reversed: A reaction with $4n$ electrons requires a Hückel system, so only the disrotatory process is allowed.

Both the frontier orbital and the Möbius–Hückel methods can also be applied to the cyclohexadiene: 1,3,5-triene reaction;[479] in either case the predicted result is that for the thermal process, only the disrotatory pathway is allowed, and for the photochemical process, only the conrotatory. For example, for a 1,3,5-triene, the symmetry of the HOMO is

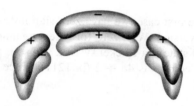

In the thermal cleavage of cyclohexadienes, then, the positive lobes must lie on the same side of the plane, requiring disrotatory motion:

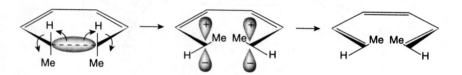

Disrotatory motion is also necessary for the reverse reaction, in order that the orbitals that overlap may be of the same sign:

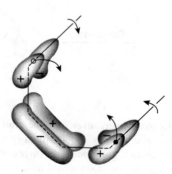

[479]For a discussion of the transition structures and energy, see Zora, M. *J. Org. Chem.* **2004**, *69*, 1940.

All these directions are reversed for photochemical processes, because in each case a higher orbital, with inverted symmetry, is occupied.

In the Möbius–Hückel approach, diagrams similar to Fig. 18.4 can be drawn for this case. Here too, the disrotatory pathway is a Hückel system and the conrotatory pathway a Möbius system, but since six electrons are now involved, the thermal reaction follows the Hückel pathway and the photochemical reaction the Möbius pathway.

In the most general case, there are four possible products that can arise from a given cyclobutene or cyclohexadiene: two from the conrotatory and two from the disrotatory pathway. For example, conrotatory ring opening of **95** gives either **96** or **97**, while disrotatory opening gives either **98** or **99**. The orbital-symmetry rules tell us when a given reaction will operate by the conrotatory and when by the disrotatory mode, but they do not say which of the two possible conrotatory or disrotatory pathways will be followed. It is often possible,

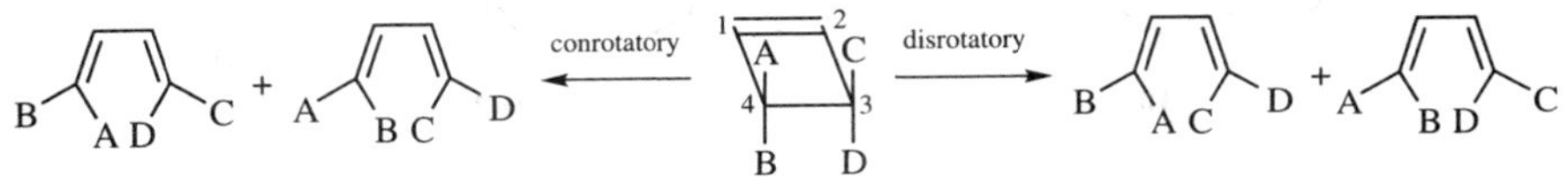

however, to make such predictions on steric grounds. For example, in the opening of **95** by the disrotatory pathway, **98** arises when groups A and C swing in toward each other (clockwise motion around C-4, counterclockwise around C-3), while **99** is formed when groups B and D swing in and A and C swing out (clockwise motion around C-3, counterclockwise around C-4). We therefore predict that when A and C are larger than B and D, the predominant or exclusive product will be **99**, rather than **98**. Predictions of this kind have largely been borne out.[480] There is evidence, however, that steric effects[481] are not the only factor, and that electronic effects also play a role, which may be even greater.[482] An electron-donating group stabilizes the transition state when it rotates *outward*, because it mixes with the LUMO; if it rotates *inward*, it mixes with the HOMO, destabilizing the transition state.[483] The compound 3-formylcyclobutene provided a test. Steric factors would cause the CHO

[480]For example, see Baldwin, J.E.; Krueger, S.M. *J. Am. Chem. Soc.* **1969**, *91*, 6444; Spangler, C.W.; Hennis, R.P. *J. Chem. Soc., Chem. Commun.* **1972**, 24; Gesche, P.; Klinger, F.; Riesen, A.; Tschamber, T.; Zehnder, M.; Streith, J. *Helv. Chim. Acta* **1987**, *70*, 2087.

[481]Leigh, W.J.; Postigo, J.A. *J. Am. Chem. Soc.* **1995**, *117*, 1688.

[482]Kirmse, W.; Rondan, N.G.; Houk, K.N. *J. Am. Chem. Soc.* **1984**, *106*, 7989; Dolbier, Jr., W.R.; Gray, T.A.; Keaffaber, J.J.; Celewicz, L.; Koroniak, H. *J. Am. Chem. Soc.* **1990**, *112*, 363; Hayes, R.; Ingham, S.; Saengchantara, S.T.; Wallace, T.W. *Tetrahedron Lett.* **1991**, *32*, 2953.

[483]For theoretical studies, see Buda, A.B.; Wang, Y.; Houk, K.N. *J. Org. Chem.* **1989**, *54*, 2264; Kallel, E.A.; Wang, Y.; Spellmeyer, D.C.; Houk, K.N. *J. Am. Chem. Soc.* **1990**, *112*, 6759.

(an electron-withdrawing group) to rotate outward; electronic effects would cause it to rotate inward. The experiment showed inward rotation.[484]

100

101   **102** $R = C_9H_{17}$

Cyclohexadienes are of course 1,3-dienes, and in certain cases it is possible to convert them to cyclobutenes instead of to 1,3,5-trienes.[485] An interesting example is found in the pyrocalciferols. Photolysis of the syn isomer **100** (or of the other syn isomer, not shown) leads to the corresponding cyclobutene,[486] while photolysis of the anti isomers (one of them is **101**) gives the ring-opened 1,3,5-triene, **102**. This difference in behavior is at first sight remarkable, but is easily explained by the orbital-symmetry rules. Photochemical ring opening to a 1,3,5-triene must be conrotatory. If **100** were to react by this pathway, the product would be the triene **102**, but this compound would have to contain a *trans*-cyclohexene ring (either the methyl group or the hydrogen would have to be directed inside the ring). On the other hand, photochemical conversion to a cyclobutene must be disrotatory, but if **101** were to give this reaction, the product would have to have a trans-fused ring junction. Compounds with such ring junctions are known (p. 188), but are very strained. Stable *trans*-cyclohexenes are unknown (p. 226). Thus, **100** and **101** give the products they do owing to a combination of orbital-symmetry rules and steric influences.

[484]Rudolf, K.; Spellmeyer, D.C.; Houk, K.N. *J. Org. Chem.* **1987**, *52*, 3708; Piers, E.; Lu, Y.-F. *J. Org. Chem.* **1989**, *54*, 2267.

[485]For a discussion of the factors favoring either direction, see Dauben, W.G.; Kellogg, M.S.; Seeman, J.I.; Vietmeyer, N.D.; Wendschuh, P.H. *Pure Appl. Chem.* **1973**, *33*, 197.

[486]Dauben, W.G.; Fonken, G.J. *J. Am. Chem. Soc.* **1959**, *81*, 4060. This was the first reported example of the conversion of a 1,3-diene to a cyclobutene.

A variation of this process is the *Bergmann cyclization*,[487] where an ene-diyne cyclizes to a biradical (**103**) and then aromatizes as shown.

103

Simply heating the en-diyne will usually lead to aromatization via this pathway.[488] Quinones can be formed via Bergman cyclization[489] and there are other synthetic applications.[490] The role of vinyl substitution has been examined.[491] An aza-Bergman cyclization is known.[492]

104 **105** **106**

The 1,3-diene-cyclobutene interconversion can even be applied to benzene rings. For example,[493] photolysis of 1,2,4-tri-*tert*-butylbenzene (**104**) gives

[487]Bergman, R.G. *Accts. Chem. Res.* **1973**, *6*, 25; Darby, N.; Kim, C.U.; Shelton, K.W.; Takada, S.; Masamune, S. *J. Chem. Soc. (D)*, **1971**, *23*, 1516; Adam, W.; Krebs, O. *Chem. Rev.* **2003**, *103*, 4131. For a discussion of electronic and stereoelectronic effects see Pourde II, G.W.; Warner, P.M.; Parrish, D.A.; Jones, G.B. *J. Org. Chem.* **2002**, *67*, 5369; Jones, G.B.; Wright, J.M.; Hynd, G.; Wyatt, J.K.; Warner, P.M.; Huber, R.S.; Li, A.; Kilgore, M.W.; Sticca, R.P.; Pollenz, R.S. *J. Org. Chem.* **2002**, *67*, 5727. For polar effects, see Schmittel, M.; Kiau, S. *Chem. Lett*, **1995**, 953; Grissom, J.W.; Calkins, T.L.; McMillen, H.A.; Jiang, Y. *J. Org. Chem.* **1994**, *59*, 5833. For free-energy relationships see Choy, N.; Kim, C.-S.; Ballestero, C.; Artigas, L.; Diez, C.; Lichtenberger, F.; Shapiro, J.; Russell, K.C. *Tetrahedron Lett.* **2000**, *41*, 6955.
[488]For examples, see Grissom, J.W.; Klingberg, D. *Tetrahedron Lett.* **1995**, *36*, 6607; Danheiser, R.L.; Gould, A.E.; de la Pradilla, R.F.; Helgason, A.L. *J. Org. Chem.* **1994**, *59*, 5514; Grissom, J.W.; Calkins, T.L.; McMillen, H.A. *J. Org. Chem.* **1993**, *58*, 6556; Tanaka, H.; Yamada, H.; Matsuda, A.; Takahashi, T. *Synlett* **1997**, 381.
[489]Jones, G.B.; Warner, P.M. *J. Org. Chem.* **2001**, *66*, 8669.
[490]Bowles, D.M.; Palmer, G.J.; Landis, C.A.; Scott, J.L.; Anthony, J.E. *Tetrahedron* **2001**, *57*, 3753.
[491]Jones, G.B.; Warner, P.M. *J. Am. Chem. Soc.* **2001**, *123*, 2134.
[492]Feng, L.; Kumar, D.; Kerwin, S.M. *J. Org. Chem.* **2003**, *68*, 2234.
[493]Unsubstituted Dewar benzene has been obtained, along with other photoproducts, by photolysis of benzene: Ward, H.R.; Wishnok, J.S. *J. Am. Chem. Soc.* **1968**, *90*, 1085; Bryce-Smith, D.; Gilbert, A.; Robinson, D.A. *Angew. Chem. Int. Ed.* **1971**, *10*, 745. For other examples, see Arnett, E.M.; Bollinger, J.M. *Tetrahedron Lett.* **1964**, 3803; Camaggi, G.; Gozzo, F.; Cevidalli, G. *Chem. Commun.* **1966**, 313; Haller, I. *J. Am. Chem. Soc.* **1966**, *88*, 2070; *J. Chem. Phys.* **1967**, *47*, 1111; Barlow, M.G.; Haszeldine, R.N.; Hubbard, R. *Chem. Commun.* **1969**, 202; Lemal, D.M.; Staros, J.V.; Austel, V. *J. Am. Chem. Soc.* **1969**, *91*, 3373.

1,2,5-tri-*tert*-butyl[2.2.0]hexadiene (**105**, a Dewar benzene).[494] The reaction owes its success to the fact that once **105** is formed, it cannot, under the conditions used, revert to **104** by either a thermal or a photochemical route. The orbital-symmetry rules prohibit thermal conversion of **105** to **104** by a pericyclic mechanism, because thermal conversion of a cyclobutene to a 1,3-diene must be conrotatory, and conrotatory reaction of **105** would result in a 1,3,5-cyclohexatriene containing one trans double bond (**106**), which is of course too strained to exist. Compound **105** cannot revert to **104** by a photochemical pathway either, because light of the frequency used to excite **104** would not be absorbed by **105**. This is thus another example of a molecule that owes its stability to the orbital-symmetry rules (see p. 1232). Pyrolysis of **105** does give **104** , probably by a diradical mechanism.[495] In the case of **107** and **108**, the Dewar benzene is actually more stable than the benzene. Compound **107** rearranges to **108** in 90% yield at 120°C.[496] In this case, thermolysis of the benzene gives the Dewar benzene (rather than the reverse), because of the strain of four adjacent *tert*-butyl groups on the ring.

A number of electrocyclic reactions have been carried out with systems of other sizes, for example, conversion of the 1,3,5,7-octatetraene **109** to the cyclooctatriene **110**.[497] The stereochemistry of these reactions can be predicted in a

[494]Wilzbach, K.E.; Kaplan, L. *J. Am. Chem. Soc.* **1965**, *87*, 4004; van Tamelen, E.E.; Pappas, S.P.; Kirk, K.L. *J. Am. Chem. Soc.* **1971**, *93*, 6092; van Tamelen, E.E. *Acc. Chem. Res.* **1972**, *5*, 186. As mentioned on p. $$$ (Lemal, D.M.; Lokensgard, J.P. *J. Am. Chem. Soc.* **1966**, *88*, 5934; Schäfer, W.; Criegee, R.; Askani, R.; Grüner, H. *Angew. Chem. Int. Ed.* **1967**, *6*, 78), Dewar benzenes can be photolyzed further to give prismanes.

[495]See, for example, Oth, J.F.M. *Recl. Trav. Chim. Pays-Bas* **1968**, *87*, 1185; Adam, W.; Chang, J.C. *Int. J. Chem. Kinet.*, **1969**, *1*, 487; Lechtken, P.; Breslow, R.; Schmidt, A.H.; Turro, N.J. *J. Am. Chem. Soc.* **1973**, *95*, 3025; Wingert, H.; Irngartinger, H.; Kallfass, D.; Regitz, M. *Chem. Ber.* **1987**, *120*, 825.

[496]Maier, G.; Schneider, K. *Angew. Chem. Int. Ed.* **1980**, *19*, 1022. See also, Wingert, H.; Maas, G.; Regitz, M. *Tetrahedron* **1986**, *42*, 5341.

[497]Marvell, E.N.; Seubert, J. *J. Am. Chem. Soc.* **1967**, *89*, 3377; Huisgen, R.; Dahmen, A.; Huber, H. *J. Am. Chem. Soc.* **1967**, *89*, 7130, *Tetrahedron Lett.* **1969**, 1461; Dahmen, A.; Huber, H. *Tetrahedron Lett.* **1969**, 1465.

similar manner. The results of such predictions can be summarized according to whether the number of electrons involved in the cyclic process is of the form $4n$ or $4n + 2$ (where n is any integer including zero).

	Thermal Reaction	Photochemical Reaction
$4n$	Conrotatory	Disrotatory
$4n + 2$	Disrotatory	Conrotatory

Although the orbital-symmetry rules predict the stereochemical results in almost all cases, it is necessary to recall (p. 1210) that they only say what is allowed and what is forbidden, but the fact that a reaction is allowed does not necessarily mean that the reaction takes place, and if an allowed reaction does take place, it does not *necessarily* follow that a concerted pathway is involved, since other pathways of lower energy may be available.[498] Furthermore, a "forbidden" reaction might still be made to go, if a method of achieving its high activation energy can be found. This was, in fact, done for the cyclobutene butadiene interconversion (*cis*-3,4-dichlorocyclobutene gave the forbidden *cis,-cis*- and *trans,trans*-1,4-dichloro-1,3-butadienes, as well as the allowed cis, trans isomer) by the use of ir laser light.[499] This is a thermal reaction. The laser light excites the molecule to a higher vibrational level (p. 330), but not to a higher electronic state.

As is the case for [2+2]-cycloaddition reactions (**15-63**), certain forbidden electrocyclic reactions can be made to take place by the use of metallic catalysts.[500] An example is the silver ion-catalyzed conversion of tricyclo[4.2.0.0$^{2.5}$]octa-3,7-diene to cyclooctatetraene:[501]

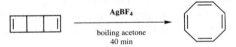

This conversion is very slow thermally (i.e., without the catalyst) because the reaction must take place by a disrotatory pathway, which is disallowed thermally.[502] In another example, the major thermal product from the barrelene anion is a

[498]For a discussion, see Baldwin, J.E.; Andrist, A.H.; Pinschmidt Jr., R.K. *Acc. Chem. Res.* **1972**, *5*, 402.
[499]Mao, C.; Presser, N.; John, L.; Moriarty, R.M.; Gordon, R.J. *J. Am. Chem. Soc.* **1981**, *103*, 2105.
[500]For a review, see Pettit, R.; Sugahara, H.; Wristers, J.; Merk, W. *Discuss. Faraday Soc.* **1969**, *47*, 71. See also, Labunskaya, V.I.; Shebaldova, A.D.; Khidekel', M.L. *Russ. Chem. Rev. 1974*, *43*, 1; Mango, F.D. *Top. Curr. Chem. 1974*, *45*, 39; *Tetrahedron Lett. 1973*, 1509; *Intra-Sci. Chem. Rep. 1972*, *6 (3)*, 171; *CHEMTECH 1971*, *1*, 758; *Adv. Catal. 1969*, *20*, 291; Mango, F.D.; Schachtschneider, J.H. *J. Am. Chem. Soc. 1971*, *93*, 1123; *1969*, *91*, 2484; van der Lugt, W.T.A.M. *Tetrahedron Lett. 1970*, 2281; Wristers, J.; Brener, L.; Pettit, R. *J. Am. Chem. Soc. 1970*, *92*, 7499.
[501]Merk, W.; Pettit, R. *J. Am. Chem. Soc. 1967*, *89*, 4788.
[502]For discussions of how these reactions take place, see Slegeir, W.; Case, R.; McKennis, J.S.; Pettit, R. *J. Am. Chem. Soc. 1974*, *96*, 287; Pinhas, A.R.; Carpenter, B.K. *J. Chem. Soc., Chem. Commun. 1980*, 15.

rearranged allyl anion that is formed by disrotatory cleavage of the cyclopropyl ring, a formally Woodward–Hoffmann-forbidden process.[503]

The ring opening of cyclopropyl cations (pp. 486, 1591) is an electrocyclic reaction and is governed by the orbital symmetry rules.[504] For this case, we invoke the rule that the s bond opens in such a way that the resulting p orbitals have the symmetry of the highest occupied orbital of the product, in this case, an allylic cation. We may recall that an allylic system has three molecular orbitals (p. 42). For the cation, with only two electrons, the highest occupied orbital is the one of the lowest energy (A). Thus, the cyclopropyl cation must undergo a

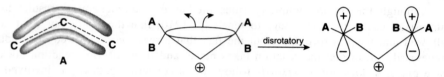

disrotatory ring opening in order to maintain the symmetry. (Note that, in contrast, ring opening of the cyclopropyl *anion* must be conrotatory,[505] since in this case it is the next orbital of the allylic system that is the highest occupied, and this has the opposite symmetry.[506]) However, it is very difficult to generate a free cyclopropyl cation (p. 487), and it is likely that in most cases, cleavage of the σ bond is concerted with departure of the leaving group in the original cyclopropyl substrate. This, of course, means that the σ bond provides anchimeric assistance to the removal of the leaving group (an S_N2-type process), and we would expect that such assistance should come from the back side. This has an important effect on the direction of ring opening. The orbital-symmetry rules require that the ring opening be disrotatory, but as we have seen, there are two disrotatory pathways and the rules do not tell us which is preferred. But the fact that the s orbital provides assistance from the backside means that the two substituents that are trans to the leaving group must move *outward*, not inward.[507] Thus, the disrotatory pathway that is followed is the one shown in B, not the one shown in C, because the former puts the electrons of the σ bond on the

[503]Leivers, M.; Tam, I.; Groves, K.; Leung, D.; Xie, Y.; Breslow, R. *Org. Lett.* **2003**, *5*, 3407.

[504]For discussions, see DePuy, C.H. *Acc. Chem. Res.* **1968**, *1*, 33; Schöllkopf, U. *Angew. Chem. Int. Ed.* **1968**, *7*, 588.

[505]For a review of ring opening of cyclopropyl anions and related reactions, see Boche, G. *Top. Curr. Chem.* **1988**, *146*, 1.

[506]For evidence that this is so, see Newcomb, M.; Ford, W.T. *J. Am. Chem. Soc.* **1974**, *96*, 2968; Boche, G.; Buckl, K.; Martens, D.; Schneider, D.R.; Wagner, H. *Chem. Ber.* **1979**, *112*, 2961; Coates, R.M.; Last, L.A. *J. Am. Chem. Soc.* **1983**, *105*, 7322. For a review of the analogous ring opening of epoxides, see Huisgen, R. *Angew. Chem. Int. Ed.* **1977**, *16*, 572.

[507]This was first proposed by DePuy, C.H.; Schnack, L.G.; Hausser, J.W.; Wiedemann, W. *J. Am. Chem. Soc.* **1965**, *87*, 4006.

side opposite that of the leaving group.[508] Strong confirmation of this picture[509] comes from acetolysis of *endo*- (**111**) and *exo*-bicyclo[3,1,0]hexyl-6-tosylate (**112**). The groups trans to the tosylate must move outward. For **111**, this means that the two hydrogens can go outside the framework of the six-membered ring, but for **112** they

111   **112**

are forced to go inside. Consequently, it is not surprising that the rate ratio for solvolysis of **111/112** was found to be $>2.5 \times 10^6$ and that at 150°C **112** did not solvolyze at all.[510] This evidence is kinetic. Unlike the cases of the cyclobutene (1,3-diene and cyclohexadiene) 1,3,5-triene interconversions, the direct product here is a cation, which is not stable but reacts with a nucleophile and loses some of its steric integrity in the process, so that much of the evidence has been of the kinetic type rather than from studies of product stereochemistry. However, it has been shown by investigations in superacids, where it is possible to keep the cations intact and to study their structures by NMR, that in all cases studied the cation that is predicted by these rules is in fact formed.[511]

OS **V**, 235, 277, 467; **VI**, 39, 145, 196, 422, 427, 862; **IX**, 180.

18-28   Conversion of One Aromatic Compound to Another

(6)*cyclo*-de-hydrogen-coupling (Overall transformation)

Stilbenes can be converted to phenanthrenes by irradiation with UV light[512] in the presence of an oxidizing agent, such as dissolved molecular oxygen, FeCl$_3$,

[508]It has been suggested that the pathway shown in **C** is possible in certain cases: Hausser, J.W.; Grubber, M.J. *J. Org. Chem.* **1972**, *37*, 2648; Hausser, J.W.; Uchic, J.T. *J. Org. Chem.* **1972**, *37*, 4087.

[509]There is much other evidence. For example, see Jefford, C.W.; Medary, R. *Tetrahedron Lett.* **1966**, 2069; Jefford, C.W.; Wojnarowski, W. *Tetrahedron Lett.* **1968**, 199; Sliwinski, W.F.; Su, T.M.; Schleyer, P.v.R. *J. Am. Chem. Soc.* **1972**, *94*, 133; Sandler, S.R. *J. Org. Chem.* **1967**, *32*, 3876; Ghosez, L.; Slinckx, G.; Glineur, M.; Hoet, P.; Laroche, P. *Tetrahedron Lett.* **1967**, 2773; Parham, W.E.; Yong, K.S. *J. Org. Chem.* **1968**, *33*, 3947; Reese, C.B.; Shaw, A. *J. Am. Chem. Soc.* **1970**, *92*, 2566; Dolbier, Jr., W.R.; Phanstiel, O. *Tetrahedron Lett.* **1988**, *29*, 53.

[510]Schöllkopf, U.; Fellenberger, K.; Patsch, M.; Schleyer, P.v.R.; Su, T.M.; Van Dine, G.W. *Tetrahedron Lett.* **1967**, 3639.

[511]Schleyer, P.v.R.; Su, T.M.; Saunders, M.; Rosenfeld, J.C. *J. Am. Chem. Soc.* **1969**, *91*, 5174.

[512]For reviews, see Mallory, F.B.; Mallory, C.W. *Org. React.* **1984**, *30*, 1; Laarhoven, W.H. *Recl. Trav. Chim. Pays-Bas* **1983**, *102*, 185, 241; Blackburn, E.V.; Timmons, C.J. *Q. Rev. Chem. Soc.* **1969**, *23*, 482; Stermitz, L.F. *Org. Photochem.* **1967**, *1*, 247. For a review of electrocyclizations of conjugated aryl olefins in general, see Laarhoven, W.H. *Org. Photochem.* **1989**, *10*, 163.

Pd–C,[513] or iodine.[514] The reaction is a photochemically allowed conrotatory[515] conversion of a 1,3,5-hexatriene to a cyclohexadiene, followed by removal of two hydrogen atoms by the oxidizing agent. The intermediate dihydrophenanthrene has been isolated.[516] The use of substrates containing heteroatoms (e.g., PhN=NPh) allows the formation of heterocyclic ring systems. The actual reacting species must be the *cis*-stilbene, but *trans*-stilbenes can often be used, because they are isomerized to the cis isomers under the reaction conditions. The reaction can be extended to the preparation of many fused aromatic systems, for example,[517]

though not all such systems give reaction.[518]

Isomerization of biphenylene to benzo[*a*]pentalene[519] is a well-known benzene ring contraction rearrangement,[520] driven by relief of strain in the four-membered ring. Related to this process is the FVP of the alternant polycyclic aromatic hydrocarbon benzo[*b*]biphenylene at 1100°C, which gives fluoranthene, a nonalternant polycyclic aromatic hydrocarbon, as the major product at 1100°C in the gas phase.[521] The mechanism used explain that this isomerization involves equilibrating diradicals of 2-phenylnaphthalene, which rearrange by the net migration of a phenyl group to give equilibrating diradicals of 1-phenylnaphthalene, one isomer of which then cyclizes to fluoranthene.

Another transformation of one aromatic compound to another is the *Stone–Wales rearrangement* of pyracyclene (**113**),[522] which is a bond-switching reaction. The rearrangement of bifluorenylidene (**114**) to dibenzo[*g,p*]chrysene (**115**) occurs at temperatures as low as 400°C and is accelerated in the presence of decomposing iodomethane, a convenient source of methyl radicals.[523] This result suggested a

[513]Rawal, V.H.; Jones, R.J.; Cava, M.P. *Tetrahedron Lett.* **1985**, *26*, 2423.

[514]For the use of iodine plus propylene oxide in the absence of air, see Liu, L.; Yang, B.; Katz, T.J.; Poindexter, M.K. *J. Org. Chem.* **1991**, *56*, 3769.

[515]Cuppen, T.J.H.M.; Laarhoven, W.H. *J. Am. Chem. Soc.* **1972**, *94*, 5914.

[516]Doyle, T.D.; Benson, W.R.; Filipescu, N. *J. Am. Chem. Soc.* **1976**, *98*, 3262.

[517]Sato, T.; Shimada, S.; Hata, K. *Bull. Chem. Soc. Jpn.* **1971**, *44*, 2484.

[518]For a discussion and lists of photocyclizing and nonphotocyclizing compounds, see Laarhoven, W.H. *Recl. Trav. Chim. Pays-Bas* **1983**, *102*, 185, 185–204.

[519]Wiersum, U.E.; Jenneskens, L.W. *Tetrahedron Lett.* **1993**, *34*, 6615; Brown, R.F.C.; Choi, N.; Coulston, K.J.; Eastwood, F.W.; Wiersum, U.E.; Jenneskens, L.W. *Tetrahedron Lett.* **1994**, *35*, 4405.

[520]Scott, LT.; Roelofs, N.H. *J. Am. Chem. Soc.* **1987**, *109*, 5461; Scott, L.T.; Roelofs, N.H. *Tetrahedron Lett.* **1988**, *29*, 6857; Anderson, M.R.; Brown, R.F.C.; Coulston, K.J.; Eastwood, F.W.; Ward, A. *Aust. J. Chem.* **1990**, *43*, 1137; Brown, R F.C.; Eastwood, F.W.; Wong, N.R. *Tetrahedron Lett.* **1993**, *34*, 3607.

[521]Preda, D.V.; Scott, L.T. *Org. Lett.* **2000**, *2*, 1489.

[522]Stone, A.J.; Wales, D.J. *Chem. Phys. Lett.* **1986**, *128*, 501.

[523]Alder, R.W.; Whittaker, G. *J. Chem. Soc., Perkin Trans. 2* **1975**, 712

radical rearrangement. This rearrangement is believed to occur by a radical-promoted mechanism consisting of a sequence of homoallyl–cyclopropylcarbinyl rearrangement steps.[524]

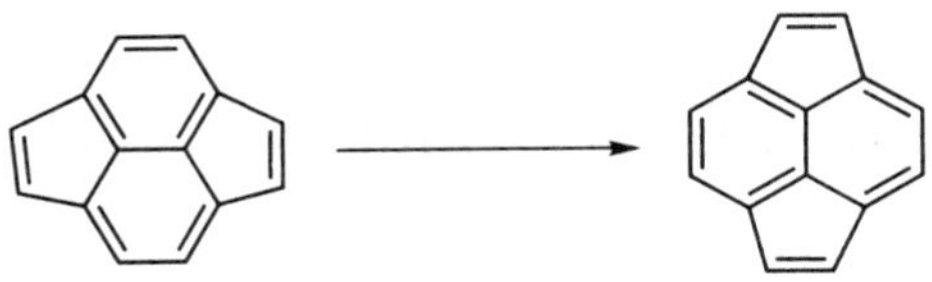

113

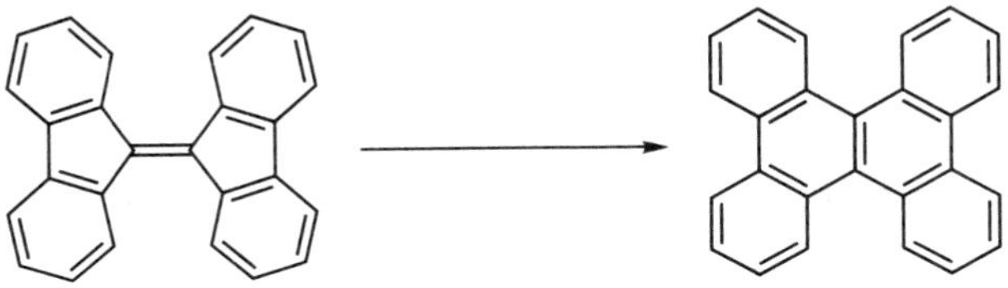

114   **115**

B. Sigmatropic Rearrangements

A sigmatropic rearrangement is defined[525] as migration, in an uncatalyzed intramolecular process, of a σ bond, adjacent to one or more π systems, to a new position in a molecule, with the π systems becoming reorganized in the process. Examples are

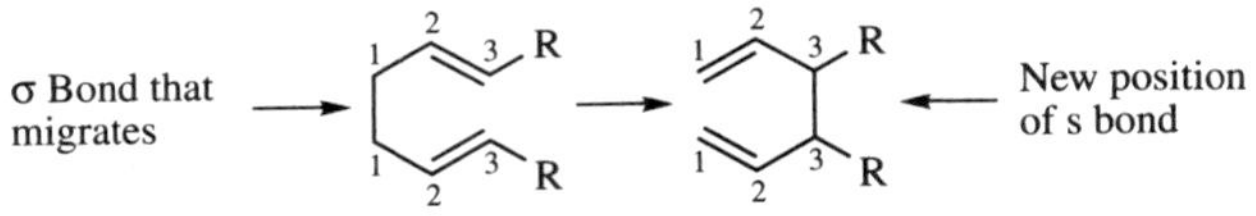

σ Bond that migrates

New position of s bond

Reaction **18-32** A [3,3]-sigmatropic rearrangement

σ Bond that migrates

New position of s bond

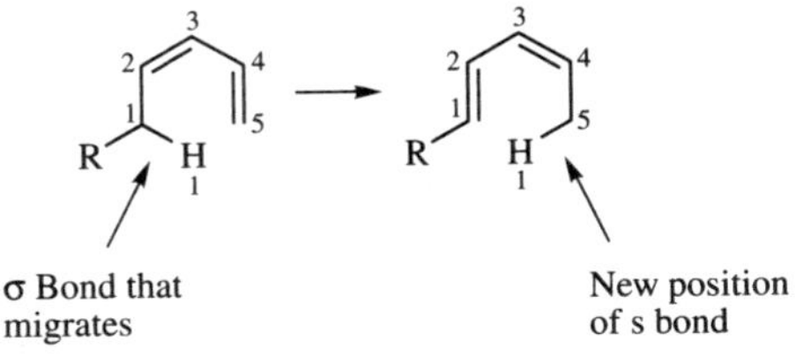

Reaction **18-39** A [1,5]-sigmatropic rearrangement

The *order* of a sigmatropic rearrangement is expressed by two numbers set in brackets: [*i,j*]. These numbers can be determined by counting the atoms over which each end of the σ bond has moved. Each of the original termini is given the number 1. Thus in the first example above, each terminus of the σ bond has

[524]Alder, R.W.; Harvey, J. N. *J. Am. Chem. Soc.* **2004**, *126*, 2490.
[525]Woodward, R.B.; Hoffmann, R. *The Conservation of Orbital Symmetry*, Academic Press, NY, *1970*, p. 114.

migrated from C-1 to C-3, so the order is [3,3]. In the second example, the carbon terminus has moved from C-1 to C-5, but the hydrogen terminus has not moved at all, so the order is [1,5].

18-29 [1,*j*]-Sigmatropic Migrations of Hydrogen

1/ → 3/Hydrogen-migration; 1/ → 5/Hydrogen-migration

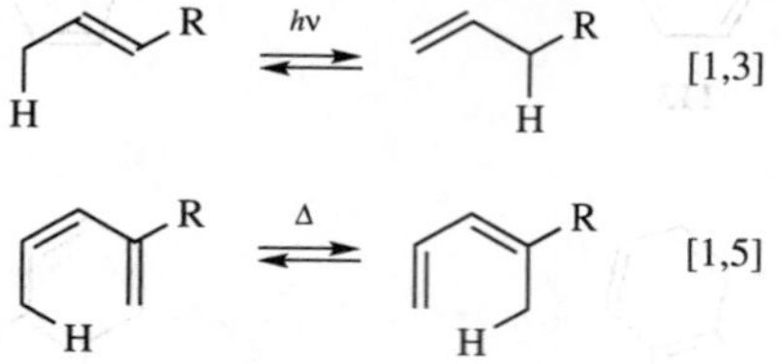

Many examples of thermal or photochemical rearrangements in which a hydrogen atom migrates from one end of a system of π bonds to the other have been reported,[526] although the reaction is subject to geometrical conditions. Isotope effects play a role in sigmatropic rearrangements, and there is evidence for a kinetic silicon isotope effect.[527] Pericyclic mechanisms are involved,[528] and the hydrogen must, in the transition state, be in contact with both ends of the chain at the same time. This means that for [1,5] and longer rearrangements, the molecule must be able to adopt the cisoid conformation. Furthermore, there are two geometrical pathways by which any sigmatropic rearrangement can take place, which we illustrate for the case of a [1,5]-sigmatropic rearrangement,[529] starting with a substrate of the form **116**, where the migration origin is an asymmetric carbon atom and U ≠ V. In one of the two pathways, the hydrogen moves along the top or bottom face of the π system. This is called *suprafacial migration*. In the other pathway, the hydrogen moves *across* the π system, from top to bottom, or vice versa. This is *antarafacial* migration. Altogether, a single isomer like **116** (different rotamers) can give four products. In a suprafacial migration, H can move across the top of the π system (as drawn above) to give the (*R*,*Z*) isomer, or it can rotate 180° and move across the bottom of the π system to give the (*S*,*E*) isomer.[530] The antarafacial migration can similarly lead to two diastereomers, in

---

[526]For a monograph, see Gajewski, J.J. *Hydrocarbon Thermal Isomerizations*, Academic Press, NY, *1981*. For reviews, see Mironov, V.A.; Fedorovich, A.D.; Akhrem, A.A. *Russ. Chem. Rev. 1981*, *50*, 666; Spangler, C.W. *Chem. Rev. 1976*, *76*, 187; DeWolfe, R.H., in Bamford, C.H.; Tipper, C.F.H. *Comprehensieve Chemical Kinetics*, Vol. 9, Elsevier, NY, *1973*, pp. 474–480; Woodward, R.B.; Hoffmann, R. *The Conservation of Orbital Symmetry*, Academic Press, NY, *1970*, pp. 114–140; Hansen, H.; Schmid, H. *Chimia, 1970*, *24*, 89; Roth, W.R. *Chimia, 1966*, *20*, 229.

[527]Lin, Y.-L.; Turos, E. *J. Am. Chem. Soc. 1999*, *121*, 856.

[528]For a discussion of catalysts that induce pericyclic rearrangements, see Moss, S.; King, B.T.; de Meijere, A.; Kozhushkov, S.I.; Eaton, P.E.; Michl, J. *Org. Lett. 2001*, *3*, 2375.

[529]Note that a [1,5]-sigmatropic rearrangement of hydrogen is also an internal ene synthesis (**15-20**).

[530]Since we are using the arbitrary designations U, V, Y, and Z, we have been arbitrary in which isomer to call (*R*,*Z*) and which to call (*S*,*E*).

this case the (*S,Z*) and (*R,E*) isomers.

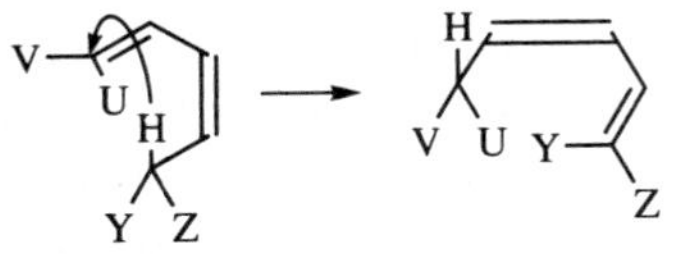

116   (*R, Z*) Isomer   Suprafacial

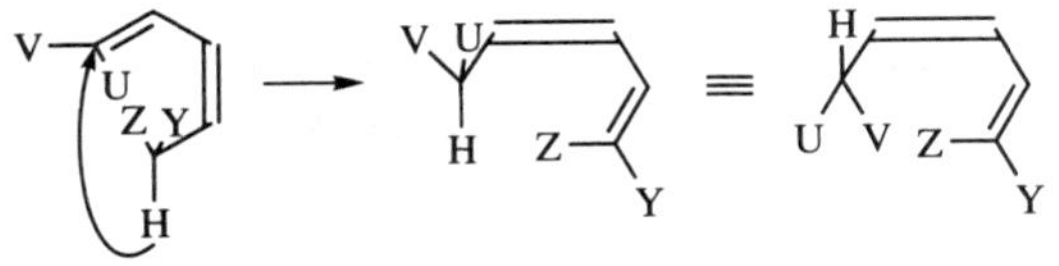

116   (*S,E*) Isomer   Suprafacial

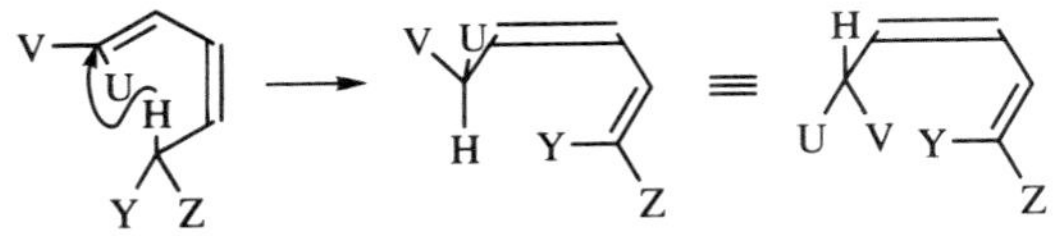

116   (*S, Z*) Isomer   Antarafacial

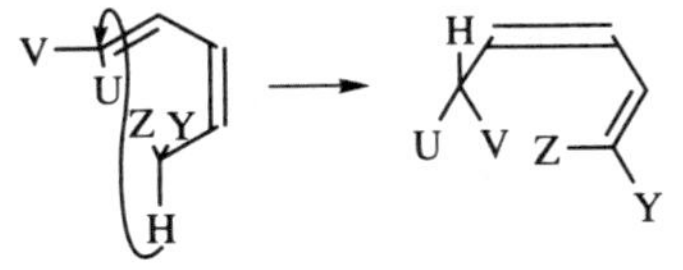

116   (*R,E*) Isomer   Antarafacial

In any given sigmatropic rearrangement, only one of the two pathways is allowed by the orbital-symmetry rules; the other is forbidden. To analyze this situation, first we use a modified frontier orbital approach.[531] We will imagine that in the transition state **C**, the migrating H atom breaks away from the rest of the system, which we may treat as if it were a free radical.

C

Imaginary
transition state
for a [1,3]-
sigmatropic
rearrangement

[531]See Woodward, R.B.; Hoffmann, R. *The Conservation of Orbital Symmetry*, Academic Press, NY, *1970*, pp. 114–140.

Note that this is not what actually takes place; we merely imagine it in order to be able to analyze the process. In a [1,3]-sigmatropic rearrangement, the imaginary transition state consists of a hydrogen atom and an allyl radical. The latter species (p. 42) has three π orbitals, but the only one that concerns us here is the HOMO which, in a thermal rearrangement is **D**. The electron of the hydrogen atom is of course in a $1s$ orbital, which has only one lobe. The rule governing sigmatropic migration of hydrogen is *the H must move from a plus to a plus or from a minus to a minus*

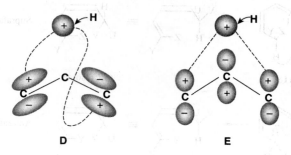

D   **E**

lobe, of the HOMO; it cannot move to a lobe of opposite sign.[532] Obviously, the only way this can happen in a thermal [1,3]-sigmatropic rearrangement is if the migration is antarafacial. Consequently, the rule predicts that antarafacial thermal [1,3]-sigmatropic rearrangements are allowed, but the suprafacial pathway is forbidden. However, in a photochemical reaction, promotion of an electron means that **E** is now the HOMO; the suprafacial pathway is now allowed and the antarafacial pathway forbidden.

A similar analysis of [1,5]-sigmatropic rearrangements shows that in this case the thermal reaction must be suprafacial and the photochemical process antarafacial. For the general case, with odd-numbered j, we can say that $[1,j]$-suprafacial migrations are allowed thermally when j is of the form $4n + 1$, and photochemically when j has the form $4n - 1$; the opposite is true for antarafacial migrations.

F   **G**

As expected, the Möbius–Hückel method leads to the same predictions. Here, we look at the basis set of orbitals shown in **F** and **G** for [1,3]- and [1,5]-rearrangements, respectively. A [1,3]-shift involves four electrons, so an allowed thermal pericyclic reaction must be a Möbius system (p. 1210) with one or an odd number

---

[532]This follows from the principle that bonds are formed only by overlap of orbitals of the same sign. Since this is a concerted reaction, the hydrogen orbital in the transition state must overlap simultaneously with one lobe from the migration origin and one from the terminus. It is obvious that both of these lobes must have the same sign.

of sign inversions. As can be seen in **F**, only an antarafacial migration can achieve this. A [1,5]-shift, with six electrons, is allowed thermally only when it is a Hückel system with zero or an even number of sign inversions; hence it requires a suprafacial migration.[533]

The actual reported results bear out this analysis. Thus a thermal [1,3] migration is allowed to take place only antarafacially, but such a transition state would be extremely strained, and thermal [1,3]-sigmatropic migrations of hydrogen are unknown.[534] On the other hand, the photochemical pathway allows suprafacial [1,3]-shifts, and a few such reactions are known, an example being the photochemical rearrangement of **117** to **118**.[535] Substituents influence the efficacy of the [1,3]-hydrogen shift.[536]

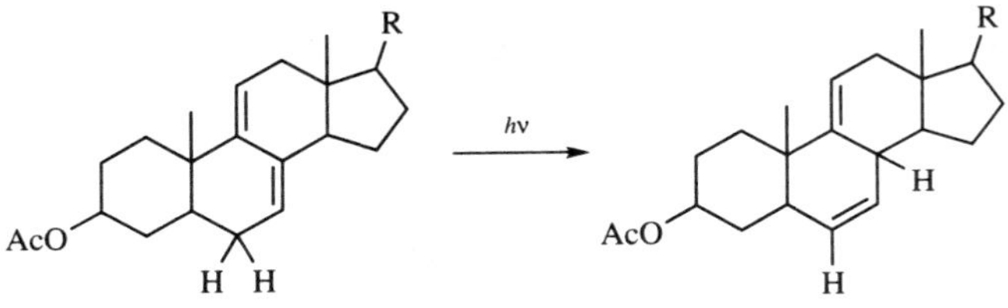

117   **118**

The situation is reversed for [1,5]-hydrogen shifts. In this case the thermal rearrangements, being suprafacial, are quite common, while photochemical rearrangements are rare.[537] Two examples of the thermal reaction are

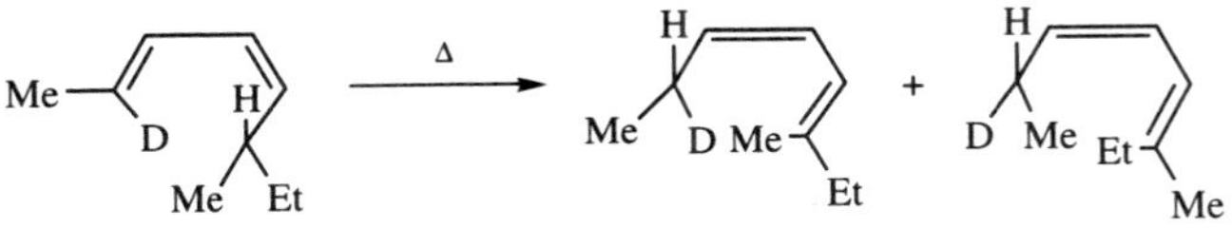

Ref. [538]

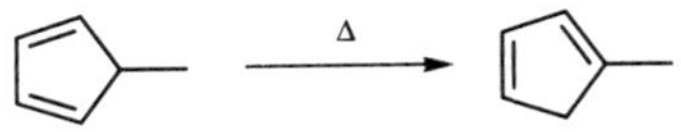

Ref. [539]

[533]For a discussion of the origins for the preference for orbital-symmetry forbidden reactions and the stereochemistry of [1,5]-sigmatropic shifts, see Kless, A.; Nendel, M.; Wilsey, S.; Houk, K.N. *J. Am. Chem. Soc.* **1999**, *121*, 4524.

[534]A possible [1,3]-migration of hydrogen has been reported. See Yeh, M.; Linder, L.; Hoffman, D.K.; Barton, T.J. *J. Am. Chem. Soc.* **1986**, *108*, 7849. See also, Pasto, D.J.; Brophy, J.E. *J. Org. Chem.* **1991**, *56*, 4554.

[535]Dauben, W.G.; Wipke, W.T. *Pure Appl. Chem.* **1964**, *9*, 539, 546. For another example, see Kropp, P.J.; Fravel, Jr., H.G.; Fields, T.R. *J. Am. Chem. Soc.* **1976**, *98*, 840.

[536]Hudson, C.E.; McAdoo, D.J. *J. Org. Chem.* **2003**, *68*, 2735.

[537]For examples of photochemical [1,5]-antarafacial reactions, see Kiefer, E.F.; Tanna, C.H. *J. Am. Chem. Soc.* **1969**, *91*, 4478; Kiefer, E.F.; Fukunaga, J.Y. *Tetrahedron Lett.* **1969**, 993; Dauben, W.G.; Poulter, C.D.; Suter, C. *J. Am. Chem. Soc.* **1970**, *92*, 7408.

[538]Roth, W.R.; König, J.; Stein, K. *Chem. Ber.* **1970**, *103*, 426.

[539]McLean, S.; Haynes, P. *Tetrahedron* **1965**, *21*, 2329. For a review of such rearrangements, see Klärner, F. *Top. Stereochem.* **1984**, *15*, 1. For a discussion of [1,5]-sigmatropic hydrogen shifts in cyclic 1,3-dienes, see Hess, Jr., B.A.; Baldwin, J.E. *J. Org. Chem.* **2002**, *67*, 6025.

Note that the first example bears out the stereochemical prediction made earlier. Only the two isomers shown were formed. In the second example, migration can continue around the ring. Migrations of this kind are called *circumambulatory rearrangements*.[540] Such migrations are known for cyclopentadiene, pyrrole, and phosphole derivatives.[541] Geminal bond participation has been observed in pentadienes,[542] the effects of phenyl substituents have been studied,[543] and the kinetics and activation parameters of [1,5] hydrogen shifts have been examined.[544] The [1,5] hydrogen shifts are also known with vinyl aziridines.[545]

The rare [1,4]-hydrogen transfer has been observed in radical cyclizations.[546] With respect to [1,7]-hydrogen shifts, the rules predict the thermal reaction to be antarafacial.[547] Unlike the case of [1,3]-shifts, the transition state is not too greatly strained, and an example of such rearrangements is the formation of **119** and **120**.[548] Photochemical [1,7]-shifts are suprafacial and, not surprisingly, many of these have been observed.[549]

The orbital symmetry rules also help us to explain, as on pp. 1232 and 1642, the unexpected stability of certain compounds. Thus, **120** could, by a thermal [1,3]-sigmatropic rearrangement, easily convert to toluene, which of course is far more stable because it has an aromatic sextet. Yet, **120** has been prepared and is stable at dry ice temperature and in dilute solutions.[550]

[540]For a review, see Childs, R.F. *Tetrahedron* **1982**, *38*, 567. See also, Minkin, V.I.; Mikhailov, I.E.; Dushenko, G.A.; Yudilevich, J.A.; Minyaev, R.M.; Zschunke, A.; Mügge, K. *J. Phys. Org. Chem.* **1991**, *4*, 31. For a study of [1,5]-sigmatropic shiftamers, see Tantillo, D.J.; Hoffmann, R. *Eur. J. Org. Chem.* **2004**, 273.

[541]Bachrach, S.M. *J. Org. Chem.* **1993**, *58*, 5414.

[542]Ikeda, H.; Ushioda, N.; Inagaki, S. *Chem. Lett.* **2001**, 166.

[543]Hayase, S.; Hrovat, D.A.; Borden, W.T. *J. Am. Chem. Soc.* **2004**, *126*, 10028.

[544]Baldwin, J.E.; Raghavan, A.S. *J. Org. Chem.* **2004**, *69*, 8128.

[545]Åhman, J.; Somfai, P.; Tanner, D. *J. Chem. Soc., Chem. Commun.* **1994**, 2785; Somfai, P.; Åhman, J. *Tetrahedron Lett.* **1995**, *36*, 1953.

[546]Journet, M.; Malacria, M. *Tetrahedron Lett.* **1992**, *33*, 1893.

[547]For a computational study that supports tunneling in thermal [1,7]-hydrogen shifts see Hess, Jr., B.A. *J. Org. Chem.* **2001**, *66*, 5897.

[548]Gurskii, M.E.; Gridnev, I.D.; Il'ichev, Y.V.; Ignatenko, A.V.; Bubnov, Y.N. *Angew. Chem. Int. Ed.* **1992**, *31*, 781; Baldwin, J.E.; Reddy, V.P. *J. Am. Chem. Soc.* **1987**, *109*, 8051; **1988**, *110*, 8223.

[549]See Murray, R.W.; Kaplan, M.L. *J. Am. Chem. Soc.* **1966**, *88*, 3527; ter Borg, A.P.; Kloosterziel, H. *Recl. Trav. Chim. Pays-Bas* **1969**, *88*, 266; Tezuka, T.; Kimura, M.; Sato, A.; Mukai, T. *Bull. Chem. Soc. Jpn.* **1970**, *43*, 1120.

[550]Bailey, W.J.; Baylouny, R.A. *J. Org. Chem.* **1962**, *27*, 3476.

Analogs of sigmatropic rearrangements in which a cyclopropane ring replaces one of the double bonds are also known, for example,[551]

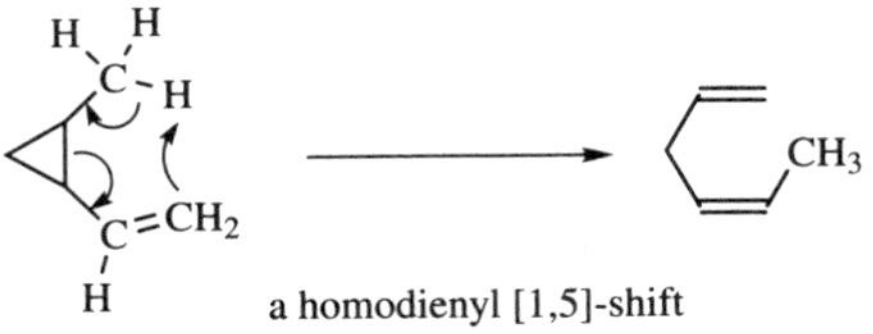

a homodienyl [1,5]-shift

The reverse reaction has also been reported.[552] 2-Vinylcycloalkanols[553] undergo an analogous reaction, as do cyclopropyl ketones (see p. 1673 for this reaction).

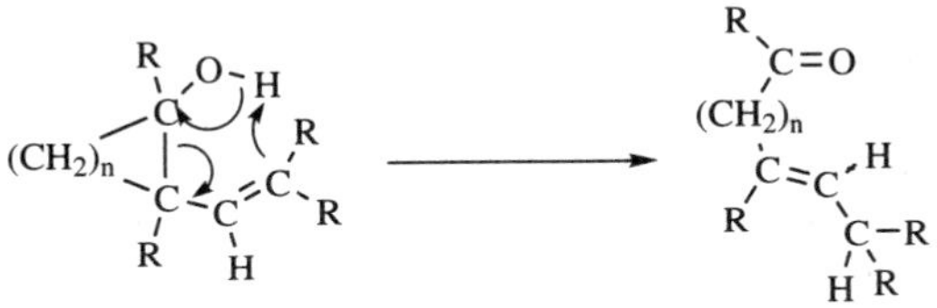

18-30 [1,j]-Sigmatropic Migrations of Carbon

[1,3] migration of alkyl

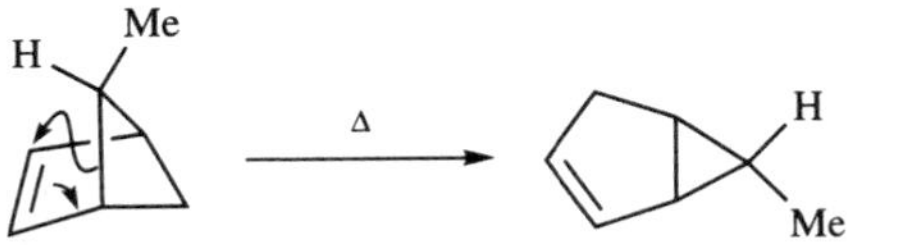

Ref. [554]

[1,5] migration of phenyl

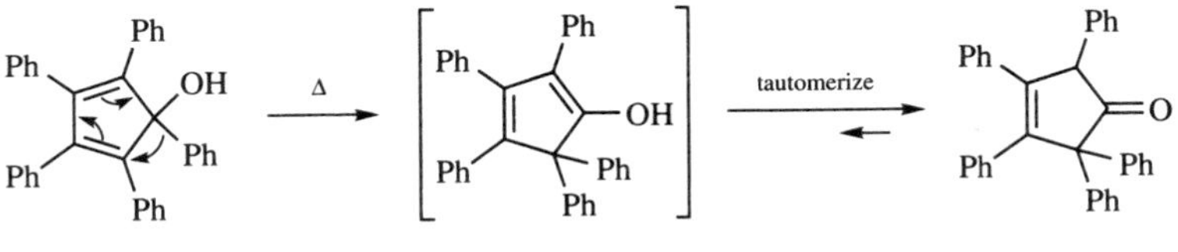

Ref. [555]

[551]Frey, H.M.; Solly, R.K. *Int. J. Chem. Kinet.*, *1969*, *1*, 473; Roth, W.R.; König, J. *Liebigs Ann. Chem. 1965*, *688*, 28; Ohloff, G. *Tetrahedron Lett. 1965*, 3795; Jorgenson, M.J.; Thacher, A.F. *Tetrahedron Lett. 1969*, 4651; Corey, E.J.; Yamamoto, H.; Herron D.K.; Achiwa, K. *J. Am. Chem. Soc. 1970*, *92*, 6635; Loncharich, R.J.; Houk, K.N. *J. Am. Chem. Soc. 1988*, *110*, 2089; Parziale, P.A.; Berson, J.A. *J. Am. Chem. Soc. 1990*, *112*, 1650; Pegg, G.G.; Meehan, G.V. *Aust. J. Chem. 1990*, *43*, 1009, 1071.
[552]Roth, W.R.; König, J. *Liebigs Ann. Chem. 1965*, *688*, 28. Also see, Grimme, W. *Chem. Ber. 1965*, *98*, 756.
[553]Arnold, R.T.; Smolinsky, G. *J. Am. Chem. Soc. 1960*, *82*, 4918; Leriverend, P.; Conia, J.M. *Tetrahedron Lett. 1969*, 2681; Conia, J.M.; Barnier, J.P. *Tetrahedron Lett. 1969*, 2679.
[554]Roth, W.R.; Friedrich, A. *Tetrahedron Lett. 1969*, 2607.
[555]Youssef, A.K.; Ogliaruso, M.A. *J. Org. Chem. 1972*, *37*, 2601.

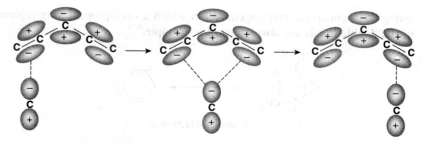

Fig. 18.5. Hypothetical orbital movement for a thermal [1,5]-sigmatropic migration of carbon. To move from one negative lobe, the migrating carbon uses only its own negative lobe, retaining its configuration.

Sigmatropic migrations of alkyl or aryl groups[556] are less common than the corresponding hydrogen migrations.[557] When they do take place, there is an important difference. Unlike a hydrogen atom, whose electron is in a $1s$ orbital with only one lobe, a carbon free radical has its odd electron in a p orbital that has *two lobes of opposite sign*. Therefore, if we draw the imaginary transition states for this case (see p. 1650), we see that in a thermal suprafacial [1,5] process (Fig. 18.5), symmetry can be conserved only if the migrating carbon moves in such a way that the lobe which was originally attached to the π system remains attached to the π system.

This can happen only if configuration is *retained within the migrating group*. On the other hand, thermal suprafacial [1,3] migration (Fig. 18.6) *can* take place if the migrating carbon switches lobes. If the migrating carbon was originally bonded by its minus lobe, it must now use its plus lobe to form the new C–C bond. Thus, configuration in the migrating group will be *inverted*. From these considerations we predict that suprafacial [1,j]-sigmatropic rearrangements in which carbon is the migrating group are always allowed, both thermally and photochemically, but that thermal [1,3] migrations will proceed with inversion and thermal [1,5]

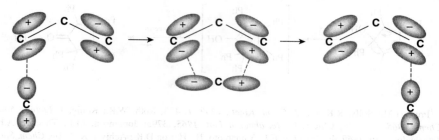

Fig. 18.6. Hypothetical orbital movement for a thermal [1,3]-sigmatropic migration of carbon. The migrating carbon moves a negative to a positive lobe, requiring it to switch its own bonding lobe from negative to positive, inverting its configuration.

[556]For reviews, see Mironov, V.A.; Fedorovich, A.D.; Akhrem, A.A. *Russ. Chem. Rev.* **1981**, *50*, 666; Spangler, C.W. *Chem. Rev.* **1976**, *76*, 187
[557]It has been shown that methyl and phenyl have lower migratory aptitudes than hydrogen in thermal sigmatropic rearrangements: Shen, K.; McEwen, W.E.; Wolf, A.P. *Tetrahedron Lett.* **1969**, 827; Miller, L.L.; Greisinger, R.; Boyer, R.F. *J. Am. Chem. Soc.* **1969**, *91*, 1578.

migrations with retention of configuration within the migrating group. More generally, we can say that suprafacial $[1,j]$ migrations of carbon in systems where
$j = 4n - 1$ proceed with inversion thermally and retention photochemically, while
systems where $j = 4n + 1$ show the opposite behavior. Where antarafacial migrations take place, all these predictions are of course reversed.

The first laboratory test of these predictions was the pyrolysis of deuterated
endo-bicyclo[3.2.0]hept-2-en-6-yl acetate (**121**), which gave the *exo*-deuterio-*exo*-
norbornyl acetate **122**.[558] Thus, as predicted by the orbital symmetry rules, this
thermal suprafacial [1,3]-sigmatropic reaction took place with complete inversion
at C-7. Similar results have been obtained in a number of other cases.[559] However,
similar studies of the pyrolysis of the parent hydrocarbon of **121**, labeled with D at
C-6 and C-7, showed that while most of the product was formed with inversion at
C-7, a significant fraction (11–29%) was formed with retention.[560] Other cases of
lack of complete inversion are also known.[561] A diradical mechanism has been
invoked to explain such cases.[562] There is strong evidence for a radical mechanism
for some [1,3]-sigmatropic rearrangements.[563] Photochemical suprafacial [1,3]
migrations of carbon have been shown to proceed with retention, as predicted.[564]
    Although allylic vinylic ethers generally undergo [3,3]-sigmatropic rearrangements
(**18-33**), they can be made to give the [1,3] kind, to give aldehydes, for example,

[558]Berson, J.A.; Nelson, G.L. *J. Am. Chem. Soc.* **1967**, *89*, 5503; Berson, J.A. *Acc. Chem. Res.* **1968**, *1*,
152.
[559]See Roth, W.R.; Friedrich, A. *Tetrahedron Lett.* **1969**, 2607; Berson, J.A. *Acc. Chem. Res.* **1972**, *5*, 406;
Bampfield, H.A.; Brook, P.R.; Hunt, K. *J. Chem. Soc., Chem. Commun.* **1976**, 146; Franzus, B.;
Scheinbaum, M.L.; Waters, D.L.; Bowlin, H.B. *J. Am. Chem. Soc.* **1976**, *98*, 1241; Klärner, F.; Adamsky,
F. *Angew. Chem. Int. Ed.* **1979**, *18*, 674.
[560]Baldwin, J.E.; Belfield, K.D. *J. Am. Chem. Soc.* **1988**, *110*, 296; Klärner, F.; Drewes, R.; Hasselmann,
D. *J. Am. Chem. Soc.* **1988**, *110*, 297.
[561]See, for example, Berson, J.A.; Holder, R.W. *J. Am. Chem. Soc.* **1973**, *95*, 2037; Pikulin, S.; Berson,
J.A. *J. Am. Chem. Soc.* **1988**, *110*, 8500.
[562]See Newman-Evans, R.H.; Carpenter, B.K. *J. Am. Chem. Soc.* **1984**, *106*, 7994; Pikulin, S.; Berson,
J.A. *J. Am. Chem. Soc.* **1988**, *110*, 8500. See also, Berson, J.A. *Chemtracts: Org. Chem.* **1989**, *2*, 213.
[563]See, for example, Bates, G.S.; Ramaswamy, S. *Can. J. Chem.* **1985**, *63*, 745; Dolbier, W.B.; Phanstiel
IV, O. *J. Am. Chem. Soc.* **1989**, *111*, 4907.
[564]Cookson, R.C.; Hudec, J.; Sharma, M. *Chem. Commun.* **1971**, 107, 108.

by treatment with $LiClO_4$ in diethyl ether.[565] In this case, the C–O bond undergoes a 1,3 migration from the O to the end vinylic carbon. When the vinylic ether is of the type $ROCR'=CH_2$, ketones RCH_2COR' are formed. There is evidence that this [1,3]-sigmatropic rearrangement is not concerted, but involves dissociation of the substrate into ions.[565]

Thermal suprafacial [1,5] migrations of carbon have been found to take place with retention,[566] but also with inversion.[567] A diradical mechanism has been suggested for the latter case.[567]

Simple nucleophilic, electrophilic, and free-radical 1,2-shifts can also be regarded as sigmatropic rearrangements (in this case, [1,2]-rearrangements). We have already (p. $$$) applied similar principles to such rearrangements to show that nucleophilic 1,2-shifts are allowed, but the other two types are forbidden unless the migrating group has some means of delocalizing the extra electron or electron pair. The mechanism of the forbidden [3s,5s]-sigmatropic shift has been examined.[568]

18-31 Conversion of Vinylcyclopropanes to Cyclopentenes

The thermal expansion of a vinylcyclopropane to a cyclopentene ring[569] is a special case of a [1,3]-sigmatropic migration of carbon, although it can also be considered an internal $[_\pi2 + _\sigma2]$-cycloaddition reaction (see **15-63**). It is known as a *vinylcyclopropane rearrangement*.[570] The reaction has been carried out on many vinylcyclopropanes bearing various substituents in the ring[571] or

---

[565]Grieco, P.A.; Clark, J.D.; Jagoe, C.T. *J. Am. Chem. Soc.* **1991**, *113*, 5488; Palani, N.; Balasubramanian, K.K. *Tetrahedron Lett.* **1995**, *36*, 9527.

[566]Boersma, M.A.M.; de Haan, J.W.; Kloosterziel, H.; van de Ven, L.J.M. *Chem. Commun.* **1970**, 1168.

[567]Klärner, F.; Yaslak, S.; Wette, M. *Chem. Ber.* **1979**, *112*, 1168; Klärner, F.; Brassel, B. *J. Am. Chem. Soc.* **1980**, *102*, 2469; Gajewski, J.J.; Gortva, A.M.; Borden, J.E. *J. Am. Chem. Soc.* **1986**, *108*, 1083; Baldwin, J.E.; Broline, B.M. *J. Am. Chem. Soc.* **1982**, *104*, 2857.

[568]Leach, A.G.; Catak, S.; Houk, K.N. *Chem. Eur. J.* **2002**, *8*, 1290.

[569]For reviews, see Baldwin, J.E. *Chem. Rev.* **2003**, *103*, 1197; Wong, H.N.C.; Hon, M.; Tse, C.; Yip, Y.; Tanko, J.; Hudlicky, T. *Chem. Rev.* **1989**, *89*, 165, see pp. 169–172; Hudlicky, T.; Kutchan, T.M.; Naqvi, S.M. *Org. React.* **1985**, *33*, 247; DeWolfe, R.H., in Bamford, C.H.; Tipper, C.F.H. *Comprehenseive Chemical Kintetics*, Vol. 9, Elseiver, NY, **1973**, pp. 470–474; Gutsche, C.D.; Redmore, D. *Carbocyclic Ring Expansion Reactions*, Academic Press, NY, **1968**, pp. 163–170.

[570]For a novel vinylcyclopropane rearrangement, see Armesto, D.; Ramos, A.; Mayoral, E.P.; Ortiz, M.J.; Agarrabeitia, A.R. *Org. Lett.* **2000**, *2*, 183.

[571]For a study of substituent effects, see McGaffin, G.; Grimm, B.; Heinecke, U.; Michaelsen, H.; de Meijere, A.; Walsh, R. *Eur. J. Org. Chem.* **2001**, 3559.

on the vinyl group and has been extended to 1,1-dicyclopropylethene[572]

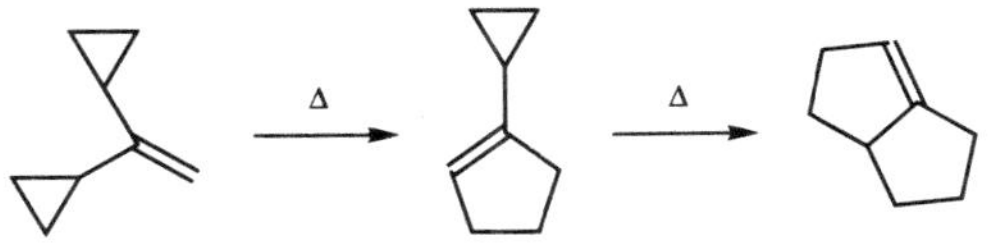

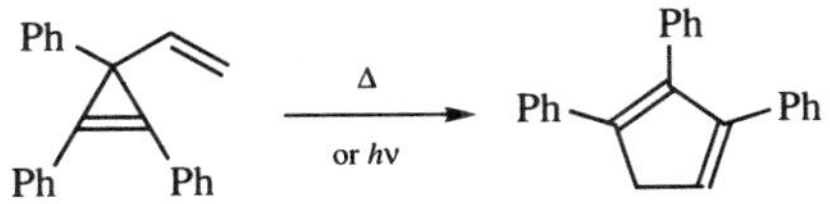

and (both thermally[573] and photochemically[574]) to vinylcyclopropenes. This rearrangement can be catalyzed by rhodium and silver compounds, and has been used to form rings.[575] Another variation converts α-trimethylsilylcyclopropyl ketones to ring-expanded ketones, such as **123**, via FVP at 550°C.[576] Flash vacuum pyrolysis of the trimethylsilyl ether of cyclopropylcarbinyl alcohols gives similar results.[577] A variation uses flash vacuum pyrolysis at 600°C to convert α-trimethylsilyloxy-α-vinyl cyclic ketones to ring expanded ketones.[578]

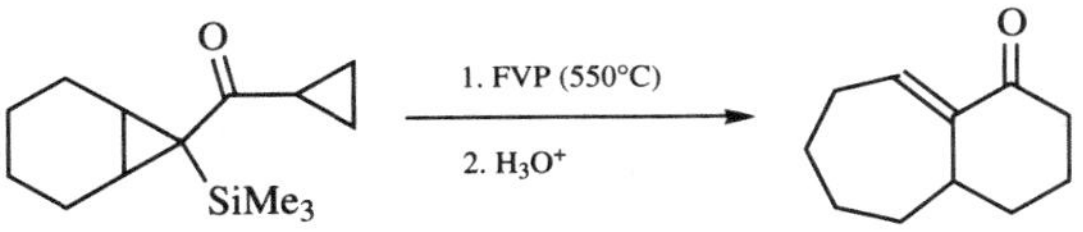

123

Various heterocyclic analogs[579] are also known, as in the rearrangement of aziridinyl amides (**124**).[580] Cyclopropyl ketones can be treated with tosylamine and a zirconium catalyst, which converts the imine formed *in situ* to a pyrroline.[581]

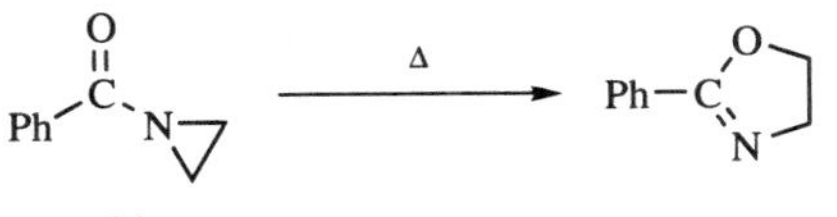

124

[572]Ketley, A.D. *Tetrahedron Lett.* *1964*, 1687; Branton, G.R.; Frey, H.M. *J. Chem. Soc. A* *1966*, 1342.

[573]Small, A.; Breslow, R. cited in Breslow, R. in de Mayo, P. *Molecular Rearrangments*, Vol. 1, Wiley, NY, *1963*, p. 236.

[574]Padwa, A.; Blacklock, T.J.; Getman, D.; Hatanaka, N.; Loza, R. *J. Org. Chem.* *1978*, *43*, 1481; Zimmerman, H.E.; Kreil, D.J. *J. Org. Chem.* *1982*, 47, 2060.

[575]Wender, P.A.; Husfeld, C.O.; Langkopf, E.; Love, J.A. *J. Am. Chem. Soc.* *1998*, *120*, 1940.

[576]Liu, H.; Shook, C.A.; Jamison, J.A.; Thiruvazhi, M.; Cohen, T. *J. Am. Chem. Soc.* *1998*, *120*, 605.

[577]Rüedi, G.; Nagel, M.; Hansen, H.-J. *Org. Lett.* *2004*, *6*, 2989.

[578]Rüedi, G.; Oberli, M.A.; Nagel, M.; Hansen, H.-J. *Org. Lett.* *2004*, *6*, 3179.

[579]For a review of a nitrogen analog, see Boeckman, Jr., R.K.; Walters, M.A. *Adv. Heterocycl. Nat. Prod. Synth.* *1990*, *1*, 1.

[580]For reviews of ring expansions of aziridines, see Heine, H.W. *Mech. Mol. Migr.* *1971*, *3*, 145; Dermer, O.C.; Ham, G.E. *Ethylenimine and Other Aziridines*, Academic Press, NY, *1969*, pp. 282–290. See also, Wong, H.N.C.; Hon, M.; Tse, C.; Yip, Y.; Tanko, J.; Hudlicky, T. *Chem. Rev.* *1989*, *89*, 165, 190–192.

[581]Shi, M.; Yang, Y.-H.; Xu, B. *Synlett* *2004*, 1622.

Two competing reactions are the homodienyl [1,5]-shift (if a suitable H is available, see **18-29**), and simple cleavage of the cyclopropane ring, leading in this case to a diene (see **18-3**).

Vinylcyclobutanes can be similarly converted to cyclohexenes,[582] but larger ring compounds do not generally give the reaction.[583] Bicyclo[2.1.0]pentane derivatives undergo this reaction, and tricyclo[4.1.0.0$^{2.5}$]heptanes rearrange to give nonconjugated cycloheptadienes.[584] Though high temperatures (as high as 500°C) are normally required for the thermal reaction, the lithium salts of 2-vinylcyclopropanols rearrange to the lithium salt of cyclopent-3-enols at 25°C.[585] Salts of 2-vinylcyclobutanols behave analogously.[586]

The reaction rate has also been greatly increased by the addition of a one-electron oxidant tris-(4-bromophenyl)aminium hexafluoroantimonate Ar$_3$N·+ SbF$_6$− (Ar = p-bromophenyl).[587] This reagent converts the substrate to a cation radical, which undergoes ring expansion much faster.[588]

The mechanisms of these ring expansions are not certain. Both concerted[589] and diradical[590] pathways have been proposed,[591] and it is possible that both pathways operate, in different systems.

For the conversion of a vinylcyclopropane to a cyclopentene in a different way, see OS **68**, 220.

[582]See, for example, Overberger, C.G.; Borchert, A.E. *J. Am. Chem. Soc.* **1960**, *82*, 1007; Gruseck, U.; Heuschmann, M. *Chem. Ber.* **1990**, *123*, 1911. The kinetics of gas-phase fragmentation of propenylmethyl cyclobutanes has been examined, see Baldwin, J.E.; Burrell, R.C. *J. Org. Chem.* **2002**, *67*, 3249. Thermal [1,3]-carbon sigmatropic rearrangements of vinylcyclobutanes have been reviewed. See Leber, P.A.; Baldwin, J.E. *Acc. Chem. Res.* **2002**, *35*, 279.

[583]For an exception, see Thies, R.W. *J. Am. Chem. Soc.* **1972**, *94*, 7074.

[584]Deak, H.L.; Stokes, S.S.; Snapper, M.L. *J. Am. Chem. Soc.* **2001**, *123*, 5152.

[585]Danheiser, R.L.; Bronson, J.J.; Okano, K. *J. Am. Chem. Soc.* **1985**, *107*, 4579.

[586]Danheiser, R.L.; Martinez-Davila, C.; Sard, H. *Tetrahedron* **1981**, *37*, 3943.

[587]Dinnocenzo, J.P.; Conlan, D.A. *J. Am. Chem. Soc.* **1988**, *110*, 2324.

[588]For a review of ring expansion of vinylcyclobutane cation radicals, see Bauld, N.L. *Tetrahedron* **1989**, *45*, 5307.

[589]For evidence favoring the concerted mechanism, see Billups, W.E.; Leavell, K.H.; Lewis, E.S.; Vanderpool, S. *J. Am. Chem. Soc.* **1973**, *95*, 8096; Berson, J.A.; Dervan, P.B.; Malherbe, R.; Jenkins, J.A. *J. Am. Chem. Soc.* **1976**, *98*, 5937; Andrews, G.D.; Baldwin, J.E. *J. Am. Chem. Soc.* **1976**, *98*, 6705, 6706; Dolbier, Jr., W.R.; Al-Sader, B.H.; Sellers, S.F.; Koroniak, H. *J. Am. Chem. Soc.* **1981**, *103*, 2138; Gajewski, J.J.; Olson, L.P. *J. Am. Chem. Soc.* **1991**, *113*, 7432.

[590]For evidence favoring the diradical mechanism, see Willcott, M.R.; Cargle, V.H. *J.Am.Chem.Soc.* **1967**, *89*, 723; Doering, W. von E.; Schmidt, E.K.G. *Tetrahedron* **1971**, *27*, 2005; Roth, W.R.; Schmidt, E.K.G. *Tetrahedron Lett.* **1971**, 3639; Simpson, J.M.; Richey Jr., H.G. *Tetrahedron Lett.* **1973**, 2545; Gilbert, J.C.; Higley, D.P. *Tetrahedron Lett.* **1973**, 2075; Caramella, P.; Huisgen, R.; Schmolke, B. *J.Am.Chem.Soc.* **1974**, *96*, 2997, 2999; Mazzocchi, P.H.; Tamburin, H.J. *J.Am.Chem. Soc.* **1975**, *97*, 555; Zimmerman, H.E.; Fleming, S.A. *J. Am. Chem. Soc.* **1983**, *105*, 622; Klumpp, G.W.; Schakel, M. *Tetrahedron Lett.* **1983**, *24*, 4595; McGaffin, G.; de Meijere, A.; Walsh, R. *Chem. Ber.* **1991**, *124*, 939. A "continuous diradical transition state" has also been proposed: Roth, W.R.; Lennartz, H.; Doering, W. von E.; Birladeanu, L.; Guyton, C.A.; Kitagawa, T. *J. Am. Chem. Soc.* **1990**, *112*, 1722, and references cited therein.

[591]For a discussion concerning whether or not this [1,3]-shift is a concerted reaction, see Gajewski, J.J.; Olson, L.P.; Willcott III, M.R. *J. Am. Chem. Soc.* **1996**, *118*, 299. For a discussion of the mechanism of this reaction, see Su, M.-D. *Tetrahedron* **1995**, *51*, 5871.

N-Cyclopropylimines undergo rearrangement to cyclic imines (pyrrolines) under photochemical conditions.[592] P-Vinyl phosphiranes (the P analog of cyclopropanes with P in the ring) under a similar rearrangement, and the mechanism has been studied.[593]

18-32   The Cope Rearrangment

(3/4/) → (1/6/)-*sigma*-Migration

Z = Ph, RCO, and so on.

When 1,5-dienes are heated, a [3,3] sigmatropic rearrangement known as the *Cope rearrangement* (not to be confused with the Cope elimination reaction, **17-9**) occurs to generate an isomeric 1,5-diene.[594] When the diene is symmetrical about the 3,4 bond, we have the unusual situation where a reaction gives a product identical with the starting material:[595]

Therefore, a Cope rearrangement can be detected only when the diene is not symmetrical about this bond. Any 1,5-diene gives the rearrangement; for example, 3-methyl-1,5-hexadiene heated to 300°C gives 1,5-heptadiene.[596] However, the reaction takes place more easily (lower temperature required) when there is a group on the 3- or 4-carbon with leads to the new double bond being substituted. The reaction is obviously reversible[597] and produces an equilibrium mixture of the two 1,5-dienes, which is richer in the thermodynamically more stable isomer. However, the equilibrium can be shifted to the right for 3-hydroxy-1,5-dienes,[598] because the product tautomerizes to the ketone or aldehyde:

[592]Campos, P.J.; Soldevilla, A.; Sampedro, D.; Rodrguez, M.A. *Org. Lett.* **2001**, *3*, 4087.

[593]Mátrai, J.; Dransfeld, A.; Veszprém, T.; Nguyen, M.T. *J. Org. Chem.* **2001**, *66*, 5671.

[594]For reviews, see Bartlett, P.A. *Tetrahedron* **1980**, *36*, 2, 28–39; Rhoads, S.J.; Raulins, N.R. *Org. React.* **1975**, *22*, 1; Smith, G.G.; Kelly, F.W. *Prog. Phys. Org. Chem.* **1971**, *8*, 75, 153–201; DeWolfe, R.H., in Bamford, C.H.; Tipper, C.F.H. *Comprehensive Chemical Kinetics*, Vol. 9, Elsevier, NY, **1973**, pp. 455–461.

[595]Note that the same holds true for [1,*j*]-sigmatropic reactions of symmetrical substrates (**18-28, 18-29**).

[596]Levy, H.; Cope, A.C. *J. Am. Chem. Soc.* **1944**, *66*, 1684.

[597]For a review of the reverse Cope cyclization, see Cooper, N.J.; Knight, D.W. *Tetrahedron* **2004**, *60*, 243.

[598]For an exception, see Elmore, S.W.; Paquette, L.A. *Tetrahedron Lett.* **1991**, *32*, 319.

The reaction of 3-hydroxy-1,5-dienes is called the *oxy-Cope rearrangement*,[599] and has proved highly useful in synthesis.[600] The oxy-Cope rearrangement is greatly accelerated (by factors of 10^{10}–10^{17}) if the alkoxide is used rather than the alcohol (the *anionic oxy-Cope rearrangement*),[601] where the direct product is the enolate ion, which is hydrolyzed to the ketone. A metal free reaction using a phosphazene base has been reported.[602] The silyloxy-Cope rearrangement has proven to be quite useful.[603] An antibody-catalyzed oxy-Cope reaction is known,[604] and the mechanism and origins of catalysis for this reaction have been studied.[605] Sulfur substitution also leads to rate enhancement of the oxy-Cope rearrangement.[606] Note that 2-oxonia Cope rearrangements have been implicated in Prins cyclization reactions (**16-54**).[607]

aza-Cope rearrangements are also known.[608] In amino-Cope rearrangements, the solvent plays a role in the regioselectivity of the reaction.[609] It has been suggested that this latter reaction does not proceed solely by a concerted [3.3]-sigmatropic rearrangement.[610]

[599]Berson, J.A.; Walsh, Jr., E.J. *J. Am. Chem. Soc.* *1968*, *90*, 4729; Warrington, J.M.; Yap, G.P.A.; Barriault, L. *Org. Lett.* *2000*, *2*, 663; Ovaska, T.V.; Roses, J.B. *Org. Lett.* *2000*, *2*, 2361. For reviews, see Paquette, L.A. *Angew. Chem. Int. Ed.* *1990*, *29*, 609; Marvell, E.N.; Whalley, W., in Patai, S. *The Chemistry of the Hydroxyl Group*, pt. 2, Wiley, NY, *1971*, pp. 738–743.

[600]For a list of references, see Larock, R.C. *Comprehensive Organic Transformations*, 2nd ed., Wiley-VCH, NY, *1999*, pp. 1306–1307.

[601]Evans, D.A.; Nelson, J.V. *J. Am. Chem. Soc.* *1980*, *102*, 774; Miyashi, T.; Hazato, A.; Mukai, T. *J. Am. Chem. Soc.* *1978*, *100*, 1008; Paquette, L.A.; Pegg, N.A.; Toops, D.; Maynard, G.D.; Rogers, R.D. *J. Am. Chem. Soc.* *1990*, *112*, 277; Gajewski, J.J.; Gee, K.R. *J. Am. Chem. Soc.* *1991*, *113*, 967. See also, Wender, P.A.; Ternansky, R.J.; Sieburth, S.M. *Tetrahedron Lett.* *1985*, *26*, 4319. For a study of isomerization of the parent substrate in the gas phase, see Schulze, S.M.; Santella, N.; Grabowski, J.J.; Lee, J.K. *J. Org. Chem.* *2001*, *66*, 7247.

[602]Mamdani, H.T.; Hartley, R.C. *Tetrahedron Lett.* *2000*, *41*, 747.

[603]For a review, see Schneider, C. *Synlett* *2001*, 1079.

[604]Braisted, A.C.; Schultz, P.G. *J. Am. Chem. Soc.* *1994*, *116*, 2211.

[605]Black, K.A.; Leach, A.G.; Kalani, Y.S.; Houk, K.N. *J. Am. Chem. Soc.* *2004*, *126*, 9695.

[606]Paquette, L.A.; Reddy, Y.R.; Vayner, G.; Houk, K.N. *J. Am. Chem. Soc.* *2000*, *122*, 10788.

[607]See Rychnovsky, S.D.; Marumoto, S.; Jaber, J.J. *Org. Lett.* *2001*, *3*, 3815.

[608]Beholz, L.G.; Stille, J.R. *J. Org. Chem.* *1993*, *58*, 5095; Sprules, T.J.; Galpin, J.D.; Macdonald, D. *Tetrahedron Lett.* *1993*, *34*, 247; Cook, G.R.; Barta, N.S.; Stille, J.R. *J. Org. Chem.* *1992*, *57*, 461. See Yadav, J.S.; Reddy, B.V.S.; Rasheed, M.A.; Kumar, H.M.S. *Synlett* *2000*, 487.

[609]Dobson, H.K.; LeBlanc, R.; Perrier, H.; Stephenson, C.; Welch, T.R.; Macdonald, D. *Tetrahedron Lett.* *1999*, *40*, 3119.

[610]Allin, S.M.; Button, M.A.C. *Tetrahedron Lett.* *1999*, *40*, 3801.

The 1,5-diene system may be inside a ring or part of an allenic system[611] (this example illustrates both of these situations):[612]

but the reaction does not take place when one of the double bonds is part of an aromatic system (e.g., 4-phenyl-1-butene).[613] When the two double bonds are in vinylic groups attached to adjacent ring positions, the product is a ring four carbons larger. This has been applied to divinylcyclopropanes and divinylcyclobutanes:[614]

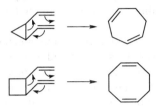

Indeed, *cis*-1,2-divinylcyclopropanes give this rearrangement so rapidly that they generally cannot be isolated at room temperature,[615] though exceptions are known.[616] When heated, 1,5-diynes are converted to 3,4-dimethylenecyclobutenes **125**.[617] A rate-determining Cope rearrangement is followed by a very rapid electrocyclic (**18-27**) reaction. The interconversion of 1,3,5-trienes and cyclohexadienes

[611]Duncan, J.A.; Azar, J.K.; Beatle, J.C.; Kennedy, S.R.; Wulf, C.M. *J. Am. Chem. Soc.* **1999**, *121*, 12029.

[612]Harris, Jr., J.F. *Tetrahedron Lett.* **1965**, 1359.

[613]See, for example, Lambert, J.B.; Fabricius, D.M.; Hoard, J.A. *J. Org. Chem.* **1979**, *44*, 1480; Marvell, E.N.; Almond, S.W. *Tetrahedron Lett.* **1979**, 2777, 2779; Newcomb, M.; Vieta, R.S. *J. Org. Chem.* **1980**, *45*, 4793. For exceptions in certain systems, see Doering, W. von E.; Bragole, R.A. *Tetrahedron* **1966**, *22*, 385; Jung, M.E.; Hudspeth, J.P. *J. Am. Chem. Soc.* **1978**, *100*, 4309; Yasuda, M.; Harano, K.; Kanematsu, K. *J. Org. Chem.* **1980**, *45*, 2368.

[614]Vogel, E.; Ott, K.H.; Gajek, K. *Liebigs Ann. Chem.* **1961**, *644*, 172. For reviews, see Wong, H.N.C.; Hon, M.; Tse, C.; Yip, Y.; Tanko, J.; Hudlicky, T. *Chem. Rev.* **1989**, *89*, 165, see pp. 172–174; Mil'vitskaya, E.M; Tarakanova, A.V.; Plate, A.F. *Russ. Chem. Rev.* **1976**, *45*, 469, see pp. 475–476.

[615]Unsubstituted *cis*-1,2-divinylcyclopropane is fairly stable at −20°C: Brown, J.M.; Golding, B.T.; Stofko, Jr., J.J. *J. Chem. Soc., Chem. Commun.* **1973**, 319; Schneider, M.P.; Rebell, J. *J. Chem. Soc., Chem. Commun.* **1975**, 283.

[616]See, for example, Brown, J.M. *Chem. Commun.* **1965**, 226; Schönleber, D. *Chem. Ber.* **1969**, *102*, 1789; Bolesov, I.G.; Ii-hsein, U.; Levina, R.Ya. *J. Org. Chem. USSR* **1970**, *6*, 1791; Schneider, M.P.; Rau, A. *J. Am. Chem. Soc.* **1979**, *101*, 4426.

[617]For reviews of Cope rearrangements involving triple bonds, see Viola, A.; Collins, J.J.; Filipp, N. *Tetrahedron* **1981**, *37*, 3765; Théron F.; Verny, M.; Vessière, R., in Patai, S. *The Chemistry of the Carbon–Carbon Triple Bond*, pt. 1, Wiley, NY, **1978**, pp. 381–445, pp. 428–430; Huntsman, W.D. *Intra-Sci. Chem. Rep.* **1972**, *6*, 151.

(in **18-27**) is very similar to the Cope rearrangement, but in **18-27**, the 3,4 bond goes from a double bond to a single bond rather than from a single bond to no bond.

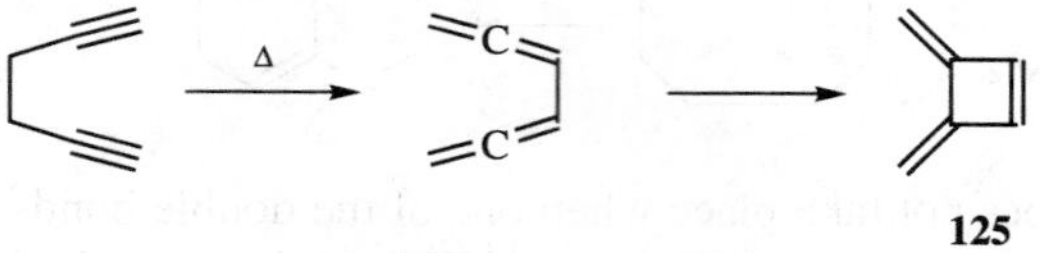

125

Like [2 + 2]-cycloadditions (p. 1220), Cope rearrangements of simple 1,5-dienes can be catalyzed by certain transition-metal compounds. For example, the addition of $PdCl_2(PhCN)_2$ causes the reaction to take place at room temperature.[618] This can be quite useful synthetically, because of the high temperatures required in the uncatalyzed process.

As we have indicated with our arrows, the mechanism of the uncatalyzed Cope rearrangement is a simple six-centered pericyclic process.[619] Since the mechanism is so simple, it has been possible to study some rather subtle points, among them the question of whether the six-membered transition state is in the boat or the chair form.[620] For the case of 3,4-dimethyl-1,5-hexadiene, it was demonstrated conclusively that the transition state is in the chair form. This was shown by the stereospecific nature of the reaction: The meso isomer gave the cis–trans product, while the (±) diastereomer gave the trans–trans diene.[621] If the transition state is in the chair form (taking the meso isomer, e.g.), one methyl must be "axial" and the other "equatorial" and the product must be the cis–trans alkene:

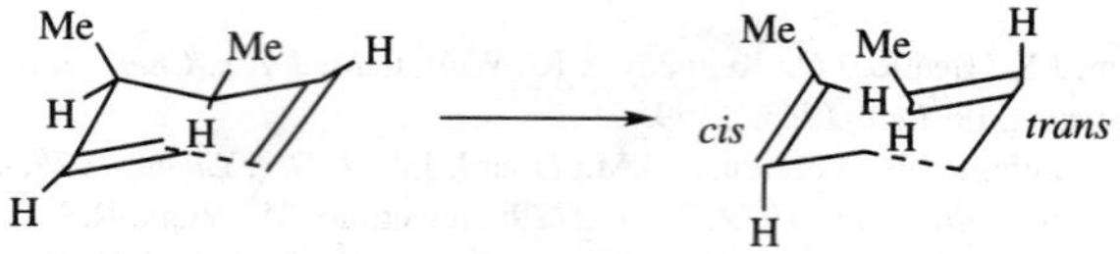

[618]Overman, L.E.; Knoll, F.M. *J. Am. Chem. Soc.* **1980**, *102*, 865; Hamilton, R.; Mitchell, T.R.B.; Rooney, J.J. *J. Chem. Soc., Chem. Commun.* **1981**, 456. For reviews of catalysis of Cope and Claisen rearrangements, see Overman, L.E. *Angew. Chem. Int. Ed.* **1984**, *23*, 579; Lutz, R.P. *Chem. Rev.* **1984**, *84*, 205. For a study of the mechanism, see Overman, L.E.; Renaldo, A.F. *J. Am. Chem. Soc.* **1990**, *112*, 3945.

[619]For a mechanistic discussion, see Poupko, R.; Zimmermann, H.; Müller, K.; Luz, Z. *J. Am. Chem. Soc.* **1996**, *118*, 7995.

[620]For a discussion showing a preference for the chair conformation, see Shea, K.J.; Stoddard, G.J.; England, W.P.; Haffner, C.D. *J. Am. Chem. Soc,* **1992**, *114*, 2635. See also, Tantillo, D.J.; Hoffmann, R. *J. Org. Chem.* **2002**, *67*, 1419.

[621]Doering, W. von E.; Roth, W.R.*Tetrahedron* **1962**, *18*, 67. See also, Hill, R.K.; Gilman, N.W. *Chem. Commun.* **1967**, 619; Goldstein, M.J.; DeCamp, M.R. *J. Am. Chem. Soc.* **1974**, *96*, 7356; Hansen, H.; Schmid, H. *Tetrahedron* **1974**, *30*, 1959; Gajewski, J.J.; Benner, C.W.; Hawkins, C.M. *J. Org. Chem.* **1987**, *52*, 5198; Paquette, L.A.; DeRussy, D.T.; Cottrell, C.E. *J. Am. Chem. Soc.* **1988**, *110*, 890.

There are two possible boat forms for the transition state of the meso isomer. One leads to a trans–trans product;

the other to a cis–cis alkene. For the ($\pm$) pair the predictions are just the opposite: There is just one boat form, and it leads to the cis–trans alkene, while one chair form ("diaxial" methyls) leads to the cis–cis product and the other ("diequatorial" methyls) predicts the trans–trans product. Thus the nature of the products obtained demonstrates that the transition state is a chair and not a boat.[622] While 3,4-dimethyl-1,5-hexadiene is free to assume either the chair or boat (it prefers the chair), other compounds are not so free. Thus 1,2-divinylcyclopropane (p. 1661) can react *only* in the boat form, demonstrating that such reactions are not impossible.[623]

Because of the nature of the transition state[624] in the pericyclic mechanism, optically active substrates with a stereogenic carbon at C-3 or C-4 transfer the chirality to the product (see p. 1673 for an example in the mechanistically similar Claisen rearrangement).[625] There are many examples of asymmetric [3,3]-sigmatropic rearrangements.[626]

[622]Preference for the chair transition state is a consequence of orbital-symmetry relationships: Hoffmann, R.; Woodward, R.B. *J. Am. Chem. Soc.* **1965**, *87*, 4389; Fukui, K.; Fujimoto, H. *Tetrahedron Lett.* **1966**, 251.

[623]For other examples of Cope rearrangements in the boat form, see Goldstein, M.J.; Benzon, M.S. *J. Am. Chem. Soc.* **1972**, *94*, 7147; Shea, K.J.; Phillips, R.B. *J. Am. Chem. Soc.* **1980**, *102*, 3156; Wiberg, K.B.; Matturro, M.; Adams, R. *J. Am. Chem. Soc.* **1981**, *103*, 1600; Gajewski, J.J.; Jiminez, J.L. *J. Am. Chem. Soc.* **1986**, *108*, 468.

[624]See Jiao, H.; Schleyer, P.v.R. *Angew. Chem. Int. Ed.* **1995**, *34*, 334; Özkan, I.; Zora, M. *J. Org. Chem.* **2003**, *68*, 9635.

[625]For a review of Cope and Claisen reactions as enantioselective syntheses, see Hill, R.K., in Morrison, J.D. *Asymmetric Synthesis*, Vol. 3, Academic Press, NY, **1984**, pp. 503–572, 503–545.

[626]For a review, see Nubbemeyer, U. *Synthesis* **2003**, 961.

Not all Cope rearrangements proceed by the cyclic six-centered mechanism.[627] Thus *cis*-1,2-divinylcyclobutane (p. 1661) rearranges smoothly to 1,5-cycloocta-diene, since the geometry is favorable. The trans isomer also gives this product, but the main product is 4-vinylcyclohexene (resulting from **18-31**). This reaction can be rationalized as proceeding by a diradical mechanism,[628] although it is pos-sible that at least part of the cyclooctadiene produced comes from a prior epimer-ization of the *trans*- to the *cis*-divinylcyclobutane followed by Cope rearrangement of the latter.[629]

It has been suggested that another type of diradical two-step mechanism may be preferred by some substrates.[630] Indeed, a nonconcerted Cope rearrangement has been reported.[631] In this pathway,[632] the 1,6 bond is formed before the 3,4 bond breaks:

This is related to the *Bergman cyclization* that was introduced in **18-27**.

It was pointed out earlier that a Cope rearrangement of the symmetrical 1,5-hexadiene gives 1,5-hexadiene. This is a *degenerate Cope rearrangement* (p. 1563). Another molecule that undergoes it is bicyclo[5.1.0]octadiene

126   **126**

[627]The diradical character of the Cope rearrangement transition state has been studied. See Staroverov, V.B.; Davidson, E.R. *J. Am. Chem. Soc.* **2000**, *122*, 186; Navarro-Vázquez, A.; Prall, M.; Schreiner, P.R. *Org. Lett.* **2004**, *6*, 2981.

[628]Hammond, G.S.; De Boer, C.D. *J. Am. Chem. Soc.* **1964**, *86*, 899; Trecker, D.J.; Henry, J.P. *J. Am. Chem. Soc.* **1964**, *86*, 902. Also see, Dolbier, Jr., W.R.; Mancini, G.J. *Tetrahedron Lett.* **1975**, 2141; Kessler, H.; Ott, W. *J. Am. Chem. Soc.* **1976**, *98*, 5014. For a discussion of diradical mechanisms in Cope rearrangements, see Berson, J.A., in de Mayo, P. *Rearrangements in Ground and Excited States*, Academic Press, NY, **1980**, pp. 358–372.

[629]See, for example, Berson, J.A.; Dervan, P.B. *J. Am. Chem. Soc.* **1972**, *94*, 8949; Baldwin, J.E.; Gilbert, K.E. *J. Am. Chem. Soc.* **1976**, *98*, 8283. For a similar result in the 1,2-divinylcyclopropane series, see Baldwin, J.E.; Ullenius, C. *J. Am. Chem. Soc.* **1984**, *96*, 1542.

[630]Doering, W. von E.; Toscano, V.G.; Beasley, G.H. *Tetrahedron* **1971**, *27*, 5299; Dewar, M.J.S.; Wade, Jr., L.E. *J. Am. Chem. Soc.* **1977**, *99*, 4417; Padwa, A.; Blacklock, T.J. *J. Am. Chem. Soc.* **1980**, *102*, 2797; Dollinger, M.; Henning, W.; Kirmse, W. *Chem. Ber.* **1982**, *115*, 2309; Kaufmann, D.; de Meijere, A. *Chem. Ber.* **1984**, *117*, 1128; Dewar, M.J.S.; Jie, C. *J. Am. Chem. Soc.* **1987**, *109*, 5893; *J. Chem. Soc., Chem. Commun.* **1989**, 98. For evidence against this view, see Gajewski, J.J. *Acc. Chem. Res.* **1980**, *13*, 142; Morokuma, K.; Borden, W.T.; Hrovat, D.A. *J. Am. Chem. Soc.* **1988**, *110*, 4474; Halevi, E.A.; Rom, R. *Isr. J. Chem.* **1989**, *29*, 311; Owens, K.A.; Berson, J.A. *J. Am. Chem. Soc.* **1990**, *112*, 5973.

[631]Roth, W.R.; Gleiter, R.; Paschmann, V.; Hackler, U.E.; Fritzsche, G.; Lange, H. *Eur. J. Org. Chem.* **1998**, 961; Roth, W.R.; Schaffers, T.; Heiber, M. *Chem. Ber.* **1992**, *125*, 739.

[632]For a report of still another mechanism, featuring a diionic variant of the diradical, see Gompper, R.; Ulrich, W. *Angew. Chem. Int. Ed.* **1976**, *15*, 299.

(**126**).[633] At room temperature, the NMR spectrum of this compound is in accord with the structure shown on the left. At 180°C, it is converted by a Cope reaction to a compound equivalent to itself. The interesting thing is that at 180°C the NMR spectrum shows that what exists is an equilibrium mixture of the two structures. That is, at this temperature the molecule rapidly (faster than 10^3 times per second) changes back and forth between the two structures. This is called *valence tautomerism* and is quite distinct from resonance, even though only electrons shift.[634] The positions of the nuclei are not the same in the two structures. Molecules like **126** that exhibit valence tautomerism (in this case, at 180°C) are said to have *fluxional* structures. It may be recalled that *cis*-1,2- divinylcyclopropane does not exist at room temperature because it rapidly rearranges to 1,4-cycloheptadiene (p. 1661), but in **126** the *cis*-divinylcyclopropane structure is frozen into the molecule in both structures. Several other compounds with this structural feature are also known. Of these, *bullvalene* (**127**) is especially interesting.

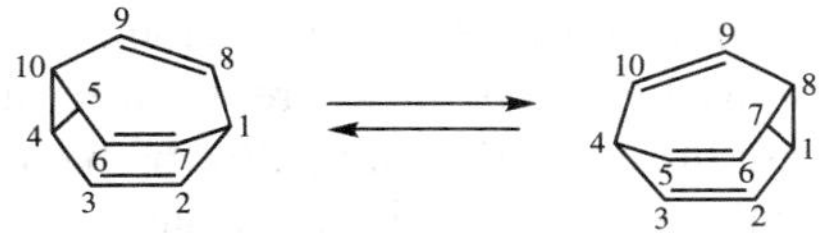

127

The Cope rearrangement shown changes the position of the cyclopropane ring from 4,5,10 to 1,7,8. But the molecule could also have undergone rearrangements to put this ring at 1,2,8 or 1,2,7. Any of these could then undergo several Cope rearrangements. In all, there are $\frac{10!}{3}$ or >1.2 million tautomeric forms, and the cyclopropane ring can be at any three carbons that are adjacent. Since each of these tautomers is equivalent to all the others, this has been called an infinitely degenerate Cope rearrangement. Bullvalene has been synthesized and its ^{1}H NMR spectrum determined.[635] At −25°C, there are two peaks with an area ratio of 6:4. This is in accord with a single non-tautomeric structure. The six are the vinylic protons and the four are the allylic ones. But at 100°C the compound shows only one NMR peak, indicating that we have here a truly unusual situation where the compound rapidly interchanges its structure among 1.2 million equivalent forms.[636] The ^{13}C NMR spectrum of bullvalene also shows

[633]Doering, W. von E.; Roth, W.R. *Tetrahedron* **1963**, *19*, 715.

[634]For reviews of valence tautomerizations, see Decock-Le Révérend, B.; Goudmand, P. *Bull. Soc. Chim. Fr. 1973*, 389; Gajewski, J.J. *Mech. Mol. Migr. 1971*, *4*, 1, see pp. 32–49; Paquette, L.A. *Angew. Chem. Int. Ed. 1971*, *10*, 11; Domareva-Mandel'shtam, T.V.; D'yakonov, I.A. *Russ. Chem. Rev. 1966*, *35*, 559, 568; Schröder, G.; Oth, J.F.M.; Merényi, R. *Angew. Chem. Int. Ed. 1965*, *4*, 752.

[635]Schröder, G. *Chem. Ber. 1964*, *97*, 3140; Merényi, R.; Oth, J.F.M.; Schröder, G. *Chem. Ber. 1964*, *97*, 3150. For a review of bullvalenes, see Schröder, G.; Oth, J.F.M. *Angew. Chem. Int. Ed. 1967*, *6*, 414.

[636]A number of azabullvalenes (**127** containing heterocyclic nitrogen) have been synthesized. They also have fluxional structures when heated, though with fewer tautomeric forms than bullvalene itself: Paquette, L.A.; Malpass, J.R.; Krow, G.R.; Barton, T.J. *J. Am. Chem. Soc. 1969*, *91*, 5296.

only one peak at 100°C.[637]

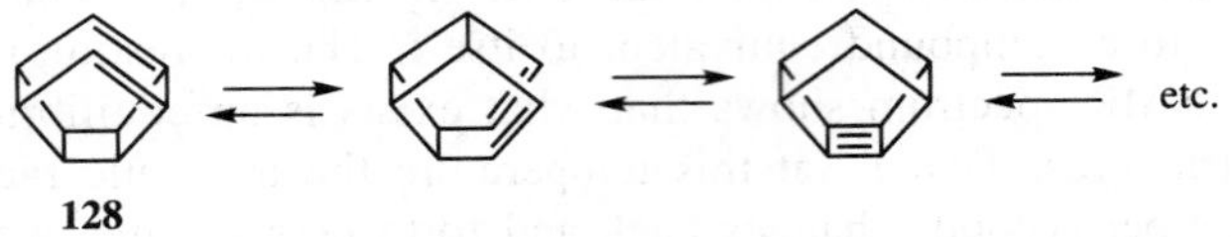

128

Another compound for which degenerate Cope rearrangements result in equivalence for all the carbons is *hypostrophene* (**128**).[638] In the case of the compound *barbaralane* (**129**)[639] (bullvalene in which one CH=CH has been replaced by a CH$_2$):

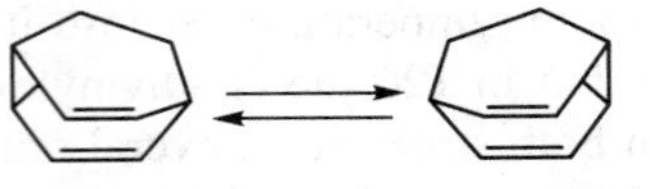

129 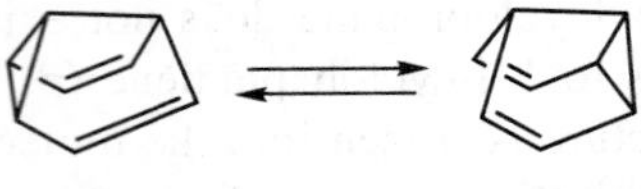   **130**

there are only 2 equivalent tautomers.[640] However, NMR spectra indicate that even at room temperature a rapid interchange of both tautomers is present, although by about −100°C this has slowed to the point where the spectrum is in accord with a single structure. In the case of *semibullvalene* (**130**) (barbaralane in which the CH$_2$ has been removed), not only is there a rapid interchange at room temperature, but even at −110°C.[641] Compound **130** has the lowest energy barrier of any known compound capable of undergoing the Cope rearrangement.[642]

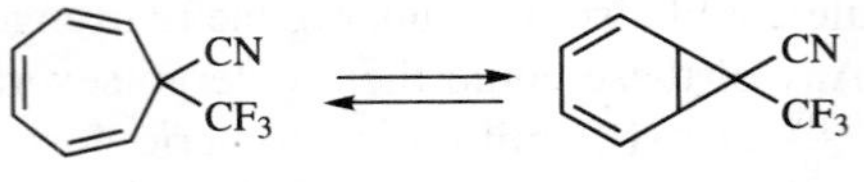

131   **132** 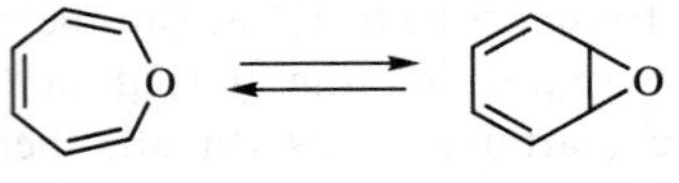   Oxepin   Benzene oxide

[637]Oth, J.F.M.; Müllen, K.; Gilles, J.; Schröder, G. *Helv. Chim. Acta* **1974**, *57*, 1415; Nakanishi, H.; Yamamoto, O. *Tetrahedron Lett.* **1974**, 1803; Günther, H.; Ulmen, J. *Tetrahedron* **1974**, *30*, 3781. For deuterium nmr spectra, see Poupko, R.; Zimmermann, H.; Luz, Z. *J. Am. Chem. Soc.* **1984**, *106*, 5391. For a crystal structure study, see Luger, P.; Buschmann, J.; McMullan, R.K.; Ruble, J.R.; Matias, P.; Jeffrey, G.A. *J. Am. Chem. Soc.* **1986**, *108*, 7825.

[638]McKennis, J.S.; Brener, L.; Ward, J.S.; Pettit, R. *J. Am. Chem. Soc.* **1971**, *93*, 4957; Paquette, L.A.; Davis, R.F.; James, D.R. *Tetrahedron Lett.* **1974**, 1615.

[639]For a study of sigmatropic shiftamers in extended barbaralanes, see Tantillo, D.J.; Hoffmann, R.; Houk, K.N.; Warner, P.M.; Brown, E.C.; Henze, D.K. *J. Am. Chem. Soc.* **2004**, *126*, 4256.

[640]Barbaralane was synthesized by Biethan, U.; Klusacek, H.; Musso, H. *Angew. Chem. Int. Ed.* **1967**, *6*, 176; by Tsuruta, H.; Kurabayashi, K.; Mukai, T. *Tetrahedron Lett.* **1965**, 3775; by Doering, W. von E.; Ferrier, B.M.; Fossel, E.T.; Hartenstein, J.H.; Jones Jr., M.; Klumpp, G.W.; Rubin, R.M.; Saunders, M. *Tetrahedron* **1967**, *23*, 3943; and by Henkel, J.G.; Hane, J.T. *J. Org. Chem.* **1983**, *48*, 3858.

[641]Meinwald, J.; Schmidt, D. *J. Am. Chem. Soc.* **1969**, *91*, 5877; Zimmerman, H.E.; Binkley, R.W.; Givens, R.S.; Grunewald, G.L.; Sherwin, M.A. *J. Am. Chem. Soc.* **1969**, *91*, 3316.

[642]Cheng, A.K.; Anet, F.A.L.; Mioduski, J.; Meinwald, J. *J. Am. Chem. Soc.* **1974**, *96*, 2887; Moskau, D.; Aydin, R.; Leber, W.; Günther, H.; Quast, H.; Martin, H.-D.; Hassenrück, K.; Miller, L.S.; Grohmann, K. *Chem. Ber.* **1989**, *122*, 925. For a discussion concerning whether or not semibullvalenes are homoaromatic, see Williams, R.V.; Gadgil, V.R.; Chauhan, K.; Jackman, L.M.; Fernandes, E. *J. Org. Chem.* **1998**, *63*, 3302.

The molecules taking part in a valence tautomerization need not be equivalent. Thus, NMR spectra indicate that a true valence tautomerization exists at room temperature between the cycloheptatriene **131** and the norcaradiene **132**.[643] In this case, one isomer (**132**) has the *cis*-1,2-divinylcyclopropane structure, while the other does not. In an analogous interconversion, benzene oxide[644] and oxepin exist in a tautomeric equilibrium at room temperature.[645]

Bullvalene and hypostrophene are members of a group of compounds all of whose formulas can be expressed by the symbol $(CH)_{10}$.[646] Many other members of this group are known. Similar groups of $(CH)_n$ compounds exist for other even-numbered values of "n".[646] For example, there are 20 possible $(CH)_8$[647] compounds,[648] and five possible $(CH)_6$ compounds,[649] all of which are known: benzene, prismane (p. 220), Dewar benzene (p. 1641), bicyclopropenyl,[650] and benzvalene.[651]

An interesting example of a valence tautomerism is the case of 1,2,3-tri-*tert*-butylcyclobutadiene (p. 74). There are two isomers, both rectangular, and ^{13}C NMR spectra show that they exist in a dynamic equilibrium, even at $-185°C$.[652]

[643]Ciganek, E. *J. Am. Chem. Soc.* **1965**, 87, 1149. For other examples of norcaradiene–cycloheptatriene valence tautomerizations, see Görlitz, M.; Günther, H. *Tetrahedron* **1969**, 25, 4467; Ciganek, E. *J. Am. Chem. Soc.* **1965**, 93, 2207; Dürr.; Kober, H. *Chem. Ber.* **1973**, 106, 1565; Betz, W.; Daub, J. *Chem. Ber.* **1974**, 107, 2095; Maas, G.; Regitz, M. *Chem. Ber.* **1976**, 109, 2039; Warner, P.M.; Lu, S. *J. Am. Chem. Soc.* **1980**, 102, 331; Neidlein, R.; Radke, C.M. *Helv. Chim. Acta* **1983**, 66, 2626; Takeuchi, K.; Kitagawa, T.; Ueda, A.; Senzaki, Y.; Okamoto, K. *Tetrahedron* **1985**, 41, 5455.

[644]For a review of arene oxides, see Shirwaiker, G.S.; Bhatt, M.V. *Adv. Heterocycl. Chem.* **1984**, 37, 67.

[645]For reviews, see Maier, G. *Angew. Chem. Int. Ed.* **1967**, 6, 402; Vogel, E.; Günther, H. *Angew. Chem. Int. Ed.* **1967**, 6, 385; Vogel, E. *Pure Appl. Chem.* **1969**, 20, 237. See also, Boyd, D.R.; Stubbs, M.E. *J. Am. Chem. Soc.* **1983**, 105, 2554.

[646]For reviews of rearrangements and interconversions of $(CH)_n$ compounds, see Balaban, A.T.; Banciu, M. *J. Chem. Educ.* **1984**, 61, 766; Greenberg, A.; Liebman, J.F. *Strained Organic Molecules*, Academic Press, NY, **1978**, pp. 203–215; Scott, L.T.; Jones, Jr., M. *Chem. Rev.* **1972**, 72, 181. See also, Maier, G.; Wiegand, N.H.; Baum, S.; Wüllner, R. *Chem. Ber.* **1989**, 122, 781.

[647]For a review of strain in $(CH)_8$ compounds, see Hassenrück, K.; Martin, H.; Walsh, R. *Chem. Rev.* **1989**, 89, 1125.

[648]The structures of all possible $(CH)_n$ compounds, for $n = 4, 6, 8,$ and 10, are shown in Balaban, A.T; Banziu, M. *J. Chem. Educ.* **1984**, 61, 766. For a review of $(CH)_{12}$ compounds, see Banciu, M.; Popa, C.; Balaban, A.T. *Chem. Scr.*, **1984**, 24, 28.

[649]For reviews of valence isomers of benzene and some related compounds, see Kobayashi, Y.; Kumadaki, I. *Top. Curr. Chem.* **1984**, 123, 103; Bickelhaupt, F.; de Wolf, W.H. *Recl. Trav. Chim. Pays-Bas* **1988**, 107, 459.

[650]For a study of how this compound isomerizes to benzene, see Davis, J.H.; Shea, K.J.; Bergman, R.G. *J. Am. Chem. Soc.* **1977**, 99, 1499.

[651]For reviews of benzvalenes, see Christl, M. *Angew. Chem. Int. Ed.* **1981**, 20, 529; Burger, U. *Chimia*, **1979**, 147.

[652]Maier, G.; Kalinowski, H.; Euler, K. *Angew. Chem. Int. Ed.* **1982**, 21, 693.

18-33 The Claisen Rearrangement[653]

Allylic aryl ethers, when heated, rearrange to *o*-allylphenols in a reaction called the *Claisen rearrangement*.[654] If both ortho positions are filled, the allylic group migrates to the para position (this is often called the *para-Claisen rearrangement*).[655] There is no reaction when the para and both ortho positions are filled. Migration to the meta position has not been observed. In the ortho migration, the allylic group always undergoes an allylic shift. That is, as shown above, a substituent α to the oxygen is now γ to the ring (and vice versa). On the other hand, in the para migration there is never an allylic shift: The allylic group is found exactly as it was in the original ether. Compounds with propargylic groups (i.e., groups with a triple bond in the appropriate position) do not generally give the corresponding products.

The mechanism is a concerted pericyclic [3,3]-sigmatropic rearrangement[656] and accounts for all these facts. For the ortho rearrangement:

Evidence is the lack of a catalyst, the fact that the reaction is first order in the ether, the absence of crossover products when mixtures are heated, and the presence of the allylic shift, which is required by this mechanism. A *retro*-Claisen rearrangement is known and its mechanism has been examined.[657] The allylic

[653]For a reiview of the Claisen rearrangment since about 1910, see Castro, A.M.M. *Chem. Rev.* **2004**, *104*, 2939.

[654]For reviews, see Fleming, I. *Pericyclic Reactions*, Oxford University Press, Oxford, **1999**, pp. 71–83; Moody, C.J. *Adv. Heterocycl. Chem.* **1987**, *42*, 203; Bartlett, P.A. *Tetrahedron* **1980**, *36*, 2, see pp. 28–39; Ziegler, F.E. *Acc. Chem. Res.* **1977**, *10*, 227; Bennett, G.B. *Synthesis* **1977**, 589; Rhoads, S.J.; Raulins, N.R. *Org. React.* **1975**, *22*, 1; Shine, H.J. *Aromatic Rearrangements*; Elsevier, NY, **1969**, pp. 89–120; Smith, G.G.; Kelly, F.W. *Prog. Phys. Org. Chem.* **1971**, *8*, 75, 153–201; Hansen, H.; Schmid, H. *Chimia*, **1970**, *24*, 89, *Chem. Br.* **1969**, *5*, 111; Jefferson, A.; Scheinmann, F. *Q. Rev. Chem. Soc.* **1968**, *22*, 391; Thyagarajan, B.S. *Adv. Heterocycl. Chem.* **1967**, *8*, 143; Dalrymplem D.L.; Kruger, T.L.; White, W.N., in Patai *The Chemistry of the Ether Linkage*, Wiley, NY, **1967**, pp. 635–660.

[655]For a discussion of regioselectivity, see Gozzo, F.C.; Fernandes, S.A.; Rodrigues, D.C.; Eberlin, M.N.; Marsaioli, A.J. *J. Org. Chem.* **2003**, *68*, 5493.

[656]For isotope effect evidence regarding the nature of the concerted transition state, see McMichael, K.D.; Korver, G.L. *J. Am. Chem. Soc.* **1979**, *101*, 2746; Gajewski, J.J.; Conrad, N.D. *J. Am. Chem. Soc.* **1979**, *101*, 2747; Kupczyk-Subotkowska, L.; Saunders, Jr., W.H.; Shine, H.J. *J. Am. Chem. Soc.* **1988**, *110*, 7153.

[657]Boeckman, Jr., R.K.; Shair, M.D.; Vargas, J.R.; Stolz, L.A. *J. Org. Chem.* **1993**, *58*, 1295.

shift for the ortho rearrangement (and the absence of one for the para) has been demonstrated by ^{14}C labeling, even when no substituents are present. Studies of the transition-state geometry have shown that, like the Cope rearrangement, the Claisen rearrangement usually prefers a chair-like transition state.[658] When the ortho positions have no hydrogen, a second [3,3]-sigmatropic migration (a Cope reaction) follows:

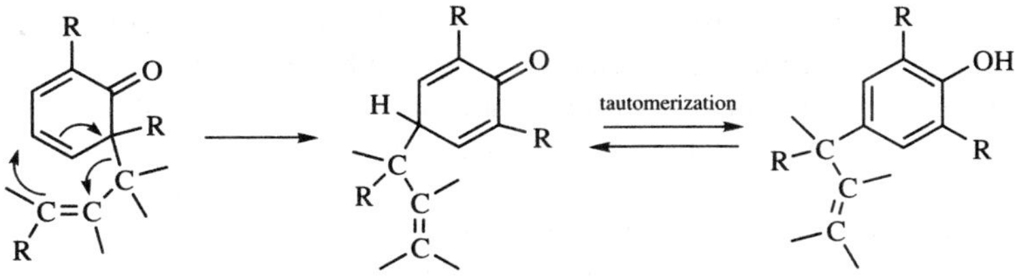

133

and the migrating group is restored to its original structure. Intermediates of structure **133** have been trapped by means of a Diels–Alder reaction.[659] The rearrangement of aryl allyl ethers is facilitated by Ag–KI in hot acetic acid,[660] and by AlMe$_3$ in water.[661] A solid-phase reaction of polymer-bound substrate undergoes the Claisen rearrangement with microwave irradiation.[662]

   Allylic ethers of enols (allylic vinylic ethers) also undergo the Claisen rearrangement;[663] in fact, it was discovered with these compounds first:[664]

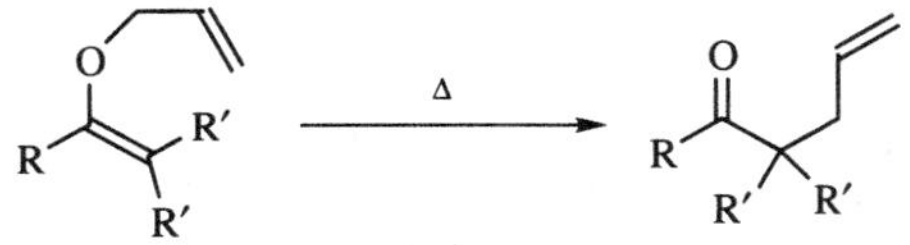

In these cases of course, the final tautomerization does not take place even when R' = H, since there is no aromaticity to restore, and ketones are more stable than enols.[665] Catalytic Claisen rearrangements of allyl vinyl ethers are well known.[666]

[658]Wunderli, A.; Winkler, T.; Hansen, H. *Helv. Chim. Acta* **1977**, *60*, 2436; Copley, S.D.; Knowles, J.R. *J. Am. Chem. Soc.* **1985**, *107*, 5306. Also see, Yoo, H.Y.; Houk, K.N. *J. Am. Chem. Soc.* **1994**, *116*, 12047; Kupczyk-Subotkowska, L.; Saunders, Jr., W.H.; Shine, H.J.; Subotkowski, W. *J. Am. Chem. Soc.* **1993**, *115*, 5957; Kupczyk-Subotkowska, L.; Subotkowski, W.; Saunders, Jr., W.H.; Shine, H.J.; *J. Am. Chem. Soc.* **1992**, *114*, 3441.
[659]Conroy, H.; Firestone, R.A. *J. Am. Chem. Soc.* **1956**, *78*, 2290.
[660]Sharghi, H.; Aghapour, G. *J. Org. Chem.* **2000**, *65*, 2813.
[661]Wipf, P.; Ribe, S. *Org. Lett.* **2001**, *3*, 1503.
[662]Kumar, H.M.S.; Anjaneyulu, S.; Reddy, B.V.S.; Yadav, J.C. *Synlett* **2000**, 1129.
[663]For a review, see Ziegler, F.E. *Chem. Rev.* **1988**, *88*, 1423.
[664]Claisen, L. *Berchte.* **1912**, *45*, 3157.
[665]However, it has proved possible to reverse the reaction, with a Lewis acid catalyst. See Boeckman Jr., R.K.; Flann, C.J.; Poss, K.M. *J. Am. Chem. Soc.* **1985**, *107*, 4359.
[666]For a review, see Hiersemann, M.; Abraham, L. *Eur. J. Org. Chem.* **2002**, 1461.

The use of water as solvent accelerates the reaction.[667] A microwave induced reaction on silica gel is known[668]. The mechanism is similar to that with allylic aryl ethers.[669] Allyl allene ethers undergo a Claisen rearrangement when heated in DMF to give the expected diene with a conjugated aldehyde unit.[670] Butenolides with a β-allylic ether unit undergo Claisen rearrangement–Conia reaction[671] cascade to give an oxaspiro heptane with β-keto lactone comprising the five-membered ring.[672] Allylic esters of β-keto acids undergo a Claisen rearrangement in what is known as the *Carroll rearrangement*[673] (also called the *Kimel–Cope rearrangement*[674]), and the reaction can be catalyzed by a ruthenium complex.[675]

It is possible to treat ketones with allyl alcohol and an acid catalyst to give γ,δ-unsaturated ketones directly, presumably by initial formation of the vinylic ethers, and then Claisen rearrangement.[676] In an analogous procedure, the enolates (**134**) of allylic esters [formed by treatment of the esters with lithium isopropylcyclohexylamide (LICA)] rearrange to γ,δ-unsaturated acids.[677] Allylic alcohols can be treated with a catalytic amount of mercuric acetate, and in the presence of an excess of allyl vinyl ethers give an alkene–aldehyde via a Claisen rearrangement.[678]

134

[667]Grieco, P.A.; Brandes, E.B.; McCann, S.; Clark, J.D. *J. Org. Chem.* **1989**, *54*, 5849. The effect of water on the transition state has been examined; see Guest, J.M.; Craw, J.S.; Vincent, M.A.; Hillier, I.H. *J. Chem. Soc. Perkin Trans. 2* **1997**, 71; Sehgal, A.; Shao, L.; Gao, J. *J. Am. Chem. Soc.* **1995**, *117*, 11337.

[668]Kotha, S.; Mandal, K.; Deb, A.C.; Banerjee, S. *Tetrahedron Lett.* **2004**, *45*, 9603.

[669]For discussions of the transition state, see Gajewski, J.J.; Jurayj, J.; Kimbrough, D.R.; Gande, M.E.; Ganem, B.; Carpenter, B.K. *J. Am. Chem. Soc.* **1987**, *109*, 1170. For MO calculations, see Vance, R.L.; Rondan, N.G.; Houk, K.N.; Jensen, F.; Borden, W.T.; Komornicki, A.; Wimmer, E. *J. Am. Chem. Soc.* **1988**, *110*, 2314; Dewar, M.J.S.; Jie, C. *J. Am. Chem. Soc.* **1989**, *111*, 511.

[670]Parsons, P.J.; Thomson, P.; Taylor, A.; Sparks, T. *Org. Lett.* **2000**, *2*, 571.

[671]For a review of the Conia-ene reaction, see Conia, J.M.; Le Perchec, P. *Synthesis* **1975**, 1.

[672]Schobert, R.; Siegfried, S.; Gordon, G.; Nieuwenhuyzen, M.; Allenmark, S. *Eur. J. Org. Chem.* **2001**, 1951.

[673]Carroll, M.F. *J. Chem. Soc.* **1940**, 704, 1266; Carroll, M.F. *J. Chem. Soc.* **1941**, 507; Ziegler, F.E. *Chem. Rev.* **1988**, *88*, 1423.

[674]Kimel, W.; Cope, A.C. *J. Am. Chem. Soc.* **1943**, *65*, 1992.

[675]Burger, E.C.; Tunge, J.A. *Org. Lett.* **2004**, *6*, 2603.

[676]Lorette, N.B. *J. Org. Chem.* **1961**, *26*, 4855. See also, Saucy, G.; Marbet, R. *Helv. Chim. Acta* **1967**, *50*, 2091; Marbet, R.; Saucy, G. *Helv. Chim. Acta* **1967**, *50*, 2095; Thomas, A.F. *J. Am. Chem. Soc.* **1969**, *91*, 3281; Johnson, W.S.; Werthemann, L.; Bartlett, W.R.; Brocksom, T.J.; Li, T.; Faulkner, D.J.; Petersen, M.R. *J. Am. Chem. Soc.* **1970**, *92*, 741; Pitteloud, R.; Petrzilka, M. *Helv. Chim. Acta* **1979**, *62*, 1319; Daub, G.W.; Sanchez, M.G.; Cromer, R.A.; Gibson, L.L. *J. Org. Chem.* **1982**, *47*, 743; Bartlett, P.A.; Tanzella, D.J.; Barstow, J.F. *J. Org. Chem.* **1982**, *47*, 3941.

[677]Ireland, R.E.; Mueller, R.H.; Willard, A.K. *J. Am. Chem. Soc.* **1976**, *98*, 2868; Gajewski, J.J.; Emrani, J. *J. Am. Chem. Soc.* **1984**, *106*, 5733; Cameron A.G.; Knight, D.W. *J. Chem. Soc. Perkin Trans. 1* **1986**, 161. See also, Wilcox, C.S.; Babston, R.E. *J. Am. Chem. Soc.* **1986**, *108*, 6636.

[678]Tokuyama, H.; Makido, T.; Ueda, T.; Fukuyama, T. *Synth. Commun.* **2002**, *32*, 869.

Alternatively, the silylketene acetal $R^3R^2C=C(OSiR_3)OCH_2CH=CHR^1$ is often used instead of **134**.[677,679] This rearrangement also proceeds at room temperature. By either procedure, the reaction is called the *Ireland–Claisen rearrangement*.[680] Note the presence of the negative charge in **134**. As with the oxy-Cope rearrangement (in **18-34**), negative charges generally accelerate the Claisen reaction,[681] although the extent of the acceleration can depend on the identity of the positive counterion.[682] The reaction proceeds with good syn selectivity in many cases.[683] The Ireland–Claisen rearrangement has been made enantioselective by converting **134** to an enol borinate in which the boron is attached to a chiral group.[684] The Ireland–Claisen rearrangement can be done with amide derivatives also.[685]

A number of expected analogs of the Claisen rearrangement are known, for example, rearrangement of $ArNHCH_2CH=CH_2$,[686] of *N*-allylic enamines $(R_2C=CRNRCR_2CR=CR_2)$,[687] of allylic imino esters, $RC(OCH_2CH=CH_2)=NR$[688] (these have often been rearranged with transition-metal catalysts[689]), and of $RCH=NRCHRCH_2CH=CH_2$. These rearrangements of nitrogen-containing compounds can be called *aza-Claisen rearrangements*,[690] but are often called

[679]Ireland, R.E.; Wipf, P.; Armstrong III, J.D. *J. Org. Chem.* **1991**, *56*, 650.

[680]For a recent example, see Dell, C.P.; Khan, K.M.; Knight, D.W. *J. Chem. Soc. Perkin Trans. 1* **1994**, 341. For a review, see Chai, Y.; Hong, S.-p.; Lindsay, H.A.; McFarland, C.; McIntosh, M.C. *Tetrahedron* **2002**, *58*, 2905.

[681]See, for example, Denmark, S.E.; Harmata, M.A.; White, K.S. *J. Am. Chem. Soc.* **1989**, *111*, 8878.

[682]Koreeda, M.; Luengo, J.I. *J. Am. Chem. Soc.* **1985**, *107*, 5572; Kirchner, J.J.; Pratt, D.V.; Hopkins, P.B. *Tetrahedron Lett.* **1988**, *29*, 4229.

[683]Mohamed, M.; Brook, M.A. *Tetrahedron Lett.* **2001**, *42*, 191. For a discussion of boat or chair preferences, see Khaledy, M.M.; Kalani, M.Y.S.; Khuong, K.S.; Houk, K.N.; Aviyente, V.; Neier, R.; Soldermann, N.; Velker, J. *J. Org. Chem.* **2003**, *68*, 572.

[684]Corey, E.J.; Lee, D. *J. Am. Chem. Soc.* **1991**, *113*, 4026.

[685]Tsunoda, T.; Tatsuki, S.; Shiraishi, Y.; Akasaka, M.; Itô, S. *Tetrahedron Lett.* **1993**, *34*, 3297. Also see, Walters, M.A.; Hoem, A.B.; Arcand, H.R.; Hegeman, A.D.; McDonough, C.S. *Tetrahedron Lett.* **1993**, *34*, 1453.

[686]Marcinkiewicz, S.; Green, J.; Mamalis, P. *Tetrahedron* **1961**, *14*, 208; Inada, S.; Ikado, S.; Okazaki, M. *Chem. Lett.* **1973**, 1213; Schmid, M.; Hansen, H.; Schmid, H. *Helv. Chim. Acta* **1973**, *56*, 105; Jolidon, S.; Hansen, H. *Helv. Chim. Acta* **1977**, *60*, 978.

[687]Ficini, J.; Barbara, C. *Tetrahedron Lett.* **1966**, 6425; Ireland, R.E.; Willard, A.K. *J. Org. Chem.* **1974**, *39*, 421; Hill, R.K.; Khatri, H.N. *Tetrahedron Lett.* **1978**, 4337; Anderson, J.C.; Flaherty, A.; Swarbrick, M.E. *J. Org. Chem.* **2000**, *65*, 9152. For the reverse of this rearrangement, see Wu, P.; Fowler, F.W. *J. Org. Chem.* **1988**, *53*, 5998.

[688]For examples, see Synerholm, M.E.; Gilman, N.W.; Morgan, J.W.; Hill, R.K. *J. Org. Chem.* **1968**, *33*, 1111; Black, D.S.; Eastwood, F.W.; Okraglik, R.; Poynton, A.J.; Wade, A.M.; Welker, C.H. *Aust. J. Chem.* **1972**, *25*, 1483; Overman, L.E. *J. Am. Chem. Soc.* **1974**, *96*, 597; Metz, P.; Mues, C. *Tetrahedron* **1988**, *44*, 6841. See Gradl, S.N.; Kennedy-Smith, J.J.; Kim, J.; Trauner, D. *Synlett* **2002**, 411.

[689]See Schenck, T.G.; Bosnich, B. *J. Am. Chem. Soc.* **1985**, *107*, 2058, and references cited therein. Palladium catalyzed: Jiang, Y.; Longmire, J.M.; Zhang, X. *Tetrahedron Lett.* **1999**, *40*, 1449; Donde, Y.; Overman, L.E. *J. Am. Chem. Soc.* **1999**, *121*, 2933; Anderson, C.E.; Overman, L.E. *J. Am. Chem. Soc.* **2003**, *125*, 12412.

[690]See Majumdar, K.C.; Samanta, S.K. *Tetrahedron* **2001**, *57*, 4955; Kirsch, S.F.; Overman, L.F.; Watson, M.P. *J. Org. Chem.* **2004**, *69*, 8101.

aza-Cope rearrangements[691] as described in **18-34**. However, a palladium catalyzed aza-Claisen has been reported.[692] A so-called amine-Claisen rearrangement was reported for *N*-allyl indoles, when heated in the presence of $BF_3\cdot OEt_2$.[693] An *azo-Cope* rearrangement: $CH_2{=}CHCR_2^1$ CR_2^2 $N{=}NAr \rightarrow R_2^1$ $C{=}CHCH_2NArN{=}CR_2^2$ has been reported.[694] Propargylic vinylic compounds give allenic aldehydes, ketones, esters, or amides:[695]

The conversion of allylic aryl thioethers $ArSCH_2CH{=}CH_2$ to *o*-allylic thiophenols is not feasible, because the latter are not stable,[696] but react to give bicyclic compounds.[697] However, many allylic vinylic sulfides do give the rearrangement (the *thio-Claisen rearrangement*).[698] Allylic vinylic sulfones, for example, $H_2C{=}CRCH_2{-}SO_2{-}CH{=}CH_2$, rearrange, when heated in the presence of ethanol and pyridine, to unsaturated sulfonate salts $CH_2{=}CRCH_2CH_2CH_2SO_3^-$, produced by reaction of the reagents with the unstable sulfene intermediates $CH_2{=}CRCH_2CH_2CH{=}SO_2$.[699] Allylic vinylic sulfoxides rapidly rearrange at room temperature or below.[700]

As mentioned for the Ireland–Claisen rearrangement, asymmetric Claisen rearrangement reactions are well known.[701] Chiral Lewis acids have been designed for

[691]For a review, see Przheval'skii, N.M.; Grandberg, I.I. *Russ. Chem. Rev.* **1987**, *56*, 477. For reviews of [3,3]-sigmatropic rearrangements with heteroatoms present, see Blechert, S. *Synthesis* **1989**, 71; Winterfeldt, E. *Fortschr. Chem. Forsch.* **1970**, *16*, 75. For a review of [3,3]-rearrangements of iminium salts, see Heimgartner, H.; Hansen, H.; Schmid, H. *Adv. Org. Chem.* **1979**, *9*, pt. 2, 655.

[692]Uozumi, Y.; Kato, K.; Hayashi, T. *Tetrahedron Asymmetry*, **1998**, *9*, 1065; Mehmandoust, M.; Petit, Y.; Larchevêque, M. *Tetrahedron Lett.* **1992**, *33*, 4313. For a 3-aza-Claisen rearrangement, see Gilbert, J.C.; Cousins, K.R. *Tetrahedron* **1994**, *50*, 10671.

[693]Anderson, W.K.; Lai, G. *Synthesis* **1995**, 1287.

[694]Mitsuhashi, T. *J. Am. Chem. Soc.* **1986**, *108*, 2400.

[695]For reviews of Claisen rearrangements involving triple bonds, see Schuster, H.F.; Coppola, G.M. *Allenes in Organic Synthesis*, Wiley, NY, **1984**, pp. 337–343; Viola, A.; Collins, J.J.; Filipp, N. *Tetrahedron* **1981**, *37*, 3765; Théron F.; Verny, M.; Vessière, R., in Patai, S. *The Chemistry of the Carbon–Carbon Triple Bond*, pt. 1, Wiley, NY, **1978**, pp. 421–428. See also, Henderson, M.A.; Heathcock, C.H. *J. Org. Chem.* **1988**, *53*, 4736.

[696]They have been trapped: See, for example, Mortensen, J.Z.; Hedegaard, B.; Lawesson, S. *Tetrahedron* **1971**, *27*, 3831; Kwart, H.; Schwartz, J.L. *J. Org. Chem.* **1974**, *39*, 1575.

[697]Meyers, C.Y.; Rinaldi, C.; Banoli, L. *J. Org. Chem.* **1963**, *28*, 2440; Kwart, H.; Cohen, M.H. *J. Org. Chem.* **1967**, *32*, 3135; *Chem. Commun.* **1968**, 319; Makisumi, Y.; Murabayashi, A. *Tetrahedron Lett.* **1969**, 1971, 2449.

[698]For a review, see Majumdar, K.C.; Ghosh, S.; Ghosh, M. *Tetrahedron* **2003**, *59*, 7251.

[699]King, J.F.; Harding, D.R.K. *J. Am. Chem. Soc.* **1976**, *98*, 3312.

[700]Block, E.; Ahmad, S. *J. Am. Chem. Soc.* **1985**, *107*, 6731.

[701]For example, see Zumpe, F.L.; Kazmaier, U. *Synlett* **1998**, 434; Ito, H.; Sato, A.; Taguchi, T. *Tetrahedron Lett.* **1997**, *38*, 4815; Kazmaier, U.; Krebs, A. *Angew. Chem. Int. Ed.* **1995**, *34*, 2012. For asymmetric induction in the thio-Claisen rearrangement, see Reddy, K.V.; Rajappa, S. *Tetrahedron Lett.* **1992**, *33*, 7957.

this purpose.[702] In general, asymmetric [3,3]-sigmatropic rearrangements are well known.[703]

Ethers with an alkyl group in the γ position (ArO–C–C=C–R systems) sometimes give abnormal products, with the β carbon becoming attached to the ring:[704]

Normal product   Abnormal product

It has been established that these abnormal products do not arise directly from the starting ether, but are formed by a further rearrangement of the normal product:[705]

This rearrangement, which has been called an *enolene rearrangement*, a *homo-dienyl [1,5]-sigmatropic hydrogen shift* (see **18-29**), and a [1,5]-*homosigmatropic rearrangement*, involves a shift of three electron pairs over *seven* atoms. It has been found that this "abnormal" Claisen rearrangement is general and can interconvert the enol forms of systems of the types **135** and **136** through the cyclopropane intermediate **137**.[706]

135   **136**   **137**

A = H, R, Ar, OR, and so on
B = H, R, Ar, COR, COAr, COOR, and so on

---

[702]Maruoka, K.; Saito, S.; Yamamoto, J. *J. Am. Chem. Soc.* **1995**, *117*, 1165. See Sharma, G.V.M.; Ilangovan, A.; Sreevivas, P.; Mahalingam, A.K. *Synlett* **2000**, 615. **Yb**: Hiersemann, M.; Abraham, L. *Org. Lett.* **2001**, *3*, 49. **Rh**: Miller, S.P.; Morken, J.P. *Org. Lett.* **2002**, *4*, 2743.
[703]For a review, see Enders, D.; Knopp, M.; Schiffers, R. *Tetrahedron Asymmetry*, **1996**, *7*, 1847.
[704]For reviews of these abnormal Claisen rearrangements, see Hansen, H. *Mech. Mol. Migr.* **1971**, *3*, 177; Marvell, E.N.; Whalley, W., in Patai, S. *The Chemistry of the Hydroxyl Group*, pt. 2, Wiley, NY, **1971**, pp. 743–750.
[705]Habich, A.; Barner, R.; Roberts, R.; Schmid, H. *Helv. Chim. Acta* **1962**, *45*, 1943; Lauer, W.M.; Johnson, T.A. *J. Org. Chem.* **1963**, *28*, 2913; Fráter, G.; Schmid, H. *Helv. Chim. Acta* **1966**, *49*, 1957; Marvell, E.N.; Schatz, B. *Tetrahedron Lett.* **1967**, 67.
[706]Watson, J.M.; Irvine, J.L.; Roberts, R.M. *J. Am. Chem. Soc.* **1973**, *95*, 3348.

Since the Claisen rearrangement mechanism does not involve ions, it should not be greatly dependent on the presence or absence of substituent groups on the ring.[707] This is the case. Electron-donating groups increase the rate and electron-withdrawing groups decrease it, but the effect is small, with the *p*-amino compound reacting only ~10–20 times faster than the *p*-nitro compound.[708] However, solvent effects[709] are greater: Rates varied over a 300-fold range when the reaction was run in 17 different solvents.[710] An especially good solvent is trifluoroacetic acid, in which the reaction can be carried out at room temperature.[711] Most Claisen rearrangements are performed without a catalyst, but $AlCl_3$ or BF_3 are sometimes used.[712] In this case, it may become a Friedel–Crafts reaction, with the mechanism no longer cyclic,[713] and ortho, meta, and para products may be obtained.

OS **III**, 418; **V**, 25; **VI**, 298, 491, 507, 584, 606; **VII**, 177; **VIII**, 251, 536.

18-34   The Fischer Indole Synthesis

When arylhydrazones of aldehydes or ketones are treated with a catalyst, elimination of ammonia takes place and an indole is formed, in the *Fischer indole synthesis*.[714] Zinc chloride is the catalyst most frequently employed, but dozens of others, including other metal halides, proton and Lewis acids, and certain transition

---

[707]However, there are substituent effects, see Aviyente, V.; Yoo, H.Y.; Houk, K.N. *J. Org. Chem.* **1997**, *62*, 6121.

[708]Goering, H.L.; Jacobson, R.R. *J. Am. Chem. Soc.* **1958**, *80*, 3277; White, W.N.; Gwynn, D.; Schlitt, R.; Girard, C.; Fife, W.K. *J. Am. Chem. Soc.* **1958**, *80*, 3271; White, W.N.; Slater, C.D. *J. Org. Chem.* **1962**, *27*, 2908; Zahl, G.; Kosbahn, W.; Kresze, G. *Liebigs Ann. Chem.* **1975**, 1733. See also, Desimoni, G.; Faita, G.; Gamba, A.; Righetti, P.P.; Tacconi, G.; Toma, L. *Tetrahedron* **1990**, *46*, 2165; Gajewski, J.J.; Gee, K.R.; Jurayj, J. *J. Org. Chem.* **1990**, *55*, 1813.

[709]For a discussion of the role played by solvent and substituents, see Gajewski, J.J. *Acc. Chem. Res.* **1997**, *30*, 219. For solvent effects, see Davidson, M.M.; Hillier, I.H.; Hall, R.J.; Burton, N.A. *J. Am. Chem. Soc.* **1994**, *116*, 9294.

[710]White, W.N.; Wolfarth, E.F. *J. Org. Chem.* **1970**, *35*, 2196. See also Brandes, E.; Greico, P.A.; Gajewski, J.J. *J. Org. Chem.* **1989**, *54*, 515.

[711]Svanholm, U.; Parker, V.D. *J. Chem. Soc. Perkin Trans. 2* **1974**, 169.

[712]For a review, see Lutz, R.P. *Chem. Rev.* **1984**, *84*, 205.

[713]For example, crossover experiments have demonstrated that the $ZnCl_2$-catalyzed reaction is intermolecular: Yagodin, V.G.; Bunina-Krivorukova, L.I.; Bal'yan, Kh.V. *J. Org. Chem. USSR* **1971**, *7*, 1491.

[714]For a monograph, see Robinson, B. *The Fischer Indole Synthesis*, Wiley, NY, **1983**. For reviews, see Grandberg, I.I.; Sorokin, V.I. *Russ. Chem. Rev.* **1974**, *43*, 115; Shine, H.J. *Aromatic Rearrangements*, Elsevier, NY, **1969**, pp. 190–207; Sundberg, R.J. *The Chemistry of Indoles*, Academic Press, NY, **1970**, pp. 142–163; Robinson, B. *Chem. Rev.* **1969**, *69*, 227. For reviews of some abnormal Fischer indole syntheses, see Ishii, H. *Acc. Chem. Res.* **1981**, *14*, 275; Fusco, R.; Sannicolo, F. *Tetrahedron* **1980**, *36*, 161.

metals have also been used. Microwave irradiation has been used to facilitate this reaction.[715] The reaction has been done using an $AlCl_3$ complex as an ionic liquid,[716] and solid-phase Fischer-indole syntheses are known.[717] Aniline derivatives react with α-diazoketones, in the presence of a rhodium catalyst, to give indoles as well.[718] Arylhydrazones are easily prepared by the treatment of aldehydes or ketones with phenylhydrazine (**16-2**) or by aliphatic diazonium coupling (**12-7**). However, it is not necessary to isolate the arylhydrazone. The aldehyde or ketone can be treated with a mixture of phenylhydrazine and the catalyst; this is now common practice. In order to obtain an indole, the aldehyde or ketone must be of the form $RCOCH_2R'$ (R = alkyl, aryl, or hydrogen). Vinyl ethers, such as dihydrofuran, serves as an aldehyde surrogate when treated with phenylhydrazine and a catalytic amount of aqueous sulfuric acid to give an 3-substituted indole.[719]

At first glance, the reaction does not seem to be a rearrangement. However, the key step of the mechanism[720] is a [3,3]-sigmatropic rearrangement:[721]

There is much evidence for this mechanism, for example, (*1*) the isolation of **142**,[722] (*2*) the detection of **141** by ^{13}C and ^{15}N NMR,[723] (*3*) the isolation of side products that could only have come from **140**,[724] and (*4*) ^{15}N labeling experiments that showed

[715]Abramovitch, R.A.; Bulman, A. *Synlett* **1992**, 795; Lipińska, T.; Guibé-Jampel, E.; Petit, A.; Loupy, A. *Synth. Commun.* **1999**, *29*, 1349.

[716]In $AlCl_3$–*N*-butylpyridinium: Rebeiro, G.L.O.; Khadilkar, B.M. *Synthesis* **2001**, 370.

[717]Rosenbaum, C.; Katzka, C.; Marzinzik, A.; Waldmann, H. *Chem. Commun.* **2003**, 1822.

[718]Moody, C.J.; Swann, E. *Synlett* **1998**, 135.

[719]Campos, K.R.; Woo, J.C.S.; Lee, S.; Tillyer, R.D. *Org. Lett.* **2004**, *6*, 79.

[720]For a mechanistic study, see Hughes, D.L.; Zhao, D. *J. Org. Chem.* **1993**, *58*, 228.

[721]This mechanism was proposed by Robinson, G.M.; Robinson, R. *J. Chem. Soc.* **1918**, *113*, 639.

[722]Southwick, P.L.; Vida, J.A.; Fitzgerald, B.M.; Lee, S.K. *J. Org. Chem.* **1968**, *33*, 2051; Forrest, T.P.; Chen, F.M.F. *J. Chem. Soc., Chem. Commun.* **1972**, 1067.

[723]Douglas, A.W. *J. Am. Chem. Soc.* **1978**, *100*, 6463; **1979**, *101*, 5676.

[724]Bajwa, G.S.; Brown, R.K. *Can. J. Chem.* **1969**, *47*, 785; **1970**, *48*, 2293, and references cited therein.

that it was the nitrogen farther from the ring that is eliminated as ammonia.[725] The main function of the catalyst seems to be to speed the conversion of **138** to **139**. The reaction can be performed without a catalyst.

There are alternative methods to produce indoles. Acetophenone reacts with 2-chloro nitrobenzene derivatives in the presence of a phenol and a palladium catalyst to give an indole.[726]

OS **III**, 725; **IV**, 884. Also see, OS **IV**, 657.

18-35 [2,3]-Sigmatropic Rearrangements

(2/S-3/) → (1/5/)-sigma-Migration

Sulfur ylids bearing an allylic group are converted on heating to unsaturated sulfides.[727] This is a concerted [2,3]-sigmatropic rearrangement[728] and has also been demonstrated for the analogous cases of nitrogen ylids[729] and the conjugate bases of allylic ethers (in the last case it is called the *[2,3]-Wittig rearrangement*).[730] It has been argued that the [2,3]-Wittig rearrangement demands severe deformation of the molecule in order to proceed.[731] The SmI_2 compound has been shown to induce

[725]Clausius, K.; Weisser, H.R. *Helv. Chim. Acta* **1952**, *35*, 400.

[726]Rutherford, J.L.; Rainka, M.P.; Buchwald, S.L. *J. Am. Chem. Soc.* **2002**, *124*, 15168.

[727]For example, see Blackburn, G.M.; Ollis, W.D.; Plackett, J.D.; Smith, C.; Sutherland, I.O. *Chem. Commun.* **1968**, 186; Trost, B.M.; LaRochelle, R. *Tetrahedron Lett.* **1968**, 3327; Baldwin, J.E.; Hackler, R.E.; Kelly, D.P. *Chem. Commun.* **1968**, 537, 538, 1083; Bates, R.B.; Feld, D. *Tetrahedron Lett.* **1968**, 417; Kirmse, W.; Kapps, M. *Chem. Ber.* **1968**, *101*, 994, 1004; Biellmann, J.F.; Ducep, J.B. *Tetrahedron Lett.* **1971**, 33; Ceré, V.; Paolucci, C.; Pollicino, S.; Sandri, E.; Fava, A. *J. Org. Chem.* **1981**, *46*, 3315; Kido, F.; Sinha, S.C.; Abiko, T.; Yoshikoshi, A. *Tetrahedron Lett.* **1989**, *30*, 1575. For a review as applied to ring expansions, see Vedejs, E. *Acc. Chem. Res.* **1984**, *17*, 358.

[728]For a review of the stereochemistry of these reactions, see Hoffmann, R.W. *Angew. Chem. Int. Ed.* **1979**, *18*, 563.

[729]For example, see Jemison, R.W.; Ollis, W.D. *Chem. Commun.* **1969**, 294; Rautenstrauch, V. *Helv. Chim. Acta* **1972**, *55*, 2233; Mageswaran, S.; Ollis, W.D.; Sutherland, I.O.; Thebtaranonth, Y. *J. Chem. Soc., Chem. Commun.* **1973**, 651; Ollis, W.D.; Sutherland, I.O.; Thebtaranonth, Y. *J. Chem. Soc., Chem. Commun.* **1973**, 657; Mander, L.N.; Turner, J.V. *J. Org. Chem.* **1973**, *38*, 2915; Stévenart-De Mesmaeker, N.; Merényi, R.; Viehe, H.G. *Tetrahedron Lett.* **1987**, *28*, 2591; Honda, K.; Inoue, S.; Sato, K. *J. Am. Chem. Soc.* **1990**, *112*, 1999.

[730]See, for example, Makisumi, Y.; Notzumoto, S. *Tetrahedron Lett.* **1966**, 6393; Schöllkopf, U.; Fellenberger, K.; Rizk, M. *Liebigs Ann. Chem.* **1970**, *734*, 106; Rautenstrauch, V. *Chem. Commun.* **1970**, 4. For a review, see Nakai, T.; Mikami, K. *Chem. Rev.* **1986**, *86*, 885. For a list of references, see Larock, R.C. *Comprehensive Organic Transformations*, 2nd ed., Wiley-VCH, NY, **1999**, pp. 1063–1067.

[731]You, Z.; Koreeda, M. *Tetrahedron Lett.* **1993**, *34*, 2597.

a [2,3]-Wittig rearrangement.[732] The reaction has been extended to certain other systems,[733] even to an all-carbon system.[734]

Since the reactions involve migration of an allylic group from a sulfur, nitrogen, or oxygen atom to an adjacent negatively charged carbon atom, they are special cases of the Stevens or Wittig rearrangements (**18-21**, **18-22**). However, in this case the migrating group *must* be allylic (in **18-21** and **18-22** other groups can also migrate). Thus, when the migrating group is allylic, there are two possible pathways: (*1*) the radical-ion or ion-pair mechanisms (**18-21**, **18-22**) and (*2*) the concerted pericyclic [2,3]-sigmatropic rearrangement. These can easily be told apart, since the latter always involves an allylic shift (as in the Claisen rearrangement), while the former pathway does not.

Of these reactions, the [2,3]-Wittig rearrangement in particular has often been used as a means of transferring chirality. The product of this reaction has potential stereogenic centers at C-3 and C-4 (if $R^5 \neq R^6$), and if the starting ether is optically active because of a stereogenic center at C-1, the product may be optically active as well. Many examples are known in which an optically active ether was converted to a product that was optically active because of chirality at C-3, C-4, or both.[735] If a

[732]Kunishima, M.; Hioki, K.; Kono, K.; Kato, A.; Tani, S. *J. Org. Chem.* **1997**, *62*, 7542. Also see, Hioki, K.; Kono, K.; Tani, S.; Kunishima, M. *Tetrahedron Lett.* **1998**, *39*, 5229. For an enantioselective [2,3]-Wittig rearrangment, see Fujimoto, K.; Nakai, T. *Tetrahedron Lett.* **1994**, *35*, 5019.

[733]See, for example, Baldwin, J.E.; Brown, J.E.; Höfle, G. *J. Am. Chem. Soc.* **1971**, *93*, 788; Yamamoto, Y.; Oda, J.; Inouye, Y. *J. Chem. Soc., Chem. Commun.* **1973**, 848; Ranganathan, S.; Ranganathan, D.; Sidhu, R.S.; Mehrotra, A.K. *Tetrahedron Lett.* **1973**, 3577; Murata, Y.; Nakai, T. *Chem. Lett.* **1990**, 2069. For reviews with respect to selenium compounds, see Reich, H.J., in Liotta, D.C. *Organoselenium Chemistry*, Wiley, NY, **1987**, pp. 365–393; Reich, H.J., in Trahanovsky, W.S. *Oxidation in Organic Chemistry*, pt. C, Academic Press, NY, **1978**, pp. 102–111.

[734]Baldwin, J.E.; Urban, F.J. *Chem. Commun.* **1970**, 165.

[735]For reviews of stereochemistry in this reaction, see Mikami, K.; Nakai, T. *Synthesis* **1991**, 594; Nakai, T.; Mikami, K. *Chem. Rev.* **1986**, *86*, 885, 888–895. See also, Nakai, T.; Nakai, E. *Tetrahedron Lett.* **1988**, *29*, 4587; Balestra, M.; Kallmerten, J. *Tetrahedron Lett.* **1988**, *29*, 6901; Brückner, R. *Chem. Ber.* **1989**, *122*, 193, 703; Scheuplein, S.W.; Kusche, A.; Brückner, R.; Harms, K. *Chem. Ber.* **1990**, *123*, 917; Wu, Y.; Houk, K.N.; Marshall, J.A. *J. Org. Chem.* **1990**, *55*, 1421; Marshall, J.A.; Wang, X. *J. Org. Chem.* **1990**, *55*, 2995.

suitable stereogenic center is present in R^1 (or if a functional group in R^1 can be so converted), then stereocontrol over three contiguous stereogenic centers can be achieved. Stereocontrol of the new double bond (E or Z) has also been accomplished.

If an OR or SR group is attached to the negative carbon, the reaction becomes a method for the preparation of β,γ-unsaturated aldehydes, because the product is easily hydrolyzed.[736]

Another [2,3]-sigmatropic rearrangement converts allylic sulfoxides to allylically rearranged alcohols by treatment with a thiophilic reagent, such as trimethyl phosphite.[737] This is often called the *Mislow–Evans rearrangement*. In this case, the migration is from sulfur to oxygen. [2,3]-Oxygen-to-sulfur migrations are also known.[738] The Sommelet–Hauser rearrangement (**13-31**) is also a [2,3]-sigmatropic rearrangement.

OS **VIII**, 427.

18-36   The Benzidine Rearrangement

Hydrazobenzene   **143**

[736]Huynh, C.; Julia, S.; Lorne, R.; Michelot, D. *Bull. Soc. Chim. Fr.* **1972**, 4057.

[737]Tang, R.; Mislow, K. *J. Am. Chem. Soc.* **1970**, *92*, 2100; Evans, D.A.; Andrews, G.C. *Acc. Chem. Res.* **1974**, *7*, 147; Hoffmann, R.W. *Angew. Chemie. Int. Ed., Engl.*, **1979**, *18*, 563; Sato, T.; Otera, J.; Nozaki, H. *J. Org. Chem.* **1989**, *54*, 2779; Bickart, P.; Carson, F.W.; Jacobus, J.; Miller, E.G.; Mislow, K. *J. Am. Chem. Soc.* **1968**, *90*, 4869.

[738]Braverman, S.; Mechoulam, H. *Isr. J. Chem.* **1967**, *5*, 71, Braverman, S.; Stabinsky, Y. *Chem. Commun.* **1967**, 270; Rautenstrauch, V. *Chem. Commun.* **1970**, 526; Smith, G.; Stirling, C.J.M. *J. Chem. Soc. C* **1971**, 1530; Tamaru, Y.; Nagao, K.; Bando, T.; Yoshida, Z. *J. Org. Chem.* **1990**, *55*, 1823.

When hydrazobenzene is treated with acids, it rearranges to give ~70% 4,4'-diaminobiphenyl (**143**, benzidine) and ~30% 2,4'-diaminobiphenyl. This reaction is called the *benzidine rearrangement* and is general for N,N'-diarylhydrazines.[739] Usually, the major product is the 4,4'-diaminobiaryl, but four other products may also be produced. These are the 2,4'-diaminobiaryl, already referred to, the 2,2'-diaminobiaryl, and the *o*- and *p*-arylaminoanilines (called *semidines*). The 2,2'- and *p*-arylaminoaniline compounds are formed less often and in smaller amounts than the other two side products. Usually, the 4,4'-diaminobiaryl predominates, except when one or both para positions of the diarylhydrazine are occupied. However, the 4,4'-diamine may still be produced even if the para positions are occupied. If SO_3H, COOH, or Cl (but not R, Ar, or NR_2) is present in the para position, it may be ejected. With dinaphthylhydrazines, the major products are not the 4,4'-diaminobinaphthyls, but the 2,2' isomers. Another side reaction is disproportionation to $ArNH_2$ and ArN=NAr. For example, p,p'-$PhC_6H_4NHNHC_6H_4Ph$ gives 88% disproportionation products at 25°C.[740]

The mechanism has been exhaustively studied and several mechanisms have been proposed.[741] At one time, it was believed that NHAr broke away from ArNHNHAr and became attached to the para position to give the semidine, which then went on to product. The fact that semidines could be isolated lent this argument support, as did the fact that this would be analogous to the rearrangements considered in Chapter 11 (**11-28–11-32**). However, this theory was killed when it was discovered that semidines could not be converted to benzidines under the reaction conditions. Cleavage into two independent pieces (either ions or radicals) has been ruled out by many types of crossover experiments, which always showed that the two rings of the starting material are in the product; that is, ArNHNHAr' gives no molecules (of any of the five products) containing two Ar groups or two Ar' groups, and mixtures of ArNHNHAr and Ar'NHNHAr' give no molecules containing both Ar and Ar'. An important discovery was the fact that, although the reaction is always first order in substrate, it can be either

---

[739]For reviews, see, in Patai, S. *The Chemistry of the Hydrazo, Azo, and Azoxy Groups*, pt. 2, Wiley, NY, *1975*, the reviews by Cox, R.A.; Buncel, E. pp. 775–807; Koga, G.; Koga, N.; Anselme, J. pp. 914–921; Williams, D.L.H., in Bamford, C.H.; Tipper, C.F.H. *Comprehensive Chemical Kinetics*, Vol. 9, Elsevier, NY, *1973*, Vol. 13, 1972, pp. 437–448; Shine, H.J. *Mech. Mol. Migr.* *1969*, 2, 191; *Aromatic Rearrangements*, Elsevier, NY, *1969*, pp. 126–179; Banthorpe, D.V. *Top. Carbocyclic Chem.* *1969*, *1*, 1; Lukashevich, V.O. *Russ. Chem. Rev.* *1967*, 36, 895.

[740]Shine, H.J.; Stanley, J.P. *J. Org. Chem.* *1967*, *32*, 905. For investigations of the mechanism of the disproportionation reactions, see Rhee, E.S.; Shine, H.J. *J. Am. Chem. Soc.* *1986*, *108*, 1000; *1987*, *109*, 5052.

[741]For a history of the mechanistic investigations and controversies, see Shine, H.J. *J. Phys. Org. Chem.* *1989*, 2, 491.

first[742] or second[743] order in [H$^+$]. With some substrates the reaction is entirely first order in [H$^+$], while with others it is entirely second order in [H$^+$], regardless of the acidity. With still other substrates, the reaction is first order in [H$^+$] at low acidities and second order at higher acidities. With the latter substrates fractional orders can often be observed,[744] because at intermediate acidities, both processes take place simultaneously. These kinetic results seem to indicate that the actual reacting species can be either the monoprotonated substrate ArNHNH$_2$Ar or the diprotonated ArNH$_2$NH$_2$Ar.

Most of the proposed mechanisms[745] attempted to show how all five products could be produced by variations of a single process. An important breakthrough was the discovery that the two main products are formed in entirely different ways, as shown by isotope-effect studies.[746] When the reaction was run with hydrazobenzene labeled with ^{15}N at both nitrogen atoms, the isotope effect was 1.022 for formation of **143**, but 1.063 for formation of 2,4'-diaminobiphenyl. This showed that the N–N bond is broken in the rate-determining step in both cases, but the steps themselves are obviously different. When the reaction was run with hydrazobenzene labeled with ^{14}C at a para position, there was an isotope effect of 1.028 for formation of **143**, but essentially no isotope effect (1.001) for formation of 2,4'-diaminobiphenyl. This can only mean that for **143** formation of the new C–C bond *and* breaking of the N–N bond both take place in the rate-determining step; in other words, the mechanism is concerted. The following [5.5]-sigmatropic rearrangement accounts for this:[745,747]

$$\bullet = {}^{14}C \qquad \textbf{144}$$

[742]Banthorpe, D.V.; Hughes, E.D.; Ingold, C.K. *J. Chem. Soc.* **1962**, 2386, 2402, 2407, 2413, 2418, 2429; Shine, H.J.; Chamness, J.T. *J. Org. Chem.* **1963**, *28*, 1232; Banthorpe, D.V.; O'Sullivan, M. *J. Chem. Soc. B* **1968**, 627.

[743]Hammond, G.S.; Shine, H.J. *J. Am. Chem. Soc.* **1950**, *72*, 220; Banthorpe, D.V.; Cooper, A. *J. Chem. Soc. B* **1968**, 618; Banthorpe, D.V.; Cooper, A.; O'Sullivan, M. *J. Chem. Soc. B* **1971**, 2054.

[744]Carlin, R.B.; Odioso, R.C. *J. Am. Chem. Soc.* **1954**, *76*, 100; Banthorpe, D.V.; Ingold, C.K.; Roy, J. *J. Chem. Soc. B* **1968**, 64; Banthorpe, D.V.; Ingold, C.K.; O'Sullivan, M. *J. Chem. Soc. B* **1968**, 624.

[745]For example, see the "polar-transition-state mechanism:" Banthorpe, D.V.; Hughes, E.D.; Ingold, C.K. *J. Chem. Soc.* **1964**, 2864, and the "π-complex mechanism:" Dewar, M.J.S., in de Mayo, P. *Molecular Rearrangments*, Vol. 1, Wiley, NY, **1963**, pp. 323–344.

[746]Shine, H.J.; Zmuda, H.; Park, K.H.; Kwart, H.; Horgan, A.J.; Collins, C.; Maxwell, B.E. *J. Am. Chem. Soc.* **1981**, *103*, 955; Shine, H.J.; Zmuda, H.; Park, K.H.; Kwart, H.; Horgan, A.J.; Brechbiel, M. *J. Am. Chem. Soc.* **1982**, *104*, 2501.

[747]This step was also part of the "polar-transition-state mechanism".

The diion **144** was obtained as a stable species in super acid solution at $-78°C$ by treatment of hydrazobenzene with $FSO_3H–SO_2$ (SO_2ClF).[748] Though the results just given were obtained with hydrazobenzene, which reacts by the diprotonated pathway, monoprotonated substrates have been found to react by the same [5,5]-sigmatropic mechanism.[749] Some of the other rearrangements in this section are also sigmatropic. Thus, formation of the *p*-semidine takes place by a [1,5]-sigmatropic rearrangement,[750] and the conversion of 2,2'-hydrazonaphthalene to 2,2'-diamino-1,1'-binaphthyl by a [3,3]-sigmatropic rearrangement.[751]

2,4'-Diaminobiphenyl is formed by a completely different mechanism, though the details are not known. There is rate-determining breaking of the N–N bond, but the C–C bond is not formed during this step.[752] The formation of the *o*-semidine also takes place by a nonconcerted pathway.[753] Under certain conditions, benzidine rearrangements have been found to go through radical cations.[754]

C. Other Cyclic Rearrangements

18-37   Metathesis of Alkenes (Alkene or Olefin Metathesis)[755]

Alkene metathesis

$$CH_3CH=CHCH_2CH_3 \quad \xrightarrow[\text{WCl}_6\text{–EtOH}]{\text{EtAlCl}_2} \quad CH_3CH=CHCH_3 \;+\; CH_3CH_2CH=CHCH_2CH_3$$

When alkenes are treated with certain catalysts they are converted to other alkenes in a reaction in which the alkylidene groups $(R^1R^2C=)$ have become interchanged by a process schematically illustrated by the equation:

---

[748]Olah, G.A.; Dunne, K.; Kelly, D.P.; Mo, Y.K. *J. Am. Chem. Soc.* **1972**, *94*, 7438.

[749]Shine, H.J.; Park, K.H.; Brownawell, M.L.; San Filippo, Jr., J. *J. Am. Chem. Soc.* **1984**, *106*, 7077.

[750]Heesing, A.; Schinke, U. *Chem. Ber.* **1977**, *110*, 3319; Shine, H.J.; Zmuda, H.; Kwart, H.; Horgan, A.G.; Brechbiel, M. *J. Am. Chem. Soc.* **1982**, *104*, 5181.

[751]Shine, H.J.; Gruszecka, E.; Subotkowski, W.; Brownawell, M.; San Filippo, Jr., J. *J. Am. Chem. Soc.* **1985**, *107*, 3218.

[752]See Rhee, E.S.; Shine, H.J. *J. Am. Chem. Soc.* **1986**, *108*, 1000; **1987**, *109*, 5052.

[753]Rhee, E.S.; Shine, H.J. *J. Org. Chem.* **1987**, *52*, 5633.

[754]See, for example, Nojima, M.; Ando, T.; Tokura, N. *J. Chem. Soc. Perkin Trans. 1* **1976**, 1504.

[755]For reviews, see Grubbs, R.H. *Tetrahedron* **2004**, *60*, 7117; Wakamatsu, H.; Blechert, S. *Angew. Chem. Int. Ed.* **2002**, *41*, 2403; Schrock, R.R.; Hoveyda, A.H. *Angew. Chem. Int. Ed.* **2003**, *42*, 4592.

$$R^1R^1 \quad R^2R^2 \quad \rightleftharpoons \quad R^1 \quad R^2 \quad R^1 \quad R^2$$

The reaction is called *metathesis* of alkenes or *alkene metathesis* (*olefin metathesis*).[756] In the example shown above, 2-pentene (either cis, trans, or a cis–trans mixture) is converted to a mixture of ∼50% 2-pentene, 25% 2-butene, and 25% 3-hexene. The reaction is reversible[757] and the alkene starting material and products exist in an equilibrium, so the same mixture can be obtained by starting with equimolar quantities of 2-butene and 3-hexene.[758] In general, the reaction can be applied to a single unsymmetrical alkene, giving a mixture of itself and two other alkenes, or to a mixture of two alkenes, in which case the number of different molecules in the product depends on the symmetry of the reactants. As in the case above, a mixture of $R^1R^2C=CR^1R^2$ and $R^3R^4C=CR^3R^4$ gives rise to only one new alkene ($R^1R^2C=CR^3R^4$), while in the most general case, a mixture of $R^1R^2C=CR^3R^4$ and $R^5R^6C=CR^7R^8$ gives a mixture of 10 alkenes: the original $2 + 8$ new ones. In early work, tungsten, molybdenum,[759] or rhenium complexes were used, and with simple alkenes the proportions of products are generally statistical,[760] which limited the synthetic utility of the reaction

---

[756]For monographs, see Drăguțn, V.; Balaban, A.T.; Dimonie, M. *Olefin Metathesis and Ring-Opening Polymerization of Cyclo-Olefins*, Wiley, NY, *1985*; Ivin, K.J. *Olefin Metathesis*, Academic Press, NY, *1983*. For reviews, see Feast, W.J.; Gibson, V.C., in Hartley, F.R. *The Chemistry of the Metal-Carbon Bond*, Vol. 5, Wiley, NY, 1989, pp. 199–228; Streck, R. *CHEMTECH 1989* , 498; Schrock, R.R. *J. Organomet. Chem. 1986*, *300*, 249; Grubbs, R.H., in Wilkinson, G. *Comprehensive Organometallic Chemistry*, Vol. 8, Pergamon, Elmsford, NY, *1982*, pp. 499–551; Basset, J.M.; Leconte, M. *CHEMTECH 1980*, 762; Banks, R.L. *CHEMTECH 1979*, 494; *Fortschr. Chem. Forsch. 1972*, *25*, 39; Calderon N.; Lawrence, J.P.; Ofstead, E.A. *Adv. Organomet. Chem. 1979*, *17*, 449; Grubbs, R.H. *Prog. Inorg. Chem. 1978*, *24*, 1; Calderon N., in Patai, S. *The Chemistry of Functional Groups: Supplement A* pt. 2, Wiley, NY, *1977*, pp. 913–964; *Acc. Chem. Res. 1972*, *5*, 127; Katz, T.J. *Adv. Organomet. Chem. 1977*, *16*, 283; Haines, R.J.; Leigh, G.J. *Chem. Soc. Rev. 1975*, *4*, 155; Hocks, L. *Bull. Soc. Chim. Fr. 1975*, 1893; Mol, J.C.; Moulijn, J.A. *Adv. Catal. 1974*, *24*, 131; Hughes, W.B. *Organomet. Chem. Synth. 1972*, *1*, 341; Khidekel', M.L.; Shebaldova, A.D.; Kalechits, I.V. *Russ. Chem. Rev. 1971*, *40*, 669; Bailey, G.C. *Catal. Rev. 1969*, *3*, 37.

[757]Smith III, A.B.; Adams, C.M.; Kozmin, S.A. *J. Am. Chem. Soc. 2001*, *123*, 990.

[758]Calderon N.; Chen, H.Y.; Scott, K.W. *Tetrahedron Lett. 1967*, 3327; Wang, J.; Menapace, H.R. *J. Org. Chem. 1968*, *33*, 3794; Hughes, W.B. *J. Am. Chem. Soc. 1970*, *92*, 532.

[759]For an example, see Crowe, W.E.; Zhang, Z.J. *J. Am. Chem. Soc. 1993*, *115*, 10998.; Fu, G.C.; Grubbs, R.H. *J. Am. Chem. Soc. 1993*, *115*, 3800.

[760]Calderon N.; Ofstead, E.A.; Ward, J.P.; Judy, W.A.; Scott, K.W. *J. Am. Chem. Soc. 1968*, *90*, 4133.

since the yield of any one product is low. However, in some cases one alkene may be more or less thermodynamically stable than the rest, so that the proportions are not statistical. Furthermore, it may be possible to shift the equilibrium. For example, 2-methyl-1-butene gives rise to ethylene and 3,4-dimethyl-3-hexene. By allowing the gaseous ethylene to escape, the yield of 3,4-dimethyl-3-hexene can be raised to 95%.[761]

145   **146**   **147**

The development of new catalysts have revolutionized this reaction, making it one of the most important methods available for synthesis. Tailoring the substrate to include two terminal alkenes leads to ethylene as a product, whose escape from the reaction drives the equilibrium to product. Many catalysts, both homogeneous[762] and heterogeneous,[763] have been used for this reaction. Although there are several examples of the former, ruthenium complexes are the most important,[764] while among the latter are oxides of Mo, W, and Re deposited on alumina or silica gel.[765] The major breakthrough in these catalysts was the development of catalysts that are relatively air stable. The three most used catalysts are carbene complexes **145**[766] and **146**[767] (Grubbs catalysts I and II, respectively), and **147** (the Shrock catalyst).[768] Catalyst **146** can be generated *in situ* from air stable

[761]Knoche, H. Ger. Pat.(Offen.) 2024835, 1970 [*Chem. Abstr.*, *1971*, 74, 44118b]. See also Chevalier, P.; Sinou, D.; Descotes, G. *Bull. Soc. Chim. Fr. 1976*, 2254; Bespalova, N.B.; Babich, E.D.; Vdovin, V.M.; Nametkin, N.S. *Dokl. Chem. 1975*, *225*, 668; Ichikawa, K.; Fukuzumi, K. *J. Org. Chem. 1976*, *41*, 2633; Baker, R.; Crimmin, M.J. *Tetrahedron Lett. 1977*, 441.

[762]First reported by Calderon N.; Chen, H.Y.; Scott, K.W. *Tetrahedron Lett. 1967*, 3327. For a lengthy list, see Hughes, W.B. *Organomet. Chem. Synth. 1972*, *1*, 341, see pp. 362–368. For a homogeneous rhenium catalyst, see Toreki, R.; Schrock, R.R. *J. Am. Chem. Soc. 1990*, *112*, 2448.

[763]First reported by Banks, R.L.; Bailey, G.C. *Ind. Eng. Chem. Prod. Res. Dev., 1964*, *3*, 170. See also, Banks, R.L. *CHEMTECH 1986*, 112.

[764]Gilbertson, S.R.; Hoge, G.S.; Genov, D.G. *J. Org. Chem. 1998*, *63*, 10077; Maier, M.E.; Bugl, M. *Synlett 1998*, 1390; Stefinovic, M.; Snieckus, V. *J. Org. Chem. 1998*, *63*, 2808.

[765]For a list of heterogeneous catalysts, see Banks, R.L. *Fortschr. Chem. Forsch. 1972*, *25*, 39, 41–46.

[766]Schwab, P.; Grubbs, R.H.; Ziller, J.W. *J. Am. Chem. Soc. 1996*, *118*, 100.

[767]Scholl, M.; Ding, S.; Lee, C.W.; Grubbs, R.H. *Org. Lett. 1999*, *1*, 953.

[768]Bazan, G.C.; Oskam, J.H.; Cho, H.-N.; Park, L.Y.; Schrock, R.R. *J. Am. Chem. Soc. 1991*, *113*, 6899, and references cited therein.

precursors. [769] Recyclable catalyst have been developed,[770] and the reaction has been done in ionic liquids,[771] as well as supercritical CO_2[772] (p. 414). Micro-wave-induced ring-closing metathesis reactions are known.[773] Polymer-bound ruthenium catalysts[774] and molybdenum catalysts[775] have been used, and the **146** has been immobilized on polyethylene glycol, PEG).[776] Efficient methods have been developed for the removal of ruthenium by-products from metathesis reactions.[777] By choice of the proper catalyst, the reaction has been applied to term-inal and internal alkenes, straight chain or branched. The effect of substitution on the ease of reaction is $CH_2 => RCH_2CH => R_2CHCH => R_2C=$.[778] Note that isomerization of the C=C unit can occur after metathesis.[779] Cross-metath-esis[780,781] (or symmetrical homo-metathesis[782]) of alkenes to give new alkenes

[769]Louie, J.; Grubbs, R.H. *Angew. Chem. Int. Ed.* **2001**, *40*, 247.

[770]Kingsbury, J.S.; Harrity, J.P.A.; Bonitatebus Jr., P.J.; Hoveyda, A.H. *J. Am. Chem. Soc.* **1999**, *121*, 791.

[771]In bmim PF$_6$, 1-butyl-3-methylimidazolium hexafluorophosphate: Buijsman, R.C.; van Vuuren, E.; Sterrenburg, J.G. *Org. Lett.* **2001**, *3*, 3785. See Clavier, H.; Audic, N.; Mauduit, M.; Guillemin, J.-C. *Chem. Commun.* **2004**, 2282.

[772]Fürstner, A.; Ackerman, L.; Beck, K.; Hori, H.; Koch, D.; Langemann, K.; Liebl, M.; Six, C.; Leitner, W. *J. Am. Chem. Soc.* **2001**, *123*, 9000.

[773]Grabacia, S.; Desai, B.; Lavastre, O.; Kappe, C.O. *J. Org. Chem.* **2003**, *68*, 9136; Mayo, K.G.; Nearhoof, E.H.; Kiddle, J.J. *Org. Lett.* **2002**, *4*, 1567; Balan, D.; Adolfsson, H. *Tetrahedron Lett.* **2004**, *45*, 3089. For a solvent-free microwave-induced reaction, see Thanh, G.V.; Loupy, A. *Tetrahedron Lett.* **2003**, *44*, 9091.

[774]Yao, Q. *Angew. Chem. Int. Ed.* **2000**, *39*, 3896; Schürer, S.C.; Gessler, S.; Buschmann, N.; Blechert, S. *Angew. Chem. Int. Ed.* **2000**, *39*, 3898.

[775]Hultzsch, K.C.; Jernelius, J.A.; Hoveyda, A.H.; Schrock, R.R. *Angew. Chem. Int. Ed.* **2002**, *41*, 589.

[776]A recyclable catalyst, see Yao, Q.; Motta, A.R. *Tetrahedron Lett.* **2004**, *45*, 2447.

[777]Ahn, Y.M.; Yang, K.; Georg, G.I. *Org. Lett.* **2001**, *3*, 1411; Cho, J.H.; Kim, B.M. *Org. Lett.* **2003**, *5*, 531. A scavenger resin has been developed, see Westhus, M.; Gonthier, E.; Brohm, D.; Breinbauer, R. *Tetrahedron Lett.* **2004**, *45*, 3141.

[778]For an explanation for this order, see McGinnis, J.; Katz, T.J.; Hurwitz, S. *J. Am. Chem. Soc.* **1976**, *98*, 605; Casey, C.J.; Tuinstra, H.E.; Saeman, M.C. *J. Am. Chem. Soc.* **1976**, *98*, 608. A model for selectivity has been proposed, see Chatterjee, A.K.; Choi, T.-L.; Sanders, D.P.; Grubbs, R.H. *J. Am. Chem. Soc.* **2003**, *125*, 11360.

[779]For example, see Schmidt, B. *J. Org. Chem.* **2004**, *69*, 7672; Sutton, A.E.; Seigal, B.A.; Finnegan, D.F.; Snapper, M.L. *J. Am. Chem. Soc.* **2002**, *124*, 13390.

[780]See La, D.S.; Sattely, E.S.; Ford, J.G.; Schrock, R.R.; Hoveyda, A.H. *J. Am. Chem. Soc.* **2001**, *123*, 7767.

[781]Chatterjee, A.K.; Grubbs, R.H. *Org. Lett.* **1999**, *1*, 1751; Chatterjee, A.K.; Morgan, J.P.; Scholl, M.; Grubbs, R.H. *J. Am. Chem. Soc.* **2000**, *122*, 3783; Fassina, V.; Ramminger, C.; Seferin, M.; Monteiro, A.L. *Tetrahedron* **2000**, *56*, 7403; Randl, S.; Buschmann, N.; Connon, S.J.; Blechert, S. *Synlett* **2001**, 1547; Grela, K.; Bieniek, M. *Tetrahedron Lett.* **2001**, *42*, 6425; Choi, T.-L.; Chatterjee, A.K.; Grubbs, R.H. *Angew. Chem. Int. Ed.* **2001**, *40*, 1277; Arjona, O.; Csákÿ, A.G.; Medel, R.; Plumet, J. *J. Org. Chem.* **2002**, *67*, 1380; Chatterjee, A.K.; Sanders, D.P.; Grubbs, R.H. *Org. Lett.* **2002**, *4*, 1939; Hansen, E.C.; Lee, D. *Org. Lett.* **2004**, *6*, 2035; BouzBouz, S.; Simmons, R.; Cossy, J. *Org. Lett.* **2004**, *6*, 3465.

[782]Blanco, O.M.; Castedo, L. *Synlett* **1999**, 557.

can be accomplished with the modern metathesis catalysts. Monosubstiuted alkenes react faster than disubstituted alkenes.[783] A double metathesis reaction of a diene (also called domino metathesis[784] or tandem metathesis[785]) with conjugated aldehydes has been reported,[786] and a triple-metathesis was reported to for a dihydropyran with two dihydropyran substituents.[787] Cross-metathesis of a terminal alkyne and a terminal alkenes (en-ynes)[788] to give a diene has also been reported.[789] Cross-metathesis of vinylcyclopropanes leads to an alkene with two cyclopropyl substituents.[790] Vinylcyclopropane-alkyne metathesis reactions have been reported.[791] Cyclic alkenes can be opened, usually with polymerization using metathesis catalysts. Ring-opening metathesis generates dienes from cyclic alkenes.[792] Allenes undergo a metathesis reaction to give symmetrical allenes.[793] The Grubbs catalyst is compatible with forming cyclic alkenes by ring-closing metathesis followed by treatment with hydrogen to give the saturated cyclic compound.[794] An interesting variation reacts an α,ω-diene with a cyclic alkene. The combination of ring-opening metathesis and ring-closing cross-metathesis leads to ring expansion to give a macrocyclic nonconjugated diene.[795]

Dienes can react intermolecularly or intramolecularly.[796] Intramolecular reactions generate rings, usually alkenes or dienes. Alkene metathesis can be

---

[783]For an example with a styrene derivative versus a terminal alkene in the same molecule, see Lautens, M.; Maddess, M.L. *Org. Lett.* **2004**, *6*, 1883.

[784]Rückert, A.; Eisele, D.; Blechert, S. *Tetrahedron Lett.* **2001**, *42*, 5245.

[785]Choi, T.-L.; Grubbs, R.H. *Chem. Commun.* **2001**, *26* 48.

[786]BouzBouz, S.; Cossy, J. *Org. Lett.* **2001**, *3*, 1451; van Otterlo, W.A.L.; Ngidi, E.L.; de Koning, C.D.; Fernandes, M.A. *Tetrahedron Lett.* **2004**, *45*, 659.

[787]Sundararajan, G.; Prabagaran, N.; Varghese, B. *Org. Lett.* **2001**, *3*, 1973.

[788]For a discussion of (*Z/E*) selectivity and substituent effects, see Kang, B.; Lee, J.M.; Kwak, J.; Lee, Y.S.; Chang, S. *J. Org. Chem.* **2004**, *69*, 7661. For a review, see Diver, S.T.; Giessert, A.J. *Chem. Rev.* **2004**, *104*, 1317.

[789]For a review, see Poulsen, C.S.; Madsen, R. *Synthesis* **2003**, 1. See Stragies, R.; Voigtmann, U.; Blechert, S. *Tetrahedron Lett.* **2000**, *41*, 5465; Yao, Q. *Org. Lett.* **2001**, *3*, 2069; Lee, H.-Y.; Kim, B.G.; Snapper, M.L. *Org. Lett.* **2003**, *5*, 1855; Giessert, A.J.; Brazis, N.J.; Diver, S.T. *Org. Lett.* **2003**, *5*, 3819; Kim, M.; Park, S.; Maifeld, S.V.; Lee, D. *J. Am. Chem. Soc.* **2004**, *126*, 10242; Tonogaki, K.; Mori, M. *Tetrahedron Lett.* **2002**, *43*, 2235. See also, Kang, B.; Kim, D.-h.; Do, Y.; Chang, S. *Org. Lett.* **2003**, *5*, 3041.

[790]Verbicky, C.A.; Zercher, C.K. *Tetrahedron Lett.* **2000**, *41*, 8723.

[791]López, F.; Delgado, A.; Rodríguez, J.R.; Castedo, L.; Mascareñas, J.L. *J. Am. Chem. Soc.* **2004**, *126*, 10262.

[792]See La, D.S.; Ford, J.G.; Sattely, E.S.; Bonitatebus, P.J.; Schrock, R.R.; Hoveyda, A.H. *J. Am. Chem. Soc.* **1999**, *121*, 11603; Wright, D.L.; Usher, L.C.; Estrella-Jimenez, M. *Org. Lett.* **2001**, *3*, 4275; Randl, S.; Connon, S.J.; Blechert, S. *Chem. Commun.* **2001**, 1796; Morgan, J.P.; Morrill, C.; Grubbs, R.H. *Org. Lett.* **2002**, *4*, 67.

[793]Ahmed, M.; Arnauld, T.; Barrett, A.G.M.; Braddock, D.C.; Flack, K.; Procopiou, P.A. *Org. Lett.* **2000**, *2*, 551.

[794]Louie, J.; Bielawski, C.W.; Grubbs, R.H. *J. Am. Chem. Soc.* **2001**, *123*, 11312.

[795]Lee, C.W.; Choi, T.-L.; Grubbs, R.H. *J. Am. Chem. Soc.* **2002**, *124*, 3224.

[796]Kroll, W.R.; Doyle, G. *Chem. Commun.* **1971**, 839. For a review see Grubbs, R.H.; Miller, S.J.; Fu, G.C. *Acc. Chem. Res.* **1995**, *28*, 446.

used to form very large rings, including 21-membered lactone rings.[797] Diynes can also react intramolecularly to give large-ring alkynes.[798] Metathesis with vinyl-cyclopropyl-alkynes is also known, producing a ring expanded product (see **148**).[799]

148

The synthetic importance of ring-closing and ring-opening metathesis reactions has led to the development of several new catalysts.[800] Catalysts have been developed that are compatible with both water and methanol.[801] The reaction is compatible with the presence of other functional groups,[802] such as other alkene units,[803] carbonyl units,[804] the alkene unit of conjugated esters,[805] butenolides[806] and other lactones,[807] amines,[808] amides,[809] sulfones,[810] phosphine oxides,[811] sulfonate esters,[812] and sulfonamides[813]

[797]Fürstner, A.; Langemann, K. *J. Org. Chem.* **1996**, *61*, 3942. Also see, Goldring, W.P.D.; Hodder, A.S.; Weiler, L. *Tetrahedron Lett.* **1998**, *39*, 4955; Ghosh, A.K.; Hussain, K.A. *Tetrahedron Lett.* **1998**, *39*, 1881.

[798]Chen, F.-E.; Kuang, Y.-Y.; Dai, H.-F.; Lu, L.; Huo, M. *Synthesis* **2003**, 2629.

[799]Wender, P.A.; Sperandio, D. *J. Org.Chem.* **1998**, *63*, 4164.

[800]Schrock, R.R.; Hoveyda, A.H. *Angew. Chem. Int. Ed.* **2003**. *42*, 4592; Garber, S.B.; Kingsbury, J.S.; Gray, B.L.; Hoveyda, A.H. *J. Am. Chem. Soc.* **2000**, *122*, 8168; Grela, K.; Kim, M. *Eur. J. Org. Chem.* **2003**, 963; Conon, S.J.; Dunne, A.M.; Blechert, S. *Angew. Chem. Int. Ed.* **2002**, *41*, 3835; Zhang, W.; Kraft, S.; Moore, J.S. *J. Am. Chem. Soc.* **2004**, *126*, 329; Aggarwal, V.K.; Alonso, E.; Fang, G.; Ferrara, M.; Hynd, G.; Porcelloni, M. *Angew. Chem. Int. Ed.* **2001**, *40*, 1433. Also see, references cited therein.

[801]Kirkland, T.A.; Lynn, D.M.; Grubbs, R.H. *J. Org. Chem.* **1998**, *63*, 9904.

[802]Oxygen and nitrogen-containing heterocycles can be prepared. For a review, see Deiter, S.A.; Martin, S.F. *Chem. Rev.* **2004**, *104*, 2199.

[803]Takahashi, T.; Kotora, M.; Kasai, K. *J. Chem. Soc., Chem. Commun.* **1994**, 2693.

[804]Schneider, M.F.; Junga, H.; Blechert, S. *Tetrahedron* **1995**, *51*, 13003; Junga, H.; Blechert, S. *Tetrahedron Lett.* **1993**, *34*, 3731; Llebaria, A.; Camps, F.; Moretó, J.M. *Tetrahedron Lett.* **1992**, *33*, 3683.

[805]Lee, C.W.; Grubbs, R.H. *J. Org. Chem.* **2001**, *66*, 7155.

[806]Paquette, L.A.; Méndez-Andino, J. *Tetrahedron Lett.* **1999**, *40*, 4301.

[807]Brimble, M.A.; Trzoss, M. *Tetrahedron* **2004**, *60*, 5613.

[808]Wright, D.L.; Schulte II, J.P.; Page, M.A. *Org. Lett.* **2000**, *2*, 1847; Dolman, S.J.; Sattely, E.S.; Hoveyda, A.H.; Schrock, R.R. *J. Am. Chem. Soc.* **2002**, *124*, 6991.

[809]Vo-Thanh, G.; Boucard, V.; Sauriat-Dorizon, H.; Guibé, F. *Synlett* **2001**, 37; Ma, S.; Ni, B.; Liang, Z. *J. Org. Chem.* **2004**, *69*, 6305.

[810]Yao, Q. *Org. Lett.* **2002**, *4*, 427.

[811]Demchuk, O.M.; Pietrusiewicz, K.M.; Michrowska, A.; Grela, K. *Org. Lett.* **2003**, *5*, 3217.

[812]LeFlohic, A. ;Meyer, C.; Cossy, J.; Desmurs, J.-R.; Galland, J.-C. *Synlett* **2003**, 667.

[813]Hanson, P.R.; Probst, D.A.; Robinson, R.E.; Yau, M. *Tetrahedron Lett.* **1999**, *40*, 4761; Kinderman, S.S.; Van Maarseveen, J.H.; Schoemaker, H.E.; Hiemstra, H.; Rutjes, F.P.J.T. *Org. Lett.* **2001**, *3*, 2045.

(see **149**).[814] Ether groups,[815] including vinyl ethers,[816] vinyl halides,[817] vinyl silanes,[818] vinyl sulfones,[819] allylic ethers,[820] and thioethers[821] are also compatible. Asymmetric ring-closing metathesis reactions have been reported.[822] Asymmetric ring-opening metathesis has also been reported.[823]

149

Two cyclic alkenes react to give dimeric dienes,[824] for example,

However, the products can then react with additional monomers and with each other, so that polymers are generally produced, and the cyclic dienes are obtained only in low yield. The reaction between a cyclic and a linear alkene can give an ring-opened diene:[825]

---

[814]Fürstner, A.; Picquet, M.; Bruneau, C.; Dixneuf, P.H. *Chem. Commun.* **1998**, 1315; Maier, M.E.; Lapeva, T. *Synlett* **1998**, 891; Mori, M.; Sakakibara, N.; Kinoshita, A. *J. Org. Chem.* **1998**, *63*, 6082; O'Mahony, D.J.R.; Belanger, D.B.; Livinghouse, T. *Synlett* **1998**, 443; Visser, M.S.; Heron N.M.; Didiuk, M.T.; Sagal, J.F.; Hoveyda, A.H. *J. Am. Chem. Soc.* **1996**, *118*, 4291.

[815]Edwards, S.D.; Lewis, T.; Taylor, R.J.K. *Tetrahedron Lett.* **1999**, *40*, 4267.

[816]Sturino, C.F.; Wong, J.C.Y. *Tetrahedron Lett.* **1998**, *39*, 9623; Rainier, J.D.; Cox, J.M.; Allwein, S.P. *Tetrahedron Lett.* **2001**, *42*, 179.

[817]Chao, W.; Weinreb, S.M. *Org. Lett.* **2003**, *5*, 2505.

[818]Schuman, M.; Gouverneur, V. *Tetrahedron Lett.* **2002**, *43*, 3513.

[819]Kim, S.; Lim, C.J. *Angew. Chem. Int. Ed.* **2002**, *41*, 3265.

[820]Delgado, M.; Martín, J.D. *Tetrahedron Lett.* **1997**, *38*, 6299; Miller, S.J.; Kim, S.-H.; Chen, Z.-R.; Grubbs, R.H. *J. Am. Chem. Soc.* **1995**, *117*, 2108.

[821]Leconte, M.; Pagano, S.; Mutch, A.; Lefebvre, F.; Basset, J.M. *Bull. Soc. Chim. Fr.* **1995**, *132*, 1069.

[822]Cefalo, D.R.; Kiely, A.F.; Wuchrer, M.; Jamieson, J.Y. ; Schrock, R.R.; Hoveyda, A.H. *J. Am. Chem. Soc.* **2001**, *123*, 3139.

[823]Gillingham, D.G.; Kataoka, O.; Garber, S.B.; Hoveyda, A.H. *J. Am. Chem. Soc.* **2004**, *126*, 12288.

[824]Calderon N.; Ofstead, E.A.; Judy, W.A. *J. Polym. Sci. Part A-1 1967*, *5*, 2209; Wasserman, E.; Ben-Efraim, D.A.; Wolovsky, R. *J. Am. Chem. Soc.* **1968**, *90*, 3286; Wolovsky, R.; Nir, Z. *Synthesis* **1972**, 134.

[825]Wasserman, E.; Ben-Efraim, D.A.; Wolovsky, R. *J. Am. Chem. Soc.* **1968**, *90*, 3286; Ray, G.C.; Crain, D.L. Fr. Pat. 1511381, 1968 [*Chem. Abstr.*, *1969*, *70*, 114580q]; Mango, F.D. U.S. Pat. 3424811, 1969 [*Chem. Abstr.*, *1969*, *70*, 106042a]; Rossi, R.; Diversi, P.; Lucherini, A.; Porri, L. *Tetrahedron Lett.* **1974**, 879; Lal, J.; Smith, R.R. *J. Org. Chem.* **1975**, *40*, 775.

Alkenes containing functional groups[826] do not give the reaction with most of the common catalysts, but some success has been reported with WCl_6–$SnMe_4$[827] and with certain other catalysts.

The reaction has also been applied to internal triple bonds:[828]

$$2\,RC\equiv CR' \rightleftharpoons RC\equiv CR + R'C\equiv CR'$$

but it has not been successful for terminal triple bonds,[829] although as noted above, molecules with a terminal alkene and a terminal alkyne react quite well. Ring-closing metathesis of alkene–alkynes leads to a cyclic alkene with a pendant vinyl unit (a diene).[830] Intramolecular reactions of a double bond with a triple bond are known[831] and a tetracyclic tetraene has been prepared from a poly-yne-diene.[832]

The generally accepted mechanism is a chain mechanism,[833] involving the intervention of a metal–carbene complex (**150** and **151**)[834] and a four-membered ring

[826]For a review, see Mol, J.C. *CHEMTECH* **1983**, 250. See also, Bosma, R.H.A.; van den Aardweg, G.C.N.; Mol, J.C. *J. Organomet. Chem.* **1983**, *255*, 159; **1985**, *280*, 115; Xiaoding, X.; Mol, J.C. *J. Chem. Soc., Chem. Commun.* **1985**, 631; Crisp, C.T.; Collis, M.P. *Aust. J. Chem.* **1988**, *41*, 935.

[827]First shown by van Dam, P.B.; Mittelmeijer, M.C.; Boelhouwer, C. *J. Chem. Soc., Chem. Commun.* **1972**, 1221.

[828]Pennella, F.; Banks, R.L.; Bailey, G.C. *Chem. Commun.* **1968**, 1548; Villemin, D.; Cadiot, P. *Tetrahedron Lett.* **1982**, *23*, 5139; McCullough, L.G.; Schrock, R.R. *J. Am. Chem. Soc.* **1984**, *106*, 4067; Fürstner, A.; Mathes, C. *Org. Lett.* **2001**, *3*, 221; Fürstner, A.; Mathes, C.; Lehmann, C.W. *J. Am. Chem. Soc.* **1999**, *121*, 9453; Fürstner, A.; Guth, O.; Rumbo, A.; Seidel, G. *J. Am. Chem. Soc.* **1999**, *121*, 11108; Brizius, G.; Bunz, U.H.F. *Org. Lett.* **2002**, *4*, 2829; Grela, K.; Ignatonska, J. *Org. Lett.* **2002**, *4*, 3747. For a review, see Tamao, K.; Kobayashi, K.; Ito, Y. *Synlett* **1992**, 539.

[829]McCullough, L.G.; Listemann, M.L.; Schrock, R.R.; Churchill, M.R.; Ziller, J.W. *J. Am. Chem. Soc.* **1983**, *105*, 6729.

[830]Mori, M.; Kitamura, T.; Sakakibara, N.; Sato, Y. *Org. Lett.* **2000**, *2*, 543; Kitamura, T.; Mori, M. *Org. Lett.* **2001**, *3*, 1161.

[831]Trost, B.M.; Trost, M.K. *J. Am. Chem. Soc.* **1991**, *113*, 1850; Gilbertson, S.R.; Hoge, G.S. *Tetrahedron Lett.* **1998**, *39*, 2075.

[832]Zuercher, W.J.; Scholl, M.; Grubbs, R.H. *J. Org. Chem.* **1998**, *63*, 4291.

[833]For a discussion of the mechanism of ring-closing meththesis, see Sanford, M.S.; Ulman, M.; Grubbs, R.H. *J. Am. Chem. Soc.* **2001**, *123*, 749; Sanford, M.S.; Love, J.A.; Grubbs, R.H. *J. Am. Chem. Soc.* **2001**, *123*, 6543; Cavallo, L. *J. Am. Chem. Soc.* **2002**, *124*, 8965; Adlhart, C.; Chen, P. *J. Am. Chem. Soc.* **2004**, *126*, 3496.

[834]For a review of these complexes and their role in this reaction, see Crabtree, R.H. *The Organometallic Chemistry of the Transition Metals*, Wiley, NY, **1988**, pp. 244–267.

containing a metal[835] (**152–155**).[836] In the cross-metathesis reaction shown as an example, $R_2C=CR_2$ reacts with $R^1_2C=CR^1_2$ in the presence of a metal catalyst, M. Initial reaction with the catalyst leads to the two expected metal carbenes, **150** and **151**. Metal carbene **151** can react with both alkenes to form metallocyclobutanes **152** and **153**. Each of these intermediates loses the metal to form the alkenes, the product of metathesis $R_2C=CR^1_2$ and the one of the original alkenes. In a likewise manner, **150** reacts with each alkene to form metallocyclobutanes **154** and **155**, which decomposes to $R_2C=CR_2$ and the metathesis product.

[835]For reviews of metallocycles, see Collman, J.C.; Hegedus, L.S.; Norton, J.R.; Finke, R.G. *Principles and Applications of Organotransition Metal Chemistry*, 2nd ed., University Science Books, Mill Valley, CA; *1987*, pp. 459–520; Lindner, E. *Adv. Heterocycl. Chem. 1986*, *39*, 237.

[836]For reviews of the mechanism, see Grubbs, R.H. *Prog. Inorg. Chem. 1978*, *24*, 1; Katz, T.J. *Adv. Organomet. Chem. 1977*, *16*, 283; Calderon N.; Ofstead, E.A.; Judy, W.A. *Angew. Chem. Int. Ed. 1976*, *15*, 401. See also McLain, S.J.; Wood, C.D.; Schrock, R.R. *J. Am. Chem. Soc. 1977*, *99*, 3519; Casey, C.P.; Polichnowski, S.W. *J. Am. Chem. Soc. 1977*, *99*, 6097; Mango, F.D. *J. Am. Chem. Soc. 1977*, *99*, 6117; Stevens, A.E.; Beauchamp, J.L. *J. Am. Chem. Soc. 1979*, *101*, 6449; Lee, J.B.; Ott, K.C.; Grubbs, R.H. *J. Am. Chem. Soc. 1982*, *104*, 7491; Levisalles, J.; Rudler, H.; Villemin, D. *J. Organomet. Chem. 1980*, *193*, 235; Iwasawa, Y.; Hamamura, H. *J. Chem. Soc., Chem. Commun. 1983*, 130; Rappé, A.K.; Upton, T.H. *Organometallics, 1984*, *3*, 1440; Kress, J.; Osborn, J.A.; Greene, R.M.E.; Ivin, K.J.; Rooney, J.J. *J. Am. Chem. Soc. 1987*, *109*, 899; Feldman, J.; Davis, W.M.; Schrock, R.R. *Organometallics, 1989*, 8, 2266.

OS **80**, 85; **81**, 1.

18-38   Metal-Ion-Catalyzed σ-Bond Rearrangements

Many highly strained cage molecules undergo rearrangement when treated with metallic ions, such as Ag⁺, Rh(I), or Pd(II).[837] The bond rearrangements observed can be formally classified into two main types: (*1*) [2+2]-ring

openings of cyclobutanes and (*2*) conversion of a bicyclo[2.2.0] system to a bicyclopropyl system. The molecule cubane supplies an example of each type (see above). Treatment with Rh(I) complexes converts cubane to tricyclo[4.2.0.0²·⁵]octa-3,7-diene (**156**),[838] an example of type 1, while Ag⁺ or Pd(II) causes the second type of reaction, producing cuneane.[839] Other examples are

---

[837]For reviews, see Halpern, J., in Wender, I.; Pino, P. *Organic Syntheses via Metal Carbonyls*, Vol. 2, Wiley, NY, *1977*, pp. 705–721; Bishop III, K.C. *Chem. Rev. 1976, 76*, 461; Cardin, D.J.; Cetinkaya, B.; Doyle, M.J.; Lappert, M.F. *Chem. Soc. Rev. 1973, 2*, 99, 132–139; Paquette, L.A. *Synthesis 1975*, 347; *Acc. Chem. Res. 1971, 4*, 280.

[838]Eaton, P.E.; Chakraborty, U.R. *J. Am. Chem. Soc. 1978, 100*, 3634.

[839]Cassar, L.; Eaton, P.E.; Halpern, J. *J. Am. Chem. Soc. 1970, 92*, 6336.

[840]Gassman, P.G.; Atkins, T.J. *J. Am. Chem. Soc. 1971, 93*, 4579; *1972, 94*, 7748; Sakai, M.; Westberg, H.H.; Yamaguchi, H.; Masamune, S. *J. Am. Chem. Soc. 1972, 93*, 4611; Paquette, L.A.; Wilson, S.E.; Henzel, R.P. *J. Am. Chem. Soc. 1972, 94*, 7771.

Type 2 [841]

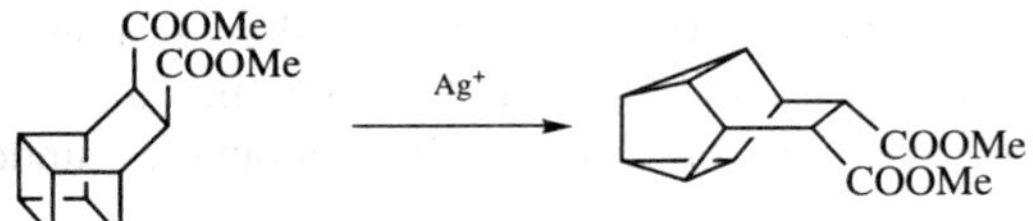

Ref. [842]

159

159 is the 9,10-dicarbomethyoxy derivative of *snoutane* (pentacyclo[3.3.2.0^{2,4}.0^{3,7}.0^{6,8}] decane).

The mechanisms of these reactions are not completely understood, although relief of strain undoubtedly supplies the driving force. The reactions are thermally forbidden by the orbital-symmetry rules, and the role of the catalyst is to provide low-energy pathways so that the reactions can take place. The type 1 reactions are the reverse of the catalyzed [2 + 2] ring closures discussed at **15-63**. The following mechanism, in which Ag$^+$ attacks one of the edge bonds, has been suggested for the conversion of **157** to **158**.[843]

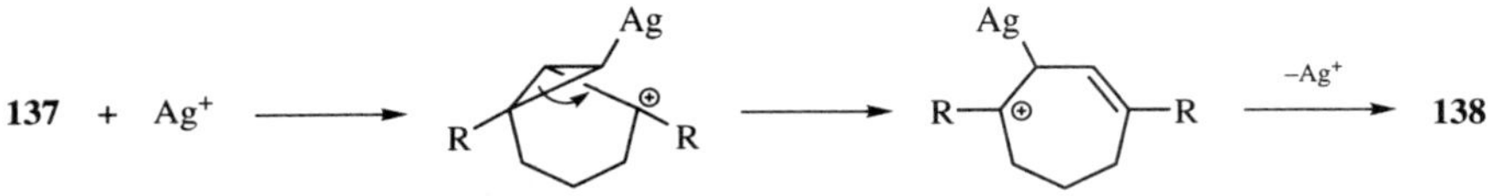

137 + Ag$^+$ ⟶ ⟶ $\xrightarrow{-\text{Ag}^+}$ **138**

Simpler bicyclobutanes can also be converted to dienes, but in this case the products usually result from cleavage of the central bond and one of the edge bonds.[844] For example, treatment of **160** with AgBF$_4$,[845]

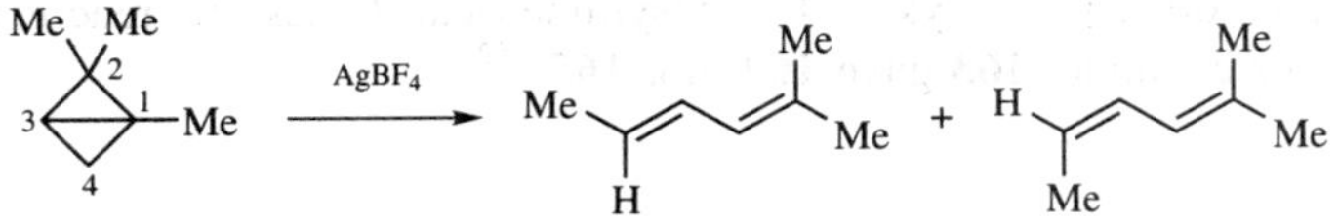

160

---

[841]The starting compound here is a derivative of basketane, or 1,8-bishomocubane. For a review of homo-, bishomo-, and trishomocubanes, see Marchand, A.P. *Chem. Rev.* **1989**, *89*, 1011.

[842]See, for example, Furstoss, R.; Lehn, J.M. *Bull. Soc. Chim. Fr.* **1966**, 2497; Dauben, W.G.; Kielbania Jr., A.J. *J. Am. Chem. Soc.* **1971**, *93*, 7345; Paquette, L.A.; Beckley, R.S.; Farnham, W.B. *J. Am. Chem. Soc.* **1975**, *97*, 1089.

[843]Gassman, P.G.; Atkins, T.J. *J. Am. Chem. Soc.* **1971**, *93*, 4579; Sakai, M.; Westberg, H.H.; Yamaguchi, H.; Masamune, S. *J. Am. Chem. Soc.* **1972**, *93*, 4611.

[844]Compound **157** can also be cleaved in this manner, giving a 3-methylenecyclohexene. See, for example, Dauben, W.G.; Kielbania Jr., A.J. *J. Am. Chem. Soc.* **1972**, *94*, 3669; Gassman, P.G.; Reitz, R.R. *J. Am. Chem. Soc.* **1973**, *95*, 3057; Paquette, L.A.; Zon, G. *J. Am. Chem. Soc.* **1974**, *96*, 203, 224.

[845]Paquette, L.A.; Henzel, R.P.; Wilson, S.E. *J. Am. Chem. Soc.* **1971**, *93*, 2335.

or $[(\pi\text{-allyl})PdCl]_2$[846] gives a mixture of the two dienes shown, resulting from a formal cleavage of the C_1–C_3 and C_1–C_2 bonds (note that a hydride shift has taken place). Dienes can also be converted to bicyclobutanes under photochemical conditions.[847]

18-39   The Di-π-methane and Related Rearrangements

Di-π-methane rearrangement

1,4-Dienes carrying alkyl or aryl substituents on C-3[848] can be photochemically rearranged to vinylcyclopropanes in a reaction called the *di-π-methane rearrangement*.[849] An example is conversion of **161** to **162**.[850] For most

161   **162**

1,4-dienes it is only the singlet excited states that give the reaction; triplet states generally take other pathways.[851] For unsymmetrical dienes, the reaction is regioselective. For example, **163** gave **164**, not **165**:[852]

[846]Gassman, P.G.; Meyer, R.G.; Williams, F.J. *Chem. Commun.* **1971**, 842.

[847]Garavelli, M.; Frabboni, B.; Fato, M.; Celani, P.; Bernardi, F.; Robb, M.A.; Olivucci, M. *J. Am. Chem. Soc.* **1999**, *121*, 1537.

[848]Zimmerman, H.E.; Pincock, J.A. *J. Am. Chem. Soc.* **1973**, *95*, 2957.

[849]For reviews, see Zimmerman, H.E. *Org. Photochem.* **1991**, *11*, 1; Zimmerman, H.E., in de Mayo, P. *Rearrangements in Ground and Excited States*, Vol. 3, Academic Press, NY, **1980**, pp. 131–166; Hixson, S.S.; Mariano, P.S.; Zimmerman, H.E. *Chem. Rev.* **1973**, *73*, 531. See also: Roth, W.R.; Wildt, H.; Schlemenat, A. *Eur. J. Org. Chem.* **2001**, 4081.

[850]Zimmerman, H.E.; Hackett, P.; Juers, D.F.; McCall, J.M.; Schröder, B. *J. Am. Chem. Soc.* **1971**, *93*, 3653.

[851]However, some substrates, generally rigid bicyclic molecules, (e.g., barrelene, p. 152, which is converted to semi-bullvalene) give the di-π-methane rearrangement only from triplet states.

[852]Zimmerman, H.E.; Baum, A.A. *J. Am. Chem. Soc.* **1971**, *93*, 3646. See also, Zimmerman, H.E.; Welter, T.R. *J. Am. Chem. Soc.* **1978**, *100*, 4131; Alexander, D.W.; Pratt, A.C.; Rowley, D.H.; Tipping, A.E. *J. Chem. Soc., Chem. Commun.* **1978**, 101; Paquette, L.A.; Bay, E.; Ku, A.Y.; Rondan, N.G.; Houk, K.N. *J. Org. Chem.* **1982**, *47*, 422.

163   **164**   **165** Not formed

The mechanism can be described by the diradical pathway given[853] (the C-3 substituents act to stabilize the radical), though the species shown are not necessarily intermediates, but may be transition states. It has been shown, for the case of certain substituted substrates, that configuration is retained at C-1 and C-5 and inverted at C-3.[854]

The reaction has been extended to allylic benzenes[855] (in this case C-3 substituents are not required), to β,γ-unsaturated ketones[856] (the latter reaction, which is called the *oxa-di-π-methane rearrangement*,[857] generally occurs only from the triplet state), to β,γ-unsaturated imines,[858] and to triple-bond systems.[859]

[853]See Zimmerman, H.E.; Little, R.D. *J. Am. Chem. Soc.* **1974**, *96*, 5143; Zimmerman, H.E.; Boettcher, R.J.; Buehler, N.E.; Keck, G.E. *J. Am. Chem. Soc.* **1975**, *97*, 5635. For an argument against the intermediacy of the •CH$_2$–cyclopropyl–CH$_2$• intermediate, see Adam, W.; De Lucchi, O.; Dörr, M. *J. Am. Chem. Soc.* **1989**, *111*, 5209.

[854]Zimmerman, H.E.; Robbins, J.D.; McKelvey, R.D.; Samuel, C.J.; Sousa, L.R. *J. Am. Chem. Soc.* **1989**, *111*, 5209.

[855]For example, see Griffin, G.W.; Covell, J.; Petterson, R.C.; Dodson, R.M.; Klose, G. *J. Am. Chem. Soc.* **1965**, *87*, 1410; Hixson, S.S. *J. Am. Chem. Soc.* **1972**, *94*, 2507; Cookson, R.C.; Ferreira, A.B.; Salisbury, K. *J. Chem. Soc., Chem. Commun.* **1974**, 665; Fasel, J.; Hansen, H. *Chimia*, **1982**, *36*, 193; Paquette, L.A.; Bay, E. *J. Am. Chem. Soc.* **1984**, *106*, 6693; Zimmerman, H.E.; Swafford, R.L. *J. Org. Chem.* **1984**, *49*, 3069.

[856]For reviews of photochemical rearrangements of unsaturated ketones, see Schuster, D.I., in de Mayo, P. *Rearrangements in Ground and Excited States*, Vol. 3, Academic Press, NY, **1980**, pp. 167–279; Houk, K.N. *Chem. Rev.* **1976**, *76*, 1; Schaffner, K. *Tetrahedron* **1976**, *32*, 641; Dauben, W.G.; Lodder, G.; Ipaktschi, J. *Top. Curr. Chem.* **1975**, *54*, 73.

[857]For a review, see Demuth, M. *Org. Photochem.* **1991**, *11*, 37.

[858]See Armesto, D.; Horspool, W.M.; Langa, F.; Ramos, A. *J. Chem. Soc. Perkin Trans. 1* **1991**, 223.

[859]See Griffin, G.W.; Chihal, D.M.; Perreten, J.; Bhacca, N.S. *J. Org. Chem.* **1976**, *41*, 3931.

When photolyzed, 2,5-cyclohexadienones can undergo a number of different reactions, one of which is formally the same as the di-π-methane rearrangement.[860] In this reaction, photolysis of the substrate **166** gives the bicyclo[3.1.0]hexenone (**171**). Although the reaction is formally the same (note the conversion of **161** to **162**

above), the mechanism is different from that of the di-π-methane rearrangement, because irradiation of a ketone can cause an $n \to \pi^*$ transition, which is of course not possible for a diene lacking a carbonyl group. The mechanism[861] in this case has been formulated as proceeding through the excited triplet states **168** and **169**. In step 1, the molecule undergoes an $n \to \pi^*$ excitation to the singlet species **167**, which cross to the triplet **168**. Step 3 is a rearrangement from one excited state to another. Step 4 is a $\pi^* \to n$ electron demotion (an intersystem crossing from $T_1 \to S_0$, see p. 339). The conversion of **170** to **171** consists of two 1,2 alkyl migrations (a one-step process would be a 1,3-migration of alkyl to a carbocation center, see p. $$$): The old $C_6–C_5$ bond becomes the new $C_6–C_4$ bond and the old $C_6–C_1$ bond becomes the new $C_6–C_5$ bond.[862]

2,4-Cyclohexadienones also undergo photochemical rearrangements, but the products are different, generally involving ring opening.[863]

[860]For reviews of the photochemistry of 2,5-cyclohexadienones and related compounds, see Schaffner, K.; Demuth, M., in de Mayo, P. *Rearrangements in Ground and Excited States*, Vol. 3, Academic Press, NY, *1980*, pp. 281–348; Zimmerman, H.E. *Angew. Chem. Int. Ed. 1969*, 8, 1; Kropp, P.J. *Org. Photochem. 1967*, 1, 1; Schaffner, K. *Adv. Photochem. 1966*, 4, 81. For synthetic use, see Schultz, A.G.; Lavieri, F.P.; Macielag, M.; Plummer, M. *J. Am. Chem. Soc. 1987*, 109, 3991, and references cited therein.

[861]Schuster, D.I. *Acc. Chem. Res. 1978*, 11, 65; Zimmerman, H.E.; Pasteris, R.J. *J. Org. Chem. 1980*, 45, 4864, 4876; Schuster, D.I.; Liu, K. *Tetrahedron 1981*, 37, 3329.

[862]Zimmerman, H.E.; Crumine, D.S.; Döpp, D.; Huyffer, P.S. *J. Am. Chem. Soc. 1969*, 91, 434.

[863]For reviews, see Schaffner, K.; Demuth, M., in de Mayo, P. *Rearrangements in Ground and Excited States*, Vol. 3, Academic Press, NY, *1980*, p. 281; Quinkert, G. *Angew. Chem. Int. Ed. 1972*, 11, 1072; Kropp, P.J. *Org. Photochem. 1967*, 1, 1.

18-40   The Hofmann–Löffler and Related Reactions

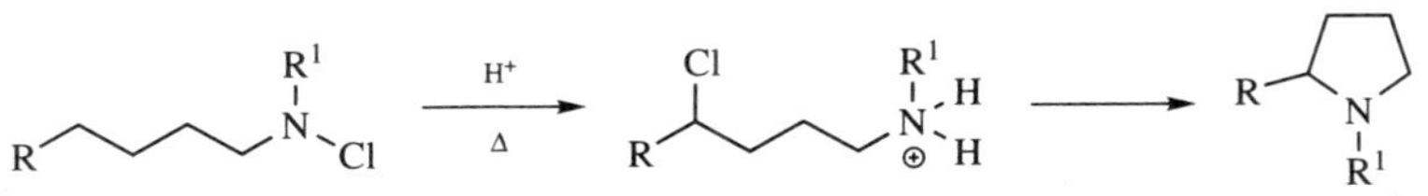

A common feature of the reactions in this section[864] is that they serve to introduce functionality at a position remote from functional groups already present. As such, they have proved very useful in synthesizing many compounds, especially in the steroid field (see also, **19-2** and **19-17**). When N-haloamines in which one alkyl group has a hydrogen in the 4 or 5 position are heated with sulfuric acid, pyrrolidines, or piperidines are formed, in a reaction known as the *Hofmann–Löffler reaction* (also called the *Hofmann–Löffler–Freytag reaction*).[865] The R′ group is normally alkyl, but the reaction has been extended to R′ = H by the use of concentrated sulfuric acid solution and ferrous salts.[866] The first step of the reaction is a rearrangement, with the halogen migrating from the nitrogen to the 4 or 5 position of the alkyl group. It is possible to isolate the resulting haloamine salt, but usually this is not done, and the second step, the ring closure (**10-31**), takes place. Though the reaction is most often induced by heat, this is not necessary, and irradiation and chemical initiators (e.g., peroxides) have been used instead. The mechanism is of a free-radical type, with the main step involving an internal hydrogen abstraction.[867]

Initiation

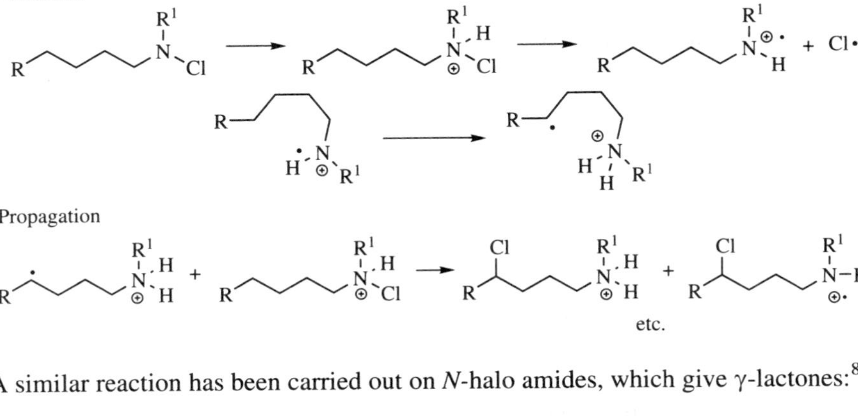

Propagation

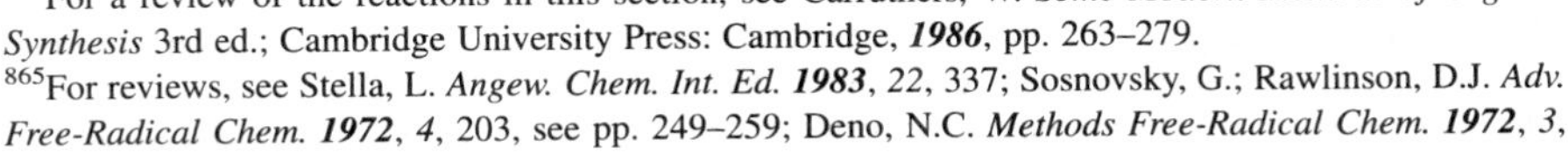

etc.

A similar reaction has been carried out on N-halo amides, which give γ-lactones:[868]

---

[864]For a review of the reactions in this section, see Carruthers, W. *Some Modern Methods of Organic Synthesis* 3rd ed.; Cambridge University Press: Cambridge, *1986*, pp. 263–279.

[865]For reviews, see Stella, L. *Angew. Chem. Int. Ed. 1983*, 22, 337; Sosnovsky, G.; Rawlinson, D.J. *Adv. Free-Radical Chem. 1972*, 4, 203, see pp. 249–259; Deno, N.C. *Methods Free-Radical Chem. 1972*, 3, 135, see pp. 136–143.

[866]Schmitz, E.; Murawski, D. *Chem. Ber. 1966*, 99, 1493.

[867]Wawzonek, S.; Thelan, P.J. *J. Am. Chem. Soc. 1950*, 72, 2118.

[868]Barton, D.H.R.; Beckwith, A.L.J.; Goosen, A. *J. Chem. Soc. 1965*, 181; Petterson, R.C.; Wambsgans, A. *J. Am. Chem. Soc. 1964*, 86, 1648; Neale, R.S.; Marcus, N.L.; Schepers, R.G. *J. Am. Chem. Soc. 1966*, 88, 3051. For a review of N-halo amide rearrangements, see Neale, R.S. *Synthesis 1971*, 1.

Another related reaction is the *Barton reaction*,[869] by which a methyl group in the ∂ position to an OH group can be oxidized to a CHO group. The alcohol is first converted to the nitrite ester. Photolysis of the nitrite results in conversion of the nitrite group to the OH group and nitrosation of the methyl group. Hydrolysis of the oxime tautomer gives the aldehyde, for example,[870]

This reaction takes place only when the methyl group is in a favorable steric position.[871] The mechanism is similar to that of the Hofmann–Löffler reaction.[872]

[869]For reviews, see Hesse, R.H. *Adv. Free-Radical Chem.* **1969**, *3*, 83; Barton, D.H.R. *Pure Appl. Chem.* **1968**, *16*, 1.

[870]Barton, D.H.R.; Beaton, J.M. *J. Am. Chem. Soc.* **1961**, *83*, 4083. Also see, Barton, D.H.R.; Beaton, J.M.; Geller, L.E.; Pechet, M.M. *J. Am. Chem. Soc.* **1960**, *82*, 2640.

[871]For a discussion of which positions are favorable, see Burke, S.D.; Silks III, L.A.; Strickland, S.M.S. *Tetrahedron Lett.* **1988**, *29*, 2761.

[872]Kabasakalian, P.; Townley, E.R. *J. Am. Chem. Soc.* **1962**, *84*, 2711; Akhtar, M.; Barton, D.H.R.; Sammes, P.G. *J. Am. Chem. Soc.* **1965**, *87*, 4601. See also, Nickon, A.; Ferguson, R.; Bosch, A.; Iwadare, T. *J. Am. Chem. Soc.* **1977**, *99*, 4518; Barton, D.H.R.; Hesse, R.H.; Pechet, M.M.; Smith, L.C. *J. Chem. Soc. Perkin Trans. 1* **1979**, 1159; Green, M.M.; Boyle, B.A.; Vairamani, M.; Mukhopadhyay, T.; Saunders, Jr., W.H.; Bowen, P.; Allinger, N.L. *J. Am. Chem. Soc.* **1986**, *108*, 2381.

This is one of the few known methods for effecting substitution at an angular methyl group. Not only CH_3 groups, but also alkyl groups of the form RCH_2 and R_2CH can give the Barton reaction if the geometry of the system is favorable. An RCH_2 group is converted to the oxime $R(C{=}NOH)$ (which is hydrolyzable to a ketone) or to a nitroso dimer, while an R_2CH group gives a nitroso compound $R_2C(NO)$. With very few exceptions, the only carbons that become nitrosated are those in the position δ to the original OH group, indicating that a six-membered transition state is necessary for the hydrogen abstraction.[873]

OS **III**, 159.

D. Noncyclic Rearrangements

18-41   Hydride Shifts

The above is a typical example of a transannular hydride shift. The 1,2-diol is formed by a normal epoxide hydrolysis reaction (**10-7**). For a discussion of 1,3 and longer hydride shifts (see p. 1572).

18-42   The Chapman Rearrangement

1/O → 3/N-Aryl-migration

In the *Chapman rearrangement,* N,N-diaryl amides are formed when aryl imino esters are heated.[874] Best yields are obtained in refluxing tetraethylene glycol dimethyl ether (tetraglyme),[875] although the reaction can also be carried out without any solvent at all. Many groups may be present in the rings, for example, alkyl, halo, OR, CN, and COOR. Aryl migrates best when it contains electron-withdrawing groups. On the other hand, electron-withdrawing groups in Ar^2 or Ar^3 decrease the reactivity. The products can be hydrolyzed to diarylamines, and

[873]For a discussion, see Nickon, A.; Ferguson, R.; Bosch, A.; Iwadare, T. *J. Am. Chem. Soc.* **1977**, *99*, 4518.
[874]For reviews, see Schulenberg, J.W.; Archer, S. *Org. React.* **1965**, *14*, 1; McCarty, C.G., in Patai, S. *The Chemistry of the Carbon-Nitrogen Double Bond*, Wiley, NY, **1970**, pp. 439–447; McCarty, C.G.; Garner, L.A., in Patai, S. *The Chemistry of Amidines and Imidates*, Wiley, NY, **1975**, pp. 189–240. For a review of 1,3 migrations of R in general, see Landis, P.S. *Mech. Mol. Migr.* **1969**, *2*, 43.
[875]Wheeler, O.H.; Roman, F.; Santiago, M.V.; Quiles, F. *Can. J. Chem.* **1969**, *47*, 503.

this is a method for preparing these compounds. The mechanism probably involves an intramolecular[876] aromatic nucleophilic substitution, resulting in a 1,3 oxygen-to-nitrogen shift. Aryl imino esters can be prepared from *N*-aryl amides by reaction with PCl_5, followed by treatment of the resulting imino chloride with an aroxide ion.[877]

Imino esters with any or all of the three groups being alkyl also rearrange, but they require catalysis by H_2SO_4 or a trace of methyl iodide or methyl sulfate.[878] The mechanism is different, involving an intermolecular process.[879] This is also true for derivatives for formamide ($Ar^2 = H$).

18-43 The Wallach Rearrangement

The conversion of azoxy compounds, on acid treatment, to *p*-hydroxy azo compounds (or sometimes the *o*-hydroxy isomers[880]) is called the *Wallach rearrangement*.[881] When both para positions are occupied, the *o*-hydroxy product may be

---

[876]For evidence for the intramolecular character of the reaction, see Wiberg, K.B.; Rowland, B.I. *J. Am. Chem. Soc.* **1955**, *77*, 2205; Wheeler, O.H.; Roman, F.; Rosado, O. *J. Org. Chem.* **1969**, *34*, 966; Kimura, M. *J. Chem. Soc. Perkin Trans. 2* **1987**, 205.

[877]For a review of the formation and reactions of imino chlorides, see Bonnett, R., in Patai, S. *The Chemistry of the Carbon–Nitrogen Double Bond*, Wiley, NY, **1970**, pp. 597–662.

[878]Landis, P.S. *Mech. Mol. Migr.* **1969**, *2*, 43.

[879]See Challis, B.C.; Frenkel, A.D. *J. Chem. Soc. Perkin Trans. 2* **1978**, 192.

[880]For example, see Dolenko, A.; Buncel, E. *Can. J. Chem.* **1974**, *52*, 623; Yamamoto, J.; Nishigaki, Y.; Umezu, M.; Matsuura, T. *Tetrahedron* **1980**, *36*, 3177.

[881]For reviews, see Buncel, E. *Mech. Mol. Migr.* **1968**, *1*, 61; Shine, H.J. *Aromatic Rearrangements*, Elsevier, NY, **1969**, pp. 272–284, 357–359; Cox, R.A.; Buncel, E., in Patai, S. *The Chemistry of the Hydrazo, Azo, and Azoxy Groups*, pt. 2, Wiley, NY, **1975**, pp. 808–837.

obtained, but ipso substitution at one of the para positions is also possible.[882] Although the mechanism[883] is not completely settled, the following facts are known: (*1*) The para rearrangement is intermolecular.[884] (*2*) When the reaction was carried out with an azoxy compound in which the N–O nitrogen was labeled with ^{15}N, *both* nitrogens of the product carried the label equally,[885] demonstrating that the oxygen did not have a preference for migration to either the near or the far ring. This shows that there is a symmetrical intermediate. (*3*) Kinetic studies show that two protons are normally required for the reaction.[886] The following mechanism,[887] involving the symmetrical intermediate **175**, has been proposed to explain the facts.[888]

It has proved possible to obtain **174** and **175** as stable species in super acid solutions.[748] Another mechanism, involving an intermediate with only one positive charge, has been proposed for certain substrates at low acidities.[889]

A photochemical Wallach rearrangement[890] is also known: The product is the *o*-hydroxy azo compound, the OH group is found in the farther ring, and the rearrangement is intramolecular.[891]

[882]See, for example, Shimao, I.; Oae, S. *Bull. Chem. Soc. Jpn.* **1983**, *56*, 643.

[883]For reviews, see Furin, G.G. *Russ. Chem. Rev.* **1987**, *56*, 532; Williams, D.L.H.; Buncel, E. *Isot. Org. Chem.* **1980**, *5*, 184; Buncel, E. *Acc. Chem. Res.* **1975**, *8*, 132.

[884]See, for example, Oae, S.; Fukumoto, T.; Yamagami, M. *Bull. Chem. Soc. Jpn.* **1963**, *36*, 601.

[885]Shemyakin, M.M.; Maimind, V.I.; Vaichunaite, B.K. *Chem. Ind.* (*London*) **1958**, 755; *Bull. Acad. Sci. USSR Div. Chem. Sci.* **1960**, 808. Also see Behr, L.C.; Hendley, E.C. *J. Org. Chem.* **1966**, *31*, 2715.

[886]Buncel, E.; Lawton, B.T. *Chem. Ind.* (*London*) **1963**, 1835; Hahn, C.S.; Lee, K.W.; Jaffé, H.H. *J. Am. Chem. Soc.* **1967**, *89*, 4975; Cox, R.A. *J. Am. Chem. Soc.* **1974**, *96*, 1059.

[887]Buncel, E.; Strachan, W.M.J. *Can. J. Chem.* **1970**, *48*, 377; Cox, R.A. *J. Am. Chem. Soc.* **1974**, *96*, 1059; Buncel, E.; Keum, S. *J. Chem. Soc., Chem. Commun.* **1983**, 578.

[888]For other proposed mechanisms, see Shemyakin, M.M.; Agadzhanyan, Ts.E.; Maimind, V.I.; Kudryavtsev, R.V. *Bull. Acad. Sci. USSR Div. Chem. Sci.* **1963**, 1216; Hahn, C.S.; Lee, K.W.; Jaffé, H.H. *J. Am. Chem. Soc.* **1967**, *89*, 4975; Hendley, E.C.; Duffey, D. *J. Org. Chem.* **1970**, *35*, 3579.

[889]Cox, R.A.; Dolenko, A.; Buncel, E. *J. Chem. Soc. Perkin Trans. 2* **1975**, 471; Cox, R.A.; Buncel, E. *J. Am. Chem. Soc.* **1975**, *97*, 1871.

[890]For a thermal rearrangement (no catalyst), see Shimao, I.; Hashidzume, H. *Bull. Chem. Soc. Jpn.* **1976**, *49*, 754.

[891]For discussions of the mechanism of the photochemical reaction, see Goon, D.J.W.; Murray, N.G.; Schoch, J.; Bunce, N.J. *Can. J. Chem.* **1973**, *51*, 3827; Squire, R.H.; Jaffé, H.H. *J. Am. Chem. Soc.* **1973**, *95*, 8188; Shine, H.J.; Subotkowski, W.; Gruszecka, E. *Can. J. Chem.* **1986**, *64*, 1108.

18-44 Dyotropic Rearrangements

1/C-Trialkylsilyl,2/O-trialkylsilyl-interchange

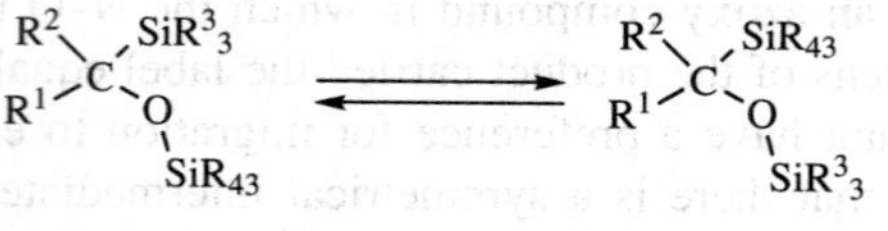

$$R^2-\underset{\underset{\displaystyle SiR_{43}}{\overset{\displaystyle |}{C}}}{\overset{\displaystyle SiR^3_3}{\overset{\displaystyle |}{\diagdown}}} O \rightleftharpoons R^2-\underset{\underset{\displaystyle SiR^3_3}{\overset{\displaystyle |}{C}}}{\overset{\displaystyle SiR_{43}}{\overset{\displaystyle |}{\diagdown}}} O$$

A *dyotropic rearrangement*[892] is an uncatalyzed process in which two σ bonds simultaneously migrate intramolecularly.[893] There are two types. The above is an example of type 1, which consists of reactions in which the two σ bonds interchange positions. In type 2, the two σ bonds do not interchange positions. An example is

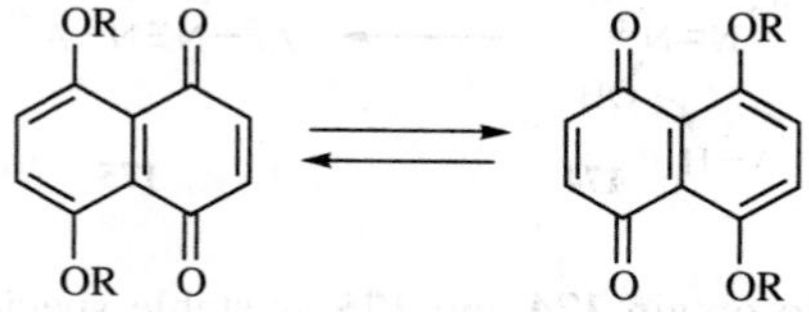

Some other examples are

Type 1

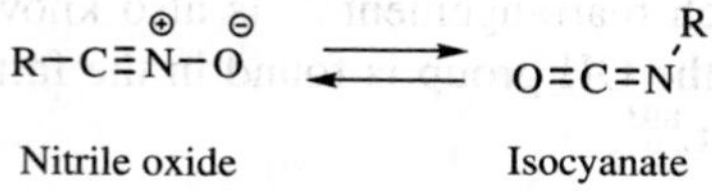

$$R-C\overset{\oplus}{\equiv}N-\overset{\ominus}{O} \rightleftharpoons O=C=N\overset{R}{\diagup}$$ Ref. [894]

Nitrile oxide   Isocyanate

Type 1

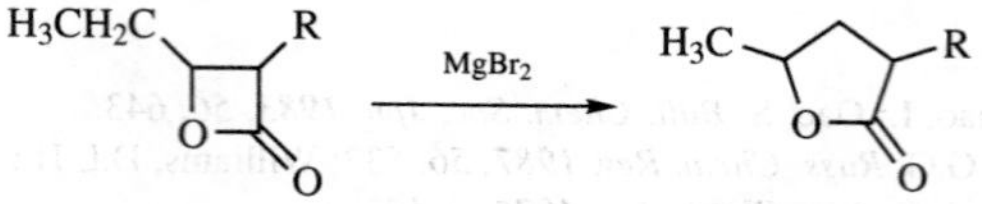

Ref. [895]

Type 2

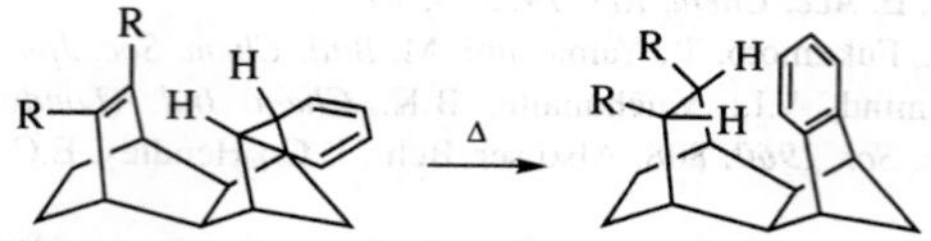

Ref. [896]

A useful type 1 example is the *Brook rearrangement*,[897] a stereospecific intramolecular migration of silicon from carbon to oxygen that occurs for

---

[892]Reetz, M.T. *Angew. Chem. Int. Ed.* **1972**, *11*, 129, 130.

[893]For reviews, see Minkin, V.I.; Olekhnovich, L.P.; Zhdanov, Yu.A. *Molecular Design of Tautomeric Compounds*, D. Reidel Publishing Co., Dordrecht, **1988**, pp. 221–246; Minkin, V.I. *Sov. Sci. Rev. Sect. B* **1985**, *7*, 51; Reetz, M.T. *Adv. Organomet. Chem.* **1977**, *16*, 33. Also see Mackenzie, K.; Gravaatt, E.C.; Gregory, R.J.; Howard, J.A.K.; Maher, J.P. *Tetrahedron Lett.* **1992**, *33*, 5629.

[894]See, for example, Taylor, G.A. *J. Chem. Soc. Perkin Trans. 1* **1985**, 1181.

[895]See Black, T.H.; Hall, J.A.; Sheu, R.G. *J. Org. Chem.* **1988**, *53*, 2371; Black, T.H.; Fields, J.D. *Synth. Commun.* **1988**, *18*, 125.

[896]See Mackenzie, K.; Proctor, G.; Woodnutt, D.J. *Tetrahedron* **1987**, *43*, 5981, and references cited therein.

[897]For a review, see Moser, W.H. *Tetrahedron* **2001**, *57*, 2065.

(α-hydroxybenzyl)trialkylsilanes (**176**) in the presence of a catalytic amount of base.[898] Formation of a Si–O bond rather than the Si–C bond drives the rearrangement, which is believed to proceed via formation of **177**, and does proceed with inversion of configuration at carbon and retention of configuration at silicon.[899] A reverse Brook rearrangement is also known.[900] The reaction has been extended to other systems. A homo-Brook rearrangement has also been reported.[901] Another variation is the aza-Brook rearrangement of α-silylallyl)amines.[902] The Brook rearrangement has been used in synthesis involving silyl dithianes.[903] A Brook rearrangement mediated [6 + 2]-annulation has been used for the construction of eight-membered carbocycles.[904]

The Brook rearrangement has been used in two important synthetic applications, a multicomponent coupling protocol initiated by a Brook rearrangement involving silyl dithianes as mentioned, and anion relay chemistry (ARC) involving a Brook rearrangement. An example of the former is the conversion of the 2-silyl dithiane **178** to the anion with *tert*-butyllithium followed by ring opening of an epoxide to give **179**.[905] Treatment with HMPA triggers a solvent-controlled Brook rearrangement that gives a new dithiane anion (**180**), which then reacts with a different epoxide to give the final product, **181**. An example of the anion relay chemistry treats dithiane (**182**) with *n*-butyllithium, and then **183** to give **184**.[906] Subsequent treatment with a variety of electrophiles, such as allyl bromide, in HMPA, leads to **185** via a Brook rearrangement, and then alkylation of the resultant dithian anion. This reaction can be initiated by nucleophiles other

[898]Brook, A.G. *Acc. Chem. Res.* **1974**, *7*, 77; Brook, A.G.; Bassendale, A.R., in de Mayo, P. *Rearrangements in Ground and Excited States*, Vol. 2, Academic Press, NY, **1980**, pp. 149–227.
[899]Brook, A. G.; Pascoe, J. D. *J. Am. Chem. Soc.* **1971** *93*, 6224.
[900]Wright, A.: West, R. *J. Am. Chem. Soc.* **1974**, *96*,3214; Wright, A.; West, R. *J. Am. Chem. Soc.* **1974**, *96*, 3227; Linderman, J.J.; Ghannam, A. *J. Am. Chem. Soc.* **1990**, *112*, 2392.
[901]Wilson, S.R.; Georgiadis, G.M. *J. Org. Chem.* **1983**, *48*, 4143.
[902]Honda, T.; Mori, M. *J. Org. Chem.* **1996**, *61*, 1196.
[903]For examples, see Smith III, A.B.; Adams, C. M. *Acc. Chem. Res.* **2004**, *37*, 365; Smith III, A.B.; Kim, D.-S. *Org. Lett.* **2005**, *7*, 3247.
[904]Takeda, K.; Haraguchi, H.; Okamoto, Y. *Org. Lett.* **2003**, *5*, 3705; Sawada, Y.; Sasaki, M.; Takeda, K. *Org. Lett.* **2004**, *6*, 2277.
[905]Smith III, A.B.; Boldi, A.M. *J. Am. Chem. Soc.* **1997**, *119*, 6925; Smith III, A.B.; Pitram, S.M. ; Boldi, A.M.; Gaunt, M.J.; Sfouggatakis, C.; Moser, W.H. *J. Am. Chem. Soc.* **2003**, *125*, 14435.
[906]Smith III, A.B.; Xian, M. *J. Am. Chem. Soc.* **2006**, *128*, 66.

than dithiane anion. Organocuprates can be used, and the anion stabilizing group can be a nitrile.[907]

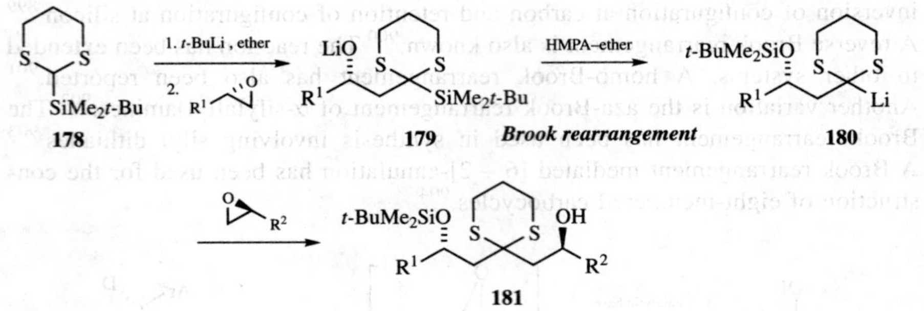

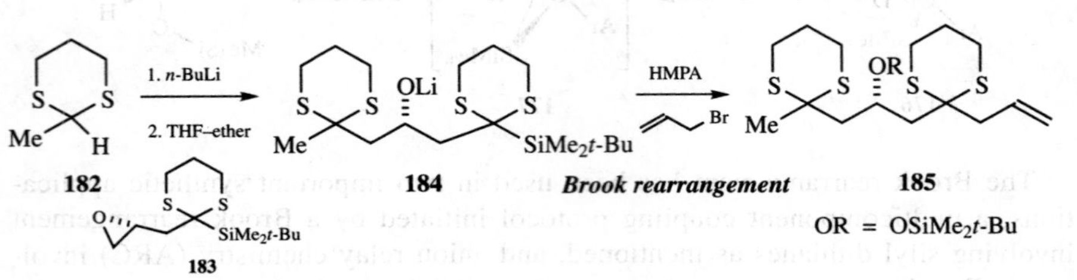

1. t-BuLi , ether
2.
SiMe2t-Bu
178
LiO
SiMe2t-Bu
179
HMPA–ether
Brook rearrangement
t-BuMe2SiO
Li
180
t-BuMe2SiO
OH
181

Me
H
182
1. n-BuLi
2. THF–ether
SiMe2t-Bu
183
OLi
Me
SiMe2t-Bu
184
HMPA
Br
Brook rearrangement
OR
Me
185
OR = OSiMe2t-Bu

[907]Private communication, Professor Amos B. Smith III, University Pennsylvania.

CHAPTER 19

Oxidations and Reductions

First, we must examine what we mean when we speak of oxidation and reduction. Inorganic chemists define oxidation in two ways: loss of electrons and increase in oxidation number. In organic chemistry, these definitions, while still technically correct, are not easy to apply. While electrons are directly transferred in some organic oxidations and reductions, the mechanisms of most of these reactions do not involve a direct electron transfer. As for oxidation number, while this is easy to apply in some cases, (e.g., the oxidation number of carbon in CH_4 is -4), in most cases attempts to apply the concept lead to fractional values or to apparent absurdities. Thus carbon in propane has an oxidation number of -2.67 and in butane of -2.5, although organic chemists seldom think of these two compounds as being in different oxidation states. An improvement could be made by assigning different oxidation states to different carbon atoms in a molecule, depending on what is bonded to them (e.g., the two carbons in acetic acid are obviously in different oxidation states), but for this a whole set of arbitrary assumptions would be required, since the oxidation number of an atom in a molecule is assigned on the basis of the oxidation numbers of the atoms attached to it. There would seem little to be gained by such a procedure. The practice in organic chemistry has been to set up a series of functional groups, in a qualitative way, arranged in order of increasing oxidation state, and then to define oxidation as *the conversion of a functional group in a molecule from one category to a higher one*. Reduction is the opposite. For the simple functional groups this series is shown in Table 19.1.[1] Note that this classification applies only to a single carbon atom or to two adjacent carbon atoms. Thus 1,3-dichloropropane is in the same oxidation state as chloromethane, but 1,2-dichloropropane is in a higher one. Obviously, such distinctions are somewhat arbitrary, and if we attempt to carry them too far, we will find ourselves painted into a corner. Nevertheless, the basic idea has served organic chemistry well. Note that

[1]For more extensive tables, with subclassifications, see Soloveichik, S.; Krakauer, H. *J. Chem. Educ.* *1966*, *43*, 532.

March's Advanced Organic Chemistry: Reactions, Mechanisms, and Structure, Sixth Edition, by Michael B. Smith and Jerry March
Copyright © 2007 John Wiley & Sons, Inc.

TABLE 19.1. Categories of Simple Functional Groups Arranged According to Oxidation State[a]

RH	$\underset{/}{\overset{\backslash}{C}}=\underset{\backslash}{\overset{/}{C}}$	$-C\equiv C-$	$\underset{R}{\overset{O}{\underset{\|}{C}}}\underset{OH}{}$	CO_2
	ROH		$\underset{R}{\overset{O}{\underset{\|}{C}}}\underset{R}{}$	CCl_4
	RCl			
	RNH_2		$\underset{R}{\overset{O}{\underset{\|}{C}}}\underset{NH_2}{}$	

and so on

$\underset{/}{\overset{\backslash}{C}}\underset{Cl}{\overset{Cl}{}}$ $\underset{/}{\overset{Cl}{C}}\underset{Cl}{\overset{Cl}{}}$

and so on

$\underset{Cl}{\overset{Cl}{C}}\underset{Cl}{\overset{/}{C}}$

and so on

$\underset{HO}{\overset{/}{C}}\underset{OH}{\overset{/}{C}}$

and so on

Approximate Oxidation Number

| −4 | −2 | 0 | +2 | +4 |

[a] Oxidation is the conversion of a functional group in a molecule to a higher category; reduction is conversion to a lower one. Conversions within a category are neither oxidations nor reductions. The numbers given at the bottom are only approximations.

conversion of any compound to another in the same category is not an oxidation or a reduction. Most oxidations in organic chemistry involve a gain of oxygen and/or a loss of hydrogen (Lavoisier's original definition of oxidation). The reverse is true for reductions.

Of course, there is no oxidation without a concurrent reduction. However, we classify reactions as oxidations or reductions depending on whether the *organic compound* is oxidized or reduced. In some cases, both the oxidant and reductant are organic; those reactions are treated separately at the end of the chapter.

MECHANISMS

Noted that our definition of oxidation has nothing to do with mechanism. Thus the conversion of bromomethane to methanol with KOH (**10-1**) and to methane with $LiAlH_4$ (**19-53**) have the same S_N2 mechanisms, but one is a reduction (according to our definition) and the other is not. It is impractical to consider the mechanisms of oxidation and reduction reactions in broad categories in this chapter as we have done for the reactions considered in Chapters 10–18.[2] The main reason is that the mechanisms are too diverse, and this in turn is because the bond changes are too different. For example, in Chapter 15, most reactions involved the bond change C=C → W—C—C—Y yet a relatively few mechanisms covered those reactions. But for oxidations and reductions the bond changes are far more diverse. Another reason is that the mechanism of a given oxidation or reduction reaction can vary greatly with the oxidizing or reducing agent employed. Very often the mechanism has been studied intensively for only one or a few of many possible agents.

Although we do not cover oxidation and reduction mechanisms in the same way as we have covered other mechanisms, it is still possible to list a few broad mechanistic categories. In doing this, we follow the scheme of Wiberg.[3]

1. *Direct Electron Transfer.*[4] We have already met some reactions in which the reduction is a direct gain of electrons or the oxidation a direct loss of them. An example is the Birch reduction (**15-13**), where sodium directly transfers an electron to an aromatic ring. An example from this chapter is found in the bimolecular reduction of ketones (**19-76**), where again it is a metal that supplies the electrons. This kind of mechanism is found largely in three types of reaction:[5] (a) the oxidation or reduction of a free radical (oxidation to a positive or reduction to a negative ion), (b) the oxidation of a negative ion or the reduction of a positive ion to a comparatively stable free radical, and (c) electrolytic oxidations or reductions (an example is the Kolbe reaction, **14-29**). An important example of (b) is oxidation of amines and phenolate ions:

[2]For monographs on oxidation mechanisms, see Bamford, C.H.; Tipper, C.F.H. *Comprehensive Chemical Kinetics*, Vol. 16, Elsevier, NY, *1980*; *Oxidation in Organic Chemistry*, Academic Press, NY, pt. A [Wiberg, K.B.], *1965*, pts. B, C, and D [Trahanovsky, W.S.], *1973, 1978, 1982*; Waters, W.A. *Mechanisms of Oxidation of Organic Compounds*, Wiley, NY, *1964*; Stewart, R. *Oxidation Mechanisms*, W. A. Benjamin, NY, *1964*. For a review, see Stewart, R. *Isot. Org. Chem. 1976, 2,* 271.

[3]Wiberg, K.B. *Surv. Prog. Chem. 1963, 1,* 211.

[4]For a monograph on direct electron-transfer mechanisms, see Eberson, L. *Electron Transfer Reactions in Organic Chemistry*, Springer, NY, *1987*. For a review, see Eberson, L. *Adv. Phys. Org. Chem. 1982, 18,* 79. For a review of multistage electron-transfer mechanisms, see Deuchert, K.; Hünig, S. *Angew. Chem. Int. Ed. 1978, 17,* 875.

[5]Littler, J.S.; Sayce, I.G. *J. Chem. Soc. 1964,* 2545.

These reactions occur easily because of the relative stability of the radicals involved.[6] The single electron-transfer mechanism (SET), which we have met several times (e.g., p. 264) is an important case.

2. *Hydride Transfer.*[7] In some reactions, a hydride ion is transferred to or from the substrate. The reduction of epoxides with LiAlH$_4$ is an example (**19-35**). Another is the Cannizzaro reaction (**19-81**). Reactions in which a carbocation abstracts a hydride ion belong in this category:[8]

$$R^+ + R'H \longrightarrow RH + R'^+$$

3. *Hydrogen-Atom Transfer.* Many oxidation and reduction reactions are free-radical substitutions and involve the transfer of a hydrogen atom. For example, one of the two main propagation steps of **14-1** involves abstraction of hydrogen:

$$RH + Cl\cdot \longrightarrow R\cdot + HCl$$

This is the case for many of the reactions of Chapter 14.

4. *Formation of Ester Intermediates.* A number of oxidations involve the formation of an ester intermediate (usually of an inorganic acid), and then the cleavage of this intermediate:

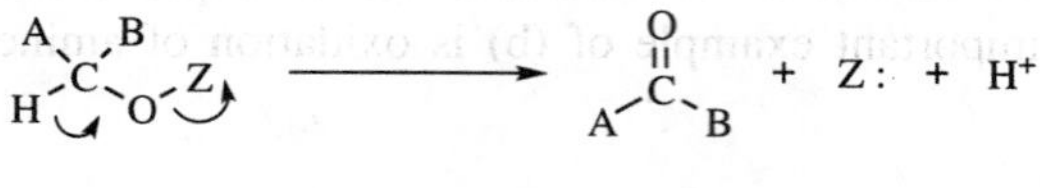

Z is usually CrO$_3$H, MnO$_3$, or a similar inorganic acid moiety. One example of this mechanism will be seen in **19-23**, where A was an alkyl or aryl group, B was OH, and Z was CrO$_3$H. Another is the oxidation of a secondary alcohol to a ketone (**19-3**), where A and B are alkyl or aryl groups and Z is also CrO$_3$H. In the lead tetraacetate oxidation of glycols (**19-7**) the mechanism also follows this pattern, but the positive leaving group is carbon instead of hydrogen. Note that the cleavage shown is an example of an E2 elimination.

5. *Displacement Mechanisms.* In these reactions, the organic substrate uses its electrons to cause displacement on an electrophilic oxidizing agent. One example is the addition of bromine to an alkene (**15-39**).

---

[6]For a review of the oxidation of phenols, see Mihailović, M.Lj.; Čeković, Z., in Patai, S. *The Chemistry of the Hydroxyl Group*, pt. 1, Wiley, NY, *1971*, pp. 505–592.

[7]For a review, see Watt, C.I.F. *Adv. Phys. Org. Chem.* **1988**, *24*, 57.

[8]For a review of these reactions, see Nenitzescu, C.D., in Olah, G.A.; Schleyer, P.V.R. *Carbonium Ions*, Vol. 2, Wiley, NY, *1970*, pp. 463–520.

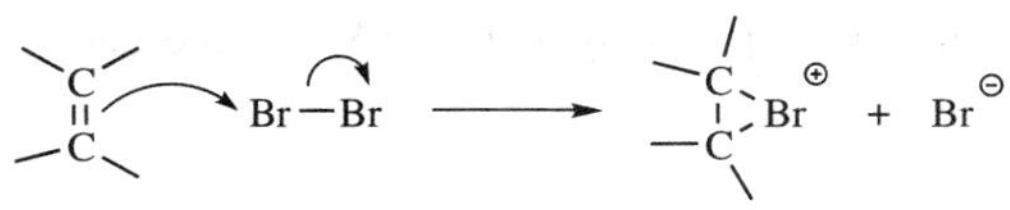

An example from this chapter is found in **19-29**:

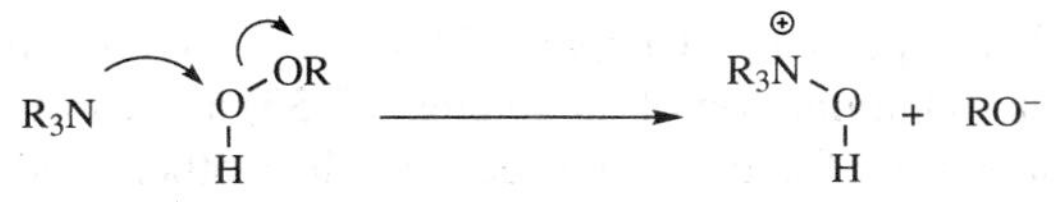

6. *Addition–Elimination Mechanisms.* In the reaction between α,β-unsaturated ketones and alkaline peroxide (**15-50**), the oxidizing agent adds to the substrate and then part of it is lost:

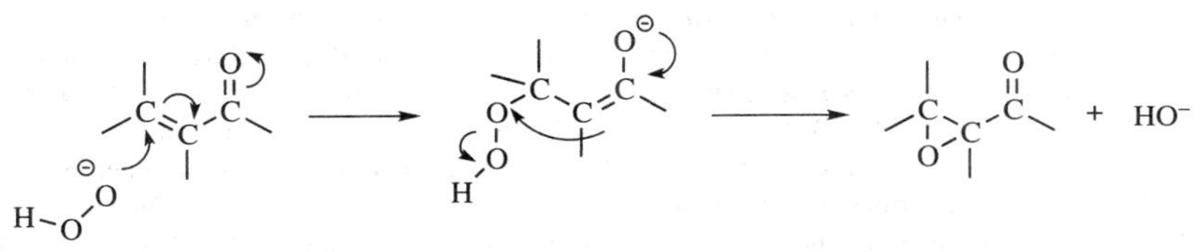

In this case, the oxygen of the oxidizing agent is in oxidation state -1 and the hydroxide ion departs with its oxygen in the -2 state, so it is reduced and the substrate oxidized. There are several reactions that follow this pattern of addition of an oxidizing agent and the loss of part of the agent, usually in a different oxidation state. Another example is the oxidation of ketones with SeO_2 (**19-17**). This reaction is also an example of category 4, since it involves formation and E2 cleavage of an ester. This example shows that these six categories are not mutually exclusive.

REACTIONS

In this chapter, the reactions are classified by the type of bond change occurring to the organic substrate, in conformity with our practice in the other chapters.[9] This means that there is no discussion in any one place of the use of a particular oxidizing or reducing agent, for example, acid dichromate or $LiAlH_4$ (except for a discussion of selectivity of reducing agents, p. 1787). Some oxidizing or reducing agents are fairly specific in their action, attacking only one or a few types of substrate.

---

[9]For a table of oxidation and reduction reactions, and the oxidizing and reducing agents for each, see Hudlický, M. *J. Chem. Educ.* **1977**, *54*, 100.

Others, like acid dichromate, permanganate, LiAlH₄, and catalytic hydrogenation, are much more versatile.[10,9,11]

OXIDATIONS[11,2]

In some cases, oxidations have been placed in another chapter. The oxidation of an alkene to a diol (**15-48**), and aromatic compound to a diol (**15-49**), or oxidation to an epoxide (**15-50**) are placed in Chapter 15, for consistency with the concept of addition to a π-bond. Diamination of an alkene (**15-53**) and formation of aziridines (**15-54**) are in Chapter 15 for the same reason. Most other oxidations have been placed here. The reactions in this section are classified into groups depending on

[10]For books on certain oxidizing agents, see Mijs, W.J.; de Jonge, C.R.J.I. *Organic Synthesis by Oxidation with Metal Compounds*, Plenum, NY, *1986*; Cainelli, G.; Cardillo, G. *Chromium Oxidations in Organic Chemistry*, Springer, NY, *1984*; Arndt, D. *Manganese Compounds as Oxidizing Agents in Organic Chemistry*, Open Court Publishing Company, La Salle, IL, *1981*; Lee, D.G. *The Oxidation of Organic Compounds by Permanganate Ion and Hexavalent Chromium*, Open Court Publishing Company, La Salle, IL, *1980*. For some reviews, see Curci, R. *Adv. Oxygenated Processes 1990*, 2, 1 (dioxiranes); Adam, W.; Curci, R.; Edwards, J.O. *Acc. Chem. Res. 1989*, 22, 205 (dioxiranes); Murray, R.W. *Chem. Rev. 1989*, 89, 1187; *Mol. Struc. Energ. 1988*, 5, 311 (dioxiranes); Kafafi, S.A.; Martinez, R.I.; Herron J.T. *Mol. Struc. Energ. 1988*, 5, 283 (dioxiranes); Krief, A.; Hevesi, L. *Organoselenium Chemistry I*; Springer, NY, *1988*, pp. 76–103 (seleninic anhydrides and acids); Ley, S.V., in Liotta, D.C. *Organoselenium Chemistry*, Wiley, NY, *1987*, pp. 163–206 (seleninic anhydrides and acids); Barton, D.H.R.; Finet, J. *Pure Appl. Chem. 1987*, 59, 937 [bismuth(V)]; Fatiadi, A.J. *Synthesis 1987*, 85 (KMnO4); Rubottom, G.M., in Trahanovsky, W.S. *Oxidation in Organic Chemistry*, pt. D, Academic Press, NY, *1982*, pp. 1–145 (lead tetraacetate); Fatiadi, A.J., in Pizey, J.S. *Synthetic Reagents*, Vol. 4, Wiley, NY, *1981*, pp. 147–335; *Synthesis 1974*, 229 (HIO₄); Fatiadi, A.J. *Synthesis 1976*, 65, 133 (MnO₂); Ogata, Y., in Trahanovsky, W.S. *Oxidation in Organic Chemistry*, pt. C, Academic Press, NY, *1978*, pp. 295–342 (nitric acid and nitrogen oxides); McKillop, A. *Pure Appl. Chem. 1975*, 43, 463 (thallium nitrate); Pizey, J.S. *Synthetic Reagents*, Vol. 2, Wiley, NY, *1974*, pp. 143–174 (MnO₂); George, M.V.; Balachandran, K.S. *Chem. Rev. 1975*, 75, 491 (nickel peroxide); Courtney, J.L.; Swansborough, K.F. *Rev. Pure Appl. Chem. 1972*, 22, 47 (ruthenium tetroxide); Ho, T.L. *Synthesis 1973*, 347 (ceric ion); Aylward, J.B. *Q. Rev. Chem. Soc. 1971*, 25, 407 (lead tetraacetate); Meth-Cohn, O.; Suschitzky, H. *Chem. Ind. (London) 1969*, 443 (MnO₂); Sklarz, B. *Q. Rev. Chem. Soc. 1967*, 21, 3 (HIO₄); Korshunov, S.P.; Vereshchagin, L.I. *Russ. Chem. Rev. 1966*, 35, 942 (MnO₂); Weinberg, N.L.; Weinberg, H.R. *Chem. Rev. 1968*, 68, 449 (electrochemical oxidation). For reviews of the behavior of certain reducing agents, see Keefer, L.K.; Lunn, G. *Chem. Rev. 1989*, 89, 459 (Ni–Al alloy); Málek, J. *Org. React. 1988*, 36, 249; *1985*, 34, 1–317 (metal alkoxyaluminum hydrides); Alpatova, N.M.; Zabusova, S.E.; Tomilov, A.P. *Russ. Chem. Rev. 1986*, 55, 99 (solvated electrons generated electrochemically); Caubère, P. *Angew. Chem. Int. Ed. 1983*, 22, 599 (modified sodium hydride); Nagai, Y. *Org. Prep. Proced. Int. 1980*, 12, 13 (hydrosilanes); Pizey, J.S. *Synthetic Reagents*, Vol. 1, Wiley, NY, *1974*, pp. 101–294 (LiAlH₄); Winterfeldt, E. *Synthesis 1975*, 617 (diisobutylaluminum hydride and triisobutylaluminum); Hückel, W. *Fortschr. Chem. Forsch. 1966*, 6, 197 (metals in ammonia or amines). For books on reductions with metal hydrides, see Seyden-Penne, J. *Reductions by the Alumino- and Borohydrides*, VCH, NY, *1991*; Štrouf, O.; Čásenský, B.; Kubánek, V. *Sodium Dihydrido-bis(2-methoxyethoxo)aluminate (SDMA)*, Elsevier, NY, *1985*; Hajós, A. *Complex Hydrides*, Elsevier, NY, *1979*. Also see, House, H.O. *Modern Synthetic Reactions*, 2nd ed., W. A. Benjamin, NY, *1972*.
[11]For books on oxidation reactions, see Hudlický, M. *Oxidations in Organic Chemistry*, American Chemical Society, Washington, DC, *1990*; Haines, A.H. *Methods for the Oxidation of Organic Compounds*, 2 vols., Academic Press, NY, *1985, 1988* [The first volume pertains to hydrocarbon substrates; the second mostly to oxygen- and nitrogen-containing substrates]; Chinn, L.J. *Selection of Oxidants in Synthesis*, Marcel Dekker, NY, *1971*; Augustine, R.L.; Trecker, D.J. *Oxidation*, 2 vols., Marcel Dekker, NY, *1969, 1971*.

the type of bond change involved. These groups are (A) eliminations of hydrogen, (B) oxidations involving cleavage of carbon–carbon bonds, (C) reactions involving replacement of hydrogen by oxygen, (D) reactions in which oxygen is added to the substrate, and (E) oxidative coupling.

A. Eliminations of Hydrogen

19-1 Aromatization of Six-Membered Rings

Hexahydro-terelimination

Six-membered alicyclic rings can be aromatized in a number of ways.[12] Aromatization is accomplished most easily if there are already one or two double bonds in the ring or if the ring is fused to an aromatic ring. The reaction can also be applied to heterocyclic five - and six-membered rings. Many groups may be present on the ring without interference, and even *gem*-dialkyl substitution does not always prevent the reaction: In such cases, one alkyl group often migrates or is eliminated. However, more drastic conditions are usually required for this. In some cases OH and COOH groups are lost from the ring. Cyclic ketones are converted to phenols. Seven-membered and larger rings are often isomerized to six-membered aromatic rings, although this is not the case for partially hydrogenated azulene systems (which are frequently found in Nature); these are converted to azulenes.

There are three types of reagents most frequently used to effect aromatization.

1. Hydrogenation catalysts,[13] such as platinum, palladium,[14] and nickel. In this case, the reaction is the reverse of double-bond hydrogenation (**15-11** and **15-15**), and presumably the mechanism is also the reverse, although not much is known.[15] Cyclohexene has been detected as an intermediate in the conversion of cyclohexane to benzene, using Pt.[16] The substrate is heated with the catalyst at ~ 300–350°C. The reactions can often be carried out under milder

---

[12]For reviews, see Haines, A.H. *Methods for the Oxidation of Organic Compounds*, Academic Press, NY, **1985**, pp. 16–22, 217–222; Fu, P.P.; Harvey, R.G. *Chem. Rev.* **1978**, *78*, 317; Valenta, Z., in Bentley, K.W.; Kirby, G.W. *Elucidation of Chemical Structures by Physical and Chemical Methods* (Vol. 4 of Weissberger, A. *Techniques of Chemistry*), 2nd ed., pt. 2, Wiley, NY, **1973**, pp. 1–76; House, H.O. *Modern Synthetic Reactions*, 2nd ed., W.A. Benjamin, NY, **1972**, pp. 34–44.

[13]For a review, see Rylander, P.N. *Organic Synthesis with Noble Metal Catalysts*, Academic Press, NY, **1973**, pp. 1–59.

[14]Ishikawa, T.; Uedo, E.; Tani, R.; Saito, S. *J. Org. Chem.* **2001**, *66*, 186; Cossy, J.; Belotti, D. *Org. Lett.* **2002**, *4*, 2557.

[15]For a discussion, see Tsai, M.; Friend, C.M.; Muetterties, E.L. *J. Am. Chem. Soc.* **1982**, *104*, 2539. See also, Augustine, R.L.; Thompson, M.M. *J. Org. Chem.* **1987**, *52*, 1911.

[16]Land, D.P.; Pettiette-Hall, C.L.; McIver, Jr., R.T.; Hemminger, J.C. *J. Am. Chem. Soc.* **1989**, *111*, 5970.

conditions if a hydrogen acceptor, such as maleic acid, cyclohexene, or benzene, is present to remove hydrogen as it is formed. The acceptor is reduced to the saturated compound. Other transition metals can be used, including $TiCl_4$-NEt_3.[17] It has been reported that dehydrogenation of 1-methylcyclohexene-1-^{13}C over an alumina catalyst gave toluene with the label partially scrambled throughout the aromatic ring.[18] For polycylic systems, heating with oxygen on activated carbon generates the aromatic compound, as in the conversion of dihydroanthracene to anthracene.[19]

2. The elements sulfur and selenium, which combine with the hydrogen evolved to give, respectively, H_2S and H_2Se. Little is known about this mechanism either.[20]

3. Quinones,[21] which become reduced to the corresponding hydroquinones. Two important quinones often used for aromatizations are chloranil (2,3,5,6-tetrachloro-1,4-benzoquinone) and DDQ (2,3-dichloro-5,6-dicyano-1,4-benzoquinone).[22] The latter is more reactive and can be used in cases where the substrate is difficult to dehydrogenate. It is likely that the mechanism involves a transfer of hydride to the quinone oxygen, followed by the transfer of a proton to the phenolate ion:[23,21]

Among other reagents[24] that have been used for aromatization of six-membered rings are atmospheric oxygen, MnO_2,[25] $KMnO_4$-Al_2O_3,[26] SeO_2, various strong bases,[27]

[17]Srinivas, G.; Periasamy, M. *Tetrahedron Lett.* **2002**, *43*, 2785.

[18]Marshall, J.L.; Miiller, D.E.; Ihrig, A.M. *Tetrahedron Lett.* **1973**, 3491.

[19]Nakamichi, N.; Kawabata, H.; Hiyashi, M. *J. Org. Chem.* **2003**, *68*, 8272.

[20]House, H.O.; Orchin, M. *J. Am. Chem. Soc.* **1960**, *82*, 639; Silverwood, H.A.; Orchin, M. *J. Org. Chem.* **1962**, *27*, 3401.

[21]For reviews, see Becker, H.; Turner, A.B., in Patai, S.; Rappoport, Z. *The Chemistry of the Quinonoid Compounds*, Vol. 2, pt. 2, Wiley, NY, **1988**, pp. 1351–1384; Becker, H., in Patai, S. *The Chemistry of the Quinonoid Compounds*, Vol. 1, pt. 1, Wiley, NY, **1974**, pp. 335–423.

[22]For reviews of DDQ, see Turner, A.B., in Pizey, J.S. *Synthetic Reagents*, Vol. 3, Wiley, NY, **1977**, pp. 193–225; Walker, D.; Hiebert, J.D. *Chem. Rev.* **1967**, *67*, 153.

[23]Braude, E.A.; Jackman, L.M.; Linstead, R.P.; Lowe, G. *J. Chem. Soc.* **1960**, 3123, 3133; Trost, B.M. *J. Am. Chem. Soc.* **1967**, *89*, 1847. See also, Stoos, F.; Roč ek, J. *J. Am. Chem. Soc.* **1972**, *94*, 2719; Hashish, Z.M.; Hoodless, I.M. *Can. J. Chem.* **1976**, *54*, 2261; Müller, P.; Joly, D.; Mermoud, F. *Helv. Chim. Acta* **1984**, *67*, 105; Radtke, R.; Hintze, H.; Rösler, K.; Heesing, A. *Chem. Ber.* **1990**, *123*, 627. Also see, Höfler, C.; Rüchardt, C. *Liebigs Ann. Chem.* **1996**, 183.

[24]For a list of reagents, with references, see Larock, R.C. *Comprehensive Organic Transformations*, 2nd ed., Wiley-VCH, NY, **1999**, pp. 187–191.

[25]See, for example, Leffingwell, J.C.; Bluhm, H.J. *Chem. Commun.* **1969**, 1151.

[26]McBride, C.M.; Chrisman, W.; Harris, C.E.; Singaram, B. *Tetrahedron Lett.* **1999**, *40*, 45.

[27]For a review, see Pines, H.; Stalick, W.M. *Base-Catalyzed Reactions of Hydrocarbons and Related Compounds*, Academic Press, NY, **1977**, pp. 483–503. See also, Reetz, M.T.; Eibach, F. *Liebigs Ann. Chem.* **1978**, 1598; Trost, B.M.; Rigby, J.H. *Tetrahedron Lett.* **1978**, 1667.

chromic acid,[28] H_2SO_4 and a ruthenium catalyst,[29] and SeO_2 on P_2O_5/Me$_3$SiOSiMe$_3$.[30] The last-mentioned reagent also dehydrogenates cyclopentanes to cyclopentadienes. In some instances, the hydrogen is not released as H_2 or transferred to an external oxidizing agent, but instead serves to reduce another molecule of substrate. This is a disproportionation reaction and can be illustrated by the conversion of cyclohexene to cyclohexane and benzene. Quinones react with allylic silanes and a Bi(OTf)$_3$ catalyst to give 2-allyl hydroquinone.[31] Similar reaction with acetic anhydride rather than an allylic silane leads to a 2-acetoxy hydroquinone.[32]

Heteroatom rings, as found in quinoline derivatives, for example, can be generated from amino-ketones with [hydroxy(tosyloxy)iodo]benzene and perchloric acid[33] or with NaHSO$_4$–Na$_2$Cr$_2$O$_7$ on wet silica.[34] Dihydropyridines are converted to pyridines with NaNO$_2$–oxalic acid and wet silica[35] BaMnO$_4$,[36] FeCl$_3$–acetic acid,[37] Mg(HSO$_4$)$_2$–NaNO$_2$,[38] NO$^+$-18-crown-6-H(NO$_3$)$_2^-$,[39] or with nicotinium dichromate.[40] Cyclic imines are converted to pyridine derivatives with NCS, and then excess sodium methoxide.[41]

Note that hydrogenolysis of cyclohexane leads to *n*-hexane with hydrogen and an iridium catalyst.[42]

OS **II**, 214, 423; **III**, 310, 358, 729, 807; **IV**, 536; **VI**, 731. Also see, OS **III**, 329.

19-2   Dehydrogenations Yielding Carbon–Carbon Double Bonds

Dihydro-elimination

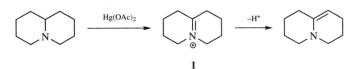

1

[28]Müller, P.; Pautex, N.; Hagemann, H. *Chimia* **1988**, *42*, 414.
[29]Tanaka, H.; Ikeno, T.; Yamada, T. *Synlett* **2003**, 576.
[30]Lee, J.G.; Kim, K.C. *Tetrahedron Lett.* **1992**, *33*, 6363.
[31]Yadav, J.S.; Reddy, B.V.S.; Swamy, T. *Tetrahedron Lett.* **2003**, *44*, 4861.
[32]Yadav, J.S.; Reddy, B.V.S.; Swamy, T.; Rao, K.R. *Tetrahedron Lett.* **2004**, *45*, 6037.
[33]Varma, R.S.; Kumar, D. *Tetrahedron Lett.* **1998**, *39*, 9113.
[34]Damavandi, J.A.; Zolfigol, M.A.; Karami, B. *Synth. Commun.* **2001**, *31*, 3183.
[35]Zolfigol, M.A.; Kiany-Borazjani, M.; Sadeghi, M.M.; Mohammadpoor-Baltork, I.; Memarian, H.R. *Synth. Comm.* **2000**, *30*, 551.
[36]Memarian, H.R.; Sadeghi, M.M.; Momeni, A.R. *Synth. Commun.* **2001**, *31*, 2241.
[37]Lu, J.; Bai, Y.; Wang, Z.; Yang, B.Q.; Li, W. *Synth. Commun.* **2001**, *31*, 2625.
[38]Zolfigol, M.A.; Kiany-Borazjani, M.; Sadeghi, M.M.; Mohammadpoor-Baltork, I.; Memarian, H.R. *Synth. Commun.* **2000**, *30*, 3919.
[39]Zolfigol, M.A.; Zebarjadian, M.H.; Sadeghi, M.M.; Mohammadpoor-Baltork, I. Memarian, H.R.; Shamsipur, M. *Synth. Commun.* **2001**, *31*, 929.
[40]Sadeghi, M.M.; Mohammadpoor-Baltork, I.; Memarian, H.R.; Sobhani, S. *Synth. Commun.* **2000**, *30*, 1661.
[41]DeKimpe, N.; Keppens, M.; Fonck, G. *Chem. Commun.* **1996**, 635.
[42]Locatelli, F.; Candy, J.-P.; Didillon, B.; Niccolai, G.P.; Uzio, D.; Basset, J.-M. *J. Am. Chem. Soc.* **2001**, *123*, 1658.

Dehydrogenation of an aliphatic compound to give a double bond in a specific location is not usually a feasible process, although industrially mixtures of alkenes are obtained in this way from mixtures of alkanes (generally by heating with chromia–alumina catalysts). There are, however, some notable exceptions. Heating cyclooctane with an iridium catalyst leads to cyclooctene.[43] Treating alkenes that have an allylic hydrogen with $CrCl_2$ converts them to allenes.[44] It is not surprising, however, that most of the exceptions generally involve cases where the new double bond can be in conjugation with a double bond or with an unshared pair of electrons already present.[45] One example is the synthesis developed by Leonard and co-workers,[46] in which tertiary amines give enamines (**10-69**) when treated with mercuric acetate[47] (see the example above). In this case the initial product is the iminium ion **1** which loses a proton to give the enamine. In another example, the oxidizing agent SeO_2 can in certain cases convert a carbonyl compound to an α,β-unsaturated carbonyl compound by removing H_2[48] (though this reagent more often gives **19-17**). This reaction has been most often applied in the steroid series, an example being formation of **2** from **3**.[49] In a similar manner, Hünig's base, diisopropylethylamine, was converted to the enamine *N,N*-diisopropyl-*N*-vinylamine by heating with an iridium catalyst.[50]

Similarly, SeO_2 has been used to dehydrogenate 1,4-diketones[51] ($RCOCH_2CH_2$-$COR \rightarrow RCOCH{=}CHCOR$) and 1,2-diarylalkanes ($ArCH_2CH_2Ar \rightarrow ArCH{=}CHAr$). These conversions can also be carried out by certain quinones, most notably DDQ (see **19-1**).[22] Ketones have been converted to conjugated ketones with

[43]Göttker-Schnetmann, I.; White, P.; Brookhart, M. *J. Am. Chem. Soc.* **2004**, *126*, 1804.

[44]Takai, K.; Kokumai, R.; Toshikawa, S. *Synlett* **2002**, 1164.

[45]For a review, see Haines, A.J. *Methods for the Oxidation of Organic Compounds*, Vol. 1, Academic Press, NY, **1985**, pp. 6–16, 206–216. For lists of examples, with references, see Larock, R.C. *Comprehensive Organic Transformations*, 2nd ed., Wiley-VCH, NY, **1999**, pp. 251–256.

[46]For example, see Leonard, N.J.; Musker, W.K. *J. Am. Chem. Soc.* **1959**, *81*, 5631; **1960**, *82*, 5148.

[47]For reviews, see Haynes, L.W.; Cook, A.G., in Cook, A.G. *Enamines*, 2nd ed. Marcel Dekker, NY, **1988**, pp. 103–163; Lee, D.G., in Augustine, R.L.; Trecker, D.J. *Oxidation*, Vol. 1, Marcel Dekker, NY, **1969**, pp. 102–107.

[48]For reviews, see Back, T.G., in Patai, S. *The Chemistry of Organic Selenium and Tellurium Compounds*, pt. 2, Wiley, NY, **1987**, pp. 91–213, 110–114; Jerussi, R.A. *Sel. Org. Transform.* **1970**, *1*, 301, see pp. 315–321.

[49]Bernstein, S.; Littell, R. *J. Am. Chem. Soc.* **1960**, *82*, 1235.

[50]Zhang, X.; Fried, A.; Knapp,S.; Goldman, A.S. *Chem. Commun.* **2003**, 2060.

[51]For example, see Barnes, C.S.; Barton, D.H.R. *J. Chem. Soc.* **1953**, 1419.

Ph(S=O)OMe and KH,[52] and also with (pyridyl)S(=O)OMe/KH, and then CuSO$_4$.[53] Silyl enol ethers also give the conjugated ketone upon treatment with ceric ammonium nitrate in DMF[54] or with Pd(OAc)$_2$/NaOAc/O$_2$.[55] Simple aldehydes and ketones have been dehydrogenated (e.g., cyclopentanone → cyclopentenone) by PdCl$_2$,[56] by FeCl$_3$,[57] and by benzeneseleninic anhydride[58] (this reagent also dehydrogenates lactones in a similar manner), among other reagents.

In an indirect method of achieving this conversion, the silyl enol ether of a simple ketone is treated with DDQ[59] or with triphenylmethyl cation[60] (for another indirect method, see **17-12**). Simple linear alkanes have been converted to alkenes by treatment with certain transition-metal compounds.[61]

An entirely different approach to specific dehydrogenation has been reported by R. Breslow[62] and by J.E. Baldwin.[63] By means of this approach it was possible, for example, to convert 3α-cholestanol (**4**) to 5α-cholest-14-en-3α-ol (**5**), thus introducing a double bond at a specific site remote from any functional group.[64] This was

[52]Resek, J.E.; Meyers, A.I. *Tetrahedron Lett.* **1995**, *36*, 7051.

[53]Trost, B.M.; Parquette, J.R. *J. Org. Chem.* **1993**, *58*, 1579.

[54]Evans, P.A.; Longmire, J.M.; Modi, D.P. *Tetrahedron Lett.* **1995**, *36*, 3985.

[55]Larock, R.C.; Hightower, T.R.; Kraus, G.A.; Hahn, P.; Zheng, O. *Tetrahedron Lett.* **1995**, *36*, 2423.

[56]Bierling, B.; Kirschke, K.; Oberender, H.; Schultz, M. *J. Prakt. Chem.* **1972**, *314*, 170; Kirschke, K.; Müller, H.; Timm, D. *J. Prakt. Chem.* **1975**, *317*, 807; Mincione, E.; Ortaggi, G.; Sirna, A. *Synthesis* **1977**, 773; Mukaiyama, T.; Ohshima, M.; Nakatsuka, T. *Chem. Lett.* **1983**, 1207. See also, Heck, R.F. *Palladium Reagents in Organic Synthesis*, Academic Press, NY, **1985**, pp. 103–110.

[57]Cardinale, G.; Laan, J.A.M.; Russell, S.W.; Ward, J.P. *Recl. Trav. Chim. Pays-Bas* **1982**, *101*, 199.

[58]Barton, D.H.R.; Hui, R.A.H.F.; Ley, S.V.; Williams, D.J. *J. Chem. Soc. Perkin Trans. 1* **1982**, 1919; Barton, D.H.R.; Godfrey, C.R.A.; Morzycki, J.W.; Motherwell, W.B.; Ley, S.V. *J. Chem. Soc. Perkin Trans. 1* **1982**, 1947.

[59]Jung, M.E.; Pan, Y.; Rathke, M.W.; Sullivan, D.F.; Woodbury, R.P. *J. Org. Chem.* **1977**, *42*, 3961.

[60]Ryu, I.; Murai, S.; Hatayama, Y.; Sonoda, N. *Tetrahedron Lett.* **1978**, 3455. For another method, which can also be applied to enol acetates, see Tsuji, J.; Minami, I.; Shimizu, I. *Tetrahedron Lett.* **1983**, *24*, 5635, 5639.

[61]See Burchard, T.; Felkin, H. *Nouv. J. Chim.* **1986**, *10*, 673; Burk, M.J.; Crabtree, R.H. *J. Am. Chem. Soc.* **1987**, *109*, 8025; Renneke, R.F.; Hill, C.L. *New J. Chem.* **1987**, *11*, 763; *Angew. Chem. Int. Ed.* **1988**, *27*, 1526; *J. Am. Chem. Soc.* **1988**, *110*, 5461; Maguire, J.A.; Boese, W.T.; Goldman, A.S. *J. Am. Chem. Soc.* **1989**, *111*, 7088; Sakakura, T.; Ishida, K.; Tanaka, M. *Chem. Lett.* **1990**, 585; and references cited therein.

[62]Breslow, R.; Baldwin, S.W. *J. Am. Chem. Soc.* **1970**, *92*, 732. For reviews, see Breslow, R. *Chemtracts: Org. Chem.* **1988**, *1*, 333; *Acc. Chem. Res.* **1980**, *13*, 170; *Isr. J. Chem.* **1979**, *18*, 187; *Chem. Soc. Rev.* **1972**, *1*, 553.

[63]Baldwin, J.E.; Bhatnagar, A.K.; Harper, R.W. *Chem. Commun.* **1970**, 659.

[64]For other methods of introducting a remote double bond, see Čeković, Z.; Cvetković, M. *Tetrahedron Lett.* **1982**, *23*, 3791; Czekay, G.; Drewello, T.; Schwarz, H. *J. Am. Chem. Soc.* **1989**, *111*, 4561. See also, Bégué, J. *J. Org. Chem.* **1982**, *47*, 4268; Nagata, R.; Saito, I. *Synlett* **1990**, 291.

accomplished by conversion of **4** to the ester **6**, followed by irradiation of **6**, which gave 55% **8**, which was then

6   **7**   **8**

hydrolyzed to **5**. The radiation excites the benzophenone portion of **6** (p. $$$), which then abstracts hydrogen from the 14 position to give the diradical **7**, which undergoes another internal abstraction to give **8**. In other cases, diradicals like **7** can close to a macrocyclic lactone (**19-17**). In an alternate approach,[65] a 9(11) double bond was introduced into a steroid nucleus by reaction of the *m*-iodo ester **9** with PhICl₂ and uv light, which results in hydrogen being abstracted regioselectively from the 9 position, resulting in chlorination at that position. Dehydrohalogenation of **10** gives the 9(11)-unsaturated steroid **11**. In contrast, use of the para isomer of **7** results in chlorination at the 14 position and loss of HCl gives the 14-unsaturated steroid. These reactions are among the very few ways to introduce functionality at a specific site remote from any functional group (see also, **19-17**).

9   **10**   **11**

Certain 1,2-diarylalkenes ArCH=CHAr' have been converted to the corresponding alkynes ArC≡CAr' by treatment with *t*-BuOK in DMF.[66] Dihydroindoles are converted to indoles with *N,N',N''*-trichloro-1,3,5-triazin-2,4,6-trione and DBU.[67]

[65]Breslow, R.; Corcoran, R.J.; Snider, B.B.; Doll, R.J.; Khanna, P.L.; Kaleya, R. *J. Am. Chem. Soc.* **1977**, *99*, 905. For related approaches, see Wolner, D. *Tetrahedron Lett.* **1979**, 4613; Breslow, R.; Brandl, M.; Hunger, J.; Adams, A.D. *J. Am. Chem. Soc.* **1987**, *109*, 3799; Batr, R.; Breslow, R. *Tetrahedron Lett.* **1989**, *30*, 535; Orito, K.; Ohto, M.; Suginome, H. *J. Chem. Soc. Chem. Commun.* **1990**, 1076.

[66]Akiyama, S.; Nakatsuji, S.; Nomura, K.; Matsuda, K.; Nakashima, K. *J. Chem. Soc. Chem. Commun.* **1991**, 948.

[67]Tilstam, U.; Harre, M.; Heckrodt, T.; Weinmann, H. *Tetrahedron Lett.* **2001**, *42*, 5385.

A different kind of dehydrogenation was used in the final step of Paquette's synthesis of dodecahedrane:[68]

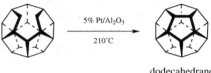

dodecahedrane

OS **V**, 428, **VII**, 4, 473.

19-3 Oxidation or Dehydrogenation of Alcohols to Aldehydes and Ketones

C,O-**Dihydro-elimination**

$$RCH_2OH \xrightarrow[\text{chromite}]{\text{copper}} RCHO$$

$$RCHOHR' \xrightarrow[\text{H}_2\text{SO}_4]{\text{K}_2\text{Cr}_2\text{O}_7} RCOR'$$

Primary alcohols can be converted to aldehydes and secondary alcohols to ketones in seven main ways:[69]

1. *With Strong Oxidizing Agents.*[70] Secondary alcohols are easily oxidized to ketones by acid dichromate[71] at room temperature or slightly above. Many

---

[68]Paquette, L.A.; Weber, J.C.; Kobayashi, T.; Miyahara, Y. *J. Am. Chem. Soc.* **1988**, *110*, 8591. For a monograph on dodecahedrane and related compounds, see Paquette, L.A.; Doherty, A.M. *Polyquinane Chemistry*; Springer, NY, **1987**. For reviews, see, in Olah, G.A. *Cage Hydrocarbons*, Wiley, NY, **1990**, the reviews by Paquette, L.A. pp. 313–352, and by Fessner, W.; Prinzbach, H. pp. 353–405; Paquette, L.A. *Chem. Rev.* **1989**, *89*, 1051; *Top. Curr. Chem.* **1984**, *119*, 1, in Lindberg, T. *Strategies and Tactics in Organic Synthesis*, Academic Press, NY, **1984**, pp. 175–200.

[69]For reviews, see Hudlický, M. *Oxidations in Organic Chemistry*, American Chemical Society, Washington, DC, **1990**, pp. 114–126, 132–149; Haines, A.M. *Methods for the Oxidation of Organic Compounds*, Vol. 2, Academic Press, NY, **1988**, pp. 5–148, 326–390; Müller, P., in Patai, S. *The Chemistry of Functional Groups, Supplement E*, Wiley, NY, **1980**, pp. 469–538; Cullis, C.F.; Fish, A., in Patai, S. *The Chemistry of the Carbonyl Group*, Vol. 1, Wiley, NY, **1966**, pp. 129–157. For a lengthy list of reagents, with references, see Larock, R.C. *Comprehensive Organic Transformations*, 2nd ed., Wiley-VCH, NY, **1999**, pp. 1234–1250.

[70]For thorough discussions, see Lee, D.G., in Augustine, R.L.; Trecker, D.J. *Oxidation*, Vol. 2, Marcel Dekker, NY, **1971**, pp. 56–81; and (with respect to chromium and manganese reagents) House, H.O. *Modern Synthetic Reactions*, 2nd ed., W.A. Benjamin, NY, **1972**, pp. 257–273.

[71]Various forms of H2CrO4 and of CrO3 are used for this reaction. For a review, see Cainelli, G.; Cardillo, G. *Chromium Oxidations in Organic Chemistry*, Open Court Publishers Co., La Salle, IL, **1981**, pp. 118–216. For discussions, see Fieser, L.F.; Fieser, M. *Reagents for Organic Synthesis*, Vol. 1, Wiley, NY, **1967**, pp. 142–147, 1059–1064, and subsequent volumes in this series.

other strong oxidizing agents (KMnO$_4$,[72] ruthenium tetroxide,[73] etc.) have also been employed. A solution of chromic acid and sulfuric acid in water is known as the *Jones reagent*.[74] When secondary alcohols are dissolved in acetone, titration with the Jones reagent oxidizes them to ketones rapidly and in high yield without disturbing any double or triple bonds that may be present (see **19-10**) and without epimerizing an adjacent stereogenic center.[75] The Jones reagent can also oxidize primary allylic alcohols to the corresponding aldehydes,[76] although overoxidation to the carboxylic acid is a problem. Oxidative cleavage of primary alcohols has been observed in the presence of molecular sieve 3 Å.[77] Indeed, for the oxidation of allylic alcohols three other Cr(VI) reagents are commonly used,[78] dipyridine Cr(VI) oxide (*Collins' reagent*),[79] pyridinium chlorochromate (PCC),[80] and pyridinium dichromate (PDC).[81] The PCC is somewhat acidic, and acid-catalyzed rearrangements have been observed.[82] A variety of amines and diamines have been converted to tetraalkylammonium halochromates or dichromates. Examples include *N*-benzyl 1,4-diazabicyclo[2.2.2]octane ammonium dichromate with microwave irradiation,[83] γ-picolinium chlorochromate,[84] and quinolinium

[72]For oxidation with KMnO$_4$ on alumina with no solvent, see Hajipour, A.R.; Mallakpour, S.E.; Imanzadeh, G. *Chem. Lett.* **1999**, 99. For oxidation with silica-supported KMnO4, see Takemoto, T.; Yasuda, K.; Ley, S.V. *Synlett* **2001**, 1555. For oxidation in the ionic liquid bmim BF4, 1-butyl-3-methylimidazolium tetrafluoroborate: Kumar, A.; Jain, N.; Chauhan, S.M.S. *Synth. Commun.* **2004**, *34*, 2835.

[73]For a review, see Lee, D.G.; van den Engh, M. in Trahanovsky, W.S. *Oxidation in Organic Chemistry*, pt. B Academic Press, NY, *1973*, pp. 197–222.

[74]Bowden, K.; Heilbron I.M.; Jones, E.R.H.; Weedon, B.C.L. *J. Chem. Soc.* **1946**, 39; Bowers, A.; Halsall, T.G.; Jones, E.R.H.; Lemin, A.J. *J. Chem. Soc.* **1953**, 2548. Also see, Scott, S.L.; Bakac, A.; Espenson, J.H. *J. Am. Chem. Soc.* **1992**, *114*, 4605. For an oxidation with Jones reagent on silica in dichloromethane, see Ali, M.H.; Wiggin, C.J. *Synth. Commun.* **2001**, *31*, 1389; Ali, M.H.; Wiggin, C.J. *Synth. Commun.* *2001*, *31*, 3383.

[75]For example, see Djerassi, C.; Hart, P.A.; Warawa, E.J. *J. Am. Chem. Soc.* *1964*, *86*, 78.

[76]Harding, K.E.; May, L.M.; Dick, K.F. *J. Org. Chem.* *1975*, *40*, 1664.

[77]Fernandes, R.A.; Kumar, P. *Tetrahedron Lett.* *2003*, *44*, 1275.

[78]For a comparative study of Jones's, Collins's, and Corey's reagents, see Warrener, R.N.; Lee, T.S.; Russell, R.A.; Paddon-Row, M.N. *Aust. J. Chem.* *1978*, *31*, 1113.

[79]Collins, J.C.; Hess, W.W.; Frank, F.J. *Tetrahedron Lett.* *1968*, 3363; Ratcliffe, R.; Rodehorst, R. *J. Org. Chem.* *1970*, *35*, 4000; Stensiö, K. *Acta Chem. Scand.* *1971*, *25*, 1125; Collins, J.C.; Hess, W.W. *Org. Synth. VI*, 644; Sharpless, K.B.; Akashi, K. *J. Am. Chem. Soc.* *1975*, *97*, 5927.

[80]Corey, E.J.; Suggs, J.W. *Tetrahedron Lett.* *1975*, 2647. For reviews of this and related reagents, see Luzzio, F.A.; Guziec, Jr., F.S. *Org. Prep. Proceed. Int.* *1988*, *20*, 533; Piancatelli, G.; Scettri, A.; D'Auria, M. *Synthesis 1982*, 245. For an improved method of preparing this reagent, see Agarwal, S.; Tiwari, H.P.; Sharma, J.P. *Tetrahedron 1990*, *46*, 4417. For a PCC oxidation with no solvent, see Salehi, P.; Firouzabadi, H.; Farrokhi, A.; Gholizadeh, M. *Synthesis 2001*, 2273.

[81]Coates, W.M.; Corrigan, J.R. *Chem. Ind. (London) 1969*, 1594; Corey, E.J.; Schmidt, G. *Tetrahedron Lett. 1979*, 399; Czernecki, S.; Georgoulis, C.; Stevens, C.L.; Vijayakumaran, K. *Tetrahedron Lett. 1985*, *26*, 1699.

[82]See Ren, S.-K.; Wang, F.; Dou, H.-N.; Fan, C.-A.; He, L.; Song, Z.-L.; Xia, W.-J.; Li, D-R.; Jia, Y.-X.; Li, X.; Tu, Y.-Q. *Synthesis 2001*, 2384.

[83]Hajipour, A.R.; Mallakpour, S.E.; Khoee, S. *Synlett 2000*, 740.

[84]Khodaei, M.M.; Salehi, P.; Goodarzi, M. *Synth. Commun. 2001*, *31*, 1253.

fluorochromate.[85] benzyltriphenylphosphonium chlorochromate has been used in a similar manner.[86] Ammonium dichromate with HIO_3 on wet silica gel[87] or ammonium chlorochromate on Montmorillonite K10[88] have also been used. The MnO_2[89] reagent is also a fairly specific reagent for oxidation of allylic and benzylic OH groups in preference to aliphatic substrates. For acid-sensitive compounds, CrO_3 in HMPA[90] or trimethylsilyl chromates[91] can be used. Benzylic alcohols are oxidized to aldehydes with $BaCr_2O_7$ in acetonitrile.[92] Both CrO_3[93] and MnO_2[94] have been used to oxidized primary and benzylic alcohols, respectively, under solvent-free conditions. A catalytic mixture of *N*-hydroxyphthalimide, $Co(OAc)_2$ and mcpba oxidizes secondary alcohols to ketones.[95] Chromium trioxide with aqueous *tert*-butylhydroperoxide oxidizes benzylic alcohols with microwave irradiation.[96] Oxidizing agents have been supported on a polymer,[97] including chromic acid[98] and permanganate,[99] as well as poly[vinyl(pyridinium fluorochromate)].[100] Microwave induced oxidation of benzylic alcohols was reported using zeolite-supported ferric nitrate.[101] Microwave irradiation of CrO_3 with various co-reagents oxidizes alcohols.[102] Phase-transfer catalysis has also been used with permanganate,[103] chromic acid,[104] and

[85]Rajkumar, G.A.; Arabindoo, B.; Murugesan, V. *Synth. Commun.* **1999**, *29*, 2105.

[86]Hajipour, A.R.; Mallakpour, S.E.; Backnejad, H. *Synth. Commun.* **2000**, *30*, 3855.

[87]Shirini, F.; Zolfigol, M.A.; Azadbar, M.R. *Russ. J. Org. Chem.* **2001**, *37*, 1600.

[88]Heravi, M.M.; Kiakojoori, R.; Tabar-Hydar, K. *Monat. Chem.* **1999**, *130*, 581.

[89]For the use of MnO_2 on silica gel with microwave irradiation, see Varma, R.S.; Saini, R.K.; Dahiya, R. *Tetrahedron Lett.* **1997**, *38*, 7823. For an example on bentonite clay with microwave irradiation, see Martinez, L.A.; García, O.; Delgado, F.; Alvarez, C.; Patiño, R. *Tetrahedron Lett.* **1993**, *34*, 5293.

[90]Cardillo, G.; Orena, M.; Sandri, S. *Synthesis* **1976**, 394.

[91]Moiseenkov, A.M.; Cheskis, B.A.; Veselovskii, A.B.; Veselovskii, V.V.; Romanovich, A.Ya.; Chizhov, B.A. *J. Org. Chem. USSR* **1987**, *23*, 1646.

[92]Mottaghinejad, E.; Shaafi, E.; Ghasemzadeh, Z. *Tetrahedron Lett.* **2004**, *45*, 8823.

[93]Lou, J.-D.; Xu, Z.-N. *Tetrahedron Lett.* **2002**, *43*, 6095.

[94]Lou, J.D.; Xu, Z.-N. *Tetrahedron Lett.* **2002**, *43*, 6149.

[95]Iwahama, T.; Yoshino, Y.; Keitoku, T.; Sakaguchi, S.; Ishii, Y. *J. Org. Chem.* **2000**, *65*, 6502.

[96]Singh, J.; Sharma, M.; Chhibber, M.; Kaur, J.; Kad, G.L. *Synth. Commun.* **2000**, *30*, 3941.

[97]For a review of oxidations and other reactions with supported reagents, see McKillop, A.; Young, D.W. *Synthesis* **1979**, 401.

[98]Cainelli, G.; Cardillo, G.; Orena, M.; Sandri, S. *J. Am. Chem. Soc.* **1976**, *98*, 6737; Santaniello, E.; Ponti, F.; Manzocchi, A. *Synthesis* **1978**, 534. See also, San Filippo, Jr., J.; Chern, C. *J. Org. Chem.* **1977**, *42*, 2182.

[99]Regen, S.L.; Koteel, C. *J. Am. Chem. Soc.* **1977**, *99*, 3837; Noureldin, N.A.; Lee, D.G. *Tetrahedron Lett.* **1981**, *22*, 4889. See also, Menger, F.M.; Lee, C. *J. Org. Chem.* **1979**, *44*, 3446.

[100]Srinivasan, R.; Balasubramanian, K. *Synth. Commun.* **2000**, *30*, 4397.

[101]Heravi, M.M.; Ajami, D.; Aghapoor, K.; Ghassemzadeh, M. *Chem. Commun.* **1999**, 833.

[102]With TMS-O-TMS: Heravi, M.M.; Ajami, D.; Tabar-Hydar, K. *Synth. Commun.* **1999**, *29*, 163. With HY zeolite: Mirza-Ayhayan, M.; Heravi, M.M. *Synth. Commun.* **1999**, *29*, 785.

[103]For a review of phase-transfer assisted permanganate oxidations, see Lee, D.G., in Trahanovsky, W.S. *Oxidation in Organic Chemistry*, pt. D Academic Press, NY, **1982**, pp. 147–206.

[104]See, for example, Hutchins, R.O.; Natale, N.R.; Cook, W.J. *Tetrahedron Lett.* **1977**, 4167; Landini, D.; Montanari, F.; Rolla, F. *Synthesis* **1979**, 134; Pletcher, D.; Tait, S.J.D. *J. Chem. Soc. Perkin Trans. 2*, **1979**, 788.

ruthenium tetroxide.[105] Phase-transfer catalysis is particularly useful because the oxidizing agents are insoluble in most organic solvents, while the substrates are generally insoluble in water (see p. 508). Ultrasound has been used for $KMnO_4$ oxidations.[106] A catalytic amount of $Cr(acac)_3$ in conjunction with H_5IO_5 oxidizes benzylic alcohols to aldehydes.[107]

Most of these oxidizing agents have also been used to convert primary alcohols to aldehydes, but precautions must be taken that the aldehyde is not further oxidized to the carboxylic acid (**19-22**).[108] When powerful oxidants, such as chromic acid, are used, one way to halt oxidation is by distillation of the aldehyde as it is formed. The following are among the oxidizing agents that have been used to convert at least some primary alcohols to aldehydes:[109] Collins' reagent, pyridinium chlorochromate and pyridinium dichromate, pyridinium dichromate, $Na_2Cr_2O_7$ in water,[110] $K_2Cr_2O_7$ in DMF at 100°C,[111] CrO_3 on silica gel,[112] wet CrO_3 on alumina with microwave irradiation,[113] $MeReO_3$,[114] HNO_3 with a $Yb(OTf)_3$ catalyst,[115] $FeBr_3$-H_2O_2,[116] a catalytic amount of $AuSiO_2$,[117] cerium (IV) immobilized on silica with $NaBrO_3$,[118] a bismuth catalyst,[119] O_2 with transition metal

[105]Morris, Jr., P.E.; Kiely, D.E. *J. Org. Chem.* **1987**, *52*, 1149.

[106]Yamawaki, J.; Sumi, S.; Ando, T.; Hanfusa, T. *Chem. Lett.* **1983**, 379.

[107]Xu, L.; Trudell, M.L. *Tetrahedron Lett.* **2003**, *44*, 2553.

[108]Though ketones are much less susceptible to further oxidation than aldehydes, such oxidation is possible (**19-8**), and care must be taken to avoid it, usually by controlling the temperature and/or the oxidizing agent.

[109]For some other reagents, not mentioned here, see Kaneda, K.; Kawanishi, Y.; Teranishi, S. *Chem. Lett.* **1984**, 1481; Semmelhack, M.F.; Schmid, C.R.; Cortés, D.A.; Chou, C.S. *J. Am. Chem. Soc.* **1984**, *106*, 3374; Cameron R.E.; Bocarsly, A.B. *J. Am. Chem. Soc.* **1985**, *107*, 6116; Anelli, P.L.; Biffi, C.; Montanari, F.; Quici, S. *J. Org. Chem.* **1987**, *52*, 2559; Bilgrien, C.; Davis, S.; Drago, R.S. *J. Am. Chem. Soc.* **1987**, *109*, 3786; Nishiguchi, T.; Asano, F. *J. Org. Chem.* **1989**, *54*, 1531. See also, Larock, R.C. *Comprehensive Organic Transformations*, 2nd ed., Wiley-VCH, NY, *1999*, pp. 1234–1250.

[110]Lee, D.G.; Spitzer, U.A. *J. Org. Chem.* **1970**, *35*, 3589. See also, Rao, Y.S.; Filler, R. *J. Org. Chem. 1974*, *39*, 3304; Lou, J. *Synth. Commun.* **1989**, *19*, 1841; *Chem. Ind. (London) 1989*, 312.

[111]Lou, J.-D.; Lu, L.-H. *Synth. Commun.* **1997**, *27*, 3701.

[112]Khadilkar, B.; Chitnavis, A.; Khare, A. *Synth. Commun.* **1996**, *26*, 205.

[113]Varma, R.S.; Saini, R.K. *Tetrahedron Lett.* **1998**, *39*, 1481.

[114]Divalentin, C.; Gandolfi, R.; Gisdakis, P.; Rösch, N. *J. Am. Chem. Soc.* **2001**, *123*, 2365; Jain, S.L.; Sharma, V.B.; Sain, B. *Tetrahedron Lett.* **2004**, *45*, 1233.

[115]Barrett, A.G.M.; Braddock, D.C.; McKinnell, R.M.; Waller, F.J. *Synlett 1999*, 1489.

[116]Martín, S.E.; Garrone, A. *Tetrahedron Lett.* **2003**, *44*, 549.

[117]Biella, S.; Rossi, M. *Chem. Commun.* **2003**, 378.

[118]Al-Haq, N.; Sullivan, A.C.; Wilson, J.R.H. *Tetrahedron Lett.* **2003**, *44*, 769.

[119]Matano, Y.; Nomura, H. *J. Am. Chem. Soc.* **2001**, *123*, 6443; Banik, B.K.; Ghatak, A.; Venkatraman, M.S.; Becker, I.F. *Synth. Commun.* **2000**, *30*, 2701.

catalysts,[120] RuO_2 with a zeolite catalyst,[121] and CuCl–phenanthroline.[122]

Tetrapropylammonium perruthenate (Pr_4N^+ RuO_4^-; also called TPAP; the *Ley reagent*)[123] has become an important oxidizing agent. This reagent has been bound to a polymer.[124] In the presence of molecular oxygen, it is catalytic in TPAP.[125] A polymer-bound morpholine *N*-oxide has been used in conjunction with a catalytic amount of TPAP.[126] Propargylic alcohols are oxidized to propargylic aldehydes with $TiCl_4/NEt_3$.[127] Methods have been developed for recovery of the catalyst and reuse of TPAP.[128]

Most of these reagents also oxidize secondary alcohols to ketones. Reagents that can be used specifically to oxidize a secondary OH group in the presence of a primary OH group[129] are H_2O_2–ammonium molybdate,[130] $NaBrO_3$–CAN,[131] and NaOCl in HOAc,[132] while $RuCl_2(PPh_3)_3$–benzene,[133]

[120] A combination of **OsO_4/CuCl catalysts**: Coleman, K.S.; Coppe, M.; Thomas, C.; Osborn, J.A. *Tetrahedron Lett.* **1999**, *40*, 3723. **A Cu catalyst**: Lipshutz, B.H.; Shimizu, H. *Angew. Chem. Int. Ed.* **2004**, *43*, 2228. **A Co–salen catalyst**: Fernández, I.; Pedro, J.R.; Roselló, A.L.; Ruiz, R.; Castro, I.; Ottenwaelder, X.; Journaux, Y. *Eur. J. Org. Chem.* **2001**, 1235. **A Co catalyst**: Minisci, F.; Punta, C.; Recupero, F.; Fontana, F.; Pedulli, G.F. *Chem. Commun.* **2002**, 688. **A Ru catalyst**: Lee, M.; Chang, S. *Tetrahedron Lett.* **2000**, *41*, 7507; Choi, E.; Lee, C.; Na, Y.; Chang, S. *Org. Lett.* **2002**, *4*, 2369; Wolfson, A.; Wuyts, S.; DeVos, D.E.; Vankelecom, I.F.J.; Jacobs, P.A. *Tetrahedron Lett.* **2002**, *43*, 8107; Yamaguchi, K.; Mizuno, N. *Angew. Chem. Int. Ed.* **2002**, *41*, 4538. **A V catalyst**: Maeda, Y.; Kakiuchi, N.; Matsumura, S.; Nishimura, T.; Kawamura, T.; Uemura, S. *J. Org. Chem.* **2002**, *67*, 6718. **A Pd catalyst**: Mori, K.; Hara, T.; Mizugaki, T.; Ebitani, K.; Kaneda, K. *J. Am. Chem. Soc.* **2004**, *126*, 10657; Schultz, M.J.; Park, C.C.; Sigman, M.S. *Chem. Commun.* **2002**, 3034; Jensen, D.R.; Schultz, M.J.; Mueller, J.A.; Sigman, M.S. *Angew. Chem. Int. Ed.* **2003**, *42*, 3810; Nishimura, T.; Onoue, T.; Ohe, K.; Uemura, S. *J. Org. Chem.* **1999**, *64*, 6750. **A Mo catalyst**: Velusamy, S.; Ahamed, M.; Punniyamurthy, T. *Org. Lett.* **2004**, *6*, 4821. For a review of aerobic oxidation of alcohols, see Zhan, B.-Z.; Thompson, A. *Tetrahedron* **2004**, *60*, 2917; Mallat, T.; Baiker, A. *Chem. Rev.* **2004**, *104*, 3037, and see Uma, R.; Crévisy, C.; Grée, R. *Chem. Rev.* **2003**, *103*, 27.

[121] Zhan, B.-Z.; White, M.A.; Sham, T.-K.; Pincock, J.A.; Doucet, R.J.; Rao, K.V.R.; Robertson, K.N.; Cameron, T.S. *J. Am. Chem. Soc.* **2003**, *125*, 2195.

[122] Markó, I.E.; Giles, P.R.; Tsukazaki, M.; Chellé-Regnaut, I.; Gautier, A.; Brown, S.M.; Urch, C.J. *J. Org. Chem.* **1999**, *64*, 2433.

[123] Griffith, W.P.; Ley, S.V.; Whitcombe, G.P.; White, A.D. *J. Chem. Soc. Chem. Commun.* **1987**, 1625; Griffith, W.P.; Ley, S.V. *Aldrichimica Acta* **1990**, *23*, 13; Markó, I.E.; Giles, P.R.; Tsukazaki, M.; Chellé-Regnaut, I.; Urch, C.J.; Brown, S.M. *J.Am. Chem. Soc.* **1997**, *119*, 12661.

[124] Hinzen, B.; Lenz, R.; Ley, S.V. *Synthesis* **1998**, 977

[125] Lenz, R.; Ley, S.V. *J. Chem. Soc. Perkin Trans. 1* **1997**, 3291.

[126] Brown, D.S.; Kerr, W.J.; Lindsay, D.M.; Pike, K.G.; Ratcliffe, P.D. *Synlett* **2001**, 1257.

[127] Han, Z.; Shinokubo, H.; Oshima, K. *Synlett* **2001**, 1421.

[128] Ley, S.V.; Ramarao, C.; Smith, M.D. *Chem. Commun.* **2001**, 2278.

[129] For other methods, see Jung, M.E.; Brown, R.W. *Tetrahedron Lett.* **1978**, 2771; Kaneda, K.; Kawanishi, Y.; Jitsukawa, K.; Teranishi, S. *Tetrahedron Lett.* **1983**, *24*, 5009; Siedlecka, R.; Skarżewski, J.; Młochowski, J. *Tetrahedron Lett.* **1990**, *31*, 2177. For a review, see Arterburn, J.B. *Tetrahedron* **2001**, *57*, 9765.

[130] Trost, B.M.; Masuyama, Y. *Isr. J. Chem.* **1984**, *24*, 134. For a method involving H_2O_2 and another catalyst, see Sakata, Y.; Ishii, Y. *J. Org. Chem.* **1991**, *56*, 6233.

[131] Tomioka, H.; Oshima, K.; Nozaki, H. *Tetrahedron Lett.* **1982**, *23*, 539.

[132] Stevens, R.V.; Chapman, K.T.; Stubbs, C.A.; Tam, W.W.; Albizati, K.F. *Tetrahedron Lett.* **1982**, *23*, 4647.

[133] Tomioka, H.; Takai, K.; Oshima, K.; Nozaki, H. *Tetrahedron Lett.* **1981**, *22*, 1605.

osmium tetroxide,[134] and $Br_2Ni(OBz)_2$[135] oxidize primary OH groups in the presence of a secondary OH group.[136] Benzylic and allylic alcohols have been selectively oxidized to the aldehydes in the presence of saturated alcohols by the use of potassium manganate $KMnO_4$ under phase-transfer conditions.[137] On the other hand, Fremy's salt (see **19-4**) selectively oxidizes benzylic alcohols and not allylic or saturated ones.[138] Certain zirconocene complexes can selectively oxidize only one OH group of a diol, even if both are primary.[139] α-Hydroxy ketones are oxidized to 1,2-diketones with $Bi(NO_3)_3$ and a $Cu(OAc)_2$ catalyst,[140] ferric chloride (solid state),[141] or O_2 and a vanadium catalyst.[142] Tetrabutylammonium periodate oxidizes primary alcohols to aldehydes,[143] as does benzyltriphenylphosphonium periodate.[144] α-Hydroxyl phosphonate esters are oxidized to the α-keto phosphonate ester with zinc dichromate, without solvent[145] or with CrO_3 on alumina with microwave irradiation.[146]

O-Trimethylsilyl ethers of benzylic alcohols are oxidized to the corresponding aldehyde with CrO_3 on wet alumina.[147] Treatment with $MnO_2/AlCl_3$ leads to similar oxidation,[148] as does $NaBrO_3$ in aq. MeCN[149] or K_2FeO_4 on clay.[150] Oxidation of trimethylsilyl ethers with O_2, a catalytic amount of N-hydroxyphthalimide and a cobalt catalyst give an aldehyde.[151] Microwave irradiation with $BiCl_3$ oxidizes benzylic TMS ethers to the aldehyde.[152] Microwave irradiation on zeolite supported ferric nitrate has been used.[153] O-Tetrahydropyranyl ethers (O-THP) have been oxidized to the aldehyde with ferric nitrate on zeolites.[154]

[134]Maione, A.M.; Romeo, A. *Synthesis* **1984**, 955.

[135]Doyle, M.P.; Dow, R.L.; Bagheri, V.; Patrie, W.J. *J. Org. Chem.* **1983**, 48, 476.

[136]For a list of references to the selective oxidation of various types of alcohol, see Kulkarni, M.G.; Mathew T.S. *Tetrahedron Lett.* **1990**, 31, 4497.

[137]Kim, K.S.; Chung, S.; Cho, I.H.; Hahn, C.S. *Tetrahedron Lett.* **1989**, 30, 2559. See also, Kim, K.S.; Song, Y.H.; Lee, N.H.; Hahn, C.S. *Tetrahedron Lett.* **1986**, 27, 2875.

[138]Morey, J.; Dzielenziak, A.; Saá, J.M. *Chem. Lett.* **1985**, 263.

[139]Nakano, T.; Terada, T.; Ishii, Y.; Ogawa, M. *Synthesis* **1986**, 774.

[140]Tymonko, S.A; Nattier, B.A.; Mohan, R.S. *Tetrahedron Lett.* **1999**, 40, 7657.

[141]Zhou, Y.-M.; Ye, X.-R.; Xin, X.-Q. *Synth. Commun.* **1999**, 29, 2229.

[142]Kirahara, M.; Ochiai, Yy.; Takizawa, S.; Takahata, H.; Nemoto, H. *Chem. Commun.* **1999**, 1387.

[143]Friedrich, H.B.; Khan, F.; Singh, N.; van Staden, M. *Synlett* **2001**, 869.

[144]Hajipour, A.R.; Mallakpour, S.E.; Samimi, H.A. *Synlett* **2001**, 1735.

[145]Firouzabadi, H.; Iranpoor, N.; Sobhani, S.; Sardarian, A.-R. *Tetrahedron Lett.* **2001**, 42, 4369.

[146]Kaboudin, B.; Nazari, R. *Synth. Commun.* **2001**, 31, 2245.

[147]Heravi, M.M.; Ajami, D.; Ghassemzadeh, M. *Synth. Commun.* **1999**, 29, 781. See also, Heravi, M.M.; Ajami, D.; Tabar-Hydar, K. *Synth. Commun.* **1999**, 29, 1009; Mojtahedi, M.M.; Saidi, M.R.; Bolourtchian, M.; Heravi, M.M. *Synth. Commun.* **1999**, 29, 3283.

[148]Firouzabadi, H.; Etemadi, S.; Karimi, B.; Jarrahpour, A.A. *Synth. Commun.* **1999**, 29, 4333.

[149]Shaabani, A.; Karimi, A.-R. *Synth. Commun.* **2001**, 31, 759.

[150]Tajbakhsh, M.; Heravi, M.M.; Habibzadeh, S.; Ghassemzadeh, M. *J. Chem. Res. (S)* **2001**, 39.

[151]Karimi, B.; Rajabi, J. *Org. Lett.* **2004**, 6, 2841.

[152]Hajipour, A.R.; Mallakpour, S.E.; Baltork, I.M.; Adibi, H. *Synth. Commun.* **2001**, 31, 1625.

[153]Heravi, M.M.; Ajami, D.; Ghassemzadeh, M.; Tabar-Hydar, K. *Synth. Commun.* **2001**, 31, 2097.

[154]Mohajerani, B.; Heravi, M.M.; Ajami, D. *Monat. Chem.* **2001**, 132, 871.

2. *The Oppenauer Oxidation.* When a ketone in the presence of an aluminum alkoxide is used as the oxidizing agent (it is reduced to a secondary alcohol), the reaction is known as the *Oppenauer oxidation.*[155] This is the reverse of the Meerwein–Ponndorf–Verley reaction (**19-36**) and the mechanism is also the reverse. The ketones most commonly used are acetone, butanone, and cyclohexanone. The most common base is aluminum *tert*-butoxide. The chief advantage of the method is its high selectivity. Although the method is most often used for the preparation of ketones, it has also been used for aldehydes. An iridium catalyst[156] has been developed for the Oppenauer oxidation, and also a water-soluble iridium catalyst[157] An uncatalyzed reaction under supercritical conditions was reported.[158]

3. *With DMSO-Based Reagents.* An alcohol is treated with DMSO, DCC,[159] and anhydrous phosphoric acid[160] in what is called *Moffatt oxidation.* In this way, a primary alcohol can be converted to the aldehyde with no carboxylic acid being produced. The strong acid conditions are sometimes a problem, and complete removal of the dicyclohexylurea by-product can be difficult. The use of oxalyl chloride and DMSO at low temperature, the Swern oxidation,[161] is generally more practical and widely used. Maintaining the low reaction temperature is essential in this reaction however, since the reagent generated *in situ* decomposes at temperatures significantly below ambient.

Similar oxidation of alcohols has been carried out with DMSO and other reagents[162] in place of DCC: acetic anhydride,[163] SO3–pyridine–triethyla-mine,[164] trifluoroacetic anhydride,[165] tosyl chloride,[166] Ph3P+Br−,[167]

[155]For a review, see Djerassi, C. *Org. React.* **1951**, *6*, 207. For the use of new catalyst,s see Akamanchi, K.G.; Chaudhari, B.A. *Tetrahedron Lett.* **1997**, *38*, 6925; Ooi, T.; Miura, T.; Itagaki, Y.; Ichikawa, H.; Maruoka, K. *Synthesis* **2002**, 279.
[156]Suzuki, T.; Morita, K.; Tsuchida, M.; Hiroi, K. *J. Org. Chem.* **2003**, *68*, 1601.
[157]Ajjou, A.N. *Tetrahedron Lett.* **2001**, *42*, 13.
[158]Sominsky, L.; Rozental, E.; Gottlieb, H.; Gedanken, A.; Hoz, S. *J. Org. Chem.* **2004**, *69*, 1492.
[159]The DCC is converted to dicyclohexylurea, which in some cases is difficult to separate from the product. One way to avoid this problem is to use a carbodiimide linked to an insoluble polymer: Weinshenker, N.M.; Shen, C. *Tetrahedron Lett.* **1972**, 3285.
[160]Pfitzner, K.E.; Moffatt, J.G. *J. Am. Chem. Soc.* **1965**, *87*, 5661, 5670; Fenselau, A.H.; Moffatt, J.G. *J. Am. Chem. Soc.* **1966**, *88*, 1762; Albright, J.D.; Goldman, L. *J. Org. Chem.* **1965**, *30*, 1107.
[161]Omura, K.; Swern, D. *Tetrahedron* **1978**, *34*, 1651. See also, Marx, M.; Tidwell, T.T. *J. Org. Chem.* **1984**, *49*, 788. For a modification of the Swern oxidation, see Liu, Y.; Vederas, J.C. *J. Org. Chem.* **1996**, *61*, 7856. For the effect of bases, see Chrisman, W.; Singaram, B. *Tetrahedron Lett.* **1997**, *38*, 2053. For an odorless Swern oxidation, see Ohsugi, S.-i.; Nishide, K.; Oono, K.;Okuyama, K.; Fudesaka, M.; Kodama, S.; Node, M. *Tetrahedron* **2003**, *59*, 8393.
[162]For a review of activated DMSO reagents and their use in this reaction, see Mancuso, A.J.; Swern, D. *Synthesis* **1981**, 165.
[163]Albright, J.D.; Goldman, L. *J. Am. Chem. Soc.* **1967**, *89*, 2416.
[164]Parikh, J.R.; Doering, W. von E. *J. Am. Chem. Soc.* **1967**, *89*, 5507.
[165]Huang, S.L.; Omura, K.; Swern, D. *Synthesis* **1978**, 297.
[166]Albright, J.D. *J. Org. Chem.* **1974**, *39*, 1977.
[167]Bisai, A.; Chandrasekhar, M.; Singh, V.K. *Tetrahedron Lett.* **2002**, *43*, 8355.

P_2O_5-Et_3N,[168] trichloromethyl chloroformate,[169] trimethylamine N-oxide,[170] 2,4,6-trichlorotriazine,[171] a molybdenum catalyst and O_2,[172] KI and $NaHCO_3$,[173] and methanesulfonic anhydride.[517] Dimethyl sulfoxide in 48% HBr oxidizes benzylic alcohols the aryl aldehydes.[174] Note that Swern oxidation of molecules having alcohol moieties, as well as a disulfide, leads to the ketone without oxidation of the sulfur.[175] Sulfoxides other than DMSO can be used in conjunction with oxalyl chloride for the oxidation of alcohols,[176] including fluorinated sulfoxides[177] and a polymer-bound sulfoxide.[178]

4. *TEMPO and Related Reagents.* The nitroxyl radical TEMPO has been used in conjunction with coreagents, including mcpba[179] Br_2–$NaNO_2$,[180] O_2 with transition-metal catalysts,[181] CuBr•SMe_2–$C_8F_{17}Br$,[182] $CuBr_2$(bpy)-air (bpy=2, 2'-bipyridyl),[183] Oxone®,[184] CuBr•SMe_2 in perfluorous solvents,[185] 2,4,6-trichlorotriazine,[186] enzymes,[187] H_5IO_6,[188] and NCS.[189] Silica-supported TEMPO,[190] polymer-bound TEMPO,[191] and PEG–TEMPO[192] (where PEG is polyethylene glycol) have been used. The TEMPO compound has also been used with a polymer-bound hypervalent iodine reagent[193] (see below). A catalytic reaction using 5% TEMPO and 5% CuCl with O_2 in an

[168]Taber, D.F.; Amedio, Jr., J.C.; Jung, K. *J. Org. Chem.* **1987**, *52*, 5621.

[169]Takano, S.; Inomata, K.; Tomita, S.; Yanase, M.; Samizu, K.; Ogasawara, K. *Tetrahedron Lett.* **1988**, *29*, 6619.

[170]Godfrey, A.G.; Ganem, B. *Tetrahedron Lett.* **1990**, *31*, 4825.

[171]DeLuca, L.; Giacomelli, G.; Porcheddu, A. *J. Org. Chem.* **2001**, *66*, 7907.

[172]Khenkin, A.M.; Neumann, R. *J. Org. Chem.* **2002**, *67*, 7075.

[173]Bauer, D.P.; Macomber, R.S. *J. Org. Chem.* **1975**, *40*, 1990.

[174]Li, C.; Xu, Y.;Lu, M.; Zhao, Z.; Liu, L.; Zhao, Z.; Cui, Y.; Zheng, P.; Ji, X.; Gao, G. *Synlett* **2002**, 2041.

[175]Fang, X.; Bandarage, U.K.; Wang, T.; Schroeder, J.D.; Garvery, D.S. *J. Org. Chem.* **2001**, *66*, 4019.

[176]Nishida, K.; Ohsugi, S.-I.; Fudesaka, M.; Kodama, S.; Node, M. *Tetrahedron Lett.* **2002**, *43*, 5177.

[177]Crich, D.; Neelamkavil, S. *Tetrahedron* **2002**, *58*, 3865.

[178]Choi, M.K.W.C.; Toy, P.H. *Tetrahedron* **2003**, *59*, 7171.

[179]Rychnovsky, S.D.; Vaidyanathan, R. *J. Org. Chem.* **1999**, *64*, 310.

[180]Liu, R.; Liang, X.; Dong, C.; Hu, X. *J. Am. Chem. Soc.* **2004**, *126*, 4112.

[181]**Mo**: Ben-Daniel, R.; Alssteers, P.; Neumann, R. *J. Org. Chem.* **2001**, *66*, 8650. **Ru**: Dijksman, A.; Marino-González, A.; Payeras, A.M.; Arends, I.W.C.E.; Sheldon, R.A. *J. Am. Chem. Soc.* **2001**, *123*, 6826; Dijksman, A.; Arends, I.W.C.E.; Sheldon, R.A. *Chem. Commun.* **1999**, 1591. **Mn/Co**: Cecchetto, A.; Fontana, F.; Minisci, F.; Recupero, F. *Tetrahedron Lett.* **2001**, *42*, 6651.

[182]Ragagnin, G.; Betzemeier, B.; Quici, S.; Knochel, P. *Tetrahedron* **2002**, *58*, 3985.

[183]Gamez, P.; Arends, I.W.C.E.; Reedijk, J.; Sheldon, R.A. *Chem. Commun.* **2003**, 2414.

[184]Bolm, C.; Magnus, A.S.; Hildebrand, J.P. *Org. Lett.* **2000**, *2*, 1173; Koo, B.-S.; Lee, C.K.; Lee, K.-J. *Synth. Commun.* **2002**, *32*, 2115.

[185]Betzemeier, B.; Cavazzini, M.; Quici, S.; Knochel, P. *Tetrahedron Lett.* **2000**, *41*, 4343.

[186]DeLuca, L.; Giacomelli, G.; Porcheddu, A. *Org. Lett.* **2001**, *3*, 3041.

[187]Fabbrini, M.; Galli, C.; Gentili, P.; Macchitella, D. *Tetrahedron Lett.* **2001**, *42*, 7551.

[188]Kim, S.S.; Nehru, K. *Synlett* **2002**, 616.

[189]Einhorn, J.; Einhorn, C.; Ratajczak, F.; Pierre, J.-L. *J. Org. Chem.* **1996**, *61*, 7452.

[190]Bolm, C.; Fey, T. *Chem. Commun.* **1999**, 1795.

[191]Fey, T.; Fischer, H.; Bachmann, S.; Albert, K.; Bolm, C. *J. Org. Chem.* **2002**, *66*, 8154.

[192]Ferreira, P.; Phillips, E.; Rippon, D.; Tsang, S.C.; Hayes, W. *J. Org. Chem.* **2004**, *69*, 6851.

[193]Sakuratani, K.; Togo, H. *Synthesis* **2003**, 21.

ionic liquid oxidizes benzylic alcohols to the corresponding aldehyde.[194] Other nitroxyl radical oxidizing agents are known.[195] A related oxidizing agent is oxoammonium salt **12** (*Bobbitt's reagent*), a stable and nonhygroscopic salt that oxidizes primary and secondary alcohols in dichloromethane.[196]

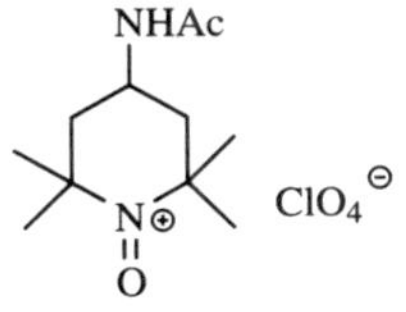

12

5. *With Hypervalent Iodine Reagents.*[197] Treatment of 2-iodobenzoic acid with KBrO$_3$ in sulfuric acid and heating the resulting product to 100°C with acetic anhydride and acetic acid gives hypervalent iodine reagent **13**, the so-called *Dess–Martin Periodinane.*[198] This reagent reacts with alcohols at ambient temperature to give the corresponding aldehyde or ketone.[199] The reaction is accelerated by water[200] and a water-soluble periodinane (*o*-iodoxybenzoic acid, **14**) has been prepared that oxidized allylic alcohols to conjugated aldehydes.[201]

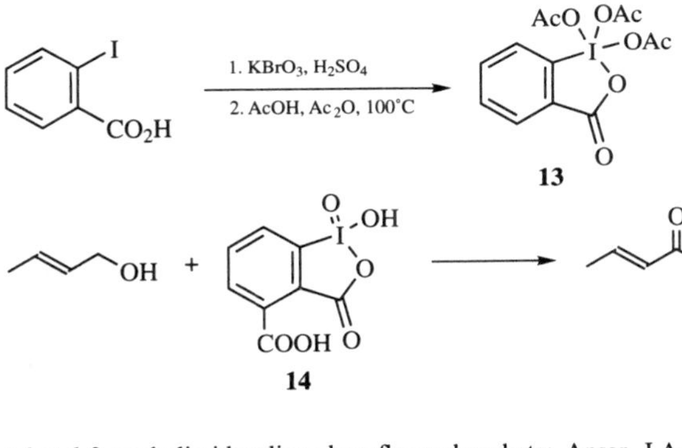

13

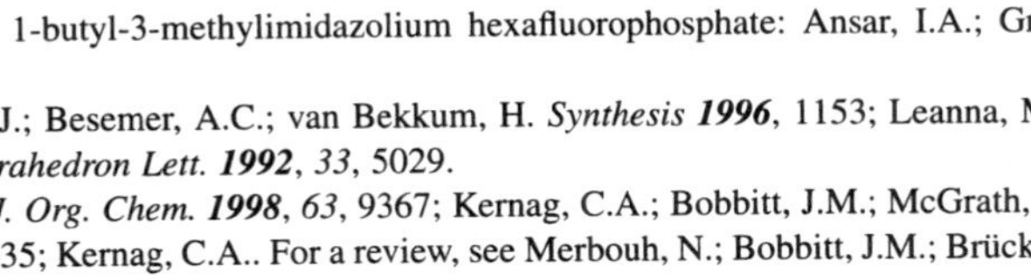

14

[194]In bmim PF$_6$, 1-butyl-3-methylimidazolium hexafluorophosphate: Ansar, I.A.; Gree, R. *Org. Lett.* *2002*, *4*, 1507.

[195]de Nooy, A.E.J.; Besemer, A.C.; van Bekkum, H. *Synthesis* *1996*, 1153; Leanna, M.R.; Sowin, T.J.; Morton, H.E. *Tetrahedron Lett.* *1992*, *33*, 5029.

[196]Bobbitt, J.M. *J. Org. Chem.* *1998*, *63*, 9367; Kernag, C.A.; Bobbitt, J.M.; McGrath, D.V. *Tetrahedron Lett.* *1999*, *40*, 1635; Kernag, C.A.. For a review, see Merbouh, N.; Bobbitt, J.M.; Brückner, C. *Org. Prep. Proceed. Int.* *2004*, *36*, 1.

[197]For a review of hypervalent iodine compounds, see Wirth, T.; Hirt, U.S. *Synthesis* *1999*, 1271.

[198]Dess, D.B.; Martin, J.C. *J. Org. Chem.* *1983*, *48*, 4155; Dess, D.B.; Martin, J.C. *J. Am. Chem. Soc.* *1991*, *113*, 7277. For a synthesis of the requisite precursor, see Frigerio, M.; Santagostino, M.; Sputore, S. *J. Org. Chem.* *1999*, *64*, 4537.

[199]For example, see Frigerio, M.; Santagostino, M. *Tetrahedron Lett.* *1994*, *35*, 8019.

[200]Meyer, S.D.; Schreiber, S.L. *J. Org. Chem.* *1994*, *59*, 7549. In aqueous β-cyclodextrin-acetone solution, see Surendra, K.; Krishnaveni, N.S.; Reddy, M.A.; Nageswar, Y.V.D.; Rao, K.R. *J. Org. Chem.* *2003*, *68*, 2058.

[201]Thottumkara, A.P.; Vinod, T.K. *Tetrahedron Lett* *2001*, *43*, 569.

The reagent has an indefinite shelf-life in a sealed container, but hydrolysis occurs upon long-term exposure to atmospheric moisture. A note of CAUTION! The Dess–Martin reagent can be shock sensitive under some conditions and explode >200°C.[202] Other hypervalent iodine oxidizing reagents are known,[203] including PhI(OAc)$_2$/TEMPO,[204] PhI(OAc)$_2$–chromium salen,[205] stabilized iodoxybenzoic acid,[206] and PhI(OAc)$_2$ supported on alumina with microwave irradiation.[207] Microwave irradiation of benzylic alcohols with PhI(OH)OTs gave the corresponding aldehyde.[208] Hypervalent iodine compounds have been used in ionic liquids.[209] Heating benzylic alcohols with o-iodoxybenzoic acid under solvent-free conditions gave the aldehyde.[210] Cyclopropylcarbinyl alcohols are oxidized to the corresponding cyclopropyl ketone or aldehyde with PhIO and a chromium–salen catalyst.[211] The Dess–Martin reagent oxidized aryl aldoximes to aryl aldehydes.[212]

6. *By Catalytic Dehydrogenation.* For the conversion of primary alcohols to aldehydes, dehydrogenation catalysts have the advantage over strong oxidizing agents that further oxidation to the carboxylic acid is prevented. Copper chromite is the agent most often used, but other catalysts (e.g., silver and copper) have also been employed. Many ketones were prepared in this manner. Catalytic dehydrogenation is more often used industrially than as a laboratory method. However, procedures using copper oxide,[213] copper(II) complexes,[214] rhodium complexes,[215] ruthenium complexes,[216] Raney nickel,[217] and palladium complexes[218] (under phase-transfer conditions)[219]

[202]Plumb, J.B.; Harper, D.J. *Chem. Eng. News,* **1990**, July 16, p. 3. For an improved procedure, see Ireland, R.E.; Liu, L. *J. Org. Chem.* **1993**, *58*, 2899.

[203]Moriarty, R.M.; Prakash, O. *Accts. Chem. Res.* **1986**, *19*, 244; Moriarty, R.M.; John, L.S.; Du, P.C. *J. Chem. Soc. Chem. Commun.* **1981**, 641; Moriarty, R.M.; Gupta, S.; Hu, H.; Berenschot, D.R.; White, K.B. *J. Am. Chem. Soc.* **1981**, *103*, 686; Moriarty, R.M.; Hu, H.; Gupta, S.C. *Tetrahedron Lett.* **1981**, *22*, 1283.

[204]DeMico, A.; Margarita, R.; Parlanti, L.; Vescovi, A.; Piancatelli, G. *J. Org. Chem.* **1997**, *62*, 6974.

[205]Adam, W.; Hajra, S.; Herderich, M.; Saha-Möller, C.R. *Org. Lett.* **2000**, *2*, 2773.

[206]Khan, T.A.; Tripoli, R.; Crawford, J.T.; Martin, C.G.; Murphy, J.A. *Org. Lett.* **2003**, *5*, 2971.

[207]Varma, R.S.; Saini, R.K.; Dahiya, R. *J. Chem. Res. (S)* **1998**, 120.

[208]Lee, J.C.; Lee, J.Y.; Lee, S.J. *Tetrahedron Lett.* **2004**, *45*, 4939.

[209]In bmim Cl, 1-butyl-3-methylimidazolium chloride: Liu, Z.; Chen, Z.-C.; Zheng, Q.-C. *Org. Lett.* **2003**, *5*, 3321; Karthikeyan, G.; Perumal, P.T. *Synlett* **2003**, 2249.

[210]Moorthy, J.N.; Singhal, N.; Venkatakrishnan, P. *Tetrahedron Let.* **2004**, *45*, 5419.

[211]Adam, W.; Gelalcha, F.G.; Saha-Möller, C.R.; Stegmann, V.R. *J. Org. Chem.* **2000**, *65*, 1915.

[212]Bose, D.S.; Narsaiah, A.V. *Synth. Commun.* **1999**, *29*, 937.

[213]Sheikh, M.Y.; Eadon, G. *Tetrahedron Lett.* **1972**, 257.

[214]Muldoon, J.; Brown, S.N. *Org. Lett.* **2002**, *4*, 1043.

[215]Takahashi, M.; Oshima, K.; Matsubara, S. *Tetrahedron Lett.* **2003**, *44*, 9201.

[216]Meijer, R.H.; Ligthart, G.B.W.L.; Meuldijk, J.; Vekemans, J.A.J.M.; Hulshof, L.A.; Mills, A.M.; Kooijman, H.; Spek, A.L. *Tetrahedron* **2004**, *60*, 1065.

[217]Krafft, M.E.; Zorc, B. *J. Org. Chem.* **1986**, *51*, 5482.

[218]For a discussion of the enantioselective palladium(II) oxidation, see Mandal, S.K.; Jensen, D.R.; Pugsley, J.S.; Sigman, M.S. *J. Org. Chem.* **2003**, *68*, 4600. See also Mandal, S.K.; Sigman, M.S. *J. Org. Chem.* **2003**, *68*, 7535; Guram, A.S.; Bei, X.; Turner, H.W. *Org. Lett.* **2003**, *5*, 2485; Ganchegui, B.; Bouquillon, S.; Hénin, F.; Muzart, J. *Tetrahedron Lett.* **2002**, *43*, 6641. For a review, see Muzart, J. *Tetrahedron* **2003**, *59*, 5789.

[219]Choudary, B.M.; Reddy, N.P.; Kantam, M.L.; Jamil, Z. *Tetrahedron Lett.* **1985**, *26*, 6257.

have been reported. Allylic alcohols[220] are oxidized to the corresponding saturated aldehyde or ketone by heating with a rhodium catalyst, and benzylic alcohols are converted to the aldehyde with a rhodium catalyst.[221] Photolysis with an iron catalyst gives similar results.[222] Propargylic alcohols are oxidized by heating with a vanadium catalyst.[223] Secondary alcohols are oxidized with $Bi(NO_3)_3$ on Montmorillonite.[224]

7. *Miscellaneous Reagents.*[225] Nitric acid in dichloromethane oxidizes benzylic alcohols to the corresponding ketone.[226] Bromine is an effective oxidant, and iodine under photochemical conditions has been used.[227] Heating a 1,2-diol with NBS in CCl_4 gave the 1,2-diketone.[228] *N*-Bromosuccinimide with β-cyclodextrin oxidizes tetrahydropyranyl ethers in water.[229] Iodine has been used in conjunction with DMSO and hydrazine.[230] Sodium bromate (NaOBr) in conjunction with HCl oxidizes α-hydroxy esters to α-keto esters.[231] Enzymatic oxidations have been reported.[232] Dimethyl dioxirane[233] oxidizes benzylic alcohols to the corresponding aldehyde,[234] and dioxirane reagents are sufficiently mild that an α,β-epoxy alcohol was oxidized to the corresponding ketone, without disturbing the epoxide, using methyl trifluoromethyl dioxirane.[235] Hydrogen peroxide with urea oxidizes aryl aldehydes in formic acid.[236] Potassium monoperoxysulfate in the presence of a

[220]Tanaka, K.; Fu, G.C. *J. Org. Chem.* **2001**, *66*, 8177.
[221]Miyata, A.; Murakami, M.; Irie, R.; Katsuki, T. *Tetrahedron Lett.* **2001**, *42*, 7067; Kölle, U.; Fränzl, H. *Monat. Chem.* **2000**, *131*, 1321; Csjernyik, G.; Ell, A.H.; Fadini, L.; Pugin, B.; Bäckvall, J.-E. *J. Org. Chem.* **2002**, *67*, 1657.
[222]Cherkaoui, H.; Soufiaoui, M.; Grée, R. *Tetrahedron* **2001**, *57*, 2379.
[223]Maeda, Y.; Kakiuchi, N.; Matsumura, S.; Nishimura, T.; Uemura, S. *Tetrahedron Lett.* **2001**, *42*, 8877.
[224]Samajdar, S.; Becker, F.F.; Banik, B.K. *Synth. Commun.* **2001**, *31*, 2691.
[225]For a review of green, catalytic oxidations of alcohols, see Sheldon, R.A.; Arends, I.W.C.E.; ten Brink, G.-J.; Dijksman, A. *Acc. Chem. Res.* **2002**, *35*, 774.
[226]Strazzolini, P.; Runcio, A. *Eur. J. Org. Chem.* **2003**, 526.
[227]Itoh, A.; Kodama, T.; Masaki, Y. *Chem. Lett.* **2001**, 686.
[228]Khurana, J.M.; Kandpal, B.M. *Tetrahedron Lett.* **2003**, *44*, 4909.
[229]Narender, M.; Reddy, M.S.; Rao, K.R. *Synthesis* **2004**, 1741. See Reddy, M.S.; Narender, M.; Nageswar, Y.V.D.; Rao, K.R. *Synthesis* **2005**, 714.
[230]Gogoi, P.; Sarmah, G.K.; Konwar, D. *J. Org. Chem.* **2004**, *69*, 5153.
[231]Chang, H.S.; Woo, J.C.; Lee, K.M.; Ko, Y.K.; Moon, S.-S.; Kim, D.-W. *Synth. Commun.* **2002**, *32*, 31.
[232]*Bacilus stearothermophilus*: Fantin, G.; Fogagnolo, M.; Giovannini, P.P.; Medici, A.; Pedrini, P.; Poli, S. *Tetrahedron Lett.* **1995**, *36*, 441. *Gluconobaccter oxydans* DSM 2343: Villa, R.; Romano, A.; Gandolfi, R.; Gargo, J.V.S.; Molinari, F. *Tetrahedron Lett.* **2002**, *43*, 6059. Chloroperoxidase: Hu, S.; Dordick, J.S. *J. Org. Chem.* **2002**, *67*, 314.
[233]For a discussion of whether dioxirane oxidation is electrophilic or nucleophilic, see Deubel, D.V. *J. Org. Chem.* **2001**, *66*, 3790.
[234]Baumstark, A.L.; Kovac, F.; Vasquez, P.C. *Can. J. Chem.* **1999**, *77*, 308. For a discussion of the mechanism, see Angelis, Y.S.; Hatzakis, N.S.; Smonou, I.; Orfanoupoulos, M. *Tetrahedron Lett.* **2001**, *42*, 3753.
[235]D'Accolti, L.; Fusco, C.; Annese, C.; Rella, M.R.; Turteltaub, J.S.; Williard, P.G.; Curci, R. *J. Org. Chem.* **2004**, *69*, 8510.
[236]Balicki, R. *Synth. Commun.* **2001**, *31*, 2195.

chiral ketone oxidizes 1,2-diols to α-hydroxy ketones enantioselectively.[237] Potassium monoperoxysulfate also oxidizes secondary alcohols in the presence of O_2.[238] air in the presence of a zeolite oxidizes benzylic alcohols.[239] The reagent $Br^+(collidine)_2PF_6$ oxidizes benzylic alcohols to the corresponding aldehyde.[240] Sodium hypochlorite in acetic acid is useful for oxidizing larger amounts of secondary alcohols.[241] Calcium hypochlorite on moist alumina with microwave irradiation has been used to oxidize benzylic alcohols.[242] Chlorosulfimines, $Ar(Cl)S=N-t-Bu$, oxidize primary alcohols to aldehydes.[243] This latter reagent is generated from $ArS-NHt-Bu$ and NCS.[244] Benzylic alcohols are converted to aldehydes with DBU (p. $$$) and Ar_3BCl_2.[245] Microwave irradiation of benzylic alcohols with $Co(CO_3)_2$ on silica gel generates the aryl aldehyde.[246]

Primary and secondary alcohols can also be oxidized, indirectly, through their esters (see **19-21**). In some cases, isolation of the ester is not required and the alcohol can then be oxidized to the aldehyde or ketone in one step.

The mechanism of oxidation with acid dichromate has been intensely studied.[247] The currently accepted mechanism is essentially that proposed by Westheimer.[248] The first two steps constitute an example of category 4 (p. 1706).

[237]Adam, W.; Saha-Möller, C.R.; Zhao, C.-G. *J. Org. Chem.* **1999**, *64*, 7492.

[238]Döbler, C.; Mehltretter, G.M.; Sundermeier, U.; Eckert, M.; Militzer, H.-C.; Beller, M. *Tetrahedron Lett.* **2001**, *42*, 8447.

[239]Son, Y.-C.; Makwana, V.D.; Howell, A.R.; Suib, S.L. *Angew. Chem. Int. Ed.* **2001**, *40*, 4280.

[240]Rousseau, G.; Robin, S. *Tetrahedron Lett* **2000**, *41*, 8881.

[241]Stevens, R.V.; Chapman, K.T.; Weller, H.N. *J. Org. Chem.* **1980**, *45*, 2030. See also, Schneider, M.; Weber, J.; Faller, P. *J. Org. Chem.* **1982**, *47*, 364; Mohrig, J.R.; Nienhuis, D.M.; Linck, C.F.; van Zoeren, C.; Fox, B.G.; Mahaffy, P.G. *J. Chem. Educ.* **1985**, *62*, 519. For a reaction with aqueous NaOCl and a guanidinium salt, see Xie, H.; Zhang, S.; Duan, H. *Tetrahedron Lett.* **2004**, *45*, 2013.

[242]Mojtahedi, M.M.; Saidi, M.R.; Bolourtchian, M.; Shirzi, J.S. *Monat. Chem.* **2001**, *132*, 655.

[243]Mukaiyama, T.; Matsuo, J.-i.; Yanagisawa, M. *Chem Lett.* **2000**, 1072; Matsuo, J.-i.; Kitgawa, H.; Iida, D.; Mukaiyama, T. *Chem. Lett.* **2001**, 150.

[244]Mukaiyama, T.; Matsuo, J.-i.; Iida, D.; Kitagawa, H. *Chem. Lett.* **2001**, 846; Matsuo, J.-i.; Iida, D.; Yamanaka, H.; Mukaiyama, T. *Tetrahedron* **2003**, *59*, 6739.

[245]Mantano, Y.; Nomura, H. *Angew. Chem. Int. Ed.* **2002**, *41*, 3028.

[246]Kiasat, A.R.; Kazemi, F.; Rafati, M. *Synth. Commun.* **2003**, *33*, 601.

[247]See Müller, P. *Chimia* **1977**, *31*, 209; Wiberg, K.B., in Wiberg, K.B. *Oxidation in Organic Chemistry*, pt. A, Academic Press, NY, **1965**, pp. 142–170; Venkatasubramanian, N. *J. Sci. Ind. Res.* **1963**, *22*, 397; Waters, W.A. *Mechanisms of Oxidation of Organic Compounds*, Wiley, NY, **1964**, pp. 49–71; Stewart, R. *Oxidation Mechanisms*, W.A. Benjamin, NY, **1964**, pp. 37–48; Durand, R.; Geneste, P.; Lamaty, G.; Moreau, C.; Pomarès, O.; Roque, J.P. *Recl. Trav. Chim. Pays-Bas* **1978**, *97*, 42; Sengupta, K.K.; Samanta, T.; Basu, S.N. *Tetrahedron* **1985**, *41*, 205.

[248]Westheimer, F.H. *Chem. Rev.* **1949**, *45*, 419, see p. 434; Holloway, F.; Cohen, M.; Westheimer, F.H. *J. Am. Chem. Soc.* **1951**, *73*, 65.

$$R_2C(H)(OH) + HCrO_4^- + H^+ \rightleftharpoons R_2C(H)(OCrO_3H)$$

$$R_2C(H)(OCrO_3H) \xrightarrow{\text{base}} R_2C=O + HCO_3-[Cr(IV)] + base-H^+$$

$$R_2C(H)(OH) + Cr(IV) \longrightarrow R_2\overset{\bullet}{C}-OH + Cr(III)$$

$$R_2\overset{\bullet}{C}-OH + Cr(VI) \longrightarrow R_2C=O + Cr(V)$$

$$R_2C(H)(OH) + Cr(IV) \longrightarrow R_2C=O + Cr(III)$$

The base in the second step may be water, although it is also possible[249] that in some cases no external base is involved and that the proton is transferred directly to one of the CrO$_3$H oxygens in which case the Cr(IV) species

$$\longrightarrow R_2C=O + H_2CrO_3$$

produced would be H$_2$CrO$_3$. Part of the evidence for this mechanism was the isotope effect of ∼ 6 found on use of MeCDOHMe, showing that the a hydrogen is removed in the rate-determining step.[250] Note that, as in **19-23** the substrate is oxidized by three different oxidation states of chromium.[251]

With other oxidizing agents, mechanisms are less clear.[252] It seems certain that some oxidizing agents operate by a hydride-shift mechanism,[253] for example,

[249]Stewart, R.; Lee, D.G. *Can. J. Chem.* **1964**, *42*, 439; Awasthy, A.; Roek, J.; Moriarty, R.M. *J. Am. Chem. Soc.* **1967**, *89*, 5400; Kwart, H.; Nickle, J.H. *J. Am. Chem. Soc.* **1979**, *98*, 2881 and cited rererences; Sengupta, K.K.; Samanta, T.; Basu, S.N. *Tetrahedron* **1986**, *42*, 681. See also Müller, P.; Perlberger, J. *Helv. Chim. Acta* **1974**, *57*, 1943; Agarwal, S.; Tiwari, H.P.; Sharma, J.P. *Tetrahedron* **1990**, *46*, 1963.

[250]Westheimer, F.H.; Nicolaides, N. *J. Am. Chem. Soc.* **1949**, *71*, 25. For other evidence, see Brownell, R.; Leo, A.; Chang, Y.W.; Westheimer, F.H. *J. Am. Chem. Soc.* **1960**, *82*, 406; Roč ek, J.; Westheimer, F.H.; Eschenmoser, A.; Moldovànyi, L.; Schreiber, J. *Helv. Chim. Acta* **1962**, *45*, 2554; Lee, D.G.; Stewart, R. *J. Org. Chem.* **1967**, *32*, 2868; Wiberg, K.B.; Schäfer, H. *J. Am. Chem. Soc.* **1967**, *89*, 455; **1969**, *91*, 927, 933; Müller, P. *Helv. Chim. Acta* **1970**, *53*, 1869; **1971**, *54*, 2000, Lee, D.G.; Raptis, M. *Tetrahedron* **1973**, *29*, 1481.

[251]Rahman, M.; Roč ek, J. *J. Am. Chem. Soc.* **1971**, *93*, 5455, 5462; Doyle, M.P.; Swedo, R.J.; Roč ek, J. *J. Am. Chem. Soc.* **1973**, *95*, 8352; Wiberg, K.B.; Mukherjee, S.K. *J. Am. Chem. Soc.* **1974**, *96*, 1884, 6647.

[252]For a review, see Cockerill, A.F.; Harrison, R.G., in Patai, S. *The Chemistry of Functional Groups, Supplement A* pt. 1, Wiley, NY, **1977**, pp. 264–277.

[253]See Barter, R.M.; Littler, J.S. *J. Chem. Soc. B* **1967**, 205. For evidence that oxidation by HNO$_2$ involves a hydride shift, see Moodie, R.B.; Richards, S.N. *J. Chem. Soc. Perkin Trans. 2* **1986**, 1833; Ross, D.S.; Gu, C.; Hum, G.P.; Malhotra, R. *Int. J. Chem. Kinet.* **1986**, *18*, 1277.

dehydrogenation with triphenylmethyl cation[254] and the Oppenauer oxidation, and some by a free-radical mechanism, (e.g., oxidation with $S_2O_8^{2-}$[255] and with VO_2^+.[256]). A summary of many proposed mechanisms is given by Littler.[257]

OS **I**, 87, 211, 241, 340; **II**, 139, 541; **III**, 37, 207; **IV**, 189, 192, 195, 467, 813, 838; **V**, 242, 310, 324, 692, 852, 866; **VI**, 218, 220, 373, 644, 1033; **VII**, 102, 112, 114, 177, 258, 297; **VIII**, 43, 367, 386; **IX**, 132, 432. Also see, OS **IV**, 283; **VIII**, 363, 501.

19-4 Oxidation of Phenols and Aromatic Amines to Quinones

1/*O*,6/*O*-Dihydro-elimination

Ortho and para diols are easily oxidized to ortho- and para-quinones, respectively.[258] Either or both OH groups can be replaced by NH_2 groups to give the same products, although for the preparation of ortho-quinones only OH groups are normally satisfactory. The reaction has been successfully carried out with other groups para to OH or NH_2; halogen, OR, Me, *t*-Bu, and even H, although with the last yields are poor. Many oxidizing agents have been used: acid dichromate,[259] silver oxide, silver carbonate, lead tetraacetate, HIO_4, NBS—H_2O—H_2SO_4,[260] MnO_2 on Bentonite with microwave irradiation,[261] dimethyl dioxirane,[262] and atmospheric oxygen,[263] to name a few. Substituted phenols, such as 4-($CH_2CH_2CH_2COOH$) phenol, are oxidized with a polymer-bound hypervalent iodine reagent to give a quinone with a spirocyclic lactone unit at C-4.[264] Oxidation

[254]Bonthrone, W.; Reid, D.H. *J. Chem. Soc.* **1959**, 2773.

[255]Ball, D.L.; Crutchfield, M.M.; Edwards, J.O. *J. Org. Chem.* **1960**, *25*, 1599; Bida, G.; Curci, R.; Edwards, J.O. *Int. J. Chem. Kinet.* **1973**, *5*, 859; Snook, M.E.; Hamilton, G.A. *J. Am. Chem. Soc.* **1974**, *96*, 860; Walling, C.; Camaioni, D.M. *J. Org. Chem.* **1978**, *43*, 3266; Clerici, A.; Minisci, F.; Ogawa, K.; Surzur, J. *Tetrahedron Lett.* **1978**, 1149; Beylerian, N.M.; Khachatrian, A.G. *J. Chem. Soc. Perkin Trans. 2* **1984**, 1937.

[256]Littler, J.S.; Waters, W.A. *J. Chem. Soc.* **1959**, 4046.

[257]Littler, J.S. *J. Chem. Soc.* **1962**, 2190.

[258]For reviews, see Haines, A.H. *Methods for the Oxidation of Organic Compounds*, Vol. 2, Academic Press, NY, **1988**, pp. 305–323, 438–447; Naruta, Y.; Maruyama, K., in Patai, S.; Rappoport, Z. *The Chemistry of the Quinoid Compounds*, Vol. 2, pt. 1, Wiley, NY, **1988**, pp. 247–276; Thomson, R.H., in Patai, S. *The Chemistry of the Quinoiid Compounds*, Vol. 1, pt. 1, Wiley, NY, **1974**, pp. 112–132.

[259]For a review of this oxidation with chromium reagents, see Cainelli, G.; Cardillo, G. *Chromium Oxiations in Organic Chemistry*, Open Court Pub. Co., La Salle, IL, **1981**, pp. 92–117.

[260]Kim, D.W.; Choi, H.Y.; Lee, K.Y.; Chi, D.Y. *Org. Lett.* **2001**, *3*, 445.

[261]Gómez-Lara, J.; Gutiérrez-Perez, R.; Penieres-Carrillo, G.; López-Cortés, J.G.; Escudero-Salas, A.; Alvarez-Toledano, C. *Synth. Commun.* **2000**, *30*, 2713.

[262]Adam, W.; Schönberger, A. *Tetrahedron Lett.* **1992**, *33*, 53.

[263]For an example on activated silica gel, see Hashemi, M.M.; Beni, Y.A. *J. Chem. Res. (S)* **1998**, 138.

[264]Ley, S.V.; Thomas, A.W.; Finch, H. *J. Chem. Soc., Perkin Trans. 1* **1999**, 669.

has been done photochemically with O_2 and tetraphenylporphine.[265] A particularly effective reagent for rings with only one OH or NH_2 group is $(KSO_3)_2N—O•$ (dipotassium nitrosodisulfonate; *Fremy's salt*), which is a stable free radical.[266] Phenols, even some whose para positions are unoccupied, can be oxidized to ortho-quinones with diphenylseleninic anhydride.[267] Quinoid coupling products are obtained from substituted phenol treated with O_2, a dicopper complex, and mushroom tyrosinase.[268]

Less is known about the mechanism than is the case for **19-3**, but, as in that case, it seems to vary with the oxidizing agent. For oxidation of catechol with $NaIO_4$, it was found that the reaction conducted in $H_2^{18}O$ gave unlabeled quinone,[269] so the following mechanism[270] was proposed:

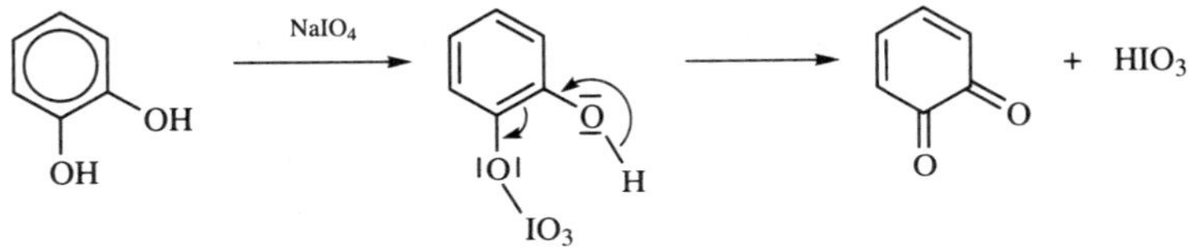

When catechol was oxidized with MnO_4^- under aprotic conditions, a semiquinone radical ion intermediate was involved.[271] For autoxidations[272] (i.e., with atmospheric oxygen) a free-radical mechanism is known to operate.[273]

OS **I**, 383, 482, 511; **II**, 175, 254, 430, 553; **III**, 663, 753; **IV**, 148; **VI**, 412, 480, 1010.

19-5   Dehydrogenation of Amines

1/1/*N*,2/2/*C*-Tetrahydro-bielimination

$$RCH_2NH_2 \longrightarrow RCN$$

Primary amines at a primary carbon can be dehydrogenated to nitriles. The reaction has been carried out with a variety of reagents, among others, lead tetraacetate,[274] NaOCl,[275]

[265]Cossy, J.; Belotti, S. *Tetrahedron Lett.* **2001**, *42*, 4329.

[266]For a review of oxidation with this salt, see Zimmer, H.; Lankin, D.C.; Horgan, S.W. *Chem. Rev. 1971*, 71, 229.

[267]Barton, D.H.R.; Brewster, A.G.; Ley, S.V.; Rosenfeld, M.N. *J. Chem. Soc. Chem. Commun.* **1976**, 985; Barton, D.H.R.; Ley, S.V., in *Further Perspectives in Organic Chemistry*, North-Holland Publishing Co., Amsterdam, The Netherlands, *1979*, pp. 53–66. For another way of accomplishing this, see Krohn, K.; Rieger, H.; Khanbabaee, K. *Chem. Ber. 1989*, *122*, 2323.

[268]Gupta, R.; Mukherjee, R. *Tetrahedron Lett.* **2000**, *41*, 7763.

[269]Adler, E.; Falkehag, I.; Smith, B. *Acta Chem. Scand. 1962*, *16*, 529.

[270]This mechanism is an example of category 4 (p. $$$).

[271]Bock, H.; Jaculi, D. *Angew. Chem. Int. Ed. 1984*, *23*, 305.

[272]For an example, see Rathore, R.; Bosch, E.; Kochi, J.K. *Tetrahedron Lett. 1994*, *35*, 1335.

[273]Sheldon, R.A.; Kochi, J.K. *Metal-Catalyzed Oxidations of Organic Compounds*, Academic Press, NY, *1981*, pp. 368–381; Walling, C. *Free Radicals in Solution*, Wiley, NY, *1957*, pp. 457–461.

[274]Stojiljković, A.; Andrejević, V.; Mihailović, M.Lj. *Tetrahedron 1967*, *23*, 721.

[275]Yamazaki, S. *Synth. Commun. 1997*, 27, 3559; Juršic, B. *J. Chem. Res. (S) 1988*, 168.

$K_2S_2O_8$/$NiSO_4$,[276] Me_3N/O/OsO_4,[277] Ru/Al_2O_3/O_2,[278] and CuCl/O_2/pyridine.[279] Several methods have been reported for the dehydrogenation of secondary amines to imines.[280] Among them[281] are treatment with (*1*) iodosylbenzene (PhIO) alone or in the presence of a ruthenium complex,[282] (*2*) DMSO and oxalyl chloride,[283] and (*3*) *t*-BuOOH and a rhenium catalyst.[284] *N*-Tosyl aziridines are converted to *N*-tosyl imines when heated with a palladium catalyst.[285] An interesting variation treats pyrrolidine with iodobenzene and a rhodium catalyst to give 2-phenylpyrroline.[286]

A reaction that involves dehydrogenation to an imine that then reacts further is the reaction of primary or secondary amines[287] with palladium black.[288] The imine initially formed by the dehydrogenation reacts with another molecule of the same or a different amine to give an aminal, which loses NH_3 or RNH_2 to give a secondary or tertiary amine. An example is the reaction between *N*-methylbenzylamine and butylmethylamine, which produces 95% *N*-methyl-*N*-butylbenzylamine.

19-6   Oxidation of Hydrazines, Hydrazones, and Hydroxylamines

1/*N*,2/*N*-Dihydro-elimination

$$\text{Ar—NH—NH—Ar} \xrightarrow{\text{NaOBr}} \text{Ar—N=N—Ar}$$

N,*N'*-Diarylhydrazines (hydrazo compounds) are oxidized to azo compounds by several oxidizing agents, including NaOBr, HgO,[289] $K_3Fe(CN)_6$ under phase-transfer

[276]Yamazaki, S.; Yamazaki, Y. *Bull. Chem. Soc. Jpn.* **1990**, *63*, 301.

[277]Gao, S.; Herzig, D.; Wang, B. *Synthesis* **2001**, 544.

[278]Yamaguchi, K.; Mizuno, N. *Angew. Chem. Int. Ed.* **2003**, *42*, 1480.

[279]Kametani, T.; Takahashi, K.; Ohsawa, T.; Ihara, M. *Synthesis* **1977**, 245; Capdevielle, P.; Lavigne, A.; Maumy, M. *Synthesis* **1989**, 453; *Tetrahedron* **1990**, 2835; Capdevielle, P.; Lavigne, A.; Sparfel, D.; Baranne-Lafont, J.; Cuong, N.K.; Maumy, M. *Tetrahedron Lett.* **1990**, *31*, 3305.

[280]For a review, see Dayagi, S.; Degani, Y., in Patai, S. *The Chemistry of the Carbon-Nitrogen Double Bond*, Wiley, NY, **1970**, pp. 117–124.

[281]For other methods, see Cornejo, J.J.; Larson, K.D.; Mendenhall, G.D. *J. Org. Chem.* **1985**, *50*, 5382; Nishinaga, A.; Yamazaki, S.; Matsuura, T. *Tetrahedron Lett.* **1988**, *29*, 4115.

[282]Müller, P.; Gilabert, D.M. *Tetrahedron* **1988**, *44*, 7171.

[283]Keirs, D.; Overton, K. *J. Chem. Soc. Chem. Commun.* **1987**, 1660.

[284]Murahashi, S.; Naot, T.; Taki, H. *J. Chem. Soc. Chem. Commun.* **1985**, 613.

[285]Wolfe, J.P.; Ney, J.E. *Org. Lett.* **2003**, *5*, 4607.

[286]Sezen, B.; Sames, D. *J. Am. Chem. Soc.* **2004**, *126*, 13244.

[287]See Larsen, J.; Jørgensen, K.A. *J. Chem. Soc. Perkin Trans. 2*, **1992**, 1213. Also see, Yamaguchi, J.; Takeda, T. *Chem. Lett.* **1992**, 1933; Yamazaki, S. *Chem. Lett.* **1992**, 823.

[288]Murahashi, S.; Yoshimura, N.; Tsumiyama, T.; Kojima, T. *J. Am. Chem. Soc.* **1983**, *105*, 5002. See also, Wilson, Jr., R.B.; Laine, R.M. *J. Am. Chem. Soc.* **1985**, *107*, 361.

[289]For a review of HgO, see Pizey, J.S. *Synthetic Reagents*, Vol. 1, Wiley, NY, **1974**, pp. 295–319.

conditions[290] or with galvinoxyl,[291] $FeCl_3$,[292] MnO_2 (this reagent yields cis-azo-benzenes),[293] $CuCl_2$, and air and NaOH.[294] The reaction is also applicable to N,N'-dialkyl- and N,N'-diacylhydrazines. Hydrazines (both alkyl and aryl) substituted on only one side also give azo compounds,[295] but these are unstable and decompose to nitrogen and the hydrocarbon:

$$Ar{-}NH{-}NH_2 \longrightarrow [\,Ar{-}N{=}NH\,] \longrightarrow ArH + N_2$$

Aniline derivatives are converted to azo compounds by heating with cetyltrimethylammonium dichromate in chloroform.[296] When hydrazones are oxidized with HgO, Ag_2O, MnO_2, lead tetraacetate, or certain other oxidizing agents, diazo compounds are obtained:[297]

$$R_2C{=}N{-}NH_2 \xrightarrow{\ HgO\ } R_2C{=}\overset{\oplus}{N}{=}\overset{\ominus}{N}$$

Hydrazones of the form $ArCH{=}NNH_2$ react with HgO in solvents, such as diglyme or ethanol, to give nitriles ArCN.[298] It is possible to oxidize dimethylhydrazones $(R{-}C{=}N{-}NMe_2)$ to the corresponding nitrile $(R{-}C{\equiv}N)$ with $MeReO_3/H_2O_2$[299] magnesium monoperoxyphthalate (MMPP),[300] or with dimethyl dioxirane.[301] Oxone® on wet alumina also converts hydrazones to nitriles with microwave irradiation.[302]

In a related reaction, primary aromatic amines have been oxidized to azo compounds by a variety of oxidizing agents, among them MnO_2, lead tetraacetate, O_2 and a base, barium permanganate,[303] and sodium perborate in acetic acid. *tert*-Butyl hydroperoxide has been used to oxidize certain primary amines to azoxy compounds.[304] Aromatic hydroxylamines $(Ar{-}NH{-}OH)$ are easily oxidized to nitroso compounds $(Ar{-}N{=}O)$, most commonly by acid dichromate.[305] Oximes of

[290]Dimroth, K.; Tüncher, W. *Synthesis* **1977**, 339.

[291]Wang, X.-Y.; Wang, Y.-L.; Li, J.-P.; Duan, Z.F.; Zhang, Z.-Y. *Synth. Commun.* **1999**, *29*, 2271.

[292]Wang, C.-L.; Wang, X.-X.; Wang, X.-Y.; Xiao, J.-P.; Wang, Y.-L. *Synth. Commun.* **1999**, *29*, 3435.

[293]Hyatt, J.A. *Tetrahedron Lett.* **1977**, 141.

[294]For a review, see Newbold, B.T., in Patai, S. *The Chemistry of the Hydrazo, Azo, and Azoxy Groups*, pt. 1, Wiley, NY, **1975**, pp. 543–557, 564–573.

[295]See Mannen, S.; Itano, H.A. *Tetrahedron* **1973**, *29*, 3497.

[296]Patel, S.; Mishra, B.K. *Tetrahedron Lett.* **2004**, *45*, 1371.

[297]For a review, see Regitz, M.; Maas, G. *Diazo Compounds*, Academic Press, NY, **1986**, pp. 233–256.

[298]Mobbs, D.B.; Suschitzky, H. *Tetrahedron Lett.* **1971**, 361.

[299]Stanković, S.; Espenson, J.H. *Chem. Commun.* **1998**, 1579.

[300]Fernández, R.; Gasch, C.; Lassaletta, J.-M.; Llera, J.-M.; Vázquez, J. *Tetrahedron Lett.* **1993**, *34*, 141.

[301]Altamura, A.; D'Accolti, L.; Detomaso, A.; Dinoi, A.; Fiorentino, M.; Fusco, C.; Curci, R. *Tetrahedron Lett.* **1998**, *39*, 2009.

[302]Ramalingam, T.; Reddy, B.V.S.; Srinivas, R.; Yadav, J.S. *Synth. Commun.* **2000**, *30*, 4507.

[303]Firouzabadi, H.; Mostafavipoor, Z. *Bull. Chem. Soc. Jpn.* **1983**, *56*, 914.

[304]Kosswig, K. *Liebigs Ann. Chem.* **1971**, *749*, 206.

[305]For a review, see Hudlický, M. *Oxidations in Organic Chemistry*, American Chemical Society, Washington, DC, **1990**, pp. 231–232.

aromatic aldehydes are converted to aryl nitriles with $InCl_3$[306] (ketoximes give a Beckmann rearrangement, **18-17**).

Nitrones, $C=N^+(R)=O^-$, are generated by the oxidation of N-hydroxyl secondary amines with 5% aq. NaOCl.[307] Secondary amines, such as dibenzylamine, can be converted to the corresponding nitrone by heating with cumyl hydroperoxide in the presence of a titanium catalyst.[308] Imines are oxidized to amides with mcpba and $BF_3 \cdot OEt_2$.[309]

OS **II**, 496; **III**, 351, 356, 375, 668; **IV**, 66, 411; **V**, 96, 160, 897; **VI**, 78, 161, 334, 392, 803, 936; **VII**, 56. Also see, OS **V**, 258. For oxidation of primary amines, see OS **V**, 341.

B. Oxidations Involving Cleavage of Carbon–Carbon Bonds[310]

19-7   Oxidative Cleavage of Glycols and Related Compounds

2/O-De-hydrogen-uncoupling

1,2-Glycols are easily cleaved under mild conditions and in good yield with periodic acid or lead tetraacetate.[311] The products are 2 equivalents of aldehyde, or 2 equivalents of ketone, or 1 equivalent of each, depending on the groups attached to the two carbons. The yields are so good that alkenes are often converted to glycols (**15-48**), and then cleaved with HIO_4 or $Pb(OAc)_4$ rather than being cleaved directly with ozone (**19-9**) or dichromate or permanganate (**19-10**). The diol can be generated and cleaved *in situ* from an alkene to give the carbonyl compounds.[312]

[306]Barman, D.C.; Thakur, A.J.; Prajapati, D.; Sandhu, J.S. *Chem. Lett.* **2000**, 1196.

[307]Cicchi, S.; Corsi, M.; Goti, A. *J. Org. Chem.* **1999**, *64*, 7243.

[308]Forcato, M.; Nugent, W.A.; Licini, G. *Tetrahedron Lett.* **2003**, *44*, 49.

[309]An, G.-i.; Rhee, H. *Synlett* **2003**, 876.

[310]For a review, see Bentley, K.W., in Bentley, K.W.; Kirby, G.W. *Elucidation of Chemical Structures by Physical and Chemical Methods* (Vol. 4 of Weissberger, A. *Techniques of Chemistry*), 2nd ed., pt. 2, Wiley, NY, **1973**, pp. 137–254.

[311]For reviews covering both reagents, see Haines, A.H. *Methods for the Oxidation of Organic Compounds*, Vol. 2, Academic Press, NY, **1988**, pp. 277–301, 432–437; House, H.O. *Modern Synthetic Reactions*, 2nd ed., W.A. Benjamin, NY, **1972**, pp. 3353–363; Perlin, A.S., in Augustine, R.L. *Oxidation*, Vol. 1, Marcel Dekker, NY, **1969**, pp. 189–212; Bunton, C.A., in Wiberg, K.B., in Wiberg, K.B. *Oxidation in Organic Chemistry*, pt. A, Academic Press, NY, **1965**, pp. 367–407. For reviews of lead tetraacetate, see Rubottom, G.M., in Trahanovsky, W.S. *Oxidation in Organic Chemistry*, pt. D, Academic Press, NY, **1982**, p. 1; Aylward, J.B. *Q. Rev. Chem. Soc.* **1971**, *25*, 407. For reviews of HIO_4, see Fatiadi, A.J. *Synthesis* **1976**, 65,133; Sklarz, B. *Q. Rev. Chem. Soc.* **1967**, *21*, 3.

[312]Yu, W.; Mei, Y.; Kang, Y.; Hua, Z.; Jin, Z. *Org. Lett.* **2004**, *6*, 3217.

A number of other oxidizing agents also give the same products, among them[313] activated MnO_2,[314] O_2 and a ruthenium catalyst,[315] PPh_3–DEAD,[316] and pyridinium chlorochromate.[317] Permanganate, dichromate, and several other oxidizing agents[318] also cleave glycols, giving carboxylic acids rather than aldehydes, but these reagents are seldom used synthetically. Electrochemical oxidation is an efficient method, and is useful not only for diols, but also for their mono- and dimethoxy derivatives.[319]

The two reagents (periodic acid and lead tetraacetate) are complementary, since periodic acid is best used in water and lead tetraacetate in organic solvents. Chiral lead carboxylates have been prepared for the oxidative cleavage of 1,2-diols.[320] When three or more OH groups are located on adjacent carbons, the middle one (or ones) is converted to formic acid.

Other compounds that contain oxygens or nitrogens on adjacent carbons undergo similar cleavage:

C–C HO NH₂	C–C HO NHR	C–C (R)H₂N NH₂(R)	C–C HO O	C–C O O
β-Amino alcohols		1,2-Diamines	α-Hydroxy aldehydes and ketones	α-Diketones a-keto aldehydes Glyoxal

Cyclic 1,2-diamines are cleaved to diketones with dimethyl dioxirane.[321] α-Diketones and α-hydroxy ketones are also cleaved by alkaline H_2O_2.[322] The HIO_4 has been used to cleave epoxides to aldehydes,[323] for example,

α-Hydroxy acids and α-keto acids are not cleaved by HIO_4, but are cleaved by $NaIO_4$ in methanol in the presence of a crown ether,[324] $Pb(OAc)_4$, alkaline H_2O_2,

[313]For a list of reagents, with references, see Larock, R.C. *Comprehensive Organic Transformations*, 2nd ed., Wiley-VCH, NY, *1999*, pp. 1250–1255.

[314]Adler, E.; Becker, H. *Acta Chem. Scand.* *1961*, *15*, 849; Ohloff, G.; Giersch, W. *Angew. Chem. Int. Ed.* *1973*, *12*, 401.

[315]Takezawa, E.; Sakaguchi, S.; Ishii, Y. *Org. Lett.* *1999*, *1*, 713.

[316]Barrero, A.F.; Alvarez-Manzaneda, E.J.; Chahboun, R. *Tetrahedron Lett.* *2000*, *41*, 1959.

[317]Cisneros, A.; Fernández, S.; Hernández, J.E. *Synth. Commun.* *1982*, *12*, 833.

[318]For a list of reagents, with references, see Larock, R.C. *Comprehensive Organic Transformations*, 2nd ed., Wiley-VCH, NY, *1999*, pp. 1650–1652.

[319]For a review, see Shono, T. *Electroorganic Chemistry as a New Tool in Organic Synthesis*, Springer, NY, *1984*, pp. 31–37. See also, Ruholl, H.; Schäfer, H.J. *Synthesis* *1988*, 54.

[320]Lena, J.I.C.; Sesenoglu, Ö.; Birlirakis, N.; Arseniyadis, S. *Tetrahedron Lett.* *2001*, *42*, 21.

[321]Gagnon, J.L.; Zajac, Jr., W.W. *Tetrahedron Lett.* *1995*, *36*, 1803.

[322]See, for example, Ogata, Y.; Sawaki, Y.; Shiroyama, M. *J. Org. Chem.* *1977*, *42*, 4061.

[323]Nagarkatti, J.P.; Ashley, K.R. *Tetrahedron Lett.* *1973*, 4599.

[324]Kore, A.R.; Sagar, A.D.; Salunkhe, M.M. *Org. Prep. Proceed. Int.* *1995*, *27*, 373.

and other reagents. These are oxidative decarboxylations. α-Hydroxy acids give aldehydes or ketones, and α-keto acids give carboxylic acids. Also see, **19-12** and **19-13**.

The mechanism of glycol oxidation with $Pb(OAc)_4$ was proposed by Criegee:[325]

This mechanism is supported by these facts: (*1*) the kinetics are second order (first order in each reactant); (*2*) added acetic acid retards the reaction (drives the equilibrium to the left); and (*3*) cis-glycols react much more rapidly than trans-glycols.[326] For periodic acid, the mechanism is similar, with the intermediate[327]

However, the cyclic-intermediate mechanism cannot account for all glycol oxidations, since some glycols that cannot form such an ester (e.g., **15**) are nevertheless cleaved by lead tetraacetate

15

[325]Criegee, R.; Kraft, L.; Rank, B. *Liebigs Ann. Chem.* **1933**, *507*, 159. For reviews, see Waters, W.A. *Mechanisms of Oxidation of Organic Compounds*, Wiley, NY, **1964**, pp. 72–81; Stewart, R. *Oxidation Mechanisms*, W.A. Benjamin, NY, **1964**, pp. 97–106.

[326]For example, see Criegee, R.; Höger, E.; Huber, G.; Kruck, P.; Marktscheffel, F.; Schellenberger, H. *Liebigs Ann. Chem.* **1956**, *599*, 81.

[327]Buist, G.J.; Bunton, C.A.; Hipperson, W.C.P. *J. Chem. Soc. B* **1971**, 2128.

(though other glycols that cannot form cyclic esters are *not* cleaved, by either reagent[328]). To account for cases like **15**, a cyclic transition state has been proposed:[326]

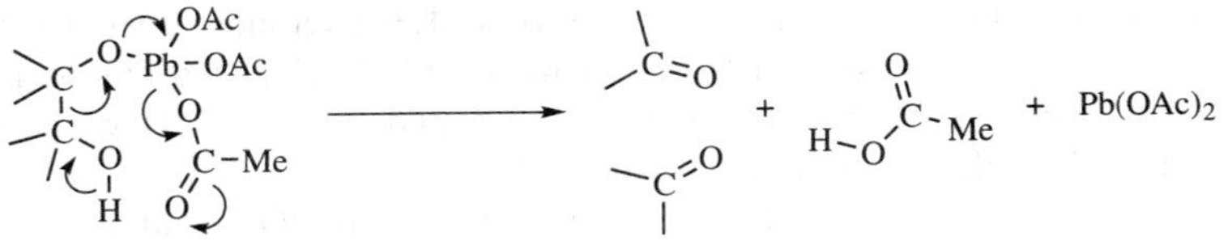

OS **IV**, 124; **VII**, 185; **VIII**, 396.

19-8   Oxidative Cleavage of Ketones, Aldehydes, and Alcohols

Cycloalkanone oxidative ring opening

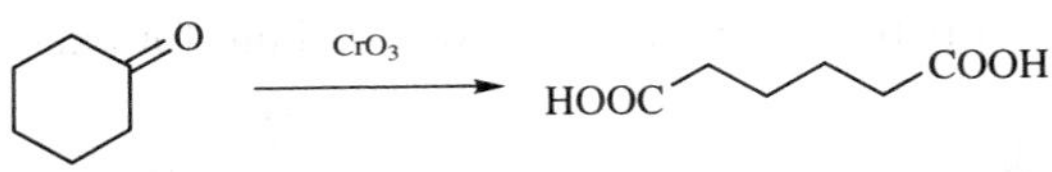

Oxidative cleavage of open-chain ketones or alcohols[329] is seldom a useful preparative procedure, not because these compounds do not undergo oxidation (they do, except for diaryl ketones), but because the result is generally a hopeless mixture. Aryl methyl ketones, such as acetophenone, however, are readily oxidized to aryl carboxylic acids with Re_2O_7 and 70% aqueous *tert*-butyl hydroperoxide.[330] Oxygen with a mixture of manganese and cobalt catalysts give similar oxidative cleavage,[331] and do hypervalent iodine compounds.[332] 1,3-Diketones, such as 1,3-diphenyl-1,3-propanedione, are oxidatively cleaved with aqueous Oxone® to give benzoic acid.[333] Noted that in the presence of benzaldehyde, aliphatic ketones are cleaved to give aliphatic carboxylic acids by treatment with $BF_3(g)$ in refluxing hexane.[334] Aldehydes, such as $PhCH_2CHO$, are cleaved to benzaldehyde with phosphonium dichromate in refluxing acetonitrile.[335]

Despite problems with acyclic ketones, the reaction is quite useful for cyclic ketones and the corresponding secondary alcohols, the dicarboxylic acid being prepared in good yield. The formation of adipic acid from cyclohexanone (shown above) is an important industrial procedure. Acid dichromate and permanganate are the most common oxidizing agents, although autoxidation (oxidation with

[328]Angyal, S.J.; Young, R.J. *J. Am. Chem. Soc.* **1959**, *81*, 5251.
[329]For a review of metal ion-catalyzed oxidative cleavage of alcohols, see Trahanovsky, W.S. *Methods Free-Radical Chem.* **1973**, *4*, 133–169. For a review of the oxidation of aldehydes and ketones, see Verter, H.S., in Zabicky, J. *The Chemistry of the Carbonyl Group*, pt. 2, Wiley, NY, **1970**, pp. 71–156.
[330]Gurunath, S.; Sudalai, A. *Synlett* **1999**, 559.
[331]Minisci, F.; Recupero, F.; Fontana, F.; Bjørsvik, H.-R.; Liguori, L. *Synlett* **2002**, 610.
[332]Lee, J.C.; Choi, J.-H.; Lee, Y.C. *Synlett* **2001**, 1563.
[333]Ashford, S.W.; Grega, K.C. *J. Org. Chem.* **2001**, *66*, 1523.
[334]Kabalka, G.W.; Li, N.-S.; Tejedor, D.; Malladi, R.R.; Gao, X.; Trotman, S. *Synth. Commun.* **1999**, *29*, 2783.
[335]Hajipour, A.R.; Mohammadpoor-Baltork, I.; Niknam, K. *Org. Prep. Proceed. Int.* **1999**, *31*, 335.

atmospheric oxygen) in alkaline solution[336] and potassium superoxide under phase-transfer conditions[337] have also been used. Other reagents include LiOCl/Chlorox[338] and MeOCO$_2$Me at 195°C.[339] Silyl-ketones have been cleaved to esters using electrolysis in alcohol solvents.[340] Cyclic 1,3-diketones are converted to α,ω-diesters with an excess of KHSO$_5$ in methanol.[341] Cyclic α-chloro ketones are cleaved to give an α,ω-functionalized compound (acetal-ester) when treated with cerium (IV) sulfate tetrahydrate and O$_2$.[342]

Cyclic ketones can also be cleaved by treatment with NOCl and an alcohol in liquid SO$_2$ to give an ω-oximinocarboxylic ester, for example,[343]

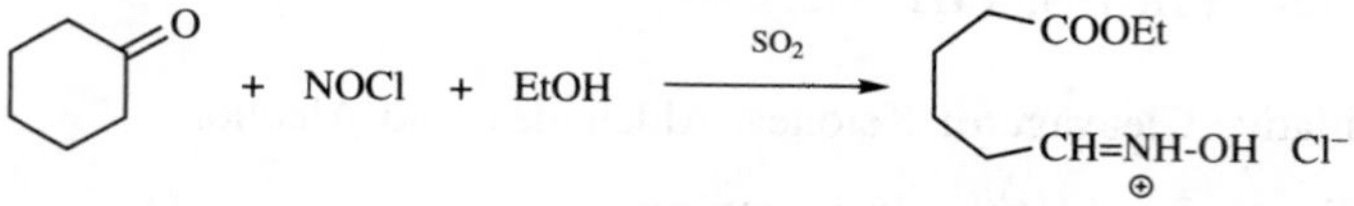

Cyclic 1,3-diketones, which exist mainly in the mono-enolic form, can be cleaved with sodium periodate with loss of one carbon, for example,[344]

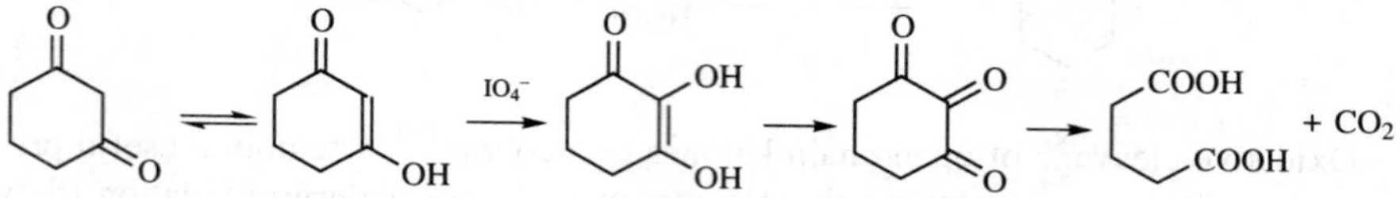

The species actually undergoing the cleavage is the triketone, so this is an example of **19-7**.

OS **I**, 18; **IV**, 19; **VI**, 690. See also, OS **VI**, 1024.

19-9 Ozonolysis

Oxo-uncoupling

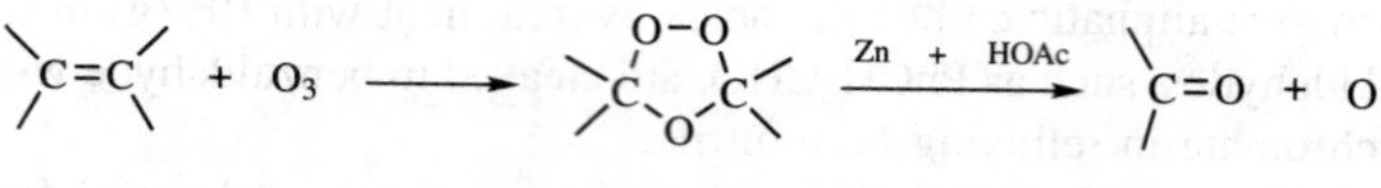

16

[336]Wallace, T.J.; Pobiner, H.; Schriesheim, A. *J. Org. Chem.* **1965**, *30*, 3768; Bjørsvik, H.-R.; Liguori, L.; González, R.R.; Merinero, J.A.V. *Tetrahedron Lett.* **2002**, *43*, 4985. See also, Osowska-Pacewicka, K.; Alper, H. *J. Org. Chem.* **1988**, *53*, 808.

[337]Lissel, M.; Dehmlow, E.V. *Tetrahedron Lett.* **1978**, 3689; Sotiriou, C.; Lee, W.; Giese, R.W. *J. Org. Chem.* **1990**, *55*, 2159.

[338]Madler, M.M.; Klucik, J.; Soell, P.S.; Brown, C.W.; Liu, S.; Berlin, K.D.; Benbrook, D.M.; Birckbichler, P.J.; Nelson, E.C. *Org. Prep. Proceed. Int.* **1998**, *30*, 230.

[339]Selva, M.; Marques, C.A.; Tundo, P. *Gazz. Chim. Ital.* **1993**, *123*, 515.

[340]Yoshida, J.; Itoh, M.; Matsunaga, S.; Isoe, S. *J. Org. Chem.* **1992**, *57*, 4877.

[341]Yan, J.; Travis, B.R.; Borhan, B. *J. Org. Chem.* **2004**, *69*, 9299.

[342]He, L.; Horiuchi, C.A. *Bull. Chem. Soc. Jpn.* **1999**, *72*, 2515.

[343]Rogić, M.M.; Vitrone, J.; Swerdloff, M.D. *J. Am. Chem. Soc.* **1977**, *99*, 1156; Moorhoff, C.M.; Paquette, L.A. *J. Org. Chem.* **1991**, *56*, 703.

[344]Wolfrom, M.L.; Bobbitt, J.M. *J. Am. Chem. Soc.* **1956**, *78*, 2489.

When compounds containing double bonds are treated with ozone, usually at low temperatures, they are converted to compounds called *ozonides* (**16**) that can be isolated but, because some of them are explosive, are more often decomposed with zinc and acetic acid, or catalytic hydrogenation to give 2 equivalents of aldehyde, or 2 equivalents of ketone, or 1 equivalent of each, depending on the groups attached to the alkene.[345] The decomposition of **16** has also been carried out with triethylamine[346] and with reducing agents, among them trimethyl phosphite,[347] thiourea,[348] and dimethyl sulfide.[349] However, ozonides can also be *oxidized* with oxygen, peroxyacids, or H_2O_2 to give ketones and/or carboxylic acids or *reduced* with $LiAlH_4$, $NaBH_4$, BH_3, or catalytic hydrogenation with excess H_2 to give 2 equivalents alcohol.[350] Ozonides can also be treated with ammonia, hydrogen, and a catalyst to give the corresponding amines,[351] or with an alcohol and anhydrous HCl to give the corresponding carboxylic esters.[352] Ozonolysis is therefore an important synthetic reaction.

A wide variety of alkenes undergo ozonolysis, including cyclic ones, where cleavage gives rise to one bifunctional product. Alkenes in which the double bond is connected to electron-donating groups react many times faster than those in which it is connected to electron-withdrawing groups.[353] The reaction has often been carried out on compounds containing more than one double bond; generally all the bonds are cleaved. In some cases, especially when bulky groups are present, conversion of the substrate to an epoxide (**15-50**) becomes an important side reaction and can be the main reaction.[354] Rearrangement is possible in some cases.[355] Ozonolysis of triple bonds[356] is less common and the reaction proceeds less easily,

[345]For monographs, see Razumovskii, S.D.; Zaikov, G.E. *Ozone and its Reactions with Organic Compounds*; Elsevier, NY, *1984*; Bailey, P.S. *Ozonation in Organic Chemistry*, 2 vols., Academic Press, NY, *1978, 1982*. For reviews, see Odinokov, V.N.; Tolstikov, G.A. *Russ. Chem. Rev. 1981*, *50*, 636; Belew, J.S., in Augustine, R.L.; Trecker, D.J. *Oxidation*, Vol. 1, Marcel Dekker, NY, *1969*, pp. 259–335; Menyailo, A.T.; Pospelov, M.V. *Russ. Chem. Rev. 1967*, *36*, 284. For a review with respect to vinylic ethers, see Kuczkowski, R.L. *Adv. Oxygenated Processes 1991*, *3*, 1. For a review with respect to haloalkenes, see Gillies, C.W.; Kuczkowski, R.L. *Isr. J. Chem. 1983*, *23*, 446.
[346]Hon, Y.-S.; Lin, S.-W.; Chen, Y.-J. *Synth. Commun. 1993*, *23*, 1543.
[347]Knowles, W.S.; Thompson, Q.E. *J. Org. Chem. 1960*, *25*, 1031.
[348]Gupta, D.; Soman, R.; Dev, S. *Tetrahedron 1982*, *38*, 3013.
[349]Pappas, J.J.; Keaveney, W.P.; Gancher, E.; Berger, M. *Tetrahedron Lett. 1966*, 4273.
[350]Sousa, J.A.; Bluhm, A.L. *J. Org. Chem. 1960*, *25*, 108; Diaper, D.G.M.; Strachan, W.M.J. *Can. J. Chem. 1967*, *45*, 33; White, R.W.; King, S.W.; O'Brien, J.L. *Tetrahedron Lett. 1971*, 3587; Flippin, L.A.; Gallagher, D.W.; Jalali-Araghi, K. *J. Org. Chem. 1989*, *54*, 1430.
[351]Diaper, D.G.M.; Mitchell, D.L. *Can. J. Chem. 1962*, *40*, 1189; Benton, F.L.; Kiess, A.A. *J. Org. Chem. 1960*, *25*, 470; Pollart, K.A.; Miller, R.E. *J. Org. Chem. 1962*, *27*, 2392; White, R.W.; King, S.W.; O'Brien, J.L. *Tetrahedron Lett. 1971*, 3591.
[352]Neumeister, J.; Keul, H.; Saxena, M.P.; Griesbaum, K. *Angew. Chem. Int. Ed. 1978*, *17*, 939. See also, Schreiber, S.L.; Claus, R.E.; Reagan, J. *Tetrahedron Lett. 1982*, *23*, 3867; Cardinale, G.; Grimmelikhuysen, J.C.; Laan, J.A.M.; Ward, J.P. *Tetrahedron 1984*, *40*, 1881.
[353]Pryor, W.A.; Giamalva, D.; Church, D.F. *J. Am. Chem. Soc. 1985*, *107*, 2793.
[354]See, for example, Bailey, P.S.; Lane, A.G. *J. Am. Chem. Soc. 1967*, *89*, 4473; Gillies, C.W. *J. Am. Chem. Soc. 1975*, *97*, 1276; Bailey, P.S.; Hwang, H.H.; Chiang, C. *J. Org. Chem. 1985*, *50*, 231.
[355]For an example, see Barrero, A.F.; Alvarez-Manzaneda, E.J.; Chahboun, R.; Cuerva, J.M.; Segovia, A. *Synlett. 2000*, 1269.
[356]For a discussion of the mechanism of ozonolysis of triple bonds, see Pryor, W.A.; Govindan, C.K.; Church, D.F. *J. Am. Chem. Soc. 1982*, *104*, 7563.

since ozone is an electrophilic agent[357] and prefers double to triple bonds (p. 1017). Compounds that contain triple bonds generally give carboxylic acids, although sometimes ozone oxidizes them to α-diketones (**19-26**). Aromatic compounds are also attacked less readily than alkenes, but have often been cleaved. Aromatic compounds behave as if the double bonds in the Kekulé structures were really there. Thus benzene gives three equivalents of glyoxal (HCOCHO), and o-xylene gives a glyoxal/MeCOCHO/MeCOCOMe ratio of 3:2:1, which shows that in this case cleavage is statistical. With polycyclic aromatic compounds the site of attack depends on the structure of the molecule and on the solvent.[358]

Although a large amount of work has been done on the mechanism of ozonization (formation of **16**), not all the details are known. The basic mechanism was formulated by Criegee.[359] The first step of the Criegee mechanism[360] is a 1,3-dipolar addition (**15-58**) of ozone to the substrate to give the "initial" or "primary" ozonide, the structure of which has been shown to be the 1,2,3-trioxolane, **17**, by microwave and other spectral methods.[361] A primary ozonide has been trapped.[362] However, **17** is highly unstable and cleaves to an aldehyde or ketone (**18**) and an intermediate,[363] which Criegee showed as a zwitterion (**19**), but which may be a diradical (**20**). This compound is usually referred to as a carbonyl oxide.[364] The carbonyl oxide (which we will represent as **19**) can then undergo various reactions, three of which lead to normal products. One is a recombination with **18**, the second

[357]See, for example, Wibaut, J.P.; Sixma, F.L.J. *Recl. Trav. Chim. Pays-Bas* **1952**, *71*, 761; Williamson, D.G.; Cvetanovi, R.J. *J. Am. Chem. Soc.* **1968**, *90*, 4248; Razumovskii, S.D.; Zaikov, G.E. *J. Org. Chem. USSR* **1972**, *8*, 468, 473; Klutsch, G.; Fliszár, S. *Can. J. Chem.* **1972**, *50*, 2841.

[358]Dobinson, F.; Bailey, P.S. *Tetrahedron Lett.* **1960** (No. 13) 14; O'Murchu, C. *Synthesis* **1989**, 880.

[359]For reviews, see Kuczkowski, R.L. *Acc. Chem. Res.* **1983**, *16*, 42; Razumovskii, S.D.; Zaikov, G.E. *Russ. Chem. Rev.* **1980**, *49*, 1163; Criegee, R. *Angew. Chem. Int. Ed.* **1975**, *14*, 745; Murray, R.W. *Acc. Chem. Res.* **1968**, *1*, 313.

[360]For a modified-Criegee mechanism, see Ponec, R.; Yuzhakov, G.; Haas, Y.; Samuni, U. *J. Org. Chem.* **1997**, *62*, 2757.

[361]Gillies, J.Z.; Gillies, C.W.; Suenram, R.D.; Lovas, F.J. *J. Am. Chem. Soc.* **1988**, *110*, 7991. See also, Criegee, R.; Schröder, G. *Chem. Ber.* **1960**, *93*, 689; Durham, L.J.; Greenwood, F.L. *J. Org. Chem.* **1968**, *33*, 1629; Bailey, P.S.; Carter, Jr., T.P.; Fischer, C.M.; Thompson, J.A. *Can. J. Chem.* **1973**, *51*, 1278; Hisatsune, I.C.; Shinoda, K.; Heicklen, J. *J. Am. Chem. Soc.* **1979**, *101*, 2524; Mile, B.; Morris, G.W.; Alcock, W.G. *J. Chem. Soc. Perkin Trans. 2* **1979**, 1644; Kohlmiller, C.K.; Andrews, L. *J. Am. Chem. Soc.* **1981**, *103*, 2578; McGarrity, J.F.; Prodolliet, J. *J. Org. Chem.* **1984**, *49*, 4465.

[362]Jung, M.E.; Davidov, P. *Org. Lett.* **2001**, *3*, 627.

[363]A Criegee intermediate has been detected for the ozonolysis of 2-butene; see Fajgar, R.; Vítek, J.; Haas, Y.; Pola, J. *Tetrahedron Lett.* **1996**, *37*, 3391.

[364]For reviews of carbonyl oxides, see Sander, W. *Angew. Chem. Int. Ed.* **1990**, *29*, 344; Brunelle, W.H. *Chem. Rev.* **1991**, *91*, 335.

a dimerization to the bis(peroxide) **21**, and the third a kind of dimerization to **22**.[365] If the first path is taken (this is normally

possible only if **15** is an aldehyde; most ketones do not do this[366]) the product is an ozonide (a 1,2,4-trioxolane),[367] and hydrolysis of the ozonide gives the normal products. If **21** is formed, hydrolysis of it gives one of the products, and, of course, **18**, which then does not undergo further reaction, is the other. Intermediate **22**, if formed, can decompose directly, as shown, to give the normal products and oxygen. In protic solvents, **19** is converted to a hydroperoxide, and these have been isolated, for example,

$$Me_2C-OMe$$
$$\overset{|}{O}OH$$

from $Me_2C=CMe_2$ in methanol. Further evidence for the mechanism is that **21** can be isolated in some cases, for example, from $Me_2C=CMe_2$. But perhaps the most impressive evidence comes from the detection of cross-products. In the Criegee mechanism, the two parts of the original alkene break apart and then recombine to form the ozonide. In the case of an unsymmetrical alkene, RCH=CHR′, there should be three ozonides:

[365]Fliszár, S.; Chyliń ska, J.B. *Can. J. Chem.* **1967**, *45*, 29; **1968**, *46*, 783.

[366]It follows that tetrasubstituted alkenes do not normally give ozonides. However, they do give the normal cleavage products (ketones) by the other pathways. For the preparation of ozonides from tetrasubstituted alkenes by ozonolysis on polyethylene, see Griesbaum, K.; Volpp, W.; Greinert, R.; Greunig, H.; Schmid, J.; Henke, H. *J. Org. Chem.* **1989**, *54*, 383.

[367]Kamata, M.; Komatsu, K.i.; Akaba, R. *Tetrahedron Lett.* **2001**, *42*, 9203. For a report of an isolable ozonide, see dos Santos, C.; de Rosso, C.R.S.; Imamura, P.M. *Synth. Commun.* **1999**, *29*, 1903.

since there are two different aldehydes **18** and two different species **19**, and these can recombine in the three ways shown. Actually *six* ozonides, corresponding to the cis and trans forms of these three, were isolated and characterized for methyl oleate.[368] Similar results have been reported for smaller alkenes, for example, 2-pentene, 4-nonene, and even 2-methyl-2-pentene.[369] The last-mentioned case is especially interesting, since it is quite plausible that this compound would cleave in only one way, so that only one ozonide (in cis and trans versions) would be found; but this is not so, and three were found for this case too. However, terminal alkenes give little or no cross-ozonide formation.[370] In general, the less alkylated end of the alkene tends to go to **18** and the other to **19**. Still other evidence[371] for the Criegee mechanism is (*1*) When $Me_2C=CMe_2$ was ozonized in the presence of

$$Me-\underset{Me}{\overset{}{C}}\overset{O-O}{\underset{O}{<}}\underset{H}{\overset{C-H}{>}}$$

23

HCHO, the ozonide **23** could be isolated;[372] (*2*) **19** prepared in an entirely different manner (photooxidation of diazo compounds), reacted with aldehydes to give ozonides;[373] and (*3*) cis- and trans-alkenes generally give the same ozonide, which would be expected if they cleave first.[374] However, this was not true for $Me_3CCH=CHCMe_3$, where the cis-alkene gave the cis-ozonide (chiefly), and the trans gave the trans.[375] The latter result is not compatible with the Criegee mechanism. Also incompatible with the Criegee mechanism was the finding that the cis/trans ratios of symmetrical (cross) ozonides obtained from *cis-* and *trans*-4-methyl-2-pentene were not the same.[376]

[368]Riezebos, G.; Grimmelikhuysen, J.C.; van Dorp, D.A. *Recl. Trav. Chim. Pays-Bas* **1963**, *82*, 1234; Privett, O.S.; Nickell, E.C. *J. Am. Oil Chem. Soc.* **1964**, *41*, 72.

[369]Loan, L.D.; Murray, R.W.; Story, P.R. *J. Am. Chem. Soc.* **1965**, *87*, 737; Lorenz, O.; Parks, C.R. *J. Org. Chem.* **1965**, *30*, 1976.

[370]Murray, R.W.; Williams, G.J. *J. Org. Chem.* **1969**, *34*, 1891.

[371]For further evidence, see Mori, M.; Nojima, M.; Kusabayashi, S. *J. Am. Chem. Soc.* **1987**, *109*, 4407; Pierrot, M.; El Idrissi, M.; Santelli, M. *Tetrahedron Lett.* **1989**, *30*, 461; Wojciechowski, B.J.; Chiang, C.; Kuczkowski, R.L. *J. Org. Chem.* **1990**, *55*, 1120; Paryzek, Z.; Martynow, J.; Swoboda, W. *J. Chem. Soc. Perkin Trans. 1* **1990**, 1220; Murray, R.W.; Morgan, M.M. *J. Org. Chem.* **1991**, *56*, 684, 6123.

[372]Even ketones can react with **19** to form ozonides, provided they are present in large excess: Criegee, R.; Korber, H. *Chem. Ber.* **1971**, *104*, 1812.

[373]Murray, R.W.; Suzui, A. *J. Am. Chem. Soc.* **1973**, *95*, 3343; Higley, D.P.; Murray, R.W. *J. Am. Chem. Soc.* **1974**, *96*, 3330.

[374]See, for example, Murray, R.W.; Williams, G.J. *J. Org. Chem.* **1969**, *34*, 1896.

[375]Schröder, G. *Chem. Ber.* **1962**, *95*, 733; Kolsaker, P. *Acta Chem. Scand. Ser. B* **1978**, *32*, 557.

[376]Murray, R.W.; Youssefyeh, R.D.; Story, P.R. *J. Am. Chem. Soc.* **1966**, *88*, 3143, 3655; Story, P.R.; Murray, R.W.; Youssefyeh, R.D. *J. Am. Chem. Soc.* **1966**, *88*, 3144. Also see, Greenwood, F.L. *J. Am. Chem. Soc.* **1966**, *88*, 3146; Choe, J.; Srinivasan, M.; Kuczkowski, R.L. *J. Am. Chem. Soc.* **1983**, *105*, 4703.

$$Me_2HC-\underset{\underset{H}{|}}{\overset{\overset{O-O}{|}}{C}}\underset{O}{}\underset{\underset{H}{|}}{\overset{}{C}}-CHMe_2 \qquad Me-\underset{\underset{H}{|}}{\overset{\overset{O-O}{|}}{C}}\underset{O}{}\underset{\underset{H}{|}}{\overset{}{C}}-CHMe_2 \qquad Me-\underset{\underset{H}{|}}{\overset{\overset{O-O}{|}}{C}}\underset{O}{}\underset{\underset{H}{|}}{\overset{}{C}}-Me$$

| | 50-50 cis–trans | 48-52 cis–trans | 38-62 cis–trans |

$$\underset{H}{\overset{Me}{\diagdown}}C=C\underset{CHMe_2}{\overset{H}{\diagup}}$$

| | 66-34 cis–trans | 49-51 cis–trans | 49-51 cis–trans |

$$\underset{Me}{\overset{H}{\diagdown}}C=C\underset{CHMe_2}{\overset{H}{\diagup}}$$

If the Criegee mechanism operated as shown above, the cis/trans ratio for each of the two cross-ozonides would have to be identical for the cis- and trans-alkenes, since in this mechanism they are completely cleaved.

The above stereochemical results have been explained[377] on the basis of the Criegee mechanism with the following refinements: (*1*) The formation of **17** is stereospecific, as expected from a 1,3-dipolar cycloaddition. (*2*) Once they are formed, **19** and **18** remain attracted to each other, much like an ion pair. (*3*) Intermediate **19** exists in syn and anti forms, which are produced in different amounts and can hold their shapes, at least for a time. This is

$$\underset{syn\text{-}\mathbf{16}}{\overset{R}{\diagdown}\overset{\oplus}{\underset{|}{C}}\overset{R'}{\diagup}} \longleftrightarrow \qquad \qquad \longleftrightarrow \underset{anti\text{-}\mathbf{16}}{}$$

plausible if we remember that a C=O canonical form contributes to the structure of **19**. (*4*) The combination of **19** and **18** is also a 1,3-dipolar cycloaddition, so configuration is retained in this step too.[378]

Evidence that the basic Criegee mechanism operates even in these cases comes from [18]O labeling experiments, making use of the fact, mentioned above, that mixed ozonides (e.g., **23**) can be isolated when an external aldehyde is added. Both the normal and modified Criegee mechanisms predict that if ^{18}O-labeled aldehyde is added to the ozonolysis mixture, the label will appear in the ether oxygen (see the reaction between **19** and **18**), and this is what is found.[379] There is evidence that the *anti*-**19** couples much more readily than the *syn*-**19**. [380]

[377]Bauld, N.L.; Thompson, J.A.; Hudson, C.E.; Bailey, P.S. *J. Am. Chem. Soc.* **1968**, *90*, 1822; Bailey, P.S.; Ferrell, T.M. *J. Am. Chem. Soc.* **1978**, *100*, 899; Keul, H.; Kuczkowski, R.L. *J. Am. Chem. Soc.* **1985**, *50*, 3371.

[378]For isotope-effect evidence that this step is concerted in some cases, see Choe, J.; Painter, M.K.; Kuczkowski, R.L. *J. Am. Chem. Soc.* **1984**, *106*, 2891. However, there is evidence that it may not always be concerted: See, for example, Murray, R.W.; Su, J. *J. Org. Chem.* **1983**, *48*, 817.

[379]Bishop, C.E.; Denson, D.D.; Story, P.R. *Tetrahedron Lett.* **1968**, 5739; Fliszár, S.; Carles, J. *J. Am. Chem. Soc.* **1969**, *91*, 2637; Gillies, C.W.; Kuczkowski, R.L. *J. Am. Chem. Soc.* **1972**, *94*, 7609; Higley, D.P.; Murray, R.W. *J. Am. Chem. Soc.* **1976**, *98*, 4526; Mazur, U.; Kuczkowski, R.L. *J. Org. Chem.* **1979**, *44*, 3185.

[380]Mile, B.; Morris, G.M. *J. Chem. Soc. Chem. Commun.* **1978**, 263.

The ozonolysis of ethylene[381] in the liquid phase (without a solvent) was shown to take place by the Criegee mechanism.[382] This reaction has been used to study the structure of the intermediate **19** or **20**. The compound dioxirane (**24**) was identified in the reaction mixture[383] at low temperatures and is probably in equilibrium with the biradical **20** (R = H). Dioxirane has been produced in solution, but it oxidatively cleaves dialkyl ethers (e.g., Et–O–Et) via a chain radical process, [384] so the choice of solvent is important.

$$H_2C \underset{O}{\overset{O}{<}} | \quad \rightleftharpoons \quad \cdot CH_2 \overset{O}{\underset{O}{\nearrow}} | \; .$$

24   **20** (R = H)

Ozonolysis in the gas phase is not generally carried out in the laboratory. However, the reaction is important because it takes place in the atmosphere and contributes to air pollution.[385] There is much evidence that the Criegee mechanism operates in the gas phase too, although the products are more complex because of other reactions that also take place.[386]

OS **V**, 489, 493; **VI**, 976; **VII**, 168; **IX**, 314. Also see OS **IV**, 554. For the preparation of ozone, see OS **III**, 673.

19-10   Oxidative Cleavage of Double Bonds and Aromatic Rings

Oxo-de-alkylidene-bisubstitution, and so on.

$$R_2C=CHR \xrightarrow{\quad CrO_3 \quad} R_2C=O \; + \; RCOOH$$

Carbon–carbon double bonds can be cleaved by many oxidizing agents,[387] the most common of which are neutral or acid permanganate and acid dichromate. The

---

[381]For a discussion of intermediates in the formation of the ozonide in this reaction, see Samuni, U.; Fraenkel, R.; Haas, Y.; Fajgar, R.; Pola, J. *J. Am. Chem. Soc.* **1996**, *118*, 3687.

[382]Fong, G.D.; Kuczkowski, R.L. *J. Am. Chem. Soc.* **1980**, *102*, 4763.

[383]Suenram, R.D.; Lovas, F.J. *J. Am. Chem. Soc.* **1978**, *100*, 5117. See, however, Ishiguro, K.; Hirano, Y.; Sawaki, Y. *J. Org. Chem.* **1988**, *53*, 5397.

[384]Ferrer, M.; Sánchez-Baeza, F.; Casas, J.; Messeguer, A. *Tetrahedron Lett.* **1994**, *35*, 2981.

[385]For a review of the mechanisms of reactions of organic compounds with ozone in the gas phase, see Atkinson, R.; Carter, W.P.L. *Chem. Rev.* **1984**, *84*, 437.

[386]See Atkinson, R.; Carter, W.P.L. *Chem. Rev.* **1984**, *84*, 437, 452–454; Kühne, H.; Forster, M.; Hulliger, J.; Ruprecht, H.; Bauder, A.; Günthard, H. *Helv. Chim. Acta* **1980**, *63*, 1971; Martinez, R.I.; Herron J.T. *J. Phys. Chem.* **1988**, *92*, 4644.

[387]For a review of the oxidation of C=C and C=N bonds, see Henry, P.M.; Lange, G.L., in Patai, S. *The Chemistry of Functional Groups, Supplement A* pt. 1, Wiley, NY, **1977**, pp. 965–1098. For a review of oxidative cleavages of C=C double bonds and aromatic rings, see Hudlický, M. *Oxidations in Organic Chemistry*, American Chemical Society, Washington, DC, **1990**, pp. 77–84, 96–98. For reviews with respect to chromium reagents, see Badanyan, Sh.O.; Minasyan, T.T.; Vardapetyan, S.K. *Russ. Chem. Rev.* **1987**, *56*, 740; Cainelli, G.; Cardillo, G. *Chromium Oxiations in Organic Chemistry*, Open Court Pub. Co., La Salle, IL, **1981**, pp. 59–92. For a list of reagents, with references, see Larock, R.C. *Comprehensive Organic Transformations*, 2nd ed., Wiley-VCH, NY, **1999**, p. 1634.

products are generally 2 equivalents of ketone, 2 equivalents of carboxylic acid, or 1 equivalent of each, depending on what groups are attached to the alkene. With ordinary solutions of permanganate or dichromate yields are generally low, and the reaction is seldom a useful synthetic method; but high yields can be obtained by oxidizing with $KMnO_4$ dissolved in benzene containing the crown ether dicyclo-hexano-18-crown-6 (see p. 120).[388] The crown ether coordinates with K^+, permitting the $KMnO_4$ to dissolve in benzene. A mixture of aq. $KMnO_4$ and $NaIO_4$ on sand is also effective.[389] Another reagent frequently used for synthetic purposes is the *Lemieux–von Rudloff reagent*: HIO_4 containing a trace of MnO_4^-.[390] The MnO_4^- is the actual oxidizing agent, being reduced to the manganate stage, and the purpose of the HIO_4 is to reoxidize the manganate back to MnO_4^-. Another reagent that behaves similarly is $NaIO_4$–ruthenium tetroxide.[391] Cyclic alkenes are cleaved to α,ω-diketones, keto-acids or dicarboxylic acids. Cyclic alkenes are cleaved to dialdehydes with $KMnO_4 \cdot CuSO_4$ in dichloromethane.[392] Hydrogen peroxide on supported heteropolyacid cleaves cyclic alkenes.[393] A combination of $RuCl_3/HIO_5$ oxidatively cleaves cyclic alkenes to dicarboxylic acids.[394]

The *Barbier–Wieland procedure* for decreasing the length of a chain by one carbon involves oxidative cleavage by acid dichromate ($NaIO_4$–ruthenium tetroxide has also been used), but this is cleavage of a 1,1-diphenyl alkene, which generally gives good yields:

$$RH_2C-COOH \xrightarrow[\text{H}^+]{\text{EtOH}} RH_2C-COOEt \xrightarrow[\text{16-29}]{\text{PhMgBr}} RH_2C-\overset{\text{Ph}}{\underset{\text{Ph}}{C}}-OH \xrightarrow[\text{17-1}]{\Delta}$$

16-64

$$RHC=\overset{\text{Ph}}{\underset{\text{Ph}}{C}} \xrightarrow[\text{NaIO}_4-\text{RuO}_4]{\text{CrO}_3 \text{ or}} R-COOH + O=\overset{\text{Ph}}{\underset{\text{Ph}}{C}}$$

Addition of a catalytic amount of OsO_4 to Jones reagent (**19-3**) leads to good yields of the carboxylic acid from simple alkenes.[395] A combination of Oxone® and OsO_4 in DMF cleaves alkenes to carboxylic acids.[396] With certain reagents, the oxidation of double bonds can be stopped at the aldehyde stage, and in these cases the products are the same as in the ozonolysis procedure. Among these reagents are

[388]Sam, D.J.; Simmons, H.E. *J. Am. Chem. Soc.* **1972**, *94*, 4024. See also, Lee, D.G.; Chang, V.S. *J. Org. Chem.* **1978**, *43*, 1532.

[389]Huang, B.; Gupton, J.T.; Hansen, K.C.; Idoux, J.P. *Synth. Commun.* **1996**, *26*, 165.

[390]Lemieux, R.U.; Rudloff, E. von *Can. J. Chem.* **1955**, *33*, 1701, 1710; Rudloff, E. von *Can. J. Chem.* **1955**, *33*, 1714; **1956**, *34*, 1413; **1965**, *43*, 1784.

[391]For a review, see Lee, D.G.; van den Engh, M., in Trahanovsky, W.S. *Oxidation in Organic Chemistry*, pt. B, Academic Press, NY, **1973**, pp. 186–192. For the use of $NaIO_4$–OsO_4, see Cainelli, G.; Contento, M.; Manescalchi, F.; Plessi, L. *Synthesis* **1989**, 47.

[392]Göksu, S.; Altudda, R.; Sütbeyaz, Y. *Synth. Commun.* **2000**, *30*, 1615.

[393]Brooks, C.D.; Huang, L.-c.; McCarron, M.; Johnstone, R.A.W. *Chem. Commun.* **1999**, 37.

[394]Griffith, W.P.; Shoair, A.G.; Suriaatmaja, M. *Synth. Commun.* **2000**, *30*, 3091.

[395]Henry, J.R.; Weinreb, S.M. *J. Org. Chem.* **1993**, *58*, 4745.

[396]Travis, B.R.; Narayan, R.S.; Borhan, B. *J. Am. Chem. Soc.* **2002**, *124*, 3824.

tert-butyl iodoxybenzene,[397] KMnO$_4$ in THF–H$_2$O,[398] and NaIO$_4$–OsO$_4$.[399] Electrolysis with LiClO$_4$ in aqueous acetonitrile also cleaves alkenes to give the aldehyde.[400] Enol ethers, RC(OR′)=CH$_2$, have been cleaved to carboxylic esters, RC(OR′)=O, by atmospheric oxygen.[401]

Cleavage of alkynes is generally rather difficult, but treatment of internal alkynes with an excess of Oxone® with a ruthenium catalyst leads to aliphatic carboxylic acids.[402]

The mechanism of oxidation probably involves in most cases the initial formation of a glycol (**15-29**) or cyclic ester,[403] and then further oxidation as in **19-7**.[404] In line with the electrophilic attack on the alkene, triple bonds are more resistant to oxidation than double bonds. Terminal triple-bond compounds can be cleaved to carboxylic acids (RC≡CH ⟶ RCOOH) with thallium(III) nitrate[405] or with [bis-(trifluoroacetoxy)iodo]pentafluorobenzene [i.e., C$_6$F$_5$I(OCOCF$_3$)$_2$],[406] among other reagents.

Aromatic rings can be cleaved with strong enough oxidizing agents. An important laboratory reagent for this purpose is ruthenium tetroxide along with a co-oxidant, such as NaIO$_4$ or NaOCl (household bleach can be used).[407] Examples[408] are the oxidation of naphthalene to phthalic acid[409] and, even more remarkably, of cyclohexylbenzene to cyclohexanecarboxylic acid[410] (note the contrast with **19-11**). The latter conversion was also accomplished with ozone.[411] Another reagent that oxidizes aromatic rings is air catalyzed by V$_2$O$_5$. The oxidations of naphthalene to phthalic anhydride and of benzene to maleic anhydride by this reagent are

[397]Ranganathan, S.; Ranganathan, D.; Singh, S.K. *Tetrahedron Lett.* **1985**, *26*, 4955.

[398]Viski, P.; Szeverényi, Z.; Simándi, L.I. *J. Org. Chem.* **1986**, *51*, 3213.

[399]Pappo, R.; Allen Jr., D.S.; Lemieux, R.U.; Johnson, W.S. *J. Org. Chem.* **1956**, *21*, 478.

[400]Maki, S.; Niwa, H.; Hirano, T. *Synlett* **1997**, 1385.

[401]Taylor, R. *J. Chem. Res. (S)* **1987**, 178. For a similar oxidation with RuO4, see Torii, S.; Inokuchi, T.; Kondo, K. *J. Org. Chem.* **1985**, *50*, 4980.

[402]Yang, D.; Chen, F.; Dong, Z.-M.; Zhang, D.-W. *J. Org. Chem.* **2004**, *69*, 2221.

[403]See, for example, Lee, D.G.; Spitzer, U.A. *J. Org. Chem.* **1976**, *41*, 3644; Lee, D.G.; Chang, V.S.; Helliwell, S. *J. Org. Chem.* **1976**, *41*, 3644, 3646.

[404]There is evidence that oxidation with Cr(VI) in aqueous acetic acid involves an epoxide intermediate: Roč ek, J.; Drozd, J.C. *J. Am. Chem. Soc.* **1970**, *92*, 6668.

[405]McKillop, A.; Oldenziel, O.H.; Swann, B.P.; Taylor, E.C.; Robey, R.L. *J. Am. Chem. Soc.* **1973**, *95*, 1296.

[406]Moriarty, R.M.; Penmasta, R.; Awasthi, A.K.; Prakash, I. *J. Org. Chem.* **1988**, *53*, 6124.

[407]Ruthenium tetroxide is an expensive reagent, but the cost can be greatly reduced by the use of an inexpensive cooxidant, such as NaOCl, the function of which is to oxidize RuO2 back to ruthenium tetroxide.

[408]For other examples, see Piatak, D.M.; Herbst, G.; Wicha, J.; Caspi, E. *J. Org. Chem.* **1969**, *34*, 116; Wolfe, S.; Hasan, S.K.; Campbell, J.R. *Chem. Commun.* **1970**, 1420; Ayres, D.C.; Hossain, A.M.M. *Chem. Commun.* **1972**, 428; Nuñez, M.T.; Martín, V.S. *J. Org. Chem.* **1990**, *55*, 1928.

[409]Spitzer, U.A.; Lee, D.G. *J. Org. Chem.* **1974**, *39*, 2468.

[410]Caputo, J.A.; Fuchs, R. *Tetrahedron Lett.* **1967**, 4729.

[411]Klein, H.; Steinmetz, A. *Tetrahedron Lett.* **1975**, 4249. For other reagents that convert an aromatic ring to COOH and leave alkyl groups untouched, see Deno, N.C.; Greigger, B.A.; Messer, L.A.; Meyer, M.D.; Stroud, S.G. *Tetrahedron Lett.* **1977**, 1703; Liotta, R.; Hoff, W.S. *J. Org. Chem.* **1980**, *45*, 2887; Chakraborti, A.K.; Ghatak, U.R. *J. Chem. Soc. Perkin Trans. 1* **1985**, 2605.

important industrial procedures.[412] *o*-Diamines have been oxidized with nickel peroxide, with lead tetraacetate,[413] and with O_2 catalyzed by CuCl:[414]

The last-named reagent also cleaves *o*-dihydroxybenzenes (catechols) to give, in the presence of MeOH, the mono-methylated dicarboxylic acids .[415]

$$\text{HOOC} - \overset{|}{C} = \overset{|}{C} - \overset{|}{C} = \overset{|}{C} - \text{COOMe}$$

Enamines ($R'_2C{=}NR_2$) are oxidatively cleaved with potassium dichromate in sulfuric acid to the ketone ($R'_2C{=}O$).[416]

OS **II**, 53, 523; **III**, 39, 234, 449; **IV**, 136, 484, 824; **V**, 393; **VI**, 662, 690; **VII**, 397; **VIII**, 377, 490; **IX**, 530. Also see, OS **II**, 551.

19-11   Oxidation of Aromatic Side Chains

Oxo,hydroxy-de-dihydro,methyl-tersubstitution

$$\text{ArR} \xrightarrow{\text{KMnO}_4} \text{ArCOOH}$$

Alkyl chains on aromatic rings can be oxidized to COOH groups by many oxidizing agents, including permanganate, nitric acid, and acid dichromate.[417] The method is most often applied to the methyl group, although longer side chains can also be cleaved. However, tertiary alkyl groups are resistant to oxidation, and when they *are* oxidized, ring cleavage usually occurs too.[418] It is usually difficult to oxidize an R group on a fused aromatic system without cleaving the ring or oxidizing it to a quinone (**19-19**). However, this has been done (e.g., 2-methylnaphthalene was converted to 2-naphthoic acid) with aqueous $Na_2Cr_2O_7$.[419] Aryl methyl groups are oxidized to aryl COOH with NaOCl in acetonitrile,[420] or with NBS in aq. NaOH under photochemical conditions.[421] Functional groups can be present anywhere on the side chain and, if in the α position, greatly increase the ease of oxidation. An exception is an a phenyl group. In such cases, the reaction stops at the diaryl ketone stage. Molecules containing aryl groups on different carbons cleave so that each

[412]For a review, see Pyatnitskii, Yu.I. *Russ. Chem. Rev.* **1976**, *45*, 762.

[413]Nakagawa, K.; Onoue, H. *Tetrahedron Lett.* **1965**, 1433; *Chem. Commun.* **1966**, 396.

[414]Kajimoto, T.; Takahashi, H.; Tsuji, J. *J. Org. Chem.* **1976**, *41*, 1389.

[415]Tsuji, J.; Takayanag, H.i *Tetrahedron* **1978**, *34*, 641; Bankston, D. *Org. Synth.* 66, 180.

[416]Harris, C.E.; Lee, L.Y.; Dorr, H.; Singaram, B. *Tetrahedron Lett.* **1995**, *36*, 2921.

[417]For many examples, see Hudlický, M. *Oxidations in Organic Chemistry*, American Chemical Society, Washington, DC, **1990**, pp. 105-109; Lee, D.G. *The Oxidation of Organic Compounds by Permanganate Ion and Hexavalent Chromium*, Open-Court Pub. Co., La Salle, IL, **1980**, pp. 43–64. For a review with chromium oxidizing agents, see Cainelli, G.; Cardillo, G. *Chromium Oxidations in Organic Chemistry*, Open Court Publishing Co., La Salle, IL, **1981**, pp. 23–33.

[418]Brandenberger, S.G.; Maas, L.W.; Dvoretzky, I. *J. Am. Chem. Soc.* **1961**, *83*, 2146.

[419]Friedman, L.; Fishel, D.L.; Shechter, H. *J. Org. Chem.* **1965**, *30*, 1453.

[420]Yamazaki, S. *Synth. Commun.* **1999**, *29*, 2211.

[421]Itoh, A.; Kodama, T.; Hashimoto, S.; Masaki, Y. *Synthesis* **2003**, 2289.

ring gets one carbon atom, as in the clevvage of the 9,10-bond of dihydrophenan-threnes **25** to **26**.

It is possible to oxidize only one alkyl group of a ring that contains more than one. The order of reactivity[422] toward most reagents is $CH_2Ar > CHR_2 > CH_2R > CH_3$.[423] Groups on the ring susceptible to oxidation (OH, NHR, NH_2, etc.) must be protected. The oxidation can be performed with oxygen, in which case it is auto-xidation, and the mechanism is like that in **14-7**, with a hydroperoxide intermediate. With this procedure it is possible to isolate ketones from $ArCH_2R$, and this is often done.[424]

The mechanism has been studied for the closely related reaction: $Ar_2CH_2 + CrO_3 \rightarrow Ar_2C=O$.[425] A deuterium isotope effect of 6.4 was found, indicating that the rate-determining step is either $Ar_2CH_2 \rightarrow Ar_2CH\bullet$ or $Ar_2CH_2 \rightarrow Ar_2CH^+$. Either way this explains why tertiary groups are not converted to COOH and why the reactivity order is $CHR_2 > CH_2R > CH_3$, as mentioned above. Both free radi-cals and carbocations exhibit this order of stability (Chapter 5). The two possibili-ties are examples of categories 2 and 3 (p. 1706). Just how the radical or the cation goes on to the product is not known.

When the alkyl group is one oxidizable to COOH (**19-11**), cupric salts are oxi-dizing agents, and the OH group is found in a position ortho to that occupied by the alkyl group.[426] This reaction is used industrially to convert toluene to phenol.

In another kind of reaction, an aromatic aldehyde ArCHO or ketone ArCOR' is con-verted to a phenol ArOH on treatment with alkaline H_2O_2,[427] but there must be an OH or NH_2 group in the ortho or para position. This is called the *Dakin reaction*.[428] The mechanism may be similar to that of the Baeyer–Villiger reaction (**18-19**):[429]

[422]Oxidation with Co(III) is an exception. The methyl group is oxidized in preference to the other alkyl groups: Onopchenko, A.; Schulz, J.G.D.; Seekircher, R. *J. Org. Chem.* **1972**, *37*, 1414.

[423]For example, see Foster, G.; Hickinbottom, W.J. *J. Chem. Soc.* **1960**, 680; Ferguson, L.N.; Wims, A.I. *J. Org. Chem.* **1960**, *25*, 668.

[424]For a review, see Pines, H.; Stalick, W.M. *Base-Catalyzed Reactions of Hydrocarbons and Related Compounds*, Academic Press, NY, *1977*, pp. 508–543.

[425]Wiberg, K.B.; Evans, R.J. *Tetrahedron* **1960**, *8*, 313.

[426]Kaeding, W.W. *J. Org. Chem.* **1961**, *26*, 3144. For a discussion, see Lee, D.G.; van den Engh, M., in Trahanovsky, W.S. *Oxidation in Organic Chemistry*, pt. B, Academic Press, NY, *1973*, pp. 91–94.

[427]For a convenient procedure, see Hocking, M.B. *Can. J. Chem.* **1973**, *51*, 2384.

[428]See Schubert, W.M.; Kintner, R.R., in Patai, S. *The Chemistry of the Carbonyl Group*, Vol. 1, Wiley, NY, *1966*, pp. 749–752.

[429]For a discussion, see Hocking, M.B.; Bhandari, K.; Shell, B.; Smyth, T.A. *J. Org. Chem.* **1982**, *47*, 4208.

27

The intermediate **27** has been isolated.[430] The reaction has been performed on aromatic aldehydes with an alkoxy group in the ring, and no OH or NH_2. In this case, acidic H_2O_2 was used.[431] The Dakin reaction has been done in ionic liquids.[432]

OS **I**, 159, 385, 392, 543; **II**, 135, 428; **III**, 334, 420, 740, 791, 820, 822; **V**, 617, 810. Also see, OS **I**, 149; **III**, 759.

19-12   Oxidative Decarboxylation
Acetoxy-de-carboxy-substitution

$$RCOOH \xrightarrow{Pb(OAc)_4} ROAc$$

Hydro-carboxyl-elimination

Carboxylic acids can be decarboxylated[433] with lead tetraacetate to give a variety of products, among them the ester ROAc (formed by replacement of COOH by an acetoxy group), the alkane RH (see **12-40**), and, if α,β hydrogen is present, the alkene formed by elimination of H and COOH, as well as numerous other products arising from rearrangements, internal cyclizations,[434] and reactions with solvent molecules. When R is tertiary, the chief product is usually the alkene, which is often obtained in good yield. High yields of alkenes can also be obtained when R is primary or secondary, in this case by the use of $Cu(OAc)_2$ along with the $Pb(OAc)_4$.[435] In the absence of $Cu(OAc)_2$, primary acids give mostly alkanes (though yields are

[430]Hocking, M.B.; Ko, M.; Smyth, T.A. *Can. J. Chem.* **1978**, *56*, 2646.

[431]Matsumoto, M.; Kobayashi, H.; Hotta, Y. *J. Org. Chem.* **1984**, *49*, 4740.

[432]In bmim PF6, 1-butyl-3-methylimidazolium hexafluorophosphate: Zambrano, J.L.; Dorta, R. *Synlett* **2003**, 1545.

[433]For reviews, see Serguchev, Yu.A.; Beletskaya, I.P. *Russ. Chem. Rev.* **1980**, *49*, 1119; Sheldon, R.A.; Kochi, J.K. *Org. React.* **1972**, *19*, 279.

[434]For examples, see Moriarty, R.M.; Walsh, H.G.; Gopal, H. *Tetrahedron Lett.* **1966**, 4363; Davies, D.I.; Waring, C. *J. Chem. Soc. C* **1968**, 1865, 2337.

[435]Bacha, J.D.; Kochi, J.K. *Tetrahedron* **1968**, *24*, 2215; Ogibin, Yu.N.; Katzin, M.I.; Nikishin, G.I. *Synthesis* **1974**, 889.

generally low) and secondary acids may give carboxylic esters or alkenes. Carboxylic esters have been obtained in good yields from some secondary acids, from β,γ-unsaturated acids, and from acids in which R is a benzylic group. Other oxidizing agents,[436] including Co(III), Ag(II), Mn(III), and Ce(IV), have also been used to effect oxidative decarboxylation.[437]

The mechanism with lead tetraacetate is generally accepted to be of the free-radical type.[438] First, there is an interchange of ester groups:

$$Pb(OAc)_4 \ + \ RCOOH \ \longrightarrow \ \underset{\textbf{28}}{Rb(OAc)_3OCOR} \quad or \quad \underset{\textbf{29}}{Pb(OAc)_2(OCOR)_2}$$

There follows a free-radical chain mechanism (shown for **28** although **29** and other lead esters can behave similarly)

$$\underset{\textbf{23}}{Pb(OAc)_3OCOR} \ \longrightarrow \ \cdot Pb(OAc)_3 \ + \ R\cdot \ + \ CO_2$$

Initiation

$$R\cdot \ + \ Pb(OAc)_3OCOR \ \longrightarrow \ R^+ \ + \ \cdot Pb(OAc)_2OCOR \ + \ OAc^-$$

$$\cdot Pb(OAc)_2OCOR \ \longrightarrow \ Pb(OAc)_2 \ + \ R\cdot \ + \ CO_2$$

Propagation

Products can then be formed either from R• or R$^+$. Primary R• abstract H from solvent molecules to give RH. R$^+$ can lose H$^+$ to give an alkene, react with HOAc to give the carboxylic ester, react with solvent molecules or with another functional group in the same molecule, or rearrange, thus accounting for the large number of possible products. The R• group can also dimerize to give RR. The effect of Cu^{2+} ions[439] is to oxidize the radicals to alkenes, thus producing good yields of alkenes from primary and secondary substrates. The Cu^{2+} ion has no effect on tertiary radicals, because these are efficiently oxidized to alkenes by lead tetraacetate.

$$H-\overset{|}{\underset{|}{C}}-\overset{|}{\underset{|}{C}}\cdot \ + \ Cu^{2+} \ \longrightarrow \ \overset{}{\underset{}{C}}=\overset{}{\underset{}{C} } \ + \ H^+ \ + \ Cu^+$$

[436]For references, see Trahanovsky, W.S.; Cramer, J.; Brixius, D.W. *J. Am. Chem. Soc.* **1974**, *96*, 1077; Kochi, J.K. *Organometallic Mechanisms and Catalysis*, Academic Press, NY, *1978*, pp. 99–106. See also, Dessau, R.M.; Heiba, E.I. *J. Org. Chem.* **1975**, *40*, 3647; Fristad, W.E.; Fry, M.A.; Klang, J.A. *J. Org. Chem.* **1983**, *48*, 3575; Barton, D.H.R.; Crich, D.; Motherwell, W.B. *J. Chem. Soc. Chem. Commun.* **1984**, 242; Toussaint, O.; Capdevielle, P.; Maumy, M. *Tetrahedron Lett.* **1984**, *25*, 3819.

[437]For another method, see Barton, D.H.R.; Bridon, D.; Zard, S.Z. *Tetrahedron* **1989**, *45*, 2615.

[438]Starnes, Jr., W.H. *J. Am. Chem. Soc.* **1964**, *86*, 5603; Davies, D.I.; Waring, C. *Chem. Commun.* **1965**, 263; Kochi, J.K.; Bacha, J.D.; Bethea III, T.W. *J. Am. Chem. Soc.* **1967**, *89*, 6538; Cantello, B.C.C.; Mellor, J.M.; Scholes, G. *J. Chem. Soc. Perkin Trans. 2*, **1974**, 348; Beckwith, A.L.J.; Cross, R.T.; Gream, G.E. *Aust. J. Chem.* **1974**, *27*, 1673, 1693.

[439]Bacha, J.D.; Kochi, J.K. *J. Org. Chem.* **1968**, *33*, 83; Kochi, J.K.; Bacha, J.D. *J. Org. Chem.* **1968**, *33*, 2746; Torssell, K. *Ark. Kemi*, **1970**, *31*, 401.

In another type of oxidative decarboxylation, arylacetic acids can be oxidized to aldehydes with one less carbon ($ArCH_2COOH \rightarrow ArCHO$) by tetrabutylammonium periodate.[440] Simple aliphatic carboxylic acids were converted to nitriles with one less carbon ($RCH_2COOH \rightarrow RC\equiv N$) by treatment with trifluoroacetic anhydride and $NaNO_2$ in F_3CCOOH.[441]

See also, **14-37**.

19-13   Bisdecarboxylation

Dicarboxy-elimination

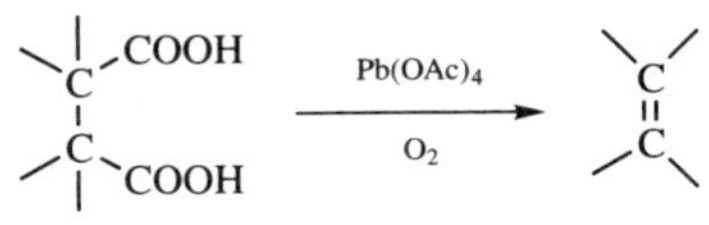

Compounds containing carboxyl groups on adjacent carbons (succinic acid derivatives) can be bisdecarboxylated with lead tetraacetate in the presence of O_2.[433] The reaction is of wide scope. The elimination is stereoselective, but not stereospecific (both *meso*- and *dl*-2,3-diphenylsuccinic acid gave *trans*-stilbene);[442] a concerted mechanism is thus unlikely. The following mechanism is not inconsistent with the data:

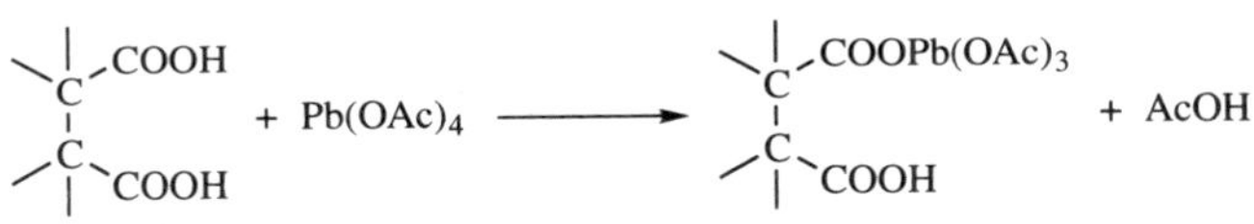

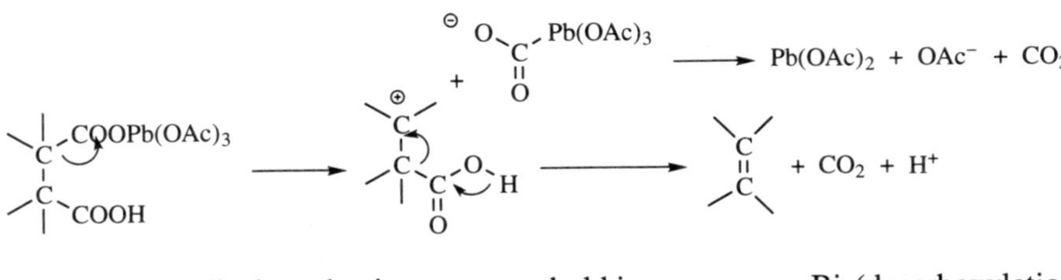

though a free-radical mechanism seems to hold in some cases. Bis(decarboxylation) of succinic acid derivatives to give alkenes[443] has also been carried out by other methods, including treatment of the corresponding anhydrides with nickel, iron,

---

[440]Santaniello, E.; Ponti, F.; Manzocchi, A. *Tetrahedron Lett.* **1980**, *21*, 2655. For other methods of accomplishing this and similar conversions, see Cohen, H.; Song, I.H.; Fager, J.H.; Deets, G.L. *J. Am. Chem. Soc.* **1967**, *89*, 4968; Wasserman, H.H.; Lipshutz, B.H. *Tetrahedron Lett.* **1975**, 4611; Kaberia, F.; Vickery, B. *J. Chem. Soc. Chem. Commun.* **1978**, 459; Doleschall, G.; Tóth, G. *Tetrahedron* **1980**, *36*, 1649.

[441]Smushkevich, Yu.I.; Usorov, M.I.; Suvorov, N.N. *J. Org. Chem. USSR* **1975**, *11*, 653.

[442]Corey, E.J.; Casanova, J. *J. Am. Chem. Soc.* **1963**, *85*, 165.

[443]For a review, see De Lucchi, O.; Modena, G. *Tetrahedron* **1984**, *40*, 2585, 2591–2608.

or rhodium complexes,[444] by decomposition of the corresponding bis(peroxye-sters),[445] and electrolytically.[446]

Compounds containing geminal carboxyl groups (disubstituted malonic acid derivatives) can also be bisdecarboxylated with lead tetraacetate,[447] *gem*-diacetates (acylals) being produced, which are easily hydrolyzable to ketones:[448]

$$
\underset{R}{\overset{R}{\diagdown}}C\underset{COOH}{\overset{COOH}{\diagup}} \xrightarrow{Pb(OAc)_4} \underset{R}{\overset{R}{\diagdown}}C\underset{OAc}{\overset{OAc}{\diagup}} \xrightarrow{hydrol.} \underset{R}{\overset{R}{\diagdown}}C=O
$$

A related reaction involves α-substituted aryl nitriles having a sufficiently acidic α hydrogen, which can be converted to ketones by oxidation with air under phase transfer conditions.[449] The nitrile is added to NaOH in benzene or DMSO containing a catalytic amount of triethylbenzylammonium chloride (TEBA).[450] This reaction could not be applied to aliphatic nitriles, but an indirect method for achieving this conversion is given in **19-60**. α-Dialkylamino nitriles can be converted to ketones, $R_2C(NMe_2)CN \rightarrow R_2C=O$, by hydrolysis with $CuSO_4$ in aqueous methanol[451] or by autoxidation in the presence of *t*-BuOK.[452]

C. Reactions Involving Replacement of Hydrogen by Heteroatoms

19-14 Hydroxylation at an Aliphatic Carbon

Hydroxylation or **Hydroxy-de-hydrogenation**

$$
R_3CH \xrightarrow[\text{silica gel}]{O_3} R_3COH
$$

Compounds containing susceptible C—H bonds can be oxidized to alcohols.[453] Nearly always, the C—H bond involved is tertiary, so the product is a tertiary alcohol. This is partly because tertiary C—H bonds are more susceptible to free-radical attack than primary and secondary bonds and partly because the reagents involved

[444]Trost, B.M.; Chen, E.N. *Tetrahedron Lett.* **1971**, 2603.

[445]Cain, E.N.; Vukov, R.; Masamune, S. *Chem. Commun.* **1969**, 98.

[446]Plieninger, H.; Lehnert, W. *Chem. Ber.* **1967**, *100*, 2427; Radlick, P.; Klem, R.; Spurlock, S.; Sims, J.J.; van Tamelen, E.E.; Whitesides, T. *Tetrahedron Lett.* **1968**, 5117; Westberg, H.H.; Dauben Jr., H.J. *Tetrahedron Lett.* **1968**, 5123. For additional references, see Fry, A.J. *Synthetic Organic Electrochemistry*, 2nd ed., Wiley, NY, **1989**, pp. 253–254.

[447]For a similar reaction with ceric ammonium nitrate, see Salomon, R.G.; Roy, S.; Salomon, R.G. *Tetrahedron Lett.* **1988**, 29, 769.

[448]Tufariello, J.J.; Kissel, W.J. *Tetrahedron Lett.* **1966**, 6145.

[449]For other methods of achieving this conversion, with references, see Larock, R.C. *Comprehensive Organic Transformations*, 2nd ed., Wiley-VCH, NY, **1999**, p. 1260.

[450]Masuyama, Y.; Ueno, Y.; Okawara, M. *Chem. Lett.* **1977**, 1439; Donetti, A.; Boniardi, O.; Ezhaya, A. *Synthesis* **1980**, 1009; Kulp, S.S.; McGee, M.J. *J. Org. Chem.* **1983**, *48*, 4097.

[451]Büchi, G.; Liang, P.H.; Wüest, H. *Tetrahedron Lett.* **1978**, 2763.

[452]Chuang, T.; Yang, C.; Chang, C.; Fang, J. *Synlett* **1990**, 733.

[453]For reviews, see Chinn, L.J. *Selection of Oxidants in Synthesis*, Marcel Dekker, NY, **1971**, pp. 7–11; Lee, D.G., in Augustine, R.L. *Oxidation*, Vol. 1, Marcel Dekker, NY, **1969**, pp. 2–6. For a monograph on all types of alkane activation, see Hill, C.L. *Activation and Functionalization of Alkanes*, Wiley, NY, **1989**.

would oxidize primary and secondary alcohols further. In the best method, the reagent is ozone and the substrate is absorbed on silica gel.[454] Yields as high as 99% have been obtained by this method. Other reagents are chromic acid,[455] potassium hydrogen persulfate $(KHSO_5)$,[456] ruthenium tetroxide (RuO_4),[457] 2,6-dichloropyridine *N*-oxide with a ruthenium catalyst,[458] thallium acetate,[459] sodium chlorite $(NaClO_2)$ with a metalloporphyrin catalyst,[460] and certain peroxybenzoic acids.[461] Alkanes and cycloalkanes have been oxidized at secondary positions, to a mixture of alcohols and trifluoroacetates, by 30% aq. H_2O_2 in trifluoroacetic acid.[462] This reagent does not oxidize the alcohols further and ketones are not found. As in the case of chlorination with *N*-haloamines and sulfuric acid (see **14-1**), the ω - 1 position is the most favored. Another reagent[463] that oxidizes secondary positions is iodosylbenzene, catalyzed by Fe^{III}–porphyrin catalysts.[464] Use of an optically active Fe^{III}–porphyrin gave enantioselective hydroxylation, with moderate ee.[465]

When chromic acid is the reagent, the mechanism is probably as follows: a Cr^{6+} species abstracts a hydrogen to give $R_3C\bullet$, which is held in a solvent cage near the resulting Cr^{5+} species. The two species then combine to give R_3COCr^{4+}, which is hydrolyzed to the alcohol. This mechanism predicts retention of configuration; this is largely observed.[466] The oxidation by permanganate also involves predominant retention of configuration, and a similar mechanism has been proposed.[467]

Treatment of double-bond compounds with selenium dioxide introduces an OH group into the allylic position (see also, **19-17**).[468] This reaction also produces conjugated aldehydes in some cases.[469] Allylic rearrangements are common. There is

[454]Cohen, Z.; Keinan, E.; Mazur, Y.; Varkony, T.H. *J. Org. Chem.* **1975**, *40*, 2141; *Org. Synth. VI*, 43; Keinan, E.; Mazur, Y. *Synthesis* **1976**, 523; McKillop, A.; Young, D.W. *Synthesis* **1979**, 401, see pp. 418–419.

[455]For a review, see Cainelli, G.; Cardillo, G. *Chromium Oxidations in Organic Chemistry*, Springer, NY, **1984**, pp. 8–23.

[456]De Poorter, B.; Ricci, M.; Meunier, B. *Tetrahedron Lett.* **1985**, *26*, 4459.

[457]Tenaglia, A.; Terranova, E.; Waegell, B. *Tetrahedron Lett.* **1989**, *30*, 5271; Bakke, J.M.; Braenden, J.E. *Acta Chem. Scand.* **1991**, *45*, 418.

[458]Ohtake, H.; Higuchi, T.; Hirobe, M. *J. Am. Chem. Soc.* **1992**, *114*, 10660.

[459]Lee, J.C.; Park, C.; Choi, Y. *Synth. Commun.* **1997**, *27*, 4079.

[460]Collman, J.P.; Tanaka, H.; Hembre, R.T.; Brauman, J.I. *J. Am. Chem. Soc.* **1990**, *112*, 3689.

[461]Schneider, H.; Müller, W. *Angew. Chem. Int. Ed.* **1982**, *21*, 146; *J. Org. Chem.* **1985**, *50*, 4609; Takaishi, N.; Fujikura, Y.; Inamoto, Y. *Synthesis* **1983**, 293; Tori, M.; Sono, M.; Asakawa, Y. *Bull. Chem. Soc. Jpn.* **1985**, *58*, 2669. See also, Querci, C.; Ricci, M. *Tetrahedron Lett.* **1990**, *31*, 1779.

[462]Deno, N.C.; Jedziniak, E.J.; Messer, L.A.; Meyer, M.D.; Stroud, S.G.; Tomezsko, E.S. *Tetrahedron* **1977**, *33*, 2503.

[463]For other procedures, see Sharma, S.N.; Sonawane, H.R.; Dev, S. *Tetrahedron* **1985**, *41*, 2483; Nam, W.; Valentine, J.S. *New J. Chem.* **1989**, *13*, 677.

[464]See Groves, J.T.; Nemo, T.E. *J. Am. Chem. Soc.* **1983**, *105*, 6243.

[465]Groves, J.T.; Viski, P. *J. Org. Chem.* **1990**, *55*, 3628.

[466]Wiberg, K.B.; Eisenthal, R. *Tetrahedron* **1964**, *20*, 1151.

[467]Wiberg, K.B.; Fox, A.S. *J. Am. Chem. Soc.* **1963**, *85*, 3487; Brauman, J.I.; Pandell, A.J. *J. Am. Chem. Soc.* **1970**, *92*, 329; Stewart, R.; Spitzer, U.A. *Can. J. Chem.* **1978**, *56*, 1273.

[468]For reviews, see Rabjohn, N. *Org. React.* **1976**, *24*, 261; Jerussi, R.A. *Sel. Org. Transform.* **1970**, *1*, 301; Trachtenberg, E.N., in Augustine, R.L. *Oxidation*, Vol. 1, Marcel Dekker, NY, **1969**, pp. 123–153.

[469]Singh, J.; Sharma, M.; Kad, G.L.; Chhabra, B.R. *J. Chem. Res. (S)* **1997**, 264.

evidence that the mechanism does not involve free radicals, but includes two pericyclic steps (A and B):[470]

The step marked A is similar to the ene synthesis (**15-23**). The step marked B is a [2,3]-sigmatropic rearrangement (see **18-35**). The reaction can also be accomplished with *tert*-butyl hydroperoxide, if SeO_2 is present in catalytic amounts (the *Sharpless method*).[471] The SeO_2 is the actual reagent; the peroxide reoxidizes the $Se(OH)_2$.[472] This method makes work-up easier, but gives significant amounts of side products when the double bond is in a ring.[473] Alkynes generally give α,α'-dihydroxylation.[474]

Ketones and carboxylic esters can be α hydroxylated by treatment of their enolate forms (prepared by adding the ketone or ester to LDA) with a molybdenum peroxide reagent (MoO_5–pyridine–HMPA) in THF–hexane at $-70°C$.[475] The reaction of ketones with $Ti(OiPr)_4$, diethyl tartrate and *tert*-butylhydroperoxide gave the α-hydroxy ketone with good enantioselectively, albeit in low yield.[476] The enolate forms of amides and esters[477] and the enamine derivatives of ketones[478] can similarly be converted to their α hydroxy derivatives by reaction with molecular oxygen. The MoO_5 method can also be applied to certain nitriles.[479] Ketones have also been α hydroxylated by treating the corresponding silyl enol ethers

[470]Arigoni, D.; Vasella, A.; Sharpless, K.B.; Jensen, H.P. *J. Am. Chem. Soc.* **1973**, *95*, 7917; Woggon, W.; Ruther, F.; Egli, H. *J. Chem. Soc. Chem. Commun.* **1980**, 706. For other mechanistic proposals, see Schaefer, J.P.; Horvath, B.; Klein, H.P. *J. Org. Chem.* **1968**, *33*, 2647; Trachtenberg, E.N.; Nelson, C.H.; Carver, J.R. *J. Org. Chem.* **1970**, *35*, 1653; Bhalerao, U.T.; Rapoport, H. *J. Am. Chem. Soc.* **1971**, *93*, 4835; Stephenson, L.M.; Speth, D.R. *J. Org. Chem.* **1979**, *44*, 4683.

[471]Umbreit, M.A.; Sharpless, K.B. *J. Am. Chem. Soc.* **1977**, *99*, 5526. See also, Uemura, S.; Fukuzawa, S.; Toshimitsu, A.; Okano, M. *Tetrahedron Lett.* **1982**, *23*, 87; Singh, J.; Sabharwal, A.; Sayal, P.K.; Chhabra, B.R. *Chem. Ind. (London)* **1989**, 533.

[472]For the use of the peroxide with O2 instead of SeO2, see Sabol, M.R.; Wiglesworth, C.; Watt, D.S. *Synth. Commun.* **1988**, *18*, 1.

[473]Warpehoski, M.A.; Chabaud, B.; Sharpless, K.B. *J. Org. Chem.* **1982**, *47*, 2897.

[474]Chabaud, B.; Sharpless, K.B. *J. Org. Chem.* **1979**, *44*, 4202.

[475]Vedejs, E.; Telschow, J.E. *J. Org. Chem.* **1976**, *41*, 740; Vedejs, E.; Larsen, S. *Org. Synth. VII*, 277; Gamboni, R.; Tamm, C. *Tetrahedron Lett.* **1986**, *27*, 3999; *Helv. Chim. Acta* **1986**, *69*, 615. See also, Anderson, J.C.; Smith, S.C. *Synlett* **1990**, 107; Hara, O.; Takizawa, J.-i.; Yamatake, T.; Makino, K.; Hamada, Y. *Tetrahedron Lett.* **1999**, *40*, 7787.

[476]Paju, A.; Kanger, T.; Pehk, T.; Lopp, M. *Tetrahedron* **2002**, *58*, 7321.

[477]Wasserman, H.H.; Lipshutz, B.H. *Tetrahedron Lett.* **1975**, 1731. For another method, see Pohmakotr, M.; Winotai, C. *Synth. Commun.* **1988**, *18*, 2141.

[478]Cuvigny, T.; Valette, G.; Larcheveque, M.; Normant, H. *J. Organomet. Chem.* **1978**, *155*, 147.

[479]Rubottom, G.M.; Gruber, J.M. *J. Org. Chem.* **1978**, *43*, 1599; Hassner, A.; Reuss, R.H.; Pinnick, H.W. *J. Org. Chem.* **1975**, *40*, 3427; Andriamialisoa, R.Z.; Langlois, N.; Langlois, Y. *Tetrahedron Lett.* **1985**, *26*, 3563; Rubottom, G.M.; Gruber, J.M.; Juve, Jr., H.D.; Charleson, D.A. *Org. Synth. VII*, 282. See also, Horiguchi, Y.; Nakamura, E.; Kuwajima, I. *Tetrahedron Lett.* **1989**, *30*, 3323.

with *m*-chloroperoxybenzoic acid,[179] or with certain other oxidizing agents.[480] When the silyl enol ethers are treated with iodosobenzene in the presence of trimethylsilyl trifluoromethyl sulfonate, the product is the α-keto triflate.[481]

Tetrahydrofuran was converted to the hemiacetal 2-hydroxytetrahydrofuran (which was relatively stable under the conditions used) by electrolysis in water.[482]

OS **IV**, 23; **VI**, 43, 946; **VII**, 263, 277, 282.

19-15   Oxidation of Methylene to OH, O₂CR, or OR

Hydroxy (or alkoxy) -de-dihydro-bisubstitution

Methyl or methylene groups α to a carbonyl can be oxidized to give α-hydroxy ketones, aldehydes, or carboxylic acid derivatives. Ketones can be α hydroxylated in good yields, without conversion to the enolates, by treatment with the hypervalent iodine reagents[483] *o*-iodosobenzoic acid[484] or phenyliodoso acetate, PhI(OAc)₂, in methanolic NaOH.[485] The latter reagent has also been used on carboxylic esters.[486] Dioxygen (O₂) and a chiral phase-transfer catalyst gave enantioselective α-hydroxylation of ketones, if the α position was tertiary.[487] Dimethyl dioxirane is quite effective for hydroxylation of 1,3-dicarbonyl compounds,[488] and O₂ with a manganese catalyst also gives hydroxylation of such compounds.[489] Oxygen with a cerium catalyst α-hydroxylates β-keto esters.[490] Ceric ammonium nitrate has been used to hydroxylate C-2 of dibenzyl malonate.[491] Methyl ketones (RCOMe) react with ammonium peroxydisulfate, (NH₄)₂S₂O₈, and a catalytic amount of diphenyl diselenide in MeOH to give α-keto acetals, RCOCH(OMe₂).[492]

[480]McCormick, J.P.; Tomasik, W.; Johnson, M.W. *Tetrahedron Lett.* **1981**, *22*, 607; Moriarty, R.M.; Prakash, O.; Duncan, M.P. *Synthesis* **1985**, 943; Iwata, C.; Takemoto, Y.; Nakamura, A.; Imanishi, T. *Tetrahedron Lett.* **1985**, *26*, 3227; Davis, F.A.; Sheppard, A.C. *J. Org. Chem.* **1987**, *52*, 954; Takai, T.; Yamada, T.; Rhode, O.; Mukaiyama, T. *Chem. Lett.* **1991**, 281.

[481]Moriarty, R.M.; Epa, W.R.; Penmasta, R.; Awasthi, A.K. *Tetrahedron Lett.* **1989**, *30*, 667.

[482]Wermeckes, B.; Beck, F.; Schulz, H. *Tetrahedron* **1987**, *43*, 577.

[483]For a review, see Moriarty, R.M.; Prakash, O. *Acc. Chem. Res.* **1986**, *19*, 244. Also see, Reddy, D.R.; Thornton, E.R. *J. Chem. Soc. Chem. Commun.* **1992**, 172.

[484]Moriarty, R.M.; Hou, K. *Tetrahedron Lett.* **1984**, *25*, 691; Moriarty, R.M.; Hou, K.; Prakash, O.; Arora, S.K. *Org. Synth.* **VII**, 263.

[485]Moriarty, R.M.; Hu, H.; Gupta, S.C. *Tetrahedron Lett.* **1981**, *22*, 1283. See Moriarty, R.M.; Berglund, B.A.; Penmasta, R. *Tetrahedron Lett.* **1992**, *33*, 6065 for reactions with PhI(O₂CCF₃)₂.

[486]Moriarty, R.M.; Hu, H. *Tetrahedron Lett.* **1981**, *22*, 2747.

[487]Masui, M.; Ando, A.; Shioiri, T. *Tetrahedron Lett.* **1988**, *29*, 2835.

[488]Adam, W.; Smerz, A.K. *Tetrahedron* **1996**, *52*, 5799. See Hull, L.A.; Budhai, L. *Tetrahedron Lett.* **1993**, *34*, 5039 for a discussion of the thermal decomposition of dimethyl dioxirane. See Murray, R.W.; Singh, M.; Jeyaraman, R. *J. Am. Chem. Soc.* **1992**, *114*, 1346 for the preparation of new dioxiranes.

[489]Christoffers, J. *J. Org. Chem.* **1999**, *64*, 7668.

[490]Christoffers, J.; Werner, T. *Synlett* **2002**, 119.

[491]Nair, V.; Nair, L.G.; Mathew, J. *Tetrahedron Lett.* **1998**, *39*, 2801.

[492]Tiecco, M.; Testaferri, L.; Tingoli, M.; Bartoli, D. *J. Org. Chem.* **1990**, *55*, 4523.

α-Acetoxylation of ketones with concurrent α-arylation occurs when ketones react with Mn(OAc)$_3$ in benzene.[493] α-Acetoxylation of ketones can occur under similar conditions without arylation.[494] α-Methyl ketones are converted to the α-acetoxy derivative under the same conditions.[495] Thallium (III) triflate converts acetophenone to α-formyloxy acetophenone.[496] α-Tosyloxy ketones are generated from acetophenone derivatives using PhI(OH)OTs.[497]

A different method for the conversion of ketones to α-hydroxy ketones consists of treating the enolate anion with a 2-sulfonyloxaziridine (e.g., **30**).[498] This is not a free-radical process; the following mechanism is likely:

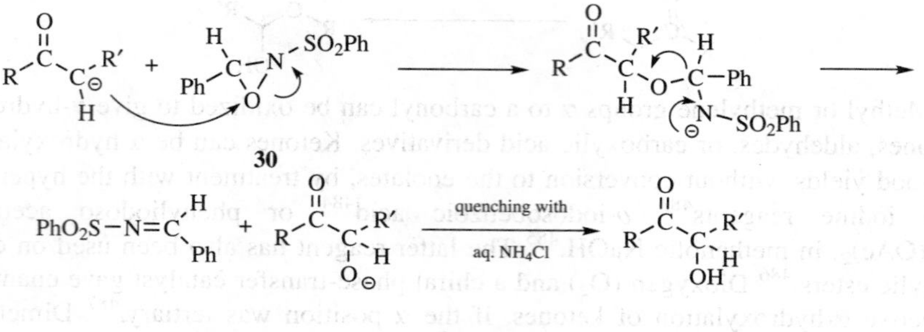

The method is also successful for carboxylic esters[351],[499] and *N,N*-disubstituted amides,[500] and can be made enantioselective by the use of a chiral oxaziridine.[501] Dimethyldioxirane also oxidizes ketones (through their enolate forms) to α-hydroxy ketones.[502] Titanium enolates can be oxidized with *tert*-butyl hydroperoxide [503] or with dimethyl dioxirane[504] and hydrolyzed with aqueous ammonium fluoride to give the α-hydroxy ketone. Ketones are converted to the α-oxamino derivative (O=C—CH$_2$- → O=C—CHONHPh) with excellent enantioselectivity using

[493]Tanyeli, C.; Özdemirhan, D.; Sezen, B. *Tetrahedron* **2002**, *58*, 9983.

[494]Tanyeli, C.; Tosun, A.; Turkut, E.; Sezen, B. *Tetrahedron* **2003**, *59*, 1055; Demir, A.S.; Reis, Ö.; Igdir, A.C. *Tetrahedron* **2004**, *60*, 3427.

[495]Tanyeli, C.; Iyigün,. *Tetrahedron* **2003**, *59*, 7135.

[496]Lee, J.C.; Jin, Y.S.; Choi, J.-H. *Chem. Commun.* **2001**, 956.

[497]Nabana, T.; Togo, H. *J. Org. Chem.* **2002**, *67*, 4362.

[498]Davis, F.A.; Vishwakarma, L.C.; Billmers, J.M.; Finn, J. *J. Org. Chem.* **1984**, *49*, 3241.

[499]For formation of α-benzyloxy lactones, see Brodsky, B.H.; DuBois, J. *Org. Lett.* **2004**, *6*, 2619.

[500]Davis, F.A.; Vishwakarma, L.C. *Tetrahedron Lett.* **1985**, *26*, 3539.

[501]Evans, D.A.; Morrissey, M.M.; Dorow, R.L. *J. Am. Chem. Soc.* **1985**, *107*, 4346; Enders, D.; Bhushan, V. *Tetrahedron Lett.* **1988**, *29*, 2437; Davis, F.A.; Sheppard, A.C.; Chen, B.; Haque, M.S. *J. Am. Chem. Soc.* **1990**, *112*, 6679; Davis, F.A.; Weismiller, M.C. *J. Org. Chem.* **1990**, *55*, 3715.

[502]Guertin, K.R.; Chan, T.H. *Tetrahedron Lett.* **1991**, *32*, 715.

[503]Schulz, M.; Kluge, R.; Schüßler, M.; Hoffmann, F. *Tetrahedron* **1995**, *51*, 3175.

[504]Adam, W.; Müller, M.; Prechtl, F. *J. Org. Chem.* **1994**, *59*, 2358.

PhN=O and L-proline[505] or (S)-proline.[506] Aldehydes undergo a similar oxidation.[507] α-Lithio sulfones have been hydroxylated with Me₃SiOOt-Bu.[508]

$$\begin{array}{c} O \\ {}^{/\!/} \\ O-C \cdot {}^{Me} \\ {}^{\backslash}_{Me} \end{array}$$ Dimethyldioxirane

α-Hydroxyketones can be generated from silyl enol ethers with a catalytic amount of MeReO₃ and H₂O₂.[509] Silyl ketene ethers are converted to α-hydroxy esters with H₂O₂ and methyl trioxorhenium.[510] The α'-position of α,β-unsaturated ketones can be selectively oxidized.[511] N-Acyl amines are converted to the α-hydroxy derivative with PhIO and a manganese–salen catalyst.[512] Note that homoallylic-type oxidation occurs when an α,α-dimethyl oxime ether is treated with PhI(OAc)₂ and a palladium catalyst in acetic acid–acetic anhydride, converting one of the methyl groups to an acetoxymethyl.[513]

Simple alkanes can be converted to esters with dialkyloxiranes. Cyclic alkanes are oxidized to alcohols with dimethyl dioxirane.[514] Cyclohexane was converted to cyclohexyl trifluoroacetate with di(trifluoromethyl) dioxirane and trifluoroacetic anhydride[515] and also with RuCl₃/MeCO₃H/CF₃CO₂H.[516] Dimethyl dioxirane converts alkanes to alcohols in some cases.[517] Adamantane is converted to adamantyl alcohol with DDQ (p. 1710) and triflic acid.[518] The mechanism of oxygen insertion into alkanes has been examined.[519]

Benzylic methylene groups are more readily oxidized to benzylic alcohols when compared to simple alkanes. Typical reagents include manganese–salen and PhIO[520] or peroxides.[521] α-Hydroxy ethers are also generated by reaction of this regents with ethers.[522] N-Benzyl phthalimide reacts with NBS, NaOAc, and acetic

[505]Hayashi, Y.; Yamaguchi, J.; Sumiya, T.; Hibino, K.; Shoji, M. *J. Org. Chem.* **2004**, *69*, 5966; Hayashi, Y.; Yamaguchi, J.; Sumiya, T.; Shoji, M. *Angew. Chem. Int. Ed.* **2004**, *43*, 1112.
[506]Bøgevig, A.; Sundén, H.; Córdova, A. *Angew. Chem. Int. Ed.* **2004**, *43*, 1109.
[507]Hayashi, Y.; Yamaguchi, J.; Hibino, K.; Shoji, M. *Tetrahedron Lett.* **2003**, *44*, 8293.
[508]Chemla, F.; Julia, M.; Uguen, D. *Bull. Soc. Chim. Fr.* **1993**, *130*, 547; **1994**, *131*, 639.
[509]Stanković, S.; Espenson, J.H. *J. Org. Chem.* **1998**, *63*, 4129.
[510]Stanković, S.; Espenson, J.H. *J. Org. Chem.* **2000**, *65*, 5528.
[511]Demir, A.S.; Jeganathan, A. *Synthesis* **1992**, 235.
[512]Punniyamurthy, T.; Katsuki, T. *Tetrahedron* **1999**, *55*, 9439.
[513]Desai, L.; Hull, K.L.; Sanford, M.S. *J. Am. Chem. Soc.* **2004**, *126*, 9542.
[514]Curci, R.; D'Accolti, L.; Fusco, C. *Tetrahedron Lett.* **2001**, *42*, 7087.
[515]Asensio, G.; Mello, R.; González-Nuñez, M.E.; Castellano, G.; Corral, J. *Angew. Chem. Int. Ed.* **1996**, *35*, 217.
[516]Murahashi, S.; Oda, Y.; Komiya, N.; Naota, T. *Tetrahedron Lett.* **1994**, *35*, 7953; Komiya, N.; Noji, S.; Murahashi, S.-I. *Chem. Commun.* **2001**, 65.
[517]Murray, R.W.; Gu, D. *J. Chem. Soc. Perkin Trans. 2* **1994**, 451.
[518]Tanemura, K.; Suzuki, T.; Nishida, Y.; Satsumabayashi, K.; Horaguchi, T. *J. Chem. Soc., Perkin Trans. 1* **2001**, 3230.
[519]Freccero, M.Gandolfi, R.; Sarzi-Amadé, M.; Rastelli, A. *Tetrahedron* **2001**, *57*, 9843.
[520]Hamada, T.; Irie, R.; Mihara, J.; Hamachi, K.; Katsuki, T. *Tetrahedron* **1998**, *54*, 10017; Hamachi, K.; Irie, R.; Katsuki, T. *Tetrahedron Lett.* **1996**, *37*, 4979.
[521]Kawasaki, K.; Tsumura, S.; Katsuki, T. *Synlett* **1995**, 1245.
[522]Miyafuji, A.; Katsuki, T. *Synlett* **1997**, 836.

acid to give N-(α-acetoxybenzyl)phthalimide.[523] Methanesulfonic acid and CuO converts ketones to α-mesyloxy (—OMs) ketones[524] and PhI(OH)OTs converts ketones to α-tosyloxy (—OTs) ketones.[525] Aryl methyl carbinols ArCH(OH)Me react with polymer-bound hypervalent iodine complexes, (polymer)–I(OH)OTs, to give a homobenzylic tosylate, ArCH(OH)CH$_2$OTs.[526] Similar oxidation to an acetoxy benzyl derivative was accomplished with PhI(OAc)$_2$ in acetic acid with a palladium catalyst,[527] and with PhI(OH)OTs in aq. DMSO.[528] With minimal water, cerium (IV) triflate converts benzylic arenes to benzylic alcohols, although the major product is the ketone when >15% of water is present.[529]

Allylic hydroxylation[530] with selenium dioxide often gives aldehydes (**19-17**), but in the presence of acetic anhydride and oxygen, SeO$_2$ converts alkenes to homoallylic acetates as the major product, C=C—C—C → C=C—C—C—OAc.[531] Allylic benzyloxylation occurs when an alkene is treated with t-BuOOCOPh and a Cu—Na zeolite,[532] a copper catalyst,[533] or with a chiral copper catalyst to give modest enantioselectivity.[534] Allylic methylene groups can be converted to ester (—CH—OCOR) derivatives in a similar manner using copper triflate.[535] Cupric acetate has also been used,[536] as well as Cu$_2$O.[537] Acyl peroxides have been used as well.[538] α-Acetoxylation of allylic alkenes can proceed with allylic rearrangement.[539]

Hydroxylation can be accomplished using enzymatic systems. In the presence of *Bacillus megaterium* and oxygen, cyclohexane is converted to cyclohexanol.[540] Allylic oxidation to an allylic alcohol was accomplished with cultured cells of *Gossypium hirsutum*.[541] Benzylic arenes are converted to the corresponding α-hydroxy compound by treatment with the enzymes of *Bacillus megaterium*, with

[523]Cho, S.-D.; Kim, H.-J.; Ahn, C.; Falck, J.R.; Shin, D.-S. *Tetrahedron Lett.* **1999**, *40*, 8215.

[524]Lee, J.C.; Choi, Y. *Tetrahedron Lett.* **1998**, *39*, 3171.

[525]Tuncay, A.; Dustman, J.A.; Fisher, G.; Tuncay, C.I.; Suslick, K.S. *Tetrahedron Lett.* **1992**, *33*, 7647.

[526]Abe, S.; Sakuratani, K.; Togo, H. *J. Org. Chem.* **2001**, *66*, 6174.

[527]Dick, A.R.; Hull, K.L.; Sanford, M.S. *J. Am. Chem. Soc.* **2004**, *126*, 2300.

[528]Xie, Y.-Y.; Chen, Z.-C. *Synth. Commun.* **2002**, *32*, 1875.

[529]Laali, K.K.; Herbert, M.; Cushnyr, B.; Bhatt, A.; Terrano, D. *J. Chem. Soc., Perkin Trans. 1* **2001**, 578.

[530]For a review, see Andrus, M.B.; Lashley, J.C. *Tetrahedron* **2002**, *58*, 845.

[531]Koltun, E.S.; Kass, S.R. *Synthesis* **2000**, 1366.

[532]Carloni, S.; Frullanti, B.; Maggi, R.; Mazzacani, A.; Bigi, F.; Sartori, G. *Tetrahedron Lett.* **2000**, *41*, 8947.

[533]LeBras, J.; Muzart, J. *Tetrahedron Lett.* **2002**, *43*, 431; LeBras, J.; Muzart, J. *Tetrahedron Asymmetry* **2003**, *14*, 1911; Fache, F.; Piva, O. *Synlett* **2002**, 2035.

[534]Lee, W.-S.; Kwong, H.-L.; Chan, H.-L.; Choi, W.-L.; Ng, L.-Y. *Tetrahedron Asymmetry* **2001**, *12*, 1007.

[535]Sekar, G.; Datta Gupta, A.; Singh, V.K. *J. Org. Chem.* **1998**, *63*, 2961; Howell, A.R.; Fan, R.; Troung, A. *Tetrahedron Lett.* **1996**, *37*, 8651; Kohmura, Y.; Katsuki, T. *Tetrahedron Lett.* **2000**, *41*, 3941.

[536]Södergren, M.J.; Andersson, P.G. *Tetrahedron Lett.* **1996**, *37*, 7577; Rispens, M.T.; Zondervan, C.; Feringa, B.L. *Tetrahedron Asymmetry*, **1995**, *6*, 661.

[537]Levina, A.; Muzart, J. *Tetrahedron Asymmetry*, **1995**, *6*, 147.

[538]Andrus, M.B.; Argade, A.B.; Chen, X.; Pamment, M.G. *Tetrahedron Lett.* **1995**, *36*, 2945; Gokhale, A.S.; Minidis, A.B.E.; Pfaltz, A. *Tetrahedron Lett.* **1995**, *36*, 1831.

[539]Chen, M.S.; White, M.C. *J. Am. Chem. Soc.* **2004**, *126*, 1346.

[540]Adam, W.; Lukacs, Z.; Saha-Möller, C.R.; Weckerle, B.; Schreier, P. *Eur. J. Org. Chem.* **2000**, 2923.

[541]Hamada, H.; Tanaka, T.; Furuya, T.; Takahata, H.; Nemoto, H. *Tetrahedron Lett.* **2001**, *42*, 909.

modest enantioselectivity.[542] Cyclic amines react with *Pseudomonas oleovorans* GPo1 to give hydroxy amines; *N*-benzylpyrrolidine is converted to 3-hydroxy *N*-benzylpyrrolidine.[543] *Sphingomonas* sp. HXN-200 gives similar results.[544] In a similar manner, lactams are converted to the corresponding 3-hydroxy lactam with *sphingomonas* sp. HXN-200.[545] *N*-Benzyl piperidine is converted to the 4-hydroxy derivative under the same conditions. [546] The reaction of tetradecanoic acid with the α-oxidase from *Pisum sativum*, in the presence of molecular oxygen, gives 2(*R*)-hydroxytetradecanoic acid with high asymmetric induction.[547]

19-16 Oxidation of Methylene to Heteroatom Functional Groups Other Than Oxygen or Carbonyl

Amino (or amido) -de-dihydro-bisubstitution

$$\underset{R\diagup\overset{\displaystyle H\diagdown\diagup H}{C}\diagdown R^1}{\ } \longrightarrow \underset{R\diagup\overset{\displaystyle H\diagdown\diagup NHR^2}{C}\diagdown R^1}{\ }$$

α-Amination or amidation of a CH unit is possible in some cases. Cyclic alkanes are converted to the *N*-alkyl *N*-tosylamine with PhI=NTs and a copper complex.[548] Benzylic CH, such as in ethylbenzene, is oxidized with PhI(OAc)$_2$ in the presence of TsNH$_2$ and a fluorinated manganese porphyrin to give the corresponding *N*-tosy-lamine, PhCHMe(NHTs).[549] Alkenes with an allylic CH react with PhI=NTs and a ruthenium catalysts to give an allylic *N*-tosylamine.[550] When an α-keto ester reacts with DEAD (diethyl azodicarboxylate) and a chiral copper complex, an α- carba-mate is formed, RCH(NHCO$_2$Et)C(=O)CO$_2$Et, with modest enantioselectivity.[551]

Similar reactions are possible, in some cases, to produce sulfur containing compounds.

Sulfo-de-dihydro-bisubstitution

$$\underset{R\diagup\overset{\displaystyle H\diagdown\diagup H}{C}\diagdown R^1}{\ } \longrightarrow \underset{R\diagup\overset{\displaystyle H\diagdown\diagup SO_3H}{C}\diagdown R^1}{\ }$$

Cyclic alkanes are converted to the corresponding alkylsulfonic acid with SO$_2$/O$_2$ and a vanadium catalyst.[552]

[542]Adam, W.; Lukacs, Z.; Harmsen, D.; Saha-Möller, C.R.; Schreier, P. *J. Org. Chem.* **2000**, *65*, 878.
[543]Li, Z.; Feiten, H.-J.; van Beilen, J.B.; Duetz, W.; Witholt, B. *Tetrahedron Asymmetry* **1999**, *10*, 1323.
[544]Li, Z.; Feiten, H.-J.; Chang, D.; Duetz, W.A.; Beilen, J.B.; Witholt, B. *J. Org. Chem.* **2001**, *66*, 8424.
[545]Chang, D.; Witholt, B.; Li, Z. *Org. Lett.* **2000**, *2*, 3949.
[546]Chang, D.; Feiten, H.-J.; Engesser, K.-H.; van Beilen, J.; Witholt, B.; Li, Z. *Org. Lett.* **2002**, *4*, 1859.
[547]Adam, W.; Boland, W.; Hartmann-Schreier, J.; Humpf, H.-U.; Lazarus, M.; Saffert, A.; Saha-Möller, C.R.; Schreier, P. *J. Am. Chem. Soc.* **1998**, *120*, 11044.
[548]Díaz-Requejo, M.M.; Belderraín, T.R.; Nicasio, M.C.; Trofimenko, S.; Pérez, P.J. *J. Am. Chem. Soc.* **2003**, *125*, 12078.
[549]Yu, X.-Q.; Huang, J.-S.; Zhou, X.-G.; Che, C.-M. *Org. Lett.* **2000**, *2*, 2233.
[550]Au, S.-M.; Huang, J.-S.; Che, C.-M.; Yu, W.-Y. *J. Org. Chem.* **2000**, *65*, 7858.
[551]Juhl, K.; Jørgensen, K.A. *J. Am. Chem. Soc.* **2002**, *124*, 2420.
[552]Ishii, Y.; Matsunaka, K.; Sakaguchi, S. *J. Am. Chem. Soc.* **200**, *122*, 7390.

19-17 Oxidation of Methylene to Carbonyl

Oxo-de-dihydro-bisubstitution

$$ \underset{H \quad H}{\overset{O}{\underset{R}{\overset{\parallel}{C}}}\overset{}{\underset{}{C}}R^1} \quad \xrightarrow{\text{SeO}_2} \quad \underset{O}{\overset{O}{\underset{R}{\overset{\parallel}{C}}}\overset{}{\underset{}{\overset{\parallel}{C}}}R^1} $$

Methyl or methylene groups α to a carbonyl can be oxidized with selenium dioxide to give, respectively, α-keto aldehydes (see **19-18**) and α-diketones.[553] The reaction can also be carried out a to an aromatic ring or to a double bond, although in the latter case, hydroxylation (see **19-14**) is the more common result. Selenium dioxide, SeO_2, is the reagent most often used, but the reaction has also been carried out with N_2O_3 and other oxidizing agents,[554] including hypervalent iodine compounds.[555] Sodium nitrite/HCl oxidizes cyclic ketones to the diketone.[556] Substrates most easily oxidized contain two aryl groups on CH_2, and these substrates can be oxidized with many oxidizing agents (see **19-11**). The benzylic position of arenes have been oxidized to alkyl aryl ketones with several oxidizing agents, including CrO_3–acetic acid,[557] the Jones reagent,[558] CrO_3 on silica,[559] pyridinium chlorochromate,[560] DDQ,[561] CrO_2Cl_2 with ultrasound,[562] $KMnO_4$ supported on MnO_2,[563] $KMnO_4$ on alumina with microwave irradiation[564] or on Montmorillonite K10 with either ultrasound or microwave irradiation,[565] $KMnO_4$/$CuSO_4$ neat[566] or with ultrasound,[567] $NaBrO_3$/CeO_2,[568] manganese–salen/PhIO,[569] tert-butylhydroperoxide and a ruthenium catalyst,[570] $Ru(OH)_x$–Al_2O_3 and O_2,[571] hydrogen peroxide with a copper catalyst,[572] as well as with SeO_2. The combination of O_2 and

[553]For reviews of oxidation by SeO_2, see Krief, A.; Hevesi, L. *Organoselenium Chemistry I*, Springer, NY, **1988**, pp. 115–180; Krongauz, E.S. *Russ. Chem. Rev.* **1977**, *46*, 59; Rabjohn, N. *Org. React.* **1976**, *24*, 261; Trachtenberg, E.N., in Augustine, R.L.; Trecker, D.J. *Oxidation*, Marcel Dekker, NY, pp. 119–187.

[554]For other methods, see Wasserman, H.H.; Ives, J.L. *J. Org. Chem.* **1978**, *43*, 3238; **1985**, *50*, 3573; Rao, D.V.; Stuber, F.A.; Ulrich, H. *J. Org. Chem.* **1979**, *44*, 456.

[555]Lee, J.C.; Park, H.-J.; Park, J.Y. *Tetrahedron Lett.* **2002**, *43*, 5661.

[556]Rüedi, G.; Oberli, M.A.; Nagel, M.; Weymuth, C.; Hansen, H.-J. *Synlett* **2004**, 2315.

[557]For example, see Harms, W.M.; Eisenbraun, E.J. *Org. Prep. Proced. Int.* **1972**, *4*, 67.

[558]Rangarajan, R.; Eisenbraun, E.J. *J. Org. Chem.* **1985**, *50*, 2435.

[559]Borkar, S.D.; Khadilkar, B.M. *Synth. Commun.* **1999**, *29*, 4295.

[560]Rathore, R.; Saxena, N.; Chandrasekaran, S. *Synth. Commun.* **1986**, *16*, 1493.

[561]Lee, H.; Harvey, R.G. *J. Org. Chem.* **1988**, *53*, 4587.

[562]Luzzio, F.A.; Moore, W.J. *J. Org. Chem.* **1993**, *58*, 512.

[563]Wei, H.-X.; Jasoni, R.L.; Shao, H.; Hu, J.; Paré, P.W. *Tetrahedron* **2004**, *60*, 11829.

[564]Oussaid, A.; Loupy, A. *J. Chem. Res. (S)* **1997**, 342.

[565]Shaabani, A.; Bazgir, A.; Teimouri, F.; Lee, D.G. *Tetrahedron Lett.* **2002**, *43*, 5165.

[566]Shaabani, A. Lee, D.G. *Tetrahedron Lett.* **2001**, *42*, 5833.

[567]Meč iarova, M.; Toma, S.; Heribanová, A. *Tetrahedron* **2000**, *56*, 8561.

[568]Shi, Q.-Z.; Wang, J.-G.; Cai, K. *Synth. Commun.* **1999**, *29*, 1177.

[569]Komiya, N.; Noji, S.; Murahashi, S.-I. *Tetrahedron Lett.* **1998**, *39*, 7921; Lee, N.H.; Lee, C.-S.; Jung, D.-S. *Tetrahedron Lett.* **1998**, *39*, 1385.

[570]Murahashi, S.-I.; Komiya, N.; Oda, Y.; Kuwabara, T.; Naota, T. *J. Org. Chem.* **2000**, *65*, 9186.

[571]Kamata, K.; Kasai, J.; Yamaguchi, K.; Mizuno, N. *Org. Lett.* **2004**, *6*, 3577.

[572]Velusamy, S.; Punniyamurthy, T. *Tetrahedron Lett.* **2003**, *44*, 8955.

mcpba oxidizes benzylic arenes to aryl ketones.[573] Note that benzyl methyl ether is oxidized to methyl benzoate with $KMnO_4$ in the presence of benzyltriethylammonium chloride.[574]

Alkenes of the form $C=C-CH_2$ (an allylic position) have been oxidized to α,β-unsaturated ketones[575] by sodium dichromate in $HOAc-Ac_2O$, by t-BuOOH and chromium compounds,[576] t-BuOOH and a palladium catalyst,[577] or a rhodium catalyst,[578] as well as electrolytically.[579] Oxygen, $MeSO_3H$ a palladium catalysts and a molybdobanadophosphate catalyst convert cyclic alkenes to saturated cyclic ketones.[580] Thallium(III) nitrate in aqueous acetic acid converts allylic alkenes to the corresponding saturated ketone, even in the presence of a primary alcohol elsewhere in the molecule.[581] The propargylic position of internal alkynes are oxidized to give propargylic ketones with an iron catalyst,[582] or with O_2/t-BuOOH in the presence of $CuCl_2 \cdot H_2O$.[583]

Cyclic amines are oxidized to lactams using a mixture of $RuCl_3$ and $NaIO_4$.[584] Lactams are also formed using $KMnO_4$ with benzyltriethylammonium chloride.[585] Tertiary amines are converted to amides[586] and cyclic tertiary amines can be converted to lactams by oxidation with Hg^{II}–EDTA complex in basic solution.[587] Lactams, which need not be N-substituted, can be converted to cyclic imides by oxidation with a hydroperoxide or peroxyacid and an Mn(II) or Mn(III) salt.[588] Lactams are oxidized to cyclic imides with oxygen and $Co(OAc)_2$ in the presence N-hydroxysuccinimide.[589]

Ethers in which at least one group is primary alkyl can be oxidized to the corresponding carboxylic esters in high yields with ruthenium tetroxide.[590] Molecular

[573]Ma, D.; Xia, C.; Tian, H. *Tetrahedron Lett.* **1999**, *40*, 8915.

[574]Markgraf, J.H.; Choi, B.Y. *Synth. Commun.* **1999**, *29*, 2405.

[575]For a review, see Muzart, J. *Bull. Soc. Chim. Fr.* **1986**, 65. For a list of reagents, with references, see Larock, R.C. *Comprehensive Organic Transformations*, 2nd ed., Wiley-VCH, NY, **1999**, pp. 1207–1210.

[576]Pearson, A.J.; Chen, Y.; Han, G.R.; Hsu, S.; Ray, T. *J. Chem. Soc. Perkin Trans. 1* **1985**, 267; Muzart, J. *Tetrahedron Lett.* **1987**, *28*, 2131; Chidambaram, N.; Chandrasekaran, S. *J. Org. Chem.* **1987**, *52*, 5048.

[577]Yu, J.-Q.; Corey, E.J. *J. Am. Chem. Soc.* **2003**, *125*, 3232.

[578]Catino, A.J.; Forslund, R.E.; Doyle, M.P. *J. Am. Chem. Soc.* **2004**, *126*, 13622.

[579]Madurro, J.M.; Chiericato Jr., G.; De Giovani, W.F.; Romero, J.R. *Tetrahedron Lett.* **1988**, *29*, 765.

[580]Kishi, A.; Higashino, T.; Sakaguchi, S.; Ishii, Y. *Tetrahedron Lett.* **2000**, *41*, 99.

[581]Ferraz, H.M.C.; Longo, Jr., L.S.; Zukerman-Schpector, J. *J. Org. Chem.* **2002**, *67*, 3518.

[582]Pérollier, C.; Sorokin, A.B. *Chem. Commun.* **2002**, 1548.

[583]Li, P.; Fong, W.M.; Chao, L.C.F.; Fung, S.H.C.; Williams, I.D. *J. Org. Chem.* **2001**, *66*, 4087.

[584]Sharma, N.K.; Ganesh, K.N. *Tetrahedron Lett.* **2004**, *45*, 1403; Zhang, X.; Schmitt, A.C.; Jiang, W. *Tetrahedron Lett.* **2001**, *42*, 5335.

[585]Markgraf, J.H.; Stickney, C.A. *J. Heterocyclic Chem.* **2000**, *37*, 109.

[586]Markgraf, J.H.; Sangani, P.K.; Finkelstein, M. *Synth. Commun.* **1995**, *27*, 1285.

[587]Wenkert, E.; Angell, E.C. *Synth. Commun.* **1988**, *18*, 1331.

[588]Doumaux Jr., A.R.; McKeon, J.E.; Trecker, D.J. *J. Am. Chem. Soc.* **1969**, *91*, 3992; Doumaux Jr., A.R.; Trecker, D.J. *J. Org. Chem.* **1970**, *35*, 2121.

[589]Minisci, F.; Punta, C.; Recupero, F.; Fontana, F.; Pedulli, G.F. *J. Org. Chem.* **2002**, *67*, 2671.

[590]Bakke, J.M.; Frøhaug, Az. *Acta Chem. Scand. B* **1995**, *49*, 615; Lee, D.G.; van den Engh, M., in Trahanovsky, W.S. *Oxidation in Organic Chemistry*, pt. B, Academic Press, NY, **1973**, pp. 222–225; Smith III, A.B.; Scarborough, Jr., R.M. *Synth. Commun.* **1980**, *10*, 205; Carlsen, P.H.J.; Katsuki, T.; Martin, V.S.; Sharpless, K.B. *J. Org. Chem.* **1981**, *46*, 3936.

oxygen with a binuclear copperII complex[591] or $PdCl_2/CuCl_2/CO$[592] also converts ethers to esters. In a variation, benzyl *tert*-butyl ethers are oxidized to benzaldehyde derivatives with NO and *N*-hydroxysuccinimide.[593] Cyclic ethers are oxidized to lactones.[594] Cyclic ethers are oxidized to lactones with CrO_3/Me_3SiONO_2.[595] Lactones are also formed from cyclic ethers with $NaBrO_3$–$KHSO_4$ in water.[596] The reaction has also been accomplished with CrO_3 in sulfuric acid,[597] and with benzyl-triethylammonium permanganate.[598]

Two mechanisms have been suggested for the reaction with SeO_2. One of these involves a selenate ester of the enol:[599]

In the other proposed mechanism,[600] the principal intermediate is α β-ketoseleninic acid

and a selenate ester is not involved.

It has proved possible to convert CH_2 to C=O groups, even if they are not near any functional groups, indirectly, by the remote oxidation method of Breslow[62] (see **19-2**). In a typical example, the keto ester **31** was irradiated to give the hydroxy lactone **32**, which was dehydrated to **33**. Ozonolysis of **33** gave the diketo ester

[591]Minakata, S.; Imai, E.; Ohshima, Y.; Inaki, K.; Ryu, I.; Komatsu, M.; Ohshiro, Y. *Chem. Lett.* **1996**, 19.

[592]Miyamoto, M.; Minami, Y.; Ukaji, Y.; Kinoshita, H.; Inomata, K. *Chem. Lett.* **1994**, 1149.

[593]Eikawa, M.; Sakaguchi, S.; Ishii, Y. *J. Org. Chem.* **1999**, *64*, 4676.

[594]For an example using titanium silicate/H_2O_2, see Sasidharan, M.; Suresh, S.; Sudalai, A. *Tetrahedron Lett.* **1995**, *36*, 9071. For an example in which a bicyclic ether was converted to a monocyclic lactone, see Ferraz, H.M.C.; Longo Jr., L.S. *Org. Lett.* **2003**, *5*, 1337.

[595]Shahi, S.P.; Gupta, A.; Pitre, S.V.; Reddy, M.V.R.; Kumareswaran, R.; Vankar, Y.D. *J. Org. Chem.* **1999**, *64*, 4509.

[596]Metsger, L.; Bittner, S. *Tetrahedron* **2000**, *56*, 1905.

[597]Henbest, H.B.; Nicholls, B. *J. Chem. Soc.* **1959**, 221, 227; Harrison, I.T.; Harrison, S. *Chem. Commun.* **1966**, 752.

[598]Schmidt, H.; Schäfer, H.J. *Angew. Chem. Int. Ed.* **1979**, *18*, 69.

[599]Corey, E.J.; Schaefer, J.P. *J. Am. Chem. Soc.* **1960**, *82*, 918.

[600]Sharpless, K.B.; Gordon, K.M. *J. Am. Chem. Soc.* **1976**, *98*, 300.

34, in which the C-14 CH$_2$ group of **31** has been oxidized to a C=O group.[601] The reaction was not completely regioselective: **34** comprised $\sim$ 60% of the product, with the remainder consisting of other compounds in which the keto group was located at C-12, C-15, and other positions along the carbon chain. Greater regioselectivity was achieved when the aromatic portion was connected to the chain at two positions.[602] In the method so far described, the reaction takes place because one portion of a molecule (the benzophenone moiety) abstracts hydrogen from another portion of the same molecule, that is, the two portions are connected by a series of covalent bonds. However, the reaction can also be carried out where the two reacting centers are actually in different molecules, providing the two molecules are held together by hydrogen bonding. For example, one of the CH$_2$ groups of n-hexadecanol monosuccinate, CH$_3$(CH$_2$)$_{14}$CH$_2$OCOCH$_2$CH$_2$COOH, was oxidized to a C=O group by applying the above procedure to a mixture of it and benzophenone-4-carboxylic acid p-PhCOC$_6$H$_4$COOH in CCl$_4$.[603]

Other remote oxidations[604] have also been reported. Among these are conversion of aryl ketones ArCO(CH$_2$)$_3$R to 1,4-diketones ArCO(CH$_2$)$_2$COR by photoirradiation in the presence of such oxidizing agents as K$_2$Cr$_2$O$_7$ or KMnO$_4$,[605] and conversion of alkyl ketones, RCO(CH$_2$)$_3$R', to 1,3- and 1,4-diketones with Na$_2$S$_2$O$_8$ and FeSO$_4$.[606] 2-Octanol was oxidized to give 2-propyl-5-methyl γ-butyrolactone with lead tetraacetate in a CO atmosphere.[607]

[601]Breslow, R.; Rothbard, J.; Herman, F.; Rodriguez, M.L. *J. Am. Chem. Soc.* **1978**, *100*, 1213.

[602]Breslow, R.; Rajagopalan, R.; Schwarz, J. *J. Am. Chem. Soc.* **1981**, *103*, 2905.

[603]Breslow, R.; Scholl, P.C. *J. Am. Chem. Soc.* **1971**, *93*, 2331. See also, Breslow, R.; Heyer, D. *Tetrahedron Lett.* **1983**, *24*, 5039.

[604]See also Beckwith, A.L.J.; Duong, T. *J. Chem. Soc. Chem. Commun.* **1978**, 413.

[605]Mitani, M.; Tamada, M.; Uehara, S.; Koyama, K. *Tetrahedron Lett.* **1984**, *25*, 2805. For an alternative photochemical procedure, see Negele, S.; Wieser, K.; Severin, T. *J. Org. Chem.* **1998**, *63*, 1138.

[606]Nikishin, G.I.; Troyansky, E.I.; Lazareva, M.I. *Tetrahedron Lett.* **1984**, *25*, 4987.

[607]Tsunoi, S.; Ryu, I.; Okuda, T.; Tanaka, M.; Komatsu, M.; Sonoda, N. *J. Am. Chem. Soc.* **1998**, *120*, 8692. Also see, Tsunoi, S.; Ryu, I.; Sonoda, N. *J. Am. Chem. Soc.* **1994**, *116*, 5473.

It is possible to perform the conversion $CH_2 \rightarrow C=O$ on an alkane, with no functional groups at all, although the most success has been achieved with substrates in which all CH_2 groups are equivalent, such as unsubstituted cycloalkanes. One method uses H_2O_2 and bis(picolinato)iron(II). Hydrogen peroxide and trifluoroacetic acid has also been used for oxidation of alkanes.[608] With this method, cyclohexane was converted with 72% efficiency to give 95% cyclohexanone and 5% cyclohexanol.[609] This was also accomplished with $BaRu(O)_2(OH)_3$.[610] The same type of conversion, with lower yields (20–30%), has been achieved with the *Gif system*.[611] There are several variations. One consists of pyridine–acetic acid, with H_2O_2 as oxidizing agent and tris(picolinato)iron(III) as catalyst.[612] Other Gif systems use O_2 as oxidizing agent and zinc as a reductant.[613] The selectivity of the Gif systems toward alkyl carbons is $CH_2 > CH \geq CH_3$, which is unusual, and shows that a simple free-radical mechanism (see p. 942) is not involved.[614] Another reagent that can oxidize the CH_2 of an alkane is methyl(trifluoromethyl)dioxirane, but this produces CH—OH more often than C=O (see **19-14**; **19-15**).[615] Simple unfunctionalized alkanes are oxidized to esters when treated with $CBr_4/2$ $AlBr_3$ and CO, but in very low yield.[616] Cyclic alkanes are oxidized to a mixture of the alcohol and the ketone with $PhI(OAc)_2$ and a manganese complex in an ionic liquid.[617] Oxidation of cyclic alkanes to cyclic ketones was accomplished using a ruthenium catalyst.[618]

OS **I**, 266; **II**, 509; **III**, 1, 420, 438; **IV**, 189, 229, 579; **VI**, 48; **IX**, 396. Also see, OS **IV**, 23.

[608]Camaioni, D.M.; Bays, J.T.; Shaw, W.J.; Linehan, J.C.; Birnbaum, J.C. *J. Org. Chem.* **2001**, *66*, 789.

[609]Sheu, C.; Richert, S.A.; Cofré, P.; Ross Jr., B.; Sobkowiak, A.; Sawyer, D.T.; Kanofsky, J.R. *J. Am. Chem. Soc.* **1990**, *112*, 1936. See also, Sheu, C.; Sobkowiak, A.; Jeon, S.; Sawyer, D.T. *J. Am. Chem. Soc.* **1990**, *112*, 879; Tung, H.; Sawyer, D.T. *J. Am. Chem. Soc.* **1990**, *112*, 8214.

[610]Lau, T.-C.; Mak, C.-K. *J. Chem. Soc. Chem. Commun.* **1993**, 766.

[611]Named for Gif-sur-Yvette, France, where it was discovered. See Schuchardt, U.; Jannini, M.J.D.M.; Richens, D.T.; Guerreiro, M.C.; Spinacé, E.V. *Tetrahedron* **2001**, *57*, 2685.

[612]About-Jaudet, E.; Barton, D.H.R.; Csuhai, E.; Ozbalik, N. *Tetrahedron Lett.* **1990**, *31*, 1657. Also see, Minisci, F.; Fontana, F.; Araneo, S.; Recupero, F. *Tetrahedron Lett.* **1994**, *35*, 3759; Barton, D.H.R.; Bévière, S.D.; Chavasiri, W.; Doller, D.; Hu, B. *Tetrahedron Lett.* **1992**, *33*, 5473. For a review of the mechanism, see Barton, D.H.R. *Chem. Soc. Rev.* **1996**, *25*, 237.

[613]See Barton, D.H.R.; Csuhai, E.; Ozbalik, N. *Tetrahedron* **1990**, *46*, 3743, and references cited therein.

[614]Barton, D.H.R.; Csuhai, E.; Doller, D.; Ozbalik, N.; Senglet, N. *Tetrahedron Lett.* **1990**, *31*, 3097. For mechanistic studies, see Barton, D.H.R.; Doller, D.; Geletii, Y.V. *Tetrahedron Lett.* **1991**, *32*, 3911, and references cited therein; Knight, C.; Perkins, M.J. *J. Chem. Soc. Chem. Commun.* **1991**, 925. Also see, Minisci, F.; Fontana, F. *Tetrahedron Lett.* **1994**, *35*, 1427; Barton, D.H.R.; Hill, D.R. *Tetrahedron Lett.* **1994**, *35*, 1431.

[615]Mello, R.; Fiorentino, M.; Fusco, C.; Curci, R. *J. Am. Chem. Soc.* **1989**, *111*, 6749; D'Accolti, L.; Dinoi, A.; Fusco, C.; Russo, A.; Curci, R. *J. Org. Chem.* **2003**, *68*, 7806.

[616]Akhrem, I.; Orlinkov, A.; Afanas'eva, L.; Petrovskii, P.; Vitt, S. *Tetrahedron Lett.* **1999**, *40*, 5897.

[617]In bmim PF6, 1-butyl-3-methylimidazolium hexafluorophosphate: Li, Z.; Xiu, C.-G.; Xu, C.-Z. *Tetrahedron Lett.* **2003**, *44*, 9229.

[618]Che, C.-M.; Cheng, K.-W.; Chan, M.C.W.; Lau, T.-C.; Mak, C.-K. *J. Org. Chem.* **2000**, *65*, 7996.

19-18   Oxidation of Arylmethanes to Aldehydes

Oxo-de-dihydro-bisubstitution

$$ArCH_3 \xrightarrow{\quad CrO_2Cl_2 \quad} ArCHO$$

Methyl groups on an aromatic ring can be oxidized to the aldehyde stage by several oxidizing agents. The reaction is a special case of **19-17**. When the reagent is chromyl chloride (CrO_2Cl_2), the reaction is called the *Étard reaction*[619] and the yields are high.[620] Another oxidizing agent is a mixture of CrO_3 and Ac_2O. In this case, the reaction stops at the aldehyde stage because the initial product is $ArCH(OAc)_2$ (an acylal), which is resistant to further oxidation. Hydrolysis of the acylal gives the aldehyde.

Among other oxidizing agents[621] that have been used to accomplish the conversion of $ArCH_3$ to $ArCHO$ are ceric ammonium nitrate,[622] ceric trifluoroacetate,[623] hypervalent iodoso compounds (see **19-3**),[624] urea–H_2O_2 with micrwoave irradiation,[625] and silver(II) oxide.[626] Oxidation of $ArCH_3$ to carboxylic acids is considered at **19-11**.

Conversion of $ArCH_3$ to $ArCHO$ can also be achieved indirectly by bromination to give $ArCHBr_2$ (**14-1**), followed by hydrolysis (**10-2**).

The mechanism of the Étard reaction is not completely known.[627] An insoluble complex is formed on addition of the reagents, which is hydrolyzed to the aldehyde. The complex is probably a kind of acylal, but the identity of the structure is not fully settled, although many proposals have been made as to its structure and as to how it is hydrolyzed.

$$\begin{array}{c} \qquad\quad O-CrCl_2OH \\ \qquad\quad / \\ Ph-C-H \\ \qquad\quad \backslash \\ \qquad\quad O-CrCl_2OH \end{array}$$

35

It is known that $ArCH_2Cl$ is not an intermediate (see **19-20**), since it reacts only very slowly with chromyl chloride. Magnetic susceptibility measurements[628]

---

[619]The name Étard reaction is often applied to any oxidation with chromyl chloride, for example, oxidation of glycols (**19-7**), alkenes (**19-10**), and so on.

[620]For a review, see Hartford, W.H.; Darrin, M. *Chem. Rev.* **1958**, *58*, 1, see pp. 25–53.

[621]For a review of the use of oxidizing agents that are regenerated electrochemically, see Steckhan, E. *Top. Curr. Chem.* **1987**, *142*, 1; 12–17.

[622]Trahanovsky, W.S.; Young, L.B. *J. Org. Chem.* **1966**, *31*, 2033; Radhakrishna Murti, P.S.; Pati, S.C. *Chem. Ind. (London)* **1967**, 702; Syper, L. *Tetrahedron Lett.* **1967**, 4193. For oxidation with ceric ammonium nitrate and KBrO3, see Ganin, E.; Amer, I. *Synth. Commun.* **1995**, *25*, 3149.

[623]Marrocco, M.; Brilmyer, G. *J. Org. Chem.* **1983**, *48*, 1487. See also, Kreh, R.P.; Spotnitz, R.M.; Lundquist, J.T. *J. Org. Chem.* **1989**, *54*, 1526.

[624]Nicolaou, K.C.; Baran, P.S.; Zhong, Y.-L. *J. Am. Chem. Soc.* **2001**, *123*, 3183.

[625]Paul, S.; Nanda, P.; Gupta, R. *Synlett* **2004**, 531.

[626]Syper, L. *Tetrahedron Lett.* **1967**, 4193.

[627]For a review, see Nenitzescu, C.D. *Bull. Soc. Chim. Fr.* **1968**, 1349.

[628]Wheeler, O.H. *Can. J. Chem.* **1960**, *38*, 2137. See also, Makhija, R.C.; Stairs, R.A. *Can. J. Chem.* **1968**, *46*, 1255.

indicate that the complex from toluene is **35**, a structure first proposed by Étard. According to this proposal, the reaction stops after only two hydrogens have been replaced because of the insolubility of **35**. There is a disagreement on how **35** is formed, assuming that the complex has this structure. Both an ionic[629] and a free-radical[630] process have been proposed. An entirely different structure for the complex was proposed by Nenitzescu and co-workers.[631] On the basis of esr studies, they proposed that the complex is $PhCH_2OCrCl_2OCrOCl_2OH$, which is isomeric with **35**. However, this view has been challenged by Wiberg and Eisenthal,[336] who interpret the esr result as being in accord with **35**. Still another proposal is that the complex is composed of benzaldehyde coordinated with reduced chromyl chloride.[632]

OS **II**, 441; **III**, 641; **IV**, 31, 713.

19-19   Oxidation of Aromatic Hydrocarbons to Quinones

Arene-quinone transformation

Condensed aromatic systems (including naphthalenes) can be directly oxidized to quinones by various oxidizing agents.[258,633] Yields are generally not high, although good yields have been reported with ceric ammonium sulfate.[634] Benzene cannot be so oxidized by strong oxidizing agents, but can be electrolytically oxidized to benzoquinone.[635] Naphthalene derivatives, however, are oxidized to naphthoquinones with H_5IO_6 and CrO_3.[636] 1,4-Dimethoxy aromatic compounds are oxidized to para-quinones with an excess of CoF_3 in water–dioxane.[637]

OS **IV**, 698, 757. Also see, OS **II**, 554.

[629]Stairs, R.A. *Can. J. Chem.* **1964**, *42*, 550.

[630]Wiberg, K.B.; Eisenthal, R. *Tetrahedron* **1964**, *20*, 1151. See also, Gragerov, I.P.; Ponomarchuk, M.P. *J. Org. Chem. USSR* **1969**, *6*, 1125.

[631]Necşoiu, I.; Przemetchi, V.; Ghenciulescu, A.; Rentea, C.N.; Nenitzescu, C.D. *Tetrahedron* **1966**, *22*, 3037.

[632]Duffin, H.C.; Tucker, R.B. *Chem. Ind. (London)* **1966**, 1262; *Tetrahedron* **1968**, *24*, 6999.

[633]For reviews, see Naruta, Y.; Maruyama, K., in Patai, S.; Rappoport, Z. *The Chemistry of the Quinoid Compounds*, Vol. 2, pt. 1, Wiley, NY, **1988**, pp. 242–247; Hudlický, M. *Oxidations in Organic Chemistry*, American Chemical Society, Washington, DC, **1990**, pp. 94–96; Haines, A.H. *Methods for the Oxidation of Organic Compounds*, Vol. 1, Academic Press, NY, **1985**, pp. 182–185, 358–360; Thomson, R.H., in Patai, S. *The Chemistry of the Quinoid Compounds*, Vol. 1, pt. 1, Wiley, NY, **1974**, pp. 132–134. See also, Sket, B.; Zupan, M. *Synth. Commun.* **1990**, *20*, 933.

[634]Periasamy, M.; Bhatt, M.V. *Synthesis* **1977**, 330; Balanikas, G.; Hussain, N.; Amin, S.; Hecht, S.S. *J. Org. Chem.* **1988**, *53*, 1007.

[635]See, for example, Ito, S.; Katayama, R.; Kunai, A.; Sasaki, K. *Tetrahedron Lett.* **1989**, *30*, 205.

[636]Yamazaki, S. *Tetrahedron Lett.* **2001**, *42*, 3355.

[637]Tomatsu, A.; Takemura, S.; Hashimoto, K.; Nakata, M. *Synlett* **1999**, 1474.

19-20 Oxidation of Primary Halides and Esters of Primary Alcohols to Aldehydes[638]

Oxo-de-hydro, halo-bisubstitution

$$RCH_2Cl \xrightarrow{\text{Me}_2\text{SO}} RCHO$$

Primary alkyl halides (chlorides, bromides, and iodides) can be oxidized to aldehydes easily and in good yields with dimethyl sulfoxide,[639] in what has been called the *Kornblum reaction*. In Kornblum's original work, the reaction of α-halo ketones with DMSO at elevated temperatures gave good yields of the corresponding glyoxal (an α-keto-aldehyde).[640] If the glyoxal could be removed from the reaction medium by distillation as it was formed, the reaction was very efficient. In many cases, it was difficult to isolate high boiling glyoxals from DMSO. Primary and secondary[641] alkyl iodides or tosylates[642] can be converted to aldehydes or ketones, although they are much less reactive than α-halo ketones. Epoxides[643] have been used to give α-hydroxy ketones or aldehydes.[644] The reaction with tosyl esters is an indirect way of oxidizing primary alcohols to aldehydes (**19-3**). Primary chlorides with DMSO, NaBr, and ZnO give the corresponding aldehyde when heated to 140°C.[645] Primary allylic bromides with a cyano group on the C=C unit are converted to conjugated α-cyano aldehydes with DMSO and NaHCO$_3$ at room temperature.[646]

---

[638]For reviews of the reactions in this section, see Tidwell, T.T. *Org. React.* **1990**, *39*, 297; *Synthesis* **1990**, 857; Haines, A.H. *Methods for the Oxidation of Organic Compounds*, Vol. 2, Academic Press, NY, **1988**, pp. 171–181, 402–406; Durst, T. *Adv. Org. Chem.* **1969**, *6*, 285, see pp. 343–356; Epstein, W.W.; Sweat, F.W. *Chem. Rev.* **1967**, *67*, 247; Moffatt, J.G., in Augustine, R.L.; Trecker, D.J. *Oxidation*, Vol. 2, Marcel Dekker, NY, **1971**, pp. 1–64. For a list of reagents, with references, see Larock, R.C. *Comprehensive Organic Transformations*, 2nd ed., Wiley-VCH, NY, **1999**, pp. 1222–1225.

[639]Nace, H.R.; Monagle, J.J. *J. Org. Chem.* **1959**, *24*, 1792; Kornblum, N.; Jones, W.J.; Anderson, G.J. *J. Am. Chem. Soc.* **1959**, *81*, 4113. This reaction is promoted by microwave irradiation; see Villemin, D.; Hammadi, M. *Synth. Commun.* **1995**, *25*, 3141.

[640]Kornblum, N.; Powers, J.W.; Anderson, G.J.; Jones, W.J.; Larson, H.O.; Levand, O.; Weaver, W.M. *J. Am. Chem. Soc.* **1957**, *79*, 6562.

[641]Baizer, M.M. *J. Org. Chem.*, **1960**, *25*, 670.

[642]Kornblum, N.; Jones, W.J.; Anderson, G.J. *J. Am. Chem. Soc.* **1959**, *81*, 4113.

[643]Epoxides can be converted to α-halo ketones by treatment with bromodimethylsulfonium bromide: Olah, G.A.; Vankar, Y.D.; Arvanaghi, M. *Tetrahedron Lett.* **1979**, 3653.

[644]Cohen, T.; Tsuji, T. *J. Org. Chem.* **1961**, *26*, 1681; Tsuji, T. *Tetrahedron Lett.* **1966**, 2413; Santosusso, T.M.; Swern, D. *Tetrahedron Lett.* **1968**, 4261; *J. Org. Chem.* **1975**, *40*, 2764.

[645]Guo, Z.; Sawyer, R.; Prakash, I. *Synth. Commun.* **2001**, *31*, 667; Guo, Z.; Sawyer, R.; Prakash, I. *Synth. Commun.* **2001**, *31*, 3395.

[646]Ravichandran, S. *Synth. Commun.* **2001**, *31*, 2185.

The mechanism of these DMSO oxidations is probably as follows:[647]

although in some cases the base abstracts a proton directly from the carbon being oxidized, in which case the ylid **37** is not an intermediate. Alkoxysulfonium salts (**36**) have been isolated.[648] This mechanism predicts that secondary compounds should be oxidizable to ketones, and this is the case. In a related procedure for the oxidation of alcohols, the intermediate **36**[649] is formed without the use of DMSO by treating the substrate with a complex generated from chlorine or NCS and dimethyl sulfide.[650]

Another way to oxidize primary alkyl halides to aldehydes is by the use of hexamethylenetetramine followed by water. However, this reaction, called the *Sommelet reaction*,[651] is limited to benzylic halides. The reaction is seldom useful when the R in RCH_2Cl is alkyl. The first part of the reaction is conversion to the amine $ArCH_2NH_2$, which can be isolated. Reaction of the amine with excess hexamethylenetetramine gives the aldehyde. It is this last step that is the actual Sommelet reaction, although the entire process can be conducted without isolation of intermediates. Once the amine is formed, it is converted to an imine ($ArCH_2N=CH_2$) with formaldehyde liberated from the reagent. The key step then follows: transfer of hydrogen from another mole of the arylamine to the imine. This last imine is then hydrolyzed by water to the aldehyde. Alternatively, the benzylamine may transfer hydrogen directly to hexamethylenetetramine. Another method that converts secondary bromides to ketones heads the bromide with $NaIO_4$ in DMF.[652]

Another reagent that convert benzylic halides to aldehydes is pyridine followed by *p*-nitrosodimethylaniline and then water, called the *Kröhnke reaction*. Primary halides and tosylates have been oxidized to aldehydes by trimethylamine *N*-oxide,[653] and by pyridine *N*-oxide with microwave irradiation.[654]

[647]Pfitzner, K.E.; Moffatt, J.G. *J. Am. Chem. Soc.* **1965**, *87*, 5661; Johnson, C.R.; Phillips, W.G. *J. Org. Chem.* **1967**, *32*, 1926; Torssell, K. *Acta Chem. Scand.* **1967**, *21*, 1.

[648]Torssell, K. *Tetrahedron Lett.* **1966**, 4445; Johnson, C.R.; Phillips, W.G. *J. Org. Chem.* **1967**, *32*, 1926; Khuddus, M.A.; Swern, D. *J. Am. Chem. Soc.* **1973**, *95*, 8393.

[649]It has been suggested that in the DCC reaction, **36** is not involved, but the ylid **37** is formed directly from a precursor containing DCC and DMSO: Torssell, K. *Tetrahedron Lett.* **1966**, 4445; Moffatt, J.G. *J. Org. Chem.* **1971**, *36*, 1909.

[650]Vilsmaier, E.; Sprügel, W. *Liebigs Ann. Chem.* **1971**, *747*, 151; Corey, E.J.; Kim, C.U. *J. Am. Chem. Soc.* **1972**, *94*, 7586; *J. Org. Chem.* **1973**, *38*, 1233; McCormick, J.P. *Tetrahedron Lett.* **1974**, 1701; Katayama, S.; Fukuda, K.; Watanabe, T.; Yamauchi, M. *Synthesis* **1988**, 178.

[651]For a review, see Angyal, S.J. *Org. React.* **1954**, *8*, 197.

[652]Das, S.; Panigrahi, A.K.; Maikap, G.C. *Tetrahedron Lett.* **2003**, *44*, 1375.

[653]Franzen, V.; Otto, S. *Chem. Ber.* **1961**, *94*, 1360. For the use of other amine oxides, see Suzuki, S.; Onishi, T.; Fujita, Y.; Misawa, H.; Otera, J. *Bull. Chem. Soc. Jpn.* **1986**, *59*, 3287.

[654]Barbry, D.; Champagne, P. *Tetrahedron Lett.* **1996**, *37*, 7725.

In a clearly related reaction, benzylic bromides are oxidized to aryl carboxylic acids by photolysis in acetone in the presence of mesoporous silica.[655]

OS **II**, 336: **III**, 811; **IV**, 690, 918, 932; **V**, 242, 668, 825, 852, 872. Also see, OS **V**, 689; **VI**, 218.

19-21   Oxidation of Amines or Nitro Compounds to Aldehydes, Ketones, or Dihalides

Oxo-de-hydro, amino-bisubstitution (overall transformation)

Primary aliphatic amines can be oxidized to aldehydes or ketones.[656] Other reagents used[657] have been *N*-bromoacetamide[658] (for benzylic amines), 3,5-di-*tert*-butyl-1,2-benzoquinone,[659] and aqueous NaOCl with phase-transfer catalysts.[660] Benzylic amine salts PhCHRNR$_2'$ H$^+$ Cl$^-$ (R,R$'$ = H or alkyl) give benzaldehydes or aryl ketones when heated in DMSO.[661] Several indirect methods for achieving the conversion RR$'$CHNH$_2$ → RR$'$C=O (R$'$ = alkyl, aryl, or H) have been reported.[662]

Primary, secondary, and tertiary aliphatic amines have been cleaved to give aldehydes, ketones, or carboxylic acids with aqueous bromine[663] and with neutral permanganate.[664] The other product of this reaction is the amine with one less alkyl group. In a different type of procedure, primary alkyl primary amines can be converted to *gem*-dihalides [RCH$_2$NH$_2$ → RCHX$_2$ (X = Br or Cl)] by treatment with an alkyl nitrite and the anhydrous copper(I) halide.[665]

Primary and secondary aliphatic nitro compounds have been oxidized to aldehydes and ketones, respectively (RR$'$CHNO$_2$ → RR$'$C=O) with sodium chlorite

[655]Itoh, A.; Kodama, T.; Inagaki, S.; Masaki, Y. *Org. Lett.* **2000**, *2*, 2455.

[656]For a review, see Haines, A.H. *Methods for the Oxidation of Organic Compounds*, Vol. 2, Academic Press, NY, **1988**, pp. 200–220, 411–415.

[657]For lists of reagents, with references, see Larock, R.C. *Comprehensive Organic Transformations*, 2nd ed., Wiley-VCH, NY, **1999**, pp. 1225–1227; Hudlický, M. *Oxidations in Organic Chemistry*, American Chemical Society, Washington, DC, **1990**, p. 240.

[658]Banerji, K.K. *Bull. Chem. Soc. Jpn.* **1988**, *61*, 3717.

[659]Corey, E.J.; Achiwa, K. *J. Am. Chem. Soc.* **1969**, *91*, 1429. For a study of the mechanism, see Klein, R.F.X.; Bargas, L.M.; Horak, V. *J. Org. Chem.* **1988**, *53*, 5994.

[660]Lee, G.A.; Freedman, H.H. *Tetrahedron Lett.* **1976**, 1641.

[661]Traynelis, V.J.; Ode, R.H. *J. Org. Chem.* **1970**, *35*, 2207. For other methods, see Takabe, K.; Yamada, T. *Chem. Ind. (London)* **1982**, 959; Azran, J.; Buchman, O.; Pri-Bar, I. *Bull. Soc. Chim. Belg.* **1990**, *99*, 345.

[662]See, for example, Dinizo, S.E.; Watt, D.S. *J. Am. Chem. Soc.* **1975**, *97*, 6900; Black, D.S.; Blackman, N.A. *Aust. J. Chem.* **1975**, *28*, 2547; Scully, Jr., F.E.; Davis, R.C. *J. Org. Chem.* **1978**, *43*, 1467; Doleschall, G. *Tetrahedron Lett.* **1978**, 2131; Babler, J.H.; Invergo, B.J. *J. Org. Chem.* **1981**, *46*, 1937.

[663]Deno, N.C.; Fruit, Jr., R.E. *J. Am. Chem. Soc.* **1968**, *90*, 3502.

[664]Rawalay, S.S.; Shechter, H. *J. Org. Chem.* **1967**, *32*, 3129. For another procedure, see Monković, I.; Wong, H.; Bachand, C. *Synthesis* **1985**, 770.

[665]Doyle, M.P.; Siegfried, B. *J. Chem. Soc. Chem. Commun.* **1976**, 433.

under phase transfer conditions,[666] tetrapropylammonium perruthenate (TPAP),[667] Oxone®,[668] as well as with other reagents.[669] Vinyl nitro compounds were converted to α-alkylated ketones, with good enantioselectivity, using R_2Zn, a chiral copper catalyst followed by hydrolysis with 20% aqueous sulfuric acid.[670]

19-22   Oxidation of Primary Alcohols to Carboxylic Acids or Carboxylic Esters

Oxo-de-dihydro-bisubstitution

$$RCH_2OH \xrightarrow{\ CrO_3\ } RCOOH$$

Primary alcohols can be oxidized to carboxylic acids by many strong oxidizing agents including chromic acid, permanganate,[671] and nitric acid.[672] Other reagents include H_5IO_6/CrO_3.[673] The reaction can be looked on as a combination of **19-3** and **19-23**. When acidic conditions are used, a considerable amount of carboxylic ester $RCOOCH_2R$ is often isolated, although this is probably not formed by a combination of the acid with unreacted alcohol, but by a combination of intermediate aldehyde with unreacted alcohol to give an acetal or hemiacetal, which is oxidized to the ester.[674] Aliphatic primary alcohols are converted to the carboxylic acid with 30% aq. H_2O_2, tetrabutylammonium hydrogen sulfate and a tungsten catalyst with microwave irradiation.[675] Oxone® in DMF also converts aliphatic aldehydes to the corresponding carboxylic acid.[676] Benzylic alcohols are oxidized to benzoic acid derivatives by treatment first with TEMPO[677] (p. 274), and then $NaClO_2$.[678] A combination of $NaClO_2$ and NaH_2PO_4 in aq. DMSO oxidizes aldehydes to acids even in the presence of a disulfide

[666]Ballini, R.; Petrini, M. *Tetrahedron Lett.* **1989**, *30*, 5329.

[667]Tokunaga, Y.; Ihara, M.; Fukumoto, K. *J. Chem. Soc. Perkin Trans. 1* **1997**, 207.

[668]Ceccherelli, P.; Curini, M.; Marcotullio, M.C.; Epifano, F.; Rosati, O. *Synth. Commun.* **1998**, *28*, 3057.

[669]For a list of reagents, with references, see Larock, R.C. *Comprehensive Organic Transformations*, 2nd ed., Wiley-VCH, NY, **1999**, pp. 1227–1228.

[670]Luchaco-Cullis, C.A.; Hoveyda, A.H. *J. Am. Chem. Soc.* **2002**, *124*, 8192.

[671]For a discussion of the mechanism of this oxidation, see Rankin, K.N.; Liu, Q.; Hendry, J.; Yee, H.; Noureldin, N.A.; Lee, D.G. *Tetrahedron Lett.* **1998**, *39*, 1095.

[672]For reviews, see Hudlický, M. *Oxidations in Organic Chemistry*, American Chemical Society, Washington, DC, **1990**, pp. 127–132; Haines, A.H. *Methods for the Oxidation of Organic Compounds*, Vol. 2, Academic Press, NY, **1988**, 148–165, 391–401. For a list of reagents, with references, see Larock, R.C. *Comprehensive Organic Transformations*, 2nd ed., Wiley-VCH, NY, **1999**, pp. 1646–1650.

[673]Zhao, M.; Li, J.; Song, Z.; Desmond, R.; Tschaen, D.M.; Grabowski, E.J.J.; Reider, P.J. *Tetrahedron Lett.* **1998**, *39*, 5323

[674]Craig, J.C.; Horning, E.C. *J. Org. Chem.* **1960**, *25*, 2098. See also, Berthon, B.; Forestiere, A.; Leleu, G.; Sillion, B. *Tetrahedron Lett.* **1981**, *22*, 4073; Nwaukwa, S.O.; Keehn, P.M. *Tetrahedron Lett.* **1982**, *23*, 35.

[675]Bogdał, D.; Łukasiewicz, M. *Synlett* **2000**, 143.

[676]Travis, B.R.; Sivakumar, M.; Hollist, G.O.; Borhan, B. *Org. Lett.* **2003**, *5*, 1031.

[677]For other oxidations of this type utilizing TEMPO, see DeLuca, L.; Giacomelli, G.; Masala, S.; Porcheddu, A. *J. Org. Chem.* **2003**, *68*, 4999. For a reaction using polymer-bound TEMPO, see Yasuda, K.; Ley, S.V. *J. Chem. Soc., Perkin Trans. 1* **2002**, 1024.

[678]Zhao, M.; Li, J.; Mano, E.; Song, Z.; Tschaen, D.M.; Grabowski, E.J.J.; Reider, P.J. *J. Org. Chem.* **1999**, *64*, 2564.

elsewhere in the molecule.[679] Similar oxidation to the acid occurred with $NaIO_4/RuCl_3$ in aqueous acetonitrile,[680] 30% aq. H_2O_2, and a cobalt–salen catalyst,[681] or oxygen on alumina with microwave irradiation.[682] Aliphatic alcohols are converted to a symmetrical ester ($RCH_2OH \rightarrow RCOOCH_2R$) by oxidation with PCC on aluminum without solvent.[683] Oxone in aqueous methanol also converts aryl aldehydes to the corresponding ester.[684] Allylic alcohols are converted to conjugated esters with MnO_2, NaCN in methanol–acetic acid.[685] Primary alcohols are oxidized to the methyl ester with trichloroisocyanuric acid in methanol.[686] This reagent also converts diols to lactones.

Primary alcohols RCH_2OH can be directly oxidized to acyl fluorides RCOF with cesium fluoroxysulfate.[687] Lactones can be prepared by oxidizing diols in which at least one OH is primary,[688] and addition of a chiral additive, such as sparteine, leads to lactones with high asymmetric induction.[689] 2-(3-Hydroxypropyl)aniline was oxidized to an acyl derivative that cyclized to give a lactam when heated with a rhodium catalyst.[690]

Primary alkyl ethers can be selectively cleaved to carboxylic acids by aq. Br_2 ($RCH_2OR' \rightarrow RCOOH$).[691] Secondary allylic alcohols are converted to ketones with 70% *tert*-butylhydroperoxide with a CrO_3 catalyst.[692]

OS **I**, 138, 168; **IV**, 499, 677; **V**, 580; **VII**, 406; **IX**, 462; **81**, 195. Also see, OS **III**, 745.

19-23   Oxidation of Aldehydes to Carboxylic Acids

Hydroxylation or **Hydroxy-de-hydrogenation**

[679]Fang, X.; Bandarage, U.P.; Wang, T.; Schroeder, J.D.; Garvey, D.S. *Synlett* **2003**, 489.

[680]Prashad, M.; Lu, Y.; Kim, H.-Y.; Hu, B.; Repic, O.; Blacklock, T.J. *Synth. Commun.* **1999**, *29*, 2937.

[681]Das, S.; Punniyamurthy, T. *Tetrahedron Lett.* **2003**, *44*, 6033.

[682]Reddy, D.S.; Reddy, P.P.; Reddy, P.S.N. *Synth. Commun.* **1999**, *29*, 2949.

[683]Bhar, S.; Chaudjuri, S.K. *Tetrahedron* **2003**, *59*, 3493.

[684]Koo, B.-S.; Kim, E.-H.; Lee, K.-J. *Synth. Commun.* **2002**, *32*, 2275.

[685]Foot, J.S.; Kanno, H.; Giblin, G.M.P.; Taylor, R.J.K. *Synlett* **2002**, 1293.

[686]Hiegel, G.A.; Gilley, C.B. *Synth. Commun.* **2003**, *33*, 2003.

[687]Stavber, S.; Planinsek, Z.; Zupan, M. *Tetrahedron Lett.* **1989**, *30*, 6095.

[688]For examples of the preparation of lactones by oxidation of diols, see Jefford, C.W.; Wang, Y. *J. Chem. Soc. Chem. Commun.* **1988**, 634; Jones, J.B.; Hirano, M.; Yakabe, S.; Morimoto, T. *Synth. Commun.* **1998**, *28*, 123; Suzuki, T.; Morita, K.; Tsuchida, M.; Hiroi, K. *Org. Lett.* **2002**, *4*, 2361; Hansen, T.M.; Florence, G.J.; Lugo-Mas, P.; Chen, J.; Abrams, J.N.; Forsynth, C.J. *Tetrahedron Lett.* **2003**, *44*, 57; Suzuki, T.; Morita, K.; Matsuo, Y.; Hiroi, K. *Tetrahedron Lett.* **2003**, *44*, 2003. For a list of reagents used to effect this conversion, with references, see Larock, R.C. *Comprehensive Organic Transformations*, 2nd ed., Wiley-VCH, NY, **1999**, pp. 1650–1652.

[689]Yanagisawa, Y.; Kashiwagi, Y.; Kurashima, F.; Anzai, J.; Osa, T.; Bobbitt, J.M. *Chem. Lett.* **1996**, 1043.

[690]Fujita, K.-i.; Takahashi, Y.; Owaki, M.; Yamamoto, K.; Yamaguchi, R. *Org. Lett.* **2004**, *6*, 2785.

[691]Although these references refer to oxidation of alkyl ethers to ketones, oxidation to carboxylic acids is also possible. See Deno, N.C.; Potter, N.H. *J. Am. Chem. Soc.* **1967**, *89*, 3550, 3555. See also, Miller, L.L.; Wolf, J.F.; Mayeda, E.A. *J. Am. Chem. Soc.* **1971**, *93*, 3306; Saigo, K.; Morikawa, A.; Mukaiyama, T. *Chem. Lett.* **1975**, 145; Olah, G.A.; Gupta, B.G.B.; Fung, A.P. *Synthesis* **1980**, 897.

[692]Chandrasekhar, S.; Mohanty, P.K.; Ramachander, T. *Synlett* **1999**, 1063.

Oxidation of aldehydes-to-carboxylic acids is quite common[693] and has been carried out with many oxidizing agents, the most popular of which is permanganate in acid, basic, or neutral solution.[694] Chromic acid,[695] bromine, and Oxone®,[696] are other reagents frequently employed. Bromate exchange resin in refluxing acetone oxidizes aryl aldehydes-to aryl-carboxylic acids.[697] Silver oxide is a fairly specific oxidizing agent for aldehydes and does not readily attack other groups. Benedict's and Fehling's solutions oxidize aldehydes,[698] and there is a test for aldehydes that depends on this reaction, but the method is seldom used for preparative purposes. In any case, it gives very poor results with aromatic aldehydes. α,β-Unsaturated aldehydes can be oxidized by sodium chlorite without disturbing the double bond.[699] Aldehydes are also oxidized to carboxylic acids by atmospheric oxygen, but the actual direct oxidation product in this case is the peroxy acid RCO_3H,[700] which with another molecule of aldehyde, disproportionates to give two molecules of acid (see **14-7**).[701] Aryl aldehydes are converted to the corresponding aryl carboxylic ester with hydrogen peroxide and a V_2O_5 catalyst[702] or a titanosilicate[703] in an alcohol solvent. Heating an α-bromoaldhyde with an alcohol and a triazolium carbene leads to the corresponding ester.[704] N-Bromophthalimide and mercuric

[693]For reviews, see Haines, A.H. *Methods for the Oxidation of Organic Compounds*, Academic Press, NY, *1988*, pp. 241–263, 423–428; Chinn, L.J. *Selection of Oxidants in Synthesis*, Marcel Dekker, NY, *1971*, pp. 63–70; Lee, D.G., in Augustine, R.L. *Oxidataion*, Vol. 1, Marcel Dekker, NY, *1969*, pp. 81–86.

[694]For lists of some of the oxidizing agents used, with references, see Hudlický, M. *Oxidations in Organic Chemistry*, American Chemical Society, Washington, DC, *1990*, pp. 174–180; Larock, R.C. *Comprehensive Organic Transformations*, 2nd ed., Wiley-VCH, NY, *1999*, pp. 1653–1661; Srivastava, R.G.; Venkataramani, P.S. *Synth. Commun. 1988*, *18*, 2193. See also, Haines, A.H. *Methods for the Oxidation of Organic Compounds*, Academic Press, NY, *1988*.

[695]For a review, see Cainelli, G.; Cardillo, G. *Chromium Oxidations in Organic Chemistry*, Springer, NY, *1984*, pp. 217–225.

[696]Webb, K.S.; Ruszkay, S.J. *Tetrahedron 1998*, *54*, 401.

[697]Chetri, A.B.; Kalita, B.; Das, P.J. *Synth. Commun. 2000*, *30*, 3317.

[698]For a review, see Nigh, W.G., in Trahanovsky, W.S. *Oxidation in Organic Chemistry*, pt. B, Academic Press, NY, *1973*, pp. 31–34.

[699]Bal, B.S.; Childers Jr., W.E.; Pinnick, H.W. *Tetrahedron 1981*, *37*, 2091; Dalcanale, E.; Montanari, F. *J. Org. Chem. 1986*, *51*, 567. See also Bayle, J.P.; Perez, F.; Courtieu, J. *Bull. Soc. Chim. Fr. 1990*, 565.

[700]For a review of the preparation of peroxy acids by this and other methods, see Swern, D., in Swern, D. *Organic Peroxides*, Vol. 1, Wiley, NY, *1970*, pp. 313–516.

[701]For reviews of the autoxidation of aldehydes, see Vardanyan, I.A.; Nalbandyan, A.B. *Russ. Chem. Rev. 1985*, *54*, 532 (gas phase); Sajus, L.; Sérée de Roch, I., in Bamford, C.H., Tipper, C.F.H. *Comprehensive Chemical Kinetics*, Vol. 16, Elsevier, NY, *1980*, pp. 89–124 (liquid phase); Maslov, S.A.; Blyumberg, E.A. *Russ. Chem. Rev. 1976*, *45*, 155 (liquid phase). For a review of photochemical oxidation of aldehydes by O2, see Niclause, M.; Lemaire, J.; Letort, M. *Adv. Photochem. 1966*, *4*, 25. For a discussion of the mechanism of catalyzed atmospheric oxidation of aldehydes, see Larkin, D.R. *J. Org. Chem. 1990*, *55*, 1563.

[702]Gopinath, R.; Patel, B.K. *Org. Lett. 2000*, *2*, 577.

[703]Chavan, S.P.; Dantale, S.W.; Govande, C.A.; Venkatraman, M.S.; Praveen, C. *Synlett 2002*, 267.

[704]Reynolds, N.T.; de Alaniz, J.R.; Rovis, T. *J. Am. Chem. Soc. 2004*, *126*, 9518.

acetate oxidizes aryl aldehydes to aryl carboxylic acids in chloroform at room temperature.[705] An aldehyde can be converted to the carboxylic acid by treatment with 30% hydrogen peroxide and methyl(trioctyl)ammonium hydrogen sulfate at 90°C.[706] Aryl aldehydes are similarly oxidized by a mixture of hydrogen peroxide and selenium dioxide (SeO_2).[707] Aldehydes (RCHO) can be directly converted to carboxylic esters (RCOOR′) by treatment with Br_2 in the presence of an alcohol.[708] Polymer-bound hypervalent iodine + TEMPO oxidizes aldehydes to acids.[709]

Mechanisms of aldehyde oxidation[710] are not firmly established, but there seem to be at least two main types: a free-radical mechanism and an ionic one. In the free-radical process, the aldehydic hydrogen is abstracted to leave an acyl radical, which obtains OH from the oxidizing agent. In the ionic process, the first step is addition of a species ^-OZ to the carbonyl bond to give **38** in alkaline solution and **39** in acid or neutral solution. The aldehydic hydrogen of **38** or **39** is then lost as a proton to a base, while Z leaves with its electron pair.

$$
\underset{\text{R}}{\overset{\text{O}}{\|}}\!\!-\!\!\text{C}\!-\!\text{H} \; + \; \text{ZO}^- \longrightarrow \underset{\textbf{38}}{\text{H} - \text{C}(\text{O}^-)\text{(O}\!-\!\text{Z)} \text{R}} \xrightarrow{\;\text{B}^-\;} \underset{\text{R}}{\overset{\text{O}}{\|}}\!\!-\!\!\text{C}\!-\!\text{O}^\ominus \; + \; \text{B}\!-\!\text{H} \; + \; \text{Z}^-
$$

$$
\underset{\text{R}}{\overset{\text{O}}{\|}}\!\!-\!\!\text{C}\!-\!\text{H} \; + \; \text{ZOH} \longrightarrow \underset{\textbf{39}}{\text{H} - \text{C}(\text{O})\text{(O}\!-\!\text{Z)}\text{R H}} \xrightarrow{\;\text{B}^-\;} \underset{\text{R}}{\overset{\text{O}}{\|}}\!\!-\!\!\text{C}\!-\!\text{OH} \; + \; \text{B}\!-\!\text{H} \; + \; \text{Z}^-
$$

For oxidation with acid dichromate the picture seems to be quite complex, with several processes of both types going on:[711]

[705]Anjum, A.; Srinivas, P. *Chem. Lett.* **2001**, 900.

[706]Sato, K.; Hyodo, M.; Takagi, J.; Aoki, M.; Noyori, R. *Tetrahedron Lett.* **2000**, *41*, 1439.

[707]Wójtowicz, H.; Brzą szcz, M.; Kloc, K.; M tochowski, J. *Tetrahedron* **2001**, *57*, 9743.

[708]Williams, D.R.; Klingler, F.D.; Allen, E.E.; Lichtenthaler, F.W. *Tetrahedron Lett.* **1988**, *29*, 5087; Al Neirabeyeh, M.; Pujol, M.D. *Tetrahedron Lett.* **1990**, *31*, 2273. For other methods, see Sundararaman, P.; Walker, E.C.; Djerassi, C. *Tetrahedron Lett.* **1978**, 1627; Grigg, R.; Mitchell, T.R.B.; Sutthivaiyakit, S. *Tetrahedron* **1981**, *37*, 4313; Massoui, M.; Beaupère, D.; Nadjo, L.; Uzan, R. *J. Organomet. Chem.* **1983**, *259*, 345; O'Connor, B.; Just, G. *Tetrahedron Lett.* **1987**, *28*, 3235; McDonald, C.; Holcomb, H.; Kennedy, K.; Kirkpatrick, E.; Leathers, T.; Vanemon, P. *J. Org. Chem.* **1989**, *54*, 1212. For a list of reagents, with references, see Larock, R.C. *Comprehensive Organic Transformations*, 2nd ed., Wiley-VCH, NY, **1999**, pp. 1661–1669.

[709]Tashino, Y.; Togo, H. *Synlett* **2004**, 2010.

[710]For a review, see Roček, J., in Patai, S. *The Chemistry of the Carbonyl Group*, Vol. 1, Wiley, NY, **1966**, pp. 461–505.

[711]Wiberg, K.B.; Szeimies, G. *J. Am. Chem. Soc.* **1974**, *96*, 1889. See also, Roček, J.; Ng, C. *J. Am. Chem. Soc.* **1974**, *96*, 1522, 2840; Sen Gupta, S.; Dey, S.; Sen Gupta, K.K. *Tetrahedron* **1990**, *46*, 2431.

Step 1
$$R\text{-}\underset{H}{\overset{O}{C}} + H_2CrO_4 \rightleftharpoons \underset{R}{\overset{H}{\underset{OH}{C}}}\text{O-CrO}_3H$$

Step 2
$$\underset{R}{\overset{H}{\underset{OH}{C}}}\text{O-CrO}_3H \xrightarrow{B^-} R\text{-}\underset{OH}{\overset{O}{C}} + B\text{-}H + Cr(IV)$$

Step 3
$$R\text{-}\underset{H}{\overset{O}{C}} + Cr(VI) \longrightarrow R\text{-}\overset{O}{C}\cdot$$

Step 4
$$\underset{R}{\overset{OH}{\underset{OH}{C}}} + Cr(IV) \longrightarrow \underset{R\cdot}{\overset{OH}{\underset{OH}{C}}} \xrightarrow{-H_2O} R\text{-}\overset{O}{C}\cdot + Cr(III)$$

Step 5
$$R\text{-}\overset{O}{C}\cdot + H_2CrO_4 \longrightarrow R\text{-}\underset{OH}{\overset{O}{C}} + Cr(V)$$

Step 6
$$R\text{-}\underset{H}{\overset{O}{C}} + Cr(V) \rightleftharpoons \underset{R}{\overset{H}{\underset{OH}{C}}}\text{O-Cr(V)}$$

Step 7
$$\underset{R}{\overset{H}{\underset{OH}{C}}}\text{O-Cr(V)} \xrightarrow{B^-} \underset{R}{\overset{O}{\underset{OH}{C}}} + B\text{-}H + Cr(III)$$

Steps 1 and 2 constitute an oxidation by the ionic pathway by Cr(VI), and steps 6 and 7 a similar oxidation by Cr(V), which is produced by an electron-transfer process. Either Cr(VI) (step 3) or Cr(IV) (step 4) [Cr(IV) is produced in step 2] may abstract a hydrogen and the resulting acyl radical is converted to carboxylic acid in step 5. Thus, chromium in three oxidation states is instrumental in oxidizing aldehydes. Still another possible process has been proposed in which the chromic acid ester decomposes as follows:[712]

$$R\text{-}C \longrightarrow R\text{-}C\underset{O}{\overset{O\text{-}H}{\parallel}} + Cr(OH)_2O$$

The mechanism with permanganate is less well known, but an ionic mechanism has been proposed[713] for neutral and acid permanganate, similar to steps 1 and 2 for dichromate:

$$R\text{-}\underset{H}{\overset{O}{C}} + HMnO_4 \longrightarrow \underset{R}{\overset{H}{\underset{OH}{C}}}\text{O-MnO}_3 \xrightarrow{B^-} R\text{-}\underset{OH}{\overset{O}{C}} + B\text{-}H + MnO_3^-$$

[712]See Roček, J.; Ng, C. *J. Org. Chem.* **1973**, *38*, 3348.
[713]See, for example, Freeman, F.; Lin, D.K.; Moore, G.R. *J. Org. Chem.* **1982**, *47*, 56; Jain, A.L.; Banerji, K.K. *J. Chem. Res. (S)* **1983**, 60.

For alkaline permanganate, the following mechanism has been proposed:[714]

$$
\begin{array}{c}
\underset{R}{\overset{O}{\underset{H}{C}}} \xrightarrow{OH^-} \underset{R}{\overset{H}{\underset{OH}{C}}} \overset{O^{\ominus}}{} \xrightarrow[slow]{MnO_4^-} \underset{R}{\overset{O}{\underset{OH}{C}}} + HMnO_4^{2-} \longrightarrow \underset{R}{\overset{O}{\underset{O\,\ominus}{C}}} + H_2O + MnO_3^-
\end{array}
$$

$$
Mn(V) + Mn(VII) \longrightarrow 2Mn(VI)
$$

OS **I**, 166; **II**, 302, 315, 538; **III**, 745; **IV**, 302, 493, 499, 919, 972, 974.

The conversion of thioketones to sulfines ($R_2C{=}S{=}O$) is difficult to categorize into the sections available, and it placed after oxidation of ketones and aldehydes. The reaction of a thioketone with hydrogen peroxide and a catalytic amount of MTO (methyl trioxorhenium) gives the sulfine.[715]

19-24   Oxidation of Carboxylic Acids to Peroxy Acids

Peroxy-de-hydroxy-substitution

$$
\underset{R}{\overset{O}{\underset{OH}{C}}} + HOOH \xrightleftharpoons{H^+} \underset{R}{\overset{O}{\underset{O^-O^-H}{C}}} + H_2O
$$

The oxidation of carboxylic acids with H_2O_2 and an acid catalyst is the best general method for the preparation of peroxy acids.[716] A mixture of $Me_2C(OMe)OOH$ and DCC has also been used.[717] The most common catalyst for aliphatic R is concentrated sulfuric acid. The reaction is an equilibrium and is driven to the right by removal of water or by the use of excess reagents. For aromatic R, the best catalyst is methanesulfonic acid, which is also used as the solvent.

D. Reactions in Which Oxygen is Added to the Substrate

19-25   Oxidation of Alkenes to Aldehydes and Ketones

1/Oxo-(1/ → 2/hydro)-*migro*-attachment

$$
\underset{/}{\overset{\backslash}{C}}{=}\underset{H}{\overset{/}{C}} \xrightarrow[H_2O]{PdCl_2} \underset{H}{\overset{\backslash}{C}}{-}\underset{\overset{\|}{O}}{C}{/}
$$

[714]Freeman, F.; Brant, J.B.; Hester, N.B.; Kamego, A.A.; Kasner, M.L.; McLaughlin, T.G.; Paul, E.W. *J. Org. Chem.* **1970**, *35*, 982.
[715]Huang, R.; Espenson, J.H. *J. Org. Chem.* **1999**, *64*, 6935.
[716]For a review of the preparation of peroxy acids, see Swern, D., in Swern, D. *Organic Peroxides*, Vol. 1, Wiley, NY, **1970**, pp. 313–516.
[717]Dussault, P.; Sahli, A. *J. Org. Chem.* **1992**, *57*, 1009.

Monosubstituted and 1,2-disubstituted alkenes can be oxidized to aldehydes and ketones by palladium chloride and similar salts of noble metals.[718] 1,1-Disubstituted alkenes generally give poor results. The reaction is used industrially to prepare acetaldehyde from ethylene (the *Wacker process*), but it is also suitable for laboratory preparations. The palladium chloride is reduced to palladium. Because the reagent is expensive, the reaction is usually carried out with a co-oxidant, most often $CuCl_2$, whose function is to reoxidize the Pd to Pd(II). The $CuCl_2$ is reduced to Cu(I), which itself is reoxidized to Cu(II) by air, so that atmospheric oxygen is the only oxidizing agent actually used up. Many other co-oxidants have been tried, among them O_3, Fe^{3+}, and PbO_2. Terminal alkenes are oxidized to methyl ketones with O_2 and a palladium catalyst with 20% pyridine in 2-propanol.[719] *tert*-Butylhydroperoxide in bromoperfluorooctane–benzene oxidizes styrene to acetophenone in a Wacker-type process.[720] The principal product is an aldehyde only from ethylene: With other alkenes Markovnikov's rule is followed, and ketones are formed predominantly.

The generally accepted mechanism involves π complexes of palladium.[721]

This mechanism accounts for the fact, established by deuterium labeling, that the four hydrogens of the acetaldehyde all come from the original ethylene and none from the solvent.

[718]For a monograph, see Henry, P.M. *Palladium Catalyzed Oxidation of Hydrocarbons*, D. Reidel Publishing Co., Dordrecht, *1980*. For reviews, see Tsuji, J. *Organic Synthesis with Palladium Compounds*, Springer, NY, *1980*, pp. 6–12; *Synthesis 1990*, 739; *1984*, 369; *Adv. Org. Chem. 1969*, 6, 109, see pp. 119–131; Heck, R.F. *Palladium Reagents in Organic Syntheses*, Academic Press, NY, *1985*, pp. 59–80; Sheldon, R.A.; Kochi, J.K. *Metal-Catalyzed Oxidations of Organic Compounds*, Academic Press, NY, *1981*, pp. 189–193, 299–303; Henry, P.M. *Adv. Organomet. Chem. 1975*, 13, 363, see pp. 378–388; Jira, R.; Freiesleben, W. *Organomet. React. 1972*, 3, 1, pp. 1–44; Khan, M.M.T.; Martell, A.E. *Homogeneous Catalysis by Metal Complexes*, Vol. 2, Academic Press, NY, *1974*, pp. 77–91; Hüttel, R. *Synthesis 1970*, 225, see pp. 225–236; Aguiló, A. *Adv. Organomet. Chem. 1967*, 5, 321; Bird, C.W. *Transition Metal Intermediates in Organic Synthesis*, Academic Press, NY, *1967*, pp. 88–111.
[719]Nishimura, T.; Kakiuchi, N.; Onoue, T.; Ohe, K.; Uemura, S. *J. Chem. Soc., Perkin Trans. 1 2000*, 1915.
[720]Betzemeier, B.; Lhermitte, F.; Knochel, P. *Tetrahedron Lett. 1998*, 39, 6667.
[721]Henry, P.M. *J. Am. Chem. Soc. 1972*, 94, 4437; Jira, R.; Sedlmeier, J.; Smidt, J. *Liebigs Ann. Chem. 1966*, 693, 99; Hosokawa, T.; Maitlis, P.M. *J. Am. Chem. Soc. 1973*, 95, 4924; Moiseev, I.I.; Levanda, O.G.; Vargaftik, M.N. *J. Am. Chem. Soc. 1974*, 96, 1003; Bäckvall, J.; Åkermark, B.; Ljunggren, S.O. *J. Am. Chem. Soc. 1979*, 101, 2411; Zaw, K.; Henry, P.M. *J. Org. Chem. 1990*, 55, 1842.

Similar reactions have been carried out with other oxidizing agents. An example involving migration of an alkyl group instead of hydrogen is oxidation of $Me_2C=CMe_2$ with peroxytrifluoroacetic acid-boron trifluoride to give Me_3COMe (pinacolone).[722] This reaction consists of epoxidation (**15-50**) followed by pinacol rearrangement of the epoxide (**18-2**). A migration is also involved in the conversion of $ArCH=CHCH_3$ to $ArCH(CH_3)CHO$ by treatment with I_2-Ag_2O in aqueous dioxane.[723]

Other reagents used have been $Pb(OAc)_4$–F_3CCOOH[724] (e.g., $PhCH=CH_2$ → $PhCH_2CHO$), H_2O_2 and a Pd catalyst,[725] H_2O–$PdCl_2$–polyethylene glycol,[726] CrO_3–H_2SO_4–Hg(II) salts,[727] and $Hg(OAc)_2$ followed by $PdCl_2$.[728] The reaction has also been accomplished electrochemically.[729] Terminal alkenes react with ceric ammonium nitrate in methanol to give α-methoxy ketones.[730]

Alkenes have also been converted to more highly oxidized products. Examples are (*1*) treatment with $KMnO_4$ in aqueous acetone containing acetic acid gives α-hydroxy ketones.[731] (*2*) 1,2-Disubstituted and trisubstituted alkenes give α-chloro ketones when oxidized with chromyl chloride in acetone: $RCH=CR'R^2$ → $RCOCClR'R^2$.[732] (*3*) α-Iodo ketones can be prepared by treating alkenes with bis-(*sym*-collidine)iodine(I) tetrafluoroborate.[733] (*4*) potassium permanganate in acetic anhydride oxidizes large-ring cycloalkenes to 1,2-diketones.[734]

Enol ethers are oxidized to carboxylic esters ($RCH=CHOR'$ → RCH_2COOR') with PCC[735] and enamines to α-amino ketones ($R^1CH=CR_2NR$ → R^1COCR_2NR) with *N*-sulfonyloxaziridines.[736] Enamines ($R^1R^4C=CR^2NR_2^3$, $R^4 \neq H$) do not give these products, but lose the amino group to give α-hydroxy ketones, $R^1R^4C(OH)$-COR^2.[736] Carboxylic acids can be prepared from terminal alkynes ($RC\equiv CH$ →

[722]Hart, H.; Lerner, L.R. *J. Org. Chem.* **1967**, *32*, 2669.

[723]Kikuchi, H.; Kogure, K.; Toyoda, M. *Chem. Lett.* **1984**, 341.

[724]Lethbridge, A.; Norman, R.O.C.; Thomas, C.B. *J. Chem. Soc. Perkin Trans. 1* **1973**, 35.

[725]Roussel, M.; Mimoun, H. *J. Org. Chem.* **1980**, *45*, 5387.

[726]Alper, H.; Januszkiewicz, K.; Smith, D.J.H. *Tetrahedron Lett.* **1985**, *26*, 2263.

[727]Rogers, H.R.; McDermott, J.X.; Whitesides, G.M. *J. Org. Chem.* **1975**, *40*, 3577.

[728]Rodeheaver, G.T.; Hunt, D.F. *Chem. Commun.* **1971**, 818. See also, Hunt, D.F.; Rodeheaver, G.T. *Tetrahedron Lett.* **1972**, 3595.

[729]See Tsuji, J.; Minato, M. *Tetrahedron Lett.* **1987**, *28*, 3683.

[730]Nair, V.; Nair, L.G.; Panicker, S.B.; Sheeba, V.; Augustine, A. *Chem. Lett.* **2000**, 584.

[731]Srinivasan, N.S.; Lee, D.G. *Synthesis* **1979**, 520. See also, Baskaran, S.; Das, J.; Chandrasekaran, S. *J. Org. Chem.* **1989**, *54*, 5182.

[732]Sharpless, K.B.; Teranishi, A.Y. *J. Org. Chem.* **1973**, *38*, 185. See also, Cardillo, G.; Shimizu, M. *J. Org. Chem.* **1978**, *42*, 4268; D'Ascoli, R.; D'Auria, M.; Nucciarelli, L.; Piancatelli, G.; Scettri, A. *Tetrahedron Lett.* **1980**, *21*, 4521; Kageyama, T.; Tobito, Y.; Katoh, A.; Ueno, Y.; Okawara, M. *Chem. Lett.* **1983**, 1481; Lee, J.G.; Ha, D.S. *Tetrahedron Lett.* **1989**, *30*, 193.

[733]Evans, R.D.; Schauble, J.H. *Synthesis* **1986**, 727.

[734]Jensen, H.P.; Sharpless, K.B. *J. Org. Chem.* **1974**, *39*, 2314.

[735]Piancatelli, G.; Scettri, A.; D'Auria, M. *Tetrahedron Lett.* **1977**, 3483. When $R^1CR^2C=CR^3OR^4$ are used, cleavage of the double bond takes place instead: Baskaran, S.; Islam, I.; Raghavan, M.; Chandrasekaran, S. *Chem. Lett.* **1987**, 1175.

[736]Davis, F.A.; Sheppard, A.C. *Tetrahedron Lett.* **1988**, *29*, 4365.

RCH$_2$COOH) by conversion of the alkyne to its phenylthio ether (RC≡CSPh) and treatment of this with HgSO$_4$ in HOAc–H$_2$SO$_4$.[737]

OS VI, 1028; VII, 137; VIII, 208.

19-26 The Oxidation of Alkynes to α-Diketones

Dioxo-biaddition

$$R-C≡C-R^1 \xrightarrow[\text{tetroxide}]{\text{ruthenium}} \underset{\underset{O}{\parallel}}{R}\overset{\overset{O}{\parallel}}{C}-\overset{}{C}-R^1$$

Internal alkynes have been oxidized[738] to α-diketones by several oxidizing agents,[739] including neutral KMnO$_4$,[740] bis(trifluoroacetoxy)iodobenzene,[741] NaIO$_4$–RuO$_2$,[742] I$_2$–DMSO,[743] MeReO$_3$/H$_2$O$_2$,[744] as well as by electrooxidation.[745] A ruthenium complex with a small amount of trifluoroacetic acid converts internal alkynes to the α-diketone.[746] Ozone generally oxidizes triple-bond compounds to carboxylic acids (19-9), but α-diketones are sometimes obtained instead. Selenium dioxide (SeO$_2$) with a small amount of H$_2$SO$_4$ oxidizes alkynes to α-diketones as well as arylacetylenes to α-keto acids (ArC≡CH → ArCOCOOH).[747]

19-27 Oxidation of Amines to Nitroso Compounds and Hydroxylamines and Related

N-Oxo-de-dihydro-bisubstitution

$$ArNH_2 \xrightarrow{H_2SO_5} Ar-N=O$$

[737]Abrams. S. R. *Can. J. Chem.* **1983**, *61*, 2423.

[738]For a review of this reaction, see Haines, A.H. *Methods for the Oxidation of Organic Compounds*, Vol. 1, Academic Press, NY, **1985**, pp. 153–162, 332-338. For a review of oxidations of triple bonds in general, see Simándi, L.I., in Patai, S.; Rappoport, Z. *The Chemistry of Functional Groups, Supplement C*, pt. 1, Wiley, NY, **1983**, pp. 513–570.

[739]For a list of reagents, with references, see Hudlický, M. *Oxidations in Organic Chemistry*, American Chemical Society, Washington, DC, **1990**, p. 92.

[740]Khan, N.A.; Newman, M.S. *J. Org. Chem.* **1952**, *17*, 1063; Lee, D.G.; Lee, E.J.; Chandler, W.D. *J. Org. Chem.* **1985**, *50*, 4306; Tatlock, J.H. *J. Org. Chem.* **1995**, *60*, 6221.

[741]Vasil'eva, V.P.; Khalfina, I.L.; Karpitskaya, L.G.; Merkushev, E.B. *J. Org. Chem. USSR* **1987**, *23*, 1967.

[742]Zibuck, R.; Seebach, D. *Helv. Chim. Acta* **1988**, *71*, 237.

[743]Yusybov, M.S.; Filimonov, V.D. *Synthesis* **1991**, 131.

[744]Zhu, Z.; Espenson, J.H. *J. Org. Chem.* **1995**, *60*, 7728.

[745]Torii, S.; Inokuchi, T.; Hirata, Y. *Synthesis* **1987**, 377.

[746]Che, C.-M.; Yu, W.-Y.; Chan, P.-M.; Cheng, W.-C.; Peng, S.-M.; Lau, K.-C.; Li, W.-K. *J. Am. Chem. Soc.* **2000**, *122*, 11380.

[747]Sonoda, N.; Yamamoto, Y.; Murai, S.; Tsutsumi, S. *Chem. Lett.* **1972**, 229.

Primary aromatic amines can be oxidized[748] to nitroso compounds. Most often the conversion is accomplished by Caro's acid (H_2SO_5) or with H_2O_2 in HOAc.[749] Hydroxylamines, which are probably intermediates in most cases, can sometimes be isolated, but under the reaction conditions are generally oxidized to the nitroso compounds. Primary aliphatic amines can be oxidized in this manner, but the nitroso compound is stable only if there is no α hydrogen. If there is an α hydrogen, the compound tautomerizes to the oxime.[750] Among the reagents used for this oxidation are sodium perborate[751] H_2O_2 with a titanium complex,[752] HOF generated *in situ*,[753] and Na_2WO_4/H_2O_2.[754] The mechanism with H_2SO_5 has been postulated to be an example of category 5 (p. 1706).[755]

Secondary amines, R_2NH, are oxidized to hydroxylamines (R_2NHOH) which are resistant to further oxidation, by dimethyldioxirane[756] and by benzoyl peroxide and Na_2HPO_4.[757] Oxone® on silica also oxidizes secondary alcohols to the hydroxylamine.[758] Hydroxylamines are formed when secondary amines react with the enzyme cyclohexanone monooxygenase.[759] Carbamates, such as *N*-Boc amines, are converted tot he *N*-hydroxy compound with bis(trifluoromethyl)-dioxirane.[760] Note that secondary alcohols can be converted to nitrones with aq. H_2O_2 and a phosphotungstate polymer complex, presumably via an hydroxylamine (see **19-28**) formed *in situ*.[761] Dialkylamiens are oxidized to the *N*-nitroso compound with N_2O_2 on polyvinylpyrrolidinone.[762]

OS **III**, 334; **VIII**, 93; **80**, 207.

[748]For reviews on the oxidation of amines, see Rosenblatt, D.H.; Burrows, E.P., in Patai, S. *The Chemistry of Functional Groups, Supplement F*, pt. 2, Wiley, NY, *1982*, pp. 1085–1149; Challis, B.C.; Butler, A.R., in Patai, S. *The Chemistry of the Amino Group*, Wiley, NY, *1968*, pp. 320–338. For reviews confined to primary aromatic amines, see Hedayatullah, M. *Bull. Soc. Chim. Fr. 1972*, 2957; Surville, R. De; Jozefowicz, M.; Buvet, R. *Ann. Chim. (Paris) 1967, [14] 2*, 149.

[749]Holmes, R.R.; Bayer, R.P. *J. Am. Chem. Soc. 1960, 82*, 3454.

[750]For example, see Kahr, K.; Berther, C. *Chem. Ber. 1960, 93*, 132.

[751]Zajac Jr., W.W.; Darcy, M.G.; Subong, A.P.; Buzby, J.H. *Tetrahedron Lett. 1989, 30*, 6495.

[752]Dewkar, G.K.; Nikalje, M.D.; Ali, I.S.; Paraskar, A.S.; Jagtap, H.S.; Sadalai, A. *Angew. Chem. Int. Ed. 2001, 40*, 405.

[753]Dirk, S.M.; Mickelson, E.T.; Henderson, J.C.; Tour, J.M. *Org. Lett. 2002, 2*, 3405.

[754]Corey, E.J.; Gross, A.W. *Org. Synth. 65*, 166.

[755]Gragerov, I.P.; Levit, A.F. *J. Gen Chem. USSR 1960, 30*, 3690.

[756]Murray, R.W.; Singh, M. *Synth. Commun. 1989, 19*, 3509. This reagent also oxidizes primary amines to hydroxylamines: Wittman, M.D.; Halcomb, R.L.; Danishefsky, S.J. *J. Org. Chem. 1990, 55*, 1981.

[757]Biloski, A.J.; Ganem, B. *Synthesis 1983*, 537.

[758]Fields, J.D.; Kropp, P.J. *J. Org. Chem. 2000, 65*, 5937.

[759]Colonna, S.; Pironti, V.; Carrea, G.; Pasta, P.; Zambianchi, F. *Tetahedron 2004, 60*, 569.

[760]Detomaso, A.; Curci, R. *Tetrahedron Lett. 2001, 42*, 755.

[761]Yamada, Y.M.A.; Tabata, H.; Takahashi, H.; Ikegami, S. *Synlett 2002*, 2031.

[762]Iranpoor, N.; Firouzabadi, H.; Pourali, A.R. *Synthesis 2003*, 1591.

19-28   Oxidation of Primary Amines, Oximes, Azides, Isocyanates, or Nitroso Compounds to Nitro Compounds

$$R_3CNH_2 \xrightarrow{KMnO_4} R_3CNO_2$$

$$R_2C=NOH \xrightarrow{F_3CCOOOH} R_2CHNO_2$$

Tertiary alkyl primary amines can be oxidized to nitro compounds in excellent yields with $KMnO_4$.[763] This type of nitro compound is not easily prepared in other ways. All classes of primary amine (including primary, secondary, and tertiary alkyl, as well as aryl) are oxidized to nitro compounds in high yields with dimethyl-dioxirane.[764] Other reagents that oxidize various types of primary amines to nitro compounds are dry ozone,[765] various peroxyacids,[766] $MeReO_3/H_2O_2$,[767] Oxone®,[768] *tert*-butyl hydroperoxide in the presence of certain molybdenum and vanadium compounds,[769] and sodium perborate.[770]

Dimethyldioxirane in wet acetone oxidizes isocyanates to nitro compounds ($RNCO \rightarrow RNO_2$).[771] Oximes can be oxidized to nitro compounds with peroxytri-fluoroacetic acid, or Oxone®,[772] sodium perborate,[773] among other ways.[763] Secondary hydroxylamines are also oxidized to nitrones with MnO_2 in dichloromethane.[774] Primary and secondary alkyl azides have been converted to nitro compounds by treatment with Ph_3P followed by ozone.[775] Aromatic nitroso compounds are easily oxidized to nitro compounds by many oxidizing agents.[776]

OS **III**, 334; **V**, 367, 845; **VI**, 803; **81**, 204.

[763]Larson, H.O., in Feuer, H. *The Chemistry of the Nitro and Nitroso Groups*, Vol. 1, Wiley, NY, *1969*, pp. 306–310. See also, Barnes, M.W.; Patterson, J.M. *J. Org. Chem.* *1976*, *41*, 733. For reviews of oxidations of nitrogen compounds, see Butler, R.N. *Chem. Rev.* *1984*, *84*, 249; Boyer, J.H. *Chem. Rev.* *1980*, *80*, 495.

[764]Murray, R.W.; Rajadhyaksha, S.N.; Mohan, L. *J. Org. Chem.* *1989*, *54*, 5783. See also, Zabrowski, D.L.; Moorman, A.E.; Beck Jr., K.R. *Tetrahedron Lett.* *1988*, *29*, 4501.

[765]Keinan, E.; Mazur, Y. *J. Org. Chem.* *1977*, *42* 844; Bachman, G.B.; Strawn, K.G. *J. Org. Chem.* *1968*, *33*, 313.

[766]Emmons, W.D. *J. Am. Chem. Soc.* *1957*, *79*, 5528; Gilbert, K.E.; Borden, W.T. *J. Org. Chem.* *1979*, *44*, 659.

[767]Murray, R.W.; Iyanar, K.; Chen, J.; Wearing, J.T. *Tetrahedron Lett.* *1996*, *37*, 805; Cardona, F.; Soldaini, G.; Goti, A. *Synlett* *2004*, 1553.

[768]Webb, K.S.; Seneviratne, V. *Tetrahedron Lett.* *1995*, *36*, 2377.

[769]Howe, G.R.; Hiatt, R.R. *J. Org. Chem.* *1970*, *35*, 4007. See also, Nielsen, A.T.; Atkins, R.L.; Norris, W.P.; Coon, C.L.; Sitzmann, M.E. *J. Org. Chem.* *1980*, *45*, 2341.

[770]McKillop, A.; Tarbin, J.A. *Tetrahedron* *1987*, *43*, 1753.

[771]Eaton, P.E.; Wicks, G.E. *J. Org. Chem.* *1988*, *53*, 5353.

[772]Bose, D.S.; Vanajatha, G. *Synth. Commun.* *1998*, *28*, 4531.

[773]Olah, G.A.; Ramaiah, P.; Lee, G.K.; Prakash, G.K.S. *Synlett* *1992*, 337.

[774]Cicchi, S.; Marradi, M.; Goti, A.; Brandi, A. *Tetrahedron Lett.* *2001*, *42*, 6503.

[775]Corey, E.J.; Samuelsson, B.; Luzzio, F.A. *J. Am. Chem. Soc.* *1984*, *106*, 3682.

[776]See Boyer, J.H., in Feuer, H. *The Chemistry of the Nitro and Nitroso Groups*, Vol. 1, Wiley, NY, *1969*, pp. 264–265.

19-29   Oxidation of Tertiary Amines to Amine Oxides

N-Oxygen-attachment

$$R_3N \xrightarrow{H_2O_2} \overset{\oplus}{R_3N} \overset{\ominus}{-O}$$

Tertiary amines can be converted to amine oxides by oxidation. Hydrogen peroxide is often used, but peroxyacids are also important reagents for this purpose. Pyridine and its derivatives are oxidized by peroxyacids[777] rather than hydrogen peroxide. Note, however, that urea–H_2O_2 in formic acid does indeed oxidize pyridine.[778] In the attack by hydrogen peroxide there is first formed a trialkylammonium peroxide, a hydrogen-bonded complex represented as $R_3N \cdot H_2O_2$, which can be isolated.[779] The decomposition of this complex probably involves an attack by the OH moiety of the H_2O_2. Oxidation with Caro's acid has been shown to proceed in this manner:[780]

This mechanism is the same as that of **19-27**; the products differ only because tertiary amine oxides cannot be further oxidized. The mechanism with other peroxyacids is probably the same. A green procedure for oxidation of tertiary amines has been developed, using a Mg–Al complex with aq. hydrogen peroxide.[781]

An alternative oxidation using O_2 and a RuCl$_3$ catalyst converted pyridine to pyridine N-oxide.[782] Bromamine-T and RuCl$_3$ in aq. acetonitrile also oxidizes pyridine to the N-oxide.[783] Tertiary amines are oxidized to the N-oxide with O_2 and Fe_2O_3 in the presence of an aliphatic aldehyde.[784] Oxygen and a cobalt–Schiff base complex also oxidzes tertiary amines, including pyridine.[785]

It is noted that azo compounds can be oxidized to azoxy compounds by peroxyacids[786] or by hydroperoxides and molybdenum complexes.[787]

Analogous to the oxidation of tertiary amines, tertiary phosphines are oxidized to phosphine oxides, ($R_3P{=}O$). Triphenylphosphine is converted to triphenylphosphine

[777]For reviews, see Albini, A.; Pietra, S. *Heterocyclic N-Oxides*; CRC Press: Boca Raton, FL, *1991*, pp. 31–41; Katritzky, A.R.; Lagowski, J.M. *Chemistry of the Heterocyclic N-Oxides*, Academic Press, NY, *1971*, pp. 21–72, 539–542.

[778]Balicki, R.; Goliski, J. *Synth. Commun. 2000, 30*, 1529.

[779]Oswald, A.A.; Guertin, D.L. *J. Org. Chem. 1963, 28*, 651.

[780]Ogata, Y.; Tabushi, I. *Bull. Chem. Soc. Jpn. 1958, 31*, 969.

[781]Choudary, B.M.; Bharathi, B.; Reddy, Ch.V.; Kantam, M.L.; Raghavan, K.V. *Chem. Commun. 2001*, 1736.

[782]Jain, S.L.; Sain, B. *Chem. Commun. 2002*, 1040.

[783]Sharma, V.B.; Jain, S.L.; Sain, B. *Tetrahedron Lett. 2004, 45*, 4281.

[784]Wang, F.; Zhang, H.; Song, G.; Lu, X. *Synth. Commun. 1999, 29*, 11.

[785]Jain, S.L.; Sain, B. *Angew. Chem. Int. Ed. 2003, 42*, 1265.

[786]For reviews, see Yandovskii, V.N.; Gidaspov, B.V.; Tselinskii, I.V. *Russ. Chem. Rev. 1981, 50*, 164; Newbold, B.T., in Patai, S. *The Chemistry of the Hydrazo, Azo, and Azoxy Groups*, pt. 1, Wiley, NY, *1975*, pp. 557–563, 573–593.

[787]Johnson, N.A.; Gould, E.S. *J. Org. Chem. 1974, 39*, 407. For a mechanistic discussion, see Mitsuhashi, T.; Simamura, O.; Tezuka, Y. *Chem. Commun. 1970*, 1300.

oxide with N_2O at $100°C$, for example. Triphenylphosphine is also oxidized with PhIO on Montmorillonite K10.[788] *tert*-Butylhydroperoxide oxides $Ph_3 \rightarrow BH_3$ to $Ph_3P{=}O$.[789]

OS **IV**, 612, 704, 828; **VI**, 342, 501; **VIII**, 87.

19-30 Oxidation of Thiols and Other Sulfur Compounds to Sulfonic Acids

Thiol-sulfonic acid oxidation

$$RSH \xrightarrow{HNO_3} RSO_3H$$

Thiols, sulfoxides, sulfones, disulfides,[790] and other sulfur compounds can be oxidized to sulfonic acids with many oxidizing agents, but for synthetic purposes the reaction is most important for thiols.[791] Among oxidizing agents used are boiling nitric acid, barium permanganate, and dimethyl dioxirane.[792] Autoxidation (oxidation by atmospheric oxygen) can be accomplished in basic solution.[793] Oxidation of thiols with chlorine and water gives sulfonyl chlorides directly.[794] Thiols can also be oxidized to disulfides (**19-34**).

OS **II**, 471; **III**, 226. Also see, OS **V**, 1070.

19-31 Oxidation of Thioethers to Sulfoxides and Sulfones

S-Oxygen-attachment

$$R{-}S{-}R \xrightarrow{H_2O_2} \underset{R{-}\overset{O}{\overset{\|}{S}}{-}R}{} \xrightarrow{KMnO_4} \underset{R{-}\overset{O}{\underset{O}{\overset{\|}{S}}}{-}R}{}$$

Thioethers can be oxidized to sulfoxides by 1 equivalent of 30% H_2O_2 or by many other oxidizing agents,[795] including H_2O_2–flavin catalyst,[796] H_2O_2 and a

[788]Mielniczak, G.; Łopusiń ski, A. *Synlett* **2001**, 505.

[789]Uziel, J.; Darcel, C.; Moulin, D.; Bauduin, C.; Juge, S. *Tetrahedron Asymmetry* **2001**, *12*, 1441.

[790]For a review of the oxidation of disulfides, see Savige, W.E.; Maclaren, J.A., in Kharasch, N.; Meyers, C.Y. *Organic Sulfur Compounds*, Vol. 2; pp. 367–402, Pergamon, NY, *1966*.

[791]For a general review of the oxidation of thiols, see Capozzi, G.; Modena, G., in Patai, S. *The Chemistry of the Thiol Group*, pt. 2, Wiley, NY, *1974*, pp. 785–839. For a review specifically on the oxidation to sulfonic acids, see Gilbert, E.E. *Sulfonation and Related Reactions*, Wiley, NY, *1965*, pp. 217–239.

[792]Gu, D.; Harpp, D.N. *Tetrahedron Lett.* **1993**, *34*, 67.

[793]Wallace, T.J.; Schriesheim, A. *Tetrahedron* **1965**, *21*, 2271.

[794]For a review, see Gilbert, E.E. *Sulfonation and Related Reactions*, Wiley, NY, *1965*, pp. 202–214.

[795]For reviews, see Hudlický, M. *Oxidations in Organic Chemistry*, American Chemical Society, Washington, DC *1990*, pp. 252–263; Drabowicz, J.; Kiełbasinski, P.; Mikołajczyk, M., in Patai, S.; Rappoport, Z.; Stirling, C. *The Chemistry of Sulphones and Sulphoxides*, Wiley, NY, *1988*, pp. 233–378, pp. 235–255; Madesclaire, M. *Tetrahedron* **1986**, *42*, 5459; Block, E., in Patai, S. *The Chemistry of Functional Groups, Supplement E*, pt. 1, Wiley, NY, *1980*, pp. 539–608. For reviews on methods of synthesis of sulfoxides, see Drabowicz, J.; Mikołajczyk, M. *Org. Prep. Proced. Int.* **1982**, *14*, 45; Oae, S., in Oae, S. *The Organic Chemistry of Sulfur*, Plenum, NY, *1977*, pp. 385–390. For a review with respect to enzymic oxidation, see Holland, H.L. *Chem. Rev.* **1988**, *88*, 473.

[796]Lindén, A.A.; Krüger, L.; Bäckvall, J.-E. *J. Org. Chem.* **2003**, *68*, 5890.

Sc(OTf)$_3$ catalyst,[797] NaIO$_4$,[798] dioxiranes,[799] MeReO$_3$/H$_2$O$_2$,[800] O$_2$ and a ceric ammonium nitrate catalyst,[801] trichloroisocyanuric acid,[802] BnPh$_3$P HSO$_5$,[803] KO$_2$/Me$_3$SiCl,[804] Fe(NO$_3$)$_3$/FeBr$_3$/air,[805] singlet oxygen on MB–Bentonite composite,[806] MnO$_2$ with a H$_2$SO$_4$/SiO$_2$ catalyst,[807] hexamethylene triamine-Br$_2$ with CHCl$_3$–H$_2$O,[808] sodium perborate,[770] H$_5$IO$_6$/FeCl$_3$,[809] hypervalent iodine compounds,[810] and peroxyacids.[811] Sulfoxides can be further oxidized to sulfones by another equivalent of H$_2$O$_2$, KMnO$_4$, sodium perborate, or a number of other agents. If enough oxidizing agent is present, thioethers can be directly converted to sulfones without isolation of the sulfoxides.[812] Thioethers can be oxidized directly to the sulfone by treatment with excess NaOCl[813] tetramethylperruthenate (TPAP,)[814] H$_2$O$_2$ and an iron catalyst,[815] H$_2$O$_2$ and 10% Na$_2$WO$_4$,[816] H$_2$O$_2$/AcOH/ MgSO$_4$,[817] urea–H$_2$O$_2$,[818] peroxy monosulfate and a manganese catalyst,[819] or with NaIO$_4$/catalytic RuCl$_3$.[820]

These reactions give high yields, and many functional groups do not interfere.[821] As with tertiary amines (**19-29**), racemic thioethers can be kinetically resolved by

---

[797]Matteucci, M.; Bhalay, G.; Bradley, M. *Org.Lett.* **2003**, *5*, 235.

[798]Leonard, N.J.; Johnson, C.R. *J. Org. Chem.* **1962**, *27*, 282; Hiskey, R.G.; Harpold, M.A. *J. Org. Chem.* **1967**, *32*, 3191. For oxidation using NaI4 on silica gel with microwave irradiation, see Varma, R.S.; Saini, R.K.; Meshram, H.M. *Tetrahedron Lett.* **1997**, *38*, 6525.

[799]Colonna, S.; Gaggero, N. *Tetrahedron Lett.* **1989**, *30*, 6233. For a discussion of the mechanism, see González-Núñez, M.E.; Mello, R.; Royo, J.; Ríos, J.V.; Asensio, G. *J. Am. Chem. Soc.* **2002**, *124*, 9154.

[800]Yamazaki, S. *Bull. Chem. Soc. Jpn.* **1996**, *69*, 2955. A combination of H2O2 and Na2WO2 gives oxidation to the sulfone, see Choi, S.; Yang, J.-D.; Ji, M.; Choi, H.; Kee, M.; Ahn, K.-H.; Byeon, S.-H.; Baik, W.; Koo, S. *J. Org. Chem.* **2001**, *66*, 8192.

[801]Riley, D.P.; Smith, M.R.; Correa, P.E. *J. Am. Chem. Soc.* **1988**, *110*, 177.

[802]Zhong, P.; Guo, M.-P.; Huang, N.-P. *Synth. Commun.* **2002**, *32*, 175.

[803]Hajipour, A.R.; Mallakpour, S.E.; Adibi, H. *J. Org. Chem.* **2002**, *67*, 8666.

[804]Chen, Y.-J.; Huang, Y.-P. *Tetrahedron Lett.* **2000**, *41*, 5233.

[805]Martín, S.E.; Rossi, L.I. *Tetrahedron Lett.* **2001**, *42*, 7147.

[806]Madhavan, D.; Pitchumani, K. *Tetrahedron* **2001**, *57*, 8391.

[807]Firouzabadi, H.; Abbassi, M. *Synth. Commun.* **1999**, *129*, 1485.

[808]Shaabani, A.; Teimouri, M.B.; Safaei, H.R. *Synth. Commun.* **2000**, *30*, 265.

[809]Kim, S.S.; Nehru, K.; Kim, S.S.; Kim, D.W.; Jung, H.C. *Synthesis* **2002**, 2484.

[810]Shukla, V.G.; Salgaonkar, P.D.; Akamanchi, K.G. *J. Org. Chem.* **2003**, *68*, 5422.

[811]For lists of some of the many oxidizing agents used in this reaction, see Ref. 672 and Block, E. *Reactions of Organosulfur Compounds*, Academic Press, NY, **1978**, p. 16.

[812]For a review, see Schank, K., in Patai, S.; Rappoport, Z.; Stirling, C. *The Chemistry of Sulphones and Sulphoxides*, Wiley, NY, **1988**, pp. 165–231, 205–213.

[813]Khurana, J.M.; Panda, A.K.; Ragi, A.; Gogia, A. *Org. Prep. Proceed. Int.* **1996**, *28*, 234.

[814]Guertin, K.R.; Kende, A.S. *Tetrahedron Lett.* **1993**, *34*, 5369.

[815]Margues, A.; Marin, M.; Ruasse, M.-F. *J. Org. Chem.* **2001**, *66*, 7588.

[816]Sato, K.; Hyodo, M.; Aoki, M.; Zheng, X.-Q.; Noyori, R. *Tetrahedron* **2001**, *57*, 2469.

[817]Makosza, M.; Surowiec, M. *Org. Prep. Proceed. Int.* **2003**, *35*, 412.

[818]Balicki, R. *Synth. Commun.* **1999**, *29*, 2235.

[819]Iranpoor, N.; Mohajer, D.; Rezaeifard, A.-R. *Tetrahedron Lett.* **2004**, *45*, 3811.

[820]Su, W. *Tetrahedron Lett.* **1994**, *35*, 4955.

[821]For a review of the oxidation of α-halo sulfides, see Venier, C.G.; Barager III, H.J. *Org. Prep. Proceed. Int.* **1974**, *6*, 77, pp. 85-86.

oxidation to sulfoxides with an optically active reagent, and this has often been done.[822] In addition, the use of chiral additives in conjunction with various oxidizing agents leads to chiral nonracemic sulfoxide with good-to-excellent enantioselectivity.[823] Asymmetric oxidation using bacterial monooxygenases is known,[824] and horseradish peroxidase gives modest enantioselectivity.[825] Chiral sulfur reagents are also known.[826] Selenides (R_2Se) can be oxidized to selenoxides and selenones.[827] It is possible to oxidize a thioether to a sulfoxide in the presence of an alcohol moiety using MnO_2/HCl.[828] Alkyl disulfides give oxidation of one sulfur to give a (RS—S(=O)R compound with good enantioselectivity when using aqueous hydrogen peroxide, a catalytic amount of a vanadium catalyst and a chiral Schiff base ligand.[829] N-Sulfonyloxaziridines can be used to oxidize sulfides to sulfoxides.[830]

When the oxidizing agent is a peroxide, the mechanism[831] of oxidation to the sulfoxide is similar to that of **19-29**.[832]

[822]For reviews, see Kagan, H.B.; Rebiere, F. *Synlett* **1990**, 643; Drabowicz, J.; Kiebasinski, P.; Mikołajczyk, M. *Org. Prep. Proceed. Int.* **1982**, 14, 45, see p. 288.

[823]For example, see Donnoli, M.I.; Superchi, S.; Rosini, C. *J. Org. Chem.* **1998**, 63, 9392; Brunel, J.-M.; Kagan, H.B. *Synlett* **1996**, 404; Brunel, J.-M.; Diter, P.; Deutsch, M.; Kagan, H.B. *J. Org. Chem.* **1995**, 60, 8086; Davis, F.A.; Reddy, R.T.; Han, W.; Carroll, P.J. *J. Am. Chem. Soc.* **1992**, 114, 1428; Palucki, M.; Hanson, P.; Jacobsen, E.N. *Tetrahedron Lett.* **1992**, 33, 7111; Sandrinelli, F.; Perrio, S.; Beslin, P. *Org. Lett.* **1999**, 1, 1177; Tokunaga, M.; Ota, M.; Haga, M.-a.; Wakatsuki, Y. *Tetrahedron Lett.* **2001**, 42, 3865; Massa, A.; Lattanzi, A.; Siniscalchi, F.R.; Scettri, A. *Tetrahedron Asymmetry* **2001**, 12, 2775; Sun, J.; Zhu, C.; Dai, Z.; Yang, M.; Pan, Y.; Hu, H. *J. Org. Chem.* **2004**, 69, 8500; Krief, A.; Lonez, F. *Tetrahedron Lett.* **2002**, 43, 6255; Massa, A.; Sinissalchi, F.R.; Bugatti, V.; Lattanzi, A.; Scettri, A. *Tetrahederon Asymmetry* **2002**, 13, 1277; Barbarini, A.; Maggi, R.; Muratori, M.; Sartori, G.; Sartorio, R. *Tetrahedron Asymmetry* **2004**, 15, 2467; Ohta, C.; Shimizu, H.; Kondo, A.; Katsuki, T. *Synlett* **2002**, 161.

[824]Colonna, S.; Gaggero, N.; Pasta, P.; Ottolina, G. *Chem. Commun.* **1996**, 2303; Pasta, P.; Carrea, G.; Holland, H.L.; Dallavalle, S. *Tetrahedron Asymmetry*, **1995**, 6, 933.

[825]Ozaki, S.-i.; Watanabe, S.; Hayasaka, S.; Konuma, M. *Chem. Commun.* **2001**, 1654.

[826]Mikołajczyk, M.; Drabowicz, J.; Kiełbasiński, P. *Chiral Sulfur Reagents*, CRC Press, Boca Raton, FL, **1997**.

[827]See Reich, H.J., in Trahanovsky, W.S. *Oxidations in Organic Chemistry*, pt. C, Academic Press, NY, **1978**, pp. 7–13; Davis, F.A.; Stringer, O.D.; Billmers, J.M. *Tetrahedron Lett.* **1983**, 24, 1213; Kobayashi, M.; Ohkubo, H.; Shimizu, T. *Bull. Chem. Soc. Jpn.* **1986**, 59, 503.

[828]Gabbi, C.; Ghelfi, F.; Grandi, R. *Synth. Commun.* **1997**, 27, 2857.

[829]Blum, S.A.; Bergman, R.G.; Ellman, J.A. *J. Org. Chem.* **2003**, 68, 150.

[830]For a review of N-sulfonyloxaziridines, see: Davis, F.A.; Sheppard, A.C. *Tetrahedron* **1989**, 45, 5703. For the use of trifluoromethyl substituted N-phosphinoyloxaziridines, see Jennings, W.B.; O'Shea, J.H.; Schweppe, A. *Tetrahedron Lett.* **2001**, 42, 101.

[831]For discussions of the mechanism with various other agents, see Rajasekaran, K.; Baskaran, T.; Gnanasekaran, C. *J. Chem. Soc. Perkin Trans. 2* **1984**, 1183; Srinivasan, C.; Chellamani, A.; Rajagopal, S. *J. Org. Chem.* **1985**, 50, 1201; Agarwal, A.; Bhatt, P.; Banerji, K.K. *J. Phys. Org. Chem.* **1990**, 3, 174; Lee, D.G.; Chen, T. *J. Org. Chem.* **1991**, 56, 5346.

[832]Modena, G.; Todesco, P.E. *J. Chem. Soc.* **1962**, 4920, and references cited therein.

The second oxidation, which is normally slower than the first[833] (which is why sulfoxides are so easily isolable), has the same mechanism in neutral or acid solution, but in basic solution it has been shown that the conjugate base of the peroxy compound ($R'OO^-$) also attacks the SO group as a nucleophile:[834]

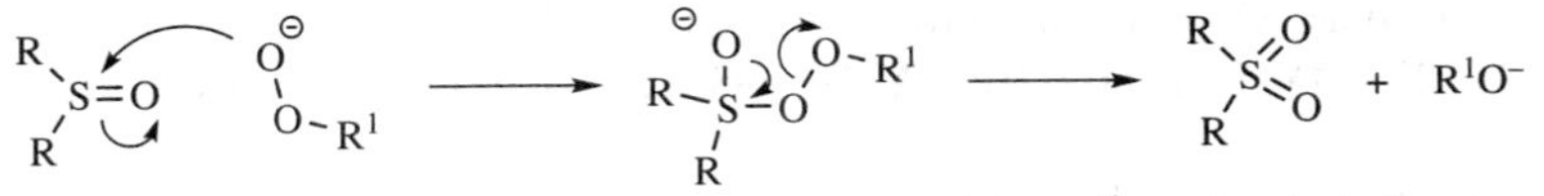

OS **V**, 791; **VI**, 403, 404, 482; **VII**, 453, 491; **VIII**, 464, 543; **IX**, 63; **80**, 190. Also see, OS **V**, 723; **VI**, 23.

E. Oxidative Coupling

19-32 Coupling Involving Carbanions

De-hydro,chloro-coupling

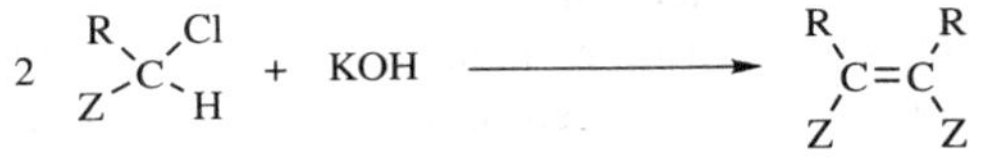

Alkyl halides with an electron-withdrawing group on the halogen-bearing carbon can be dimerized to alkenes by treatment with bases. The Z group may be nitro, aryl, and so on. It is likely that in most cases the mechanism[835] involves nucleophilic substitution followed by elimination[836] (illustrated for benzyl chloride):

$$PhCH_2Cl \xrightarrow{\text{base}} Ph\overset{\ominus}{C}HCl \xrightarrow[S_N]{PhCH_2Cl} PhCHClCH_2Ph \xrightarrow{-HCl} PhCH{=}CHPh$$

α,α-Dibromotoluenes ($ArCHBr_2$) give tolanes $ArC{\equiv}CAr$), by debromination of the intermediates $ArCBr{=}CBrAr$.[837] In a related reaction, diarylmethane dihalides

[833]There are some reagents that oxidize sulfoxides in preference to sulfides, for example, NaMnO4: see Henbest, H.B.; Khan, S.A. *Chem. Commun.* ***1968***, 1036.

[834]Curci, R.; Di Furia, F.; Modena, G. *J. Chem. Soc. Perkin Trans. 2* ***1978***, 603, and references cited therein. See also, Oae, S.; Takata, T. *Tetrahedron Lett.* ***1980***, *21*, 3213; Akasaka, T.; Ando, W. *J. Chem. Soc. Chem. Commun.* ***1983***, 1203.

[835]For discussion, see Saunders, Jr., W.H.; Cockerill, A.F. *Mechanisms of Elimination Reactions*, Wiley, NY, ***1973***, pp. 548–554.

[836]For example, see Hauser, C.R.; Brasen, W.R.; Skell, P.S.; Kantor, S.W.; Brodhag, A.E. *J. Am. Chem. Soc.* ***1956***, *78*, 1653; Hoeg, D.F.; Lusk, D.I. *J. Organomet. Chem.* ***1966***, *5*, 1; Reisdorf, D.; Normant, H. *Organomet. Chem. Synth.* ***1972***, *1*, 375; Hanna, S.B.; Wideman, L.G. *Chem. Ind. (London)* ***1968***, 486. In some cases, a radical anion chain mechanism can take place: Bethell, D.; Bird, R. *J. Chem. Soc. Perkin Trans. 2* ***1977***, 1856.

[837]Vernigor, E.M.; Shalaev, V.K.; Luk'yanets, E.A. *J. Org. Chem. USSR* ***1981***, *17*, 317.

(Ar_2CX_2) have been dimerized to tetraaryl alkenes ($Ar_2C=CAr_2$) with copper,[838] and with iron(II) oxalate dihydrate.[839]

A somewhat different type of coupling is observed when salts of β-keto esters, arylacetonitriles ($ArCH_2CN$), and other compounds of the form ZCH_2Z' are treated with an oxidizing agent, such as iodine,[840] or Cu(II) salts.[841] Arylmethanesulfonyl chlorides ($ArCH_2SO_2Cl$) couple to give $ArCH=CHAr$ when treated with Et_3N.[842]

OS **II**, 273; **IV**, 372, 869, 914; **VIII**, 298. Also see, OS **I**, 46; **IV**, 877.

19-33 Dimerization of Silyl Enol Ethers or of Lithium Enolates

3/O-De-trimethylsilyl-1/C-coupling

40

Silyl enol ethers can be dimerized to symmetrical 1,4-diketones by treatment with Ag_2O in DMSO or certain other polar aprotic solvents.[843] The reaction has been performed with R^2, R^3 = hydrogen or alkyl, although best yields are obtained when $R^2 = R^3 = H$. In certain cases, unsymmetrical 1,4-diketones have been prepared by using a mixture of two silyl enol ethers. Other reagents that have been used to achieve either symmetrical or cross-coupled products are iodosobenzene–BF_3–Et_2O,[844] ceric ammonium nitrate,[845] and lead tetraacetate.[846] If R^1 = OR (in which case the substrate is a ketene silyl acetal), dimerization with $TiCl_4$ leads to a dialkyl succinate (**40**, R^1 = OR).[847]

In a similar reaction, lithium enolates, $RC(OLi)=CH_2$, were dimerized to 1,4-diketones ($RCOCH_2CH_2COR$) with $CuCl_2$, $FeCl_3$, or copper(II) triflate, in a non-protic solvent.[848]

[838]Buckles, R.E.; Matlack, G.M. *Org. Synth.* **IV**, 914.

[839]Khurana, J.M.; Maikap, G.C.; Mehta, S. *Synthesis* **1990**, 731.

[840]See, for example, Kaiser, E.M. *J. Am. Chem. Soc.* **1967**, *89*, 3659; Belletire, J.L.; Spletzer, E.G.; Pinhas, A.R. *Tetrahedron Lett.* **1984**, *25*, 5969; Mignani, S.; Lahousse, F.; Merényi, R.; Janousek, Z.; Viehe, H.G. *Tetrahedron Lett.* **1985**, *26*, 4607; Aurell, M.J.; Gil, S.; Tortajada, A.; Mestres, R. *Synthesis* **1990**, 317.

[841]Rathke, M.W.; Lindert, A. *J. Am. Chem. Soc.* **1971**, *93*, 4605; Baudin, J.; Julia, M.; Rolando, C.; Verpeaux, J. *Bull. Soc. Chim. Fr.* **1987**, 493.

[842]King, J.F.; Durst, T. *Tetrahedron Lett.* **1963**, 585; King, J.F.; Harding, D.R.K. *Can. J. Chem.* **1976**, *54*, 2652; Nakayama, J.; Tanuma, M.; Honda, Y.; Hoshino, M. *Tetrahedron Lett.* **1984**, *25*, 4553.

[843]Ito, Y.; Konoike, T.; Saegusa, T. *J. Am. Chem. Soc.* **1975**, *97*, 649.

[844]Moriarty, R.; Prakash, O.; Duncan, M.P. *J. Chem. Soc. Perkin Trans. 1* **1987**, 559.

[845]Baiocchi, E.; Casu, A.; Ruzziconi, R. *Tetrahedron Lett.* **1989**, *30*, 3707.

[846]Moriarty, R.M.; Penmasta, R.; Prakash, I. *Tetrahedron Lett.* **1987**, *28*, 873.

[847]Inaba, S.; Ojima, I. *Tetrahedron Lett.* **1977**, 2009. See also, Totten, G.E.; Wenke, G.; Rhodes, Y.E. *Synth. Commun.* **1985**, *15*, 291, 301.

[848]Ito, Y.; Konoike, T.; Harada, T.; Saegusa, T. *J. Am. Chem. Soc.* **1977**, *99*, 1487; Kobayashi, Y.; Taguchi, T.; Tokuno, E. *Tetrahedron Lett.* **1977**, 3741; Frazier Jr., R.H.; Harlow, R.L. *J. Org. Chem.* **1980**, *45*, 5408.

OS **VIII**, 467.

19-34   Oxidation of Thiols to Disulfides

S-**De-hydrogen-coupling**

$$2\ \text{RSH} \xrightarrow{\text{H}_2\text{O}_2} \text{RSSR}$$

Thiols are easily oxidized to disulfides.[849] Hydrogen peroxide is the most common reagent,[850] but many oxidizing agents give the reaction, among them KMnO$_4$/CuSO$_4$,[851] Me$_2$SO–I$_2$,[852] Br$_2$ under phase-transfer conditions,[853] Br$_2$ on hydrated silica,[854] sodium perborate,[855] NaI/air,[856] *t*-BuOOH/VO(acac)$_2$,[857] SmI$_2$,[858] PPh$_3$ with a rhodium catalyst,[859] dibromohydantoin,[860] cetyltrimethylammonium dichromate,[861] and NO. It can also be done electrochemically.[862] Hydrogen peroxide 30% in hexafluoroisopropanol converts thiols to disulfides,[863] on Clayan with microwave irradiation,[864] and solventless reactions on MnO$_2$,[865] PCC (p. 1716)[866] or SO$_2$Cl$_2$[867] are also effective. However, strong oxidizing agents may give **19-26**. Even the oxygen in the air oxidizes thiols on standing, if a small amount of base is present. The reaction is reversible (see **19-75**), and the interconversion between cysteine and cystine is an important one in biochemistry.

[849]For a review, see Capozzi, G.; Modena, G., in Patai, S. *The Chemistry of the Thiol Group*, pt. 2, Wiley, NY, *1974*, pp. 785–839. For a list of reagents, with references, see Block, E. *Reactions of Organosulfur Compounds*, Academic Press, NY, *1978*.

[850]It has been pointed out that, nevertheless, H2O2 is not a very good reagent for this reaction, since it gives sulfonic acids (**19-30**) as well as disulfides: Evans, B.J.; Doi, J.T.; Musker, W.K. *J. Org. Chem. 1990*, *55*, 2337.

[851]Noureldin, N.A.; Caldwell, M.; Hendry, J.; Lee, D.G. *Synthesis 1998*, 1587.

[852]Aida, T.; Akasaka, T.; Furukawa, N.; Oae, S. *Bull. Chem. Soc. Jpn. 1976*, *49*, 1441. See also, Fristad, W.E.; Peterson, J.R. *Synth. Commun. 1985*, *15*, 1.

[853]Drabowicz, J.; Mikołajczyk, M. *Synthesis 1980*, 32.

[854]Ali, M.H.; McDermott, M. *Tetrahedron Lett. 2002*, *43*, 6271.

[855]McKillop, A.; Koyunçu, D. *Tetrahedron Lett. 1990*, *31*, 5007.

[856]Iranpoor, N.; Zeynizadeh, B. *Synthesis 1999*, 49.

[857]Raghavan, S.; Rajender, A.; Joseph, S.C.; Rasheed, M.A. *Synth. Commun. 2001*, *31*, 1477.

[858]Zhan, Z.-P.; Lang, K.; Liu, F.; Hu, L.-m. *Synth. Commun. 2004*, *34*, 3203.

[859]Tanaka, K.; Ajiki, K. *Tetrahedron Lett. 2004*, *45*, 25.

[860]Khazaei, A.; Zolfigol, M.A.; Rostami, A. *Synthesis 2004*, 2959.

[861]Patel, S.; Mishra, B.K. *Tetrahedron Lett. 2004*, *45*, 1371. See also Tajbakhsh, M.; Hosseinzadeh, R.; Shakoori, A. *Tetrahedron Lett. 2004*, *45*, 1889.

[862]See, for example, Leite, S.L.S.; Pardini, V.L.; Viertler, H. *Synth. Commun. 1990*, *20*, 393. For a review, see Shono, T. *Electroorganic Chemistry as a New Tool in Organic Synthesis*, Springer, NY, *1984*, pp. 38–43.

[863]Kesavan, V.; Bonnet-Delpon, D.; Bégué, J.-P. *Synthesis 2000*, 223.

[864]Meshram, H.M.; Bandyopadhyay, A.; Reddy, G.S.; Yadav, J.S. *Synth. Commun. 2000*, *30*, 701.

[865]Firouzabadi, H.; Abbassi, M.; Karimi, B. *Synth. Commun. 1999*, *129*, 2527.

[866]Salehi, P.; Farrokhi, A.; Gholizadeh, M. *Synth. Commun. 2001*, *31*, 2777.

[867]Leino, R.; Lönnqvist, J.-E. *Tetrahedron Lett. 2004*, *45*, 8489.

The mechanism has been studied for several oxidizing agents and varies with the agent.[868] For oxygen it is[869]

$$RSH + B^- \rightleftharpoons RS^- + BH$$

$$RS^- + O_2 \longrightarrow RS\cdot + \cdot O_2^-$$

$$RS^- + \cdot O_2^- \longrightarrow RS\cdot + O_2{}^{2-}$$

$$2\ O_2{}^{2-} + 2\ BH \longrightarrow 2\ OH^- + 2\ B^- + O_2$$

With respect to the sulfur, this mechanism is similar to that of **14-16**, involving as it does loss of a proton, oxidation to a free radical, and radical coupling.

Unsymmetrical disulfides can be prepared[870] by treatment of a thiol RSH with diethyl azodicarboxylate EtOOCN=NCOOEt to give an adduct, to which another thiol R′SH is then added, producing the disulfide RSSR′.[871]

OS **III**, 86, 116.

REDUCTIONS

For the most part, reductions have been grouped into this chapter, with a few notable exceptions. Catalytic hydrogenation of alkenes and alkynes in **15-11** and **15-12**, hydrogenation of aromatic rings in **15-13** and reductive cleavage of cyclopropanes in **15-15** were placed in Chapter 15 to coincide with addition reactions, and protonolysis of alkyl boranes in **15-16** was placed there also for continuity. In general, reductions of functional groups encompass a variety of reaction types. The reactions in this section are classified into groups depending on the type of bond change involved. These groups are (*1*) attack at carbon (C—O and C=O), (*2*) attack at non-carbonyl multiple bonds to heteroatoms, (*3*) reactions in which a heteroatom is removed from the substrate, (*4*) reduction with cleavage, (*5*) reductive coupling, and (*6*) reactions in which an organic substrate is both oxidized and reduced. Most of the reagents in this section are metal hydrides, metals with an acid or a protic solvent, hydrogen gas with a catalyst, and so on. Other reducing agents are available, and will be introduced in the appropriate section. Note that plants can be used as reducing agents.[872]

[868]See Tarbell, D.S. in Kharasch, N. *Organic Sulfur Compounds*, Pergamon, Elmsford, NY, *1961*, pp. 97–102.

[869]Wallace, T.J.; Schriesheim, A.; Bartok, W. *J. Org. Chem.* *1963*, *28*, 1311.

[870]Mukaiyama, T.; Takahashi, K. *Tetrahedron Lett.* *1968*, 5907.

[871]For other methods, see Boustany, K.S.; Sullivan, A.B. *Tetrahedron Lett.* *1970*, 3547; Harpp, D.N.; Ash, D.K.; Back, T.G.; Gleason, J.G.; Orwig, B.A.; VanHorn, W.F.; Snyder, J.P. *Tetrahedron Lett.* *1970*, 3551; Oae, S.; Fukushima, D.; Kim, Y.H. *J. Chem. Soc. Chem. Commun.* *1977*, 407.

[872]Bruni, R.; Fantin, G.; Medici, A.; Pedrini, P.; Sacchetti, G. *Tetrahedron Lett.* *2002*, *43*, 3377.

TABLE 19.2. The Ease of Reduction of Various Functional Groups Toward Catalytic Hydrogenation.[876]

Reaction	Substrate[a]	Product	
19-39	RCOCl	RCHO	Easiest
19-45	RNO_2	RNH_2	
15-11	$RC{\equiv}CR$	RCH=CHR	
19-36	RCHO	RCH_2OH	
15-11	RCH=CHR	RCH_2CH_2R	
19-36	RCOR	RCHOHR	
19-56	$ArCH_2OR$	$ArCH_3$ + ROH	
19-43	$RC{\equiv}N$	RCH_2NH_2	
15-14			
19-38	RCOOR'	RCH_2OH + R'OH	
19-64	RCOHNR'	RCH_2NHR	
15-13			Most difficult
19-37	$RCOO^-$		Inert

[a]The groups are listed in approximate order of ease of reduction.

Selectivity[873]

It is often necessary to reduce one group in a molecule without affecting another reducible group. It is usually possible to find a reducing agent that will do this. The most common broad-spectrum reducing agents are the metal hydrides[874] and hydrogen (with a catalyst).[875] Many different metal-hydride systems and hydrogenation catalysts have been investigated in order to find conditions under which a given group will be reduced chemoselectively. Tables 19.2–19.4 list the reactivity of various functional groups toward catalytic hydrogenation, $LiAlH_4$, and BH_3, respectively.[876]

[873]For monographs on reductions in general, see Hudlický, M. *Reductions in Organic Chemistry*, Wiley, NY, *1984*; Augustine, R.L. *Reduction*, Marcel Dekker, NY, *1968*. For a review, see Candlin, J.P.; Rennie, R.A.C., in Bentley, K.W.; Kirby, G.W. *Elucidation of Chemical Structures by Physical and Chemical Methods* (Vol. 4 of Weissberger, A. *Techniques of Chemistry*), 2nd ed., pt. 2, Wiley, NY, *1973*, pp. 77–135.
[874]For discussions of selectivity with metal hydride reducing agents, see Brown, H.C.; Krishnamurthy, S. *Tetrahedron 1979*, *35*, 567; Walker, E.R.H. *Chem. Soc. Rev. 1976*, *5*, 23; Brown, H.C. *Boranes in Organic Chemistry*, Cornell University Press, Ithaca, NY, *1972*, pp. 209–251; Rerick, M.N., in Augustine, R.L. *Reduction*, Marcel Dekker, NY, *1968*. For books, see, in Ref. 10, the works by Seyden-Penne, J.; Strouf, O. et al., and Hajós, A.
[875]For a discussion of catalyst selectivity for hydrogenations, see Rylander, P.N. *Aldrichimica Acta 1979*, *12*, 53. See also, Rylander, P.N. *Hydrogenation Methods*, Academic Press, NY, *1985*.
[876]Table 19.2 is from House, H.O. *Modern Synthetic Reactions*, 2nd ed., W.A. Benjamin, NY, *1972*, p. 9. Tables 19.3 and 19.4 are from Brown, H.C. *Boranes in Organic Chemistry*, Cornell University Press, Ithaca, NY, *1972*, pp. 213 and 232, respectively.

TABLE 19.3. The Ease of Reduction of Various Functional Groups with LiAlH₄ in Ether[876]

Reaction	Substrate[a]	Product	
19-36	RCHO	RCH₂OH	Easiest
19-36	RCOR	RCHOHR	
19-63	RCOCl	RCH₂OH	
19-38	Lactone	Diol	
19-35	$\overset{H}{\underset{R}{\diagdown}}C \overset{}{\underset{O}{-}}C\overset{H}{\underset{R}{\diagup}}$	RCH₂CHOHR	
19-38	RCOOR'	RCH₂OH + R'OH	
19-37	RCOOH	RCH₂OH	
19-37	RCOO⁻	RCH₂OH	
19-64	RCONR'₂	RCH₂NR'₂	
19-43	RC≡N	RCH₂NH₂	
19-45	RNO₂	RNH₂	
19-80	ArNO₂	ArN=NAr	Most difficult
15-11	RCH=CHR⁻	Inert	

[a]However, LiAlH₄ is a very powerful reagent, and much less chemoselectivity is possible here than with most of the other metal hydrides.

Table 19.5 shows which groups can be reduced by catalytic hydrogenation and various metal hydrides.[877] Of course, the tables cannot be exact, because the nature of R and the reaction conditions obviously affect reactivity. Nevertheless, the tables do give a fairly good indication of which reagents reduce which

TABLE 19.4. The Ease of Reduction of Various Functional Groups With Borane[876]

Reaction	Substrate[a]	Product	
19-37	RCOOH	RCH₂OH	Easiest
15-16	RCH=CHR	(RCH₂CHR)₃B	
19-36	RCOR	RCHOHR	
19-43	RCN	RCH₂NH₂	
19-35	$\overset{H}{\underset{R}{\diagdown}}C \overset{}{\underset{O}{-}}C\overset{H}{\underset{R}{\diagup}}$	RCH₂CHOHR	
19-38	RCOOR'	RCH₂OH + R'OH	Most difficult
19-39,19-63	RCOCl	Inert	

[a]It is evident that this reagent and LiAlH₄ (Table 19.3) complement each other.

[877]The first 10 columns are from Brown, H.C.; Krishnamurthy, S. *Tetrahedron* **1979**, *35*, 567, p. 604. The column on (*i*-Bu)2AlH is from Yoon, N.M.; Gyoung, Y.S. *J. Org. Chem.* **1985**, *50*, 2443; the one on NaAlEt2H2 from Stinson, S.R. *Chem. Eng. News, Nov. 3,* **1980**, *58*, No. 44, 19; and the one on LiBEt3H from Brown, H.C.; Kim, S.C.; Krishnamurthy, S. *J. Org. Chem.* **1980**, *45*, 1. For similar tables that show additional reducting agents, see Pelter, A.; Smith, K.; Brown, H.C. *Borane Reagents*, Academic Press, NY, **1988**, p. 129; Hajós, A. *Complex Hydrides*, Elsevier, NY, **1979**, pp. 16–17. For tables showing which agents reduce a wide variety of functional groups, see Hudlický, M. *Reductions in Organic Chemistry*, Wiley, NY, **1984**, pp. 177–200.

groups.[878] Lithium aluminium hydride is a very powerful and unselective reagent.[879] Consequently, other metal hydrides are generally used when chemoselectivity is required. As will be seen on p. 1794, a number of less reactive (and more selective) reagents have been prepared by replacing some of the hydrogens of $LiAlH_4$ with alkoxy groups (by treatment of $LiAlH_4$ with ROH).[880] Most of the metal hydrides are nucleophilic reagents and attack the carbon atom of a carbon-hetero single or multiple bond. Another useful reagent is $LiAlHSeH$.[881] However, BH_3[882,883] and AlH_3[884] are electrophiles (Lewis acids) and attack the heteroatom. This accounts for the different patterns of selectivity shown in the tables.

TABLE 19.5. Reactivity of Various Functional Groups With Some Metal Hydrides and Toward Catalytic Hydrogenation.[872]

Reaction[a]	A	B	C	D[374]	E[885]	F[886]	G	H	I	J[887]	K[888]	L	M	N
19-36 RCHO ⟶ RCH₂OH	+	+	+	+	+	+	+	+	+	+	+	+	+	+
19-36 RCOR ⟶ RCHOHR	+	+	+	+	+	+	+	+	+	+	+	+	+	+
19-39 ⟶ RCHO														
19-63 RCOCl ↗ ↘ RCH₂OH	+[889]	+	+	−	−	+	+	+	+	+	+	+	+	+
19-63 lactone ⟶ diol	−	+	+	+	+	+	±	+	+	+	+	+	+	+
19-35 epoxide ⟶ alcohol	−	+	+	+	±	±	±	+	+	+	+	+	+	+
19-38 RCOOR′ ⟶ RCH₂OH +R′OH	−	+	+	±	−	±	±	+	+	+	+	+	+	+
19-37 RCOOH ⟶ RCH₂OH	−	−	+	+	−	±	−	+	+	+	−	+	+	−

(Continued)

[878]See also, the table in Hudlický, M. *J. Chem. Educ.* **1977**, *54*, 100.

[879]For a review of LiAlH4, see Pizey, J.S. *Synthetic Reagents*, Vol. 1, Wiley, NY, **1974**, pp. 101–194.

[880]For reviews of reductions by these reagents, see Málek, J. *J. Org. Chem.* **1988**, *36*, 249; **1985**, *34*, 1; Málek, J.; Č erny, M. *Synthesis* **1972**, 217.

[881]Ishihara, H.; Koketsu, M.; Fukuta, Y.; Nada, F. *J. Am. Chem. Soc.* **2001**, *123*, 8408.

[882]See Brown, H.C.; Heim, P.; Yoon, N.M. *J. Am. Chem. Soc.* **1970**, *92*, 1637; Cragg, G.M.L. *Organoboranes in Organic Synthesis*, Marcel Dekker, NY, **1973**, pp. 319–371. For reviews of reductions with BH3, see Wade, R.C. *J. Mol. Catal.* **1983**, *18*, 273 (BH3 and a catalyst); Lane, C.F. *Chem. Rev.* **1976**, *76*, 773; *Aldrichimica Acta* **1977**, *10*, 41; Brown, H.C.; Krishnamurthy, S. *Aldrichimica Acta* **1979**, *12*, 3. For reviews of reduction with borane derivatives, see Pelter, A.; Smith, K.; Brown, H.C. *Borane Reagents*, Academic Press, NY, **1988**, pp. 125–164; Pelter, A. *Chem. Ind. (London)* **1976**, 888.

[883]Reacts with solvent, reduced in aprotic solvents.

[884]Reduced to aldehyde (**19-44**)

[885]Brown, H.C.; Bigley, D.B.; Arora, S.K.; Yoon, N.M. *J. Am. Chem. Soc.* **1970**, *92*, 7161. For reductions with thexylborane, see Brown, H.C.; Heim, P.; Yoon, N.M. *J. Org. Chem.* **1972**, *37*, 2942.

[886]Brown, H.C.; Krishnamurthy, S.; Yoon, N.M. *J. Org. Chem.* **1976**, *41*, 1778.

[887]See Yoon, N.M.; Brown, H.C. *J. Am. Chem. Soc.* **1968**, *90*, 2927.

[888]Brown, H.C.; Kim, S.C.; Krishnamurthy, S. *J. Org. Chem.* **1980**, *45*, 1. For a review of the synthesis of alkylsubstituted borohydrides, see Brown, H.C.; Singaram, B.; Singaram, S. *J. Organomet. Chem.* **1982**, *239*, 43.

[889]See Brown, H.C.; Heim, P.; Yoon, N.M. *J. Am. Chem. Soc.* **1970**, *92*, 1637; Cragg, G.M.L. *Organoboranes in Organic Synthesis*, Marcel Dekker, NY, **1973**, pp. 319–371. For reviews of reductions with BH3, see Wade, R.C. *J. Mol. Catal.*, **1983**, *18*, 273 (BH3 and a catalyst); Lane, C.F. *Chem. Rev.* **1976**, *76*, 773; *Aldrichimica Acta* **1977**, *10*, 41; Brown, H.C.; Krishnamurthy, S. *Aldrichimica Acta* **1979**, *12*, 3. For reviews of reduction with borane derivatives, see Pelter, A.; Smith, K.; Brown, H.C. *Borane Reagents*, Academic Press, NY, **1988**, pp. 125–164; Pelter, A. *Chem. Ind. (London)* **1976**, 888.

TABLE 19.5. (*Continued*)

Reaction[a]	A	B	C	D[374]	E[885]	F[886]	G	H	I	J[887]	K[888]	L	M	N
19-37 RCOO⁻ ⟶ RCH₂OH	−	−	+	−	−	−	−	+	+	+	−	−	−	
19-64 ⟶ RCH₂NR₂′ RCNR₂′ ↗														
19-41 ↘ RCHO	−	−	−	+	+	+	−	+	+	+	+	+	+	+
19-43 RC≡N ⟶ RCH₂NH₂	−	−	−	+	−	±	−	+	+	+	±	+[384]	+	+
19-45 ⟶ RCH₂NH₂′ RCONR₂′ ↗														
19-80 ↘ RCHO	−	−	−	−	−	−	−	+	+	−	−	+[890]	+	+
15-11 RCH=CHR ⟶ RCH₂CH₂R	−	−	−	+	+	+	−	−	−	−	−	+	−	+

A = NaBH₄ in EtOH. B = NaBH₄ + LiCl in diglyme. C = NaBH₄ + AlCl₃ in diglyme. D = BH₃–THF. E = bis-3-methyl-2-butylborane (disiamylborane) in THF. F = 9-BBN. G = LiAlH(O*t*-Bu)₃ in THF. H = LiAlH(OMe)₃ in THF. I = LiAlH₄ in ether. J = AlH₃ in THF. K = LiBEt₃H. L = (*i*Bu)₂AlH [DIBALH]. M = NaAlEt₂H₂. N = catalytic hydrogenation.

19-53   RX + LiAlH₄ ⟶ RH

19-57   R–OSO₂R′ + LiAlH₄ ⟶ RH

19-35

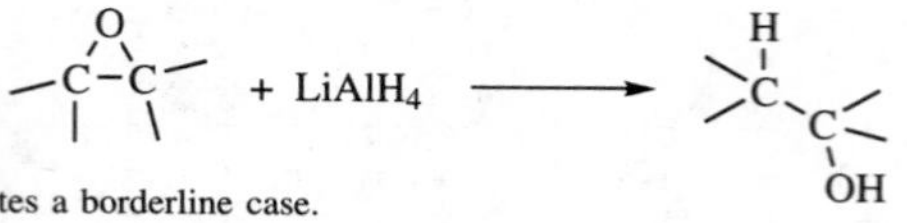

±indicates a borderline case.

A. Attack at Carbon (C–O and C=O)

19-35   Reduction of Epoxides

(3)*OC-seco*-Hydro-de-alkoxylation

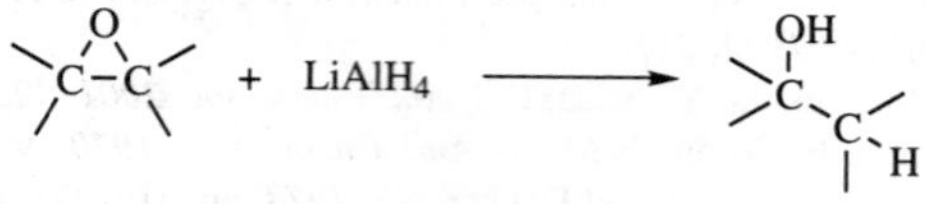

Reduction of epoxides is a special case of **19-56** and is easily carried out.[891] The most common reagent is LiAlH₄,[892] which reacts by the S_N2-type mechanism, giving inversion of configuration. An epoxide on a substituted cyclohexane ring cleaves in such a direction as to give an axial alcohol. As expected for an S_N2 mechanism, cleavage usually occurs so that a tertiary alcohol is formed if possible. If not, a secondary alcohol is preferred. However, for certain substrates, the epoxide ring can be opened the other way by reduction with NaBH₄–ZrCl₄,[893] Pd/C and

[890]Reduced to hydroxylamine (**19-46**).

[891]For a list of reagents, with references, see Larock, R.C. *Comprehensive Organic Transformations*, 2nd ed., Wiley-VCH, NY, *1999*, pp. 1019–1027.

[892]See Healy, E.F.; Lewis, J.D.; Minniear, A.B. *Tetrahedron Lett.* **1994**, *35*, 6647 for a discussion of the LiAlH4 reduction of unsaturated cyclic epoxides.

[893]Laxmi, Y.R.S.; Iyengar, D.S. *Synth. Commun.* **1997**, *27*, 1731 (addition of L-proline to this reaction leads to moderate asymmetric induction).

$HCOONH_4$,[894] $SiO_2-Zn(BH_4)_2$,[895] or with BH_3 in THF.[896] The reaction has also been carried out with other reagents, for example, sodium amalgam in EtOH, Li in ethylenediamine,[897] $Bu_3SnH-NaI$,[898] and by catalytic hydrogenolysis.[899] Chemoselective and regioselective ring opening (e.g., of allylic epoxides and of epoxy ketones and esters) has been achieved with SmI_2,[900] $HCOOH-NEt_3$ and a palladium catalyst,[901] and sodium bis(2-methoxyethoxy)aluminum hydride (Red-Al).[902] Highly hindered epoxides can be conveniently reduced, without rearrangement, with lithium triethylborohydride.[903]

Epoxy ketones are selectively reduced with lithium naphthalenide[904] or Cp_2TiCl in THF/MeOH[905] to the β-hydroxyketone. Other reduction methods can lead to the epoxy alcohol (see p.$$$). Reduction of epoxy amides with SmI_2 in methanol gave the α-hydroxyamide.[906]

Epi-sulfides can be reduced to give the alkene using Bu_3SnH in the presence of BEt_3.[907]

Epoxides can be reductively halogenated (the product is the alkyl bromide or iodide rather than the alcohol) with $Me_3SiCl-NaX-(Me_2SiH)_2O$ (1,1,3,3-tetramethyldisiloxane).[908]

The usual product of epoxide reductions is the alcohol, but epoxides are reduced all the way to the alkane by titanocene dichloride[909] and by $Et_3SiH-BH_3$.[910]

[894]Dragovich, P.S.; Prins, T.J.; Zhou, R. *J. Org. Chem.* **1995**, *60*, 4922. For reduction with a palladium catalyst in formic acid see Ley, S.V.; Mitchell, C.; Pears, D.; Ramarao, C.; Yu, J.Q.; Zhou, W. *Org. Lett.* **2003**, *5*, 4665.

[895]Ranu, B.C.; Das, A.R. *J. Chem. Soc. Perkin Trans. 1* **1992**, 1881.

[896]For a review of epoxide reduction with BH3, see Cragg, G.M.L. *Organoboranes in Organic Synthesis*, Marcel Dekker, NY, **1973**, pp. 345–348. See also Yamamoto, Y.; Toi, H.; Sonoda, A.; Murahashi, S. *J. Chem. Soc. Chem. Commun.* **1976**, 672.

[897]Brown, H.C.; Ikegami, S.; Kawakami, J.H. *J. Org. Chem.* **1970**, *35*, 3243.

[898]Bonini, C.; Di Fabio, R. *Tetrahedron Lett.* **1988**, *29*, 819.

[899]For a review, see Rylander, P.N. *Catalytic Hydrogenation over Platinum Metals*, Academic Press, NY, **1967**, pp. 478–485. See Oshima, M.; Yamazaki, H.; Shimizu, I.; Nizar, M.; Tsuji, J. *J. Am. Chem. Soc.* **1989**, *111*, 6280.

[900]Molander, G.A.; La Belle, B.E.; Hahn, G. *J. Org. Chem.* **1986**, *51*, 5259; Otsubo, K.; Inanaga, J.; Yamaguchi, M. *Tetrahedron Lett.* **1987**, *28*, 4437. See also, Miyashita, M.; Hoshino, M.; Suzuki, T.; Yoshikoshi, A. *Chem. Lett.* **1988**, 507.

[901]Noguchi, Y.; Yamada, T.; Uchiro, H.; Kobayashi, S. *Tetrahedron Lett.* **2000**, *41*, 7493, 7499.

[902]Gao, Y.; Sharpless, K.B. *J. Org. Chem.* **1988**, *53*, 4081.

[903]Krishnamurthy, S.; Schubert, R.M.; Brown, H.C. *J. Am. Chem. Soc.* **1973**, *95*, 8486.

[904]Jankowska, R.; Liu, H.-J.; Mhehe, G.L. *Chem. Commun.* **1999**, 1581.

[905]Hardouin, C.; Chevallier, F.; Rousseau, B.; Doris, E. *J. Org. Chem.* **2001**, *66*, 1046.

[906]Concellón, J.M.; Bardales, E. *Org. Lett.* **2003**, *5*, 4783.

[907]Uenishi, J.; Kubo, Y. *Tetrahedron Lett.* **1994**, *35*, 6697.

[908]Aizpurua, J.M.; Palomo, C. *Tetrahedron Lett.* **1984**, *25*, 3123.

[909]van Tamelen, E.E.; Gladys, J.A. *J. Am. Chem. Soc.* **1974**, *96*, 5290.

[910]Fry, J.L.; Mraz, T.J. *Tetrahedron Lett.* **1979**, 849.

19-36 Reduction of Aldehydes and Ketones to Alcohols[911]

C,O-**Dihydro-addition**

$$\underset{C}{\overset{O}{\parallel}} + \text{LiAlH}_4 \xrightarrow{\hspace{1cm}} \xrightarrow{\text{H}^+} \underset{C}{\overset{\text{H} \quad \text{OH}}{}}$$

Aldehydes can be reduced to primary alcohols, and ketones to secondary alcohols, by a number of reducing agents,[912] of which LiAlH$_4$ and other metallic hydrides are the most commonly used.[913] These reagents have two main advantages over many other reducing agents: They do not reduce carbon–carbon double or triple bonds (with the exception of propargylic alcohols),[914] and with LiAlH$_4$ all four hydrogens are usable for reduction. The reaction is broad and general. Lithium aluminum hydride easily reduces aliphatic, aromatic, alicyclic, and heterocyclic aldehydes, containing double or triple bonds and/or nonreducible groups, such as NR$_3$, OH, OR, and F. If the molecule contains a group reducible by LiAlH$_4$ (e.g., NO$_2$, CN, COOR), then it is also reduced. Since LiAlH$_4$ reacts readily with water and alcohols, these compounds must be excluded. Common solvents are ether and THF. The compound NaBH$_4$ has a similar scope, but is more selective and so may be used with NO$_2$, Cl, COOR, CN, and so on in the molecule. Another advantage of NaBH$_4$ is that it can be used in water or alcoholic solvents and so reduces compounds, such as sugars that are not soluble in ethers.[915] Other solvents can be used with some modification of the borohydride. For example, butyltriphenylphosphonium borohydride reduces aldehydes to alcohols in dichloromethane.[916] A polymer-bound phase-transfer material with NaBH$_4$ in wet THF has also been used.[917] Sodium borohydride on alumina, under microwave irradiation, is also an effective reagent.[918] Sodium borohydride has been used on silica gel.[919] The scope of these reagents with ketones is similar to that with aldehydes. Lithium aluminum hydride reduces even sterically hindered ketones.

[911]See Smith, M.B. *Organic Synthesis*, 2nd ed., McGraw-Hill, NY, *2001*, pp. 306–368.

[912]For a review, see Hudlický, M. *Reductions in Organic Chemistry*, Ellis Horwood, Chichester, *1984*, pp. 96–129. For a list of reagents, with references, see Larock, R.C. *Comprehensive Organic Transformations*, 2nd ed., Wiley-VCH, NY, *1999*, pp. 1075–1113.

[913]For books on metal hydrides, see Abdel-Magid, A.F., Ed., *Reductions in Organic Synthesis*, American Chemical Society, Washington, DC, *1996*; Seyden-Penne, J. *Reductions by the Alumino- and Borohydrides*, VCH, NY, *1991*; Hajos, A. *Complex Hydrides*, Elsevier, NY, *1979*. For reviews, see House, H.O. *Modern Synthetic Reactions*, 2nd ed., W.A. Benjamin, NY, *1972*, pp. 49–71; Wheeler, O.H., in Patai, S. *The Chemistry of the Carbonyl Group*, pt. 1, Wiley, NY, *1966*, pp. 507–566.

[914]See Meta, C.T.; Koide, K. *Org. Lett.* *2004*, *6*, 1785; Naka, T.; Koide, K. *Tetrahedron Lett.* *2003*, *44*, 443.

[915]The compound NaBH4 reduces solid ketones in the absence of any solvent (by mixing the powders): Toda, F.; Kiyoshige, K.; Yagi, M. *Angew. Chem. Int. Ed.* *1989*, *28*, 320.

[916]Hajipour, A.R.; Mallakpour, S.E. *Synth. Commun.* *2001*, *31*, 1177.

[917]Tamami, B.; Mahdavi, H. *Tetrahedron* *2003*, *59*, 821.

[918]Varma, R.S.; Saini, R.K. *Tetrahedron Lett.* *1997*, *38*, 4337.

[919]Yakabe, S.; Hirano, M.; Morimoto, T. *Synth. Commun.* *1999*, *29*, 295; Liu, W.-y.; Xu, Q.-h.; Ma, Y.-x. *Org. Prep. Proceed. Int.* *2000*, *32*, 596.

The double bonds that are generally not affected by metallic hydrides may be isolated or conjugated, but double bonds that are conjugated with the C=O group may or may not be reduced, depending on the substrate, reagent, and reaction conditions.[920] Some reagents that reduce only the C=O bonds of α,β-unsaturated aldehydes and ketones are AlH_3,[921] $NaBH_4$, or $LiAlH_4$ in the presence of lanthanide salts,[922] cobalt complexes,[923] nickel compounds,[924] I_2,[925] $NaBH_3(OAc)$,[926] $Zn(BH_4)_2$[927] on Y-zeolite,[928] and Et_3SiH.,[929] Also, both $LiAlH_4$[930] and $NaBH_4$[931] predominantly reduce only the C=O bonds of C=C–C=O systems in most cases, although substantial amounts of fully saturated alcohols have been found in some cases[930] (**15-14**). For some reagents that reduce only the C=C bonds of conjugated aldehydes and ketones, see **15-11**. A mixture of $InCl_3$ and $NaBH_4$ reduced both the C=C and C=O units of conjugated ketones.[932]

When a functional group is selectively attacked in the presence of a different functional group, the reaction is said to be *chemoselective*.[933] A number of reagents have been found to reduce aldehydes much faster than ketones. Among these[934] are sodium triacetoxyborohydride[935] ($NaBH_4$–HCOOH),[936] zinc borohydride in THF,[937] bis-(isopropoxytitanium borohydride),[938] a complex of $LiAlH_4$ and N-methyl-2-pyrrolidinone (of particular interest since it is stable in air and to heating),[939] and Raney nickel.[940] On

[920]For a review of the reduction of α,β-unsaturated carbonyl compounds, see Keinan, E.; Greenspoon, N., in Patai, S.; Rappoport, Z. *The Chemistry of Enones*, pt. 2, Wiley, NY, *1989*, pp. 923–1022.

[921]Jorgenson, M.J. *Tetrahedron Lett.* *1962*, 559; Dilling, W.L.; Plepys, R.A. *J. Org. Chem.* *1970*, *35*, 2971.

[922]Gemal, A.L.; Luche, J. *J. Am. Chem. Soc.* *1981*, *103*, 5454; Fukuzawa, S.; Fujinami, T.; Yamauchi, S.; Sakai, S. *J. Chem. Soc. Perkin Trans. 1* *1986*, 1929. See also Chênevert, R.; Ampleman, G. *Chem. Lett.* *1985*, 1489; Varma, R.S.; Kabalka, G.W. *Synth. Commun.* *1985*, *15*, 985.

[923]Ohtsuka, Y.; Koyasu, K.; Ikeno, T.; Yamada, T. *Org. Lett.* *2001*, *3*, 2543.

[924]Khurana, J.M.; Chauhan, S. *Synth. Commun.* *2001*, *31*, 3485.

[925]Singh, J.; Kaur, I.; Kaur, J.; Bhalla, A.; Kad, G.L. *Synth. Commun.* *2003*, *33*, 191.

[926]Nutaitis, C.F.; Bernardo, J.E. *J. Org. Chem.* *1989*, *54*, 5629.

[927]For a review of the reactivity of this reagent, see Ranu, B. *Synlett* *1993*, 885.

[928]Sreekumar, R.; Padmakumar, R.; Rugmini, P. *Tetrahedron Lett.* *1998*, *39*, 5151.

[929]Ojima, I.; Kogure, T. *Organometallics* *1982*, *1*, 1390.

[930]Johnson, M.R.; Rickborn, B. *J. Org. Chem.* *1970*, *35*, 1041.

[931]Chaikin, S.W.; Brown, W.G. *J. Am. Chem. Soc.* *1949*, *71*, 122.

[932]Ranu, B.C.; Samanta, S. *Tetrahedron* *2003*, *59*, 7901.

[933]See Luibrand, R.T.; Taigounov, I.R.; Taigounov, A.A. *J. Org. Chem.* *2001*, *66*, 7254.

[934]For some others (not all of them metal hydrides), see Hutchins, R.O.; Kandasamy, D. *J. Am. Chem. Soc.* *1973*, *95*, 6131; Risbood, P.A.; Ruthven, D.M. *J. Org. Chem.* *1979*, *44*, 3969; Babler, J.H.; Invergo, B.J. *Tetrahedron Lett.* *1981*, *22*, 621; Fleet, G.W.J.; Harding, P.J.C. *Tetrahedron Lett.* *1981*, *22*, 675; Yamaguchi, S.; Kabuto, K.; Yasuhara, F. *Chem. Lett.* *1981*, 461; Kim, S.; Kang, H.J.; Yang, S. *Tetrahedron Lett.* *1984*, *25*, 2985; Kamitori, Y.; Hojo, M.; Masuda, R.; Yamamoto, M. *Chem. Lett.* *1985*, 253; Borbaruah, M.; Barua, N.C.; Sharma, R.P. *Tetrahedron Lett.* *1987*, *28*, 5741.

[935]Gribble, G.W.; Ferguson, D.C. *J. Chem. Soc. Chem. Commun.* *1975*, 535. See also, Nutaitis, C.F.; Gribble, G.W. *Tetrahedron Lett.* *1983*, *24*, 4287.

[936]Blanton, J.R. *Synth. Commun.* *1997*, *27*, 2093.

[937]Ranu, B.C.; Chakraborty, R. *Tetrahedron Lett.* *1990*, *31*, 7663; See Ranu, B. *Synlett* *1993*, 885.

[938]Ravikumar, K.S.; Chandrasekaran, S. *Tetrahedron* *1996*, *52*, 9137.

[939]Fuller, J.C.; Stangeland, E.L.; Jackson, T.C.; Singaram, B. *Tetrahedron Lett.* *1994*, *35*, 1515. See also, Mogali, S.; Darville, K.; Pratt, L.M. *J. Org. Chem.* *2001*, *66*, 2368.

[940]Barrero, A.F.; Alvarez-Manzaneda, E.J.; Chahboun, R.; Meneses, R. *Synlett* *2000*, 197.

the other hand, ketones can be chemoselectively reduced in the presence of aldehydes with $NaBH_4$ in aq. EtOH at $-15°C$ in the presence of cerium trichloride $CeCl_3$.[941] The reagent lithium n-dihydropyridylaluminum hydride reduces diaryl ketones much better than dialkyl or alkyl aryl ketones.[942] Most other hydrides reduce diaryl ketones more slowly than other types of ketones. Saturated ketones can be reduced in the presence of α,β-unsaturated ketones with $NaBH_4$-50% $MeOH—CH_2Cl_2$ at $-78°C$[943] and with zinc borohydride.[944]

In general, $NaBH_4$ reduces carbonyl compounds in this order: aldehydes > α,β-unsaturated aldehydes > ketones > α,β-unsaturated ketones, and a carbonyl group of one type can be selectively reduced in the presence of a carbonyl group of a less reactive type.[945] A number of reagents will preferentially reduce the less sterically hindered of two carbonyl compounds, but by the use of DIBALH in the presence of the Lewis acid methylaluminum bis(2,16-di-tert-butyl-4-methylphenoxide), it was possible selectively to reduce the *more hindered* of a mixture of two ketones.[946] It is obvious that reagents can often be found to reduce one kind of carbonyl function in the presence of another.[947] For a discussion of selectivity in reduction reactions, see p. 1787. A syn-selective reduction of β-hydroxy ketones was achieved using $(iPrO)_2TiBH_4$.[948]

Quinones are reduced to hydroquinones by $LiAlH_4$, $SnCl_2—HCl$, or sodium hydrosulfite ($Na_2S_2O_4$), as well as by other reducing agents.

The reagent lithium tri-sec-butylborohydride $LiBH(sec-Bu)_3$ (L-Selectride) reduces cyclic and bicyclic ketones in a highly stereoselective manner,.[949] For example, 2-methylcyclohexanone gave cis-2-methylcyclohexanol with an isomeric purity >99%. Both L-Selectride and the potassium salt (K-Selectride) reduce carbonyls in cyclic and acyclic molecules with high diastereoselectivity.[950] The more usual reagents, for example, $LiAlH_4$, $NaBH_4$, reduce relatively unhindered cyclic ketones either with little or no stereoselectivity[951] or give predominant formation of the more stable isomer (axial attack).[952] Mixed reagents, such as

[941]See Gemal, A.L.; Luche, J. *Tetrahedron Lett.* **1981**, *22*, 4077; Li, K.; Hamann, L.G.; Koreeda, M. *Tetrahedron Lett.* **1992**, *33*, 6569.

[942]Lansbury, P.T.; Peterson, J.O. *J. Am. Chem. Soc.* **1962**, *84*, 1756.

[943]Ward, D.E.; Rhee, C.K.; Zoghaib, W.M. *Tetrahedron Lett.* **1988**, *29*, 517.

[944]Sarkar, D.C.; Das, A.R.; Ranu, B.C. *J. Org. Chem.* **1990**, *55*, 5799.

[945]Ward, D.E.; Rhee, C.K. *Can. J. Chem.* **1989**, *67*, 1206.

[946]Maruoka, K.; Araki, Y.; Yamamoto, H. *J. Am. Chem. Soc.* **1988**, *110*, 2650.

[947]For lists of some of these chemoselective reagents, with references, see Larock, R.C. *Comprehensive Organic Transformations*, 2nd ed., Wiley-VCH, NY, **1999**, pp. 1089–1092, and references given in Ward, D.E.; Rhee, C.K. *Can. J. Chem.* **1989**, *67*, 1206.

[948]Ravikumar, K.S.; Sinha, S.; Chandrasekaran, S. *J. Org. Chem.* **1999**, *64*, 5841.

[949]Brown, H.C.; Krishnamurthy, S. *J. Am. Chem. Soc.* **1972**, *94*, 7159; Krishnamurthy, S.; Brown, H.C. *J. Am. Chem. Soc.* **1976**, *98*, 3383.

[950]K-Selectride: Lawson, E.C.; Zhang, H.-C.; Maryanoff, B.E. *Tetrahedron Lett.* **1999**, *40*, 593.

[951]For reviews of the stereochemistry and mechanism, see Caro, B.; Boyer, B.; Lamaty, G.; Jaouen, G. *Bull. Soc. Chim. Fr.* **1983**, II-281; Boone, J.R.; Ashby, E.C. *Top. Stereochem.* **1979**, *11*, 53; Wigfield, D.C. *Tetrahedron* **1979**, *35*, 449. For a review of stereoselective synthesis of amino alcohols by this method, see Tramontini, M. *Synthesis* **1982**, 605.

[952]For a discussion of why this isomer is predominantly formed, see Mukherjee, D.; Wu, Y.; Fronczek, F.R.; Houk, K.N. *J. Am. Chem. Soc.* **1988**, *110*, 3328.

$LiBH_3[N(C_3H_7)_2]$, gives high selectivity for axial attack.[953] Reduction of cyclohexanone derivatives with the very hindered $LiAlH(CEt_2CMe_3)_3$ gave primarily the cis-alcohol.[954] Cyclohexanones that have a large degree of steric hindrance near the carbonyl group usually give predominant formation of the less stable alcohol, even with $LiAlH_4$ and $NaBH_4$.

Other reagents reduce aldehydes and ketones to alcohols,[955] including:

1. *Hydrogen and a Catalyst.*[956] The most common catalysts are platinum and ruthenium, but homogeneous catalysts have also been used,[957] including copper on silica gel[958] and a ruthenium catalyst on mesoporous silica.[959] Before the discovery of the metal hydrides this was one of the most common ways of effecting this reduction, but it suffers from the fact that C=C, C≡C, C=N, and C≡N bonds are more susceptible to attack than C=O bonds.[960] For aromatic aldehydes and ketones, reduction to the hydrocarbon (**19-61**) is a side reaction, stemming from hydrogenolysis of the alcohol initially produced (**19-54**).

2. *Sodium in Ethanol.*[961] This is called the *Bouveault–Blanc procedure* and was more popular for the reduction of carboxylic esters (**19-38**) than of aldehydes or ketones before the discovery of $LiAlH_4$.

For the reaction with sodium in ethanol the following mechanism[962] has been suggested:[963]

[953]Harrison, J.; Fuller, J.C.; Goralski, C.T.; Singaram, B. *Tetrahedron Lett.* **1994**, *35*, 5201.
[954]Boireau, G.; Deberly, A.; Toneva, R. *Synlett* **1993**, 585. In this study, reduction with LiAlH(Ot-Bu)3 was shown to give primarily the trans-alcohol.
[955]This can also be done electrochemically. For a review, see Feoktistov, L.G.; Lund, H., in Baizer, M.M.; Lund, H. *Organic Electochemistry*, Marcel Dekker, NY, **1983**, pp. 315–358, 315–326. See also, Coche, L.; Moutet, J. *J. Am. Chem. Soc.* **1987**, *109*, 6887.
[956]For reviews, see Abdel-Magid, A.F., Ed., *Reductions in Organic Synthesis*, American Chemical Society Washington, DC, **1996**, pp. 31–50; Parker, D., in Hartley, F.R. *The Chemistry of the Metal-Carbon Bond*, Vol. 4, Wiley, NY, **1987**, pp. 979–1047; Tanaka, K., in Červený, l. *Catalytic Hydrogenation*, Elsevier, NY, **1986**, pp. 79–104; Rylander, P.N. *Hydrogenation Methods*, Academic Press, NY, **1985**, pp. 66–77; Rylander, P.N. *Catalytic Hydrogenation over Platinum Metals*, Academic Press, NY., **1967**, pp. 238–290.
[957]For a review, see Heck, R.F. *Organotransition Metal Chemistry*, Academic Press, NY, **1974**, pp. 65–70.
[958]Ravasio, N.; Psaro, R.; Zaccheria, F. *Tetrahedron Lett.* **2002**, *43*, 3943.
[959]Kesanli, B.; Lin, W. *Chem. Commun.* **2004**, 2284.
[960]For catalysts that allow hydrogenation of only the C=O bond of α,β-unsaturated aldehydes, see Galvagno, S.; Poltarzewski, Z.; Donato, A.; Neri, G.; Pietropaolo, R. *J. Chem. Soc. Chem. Commun.* **1986**, 1729; Farnetti, E.; Pesce, M.; Kaspar, J.; Spogliarich, R.; Graziani, M. *J. Chem. Soc. Chem. Commun.* **1986**, 746; Narasimhan, C.S.; Deshpande, V.M.; Ramnarayan, K. *J. Chem. Soc. Chem. Commun.* **1988**, 99.
[961]For a discussion, see House, H.O. *Modern Synthetic Reactions*, 2nd ed., W.A. Benjamin, NY, **1972**, pp. 152–160.
[962]For reviews of the mechanisms of these reactions, see Pradhan, S.K. *Tetrahedron* **1986**, *42*, 6351; Huffman, J.W. *Acc. Chem. Res.* **1983**, *16*, 399. For discussions of the mechanism in the absence of protic solvents, see Huffman, J.W.; Liao, W.; Wallace, R.H. *Tetrahedron Lett.* **1987**, *28*, 3315; Rautenstrauch, V. *Tetrahedron* **1988**, *44*, 1613; Song, W.M.; Dewald, R.R. *J. Chem. Soc. Perkin Trans.* 2 **1989**, 269. For a review of the stereochemistry of these reactions in liquid NH3, see Rassat, A. *Pure Appl. Chem.* **1977**, *49*, 1049.
[963]House, H.O. *Modern Synthetic Reactions*, 2nd ed., W.A. Benjamin, NY, **1972**, p. 151. See, however, Giordano, C.; Perdoncin, G.; Castaldi, G. *Angew. Chem. Int. Ed.* **1985**, *24*, 499.

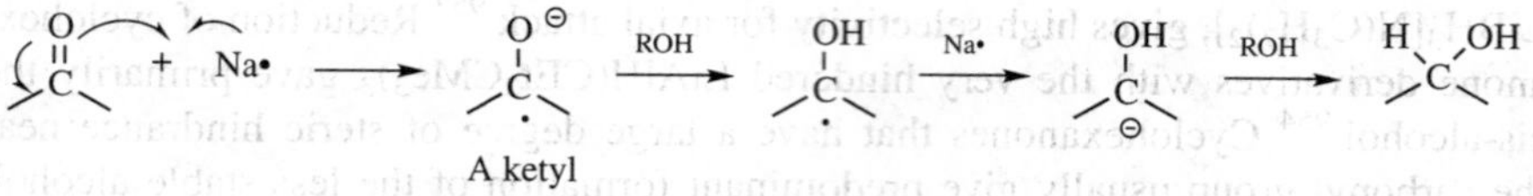

A ketyl

The ketyl intermediate can be isolated.[964]

3. *Isopropyl Alcohol and Aluminum Isopropoxide.* This is called the *Meerwein–Ponndorf–Verley reduction.*[965] It is reversible, and the reverse reaction is known as the *Oppenauer oxidation* (see **19-3**):

The equilibrium is shifted by removal of the acetone by distillation. There is a report of the reduction of benzaldehyde to benzyl alcohol by heating with Z-propanol at 225°C for 1 day.[966] The reaction takes place under very mild conditions and is highly specific for aldehydes and ketones, so that C=C bonds (including those conjugated with the C=O bonds) and many other functional groups can be present without themselves being reduced.[967] This includes acetals, so that one of two carbonyl groups in a molecule can be specifically reduced if the other is first converted to an acetal. β-Keto esters, β-diketones, and other ketones and aldehydes with a relatively high enol content do not give this reaction. A SmI_3-assisted version of this reduction has been reported.[968] Zeolites have been used as a medium for this reduction.[969] This reduction can be done catalytically[970] and an aluminum-free, zirconium zeolite catalyst has been developed.[971] A combination of Z-propanol with BINOL and $AlMe_3$ leads to reduction of α-chloroketones to the chlorohydrin with good enantioselectivity.[972] Microwave irradiation of a ketone with Z-propanol, KOH, and activated alumina gives good yields of the alcohol.[973]

---

[964]For example, see Rautenstrauch, V.; Geoffroy, M. *J. Am. Chem. Soc.* **1976**, *98*, 5035; **1977**, *99*, 6280.

[965]For other catalysts, see Akamanchi, K.G.; Noorani, V.R. *Tetrahedron Lett.* **1995**, *36*, 5085; Akamanchi, K.G.; Varalakshmy, N.R. *Tetrahedron Lett.* **1995**, *36*, 3571; Maruoka, K.; Saito, S.; Concepcion, A.B.; Yamamoto, H. *J. Am. Chem. Soc.* **1993**, *115*, 1183. For a microwave-induced version of this reaction, see Barbry, D.; Torchy, S. *Tetrahedron Lett.* **1997**, *38*, 2959.

[966]Bagnell, L.; Strauss, C.R. *Chem. Commun.* **1999**, 287.

[967]Diisobornyloxyaluminum isopropoxide gives higher yields under milder conditions than aluminum isopropoxide: Hutton, J. *Synth. Commun.* **1979**, *9*, 483. For other substitutes for aluminum isopropoxide, see Namy, J.L.; Souppe, J.; Collin, J.; Kagan, H.B. *J. Org. Chem.* **1984**, *49*, 2045; Okano, T.; Matsuoka, M.; Konishi, H.; Kiji, J. *Chem. Lett.* **1987**, 181.

[968]Evans, D.A.; Nelson, S.G.; Gagné, M.R.; Muci, A.R. *J. Am. Chem. Soc.* **1993**, *115*, 9800.

[969]Corma, A.; Domine, M.E.; Nemeth, L.; Valencia, S. *J. Am. Chem. Soc.* **2002**, *124*, 3194.

[970]Campbell, E.J.; Zhou, H.; Nguyen, S.T. *Org. Lett.* **2001**, *3*, 2391. See Albrecht, M.; Crabtree, R.H.; Mata, J.; Peris, E. *Chem. Commun.* **2002**, 32.

[971]Zhu, Y.; Chuah, G.; Jaenicke, S. *Chem. Commun.* **2003**, 2734.

[972]Campbell, E.J.; Zhou, H.; Nguyen, S.T. *Angew. Chem. Int. Ed.* **2002**, *41*, 1020.

[973]Kazemi, F.; Kiasat, A.R. *Synth. Commun.* **2002**, *32*, 2255.

The Meerwein–Ponndorf–Verley reaction usually[974] involves a cyclic transition state:[975]

but in some cases 2 equivalents of aluminum alkoxide are involved: one attacking the carbon and the other the oxygen, a conclusion that stems from the finding that in these cases the reaction was 1.5 order in alkoxide.[976] Although, for simplicity, we have shown the alkoxide as a monomer, it actually exists as trimers and tetramers, and it is these that react.[977]

4. *Metal Reductions.* A single carbonyl group of an α-diketone can be reduced (to give an α-hydroxy ketone) by heating with zinc powder in aq. DMF[978] or zinc in methanol in the presence of benzyltriethylammonium chloride.[979] This has also been accomplished with aq. VCl_2[980] and with $Zn-ZnCl_2-EtOH$.[981] Aluminum and NaOH in aqueous methanol reduces ketones.[982] β-Hydroxy ketones are reduced with good anti-selectivity using an excess of SmI_2 in water,[983] and other ketones or aldehydes are reduced with SmI_2[984] in aq. THF,[985] in 2-propanol,[986] or methanol.[987] Other metals can be used, including $FeCl_3/Zn$ in aq. DMF[988] or DME/MeOH.[989] 1,2-Diketones were reduced to the α-hydroxy ketone with TiI_4 in acetonitrile, followed by hydrolysis.[990] Ammonia and aq. $TiCl_3$ in methanol reduces ketones.[991]

[974]It has been that shown in some cases reduction with metal alkoxides, including aluminum isopropoxide, involves free-radical intermediates (SET mechanism): Screttas, C.G.; Cazianis, C.T. *Tetrahedron* **1978**, *34*, 933; Nasipuri, D.; Gupta, M.D.; Banerjee, S. *Tetrahedron Lett.* **1984**, *25*, 5551; Ashby, E.C.; Argyropoulos, J.N. *J. Org. Chem.* **1986**, *51*, 3593; Yamataka, H.; Hanafusa, T. *Chem. Lett.* **1987**, 643.

[975]See, for example, Shiner, Jr., V.J.; Whittaker, D. *J. Am. Chem. Soc.* **1963**, *85*, 2337; Warnhoff, E.W.; Reynolds-Warnhoff, P.; Wong, M.Y.H. *J. Am. Chem. Soc.* **1980**, *102*, 5956.

[976]Moulton, W.N.; Van Atta, R.E.; Ruch, R.R. *J. Org. Chem.* **1961**, *26*, 290.

[977]Williams, E.D.; Krieger, K.A.; Day, A.R. *J. Am. Chem. Soc.* **1953**, *75*, 2404; Shiner, Jr., V.J.; Whittaker, D. *J. Am. Chem. Soc.* **1969**, *91*, 394.

[978]Kreiser, W. *Liebigs Ann. Chem.* **1971**, *745*, 164.

[979]Kardile, G.B.; Desai, D.G.; Swami, S.S. *Synth. Commun.* **1999**, *29*, 2129.

[980]Ho, T.; Olah, G.A. *Synthesis* **1976**, 815.

[981]Toda, F.; Tanaka, K.; Tange, H. *J. Chem. Soc. Perkin Trans. 1* **1989**, 1555.

[982]Bhar, S.; Guha, S. *Tetrahedron Lett.* **2004**, *45*, 3775.

[983]Keck, G.E.; Wager, C.A.; Sell, T.; Wager, T.T. *J. Org. Chem.* **1999**, *64*, 2172.

[984]See Prasad, E.; Flowers II, R.A. *J. Am. Chem. Soc.* **2002**, *124*, 6895.

[985]Fukuzawa, S.-i.; Miura, M.; Matsuzawa, H. *Tetrahedron Lett.* **2000**, *42*, 4167; Dahlén, A.; Hilmersson, G. *Tetrahedron Lett.* **2002**, *43*, 7197.

[986]Fukuzawa, S.-i.; Nakano, N.; Saitoh, T. *Eur. J. Org. Chem.* **2004**, 2863.

[987]Keck, G.E.; Wager, C.A. *Org. Lett.* **2000**, *2*, 2307.

[988]Sadavarte, V.S.; Swami, S.S.; Desai, D.G. *Synth. Commun.* **1998**, *28*, 1139.

[989]Chopade, P.R.; Davis, T.A.; Prasad, E.; Flowers II, R.A. *Org. Lett.* **2004**, *6*, 2685.

[990]Hayakawa, R.; Sahara, T.; Shimizu, M. *Tetrahedron Lett.* **2000**, *41*, 7939.

[991]Clerici, A.; Pastori, N.; Porta, O. *Eur. J. Org. Chem.* **2001**, 2235.

5. *Boranes.* Borane (BH$_3$) and substituted boranes reduce aldehydes and ketones in a manner similar to their addition to C=C bonds (**15-16**).[992] That is, the boron adds to the oxygen and the hydrogen to the carbon:[993]

The borate is then hydrolyzed to the alcohol. Both 9-BBN[994] (p. 1077) and BH$_3$—Me$_2$S[995] reduce only the C=O group of conjugated aldehydes and ketones. A variety of alkylboranes can be used for reduction.[996] Borane reduction of a titanium complex of a 1,3-diketone gives the syn-diol.[997] Reduction occurs with B$_{10}$H$_{14}$ with CeCl$_3$,[998] Alane (AlH$_3$) derivatives can also be used, including diisobutylaluminum hydride.[999] Tributylborane in ionic solvents reduces aldehydes to alcohols.[1000]

6. *Tin Hydrides.* Tributyltin hydride reduces aldehydes to primary alcohols by simply heating in methanol.[1001] A mixture of Bu$_3$SnH and phenylboronic acid (p. 815) reduces aldehydes in dichloromethane.[1002] Reduction of ketones was achieved with Bu$_2$SnH$_2$ and a palladium catalyst.[1003] Using triaryltin hydrides with BF$_3$•OEt$_2$, where aryl is 2,6-diphenylbenzyl, selective reduction of aliphatic aldehydes in the presence of a conjugated aldehyde was achieved.[1004]

7. *Cannizzaro Reaction.* In the Cannizzaro reaction (**19-81**), aldehydes without an α hydrogen are reduced to alcohols.

8. *Silanes.* In the presence of bases, certain silanes can selectively reduce carbonyls. Epoxy-ketones are reduced to epoxy-alcohols, for example, with

[992]For a review, see Cragg, G.M.L. *Organoboranes in Organic Synthesis*, Marcel Dekker, NY, *1973*, pp. 324–335. See Cha, J.S.; Moon, S.J.; Park, J.H. *J. Org. Chem.* *2001*, *66*, 7514.

[993]Brown, H.C.; Subba Rao, B.C. *J. Am. Chem. Soc.* *1960*, *82*, 681; Brown, H.C.; Korytnyk, W. *J. Am. Chem. Soc.* *1960*, *82*, 3866.

[994]Krishnamurthy, S.; Brown, H.C. *J. Org. Chem.* *1975*, *40*, 1864; Lane, C.F. *Aldrichimica Acta 1976*, *9*, 31.

[995]Mincione, E. *J. Org. Chem.* *1978*, *43*, 1829.

[996]Smith, K.; El-Hiti, G.A.; Hou, D.; De Boos, G.A. *J. Chem. Soc. Perkin Trans. 1 1999*, 2807.

[997]Bartoli, G.; Bosco, M.; Bellucci, M.C.; Daplozzo, R., Marcantoni, E.; Sambri, L. *Org. Lett. 2000*, *2*, 45.

[998]Bae, J.W.; Lee, S.H.; Jung, Y.J.; Yoon, C.-O.M.; Yoon, C.M. *Tetrahedron Lett.* *2001*, *42*, 2137.

[999]Nakamura, S.; Kuroyanagi, M.; Watanabe, Y. Toru, T. *J. Chem. Soc. Perkin Trans. 1 2000*, 3143.

[1000]In bmim PF$_6$, 1-butyl-3-methylimidazoliium hexafluorophosphate and in emim PF$_6$, 1-ethyl-3-methylimidazolium hexafluorophosphate: Kabalka, G.W.; Malladi, R.R. *Chem. Commun. 2000*, 2191.

[1001]Kamiura, K.; Wada, M. *Tetrahedron Lett.* *1999*, *40*, 9059; Fung, N.Y.M.; de Mayo, P.; Schauble, J.H.; Weedon, A.C. *J. Org. Chem.* *1978*, *43*, 3977; Shibata, I.; Yoshida, T.; Baba, A.; Matsuda, H. *Chem. Lett.* *1989*, 619; Adams, C.M.; Schemenaur, J.E. *Synth. Commun.* *1990*, *20*, 2359. For a review, see Kuivila, H.G. *Synthesis 1970*, 499.

[1002]Yu, H.; Wang, B. *Synth. Commun.* *2001*, *31*, 2719.

[1003]Kamiya, I.; Ogawa, A. *Tetrahedron Lett.* *2002*, *43*, 1701.

[1004]Sasaki, K.; Komatsu, N.; Shivakawa, S.; Maruoka, K. *Synlett 2002*, 575.

(MeO)$_3$SiH and LiOMe.[1005] Controlling temperature and solvent leads to different ratios of syn- and anti- products.[1006] Silanes reduce ketones in the presence of BF$_3$•OEt$_2$[1007] and transition-metal compounds catalyze this reduction.[1008] Ketones are reduced with Cl$_3$SiH in the presence of pyrrolidine carboxaldehyde[1009] or under photochemical conditions.[1010] Polymethylhydrosiloxane with tetrabutylammonium fluoride reduces α-amino ketones to give the syn-amino alcohol.[1011]

9. *Ammonium Formates.* Sodium formate and trialkylammonium formates can be used to reduce aldehydes and ketones to the corresponding alcohol. Decanal was reduced to decanol, for example, using sodium formate in *N*-methyl-2-pyrrolidinone as a solvent.[1012] A mixture of formic acid and ethyl magnesium bromide was used to reduce decanal to decanol in 70% yield.[1013]

Unsymmetrical ketones are prochiral (p. 193); that is, reduction creates a new stereogenic center:

Much effort has been put into finding optically active reducing agents that will produce one enantiomer of the alcohol enantioselectively, and considerable success has been achieved,.[1014] Each reagent tends to show a specificity for certain types of ketones.[1015] H.C. Brown and co-workers[1016] reduced various types of ketone with a number of reducing agents. These workers also determined the relative effectiveness of various reagents for reduction of eight other types of ketone, including heterocyclic, aralkyl, β-keto esters, β-keto acids,[1017] and so on.[1016] In most cases, good enantioselectivity can be obtained with the proper reagent.[1018] Substituents that are

---

[1005]Hojo, M.; Fujii, A.; Murakami, C.; Aihara, H.; Hosomi, A. *Tetrahedron Lett.* **1995**, *36*, 571.

[1006]See Yamamoto, Y.; Matsuoka, K.; Nemoto, H. *J. Am. Chem. Soc.* **1988**, *110*, 4475.

[1007]Smonou, I. *Tetrahedron Lett.* **1994**, *35*, 2071.

[1008]Schmidt, T. *Tetrahedron Lett.* **1994**, *35*, 3513.

[1009]Iwasaki, F.; Onomura, O.; Mishima, K.; Maki, T.; Matsumura, Y. *Tetrahedron Lett.* **1999**, *40*, 7507.

[1010]Enholm, E.J.; Schulte II, J.P. *J. Org. Chem.* **1999**, *64*, 2610.

[1011]Nadkarni, D.; Hallissey, J.; Mojica, C. *J. Org. Chem.* **2003**, *68*, 594.

[1012]Babler, J.H.; Sarussi, S.J. *J. Org. Chem.* **1981**, *46*, 3367.

[1013]Babler, J.H.; Invergo, B.J. *Tetrahedron Lett.* **1981**, *22*, 621.

[1014]For reviews, see Singh, V.K. *Synthesis* **1992**, 605; Midland, M.M. *Chem. Rev.* **1989**, *89*, 1553; Nógrádi, M. *Stereoselective Synthesis*, VCH, NY, **1986**, pp. 105–130; in Morrison, J.D. *Asymmetric Synthesis*, Academic Press, NY, **1983**, the articles by Midland, M.M. Vol. 2, pp. 45–69, and Grandbois, E.R.; Howard, S.I.; Morrison, J.D. Vol. 2, pp. 71–90; Haubenstock, H. *Top. Stereochem.* **1983**, *14*, 231.

[1015]For a list of many of these reducing agents, with references, see Larock, R.C. *Comprehensive Organic Transformations*, 2nd ed., Wiley-VCH, NY, **1999**, pp. 1097–1111.

[1016]Brown, H.C.; Park, W.S.; Cho, B.T.; Ramachandran, P.V. *J. Org. Chem.* **1987**, *52*, 5406.

[1017]Wang, Z.; La, B.; Fortunak, J.M.; Meng, X.-J.; Kabalka, G.W. *Tetrahedron Lett.* **1998**, *39*, 5501.

[1018]See Brown, H.C.; Ramachandran, P.V.; Weissman, S.A.; Swaminathan, S. *J. Org. Chem.* **1990**, *55*, 6328; Rama Rao, A.V.; Gurjar, M.K.; Sharma, P.A.; Kaiwar, V. *Tetrahedron Lett.* **1990**, *31*, 2341; Midland, M.M.; Kazubski, A.; Woodling, R.E. *J. Org. Chem.* **1991**, *56*, 1068.

remote to the carbonyl group can play a role in facial selectivity of the reduction.[1019] Successful asymmetric reductions have been achieved with biologically derived reducing agents,[1020] such as baker's yeast,[1021] enzymes from other organisms,[1022] or with biocatalysts.[1023] Immobilized bakers yeast has been used in an ionic liquid.[1024]

Asymmetric reduction with very high enantioselectivity has also been achieved with achiral reducing agents and optically active catalysts.[1025] Two approaches are represented by (1) homogeneous catalytic hydrogenation with the catalyst 2,2'-bis(diphenylphosphino)-1,1'-binaphthyl-ruthenium acetate, BINAP—Ru(OAc)$_2$,[1026] which reduces

---

[1019]Kaselj, M.; Gonikberg, E.M.; le Noble, W.J. *J. Org. Chem.* **1998**, *63*, 3218.

[1020]For a review, see Sih, C.J.; Chen, C. *Angew. Chem. Int. Ed.* **1984**, *23*, 570.

[1021]See, for example, Fujisawa, T.; Hayashi, H.; Kishioka, Y. *Chem. Lett.* **1987**, 129; Nakamura, K.; Kawai, Y.; Ohno, A. *Tetrahedron Lett.* **1990**, *31*, 267; Spiliotis, V.; Papahatjis, D.; Ragoussis, N. *Tetrahedron Lett.* **1990**, *31*, 1615; Ishihara, K.; Sakai, T.; Tsuboi, S.; Utaka, M. *Tetrahedron Lett.* **1994**, *35*, 4569; Tsuboi, S.; Furutani, H.; Ansari, M.H.; Sakai, T.; Utaka, M.; Takeda, A. *J. Org. Chem.* **1993**, *58*, 486; Hayakawa, R.; Nozawa, K.; Kimura, K.; Shimizu, M. *Tetrahedron* **1999**, *55*, 7519; Kreutz, O.C.; Segura, R.C.M.; Rodrigues, J.A.R.; Moran, P.J.S. *Tetrahedron Asymmetry* **2000**, *11*, 2107; Johns, M.K.; Smallridge, A.J.; Trewhella, M.A. *Tetrahedron Lett.* **2001**, *42*, 4261; Attolini, M.; Bouguir, F.; Iacazio, F.; Peiffer, G.; Maffei, M. *Tetrahedron* **2001**, *57*, 537; Wei, Z.-L.; Li, Z.-Y.; Lin, G.-Q. *Tetrahedron Asymmetry* **2001**, *12*, 229. For reduction with designer yeast, see Chmur zyński, L *J. Heterocyclic Chem.* **2000**, *37*, 71.

[1022]See Wei, Z.-L.; Li, Z.-Y.; Lin, G.-Q. *Tetrahedron* **1998**, *54*, 13059; Guarna, A.; Occhiato, E.G.; Spinetti, L.M.; Vallecchi, M.E.; Scarpi, D. *Tetrahedron* **1995**, *51*, 1775; Medson, C.; Smallridge, A.J.; Trewhella, M.A. *Tetrahedron Asymmetry,* **1997**, *8*, 1049; Nakamura, K.; Inoue, Y.; Ohno, A. *Tetrahedron Lett.* **1995**, *36*, 265; Casy, G.; Lee, T.V.; Lovell, H. *Tetrahedron Lett.* **1992**, *33*, 817; Heiss, C.; Phillips, R.S. *J. Chem. Soc. Perkin Trans. 1* **2000**, 2821; Gotor, V.; Rebolledo, F.; Liz, R. *Tetrahedron Asymmetry* **2001**, *12*, 513; Hage, A.; Petra, D.G.I.; Field, J.A.; Schipper, D.; Wijnberg, J.B.P.A.; Kamer, P.C.J.; Reek, J.N.H.; van Leeuwen, P.W.N.M.; Wever, R.; Schoemaker, H.E. *Tetrahedron Asymmetry* **2001**, *12*, 1025; Yasohara, Y.; Kizaki, N.; Hasegawa, J.; Wada, M.; Kataoka, M.; Shimizu, S. *Tetrahedron Asymmetry* **2001**, *12*, 1713; Tsujigami, T.; Sugai, T.; Ohta, H. *Tetrahedron Asymmetry* **2001**, *12*, 2543; Yadav, J.S.; Nanda, S.; Reddy, P.T.; Rao, A.B. *J. Org. Chem.* **2002**, *67*, 3900; Stampfer, W.; Kosjek, B.; Faber, K.; Kroutil, W. *J. Org. Chem.* **2003**, *68*, 402; Gröger, H.; Hummel, W.; Buchholz, S.; Drauz, K.; Nguyen, T.V.; Rollmann, C.; Hüsken, H.; Abokitse, K. *Org. Lett.* **2003**, *5*, 173; Matsuda, T.; Nakajima, Y.; Harada, T.; Nakamura, K. *Tetrahedron Asymmetry* **2002**, *13*, 971; Nakamura, K.; Yamanaka, R. *Tetrahedron Asymmetry* **2002**, *13*, 2529, and references cited therein; Carballeira, J.D.; Álvarez, E.; Campillo, M.; Pardo, L.; Sinisterra, J.V. *Tetrahedron Asymmetry* **2004**, *15*, 951; Shkmoda, K.; Kubota, N.; Hamada, H.; Kaji, M.; Hirata, T. *Tetrahedron Asymmetry* **2004**, *15*, 1677; Salvi, N.A.; Chattopadhyay, S. *Tetrahedron Asymmetry* **2004**, *15*, 3397. For enzymatic reduction of thio ketones, see Nielsen, J.K.; Madsen, J.Ø. *Tetrahedron Asymmetry* **1994**, *5*, 403.

[1023]For a review, see Nakamura, K.; Yamanaka, R.; Matsuda, T.; Harada, T. *Tetrahedron Asymmetry* **2003**, *14*, 2659.

[1024]In bmim PF6, 1-butyl-3-methylimidazolium hexafluorophosphate: Howarth, J.; James, P.; Dai, J. *Tetrahedron Lett.* **2001**, *42*, 7517.

[1025]See Smith, M.B. *Organic Synthesis*, 2nd ed., McGraw-Hill, NY, **2001**, pp. 343–359.

[1026]For reviews of BINAP, see Noyori, R. *Science* **1990**, *248*, 1194; Noyori, R.; Takaya, H. *Acc. Chem. Res.* **1990**, *23*, 345. For the synthesis of BINAP, see Takaya, H.; Akutagawa, S.; Noyori, R. *Org. Synth.* 67, 20.

(R)(+)-BINAP

41

β-keto esters with high enantioselectivity.[1027] A variety of chiral additives and/or ligands have been used with catalytic hydrogenation reactions, and many functional groups can be tolerated.[1028] Asymmetric catalytic hydrogenation has been done in ionic liquids.[1029] A second approach is reduction with BH_3–THF or catecholborane,[1030] using an oxazaborolidine **41** (R = H, Me, or *n*-Bu; Ar = Ph or β-naphthyl)[1031] or other chiral compounds[1032] as a catalyst. Both a polymer-bound oxazaborolidine[1033] and a

[1027]Noyori, R.; Ohkuma, T.; Kitamura, M.; Takaya, H.; Sayo, N.; Kumobayashi, H.; Akutagawa, S. *J. Am. Chem. Soc.* **1987**, *109*, 5856; Taber, D.F.; Silverberg, L.J. *Tetrahedron Lett.* **1991**, *32*, 4227. See also, Kitamura, M.; Ohkuma, T.; Inoue, S.; Sayo, N.; Kumobayashi, H.; Akutagawa, S.; Ohta, T.; Takaya, H.; Noyori, R. *J. Am. Chem. Soc.* **1988**, *110*, 629.
[1028]Alonso, D.A.; Guijarro, D.; Pinho, P.; Temme, O.; Andersson, P.G. *J. Org. Chem.* **1998**, *63*, 2749; Le Blond, C.; Wang, J.; Liu, J.; Andrews, A.T.; Sun, Y.-K. *J. Am. Chem. Soc.* **1999**, *121*, 4920; ter Halle, R.; Colasson, B.; Schulz, E.; Spagnol, M.; Lemaire, M. *Tetrahedron Lett.* **2000**, *41*, 643; ter Halle, R.; Schulz, E.; Spagnol, M.; Lemaire, M. *Synlett* **2000**, 680; Ohkuma, T.; Ishii, D.; Takeno, H.; Noyori, R. *J. Am. Chem. Soc.* **2000**, *122*, 6510; Burk, M.J.; Hems, W.; Herzberg, D.; Malan, C.; Zanotti-Gerosa, A. *Org. Lett.* **2000**, *2*, 4173; Wu, J.; Chen, H.; Zhou, Z.-Y.; Yueng, C.H.; Chan, A.S.C. *Synlett* **2001**, 1050; Madec, J.; Pfister, X.; Phansavath, P.; Ratovelomanana-Vidal, V.; Genêt, J.-P. *Tetrahedron* **2001**, *57*, 2563; Ohkuma, T.; Hattori, T.; Ooka, H.; Inoue, T.; Noyori, R. *Org. Lett.* **2004**, *6*, 2681; Xie, J.-H.; Wang, L.-X.; Fu, Y.; Zhu, S.-F.; Fan, B.-M.; Duan, H.-F.; Zhou, Q.-L. *J. Am. Chem. Soc.* **2003**, *125*, 4404; Raja, R.; Thomas, J.M.; Jones, M.D.; Johnson, B.F.G.; Vaushan, D.E.W. *J. Am. Chem. Soc.* **2003**, *125*, 14982; Lei, A.; Wu, S.; He, M.; Zhang, X. *J. Am. Chem. Soc.* **2004**, *126*, 1626; Sun, Y.; Wan, X.; Guo, M.; Wang, D.; Dong, X.; Pan, Y.; Zhang, Z. *Tetrahedron Asymmetry* **2004**, *15*, 2185. For a discussion of the mechanism, see Sandoval, C.A.; Ohkuma, T.; Muñiz, K.; Noyori, R. *J. Am. Chem. Soc.* **2003**, *125*, 13490.
[1029]In bmim BF4, 1-butyl-3-methylimidazolium tetrafluoroborate: Ngo, H.L.; Hu, A.; Lin, W. *Chem. Commun.* **2003**, 1912.
[1030]For an example using catecholborane and a chiral gallium complex, see Ford, A.; Woodward, S. *Angew. Chem. Int. Ed.* **1999**, *38*, 335.
[1031]Corey, E.J.; Bakshi, R.K. *Tetrahedron Lett.* **1990**, *31*, 611; Puigjaner, C.; Vidal-Ferran, A.; Moyano, A.; Pericàs, M.A.; Riera, A. *J. Org. Chem.* **1999**, *64*, 7902; Yadav, J.S.; Reddy, P.T.; Hashim, S.R. *Synlett* **2000**, 1049; Li, X.; Yeung, C.-h.; Chan, A.S.C.; Yang, T.-K. *Tetrahedron Asymmetry* **1999**, *10*, 759; Cho, B.T.; Chun, Y.S. *J. Chem. Soc. Perkin Trans. 1* **1999**, 2095; Santhi, V.; Rao, J.M. *Tetrahedron Asymmetry* **2000**, *11*, 3553; Jones, S.; Atherton, J.C.C. *Tetrahedron Asymmetry* **2000**, *11*, 4543; Cho, B.T.; Kim, D.J. *Tetrahedron Asymmetry* **2001**, *12*, 2043; Jiang, B.; Feng, Y.; Hang, J.-F. *Tetrahedron Asymmetry* **2001**, *12*, 2323; Gilmore, N.J.; Jones, S.; Muldowney, M.P. *Org. Lett.* **2004**, *6*, 2805; Huertas, R.E.; Corella, J.A.; Soderquist, J.A. *Tetrahedron Lett.* **2003**, *44*, 4435.
[1032]See Hong, Y.; Gao, Y.; Nie, X.; Zepp, C.M. *Tetrahedron Lett.* **1994**, *35*, 6631; Quallich, G.J.; Woodall, T.M. *Tetrahedron Lett.* **1993**, *34*, 4145; Brunel, J.M.; Legrand, O.; Buono, G. *Eur. J. Org. Chem.* **2000**, 3313; Ford, A.; Woodward, S. *Synth. Commun.* **1999**, *29*, 189; Calmes, M.; Escale, F. *Synth. Commun.* **1999**, *29*, 1341; Kawanami, Y.; Murao, S.; Ohga, T.; Kobayashi, N. *Tetrahedron* **2003**, *59*, 8411; Basaviah, D.; Reddy, G.J.; Chandrashekar, V. *Tetrahedron Asymmetry* **2004**, *15*, 47; Zhang, Y.-X.; Du, D.-M.; Chen, X.; Lü, S.-F.; Hua, W.-T. *Tetrahedron Asymmetry* **2004**, *15*, 177.
[1033]Price, M.D.; Sui, J.K.; Kurth, M.J.; Schore, N.E. *J. Org. Chem.* **2002**, *67*, 8086.

dendritic chiral catalyst has been used in conjunction with borane,[1034] as well as other chiral additives can be used.[1035]

A third important method is the combination of LiAlH$_4$ or NaBH$_4$ with a chiral ligand, often in the presence of a transition-metal complex.[1036] Examples include LiBH$_4$/NiCl$_2$ and a chiral amino alcohol,[1037] NaBH$_4$ with chiral Lewis acid complexes,[1038] or NaBH$_4$/Me$_3$SiCl and a chiral ligand.[1039] A mixture of NaBH$_4$ and Me$_3$SiCl with a catalytic amount of a chiral, polymer-bound sulfonamide leads to asymmetric reduction.[1040]

Enantioselective reduction is possible with the other methods mentioned above. Reduction with silanes and transition-metal catalysts, such as ruthenium compounds, is also very effective.[1041] This method gives high enantioselectivity with various types of ketone, especially α,β-unsaturated ketones. Chiral ruthenium catalysts have been used with triethylammonium formate for the enantioselective reduction.[1042] A ruthenium catalyst with a polymer-supported chiral ligand has been used with Bu$_4$NBr and HCO$_2$Na in water.[1043] Chiral additives mixed with surfactants have been used with sodium formate.[1044] Enantioselective reduction was observed with PhSiH$_3$ and copper compounds with a chiral ligand,[1045] with a mixture of ruthenium and silver catalysts,[1046] or with Mn(dpm)$_3$ and oxygen (dpm = diphenylmethylene).[1047] Asymmetric reduction was achieved using an

[1034]Bolm, C.; Derrien, N.; Seger, A. *Chem. Commun.* **1999**, 2087.

[1035]Yanagi, T.; Kikuchi, K.; Takeuchi, H.; Ishikawa, T.; Nishimura, T.; Kamijo, T. *Chem. Lett.* **1999**, 1203; Hu, J.-b.; Zhao, G.; Yang, G.-s.; Ding, Z.-d. *J. Org. Chem.* **2001**, *66*, 303; Zhou, H.; Lü, S.; Xie, R.; Chan, A.S.C.; Yang, T.-K. *Tetrahedron Lett.* **2001**, *42*, 1107; Basavaiah, D.; Reddy, G.J.; Chandrashekar, V. *Tetrahedron Asymmetry* **2001**, *12*, 685.

[1036]For a review, see Daverio, P.; Zanda, M. *Tetrahedron Asymmetry* **2001**, *12*, 2225.

[1037]Molvinger, K.; Lopez, M.; Court, J. *Tetrahedron Lett.* **1999**, *40*, 8375.

[1038]Nozaki, K.; Kobori, K.; Uemura, T.; Tsutsumi, T.; Takaya, H.; Hiyama, T. *Bull. Chem. Soc. Jpn.* **1999**, *72*, 1109.

[1039]Jiang, B.; Feng, Y.; Zheng, J. *Tetrahedron Lett.* **2000**, *41*, 10281.

[1040]Zhao, G.; Hu, J.-b.; Qian, Z.-s.; Yin, X.-x. *Tetrahedron Asymmetry* **2002**, *13*, 2095.

[1041]Hayashi, T.; Hayashi, C.; Uozumi, Y. *Tetrahedron Asymmetry*, **1995**, *6*, 2503.

[1042]Koike, T.; Murata, K.; Ikariya, T. *Org. Lett.* **2000**, *2*, 3833; Okano, K.; Murata, K.; Ikariya, T. *Tetrahedron Lett.* **2000**, *41*, 9277; Cossy, J.; Eustache, F.; Dalko, P.I. *Tetrahedron Lett.* **2001**, *42*, 5005; Rhyoo, H.Y.; Yoon, Y.-A.; Park, H.-J.; Chung, Y.K. *Tetrahedron Lett.* **2000**, *42*, 5045; Chen, Y.-C.; Wu, T.-F.; Deng, J.-G.; Liu, H.; Jiang, Y.-Z.; Choi, M.C.K.; Chan, A.S.C. *Chem. Commun.* **2001**, 1488; Liu, P.N.; Gu, P.M.; Wang, F.; Tu, Y.Q. *Org. Lett.* **2004**, *6*, 169; Wu, X.; Li, X.; Hems, W.; King, F.; Xiao, J. *Org. Biomol. Chem.* **2004**, *2*, 1818; Schlatter, A.; Kundu, M.K.; Woggon, W.-D. *Angew. Chem. Int. Ed.* **2004**, *43*, 6731; Hannedouche, J.; Kenny, J.A.; Walsgrove, J.; Wills, M. *Synlett* **2002**, 263.

[1043]Liu, P.N.; Deng, J.G.; Tu, Y.Q.; Wang, S.H. *Chem. Commun.* **2004**, 2070.

[1044]Rhyoo, H.Y.; Park, H.-J.; Suh, W.H.; Chung, Y.K. *Tetrahedron Lett.* **2002**, *43*, 269.

[1045]Sirol, S.; Courmarcel, J.; Mostefai, N.; Riant, O. *Org. Lett.* **2001**, *3*, 4111; Lipshutz, B.H.; Lower, A.; Noson, K. *Org. Lett.* **2002**, *4*, 4045; Lipshutz, B.H.; Noson, K.; Chrisman, W.; Lower, A. *J. Am. Chem. Soc.* **2003**, *125*, 8779.

[1046]Gade, L.H.; César, V.; Bellemin-Laponnaz, S. *Angew. Chem. Int. Ed.* **2004**, *43*, 1014.

[1047]Cecchetto, A.; Fontana, F.; Minisci, F.; Recupero, F. *Tetrahedron Lett.* **2001**, *42*, 6651.

alkoxide or hydroxide base with a chiral rhodium,[1048] ruthenium,[1049] or iridium complex.[1050] A chiral samarium complex has been used in conjunction with Z-propanol.[1051] Chiral mercapto alcohols have also been used for asymmetric reduction.[1052]

Enantioselective reduction is not possible for aldehydes, since the products are primary alcohols in which the reduced carbon is not chiral, but deuterated aldehydes RCDO give a chiral product, and these have been reduced enantioselectively with B-(3-pinanyl)-9-borabicyclo[3.3.1]nonane (Alpine-Borane) with almost complete optical purity.[1053] Other chiral boranes can be used to reduce aldehydes or ketones.[1054]

In the above cases, an optically active reducing agent or catalyst interacts with a prochiral substrate. Asymmetric reduction of ketones has also been achieved with an achiral reducing agent, if the ketone is complexed to an optically active transition-metal Lewis acid.[1055]

Diastereomers

There are other stereochemical aspects to the reduction of aldehydes and ketones. If there is a stereogenic center α to the carbonyl group,[1056] even an achiral reducing agent can give more of one diastereomer than of the other. Such

---

[1048]Murata, K.; Ikariya, T.; Noyori, R. *J. Org. Chem.* **1999**, *64*, 2186.

[1049]With Yb(OTf)3 as a co-reagent, see Matsunaga, H.; Yoshioka, N.; Kunieda, T. *Tetrahedron Lett.* **2001**, *42*, 8857. With microwave irradiation, see Lutsenko, S.; Moberg, C. *Tetrahedron Asymmetry* **2001**, *12*, 2529.

[1050]Maillard, D.; Nguefack, C.; Pozzi, G.; Quici, S.; Valad, B.; Sinou, D. *Tetrahedron Asymmetry* **2000**, *11*, 2881.

[1051]Ohno, K.; Kataoka, Y.; Mashima, K. *Org. Lett.* **2004**, *6*, 4695.

[1052]Yang, T.-K.; Lee, D.-S. *Tetrahedron Asymmetry* **1999**, *10*, 405.

[1053]Midland, M.M.; Greer, S.; Tramontano, A.; Zderic, S.A. *J. Am. Chem. Soc.* **1979**, *101*, 2352. See also, Noyori, R.; Tomino, I.; Tanimoto, Y. *J. Am. Chem. Soc.* **1979**, *101*, 3129; Brown, H.C.; Jadhav, P.K.; Mandal, A.K. *Tetrahedron* **1981**, *37*, 3547; Midland, M.M.; Zderic, S.A. *J. Am. Chem. Soc.* **1982**, *104*, 525.

[1054]Wang, Z.; Zhao, C.; Pierce, M.E.; Fortunak, J.M. *Tetrahedron Asymmetry* **1999**, *10*, 225; Ramachandran, P.V.; Pitre, S.; Brown, H.C. *J. Org. Chem.* **2002**, *67*, 5315. For a discussion of the sources of stereoselectivity, see Rogic, M.M. *J. Org. Chem.* **2000**, *65*, 6868; Xu, J.; Wei, T.; Zhang, Q. *J. Org. Chem.* **2004**, *69*, 6860.

[1055]Dalton, D.M.; Gladysz, J.A. *J. Organomet. Chem.* **1989**, *370*, C17.

[1056]In theory, the chiral center can be anywhere in the molecule, but in practice, reasonable diastereoselectivity is most often achieved when it is in the α position. For examples of high diastereoselectivity when the chiral center is further away, especially in reduction of β-hydroxy ketones, see Narasaka, K.; Pai, F. *Tetrahedron* **1984**, *40*, 2233; Hassine, B.B.; Gorsane, M.; Pecher, J.; Martin, R.H. *Bull. Soc. Chim. Belg.* **1985**, *94*, 597; Bloch, R.; Gilbert, L.; Girard, C. *Tetrahedron Lett.* **1988**, *53*, 1021; Evans, D.A.; Chapman, K.T.; Carreira, E.M. *J. Am. Chem. Soc.* **1988**, *110*, 3560.

diastereoselective reductions have been carried out with considerable success.[1057] In most such cases Cram's rule (p. 168) is followed, but exceptions are known.[1058]

With most reagents there is an initial attack on the carbon of the carbonyl group by a hydride equivalent (H⁻) although with BH_3[1059] the initial attack is on the oxygen. Detailed mechanisms are not known in most cases.[1060] With tetrahydroaluminate or borohydride compounds, the attacking species is the AlH_4^- (or BH_4^-) ion, which, in effect, transfers H⁻ to the carbon. The following mechanism has been proposed for LiAlH₄:[1061]

S = solvent molecules

42

Evidence that the cation plays an essential role, at least in some cases, is that when the Li⁺ was effectively removed from LiAlH₄ (by the addition of a crown ether), the reaction did not take place.[1062] The complex **42** must now be hydrolyzed to the alcohol. For NaBH₄, the Na⁺ does not seem to participate in the transition state, but kinetic evidence shows that an OR group from the solvent does participate and remains attached to the boron:[1063]

Free H⁻ cannot be the attacking entity in most reductions with boron or aluminum hydrides because the reactions are frequently sensitive to the size of the MH_4^- [or MR_mHn- or $M(OR)_mH_n^-$ etc.].

[1057]For reviews, see Nógrádi, M. *Stereoselective Synthesis*, VCH, NY, *1986*, pp. 131–148; Oishi, T.; Nakata, T. *Acc. Chem. Res.* *1984*, *17*, 338.

[1058]One study showed that the Cram's rule product predominates with metal hydride reducing agents, but the other product with Bouveault-Blanc and dissolving metal reductions: Yamamoto, Y.; Matsuoka, K.; Nemoto, H. *J. Am. Chem. Soc.* *1988*, *110*, 4475.

[1059]For a discussion of the mechanism with boranes, see Brown, H.C.; Wang, K.K.; Chandrasekharan, J. *J. Am. Chem. Soc.* *1983*, *105*, 2340.

[1060]For reviews of the stereochemistry and mechanism, see Caro, B.; Boyer, B.; Lamaty, G.; Jaouen, G. *Bull. Soc. Chim. Fr.* *1983*, II-281; Boone, J.R.; Ashby, E.C. *Top. Stereochem.* *1979*, *11*, 53; Wigfield, D.C. *Tetrahedron* *1979*, *35*, 449.

[1061]Ashby, E.C.; Boone, J.R. *J. Am. Chem. Soc.* *1976*, *98*, 5524.

[1062]Pierre, J.; Handel, H. *Tetrahedron Lett.* *1974*, 2317. See also Loupy, A.; Seyden-Penne, J.; Tchoubar, B. *Tetrahedron Lett.* *1976*, 1677; Ashby, E.C.; Boone, J.R. *J. Am. Chem. Soc.* *1976*, *98*, 5524.

[1063]Wigfield, D.C.; Gowland, F.W. *J. Org. Chem.* *1977*, *42*, 1108; *Tetrahedron Lett.* *1976*, 3373. See however Adams, C.; Gold, V.; Reuben, D.M.E. *J. Chem. Soc. Chem. Commun.* *1977*, 182; *J. Chem. Soc. Perkin Trans. 2* *1977*, 1466, 1472; Kayser, M.M.; Eliev, S.; Eisenstein, O. *Tetrahedron Lett.* *1983*, *24*, 1015.

The question of whether the initial complex in the LiAlH$_4$ reduction (**42**, which can be written as $\text{H}-\overset{|}{\underset{|}{\text{C}}}-\text{O}\overset{\ominus}{\text{A}}\text{lH}_3 = \textbf{43}$) can reduce another carbonyl to give $\text{H}-\overset{|}{\underset{|}{\text{C}}}\cdot\text{O}_2\overset{\ominus}{\text{A}}\text{lH}_4$ and so on has been controversial. It has been shown[1064] that this is probably not the case but that, more likely, **43** disproportionates to $(\text{H}-\overset{|}{\underset{|}{\text{C}}}-\text{O})_4\text{Al}^{\ominus}$ and AlH_4^{-}, which is the only attacking species. Disproportionation has also been reported in the NaBH$_4$ reaction.[1065]

Aluminate, **43**, is essentially LiAlH$_4$ with one of the hydrogens replaced by an alkoxy group, that is, LiAlH$_3$OR. The fact that **43** and other alkoxy derivatives of LiAlH$_4$ are less reactive than LiAlH$_4$ itself has led to the use of such compounds as reducing agents that are less reactive and more selective than LiAlH$_4$.[1066] We have already met some of these, for example, LiAlH(O$-t$-Bu)$_3$ (reactions **19-39–19-41**; see also, Table 19.5). As an example of chemoselectivity in this reaction it may be mentioned that LiAlH(O-t-Bu)$_3$ has been used to reduce only the keto group in a molecule containing both keto and carboxylic ester groups.[1067] However, the use of such reagents is sometimes complicated by the disproportionation mentioned above, which may cause LiAlH$_4$ to be the active species, even if the reagent is an alkoxy derivative. Another highly selective reagent (reducing aldehydes and ketones, but not other functional groups), which does not disproportionate, is potassium triisopropoxyborohydride.[1068]

The mechanism of catalytic hydrogenation of aldehydes and ketones is probably similar to that of reaction **15-11**, although not much is known about it.[1069]

For other reduction reactions of aldehydes and ketones (see **19-61**, **19-76**, and **19-81**).

OS **I**, 90, 304, 554; **II**, 317, 545, 598; **III**, 286; **IV**, 15, 25, 216, 660; **V**, 175, 294, 595, 692; **VI**, 215, 769, 887; **VII**, 129, 215, 241, 402, 417; **VIII**, 302, 312, 326, 527; **IX**, 58, 362, 676.

19-37   Reduction of Carboxylic Acids to Alcohols

Dihydro-de-oxo-bisubstitution

$$\text{RCOOH} \xrightarrow{\text{LiAlH}_4} \text{RCH}_2\text{OH}$$

[1064]Haubenstock, H.; Eliel, E.L. *J. Am. Chem. Soc.* **1962**, *84*, 2363; Malmvik, A.; Obenius, U.; Henriksson, U. *J. Chem. Soc. Perkin Trans. 2* **1986**, 1899, 1905.

[1065]Malmvik, A.; Obenius, U.; Henriksson, U. *J. Org. Chem.* **1988**, *53*, 221.

[1066]For reviews of reductions with alkoxyaluminum hydrides, see Málek, J. *Org. React.* **1988**, *36*, 249; **1985**, *34*, 1; Málek, J.; Č erný, M. *Synthesis* **1972**, 217.

[1067]Levine, S.G.; Eudy, N.H. *J. Org. Chem.* **1970**, *35*, 549; Heusler, K.; Wieland, P.; Meystre, C. *Org. Synth.* **V**, 692.

[1068]Brown, C.A.; Krishnamurthy, S.; Kim, S.C. *J. Chem. Soc. Chem. Commun.* **1973**, 391.

[1069]For a review of the mechanism of gas-phase hydrogenation, see Pavlenko, N.V. *Russ. Chem. Rev.* **1989**, *58*, 453.

Carboxylic acids are easily reduced to primary alcohols by LiAlH$_4$.[1070] The reaction does not stop at the aldehyde stage (but see **19-40**). The conditions are particularly mild, the reduction proceeding quite well at room temperature. Other hydrides have also been used,[1071] but not NaBH$_4$ (see Table 19.5).[1072] Note, however, that complexion of the carboxylic acid with cyanuric chloride (2,4,6-trichlorotriazine) also smooth reduction to the alcohol.[1073] A combination of NaBH$_4$ and an arylboronic acid (p. 815) is also effective.[1074] Benzyltriethylammonium borohydride is dichloromethane also reduces carboxylic acids to the alcohol.[1075] Catalytic hydrogenation is also generally ineffective.[1076] Borane is particularly good for carboxyl groups (Table 19.4) and permits selective reduction of them in the presence of many other groups (although the reaction with double bonds takes place at about the same rate in ether solvents).[1077] Borane also reduces carboxylic acid salts.[1078] Aluminum hydride reduces COOH groups without affecting carbon–halogen bonds in the same molecule. The reduction has also been carried out with SmI$_2$ in basic media[1079] or aq. H$_3$PO$_4$,[1080] or simply with SmI$_2$ in water.[1081] A mixture of NaBH$_4$ and I$_2$ has been used to reduced amino acids to amino alcohols.[1082]

OS **III**, 60; **VII**, 221; 530; **VIII**, 26, 434, 528.

19-38 Reduction of Carboxylic Esters to Alcohols

Dihydro,hydroxy-de-oxo,alkoxy-tersubstitution

$$\text{RCOOR}' \xrightarrow{\text{LiAlH}_4} \text{RCH}_2\text{OH} + \text{R}'\text{OH}$$

[1070]For a review, see Gaylord, N.G. *Reduction with Complex Metal Hydrides*, Wiley, NY, *1956*, pp. 322–373.

[1071]For a list of reagents, with references, see Larock, R.C. *Comprehensive Organic Transformations*, 2nd ed., Wiley-VCH, NY, *1999*, pp. 1114–1116. Zinc borohydride has also been used; see Narashimhan, S.; Madhavan, S.; Prasad, K.G. *J. Org. Chem. 1995*, *60*, 5314.

[1072]NaBH4 in the presence of Me2N=CHCl$^+$ Cl$^-$ reduces carboxylic acids to primary alcohols chemoselectively in the presence of halide, ester, and nitrile groups: Fujisawa, T.; Mori, T.; Sato, T. *Chem. Lett. 1983*, 835.

[1073]Falorni, M.; Porcheddu, A.; Taddei, M. *Tetrahedron Lett. 1999*, *40*, 4395.

[1074]Tale, R.H.; Patil, K.M.; Dapurkar, S.E. *Tetrahedron Lett. 2003*, *44*, 3427.

[1075]Narashimhan, S.; Swarnalakshmi, S.; Balakumar, R. *Synth. Commun. 2000*, *30*, 941.

[1076]See Rylander, P.N. *Hydrogenation Methods*, Academic Press, NY, *1985*, pp. 78–79.

[1077]Brown, H.C.; Korytnyk, W. *J. Am. Chem. Soc. 1960*, *82*, 3866; Batrakov, S.G.; Bergel'son, L.D. *Bull. Acad. Sci. USSR Div. Chem. Sci. 1965*, 348; Pelter, A.; Hutchings, M.G.; Levitt, T.E.; Smith, K. *Chem. Commun. 1970*, 347; Brown, H.C.; Stocky, T.P. *J. Am. Chem. Soc. 1977*, *99*, 8218; Chen, M.H.; Kiesten, E.I.S.; Magano, J.; Rodriguez, D.; Sexton, K.E.; Zhang, J.; Lee, H.T. *Org. Prep. Proceed. Int. 2002*, *34*, 665.

[1078]Yoon, N.M.; Cho, B.T. *Tetrahedron Lett. 1982*, *23*, 2475.

[1079]Kamochi, Y.; Kudo, T. *Bull. Chem. Soc. Jpn. 1992*, *65*, 3049.

[1080]Kamochi, Y.; Kudo, T. *Tetrahedron 1992*, *48*, 4301.

[1081]Kamochi, Y.; Kudo, T. *Chem. Lett. 1993*, 1495.

[1082]McKennon, M.J.; Meyers, A.I.; Drauz, K.; Schwarm, M. *J. Org. Chem. 1993*, *58*, 3568.

Lithium aluminum hydride reduces carboxylic esters to give 2 equivalents of alcohol.[1083] The reaction is of wide scope and has been used to reduce many esters. Where the interest is in obtaining R'OH, this is a method that is often a working equivalent of "hydrolyzing" esters. Lactones yield diols. Among the reagents that give the same products[1084] are DIBALH, lithium triethylborohydride, LiAl-H(Ot-Bu)$_3$,[1085] and BH$_3$–SMe$_2$ in refluxing THF.[1086] Although NaBH$_4$ reduces phenolic esters, especially those containing electron-withdrawing groups,[1087] its reaction with other esters is usually so slow that it is not the reagent of choice (exceptions are known[1088]), and it is generally possible to reduce an aldehyde or ketone without reducing an ester function in the same molecule. Note that NaBH$_4$ in DMF–MeOH reduces aryl carboxylic esters to benzylic alcohols,[1089] and NaBH$_4$–LiCl with microwave irradiation also reduces esters to primary alcohols.[1090] However, NaBH$_4$ reduces esters in the presence of certain compounds (see Table 19.5).[1091] Carboxylic esters can also be reduced to alcohols by hydrogenation over copper chromite catalysts,[1092] although high pressures and temperatures are required. Ester functions generally survive low-pressure catalytic hydrogenations. Before the discovery of LiAlH$_4$, the most common way of carrying out the reaction was with sodium in ethanol, a method known as the *Bouveault–Blanc procedure*. This procedure is still sometimes used where selectivity is necessary (see also, **19-62**, **19-65**, and **19-59**).

Silanes, such as Ph$_2$SiH$_2$, with a catalytic amount of triphenylphosphine and a rhodium catalyst reduced esters to primary alcohols.[1093] Aliphatic silanes such as EtMe$_2$SiH, also reduced esters with a ruthenium catalyst.[1094]

OS **II**, 154, 325, 372, 468; **III**, 671; **IV**, 834; **VI**, 781; **VII**, 356; **VIII**, 155; **IX**, 251.

[1083]For a review, see Gaylord, N.G. *Reduction with Complex Metal Hydrides*, Wiley, NY, *1956*, pp. 391–531.
[1084]For a list of reagents, with references, see Larock, R.C. *Comprehensive Organic Transformations*, 2nd ed., Wiley-VCH, NY, *1999*, pp. 1116–1120.
[1085]Ayers, T.A. *Tetrahedron Lett.* *1999*, *40*, 5467.
[1086]Brown, H.C.; Choi, Y.M. *Synthesis* *1981*, 439; Brown, H.C.; Choi,Y.M.; Narasimhan, S. *J. Org. Chem.* *1982*, *47*, 3153.
[1087]Takahashi, S.; Cohen, L.A. *J. Org. Chem.* *1970*, *35*, 1505.
[1088]For example, see Brown, M.S.; Rapoport, H. *J. Org. Chem.* *1963*, *28*, 3261; Bianco, A.; Passacantilli, P.; Righi, G. *Synth. Commun.* *1988*, *18*, 1765; Boechat, N.; da Costa, J.C.S.; Mendonç a, J.de S.; de Oliveira, P.S.M.; DeSouza, M.V.N. *Tetrahedron Lett.* *2004*, *45*, 6021.
[1089]Zanka, A.; Ohmori, H.; Okamoto, T. *Synlett* *1999*, 1636.
[1090]Feng, J.-C.; Liu, B.; Dai, L.; Yang, X.-L.; Tu, S.-J. *Synth. Commun.* *2001*, *31*, 1875.
[1091]See also Kikugawa, Y. *Chem. Lett.* *1975*, 1029; Santaniello, E.; Ferraboschi, P.; Sozzani, P. *J. Org. Chem.* *1981*, *46*, 4584; Brown, H.C.; Narasimhan, S.; Choi,Y.M. *J. Org. Chem.* *1982*, *47*, 4702; Soai, K.; Oyamada, H.; Takase, M.; Ookawa, A. *Bull. Chem. Soc. Jpn.* *1984*, *57*, 1948; Guida, W.C.; Entreken, E.E.; Guida, W.C. *J. Org. Chem.* *1984*, *49*, 3024.
[1092]For a review, see Adkins, H. *Org. React.* *1954*, *8*, 1.
[1093]Ohta, T.; Kamiya, M.; Kusui, K.; Michibata, T.; Nobutomo, M.; Furukawa, I. *Tetrahedron Lett.* *1999*, *40*, 6963.
[1094]Matsubara, K.; Iura, T.; Maki, T.; Nagashima, H. *J. Org. Chem.* *2002*, *67*, 4985.

19-39   Reduction of Acyl Halides

Hydro-de-halogenation or Dehalogenation

$$RCOCl \xrightarrow[-78°C]{LiAlH(O\textit{t}-Bu)_3} RCHO$$

Acyl halides can be reduced to aldehydes[1095] by treatment with lithium tri-*tert*-butoxyaluminum hydride in diglyme at $-78°C$.[1096] The R group may be alkyl or aryl and may contain many types of substituents, including NO_2, CN, and EtOOC groups. The reaction stops at the aldehyde stage because steric hindrance prevents further reduction under these conditions. Acyl halides can also be reduced to aldehydes by hydrogenolysis with palladium-on-barium sulfate as catalyst. This is called the *Rosenmund reduction*.[1097] A more convenient hydrogenolysis procedure involves palladium-on-charcoal as the catalyst, with ethyldiisopropylamine as acceptor of the liberated HCl and acetone as the solvent.[1098] The reduction of acyl halides to aldehydes has also been carried out[1099] with Bu_3SnH,[1100] with the $InCl_3$-catalyzed reaction with Bu_3SnH,[1101] with $NaBH_4$ in a mixture of DMF and THF,[1102] and with formic acid/NH_4OH.[1103] In some of these cases, the mechanisms are free-radical. There are several indirect methods for the conversion of acyl halides to aldehydes, most of them involving prior conversion of the halides to certain types of amides (see **19-41**). There is also a method in which the COOH group is replaced by a completely different CHO group (**16-87**).

OS **III**, 551, 627; **VI**, 529, 1007. Also see, OS **III**, 818; **VI**, 312.

[1095]For a review of the formation of aldehydes from acid derivatives, see Fuson, R.C., in Patai, S. *The Chemistry of the Carbonyl Group*, Vol. 1, Wiley, NY, *1966*, pp. 211–232. For a review of the reduction of acyl halides, see Wheeler, O.H., in Patai, S. *The Chemistry of Acyl Halides*, Wiley, NY, *1972*, pp. 231–251.

[1096]Cha, J.S.; Brown, H.C. *J. Org. Chem.* **1993**, *58*, 4732, and references cited therein.

[1097]For a review, see Rylander, P.N. *Catalytic Hydrogenation Over Platinum Metals*, Academic Press, NY, *1967*, pp. 398–404. For a discussion of the Pt catalyst, see Maier, W.F.; Chettle, S.J.; Rai, R.S.; Thomas, G. *J. Am. Chem. Soc.* **1986**, *108*, 2608.

[1098]Peters, J.A.; van Bekkum, H. *Recl. Trav. Chim. Pays-Bas* **1971**, *90*, 1323; **1981**, *100*, 21. See also, Burgstahler, A.W.; Weigel, L.O.; Shaefer, C.G. *Synthesis* **1976**, 767.

[1099]For some other methods, see Wagenknecht, J.H. *J. Org. Chem.* **1972**, *37*, 1513; Smith, D.G.; Smith, D.J.H. *J. Chem. Soc. Chem. Commun.* **1975**, 459; Leblanc, J.C.; Moise, C.; Tirouflet, J. *J. Organomet. Chem.* **1985**, *292*, 225; Corriu, R.J.P.; Lanneau, G.F.; Perrot, M. *Tetrahedron Lett.* **1988**, *29*, 1271. For a list of reagents, with references, see Larock, R.C. *Comprehensive Organic Transformations*, 2nd ed., Wiley-VCH, NY, *1999*, pp. 1265–1266.

[1100]Kuivila, H.G. *J. Org. Chem.* **1960**, *25*, 284; Walsh, Jr., E.J.; Stoneberg, R.L.; Yorke, M.; Kuivila, H.G. *J. Org. Chem.* **1969**, *34*, 1156; Four, P.; Guibe, F. *J. Org. Chem.* **1981**, *46*, 4439; Lusztyk, J.; Lusztyk, E.; Maillard, B.; Ingold, K.U. *J. Am. Chem. Soc.* **1984**, *106*, 2923.

[1101]Inoue, K.; Yasuda, M.; Shibata, I.; Baba, A. *Tetrahedron Lett.* **2000**, *41*, 113.

[1102]Babler, J.H. *Synth. Commun.* **1982**, *12*, 839. For the use of NaBH4 and metal ions, see Entwistle, I.D.; Boehm, P.; Johnstone, R.A.W.; Telford, R.P. *J. Chem. Soc. Perkin Trans. 1* **1980**, 27.

[1103]Shamsuddin, K.M.; Zubairi, Md.O.; Musharraf, M.A. *Tetrahedron Lett.* **1998**, *39*, 8153.

19-40   Reduction of Carboxylic Acids, Esters, and Anhydrides to Aldehydes[1104]

Hydro-de-hydroxylation or Dehydroxylation (overall transformation)

$$RCOOH \longrightarrow RCHO$$

$$RCOOR' \longrightarrow RCHO$$

With most reducing agents, reduction of carboxylic acids generally gives the primary alcohol (**19-37**) and the isolation of aldehydes is not feasible. However, simple straight-chain carboxylic acids have been reduced to aldehydes[1105] by treatment with Li in $MeNH_2$ or NH_3 followed by hydrolysis of the resulting imine,[1106] with

$$RCOOH \xrightarrow[MeNH_2]{Li} RCH{=}N{-}Me \xrightarrow{H_2O} RCHO$$

with thexylchloro(or bromo)borane-Me_2S[1107] (see **15-16** for the thexyl group), $Me_2N{=}CHCl^+ \ Cl^-$ in pyridine,[1108] and with diaminoaluminum hydrides.[1109] Benzoic acid derivatives were reduced to benzaldehyde derivatives with NaH_2PO_2 and a diacylperoxide and a palladium catalyst.[1110] Caproic and isovaleric acids have been reduced to aldehydes in 50% yields or better with DIBALH (i-Bu_2AlH) at -75 to $-70°C$.[1111] Carboxylic acids can be reduced directly on Claycop–H_2O_2 using microwave irradiation.[1112]

Carboxylic esters have been reduced to aldehydes with DIBALH at $-70°C$, with diaminoaluminum hydrides,[1113] with $LiAlH_4$–Et_2NH,[1114] and for phenolic esters with $LiAlH(O{-}t{-}Bu)_3$ at $0°C$.[1115] Aldehydes have also been prepared by reducing ethyl thiol esters (RCOSEt) with Et_3SiH and a Pd–C catalyst.[1116] Pretreatment of

---

[1104]For a review, see Cha, J.S. *Org. Prep. Proceed. Int.* **1989**, *21*, 451.

[1105]For other reagents, see Lanneau, G.F.; Perrot, M. *Tetrahedron Lett.* **1987**, *28*, 3941; Cha, J.S.; Kim, J.E.; Yoon, M.S.; Kim, Y.S. *Tetrahedron Lett.* **1987**, *28*, 6231. See also, the lists, in Larock, R.C. *Comprehensive Organic Transformations*, 2nd ed., Wiley-VCH, NY, **1999**, pp. 1265–1268.

[1106]Bedenbaugh, A.O.; Bedenbaugh, J.H.; Bergin, W.A.; Adkins, J.D. *J. Am. Chem. Soc.* **1970**, *92*, 5774.

[1107]Chloro - see Brown, H.C.; Cha, J.S.; Yoon, N.M.; Nazer, B. *J. Org. Chem.* **1987**, *52*, 5400; Bromo, see Cha, J.S.; Kim, J.E.; Lee, K.W. *J. Org. Chem.* **1987**, *52*, 5030.

[1108]Fujisawa, T.; Mori, T.; Tsuge, S.; Sato, T. *Tetrahedron Lett.* **1983**, *24*, 1543.

[1109]Muraki, M.; Mukaiyama, T. *Chem. Lett.* **1974**, 1447; **1975**, 215; Cha, J.S.; Kim, J.M.; Jeoung, M.K.; Kwon, O.O.; Kim, E.J. *Org. Prep. Proceed. Int.* **1995**, *27*, 95.

[1110]Gooßen, L.J.; Ghosh, K. *Chem. Commun.* **2002**, 836.

[1111]Zakharkin, L.I.; Sorokina, L.P. *J. Gen. Chem. USSR* **1967**, *37*, 525.

[1112]Varma, R.S.; Dahiya, R. *Tetrahedron Lett.* **1998**, *39*, 1307.

[1113]Muraki, M.; Mukaiyama, T. *Chem. Lett.* **1974**, 1447; **1975**, 215; Cha, J.S.; Kim, J.M.; Jeoung, M.K.; Kwon, O.O.; Kim, E.J. *Org. Prep. Proceed. Int.* **1995**, *27*, 95.

[1114]Cha, J.S.; Kwon, S.S. *J. Org. Chem.* **1987**, *52*, 5486.

[1115]Zakharkin, L.I.; Khorlina, I.M. *Tetrahedron Lett.* **1962**, 619, *Bull. Acad. Sci. USSR Div. Chem. Sci.* **1963**, 288; **1964**, 435; Zakharkin, L.I.; Gavrilenko, V.V.; Maslin, D.N.; Khorlina, I.M. *Tetrahedron Lett.* **1963**, 2087; Zakharkin, L.I.; Gavrilenko, V.V.; Maslin, D.N. *Bull. Acad. Sci. USSR Div. Chem. Sci.* **1964**, 867; Weissman, P.M.; Brown, H.C. *J. Org. Chem.* **1966**, *31*, 283.

[1116]Fukuyama, T.; Lin, S.; Li, L. *J. Am. Chem. Soc.* **1990**, *112*, 7050.

the acid with Me_3SiCl followed by reduction with DIBALH also gives the aldehyde.[1117] Thioesters have been reduced to the aldehyde with lithium metal in THF at $-78°C$, followed by quenching with methanol.[1118]

Anhydrides, both aliphatic and aromatic, as well as mixed anhydrides of carboxylic and carbonic acids, have been reduced to aldehydes in moderate yields with disodium tetracarbonylferrate, $Na_2Fe(CO)_4$.[1119] Heating a carboxylic acid, presumably to form the anhydride, and then reaction with Na/EtOH leads to the aldehyde.[1120]

Acid chlorides are reduced to aldehydes with Bu_3SnH and a nickel catalyst.[1121] Also see, **19-62** and **19-38**.

OS **VI**, 312; **VIII**, 241, 498.

19-41   Reduction of Amides to Aldehydes

Hydro-de-dialkylamino-substitution

$$RCONR_2' + LiAlH_4 \longrightarrow RCHO + NHR_2'$$

N,N-Disubstituted amides can be reduced to amines with $LiAlH_4$ (see **19-64**), but also to aldehydes.[1122] Keeping the amide in excess gives the aldehyde rather than the amine. Sometimes it is not possible to prevent further reduction and primary alcohols are obtained instead. Other reagents[1123] that give good yields of aldehydes are DIBALH,[1124] $LiAlH(O-t-Bu)_3$, diaminoaluminum hydrides,[1125] disiamylborane (see **15-16** for the disiamyl group),[1126] and $Cp_2Zr(H)Cl$.[1127]

Aldehydes have been prepared from carboxylic acids or acyl halides by first converting them to certain types of amides that are easily reducible. There are several examples:[1128]

[1117]Chandrasekhar, S.; Kumar, M.S.; Muralidhar, B. *Tetrahedron Lett.* **1998**, *39*, 909.

[1118]Penn, J.H.; Owens, W.H. *Tetrahedron Lett.* **1992**, *33*, 3737.

[1119]Watanabe, Y.; Yamashita, M.; Mitsudo, T.; Igami, M.; Takegami, Y. *Bull. Chem. Soc. Jpn.* **1975**, *48*, 2490; Watanabe, Y.; Yamashita, M.; Mitsudo, T.; Igami, M.; Tomi, K.; Takegami, Y. *Tetrahedron Lett.* **1975**, 1063.

[1120]Shi, Z.; Gu, H. *Synth. Commun.* **1997**, *27*, 2701.

[1121]Malanga, C.; Mannucci, S.; Lardicci, L. *Tetrahedron Lett.* **1997**, *38*, 8093.

[1122]For a review, see Fuson, R.C., in Patai, S. *The Chemistry of the Carbonyl Group*, Vol. 1, Wiley, NY, **1966**, pp. 220–225.

[1123]For a list of reagents, with references, see Larock, R.C. *Comprehensive Organic Transformations*, 2nd ed., Wiley-VCH, NY, **1999**, pp.1269–1271.

[1124]Zakharkin, L.I.; Khorlina, I.M. *Bull. Acad. Sci. USSR Div. Chem. Sci.* **1959**, 2046.

[1125]Muraki, M.; Mukaiyama, T. *Chem. Lett.* **1975**, 875.

[1126]Godjoian, G.; Singaram, B. *Tetrahedron Lett.* **1997**, *38*, 1717.

[1127]White, J.M.; Tunoori, A.R.; Georg, G.I. *J. Am. Chem. Soc.* **2000**, *122*, 11995.

[1128]For other examples, see Doleschall, G. *Tetrahedron* **1976**, *32*, 2549; Atta-ur-Rahman; Basha, A. *J. Chem. Soc. Chem. Commun.* **1976**, 594; Izawa, T.; Mukaiyama, T. *Bull. Chem. Soc. Jpn.* **1979**, *52*, 555; Craig, J.C.; Ekwurieb, N.N.; Fu, C.C.; Walker, K.A.M. *Synthesis* **1981**, 303.

1. *Reissert Compounds.*[1129] Compounds such as **44** are prepared from the acyl halide by treatment with quinoline and cyanide ion. Treatment of **44** with sulfuric acid gives the corresponding aldehyde.

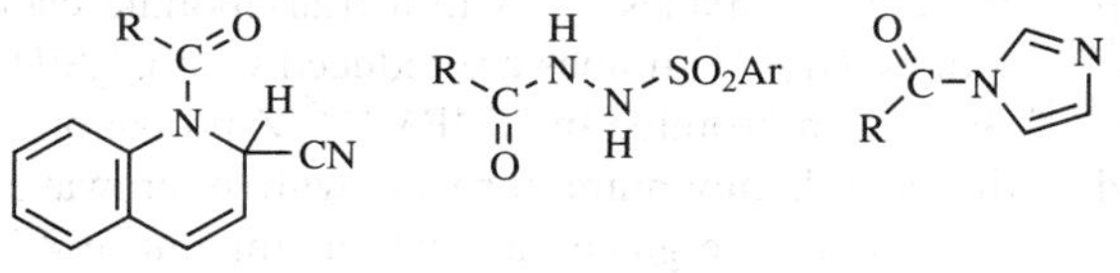

44	**45**	**46**

2. *Acyl Sulfonylhydrazides.* Compounds such as **45** are cleaved with base to give aldehydes. This is known as the *McFadyen–Stevens reduction* and is applicable only to aromatic aldehydes or aliphatic aldehydes with no a hydrogen.[1130] RCON=NH (see **19-67**) has been proposed as an intermediate in this reaction.[1131]

3. *Imidazoles.* Compounds **46**[1132] can be reduced to aldehydes with LiAlH$_4$.

4. *See Also the Sonn–Müller Method.* (**19-44**).

OS **VIII**, 68. See OS **IV**, 641, **VI**, 115 for the preparation of Reissert compounds.

B. Attack at Non-Carbonyl Multiple-Bonded Heteroatoms

19-42   Reduction of the Carbon–Nitrogen Double Bond

C,N-**Dihydro-addition**

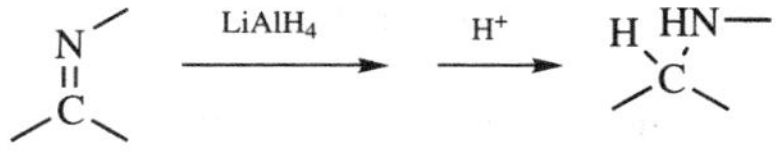

Imines and Schiff bases,[1133] hydrazones,[1134] and other C=N compounds can be reduced with LiAlH$_4$, NaBH$_4$,[1135] Na—EtOH, hydrogen and a catalyst, as well as

---

[1129]For reviews of Reissert compounds, see Popp, F.D.; Uff, B.C. *Heterocycles* **1985**, *23*, 731; Popp, F.D. *Bull. Soc. Chim. Belg.* **1981**, *90*, 609; *Adv. Heterocycl. Chem.* **1979**, *24*, 187; *1968*, *9*, 1. See Bridge, A.W.; Hursthouse, M.B.; Lehmann, C.W.; Lythgoe, D.J.; Newton, C.G. *J. Chem. Soc. Perkin Trans. 1* **1993**, 1839 for isoquinoline Reissert salts.

[1130]Babad, H.; Herbert, W.; Stiles, A.W. *Tetrahedron Lett.* **1966**, 2927; Dudman, C.C.; Grice, P.; Reese, C.B. *Tetrahedron Lett.* **1980**, *21*, 4645.

[1131]For discussions, see Cacchi, S.; Paolucci, G. *Gazz. Chem. Ital.* **1974**, *104*, 221; Matin, S.B.; Craig, J.C.; Chan, R.P.K. *J. Org. Chem.* **1974**, *39*, 2285.

[1132]For a review, see Staab, H.A.; Rohr, W. *Newer Methods Prep. Org. Chem.* **1968**, *5*, 61.

[1133]See Ranu, B.C.; Sarkar, A.; Majee, A. *J. Org. Chem.* **1997**, *62*, 1841; Verdaguer, X.; Lange, U.E.W.; Buchwald, S.L. *Angew. Chem. Int. Ed.* **1998**, *37*, 1103; Amin, Sk.R.; Crowe, W.E. *Tetrahedron Lett.* **1997**, *38*, 7487; Vetter, A.H.; Berkessel, A. *Synthesis* **1995**, 419.

[1134]For an enantioselective reduction of hydrazone derivatives, see Burk, M.J.; Feaster, J.E. *J. Am. Chem. Soc.* **1992**, *114*, 6266.

[1135]Bhattacharyya, S.; Neidigh, K.A.; Avery, M.A.; Williamson, J.S. *Synlett* **1999**, 1781.

with other reducing agents.[1136] A mixture of Sm/I_2[1137] or In/NH_4Cl[1138] reduces imines. Reduction with Bu_2SnClH in HMPA has been shown to be chemoselective for imines.[1139] Iminium salts are also reduced by $LiAlH_4$, although here there is no "addition" to the nitrogen:[1140] Silanes[1141] with a triarylborane catalyst reduces N-sulfonyl imines[1142] as does TiI_4.[1143] Imines are reduced with Cl_3SiH and pyrrolidine carboxaldehyde,[1144] Samarium bromide in HMPA,[1145] Z-propanol with a ruthenium catalyst,[1146] and with triethylammonium formate with microwave irradiation.[1147] Oximes are reduced with hydrogen gas an a catalytic amount of 48% HBr.[1148]

Oximes are generally reduced to amines (**19-48**),[1149] but simple reduction to give hydroxylamines can be accomplished with borane[1150] or sodium cyanoborohydride.[1151] Oxime O-ethers are reduced with Bu_3SnH and $BF_3 \cdot OEt_2$.[1152] Diazo compounds (ArN=NAr) are reductively cleaved to aniline derivatives with Zn and ammonium formate in methanol.[1153]

[1136]For a review, see Harada, K., in Patai, S. *The Chemistry of the Carbon–Nitrogen Double Bond*, Wiley, NY, *1970*, pp. 276–293. For a review with respect to catalytic hydrogenation, see Rylander, P.N. *Catalytic Hydrogenation over Platinum Metals*, Academic Press, NY, *1967*, pp. 123–138.

[1137]Banik, B.K.; Zegrocka, O.; Banik, I.; Hackfeld, L.; Becker, F.F. *Tetrahedron Lett.* *1999*, *40*, 6731.

[1138]Banik, B.K.; Hackfeld, L.; Becker, F.F. *Synth. Commun.* *22001*, *31*, 1581.

[1139]Shibata, I.; Moriuchi-Kawakami, T.; Tanizawa, D.; Suwa, T.; Sugiyama, E.; Matsuda, H.; Baba, A. *J. Org. Chem.* *1998*, *63*, 383.

[1140]For a review of nucleophilic addition to iminium salts, see Paukstelis, J.V.; Cook, A.G. in Cook, A.G. *Enamines*, 2nd ed., Marcel Dekker, NY, *1988*, pp. 275–356.

[1141]For a discussion of noncovalent interactions in the reduction of imines, see Malkov, A.V.; Mariani, A.; MacDougall, K.N.; Kočovský, P. *Org. Lett.* *2004*, *6*, 2253.

[1142]Blackwell, J.M.; Sonmor, E.R.; Scoccitti, T.; Piers, W.E. *Org. Lett.* *2000*, *2*, 3921.

[1143]Shimizu, M.; Sahara, T.; Hayakawa, R. *Chem. Lett.* *2001*, 792.

[1144]Iwasaki, F.; Onomura, O.; Mishima, K.; Kanematsu, T.; Maki, T.; Matsumura, Y. *Tetrahedron Lett.* *2001*, *42*, 2525.

[1145]Knettle, B.W.; Flowers II, R.A. *Org. Lett.* *2001*, *3*, 2321.

[1146]Samec, J.S.M.; Bäckvall, J.-E. *Chem. Eur. J.* *2002*, *8*, 2955.

[1147]Moghaddam, F.M.; Khakshoor, O.; Ghaffarzadeh, M. *J. Chem. Res. (S)* *2001*, 525.

[1148]Davies, I.W.; Taylor, M., Marcoux, J.-F.; Matty, L.; Wu, J.; Hughes, D.; Reider, P.J. *Tetrahedron Lett.* *2000*, *41*, 8021.

[1149]For examples, see Bolm, C.; Felder, M. *Synlett* *1994*, 655; Williams, D.R.; Osterhout, M.H.; Reddy, J.P. *Tetrahedron Lett.* *1993*, *34*, 3271.

[1150]Feuer, H.; Vincent Jr., B.F.; Bartlett, R.S. *J. Org. Chem.* *1965*, *30*, 2877; Kawase, M.; Kikugawa, Y. *J. Chem. Soc. Perkin Trans. 1* *1979*, 643.

[1151]For reviews of $NaBH_3CN$, see Hutchins, R.O.; Natale, N.R. *Org. Prep. Proced. Int.* *1979*, *11*, 201; Lane, C.F. *Synthesis* *1975*, 135.

[1152]Ueda, M.; Miyabe, H.; Namba, M.; Nakabayashi, T.; Naito, T. *Tetrahedron Lett.* *2002*, *43*, 4369.

[1153]Gowda, S.; Abiraj, K.; Gowda, D.C. *Tetrahedron Lett.* *2002*, *43*, 1329.

Reduction of imines has been carried out enantioselectively.[1154] Catalytic hydrogenation with a chiral iridium[1155] or palladium[1156] catalyst has been used. Catalytic hydrogenation of iminium salts with a chiral ruthenium catalyst gives the amine.[1157] In a related reaction, enamines were reduced by hydrogenation over a chiral rhodium catalyst.[1158] An ammonium formate with a chiral ruthenium complex was used with imines.[1159] Hydrogenation of oximes with Pd/C and a nickel complex gives the imine, and in the presence of a lipase and ethyl acetate the final product was an acetamide, formed with high enantioselectivity.[1160] Conjugated N-sulfonyl imines are reduced to the conjugated sulfonamide with good enantioselectivity using a chiral rhodium catalyst in the presence of LiF and PhSnMe$_3$.[1161] Phosphinyl imines, $R_2C=N-P(=O)Ar_2$, are reduced with high enantioselectivity using a chiral copper catalyst.[1162] Silanes, such as PhSiH$_3$, can be used for the reduction of imines, and in the presence of a chiral titanium catalyst the resulting amine was formed with excellent enantioselectivity.[1163]

Isocyanates have been catalytically hydrogenated to N-substituted formamides: $RNCO \rightarrow R-NH-CHO$.[1164] Isothiocyanates were reduced to thioformamides with SmI$_2$ in HMPA/t-BuOH.[1165]

OS **III**, 328, 827; **VI**, 905; **VIII**, 110, 568. Also see, OS **IV**, 283.

19-43 The Reduction of Nitriles to Amines

CC,NN-**Tetrahydro-biaddition**

$$R-C \equiv N + LiAlH_4 \longrightarrow R-CH_2-NH_2$$

Nitriles can be reduced to primary amines with many reducing agents,[1166] including LiAlH$_4$, and H$_3$B•SMe$_2$•.[1167] The reagent NaBH$_4$ does not generally

---

[1154]See Denmark, S.E.; Nakajima, N.; Nicaise, O. J.-C. *J. Am. Chem. Soc.* **1994**, *116*, 8797; Fuller, J.C.; Belisle, C.M.; Goralski, C.T.; Singaram, B. *Tetrahedron Lett.* **1994**, *35*, 5389; Willoughby, C.A.; Buchwald, S.L. *J. Org. Chem.* **1993**, *58*, 7627; *J. Am. Chem. Soc.* **1992**, *114*, 7562; Kawate, T.; Nakagawa, M.; Kakikawa, T.; Hino, T. *Tetrahedron Asymmetry* **1992**, *3*, 227. For a review of asymmetric reductions involving the C=N unit, see Zhu, Q.-C.; Hutchins, R.O. *Org. Prep. Proceed. Int.* **1994**, *26*, 193.
[1155]Kainz, S.; Brinkmann, A.; Leitner, W.; Pfaltz, A. *J. Am. Chem. Soc.* **1999**, *121*, 6421; Xiao, D.; Zhang, X. *Angew. Chem. Int. Ed.* **2001**, *40*, 3425; Trifonova, A.; Diesen, J.S.; Chapman, C.J.; Andersson, P.G. *Org. Lett.* **2004**, *6*, 3825.
[1156]Abe, H.; Amii, H.; Uneyama, K. *Org. Lett.* **2001**, *3*, 313.
[1157]Magee, M.P.; Norton, J.R. *J. Am. Chem. Soc.* **2001**, *123*, 1778.
[1158]Tararov, V.I.; Kadyrov, R.; Riermeier, T.H.; Holz, J.; Börner, A. *Tetrahedron Lett.* **2000**, *41*, 2351.
[1159]Mao, J.; Baker, D.C. *Org. Lett.* **1999**, *1*, 841.
[1160]Choi, Y.K.; Kim, M.J.; Ahn, Y.; Kim, M.-J. *Org. Lett.* **2001**, *3*, 4099.
[1161]Hayashi, T.; Ishigedani, M. *Tetrahedron* **2001**, *57*, 2589.
[1162]Lipshutz, B.H.; Shimizu, H. *Angew. Chem. Int. Ed.* **2004**, *43*, 2228.
[1163]Hansen, M.C.; Buchwald, S.L. *Org. Lett.* **2000**, *2*, 713.
[1164]Howell, H.G. *Synth. Commun.* **1983**, *13*, 635.
[1165]Park, H.S.; Lee, I.S.; Kim, Y.H. *Chem. Commun.* **1996**, 1805.
[1166]For a review, see Rabinovitz, M., in Rappoport, Z. *The Chemistry of the Cyano Group*, Wiley, NY, **1970**, pp. 307–340. For a list of reagents, with references, see Larock, R.C. *Comprehensive Organic Transformations*, 2nd ed., Wiley-VCH, NY, **1999**, pp. 875–878.
[1167]See Brown, H.C.; Choi, Y.M.; Narasimhan, S. *Synthesis* **1981**, 605.

reduce nitriles except in alcoholic solvents with a catlayst, such as $CoCl_2$,[1168] $NiCl_2$,[1169] or Raney nickel.[1170] A mixture of $NaBH_4/NiCl_2$ in acetic anhydride reduces the nitrile to the amine, which is trapped as the acetamide.[1171] Lithium dimethylamino- borohydride ($LiBH_3NMe_2$) reduces aryl nitriles to the corresponding benzylamines.[1172]

The reduction of nitriles is of wide scope and has been applied to many nitriles. When catalytic hydrogenation is used, secondary amines, $(RCH_2)_2NH$, are often side products.[1173] These can be avoided by adding a compound, such as acetic anhydride, which removes the primary amine as soon as it is formed,[1174] or by the use of excess ammonia to drive the equilibria backward.[1175] Sponge nickel[1176] or nickel on silica gel[1177] have been used for the catalytic hydrogenation of aryl nitriles to amines.

Attempts to stop with the addition with only 1 equivalent of hydrogen, have failed that is, to convert the nitrile to an imine, except where the imine is subsequently hydrolyzed (**19-44**).

N-Alkylnitrilium ions are reduced to secondary amines by $NaBH_4$.[1178]

$$RCN \xrightarrow{R'_3O^+BF_4^-} R-C{\equiv}N^{\oplus}-R' \xrightarrow[\text{diglyme}]{NaBH_4} RCH_2-NH-R'$$

Since nitrilium salts can be prepared by treatment of nitriles with trialkyloxonium salts (see **16-8**), this is a method for the conversion of nitriles to secondary amines.

Note that the related compounds, the isonitriles ($R-^{\oplus}N{\equiv}C^{\ominus}$, also called isocyanides) have been reduced to N-methylamines with $LiAlH_4$, as well as with other reducing agents.

OS **III**, 229, 358, 720; **VI**, 223.

[1168]Satoh, T.; Suzuki, S. *Tetrahedron Lett.* **1969**, 4555. For a discussion of the mechanism, see Heinzman, S.W.; Ganem, B. *J. Am. Chem. Soc.* **1982**, *104*, 6801.

[1169]Khurana, J.M.; Kukreja, G. *Synth. Commun.* **2002**, *32*, 1265.

[1170]Egli, R.A. *Helv. Chim. Acta* **1970**, *53*, 47.

[1171]Caddick, S.; de K. Haynes, A.K.; Judd, D.B.; Williams, M.R.V. *Tetrahedron Lett.* **2000**, *41*, 3513.

[1172]Thomas, S.; Collins, C.J.; Cuzens, J.R.; Spieciarich, D.; Goralski, C.T.; Singaram, B. *J. Org. Chem.* **2001**, *66*, 1999.

[1173]For a method of making secondary amines the main products, see Galán, A.; de Mendoza, J.; Prados, P.; Rojo, J.; Echavarren, A.M. *J. Org. Chem.* **1991**, *56*, 452.

[1174]For example, see Carothers, W.H.; Jones, G.A. *J. Am. Chem. Soc.* **1925**, *47*, 3051; Gould, F.E.; Johnson, G.S.; Ferris, A.F. *J. Org. Chem.* **1960**, *25*, 1658.

[1175]For example, see Freifelder, M. *J. Am. Chem. Soc.* **1960**, *82*, 2386.

[1176]Tanaka, K.; Nagasawa, M.; Kasuga, Y.; Sakamura, H.; Takuma, Y.; Iwatani, K. *Tetrahedron Lett.* **1999**, *40*, 5885.

[1177]Takamizawa, S.; Wakasa, N.; Fuchikami, T. *Synlett* **2001**, 1623.

[1178]Borch, R.F. *Chem. Commun.* **1968**, 442.

19-44   The Reduction of Nitriles to Aldehydes

Hydro,oxy-de-nitrilo-tersubstitution

$$R-C{\equiv}N \xrightarrow[\text{2. hydrolysis}]{\text{1. HCl, SnCl}_2} RCH{=}O$$

There are two principal methods for the reduction of nitriles to aldehydes.[1179] In one of these, known as the *Stephen reduction*, the nitrile is treated with HCl to form an iminium salt, **47**.

$$RCCl = \overset{\oplus}{N}H_2 \quad \overset{\ominus}{Cl}$$
47

Iminium salt **47** is reduced with anhydrous $SnCl_2$ to $RCH{=}NH$, which precipitates as a complex with $SnCl_4$ and is then hydrolyzed (**16-2**) to the aldehyde. The Stephen reduction is most successful when R is aromatic, but it can be done for aliphatic R up to about six carbons.[1180] It is also possible to prepare **47** in a different way, by treating ArCONHPh with PCl_5, which can then be converted to the aldehyde. This is known as the *Sonn–Müller method*. Aqueous formic acid in the presence of PtO_2, followed by treatment with aqueous acid, converts aryl nitriles to aryl aldehydes.[1181]

The other way of reducing nitriles to aldehydes involves using a metal hydride reducing agent to add 1 equivalent of hydrogen and hydrolysis, *in situ*, of the resulting imine (which is undoubtedly coordinated to the metal). This has been carried out with $LiAlH_4$, $LiAlH(OEt)_3$,[1182] $LiAlH(NR_2)_3$,[1183] and DIBALH.[1184] The metal hydride method is useful for aliphatic and aromatic nitriles.

OS **III**, 626, 818; **VI**, 631.

19-45   Reduction of Nitro Compounds to Amines

$$RNO_2 \xrightarrow[\text{HCl}]{\text{Zn}} RNH_2$$

Both aliphatic[1185] and aromatic nitro compounds can be reduced to amines, although the reaction has been applied much more often to aromatic nitro

---

[1179]For a review, see Rabinovitz, M., in Rappoport, Z. *The Chemistry of the Cyano Group*, Wiley, NY, *1970*, p. 307. For a list of reagents, with references, see Larock, R.C. *Comprehensive Organic Transformations*, 2nd ed., Wiley-VCH, NY, *1999*, pp. 1271–1272.

[1180]Zil'berman, E.N.; Pyryalova, P.S. *J. Gen. Chem. USSR* *1963*, *33*, 3348.

[1181]Xi, F.; Kamal, F.; Schenerman, M.A. *Tetrahedron Lett.* *2002*, *43*, 1395.

[1182]Brown, H.C.; Shoaf, C.J. *J. Am. Chem. Soc.* *1964*, *86*, 1079. For a review of reductions with this and related reagents, see Málek, J. *Org. React.* *1988*, *36*, 249, see pp. 287–289, 438–448.

[1183]Cha, J.S.; Lee, S.E.; Lee, H.S. *Org. Prep. Proceed. Int.* *1992*, *24*, 331. Also see, Cha, J.S.; Jeoung, M.K.; Kim, J.M.; Kwon, O.O.; Lee, J.C. *Org. Prep. Proceed. Int.* *1994*, *26*, 583.

[1184]Miller, A.E.G.; Biss, J.W.; Schwartzman, L.H. *J. Org. Chem.* *1959*, *24*, 627; Marshall, J.A.; Andersen, N.H.; Schlicher, J.W. *J. Org. Chem.* *1970*, *35*, 858.

[1185]For a review of selective reduction of aliphatic nitro compounds without disturbance of other functional groups, see Ioffe, S.L.; Tartakovskii, V.A.; Novikov, S.S. *Russ. Chem. Rev.* *1966*, *35*, 19.

compounds, owing to their greater availability. Many reducing agents have been used to reduce aromatic nitro compounds, the most common being Zn, Sn, or Fe (or sometimes other metals) and acid, and catalytic hydrogenation.[1186] Indium metal in aqueous ethanol with ammonium chloride[1187] or with water in aq. THF[1188] also reduces aromatic nitro compounds to the corresponding aniline derivative. Indium metal in methanol, with acetic anhydride and acetic acid, converts aromatic nitro compounds to the acetanilide.[1189] Samarium and a catalytic amount of iodine also accomplishes this reduction,[1190] as does Sm with a bipyridinium dibromide in methanol.[1191] Samarium metal in methanol with ultrasound also reduces aryl nitro compounds.[1192] Sodium sulfide (NaHS) on alumina with microwave irradiation reduces aryl nitro compounds to aniline derivatives.[1193] A mild reduction uses Al(Hg) in aq. THF with ultrasound.[1194] An Al/NiCl$_2$ reagent was used to reduced the nitro group of a polymer-bound $CH_2OCH_2C_6H_4NO_2$ moiety.[1195] Some other reagents used[1196] were Et$_3$SiH/RhCl(PPh$_3$)$_3$,[1197] AlH$_3$–AlCl$_3$, Mn with CrCl$_2$,[1198] nanoparticulate iron in water at 210°C,[1199] formic acid and Pd–C[1200] for formic acid with Raney nickel in methanol,[1201] and sulfides, such as NaHS, (NH$_4$)$_2$S, or polysulfides. The reaction with sulfides or polysulfides is called the *Zinin reduction*.[1202] Amines are also the products when nitro compounds, both alkyl and aryl, are reduced with HCOONH$_4$–Pd–C.[1203] Many other functional groups (e.g., COOH, COOR, CN, amide) are not affected by this reagent (although ketones are reduced, see **19-33**). With optically active alkyl substrates this method gives

[1186]For reviews, see Rylander, P.N. *Hydrogenation Methods*, Academic Press, NY, *1985*, pp. 104–116, *Catalytic Hydrogenation over Platinum Metals*, Academic Press, NY, *1967*, pp. 168–202. See Deshpande, R.M.; Mahajan, A.N.; Diwakar, M.M.; Ozarde, P.S.; Chaudhari, R.V. *J. Org. Chem.* *2004*, *69*, 4835; Wu, G.; Huang, M.; Richards, M.; Poirer, M.; Wen, X.; Draper, R.W. *Synthesis* *2003*, 1657.

[1187]Moody, C.J.; Pitts, M.R. *Synlett* *1998*, 1028; Banik, B.K.; Suhendra, M.; Banik, I.; Becker, F.F. *Synth. Commun.* *2000*, *30*, 3745.

[1188]Lee, J.G.; Choi, K.I.; Koh, H.Y.; Kim, Y.; Kang, Y.; Cho, Y.S. *Synthesis* *2001*, 81.

[1189]Kim, B.H.; Han, R.; Piao, F.; Jun, Y.M.; Baik, W.; Lee, B.M. *Tetrahedron Lett.* *2003*, *44*, 77.

[1190]Banik, B.K.; Mukhopadhyay, C.; Venkatraman, M.S.; Becker, F.F. *Tetrahedron Lett.* *1998*, *39*, 7243; Wang, L.; Zhou, L.; Zhang, Y. *Synlett* *1999*, 1065.

[1191]Yu, C.; Liu, B.; Hu, L. *J. Org. Chem.* *2001*, *66*, 919.

[1192]Basu, M.K.; Becker, F.F.; Banik, B.K. *Tetrahedron Lett.* *2000*, *41*, 5603.

[1193]Kanth, S.R.; Reddy, G.V.; Rao, V.V.V.N.S.R.; Maitraie, P.; Narsaiah, B.; Rao, P.S. *Synth. Commun.* *2002*, *32*, 2849.

[1194]Fitch, R.W.; Luzzio, F.A. *Tetrahedron Lett.* *1994*, *35*, 6013.

[1195]Kamal, A.; Reddy, K.L.; Devaiah, V.; Reddy, G.S.K. *Tetrahedron Lett.* *2003*, *44*, 4741.

[1196]For a list of reagents, with references, see Larock, R.C. *Comprehensive Organic Transformations*, 2nd ed., Wiley-VCH, NY, *1999*, pp. 821–828.

[1197]Brinkman, H.R. *Synth. Commun.* *1996*, *26*, 973.

[1198]Hari, A.; Miller, B.L. *Angew. Chem. Int. Ed.* *1999*, *38*, 2777.

[1199]Wang, L.; Li, P.; Wu, Z.; Yan, J.; Wang, M.; Ding, Y. *Synthesis* *2003*, 2001.

[1200]Entwistle, I.D.; Jackson, A.E.; Johnstone, R.A.W.; Telford, R.P. *J. Chem. Soc. Perkin Trans. 1* *1977*, 443. See also, Terpko, M.O.; Heck, R.F. *J. Org. Chem.* *1980*, *45*, 4992; Babler, J.H.; Sarussi, S.J. *Synth. Commun.* *1981*, *11*, 925.

[1201]Gowda, D.C.; Gowda, A.S.P.; Baba, A.R.; Gowda, S. *Synth. Commun.* *2000*, *30*, 2889.

[1202]For a review of the Zinin reduction, see Porter, H.K. *Org. React.* *1973*, *20*, 455.

[1203]Ram, S.; Ehrenkaufer, R.E. *Tetrahedron Lett.* *1984*, *25*, 3415.

retention of configuration.[1204] Ammonium formate in methanol reduces aromatic nitro compounds.[1205] Lithium aluminum hydride reduces aliphatic nitro compounds to amines, but with aromatic nitro compounds the products with this reagent are azo compounds (**19-80**). Most metal hydrides, including $NaBH_4$ and BH_3, do not reduce nitro groups at all, although both aliphatic and aromatic nitro compounds have been reduced to amines with $NaBH_4$ and various catalysts, such as $NiCl_2$ or $CoCl_2$[1206] phthalocyanine iron (II),[1207] and $ZrCl_4$.[1208] Borohydride exchange resin in the presence of $Ni(OAc)_2$, however, gives the amine.[1209] Treatment of aromatic nitro compounds with $NaBH_4$ alone has resulted in reduction of the *ring* to a cyclohexane ring with the nitro group still intact[1210] or in cleavage of the nitro group from the ring.[1211] With $(NH_4)_2S$ or other sulfides or polysulfides it is often possible to reduce just one of two or three nitro groups on an aromatic ring or on two different rings in one molecule.[1212] The nitro groups of *N*-nitro compounds can also be reduced to amino groups, for example, nitrourea $NH_2CONHNO_2$ gives semicarbazide $NH_2CONHNH_2$. Bakers yeast reduces aromatic nitro compounds to aniline derivatives.[1213] A combination of $NaH_2PO_2/FeSO_4$ with microwave irradiation reduces aromatic nitro compounds to aniline derivatives.[1214] Hydrazine on alumina, with $FeCl_3$ and microwave irradiation accomplishes this reduction.[1215] Hydrazine–formic acid with Raney nickel in methanol reduces aromatic nitro compounds.[1216] Heating aromatic nitro compounds with 57% HI reduces the nitro group to the amino group.[1217]

With some reducing agents, especially with aromatic nitro compounds, the reduction can be stopped at an intermediate stage, and hydroxylamines (**19-46**), hydrazobenzenes, azobenzenes (**19-80**), and azoxybenzenes (**19-79**) can be obtained in this manner. However, nitroso compounds, which are often postulated as intermediates, are too reactive to be isolated, if indeed they are intermediates. Reduction by metals in mineral acids cannot be stopped, but always produces the amine.

[1204]Barrett, A.G.M.; Spilling, C.D. *Tetrahedron Lett.* **1988**, *29*, 5733.

[1205]Gowda, D.C.; Mahesh, B. *Synth. Commun.* **2000**, *30*, 3639.

[1206]See, for example, Osby, J.O.; Ganem, B. *Tetrahedron Lett.* **1985**, *26*, 6413; Petrini, M.; Ballini, R.; Rosini, G. *Synthesis* **1987**, 713; He, Y.; Zhao, H.; Pan, X.; Wang, S. *Synth. Commun.* **1989**, *19*, 3047. See also, references cited therein.

[1207]Wilkinson, H.S.; Tanoury, G.J.; Wald, S.A.; Senanayake, C.H. *Tetrahedron Lett.* **2001**, *42*, 167.

[1208]Chary, K.P.; Ram, S.R.; Iyengar, D.S. *Synlett* **2000**, 683.

[1209]Yoon, N.M.; Choi, J. *Synlett* **1993**, 135.

[1210]Severin, T.; Schmitz, R. *Chem. Ber.* **1962**, *95*, 1417; Severin, T.; Adam, M. *Chem. Ber.* **1963**, *96*, 448.

[1211]Kaplan, L.A. *J. Am. Chem. Soc.* **1964**, *86*, 740. See also, Swanwick, M.G.; Waters, W.A. *Chem. Commun.* **1970**, 63.

[1212]This result has also been achieved by hydrogenation with certain catalysts [Lyle, R.E.; LaMattina, J.L. *Synthesis* **1974**, 726; Knifton, J.F. *J. Org. Chem.* **1976**, *41*, 1200; Ono, A.; Terasaki, S.; Tsuruoka, Y. *Chem. Ind. (London) 1983*, 477], and with hydrazine hydrate and Raney nickel: Ayyangar, N.R.; Kalkote, U.R.; Lugad, A.G.; Nikrad, P.V.; Sharma, V.K. *Bull. Chem. Soc. Jpn.* **1983**, *56*, 3159.

[1213]Baik, W.; Han, J.L.; Lee, K.C.; Lee, N.H.; Kim, B.H.; Hahn, J.-T. *Tetrahedron Lett.* **1994**, *35*, 3965.

[1214]Meshram, H.M.; Ganesh, Y.S.S.; Sekhar, K.C.; Yadav, J.S. *Synlett* **2000**, 993.

[1215]Vass, A.; Dudás, J.; Tóth, J.; Varma, R.S. *Tetrahedron Lett.* **2001**, *42*, 5347.

[1216]Gowda, S.; Gowda, D.C. *Tetrahedron* **2002**, *58*, 2211.

[1217]Kumar, J.S.D.; Ho, M.M.; Toyokuni, T. *Tetrahedron Lett.* **2001**, *42*, 5601.

The mechanisms of these reductions have not been studied much, although it is usually presumed that, at least with some reducing agents, nitroso compounds and hydroxylamines are intermediates. Both of these types of compounds give amines when exposed to most of these reducing agents (**19-47**), and hydroxylamines can be isolated (**19-46**). With metals and acid the following path has been suggested:[1218]

$$
\text{Ar—N}^{\oplus}(=O)(O^{\ominus}) \xrightarrow{\text{metal}} \text{Ar—N}^{\cdot\oplus}(O^{\ominus})(O^{\ominus}) \xrightarrow{H^+} \text{Ar—N}^{\cdot\oplus}(O^{\ominus})(O\text{—}H) \xrightarrow{\text{metal}} \text{Ar—N}(O^{\ominus})(O\text{—}H) \longrightarrow
$$

$$
\text{Ar—N}=O \xrightarrow{\text{metal}} \text{Ar—N—O}^{\ominus} \xrightarrow{H^+} \text{Ar—N}^{\cdot}\text{—O—H} \xrightarrow{\text{metal}} \text{Ar—N}^{\ominus}\text{—O—H}
$$

$$
\xrightarrow{H^+} \text{Ar—N(H)—O—H} \xrightarrow[H^+]{\text{metal}} \text{Ar—N(H)—H}
$$

Certain aromatic nitroso compounds (Ar—NO) can be obtained in good yields by irradiation of the corresponding nitro compounds in 0.1 M aq. KCN with uv light.[1219] The reaction has also been performed electrochemically.[1220] When nitro compounds are treated with most reducing agents, nitroso compounds are either not formed or react further under the reaction conditions and cannot be isolated.

Reductive alkylation of aromatic nitro compounds is possible. The reaction of nitrobenzene with allylic or benzyl halides in the presence of an excess of tin metal in methanol, leads to the N,N-diallyl or dibenzyl aniline.[1221] A similar reaction occurs with nitrobenzene, allyl bromide, and indium metal in aq. acetonitrile.[1222]

OS **I**, 52, 240, 455, 485; **II**, 130, 160, 175, 254, 447, 471, 501, 617; **III**, 56, 59, 63, 69, 73, 82, 86, 239, 242, 453; **IV**, 31, 357; **V**, 30, 346, 552, 567, 829, 1067, 1130; **81**, 188.

19-46   Reduction of Nitro Compounds to Hydroxylamines

$$\text{ArNO}_2 \xrightarrow[\text{H}_2\text{O}]{\text{Zn}} \text{ArNHOH}$$

When aromatic nitro compounds are reduced with zinc and water under neutral conditions,[1223] hydroxylamines are formed. Among other reagents used for this

[1218]House, H.O. *Modern Synthetic Reactions*, 2nd ed., W.A. Benjamin, NY, *1972*, p. 211.

[1219]Petersen, W.C.; Letsinger, R.L. *Tetrahedron Lett.* *1971*, 2197; Vink, J.A.J.; Cornelisse, J.; Havinga, E. *Recl. Trav. Chim. Pays-Bas* *1971*, 90, 1333.

[1220]Lamoureux, C.; Moinet, C. *Bull. Soc. Chim. Fr.* *1988*, 59.

[1221]Bieber, L.W.; da Costa, R.C.; da Silva, M.F. *Tetrahedron Lett.* *2000*, 41, 4827.

[1222]Kang, K.H.; Choi, K.I.; Koh, H.Y.; Kim, Y.; Chung, B.Y.; Cho, Y.S. *Synth. Commun.* *2001*, 31, 2277.

[1223]For some other methods of accomplishing this conversion, see Rondestvedt Jr., C.S.; Johnson, T.A. *Synthesis* *1977*, 850; Entwistle, I.D.; Gilkerson, T.; Johnstone, R.A.W.; Telford, R.P. *Tetrahedron* *1978*, 34, 213.

purpose have been SmI$_2$,[1224] N$_2$H$_4$–Rh–C,[1225] and KBH$_4$/BiCl$_3$.[1226] Borane in THF reduces aliphatic nitro enolate anions to hydroxylamines:[1227]

$$\underset{R}{\overset{R}{\underset{\ominus}{\underset{|}{C}}}}\text{—NO}_2 \quad \xrightarrow{\text{BF}_3\text{–THF}} \quad \underset{R}{\overset{R}{\underset{H}{\underset{|}{C}}}}\text{—NHOH}$$

Nitro compounds have been reduced electrochemically, to hydroxylamines, as well as to other products.[1228]

OS **I**, 445; **III**, 668; **IV**, 148; **VI**, 803; **VIII**, 16.

19-47 Reduction of Nitroso Compounds and Hydroxylamines to Amines

N-Dihydro-de-oxo-bisubstitution

$$\text{RNO} \quad \xrightarrow[\text{HCl}]{\text{Zn}} \quad \text{RNH}_2$$

N-Hydro-de-hydroxylation or *N*-Dehydroxylation

$$\text{RNHOH} \quad \xrightarrow[\text{HCl}]{\text{Zn}} \quad \text{RNH}_2$$

Nitroso compounds and hydroxylamines can be reduced to amines by the same reagents that reduce nitro compounds (**19-45**). Reaction with CuCl, and then phenylboronic acid (p. 815), also reduces nitroso compounds to the amine.[1229] A hydroxylamine can be reduced to the amine with CS$_2$ in acetonitrile.[1230] Indium metal in EtOH/aq. NH$_4$Cl reduces hydroxylamines to the amine.[1231] *N*-Nitroso compounds are similarly reduced to hydrazinesm R$_2$N–NO → R$_2$N–NH$_2$.[1232]

OS **I**, 511; **II**, 33, 202, 211, 418; **III**, 91; **IV**, 247. See also, OS **VIII**, 93.

19-48 Reduction of Oximes to Primary Amines or Aziridines

$$\underset{R}{\overset{\text{N—OH}}{\underset{\|}{\underset{C}{}}}}\underset{R^1}{}\quad \xrightarrow{\text{LiAlH}_4} \quad \underset{R}{\overset{H}{\underset{R^1}{\underset{|}{C}}}}\text{—NH}_2$$

[1224]Kende, A.S.; Mendoza, J.S. *Tetrahedron Lett.* **1991**, *32*, 1699.

[1225]Oxley, P.W.; Adger, B.M.; Sasse, M.J.; Forth, M.A. *Org. Synth.* **67**, 187.

[1226]Ren, P.D-D.; Pan, X.-W.; Jin, Q.-H.; Yao, Z.-P. *Synth. Commun.* **1997**, *27*, 3497.

[1227]Feuer, H.; Bartlett, R.S.; Vincent Jr., B.F.; Anderson, R.S. *J. Org. Chem.* **1965**, *31*, 2880.

[1228]For reviews of the electroreduction of nitro compounds, see Fry, A.J. *Synthetic Organic Electrochemistry*, 2nd ed., Wiley, NY, **1989**, pp. 188–198; Lund, H. in Baizer; Lund *Organic Electrochemistry*, Marcel Dekker, NY, **1983**, pp. 285–313.

[1229]Yu, Y.; Srogl, J.; Liebeskind, L.S. *Org. Lett.* **2004**, *6*, 2631.

[1230]Schwartz, M.A..; Gu, J.; Hu, X. *Tetrahedron Lett.* **1992**, *33*, 1687.

[1231]Cicchi, S.; Bonanni, M.; Cardona, F.; Revuelta, J.; Goti, A. *Org. Lett.* **2003**, *5*, 1773.

[1232]For examples of this reduction, accomplished with titanium reagents, see Entwistle, I.D.; Johnstone, R.A.W.; Wilby, A.H. *Tetrahedron* **1982**, *38*, 419; Lunn, G.; Sansone, E.B.; Keefer, L.K. *J. Org. Chem.* **1984**, *49*, 3470.

Both aldoximes and ketoximes can be reduced to primary amines with $LiAlH_4$. The reaction is slower than with ketones, so that, for example, $PhCOCH=NOH$ gave 34% $PhCHOHCH=NOH$.[1233] Among other reducing agents that give this reduction[1234] are zinc and acetic acid, BH_3,[1235] $NaBH_3CN-TiCl_3$,[1236] polymethylhydrosiloxane (PMHS) with Pd-C,[1237] and sodium and an alcohol.[1238] Catalytic hydrogenation is also effective.[1239] The reduction has been performed enantioselectively with Baker's yeast[1240] and with Ph_2SiH_2 and an optically active rhodium complex catalyst.[1241] Reduction of oximes with indium metal in acetic anhydride/acetic acid–THF leads to the acetamide.[1242] Oxime O-ethers are reduced to the amine with modest enantioselectivity using a chiral oxazaboroline.[1243]

When the reducing agent is DIBALH, the product is a secondary amine, arising from a rearrangement:[1244]

$$\underset{R}{\overset{}{}}\overset{N\text{-OH}}{\underset{C}{\parallel}}R^1 \quad \xrightarrow{\;t\text{-Bu}_2\text{AlH}\;} \quad R\overset{H}{\underset{}{\;N\;}}CH_2\text{-}R^1$$

With certain oximes (e.g., those of the type $ArCH_2CR=NOH$), treatment with $LiAlH_4$ gives aziridines,[1245] for example,

$$Ph\smallsetminus\overset{N\text{-OH}}{\underset{C}{\parallel}}\smallsetminus Ph \quad \xrightarrow[\text{THF}]{\;LiAlH_4\;} \quad$$

Hydrazones, arylhydrazones, and semicarbazones can also be reduced to amines with various reducing agents, including $Zn-HCl$ and H_2 and Raney nickel.

[1233]Felkin, H. *C. R. Acad. Sci.* **1950**, *230*, 304.

[1234]For a list of reagents, with references, see Larock, R.C. *Comprehensive Organic Transformations*, 2nd ed., Wiley-VCH, NY, **1999**, pp. 845–846.

[1235]Feuer, H.; Braunstein, D.M. *J. Org. Chem.* **1969**, *34*, 1817.

[1236]Leeds, J.P.; Kirst, H.A. *Synth. Commun.* **1988**, *18*, 777.

[1237]Chandrasekhar, S.; Reddy, M.V.; Chandraiah, L. *Synlett* **2000**, 1351.

[1238]For example, see Sugden, J.K.; Patel, J.J.B. *Chem. Ind. (London)* **1972**, 683.

[1239]For a review, see Rylander, P.N. *Catalytic Hydrogenation over Platinum Metals*, Academic Press, NY, **1967**, pp. 139–159.

[1240]Gibbs, D.E.; Barnes, D. *Tetrahedron Lett.* **1990**, *31*, 5555.

[1241]Brunner, H.; Becker, R.; Gauder, S. *Organometallics* **1986**, *5*, 739; Takei, I.; Nishibayashi, Y.; Ishii, Y.; Mizobe, Y.; Uemura, S.; Hidai, M. *Chem. Commun.* **2001**, 2360.

[1242]Harrison, J.R.; Moody, C.J.; Pitts, M.R. *Synlett* **2000**, 1601.

[1243]Fontaine, E.; Namane, C.; Meneyrol, J.; Geslin, M.; Serva, L.; Russey, E.; Tissandié, S.; Maftouh, M.; Roger, P. *Tetrahedron Asymmetry* **2001**, *12*, 2185.

[1244]Sasatani, S.; Miyazaki, T.; Maruoka, K.; Yamamoto, H. *Tetrahedron Lett.* **1983**, *24*, 4711; Graham, S.H.; Williams, A.J.S. *Tetrahedron* **1965**, *21*, 3263.

[1245]For a review, see Kotera, K.; Kitahonoki, K. *Org. Prep. Proced.* **1969**, *1*, 305. For examples, see Tatchell, A.R. *J. Chem. Soc. Perkin Trans. 1* **1974**, 1294; Ferrero, L.; Rouillard, M.; Decouzon, M.; Azzaro, M. *Tetrahedron Lett.* **1974**, 131; Diab, Y.; Laurent, A.; Mison, P. *Tetrahedron Lett.* **1974**, 1605.

Oximes have been reduced in a different way, to give imines (RR'C=NOH →
RR'C=NH), which are generally unstable but which can be trapped to give
useful products. Among reagents used for this purpose have been Bu$_3$P—SPh$_2$[1246]
and Ru$_3$(CO)$_{12}$.[1247] Oximes can also be reduced to hydroxylamines (**19-42**).
Nitrones have been reduced to imines using AlCl$_3$•6 H$_2$O/KI followed by
Na$_2$S$_2$O$_3$—H$_2$O.[1248]

OS **II**, 318; **III**, 513; **V**, 32, 83, 373, 376.

19-49   Reduction of Aliphatic Nitro Compounds to Oximes or Nitriles

$$RCH_2NO_2 \xrightarrow[\text{HOAc}]{\text{Zn}} RCH=NOH$$

Nitro compounds that contain an α hydrogen can be reduced to oximes with zinc
dust in acetic acid[1249] or with other reagents, among them CS$_2$—NEt$_3$,[1250]
CrCl$_2$,[1251] and (for α-nitro sulfones) NaNO$_2$.[1252] α-Nitro alkenes have been converted to oximes

$$-\overset{|}{\underset{|}{C}}=\overset{|}{\underset{|}{C}}-NO_2 \longrightarrow -\overset{|}{\underset{|}{C}}H-\overset{|}{\underset{|}{C}}=NOH$$

with sodium hypophosphite, indium with aq. NH$_4$Cl/MeOH,[1253] and with
Pb—HOAc—DMF, as well as with certain other reagents.[1254]

$$RCH_2NO_2 \xrightarrow{\text{NaBH}_2\text{S}_3} RC\equiv N$$

Primary aliphatic nitro compounds can be reduced to aliphatic nitriles with
sodium dihydro(trithio)borate[1087] or with t-BuN≡C/BuN=C=O.[1255] Secondary
compounds give mostly ketones (e.g., nitrocyclohexane gave 45% cyclohexanone,
30% cyclohexanone oxime, and 19% N-cyclohexylhydroxylamine). Tertiary aliphatic nitro compounds do not react with this reagent (see also, **19-45**).

OS **IV**, 932.

[1246]Barton, D.H.R.; Motherwell, W.B.; Simon, E.S.; Zard, S.Z. *J. Chem. Soc. Chem. Commun.* **1984**, 337.
[1247]Akazome, M.; Tsuji, Y.; Watanabe, Y. *Chem. Lett.* **1990**, 635.
[1248]Boruah, M.; Konwar, D. *Synlett* **2001**, 795.
[1249]Johnson, K.; Degering, E.F. *J. Am. Chem. Soc.* **1939**, *61*, 3194.
[1250]Barton, D.H.R.; Fernandez, I.; Richard, C.S.; Zard, S.Z. *Tetrahedron* **1987**, *43*, 551; Albanese, D.;
Landini, D.; Penso, M. *Synthesis* **1990**, 333.
[1251]Hanson, J.R. *Synthesis* **1974**, 1, pp. 7-8.
[1252]Zeilstra, J.J.; Engberts, J.B.F.N. *Synthesis* **1974**, 49.
[1253]Yadav, J.S.; Subba Reddy, B.V.; Srinivas, R.; Ramalingam, T. *Synlett* **2000**, 1447.
[1254]See Kabalka, G.W.; Pace, E.D.; Wadgaonkar, P.P. *Synth. Commun.* **1990**, *20*, 2453; Sera, A.;
Yamauchi, H.; Yamada, H.; Itoh, K. *Synlett* **1990**, 477.
[1255]El Kaim, L.; Gacon, A. *Tetrahedron Lett.* **1997**, *38*, 3391.

19-50   Reduction of Azides to Primary Amines

N-Dihydro-de-diazo-bisubstitution

$$RN_3 \xrightarrow{\text{LiAlH}_4} RNH_2$$

Azides are easily reduced to primary amines by $LiAlH_4$, as well as by a number of other reducing agents,[1256] including $NaBH_4$, $NaBH_4/LiCl$,[1257] $NaBH_4/CoCl_2/H_2O$,[1258] $NaBH_4/ZrCl_4$,[1259] $BHCl_2 \cdot SMe_2$,[1260] H_2 and a catalyst, $Bu_3SnH/PhSiH_3/AIBN$,[1261] Mg or Ca in MeOH,[1262] $Sm/NiCl_2$,[1263] $Zn-FeCl_3/EtOH$,[1264] $Zn/NH_4Cl/aq.$ EtOH,[1265] $FeCl_3/NaI$,[1266] $FeSO_4/NH_3/MeOH$,[1267] baker's yeast,[1268] Sm/I_2,[1269] Indium metal in EtOH,[1270] $LiMe_2NBH_3$,[1271] and tin complexes prepared from $SnCl_2$ or $Sn(SR)_2$.[1272] Reaction with PPh_3 leads to a phosphazide, $Ph_3P=N-N=N-R$, which loses nitrogen in what is called the *Staudinger reaction*[1273] a method to prepare phosphazo compounds, but in this case leads to reduction. Alkylation is possible, and the reaction of an alkyl azide with PMe_3, and then an excess of iodomethane leads to the N-methylated amine.[1274] This reaction, combined with $RX \rightarrow RN_3$ (**10-43**), is an important way of converting alkyl halides RX to primary amines RNH_2; in some cases the two procedures have been combined into one laboratory step.[1275] Sulfonyl azides RSO_2N_3

[1256]For a review, see Scriven, E.F.V.; Turnbull, K. *Chem. Rev.* **1988**, *88*, 297, see pp. 321–327. For lists of reagents, with references, see Larock, R.C. *Comprehensive Organic Transformations*, 2nd ed., Wiley-VCH, NY, **1999**, pp. 815–820; Rolla, F. *J. Org. Chem.* **1982**, *47*, 4327.

[1257]Ram, S.R.; Chary, K.P.; Iyengar, D.S. *Synth. Commun.* **2000**, *30*, 4495.

[1258]Fringuelli, F.; Pizzo, F.; Vaccaro, L. *Synthesis* **2000**, 646.

[1259]Chary, K.P.; Ram, S.R.; Salahuddin, S.; Iyengar, D.S. *Synth. Commun.* **2000**, *30*, 3559.

[1260]Salunkhe, A.M.; Ramachandran, P.V.; Brown, H.C. *Tetrahedron* **2002**, *58*, 10059.

[1261]Hays, D.S.; Fu, G.C. *J. Org. Chem.* **1998**, *63*, 2796.

[1262]Maiti, S.N.; Spevak, P.; Narender Reddy, A.V. *Synth. Commun.* **1988**, *18*, 1201.

[1263]Wu, H.; Chen, R.; Zhang, Y. *Synth. Commun.* **2002**, *32*, 189.

[1264]Pathak, D.; Laskar, D.D.; Prajapati, D.; Sandhu, J.S. *Chem. Lett.* **2000**, 816.

[1265]Lin, W.; Zhang, X.; He, Z.; Jin, Y.; Gong, L.; Mi, A. *Synth. Commun.* **2002**, *32*, 3279.

[1266]Kamal, A.; Ramana, K.V.; Ankati, H.B.; Ramana, A.V. *Tetrahedron Lett.* **2002**, *43*, 6861.

[1267]Kamal, A.; Laxman, E.; Arifuddin, M. *Tetrahedron Lett.* **2000**, *41*, 7743.

[1268]Kamal, A.; Damayanthi, Y.; Reddy, B.S.N.; Lakminarayana, B.; Reddy, B.S.P. *Chem. Commun.* **1997**, 1015; Baruah, M.; Boruah, A.; Prajapati, D.; Sandhu, J.S. *Synlett* **1996**, 1193.

[1269]Huang, Y.; Zhang, Y.; Wang, Y. *Tetrahedron Lett.* **1997**, *38*, 1065.

[1270]Reddy, G.V.; Rao, G.V.; Iyengar, D.S. *Tetrahedron Lett.* **1999**, *40*, 3937.

[1271]Alvarez, S.G.; Fisher, G.B.; Singaram, B. *Tetrahedron Lett.* **1995**, *36*, 2567.

[1272]Bartra, M.; Romea, P.; Urpí, F.; Vilarrasa, J. *Tetrahedron* **1990**, *46*, 587. See also, Bosch, I.; Costa, A.M.; Martín, M.; Urpí, F.; Vilarrasa, J. *Org. Lett.* **2000**, *2*, 397.

[1273]First reported by Staudinger, H.; Meyer, J. *Helv. Chim. Acta* **1919**, *2*, 635. For a review, see Golobov, Y.G.; Zhmurova, I.N.; Kasukhin, L.F. *Tetrahedron* **1981**, *37*, 437. For a discussion of the mechanism, see Tian, W.Q.; Wang, Y.A. *J. Org. Chem.* **2004**, *69*, 4299. For a modification that leads to β-lactams, see Krishnaswamy, D.; Bhawal, B.M.; Deshmukh, A.R.A.S. *Tetrahedron Lett.* **2000**, *41*, 417; Wack, H.; Drury III, W.J.; Taggi, A.E.; Ferraris, D.; Lectka, T. *Org. Lett.* **1999**, *1*, 1985.

[1274]Kato, H.; Ohmori, K.; Suzuki, K. *Synlett* **2001**, 1003.

[1275]See, for example, Koziara, A.; Osowska-Pacewicka, K.; Zawadzki, S.; Zwierzak, A. *Synthesis* **1985**, 202; **1987**, 487. The reactions **10-48**, **10-43**, and **19-50** have also been accomplished in one laboratory step: Koziara, A. *J. Chem. Res. (S)* **1989**, 296.

have been reduced to sulfonamides RSO_2NH_2 by irradiation in isopropyl alcohol[1276] and with NaH.[1277]

OS **V**, 586; **VII**, 433.

19-51 Reduction of Miscellaneous Nitrogen Compounds

Isocyanate–methylamine transformation	$R-N=C=O \xrightarrow{\text{LiAlH}_4}$	$R-NH-CH_3$
Isothiocyanate–methylamine transformation	$R-N=C=S \xrightarrow{\text{LiAlH}_4}$	$R-NH-CH_3$
***N,N*-Dihydro-addition**	$Ar-N=N-Ar \xrightarrow[\text{catalyst}]{\text{H}_2}$	$Ar-NH-NH-Ar$
Diazonium–arylhydrazone reduction	$ArN_2{}^+\ Cl^- \xrightarrow{\text{Na}_2\text{SO}_3}$	$ArNHNH_2$
***N*-Hydro-de-nitroso-substitution**	$R_2N-NO \xrightarrow[\text{Ni}]{\text{H}_2}$	R_2NH

Isocyanates and isothiocyanates are reduced to methylamines on treatment with LiAlH$_4$. LiAlH$_4$ does not usually reduce azo compounds[1278] (indeed these are the products from LiAlH$_4$ reduction of nitro compounds, **19-80**), but these can be reduced to hydrazo compounds by catalytic hydrogenation or with diimide[1279] (see **15-11**). Diazonium salts are reduced to hydrazines by sodium sulfite. This reaction probably has a nucleophilic mechanism.[1280]

The initial product is a salt of hydrazinesulfonic acid, which is converted to the hydrazine by acid treatment. Diazonium salts can also be reduced to arenes (**19-69**). *N*-Nitrosoamines can be denitrosated to secondary amines by a number of reducing agents, including H$_2$ and a catalyst,[1281] BF$_3$—THF—NaHCO$_3$,[1282] and NaBH$_4$—TiCl$_4$,[1283] as well as by hydrolysis.[1284]

[1276]Reagen, M.T.; Nickon, A. *J. Am. Chem. Soc.* **1968**, *90*, 4096.

[1277]Lee, Y.; Closson, W.D. *Tetrahedron Lett.* **1974**, 381.

[1278]For a review see Newbold, B.T., in Patai, S. *The Chemistry of the Hydrazo, Azo, and Azoxy Groups*, pt. 2, Wiley, NY, **1975**, pp. 601, 604–614.

[1279]For example, see Ioffe, B.V.; Sergeeva, Z.I.; Dumpis,Yu.Ya. *J. Org. Chem. USSR* **1969**, *5*, 1683.

[1280]Huisgen, R.; Lux, R. *Chem. Ber.* **1960**, *93*, 540.

[1281]Enders, D.; Hassel, T.; Pieter, R.; Renger, B.; Seebach, D. *Synthesis* **1976**, 548.

[1282]Jeyaraman, R.; Ravindran, T. *Tetrahedron Lett.* **1990**, *31*, 2787.

[1283]Kano, S.; Tanaka, Y.; Sugino, E.; Shibuya, S.; Hibino, S. *Synthesis* **1980**, 741.

[1284]Fridman, A.L.; Mukhametshin, F.M.; Novikov, S.S. *Russ. Chem. Rev.* **1971**, *40*, 34, pp. 41–42.

$$\text{RCN} + \text{[limonene structure]} \xrightarrow{\text{Pd-C}} \text{RCH}_3 + \text{[structure]}$$

Limonene

A cyano group can be reduced to a methyl group by treatment with a terpene, such as limonene (which acts as reducing agent) in the presence of palladium–charcoal.[1285] Hydrogen gas (H_2) is also effective,[1286] although higher temperatures are required. The group R may be alkyl or aryl. Cyano groups CN have also been reduced to CH_2OH, in the vapor phase, with 2-propanol and zirconium oxide.[1287]

Aryl nitro compounds are reduced to diaryl hydrazines with Al—KOH in methanol.[1288]

OS **I**, 442; **III**, 475. Also see, OS **V**, 43.

C. Reactions in Which a Heteroatom Is Removed from the Substrate

19-52 Reduction of Silanes to Methylene Compounds
Si-Hydrogen-uncoupling

$$\text{R–SiR}'_3 \longrightarrow \text{R–H}$$

In certain cases, the C—Si bond of silanes can be converted to C—H. α-Silyl esters are reduced to esters with mercuric acetate and tetrabutylammonium fluoride, for example.[1289]

19-53 Reduction of Alkyl Halides
Hydro-de-halogenation or Dehalogenation

$$\text{RX} \longrightarrow \text{RH}$$

This type of reduction can be accomplished with many reducing agents.[1290] A powerful, but highly useful reagent is $LiAlH_4$,[1291] which reduces almost all types of

[1285]Kindler, K.; Lührs, K. *Chem. Ber.* **1966**, 99, 227; *Liebigs Ann. Chem.* **1967**, 707, 26.

[1286]See also Andrade, J.G.; Maier, W.F.; Zapf, L.; Schleyer, P.v.R. *Synthesis* **1980**, 802; Brown, G.R.; Foubister, A.J. *Synthesis* **1982**, 1036.

[1287]Takahashi, K.; Shibagaki, M.; Matsushita, H. *Chem. Lett.* **1990**, 311.

[1288]Khurana, J.M.; Singh, S. *J. Chem. Soc., Perkin Trans. 1* **1999**, 1893.

[1289]Poliskie, G.M.; Mader, M.M.; van Well, R. *Tetrahedron Lett.* **1999**, 40, 589.

[1290]For reviews, see Hudlický, M. *Reductions in Organic Chemistry*, Ellis Horwood, Chichester, **1984**, pp. 62–67, 181; Pinder, A.R. *Synthesis* **1980**, 425. For a list of reagents, see Larock, R.C. *Comprehensive Organic Transformations*, 2nd ed., Wiley-VCH, NY, **1999**, pp. 29–39.

[1291]For a review of LiAlH4, see Pizey, J.S. *Synthetic reagents*, Vol. 1, Wiley, NY, **1974**, pp. 101–294. For monographs on complex metal hydrides, see Seyden-Penne, J. *Reductions by the Alumino- and Borohydrides*, VCH, NY, **1991**; Hajós, A. *Complex Hydrides*, Elsevier, NY, **1979**.

alkyl halide, including vinylic, bridgehead, and cyclopropyl halides.[1292] Reduction with lithium aluminum deuteride serves to introduce deuterium into organic compounds. An even more powerful reducing agent, is lithium triethylborohydride (LiEt$_3$BH), which rapidly reduces primary, secondary, allylic, benzylic, and neopentyl halides, but not tertiary (these give elimination) or aryl halides.[1293] Another powerful reagent, which reduces primary, secondary, tertiary, allylic, vinylic, aryl, and neopentyl halides, is a complex formed from lithium trimethoxyaluminum hydride, LiAlH(OMe)$_3$, and CuI.[1294] A milder reducing agent is NaBH$_4$ in a dipolar aprotic solvent, such as Me$_2$SO, DMF, or sulfolane,[1295] which at room temperature or above reduces primary, secondary, and some tertiary[1296] halides in good yield without affecting other functional groups that would be reduced by LiAlH$_4$, for example, COOH, COOR, CN.[1297] A mixture of NaBH$_4$ and InCl$_3$ efficiently reduces secondary bromides.[1298] Borohydride exchange resin is also an effective reducing agent in the presence of metal catalysts, such as Ni(OAc)$_2$,[1299] and Bu$_4$NBH$_4$, is also effective.[1300]

Other reducing agents[1301] include zinc (with acid or base), SnCl$_2$, SmI$_2$–THF–HMPA,[1302] and Et$_3$SiH in the presence of AlCl$_3$.[1303] Diethyl phosphonate–Et$_3$N,[1304] phosphorus tris(dimethylamide) (Me$_2$N)$_3$P,[1305] and organotin hydrides R$_n$SnH$_{4-n}$[1306] (chiefly Bu$_3$SnH) usually used in conjunction with a radical

[1292]Jefford, C.W.; Kirkpatrick, D.; Delay, F. *J. Am. Chem. Soc.* *1972*, *94*, 8905; Krishnamurthy, S.; Brown, H.C. *J. Org. Chem.* *1982*, *47*, 276.

[1293]Krishnamurthy, S.; Brown, H.C. *J. Org. Chem.* *1980*, *45*, 849; *1983*, *48*, 3085.

[1294]Masamune, S.; Rossy, P.A.; Bates, G.S. *J. Am. Chem. Soc.* *1973*, *95*, 6452; Masamune, S.; Bates, G.S.; Georghiou, P.E. *J. Am. Chem. Soc.* *1974*, *96*, 3686.

[1295]Bell, H.M.; Vanderslice, C.W.; Spehar, A. *J. Org. Chem.* *1969*, *34*, 3923; Hutchins, R.O.; Hoke, D.; Keogh, J.; Koharski, D. *Tetrahedron Lett.* *1969*, 3495; Vol'pin, M.E.; Dvolaitzky, M.; Levitin, I. *Bull. Soc. Chim. Fr.* *1970*, 1526; Hutchins, R.O.; Kandasamy, D.; Dux III, F.; Maryanoff, C.A.; Rotstein, D.; Goldsmith, B.; Burgoyne, W.; Cistone, F.; Dalessandro, J.; Puglis, J. *J. Org. Chem.* *1978*, *43*, 2259.

[1296]Hutchins, R.O.; Bertsch, R.J.; Hoke, D. *J. Org. Chem.* *1971*, *36*, 1568.

[1297]For the use of NaBH$_4$ under phase-transfer conditions, see Bergbreiter, D.E.; Blanton, J.R. *J. Org. Chem.* *1987*, *52*, 472.

[1298]Inoue, K.; Sawada, A.; Shibata, I.; Baba, A. *J. Am. Chem. Soc.* *2002*, *124*, 906.

[1299]Yoon, N.M.; Lee, H.J.; Ahn, J.H.; Choi, J. *J. Org. Chem.* *1994*, *59*, 4687.

[1300]Narasimhan, S.; Swarnalakshmi, S.; Balakumar, R.; Velmathi, S. *Synth. Commun.* *1999*, *29*, 685.

[1301]For some other reducing agents, not mentioned here, see Akiba, K.; Shimizu, A.; Ohnari, H.; Ohkata, K. *Tetrahedron Lett.* *1985*, *26*, 3211; Kim, S.; Yi, K.Y. *Bull. Chem. Soc. Jpn.* *1985*, *58*, 789; Cole, S.J.; Kirwan, J.N.; Roberts, B.P.; Willis, C.R. *J. Chem. Soc. Perkin Trans. 1* *1991*, 103; Hudlický, M. *Reductions in Organic Chemistry*, Ellis Horwood, Chichester, *1984*, pp. 62–67, 181, Pinder, A.R. *Synthesis* *1980*, 425. For a list of reagents, see Larock, R.C. *Comprehensive Organic Transformations*, 2nd ed., Wiley-VCH, NY, *1999*, pp. 29–39.

[1302]For discussions of mechansims related to SmI$_2$ reduction of halides. Inanaga, J.; Ishikawa, M.; Yamaguchi, M. *Chem. Lett.* *1987*, 1485; Shabangi, M.; Kuhlman, M.L.; Flowers II, R.A. *Org. Lett.* *1999*, *1*, 2133. See also, Molander, G.A.; Hahn, G. *J. Org. Chem.* *1986*, *51*, 1135. See Ogawa, A.; Ohya, S.; Hirao, T. *Chem. Lett.* *1997*, 275 for reduction with SmI$_2$/hv.

[1303]Doyle, M.P.; McOsker, C.C.; West, C.T. *J. Org. Chem.* *1976*, *41*, 1393; Parnes, Z.N.; Romanova, V.S.; Vol'pin, M.E. *J. Org. Chem. USSR* *1988*, *24*, 254.

[1304]Hirao, T.; Kohno, S.; Ohshiro, Y.; Agawa, T. *Bull. Chem. Soc. Jpn.* *1983*, *56*, 1881.

[1305]Downie, I.M.; Lee, J.B. *Tetrahedron Lett.* *1968*, 4951.

[1306]Seyferth, D.; Yamazaki, H.; Alleston, D.L. *J. Org. Chem.* *1963*, *28*, 703. For a novel trialkyltin hydride, see Gastaldi, S.; Stein, D. *Tetrahedron Lett.* *2002*, *43*, 4309.

initiator, such as AIBN.[1307] Tributyltin hydride can be used in conjunction with transition-metal salts, such as InCl$_3$.[1308] The organotin hydride (MeOCH$_2$-CH$_2$OCH$_2$CH$_2$CH$_2$)$_3$SnH reduces alkyl halides and is water soluble, unlike Bu$_3$SnH.[1309] In a related area, silylated cyclohexadienes have been used with AIBN as radical-chain reducing reagents, effective for tertiary halides.[1310] Other transition metal-based reducing agents include NiCl$_2$,[1311] Ni(OAc)$_2$/Al(acac)$_3$/ NaH.[1312] Raney nickel in Z-propanol reduces primary iodides in the presence of a lactone moiety.[1313] Aluminum amalgam efficiently reduced an iodohydrin to the alcohol.[1314] A polymer-bound dialkyltin halide has been used in conjunction with NaBH$_4$ to reduce alkyl bromides.[1315]

Reduction, especially of bromides and iodides, can also be effected by catalytic hydrogenation,[1316] and electrochemically.[1317] Raney nickel by itself can reduce alkyl halides.[1318] A good reducing agent for the removal of all halogen atoms in a polyhalo compound (including vinylic, allylic, geminal, and even bridgehead halogens) is lithium[1319] or sodium[1320] and t-BuOH in THF. Propargylic halides can often be reduced with allylic rearrangement to give allenes.[1321]

$$ \underset{\underset{X}{\overset{|}{\underset{R}{\overset{R}{|}}}}{R-C-C{\equiv}C-H} \xrightarrow{\text{LiAlH}_4} \underset{\underset{H}{R}}{\overset{R}{\underset{}{}}}C=C=C\overset{H}{\underset{H}{}} $$

The choice of a reducing agent usually depends on what other functional groups are present. Each reducing agent reduces certain groups and not others. This type of

[1307]For reviews of organotin hydrides, see Neumann, W.P. *Synthesis* **1987**, 665; Kuivila, H.G. *Synthesis* **1970**, 499, *Acc. Chem. Res.* **1968**, *1*, 299. Tributyltin hydride also reduces vinyl halides in the prescence of a palladium catalyst. See Uenishi, J.; Kawahama, R.; Shiga, Y.; Yonemitsu, O.; Tsuji, J. *Tetrahedron Lett.* **1996**, *37*, 6759.

[1308]Inoue, K.; Sawada, A.; Shibata, I.; Baba, A. *Tetrahedron Lett.* **2001**, *42*, 4661; Hayashi, N.; Shibata, I.; Baba, A. *Org. Lett.* **2004**, *6*, 4981.

[1309]Light, J.; Breslow, R. *Tetrahedron Lett.* **1990**, *31*, 2957.

[1310]Studer, A.; Amrein, S.; Schleth, F.; Schulte, T.; Walton, J.C. *J. Am. Chem. Soc.* **2003**, *125*, 5726.

[1311]Alonso, F.; Radivoy, G.; Yus, M. *Tetrahedron* **1999**, *55*, 4441.

[1312]Massicot, F.; Schneider, R.; Fort, Y.; Illy-Cherry, S.; Tillement, O. *Tetrahedron* **2000**, *56*, 4765.

[1313]Mebane, R.C.; Grimes, K.D.; Jenkins, S.R.; Deardorff, J.D.; Gross, B.H. *Synth. Commun.* **2002**, *32*, 2049.

[1314]Wang, Y.-C.; Yan, T.-H. *Chem. Commun.* **2000**, 545.

[1315]Enholm, E.J.; Schulte II, J.P. *Org. Lett.* **1999**, *1*, 1275.

[1316]For a discussion, see Rylander, P.N. *Hydrogenation Methods*, Academic Press, NY, **1985**. See also, Kantam, M.L.; Rahman, A.; Bandyopadhyay, T.; Haritha, Y. *Synth. Commun.* **1999**, *29*, 691.

[1317]For reviews, see Fry, A.J. *Synthetic Organic Electrochemistry*, 2nd ed., Wiley, NY, **1989**, pp. 136–151; Feoktistov, L.G., in Baizer, M.M.; Lund, H. *Organic Electrochemistry*, Marcel Dekker, NY, **1983**, pp. 259–284.

[1318]For an example see Marquié, J.; Laporterie, A.; Dubac, J.; Roques, N. *Synlett* **2001**, 493.

[1319]For example, see Gassman, P.G.; Pape, P.G. *J. Org. Chem.* **1964**, *29*, 160; Fieser, L.F.; Sachs, D.H. *J. Org. Chem.* **1964**, *29*, 1113; Berkowitz, D.B. *Synthesis* **1990**, 649.

[1320]For example, see Gassman, P.G.; Aue, D.H.; Patton, D.S. *J. Am. Chem. Soc.* **1968**, *90*, 7271; Gassman, P.G.; Marshall, J.L. *Org. Synth. V*, 424.

[1321]For examples, see Crandall, J.K.; Keyton, D.J.; Kohne, J. *J. Org. Chem.* **1968**, *33*, 3655; Claesson, A.; Olsson, L. *J. Am. Chem. Soc.* **1979**, *101*, 7302.

selectivity is called *chemoselectivity*. A chemoselective reagent is one that reacts with one functional group (e.g., halide), but not another (e.g., C=O). For example, there are several reagents that reduce only the halogen of α-halo ketones, leaving the carbonyl group intact.[1322] Among them are polymer-supported triphenylphosphine,[1323] decaborane with 10% Pd/C,[1324] Bi in aq. THF[1325] or In metal in water,[1326] and *i*-Bu$_2$AlH–SnCl$_2$.[1327] In a similar chemoselective reaction, the halogen in α-haloimines has been reduced with SnCl$_2$/MeOH without reducing the C=N bond.[1328]

Both NaBH$_3$CN–SnCl$_2$.[1329] and the *n*-butyllithium ate complex of B-*n*-butyl-9-BBN[1330] (see p. 1077) reduce tertiary alkyl, benzylic, and allylic halides, but do not react with primary or secondary alkyl or aryl halides. Another highly selective reagent, in this case for primary and secondary iodo and bromo groups, is sodium cyanoborohydride, NaBH$_3$CN, in HMPA.[1331] Most of the reducing agents mentioned reduce chlorides, bromides, and iodides, but organotin hydrides also reduce fluorides.[1332] See p. 1787 for a discussion of selectivity in reduction reactions.

Vinyl halides can be reduced to the corresponding alkene is some cases.[1333] As mentioned above, electrochemical reduction of aryl and vinyl halides is well known.[1334] When vinyl dibromides, such as RCH=CBr$_2$, are treated with (MeO)$_2$P(=O)H and triethylamine, for example, the product is the vinyl bromide RCH=HBr.[1335] Indium in ethanol accomplishes the same transformation.[1336] Similar reduction occurs when vinyl diiodides are treated with Zn–Cu in acetic acid.[1337]

[1322]For a review of reductive dehalogenation of polyhalo ketones, see Noyori, R.; Hayakawa, Y. *Org. React.* **1983**, *29*, 163.

[1323]Dhuru, S.P.; Padiya, K.J.; Salunkhe, M.M. *J. Chem. Res. (S)* **1998**, 56.

[1324]Lee, S.H.; Jung, Y.J.; Cho, Y.J.; Yoon, C.-O.M.; Hwang, H.-J.; Yoon, C.M. *Synth. Commun.* **2001**, *31*, 2251.

[1325]Ren, P.-D.; Hin, Q.-H.; Yao, Z.-P. *Synth. Commun.* **1997**, *27*, 2577.

[1326]Park, L.; Keum, G.; Kang, S.B.; Kim, K.S.; Kim, Y. *J. Chem. Soc. Perkin Trans. 1* **2000**, 4462.

[1327]Oriyama, T.; Mukaiyama, T. *Chem. Lett.* **1984**, 2069.

[1328]Aelterman, W.; Eeckhaut, A.; De Kimpe, N. *Synlett* **2000**, 1283.

[1329]Kim, S.; Ko, J.S. *Synth. Commun.* **1985**, *15*, 603.

[1330]Toi, H.; Yamamoto, Y.; Sonoda, A.; Murahashi, S. *Tetrahedron* **1981**, *37*, 2261.

[1331]Hutchins, R.O.; Kandasamy, D.; Maryanoff, C.A.; Masilamani, D.; Maryanoff, B.E. *J. Org. Chem.* **1977**, *42*, 82.

[1332]Fluorides can also be reduced by a solution of K and dicyclohexano-18-crown-6 in toluene or diglyme: Ohsawa, T.; Takagaki, T.; Haneda, A.; Oishi, T. *Tetrahedron Lett.* **1981**, *22*, 2583. See also, Brandänge, S.; Dahlman, O.; Ölund, J. *Acta Chem. Scand. Ser. B* **1983**, *37*, 141.

[1333]For a general discussion that includes reduction of vinyl halides with tin compounds, see Curran, D.P. *Synthesis* **1988**, 417, 489.

[1334]Fry, A.; Mitnick, M.A.; Reed, R.G. *J. Org. Chem.* **1970**, *35*, 1232; Bhuvaneswari, N.; Venkatachalam, C.S.; Balasubramanian, K.K. *Tetrahedron Lett.* **1992**, *33*, 1499; Urove, G.A.; Peters, D.G.; Mubarak, M.S. *J. Org. Chem.* **1992**, *57*, 786; Miller, L.L.; Rienkena, E. *J. Org. Chem.* **1969**, *34*, 3359; Fry, A.J.; Mitnick, M.A. *J. Am. Chem. Soc.* **1969**, *91*, 6207.

[1335]Abbas, S.; Hayes, C.J.; Worden, S. *Tetrahedron Lett.* **2000**, *41*, 3215.

[1336]Ranu, B.C.; Samanta, S.; Guchhait, S.K. *J. Org. Chem.* **2001**, *66*, 4102.

[1337]Kdota, I.; Ueno, H.; Ohno, A.; Yamamoto, Y. *Tetrhaedron Lett.* **2003**, *44*, 8645.

With LiAlH$_4$ and most other metallic hydrides, the mechanism usually consists of simple nucleophilic substitution with attack by hydride ion that may or may not be completely free. The mechanism is S$_N$2 rather than S$_N$1, since primary halides react better than secondary or tertiary (tertiary generally give alkenes or do not react at all) and since Walden inversion has been demonstrated. However, rearrangements found in the reduction of bicyclic tosylates with LiAlH$_4$ indicate that the S$_N$1 mechanism can take place.[1338] There is evidence that LiAlH$_4$ and other metal hydrides can also reduce halides by an SET mechanism,[1339] especially those, such as vinylic,[1340] cyclopropyl,[1341] or bridgehead halides, that are resistant to nucleophilic substitution. Reduction of halides by NaBH$_4$ in 80% aqueous diglyme[1342] and by BH$_3$ in nitromethane[1343] takes place by an S$_N$1 mechanism. It is known that NaBH$_4$ in sulfolane reduces tertiary halides possessing a β-hydrogen by an elimination-addition mechanism.[1344]

The mechanism for reduction of alkyl halides is not always nucleophilic substitution. For example, reductions with organotin hydrides generally[1345] take place by free-radical mechanisms,[1346] as do those with Fe(CO)$_5$. Alkyl halides, including fluorides and polyhalides, can be reduced with magnesium and a secondary or tertiary alcohol (most often 2-propanol).[1347] This is actually an example of the occurrence in one step of the sequence:

$$RX \longrightarrow RMgX \longrightarrow H^+RH$$

More often the process is carried out in two separate steps (**12-36** and **12-22**).

OS **I**, 357, 358, 548; **II**, 320, 393; **V**, 424; **VI**, 142, 376, 731; **VIII**, 82. See also, OS **VIII**, 583.

[1338]Appleton, R.A.; Fairlie, J.C.; McCrindle, R. *Chem. Commun.* **1967**, 690; Kraus, W.; Chassin, C. *Tetrahedron Lett.* **1970**, 1443. See Omoto, M.; Kato, N.; Sogon, T.; Mori, A. *Tetrahedron Lett.* **2001**, 42, 939.

[1339]Singh, P.R.; Khurana, J.M.; Nigam, A. *Tetrahedron Lett.* **1981**, 22, 2901; Srivastava, S.; le Noble, W.J. *Tetrahedron Lett.* **1984**, 25, 4871; Ashby, E.C.; Pham, T.N. *J. Org. Chem.* **1986**, 51, 3598; Hatem, J.; Meslem, J.M.; Waegell, B. *Tetrahedron Lett.* **1986**, 27, 3723; Ashby, E.C.; Deshpande, A.K. *J. Org. Chem.* **1994**, 59, 3798; Ashby, E.C.; Welder, C.; Doctorovich, F. *Tetrahedron Lett.* **1993**, 34, 7235. See, however, Hirabe, T.; Takagi, M.; Muraoka, K.; Nojima, M.; Kusabayashi, S. *J. Org. Chem.* **1985**, 50, 1797; Park, S.; Chung, S.; Newcomb, M. *J. Org. Chem.* **1987**, 52, 3275.

[1340]Chung, S. *J. Org. Chem.* **1980**, 45, 3513.

[1341]McKinney, M.A.; Anderson, S.W.; Keyes, M.; Schmidt, R. *Tetrahedron Lett.* **1982**, 23, 3443; Hatem, J.; Waegell, B. *Tetrahedron* **1990**, 46, 2789.

[1342]Bell, H.M.; Brown, H.C. *J. Am. Chem. Soc.* **1966**, 88, 1473.

[1343]Matsumura, S.; Tokura, N. *Tetrahedron Lett.* **1969**, 363.

[1344]Jacobus, J. *Chem. Commun.* **1970**, 338; Hutchins, R.O.; Bertsch, R.J.; Hoke, D. *J. Org. Chem.* **1971**, 36, 1568.

[1345]For an exception, see Carey, F.A.; Tramper, H.S. *Tetrahedron Lett.* **1969**, 1645.

[1346]Menapace, L.W.; Kuivila, H.G. *J. Am. Chem. Soc.* **1964**, 86, 3047; Tanner, D.D.; Singh, H.K. *J. Org. Chem.* **1986**, 51, 5182.

[1347]Bryce-Smith, D.; Wakefield, B.J.; Blues, E.T. *Proc. Chem. Soc.* **1963**, 219.

19-54   Reduction of Alcohols[1348]

Hydro-de-hydroxylation or Dehydroxylation

$$ROH + H_2 \xrightarrow{\text{catalyst}} RH$$

The hydroxyl groups of most alcohols can seldom be cleaved by catalytic hydrogenation and alcohols are often used as solvents for hydrogenation of other compounds. However, benzyl-type alcohols undergo the reaction readily and have often been reduced.[1349] Diaryl and triarylcarbinols are similarly easy to reduce and this has been accomplished with $LiAlH_4–AlCl_3$,[1350] with $NaBH_4$ in F_3CCOOH,[1351] and with iodine, water, and red phosphorus (OS **I**, 224). Other reagents have been used,[1352] among them PPh_3/diethyl-azo-dicarboxylate and arylsulfonyl hydrazine,[1353] PPh_3 and electrolysis,[1354] $Me_3SiCl–MeI–MeCN$,[1355] $Me_3SiCl–NaI$,[1356] $Et_3SiH–BF_3$,[1357] $SmI_2–THF–HMPA$,[1358] and tin and HCl. The reduction of secondary alcohols was accomplished using Ph_2SiClH and $InCl_3$.[1359] 1,3-Diols are especially susceptible to hydrogenolysis. Tertiary alcohols can be reduced by catalytic hydrogenolysis when the catalyst is Raney nickel.[1360] Allylic alcohols (and ethers and acetates) can be reduced (often with accompanying allylic rearrangement) with Zn amalgam and HCl, as well as with certain other reagents.[1361] α-Acetylenic alcohols are converted to alkynes by reduction of their cobalt carbonyl complexes with $NaBH_4$ and CF_3COOH.[1362] Reagents that reduce the OH group

---

[1348]For a review, see Müller, P., in Patai, S. *The Chemistry of Functional Groups, Supplement E*, pt. 1, Wiley, NY, *1980*, pp. 515–522.

[1349]For reviews, see Rylander, P.N. *Hydrogenation Methods*, Academic Press, NY, *1985*, pp. 157–163, *Catalytic Hydrogenation over Platinum Metals*, Academic Press, NY, *1967*, pp. 449–468. For a review of the stereochemistry of hydrogenolysis, see Klabunovskii, E.I. *Russ. Chem. Rev.* *1966*, *35*, 546.

[1350]Blackwell, J.; Hickinbottom, W.J. *J. Chem. Soc.* *1961*, 1405; Avendaño, C.; de Diego, C.; Elguero, J. *Monatsh. Chem.* *1990*, *121*, 649.

[1351]For a review, see Gribble, G.W.; Nutaitis, C.F. *Org. Prep. Proced. Int.* *1985*, *17*, 317. Also see, Nutaitis, C.F.; Bernardo, J.E. *Synth. Commun.* *1990*, *20*, 487.

[1352]For a list of reagents, with references, see Larock, R.C. *Comprehensive Organic Transformations*, 2nd ed., Wiley-VCH, NY, *1999*, pp. 44–46.

[1353]Myers, A.G.; Movassaghi, M.; Zheng, B. *J. Am. Chem. Soc.* *1997*, *119*, 8572.

[1354]Maeda, H.; Maki, T.; Eguchi, K.; Koide, T.; Ohmori, H. *Tetrahedron Lett.* *1994*, *35*, 4129.

[1355]Sakai, T.; Miyata, K.; Utaka, M.; Takeda, A. *Tetrahedron Lett.* *1987*, *28*, 3817.

[1356]Cain, G.A.; Holler, E.R. *Chem. Commun.* *2001*, 1168.

[1357]Orfanopoulos, M.; Smonou, I. *Synth. Commun.* *1988*, *18*, 833; Smonou, I.; Orfanopoulos, M. *Tetrahedron Lett.* *1988*, *29*, 5793. See Wustrow, D.J.; Smith III, W.J.; Wise, L.D. *Tetrahedron Lett.* *1994*, *35*, 61 for reduction with $Et_3SiH/LiClO_4$.

[1358]Kusuda, K.; Inanaga, J.; Yamaguchi, M. *Tetrahedron Lett.* *1989*, *30*, 2945.

[1359]Yasuda, M.; Onishi, Y.; Ueba, M.; Miyai, T.; Baba, A. *J. Org. Chem.* *2001*, *66*, 7741.

[1360]Krafft, M.E.; Crooks III, W.J. *J. Org. Chem.* *1988*, *53*, 432. For another catalyst, see Parnes, Z.N.; Shaapuni, D.Kh.; Kalinkin, M.I.; Kursanov, D.N. *Bull. Acad. Sci. USSR Div. Chem. Sci.* *1974*, *23*, 1592.

[1361]For discussion, see Elphimoff-Felkin, I.; Sarda, P. *Org. Synth.* **VI**, 769; *Tetrahedron* *1977*, *33*, 511. For another reagent, see Lee, J.; Alper, H. *Tetrahedron Lett.* *1990*, *31*, 4101.

[1362]Nicholas, K.M.; Siegel, J. *J. Am. Chem. Soc.* *1985*, *107*, 4999.

of α-hydroxy ketones without affecting the C=O group include lithium diphenyl-phosphide, Ph_2PLi,[1363] red phosphorus–iodine,[1364] and Me_3SiI.[1365]

Alcohols can also be reduced indirectly by conversion to a sulfonate and reduction of that compound (**19-57**). The two reactions can be carried out without isolation of the sulfonate if the alcohol is treated with pyridine–SO_3 in THF, and $LiAlH_4$ then added.[1366] Another indirect reduction that can be done in one step involves treatment of the alcohol (primary, secondary, or benzylic) with NaI, Zn, and Me_3SiCl.[1367] In this case, the alcohol is first converted to the iodide, which is reduced. For other indirect reductions of OH, see **19-59**.

The mechanisms of most alcohol reductions are obscure.[1368] Hydrogenolysis of benzyl alcohols can give inversion or retention of configuration, depending on the catalyst.[1369] The mechanism of electroreduction of allylic alcohols in acidic aqueous media has been examined.[1370]

Note that tertiary benzylic alcohols are cleaved to give the aromatic compound $[ArC(OH)Ar'_2 \rightarrow Ar-H]$ by heating with cesium carbonate and $Pd(OAc)_2$.[1371]

OS **I**, 224; **IV**, 25, 218, 482; **V**, 339; **VI**, 769.

19-55   Reduction of Phenolic and Other Hydroxyaryl Compunds

Hydro-de-hydroxylation or Dehydroxylation, etc.

$$ArOH \xrightarrow{\text{Zn}} ArH$$

Oxygenated compounds, such as phenols, phenolic esters, and ethers, can be reduced.[1372] Phenols can be reduced by distillation over zinc dust or with HI and red phosphorus, but these methods are quite poor and are seldom feasible. Catalytic hydrogenation has also been used, but the corresponding cyclohexanol (see **15-13**) is a side product.[1373]

[1363]Leone-Bay, A. *J. Org. Chem.* **1986**, *51*, 2378.

[1364]Ho, T.L.; Wong, C.M. *Synthesis* **1975**, 161.

[1365]Ho, T.L. *Synth. Commun.* **1979**, *9*, 665.

[1366]Corey, E.J.; Achiwa, K. *J. Org. Chem.* **1969**, *34*, 3667.

[1367]Morita, T.; Okamoto, Y.; Sakurai, H. *Synthesis* **1981**, 32.

[1368]For discussions of the mechanisms of the hydrogenolysis of benzyl alcohols, see Khan, A.M.; McQuillin, F.J.; Jardine, I. *Tetrahedron Lett.* **1966**, 2649; *J. Chem. Soc. C* **1967**, 136; Garbisch, Jr., E.W.; Schreader, L.; Frankel, J.J. *J. Am. Chem. Soc.* **1967**, *89*, 4233; Mitsui, S.; Imaizumi, S.; Esashi, Y. *Bull. Chem. Soc. Jpn.* **1970**, *43*, 2143.

[1369]Mitsui, S.; Kudo, Y.; Kobayashi, M. *Tetrahedron* **1969**, *25*, 1921; Mitsui, S.; Imaizumi, S.; Esashi, Y. *Bull. Chem. Soc. Jpn.* **1970**, *43*, 2143.

[1370]Shukun, H.; Yougun, S.; Jindong, Z.; Jian, S. *J. Org. Chem.* **2001**, *66*, 4487.

[1371]Terao, Y.; Nomoto, M.; Satoh, T.; Miura, M.; Nomura, M. *J. Org. Chem.* **2004**, *69*, 6942.

[1372]For a list of reagents, with references, see Larock, R.C. *Comprehensive Organic Transformations*, 2nd ed., Wiley-VCH, NY, **1999**, pp. 44–52ff.

[1373]Shuikin, N.I.; Erivanskaya, L.A. *Russ. Chem. Rev.* **1960**, *29*, 309, see pp. 313–315. See also, Bagnell, L.J.; Jeffery, E.A. *Aust. J. Chem.* **1981**, *34*, 697.

Much better results have been obtained by conversion of phenols to certain esters or ethers and reduction of the latter:

$$ArOSO_2CF_3 \xrightarrow[\substack{Pd(OAc)_2,\ Ph_3P \\ DMF}]{HCOOH,\ Et_3N} ArH \qquad Ref.[1374]$$

$$ArOTs + NaBH_4/NiCl_2 \longrightarrow ArH \qquad Ref.[1375]$$

$$\underset{\substack{| \\ OEt}}{\overset{\overset{O}{\|}}{ArO{-}P{-}OEt}} \xrightarrow[THF]{Ti} Ar\text{-}H \qquad Ref.[1376]$$

OS **VI**, 150. See also, OS **VII**, 476.

19-56   Replacement of Alkoxyl by Hydrogen

Hydro-de-alkoxylation or Dealkoxylation

$$R{-}O{-}R' \longrightarrow R{-}H + R'{-}H$$

R, R' = allyl, aryl, vinyl, benzylic

Simple ethers are not normally cleaved by reducing agents, although such cleavage has sometimes been reported[1377] (e.g., THF treated with $LiAlH_4{-}AlCl_3$[1378] or with a mixture of $LiAlH(O{-}t\text{-}Bu)_3$ and Et_3B[1379] gave 1-butanol; the latter reagent also cleaves methyl alkyl ethers).[1380] Certain types of ethers can be cleaved quite well by reducing agents.[1381] Among these are allyl aryl,[1382] vinyl aryl,[1383] benzylic ethers,[1349,1384] and anisole[1385] (for epoxides, see **19-35**). 7-Oxobicyclo[2.2.1]heptanes

[1374]Cacchi, S.; Ciattini. P.G.; Morera, E.; Ortar, G. *Tetrahedron Lett.* **1986**, *27*, 5541. See also, Peterson, G.A.; Kunng, F.; McCallum, J.S.; Wulff, W.D. *Tetrahedron Lett.* **1987**, *28*, 1381; Chen, Q.; He, Y. *Synthesis* **1988**, 896; Cabri, W.; De Bernardinis, S.; Francalanci, F.; Penco, S. *J. Org. Chem.* **1990**, *55*, 350.
[1375]Wang, F.; Chiba, K.; Tada, M. *J. Chem. Soc. Perkin Trans. 1* **1992**, 1897.
[1376]Welch, S.C.; Walters, M.E. *J. Org. Chem.* **1978**, *43*, 4797. See also, Rossi, R.A.; Bunnett, J.F. *J. Org. Chem.* **1973**, *38*, 2314.
[1377]Ranu, B.C.; Bhar, S. *Org. Prep. Proceed. Int.* **1996**, *28*, 371.
[1378]Bailey, W.J.; Marktscheffel, F. *J. Org. Chem.* **1960**, *25*, 1797.
[1379]Krishnamurthy, S.; Brown, H.C. *J. Org. Chem.* **1979**, *44*, 3678.
[1380]For a review of ether reduction, see Müller, P., in Patai, S. *The Chemistry of Functional Groups, Supplement E*, pt. 1, Wiley, NY, **1980**, pp. 522–528.
[1381]For a list of reagents, with references, see Larock, R.C. *Comprehensive Organic Transformations*, 2nd ed., Wiley-VCH, NY, **1999**, pp. 1013–1019.
[1382]Tankguchi, T.; Ogasawara, K. *Angew. Chem. Int. Ed.* **1998**, *37*, 1136; Rao, G.V.; Reddy, D.S.; Mohan, G.H.; Iyengar, D.S. *Synth. Commun.* **2000**, *30*, 3565.
[1383]Tweedie, V.L.; Barron B.G. *J. Org. Chem.* **1960**, *25*, 2023. See also, Hutchins, R.O.; Learn, K. *J. Org. Chem.* **1982**, *47*, 4380.
[1384]Bouzide, A.; Sauvé, G. *Synlett* **1997**, 1153; Thomas, R.M.; Mohan, G.H.; Iyengar, D.S. *Tetrahedron Lett.* **1997**, *38*, 4721; Shi, L.; Xia, W.J.; Zhang, F.M.; Tu, Y.Q. *Synlett* **2002**, 1505. See also Olivero, S.; Duñach, E. *Tetrahedron Lett.* **1997**, *38*, 6193.
[1385]Majetich, G.; Zhang, Y.; Wheless, K. *Tetrahedron Lett.* **1994**, *35*, 8727.

can be reductively cleaved with DIBAL and nickel catalysts.[1386] α-Methoxy ketones are demethoxylated (O=C–COMe → O=C–CH) with SmI$_2$.[1387]

Acetals and ketals are resistant to LiAlH$_4$ and similar hydrides, and carbonyl groups are often converted to acetals or ketals for protection (**16-5**). However, a combination of LiAlH$_4$ and AlCl$_3$[1388] does reduce acetals and ketals, removing one group, as shown above.[1389] The actual reducing agents in this case are primarily chloroaluminum hydride (AlH$_2$Cl) and dichloroaluminum hydride (AlHCl$_2$), which are formed from the reagents.[1390] This conversion can also be accomplished with DIBALH,[1391] as well as with other reagents.[1392] Ortho esters are easily reduced to acetals by LiAlH$_4$ alone, offering a route to aldehydes, which are easily prepared by hydrolysis of the acetals (**10-6**). Mixed ketals [R(OMe)OR′] can be demethoxylated (to give RHOR′) with Bn$_3$SnCl/NaCHBH$_3$ in the presence of AIBN.[1393]

OS **III**, 693; **IV**, 798; **V**, 303. Also see, OS **III**, 742; **VII**, 386.

19-57 Reduction of Tosylates and Similar Compounds

Hydro-de-sulfonyloxy-substitution

$$\text{RCH}_2\text{OTs} + \text{LiAlH}_4 \longrightarrow \text{RCH}_3$$

Tosylates and other sulfonates can be reduced[1394] with LiAlH$_4$,[1395] with NaBH$_4$ in a dipolar aprotic solvent,[1396] with LiEt$_3$BH, with i-Bu$_2$AlH (DIBALH),[1397] or with Bu$_3$SnH–NaI.[1398] The scope of the reaction seems to be similar to that of **19-53**.

[1386]Lautens, M.; Chiu, P.; Ma, S.; Rovis, T. *J. Am. Chem.Soc.* **1995**, *117*, 532.

[1387]Mikami, K.; Yamaoka, M.; Yoshida, A. *Synlett* **1998**, 607.

[1388]For a review of reductions by metal hydride–Lewis acid combinations, see Rerick, M.N., in Augustine, R.L. *Reduction*, Marcel Dekker, NY, **1968**, pp. 1–94.

[1389]Eliel, E.L.; Badding, V.G.; Rerick, M.N. *J. Am. Chem. Soc.* **1962**, *84*, 2371.

[1390]Ashby, E.C.; Prather, J. *J. Am. Chem. Soc.* **1966**, *88*, 729; Diner, U.E.; Davis, H.A.; Brown, R.K. *Can. J. Chem.* **1967**, *45*, 207.

[1391]See, for example, Zakharkin, L.I.; Khorlina, I.M. *Bull. Acad. Sci. USSR Div. Chem. Sci.* **1959**, 2156; Takano, S.; Akiyama, M.; Sato, S.; Ogasawara, K. *Chem. Lett.* **1983**, 1593.

[1392]For other reagents that accomplish this conversion, see Kotsuki, H.; Ushio, Y.; Yoshimura, N.; Ochi, M. *J. Org. Chem.* **1987**, *52*, 2594; Hojo, M.; Ushioda, N.; Hosomi, A. *Tetrahedron Lett.* **2004**, *45*, 4499; Larock, R.C. *Comprehensive Organic Transformations*, 2nd ed., Wiley-VCH, NY, **1999**, pp. 931–942.

[1393]Srikrishna, A.; Viswajanani, R. *Synlett* **1995**, 95.

[1394]For a list of substrate types and reagents, with references, see Larock, R.C. *Comprehensive Organic Transformations*, 2nd ed., Wiley-VCH, NY, **1999**, pp. 46–52.

[1395]For examples, see Dimitriadis, E.; Massy-Westropp, R.A. *Aust. J. Chem.* **1982**, *35*, 1895; Goodenough, K.M.; Moran, W.J.; Raubo, P.; Harrity, J.P.A. *J. Org. Chem.* **2005**, *70*, 207.

[1396]Hutchins, R.O.; Hoke, D.; Keogh, J.; Koharski, D. *Tetrahedron Lett.* **1969**, 3495.

[1397]Janssen, C.G.M.; Hendriks, A.H.M.; Godefroi, E.F. *Recl. Trav. Chim. Pays-Bas* **1984**, *103*, 220.

[1398]Ueno, Y.; Tanaka, C.; Okawara, M. *Chem. Lett.* **1983**, 795.

When the reagent is LiAlH$_4$, alkyl tosylates are reduced more rapidly than iodides or bromides if the solvent is Et$_2$O, but the order is reversed in diglyme.[1399] The reactivity difference is great enough so that a tosylate function can be reduced in the presence of a halide and vice versa. Tertiary allylcyclopropyl tosylates have been reduced with BuZnCl and a palladium catalyst.[1400]

OS **VI**, 376, 762; **VIII**, 126. See also, OS **VII**, 66.

19-58 Hydrogenolysis of esters (Barton–McCombie Reaction)

Hydro-de-thioacetoxylation

Alcohols can readily be converted to carbonate and thiocarbonate derivatives. Under radical conditions,[1401] using *azobis*-isobutyronitrile (AIBN, p. 935) and Bu$_3$SnH, the carbonate or thiocarbonate unit is reduced and replaced with hydrogen. The overall process is reduction of the ROH unit to RH. This is called the *Barton–McCombie reaction*.[1402] When R is cyclododecane (OCSOR), for example, this reduction yields the parent cyclododecane in 76% yield.[1403] When R = cyclododecane (OCSSMe), treatment with Bu$_3$P=O and AIBN, gives the alkane in 94% yield.[1404] Both PhSiH$_3$/AIBN[1405] and PhSiH$_2$—BEt$_3$•O$_2$ can be used.[1406] This reaction can be catalytic in Bu$_3$SnH.[1407] Variations include reduction of ROCSNHPh derivatives using Ph$_3$SiH/BEt$_3$.[1408]

19-59 Reductive Cleavage of Carboxylic Esters

Hydro-de-acyloxylation or Deacyloxylation

[1399]Krishnamurthy, S. *J. Org. Chem.* **1980**, *45*, 2550.

[1400]Ollivier, J.; Piras, P.P.; Stolle, A.; Aufranc, P.; de Meijere, A.; Salaün, J. *Tetrahedron Lett.* **1992**, *33*, 3307.

[1401]Barton, D.H.R.; Jaszberenyi, J.Cs.; Tang, D. *Tetrahedron Lett.* **1993**, *34*, 3381.

[1402]Barton, D.H.R.; McCombie, S.W. *J. Chem. Soc. Perkin Trans. 1* **1975**, 1574; Robins, M.J.; Wilson, J.S.; Hansske, F. *J. Am. Chem. Soc.* **1983**, *105*, 4059.

[1403]Jang, D.O.; Cho, D.H.; Kim, J. *Synth. Commun*, **1998**, *28*, 3559. Also see Gimisis, T.; Ballestri, M.; Ferreri, C.; Chatgilialoglu, C.; Boukherroub, R.; Manuel, G. *Tetrahedron Lett.* **1995**, *36*, 3897; Crimmins, M.T.; Dudek, C.M.; Cheung, A.W-H. *Tetrahedron Lett.* **1992**, *33*, 181.

[1404]Jang, D.O.; Cho, D.H.; Barton, D.H.R. *Synlett* **1998**, 39; Barton, D.H.R.; Parekh, S.I.; Tse, C.-L. *Tetrahedron Lett.* **1993**, *34*, 2733.

[1405]Barton, D.H.R.; Jang, D.O.; Jaszberenyi, J.Cs. *Tetrahedron* **1993**, *49*, 2793.

[1406]Barton, D.H.R.; Jang, D.O.; Jaszberenyi, J.Cs. *Tetrahedron* **1993**, *49*, 7193.

[1407]Lopez, R.M.; Hays, D.S.; Fu, G.C. *J. Am. Chem. Soc.* **1997**, *119*, 6949.

[1408]Oba, M.; Nishiyama, K. *Tetrahedron* **1994**, *50*, 10193.

The alkyl group R of certain carboxylic esters can be reduced to RH^{1409} by treatment with lithium in ethylamine.[1410] The reaction is successful when R is a tertiary or a sterically hindered secondary alkyl group. A free-radical mechanism is likely.[1411] Similar reduction, also by a free-radical mechanism, has been reported with sodium in HMPA—t-BuOH.[1412] In the latter case, tertiary R groups give high yields of RH, but primary and secondary R are converted to a mixture of RH and ROH. Both of these methods provide an indirect method of accomplishing **19-54** for tertiary R.[1413] The same thing can be done for primary and secondary R by treating alkyl chloroformates, ROCOCl, with tri-n-propylsilane in the presence of *tert*-butylperoxide[1414] and by treating thiono ethers ROC(=S)W (where W can be OAr or other groups) with Ph_2SiH_2[1415] or Ph_3SiH[1416] and a free-radical initiator. Allylic acetates can be reduced with $NaBH_4$ and a palladium complex,[1417] and with $SmI_2Pd(0)$.[1418] The last reagent converts propargylic acetates to allenes $R^1C{\equiv}C{-}CR^2R^3OAc \rightarrow R^1CH{=}C{=}CR^2R^3$.[1418] For other carboxylic ester reductions, see **19-62**, **19-38**, and **19-65**.

Note that acid chlorides can be reduced (R—COCl $\rightarrow$ R—H) using $(Me_3Si)_3SiH/$ AIBN.[1419]

OS **VII**, 139.

19-60 Reduction of Hydroperoxides and Peroxides

$$R{-}O{-}O{-}H \xrightarrow{\text{LiAlH}_4} ROH$$

Hydroperoxides can be reduced to alcohols with $LiAlH_4$ or Ph_3P^{1420} or by catalytic hydrogenation. This functional group is very susceptible to catalytic

[1409]For a review of some of the reactions in this section and some others, see Hartwig, W. *Tetrahedron* **1983**, *39*, 2609.

[1410]Barrett, A.G.M.; Godfrey, C.R.A.; Hollinshead, D.M.; Prokopiou, P.A.; Barton, D.H.R.; Boar, R.B.; Joukhadar, L.; McGhie, J.F.; Misra, S.C. *J. Chem. Soc. Perkin Trans. 1* **1981**, 1501. See Garst, M.E.; Dolby, L.J.; Esfandiari, S.; Fedoruk, N.A.; Chamberlain, N.C.; Avey, A.A. *J. Org. Chem.* **2000**, *65*, 7098.

[1411]Barrett, A.G.M.; Prokopiou, P.A.; Barton, D.H.R.; Boar, R.B.; McGhie, J.F. *J. Chem. Soc. Chem. Commun.* **1979**, 1173.

[1412]Deshayes, H.; Pete, J. *Can. J. Chem.* **1984**, *62*, 2063.

[1413]Also see Barton, D.H.R.; Crich, D. *J. Chem. Soc. Perkin Trans. 1* **1986**, 1603.

[1414]Jackson, R.A.; Malek, F. *J. Chem. Soc. Perkin Trans. 1* **1980**, 1207.

[1415]See Barton, D.H.R.; Jang, D.O.; Jaszberenyi, J.C. *Tetrahedron Lett.* **1990**, *31*, 4681, and references cited therein. For similar methods, see Nozaki, K.; Oshima, K.; Utimoto, K. *Bull. Chem. Soc. Jpn.* **1990**, *63*, 2578; Kirwan, J.N.; Roberts, B.P.; Willis, C.R. *Tetrahedron Lett.* **1990**, *31*, 5093.

[1416]Oba, M.; Nishiyama, K. *Synthesis* **1994**, 624.

[1417]Hutchins, R.O.; Learn, K.; Fulton, R.P. *Tetrahedron Lett.* **1980**, *21*, 27. See also Ipaktschi, J. *Chem. Ber.* **1984**, *117*, 3320.

[1418]Tabuchi, T.; Inanaga, J.; Yamaguchi, M. *Tetrahedron Lett.* **1986**, *27*, 601, 5237. See also Kusuda, K.; Inanaga, J.; Yamaguchi, M. *Tetrahedron Lett.* **1989**, *30*, 2945.

[1419]Ballestri, M.; Chatgilialoglu, C.; Cardi, N.; Sommazzi, A. *Tetrahedron Lett.* **1992**, *33*, 1787.

[1420]For a review, see Rowley, A.G., in Cadogan, J.I.G. *Organophosphorus Reagents in Organic Synthesis*, Academic Press, NY, **1979**, pp. 318–320.

hydrogenation, as shown by the fact that a double bond may be present in the same molecule without being reduced.[1421]

$$R \overset{H}{\underset{R^1}{\diagdown}} \overset{}{C} \overset{}{\diagup} \underset{CN}{} \xrightarrow[-78°C]{LiN(i\text{-}Pr)_2} \left[R \overset{\ominus}{\underset{R^1}{\diagdown}} \overset{}{C} - CN \right] \xrightarrow[-78°C]{O_2} R \overset{O-O^{\ominus}}{\underset{R^1}{\diagdown}} \overset{}{C} \overset{}{\diagup} \underset{CN}{} \xrightarrow[2. \, Sn^{2+}]{1. \, H^+} R \overset{OH}{\underset{R^1}{\diagdown}} \overset{}{C} \overset{}{\diagup} \underset{CN}{} \xrightarrow{OH^-} R \overset{}{\underset{R^1}{\diagdown}} C = O$$

The reaction is an important step in a method for the oxidative decyanation of nitriles containing an α hydrogen.[1422] The nitrile is first converted to the α-hydroperoxy nitrile by treatment with base at −78°C followed by O_2. The hydroperoxy nitrile is then reduced to the cyanohydrin, which is cleaved (the reverse of **16-52**) to the corresponding ketone. The method is not successful for the preparation of aldehydes ($R' = H$).

Peroxides are cleaved to 2 equivalents of alcohols by $LiAlH_4$, Mg/MeOH,[1423] or by catalytic hydrogenation. Peroxides can be reduced to ethers with $P(OEt)_3$.[1424] In a similar reaction, disulfides ($RSSR'$) can be converted to sulfides RSR' by treatment with tris(diethylamino)phosphine, $(Et_2N)_3P$.[1425]

OS **VI**, 130.

19-61   Reduction of Carbonyl to Methylene in Aldehydes and Ketones

Dihydro-de-oxo-bisubstitution

$$R \overset{O}{\underset{R^1}{\diagdown}} \overset{\|}{C} \xrightarrow[HCl]{Zn\text{-}Hg} R \overset{H \quad H}{\underset{R^1}{\diagdown}} \overset{}{C} $$

There are various ways of reducing the C=O group of aldehydes and ketones to CH_2.[1426] The two oldest, but still very popular, methods are the *Clemmensen reduction*[1427] and the *Wolff–Kishner reduction*. The Clemmensen reduction consists of heating the aldehyde or ketone with zinc amalgam and aq. HCl.[1428] Ketones are reduced more often than aldehydes. In the Wolff–Kishner reduction,[1429] the aldehyde or ketone is heated with hydrazine hydrate and a base (usually NaOH

[1421]Rebeller, M.; Clément, G. *Bull. Soc. Chim. Fr.* ***1964***, 1302.

[1422]Freerksen, R.W.; Selikson, S.J.; Wroble, R.R.; Kyler, K.S.; Watt, D.S. *J. Org. Chem.* ***1983***, *48*, 4087. This paper also reports several other methods for achieving this conversion.

[1423]Dai, P.; Dussault, P.H.; Trullinger, T.K. *J. Org. Chem.* ***2004***, *69*, 2851.

[1424]Horner, L.; Jurgeleit, W. *Liebigs Ann. Chem.* ***1955***, *591*, 138. See also, Rowley, A.G., in Cadogan, J.I.G. *Organophosphorus Reagents in Organic Synthesis*, Academic Press, NY, ***1979***, pp. 320–322.

[1425]Harpp, D.N.; Gleason, J.G. *J. Am. Chem. Soc.* ***1971***, *93*, 2437. For another method, see Comasseto, J.V.; Lang, E.S.; Ferreira, J.T.B.; Simonelli, F.; Correi, V.R. *J. Organomet. Chem.* ***1987***, *334*, 329.

[1426]For a review, see Reusch, W. in Augustine, R.L. *Reduction*, Marcel Dekker, NY, ***1968***, pp. 171–211.

[1427]Fragmentation reactions sometimes accompany Clemmenson reduction. See Bailey, K.E.; Davis, B.R. *Aust. J. Chem.* ***1995***, *48*, 1827. Also see Rosnati, V. *Tetrahedron Lett.* ***1992***, *33*, 4791.

[1428]For a review, see Vedejs, E. *Org. React.* ***1975***, *22*, 401. For a discussion of experimental conditions, see Fieser, L.F.; Fieser, M. *Reagents for Organic Synthesis*, Vol. 1, Wiley, NY, ***1967***, pp. 1287–1289.

[1429]For a review, see Todd, D. *Org. React.* ***1948***, *4*, 378.

or KOH). The *Huang–Minlon modification*[1430] of the Wolff–Kishner reaction, in which the reaction is carried out in refluxing diethylene glycol, has completely replaced the original procedure. A microwave-assisted Huang–Minlon procedure has been reported.[1431] The reaction can also be carried out under more moderate conditions (room temperature) in DMSO with potassium *tert*-butoxide as base.[1432] A new modification of the reduction treats a ketone with hydrazine in toluene with microwave irradiation, and subsequent reaction with KOH with microwave irradiation completes the Wolff–Kishner reduction.[1433] The Wolff–Kishner reaction can also be applied to the semicarbazones of aldehydes or ketones. The Clemmensen reduction is usually easier to perform, but it fails for acid-sensitive and high-molecular-weight substrates. For these cases, the Wolff–Kishner reduction is quite useful. For high-molecular-weight substrates, a modified Clemmensen reduction, using activated zinc and gaseous HCl in an organic solvent, such as ether or acetic anhydride, has proved successful.[1434] The Clemmensen and Wolff–Kishner reactions are complementary, since the former uses acidic and the latter basic conditions.

Both methods are fairly specific for aldehydes and ketones and can be carried out with many other functional groups present. However, certain types of aldehydes and ketones do not give normal reduction products. Under Clemmensen conditions,[1435] α-hydroxy ketones give either ketones (hydrogenolysis of the OH, **19-54**) or alkenes, and 1,3-diones usually undergo rearrangement (e.g., MeCOCH$_2$-COMe → MeCOCHMe$_2$).[1436] Neither method is suitable for α,β-unsaturated ketones. These give pyrazolines[1437] under Wolff–Kishner conditions, while under Clemmensen conditions both groups of these molecules may be reduced or if only one group is reduced, it is the C=C bond.[1438] Sterically hindered ketones are resistant to both the Clemmensen and Huang–Minlon procedures, but can be reduced by vigorous treatment with anhydrous hydrazine.[1439] In the Clemmensen reduction, pinacols (**19-76**) are often side products.

Other reagents have also been used to reduce the C=O of aldehydes and ketones to CH$_2$.[1440] Among these are Me$_3$SiCl followed by Et$_3$SiH/TiCl$_4$,[1441] Ni(OAc)$_2$ on borohydride exchange resin,[1442] Et$_3$SiH on pyridinium poly(hydrogen fluoride),

[1430]Huang-Minlon *J. Am. Chem. Soc.* **1946**, *68*, 2487; **1949**, *71*, 3301.

[1431]Jaisankar, P.; Pal, B.; Giri, V.S. *Synth. Commun.* **2002**, *32*, 2569.

[1432]Cram, D.J.; Sahyun, M.R.V.; Knox, G.R. *J. Am. Chem. Soc.* **1962**, *84*, 1734.

[1433]Gadhwal, S.; Baruah, M.; Sandhu, J.S. *Synlett* **1999**, 1573.

[1434]Toda, M.; Hayashi, M.; Hirata, Y.; Yamamura, S. *Bull. Chem. Soc. Jpn.* **1972**, *45*, 264.

[1435]For a review of Clemmensen reduction of diketones and unsaturated ketones, see Buchanan, J.G.S.; Woodgate, P.D. *Q. Rev. Chem. Soc.* **1969**, *23*, 522.

[1436]Cusack, N.J.; Davis, B.R. *J. Org. Chem.* **1965**, *30*, 2062; Wenkert, E.; Kariv, E. *Chem. Commun.* **1965**, 570; Galton, S.A.; Kalafer, M.; Beringer, F.M. *J. Org. Chem.* **1970**, *35*, 1.

[1437]Pyrazolines can be converted to cyclopropanes; see **17-34**.

[1438]See, however, Banerjee, A.K.; Alvárez, J.; Santana, M.; Carrasco, M.C. *Tetrahedron* **1986**, *42*, 6615.

[1439]Barton, D.H.R.; Ives, D.A.J.; Thomas, B.R. *J. Chem. Soc.* **1955**, 2056.

[1440]For a list, with references, see Larock, R.C. *Comprehensive Organic Transformations*, 2nd ed., Wiley-VCH, NY, **1999**, pp. 61–66.

[1441]Yato, M.; Homma, K.; Ishida, A. *Heterocycles* **1995**, *41*, 17.

[1442]Bandgar, B.P.; Nikat, S.M.; Wadgaonkar, P.P. *Synth. Commun.* **1995**, *25*, 863.

[PPHF],[1443] and, for aryl ketones (ArCOR and ArCOAr), $NaBH_4-F_3CCOOH$,[1444] $NaBH_4-AlCl_3$,[1445] $NaBH_3CN$ in THF—aq. HCl,[1446] Ni—Al in H_2O,[1447] HCOO-NH_4-Pd-C,[1448] $H_3PO_2/AcOH$ and an I_2 catalyst,[1449] or trialkylsilanes in F_3CC-OOH.[1450] Silanes, such as Et_3SiH and a triarylborane catalyst, reduce aliphatic aldehydes to the alkane, $-CHO \rightarrow -CH_3$.[1451] Chlorosilanes, such as Me_2SiClH, with an $InCl_3$ catalyst reduced ketones to the methylene compound.[1452] Polymethyl-hydroxysiloxane and a triarylborane catalyst deoxygenates ketones.[1453] Most of these reagents also reduce aryl aldehydes (ArCHO) to methylbenzenes (ArCH$_3$).[1454] Aliphatic aldehydes (RCHO) can be reduced to RCH_3 with titanocene dichloride, $(C_5H_5)_2TiCl_2$.[1455] One carbonyl group of 1,2-diketones can be selectively reduced by H_2S with an amine catalyst[1456] or by HI in refluxing acetic acid.[1457] One carbo-nyl group of quinones, such as **48**, can be reduced with copper and sulfuric acid or with tin and HCl.[1458] One carbonyl group of 1,3-diketones was selectively reduced by catalytic hydrogenolysis.[1459] Catalytic hydrogenation at 170°C with Pt/K10 removes oxygen from the molecule.[1460] Simply heating a ketone in supercritical Z-propanol reduces the ketone to the methylene compound.[1461]

48

[1443]Olah, G.A.; Wang, Q.; Prakash, G.K.S. *Synlett* **1992**, 647.

[1444]Gribble, G.W.; Nutaitis, C.F. *Org. Prep. Proced. Int.* **1985**, *17*, 317.

[1445]Ono, A.; Suzuki, N.; Kamimura, J. *Synthesis* **1987**, 736.

[1446]Pashkovsky, F.S.; Lokot, I.P.; Lakhvich, F.A. *Synlett* **2001**, 1391.

[1447]Ishimoto, K.; Mitoma, Y.; Negashima, S.; Tashiro, H.; Prakash, G.K.S.; Olah, G.A.; Tahshiro, M. *Chem. Commun.* **2003**, 514.

[1448]Ram, S.; Spicer, L.D. *Tetrahedron Lett.* **1988**, *29*, 3741.

[1449]Hicks, L.D.; Han, J.K.; Fry, A.J. *Tetrahedron Lett.* **2000**, *41*, 7817; Gordon, P.E.; Fry, A.J. *Tetrahedron Lett.* **2001**, *42*, 831.

[1450]Kursanov, D.N.; Parnes, Z.N.; Loim, N.M. *Bull. Acad. Sci. USSR Div. Chem. Sci.* **1966**, 1245; West, C.T.; Donnelly, S.J.; Kooistra, D.A.; Doyle, M.P. *J. Org. Chem.* **1973**, *38*, 2675. See also, Fry, J.L.; Orfanopoulos, M.; Adlington, M.G.; Dittman, Jr., W.R.; Silverman, S.B. *J. Org. Chem.* **1978**, *43*, 374; Olah, G.A.; Arvanaghi, M.; Ohannesian, L. *Synthesis* **1986**, 770.

[1451]Gevorgyan, V.; Rubin, M.; Liu, J.-X.; Yamamoto, Y. *J. Org. Chem.* **2001**, *66*, 1672.

[1452]Miyai, T.; Ueba, M.; Baba, A. *Synlett* **1999**, 182.

[1453]Chandrasekar, S.; Reddy, Ch.R.; Babu, B.N. *J. Org. Chem.* **2002**, *67*, 9080.

[1454]See, for example, Hall, S.S.; Bartels, A.P.; Engman, A.M. *J. Org. Chem.* **1972**, *37*, 760; Kursanov, D.N.; Parnes, Z.N.; Loim, N.M.; Bakalova, G.V. *Doklad. Chem.* **1968**, *179*, 328; Zahalka, H.A.; Alper, H. *Organometallics* **1986**, *5*, 1909.

[1455]van Tamelen, E.E.; Gladys, J.A. *J. Am. Chem. Soc.* **1974**, *96*, 5290.

[1456]Mayer, R.; Hiller, G.; Nitzschke, M.; Jentzsch, J. *Angew. Chem. Int. Ed.* **1963**, *2*, 370.

[1457]Reusch, W.; LeMahieu, R. *J. Am. Chem. Soc.* **1964**, *86*, 3068.

[1458]Meyer, K.H. *Org. Synth. I*, 60; Macleod, L.C.; Allen, C.F.H. *Org. Synth. II*, 62.

[1459]Cormier, R.A.; McCauley, M.D. *Synth. Commun.* **1988**, *18*, 675.

[1460]Török, B.; London, G. Bartók, M. *Synlett* **2000**, 631.

[1461]Hatano, B.; Tagaya, H. *Tetraehedron Lett.* **2003**, *44*, 6331.

An indirect method of accomplishing the reaction is reduction of tosylhydrazones ($R_2C=N-NHTs$) to R_2CH_2 with $NaBH_4$, BH_3, catecholborane, bis(benzyloxy)borane, or $NaBH_3CN$. The reduction of α,β-unsaturated tosylhydrazones with $NaBH_3CN$, with $NaBH_4$-HOAc, or with catecholborane proceeds with migration of the double bond to the position formerly occupied by the carbonyl carbon, even if this removes the double bond from conjugation with an aromatic ring,[1462] for example,

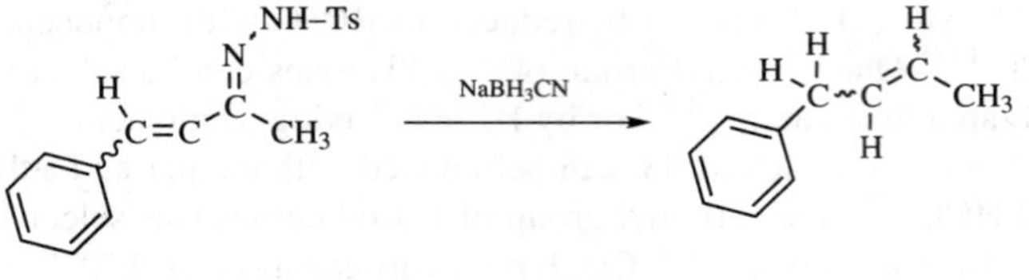

A cyclic mechanism is apparently involved:

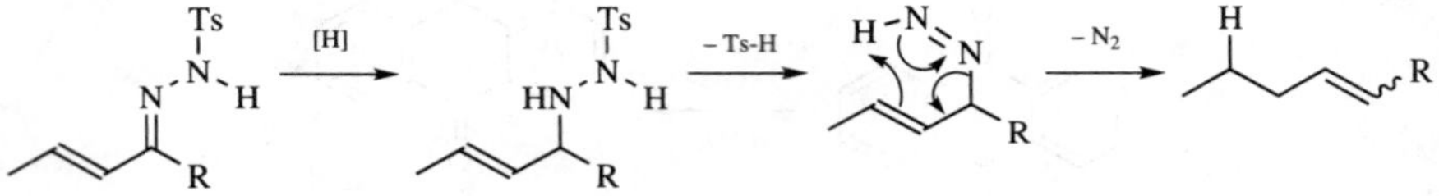

Another indirect method is conversion of the aldehyde or ketone to a dithioacetal or ketal, and desulfurization of using Raney nickel or another reagent (**14-27**).

It is interesting to see that amines can be deaminated to give the corresponding methylene compounds with low-valent titanium ($TiCl_3/Li/THF$).[1463]

The first step in the mechanism[1464] of the Wolff–Kishner reaction consists of formation of the hydrazone (**16-14**). It is this species that undergoes reduction in the presence of base, most likely in the following manner:

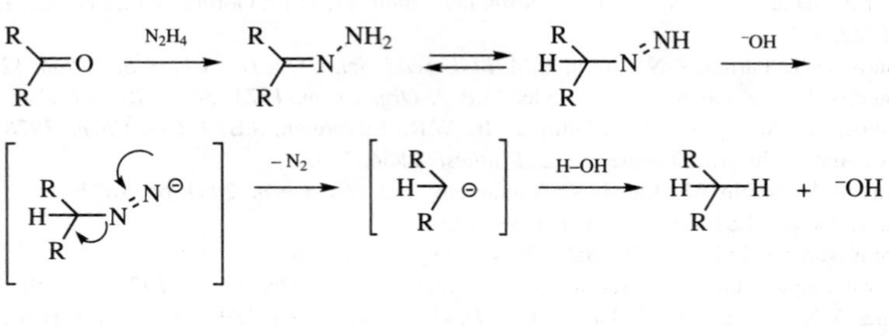

[1462]Kabalka, G.W.; Yang, D.T.C.; Baker, Jr., J.D. *J. Org. Chem.* **1976**, *41*, 574; Taylor, E.J.; Djerassi, C. *J. Am. Chem. Soc.* **1976**, *98*, 2275; Hutchins, R.O.; Natale, N.R. *J. Org. Chem.* **1978**, *43*, 2299; Greene, A.E. *Tetrahedron Lett.* **1979**, 63.

[1463]Talukdar, S.; Banerji, A. *Synth. Commun,* **1996**, *26*, 1051.

[1464]For a review of the mechanism, see Szmant, H.H. *Angew. Chem. Int. Ed.* **1968**, *7*, 120. Also see, Taber, D.F.; Stachel, S.J. *Tetrahedron Lett.* **1992**, *33*, 903.

Not much is known about the mechanism of the Clemmensen reduction. Several mechanisms have been proposed,[1465] including one going through a zinc–carbene intermediate.[1466] One thing reasonably certain is that the corresponding alcohol is not an intermediate, since alcohols prepared in other ways fail to give the reaction. Note that the alcohol is not an intermediate in the Wolff–Kishner reduction either.

OS **I**, 60; **II**, 62, 499; **III**, 410, 444, 513, 786; **IV**, 203, 510; **V**, 533, 747; **VI**, 62, 293, 919; **VII**, 393. Also see, OS **IV**, 218; **VII**, 18.

19-62   Reduction of Carboxylic Esters to Ethers

Dihydro-de-oxo-bisubstitution

$$\text{RCOOR}' \xrightarrow[\text{LiAlH}_4]{\text{BF}_3\text{–etherate}} \text{RCH}_2\text{OR}'$$

Carboxylic esters and lactones have been reduced to ethers, although 2 equivalents of alcohol are more commonly obtained (**19-38**). Reduction to ethers has been accomplished with a reagent prepared from BF_3–etherate and either $LiAlH_4$, $LiBH_4$, or $NaBH_4$,[1467] with trichlorosilane and uv light,[1468] and with catalytic hydrogenation. The reaction with the BF_3 reagent apparently succeeds with secondary R′, but not with primary R′, which give **19-38**. Lactones give cyclic ethers.[1469] Acyloxy groups are reduced by cleavage of the C—C=O bond, R(Ar)COO—C → C—H) with an excess of Ph_2SiH_2 and di-*tert*-butyl peroxide.[1470] Esters are reduced to ethers using Et_3SiH and $TiCl_4$.[1471] Lactones are converted to cyclic ethers by treatment with Cp_2TiCl_2 followed by Et_3SiH on Amberlyst 15.[1472]

Thiono esters RCSOR′ can be reduced to ethers RCH₂OR′ with Raney nickel (**14-27**).[1473] Reaction of thio esters, such as C—OC(=O)Ph with Ph_2SiH_2 and Ph_3SnH with BEt_3, followed by AIBN (p. 935) leads to reduction of the C=S unit to give an ether.[1474] Since the thiono esters can be prepared from carboxylic

[1465]See, for example, Horner, L.; Schmitt, E. *Liebigs Ann. Chem. 1978*, 1617; Poutsma, M.L.; Wolthius, E. *J. Org. Chem. 1959, 24*, 875; Nakabayashi, T. *J. Am. Chem. Soc. 1960, 82*, 3900, 3906; Di Vona, M.L.; Rosnati, V. *J. Org. Chem. 1991, 56*, 4269.

[1466]Burdon, J.; Price, R.C. *J. Chem. Soc. Chem. Commun. 1986*, 893.

[1467]Pettit, G.R.; Green, B.; Kasturi, T.R.; Ghatak, U.R. *Tetrahedron 1962, 18*, 953; Ager, D.J.; Sutherland, I.O. *J. Chem. Soc. Chem. Commun. 1982*, 248. See also, Dias, J.R.; Pettit, G.R. *J. Org. Chem. 1971, 36*, 3485.

[1468]Nagata, Y.; Dohmaru, T.; Tsurugi, J. *J. Org. Chem. 1973, 38*, 795; Baldwin, S.W.; Haut, S.A. *J. Org. Chem. 1975, 40*, 3885. See also, Kraus, G.A.; Frazier, K.A.; Roth, B.D.; Taschner, M.J.; Neuenschwander, K. *J. Org. Chem. 1981, 46*, 2417.

[1469]See, for example, Pettit, G.R.; Kasturi, T.R.; Green, B.; Knight, J.C. *J. Org. Chem. 1961, 26*, 4773; Edward, J.T.; Ferland, J.M. *Chem. Ind. (London) 1964*, 975.

[1470]Kim, J.-G.; Cho, D.H.; Jang, D.O. *Tetrahedron Lett. 2004, 45*, 3031; Jiang, D.O.; Kim, J.; Cho, D.H.; Chung, C.-M. *Tetrahedron Lett. 2001, 42*, 1073.

[1471]Yato, M.; Homma, K.; Ishida, A. *Tetrahedron 2001, 57*, 5353.

[1472]Hansen, M.C.; Verdaguer, X.; Buchwald, S.L. *J. Org. Chem. 1998, 63*, 2360.

[1473]Baxter, S.L.; Bradshaw, J.S. *J. Org. Chem. 1981, 46*, 831.

[1474]Jang, D.O.; Song, S.H. *Synlett 2000*, 811; Jang, D.O.; Song, S.H.; Cho, D.H. *Tetrahedron 1999, 55*, 3479.

esters (**16-11**), this provides an indirect method for the conversion of carboxylic esters to ethers. Thiol esters (RCOSR') have been reduced to thioethers (RCH$_2$SR').[1475]

See also, **19-65**, **19-59**.

19-63 Reduction of Cyclic Anhydrides to Lactones and Acid Derivatives to Alcohols

Dihydro-de-oxo-bisubstitution

Cyclic anhydrides can give lactones if reduced with Zn—HOAc, with hydrogen and platinum or RuCl$_2$(Ph$_3$P)$_3$,[1476] with NaBH$_4$,[1477] or even with LiAlH$_4$, although with the last-mentioned reagent diols are the more usual product. With a BINOL–AlHOEt complex, however, reduction to the lactone proceeds smoothly.[1478] With some reagents the reaction can be accomplished regioselectively, that is, only a specific one of the two C=O groups of an unsymmetrical anhydride is reduced.[1479] Open-chain anhydrides either are not reduced at all (e.g., with NaBH$_4$) or give 2 equivalents of alcohol. The LiAlH$_4$ usually reduces open-chain anhydrides to give 2 equivalents of alcohol. With cyclic anhydrides the reaction with LiAlH$_4$ can be controlled to give either diols or lactones.[1480] The NaBH$_4$ in THF, with dropwise addition of methanol, reduces open-chain anhydrides to 1 equivalent of primary alcohol and 1 equivalent of carboxylic acid.[1481]

Acyl halides are reduced[1482] to alcohols by LiAlH$_4$ or NaBH$_4$, as well as by other metal hydrides (Table 19.5), but not by borane.

In general, reduction of amides to alcohols is difficult. More commonly the amide is reduced to an amine. An exception uses LiH$_2$NBH$_3$ to give the alcohol.[1483] Reduction with sodium metal in propanol also gives the alcohol.[1484] Acyl

[1475]Eliel, E.L.; Daignault, R.A. *J. Org. Chem.* **1964**, *29*, 1630; Bublitz, D.E. *J. Org. Chem.* **1967**, *32*, 1630.

[1476]Lyons, J.E. *J. Chem. Soc. Chem. Commun.* **1975**, 412; Morand, P.; Kayser, M.M. *J. Chem. Soc. Chem. Commun.* **1976**, 314. See also Hara, Y.; Wada, K. *Chem. Lett.* **1991**, 553.

[1477]Bailey, D.M.; Johnson, R.E. *J. Org. Chem.* **1970**, *35*, 3574.

[1478]Matsuki, K.; Inoue, H.; Takeda, M. *Tetrahedron Lett.* **1993**, *34*, 1167.

[1479]See, for example, Kayser, M.M.; Salvador, J.; Morand, P. *Can. J. Chem.* **1983**, *61*, 439; Ikariya, T.; Osakada, K.; Ishii, Y.; Osawa, S.; Saburi, M.; Yoshikawa, S. *Bull. Chem. Soc. Jpn.* **1984**, *57*, 897; Soucy, C.; Favreau, D.; Kayser, M.M. *J. Org. Chem.* **1987**, *52*, 129.

[1480]Bloomfield, J.J.; Lee, S.L. *J. Org. Chem.* **1967**, *32*, 3919.

[1481]Soai, K.; Yokoyama, S.; Mochida, K. *Synthesis* **1987**, 647.

[1482]For a review of the reduction of acyl halides, see Wheeler, O.H., in Patai, S. *The Chemistry of Acyl Halides*, Wiley, NY, **1972**, pp. 231–251. For a list of reagents, with references, see Larock, R.C. *Comprehensive Organic Transformations*, 2nd ed., Wiley-VCH, NY, **1999**, pp. 1263–1264.

[1483]Myers, A.G.; Yang, B.H.; Kopecky, D.J. *Tetrahedron Lett.* **1996**, *37*, 3623.

[1484]Moody, H.M.; Kaptein, B.; Broxterman, Q.B.; Boesten, W.H.J.; Kamphuis, J. *Tetrahedron Lett.* **1994**, *35*, 1777.

imidazoles are also reduced to the corresponding alcohol with NaBH$_4$ in aq. HCl.[1485]

There are no *Organic Syntheses* references, but see OS **II**, 526, for a related reaction. See OS **VI**, 482 for reduction to alcohols and OS **IV**, 271 for reduction of acyl halides.

19-64 Reduction of Amides to Amines

Dihydro-deoxo-bisubstitution

$$RCONH_2 \xrightarrow{\text{LiAlH}_4} RCH_2NH_2$$

Amides can be reduced[1486] to amines with LiAlH$_4$ or by catalytic hydrogenation, but high temperatures and pressures are usually required for the latter. Even with LiAlH$_4$, the reaction is more difficult than the reduction of most other functional groups, and other groups often can be reduced without disturbing an amide function. Although NaBH$_4$ by itself does not reduce amides, it does so in the presence of certain other reagents[1487] including iodine.[1488] Lithium borohydride reduces acetamides.[1489] Substituted amides can be reduced with these powerful reagents; secondary amides to secondary amine and tertiary amides to tertiary amines. Borane[1490] and sodium in 1-propanol[1491] are good reducing agents for all three types of amides. Another reagent that reduces disubstituted amides to amines is trichlorosilane.[1492] Other silanes, such as Et$_3$SiH in the presence of a rhenium catalyst, reduce amides to amines.[1493] Sodium (dimethylamino)borohydride reduces unsubstituted and disubstituted, but not monosubstituted amides.[1494] Electrolytic reduction of carbamates to give an amine are possible.[1495]

[1485]Sharma, R.; Voynov, G.H.; Ovaska, T.V.; Marquez, V.E. *Synlett* **1995**, 839.

[1486]For a review, see Challis, B.C.; Challis, J.A., in Zabicky, J. *The Chemistry of Amides*, Wiley, NY, *1970*, pp. 795–801. For a review of the reduction of amides, lactams, and imides with metallic hydrides, see Gaylord, N.G. *Reduction with Complex Metal Hydrides*, Wiley, NY, *1956*, p. 544. For a list of reagents, with references, see Larock, R.C. *Comprehensive Organic Transformations*, 2nd ed., Wiley-VCH, NY, *1999*, pp. 869–872.

[1487]See, for example, Satoh, T.; Suzuki, S.; Suzuki, Y.; Miyaji, Y.; Imai, Z. *Tetrahedron Lett.* **1969**, 4555; Rahman, A.; Basha, A.; Waheed, N.; Ahmed, S. *Tetrahedron Lett.* **1976**, 219; Kuehne, M.E.; Shannon, P.J. *J. Org. Chem.* **1977**, *42*, 2082; Wann, S.R.; Thorsen, P.T.; Kreevoy, M.M. *J. Org. Chem.* **1981**, *46*, 2579; Mandal, S.B.; Giri, V.S.; Pakrashi, S.C. *Synthesis* **1987**, 1128; Akabori, S.; Takanohashi, Y. *Chem. Lett.* *1990*, 251.

[1488]Prasad, A.S.B.; Kanth, J.V.B.; Periasamy, M. *Tetrahedron* **1992**, *48*, 4623.

[1489]Tanaka, H.; Ogasawara, K. *Tetrahedron Lett.* **2002**, *43*, 4417.

[1490]Brown, H.C.; Narasimhan, S.; Choi, Y.M. *Synthesis* **1981**, 441, 996; Krishnamurthy, S. *Tetrahedron Lett.* **1982**, *23*, 3315; Bonnat, M.; Hercourt, A.; Le Corre, M. *Synth. Commun.* **1991**, *21*, 1579.

[1491]Bhandari, K.; Sharma, V.L.; Chatterjee, S.K. *Chem. Ind. (London)* **1990**, 547.

[1492]Nagata, Y.; Dohmaru, T.; Tsurugi, J. *Chem. Lett.* **1972**, 989. See also, Benkeser, R.A.; Li, G.S.; Mozdzen, E.C. *J. Organomet. Chem.* **1979**, *178*, 21.

[1493]Igarashi, M.; Fuchikami, T. *Tetrahedron Lett.* **2001**, *42*, 1945.

[1494]Hutchins, R.O.; Learn, K.; El-Telbany, F.; Stercho, Y.P. *J. Org. Chem.* **1984**, *49*, 2438.

[1495]Franco, D.; Duñach, E. *Tetrahedron Lett.* **2000**, *41*, 7333.

With some RCONR, LiAlH$_4$ causes cleavage, and the aldehyde (**10-41**) or alcohol is obtained. Lithium triethylborohydride produces the alcohol with most *N,N*-disubstituted amides, but not with unsubstituted or *N*-substituted amides.[1496] Lactams are reduced to cyclic amines in high yields with LiAlH$_4$, although cleavage sometimes occurs here too. A mixture of LiBHEt$_3$/Et$_3$SiH is also effective.[1497] Lactams are also reduced to cyclic amines with 9-BBN[1498] (p. 1077) or LiBH$_3$NMe$_2$.[1499] Imides are generally reduced on both sides,[1500] although it is sometimes possible to stop with just one. Both cyclic and acyclic imides have been reduced in this manner, although with acyclic imides cleavage is often obtained, for example,[1501]

$$PhN(COMe)_2 \longrightarrow PhNHEt$$

Acyl sulfonamides have been reduced (RCONHSO$_2$Ph → RCH$_2$NHSO$_2$Ph) with BH$_3$–SMe$_2$[1502] and with SmI$_2$/DMPU.[1503]

OS **IV**, 339, 354, 564; **VI**, 382; **VII**, 41.

19-65   Reduction of Carboxylic Acids and Esters to Alkanes
Trihydro-de-alkoxy,oxo-tersubstitution, and so on.

$$RCOOR' \xrightarrow{(C_5H_5)_2TiCl_2} RCH_3 + R'OH$$

The reagent titanocene dichloride reduces carboxylic esters in a different manner from that of **19-59**, **19-62**, or **19-38**. The products are the alkane RCH$_3$ and the alcohol R'OH.[909] The mechanism probably involves an alkene intermediate. Aromatic acids can be reduced to methylbenzenes by a procedure involving refluxing first with trichlorosilane in MeCN, then with tripropylamine added, and finally with KOH and MeOH (after removal of the MeCN).[1504] The following sequence has been suggested:[1504]

$$ArCOOH \xrightarrow{SiHCl_3} (ArCO)_2O \xrightarrow[R_3N]{SiHCl_3} ArCH_2SiCl_3 \xrightarrow[MeOH]{KOH} ArCH_3$$

Esters of aromatic acids are not reduced by this procedure, so an aromatic COOH group can be reduced in the presence of a COOR' group.[1505] However, it is also

[1496]Brown, H.C.; Kim, S.C. *Synthesis* **1977**, 635.

[1497]Pedregal, C.; Ezquerra, J.; Escribano, A.; Carreño, M.C.; García Ruano, J.L.G. *Tetrahedron Lett.* **1994**, *35*, 2053.

[1498]Colllins, C.J.; Lanz, M.; Singaram, B. *Tetrahedron Lett.* **1999**, *40*, 3673.

[1499]Flaniken, J.M.; Collins, C.J.; Lanz, M.; Singaram, B. *Org. Lett.* **1999**, *1*, 799.

[1500]For a reduction with borane•THF, see Akula, M.R.; Kabalka, G.W. *Org. Prep. Proceed. Int.* **1999**, *31*, 214.

[1501]Witkop, B.; Patrick, J.B. *J. Am. Chem. Soc.* **1952**, *74*, 3861.

[1502]Belletire, J.L.; Fry, D.F. *Synth. Commun.* **1988**, *18*, 29.

[1503]Vedejs, E.; Lin, S. *J. Org. Chem.* **1994**, *59*, 1602.

[1504]Benkeser, R.A.; Foley, K.M.; Gaul, J.M.; Li, G.S. *J. Am. Chem. Soc.* **1970**, *92*, 3232.

[1505]Benkeser, R.A.; Ehler, D.F. *J. Org. Chem.* **1973**, *38*, 3660.

possible to reduce aromatic ester groups, by a variation of the trichlorosilane procedure.[1506] Both *o*- and *p*-hydroxybenzoic acids and their esters have been reduced
to cresols $HOC_6H_4CH_3$ with sodium bis(2-methoxyethoxy)aluminum hydride,
$NaAlH_2(OC_2H_4OMe)_2$ (Red-Al).[1507] Heating a 2-pyridylbenzyl ester with ammonium formate and a rutheniumc atlyst leads to reduction of the CH_2COO unit to the
the alkane.[1508]

Carboxylic acids can also be converted to alkanes, indirectly,[1509] by reduction of
the corresponding tosylhydrazides $RCONHNH_2$ with $LiAlH_4$ or borane.[1510]

OS **VI**, 747.

19-66   Hydrogenolysis of Nitriles

Hydro-de-cyanation

$$R-CN \longrightarrow R-H$$

This transformation is not common, but given the proliferation of nitriles in organic
chemistry, it is potentially quite useful. In the presence of mercuric compounds, tertiary nitriles can be reduced to the hydrocarbon with sodium cyanoborohydride.[1511]
gem-Dinitriles can be reduced to the corresponding mononitrile with SmI_2.[1512]

19-67   Reduction of the C—N Bond

Hydro-de-amination or Deamination

$$RNH_2 \longrightarrow RH$$

Benzylic amines are particularly susceptible to hydrogenolysis by catalytic
hydrogenation[1513] or dissolving metal reduction.[1514] Note that the Wolff–Kishner
reduction in **19-61** involved formation of a hydrazone and deprotonation by base
led to loss of nitrogen and reduction. Ceric ammonium nitrate in aqueous acetonitrile has also been shown to reductively cleave the *N*-benzyl group.[1515] Primary
amines have been reduced to RH with hydroxylamine-*O*-sulfonic acid and

[1506]Benkeser, R.A.; Mozdzen, E.C.; Muth, C.L. *J. Org. Chem.* **1979**, *44*, 2185.

[1507]Černý, M.; Málek, J. *Collect. Czech. Chem. Commun.* **1970**, *35*, 2030.

[1508]Chatani, N.; Tatamidani, H.; Ie, Y.; Kakiuchi, F.; Murai, S. *J. Am. Chem. Soc.* **2001**, *123*, 4849.

[1509]For another indirect method, which can also be applied to acid derivatives, see Degani, I.; Fochi, R. *J. Chem. Soc. Perkin Trans. 1* **1978**, 1133. For a direct method, see Le Deit, H.; Cron S.; Le Corre, M. *Tetrahedron Lett.* **1991**, *32*, 2759.

[1510]Attanasi, O.; Caglioti, L.; Gasparrini, F.; Misiti, D. *Tetrahedron* **1975**, *31*, 341, and references cited therein.

[1511]Sassaman, M.B. *Tetrahedron* **1996**, *52*, 10835.

[1512]Kang, H.-Y.; Hong, W.S.; Cho, Y.S.; Koh, H.Y. *Tetrahedron Lett.* **1995**, *36*, 7661.

[1513]Hartung, W.H.; Simonoff, R. *Org. React.* **1953**, *7*, 263.

[1514]du Vigneaud, V.; Behrens, O.K. *J. Biol. Chem.* **1937**, *117*, 27.

[1515]Bull, S.D.; Davies, S.G.; Fenton, G.; Mulvaney, A.W.; Prasad, R.S.; Smith, A.D. *J. Chem. Soc. Perkin Trans. 1* **2000**, 3765.

aq. NaOH to give the hydrocarbon, nitrogen gas, and the sulfate anion.[1516] It is postulated that $R-N=N-H$ is an intermediate that decomposes to the carbocation. The reaction has also been accomplished with difluoroamine HNF_2;[1517] the same intermediates are postulated in this case. Treatment of aniline with 20 equivalents of NO gave benzene.[1518] An indirect means of achieving the same result is the conversion of the primary amine to the sulfonamide, $RNHSO_2R'$ (**16-102**), and treatment of this with NH_2OSO_2OH[1519] or NaOH, and then NH_2Cl.[1520] Tosylaziridines derived from terminal alkenes are reduced to the corresponding primary tosylamine with polymethylhydrosiloxane/Pd–C.[1521]

Other indirect methods involve reduction of N,N-ditosylates (p. 497) with $NaBH_4$ in HMPA[1522] and modifications of the Katritzky pyrylium–pyridinium method.[1523] Allylic and benzylic amines[1349] can be reduced by catalytic hydrogenolysis. Aziridines can be reductively opened with SmI_2[1524] or with Bu_3SnH and AIBN.[1525] The C–N bond of enamines is reductively cleaved to give an alkene with alane (AlH_3).[1526]

and with 9-BBN (p. 1077) or borane methyl sulfide (BMS).[1527] Since enamines can be prepared from ketones (**16-13**), this is a way of converting ketones to alkenes. In the latter case, BMS gives retention of configuration [an (E) isomer gives the (E) product], while 9-BBN gives the other isomer.[1527] Diazo ketones are reduced to methyl ketones by HI: $RCOCHN_2 + HI \rightarrow RCOCH_3$.[1528]

Quaternary ammonium salts can be cleaved with $LiAlH_4$, $R_4N^+ + LiAlH_4 \rightarrow R_3N + R^-$, as can quaternary phosphonium salts R_4P^+. Other reducing agents have also been used, for example, lithium triethylborohydride (which preferentially cleaves methyl groups)[1529] and sodium in liquid ammonia. When quaternary salts

[1516]Doldouras, G.A.; Kollonitsch, J. *J. Am. Chem. Soc.* **1978**, *100*, 341.

[1517]Bumgardner, C.L.; Martin, K.J.; Freeman, J.P. *J. Am. Chem. Soc.* **1963**, *85*, 97.

[1518]Itoh, T.; Matsuya, Y.; Nagata, K.; Ohsawa, A. *Tetrahedron Lett.* **1996**, *37*, 4165.

[1519]Nickon, A.; Hill, R.H. *J. Am. Chem. Soc.* **1964**, *86*, 1152.

[1520]Guziec Jr., F.S.; Wei, D. *J. Org. Chem.* **1992**, *57*, 3772.

[1521]Chandrasekhar, S.; Ahmed, M. *Tetrahedron Lett.* **1999**, *40*, 9325.

[1522]Hutchins, R.O.; Cistone, F.; Goldsmith, B.; Heuman, P. *J. Org. Chem.* **1975**, *40*, 2018.

[1523]See Katritzky, A.R.; Bravo-Borja, S.; El-Mowafy, A.M.; Lopez-Rodriguez, G. *J. Chem. Soc. Perkin Trans. 1* **1984**, 1671.

[1524]Molander, G.A.; Stengel, P.J. *Tetrahedron,* **1997**, *53*, 8887.

[1525]Schwan, A.L.; Refvik, M.D. *Tetrahedron Lett.* **1993**, *34*, 4901.

[1526]Coulter, J.M.; Lewis, J.W.; Lynch, P.P. *Tetrahedron* **1968**, *24*, 4489.

[1527]Singaram, B.; Goralski, C.T.; Rangaishenvi, M.V.; Brown, H.C. *J. Am. Chem. Soc.* **1989**, *111*, 384.

[1528]For example, see Pojer, P.M.; Ritchie, E.; Taylor, W.C. *Aust. J. Chem.* **1968**, *21*, 1375.

[1529]Cooke Jr., M.P.; Parlman, R.M. *J. Org. Chem.* **1975**, *40*, 531.

are reduced with sodium amalgam in water, the reaction is known as the *Emde reduction*. However, this reagent is not applicable to the cleavage of ammonium salts with four *saturated* alkyl groups. Of course, aziridines[899] can be reduced in the same way as epoxides (**19-35**).

Nitro compounds, RNO_2, can be reduced to RH[1530] by sodium methylmercaptide, CH_3SNa, in an aprotic solvent[1531] or by Bu_3SnH.[1532] Both reactions have free-radical mechanisms.[1533] Tertiary nitro compounds can be reduced to RH by NaHTe.[1534] Hydrogenolysis with a Pt catalyst in the gas phase has been reported to reduce nitro compounds, as well as primary and secondary amines.[1535] The nitro group of aromatic nitro compounds has been removed with sodium borohydride.[1536] This reaction involves an addition–elimination mechanism. Reduction of the C—N bond on aromatic amines with Li metal in THF generates the aryl compounds.[1537] Sodium nitrite, sodium bisulfite in EtOH/water/acetic acid does a similar reduction.[1538] Conversion of the aniline derivative to the methanesulfonamide and subsequent treatment with NaH and NH_2Cl gives the same result.[1539] The Bu_3SnH reagent also reduces isocyanides, RNC (prepared from RNH_2 by formylation followed by **17-31**), to RH,[1540] a reaction that can also be accomplished with Li or Na in liquid NH_3,[1541] or with K and a crown ether in toluene.[1542] α-Nitro ketones can be reduced to ketones with $Na_2S_2O_4$—Et_3SiH in $HMPA$—H_2O.[1543]

OS **III**, 148; **IV**, 508; **VIII**, 152.

[1530]For a method of reducing allylic nitro groups, see Ono, N.; Hamamoto, I.; Kamimura, A.; Kaji, A. *J. Org. Chem.* **1986**, *51*, 3734.

[1531]Kornblum, N.; Carlson, S.C.; Smith, R.G. *J. Am. Chem. Soc.* **1979**, *101*, 647; Kornblum, N.; Widmer, J.; Carlson, S.C. *J. Am. Chem. Soc.* **1979**, *101*, 658.

[1532]For reviews, see Ono, N., in Feuer, H.; Nielsen, A.T. *Nitro Compounds; Recent Advances in Synthesis and Chemistry*, VCH, NY, **1990**, pp. 1–135, 1–45; Rosini, G.; Ballini, R. *Synthesis* **1988**, 833, see pp. 835–837; Ono, N.; Kaji, A. *Synthesis* **1986**, 693. For discussions of the mechanism, see Korth, H.; Sustmann, R.; Dupuis, J.; Geise, B. *Chem. Ber.* **1987**, *120*, 1197; Kamimura, A.; Ono, N. *Bull. Chem. Soc. Jpn.* **1988**, *61*, 3629.

[1533]For a discussion of the mechanism with Bu_3SnH, see Tanner, D.D.; Harrison, D.J.; Chen, J.; Kharrat, A.; Wayner, D.D.M.; Griller, D.; McPhee, D.J. *J. Org. Chem.* **1990**, *55*, 3321. If an α substituent is present, it may be reduced instead of the NO_2. For a mechanistic discussion, see Bowman, W.R.; Crosby, D.; Westlake, P.J. *J. Chem. Soc. Perkin Trans. 2* **1991**, 73.

[1534]Suzuki, H.; Takaoka, K.; Osuka, A. *Bull. Chem. Soc. Jpn.* **1985**, *58*, 1067.

[1535]Guttieri, M.J.; Maier, W.F. *J. Org. Chem.* **1984**, *49*, 2875.

[1536]Severin, T.; Schmitz, R.; Temme, H. *Chem. Ber.* **1963**, *96*, 2499; Kniel, P. *Helv. Chim. Acta* **1968**, *51*, 371. For another method, see Ono, N.; Tamura, R.; Kaji, A. *J. Am. Chem. Soc.* **1983**, *105*, 4017.

[1537]Azzena, U.; Dessanti, F.; Melloni, G.; Pisano, L. *Tetrahedron Lett.* **1999**, *40*, 8291.

[1538]Geoffroy, O.J.; Morinelli, T.A.; Meier, G.B. *Tetrahedron Lett.* **2001**, *42*, 5367.

[1539]Wang, Y.; Guziec, Jr., F.S. *J. Org. Chem.* **2001**, *66*, 8293.

[1540]Barton, D.H.R.; Bringmann, G.; Motherwell, W.B. *Synthesis* **1980**, 68.

[1541]See Niznik, G.E.; Walborsky, H.M. *J. Org. Chem.* **1978**, *43*, 2396; Yadav, J.S.; Reddy, P.S.; Joshi, B.V. *Tetrahedron Lett.* **1988**, *44*, 7243.

[1542]Ohsawa, T.; Mitsuda, N.; Nezu, J.; Oishi, T. *Tetrahedron Lett.* **1989**, *30*, 845.

[1543]Kamimura, A.; Kurata, K.; Ono, N. *Tetrahedron Lett.* **1989**, *30*, 4819.

19-68 Reduction of Amine Oxides and Azoxy Compounds

N-Oxygen-detachment

$$R-\overset{\overset{\displaystyle R}{|}}{\underset{\underset{\displaystyle R}{|}}{N}}{}^{\oplus}-O^{\ominus} \xrightarrow{\ PPh_3\ } R-\overset{\overset{\displaystyle R}{|}}{\underset{\underset{\displaystyle R}{|}}{N}}$$

$$\underset{Ar}{\overset{Ar}{>}}N=\overset{\oplus}{N}\underset{O^{\ominus}}{\overset{\diagup}{\diagdown}} \xrightarrow{\ PPh_3\ } \underset{Ar}{\overset{Ar}{>}}N=N\diagdown_{Ar}$$

Amine oxides[1544] and azoxy compounds (both alkyl and aryl)[1545] can be reduced practically quantitatively with triphenylphosphine.[1546] Other reducing agents, for example, LiAlH$_4$, NaBH$_4$/LiCl,[1547] H$_2$—Ni, PCl$_3$, TiCl$_3$,[1548] Ga/H$_2$O,[1549] In/TiCl$_4$,[1550] LiAlH$_4$/TiCl$_4$, or SbCl$_2$ have also been used. Indium metal with aqueous ammonium chloride in methanol gives good yields of pyridine from pyridine *N*-oxide.[1551] Similar results are obtained using ammonium formate and Raney nickel[1552] or zinc.[1553] Indium (III) chloride has been used for the reduction of quinoline *N*-oxide to quinoline.[1554] Polymethylhydrosiloxane with Pd—C is also an effective reducing agent for amino oxides.[1555] Nitrile oxides[1556] (R—C≡N$^+$—O$^-$) can be reduced to nitriles with trialkylphosphines,[1557] and isocyanates (RNCO) to isocyanides (RNC) with Cl$_3$SiH—Et$_3$N.[1558]

Analogous to amino *N*-oxides, phosphine oxides (R$_3$P=O) are reduced to phosphines (R$_3$P). Treatment of a phosphine oxide with MeOTf followed by reduced

[1544]For reviews of the reduction of heterocyclic amine oxides, see Albini, A.; Pietra, S. *Heterocyclic N-Oxides*, CRC Press, Boca Raton, FL, *1991*, pp. 120–134; Katritzky, A.R.; Lagowski, J.M. *Chemistry of the Heterocyclic N-Oxides*, Academic Press, NY, *1971*, pp. 166–231.

[1545]For a review, see Newbold, B.T., in Patai, S. *The Chemistry of the Hydrazo, Azo, and Azoxy Groups*, pt. 2, Wiley, NY, *1975*, pp. 602–603, 614–624.

[1546]For a review, see Rowley, A.G., in Cadogan, J.I.G. *Organophosphorus Reagents in Organic Synthesis*, Academic Press, NY, *1979*, pp. 295–350.

[1547]Ram, S.R.; Chary, K.P.; Iyengar, D.S. *Synth. Commun.* **2000**, *30*, 3511.

[1548]Kuz'min, S.V.; Mizhiritskii, M.D.; Kogan, L.M. *J. Org. Chem. USSR* **1989**, *25*, 596.

[1549]Han, J.H.; Choi, K.I.; Kim, J.H.; Yoo, B.W. *Synth. Commun.* **2004**, *34*, 3197.

[1550]Yoo, B.W.; Choi, K.H.; Choi, K.I.; Kim, J.H. *Synth. Commun.* **2003**, *33*, 4185.

[1551]Yadav, J.S.; Reddy, B.V.S.; Reddy, M.M. *Tetrahedron Lett.* **200**, *41*, 2663.

[1552]Balicki, R.; Maciejewski, G. *Synth. Commun.* **2002**, *32*, 1681.

[1553]Balicki, R.; Cybulski, M.; Maciejewski, G. *Synth. Commun.* **2003**, *33*, 4137.

[1554]Ilias, Md.; Barman, D.C.; Prajapati, D.; Sandhu, J.S. *Tetrahedron Lett.* **2002**, *43*, 1877.

[1555]Chandrasekhar, S.; Reddy, Ch.R.; Rao, R.J.; Rao, J.M. *Synlett* **2002**, 349.

[1556]For reviews of the chemistry of nitrile oxides, see Torssell, K.B.G. *Nitrile Oxides, Nitrones, and Nitronates in Organic Synthesis*, VCH, NY, *1988*, pp. 55–74; Grundmann, C. *Fortschr. Chem. Forsch.* *1966*, *7*, 62.

[1557]Grundmann, C.; Frommeld, H.D. *J. Org. Chem.* **1965**, *30*, 2077.

[1558]Baldwin, J.E.; Derome, A.E.; Riordan, P.D. *Tetrahedron* **1983**, *39*, 2989.

with LiAlH$_4$ gives the phosphine.[1559] Chiral phosphine oxides are reduced to the phosphine with excellent enantioselectivity using PPh$_3$ and Cl$_3$SiH.[1560]

OS **IV**, 166. See also, OS **VIII**, 57.

19-69   Replacement of the Diazonium Group by Hydrogen

Dediazoniation or **Hydro-de-diazoniation**

$$ArN_2^+ + H_3PO_2 \longrightarrow ArH$$

Reduction of the diazonium group (*dediazoniation*) provides an indirect method for the removal of an amino group from an aromatic ring.[1561] The best and most common way of accomplishing this is by use of hypophosphorous acid H$_3$PO$_2$, although many other reducing agents[1562] have been used, among them ethanol, HMPA,[1563] thiophenol,[1564] and sodium stannite. Ethanol was the earliest reagent used, and it frequently gives good yields, but often ethers (ArOEt) are side products. When H$_3$PO$_2$ is used, 5–15 equivalents of this reagent are required per equivalent of substrate. Diazonium salts can be reduced in nonaqueous media by several methods, including treatment with Bu$_3$SnH or Et$_3$SiH in ethers or MeCN[1565] and by isolation as the BF$_4^-$ salt and reduction of this with NaBH$_4$ in DMF.[1566] Aromatic amines can be deaminated (ArNH$_2 \longrightarrow$ ArH) in one laboratory step by treatment with an alkyl nitrite in DMF[1567] or boiling THF.[1568] The corresponding diazonium salt is an intermediate.

Not many investigations of the mechanism have been carried out. It is generally assumed that the reaction of diazonium salts with ethanol to produce ethers takes place by an ionic (S$_N$1) mechanism while the reduction to ArH proceeds by a free-radical process.[1569] The reduction with H$_3$PO$_2$ is also believed to have a free-radical mechanism.[1570] In the reduction with NaBH$_4$, an aryldiazene intermediate

[1559]Imamoto, T.; Kikuchi, S.-i.; Miura, T.; Wada, Y. *Org. Lett.* **2001**, *3*, 87.

[1560]Wu, H.-C.; Yu, J.-Q.; Spencer, J.B. *Org. Lett.* **2004**, *6*, 4675.

[1561]For a review, see Zollinger, H., in Patai, S.; Rappoport, Z. *The Chemistry of Functinal Groups, Supplement C* pt. 1, Wiley, NY, **1983**, pp. 603–669.

[1562]For lists of some of these, with references, see Larock, R.C. *Comprehensive Organic Transformations*, 2nd ed., Wiley-VCH, NY, **1999**, pp. 39–41; Tröndlin, F.; Rüchardt, C. *Chem. Ber.* **1977**, *110*, 2494.

[1563]Shono, T.; Matsumura, Y.; Tsubata, K. *Chem. Lett.* **1979**, 1051.

[1564]For a list of some of these, with references, see Korzeniowski, S.H.; Blum, L.; Gokel, G.W. *J. Org. Chem.* **1977**, *42*, 1469.

[1565]Nakayama, J.; Yoshida, M.; Simamura, O. *Tetrahedron* **1970**, *26*, 4609.

[1566]Hendrickson, J.B. *J. Am. Chem. Soc.* **1961**, *83*, 1251. See also, Threadgill, M.D.; Gledhill, A.P. *J. Chem. Soc. Perkin Trans. 1* **1986**, 873.

[1567]Doyle, M.P.; Dellaria, Jr., J.F.; Siegfried, B.; Bishop, S.W. *J. Org. Chem.* **1977**, *42*, 3494.

[1568]Cadogan, J.I.G.; Molina, G.A. *J. Chem. Soc. Perkin Trans. 1* **1973**, 541.

[1569]For examples, see DeTar, D.F.; Kosuge, T. *J. Am. Chem. Soc.* **1958**, *80*, 6072; Lewis, E.S.; Chambers, D.J. *J. Am. Chem. Soc.* **1971**, *93*, 3267; Broxton, T.J.; Bunnett, J.F.; Paik, C.H. *J. Org. Chem.* **1977**, *42*, 643.

[1570]See, for example, Kornblum, N.; Cooper, G.D.; Taylor, J.E. *J. Am. Chem. Soc.* **1950**, *72*, 3013; Beckwith, A.L.J. *Aust. J. Chem.* **1972**, *25*, 1887; Levit, A.F.; Kiprianova, L.A.; Gragerov, I.P. *J. Org. Chem. USSR* **1975**, *11*, 2395.

(ArN=NH) has been demonstrated,[1571] arising from nucleophilic attack by BH_4^- on the β nitrogen. Such diazenes can be obtained as moderately stable (half-life of several hours) species in solution.[1572] It is not entirely clear how the aryldiazene decomposes, but there are indications that either the aryl radical AR• or the corresponding anion Ar⁻ may be involved.[1573]

An important use of the dediazoniation reaction is to remove an amino group after it has been used to direct one or more other groups to ortho and para positions. For example, the compound 1,3,5-tribromobenzene cannot be prepared by direct bromination of benzene because the bromo group is ortho–para-directing; however, this compound is easily prepared by the following sequence:

$$C_6H_6 \xrightarrow[\substack{H_2SO_4 \\ 11\text{-}2}]{HNO_3} PhNO_2 \xrightarrow[\substack{19\text{-}45}]{Sn \cdot HCl} PhNH_2 \xrightarrow[\substack{11\text{-}10}]{3Br_2} \text{(tribromoaniline)} \xrightarrow[\substack{13\text{-}19}]{HONO} \text{(diazonium salt)} \xrightarrow[H_3PO_2]{} \text{(1,3,5-tribromobenzene)}$$

Many other compounds that would otherwise be difficult to prepare are easily synthesized with the aid of the dediazoniation reaction.

Unwanted dediazoniation can be suppressed by using hexasulfonated calix[6]arenes (see p. 122).[1574]

OS I, 133, 415; II, 353, 592; III, 295; IV, 947; VI, 334.

19-70 Desulfurization

Hydro-de-thio-substitution, and so on

$$RSH \longrightarrow RH$$
$$RSR' \longrightarrow RH + R'H$$
$$RS(O)_nR' \longrightarrow RH + R'H$$

Thiols and thioethers,[1575] both alkyl and aryl, can be desulfurized by hydrogenolysis with Raney nickel.[1576] The hydrogen is usually not applied externally, since

[1571]König, E.; Musso, H.; Záhorszky, U.I. *Angew. Chem. Int. Ed.* **1972**, *11*, 45; McKenna, C.E.; Traylor, T.G. *J. Am. Chem. Soc.* **1971**, *93*, 2313.

[1572]Huang, P.C.; Kosower, E.M. *J. Am. Chem. Soc.* **1968**, *90*, 2354, 2362, 2367; Smith III, M.R.; Hillhouse, G.L. *J. Am. Chem. Soc.* **1988**, *110*, 4066.

[1573]Rieker, A.; Niederer, P.; Leibfritz, D. *Tetrahedron Lett.* **1969**, 4287; Kosower, E.M.; Huang, P.C.; Tsuji, T. *J. Am. Chem. Soc.* **1969**, *91*, 2325; König, E.; Musso, H.; Záhorszky, U.I. König, E.; Musso, H.; Záhorszky, U.I. *Angew. Chem. Int. Ed.* **1972**, *11*, 45; McKenna, C.E.; Traylor, T.G. *J. Am. Chem. Soc.* **1971**, *93*, 2313.; Broxton, T.J.; McLeish, M.J. *Aust. J. Chem.* **1983**, *36*, 1031.

[1574]Shinkai, S.; Mori, S.; Araki, K.; Manabe, O. *Bull. Chem. Soc. Jpn.* **1987**, *60*, 3679.

[1575]For a review of the reduction of thioethers, see Block, E., in Patai, S. *The Chemistry of Functional Groups, Supplement E*, pt. 1, Wiley, NY, **1980**, pp. 585–600.

[1576]For reviews, see Belen'kii, L.I., in Belen'kii, L.I. *Chemistry of Organosulfur Compounds*, Ellis Horwood, Chichester, **1990**, pp. 193–228; Pettit, G.R.; van Tamelen, E.E. *Org. React.* **1962**, *12*, 356; Hauptmann, H.; Walter, W.F. *Chem. Rev.* **1962**, *62*, 347.

Raney nickel already contains enough hydrogen for the reaction. Other sulfur compounds can be similarly desulfurized, including disulfides, thiono esters,[1577] thioamides, sulfoxides, and thioacetals.[1578] Reduction of thioacetals is an indirect way of accomplishing reduction of a carbonyl to a methylene group (see **19-61**), and it can also give the alkene if a hydrogen is present.[1579] In most of the examples given, R can also be aryl. Other reagents[1580] have also been used.[1581]

Lithium aluminum hydride reduces most sulfur compounds with cleavage of the C—S bond, including thiols.[1582] Thioesters can be reduced with Ni_2B (from $NiBr_2$/ $NaBH_4$).[1583] β-Ketosulfones are reduced with $TiCl_4$–Zn,[1584] $TiCl_4$–Sm,[1585] or Bu_3SnCl–$NaCNBH_3$/AIBN.[1586]

An important special case of RSR reduction is desulfurization of thiophene derivatives. This proceeds with concomitant reduction of the double bonds. Many compounds have been made by alkylation of thiophene to **49**, followed by reduction to give **50**.

Thiophenes can also be desulfurized to alkenes ($RCH_2CH=CHCH_2R'$ from **49**) with a nickel boride catalyst prepared from nickel(II) chloride and $NaBH_4$ in methanol.[1587] It is possible to reduce just one SR group of a dithioacetal by treatment with borane–pyridine in trifluoroacetic acid or in CH_2Cl_2 in the presence of $AlCl_3$.[1588] Phenyl selenides RSePh can be reduced to RH with Ph_3SnH[1589] and with nickel boride.[1590] Cleavage of the C—Se bond can also be achieved with SmI_2.[1591]

[1577]See Baxter, S.L.; Bradshaw, J.S. *J. Org. Chem.* **1981**, *46*, 831.

[1578]For desulfurization of the mixed acetal PhCHC(OBu)SPh to $PhCH_2OBu$ see Nakata, D.; Kusaka, C.; Tani, S.; Kunishima, M. *Tetrahedron Lett.* **2001**, *42*, 415.

[1579]Fishman, J.; Torigoe, M.; Guzik, H. *J. Org. Chem.* **1963**, *28*, 1443.

[1580]For lists of reagents, with references, see Larock, R.C. *Comprehensive Organic Transformations*, 2nd ed., Wiley-VCH, NY, **1999**, pp. 53–60. For a review with respect to transition-metal reagents, see Luh, T.; Ni, Z. *Synthesis* **1990**, 89. For some very efficient nickel-containing reagents, see Becker, S.; Fort, Y.; Vanderesse, R.; Caubère, P. *J. Org. Chem.* **1989**, *54*, 4848.

[1581]For example, diphosphorus tetraiodide by Suzuki, H.; Tani, H.; Takeuchi, S. *Bull. Chem. Soc. Jpn.* **1985**, *58*, 2421; Shigemasa, Y.; Ogawa, M.; Sashiwa, H.; Saimoto, H. *Tetrahedron Lett.* **1989**, *30*, 1277; $NiBr_2$-Ph_3P-LiAlH$_4$ by Ho, K.M.; Lam, C.H.; Luh, T. *J. Org. Chem.* **1989**, *54*, 4474.

[1582]Smith, M.B.; Wolinsky, J. *J. Chem. Soc. Perkin Trans. 2* **1998**, 1431.

[1583]Back, T.G.; Baron D.L.; Yang, K. *J. Org. Chem.* **1993**, *58*, 2407.

[1584]Guo, H.; Ye, S.; Wang, J.; Zhang, Y. *J. Chem. Res. (S)* **1997**, 114.

[1585]Wang, J.; Zhang, Y. *Synth. Commun.* **1996**, *26*, 1931.

[1586]Giovannini, R.; Petrini, M. *Synlett* **1995**, 973.

[1587]Schut, J.; Engberts, J.B.F.N.; Wynberg, H. *Synth. Commun.* **1972**, *2*, 415.

[1588]Kikugawa, Y. *J. Chem. Soc. Perkin Trans. 1* **1984**, 609.

[1589]Clive, D.L.J.; Chittattu, G.; Wong, C.K. *J. Chem. Soc. Chem. Commun.* **1978**, 41.

[1590]Back, T.G. *J. Chem. Soc. Chem. Commun.* **1984**, 1417.

[1591]Ogawa, A.; Ohya, S.; Doi, M.; Sumino, Y.; Sonoda, N.; Hirao, T. *Tetrahedron Lett.* **1998**, *39*, 6341.

The exact mechanism of the Raney nickel reactions are still in doubt, although they are probably of the free-radical type.[1592] It has been shown that reduction of thiophene proceeds through butadiene and butene, not through 1-butanethiol or other sulfur compounds, that is, the sulfur is removed before the double bonds are reduced. This was demonstrated by isolation of the alkenes and the failure to isolate any potential sulfur-containing intermediates.[1593]

Sulfonamides are reduced to the corresponding amine by heating with Me_3SiCl and NaI.[1594]

OS **IV**, 638; **V**, 419; **VI**, 109, 581, 601. See also, OS **VII**, 124, 476.

19-71 Reduction of Sulfonyl Halides and Sulfonic Acids to Thiols or Disulfides

$$RSO_2Cl \xrightarrow{\text{LiAlH}_4} RSH$$

Thiols can be prepared by the reduction of sulfonyl halides[1595] with $LiAlH_4$. Usually, the reaction is carried out on aromatic sulfonyl chlorides. Zinc and acetic acid, and HI, also give the reduction. Another reagent for this reduction is Me_2SiCl_2 and Zn with dimethyl acetamide.[1596] Sulfonic acids have been reduced to thiols with a mixture of triphenylphosphine and either I_2 or a diaryl disulfide.[1597] For the reduction of sulfonyl chlorides to sulfinic acids, see **16-104**.

Disulfides RSSR can also be produced.[1598] Other sulfonic acid derivatives can be converted to disulfides. Esters, such as PhSAc, are converted to disulfides PhS—SPh with Clayan and microwave irradiation.[1599] Thiobenzoate derivatives PhSBz are similarly converted to PhS—SPh with SmI_2.[1600] In a similar manner, RS—SO_3Na is converted to RS—SR when heated with samarium metal in water.[1601]

OS **I**, 504; **IV**, 695; **V**, 843.

[1592]For a review, see Bonner, W.A.; Grimm, R.A., in Kharasch, N.; Meyers, C.Y. *The Chemistry of Organic Sulfur Compounds*, Vol. 2, Pergamon, NY, *1966*, pp. 35–71, 410–413. For a review of the mechanism of desulfurization on molybdenum surfaces, see Friend, C.M.; Roberts, J.T. *Acc. Chem. Res.* *1988*, *21*, 394.

[1593]Owens, P.J.; Ahmberg, C.H. *Can. J. Chem.* *1962*, *40*, 941.

[1594]Sabitha, G.; Reddy, B.V.S.; Abraham, S.; Yadav, J.S. *Tetrahedron Lett.* *1999*, *40*, 1569.

[1595]For a review, see Wardell, J.L., in Patai, S. *The Chemistry of the Thiol Group*, pt. 2, Wiley, NY, *1974*, pp. 216–220.

[1596]Uchiro, H.; Kobayashi, S. *Tetrahedron Lett.* *1999*, *40*, 3179.

[1597]Oae, S.; Togo, H. *Bull. Chem. Soc. Jpn.* *1983*, *56*, 3802; *1984*, *57*, 232.

[1598]For example, see Alper, H. *Angew. Chem. Int. Ed.* *1969*, *8*, 677; Chan, T.H.; Montillier, J.P.; Van Horn, W.F.; Harpp, D.N. *J. Am. Chem. Soc.* *1970*, *92*, 7224. See also, Olah, G.A.; Narang, S.C.; Field, L.D.; Karpeles, R. *J. Org. Chem.* *1981*, *46*, 2408; Oae, S.; Togo, H. *Bull. Chem. Soc. Jpn.* *1983*, *56*, 3813; Suzuki, H.; Tani, H.; Osuka, A. *Chem. Lett.* *1984*, 139; Babu, J.R.; Bhatt, M.V. *Tetrahedron Lett.* *1986*, 27, 1073; Narayana, C.; Padmanabhan, S.; Kabalka, G.W. *Synlett* *1991*, 125.

[1599]Meshram, H.M.; Bandyopadhyay, A.; Reddy, G.S.; Yadav, J.S. *Synth. Commun.* *1999*, *29*, 2705.

[1600]Yoo, B.W.; Baek, H.S.; Keum, S.R.; Yoon, C.M.; Nam. G.S.; Kim, S.H.; Kim, J.H. *Synth. Commun.* *2000*, *30*, 4317.

[1601]Wang, L.; Li, P.; Zhou, L. *Tetrahedron Lett.* *2002*, *43*, 8141.

19-72   Reduction of Sulfoxides and Sulfones

S-Oxygen-detachment

$$R^{S}_{R}^{O} \xrightarrow{\text{LiAlH}_4} R^{S}_{R}$$

Sulfoxides can be reduced to sulfides by many reagents,[1602] among them Ph_3P,[1603] $LiAlH_4$, HI, Bu_3SnH,[1604] $MeSiCl_3$–NaI,[1605] H_2–Pd–C,[1606] $NaBH_4$–$NiCl_2$,[1607] $NaBH_4/I_2$,[1608] catecholborane,[1609] TiI_4,[1610] $TiCl_4/In$,[1611] Cp_2TiCl_2/In,[1612] Sm/ methanolic NH_4Cl with ultrasound,[1613] $(EtO)_2PCl/NEt_3$,[1614] and $SiO_2/SOCl_2$. Sulfones, however, are usually stable to reducing agents, although they have been reduced to sulfides with DIBALH, $(iBu)_2AlH$.[1615] A less general reagent is $LiAlH_4$, which reduces some sulfones to sulfides, but not others.[1616] Heating sulfoxides with 2,6-dihydroxypyridine gives the corresponding sulfide.[1617] Both sulfoxides and sulfones can be reduced by heating with sulfur (which is oxidized to SO_2), although the reaction with sulfoxides proceeds at a lower temperature. It has been shown by using substrate labeled with ^{35}S that sulfoxides simply give up the oxygen to the sulfur, but that the reaction with sulfones is more complex, since $\sim 75\%$ of the original radioactivity of the sulfone is lost.[1618] This indicates that most of the sulfur in the sulfide product comes in this case from the *reagent*. There is no direct general

[1602]For reviews, see Kukushkin, V.Yu. *Russ. Chem. Rev.* ***1990***, *59*, 844; Madesclaire, M. *Tetrahedron* ***1988***, *44*, 6537; Drabowicz, J.; Togo, H.; Mikołajczyk, M.; Oae, S. *Org. Prep. Proced. Int.* ***1984***, *16*, 171; Drabowicz, J.; Numata, T.; Oae, S. *Org. Prep. Proced. Int.* ***1977***, *9*, 63. For a list of reagents, with references, see Block, E. *Reactions of Organosulfur Compounds*, Academic Press, NY, ***1978***.

[1603]For a review, see Rowley, A.G., in Cadogan, J.I.G. *Organophosphorus Reagents in Organic Synthesis*, Academic Press, NY, ***1979***, pp. 301–304.

[1604]Kozuka, S.; Furumai; S.; Akasaka, T.; Oae, S. *Chem. Ind. (London)* ***1974***, 496.

[1605]Olah, G.A.; Husain, A.; Singh, B.P.; Mehrotra, A.K. *J. Org. Chem.* ***1983***, *48*, 3667. See also, Schmidt, A.H. *Russ Chem. Ber.* ***1981***, *114*, 822.

[1606]Ogura, K.; Yamashita, M.; Tsuchihashi, G. *Synthesis* ***1975***, 385.

[1607]Khurana, J.M.; Ray, A.; Singh, S. *Tetrahedron Lett.* ***1998***, *39*, 3829.

[1608]Karimi, B.; Zareyee, D. *Synthesis* ***2003***, 335.

[1609]Harrison, D.J.; Tam, N.C.; Vogels, C.M.; Langler, R.F.; Baker, R.T.; Decken, A.; Westcott, S.A. *Tetrahedron Lett.* ***2004***, *45*, 8493.

[1610]Shimizu, M.; Shibuya, K.; Hayakawa, R. *Synlett* ***2000***, 1437.

[1611]Yoo, B.W.; Choi, K.H.; Kim, D.Y.; Choi, K.I.; Kim, J.H. *Synth. Commun.* ***2003***, 33, 53.

[1612]Yoo, B.W.; Choi, K.H.; Lee, S.J.; Yoon, C.M.; Kim, S.H.; Kim, J.H. *Synth. Commun.* ***2002***, *32*, 63.

[1613]Yadav, J.S.; Subba Reddy, B.V.; Srinivas, C.; Srihari, P. *Synlett* ***2001***, 854.

[1614]Jie, Z.; Rammoorty, V.; Fischer, B. *J. Org. Chem.* ***2002***, *67*, 711.

[1615]Gardner, J.N.; Kaiser, S.; Krubiner, A.; Lucas, H. *Can. J. Chem.* ***1973***, *51*, 1419.

[1616]Bordwell, F.G.; McKellin, W.H. *J. Am. Chem. Soc.* ***1951***, *73*, 2251; Whitney, T.A.; Cram, D.J. *J. Org. Chem.* ***1970***, *35*, 3964; Weber, W.P.; Stromquist, P.; Ito, T.I. *Tetrahedron Lett.* ***1974***, 2595.

[1617]Miller, S.J.; Collier, T.R.; Wu, W. *Tetrahedron Lett.* ***2000***, *41*, 3781.

[1618]Kiso, S.; Oae, S. *Bull. Chem. Soc. Jpn.* ***1967***, *40*, 1722. See also, Oae, S.; Nakai, M.; Tsuchida, Y.; Furukawa, N. *Bull. Chem. Soc. Jpn.* ***1971***, *44*, 445.

method for the reduction of sulfones to sulfoxides, but an indirect method has been reported.[1619] Selenoxides can be reduced to selenides with a number of reagents.[1620]

OS IX, 446

D. Reduction With Cleavage

19-73   de-Alkylation of Amines and Amides

$$R_2'N-R \longrightarrow R_2'N-H$$

Certain amines can be dealkylated, usually under reductive conditions. N-Allyl amines, $R_2N-CH_2CH=CH_2$ are converted to the corresponding amine, R_2N-H, with Dibal/NiCl$_2$dppp,[1621] and with Pd(dba)$_2$dppb.[1622] A mixture of TiCl$_3$ and Li converts N-benzylamines to the amine ($R_2NCH_2Ph \rightarrow R_2NH$).[1623] In the case of N,N-dimethyl amines, RuCl$_3$ and H$_2$O$_2$ demethylate the amine (ArNMe$_2 \rightarrow$ ArNHMe).[1624] Tribenzylamines are dealkylated to give the dibenzylamine with ceric ammonium nitrate in aqueous acetonitrile.[1625] N-Benzyl indoles are cleaved to indoles with O$_2$, DMSO/KOt-Bu[1626] or with tetrabutylammonium fluoride.[1627]

The process is not limited to amines. Amides can also be dealkylated. N-Benzyl amides are debenzylated in the presence of NBS and AIBN.[1628]

N-Alkyl sulfonamides are dealkylated with PhI(OAc)$_2$ and I$_2$ with ultrasound to give a primary sulfonamide.[1629] Similar results are obtained with H$_5$IO$_6$ and a chromium catalyst.[1630] *tert*-Butyl sulfonamides are cleaved to the primary sulfonamide with BCl$_3$.[1631]

[1619]Still, I.W.J.; Ablenas, F.J. *J. Org. Chem.* **1983**, *48*, 1617.

[1620]See, for example, Sakaki, K.; Oae, S. *Chem. Lett.* **1977**, 1003; Still, I.W.J.; Hasan, S.K.; Turnbull, K. *Can. J. Chem.* **1978**, *56*, 1423; Denis, J.N.; Krief, A. *J. Chem. Soc. Chem. Commun.* **1980**, 544.

[1621]Taniguchi, T.; Ogasawara, K. *Tetrahedron Lett.* **1998**, *39*, 4679.

[1622]Lemaire-Audoire, S.; Savignac, M.; Dupuis, C.; Genêt, J.-P. *Bull. Soc. Chim. Fr.* **1995**, *132*, 1157; Lemaire-Audoire, S.; Savignac, M.; Genêt, J.-P.; Bernard, J.-M. *Tetrahedron Lett.* **1995**, *36*, 1267.

[1623]Talukdar, S.; Banerji, A. *Synth. Commun.* **1995**, *25*, 813.

[1624]Murahashi, S.-I.; Naota, T.; Miyaguchi, N.; Nakato, T. *Tetrahedron Lett.* **1992**, *33*, 6991.

[1625]Bull, S.D.; Davies, S.G.; Mulvaney, A.W.; Prasad, R.S.; Smith, A.D.; Fenton, G. *Chem. Commun.* **2000**, 337.

[1626]Haddach, A.A.; Kelleman, A.; Deaton-Rewoliwski, M.V. *Tetrahedron Lett.* **2002**, *43*, 399.

[1627]Routier, S.; Saugé, L.; Ayerbe, N,; Couderet, G.; Mérour, J.-Y. *Tetrahedron Lett.* **2002**, *43*, 589. For a related debenzylation see Meng, G.; He, Y.-P.; Chen, F.-E. *Synth. Commun.* **2003**, *33*, 2593.

[1628]Baker, S.R.; Parsons, A.F.; Wilson, M. *Tetrahedron Lett.* **1998**, *39*, 331.

[1629]Katohgi, M.; Yokoyama, M.; Togo, H. *Synlett* **2000**, 1055; Katohgi, M.; Togo, H. *Tetrahedron* **2001**, *57*, 7481.

[1630]Xu, L.; Zhang, S.; Trudell, M.L. *Synlett* **2004**, 1901.

[1631]Wan, Y.; Wu, X.; Kannan, M.A.; Alterman, M. *Tetrahedron Lett.* **2003**, *44*, 4523.

19-74   Reduction of Azo, Azoxy, and Hydrazo Compounds to Amines

$$
\left.
\begin{array}{c}
Ar\diagdown N{=}N\diagup ^{Ar} \\[2pt]
Ar\diagdown N{=}N\overset{\oplus}{\underset{O_{\ominus}}{\diagup}}{}^{Ar} \\[2pt]
Ar\diagdown \overset{H}{\underset{}{N}}\diagdown \underset{H}{N}\diagup ^{Ar}
\end{array}
\right\}
\quad \xrightarrow[\text{HCl}]{\text{Zn}} \quad 2\ ArNH_2
$$

Azo, azoxy, and hydrazo compounds can all be reduced to amines.[1632] Metals (notably zinc) and acids, and $Na_2S_2O_4$, are frequently used as reducing agents, and Bu_3SnH with a copper catalyst has been used.[1633] Borane reduces azo compounds to amines, although it does not reduce nitro compounds.[1634] $LiAlH_4$ does not reduce hydrazo compounds or azo compounds, although with the latter, hydrazo compounds are sometimes isolated. With azoxy compounds, $LiAlH_4$ gives only azo compounds (**19-68**). Noted that azo compounds are reduced to the hydrazine by reaction with hydrazine hydrate in ethanol.[1635]

OS **I**, 49; **II**, 35, 39; **III**, 360; **X**, 327. Also see, OS **II**, 290.

19-75   Reduction of Disulfides to Thiols

S-Hydrogen-uncoupling

$$
RSSR \quad \xrightarrow[\text{H}^+]{\text{Zn}} \quad 2\ RSH
$$

Disulfides can be reduced to thiols by mild reducing agents,[1636] such as zinc and dilute acid, In and $NH_4Cl/EtOH$,[1637] or Ph_3P and H_2O.[1638] The reaction can also be accomplished simply by heating with alkali.[1639] Among other reagents used have been $LiAlH_4$, $NaBH_4/ZrCl_4$,[1640] Mg/MeOH,[1641] $KBH(O{-}iPr)_3$,[1642] and hydrazine or substituted hydrazines.[1643]

[1632]For a review, see Newbold, B.T., in Patai, S. *The Chemistry of Hydrazo, Azo, and azoxy Groups*, pt. 2, Wiley, NY, *1975*, pp. 629–637.

[1633]Tan, Z.; Qu, Z.; Chen, B.; Wang, J. *Tetrahedron 2000*, *56*, 7457.

[1634]Brown, H.C.; Subba Rao, B.C. *J. Am. Chem. Soc. 1960*, *82*, 681.

[1635]Zhang, C.-R.; Wang, Y.-L. *Synth. Commun. 2003*, *33*, 4205.

[1636]For a review, see Wardell, J.L., in Patai, S. *The Chemistry of the Thiol Group*, pt. 2, Wiley, NY, *1974*, pp. 220–229.

[1637]Reddy, G.V.S.; Rao, G.V.; Iyengar, D.S. *Synth. Commun. 2000*, *30*, 859.

[1638]Overman, L.E.; Smoot, J.; Overman, J.D. *Synthesis 1974*, 59.

[1639]For discussions, see Danehy, J.P.; Hunter, W.E. *J. Org. Chem. 1967*, *32*, 2047.

[1640]Chary, K.P.; Rajaram, S.; Iyengar, D.S. *Synth. Commun. 2000*, *30*, 3905.

[1641]Sridhar, M.; Vadivel, S.K.; Bhalerao, U.T. *Synth. Commun. 1997*, *27*, 1347.

[1642]Brown, H.C.; Nazer, B.; Cha, J.S. *Synthesis 1984*, 498.

[1643]Maiti, S.N.; Spevak, P.; Singh, M.P.; Micetich, R.G.; Narender Reddy, A.V. *Synth. Commun. 1988*, *18*, 575.

Aryl diselenides are similarly cleaved to selenols (ArSeH) with Cp_2TiH followed by $Ph_2I^+X^-$.[1644]

OS **II**, 580. Also see, OS **IV**, 295.

E. Reductive Coupling

19-76   Bimolecular Reduction of Aldehydes and Ketones to 1,2-Diols and Imines to 1,2-Diamines

2/O-Hydrogen-coupling and 2/N-Hydrogen-coupling

1,2-Diols (pinacols) can be synthesized by reduction of aldehydes and ketones with active metals, such as sodium, magnesium, or aluminum.[1645] Aromatic ketones give better yields than aliphatic ones. The use of a $Mg-MgI_2$ mixture has been called the *Gomberg–Bachmann pinacol synthesis*.[1646] As with a number of other reactions involving sodium, there is a direct electron transfer here, converting the ketone or aldehyde to a ketyl, which dimerizes.

Other reagents have been used,[1647] including Sm,[1648] Yb,[1649] $Yb-Me_3SiCl$,[1650] $InCl_3$ catalyst with Mg,[1651] $Al/TiCl_3$,[1652] $VOCl_3$ catalyst with Me_3SiCl,[1653]

[1644]Huang, X.; Wu, L.-L.; Xu, X.-H. *Synth. Commun.* **2001**, *31*, 1871.

[1645]For efficient methods, see Schreibmann, A.A.P. *Tetrahedron Lett.* **1970**, 4271; Fürstner, A.; Csuk, R.; Rohrer, C.; Weidmann, H. *J. Chem. Soc. Perkin Trans. 1* **1988**, 1729. For an ultrasound promoted reaction with aluminum, see Bian, Y.-J.; Liu, S.-M.; Li, J.-T.; Li, T.-S. *Synth. Commun.* **2002**, *32*, 1169.

[1646]For an ultrasound promoted reaction, see Li, J.-T.; Bian, Y.-J.; Zang, H.-J.; Li, T.-s. *Synth. Commun.* **2002**, *32*, 547.

[1647]For a list of reagents, with references, see Larock, R.C. *Comprehensive Organic Transformations*, 2nd ed., Wiley-VCH, NY, **1999**, pp.1111–1114.

[1648]Ghatak, A.; Becker, F.F.; Banik, B.K. *Tetrahedron Lett.* **2000**, *41*, 3793; Talukdar, S.; Fang, J.-M. *J. Org. Chem.* **2001**, *66*, 330; Yu, M.; Zhang, Y. *Org. Prep. Proceed. Int.* **2001**, *33*, 187; Hélion, F.; Lannou, M.-I.; Namy, J.-L. *Tetrahedron Lett.* **2003**, *44*, 5507.

[1649]Hou, Z.; Takamine, K.; Fujiwara, Y.; Taniguchi, K. *Chem. Lett.* **1987**, 2061.

[1650]Ogawa, A.; Takeuchi, H.; Hirao, T. *Tetrahedron Lett.* **1999**, *40*, 7113.

[1651]Mori, K.; Ohtaka, S.; Uemura, S. *Bull. Chem. Soc. Jpn.* **2001**, *74*, 1497.

[1652]Li, J.-T.; Lin, Z.-P.; Qi, N.; Li, T.-S. *Synth. Commun.* **2004**, *34*, 4339.

[1653]Hirao, T.; Hatano, B.; Imamoto, Y.; Ogawa, A. *J. Org. Chem.* **1999**, *64*, 7665.

activated Mn,[1654] In with ultrasound,[1655] Zn,[1656] and a reagent prepared from TiCl$_4$[1657] and Mg amalgam[1658] (a low-valent titanium reagent;[1659] see **19-76**). A mixture of TiCl$_4$[1660] or TiCl$_2$[1661] and Zn can also be used. Unsymmetrical coupling between two different aldehydes has been achieved by the use of a vanadium complex,[1662] while TiCl$_3$ in aqueous solution has been used to couple two different ketones.[1663] Two aldehydes have also been coupled using magnesium in water.[1664] Coupling leads to a mixture of syn- and anti-diols. "Syn-selective" reagents are Cp$_2$TiCl$_2$/Mn,[1665] TiCl$_4$/Bu$_4$I,[1666] TiI$_4$,[1667] TiBr$_2$+Cu,[1668] and NbCl$_3$.[1669] With SmI$_2$,[1670] coupling in the presence of a primary alkyl iodide leads to acyl addition using an excess of HMPA, but pinacol coupling with an excess of LiBr.[1671] "Anti-selective" coupling reactions are also known: Ti–salen,[1672] Mg with a NiCl$_2$ catalyst,[1673] Sm/SmCl$_3$,[1674] and TiCl$_4$(thf)$_2$ with a chiral Schiff base.[1675] Aryl

[1654]Rieke, R.D.; Kim, S.-H. *J. Org. Chem.* **1998**, *63*, 5235. For a reaction using Mn, Me$_3$SiCl and a cerium catalyst, see Groth, U.; Jeske, M. *Synlett* **2001**, 129.

[1655]Lim, H.J.; Keum, G.; Kang, S.B.; Chung, B.Y.; Kim, Y. *Tetrahedron Lett.* **1998**, *39*, 4367.

[1656]Hekmatshoar, R.; Yavari, I.; Beheshtiha, Y.S.; Heravi, M.M. *Monat. Chem.* **2001**, *132*, 689.

[1657]For a discussion of pinacol versus reduction with this reagent, and mechanistic considerations, see Clerici, A.; Pastori, N.; Porta, O. *Tetrahaedron Lett.* **2004**, *45*, 1825.

[1658]Corey, E.J.; Danheiser, R.L.; Chandrasekaran, S. *J. Org. Chem.* **1976**, *41*, 260; Pons, J.; Zahra, J.; Santelli, M. *Tetrahedron Lett.* **1981**, *22*, 3965. For some other titanium-containing reagents, see Clerici, A.; Porta, O. *J. Org. Chem.* **1985**, *50*, 76; Handa, Y.; Inanaga, J. *Tetrahedron Lett.* **1987**, *28*, 5717. For a review of such coupling with Ti and V halides, see Lai, Y. *Org. Prep. Proceed. Int.* **1980**, *12*, 363.

[1659]For a discussion of the mechanism, see Hashimoto, Y.; Mizuno, U.; Matsuoka, H.; Miyahara, T.; Takakura, M.; Yoshimoto, M.; Oshima, K.; Utimoto, K.; Matsubara, S. *J. Am. Chem Soc.* **2001**, *123*, 1503.

[1660]Li, T.; Cui, W.; Liu, J.; Zhao, J.; Wang, Z. *Chem. Commun.* **2000**, 139.

[1661]Kagayama, A.; Igarashi, K.; Mukaiyama, T. *Can. J. Chem.* **2000**, *78*, 657.

[1662]Freudenberger, J.H.; Konradi, A.W.; Pedersen, S.F. *J. Am. Chem. Soc.* **1989**, *111*, 8014.

[1663]Clerici, A.; Porta, O. *J. Org. Chem.* **1982**, *47*, 2852; *Tetrahedron* **1983**, *39*, 1239. For some other unsymmetrical couplings, see Hou, Z.; Takamine, K.; Aoki, O.; Shiraishi, H.; Fujiwara, Y.; Taniguchi, H. *J. Chem. Soc. Chem. Commun.* **1988**, 668; Delair, P.; Luche, J. *J. Chem. Soc. Chem. Commun.* **1989**, 398; Takahara, P.M.; Freudenberger, J.H.; Konradi, A.W.; Pedersen, S.F. *Tetrahedron Lett.* **1989**, *30*, 7177.

[1664]Zhang, W.-C.; Li, C.-J. *J. Chem. Soc. Perkin Trans. 1* **1998**, 3131.

[1665]Gansäuer, A.; Bauer, D. *Eur. J. Org. Chem.* **1998**, 2673. Also see, Barden, M.C.; Schwartz, J. *J. Am. Chem. Soc.* **1996**, *118*, 5484; Gansäuer, A. *Chem. Commun.* **1997**, 457; Gansäuer, A. *Synlett* **1997**, 363; Clerici, A.; Clerici, L.; Porta, O. *Tetrahedron Lett.* **1996**, *37*, 3035.

[1666]Tsuritani, T.; Ito, S.; Shinokubo, H.; Oshima, K. *J. Org. Chem.* **2000**, 65, 5066.

[1667]Hayakawa, R.; Shimizu, M. *Chem. Lett.* **2000**, 724. For a syn-selective coupling with conjugated aldehydes, see Shimizu, M.; Goto, H.; Hayakawa, R. *Org. Lett.* **2002**, *4*, 4097.

[1668]Mukaiyama, T.; Yoshimura, N.; Igarashi, K.; Kagayama, A. *Tetrahedron* **2001**, *57*, 2499.

[1669]Szymoniak, J.; Besançon, J.; Moïse, C. *Tetrahedron* **1994**, *50*, 2841.

[1670]Namy, J.L.; Souppe, J.; Kagan, H.B. *Tetrahedron Lett.* **1983**, *24*, 765; Nomura, R.; Matsuno, T.; Endo, T. *J. Am. Chem. Soc,* **1996**, *118*, 11666; Honda, T.; Katoh, M. *Chem. Commun.* **1997**, 369; Shiue, J.-S.; Lin, C.-C.; Fang, J.-M. *Tetrahedron Lett.* **1993**, *34*, 335. Also see Yamashita, M.; Okuyama, K.; Kawasaki, I.; Ohta, S. *Tetrahedron Lett.* **1996**, *37*, 7755.

[1671]Miller, R.S.; Sealy, J.M.; Shabangi, M.; Kuhlman, M.L.; Fuchs, J.R.; Flowers II, R.A. *J. Am. Chem. Soc.* **2000**, *122*, 7718.

[1672]Chatterjee, A.; Bennur, T.H.; Joshi, N.N. *J. Org. Chem.* **2003**, *68*, 5668.

[1673]Shi, L.; Fan, C.-A.; Tu, Y.-Q.; Wang, M.; Zhang, F.-M. *Tetrahedron* **2004**, *60*, 2851.

[1674]Matsukawa, S.; Hinakubo, Y. *Org. Lett.* **2003**, *5*, 1221.

[1675]Li, Y.-G.; Tian, Q.-S.; Zhao, J.; Feng, Y.; Li, M.-J.; You, T.-P. *Tetrahedron Asymmetry* **2004**, *15*, 1707.

aldehydes are coupled to give the bis-trimethylsilyl ether using Mn, Me₃SiCl, and Cp₂TiCl₂.[1676]

A crossed-pinacol coupling was reported using Et₂Zn and with a BINOL catalyst gave good enantioselectivity.[1677] A combination of Mg and Me₃SiCl was also used to a crossed-pinacol.[1678]

Intramolecular pinacol coupling reactions are known, giving cyclic 1,2-diols.[1679] Dialdehydes have been cyclized by reaction with TiCl₃ to give cyclic 1,2-diols in good yield.[1680] A radical-induced coupling of an α,ω-dialdehyde led to cis-1,2-cyclopentanediol when treated with Bu₃SnH and AIBN.[1681] or induced photochemically.[1682]

Chiral additives with pinacol couplings lead to formation of a diol with moderate to good enantioselectivity.[1683] Chiral metal complexes in conjunction with a metal leads to diol formation with good enantioselectivity.[1684]

A variation of the pinacol coupling treats acyl nitriles with indium metal and ultrasound to give a 1,2-diketone.[1685] Another variation couples acetals to give 1,2-diols.[1686]

The dimerization of ketones to 1,2-diols can also be accomplished photochemically; indeed, this is one of the most common photochemical reactions.[1687] The substrate, which is usually a diaryl or aryl alkyl ketone (though a few aromatic aldehydes and dialkyl ketones have been dimerized), is irradiated with UV light in the presence of a hydrogen donor, such as isopropyl alcohol, toluene, or an amine.[1688]

[1676]Dunlap, M.S.; Nicholas, K.M. *Synth. Commun.* **1999**, *29*, 1097.

[1677]Kumagai, N.; Matsunaga, S.; Kinoshita, T.; Harada, S.; Okada, S.; Sakamoto, S.; Yamaguchi, K.; Shibasaki, M. *J. Am. Chem. Soc.* **2003**, *125*, 2169.

[1678]Maekawa, H.; Yamamoto, Y.; Shimada, H.; Yonemura, K.; Nishiguchi, I. *Tetraheron Lett.* **2004**, *45*, 3869.

[1679]**With a Ti catalyst + Zn**: Yamamoto, Y.; Hattori, R.; Itoh, K. *Chem. Commun.* **1999**, 825; Yamamoto, Y.; Hattori, R.; Miwa, T.; Nakagai, Y.-I.; Kubota, T.; Yamamoto, C.; Okamoto, Y.; Itoh, K. *J. Org. Chem.* **2001**, *66*, 3865. **With SmI₂/t-BuOH**: Handa, S.; Kachala, M.S.; Lowe, S.R. *Tetrahedron Lett.* **2004**, *45*, 253.

[1680]McMurry, J.E.; Rico, J.G. *Tetrahedron Lett.* **1989**, *30*, 1169. For the stereochemistry of this coupling, see McMurry, J.E.; Siemers, N.O. *Tetrahedron Lett.* **1993**, *34*, 7891. For other cyclization reactions of dialdehydes and ketoaldehydes, see Molander, G.A.; Kenny, C. *J. Am. Chem. Soc.* **1989**, *111*, 8236; Raw, A.S.; Pedersen, S.F. *J. Org. Chem.* **1991**, *56*, 830; Chiara, J.L.; Cabri, W.; Hanessian, S. *Tetrahedron Lett.* **1991**, *32*, 1125.

[1681]Hays, D.S.; Fu, G.C. *J. Org. Chem.* **1998**, *63*, 6375.

[1682]Hays, D.S.; Fu. G.C. *J. Am. Chem. Soc.* **1995**, *117*, 7283.

[1683]Enders, D.; Ullrich, E.C. *Tetrahedron Asymmetry* **2000**, *11*, 3861.

[1684]See Bensari, A.; Renaud, J.-L.; Riant, O. *Org. Lett.* **2001**, *3*, 3863; Takenaka, N.; Xia, G.; Yamamoto, H. *J. Am. Chem. Soc.* **2004**, *126*, 13198.

[1685]Baek, H.S. et al. *Tetrahedron Lett.* **2000**, *41*, 8097.

[1686]Studer, A.; Curran, D.P. *Synlett* **1996**, 255.

[1687]For reviews, see Schönberg, A. *Preparative Organic Photochemistry*, Springer, NY, **1968**, pp. 203–217; Neckers, D.C. *Mechanistic Organic Photochemistry*, Reinhold, NY, **1967**, pp. 163–177; Calvert, J.G.; Pitts Jr., J.N. *Photochemistry*, Wiley, NY, **1966**, pp. 532–536; Turro, N.J. *Modern Molecular Photochemistry*, W.A. Benjamin, NY, **1978**, pp. 363–385; Kan, R.O. *Organic Photochemistry*, McGraw-Hill, NY, **1966**, pp. 222–229.

[1688]For a review of amines as hydrogen donors in this reaction, see Cohen, S.G.; Parola, A.; Parsons, Jr., G.H. *Chem. Rev.* **1973**, *73*, 141.

In the case of benzophenone, irradiated in the presence of 2-propanol, the ketone molecule initially undergoes $n \rightarrow \pi^*$ excitation, and the singlet species thus formed crosses to the T_1 state with a very high efficiency.

The T_1 species abstracts hydrogen from the alcohol (p. 347), and then dimerizes. The *i*PrO• radical, which is formed by this process, donates H• to another molecule of ground-state benzophenone, producing acetone and another molecule of **51**. This mechanism[1689] predicts that the quantum yield for the disappearance of benzophenone should be 2, since each quantum of light results in the conversion of 2 equivalents of benzophenone to **51**. Under favorable experimental conditions, the observed quantum yield does approach 2. Benzophenone abstracts hydrogen with very high efficiency. Other aromatic ketones are dimerized with lower quantum yields, and some (e.g., *p*-aminobenzophenone, *o*-methylacetophenone) cannot be dimerized at all in 2-propanol (although *p*-aminobenzophenone, e.g., can be dimerized in cyclohexane[1690]). The reaction has also been carried out electrochemically.[1691]

A coupling reaction similar to pinacol coupling has been used with imines, which dimerize to give 1,2-diamines. A number of reagents have been used, including treatment with TiCl$_4$–Mg,[1692] In/aq. EtOH,[1693] Zn/aq. NaOH,[1694] Cp$_2$VCl$_2$/Zn/

[1689]For some of the evidence for this mechanism, see Pitts, Jr., J.N.; Letsinger, R.L.; Taylor, R.; Patterson, S.; Recktenwald, G.; Martin, R.B. *J. Am. Chem. Soc.* **1959**, *81*, 1068; Moore, W.M.; Hammond, G.S.; Foss, R.P. *J. Am. Chem. Soc.* **1961**, *83*, 2789; Huyser, E.S.; Neckers, D.C. *J. Am. Chem. Soc.* **1963**, *85*, 3641.

[1690]Porter, G.; Suppan, P. *Proc. Chem. Soc.* **1964**, 191.

[1691]Elinson, M.N.; Feducovich, S.K.; Dorofeev, A.S.; Vereshchagin, A.N.; Nikishin, G.I. *Tetrahedron* **2000**, *56*, 9999. For reviews, see Fry, A.J. *Synthetic Organic Electrochemistry*, 2nd ed., Wiley, NY, **1989**, pp. 174–180; Shono, T. *Electroorganic Chemistry as a New Tool in Organic Synthesis*, Springer, NY, **1984**, pp. 137–140; Baizer, M.M.; Petrovich, J.P. *Prog. Phys. Org. Chem.* **1970**, *7*, 189. For a review of electrolytic reductive coupling, see Baizer, M.M. in Baizer, M.M.; Lund, H. *Organic Electrochemistry*, Marcel Dekker, NY, **1983**, pp. 639–689.

[1692]Betschart, C.; Schmidt, B.; Seebach, D. *Helv. Chim. Acta* **1988**, *71*, 1999; Mangeney, P.; Tejero, T.; Alexakis, A.; Grosjean, F.; Normant, J. *Synthesis* **1988**, 255; Alexakis, A.; Aujard, I.; Mangeney, P. *Synlett* **1998**, 873, 875.

[1693]Kalyanam, N.; Rao, G.V. *Tetrahedron Lett.* **1993**, *34*, 1647.

[1694]Dutta, M.P.; Baruah, B.; Boruah, A.; Prajapati, D.; Sandu, J.S. *Synlett* **1998**, 857.

PhMe$_2$SiCl,[1695] Et$_2$AlCl,[1696] SmI$_2$,[1697] and (for silylated imines) NbCl$_4$(thf)$_2$.[1698] When electroreduction was used, it was even possible to obtain cross-products, by coupling a ketone to an O-methyl oxime:[1699] O-Methyl oxime ethers are coupled to give 1,2-diamines using Zn and TiCl$_4$.[1700] Aldehydes are converted to 1,2-diamines by treatment with TMS$_2$NH, NaH, and Li metal in 5 M LiClO$_4$ in ether, with sonication.[1701] Hemiaminals are coupled to give 1,2-diamines with TiI$_4$/Zn.[1702] Amides are converted to 1,2-diamines with Cp$_2$TiF$_2$ and PhMeSiH$_2$.[1703] Samarium(II) iodide was used to couple iminium salts, giving the 1,2-diamine.[1704] Ketones can be treated with Yb, and then an imine to give amino alcohols.[1705]

The N-methoxyamino alcohol could then be reduced to the amino alcohol.[1699] A photochemical coupling has also been reported.[1706] A variation of this reaction treats an imine with Yb in THF/HMPA and then an aldehyde to give a 1,2-bis(imine).[1707]

OS **I**, 459; **II**, 71; **X**, 312; **81**, 26.

19-77   Bimolecular Reduction of Aldehydes or Ketones to Alkenes

De-oxygen-coupling

Aldehydes and ketones, both aromatic and aliphatic (including cyclic ketones), can be converted in high yields to dimeric alkenes by treatment low valent titanium,[1708] initially generated with TiCl$_3$ and a zinc–copper couple.[1709] This is called

[1695]Hatano, B.; Ogawa, A.; Hirao, T. *J. Org. Chem.* **1998**, *63*, 9421.

[1696]This reaction proceeds with N-ethylation. See Shimizu, M.;Niwa, Y. *Tetrahedron Lett.* **2001**, *42*, 2829.

[1697]Enholm, E.J.; Forbes, D.C.; Holub, D.P. *Synth. Commun.* **1990**, *20*, 981; Imamoto, T.; Nishimura, S. *Chem. Lett.* **1990**, 1141; Zhong, Y.-W.; Izumi, K.; Xu, M.-H.; Lin, G.-Q. *Org. Lett.* **2004**, *6*, 4747.

[1698]Roskamp, E.J.; Pedersen, S.F. *J. Am. Chem. Soc.* **1987**, *109*, 3152.

[1699]Shono, T.; Kise, N.; Fujimoto, T. *Tetrahedron Lett.* **1991**, *32*, 525.

[1700]Kise, N.; Ueda, N. *Tetrahedron Lett.* **2001**, *42*, 2365.

[1701]Mojtahedi, M.M.; Saidi, M.R.; Shirzi, J.S.; Bolourtchian, M. *Synth. Commun.* **2001**, *31*, 3587.

[1702]Yoshimura, N.; Mukaiyama, T. *Chem. Lett.* **2001**, 1334.

[1703]Selvakumar, K.; Harrod, J.F. *Angew. Chem. Int. Ed.* **2001**, *40*, 2129.

[1704]Kim, M.; Knettle, B.W.; Dahlén, A.; Hilmersson, G.; Flowers III, R.A. *Tetrahedron* **2003**, *59*, 10397.

[1705]Su, W.; Yang, B. *Synth. Commun.* **2003**, *33*, 2613.

[1706]Campos, P.J.; Arranz, J.; Rodríguez, M.A. *Tetrahedron* **2000**, *56*, 7285; Ortega, M.; Rodríguez, M.A.; Campos, P.J. *Tetrahedron* **2004**, *60*, 6475.

[1707]Jin, W.; Makioka, Y.; Kitamura, T.; Fujiwara, Y. *J. Org. Chem.* **2001**, *66*, 514.

[1708]For a highly active reagent see Rele, S.; Chattopadhyay, S.; Nayak, S.K. *Tetrahedron Lett.* **2001**, *42*, 9093.

[1709]McMurry, J.E.; Fleming, M.P.; Kees, K.L.; Krepski, L.R. *J. Org. Chem.* **1978**, *43*, 3255. For an optimized procedure, see McMurry, J.E.; Lectka, T.; Rico, J.G. *J. Org. Chem.* **1989**, *54*, 3748.

the *McMurry reaction*.[1710] The reagent produced in this way is called a *low-valent titanium reagent*, and the reaction has also been accomplished[1711] with low-valent titanium reagents prepared in other ways, for example, from Mg and a $TiCl_3$–THF complex,[1712] from $TiCl_4$ and Zn or Mg,[1713] from $TiCl_3$ and $LiAlH_4$,[1714] from $TiCl_3$ and lamellar potassium graphite,[1715] from $TiCl_3$ and K or Li,[1716] as well as with $ZnMe_3SiCl$[1717] and with certain compounds prepared from WCl_6 and either lithium, lithium iodide, $LiAlH_4$, or an alkyllithium[1718] (see **17-18**). The reaction has been used to convert dialdehydes and diketones to cycloalkenes.[1719] Rings of 3–16 and 22 members have been closed in this way, for example,[1720]

The same reaction on a keto ester gives a cycloalkanone.[1721]

[1710]For reviews, see McMurry, J.E. *Chem. Rev.* **1989**, *89*, 1513; *Acc. Chem. Res.* **1983**, *16*, 405; Lenoir, D. *Synthesis* **1989**, 883; Betschart, C.; Seebach, D. *Chimia* **1989**, *43*, 39; Lai, Y. *Org. Prep. Proceed. Int.* **1980**, *12*, 363. For related reviews, see Kahn, B.E.; Rieke, R.D. *Chem. Rev.* **1988**, *88*, 733; Pons, J.; Santelli, M. *Tetrahedron* **1988**, *44*, 4295. For the stereochemistry associated with this reaction, see Andersson, P.G. *Tetrahedron Lett.* **1994**, *35*, 2609.

[1711]For a list of reagents, with references, see Larock, R.C. *Comprehensive Organic Transformations*, 2nd ed., Wiley-VCH, NY, **1999**, pp. 305–308.

[1712]Tyrlik, S.; Wolochowicz, I. *Bull. Soc. Chim. Fr.* **1973**, 2147.

[1713]Mukaiyama, T.; Sato, T.; Hanna, J. *Chem. Lett.* **1973**, 1041; Lenoir, D. *Synthesis* **1977**, 553; Lenoir, D.; Burghard, H. *J. Chem. Res. (S)* **1980**, 396; Carroll, A.R.; Taylor, W.C. *Aust. J. Chem.* **1990**, *43*, 1439.

[1714]McMurry, J.E.; Fleming, M.P. *J. Am. Chem. Soc.* **1974**, *96*, 4708; Dams, R.; Malinowski, M.; Geise, H.J. *Bull. Soc. Chim. Belg.* **1982**, *91*, 149, 311; Bottino, F.A.; Finocchiaro, P.; Libertini, E.; Reale, A.; Recca, A. *J. Chem. Soc. Perkin Trans. 2* **1982**, 77. This reagent has been reported to give capricious results; see McMurry, J.E.; Fleming, M.P. *J. Org. Chem.* **1976**, *41*, 896.

[1715]Fürstner, A.; Weidmann, H. *Synthesis* **1987**, 1071.

[1716]McMurry, J.E.; Fleming, M.P. *J. Org. Chem.* **1976**, *41*, 896; Richardson, W.H. *Synth. Commun.* **1981**, *11*, 895; Rele, S.; Talukdar, S.; Banerji, A.; Chattopadhyay, S. *J. Org. Chem.* **2001**, *66*, 2990.

[1717]Banerjee, A.K.; Sulbaran de Carrasco, M.C.; Frydrych-Houge, C.S.V.; Motherwell, W.B. *J. Chem. Soc. Chem. Commun.* **1986**, 1803.

[1718]Sharpless, K.B.; Umbreit, M.A.; Nieh, M.T.; Flood, T.C. *J. Am. Chem. Soc.* **1972**, *94*, 6538; Fujiwara, Y.; Ishikawa, R.; Akiyama, F.; Teranishi, S. *J. Org. Chem.* **1978**, *43*, 2477; Dams, R.; Malinowski, M.; Geise, H.J. *Bull. Soc. Chim. Belg.* **1982**, *19*, 149, 311. See also, Petit, M.; Mortreux, A.; Petit, F. *J. Chem. Soc. Chem. Commun.* **1984**, 341; Chisholm, M.H.; Klang, J.A. *J. Am. Chem. Soc.* **1989**, *111*, 2324.

[1719]Baumstark, A.L.; Bechara, E.J.H.; Semigran, M.J. *Tetrahedron Lett.* **1976**, 3265; McMurry, J.E.; Fleming, M.P.; Kees, K.L.; Krepski, L.R. *J. Org. Chem.* **1978**, *43*, 3255.

[1720]Baumstark, A.L.; McCloskey, C.J.; Witt, K.E. *J. Org. Chem.* **1978**, *43*, 3609.

[1721]McMurry, J.E.; Miller, D.D. *J. Am. Chem. Soc.* **1983**, *105*, 1660.

Indoles have been prepared form ortho-acyl amides with Ti(powder) and Me₃SiCl[1722] or with TiCl₃—C₈K.[1723] Benzofurans have been prepared by a closely related reaction.[1724]

Unsymmetrical alkenes can be prepared from a mixture of two ketones in a cross-coupling reaction, if one is in excess.[1725] An aldehyde and a ketone were cross-coupled using Yb(OTf)₃, for example.[1726] The mechanism consists of initial coupling of two radical species to give a 1,2-dioxygen compound (a titanium pinacolate), which is then deoxygenated.[1727]

OS **VII**, 1.

19-78 Acyloin Ester Condensation

$$2 \quad \underset{R}{\overset{O}{\underset{\|}{C}}} \text{—OR}^1 \quad \xrightarrow[\text{xylene}]{\text{Na}} \quad \underset{\text{NaO}}{\overset{R}{\underset{\|}{C}}} = \underset{\text{ONa}}{\overset{R}{\underset{\|}{C}}} \quad \xrightarrow{\text{H}_2\text{O}} \quad \underset{H}{\overset{R}{\underset{|}{C}}} \underset{\|}{\overset{OH}{\underset{O}{C}}} \text{—R}$$

52

When carboxylic esters are heated with sodium in refluxing ether or benzene, a bimolecular reduction takes place, and the product is an α-hydroxy ketone (called an acyloin).[1728] The reaction, called the *acyloin ester condensation*,[1729] is quite successful when R is alkyl. Acyloins with long chains have been prepared in this way, for example, R = C₁₇H₃₅, but for high-molecular-weight esters, toluene or xylene is used as the solvent. Modifications to this procedure have been reported, including an ultrasound-promoted acyloin condensation in ether,[1730] which imporvied the yields of four-, five-, and six-membered rings, and Olah's procedure, which was also done in ether.[1731]

The acyloin condensation has been used with great success, in boiling xylene, to prepare cyclic acyloins from diesters.[1732] The yields are 50–60% for the preparation

[1722]Fürstner, A.; Hupperts, A. *J. Am. Chem. Soc.* **1995**, *117*, 4468.

[1723]Fürstner, A.; Hupperts, A.; Ptock, A.; Janssen, E. *J. Org. Chem.* **1994**, *59*, 5215.

[1724]Fürstner, A.; Jumbam, D.N. *Tetrahedron* **1992**, *48*, 5991.

[1725]McMurry, J.E.; Fleming, M.P.; Kees, K.L.; Krepski, L.R. *J. Org. Chem.* **1978**, *43*, 3255; Nishida, S.; Kataoka, F. *J. Org. Chem.* **1978**, *43*, 1612; Coe, P.L.; Scriven, C.E. *J. Chem. Soc. Perkin Trans. 1* **1986**, 475; Chisholm, M.H.; Klang, J.A. *J. Am. Chem. Soc.* **1989**, *111*, 2324.

[1726]Curini, M.; Epifano, F.; Maltese, F., Marcotullio, M.C. *Eur. J. Org. Chem.* **2003**, 1631.

[1727]McMurry, J.E.; Fleming, M.P.; Kees, K.L.; Krepski, L.R. *J. Org. Chem.* **1978**, *43*, 3255; Dams, R.; Malinowski, M.; Westdorp, I.; Geise, H.Y. *J. Org. Chem.* **1982**, *47*, 248. See Villiers, C.; Ephritikhine, M. *Angew. Chem,. Int. Ed.* **1997**, *36*, 2380; Stahl, M.; Pindur, U.; Frenking, G. *Angew. Chem. Int. Ed.* **1997**, *36*, 2234.

[1728]For a review, see Bloomfield, J.J.; Owsley, D.C.; Nelke, J.M. *Org. React.* **1976**, *23*, 259. For a list of reactions, with references, see Larock, R.C. *Comprehensive Organic Transformations*, 2nd ed., Wiley-VCH, NY, **1999**, pp. 1313–1315.

[1729]For reaction with tethered diesters, see Daynard, T.S.; Eby, P.S.; Hutchinson, J.H. *Can. J. Chem.* **1993**, *71*, 1022.

[1730]Fadel, A.; Canet, J,-L.; Salaün, J. *Synlett* **1990**, 89.

[1731]Olah, G.A.; Wu, A. *Synthesis* **1991**, 1177.

[1732]For a review of cyclizations by means of the acyloin condensation, see Finley, K.T. *Chem. Rev.* **1964**, *64*, 573.

of 6- and 7-membered rings, 30–40% for 8- and 19-membered, and 60–95% for rings of 10–20 members. Even larger rings have been closed in this manner. This is one of the best ways of closing rings of 10 members or more. The reaction has been used to close 4-membered rings,[1733] although this is generally not successful. The presence of double or triple bonds does not interfere.[1734] Even a benzene ring can be present, and many paracyclophane derivatives (**53**) with $n = 9$ or more have been synthesized in this manner.[1735]

53   **54**

Yields in the acyloin condensation can be improved by running the reaction in the presence of chlorotrimethylsilane Me_3SiCl, in which case the dianion **52** is converted to the bis silyl enol ether **54**, which can be isolated and subsequently hydrolyzed to the acyloin with aqueous acid.[1736] This is now the standard way to conduct the acyloin condensation. Among other things, this method inhibits the Dieckmann condensation[1737] (**16-85**), which otherwise competes with the acyloin condensation when a five-, six-, or seven-membered ring can be closed (note that the ring formed by a Dieckmann condensation is always one carbon atom smaller than that formed by an acyloin condensation of the same substrate). The Me_3SiCl method is especially good for the closing of four-membered rings.[1738]

The mechanism is not known with certainty, but it is usually presumed that the diketone RCOCOR is an intermediate,[1739] since small amounts of it are usually isolated as side products, and when it is resistant to reduction (e.g., t-Bu—COCO—t-Bu), it is the major product. A possible sequence (analogous to that of **19-76**) is

[1733]Cope, A.C.; Herrick, E.C. *J. Am. Chem. Soc.* **1950**, *72*, 983; Bloomfield, J.J.; Irelan, J.R.S. *J. Org. Chem.* **1966**, *31*, 2017.

[1734]Cram, D.J.; Gaston, L.K. *J. Am. Chem. Soc.* **1960**, *82*, 6386.

[1735]For a review, see Cram, D.J. *Rec. Chem. Prog.*, **1959**, *20*, 71.

[1736]Schräpler, U.; Rühlmann, K. *Chem. Ber.* **1964**, *97*, 1383. For a review of the Me_3SiCl method, see Rühlmann, K. *Synthesis* **1971**, 236.

[1737]Bloomfield, J.J. *Tetrahedron Lett.* **1968**, 591.

[1738]Gream, G.E.; Worthley, S. *Tetrahedron Lett.* **1968**, 3319; Wynberg, H.; Reiffers, S.; Strating, J. *Recl. Trav. Chim. Pays-Bas* **1970**, *89*, 982; Bloomfield, J.J.; Martin, R.A.; Nelke, J.M. *J. Chem. Soc. Chem. Commun.* **1972**, 96.

[1739]Another mechanism, involving addition of the ketyl to another molecule of ester (rather than a dimerization of two ketyl radicals), in which a diketone is not an intermediate, has been proposed: Bloomfield, J.J.; Owsley, D.C.; Ainsworth, C.; Robertson, R.E. *J. Org. Chem.* **1975**, *40*, 393.

A large surface area for the sodium is usually required for good results in this coupling, consistent with a surface reaction. In order to account for the ready formation of large rings, which means that the two ends of the chain must approach each other even although this is conformationally unfavorable for long chains, it may be postulated that the two ends become attached to nearby sites on the surface[1740] of the sodium. Although high dilution techniques are not always necessary, effective stirring (high speed stirrer at 2000–2500 rpm) is usually required to generate "sodium sand". Highly pure sodium gives poorer results, and a small percentage of potassium is important. Up to 50% potassium (1:1 Na/K)[1741] has been used in acyloin condensations.

In a related reaction, aromatic carboxylic acids were condensed to α-diketones (2ArCOOH → ArCOCOAr) on treatment with excess Li in dry THF in the presence of ultrasound.[1742]

The acyloin condensation was used in an ingenious manner to prepare the first reported catenane (see p. 131).[1743] This synthesis of a catenane produced only a small yield and relied on chance for threading the molecules before ring closure.

OS **II**, 114; **IV**, 840; **VI**, 167.

19-79 Reduction of Nitro to Azoxy Compounds

Nitro-azoxy reductive transformation

$$2\,ArNO_2 \xrightarrow{\;Na_3AsO_3\;} \overset{\ominus}{O}\!\!-\!\!\overset{Ar}{\underset{Ar}{\overset{\oplus}{N}=N}}$$

Azoxy compounds can be obtained from nitro compounds with certain reducing agents, notably sodium arsenite, sodium ethoxide, NaTeH,[1744] NaBH4–PhTe-TePh,[1745] and glucose. The most probable mechanism with most reagents is that one molecule of nitro compound is reduced to a nitroso compound and another to a hydroxylamine (**19-46**), and these combine (**12-51**). The combination step is rapid compared to the reduction process.[1746] Nitroso compounds can be reduced to azoxy compounds with triethyl phosphite or triphenylphosphine[1747] or with an alkaline aqueous solution of an alcohol.[1748]

OS **II**, 57.

[1740]For the preparation of high-surface sodium, see Makosza, M.; Grela, K. *Synlett* **1997**, 267.

[1741]Vogel, I.A. *A Textbook of Practical Organic Chemistry*, 3rd ed, Wiley, NY, **1966**, p. 856.

[1742]Karaman, R.; Fry, J.L. *Tetrahedron Lett.* **1989**, *30*, 6267.

[1743]For reviews of the synthesis of catenanes, see Sauvage, J. *Acc. Chem. Res.* **1990**, *23*, 319; *Nouv. J. Chim.* **1985**, *9*, 299; Dietrich-Buchecker, C.O.; Sauvage, J. *Chem. Rev.* **1987**, *87*, 795.

[1744]Osuka, A.; Shimizu, H.; Suzuki, H. *Chem. Lett.* **1983**, 1373.

[1745]Ohe, K.; Uemura, S.; Sugita, N.; Masuda, H.; Taga, T. *J. Org. Chem.* **1989**, *54*, 4169.

[1746]Ogata, Y.; Mibae, J. *J. Org. Chem.* **1962**, *27*, 2048.

[1747]Bunyan, P.J.; Cadogan, J.I.G. *J. Chem. Soc.* **1963**, 42.

[1748]See, for example, Hutton, J.; Waters, W.A. *J. Chem. Soc. B* **1968**, 191. See also, Porta, F.; Pizzotti, M.; Cenini, S. *J. Organomet. Chem.* **1981**, *222*, 279.

19-80 Reduction of Nitro to Azo Compounds

N-De-bisoxygen-coupling

$$2\ ArNO_2 \xrightarrow{LiAlH_4} Ar-N=N-Ar$$

Nitro compounds can be reduced to azo compounds with various reducing agents, of which $LiAlH_4$ and zinc and alkali are the most common. A combination of triethylammonium formate and lead in methanol is also effective.[1749] With many of these reagents, slight differences in conditions can lead either to the azo or azoxy (**19-79**) compound. By analogy to **19-79**, this reaction may be looked on as a combination of $ArN=O$ and $ArNH_2$ (**13-24**). However, when the reducing agent was $NaBH_4$,[1750] it was shown that azoxy compounds were intermediates. Nitroso compounds can be reduced to azo compounds with $LiAlH_4$. Dicarborane, with a catalytic amount of acetic acid, reduces aromatic nitro compounds to the amine.[1751]

Nitro compounds can be further reduced to hydrazo compounds with zinc and sodium hydroxide, with hydrazine hydrate and Raney nickel,[1752] or with $LiAlH_4$ mixed with a metal chloride such as $TiCl_4$ or VCl_3.[1753] The reduction has also been accomplished electrochemically.

OS **III**, 103.

F. Reactions in Which an Organic Substrate is Both Oxidized and Reduced

Some reactions that belong in this category have been considered in earlier chapters. Among these are the Tollens' condensation (**16-43**), the benzil–benzilic acid rearrangement (**18-6**), and the Wallach rearrangement (**18-43**).

19-81 The Cannizzaro Reaction

Cannizzaro Aldehyde Disproportionation

$$2\ ArCHO \xrightarrow{NaOH} ArCH_2OH\ +\ ArCOO^-$$

Aromatic aldehydes, and aliphatic ones with no a hydrogen, give the *Cannizzaro reaction* when treated with NaOH or other strong bases.[1754] In this reaction, one molecule of aldehyde oxidizes another to the acid and is itself reduced to the primary alcohol. Aldehydes with an α-hydrogen do not give the reaction, because when these compounds are treated with base the aldol reaction (**16-34**) is much faster.[1755] Normally, the best yield of acid or alcohol is 50% each, but this can

[1749]Srinavasa, G.R.; Abiraj, K.; Gowda, D.C. *Tetrahedron Lett.* **2003**, *44*, 5835.
[1750]Hutchins, R.O.; Lamson, D.W.; Rufa, L.; Milewski, C.; Maryanoff, B. *J. Org. Chem.* **1971**, *36*, 803.
[1751]Bae, J.W.; Cho, Y.J.; Lee, S.H.; Yoon, C.M. *Tetrahedron Lett.* **2000**, *41*, 175.
[1752]Furst, A.; Moore, R.E. *J. Am. Chem. Soc.* **1957**, *79*, 5492.
[1753]Olah, G.A. *J. Am. Chem. Soc.* **1959**, *81*, 3165.
[1754]For a review, see Geissman, T.A. *Org. React.* **1944**, *2*, 94.
[1755]An exception is cyclopropanecarboxaldehyde: van der Maeden, F.P.B.; Steinberg, H.; de Boer, T.J. *Recl. Trav. Chim. Pays-Bas* **1972**, *91*, 221.

be altered in certain cases. Solvent-free reactions are known.[1756] On the other hand, high yields of alcohol can be obtained from almost any aldehyde by running the reaction in the presence of formaldehyde.[1757] In this case, the formaldehyde reduces the aldehyde to alcohol and is itself oxidized to formic acid. In such a case, where the oxidant aldehyde differs from the reductant aldehyde, the reaction is called the *crossed-Cannizzaro reaction*.[1758] The Tollens' condensation (**16-43**) includes a crossed-Cannizzaro reaction as its last step. A Cannizzaro reaction run on 1,4-dialdehydes (note that α hydrogens are present here) with a rhodium catalyst gives ring closure, for example,[1759]

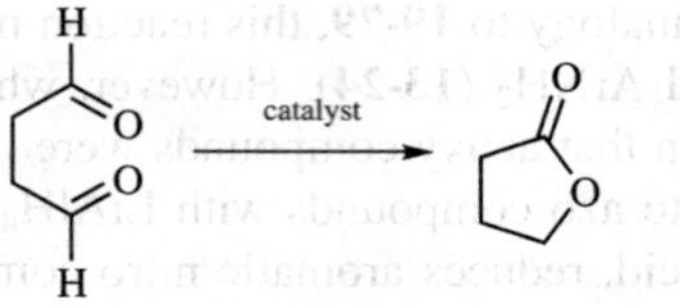

The product is the lactone derived from the hydroxy acid that would result from a normal Cannizzaro reaction. Chiral additives have been used, but with bis(oxazolidine) derivatives the reaction proceeded with poor enantioselectivity.[1760]

α-Keto aldehydes give internal Cannizzaro reactions:

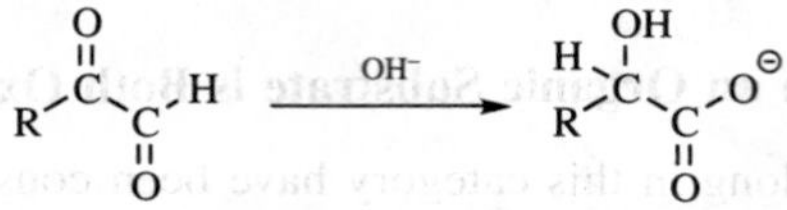

This product is also obtained on alkaline hydrolysis of compounds of the formula $RCOCHX_2$. Similar reactions have been performed on α-keto acetals[1761] and γ-keto aldehydes.

The mechanism[1762] of the Cannizzaro reaction[1763] involves a hydride shift (an example of mechanism type 2, p. 1706). First ⁻OH adds to the C=O to give **55**, which may lose a proton in the basic solution to give the diion **56**.

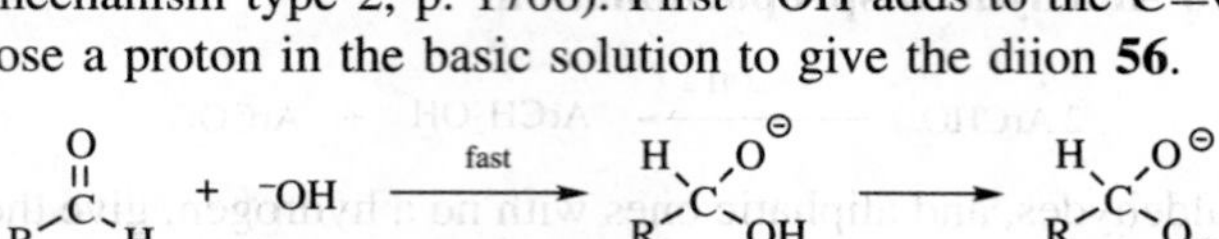

[1756]Yoshizawa, K.; Toyota, S.; Toda, F. *Tetrahedron Lett.* **2001**, *42*, 7983.

[1757]For an example using microwave irradiation, see Thakuria, J.A.; Baruah, M.; Sandhu, J.S. *Chem. Lett.* **1999**, 995.

[1758]For a microwave assisted crossed Cannizzaro reaction, see Varma, R.S.; Naicker, K.P.; Liesen, P.J. *Tetrahedron Lett.* **1998**, *39*, 8437. See Reddy, B.V.S.; Srinivas, R.; Yadav, J.S.; Ramalingam, T. *Synth. Commun.* **2002**, *32*, 219.

[1759]Bergens, S.H.; Fairlie, D.P.; Bosnich, B. *Organometallics* **1990**, *9*, 566.

[1760]Russell, A.E.; Miller, S.P.; Morken, J.P. *J. Org. Chem.* **2000**, *65*, 8381.

[1761]Thompson, J.E. *J. Org. Chem.* **1967**, *32*, 3947.

[1762]For evidence that an SET pathway may intervene, see Ashby, E.C.; Coleman III, D.T.; Gamasa, M.P. *J. Org. Chem.* **1987**, *52*, 4079; Fuentes, A.; Marinas, J.M.; Sinisterra, J.V. *Tetrahedron Lett.* **1987**, *28*, 2947.

[1763]See for example, Swain, C.G.; Powell, A.L.; Sheppard, W.A.; Morgan, C.R. *J. Am. Chem. Soc.* **1979**, *101*, 3576; Watt, C.I.F. *Adv. Phys. Org. Chem.* **1988**, *24*, 57, 81–86.

The strong electron-donating character of O$^-$ greatly facilitates the ability of the aldehydic hydrogen to leave with its electron pair. Of course, this effect is even stronger in **56**. Hydride is transferred to another molecule of aldehyde. The hydride can come from **55** or **56**:

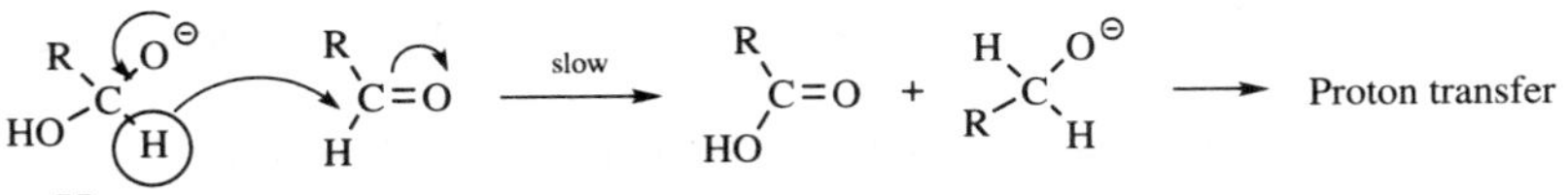

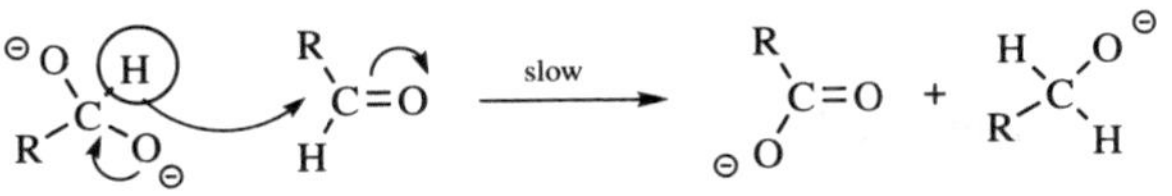

If the hydride ion comes from **55**, the final step is a rapid proton transfer. In the other case, the acid salt is formed directly, and the alkoxide ion acquires a proton from the solvent. Evidence for this mechanism is (*1*) The reaction can be first order in base and second order in substrate (thus going through **55**) or, at higher base concentrations, second order in each (going through **56**); and (*2*) when the reaction was run in D$_2$O, the recovered alcohol contained no α deuterium,[1764] indicating that the hydrogen comes from another equivalent of aldehyde and not from the medium.[1765]

OS **I**, 276; **II**, 590; **III**, 538; **IV**, 110.

19-82   The Tishchenko Reaction

Tishchenko aldehyde-ester disproportionation

$$2 \text{ ArCHO} \xrightarrow{\text{Al(OEt)}_3} \text{ROOCH}_2\text{R}$$

When aldehydes, with or without a hydrogen, are treated with aluminum ethoxide, one molecule is oxidized and another reduced, as in **19-81**, but here they are found as the ester. The process is called the *Tishchenko reaction*. Crossed-Tishchenko reactions are also possible. With more strongly basic alkoxides, such as magnesium or sodium alkoxides, aldehydes with an a hydrogen give the aldol reaction. Treatment of a dialdehyde, such as phthalic dicarboxaldehyde (phthalaldehyde) with CaO, leads to a lactone.[1766] Like **19-81**, this reaction has a mechanism that

---

[1764]Fredenhagen, H.; Bonhoeffer, K.F. *Z. Phys. Chem. Abt. A* ***1938***, *181*, 379; Hauser, C.R.; Hamrick, Jr., P.J.; Stewart, A.T. *J. Org. Chem.* ***1956***, *21*, 260.

[1765]When the reaction was run at 100°C, in MeOH—H$_2$O, isotopic exchange was observed (the product from PhCDO had lost some of its deuterium): Swain, C.G.; Powell, A.L.; Lynch, T.J.; Alpha, S.R.; Dunlap, R.P. *J. Am. Chem. Soc.* ***1979***, *101*, 3584. Side reactions were postulated to account for the loss of deuterium. See, however, Chung, S. *J. Chem. Soc. Chem. Commun.* ***1982***, 480.

[1766]Seki, T.; Hattori, H. *Chem. Commun.* ***2001***, 2510.

involves hydride transfer.[1767] The Tishchenko reaction can also be catalyzed[1768] by ruthenium complexes,[1769] by Cp_2ZrH_2[1770] or $BuTi(OiPr)_4Li$,[1771] and, for aromatic aldehydes, by disodium tetracarbonylferrate, $Na_2Fe(CO)_4$.[1772] Both CaO (noted above) and SrO have been used as catalysts.[1773] A bisphenylenedioxy bis-(aluminum) catalyst has been used to convert aliphatic aldehydes to the corresponding ester.[1774] The bis $Al(OiPr)_2$ derivative of catechol has also been used as a catalyst.[1775]

A Tishchenko–aldol-transfer reaction was reported using β-hydroxy ketones and an aldehydes with an $AlMe_3$ catalyst, giving a mono acyl diol.[1776]
OS I, 104.

19-83 The Pummerer Rearrangement[1777]

Pummerer methyl sulfoxide rearrangement

When sulfoxides bearing an a hydrogen are treated with acetic anhydride, the product is an α-acetoxy sulfide. This is one example of *the Pummerer rearrangement*, in which the sulfur is reduced while an adjacent carbon is oxidized.[1778] The product is readily hydrolyzed (**10-6**) to the aldehyde R_2CHO.[1779] Besides acetic anhydride, other anhydrides and acyl halides give similar products. Inorganic acids, such as HCl, also give the reaction, and $RSOCH_2R'$ can be converted to $RSCHClR'$ in this way. Sulfoxides can also be converted to α-halo sulfides[1780] by other

---

[1767]See, for example, Zakharkin, L.I.; Sorokina, L.P. *J. Gen. Chem. USSR* **1967**, *37*, 525; Saegusa, T.; Ueshima, T.; Kitagawa, S. *Bull. Chem. Soc. Jpn.* **1969**, *42*, 248; Ogata, Y.; Kishi, I. *Tetrahedron* **1969**, *25*, 929.

[1768]For a list of reagents, with references, see Larock, R.C. *Comprehensive Organic Transformations*, 2nd ed., Wiley-VCH, NY, **1999**, pp. 1653–1655.

[1769]Ito, T.; Horino, H.; Koshiro,Y.; Yamamoto, A. *Bull. Chem. Soc. Jpn.* **1982**, *55*, 504.

[1770]DeMico, A.; Margarita, R.; Parlanti, L.; Vescovi, A.; Piancatelli, G. *J. Org. Chem.* **1997**, *62*, 6974.

[1771]Mahrwald, R.; Costisella, B. *Synthesis* **1996**, 1087.

[1772]Yamashita, A.; Watanabe, Y.; Mitsudo, T.; Takegami, Y. *Bull. Chem. Soc. Jpn.* **1976**, *49*, 3597.

[1773]Seki, T.; Akutsu, K.; Hattori, H. *Chem. Commun.* **2001**, 1000.

[1774]Ooi, T.; Miura, T.; Takaya, K.; Maruoka, K. *Tetrahedron Lett.* **1999**, *40*, 7695.

[1775]Simpura, I.; Nevalainen, V. *Tetrahedron* **2001**, *57*, 9867.

[1776]Mascarenhas, C.M.; Duffey, M.O.; Liu, S.-Y.; Morken, J.P. *Org. Lett.* **1999**, *1*, 1427; Simpura, I.; Nevalainen, V. *Tetrahedron Lett.* **2001**, *42*, 3905; Cavazzini, M.; Pozzi, G.; Quici, S.; Maillard, D.; Sinou, D. *Chem. Commun.* **2001**, 1220.

[1777]For a review of the Pummerer reaction for the synthesis of heterocyclic compounds, see Bur, S.K.; Padwa, A. *Chem. Rev.* **2004**, *104*, 2401.

[1778]For reviews, see De Lucchi, O.; Miotti, U.; Modena, G. *Org. React.* **1991**, *40*, 157; Warren, S. *Chem. Ind. (London)* **1980**, 824; Oae, S.; Numata, T. *Isot. Org. Chem.* **1980**, *5*, 45; Block, E. *Reactions of Organosulfur Compounds*, Academic Press, NY, **1978**, pp. 154–162.

[1779]See, for example, Sugihara, H.; Tanikaga, R.; Kaji, A. *Synthesis* **1978**, 881.

[1780]For a review of α-chloro sulfides, see Dilworth, B.M.; McKervey, M.A. *Tetrahedron* **1986**, *42*, 3731.

reagents, including sulfuryl chloride, NBS, and NCS. Enantioselective Pummerer rearrangements are known.[1781] Uncatalyzed thermal rearrangements are also known.[1782]

The following four-step mechanism has been proposed for the reaction between acetic anhydride and DMSO:[1783]

57

For DMSO and acetic anhydride, step 4 is intermolecular, as shown by [18]O isotopic labeling studies.[1784] With other substrates, however, step 4 can be inter- or intramolecular, depending on the structure of the sulfoxide.[1785] Depending on the substrate and reagent, any of the first three steps can be rate determining. In the case of Me_2SO treated with $(F_3CCO)_2O$, the intermediate corresponding to 57^{1786} could be isolated at low temperature, and on warming gave the expected product.[1787] There is much other evidence for this mechanism.[1788]

A sila-Pummerer rearrangement has been reported.[1789]

19-84 The Willgerodt Reaction

Willgerodt carbonyl transformation

$$ArCOCH_3 \xrightarrow{(NH_4)_2S_x} ArCH_2CONH_2 + ArCH_2COO^- NH_4^+$$

[1781]Kita, Y.; Shibata, N.; Kawano, N.; Tohjo, T.; Fujimori, C.; Matsumoto, K. *Tetrahedron Lett.* **1995**, *36*, 115; Kita, Y.; Shibata, N.; Fukui, S.; Fujita, S. *Tetrahedron Lett.* **1994**, *35*, 9733; Kita, Y.; Shibata, N.; Kawano, N.; Fukui, S.; Fujimori, C. *Tetrahedron Lett.* **1994**, *35*, 3575; Kita, Y.; Shibata, N.; Yoshida, N. *Tetrahedron Lett.* **1993**, *34*, 4063.

[1782]Wladislaw, B.; Marzorati, L.; Biaggio, F.C. *J. Org. Chem.* **1993**, *58*, 6132.

[1783]See, for example, Numata, T.; Itoh, O.; Yoshimura, T.; Oae, S. *Bull. Chem. Soc. Jpn.* **1983**, *56*, 257; Kita, Y.; Shibata, N.; Yoshida, N.; Fukui, S.; Fujimori, C. *Tetrahedron Lett.* **1994**, *35*, 2569.

[1784]Oae, S.; Kitao, T.; Kawamura, S.; Kitaoka, Y. *Tetrahedron* **1963**, *19*, 817.

[1785]See, for example, Itoh, O.; Numata, T.; Yoshimura, T.; Oae, S. *Bull. Chem. Soc. Jpn.* **1983**, *56*, 266; Oae, S.; Itoh, O.; Numata, T.; Yoshimura, T. *Bull. Chem. Soc. Jpn.* **1983**, *56*, 270.

[1786]For a review of sulfur-containing cations, see Marino, J.P. *Top. Sulfur Chem.* **1976**, *1*, 1.

[1787]Sharma, A.K.; Swern, D. *Tetrahedron Lett.* **1974**, 1503.

[1788]See Block, E. *Reactions of Organosulfur Compounds*, Academic Press, NY, **1978**, pp. 154–156; Oae, S.; Numata, T. *Isot. Org. Chem.* **1980**, *5*, 45, 48; Wolfe, S.; Kazmaier, P.M. *Can. J. Chem.* **1979**, *57*, 2388, 2397; Russell, G.A.; Mikol, G.J. *Mech. Mol. Migr.* **1968**, *1*, 157.

[1789]Kirpichenko, S.V.; Suslova, E.N.; Albanov, A.I.; Shainyan, B.A. *Tetrahedron Lett.* **1999**, *40*, 185.

In the *Willgerodt reaction*, a straight- or branched-chain aryl alkyl ketone is converted to the amide and/or the ammonium salt of the acid by heating with ammonium polysulfide.[1790] The carbonyl group of the product is always at the end of the chain. Thus $ArCOCH_2CH_3$ gives the amide and the salt of $ArCH_2CH_2COOH$, and $ArCOCH_2CH_2CH_3$ gives derivatives of $ArCH_2CH_2CH_2COOH$. However, yields sharply decrease with increasing length of chain. The reaction has also been carried out on vinylic and ethynyl aromatic compounds and on aliphatic ketones, but yields are usually lower in these cases. Unlike the Pummerer rearrangement (**19-83**), which involves transposition of an oxygen from S to C, the Willgerodt reaction involves oxygen migration *and* oxidation of the organic species. The use of sulfur and a dry primary or secondary amine (or ammonia), as the reagent is called the *Kindler modification* of the Willgerodt reaction.[1791] The product in this case is $Ar(CH_2)_nCSNR_2$,[1792] which can be hydrolyzed to the acid. Particularly good results are obtained with morpholine as the amine. For volatile amines, the HCl salts can be used instead, with NaOAc in DMF at 100°C.[1793] Dimethylamine has also been used in the form of dimethylammonium dimethylcarbamate, Me_2NCOO^- $Me_2NH_2^+$.[1794] The Kindler modification has also been applied to aliphatic ketones.[1795] Thioamides have been prepared from ketones in a base-catalyzed reaction.[1796]

Alkyl aryl ketones can be converted to arylacetic acid derivatives in an entirely different manner. The reaction consists of treatment of the substrate with silver nitrate and I_2 or Br_2,[1797] or with thallium nitrate, MeOH, and trimethyl orthoformate adsorbed on Montmorillonite K10, an acidic clay.[1798]

The mechanism of the Willgerodt reaction is not completely known, but some conceivable mechanisms can be excluded. Thus, one might suppose that the alkyl group becomes completely detached from the ring, and then attacks it with its other

[1790]For a review, see Brown, E.V. *Synthesis* **1975**, 358.
[1791]For a review, see Mayer, R., in Oae, S. *The Organic Chemistry of Sulfur*, Plenum, NY, **1977**, pp. 58–63. For a study of the optimum conditions for this reaction, see Lundstedt, T.; Carlson, R.; Shabana, R. *Acta Chem. Scand. Ser. B* **1987**, *41*, 157, and other papers in this series. See also, Carlson, R.; Lundstedt, T. *Acta Chem. Scand. Ser. B* **1987**, *41*, 164; Kanyonyo, M.R.; Gozzo, A.; Lambert, D.M.; Lesieur, D.; Poupaert, J.H. *Bull. Soc. Chim. Belg.* **1997**, *106*, 39.
[1792]The reaction between ketones, sulfur, and ammonia can also lead to heterocyclic compounds. For a review, see Asinger, F.; Offermanns, H. *Angew. Chem. Int. Ed.* **1967**, *6*, 907.
[1793]Amupitan, J.O. *Synthesis* **1983**, 730.
[1794]Schroth, W.; Andersch, J. *Synthesis* **1989**, 202.
[1795]See Dutron-Woitrin, F.; Merényi, R.; Viehe, H.G. *Synthesis* **1985**, 77.
[1796]For a review, see Poupaert, J.H.; Bouinidane, K.; Renard, M.; Lambert, D.; Isa, M. *Org. Prep. Proceed. Int.* **2001**, *33*, 335.
[1797]Higgins, S.D.; Thomas, C.B. *J. Chem. Soc. Perkin Trans. 1* **1982**, 235. See also, Higgins, S.D.; Thomas, C.B. *J. Chem. Soc. Perkin Trans. 1* **1983**, 1483.
[1798]Taylor, E.C.; Conley, R.A.; Katz, A.H.; McKillop, A. *J. Org. Chem.* **1984**, *49*, 3840.

end. However, this possibility is ruled out by experiments such as the following: When isobutyl phenyl ketone (**58**) is subjected to the Willgerodt reaction, the product is **59**, not **60**, which would arise if the end carbon of the ketone became bonded to the ring in the product:[1799]

This also excludes a cyclic-intermediate mechanism similar to that of the Claisen rearrangement (**18-33**). Another important fact is that the reaction is successful for singly branched side chains, such as **58**, but not for doubly branched side chains, as in PhCOCMe$_3$.[1799] Still another piece of evidence is that compounds oxygenated along the chain give the same products; thus PhCOCH$_2$CH$_3$, PhCH$_2$COMe, and PhCH$_2$CH$_2$CHO all give PhCH$_2$CH$_2$CONH$_2$.[1800] All these facts point to a mechanism consisting of consecutive oxidations and reductions along the chain, although just what form these take is not certain. Initial reduction to the hydrocarbon can be ruled out, since alkylbenzenes do not give the reaction. In certain cases, imines[1801] or enamines[1802] have been isolated from primary and secondary amines, respectively, and these have been shown to give the normal products, leading to the suggestion that they may be reaction intermediates.

[1799]King, J.A.; McMillan, F.H. *J. Am. Chem. Soc.* **1946**, *68*, 632.

[1800]For an example of this type of behavior, see Asinger, F.; Saus, A.; Mayer, A. *Monatsh. Chem.* **1967**, *98*, 825.

[1801]Asinger, F.; Halcour, K. *Monatsh. Chem.* **1964**, *95*, 24. See also, Nakova, E.P.; Tolkachev, O.N.; Evstigneeva, R.P. *J. Org. Chem. USSR* **1975**, *11*, 2660.

[1802]Mayer, R., in Janssen, M.J. *Organosulfur Chemistry*, Wiley, NY, **1967**, pp. 229–232.

APPENDIX A

The Literature of Organic Chemistry

All discoveries in the laboratory must be published somewhere if the information is to be made generally available. A new experimental result that is not published might as well not have been obtained, insofar as it benefits the entire chemical world. The total body of chemical knowledge (called *the literature*) is located on the combined shelves of all the chemical libraries in the world. Anyone who wishes to learn whether the answer to any chemical question is known, and, if so, what the answer is, has only to turn to the contents of these shelves. Indeed, the very expressions "is known," "has been done," and so on, really mean "has been published." To the uninitiated, the contents of the shelves may appear formidably large, but fortunately the process of extracting information from the literature of organic chemistry is usually not difficult. In this appendix, we will examine the literature of organic chemistry.[1] It is quite clear that The Literature can be divided into two broad categories: primary sources and secondary sources. A *primary source* publishes the original results of laboratory investigations. Books, indexes, and other publications that cover material that has previously been published in primary sources are called *secondary sources*. It is because of the excellence of the secondary sources in organic chemistry (especially *Chemical Abstracts*™, SciFinder®, and Beilstein) that literature searching is comparatively not difficult. The two chief kinds of primary source are journals and patents. There are several types of secondary source.

---

[1]For books on the chemical literature, see Wolman, Y. *Chemical Information*, 2nd ed., Wiley, NY, *1988*; Maizell, R.E. *How to Find Chemical Information*, 2nd ed., Wiley, NY, *1987*; Mellon, M.G. *Chemical Publications*, 5th ed., McGraw-Hill, NY, *1982*; Skolnik, H. *The Literature Matrix of Chemistry*, Wiley, NY, *1982*; Antony, A. *Guide to Basic Information Sources in Chemistry*, Jeffrey Norton Publishers, NY, *1979*; Bottle, R.T. *Use of the Chemical Literature*, Butterworth, London, *1979*; Woodburn, H.M. *Using the Chemical Literature*, Marcel Dekker, NY, *1974*. For a three-part article on the literature of organic chemistry, see Hancock, J.E.H. *J. Chem. Educ. 1968*, *45*, 193–199, 260–266, 336–339.

March's Advanced Organic Chemistry: Reactions, Mechanisms, and Structure, Sixth Edition, by Michael B. Smith and Jerry March
Copyright © 2007 John Wiley & Sons, Inc.

1870

PRIMARY SOURCES

Journals

For well over 100 years, nearly all new work in organic chemistry (except for that disclosed in patents) has been published in journals. There are thousands of journals that publish chemical papers, in many countries and in many languages. Some print papers covering all fields of science; some are restricted to chemistry; some to organic chemistry; and some are still more specialized. Fortunately for the sanity of organic chemists, the vast majority of important papers in "pure" organic chemistry (as opposed to "applied") are published in relatively few journals, perhaps 50 or fewer. The concept of "pure" organic chemistry is not as useful because organic chemistry is now important in many areas. Literature that is important to an organic chemist is found in journals and patents that focus on bioorganic, organometallic, materials science, separation science, medicinal chemistry, pharmaceutical sciences, and medicine to name a few. The reader is therefore cautioned that the journals listed in this section have organic chemistry as their primary focus, but are by no means the only sources of information concerning organic chemistry. The literature is vast and many journals are published weekly, and some semi-monthly.

In addition to ordinary papers, there are two other types of publications in which original work is reported: *notes* and *communications*. A note is a brief paper, often without a summary (nearly all papers are published with summaries or abstracts prepared by the author). Otherwise, a note is similar to a paper.[2] Communications (also called *letters*) are also brief and usually without summaries (though some journals now publish summaries along with their communications, a welcome trend). However, communications differ from notes and papers in three respects:

1. They are brief, not because the work is of small scope, but because they are condensed. Usually, they include only the most important experimental details or none at all.

2. They are often of immediate significance. Journals that publish communications make every effort to have them appear as soon as possible after they are received. Some papers and notes are of great importance, and some are of lesser importance, but all communications are supposed to be of high importance. With modern computer technology, communications can often be published in a matter of weeks, and the on-line version (e.g., the American Chemical Society ASAP papers) can be found before the print version appears.

3. Communications are preliminary reports, and the material in them may be republished as papers at a later date, in contrast to the material in papers and notes, which cannot be republished.

---

[2] In some journals, notes are called "short communications," an unfortunate practice, because they are not communications as that term is defined in the text.

Although papers (we use the term in its general sense, to cover notes and communications also) are published in many languages, the English-speaking chemist is in a fairly fortunate position. At present well over half of the important papers in organic chemistry are published in English. Not only are American, British, and British Commonwealth journals published almost entirely in English, but so are many others around the world. There are predominantly English-language journals published in Japan, Italy, Czechoslovakia, Sweden, The Netherlands, Israel, and other countries. In a reorganization, six prominent European journals (*Chemische Berichte, Liebigs Annalen der Chemie, Bulletin de la Société Chimique de France, Bulletin des Sociétés Chimique Belges, Recueil des Travaux Chimiques des Pays-Bas, and Gazzetta Chimica Italiana*) have been discontinued. In their place is the *European Journal of Organic Chemistry, published in English*. Most of the articles published in other languages have summaries printed in English also. For many years, the second most important language (in terms of the number of organic chemical papers published) was Russian, and most of these papers are available in English translation, although in most cases, 6 months to 1 year later. With the political changes in Russia, however, many of these Journals have been modified or discontinued. Important papers have been published in German and French for >200 years, and these are generally not available in translation, so that the organic chemist should have at least a reading knowledge of these languages. In recent years, however, fewer papers in French or German have appeared without an English translation, such as in the journal *Angewandte Chemie*, which in 1962 became available in English under the title *Angewandte Chemie International Edition in English*. Of course, a reading knowledge of French and German (especially German) is critical for the older literature. Before ~1920, more than one-half of the important chemical papers were in these languages. It must be realized that the original literature is never obsolete. With the rise of China in the scientific community, journals are published in Chinese, and there are journals published in Japanese. Work by Chinese and Japanese scientists regularly appears in English-language journals, however. Secondary sources become superseded or outdated, but nineteenth century journals are found in most chemical libraries and are still consulted. Table A.1 presents a list of the more important current journals that publish original papers[3] and communications in organic chemistry. Some of them also publish review articles, book reviews, and other material. Changes in journal title have not been infrequent; footnotes to the table indicate some of the more important, but some of the other journals listed have also undergone title changes. In 1999, the *Journal of Organic Chemistry* stopped publishing communications, and these are now published in a new journal, *Organic Letters*.

The primary literature has grown so much in recent years that attempts have been made to reduce the volume. One such attempt is the *Journal of Chemical Research*, begun in 1977. The main section of this journal, called the "Synopses," publishes synopses, which are essentially long abstracts, with references. The full texts of most of the papers are published only in microfiche and miniprint versions.

---

[3]In Table A.1, notes are counted as papers.

TABLE A.1. A List of the More Important *Current* Journals That Currently Publish Original Papers in Organic Chemistry[a]

No.	Name	Papers or Communications	Issues per Year
1	**Angew**andte **Chem**ie (1887)[4]	C[5]	12
2	**Angew**andte **Chem**ie International Edition (1962)[6]	C[7]	48
3	**Aust**ralian **J**ournal of **Chem**istry (1948)	P	12
4	**Bioorg**anic **Chem**istry (1971)	P[5]	4
5	**Bioorg**anic & **Med**icinal **Chem**istry **Lett**ers (1991)	C	12
6	**Bull**etin of the **Chem**ical **Soc**iety of **Jap**an (1926)	P	12
7	**Can**adian **J**ournal of **Chem**istry (1929)	P,C	12
8	**Carbohydr**ate **Res**earch (1965)	P,C	22
9	**Chem**istry, a **Eur**opean **Journal** (1995)	P	24
10	**Chem**istry, an **Asian J**ournal (2006)	New	New
11	**Chem**istry and **Ind**ustry (**London**) (1923)	C	24
12	**Chem**istry **Lett**ers (1972)	C	12
13	**Chimia** (1947)	C[5]	12
14	**Collec**tion of **Czech**oslovak **Chem**ical **Commun**ications (1929)	P	12
15	**Dokl**ady **Chem**istry (1922)[4]	C	12
17	**Eur**opean **J**ournal of **Org**anic **Chem**istry (1998)	P	12
18	**Helv**etica **Chim**ica **Acta** (1918)	P	8
19	**Heteroat**om **Chem**istry (1990)	P	6
20	**Heterocycles** (1973)	C[5]	12
21	**Ind**ian **J**ournal of **Chem**istry (Section **B**)	P	12
22	**Int**ernational **J**ournal of chemical **Kinet**ics (1969)	P	12
23	**Isr**ael **J**ournal of **Chem**istry (1963)	P[8]	4
23	**J**ournal of the **Am**erican **Chem**ical **Soc**iety (1879)	P,C	52
25	**J**ournal of **Carbohydr**ate **Chem**istry (1981)	P,C	6
26	**J**ournal of **Chem**ical **Res**earch, **Synop**ses (1977)	P	12
27	**Chem**ical **Commun**ications (1965)	C	24
28	**J**ournal of **Comb**inatorial **Chem**istry (2000)	P,C	6
29	**J**ournal of **Comput**ational **Chem**istry (1979)	P	16
30	**J**ournal of **Fluor**ine **Chem**istry (1971)	P,C	12

[4]These journals are available in English translation.
[5]These journals also publish review articles regularly.
[6]These journals are available in English translation.
[7]These journals also publish review articles regularly.
[8]Each issue of this journal is devoted to a specific topic.

TABLE A.1 (*Continued*)

No.	Name	Papers or Communications	Issues per Year
31	Journal of **Heterocyclic Chem**istry (1964)	P,C	12
32	Journal of the **Indian** Chemical **Soci**ety (1924)	P	12
33	Journal of **Lipid Res**earch (1959)	P	12
34	Journal of **Medicinal Chem**istry (1958)	P,C	12
35	Journal of **Molecular Struct**ure (1967)	P,C	16
36	Journal of **Organomet**allic **Chem**istry (1963)	P,C	48
37	Journal of **Organic Chem**istry (1936)	P,C	26
38	Journal of **Photochem**istry and **Photobiol**ogy, A: Chemistry (1972)	P	12
39	Journal of **Physical Organic Chem**istry (1988)	P	12
40	Journal of Polymer Science Part A (1962)	P	24
41	Journal für **Praktische Chem**ie (1834)	P	6
42	**Macromolecules** (1968)	P,C	26
43	**Liebigs Ann**alen der **Chem**ie (1832)	P	12
44	**Mendeleev Commun**ications (1991)	C	8
45	**Monats**hefte für **Chem**ie (1870)	P	12
46	**New J**ournal of **Chem**istry (1977)[9]	P	11
47	**Organometallics** (1982)	P,C	12
48	**Organic** and **Biomolecular Chem**istry (2003)	P	24
49	**Organic Lett**ers (1999)	C	12
50	**Organic Mass Spectrom**etry (1968)	PC	12
51	**Organic Prep**arations and **Proced**ures International (1969)	P[5]	6
52	**Organic Process Res**earch & **Development** (1997)	P	6
53	**Photochem**istry and **Photobiol**ogy (1962)	P[5]	12
54	**Polish J**ournal of **Chem**istry (1921)[9]	PC	12
55	**Pure** and **Applied Chem**istry (1960)	[10]	12
56	**Res**earch on Chemical **Intermed**iates (1973)[11]	P[5]	6
57	**Russian J**ournal of **Org**anic **Chem**istry (1984)	P,C	12
58	**Sulfur Lett**ers (1982)	C	6
59	**Synlett** (1989)	C[5]	12
60	**Synthetic Commun**ications (1971)	C	22
61	**Synthesis** (1969)	P[5]	12
62	**Tetrahedron** (1958)	P[5]	52
63	**Tetrahedron: Asymmetry** (1990)	PC	12
64	**Tetrahedron Lett**ers (1959)	C	52

[a]These Journals are listed in alphabetical order of *Chemical Abstracts*™ abbreviations, Which Are Indicated in Boldface. Also given are the year of founding, number of issues per year as of 1998, and whether the journal primarily publishes papers (P), communications (C), or both.

[9]Before 1978 this journal was called *Roczniki Chemii*.

[10]*Pure Appl. Chem.* publishes IUPAC reports and lectures given at IUPAC meetings.

[11]Before 1989 this journal was called *Reviews of Chemical Intermediates*.

For some years, the American Chemical Society journals, including *J. Am. Chem. Soc.* and *J. Org. Chem.*, have provided supplementary material for some of their papers. This material is available on-line, and for older literature from the Microforms and Back Issues Office at the ACS Washington, DC, office, either on microfiche or as a photocopy. Many other journals now offer supplementary material, generally experimental procedures, supporting spectral data and crystallographic data. These practices have not yet succeeded in substantially reducing the total volume of the world's primary chemical literature.

Patents

In many countries, including the United States, it is possible to patent a new compound or a new method for making a known compound (either laboratory or industrial procedures), as long as the compounds are useful. It comes as a surprise to many to learn that a substantial proportion of the patents granted (perhaps 20–30%) have been chemical patents. Chemical patents are part of the chemical literature, and both U.S. and foreign patents are regularly abstracted by *Chemical Abstracts*™. In addition to learning about the contents of patents from this source, chemists may consult the *Official Gazette* of the U.S. Patent Office, which, published weekly and available in many libraries, lists titles of all patents issued that week. Bound volumes of all U.S. patents are kept in a number of large libraries, including the New York Public Library, which also has an extensive collection of foreign patents. Photocopies of any U.S. patent and most foreign patents can be obtained at low cost from the U.S. Patent and Trademark Office, Washington, DC, 20231. Many patents can now be obtained on-line as well. In addition, *Chemical Abstracts*™ lists, in the introduction to the first issue of each volume, instructions for obtaining patents from 26 countries. Patents are also available via SciFinder®.

Although patents are often very useful to the laboratory chemist, and no literature search is complete that neglects relevant patents, as a rule they are not as reliable as papers. There are two reasons for this:

1. It is in the interest of the inventor to claim as much as possible. Therefore, they may, for example, actually have carried out a reaction with ethanol and with 1-propanol, but will claim all primary alcohols, and perhaps even secondary and tertiary alcohols, glycols, and phenols. An investigator repeating the reaction on an alcohol that the inventor did not use may find that the reaction gives no yield at all. In general, it is safest to duplicate the actual examples given, of which most chemical patents contain one or more.

2. Although legally a patent gives an inventor a monopoly, any alleged infringements must be protected in court, and this may cost a good deal of money. Therefore some patents are written so that certain essential details are concealed or entirely omitted. This practice is not exactly cricket, because a patent is supposed to be a full disclosure, but patent attorneys are generally

skilled in the art of writing patents, and procedures given are not always sufficient to duplicate the results.

Fortunately, the above statements do not apply to all chemical patents: many make full disclosures and claim only what was actually done. It must also be pointed out that it is not always possible to duplicate the work reported in every paper in a journal due to the use of proprietary catalysts, equipment, or procedures. Note, however, that some work is not published or patented, but rather maintained within the company as a trade secret. Such work is not, of course, available to the public.

SECONDARY SOURCES

Journal articles and patents contain virtually all of the original work in organic chemistry. However, if therewere no indexes, abstracts, review articles, and other secondary sources, the literature would be unusable, because it is so vast that no one could hope to find anything in particular. Fortunately, the secondary sources are excellent. There are various kinds and the categories tend to merge. Our classification is somewhat arbitrary.

Listings of Titles

The profusion of original papers is so great that publications that merely list the titles of current papers find much use. Such lists are primarily methods of alerting the chemist to useful papers published in journals that they do not normally read. This approach, using print versions containing lists or journals and articles, is not used much nowadays. Most journals are available on-line, and many have the original papers, with supplemental material, as HTML and pdf[12] documents. The pdf document can be downloaded to the searcher's desktop, usually for a fee, and is most convenient. *Chemical Abstracts*™ was originally available on-line as *CAS ONLINE*. CAS ONLINE was discontinued when STN® (a service that CAS operates in cooperation with the German organization FIZ Karlsruhe, see below) was introduced in 1984. All of the *Chemical Abstracts*™ literature can be accessed using CAS databases online. Many libraries and companies pay the appropriate fees, so accessing the journals is usually quite easy via those organizations. Search engines allow one to quickly scan an enormous amount of literature from office or home. in addition, most browsers have on-line searching capabilities via various search engines, and simply typing in an author, a topic, a chemical, or a few key words can lead to important articles or information. Some of the important on-line technology will be discussed below. However, some of the available resources[13] include Specialty Citation Indexes, Science Citation Index Expanded™, Web of Science®,

---

[12]Adobe Acrobat© files.
[13]See http://scientific.thomson.com/products/categories/citation/

Science Citation Index®, ISI ProceedingsSM, Reaction Citation Index™, and the Derwent Innovations IndexSM. Science Citation Index® is available on STN®, where it is called SciSearch. We will begin with the older print-versions for chemical searches.

A print-version "title" publication covering the whole of chemistry is *Current Contents Physical, Chemical & Earth Sciences*,[14] which began in 1967 and appears weekly, contains the contents pages of all issues of ~800 journals in chemistry, physics, earth sciences, mathematics, and allied sciences. Each issue contains an index of important words taken from the titles of the papers listed in that issue, and an author index, which, however, lists only the first-named author of each paper. The author's address is also given, so that one may write for reprints. An on-line service is available called Current Contents Connect® is a multidisciplinary Web resource providing access to complete bibliographic information from >8000 of the world's leading scholarly journals and >2000 books.[15]

CAS databases on-line described below allows one to search a variety of databases, including journal titles.

Abstracts

Listings of titles are valuable, as far as they go, but they do not tell what is in the paper, beyond the implications carried by the titles. Most current journals contain a graphic abstract as well as a title and a brief print description of the research. The graphical abstract is extremely useful for scanning the literature presented in a journal, and both the print and graphical abstracts are available on-line for most journals.

From the earliest days of organic chemistry, abstracts of papers have been widely available, often as sections of journals whose principal interests lay elsewhere.[16] At the present time there are only two publications entirely devoted to abstracts covering the whole field of chemistry. One of these, *Referativnyi Zhurnal, Khimiya*, which began in 1953, is published in Russian and is chiefly of interest to Russian-speaking chemists. The other is *Chemical Abstracts*™ (CA). This publication, which appears weekly, prints abstracts in English of virtually every paper containing original work in pure or applied chemistry published anywhere in the world.[17] More than 9500 currently published journals are covered, in many languages. In addition, *CA* publishes abstracts of every patent of chemical interest from 50 national and international patent offices. *Chemical Abstracts*™ lists and indexes, but does not abstract review articles and books. The abstracts currently appear in 80 sections, of which sections 21–34 are devoted to organic chemistry, under such

---

[14]Title pages of organic chemistry journals are also carried by *Current Contents Life Sciences*, which is a similar publication covering biochemistry and medicine.
[15]http://scientific.thomson.com/products/ccc/
[16]For example, *Chem. Ind. (London)* publishes abstracts of papers that appear in other journals. In the past, journals, such as *J. Am. Chem. Soc., J. Chem Soc.,* and *Berchti* also did so.
[17]For a guide to the use of *CA*, see Schulz, H. *From CA to CAS ONLINE*; VCH: NY, *1988*.

headings as Alicyclic Compounds, Alkaloids, Physical Organic Chemistry, Hetero-cyclic Compounds (One Heteroatom), and so on. Each abstract of a paper begins with a heading that gives (*1*) the abstract number;[18] (*2*) the title of the paper; (*3*) the authors' names as fully as given in the paper; (*4*) the authors' address; (*5*) the abbreviated name of the journal (see Table A.1);[19] (*6*) the year, volume, issue, and page numbers; and (*7*) the language of the paper. In earlier years, *CA* gave the language only if it differed from the language of the journal title. Abstracts of patents begin with the abstract number, title, inventor and company (if any), patent number, patent class number, date patent issued, country of priority, patent application number, date patent applied for, and number of pages in the patent. The body of the abstract is a concise summary of the information in the paper. For many common journals the author's summary (if there is one) is used in *CA* as it appears in the original paper, with perhaps some editing and additional information. Each issue of *CA* contains an author index, a patent index, and an index of keywords taken from the titles and the texts or contexts of the abstracts. The patent index lists all patents in order of number. The same compound or method is often patented in several countries. *Chemical Abstracts* will abstract only the first patent, but it does list the patent numbers of the duplicated patents in the patent index along with all previous patent numbers that correspond to it. Before 1981, there were separate Patent Number Indexes and Patent Concordances (the latter began in 1963).

At the end of each section of *CA*, there is a list of cross-references to related papers in other sections.

Chemical Abstracts[TM] is, of course, highly used for "current awareness"; it allows one to read, in one place, abstracts of virtually all new work in chemistry, though its large size puts a limit on the extent of this type of usefulness.[20] *Chemical Abstracts*[TM] is even more useful as a repository of chemical information, a place for finding out what was done in the past. This value stems from the excellent indexes, which enable the chemist in most cases to ascertain quickly where information is located. From the time of its founding in 1907 until 1961, *CA* published annual indexes. Since 1962 there are two volumes published each year, and a separate index is issued for each volume. For each volume there is an index of subjects, authors, formulas, and patent numbers. Beginning in 1972, the subject index has been issued in two parts, a chemical substance index and a general subject index, which includes all entries that are not the names of single chemical substances. However, the indexes to each volume become essentially superseded as collective indexes are issued. The first collective indexes are 10-year (decennial) indexes, but the volume of information has made 5-year indexes necessary since 1956. Collective indexes published to date are shown in Table A.2. Thus a user of the indexes at the time of this writing would consult the collective indexes through 1995 and the semiannual indexes thereafter. The 14th collective index (covering

[18]Beginning in 1967. See p. 1880.
[19]These abbreviations are changed from time to time. Therefore the reader may notice inconsistencies.
[20]It is possible to subscribe to *CA Selects,* which provides copies of all abstracts within various narrow fields, such as organofluorine chemistry, organic reaction mechanisms, and organic stereochemistry.

TABLE A.2. *CA* **Collective Indexes Published**

Collective Index	Subject and General Subject	Chemical Substance	Author	Formula	Patent
1	1907–1916		1907–1916		
2	1917–1926		1917–1926		1907–1936
3	1927–1936		1927–1936	1920–1946	
4	1937–1946		1937–1946		1937–1946
5	1947–1956		1947–1956	1947–1956	1947–1956
6	1957–1961		1957–1961	1957–1961	1957–1961
7	1962–1966		1962–1966	1962–1966	1962–1966
8	1967–1971		1967–1971	1967–1971	1967–1971
9	1972–1976	1972–1976	1972–1976	1972–1976	1972–1976
10	1977–1981	1977–1981	1977–1981	1977–1981	1977–1981
11	1982–1986	1982–1986	1982–1986	1982–1986	1982–1986
12	1987–1991	1987–1991	1987–1991	1987–1991	1987–1991
13	1992–1996	1992–1996	1992–1996	1992–1996	1992–1996
14	1997–2001	1997–2001	1997–2001	1997–2001	1997–2001
15	2002–2006	2002–2006	2002–2006	2002–2006	2002–2006

1997–2001) appeared in 2002. The 15th Collective index will be issued only in CD ROM, as well as any subsequent collective, presumably.

Beginning with the eighth collective index period, *CA* has published an *Index Guide*. This publication gives structural formulas and/or alternate names for thousands of compounds, as well as many other cross-references. It is designed to help the user efficiently and rapidly to find *CA* references to subjects of interest in the general subject, formula, and chemical substance indexes. Each collective index contains its own *Index Guide*. A new *Index Guide* is issued every 18 months. The *Index Guide* is necessary because the *CA* general subject index is a "controlled index," meaning it restricts its entries only to certain terms. For example, anyone who looks for the term "refraction" in the general subject index will not find it. The *Index Guide* includes this term, and directs the reader to "Electromagnetic wave, refraction of," "Sound and ultrasound, refraction of," and other terms, all of which will be found in the general subject index. Similarly, the chemical substance index usually lists a compound only under one name, the approved *CA* name. Trivial and other names will be found in the *Index Guide*. For example, the term "methyl carbonate" is not in the chemical substance index, but the *Index Guide* does have this term, and tells us to look for it in the chemical substance index under the headings "carbonic acid, esters, dimethyl ester" (for Me_2CO_3) and "carbonic acid, esters, monomethyl ester" (for $MeHCO_3$). Furthermore, the *Index Guide* gives terms related to the chosen term, helping users to broaden a search. For example, one who looks for "Atomic orbital" in the *Index Guide* will find the terms "Energy Level," "Molecular orbital," "Atomic integral," and "Exchange, quantum mechanical, integrals for," all of which are controlled index terms.

Along with each index (annual, semiannual, or collective) appears an index of ring systems. This valuable index enables the user to ascertain immediately if any ring system appears in the corresponding subject or chemical substance index and under what names. For example, someone wishing to determine whether any compounds containing this ring system

Benz(*h*)isoquinoline

are reported in the 1982–1986 collective index (even if they did not know the name) would locate, under the heading "3-ring systems," the listing **6, 6, 6** (since the compound has three rings of six members each), under which they would find the sublisting $C_5N-C_6-C_6$ (since one ring contains five carbons and a nitrogen while the others are all-carbon), under which is listed the name benz(*h*)isoquinoline, as well as the names of 30 other systems $C_5N-C_6-C_6$. A search of the chemical substance index under these names will give all references to these ring systems that appeared in *CA* from 1982 to 1986.

Before 1967, *CA* used a two-column page, with each column separately numbered. A row of letters from *a* to *h* appeared down the center of the page. These letters are for the guidance of the user. Thus an entry 7337*b* refers to the *b* section of column 7337. In early years, superscript numbers (e.g., 4327^5), were used in a similar manner. In very early years, these numbers were not printed on the page at all, though they are given in the decennial indexes, so that the user must mentally divide the page into nine parts. Beginning with 1967, abstracts are individually numbered and column numbers are discarded. Therefore, beginning with 1967, index entries give abstract number rather than column number. The abstract numbers are followed by a letter that serves as a check character to prevent miscopying errors in computer handling. To use the *CA* general subject, chemical substance, and formula indexes intelligently requires practice, and the student should become familiar with representative volumes of these indexes and with the introductory sections to them, as well as with the *Index Guides*.

In the *CA* formula indexes, formulas are listed in order of (*1*) number of carbon atoms; (*2*) number of hydrogen atoms; (*3*) other elements in alphabetic order. Thus, all C_3 compounds are listed before any C_4 compound; all C_5H_7 compounds before any C_5H_8 compound; $C_7H_{11}Br$ before $C_7H_{11}N$; $C_9H_6N_4S$ before C_9H_6O, and so on. Deuterium and tritium are represented by D and T and treated alphabetically, for example, C_2H_5DO after C_2H_5Cl and before C_2H_5F or C_2H_6.

Since 1965, *CA* has assigned a Registry Number to each unique chemical substance; CAS Registry Number®. This is a number of the form [*766-51-8*] that remains invariant, no matter what names are used in the literature. More than

10 million numbers have already been assigned and thousands are added each week. At one time, all numbers published with the *CA* preferred names appeared in a multivolume "Registry Handbook," but this work was discontinued. The CAS Registry contains >27 million organic and inorganic substances and >57 million sequences.[21] For a discussion of online searching see pp. 1901–1905.

There were a number of earlier abstracting publications now defunct. The most important are *Chemisches Zentralblatt* and *British Abstracts*. These publications are still valuable because they began before *CA* and can therefore supply abstracts for papers that appeared before 1907. Furthermore, even for papers published after 1907, *Zentralblatt* and *British Abstracts* are often more detailed. *Zentralblatt* was published, under various names, from 1830 to 1969.[22] *British Abstracts* was a separate publication from 1926 to 1953, but earlier abstracts from this source are available in the *Journal of the Chemical Society* from 1871 to 1925.

Beilstein

This publication is so important to organic chemistry that it deserves a section by itself. Beilstein's *Handbuch der Organischen Chemie*, usually referred to as *Beilstein*, lists all the known organic compounds reported in the literature during its period of coverage. We will first describe the print version, particularly important for older literature. This discussion will be followed by a description of the on-line version of Beilstein – Crossfire Beilstein. For each compound are given: all names; the molecular formula; the structural formula; all methods of preparation (briefly, e.g., "by refluxing 1-butanol with NaBr and sulfuric acid"); physical constants, such as melting point and refractive index; other physical properties; chemical properties including reactions; occurrence in nature (i.e., which species it was isolated from); biological properties, if any; derivatives with melting points; analytical data, and any other information that has been reported in the literature.[23] Equally important, for every piece of information, a reference is given to the original literature. Furthermore, the data in *Beilstein* have been critically evaluated. That is, all information is carefully researched and documented, and duplicate and erroneous results are eliminated. Some compounds are discussed in two or three lines and others require several pages. The value of such a work should be obvious.

The first three editions of *Beilstein* are obsolete. The fourth edition (*vierte Auflage*) covers the literature from its beginnings through 1909. This edition, called *das Hauptwerk*, consists of 27 volumes. The compounds are arranged in order of a

---

[21]This total is updated daily. See http://www.cas.org/cgi-bin/regreport.pl.
[22]An "obituary" of *Zentralblatt* by Weiske, which gives its history and statistical data about its abstracts and indexes, was published in the April 1973 issue of *Chem. Ber.* (pp. I–XVI).
[23]For a discussion of how data are processed for inclusion in *Beilstein*, see Luckenbach, R.; Ecker, R.; Sunkel, J. *Angew. Chem. Int. Ed.* **1981**, *20*, 841.

system too elaborate to discuss fully here.[24] The compounds are divided into three divisions that are further subdivided into "systems":

Division	Volumes	System Numbers
I. Acyclic Compounds	1–4	1–499
II. Carbocyclic Compounds	5–16	450–2359
III. Heterocyclic Compounds	17–27	2360–4720

Das Hauptwerk is still the basis of *Beilstein* and has not been superseded. The later literature is covered by supplements that have been arranged to parallel *das Hauptwerk*. The same system is used, so that the compounds are treated in the same order. The first supplement (*erstes Ergänzungswerk*) covers 1910–1919; the second supplement (*zweites Ergänzungwerk*) covers 1920–1929; the third supplement (*drittes Ergänzungswerk*) covers 1930–1949; the fourth supplement (*viertes Ergänzungswerk*) covers 1950–1959, and the fifth supplement covers 1960-1979. Like *das Hauptwerk*, each supplement contains 27 volumes,[25] except that supplements 3 and 4 are combined for Vols. 17–27, so that for these volumes the combined third and fourth supplement covers the years 1930–1959. Each supplement has been divided into volumes in the same way as *das Hauptwerk*, and, for example, compounds found in Vol. 3, system number 199 of *das Hauptwerk* will also be found in Vol. 3, system number 199 of each supplement. To make cross-referencing even easier, each supplement gives, for each compound, the page numbers at which the same compound can be found in the earlier books. Thus, on page 554 of Vol. 6 of the fourth supplement, under the listing phenetole are found the symbols (H 140; E I 80; E II 142; E III 545) indicating that earlier information on phenetole is given on page 140 of Vol. 6 of *das Hauptwerk*, on page 80 of the first, page 142 of the second, and page 545 of the third supplement. Furthermore, each page of the supplements contains, at the top center, the corresponding page numbers of *das Hauptwerk*. Since the same systematic order is followed in all six series, location of a compound in any one series gives its location in the other five. If a compound is found, for example, in Vol. 5 of *das Hauptwerk*, one has but to note the page number and scan Vol. 5 of each supplement until that number appears in the top center of the page (the same number often covers several pages). Of course, many compounds are found in only one, two, three, four, or five of the series, since no work may have been published on that compound during a

---

[24]For descriptions of the *Beilstein* system and directions for using it, see Sunkel, J.; Hoffmann, E.; Luckenbach, R. *J. Chem. Educ.* **1981**, *58*, 982; Luckenbach, R. *CHEMTECH* **1979**, 612. The Beilstein Institute has also published two English-language guides to the system. One, available free, is *How to Use Beilstein*; Beilstein Institute: Frankfurt/Main, *1979*. The other is by Weissbach, O. *A Manual for the Use of Beilstein's Handbuch der Organischen Chemie*, Springer, NY, *1976*. An older work, which many students will find easier to follow, is by Huntress, E.H. *A Brief Introduction to the Use of Beilstein's Handbuch der Organischen Chemie*, 2nd ed., Wiley, NY, *1938*.

[25]In some cases, to keep the system parallel and to avoid books that are too big or too small, volumes are issued in two or more parts, and, in other cases, two volumes are bound as one.

particular period covered. From *das Hauptwerk* to the fourth supplement, *Beilstein* is in German, though it is not difficult to read since most of the words are the names of compounds (a *Beilstein* German–English Dictionary, available free from the publisher, is in many libraries). For the fifth supplement (covering 1960–1979), which is in English, publication of Division III began before the earlier divisions. At the time of this writing, Vols. 17–22 (totaling 70 separate parts exclusive of index volumes) of this supplement have been published, as well as a combined index for Volumes 17–19. This index covers only the fifth supplement. The subject portion of this index, which lists compound names only, gives these names in English.

Volumes 28 and 29 of *Beilstein* are subject and formula indexes, respectively. The most recent complete edition of these volumes is part of the second supplement and covers only *das Hauptwerk* and the first two supplements (though complete indexes covering *das Hauptwerk* and the first four supplements have been announced to appear in the next few years). For Vol. 1, there is a cumulative subject and a cumulative formula index, which combine *das Hauptwerk* and the first four supplements.[26] Similar index volumes, covering all four supplements, have been issued for the other volumes, 2–27. Some of these are combined (e.g., 2–3, 12–14, and 23–25). For English-speaking chemists (and probably for many German-speaking chemists) the formula indexes are more convenient. Of course (except for the fifth supplement indexes), one must still know some German, because most formula listings contain the names of many isomers. If a compound is found only in *das Hauptwerk*, the index listing is merely the volume and page numbers (e.g., **1**, 501). Roman numbers are used to indicate the supplements, for example, **26**, 15, I 5, II 7. Thus the subject and formula indexes lead at once to locations in *das Hauptwerk* and the first four supplements. The *Beilstein* formula indexes are constructed the same way as the *CA* indexes (p. 1880).

There is also a fourth division of *Beilstein* (systems 4721–4877) that covers natural products of uncertain structure: rubbers, sugars, and so on. These are treated in Vols. 30 and 31, which do not go beyond 1935, and which are covered in the collective indexes. These volumes will not be updated. All such compounds are now included in the regular *Beilstein* volumes.

Like *CA*, *Beilstein* is available on-line, with the useful search engine *CrossFire Beilstein*. The *Beilstein* database is one of the best databases for organic chemistry, particularly synthetic organic chemistry. It contains information on ∼9.4 million organic substances that have been fully characterized, as well as 9.8 million reactions. Information available includes:[27]

- Structure and reaction diagrams
- Identification information (name, formula, registry numbers, etc.)

[26]Most page number entries in the combined indexes contain a letter, for example, $CHBr_2Cl$ 67f, II 33a, III 87d, IV, 81. These letters tell where on the page to find the compound and are useful because the names given in the index are not necessarily those used in the earlier series. The letter "a" means the compound is the first on its page, "b" is the second, and so on. No letters are given for the fourth supplement.
[27]http://gethelp.library.upenn.edu/guides/tutorials/scitech/beilstein/BCWhat.html

- Isolation and purification information
- Derivatives
- Physical properties, including
- Structure and energy parameters
- State of matter and change of state information
- Transport phenomena
- Optical properties
- Spectra (UV, VIS, MS, NMR, IR, ESR, X-ray, etc.)
- Electrical and magnetic properties
- Electrochemistry
- Behavior of liquid/solid, liquid/liquid, and liquid/vapor systems
- Solution behavior
- Energy data
- Optical data
- Boundary surface phenomena
- Adsorption
- Association
- Pharmacological and ecological data

Not all compounds contain all different types of data. Some data, often those quantities that can be described by a number, word, or short phrase, are given explicitly in the record for the compound. Others simply have a literature reference. The factual data, of which there are 260 million, are only experimental data, scientifically measured.

Coverage of the Literature and Search Capabilities

Beilstein references the chemical literature published in ~175 prestigious organic chemistry journals[28] from 1771 to date. It is updated quarterly. The database may be searched using structures and/or text (property information, chemical name, molecular formula)." Figure A. 1 shows the on-line page after a structure has been drawn using the drawing tools provided with the program, used by Crossfire to begin a session. Figure A. 2 shows an on-line page that displays data obtained after a structure search. The variety of tools available for searching is much more extensive than the ones shown in Figs. A.1 and A.2. One can use the structure editor to draw a specific structure, and it is possible to draw two or more structures, label them as reactant and/or product and search for a chemical reactions or transformation. A variety of data can be accessed for a compound or a reaction, and the output includes the chemical structures, references to the primary literature, physical, and spectral properties where available.

[28]See http://www.mdl.com/products/pdfs/BSJournals.pdf.

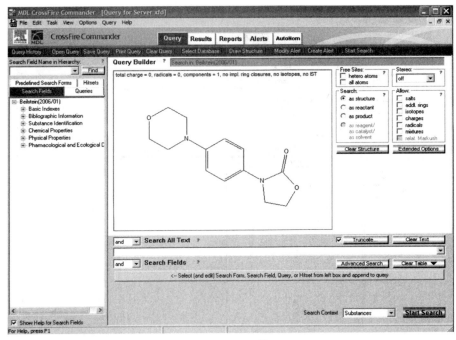

Fig. A.1. Structure Search for CrossFire Beilstein[29] Beilstein Data: Copyright (c) 1988–2006, Beilstein Institut zur Foerderung der Chemischen Wissenschaften licensed to Beilstein GmbH and MDL Information Systems GmbH. All rights reserved.

Tables of Information

In addition to *Beilstein*, there are many other reference works in organic chemistry that are essentially compilations of data. These books are very useful and often save the research worker a great deal of time. In this section we discuss some of the more important of such works.

1. The sixth edition of *Heilbron's Dictionary of Organic Compounds*, J. Buckingham, Ed., 9 vols., Chapman and Hall, London, *1996*, contains brief listings of >150,000 organic compounds, giving names, structural formulas, physical properties, and derivatives, with references. For many entries additional data concerning occurrence, biological activity, and toxicity hazard information are also given. The arrangement is alphabetical. The dictionary contains indexes of names, formulas, heteroatoms, and *CA* Registry Numbers. Annual supplements, with cumulative indexes, have appeared since 1983. A similar work, devoted to organometallic compounds, is the 2nd edition of the *Dictionary of Organometallic Compounds*, 6 vols. in its 5th supplement,

---

[29]See http://www.mdl.com/products/knowledge/crossfire_commander/.

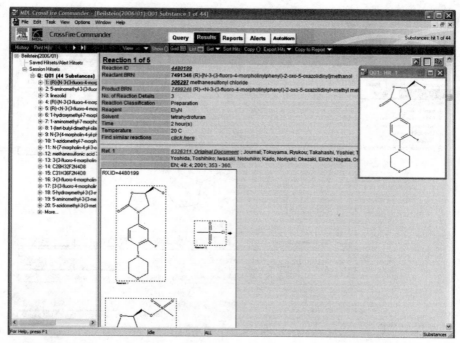

Fig. A.2. Preparation Data from CrossFire Beilstein.[30] Beilstein Data: Copyright (c) 1988–2006, Beilstein Institut zur Foerderung der Chemischen Wissenschaften licensed to Beilstein GmbH and MDL Information Systems GmbH. All rights reserved.

published by Chapman and Hall in *1989*. Another, *Dictionary of Steroids*, 2 vols., 1991, is also published by Chapman and Hall.

2. A multivolume compendium of physical data is Landolt-Börnstein's *Zahlenwerte und Funktionen aus Physik, Chemie, Astronomie, Geophysik, und Technik*, 6th ed., Springer, Berlin, **1950**–. There is also a "New Series," for which the volumes are given the English title *Numerical Data and Functional Relationships in Science and Technology*, as well as the German title. This compendium, which is not yet complete, lists a great deal of data, some of which are of interest to organic chemists (e.g., indexes of refraction, heats of combustion, optical rotations, and spectral data). Literature references are given for all data.

3. *The Handbook of Chemistry and Physics*, CRC Press, Boca Raton, FL (called the "rubber handbook"), which is revised annually (87th ed., *2006–2007*), is a valuable repository of data quickly found. For organic chemists the most important table is "Physical Constants of Organic Compounds,"

[30]See http://www.mdl.com/products/knowledge/crossfire_beilstein/index.jsp. Also see, http://chemistry.library.wisc.edu/beilstein/quickguide6.htm

which lists names, formulas, color, solubilities, and physical properties of thousands of compounds. However, there are many other useful tables. A similar work is *Lange's Handbook of Chemistry*, 16th ed., McGraw-Hill, New York, *2004*. Another such handbook, but restricted to data of interest to organic chemists, is *Dean's Handbook of Organic Chemistry*, 2nd ed., McGraw-Hill, New York, *2003*. This book also contains a long table of "Physical Constants of Organic Compounds," and has much other information including tables of thermodynamic properties, spectral peaks, pK_a values, bond distances, and dipole moments.

4. A list of most of the known natural compounds (e.g., terpenes, alkaloids, carbohydrates), to which structures have been assigned, along with structural formulas, melting points, optical rotations, and references, is provided in Devon and Scott, *Handbook of Naturally Occurring Compounds*, 3 vols., Academic Press, New York, *1972*.

5. Dreisbach, *Physical Properties of Chemical Compounds*, Advances in Chemistry Series Nos. 15, 22, 29, American Chemical Society, Washington, DC, *1955–1961* lists many physical properties of >1000 organic compounds.

6. Physical properties of thousands of organometallic compounds, with references, are collected in five large compendia: the *Dictionary of Organometallic Compounds*, mentioned under item 1, above; Dub, *Organometallic Compounds*, 2nd ed., 3 vols. with supplements and index, Springer, New York, *1966–1975*; Hagihara, Kumada, and Okawara, *Handbook of Organometallic Compounds*, W. A. Benjamin, New York, 1968; and Kaufman, *Handbook of Organometallic Compounds*, Van Nostrand, Princeton, NJ, 1961; *Comprehensive Organometallic Chemistry II*, 14 Vols, Pergamon, 1995.

7. The *Merck Index*, 13th ed., Merck and Company, Rahway, NJ, *2001* is a good source of information about chemicals of medicinal importance. Many drugs are given three types of name: chemical *name* (which is the name an organic chemist would give it; of course, there may well be more than one); *generic name*, which must be placed on all containers of the drug; and *trade names,* which are different for each company that markets the drug. For example, the generic name for 1-(4-chlorobenzhydryl)-4-methylpiperazine is chlorcyclazine. Among the trade names for this drug, which is an antihistamine, are Trihistan, Perazyl, and Alergicide. The *Merck Index* is especially valuable because it gives all known names of all three types for each compound and the names are cross-indexed. Also given, for each compound, are the structural formula, *CA* preferred name and Registry Number, physical properties, medicinal and other uses, toxicity indications, and references to methods of synthesis. There are indexes of formulas and Registry Numbers, and miscellaneous tables. The 10th edition of the *Merck Index* (1983) also includes a lengthy list of organic name reactions, with references. Although the 11th edition omitted this list, it was restored in the 12th and subsequent editions. Note that the Merck Index Online™ is also available.

8. There are two publications that list properties of azeotropic mixtures. Timmermans, *The Physicochemical Constants of Binary Systems in Concentrated Solutions*, 4 vols., Interscience, New York, *1959–1960*, is by far the more comprehensive. The other is *Azeotropic Data*, 2 vols., Advances in Chemistry Series Nos. 6 and 35, American chemical Society, Washington, DC, *1952*, *1962*.

9. Thousands of dipole moments, with references, are collected in McClellan, *Tables of Experimental Dipole Moments*, Vol. 1, W.H. Freeman, San Francisco, CA, *1963*; Vol. 2, Rahara Enterprises, El Cerrita, CA, *1974*.

10. *Tables of Interatomic Distances and Configurations in Molecules and Ions*, London chemical Society Special publication No. 11, 1958, and its supplement, Special publication No. 18, *1965*, include bond distances and angles for hundreds of compounds, along with references.

11. The *Ring Systems Handbook*, published in *1988* by the Chemical Abstracts Service, provides the names and formulas of ring and cage systems that have been published in *CA*. The ring systems are listed under a system essentially the same as that used for the *CA* index of ring systems (p. $$$). Each entry gives the *CA* index name and Registry Number for that ring system. In many cases, a *CA* reference is also given. There is a separate Formula Index (for the parent ring systems) and a Ring Name Index. Cumulative supplements are issued twice a year. The *Ring Systems Handbook* supersedes earlier publications called *The Parent Compound Handbook* and *The Ring Index*.

12. The Sadtler Research Laboratories publish large collections of IR, UV, NMR and other spectra, in loose-leaf form. Indexes are available.

13. Infrared, UV, NMR, Raman, and mass spectral data, as well as melting-point, boiling-point, solubility, density, and other data for >30,000 organic compounds are collected in the *CRC Handbook of Data on Organic Compounds*, 2nd ed., 9 vols., CRC Press, Boca Raton, FL, *1988*, edited by Weast and Grasselli. It differs from the Sadtler collection in that the data are given in tabular form (lists of peaks) rather than reproduction of the actual spectra, but this book has the advantage that all the spectral and physical data for a given compound appear at one place. References are given to the Sadtler and other collections of spectra. Volumes 7–9 contain indexes of spectral peaks for IR, UV, ^{1}H NMR, ^{13}C NMR, mass, and Raman spectra, as well as indexes of other names, molecular formulas, molecular weights, and physical constants. Annual updates began appearing in *1990* (the first one is called Vol. 10).

14. The *Aldrich Library of Infrared Spectra*, 3rd ed., Aldrich chemical Company, Milwaukee, WI, *1981*, by Pouchert contains >12,000 IR spectra so arranged that the user can readily see the change that takes place in a given spectrum when a slight change is made in the structure of a molecule. The same company also publishes the *Aldrich Library of FT–IR Spectra* and the *Aldrich Library of NMR Spectra*, both also by Pouchert. A similar volume, which has IR and Raman spectra of ~1000 compounds, is *Raman/Infrared Atlas of Organic Compounds*, 2nd ed., VCH, New York, *1989*, by Schrader.

15. An extensive list of visible and UV peaks is given in *Organic Electronic Spectral Data*, Wiley, New York. Twenty-six volumes have appeared so far, covering the literature through *1984*.

16. A collection of 500 ^{13}C NMR spectra is found in Johnson and Jankowski, *Carbon-13 NMR Spectra*, Wiley, New York, *1972*.

Reviews

A review article is an intensive survey of a rather narrow field; for example, the titles of some recent reviews are "Desulfonation Reactions: Recent Developments,"[31] "Pyrrolizidine and Indolizidine Syntheses Involving 1,3-Dipolar Cycloaddition,"[32] and "From Corrin Chemistry to Asymmetry Catalysis—A Personal Account."[33] A good review article is of enormous value, because it is a thorough survey of all the work done in the field under discussion. Review articles are printed in review journals and in certain books. The most important review journals in organic chemistry (though most are not exclusively devoted to organic chemistry) are shown in Table A.3. Some of the journals listed in Table A.1, for

TABLE A.3. Review Journals, With Year of Founding and Issues per Year as of 1998

Journal	Issues
Accounts of Chemical Research (1968)	12
Aldrichimica Acta (1968)	4
Angewandte Chemie (1888) and its English Translation:	12
Angewandte Chemie, International Edition (1962)	12
Chemical Reviews (1924)	8
Chemical Society Reviews (1947)[34]	4
Heterocycles (1973)	12
Natural Product Reports (1984)	6
Organic Preparations and Procedures International (1969)	6
Soviet Scientific Reviews, Section B, Chemistry Reviews (1979) Irreg.	
Sulfur Reports (1980)	6
Synlett (1989)	12
Synthesis (1969)	12
Tetrahedron (1958)	52
Topics in Current Chemistry (1949)[35]	Irreg.
Uspekhi Khimii (1932)	12
and its English translation: Russian chemical Reviews (1960)	12

[31]Nájera, C.; Yus, M. *Tetrahedron* *1999*, *55*, 10547.
[32]Broggini, G.; Zecchi, G. *Synthesis* *1999*, 905.
[33]Pfaltz, A. *Synlett* *1999*, 835.
[34]Successor to *Quarterly Reviews* (abbreviated as *Q. Rev., Chem. Soc.*).
[35]Formerly called *Fortschritte der Chemischen Forschung*.

example, *Synlett, Tetrahedron, Synthesis, Organic Preparations and Proce-dures International,* and *J. Organomet. Chem.* also publish occasional review articles.

There are several open-ended serial publications that are similar in content to the review journals, but are published irregularly (seldom more often than once a year) and are hardbound. Some of these publish reviews in all fields of chemistry; some cover only organic chemistry; some specialize further. The coverage is indicated by the titles. Table A.4 shows some of the more important such publications, with *CA* abbreviations.

There are several publications that provide listings of review articles in organic chemistry. The most important is the *J. Org. Chem.,* which began to list review arti-cles in 1978 (the first list is at *J. Org. Chem., 43,* 3085), suspended the listings in 1985, and resumed them in 1990 (at *J. Org. Chem., 55,* 398). These lists, which appear about four times a year, give the titles and reference sources of virtually

TABLE A.4. Irregularly Published Serial Publications

Advances in **Carbo**cation **Chem**istry
Advances in **Carbohydr**ate **Chem**istry and **Biochem**istry
Advances in **Cata**lysis
Advances in **Cycloadd**ition
Advances in **Free Radical Chem**istry
Advances in **Heterocyclic Chem**istry
Advances in **Metal-Organic Chem**istry
Advances in **Molecular Mod**eling
Advances in **Organometallic Chem**istry
Advances in **Oxygenated Processes**
Advances in **Photochem**istry
Advances in physical **Organic Chem**istry
Advances in **Protein Chem**istry
Advances in **Theor**etically **Interesting Mol**ecules
Fluorine Chemistry Reviews
Fortshritte der Chemie Organischer Naturstoffe
Isotopes in **Organic Chem**istry
Molecular Structure and **Energetics**
Organic Photochemistry
Organometallic Reactions
Organic Reactions
Organic Synthesis: Theory and **Appl**ications
Progress in Heterocyclic Chemistry
Progress in Macrocyclic Chemistry
Progress in physical **Organic Chem**istry
Reactive Intermediates (Plenum)
Reactive **Intermediates** (Wiley)
Survey of **Progress in Chem**istry
Topics in physical **Organometallic Chem**istry
Topics in Stereochemistry

all review articles in the field of organic chemistry that have appeared in the preceding 3 months, including those in the review journals and serials mentioned above, as well as those in monographs and treatises. There is also a listing of new monographs on a single subject. Each list includes a subject index.

Another publication is the *Index of Reviews in Organic Chemistry*, complied by Lewis, Chemical Society, London, a classified listing of review articles. The first volume, published in 1971, lists reviews from ~1960 (in some cases much earlier) to ≈ 1970 in alphabetical order of topic. Thus four reviews are listed under "Knoevenagel condensation," five under "Inclusion compounds," and one under "Vinyl ketones." There is no index. A second volume (1977) covers the literature to 1976. Annual or biannual supplements appeared from 1979 until the publication was terminated in 1985. Classified lists of review articles on organometallic chemistry are found in articles by Smith and Walton[36] and by Bruce.[37] A similar list for heterocyclic chemistry is found in articles by Katritzky and others.[38] See also the discussion of the *Index of Scientific Reviews*, p. 1908.

Annual Reviews

The review articles discussed in the previous section are each devoted to a narrow topic covering the work done in that area over a period of years. An annual review is a publication that covers a broad area, but limits the period covered, usually to 1 or 2 years.

1. The oldest annual review publication still publishing is *Annual Reports on the Progress of Chemistry*, published by the Royal Society of Chemistry (formerly the chemical Society), which began in 1905 and which covers the whole field of chemistry. Since 1967 it has been divided into sections. Organic chemistry is found in Section B.

2. Because the number of papers in chemistry has become so large, the Royal Society of Chemistry publishes annual-review-type volumes of smaller scope, called *Specialist Periodical Reports*. Among those of interest to organic chemists are "Carbohydrate Chemistry" (Vol. 22 covers 1988); "Photochemistry" (Vol. 21 covers 1988–1989); and "General and Synthetic Methods," (Vol. 12 covers 1987).

3. *Organic Reaction Mechanisms*, published by Wiley, New York, is an annual survey that covers the latest developments in the field of mechanisms. The first volume, covering 1965, appeared in 1966.

4. There are two annual reviews devoted to progress in organic synthesis. Theilheimer, *Synthetic Methods of Organic Chemistry*, S. Karger Verlag,

[36]Smith, J.D.; Walton, D.R.M. *Adv. Organomet. Chem.* **1975**, *13*, 453.

[37]Bruce, M.IK. *Adv. Organomet. Chem.* **1972**, *10*, 273, **1973**, *11*, 447, **1974**, *12*, 380.

[38]Belen'kii, L.I. *Adv. Heterocyclic Chem.* **1988**, *44*, 269; Katritzky, A.R.; Jones, P.M. *Adv. Heterocyclic Chem.* **1979**, *25*, 303; Katritzky, A.R.; Weeds, S.M. *Adv. Heterocyclic Chem.* **1966**, *7*, 225.

Basel, is an annual compilation, beginning in 1946, of new methods for the synthesis of organic compounds, arranged according to a system based on bond closing and bond breaking reactions. Equations, brief procedures, yields, and literature references are given. Volume 44 was issued in 1990. Volumes 3 and 4 are available only in German, but all the rest are in English. There is an index to each volume. Cumulative indexes appear in every fifth volume. Beginning with Vol. 8, each volume includes a short summary of trends in synthetic organic chemistry. A more recent series is *Annual Reports in Organic Synthesis*, Academic Press, New York, which has covered the literature of each year since 1970. Equations are listed with yields and references according to a fairly simple system.

5. The *Journal Of Organometallic Chemistry* several times a year publishes annual surveys arranged according to metallic element. For example, Vol. 404, published in February 1991, contains annual surveys for 1989 of organic compounds containing Sb, Bi, and Fe, and the use of transition metals in organic synthesis, and surveys for 1988 covering B, Ru, and Os.

Awareness Services

Besides the annual reviews and the title and abstract services previously mentioned, there exist a number of publications designed to keep readers aware of new developments in organic chemistry or in specific areas of it.

1. *Chemtracts: Organic Chemistry* is a bimonthly periodical, begun in 1988, that prints abstracts of certain recently published papers (those that the editors consider most important), with commentaries on these papers by distinguished organic chemists. Important current research in bioorganic, organometallic, synthesis, physical-organic and theoretical chemistry, and pharmaceutical–medicinal chemistry is covered in each issue, giving readers updates on the newest trends and developments in organic chemistry by summarizing and commenting on current and past research..

2. The Institute for Scientific Information (ISI), besides publishing *Current Contents* (p. 1877) and the *Science Citation Index* (p. 1877), also publishes *Index Chemicus* (formerly called *Current Abstracts of Chemistry and Index Chemicus*). This publication, begun in 1960 and appearing weekly, is devoted to printing structural formulas of all new compounds appearing in >100 journals, along with equations to show how they were synthesized and an author's summary of the work. Each issue contains five indexes: author, journal, biological activity, labeled compounds, and intermediates that were not isolated. These indexes are cumulated annually.

3. Theilheimer and the *Annual Reports on Organic Synthesis*, mentioned in the previous section, list new synthetic methods once a year. There are several publications that do this monthly. Among these are *Current Chemical Reactions* (begun in 1979 and published by ISI), *Journal of Synthetic*

Methods (begun in 1975 and published by Derwent publications), and *Methods in Organic Synthesis*, begun in 1984 and published by the Royal Society of Chemistry. *Methods in Organic Synthesis* also lists books and review articles pertaining to organic synthesis.

4. *Natural Product Updates*, a monthly publication begun in 1987 and published by the Royal Society of Chemistry, lists recent results in the chemistry of natural products, along with structural formulas. It covers new compounds, structure determinations, new properties and total syntheses, among other topics.

General Treatises

There are a number of large-scale multivolume treatises that cover the whole field of organic chemistry or large areas of it.

1. *Rodd's Chemistry of Carbon Compounds*, edited by Coffey, Elsevier, Amsterdam, The Netherlands, is a treatise consisting of five main volumes, each of which contains several parts. publication began in 1964 and is not yet complete. The organization is not greatly different from most textbooks, but the coverage is much broader and deeper. Supplements to many of the volumes have appeared. An earlier edition, called *Chemistry of Carbon Compounds*, edited by Rodd, was published in 10 parts from 1951 to 1962.

2. Houben–Weyl's, *Methoden der Organischen Chemie*, Georg Thieme Verlag, Stuttgart, is a major treatise in German devoted to laboratory methods. The fourth edition, which was begun in 1952 and consists of 20 volumes, most of them in several parts, is edited by E. Muller. The series includes supplementary volumes. The first four volumes contain general laboratory methods, analytical methods, physical methods, and general chemical methods. The later volumes are devoted to the synthesis of specific types of compounds, for example, hydrocarbons, oxygen compounds, and nitrogen compounds. Beginning in 1990 parts of the series have appeared in English.

3. *Comprehensive Organic Chemistry*, Pergamon, Elmsford, NY, *1979*, is a six-volume treatise on the synthesis and reactions of organic compounds. The first three volumes cover the various functional groups, Vol. 4, heterocyclic compounds, and Vol. 5, biological compounds, such as proteins, carbohydrates, and lipids. Probably the most useful volume is Vol. 6, which contains formula, subject, and author indexes, as well as indexes of reactions and reagents. The last two of these not only refer to pages within the treatise, but directly give references to review articles and original papers. For example, on p. 1129, under "Chromic acid-sulphuric acid (Jones reagent), oxidation, alcohols," are listed 13 references to original papers. Several similar treatises, including the nine-volume *Comprehensive Organometallic Chemistry* (1982), the eight volume *Comprehensive Heterocyclic Chemistry* (1984), and the six volume *Comprehensive Medicinal Chemistry* (1989) are also published by Pergamon. The indexes to these works also include references.

4. A major treatise devoted to experimental methods of chemistry is *Techniques of Chemistry*, edited first by Weissberger and then by Saunders, Wiley, New York. This publication, which began in 1970, so far consists of 21 volumes, most of them in several parts, covering such topics as electrochemical and spectral methods, kinetic methods, photochromism, and organic solvents. *Techniques of Chemistry* is a successor to an earlier series, called *Techniques of Organic Chemistry*, which appeared in 14 volumes, some of them in more than one edition, from 1945 to 1969.

5. *Comprehensive Chemical Kinetics*, edited by Bamford and Tipper, 1969–, Elsevier, Amsterdam, The Netherlands, is a multivolume treatise covering the area of reaction kinetics. Six of these volumes (not all published at the time of writing) deal with the kinetics and mechanisms of organic reactions in a thorough and comprehensive manner.

6. Three multivolume treatises that cover specific areas are Elderfield, *Heterocyclic Compounds*, Wiley, New York, 1950-; Manske and Holmes, *The Alkaloids*, Academic Press, New York, 1950-; and Simonson, Owen, Barton, and Ross, *The Terpenes*, Cambridge University Press, London, 1947–1957.

7. *Encyclopedia of Reagents for Organic Synthesis*, edited by Paquette, Wiley, New York, was published in 1995. It is an eight volume, alphabetic listing of reagents used in organic chemistry with descriptions of the preparation, use and chemistry, with references. Each reagent was researched by organic chemists active in research, who contributed to the total publication.

This work is available on-line as **eEROS**. The *Encyclopedia of Reagents for Organic Synthesis*, e-EROS, provides updated information on ~3800 reagents with a database of close to 50,000 reactions. Each reagent entry includes properties such as physical data, solubility, form supplied in, purification, and preparative methods; examples of use in reactions; and literature references. Search options include: name, CAS number, structure and reaction.

8. *Comprehensive Organic Synthesis*, edited by Trost and Fleming, Pergamon, was published in 1991. It is a nine volume compilation.

9. *Comprehensive Organic Functional Group Transformations*, edited by Katritzky, Meth-Cohn, and Rees, Pergamon, was published in 1995. It is a seven volume compilation.

Monographs and Treatises on Specific Areas

Organic chemistry is blessed with a large number of books devoted to a thorough coverage of a specific area. Many of these are essentially very long review articles, differing from ordinary review articles only in size and scope. Some of the books are by a single author, and others have chapters by different authors but all are carefully planned to cover a specific area. Many of these books have been referred to in footnotes in appropriate places in this book. There have been several series of monographs, one of which is worth special mention: *The Chemistry of Functional Groups*, under the general editorship of Patai, published by

Wiley, New York. Each volume deals with the preparation, reactions, and physical and chemical properties of compounds containing a given functional group. Volumes covering >20 functional groups have appeared so far, including books on alkenes, cyano compounds, amines, carboxylic acids and esters, and quinones.

Textbooks

There are many excellent textbooks in the field of organic chemistry. We restrict ourselves to listing only a few of those published, mostly since 1985. Some of these are first-year texts and some are advanced (advanced texts generally give references; first-year texts do not, though they may give general bibliographies, suggestions for further reading, etc.); some cover the whole field, and others cover reactions, structure, and/or mechanism only. All the books listed here are not only good textbooks and the advanced books are valuable reference books for graduate students and practicing chemists.

Bruckner, *Advanced Organic Chemistry: Reaction Mechanisms*, Academic Press, NY *2001*.

Bruice, *Organic Chemistry*, 4th ed., Prentice-Hall, NJ, *2004*.

Carey, *Organic Chemistry*, 6th ed., McGraw-Hill, NY, *2006*.

Carey and Sundberg, *Advanced Organic Chemistry: Structure and Mechanisms (Part A)*, 4th ed., Springer, *2004*.

Carey and Sundberg, *Advanced Organic Chemistry: Structure and Mechanisms (Part B)*, 4th ed., Springer, *2001*.

Carruthers and Coldham, *Some Modern Methods of Organic Synthesis*, 4th ed., Cambridge University Press, Cambridge, *2004*.

Ege, *Organic Chemistry: Structure and Reactivity*, 5th ed., D.C. Houghton Mifflin, Boston, *2003*.

Fox and Whitesell, *Organic Chemistry*, 3rd ed, Jones and Bartlett, Sudbury, MA, *2004*.

Grossman, *The Art of Writing Reasonable Organic Reaction Mechanisms*, 2nd ed, Springer, *2005*.

House, *Modern Synthetic Reactions*, 2nd ed., W. A. Benjamin, New York, *1972*.

Ingold, *Structure and Mechanism in Organic Chemistry*, 2nd ed., Cornell University Press, Ithaca, NY, *1969*.

Isaacs, *Physical Organic Chemistry*, Wiley, NY, *1987*.

Jones, *Organic Chemistry*, 3rd ed. W.W. Norton, NY, *2004*.

Loudon, *Organic Chemistry*, 4th ed., Oxford University Press, Cambridge, *2001*.

Lowry and Richardson, *Mechanism and Theory in Organic Chemistry*, 3rd ed., Harper and Row, New York, *1987*.

McMurry, *Organic Chemistry*, 6th ed., Brooks/Cole, Monterey CA, *2003*.

Maskill, *The Physical Basis of Organic Chemistry*, Oxford University Press, Oxford, *1985*.

Morrison, Boyd and Boyd, *Organic Chemistry*, 6th ed., Benjamin Cummings, CA, *1992*.

Mundy, Ellerd, and Favaloro, Jr., *Name Reactions and Reagents in Organic Synthesis*, 2nd ed, Wiley, Hoboken, ND, *2005*.

Ritchie, *Physical Organic Chemistry*, 2nd ed., Marcel Dekker, NY, *1989*.

Solomons and Fryhle, *Organic Chemistry*, 8th ed., Wiley, NY, *2003*.

Smith, *Organic Synthesis*, 2nd edition McGraw-Hill, NY, *2000*.

Streitwieser, Heathcock, and Kosower, *Introductory Organic Chemistry*, 4th ed., Prentice-Hall, Saddle River, NJ, *1998*.

Sykes, *A Guidebook to Mechanism in Organic Chemistry*, 6th ed., Longmans Scientific and Technical, Essex, *1986*.

Vollhardt and Schore, *Organic Chemistry*, 4th ed., W.H. Freeman, New York, *2002*.

Wade, *Organic Chemistry*, 4th ed., Prentice-Hall, Upper Saddle River, NJ, *1999*.

Other Books

In this section, we mention several books that do not fit conveniently into the previous categories. All but the last have to do with laboratory synthesis.

1. *Organic Syntheses*, published by Wiley, New York is a collection of procedures for the preparation of specific compounds. The thin annual volumes have appeared each year since 1921. For the first 59 volumes, the procedures for each 10- (or 9-) year period are collected in cumulative volumes. Beginning with Vol. 60, the cumulative volumes cover five-year periods. The cumulative volumes published so far are

Annual Volumes	Collective Volumes
1–9	I
10–19	II
20–29	III
30–39	IV
40–49	V
50–59	VI
60–64	VII
65–69	VIII
70–74	IX
75–80	X

The advantage of the procedures in *Organic Syntheses*, compared with those found in original journals, is that these procedures are *tested*. Each preparation is carried out first by its author, and then by a member of the *Organic Syntheses* editorial board, and only if the yield is essentially duplicated is the procedure published. While it is possible to repeat most procedures given in journals, this is not always the case. All *Organic Syntheses* preparations are noted in *Beilstein* and in *CA*. In order to locate a given reaction in *Organic Syntheses*, the reader may use the OS references given in the present volume (through OS **69**); the indexes in *Organic Syntheses* itself; Shriner and Shriner, *Organic Syntheses* Collective Volumes I, II, III, IV, V *Cumulative Indices*, Wiley, New York, 1976, or Sugasawa and Nakai; Reaction Index of *Organic Syntheses*, Wiley, New York, 1967 (through OS **45**). Another book classifies virtually all the reactions in *Organic Syntheses* (collective vols. I–VII and annual vols. 65–68) into 11 categories: annulation, rearrangement, oxidation, reduction, addition, elimination, substitution, C—C bond formation, cleavage, protection–deprotection, and miscellaneous. This is *Organic Syntheses: Reaction Guide*, by Liotta and Volmer, published by Wiley, New York, in 1991. Some of the categories are subdivided further, and some reactions are listed in more than one category. What is given under each entry are the equation and the volume and page reference to *Organic Syntheses*.

2. Volume 1 of *Reagents for Organic Synthesis*, by Fieser and Fieser, Wiley, New York, 1967, is a 1457–page volume that discusses, in separate sections, some 1120 reagents and catalysts. It tells how each reagent is used in organic synthesis (with references) and, for each, tells which companies sell it, or how to prepare it, or both. The listing is alphabetical. Eighteen additional volumes have so far been published, which continue the format of Vol. 1 and add more recent material. A cumulative index for Vols. 1–12, by Smith and Fieser, was published in 1990. A complete cumulative index for Vol. 1–22 was published in 2005, by M.B. Smith. The series included Vol. 1–18 with Mary Fieser. After the death of Mary Fieser, the series was resumed by T. -L. Ho, and now includes Vol. 19–22.

3. *Comprehensive Organic Transformations*, 2nd ed. by Larock, Wiley-VCH, New York, *1999*, has been frequently referred to in footnotes in Part 2 of this book. This compendium is devoted to listings of methods for the conversion of one functional group into another, and covers the literature through 1987. It is divided into nine sections covering the preparation of alkanes and arenes, alkenes, alkynes, halides, amines, ethers, alcohols and phenols, aldehydes and ketones, and nitriles, carboxylic acids and derivatives. Within each section are given many methods for synthesizing the given type of compound, arranged in a logical system. A schematic equation is given for each method, and then a list of references (without author names, to save space) for locating examples of the use of that method. When different reagents are used for the same functional group transformation, the particular reagent is shown for each reference. There is a 164-page index of group transformations. The 2nd edition has only recently been published and is *not* referenced in this edition, and a CD ROM version is now available.

4. *Survey of Organic Synthesis*, by Buehler and Pearson, Wiley, New York, 2 vols., *1970, 1977*, discusses hundreds of reactions used to prepare the principal types of organic compounds. The arrangement is by chapters, each covering a functional group, (e.g., ketones, acyl halides, amines). Each reaction is thoroughly discussed and brief synthetic procedures are given. There are many references.

5. A similar publication is Sandler and Karo, *Organic Functional Group Preparations*, 2nd ed., 3 vols., Academic Press, New York, *1983–1989*. This publication covers more functional groups than Buehler and Pearson.

6. *Compendium of Organic Synthetic Methods*, Wiley, New York, contains equations describing the preparation of thousands of monofunctional and difunctional compounds with references. Eleven volumes have been published so far (vol. 1–2, edited by Harrison and Harrison; Vol. 3, edited by Hegedus and Wade; Vol. 4–5, edited by Wade; Vol. 6–11 edited by Smith). Volume 12 will appear in 2007.

7. *The Vocabulary of Organic Chemistry*, by Orchin, Kaplan, Macomber, Wilson, and Zimmer, Wiley, New York, *1980*, presents definitions of >1000 terms used in many branches of organic chemistry, including stereochemistry, thermodynamics, wave mechanics, natural products, and fossil fuels. There are also lists of classes of organic compounds, types of mechanism, and name reactions (with mechanisms). The arrangement is topical rather than alphabetical, but there is a good index. *Compendium of Chemical Terminology*, by Gold, Loening, McNaught, and Sehmi (the "Gold book"), published by Blackwell Scientific publications, Oxford, in 1987, is an official IUPAC list of definitions of terms in several areas of chemistry, including organic.

LITERATURE SEARCHING

Until recently searching the chemical literature meant looking only at printed materials (some of which might be on microfilm or microfiche). Now, however, much of the literature can be searched online, including some of the most important. Whether the search is online or uses only the printed material, there are two basic types of search, (*1*) searches for information about one or more specific compounds or classes of compounds, and (*2*) other types of searches. First, we will discuss searches using only printed materials, and then on-line searching.[39]

Literature Searching Using Printed Materials

Searching for Specific Compounds. Organic chemists often need to know if a compound has ever been prepared and if so, how, and/or they may be seeking a melting point, an ir spectrum, or some other property. Someone who wants all

[39]For a monograph that covers both online searching and searching using printed materials, see Wiggins, G. *Chemical Information Sources*, McGraw-Hill: NY, *1991*.

the information that has ever been published on any compound begins by consulting the formula indexes in *Beilstein* (p. 1883). At this time there are two ways to do this. (*1*) The formula index to the second supplement (Vol. 29, see p. $$$) will quickly show whether the compound is mentioned in the literature through 1929. If it is there, the searcher turns to the pages indicated, where all methods used to prepare the compound are given, as well as all physical properties, with references. Use of the page heading method described on p. 1882 will then show the locations, if any, in the third and later supplements. (*2*) If one has an idea which volume of *Beilstein* the compound is in (and the tables of contents at the front of the volumes may help), one may search the cumulative index for that volume. If not sure, one may consult several indexes. One of these two procedures will locate all compounds mentioned in the literature through 1959. If the compound is heterocyclic, it may be in the fifth supplement. If it is in Vols. 17–19 (or in a later volume whose index has been published), the corresponding indexes may be consulted. If not, the page heading method will find it, if it was reported before 1960.[40] There is a way by which all of the above can be avoided. A computer program, called SANDRA (available from the *Beilstein* publisher), allows the user to find the *Beilstein* location by using a mouse to draw the structural formula of the compound sought. At this point, the investigator will know (*1*) all information published through 1959 or 1979,[34] or (*2*) that the compound is not mentioned in the literature through 1959 or 1979.[41] In some cases, scrutiny of *Beilstein* will be sufficient, perhaps if only a boiling point or a refractive index is required. In other cases, especially where specific laboratory directions are needed, the investigator will have to turn to the original papers.

To carry the search past 1959 (or 1979), the chemist next turns to the collective formula indexes of *Chemical Abstracts*[TM]: 1957–1961; 1962–1966; 1967–1971; 1972–1976; 1977–1981; 1982–1986; 1987–1991; 1992–1996; 1997–2002, and such later collective indexes as have appeared; and the semiannual indexes thereafter. If a given formula index contains only a few references to the compound in question, the pages or abstract numbers will be given directly in the formula index. However, if there are many references, the reader will be directed to see the chemical substance index or (before 1972) the subject index for the same period; and here the number of page or abstract numbers may be very large indeed. Fortunately, numerous subheadings are given, and these often help the user to narrow the search to the more promising entries. Nevertheless, one will undoubtedly turn to many abstracts that do not prove to be helpful. In many cases, the information in the abstracts will be sufficient. If not, the original references must be consulted. In some cases (the index entry is marked by an asterisk or a double asterisk), the

---

[40]Compounds newly reported in the fifth supplement that are in a volume whose index has not yet been published will not be found by this procedure. To find them in *Beilstein* it is necessary to know something about the system (see Ref. 25), but they may also be found by consulting *CA* indexes beginning with the sixth collective index, or by using *Beilstein* on-line.

[41]For those heterocyclic compounds that would naturally belong to a volume for which the fifth supplement has been published.

compound is not mentioned in the abstract, though it is in the original paper or patent. Incidentally, all entries in the *CA* indexes that refer to patents are prefixed by the letter P. Since 1967, the prefixes B and R have also been used, to signify books and reviews, respectively.

By the procedure outlined above, all information regarding a specific compound that has been published up to about 1 year before the search can be found by a procedure that is always straightforward and that in many cases is rapid (if the compound has been reported only a few times). Equally important, if the compound has not been reported, the investigator will know that, too. It should be pointed out that for common compounds, such as benzene, ether, acetone, trivial mentions in the literature are not indexed (so they will not be found by this procedure), only significant ones. Thus, if acetone is converted to another compound, an index entry will be found, but not if it is used as a solvent or an eluant in a common procedure.

The best way to learn if a compound is mentioned in the literature after the period covered by the latest semiannual formula index of *CA* is to use the online services. However, if one lacks access to these, one may consult the keyword index at the end of each issue of *CA*. In these cases, of course, it is necessary to know what name might be used for the compound. The name is not necessary for *Index Chemicus*; one consults the formula indexes. However, these methods are far from complete. *Index Chemicus* lists primarily new compounds, those that would not have been found in the earlier search. The keyword indexes in *CA* are more complete, being based on internal subject matter, as well as title, but they are by no means exhaustive. Furthermore, all three of these publications lag some distance behind the original journals. To locate all references to a compound after the period covered by the latest semiannual formula index of *CA*, it is necessary to use *CAS* on-line.

The complete procedure described above may not be necessary in all cases. Often all the information one needs about a compound will be found in one of the handbooks (p. 1887), in the *Dictionary of Organic Compounds* (p. 1885), or in one of the other compendia listed in this chapter, most of which give references to the original literature.

Other Searches.[42] There is no definite procedure for making other literature searches using only printed materials. Any chemist who wishes to learn all that is known about the mechanism of the reaction between aldehydes and HCN, or which compounds of the general formula Ar_3CR have been prepared, or which are the best catalysts for Friedel–Crafts acylation of naphthalene derivatives with anhydrides, or where the group $-C(NH_2)=N-$ absorbs in the IR, is dependent on their ingenuity and knowledge of the literature. If a specific piece of information is needed, it may be possible to find it in one of the compendia mentioned previously. If the topic is more general, the best procedure is often to begin by consulting one or more monographs, treatises, or textbooks that will give general

---

[42]This discussion is necessarily short. For much more extensive discussions, consult the books in Refs. 1 and 17.

background information and often provide references to review articles and original papers. In many cases, this is sufficient, but when a complete search is required, it is necessary to consult the *CA* subject and/or chemical substance indexes, where the ingenuity of the investigator is most required, for now it must be decided which words to look under. If one is interested in the mechanism of the reaction between aldehydes and HCN, one might look under "aldehydes," or "hydrogen cyanide," or even under "acetaldehyde" or "benzaldehyde," and so on, but then the search is likely to prove long. A better choice in this case would be "cyanohydrin," since these are the normal products and references there would be fewer. It would be a waste of time to look under "mechanism." In any case, many of the abstracts would not prove helpful. Literature searching of this kind is necessarily a wasteful process. Of course, the searcher would not consult the *CA* annual indexes, but only the collective indexes as far as they go and the semiannual indexes thereafter. If it is necessary to search before 1907 (and even before 1920, since *CA* was not very complete from 1907 to ~1920), recourse may be made to *Chemisches Zentralblatt* and the abstracts in the *Journal of the Chemical Society*.

Literature Searching Online[39]

Most of the *Chemical Abstracts*™ literature can be accessed using CAS databases on-line. The CAS registry[43] is the largest and most current database of chemical substance information in the world containing >26 million organic and inorganic substances and 56 million sequences. CAS is a team of scientists who provide digital information environment for scientific research and discovery and provides pathways to published research in the world's journal and patent literature back to the beginning of the twentieth century. Since 1907, CAS has indexed and summarized chemistry-related articles from >40,000 scientific journals, in addition to patents, conference proceedings and other documents pertinent to chemistry, life sciences and many other fields. In total, abstracts for >24 million documents are accessible online through CAS. Through the printed CA, CA on CD, STN, the CAS files distributed through licensed vendors, the SciFinder® and SciFinder Scholar™ desktop research tools, and the STN® Easy or STN® on the Web services, data produced by CAS is accessible to virtually any scientific researcher worldwide in industry, governmental research institutions, and academia.

Substance identification is a special strength of CAS. It is widely known as the CAS Registry, the largest substance identification system in existence. When a chemical substance, newly encountered in the literature, is processed by CAS, its molecular structure diagram, systematic chemical name, molecular formula, and other identifying information are added to the Registry and it is assigned a unique CAS Registry Number®. Registry now contains records for >26 million organic and inorganic substances and >56 million sequences.

The CAS REGISTRY contains >27 million organic and inorganic substances and >57 million sequences. A current count of substance records like this is

---

[43]http://www.cas.org/about.html

updated daily on the CAS web site.[44] An important piece of information that assists in such a search is the CAS Registry Number®. Each substance in REGISTRY is identified by a unique numeric identifier called a CAS Registry Number®.[45] "The CAS Registry Number® is a unique number assigned to a chemical by the Chemical Abstracts Service.[46] A fairly large collection of CAS numbers, with links to safety data for many chemicals, can be found at the listing of chemicals by CAS number at the Safety Home Page of the Physical and Theoretical Chemistry Laboratory at Oxford University."[47] The CAS Registry Number® is a unique numeric identifier that designates only one substance, has no chemical significance, and is a link to finding information about a specific chemical substance. A CAS Registry Number® includes up to nine digits that are separated into three groups by hyphens. The first part of the number, starting from the left, has up to six digits; the second part has two digits. The final part consists of a single check digit.[48]

CAS is located in Columbus, Ohio and is a division of the American Chemical Society. CAS can be contacted at Chemical Abstracts Service, 2540 Olentangy River Road, P.O. Box 3012, Columbus, Ohio 43210 (E-mail: help@cas.org).

Online searching means using a computer terminal to search a *database*. Although databases in chemistry are available from several organizations, STN® International (The Scientific & Technical Information Network) is important because it is comprehensive and available in many countries. STN® has dozens of databases, including many that cover chemistry and chemical engineering. To access these databases a chemistry department, a library, or an individual subscribes to STN® (for a nominal fee), and receives code numbers that will permit access to the system. Then all one needs is a computer and a modem. STN® charges for each use, depending on which databases are used, for how long, and what kind of information is requested. One of the nice features of STN® is that the same command language is used for all databases, so when one has mastered the language for one database, one can use it for all the others. In this section, we will discuss literature searching using *CAS* databases online, which is one of the databases available from STN. One thing that must be remembered is that *CAS databases online* is complete only from 1967 to the present,[49] so that searches for earlier abstracts must use the printed volumes. However, for the period since 1967, not only is online searching a great deal faster than searching the printed *CA*, but, as we will see, one can do kinds of searches online that are simply not possible using only the printed volumes. Furthermore, the on-line files are updated every two weeks, so that one will find all the abstracts on-line well past the appearance of the latest semiannual indexes, often even before the library has received the latest

[44] http://www.cas.org/cgi-bin/regreport.pl
[45] http://www.cas.org/EO/regsys.html
[46] http://www.cas.org/
[47] http://ptcl.chem.ox.ac.uk/MSDS/glossary/casnumber.html
[48] http://www.cas.org/EO/checkdig.html
[49] There is also a file called CAOLD that has some papers earlier than 1967.

weekly printed issue of *CA*. The *CAS databases online* is extremely flexible; one can search in a great many ways. It is beyond the scope of this book to discuss the system in detail (*CA* conducts workshops on its use), but even with the few commands we will give here, a user can often find all that he or she is looking for. The *CAS databases online* has two major files, *the CA File* and *the Registry File*.[50] T

SciFinder®: A Search Tool for Exploring CAS Databases[51]

Tutorials are available to help use SciFinder®,[52] which is a searching tool and not the CAS database. SciFinder® can search a research topic[53] or a compound can be searched by structure.[54] Another search engine is available, different from SciFinder®, known as STN Express® with Discover!,[55] and this searching tool can also be used to search CAS databases.[56] The Analysis Edition of STN Express® with Discover! allows one to search, analyze, visualize, and discover sci-tech information by the ability to create a table for substance analysis that identifies the common substructure for an answer set of structurally related substances,; Group related author–inventor names and company names for better analysis and visualization results; Analyze and tabulate data from single- or multi-file search results, and create a data table and 3D chart; Save an answer set from databases such as CAplusSM, PCTFULL, and USPATFULL with the Save for STN® AnaVist Wizard, and then import and open it in STN® AnaVistTM; Create an interactive spreadsheet from all or only hit CAS Registry Number®s and their corresponding CAS Roles through the CAS Registry Number®; Upload lengthy genetic sequences automatically for searching in DGENE and PCTGEN via the Upload Query Wizard. STN Express® was developed in collaboration with Hampden Data Services. Note that the CAplus database makes it possible to see not only what a paper has cited, but also what papers it been cited by.

To illustrate how STN® is used, an on-line tutorial is available.[57] A few on-line windows from a SciFinder® search are provided to illustrate how searches can be done. This presentation is by no means complete or intended as an alternative to the actual tutorial. Indeed, one could *not* use SciFinder® properly after simply reading this discussion. The intent is to illustrate some features that are available and to present an overview of the use of this important tool.

---

[50]There is also a file, LCA, which is used for learning the system. It includes only a small fraction of the papers in the CA File, and is not updated. There is no charge for using the LCA File, except for a small hourly fee.
[51]http://www.cas.org/. For a tutorial, see http://www.cas.org/SCIFINDER/citation.html.
[52]http://www.cas.org/SCIFINDER/SCHOLAR/interact/
[53]http://www.cas.org/SCIFINDER/SCHOLAR/page2a.html
[54]http://www.cas.org/SCIFINDER/SCHOLAR/scholstruc.html
[55]http://www.cas.org/ONLINE/STN/discover.html
[56]http://www.cas.org/stn.html
[57]http://www.cas.org/ONLINE/STN/expressmac.pdf

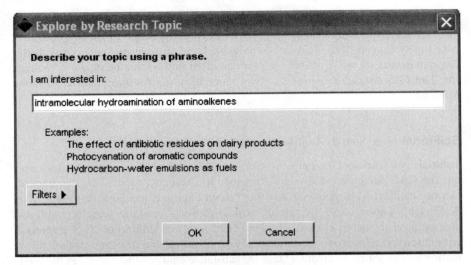

Fig. A.3. Explore by research topic.

Using SciFinder®, a search can be done in one of several different ways. In one example, a search is done by research topic:[58] The example shown in Fig. A.3 shows a search for intramolecular hydroamination of aminoalkenes. To begin, click Explore by research topic and enter the appropriate information.

It is possible to use filters (see Fig. A.3) in order to refine the search by year, document type (journal, patent, review, etc.), author, or company. A window is returned containing references categorized by their relationship to the search phrase, as shown in Fig. A.4. One then simply checks those reference lists that appear closest to the area of interest.

After clicking on "get references," a screen is returned (Fig. A.5) that has the original references as shown in the window. One is given the option of refining this list further, and for each reference, most browsers allow viewing of the abstract or the full references as an HTML or a pdf file. Two icons appear beside each reference in the list: a microscope, which can be clicked on to see the abstract, and an document icon that can be clicked on to see the full-text, original article on the publisher's web-site. This feature is an alternative to document delivery in the traditional sense.

Other examples of typical searches allowed by SciFinder® include search by authors name, as with Professor K. Barry Sharpless shown in Fig. A.6. Search by structure is also possible, such as the one shown in Fig. A.7, and the program provides drawing tools. SciFinder® then searches to finds matches based on structure or reaction.

It is also possible to using the drawing tools to search by reaction and reaction type. SciFinder® returns reaction information, such as that shown in Fig. A. 8. In

[58]http://www.cas.org/SCIFINDER/topic.html

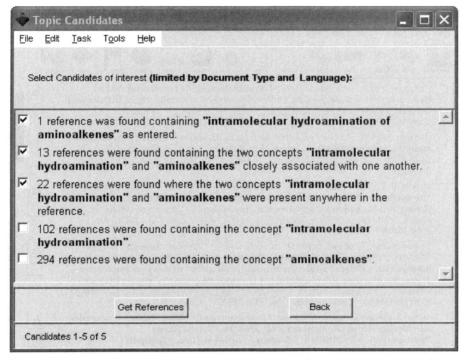

Fig. A.4. Selection of candidates of interest for search by research topic.

all cases, journal articles and/or patents are returned that provide direct access to the literature of interest.

Science Citation Index

As seen in the SciFinder® search tutorials, it is possible to track papers that have cited a particular article or author. A publication that can greatly facilitate literature searching is *Science Citation Index (SCI)*, begun in 1961. This publication, which is quite different from any other mentioned in this chapter, gives a list of all papers in a given year that have cited a given paper, patent, or book. Its utility lies in the fact that it enables the user to search *forward* from a given paper or patent, rather than backward, as is usually the case. For example, suppose a chemist is familiar with a paper by Jencks and Gilchrist (*J. Am. Chem. Soc.,* *1968*, *90*, 2622) entitled "Nonlinear Structure–Reactivity Correlations. The Reactivity of Nucleophilic Reagents toward Esters." The chemist is easily able to begin a search for earlier papers by using references supplied in this paper, and can then go further backward with the aid of references in those papers, and so on. But for obvious reasons, the paper itself supplies no way to locate *later* papers.

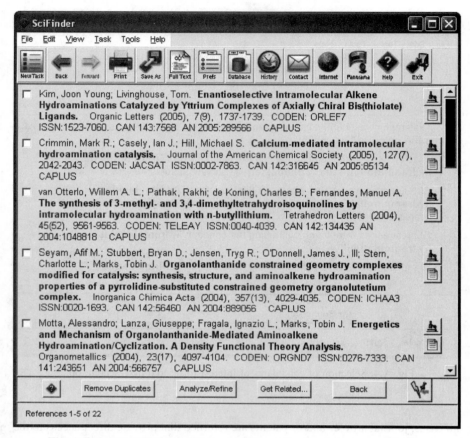

Fig. A.5. Original literature references returned for search by research topic.

The *SCI* is designed to make up for this gap. The citation index of *SCI* lists all papers, patents, or books cited in a given year or 2-month period (by first author only) and then gives a list of papers that have done the citing. The index is published bimonthly and cumulated annually. For example, column 43,901 of the 1989 citation index shows that the Jencks paper mentioned above was cited as a footnote in 16 papers published in 1989. It is reasonable to assume that most of the papers that cited the Jencks paper were on closely related subjects. For each of the 16 papers are listed the first author, journal abbreviation, volume and page numbers, and year. In a similar manner, if one consulted *SCI* for all the years from 1968 on, one would have a complete list of papers that cited that paper. One could obviously broaden the search by then consulting *SCI* (from 1989 on) for papers that cited these 16 papers, and so on. Papers, patents, or books listed, for example, in the 1989 *SCI* may go back many years (e.g., papers published by Einstein in 1905 and 1906 are included). The only

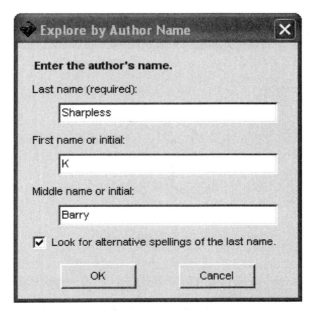

Fig. A.6. Screen shot for beginning a search by author.

requirement is that a paper published in 1989 (or late 1988) has mentioned the earlier paper in a footnote. The arrangement of cited papers or books is alphabetical by cited first author and then by cited year. Cited patents are listed in a separate table, in order of patent number, though the inventor and country are also given.

The *SCI* covers ~3200 journals in the physical and biological sciences, as well as in medicine, agriculture, and technology. In addition to the citation index, each bimonthly and annual *SCI* also includes three other indexes. One of these, called *Source Index*, is similar to the *CA* author index. It lists the titles, journal abbreviations, volume, issue, page numbers, and year of all papers published by a given author during that 2-month period or year. All authors are listed; not just first authors. The second, called the *Corporate Index*, lists all publications that have been published from a given institution during that period, by first author. Thus, the corporate index for 1989 lists 63 papers by 45 different first authors emanating from the Department of Chemistry of Rutgers University, New Brunswick, NJ. The main section of the corporate index (the Geographic Section) lists institutions by country or (for the U.S.) by state. There is also an Organization Section, which lists the names of institutions alphabetically, and for each gives the location, so it can be found in the geographic section. The third index included in *SCI* is the *Permuterm*[59] *Subject Index*. This index alphabetically lists every significant word in the titles of

[59]Registered trade name.

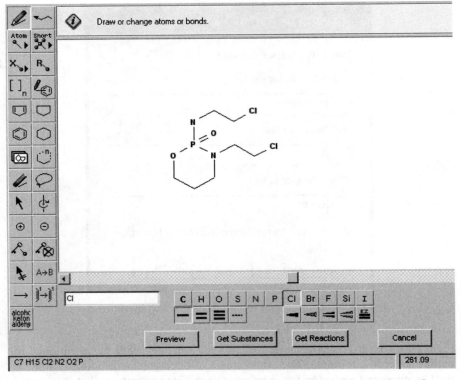

Fig. A.7. Screen shot for beginning a search by structure, using the drawing tools.

all papers published in that year or bimonthly period, paired with all other significant words in the same title. Thus, for example, a title with seven significant words appears at 42 separate places in the index. Each of the seven words appears six times as the main word, each time paired with a different word as the coword. The user is then led to the *Source Index*, where the full reference is given. The *SCI* is also available on-line (though not through STN) and on CD ROM disks. A version of *SCI* that is restricted to chemistry, but also includes searchable abstracts, is available only in the CD ROM format.

The publishers of *SCI* also produce another publication, called *Index to Scientific Reviews*, that appears semiannually. This publication, which began in 1974, is very similar to *SCI*, but confines itself to listing citations to review articles. The citations come from ~2500 journals in the same general areas as are covered by *SCI*. The review articles cited appeared in ~215 review journals and books, as well as in those journals that publish occasional review articles. Like *SCI*, the *Index to Scientific Reviews* contains citation, source, corporate, and Permuterm indexes. It also contains a "Research Front Specialty Index," which classifies reviews by subject.

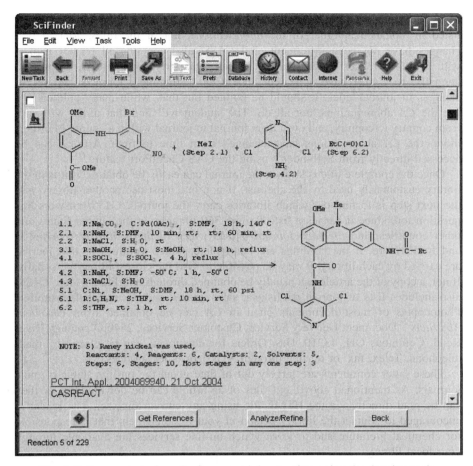

Fig. A.8. Screen shot of results for a search by reaction, using the drawing tools.

How to Locate Journal Articles

Having obtained a reference from various sources or searches, one often needs to consult the original journal (the location of patents is discussed on p. $$$). The first step is to ascertain the full name of the journal, since it is the abbreviation that is generally given. Of course, everyone should be familiar with the abbreviations of the very important journals, such as *J. Org. Chem.*, *Angew. Chem. Int. Ed.*, and so on, but references are often found to journals whose titles are not at all familiar (e.g., *K. Skogs Lantbruksakad. Tidskr.* or *Nauchn. Tr. Mosk. Lesotekh. Inst.*). In such cases, one consults the *Chemical Abstracts Service Source Index (CASSI)*, with the most recent abbreviations in bold print. *CASSI* is available in a 1907–2004 cumulative, containing information 80,000 serial and nonserial publications. *CASSI* also lists journals covered by *Chemisches Zentralblatt* and its predecessors from 1830 to 1969, and journals cited in *Beilstein* before 1907. The journals are listed in alphabetical order of the *abbreviations*, not of the titles. Journal title changes have not been infrequent, and *CASSI* also contains all former names, with cross-references to the current names. Quarterly supplements, cumulated annually, to *CASSI* have appeared since 1990 listing new journals and recent changes in journal titles. It should be pointed out that, while many publications use the *CA* abbreviations, not all do. The student will find that usage will vary from country to country, and even from journal to journal within a country. Furthermore, the *CA* abbreviations have changed from time to time. Articles can be accessed directly from SciFinder® using the CAS ChemPort feature.[60]

Once the complete title is known, the journal can easily be obtained if it is in the library customarily used by the chemist. If not, one must use another library, and the next step is to find out which libraries carry the journal. *CASSI* answers this question too, since it carries a list of some 360 libraries in the United States and other countries, and *for each journal it tells which of these libraries carries it*, and furthermore, if the holdings are incomplete, which volumes of that journal are carried by each library. It may be possible to visit the closest library personally. If not, a copy of the article can usually be obtained through interlibrary loan. *CASSI* also includes lists of journal publishers, sales agents, and document depositories. Photocopies of most documents cited in *CA* can be obtained from *Chemical Abstracts*™ Document Delivery Service, Customer Services, 2540 Olentangy River Road, Columbus OH, 43210, U.S. Orders for documents can be placed by mail, telephone, Telex, fax, or on-line through STN® or other services.

These latter comments are largely out of date given the on-line status of most journals. As mentioned above, pdf files of an article can be downloaded, or they can be read directly via the HTML file using any current browser. The reader is encouraged to contact the library person in your establishment that is responsible for chemical literature and to learn which on-line services are available through your local library.

---

[60]See http://www.cas.org/chemport/index.html

Classification of Reactions by Type of Compounds Synthesized

March's Advanced Organic Chemistry: Reactions, Mechanisms, and Structure, Sixth Edition, by Michael B. Smith and Jerry March
Copyright © 2007 John Wiley & Sons, Inc.

■ AUTHOR INDEX

Entries in this index refer to chapter number (**boldface**) and reference number.

Grohmann, K., 02-0272, 18-0642
Gron, L.U., 13-0403
Grondin, J., 12-0599
Grondin, R., 15-0384
Gronert, S., 05-0106, 05-0150 10-0267, 10-0307
Gronheid, R., 05-0065, 12-0484
Gronowitz, S., 12-0512
Gros, C., 04-0096
Gros, E.G., 10-0540
Gros, G., 15-0375
Gros, P., 13-0647
Grosjean, D., 15-0068
Grosjean, F., 19-1692
Grosjean, M., 04-0268
Gross, A.W., 10-1652, 12-0313, 15-1921, 19-0754
Gross, B., 12-0605
Gross, B.H., 19-1313
Gross, E., 16-1797, 16-1816, 16-2133
Gross, E.K.M., 10-1208
Gross, G., 04-0476, 04-0486
Gross, H., 10-0292, 11-0488, 11-0489
Gross, K.C., 08-0175
Gross, M.F., 15-0429
Gross, M.L., 11-0031
Gross, P.H., 16-0673
Gross, T., 16-0344
Grosse, D., 16-1889
Grossert, J.S., 05-0144
Grossi, L., 05-0215, 13-0011, 15-1167
Grossman, N., 18-0186
Grossman, R.B., 15-0773, 15-1029, 15-1041
Grosu, I., 02-0428, 04-0076
Grotemeier, G., 10-1640
Groth, P., 02-0346, 03-0107
Groth, U., 16-0576, 19-1654
Groth, U.M., 15-0878
Grotjahn, D.B., 08-0197, 15-0213
Groundwater, P.W., 15-1940
Grout, A., 17-0061
Groutas, W.C., 16-1425
Grove, D.I., 02-0362
Grove, D.M., 10-1438
Grovenstein Jr., E., 05-0145, 11-0305, 11-0620, 12-0688, 18-0094, 18-0120, 18-0121, 18-0122
Grover, R., 04-0304
Grover, V., 10-0720
Grover, V.K., 10-0606
Groves, E.E., 12-0302
Groves, J.K., 15-1648

Groves, J.T., 02-0197, 14-0082, 15-1892, 19-0464, 19-0465
Groves, K., 18-0503
Groves, L.G., 02-0368
Groves, P.T., 14-0288
Grozinger, K., 11-0398
Grubber, M.J., 18-0508
Grubbs, E.J., 01-0041
Grubbs, R., 02-0226
Grubbs, R.H., 15-2201, 16-1002, 16-1347, 16-1348, 18-0755, 18-0756, 18-0759, 18-0766, 18-0767, 18-0769, 18-0778, 18-0781, 18-0785, 18-0792, 18-0794, 18-0795, 18-0796, 18-0801, 18-0805, 18-0820, 18-0832, 18-0833, 18-0836
Gruber, A.S., 13-0397
Gruber, J.M., 19-0479
Gruber, W., 18-0011
Grubmüller, P., 12-0687
Gruen, L.C., 11-0110
Gruetzmacher, G., 11-0520
Grugier, J., 16-1999
Grünanger, P., 15-1932
Gründemann, E., 04-0039
Grundke, H., 18-0420
Grundmann, C., 19-1556, 19-1557
Grundon, M.F., 13-0773, 17-0446
Grundy, J., 10-0718
Grüner, H., 15-2255, 18-0494
Grunewald, G.L., 10-0993, 18-0641
Gruntz, U., 10-0725, 10-0850
Grunwald, E., 03-0009, 06-0004, 06-0043, 08-0220, 09-0016, 10-0282, 10-0422, 10-0427, 16-0156, 16-0157
Grunwell, J.R., 15-1171
Gruseck, U., 18-0582
Gruszecka, E., 18-0751, 18-0891
Grutladauria, M., 15-0224, 15-0439
Grutzner, J.B., 05-0126, 05-0134, 12-0102, 12-0208, 15-1512
Gruzdneva, V.N., 09-0034, 18-0044
Grynszpan, F., 03-0124
Grzyb, J.A., 16-1803
Gschrei, C.J., 15-0331
Gschwend, H.W., 10-1284, 12-0388, 13-0641
Gschwind, R.M., 08-0088
Gstöttmayr, C.W.K., 10-1303
Gu, C., 19-0253

Gu, D., 19-0517, 19-0792
Gu, H., 19-1120
Gu, J., 19-1230
Gu, P.M., 19-1042
Gu, S., 15-0867
Guan, X.-P., 16-1768
Guanti, G., 12-0230, 16-1571, 16-2146
Guardeño, R., 15-0454
Guari, Y., 13-0177
Guarna, A., 10-1414, 12-0283, 16-0687, 19-1022
Guarnieri, A., 05-0241, 13-0060
Guay, D., 15-0783, 16-1372, 16-2180
Gubelt, C., 12-0480
Guchhait, S.K., 13-0386, 15-0480, 15-0582, 17-0372, 19-1336
Gudipati, M.S., 16-2090
Gudkov, B.S., 15-0447, 15-0450
Gudkova, A.S., 18-0250
Gudkova, I.P., 04-0169, 04-0209
Gudmundsdottir, A.D., 05-0402
Gudmundsson, B.O., 05-0177, 05-0181
Gudnason, P.I., 04-0371
Guella, F., 10-0242
Guemas, J., 16-0209, 15-0024, 15-0268, 15-0290
Guerin, D.J., 15-1318
Guérin, B., 15-1245, 15-1281
Guérin, C., 04-0026
Guerra, F.M., 14-0177, 15-0284
Guerra, I.M., 15-1710
Guerra, M., 01-0110, 15-1219
Guerreiro, M.C., 19-0611
Guerrero, A., 10-1089, 10-1798, 15-1545
Guerrieri, F., 10-1841, 15-1381
Guertin, D.L., 19-0779
Guertin, K.R., 19-0502, 19-0814
Guest, J.M., 18-0667
Guette, C., 11-0226
Guette, J., 11-0261
Guetté, J., 04-0209, 04-0214, 04-0214
Gugel, H., 04-0472
Guggenberger, L.J., 05-0159
Guggisberg, A., 16-1858
Guha, S., 19-0982
Guibe, F., 10-0340, 10-0489, 19-1100
Guibé, F., 18-0809
Guibé-Jampel, E., 18-0362, 18-0715
Guida, A.R., 10-0727
Guida, W.C., 10-0727, 19-1091
Guihéneuf, G., 03-0015

Mehmedbasich, E., **16**-1469
Mehrdad, M., **10**-0928
Mehrotra, **19**-1605
Mehrotra, A., **10**-0808
Mehrotra, A.K., **10**-1089, **15**-2185, **16**-0164, **16**-1523, **18**-0733
Mehrsheikh-Mohammadi, M.E., **05**-0245, **10**-0306, **14**-0089
Mehta, A.S., **17**-0248
Mehta, G., **02**-0197, **02**-0360, **10**-0188, **10**-0693, **16**-0007, **16**-0008
Mehta, S., **10**-1199, **19**-0839
Mei, Y., **19**-0312
Meidar, D., **10**-1147, **11**-0636, **15**-0207
Meidine, M.F., **10**-0995
Meienhofer, J., **16**-1797, **16**-1816, **16**-2133
Meier, G.B., **19**-1538
Meier, G.P., **16**-1301
Meier, H., **02**-0066, **04**-0471, **04**-0472, **12**-0646, **15**-2047, **17**-0502, **18**-0211
Meier, M., **18**-0404
Meier, M.S., **02**-0366
Meier, T., **10**-1456
Meiere, S.H., **15**-2074
Meignein, C., **16**-0716
Meijer, A., **15**-2030
Meijer, E.W., **07**-0061, **11**-0480, **12**-0743
Meijer, J., **10**-0230, **12**-0500, **15**-0951
Meijer, R.H., **19**-0216
Meijide, F., **12**-0703
Meinecke, A., **10**-1536
Meinema, H.A., **15**-0522
Meintzer, C.P., **14**-0158, **14**-0188
Meinwald, J., **04**-0232, **04**-0435, **12**-0237, **15**-1618, **17**-0269, **18**-0641, **18**-0642
Meinwald, Y.C., **04**-0078, **04**-0435, **15**-1618
Meisenheimer, J., **13**-0006
Meisinger, R.H., **02**-0255
Meislich, H., **10**-0224
Meister, A., **16**-2136
Meister, H., **17**-0252, **17**-0311
Meisters, A., **10**-1506, **15**-1146, **16**-0676, **16**-1974
Meiswinkel, A., **15**-0413
Meixner, J., **15**-1214
Mekhalfia, A., **14**-0194, **14**-0274
Mekhtiev, S.D., **11**-0407
Mekhtieva, V.Z., **15**-0261
Mekley, N., **05**-0394

Melamed, D.B., **12**-0699
Melamed, M., **12**-0449
Melamed, U., **05**-0135
Melander, L., **06**-0059, **11**-0006
Melby, L.R., **03**-0097
Melchiorre, P., **15**-0993, **16**-1434
Mele, G., **18**-0123
Melega, W.P., **02**-0172
Melera, A., **02**-0273
Meli, A., **15**-0279, **15**-2406
Melikyan, G.G., **15**-1650
Melikyan, V.R., **16**-1462
Mello, R., **19**-0515, **19**-0615, **19**-0799
Mellon, M.G., appendixA-0001
Melloni, A., **10**-1469
Melloni, G., **10**-0258, **11**-0376, **15**-0034, **15**-0079, **15**-0082, **15**-1639, **19**-1537
Mellor, D.P., **04**-0007
Mellor, J.H., **14**-0266
Mellor, J.M., **04**-0240, **15**-1851, **15**-1917, **15**-2117, **19**-0438
Melman, A., **16**-1663
Melnyk, O., **12**-0103
Melpolder, J.B., **02**-0047
Melpolder, J.P., **13**-0423
Melville, M.G., **11**-0481
Melvin Jr., L.S., **02**-0089, **16**-1376, **16**-1402
Memarian, H.R., **19**-0035, **19**-0036, **19**-0038, **19**-0039, **19**-0040
Menapace, H.R., **18**-0758
Menapace, L.W., **19**-1346
Menashe, N., **15**-0220
Mencarelli, P., **04**-0052, **05**-0241
Menche, D., **16**-0647
Mende, U., **02**-0213
Mendenhall, G.D., **14**-0232, **19**-0281
Mendez, R., **04**-0372
Méndez, F., **08**-0158
Méndez, M., **10**-1472, **13**-0311, **15**-0776, **15**-0818, **15**-0820
Méndez-Andino, J., **18**-0806
Mendonça, G.F., **15**-0178
Mendonça, J.de S., **19**-1088
Mendonça, P., **13**-0194
Mendoza, J.S., **19**-1224
Menegheli, P., **11**-0495
Menendez, M., **16**-0135
Menéndez, J.C., **10**-0990, **15**-2144
Menes-Arzate, M., **15**-1244
Meneses, R., **15**-0578, **16**-1618, **19**-0940

Meneyrol, J., **19**-1243
Menezes, P.H., **10**-0828, **15**-0944, **15**-1325
Menezes, R.F., **15**-2061
Meng, G., **16**-0323, **19**-1627
Meng, Q., **12**-0100, **12**-0101, **16**-0762
Meng, W.-D., **15**-1671
Meng, X.-J., **19**-1017
Meng, Y., **15**-1158
Meng, Z., **13**-0040
Menge, W.M.P.B., **16**-1249
Mengel, W., **16**-1226
Mengenhauser, J.V., **16**-2070
Menger, F.M., **03**-0143, **04**-0310, **04**-0435, **06**-0010, **10**-0176, **16**-0028, **16**-1570, **16**-1593, **19**-0099
Menges, F., **15**-0425, **15**-0605
Mengler, H., **16**-2047
Menichetti, S., **15**-1607, **15**-1845
Mennen, S., **15**-1368, **16**-0724
Menon, B., **05**-0110, **05**-0130, **08**-0065, **08**-0066
Menon, B.C., **12**-0639
Menoret, G., **18**-0143
Menyailo, A.T., **19**-0345
Menzel, K., **12**-0299
Menzer, S., **03**-0184
Meot-Ner, M., **03**-0001, **05**-0026, **08**-0191, **08**-0223
Méou, A., **15**-0999, **15**-1923
Meracz, I., **15**-2042
Merbouh, N., **19**-0196
Mercier, F., **10**-1454
Mereddy, A.R., **12**-0485
Meredith, C.C., **02**-0134
Mereiter, K., **16**-0687
Merényi, R., **04**-0440, **05**-0243, **05**-0245, **15**-2182, **15**-2222, **15**-2359, **17**-0358, **18**-0634, **18**-0635, **18**-0729, **19**-0840, **19**-1795
Merer, A.J., **07**-0014
Meresz, O., **16**-1468
Mereu, A., **16**-0777, **16**-0803
Mergelsberg, I., **14**-0455
Merger, F., **16**-1649
Merinero, J.A.V., **19**-0336
Merino, P., **16**-0929
Merk, W., **18**-0500, **18**-0501
Merkel, A., **10**-0344
Merkert, J., **15**-1298
Merkley, J.H., **15**-0861, **15**-0862
Merkushev, E.B., **11**-0282, **12**-0522, **13**-0262, **19**-0741
Merle, D., **16**-2132
Merlic, C.A., **10**-1485, **13**-0361, **15**-2266

The vast use of transition metal catalysts led to limited citations of individual metals. In many cases, the term metal catalysis, or catalysts, metal is used as a general heading for several different metals. The term organometallics is used in many cases to include Grignard reagents or organolithium reagents.

March's Advanced Organic Chemistry: Reactions, Mechanisms, and Structure, Sixth Edition, by Michael B. Smith and Jerry March
Copyright © 2007 John Wiley & Sons, Inc.

hydride, tributyltin, and AIBN,
reduction of
thiocarbonates, 1833
and phenylboronic acid,
reduction of aldehydes,
1798
and radical cyclization, 1125,
1126
and reduction of conjugated
esters, 1072
reduction of aldehydes,
1798
with radicals, 0937
hydrides, alkoxy, and reduction
of nitriles, 1815
hydrides, metal, see metal
hydrides
metal, reduction of boranes,
1628
reduction of aldehydes or
ketones, 1792–1805
reduction of ketones, and
Cram's rule, 1804
tin, reduction of aldehydes or
ketones, 1798
tri-*tert*-butoxy, lithium,
reduction of acyl halides,
1808
hydroamination of alkenes,
enantioselectivity, 1181
hydroamination, of alkenes,
metal catalyzed, 1181
hydroboration, 1075–1082
and amination of alkenes,
1050
and enantioselective
synthesis, 1080
and ether solvents, 1075
and functional groups,
1079
and hydration of alkenes,
1034
and oxidation of boranes,
1079
and reduction of alkenes,
1063
and reduction of alkynes to
alkenes, 1064
and steric effects, 1018
and synthetic
transformations, 1079
and Wilkinson's catalyst,
1082
cyclic, with dienes, 1081
enantioselective, 1079, 1082
mechanism, 1078
of alkynes or dienes, 1081
of norbornene, 1023
hydrobromic acid, see HBr
hydrocarbons, acidity, 0251

alternant and nonalternant,
0069
amination, 1757
and chirality, 0152
and guest-host interactions,
0127
and the Kolbe reaction, 0993
bond energy, 0028
by deamination, reagents,
1843, 1844
by decarboxylation, 0835
by dehalogenation of alkyl
halides, reagents, 1824
by dehydroxylation of
alcohols, reagents, 1829
by desilylation of silanes,
1824
by hydrogenation of
thiophenes, 0991
by metal catalyzed
decarbonylation of
aldehydes, 0997
by reduction of alcohols,
reagents, 1829
by reduction of alkyl halides,
reagents, 1824
by reduction of amines,
reagents, 1843, 1844
by reduction of carboxylic
acids, 1843
by reduction of ethers,
reagents, 1831
by reduction of nitriles, 1843
by reduction of silanes, 1824
by reduction of sulfones,
sulfoxides, thiols,
thioethers, 1848, 1849
coupling of boranes, 0989
coupling of organocuprates,
0988
deamination of amines,
indirect methods, 1844
dipole moments, 0096
halosulfonation, 0973
hydroxylation, 1750, 1751
metal catalyzed coupling of
Grignard reagents, 0987
metal mediated
carbonylation, 0824
oxidation to alcohols, 1750
oxidation to amines, 1757
oxidation to hydroperoxides,
0967
oxidation to peroxides, 0971
oxidation to sulfonic acids,
1757
photochemical reaction with
nitriles, 0975
radical coupling of
carboxylate salts, 0992

reagents, for oxidation to
alcohols, 1751
reduction of aldehydes or
ketones, 1835
reduction of thiocarbonates,
1833
reductive cleavage of esters,
1833
remote oxidation, 1761,
1762
with acylperoxides, 0971
with halogen and sulfur
dioxide, 0973
with hydroperoxides, 0971
with oxygen, 0967
with sulfur dichloride, 0974
hydrocarboxylation, and nickel
carbonyl, 1139
and the Koch reaction, 1137
metal catalyzed, mechanism,
1138
of alkenes, 1136, 1138
of alkynes, 1137, 1138
hydrochloric acid, see HCl
hydrocyanic acid, see HCN
hydroformylation, and dicobalt
octacarbonyl, 1146, 1147
and the Tischenko reaction,
1146
hydrogen abstraction, see
abstraction
hydrogen atom abstraction,
0347
hydrogen atom transfer, and
oxidation-reduction,
1706
hydrogen atoms, and the
Schrödinger equation,
0004
and wave functions, 0004
transannular shifts, 1578
hydrogen bonding, 0106–0114
and acidity, 0385
and molecular recognition,
0124
and pK, 0385
and proton transfer, 0367
hydrogen bonds, funtional
groups and characteristics,
0106–0109, 0112, 0113
and enol stabilization, 0772
and guest-host interactions,
0124
and X-ray crystallography,
0109
hydrogen cyanide, see HCN
hydrogen exchange, and
cyclopropenyl derivatives,
0077
coupling of alkenes, 0802